AMERICA IN TWO CENTURIES:
An Inventory

This is a volume in the Arno Press collection

AMERICA IN TWO CENTURIES: An Inventory

Advisory Editor

DANIEL J. BOORSTIN

See last pages of this volume for a complete list of titles

SUBJECT-MATTER INDEX

OF

PATENTS FOR INVENTIONS

ISSUED BY THE

UNITED STATES PATENT OFFICE

FROM

1790 TO 1873, INCLUSIVE

VOLUME I

ARNO PRESS
A New York Times Company
1976

Editorial Supervision: ANDREA HICKS

Reprint Edition 1976 by Arno Press Inc.

Reprinted from a copy in
The New York Public Library

AMERICA IN TWO CENTURIES: An Inventory
ISBN for complete set: 0-405-07666-5
See last pages of this volume for titles.

Manufactured in the United States of America

Library of Congress Cataloging in Publication Data

United States. Patent Office.
Subject-matter index of patents for inventions issued by the United States Patent Office from 1790 to 1873, inclusive.

(America in two centuries, an inventory)
On spine: Patents for inventions: 1790-1873.
Reprint of the 1874 ed. published by Government Printing Office, Washington.
1. Patents--United States--Indexes. I. Leggett, Mortimer Dormer, 1821-1896. II. Title. III. Title: Patents for inventions: 1790-1873. IV. Series.
T223.D7A45 1976 608'.7'73 75-24110
ISBN 0-405-07737-8

SUBJECT-MATTER INDEX

OF

PATENTS FOR INVENTIONS

ISSUED BY THE

UNITED STATES PATENT OFFICE

FROM

1790 TO 1873, INCLUSIVE.

VOLUME I.

COMPILED AND PUBLISHED UNDER THE DIRECTION OF

M. D. LEGGETT,

COMMISSIONER OF PATENTS.

WASHINGTON:
GOVERNMENT PRINTING OFFICE.
1874.

INDEX OF PATENTS

ISSUED FROM

THE UNITED STATES PATENT OFFICE

FROM 1790 TO 1873, INCLUSIVE.

Invention.	Inventor.	Residence.	Date.	No.
A.				
Abdominal and back supporter and truss combined.	R. J. Cundiff	Lynchburgh, Va	June 7, 1870	103, 992
Abdominal and spinal supporter	J. C. Zachos	New York, N. Y	July 16, 1872	129, 202
Abdominal and uterine supporter	J. A. Campbell	Lima, N. Y	Apr. 10, 1841	2, 041
Abdominal and uterine supporter	E. J. Harding	Saint Louis, Mo	Mar. 24, 1868	75, 902
Abdominal and uterine supporter combined	Z. Waters	Bloomington, Ill	July 27, 1869	93, 144
Abdominal supporter	M. G. Briggs	Boston, Mass	Feb. 6, 1872	123, 326
Abdominal supporter	H. B. Conant	Geneva, Wis	Aug. 2, 1853	9, 896
Abdominal supporter	E. Dexter	Quincy, Ill	June 5, 1866	55, 272
Abdominal supporter	M. Faloon	Bloomington, Ill	June 16, 1868	78, 946
Abdominal supporter	S. S. Fitch	New York, N. Y	Mar. 19, 1850	7, 186
Abdominal supporter	L. D. Fleming	Newark, N. J	Dec. 31, 1845	4, 336
Abdominal supporter	J. Funkhouser	Rockingham County, Va	Oct. 6, 1868	82, 702
Abdominal supporter	E. K. Gale	New York, N. Y	Mar. 14, 1846	4, 425
Abdominal supporter	B. A. Grover	Momence, Ill	July 5, 1859	24, 630
Abdominal supporter	J. W. Gurley	Petersburgh, Va	July 22, 1873	141, 137
Abdominal supporter	E. J. Harding	Saint Louis, Mo	Feb. 2, 1869	86, 539
Abdominal supporter	E. J. Harding	Saint Louis, Mo	July 15, 1873	140, 778
Abdominal supporter	W. Henderson and J. Greenawalt.	Pittsburgh, Pa	June 25, 1867	66, 022
Abdominal supporter	S. L. Hockert	Milwaukee, Wis	Sept. 8, 1868	82, 001
Abdominal supporter	J. W. Hood	Mount Sterling, N. Y	Dec. 11, 1847	5, 395
Abdominal supporter	I. J. H. Howard	San Francisco, Cal	June 24, 1862	35, 683
Abdominal supporter	H. R. and G. W. Hubbard	Middletown, Conn	Apr. 17, 1849	6, 353
Abdominal supporter	A. F. Jennings	Sherman, N. Y	Aug. 25, 1868	81, 510
Abdominal supporter	A. T. Kirk	Chicago, Ill	May 28, 1872	127, 352
Abdominal supporter	M. L. Knapp	Painesville, Ohio	Jan. 28, 1851	7, 916
Abdominal supporter	A. I. Lonsbury	Somerville, Tenn	Nov. 11, 1851	8, 515
Abdominal supporter	M. M. Merrill	Boston, Mass	July 8, 1873	140, 585
Abdominal supporter	J. M. Milligan	New Albany, Ind	Feb. 10, 1857	16, 602
Abdominal supporter	J. A. Morrell	Chicago, Ill	Mar. 17, 1868	75, 695
Abdominal supporter	M. W. O'Meara	New York, N. Y	Mar. 12, 1850	7, 175
Abdominal supporter	J. L. Porter	Kirksville, Mo	May 25, 1869	90, 389
Abdominal supporter	M. F. Potter	Kaneville, Ill	May 18, 1869	90, 191
Abdominal supporter	J. V. Richardson	Milwaukee, Wis	Aug. 29, 1871	118, 645
Abdominal supporter	H. H. Thompson	Washington, D. C	Nov. 22, 1864	45, 193
Abdominal supporter	J. Thompson	Milwaukee, Wis	Nov. 19, 1867	71, 246
Abdominal supporter	A. B. Weaver	Carthage, Ind	Sept. 27, 1859	25, 597
Abdominal supporter	J. White, S. N. Marsh, and H. Smith.	Canajoharie, N. Y	Sept. 28, 1843	3, 289
Abdominal supporter	W. M. Young	Trempealeau County, Wis	Dec. 24, 1867	72, 710
Abdominal supporter and corsets combined	S. A. Moody	New York, N. Y	May 3, 1864	42, 591
Abrading and polishing wheel	E. C. Merrill	Charleston, Vt	Dec. 7, 1869	97, 671
Abrasion, Form of rubbing surfaces for regulating.	C. Schiele	Frankfort, Germany	May 21, 1850	7, 385
Abrasive powder	J. Russell	Bath, Me	Jan. 22, 1867	61, 363
Accelerating washer	M. Cass	Caroline, N. Y	Aug. 29, 1827	......
Accelerating wheel-head	D. H. Tuttle	Williamson, N. Y	May 17, 1824	......
Accordeon	A. Faas	Philadelphia, Pa	Aug. 12, 1856	15, 511
Accordeon	E. Prics	New York, N. Y	June 21, 1864	43, 226
Accordeon	C. F. Zimmerman	Philadelphia, Pa	July 10, 1866	56, 319
Accordeon	C. F. Zimmermann	Philadelphia, Pa	Aug. 2, 1870	106, 018
Accordeon, &c	F. Goetze and D. Müller	New York, N. Y	Oct. 28, 1873	144, 025
Accordeon, Construction of	A. Faas	Philadelphia, Pa	June 13, 1854	11, 062
Accordeon, Method of tuning	E. A. Robbins	Rochester, N. Y	June 12, 1855	13, 044
Accordeon, Musical notation for	C. F. Zimmermann	Philadelphia, Pa	Jan. 3, 1871	110, 719
Accordeon-strap	C. F. Zimmermann	Philadelphia, Pa	Jan. 29, 1867	61, 698
Accordeon, Valve of	C. M. Zimmermann	Philadelphia, Pa	July 22, 1856	15, 401
Accordeons, &c., Tuning reeds of	M. and N. Schneider	New Orleans, La	Sept. 27, 1845	4, 212
Accoucheur's chair	N. W. Smith	Shutesbury, Mass	Oct. 16, 1849	6, 795
Account-indicator, Commercial	C. Brunschwiler	New York, N. Y	June 29, 1869	91, 905
Accouterments, Cavalry	J. K. Mizner	Detroit, Mich	Jan. 16, 1866	52, 064
Accouterments, Mode of slinging	W. D. Mann	Detroit, Mich	Dec. 8, 1863	40, 849
Acetate of copper or verdigris, Manufacture of	S. Dempsey	New York, N. Y	Feb. 4, 1813	......
Acid, and in the application of the same for various useful purposes, Production and manufacture of carbonic.	S. Stevens	New York, N. Y	Aug. 27, 1867	68, 321
Acid and other liquids, Apparatus for concentrating sulphuric.	J. Hughes	Brooklyn, N. Y	May 28, 1867	65, 227
Acid and other vessels, Cap to prevent the bursting of carbonic.	J. Matthews	New York, N. Y	Apr. 22, 1873	138, 171
Acid and paint from material used to purify gas, Manufacture of.	J. Hughes	Stapleton, N. Y	May 28, 1872	127, 350
Acid and phosphates, Manufacture of phosphoric.	T. A. Genth	Philadelphia, Pa	Aug. 2, 1859	24, 931
Acid and water-proof composition for coating cloth.	H. W. Johns	New York, N. Y	May 10, 1870	102, 824

Index of patents issued from the United States Patent Office from 1790 *to* 1873, *inclusive*—**Continued.**

Invention.	Inventor.	Residence.	Date.	No.
Acid, Apparatus for concentrating sulphuric	D. Ashworth and R. Eaton	Woburn, Mass	May 26, 1868	78,352
Acid, Apparatus for generating carbonic	W. H. Bate	East Somerville, Mass	Dec. 20, 1870	110,190
Acid, Apparatus for generating carbonic	B. Bates	Baltimore, Md	Apr. 5, 1870	101,415
Acid, Apparatus for generating carbonic	L. Kimball	Bradford, Mass	June 11, 1872	127,893
Acid, Apparatus for making nitric	P. O'Reilly	Providence, R. I	Mar. 24, 1857	16,879
Acid, Apparatus for making sulphuric	S. Ravenel	Charleston, S. C	Apr. 25, 1871	114,042
Acid, Apparatus for producing sulphurous	M. Hatschek	Pesth, Hungary	Mar. 22, 1870	101,011
Acid as a substitute for other solid acids, Preparing phosphoric.	E. N. Horsford	Cambridge, Mass	Apr. 22, 1856	14,722
Acid by lime decomposition, Margaric, stearic, and oleic.	H. Seybert and L. Vanuxem	Philadelphia, Pa	Aug. 16, 1830	....
Acid compound for use in baking and cooking	J. E. Lauer	New York, N. Y	Feb. 19, 1867	62,277
Acid, Concentrating apparatus for sulphuric	W. T. Clough	Newark, N. J	July 1, 1856	15,222
Acid concentrating-pan, Sulphuric	P. Marcelin and J. Saunders	Greenpoint, N. Y	Apr. 28, 1868	77,202
Acid, Concentrating sulphuric	D. Ashworth and R. B. Eaton	Woburn, Mass	Mar. 19, 1867	62,919
Acid, Concentrating sulphuric	J. D. Loftus	Chelsea, Mass	Jan. 1, 1867	60,759
Acid for dyeing woolens, &c., Mixing	H. Young	Rochester, N. Y	Nov. 19, 1833	....
Acid for extinguishing fire, Mode of collecting and storing carbonic.	E. Thayer	Worcester, Mass	Nov. 16, 1869	97,001
Acid from carboys, Apparatus for discharging	W. Gee	New York, N. Y	Oct. 24, 1871	120,259
Acid from oil-refineries, Mode of recovering spent	L. S. Fales	New York, N. Y	Nov. 23, 1869	97,182
Acid from petroleum-refinery, Mode of utilizing the waste.	R. M. Smith	Baltimore, Md	May 17, 1864	42,803
Acid generator, fountain, and other vessels, Carbonic.	J. Matthews	New York, N. Y	May 13, 1873	138,908
Acid in liquids, Apparatus for ascertaining the amount of.	H. Twitchell	Cincinnati, Ohio	Feb. 15, 1870	99,976
Acid, Manufacture of acetic	C. J. T. Burcey	Black Rock, Conn	Sept. 12, 1871	118,788
Acid, Manufacture of acetic	A. and M. Pirz	East New York, N. Y	Mar. 2, 1869	87,365
Acid, Manufacture of boracic	F. Gutzkow	San Francisco, Cal	Mar. 25, 1873	137,072
Acid, Manufacture of citric	W. K. Johnston	Memphis, Tenn	May 24, 1870	103,340
Acid, Manufacture of nitric	H. M. Baker	Williamsburgh, N. Y	Oct. 11, 1870	108,090
Acid, Manufacture of pyroligneous	M. E. Converse and A. T. Atherton.	Rindge, N. H., and Lowell, Mass.	May 26, 1868	78,264
Acid, Manufacture of pyroligneous	A. H. Emery	New York, N. Y	Aug. 8, 1865	49,247
Acid, Manufacture of solid fatty	L. A. De Milly	Paris, France	Sept. 8, 1868	81,884
Acid, Manufacture of stearic	E. De Bassano and A. Brudeun	Brussels, Belgium	Aug. 27, 1861	33,135
Acid, Manufacture of stearic	F. C. A. Bock	Copenhagen, Denmark	June 29, 1869	92,004
Acid, Manufacture of sulphuric	L. Chandor	St. Petersburg, Russia	May 31, 1864	42,985
Acid, Manufacture of sulphuric	D. E. Contaret	Roxbury, Mass	June 13, 1854	11,050
Acid, Manufacture of sulphuric	C. Hinrichs	New York, N. Y	Sept. 7, 1852	9,249
Acid, Manufacture of sulphuric	D. Jackson	Walworth, England	Aug. 13, 1872	130,432
Acid, Manufacture of sulphuric	A. Monnier	Camden, N. J	Aug. 11, 1857	17,976
Acid, Manufacture of sulphuric	J. Saunders	Brooklyn, N. Y	Nov. 25, 1873	144,928
Acid, Manufacture of sulphuric	J. Smith and J. R. Savage	Philadelphia, Pa	Feb. 16, 1864	41,647
Acid, Manufacture of sulphuric	A. H. Tait	New York, N. Y	Feb. 9, 1869	86,881
Acid, Manufacture of sulphuric	E. Thomson and W. H. Greene	Philadelphia, Pa	Sept. 23, 1873	143,202
Acid, Manufacture of sulphuric and hydrochloric	H. M. Baker	Washington, D. C	Jan. 26, 1869	86,200
Acid, Manufacture of sulphurous	N. P. Akin	Philmont, N. Y	May 21, 1872	127,008
Acid, Manufacture of sulphurous	P. Marcelin	New York, N. Y	Feb. 13, 1872	123,713
Acid, Manufacture of sulphurous	P. Marcelin and E. Ende	New Orleans, La	June 12, 1860	28,678
Acid, Method of manufacturing sulphuric	J. Hargreaves	Paterson, N. J	Aug. 24, 1839	1,303
Acid, Method of purifying acetic	A. A. Fesquet	Marseilles, France	Aug. 30, 1864	44,053
Acid, Obtaining pure sulphurous	J. Albrecht	New Orleans, La	June 29, 1858	20,755
Acid on sheet-iron, Method of neutralizing	E. A. Harvey	Wilmington, Del	May 8, 1866	54,538
Acid, Pan for concentrating sulphuric	P. Marcelin and J. Saunders	Greenpoint, N. Y	May 12, 1868	77,826
Acid phosphate of lime, Method of preparing	E. N. Horsford	Cambridge, Mass	Apr. 14, 1868	76,763
Acid, Process for making sulphuric	B. Bell	Boston, Mass	Nov. 7, 1811	....
Acid, Process for making sulphuric	H. Helland	Westfield, Mass	Dec. 27, 1859	26,588
Acid, Purifying acetic	J. F. Cavarly	Flushing, N. Y	Dec. 5, 1871	121,586
Acid, Purifying acetic	T. L. Olden	Brooklyn, N. Y	Mar. 8, 1870	100,553
Acid, Purifying pyroligneous	C. F. Binder	Philadelphia, Pa	June 20, 1871	116,142
Acid, Purifying pyroligneous or acetic	C. C. Parsons	New York, N. Y	Feb. 23, 1869	87,193
Acid to cane-juice, Method of applying sulphurous	T. Byrne	Baton Rouge, La	Sept. 4, 1860	29,860
Acid used in refining petroleum, Process of recovering the.	R. G. Loftus	Chelsea, Mass	June 14, 1864	43,157
Acids, Apparatus for manufacturing fatty	M. Werk	Cincinnati, Ohio	Oct. 5, 1858	21,711
Acids, Evaporating and concentrating sulphuric and other.	W. T. Clough	Newark, N. J	June 18, 1872	127,957
Acids from leather, Removing	M. W. Fry	Guyandotte, W. Va	Nov. 4, 1873	144,328
Acids, Mode of preparing vegetable	N. S. Allison and B. Kugler	Philadelphia, Pa	June 5, 1812	....
Acids, Preparing wooden vessels for holding	W. Archdeacon	Chicago, Ill	July 16, 1872	129,204
Acids, Process for purifying pyroligneous and acetic.	C. C. Parsons	New York, N. Y	Nov. 9, 1869	96,721
Acids, Receivers or carboys for the manufacture of muriatic and other.	A. Baumgarten	New York, N. Y	Aug. 3, 1869	93,270
Acids, Separating oleic and stearic	J. S. Gwynne	Pittsburgh, Pa	Sept. 3, 1846	4,735
Acids, Treatment of fatty	J. C. Appenzeller	Cincinnati, Ohio	Jan. 25, 1859	22,691
Acoustic apparatus	D. D. Stelle	New Brunswick, N. J	Feb. 14, 1860	27,165
Acoustic auricle	E. G. Hyde	Irvington, N. J	Jan. 27, 1857	16,485
Acoustic drum	S. Sawyer	Boston, Mass	Oct. 24, 1834	....
Acupuncture instrument	A. R. Brown	Litchfield, Mich	Jan. 1, 1867	60,917
Acupuncture instrument	A. R. Brown	Albion, Mich	Dec. 15, 1868	84,854
Acupuncture instrument	G. Herrick	Albion, Mich	Apr. 26, 1870	102,262
Acupuncture instrument, Mold for making	A. R. Brown	Albion, Mich	Oct. 6, 1868	82,686
Adding and registering numbers, Instrument for	N. Spofford and C. Corliss	Haverhill, Mass	June 23, 1868	79,272
Adding and subtracting register	H. A. House	Bridgeport, Conn	Nov. 29, 1870	109,619
Adding figures, Instrument for	C. Corliss	Haverhill, Mass	June 23, 1868	79,272
Adding numbers, Instrument for	J. B. Newbrough	Saint Louis, Mo	June 21, 1859	24,481
Adding-machine	I. W. Arndt	Green Bay, Wis	Apr. 12, 1859	23,537
Adding-machine	J. Ballou	Cincinnati, Ohio	Mar. 13, 1860	27,418
Adding-machine	B. B. Brown	Delaware, Ohio	Oct. 19, 1869	95,876
Adding-machine	J. T. Campbell	Rockville, Ind	Aug. 9, 1859	24,990
Adding-machine	G. W. Chapin	Brooklyn, N. Y	Feb. 8, 1870	99,513
Adding-machine	G. W. Chapin	Brooklyn, N. Y	Sept. 6, 1870	106,999
Adding-machine	G. B. Fowler	Chicago, Ill	July 14, 1863	39,222
Adding-machine	J. Groesbeck	Philadelphia, Pa	Mar. 1, 1870	100,288

Index of patents issued from the United States Patent Office from 1790 *to* 1873, *inclusive*—Continued.

Invention.	Inventor.	Residence.	Date.	No.
Adding-machine	E. M. Hamilton	New York, N. Y	July 18, 1871	117, 169
Adding-machine	J. Harris, jr.	Roxbury, Mass	Jan. 1, 1861	31, 016
Adding-machine	A. L. Hatfield	Lewisburgh, Pa	Sept. 26, 1854	11, 726
Adding-machine	I. G. Hubbs	New York, N. Y	Aug. 19, 1856	15, 565
Adding-machine	M. C. Jeffers	New York, N. Y	Sept. 29, 1863	40, 105
Adding-machine	D. Kohler	Sunbury, Pa	June 6, 1834	....
Adding-machine	F. T. Leilich	Frederick, Md	Aug. 23, 1870	106, 701
Adding-machine	G. Lindervos	Point Arena, Cal	June 24, 1873	140, 146
Adding-machine	D. R. Nelson	Jackson, Ohio	Apr. 24, 1860	28, 006
Adding-machine	J. B. Newbrough	Saint Louis, Mo	Sept. 28, 1858	21, 621
Adding-machine	N. Ockerland	New York, N. Y	July 26, 1870	105, 717
Adding-machine	V. Parks	Fort Wayne, Ind	Mar. 5, 1867	62, 677
Adding-machine	H. Parmelee	Philadelphia, Pa	Jan. 25, 1870	99, 226
Adding-machine	S. Pool	Chapel Hill, N. C	Sept. 23, 1873	143, 184
Adding-machine	T. Rossiter	New Haven, Conn	Aug. 3, 1869	93, 350
Adding-machine	N. S. Saxton	Riverhead, N. Y	Nov. 13, 1855	13, 800
Adding-machine	C. E. Spear	Gardiner, Me	Aug. 30, 1870	106, 881
Adding-machine	A. M. Stephenson	Manteno, Ill	Mar. 25, 1873	137, 107
Adding-machine	T. T. Strode	Mortonville, Pa	Oct. 2, 1860	30, 264
Adding-machine	T. T. Strode	Mortonville, Pa	Aug. 1, 1865	49, 168
Adding-machine	E. A. Swain	New York, N. Y	Jan. 11, 1870	98, 720
Adding-machine	F. F. Warner	Chicago, Ill	Dec. 27, 1870	110, 520
Adding-machine	C. H. Webb	New York, N. Y	Mar. 10, 1868	75, 392
Adding-machine	C. Winter	Piqua, Ohio	Apr. 12, 1859	23, 637
Addometer	J. Burns	New York, N. Y	Aug. 24, 1858	21, 243
Addometer	L. N. Nutz	Alton, Ill	Aug. 17, 1858	21, 236
Address-case for railway-car	S. W. Downey	Piedmont, W. Va	Jan. 28, 1868	73, 881
Address printing-machine	H. Julien	Ottawa, Canada	July 28, 1868	80, 285
Addresses, &c., Machine for printing	J. A. Barrington	Fredericktown, Ohio	June 14, 1859	24, 364
Addresses on newspapers, &c., Apparatus for printing.	J. A. Campbell	Georgetown, Canada	Jan. 17, 1860	26, 831
Addresses on newspapers, &c., Apparatus for printing.	G. Hutchison	Allegheny, Pa	Sept. 6, 1859	25, 337
Addresses on newspapers, &c., Apparatus for printing.	C. K. Marshall	Vicksburgh, Miss	Nov. 1, 1859	25, 974
Addresses on newspapers, Machine for printing	N. Bowlus	Middletown, Md	May 1, 1860	28, 059
Addresses on newspapers, Machine for printing	S. D. Carpenter	Madison, Wis	May 5, 1857	17, 194
Addresses on newspapers, Machine for printing	E. P. Day	New York, N. Y	June 6, 1854	11, 022
Addresses on newspapers, Machine for printing	D. B. Tiffany and S. W. Soule	Xenia, Ohio, and Milwaukee, Wis.	Mar. 20, 1860	27, 580
Addresses on newspapers, &c., Machine for printing.	J. A. Campbell	Buffalo, N. Y	Jan. 20, 1863	37, 432
Addresses on newspapers, &c., Machine for printing.	R. W. and D. Davis	Yellow Springs, Ohio	Sept. 6, 1859	25, 319
Addresses on newspapers, &c., Machine for printing.	D. Fuller	Cherry Valley, Ill	Sept. 15, 1863	39, 913
Addresses on newspapers, &c., Machine for printing.	J. C. Gaither	Somerset, Pa	Apr. 7, 1868	76, 433
Addresses on newspapers, &c., Machine for printing.	J. Lord	Pawtucket, Mass	Sept. 7, 1858	21, 429
Addresses on newspapers, &c., Machine for printing.	A. H. Nordyke	Richmond, Ind	Mar. 1, 1859	23, 107
Addresses on newspapers, &c., Machine for printing.	M. and C. Peck and R. W. Wright.	New Haven and Orange, Conn.	Jan. 12, 1864	41, 234
Addresses on newspapers, &c., Printing	H. Moeser	Pittsburgh, Pa	June 24, 1851	8, 175
Addresses on newspapers, &c., Printing	S. W. Soule	Cincinnati, Ohio	Oct. 2, 1860	30, 259
Addressing letters	J. S. Brown	Washington, D. C	Sept. 9, 1862	36, 393
Addressing-machine	W. H. Clague and R. B. Randall	Rochester, N. Y	Dec. 19, 1871	121, 931
Addressing-machine	F. A. Darling	Fayetteville, N. Y	Apr. 8, 1873	137, 599
Addressing-machine	G. A. Davison	Montana, Iowa	July 4, 1871	116, 567
Addressing-machine	R. Dick	Buffalo, N. Y	May 7, 1867	64, 502
Addressing-machine	G. Gibbons	Meriden, Conn	Aug. 10, 1869	93, 527
Addressing-machine	S. Holton	Middlebury, Vt	Feb. 25, 1873	136, 323
Addressing-machine	J. McFatrick	Lena, Ill	Oct. 4, 1870	108, 038
Addressing-machine	J. K. Rukenbrod	Salem, Ohio	Dec. 12, 1871	121, 900
Addressing-machine	N. E. and G. W. Warren	Cleveland, Ohio, and Hillsdale, Mich.	Apr. 4, 1865	47, 142
Addressing-machine, Feed-rack for	W. H. Henderson and W. H. Snider.	Lena, Ill	July 13, 1869	92, 444
Addressing-machine, Newspaper	H. A. Gage	Manchester, N. H	July 9, 1861	32, 763
Addressing-machine, Newspaper	C. K. Marshall	New Orleans, La	Feb. 9, 1869	86, 680
Addressing-machine, Newspaper	P. O'Conner	Youngstown, Ohio	Apr. 25, 1871	114, 032
Addressing newspapers, &c., Apparatus for	W. M. Doty	New York, N. Y	Jan 26, 1864	41, 369
Addressing newspapers, &c., Machine for	J. Battey	Honeoye Falls, N. Y	Jan. 17, 1860	26, 827
Addressing newspapers, &c., Machine for	G. Schuh	Madison, Ind	Apr. 26, 1859	23, 787
Addressing newspapers, &c., Printing-press for	G. Henderson	Allegheny, Pa	Sept. 6, 1859	25, 363
Adhesive material	A. J. Russell	New York, N. Y	Apr. 2, 1861	31, 909
Adjustable bit	F. Jonas	Burlington, Iowa	Apr. 1, 1873	137, 373
Adjustable box for arbors, &c	B. D. and H. E. Kay	Fall River, Mass	Nov. 10, 1868	83, 859
Adjustable bracket	J. H. Davis	Chillicothe, Mo	July 13, 1869	92, 519
Adjustable bracket	J. B. Morrison	Brooklyn, N. Y	Aug. 26, 1873	142, 263
Adjustable chair	A. Chase	North Weare, N. H	June 11, 1861	32, 511
Adjustable chair	R. A. Thompson	Beaver Falls, Pa	Dec. 3, 1872	133, 550
Adjustable chair	T. Weaver	Harrisburgh, Pa	Jan. 10, 1865	45, 887
Adjustable clamp	W. G. Floyd	Brooklyn, E. D., N. Y	Mar. 6, 1866	52, 983
Adjustable clamp	C. S. Meeker	New Haven, Conn	June 1, 1869	90, 768
Adjustable gage	R. E. Jones	New York, N. Y	Jan. 26, 1869	86, 308
Adjustable hanger	R. A. Stratton	Philadelphia, Pa	May 12, 1863	38, 539
Adjustable miter	P. A. Snyder	Jersey City, N. J	Dec. 18, 1866	60, 646
Adjustable press	N. C. Stiles	Middletown, Conn	Nov. 24, 1868	84, 313
Adjustable seat	I. Cook	Saint Louis, Mo	Mar. 3, 1868	74, 993
Adjustable seat	B. N. Hemenway	Rockland, Me	July 16, 1872	129, 560
Adjustable seat	L. Postawka	Cambridge, Mass	Apr. 4, 1871	113, 342
Adjustable seat-fastener	W. T. Thornton	Belleville, Mich	Dec. 3, 1867	71, 821
Adjustable spring	A. Roff	Southport, Conn	July 7, 1868	79, 778

Index of patents issued from the United States Patent Office from 1790 *to* 1873, *inclusive*—Continued.

Invention.	Inventor.	Residence.	Date.	No.
Adjustable table	O. C. Dodge	Brooklyn, N. Y	Nov. 20, 1860	30,671
Adjustable wrench	A. Hotchkiss	Sharon, Conn	May 4, 1852	8,922
Adjustable wrench	J. Magee	Usquepaugh, R. I	Dec. 28, 1869	98,393
Adjustable wrench	A. Sedgwick	Poughkeepsie, N. Y	Sept. 25, 1866	58,306
Advertiser, Balloon	E. L. Moodie	New York, N. Y	Mar. 18, 1873	136,930
Advertising-album	W. S. Gavan	Savannah, Ga	Dec. 19, 1871	122,005
Advertising-album	G. W. Hawes	New York, N. Y	Oct. 24, 1871	120,260
Advertising-apparatus	A. Berthoud	New York, N. Y	Nov. 26, 1861	33,772
Advertising-apparatus	W. Hebdon	New York, N. Y	Nov. 15, 1870	109,208
Advertising-apparatus	J. A. Royce	Lee, Mass	July 9, 1867	66,637
Advertising-apparatus, Panoramic	J. Berthoud	Paris, France	Nov. 18, 1862	36,937
Advertising-box to lamp and other posts, Clamp for securing.	P. McCosker	Pittsburgh, Pa	Sept. 30, 1873	143,365
Advertising-cabinet, Electric	H. W. McAllister	Chicago, Ill	May 30, 1871	115,494
Advertising-calendar	J. D. Parsons	Albany, N. Y	May 11, 1869	90,017
Advertising-carriage	W. C. Harris, A. R. Roseman, and H. B. Hutchins.	Philadelphia, Pa	Feb. 18, 1868	74,530
Advertising-device	I. Beziegler	Philadelphia, Pa	Apr. 30, 1872	126,367
Advertising-device	H. H. Browne	Mount Vernon, N. Y	June 16, 1868	78,923
Advertising-device	A. Davis	Lowell, Mass	Oct. 8, 1867	69,637
Advertising-device	J. E. Emery	Albany, N. Y	Apr. 4, 1871	113,409
Advertising-device	C. Peabody and P. H. Delancy	Detroit, Mich	Dec. 8, 1868	84,707
Advertising-device	R. F. Rankin	Columbus, Ohio	June 29, 1869	92,137
Advertising-device	W. Raphael	Saint Louis, Mo	July 30, 1872	130,070
Advertising-device	W. H. Reiff	Philadelphia, Pa	Apr. 25, 1871	114,198
Advertising-device	W. S. Webb	Providence, R. I	Feb. 23, 1869	87,228
Advertising-directory	B. F. Stillwell	San Francisco, Cal	Nov. 17, 1863	40,648
Advertising-frame	D. G. Howell	Danby, N. Y	Feb. 8, 1870	99,675
Advertising-frame	B. S. Moulton	Danvers, Mass	Oct. 15, 1872	132,314
Advertising-frame	W. H. Sadler and J. M. Drysdale	Baltimore, Md	Apr. 13, 1869	88,910
Advertising-frame	J. Sonnedecker	Cincinnati, Ohio	May 5, 1868	77,545
Advertising-frame, Electro-magnetic	J. Brooks	Boston, Mass	Dec. 14, 1869	97,756
Advertising-lamp	E. Boesch	San Francisco, Cal	Mar. 21, 1871	112,890
Advertising-lantern	T. L. Wright	New York, N. Y	Dec. 12, 1871	121,923
Advertising-machine	I. W. Sylvester	New York, N. Y	June 25, 1867	66,055
Advertising-medium, Ornamental	J. O. Belknap	New Orleans, La	July 11, 1871	116,799
Advertising, Mode of	E. Wiebé	Brooklyn, N. Y	Nov. 15, 1859	26,136
Adz	P. H. Bradley	Portland, Me	Aug. 24, 1869	93,957
Æolian-attachment	G. W. Ingalls	Concord, N. H	Dec. 23, 1851	8,608
Æolian-attachment, &c., Manner of adjusting the pitch of reeds for.	C. Horst	New Orleans, La	Sept. 27, 1845	4,210
Aërated liquids, Draft-pipe for	O. F. Stedman	Westfield, N. Y	Jan. 7, 1873	134,710
Aërated liquids, Mode of preparing	P. H. Vander Weyde	Philadelphia, Pa	Feb. 4, 1868	74,175
Aërated water, Manufacture of	G. McCoy	New York, N. Y	Apr. 23, 1867	64,019
Aërating and mixing substances, Apparatus for	E. L. Pratt	Boston, Mass	Nov. 6, 1866	59,449
Aërating liquid, Apparatus for	E. L. Pratt	Boston, Mass	Sept. 10, 1867	68,788
Aërating liquid, Apparatus for	T. Warker	New York, N. Y	Apr. 8, 1862	34,916
Aërating water	C. D. Simons and J. J. Riondel	Charleston, S. C	Aug. 16, 1810	....
Aërator, Flame	C. L. Browne	Washington, D. C	May 12, 1868	77,799
Aërial car	O. Abbruzzo	St. Margherita, Italy	July 21, 1868	80,107
Aërial car	I. W. Forbes	La Porte, Ind	July 16, 1872	129,401
Aërial car	F. Just and A. Koellener	Buffalo, N. Y	Oct. 10, 1865	50,365
Aërial car	A. P. Keith	Easton, Mass	Mar. 1, 1870	100,415
Aërial car	E. Oakes	Richmond, Ind	Aug. 30, 1870	106,862
Aërial carriage and way	D. Towse	Pittsburgh, Pa	Dec. 10, 1867	71,921
Aërial carriage and way	D. Towse	Pittsburgh, Pa	Dec. 10, 1867	71,922
Aërial carriage and way	D. Towse	Pittsburgh, Pa	Dec. 10, 1867	71,923
Aërial machine	L. C. Crowell	West Dennis, Mass	June 3, 1862	35,437
Aërial machine	J. A. Elston	Elston Station, Mo	Aug. 13, 1867	67,739
Aërial machine	A. Kinsella	Cascades, Wash	June 3, 1862	35,453
Aërial navigator	T. Moy and R. E. Shill	London and Mile End, England.	Nov. 26, 1872	133,381
Aërial navigator	Z. Stone	Kinsmans, Ohio	May 12, 1868	77,850
Aërial railway	J. A. A. Fountaine	New York, N. Y	Feb. 5, 1867	61,824
Aërial railway	R. Montgomery	New York, N. Y	Sept. 11, 1866	57,949
Aërial ship	M. Braun	Cape Vincent, N. Y	Feb. 8, 1870	99,629
Aërial transit, Method of	R. A. Cheesebrough	New York, N. Y	Feb. 1, 1870	99,406
Aërostat	S. Andrews	Perth Amboy, N. J	July 5, 1864	43,449
Aërostats or balloons of various forms, Manner of directing the course of.	M. Muzzi	Bologna, Italy	Oct. 16, 1844	3,799
Aërostatic jack	A. V. Ojeda	San Francisco, Cal	Aug. 29, 1871	118,475
Afghan	D. Bickford	Boston, Mass	July 21, 1868	80,122
Agare Americana, Machine for dressing the leaves of.	E. J. y Patrullo	Merida, Mexico	Apr. 23, 1861	32,143
Agare-plant, Machine for dressing the leaves of the.	E. J. y Patrullo	Merida, Mexico	Mar. 5, 1861	31,616
Age to wine, Process of imparting	J. Searle	San Francisco, Cal	July 11, 1865	48,728
Ageing alcoholic liquors and producing vinegar, Apparatus for.	R. D. Turner	New York, N. Y	Mar. 4, 1873	136,470
Ageing alcoholic liquor	H. Purdy	Burlington, Iowa	Oct. 18, 1870	108,388
Ageing alcoholic liquors, Mode of	J. L. Martin	Baltimore, Md	May 21, 1867	64,990
Ageing and mixing liquors, Apparatus for	S. C. Bruce	New York, N. Y	Jan. 9, 1872	122,513
Ageing and purifying spirits	J. M. Crafts	Boston, Mass	July 28, 1868	80,459
Ageing and refining wine and liquor, Apparatus for.	R. D. Turner	New York, N. Y	Sept. 24, 1867	69,275
Ageing-apparatus for whisky and other spirits	J. P. Greeley	Boston, Mass	Nov. 29, 1870	109,611
Ageing-apparatus for whisky and other spirits	W. P. Martin	Millersburgh, Ky	Oct. 12, 1869	95,703
Ageing liquors and spirits and for producing aromatic ethers, Process for.	C. L. Fleischmann	Washington, D. C	May 4, 1869	89,748
Ageing liquors, Apparatus for	R. D. Turner	New York, N. Y	July 16, 1872	129,440
Ageing liquors, Apparatus for mixing and	S. C. Bruce	New York, N. Y	Dec. 28, 1869	98,226
Ageing spirits	A. Caldwell	Lexington, Ky	Sept. 27, 1870	107,658
Ageing spirits, Apparatus for	E. L. Morse	Saint Louis, Mo	July 13, 1869	92,633
Ageing spirits, Apparatus for	P. M. Papin	Saint Louis, Mo	May 18, 1869	90,120
Ageing spirituous liquors, Apparatus for	R. D. Turner	New York, N. Y	Oct. 19, 1869	96,056
Ageing whisky and other spirits, Apparatus for	J. Peiffer and S. Richards	Valonia, Pa	Mar. 7, 1871	112,485
Ageing wine	F. Haeck	Brussels, Belgium	Dec. 4, 1866	60,179
Ageing wine and liquor, Apparatus for	A. Luquet and P. Huerne	San Francisco, Cal	May 14, 1872	126,722

Index of patents issued from the United States Patent Office from 1790 *to* 1873, *inclusive*—Continued.

Invention.	Inventor.	Residence.	Date.	No.
Ageing wine and spirits, Apparatus for	S. C. Bruce	New York, N. Y	Feb. 9, 1869	86, 640
Ageing wines and liquors	A. Luquet	San Francisco, Cal	Mar. 11, 1873	136, 741
Agitating and heating substances, Apparatus for	G. Nebeker	Wilmington, Del	June 26, 1866	55, 886
Agitating and kneading substances, Apparatus for.	W. Adamson	Philadelphia, Pa	Apr. 18, 1865	47, 264
Agricultural boiler	C. M. Cloud	Grinnell, Iowa	Sept. 2, 1873	142, 326
Agricultural boiler	F. Farquhar	Richmond, Ind	May 3, 1870	102, 671
Agricultural boiler	J. H. Haviland, G. W. Cronk, and J. F. Antisdel.	Janesville, Wis	Nov. 19, 1872	133, 097
Agricultural boiler	P. H. Inman and C. B. Withington.	Janesville, Wis	Jan. 16, 1872	122, 723
Agricultural boiler	H. A. Mears	Rockford, Ill	June 24, 1873	140, 213
Agricultural boiler	H. A. Mears	Rockford, Ill	Dec. 16, 1873	145, 517
Agricultural boiler	J. Murdock	South Carver, Mass	Nov. 14, 1871	120, 894
Agricultural boiler	D. R. Prindle	East Bethany, N. Y	July 7, 1868	79, 685
Agricultural boiler and steamer	R. S. Hazen, sr	Calmus, Iowa	Sept. 3, 1872	130, 996
Agricultural digger	A. L. Kennedy	Philadelphia, Pa	Nov. 8, 1870	109, 019
Agricultural fork	T. Beale	New Milford, Ill	Mar. 5, 1867	62, 594
Agricultural fork	C., A., and C. N. Clow	Port Byron, N. Y	Dec. 8, 1857	18, 804
Agricultural forks, shovels, and hoes, Handle of	R. M. Hine	Throopsville, N. Y	Oct. 28, 1856	15, 976
Agricultural implements, Blade for	W. Scott	Floyd Court-House, Va	Dec. 30, 1873	146, 026
Agricultural implements, Combined	L. Lehmann	Monee, Ill	Nov. 15, 1870	109, 226
Agricultural implements, Combined	C. R. Rand	Dubuque, Iowa	Feb. 14, 1871	111, 871
Agricultural implements, Disk for	E. T. Bussell	Indianapolis, Ind	June 11, 1872	127, 677
Agricultural implements mounted on wheels, Leveling attachment to.	B. F. Cook	Olema, Cal	Nov. 12, 1867	70, 804
Agricultural machine	H. Cowing	New Orleans, La	Aug. 4, 1868	80, 795
Agricultural steam-apparatus	E. C. Bellinger	Barnwell District, S. C	Nov. 19, 1833	....
Agricultural steam-boiler	L. S. Robbins	New York, N. Y	May 26, 1868	78, 235
Agriculture	V. Pardee	Trenton, N. Y	Feb. 6, 1829	....
Air and gas, Apparatus for mixing	T. C. Hopper	Philadelphia, Pa	Dec. 12, 1871	121, 719
Air and gas engine	A. H. De Villeneuve	Paris, France	Mar. 19, 1872	124, 671
Air and gas engine	D. Dick	Meadville, Pa	Apr. 9, 1867	63, 619
Air and gas engine	J. F. Haskins	Fitchburgh, Mass	July 16, 1872	129, 337
Air and gas engine	O. Trossin	Berlin, Germany	July 22, 1873	141, 189
Air and liquid cooler	D. E. Somes	Washington, D. C	Mar. 29, 1870	101, 392
Air and making ice, Apparatus for cooling	T. D. Kingan	Indianapolis, Ind	Dec. 16, 1873	145, 659
Air and other substances, Cooling	D. E. Somes	Washington, D. C	Nov. 12, 1867	70, 909
Air and steam engine	F. B. Blanchard	Waterville, Me	July 10, 1855	13, 209
Air and steam engine, Combined	O. M. Stillman	Westerly, R. I	Aug. 9, 1864	43, 803
Air and steam for actuating engine, Process for mixing.	W. M. Storm	New York, N. Y	Apr. 5, 1853	9, 654
Air and steam jet to promote combustion	G. M. Copeland	Brooklyn, N. Y	Feb. 26, 1867	62, 397
Air and supplying boilers therewith, Heating	G. E. Hibbard	Fond du Lac, Wis	Dec. 16, 1873	145, 568
Air and water warming apparatus for dwellings	L. C. St. John	Buffalo, N. Y	Oct. 7, 1851	8, 413
Air, Apparatus for compressing	W. Arthur	Brooklyn, N. Y	July 25, 1865	48, 886
Air, Apparatus for compressing	J. B. J. Mignon and S. H. Rouart.	Paris, France	Mar. 19, 1867	63, 075
Air, Apparatus for navigating the	C. McDermott	Monticello, Ark	Nov. 12, 1872	133, 046
Air, Apparatus for navigating the	W. F. Quimby	Stanton, Del	Nov. 26, 1861	33, 797
Air, Apparatus for purifying and cooling	A. S. Lyman	New York, N. Y	Jan. 19, 1864	41, 309
Air, Apparatus for removing dust and gas from	J. D. Whelpley and J. J. Storer	Boston, Mass	Mar. 6, 1866	53, 068
Air, Apparatus for supplying a continuous flow of	B. Rouquayrol	Paris, France	Mar. 20, 1866	53, 385
Air, Apparatus for transmitting power by the medium of.	H. Call	Concord, N. H	Aug. 24, 1869	93, 964
Air, Apparatus for transmitting power by the medium of.	H. Call	Concord, N. H	Jan. 18, 1870	98, 846
Air apparatus, Fresh	J. McNeven	New York, N. Y	Sept. 14, 1869	94, 905
Air bath, Compressed	P. T. Ware	Toronto, Canada	May 8, 1866	54, 655
Air bath, Hot and cold	S. M. Landis	Philadelphia, Pa	May 14, 1867	64, 677
Air-blast engine	J. Grimm	Saint Louis, Mo	Oct. 26, 1869	96, 223
Air-brake	T. Luce	Richmondville, Mich	Sept. 10, 1872	131, 286
Air-brake and signal, Steam-power	G. Westinghouse, jr	Pittsburgh, Pa	Mar. 5, 1872	124, 404
Air-brake for car	L. H. Dwelley	Dorchester, Mass	Nov. 28, 1865	51, 158
Air-brake for railway-car	C. Fogelberg	Boston, Mass	May 28, 1872	127, 332
Air-brake for railway-car	T. O. Ward	Paw Paw, Mich	Jan. 30, 1872	123, 312
Air-brake, Locomotive	G. Westinghouse, jr	Pittsburgh, Pa	Oct. 28, 1873	144, 005
Air-brake, Steam	G. Westinghouse, jr	Pittsburgh, Pa	Mar. 5, 1872	124, 405
Air-brake, Steam-power	H. L. McAvoy	Baltimore, Md	Apr. 29, 1873	138, 339
Air-brake, Steam-power	G. Westinghouse, jr	Pittsburgh, Pa	Jan. 23, 1872	123, 067
Air by exhaust-steam, Method of heating	A. C. Fletcher	New York, N. Y	May 19, 1863	38, 637
Air by steam, Apparatus for heating	J. Hollingsworth	Chicago, Ill	June 5, 1860	28, 578
Air-chamber	R. Creuzbaur	Travis County, Texas	June 18, 1861	32, 595
Air-compressing apparatus	B. T. Babbitt	New York, N. Y	May 17, 1870	103, 121
Air-compressing apparatus	B. T. Babbitt	New York, N. Y	Nov. 12, 1872	133, 004
Air compressing apparatus	F. S. Dumont	New York, N. Y	Apr. 27, 1869	89, 390
Air-compressing apparatus	G. D. Emerson	New York, N. Y	Aug. 20, 1872	130, 627
Air-compressing apparatus	J. Ericsson	New York, N. Y	Dec. 30, 1873	146, 055
Air-compressing apparatus	R. S. Pardée	San Diego, Cal	Oct. 14, 1873	143, 634
Air-compressing apparatus	J. S. Patric	Victor, N. Y	Apr. 18, 1865	47, 328
Air-compressing apparatus	J. S. Patric	Rochester, N. Y	Oct. 17, 1871	120, 094
Air-compressing apparatus	J. B. Waring	South Norwalk, Conn	July 16, 1872	129, 631
Air-compressing apparatus, Hydraulic	M. Hey	Philadelphia, Pa	Apr. 26, 1870	102, 397
Air-compressing apparatus, Hydraulic	C. Moore	Jersey City, N. J	July 1, 1873	140, 524
Air-compressing apparatus, Hydraulic	W. D. Seal	Washington, D. C	June 14, 1870	104, 362
Air-compressing engine	J. F. Haskins	Fitchburgh, Mass	Aug. 6, 1872	130, 296
Air-compressor	H. H. Day	New York, N. Y	Nov. 11, 1873	144, 390
Air-compressor	H. P. Fairfield	Boston, Mass	Sept. 2, 1873	142, 452
Air-compressor, Hydraulic	M. Hey	Philadelphia, Pa	July 25, 1871	117, 285
Air-compressor, Hydraulic	W. E. Prall	Washington, D. C	Nov. 7, 1871	120, 597
Air-compressor or blower	L. Chase	Portland, Me	Aug. 12, 1873	141, 762
Air-compressor or blower	L. Chase	Portland, Me	Sept. 30, 1873	143, 329
Air-condensing and storing by the pressure of water, Apparatus for.	J. Cochrane	New York, N. Y	July 23, 1872	129, 791
Air-condensing apparatus	H. J. Bailey	Pittsburgh, Pa	Jan. 14, 1868	73, 283
Air-condensing apparatus	H. Moore	Bangall, N. Y	Mar. 3, 1868	75, 042
Air-condensing apparatus	J. S. Patric	Rochester, N. Y	Oct. 17, 1871	120, 095
Air-cooling and ice-manufacturing machine	J. Kraffert	Hoboken, N. J	Dec. 27, 1870	110, 573

Index of patents issued from the United States Patent Office from 1790 *to* 1873, *inclusive*—Continued.

Invention.	Inventor.	Residence.	Date.	No.
Air-cooling apparatus	T. D. Kingan	Indianapolis, Ind	July 1, 1873	140, 375
Air-cooling apparatus	N. S. Shaler	Newport, Ky	May 30, 1865	47, 991
Air-cooling process	E. H. Grant	Washington, D. C	Feb. 16, 1869	87, 041
Air-cooling ventilator	F. Villard	Mount Eaton, Ohio	May 29, 1866	55, 180
Air-cushion for doors	J. W. Wetmore	Erie, Pa	Sept. 2, 1873	142, 540
Air cylinder, Compressed	G. W. W. Goodwyn	New Orleans, La	Mar. 1, 1870	100, 282
Air-draft, Subignis	S. Randall	Providence, R. I	Mar. 12, 1811	
Air-engine	W. Alworth	Scranton, Pa	May 28, 1872	127, 137
Air-engine	J. B. Atwater	Chicago, Ill	Mar. 13, 1866	53, 097
Air-engine	D. Bickford	Boston, Mass	June 6, 1865	48, 043
Air-engine	J. R. Cameron	Pittsburgh, Pa	Nov. 12, 1867	70, 800
Air-engine	P. Chick	Taunton, Mass	Dec. 18, 1866	60, 474
Air-engine	W. Denkmann	Washington, D. C	Dec. 16, 1862	37, 155
Air-engine	J. Ericsson	New York, N. Y	Nov. 4, 1851	8, 481
Air-engine	J. Ericsson	New York, N. Y	July 31, 1855	13, 348
Air-engine	J. Ericsson	New York, N. Y	Apr. 15, 1856	14, 690
Air-engine	J. Ericsson	New York, N. Y	Dec. 14, 1858	22, 287
Air-engine	J. Ericsson	New York, N. Y	Oct. 9, 1860	30, 306
Air-engine	E. Langen and N. A. Otto	Cologne, Prussia	Aug. 13, 1867	67, 659
Air-engine	F. J. Laubereau	Paris, France	Apr. 10, 1849	6, 301
Air-engine	J. J. E. Lenoir	Paris, France	Mar. 19, 1861	31, 722
Air-engine	R. Lord	Pawtucket, Mass	Sept. 17, 1861	33, 308
Air-engine	A. S. Lyman	New York, N. Y	Feb. 28, 1854	10, 576
Air-engine	T. McDonough	Middletown, Conn	Sept. 23, 1856	15, 771
Air-engine	H. Messer	Roxbury, Mass	Jan. 6, 1863	37, 299
Air-engine	D. Myers	Chicago, Ill	Nov. 15, 1870	109, 338
Air-engine	J. R. Napier and W. J. M. Rankine.	Glasgow and Govan Parish, Great Britain.	Sept. 19, 1854	11, 696
Air-engine	H. Norman and C. F. Dietrich	New Orleans, La	Feb. 25, 1873	136, 259
Air-engine	H. M. Paine	Worcester, Mass	Nov. 30, 1858	22, 219
Air-engine	J. R. Peters	New York, N. Y	Nov. 18, 1862	36, 964
Air-engine	B. F. Rice	Clinton, Mass	Apr. 5, 1859	23, 495
Air-engine	A. K. Rider	New York, N. Y	Jan. 17, 1871	111, 088
Air-engine	A. K. Rider	New York, N. Y	Oct. 24, 1871	120, 325
Air-engine	P. Shaw	East Abington, Mass	May 2, 1854	10, 868
Air-engine	O. M. Stillman	Stonington, Conn	June 26, 1860	28, 910
Air-engine	O. M. Stillman	Stonington, Conn	June 26, 1860	28, 911
Air-engine	S. Wilcox, jr	Westerly, R. I	May 3, 1859	23, 876
Air-engine	S. Wilcox, jr	Westerly, R. I	Nov. 20, 1860	30, 700
Air-engine	S. Wilcox, jr	Westerly, R. I	Nov. 20, 1860	30, 701
Air-engine	S. Wilcox, jr	Westerly, R. I	Sept. 19, 1865	50, 061
Air-engine	A. O. Wilcox	Philadelphia, Pa	July 19, 1853	9, 871
Air-engine	J. A. Woodbury, J. Merrill, and G. Patten.	Winchester, Boston, and Charlestown, Mass.	May 17, 1853	9, 739
Air-engine	J. A. Woodbury, J. Merrill, and G. Patten.	Winchester and Boston, Mass.	Oct. 4, 1853	10, 081
Air engine and motor	P. A. Ensign	Adrian, Mich	Oct. 2, 1866	58, 397
Air engine, Compressed	D. Bickford	Westerly, R. I	May 15, 1860	28, 248
Air engine, Compressed	A. M. Smith	Chicago, Ill	Aug. 8, 1871	117, 825
Air engine, Compressed	W. M. Storm	Troy, N. Y	Sept. 23, 1851	8, 380
Air engine, Compressed	W. C. Turnbull	Baltimore, Md	Apr. 17, 1860	27, 938
Air engine, Condensed	A. Parsey	London, England	July 31, 1847	5, 205
Air-engine or fan-blower	G. C. Hawkins	Boston, Mass	Jan. 10, 1871	110, 915
Air-engines, Air-compressing apparatus for	J. McLeish	Philadelphia, Pa	Dec. 10, 1872	133, 713
Air-engines, Method of distributing the air over the heating and cooling surfaces of.	E. Buckup	New York, N. Y	July 2, 1850	7, 470
Air-flue register	W. L. McDowell	Philadelphia, Pa	Feb. 8, 1870	99, 692
Air flue with the chimney-pipes, Descending	J. Macomber	Greenwich, N. Y	Nov. 19, 1811	
Air for illuminating purposes, Apparatus for carbonizing.	W. A. Simonds	Boston, Mass	June 21, 1864	43, 264
Air for illuminating purposes, Apparatus for carbonizing.	W. A. Simonds	Boston, Mass	Oct. 4, 1864	44, 560
Air for preserving animal and vegetable substances, Mode of drying and purifying.	E. D. Brainard	Albany, N. Y	July 16, 1867	66, 786
Air for use in the arts, Heating	J. A. Morrell	New York, N. Y	Sept 2, 1873	142, 497
Air from cases, Apparatus for expelling	D. Clavidge	Indianapolis, Ind	May 15, 1866	54, 686
Air from liquors on tap, Method of excluding	A. F. Boyd	Muskingum County, Ohio	Apr. 21, 1857	17, 073
Air-furnace for melting ores and refining and heating iron with stone-coal.	T. Gregg	Connellsville, Pa	Mar. 16, 1814	
Air, gas, &c., Apparatus for compressing	S. Bidwell	Philadelphia, Pa	Nov. 8, 1864	44, 930
Air-heater	W. Sage	Durham, Conn	Feb. 20, 1855	12, 425
Air heater and cooler	L. Perkins	London, England	July 10, 1866	56, 338
Air-heater and steam-condenser, Compound	B. F. Sturtevant	Jamaica Plain, Mass	Feb. 22, 1870	100, 240
Air-heater and steam-condenser, Compound	B. F. Sturtevant	Jamaica Plain, Mass	Feb. 22, 1870	100, 241
Air-heater and steam-condenser, Compound	B. F. Sturtevant	Jamaica Plain, Mass	Feb. 22, 1870	100, 242
Air-heater, Tubular	R. R. Hawley	Normal, Ill	Sept. 1, 1868	81, 781
Air-heating apparatus, Compressed	H. Bushnell	New Haven, Conn	July 1, 1873	140, 466
Air-heating furnace	J. Barker	Baltimore, Md	July 7, 1846	4, 622
Air-heating furnace	J. Barker	Baltimore, Md	Feb. 8, 1848	5, 436
Air-heating furnace	E. Barrow	New York, N. Y	Dec. 11, 1845	4, 301
Air-heating furnace	R. L. Bate	Adrian, Mich	Apr. 1, 1862	34, 856
Air-heating furnace	J. Bradley	Saint Albans, Vt	June 24, 1844	3, 636
Air-heating furnace	W. Bryent	Boston, Mass	Oct. 24, 1854	11, 847
Air-heating furnace	H. Bushnell	Hartford, Conn	Mar. 27, 1849	6, 238
Air-heating furnace	P. H. Carman	Brooklyn, N. Y	June 29, 1869	91, 908
Air-heating furnace	G. Chilson	Boston, Mass	Nov. 19, 1850	7, 780
Air-heating furnace	R. T. Crane	Chicago, Ill	Mar. 20, 1860	27, 530
Air-heating furnace	D. Culver	Hartford, Conn	May 15, 1847	5, 114
Air-heating furnace	H. A. Engels	Cincinnati, Ohio	Feb. 12, 1850	7, 082
Air-heating furnace	C. Fowler	Hartford, Conn	July 10, 1829	
Air-heating furnace	G. Fox	Hartford, Conn	Mar. 10, 1843	2, 994
Air-heating furnace	J. D. Green and E. Ivers	Philadelphia, Pa	Jan. 6, 1857	16, 348
Air-heating furnace	W. H. Harris	Grand Rapids, Mich	Sept. 22, 1863	40, 033
Air-heating furnace	C. R. Harvey	New York, N. Y	Jan. 24, 1854	10, 447
Air-heating furnace	J. P. Hayes	Philadelphia, Pa	June 22, 1858	20, 640
Air-heating furnace	W. Hickok	New York, N. Y	May 15, 1847	5, 118
Air-heating furnace	T. D. Ingersoll	Monroe, Mich	Mar. 30, 1858	19, 781
Air-heating furnace	A. Janes	New York, N. Y	Aug. 9, 1845	4, 140

Index of patents issued from the United States Patent Office from 1790 *to* 1873, *inclusive*—Continued.

Invention.	Inventor.	Residence.	Date.	No.
Air-heating furnace	J. Liddle	New York, N. Y	Aug. 26, 1856	15, 613
Air-heating furnace	J. MacGregor, jr	Wilton, N. Y	Mar. 5, 1850	7, 143
Air-heating furnace	E. D. Norcross	Augusta, Me	Feb. 24, 1852	8, 758
Air-heating furnace	J. S. Sumner	Newton, Mass	Mar. 5, 1872	124, 229
Air-heating furnace	J. H. Sutton	Honesdale, Pa	Mar. 13, 1855	12, 533
Air-heating furnace	J. M. Thatcher	Bergen, N. J	Nov. 19, 1867	71, 244
Air-heating furnace	O. Tiffany	New York, N. Y	Mar. 20, 1849	6, 205
Air-heating furnace	G. W. Wilson	Chelsea, Mass	Feb. 28, 1865	46, 601
Air-heating furnace	G. W. Wilson	Chelsea, Mass	June 11, 1867	65, 784
Air-heating furnace	J. M. Wilson	Philadelphia, Pa	Sept. 2, 1873	142, 363
Air-heating pipe, Connecting the joints of	J. Young	Franklin Furnace, Ohio	Mar. 7, 1854	10, 617
Air in dwellings, &c., Mode of securing pure and wholesome.	D. E. Somes	Washington, D. C	Oct. 11, 1864	44, 671
Air in heated apartments, Apparatus to effect the hydration of.	P. J. Schopp	Louisville, Ky	May 31, 1870	103, 665
Air, Keeping air-springs supplied with	S. G. Randall	Middlebury, Vt	Apr. 5, 1859	23, 497
Air, Machine for blowing uniform currents of	J. Griffin	Harpersfield, N. Y	Mar. 1, 1859	23, 084
Air, Moistening, cooling, and warming	D. E. Somes	Washington, D. C	Feb. 5, 1867	61, 886
Air-navigating apparatus	T. Green	New York, N. Y	Aug. 12, 1873	141, 785
Air of rooms with antiseptic vapors, Apparatus for impregnating the.	A. J. Sax	Paris, France	June 20, 1865	48, 352
Air or gas engine	P. Shearer	Reading, Pa	Sept. 3, 1861	33, 215
Air, Paper reservoir for compressed	C. W. Wailey	New Orleans, La	Sept. 1, 1868	81, 713
Air-pipe, Furnace	J. P. Dawson	Des Moines, Iowa	July 9, 1872	128, 864
Air-power	S. Morey	Philadelphia, Pa	Apr. 8, 1812	......
Air, Process for cooling	D. E. Somes	Washington, D. C	Oct. 3, 1865	50, 286
Air-purifying apparatus in the manufacture of fertilizers.	J. E. Cox	Cincinnati, Ohio	Mar. 29, 1870	101, 353
Air-register	H. F. Hayden	Washington, D. C	Dec. 19, 1871	121, 942
Air-register	E. A. Tuttle	Brooklyn, N. Y	Nov. 14, 1871	120, 912
Air-spring	G. M. Alsop	Philadelphia, Pa	Sept. 23, 1862	36, 498
Air-spring	J. Corriston	Sandusky, Ohio	Oct. 13, 1868	83, 043
Air-spring for railway-cars	P. G. Gardiner	New York, N. Y	June 1, 1869	90, 657
Air spring for railway-cars, &c., Compressed	L. Bissell	Newark, N. J	Oct. 11, 1841	2, 307
Air-springs, Automatic vent-opener for	J. Walther	Brooklyn, N. Y	May 27, 1873	139, 346
Air-supplying apparatus	H. W. Adams	Philadelphia, Pa	Dec. 27, 1870	110, 414
Air-tight box, case, &c	J. G. Staunton	Buffalo, N. Y	Jan. 3, 1865	45, 766
Air-tight can	C. Barry	Philadelphia, Pa	Aug. 18, 1868	81, 243
Air-tight can	W. J. Gordon	Philadelphia, Pa	Mar. 16, 1869	87, 840
Air-tight jar	G. M. Ramsey	New York, N. Y	Nov. 19, 1867	71, 215
Air-tight vessel, Constructing	T. C. Taylor	Philadelphia, Pa	Feb. 27, 1866	52, 909
Air to air-chambers, Mode of supplying	P. H. Vander Weyde	Philadelphia, Pa	Aug. 21, 1866	57, 412
Air to be used as a motive-power, Apparatus for heating and cooling.	W. Hidden and J. Reeves	New York, N. Y	Aug. 10, 1858	21, 133
Air to furnaces, Applying heated	J. Silsbe	Tyrone, N. Y	Aug. 8, 1837	343
Air-trap for steam and other enginery	G. W. Blake	New York, N. Y	July 13, 1869	92, 571
Air, Treating diseases by condensed	O. Stone	Rochester, N. Y	Oct. 24, 1865	50, 641
Air with gasoline, Apparatus for charging	B. Pickering	Milton, Ohio	Dec. 11, 1866	60, 417
Air with hydrocarbon vapors, Apparatus for charging.	H. M. Paine	Newark, N. J	Feb. 23, 1869	87, 192
Air with hydrocarbon vapors, Machine for charging.	H. B. Myer	Cleveland, Ohio	Feb. 27, 1866	52, 876

Alarm:
See Bell alarm.
Blast-furnace alarm.
Box-alarm.
Burglar-alarm.
Clock-alarm.
Door-bell alarm.
Electric-alarm.
Electric fire and burglar alarm.
Electro-magnetic alarm.
Electro-magnetic house-alarm.
Electro-magnetic temperature-alarm.
Fire-alarm.
Fog-alarm.
Hat-body-machine alarm.
High-water alarm.
Hotel-alarm.
House-alarm.
Leak-alarm.
Letter-box alarm.
Lock-alarm.
Locomotive-engine alarm.
Low-water alarm.
Milk-boiler alarm.
Money-drawer alarm.
Nautical alarm.
Overflow-alarm.
Pickpocket-alarm.
Pocket-alarm.
Portable alarm.
Portable-box alarm.
Portable burglar-alarm.
Pressure-alarm.
Railway-alarm.
Railway-switch alarm.
Railway-telegraph alarm.
Steam-alarm.
Steam-boiler alarm.
Steam-pressure alarm.
Switch-alarm.
Target-alarm.
Thill-alarm.
Tidal alarm.
Till-alarm.
Till-lock alarm.
Time-alarm.
Vessel-alarm.

Index of patents issued from the United States Patent Office from 1790 to 1873, inclusive—Continued.

Invention.	Inventor.	Residence.	Date.	No.
Alarm-attachment	J. H. Thorp	New York, N. Y	Dec. 6, 1870	109, 971
Alarm-gage for steam-boiler	J. Whitmore	Lowell, Mass	July 27, 1858	21, 040
Alarm-gage for steam-boiler, Steam and water	G. W. Grader and B. F. Cowan	Memphis, Tenn	Dec. 14, 1858	22, 287
Alarm-lock	S. A. Andrews	Farmers' Valley, Wis	Sept. 18, 1866	58, 043
Alarm-lock	F. Brewster	Cleveland, Ohio	Mar. 16, 1869	87, 905
Alarm-lock	G. N. Bruster	Factoryville, N. Y	May 19, 1863	38, 556
Alarm-lock	J. Cone	Yellow Springs, Ohio	June 24, 1856	15, 168
Alarm-lock	A. B. Crane	Newton, Mass	July 16, 1872	129, 213
Alarm-lock	C. Fleischel	New York, N. Y	June 15, 1852	9, 019
Alarm-lock	E. Herbster	Chicago, Ill	May 4, 1869	89, 762
Alarm-lock	H. L. Hervey	Windsor, Conn	Feb. 9, 1858	19, 295
Alarm-lock	H. L. Hervey	Windsor, Conn	Apr. 13, 1858	19, 926
Alarm-lock	B. F. Irvine and T. A. Hitchcock.	North La Crosse, Wis	Dec. 7, 1869	97, 513
Alarm-lock	H. Lockwood	New York, N. Y	Apr. 12, 1859	23, 591
Alarm-lock	C. E. Pierce	New York, N. Y	Sept. 26, 1871	119, 403
Alarm-lock	J. S. and R. Porter	Waterford, N. Y	Mar. 5, 1867	62, 683
Alarm-lock	C. Schnepf	Marietta, Ohio	Oct. 31, 1871	120, 541
Alarm-lock	N. Seubert	Syracuse, N. Y	June 1, 1869	90, 962
Alarm-lock	N. Seubert	Syracuse, N. Y	Aug. 24, 1869	94, 137
Alarm-lock	T. P. Sink	Fairton, N. J	Dec. 6, 1870	109, 954
Alarm-lock	G. J. Swingle	Davenport, Iowa	Jan. 30, 1872	123, 305
Alarm-lock	J. T. Taylor	Newnan, Ga	Oct. 17, 1871	120, 125
Alarm-lock	J. H. Thorp	New York, N. Y	Mar. 21, 1871	112, 987
Alarm-lock	S. I. Trask	Guilford Centre, N. Y	Feb. 5, 1856	14, 209
Alarm-lock	O. D. Warner	Bristol, Conn	Jan. 19, 1869	86, 116
Alarm-lock	J. W. Wells	Pittsburgh, Pa	Sept. 7, 1858	21, 457
Alarm-lock	J. Ziegler	Baltimore, Md	Apr. 10, 1860	27, 856
Alarm-lock, Burglar	N. Cheek	Chapel Hill, N. C	Aug. 4, 1868	80, 599
Alarm-lock, Burglar	D. Edwards	McConnellsville, Ohio	Aug. 12, 1840	1, 722
Alarm-lock, Burglar	S. T. Heiminway	Saratoga Springs, N. Y	Jan. 7, 1868	73, 182
Alarm-lock, Burglar	A. Isensee	Indianapolis, Ind	Jan. 28, 1868	73, 897
Alarm-lock, Burglar	G. Jacobs	Washington, D. C	June 25, 1867	66, 090
Alarm-lock, Electrical	F. Girard	Havana, Cuba	Mar. 14, 1871	112, 585
Alarm-lock for doors	A. Iske	Lancaster, Pa	Jan. 7, 1868	73, 013
Alarm-lock for money-drawers	G. Kimball	Cleveland, Ohio	Oct. 18, 1870	108, 363
Alarm-lock for tills	D. K. Miller	Bernville, Pa	Apr. 16, 1867	63, 922
Alarm-lock for tills	C. Tucker	Bloomington, Ill	May 7, 1867	64, 598
Album	A. Hathaway	Charlestown, Mass	Aug. 6, 1867	67, 540
Album-case	C. E. Rankin	New York, N. Y	May 6, 1862	35, 173
Album-clip	C. Weil	New York, N. Y	Mar. 29, 1864	42, 134
Album, Construction of	J. D. Mets	Dubuque, Iowa	Apr. 4 1865	47, 120
Album, Machine for cutting the front edge of photographic.	S. D. Burlock	Philadelphia, Pa	Dec. 6, 1864	45, 314
Album, Photograph	J. F. Tapley	Springfield, Mass	Aug. 9 1870	106, 232
Album, Photographic	J. W. Beackley	Philadelphia, Pa	May 31 1864	42, 922
Album, Photographic	E. D. Griggs	Waterbury, Conn	May 26, 1862	35, 310
Album, Photographic	F. R. Grunel	Geneva, Switzerland	May 14, 1861	32, 287
Album, Photographic	W. Matthews	New York, N. Y	July 29, 1862	36, 011
Album, Photographic	E. Maynard	Washington, D. C	Aug. 2, 1864	43, 696
Album, Photographic	J. D. Mets	Dubuque, Iowa	July 21, 1863	39, 300
Album, Photographic	J. D. Mets	Dubuque, Iowa	Nov. 24, 1863	40, 702
Album, Photographic	C. E. Prétat	New York, N. Y	Nov. 1, 1864	44, 890
Album, Photographic	J. F. Tapley	Springfield, Mass	Mar. 16, 1869	87, 804
Album, Photographic	R. Van Velthoven	Philadelphia, Pa	Oct. 30, 1866	59, 323
Album, Photographic	R. Van Velthoven and J. H. Hazzard.	Philadelphia, Pa	Oct. 17, 1865	50, 521
Album, Porcelain and card	J. C. Spooner	Springfield, Mass	Mar. 20, 1866	53, 354
Album, Stereoscope and photographic	J. Q. A. Tresize	Zanesville, Ohio	Nov. 17, 1863	40, 654
Albumen and prussiate of potash from blood, Manufacture of.	A. H. Hirsh	Chicago, Ill	Nov. 28, 1865	51, 181
Albumen from blood, Process of purifying and decoloring.	P. Jacques	Paris, France	Dec. 21, 1869	98, 165
Albumen, Manufacture of	G. Bourgade	New York, N. Y	Sept 6, 1870	107, 158
Albumen, Preparation of	G. Bourgade	New York, N. Y	Aug 9, 1870	107, 112
Albumen, Process for manufacturing	J. M. Fuchs	New York, N. Y	Dec 24, 1867	72, 625
Alcohol and other pure distillates, Manufacture of	J. F. Collins	New York, N. Y	Jan. 1, 1867	60, 835
Alcohol and other spirits, Manufacture of	J. F. Collins	New York, N. Y	Nov 20, 1866	59, 891
Alcohol and spirits, Purifying	C. C. Parsons	New York, N. Y	June 14, 1870	104, 343
Alcohol, Apparatus for heating by vapor of	T. K. Anderson	Painted Post, N. Y	Feb 13, 1849	6, 107
Alcohol, Apparatus for purifying	C. J. Falkmann	London, Great Britain	Nov. 7, 1865	50, 884
Alcohol, Apparatus for rectifying	A. A. Foubert and J. G. Béquet	St. Helier, England, and Paris, France.	Oct. 24, 1865	50, 668
Alcohol, Apparatus for rectifying	H. Lamotte	London, England	Oct. 23, 1866	59, 149
Alcohol, Apparatus for rectifying	M. Thompson	Saint Louis, Mo	Aug 9, 1864	43, 805
Alcohol, &c., Distillation of	A. Fries	Cincinnati, Ohio	July 11, 1865	48, 711
Alcohol for burning-fluid, Carbonated	S. Casey	Lebanon, Me	Mar 17, 1834	....
Alcohol from apples, Extracting	A. Wolcott	East Bloomfield, N. Y	Oct. 6, 1835	....
Alcohol from common spirits by steam, Extracting	A. and N. Wolcott	Bloomfield, N. Y	Mar 19, 1827	....
Alcohol from olefiant gas, Mode of manufacturing	E. A. Cotelle	Paris, France	Feb 23, 1864	41, 685
Alcohol from water and other heavier fluids, Separating.	B. F. Greenough	Cincinnati, Ohio	Dec. 20, 1853	10, 349
Alcohol from whisky, Apparatus for separating	A. V. H. Webb	New York, N. Y	Aug. 28, 1841	2, 234
Alcohol, Method of treating grain for the manufacture of.	W. M. Watson	Tonica, Ill	Dec. 5, 1865	51, 369
Alcohol, Process for purifying	L. Atwood	Boston, Mass	Aug. 23, 1853	9, 951
Alcohol, &c., Process of purifying	C. C. Parsons	New York, N. Y	July 13, 1869	92, 640
Alcohol, &c., Process of treating maize, barley, and other cereals for the manufacture of.	A. Fleischmann	Olmütz, Moravia, Austria	July 12, 1864	43, 555
Alcoholic and other spirits, Apparatus for treatment of.	R. D. Turner	New York, N. Y	Feb. 14, 1871	111, 791
Alcoholic liquid, Apparatus for concentrating and analyzing.	F. Haeck	Brussels, Belgium	Nov. 8, 1864	45, 002
Alcoholic liquid, Process of purifying and refining	F. Schleifer	San Francisco, Cal	July 27, 1869	93, 127
Alcoholic spirits from tomatoes, Manufacture of	J. S. Williams	Cinnaminson, N. J	Nov. 24, 1868	84, 455
Alcoholic spirits, Manufacture of	J. Neely and S. Allen	Buckingham County, Va	May 28, 1867	65, 108
Alcoholic spirits, Manufacture of	I. J. Rolfe and J. Rogers	Nevada City, Cal	Feb. 6, 1872	122, 423
Alcoholic spirits, Manufacture of	M. Thompson	Saint Louis, Mo	June 28, 1864	43, 348

Index of patents issued from the United States Patent Office from 1790 *to* 1873, *inclusive*—Continued.

Index of patents issued from the United States Patent Office from 1790 *to* 1873, *inclusive*—Continued.

Invention.	Inventor.	Residence.	Date.	No.
Alloy of metals in forming water-pipes, &c., Art of ufacturing and uniting.	W. A. Shaw	New York, N. Y	Dec. 31, 1867	72, 919
Alloy of nickel, zinc, and copper	H. W. Wright	Glastenbury, Conn	Aug. 5, 1873	141, 530
Alloy or bell-metal	H. L. Macker	Boston, Mass	Aug. 27, 1872	130, 814
Alloy or bronze	G. M. Levi and C. M. Küzel	Brussels and Liege, Belgium	May 23, 1871	115, 220
Alloy or metallic compound for bearings, &c	C. Adams	Pittsburgh, Pa	Nov. 5, 1872	132, 744
Alloy to imitate precious metals, Metallic	L. Sibert	Staunton, Va	Mar. 15, 1870	100, 937
Alloy to imitate silver, &c	H. L. Macker	Boston, Mass	Jan. 23, 1872	122, 901
Alloy to imitate silver	A. Schmitte and H. A. Levallois	Paris, France	Nov. 19, 1867	71, 072
Alloy to resemble silver	A. E. P. Baudvin	Paris, France	Sept. 16, 1873	142, 760
Alloys of aluminium and vulcanite, Combination of	N. C. Fowler	Yarmouth Port, Mass	Feb. 14, 1865	46, 347
Alloying copper, iron, &c., by cementation, Mode of	M. Sorel	Paris, France	Sept. 17, 1838	924
Allumettes, Machine for making paper	B. B. Lehman	Lebanon, Pa	Mar. 26, 1867	63, 265
Almanac, Perpetual	W. Gibson	Lanark, Scotland	Oct. 10, 1865	50, 430
Almond-peeling machine	H. Wathew	Philadelphia, Pa	Oct. 30, 1866	59, 324
Alphabet block	S. L. Hill	Williamsburgh, N. Y	Nov. 13, 1866	59, 603
Alphabet-block	E. H. Muldaur	Dover, Del	July 8, 1873	147, 722
Alphabet-frame	E. B. Nourse	Eaton, Ohio	Oct. 2, 1866	58, 463
Alphabeticon	J. H. R. Reffelt	Hoboken, N. J	Jan. 2, 1866	51, 864
Altiscope	J. Clark	Philadelphia, Pa	Mar. 13, 1866	53, 115
Altitude-instrument	H. Colby	New York, N. Y	Feb. 22, 1870	100, 119
Altitude of the sun, Instrument for taking	F. Yeiser	Lexington, Ky	Feb. 8, 1859	22, 913
Alum and fertilizers from mineral phosphates, Manufacture of.	P. Spence	Newton Heath, Manchester, Great Britain.	Dec. 13, 1870	110, 084
Alum from lignite, Mode of making	G. Troost	Philadelphia, Pa	Mar. 3, 1817	......
Alum-making, Preparing clay for	H. D. Pochin	Salford, England	Oct. 21, 1856	15, 934
Alum, Porous	H. Pemberton	Allegheny City, Pa	Oct. 6, 1868	82, 747
Alum, Process for manufacturing	J. H. Wurtz	New York, N. Y	Oct. 22, 1850	7, 737
Alumina, alum, and other aluminous substances, Manufacture of sulphate of.	H. Pemberton	Allegheny City, Pa	Jan. 1, 1867	60, 780
Alumina, Manufacture of acetate of	G. T. Lewis	Philadelphia, Pa	Oct. 30, 1866	59, 238
Alumina, Manufacture of sulphate of	H. Pemberton	Allegheny City, Pa	May 19, 1868	78, 005
Alumina, Manufacture of sulpho-acetate of	G. T. Lewis	Philadelphia, Pa	Oct. 30, 1866	59, 239
Alumina, Process of manufacturing sulphate of	M. J. Funcke	Eichelskamp, Prussia	Jan. 23, 1841	1, 945
Alumina, Treatment of sulphate of	A. A. Croll	London, England	May 28, 1867	65, 175
Aluminium and calomel, Manufacture of	L. T. Corbelli and V. Raitti	Florence, Tuscany, and Modena.	Oct. 26, 1858	21, 923
Aluminium, Apparatus for and method of casting	J. B. Bean	Baltimore, Md	Sept. 3, 1867	68, 548
Aluminium, Electro-deposition of	J. A. Jeançon	Newport, Ky	July 15, 1873	140, 924
Aluminium, Preparation of	L. T. Corbelli and V. Raitti	Florence, Tuscany, and Modena.	Oct. 26, 1858	21, 922
Aluminium with vulcanite and other materials, Combining.	N. C. Fowler	Yarmouth, Mass	Feb. 7, 1865	46, 230
Amalgam and mercury from ore-pulp, Machine for collecting.	Z. Wheeler	San Francisco, Cal	July 14, 1863	39, 251
Amalgam, Device for straining gold and silver	T. Varney	San Francisco, Cal	Mar. 18, 1862	34, 708
Amalgam for coating harness-trimmings, &c	H. L. Macker	Boston, Mass	Feb. 13, 1872	123, 712
Amalgam, Process for refining	W. Brückner	San Francisco, Cal	Mar. 21, 1865	46, 875
Amalgam, Separating and collecting gold and silver.	S. W. Wood	Cornwall, N. Y	Feb. 9, 1864	41, 565
Amalgamating and collecting gold and silver	H. W. Adams and W. S. Worthington.	New York and Newtown, N. Y.	Feb. 16, 1864	41, 588
Amalgamating and collecting gold and silver from ore, Apparatus for.	W. P. Parrott and J. J. Bordman.	Boston, Mass	June 11, 1867	65, 593
Amalgamating-apparatus	W. H. Butler	Chicago, Ill	Sept. 19, 1865	49, 975
Amalgamating-apparatus	J. N. Phelps	New York, N. Y	Oct. 18, 1864	44, 775
Amalgamating-apparatus, Method of elevating the mercury in.	O. Hofmann	Ellsworth, Nev	July 2, 1872	128, 623
Amalgamating gold and silver	A. K. Eaton	New York, N. Y	Apr. 4, 1854	10, 734
Amalgamating gold and silver	S. Longman	Brooklyn, N. Y	Aug. 17, 1858	21, 204
Amalgamating gold and silver	I. M. Phelps	Chicago, Ill	July 11, 1871	116, 865
Amalgamating gold and silver	G. B. Simpson	Washington, D. C	Feb. 6, 1866	52, 456
Amalgamating gold and silver, Apparatus for	J. S. Diltz	Mount Ophir, Cal	Mar. 1, 1864	41, 763
Amalgamating gold and silver, Apparatus for	R. McCully	Philadelphia, Pa	Feb. 23, 1869	87, 276
Amalgamating gold and silver, Apparatus for	J. Oliver	Ophir, Cal	Nov. 12, 1872	132, 919
Amalgamating gold and silver, Apparatus for	H. H. Scoville and P. W. Gates	Chicago, Ill	Sept. 20, 1864	44, 343
Amalgamating gold and silver, Apparatus for	J. J. Storer	Boston, Mass	Nov. 1, 1864	44, 898
Amalgamating gold and silver, Machine for	A. B. Crosby	Boston, Mass	Feb. 28, 1865	46, 546
Amalgamating gold and silver, Machine for	J. C. Dickey	Saratoga Springs, N. Y	Nov. 22, 1864	45, 144
Amalgamating gold and silver, Machine for	J. Hoeniger	New York, N. Y	July 15, 1862	35, 876
Amalgamating gold and silver, Machine for	T. Varney	San Francisco, Cal	July 15, 1862	35, 904
Amalgamating gold and silver, Process of	A. B. Paul and J. L. Wood	San Francisco and Independence, Cal.	Oct. 19, 1869	95, 931
Amalgamating gold and silver, Process of	J. Tunbridge	Newark, N. J	Dec. 16, 1873	145, 603
Amalgamating gold and silver, Process of	J. N. Wyckoff	New York, N. Y	Oct. 3, 1865	50, 296
Amalgamating gold and silver ores, Apparatus for	L. E. Rivot	Paris, France;	Jan. 28, 1868	73, 839
Amalgamating gold and silver ores, Process of	P. T. G. Stockman	Brooklyn, N. Y	Dec. 5, 1871	121, 554
Amalgamating gold and silver with lead, Apparatus for.	H. L. Fulton	Chicago, Ill	June 5, 1866	55, 273
Amalgamating gold in tailings, Apparatus for	T. A. Pratt	Marysville, Cal	Nov. 28, 1871	121, 296
Amalgamating gold, Machine for	F. N. Du Bois	Chicago, Ill	Apr. 15, 1862	34, 948
Amalgamating gold, &c., Process for	F. N. Du Bois	Black Hawk, Colo	Apr. 3, 1866	53, 590
Amalgamating gold, Process for	C. C. Knowles	Lowell, Mass	Aug. 6, 1850	7, 546
Amalgamating gold, &c., shaking and rocking table for.	A. Behr and W. J. Ward	Black Hawk, Colo	Apr. 25, 1865	47, 384
Amalgamating gold with mercury	M. Foreman and J. R. Mathewson.	Philadelphia, Pa	Sept. 18, 1866	58, 088
Amalgamating-machine	M. B. Dodge	Black Hawk Point, Colo	May 3, 1864	42, 568
Amalgamating-machine	E. Heath	San Francisco, Cal	Feb. 17, 1863	37, 688
Amalgamating machine, Gold and silver	Z. Wheeler	San Francisco, Cal	Dec. 8, 1863	40, 874
Amalgamating ores, Apparatus for	A. Bassett	Virginia, City, Nev	Jan. 22, 1867	61, 383
Amalgamating ores, Apparatus for	J. Tunbridge	Newark, N. J	Dec. 12, 1871	121, 827
Amalgamating ores, &c., Apparatus for	G. D. Wyckoff	Oil City, Pa	Mar. 19, 1872	124, 873
Amalgamating ores of silver, Process of	W. R. Frink	Virginia City, Nev	Aug. 30, 1864	43, 983
Amalgamating ores of the precious metals, Method of	F. Libben	New York, N. Y	July 2, 1861	32, 686
Amalgamating ores, Process for	W. F. Stewart	Austin, Nev	Dec. 20, 1864	45, 534
Amalgamating-pan	S. W. Bullock	Elizabeth, N. J	June 27, 1865	48, 480

Index of patents issued from the United States Patent Office from 1790 *to* 1873, *inclusive*—Continued.

Invention.	Inventor.	Residence.	Date.	No.
Amalgamating-pan	I. S Parke	Virginia City, Nev	Oct. 14, 1873	143, 635
Amalgamating-pan, &c., Composition for coating	C. H. Golding	Virginia City, Nev	Jan. 16, 1866	52, 040
Amalgamating-pan for gold and silver ores	I. S. Parke	Virginia City, Nev	Apr. 18, 1871	113, 791
Amalgamating-riffle	J. S. Briggs	Michigan Bluffs, Cal	Mar. 1, 1859	23, 072
Amalgamating the precious metals	J. B. Beers	San Francisco, Cal	Aug. 20, 1867	67, 842
Amalgamating the precious metals	H. J. Smith	Boston, Mass	July 9, 1867	66, 529
Amalgamating the precious metals	L. R. Streeter	Lowell, Mass	May 29, 1855	12, 988
Amalgating the precious metals and preventing the loss of mercury.	J. S. Phillips	San Francisco, Cal	May 16, 1871	114, 848
Amalgamating the precious metals, Apparatus for	J. B. Atwater	Chicago, Ill	Dec. 15, 1863	40, 894
Amalgamating the precious metals, Apparatus for	E. J. Fraser	San Francisco, Cal	Apr. 30, 1872	126, 196
Amalgamating the precious metals, Machine for	E. Coleman	San Francisco, Cal	Aug. 18, 1863	39, 550
Amalgamating the precious metals, Mode of	C. A. Seely	New York, N. Y	May 2, 1865	47, 577
Amalgamating the precious metals, Polishing ore to aid in.	I. W. Forbes	LaPorte, Ind	May 30, 1871	115, 293
Amalgamation, Cleaning pulverized ore so as to aid.	I. W. Forbes	LaPorte, Ind	May 30, 1871	115, 294
Amalgamation, Composition for preparing gold and silver ores for.	W. Gluyas	San Francisco, Cal	Oct. 11, 1859	25, 787
Amalgamation of gold and silver	A. B. Crosby	Greene, Me	Sept. 26, 1871	119, 264
Amalgamation, Preparing ores for	A. B. Paul	San Francisco, Cal	Nov. 19, 1872	133, 172
Amalgamator	S. F. Ambler	Brooklyn, N. Y	Feb. 4, 1862	34, 286
Amalgamator	J. B. Atwater	Chicago, Ill	July 25, 1865	48, 887
Amalgamator	J. and E. W. Barker	Baltimore, Md	Jan. 18, 1859	22, 616
Amalgamator	J. A. Bertola	New York, N. Y	Oct. 20, 1857	18, 485
Amalgamator	J. C. Brewster	New York, N. Y	Feb. 8, 1870	99, 630
Amalgamator	J. A. Brock	Chicago, Ill	May 1, 1860	28, 136
Amalgamator	E. Brown	Chicago, Ill	Oct. 23, 1866	58, 981
Amalgamator	G. W. Carter	San Francisco, Cal	Feb. 7, 1860	27, 035
Amalgamator	F. P. Cavanah	Pioneer Mills, N. C	May 3, 1859	23, 881
Amalgamator	S. F. Charles	Dahlonega, Ga	Sept. 25, 1866	58, 335
Amalgamator	T. J. Chubb	Brooklyn, N. Y	Aug. 8, 1865	49, 232
Amalgamator	T. J. Chubb	Brooklyn, N. Y	Aug. 6, 1867	67, 498
Amalgamator	A. M. Church	Augusta, Ga	Jan. 25, 1859	22, 704
Amalgamator	M. H. Collins	Chelsea, Mass	Nov. 21, 1865	51, 021
Amalgamator	F. W. Crosby	New York, N. Y	Mar. 26, 1867	63, 221
Amalgamator	J. Curtis	New York, N. Y	Dec. 28, 1835	......
Amalgamator	J. Curtis	New York, N. Y	Dec. 28, 1835	......
Amalgamator	J. Curtis	New York, N. Y	Dec. 28, 1835	......
Amalgamator	G. S. Curtis and T. Tripp	Chicago, Ill	Apr. 17, 1866	54, 062
Amalgamator	A. G. Day	New York, N. Y	Sept. 26, 1865	50, 101
Amalgamator	L. Eddleblute	Garden Valley, Cal	Dec. 27, 1859	26, 576
Amalgamator	J. W. Evans	New York, N. Y	June 3, 1856	15, 037
Amalgamator	G. B. Field	New York, N. Y	Aug. 27, 1867	68, 175
Amalgamator	A. L. Fleury	New York, N. Y	Jan. 14, 1868	73, 241
Amalgamator	J. B. Forrissier	New York, N. Y	June 25, 1867	66, 142
Amalgamator	S. Fountain	Silver City, Nev	Apr. 28, 1868	77, 183
Amalgamator	W. M. Fuller	Chicago, Ill	Apr. 10, 1866	53, 805
Amalgamator	S. Gardiner, jr	New York, N. Y	Oct. 9, 1855	13, 645
Amalgamator	H. A. Gaston	Austin, Nev	Mar. 27, 1866	53, 435
Amalgamator	H. A. Gaston	Nevada City, Cal	Sept. 3, 1867	68, 359
Amalgamator	P. W. Gates	Chicago, Ill	Oct. 24, 1865	50, 572
Amalgamator	W. Gluyas	San Francisco, Cal	Aug. 7, 1860	29, 545
Amalgamator	A. W. Hall	New York, N. Y	Sept. 6, 1864	44, 140
Amalgamator	A. W. Hall	New York, N. Y	Feb. 28, 1865	46, 560
Amalgamator	I. T. Halsted	Fredonia, N. Y	July 25, 1865	48, 930
Amalgamator	H. Halvorson	North Cambridge, Mass	Oct. 17, 1865	50, 534
Amalgamator	E. Hamilton	Chicago, Ill	Feb. 27, 1866	52, 846
Amalgamator	T. Hansbrow	Sacramento, Cal	Oct. 27, 1863	40, 406
Amalgamator	K. Hazen	Brooklyn, N. Y	Sept. 6, 1859	25, 333
Amalgamator	F. G. Hesse	San Francisco, Cal	Oct. 8, 1867	69, 564
Amalgamator	F. G. Hesse	San Francisco, Cal	Nov. 12, 1867	70, 839
Amalgamator	J. M. Hill	Angel's Camp, Cal	Jan. 1, 1861	31, 063
Amalgamator	H. L. Hopkins	San Francisco, Cal	Sept, 27, 1864	44, 422
Amalgamator	A. Horn	Silver City, Nev	Nov. 19, 1867	71, 011
Amalgamator	A. Horn	Silver City, Nev	Aug. 4, 1868	80, 739
Amalgamator	R. W. Howard	Warwick, R. I	June 18, 1867	65, 912
Amalgamator	W. H. Howland	San Francisco, Cal	Oct. 25, 1859	25, 933
Amalgamator	W. W. Hubbell	Philadelphia, Pa	Oct. 8, 1867	69, 672
Amalgamator	A. Hunter	San Francisco, Cal	May 7, 1867	64, 534
Amalgamator	S. Johnson	San Francisco, Cal	Apr. 13, 1869	88, 963
Amalgamator	W. Kendrick	New York, N. Y	May 29, 1866	55, 114
Amalgamator	E. N. Kent	New York, N. Y	Dec. 4, 1855	13, 879
Amalgamator	J. Kenyon	Black Hawk, Colo	July 19, 1864	43, 589
Amalgamator	I. W. Knox	San Francisco, Cal	Apr. 24, 1860	27, 990
Amalgamator	G. C. Langtry and G. Emmett	Gold Hill, Nev	Feb. 20, 1872	123, 918
Amalgamator	G. A. Mariner and J. Kune	Chicago, Ill	Oct. 6, 1868	82, 730
Amalgamator	L. G. Marshall	San Francisco, Cal	Jan. 7, 1862	34, 115
Amalgamator	F. Maxson	San Francisco, Cal	Oct. 18, 1859	25, 840
Amalgamator	G. E. Mills	New York, N. Y	June 26, 1860	28, 885
Amalgamator	F. Morris	San Francisco, Cal	Mar. 25, 1873	137, 088
Amalgamator	G. M. Norton	San Francisco, Cal	Sept. 18, 1860	30, 085
Amalgamator	A. F. W. Partz	Wurtsborough, N. Y	July 14, 1863	39, 240
Amalgamator	A. B. Paul	Nevada, Cal	Aug. 27, 1861	33, 159
Amalgamator	C. C. Peck	Black Hawk, Colo	Feb. 21, 1865	46, 492
Amalgamator	C. C. Peck	Black Hawk, Colo	Nov. 21, 1865	51, 079
Amalgamator	W. Robbins and E. Swasey	Hindsdale, Ill., and Buckport, Me.	Apr. 21, 1868	77, 097
Amalgamator	J. A. Robinson, jr	San Francisco, Cal	Jan. 22, 1867	61, 465
Amalgamator	D. E. Rose	Cincinnati, Ohio	Jan. 15, 1867	61, 262
Amalgamator	H. P. Russ	San Francisco, Cal	Mar. 22, 1859	23, 313
Amalgamator	H. H. Scoville, jr	Chicago, Ill	May 30, 1865	48, 030
Amalgamator	A. J. Senatz and G. W. Knowlton.	Sacramento, Cal	Apr. 30, 1867	64, 258
Amalgamator	G. B. Simpson	Washington, D. C	Apr. 17, 1866	54, 028
Amalgamator	R. and W. T. Smith	Carondelet, Mo	Apr. 7, 1868	76, 538
Amalgamator	L. Solomon	New York, N. Y	Dec. 7, 1858	22, 245
Amalgamator	J. T. Staats	New York, N. Y	Mar. 13, 1866	53, 194

Index of patents issued from the United States Patent Office from 1790 *to* 1873, *inclusive*—Continued.

Invention.	Inventor.	Residence.	Date.	No.
Amalgamator	S. Standish	Pacheco, Cal	Jan. 15, 1867	61,274
Amalgamator	A. M. Stetson	San Francisco, Cal	Aug. 26, 1856	15,619
Amalgamator	G. Stevens	San Francisco, Cal	June 29, 1869	91,878
Amalgamator	C. C. Stevenson	Gold Hill, Nev	Apr. 19, 1870	102,177
Amalgamator	W. L. Strong	San Francisco, Cal	May 7, 1867	64,458
Amalgamator	S. G. Sturges	Newark, N. J	Sept. 18, 1866	58,149
Amalgamator	A. Swazey	Chicago, Ill	Nov. 5, 1867	70,642
Amalgamator	J. Thomson	Gibsonville, Cal	May 25, 1869	90,409
Amalgamator	T. Tripp	Chicago, Ill	Apr. 17, 1866	54,074
Amalgamator	T. Tripp and G. S. Curtis	Chicago, Ill	Apr. 17, 1866	54,075
Amalgamator	T. Tripp and G. S. Curtis	Chicago, Ill	Apr. 17, 1866	54,076
Amalgamator	T. Varney	San Francisco, Cal	July 12, 1864	43,535
Amalgamator	Z. Wheeler	San Francisco, Cal	Nov. 29, 1864	45,289
Amalgamator	J. D. Whelpley and J. J. Storer	Boston, Mass	Sept. 11, 1866	58,010
Amalgamator	J. White	Bangor, Me	Aug. 28, 1860	29,837
Amalgamator	S. E. Woodworth	Murphy's, Cal	Apr. 9, 1861	32,022
Amalgamator	S. E. Woodworth and J. S. Wethered.	San Francisco, Cal	Nov. 27, 1860	30,781
Amalgamator	J. N. Wickoff and T. M. Fell	Brooklyn, N. Y., and Melvin Mines, Va.	Nov. 27, 1860	30,783
Amalgamator and concentrator	G. Johnston and E. G. Smith	Auburn, Cal	July 9, 1867	66,499
Amalgamator and ore-concentrator	J. Scott	San Francisco, Cal	May 24, 1870	103,377
Amalgamator and ore-crusher	J. Burrell	Central City, Colo	Jan. 28, 1862	34,238
Amalgamator and ore-mill	A. B. Paul	Silver City, Nev	May 13, 1862	35,253
Amalgamator, Cylindrical	T. M. Fell	New York, N. Y	Jan. 1, 1867	60,709
Amalgamator, Electrical	J. H. Rae	Syracuse, N. Y	Oct. 13, 1868	83,091
Amalgamator for collecting gold and silver	J. A. Bertola	New York, N. Y	July 29, 1862	35,987
Amalgamator for collecting gold and silver	J. S. and H. Curtis	Chicago, Ill	Feb. 23, 1869	87,248
Amalgamator for gold and silver	W. H. Deviar	Valley Township, Mo	Mar. 18, 1862	34,673
Amalgamator for gold and silver	P. G. Gardiner	New York, N. Y	Oct. 4, 1864	44,525
Amalgamator for gold and silver	W. A. Palmer	San Francisco, Cal	Nov. 18, 1862	36,963
Amalgamator for gold and silver	T. Varney	San Francisco, Cal	Dec. 16, 1862	37,185
Amalgamator for gold and silver, Electro-magnetic	A. B. Paul	San Francisco, Cal	June 1, 1869	90,777
Amalgamator for gold and silver, Voltaic	J. H. Rae	Syracuse, N. Y	Feb. 20, 1872	123,932
Amalgamator for ores of gold and silver	W. M. Fuller	Chicago, Ill	Dec. 24, 1867	72,626
Amalgamator, Gold	W. Ball	Chicopee, Mass	Sept. 9, 1851	8,344
Amalgamator, Gold	H. Brevoort	San Francisco, Cal	Aug. 30, 1859	25,242
Amalgamator, Gold	R. H. Collyer	New York, N. Y	June 6, 1854	11,034
Amalgamator, Gold	A. K. Eaton	New York, N. Y	Nov. 6, 1860	30,567
Amalgamator, Gold	J. H. Fisher	Placerville, Cal	Feb. 2, 1858	19,246
Amalgamator, Gold	D. Leibee	Middletown, Ohio	Jan. 1, 1856	14,023
Amalgamator, Gold	A. S. Wright	San Francisco, Cal	Aug. 15, 1854	11,541
Amalgamator, Gold	J. N. Wickoff and T. M. Fell	Brooklyn, N. Y., and Orange Mines, Va.	July 26, 1859	24,902
Amalgamator, Gold and silver	H. W. Adams	New York, N. Y	Nov. 29, 1864	45,214
Amalgamator, Gold and silver	H. Bohlhoff	Burlington, Iowa	Mar. 7, 1865	46,632
Amalgamator, Gold and silver	M. Laflin	Chicago, Ill	Mar. 5, 1867	62,639
Amalgamator, Gold and silver	H. Pietsch	New York, N. Y	Jan. 7, 1862	34,072
Amalgamator, Ore	J. M. Beath	San Francisco, Cal	Dec. 20, 1864	45,468
Amalgamator, Re-immersing	J. R. Miller	Fredericksburgh, Va	July 2, 1850	7,478
Amalgamator, Shoe for	J. H. Bullock	Gold Hill, Nev	May 12, 1868	77,868
Ambrotypes, Method of mounting	J. S. McClure	Mobile, Ala	Nov. 1, 1859	25,977
Ambulance	G. W. Arnold	Morgantown, W. Va	Apr. 5, 1864	42,152
Ambulance	C. Britain	Saint Joseph, Mich	Aug. 11, 1863	39,460
Ambulance	J. M. Hayward	Boston, Mass	May 16, 1865	47,719
Ambulance	B. Howard	New York, N. Y	June 27, 1865	48,404
Ambulance	E. R. McKean	Washington, D. C	Oct. 11, 1864	44,643
Ambulance	M. Pinner	New York, N. Y	Jan. 27, 1863	37,518
Ambulance	D. H. Rucker, J. E. Allen, and J. S. Smith.	Washington, D. C	Nov. 6, 1866	59,459
Ambulance	W. Slatter	Allegheny City, Pa	May 30, 1865	47,992
Ambulance	A. W. Süs	New York, N. Y	Aug. 18, 1863	39,595
Ammonia-engine	C. Fellier	Paris, France	Dec. 12, 1871	121,909
Ammonia-gas engine	E. Lamm	New Orleans, La	Mar. 12, 1872	124,495
Ammonia, Manufacture of	A. Paraf	Thann, France	Aug. 6, 1867	67,447
Ammonia, Manufacture of carbonate of	S. R. Divine	New York, N. Y	Jan. 26, 1869	86,142
Ammonia, sulphur, and other products from gas-lime, Preparation of.	R. J. Everett	Bridgeport, Conn	June 4, 1872	127,470
Ammoniacal and other liquids, Apparatus for evaporating.	L. S. Fales	New York, N. Y	July 27, 1869	93,072
Ammoniacal-gas engine	E. Lamm	New Orleans, La	July 19, 1870	105,581
Ammoniacal-gas engine	W. H. Laubach	Philadelphia, Pa	May 28, 1872	127,250
Ammoniacal-gas engine	W. H. Smith and F. V. De Coppet	New Orleans, La	July 29, 1873	141,242
Ammoniacal-gas-engine feeder	E. Lamm	New Orleans, La	Dec. 5, 1871	121,527
Ammunition box, Military	J. J. Hirschbuhl	Louisville, Ky	Feb. 18, 1862	34,423
Ammunition for fire-arms, Fixed	G. Conover	Middletown, Conn	Feb. 23, 1864	41,684
Ammunition holder for cartridge-box, Fixed	J. W. Cochran	New York, N. Y	Mar. 17, 1868	75,625
Ammunition holder for cartridge-box, Fixed	J. W. Cochran	New York, N. Y	Mar. 17, 1868	75,626
Ammunition in chests and boxes, Packing	F. L. Hagadorn	Baltimore, Md	Nov. 26, 1867	71,298
Amputating-apparatus	G. W. Griswold	Carbondale, Pa	Jan. 17, 1854	10,435
Amulet, Metal	E. West	....	May 19, 1809	....
Anæsthetics, Instrument for administration of	O. Willson	Aurora, Ill	Dec. 10, 1867	71,934
Anchor	W. J. Armstrong and C. Browne	Brooklyn, N. Y	Nov. 5, 1867	70,498
Anchor	A. B. Babbitt	Taunton, Mass	May 23, 1871	115,011
Anchor	E. I. Barlow	San Francisco, Cal	Nov. 30, 1869	97,342
Anchor	A. Bradford	Raynham, Mass	Mar. 1, 1864	41,754
Anchor	C. F. Brown	Warren, R. I	May 22, 1866	54,847
Anchor	C. F. Brown	Warren, R. I	Jan. 26, 1869	86,206
Anchor	C. A. Chamberlin	Allegheny City, Pa	Apr. 19, 1864	42,346
Anchor	C. A. Chamberlin	Pittsburgh, Pa	July 11, 1871	116,806
Anchor	C. A. Chamberlin	Philadelphia, Pa	Sept. 16, 1873	142,769
Anchor	G. Coffin	Jamaica Plains, Mass	June 27, 1865	48,370
Anchor	T. L. Dalton	New York, N. Y	Jan. 6, 1857	16,356
Anchor	P. Dinzey	Bartholomew, West Indies	June 12, 1866	55,593
Anchor	J. D. Greene	Cambridge, Mass	Apr. 13, 1869	88,781
Anchor	R. V. D. Guinon	Brooklyn, N. Y	Apr. 3, 1855	12,622
Anchor	J. W. Habberley	South Malden, Mass	Oct. 13, 1868	82,940
Anchor	Z. Hall	Dennis, Mass	Nov. 25, 1873	144,901

Index of patents issued from the United States Patent Office from 1790 *to* 1873, *inclusive*—Continued.

Invention.	Inventor.	Residence.	Date.	No.
Anchor	F. Howes	Boston, Mass	Dec. 22, 1868	85, 228
Anchor	W. M. Hughes	San Francisco, Cal	Feb. 28, 1871	112, 247
Anchor	C. T. Julius	Philadelphia, Pa	Jan. 1, 1867	60, 903
Anchor	W. A. Kentish	New York, N. Y	May 2, 1848	5, 547
Anchor	F. J. Latham	New York, N. Y	Aug. 21, 1866	57, 339
Anchor	A. H. Law	San Francisco, Cal	May 17, 1870	103, 205
Anchor	G. A. Lloyd and C. A. Stewart	San Francisco, Cal	July 9, 1867	66, 665
Anchor	G. A. Lloyd and C. A. Stewart	San Francisco, Cal	Oct. 12, 1869	95, 698
Anchor	C. E. Marshall	Digby, Nova Scotia	Oct. 17, 1865	50, 542
Anchor	F. Martin	Marseilles, France	Mar. 19, 1861	31, 726
Anchor	E. P. McCarthy and J. Johnston.	San Francisco, Cal	Dec. 1, 1868	84, 565
Anchor	E. R. C. Morgan	Mumbles, South Wales	June 21, 1864	43, 273
Anchor	J. W. Morgan	Saltney, Great Britain	Apr. 20, 1869	89, 232
Anchor	G. Norton	Boston, Mass	Feb. 23, 1869	87, 109
Anchor	G. C. Pattison	Baltimore, Md	Jan. 24, 1871	111, 144
Anchor	G. C. Pattison	Baltimore, Md	Jan. 24, 1871	111, 145
Anchor	G. C. Pattison	Baltimore, Md	Sept. 18, 1866	58, 128
Anchor	G. C. Pattison	Baltimore, Md	Feb. 2, 1869	86, 439
Anchor	G. C. Pattison	Baltimore, Md	May 3, 1870	102, 583
Anchor	G. C. Pattison	Baltimore, Md	May 3, 1870	102, 584
Anchor	D. C. Pierce	Clayton, N. Y	Jan. 7, 1868	73, 035
Anchor	D. C. Pierce	Clayton, N. Y	Jan. 7, 1868	73, 036
Anchor	W. H. Porter	Great Britain	Mar. 18, 1842	2, 497
Anchor	C. W. Roeden	San Francisco, Cal	Nov. 24, 1868	84, 304
Anchor	J. W. Russell and D. Joline	Tottenville, N. Y	Jan. 28, 1868	73, 758
Anchor	W. W. Smith	Newcastle-on-Tyne, England.	Feb. 4, 1873	135, 450
Anchor	E. Snell	London, Great Britain	Aug. 29, 1869	49, 688
Anchor	W. Williams	Saint Louis, Mo	Mar. 16, 1858	19, 659
Anchor	F. Wittram	San Francisco, Cal	June 9, 1868	78, 852
Anchor	F. Wittram	San Francisco, Cal	Apr. 27, 1869	89, 454
Anchor and life-preserver, Combined floating	J. Humphries	Washington, D. C	Mar. 16, 1858	19, 638
Anchor, Apparatus for working	A. Campbell	Jersey City, N. J	July 3, 1860	28, 965
Anchor-ball	H. W. Harkness	Bristol, Conn	Aug. 24, 1858	21, 298
Anchor-cable of iron-clad vessels, Means for the protection of.	I. Newton	New York, N. Y	Feb. 9, 1864	41, 530
Anchor, Cast-iron	J. S. Stoddard	Palmyra, N. Y	July 2, 1836	....
Anchor, Cat-block for freeing a ship's	G. W. Duncan	Bath, Me	Apr. 28, 1863	38, 356
Anchor, Compound	S. N. Miller	West Roxbury, Mass	June 29, 1852	9, 076
Anchor, Floating	G. L. Baker	Astoria, N. Y	Feb. 12, 1867	61, 983
Anchor, Floating drag or	A. F. Lewis	Shopiere, Wis	Aug. 22, 1854	11, 555
Anchor for animal	P. H. Raiford	Houston, Tex	June 8, 1869	91, 042
Anchor, Grappling	A. S. and T. J. Trafton	Portsmouth, N. H	Mar. 30, 1869	88, 346
Anchor shackle, Second	G. Gilmour	Chelsea, Mass	Mar. 10, 1857	16, 821
Anchor shackle, Second	T. Leavitt	Malden, Mass	Jan. 3, 1860	26, 684
Anchor, Ship's	N. P. Isaacs and J. Raisbach	New York, N. Y	July 5, 1845	4, 096
Anchor-stock	J. L. Hanly	San Francisco, Cal	May 14, 1867	64, 762
Anchor-stopper	G. H. Babcock	Providence, R. I	Aug. 7, 1866	56, 873
Anchor-stopper, Cat-head	W. H. Barker	Windsor, Nova Scotia	Dec. 13, 1870	109, 998
Anchor-stopper, Cat-head	P. H. Jackson	New York, N. Y	Nov. 27, 1855	13, 847
Anchor-tripper	T. L. Baylies	Richmond, Ind	Apr. 19, 1859	23, 654
Anchor-tripper	S. R. Bryant	New York, N. Y	Feb. 27, 1855	12, 435
Anchor-tripper	E. Davidson	Providence, R. I	Jan. 23, 1866	52, 143
Anchor-tripper	G. Gibson	Port Richmond, N. Y	Dec. 5, 1865	51, 305
Anchor-tripper	B. H. Heitmann	Hoboken, N. J	May 16, 1865	47, 729
Anchor-tripper	H. Higgins	Orleans, Mass	Jan. 10, 1860	26, 765
Anchor-tripper	J. B. Holmes	New York, N. Y	Apr. 28, 1857	17, 182
Anchor-tripper	J. B. Hopkins	Dennis Port, Mass	Apr. 6, 1869	88, 571
Anchor-tripper	W. Stacey	Kittery, Me	Dec. 27, 1864	45, 649
Anchor, Water	N. W. Wheeler	Brooklyn, N. Y	June 11, 1867	65, 625
Anchor-well and anchor	R. and T. Winans	Baltimore, Md	Jan. 29, 1861	31, 276
Anchors, Making and fagoting	J. Tucker and J. Judge	Washington, D. C	Nov. 19, 1829	....
Anchors, Means for casting	J. Evans, jr	Millbridge, Me	Aug. 13, 1867	67, 741
Anchors, Mode of recovering lost	C. B. Whittemore	Boston, Mass	Sept. 3, 1867	68, 402
Anchoring stationary machinery	S. Boone	Le Gros, Ind	Aug. 20, 1867	67, 943
Andiron	J. T. Dee and I. Murray	Fredericktown, Mo	Apr. 30, 1872	126, 273
Andiron	W. Dexter	Winchester, Conn	May 17, 1814	....
Andiron	J. B. Logan	Blountville, Tenn	Mar. 27, 1860	27, 640
Andiron	W. Pye	New York, N. Y	Nov. 16, 1821	....
Andiron	J. B. Sargent	New Britain, Conn	Aug. 17, 1858	21, 218
Andiron	J. Stickney	Baltimore, Md	May 4, 1805	....
Andiron and fire-place	F. Passy	Paris, France	June 23, 1868	79, 253
Andiron and fire-place	J. Sweet	Berkshire, Mass	Feb. 26, 1811	....
Andiron-bars	J. Cochran	Batavia, N. Y	July 1, 1836	....
Andirons, Construction of	J. R. Remington	Lowndes County, Ala	May 16, 1846	4, 523
Andirons, Construction of	E. Smylie	New York, N. Y	July 12, 1843	3, 170
Andirons, Construction of feet of brass	J. Griffiths	New York, N. Y	Mar. 15, 1827	....
Andirons, Manufacturing	W. Wilson	Greenfield, Mass	May 16, 1832	....
Andirons, Pedestal feet for	E. Smylie	New York, N. Y	Feb. 1, 1827	....
Andirons, Preparing and finishing	E. Smylie	New York, N. Y	Feb. 22, 1827	....
Anemometer	G. R. Stuntz	Superior, Wis	Feb. 4, 1862	34, 321
Anemometer, Electrical	F. S. Baldwin	Saint Louis, Mo	July 1, 1873	140, 340
Angle-instrument	B. Dearborn	Boston, Mass	Apr. 29, 1808	....
Angle-iron cutter	A. McGuffie	Buffalo, N. Y	Dec. 9, 1873	145, 437
Angle-iron, Machine for bending	D. Bell	Buffalo, N. Y	Dec. 15, 1863	40, 899
Angle-protractor	H. Taylor	United States Navy	Dec. 20, 1864	45, 535
Angles, Gage for determining	E. C. C. Kellogg	Hartford, Conn	Nov. 13, 1866	59, 721
Angler's combined float and sinker	J. W. Hoard	Providence, R. I	Dec. 12, 1854	12, 060
Angler's float	J. A. Terrell	Bloomfield, Ky	Feb. 2, 1869	86, 609
Angler's reel	W. H. Bradley	New York, N. Y	Sept. 22, 1868	82, 377
Angling-fly	J. Mullaly	New York, N. Y	May 20, 1873	139, 180
Angling-reel	C. L. Noe	Bergen Point, N. J	Jan. 28, 1873	135, 283
Angulometer, Plane	E. Thayer	Worcester, Mass	Aug. 26, 1862	36, 312
Aniline colors	G. E. C. Delaire	Paris, France	July 30, 1861	32, 965
Aniline colors, Compound of	E. Zinssmann	New York, N. Y	Mar. 24, 1868	76, 031
Aniline colors for dyeing and printing, Method of preparing.	A. S. L. Leonhardt	Berlin, Prussia	Sept. 12, 1865	49, 958

Index of patents issued from the United States Patent Office from 1790 *to* 1873, *inclusive*—Continued.

Invention.	Inventor.	Residence.	Date.	No.
Aniline colors, Manufacture of	P. Monnet	Lyons, France	Sept. 2, 1862	36, 357
Aniline colors, Preparation of	J. Renard	Lyons, France	July 31, 1860	29, 434
Aniline colors, Producing	P. Monnet	Lyons, France	Sept. 2, 1862	36, 356
Aniline red, Manufacture of	C. and A. Clemm	Philadelphia, Pa., and Manheim, Baden.	Oct. 10, 1865	50, 335
Anilines, Manufacture of coloring-matter from	P. Chevalier	Lyons, France	Mar. 13, 1866	53, 241
Animal and bird trap	A. T. Latta	Camden, S. C	July 14, 1868	79, 987
Animal and bird trap	G. C. Stamper	Pella, Iowa	June 23, 1868	79, 154
Animal and game trap	J. H. Richardson	Westport, Mo	May 2, 1871	114, 343
Animal and insect trap	W. W. Hannah	Hudson, N. Y	July 29, 1873	141, 346
Animal and vegetable substances, Apparatus for drying and curing.	M. J. Stein	New York, N. Y	Mar. 26, 1872	124, 982
Animal and vegetable substances, Apparatus for packing.	F. Stabler	Baltimore, Md	Mar. 20, 1866	53, 355
Animal and vegetable substances for curing meat, tanning, &c., Apparatus for decomposing.	D. Bruce	Rossville, N. Y	Jan. 22, 1867	61, 315
Animal-cage trap	S. W. Rice	Roseburgh, Oreg	Apr. 15, 1873	137, 864
Animal-cage, Wheel for	G. R. Peckham	Worcester, Mass	Sept. 1, 1868	81, 678
Animal-carcass scraping-machine	R. C. Tompkins	New York, N. Y	Nov. 11, 1873	144, 579
Animal-cleaning machine	C. D. Wheeler	New York, N. Y	Nov. 29, 1859	26, 313
Animal deposits in streets, Apparatus for preventing.	E. Berlinger	New York, N. Y	Nov. 12, 1872	133, 007
Animal-dosing instrument	O. Barker	Brooklyn, N. Y	Mar. 22, 1870	100, 967
Animal-fats, Treating	H. Mége	Paris, France	Dec. 30, 1873	146, 012
Animal-fetter	M. E. Burlingame	Willett, N. Y	July 29, 1862	35, 991
Animal-fetter	M. E. Burlingame	Willett, N. Y	Aug. 1, 1865	49, 081
Animal-gag	W. H. H. Hallock	Mattituck, N. Y	Oct. 17, 1871	120, 062
Animal-marking instrument	J. S. Bodle	Mechlenburg, N. Y	Mar. 5, 1867	62, 596
Animal matter, Apparatus for treating and drying	W. C. Marshall	New York, N. Y	Apr. 8, 1873	137, 699
Animal matter, Drying and disintegrating	C. C. Coe	Hartford, Conn	Aug. 19, 1873	141, 853
Animal matter, Machine for cutting and grinding	A. Smith	Cincinnati, Ohio	Nov. 26, 1867	71, 544
Animal matter, oil, &c., Drying and deodorizing	M. J. Stein	New York, N. Y	June 4, 1872	127, 670
Animal matter, Process and apparatus for treating and drying.	C. G. Bruce and M. J. Stein	New York, N. Y	Oct. 15, 1872	132, 243
Animal matter, Process and apparatus for treatment of.	M. J. Stein	New York, N. Y	Nov. 5, 1872	132, 873
Animal matters, Mode of salting	C. Payne	South Lambeth, England	Sept. 11, 1841	2, 247
Animal or other power to propel machinery, Constructing machine for applying.	I. F. Moore	Falmouth, Va	May 4, 1841	2, 073
Animal-power	J. J. Adgate	Liberty, N. Y	Apr. 28, 1868	77, 236
Animal-power	H. Bolton	Elizabethtown, Canada	Jan. 23, 1872	122, 907
Animal-power	H. F. Carpenter	Polo, Ill	Sept. 24, 1872	131, 498
Animal-power	J. R. Deyo	Sterling, Ill	Apr. 9, 1872	125, 444
Animal-power	J. C. Gentry	Philadelphia, Pa	Dec. 14, 1830	....
Animal-power	J. B. Hall	Cheshire, N. Y	Nov. 23, 1869	97, 193
Animal-power	W. Patterson	Oneonta, N. Y	Aug. 8, 1871	117, 922
Animal-power	C. M. and G. Richards	Harpersville, N. Y	May 9, 1865	47, 659
Animal-power	H. Rosameyer, jr	Allegheny City, Pa	June 11, 1872	127, 800
Animal-power	T. Starr	New Lisbon, Ohio	Dec. 1, 1868	84, 590
Animal-power	M. G. Wood	Church Corners, Mich	Oct. 24, 1871	120, 357
Animal-power, Adjustable	N. Potter	East Troy, Pa	Feb. 28, 1871	112, 179
Animal-power apparatus	O. M. Brock	Monroeton, Pa	Jan. 10, 1871	110, 892
Animal-power, Endless chain for	G. N. Palmer	Greene, N. Y	June 3, 1873	139, 604
Animal-power, Machine for applying	J. Dearborn	Seabrook, N. H	May 21, 1867	64, 951
Animal-power, Propelling machinery by	S. Cooper	Augusta County, Va	Apr. 12, 1830	....
Animal-power to mechanical purposes, Applying	W. E. Arnold	Chatham, Conn	May 7, 1831	....
Animal-seizing device	D. Fasig	Rowsburg, Ohio	May 11, 1869	89, 922
Animal-shearing apparatus	C. A. J. Lengelée	Paris, France	Mar. 25, 1873	137, 220
Animal-shoeing stock	J. Sinclair	Davenport, Iowa	Nov. 11, 1862	36, 920
Animal substances, Separating fatty matter from	H. A. Amelung	Chicago, Ill	Jan. 27, 1863	37, 482
Animal-tether	T. N. Wheeler	Blue Earth City, Minn	July 5, 1870	105, 018
Animal-trap	A. J. Adams	Portland, Oreg	Apr. 28, 1868	77, 157
Animal-trap	H. Adams	Erie County, Pa	Aug. 10, 1833	....
Animal-trap	S. L. Allen	Cinnaminson, N. J	July 16, 1872	129, 203
Animal-trap	H. L. Anderson	Smithville, Ind	Apr. 9, 1867	63, 603
Animal-trap	C. Angle	Hazel Green, Mich	June 28, 1870	104, 684
Animal-trap	J. Annis	Galva, Ill	Oct. 18, 1864	44, 763
Animal-trap	S. M. Armstead	Grand Haven, Mich	Dec. 31, 1867	72, 713
Animal-trap	H. H. C. Arnold	Burlington, Kans	Feb. 27, 1872	124, 027
Animal-trap	H. H. C. Arnold	Nicholasville, Ky	Dec. 2, 1873	145, 082
Animal-trap	S. Arnold	Silver Springs, Tenn	June 21, 1870	104, 406
Animal-trap	S. Arnold	Silver Springs, Tenn	Aug. 2, 1870	105, 880
Animal-trap	S. Ayres	Worcester, Mass	Jan. 12, 1869	85, 780
Animal-trap	H. J. Baddeley	Napa, Cal	Dec. 3, 1872	133, 557
Animal-trap	L. V. Badger	Chicago, Ill	July 9, 1867	66, 550
Animal-trap	J. W. Bagby	Northcutt's Store, Ky	Aug. 31, 1869	94, 173
Animal-trap	H. Bagley	Mechanicsville, Iowa	Mar. 29, 1864	42, 063
Animal-trap	F. P. Baker	Boston, Mass	Sept. 7, 1869	94, 464
Animal-trap	L. J. Baker	East Machias, Me	Dec. 5, 1865	51, 283
Animal-trap	W. Ball	Oregon, Mo	Aug. 9, 1870	106, 107
Animal-trap	G. T. Barker	Pittsfield, Mass	May 12, 1863	38, 458
Animal-trap	G. Barr	Clatskanie, Oreg	Nov. 12, 1872	132, 948
Animal-trap	T. J. Belford	Worthington, Ohio	Feb. 16, 1869	87, 022
Animal-trap	H. Belmer	Cincinnati, Ohio	Aug 13, 1867	67, 709
Animal-trap	J. Biddle	Edinburgh, Pa	Aug. 29, 1871	118, 581
Animal-trap	W. Biddle	Pittsburgh, Pa	July 12, 1839	1, 238
Animal-trap	W. J. and H. L. Biddle	New Madison, Ohio	Feb. 18, 1868	74, 658
Animal-trap	T. Bingham	Stockport, Ohio	Jan. 5, 1869	85, 634
Animal-trap	A. S. Blake	Waterbury, Conn	Apr. 26, 1859	23, 750
Animal-trap	J. Blume	Mount Pleasant, Md	Dec. 14, 1869	97, 755
Animal-trap	J. L. Brabyn	New York, N. Y	Apr. 6, 1858	19, 825
Animal-trap	L. B. Bradley	Watertown, Conn	Aug. 28, 1855	13, 483
Animal-trap	J. W. Bradway	Akron, Ind	Mar. 31, 1863	38, 024
Animal-trap	J. Branch	Mooers, N. Y	Sept. 14, 1869	94, 708
Animal-trap	A. C. Briant	Lafayette, Ind	Feb. 25, 1868	74, 886
Animal-trap	W. W. Brigg	Home, Tenn	June 23, 1871	116, 017
Animal-trap	A. Brown	Bridgeport, Oreg	Dec. 21, 1869	98, 023
Animal-trap	E. Brown	Wayne, Mich	May 4, 1869	89, 626

Index of patents issued from the United States Patent Office from 1790 *to* 1873, *inclusive.*—**Continued.**

Invention.	Inventor.	Residence.	Date.	No.
Animal-trap	G. W. Brown	Sacramento, Cal	Dec. 3, 1867	71, 693
Animal-trap	M. D. Brown	Newburgh, Tenn	Nov. 28, 1871	121, 324
Animal-trap	D. R. Bruton	Thomasville, N. C	June 29, 1869	92, 010
Animal-trap	A. Burnham	Montague, Mass	May 19, 1863	38, 559
Animal-trap	M. Butler	Vernon, Ind	Feb. 25, 1868	74, 887
Animal-trap	L. W. Buxton	Nashua, N. H	Apr. 24, 1860	27, 967
Animal-trap	J. Caffrey	Paradise Township, Pa	Jan. 2, 1855	12, 125
Animal-trap	A. Campbell	Oxford, Ind	Sept. 1, 1868	81, 596
Animal-trap	W. H. Campbell	Brooklyn, N. Y	July 10, 1866	56, 175
Animal-trap	C. R. Capps	Illiopolis, Ill	Feb. 25, 1868	74, 889
Animal-trap	H. C. Case	Pekin, Ill	Mar. 8, 1870	100, 599
Animal-trap	W. F. Caswell	Raynham, Mass	July 25, 1865	48, 905
Animal-trap	R. Chadwick	Nantucket, Mass	July 30, 1861	32, 961
Animal-trap	La F. C. Chamberlin	Osawatomie, Kan	Mar. 12, 1872	124, 417
Animal-trap	J. W. Churchill	Pittston, Pa	Aug. 15, 1865	49, 378
Animal-trap	J. W. Churchill	Pittston, Pa	Nov. 26, 1867	71, 364
Animal-trap	G. E. Clarke	Racine, Wis	Dec. 26, 1865	51, 777
Animal-trap	G. E. Clarke	Racine, Wis	Dec. 4, 1866	60, 143
Animal-trap	G. E. Clarke	Racine, Wis	Aug. 27, 1867	68, 046
Animal-trap	W. J. Clarkson	Gourdin's Depot, N. E. R. R., S. C.	Feb. 1, 1870	99, 293
Animal-trap	J. J. Cline	High Hill, Ohio	Apr. 14, 1868	76, 713
Animal-trap	D. Cole	Orwell, Pa	Aug. 28, 1866	57, 480
Animal-trap	W. F. Collier	Worcester, Mass	Feb. 15, 1870	99, 995
Animal-trap	I. N. Connell	Spencer's Station, Ohio	Aug. 20, 1867	67, 959
Animal-trap	J. Coombe	San José, Cal	Feb. 6, 1872	123, 381
Animal-trap	J. F. Coppock	Dexter, Iowa	Oct. 10, 1871	119, 826
Animal-trap	J. Cosolowsky	Titusville, Pa	Nov. 8, 1870	109, 112
Animal-trap	G. G. Cottrell	Sharon, Conn	June 5, 1866	55, 245
Animal-trap	H. H. Cottrill	Vinton Station, Ohio	Mar. 9, 1869	87, 548
Animal-trap	W. Cover	Jenner's Cross Roads, Pa	Jan. 7, 1868	72, 979
Animal-trap	R. S. Craig	Cincinnati, Ohio	Nov. 7, 1854	11, 887
Animal-trap	E. H. Crane	Jonesville, Mich	July 5, 1864	43, 396
Animal-trap	J. Curtis	Saint Charles, Minn	Oct. 15, 1867	69, 777
Animal-trap	E. A. C. Da Silva	New York, N. Y	Feb. 11, 1873	135, 732
Animal-trap	A. W. Davis	South Haven, Mich	Nov. 19, 1872	133, 082
Animal-trap	G. H. Davis	Patten, Me	Jan. 30, 1872	123, 089
Animal-trap	W. H. Davis	Lexington, Ind	Dec. 24, 1867	72, 461
Animal-trap	J. M. Dearborn	Boston, Mass	Mar. 3, 1868	74, 998
Animal-trap	H. D. Deming and P. G. Walker	Delmar, Pa	Mar. 5, 1861	31, 651
Animal-trap	J. Dildine	Limestoneville, Pa	Dec. 10, 1872	133, 836
Animal-trap	S. F. Dimock	Spencer, Ohio	Nov. 27, 1866	59, 980
Animal-trap	M. W. Drake	Owasso, Mich	Aug. 3, 1869	93, 185
Animal-trap	U. H. Duparck	Albion, Mich	Sept. 13, 1864	44, 168
Animal-trap	A. J. Eddy and J. B. Wilber	Winnesheik County and Howard County, Iowa	Dec. 13, 1864	45, 399
Animal-trap	A. Edwards	Chicago, Ill	Sept. 5, 1865	49, 735
Animal-trap	A. Ellis and O. Albertson	Salem, Ind	July 31, 1866	56, 735
Animal trap	A. Ellis and O. Albertson	Salem, Ind	Dec. 24, 1867	72, 470
Animal-trap	J. W. Ells	Pittsburgh, Pa	Apr. 13, 1869	88, 778
Animal-trap	J. H. Elward	Polo, Ill	Jan. 9, 1866	51, 933
Animal-trap	J. P. Emswiler	Knightstown, Ind	June 25, 1867	66, 007
Animal-trap	S. F. Estell	Richmond, Ind	Oct. 16, 1866	58, 795
Animal-trap	S. F. Estell	Richmond, Ind	Nov. 20, 1866	59, 762
Animal-trap	S. F. Estell	Richmond, Ind	Aug. 18, 1868	81, 265
Animal-trap	S. F. Estell	Chicago, Ill	Feb. 4, 1873	135, 534
Animal-trap	E. B. Everitt	West Meriden, Conn	June 14, 1870	104, 293
Animal-trap	O. S. Ewing	Rome, Tenn	Nov. 14, 1871	120, 953
Animal-trap	J. B. Fairchild	Covington, Ky	June 12, 1866	55, 478
Animal-trap	L. Faris	Princetown, Ohio	Feb. 23, 1869	87, 159
Animal-trap	T. Fell	New York, N. Y	Oct. 3, 1871	119, 592
Animal-trap	W. A. Fenn	Rochester, N. Y	Apr. 27, 1869	89, 300
Animal-trap	A. C. Flanders	Owatonna, Minn	June 15, 1869	91, 427
Animal-trap	J. M. Flautt	Reedsburgh, Wis	Dec. 27, 1864	45, 595
Animal-trap	F. Flora	Pierce, Ohio	Jan. 28, 1873	135, 268
Animal-trap	H. Foust	Mill Village, Pa	Oct. 3, 1871	119, 594
Animal-trap	F. Fox	Meadville, Pa	June 21, 1870	104, 570
Animal-trap	A. A. Fradenburg	Nevada City, Cal	May 22, 1866	54, 885
Animal-trap	D. M. Francisco	Three Rivers, Mich	Dec. 5, 1871	121, 608
Animal-trap	S. Frisbie and H. C. Hart	Unionville, Conn	Apr. 9, 1872	125, 449
Animal-trap	A. Frost	Seymour, Ind	June 11, 1867	65, 661
Animal-trap	H. S. Frost	Watertown, Conn	Oct. 31, 1871	120, 375
Animal-trap	C. G. Frushour	La Gro, Ind	Oct. 19, 1869	95, 893
Animal-trap	S. Gibson	Martic Township, Pa	July 13, 1858	20, 873
Animal-trap	W. S. Gitchell	Peru, Ind	Feb. 28, 1865	46, 557
Animal-trap	J. Gould	Clinton, Pa	June 3, 1873	139, 572
Animal-trap	J. W. Greene	Chillicothe, Mo	Jan. 7, 1873	134, 539
Animal-trap	H. Hackman, jr	Pequea, Pa	June 16, 1857	17, 570
Animal-trap	J. Haddleton	Rochester, N. Y	Sept. 24, 1867	69, 089
Animal-trap	J. H. Hair	Galesburgh, Ill	June 17, 1873	140, 036
Animal-trap	W. R. Hampton	Fairfield, Ill	Nov. 29, 1870	109, 614
Animal-trap	B. F. Hancock	Monroe, Wis	Apr. 28, 1868	77, 372
Animal-trap	G. Haneline	Akron, Ohio	Aug. 10, 1869	93, 533
Animal-trap	G. R. Harding	Manchester, Va	Oct. 17, 1871	120, 063
Animal-trap	B. and D. H. Harnish	Lancaster and Pequea, Pa	Apr. 2, 1872	125, 290
Animal-trap	G. Hart	Granger, Ohio	Sept. 1, 1857	18, 092
Animal-trap	G. L. Hart	New Britain, Conn	Jan. 17, 1871	110, 973
Animal-trap	G. L. and H. C. Hart	Unionville, Conn	Oct. 18, 1864	44, 721
Animal-trap	S. H. and H. C. Hart	Unionville, Conn	Feb. 14, 1871	111, 641
Animal-trap	D. Harwood	Dutch Flat, Cal	July 20, 1869	92, 823
Animal-trap	E. E. Haughwout	New York, N. Y	Apr. 5, 1870	101, 459
Animal-trap	E. Hause and L. Kelley	Tecumseh, Mich	July 18, 1871	117, 072
Animal-trap	C. Henert	Washington, D. C	June 22, 1869	91, 744
Animal-trap	J. Herr	Carbondale, Ill	June 8, 1869	91, 131
Animal-trap	N. T. Hersch	Round Hill, Pa	June 7, 1870	104, 031
Animal-trap	J. M. Hill	Cisne, Ill	Feb. 20, 1872	123, 779
Animal-trap	T. M. Hill	Richmond, Ind	Nov. 3, 1868	83, 633
Animal-trap	S. Hoke	Union City, Ind	Oct. 29, 1867	70, 336

Index of patents issued from the United States Patent Office from 1790 *to* 1873, *inclusive*—Continued.

Invention.	Inventor.	Residence.	Date.	No.
Animal-trap	J. W. Hollingsworth	Seymour, Ind	Oct. 16, 1866	58, 826
Animal-trap	J. W. Hollingsworth	Salem, Ind	May 21, 1867	64, 869
Animal-trap	A. A. Hotchkiss	Sharon, Conn	Feb. 14, 1871	111, 745
Animal-trap	J. W. F. How	Douglas County, Oreg	Mar. 28, 1871	113, 168
Animal-trap	J. W. F. How	Canyonville, Oreg	Mar. 19, 1872	124, 821
Animal-trap	N. S. Howell	Tualitin, Oreg	Nov. 14, 1871	120, 877
Animal-trap	S. Huffman	Westfield, Ill	Feb. 25, 1868	74, 759
Animal-trap	J. Hughson	Buffalo, N. Y	Sept. 7, 1869	94, 496
Animal-trap	W. Huntington	Howell, Mich	Dec. 17, 1867	72, 396
Animal-trap	B. Illingworth	Freeport, Ill	Oct. 22, 1867	70, 093
Animal-trap	L. E. Ingersoll	Columbus, Pa	June 11, 1872	127, 887
Animal-trap	G. Irwin	Elizabethtown, Ky	July 9, 1867	66, 592
Animal-trap	W. B. Jarvis	Washington, N. C	Feb. 23, 1869	87, 173
Animal-trap	C. Jillson	Worcester, Mass	Nov. 16, 1858	22, 078
Animal-trap	S. F. Jones	Saint Paul, Ind	May 1, 1866	54, 361
Animal-trap	P. W. King	Lowville, N. Y	May 5, 1868	77, 622
Animal-trap	A. Kinkead	Greenwich, Ohio	June 7, 1864	43, 030
Animal-trap	G. W. Kintz	Rochester, N. Y	Mar. 16, 1869	87, 945
Animal-trap	T. B. Kirby	Flowerfield, Mich	Sept. 8, 1868	82, 008
Animal-trap	S. Knight	Adel, Iowa	Dec. 12, 1865	51, 463
Animal-trap	J. Kohler	Cincinnati, Ohio	July 7, 1868	79, 577
Animal-trap	J. O. Kopas	Washington, D. C	Apr. 19, 1870	102, 134
Animal-trap	J. O. Kopas and G. W. Bauer	Washington, D. C	Apr. 19, 1870	102, 133
Animal-trap	J. Krummenauer	New York, N. Y	Mar. 4, 1873	136, 378
Animal-trap	G. F. Lampkin	Georgetown, Ky	July 9, 1872	128, 802
Animal-trap	W. S. Lawrence	Winchester, Ky	Jan. 21, 1873	135, 134
Animal-trap	D. A. Leach	Bloomington, Ind	Sept. 23, 1873	143, 169
Animal-trap	H. Lee	Oberlin, Ohio	Feb. 12, 1867	61, 941
Animal-trap	J. A. Lee	Chattanooga, Tenn	Nov. 3, 1868	83, 642
Animal-trap	W. W. Leech	Pittsburgh, Pa	Feb. 25, 1868	74, 762
Animal-trap	H. S. Lesher	Galesburgh, Ill	Sept. 1, 1868	81, 798
Animal-trap	L. Lewis	Tuskegee, Ala	Feb. 13, 1872	123, 633
Animal-trap	J. Liming	Philadelphia, Pa	Dec. 22, 1863	41, 012
Animal-trap	W. D. Lindsley	Eudora, Kans	June 14, 1870	104, 172
Animal-trap	N. M. Linton	Wilmington, Ohio	Nov. 19, 1867	71, 188
Animal-trap	C. Long	Newark, Ohio	May 23, 1871	115, 222
Animal-trap	J. H. Lord	San Francisco, Cal	Apr. 26, 1870	102, 411
Animal-trap	A. M. Lovett	Zanesville, Ohio	Mar. 18, 1873	136, 842
Animal-trap	J. N. Lowrance	Montezuma, Ill	July 16, 1872	129, 415
Animal-trap	N. B. Lucas	Jersey County, Ill	Apr. 11, 1848	5, 508
Animal-trap	W. Luker	Kalamazoo, Mich	Nov. 16, 1869	96, 821
Animal-trap	H. F. Lushbaugh and O. Z. Hurd	Mount Pulaski, Ill	Nov. 24, 1868	84, 291
Animal-trap	C. C. Lyman	Edenborough, Pa	Dec. 17, 1867	72, 211
Animal-trap	C. C. Lyman	Edenborough, Pa	Mar. 24, 1868	75, 777
Animal-trap	M. W. Lyman	Chicago, Ill	Mar. 15, 1870	100, 779
Animal-trap	M. W. Lyman	Chicago, Ill	Mar. 19, 1872	124, 838
Animal-trap	J. Manchester	New York, N. Y	Feb. 4, 1868	74, 105
Animal-trap	J. Manchester	New York, N. Y	Oct. 19, 1869	95, 920
Animal-trap	H. Mansfield	Warsaw, Ind	Aug. 20, 1867	67, 893
Animal-trap	T. E. Marable	Petersburgh, Va	Nov. 22, 1870	109, 530
Animal-trap	M. B. Marshall	Draw Bridge, Md	Mar. 10, 1868	75, 438
Animal-trap	D. J. Martin	Covington, Ohio	July 23, 1867	66, 980
Animal-trap	A. A. Mattern	Morrison, Ill	June 25, 1872	128, 233
Animal-trap	A. J. Matthews	Hartwell, Ga	June 10, 1873	139, 800
Animal-trap	R. B. May	Murfreesborough, Tenn	Feb. 25, 1873	136, 330
Animal-trap	J. C. McClamroch	Edina, Mo	May 26, 1868	78, 306
Animal-trap	W. McClure	Sinking Spring, Ohio	Oct. 13, 1868	83, 074
Animal-trap	C. Melone	Lawrence, Kans	Mar. 13, 1866	53, 164
Animal-trap	J. Merchen	Brookville, Ind	June 17, 1873	139, 967
Animal-trap	J. B. Merriman and G. B. Lewis	Plantsville, Conn	Nov. 8, 1870	109, 035
Animal-trap	G. W. Merritt and H. S. Gibbs	Norwalk, Conn	Jan. 26, 1869	86, 175
Animal-trap	O. Metcalf	Salem, Ind	Jan. 14, 1868	73, 254
Animal-trap	L. Meyer	Saint Louis, Mo	June 11, 1872	127, 701
Animal-trap	J. H. Miller	Vernon, Ind	Jan. 21, 1868	73, 625
Animal-trap	W. Miller	Chicopee, Mass	Feb. 25, 1868	74, 842
Animal-trap	J. H. Mooney and G. A. Lloyd	San Francisco, Cal	Sept. 26, 1871	119, 237
Animal-trap	L. Moore	Baraboo, Wis	July 13, 1869	92, 631
Animal-trap	W. Morris	Elkhart City, Ill	Apr. 9, 1867	63, 651
Animal-trap	J. Nampel	Freeport, Ill	Nov. 16, 1869	96, 946
Animal-trap	W. H. Newby	Seymour, Ind	May 1, 1866	54, 393
Animal-trap	J. Newell	Ypsilanti, Mich	Feb. 27, 1872	124, 149
Animal-trap	A. Newton	Darby Creek, Ohio	Aug. 30, 1870	106, 858
Animal-trap	G. A. Norman	Baltimore, Md	May 20, 1873	139, 182
Animal-trap	H. S. North and J. O. Couch	Middletown, Conn	June 28, 1859	24, 573
Animal-trap	H. Ogborn	Richmond, Ind	Nov. 19, 1867	71, 206
Animal-trap	E. Oliver	New York, N. Y	Dec. 13, 1870	110, 065
Animal-trap	E. Oliver	New York, N. Y	Feb. 14, 1871	111, 771
Animal-trap	D. J. Owen	Springville, Pa	Apr. 27, 1869	89, 429
Animal-trap	D. J. Owen	Springville, Pa	May 31, 1870	103, 767
Animal-trap	D. J. Owen	Springville, Pa	Apr. 29, 1873	138, 275
Animal-trap	G. W. Pagett	Adams Township, Ind	May 2, 1865	47, 563
Animal-trap	G. J. Parham	Harrodsburgh, Ind	Nov. 26, 1867	71, 321
Animal-trap	J. C. Parish	Petersburgh, Va	Feb. 7, 1871	111, 673
Animal-trap	W. Patterson	Waukegan, Ill	July 16, 1872	129, 295
Animal-trap	H. Pattison	Duck Creek, Ill	June 15, 1869	91, 363
Animal-trap	R. L. Payne	Halifax, Va	Nov. 2, 1858	21, 978
Animal-trap	J. D. Pell	New York, N. Y	Mar. 21, 1871	112, 844
Animal-trap	S. Pence	Eaton, Ohio	Oct. 8, 1867	69, 695
Animal-trap	L. V. Percival and J. Link	United States Army	Sept. 20, 1870	107, 532
Animal-trap	D. N. Phelps	San Leandro, Cal	Sept. 25, 1866	58, 2[illegible]6
Animal-trap	T. S. Phillips	Casadaga, N. Y	Oct. 8, 1867	69, 583
Animal-trap	O. Pier	Harmony, N. Y	June 27, 1854	11, 175
Animal-trap	D. Pittman	Fort Madison, Iowa	June 4, 1861	32, 486
Animal-trap	C. Polley	Shelbyville, Tenn	Sept. 7, 1869	94, 642
Animal-trap	H. Polley	San Francisco, Cal	Dec. 5, 1871	121, 468
Animal-trap	H. W. Prouty	Boston, Mass	Sept. 15, 1868	82, 246
Animal-trap	J. Quigley	Winona, Minn	May 28, 1861	32, 434
Animal-trap	S. S. Rain	Lowville, N. Y	Sept. 10, 1867	68, 651

Index of patents issued from the United States Patent Office from 1790 *to* 1873, *inclusive*—Continued.

Invention.	Inventor.	Residence.	Date.	No.
Animal-trap	S. S. Rain	Lowville, N. Y	July 7, 1868	79, 686
Animal-trap	N. Rasmussen	Chicago, Ill	Oct. 3, 1871	119, 645
Animal-trap	S. Reed	Whitestown, Pa	Jan. 14, 1868	73, 384
Animal-trap	W. N. Reed	Washington, D. C	Sept. 21, 1869	94, 975
Animal-trap	E. Reichard	Washington, Mo	Oct. 13, 1868	83, 094
Animal-trap	J. H. Reisinger	Vinton, Ohio	May 5, 1868	77, 657
Animal-trap	F. Reuthe	Hartford, Conn	May 12, 1857	17, 297
Animal-trap	F. Reuthe	Hartford, Conn	Aug. 24, 1858	21, 302
Animal-trap	J. H. Richardson	Westport, Mo	Sept. 13, 1870	107, 292
Animal-trap	T. L. Rivers	Saint Louis, Mo	May 19, 1868	78, 015
Animal-trap	J. Rollins	Kingston, Tenn	Mar. 26, 1872	124, 976
Animal-trap	B. F. Sanford	Galesburgh, Ill	Dec. 5, 1865	51, 356
Animal-trap	J. Sebroy	Fortville, Ind	Nov. 5, 1867	70, 624
Animal-trap	H. Seehausen	Memphis, Tenn	Jan. 25, 1870	99, 245
Animal-trap	T. Shailer	Haddam, Conn	May 10, 1844	3, 580
Animal-trap	J. Sherman	New Oxford, Pa	Sept. 1, 1868	81, 829
Animal-trap	T. Silliman	Three Rivers, Mich	Nov. 26, 1867	71, 543
Animal-trap	T. Silliman	Three Rivers, Mich	Apr. 27, 1869	89, 352
Animal-trap	R. Simpson	Port Jefferson, Ohio	Oct. 29, 1867	70, 370
Animal-trap	J. A. Sinclair	Woodsfield, Ohio	Feb. 18, 1868	74, 617
Animal-trap	G. Slusser	Hillsborough, Ohio	Aug. 14, 1860	29, 627
Animal-trap	B. C. Smith	Pekin, Ill	July 11, 1871	117, 005
Animal-trap	B. F. Smith	Philadelphia, Pa	Oct. 14, 1873	143, 727
Animal-trap	E. B. Smith	Marietta, Ohio	June 30, 1868	79, 506
Animal-trap	O. R. Smith	Elgin, Minn	Aug. 11, 1868	81, 023
Animal-trap	V. O. and J. R Spencer	Mansfield, Pa	Feb. 7, 1860	27, 080
Animal-trap	E. Sprague and G. C. Belt	Bridgeton, Ind	June 28, 1870	104, 896
Animal-trap	W. A. Stack	Hillsborough, Md	Nov. 10, 1868	84, 013
Animal-trap	W. L. Starr	Columbus, Ohio	Mar. 31, 1868	76, 289
Animal-trap	J. J. St. Ledger	Philadelphia, Pa	Mar. 30, 1869	88, 526
Animal-trap	A. Storm	Brooklyn, N. Y	Feb. 5, 1867	61, 889
Animal-trap	J. N. Stow and R. Loop	Camden, Ohio	Apr. 25, 1871	114, 062
Animal-trap	Z. Swope	Lancaster, Pa	Aug. 16, 1859	25, 224
Animal-trap	J. Teed	Tompkins, N. Y	Apr. 16, 1867	63, 961
Animal-trap	A. C. Thomas	Camp Charlotte, Ohio	Feb. 11, 1868	74, 256
Animal-trap	J. S. Thompson	Sycamore, Ill	Feb. 11, 1868	74, 451
Animal-trap	N. S. Thompson	Germantown, Ohio	Jan. 21, 1868	73, 475
Animal-trap	W. Tinsley	New York, N. Y	Feb. 7, 1860	27, 083
Animal-trap	R. Tompkins	Clarksville, Tenn	Nov. 23, 1869	97, 248
Animal-trap	J. Trainer	Rural Dale, Ohio	Oct. 22, 1867	70, 133
Animal-trap	J. Trainer	Vinton Station, Ohio	May 19, 1868	78, 029
Animal-trap	C. S. Trevitt	Washington, D. C	Oct. 22, 1867	70, 134
Animal-trap	R. M. Turner	Woodland, Mich	Sept. 7, 1858	21, 454
Animal-trap	J. L. Tusten	Winona, Miss	Nov. 9, 1869	96, 744
Animal-trap	T. B. Van Pelt	Westport, Mo	July 20, 1869	92, 911
Animal-trap	C. B. Veronee	Athens, Ga	Apr. 4, 1871	113, 371
Animal-trap	G. S. Walker	Erie, Pa	Aug. 22, 1871	118, 309
Animal-trap	S. Ward	Richmond, Ill	Sept. 24, 1867	69, 279
Animal-trap	A. L. Waring	Coshocton, Ohio	Oct. 13, 1868	83, 011
Animal-trap	A. Warner	Cleveland, Ohio	Sept. 23, 1862	36, 543
Animal-trap	J. Westcott	Patchogue, N. Y	Oct. 15, 1867	69, 878
Animal-trap	L. Wetmore	Tioga County, Pa	July 12, 1859	24, 771
Animal-trap	J. Wheelock	San Francisco, Cal	Mar. 7, 1865	46, 741
Animal-trap	J. P. Wigal	Henderson, Ky	Jan. 14, 1868	73, 418
Animal-trap	H. Y. Wildey	Philadelphia, Pa	Oct. 2, 1860	30, 269
Animal-trap	J. M. Wilkinson	Bloomington, Tenn	Sept. 2, 1873	142, 362
Animal-trap	W. T. Williams	New York, N. Y	Feb. 19, 1861	31, 504
Animal-trap	J. P. Wilson	Frankfort, N. Y	Jan. 31, 1860	27, 017
Animal-trap	G. Wolf	Williamsport, Md	Apr. 24, 1866	54, 241
Animal-trap	J. J. Wood	North Manchester, Ind	Sept. 27, 1870	107, 844
Animal-trap	R. E. Wood	Santa Cruz, Cal	Nov. 29, 1870	109, 789
Animal-trap	W. J. Woodside	Zanesville, Ohio	Apr. 28, 1868	77, 341
Animal-trap	W. Wright	Philadelphia, Pa	June 19, 1860	28, 820
Animal-trap	H. D. Wrightson	Queenstown, Md	Mar. 7, 1871	112, 403
Animal-trap	C. Zaiser	Newark, N. J	Feb. 11, 1868	74, 264
Animal-trap adjustable platform	J. Thomas	West Chester, Pa	June 26, 1849	6, 554
Animal-trap and seed-safe combined	S. V. Greer	Rocky Hill Station, Ky	July 15, 1873	140, 773
Animal-trap, Construction of	M. H. Biddle	Mount Carmel, Ill	Oct. 5, 1858	21, 647
Animal-trap, Construction of	E. Hill	Cincinnati, Ohio	Oct. 5, 1858	21, 676
Animal-trap, Device for setting	I. Miller	Bryan, Ohio	Sept. 10, 1867	68, 776
Animal-trap, Manufacture of	C. Jillson	Worcester, Mass	Jan. 6, 1857	16, 335
Animal-trap, Self-setting	H. B. Myers	Schoolcraft, Mich	Feb. 14, 1865	46, 379
Animal-trap wheel	W. F. Collier	Worcester, Mass	June 13, 1871	115, 933
Animal-trap wheel	B. B. and J. R. Hill	Worcester, Mass	July 4, 1871	116, 592
Animals against flies, Lotion for protecting	J. Greene	Providence, R. I	Oct. 24, 1871	120, 191
Animals, Apparatus for relieving choked	G. Clump	Hamden, Conn	Apr. 21, 1868	76, 998
Animals, Apparatus for taming wild	P. R. Sanderson	Caledonia, N. Y	Dec. 10, 1867	71, 914
Animals, Brand for marking	H. Thompson	Hector, N. Y	Apr. 30, 1867	64, 263
Animals, Device for catching	W. L. Hopper	Monmouth, Ill	Aug. 27, 1867	68, 197
Animals, Device for catching and holding domestic	H. V. Van Etten	Auburn, N. Y	Dec. 29, 1868	85, 413
Animals from railways, Apparatus for removing	L. Montgilion	Elk Ridge Landing, Md	Feb. 13, 1849	6, 113
Animals from the heat of the sun, Mode of protecting.	C. Elveena	New York, N. Y	Sept. 18, 1866	58, 081
Animals, Instrument for administering balls to	T. H. Bex	Syracuse, N. Y	June 17, 1873	139, 856
Animals, Slings for raising	F. Hohorst	New York, N. Y	Feb. 18, 1873	136, 059
Ankle-brace	J. S. Niswander	Oakland, Cal	Mar. 21, 1871	112, 952
Ankle-brace	S. B. Sherer	Aurora, Ill	Apr. 7, 1868	76, 353
Ankle-brace joint	E. E. Howe	Boston, Mass	Aug. 20, 1872	130, 639
Ankle or knee guard	H. A. Hall	Boston, Mass	May 12, 1868	77, 728
Ankle-supporter	R. Cunningham	Chicago, Ill	Mar. 20, 1866	53, 276
Ankle supporter and filler	S. Silberschmidt	Baltimore, Md	Oct. 7, 1873	143, 537
Ankles, Surgical apparatus for fractured or injured	G. W. Yerger	Philadelphia, Pa	Mar. 20, 1849	6, 214
Annealing and hardening metals, Process for	J. N. Lauth	Pittsburgh, Pa	Oct. 29, 1872	132, 675
Annealing and swaging castings, Method of	E. B. Wilson	Westminster, England	Sept. 17, 1861	33, 315
Annealing-apparatus	J. Worcester	Newport, Ky	Sept. 25, 1860	30, 174
Annealing-box	J. C. Lewis	Sharpsburgh, Pa	May 1, 1866	54, 376
Annealing car-wheels	A. L. Mowry	Cincinnati, Ohio	May 7, 1861	32, 252
Annealing cut nails	J. McCarty	Reading, Pa	June 11, 1861	32, 525

Index of patents issued from the United States Patent Office from 1790 *to* 1873, *inclusive*—Continued.

Invention.	Inventor.	Residence.	Date.	No.
Annealing-furnace	E. Bennett	Philadelphia, Pa	July 11, 1865	48,761
Annealing-furnace	J. J. Eagleton	New York, N. Y	May 20, 1856	14,908
Annealing-furnace	T. F. Hammer	Branford, Conn	Jan. 9, 1872	122,604
Annealing-furnace	H. W. Moore	Bridgeport, Conn	Oct. 9, 1866	58,666
Annealing-furnace	W. R. Thomas	Catasaqua, Pa	Jan. 14, 1868	73,407
Annealing-furnace for sheet-iron, &c	J. Malone	Temperanceville, Pa	May 22, 1866	55,014
Annealing-furnace, Nail and tack	E. G. Paull	Fair Haven, Mass	Apr. 13, 1869	88,805
Annealing-furnace, Plowshare	W. M. Watson	Tonica, Ill	Feb. 6, 1872	123,528
Annealing hollow iron-ware	D. Stuart	Philadelphia, Pa	Sept. 27, 1853	10,057
Annealing metal, Process and apparatus for	J. M. Bottum	New York, N. Y	Nov. 16, 1869	96,874
Annealing pot and saucer	J. Hibell	Nechells, England	July 7, 1868	79,759
Annotto, Solution of	A. Macphail	Jersey City, N. J	June 7, 1864	43,034
Annunciator	J. S. Birch	New York, N. Y	Dec. 1, 1868	84,474
Annunciator	J. Capron	New York, N. Y	Dec. 27, 1870	110,547
Annunciator	M. L. Deering	New York, N. Y	Jan. 2, 1866	51,809
Annunciator	C. H. Greenleaf	Franconia, N. H	July 9, 1872	128,724
Annunciator	C. H. Greenleaf	Franconia, N. H	Aug. 19, 1873	141,927
Annunciator	H. Gross and G. S. Zingling	Tiffin, Ohio	Nov. 5, 1867	70,552
Annunciator	H. Horsfall	New York, N. Y	Oct. 31, 1856	50,709
Annunciator	S. F. Nichols	Trappe, Md	May 28, 1872	127,185
Annunciator	A. Rankin	Philadelphia, Pa	Feb. 17, 1863	37,706
Annunciator and fire-alarm, Hotel	E. A. Hill	Chicago, Ill	Apr. 25, 1871	114,007
Annunciator, Electric	J. B. Shannon	Philadelphia, Pa	June 11, 1872	127,931
Annunciator, Electric	G. W. Shawk	Cleveland, Ohio	June 10, 1873	139,826
Annunciator, Electric	A. Storer	Cleveland, Ohio	Aug. 19, 1873	141,898
Annunciator, Electric hotel	W. W. Foot	Saybrook, Ohio	June 24, 1873	140,129
Annunciator, Electric vote	T. B. Doolittle	Bridgeport, Conn	Oct. 14, 1873	143,679
Annunciator, Electro-magnetic	C. S. Bulkley	New York, N. Y	Nov. 15, 1853	10,226
Annunciator, Electro-magnetic	J. Capron	New York, N. Y	Aug. 15, 1871	117,979
Annunciator, Electro-magnetic	C. E. Chinnock	New York, N. Y	Feb. 20, 1872	123,808
Annunciator, Electro-magnetic	E. Gray	Chicago, Ill	Aug. 22, 1871	118,231
Annunciator, Electro-magnetic	W. Humans	Cambridge, Mass	Mar. 4, 1873	136,518
Annunciator, Electro-magnetic	W. N. McInnis	Northumberland, Pa	Mar. 29, 1870	101,372
Annunciator, Electro-magnetic	G. W. Shawk and A. Storer	Cleveland, Ohio	Mar. 4, 1873	136,465
Annunciator for hotels, &c., Electro-magnetic	O. Hagendorf	New York, N. Y	Dec. 17, 1872	134,053
Annunciator for railway-carriage	M. H. Ford	Boston, Mass	May 8, 1849	6,442
Annunciator for signals in hotels, &c., Electro-magnetic.	C. S. Bulkley	Macon, Ga	Oct. 29, 1850	7,739
Annunciator, Hotel	J. Bale	Buffalo, N. Y	Feb. 27, 1855	12,433
Annunciator, Hotel	J. H. H. Bennett	Hunt's Hollow, N. Y	Dec. 17, 1861	33,926
Annunciator, Hotel	W. H. Hale	Worcester, Mass	Apr. 22, 1856	14,719
Annunciator, Hotel	W. Horsfall	New York, N. Y	Oct. 4, 1853	10,071
Annunciator, Hotel	H. B. Porter	Chicago, Ill	Nov. 19, 1867	71,214
Annunciator, Hotel	L. J. Vansands	Chicago, Ill	Oct. 5, 1869	95,620
Annunciator, Hotel electrical	H. B. Porter	Chicago, Ill	Dec. 28, 1869	98,296
Annunciator, Magnetic	J. Blackie	Washington, D. C	July 26, 1864	43,633
Annunciator or bell-telegraph	J. Garvey	New York, N. Y	Nov. 19, 1850	7,783
Annunciators, Electro-magnetic hotel	G. B. Scott	Brooklyn, N. Y	Jan. 9, 1872	122,664
Annunciator, Speaking-tube	R. May	New York, N. Y	July 2, 1872	128,552
Annunciator, Speaking-tube	D. Rousseau	New York, N. Y	June 10, 1873	139,818
Ant-destroying composition	J. D. Dennis	Gilroy, Cal	June 8, 1869	91,094
Ant-destroying process	J. J. Dulany and H. Dreyer	Oakland, Tex	Oct. 28, 1873	144,075
Ant-obstructing device	E. Rooks	Trenton, Tenn	Feb. 14, 1871	111,875
Ant-trap	T. G. Ames	Kosse, Tex	Apr. 1, 1873	137,336
Ant-trap	G. W. Cottingham and J. S. Menefee.	Texana, Tex	Aug. 7, 1860	29,468
Ant-trap	L. Rubarth	Davilla, Tex	July 15, 1873	140,954
Anthracene, Apparatus for preparing	J. C. F. Cheever	New York, N. Y	Sept. 17, 1872	131,393
Anthracene, Apparatus for preparing	H. J. Fenner and F. Versmann.	Greenwich and London, England.	Apr. 30, 1872	126,277
Anthracene, Manufacture of	H. J. Fenner and F. Versmann.	Greenwich and London, England.	Aug. 27, 1872	130,909
Anthracite and other fuel, Promoting combustion of.	P. Davis	Baltimore, Md	July 29, 1834	....
Anti-attrition compound	E. Mott	Philadelphia, Pa	Oct. 4, 1817	....
Anti-friction axle and journal-box	J. H. Carkeet	Montgomery, Ala	Dec. 31, 1867	72,796
Anti-friction box	J. L. Dutton, sr	Philadelphia, Pa	Dec. 20, 1859	26,483
Anti-friction box	D. A. Morris	Pittsburgh, Pa	Mar. 21, 1854	10,676
Anti friction box	G. T. Parry	Spring Garden, Pa	Aug. 2, 1853	9,912
Anti-friction box	J. L. Parry	Philadelphia, Pa	Nov. 9, 1869	96,720
Anti-friction box and axle	J. Harris, jr	Boston, Mass	Feb. 22, 1848	5,452
Anti-friction box for axles, &c	W. Rowan	Belfast, Ireland	Oct. 9, 1845	4,226
Anti-friction box for hubs and rotary bearings	E. Fisk and J. C. Green	Fayette, Me	Nov. 14, 1835	....
Anti-friction box for shafting	J. McIlvain	Churchville, Md	Dec. 29, 1868	85,393
Anti-friction press	A. H. Emery	New York, N. Y	Sept. 11, 1860	29,900
Anti-friction roller for shafting	W. E. Wilcox	Peoria, Ill	June 22, 1869	91,695
Anti-friction washer	U. H. Reed, J. Lake, and L. Sisson.	North Easton, Mass	Nov. 17, 1868	84,067
Anti-friction wheel for belt-gearing	D. Eldredge	Philadelphia, Pa	Aug. 29, 1865	49,614
Anti-stream boat	W. Wadsworth	Hartford, Conn	July 17, 1806	....
Anvil	O. and S. E. Brigham	Fitchburg, Mass	Feb. 3, 1857	16,525
Anvil	L. Kirkup	Brooklyn, N. Y	Jan. 1, 1867	60,745
Anvil	C. Peters and W. Fetter	Trenton, N. J., and Bucks County, Pa.	May 4, 1852	8,925
Anvil	C. H. Schadt	New York, N. Y	Feb. 12, 1861	31,410
Anvil and hammer	C. Josselyn	Pembroke, Mass	Oct. 9, 1816	....
Anvil and vise combined	J. D. Barton, F. S. Rogers, and D. Fisher.	Kalamazoo, Mich	July 17, 1866	56,346
Anvil and vise combined	R. D. Chandler	Fairhaven, N. J	June 29, 1869	91,909
Anvil, Apparatus for polishing	M. Fisher and J. Norris	Trenton, N. J	Oct. 4, 1853	10,066
Anvil-block	S. Van Tiers	Hanover Iron Works, Pa	Mar. 19, 1836	....
Anvil, Centering	J. Adt	Waterbury, Conn	July 7, 1863	39,193
Anvil-clamp for holding ties while being upset by hand-forging.	N. P. Quick	Carmel, N. Y	July 20, 1869	92,881
Anvil-cutter	V. A. Dunn	West Peru, Me	Mar. 10, 1868	75,395
Anvil for forming horseshoe-calks	R. Saylor and E. T. Rhodes	Marshall, Mich	Dec. 1, 1868	84,583
Anvil for making horseshoes	A. S. Wilkinson	Pawtucket, R. I	July 10, 1866	56,311
Anvil for repairing T-rails, Vise	S. Mason and E. M. Davis	Michigan City, Ind	Apr. 6, 1858	19,861
Anvil for swaging horse shoe-calks	P. Badore	Montpelier, Vt	Dec. 18, 1866	60,610

Index of patents issued from the United States Patent Office from 1790 *to* 1873, *inclusive*—Continued.

Invention.	Inventor.	Residence.	Date.	No.
Anvil-last	D. Bainbridge	Philadelphia, Pa	Oct. 11, 1870	108, 231
Anvil-making machine	J. Taylor	Shade Gap, Pa	Jan. 31, 1844	3, 418
Anvil on which to rivet trunks	W. D. Burnett	Newark, N. J	Dec. 18, 1866	60, 617
Anvils and the top and bottom parts of hammers, &c., Manufacture of.	D. Foster	Sheffield, England	Oct. 5, 1869	95, 577
Apiary	W. Faulkner	Vevay, Ind	Feb. 4, 1868	74, 065
Apparel, Machine for incising button-holes and embossing and printing articles of.	S. S. Stone	Troy, N. Y	Aug. 23, 1864	43, 932
Apparel, &c., Manufacture of seamless felt wearing.	S. M. Perkins	Springfield, Pa	Jan. 25, 1853	9, 557
Apparel, Means for stiffening articles of wearing	J. Sloan	Philadelphia, Pa	Feb. 11, 1868	74, 250
Apparel, Perforated wearing	H. E. Smith	New York, N. Y	Nov. 2, 1869	96, 358
Apparel, Safety-attachment for pocket of	A. Arnemann	Guttenberg, Iowa	Sept. 18, 1868	81, 862
Apparel, shoes, &c., Fastening wearing	R. Judson and W. H. Lynch	Mattcawan, N. Y	Mar. 24, 1868	75, 924
Apparel, Waist-band for wearing	Z. Wolfsbruck	New York, N. Y	Apr. 14, 1868	76, 871
Apple and other fruit, Grinding	B. Churchell	Buckfield, Me	Dec. 9, 1825	....
Apple and other fruit, Machine for stringing dried	S. T. Sanford	Fall River, Mass	Feb. 11, 1863	37, 708
Apple and vegetable knife	A. Kimber	Muncie, Ind	June 30, 1872	129, 966
Apple-bin	S. S. and G. W. Cole	Canton, Ill	Aug. 30, 1864	43, 973
Apple-butter stirrer	G. W. Collins	West Lebanon, Pa	Aug. 26, 1873	142, 208
Apple cutter and corer	J. Wroten	Salisbury, Md	Sept. 26, 1865	50, 197
Apple cutting and coring machine	C. Darling	Utica, N. Y	July 1, 1856	15, 224
Apple cutting and coring machine	A. F. Ledbetter	Westminster, N. C	Mar. 1, 1859	23, 095
Apple-cutting machine	A. Glendening	Loudoun County, Va	Sept. 9, 1824	....
Apple gathering and cleansing machine	E. Wells	Duxbury, Vt	Dec. 6, 1831	....
Apple-gathering device	H. F. Wadhams	South Dansville, N. Y	Apr. 2, 1867	63, 584
Apple-gathering instrument	E. Tyler	Hancock, Ill	Nov. 7, 1865	50, 859
Apple-grinder	J. R. Dean	Augusta Township, Ohio	Apr. 5, 1832	....
Apple-grinder	C. F. D. Jones, jr	New Hartford, N. Y	Aug. 13, 1872	130, 505
Apple-grinder and corn-sheller	E. Harris, J. Newton, W. Webster, and J. W. Dart.	Truxton, N. Y	Mar. 10, 1832	....
Apple-grinding machine	R. Butterworth	Trenton, N. J	Sept. 5, 1865	49, 714
Apple-grinding machine	I. T. Carpenter	Martin's Ferry, Ohio	Jan. 24, 1860	26, 892
Apple-grinding machine	D. D. Demarest	New Milford, N. Y	Jan. 14, 1829	....
Apple-grinding machine	C. H. Weeks	Paris, N. Y	Apr. 9, 1827	....
Apple-grinding machine	T. I. Wells	New York	Aug. 25, 1840	1, 739
Apple-grinding mill	W. O. Hickok	Harrisburgh, Pa	Nov. 20, 1855	13, 839
Apple-grinding mill	F. B. Hunt	Westfield, Ind	July 26, 1853	9, 876
Apple-grinding mill	L. McKee	Hagerstown, Md	Feb. 26. 1861	31, 553
Apple-grinding mill, Hand	R. P. Clark	Johnstown, N. Y	Sept. 13, 1859	25, 385
Apple grinding or grating machine	J. Farnham	Tioga, N. Y	June 28, 1825	....
Apple-mill	S. Keeler	Lancaster, Pa	Jan. 27, 1863	37, 510
Apple-mill	G. S. Rust	Chester, Ill	Feb. 23, 1864	41, 723
Apple-mill, &c., Convertible	G. S. Rust	Chester, Ill	Aug. 12, 1862	36, 171
Apple-mill, Grinding-cylinder for	J. Shaefer	Lancaster, Pa	Aug. 9, 1859	25, 050
Apple-packing machine	H. B. Gill	Ogden, N. Y	May 6, 1862	35, 199
Apple-parer	W. A. C. Oaks	Reading, Pa	July 22, 1873	141, 070
Apple-pulping machine	G. Clayton	Marshallton, Pa	June 6, 1871	115, 705
Apple-quarterer	C. E. Billings	Warren, Vt	Dec. 8, 1868	84, 791
Apple-rake	E. Hart	Berlin, Conn	Dec. 31, 1812	....
Apple-sauce	A. R. Davis	Cambridge, Mass	Feb. 9, 1869	86, 649
Apple-seed gatherer	R. Brusie	Cleveland, Ohio	July 2, 1867	66, 211
Apple steaming and mashing machine	J. Dimm	Greenwood, Pa	Dec. 1, 1837	499
Apples and shelling corn, Grinding	M. Morehouse	Butler County, Ohio	Jan. 23, 1834	....
Apples, Apparatus for expressing juice from	G. R. Burt	Perry, N. Y	Jan. 23, 1872	122, 878
Apples, corn, bark, &c., Grinding	G. Parmalee	Reading, Conn	Mar. 4, 1830	....
Apples, Cutting and coring	G. C. Wright	Le Roy, Ohio	June 26, 1860	28, 935
Apples, Drying and preserving	M. P. Smith	Baltimore, Md	Nov. 8, 1870	109, 068
Apples, Grinding	A. Dean	Jerusalem, N. Y	Mar. 15, 1859	23, 231
Apples, Machine for gathering	S. Laning	Camden, N. J	Nov. 1, 1830	....
Approach-gate	S. H. Cole	East Enterprise, Ind	Aug. 16, 1870	106, 330
Approach-opening gate	R. Brinkerhoff	Mansfield, Ohio	May 28, 1861	32, 407
Approach-opening gate	S. Cole	Rochester, N. Y	May 5, 1857	17, 202
Approach-opening gate	W. T. Cole	Reed, Ohio	Sept. 24, 1861	33, 338
Approach-opening gate	R. E. House	Binghamton, N. Y	Jan. 13, 1857	16, 386
Approach-opening gate	C. A. Howard	Pontiac, Mich	Sept. 29, 1857	18, 283
Approach-opening gate	G. W. McGill	Buffalo, N. Y	May 19, 1857	17, 335
Approach-opening gate	W. G. Philips	Newport, Del	Dec. 9, 1856	16, 187
Approach-opening gate	E. Woodruff	Elizabethtown, N. J	Apr. 3, 1855	12, 652
Aquaria	E. D. Davis	Brooklyn, N. Y	Nov. 9, 1858	22, 019
Aquaria, Construction of	J. Chilcott and J. Scrimgeour	Brooklyn, N. Y	Oct. 5, 1858	21, 719
Aquarium	J. A. Cutting	Boston, Mass	Mar. 12, 1861	31, 657
Aquarium	H. Shlarbaum	New York, N. Y	Jan. 1, 1861	31, 040
Aqueduct	J. Osborn	Mount Carmel, Conn	Sept. 4, 1860	29, 906
Aqueduct-coupling	J. Aldrich	Lake Village, N. H	Jan. 16, 1866	52, 010
Aqueduct, Earthen or clay tube for	J. Ramsey	Burke, Vt	Apr. 4, 1810	....
Aqueduct formed of water-proof lime, Water or gas	J. M. Benham	Bridgewater, N. Y	Oct. 1, 1830	....
Aqueduct or water-conductor, Elevating	J. Clapp	Greenfield, Mass	Nov. 16, 1811	....
Aqueduct-pipe	T. B. Armistead	Bloomfield, N. Y	Apr. 15, 1830	....
Aqueduct-pipe, Composition for	G. Myers	Bridgewater, N. Y	Mar. 28, 1844	3, 514
Aqueduct-pipe, Making	T. B. Robbins	Stockbridge, Mass	Mar. 19, 1821	....
Aqueduct supply-pipe, Construction of	J. H. Thorndike	Boston, Mass	July 1, 1844	3, 650
Aqueducts of water-proof lime, Making	J. M. Benham	Bridgewater, N. Y	Aug. 29, 1827	....
Arastra	J. C. Davis	Alameda County, Cal	Apr. 8, 1862	34, 881
Arastra	S. E. Woodworth and J. S. Wethered.	Murphy's and San Francisco, Cal.	June 11, 1861	32, 548
Arbor, Expansion	H. C. Taylor	Marquette, Mich	June 25, 1872	128, 260
Arbor, Grape, &c	J. O. Altick	Dayton, Ohio	Mar. 19, 1867	62, 989
Arbor or fence-post	J. P. Dorman	Galesburgh, Ill	Nov. 30, 1869	97, 367
Arch, Construction of	F. Alsip	North McGregor, Iowa	Nov. 7, 1871	120, 608
Arch-girder	J. Bevan	New York, N. Y	May 21, 1850	7, 374
Arch, tunnel, &c., Construction of	G. T. Lape	Summit, N. Y	Sept. 1, 1868	81, 797
Arched structure	C. Henderson	London, England	Apr. 18, 1871	113, 881
Archil, Preparation of	J. Eberhardt	Philadelphia County, Pa	June 27, 1854	11, 194
Archil, Preparation of	L. Jarosson	New York, N. Y	June 15, 1852	9, 027
Architecture, Naval	J. Bowdlear	Roxbury, Mass	Oct. 20, 1863	40, 321
Architecture, Naval	C. Hoxie	New York	Feb. 11, 1834	....
Area of irregular figures, Mode of finding the	T. Wood	Smithfield, Ohio	July 22, 1839	1, 256
Arithmetic, Art of teaching	O. H. Shaw	Richmond, Va	Apr. 16, 1831	....

Index of patents issued from the United States Patent Office from 1790 *to* 1873, *inclusive*—Continued.

Invention.	Inventor.	Residence.	Date.	No.
Arithmetical frame	H. K. Bugbee	New York, N. Y	July 12, 1864	43, 545
Arithmetical frame	E. T. Curtis	Calumet, Mich	Aug. 26, 1873	142, 151
Arithmetical sum setter	A. W. Price	Detroit, Mich	Apr. 23, 1872	126, 123
Arithmometer	T. Hill	Waltham, Mass	Nov. 24, 1857	18, 692
Arithmometer for adding	O. L. Castle	Upper Alton, Ill	Nov. 24, 1857	18, 675
Arithmometer for addition	O. L. Castle	Upper Alton, Ill	Nov. 2, 1858	21, 941
Ark, Safety	W. Hollins	Baltimore, Md	May 4, 1824	....
Arm and hand, Artificial	B. F. Palmer	Philadelphia, Pa	Jan. 11, 1859	22, 576
Arm and hand, Artificial	T. Uren	New York, N. Y	Jan. 31, 1865	46, 158
Arm and hand, Artificial	T. Uren	New York, N. Y	Jan. 31, 1865	46, 159
Arm, Artificial	J. Condell	Morristown, N. Y	July 11, 1865	48, 659
Arm, Artificial	E. Cotly	Washington, D. C	Oct. 27, 1863	40, 397
Arm, Artificial	H. A. Kimball and A. J. Lawrence.	Philadelphia, Pa	May 23, 1865	47, 835
Arm, Artificial	J. H. Koeller	New York, N. Y	July 19, 1864	43, 590
Arm, Artificial	J. H. Koeller	New York, N. Y	Oct. 11, 1864	44, 638
Arm, Artificial	D. W. Kolbe	Philadelphia, Pa	Nov. 15, 1864	45, 052
Arm, Artificial	M. Lincoln	Malden, Mass	Aug. 11, 1863	39, 487
Arm, Artificial	A. McOmber	Schenectady, N. Y	Jan. 1, 1867	60, 921
Arm, Artificial	J. Peterson	Canoga, N. Y	Mar. 7, 1865	46, 696
Arm, Artificial	E. Spellerberg	Philadelphia, Pa	Apr. 26, 1864	42, 515
Arm, Artificial	E. Spellerberg	Philadelphia, Pa	Nov. 28, 1865	51, 238
Arm, Artificial	I. Stoffel	Washington, D. C	Jan. 10, 1865	45, 876
Arm, Artificial	I. Stoffel	Washington, D. C	Aug. 28, 1866	57, 594
Arm, Artificial	T. Uren	New York, N. Y	May 30, 1865	48, 002
Arm-chair, Folding	H. S. Golightly and C. S. Twitchell.	New Haven, Conn	June 28, 1864	43, 366
Arm-chair, Folding	W. C. Goodwin	Hamden, Conn	Jan. 21, 1862	34, 204
Arm-rest and paper-cutter, Combined	C. B. Dickinson	Brooklyn, N. Y	May 26, 1868	78, 192
Arm-rest, Compositor's	C. L. Alexander	Washington, D. C	May 28, 1872	127, 208
Arm-rest, Penman's	J. B. Withey	Lexington, Mich	May 26, 1868	78, 250
Arm-support for keyed instrument	L. Buchbuger	Chicago, Ill	Nov. 29, 1870	109, 582
Arms, Slinging	O. E. Woods	Philadelphia, Pa	May 15, 1866	54, 807
Arms-supporter for riflemen	S. Kinman	Humboldt, Cal	Feb. 14, 1865	46, 365
Arms-supporter, India-rubber	H. Greentree	Baltimore, Md	Dec. 12, 1871	121, 868
Armillary sphere	H. Bryant	Hartford, Conn	Sept. 10, 1872	131, 148
Armlet	A. S. Potter	Providence, R. I	Apr. 9, 1872	125, 407
Arm-pit shield	W. E. Beames	New York, N. Y	Oct. 22, 1872	132, 348
Arm-pit shield	J. Sibley	New York, N. Y	Sept. 16, 1873	142, 875
Armor, Defensive	C. H. Hudson	Roxbury, Mass	Sept. 27, 1864	44, 426
Armor for marine and other batteries, Defensive	J. B Eads	Saint Louis, Mo	July 14, 1863	39, 218
Armor for ship	T. Whitby	Lambeth, England	Sept. 3, 1867	68, 474
Armor for ship, Defensive	M. L. Callender and N. W. Northrup.	New York and Greene, N. Y.	May 27, 1862	35, 412
Armor for ships and other batteries, Defensive	C. W. S. Heaton	Belleville, Ill	Apr. 14, 1863	38, 206
Armor for ships and other batteries, Defensive	R. H. Jewett	Mount Sterling, Ill	Feb. 17, 1863	37, 695
Armor for ships and other batteries, Defensive	R. Montgomery	New York, N. Y	Feb. 10, 1863	37, 633
Armor for ships and other batteries, Defensive	W. W. W. Wood	Philadelphia, Pa	Sept. 23, 1862	36, 546
Armor for ships, batteries, &c., Defensive	G. B. Manley	Danville, Pa	Jan. 13, 1863	37, 402
Armor for ships, Construction of the defensive	W. Rumbold	Saint Louis, Mo	July 15, 1862	35, 895
Armor for ships, Metallic defensive	W. Ballard	New York, N. Y	June 24, 1862	35, 665
Armor for water and land batteries, Defensive	J. L. Jones	Saint Louis, Mo	Apr. 15, 1862	35, 001
Armor, Metallic defensive	B. B. Hotchkiss	Sharon, Conn	Aug. 12, 1862	36, 152
Armor, Naval defensive	G. M. Mowbray	Titusville, Pa	Sept. 9, 1862	36, 439
Armor plate, Defensive	E. Cox	Covington, Ky	May 27, 1862	35, 364
Armor-plate, Defensive	M. Wappich	Sacramento, Cal	Mar. 3, 1863	37, 836
Armor-plate for land or marine battery	F. P. Dimpfel	Philadelphia, Pa	Aug. 4, 1863	39, 384
Armor-plate for marine or other batteries, Means of connecting metallic.	T. Shaw	Philadelphia, Pa	May 13, 1862	35, 279
Armor-plate for ships and other batteries	B. T. Babbitt	New York, N. Y	Jan. 13, 1863	37, 380
Armor-plate for vessels	J. F. Winslow	Troy, N. Y	May 27, 1862	35, 407
Armor-plate, Ships'	H. H. Warden	New York, N. Y	Feb. 25, 1862	34, 539
Armor-plates, Means of affixing defensive	E. Brady	Philadelphia, Pa	Mar. 3, 1863	37, 807
Armor-plates, Mode of protecting	M. Bernabé	Toulon, France	Jan. 15, 1867	61, 143
Armor-plates, Pile for	J. Jeavons	Sheffield, England	Dec. 13, 1870	110, 143
Armor-plates, Press for bending ships'	E. Sauer	New York, N. Y	Mar. 24, 1863	37, 982
Armor-plates to marine batteries, Means of affixing defensive.	O. G. Stillman	Fabius, N. Y	Nov. 25, 1862	37, 013
Armor-plates to vessels, Attaching	J. Rusch	New York, N. Y	May 20, 1862	35, 353
Armor, Ships' defensive	S. D. Carpenter	Madison, Wis	May 23, 1865	47, 796
Armor to navigable vessels and water batteries, Means of attaching.	J. B. Love	Philadelphia, Pa	Oct. 22, 1861	33, 532
Armored can	W. F. Thompson	Toledo, Ohio	May 7, 1872	126, 502
Armor-plating for vessels	G. J. Gunther	London, England	Nov. 24, 1868	84, 418
Army stretcher	J. J. Smith	Philadelphia, Pa	Sept. 8, 1863	39, 840
Arrow-spring	J. B. Cleaveland	Indianapolis, Ind	Aug. 15, 1871	118, 108
Articulator	E. T. Starr	Philadelphia, Pa	May 19, 1868	78, 151
Artillery and mining blasting	T. P. Shaffner	Louisville, Ky	Dec. 18, 1866	60, 572
Artillery, Carriage for field	D. Cobb	Boston, Mass	Apr. 25, 1808	....
Artillery carriages, Wheel for flying	J. D. Murphy	Baltimore, Md	Apr. 3, 1860	27, 733
Artillery, Field	S. W. Wood	Cornwall, N. Y	May 7, 1872	126, 607
Artist's stretcher	J. E. Todd	Middletown, Conn	Sept. 18, 1866	58, 154
Artist's stretching-frame	J. F. Carroll	South Boston, Mass	July 21, 1868	80, 135
Artist's materials, Apparatus for drying	G. D. Jones	New York, N. Y	Dec. 11, 1866	60, 382
Arts, Composition of matter for various uses in the	C. L. Coombs	Washington, D. C	Aug. 24, 1869	94, 080
Asbestus and obtaining useful products therefrom, Treating.	J. S. Rosenthal	Philadelphia, Pa	Aug. 20, 1872	130, 663
Asbestus and other fibrous minerals, Treating	C. A. Stevens	New York, N. Y	Mar. 14, 1871	112, 650
Asbestus and other mineral fibers for useful purposes, Treating.	C. A. Stevens	New York, N. Y	Mar. 14, 1871	112, 649
Asbestus for the production of textile fiber, Treatment of.	J. S. Rosenthal	Philadelphia, Pa	Aug. 6, 1872	130, 245
Asbestus, Treating	J. S Rosenthal	Philadelphia, Pa	Aug. 13, 1872	130, 538
Asbestus Use and application of	J. Scott	Philadelphia, Pa	Nov. 26, 1835	....
Asbestus yarn	J. S. Rosenthal	Philadelphia, Pa	Aug. 13, 1872	130, 537
Ash-bin	W. W. Chase	Springfield, N. H	July 28, 1868	80, 451
Ash-box	G. Dunlop	Williamsburgh, N. Y	Sept. 24, 1872	131, 607
Ash-box	J. Kee and J. Sloan	Philadelphia, Pa	Apr. 24, 1866	54, 173

Index of patents issued from the United States Patent Office from 1790 *to* 1873, *inclusive*—**Continued.**

Invention.	Inventor.	Residence.	Date.	No.
Ash-box and sifter	H. D. Rogers	Cincinnati, Ohio	Mar. 1, 1870	100, 452
Ash-hopper	P. Zimmerman	Sylvan. Pa	Aug. 3, 1869	93, 388
Ash-house	M. Hall, jr	Osborn, Ohio	Dec. 17, 1867	72, 290
Ash-leach	J. W. Kernodle and A. H. Haun.	Lebanon, Ind	June 17, 1873	139, 963
Ash-leach	D. T. Miller	Woodbury, Pa	Apr. 9, 1872	125, 607
Ash-leach	S. A. Porter	Prescott, Wis	Feb. 16, 1864	41, 638
Ash-leaching apparatus	P. Perdew and A. W. Brinkerhoff.	Syracuse, Ohio	May 20, 1856	14, 925
Ash-pan	J. A. Lawson	Troy, N. Y	June 16, 1863	38, 902
Ash-pan	C. H. Low	Cleveland, Ohio	Nov. 9, 1869	96, 711
Ash-pan cleaner, Locomotive	A. O. Denio	Wilmington, Del	Aug. 12, 1873	141, 769
Ash-pan drawer and lifter	J. Morrison, jr	Troy, N. Y	June 20, 1865	48, 299
Ash-pan, Locomotive	L. H. Dee	Grand Junction, Iowa	July 12, 1870	105, 317
Ash-pan, Locomotive	H. A. Stoddard	Springfield, Mass	Sept. 6, 1870	107, 117
Ash-screen	E. C. Jenkins, jr	Worcester, Mass	Jan. 24, 1871	111, 126
Ash-screen and coal-hod combined	T. J. Thurston	Lewiston, Me	Aug. 11, 1868	81, 037
Ash-screen and commode	J. C. Morrell	Manchester, England	Mar. 11, 1873	136, 609
Ash-shovel	R. J. S. Thompson	Washington, Pa	Mar. 15, 1870	100, 952
Ash-shovel and sifter	W. C. McGill	Cincinnati, Ohio	Feb. 23, 1869	87, 186
Ash-sifting shovel	A. M. Olds	Chicago, Ill	Sept. 5, 1865	49, 781
Ash-tub or leach	C. Roop	Middletown, Pa	July 9, 1867	66, 636
Ashes, Apparatus for leaching	E. Williams	Westfield, N. Y	Jan. 9, 1838	5[illegible]9
Ashes, Leaching	G. Clement	Canandaigua, N. Y	June 10, 1837	228
Ashes, Manufacturing	J. Bellows, jr., and E. White	....	Dec. 29, 1804	....
Ashlar-cutting	W. Boulton	Tompkinsville, N. Y	July 26, 1870	105, 633
Asparagus-plant, Utilizing the products of the	J. P. Gage and J. C. Gilbert	New York, N. Y	Oct. 7, 1862	36, 608
Asphalt pipe, Manufacture of	A. Müller	Jersey City, N. J	Oct. 21, 1873	143, 922
Asphalt, Treating	P. Barthel	Frankfort-on-the-Main, Germany.	Feb. 18, 1873	135, 879
Asphaltum, Process for burning	S. Stevens	Washington, D. C	Apr. 9, 1872	125, 497
Astronomical instrument	C. Emmanuel	Paris, France	Jan. 17, 1865	45, 954
Astronomical instrument	H. Glover	New York, N. Y	Nov. 16, 1858	22, 075
Astronomical machine	J. Swaim	Philadelphia, Pa	June 29, 1833	....
Atmosphere, Apparatus for moistening	J. G. Garland	Biddeford, Me	July 1, 1873	140, 410
Atmosphere-cooler	M. J. Kelly	Chicago, Ill	Oct. 18, 1864	44, 731
Atmosphere of apartments, Boiler for hydrating the.	P. I. Schopp	Louisville, Ky	Oct. 12, 1869	95, 735
Atmospheric engine	E. and J. Prentis	Baltimore, Md	June 22, 1824	....
Atmospheric engine	S. E. Tuttle	Evansville, Wis	Mar. 12, 1872	124, 520
Atmospheric lubricator, Self-feeding	J. Sutton	New York, N. Y	Nov. 6, 1855	13, 769
Atmospheric motive-power	W. Jones	New Albany, Ind	Mar. 5, 1872	124, 361
Atmospheric motive-power	W. Jones	New Albany, Ind	July 16, 1872	129, 567
Atmospheric propelling-engine, Talbot's	E. A. Talbot	Dublin, Ireland	June 21, 1828	....
Atmospheric regulator for stove, furnace, &c	B. Holly	Seneca Falls, N. Y	July 13, 1858	20, 919
Atmospheric steam-engine	J. Mead and M. Kitchell	Lebanon, Ohio	July 27, 1829	....
Atmospheric steam-engine, Two-cylinder	W. Willis	Charleston, S. C	Nov. 14, 1826	....
Atmospheric transportation	R. H. Gilbert	Washington, D. C	Feb. 8, 1870	99, 663
Atomizer	J. J. Essex	Newport, R. I	June 27, 1871	116, 286
Atomizer	C. P. Janes	Boston, Mass	June 6, 1871	115, 615
Atomizer	H. D. Lockwood	Charlestown, Mass	Aug. 1, 1871	117, 651
Atomizer	H. D. Lockwood	Charlestown, Mass	May 28, 1872	127, 356
Atomizer-bulb	H. D. Lockwood	Charlestown, Mass	Apr. 29, 1873	138, 416
Atomizer for administering medicine	J. Sheedy	New York, N. Y	Sept. 21, 1869	95, 152
Atomizer, Liquid	H. Kraut	Saint Louis, Mo	July 7, 1868	79, 764
Atomizer or vaporizer	J. N. Gerard	New York, N. Y	Sept. 23, 1873	143, 070
Atomizer, Steam	I. P. Leete	Philadelphia, Pa	Oct. 25, 1870	108, 710
Atomizing apparatus	A. M. Shurtleff	Boston, Mass	Mar. 16, 1869	87, 978
Atomizing apparatus for surgical use	A. M. Shurtleff	Boston, Mass	Mar. 24, 1868	75, 991
Atomizing, Drying and concentrating liquid substance by.	S. R. Percy	New York, N. Y	Apr. 9, 1872	125, 406
Atomizing, inhaling, and injecting liquids and gases, Apparatus for.	W. R. Leonard	New York, N. Y	Apr. 23, 1872	125, 965
Atomizing liquids, Apparatus for	A. M. Shurtleff	Boston, Mass	Mar. 3, 1868	75, 208
Atomizing liquids, Apparatus for	A. M. Shurtleff	Boston, Mass	May 2, 1871	114, 482
Atomizing-tube	C. H. Eccleston	Oxford, N. Y	Sept. 10, 1867	68, 614
Atomizing-tube	T. J. Holmes	Malden, Mass	Aug. 23, 1870	106, 587
Atomizing-tube	W. L. Leach	Boston, Mass	Jan. 5, 1869	85, 596
Atomizing-tube	A. M. Shurtleff	Boston, Mass	Dec. 18, 1866	60, 580
Auger	J. Blake	East Pepperell, Mass	Apr. 17, 1860	27, 946
Auger	J. Brownell	Otsego, N. Y	May 16, 1815	....
Auger	L. Colt	Niagara Falls, N. Y	Aug. 11, 1868	80, 915
Auger	R. Cook	Saratoga Springs, N. Y	June 17, 1851	8, 162
Auger	K. Curtiss	Winchester, Conn	Apr. 22, 1856	14, 752
Auger	R. French	Derby, Conn	July 20, 1843	3, 181
Auger	E. C. Gillette	Richfield, British Columbia	Mar. 14, 1865	46, 854
Auger	C. L. Griswold	Chester, Conn	May 30, 1865	47, 946
Auger	W. Hale	Champlain, N. Y	Feb. 7, 1807	....
Auger	J. M. Hathaway	New York, N. Y	Sept. 4, 1860	29, 883
Auger	T. Hofstatter, jr	New York, N. Y	Mar. 12, 1867	62, 849
Auger	R. H. Hopkins	Hinsdale, N. H	June 21, 1870	104, 457
Auger	M. Howe	New York	Apr. 2, 1814	....
Auger	C. Hoxie	....	July 12, 1804	....
Auger	W. A. Ives	New Haven, Conn	Aug. 28, 1860	29, 793
Auger	W. A. Ives	New Haven, Conn	Oct. 12, 1869	95, 803
Auger	W. A. Ives	New Haven, Conn	Feb. 7, 1871	111, 648
Auger	R. Jennings	Deep River, Conn	Jan. 30, 1855	12, 318
Auger	W. Jones	Portsmouth, Va	June 12, 1835	....
Auger	W. Jones	Portsmouth, Va	June 15, 1835	....
Auger	A. C. Kasson	Milwaukee, Wis	Jan. 15, 1867	61, 208
Auger	P. Kavanagh	Canarsie, N. Y	Sept. 10, 1872	131, 279
Auger	T. C. Keith	Valley Falls, R. I	Mar. 12, 1867	62, 754
Auger	R. M. Lafferty and E. P. Smith.	Three Rivers, Mich	Sept. 12, 1871	118, 806
Auger	E. G. Lamson	Windsor, Vt	Jan. 16, 1866	52, 056
Auger	H. T. Love	Vermillion Township, Kan	Nov. 7, 1865	50, 887
Auger	J. A. McGee	New York, N. Y	Dec. 10, 1867	72, 065
Auger	C. Monson	Moscow, Wis	June 14, 1870	104, 335
Auger	H. T. Moody	Newburyport, Mass	May 29, 1866	55, 144

Index of patents issued from the United States Patent Office from 1790 *to* 1873, *inclusive*—Continued.

Invention.	Inventor.	Residence.	Date.	No.
Auger	A. Newton, L. B. Smith, and E. Sanford.	Meriden, Conn	Mar. 27, 1847	5,036
Auger	I. T. Payne	Chester, Conn	Mar. 10, 1868	75,454
Auger	N. B. Phelps	Rochester, N. Y	Dec. 27, 1859	26,613
Auger	H. Pitcher	Fond du Lac, Wis	Aug. 29, 1871	118,552
Auger	H. Pitcher	Fond du Lac, Wis	Mar. 26, 1872	126,076
Auger	T. M. Richardson	Stockton, Me	Sept. 28, 1869	95,379
Auger	N. C. Sanford	Meriden, Conn	Sept. 23, 1862	36,534
Auger	N. C. Sanford	Meriden, Conn	June 16, 1868	79,012
Auger	J. H. Schreiner	Philadelphia, Pa	Aug. 17, 1826	....
Auger	J. Swan	Seymour, Conn	May 30, 1871	115,541
Auger	A. C. Vaughan	Philadelphia, Pa	Mar. 31, 1868	76,278
Auger	B. Walch	Frederick, Md	Jan. 17, 1871	111,099
Auger	C. Wardwell	Painesville, Ohio	Apr. 20, 1869	89,097
Auger	S. Wood	Worcester, Mass	Nov. 1, 1859	25,990
Auger and gimlet	O. Percival	East Haddam, Conn	Oct. 14, 1835	....
Auger and reamer combined	O. W. Townsend	Fond du Lac, Wis	Feb. 21, 1871	111,990
Auger, Annular	R. Stewart	Elmira, N. Y	Apr. 10, 1866	53,896
Auger-bit	A. L. Andrews	Bristol, Conn	June 21, 1870	104,404
Auger-bit	W. A. Ives	New Haven, Conn	June 1, 1869	90,755
Auger-bit	H. C. Lewis	Essex, Conn	June 1, 1869	90,759
Auger-bit	J. C. Mills	Rochester, N. Y	Feb. 21, 1871	112,065
Auger-bit	H. L. Shailer	Deep River, Conn	Mar. 16, 1869	87,796
Auger-bit	R. A. Whitmore	De Witt, Ark	Nov. 5, 1872	132,883
Auger-bit die, Construction of	R. N. Watrous	Elmira, N. Y	Aug. 8, 1871	117,838
Auger-bit, Expanding	A. Weeks	South Boston, Mass	Aug. 11, 1857	18,003
Auger-bits, Device for re-setting old	C. W. Beals	Greig, N. Y	Aug. 1, 1871	117,594
Auger-bits, Machine for bending the lips of	J. Swan	Seymour, Conn	Mar. 15, 1870	100,816
Auger-bits, Machine for forming	W. A. Ives	New Haven, Conn	Mar. 19, 1872	124,683
Auger-bits, Machine for forming	J. Swan	Seymour, Conn	July 29, 1873	141,401
Auger-bits, Machine for forming lips of	J. Swan	Seymour, Conn	June 27, 1871	116,509
Auger-bits, Machine for making	W. W. Grier and R. H. Boyd	Hutton, Pa	May 22, 1866	54,893
Auger-bits, Machine for making	J. Swan	Seymour, Conn	Apr. 21, 1868	76,955
Auger-bits, Manufacture of	J. Swan	Seymour, Conn	July 14, 1868	80,027
Auger-bits, Method of forming spur-lip for	W. A. Ives	New Haven, Conn	Feb. 28, 1871	112,255
Auger-blanks, Die for swaging the ends of	R. Jennings	Deep River, Conn	June 12, 1866	55,498
Auger, Boring and withdrawing	J. Snyder	Tuscarora, N. Y	June 1, 1830	....
Auger, Combined convex and concave	N. C. Sanford	Meriden, Conn	Mar. 27, 1849	6,221
Auger, Double-podded center-screw	E. L'Hommedieu	Saybrook, Conn	July 31, 1809	....
Auger, Earth	G. H. Baisley and G. Wilson	Hamilton, Mo	July 15, 1873	140,756
Auger, Earth	W. H. Beach and C. W. Hanson	Hamburg, Iowa	Feb. 4, 1873	135,509
Auger, Earth	E. H. Clark	Appleton, Wis	Oct. 28, 1873	143,963
Auger, Earth	W. Cole	Milan, Tenn	Nov. 11, 1873	144,510
Auger, Earth	G. G. Collins and J. A. Morrison	Philo, Ill	Jan. 14, 1873	134,734
Auger, Earth	W. T. Cooley	Chicago, Ill	Sept. 30, 1873	143,276
Auger, Earth	A. J. Dine	Xenia, Ind	May 7, 1867	64,503
Auger, Earth	X. Earle	Depere, Wis	May 21, 1872	126,941
Auger, Earth	S. Emery	Cameron, Mo	Aug. 26, 1873	142,218
Auger, Earth	W. H. Gates	Perry Township, Iowa	Oct. 1, 1872	131,872
Auger, Earth	T. C. Harris	Dresden, Iowa	Nov. 5, 1872	132,830
Auger, Earth	H. P. Haskin	Roscoe, Ill	Oct. 8, 1872	132,072
Auger, Earth	N. H. Lindley	Bridgeport, Conn	July 18, 1871	117,180
Auger, Earth	T. Orchard	Sacramento, Cal	Apr. 25, 1871	114,185
Auger, Earth	C. D. Pierce	Lawrence, Kans	Aug. 12, 1873	141,664
Auger, Earth	S. Pope	Covington County, Miss	Mar. 22, 1870	101,160
Auger, Earth	I. N. Pyle	Cameron, Mo	Nov. 12, 1872	132,980
Auger, Earth	W. H. Salyer	Corning, Iowa	Aug. 19, 1873	141,892
Auger, Earth	J. B. Smith	Memphis, Tenn	Mar. 4, 1873	136,555
Auger, Earth	A. Sorg and S. C. Bollman	Decatur, Ind	June 24, 1873	140,226
Auger, Earth	H. C. Stouffer	Canfield, Ohio	Dec. 24, 1872	134,324
Auger, Earth	H. C Stouffer	Canfield, Ohio	Mar. 18, 1873	137,036
Auger, Earth	I. M. West	Summit, Nebr	Apr. 8, 1873	137,744
Auger, Earth	B. F. White and S. R. Owen	Stewartsville, Mo	Feb. 13, 1872	123,755
Auger, Earth	J. Wilson	Cameron, Mo	Oct. 29, 1872	132,611
Auger, Earth	I. Yeazel	Clark County, Ohio	Mar. 19, 1872	124,711
Auger, Earth-boring	A. Crafts and E. Weeks	Auburn, Ohio	Nov. 20, 1849	6,880
Auger, Earth-boring	J. W. Heath	Memphis, Tenn	Mar. 31, 1868	76,188
Auger, Earth-boring	W. W. Jilz	Hamilton, Mo	Aug. 1, 1871	117,542
Auger, Earth-boring	W. W. Jilz	St. Joseph, Mo	July 16, 1872	129,283
Auger, Earth-boring	A. A. McMahen	Oxford, Miss	July 12, 1859	24,749
Auger, Earth-boring	G. Page	Baltimore, Md	May 3, 1839	1,140
Auger, Excavating	J. Buck	Bucksport, Me	Feb. 5, 1850	7,067
Auger, Expanding	C. Meyer	Fond du Lac, Wis	Mar. 29, 1859	23,381
Auger finishing-tool	R. Jennings	Deep River, Conn	July 3, 1866	56,058
Auger for barrels, Shredding	A. Kelly	Sharpsburgh, Pa	July 6, 1869	92,195
Auger for boring boxes	R. L. Priester	Souder's Station P. O., Md	Nov. 14, 1871	120,827
Auger for boring earth	J. M. Cooper	Newbern, Ala	Oct. 7, 1842	2,807
Auger for boring guns, pistols, &c	D. Pettibone	Philadelphia, Pa	Feb. 12, 1814	....
Auger for boring guns, Twisted screw	W. Holmes	Winchester, N. H	Apr. 4, 1820	....
Auger for boring hubs of wheels, Set	S. Caldwell	Windham, Conn	Nov. 20, 1826	....
Auger for boring large holes	N. J. Lampman	Coxsackie, N. Y	Apr. 8, 1835	....
Auger for boring-machine	G. Flautt	Cavetown, Md	Feb. 12, 1850	7,083
Auger for boring post-holes	C. Hoxie	Hudson, N. Y	July 26, 1824	....
Auger for boring square holes	A. S. Perrine	Louisville, Ky	Mar. 1, 1870	100,440
Auger for boring wood	G. W. Low	Chillicothe, Ohio	Feb. 9, 1864	41,517
Auger for cutting rotary tenons	G. Taylor and G. H. Burger	Worthington, Ohio	Aug. 30, 1859	25,289
Auger for wood	M. Morris	Broad Brook, Conn	Nov. 30, 1858	22,195
Auger-gage	S. L. Lyford	Portland, Me	May 29, 1866	55,132
Auger-gage	A. Stowel		Feb. 14, 1803	....
Auger-gage	W. E. Whiting	Providence, R. I	Feb. 11, 1868	74,466
Auger-gimlet, &c	W. Spangler	Harper's Ferry, Va	July 11, 1854	11,280
Auger-gimlet, &c., Single-twist	W. N. Clark	Chester, Conn	Jan. 31, 1845	3,899
Auger, Ground	D. Ring	Damariscotta, Me	Apr. 4, 1865	47,172
Auger-handle	D. W. George	Minnesota City, Minn	Sept. 28, 1869	95,220
Auger-handle	S. W. Hemenway	Lansing, Iowa	May 20, 1873	139,066
Auger-handle	T. C. Hendry	Union Point, Ga	Oct. 20, 1868	83,156
Auger-handle	G. W. Herring	Bangor, Me	Nov. 30, 1869	97,398
Auger-handle	W. A. Ives	New Haven, Conn	Oct. 11, 1870	108,267

Index of patents issued from the United States Patent Office from 1790 *to* 1873, *inclusive*—Continued.

Invention.	Inventor.	Residence.	Date.	No.
Auger-handle	W. A. Ives	New Haven, Conn	Jan. 10, 1871	110,920
Auger-handle	F. B. Pease	Ontario Centre, N. Y	Sept. 2, 1873	142,410
Auger-handle	S. T. Peat	Florence, N. J	July 13, 1869	92,541
Auger-handle	H. D. Pennoyer	Athens, N. Y	July 28, 1868	80,365
Auger-handle	L. L. Pollard	Worcester, Mass	July 23, 1861	32,890
Auger-handle	N. C. Sanford	Meriden, Conn	June 17, 1856	15,147
Auger-handle	G. Sanford	Ellenville, N. Y	Oct. 1, 1850	7,688
Auger-handle	W. W. Simrell	Great Bend, Pa	Aug. 19, 1862	36,240
Auger-handle	D. Y. Smith	Joliet, Ill	Sept. 19, 1865	50,045
Auger-handle	D. Y. Smith	Joliet, Ill	Dec. 10, 1867	72,099
Auger-handle	J. Swan	Seymour, Conn	Apr. 21, 1868	76,956
Auger-handle	J. Swan	Seymour, Conn	Dec. 14, 1869	97,830
Auger-handle	G. H. Talbot	Boston, Mass	Dec. 11, 1855	13,925
Auger-handle	A. Thayer	Malden Bridge, N. Y	Dec. 3, 1850	7,815
Auger-handle, &c	D. C. Stone	Wawarsing, N. Y	Mar. 12, 1845	3,949
Auger-handle fastening	W. N. Clark	Chester, Conn	July 28, 1857	17,868
Auger-handle fastening	G. H. Hubbard	Shelburne Falls, Mass	Mar. 6, 1855	12,484
Auger-handles and braces, Socket for	A. H. McKinley	Higginsport, Ohio	Aug. 16, 1853	9,939
Auger-handles, Shank-socket for	J. M. Horton	Albany, N. Y	July 8, 1862	35,856
Auger, Hollow	G. F. Almy	Delphos, Ohio	Aug. 12, 1873	141,745
Auger, Hollow	A. Bauman and O. O. Witherell	Toledo, Ohio	Jan. 24, 1871	111,167
Auger, Hollow	F. Beals and M. Smith	New Haven, Conn	Jan. 9, 1866	51,911
Auger, Hollow	H. T. Bean, J. C. Freeman, and D. B. Mills.	Palestine, Ill	Dec. 7, 1869	97,590
Auger, Hollow	J. H. Beauregard	Sandy Hill, N. Y	Dec. 29, 1868	85,423
Auger, Hollow	C. S. Bonney	Syracuse, N. Y	Aug. 2, 1870	105,896
Auger, Hollow	G. E. Booth	Seymour, Conn	May 7, 1867	64,478
Auger, Hollow	A. Brush	East Constable, N. Y	Jan. 7, 1868	73,162
Auger, Hollow	C. L. Campbell	Binghamton, N. Y	May 10, 1864	42,639
Auger, Hollow	W. A. Clark	Woodbridge, Conn	Feb. 2, 1869	86,364
Auger, Hollow	W. A. Clark	Woodbridge, Conn	Aug. 17, 1869	93,808
Auger, Hollow	W. A. Clark	Woodbridge, Conn	Dec. 12, 1871	121,707
Auger, Hollow	A. Conant	Pepperville, Mass	June 11, 1829	....
Auger, Hollow	C. W. Corr	Carlinville, Ill	Feb. 25, 1873	136,138
Auger, Hollow	A. F. Cushman	Hartford, Conn	Aug. 30, 1870	106,919
Auger, Hollow	J. Deming	Salem, Ohio	Dec. 3, 1872	133,570
Auger, Hollow	M. L. Edwards	Salem, Ohio	Dec. 3, 1872	133,573
Auger, Hollow	O. A. Essig	Canton, Ohio	May 6, 1873	138,623
Auger, Hollow	E. W. Fawcett, E. W. Silver, and J. Deming.	Salem, Ohio	Sept. 23, 1873	143,132
Auger, Hollow	R. Gaylord	Seymour, Conn	July 21, 1863	39,284
Auger, Hollow	D. George	Granville, Ohio	Mar. 28, 1848	5,487
Auger, Hollow	S. P. Gilbert	Racine, Wis	Apr. 2, 1861	31,883
Auger, Hollow	A. B. Hendryx	Seymour, Conn	Feb. 25, 1862	34,497
Auger, Hollow	M. Isbell	New Haven, Conn	May 7, 1867	64,423
Auger, Hollow	W. A. Ives	New Haven, Conn	Oct. 13, 1868	82,957
Auger, Hollow	S. Katz	Bossardsville, Pa	Aug. 8, 1871	117,784
Auger, Hollow	F. Kraus	Philadelphia, Pa	Mar. 31, 1868	76,205
Auger, Hollow	J. Lefeber	Cambridge City, Ind	Oct. 2, 1866	58,436
Auger, Hollow	J. McClure	Rockland, Me	Oct. 9, 1866	58,658
Auger, Hollow	G. M. Nye and A. T. Haviland	Elmira, N. Y	May 22, 1860	28,400
Auger, Hollow	J. L. Parker	Harrisonburgh, Va	Apr. 14, 1868	76,804
Auger, Hollow	H. J. Rickard	Rochester, N. Y	May 18, 1869	90,195
Auger, Hollow	J. H. Smith	Pineville, Pa	July 17, 1866	56,459
Auger, Hollow	J. M. Smith	Seymour, Conn	June 5, 1866	55,382
Auger, Hollow	G. N. Stearns	Syracuse, N. Y	Sept. 8, 1863	39,841
Auger, Hollow	G. N. Stearns	Syracuse, N. Y	Aug. 27, 1872	130,826
Auger, Hollow	J. Swan	Seymour, Conn	Sept. 19, 1871	119,096
Auger, Hollow	I. H. Van Wie	Clarksville, N. Y	July 26, 1870	105,744
Auger, Hollow	J. Ward	New York, N. Y	Jan. 1, 1867	60,972
Auger, Hollow	A. Wyckoff	Elmira, N. Y	July 12, 1859	24,773
Auger, Hollow	A. Wyckoff	Elmira, N. Y	Apr. 3, 1866	53,722
Auger, Hollow	A. Wyckoff and L. F. Stevens	Elmira, N. Y	Mar. 12, 1861	31,694
Auger, Hollow	E. Young	Amanda, Ohio	Jan. 1, 1867	60,985
Auger, Ice	W. A. Clark	New Haven, Conn	June 10, 1873	139,769
Auger-lips, Machine for forming	J. Swan	Seymour, Conn	May 20, 1873	139,091
Auger, Mining	J. Hobart	Dubuque, Iowa	May 23, 1846	4,531
Auger, Mining	P. Nichols	Boston, Mass	May 18, 1810	....
Auger or bit	E. L'Hommedieu	Saybrook, Conn	Feb. 11, 1835	....
Auger or bit, Expanding	L. H. Gibbs	Washington, D. C	July 17, 1855	13,261
Auger, Peat	N. Aubin	Montreal, Canada	Jan. 25, 1870	99,129
Auger, Post	C. Adams	Pittsburgh, Pa	Oct. 20, 1868	83,233
Auger, Post	J. Armstrong	Bucyrus, Ohio	July 13, 1869	92,504
Auger, Post	H. W. Caswell	Yarmouth, Me	Mar. 19, 1867	63,016
Auger, Post	Z. S. Cracraft	Lacon, Ill	Oct. 5, 1869	95,566
Auger, Post	J. B. Drapar	Salem, Ill	Nov. 3, 1868	83,611
Auger, Post	H. C. Partridge and J. Preston	Bainbridge, N. Y	Jan. 25, 1870	99,227
Auger, Post	G. Seeger and C. H. Shaffer	Clark's Hill, Ind	Dec. 21, 1869	98,196
Auger, Post	S. S. and J. G. Sherman	McHenry, Ill	July 6, 1869	92,214
Auger, Post	A. Vaughan	Chicago, Ill	June 15, 1869	91,387
Auger, Post-hole	A. Burton	Chicago, Ill	Apr. 14, 1868	76,598
Auger, Post-hole	A. R. Clark	Albia, N. Y	Aug. 22, 1871	118,196
Auger, Post-hole	S. W. Corbin	Bainbridge, N. Y	Mar. 2, 1869	87,470
Auger, Post-hole	I. Hart	Clarksburgh, W. Va	Aug. 31, 1869	94,412
Auger, Post-hole	S. C. Horton	Tarrytown, N. Y	Feb. 2, 1869	86,400
Auger, Post-hole	J. Killgore, G. D. Clapsaddle, and E. Smart.	Arcola, Ill	June 11, 1867	65,577
Auger, Post-hole	J. M. Leeds and J. E. Hallowell	Kokomo, Ind	Jan. 23, 1866	52,180
Auger, Post-hole	T. Luson	Sharon, Wis	June 4, 1867	65,400
Auger, Post-hole	S. McCray	Woodstock, Ill	Sept. 10, 1867	68,638
Auger, Post-hole hollow	S. H. Yocum	Tipton, Ind	Mar. 31, 1868	76,132
Auger, Pump, and spur-bits	R. French	Derby, Conn	Nov. 11, 1834	....
Auger, Screw	D. Bassett	Derby, Conn	Mar. 18, 1829	....
Auger, Screw	E. L'Hommedieu	Norwich, Conn	July 17, 1816	....
Auger, Screw	P. Williamson	New York, N. Y	Sept. 9, 1833	....
Auger, Serpentine-screw	G. Shelter	York, Pa	Sept. 10, 1829	....
Auger, Serpentine-screw	G. Shelter	York, Pa	Mar. 21, 1831	....
Auger shafts, Coupling for earth	G. W. Irwin and T. A. Cox	Champaign County, Ill	July 8, 1873	140,710

Index of patents issued from the United States Patent Office from 1790 *to* 1873, *inclusive*—Continued.

Invention.	Inventor.	Residence.	Date.	No.
Auger-shank, Method of attaching cutting-lips to.	N. S. White and A. Denio	Shaftsbury, Vt	Dec. 21, 1858	22, 394
Auger, &c., Ship	I. W. Hoagland	Jersey City, N. J	Mar. 20, 1855	12, 551
Auger, Single-twist	E. L'Hommedieu	Saybrook, Conn	Oct. 1, 1830	
Auger, Single-twist spiral-screw	N. C. Sanford	Meriden, Conn	Aug. 8, 1834	
Auger, Slotting	P. Cunningham	Eckley, Pa	Sept. 1, 1868	81, 607
Auger, Spiral-screw	A. Newton	Meriden, Conn	Aug. 8, 1834	
Auger, Spoke-tenon	W. Morehouse	Buffalo, N. Y	Jan. 14, 1862	34, 159
Auger-stock	W. T. Barnes	Buffalo, N. Y	Apr. 3, 1849	6, 256
Auger-stock	S. C. Norcross	Norway, Me	Aug. 25, 1863	39, 669
Auger-strap	S. Gore	Sheshequin, Pa	July 20, 1825	
Auger, Submarine	N. Blake	Ira, N. Y	Apr. 20, 1852	8, 882
Auger to bore square holes, &c	C. Jackson	Otsego County, N. Y	Feb. 14, 1818	
Auger, Tube	R. Thomas	Richmond County, N. C	May 29, 1819	
Auger, Tubular	J. A. Reynolds	Elmira, N. Y	Jan. 13, 1857	16, 399
Auger-twisting machine	W. L. Aldrich and W. Evans	Norwich and Seymour, Conn.	Aug. 6, 1867	67, 395
Auger, Undermining	P. Sheldon	Jamestown, N. Y	Oct. 7, 1873	143, 535
Auger, Well	E. Altman	Hamilton, Mo	Jan. 31, 1871	111, 300
Auger, Well	J. B. Christian	Hamburg, Iowa	Sept. 17, 1872	131, 427
Auger, Well	J. Y. Goode	Water Valley, Miss	Dec. 7, 1869	97, 501
Auger, Well	J. and T. J. Ingels	Atchison, Kans	Nov. 15, 1870	109, 214
Auger, Well	H. R. King	Poplar Bluff, Ark	Oct. 17, 1871	120, 076
Auger, Well	A. A. McMahen	Oxford, Miss	Apr. 13, 1869	88, 891
Auger, Well	W. L. Payne	Topeka, Kans	Sept. 23, 1873	143 181
Auger, Well	F. Spees	Tabor, Iowa	Feb. 20, 1872	123, 847
Auger, Well	W. H. Stone	Pattonsburgh, Mo	July 11, 1871	116, 883
Auger, Well	J. Wilson and G. H. Baisley	Hamilton, Mo	July 5, 1870	105, 021
Auger, Well-boring	P. Ollom	Muncie, Ind	Feb. 19, 1867	62, 216
Augers and bits, Manufacture of	J. Swan	Seymour, Conn	June 9, 1868	78, 769
Augers and boring-bits, Process for making	C. Whitehouse	Bridgtown, England	June 15, 1869	91, 503
Augers and spiral conveyers, Packing	J. Mattison	Oswego, N. Y	Nov. 2, 1869	96, 455
Augers, Casting screw	J. Carl and J. W. Heath	Grenada, Miss	Aug. 21, 1860	29, 668
Augers, Coupling-device for earth-boring	A. J. and G. Heine	Frémont, Nebr	July 16, 1872	129, 027
Augers, Coupling for earth-boring	T. Orchard	Lincoln, Cal	Dec. 6, 1870	109, 930
Auger, Device for fastening cutter of hollow	W. A. Clark	Bethany, Conn	July 12, 1859	24, 722
Augers, Die for making	E. L'Hommedieu	Chester, Conn	Aug. 29, 1854	11, 613
Augers, Die for making	E. H. Tracy	Meriden, Conn	July 4, 1865	48, 633
Augers, Die for manufacturing	S. A. Smith	Essex, Conn	Mar. 4, 1873	136, 391
Augers, Handle-fastening for	J. M. Hathaway	New York, N. Y	Aug. 21, 1860	29, 692
Augers, Machine for cutting down	E. O. and E. Carrington	Wallingford, Conn	July 2, 1867	66, 297
Augers, Machine for finishing the lips and points of.	W. P. Maxson	Elmira, N. Y	Nov. 19, 1872	133, 238
Augers, Machine for grinding and cutting down	E. O. and E. Carrington	West Meriden, Conn	Apr. 7, 1868	76, 301
Augers, Machine for making	I. Tower	Rochester, N. Y	Mar. 1, 1864	41, 822
Augers, Machine for regulating the twist and diameter of screw.	N. C. Sanford and L. P. Smith	Meriden, Conn	Apr. 10, 1849	6, 305
Augers, Machine for swaging the head of screw	R. Jennings	Deep River, Conn	July 31, 1866	56, 869
Augers, Machine for turning the lips of	R. Cook	Shelburne Falls, Mass	Mar. 27, 1855	12, 583
Augers, Machinery for manufacturing double-twist screw.	E. L'Hommedieu and R. N. Watrous.	Chester, Conn	July 24, 1838	851
Augers, Manufacture of	J. Swan	Seymour, Conn	Aug. 20, 1867	68, 012
Augers, Manufacturing screw	W. Field	Pawtucket, R. I	Apr. 15, 1840	1, 553
Augers, Means for operating earth	W. H. Beach and C. N. Hanson.	Hamburg, Iowa	Feb. 4, 1873	135, 510
Augers, Means for operating earth	J. B. Christian	Hamburg, Iowa	May 13, 1873	138, 788
Augers, Method of attaching expansible cutting-lips to.	N. Clare and J. Quigly	Malden, N. Y	Apr. 6, 1858	19, 829
Augers, Method of manufacturing	G. G. Griswold	Chester, Conn	Apr. 1, 1856	14, 561
Augers, Method of securing the cutters to the spindles of.	C. L. Barnes	New York, N. Y	Aug. 17, 1858	21, 179
Augers, Method of using the common screw	O. Stetson and W. Sebree	Georgetown, Ky	Dec. 11, 1810	
Augers, Mold for casting	W. Evans and R. E. Hayden	Seymour, Conn	Apr. 6, 1869	88, 559
Augers, Tenoning	W. A. Clark	Bethany, Conn	June 12, 1860	28, 653
Augers to handles, Attaching	E. Broad	Saint Anthony, Minn	Jan. 8, 1869	90, 919
Augers to handles, Attaching	M. S. Brooks	Chester, Conn	Oct. 28, 1851	8, 464
Augers to handles, Attaching	C. W. Cotton	Shelburne Falls, Mass	Mar. 20, 1855	12, 575
Augers to handles, Attaching	J. E. Larkin	Ballston Spa, N. Y	Nov. 19, 1850	7, 785
Augers to handles, Attaching	H. W. Olney	Allegheny, Pa	Nov. 7, 1865	50, 838
Augers to sinker for boring artesian well, Mode of uniting.	W. Morris	Kanawaha County, Va	Sept. 4, 1841	2, 243
Augers, Turning any number of	H. Branch	New York	Aug. 7, 1826	
Augers, Twisting-machine	O. Snow	Meriden, Conn	May 13, 1834	
Aural instruments	C. G. Page	Washington, D. C	Nov. 20, 1860	30, 688
Automatic brake	J. Wilkinson	Baltimore, Md	Jan. 1, 1861	31, 047
Automatic brake for machinery	T. Stebins	San Francisco, Cal	Oct. 8, 1872	132, 113
Automatic folding-gate	J. B. Mahana	Benson, Vt	Jan. 10, 1865	45, 842
Automatic gate	W. W. Burson	Rockford, Ill	Nov. 10, 1868	83, 923
Automatic gate	J. B. Cotton	Dayton, Ohio	Aug. 30, 1870	106, 917
Automatic gate	G. C. Crum	Barr's Store, Ill	July 1, 1873	140, 479
Automatic gate	L. S. Deming	Newington, Conn	Sept. 8, 1863	39, 798
Automatic gate	B. F. Dickey	Marshall, Michigan	July 13, 1869	92, 430
Automatic gate	J. S. Elkins and J. T. Green	Marquette, Wis	June 19, 1866	55, 633
Automatic gate	L. Filson	Bushnell, Ill	Apr. 29, 1873	138, 238
Automatic gate	J. W. Foster	Racine, Wis	Apr. 28, 1863	38, 304
Automatic gate	J. S. Fort	Kenton, Ohio	Aug. 19, 1873	141, 925
Automatic gate	C. E. Gillespie	Edwardsville, Ill	June 10, 1873	139, 7[illegible]2
Automatic gate	E. Harter	Dowagiac, Mich	Oct. 6, 1868	82, 831
Automatic gate	N. Long	Muncie, Indiana	Apr. 22, 1873	138, 033
Automatic gate	G. McKnight	Hebron, N. Y	Nov. 15, 1864	45, 063
Automatic gate	H. W. Miskimen	Kingston Mines, Ill	Dec. 10, 1861	33, 896
Automatic gate	T. J. Murphy	Rochester, N. Y	Dec. 1, 1868	84, 569
Automatic gate	G. W. Olbert and W. Young	Barr's Store, Ill	Oct. 21, 1873	143, 839
Automatic gate	M. Orewiler	Bucyrus, Ohio	Sept. 27, 1870	107, 801
Automatic gate	J. G. Page	Rockford, Ill	Aug. 7, 1866	56, 987
Automatic gate	J. P. Ponce	Mayfield, Cal	Apr. 29, 1873	138, 283
Automatic gate	J. H. Schenck	Chicago, Ill	June 26, 1866	55, 914
Automatic gate	P. A. Spicer and M. Crossman	Marshall, Mich	July 26, 1870	105, 735
Automatic gate	G. B. Stevenson	Upper Sandusky, Ohio	Apr. 15, 1873	137, 971
Automatic gate	J. E. Stong	Newton Brook, Canada	Aug. 19, 1873	141, 960
Automatic gate	E. Waterbury	Stamford, Conn	Nov. 12, 1861	33, 723

Index of patents issued from the United States Patent Office from 1790 *to* 1873, *inclusive*—Continued.

Invention.	Inventor.	Residence.	Date.	No.
Automatic gate	E. P. Wheeler	Corinth, Miss	Oct. 7, 1873	143, 551
Automatic gate	S. Whitaker	Bel Air, Md	May 3, 1870	102, 741
Automatic gate	W. I. Wooster	Harvard, Ill	Mar. 18, 1873	136, 892
Automatic gate	S. J. Wright	Ellsworth, N. Y	Apr. 30, 1867	64, 394
Automatic gate for railway-crossing	G. B. Pullinger	Philadelphia, Pa	Dec. 18, 1855	13, 956
Automatic hoop	L. W. Taylor	Weathersfield, Vt	June 1, 1869	90, 700
Automatic lubricator	I. Dreyfus	New York, N. Y	May 21, 1867	64, 956
Automatic lubricator	E. Von Jeinsen	Omaha, Nebr	July 5, 1870	105, 147
Automatic press	P. Hayden	Pittsburgh, Pa	Sept. 12, 1865	49, 877
Automatic register	S. Rigler	Ottawa, Ill	Dec. 10, 1867	72, 084
Automatic regulator	J. M. Osgood	Somerville, Mass	Nov. 1, 1864	44, 884
Automatic rope-skipper	J. Hoffmann	Waterbury, Conn	July 1, 1873	140, 503
Automatic switch	S. Hodkinson	Louisville, Ky	May 4, 1869	89, 578
Automatic ventilator	M. E. Mead	Darien Depot, Conn	May 18, 1869	90, 180
Automatic ventilator	B. F. Prentis	Benwood, W. Va	Oct. 11, 1870	108, 185
Auxiliary engine to be used in supplying steam-boilers with water, Manner of arranging the parts of.	J. Cochrane	Baltimore, Md	Apr. 16, 1845	4, 003
Auxiliary table	J. Blake	Scranton, Pa	Mar. 23, 1869	88, 121
Auxiliary table	E. Johnson	McLean County, Ill	Apr. 29, 1873	138, 406
Awl	S. Babbitt	Brazil, Ind	July 12, 1870	105, 161
Awl	S. Babbitt	Brazil, Ind	Aug. 1, 1871	117, 501
Awl	R. Egan	Brooklyn, N. Y	Apr. 25, 1865	47, 403
Awl	T. Kenney	Lynn, Mass	May 2, 1871	114, 446
Awl	G. K. Mellor	Woonsocket, R. I	Dec. 24, 1872	134, 214
Awl	S. E. Totten	Brooklyn, N. Y	June 18, 1867	65, 843
Awl and punch, Belt	W. J. Innis	Providence, R. I	Nov. 15, 1859	26, 107
Awl and similar tools, Handle for	D. H. Chamberlain	Boston, Mass	May 30, 1848	5, 609
Awl and tool	H. Aiken	Franklin, N. H	Apr. 13, 1858	19, 901
Awl, Belt	S. Y. Beach	Seymour, Conn	May 25, 1869	90, 485
Awl, Belt	F. I. Palmer	Youngstown, Ohio	Apr. 5, 1870	101, 655
Awl, Centering	D. A. Wilcox	Woodstock, Vt	Aug. 17, 1869	93, 937
Awl, Centering	N. Woodbury	Woodstock, Vt	June 22, 1869	91, 806
Awl-drill, &c., Setting	E. B. Bigelow and S. P. Brigham.	West Boyleston Township, Mass.	Jan. 27, 1835	....
Awl for heel-machine	H. H. Bigelow	Worcester, Mass	Mar. 28, 1871	113, 008
Awl, Grafting	D. B. Oliver	Cambria, N. Y	Nov. 22, 1870	109, 442
Awl-haft	W. Campbell	Gilsum, N. H	July 1, 1836	....
Awl-haft	D. H. Chamberlain	Boston, Mass	Apr. 3, 1849	6, 261
Awl-haft	N. S. Clement	Worcester, N. Y	Apr. 21, 1857	17, 078
Awl-haft	L. H. Farnsworth	Hudson, Mass	June 16, 1868	78, 947
Awl-haft	B. James	Worcester, Mass	Apr. 15, 1856	14, 704
Awl-haft	E. Martin	Grafton, Mass	Mar. 21, 1843	3, 014
Awl-haft	D. M. Smith	Gilsum, N. H	Oct. 25, 1832	....
Awl-haft	D. M. Smith	Gilsum, N. H	Mar. 31, 1836	....
Awl-handle	N. S. Clement	New Britain, Conn	Mar. 19, 1867	62, 938
Awl-handle	D. R. Wight	Sturbridge, Mass	Mar. 24, 1868	76, 020
Awl handle, Brad	H. L. Barker	Hartford, Conn	Feb. 6, 1866	52, 443
Awl holder and extractor, Pegging	H. Huston	Cannonsburgh, Pa	Sept. 8, 1863	39, 815
Awl, Lasting	C. K. Bradford	Lynnfield, Mass	Oct. 1, 1867	69, 395
Awl-making die	J. P. Blake	Rockville, Mass	Apr. 13, 1869	88, 637
Awl, Pegging	G. B. Paine	Montpelier, Vt	Feb. 16, 1869	87, 062
Awl, Shoemakers'	B. I. Lane	South Framingham, Mass	Apr. 17, 1860	27, 913
Awl-sockets, &c	H. Aiken	Dracut, Mass	Dec. 16, 1833	....
Awl, Tying-up	N. W. Baker	Lynn, Mass	June 22, 1869	91, 593
Awls to hafts, Method of attaching	D. Pierce	Montague, Mass	Aug. 13, 1838	882
Awning	W. Armstrong	Milwaukee, Wis	June 14, 1864	43, 081
Awning	G. H. Bancroft	Philadelphia, Pa	May 7, 1867	64, 399
Awning	J. C. Bowe	Urbana, Ohio	Nov. 13, 1866	59, 548
Awning	H. A. Bowman	Worcester, Mass	Apr. 27, 1869	89, 462
Awning	J. Boyle	New York, N. Y	Feb. 18, 1873	135, 966
Awning	S. Chace	Providence, R. I	Oct. 11, 1859	25, 723
Awning	G. W. Gerau	Brooklyn, N. Y	Apr. 15, 1873	137, 772
Awning	W. Hildebrand	Fort Wayne, Ind	Nov. 8, 1870	109, 007
Awning	E. E. Laumont	New York, N. Y	Oct. 17, 1865	50, 479
Awning	W. Matthews	Orange, N. J	June 11, 1872	127, 781
Awning	S. Miller and J. S. McClellan	Champaign County, Ohio	Nov. 12, 1867	70, 880
Awning	S. Miller and J. S. McClellan	Champaign County, Ohio	Nov. 26, 1867	71, 509
Awning	J. Sebo	Wilmington, Del	May 25, 1844	3, 594
Awning	L. G. Sert and C. L. Schurr	Baltimore, Md	Oct. 25, 1870	108, 639
Awning	H. Sykes	New York, N. Y	Oct. 7, 1873	143, 542
Awning	J. A. Pain	Clyde, N. Y	Apr. 10, 1866	53, 859
Awning	N. Poulson	Washington, D. C	Sept. 8, 1868	82, 032
Awning	R. P. Pratt	Hartford, Conn	Oct. 13, 1868	83, 090
Awning	T. G. Tyler	New York, N. Y	Sept. 29, 1868	82, 569
Awning	J. B. Watkins	New Bedford, Mass	Aug. 19, 1862	36, 251
Awning, Adjustable	L. Yenne and C. Schneider	New York, N.Y., and Newark, N. J.	Sept. 20, 1870	107, 584
Awning and fan for horses' heads	H. L. Byrd	Baltimore, Md	Sept. 13, 1870	107, 333
Awning and reflector	J. Corduan	Brooklyn, N. Y	May 2, 1865	47, 524
Awning, device for operating	A. Thalhofer	South Bend, Ind	June 28, 1870	104, 900
Awning-fixture	E. Peach	Utica, N. Y	July 31, 1860	29, 399
Awning for animals	S. Moffit	Minneapolis, Minn	Apr. 4, 1871	113, 323
Awning for car, &c	J. H. Monce	New York, N. Y	July 26, 1870	105, 829
Awning for horse and dray	J. Nelson	Cincinnati, Ohio	Sept. 25, 1855	13, 600
Awning for horse-car	M. C. Battey	Washington, D. C	Nov. 17, 1868	84, 160
Awning-frame	H. Hilliard	Brooklyn, N. Y	July 5, 1870	105, 076
Awning-frame	J. W. Loane	Baltimore, Md	May 7, 1872	126, 556
Awning-frame	H. Stephens	New York, N. Y	Aug. 26, 1873	142, 291
Awning-frame	C. Werner	Charleston, S. C	Nov. 29, 1870	109, 694
Awning-frame, Adjustable	L. Yenne	New York, N. Y	May 3, 1870	102, 642
Awning-frame for horses, attachable to harness	N. Pullman	New Oregon, Iowa	Sept. 29, 1857	18, 300
Awning frame, Shade	J. Drechsler	New York, N. Y	Dec. 3, 1872	133, 572
Awning frame, Window	J. A. Allen	Philadelphia, Pa	Apr. 29, 1873	138, 223
Awning, metal	W. O. Paisen	New York, N. Y	Apr. 27, 1858	20, 085
Awning, Shop	W. H. Bakewell	New York, N. Y	Mar. 23, 1852	8, 817
Awning-slide	J. Boyle	New York, N. Y	Oct. 29, 1872	132, 559
Awning-slide	T. F. Darcy	New York, N. Y	Feb. 27, 1872	124, 037

Index of patents issued from the United States Patent Office from 1790 *to* 1873, *inclusive*—Continued.

Invention.	Inventor.	Residence.	Date.	No.
Awning, Street-car	C. B. Turnbull	Washington, D. C	July 15, 1873	140,972
Awning-windlass	I. Walter	Cincinnati, Ohio	Feb. 27, 1872	124,178
Awning, Window	C. C. Moore	New York, N. Y	Dec. 5, 1871	121,650
Awning, Window	J. B. Wheeden	Baltimore, Md	July 6, 1869	92,239
Awnings, Painting striped	C. E. Wheeler	Boston, Mass	May 13, 1873	138,773
Ax	J. H. Beidler	Adrian, Mich	Aug. 10, 1869	93,585
Ax	D. W. Colburn	Laomi, Ill	July 9, 1867	66,563
Ax	J. Franklin	Springfield, Ohio	Nov. 2, 1869	96,416
Ax	J. W. Hilton and R. W. Green	Bradford, Pa	Sept. 1, 1868	81,635
Ax	A. H. Jumper	Sunman, Ind	Aug. 31, 1869	94,318
Ax	H. Mann	Bellefonte, Pa	June 3, 1862	35,480
Ax	W. Peabody	Orono, Me	Aug. 22, 1871	118,266
Ax	E. Quast	Freedom, Mo	Nov. 16, 1869	96,937
Ax	B. Smith	Canton, Conn	Nov. 2, 1832	-------
Ax and hook combined, Grubbing	P. W. Norris	Detroit, Mich	Feb. 15, 1870	99,784
Ax and pick, Ice	J. N. Bunnell	Unionville, Conn	Dec. 6, 1870	109,867
Ax-bit-blank bar	J. Lippincott	Pittsburgh, Pa	Dec. 22, 1868	85,110
Ax-bit-blank machine	E. Bartholomew	Mill Hall, Pa	June 20, 1871	116,011
Ax-bits, Machine for forming	L. Chapman	Collinsville, Conn	June 1, 1869	90,726
Ax-bits, Manufacture of	C. Blair	Collinsville, Conn	Aug. 17, 1869	93,666
Ax blanks, Method of making	J. W. Ells	Pittsburgh, Pa	Sept. 3, 1867	68,423
Ax, Chopping	T. Merriam	Waterloo, Wis	Apr. 12, 1864	42,303
Ax-dressing machinery	J. Mackey	Napanock, N. Y	Aug. 29, 1848	5,736
Ax-dressing machinery	E. K. Root	Collinsville, Conn	Aug. 22, 1848	5,731
Ax for wood-splitting machine	J. H. Silkman	Milwaukee, Wis	Jan. 23, 1866	52,216
Ax, Grinding	H. Mann	Bellefonte, Pa	Aug. 17, 1869	93,727
Ax, Hand	E. H. Meigs	East Berlin, Conn	Aug. 11, 1868	80,868
Ax-handle	J. M. Sears	Vandalia, Ill	May 30, 1871	115,532
Ax-handle	B. D. Stevens	Decorah, Iowa	June 2, 1868	78,550
Ax-handle fastening	J. E. Emerson	San Francisco, Cal	Apr. 10, 1860	27,784
Ax-handle guard	L. B. Hoit	Cedar Falls, Iowa	July 12, 1870	105,205
Ax-handle shield	B. Butler	Saint Johnsbury Centre, Vt.	Jan. 26, 1869	86,130
Ax-handles, Mode of securing	G. W. Simonds	Lynnfield, Mass	July 10, 1866	56,281
Ax, hatchet, &c	D. Hinman	Winchester, Conn	Nov. 2, 1832	-------
Ax-helve	W. Morehouse	Buffalo, N. Y	Apr. 11, 1865	47,214
Ax-helves, Metal cap for	A. W. Porter	Saint Johnsville, N. Y	July 17, 1860	29,192
Ax-making machine	E. F. Hurd	Johnsonville, N. Y	Aug. 14, 1866	57,142
Ax making machine	E. F. Hurd	Hoosick Falls, N. Y	Nov. 21, 1871	121,172
Ax-making machine	C. Hutchins	East Douglass, Mass	Mar. 3, 1857	16,732
Ax-making machine	J. Simmons	Cohoes, N. Y	Mar. 1, 1853	9,601
Ax-making machine	I. W. Turne	Baltimore, Md	Nov. 4, 1842	2,841
Ax-making process	J. Orelup	Ballston Spa, N. Y	June 8, 1852	9,000
Ax, Oval	A. Collins	Winsted, Conn	Oct. 25, 1832	-------
Ax, Oval	S. Hyde	Williamsburgh, Mass	Jan. 29, 1830	-------
Ax, Oval	S. Hyde	Williamsburgh, Mass	Mar. 2, 1831	-------
Ax-pole, Making	D. P. Estep	Pittsburgh, Pa	Oct. 14, 1856	15,880
Ax-poll blank	W. Bunton	Pittsburgh, Pa	Oct. 12, 1869	95,646
Ax-poll swaging-machine	R. Blake and A. Carpenter	Scranton, Pa	May 4, 1869	89,623
Ax-polls, Machine for making	R. H. Cole	Saint Louis, Mo	June 16, 1857	17,556
Ax-polls, Machine for making	G. Reynolds	Manchester, N. H	July 20, 1858	20,957
Ax-polls, Machine for punching	R. Blake	Scranton, Pa	Aug. 11, 1868	80,900
Ax-shaving machine	H. C. Reynolds	Manchester, N. H	Aug. 6, 1867	67,584
Ax, Stone	W. C. Peckham	Troy, Ohio	Oct. 29, 1872	132,540
Ax-testing machine	W. Hunt	East Douglass, Mass	Sept. 2, 1856	15,656
Axes, Attaching handles to	T. H. Tyndale	Belleville, Ill	May 25, 1869	90,412
Axes, Attaching handles to	H. C. Wooding	Wallingford, Conn	June 5, 1866	55,437
Axes, Die for forming the eyes of pick	H. L. Lowman	New York, N. Y	Nov. 27, 1866	60,022
Axes, Forging	D. Hinman	Canton, Conn	June 29, 1833	-------
Axes, Forging	E. Shaw	Canton, Conn	June 29, 1833	-------
Axes, hammers, &c., Manufacture of	F. C. Curie	Lancaster, Pa	Sept. 29, 1868	82,607
Axes, &c., Hardening	J. N. Rockwell	Napanock, N. Y	June 23, 1857	17,639
Axes, hatchets, &c., Method of punching the eyes of	E. K. Root	Collinsville, Conn	Dec. 10, 1838	1,027
Axes, hoes, scythes, pitch-forks, &c., Making	J. P. Hazard	Washington County, R. I	Aug. 9, 1833	-------
Axes, hoes, scythes, pitch-forks, &c., Making	J. F. Mackie	New York	Nov. 19, 1833	-------
Axes, hoes, scythes, pitch-forks, &c., Making	L. Olds	Oneonta, N. Y	July 9, 1833	-------
Axes, Machine for making	J. Pratt, jr	Charlemont, Mass	May 16, 1832	-------
Axes, Machine for manufacturing	D. C. Stone	Napanock, N. Y	Apr. 21, 1838	699
Axes, Machinery for hammering heads of	L. Dodge	Waterford, N. Y	May 1, 1866	54,311
Axes, Machinery for making	H. D. Morris	Baldwinsville, N. Y	Aug. 31, 1869	93,232
Axes, Making	J. L. Lewis	Pittsburgh, Pa	Mar. 13, 1866	53,155
Axes, Manufacture of	J. Lippincott	Pittsburgh, Pa	Feb. 21, 1860	27,227
Axes, Manufacture of	J. Lippincott	Pittsburgh, Pa	July 23, 1867	66,978
Axes, Manufacture of	H. C. Reynolds	Manchester, N. H	Aug. 1, 1865	49,156
Axes, Manufacturing	H. D. Morris	Baldwinsville, N. Y	Jan. 11, 1870	98,789
Axes, Manufacturing	E. K. Root	Canton, Conn	Mar. 30, 1836	-------
Axes, Manufacturing	E. Shaw and R. H. Burk	Canton, Conn	Nov. 3, 1831	-------
Axes, Manufacturing and forming the eyes of	M. D. Whipple	Douglass, Mass	Jan. 31, 1833	-------
Axes, Method of manufacturing	G. Palmer and C. W. Hubbard	Pittsburgh, Pa	Aug. 22, 1871	118,264
Axes, Method of securing helves in	H. N. and J. C. Bill	Willimantic, Conn	July 25, 1854	11,350
Axes, Mode of making	J. Alley, sr	Pise, Ind	Apr. 9, 1829	-------
Axes, picks, &c., Fastening handles to	J. Stewart	Saint Cloud, Minn	Aug. 18, 1868	81,308
Axes, &c., Sharpening	J. Shotwell		Mar. 16, 1799	-------
Axes, Shears for the manufacture of	L. Dodge	Cohoes, N. Y	Apr. 2, 1861	31,927
Axes to handles, Fastening	W. H. Livingston	New York, N. Y	Sept. 25, 1860	30,146
Axes to handles, Mode of attaching	J. Stewart	Money Creek, Minn	Oct. 29, 1867	70,284
Axes, Step or gudgeon, Vertical	J. Andrews	Dinwiddie County, Va	Aug. 11, 1817	-------

Axle:

See Anti-friction.
Car.
Carriage.
Carriage-wheel.
Divided.
Grindstone.
Iron.
Locomotive.
Mail.
Metallic.

Index of patents issued from the United States Patent Office from 1790 *to* 1873, *inclusive*—Continued.

Invention.	Inventor.	Residence.	Date.	No.
Axle—Continued.				
See Movable.				
Trussed.				
Tubular.				
Vehicle.				
Wagon.				
Wheat-drill.				
Wheel.				
Wooden				
Axle	J. Elmire	Martic Township, Pa	Aug. 18, 1868	81,079
Axle	E. Finn	Berlin, Wis	Dec. 15, 1868	84,940
Axle	S. D. Littlefield	Burlington, Wis	Oct. 1, 1867	69,350
Axle	W. Nevins	Falmouth, Me	Oct. 1, 1867	69,362
Axle	J. D. Smith	Schuylerville, N. Y	May 18, 1869	90,317
Axle	H. T. Tichenor	Fort Branch, Ind	Aug. 27, 1867	68,130
Axle	H. W. Tilton	Walpole, Mass	June 29, 1869	91,987
Axle	H. G. Weibling	Denver City, Col	Jan. 20, 1863	37,479
Axle and axle-box	D. Wigger	New York, N. Y	Nov. 12, 1867	70,928
Axle and bolster coupling	G. Brown	Carlisle, N. Y	Nov. 21, 1848	5,932
Axle and gudgeon, Mode of making boxes for	I. Babbit	Boston, Mass	July 17, 1839	1,252
Axle and its bearing	R. R. McGregor	Covington, Tenn	Apr. 19, 1870	102,025
Axle and journal-box lubricator	J. B. G. M. F. Piret	Paris, France	Aug. 2, 1870	105,977
Axle and journal, Lubricating	W. Kenworthy and J. H. Pollitt	Birmingham, Pa	June 6, 1871	115,618
Axle and other spindles, Lubricator for	W. H. Harvey	Bangor, Me	Oct. 31, 1871	120,515
Axle and shaft	W. H. Hawley	Utica, N. Y	Nov. 2, 1869	96,318
Axle, Attaching draft-pole to	E. C. Smith	Birmingham, Conn	Nov. 27, 1866	60,075
Axle bearing, Railway-car	W. E. Wilcox and T. H. Wills	Peoria and Beardstown, Ill	July 27, 1869	93,151
Axle-blanks, Tool for finishing	H. E Forrest	Cambridgeport, Mass	Jan. 14, 1873	134,742
Axle-box	S. F. Allen	Chicago, Ill	Sept. 4, 1866	57,657
Axle-box	S. Barker	Hartford, Conn	Sept. 24, 1867	69,160
Axle-box	J. Bevin	Unadilla Forks, N. Y	Apr. 8, 1856	14,639
Axle-box	W. A. Boyden	Altoona, Pa	Nov. 19, 1867	71,125
Axle-box	H. Brady	Factoryville, N. Y	Aug. 20, 1867	67,946
Axle-box	G. Brill	Philadelphia, Pa	Mar. 19, 1867	62,931
Axle-box	N. Campbell	Brooklyn, N. Y	Apr. 16, 1867	63,783
Axle-box	H. L. Castile	Memphis, Tenn	May 15, 1860	28,253
Axle-box	J. Christy	Philadelphia, Pa	Jan. 16, 1866	52,029
Axle-box	C. Cook	Winsted, Conn	Sept. 5, 1865	49,726
Axle-box	D. Dalzell	South Egremont, Mass	Aug. 16, 1870	106,469
Axle-box	D. Dalzell	South Egremont, Mass	July 25, 1871	117,390
Axle-box	L. A. Dochez	New York, N. Y	Dec. 31, 1867	72,821
Axle-box	D. H. Dotterer	Philadelphia, Pa	Apr. 3, 1866	53,588
Axle-box	D. H. Dotterer	Philadelphia, Pa	June 18, 1867	65,797
Axle-box	D. H. Dotterer	Philadelphia, Pa	May 30, 1871	115,449
Axle-box	W. B. Fahnestock	Lancaster, Pa	Mar. 30, 1858	19,762
Axle-box	A. A. Freeman	Philadelphia, Pa	Apr. 2, 1867	63,379
Axle-box	A. Goodyear, 2d	Hamden, Conn	July 4, 1871	116,702
Axle-box	E. P. Haskell	New Bedford, Mass	Nov. 17, 1868	84,185
Axle-box	J. B. Hendricks	Clayton, Ill	Sept. 14, 1869	94,823
Axle-box	J. T. Henry	New Haven, Conn	Jan. 16, 1872	122,829
Axle-box	H. Howson	Philadelphia, Pa	Nov. 2, 1858	21,998
Axle-box	J. J. Lahaye and J. E. Wooten	Reading, Pa	July 5, 1870	105,096
Axle box	B. P. Lamason	Milton, Pa	Sept. 24, 1867	69,225
Axle-box	F. Leppens	Hartford, Conn	Nov. 13, 1866	59,616
Axle-box	J. Lightner	Roxbury, Mass	Nov. 21, 1848	5,935
Axle-box	I. S. Lister	Philadelphia, Pa	June 23, 1868	79,238
Axle-box	J. S. Lister	Philadelphia, Pa	Feb. 9, 1869	86,679
Axle-box	E. Lockwood	Bordentown, N. J	Sept. 5, 1865	49,772
Axle-box	D. Metz	Washington, D. C	Nov. 2, 1869	96,337
Axle-box	M. V. Miller and G. Henry	Manchester, Pa., and Steubenville, Ohio.	Aug. 7, 1866	56,977
Axle-box	J. Montgomery	Harrisburgh, Pa	May 15, 1866	54,753
Axle-box	C. M. Oliver	Port Carbon, Pa	Nov. 13, 1866	59,632
Axle-box	B. M. Pearne and L. Coville	Oxford, N. Y	July 14, 1868	79,916
Axle-box	P. Philippi	Beardstown, Ill	Feb. 19, 1867	62,221
Axle-box	H. B. Pitner	La Porte, Ind	Dec. 10, 1867	72,079
Axle-box	W. H. Pollard	Seneca Falls, N. Y	June 4, 1867	65,428
Axle-box	W. F. Rippon	Providence, R. I	May 15, 1866	54,775
Axle-box	O. H. P. Robinson	Bellport, N. Y	Mar. 13, 1866	53,186
Axle-box	H. B. Rowley	Rushville, N. Y	Nov. 27, 1866	60,064
Axle-box	W. H. Saunders	Hastings, N. Y	Dec. 2, 1856	16,153
Axle-box	A. E. Smith	Bronxville, N. Y	Mar. 11, 1856	14,415
Axle-box	A. E. Smith	Bronxville, N. Y	Jan. 27, 1857	16,499
Axle-box	W. Stechschult	Glaudorf, Ohio	Mar. 15, 1864	41,945
Axle-box	J. Stephenson	New York, N. Y	July 25, 1865	49,005
Axle-box	L. Wakefield	Minneapolis, Minn	Mar. 23, 1869	88,242
Axle-box	S. A. Wing and L. G Johnston	Greenfield Centre, N. Y	Sept. 5, 1871	118,770
Axle-box, &c	D. Cumming	Sorrel Horse, Pa	July 27, 1858	20,991
Axle-box alarm and lubricator, Car	O. Evans	Saint Paul, Minn	Jan. 28, 1873	135,213
Axle-box and hanger	D. H. Dotterer	Philadelphia, Pa	Aug. 13, 1867	67,638
Axle-box and hub	G. Leverich	Elizabethport, N. J	Feb. 2, 1869	86,558
Axle-box and sleeve, Vehicle-wheel	W. H. Cowell	Columbus, Ohio	Aug. 12, 1873	141,699
Axle-box and spindle	J. Bower	Lafayette, Ind	Jan. 14, 1873	134,725
Axle-box, Anti-friction	P. Seyl, P. Fischer, and P. Brenner.	Chicago, Ill	July 18, 1871	117,211
Axle-box bearing	A. B. Allen	Rutland, Vt	Mar. 30, 1869	88,356
Axle box, Car	W. G. Beattie	Nine Elms, England	Dec. 17, 1872	134,026
Axle box, Car	J. E. Bering	Newburgh, N. Y	Dec. 2, 1873	145,146
Axle box, Car	J. W. Cochran	New York, N. Y	Nov. 2, 1858	21,943
Axle box, Car	W. W. Crane	Auburn, N. Y	May 13, 1873	138,794
Axle box, Car	J. C. Creed	Omaha, Nebr	Jan. 12, 1869	85,793
Axle box, Car	J. M. Dodge	Newark, N. J	Mar. 4, 1873	136,424
Axle box, Car	D. H. Dotterer	Philadelphia, Pa	Sept. 5, 1871	118,701
Axle box, Car	H. G. Downs	Seneca Falls, N. Y	Jan. 7, 1873	134,594
Axle box, Car	W. Ebbit	New York, N. Y	Mar. 18, 1862	34,674
Axle box, Car	C. T. Fay	San Francisco, Cal	Feb. 18, 1873	135,977
Axle box, Car	S. F. Gates	Cambridge, Mass	Apr. 29, 1873	138,242
Axle box, Car	J. T. Hagerty	Camp Point, Ill	Feb. 15, 1870	99,882

Index of patents issued from the United States Patent Office from 1790 *to* 1873, *inclusive*—Continued.

Invention.	Inventor.	Residence.	Date.	No.
Axle box, Car	J. Harris	Dorchester, Mass	Dec. 10, 1867	71, 873
Axle box, Car	C. A. Haskins	Chicago, Ill	May 28, 1872	127, 163
Axle box, Car	G. H. Henfield	San Francisco, Cal	Jan. 7, 1868	73, 183
Axle box, Car	H. H. Hill and G. M. Sargent	Moline, Ill	July 29, 1873	141, 350
Axle box, Car	J. Hogan	South Boston, Mass	Aug. 26, 1873	142, 104
Axle box, Car	W. B. Howe	Troy, N. Y	Oct. 21, 1873	143, 763
Axle box, Car	W. R. Hunter	Erie, Pa	Jan. 26, 1869	86, 230
Axle box, Car	J. G. Johnson	Elkton, Md	Dec. 9, 1873	145, 425
Axle box, Car	J. Kinzer	Pittsburgh, Pa	Jan. 28, 1873	135, 343
Axle box, Car	J. J. Lahaye	Reading, Pa	July 4, 1871	116, 721
Axle box, Car	G. F. Lynch	Milwaukee, Wis	Oct. 11, 1870	108, 277
Axle box, Car	W. J. Manker	Sedalia, Mo	Apr. 29, 1873	138, 265
Axle box, Car	H. E. Marchand	Pittsburgh, Pa	May 27, 1873	139, 322
Axle box, Car	W. J. L. Moulton	San Francisco, Cal	Dec. 2, 1873	145, 227
Axle box, Car	D. Muzzey	Saint Albans, Vt	Nov. 25, 1873	145, 004
Axle box, Car	A. H. Nathan and M. Thornton.	Macon, Ga	May 12, 1868	77, 831
Axle box, Car	W. G. Parr	Normal, Ill	Oct. 10, 1871	119, 877
Axle box, Car	C. Pinder and D. C. Robinson	Lowell, Mass	Oct. 29, 1867	70, 356
Axle box, Car	L. Schulze	Chicago, Ill	Oct. 14, 1873	143, 592
Axle box, Car	M. Sessions and D. Muzzey	Saint Albans, Vt	Dec. 19, 1871	122, 069
Axle box, Car	T. B. Stewart	Wethersfield, Conn	Nov. 19, 1867	71, 241
Axle box, Car	T. B. Stewart	Hartford, Conn	Mar. 5, 1872	124, 227
Axle box, Car	W. Stone	Hollidaysburgh, Pa	Nov. 26, 1867	71, 422
Axle box, Car	J. E. Uhl	Renovo, Pa	May 6, 1873	138, 717
Axle box, Car	S. Ustick	Philadelphia, Pa	Mar. 4, 1873	136, 472
Axle box, Car	S. Ustick	Philadelphia, Pa	Aug. 26, 1873	142, 302
Axle box, Car	S. Ustick	Philadelphia, Pa	Aug. 26, 1873	142, 303
Axle box, Car	J. L. Vanclain	Lafayette, Ind	July 9, 1861	32, 803
Axle box, Car	W. W. Whitaker	Gloversville, N. Y	Sept. 3, 1873	142, 752
Axle box, Car	J. Whitaker and S. S. Cook	Woonsocket, R. I	Dec. 30, 1873	146, 116
Axle box, Car	C. Williams	Adrian, Mich	Oct. 31, 1871	120, 473
Axle box, Carriage	A. G. Baker and G. M. Ennis	New Bedford, Mass	Jan. 4, 1870	98, 543
Axle box, Carriage	G. B. Durkee	Alden, N. Y	May 17, 1870	103, 161
Axle box, Carriage	L. R. Dye	Cranberry, N. J	Sept. 6, 1871	107, 170
Axle box, Carriage	E. W. Ives	Hamden, Conn	Apr. 25, 1871	114, 148
Axle box, Carriage	D. Jewett	Lynn, Mass	Apr. 19, 1870	102, 127
Axle box, Carriage	J. Jones and J. and J. Dunkerly	Paterson, N. J	Apr. 22, 1873	138, 091
Axle box, Carriage	F. B. Morse	Plantsville, Conn	Jan. 9, 1872	122, 635
Axle box, Carriage	A. E. Smith	Bronxville, N. Y	Feb. 8, 1870	99, 607
Axle box, Carriage	P. R. Stage	Greensburgh, Ind	Nov. 26, 1872	133, 494
Axle box, Carriage	W. Stechschult	Glandorf, (Ottawa Post-Office,) Ohio.	Nov. 29, 1864	45, 279
Axle box, Carriage	E. M. Stratton	New York, N. Y	Apr. 1, 1856	14, 579
Axle box, Carriage	F. Wood	Bridgeport, Conn	Oct. 24, 1865	50, 648
Axle-box cover	W. S. Auchincloss	New York, N. Y	May 15, 1866	54, 666
Axle-box cover	W. H. Fitz Gerald	Brooklyn, N. Y	Apr. 25, 1871	114, 122
Axle-box cover	F. Grinnell	Meadville, Pa	June 8, 1869	91, 115
Axle-box cover	F. K. Hain	Renova, Pa	Oct. 9, 1866	58, 640
Axle-box cover, Car	D. Cowley	Erie, Pa	Dec. 13, 1870	110, 117
Axle-box cover, Car	R. McDowell	Lambertville, N. J	Mar. 19, 1867	63, 070
Axle-box cover, Railway	E. Moyel and J. Howell	Wyandotte, Kans	May 20, 1873	139, 179
Axle-box for car-trucks	W. Loughridge	Weverton, Md	Nov. 15, 1864	45, 058
Axle-box for journal of railway-cars	O. N. French	New London, Conn	July 15, 1851	8, 220
Axle-box for locomotives, Driving	J. W. Goff	Providence, R. I	Feb. 2, 1869	86, 390
Axle-box for lubricating railway rolling-stock, &c	P. F. Aerts	London, England	Aug. 2, 1859	24, 914
Axle-box for railway-cars	D. Jewett	Lynn, Mass	May 26, 1868	78, 210
Axle-box for railway-cars	D. S. Wood	Albany, N. Y	Mar. 22, 1864	42, 037
Axle-box for railway-cars, Replaceable	W. D. Arnett	Cincinnati, Ohio,	May 27, 1856	14, 981
Axle-box for vehicles	C. H. Allen	Saint Louis, Mo	Oct. 28, 1873	143, 950
Axle-box for vehicles	S. F. Green	Croton Falls, N. Y	Sept. 29, 1863	40, 100
Axle box from pedestal, Disconnecting car	W. D. Arnett	Chicago, Ill	June 15, 1858	20, 535
Axle-box guide	R. Hitchcock	Springfield, Mass	Apr. 30, 1867	64, 316
Axle-box lid	J. Bristow	Detroit, Mich	Sept. 21, 1869	94, 943
Axle-box, Lubricating	T. S. Speakman	Camden, N. J	Apr. 12, 1870	101, 933
Axle-box lubricator	C. D. Flynt	New York, N. Y	June 18, 1872	128, 135
Axle-box lubricator	C. Meziex	New York, N. Y	June 30, 1863	39, 059
Axle-box lubricator	S. S. Putnam	Dorchester, Mass	Nov. 2, 1869	96, 348
Axle-box lubricator	C. M. Ried	Greensborough, Ala	Feb. 16, 1869	87, 069
Axle-box lubricator	J. and W. Shackleton	Cleveland, Ohio	Dec. 10, 1872	133, 895
Axle-box lubricator, Car	M. C. Hubbard	Philadelphia, Pa	Mar. 16, 1869	87, 777
Axle-box lubricator, Car	J. Trent	Brooklyn, N. Y	Mar. 12, 1872	124, 642
Axle-box lubricator, Car	S. Ustick	Philadelphia, Pa	Apr. 30, 1872	126, 349
Axle-box lubricator, Car	S. Ustick	Philadelphia, Pa	Apr. 30, 1872	126, 350
Axle-box lubricator, Car	S. Ustick	Philadelphia, Pa	Apr. 30, 1872	126, 351
Axle-box lubricator, Carriage	S. S. Putnam	Dorchester, Mass	Nov. 20, 1866	59, 779
Axle-box lubricator, Carriage and wagon	W. W. Crane	Auburn, N. Y	Nov. 26, 1872	133, 417
Axle box, Railway	C. B. Boynton	Saint Paul, Minn	Nov. 12, 1867	70, 793
Axle box, Railway	A. Higley	South Bend, Ind	Feb. 2, 1869	86, 543
Axle box, Railway	J. J. Lahaye	Reading, Pa	May 8, 1866	54, 563
Axle box, Railway	H. Rice	Concord, N. H	Apr. 3, 1860	27, 738
Axle box, Railway	S. T. Shelley	Louisville, Ky	Jan. 10, 1865	45, 870
Axle box, Railway	J. B. Sutherland	Detroit, Mich	May 16, 1871	114, 877
Axle box, Railway	J. Wardrobe, C. D. B. Fisk, J. F. Curtis, and G. Fetley.	Carlin, Nev	Dec. 13, 1870	110, 093
Axle-box, Railway-car	R. N. Allen	Cleveland, Ohio	Mar. 23, 1858	19, 741
Axle-box, Railway-car	W. S. Auchincloss	Wilmington, Del	Dec. 5, 1871	121, 572
Axle-box, Railway-car	R. Brewer	New York, N. Y	June 10, 1873	139, 655
Axle-box, Railway-car	W. Campbell	Clinton, Iowa	Aug. 17, 1869	93, 858
Axle-box, Railway-car	W. Davis and C. Macan	Elizabethport and Rahway, N. J.	June 29, 1869	91, 919
Axle-box, Railway-car	G. W. and J. C. Geisendorff	Indianapolis, Ind., and Cincinnati, Ohio.	Feb. 9, 1858	19, 290
Axle-box, Railway-car	J. Harris	Dorchester, Mass	Jan. 3, 1860	26, 673
Axle-box, Railway-car	C. W. Harvey	Buffalo, N. Y	Jan. 14, 1873	134, 801
Axle-box, Railway-car	R. Levington	Monroe, Mich	Oct. 14, 1851	8, 428
Axle-box, Railway-car	M. McCammon	Chicago, Ill	Oct. 12, 1869	95, 705
Axle-box, Railway-car	D. R. Perkinpine	Philadelphia, Pa	Mar. 18, 1856	14, 468

Index of patents issued from the United States Patent Office from 1790 to 1873, inclusive—Continued.

Invention.	Inventor.	Residence.	Date.	No.
Axle-box, Railway-car	W. W. Simrell	Great Bend, Pa	July 29, 1862	36, 033
Axle-box, Railway-car	L. Stevens	Fitchburgh, Mass	May 31, 1859	24, 245
Axle-box, Railway-car	J. Tull and S. Norris	Philadelphia, Pa	July 3, 1847	5, 180
Axle-box, Railway-car	W. E. Wilcox	Peoria, Ill	June 22, 1869	91, 804
Axle-box, Railway-car	W. E. Wilcox	Peoria, Ill	July 27, 1869	93, 150
Axle-box roller	G. W. and J. C. Geisendorff	Indianapolis, Ind., and Cincinnati, Ohio.	Feb. 6, 1855	12, 347
Axle-box, sand-band, and casing for carriage-wheel hub, Combined.	M. McNalley	Houston, Tex	Dec. 26, 1871	122, 185
Axle-box, Self-acting lubricator for	E. von Jeinsen	Omaha, Nebr	July 5, 1870	105, 146
Axle-box, Street-car	J. Stephenson	New York, N. Y	June 4, 1872	127, 524
Axle-box, Street-car	A. Wright	Saint Louis, Mo	Nov. 12, 1872	133, 073
Axle-box, Vehicle	C. Ahrenbeck	Navasota, Tex	June 11, 1872	127, 821
Axle-box, Vehicle	W. A. Clark	New Haven, Conn	Sept. 9, 1873	142, 556
Axle-box, Vehicle	C. F. Gillette	Sparta, Wis	Feb. 19, 1873	62, 126
Axle-box, Vehicle	I. Halloway	New York, N. Y	Jan. 3, 1860	26, 672
Axle-box, Vehicle	J. Reilly	Racine, Wis	Feb. 12, 1868	62, 068
Axle-box, Vehicle	O. P. Rice	New York, N. Y	Sept. 30, 1873	143, 300
Axle-box, Vehicle	J. P. Smith	Terre Haute, Ind	Mar. 11, 1873	166, 778
Axle-box, Vehicle-wheel	D. Dalzell	South Egremont, Mass	July 9, 1872	128, 861
Axle-box, Wagon	A. E. Smith	Bronxville, N. Y	Sept. 11, 1866	57, 987
Axle-box washer	W. K. Foster	Cambridgeport, Mass	Oct. 22, 1867	70, 082
Axle boxes, Attachment to car	D. A. Hopkins	Jersey City, N. J	June 10, 1873	139, 791
Axle boxes, Boring carriage	A. Broad	Salisbury, Conn	June 14, 1809	....
Axle boxes, Casting carriage	S. Williamson	Cincinnati, Ohio	July 7, 1863	39, 189
Axle-boxes, Core for casting	W. H. Hawley	Utica, N. Y	Aug. 9, 1870	106, 265
Axle-boxes, Device for making core for	H. Kellogg	Milford, Conn	Sept. 3, 1872	131, 007
Axle boxes, Elastic check for car	J. Stephenson	New York, N. Y	June 4, 1872	127, 523
Axle-boxes, Fastening cover to	J. O. Clute and P. Kinney	Albany, N. Y	Mar. 31, 1863	38, 032
Axle-boxes from carriage-hubs, Apparatus for removing.	C. Searle	Brooklyn, N. Y	Aug. 17, 1869	93, 911
Axle-boxes from sand and dust, Device for protecting.	C. D. Seys	Nokomis, Ill	Feb. 13, 1872	123, 582
Axle-boxes, Gage for setting	T. Scott	Saint Paul, Minn	Apr. 29, 1873	138, 347
Axle-boxes, Machine for forming lining for	H. Kellogg	Milford, Conn	Nov, 26, 1872	133, 450
Axle boxes, Machine for lining carriage	H. Kellogg	Milford, Conn	Nov. 26, 1872	133, 449
Axle-boxes, Machinery for dressing	W. Hamilton	Allegheny City, Pa	May 5, 1863	38, 385
Axle-boxes, Making	A. E. Smith	Bronxville, N. Y	May 8, 1860	28, 207
Axle boxes, Manufacture of carriage	W. J. Parmelee	Fort Plain, N. Y	Apr. 1, 1873	137, 475
Axle-boxes, Oil reservoir for	E. von Jeinsen and J. M. McDonald.	San Francisco, Cal	Apr. 4, 1871	113, 599
Axle-brace for carriages, Adjustable	T. O. Rogers	Elmira, N. Y	Feb. 2, 1858	19, 264
Axle-clip for carriage-work	M. Seward	New Haven, Conn	July 19, 1864	43, 607
Axle-clip trimming and straightening die	E. H. Plant	Plantsville, Conn	July 19, 1870	105, 595
Axle-clips, Mode of making	H. M. Beecher	West Meriden, Conn	Aug. 12, 1873	141, 625
Axle coupling, Car	D. F. Fetter	New York, N. Y	Jan. 11, 1870	98, 681
Axle coupling, Carriage	G. F. Smith	Plantsville, Conn	Aug. 18, 1868	81, 115
Axle-coupling, Divided	C. A. Nutting	Macon, Ga	June 4, 1872	127, 637
Axle fitting and setting apparatus	J. Cunningham	Rochester, N. Y	Nov. 8, 1870	109, 115
Axle-forging die	J. Harrington	Bridgeport, Conn	May 23, 1871	115, 196
Axle-forming die	J. Nicol	Auburn, N. Y	Jan. 18, 1870	98, 878
Axle-gage	A. J. and A. H. Beach	Linden, Mich	Jan. 28, 1868	73 867
Axle-gage	J. Birkett	Tazewell County, Ill	July 6, 1869	92, 148
Axle-gage	R. C. Kelly	Brandon, Wis	Aug. 16, 1870	106, 373
Axle-gage	G. Luedke	Princeton, Wis	Dec. 28, 1869	98, 280
Axle-gage	H. W. Spaulding	Chelsea, Vt	June 6, 1871	115, 778
Axle-gage	R. K. Vestal	Santa Cruz, Cal	Apr. 4, 1871	113, 598
Axle-gage	D. C. Witsell	Carrollton, Pa	May 30, 1871	115, 549
Axle-grease	J. J. Barrett	Chillicothe, Ohio	Oct. 6, 1868	82, 675
Axle-head	R. E. Bean	Franklin, N. H	Oct. 6, 1868	82, 680
Axle journal and box, Railway-car	O. Robirds	Saint Louis, Mo	Aug. 3, 1869	93, 348
Axle journal-lubricator, Carriage	G. W. Parsons	Harrisburgh, Pa	Nov. 27, 1866	60, 049
Axle, Lubricating	J. Ives	Mount Carmel, Conn	May 2, 1871	114, 298
Axle-lubricating compound	G. M. Denison	Essex, Conn	Nov. 7, 1871	120, 631
Axle-lubricator	G. P. Blaisdell	North Easton, Mass	Dec. 13, 1870	110, 002
Axle-lubricator	G. A. Brannan	Baltimore, Md	Feb. 4, 1873	135, 517
Axle-lubricator	J. F. Hinman	Battle Creek, Mich	Nov. 24, 1863	40, 692
Axle-lubricator	J. E. Mowerson and C. S. D. Baun.	Westwood, N. J	Apr. 9, 1872	125, 477
Axle-lubricator	E. Sawyer	Hollow Square, Ala	Mar. 29, 1870	101, 384
Axle-lubricator	E. W. Smith	Waterbury, Conn	July 23, 1872	129, 760
Axle-lubricator	M. Soellinger and H. Noetzli	Keokuk, Iowa	Feb. 20, 1872	123, 949
Axle-lubricator	J. L. Sorber	Chillicothe, Ohio	Apr. 7, 1868	76, 541
Axle-lubricator	H. Thurlow	Skaneateles, N. Y	Jan. 17, 1871	111, 095
Axle-lubricator	H. S. Weaver	Irwin Station, Pa	Aug. 1, 1871	117, 581
Axle-lubricator	J. L. Winslow	Portland, Me	Dec. 1, 1868	84, 667
Axle-lubricator	J. Worden	Normal, Ill	Sept. 1, 1868	81, 858
Axle-lubricator, and mode of attaching it to axle	L. Adams	Amherst, Mass	Feb. 18, 1868	74, 476
Axle-lubricator, Automatic railway	M. Egan	Ogdensburgh, N. Y	Nov. 6, 1855	13, 749
Axle lubricator, Car	J. Barber	Bridesburgh, Pa	Dec. 5, 1871	121, 575
Axle lubricator, Car	J. E. Bering	Newburgh, N. Y	Dec. 23, 1873	145, 834
Axle lubricator, Car	G. A. Brannan	Baltimore, Md	Aug. 13, 1872	130, 470
Axle lubricator, Car	W. P. Burrow	Norfolk, Va	Feb. 20, 1872	123, 865
Axle lubricator, Car	J. W. Cochran	New York, N. Y	May 25, 1858	20, 331
Axle lubricator, Car	J. W. Cochran	New York, N. Y	Oct. 5, 1858	21, 652
Axle lubricator, Car	J. W. Cochran	New York, N. Y	June 1, 1858	20, 406
Axle lubricator, Car	J. S. Eggleston	Auburn, N. Y	Sept. 3, 1872	131, 087
Axle lubricator, Car	T. C. Hargrave	Boston, Mass	May 3, 1864	42, 577
Axle lubricator, Car	T. C. Hargrave	Boston, Mass	May 3, 1864	42, 578
Axle lubricator, Car	B. P. La Mothe	New York, N. Y	June 7, 1864	43, 033
Axle lubricator, Car	J. Lichtenstein	Baltimore, Md	Nov. 22, 1870	109, 525
Axle lubricator, Car	T. J. Mooers	Blossburgh, Pa	Jan. 11, 1870	98, 701
Axle lubricator, Car	J. R. Morris	Houston, Tex	Jan. 7, 1873	134, 696
Axle lubricator, Car	W J. L. Moulton	San Francisco, Cal	Dec. 2, 1873	145, 226
Axle lubricator, Car	W. Painter	Baltimore, Md	Apr. 16, 1872	125, 841
Axle lubricator, Car	T. H. Paul	Frostburgh, Md	Mar. 12, 1872	124, 618
Axle lubricator, Car	J. S. Sanson	Morrisania, N. Y	Jan. 2, 1872	122, 495
Axle lubricator, Car	T. Sayles	Chicago, Ill	Sept. 20, 1870	107, 631

Index of patents issued from the United States Patent Office from 1790 *to* 1873, *inclusive*—Continued.

Invention.	Inventor.	Residence.	Date.	No.
Axle lubricator, Car	T. R. Timby	Tarrytown, N. Y	Jan. 28, 1873	135, 386
Axle lubricator, Car	S. Ustick	Philadelphia, Pa	Apr. 30, 1872	126, 352
Axle lubricator, Car	E. von Jeinsen	Omaha, Nebr	Aug. 2, 1870	106, 096
Axle lubricator, Car	R. Vose and J. M. Evans	New York, N. Y	May 28, 1872	127, 391
Axle lubricator, Car	I. P. Wendell	Philadelphia, Pa	Jan. 24, 1871	111, 287
Axle lubricator, Car	M. W. Woodruff	Camillus, N. Y	May 7, 1872	126, 508
Axle lubricator, Carriage	H. F. Phillips	Auburn, N. Y	Apr. 1, 1873	137, 479
Axle lubricator, Carriage	C. Polley	Scott, Ohio	Sept. 20, 1864	44, 338
Axle lubricator, Carriage	S. S. Putnam	Dorchester, Mass	Feb. 4, 1868	74, 134
Axle-lubricator plug	A. C. Garratt	Hanover, Mass	Jan. 16, 1855	12, 238
Axle lubricator, Railway	W. Clough	Madison, Ind	Mar. 30, 1858	19, 750
Axle lubricator, Railway-car	C. Hyatt	Buffalo, N. Y	Nov. 1, 1870	108, 789
Axle lubricator, Railway-car	I. P. Wendell	Philadelphia, Pa	Jan. 24, 1871	111, 286
Axle lubricator, Vehicle	J. S. Eggleston	Auburn, N. Y	Jan. 2, 1872	122, 374
Axle lubricator, Vehicle	L. H. Fisher	Walpole, Mass	Dec. 10, 1872	133, 769
Axle lubricator, Wagon	W. Loewenstein	Keokuk, Iowa	Apr. 29, 1873	138, 263
Axle lubricator, Wagon	C. A. Wakefield	Pittsfield, Mass	June 13, 1871	115, 914
Axle-nut and axle	C. Thomas	Boston, Mass	June 28, 1870	104, 901
Axle-nut die	A. B. Candee and L. S. Taylor	Hamden and Southington, Conn.	June 2, 1868	78, 576
Axle-nut lubricator	D. Dalzell	South Egremont, Mass	May 28, 1872	127, 313
Axle-nuts, Method of manufacturing	S. Vanstone	Providence, R. I	June 28, 1870	104, 984
Axle or journal lubricator	R. V. Laney	Cumberland, Md	Mar. 12, 1872	124, 595
Axle or shaft	G. Foster	Brooklyn, N. Y	Dec. 13, 1859	26, 425
Axle-roller	W. S. Mackintosh	Pittsburgh, Pa	June 16, 1868	78, 984
Axle-rolling machine	W. P. Porter	Pittsburgh, Pa	Mar. 10, 1868	75, 457
Axle-set	H. R. Ladd	Orwell, Ohio	Mar. 9, 1869	87, 571
Axle-shafts, &c., Coating the bearing of boxes for	W. Peters	Baltimore, Md	July 22, 1862	35, 976
Axle-skein	D. and J. Gray	Wayland, N. Y	June 24, 1862	35, 676
Axle-skein	H. C. Kochensperger	Thornville, Ohio	Jan. 2, 1872	122, 391
Axle-skein	L. Mayhew	Saratoga Springs, N. Y	Dec. 27, 1870	110, 579
Axle-skein	L. Mayhew	Greenfield, N. Y	Apr. 4, 1871	113, 681
Axle-skein	L. Mayhew	Greenfield, N. Y	Aug. 22, 1871	118, 377
Axle-skein	L. Mayhew	Rock City Falls, N. Y	Oct. 24, 1871	120, 303
Axle-skein	H. F. Phillips	Auburn, N. Y	Aug. 18, 1863	39, 617
Axle-skein	E. Rooks	Trenton, Tenn	Nov. 9, 1869	96, 620
Axle-skein	G. Schreyer	Columbus, Ohio	May 4, 1869	89, 602
Axle-skein	T. S. Sleeper	Binghamton, N. Y	Sept. 28, 1869	95, 279
Axle skein and box	G. E. Clarke	Racine, Wis	May 5, 1868	77, 585
Axle-skein, Malleable-iron	A. J. Alston	Allegheny City, Pa	Dec. 24, 1872	134, 189
Axle-skein, Reversible	A. F. Smith	Aiken, Tex	Apr. 16, 1872	125, 762
Axle-skein setter	J. Burt	Sturgis, Mich	Aug. 4, 1868	80, 594
Axle-skein, Steel	A. J. Alston	Allegheny City, Pa	Dec. 24, 1872	134, 188
Axle-skein, Truss-brace attachment for	O. Vanorman	Fond du Lac, Wis	July 16, 1872	129, 258
Axle-skein, Truss-brace for	O. Vanorman	Fond du Lac, Wis	Dec. 5, 1871	121, 692
Axle-skein mold, Machine for preparing	J. G. Holt	Chicago, Ill	June 5, 1866	55, 295
Axle-spindles to skeins, Device for fitting	C. L. Campbell	Binghamton, N. Y	Sept. 28, 1869	95, 319
Axle-spindles to skeins of wagons, Machine for fitting	C. L. Campbell	Binghamton, N. Y	Oct. 9, 1866	58, 714
Axle thimble-skein	W. D. Baughn	Milford, Mich	Nov. 26, 1867	71, 264
Axle thimble-skein	M. Ehrgott	Pittsburgh, Pa	Mar. 9, 1869	87, 652
Axle thimble-skein	J. A. Williams	Elizabeth, Ill	Oct. 8, 1867	69, 735
Axle-tourniquet	J. Strong		Jan. 29, 1801	
Axletree	C. E. Buck	Racine, Wis	Jan. 21, 1868	73, 575
Axletree	R. Haslup	Baltimore, Md	Feb. 24, 1831	
Axletree	J. W. Whitney	Buffalo, N. Y	Sept. 5, 1848	5, 753
Axletree	J. W. Wilkie	Auburn, N. Y	Mar. 26, 1867	63, 347
Axletree	J. W. Wilkie	Auburn, N. Y	Mar. 26, 1867	63, 348
Axletree-arm	D. Philips	Sharon, Pa	Feb. 3, 1852	8, 707
Axletree, hub, and spoke	G. Wilkinson	White Creek, N. Y	Dec. 6, 1831	
Axletree safety-carriage machine, Double	E. G. Fitch	Blakely, Ala	Feb. 7, 1829	
Axletrees, Diminishing friction in	B. Hinkley	Fayette, Me	Apr. 14, 1835	
Axletrees, Machine for turning	W. H. Heffley	Rochester, Ind	Aug. 16, 1870	106, 364
Axletrees, Mode and machinery for setting	T. Fessenden	Boston, Mass	Sept. 25, 1837	410
Axletrees, Reduced friction-roller	F. Rees	Clarksville, Ga	Apr. 14, 1832	
Axletrees, Setting	J. N. Arvin and S. H. Perkins	Valparaiso, Ind	July 17, 1860	29, 132
Axle-turning machine	G. S. Knight	Syracuse, N. Y	Aug. 20, 1867	67, 886
Axle-turning machine	A. F. Moore	Florence, Ind	Mar. 28, 1871	113, 078
Axle-turning machine	A. F. Moore	Florence, Ind	Apr. 1, 1873	137, 316
Axle-turning machine	R. Zeider	El Paso, Ill	Aug. 20, 1872	130, 782
Axle-washer	A. E. Smith	Bronxville, N. Y	July 24, 1855	13, 327
Axle-wheel	C. J. Crane	Burr Oak, Mich	Oct. 24, 1865	50, 655
Axles and shafts, Bearing for	W. H. Hovey	Hartford, Conn	Dec. 10, 1850	7, 819
Axles, Connecting skeins with	A. Combs	Farmington, Ohio	May 14, 1850	7, 363
Axles, Constructing	J. Montgomery	New York, N. Y	Nov. 21, 1871	121, 187
Axles, Gage for making	P. Geiser	Waynesborough, Pa	July 20, 1869	92, 817
Axles, Grease-box for	J. M. Smart	New York, N. Y	Nov. 6, 1849	6, 854
Axle, Machine for rolling shoulder on	W. Van Anden	Poughkeepsie, N. Y	Aug. 22, 1854	11, 596
Axles, Mode of lubricating	J. F. Stevenson and T. B. Hammer.	McKeesport, Pa	Mar. 18, 1862	34, 704
Axles of machinery, Anti-friction device for	N. Bailly, C. Durand, G. H. Mesnard, and Z. Poirier.	France, England, and South Lambeth, England.	Oct. 9, 1866	58, 739
Axles or journals, Mode of preventing the heating of	E. D. Murfey	New York, N. Y	Nov. 15, 1870	109, 237
Axles, Preventing friction on	T. S. and T. S. Minniss	Meadville, Pa	Feb. 15, 1859	22, 971
Axles, Splicing-bar for	J. E. Balderston	Philadelphia, Pa	Jan. 21, 1862	34, 189
Axles to vehicles, Mode of attaching	W. Gray and H. E. Porter	Hebron, Conn	Apr. 28, 1868	77, 185
B.				
Baby-chair	P. Henrichs	Erie, Pa	Jan. 7, 1873	134, 668
Baby-chair	C. Holtz	New York, N. Y	Mar. 16, 1869	87, 933
Baby chair and table	J. Kopp	Bridgeport, Conn	Dec. 11, 1866	60, 389
Baby creeper or walker	P. Hurd	Croton, Mich	Mar. 10, 1868	75, 274
Baby-holder	R. Hale	Chicago, Ill	Jan. 28, 1868	73, 801
Baby-jumper	W. Berg and M. Stephan	Canton, Ohio	June 8, 1869	90, 985
Baby-jumper	S. G. Bigelow	Silver Lake, Ind	Nov. 12, 1872	133, 008
Baby-jumper	J. H. Coldwell	Poughkeepsie, N. Y	Mar. 31, 1868	76, 053
Baby-jumper	J. A. De Brame	New York, N. Y	Apr. 16, 1861	32, 109
Baby-jumper	J. P. Faulks	Glasgow, Mo	Aug. 29, 1871	118, 599

Index of patents issued from the United States Patent Office from 1790 *to* 1873, *inclusive*—Continued.

Invention.	Inventor.	Residence.	Date.	No.
Baby-jumper	W. Gibson	Fort Wayne, Ind	June 26, 1866	55, 849
Baby-jumper	D. M. Holmes	Westchester, N. Y	Sept. 17, 1872	131, 349
Baby-jumper	A. H. Mason	Binghamton, N. Y	Sept. 10, 1867	68, 637
Baby-jumper	C. Rich	Poughkeepsie, N. Y	Oct. 13, 1868	82, 992
Baby-jumper	E. Rice	Elizabethtown, N. J	Oct. 28, 1851	8, 478
Baby-jumper	A. F. Spooner	Sterling, Ill	Apr. 22, 1873	138, 209
Baby-jumper	J. P. Thompson	Philadelphia, Pa	Oct. 2, 1866	58, 510
Baby-jumper	E. L. Warner	Oxford, Conn	July 5, 1870	105, 149
Baby-jumper	M. J. Wellman	New York, N. Y	Mar. 31, 1857	16, 942
Baby-jumper	J. E. Wells	Chicago, Ill	Aug. 13, 1872	130, 397
Baby jumper and cradle	R. Ashe	Somerville, Mass	July 23, 1867	67, 153
Baby jumper and cradle, Combined	G. H. Mellen	Chicago, Ill	June 30, 1868	79, 534
Baby jumper and rocker	B. Beach	Meriden, Conn	Sept. 28, 1869	95, 183
Baby jumper and supporter	H. Frankfurth	Utica, N. Y	Dec. 3, 1861	33, 833
Baby jumper and swing	S. T. McDougall	New York, N. Y	July 5, 1864	43, 420
Baby jumper and walker	J. H. Coldwell	New York, N. Y	Aug. 30, 1864	43, 972
Baby jumper and walker	L. O. Colvin	Newark, N. J	Sept. 3, 1872	131, 083
Baby jumper, couch, and carriage	J. S. Brown	Green Point, N. Y	Nov. 27, 1860	30, 717
Baby-swing	J. Wolf	Cleveland, Ohio	Dec. 26, 1865	51, 772
Baby-swing	J. H. Wygant and R. P. Paulison	Hackensack, N. J	May 21, 1872	127, 134
Baby-tender	J. S. Brown	New York, N. Y	May 17, 1864	42, 745
Baby-tender	A. H. Carson and A. Brown	Newport, R. I., and Troy, N. Y.	Apr. 25, 1871	113, 979
Baby-tender	L. Limerick	Louisville, Ky	May 7, 1867	64, 432
Baby-tender	S. M. Simonds	Lynn, Mass	Mar. 28, 1871	113, 103
Baby-tender	A. Wheeler	Newton, Mass	Sept. 1, 1863	39, 766
Baby-tender	C. N. Ziegler	Grafton, Ohio	Jan. 9, 1872	122, 546
Baby-tender, Locomotive	J. C. Smith	Boston, Mass	Apr. 3, 1849	6, 247
Baby traveling and toilet chair	E. Hoare	Dedham, Mass	May 16, 1871	114, 818
Baby-walker	P. W. Clark	Paw Paw, Mich	June 25, 1872	128, 561
Baby-walker	J. Erikson	Madison, Wis	Aug. 5, 1873	141, 431
Baby-walker	G. Enell	Guttenburgh, N. J	May 7, 1872	126, 452
Baby-walker	F. A. Geisler	Bristol, R. I	Sept. 22, 1868	82, 304
Baby-walker	J. C. Goulding	Trenton, N. J	Nov. 23, 1869	97, 078
Baby-walker	C. Maschmann	Watertown, N. Y	June 20, 1871	116, 073
Baby-walker	P. Pallissard	Aroma, Ill	June 12, 1866	55, 529
Baby-walker	P. H. Randolph	Leavenworth City, Kans	Nov. 30, 1869	97, 315
Baby-walker	J. Rohr	Batesville, Ind	Nov. 19, 1867	71, 220
Baby-walker	J. Thomas	Brooklyn, N. Y	Mar. 17, 1857	16, 862
Baby-walker	C. D. and J. A. Westlake	Laurel, Md	Mar. 19, 1872	124, 706
Baby walker and cradle	J. H. Brown	New York, N. Y	Jan. 2, 1866	51, 795
Baby walker and hanging chair	L. O. Colvin	Newark, N. J	July 11, 1871	116, 813
Baby walker and jumper	E. Y. Robbins	Cincinnati, Ohio	Dec. 2, 1856	16, 150
Baby-walker, Willow	J. E. Kauffman	Osborn, Ohio	June 25, 1872	128, 230
Back and abdominal supporter	J. Ford	Salem, Oreg	Mar. 31, 1868	76, 070
Back-band fastener	D. L. McGregor	Charlestown, Mass	Dec. 17, 1867	72, 312
Back-band hook	H. Beagle	Philadelphia, Pa	Nov. 29, 1870	109, 572
Back-band hook	H. Beagle, jr	Philadelphia, Pa	Oct. 3, 1871	119, 556
Back-band hook	W. McKerahan	Allegheny City, Pa	Feb. 9, 1869	86, 852
Back-band hook	J. Straus	Saint Louis, Mo	June 3, 1873	139, 524
Back-band hook	S. Ward	Princeton, Ind	Feb. 26, 1867	62, 513
Back-band hook	S. Ward	Princeton, Ind	Mar. 9, 1869	87, 737
Back-band hook and buckle	S. Ward	Princeton, Ind	Sept. 3, 1872	131, 139
Back-band hook for plow-harness	N. Warlick	Chambers County, Ala	Oct. 28, 1856	15, 993
Back-band strap, Self-adjusting and vibrating	R. Jancovins	Newark, N. J	June 1, 1858	20, 434
Back-brace	E. P. Banning, jr	New York, N. Y	Mar. 12, 1872	124, 473
Backgammon and checker board	E. O. Goodwin	Bristol, Conn	Nov. 25, 1856	16, 116
Backgammon-board	N. B. Williams	New York, N. Y	Mar. 16, 1869	87, 895
Backlash for mill-gearing	J. L. Post	Ashley, Ill	Oct. 18, 1870	108, 515
Backlash in machinery, Device for preventing	J. L. Post	Ashley, Ill	Mar. 14, 1871	112, 736
Backlash in machines driven by gearing, Means for preventing.	G. H. Babcock, J. P. Manton, and J. Boyd.	Providence, R. I	Oct. 12, 1869	95, 757
Backlash in mills, Device for preventing	A. B. Rider	Fairfield, Ill	June 25, 1872	128, 327
Backlash spring for machinery	H. W. Bachman	McLean, Ill	Oct. 22, 1872	132, 347
Back-log boiler for ranges, &c	B. Hunter	Philadelphia, Pa	Mar. 26, 1872	124, 899
Back-pad press	E. W. Harlow	Hastings, Mich	Jan. 24, 1871	111, 119
Bacon and ham box or case	M. W. Brown	New York, N. Y	Oct. 18, 1870	108, 322
Bacon, ham, and shoulders, Mode of curing	W. G. Bell	Charlestown, Mass	Nov. 30, 1869	97, 268
Badge, Campaign	H. C. Griggs	Waterbury, Conn	Oct. 27, 1868	83, 486
Badge, Fastening for policeman's	T. Kirkpatrick	New York, N. Y	Aug. 30, 1870	106, 942
Badge for hats, Illuminated	W. J. Scott	Albany, N. Y	Jan. 9, 1872	122, 665
Bag:				
See Broom-bag.				
Carpet-bag.				
Dash-board bag.				
Feed-bag.				
Floating bag.				
Fruit-gathering bag.				
Grain-bag.				
Guano-bag.				
Horse nose-bag.				
Ice-bag.				
Mail-bag.				
Packing-bag.				
Paper-bag.				
Phosphate-bag.				
Saddle-bag.				
Striking-bag.				
Tobacco-bag.				
Tobacco-packing bag.				
Traveling-bag.				
Woven bag.				
Bag and shoe-string fastener	J. H. Weeden	Waterbury, Conn	Sept. 21, 1869	94, 987
Bag-clasp	E. L. Dickey	Chicago, Ill	Feb. 18, 1868	74, 517
Bag-clasp fastener	A. H. Balch and W. D. E. Nelson.	Montreal, Canada	Apr. 18, 1871	113, 834
Bag-fastener	W. P. Brooks	Fairmount, Minn	May 1, 1866	54, 287
Bag-fastener	W. H. Brown	Worcester, Mass	Aug. 7, 1866	56, 894

Index of patents issued from the United States Patent Office from 1790 *to* 1873, *inclusive*—Continued.

Invention.	Inventor.	Residence.	Date.	No.
Bag-fastener	S. R. Bush	Bush's Mill, Ohio	Nov. 18, 1873	144, 738
Bag-fastener	L. H. Colborn	Chicago, Ill	May 5, 1868	77, 587
Bag-fastener	S. C. Dix	Neponset, Ill	Apr. 12, 1870	101, 837
Bag-fastener	F. H. Drake	Kenosha, Wis	Mar. 24, 1868	75, 877
Bag-fastener	A. C. Fletcher	New York, N. Y	July 25, 1865	48, 922
Bag-fastener	C. A. Haring	Peoria, Ill	Oct. 26, 1869	96, 229
Bag-fastener	C. J. Huntington	Rockford, Ill	Feb. 2, 1869	86, 403
Bag-fastener	W. P. Maxson	Albion, Wis	May 24, 1859	24, 133
Bag-fastener	J. Miller	Ovid, Mich	June 4, 1867	65, 499
Bag-fastener	C. H. Nye	Vineland, N. J	Oct. 27, 1868	83, 534
Bag-fastener	D. Overholtzer	Polo, Ill	Jan. 7, 1868	73, 193
Bag-fastener	S. P. Parmly	New Orleans, La	June 15, 1869	91, 362
Bag-fastener	G. H. Peacock	Fairport, N. Y	Apr. 21, 1868	76, 941
Bag-fastener	E. Romans	La Porte, Ind	Nov. 5, 1867	70, 620
Bag-fastener	G. C. Setchell and C. L. Taylor	Norwich, Conn	Dec. 26, 1871	122, 135
Bag-fastener	A. M. Wright	Safe Harbor, Pa	July 30, 1867	67, 393
Bag-fastener	W. Zeller and R. Lechner	Lebanon County and Berks County, Pa.	Sept. 24, 1867	69, 295
Bag fastener or tie	D. B. Baker	Rollersville, Ohio	June 11, 1867	65, 632
Bag-filler	W. H. and J. G. Mitchell	Circleville, Ohio	May 14, 1872	126, 732
Bag-filler	A. J. Olney	Van Buren, Ind	July 6, 1869	92, 347
Bag-filling and weighing-machine	C. A. Whelan and C. F. Wakely	Madison, Wis	Jan. 3, 1871	110, 707
Bag-frame	W. T. Mercereau	Newark, N. J	Apr. 24, 1866	54, 190
Bag-frame	A. J. Sessions	Bristol, Conn	June 12, 1866	55, 543
Bag-frame bending-machine	G. Havell	Newark, N. J	Nov. 9, 1869	96, 695
Bag, grain, &c., tie	G. W. Osborn	Parkville, Mich	Oct. 4, 1870	107, 952
Bag-holder	I. Allen	Manchester, N. Y	July 4, 1871	116, 661
Bag-holder	O. Barnett and A. D. Brooks	Dartford, Wis	Sept. 12, 1871	118, 897
Bag-holder	C. J. Barney	Edgartown, Mass	Dec. 4, 1866	60, 121
Bag-holder	B. S. Boydston	Richmond, Ind	July 9, 1867	66, 554
Bag-holder	E. Boynton	Palmyra, Wis	June 28, 1870	104, 698
Bag-holder	C. D. Brainerd	Danville, Vt	Dec. 31, 1867	72, 788
Bag-holder	J. B. Brown	Washington, D. C	Jan. 7, 1873	134, 637
Bag-holder	J. M. Burke	Dansville, N. Y	Jan. 4, 1870	98, 553
Bag-holder	S. P. Clemons	Dansville, N. Y	Jan. 5, 1869	85, 643
Bag-holder	J. N. Collins	Menasha, Wis	Aug. 31, 1869	94, 283
Bag-holder	G. E. Corbin	Saint Johns, Mich	Sept. 4, 1866	57, 683
Bag-holder	J. S. Corbin	Ann Arbor, Mich	July 25, 1865	48, 911
Bag-holder	L. Crofoot	Pavilion, N. Y	Jan. 5, 1869	85, 511
Bag-holder	L. Crofoot	Pavilion, N. Y	Feb. 22, 1870	100, 123
Bag-holder	L. Crofoot	Pavilion, N. Y	Aug. 13, 1872	130, 489
Bag-holder	D. Culver	Kingston, Pa	July 2, 1872	128, 598
Bag-holder	G. Dare	Auburn, N. Y	May 25, 1869	90, 344
Bag-holder	G. W. Dungan and W. Wasson	Genoa, Nev	Feb. 13, 1872	123, 684
Bag-holder	E. C. Fairchild	Sunderland, Mass	Mar. 3, 1868	75, 138
Bag-holder	E. A. Fisher	Morganville, N. Y	June 8, 1869	91, 004
Bag-holder	L. H. Gano	Milwaukee, Wis	Mar. 12, 1867	62, 741
Bag-holder	N. A. Geisinger	Coopersburgh, Pa	Oct. 21, 1873	143, 897
Bag-holder	E. N. Giles	Brownville, N. Y	Sept. 10, 1872	131, 267
Bag-holder	C. F. Gillett	Sparta, Wis	Apr. 10, 1866	53, 809
Bag-holder	C. F. Gillett	Sparta, Wis	July 17, 1866	56, 402
Bag-holder	F. Godfrey	Grand Rapids, Mich	Apr. 12, 1864	42, 283
Bag-holder	O. Hanks	Cincinnati, Ohio	Apr. 2, 1867	63, 383
Bag-holder	T. Harding	Springfield, Ohio	Oct. 22, 1867	69, 989
Bag-holder	A. V. Heyden	Milwaukee, Wis	Feb. 26, 1867	62, 332
Bag-holder	C. W. Hills and O. F. Woodruff	Morrison, Ill	Jan. 30, 1866	52, 288
Bag-holder	J. H. Hollidge	Washington, D. C	July 30, 1872	130, 051
Bag-holder	G. G. Hollinger	Mount Joy, Pa	Aug. 6, 1872	130, 130
Bag-holder	J. Hunter and H. M. Clayton	Ashland, Pa	May 13, 1873	138, 752
Bag-holder	E. L. Lyon	Steamburgh, N. Y	May 14, 1872	126, 723
Bag-holder	J. McPhail	Charles City, Iowa	Nov. 16, 1869	96, 941
Bag-holder	J. Melcher	Minneapolis, Minn	May 28, 1867	65, 105
Bag-holder	P. Meyers	Stoutsville, Ohio	Feb. 16, 1869	87, 059
Bag-holder	C. K. Mitchell	Greenville, Mich	July 11, 1871	116, 981
Bag-holder	L. W. Morlan	New Lisbon, Ohio	June 6, 1865	48, 083
Bag-holder	E. S. Moulton	Plymouth, Mich	July 9, 1867	66, 614
Bag-holder	J. V. H. Nott	Guilderland, N. Y	Dec. 4, 1866	60, 233
Bag-holder	A. M. Olds	Glenmont, Wis	Aug. 20, 1861	33, 097
Bag-holder	J. H. Park	White House, N. J	May 5, 1868	77, 648
Bag-holder	J. I. Peyton and C. N. S. Wallach.	Washington, D. C	Apr. 27, 1869	89, 500
Bag-holder	R. Ramsey	New Wilmington, Pa	Feb. 21, 1865	46, 498
Bag-holder	E. Reynolds	Corunna, Mich	Nov. 28, 1865	51, 220
Bag-holder	S. S. Rockwell	Lansing, Mich	Apr. 3, 1866	53, 681
Bag-holder	J. Roseborough, jr	South Hermitage, Pa	June 18, 1872	128, 073
Bag-holder	N. N. Rugg	Geneva, Ill	June 22, 1869	91, 568
Bag-holder	W. H. Smith	Sparta, Wis	July 31, 1866	56, 819
Bag-holder	W. H. Starry	Middletown, Ohio	Nov. 13, 1866	59, 677
Bag-holder	G. W. Striker	Iowa Falls, Iowa	Sept. 23, 1873	143, 194
Bag-holder	G. D. Sweigert	Martic Township, Pa	May 3, 1870	102, 616
Bag-holder	A. D. Swogger	Worth, Pa	Dec. 27, 1870	110, 511
Bag-holder	C. H. Thomas	Williamsburgh, Ohio	July 22, 1873	141, 186
Bag-holder	T. J. Trapp	Williamsport, Pa	July 4, 1871	116, 777
Bag-holder	T. J. Trapp	Williamsport, Pa	Aug. 13, 1872	130, 547
Bag-holder	L. Turner	Cedar Rapids, Iowa	July 2, 1867	66, 424
Bag-holder	I. Vance and C. Rogers	Pittsburgh, Pa	May 21, 1872	127, 122
Bag-holder	D. S. Wing	Rome, N. Y	Nov. 29, 1870	109, 788
Bag-holder	J. and T. R. Yaggy	Plainfield, Ill	Apr. 14, 1868	76, 686
Bag-holder and conveyor	C. K. Hostetter	East Donegal Township, Pa	Dec. 18, 1860	30, 943
Bag holder and elevator, Combined	G. H. Smith	Des Moines, Iowa	Apr. 14, 1863	38, 183
Bag holder and filler	A. M. Darling	Davenport, Iowa	Nov. 26, 1867	71, 286
Bag holder and filler	W. F. Lum	Waterloo, Wis	Dec. 5, 1871	121, 529
Bag holder and filler, Combined	W. F. Lum	Waterloo, Wis	Mar. 14, 1871	112, 727
Bag-holder and measure	G. E. Randall	Yaphank, N. Y	Aug. 2, 1864	43, 706
Bag-holder and scale	W. Zimmerman	Lebanon, Pa	Nov. 9, 1869	96, 759
Bag-holder and scales, Combined	W. Zimmerman	Lebanon, Pa	Mar. 7, 1871	112, 522
Bag-holder and truck, Combined	J. Hewitt	Albany, N. Y	May 28, 1872	127, 346
Bag-holder and truck, Combined	H. A. Reid	Beaver Dam, Wis	Apr. 20, 1869	89, 077

Index of patents issued from the United States Patent Office from 1790 *to* 1873, *inclusive*—Continued.

Invention.	Inventor.	Residence.	Date.	No.
Bag-holder and truck, Combined	P. C. Van Brocklin	Paris, Canada	Jan. 9, 1872	122, 685
Bag-holder, Hopper for	W. Wasson and G. W. Dungan.	Genoa, Nev	July 16, 1872	129, 633
Bag holding and filling machine	J. Agate	Cuba, N. Y	Sept. 2, 1862	36, 328
Bag-holding apparatus	J. S. Lehman	Mount Joy, Pa	Feb. 11, 1868	74, 232
Bag-holding device and truck	J. S. Lehman	Mount Joy, Pa	Sept. 15, 1868	82, 132
Bag-lock	H. Ahrend	Newark, N. J	June 3, 1873	139, 526
Bag-lock	E. L. Gaylord	Bridgeport, Conn	Aug. 6, 1872	130, 209
Bag-lock	B. Steinmetz	Paris, France	Sept. 13, 1870	107, 422
Bag-machine	N. Biedinger	Cincinnati, Ohio	July 1, 1873	140, 342
Bag-machine, Pasting-apparatus for	S. E. Pettee	Mansfield, Mass	Nov. 30, 1858	22, 199
Bag-mouth fastener	T. K. Reed	North Bridgewater, Mass	Feb. 28, 1865	46, 587
Bag-string fastener	E. B. Southwick	Mendon, Mich	Jan. 4, 1870	98, 527
Bag-string inserter	W. J. Cussen	Richmond, Va	June 4, 1872	127, 579
Bag-tie	J. W. Bates	Glencoe, Minn	Oct. 20, 1868	83, 240
Bag-tie	J. W. Blackstone	Darlington, Wis	July 2, 1872	128, 526
Bag-tie	J. Bowers	Brookville, Ill	Sept. 24, 1872	131, 593
Bag-tie	J. Brannihr and D. H. Rhodes	Hempstead and Baldwinsville, N. Y.	Mar. 11, 1873	136, 637
Bag-tie	D. Dick and O. W. Preston, jr	Corning, N. Y	Dec. 15, 1868	84, 998
Bag-tie	J. W. H. Doubler	Darlington, Wis	Apr. 16, 1872	125, 795
Bag-tie	L. H. Gano	Milwaukee, Wis	Mar. 3, 1868	75, 008
Bag-tie	L. H. Gano	New York, N. Y	Feb. 9, 1869	86, 745
Bag-tie	J. J. Gordon	Flint, Mich	Dec. 3, 1867	71, 739
Bag-tie	J. Grimes	Alexandria, Va	Oct. 29, 1867	70, 201
Bag-tie	D. B. Hall	Bucksport, Me	Apr. 2, 1867	63, 510
Bag-tie	C. P. and W. H. Markham	Rogersville, N. Y	May 30, 1871	115, 336
Bag-tie	J. C. Meloy	Hastings, Minn	Nov. 2, 1869	96, 459
Bag-tie	L. Morehouse	Barton, Wis	Mar. 5, 1867	62, 667
Bag-tie	G. Murray	Waterloo, N. Y	May 11, 1869	90, 013
Bag-tie	T. H. Russell	Lebanon, N. H	Apr. 30, 1872	126, 157
Bag-tie	E. Truslow	New York, N. Y	June 30, 1868	79, 414
Bag-turning machine	J. Martin	New York, N. Y	Mar. 18, 1873	136, 926
Bags, Clasp for fastening	W. H. Cloud, A. L. Hatfield, and C. H. Burdick.	Tremont, Ohio	Apr. 6, 1858	19, 830
Bags, Holding and filling	G. E. Randall	Yaphank, N. Y	June 13, 1865	48, 208
Bags, Machinery for turning	W. V. Gee	New Haven, Conn	Dec. 4, 1860	30, 808
Bags and sacks, Machine for holding open	J. Robinson	New Wilmington, Pa	Sept. 9, 1862	36, 423
Bags and sacks, Manufacture of	W. B. Carlock	New York, N. Y	July 3, 1849	6, 566
Bags, sacks, &c., Fastening	A. B. Wood, jr	Ovid, Mich	Nov. 1, 1864	44, 907
Bags while being filled, Machine for holding	J. H. Morris	Niles, Mich	Sept. 3, 1861	33, 209
Bagasse, Drying	G. Merrick	New Orleans, La	Apr. 10, 1845	3, 994
Bagasse-drying-machine	J. H. Dakin	Baton Rouge, La	May 21, 1850	7, 375
Bagasse-drying-machine	S. H. Gilman	Cincinnati, Ohio	Oct. 28, 1851	8, 466
Bagasse for fuel, Preparing	J. M. Frink	Coral, Ill	May 26, 1863	38, 667
Bagasse-furnace	J. Amick	Assumption Parish, La	Apr. 19, 1870	101, 968
Bagasse-furnace	A. JnChapman	Bayou Goula, La	Nov. 22, 1859	26, 162
Bagasse-furnace	F. Daunoy	Carrollton, La	Dec. 21, 1858	22, 353
Bagasse-furnace	C. A. Desobry	Plaquemine, La	Sept. 6, 1859	25, 322
Bagasse-furnace	C. A. Desobry	Plaquemine, La	June 19, 1860	28, 741
Bagasse-furnace	J. M. Frink	Coral, Ill	May 26, 1863	38, 666
Bagasse-furnace	S. H. Gilman	New Orleans, La	Dec. 4, 1855	13, 873
Bagasse-furnace	S. H. Gilman	New Orleans, La	Aug. 5, 1856	15, 481
Bagasse-furnace	S. H. Gilman	New Orleans, La	Jan. 10, 1860	26, 759
Bagasse-furnace	A. Hager and Y. Allyn	Baton Rouge and New Orleans, La.	May 6, 1856	14, 812
Bagasse-furnace	J. M. Jones and J. M. Charpentier.	New Orleans and Saint Mary's Parish, La.	Sept. 18, 1860	30, 107
Bagasse-furnace	J. M. Jones and J. M. Charpentier.	New Orleans and Pattersonville, La.	Oct. 23, 1860	30, 477
Bagasse-furnace	G. M. Longacre	New Orleans, La	July 21, 1857	17, 842
Bagasse-furnace	C. Neames	New Orleans, La	Feb. 14, 1860	27, 146
Bagasse-furnace	E. Skelly	Plaquemine, La	June 15, 1858	20, 591
Bagasse-furnace	A. Stillman	New York, N. Y	May 1, 1855	12, 807
Bagasse-furnace	M. Thompson	New York, N. Y	Dec. 15, 1857	18, 874
Bagasse-furnace	L. Tregre	Parish of Saint John the Baptist, La.	July 5, 1859	24, 675
Bagatelle	M. Redgrave	Cincinnati, Ohio	May 30, 1871	115, 357
Bagatelle game and apparatus	W. Evers	San Francisco, Cal	Dec. 12, 1871	121, 767
Bagatelle-table	W. Evers	San Francisco, Cal	Sept. 10, 1872	131, 259
Baggage-check	V. W. Blanchard	Bridport, Vt	Jan. 7, 1868	72, 965
Baggage-check	F. X. Bellerive	Plattsburgh, N. Y	July 4, 1871	116, 667
Baggage-check	M. N. Coe	Water Valley, Miss	Mar. 12, 1872	124, 418
Baggage-check	L. O. Cottle	Cedar Rapids, Iowa	Jan. 23, 1872	122, 994
Baggage-check	L. O. Cottle	Cedar Rapids, Iowa	Dec. 24, 1872	134, 257
Baggage-check	J. M. Curless	Cedar Rapids, Iowa	July 29, 1873	141, 205
Baggage-check	E. Flather	Bridgeport, Conn	Jan. 14, 1868	73, 315
Baggage-check	C. E. Ingalls	Boston, Mass	Feb. 13, 1872	123, 703
Baggage-check	J. H. McAlvin	Cedar Rapids, Iowa	May 7, 1872	126, 559
Baggage-check	G. F. Newcomb	New Haven, Conn	June 14, 1870	104, 339
Baggage-check	E. H. Paine	Louisville, Ky	Nov. 24, 1868	84, 372
Baggage-check	W. D. Richardson	Springfield, Ill	Oct. 6, 1863	40, 186
Baggage-check	H. S. Ross	Buffalo, N. Y	May 26, 1868	78, 394
Baggage-check	G. C. Thomas	Brooklyn, N. Y	Mar. 20, 1866	53, 360
Baggage-check	G. C. Thomas	Brooklyn, N. Y	Feb. 19, 1867	62, 300
Baggage-check, Folding	E. H. Graves	Chicago, Ill	Sept. 7, 1869	94, 487
Baggage-check, Railway	F. H. Furniss and F. R. Myers	Cleveland, Ohio	Jan. 20, 1863	37, 441
Baggage-director	T. M. Richardson	Searsport, Me	July 2, 1861	32, 718
Bail, Detachable tipping	J. Keith	Brooklyn, N. Y	Dec. 7, 1869	97, 519
Bail-ear	W. D. Mason and A. T. Rice	Chicago, Ill	Apr. 5, 1870	101, 483
Bail hooks or eyes, Machine for forming pot	J. Miller	Washington, Pa	Oct. 27, 1868	83, 522
Bail, Machine for forming wire	J. P. Van Bramer	Galesburgh, Ill	Sept. 19, 1871	119, 204
Bail-making machine	H. C. Wilder	Ashby, Mass	Aug. 25, 1868	81, 447
Bailing-press	P. K. Dederick	Albany, N. Y	Oct. 29, 1872	132, 639
Bailing-wheel for raising water from the hold of steam-boats, &c.	H. D. Forbes	New York, N. Y	May 26, 1843	3, 107
Bait and vegetable cutter	Z. G. Greenleaf	Bath, Me	Feb. 18, 1868	74, 682
Bait mill for fisherman	S. Hamblin	Taunton, Mass	Mar. 29, 1870	101, 260
Bait or meat cutter	N. Richardson	Gloucester, Mass	Jan. 21, 1868	73, 464

Index of patents issued from the United States Patent Office from 1790 *to* 1873, *inclusive*—Continued.

Invention.	Inventor.	Residence.	Date.	No.
Baker	C. H. Beeman, 2d	North Fairfax, Vt	Feb. 11, 1868	74, 282
Baker	J. T. Davy	Troy, N. Y	Feb. 12, 1845	3, 910
Baker and boiler, Combined	W. H. Miller	Brandenburgh, Ky	Mar. 7, 1871	112, 479
Baker and roaster, Double-reflecting	S. Hasey	Rensselaerville, N. Y	Dec. 28, 1832	....
Baker and roaster, Economical	P. Willcox	Springfield, Mass	Sept. 25, 1831	....
Baker, Coal-heating	J. D. Wheelock	Mayville, Wis	June 17, 1856	15, 155
Baker, Custard and cake	C. L. Sweatt and G. A. Huntoon	Fisherville, N. H	Jan. 17, 1871	111, 012
Baker, Metal	J. St. John	Hudson, N. Y	Sept. 30, 1834	....
Baker, Portable	J. H. Jackson	United States Army	Apr. 26, 1864	42, 535
Baker, Reflecting	L. B. Olmsted	Binghamton, N. Y	Jan. 20, 1836	....
Baker, Reflecting	W. Taintor and H. S. Orton	Porter County, Ind	Aug. 9, 1845	4, 145
Baker, Reflecting tin	W. Prescott	Boston, Mass	May 24, 1832	....
Baker, Steam	P. F. Wilcox	Springfield, Mass	Sept. 3, 1834	....
Baker, Tin	N. D. Whiten	New York, N. Y	Oct. 10, 1835	....
Baker, Tin	G. Williston	New London, Conn	Jan. 11, 1831	....
Baker's table	A. S. Maxwell	Dixon, Ill	Oct. 18, 1870	108, 373
Bake-pan	W. Beach	Philadelphia, Pa	Dec. 16, 1856	16, 226
Bake-pan	W. B. Chamberlin	Westfield, N. J	Aug. 26, 1873	142, 144
Bake-pan	J. Chase, jr	Watertown, N. Y	Mar. 17, 1868	75, 524
Bake-pan	W. A. Daggett	South Vineland, N. J	Oct. 1, 1872	131, 745
Bake-pan	R. D. McDonald	Jersey City, N. J	Nov. 26, 1872	133, 377
Bake-pan	J. G. Metsker	Logansport, Ind	Sept. 16, 1873	142, 928
Bake-pan	B. F. Miller	New York, N. Y	Oct. 8, 1861	33, 444
Bake-pan, Perforated	C. M. Cooney	Washington, D. C	July 21, 1868	80, 147
Baking and cooking, Apparatus for	W. G. Ruggles	Worcester, Mass	May 18, 1858	20, 304
Baking and roasting apparatus	T. J. T. Cummings	Fort Wayne, Ind	Mar. 5, 1872	124, 256
Baking-apparatus	J. P. Hayes	Boston, Mass	Jan. 30, 1849	6, 063
Baking-frame	T. C. Riddell	Wilmington, Del	Oct. 15, 1867	69, 939
Baking-iron	E. Skinner	Sandwich, N. H	Oct. 1, 1830	....
Baking-mold	G. Wingate	Boston, Mass	Nov. 25, 1873	144, 939
Baking or burning pan	W. Burtus	New York, N. Y	Nov. 7, 1820	....
Baking or cooking	A. Stowell	Medford, Mass	Dec. 14, 1830	....
Baking-pan	R. G. Elder	New York, N. Y	Jan. 28, 1868	73, 703
Baking-pan	A. Heminway and W. A. Daggett.	Landis Township, N. J	May 10, 1870	102, 938
Baking-pan	G. W. Mitchell	New York, N. Y	Oct. 15, 1867	69, 830
Baking-pan	W. Moran	Jersey City, N. J	June 25, 1872	128, 238
Baking-pan	S. W. Rogers	Harwich, Mass	Dec. 15, 1868	84, 908
Baking-pan	C. T. Smith	Jersey City, N. J	Nov. 11, 1873	144, 417
Baking-pan	S. West	Trenton, N. J	Nov. 13, 1866	59, 732
Baking-pan, Cover for	W. C. Davis	Cincinnati, Ohio	Nov. 3, 1863	40, 461
Baking pan, Potato	A. Jones	Lebanon, N. H	May 2, 1871	114, 444
Baking-powder	G. A. Mariner and J. Fish	Chicago, Ill	Dec. 13, 1864	45, 419
Baking-powder	J. Stowell	Charlestown, Mass	Nov. 16, 1869	96, 994
Baking powder, Bread	J. G. Copping and J. Weideman	Clinton, Iowa	Oct. 14, 1873	143, 580
Baking-powder, &c., Manufacture of acid phosphates for use in.	N. B. Rice	East Saginaw, Mich	Jan. 3, 1871	110, 680
Baking-powder, Putting up	G. F. Wilson	Providence, R. I	Sept. 5, 1871	118, 768
Baking kitchen, Steam	S. Smith	Herkimer County, N. Y	June 11, 1811	....
Balance	I. Bisbee	Richmond, Mo	Dec. 24, 1867	72, 593
Balance	H. W. Catlin	Burlington, Vt	Dec. 15, 1863	40, 906
Balance	H. A. Clum	Rochester, N. Y	Mar. 28, 1865	46, 992
Balance	B. Dearborn	Boston, Mass	Mar. 24, 1819	....
Balance	N. Griffing	New York, N. Y	Apr. 8, 1835	....
Balance	S. Harris	Philadelphia, Pa	July 7, 1863	39, 145
Balance	J. B. Hendricks	Clayton, Ill	Feb. 13, 1872	123, 564
Balance	W. S. How	Cincinnati, Ohio	Aug. 9, 1870	106, 163
Balance	C. Howlett	Hartford, Conn	May 6, 1862	35, 155
Balance	G. W. King	Georgetown, D. C	Sept. 6, 1870	107, 182
Balance	C. C. Marsh	New York, N. Y	Oct. 10, 1871	119, 871
Balance	L. A. Mátos	Philadelphia, Pa	June 2, 1868	78, 532
Balance	A. Y. McDonald	Dubuque, Iowa	Mar. 6, 1866	53, 020
Balance	G. R. Moore	Brattleborough, Vt	Jan. 6, 1844	3, 396
Balance	B. Morrill	Boscawen, N. H	May 26, 1843	3, 111
Balance	E. C. Pickering	Boston, Mass	Nov. 4, 1873	144, 286
Balance	S. A. Rogers	Geneva, N. Y	July 31, 1817	....
Balance	R. Shaler	Madison, Conn	Nov. 28, 1865	51, 226
Balance	J. Vaughn	Rutland, Vt	May 6, 1836	....
Balance, Aeromatic	J. A. Gridley	Southampton, Mass	Mar. 26, 1861	31, 802
Balance and scale-beam for weighing, Platform	J. D. Dale	Lansingburgh, N. Y	Aug. 30, 1838	899
Balance, Columbian	W. Granger	Middlebury, Vt	Sept. 8, 1810	....
Balance, Compensating spring	O. C. Squyer	West Dresden, N. Y	May 8, 1860	28, 212
Balance, Compound lever	J. B. Whetmore	Bush, N. Y	Feb. 27, 1832	....
Balance, Computing spring	S. R. P. Camp	New York, N. Y	June 7, 1870	103, 977
Balance, Constructing	D. Rogers	Auburn, N. Y	July 13, 1824	....
Balance, Construction of platform	A. Dole	Bangor, Me	July 18, 1840	1, 699
Balance, Dearborn's	S. Blaisdel	Lancaster, Ohio	Oct. 10, 1827	....
Balance, Dearborn's	U. West and D. Loring	New York, N. Y	Aug. 23, 1831	....
Balance, Double-scale	T. Fairbanks	Saint Johnsbury, Vt	Mar. 13, 1849	6, 169
Balance, Eagle	L. Jencks and H. Dexter	Killingly, Conn	Nov. 24, 1834	....
Balanced elevator	C. B. Sawyer	Fitchburgh, Mass	June 14, 1864	43, 136
Balance-elevator	C. B. Sawyer	Fitchburgh, Mass	May 24, 1870	103, 375
Balance for counter, &c	E. Hibbard	Lunenburgh, Vt	Nov. 7, 1835	....
Balance for detecting counterfeit coin	H. Maranville	Clinton, Ohio	Jan. 13, 1857	16, 390
Balance for detecting counterfeit money	F. J. Herpers	Newark, N. J	Dec. 29, 1857	18, 973
Balance for detecting spurious coin	J. Allender	New London, Conn	Nov. 27, 1855	13, 840
Balance for weighing, Platform	B. Bull	New York, N. Y	Mar. 3, 1837	133
Balance for weighing, Platform	B. Morison	Milton, Pa	Mar. 17, 1838	641
Balance-gate	W. C. Van Hoesen	Leeds, N. Y	Apr. 20, 1852	8, 894
Balance, Hydrostatic	J. O. Bandissin	Saint Louis, Mo	May 10, 1864	42, 722
Balance, Hydrostatic	S. Squire	Brooklyn, N. Y	Jan. 10, 1860	26, 795
Balance in combination with a knife, Spring	G. H. Smith	Greenwood, Iowa	Sept. 14, 1858	21, 520
Balance or fly wheel applied to machinery	J. Ordroman	....	Apr. 26, 1819	....
Balance or scale-beam	S. E. Winslow	Philadelphia, Pa	July 31, 1840	1, 708
Balance-pendulum lock	R. Gaines	Georgetown, D. C	Mar. 17, 1806	....
Balance-pendulum lock	S. Goodwin and R. Gaines	Baltimore, Md., and Georgetown, D. C.	Jan. 22, 1806	....
Balance pivot-gate	D. Brown	Hampden, Me	Apr. 26, 1870	102, 365

Index of patents issued from the United States Patent Office from 1790 to 1873, inclusive—Continued.

Invention.	Inventor.	Residence.	Date.	No.
Balance-platform	A. Bliss	Benson, Vt	June 12, 1835	....
Balance, Platform	C. Crain and E. L. Wemple	Madison, N. Y	Apr. 8, 1840	1, 546
Balance-platform	I. Gay	Dunstable, N. H	Jan. 14, 1833	....
Balance, Platform	J. Gibbs	Newark, Ohio	Sept. 12, 1846	4, 757
Balance, Platform	S. L. Hay	Boston, Mass	July 14, 1834	....
Balance, Platform	J. Horton	Madrid, N. Y	July 2, 1836	....
Balance, Platform	T. Y. Jennings	Geneva, Ohio	Nov. 10, 1841	2, 339
Balance, Platform	C. P. Ladd	Strasburgh, Vt	May 17, 1836	....
Balance-platform	J. B. Maag	New York, N. Y	Feb. 19, 1833	....
Balance, Platform	R. L. McCollum	Rochester, N. Y	Dec. 31, 1838	1, 048
Balance-platform	B. Morison	Milton, Pa	Jan. 13, 1833	....
Balance-platform	B. Morison	Milton, Pa	Dec. 31, 1833	....
Balance, Platform	J. M. Peck	Strasburgh, Vt	May 17, 1836	....
Balance, Platform	D. M. Smyth	New York, N. Y	Apr. 10, 1855	12, 698
Balance, Portable	A. Dole	Bangor, Me	Dec. 23, 1841	2, 397
Balance, Safety-valve spring	T. S. Ray and S. E. Cleveland	Buffalo, N. Y	Dec. 5, 1865	51, 350
Balance, Self-indicating	Z. W. and O. Avery	Bethany, Pa	May 12, 1857	17, 252
Balance, Spring	W. G. Barker	Detroit, Mich	Aug. 6, 1867	67, 482
Balance, Spring	E. P. Beckwith	New London, Conn	Jan. 16, 1855	12, 249
Balance, Spring	J. H. and R. H. Bull	New York, N. Y	Sept. 20, 1844	3, 752
Balance, Spring	N. Burnham	Norwalk, Ohio	July 15, 1862	35, 869
Balance, Spring	W. A. Crowell	Salisbury, Conn	June 26, 1860	28, 838
Balance, Spring	D. C. Lawrence	Cedar Falls, Iowa	Apr. 29, 1862	35, 125
Balance, Spring	J. K. O'Neil	Kingston, N. Y	May 22, 1866	54, 941
Balance, Spring	H. Saloshinsky	New York, N. Y	June 13, 1865	48, 211
Balance, Spring	W. B. Snyder	Lakeville, Conn	Feb. 28, 1865	46, 592
Balance, Steelyard	E. and T. Fairbanks	Saint Johnsbury, Vt	Feb. 10, 1837	120
Balance, Steelyard	E. and T. Fairbanks	Saint Johnsbury, Vt	Feb. 10, 1837	121
Balance, Weighing	R. Eastman	Concord, N. H	Mar. 13, 1849	6, 174
Balance, Weighing	B. Morison	Milton, Pa	Feb. 16, 1837	129
Balance, Weighing and price	A. M. Maynard	Savoy, Mass	Nov. 25, 1873	144, 914
Balance for quick weighing, Pendulum	B. Fenn	Hartford, Ohio	Mar. 29, 1853	9, 635
Balance-weight	S. Harris	Philadelphia, Pa	Dec. 4, 1866	60, 184
Balance-wheel	E. S. Pierce	Hartford, Conn	Nov. 19, 1867	71, 212
Balance-wheel	H. Randall	Scott, N. Y	Oct. 1, 1872	131, 782
Balance-wheel	F. W. Wild	Baltimore, Md	Apr. 25, 1871	114, 240
Balance-wheels, &c., Mode of hanging	J. Brinkerhoff	Auburn, N. Y	Mar. 31, 1863	38, 025
Balancing vertical reciprocating masses	W. F. Durfee	Bridgeport, Conn	Mar. 21, 1871	112, 792
Balcony, Portable	E. Balmforth	Danbury, Conn	Dec. 17, 1872	133, 962
Baldric	V. Price	New York, N. Y	Feb. 28, 1871	112, 180
Bale-band shears	A. H. Daniels	Manchester, N. H	Mar. 2, 1869	87, 328
Bale-band stretcher	L. J. Anderson	Water Valley, Miss	Feb. 6, 1872	123, 374
Bale-band stretcher	F. M. Logue	Satartia, Miss	Aug. 5, 1873	141, 447
Bale-band tightener	P. C. Ingersoll	Greenpoint, N. Y	June 25, 1867	66, 087
Bale-fastener	A. C. Fletcher	New York, N. Y	Jan. 28, 1868	73, 791
Bale-fastening	T. B. Bunting	New York, N. Y	Sept. 11, 1866	58, 018
Bale-fastening	W. M. Irvine	Montgomery, Ala	Aug. 25, 1868	81, 374
Bale fastening, Cotton	T. McIntire	Franklin Furnace, Ohio	July 17, 1860	29, 185
Bale-fastening, Wire	E. S. Lenox	New Brighton, N. Y	Dec. 21, 1869	98, 169
Bale, Hay, straw, &c	L. Dodge	Waterford, N. Y	Mar. 26, 1872	125, 031
Bale-hook	R. T. Yardley	Baltimore, Md	Sept. 19, 1871	119, 213
Bale-hoop, Cotton	J. McMurtry	Lexington, Ky	Feb. 23, 1858	19, 437
Bale-hoop fastening	R. Dillon	New York, N. Y	Feb 26, 1867	62, 400
Bale-hoop, Fastening	J. F. Milligan	Saint Louis, Mo	Mar. 13, 1866	53, 230
Bale-hoop fastening	J. Reese	Pittsburgh, Pa	Feb. 5, 1867	61, 868
Bale-hoop-straightening machine	C. Hughes	New Orleans, La	Oct. 9, 1860	30, 322
Bale-hoop tie	G. Brodie	Plum Bayou, Ark	July 11, 1871	116, 925
Bale-hoops, Clasps for cotton	A. C. Richard	Newtown, Conn	Sept. 14, 1858	21, 517
Bale-hoops, Coupling for	J. Agnew	Columbia, S. C	Aug. 31, 1858	21, 305
Bale hoops, Machine for tightening and securing compressed.	R. S. Adams	Loyd, N. Y	May 15, 1866	54, 810
Bale-hoops, Tool for fastening	E. A. Jeffery	Corning, N. Y	June 21, 1859	24, 464
Bale-label	N. C. Jones	New York, N. Y	Aug. 11, 1868	80, 967
Bale, Metallic band for binding	W. Field	Providence, R. I	Aug. 17, 1858	21, 190
Bale-raft, Cotton	T. Byrne	New York, N. Y	Nov. 28, 1865	51, 140
Bale-tag, Cotton	E. A. Locke	Boston, Mass	May 24, 1864	42, 860
Bale-tie	L. Arnold	Galveston, Tex	Nov. 26, 1872	133, 349
Bale-tie	A. Barbarin	New Orleans, La	Aug. 27, 1867	68, 149
Bale-tie	S. Brett	New York, N. Y	June 6, 1871	115, 692
Bale-tie	G. Brodie	Plum Bayou, Ark	Sept. 16, 1873	142, 894
Bale-tie	F. G. Brown	Brenham, Tex	Dec. 19, 1871	121, 988
Bale-tie	J. T. Butler	New Orleans, La	Jan. 9, 1872	122, 563
Bale-tie	L. Carter	Huntsville, Ala	Dec. 2, 1873	145, 091
Bale-tie	F. Cook	New Orleans, La	Nov. 26, 1872	133, 412
Bale-tie	F. Cook	New Orleans, La	Feb. 4, 1873	135, 526
Bale-tie	F. Cook	New Orleans, La	May 6, 1873	138, 479
Bale-tie	F. Cook	New Orleans, La	May 6, 1873	138, 480
Bale-tie	F. Cook	New Orleans, La	Sept. 16, 1873	142, 772
Bale-tie	F. Cook	New Orleans, La	Sept. 30, 1873	143, 223
Bale-tie	F. Cook	New Orleans, La	Dec. 23, 1873	145, 847
Bale-tie	J. S. Davis	Louisville, Ky	Oct. 17, 1871	120, 045
Bale-tie	D. K. Dedrick	Albany, N. Y	Feb. 11, 1873	135, 700
Bale tie	J. W. Gurley	Petersburgh, Va	Dec. 17, 1872	134, 052
Bale-tie	J. L. Haigh	New York, N. Y	Aug. 6, 1872	130, 218
Bale-tie	J. W. Hedenberg	Chicago, Ill	Apr. 8, 1873	137, 549
Bale-tie	J. Holmes and J. C. H. Slack	Manchester, England	Apr. 4, 1871	113, 518
Bale-tie	H. A. House	Bridgeport, Conn	Dec. 10, 1872	133, 858
Bale-tie	E. P. Jones	Shell Mound, Miss	June 10, 1873	139, 675
Bale-tie	W. A. Jordan	New Orleans, La	Aug. 23, 1870	106, 698
Bale-tie	W. A. Jordan	New Orleans, La	Sept. 6, 1870	107, 058
Bale-tie	H. Lampson	London, England	May 28, 1867	65, 239
Bale-tie	J. McMurtry	Lexington, Ky	Sept. 16, 1873	142, 803
Bale-tie	D. L. Miller	Madison, N. J	Apr. 9, 1872	125, 474
Bale tie	W. Parsons	Palmyra, N. Y	Jan. 23, 1872	122, 907
Bale-tie	J. E. Perkins	San Francisco, Cal	Mar. 7, 1871	112, 375
Bale-tie	M. Quin	Galveston, Tex	Feb. 18, 1873	135, 938
Bale-tie	J. F. Rusling	Lawrenceville, Pa	Oct. 17, 1871	120, 104
Bale-tie	S. O. Ryder	New York, N. Y	Apr. 2, 1867	63, 563

Index of patents issued from the United States Patent Office from 1790 *to* 1873, *inclusive*—Continued

Invention.	Inventor.	Residence.	Date.	No.
Bale-tie	J. L. Sheppard	Charleston, S. C	Aug. 22, 1871	118, 286
Bale-tie	W. C. Stiff	Birmingham, England	July 10, 1873	140, 556
Bale-tie	C. Swett	Copiah County, Miss	July 16, 1872	129, 186
Bale-tie	J. D. Van Benthuysen	New Orleans, La	Oct. 29, 1866	70, 295
Bale-tie	N. S. Walker	Liverpool, England	Oct. 15, 1872	132, 223
Bale-tie	F. Watkins	London Works, Birmingham, England.	Apr. 25, 1871	114, 236
Bale-tie and straining-lever	J. C. Coit	Cheraw, S. C	Jan. 16, 1872	122, 813
Bale-tie and straining-lever	J. C. Coit	Cheraw, S. C	July 23, 1872	129, 715
Bale-tie, Band or hoop	E. J. Beard	Saint Louis, Mo	July 15, 1873	140, 873
Bale-tie clamp	T. D. Leonard	Waco, Tex	Nov. 18, 4873	144, 776
Bale tie, Cotton	G. W. Adams	New Orleans, La	Feb. 20, 1872	123, 853
Bale tie, Cotton	A. S. Armstrong	New Orleans, La	June 10, 1873	139, 754
Bale tie, Cotton	A. Barbarin	New Orleans, La	Aug. 27, 1867	68, 148
Bale tie, Cotton	J. W. Barnum	New Orleans, La	Nov. 5, 1867	70, 503
Bale tie, Cotton	J. W. Barnum	New Orleans, La	Mar. 31, 1868	76, 141
Bale tie, Cotton	J. W. Barnum	New Orleans, La	Mar. 31, 1868	76, 142
Bale tie, Cotton	J. W. Barnum	New Orleans, La	Mar. 31, 1868	76, 143
Bale tie, Cotton	J. W. Barnum	New Orleans, La	Mar. 31, 1868	76, 144
Bale tie, Cotton	J. W. Barnum	New Orleans, La	Mar. 31, 1868	76, 145
Bale tie, Cotton	J. W. Barnum	New Orleans, La	Mar. 31, 1868	76, 146
Bale tie, Cotton	E. J. Beard	Saint Louis, Mo	July 30, 1872	129, 917
Bale tie, Cotton	G. N. Beard	Saint Louis, Mo	Aug. 13, 1867	67, 707
Bale tie, Cotton	G. N. Beard	Saint Louis, Mo	May 11, 1869	89, 844
Bale tie, Cotton	G. N. Beard	Saint Louis, Mo	Dec. 27, 1870	110, 539
Bale tie, Cotton	G. N. Beard	Saint Louis, Mo	Sept. 30, 1873	143, 319
Bale tie, Cotton	G. N. Beard	Saint Louis, Mo	Dec. 9, 1873	145, 273
Bale tie, Cotton	G. Brodie	Plum Bayou, Ark	Feb. 27, 1872	123, 976
Bale tie, Cotton	G. Brodie	Jefferson County, Ark	July 30, 1872	129, 925
Bale tie, Cotton	A. G. Buford	Water Valley, Miss	Nov. 11, 1873	144, 502
Bale tie, Cotton	W. J. Carroll	Natchez, Miss	May 16, 1871	114, 760
Bale tie, Cotton	J. S. Carson	Brook Haven, Miss	Mar. 9, 1869	87, 541
Bale tie, Cotton	W. Chambers	New Orleans, La	July 12, 1870	105, 172
Bale tie, Cotton	W. Chambers	New Orleans, La	Dec. 6, 1870	109, 804
Bale tie, Cotton	M. D. Cheek	Clarendon, Ark	Aug. 27, 1867	68, 167
Bale tie, Cotton	M. R. Clark	Columbia, S. C	Oct. 18, 1870	108, 450
Bale tie, Cotton	R. W. Cobb	Helena, Ala	July 8, 1873	140, 575
Bale tie, Cotton	B. Coleman	Louisville, Ky	Mar. 5, 1867	62, 610
Bale tie, Cotton	F. Cook	New Orleans, La	May 6, 1873	138, 481
Bale tie, Cotton	F. Cook	New Orleans, La	May 6, 1873	138, 482
Bale tie, Cotton	F. Cook	New Orleans, La	May 6, 1873	138, 483
Bale tie, Cotton	F. Cook	New Orleans, La	May 6, 1873	138, 484
Bale tie, Cotton	F. Cook	New Orleans, La	June 10, 1873	139, 777
Bale tie, Cotton	F. Cook	New Orleans, La	June 24, 1873	140, 246
Bale tie, Cotton	F. Cook	New Orleans, La	July 15, 1873	140, 766
Bale tie, Cotton	F. Cook	New Orleans, La	Aug. 19, 1873	142, 001
Bale tie, Cotton	T. Cromer	Galveston, Tex	Sept. 23, 1873	143, 124
Bale tie, Cotton	W. Crone	Galveston, Tex	Aug. 5, 1873	141, 494
Bale tie, Cotton	J. Crookes	Saint Louis, Mo	June 8, 1869	91, 091
Bale tie, Cotton	J. B. Dunn	Petersburgh, Va	Feb. 4, 1868	74, 059
Bale tie, Cotton	N. T. Edson	New Orleans, La	Sept. 17, 1867	68, 859
Bale tie, Cotton	H. Fassmann	New Orleans, La	Jan. 29, 1867	61, 527
Bale tie, Cotton	H. Fassmann	New Orleans, La	Feb. 5, 1867	61, 727
Bale tie, Cotton	H. Fassmann	New Orleans, La	Feb. 19, 1867	62, 188
Bale tie, Cotton	H. Fassmann	New Orleans, La	Mar. 3, 1868	75, 140
Bale tie, Cotton	W. D Field	Providence, R. I	May 6, 1873	138, 492
Bale tie, Cotton	W. D. Field	Providence, R. I	June 17, 1873	140, 024
Bale tie, Cotton	A. C. Fletcher	New York, N. Y	Feb. 4, 1868	74, 070
Bale tie, Cotton	E. A. Franklin	Brenham, Tex	May 14, 1872	126, 687
Bale tie, Cotton	E. Garrett	New Orleans, La	May 24. 1859	24, 112
Bale tie, Cotton	A. J. Going	Clinton, La	Apr. 1, 1873	137, 301
Bale tie, Cotton	J. M. Goldsmith	Boston, Mass	Sept. 30, 1873	143, 343
Bale tie, Cotton	R. S. Goodgion	Goodgion's Factory, S. C	Aug. 6, 1872	130, 293
Bale tie, Cotton	J. H. Gridley	Washington, D. C	Oct. 23, 1866	59, 007
Bale tie, Cotton	J. W. Hedenberg	Chicago, Ill	Mar. 10, 1868	75, 267
Bale tie, Cotton	C. G. Johnsen	New Orleans, La	July 11, 1871	116, 964
Bale tie, Cotton	E. P. Jones	Sun Flower County, Miss	Oct. 25, 1870	108, 600
Bale tie, Cotton	B. Kimball	Saint Louis, Mo	Oct. 21, 1873	143, 911
Bale tie, Cotton	J. Knight	Louisville, Ky	Feb. 12, 1867	61, 040
Bale tie, Cotton	R. G. Latting	New Orleans, La	Dec. 18, 1866	60, 528
Bale tie, Cotton	R. G. Latting	New Orleans, La	Jan. 15, 1867	61, 217
Bale tie, Cotton	R. H. Lecky	Allegheny City, Pa	Oct. 29, 1867	70. 230
Bale tie, Cotton	Z. W. Lee	Blakely, Ga	Oct. 16, 1866	58, 844
Bale tie, Cotton	Z. W. and E. D. Lee	Blakely, Ga	July 3, 1860	28, 991
Bale tie, Cotton	L. Littlejohn	New York, N. Y	Aug. 13, 1867	67, 777
Bale tie, Cotton	E. T. Mainwaring	Tipton, England	Jan. 1, 1869	90, 761
Bale tie, Cotton	M. F. Maury	New Orleans, La	May 28, 1867	65, 102
Bale tie, Cotton	O. K. McClean	New Orleans, La	Sept. 16, 1873	142, 802
Bale tie, Cotton	J. R. McClintock and J. Cumberland.	New Orleans, La	May 23, 1871	115, 225
Bale tie, Cotton	J. I. McComb	Liverpool, England	Oct. 23, 1866	59, 152
Bale tie, Cotton	J. B. McDonald	Louisville, Ky	June 8, 1869	91, 034
Bale tie, Cotton	J. McMurtry	Lexington, Ky	Oct. 2, 1866	58, 445
Bale tie, Cotton	W. McNabb	Clapton, England	May 14, 1872	126, 645
Bale tie, Cotton	H. B. Merritt	Saint Louis, Mo	Nov. 5, 1867	70, 452
Bale tie, Cotton	I. H. Merritt	Cincinnati, Ohio	Apr. 10, 1866	53, 851
Bale tie, Cotton	J. F. Milligan	Saint Louis, Mo	July 30, 1867	67, 334
Bale tie, Cotton	J. F. Milligan	Saint Louis, Mo	Sept. 24, 1867	69, 113
Bale tie, Cotton	J. F. Milligan	Saint Louis, Mo	Nov. 29. 1870	109, 748
Bale tie, Cotton	H. T. Minor, jr	Savannah, Ga	July 16, 1872	129, 157
Bale tie, Cotton	S. J. Mitchell	Saint Louis, Mo	Aug. 27, 1867	68, 225
Bale tie, Cotton	W. M. Morris	Washington County, Miss	Apr. 6, 1869	88, 727
Bale tie, Cotton	C. Mudge	New Orleans, La	Jan. 7, 1868	73, 026
Bale tie, Cotton	T. C. Oakman	Paterson, N. J	June 15, 1869	91, 361
Bale tie, Cotton	D. G. Olmstead	Vicksburgh, Miss	Mar. 23, 1858	19, 707
Bale tie, Cotton	W. Onions	Saint Louis, Mo	June 5, 1866	55, 353
Bale tie, Cotton	W. Onions	Saint Louis, Mo	Nov. 6, 1866	59, 443
Bale tie, Cotton	W. Onions	Saint Louis, Mo	May 14, 1867	64, 696

Index of patents issued from the United States Patent Office from 1790 *to* 1873, *inclusive*—Continued.

Invention.	Inventor.	Residence.	Date.	No.
Bale tie, Cotton	W. J. Orr	Charlotte, N. C	Sept. 2, 1873	142, 505
Bale tie, Cotton	C. J. Paine	Painesville, Ohio	Sept. 24, 1867	69, 120
Bale tie, Cotton	J. W. Petty	New Orleans, La	Mar. 5, 1867	62, 550
Bale tie, Cotton	J. L. Phillips	Washington County, Miss.	Jan. 7, 1868	73, 034
Bale tie, Cotton	F. Quarles	Waco, Tex	Nov. 18, 1873	144, 793
Bale tie, Cotton	W. C. Ramsay	Wadesborough, N. C	Apr. 8, 1873	137, 570
Bale tie, Cotton	J. Reese	Pittsburgh, Pa	Feb. 26, 1867	62, 365
Bale tie, Cotton	J. Reinecker	New Orleans, La	Feb. 12, 1867	62, 069
Bale tie, Cotton	G. Ricker	Covington, Ky	Mar. 12, 1867	62, 777
Bale tie, Cotton	E. S. Roberts	Columbus, Ga	Mar. 10, 1868	75, 461
Bale tie, Cotton	J. W. Rogan	Memphis, Tenn	Jan. 25, 1870	99, 239
Bale tie, Cotton	S. Rogers	Pittsburgh, Pa	Apr. 30, 1867	64, 255
Bale tie, Cotton	G. A. Seaver	New Orleans, La	Oct. 23, 1866	59, 141
Bale tie, Cotton	G. A. Seaver	New York, N. Y	May 12, 1868	77, 770
Bale tie, Cotton	D. M. Sechler	Cincinnati, Ohio	Mar. 19, 1867	63, 105
Bale tie, Cotton	J. L. Sheppard	Charleston, S. C	Oct. 15, 1867	69, 849
Bale tie, Cotton	J. L. Sheppard	Charleston, S. C	Mar. 17, 1868	75, 705
Bale tie, Cotton	J. L. Sheppard	Charleston, S. C	Feb. 27, 1872	124, 090
Bale tie, Cotton	J. A. Shone	Holly Springs, Miss	Aug. 11, 1868	81, 018
Bale tie, Cotton	D. S. Skinner	Providence, R. I	Sept. 24, 1872	131, 573
Bale tie, Cotton	D. S. Skinner	Providence, R. I	Dec. 16, 1873	145, 536
Bale tie, Cotton	W. M. Smith	Augusta, Ga	Dec. 14, 1869	97, 977
Bale tie, Cotton	W. M. Smith	Augusta, Ga	Feb. 15, 1870	99, 964
Bale tie, Cotton	J. R. Speer	Pittsburgh, Pa	Nov. 26, 1867	71, 338
Bale tie, Cotton	H. D. Starr	Texana, Tex	Sept. 2, 1873	142, 527
Bale tie, Cotton	C. Swett	Vicksburgh, Miss	Oct. 23, 1866	59, 144
Bale tie, Cotton	C. Swett	Copiah County, Miss	July 16, 1872	129, 187
Bale tie, Cotton	C. Swett	Copiah County, Miss	July 16, 1872	129, 188
Bale tie, Cotton	C. Swett	Copiah County, Miss	July 16, 1872	129, 189
Bale tie, Cotton	C. Swett	Copiah County, Miss	July 16, 1872	129, 190
Bale tie, Cotton	C. Swett	Copiah County, Miss	July 16, 1872	129, 191
Bale tie, Cotton	W. Trowbridge	New Orleans, La	Mar. 10, 1868	75, 319
Bale tie, Cotton	J. W. Truman	Macon, Ga	Sept. 24, 1867	69, 274
Bale tie, Cotton	C. Ulmer	Mobile, Ala	July 30, 1867	67, 231
Bale tie, Cotton	C. W. Wailey	New Orleans, La	Oct. 15, 1867	69, 870
Bale tie, Cotton	J. S. Wallis	New Orleans, La	May 4, 1869	89, 612
Bale tie, Cotton	L. Weil	New York, N. Y	Dec. 30, 1873	145, 037
Bale-tie fastener	R. S. Sayre	Stilesborough, Ga	Dec. 10, 1872	133, 803
Bale-tie fastener, Wire	C. Brown	New York, N. Y	May 7, 1872	126, 515
Bale-tie for cotton, Iron	C. C. Bier	New Orleans, La	June 26, 1860	28, 825
Bale-tie for cotton, Iron	W. Boyd	New Orleans, La	Aug. 30, 1859	25, 940
Bale-tie for cotton, Iron	P. Davey	Portsmouth, Ohio	Sept. 25, 1860	30, 126
Bale-tie for cotton, Iron	W. S. Loughborough	Rochester, N. Y	May 8, 1860	28, 187
Bale-tie for cotton, Iron	J. Nuttall	New Orleans, La	May 10, 1859	23, 940
Bale-tie for cotton, Iron	W. Stewart	Natchez, Miss	May 1, 1860	28, 1[illegible]0
Bale-tie for cotton, Iron	C. W. Wailey	Lexington, Ky	Mar. 27, 1860	27, 660
Bale-tie for cotton, Metallic	J. Aiken	Natchez, Miss	May 1, 1860	28, 044
Bale-tie for cotton, Metallic	T. Cook	New Orleans, La	Mar. 2, 1858	19, 490
Bale-tie for cotton, Metallic	T. Guyol	New Orleans, La	Jan. 28, 1868	73, 713
Bale-tie for cotton, &c., Wire	D. McComb	Memphis, Tenn	Aug. 13, 1872	130, 519
Bale-tie, Hoop	J. F. Milligan	Saint Louis, Mo	July 23, 1872	129, 851
Bale-tie lock	J. Wheelock	San Francisco, Cal	Nov. 2, 1869	96, 521
Bale-tie, Machine for fastening	J. Adams	New Orleans, La	June 18, 1867	65, 856
Bale-tie, Metallic	I. C. Plant	Macon, Ga	Aug. 24, 1858	21, 272
Bale-tie, Wire	O. Macdaniel	New York, N. Y	Mar. 26, 1867	63, 166
Bale-tie, Wire	E. S. Lenox	New York, N. Y	July 9, 1872	128, 803
Bale-ties, Device for forming	J. McClean	New Orleans, La	Dec. 9, 1873	145, 357
Bale-ties, Device for making	M. N. Coe	Madison Parish, La	Dec. 12, 1871	121, 756
Bale ties, Machine for adjusting cotton	M. Martin	Charlotte, N. C	May 9, 1871	114, 581
Bales already formed, Press for compressing	F. F. Cornell, jr., and E. M. Wight.	New York, N. Y	Feb. 5, 1867	61, 719
Bales and packages, Fastening band on	H. Knowles	New York, N. Y	Mar. 1, 1859	23, 092
Bales, &c., Clasp for fastening bands on cotton	H. Knowles	New York, N. Y	Aug. 16, 1859	25, 125
Bales, Fastening for cotton	J. W. Evans	New York, N. Y	Aug. 21, 1860	29, 683
Bales, Fastening for metallic bands for cotton	W. Minor	Houma, La	Oct. 27, 1857	18, 514
Bales, Fastening for metallic bands of cotton	C. J. Provost	Sardis, Ala	Sept. 29, 1857	18, 299
Bales, Fastening iron bands on cotton	C. G. Wells	Galveston, Tex	Apr. 5, 1859	23, 518
Bales, Fastening metal hoops on cotton	J. T. Butler	Natchez, Miss	Nov. 15, 1859	26, 087
Bales, Fastening of metallic band for cotton	O. C. Evans	New York, N. Y	Mar. 26, 1861	31, 848
Bales, &c., Hoop-fastening for cotton	E. Davidson	Batesville, Ark	Aug. 9, 1859	24, 995
Bales, Hoop fastening for cotton	G. J. Widrig	Memphis, Tenn	Apr. 12, 1859	23, 632
Bales, Hoop-lock for cotton	E. V. Fassman	New Orleans, La	Apr. 18, 1865	47, 288
Bales, Iron tie for cotton	G. N. Beard	Saint Louis, Mo	July 16, 1861	32, 818
Bales, Iron tie for cotton	J. J. McComb	New Orleans, La	Jan. 29, 1861	31, 252
Bales, Label for cotton	P. H. Taylor	New Orleans, La	Mar. 1, 1870	100, 467
Bales, Lock for cotton	A. P. Merrill, jr	Natchez, Miss	May 22, 1860	28, 392
Bales, Machine for preparing hay for pressing into	O. and C. Waste	Cameron, Ill	Nov. 29, 1864	45, 285
Bales, &c., Machine for tightening and securing metallic band for cotton.	G. W. Penniston	North Vernon, Ind	Aug. 31, 1858	21, 360
Bales, &c., Mark-holder for	P. Fitch	Brooklyn, N. Y	Apr. 28, 1863	38, 363
Bales, &c., Mark-holder for	H. W. Goodrich and E. A. Locke.	Chelsea and Boston, Mass.	Apr. 28, 1863	38, 365
Bales, &c., Metallic band-fastening for	A. O. Broad	Louisville, Ky	June 9, 1857	17, 485
Bales, Metallic band for cotton	R. W. Fenwick	Washington, D. C	Sept. 25, 1860	30, 133
Bales, &c., Mode of attaching labels to	E. A. Locke	Boston, Mass	Sept. 29, 1863	40, 109
Bales, Mode of securing metal hoops on cotton	J. McMurtry	Fayette County, Ky	Apr. 10, 1860	27, 866
Bales of cotton and other fibrous materials, Non-elastic band for.	D. McComb	Memphis, Tenn	June 17, 1856	15, 142
Bales of cotton, &c., Incorrodible mark or label for	H. D. Mears	Washington, D. C	Sept. 15, 1863	39, 944
Bales of goods, Machine for roping	R. Dillon	New York, N. Y	Feb. 22, 1848	5, 451
Bales of hay, Constructing	O. and C. Waste	Cameron, Ill	Nov. 29, 1864	45, 284
Bales of merchandise, Mode of fastening	T. W. Reilly	New Orleans, La	Dec. 15, 1863	40, 983
Bales, Pressing cotton, &c., into	S. A. Clemens	Granby, Conn	Sept. 3, 1850	7, 612
Bales, Securing iron bands on cotton	C. W. Pyle	Galveston, Tex	Sept. 27, 1859	25, 584
Bales, Securing metallic band on cotton	P. C. Ingersoll	Green Point, N. Y	May 18, 1858	20, 311
Bales, Tightening rope on cotton	C. Wilson	Brooklyn, N. Y	Feb. 19, 1861	31, 505
Baling cotton	C. Colahan	Alton, Ill	Jan. 29, 1867	61, 517
Baling cotton, &c	B. W. Collier	Oxford, Miss	Apr. 16, 1872	125, 724

Index of patents issued from the United States Patent Office from 1790 *to* 1873, *inclusive*—Continued.

Invention.	Inventor.	Residence.	Date.	No.
Baling cotton, Band for	H. W. Oliver, jr	Pittsburgh, Pa	Mar. 31, 1868	76, 238
Baling cotton, Hoop-lock for	F. Quant	Painesville, Ohio	Nov. 28, 1865	51, 214
Baling cotton, &c., Metallic band for	J. Downes	Handsworth, England	Nov. 7, 1871	120, 727
Baling device, Hay	L. Doty	Frankfort, Ill	Aug. 12, 1873	141, 636
Baling hay	R. Wakeman and J. L. Ballance	Port Deposit, Md	July 10, 1866	56, 294
Baling hay, &c., Machine for	E. Dorr	Rockford, Ill	July 12, 1864	43, 482
Baling hay, &c., Mode of	E. Dorr	Rockford, Ill	July 12, 1864	43, 481
Baling, Machine for cutting and preparing hay for	S. Colahan	Cleveland, Ohio	Mar. 28, 1865	46, 993
Baling manure	H. C. Babcock	Hartford, Conn	July 2, 1872	128, 454
Baling, Metallic band for	G. Brodie	Little Rock, Ark	Mar. 22, 1859	23, 291
Baling-press	J. M. Albertson	New London, Conn	Dec. 15, 1868	84, 982
Baling-press	J. M. Albertson	New London, Conn	Dec. 7, 1869	97, 471
Baling-press	J. M. Albertson	New London, Conn	June 17, 1873	139, 849
Baling-press	W. J. Arrington	Louisville, Ga	Aug. 15, 1871	117, 962
Baling-press	H. A. Ashley	Springfield, Ohio	Oct. 30, 1866	59, 309
Baling-press	S. J. Austin	Freeport, Me	Sept. 17, 1867	68, 935
Baling-press	C. J. Barney	Rockport, Ind	May 7, 1872	126, 435
Baling-press	H. W. Baumann	Memphis, Tenn	Aug. 12, 1873	141, 622
Baling-press	J. Berkeley	Washington, Tex	Jan. 12, 1869	85, 783
Baling-press	A. P. Boren	Greensborough, N. C	May 18, 1869	90, 152
Baling-press	G. Brodie	Plum Bayou, Ark	Sept. 19, 1871	119, 009
Baling-press	C. B. Brooks	Auburn, Me	June 13, 1865	48, 151
Baling-press	C. Brown and D. L. Miller	Buffalo, N. Y., and Madison, N. J.	Sept. 1, 1868	81, 745
Baling-press	M. Brunner, jr	Fremont, Ohio	Sept. 17, 1872	131, 424
Baling-press	H. K. Burnett	Poughkeepsie, N. Y	Mar. 18, 1873	136, 965
Baling-press	S. P. Cady	Hastings, Mich	Sept. 10, 1872	131, 250
Baling-press	N. Chapman	Milford, Mass	Apr. 13, 1869	88, 846
Baling-press	N. Chapman	Hopedale, Mass	Oct. 3, 1871	119, 573
Baling-press	M. D. Cheek	Clarendon, Ark	Mar. 12, 1867	62, 815
Baling-press	M. D. Cheek	Memphis, Tenn	Dec. 29, 1868	85, 365
Baling-press	M. D. Cheek	Memphis, Tenn	Aug. 23, 1870	106, 660
Baling-press	C. C. Converse	New York, N. Y	Sept. 18, 1866	58, 222
Baling-press	W. M. Conner	Burlington, Ky	July 22, 1873	141, 035
Baling-press	J. S. Cook	West Groton, Mass	Nov. 21, 1871	121, 087
Baling-press	F. F. Cornell, jr	New York, N. Y	May 23, 1865	47, 800
Baling-press	F. F. Cornell, jr	New York, N. Y	June 20, 1865	48, 261
Baling-press	F. F. Cornell, jr	New York, N. Y	Feb. 5, 1867	61, 718
Baling-press	T. J. Corning	San José, Cal	Feb. 28, 1871	112, 224
Baling-press	T. J. Corning	San José, Cal	Nov. 14, 1871	120, 942
Baling-press	W. P. Craig	Milton, Ky	July 4, 1865	48, 523
Baling-press	D. Cumming, jr	New York, N. Y	Apr. 7, 1868	76, 407
Baling-press	P. K. Dederick	Greenbush, N. Y	Apr. 13, 1869	88, 775
Baling-press	P. K. Dederick	Greenbush, N. Y	Feb. 1, 1870	99, 295
Baling-press	P. K. Dederick	Albany, N. Y	Feb. 1, 1870	99, 296
Baling-press	P. K. Dederick	Albany, N. Y	Feb. 1, 1870	99, 297
Baling-press	W. Deering	Louisville, Ky	Nov. 23, 1869	97, 170
Baling-press	L. Dodge	Waterford, N. Y	Oct. 17, 1871	120, 048
Baling-press	G. Duncan	San Francisco, Cal	Nov. 1, 1870	108, 772
Baling-press	D. Dunn	Lewisport, Ky	Sept. 1, 1868	81, 612
Baling-press	C. J. Emmett	New York, N. Y	Oct. 5, 1869	95, 447
Baling-press	D. A. Faughaenel	Kansas City, Mo	July 16, 1872	129, 119
Baling-press	G. F. Felton	De Pere, Wis	Apr. 23, 1872	125, 944
Baling-press	C. V. Fleetwood and E. W. Morton.	Vincennes, Ind	Mar. 22, 1864	42, 046
Baling-press	D. Frisbie and S. C. Goodsell	New Haven, Conn	May 8, 1866	54, 644
Baling-press	G. B. Garlinghouse	North Madison, Ind	Aug. 27, 1867	68, 062
Baling-press	C. W. Gillis	San Antonio, Tex	Nov. 20, 1866	59, 832
Baling-press	J. H. Godwin	Scotland Neck, N. C	Aug. 27, 1867	68, 182
Baling-press	W. Golding	New Orleans, La	Apr. 23, 1872	125, 890
Baling-press	J. Gorham	Bairdstown, Ga	Dec. 3, 1867	71, 740
Baling-press	J. B. Gridley	Louisville, Ky	Apr. 10, 1866	53, 917
Baling-press	G. F. Grund	Fremont, Ohio	Apr. 9, 1872	125, 454
Baling-press	T. D. Guthrie, jr	Galva, Ill	Oct. 8, 1867	69, 657
Baling-press	J. Harder	Lock Haven, Pa	July 14, 1863	39, 230
Baling-press	J. K. Harris	Allensville, Ind	Aug. 26, 1862	36, 297
Baling-press	J. K. Harris	Allensville, Ind	June 23, 1863	38, 963
Baling-press	G. W. Hart	Aurora, Ind	Nov. 21, 1865	51, 043
Baling-press	H. F. Hicks	Grand View, Ind	June 27, 1865	48, 487
Baling-press	P. Higdon	Lewisport, Ky	Aug. 24, 1869	94, 110
Baling-press	D. H. Hill	Union Springs, Ala	Jan. 12, 1869	85, 822
Baling-press	E. Hill	Hamilton, Ill	Nov. 22, 1870	109, 515
Baling-press	G. H. Hoke and J. A. Brown	Elizabeth, Ind	Oct. 15, 1867	69, 807
Baling-press	G. D. Howe	Lewisport, Ky	Dec. 11, 1866	60, 376
Baling-press	A. J. Hunt	Walla Walla, Wash	July 16, 1872	129, 477
Baling-press	F. A. Huntington and J. F. Carter.	San Francisco, Cal	May 7, 1872	126, 394
Baling-press	W. Iler	Shreveport, La	Sept. 27, 1870	107, 688
Baling-press	W. R. King	Chicago, Ill	Dec. 4, 1866	60, 196
Baling-press	W. R. King	Chicago, Ill	Apr. 16, 1872	125, 741
Baling-press	L. W. Liles	Roanoke, Ala	May 7, 1872	126, 405
Baling-press	C. Locher	Oroville, Cal	Aug. 23, 1870	106, 703
Baling-press	T. E. Marable	Petersburgh, Va	Oct. 24, 1871	120, 297
Baling-press	E. J. Marsters	Shaw's Flat, Cal	June 7, 1870	103, 907
Baling-press	A. W. Mason	Station 2, (P. and G. R. R.,) Fla.	Apr. 23, 1872	125, 971
Baling-press	D. L. Miller	Madison, N. J	July 7, 1863	39, 158
Baling-press	J. F. Milligan	Saint Louis, Mo	Oct. 13, 1868	83, 079
Baling-press	W. H. Morris	Troy, Tenn	Aug. 13, 1872	130, 524
Baling-press	B. L. and T. H. B. Myers	Washington, N. C	May 7, 1872	126, 480
Baling-press	G. W. Nutter	Santa Cruz, Cal	Aug. 2, 1870	106, 075
Baling-press	G. C. Paine	San Francisco, Cal	Apr. 4, 1865	47, 124
Baling-press	W. H. H. Peairs	Olathe, Kans	Nov. 28, 1871	121, 408
Baling-press	P. Philip and P. J. Stophilbeen	Hudson and Schodack, N. Y.	Apr. 5, 1864	42, 223
Baling-press	J. Price, jr	Petaluma, Cal	Sept. 1, 1863	39, 748
Baling-press	A. J. Purviance	Mount Zion, Iowa	July 2, 1867	66, 388
Baling-press	J. Randolph	Huntsville, Tex	Jan. 2, 1872	122, 406

Index of patents issued from the United States Patent Office from 1790 to 1873, inclusive—Continued.

Invention.	Inventor.	Residence.	Date.	No.
Baling-press	B. Roberts	Clintondale, N. Y	Feb. 23, 1864	41, 722
Baling-press	J. W. Roberts	New Monmouth, N. J	June 14, 1864	43, 132
Baling-press	C. H. Robinson	Bath, Me	Jan. 10, 1865	45, 863
Baling-press	J. B. Root	New York, N. Y	Dec. 30, 1873	146, 096
Baling-press	C. H. Schnelle	Saint Louis, Mo	May 7, 1872	126, 415
Baling-press	J. S. Schofield	Macon, Ga	Feb. 13, 1872	123, 734
Baling-press	I. S. Schuyler	New York, N. Y	May 6, 1862	35, 203
Baling-press	L. Seeberger and N. Levy	Cincinnati, Ohio	July 17, 1866	56, 456
Baling-press	G. W. Serrin	Memphis, Tenn	Apr. 21, 1868	77, 108
Baling-press	W. B. Smith	Aberdeen, Ind	July 2, 1867	66, 404
Baling-press	B. F. Stroud	Marshall, Tex	May 10, 1870	102, 879
Baling-press	E. Taylor	Memphis, Tenn	Apr. 7, 1868	76, 548
Baling-press	J. P. Taylor	Hudson City, N. J	Sept. 10, 1872	131, 232
Baling-press	J. P. Taylor and J. R. Baker	Hudson City and Jersey City, N. J.	Nov. 5, 1867	70, 649
Baling-press	C. H. Thatford	Jamaica, N. Y	Feb. 27, 1866	52, 911
Baling-press	H. H. Tift	Mystic, Conn	June 15, 1869	91, 495
Baling-press	J. D. Towner and G. J. Harris	Murfreesborough, Tenn	Oct. 24, 1871	120, 347
Baling-press	I. P. Walker	Milwaukee, Wis	July 4, 1871	116, 650
Baling-press	E. R. Wallace	Jonesville, S. C	Feb. 21, 1871	111, 993
Baling-press	F. B. Wallin	Saugatuck, Mich	May 14, 1872	126, 853
Baling-press	J. P. White	New York, N. Y	Sept. 5, 1865	49, 815
Baling-press	J. I. Williams	Meridian, Miss	Oct. 2, 1866	58, 526
Baling-press	G. Winship	Atlanta, Ga	Nov. 4, 1873	144, 245
Baling-press	P. W. Yarrell	Graysburgh, N. C	Oct. 1, 1872	131, 805
Baling-press, Beater	F. F. Cornell, jr	New York, N. Y	Feb. 5, 1867	61, 717
Baling-press, Beating-device for	L. C. Field	Galesburgh, Ill	July 4, 1865	48, 621
Baling-press, Broom-corn	W. J. Willson	Coles County, Ill	Aug. 16, 1870	106, 440
Baling-press, Construction of	F. F. Cornell, jr	New York, N. Y	Nov. 7, 1865	50, 798
Baling press, Cotton	B. D. Gullett	Amity City, La	Feb. 2, 1869	86, 532
Baling press, Cotton	W. Norman	Van Buren, Ark	Oct. 31, 1865	50, 772
Baling-press, Hand-power	P. C. Ingersoll	Green Point, N. Y	Nov. 25, 1873	144, 982
Baling press, Hand-power	W. R. Newman	Galesburgh, Ill	Nov. 29, 1870	109, 653
Baling press, Hay and cotton	E. Buel	Silver Creek, N. Y	Apr. 20, 1869	89, 022
Baling press, Hay and cotton	P. K. Dederick	Albany, N. Y	Oct. 29, 1872	132, 566
Baling-press, Horizontal	J. D. Wilber	Pleasant Plains, N. Y	May 2, 1865	47, 591
Baling short-cut hay, &c	C. Brown	Buffalo, N. Y	Aug. 27, 1867	68, 282
Baling short-cut hay and straw	S. W. Adwen	Rochester, N. Y	June 8, 1869	91, 057
Baling short-cut hay and straw	W. Hadwin	Rochester, N. Y	Feb. 20, 1872	123, 824
Baling short-cut hay or straw	C. Brown	New York, N. Y	Apr. 16, 1872	125, 786
Baling short-cut hay or straw	W. Hadwin	Rochester, N. Y	Feb. 28, 1871	112, 140
Ball:				
See Base-ball.				
Billiard-ball.				
Cannon-ball.				
Cat-ball.				
Copper ball.				
Croquet-ball.				
Foot-ball.				
Loaded ball.				
Mast-ball.				
Minie-ball.				
Musket-ball.				
Nine-pin ball.				
Pop-corn ball.				
Puddle-ball.				
Puddlers' ball.				
Rifle-ball.				
Rubber ball.				
Ball alley	J. D. Patrick	San Francisco, Cal	Oct. 8, 1867	69, 581
Ball-alley	J. D. Patrick	San Francisco, Cal	Sept. 1, 1868	81, 676
Ball-and-socket joint	R. R. and J. Craig	Nevada, Cal	Dec. 28, 1869	98, 234
Ball-and-socket joint	W. P. Haskins	Mendota, Ill	Oct. 29, 1867	70, 334
Ball-and-socket joint	E. Maynard	Washington, D. C	Jan. 26, 1869	86, 173
Ball-and-socket joint	M. W. St. John	Leonardsville, N. Y	June 30, 1868	79, 408
Ball-and-socket joint	M. W. St. John	Leonardsville, N. Y	Aug. 26, 1873	142, 293
Ball-brace	H. S. Bartholomew	Bristol, Conn	May 21, 1861	32, 347
Ball holder, Billiard-table	H. W. Collender	New York, N. Y	Oct. 7, 1873	143, 500
Ball-joint, Detachable	H. Breevort	Brooklyn, N. Y	Apr. 14, 1868	76, 706
Ball, Loaded	W. Hunt	New York, N. Y	Aug. 10, 1848	5, 701
Ball-motor, Rotating	D. A. Drasch	St. Egidi, Austria	Dec. 22, 1868	85, 220
Ball-players, Base for	E. D. Taylor	Hornellsville, N. Y	Mar. 3, 1868	75, 076
Ball rounding and polishing machine	J. L. Knowlton	Philadelphia, Pa	Aug. 1, 1865	49, 122
Ball tally-board, Base	T. L. Canary	Brownsburgh, Ind	Sept. 1, 1868	81, 598
Balls and shot by pressure, Machine for making	T. Bruff	Washington, D. C	Oct. 5, 1808	....
Balls, Apparatus for hardening copper	A. Weldon	Chicago, Ill	July 29, 1873	141, 247
Balls, &c., Machine by which the centrifugal force is controlled in throwing.	J. Martin	Louisville, Ky	Aug. 3, 1840	1, 713
Balls or mandrels, Machine for turning	W. Newsham	Philadelphia, Pa	Mar. 8, 1870	100, 654
Balls or shot, Machine for manufacturing	L. Magers	Baltimore, Md	Nov. 13, 1840	1, 855
Balls, shot, &c., Machine for throwing	R. McCarty	New York, N. Y	Dec. 31, 1838	1, 049
Balls, shot, &c., of lead, Casting	D. Pettibone	Philadelphia, Pa	Nov. 10, 1819	....
Balls, springs, &c., Manufacture of	A. R. Davis	Cambridge, Mass	June 9, 1863	38, 813
Ballast-box for vessel, Deck	T. Fowler	Richmond Valley, N. Y	Feb. 11, 1862	34, 358
Balloon	J. H. Connell	Lexington, Ky	Nov. 17, 1863	40, 608
Balloon	J. P. Gage	New York, N. Y	Mar. 8, 1859	23, 163
Balloon	M. Nelson	New York, N. Y	May 21, 1861	32, 378
Balloon	T. L. Shaw	Omaha City, Nebr	Feb. 10, 1863	37, 667
Balloon	J. J. Sherman	Albany, N. Y	Aug. 27, 1861	33, 165
Balloon	A. G. Wright	Santa Cruz, Cal	Dec. 27, 1864	45, 665
Balloon, Advertising	W. F. Browne	New York, N. Y	Nov. 11, 1873	144, 436
Balloon and appendage	H. Bell	London, England	Mar. 26, 1850	7, 207
Balloons, Arrangement for mooring and managing	J. W. Brewer	Cincinnati, Ohio	July 11, 1854	11, 248
Balloon, Federal	M. McFarland		Oct. 28, 1799	....
Balloon, Locomotive	P. Haenlein	Mainz, Germany	Aug. 27, 1872	130, 915
Balloon, Marine	C. W. Sykes	Suffield, Conn	Sept. 11, 1866	57, 996
Ballot-box	G. L. Bailey	Portland, Me	May 22, 1860	28, 339
Ballot-box	A. Cummings	New York, N. Y	May 18, 1858	20, 256

Index of patents issued from the United States Patent Office from 1790 *to* 1873, *inclusive*—Continued.

Invention.	Inventor.	Residence.	Date.	No.
Ballot-box	G. A. Frébault	Paris, France	Aug. 23, 1870	106, 682
Ballot box	J. Gamber	Petersburgh, Pa	Aug. 19, 1873	141, 872
Ballot-box	J. A. Hill	Bloomington, Iowa	Aug. 6, 1850	7, 545
Ballot-box	S. C. Jollie	New York, N. Y	Oct. 5, 1858	21, 684
Ballot-box	J. A. McPherson	Troy, N. Y	Jan. 24, 1865	46, 012
Ballot-box	J. P. Outcalt	Lancaster, Ohio	Feb. 25, 1868	74, 932
Ballot-box, Registering	J. S. Savage	Kingston, N H	Aug. 26, 1873	142, 124
Ballot-boxes, Apparatus for detecting fraud in	M. I. Shinn	Richmond, Ind	Oct. 23, 1860	30, 503
Balm	S. G. Wallis	Waterford, Pa	July 21, 1868	80, 101
Balsam, Lavender	E. Bartlette	New York, N. Y	June 29, 1833	......
Baluster-cutting machine	P. J. Frantze, M. Tillesen, and J. A. Gregersen.	Chicago, Ill	July 29, 1873	141, 214
Balusters, Construction of sheet-metal	G. Fischer	New York, N. Y	Feb. 16, 1869	86, 977
Bamboo, cane, and other fibrous plants, Mode of treating.	C. Heaton	New York, N. Y	July 2, 1867	66, 338
Bamboo-fiber, Process of preparing	L. S. Robbins and J. A. Southmayd.	New York, N. Y., and Elizabeth, N. J.	Mar. 2, 1869	87, 432
Bamboo or cane fiber, Manufacture of flock from	L. S. Robbins and J. A. Southmayd.	New York, N. Y., and Elizabeth, N. J.	Mar. 2, 1869	87, 433
Bananas and plantains, Manufacture from	J. Fry	New Orleans, La	July 27, 1869	93, 075
Band and fastener, Package	F. R. Hunt	Leavenworth, Kans	Mar. 18, 1873	137, 000
Band and skirt-hoop attachment	A. K. Young	Boston, Mass	Aug. 19, 1862	36, 253
Band, Anti-interfering	W. Hall and J. R. Clifford	Boston, Mass	Sept. 8, 1868	81, 998
Band-clasp, Elastic	J. C. Arms	Northampton, Mass	Oct. 15, 1872	132, 230
Band-clasp, Elastic	A. Scheydecker	Amsterdam, N. Y	Sept. 20, 1870	107, 550
Band-comb	E. Brown	Wappinger's Falls, N. Y	Mar. 4, 1873	136, 358
Band-comb	D. N. Ropes	Orange, N. J	Dec. 2, 1873	145, 071
Band-coupling	S. Moulton	Hartford, Conn	July 2, 1867	66, 244
Band-cutter	W. C. Barr and E. J. Hunkins	Macon City, Mo	Oct. 5, 1869	95, 414
Band-cutter	A. Zwiebel	Burlington, Wis	Apr. 27, 1869	89, 372
Band-cutter and thrashing-machine attachment	S. C. Myers and J. McCanley	McAllisterville, Pa	Aug. 29, 1871	118, 634
Band-cutter, spreader, and feeder, Automatic	W. H. H. Youngs	Waverly, Iowa	Dec. 20, 1870	110, 324
Band-cutting fork	D. Arnold	West Lodi, Ohio	Apr. 23, 1872	125, 926
Band-cutting fork	D. Arnold	Clyde, Ohio	Nov. 26, 1872	133, 396
Band-cutting knife, Straw	F. Coulon	Rockford, Ill	Mar. 31, 1868	75, 055
Band-cutting machine	W. U. Hoover	Macomb, Ill	June 20, 1865	48, 280
Band-drawer	B. W. Field	Ferrisburgh, Vt	Apr. 20, 1869	89, 136
Band, Endless	L. Binns	Bradford, England	Oct. 21, 1873	143, 743
Band-fastening	D. E. Hall	Detroit, Mich	Jan. 19, 1869	85, 929
Band-fastening, Metallic	A. Barbarin	New Orleans, La	Oct. 9, 1866	58, 574
Band for bundle	C. Perley	New York, N. Y	Aug. 15, 1865	49, 435
Band for machinery	J. H. Clifton	New Castle, Pa	May 22, 1860	28, 350
Band for machinery, Driving	C. M. Roullier	Paris, France	Aug. 5, 1862	36, 110
Band, Gripe	E. White and J. H. Warren	French Corral, Cal	Apr. 17, 1866	54, 048
Band hook, Back	C. Wack	Evansville, Ind	July 28, 1868	80, 373
Band-knife	F. Coulon	Rockford, Ill	June 20, 1871	116, 027
Band, Machine	S. J. Whitton	Coleraine, Mass	May 10, 1870	102, 898
Band-perforating machine, Metal	A. Leland	Philadelphia, Pa	Mar. 13, 1866	53, 228
Band-puller	H. M. Wait	Woodstock, Ill	July 21, 1868	80, 251
Band-tie	J. I. Peyton	Washington, D. C	July 18, 1871	117, 202
Band-tightener	G. Brodie	Jefferson County, Ark	Jan. 30, 1872	123, 228
Band-tightener	F. M. Lottridge	Portland, Me	Dec. 14, 1869	97, 784
Band-twisting machine, Jack	J. Collier	Morenci, Mich	Oct. 27, 1868	83, 465
Bands, clamps, &c., Fastening for attaching the ends of.	E. Carter	Norwalk, Conn	Sept. 28, 1869	95, 322
Bands, Clasp for metallic or other flexible	A. C. Richard	Newtown, Conn	Oct. 19, 1858	21, 848
Bands, Composition for making self-cementing	E. M. Carrington	New York, N. Y	Jan. 12, 1869	85, 790
Bands, Contracting wrought-iron	E. Cooper	Richmond, Va	Mar. 16, 1833	......
Bands, Guiding endless belt or apron	S. Sawyer	Boston, Mass	Mar. 30, 1833	......
Bands, Hoop-lock for securing the ends of metallic	P. C. Ingersoll	Green Point, N. Y	Mar. 15, 1859	23, 249
Bands, Machine for making endless	L. Binns	Bradford, England	Dec. 9, 1873	145, 386
Bands, Machine for re-rolling metallic	C. W. Chapman	Liverpool, England	Mar. 11, 1873	136, 585
Bands, Machine for trimming metallic	J. G. Merrill	Boston, Mass	Oct. 8, 1872	131, 962
Bands, &c., Making iron-keyed	E. Cooper	Richmond, Va	Sept. 14, 1826	......
Bands, &c., Method of joining sheet-metal	W. Painter	Baltimore, Md	Sept. 5, 1865	49, 782
Bands of iron, Clasp for the ends of	C. Warner	New York, N. Y	Mar. 15, 1859	23, 281
Bands or belts, Making gutta-percha	C. Hancock	Grosvenor Place, England	May 23, 1848	5, 590
Bandage	N. Jensen	Washington, D. C	Dec. 14, 1858	22, 293
Bandage for preternatural enlargements	A. R. Brown	Albion, Mich	Nov. 24, 1868	84, 255
Bandage, Obstetrical	M. Willis	Rochester, N. Y	Mar. 26, 1861	31, 843
Bandage, pad, &c., for cure or relief of hernia, &c.	R. Thompson	Columbus, Ohio	July 29, 1837	315
Bandage, Supporting	E. F. Hofmann	New York, N. Y	Mar. 16, 1869	87, 932
Bandage, Suspensory	E. Heaton	New Haven, Conn	May 14, 1867	64, 766
Bandage, Suspensory	W. H. Phelps	Boston, Mass	May 12, 1868	77, 757
Bandage, Suspensory	J. L. Taylor	New York, N. Y	Apr. 12, 1870	101, 941
Bandbox	G. H. Dickerman	Boston, Mass	Jan. 4, 1859	22, 493
Bandboxes, Machine for cutting the top and bottom of	B. Mestayer	New York, N. Y	Nov. 13, 1821	......
Bandboxes, Manufacture of	W. Tabele	New York, N. Y	Oct. 9, 1849	6, 783
Bandelore	J. L. Haven and C. Hettrick	Cincinnati, Ohio	Nov. 20, 1866	59, 745
Banding and ornamental binding, Raised	D. K. Van Vechten	New York, N. Y	Dec. 12, 1820	......
Banding, Machine	C. Lenzmann	Brooklyn, N. Y	Dec. 1, 1857	18, 753
Banjo	L. Brown	Baltimore, Md	Apr. 6, 1869	88, 541
Banjo	F. P. Dobson	New York, N. Y	Apr 6, 1869	88, 555
Banjo	G. C. Dobson and W. McDonnell.	Boston, Mass	May 25, 1869	90, 350
Banjo	H. C. Dobson	New York, N. Y	July 16, 1867	66, 810
Banjo	H. C. Dobson	New York, N. Y	Mar. 4, 1873	136, 491
Banjo	J. Field	New York, N. Y	May 11, 1869	89, 923
Banjo	F. W. Harlass	New York, N. Y	Aug. 15, 1865	49, 401
Banjo	E. B. Mansfield	Boston, Mass	May 28, 1872	127, 179
Banjo	J. Mayberger	New York, N. Y	Dec. 24, 1867	72, 517
Banjo	G. Mein	Williamsburgh, N. Y	Aug. 28, 1866	57, 540
Banjo	J. S. Stiles	Springfield, Vt	July 22, 1873	141, 182
Banjo	G. Teed	New York, N. Y	Apr. 8, 1862	34, 913
Banjo	W. B. Tilton	New York, N. Y	Apr. 24, 1866	54, 264
Banjo	S. F. Van Hagen	Albany N. Y	Oct. 18, 1859	25, 872
Bank and other lock	W. Johnson	Milwaukee, Wis	June 29, 1858	20, 716
Bank and safe lock	S. T. Bacon	Boston, Mass	July 12, 1859	24, 710

Index of patents issued from the United States Patent Office from 1790 *to* 1873, *inclusive*—Continued.

Invention.	Inventor.	Residence.	Date.	No.
Bank and safe lock	W. Hall	Boston, Mass	Oct. 16, 1845	4, 236
Bank and safe, Safety-lock for	M. R. Stephenson and O. Edwards.	Boston, Mass	July 9, 1844	3, 651
Bank-bills, &c., Clasp for	C. Perley	New York, N. Y	Aug. 9, 1864	43, 792
Bank-check	A. Man	Brooklyn, N. Y	Sept. 3, 1867	68, 448
Bank-check, &c	A. C. Paquet	Philadelphia, Pa	May 16, 1871	114, 963
Bank-check canceler	W. M. Simpson	Newark, N. J	Feb. 16, 1858	19, 384
Bank-checks, &c., Preventing and detecting the alteration of.	S. Crane	Dalton, Mass	Aug. 22, 1871	118, 204
Bank-lock	S. S. Burlingame	Warwick, R. I	Oct. 19, 1858	21, 862
Bank-lock	J. H. Crygier	New York, N. Y	Nov. 22, 1853	10, 265
Bank-lock	F. Denzler	New York, N. Y	Feb. 20, 1855	12, 403
Bank-lock	L. Derby	New York, N. Y	Nov. 2, 1858	21, 947
Bank-lock	W. Hall	Boston, Mass	June 27, 1854	11, 158
Bank-lock	A. C. Harig and D. C. Stoy	Louisville, Ky	July 25, 1854	11, 374
Bank-lock	W. Johnson	Milwaukee, Wis	Aug. 2, 1859	24, 975
Bank-lock	L. H. Miller	Providence, R. I	Oct. 5, 1858	21, 689
Bank-lock	C. G. Mueller	Charleston, S. C	Apr. 3, 1855	12, 647
Bank-lock	T. P. Murphy	New York, N. Y	May 31, 1853	9, 757
Bank-lock	H. Ritchie	Newark, N. J	Apr. 3, 1849	6, 252
Bank-lock	D. M. Smith	Springfield, Vt	Apr. 3, 1849	6, 272
Bank-lock	L. Yale, jr	Newport, N. Y	July 12, 1853	9, 853
Bank-lock	L. Yale	Newport, N. Y	Feb. 28, 1854	10, 584
Bank-lock	L. Yale	Newport, N. Y	May 22, 1855	12, 932
Bank-lock, Powder-proof	W. Hall	Boston, Mass	July 29, 1851	8, 257
Bank-note	J. M. Batchelder	Cambridge, Mass	Feb. 3, 1863	37, 561
Bank-note	J. Gibson, jr	Albany, N. Y	Oct. 3, 1871	119, 599
Bank-note	J. Murdock	New York, N. Y	Mar. 26, 1861	31, 820
Bank-note, bond, revenue-stamp, &c	G. W. Casilear	Washington, D. C	Feb. 15, 1870	99, 757
Bank-note, bond, revenue-stamp, &c	J. Duthie	New York, N. Y	Apr. 9, 1872	125, 550
Bank-note plates, Method of engraving	G. W. Casilear	Washington, D. C	May 6, 1873	138, 613
Bank-note plates, Method of engraving	G. W. Casilear	Washington, D. C	May 6, 1873	138, 614
Bank-notes against forgery, Use of type in guarding	A. Brewster	Hartford, Conn	July 15, 1816	....
Bank-notes, bills, &c., Blank for	P. Hannay	Washington, D. C	Aug. 5, 1856	15, 486
Bank-notes by lithography, Preventing counterfeit of.	F. Peabody and J. Dixon	Salem, Mass	Apr. 20, 1832	....
Bank-notes, &c., Device to prevent counterfeiting	M. C. Lea	Philadelphia, Pa	Apr. 21, 1863	38, 231
Bank-notes, Etching the end-piece of	H. S. Tanner	Philadelphia, Pa	July 1, 1815	....
Bank-notes, Formation of	G. Murray	Philadelphia, Pa	Mar. 23, 1822	....
Bank-notes from being counterfeited, Method of preventing.	C. D. Scropyan	New Haven, Conn	Jan. 8, 1856	14, 069
Bank notes, Graphic-plate for	J. Meere	Philadelphia, Pa	July 1, 1815	....
Bank-notes, Manufacture of	A. C. Carey	Lynn, Mass	Apr. 10, 1860	27, 857
Bank-notes, Manufacturing	G. T. Jones	Cincinnati, Ohio	May 15, 1866	54, 834
Bank-notes, &c., Mode of detecting counterfeited	L. Heath	Boston, Mass	July 2, 1867	66, 337
Bank-notes, Mode of making	J. P. Puglia	Philadelphia, Pa	Aug. 13, 1822	....
Bank-notes, &c., Mode of preventing the counterfeiting of.	I. Rehn	Philadelphia, Pa	Apr. 28, 1863	38, 335
Bank-notes, papers, &c., Self-cementing band for holding.	E. M. Carrington	New York, N. Y	Aug. 25, 1868	81, 339
Bank-notes, &c., Preventing the counterfeiting of	S. Beer	New York, N. Y	Dec. 1, 1868	84, 606
Bank-notes, &c., Preventing the counterfeiting of	C. D. Scropyan	New York, N. Y	June 2, 1857	17, 473
Bank-notes, Process of manufacturing	G. T. Jones	Cincinnati, Ohio	July 9, 1867	66, 500
Bank-notes, &c., Shears for cutting	S. P. Ruggles	Boston, Mass	Jan. 5, 1858	19, 046
Bank-notes to prevent counterfeiting, Mode of preparing.	G. T. Jones	Cincinnati, Ohio	May 15, 1866	54, 835
Bank-notes when cut, Table to hold	F. G. Johnson	Brooklyn, N. Y	Feb. 28, 1854	10, 575
Bank or other bills, Universal vitriolic test for making.	A. Brewster	Hartford, Conn	July 11, 1808	....
Bank, vault, and safe lock	J. Oxnard	Portland, Me	Apr. 10, 1845	3, 990
Banks, Mode of preventing forgery on	N. Sylvester	New York, N. Y	July 12, 1824	....
Banks, vaults, safes, &c., Lock for	M. R. Stephenson and O. Edwards.	Boston, Mass	Apr. 17, 1844	3, 546

Bar:

See Axle-bit blank bar.
Boring-bar.
Calk-bar.
Chain cross-bar.
Claw-bar.
Clevis-bar.
Core-bar.
Crow-bar.
Draft-bar.
Draw-bar.
Equalizing-bar.
Fire-bar.
Fish-bar.
Furnace-grate bar.
Grate-bar.
Harvester-finger bar.
Horseshoe-bar.
Insect-bar.
Iron-bar.
Iron and steel bar.
Lifting bar.
Metal bar.
Metallic bar.
Miners' bar.
Mosquito-bar.
Nut-bar.
Pinch-bar.
Railway-bar.
Rolled-bar.
Sand-bar.
Shaker-bar.
Shifting-bar.
Shutter-bar.
Sickle-bar.

Index of patents issued from the United States Patent Office from 1790 *to* 1873, *inclusive*—Continued.

Invention.	Inventor.	Residence.	Date.	No.
Bar—Continued.				
See Skylight-bar.				
Slide-bar.				
Spike-bar.				
Splinter-bar.				
Stove-bar.				
Tamping-bar.				
Tapered bar.				
Tenter-bar.				
Tie-bar.				
Toe-calk bar.				
Trace safety-bar.				
Vehicle side-bar.				
Wrench-bar.				
Bar-room register	J. McNamee	Easton, Pa	June 10, 1862	35,566
Bar-straightening machine	G. Lander	Pittsburgh, Pa	May 28, 1867	65,242
Bars and tubes, Machine for bending	A. Harris	Minneapolis, Minn	Dec. 31, 1872	134,428
Bars or plates of metal, Machinery for rolling tapering.	J. Holmes	Pittsburgh, Pa	July 18, 1865	48,868
Barbers' and dental chair	M. Leidecker	Rochester, N. Y	May 3, 1870	102,688
Barbers' and dental chair	M. Leidecker and P. Cron	Rochester N. Y	Nov. 3, 1868	83,644
Barbers' chair	A. Abel	New York, N. Y	Mar. 8, 1870	100,487
Barbers' chair	A. Abel	New York, N. Y	Mar. 19, 1872	124,713
Barbers' chair	N. W. Bonney	Lewiston, Me	Mar. 5, 1867	62,523
Barbers' chair	F. J. Coates	Cincinnati, Ohio	Dec. 3, 1872	133,566
Barbers' chair	J. N. Ewald	Frankfort, Ind	June 18, 1872	128,029
Barbers' chair	A. Gerdes and J. Reiché	New York, N. Y	Oct. 6, 1868	82,704
Barbers' chair	P. Haberstich	Dayton, Ohio	Feb. 18, 1873	135,986
Barbers' chair	C. Kaestner	Chicago, Ill	June 2, 1868	78,525
Barbers' chair	H. Perry	Bethel, Me	July 16, 1872	129,423
Barbers' chair	A. Schwaab	New York, N. Y	Feb. 18, 1873	136,009
Barbers' chair	J. and J. Stock	New York, N. Y	May 5, 1868	77,548
Barbers' chair and stool	H. Remick	Portsmouth, N. H	Sept. 19, 1865	50,032
Barbers' check-holder	G. W. Hoglen	Dayton, Ohio	Sept. 17, 1872	131,499
Barge and army boat, Portable safety	S. C. Batchelor	Cincinnati, Ohio	Jan. 20, 1841	1,935
Barge-coupling ratchet	W. W. Patterson and E. Bishop	Pittsburgh, Pa	Feb. 7, 1871	111,564
Barges, Construction of	A. Snyder	Freeport, Pa	Apr. 16, 1872	125,700
Barges in river, Steering apparatus for	M. Lytle	Allegheny, Pa	Sept. 13, 1859	25,429
Barges, Unloading coal	W. A. Wright	East Haven, Conn	May 6, 1873	138,727
Barilla from tobacco-stems, &c., Manufacturing	G. Easterly	Richmond, Va	Feb. 5, 1810	
Bark and corn grinding mill	C. Parker	Meriden, Conn	Nov. 25, 1837	484
Bark and dye-wood, Machine for cutting	H. Haight, jr., and H. White	Stamford, Conn	Apr. 11, 1825	
Bark and dye-woods, Machine for grinding	O. Pease and A. Donalds	Norfolk, Conn	Mar. 18, 1808	
Bark and grain, Mill for grinding	V. Birely	Frederick, Md	July 11, 1842	2,716
Bark and leaves of various trees, Mode of extracting filamentous matter similar to silk, cotton, &c., from the.	A. C. Vautier	Paris, France	Oct. 22, 1861	33,551
Bark and other materials, Apparatus for making extracts from.	J. W. Jones	Cumberland, Md	Apr. 14, 1868	76,775
Bark and other materials, Apparatus for leaching	G. A. Starkweather	Waymart, Pa	Mar. 16, 1869	87,984
Bark, &c., Apparatus for, and mode of, packing ground.	J. Lyon	Philadelphia, Pa	Jan. 5, 1815	
Bark, Apparatus for preparing quercitron	J. Andrews	Philadelphia, Pa	July 15, 1816	
Bark breaking and rossing machine	J. W. Burdwin and G. S. Tillinghast.	Morrisville, N. Y	Mar. 19, 1872	124,788
Bark-breaking machine	N. Sears	Hudson, N. Y	Oct. 31, 1820	
Bark-crusher	B. Irving	New York, N. Y	Dec. 22, 1868	85,172
Bark cutting and grinding machine	J. Warrell		Mar. 17, 1802	
Bark-extracts, Apparatus for manufacturing	W. Maynard	Salem, Mass	Sept. 12, 1871	118,956
Bark for exportation, Grinding and packing	J. Richardson and B. Stout	Bucks County, Pa	June 6, 1812	
Bark for tanning and other purposes, Process of extracting the strength of.	J. Brainer and W. H. Burridge.	Cleveland, Ohio	Apr. 8, 1862	34,873
Bark from tree, Tool for splitting	E. Michaels	Palermo, Me	Aug. 28, 1866	57,542
Bark from willows, &c., Machine for removing	L. A. Beardsley	South Edmeston, N. Y	June 26, 1860	28,824
Bark from wood, Mode of separating	J. Maitre	Chatillon, France	Dec. 31, 1867	72,873
Bark-grinding machine	A. Bull	Caroline, N. Y	Mar. 27, 1828	
Bark-grinding machine	M. Hurd	Augusta, N. Y	May 3, 1831	
Bark-grinding mill	B. A. Beardsley	Sangerfield, N. Y	Feb. 4, 1843	2,944
Bark-grinding mill	A. Lindsey	Canton, Me	Dec. 26, 1845	4,335
Bark-leach	S. B. Patterson	Bridgeport, Ct	Mar. 17, 1868	75,571
Bark-leaching apparatus	H. C. Crowell	Morgan, Ohio	Feb. 14, 1871	111,730
Bark-leaching apparatus	C. Korn	Wurtsborough, N. Y	Dec. 3, 1867	71,765
Bark, Machine for breaking and grinding	C. Tobey	Hudson, N. Y	May 7, 1807	
Bark, Machine for rossing	R. Healy	Swanton Falls, Vt	June 9, 1863	38,861
Bark, Machine for separating the qualities of	J. Brakeley	New York, N. Y	June 5, 1860	28,554
Bark, Machine for shaving	M. Winger	Lancaster County, Pa	Nov. 29, 1859	26,317
Bark, Manufacture of quercitron	M. Winger	Ephratah, Pa	Oct. 25, 1870	108,669
Bark-mill	C. Churchman and G. Martin, jr.	Upper Chichester, Pa	July 16, 1813	
Bark-mill	O. Coogan	Pittsfield, Mass	Oct. 24, 1871	120,246
Bark-mill	J. G. Curtis	Emporium, Pa	Apr. 19, 1870	101,984
Bark-mill	J. Elliott	Philadelphia, Pa	May 24, 1822	
Bark-mill	C. Foss	Madison, Ohio	Oct. 25, 1826	
Bark-mill	L. Gale	Berkshire County, Mass	Mar. 20, 1811	
Bark-mill	L. Gale	Lenox, Mass	May 16, 1815	
Bark-mill	J. T. Gifford	Veteran, N. Y	Feb. 27, 1832	
Bark-mill	C. H. Green and R. Montgomery	Sangerfield, N. Y	Sept. 30, 1831	
Bark-mill	J. Helenbrook	Olean, N. Y	Feb. 14, 1871	111,744
Bark-mill	L. N. Hermance	Kingston, N. Y	June 7, 1870	103,881
Bark-mill	D. Humberd and G. Downs	McConnellstown, Pa	Mar. 27, 1832	
Bark-mill	M. Hurd	Augusta, N. Y	Feb. 1, 1830	
Bark-mill	C. Korn	Wurtsborough, N. Y	Oct. 4, 1870	107,923
Bark-mill	R. Montgomery and L. W. Harris.	Sangerfield, N. Y	Aug. 12, 1840	1,714
Bark-mill	A. P. Norton and M. Owen	Sangerfield, N. Y	June 25, 1845	4,090
Bark-mill	J. Olds	Meriden, Conn	Apr. 18, 1814	
Bark-mill	G. E. Palen and F. P. Avery	Tunkhannock, Pa	Jan. 24, 1871	111,239
Bark-mill	T. W. Pryor	Philadelphia, Pa	May 21, 1805	
Bark-mill	I. Scudder	Prattsville, N. Y	Oct. 25, 1845	4,237

Index of patents issued from the United States Patent Office from 1790 *to* 1873, *inclusive*—Continued.

Invention.	Inventor.	Residence.	Date.	No.
Bark-mill	R. H. Shultis	Ellenville, N. Y	May 17, 1870	103, 246
Bark-mill	F. Stamm	East Lampeter, Pa	Jan. 31, 1871	111, 397
Bark-mill	W. Tansley	Salisbury Centre, N. Y	May 29, 1860	28, 518
Bark-mill	W. Tansley	Salisbury Centre, N. Y	Oct. 18, 1864	44, 756
Bark-mill	W. Tansley	Salisbury Centre, N. Y	Dec. 14, 1869	97, 989
Bark-mill	N. S. Thomas	Painted Post, N. Y	Aug. 8, 1865	49, 319
Bark-mill	E. and J. Trask	Sangerfield, N. Y	Nov. 22, 1821	....
Bark-mill	J. Trask, A. Seabury, and W. Young.	Sangerfield, N. Y	Apr. 17, 1833	....
Bark-mill	M. Winger	Ephratah, Pa	May 17, 1864	42, 811
Bark-mill, Cast-iron	W. Torrey	Westbrook, Me	Sept. 13, 1827	....
Bark-mill, Feed-regulator for	B. Irving	New York, N. Y	Feb. 9, 1869	86, 675
Bark, Mill for breaking and grinding tanners'	J. Montgomery	Sangerfield, N. Y	May 29, 1828	....
Bark-mill for grinding farmers' bark	M. Beecher	Remson, N. Y	Sept. 27, 1844	3, 767
Bark, Mode of grinding	J. Markley	....	July 19, 1794	....
Bark, Obtaining coloring-matter from oak	C. Hinrichs	New York, N. Y	Aug. 7, 1860	29, 489
Bark-planing machine	J. Brakeley	New York, N. Y	Mar. 19, 1861	31, 697
Bark, Preparation of quercitron	J. Elliott	Philadelphia, Pa	Aug. 20, 1822	....
Bark rosser and breaker	G. S. Tillinghast and J. W. Burdwin.	Morrisville, N. Y	Dec. 20, 1870	110, 403
Bark-rossing machine	J. Cowie	Portland, Me	Sept. 15, 1863	39, 888
Bark-rossing machine	C. Gilpin and J. T. Hill	Cumberland, Md	July 2, 1872	128, 612
Bark-rossing machine	J. Martin and G. W. Wilson	Morenci, Mich	Aug. 5, 1873	141, 448
Bark-rossing machine	B. F. Taber	Buffalo, N. Y	Jan. 27, 1863	37, 530
Bark-shaving machine	M. Winger	Ephratah, Pa	Oct. 25, 1870	108, 668
Barkers' mill	J. Rumsay	....	Aug. 26, 1791	....
Barley-fork	F. Dunn	Pulaski, N. Y	Mar. 1, 1870	100, 382
Barley-fork	M. C. Remington	Auburn, N. Y	May 14, 1867	64, 798
Barley, Machine for making pearl	A. Wulze	Saint Louis, Mo	May 3, 1859	23, 879
Barley machine, Pearl	W. Rickard	Chicago, Ill	Feb. 18, 1868	74, 595
Barley or gavel fork	D. P. Sharp	Ithaca, N. Y	May 3, 1870	102, 598
Barley-shelling machine	H. Handschy	Bridgeville, Ohio	Sept. 5, 1834	....
Barn-door fastening	D. N. Minor	Bridgewater, Mich	Oct. 30, 1866	59, 248
Barn-fork, Manufacture of	C. T. Beebe	Jackson, Mich	Dec. 19, 1871	121, 983
Barn-shovel, &c., Sheet-iron	J. Brower	Berks County, Pa	Feb. 19, 1819	....
Barn, Stock	J. Tyler	New Carlisle, Ohio	Aug. 2, 1870	106, 095
Barometer	P. Armand le comte de Fontaine-Moreau.	London, England	Aug. 20, 1846	4, 702
Barometer	H. A. Clum	Auburn, N. Y	May 29, 1860	28, 454
Barometer	H. A. Clum	Auburn, N. Y	Apr. 16, 1861	32, 050
Barometer	A. H. Emery and J. Johnson	New York, N. Y., and Saco, Me.	June 11, 1872	127, 752
Barometer	E. F. Hamann	Collikoon, N. Y	Sept. 18, 1860	30, 061
Barometer	W. R. Hopkins	Geneva, N. Y	Jan. 27, 1841	1, 951
Barometer	J. T. Large	Brooklyn, N. Y	July 16, 1861	32, 836
Barometer	G. V. Mooney	New York, N. Y	May 30, 1865	47, 971
Barometer	J. Thomson	Wayne, Ill	Feb. 9, 1864	41, 549
Barometer	T. R. Timby	Medina, N. Y	Nov. 3, 1857	18, 560
Barometer	L. Woodruff	Ann Arbor, Mich	June 5, 1860	28, 626
Barometer, Mercurial	G. Tagliabue	New York, N. Y	July 5, 1859	24, 674
Barometer, Mercurial	T. R. Timby	Worcester, Mass	Oct. 28, 1862	36, 872
Barometer, Self-recording	D. Peelor	Johnstown, Pa	Apr. 4, 1871	113, 693
Barrack, Portable	M. F. Brantingham	Sangamond County, Ill	May 12, 1857	17, 256
Barrel	E. H. Cady	Grand Rapids, Mich	May 24, 1870	103, 298
Barrel	P. H. Griswold	Grand Rapids, Mich	July 19, 1870	105, 565
Barrel	T. Hanvey	Elma, N. Y	Dec. 20, 1870	110, 232
Barrel	S. Harris	Kilburn, England	Nov. 1, 1864	44, 864
Barrel	G. M. Huntley	Grand Rapids, Mich	Aug. 30, 1870	106 827
Barrel	H. G. Porter	Grand Rapids, Mich	Apr. 26, 1870	102, 432
Barrel	H. G. Porter	Grand Rapids, Mich	Feb. 14, 1871	111, 868
Barrel	H. G. Porter	Grand Rapids, Mich	Dec. 26, 1871	122, 133
Barrel	S. Roberts	Cleveland, Ohio	May 14, 1861	32, 340
Barrel	J. Tomlinson	Goderich, Ontario, Canada	Oct. 3, 1871	119, 546
Barrel	D. H. Waters	Grand Rapids, Mich	Dec. 14, 1869	97, 840
Barrel	D. H. Waters	Grand Rapids, Mich	Dec. 14, 1869	97, 841
Barrel	J. W. Westen	New York, N. Y	Dec. 13, 1870	110, 096
Barrel	H. Willard	Grand Rapids, Mich	Apr. 12, 1870	101, 956
Barrel	H. Willard	Grand Rapids, Mich	Apr. 26, 1870	102, 346
Barrel	H. Willard	Grand Rapids, Mich	Apr. 26, 1870	102, 347
Barrel	H. Willard	Grand Rapids, Mich	May 10, 1870	103, 000
Barrel	H. Willard	Grand Rapids, Mich	Mar. 21, 1871	112, 875
Barrel	H. Willard	Grand Rapids, Mich	Feb. 25, 1873	136, 196
Barrel, Amalgamating	J. Brodie	San Francisco, Cal	July 5, 1864	43, 462
Barrel, Amalgamating	J. B. Johnson	San Francisco, Cal	July 12, 1864	43, 507
Barrel and box elevator	E. and J. E. Blake	Chicopee, Mass	Aug. 15, 1871	118, 097
Barrel and cask for containing oil, &c	A. Pelletier	Washington, D. C	Jan. 21, 1868	73, 643
Barrel and cask closing apparatus	A. Koegler	Newark, N. J	Dec. 26, 1871	122, 177
Barrel and keg	E. Johnson, jr., and G. W. Ansly.	Cleveland, Ohio	May 31, 1864	42, 949
Barrel and keg	J. Merrill	Boston, Mass	Oct. 2, 1866	58, 452
Barrel and keg making machinery	N. Goodall and L. Tainter	Watertown, N. Y	June 9, 1828	....
Barrel and other cylindrical package	H. Willard	Grand Rapids, Mich	Nov. 14, 1871	120, 920
Barrel and other vessel	J. A. Frey, J. Allen, and G. D. Smith.	Washington, D. C	Apr. 17, 1866	53, 970
Barrel and tank, Cement for coating oil	I. Waterman	London, Canada	July 4, 1871	116, 651
Barrel, Ash	T. H. Kane	New York, N. Y	Dec. 24, 1872	134, 145
Barrel beveling and crozing machine	H. A. Crossley	Cleveland, Ohio	May 24, 1870	103, 333
Barrel-carriage	W. Furley	Smithsburgh, Md	Oct. 30, 1849	6, 826
Barrel, Cast-iron	A. Putnam, jr	Saratoga Springs, N. Y	Feb. 7, 1871	111, 679
Barrel-chamfering machine	J. Tilley	West Troy, N. Y	Dec. 10, 1861	33, 908
Barrel chamfering and crozing machine	T. M. Adams and T. B. Luce	Linden, Mich	Dec. 13, 1830	110, 106
Barrel chamfering and crozing machine	A. H. Crozier and C. Carrier	Oswego, N. Y	Aug. 10, 1858	21, 117
Barrel chamfering, beveling, and howeling machine.	S. Widerman	Elysville, Md	Apr. 23, 1842	2, 583
Barrel-cleaning apparatus	I. Pfeiffer	New York, N. Y	Dec. 7, 1869	97, 688
Barrel, Core	R. Cartwright	Chicago, Ill	Apr. 26, 1870	102, 220
Barrel-cover	C. Bird	Dorchester, Mass	Nov. 12, 1867	70, 789
Barrel-cover	G. Hunt	Wawayanda, N. J	Apr. 8, 1873	137, 683

Index of patents issued from the United States Patent Office from 1790 *to* 1873, *inclusive*—Continued.

Invention.	Inventor.	Residence.	Date.	No.
Barrel-cover	H. Steiger	Washington, D. C	May 10, 1864	42, 699
Barrel creasing and beveling machine	A. H. Crozier	Oswego, N. Y	July 4, 1854	11, 211
Barrel cresset	J. F. Little	Lockport, N. Y	July 23, 1867	67, 062
Barrel crozing and chamfering machine	J. Greenwood	Rochester, N. Y	June 4, 1872	127, 478
Barrel crozing and chamfering machine	C. Murdock	Ellenville, N. Y	June 2, 1868	78, 470
Barrel crozing and chamfering machine, Tool for	H. Martin	Louisville, Ky	Feb. 1, 1859	22, 813
Barrel crozing and howeling machine	E. and B. Holmes	Buffalo, N. Y	July 28, 1868	80, 481
Barrel-crozing machine	H. De Bus and G. Johnson	Cincinnati, Ohio	Feb. 11, 1868	74, 319
Barrel-crozing machine	J. Ellis	South Brooks, Me	July 16, 1872	128, 955
Barrel-crozing machine	J. Maley	Middletown, Ohio	Dec. 1, 1868	84, 500
Barrel-crozing machine	H. Nelson	Lake Village, N. H	Apr. 5, 1870	101, 650
Barrel-crozing machine	J. Solter	Baltimore, Md	Oct. 22, 1872	132, 377
Barrel-dressing	C. Titus	Union, Me	July 14, 1868	80, 032
Barrel dressing and crozing machine	W. Brown	Saint Louis, Mo	July 29, 1873	141, 317
Barrel-dressing machine	L. Wirthlin	Saint Louis, Mo	Aug. 25, 1863	39, 696
Barrel, Drying	S. Gibbons	Binghamton, N. Y	Aug. 27, 1867	68, 181
Barrel-fastening	E. T. Gilmore	New York, N. Y	Dec. 12, 1871	121, 865
Barrel-filler	S. C. Catlin	Cleveland, Ohio	Dec. 20, 1870	110, 201
Barrel-filler	F. Stitzel	Louisville, Ky	Nov. 8, 1870	109, 071
Barrel-filler	L. H. Watson	Pittsburgh, Pa	Apr. 18, 1871	113, 954
Barrel-filler, Automatic	S. C. Catlin	Cleveland, Ohio	June 25, 1870	99, 159
Barrel-filler, Self-closing	H. A. Webber and C. Reifsnyder.	Chicago, Ill	Dec. 5, 1865	51, 372
Barrel-filling apparatus	H. A. Webber and C. Reifsnyder.	Chicago, Ill	Dec. 5, 1865	51, 370
Barrel-filling apparatus	H. A. Webber and C. Reifsnyder.	Chicago, Ill	Dec. 5, 1865	51, 371
Barrel-filling apparatus, Automatic self-closing	W. S. Payne	Petroleum Centre, Pa	Sept. 10, 1867	68, 649
Barrel-filling apparatus with whistling indicator	H. S. Phillips	Sewickley, Pa	Nov. 16, 1869	96, 933
Barrel-filling device	F. Stitzel	Louisville, Ky	Nov. 21, 1871	121, 134
Barrel-filling machine	J. L. Stewart	Philadelphia, Pa	Nov. 8, 1870	109, 151
Barrel, Flour	W. H. Towers	New York, N. Y	May 1, 1866	54, 444
Barrel for beer, &c, Construction of	M. Hawe	Albany, N. Y	Jan. 17, 1871	111, 058
Barrel for carbon-oil	B. Hackett	Pittsburgh, Pa	Jan. 19, 1864	41, 340
Barrel for cooling fluids	G. Verplaetse	New York, N. Y	Mar. 2, 1869	87, 447
Barrel for holding oil	J. A. Bassett	Salem, Mass	Apr. 18, 1865	47, 273
Barrel for holding petroleum	L. Day and H. Chapman	Buffalo, N. Y	Oct. 31, 1865	50, 694
Barrel for holding petroleum	G. W. Williamson	Gouldsborough, Pa	Aug. 1, 1865	49, 202
Barrel for holding petroleum and other oils	J. Holland	Conshohocken, Pa	Feb. 14, 1865	46, 359
Barrel for petroleum, &c	J. S. Lipps	Brooklyn, N. Y	Sept. 4, 1866	57, 738
Barrel-handle	M. S. Scofield	Stamford, Conn	Dec. 27, 1870	110, 502
Barrel-handling device for stores	G. M. Moore	Farmington, Ill	Aug. 27, 1872	130, 931
Barrel-head	J. B. Barsaloux	Saint Louis, Mo	June 27, 1871	116, 398
Barrel-head	J. A. Cook	Oswego, N. Y	Mar. 16, 1869	87, 760
Barrel head	W. W. Crooker	Waukegan, Ill	Nov. 5, 1872	132, 752
Barrel-head	R. B. De Bare	Philadelphia, Pa	Oct. 4, 1870	107, 885
Barrel-head	G. W. Gilbert	Radnor, Pa	Nov. 28, 1865	51, 260
Barrel-head	L. L. Gilliland	Dayton, Ohio	June 13, 1865	48, 168
Barrel head	S. S. Gray	Boston, Mass	July 25, 1871	117, 274
Barrel-head	A. Hanvey	Steubenville, Ohio	Aug. 1, 1871	117, 536
Barrel-head	A. Hanvey	Steubenville, Ohio	Oct. 14, 1873	143, 571
Barrel-head	G. Righter	Radnor, Pa	July 20, 1869	92, 885
Barrel-head	J. McCammon	Dayton, Ohio	Jan. 23, 1866	52, 182
Barrel-head	J. W. Taylor	Ashland, Va	Aug. 20, 1872	130, 600
Barrel-head	C. W. Saladee	Saint Catharine's, Canada	Apr. 16, 1872	125, 694
Barrel-head	J. W. Taylor	Ashland, Va	June 17, 1873	140, 095
Barrel-head	M. L. Thompson	Flemington, N. J	Oct. 22, 1867	70, 050
Barrel-head	M. L. Thompson	Flemington, N. J	Aug. 11, 1868	80, 886
Barrel-head, Adjustable	P. Rink and J. Docherty	Wertsville, N. J	July 14, 1868	79, 919
Barrel-head, Adjustable	A. C. Yawger	Newark, N. J	Aug. 4, 1868	80, 582
Barrel head and tap	G. Weaver and H. N. Allen	Boston, Mass	Dec. 22, 1868	85, 260
Barrel-head-centering device	J. J. Ralya	Cleveland, Ohio	Aug. 12, 1873	141, 816
Barrel-head circling and beveling machine	W. H. Doane	Cincinnati, Ohio	June 24, 1862	35, 674
Barrel-head-cutter	P. Chase	Peoria, Ill	Jan. 11, 1870	98, 666
Barrel-head cutting and beveling tool	W. Watkins	Crete, Ill	May 12, 1863	38, 520
Barrel-head-cutting machine	A. H. Crozier	Oswego, Ill	Mar. 20, 1855	12, 543
Barrel-head-cutting machine	W. Manning	Rouse's Point, N. Y	Mar. 2, 1858	19, 509
Barrel-head-cutting machine	J. H. Mattison	Scriba, N. Y	Mar. 2, 1858	19, 510
Barrel-head-cutting machine	J. B. Stanhope	Philadelphia, Pa	Mar. 26, 1872	124, 917
Barrel-head-cutting machine	C. R. Tompkins	Rochester, N. Y	Feb. 21, 1865	46, 508
Barrel-head-cutting machine	W. L. Young	Muscatine, Iowa	Feb. 20, 1855	12, 452
Barrel-head-cutting machinery	C. B. Hutchinson	Waterloo, N. Y	Feb. 1, 1853	9, 565
Barrel-head-dressing machine	W. Trapp	Elmira, N. Y	Aug. 23, 1864	43, 933
Barrel-head-dressing machine	P. Welch	Oswego, N. Y	Jan. 12, 1864	41, 248
Barrel-head machine	A. Benster	Detroit, Mich	Jan. 26, 1864	41, 356
Barrel-head machine	J. B Dougherty	Rochester, N. Y	Sept. 17, 1867	68, 856
Barrel-head machine	B. Fitch	Mooers, N. Y	May 18, 1858	20, 261
Barrel-head machine	C. B. Hutchinson	Auburn, N. Y	Sept. 11, 1860	30, 024
Barrel-head machine	H. L. McNish	Lowell, Mass	Oct. 9, 1860	30, 378
Barrel-head machine	W. Mickle	Oneonta, N. Y	May 31, 1870	103, 762
Barrel-head machine	J. S. Thompson	Glen's Falls, N. Y	Feb. 20, 1866	52, 772
Barrel-head machinery, Method of connecting the beveling-knives in circular cutting.	W. Bevard	Muscatine, Iowa	Jan. 12, 1858	19, 066
Barrel-head-making machine	A. Cutter	Boston, Mass	Oct. 11, 1870	108, 115
Barrel-head-manufacturing machine	L. B. Batcheller	Rochester, N. Y	Apr. 3, 1860	27, 682
Barrel-head-rounding machine	O. Redmond	Rochester, N. Y	Dec. 1, 1868	84, 509
Barrel-head-sawing machine	P. I. Steere	Cheshire, Mass	May 3, 1853	9, 699
Barrel-head-turning machine	J. J. Ralya	Cleveland, Ohio	Apr. 18, 1871	113, 795
Barrel-heads, Apparatus for rounding and beveling	J. P. Heacock	Marlborough, Ohio	Mar. 7, 1854	10, 594
Barrel-heads, Charring	J. D. Copenhaver	Martinsburgh, W. Va	Sept. 13, 1370	107, 339
Barrel-heads, Cutter for	W. H. Bennette	Utica, N. Y	Oct. 31, 1865	50, 677
Barrel-heads, Machine for chamfering	J. Greenwood	Rochester, N. Y	June 21, 1859	24, 458
Barrel-heads, Machine for cutting	A. H. Crozier	Oswego, N. Y	July 13, 1858	20, 864
Barrel-heads, Machine for cutting both bevels simultaneously on.	A. D. Stewart	Bennington, Vt	July 20, 1858	20, 962
Barrel-heads, Machine for forming	P. H. Lawler and J. B. Dougherty.	Spencerport and Rochester, N. Y.	Aug. 27, 1861	33, 153
Barrel-heads, Machine for making	E. Greenlee	Summerhill, Pa	Aug. 30, 1864	43, 985

Index of patents issued from the United States Patent Office from 1790 *to* 1873, *inclusive*—Continued.

Invention.	Inventor.	Residence.	Date.	No.
Barrel-heads, Machine for making	J. Greenwood	Rochester, N. Y	Mar. 7, 1865	46, 661
Barrel-heads, Machine for making	G. Schloemer	Greenfield, Wis	Jan. 30, 1872	123, 201
Barrel-heads, Machine for manufacturing	N. W. Robinson	Keesville, N. Y	May 6, 1856	14, 829
Barrel-heads, Machine for planing	G. R. Hay	Cleveland, Ohio	Jan. 16, 1872	122, 828
Barrel-heads, Machinery for dressing	T. Shepard	Oswegatchie, N. Y	Nov. 27, 1849	6, 912
Barrel-heads, Making	J. L. Kilgore	Cleveland, Ohio	Nov. 18, 1873	144, 771
Barrel-heads, Method of securing	G. W. Banker	Medford, Mass	June 12, 1860	28, 712
Barrel-heads, Method of securing	J. T. Tomkins	New York, N. Y	Jan. 3, 1865	45, 771
Barrel-heads, shingles, &c., Machine for sawing	J. B. Dougherty	Rochester, N. Y	Mar. 10, 1863	37, 885
Barrel-heads, Submerging	N. B. Cleveland	Waupun, Wis	Feb. 21, 1860	27, 256
Barrel-header	C. B. Pettengill	Hebron, Me	Apr. 8, 1873	137, 567
Barrel-header	L. S. Thompson	Spencerport, N. Y	July 16, 1872	129, 375
Barrel-header	C. F. Weaver and J. W. Goss	Rochester, N. Y	July 30, 1872	130, 090
Barrel heading, circling, and beveling machine	R. Grotz	Chicago, Ill	Jan. 14, 1868	73, 321
Barrel-headings, Machine for jointing	W. Trapp	Elmira, N. Y	Aug. 23, 1864	43, 935
Barrel-headings, Machine for sawing	C. J. Holman	Chicago, Ill	Dec. 24, 1867	72, 639
Barrel-heater	A. E. Salisbury and S. S. Steel	Martin, Ohio	May 6, 1873	138, 531
Barrel-hoop	J. B. Dougherty	Rochester, N. Y	Sept. 1, 1863	39, 720
Barrel-hoop	J. B. Dougherty	Rochester, N. Y	Oct. 13, 1863	40, 250
Barrel-hoop	H. Ogborn	Richmond, Ind	Nov. 25, 1873	144, 920
Barrel-hoop	H. W. C. Tweddle	Allegheny, Pa	Oct. 23, 1866	59, 097
Barrel-hoop	H. Willard	Grand Rapids, Mich	July 15, 1873	140, 862
Barrel-hoop-chamfering device	H. Pelsue	East Wallingford, Vt	May 17, 1864	42, 790
Barrel-hoop-crimping machine	M. Reed	Rochester, N. Y	May 3, 1864	42, 597
Barrel-hoop-cutting machine	J. Dobbins	Litchfield, Mich	July 21, 1868	80, 154
Barrel-hoop-cutting machine	A. Goodyear	Springport, Mich	Aug. 14, 1866	57, 120
Barrel-hoop-dressing machine	A. McAlpine	Pittston, Pa	Jan. 15, 1867	61, 226
Barrel-hoop lock	J. Chase	Orange, Mass	Feb. 11, 1868	74, 305
Barrel-hoop loosener and nail extractor	C. F. Dean	Saint Johnsbury, Vt	Jan. 26, 1869	86, 285
Barrel-hoop-sawing machine	G. H. Shearer	Bay City, Mich	Dec. 24, 1867	72, 551
Barrel-hoops, Box for steaming	J. L. Gage	Rochester, N. Y	Apr. 23, 1872	126, 047
Barrel-hoops, Machine for dressing	J. T. Forsyth	Wheeling, W. Va	Sept. 29, 1868	82, 613
Barrel-hoops, Machine for making splints for	J. B. Dougherty	Rochester, N. Y	Feb. 23, 1864	41, 688
Barrel-hoops, Metal clasp for	D. M. Lawrence	Washington, D. C	Mar. 19, 1867	62, 964
Barrel-hoops, Staple for securing ends of	H. Ogborn	Richmond, Indiana	July 1, 1873	140, 430
Barrel, keg, and pail, Combined	G. W. Banker	New York, N. Y	June 13, 1871	115, 807
Barrel-lifter	L. H. Goff	Saint Albans, Vt	Sept. 4, 1866	57, 819
Barrel-lining composition	B. H. Howell	New York, N. Y	May 23, 1865	47, 825
Barrel-lining composition	H. W. C. Tweddle	Pittsburgh, Pa	Aug. 7, 1866	57, 016
Barrel lining, Oil	W. R. Bree	Pottsville, Pa	Aug. 20, 1867	67, 947
Barrel lining, Oil	M. G. Huntley	Grand Rapids, Mich	Nov. 21, 1871	121, 106
Barrel-machine	R. W. George	Richmond, Me	June 7, 1864	43, 074
Barrel-machine	P. H. Lawler	Rochester, N. Y	Oct. 9, 1866	58, 729
Barrel-machine	W. R. and E. Middleton	Cleveland, Ohio	May 2, 1871	114, 321
Barrel-machine	A. Wyckoff	Elmira, N. Y	Dec. 19, 1865	51, 643
Barrel, Machine for compressing the cylinder of cask upon its head to form a tight.	W. Reid	West Arlington, Vt	June 5, 1866	55, 363
Barrel-machinery	S. Andrews	Perth Amboy, N. J	Feb. 12, 1850	7, 077
Barrel-machinery	H. Baker	Cortland, N. Y	Oct. 16, 1866	58, 754
Barrel-machinery	H. S. Higgens	Graham, Ind	Nov. 7, 1854	11, 894
Barrel-machinery	M. T. Kennedy	New Brighton, Pa	Aug. 6, 1872	130, 224
Barrel-machinery	T. Molinier	New Orleans, La	Feb. 10, 1847	4, 966
Barrel-machinery	R. Murdock	Rochester, N. Y	June 12, 1849	6, 523
Barrel-machinery	G. W. Pierce	Holly, N. Y	Oct. 24, 1865	50, 625
Barrel-machinery	W. Trapp, jr	Dryden, N. Y	Oct. 1, 1845	4, 218
Barrel-making	L. Raymond	Rockland, Del	Jan. 21, 1868	73, 461
Barrel-making machine	W. and M. Adams	Ogden, N. Y	Jan. 24, 1828	....
Barrel-making machine	S. J. Arnold and A. F. Clark	Raymondsville, N. Y	Sept. 1, 1868	81, 575
Barrel-making machine	H. Baker	McLean, N. Y	July 30, 1844	3, 683
Barrel-making machine	W. Brown	Saint Louis, Mo	Aug. 17, 1869	93, 855
Barrel making machine	W. Brown	Saint Louis, Mo	July 9, 1872	128, 786
Barrel-making machine	S. Roberts	Cleveland, Ohio	Apr. 1, 1862	34, 847
Barrel-making machinery	I. Crossett	East Bennington, Vt	July 1, 1844	3, 648
Barrel-making machinery	H. P. Hall	Akron, N. Y	July 15, 1873	140, 776
Barrel-making machinery	G. W. Livermore	Cambridgeport, Mass	Mar. 21, 1854	10, 680
Barrel, Metallic	J. I. Bard	New Orleans, La	Jan. 16, 1872	122, 796
Barrel, Metallic	J. A. Frey	New York, N. Y	June 11, 1867	65, 741
Barrel, Metallic oil	E. Parker	Philadelphia, Pa	July 19, 1864	43, 625
Barrel, Oil	R. N. Allen	Cleveland, Ohio	Jan. 20, 1863	37, 427
Barrel-opening tool	D. Snedeker	New York, N. Y	May 8, 1860	28, 208
Barrel-opening tool	T. J. Phillips	Washington, D. C	Jan. 28, 1868	73, 923
Barrel or cask	C. H. Carver	Taunton, Mass	Sept. 11, 1866	57, 863
Barrel or cask	C. T. Provost	New York, N. Y	Aug. 27, 1867	68, 234
Barrel or cask	S. Shea	Corry, Pa	Jan. 29, 1867	61, 690
Barrel or kibble, Lime	F. K. Winsor	Hillsdale, Mich	June 7, 1870	103, 953
Barrel-packer	T. Burns	Williamsburg, N. Y	Apr. 11, 1865	47, 253
Barrel-packer	W. H. Glasgow	New York, N. Y	Mar. 15, 1859	23, 243
Barrel-pitching apparatus	J. W. Brady	Baltimore, Md	Dec. 20, 1870	110, 193
Barrel-pitching apparatus	A. A. C. Klaucke	Washington, D. C	Dec. 20, 1870	110, 373
Barrel-pitching apparatus	R. Rosochacki	Cleveland, Ohio	Mar. 22, 1870	101, 166
Barrel-pitching apparatus	L. Schulze	Louisville, Ky	Aug. 30, 1870	106, 964
Barrel-pitching apparatus	G. Sichler	New York, N. Y	June 17, 1873	139, 976
Barrel-pitching apparatus	W. C. Vollmer and R. Rosochaiki.	Cuyahoga County, Ohio	July 16, 1872	128, 987
Barrel-pitching furnace	H. Shlandeman	Decatur, Ill	June 25, 1872	128, 329
Barrel-pitching machine	D. Myers	Baltimore, Md	Dec. 26, 1871	122, 128
Barrel-pitching machine	A. de Witzleben	Washington, D. C	Aug. 8, 1871	117, 956
Barrel-roller	S. G. Hill	Muscatine, Iowa	May 28, 1872	127, 236
Barrel-rolling apparatus	L. L. Hyatt and A. G. Hüpfel	New York and Morrisania, N. Y.	Apr. 18, 1871	113, 771
Barrel rolling device	H. W. Stephenson	Cincinnati, Ohio	Apr. 25, 1865	47, 465
Barrel-rolling tongs	M. W. Ingle	Indianapolis, Ind	Mar. 7, 1871	112, 462
Barrel-setting-up apparatus	W. B. Elliott	Corning, N. Y	Mar. 29, 1870	101, 242
Barrel-stand and keg-holder	C. N. Cass	Winchester, Ill	July 8, 1873	140, 573
Barrel-stand, Movable	P. J. Skinner	Oswego, N. Y	Jan. 14, 1868	73, 395
Barrel-stave	C. Murdock	Hartford, Conn	June 1, 1869	90, 679
Barrel stave-dressing machine	J. J. Raiya	Allegheny City, Pa	Nov. 26, 1867	71, 326
Barrel-stave jointer	M. Randolph	Saint Louis, Mo	July 16, 1867	66, 737

Index of patents issued from the United States Patent Office from 1790 *to* 1873, *inclusive*—Continued.

Invention.	Inventor.	Residence.	Date.	No.
Barrel staves and heads, Machine for shaving, jointing, and forming.	B. Langdon and W. Mowry	Washington County, N. Y	June 20, 1811	....
Barrel-support	T. W. Claussen	Mars Bluff, S. C	Mar. 4, 1873	136, 416
Barrel, tank, and cask for transporting acid and petroleum.	J. B. Davenport	New York, N. Y	Aug. 29, 1871	118, 438
Barrel-tapping	W. Boynton	Auburn, N. Y	Aug. 7, 1866	56, 889
Barrel tapping-piece	A. Zinsser	New York, N. Y	Mar. 24, 1868	76, 030
Barrel-tightening strap	S. Macferren	Philadelphia, Pa	Dec. 20, 1870	110, 255
Barrel-transporter	J. Griffing	Ipswich, Mass	July 1, 1873	140, 364
Barrel-trussing machine	J. Reid	Allegheny City, Pa	Oct. 29, 1872	132, 601
Barrel-trussing machine	P. Welch	Saint Louis, Mo	Dec. 6, 1870	109, 854
Barrel-vent	R. C. Fleming	Philadelphia, Pa	June 2, 1868	78, 447
Barrel, Ventilated flour	T. Pearsall	Smithborough, N. Y	June 27, 1854	11, 200
Barrel-washer	A. Muntzenberger	Kenosha, Wis	Nov. 25, 1873	145, 003
Barrel-washing machine	J. Peacock	Rockford, Ill	Mar. 13, 1866	53, 177
Barrel-washing machine	J. Peacock	Rockford, Ill	Aug. 27, 1867	68, 311
Barrel-washing machine	T. Reynolds	Yonkers, N. Y	May 24, 1870	103, 371
Barrels and casks, Apparatus for pitching beer and other.	W. Vogt	Louisville, Ky	July 11, 1871	116, 896
Barrels and casks, Coating for oil	S. H. Titus	Saint Louis, Mo	Nov. 25, 1862	37, 018
Barrels and casks during the process of pitching, Apparatus for agitating and cooling.	D. Cammerer	Cincinnati, Ohio	June 21, 1870	104, 421
Barrels and firkins, Adjustable head for	H. D. Rumsey	Homer, N. Y	Feb. 4, 1868	74, 150
Barrels and casks, Machine for distributing pitch in	B. J. Stukenborg	Cincinnati, Ohio	July 4, 1871	116, 772
Barrels and packages, Construction of	C. Green	Wilmington, Del	Mar. 1, 1870	100, 286
Barrels, and placing them in tiers or rows, Machine for hoisting.	G. B. Vroom and S. Kinzie	Jersey City, N. J., and Brooklyn, N. Y.	Jan. 22, 1861	31, 197
Barrels, Apparatus for aerating beer	J. Metzger	East Cambridge, Mass	May 16, 1871	114, 839
Barrels, Apparatus for lifting	J. S. Brewer	Chicago, Ill	Feb. 28, 1871	112, 214
Barrels, casks, &c., Apparatus for pitching and coating.	L. Shulze	Baltimore, Md	Dec. 6, 1870	109, 948
Barrels, Apparatus for pitching beer, &c	H. Reutti and P. Winklehaus	Hamilton, Ohio	Apr. 19, 1870	102, 156
Barrels, &c., Apparatus for rolling and handling	S. Longley	Cincinnati, Ohio	June 3, 1856	15, 042
Barrels, &c., Apparatus for steaming and cleaning	A. K. Lee	Galveston, Tex	May 3, 1873	138, 810
Barrels, &c., Applying solution to the interior and exterior of oil.	J. O. Woodruff	Auburn, N. Y	Oct. 24, 1865	50, 649
Barrels, Automatic plug for	T. Windle and J. H. Dorst	New Albany, Ind	Jan. 26, 1869	86, 196
Barrels, boxes, &c., Machine for lifting and carrying.	J. W. Babcock	Chicopee, Mass	Feb. 18, 1873	135, 958
Barrels, &c., Bush for	D. F. Fetter	New York, N. Y	June 22, 1869	91, 618
Barrels, Bushing for	D. F. Fetter	New York, N. Y	Feb 19, 1867	62, 189
Barrels, casks, &c., Apparatus for pitching	L. Schulze	Baltimore, Md	May 30, 1871	115, 530
Barrels, casks, &c., Machine for pitching	G. Meyer	Quincy, Ill	Jan. 21, 1873	135, 139
Barrels, casks, &c., Metallic hoop for	W. Wilson, jr	Wilmington, Del	June 27, 1865	48, 497
Barrels, Cleansing and purifying beer	W. Harvey	Waterville, N. Y	Apr. 2, 1872	125, 193
Barrels, Composition for coating oil	D. Ahl	Newville, Pa	Dec. 6, 1864	45, 379
Barrels, Composition for coating oil	J. G. Thompson	Carbondale, Pa	Mar. 7, 1865	46, 734
Barrels, Composition for lining	A. C. Dunn	New York, N. Y	May 29, 1866	55, 073
Barrels, Composition for lining	L. Held	Harlem, N. Y	June 27, 1865	48, 398
Barrels, &c., Composition for lining oil	J. Baur	Brooklyn, N. Y	June 6, 1865	48, 041
Barrels, &c., Composition for lining oil	F. Becker	Scranton, Pa	Mar. 10, 1863	37, 848
Barrels, Composition for lining oil	W. Budd and J. L. Husband	Philadelphia, Pa	Mar. 7, 1865	46, 636
Barrels, Composition for lining oil	H. Loewenberg	New York, N. Y	Apr. 25, 1865	47, 496
Barrels, Composition for lining oil	J. P. Schenck, jr	Matteawan, N. Y	Jan. 1, 1867	60, 944
Barrels, Composition for lining oil	G. W. Williamson	Goldsborough, Pa	July 12, 1864	43, 552
Barrels, Composition for lining petroleum	J. Fox	Philadelphia, Pa	May 23, 1865	47, 810
Barrels, Composition for lining petroleum	R. Price	Jersey City, N. J	Aug. 1, 1865	49, 152
Barrels, Connection for floating	R. W. Park	Pittsburgh, Pa	May 3, 1864	42, 594
Barrels, Construction of	G. St. George, jr	New York, N. Y	Aug. 27, 1867	68, 322
Barrels, Cresset for heating	J. S. Thompson and M. J. Seymour.	Glen's Falls, N. Y	Mar. 29, 1859	23, 410
Barrels, Device for tilting	J. C. Curran	Philadelphia, Pa	Aug. 3, 1869	93, 180
Barrels, Elastic cement for lining petroleum	J. J. K. Boote and W. A. Gibson	Cleveland, Ohio	Apr. 4, 1871	113, 485
Barrels, &c., Fermentation guard for	A. P. Gardner	Moscow, Pa	Dec. 23, 1873	145, 860
Barrels, &c., for containing petroleum, Composition for lining.	H. Preuss	New York, N. Y	Jan. 10, 1865	45, 857
Barrels for holding oil, &c., Process for lining	J. Baur	New York, N. Y	Mar. 28, 1865	46, 988
Barrels for holding petroleum, Composition for lining.	H. Loewenberg	New York, N. Y	Jan. 24, 1865	46, 060
Barrels for petroleum, &c., Composition for lining	L. Francis	New York, N. Y	Feb. 28, 1865	46, 554
Barrels from bursting, Safety-plug to prevent	R. Bridge	Paterson, N. J	Mar. 6, 1866	52, 957
Barrels from leaking, Method of preventing oil	D. Ahl	Newville, Pa	Jan. 17, 1865	45, 902
Barrels, &c., from leaking, Method of preventing oil	G. T. Parry and W. S. Warner	Philadelphia, Pa	Jan. 10, 1865	45, 855
Barrels, Gage for filling	W. H. Noyes	Boston, Mass	June 5, 1860	28, 637
Barrels, Head-lining for	G. A. Reed	New York, N. Y	Mar. 11, 1873	136, 763
Barrels impervious to oil, &c., Composition for rendering.	G. R. Percy	New York, N. Y	Jan. 24, 1865	46, 021
Barrels impervious to petroleum, Process of rendering.	J. Baur	Brooklyn, N. Y	Aug. 8, 1865	49, 215
Barrels, &c., Implement for rolling, guiding, and controlling.	B. V. Tamplin	Peoria, Ill	Apr. 22, 1873	138, 055
Barrels, kegs, &c., Tool to be used in the manufacture of.	W. G. Burr	Mount Pleasant, N. Y	Oct. 12, 1839	1, 367
Barrels, &c., Lining	G. Arnd	New York, N. Y	June 21, 1864	43, 172
Barrels, Lining for oil	R. V. Jones	Canton, Ohio	Aug. 21, 1866	57, 333
Barrels, Lining for petroleum	W. H. Stone	Brooklyn, N. Y	May 2, 1865	47, 581
Barrels, &c., Lining petroleum	T. O. Oliver	New York, N. Y	Apr. 18, 1865	47, 363
Barrels, Lock for securing stamp upon	J. L. Harley	Baltimore, Md	Sept. 6, 1870	107, 176
Barrels, Machine for chamfering and crozing	J. H. Mattison	Scriba, N. Y	Oct. 12, 1858	21, 769
Barrels, Machine for chamfering and crozing	P. Welch	Oswego, N. Y	Jan. 12, 1864	41, 249
Barrels, Machine for crozing and chamfering	H. Littlejohn	Troy, N. Y	Mar. 1, 1859	23, 097
Barrels, Machine for crozing, chamfering, and beveling.	W. M. Arnall	Sperryville, Va	Oct. 5, 1858	21, 718
Barrels, Machine for cutting the locks in hoops for	F. C. La Riviere	Minneapolis, Minn	June 4, 1867	65, 491
Barrels, Machine for finishing the inside of	E. Greenlee	Summer Hill, Pa	Dec. 4, 1860	30, 811
Barrels, Machine for forcing hoops on	J. Greenwood	Rochester, N. Y	Dec. 30, 1873	145, 942
Barrels, &c., Machine for forming	J. Rees	Elk Horn Grove, Ill	Oct. 5, 1858	21, 725

Index of patents issued from the United States Patent Office from 1790 to 1873, inclusive—Continued.

Invention.	Inventor.	Residence.	Date.	No.
Barrels, Machine for leveling the staves in	E. and B. Holmes	Buffalo, N. Y	July 28, 1868	80, 482
Barrels, Machine for making plugs for	L. H. Dwelley	Dorchester, Mass	Nov. 13, 1866	59, 567
Barrels, Machine for sawing heading for	P. Welch	Oswego, N. Y	Apr. 5, 1864	42, 244
Barrels, Machine for setting staves in	C. B. Hutchinson	Auburn, N. Y	Mar. 19, 1867	63, 052
Barrels, Machine for setting up	A. Cutter	Boston, Mass	Oct. 11, 1870	108, 113
Barrels, Machine for setting up	A. G. Mack	Rochester, N. Y	Jan. 1, 1861	31, 026
Barrels, &c., Machine to be used in combination with improved iron truss-hoop in the manufacture of.	J. H. Bruner and R. H. Thompson.	Mitford and Granville, Ohio	Nov. 12, 1842	2, 852
Barrels, Machinery for making	C. Ruggles	Huron, Ohio	Nov. 18, 1873	144, 635
Barrels, Making	P. Werum	Berlin, Ohio	Aug. 25, 1863	39, 694
Barrels, Manufacturing	J. Cole and S. Miller	Washington County, Ohio	Sept. 29, 1831	....
Barrels, Manufacturing oil	J. Squier	Salina, N. Y	Jan. 21, 1835	....
Barrels, Manufacturing oil	B. Springer	Henry County, Ind	Apr. 27, 1832	....
Barrels, Mechanism for heading	J. Griffing	Ipswich, Mass	Mar. 11, 1873	136, 595
Barrels, Metallic head for	S. Lewis	Rochester, N. Y	Dec. 20, 1864	45, 507
Barrels, Method of boring gun	H. Peeler	Boston, Mass	Feb. 6, 1849	6, 088
Barrels, Method of making	S. Roberts	Cleveland, Ohio	Oct. 16, 1860	30, 425
Barrels, Method of making	S. Roberts	Cleveland, Ohio	Feb. 26, 1861	31, 558
Barrels, Method of securing bush for bung to	T. Summerfield	New York, N. Y	May 23, 1865	47, 877
Barrels, Mode of chamfering and crozing	J. H. Mattison	Scriba, N. Y	Oct. 3, 1857	18, 549
Barrels, Mode of manufacturing	G. W. Banker	Medford, Mass	Oct. 11, 1859	25, 784
Barrels, Mode of pitching	J. F. T. Halbeck and M. Gottfried.	Chicago, Ill	May 3, 1864	42, 580
Barrels, Mode of repairing	E. W. Gillman	Hunter's Point, N. Y	June 30, 1868	79, 340
Barrels of oil, &c., Instrument for ascertaining the amount of water, &c., in.	G. Tagliabue	New York, N. Y	May 5, 1863	38, 427
Barrels, Pitching	W. Anheuser	Saint Louis, Mo	Apr. 4, 1871	113, 480
Barrels, Pitching	J. P. Benoit	Detroit, Mich	Oct. 26, 1869	96, 076
Barrels, Raising liquid from	E. Cook	Rochester, N. Y	Apr. 30, 1872	126, 267
Barrels, Self-acting tilt for	I. Hudson and W. Winshall	Stockport, Great Britain	Sept. 27, 1870	107, 779
Barrels, &c., Skid for elevating and lowering	C. Brosius	Hancock, Md	Feb. 18, 1868	74, 493
Barrels, Skid for supporting	W. W. Doane and W. P. Burr	Brewer, Me	Oct. 16, 1866	58, 787
Barrels to contain petroleum, coal-oil, &c., Preparing.	L. S. Robbins	New York, N. Y	May 3, 1864	42, 598
Barrels to hold oil, petroleum, &c., Preparing	H. Wurtz	New York, N. Y	Feb. 21, 1865	46, 518
Barrels to render them oil-tight, Coating	A. H. Hook and J. H. Darlington	New York, N. Y	June 7, 1864	43, 025
Barrels water and oil tight, Composition for making	T. Schumann and C. G. Frash	New York, N. Y	July 5, 1864	43, 433
Barrels with glue, Process of lining oil	E. W. Leggett	New York, N. Y	Oct. 21, 1873	143, 770
Barrels with sheet-metal, Mode of lining	T. R. Cook	Saratoga Springs, N. Y	Nov. 5, 1867	70, 414
Barrow and cultivator, Combined wheel	L. Trefftz and G. H. Shimpert	Pinckneyville, Ill	Jan. 5, 1869	85, 709
Barrow and garden-plow, Combined wheel	J. D. O'Callaghan	Calhoun, Ga	Oct. 12, 1869	95, 718
Barrow and hand-plow combined, Wheel	B. G. Fitzhugh	Frederick, Md	Oct. 17, 1871	119, 973
Barrow and step-ladder combined, Wheel	H. J. Evans	Christieville, Canada	Sept. 26, 1871	119, 339
Barrows and trucks, Tubular frame of wheel	B. W. Tuthill	New York, N. Y	Dec. 12, 1871	121, 916
Barrow and turnip-drill, Combined	A. M. Nowland	Olivet, Mich	Nov. 8, 1870	109, 041
Barrow, Coal	P. K. Dederick	Albany, N. Y	Apr. 26, 1870	102, 376
Barrow, Construction of wheel	N. Tufts, 2d	Charlestown, Mass	Sept. 13, 1824	....
Barrow, Drill	G. Colby	Fayetteville, Pa	June 12, 1849	6, 520
Barrow, Dumping wheel	G. H. Kanmacher	Chicago, Ill	May 21, 1872	127, 065
Barrow frame, Wheel	B. W. Tuthill	Oregon City, Oreg	Jan. 3, 1871	110, 698
Barrow, Hand	M. S. Schofield	Stamford, Conn	Jan. 18, 1870	99, 009
Barrow, Horse wheel	J. Rowley	Pittsford, N. Y	Nov. 1, 1825	....
Barrow trays, Machine for making wheel	E. Breed	Jamestown, N. Y	Aug. 6, 1872	130, 273
Barrow, Truck	J. J. Richardson	Marion, S. C	Jan. 21, 1873	135, 157
Barrow, Wheel	N. Cronkite	Milford, N. Y	May 24, 1870	103, 430
Barrow, Wheel	D. B. Ellis	Ypsilanti, Mich	Nov. 12, 1867	70, 821
Barrow, Wheel	J. Ennis	Columbus, Ga	Oct. 18, 1870	108, 341
Barrow, Wheel	J. Gehr	Clear Spring, Md	Nov. 7, 1871	120, 736
Barrow, Wheel	C. R. Hight	Geneva, Ill	Aug. 6, 1872	130, 129
Barrow, Wheel	C. C. Johnson	Springfield, Vt	Jan. 19, 1869	86, 077
Barrow, Wheel	J. M. and J. L. Jones	Paris and Lexington, Ky	Oct. 21, 1873	143, 767
Barrow, Wheel	H. Lawrence	New York, N. Y	July 18, 1871	117, 085
Barrow, Wheel	H. Lawrence	New York, N. Y	July 18, 1871	117, 086
Barrow, Wheel	E. B. Marshall	Atlanta, Ga	Oct. 20, 1868	83, 189
Barrow, Wheel	W. McKibbin	San Francisco, Cal	Oct. 24, 1871	120, 159
Barrow, Wheel	W. F. Newcombe	Cleveland, Ohio	May 19, 1868	78, 003
Barrow, Wheel	P. Noling	Woodside, Wis	Oct. 12, 1869	95, 717
Barrow, Wheel	C. and C. Nutting, jr	San Francisco, Cal	Nov. 19, 1872	133, 170
Barrow, Wheel	F. Parkerson	Philadelphia, Pa	Mar. 31, 1868	76, 241
Barrow, Wheel	R. M. Reynolds	East Saginaw, Mich	Oct. 4, 1870	107, 053
Barrow, Wheel	N. C. Sanford	Meriden, Conn	June 6, 1865	48, 101
Barrow, Wheel	J. B. Schneider	Kankakee, Ill	June 10, 1873	139, 823
Barrow, Wheel	B. W. Tuthill	Oregon City, Oreg	Sept. 28, 1869	95, 288
Barrow, Wheel	J. J. Van Kersen	Kalamazoo, Mich	Dec. 27, 1864	45, 658
Barrow, Wheel	J. G. Weir	Pittsburgh, Pa	Dec. 28, 1869	98, 324
Baryta, Manufacture of	C. M. T. Du Motay	Paris, France	Nov. 11, 1873	144, 517
Baryta, Manufacture of artificial sulphates of	J. Philip	Hamburg, Germany	Nov. 7, 1871	120, 771
Barytes, Process of preparing sulphate of	W. M. Page and E. B. Krausse	Saint Louis, Mo	Sept. 15, 1868	82, 154
Base-ball, Apparatus for playing parlor	F. C. Lebring	Hoboken, N. J	Feb. 4, 1868	74, 154
Base-ball, India-rubber	H. A. Alden	Matteawan, N. Y	July 7, 1868	79, 719
Base-ball table	W. Buckley	New York, N. Y	Aug. 20, 1867	67, 951
Base-balls, Manufacture of	H. A. Alden	Matteawan, N. Y	Dec. 17, 1867	72, 355
Basin	H. H. Craigie	New York, N. Y	Mar. 12, 1867	62, 734
Basin, Overflow	H. Boyd	New York, N. Y	Mar. 12, 1867	62, 812
Basin-plug	W. S. Carr	New York, N. Y	June 4, 1867	65, 343
Basin-trap	H. H. Craigie	New York, N. Y	Dec. 7, 1869	97, 482
Basin with waste-pipe attachment	G. C. Miller and J. H. Coates	Goshen and New York, N. Y	Mar. 15, 1870	100, 782
Basins, Means for draining	L. R. Comstock	Philadelphia, Pa	July 8, 1873	140, 577
Basins, Press for forming metal	G. Murray	Cambridge, Mass	Dec. 5, 1865	51, 342
Basket	E. Beecher	Plymouth, Conn	Aug. 30, 1864	44, 039
Basket	L. W. Beecher	Westville, Conn	Sept. 28, 1869	95, 185
Basket	W. H. Carpenter	New York, N. Y	Aug. 24, 1869	94, 075
Basket	E. B. Cole	Huntington, Mass	Dec. 12, 1871	121, 758
Basket	J. Graham	Vassar, Mich	June 18, 1872	128, 036
Basket	W. C. Higgins	North Blandford, Mass	Apr. 1, 1873	137, 368
Basket	H. C. Jones	Dowagiac, Mich	Dec. 11, 1866	60, 383

Index of patents issued from the United States Patent Office from 1790 *to* 1873, *inclusive*—Continued.

Invention.	Inventor.	Residence.	Date.	No.
Basket	H. C. Jones	Dowagiac, Mich	Aug. 11, 1868	80, 825
Basket	H. C. Jones	Dowagiac, Mich	Aug. 6, 1872	130, 135
Basket	H. C. Jones	Dowagiac, Mich	Aug. 6, 1872	130, 136
Basket	H. C. Jones	Dowagiac, Mich	Aug. 6, 1872	130, 137
Basket	H. C. Jones	Dowagiac, Mich	Aug. 6, 1872	130, 138
Basket	H. C. Jones	Dowagiac, Mich	Aug. 6, 1872	130, 139
Basket	H. C. Jones	Dowagiac, Mich	Dec. 17, 1872	133, 941
Basket	H. C. Jones	Dowagiac, Mich	Sept. 2, 1873	142, 343
Basket	E. B. Lyman	Waterbury, Conn	Sept. 5, 1865	49, 774
Basket	L. Marble	Vassar, Mich	Jan. 7, 1862	34, 088
Basket	S. P. P. Miller	Beaver, Pa	May 5, 1868	77, 516
Basket	J. R. Park	Marlborough, N. Y	May 1, 1860	28, 105
Basket	S. I. Russell	Chicago, Ill	Dec. 10, 1872	133, 892
Basket	S. M. Sherman	Fort Dodge, Iowa	Aug. 20, 1861	33, 106
Basket	T. R. Sherry	Newark, N. J	July 4, 1865	48, 595
Basket	H. E. Tower	Worthington, Mass	Mar. 5, 1872	124, 231
Basket	H. E. Tower	Worthington, Mass	Mar. 5, 1872	124, 232
Basket	B. F. Tuthill	Chicago, Ill	June 27, 1871	116, 374
Basket	A. Van Riper	Washington Tp., N. J	Aug. 21, 1847	5, 247
Basket	G. Z. and W. A. Van Riper	Lagrange, Mich	Feb. 27, 1866	52, 914
Basket	R. B. Wheeler	Dowagiac, Mich	Oct. 29, 1872	132, 703
Basket	S. H. Wheeler	Niles, Mich	July 22, 1873	141, 023
Basket-and bird-cage, Combined hanging	S. Vanstone	Providence, R. I	Apr. 15, 1873	137, 980
Basket braiding block	G. O. Gillette	Streetsborough, Ohio	Aug. 20, 1872	130, 711
Basket' Clothes	L. and A. C. Carman	McCoy's Station, Ohio	Sept. 20, 1870	107, 446
Basket, Cotton	R. L. Myers	Washington, N. C	Dec. 21, 1869	98, 091
Basket, Cover for fruit	J. C. Ingham and C. Colby	Benton Harbor, Mich	July 22, 1873	141, 142
Basket, Extension	A. Norman	Rochester, N. Y	Mar. 30, 1869	88, 327
Basket-finishing machine	F. H. Brown	Chicago, Ill	Oct. 29, 1867	70, 318
Basket, Flower	G. Gunther	New York, N. Y	June 13, 1865	48, 171
Basket, Folding	A. M. Olds	New York, N. Y	Jan. 28, 1868	73, 749
Basket for berries	S. R. Wilmot	Brooklyn, N. Y	May 14, 1861	32, 324
Basket for feeding tarred corn-cobs to furnace	E. Bimm	Dayton, Ohio	May 28, 1867	65, 157
Basket for grinding tile	P. C. Reniers	Pittsburgh, Pa	Sept. 21, 1869	94, 977
Basket for house-plant	A. P. Eastman	Washington, D. C	Mar. 7, 1871	112, 433
Basket for railway-car, Bracket	J. L. Howard	Hartford, Conn	May 25, 1869	90, 543
Basket for tile-grinder	P. C. Reniers	Pittsburgh, Pa	Dec. 13, 1870	110, 072
Basket-form	A. F. Scow	Chicago, Ill	May 13, 1873	138, 765
Basket-form	A. F. Scow	Chicago, Ill	May 20, 1873	139, 025
Basket, Form-block for	W. H. Carpenter	New York, N. Y	Aug. 11, 1868	80, 910
Basket-former	H. C. Jones	Dowagiac, Mich	Mar. 28, 1871	113, 058
Basket-forming	T. and J. Churchill	Detroit, Mich	Dec. 27, 1864	45, 585
Basket, Fruit	L. Carpenter	Saint Joseph, Mich	May 3, 1870	102, 488
Basket, Grain	H. C. Jones	Dowagiac, Mich	Apr. 15, 1873	137, 845
Basket, Grain	H. C. Jones	Dowagiac, Mich	Apr. 15, 1873	137, 846
Basket, Grain	H. C. Jones	Dowagiac, Mich	Sept. 2, 1873	142, 341
Basket, Grain	H. C. Jones	Dowagiac, Mich	Sept. 2, 1873	142, 342
Basket, Grain	H. C. Jones	Dowagiac, Mich	Dec. 16, 1873	145, 573
Basket-handle, Fastening	A. Faas	Philadelphia, Pa	June 23, 1857	17, 621
Basket-handle, Metallic	O. A. North	New Britain, Conn	Nov. 19, 1872	133, 168
Basket, Hanging	C. C. Hibbert	Lynn, Mass	Mar. 19, 1872	124, 817
Basket, Hanging	J. H. O'Neil	Cleveland, Ohio	Apr. 12, 1870	101, 907
Basket, Hanging	D. Sherwood	Lowell, Mass	Nov. 16, 1869	96, 977
Basket, Lady's work	S. Ainsworth	Saratoga Springs, N. Y	Dec. 27, 1864	45, 571
Basket, Lady's work	R. V. Jones	Canton, Ohio	May 25, 1869	90, 553
Basket-machine	M. Shaffer	La Fayette, Ind	Dec. 28, 1869	98, 424
Basket-machine	G. Storer	New Britain, Conn	Feb. 12, 1867	62, 085
Basket-machine	J. B. Sweetland and E. C. Goodrich.	Pontiac, Mich	Dec. 12, 1865	51, 491
Basket-making machine	F. H. Brown	Chicago, Ill	Oct. 22, 1867	70, 072
Basket-making machine	J. Moore and S. W. Day	Oak Creek, Wis	Jan. 14, 1873	134, 920
Basket, Metallic	A. Schenck	Chicago, Ill	Aug. 4, 1863	39, 426
Basket or dish, Table	J. Gibson, jr	Albany, N. Y	Sept. 19, 1871	119, 086
Basket, Peach	H. Carpenter	Brooklyn, N. Y	Oct. 27, 1868	83, 459
Basket-protector	P. Eley	New York, N. Y	Apr. 4, 1865	47, 096
Basket, Reticule wicker	J. Venet	New York, N. Y	Dec. 14, 1869	97, 836
Basket, Reticule wicker	J. Venet	New York, N. Y	Apr. 16, 1872	125, 706
Basket, Silver-plate cake and fruit	R. Gleason, jr	Dorchester, Mass	June 17, 1856	15, 127
Basket-splint	I. F. Wilcox	Streetsborough, Ohio	Nov. 28, 1871	121, 304
Basket-splint machine	A. and G. Van Riper	Bergen County, N. J	Oct. 2, 1847	5, 319
Basket-splints, Machine for riving	W. I. Horton	La Grange, Ala	Nov. 29, 1859	26, 268
Basket, Stave	H. C. Jones	Dowagiac, Mich	Mar. 28, 1871	113, 059
Basket-stuff-cutting machine	C. Jordan	Wrentham, Mass	Mar. 22, 1870	101, 021
Basket, Trout	P. R. Ridgway	Boston, Mass	July 16, 1872	129, 592
Basket, Water-proof	G. M. Allerton	Dover Plains, N. Y	Oct. 22, 1872	132, 428
Basket-weaving machine	F. H. Brown	Chicago, Ill	Sept. 10, 1867	68, 695
Basket-weaving machine	F. H. Brown	Chicago, Ill	Oct. 1, 1867	69, 309
Basket, Weighing	C. Renne and P. Landenberger	New York, N. Y	Nov. 23, 1869	97, 119
Basket, Wine	J. Roussillon	Epernay, France	Mar. 21, 1871	112, 852
Basket, Wire	D. Sherwood	Lowell, Mass	Feb. 18, 1873	135, 943
Basket, Work	T. V. Hobbs	Chicago, Ill	June 3, 1873	139, 508
Baskets, Apparatus for casting metal for metallic	W. H. Trissler	Dunkirk, N. Y	July 9, 1872	128, 928
Baskets, Construction of	J. A. H. Ellis	Springfield, Vt	Apr. 7, 1857	16, 970
Baskets for catching eels, Method of making	J. Downs	Belleport, N. Y	Apr. 25, 1843	3, 056
Baskets, Form-block for shaping	H. Carpenter	New York, N. Y	Dec. 31, 1867	72, 797
Baskets, Form or mold on which wooden slats, &c., are made into.	J. A. H. Ellis	Springfield, Vt	Mar. 31, 1857	16, 953
Baskets, Machine for braiding open-work	F. H. Brown	Chicago, Ill	Oct. 29, 1867	70, 160
Baskets, Machine for cutting material for	L. J. Bridgeman	Rock Stream, N. Y	Sept. 27, 1870	107, 654
Baskets, Machine for dressing willow for	E. K. Root	Hartford, Conn	Oct. 16, 1866	58, 958
Baskets, Machine for forming	E. S. Jeffery	New Haven, Conn	July 18, 1865	48, 869
Baskets, Machinery for forming	J. W. Millet	Bachelorsville, N. Y	May 2, 1865	47, 602
Baskets, Making splits for	J. S. Borden	Shrewsbury, N. J	June 29, 1833	....
Baskets, Manufacture of stave	H. C. Jones	Dowagiac, Mich	Mar. 15, 1870	100, 902
Baskets, Means for manufacturing	E. Bredt	New York, N. Y	June 20, 1865	48, 255
Baskets, Means of manufacturing	J. D. and J. T. Shuler	Lockport, N. Y	May 13, 1862	35, 265
Baskets, Method of manufacturing	L. Marble	Vassar, Mich	July 5, 1859	24, 699
Baskets, Preparing moss for ornamental	J. W. Shiveley	Saratoga Springs, N. Y	Apr. 30, 1873	126, 237

Index of patents issued from the United States Patent Office from 1790 *to* 1873, *inclusive*—Continued.

Invention.	Inventor.	Residence.	Date.	No.
Baskets, Tool for manufacturing splint	A. Baker	Templeton, Mass	Feb. 2, 1858	19, 229
Bat for cricket, &c	P. Caminoni	New York, N. Y	May 24, 1864	42, 834
Bat, Spring	G. W. Hill	Deep River, Conn	Oct. 30, 1866	59, 313
Bath:				
See Air-bath.				
Bird-bath.				
Combination bath.				
Compressed-air bath.				
Cot-bath.				
Cot elastic bath.				
Electrical bath.				
Electro-magnetic bath.				
Foot-bath.				
Galvanic bath.				
Immersion-bath.				
Medicated bath.				
Medicated shampoo bath.				
Mineral bath.				
Parlor bath.				
Photographic bath.				
Portable bath				
Shower-bath.				
Steam-bath.				
Swimming-bath.				
Vapor-bath.				
Ventilated bath.				
Warm bath.				
Zincing and tinning bath.				
Bath and wash-stand	R. Cooper	Georgetown, D. C	Jan. 31, 1871	111, 321
Bath-boiler	E. H. Chapman and T. M. Hammett.	Philadelphia, Pa	Dec. 31, 1867	72, 798
Bath-chamber	G. F. Foote	Middletown, N. Y	Dec. 10, 1872	133, 702
Bath, Confluent pipe for	H. Jones	Philadelphia, Pa	Apr. 12, 1870	101, 884
Bath-house, Portable	J. Gourlay	Escanaba, Mich	May 5, 1868	77, 480
Bath, Portable	E. J. Knowlton	Ann Arbor, Mich	Mar. 1, 1870	100, 297
Bath-room	H. G. Covel	Brooklyn, N. Y	Dec. 24, 1872	134, 258
Bath-room rack	M. A. H. Saurman	Philadelphia, Pa	Nov. 10, 1868	83, 884
Bath-seat, Portable	A. P. Young	Providence, R. I	Aug. 16, 1870	106, 445
Bath-tub	L. Bagger	Washington, D. C	June 17, 1873	139, 995
Bath-tub	A. Barrows	Philadelphia, Pa	Oct. 12, 1869	95, 633
Bath-tub	C. A. Blessing	Philadelphia, Pa	Aug. 31, 1869	94, 385
Bath-tub	A. Brady	New York, N. Y	June 7, 1864	43, 002
Bath-tub	A. C. Brownell	Brooklyn, N. Y	Apr. 7, 1868	76, 298
Bath-tub	J. Carroll	New York, N. Y	May 28, 1867	65, 17[illegible]
Bath-tub	J. C. Clapp	Homer, N. Y	Oct. 22, 1867	69, 968
Bath-tub	E. F. Cook	Omaha, Nebr	Feb. 9, 1864	41, 486
Bath-tub	E. J. Knowlton	Ann Arbor, Mich	Jan. 28, 1868	73, 906
Bath-tub	J. H. Mercer	New York, N. Y	Mar. 3, 1868	75, 180
Bath-tub	J. L. Mott	New York, N. Y	Sept. 27, 1853	10, 049
Bath-tub	A. R. Robb	Brooklyn, N. Y	July 9, 1872	128, 755
Bath-tub	M. A. Stevens	Hartford, Conn	June 22, 1869	91, 785
Bath-tub, Combination	A. C. Brownell	Brooklyn, N. Y	May 24, 1870	103, 293
Bath-tub eduction-tube	W. H. Walton	Philadelphia, Pa	Oct. 5, 1869	95, 542
Bath tub, Electrical	J. Kidder	New York, N. Y	Jan. 9, 1866	51, 948
Bath tub, Electro-medical	S. Russell, jr	New York, N. Y	June 10, 1873	139, 819
Bath tub, Magnetic	J. R. Anderson	New York, N. Y	June 18, 1872	128, 094
Bath-tub, Plug-socket	R. C. Scrimgeour	Brooklyn, N. Y	Apr. 8, 1873	137, 726
Bath-tub, Portable	R. McCully	Philadelphia, Pa	Mar. 23, 1869	88, 154
Bath-tub, Portable	J. P. Rider	Brooklyn, N. Y	July 6, 1869	92, 364
Bath-tub, Portable	G. Wheeler	New York, N. Y	Nov. 19, 1872	133, 276
Bath-tub, Portable India-rubber	J. W. Ramsay and D. Wilson	Grafton, Va	Sept. 4, 1860	29, 914
Bath-tubs, Flask for molding	J. Demarest	Mott Haven, N. Y	Aug. 28, 1855	13, 514
Bath-tubs, Sheet-metal lining for	C. A. Blessing	Philadelphia, Pa	July 28, 1868	80, 441
Bathing-apparatus	H. Bachmann and J. Ehrenfried.	Lancaster, Pa	Jan. 26, 1833	....
Bathing-apparatus	D. Brumley	New York, N. Y	Nov. 19, 1833	....
Bathing-apparatus	G. I. Byrd and P. Milne	New York, N. Y	June 17, 1840	1, 636
Bathing-apparatus	C. Escudier	Pattersonville, La	Jan. 18, 1859	22, 636
Bathing-apparatus	W. Hensler	Milwaukee, Wis	Mar. 1, 1864	41, 773
Bathing-apparatus	J. Ingram	New York, N. Y	Oct. 8, 1867	69, 675
Bathing-apparatus	F. Kraemer	Brooklyn, N. Y	Aug. 10, 1858	21, 138
Bathing-apparatus	L. Lefebvre	New Orleans, La	Apr. 21, 1857	17, 102
Bathing-apparatus	J. K. O'Neil	Kingston, N. Y	Aug. 11, 1857	17, 979
Bathing-apparatus	G. Scholl	New Philadelphia, Ohio	Sept. 5, 1871	118, 750
Bathing-apparatus	J. C. Schooley	Cincinnati, Ohio	Apr. 2, 1861	31, 912
Bathing-apparatus	C. Schultz and T. Warker	New York, N. Y	Oct. 17, 1865	50, 499
Bathing-apparatus	W. T. Street	Frankford, Pa	May 10, 1870	102, 878
Bathing-apparatus	C. Venn	Kastnersville, Canada	Oct. 31, 1871	120, 555
Bathing-apparatus	W. G. Young	Baltimore, Md	Dec. 16, 1845	4, 309
Bathing apparatus, Rectum	C. B. Veronee	Athens, Ga	July 11, 1871	116, 895
Bathing-apparatus, Safety	W. H. Pitt	Philadelphia, Pa	Dec. 22, 1868	85, 126
Bathing-dress	O. Morse	Concord, Mass	Feb. 23, 1869	87, 107
Bathing, Life-line for sea	W. T. Street	Frankford, Pa	Dec. 1, 1868	84, 592
Bathing-machine	L. Desens	Paris, France	Dec. 6, 1864	45, 368
Bathing-machine	W. Merritt	New York, N. Y	Feb. 1, 1814	....
Bathing vessel, Warm	D. Harrington	Centreville, Va	Oct. 4, 1813	....
Baton, Alarm	J. F. Haskins	Fitchburgh, Mass	May 30, 1871	115, 464
Baton, Police	S. Beery and J. W. McDonald	Urbana, Ohio, and Houston, Tex.	Oct. 14, 1873	143, 610
Baton, Police	C. Kast	Williamsport, Pa	Mar. 28, 1871	113, 061
Baton, Police	H. C. Richard	Pottsville, Pa	June 20, 1871	116, 097
Batter and egg stirrer and beater	H. Muth	Cincinnati, Ohio	Oct. 22, 1872	132, 483
Batter-cup	B. Wieland	Orangeville, Ill	July 25, 1865	49, 019
Batter-machine	E. Yerby	Washington, D. C	July 27, 1858	21, 042
Batter-mixer	A. Burdick	Middletown, Conn	June 18, 1872	128, 013

Index of patents issued from the United States Patent Office from 1790 *to* 1873, *inclusive*—Continued.

Invention.	Inventor.	Residence.	Date.	No.
Batter-pot	E. A. Jeffery	New York, N. Y	Feb. 7, 1871	111, 548
Battery:				
See Carbon battery.				
Centrifugal battery.				
Chain-shot battery.				
Electric battery.				
Electro battery.				
Electro-magnetic battery.				
Electro-medical battery.				
Electrotyping battery.				
Field battery.				
Floating battery.				
Fuse-firing battery.				
Galvanic battery.				
Gun battery.				
Non-freezing battery.				
Physiological battery.				
Platoon battery.				
Portable body battery.				
Sap battery.				
Self-acting battery.				
Ship's battery.				
Stamp battery.				
Stamping battery.				
Thermo-electric battery				
Traveling battery.				
Voltaic battery.				
Battery-current manipulator	J. Kidder	New York, N. Y	Nov. 7, 1871	120, 750
Battery, Galvanic	A. L. Nolf	New York, N. Y	Sept. 2, 1873	142, 502
Battery, Traveling	C. Perley	New York, N. Y	Feb. 24, 1863	37, 766
Batting and wadding	S. Baxendale	Boston, Mass	Dec. 11, 1866	60, 324
Batting-machine feed-mechanism	W. Fuzzard	Malden, Mass	Aug. 16, 1864	43, 845
Batting, Machine for making	F. W. Bloodgood	New York, N. Y	Dec. 16, 1873	145, 615
Batting or wadding, Fibrous	J. T. Stoddard	Plymouth, Mass	Feb. 23, 1864	41, 727
Battledore	M. Cregen	Chicago, Ill	Aug. 1, 1871	117, 518
Bayonet	F. W. Alexander	Baltimore, Md	Nov. 15, 1864	45, 009
Bayonet	A. Mills	United States Army	Mar. 29, 1870	101, 297
Bayonet and bayonet-fastening, Spade	F. Chillingworth and I. Merrill	Springfield, Mass	Apr. 16, 1872	125, 720
Bayonet-attachment	W. Hoffman	Washington, D. C	Apr. 18, 1865.	47, 303
Bayonet-attachment	C. Howard	New York, N. Y	May 15, 1866	54, 728
Bayonet-attachment	J. W. Neil	Brooklyn, N. Y	Jan. 7, 1873	134, 608
Bayonet-attachment	P. A. Oliver	Elizabeth, N. J	July 30, 1867	67, 210
Bayonet-shank	H. Waters	Northbridge, Mass	June 21, 1864	43, 247
Bayonet, Drill	T. Wahlfelt	Stockholm, Sweden	Aug. 1, 1871	117, 702
Bayonet-fastening	J. N. Ward	United States Army, New York, N. Y.	Dec. 15, 1857	18, 876
Bayonet fastening, Saber	C. A. McEvoy	Richmond, Va	Oct. 30, 1860	30, 539
Bayonet-guard, Removable	J. G. Ernst	York, Pa	Aug. 5, 1862	36, 081
Bayonet-scabbard	E. Rice	Cambridge, Mass	May 21, 1872	127, 104
Bayonet scabbard and guard	J. G. Ernst	York, Pa	Dec. 23, 1862	37, 222
Bayonet-scabbard, Frog for	W. Hoffman	Solano County, Cal	May 8, 1860	28, 175
Bayonet-scabbards, Machine for cutting out	H. D. Smith	New York, N. Y	June 30, 1863	39, 074
Bayonet-scabbards, Manufacture of	E. Gaylord	Chicopee, Mass	May 15, 1860	28, 269
Bayonet, Sliding	J. Jenkinson	Brooklyn, N. Y	July 1, 1862	35, 760
Bayonet-socket	B. Burton	Brooklyn, N. Y	Oct. 7, 1873	143, 495
Bayonet-socket	A. Willson	New York	June 11, 1814	....
Bayonet-sockets, Turning	J. Humphreys	Millburgh, Mass	Feb. 23, 1864	41, 704
Bayonet, Spade	S. S. Alexander	Philadelphia, Pa	June 22, 1869	91, 586
Bayonet, Spade	E. Rise	United States Army	June 22, 1869	91, 564
Bayonet, Spade	W. S. Wetmore	London, England	Dec. 26, 1871	122, 206
Bayonet, Steel scabbard for	J. E. Emerson	Trenton, N. J	Aug. 19, 1862	36, 209
Bayonets, Scabbard for trowel	F. Chillingworth	Springfield, Mass.	Dec. 1, 1868	84, 612
Bead-forming die for sheet-metal vessel	S. R. Wilmot	Bridgeport, Conn	July 10, 1866	56, 313
Bead-miters, Machine for cutting	R. F. Thompkins	New York, N. Y	July 13, 1869	92, 491
Beads, Manufacturing metallic	J. R. Wendt	Boston, Mass	Feb. 10, 1857	16, 631
Beading-machine, Sheet-metal-ware	C. E. Marchand	Alliance, Ohio	June 10, 1873	139, 681
Beading moldings, &c., Plane for	S. C. Howes	South Chatham, Mass	Nov. 3, 1863	40, 483
Beading-tool and circular shears, Combined	J. F. Flanders	Newburyport, Mass	Jan. 2, 1849	5, 994
Beam	R. J. Gatling	Indianapolis, Ind	Nov. 16, 1869	96, 793
Beam and girder	J. Montgomery	Croton Landing, N. Y	Oct. 20, 1868	83, 196
Beam and girder, Wrought-iron	A. Pollak	Philadelphia, Pa	June 5, 1855	13, 025
Beam and scale, Movable suspended	S. Wallis	....	Sept. 21, 1801	....
Beam, Composite	W. W. Lummus	Lynn, Mass	June 17, 1873	139, 902
Beam, Corrugated	R. Montgomery	New York, N. Y	Sept. 25, 1855	13, 599
Beam, Corrugated	S. J. Seely	New York, N. Y	May 10, 1864	42, 692
Beam engine, Triangular	N. W. Wheeler	Brooklyn, N. Y	Jan. 16, 1866	52, 101
Beam-engines, Parallel motion for	J. M. Thompson	Taunton, Mass	July 11, 1854	11, 252
Beam-fagot	H. T. Buffington, jr	Buffalo, N. Y	Sept. 29, 1868	82, 487
Beam, Floor	W. W. Lummus	Lynn, Mass	June 17, 1873	139, 903
Beam, Floor	W. W. Lummus	Lynn, Mass	June 17, 1873	139, 904
Beam, floor, and ceiling, Fire-proof	I. Hodgson and W. H. Brown	Indianapolis, Ind	May 30, 1871	115, 472
Beam, Hollow	J. W. Cremin	New York, N. Y	Dec. 26, 1871	122, 230
Beam, Metal	J. K. Ingalls	Brooklyn, N. Y	June 10, 1856	15, 113
Beam, Metal	J. Merrill	Boston, Mass	May 14, 1867	64, 783
Beam, Metal	R. Montgomery	New York, N. Y	Apr. 14, 1868	76, 795
Beam, Metallic	L. Holms	Paterson, N. J	Mar. 22, 1870	101, 015
Beam or girder for fire-proof structure	W. W. Hughes	Philadelphia, Pa	Jan. 17, 1871	111, 063
Beam, Plow	J. S. Hall	Manchester, Pa	Feb. 22, 1859	23, 024
Beam, Scale	A. B. Davis	Philadelphia, Pa	Feb. 26, 1861	31, 534
Beam, Sheet-metal	R. Montgomery	New York, N. Y	July 12, 1853	9, 842
Beams and columns, Connection for wrought-iron	A. Bonzano	Phœnixville, Pa	July 1, 1873	140, 455
Beams, Connecting rigidly the ends of metal	S. Nowlan	New York, N. Y	Apr. 13, 1858	19, 945
Beams, Construction of fagots for	W. W. Miller	Safe Harbor, Pa	Oct. 22, 1867	70, 011
Beams, Construction of fagots for	G. Walters and T. Shaffer	Phœnixville, Pa	Oct. 15, 1867	69, 872
Beams, Machine for shearing the ends of	J. M. Cornell	New York, N. Y	Aug. 13, 1872	130, 486
Beams, Manufacture of corrugated	R. Montgomery	New York, N. Y	May 31, 1859	24, 229
Beams or girders, Mode of preparing fagots for manufacturing wrought.	G. Walters and T. Shaffer	Phœnixville, Pa	Dec. 17, 1867	72, 246

Index of patents issued from the United States Patent Office from 1790 *to* 1873, *inclusive*—Continued.

Invention.	Inventor.	Residence.	Date.	No.
Beaming-machine	A. Hill	Hinsdale, Mass	May 6, 1805	
Beaming-machine	J. Ladd		July 17, 1801	
Bean-pot lifter and carrier	G. F. Foss	East Boston, Mass	Apr. 28, 1868	77, 182
Bean-puller	S. W. Moore	Albion, N. Y	June 25, 1867	66, 034
Bean-puller	S. R. Niles	Rawsonville, Mich	Aug. 11, 1868	80, 994
Bean-pulling machine	J. Baxter	North Greece, N. Y	May 3, 1864	42, 548
Bean pulling machine	J. Day	Murray, N. Y	Nov. 10, 1857	18, 576
Bean-pulling machine	S. R. Niles	Rawsonville, Mich	Apr. 20, 1869	89, 164
Beans, Machine for cleaning and assorting	R. Read	Brockport, N. Y	Mar. 11, 1862	34, 645
Bearing and packing	E. D. Murfey	New York, N. Y	Aug. 1, 1871	117, 558
Bearing and packing material	E. D Murfey	New York, N. Y	Dec. 27, 1870	110, 582
Bearing and packing material	E. D Murfey	New York, N. Y	Dec. 27, 1870	110, 583
Bearing and packing, Saturating fibrous materials and powdered substances for.	E. D. Murfey	New York, N. Y	Oct. 11, 1870	108, 285
Bearing, Anti-friction	A. Hall	Tiffin, Ohio	Sept. 10, 1872	131, 162
Bearing, Compensation	L. Smith	Buffalo, N. Y	Apr. 17, 1855	12, 732
Bearing, End-thrust	C. Perley	New York, N. Y	Feb. 24, 1863	37, 765
Bearing for machinery	C. B. Richards	Hartford, Conn	Aug. 1, 1871	117, 682
Bearing for machinery, Anti-friction	J. Harden	Chicago, Ill	Jan. 14, 1868	73, 245
Bearing for machinery, Lubricating	L. H. Olmstead	Stamford, Conn	Jan. 21, 1868	73, 459
Bearing for machinery, Self oiling and adjusting	T. S. Brown	Poughkeepsie, N. Y	Dec. 18, 1866	60, 616
Bearing-step and vertical shaft	F. A. Gardner	Danbury Conn	Apr. 18, 1871	113, 867
Bearing-surface and journal for machinery	J. Wharton	Philadelphia, Pa	Aug. 15, 1871	118, 086
Bearing-surfaces in machinery, Material for	J. Schieder	New York, N. Y	Dec. 13, 1870	110, 163
Bearing, Thrust	A. W. Case	South Manchester, Conn	June 2, 1868	78, 428
Bearings and journals, Material for	E. D. Murfey	New York, N. Y	Nov. 15, 1870	109, 239
Bearings, Composition for	E. D. Murfey	New York, N. Y	Nov. 1, 1870	108, 931
Bearings, Composition for coating	P. J. Kelly	New York, N. Y	Nov. 8, 1870	109, 018
Bearings for machinery, Casting	H. F. and G. S. Snyder and W. N. Jones.	Williamsport, Pa	Dec. 29, 1868	85, 341
Bearings of machinery, &c., Metallic composition for.	I. S. Hill and J. Dixon	Boston and Taunton, Mass	Jan. 31, 1845	3, 896
Bearings, &c., Rubber compound for	P. J. Kelly	New York, N. Y	June 27, 1871	116, 322
Beater and compressing press	G. Ertel	Quincy, Ill	Feb. 4, 1873	135, 533
Beater or power press, Operating	C. Nelson	Newburgh, N. Y	June 19, 1866	55, 779
Beater-press	J. Cory	Chicago, Ill	Feb. 11, 1873	135, 780
Beater-press	P. K. Dederick	Albany, N. Y	July 4, 1865	48, 619
Beater-press	S. R. Dummer	New York, N. Y	Jan. 30, 1866	52, 275
Beater-press	G. Ertel	Liberty, Ill	Dec. 31, 1867	72, 728
Beater-press	L. C. Field	Galesburgh, Ill	Aug. 1, 1865	49, 193
Beater-press	F. Frey	Liberty, Ill	Feb. 27, 1872	124, 047
Beating, bruising, and cutting substances, Mach. for	A. Britz	Philadelphia, Pa	July 18, 1816	
Bed, Air	J. R. Hamilton	Portland, Oregon	July 16, 1867	66, 706
Bed, Air	J. Scott	Philadelphia, Pa	Oct. 18, 1853	10, 139
Bed and bed-bottom	J. W. C. Peters and W. A. Le Row.	Chicago, Ill	Oct. 25, 1870	108, 622
Bed and bedstead, Invalid	S. S. Brown	Woonsocket, R. I	Mar. 2, 1869	87, 463
Bed and chair, Combination of camp	S. G. Crane	Rochester, N. Y	July 16, 1861	32, 822
Bed and chair, Combined	G. A. Keene	Lynn, Mass	July 3, 1860	29, 038
Bed and crib, Sofa	S. Chapin	Cincinnati, Ohio	Mar. 21, 1865	46, 879
Bed and cushion spring	D. E. Somes	Washington, D. C	Oct. 12, 1869	95, 848
Bed and frame, Spring	W. Z. Cooke	Hartford, Conn	Feb. 18, 1873	135, 892
Bed and knapsack, Atmospheric air	C. Linden	Eden Township, Ill	Oct. 7, 1862	36, 618
Bed and lounge spring	F. Fraps	Springfield, Mass	July 30, 1867	67, 186
Bed and musical-instrument board combined	J. McDonald	New York, N. Y	Nov. 23, 1869	97, 101
Bed and seat bottom	W. F. Clark	Mount Pleasant, Iowa	Sept. 12, 1871	118, 908
Bed and seat spring	J. P. Chamberlin	North Abington, Mass	May 16, 1871	114, 923
Bed and seat spring	J. P. Chamberlin	North Abington, Mass	Dec. 19, 1871	121, 990
Bed and seat spring, Means of securing	J. D. Eleston	Canaan, Conn	Sept. 8, 1868	81, 991
Bed and sofa	W. H. Schwalbe	New York, N. Y	Sept. 1, 1868	81, 691
Bed attachment, Invalid	N. Teal	Kendallville, Ind	Aug. 6, 1867	67, 610
Bed, berth, and car-seat spring	N. S. Whipple	Detroit, Mich	Oct. 5, 1869	95, 622
Bed-boat or life-preserver	J. Stevenson	Philadelphia, Pa	Feb. 20, 1855	12, 450
Bed bottom	J. R. Abbe	Providence, R. I	May 15, 1866	54, 658
Bed-bottom	P. Agger	Cincinnati, Ohio	Sept. 12, 1871	118, 834
Bed-bottom	I. T. Allen	New York, N. Y	Jan. 24, 1860	26, 878
Bed-bottom	J. Allen	Pughtown, Pa	Apr. 21, 1868	76, 968
Bed-bottom	J. H. Allyn	Whitesborough, N. Y	Mar. 21, 1871	112, 765
Bed-bottom	P. Anderson	Chicago, Ill	Sept. 9, 1873	142, 549
Bed bottom	C. D. Austin	Amsterdam, N. Y	Feb. 27, 1872	124, 109
Bed-bottom	A. S. Babbit	Keeseville, N. Y	Jan. 29, 1867	61, 595
Bed bottom	D. Babcock	Seneca Falls, N. Y	July 9, 1867	66, 549
Bed-bottom	H. Barber	Juneau, Wis	Dec. 23, 1862	37, 210
Bed-bottom	J. J. Baxter	Grand Rapids, Mich	Mar. 16, 1869	87, 752
Bed-bottom	S. W. Beach	Niles, Mich	May 1, 1866	54, 281
Bed-bottom	A. H. Bell	Cedar Falls, Iowa	Jan. 14, 1873	124, 839
Bed-bottom	V. Bell	Seville, Ohio	July 5, 1870	105, 028
Bed-bottom	V. Bell	Seville, Ohio	Feb. 14, 1871	111, 807
Bed-bottom	H. Benedict	Philadelphia, Pa	Jan. 16, 1872	122, 698
Bed-bottom	H. Benedict	Philadelphia, Pa	July 8, 1873	140, 571
Bed-bottom	C. E. Best	Jordan, N. Y	Feb. 19, 1867	62, 248
Bed-bottom	R. F. Billings	Portland, Me	Aug. 23, 1859	25, 170
Bed-bottom	A. Bingham	Talladega, Fla	Aug. 23, 1859	25, 171
Bed-bottom	A. Bingham	Newtonville, Mass	Nov. 12, 1867	70, 687
Bed-bottom	D. E. Bishop	Baltimore, Md	Apr. 16, 1872	125, 783
Bed-bottom	C. D. Blinn	Port Hudson, Mich	May 23, 1865	47, 789
Bed-bottom	J. Blythe	Lafayette, Ind	May 19, 1863	38, 551
Bed-bottom	P. Boesen and M. Bedessem	Kenosha, Wis	Feb. 18, 1873	135, 965
Bed-bottom	L. M. Bolles	Cooperstown, N. Y	Oct. 8, 1867	69, 617
Bed-bottom	A. T. Boon and J. H. Bell	Galesburgh, Ill	May 11, 1869	89, 912
Bed-bottom	W. Bowen	Dayton, Mich	Mar. 8, 1870	100, 491
Bed-bottom	B. R. Boynton	Keeseville, N. Y	Sept. 3, 1867	68, 551
Bed-bottom	F. S. Bradley	New Haven, Conn	Feb. 11, 1868	74, 293
Bed-bottom	C. Bradway	Maquoketa, Iowa	July 20, 1869	92, 787
Bed-bottom	W. R. Briggs	Boston, Mass	Jan. 25, 1870	99, 056
Bed-bottom	C. A. Brigham	Cleveland, Ohio	Apr. 13, 1869	88, 772
Bed-bottom	E. L. Brockett	Nelson, Ohio	June 17, 1873	140, 005
Bed-bottom	J. Bromiley	Pawtucket, R. I	Sept. 20, 1864	44, 279

Index of patents issued from the United States Patent Office from 1790 *to* 1873, *inclusive*—Continued.

Invention.	Inventor.	Residence.	Date.	No.
Bed-bottom	C. M. Brown	Aurora, Ill	Mar. 27, 1866	53, 519
Bed-bottom	J. H. Brown	Hudson, Wis	Feb. 18, 1868	74, 660
Bed-bottom	D. W. Burbank	New York, N. Y	June 12, 1866	55, 462
Bed-bottom	E. L. Bushnell	Poughkeepsie	Oct. 18, 1870	108, 447
Bed-bottom	L. W. Buxton	Nashua, N. H	May 8, 1860	28, 233
Bed-bottom	L. W. Buxton	Nashua, N. H	Aug. 7, 1860	29, 540
Bed-bottom	D. M. Bye	Roanoke, Ind	Sept. 28, 1869	95, 192
Bed-bottom	E. G. Cameron	Ashtabula, Ohio	Feb. 11, 1873	135, 774
Bed-bottom	J. B. Campbell	Cincinnati, Ohio	Feb. 26, 1867	62, 315
Bed-bottom	S. A. Canfield	Carlstadt, N. J	May 1, 1866	54, 292
Bed-bottom	F. Carré	Paris, France	Nov. 7, 1865	50, 883
Bed-bottom	E. Caswell	Lyons, N. Y	Mar. 17, 1868	75, 519
Bed-bottom	W. A. Chamberlin	Alexander, N. Y	May 26, 1868	78, 184
Bed-bottom	D. G. Chapin	Galena, Ill	Dec. 3, 1867	71, 700
Bed-bottom	A. D. Chase	Reading, Pa	Aug. 7, 1866	57, 039
Bed-bottom	P. G. Chase	Berlin, Wis	May 23, 1865	47, 797
Bed-bottom	J. Christie	Lowell, Mich	Oct. 6, 1868	82, 800
Bed-bottom	H. M. Clark	West Meriden, Conn	Apr. 23, 1867	63, 993
Bed-bottom	I. A. Clippinger and S. S. Pratt	Newton, Iowa	Feb. 8, 1870	99, 636
Bed-bottom	H. A. Coats	Wellsville, N. Y	Feb. 25, 1867	62, 395
Bed-bottom	D. C. Colby	Washington, D. C	July 9, 1867	66, 564
Bed-bottom	G. T. Comins	Lowell, Mass	Dec. 9, 1862	37, 085
Bed-bottom	H. M. Conklin	Syracuse, N. Y	Oct. 8, 1867	69, 632
Bed-bottom	W. F. Converse	Harrison, Ohio	July 2, 1861	32, 683
Bed-bottom	H. Cook and C. E. Simmons	Waukegan, Ill	Dec. 24, 1867	72, 605
Bed-bottom	W. Cooke	New York, N. Y	May 28, 1872	127, 152
Bed-bottom	H. A. Cooke	Charlestown, Mass	Dec. 12, 1865	51, 427
Bed-bottom	H. A. Cooke	Charlestown, Mass	July 9, 1867	66, 567
Bed-bottom	A. W. Cramer	Honesdale, Pa	Aug. 23, 1870	106, 666
Bed-bottom	C. Croley	Dayton, Ohio	May 28, 1867	65, 061
Bed-bottom	M. C. Cronk	Auburn, N. Y	Dec. 12, 1865	51, 429
Bed-bottom	L. M. Crosby and E. G. Cameron.	Ashtabula, Ohio	Nov. 25, 1873	144, 959
Bed-bottom	S. H. Crossman	Battle Creek, Mich	Mar. 13, 1866	53, 121
Bed-bottom	R. H. Cutter	Cleveland, Ohio	Sept. 10, 1867	68, 612
Bed-bottom	J. D. F. Dahl	Milwaukee, Wis	May 12, 1868	77, 720
Bed-bottom	J. and S. Danner	Canton, Ohio	Apr. 23, 1867	63, 998
Bed-bottom	W. F. Daugherty	Mount Pleasant, Iowa	Oct. 15, 1867	69, 779
Bed-bottom	G. B. and C. B. Davis	Freeport, Ill	Oct. 23, 1866	58, 994
Bed-bottom	J. W. Davis	Washington, D. C	Feb. 5, 1867	61, 724
Bed-bottom	J. Decamp	Cincinnati, Ohio	Sept. 28, 1869	95, 329
Bed-bottom	P. Demeure and A. Mauritz	New York, N. Y	Sept. 13, 1853	10, 010
Bed-bottom	L. L. and A. J. Deming and R. Alden.	Erie, Pa	Nov. 28, 1871	121, 279
Bed-bottom	H. Doebele	Philo, Ohio	Aug. 6, 1867	67, 420
Bed-bottom	G. H. Dow	Freeport, Ill	June 11, 1867	65, 731
Bed-bottom	J. Dreusike	Cincinnati, Ohio	Sept. 24, 1867	69, 193
Bed-bottom	J. Dreusike	New York, N. Y	Oct. 28, 1873	143, 970
Bed-bottom	F. B. Duffey	Sparta, Wis	Oct. 2, 1866	58, 394
Bed-bottom	S. Dunlap	Rome, Ga	June 8, 1869	90, 999
Bed-bottom	A. M. Dye	Clinton, Ill	Dec. 27, 1859	26, 575
Bed-bottom	W. H. Elliot	Plattsburgh, Pa	Aug. 10, 1858	21, 123
Bed-bottom	T. Falloon	Lyons, N. Y	May 22, 1866	54, 877
Bed-bottom	M. Falloon	Bloomington, Ill	Mar. 21, 1871	112, 912
Bed-bottom	J. S. Farrington	Wilwaukee, Wis	June 11, 1867	65, 556
Bed-bottom	H. E. Ficket and J. W. Summers.	Glen's Falls, N. Y	Nov. 22, 1859	26, 228
Bed-bottom	E. S. Field	Hartford, Conn	Sept. 10, 1872	131, 158
Bed-bottom	D. Fitzgerald	New York, N. Y	Apr. 16, 1867	63, 790
Bed-bottom	J. P. Flanders and S. K. Wells	Burlington, Vt	Oct. 29, 1867	70, 330
Bed-bottom	J. H. Fletcher	Lowell, Mass	May 22, 1866	54, 879
Bed-bottom	J. Flinn	Philadelphia, Pa	July 31, 1866	56, 740
Bed-bottom	H. A. and A. Follett	Smithfield, R. I	Aug. 6, 1867	67, 521
Bed-bottom	D. Frankfoder	Wakarusa, Ind	July 1, 1873	140, 492
Bed-bottom	D. Frankfoder and G. W. McGeorge.	Wakarusa, Ind	Nov. 19, 1872	133, 090
Bed-bottom	S. B. Freeman	Burlington, Iowa	Oct. 18, 1870	108, 470
Bed-bottom	S. B. Freeman	Fort Wayne, Ind	Nov. 18, 1873	144, 754
Bed-bottom	G. Frey	New York, N. Y	Dec. 8, 1863	40, 831
Bed-bottom	F. S. Frost	West Cambridge, Mass	July 9, 1867	66, 481
Bed-bottom	J. C. Fry	Sidney, Ohio	Feb. 25, 1868	74, 812
Bed-bottom	E. K. Garretson and W. Fulghum.	Winchester, Ind	May 6, 1873	138, 630
Bed-bottom	C. Gense	Fouilloy, France	Apr. 15, 1873	137, 909
Bed-bottom	G. L. Gerard	New Haven, Conn	Mar. 26, 1867	63, 153
Bed-bottom	G. L. Gerard	New Haven, Conn	Sept. 15, 1868	82, 218
Bed-bottom	E. and O. W. Gibbs	Richland Centre, Wis	June 9, 1868	78, 732
Bed-bottom	C. F. Gillette	Sparta, Wis	Apr. 2, 1867	63, 500
Bed-bottom	E. T. Gilmore	Westborough, Mass	May 14, 1872	126, 801
Bed-bottom	S. Gissinger	Pittsburgh, Pa	Oct. 3, 1871	119, 600
Bed-bottom	S. Gissinger	Pittsburgh, Pa	Mar. 5, 1872	124, 205
Bed-bottom	W. Graham	Sharpsburgh, Pa	Nov. 19, 1872	133, 093
Bed-bottom	J. S. Grant	Sidney Centre, Me	Sept. 3, 1867	68, 499
Bed-bottom	B. Gregg	Bennington, Vt	Aug. 11, 1868	80, 944
Bed-bottom	B. Gregg	Bennington, Vt	Oct. 4, 1870	108, 018
Bed-bottom	F. C. Gridley	Hudson, Wis	Jan. 28, 1868	73, 799
Bed-bottom	B. Griffin	Lawrence, Mass	June 8, 1858	20, 486
Bed-bottom	B. Griffin	Lawrence, Mass	Aug. 13, 1867	67, 647
Bed-bottom	F. C. Hagen	Cuba, N. Y	June 29, 1869	91, 931
Bed-bottom	H. J. Hale	Indianapolis, Ind	Oct. 6, 1868	82, 829
Bed-bottom	C. H. Hall	Binghamton, N. Y	May 21, 1867	64, 973
Bed-bottom	R. L. Hall	Lowell, Mass	Sept. 26, 1865	50, 118
Bed-bottom	R. L. Hall	Lowell, Mass	Oct. 2, 1866	58, 545
Bed-bsttom	W. M. Hamilton	Wenona, Ill	Sept. 29, 1868	82, 521
Bed-bottom	J. W. Hampton	Mount Pleasant, Iowa	Oct. 24, 1871	120, 271
Bed-bottom	J. W. Hampton	Mount Pleasant, Iowa	Sept. 17, 1872	131, 438
Bed-bottom	P. C. Hard	Wadsworth, Ohio	Feb. 25, 1873	136, 158

Index of patents issued from the United States Patent Office from 1790 *to* 1873, *inclusive*—Continued.

Invention.	Inventor.	Residence.	Date.	No.
Bed-bottom	C. H. Hardy	Charlestown, Mass	May 7, 1867	64, 528
Bed-bottom	J. Harper	Hillsborough, Iowa	Sept. 17, 1872	131, 397
Bed-bottom	J. J. Harris	Pewamo, Mich	Mar. 2, 1869	87, 489
Bed-bottom	H. P. Hart	New Woodstock, N. Y	Apr. 12, 1859	23, 573
Bed-bottom	W. J. Haswell	Waverly, N. Y	Dec. 11, 1866	60, 368
Bed-bottom	R. Hatch	Strafford, Vt	Aug. 30, 1859	25, 263
Bed-bottom	L. K. Hawes	Whitewater, Wis	Mar. 31, 1868	76, 077
Bed-bottom	A. R. Henderson and J. Ford	Andover, N. Y	May 22, 1866	54, 902
Bed-bottom	P. Hinkle	San Francisco, Cal	July 21, 1868	80, 176
Bed-bottom	A. Hitchcock	Wayne County, Mich	May 8, 1866	54, 545
Bed-bottom	B. Hitchcock	Waukegan, Ill	Apr. 14, 1868	76, 633
Bed-bottom	S. Hobbs	Wilmot, Ohio	Sept. 12, 1871	118, 804
Bed-bottom	E. C. Holden and E. L. Brockett	Owatonna, Minn	May 18, 1869	90, 171
Bed-bottom	E. F. Holloway and J. W. Hudelson.	Knightstown, Ind	Sept. 4, 1866	57, 720
Bed-bottom	B. Holmes	New York, N. Y	Feb. 18, 1873	135, 989
Bed-bottom	E. Hopkins	Newaygo, Mich	June 14, 1870	104, 310
Bed-bottom	D. Hoskins	Philadelphia, Pa	Sept. 13, 1870	107, 261
Bed-bottom	T. Howe	Cambridgeport, Mass	June 13, 1854	11, 054
Bed-bottom	Z. Howe	Lowell, Mich	Aug. 27, 1867	68, 078
Bed-bottom	E. Hoyt	Stamford, Conn	Nov. 25, 1873	144, 906
Bed-bottom	B. R. Hundley	Lynchburgh, Ohio	June 13, 1854	11, 066
Bed-bottom	M. A. Hunt	Cincinnati, Ohio	Apr. 29, 1873	138, 252
Bed-bottom	W. R. S. Hunter	Blackberry Station, Ill	June 16, 1868	78, 876
Bed-bottom	P. C. Ingersoll	Green Point, N. Y	Oct. 31, 1865	50, 713
Bed-bottom	W. B. Ingersoll	New York, N. Y	July 2, 1867	66, 350
Bed-bottom	H. L. Isham	Plattsburgh, N. Y	Oct. 1, 1867	69, 441
Bed-bottom	R. V. Jenks and W. A. Miller	Paterson, N. J	June 13, 1871	115, 863
Bed-bottom	S. C. Jennings	Wautoma, Wis	Aug. 6, 1867	67, 554
Bed-bottom	C. Johnson	Chicago, Ill	Nov. 7, 1871	120, 587
Bed-bottom	F. G. Johnson	Brooklyn, N. Y	Jan. 10, 1865	45, 832
Bed-bottom	W. Jones	Berlin, Wis	May 15, 1866	54, 737
Bed-bottom	B. D. Joslin and R. A. Newhall	North Brownville, Mich	May 12, 1868	77, 738
Bed-bottom	W. B. Judson	Poughkeepsie, N. Y	Jan. 10, 1871	110, 854
Bed-bottom	J. F. Keeler	Cleveland, Ohio	May 12, 1857	17, 282
Bed-bottom	C. A. Kellogg	Elyria, Ohio	July 24, 1866	56, 570
Bed-bottom	A. W. Kendrick	Xenia, Ohio	Mar. 24, 1868	75, 926
Bed-bottom	M. Kohn	Hartford, Conn	Nov. 21, 1871	121, 111
Bed-bottom	S. E. Lanphear and H. H. Blair	Brunswick, Ohio	June 19, 1866	55, 678
Bed-bottom	S. L. Leach	Corry, Pa	June 24, 1873	140, 284
Bed-bottom	F. Leadbetter	Plymouth, Mich	Nov. 5, 1867	70, 583
Bed-bottom	R. Leavitt	Cambridge, Mass	Aug. 24, 1858	21, 263
Bed-bottom	G. D. Leonard	Chicago, Ill	Oct. 10, 1871	119, 862
Bed-bottom	G. D. Leonard	Chicago, Ill	May 21, 1872	126, 970
Bed-bottom	C. P. Loeser	Hartford, Conn	Jan. 1, 1867	60, 913
Bed-bottom	R. O. Lowrey	Salem, N. Y	Mar. 16, 1869	87, 948
Bed-bottom	J. Lynn and J. R. Crowell	Seneca Falls, N. Y	May 8, 1866	54, 567
Bed-bottom	J. I. Mabbett	Titusville, Pa	Nov. 5, 1867	70, 448
Bed-bottom	J. I. Mabbett	Titusville, Pa	Apr. 14, 1868	76, 785
Bed-bottom	D. P. Mahan	Antioch, Cal	Dec. 17, 1872	134, 079
Bed-bottom	D. Manuel	Boston, Mass	Feb. 19, 1867	62, 213
Bed-bottom	D. Manuel	Boston, Mass	Mar. 19, 1867	63, 066
Bed-bottom	J. S. Martin	Mount Pleasant, Iowa	July 16, 1872	129, 152
Bed-bottom	A. Matson	Blockville, N. Y	June 29, 1869	92, 074
Bed-bottom	N. Maxey	Plymouth, Ind	Jan. 21, 1868	73, 622
Bed-bottom	A. and W. P. McBride	Lowell, Mich	Aug. 10, 1869	93, 632
Bed bottom	S. McDonald	Cincinnati, Ohio	Aug. 6, 1867	67, 442
Bed-bottom	A. D. McMaster	Rochester, N. Y	Dec. 5, 1871	121, 647
Bed-bottom	B. F. S. Monroe	Utica, N. Y	Jan. 25, 1859	22, 742
Bed-bottom	T. B. Moore	Bridesburgh, Pa	Mar. 12, 1867	62, 871
Bed-bottom	T. B. Moore and G. De Bow	Bridesburgh, Pa	June 9, 1868	78, 753
Bed-bottom	M. Morehouse	Boscobel, Wis	Mar. 29, 1870	101, 376
Bed-bottom	T. H. B. Morehouse	Lansing, Mich	Dec. 19, 1871	122, 048
Bed-bottom	A. W. Morse	Eaton, N. Y	Feb. 22, 1859	23, 064
Bed-bottom	L. Mudge	Springfield, Ohio	Nov. 26, 1867	71, 504
Bed-bottom	M. M. Murray	Cincinnati, Ohio	Mar. 19, 1872	124, 847
Bed-bottom	R. B. Nevens	Lowell, Mass	Sept. 10, 1867	68, 645
Bed-bottom	J. N. Newell	Adrian, Mich	May 17, 1870	103, 226
Bed-bottom	L. L. Newman	East Saginaw, Mich	Aug. 2, 1870	106, 074
Bed-bottom	J. M. Noble	Delhi, Iowa	May 25, 1858	20, 362
Bed-bottom	A. W. Obermann	Chicago, Ill	Aug. 27, 1872	130, 858
Bed-bottom	T. L. Odell and J. C. Hudson	Iowa Falls, Iowa	Feb. 11, 1873	135, 724
Bed-bottom	H. Ogborn and A. W. Kendrick	Richmond, Ind., and Brooklyn, N. Y.	Feb. 11, 1873	135, 725
Bed-bottom	H. Ogborn and A. W. Kendrick	Richmond, Ind., and Brooklyn, N. Y.	Feb. 11, 1873	135, 835
Bed-bottom	H. Ogborn and A. W. Kendrick	Richmond, Ind., and Brooklyn, N. Y.	Feb. 18, 1873	135, 933
Bed-bottom	H. Ogborn and A. W. Kendrick	Richmond, Ind., and Brooklyn, N. Y.	Mar. 4, 1873	136, 382
Bed-bottom	O. S. Osgood	Burlington, Iowa	Dec. 28, 1869	98, 401
Bed-bottom	O. S. Osgood	Burlington, Iowa	Aug. 30, 1870	106, 954
Bed-bottom	O. S. Osgood	Burlington, Iowa	May 9, 1871	114, 704
Bed-bottom	O. S. Osgood	Mount Pleasant, Iowa	Oct. 24, 1871	120, 315
Bed-bottom	H. H. Palmer	Rockford, Ill	Jan. 2, 1866	51, 856
Bed-bottom	H. H. Palmer	Rockford, Ill	Jan. 15, 1867	61, 242
Bed-bottom	B. Partello	Detroit, Mich	July 20, 1869	92, 878
Bed-bottom	J. D. Patton	Davenport, Iowa	Aug. 28, 1866	57, 560
Bed-bottom	E. M. Payne	Waverly, N. Y	Aug. 14, 1866	57, 181
Bed-bottom	T. Payne	Grand Rapids, Mich	June 15, 1869	91, 259
Bed-bottom	S. Pearson	Cincinnati, Ohio	Aug. 15, 1865	49, 434
Bed-bottom	C. Penfield	New Britain, Conn	Feb. 4, 1873	135, 580
Bed-bottom	S. E. Pettee	Bethlehem, Pa	Feb. 5, 1867	61, 757
Bed-bottom	J. Potter	Portland, Me	Feb. 4, 1868	73, 997
Bed-bottom	J. Potter	Portland, Me	Jan. 4, 1870	98, 520
Bed-bottom	J. H. Power	Burlington, Iowa	Jan. 23, 1872	123, 046
Bed-bottom	J. D. Pratt	Cleveland, Ohio	Nov. 2, 1869	96, 346
Bed-bottom	A. M. Pugh	Bucyrus, Ohio	Jan. 19, 1869	85, 960

Index of patents issued from the United States Patent Office from 1790 *to* 1873, *inclusive*—**Continued.**

Invention.	Inventor.	Residence.	Date.	No.
Bed-bottom	E. W. Quincy	Peoria, Ill	July 1, 1873	140, 387
Bed-bottom	R. Rakestraw	Wyoming, Ill	Feb. 11, 1873	135, 844
Bed-bottom	H. B. Ramsey	Rockville, Ind	July 2, 1872	128, 562
Bed-bottom	T. Raser	Genesee, Ill	Oct. 1, 1867	69, 366
Bed-bottom	C. D Read	Burlington, Vt	July 2, 1867	66, 390
Bed-bottom	C. D. Read	Burlington, Vt	May 18, 1869	90, 303
Bed-bottom	E. P. Read	Chicago, Ill	Dec. 12, 1871	121, 728
Bed-bottom	E. P. Read	Chicago, Ill	Dec. 12, 1871	121, 729
Bed-bottom	C. V. Reeder	San José, Cal	July 1, 1873	140, 540
Bed-bottom	J. Reynolds, jr	Cincinnati, Ohio	Aug. 20, 1872	130, 746
Bed-bottom	G. Richardson	Milwaukee, Wis	Aug. 23, 1870	106, 726
Bed-bottom	G. Richardson	Milwaukee, Wis	Sept. 5, 1871	118, 745
Bed-bottom	J. Rickard and J. Cook	Philadelphia, Pa	Mar. 19, 1867	62, 973
Bed-bottom	E. R. Rison	Kinmundy, Ill	Sept. 4, 1866	57, 770
Bed-bottom	G. W. Robbins	Fond du Lac, Wis	Aug. 6, 1867	67, 585
Bed-bottom	G. W. Robinson	Galesburgh, Ill	June 27, 1871	116, 354
Bed-bottom	W. W. Robinson	Ripon, Wis	Jan. 2, 1866	51, 866
Bed-bottom	O. A. A. Rouillion	New York, N. Y	Nov. 5, 1861	33, 685
Bed-bottom	J. Rouse	Dowagiac, Mich	May 15, 1866	54, 779
Bed-bottom	C. Ruprecht	Cleveland, Ohio	July 30, 1872	130, 077
Bed-bottom	F. Russell	Otselic, N. Y	Sept. 14, 1858	21, 519
Bed-bottom	R. S. Sanborn	Ripon, Wis	Sept. 26, 1865	50, 168
Bed-bottom	R. S. Sanborn	Ripon, Wis	Oct. 9, 1866	58, 679
Bed-bottom	C. H. Sawyer	Hollis, Me	Dec. 19, 1865	51, 624
Bed-bottom	F. Schimming	Philadelphia, Pa	Apr. 10, 1866	53, 888
Bed-bottom	A. Schlingman	Alexandria, Ohio	Jan. 7, 1868	73, 052
Bed-bottom	C. A. Scott	Aurora, Ill	Mar. 27, 1866	53, 492
Bed-bottom	D. A. Scott	Cincinnati, Ohio	Apr. 9, 1872	125, 490
Bed-bottom	D. A. Scott	Cincinnati, Ohio	Sept. 3, 1872	131, 122
Bed-bottom	D. A. Scott and J. E. Burdge	Cincinnati, Ohio	Nov. 19, 1867	71, 225
Bed-bottom	W. H. Sears	Cheshire, Conn	Sept. 23, 1873	143, 190
Bed-bottom	H. G. Seekins	Elyria, Ohio	July 10, 1866	56, 278
Bed-bottom	G. W. Seidler	Hartford, Conn	Oct. 31, 1865	50, 738
Bed-bottom	T. J. Sheears	Ypsilanti, Mich	July 15, 1873	140, 955
Bed-bottom	I. N. Sheets	Jefferson, Ohio	July 28, 1868	80, 312
Bed-bottom	J. H. Sherman	Galesburgh, Ill	Nov. 7, 1871	120, 599
Bed-bottom	C. E. Simmons and H. Cook	Waukegan, Ill	Dec. 3, 1867	71, 654
Bed-bottom	P. P. Simmons	Davenport, Iowa	Apr. 21, 1868	77, 113
Bed-bottom	W. W. Skaats	Cincinnati, Ohio	Sept. 23, 1873	143, 035
Bed-bottom	E. Small	Dennisport, Mass	July 30, 1867	67, 362
Bed-bottom	E. Smith, jr., and A. Chase	Worcester, Mass	Aug. 21, 1866	57, 395
Bed-bottom	H. T. Smith	Washington, D. C	Feb. 6, 1872	123, 520
Bed-bottom	H. T. Smith	Washington, D. C	May 28, 1872	127, 278
Bed-bottom	J. G. Smith	Battle Creek, Mich	Aug. 7, 1866	57, 003
Bed-bottom	J. J. Smith	Philadelphia, Pa	Oct. 13, 1868	83, 001
Bed-bottom	J. L. Smith	Lowell, Mass	Jan. 29, 1861	31, 264
Bed-bottom	N. Smith and E. S. Field	Hartford, Conn	Sept. 10, 1872	131, 187
Bed-bottom	R. A. Smith	East Weare, N. H	Dec. 5, 1871	121, 473
Bed-bottom	S. A. Smith	Monroe, Wis	Dec. 3, 1867	71, 804
Bed-bottom	W. C. Smith	Warrensburgh, Mo	Oct. 19, 1869	95, 945
Bed-bottom	S. L. Southard	Rock Island, Ill	Feb. 4, 1868	74, 162
Bed-bottom	C. B. Spencer	Hastings, Minn	Sept. 3, 1872	131, 034
Bed-bottom	C. B. Spencer	Hastings, Minn	Nov. 25, 1873	145, 023
Bed-bottom	J. Sperry	Charleston, Ill	Mar. 29, 1870	101, 326
Bed-bottom	D. Spore	Sharon, Wis	Nov. 6, 1866	59, 470
Bed-bottom	O. W. Stanford	Lebanon, Ohio	Jan. 12, 1869	85, 866
Bed-bottom	W. Stenger and A. Beyrnheimer.	Jefferson, Ohio	July 2, 1867	66, 410
Bed-bottom	W. Stickney	Lockport, N. Y	Aug. 6, 1867	67, 604
Bed-bottom	W. Stickney	Lockport, N. Y	Sept. 17, 1867	68, 912
Bed-bottom	W. H. Tambling	Berlin, Wis	Oct. 16, 1860	30, 432
Bed-bottom	W. H. Tambling	Mazo Manie, Wis	Apr. 14, 1868	76, 674
Bed-bottom	J. V. Taylor	La Cygne, Kans	July 15, 1873	140, 856
Bed-bottom	P. Taylor and W. A. Gove	Pawtucket, R. I., and Charlestown, Mass.	June 5, 1866	55, 390
Bed-bottom	D. M. Thomas	Dowagiac, Mich	Feb. 5, 1867	61, 775
Bed-bottom	L. Thomas and J. A. Kurtz	Pittsburgh, Pa	Jan. 23, 1872	123, 062
Bed-bottom	E. C. Thompson	New York, N. Y	Jan. 31, 1871	111, 402
Bed-bottom	T. P. Thompson	Charlestown, Mass	June 18, 1867	65, 364
Bed-bottom	L. B. Tinkham	Lawrence, Mass	Apr. 26, 1859	23, 813
Bed-bottom	J. Tinney	Westfield, N. Y	Apr. 4, 1871	113, 596
Bed-bottom	J. E. Todd	Middletown, Conn	Jan. 9, 1866	51, 981
Bed-bottom	A. M. Tomb	Lyons, N. Y	Jan. 2, 1866	51, 882
Bed-bottom	A. M. Tomb	Lyons, N. Y	June 25, 1867	66, 057
Bed-bottom	M. B. Towslee	Pewamo, Mich	June 20, 1871	116, 237
Bed-bottom	W. H. Trissler	Dunkirk, N. Y	Oct. 22, 1872	132, 422
Bed-bottom	C. B. Tucker	Cambridgeport, Mass	May 16, 1865	47, 757
Bed-bottom	C. B. Tucker and L. S. Babbitt.	Chicago, Ill	Aug. 20, 1867	67, 930
Bed-bottom	E. W. Tucker	Lowell, Mich	Apr. 3, 1866	53, 706
Bed-bottom	H. Tucker	Newton, Mass	Sept. 9, 1862	36, 429
Bed-bottom	A. Turnbull	New Britain, Conn	June 25, 1872	128, 438
Bed-bottom	A. Turnbull	New Britain, Conn	Apr. 1, 1873	137, 511
Bed-bottom	A. Turnbull and R. L. Webb	New Britain, Conn	Apr. 9, 1872	125, 421
Bed-bottom	P. Ulmer	Charlestown, Mass	Oct. 2, 1860	30, 267
Bed-bottom	C. Valkmar	New York, N. Y	Dec. 6, 1870	109, 852
Bed-bottom	C. Van Deusen	Clarksville, N. Y	Oct. 10, 1871	119, 727
Bed-bottom	B C. Vanduzen	Cincinnati, Ohio	Mar. 18, 1873	137, 039
Bed-bottom	W. W. Wait	Richmond, Ind	Mar. 8, 1870	100, 569
Bed-bottom	H. B. Walbridge	Toledo, Ohio	Mar. 10, 1868	75, 321
Bed-bottom	C. Walker	Chester, Vt	Apr. 28, 1868	77, 339
Bed-bottom	J. C. Walker and W. Lapish	Burlington, Iowa	Nov. 28, 1871	121, 437
Bed-bottom	C. D. Walkes	Elyria, Ohio	Dec. 14, 1869	97, 997
Bed-bottom	W. Weaver	Phœnixville, Pa	Feb. 5, 1867	61, 901
Bed-bottom	P. W. Webster and W. H. Prescott.	Concord, N. H	July 30, 1867	67, 238
Bed-bottom	C. Weed	Boston, Mass	May 21, 1867	65, 028
Bed-bottom	F. Wellhouse	Leavenworth, Kans	July 16, 1872	129, 074
Bed-bottom	A. S. Wells	New Britain, Conn	Apr. 1, 1873	137, 398

Index of patents issued from the United States Patent Office from 1790 *to* 1873, *inclusive*—Continued.

Invention.	Inventor.	Residence.	Date.	No.
Bed-bottom	A. West	Burlington, Iowa	Nov. 3, 1868	83, 809
Bed-bottom	C. W. White	Cincinnati, Ohio	June 18, 1867	65, 975
Bed-bottom	G. Widdicomb	Grand Rapids, Mich	Nov. 9, 1869	96, 750
Bed-bottom	N. O. Wilcox	Omaha, Nebr	May 13, 1873	138, 971
Bed-bottom	J. W. Wilder	New York, N. Y	Nov. 5, 1867	70, 663
Bed-bottom	J. D. Wilkinson	Plattsburgh, N. Y	Mar. 24, 1868	75, 820
Bed-bottom	D. S. Williams	Coldwater, Mich	Dec. 10, 1867	71, 933
Bed-bottom	H. W. Williams	Stowe, Vt	Nov. 5, 1867	70, 489
Bed-bottom	N. J. Willis	Waltham, Mass	Jan. 22, 1867	61, 374
Bed-bottom	J. E. Wilsey and D. Forbes	Chicago, Ill	Aug. 7, 1866	57, 031
Bed-bottom	L. Wilson	Ovid, N. Y	Nov. 26, 1867	71, 352
Bed-bottom	D. Winder	Cincinnati, Ohio	Oct. 11, 1859	25, 780
Bed-bottom	S. J. Wingate	Decatur, Ill	Jan. 7, 1868	73, 218
Bed-bottom	W. Workman and C. F. Swain	Ripon, Wis	Aug. 1, 1865	49, 189
Bed-bottom	E. Yeoman	Waukegan, Ill	Aug. 13, 1867	67, 698
Bed-bottom, Adjustable	E. Hoag	Coxsackie, N. Y	May 25, 1869	90, 449
Bed-bottom and chair spring	A. H. Knapp	Newton Centre, Mass	Jan. 28, 1868	73, 905
Bed-bottom and mattress spring fastener	J. S. Barnum	Topeka, Kans	Apr. 6, 1869	88, 602
Bed-bottom and seat	E. C. Cross	Lime Rock, Conn	Jan. 8, 1867	61, 001
Bed-bottom and sofa, Spring	W. D. Adams	Poughkeepsie, N. Y	Mar. 26, 1872	124, 927
Bed-bottom connection	F. Metz	Lyons, N. Y	Apr. 10, 1866	53, 853
Bed-bottom, Folding	F. C. Payne	New York, N. Y	Mar. 29, 1864	42, 108
Bed-bottom frame	J. W. C. Peters	Chicago, Ill	July 2, 1872	128, 504
Bed-bottom frame	J. E. Whittlesey and J. W. C. Peters.	Chicago, Ill	Oct. 15, 1872	132, 226
Bed bottom, Invalid	J. D. Brunner	Doylestown, Pa	Dec. 31, 1872	134, 460
Bed bottom, Invalid	C. A. Tousley	Battle Creek, Mich	Nov. 26, 1872	133, 502
Bed bottom, Invalid	C. A. Tousley	Battle Creek, Mich	Mar. 4, 1873	136, 562
Bed bottom, Invalid	S. M. Wilkes	Staunton, Va	Sept. 17, 1872	131, 488
Bed-bottom or cushion, Spring	H. E. Maker	Newton, (Upper Falls,) Mass.	Apr. 27, 1869	89, 321
Bed-bottom, Paper	J. B. Crane	Dalton, Mass	Jan. 17, 1871	110, 958
Bed-bottom, seat, &c., Spring	C. Rich	Poughkeepsie, N. Y	Feb 21, 1871	111, 974
Bed-bottom slat	J. N. Dennett	Bath, Me	Aug. 21, 1860	29, 679
Bed-bottom, sofa, and chair-seat, Spring	C. Rich	Poughkeepsie, N. Y	June 27, 1871	116, 529
Bed-bottom, Spring	E. Adler	New York, N. Y	Dec. 23, 1856	16, 310
Bed-bottom, Spring	J. H. Almond	Louisville, Ky	Jan. 26, 1869	86, 120
Bed-bottom, Spring	C. L. Ames and A. H. Frost	Chicago, Ill	Oct. 7, 1873	143, 397
Bed-bottom, Spring	S. B. Andrews	Chicago, Ill	Apr. 2, 1872	125, 259
Bed-bottom, Spring	C. F. Baade	Brooklyn, N. Y	Aug. 16, 1870	106, 316
Bed-bottom, Spring	H. D. W. Bailey	Sterling, Ill	Sept. 13, 1870	107, 213
Bed-bottom, Spring	J. Bailey and J. Decamp	Cincinnati, Ohio	Apr. 3, 1860	27, 680
Bed-bottom, Spring	F. P. Baldwin and C. T. Segar	Utica, N. Y	Oct. 10, 1871	119, 688
Bed-bottom, Spring	J. Barnes	New York, N. Y	Aug. 2, 1864	43, 662
Bed-bottom, Spring	J. Barnes	New York, N. Y	Feb. 20, 1866	52, 658
Bed-bottom, Spring	J. Barnes	New York, N. Y	July 17, 1866	56, 484
Bed-bottom, Spring	W. W. Bartlett	Portland, Me	June 21, 1870	104, 538
Bed-bottom, Spring	L. M. Bates	Cleveland, Ohio	Apr. 4, 1871	113, 385
Bed-bottom, Spring	A. B. Batey	Binghamton, N. Y	Dec. 10, 1867	71, 840
Bed-bottom, Spring	B. F. Bennett	Lockport, N. Y	Nov. 13, 1866	59, 546
Bed-bottom, Spring	E. R. Benton	Cleveland, Ohio	May 24, 1859	24, 091
Bed-bottom, Spring	C. H. Berry	East Somerville, Mass	Dec. 28, 1869	98, 339
Bed-bottom, Spring	G. Bevis	Rochester, N. Y	Jan. 26, 1864	41, 358
Bed-bottom, Spring	C. Bigeon	Cincinnati, Ohio	Jan. 16, 1872	122, 798
Bed-bottom, Spring	C. C. Bisbee	Rochester, N. Y	Aug. 2, 1859	24, 924
Bed-bottom, Spring	H. E. Bissell	Hartford, Conn	Sept. 14, 1869	94, 701
Bed-bottom, Spring	H. H. Blair and S. Barry	Brunswick, Ohio	May 6, 1873	133, 472
Bed-bottom, Spring	A. M. Blake	Canton, Ohio	Jan. 28, 1868	73, 770
Bed-bottom, Spring	O. Blake	Peru, Ind	May 31, 1870	103, 548
Bed-bottom, Spring	S. F. Bouton and N. P. Ames	Chicago, Ill	Feb. 25, 1868	74, 747
Bed-bottom, Spring	J. B. Bowditch	New Haven, Conn	Apr. 4, 1865	47, 086
Bed-bottom, Spring	L. R. Bradbury	Charlestown, Mass	Apr. 7, 1868	76, 296
Bed-bottom, Spring	C. B. Bristol	New Haven, Conn	June 12, 1866	55, 574
Bed-bottom, Spring	D. C. Bronson	Great Bend, Pa	May 24, 1870	103, 425
Bed-bottom, Spring	S. M. Brooks	Memphis, Tenn	Aug. 2, 1870	106, 028
Bed-bottom, Spring	S. M. Brooks	Burlington, Conn	Aug. 27, 1872	130, 788
Bed-bottom, Spring	G. A. Brown	Kalamazoo, Mich	Oct. 6, 1868	82, 687
Bed-bottom, Spring	G. W. Brown	Buffalo, N. Y	Sept. 16, 1873	142, 895
Bed-bottom, Spring	G. Brownlee	Princeton, Ind	Aug. 15, 1871	117, 976
Bed-bottom, Spring	G. Brownlee	Princeton, Ind	Feb. 27, 1872	124, 032
Bed-bottom, Spring	J. N. Bull	Springfield, Mass	Apr. 19, 1870	102, 089
Bed-bottom, Spring	G. E. Burt	Harvard, Mass	July 9, 1872	128, 704
Bed-bottom, Spring	I. Bush	Indianola, Iowa	Feb. 18, 1873	135, 885
Bed-bottom, Spring	M. Cahill	Kalamazoo, Mich	July 26, 1870	105, 636
Bed-bottom, Spring	J. B. Campbell	Cincinnati, Ohio	Mar. 1, 1870	100, 256
Bed-bottom, Spring	G. O. Capen	Providence, R. I	May 17, 1870	103, 138
Bed-bottom, Spring	A. H. Ceiley	Springfield, Mass	Jan. 25, 1870	99, 058
Bed-bottom, Spring	C. L. Chadeayne	Yonkers, N. Y	Oct. 5, 1869	95, 564
Bed-bottom, Spring	N. W. Clark	Clarkston, Mich	May 14, 1872	126, 675
Bed-bottom, Spring	A. Cole	Lockport, N. Y	July 17, 1866	56, 373
Bed-bottom, Spring	A. Cole	Mishawaka, Ind	Dec. 28, 1869	98, 230
Bed-bottom, Spring	A. Cole	Manamuskin, N. J	Mar. 7, 1871	112, 323
Bed-bottom, Spring	J. Collins	Grand Rapids, Mich	June 29, 1869	91, 913
Bed-bottom, Spring	W. N. Cook	Grand Rapids, Mich	Oct. 19, 1869	95, 880
Bed-bottom, Spring	W. Z. Cooke	New Britain, Conn	Oct. 1, 1872	131, 743
Bed-bottom, Spring	J. L. Cooper and E. A. Monroe	Elmira, N. Y	June 1, 1869	90, 820
Bed-bottom, Spring	J. Coover	Chambersburgh, Pa	Feb. 23, 1858	19, 410
Bed-bottom, Spring	D. V. Crandall	Chicago, Ill	July 11, 1871	116, 815
Bed-bottom, Spring	W. S. Crippen	Grand Rapids, Mich	Apr. 26, 1870	102, 231
Bed-bottom, Spring	A. C. Crondal	New York, N. Y	Feb. 9, 1864	41, 488
Bed-bottom, Spring	L. Cutler	Springfield, Mass	July 5, 1870	104, 938
Bed-bottom, Spring	J. and S. Danner	Canton, Ohio	June 14, 1864	43, 104
Bed-bottom, Spring	J. and S. Danner	Canton, Ohio	Sept. 11, 1866	57, 875
Bed-bottom, Spring	G. W. Dow	Lynn, Mass	June 16, 1857	17, 605
Bed-bottom, Spring	J. F. Duffy and W. P. Frailey	Chicago, Ill	May 17, 1870	103, 030
Bed-bottom, Spring	C. H. Dunks	Detroit, Mich	Feb. 5, 1870	99, 765
Bed-bottom, Spring	C. H. Dunks	New York, N. Y	Sept. 2, 1873	142, 377

Index of patents issued from the United States Patent Office from 1790 *to* 1873, *inclusive*—Continued.

Invention.	Inventor.	Residence.	Date.	No.
Bed-bottom, Spring	S. Dunlap	Rome, Ga	Mar. 4, 1873	136, 494
Bed-bottom, Spring	B. F. Ells	Dayton, Ohio	Mar. 1, 1870	100, 385
Bed-bottom, Spring	W. J. Emens	Louisville, Ky	Feb. 18, 1868	74, 675
Bed-bottom, Spring	B. C. English and F. Fraps	Springfield, Mass	Apr. 7, 1868	76, 425
Bed-bottom, Spring	E. E. Everitt	Philadelphia, Pa	Sept. 24, 1867	69, 198
Bed-bottom, Spring	M. Faloon	Bloomington, Ill	Apr. 5, 1870	101, 600
Bed-bottom, Spring	A. H. Fatzinger	New York, N. Y	Sept. 13, 1870	107, 348
Bed-bottom, Spring	J. A. Fegan	Brooklyn, N. Y	Mar. 2, 1869	87, 403
Bed-bottom, Spring	E. S. Field	Hartford, Conn	May 6, 1873	138, 490
Bed-bottom, Spring	E. S. Field	Hartford, Conn	May 6, 1873	138, 491
Bed-bottom, Spring	C. D. Flynt	New York, N. Y	Sept. 24, 1872	131, 675
Bed-bottom, Spring	E. Foster	Hartford, Conn	Apr. 13, 1858	19, 922
Bed-bottom, Spring	F. B. Franklin	Appleton, Wis	Mar. 4, 1862	34, 570
Bed-bottom, Spring	A. Frazee and L. W. Smith	Canandaigua, N. Y	July 31, 1866	56, 741
Bed-bottom, Spring	M. Freeman	Chicago, Ill	Aug. 20, 1872	130, 574
Bed-bottom, Spring	T. J. Gaffney and C. H. Dunks	Detroit, Mich	Oct. 27, 1868	83, 373
Bed-bottom, Spring	C. Gammel	Utica, N. Y	June 3, 1873	139, 567
Bed-bottom, Spring	J. C. Gaston	Cincinnati, Ohio	Feb. 16, 1869	86, 916
Bed-bottom, Spring	G. L. Gerard	New Haven, Conn	May 4, 1869	89, 651
Bed-bottom, Spring	J. E. Gillespie	Trenton, N. J	Dec. 20, 1864	45, 487
Bed-bottom, Spring	C. Glenn	Allegheny City, Pa	Dec. 6, 1870	109, 889
Bed-bottom, Spring	R. A. Goodyear	Binghamton, N. Y	June 22, 1869	91, 738
Bed-bottom, Spring	L. Granger	Memphis, Mich	June 16, 1868	78, 956
Bed-bottom, Spring	G. C. Grut	Milwaukee, Wis	Sept. 21, 1869	95, 105
Bed-bottom, Spring	G. C. Grut	Chicago, Ill	Oct. 3, 1871	119, 602
Bed-bottom, Spring	S. M. Guest	Chicago, Ill	Mar. 12, 1872	124, 436
Bed-bottom, Spring	C. Hacker	Euphemia, Ohio	Oct. 27, 1868	83, 489
Bed-bottom, Spring	H. Hard	Akron, Ohio	Nov. 21, 1871	121, 169
Bed-bottom, Spring	H. Hard	Akron, Ohio	July 30, 1872	130, 043
Bed-bottom, Spring	G. W. Hardy	Grand Rapids, Mich	Sept. 10, 1872	131, 165
Bed-bottom, Spring	P. J. Harvey	Chicago, Ill	Feb. 11, 1868	74, 353
Bed-bottom, Spring	W. W. Hawk	Detroit, Mich	Dec. 31, 1872	134, 430
Bed-bottom, Spring	W. Haworth	Canton, Ohio	June 1, 1869	90, 840
Bed-bottom, Spring	C. E. Hendrick	Chicopee, Mass	July 16, 1872	128, 960
Bed-bottom, Spring	E. Hennessey	Waterville, Me	Oct. 1, 1861	33, 417
Bed-bottom, Spring	H. A. Hight, sr	Fort Wayne, Ind	Dec. 30, 1873	146, 072
Bed-bottom, Spring	E. B. Hill	Grand Rapids, Mich	Feb. 7, 1871	111, 646
Bed-bottom, Spring	E. B. Hill and E. Van Valkenburg.	Grand Rapids, Mich	Sept. 2, 1873	142, 391
Bed-bottom, Spring	F. D. Hill	Poughkeepsie, N. Y	July 9, 1872	128, 867
Bed-bottom, Spring	P. Hill	Millport, N. Y	Sept. 12, 1871	118, 856
Bed-bottom, Spring	S. H. Hine	Sedan, Ind	Feb. 27, 1872	123, 994
Bed-bottom, Spring	C. and M. Hogeboom and L. Van Vleck.	Winslow, Ill	Aug. 31, 1869	94, 207
Bed-bottom, Spring	R. Holden	New York, N. Y	Jan. 3, 1860	26, 675
Bed-bottom, Spring	O. Howe	Cambridgeport, Mass	Nov. 19, 1861	33, 741
Bed-bottom, Spring	T. Howe	Cambridgeport, Mass	Mar. 1, 1870	100, 408
Bed-bottom, Spring	H. F. Howell	Mount Pleasant, Iowa	June 18, 1872	128, 043
Bed-bottom, Spring	W. C. Hubbard	Hubbardston, Mich	July 4, 1871	116, 712
Bed-bottom, Spring	L. Hull	Charlestown, Mass	May 21, 1872	127, 057
Bed-bottom, Spring	L. Hull	Charlestown, Mass	Sept. 17, 1872	131, 447
Bed-bottom, Spring	F. A. Huntington	San Francisco, Cal	Dec. 17, 1867	72, 202
Bed-bottom, Spring	H. L. Huntington	Chicago, Ill	Nov. 19, 1867	71, 014
Bed-bottom, Spring	J. Johnson	Hartford, Conn	July 9, 1872	128, 886
Bed-bottom, Spring	T. W. Johnston	Richmond, Me	Mar. 8, 1870	100, 635
Bed-bottom, Spring	J. S. Judson	Austin, Minn	Nov. 4, 1873	144, 338
Bed-bottom, Spring	W. B. Judson	Poughkeepsie, N. Y	Dec. 27, 1870	110, 471
Bed-bottom, Spring	J. B. Kelley and N. P. Kingsley.	Brandon, Vt	Oct. 13, 1868	82, 960
Bed-bottom, Spring	E. S. Kimball	Springfield, Mass	Dec. 21, 1869	98, 069
Bed-bottom, Spring	B. F. Kingman and M. V. B. Shepard.	Chicago, Ill	July 14, 1868	79, 839
Bed-bottom, Spring	S. P. Kittle	Brooklyn, N. Y	Nov. 21, 1865	51, 061
Bed-bottom, Spring	A. Kneppler	East New York, N. Y	July 27, 1869	93, 096
Bed-bottom, Spring	J. P. Knowles	Lockport, N. Y	July 1, 1862	35, 763
Bed-bottom, Spring	S. Landgraf	New York, N. Y	Oct. 29, 1872	132, 673
Bed-bottom, Spring	H. Lathrop	Utica, N. Y	Nov. 8, 1864	44, 962
Bed-bottom, Spring	G. Lightfoot	Elgin, Ill	Jan. 2, 1872	122, 325
Bed-bottom, Spring	T. Linfoot	Cincinnati, Ohio	Feb. 6, 1866	52, 426
Bed-bottom, Spring	S. Logan	Greenville, Pa	Jan. 9, 1872	122, 618
Bed-bottom, Spring	G. W. Loomis	Hartford, Conn	Mar. 22, 1870	101, 029
Bed-bottom, Spring	G. E. Lord	Utica, N. Y	Jan. 31, 1865	46, 119
Bed-bottom, Spring	J. E. Lord	Quincy, Ill	Feb. 28, 1871	112, 158
Bed-bottom, Spring	R. O. Lowrey	Saratoga Springs, N. Y	Feb. 11, 1868	74, 388
Bed-bottom, Spring	H. E. Maker	South Framingham, Mass	Nov. 5, 1872	132, 844
Bed-bottom, Spring	C. F. Manuel and D. M. Leonard	La Crosse, Wis	May 16, 1871	114, 355
Bed-bottom, Spring	D. Manuel	Boston, Mass	Jan. 26, 1869	86, 169
Bed-bottom, Spring	F. N. Marvin	Birmingham, Conn	Sept. 3, 1872	131, 107
Bed-bottom, Spring	W. D. Mason, C. H. Jacobus, and R. Millen.	New York, N. Y., and Newark, N. J.	Sept. 24, 1872	131, 621
Bed-bottom, Spring	F. B. Mattson	Rockford, Ill	Feb. 23, 1869	87, 184
Bed-bottom, Spring	E. W. Maxson	Scranton, Pa	Jan. 31, 1871	111, 362
Bed-bottom, Spring	W. McArthur	Philadelphia, Pa	June 7, 1870	104, 048
Bed-bottom, Spring	J. J. McCormick	Paterson, N. J	Apr. 30, 1861	30, 235
Bed-bottom, Spring	S. B. McCracken	Detroit, Mich	May 26, 1868	78, 223
Bed-bottom, Spring	W. C. McGill	Cincinnati, Ohio	Nov. 18, 1873	144, 688
Bed-bottom, Spring	R. W. McIntyre	Oregon, Wis	Nov. 5, 1872	132, 728
Bed-bottom, Spring	D. McMurchy	Jeffersonville, Ind	May 21, 1872	126, 977
Bed-bottom, Spring	W. H. Meriwether	New Braunfels, Tex	Aug. 8, 1854	11, 484
Bed-bottom, Spring	A. E. Miller	Arcadia, Ohio	Nov. 26, 1872	133, 329
Bed-bottom, Spring	J. Moore	Tarentum, Pa	Aug. 24, 1869	94, 020
Bed-bottom, Spring	G. W. Morrill	Sterling, Ill	July 20, 1869	92, 740
Bed-bottom, Spring	W. A. Morse and D. S. Bean	Boston, Mass.	June 19, 1860	28, 766
Bed-bottom, Spring	C. W. Mutell	Springfield, Mass	May 31, 1870	103, 646
Bed-bottom, Spring	A. W. Newell	Bradford, Pa	Jan. 28, 1868	73, 744
Bed-bottom, Spring	C. W. Northrup	Utica, N. Y	Sept. 30, 1873	143, 373
Bed-bottom, Spring	H Ogborn and A. W. Kendrick	Richmond, Ind., and Brooklyn, N. Y.	Jan. 23, 1872	122, 906
Bed-bottom, Spring	M. Ohmer	Dayton, Ohio	Apr. 27, 1869	89, 494

Index of patents issued from the United States Patent Office from 1790 *to* 1873, *inclusive*—Continued.

Invention.	Inventor.	Residence.	Date.	No.
Bed-bottom, Spring	B. J. Oleff	Milwaukee, Wis	Dec. 3, 1867	71,634
Bed-bottom, Spring	B. H. Otis	Havana, Ill	Jan. 9, 1872	122,645
Bed-bottom, Spring	W. Owen and S. Harter	Pierceton, Ill	Mar. 12, 1872	124,507
Bed-bottom, Spring	M. Le Page	Woodhaven, N. Y	Aug. 13, 1867	67,776
Bed-bottom, Spring	H. H. Palmer	Rockford, Ill	Nov. 24, 1868	84,300
Bed-bottom, Spring	E. Parker	Plymouth, Conn	Apr. 23, 1872	126,081
Bed-bottom, Spring	L. E. Payne	Disco, Mich	Feb. 9, 1864	41,533
Bed-bottom, Spring	J. F. Peck	Springfield, Mass	May 25, 1869	90,384
Bed-bottom, Spring	E. Perry	Hopkinton, Mass	Aug. 24, 1869	94,025
Bed-bottom, Spring	C. B. Pickett	Delaware, Ohio	Sept. 10, 1872	131,180
Bed-bottom, Spring	M. Pierce	Winona, Minn	Oct. 29, 1867	70,355
Bed-bottom, Spring	J. Pollitt	Edon, Ohio	Aug. 12, 1873	141,726
Bed-bottom, Spring	J. Potter	Portland, Me	July 27, 1869	93,118
Bed-bottom, Spring	J. H. Potts	Ottumwa, Iowa	Mar. 22, 1870	101,161
Bed-bottom, Spring	S. Puffer	Oxford, N. Y	July 9, 1872	128,905
Bed-bottom, Spring	J. H. Quackenbush	Luddington, Mich	Dec. 20, 1870	110,284
Bed-bottom, Spring	J. Ralston	Mansfield, Ohio	Nov. 12, 1872	132,982
Bed-bottom, Spring	B. Rear	Toronto, Canada	Jan. 28, 1868	73,755
Bed-bottom, Spring	S. H. Reeves	New York, N. Y	May 6, 1873	138,529
Bed-bottom, Spring	G. Reneky and S. Kiess	Edgerton, Ohio	Nov. 17, 1868	84,069
Bed-bottom, Spring	M. A. Richardson	Sherman, N. Y	May 14, 1872	126,743
Bed-bottom, Spring	G. B. Richmond	Charlestown, Mass	May 31, 1870	103,658
Bed-bottom, Spring	G. W. Robbins	Dubuque, Iowa	May 9, 1871	114,712
Bed-bottom, Spring	M. Roberts	Saint Paul, Minn	July 24, 1866	56,678
Bed-bottom, Spring	M. Rosenberg and H. Scheuerle	New York, N. Y	Nov. 13, 1860	30,635
Bed-bottom, Spring	H. Russell and M. S. Fuller	Nashville, Mich	Dec. 1, 1868	84,581
Bed-bottom, Spring	I. M. Russell	Lewiston, Me	Dec. 7, 1869	97,706
Bed-bottom, Spring	W. P. Sadler	Springfield, Ill	May 7, 1872	126,579
Bed-bottom, Spring	G. E. Safford	New York, N. Y	June 15, 1858	20,609
Bed-bottom, Spring	A. L. Sawyer and W. Baldwin	Detroit, Mich	May 26, 1868	78,397
Bed-bottom, Spring	G. Schott	New York, N. Y	Nov. 17, 1863	40,645
Bed-bottom, Spring	C. Schroder	New York, N. Y	Oct. 14, 1856	15,900
Bed-bottom, Spring	C. Schroder	New York, N Y	Feb. 23, 1858	19,473
Bed-bottom, Spring	D. A. Scott	Cincinnati, Ohio	July 6, 1869	92,375
Bed-bottom, Spring	D. A. Scott	Cincinnati, Ohio	June 21, 1870	104,504
Bed-bottom, Spring	D. A. Scott and J. E. Burdge	Cincinnati, Ohio	Feb. 4, 1868	74,153
Bed-bottom, Spring	W. S. Seaman	Milwaukee, Wis	Aug. 20, 1872	130,753
Bed-bottom, Spring	J. L. Secomb	Detroit, Mich	Feb. 15, 1870	99,790
Bed-bottom, Spring	J. L. Secomb	Chicago, Ill	June 18, 1872	128,175
Bed-bottom, Spring	J. L. Secomb	Chicago, Ill	Jan. 7, 1873	134,614
Bed-bottom, Spring	D. N. Sellig	New York, N. Y	Apr. 5, 1870	101,526
Bed-bottom, Spring	I. A. Sergeant	Springfield, Ohio	Apr. 26, 1859	23,786
Bed-bottom, Spring	M. Shepard	Chicago, Ill	July 26, 1870	105,853
Bed-bottom, Spring	E. G. Sherman	Forestville, N. Y	Apr. 29, 1873	138,348
Bed-bottom, Spring	P. Sisson	Brant, N. Y	Aug. 25, 1868	81,422
Bed-bottom, Spring	C. I. Skow	Racine, Wis	May 17, 1870	103,249
Bed-bottom, Spring	H. Smith and J. Potter	Groveton, N. H., and Portland, Me.	Feb. 21, 1871	112,085
Bed-bottom, Spring	H. T. Smith	Washington, D. C	Oct. 6, 1857	18,357
Bed-bottom, Spring	G. Speckner	Madison, Wis	May 9, 1871	114,618
Bed-bottom, Spring	W. D. Spencer and E. Parker	New Britain, Conn	Apr. 1, 1873	137,498
Bed-bottom, Spring	W. Starke	Chicago, Ill	Apr. 29, 1873	138,446
Bed-bottom, Spring	E. Steinel	Amsterdam, N. Y	Dec. 3, 1867	71,808
Bed-bottom, Spring	J. Stengel	Croton, Mich	Dec. 7, 1869	97,720
Bed-bottom, Spring	C. S. Stevens	Portland, Me	May 24, 1870	103,521
Bed-bottom, Spring	J. Stevens	Lowell, Mass	Apr. 23, 1861	32,165
Bed-bottom, Spring	P. Stovall	Newman, Ga	Oct. 25, 1870	108,650
Bed-bottom, Spring	R. Tattershall	Beloit, Wis	Mar. 10, 1868	75,490
Bed-bottom, Spring	R. Tattershall	Beloit, Wis	Apr. 19, 1870	102,182
Bed-bottom, Spring	J. C. Taylor	Ann Arbor, Mich	May 19, 1868	78,026
Bed-bottom, Spring	H. L. Thistle	New York, N. Y	Jan. 22, 1861	31,196
Bed-bottom, Spring	S. A. Thompson and C. T. Kendrick.	Manchester, Mich., and Brooklyn, N. Y.	Apr. 22, 1873	138,056
Bed-bottom, Spring	J. D. Tifft	Cuyahoga Falls, Ohio	Jan. 14, 1868	73,268
Bed-bottom, Spring	L. L. Tingley	Pawtucket, R. I	May 1, 1866	54,443
Bed-bottom, Spring	C. H. Triphagen	Pewamo, Mich	Aug. 8, 1871	117,830
Bed-bottom, Spring	H. Tucker	Cambridgeport, Mass	July 3, 1855	13,188
Bed-bottom, Spring	F. Tyler	Cleveland, Ohio	Aug. 9, 1859	25,063
Bed-bottom, Spring	S. E. Tyler	Beloit, Wis	July 7, 1868	79,705
Bed-bottom, Spring	P. Ulmer	New York, N. Y	Oct. 4, 1859	25,689
Bed-bottom, Spring	M. and L. Van Vleck	Monroe, Wis	Apr. 26, 1870	102,337
Bed-bottom, Spring	A. Walker	New Haven, Conn	Dec. 31, 1867	72,943
Bed-bottom, Spring	N. Warlick	Chambers Court-House, Ala	Nov. 16, 1858	22,098
Bed-bottom, Spring	D. P. Webster	Upper Gilmanton, N. H	Aug. 14, 1866	57,229
Bed-bottom, Spring	D. P. Webster and H. W. Ladd	Philadelphia, Pa	Apr. 23, 1867	64,175
Bed-bottom, Spring	D. P. Webster and H. W. Ladd	Upper Gilmanton, N. H., and Philadelphia, Pa.	Sept. 10, 1867	68,818
Bed-bottom, Spring	O. H. Weed	Charlestown, Mass	Aug. 27, 1867	68,268
Bed-bottom, Spring	W. Wells	Salem, Mass	Mar. 8, 1870	100,698
Bed-bottom, Spring	G. Widdicomb	Grand Rapids, Mich	Dec. 17, 1867	72,350
Bed-bottom, Spring	N. J. Willis	Boston, Mass	Feb. 27, 1866	52,937
Bed-bottom, Spring	N. J. Willis	Boston, Mass	June 23, 1868	79,285
Bed-bottom, Spring	J. E. Wilsey and D. Forbes	Chicago, Ill., and Scotland, Great Britain.	July 31, 1866	56,844
Bed-bottom, Spring	G. Wilson	Chicago, Ill	Oct. 3, 1871	119,487
Bed-bottom, Spring	S. Winslow	Charlestown, Mass	Jan. 19, 1869	85,980
Bed-bottom, Spring	W. H. Woodworth	Pewamo, Mich	Oct. 17, 1871	120,142
Bed-bottom, Spring	E. E. Worden and H. Wilms	Brandon, Vt	Feb. 25, 1868	74,872
Bed-bottom, Spring	A. W. Wright	Bunker Hill, Ill	Apr. 7, 1868	76,575
Bed-bottom, Spring	E. L. Wright	Sterling, Ill	June 20, 1871	116,250
Bed-bottom spring	W. Wright	New York, N. Y	Dec. 19, 1854	12,111
Bed-bottom spring	L. M. Bates	Newark, Ohio	Apr. 23, 1867	63,985
Bed-bottom spring	L. M. Bates	Jackson, Mich	Oct. 5, 1869	95,553
Bed-bottom spring	T. A. Carl	Nashville, Tenn	Nov. 8, 1870	108,966
Bed-bottom spring	J. P. Chamberlin	Abington, Mass	Dec. 21, 1869	98,146
Bed-bottom spring	H. R. Crampton	Lockport, N. Y	Sept. 18, 1860	30,049
Bed-bottom spring	W. C. T. Davidson	Hannibal, Mo	Feb. 6, 1872	123,335
Bed-bottom spring	M. W. Farber	Mount Pleasant, Iowa	May 23, 1871	115,043

Index of patents issued from the United States Patent Office from 1790 *to* 1873, *inclusive*—**Continued.**

Invention.	Inventor.	Residence.	Date.	No.
Bed-bottom spring	M. Flanigan	Detroit, Mich	Mar. 23, 1869	88,153
Bed-bottom spring	O. F. A. Faulkner	Mount Pleasant, Iowa	July 16, 1872	129,218
Bed-bottom spring	J. Fox	Albion, Mich	Nov. 9, 1869	96,686
Bed-bottom spring	J. P. Haskins	Saratoga Springs, N. Y	Aug. 20, 1872	130,579
Bed-bottom spring	H. Ingraham	Naples, N. Y	Nov. 22, 1870	109,413
Bed-bottom spring	J. Johnson	Northampton County, N. C	July 16, 1867	66,849
Bed-bottom spring	P. W. Kniskern	Fort Smith, Ark	Nov. 9, 1869	96,706
Bed-bottom spring	G. Koenig	Plymouth, Mich	Jan. 7, 1868	73,188
Bed-bottom spring	F. F. Lahm	Chicago, Ill	Nov. 2, 1869	96,446
Bed-bottom spring	J. M. Losie	Indianapolis, Ind	Sept. 8, 1868	82,012
Bed-bottom spring	A. McDaniel	Dubuque, Iowa	Sept. 8, 1868	81,923
Bed-bottom spring	E. D. Merriam and S. Aldrich	La Grange, Ohio	Aug. 27, 1867	68,098
Bed-bottom spring	W. H. Miller	Hannibal, Mo	Jan. 30, 1872	123,279
Bed-bottom spring	H. Ogborn and A. W. Kendrick	Richmond, Ind., and Brooklyn, N. Y.	Mar. 11, 1873	136,665
Bed-bottom spring	B. Partello	Detroit, Mich	Mar. 23, 1869	88,070
Bed-bottom spring	J. F. Peck	Springfield, Mass	Nov. 22, 1870	109,445
Bed-bottom spring	D. Punchies	Plymouth, Mich	Nov. 13, 1866	59,649
Bed-bottom spring	H. M. Scott	Portland, Me	Nov. 22, 1859	26,235
Bed-bottom spring	H. G. Seekins	Elyria, Ohio	Mar. 26, 1867	63,313
Bed-bottom spring	A. E. Thayer	Philadelphia, Pa	Apr. 25, 1871	114,225
Bed-bottom spring	J. A. Van Wert and A. Crooley	Elmira, N. Y	July 15, 1873	140,975
Bed-bottom, Spring-slat	J. S. Vanhorn and W. H. Pack	Jersey City, N. J	Aug. 18, 1868	81,120
Bed-bottom, Woven-wire	G. Richardson	Milwaukee, Wis	Aug. 29, 1871	118,644
Bed-bottoms, Foundation for spring	D. Manuel	Dedham, Mass	Nov. 30, 1869	97,306
Bed-bottoms, Means of securing springs to slats of	Z. S. Cracraft	Lacon, Ill	Nov. 17, 1868	84,092
Bed-bottoms, seats, &c., Connection of springs for	C. Rich	Poughkeepsie, N. Y	May 10, 1870	102,867
Bed-bottoms, Spring-fastener for	D. A. Scott	Cincinnati, Ohio	Apr. 22, 1873	138,203
Bed-bottoms, Spring-holder for	R. Lapham	New York, N. Y	Nov. 20, 1868	59,799
Bed-bottoms, Spring-slat for	J. M. French	East Cambridge, Mass	Dec. 12, 1865	51,433
Bed-brace, Spring	H. D. Bolt	Elmira, N. Y	May 18, 1869	90,229
Bed-bug destroyer	B. Garrand	Maryville, Tenn	Mar. 12, 1836	....
Bed-bug destroyer	R. Sealy	New York, N. Y	June 20, 1823	....
Bed-bug destroying compound	E. Hooper	Diamond Springs, Cal	Oct. 8, 1872	132,080
Bed-bug exterminator	C. L. Fewell	Corsicana, Tex	June 3, 1873	139,562
Bed-bug trap	J. A. Clark	Georgetown, D. C	June 13, 1829	....
Bed-bug trap	E. B. Lake	Bridgeport, N. J	Aug. 21, 1866	57,338
Bed-bug trap	W. Tapper	New York, N. Y	Dec. 12, 1865	51,493
Bed-bug trap for bedsteads	C. Legab	Pittsburgh, Pa	Apr. 15, 1873	137,936
Bed-bugs and other vermin, Compound for destroying.	P. Seebald	New York, N. Y	June 23, 1868	79,151
Bed-bugs by steam, Destroying	T. Miller	Newburgh, N. Y	Nov. 19, 1833	....
Bed-bugs by steam, Destroying	B. Overman	Greensborough, N. C	June 22, 1832	....
Bed, Camp	W. S. G. Baker	Baltimore, Md	Dec. 17, 1861	33,924
Bed, Camp	L. Hutchins	Norwich, Conn	Oct. 28, 1862	36,783
Bed, Canopy	I. E. Palmer	Montville, Conn	Jan. 31, 1860	27,007
Bed, chair, &c., Spring-bottom for	G. W. Griswold	Abington, Pa	Oct. 29, 1861	33,622
Bed, Child's	M. E. J. Marr	Jefferson, La	Apr. 7, 1868	76,336
Bed-clothes clamp	W. Fisher	Ripon, Wis	Dec. 11, 1866	60,351
Bed-clothes clamp	A. Storm	Rutland, Vt	Sept. 10, 1867	68,668
Bed-clothes clasp	F. A. Rockwell	Ridgefield, Conn	Feb. 12, 1850	7,088
Bed-clothes, Device for securing	G. Inwood	San Francisco, Cal	Mar. 30, 1869	88,298
Bed-clothes, Frame for supporting	G. A. McLane	Chicago, Ill	Mar. 5, 1872	124,213
Bed-clothes holder	M. D. Brooks	Albany, N. Y	Oct. 15, 1867	69,898
Bed-clothes holder	J. B. Munson	Bailey Hollow, Pa	June 9, 1868	78,756
Bed-clothes retainer	J. Brikenhead	Canton, Mass	June 20, 1871	116,013
Bed-clothes retainer, Child's	M. L. Thompson	Brooklyn, N. Y	Nov. 28, 1865	51,270
Bed, Conical spring	W. J. and A. E. Lyman	East Hampton, Mass	Sept. 18, 1835	....
Bed-cord	S. Albro	Buffalo, N. Y	July 3, 1860	28,950
Bed-cord tightener	A. B. Stroup	Waldron, Ind	Apr. 10, 1860	27,844
Bed, Cot	C. W. Irwin	Saint Louis, Mo	Oct. 22, 1861	33,527
Bed, Cot	L. B. Morse	Athol, Mass	Feb. 4, 1873	135,573
Bed-cushion and other upholstering	H. H. Barnes	Hoosick Falls, N. Y	Jan. 10, 1871	110,820
Bed, Elastic-spring	A. G. Hull	New York, N. Y	Oct. 1, 1831	....
Bed elevator, Invalid	D. S. Dunning	New York, N. Y	June 16, 1857	17,558
Bed, Inclosed folding spring	L. Derome	San Francisco, Cal	Jan. 18, 1870	98,932
Bed, Extension	F. Menzer	San Francisco, Cal	Apr. 5, 1870	101,489
Bed, Folding	F. C. Payne	New York, N. Y	Apr. 28, 1868	77,311
Bed, Folding	L. Whitehead, sr	Brooklyn, N. Y	Mar. 28, 1871	113,231
Bed, Folding	W. Wright	Bloomfield, N. J	Nov. 14, 1871	120,924
Bed, Folding box	A. G. Bayles	New York, N. Y	Nov. 19, 1872	133,137
Bed for ships and hospitals	B. F. McAlhatten	New York, N. Y	Mar. 4, 1862	34,583
Bed, Fracture	M. M. Latta	Goshen, Ind	Aug. 21, 1866	57,340
Bed-frame	S. Springer	Chicago, Ill	Mar. 29, 1870	101,327
Bed-frame and attachment	J. M. Farnham	Hartford, Conn	July 16, 1872	129,121
Bed frame, Military and civic	T. S. Lambert	Peekskill, N. Y	Apr. 15, 1862	34,971
Bed foundation	W. A. N. Long	Fisherville, N. H	Mar. 26, 1861	31,850
Bed, Hospital	H. Conway	Dayton, Ohio	Feb. 18, 1868	74,669
Bed, Hospital	E. Gray	Cuyahoga Falls, Ohio	Apr. 14, 1868	76,747
Bed, Hospital	I. Waller	Cleveland, Ohio	Mar. 5, 1872	124,398
Bed, Invalid	G. H. Clark	Pontiac, Mich	Sept. 15, 1857	18,189
Bed, Invalid	J. Crosby	Manchester, N. H	Aug. 23, 1864	43,900
Bed, Invalid	S. D. Hopkins	Staunton, Va	Aug. 24, 1852	9,215
Bed, Invalid	R. H. Mathews	Painesville, Ohio	Feb. 28, 1865	46,569
Bed, Invalid	J. N. Morrison	Philadelphia, Pa	Sept. 22, 1863	40,056
Bed-key, Ratchet	W. M. Gray	Brooklyn, N. Y	Dec. 10, 1867	71,871
Bed-lounge	A. F. Benten	San Francisco, Cal	Sept. 5, 1871	118,677
Bed-lounge	E. and F. Boese and A. Neuberger.	Chicago, Ill	Jan. 5, 1869	85,560
Bed-lounge	J. Brader	New York, N. Y	Mar. 28, 1871	113,012
Bed-lounge	H. S. Carter	Chicago, Ill	Mar. 1, 1870	100,257
Bed-lounge	H. Closterman	Cincinnati, Ohio	June 11, 1872	127,741
Bed-lounge	J. L. Cox	Manchester, N. H	Sept. 8, 1868	81,880
Bed-lounge	R. H. Cutter	Cleveland, Ohio	Jan. 29, 1867	61,610
Bed-lounge	G. Hartzell	Philadelphia, Pa	Apr. 8, 1873	137,677
Bed-lounge	G. Hartzell and J. P. Reifsneider.	Philadelphia, Pa	June 6, 1871	115,732
Bed-lounge	C. H. Hildreth	Chicago, Ill	June 11, 1872	127,693

Index of patents issued from the United States Patent Office from 1790 *to* 1873, *inclusive*—Continued.

Index fo patents issued from the United States Patent Office from 1790 *to* 1873, *inclusive*—Continued.

Invention.	Inventor.	Residence.	Date.	No.
Bed-spring attachment	A. Smith	Boston, Mass	Oct. 15, 1872	132, 328
Bed-spring fastening	D. Manuel	Boston, Mass	Mar. 5, 1867	62, 653
Bed-spring guide	A. C. Stich	Kalamazoo, Mich	Oct. 22, 1867	70, 041
Bed-springs, Device for forming the eyes of	N. B. White	South Dedham, Mass	Oct. 23, 1866	59, 145
Bed-springs, Device for forming spiral	E. L. Bushnell	Poughkeepsie, N. Y	Aug. 8, 1871	117, 860
Bed-springs, Machine for coiling	M. Van Pleck	Albany, N. Y	Mar. 21, 1871	112, 868
Bed ventilating, cooling, or warming device	D. E. Somes	Washington, D. C	Feb. 8, 1870	99, 722
Bed, Wardrobe	F. C. Payne and A. Reid	New York, N. Y	Apr. 12, 1859	23, 604
Beds, cushions, &c., Filling for	G. C. Barney	Chicago, Ill	June 2, 1868	78, 412
Beds, Feathers for	G. Schott	New York, N. Y	Sept. 13, 1864	44, 223
Beds, Head-rest for	B. F. Walton	Jersey City, N. J	Mar. 4, 1873	136, 395
Beds, Instrument for stirring straw and husk	C. A. Richardson	Waterville, Me	Apr. 1, 1856	14, 574
Beds, Manufacturing wood to be used as a substitute for curled hair in stuffing.	W. Baker	Utica, N. Y	May 30, 1842	2, 654
Beds, mattresses, &c., Filling for	J. C. Wightman	Newton, Mass	Dec. 2, 1873	145, 140
Beds, Means of ventilating, cooling, and warming.	D. E. Somes	Washington, D. C	Nov. 16, 1869	96, 989
Beds, Mode of warming	P. Lannay	Baltimore, Md	Nov. 11, 1823	....
Beds, Mosquito canopy for	W. B. Ellis	Boston, Mass	Feb. 23, 1869	87, 154
Beds or lounges, Converting the backs of car-seats into	H. B. Myer	Buffalo, N. Y	Sept. 19, 1854	11, 699
Beds, Palm-leaf for stuffing	J. C. Smith	Cambridgeport, Mass	Apr. 3, 1835	....
Beds, seats, &c., Expanding bottom for	E. K. Garretson	Ottawa, Ill	Aug. 25, 1868	81, 358
Beds, sofas, &c., Spring foundation for	J. H. Crane	Charlestown, Mass	Feb. 7, 1860	27, 037
Beds to tents, Attaching	E. Lynch	Buffalo, N. Y	Jan. 28, 1862	34, 258
Bedstead	W. R. Bacall	Boston, Mass	Aug. 6, 1867	67, 397
Bedstead	T. B. Baldwin	Marshall, Tex	Aug. 8, 1871	117, 724
Bedstead	J. H. Barth	Indianapolis, Ind	Jan. 10, 1854	10, 418
Bedstead	W. Bell	Lexington, Ky	Nov. 7, 1826	....
Bedstead	R. Belt	Centreville, Ky	Feb. 22, 1831	....
Bedstead	J. H. Belter	New York, N. Y	Aug. 19, 1856	15, 552
Bedstead	B. F. Berry	Utica, N. Y	Apr. 22, 1835	....
Bedstead	G. Beurer	Brooklyn, N. Y	Feb. 11, 1868	74, 193
Bedstead	W. H. Bramble	Springfield, Ohio	Apr. 19, 1859	23, 659
Bedstead	W. E. Briggs	Sparta, Wis	Aug. 13, 1872	130, 355
Bedstead	S. Caro	Washington, D. C	Dec. 31, 1872	134, 359
Bedstead	J. C. Clime	Philadelphia, Pa	Mar. 17, 1868	75, 527
Bedstead	J. C. Clime	Philadelphia, Pa	June 9, 1868	78, 647
Bedstead	G. G. Cochran	Brooklyn, N. Y	May 14, 1867	64, 632
Bedstead	E. Coleman	New Bedford, Mass	Oct. 31, 1834	....
Bedstead	N. Colver	Boston, Mass	Apr. 24, 1849	6, 386
Bedstead	I. Cooper	Baltimore, Md	Feb. 22, 1825	....
Bedstead	J. R. Coxe	Philadelphia, Pa	Oct. 16, 1812	....
Bedstead	H. B. Coyle	Philadelphia, Pa	May 28, 1872	127, 312
Bedstead	C. W. Curtiss	New Haven, Conn	Dec. 12, 1842	2, 881
Bedstead	S. Davis	Catskill, N. Y	Dec. 29, 1826	....
Bedstead	W. Deckman	Canton, Ohio	Aug. 21, 1860	29, 678
Bedstead	I. Deyo	Naples, N. Y	June 6, 1871	115, 716
Bedstead	A. P. Dorman	Galesburgh, Ill	Nov. 14, 1865	50, 920
Bedstead	D. E. Dugan	Springville, Pa	Dec. 7, 1869	97, 619
Bedstead	E. Eaton	Cincinnati, Ohio	Sept. 27, 1859	25, 563
Bedstead	I. Eaton	Mount Gilead, Va	Dec. 31, 1833	....
Bedstead	S. Espach	Cincinnati, Ohio	Dec. 27, 1859	26, 577
Bedstead	D. M. Estey	Brattleborough, Vt	Nov. 16, 1869	96, 787
Bedstead	H. H. Evarts	Chicago, Ill	Feb. 22, 1870	100, 133
Bedstead	F. G. Ford	Bridgeton, N. J	Mar. 25, 1873	137, 191
Bedstead	A. Foster	New York, N. Y	July 2, 1861	32, 687
Bedstead	W. Gambel	Baltimore, Md	Aug. 25, 1831	....
Bedstead	C. H. Gould	Concord, N. H	Sept. 2, 1856	15, 648
Bedstead	J. R. Guy	Springfield, Ohio	Oct. 11, 1859	25, 732
Bedstead	T. Q. Hall	Indianapolis, Ind	June 14, 1870	104, 143
Bedstead	J. Harding	Detroit, Mich	Sept. 28, 1869	95, 223
Bedstead	J. Hart	Lebanon, Ky	Mar. 3, 1840	1, 504
Bedstead	J. Henfrey	United States Army	Sept. 23, 1862	36, 519
Bedstead	F. Hesz	Cincinnati, Ohio	June 12, 1855	13, 034
Bedstead	P. Hinds	Kendall's Mill's, Me	June 16, 1857	17, 569
Bedstead	B. Hinkley	Utica, N. Y	July 5, 1848	5, 656
Bedstead	B. Hinkley	Troy, N. Y	Dec. 25, 1849	6, 972
Bedstead	B. Hinkley	Troy, N. Y	Dec. 11, 1855	13, 9[illegible]9
Bedstead	B. Hinkley	Troy, N. Y	July 26, 1859	24, 873
Bedstead	J. Hoert	New York, N. Y	May 28, 1867	65, 222
Bedstead	J. Horner	New Brunswick, N. J	July 30, 1867	67, 304
Bedstead	E. Howe, jr	Brooklyn, N. Y	Aug. 26, 1856	15, 609
Bedstead	T. Howe	Cambridgeport, Mass	July 17, 1855	13, 265
Bedstead	S. Huddleston	Cottage Grove, Ind	July 1, 1856	15, 235
Bedstead	B. Humphreville and J. King	Morristown, N. J	Dec, 17, 1834	....
Bedstead	W. Huntress	South Berwick, Me	June 10, 1856	15, 076
Bedstead	S. Hyde	Arcadia, N. Y	May 2, 1828	....
Bedstead	H. Ingraham	Naples, N. Y	Aug. 3, 1869	93, 305
Bedstead	R. Jenkins	Covington, Ky	June 5, 1860	28, 582
Bedstead	D. H. Jennings and J. Bounds	Bridgeport, Conn	June 22, 1869	91, 749
Bedstead	G. A. Jeremiah	New York, N. Y	Mar. 14, 1871	112, 716
Bedstead	W. B. Johns	United States Army	June 1, 1858	20, 435
Bedstead	L. J. Johnson	Cleveland, Ohio	July 25, 1871	117, 424
Bedstead	B. Judd	New London, Conn	May 25, 1832	....
Bedstead	C. Kniseley	Meadville, Pa	June 16, 1836	....
Bedstead	L. C. Kuhn		Aug. 10, 1791	....
Bedstead	M. Lally	East Palestine, Ohio	May 30, 1871	115, 485
Bedstead	N. Lamphear	Monmouth, Ill	June 29, 1858	20, 723
Bedstead	J. Leigh	Edgefield, S. C	Aug. 7, 1860	29, 498
Bedstead	J. Maguire	Philadelphia, Pa	Jan. 15, 1836	....
Bedstead	D. Manuel	Lancaster, Pa	Apr. 14, 1863	38, 202
Bedstead	R. Martin	Chicago, Ill	Feb. 23, 1869	87, 273
Bedstead	R. Maxwell	Tucker County, Va	Oct. 19, 1858	21, 841
Bedstead	J. McClintic	Chambersburgh, Pa	Nov. 2, 1825	....
Bedstead	S. McQuerns and B. M. Lyon	Abbeville, S. C	Mar. 8, 1859	23, 222
Bedstead	J. C. Merritt	West Point, N. Y	June 14, 1870	104, 182
Bedstead	C. Messenger	Warren, Ohio	July 26, 1859	24, 880
Bedstead	R. Miller	Glasgow, Ky	May 23, 1834	....
Bedstead	A. Moon	Bristol, Wis	July 1, 1856	15, 249

Index of patents issued from the United States Patent Office from 1790 *to* 1873, *inclusive*—Continued.

Invention.	Inventor.	Residence.	Date.	No.
Bedstead	E. Morris	Burlington, N. J	Dec. 9, 1873	145, 305
Bedstead	J. Morris	Derby, Conn	July 8, 1843	3, 162
Bedstead	C. F. Nahmmacher	Pittsburgh, Pa	June 13, 1871	115, 836
Bedstead	L. Newcomb, jr	New Bedford, Mass	Nov. 11, 1851	8, 516
Bedstead	S. Oberholzer	Wheaton, Ill	Apr. 30, 1872	126, 226
Bedstead	H. D. Olds	Cedar Rapids, Iowa	May 13, 1873	138, 756
Bedstead	D. O'Leary	Hubbard, Ohio	Aug. 20, 1872	130, 655
Bedstead	H. Pace, sr	Cincinnati, Ohio	Dec. 10, 1846	4, 880
Bedstead	I. Pedrick	Bridgeton, N. J	May 21, 1867	64, 901
Bedstead	J. E. Pencille	Lockport, N. Y	July 13, 1869	92, 643
Bedstead	C. Perley	New York, N. Y	Feb. 27, 1866	52, 879
Bedstead	D. Powles	Baltimore, Md	Oct. 31, 1821	....
Bedstead	D. U. Pratt	Cleveland, Ohio	Feb. 3, 1863	37, 589
Bedstead	P. Prettyman	Georgetown, Del	Mar. 31, 1834	....
Bedstead	P. Prettyman	Georgetown, Del	Sept. 26, 1835	....
Bedstead	B. S. Pringle	Barnesville, Ga	Sept. 4, 1860	29, 913
Bedstead	H. B. Ramsey	Rockville, Ind	Oct. 25, 1870	108, 729
Bedstead	D. C. Raub	Saint Louis, Mo	May 6, 1873	138, 582
Bedstead	H. Reisinger	York, Pa	Aug. 18, 1832	....
Bedstead	L. W. Roath	Lexington, Ohio	Aug. 4, 1868	80, 668
Bedstead	R. Roberts	Saint Paul, Ind	Nov. 5, 1867	70, 618
Bedstead	E. Rogers and M. Pearson	Essex County, Mass	June 11, 1829	....
Bedstead	A. Rothwell	Washington, D. C	July 2, 1867	66, 398
Bedstead	I. Russell	Dedham, Mass	Sept. 16, 1851	8, 358
Bedstead	H. W. Sabin	Canandaigua, N. Y	July 1, 1851	8, 190
Bedstead	D. W. Smead	Peru, Ill	Oct. 19, 1852	9, 347
Bedstead	J. J. Smith and J. H. Pugh	Philadelphia, Pa	Aug. 26, 1856	15, 621
Bedstead	G. Snowden	Hudson, N. J	Feb. 14, 1871	111, 785
Bedstead	N. W. Speers	Cincinnati, Ohio	Apr. 27, 1858	20, 097
Bedstead	S. Springer	Chicago, Ill	Dec. 12, 1871	121, 734
Bedstead	A. Stark	Topeka, Kans	Aug. 22, 1865	49, 568
Bedstead	W. St. Charles	Fairmont, Va	June 8, 1858	20, 518
Bedstead	W. B. Stewart	Brooklyn, N. Y	Aug. 27, 1867	68, 255
Bedstead	W. L. Thomas	Middlebury, Ohio	Apr. 7, 1868	76, 553
Bedstead	P. Thompson	Springfield, Ohio	Apr. 5, 1859	23, 513
Bedstead	W. S. Todd	Mechanicsville, Iowa	Sept. 14, 1858	21, 527
Bedstead	C. Van Noy	Lexington, Ky	Jan. 10, 1832	....
Bedstead	G. Vaughan, jr	Philadelphia, Pa	Sept. 18, 1833	....
Bedstead	S. Walker	Boston, Mass	Mar. 20, 1860	27, 585
Bedstead	T. Wall	Jones' Station, Ohio	June 5, 1860	28, 621
Bedstead	W. M. Ward and P. Beunage	Eureka, Ill	Sept. 22, 1868	82, 457
Bedstead	C. A. Warner	Bristol, Conn	June 29, 1858	20, 750
Bedstead	W. White	Portsmouth, Va	Sept. 25, 1855	13, 607
Bedstead	H. K. Whitner	Philadelphia, Pa	Mar. 9, 1869	87, 742
Bedstead	S. Willard	Cincinnati, Ohio	Aug. 1, 1854	11, 449
Bedstead	P. Williamson	Baltimore, Md	Dec. 6, 1821	....
Bedstead	P. Williamson	Baltimore, Md	Oct. 17, 1822	....
Bedstead	D. Witt	Hubbardston, Mass	Feb. 9, 1869	86, 892
Bedstead	B. F. Woodside	Atlanta, Ga	Aug. 20, 1867	68, 023
Bedstead	W. Zaiser	Cincinnati, Ohio	Nov. 19, 1850	7, 791
Bedstead	J. P. Zeller	South Bend, Ind	Apr. 9, 1872	125, 653
Bedstead, Adjustable	W. O. Reid	Vienna, N. C	Mar. 1, 1870	100, 324
Bedstead, Adjustable bolster for	J. Hofmann	Chicago, Ill	May 16, 1871	114, 819
Bedstead, Alarm	J. C. House	Lowville, N. Y	July 17, 1855	13, 263
Bedstead and bed-bottom	F. G. Ford	Baltimore, Md	Feb. 14, 1871	111, 832
Bedstead and bed-bottom	E. Krieghoff	Rochester, N. Y	July 9, 1867	66, 598
Bedstead and bed for the sick and lame	J. Baily	Philadelphia, Pa	Mar. 18, 1825	....
Bedstead and cot	V. L. Baker	Richmond, Va	Aug. 2, 1870	105, 885
Bedstead and crib, Folding	R. S. Titcomb	Gloversville, N. Y	Dec. 10, 1867	72, 122
Bedstead and crib, Folding	R. S. Titcomb	Gloversville, N. Y	Apr. 28, 1868	77, 334
Bedstead and lounge	J. Dourson	Columbus, Ohio	Dec. 28, 1869	98, 359
Bedstead and lounge	R. Jewell	New York, N. Y	Apr. 12, 1870	101, 882
Bedstead and mattress	E. Cherrington	Boston, Mass	Nov. 23, 1835	....
Bedstead and quilting-frame	J. Park	Joliet, Ill	Sept. 8, 1868	81, 936
Bedstead and sacking-bottom, Portable	S. Whitmarsh	Northampton, Mass	Oct. 8, 1850	7, 712
Bedstead and sofa or chair	J. S. McCurdy	New York, N. Y	June 29, 1869	92, 075
Bedstead and spring bed-bottom	W. McArthur	Philadelphia, Pa	Jan. 31, 1871	111, 364
Bedstead and table	G. Wilson	Chicago, Ill	Sept. 20, 1870	107, 580
Bedstead and table combined	O. Lafreniere	New York, N. Y	Sept. 15, 1863	39, 933
Bedstead and table fastener	J. M. Baird	Wheeling, W. Va	May 3, 1870	102, 473
Bedstead and trunk combined	F. Boissard and S. Conrath	New York, N. Y	Aug. 27, 1861	33, 128
Bedstead and wardrobe combined	J. A. Morgan	Bloomfield, Iowa	Mar. 12, 1872	124, 609
Bedstead, Barrack or hospital	C. S. Snead	Louisville, Ky	Dec. 14, 1869	97, 981
Bedstead, Bisected	P. Breasted	Green County, N. Y	Feb. 21, 1826	....
Bedstead, Book-case	E. Putnam	Boston, Mass	Sept. 1, 1868	81, 687
Bedstead-bottom	S. E. Hartwell	New York, N. Y	Oct. 26, 1858	21, 878
Bedstead-bottom	G. Schott	New York, N. Y	Dec. 26, 1865	51, 755
Bedstead-bottom	J. C. Stuck	Boston, Mass	Aug. 2, 1864	43, 719
Bedstead-bottom, Spring	G. Schott and J. Loudon	New York, N. Y	June 7, 1859	24, 333
Bedstead, Bureau	D. Arnaud	Boston, Mass	Dec. 1, 1868	84, 466
Bedstead, Bureau	W. F. Browne	New York, N. Y	Aug. 27, 1872	130, 789
Bedstead, Bureau	L. Derome	San Francisco, Cal	Mar. 31, 1863	38, 035
Bedstead, Bureau	F. Hoffman	New York, N. Y	Oct. 26, 1858	21, 926
Bedstead, Bureau	H. W. Kingman	New York, N. Y	Oct. 12, 1844	3, 792
Bedstead, Bureau	A. Parker	New York, N. Y	July 21, 1868	80, 208
Bedstead, Bureau	F. C. Payne	New York, N. Y	Nov. 1, 1864	44, 887
Bedstead, Bureau	E. Shackford and D. Arnaud	Boston, Mass	Apr. 14, 1868	76, 664
Bedstead, Bureau	M. Sulzbacher	New York, N. Y	Feb. 18, 1868	74, 628
Bedstead, Bureau	H. L. Thistle	New York, N. Y	May 10, 1859	23, 960
Bedstead, Bureau	D. Trefry	Boston, Mass	June 22, 1869	91, 688
Bedstead, Bureau	E. Whitney	Boston, Mass	July 14, 1857	17, 818
Bedstead, bureau, and stand combined	A. and L. A. Parker	Girard, Kans	Apr. 4, 1871	113, 692
Bedstead, Cabinet	S. S. Burr	Dedham, Mass	Oct. 25, 1869	96, 196
Bedstead, Cabinet	S. S. Burr	Boston, Mass	Sept. 26, 1871	119, 312
Bedstead, Cabinet	W. H. Dutton	Philadelphia, Pa	Apr. 7, 1868	76, 423
Bedstead, Cabinet	M. Crosby	Boston, Mass	June 18, 1872	127, 962
Bedstead, Cabinet	S. C. Maine	Boston, Mass	June 15, 1869	91, 244
Bedstead, Cabinet	A. J. Roberts	Boston, Mass	July 26, 1870	105, 850

Index of patents issued from the United States Patent Office from 1790 *to* 1873, *inclusive*—Continued.

Invention.	Inventor.	Residence.	Date.	No.
Bedstead, Cabinet	D. T. Robinson	Boston, Mass	Aug. 10, 1869	93, 556
Bedstead, Cabinet	W. H. and L. Young	Boston, Mass	July 7, 1868	79, 622
Bedstead, Camp	E. J. Everickx	Paris, France	Mar. 1, 1870	100, 386
Bedstead, Camp	C. Festensen, H. Iverson, and C. J. Skow.	Racine, Wis	July 11, 1865	48, 671
Bedstead, Camp	W. A. Mauran	Boston, Mass	Aug. 20, 1861	33, 124
Bedstead, Camp	A. D. McCoy	Alexandria, La	Dec. 27, 1870	110, 485
Bedstead, Camp	A. Polino	Paris, France	Dec. 10, 1861	33, 901
Bedstead, Camp	W. C. Shaw and J. Stalcup	Philadelphia, Pa., and Wilmington, Del.	Sept. 10, 1850	7, 633
Bedstead, Camp	J. White	Cleveland, Ohio	Nov. 12, 1861	33, 724
Bedstead, Canopy	A. M. Rodgers	Brooklyn, N. Y	Mar. 22, 1870	101, 165
Bedstead, chair, secretary, and wardrobe	W. Reckards	New York, N. Y	May 12, 1868	77, 841
Bedstead-clamp	T. B. Gregory	Champaign, Ill	Sept. 27, 1870	107, 774
Bedstead combined with other furniture, Wardrobe	H. R. and J. L. Plimpton	Hampton County, Mass	Mar. 25, 1856	14, 514
Bedstead, Cot	S. Clark	New York, N. Y	Mar. 23, 1836	....
Bedstead, Cot	P. Williamson	Baltimore, Md	Nov. 11, 1830	....
Bedstead cord-pin	J. T. Bever	Hainesville, Mo	July 19, 1859	24, 788
Bedstead crib-attachment	E. F. Bryan	Savannah, Ga	Jan. 14, 1873	134, 847
Bedstead crib-attachment	F. Ryder	Great Falls, N. H	Aug. 13, 1872	130, 443
Bedstead crib-attachment	B. Snyder	Philadelphia, Pa	June 10, 1873	139, 829
Bedstead crib-attachment	M. W. Tarbell and J. H. P. Inslee.	Jersey City, N. J	Feb. 25, 1873	136, 280
Bedstead crib-attachment	H. R. Tracy	New York, N. Y	Feb. 25, 1868	74, 865
Bedstead crib-attachment	J. H. L. Wilson	Auburn, Kans	Apr. 4, 1871	113, 382
Bedstead, Double-tenon draw-key	S. Cooper	Chambersburgh, Pa	Nov. 5, 1825	....
Bedstead drapery fastening	G. W. Watrous	Hartford, Conn	July 19, 1859	24, 837
Bedstead drapery fastener or suspender	R. B. Pullan	Cincinnati, Ohio	June 18, 1861	32, 585
Bedstead, Drum-sacking	H. Wilbur	Newburyport, Mass	June 16, 1828	....
Bedstead, Easy-draft	P. T. Walcott	Monmouth County, N. J	Jan. 6, 1812	....
Bedstead, Elevated	D. Burnett	Bedford Station, N. Y	Aug. 6, 1867	67, 494
Bedstead, Expanding	N. Cross	New York, N. Y	Dec. 16, 1862	37, 153
Bedstead, Extension	C. S. Debow	New York, N. Y	Apr. 18, 1846	4, 463
Bedstead, Extension	J. Holzman	New York, N. Y	July 23, 1867	67, 055
Bedstead, Extension	A. Iske	Lancaster, Pa	Nov. 10, 1863	40, 564
Bedstead, Extension	F. Menzer	San Francisco, Cal	Feb. 1, 1870	99, 455
Bedstead, Fan ventilating	F. Moore	Panola, Miss	Oct. 23, 1860	30, 492
Bedstead-fastener	D. Babcock	Seneca Falls, N. Y	Feb. 3, 1863	37, 560
Bedstead-fastener	J. C. Cline	Philadelphia, Pa	May 5, 1868	77, 457
Bedstead-fastener	J. Doering	Philadelphia, Pa	Apr. 6, 1869	88, 617
Bedstead-fastener	E. S. Earley	Philadelphia, Pa	Apr. 20, 1869	89, 209
Bedstead-fastener	J. Janeway	Indianapolis, Ind	Aug. 4, 1868	80, 743
Bedstead-fastener	A. B. Sheaffer	Ephrata, Pa	May 24, 1870	103, 511
Bedstead-fastener	H. Swincford	Mifflinburgh, Pa	Jan. 18, 1870	99, 030
Bedstead-fastener	H. S. Wing	Plattsburgh, N. Y	May 18, 1869	90, 329
Bedstead-fastening	J. W. Adams	Boston, Mass	Sept. 10, 1846	4, 748
Bedstead-fastening	E. H. Badger	Petersburgh, Va	Apr. 3, 1826	....
Bedstead-fastening	J. M. Baird	Wheeling, W. Va	Oct. 7, 1873	143, 431
Bedstead-fastening	G. W. Baker	Cochranton, Pa	May 24, 1859	24, 085
Bedstead-fastening	D. Ball	Sandy Hill, N. Y	Apr. 16, 1842	2, 562
Bedstead-fastening	D. Ball	Albany, N. Y	Apr. 10, 1847	5, 060
Bedstead-fastening	E. R. Ball	Kalamazoo, Mich	Apr. 4, 1854	10, 731
Bedstead-fastening	E. T. Barlow	San Francisco, Cal	Aug. 29, 1871	118, 424
Bedstead-fastening	J. Barnes	Burlington, Vt	Apr. 20, 1869	89, 116
Bedstead-fastening	W. S. Bartle	Newark, N. Y	Jan. 29, 1867	61, 504
Bedstead-fastening	C. E. Bander	Cleveland, Ohio	Jan. 18, 1853	9, 541
Bedstead-fastening	G. W. Baynes, T. Hinty, and M. Jackson.	Glenville, Va	Aug. 16, 1853	9, 931
Bedstead-fastening	A. Bechtol	Berkeley Springs, Va	Oct. 9, 1860	30, 286
Bedstead-fastening	W. Bell	Lexington, Ky	Feb. 15, 1838	606
Bedstead-fastening	J. Benjamen	Naples, N. Y	July 26, 1870	105, 771
Bedstead-fastening	J. Bodefer	Cincinnati, Ohio	Apr. 10, 1855	12, 693
Bedstead-fastening	L. G. Bradford	Plymouth, Mass	Sept. 3, 1872	130, 973
Bedstead-fastening	L. G. Bradford	Plymouth, Mass	Feb. 25, 1873	136, 301
Bedstead-fastening	C. Bradway	Maquoketa, Iowa	Nov. 30, 1869	97, 270
Bedstead-fastening	H. Branch	New York, N. Y	Apr. 18, 1832	....
Bedstead-fastening	J. Brooke	Baltimore, Ohio	May 15, 1849	6, 451
Bedstead-fastening	G. Burket	Croghan, Ohio	June 8, 1858	20, 478
Bedstead-fastening	L. W. Buxton	Nashua, N. H	Apr. 5, 1859	23, 525
Bedstead-fastening	W. H. Carter	Candor, N. Y	Sept. 6, 1870	106, 997
Bedstead-fastening	A. N. and A. Case	Gustavus, Ohio	Mar. 1, 1853	9, 598
Bedstead-fastening	W. Clark	Weymouth, Ohio	Mar. 9, 1858	19, 544
Bedstead-fastening	C. S. Comins	Lowell, Mass	Jan. 3, 1871	110 632
Bedstead-fastening	C. C. Coolidge	Boston, Mass	May 7, 1850	7, 339
Bedstead-fastening	A. S. Dalbey	Richmond, Ind	Feb. 16, 1869	86, 909
Bedstead-fastening	J. Doering	Philadelphia, Pa	Sept. 13, 1864	44, 167
Bedstead-fastening	J. Drayton	Buffalo, N. Y	Sept. 26, 1854	11, 722
Bedstead-fastening	M. Elder	Mansfield, Ohio	Jan. 22, 1850	7, 035
Bedstead-fastening	W. H. Elliot	New York, N. Y	Nov. 20, 1866	59, 794
Bedstead-fastening	W. H. Elliot	New York, N. Y	Nov. 20, 1866	59, 795
Bedstead-fastening	W. H. Elliot	New York, N. Y	Nov. 20, 1866	59, 825
Bedstead fastening	W. H. Elliot	New York, N. Y	Aug. 3, 1869	93, 188
Bedstead-fastening	J. W. Elstun	Indianapolis, Ind	May 28, 1872	127, 226
Bedstead-fastening	H. C. Ernst	Vandalia, Ill	Feb. 23, 1841	1, 991
Bedstead-fastening	B. Essig	Sacramento, Cal	Mar. 10, 1868	75, 254
Bedstead-fastening	E. E. Everitt	Philadelphia, Pa	Mar. 29, 1859	23, 359
Bedstead-fastening	S. Fahs and A. H. Lochman	York, Pa	Oct. 30, 1847	5, 342
Bedstead-fastening	C. H. Fessenden	Candor, N. Y	Mar. 22, 1870	100. 993
Bedstead-fastening	J. R. Finley and L. Martin	Delphi, Ind	Apr. 16, 1872	125, 801
Bedstead-fastening	J. Fowler	Pittsburgh, Pa	July 11, 1842	2, 715
Bedstead-fastening	S. A. Frayer	Coxsackie, N. Y	June 6, 1871	115, 725
Bedstead-fastening	P. H. Frelinghousen	Jonestown, Pa	Feb. 19, 1867	62, 262
Bedstead-fastening	E. I. Gazzan	Pittsburgh, Pa	Jan. 23, 1846	4, 360
Bedstead fastening	C. M. Gilbert	Philadelphia, Pa	Apr. 17, 1866	53, 972
Bedstead-fastening	C. M. Gilbert	Philadelphia, Pa	Sept. 22, 1868	82, 305
Bedstead-fastening	C. M. Gilbert	Philadelphia, Pa	May 14, 1872	126, 630
Bedstead-fastening	E. F. Gilbert	Lyons, N. Y	Oct. 18, 1870	108, 472
Bedstead-fastening	S. Goetz	Reed's Mills, Ohio	Feb. 7, 1871	111, 528

Index of patents issued from the United States Patent Office from 1790 *to* 1873, *inclusive*—Continued.

Invention.	Inventor.	Residence.	Date.	No.
Bedstead-fastening	E. G. Gory	Cincinnati, Ohio	Mar. 19, 1872	124, 735
Bedstead-fastening	A. Grillet	Philadelphia, Pa	July 26, 1870	105, 798
Bedstead-fastening	H. Gross	Hoboken, N. J	Jan. 2, 1872	122, 457
Bedstead-fastening	J. Guild	Cincinnati, Ohio	Dec. 20, 1843	3, 387
Bedstead-fastening	S. Harris	Philadelphia, Pa	Dec. 23, 1856	16, 276
Bedstead-fastening	J. Haslet and C. Devitt	Irville, Ohio	May 9, 1846	4, 508
Bedstead-fastening	J. C. Helme	Wilkesbarre, Pa	June 27, 1848	5, 649
Bedstead-fastening	B. Hinkley	Utica, N. Y	Dec. 9, 1846	4, 878
Bedstead-fastening	W. Hinman	Elkhart, Ind	Apr. 15, 1856	14, 660
Bedstead-fastening	E. G. Hopkins	Pen Yan, N. Y	Nov. 15, 1859	26, 104
Bedstead-fastening	S. Hovey	Painesville, Ohio	Aug. 28, 1849	6, 674
Bedstead-fastening	R. Hubbard	Milton, Ind	May 15, 1860	28, 276
Bedstead-fastening	J. C. and F. J. Jackson	Rochester, N. Y., and Danbury, Conn.	May 5, 1868	77, 491
Bedstead-fastening	L. L. Jackson	Paterson, N. J	Aug. 13, 1867	67, 766
Bedstead-fastening	J. C. Jeffries	Mount Vernon, Ind	June 28, 1859	24, 562
Bedstead-fastening	J. Johnson	Geneseo, N. Y	June 29, 1852	9, 073
Bedstead-fastening	W. Johnston	Appleton, Wis	Mar. 3, 1868	75, 025
Bedstead-fastening	W. Johnston	Appleton, Wis	Dec. 15, 1868	84, 949
Bedstead-fastening	J. A. and A. F. Jones	Lexington, Ky	Aug. 1, 1848	5, 689
Bedstead-fastening	S. Jones	San Francisco, Cal	June 11, 1872	127, 694
Bedstead-fastening	J. R. Kain and S. Lewis	Tiffin, Ohio	May 6, 1851	8, 076
Bedstead-fastening	L. Kent	Dorset, Vt	Sept. 12, 1838	911
Bedstead-fastening	J. Knipe	New York, N. Y	Jan. 8, 1842	2, 415
Bedstead-fastening	J. Lemman	Cincinnati, Ohio	Oct. 2, 1866	58, 437
Bedstead-fastening	S. Lewis	Tiffin, Ohio	June 27, 1848	5, 650
Bedstead-fastening	S. Lewis	Tiffin, Ohio	Nov. 18, 1856	16, 093
Bedstead-fastening	S. Lewis	Tiffin, Ohio	Dec. 13, 1864	45, 417
Bedstead-fastening	J. Maguire	Trenton, N. J	Feb. 19, 1867	62, 212
Bedstead-fastening	J. F. Mancha	Ridgely, Md	Jan. 28, 1873	135, 279
Bedstead-fastening	P. Maulding and J. W. Fraley	Marshall, Tex	Aug. 16, 1870	106, 382
Bedstead-fastening	I. M. May	Anderson, Ind	Sept. 14, 1858	21, 511
Bedstead-fastening	L. May	Columbus, Ga	May 28, 1872	127, 255
Bedstead-fastening	P. McIntyre	Norwich, Conn	May 25, 1869	90, 568
Bedstead-fastening	I. McLaughlin	Sunderland, Vt	Apr. 28, 1836	......
Bedstead-fastening	W. H. McPherson	Nashville, Tenn	Dec. 12, 1871	121, 726
Bedstead-fastening	W. E. Merrill and T. Tupper	Nashua, N. H	Dec. 13, 1853	10, 313
Bedstead-fastening	E. Mets and W. B. Geddes	Rochester, N. Y	Nov. 22, 1870	109, 538
Bedstead-fastening	P. Miles	New Haven, Conn	May 15, 1860	28, 326
Bedstead-fastening	H. Miller	South Bend, Ind	Aug. 7, 1849	6, 630
Bedstead-fastening	J. E. Milliken	Bridgeton, Me	Mar. 10, 1868	75, 445
Bedstead-fastening	T. W. Moore	New York, N. Y	Nov. 14, 1871	120, 893
Bedstead-fastening	T. W. Moore	New York, N. Y	June 11, 1872	127, 786
Bedstead-fastening	J. Morrison	McArthurstown, Ohio	Oct. 29, 1850	7, 743
Bedstead-fastening	J. Moulton	Ossipee, N. H	Dec. 11, 1849	6, 940
Bedstead-fastening	I. W. Moyer	Utica, N. Y	May 15, 1847	5, 117
Bedstead fastening	H. B. Nash	Kingsbury, N. Y	Dec. 4, 1847	5, 387
Bedstead-fastening	A. S. Newnouse	Richmond County, Ga	June 15, 1852	9, 034
Bedstead-fastening	T. O'Keefe	Appleton, Wis	June 1, 1869	90, 776
Bedstead-fastening	J. P. Owen	Norwalk, Ohio	Dec. 11, 1849	6, 944
Bedstead-fastening	C. H. Parker	New Geneva, Pa	June 11, 1850	7, 428
Bedstead-fastening	M. Pechman	New York, N. Y	May 14, 1867	64, 699
Bedstead-fastening	G. Porter	Cincinnati, Ohio	Dec. 30, 1833	......
Bedstead-fastening	P. Post	New Haven, Conn	June 12, 1838	780
Bedstead-fastening	R. M. Price	Leesville, Ohio	Apr. 18, 1871	113, 925
Bedstead-fastening	W. H. Price	Philadelphia, Pa	Jan. 3, 1854	10, 378
Bedstead-fastening	C. D. Purdy	La Porte, Ind	Jan. 1, 1869	90, 780
Bedstead-fastening	C. D. Purdy	La Porte, Ind	Sept. 27, 1870	107, 812
Bedstead fastening	R. Ramsey	Wilmington, Pa	Feb. 19, 1850	7, 107
Bedstead-fastening	A. H. Rennie	Binghamton, N. Y	Mar. 19, 1867	63, 095
Bedstead-fastening	J. Richman	Lancaster, Ohio	Dec. 28, 1832	......
Bedstead-fastening	O. Robinson	Rochester, N. Y	Dec. 28, 1858	22, 456
Bedstead-fastening	A. Roda	Rochester, N. Y	Jan. 24, 1860	26, 924
Bedstead-fastening	J. Rodefer	Cincinnati, Ohio	Dec. 18, 1839	1, 431
Bedstead-fastening	W. H. Sabin	New Milford, Pa	May 12, 1840	1, 603
Bedstead-fastening	J. D. Sanborn	Bennington, N. Y	Apr. 3, 1849	6, 275
Bedstead-fastening	J. C. Santee	Hughesville, Pa	Jan. 24, 1865	46, 029
Bedstead-fastening	J. Schepler	Lambertville, N. J	Feb. 14, 1871	111, 878
Bedstead-fastening	A. Schlingman	West Alexandria, Ohio	Apr. 30, 1867	64, 257
Bedstead-fastening	I. A. Sergeant	Hamilton, Ohio	July 13, 1852	9, 122
Bedstead-fastening	W. Shaw	Clarion, Pa	Mar. 30, 1852	8, 845
Bedstead-fastening	W. Shaw	Clarion, Pa	Aug. 17, 1852	9, 203
Bedstead-fastening	J. Simpson	Newark, Ohio	Apr. 2, 1872	125, 342
Bedstead-fastening	H. T. Smith	Washington, D. C	Aug. 14, 1860	29, 628
Bedstead-fastening	I. Smith	Chagrin Falls, Ohio	Oct. 25, 1845	4, 243
Bedstead-fastening	J. J. Smith	Philadelphia, Pa	Mar. 29, 1864	42, 120
Bedstead-fastening	W. Stevens	Tarentum, Pa	Dec. 7, 1869	97, 722
Bedstead-fastening	J. M. Stewart	Newark, N. J	Aug. 6, 1872	130, 328
Bedstead-fastening	R. H. St. John	Columbus, Ohio	July 11, 1854	11, 305
Bedstead-fastening	D. Hotlemeyer	Hancock, Md	May 29, 1849	6, 483
Bedstead-fastening	G. Sugg and W. Metz	Chicago, Ill	Apr. 23, 1867	64, 048
Bedstead-fastening	G. Sugg and W. Metz	Chicago, Ill	Mar. 2, 1869	87, 380
Bedstead-fastening	W. Swift and W. Ottiwell	New Bedford, Mass	Sept. 11, 1828	......
Bedstead-fastening	E. S. Taylor	Cleveland, Ohio	Mar. 1, 1853	9, 604
Bedstead fastening	J. Taylor	Macon, Ga	Sept. 4, 1849	6, 687
Bedstead-fastening	J. C. Turner	Newark, N. J	June 7, 1870	104, 082
Bedstead-fastening	J. H. Waite	Orange, Mass	Sept. 6, 1870	107, 131
Bedstead-fastening	J. B. Wardwell	Georgetown, D. C	June 18, 1867	65, 970
Bedstead-fastening	C. Wolf	Rochester, N. Y	June 10, 1873	139, 844
Bedstead-fastening	C. Wolf	Rochester, N. Y	Sept. 16, 1873	142, 973
Bedstead-fastening	C. Wolf	Rochester, N. Y	Dec. 2, 1873	145, 266
Bedstead-fastening	E. S. Wright	Buffalo, N. Y	July 6, 1858	20, 839
Bedstead-fastening	J. W. Yothers	Spruce Grove, Pa	Oct. 30, 1855	13, 736
Bedstead-fastening	N. Zins	Evansville, Ind	Nov. 30, 1869	97, 468
Bedstead, Fastening of cast-iron	A. C. Semple	New York, N. Y	Feb. 23, 1858	19, 451
Bedstead-fastening, Plug	J. J. Smith	Philadelphia, Pa	June 7, 1859	24, 361
Bedstead, Folding	C. P. Alling, jr	Sylvan, Wis	Nov. 24, 1868	84, 337
Bedstead, Folding	D. Bach and K. Krenkel	New York, N. Y	Feb. 28, 1860	27, 263

Index of patents issued from the United States Patent Office from 1790 *to* 1873, *inclusive*—Continued.

Invention.	Inventor.	Residence.	Date.	No.
Bedstead, Folding	V. Baker	Weedsport, N. Y	July 11, 1848	5, 662
Bedstead, Folding	A. G. Bayles, J. W. H. Carroll, and Y. D. Miner.	New York and Williamsburgh, N. Y.	May 13, 1873	138, 840
Bedstead, Folding	J. Binder	Chelsea, Mass	Jan. 15, 1850	7, 014
Bedstead, Folding	T. B. Bleecker	New York, N. Y	Apr. 17, 1847	5, 072
Bedstead, Folding	W. H. Buel	Laughlintown, Pa	Feb. 15, 1870	99, 755
Bedstead, Folding	S. S. Burr	Dedham, Mass	Feb. 18, 1868	74, 494
Bedstead, Folding	E. Cotty	Brooklyn, N. Y	May 1, 1860	28, 127
Bedstead, Folding	M. Crosby	Boston, Mass	Oct. 18, 1870	108, 457
Bedstead, Folding	J. H. Durand	Kalamazoo, Mich	Oct. 3, 1865	50, 300
Bedstead, Folding	A. M. Dye	Clinton, Ill	Oct. 9, 1860	30, 304
Bedstead, Folding	T. G. Gifford	New York, N. Y	Feb. 15, 1870	99, 847
Bedstead, Folding	H. Gill	Dedham, Mass	Apr. 22, 1862	35, 018
Bedstead, Folding	S. Gillespie	New York, N. Y	Feb. 28, 1860	27, 284
Bedstead, Folding	M. B. Goodell	Worcester, Mass	Mar. 23, 1869	88, 158
Bedstead, Folding	J. C. Fall and A. C. Richards	Cincinnati, Ohio	July 5, 1870	105, 069
Bedstead, Folding	T. B. Harkins	Bristol, Pa	May 11, 1869	89, 993
Bedstead, Folding	H. H. Hill	Pontiac, Ill	Oct. 29, 1872	132, 576
Bedstead, Folding	B. Hinkley	Troy, N. Y	Feb. 17, 1857	16, 647
Bedstead, Folding	J. Housiaux	Washington, D. C	June 9, 1863	38, 827
Bedstead, Folding	T. Howe	Cambridgeport, Mass	Apr. 3, 1866	53, 620
Bedstead, Folding	J. A. Johnston	Antrim, Ohio	May 12, 1857	17, 281
Bedstead, Folding	J. L. Killgore	Wilmington, Del	Apr. 8, 1873	137, 690
Bedstead, Folding	J. P. Koch	New York, N. Y	Dec. 27, 1859	26, 598
Bedstead, Folding	K. Kunkel	New York, N. Y	Feb. 21, 1871	112, 053
Bedstead, Folding	W. C. Lutz	Jacob's Church, Va	Feb. 28, 1860	27, 299
Bedstead, Folding	S. H. Mapes	Almond, N. Y	Oct. 1, 1867	69, 457
Bedstead, Folding	J. Müller	Philadelphia, Pa	Apr. 13, 1869	88, 977
Bedstead, Folding	J. Müller	Philadelphia, Pa	Dec. 7, 1869	97, 545
Bedstead, Folding	F. C. Payne	New York, N. Y	Dec. 24, 1861	34, 009
Bedstead, Folding	J. L. Roberts	Dorchester, Mass	July 26, 1864	43, 656
Bedstead, Folding	G. D. Sargent	Boston, Mass	Dec. 11, 1860	30, 888
Bedstead, Folding	G. Sickels	Boston, Mass	Nov. 2, 1869	96, 356
Bedstead, Folding	W. Stoddard	Hingham, Mass	Apr. 3, 1855	12, 656
Bedstead, Folding	J. Sutter	New York, N. Y	May 3, 1864	42, 601
Bedstead, Folding	T. S. Thorson	Chicago, Ill	Dec. 17, 1872	134, 111
Bedstead, Folding	J. Turner	Cambridgeport, Mass	Jan. 3, 1871	110, 697
Bedstead, Folding	J. B. Wickersham	New York, N. Y	June 2, 1857	17, 460
Bedstead, Folding and cot	E. M. Payne	Waverly, N. Y	Jan. 29, 1867	61, 681
Bedstead, Folding bureau	D. B. Maynard	Worcester, Mass	Dec. 22, 1868	85, 232
Bedstead, Folding bureau or wardrobe	A. E. Botter	New York, N. Y	Dec. 20, 1853	10, 351
Bedstead, Folding case	E. Whitney	Lynn, Mass	Sept. 18, 1860	30, 101
Bedstead, Folding iron	H. F. Vandenhove	New York, N. Y	Nov. 3, 1857	18, 565
Bedstead, Folding metallic	W. Gardner	New York, N. Y	Dec. 9, 1862	37, 094
Bedstead for the sick	N. Richardson	Boston, Mass	Sept. 9, 1835	....
Bedstead for the sick	A. Smith	Milo, N. Y	July 12, 1830	....
Bedstead for the sick and wounded	D. Anthony	Adams, Mass	Sept. 15, 1834	....
Bedstead for the sick and wounded	W. S Wolf	New York, N. Y	July 15, 1834	....
Bedstead for the sick and wounded	W. Wooliey	New York, N. Y	Aug. 30, 1834	....
Bedstead for the sick, Elevating	D. Bancroft	Grafton, Vt.	Oct. 12, 1830	....
Bedstead, Fracture	E. Daniels	Owego, N. Y	Apr. 9, 1861	31, 958
Bedstead-frame	J. M. Farnham	Hartford, Conn	Nov. 30, 1869	97, 375
Bedstead, Hammock	I. M. Kilner	Chester, Great Britain	June 11, 1872	127, 892
Bedstead, Hospital	J. D. Cochran	Milford, N. H	Sept. 27, 1864	44, 399
Bedstead, Hospital	A. Iske	Lancaster, Pa	Aug. 15, 1865	49, 412
Bedstead, Hospital	A. Iske	Lancaster, Pa	Apr. 28, 1868	77, 193
Bedstead, Hospital	J. Sebo	Wilmington, Del	July 1, 1862	35, 782
Bedstead, Hospital	J. Sebo	Wilmington, Del	May 26, 1863	38, 700
Bedstead, Invalid	T. Arnold	Mobile, Ala	May 29, 1855	12, 940
Bedstead, Invalid	C. S. Baker	Manchester, N. H	Oct. 22, 1867	69, 958
Bedstead, Invalid	I. F. Baker	West Yarmouth, Mass	Sept. 2, 1862	36, 332
Bedstead, Invalid	A. W. Barker	Suffolk County, Mass	Mar. 5, 1850	7, 131
Bedstead, Invalid	H. N. Blair	Troy, Pa	Dec. 31, 1872	134, 456
Bedstead, Invalid	A. Bray	Chicopee, Mass	June 3, 1873	139, 493
Bedstead, Invalid	I. Buckman	South Woodstock, Vt	Apr. 17, 1849	6, 314
Bedstead, Invalid	A. W. Chase	Ann Arbor, Mich	May 15, 1860	28, 254
Bedstead, Invalid	C. M. Clinton and E. J. Morgan.	Ithaca, N. Y	Apr. 8, 1873	137, 596
Bedstead, Invalid	H. Cordes	Belleville, N. J	July 17, 1866	56, 376
Bedstead, Invalid	O. G. Cosby	Richmond, Va	Apr. 15, 1873	137, 828
Bedstead, Invalid	E. Daniels	Union, N. Y	May 29, 1855	12, 944
Bedstead, Invalid	B. Eastman	Philadelphia, Pa	July 17, 1855	13, 253
Bedstead, Invalid	I. T. Forbes	Coburg, Canada	Aug. 29, 1854	11, 605
Bedstead, Invalid	F. G. Ford	Washington, D. C	June 13, 1871	115, 948
Bedstead, Invalid	O. P. Furman	Addison, N. Y	Nov. 22, 1864	45, 151
Bedstead, Invalid	J. H. Gibson	McDonough, N. Y	Mar. 6, 1866	52, 988
Bedstead, Invalid	S. Grantz	Beaver Creek, Md	Jan. 3, 1860	26, 666
Bedstead, Invalid	G. W. Grote	Saint Catharine's, Canada.	Apr. 22, 1873	138, 083
Bedstead, Invalid	W. A. Guyer	New York, N Y	July 10, 1866	56, 211
Bedstead, Invalid	P. W. Hardwick	Wayne County, Ind	May 8, 1866	54, 533
Bedstead, Invalid	W. Heath	Bath, Me	Nov. 13, 1866	59, 600
Bedstead, Invalid	W. Heath	Bath, Me	Jan. 7, 1868	73, 008
Bedstead, Invalid	W. Heath	Bath, Me	Mar. 10, 1868	75, 265
Bedstead, Invalid	W. Heath	Bath, Me	May 23, 1871	115, 054
Bedstead, Invalid	J. R. Hill	Goshen, Ohio	Mar. 22, 1864	41, 996
Bedstead, Invalid	W. S. Hill	Manchester, N. H	Mar. 27, 1866	53, 533
Bedstead, Invalid	E. Hutchinson	Manchester, N. H	June 12, 1866	55, 496
Bedstead, Invalid	L. J. Johnson	Norwich, Conn	May 11, 1869	89, 868
Bedstead, Invalid	M. Johnson	Colebrook, N. H	Oct. 27, 1863	40, 413
Bedstead, Invalid	S. P. Johnson	Portland, Me	Nov. 12, 1867	70, 722
Bedstead, Invalid	J. Karney	Cincinnati, Ohio	Oct. 23, 1849	6, 814
Bedstead, Invalid	C. G. Kuhn	New York, N. Y	May 21, 1872	127, 070
Bedstead, Invalid	I. H. Latourandais	Flint Hill, Va	May 1, 1847	5, 100
Bedstead, Invalid	D. D. Marsh	Sunapee, N. H	June 16, 1868	78, 986
Bedstead, Invalid	J. Massey	New York, N. Y	Sept. 17, 1867	68, 892
Bedstead, Invalid	T. A. McFarland	Erie, Pa	July 25, 1871	117, 440
Bedstead, Invalid	D. Merrill	Worcester, Mass	Feb. 9, 1864	41, 526
Bedstead, Invalid	G. Miller	Fremont, Ohio	Feb. 2, 1858	19, 254

Index of patents issued from the United States Patent Office from 1790 *to* 1873, *inclusive*—Continued.

Invention.	Inventor.	Residence.	Date.	No.
Bedstead, Invalid	J. H. Oerter	New York, N. Y	Feb. 7, 1871	111, 561
Bedstead, Invalid	M. A. Ormsbee	Fair Haven, Vt	July 11, 1871	116, 861
Bedstead, Invalid	J. Parker	Liverpool, England	June 15, 1858	20, 580
Bedstead, Invalid	B. Pollard	Dutch Flat, Cal	Oct. 1, 1872	131, 903
Bedstead, Invalid	S. Puffer	Oxford, N. Y	Mar. 6, 1866	53, 041
Bedstead, Invalid	W. O. Reid	Vienna, N. C	Jan. 31, 1871	111, 380
Bedstead, Invalid	J. Robinson	Madison Station, Miss	July 8, 1873	140, 646
Bedstead, Invalid	W. W. Rowles and A. J. Russell	Baltimore, Md	Mar. 24, 1868	75, 983
Bedstead, Invalid	A. J. Russell	Baltimore, Md	June 28, 1870	104, 774
Bedstead, Invalid	A. J. Russell	Baltimore, Md	May 23, 1871	115, 108
Bedstead, Invalid	H. A. Scott	Winchester, N. H	Apr. 9, 1872	125, 491
Bedstead, Invalid	H. A. Scott	Winchester, N. H	July 23, 1872	129, 754
Bedstead, Invalid	H. A. Scott	Winchester, N. H	Dec. 17, 1872	134, 008
Bedstead, Invalid	J. Sebo	Wilmington, Del	Sept. 30, 1862	36, 578
Bedstead, Invalid	H. O. Sheidley	Republic, Ohio	Oct. 11, 1859	25, 766
Bedstead, Invalid	F. H. Smith and W. F. Wood	North Hebron, N. Y	July 6, 1869	92, 380
Bedstead, Invalid	G. Smith	Stratford, Canada	May 30, 1870	115, 371
Bedstead, Invalid	W. Swift	Brooklyn, N. Y	Feb. 14, 1860	27, 169
Bedstead, Invalid	W. Swift	Brooklyn, N. Y	Nov. 24, 1863	40, 716
Bedstead, Invalid	C. D. Van Allen	New York, N. Y	Apr. 25, 1854	10, 840
Bedstead, Invalid	F. M. Webster	Newport, Ky	May 8, 1849	6, 433
Bedstead, Invalid	C. Woolley	New York, N. Y	Apr. 16, 1842	2, 567
Bedstead, Invalid and fracture	T. McIlroy	New York, N. Y	Nov. 19, 1867	71, 196
Bedstead, Iron	J. Bayston and G. W. Nicholson	Chicago and Naperville, Ill.	Mar. 30, 1869	88, 435
Bedstead, Iron	J. M. French	East Cambridge, Mass	Dec. 16, 1862	37, 163
Bedstead, Iron folding	J. Biberthaler	New York, N. Y	July 26, 1859	24, 853*
Bedstead-joint	L. G. Bradford	Plymouth, Mass	Dec. 6, 1870	109, 799
Bedstead-key	J. Blake and D. Cushing	Providence, R. I	Dec. 14, 1830	....
Bedstead-key	H. Stein	New York, N. Y	Aug. 20, 1872	130, 671
Bedstead, Lounge	B. Cerf	New York, N. Y	Jan. 2, 1866	51, 802
Bedstead, Lounge	F. C. Payne	New York, N. Y	Aug. 15, 1871	118, 047
Bedstead, Lounge	H. H. Reichert	New York, N. Y	July 9, 1872	128, 908
Bedstead, lounge, and chair	S. A. Skinner	Bristol, Vt	Sept. 9, 1862	36, 441
Bedstead-machine	A. Stedman	Pittsford, N. Y	Jan. 16, 1835	....
Bedstead, Metallic	J. Carey	Brooklyn, N. Y	May 27, 1862	35, 414
Bedstead, Metallic	M. Lefferts	New York, N. Y	Apr. 15, 1856	14, 668
Bedstead, Metallic portable	J. Homer	New York, N. Y	Jan. 28, 1818	....
Bedstead, Metamorphosic alleviator	J. Lowe	Vienna, N. Y	Oct. 25, 1830	....
Bedstead, Metamorphosic alleviator	J. Lowe	Whitesborough, N. Y	Nov. 21, 1831	....
Bedstead or chair, Hospital	W. Woolley	New York, N. Y	Sept. 28, 1831	....
Bedstead or cot	H. W. Ladd	Chelsea, Mass	Sept. 20, 1870	107, 509
Bedstead, Parlor	F. E Coffin	Boston, Mass	Jan. 24, 1871	111, 108
Bedstead, Parlor	F. E. Coffin	Boston, Mass	June 3, 1873	139, 434
Bedstead, Parlor	M. Crosby	Boston, Mass	Nov. 10, 1868	83, 936
Bedstead, Parlor	M. Crosby	Boston, Mass	Oct. 4, 1870	107, 881
Bedstead, Parlor	J. A. Morgan	Bloomfield, Iowa	Aug. 31, 1869	94, 431
Bedstead, Parturient invalid	J. N. Robbins	Troy, Mo	Oct. 20, 1831	....
Bedstead, Portable	F. Cotton	Brooklyn, N. Y	Jan. 25, 1859	22, 706
Bedstead, Portable	J. Daley	Baltimore, Md	May 14, 1827	....
Bedstead, Portable	S. Willard	Cincinnati, Ohio	July 25, 1854	11, 398
Bedstead, Portable cot	W. C. Betts	New York, N. Y	May 11, 1852	8, 934
Bedstead, Portable cot	A. McDonough	Philadelphia, Pa	July 10, 1849	6, 588
Bedstead, Portable folding	H. W. Eastman	Baltimore, Md	Apr. 9, 1861	31, 962
Bedstead, Portable folding-chair	G. H. Cottam	Hampton Road, England	Apr. 25, 1854	10, 825
Bedstead, Portable invalid	Z. C. Favor	Chicago, Ill	Apr. 20, 1858	19, 987
Bedstead put together with right and left screws	T. B. Smith	Troy, N. Y	Oct. 11, 1824	....
Bedstead-rail	C. Robinson	Cambridgeport, Mass	Apr. 27, 1858	20, 092
Bedstead-rail attachment	J. Shaefer	Lancaster, Pa	May 6, 1862	35, 182
Bedstead-rail fastening	J. P. Allen	Manchester, Mass	Nov. 3, 1841	2, 326
Bedstead-rails, Apparatus for cutting screws on	J. R. Kain	Tiffin, Ohio	Nov. 22, 1853	10, 254
Bedstead-rails, Apparatus for cutting screws on	H. Smith	Norwalk, Ohio	Nov. 22, 1853	10, 261
Bedstead-rails, Machine for cutting screws on	H. Gross and W. Campbell	Tiffin City, Ohio	Apr. 1, 1851	8, 019
Bedstead rails, Machine for cutting screws on	J. Lindly	Cynthiana, Ky	June 20, 1838	797
Bedstead-rails, Machine for cutting screws on	J. P. Owens	Norwalk, Ohio	Nov. 22, 1853	10, 257
Bedstead-rails, Manner of cutting tenons and boring holes in.	T. Cole	Greensburgh, Ind	Nov. 12, 1841	2, 356
Bedstead, Revolving rail and round tenon	G. Post	Auburn, N. Y	Oct. 11, 1828	....
Bedstead, sacking-bottom, &c	D. Powles	Baltimore, Md	Jan. 26, 1827	....
Bedstead, Sacking-bottom for	I. Cooper	Johnstown, Pa	Oct. 7, 1844	3, 778
Bedstead, Secret	W. Woolley	New York, N. Y	May 17, 1830	....
Bedstead, Secret or sofa	W. Woolley	New York, N. Y	Oct. 3, 1831	....
Bedstead, Secretary	E. E. and M. G. Briggs	Boston, Mass	Feb. 16, 1869	86, 901
Bedstead, Secretary	G. Gage	Kendall's Mills, Me	July 24, 1860	29, 265
Bedstead, Sectional	C. Page	North Danvers, Mass	Dec. 13, 1853	10, 315
Bedstead, seat, &c	C. T. Frost	Medfield, Mass	June 8, 1869	90, 941
Bedstead-slat	A. S. and O. H. Drisko	Boston, Mass	July 5, 1870	104, 944
Bedstead-slat	S. Hickock	Buffalo, N. Y	Dec. 22, 1857	18, 902
Bedstead-slat	T. Howe	Cambridgeport, Mass	Sept. 27, 1859	25, 569
Bedstead-slat	O. H. Weed	Charlestown, Mass	Feb. 11, 1868	74, 262
Bedstead-slat	D. Wehrle	Williamsville, N. Y	Sept. 10, 1872	131, 198
Bedstead-slat holder	F. G. Ford	Philadelphia, Pa	Nov. 7, 1871	120, 734
Bedstead-slats, Elastic loop for suspending	C Robinson	Cambridgeport, Mass	June 30, 1857	17, 695
Bedstead, Sliding sofa	G. Wode	New York, N. Y	June 12, 1838	782
Bedstead so constructed as to be easily removable in case of fire, &c.	W. Gilman and W. Jackson		Mar. 19, 1804	....
Bedstead, Sofa	J. J. Anderson	Boston, Mass	Feb. 2, 1869	86, 344
Bedstead, Sofa	W. P. Barclay	Chicago, Ill	Oct. 13, 1868	83, 025
Bedstead, Sofa	J. Beiersdorf	Chicago, Ill	July 1, 1873	140, 453
Bedstead, Sofa	C. H. Berry	East Somerville, Mass	Dec. 10, 1867	71, 957
Bedstead, Sofa	F. Boeger	Philadelphia, Pa	Jan. 16, 1866	52, 020
Bedstead, Sofa	K. Borren	New York, N. Y	Apr. 26, 1859	23, 806
Bedstead, Sofa	E. B. Bowditch	New Haven, Conn	July 24, 1849	6, 607
Bedstead, Sofa	E. B. Bowditch	New Haven, Conn	Feb. 26, 1850	7, 116
Bedstead, Sofa	J. Brada	New York, N. Y	Jan. 14, 1873	134, 843
Bedstead, Sofa	E. Brady	New York, N. Y	May 2, 1865	47, 512
Bedstead, Sofa	J. C. Butler	Albany, N. Y	May 10, 1870	102, 769
Bedstead, Sofa	W. Clowes	Hartwood, N. Y	June 24, 1843	3, 140

Index of patents issued from the United States Patent Office from 1790 *to* 1873, *inclusive*—Continued.

Invention.	Inventor.	Residence.	Date.	No.
Bedstead, Sofa	F. Cotton	Brooklyn, N. Y	Feb. 23, 1864	41, 682
Bedstead, Sofa	J. Dourson	Columbus, Ohio	Mar. 23, 1869	88, 146
Bedstead, Sofa	J. Dourson	Columbus, Ohio	Feb. 13, 1872	123, 620
Bedstead, Sofa	J. C. Emery	Concord, N. H	June 20, 1848	5, 641
Bedstead, Sofa	A. D. Farrell	New York, N. Y	May 2, 1843	3, 072
Bedstead, Sofa	W. Farson	Philadelphia, Pa	Jan. 18, 1870	98, 858
Bedstead, Sofa	W. Farson	Philadelphia, Pa	Oct. 31, 1871	120, 374
Bedstead, Sofa	W. Farson	Philadelphia, Pa	Oct. 22, 1872	132, 392
Bedstead, Sofa	S. S. Gilliland and J. R. Wagoner.	Dayton, Ohio	May 17, 1853	9, 726
Bedstead, Sofa	N. M. Graw	New York, N. Y	Dec. 10, 1838	1, 030
Bedstead, Sofa	C. F. Grundin	Boston, Mass	June 17, 1873	139, 952
Bedstead, Sofa	J. T. Hammitt	Philadelphia, Pa	Mar. 16, 1852	8, 800
Bedstead, Sofa	J. B. Harlow	Portland, Me	Feb. 11, 1873	135, 643
Bedstead, Sofa	D. B. Hubbard	Wheeling, W. Va	Aug. 22, 1871	118, 368
Bedstead, Sofa	C. F. H. Huff	New York, N. Y	May 9, 1871	114, 684
Bedstead, Sofa	J. Irwin	Philadelphia, Pa	May 11, 1858	20, 206
Bedstead, Sofa	F. Keller	New York, N. Y	Aug. 1, 1865	49, 120
Bedstead, Sofa	J. Kena	Brooklyn, N. Y	Nov. 14, 1865	50, 939
Bedstead, Sofa	J. S. Kenyon	Webster City, Iowa	July 3, 1866	56, 064
Bedstead, Sofa	P. J. Larson	Chicago, Ill	Dec. 16, 1873	145, 508
Bedstead, Sofa	E. Lord	Portland, Me	Dec. 24, 1872	134, 210
Bedstead, Sofa	E. Lord	Portland, Me	Dec. 23, 1873	145, 741
Bedstead, Sofa	E. Lovell	Saint Louis, Mo	Aug. 26, 1873	142, 249
Bedstead, Sofa	M. Sulzbacher	New York, N. Y	Mar. 1, 1870	100, 464
Bedstead, Sofa	T. J. Magee	Cincinnati, Ohio	Sept. 8, 1863	39, 863
Bedstead, Sofa	T. J. Magee	Cincinnati, Ohio	Dec. 8, 1863	40, 848
Bedstead, Sofa	C. F. Martine	Boston, Mass	June 6, 1854	11, 026
Bedstead, Sofa	C. F. Martine	Boston, Mass	Sept. 3, 1867	68, 373
Bedstead, Sofa	M. K. Maximilian	New York, N. Y	Sept. 1, 1868	81, 659
Bedstead, Sofa	J. M. Meschutt	New York, N. Y	July 23, 1841	2, 185
Bedstead, Sofa	S. Millet	New York, N. Y	Aug. 8, 1854	11, 488
Bedstead, Sofa	J. Needham	New York, N. Y	Aug. 20, 1846	4, 703
Bedstead, Sofa	F. C. Payne	New York, N. Y	June 23, 1863	38, 978
Bedstead, Sofa	J. F. C. Peikhardt	New York, N. Y	Feb. 21, 1860	27, 234
Bedstead, Sofa	J. R. Penniman	Boston, Mass	Aug. 22, 1827	......
Bedstead, Sofa	J. F. C. Pickhardt	New York, N. Y	May 8, 1866	54, 592
Bedstead, Sofa	J. Pratt, 3d	Hartford, Conn	Apr. 1, 1842	2, 530
Bedstead, Sofa	H. H. Reichert	New York, N. Y	Apr. 23, 1872	126, 088
Bedstead, Sofa	J. A. Robson	New York, N. Y	Nov. 20, 1849	6, 893
Bedstead, Sofa	S. R. Roscoe	Obion County, Tenn	May 14, 1867	64, 799
Bedstead, Sofa	S. R. Roscoe	Obion, Tenn	Sept. 8, 1868	81, 949
Bedstead, Sofa	H. S. Rose	Bath, N. Y	Mar. 13, 1866	53, 188
Bedstead, Sofa	C. Rossow	Chicago, Ill	May 13, 1873	138, 819
Bedstead, Sofa	R. Scarritt	Saint Louis, Mo	Oct. 8, 1850	7, 707
Bedstead, Sofa	L. Schaefer	Cleveland, Ohio	Sept. 3, 1867	68, 389
Bedstead, Sofa	J. Schafer	New York, N. Y	Mar. 19, 1872	124, 765
Bedstead, Sofa	A. Schwaab	New York, N. Y	Sept. 7, 1869	94, 656
Bedstead, Sofa	G. Sickels	Middletown, Conn	Dec. 4, 1844	3, 845
Bedstead, Sofa	B. L. Southack	New York, N. Y	Oct. 20, 1868	83, 333
Bedstead, Sofa	J. K. Stockton	New York, N. Y	Feb. 25, 1873	136, 276
Bedstead, Sofa	C. Street	Cincinnati, Ohio	Apr. 1, 1873	137, 5[illegible]5
Bedstead, Sofa	C. F. Tauchert	Cincinnati, Ohio	Dec. 17, 1872	134, 016
Bedstead, Sofa	W. H. Tendler	Cambridge, Mass	May 31, 1859	24, 267
Bedstead, Sofa	W. H. Tendler	Cambridge, Mass	Apr. 10, 1860	27, 871
Bedstead, Sofa	W. H. Tendler and J. F. Moeshlin.	Cambridge, Mass	Aug. 28, 1860	29, 832
Bedstead, Sofa	W. Walcutt	New York, N. Y	Mar. 26, 1872	124, 990
Bedstead, Sofa	A. Walker	New Haven, Conn	June 29, 1852	9, 083
Bedstead, Sofa	A. G. Warren	Norwich, Conn	May 28, 1850	7, 404
Bedstead, Sofa	J. Werner	New York, N. Y	Apr. 7, 1868	76, 570
Bedstead, Sofa	S. M. and S. A. Winn	San Francisco, Cal	Oct. 7, 1873	143, 553
Bedstead, sofa, and settee	F. Breckels	New York, N. Y	Apr. 7, 1834	......
Bedstead, sofa, settee, &c	J. P. Allen	Manchester, Mass	May 19, 1832	......
Bedstead, Spring	J. Böhmer	Saint Louis, Mo	Apr. 20, 1869	89, 119
Bedstead, Spring	W. H. Kimball and A. J. French	Lynn, Mass	Sept. 2, 1856	15, 658
Bedstead, Spring	N. Houghton	New York, N. Y	Feb. 25, 1862	34, 500
Bedstead, Spring	R. Lapham	New York, N. Y	June 19, 1866	55, 680
Bedstead, Spring	D. R. Lightner	Bucyrus, Ohio	Sept. 4, 1860	29, 892
Bedstead, Spring	J. B. McClanathan	Horicon, Wis	Feb. 13, 1866	52, 585
Bedstead, Spring	N. M. Phillips	New York, N. Y	Mar. 16, 1858	19, 649
Bedstead, Spring	C. F. Spencer	Rochester, N. Y	Mar. 29, 1859	23, 404
Bedstead, Spring	S. H. Tift	Morrisville, Vt	Oct. 16, 1866	58, 914
Bedstead, Spring	W. Woods and E. Smith	Worcester, Mass	May 2, 1865	47, 594
Bedstead-spring	E. F. Dunaway	Cincinnati, Ohio	Aug. 22, 1871	118, 351
Bedstead-spring	E. F. Dunaway	Cincinnati, Ohio	Aug. 22, 1871	118, 352
Bedstead-spring	D. Manuel	Newton, Mass	Dec. 12, 1865	51, 516
Bedstead, Spring-bottom for	A. Foote	North Blandford, Mass	June 24, 1856	15, 209
Bedstead, Spring-bottom for	R. Hatch	Strafford, Vt	Dec. 29, 1857	18, 968
Bedstead, Spring cot	F. C. Hall	Baltimore, Md	Apr. 9, 1872	125, 389
Bedstead, Steam spring	C. T. Young	North Chelmsford, Mass	Dec. 29, 1857	18, 008
Bedstead, Tent	A. D. McCoy	New Orleans, La	Feb. 12, 1867	62, 049
Bedstead-tester	F. Layaux	Monroe, La	Aug. 30, 1870	106, 842
Bedstead, Toilet	G. V. Leicester	Boston, Mass	Nov. 9, 1869	96, 599
Bedstead, Trunk	R. C. Dubois	Washington, D. C	Apr. 19, 1870	102, 098
Bedstead, Ventilating	H. W. Henley	New York, N. Y	Jan. 24, 1860	26, 946
Bedstead-vetilating attachment	J. H. Martin	Hartford, N. Y	Aug. 16, 1864	43, 857
Bedstead, Wardrobe	R. M. Austin	Philadelphia, Pa	June 4, 1872	127, 546
Bedstead, Wardrobe	W. R. Bagnall	Chelsea, Mass	Oct. 1, 1867	69, 304
Bedstead, Wardrobe	C. L. Barritt	Brooklyn, N. Y	June 24, 1873	140, 109
Bedstead, Wardrobe	W. Berg	New York, N. Y	Feb. 28, 1860	27, 329
Bedstead, Wardrobe	S. S. Burr	Boston, Mass	Aug. 30, 1870	102, 776
Bedstead, Wardrobe	S. S. Burr and L. Pierce	Dedham and Charlestown, Mass.	Sept. 7, 1869	94, 560
Bedstead, Wardrobe	J. Byrnes	Chicago, Ill	May 6, 1873	138, 610
Bedstead, Wardrobe	A. Dietz	New Orleans, La	Dec. 13, 1864	45, 393
Bedstead, Wardrobe	J. F. Dodge	Nashua, N. H	July 19, 1870	105, 556
Bedstead, Wardrobe	J. Eby	Muncie, Ind	Apr. 11, 1865	47, 197

Index of patents issued from the United States Patent Office from 1790 to 1873, inclusive—Continued.

Invention.	Inventor.	Residence.	Date.	No.
Bedstead, Wardrobe	Z. C. Favor	Boston, Mass	Apr. 2, 1838	668
Bedstead, Wardrobe	T. L. Fortune	Weston, Mo	July 5, 1864	43, 4 5
Bedstead, Wardrobe	R. W. Frost	New York, N. Y	Apr. 1, 1873	137, 3 6
Bedstead, Wardrobe	E. Hunter and D. Van Sickle	Chicago, Ill	May 27, 1873	139, 316
Bedstead, Wardrobe	S. C. Maine	Boston, Mass	Mar. 19, 1867	63, 065
Bedstead, Wardrobe	W. H. Pease	Dayton, Ohio	Oct. 6, 1863	40, 184
Bedstead, Wardrobe	J. F. C. Pickhardt	New York, N. Y	Aug. 13, 1867	67, 797
Bedstead, Wardrobe	C. Robbins	Chicago, Ill	Feb. 23, 1858	19, 449
Bedstead, Wardrobe	J. Smith	Chicago, Ill	Apr. 8, 1873	137, 628
Bedstead, Wardrobe	A. A. Young	Boston, Mass	Oct. 27, 1868	83, 432
Bedstead, Wardrobe or bureau	J. S. McCurdy	New York, N. Y	Apr. 14, 1857	17, 047
Bedstead, Windlass	C. Adams	Boston, Mass	Feb. 26, 1825	....
Bedstead, Windlass	L. Foster	Boston, Mass	Feb. 16, 1824	....
Bedstead, Windlass	L. Foster	Boston, Mass	June 25, 1824	....
Bedstead, Windlass	T. Lamb	Washington, D. C	Nov. 13, 1840	1, 836
Bedsteads and attaching and stretching sackings, Manner of fastening.	M. Gregg	Philadelphia, Pa	Mar. 25, 1840	1, 526
Bedsteads, Attaching curtain-posts to	T. Early	Williamson County, Tenn	Jan. 8, 1839	1, 062
Bedsteads, Attaching cutters for cutting screws on rails of.	J. Zimmer	Tiffin, Ohio	Dec. 23, 1851	8, 6 0
Bedsteads, Bed-bug protector for	S. Clarke	New York, N. Y	July 3, 1866	56, 0 8
Bedsteads, Bevel-grooved and dovetail joints for	J. Mitchel	Harrisburgh, Pa	Jan. 11, 1826	....
Bedsteads, Boring and cutting screws on	J. Guild	Cincinnati, Ohio	Oct. 29, 1846	4, 829
Bedsteads, Constructing	C. J. Fountain, J. F. Adams, and G. F. Hillyer.	New York, N. Y	Apr. 20, 1839	1, 129
Bedsteads, Constructing	S. P. Smith	Salina, N. Y	Oct. 26, 1838	990
Bedsteads, Constructing and cording	M. Engle	Easton, Pa	Apr. 24, 1840	1, 562
Bedsteads, Construction of	O. R. Herbert	Mount Pleasant, N. J	July 14, 1824	....
Bedsteads, Construction of	J. F. Hollister	Plano, Ill	May 2, 1871	114, 439
Bedsteads, Cutting screws on rails of	S. Lewis	Tiffin, Ohio	Sept. 26, 1848	5, 812
Bedsteads, Machine for cutting screws of screw	P. Williams	Towanda, Pa	Feb. 16, 1837	126
Bedsteads, Machine for cutting screws on	O. Parkhurst and D. Bullock	Cohoes, N. Y	July 10, 1855	13, 225
Bedsteads, Machine for cutting screws on posts and rails of.	O. Thornley	Lebanon, Ind	Oct. 7, 1851	8, 414
Bedsteads, Machine for cutting screws on rails and posts of.	J. P. Owen	Norwalk, Ohio	Mar. 16, 1852	8, 806
Bedsteads, Machine for cutting screws on rails of	S. Lewis	Rochester, N. Y	Oct. 7, 1851	8, 416
Bedsteads, Machine for cutting screws on rails of	J. Lindley	Cynthiana, Ky	Oct. 10, 1840	1, 816
Bedsteads, Machine for cutting screws on rails of	J. Thompson	Cynthiana, Ky	July 29, 1841	2, 1 8
Bedsteads, Machinery for cutting screws in	J. Garside and H. J. Betjeman	Harrison, Ohio	Aug. 28, 1849	6, 668
Bedsteads, Machinery for cutting screws in the posts and rails of.	W. F. Converse, R. H. Penny, and R. S. Hannaford.	Harrison, Ohio	Dec. 31, 1844	3, 872
Bedsteads, Machinery for cutting screws on rails of	W. F. Converse and J. Burdge	Cincinnati, Ohio	Apr. 24, 1849	6, 387
Bedsteads, Machinery for cutting screws on rails of	S. Lewis	Tiffin, Ohio	Apr. 9, 1850	7, 264
Bedsteads, Making and stretching the sacking of	W. S. Anderson	Shelbyville, Tenn	June 4, 1838	765
Bedsteads, Mortise and tenon for	W. H. Elliot	New York, N. Y	Apr. 14, 1868	76, 611
Bedsteads, sofas, chairs, &c., Raising or inclining the various parts of.	E. Cherrington	Boston, Mass	Feb. 9, 1839	1, 078
Bedsteads, Spiral spring for	W. L. Beardsley	Binghamton, N. Y	Nov. 29, 1870	109, 573
Bedsteads, Table attachment for	D. Bull	Amboy, Ill	Sept. 17, 1867	68, 948
Bedsteads, Table attachment for	E. D. W. Hatch	Chicago, Ill	Apr. 20, 1869	89, 146
Bedsteads, Tester-frame for	J. C. Davis	Montgomery, Ala	Dec. 31, 1867	72, 814
Bedsteads, Tightening sacking-bottoms of	B. Bosworth	Fayette County, Ky	Dec. 14, 1839	1, 429
Bedsteads, Tightening sacking-bottoms of	J. Hart	Nicholasville, Ky	June 21, 1839	1, 185
Bedsteads to appartments, Application of	J. B. Brown	Boston, Mass	Oct. 9, 1866	58, 590
Bee-feeder	J. M. Beebe	Casadaga, N. Y	Mar. 26, 1867	63, 135
Bee-feeder	P. Miller, jr	Fredonia, N. Y	Mar. 23, 1869	88, 158
Bee-feeder	E. J. Peck	Linden, N. J	Apr. 30, 1872	126, 229
Bee-feeder	C. C. Van Deusen	Sprout Brook, N. Y	July 5, 1870	105, 144
Bee-feeding apparatus	E. Buel	Silver Creek, N. Y	Jan. 29, 1867	61, 514
Bee-feeding device	W. Brown	Shelbyville, Ind	June 21, 1859	24, 516
Bee-hive	D. L. Addair	Hawesville, Ky	Aug. 27, 1867	68, 141
Bee-hive	L. Adams	Mason City, Ill	Dec. 9, 1873	145, 324
Bee-hive	G. C. Aiken	Nashua, N. H	Feb. 15, 1859	22, 924
Bee-hive	L. W. and S. W. Albee	South Charlestown, N. H	Aug. 21, 1866	57, 272
Bee-hive	C. C. Aldrich	Morristown, Minn	Apr. 9, 1872	125, 427
Bee-hive	T. R. Allen	Syracuse, N. Y	Dec. 8, 1868	84, 788
Bee-hive	H. Alley	Wenham, Mass	June 18, 1872	128, 005
Bee-hive	W. J. Andrews	Columbia, Tenn	June 11, 1867	65, 629
Bee-hive	J. H. Andrews	Almont, Mich	Aug. 25, 1863	39, 622
Bee-hive	E. S. Armstrong	Jerseyville, Ill	Dec. 12, 1871	121, 837
Bee hive	D. Arndt	West Middletown, Pa	May 9, 1846	4, 502
Bee-hive	D. Arndt	Zanesville, Ohio	Aug. 21, 1860	29, 6 8
Bee-hive	W. B. Arndt	Nova, Ohio	Nov. 5, 1872	132, 787
Bee-hive	D. J. Arnold	Brownville, Nebr	Sept. 24, 1872	131, 589
Bee hive	R. Arnold	Suffolk, Va	Dec. 20, 1870	110, 332
Bee-hive	R. Arnold	Suffolk, Va	Sept. 12, 1871	118, 894
Bee hive	T. Atkinson	Memphis, Tenn	May 11, 1869	89, 841
Bee-hive	T. Atkinson	Memphis, Tenn	Aug. 10, 1869	93, 395
Bee-hive	E. S. Bacon	Albion, N. Y	Aug. 28, 1860	29, 753
Bee-hive	A. C. Badgley	Earleville, Ill	Oct. 29, 1867	70, 313
Bee-hive	C. W. Banks	Baltimore, Md	Oct. 1, 1872	131, 7 0
Bee-hive	S. D. Barber and J. Wolf	Mattoon, Ill	Jan. 4, 1870	98, 544
Bee-hive	E. Bartholomew	Cleveland, Ohio	Nov. 22, 1859	26, 154
Bee-hive	N. P. Bassett	Fulton, N. Y	Aug. 4, 1868	80, 587
Bee-hive	J. H. Bassler	Pine Grove, Pa	June 8, 1869	90, 984
Bee-hive	H. A. Bathurst	Clearfield, Pa	Dec. 12, 1871	121, 840
Bee-hive	H. Baughman	Columbus, Ohio	May 26, 1868	78, 253
Bee-hive	E. B. Beach	West Meriden, Conn	Apr. 21, 1868	76, 976
Bee hive	D. S. Bear	Toledo, Iowa	Feb. 18, 1868	74, 487
Bee-hive	C. Beard	Waynesborough, Va	Jan. 16, 1872	122, 697
Bee-hive	E. Beard	New Sharon, Me	Feb. 25, 1836	....
Bee-hive	J. M. Beebe	Casadaga, N. Y	Nov. 12, 1867	70, 782
Bee-hive	J. M. Beebe	Casadaga, N. Y	Aug. 6, 1872	130, 101
Bee-hive	J. Bebe	Carrolltown, Pa	Nov. 8, 1870	108, 900
Bee-hive	B. Benton	Weesaw, Mich	Nov. 7, 1871	120, 698
Bee-hive	H. Berix	Petersburgh, Ohio	June 15, 1869	91, 203
Bee-hive	T. F. Bingham	Gowanda, N. Y	Nov. 17, 1863	40, 000

Index of patents issued from the United States Patent Office from 1790 *to* 1873, *inclusive*—Continued.

Invention.	Inventor.	Residence.	Date.	No.
Bee-hive	T. F. Bingham	Gowanda, N. Y	Jan. 5, 1864	41, 052
Bee-hive	T. F. Bingham	Gowanda, N. Y	Aug. 9, 1864	43, 756
Bee-hive	T. F. Bingham	Gowanda, N. Y	Oct. 30, 1866	59, 169
Bee-hive	T. F. Bingham	Gowanda, N. Y	Oct. 30, 1866	59, 170
Bee-hive	J. S. Black	Bloomfield, Ky	May 15, 1860	28, 249
Bee-hive	W. Black	Harrisburgh, Pa	Jan. 26, 1869	86, 128
Bee-hive	E. Blake	Hartford County, Me	Nov. 16, 1820	
Bee-hive	A. Blood, sr	Norfolk, Va	Nov. 16, 1858	22, 059
Bee-hive	E. Booth	Springfield, Mass	Nov. 28, 1842	2, 861
Bee-hive	H. C. Boyers	Danville, Iowa	July 19, 1864	43, 564
Bee-hive	A. Bradshaw	Rantoul, Ill	Jan. 31, 1871	111, 310
Bee-hive	J. Bradt	La Porte, Ind	Nov. 18, 1862	36, 938
Bee-hive	J. Bradt	La Porte, Ind	July 3, 1866	55, 998
Bee-hive	N. Brasher	Green's Fork, Ind	July 31, 1860	29, 356
Bee-hive	F. Brewer	Waynesville, Mo	Mar. 7, 1871	112, 317
Bee-hive	L. C. Bristol and C. T. Alverson	Victor, N. Y	Oct. 9, 1866	58, 585
Bee-hive	J. F. Broomfield	Richmond, Ky	May 16, 1871	114, 912
Bee-hive	G. A. Brown and F. A. McCallen.	Russellville, Ky	Jan. 10, 1871	110, 894
Bee-hive	J. S. Brown	Washington, D. C	July 29, 1856	15, 457
Bee-hive	L. Brown	Pontiac, N. Y	Oct. 20, 1863	40, 323
Bee-hive	P. Brown	Taylorville, Ill	Sept. 10, 1872	131, 247
Bee-hive	A. M. Brubaker and B. R. Witmer.	Millersville, Pa	Mar. 11, 1873	136, 642
Bee-hive	S. R. Bryant	Waterford, Pa	Oct. 2, 1860	30, 202
Bee-hive	S. R. Bryant	Waterford, Pa	Feb. 12, 1861	31, 369
Bee-hive	W. Bryant	Nashville, Tenn	Dec. 21, 1842	2, 890
Bee-hive	B. F. Bucklin	Cuyahoga Falls, Ohio	Jan. 23, 1872	122, 990
Bee-hive	R. Bullard	Litchfield, Mich	June 4, 1861	32, 463
Bee-hive	D. S. Burget	Martinsburgh Borough, Pa	July 5, 1870	105, 037
Bee-hive	J. Burnham	York, Mich	Nov. 7, 1871	120, 709
Bee-hive	T. Burris	Fieldon, Ill	Aug. 15, 1865	49, 375
Bee-hive	H. Burton	Richview, Ill	Sept. 1, 1868	81, 747
Bee-hive	R. P. Buttles	Mansfield, Pa	Feb. 9, 1869	86, 729
Bee-hive	G. Calvert	Upperville, Va	Nov. 1, 1853	10, 196
Bee-hive	G. Calvert	Upperville, Va	June 22, 1869	91, 712
Bee-hive	D. M. Calvin	East Fallowfield, Pa	Apr. 8, 1873	137, 530
Bee-hive	D. M. Calvin	East Fallowfield, Pa	Dec. 2, 1873	145, 048
Bee-hive	J. A. Cameron	Memphis, Tenn	Feb. 9, 1869	86, 643
Bee-hive	P. Campbell	Carrolltown, Pa	Dec. 14, 1869	97, 759
Bee-hive	A. Canniff	Fort Scott, Kans	Oct. 8, 1872	132, 052
Bee-hive	H. F. Carpenter	Greencastle, Pa	June 7, 1870	103, 978
Bee-hive	J. I. Cassel and W. Quin	Eaton, Ohio	Apr. 14, 1868	76, 602
Bee-hive	J. Chase	Ripley, Ohio	July 21, 1868	80, 138
Bee-hive	E. Y. Chevalier	Fort Wayne, Ind	Aug. 5, 1862	36, 070
Bee-hive	J. B. Child	Lee Centre, Ill	Apr. 10, 1866	53, 785
Bee-hive	W. R. Clark	Piqua, Ohio	Oct. 18, 1864	44, 707
Bee-hive	W. R. Clark	Piqua, Ohio	Nov. 7, 1871	120, 624
Bee-hive	W. R. Clark	Piqua, Ohio	Nov. 26, 1872	133, 355
Bee-hive	G. H. Clarke	East Washington, N. H	Jan. 8, 1856	14, 051
Bee-hive	G. H. Clarke	East Washington, N. H	Mar. 1, 1859	23, 078
Bee-hive	A. Claypool	Weston, Ohio	Jan. 24, 1871	111, 107
Bee-hive	H. Clipp	Orange, Ind	Apr. 2, 1867	63, 472
Bee-hive	A. Clow	Waterford, Pa	Apr. 16, 1861	32, 101
Bee-hive	J. Coats	Camden, Ohio	Oct. 29, 1867	70, 167
Bee-hive	A. F. Cobb	Chapel Hill, Mo	Apr. 21, 1868	76, 999
Bee-hive	T. S. Collins and H. Senseman	Tremont, Ohio	Nov. 14, 1871	120, 858
Bee-hive	D. Collom	Tallmadge, Ohio	Aug. 31, 1869	94, 186
Bee-hive	A. Colton	Pittsfield, Vt	Dec. 31, 1845	4, 343
Bee-hive	O. Colvin	Belvidere, Ill	Jan. 15, 1867	61, 164
Bee-hive	R. Colvin	Baltimore, Md	Dec. 3, 1861	33, 826
Bee-hive	P. Compton	Sullivansville, N. Y	May 19, 1868	78, 062
Bee-hive	A. V. Conklin	Bennington, Ohio	Oct. 20, 1868	83, 257
Bee-hive	G. E. Conwell	Knoxville, Iowa	Dec. 31, 1867	72, 805
Bee-hive	H. B. Cooper	Memphis, Tenn	Feb. 22, 1870	100, 015
Bee-hive	S. and J. D. Cope	Damascusville, Ohio	Feb. 12, 1844	3, 439
Bee-hive	E. Corner	Columbus, Ohio	Dec. 24, 1861	34, 034
Bee-hive	W. Courtney	Richview, Ill	Dec. 31, 1867	72, 808
Bee-hive	W. Courtney	Richview, Ill	May 10, 1870	102, 917
Bee-hive	E. Cox	Monroe, Wis	July 16, 1867	66, 680
Bee-hive	J. H. Crandell	Upper Marlborough, Md	Apr. 20, 1869	89, 129
Bee-hive	S. B. Cranford	Upper Marlborough, Md	Aug. 10, 1869	93, 418
Bee-hive	L. H. Critchfield	Shreeve, Ohio	Oct. 5, 1869	95, 436
Bee-hive	W. F. Cunningham	Middletown, Ky	July 16, 1872	129, 464
Bee-hive	S. Cuplin	Iowa Falls, Iowa	July 28, 1868	80, 395
Bee-hive	J. Croner	Cross Creek Village, Pa	June 29, 1869	92, 018
Bee-hive	S. Cuplin	Iowa Falls, Iowa	Nov. 24, 1868	84, 415
Bee-hive	S. Cuplin	Iowa Falls, Iowa	Jan. 31, 1871	111, 434
Bee-hive	S. Cuplin	Iowa Falls, Iowa	Nov. 14, 1871	120, 944
Bee-hive	J. A. Cutting	Haverhill, N. H	June 24, 1844	3, 638
Bee-hive	E. J. Danavan	Indianapolis, Ind	Nov. 1, 1870	108, 893
Bee-hive	R. Daniels and G. P. Cobb	Woodstock, Vt	Mar. 12, 1861	31, 658
Bee-hive	J. W. Davidson	Rome, Ohio	Sept. 17, 1842	2, 777
Bee-hive	E. Davis	Noblesville, Ind	Sept. 6, 1870	107, 012
Bee-hive	O. Davis	Newton, Iowa	July 7, 1868	79, 734
Bee-hive	S. Davis	Claremont, N. H	July 26, 1853	9, 883
Bee hive	S. T. Davis	Millersville, Pa	Apr. 15, 1873	137, 831
Bee-hive	A. Decker	Fairfield County, Ohio	July 30, 1845	4, 129
Bee-hive	L. Defenbaugh	Kokomo, Ind	Dec. 19, 1865	51, 566
Bee-hive	J. H. Dennis	Boston, Mass	Jan. 24, 1854	10, 445
Bee-hive	J. Detrick and W. B. Niven	Bellefontaine, Ohio	June 13, 1871	115, 941
Bee-hive	A. Deweese	Oak Mills, Kans	Apr. 1, 1873	137, 348
Bee-hive	A. F. Dickey	Benford's Store, Pa	Dec. 5, 1871	121, 597
Bee-hive	W. M. Dickinson	Goshen, Ind	May 26, 1863	38, 660
Bee-hive	E. W. Diefendorf	Moniteau, Mo	Aug. 29, 1871	118, 520
Bee-hive	L. J. Diehl	Butler, Ind	Dec. 16, 1873	145, 634
Bee-hive	W. W. Dodson and J. B. Bray	Lynnville, Tenn	Dec. 19, 1871	121, 999
Bee-hive	G. Dofler	Frederick, Md	Mar. 4, 1843	2, 986

Index of patents issued from the United States Patent Office from 1790 *to* 1873, *inclusive*—Continued.

Invention.	Inventor.	Residence.	Date.	No.
Bee-hive	J. Donnel	Davenport, Iowa	Feb. 24, 1863	37, 741
Bee-hive	D. Dougal and W. Truxal	Butler, Pa	Sept. 3, 1861	33, 189
Bee-hive	J. A. Douglass	Altoona, Pa	Nov. 8, 1870	108, 984
Bee-hive	B. Douthett	Pittsburgh, Pa	Nov. 17, 1868	84, 052
Bee-hive	J. A. Dugdale	Selma, Ohio	July 31, 1849	6, 622
Bee-hive	H. N. Dunham and B. Addington.	Centreville, Ind	Sept. 7, 1869	94, 583
Bee-hive	A. P. Durant	Athens, Ohio	Dec. 8, 1868	84, 805
Bee-hive	J. and H. A. Earhart	Campbellstown, Pa	Nov. 8, 1864	44, 943
Bee-hive	M. Easley	Rush, Ill	June 1, 1869	90, 735
Bee-hive	G. Eason	Lyons, N. Y	Sept. 22, 1868	82, 298
Bee-hive	J. R. East	Fincastle, Tenn	Nov. 18, 1873	144, 750
Bee-hive	H. Eddy	North Bridgewater, Mass	Sept. 19, 1854	11, 691
Bee-hive	D. Edge	Saint Mary's, Ill	Oct. 27, 1868	83, 479
Bee-hive	J. C. Edwards	Cattleville, Mo	Sept. 26, 1871	119, 336
Bee-hive	W. T. Eisenhart	Doylestown, Pa	Nov. 13, 1866	59, 569
Bee-hive	W. A. Elam	Milan, Tenn	Nov. 23, 1869	97, 067
Bee-hive	A. E. Ellis	Friendsville, Ill	Oct. 9, 1866	58, 624
Bee-hive	W. J. Elvin	North Madison, Ind	June 30, 1868	79, 454
Bee-hive	C. Embrey	Williamsport, Md	Oct. 11, 1870	108, 244
Bee-hive	R. G. Emerson	Fair Haven, Ill	May 12, 1863	38, 475
Bee-hive	M. Engel	Easton, Pa	Apr. 15, 1840	1, 555
Bee-hive	T. S. Engledow	Cedar Falls, Iowa	Jan. 19, 1869	86, 063
Bee-hive	T. S. Engledow	Cedar Falls, Iowa	Jan. 10, 1871	110, 909
Bee-hive	D. H. Farnam	Litchfield, Conn	May 20, 1842	2, 633
Bee-hive	H. A. Farnam	South Bend, Ind	Dec. 26, 1871	122, 242
Bee-hive	I. B. Farquhar	Bloody Run, Pa	June 1, 1869	90, 649
Bee-hive	O. Field	Independence, Iowa	Sept. 15, 1868	82, 101
Bee-hive	J. T. Fife	Tyner City, Ind	Dec. 8, 1868	84, 812
Bee-hive	H. Filson	Monongahela City, Pa	July 20, 1869	92, 947
Bee-hive	S. Fink	Lindsey, Ohio	Feb. 6, 1872	123, 470
Bee-hive	J. E. Finley	Memphis, Tenn	Nov. 6, 1866	59, 377
Bee-hive	J. E. Finley	Memphis, Tenn	May 4, 1869	89, 644
Bee-hive	J. E. Finley	Enon, Ohio	June 15, 1869	91, 321
Bee-hive	C. Finn	Des Moines, Iowa	May 3, 1870	102, 526
Bee-hive	C. Finn	Des Moines, Iowa	Apr. 15, 1873	137, 834
Bee-hive	G. J. Flansburgh	Bethlehem, N. Y	Feb. 16, 1869	86, 913
Bee-hive	W. A. Flanders	Sharon, Vt	Oct. 25, 1853	10, 152
Bee-hive	W. A. Flanders	Cleveland, Ohio	Mar. 6, 1860	27, 408
Bee-hive	W. A. Flanders	Shelby, Ohio	July 14, 1863	39, 221
Bee-hive	W. A. Flanders	Shelby, Ohio	Apr. 5, 1864	42, 181
Bee-hive	W. A. Flanders	Shelby, Ohio	May 7, 1867	64, 515
Bee-hive	W. A. Flanders	Shelby, Ohio	May 5, 1868	77, 473
Bee-hive	H. H. Flick	Lavansville, Pa	Sept. 21, 1869	95, 100
Bee-hive	E. A. Floyd	Macomb, Ill	June 4, 1867	65, 367
Bee-hive	M. D. Fogel	Alpha, Ohio	Feb. 25, 1868	74, 810
Bee-hive	S. P. Forgy	Allensville, Ky	Aug. 4, 1868	80, 716
Bee-hive	T. H. Forster	Indianapolis, Ind	June 7, 1870	104, 011
Bee-hive	T. A. Frakes	Middletown, Ill	May 17, 1870	103, 166
Bee-hive	A. Francis and I. Carlisle	Chandlersville, Ohio	Nov. 28, 1842	2, 862
Bee-hive	A. H. Frank	Buffalo, N. Y	Dec. 27, 1864	45, 598
Bee-hive	E. P. French	Nashua, N. H	May 11, 1858	20, 202
Bee-hive	J. J. and A. J. Frey	Hook's Point, Iowa	Mar. 23, 1869	88, 157
Bee-hive	J. D. Fulkerson	Unity, Ohio	July 1, 1844	3, 646
Bee-hive	A. Fuller	Plymouth, Ind	May 1, 1866	54, 325
Bee-hive	P. J. Furlong	Galen, N. Y	June 1, 1858	20, 417
Bee-hive	G. G. Gabrion	Olive, Mich	Nov. 8, 1870	108, 997
Bee-hive	J. C. Gaston	Cincinnati, Ohio	Nov. 10, 1868	83, 952
Bee-hive	J. C. Gaston	Cincinnati, Ohio	Feb. 16, 1869	86, 915
Bee-hive	J. Gatschets	York Township, Ohio	Nov. 19, 1867	70, 989
Bee-hive	A. W. Geaheart	Beallsville, Ohio	Feb. 14, 1860	27, 122
Bee-hive	D. Gebhart	Sallamonia, Ind	Apr. 18, 1871	113, 870
Bee-hive	G. Gebhart	Lebanon, Pa	Feb. 20, 1843	2, 967
Bee-hive	E. Gerry	Garden City, Minn	Apr. 1, 1873	137, 433
Bee-hive	A. Gilmore	Wayne, Me	June 5, 1849	6, 494
Bee-hive	R. Gipson	Shelby, Ohio	May 31, 1864	42, 995
Bee-hive	J. H. Gisler	Saint Louis, Mo	Feb. 28, 1871	112, 139
Bee-hive	J. W. Gladding	Normal, Ill	Dec. 6, 1870	109, 888
Bee-hive	J. W. Gladding	Normal, Ill	Jan. 9, 1872	122, 598
Bee-hive	L. L. Goodwin	Toronto, Ind	Nov. 15, 1870	109, 201
Bee-hive	J. Gould	Grinnell, Iowa	Jan. 11, 1870	98, 761
Bee-hive	F. Grabbe	North Topeka, Kans	Sept. 24, 1872	131, 610
Bee-hive	G. Graffham	Lawrenceville, Ill	July 24, 1866	56, 551
Bee-hive	M. Graham	Coshocton, Ohio	July 27, 1869	93, 081
Bee-hive	J. H. Graves	Rochester, N. Y	Feb. 10, 1863	37, 622
Bee-hive	A. Gray	Reiley, Ohio	Nov. 26, 1867	71, 381
Bee-hive	D. S. Gray	Onarga, Ill	July 11, 1865	48, 768
Bee-hive	J. C. Gray	Frankfort, Ind	Feb. 19, 1861	31, 454
Bee-hive	W. Green	Kinzua, Pa	Sept. 1, 1863	39, 729
Bee-hive	S. V. Greer	Glasgow, Ky	June 4, 1872	127, 479
Bee-hive	H. Grems	Westmoreland, N. Y	Mar. 8, 1870	100, 524
Bee-hive	D. M. Griffith	Alum Bank, Pa	June 13, 1871	115, 843
Bee-hive	D. L. Grover	Groton, N. Y	July 1, 1873	140, 497
Bee-hive	E. Grumman	Norwalk, Conn	May 26, 1842	2, 645
Bee-hive	J. A. Gruver	West Union, Iowa	July 7, 1863	39, 142
Bee-hive	D. M. Gunn and C. L. Cain	Oskaloosa, Iowa	Oct. 15, 1861	33, 484
Bee-hive	D. M. Gunn and C. L. Cain	Oskaloosa, Iowa	Dec. 2, 1862	37, 039
Bee-hive	H. Gushee and J. G. Dawes	San Francisco, Cal	Oct. 11, 1859	25, 731
Bee-hive	M. Guthrie	Clifton, Iowa	Nov. 13, 1866	59, 593
Bee-hive	W. H. Hall	Wallingford, Conn	Dec. 27, 1839	1, 446
Bee-hive	H. A. Hannum	Cazenovia, N. Y	Feb. 3, 1863	37, 576
Bee-hive	H. A. Hannum	Cazenovia, N. Y	Nov. 8, 1864	44, 949
Bee-hive	J. S. Harbison	Sacramento, Cal	Jan. 4, 1859	22, 500
Bee-hive	J. S. Harbison	Sacramento, Cal	Dec. 13, 1859	26, 431
Bee-hive	W. Harden	Chariton, Iowa	Sept. 20, 1864	44, 305
Bee-hive	H. S. Harned and F. S. Elliott	Boonesborough, Iowa	Feb. 28, 1871	112, 244
Bee-hive	J. H. Harper	Washington, D. C	Aug. 5, 1862	36, 088
Bee-hive	J. Harper	Hillsborough, Iowa	Feb. 21, 1865	46, 467

Index of patents issued from the United States Patent Office from 1790 *to* 1873, *inclusive*—Continued.

Invention.	Inventor.	Residence.	Date.	No.
Bee-hive	J. Harris	Belmont County, Ohio	Jan. 27, 1843	2, 929
Bee-hive	J. Harris	New Carlisle, Ind	May 27, 1862	35, 426
Bee-hive	E. Harrison	Springfield, Ohio	Oct. 1, 1867	69, 428
Bee-hive	A. H. Hart	Stockbridge, Wis	Aug. 13, 1867	67, 649
Bee-hive	A. H. Hart	Appleton, Wis	July 2, 1872	128, 619
Bee-hive	S. Hart	New Haven, N. Y	Mar. 26, 1845	3, 971
Bee-hive	N. D. Hartley and M. L. Morehouse.	Quincy, Ill	Feb. 12, 1861	31, 384
Bee-hive	G. R. Hartman	Fort Wayne, Ind	July 9, 1861	32, 769
Bee-hive	H. Harvey	Meriden, Ill	June 13, 1871	115, 849
Bee-hive	C. Hastings	Dowagiac, Mich	Jan. 7, 1868	73, 096
Bee-hive	B. S. and E. H. Haviland	Fort Dodge, Iowa	May 21, 1867	64, 864
Bee-hive	R. Hawkins	Beallsville, Pa	Apr. 24, 1860	27, 980
Bee-hive	T. Hawkins	Auburn, N. Y	July 10, 1866	56, 219
Bee-hive	J. Hazen	Bethlehem, N. Y	Sept. 2, 1862	36, 349
Bee-hive	J. Hazen	Bethlehem, N. Y	July 2, 1867	66, 336
Bee-hive	J. Heacock	Marlborough, Ohio	June 4, 1867	65, 381
Bee-hive	H. L. Heckman	Brooklyn, Iowa	Jan. 24, 1871	111, 204
Bee-hive	Z. R. Henck	Port Royal, Pa	Oct. 12, 1842	2, 814
Bee-hive	J. H. Hendrick	Clinton, Ill	Dec. 26, 1865	51, 716
Bee-hive	J. H. Hendrick	Clinton, Ill	Oct. 1, 1867	69, 431
Bee-hive	J. H. Hendrick	Clinton, Ill	Jan. 31, 1871	111, 451
Bee-hive	E. E. Henegan	Downsville, Wis	July 8, 1873	140, 702
Bee-hive	W. M. Henry	Leo, Ind	Aug. 1, 1871	117, 539
Bee-hive	W. Henschen	Hennepin County, Minn	Apr. 4, 1865	47, 169
Bee-hive	D. Herman	Bigler Post-Office, Pa	Dec. 5, 1865	51, 315
Bee-hive	J. M. Hicks	Indianapolis, Ind	Sept. 6, 1870	107, 047
Bee-hive	J. N. Hieronymus	Fairbury, Ill	Dec. 5, 1871	121, 618
Bee-hive	J. Hiestand	Palestine, Ill	Nov. 5, 1867	70, 436
Bee-hive	S. O. Higgason	Union City, Tenn	Apr. 23, 1872	126, 055
Bee-hive	H. Hiser	Wooster, Ohio	June 27, 1842	2, 692
Bee-hive	G. T. Hixson	Gallipolis, Ohio	Sept. 24, 1872	131, 613
Bee-hive	H. Hodgson	Fremont, Ohio	Feb. 13, 1866	52, 571
Bee-hive	A. Hogg	Rutland, Ohio	Sept. 22, 1863	40, 038
Bee-hive	I. Honeywell	Toledo, Iowa	Feb. 11, 1868	74, 362
Bee-hive	J. S. Hooton	New Carlisle, Ind	June 30, 1868	79, 471
Bee-hive	T. P. Hornbrook	Wheeling, Va	May 21, 1861	32, 367
Bee-hive	J. and W. H. Horsman	New London, Ind	July 30, 1872	130, 052
Bee-hive	W. S. Hough	Galt, Canada	Nov. 25, 1873	144, 847
Bee-hive	J. M. Hubbard	Canterbury, N. H	Feb. 5, 1836	....
Bee-hive	H. O. Hughes	Judson, Mo	Dec. 8, 1868	84, 826
Bee-hive	P. and S. Hurst	Orrville, Ohio	July 9, 1872	128, 883
Bee-hive	S. Hutchinson	North Lewisburgh, Ohio	July 2, 1867	66, 348
Bee hive	W. Hyde	Emery, Ohio	Aug. 7, 1860	29, 491
Bee-hive	S. Ide	East Shelby, N. Y	Apr. 2, 1861	31, 890
Bee-hive	C. R. Isham	Peoria, N. Y	Jan. 30, 1872	123, 104
Bee-hive	J. A. Jackson	Macon, Mich	Feb. 11, 1868	74, 369
Bee-hive	J. Jacobs	Yellow Springs, Ohio	Dec. 27, 1859	26, 591
Bee-hive	J. Jacobs	Columbus, Ohio	Sept. 25, 1860	30, 143
Bee-hive	Z. L. Jacobs	Hebron, Conn	Apr. 9, 1867	63, 641
Bee-hive	A. E. James	Point Pleasant, Va	Jan. 6, 1844	3, 398
Bee-hive	E. L. Jinnett	Vermillion County, Ill	Mar. 19, 1861	31, 719
Bee-hive	C. and A. Jones	Santa Anna, Ill	Apr. 18, 1871	113, 892
Bee-hive	E. Jones	Amsterdam, N. Y	Feb. 12, 1845	3, 911
Bee-hive	J. Jones	Galway, N. Y	May 12, 1842	2, 618
Bee-hive	R. Jones	Cedarville, Ill	Oct. 29, 1867	70, 338
Bee-hive	S. M. Judd	Danbury, Conn	Mar. 30, 1836	....
Bee-hive	J. J. Justin	Milwaukee, Wis	July 10, 1866	56, 230
Bee-hive	H. C. Keith	Ancona, Ill	Feb. 26, 1867	62, 425
Bee-hive	W. J. Kelly	Commerce, Mich	Mar. 21, 1871	112, 813
Bee-hive	S. Kelly	Washington, D. C	Dec. 8, 1857	18, 815
Bee-hive	A. Kelsey	Westport, Mo	Mar. 31, 1857	16, 926
Bee-hive	W. R. Kelsey	Big Steam Point, N. Y	May 9, 1846	4, 501
Bee-hive	L. Kennedy	Hartford, Conn	Feb. 9, 1864	41, 513
Bee-hive	W. Kennedy	Roxbury, Mass	June 21, 1864	41, 211
Bee-hive	K. P. Kidder	Burlington, Vt	Apr. 13, 1858	19, 931
Bee-hive	K. P. Kidder	Burlington, Vt	June 26, 1860	28, 871
Bee-hive	K. P. Kidder	Burlington, Vt	Mar. 17, 1863	37, 915
Bee-hive	K. P. Kidder	Burlington, Vt	May 14, 1867	64, 773
Bee-hive	H. A. King	Nevada, Ohio	Sept. 8, 1868	82, 006
Bee-hive	H. A. King	New York, N. Y	Mar. 26, 1872	124, 962
Bee-hive	H. A. King	New York, N. Y	Sept. 10, 1872	131, 168
Bee-hive	H. A. King and J. Loughmaster.	Seal, Ohio	Nov. 24, 1863	40, 725
Bee-hive	H. A. and N. H. King and F. S. Walker.	Nevada, Ohio	Oct. 10, 1865	50, 367
Bee-hive	I. King	Germantown, Ohio	Nov. 10, 1868	83, 861
Bee-hive	W. R. King	Shelbyville, Ky	May 9, 1871	114, 690
Bee-hive	E. N. Kingsley	Minneapolis, Minn	Sept. 11, 1866	57, 924
Bee-hive	W. T. Kirkpatrick	Tamaroa, Ill	Dec. 14, 1869	97, 933
Bee-hive	A. H. Klepper	Muscatine, Iowa	Sept. 26, 1871	119, 366
Bee-hive	E. J. Kline	Kirkville, Iowa	May 12, 1868	77, 892
Bee-hive	W. Kraiss	Fairview, Pa	Apr. 30, 1867	64, 334
Bee-hive	L. Kramer	Point Pleasant, Pa	May 14, 1867	64, 675
Bee-hive	E. Kretchmer	Pleasant Grove, Iowa	Nov. 13, 1866	59, 614
Bee-hive	E. Kretchmer	Pleasant Grove, Iowa	July 23, 1867	67, 123
Bee-hive	E. Kretchmer	Coburg, Iowa	May 6, 1873	138, 661
Bee-hive	J. Lamborn	Marshallton, Pa	Aug. 26, 1845	4, 168
Bee-hive	J. C. and R. R. Lander	Mazo Manie, Wis	Nov. 24, 1868	84, 286
Bee-hive	L. L. Langstroth	Philadelphia, Pa	Oct. 5, 1852	9, 300
Bee-hive	J. Lash	Mount Vernon, Nebr	Aug. 9, 1864	43, 779
Bee-hive	D. Latchaw	Barkeyville, Pa	Feb. 13, 1872	123, 632
Bee-hive	D. Latchaw	Barkeyville, Pa	Aug. 5, 1873	141, 443
Bee-hive	A. S. Layton	Yellville, Ark	Feb. 9, 1869	86, 765
Bee-hive	W. M. Lee	Rosendale, Wis	Oct. 15, 1861	33, 486
Bee-hive	W. M. Lee	Rosendale, Wis	Aug. 25, 1863	39, 660
Bee-hive	J. K. Leedy	Woodstock, Va	July 12, 1859	24, 745
Bee-hive	J. Leffel and E. Harrison	Springfield, Ohio	Sept. 27, 1870	107, 792

Index of patents issued from the United States Patent Office from 1790 *to* 1873, *inclusive*—Continued.

Invention.	Inventor.	Residence.	Date.	No.
Bee-hive	G. W. Leffingwell	Columbus, Wis	July 23, 1867	66, 977
Bee-hive	W. Lehman	Lexington, Mo	Feb. 4, 1873	135, 566
Bee-hive	V. Leonard	Springfield, Pa	June 4, 1867	65, 402
Bee-hive	V. Leonard	Springfield, Pa	Sept. 3, 1867	68, 759
Bee-hive	V. Leonard	Springfield, Pa	Aug. 16, 1870	106, 377
Bee-hive	C. P. Lloyd	Portsmouth, Ohio	Feb. 12, 1867	61, 043
Bee-hive	B. Lockrone	Somerset, Ohio	Oct. 27, 1868	83, 514
Bee-hive	J. Longgrear and J. E. Clark	Rolla, Mo	Aug. 13, 1872	130, 511
Bee-hive	D. Loofbourrow	Hillsborough, Iowa	Jan. 7, 1873	134, 687
Bee-hive	R. Lovett	Canton, Ohio	Mar. 29, 1864	42, 093
Bee-hive	J. J. Lower	Tennessee, Ill	Dec. 24, 1867	72, 651
Bee-hive	G. W. Lowry	Lavansville, Pa	Nov. 9, 1869	96, 712
Bee-hive	O. Mack	Gilsum, N. H	Apr. 22, 1835	......
Bee-hive	S. Maitland	Fort Wayne, Ind	Aug. 28, 1860	29, 799
Bee-hive	H. Markham	Henderson, Ill	Feb. 23, 1864	41, 711
Bee-hive	G .Marsh	Steamburgh, N. Y	Feb. 28, 1871	112, 262
Bee-hive	J. S. Marshall	West Greenville, Pa	Apr. 22, 1862	35, 031
Bee-hive	J. R. Martin	Martinsburgh, Ind	Dec. 17, 1867	72, 308
Bee-hive	R. Martin	Fairfield, Ohio	May 19, 1840	1, 612
Bee-hive	C. R. C. Masten and A. D. Van Vlack.	Pleasant Valley, N. Y	Jan. 19, 1869	86, 023
Bee-hive	T. H. Mathews	Rushville, Ill	Aug. 23, 1864	43, 921
Bee-hive	J. Mathius	Pemberton, Ohio	Oct. 8, 1867	69, 569
Bee-hive	J. A. McAnulty	Gilpin, Pa	Jan. 31, 1871	111, 466
Bee-hive	T. F. McCafferty	Columbus, Ohio	Mar. 30, 1869	88, 315
Bee-hive	S. McClanathan	Warren, Ill	Mar. 24, 1863	38, 012
Bee-hive	H. M. McClellan	York, Pa	Dec. 1, 1857	18, 757
Bee-hive	J. McDonald	Buffalo, Ohio	June 23, 1868	79, 243
Bee-hive	N. McGonnigle	Allegheny, Pa	July 31, 1860	29, 394
Bee-hive	C. McGrew	Bloomington, Ill	Feb. 12, 1867	62, 051
Bee-hive	S. D. McLean	Sunny Slope, Tenn	June 4, 1872	127, 498
Bee-hive	J. D. Meador	Independence, Mo	Feb. 2, 1869	86, 569
Bee-hive	J. Meese, sr	Milton, Ohio	Jan. 31, 1860	27, 004
Bee-hive	G. W. Merchant	La Porte, Ind	Sept. 25, 1866	58, 276
Bee-hive	M. Metcalf	Grand Rapids, Mich	July 30, 1861	32, 952
Bee-hive	J. F. Metherd and S. Young	Union City, Ind	Mar. 28, 1871	113, 187
Bee-hive	J. Milholland and B. Crane	Chandlersville, Ohio	Apr. 29, 1842	2, 592
Bee-hive	G. Miller	Battle Ground, Wash	Sept. 30, 1873	143, 295
Bee-hive	J. Miller	Lexington, Ohio	Nov. 6, 1843	3, 325
Bee-hive	O. Miller	Girard, Pa	Apr. 2, 1861	31, 900
Bee-hive	W. D. Miller	Mad River, Ohio	Sept. 1, 1843	3, 246
Bee-hive	N. C. Mitchell	Caledonia, Ohio	Feb. 18, 1868	74, 709
Bee-hive	J. Montgomery	Union City, Tenn	Jan. 4, 1870	98, 612
Bee-hive	F. Moore	Carrollton, Ohio	June 9, 1868	78, 813
Bee-hive	J. E. Moore	Bridgewater, Pa	Feb. 20, 1872	123, 924
Bee-hive	S. C. Moore	Pattonsburgh, Mo	Aug. 8, 1871	117, 914
Bee-hive	W. L. Moorhead and T. D. Howell.	Lanesville, Ohio	Feb. 16, 1843	2, 953
Bee-hive	S. Morrill	Dixfield, Me	Feb. 4, 1834	......
Bee-hive	S. Morrill	Dixfield, Me	Jan. 16, 1835	......
Bee-hive	M. Morton	Gallatin, Mo	Apr. 5, 1870	101, 647
Bee-hive	W. T. Mosher	Poplar Ridge, N. Y	Aug. 27, 1872	130, 932
Bee-hive	A. R. Moulton	Fall Branch, Tenn	Sept. 10, 1872	131, 296
Bee-hive	M. D. Mulford, jr	New Providence, Iowa	July 31, 1866	56, 782
Bee-hive	E. F. Mulkey and J. Case	Tamaroa, Ill	Aug. 27, 1872	130, 933
Bee-hive	A. Mutersbaugh	Lewinsville, Va	Nov. 7, 1871	120, 766
Bee-hive	P. M. Myers, J. W. Walser, and J. Spangler.	Canton, Ohio	May 7, 1867	64, 557
Bee-hive	S. C. Myers	Mount Pleasant, Pa	July 22, 1839	1, 255
Bee-hive	B. F. Nave	Fort Wayne, Ind	Sept. 29, 1868	82, 628
Bee-hive	J. Neal	Orleans, Iowa	Jan. 28, 1868	73, 915
Bee-hive	J. Newman and N. Brown	Rossville, Ill	Mar. 18, 1873	136, 933
Bee-hive	C. S. Newsom	Gallipolis, Ohio	Apr. 16, 1872	125, 837
Bee-hive	P. Nicolle	Lindsay, Canada	Mar, 28, 1871	113, 084
Bee-hive	A. Niebel	Tiffin, Ohio	Oct. 27, 1868	83, 532
Bee-hive	E. O'Connor	Philadelphia, Pa	Oct. 17, 1871	119, 991
Bee-hive	H. Ogborn	Richmond, Ind	Mar. 22, 1864	42, 017
Bee-hive	L. M. Olden	Pana, Ill	Nov. 26, 1867	71, 405
Bee-hive	O. Osmundson	Mission, Ill	Apr. 20, 1869	89, 067
Bee-hive	J. N. Outten	Caseyville, Ky	Sept. 29, 1868	82, 634
Bee-hive	J. A. Paddock and J. S. Estep	Cass County, Ill	July 6, 1869	92, 348
Bee-hive	S. E. Paine and W. Kerr	Xenia, Ill	Feb. 13, 1872	123, 724
Bee-hive	J. W. Palmer	Port Republic, Va	Nov. 1, 1859	25, 979
Bee-hive	J. W. Palmer and J. K. Leedy	Port Republic and Tom's Brook, Va.	Oct. 9, 1860	30, 341
Bee-hive	L. Parker	Winterset, Iowa	Apr. 28, 1863	38, 328
Bee-hive	E. Parks	Wheatfield, N. Y	July 22. 1845	4, 116
Bee-hive	J. L. Patourel	Chandlersville, Ohio	June 11, 1841	2, 122
Bee-hive	J. M. Patton	Tipton, Iowa	Feb. 11, 1868	74, 415
Bee-hive	C. Pawling	New Pittsburgh, Ohio	Oct. 14, 1856	15, 894
Bee-hive	J. Pearson	West Milton, Ohio	Feb. 2, 1869	86, 441
Bee-hive	R. Pearson	Appleton, Wis	Aug. 3, 1869	93, 340
Bee-hive	T. W. Peirce	Minneapolis, Minn	Mar. 27, 1866	53, 475
Bee-hive	H. Penoyer	Centralia, Ill	Feb. 28, 1865	46, 584
Bee-hive	E. Pepple	Canton, Ohio	June 24, 1843	3, 141
Bee-hive	P. O. Petersen	Oakland, Cal	May 14, 1872	126, 739
Bee-hive	E. W. Phelps	Elizabeth, N. J	Nov. 9, 1858	22, 030
Bee-hive	W. H. Pierson	West Jersey, Ill	Oct. 16, 1866	58, 881
Bee-hive	H. A. Pitts	Winthrop, Me	Sept. 25, 1841	2, 268
Bee-hive	J. F. Pool	Monroe, Wis	Aug. 18, 1868	81, 206
Bee-hive	C. H. Potter	Philadelphia, Pa	June 4, 1872	127, 513
Bee-hive	W. Powers	Youngstown, Ohio	Mar. 8, 1859	23, 192
Bee-hive	I. C. Pratt	Morton, Ill	Dec. 18, 1860	30, 984
Bee-hive	J. P. Praul	Pleasant Hill, Ill	Feb. 16, 1869	87, 003
Bee-hive	J. M. Price	Buffalo Grove, Iowa	Mar. 22, 1870	101, 039
Bee-hive	J. S. Procter	Franklin, Ky	Dec. 26, 1871	122, 279
Bee-hive	T. Prosser	Birmingham, Pa	June 8, 1858	20, 508
Bee-hive	E. D. Pugh	Fort Plain, Iowa	Sept. 12, 1871	118, 972

Index of patents issued from the United States Patent Office from 1790 *to* 1873, *inclusive*—Continued.

Invention.	Inventor.	Residence.	Date.	No.
Bee-hive	R. Ramsay	New Wilmington, Pa	Jan. 7, 1862	34, 094
Bee-hive	D. R. Read	Lawrence, Kans	Apr. 18, 1871	113, 928
Bee-hive	E. B. Redfield and E. C. Hubbard.	White's Corners and Water Valley, N. Y.	June 15, 1869	91, 266
Bee-hive	B. H. Reece	Marion, Iowa	July 15, 1862	35, 892
Bee-hive	O. P. Reeve	Tipton, Iowa	Jan. 10, 1865	45, 859
Bee-hive	O. P. Reeve and C. C. Parker	Central City, Mo	Aug. 8, 1871	117, 815
Bee-hive	J. L. Reid	Van Wert, Ohio	Aug. 12, 1862	36, 169
Bee-hive	S. B. Replogle	Martinsburgh, Pa	July 21, 1868	80, 091
Bee-hive	O. Reynolds	Webster, N. Y	Dec. 4, 1844	3, 841
Bee-hive	O. Reynolds	Webster, N. Y	July 31, 1847	5, 211
Bee-hive	O. Reynolds	Webster, N. Y	June 4, 1861	32, 484
Bee-hive	W. Reynolds and J. V. Brooks	Lexington, Ill	Mar. 28, 1871	113, 207
Bee-hive	D. Rice	Richfield, Ill	Aug. 12, 1873	141, 818
Bee-hive	J. C. Rich	Penfield, N. Y	Mar. 4, 1843	2, 988
Bee-hive	S. L. Richardson	Webster City, Iowa	Dec. 29, 1868	85, 401
Bee-hive	J. E. Richey and C. Hotchkiss	Van Wert, Ohio	Mar. 15, 1864	41, 942
Bee-hive	J. Robb	Lewistown, Pa	Dec. 31, 1845	4, 340
Bee-hive	W. H. Roberts	Campbell's Station, Tenn	June 18, 1872	128, 070
Bee-hive	H. G. Robertson	Greenville, Tenn	Jan. 29, 1856	14, 168
Bee-hive	G. A. Robinson	Mount Pulaski, Ill	July 27, 1869	93, 009
Bee-hive	J. M. Robnett	Centralia, Ill	June 1, 1869	90, 873
Bee-hive	S. Rogers and A. J. Mason	Butler, Ind	Sept. 24, 1872	131, 628
Bee-hive	S. Rogers and A. J. Mason	Butler, Ind	Aug. 26, 1873	142, 281
Bee-hive	H. F. Rohm	West Providence, Pa	Sept. 20, 1870	107, 548
Bee-hive	H. F. Rohm	West Providence, Pa	Sept. 20, 1870	107, 628
Bee-hive	J. Rohrer	Middletown, Pa	Oct. 8, 1861	33, 451
Bee-hive	A. I. Root and M. Andrews	Medina, Ohio	June 18, 1872	128, 072
Bee-hive	J. E. Ross	Mount Sidney, Va	July 8, 1842	2, 707
Bee-hive	D. Rudolph	Sugar Grove, Ohio	Mar. 3, 1868	75, 201
Bee-hive	W. A. Ruth	Wyoming, Del	Nov. 29, 1870	109, 672
Bee-hive	H. W. Sabin	Gorham, N. Y	Jan. 15, 1846	4, 355
Bee-hive	A. Sanburn	Carthage, Ohio	Mar. 26, 1845	3, 972
Bee-hive	B. D. Sanders	Holliday's Cove, Va,	Oct. 27, 1857	18, 523
Bee-hive	M. R. Sanders	Cambria Township, Pa	Nov. 14, 1871	120, 899
Bee-hive	J. D. Sanderson	Stetson, Me	Dec. 14, 1858	22, 309
Bee-hive	R. Sanford	Marion, N. Y	Nov. 23, 1869	97, 124
Bee-hive	W. K. Sawyer	Three Oaks, Mich	Aug. 1, 1871	117, 690
Bee-hive	G. C Schneider	Adrian, Mich	Oct. 6, 1868	82, 880
Bee-hive	H. Scovell and J. C. Banker	Waseca, Minn	Sept. 24, 1872	131, 632
Bee-hive	N. B. Sebring	Metamora, Ohio	May 18, 1869	90, 199
Bee-hive	B. F. See	Monroe, Ohio	Jan. 19, 1869	86, 037
Bee-hive	P. J. Severson	Knowersville, N. Y	Feb. 25, 1868	74, 854
Bee-hive	D. and A. M. Shaeffer	Centreville, Iowa	Dec. 31, 1867	72, 918
Bee-hive	H. M. Shaffer	Bucyrus, Ohio	Oct. 2, 1860	30, 255
Bee-hive	M. B. Shaw	Zionsville, Iowa	June 18, 1872	127, 991
Bee-hive	E. N. Shedd	Three Oaks, Mich	Sept. 21, 1869	95, 151
Bee-hive	T. Shields	Hillsborough, Ohio	Oct. 6, 1868	82, 757
Bee-hive	S. P. Shipley	Olena, Ohio	June 16, 1868	78, 898
Bee-hive	J. Shoe	Pleasant Hill, Ohio	Dec. 8, 1868	84, 774
Bee-hive	J. Sholl	New York, N. Y	Nov. 9, 1839	1, 402
Bee-hive	J. H. Shook	Normal, Ill	Oct. 14, 1873	143, 643
Bee-hive	S. A., F. J., and J. B. Short, and J. Kile.	Decatur, Ala	Mar. 7, 1871	112, 388
Bee-hive	W. P. Shortridge	Easton, Mo	Sept. 12, 1871	118, 822
Bee-hive	S. Shrock	New Philadelphia, Pa	Nov. 10, 1868	84, 010
Bee-hive	H. P. Simmons and A. J. King	Paterson and Hudson, N. J	Feb. 4, 1873	135, 493
Bee-hive	N. Simpson	Pomeroy, Ohio	May 21, 1867	65, 020
Bee-hive	W. Y. Singleton	Springfield, Ill	Feb. 18, 1868	74, 618
Bee-hive	L. S. Sisson	West Edmeston, N. Y	Jan. 10, 1871	110, 937
Bee-hive	G. Slusser	Hillsborough, Ohio	Aug. 13, 1867	67, 680
Bee-hive	A. J. Smith	Decorah, Iowa	Oct. 15, 1861	33, 499
Bee-hive	A. J. Smith	Wayland, Mich	Aug. 22, 1865	49, 563
Bee-hive	A. J. Smith and H. C. Reed	Decorah, Iowa	Mar. 31, 1868	76, 258
Bee-hive	G. W. Smith, N. B. Vosburgh, A. J. Kramer, and W. L. Winter.	Linn County, Iowa	Aug. 4, 1863	39, 451
Bee-hive	J. L. Smith	Liberty Centre, Ohio	Sept. 27, 1870	107, 732
Bee-hive	J. W. Smith	Iowa Point, Kans	Jan. 6, 1863	37, 363
Bee-hive	M. S. Snow	Forestville, N. Y	Mar. 24, 1868	75, 804
Bee-hive	C. E. Spaulding	Theresa, N. Y	May 11, 1869	89, 896
Bee-hive	C. J. Sperry and L. Chandler	New London, Minn	Sept. 30, 1873	143, 307
Bee-hive	G. Spinney	Saugus, Mass	Sept. 26, 1865	50, 179
Bee hive	O. Sprague	Fulton, Ill	Mar. 28, 1865	47, 047
Bee-hive	N. A. Springer	Pontiac, Ill	Nov. 22, 1870	109, 557
Bee-hive	H. Staggs	Topeka, Kans	Jan. 9, 1872	122, 542
Bee-hive	U. Stanbury	Plymouth, Ind	Oct. 20, 1868	83, 335
Bee-hive	T. Stanly	Farmington, Conn	Dec. 9, 1811	......
Bee-hive	T. Stanly	Farmington, Conn	Jan. 6, 1812	......
Bee-hive	S. Stansberry	Knoxville, Tenn	Mar. 2, 1858	19, 520
Bee-hive	R. P. Starbuck	Gallatin, Mo	May 18, 1869	90, 203
Bee hive	J. H. Starr	Middleburgh, N. Y	Nov. 6, 1866	59, 471
Bee-hive	J. B. Staunton	Ellicottsville, N. Y	May 28, 1872	127, 281
Bee-hive	J. G. Staunton	Ellicottsville, N. Y	Nov. 19, 1872	133, 128
Bee-hive	M. J. Stearnes	Galesville, Wis	Feb. 7, 1871	111, 696
Bee-hive	S. D. Stearns and G. Ellsworth.	Weston, Ohio	Sept. 12, 1871	118, 827
Bee-hive	S. Stevens	New Carlisle, Ind	May 14, 1867	64, 806
Bee-hive	S. Stevens	Terre Coupee, Ind	June 15, 1869	91, 492
Bee-hive	A. J. Sternberg	Butler, Ind	Sept. 9, 1873	142, 748
Bee-hive	H. A. Stidger	Carrollton, Ohio	Feb. 18, 1868	74, 627
Bee-hive	M. Stillwell	Manlius, N. Y	Oct. 22, 1861	33, 548
Bee-hive	J. H. Stockwell	Bronson, Mich	Oct. 21, 1873	143, 937
Bee-hive	O. Stoddard	Busti, N. Y	Aug. 15, 1848	5, 714
Bee-hive	L. M. Stoops	Grand View, Ind	Oct. 27, 1868	83, 565
Bee-hive	J. B. Strickler	Milford, Ill	Jan. 6, 1863	37, 310
Bee-hive	H. Stump	Adel, Iowa	Dec. 6, 1864	45, 378
Bee-hive	A. J. Surles	Florence, Ga	Mar. 18, 1851	7, 991
Bee-hive	C. Suydam	Lambertville, N. J	Nov. 18, 1845	4, 272

Index of patents issued from the United States Patent Office from 1790 *to* 1873, *inclusive*—Continued.

Invention.	Inventor.	Residence.	Date.	No.
Bee-hive	D. H. Swartz	Lancaster, Ohio	Jan. 24, 1871	111, 273
Bee-hive	J. Sweet	Bethlehem, Mass	Apr. 11, 1810	....
Bee-hive	J. Tallman	Clayton, Ill	Sept. 22, 1868	82, 363
Bee-hive	P. Taltavull	Washington, D. C	Oct. 26, 1858	21, 912
Bee-hive	L. and M. Taylor and E. Cox	Jordan, Wis	Feb. 13, 1866	52, 622
Bee-hive	S. Taylor	Burlington, Me	Oct. 9, 1866	58, 696
Bee-hive	H. R. Terry	Edinborough, Pa	Nov. 5, 1861	33, 668
Bee-hive	J. H. Thomas	Rochester, N. Y	July 2, 1867	66, 415
Bee-hive	J. H Thurston	Rainsborough, Ohio	Nov. 17, 1868	84, 232
Bee-hive	S. Titcomb	Farmington, Me	Apr. 10, 1849	6, 285
Bee-hive	R. S. Torrey	Bangor, Me	June 7, 1859	24, 342
Bee-hive	R. S. Torrey	Bangor, Me	Aug. 3, 1869	93, 372
Bee-hive	H. A. Tozier	Littleton, Me	Oct. 9, 1866	58, 697
Bee-hive	J. Tritt	New Richmond, Ohio	Dec. 19, 1871	122, 080
Bee-hive	S. Trumbull	Suffield, Conn	Apr. 6, 1842	2, 539
Bee-hive	H. Tuller	Ash Grove, Ill	June 2, 1868	78, 621
Bee-hive	E. B. Turnipseed	Columbia, S. C	Oct. 18, 1870	108, 536
Bee-hive	G. V. Umbaugh	Lima, Ohio	Mar. 23, 1869	88, 235
Bee-hive	T. S. Underhill	Saint Johnsville, N. Y	Dec. 13, 1859	26, 452
Bee-hive	J. H. Valentine	Sparta, Ill	July 1, 1862	35, 791
Bee-hive	J. T. Vanduzer	Tyrone, N. Y	Dec. 27, 1864	45, 657
Bee-hive	A. C. Varela	Washington, D. C	Dec. 24, 1867	72, 572
Bee-hive	F. Varin	Huntsville, Ala	Nov. 7, 1871	120, 799
Bee-hive	S. Vreeland	Cuba, N. Y	July 20, 1869	92, 913
Bee-hive	E. Walker	Indianapolis, Ind	Dec. 27, 1870	110, 517
Bee-hive	F. R. Walker	Waterford, Pa	May 7, 1861	32, 259
Bee-hive	S. H. Walker	Somerville, Tenn	May 31, 1859	24, 251
Bee-hive	W. W. Walker	Nettle Lake, Ohio	Jan. 13, 1863	37, 414
Bee-hive	N. W. Walton and J. L. Cunningham.	Pennsborough, W. Va	Sept. 2, 1873	142, 537
Bee-hive	W. Wambach	Indianapolis, Ind	Jan. 24, 1871	111, 283
Bee-hive	P. S. Ward	Millville, Iowa	Oct. 21, 1862	36, 740
Bee-hive	W. Warren	Three Oaks, Mich	May 5, 1863	38, 431
Bee-hive	J. Wash	Mount Sterling, Ill	July 3, 1866	56, 127
Bee-hive	J. Wash	Mount Sterling, Ill	Apr. 21, 1868	77, 139
Bee-hive	D. G. Watt	Lawrence, Kans	June 29, 1869	91, 885
Bee-hive	C. Webb	Wallingford, Conn	May 4, 1841	2, 079
Bee-hive	T. Webb	Buda, Ill	Jan. 12, 1869	85, 878
Bee-hive	J. M. Weeks	Salisbury, Vt	June 30, 1836	....
Bee-hive	J. M. Weeks	Salisbury, Vt	July 1, 1841	2, 151
Bee-hive	J. H. Welty	Mount Carroll, Ill	Apr. 28, 1863	38, 340
Bee-hive	G. R. West	Fairfield, Ohio	Apr. 20, 1844	3, 554
Bee-hive	W. L. West	Elmira, N. Y	Feb. 8, 1859	22, 908
Bee-hive	J. Wheeldon	Greensburgh, Ind	Jan. 3, 1871	110, 809
Bee-hive	J. Wheeldon	Greensburgh, Ind	Nov. 14, 1871	120, 918
Bee-hive	C. Wheeler	Little Valley, N. Y	June 20, 1845	4, 087
Bee-hive	C. Wheeler	Little Valley, N. Y	May 9, 1854	10, 877
Bee-hive	G. Wheeler	Little Valley, N. Y	July 3, 1849	6, 576
Bee-hive	G. T. Wheeler	Mexico, N. Y	Nov. 22, 1870	109, 477
Bee-hive	W. Whitcomb	Grafton, Vt	Sept. 19, 1846	4, 763
Bee-hive	J. C. White	Huntsville, Ala	Feb. 13, 1872	123, 654
Bee-hive	N. F. White	Mount Pleasant, Iowa	June 21, 1870	104, 521
Bee-hive	A. Whitman, jr	Londonderry, Vt	Nov. 6, 1834	....
Bee-hive	H. H. Whitney	Waterford, Pa	Apr. 30, 1861	32, 217
Bee-hive	C. Wiggins	Fayette County, Pa	Feb. 27, 1827	....
Bee-hive	J. W. Winder	Cincinnati, Ohio	May 31, 1870	103, 697
Bee-hive	T. H. Windle	Wagontown, Pa	Aug. 10, 1858	21, 163
Bee-hive	A. Wilkinson	Greensburgh, Ind	Jan. 24, 1871	111, 289
Bee-hive	C. Williams	Weston, Mo	Aug. 28, 1860	29, 848
Bee-hive	J. Williams	Bean's Station, Tenn	May 7, 1872	126, 605
Bee-hive	R. A. Williams	Colusa, Cal	Dec. 6, 1870	109, 986
Bee-hive	S. C. Williams	Shenandoah County, Va	May 14, 1823	....
Bee-hive	H. K. Wilson	Barboursville, Ky	Apr. 16, 1872	125, 776
Bee-hive	F. Wolfersberger	Plymouth, Ohio	May 12, 1842	2, 617
Bee-hive	J. Wood	Alert, Ohio	Jan. 12, 1869	85, 884
Bee-hive	A. E. Woodhull	Spencer, Mich	Dec. 26, 1871	122, 302
Bee-hive	R. H. Woodside	Coultersville, Ill	Apr. 8, 1873	137, 645
Bee-hive	F. Woodward	Sacramento City, Cal	Aug. 13, 1861	33, 059
Bee-hive	A. T. Wright	Oskaloosa, Iowa	June 16, 1863	38, 925
Bee-hive	A. T. Wright	New Vienna, Ohio	June 4, 1867	65, 522
Bee-hive	A. T. Wright	New Vienna, Ohio	Jan. 5, 1869	85, 716
Bee-hive	A. T. Wright	New Vienna, Ohio	Jan. 12, 1869	85, 885
Bee-hive	A. T. Wright	Oskaloosa, Iowa	Aug. 26, 1873	142, 318
Bee-hive	J. M. Youart	Indianapolis, Ind	Feb. 15, 1870	99, 994
Bee-hive	V. Zimmerman	Morris, Ill	Sept. 22, 1868	82, 372
Bee-hive and management of bees	A. Alley	Cincinnati, Ohio	Aug. 26, 1831	....
Bee-hive and management of bees	E. Britton	Little Falls, N. Y	Sept. 9, 1831	....
Bee-hive and management of bees	P. Munch	Putnam, Ohio	July 27, 1831	....
Bee-hive and management of bees	L. H. Parish	Brighton, N. Y	Aug. 6, 1834	....
Bee-hive for raising queen-bees	J. Davis	Indianapolis, Ind	Nov. 23, 1869	97, 055
Bee-hive honey-box	H. M. Johnson and C. M. Barker	Marshall and Albion, Mich.	Jan. 7, 1873	134, 674
Bee-hive honey-box	E. C. Lewis	Glasgow, Mo	Nov. 14, 1871	120, 885
Bee-hive miller-trap	T. L. Gray	Thomasville, Tenn	Dec. 21, 1869	98, 052
Bee-hive portal, Moth-proof	E. Beard	Salem, Iowa	Mar. 10, 1868	75, 349
Bee-hive protector	A. S. Johnson	Waupun, Wis	Oct. 27, 1868	83, 388
Bee-hive slide	E. Beard	Charlestown, Mass	Mar. 12, 1830	....
Bee-hives, Closing and opening the entrance to	G. Upham	Hebron, Ohio	May 1, 1845	4, 018
Bee-hives, Comb frame for	W. Kenyon	Crawfordsville, Ind	Feb. 6, 1872	123, 483
Bee-hives, Comb-frame for	M. Metcalf	Grand Rapids, Mich	Jan. 14, 1862	34, 157
Bee-hives, Compound to be used in	T. F. McCafferty	Forest, Ohio	Aug. 27, 1867	68, 221
Bee-hives, Entrance to	J E. Dalton and T. Stevens	New Vienna, Ohio	Apr. 9, 1850	7, 258
Bee-hives, Feed-box of	J. D. Egleston	Canaan, Conn	Jan. 27, 1857	16, 474
Bee-hives, Fumigator for	A. R. Denlinger	Gordonsville, Pa	June 17, 1873	139, 881
Bee-hives, Machine for making comb-guide for	H. A. King	New York, N. Y	May 23, 1871	115, 066
Bee-hives, Moth-fly trap for	J. D. Meador	Independence, Mo	Sept. 1, 1868	81, 808
Bee-hives, Moth-trap for	L. Gates	Pleasant Hill, Mo	Mar. 18, 1873	136, 911
Bee-hives, Moth-trap for	R. Haven	Perrysville, Ind	Sept. 24, 1861	33, 347

Index of patents issued from the United States Patent Office from 1790 *to* 1873, *inclusive*—Continued.

Invention.	Inventor.	Residence.	Date.	No.
Bee-hives, Moth-trap for	W. J. Hazen	Bethany, Pa	Nov. 12, 1861	33, 704
Bee-hives, Moth-trap for	E. W. Phelps	Newark, Ohio	Apr. 6, 1852	8, 585
Bee-hives, Surplus honey-boxes in	H. Moon and D. C. Turner	Red Creek, N. Y	Aug. 25, 1868	81, 398
Bee-hives, Swarm-indicator for	W. W. Snell	Rushford, Minn	Oct. 24, 1865	50, 639
Bee-hives, Use of slides in	N. Potter	Buffalo, N. Y	Mar. 11, 1851	7, 970
Bee-hives, Working the doors of	J. Case	Selma, Ohio	Nov. 19, 1850	7, 779
Bee-hiving device	C. Cutler	Tonawanda, N. Y	May 21, 1861	32, 357
Bee-house	D. Burbank	Lexington, Ky	Nov. 9, 1869	96, 668
Bee-house	W. Carter	Saint Louis, Mo	Dec. 15, 1868	84, 994
Bee-house	C. Decker	New Michigan, Ill	Oct. 20, 1868	83, 263
Bee-house	D. S. Gray	Onargo, Ill	Feb. 2, 1869	86, 528
Bee-house	W. Groves	Harrisburgh, Pa	June 12, 1835	
Bee-house	J. W. Wood	Alden, Ill	Nov. 29, 1870	109, 702
Bee house and hive	W. M. Simpson	Davidsburgh, Mich	Apr. 28, 1868	77, 222
Bee-houses and bee-hives and the management thereof, Construction of.	J. Searle	Hill, N. H	Jan. 20, 1838	580
Bee-keepers' protecting-mask	L. C. Huff	Sonina, Ala	July 1, 1873	140, 417
Bee-moth instrument	R. P. Buttles	Mansfield, Pa	Nov. 23, 1869	97, 161
Bee-moth trap	G. Fletcher, sr	Greensburgh, Ind	Mar. 26, 1850	7, 212
Bee-palace	L. Hamlin	Kerkersville, Ohio	July 13, 1844	3, 664
Bee-protector	J. Cory	Holden, Mo	June 28, 1870	104, 832
Bee-protector	T. D. Howell	Zanesville, Ohio	Dec. 15, 1843	3, 379
Bee-separator	J. H. Starr	Middlebush, N. Y	Aug. 8, 1865	49, 314
Bees, Composition for stupefying	A. Y. Rozenbury	Waterloo, Ind	May 23, 1871	115, 107
Bees, House for the management of	M. A. Glass	Independence, Iowa	July 4, 1871	116, 700
Bees, Mode of managing honey	F. Kelsey	Lockport, N. Y	Aug. 26, 1828	
Bees, Trellis for propagating	A. Simons	Fairfield, Iowa	Aug. 18, 1868	81, 220
Beef and vegetable cutter, Dried	W. and J. T. Bird	Flemington, N. J	Dec. 5, 1865	51, 288
Beef-cutter	B. Meeker	Brooklyn, N. Y	Oct. 21, 1815	
Beef cutter, Dried	D. W. Goble	Newark, N. J	July 9, 1850	7, 489
Beef cutter, Dried	C. J. and D. C. Holmes	Stafford Springs, Conn	Dec. 28, 1869	98, 263
Beef, Machine for cutting smoked	S. R. Kneeland		Oct. 9, 1810	
Beef, Manufacture of chipped	C. L. Tucker	Chicago, Ill	July 20, 1869	92, 907
Beef, &c., Method of preparing	B. F. Stephens	Brooklyn, N. Y	June 8, 1869	91, 176
Beef or cold-slaw shaving or cutting machine	H. Bangs	New York, N. Y	Apr. 28, 1831	
Beef, salted, Mode of making and curing	W. A. Tomlinson	New York, N. Y	Dec. 14, 1830	
Beef-spreader	F. Tesh	Johnstown, Pa	Jan. 9, 1855	12, 222
Beef-steak and other meat, Machine for mangling	R. McCorkell	Warsaw, Minn	Oct. 25, 1864	44, 811
Beef-steak chopper	E. Atkins	Monroe, La	July 16, 1872	128, 997
Beef-steak crusher	J. J. Doyle	Sharon, Conn	Nov. 14, 1865	50, 921
Beef-steak crusher	H. Thompson	Concord, N. H	Sept. 14, 1869	94, 852
Beef-steak cutter and mangler	D. C. Thompson	Ischua, N. Y	Dec. 8, 1868	84, 718
Beef-steak cutters, Die for forming	H. Thompson	Concord, N. H	June 27, 1871	116, 372
Beef-steak for broiling, Device for preparing	W. A. Clark	Woodbridge, Conn	Apr. 4, 1871	113, 260
Beef-steak for cooking, Preparing	T. G. Stagg	New York, N. Y	Oct. 22, 1850	7, 735
Beef-steak, Machine for preparing	A. M. Bond	Concord, N. H	May 2, 1871	114, 396
Beef-steak mangle	J. Locke	Lewisburgh, Pa	Aug. 29, 1871	118, 542
Beef-steak, paring apples, and sharpening knives, Machine for pressing.	B. F. Alexander	Glen Hope, Pa	Oct. 31, 1865	50, 671
Beef-steak preparer	I. C. Nichols	Union, N. Y	Aug. 6, 1867	67, 568
Beef-steak tenderer	J. S. Morris	Mosiertown, Pa	Nov. 4, 1873	144, 216
Beer	G. W. La Baw	Jersey City, N. J	Feb. 16, 1864	41, 626
Beer, alcohol, &c., cooler and condenser	J. Frageser	New York, N. Y	June 19, 1866	55, 746
Beer, ale, and other fermented liquors, Manufacture of.	J. Firmenich	Buffalo, N. Y	Sept. 14, 1869	94, 880
Beer, ale, &c., Apparatus for preserving	R. Eickemeyer	Yonkers, N. Y	Sept. 22, 1868	82, 394
Beer, ale, &c., cooler	T. Bergner and S. Zeisse	Philadelphia, Pa	May 23, 1871	115, 018
Beer, ale, &c., on draught by means of carbonic-acid gas, Preserving and delivering.	T. Ahrens	Louisville, Ky	Aug. 2, 1870	106, 019
Beer, ale, porter, &c., Composition for the manufacture of.	T. Hawks	Rochester, N. Y	June 16, 1868	78, 874
Beer, ale, &c., Process of "hopping"	W. S. Haight	Waterford, N. Y	July 16, 1867	66, 833
Beer and ale on draught, Preserving	C. E. Haynes	Boston, Mass	Nov. 8, 1870	109, 126
Beer and liquid cooler	T. Gründmann	Cleveland, Ohio	Mar. 15, 1870	100, 752
Beer and mash cooler	C. Wise and B. Loeffler	New York, N. Y	May 7, 1867	64, 614
Beer and other liquids, Cooler for	J. Chollar and C. W. Cunningham.	Washington, D. C	July 18, 1865	48, 791
Beer and other liquids, Cooler for	J. J. Märki	Richmond, Ind	Dec. 7, 1869	97, 540
Beer and other liquids, Cooler for	G. B. Turrell	New York, N. Y	Dec. 22, 1868	85, 190
Beer and other liquids on draught, Apparatus for cooling and discharging.	J. M. Heiss	Baltimore, Md	June 20, 1871	116, 053
Beer and other liquids on draught, Cooling and preserving.	J. M. Heiss	Baltimore, Md	May 30, 1871	115, 468
Beer and water cooler	C. Geenen	New Orleans, La	Feb. 6, 1872	123, 390
Beer and water cooler	F. Wagner	Philadelphia, Pa	July 9, 1872	128, 768
Beer and water cooler	J. Weinberger	New Orleans, La	Mar. 12, 1872	124, 465
Beer and wine, Cooler for preserving	F. Loesch	Buffalo, N. Y	Dec. 30, 1873	145, 955
Beer and yeast, Manufacture of	L. Pasteur	Paris, France	July 22, 1873	141, 072
Beer, Apparatus for manufacture of	G. Habich	Roxbury, Mass	June 8, 1858	20, 488
Beer, Apparatus for preserving	C. Pholmann	Lanesville, Ky	May 24, 1870	103, 498
Beer, Apparatus for preserving and forcing	E. C. Krausnick	Saint Louis, Mo	Nov. 25, 1873	144, 987
Beer, Apparatus for raising	J. L. Treat	New York, N. Y	Apr. 13, 1869	88, 927
Beer, Brewing	C. Abresch	New York, N. Y	Nov. 23, 1869	97, 143
Beer, Brewing spruce	W. Bezeau	Philadelphia, Pa	June 13, 1815	
Beer, Composition for making	S. Hinds	Montrose, Pa	May 11, 1831	
Beer, &c., Compound for clarifying	C. Hefft	Pekin, Ill	July 23, 1872	129, 732
Beer-cooler	J. Agate	Pittsford, N. Y	July 14, 1868	79, 931
Beer-cooler	J. Agate	Pittsford, N. Y	June 1, 1869	90, 716
Beer-cooler	J. Boyle	Roxbury, Mass	May 17, 1859	24, 004
Beer-cooler	D. Cammerer	Cincinnati, Ohio	Mar. 19, 1867	63, 013
Beer-cooler	D. Cammerer	Cincinnati, Ohio	Sept. 29, 1868	82, 593
Beer-cooler	J. Chandless	New York, N. Y	Feb. 11, 1873	135, 683
Beer-cooler	H. C. Dart	New York, N. Y	May 18, 1869	90, 242
Beer-cooler	H. C. Dart	New York, N. Y	July 13, 1869	92, 429
Beer-cooler	R. Dreher	Milwaukee, Wis	Apr. 16, 1872	125, 668
Beer-cooler	H. Ellerbrock and C. Mahler	Baltimore, Md	Feb. 6, 1872	123, 465
Beer-cooler	J. Fallows	Philadelphia, Pa	Nov. 3, 1868	93, 702
Beer-cooler	A. Fischer	New York, N. Y	Mar. 20, 1866	53, 285

Invention.	Inventor.	Residence.	Date.	No.
Beer-cooler	A. Foubert	Buffalo, N. Y	Jan. 9, 1872	122, 592
Beer-cooler	J. Frageser	New York, N. Y	Dec. 17, 1861	33, 967
Beer-cooler	G. Fuchs and J. Luigart	Logansport, Ind	June 9, 1868	78, 659
Beer-cooler	J. Geemen	Chicago, Ill	Nov. 10, 1868	83, 953
Beer-cooler	W. Gee	New York, N. Y	Aug. 23, 1870	106, 686
Beer-cooler	J. Gimlich	Pittsfield Mass	May 27, 1873	139, 310
Beer-cooler	P. H. Griffin	Albany, N. Y	Dec. 31, 1867	72, 731
Beer-cooler	V. Haeffner	Dobb's Ferry, N. Y	Jan. 20, 1863	37, 443
Beer-cooler	G. M. Haszinger	Vicksburgh, Miss	June 13, 1871	115, 850
Beer-cooler	C. W. Haug	New York, N. Y	Jan. 24, 1871	111, 203
Beer-cooler	H. Heimerle	Buffalo, N. Y	Dec. 31, 1867	72, 733
Beer-cooler	J. Herget	Saint Louis, Mo	Sept. 11, 1866	57, 900
Beer-cooler	N. Hiemenz	Buffalo, N. Y	May 12, 1868	77, 732
Beer-cooler	A. Hitscherich	Milwaukee, Wis	Nov. 9, 1869	96, 590
Beer-cooler	J. Hoefer	New York, N. Y	May 23, 1865	47, 823
Beer-cooler	C. Jones	Brooklyn, N. Y	May 24, 1859	24, 124
Beer-cooler	E. Kraft	Tyrone, Pa	Dec. 24, 1872	134, 208
Beer-cooler	J. M. Otto	Brooklyn, N. Y	June 4, 1872	127, 639
Beer-cooler	A. Pfund	New York, N. Y	June 17, 1873	140, 073
Beer-cooler	H. Pietsch and M. Walter	Milwaukee, Wis	Apr. 6, 1869	88, 584
Beer-cooler	A. D. Puffer	Somerville, Mass	Nov. 24, 1868	84, 439
Beer-cooler	T. L. Rankin and C. W. Grassmuck.	New Richmond, Ohio, and Peru, Ill.	Dec. 29, 1868	85, 477
Beer-cooler	H. Rensch	Quincy, Ill	Nov. 27, 1866	60, 059
Beer-cooler	A. F. Rick	Chicago, Ill	Apr. 22, 1873	138, 195
Beer-cooler	C. L. Ridgway	Boston, Mass	Aug. 29, 1871	118, 646
Beer-cooler	W. Rose	Saint Louis, Mo	Apr. 3, 1866	53, 682
Beer-cooler	D. Sager	Albany, N. Y	June 28, 1864	43, 342
Beer-cooler	A. W. and J. Sangster	Buffalo, N. Y	Apr. 16, 1867	63, 947
Beer-cooler	F. Schmidt	Milwaukee, Wis	Apr. 29, 1873	138, 289
Beer-cooler	L. Schulze	Louisville, Ky	Oct. 13, 1868	83, 099
Beer-cooler	P. Schweikhart	Buffalo, N. Y	May 7, 1867	64, 452
Beer-cooler	H. Shlaudeman	Decatur, Ill	May 5, 1868	77, 664
Beer-cooler	H. Steubing	New York, N. Y	Oct. 6, 1863	40, 200
Beer-cooler	F. Streubel and A. Roos	New York, N. Y	Oct. 22, 1861	33, 549
Beer-cooler	M. Tschirgi and L. Kammüller	Dubuque, Iowa	Aug. 14, 1866	57, 221
Beer-cooler	G. B. Turrell	New York, N. Y	July 16, 1861	32, 845
Beer-cooler	F. Uhrland	Buffalo, N. Y	Sept. 19, 1865	50, 055
Beer-cooler	G. Waters	Cincinnati, Ohio	Sept. 19, 1865	50, 056
Beer-cooler	G. Winter	Buffalo, N. Y	Oct. 27, 1868	83, 431
Beer-cooler	J. Yates and E. Deuell	Brooklyn, N. Y	Nov. 17, 1868	84, 244
Beer-cooler for beer on draught	H. Sell	New York, N. Y	Mar. 14, 1871	112, 742
Beer-cooler, Portable	E. C. Bundy	Oneonta, N. Y	Mar. 15, 1870	100, 852
Beer-cooler tap-holder attachment	J. H. Fisher	Chicago, Ill	Sept. 23, 1873	143, 068
Beer-drawing attachments for casks, &c	S. Marks	San Francisco, Cal	Oct. 28, 1873	144, 119
Beer-elevator	A. H. Ladner and T. F. Fenlin	Philadelphia, Pa	June 28, 1870	104, 742
Beer, Factitious	F. Luedke	New York, N. Y	Sept. 25, 1860	30, 147
Beer, Fermenting-tun for	A. Hammer	Philadelphia, Pa	Jan. 2, 1855	12, 160
Beer-forcing apparatus	W. H. Otto and P. Korper	Tremont, Pa	Feb. 13, 1872	123, 639
Beer-fountain, Portable	D. Gay	Bath, Me	Apr. 24, 1849	6, 393
Beer from casks, Hydro-pneumatic apparatus for raising.	R. Sealy	New York, N. Y	Aug. 31, 1844	3, 725
Beer from barrels, Apparatus for forcing	J. Devlin	Brooklyn, N. Y	May 28, 1872	127, 156
Beer from barrels, Apparatus for forcing	J. Devlin	Brooklyn, N. Y	Nov. 19, 1872	133, 084
Beer, &c., from barrels, Apparatus for forcing	J. Devlin	Brooklyn, N. Y	June 17, 1873	139, 882
Beer from malt and Indian meal, Manufacture of	J. Singer	Chicago, Ill	Aug. 25, 1863	39, 685
Beer-glass, &c., Froth-arrester for	J. Winkler	Hudson City, N. J	Nov. 17, 1868	84, 242
Beer-hopping apparatus	W. S. Haight	Waterford, N. Y	June 30, 1868	79, 342
Beer, Machine for mashing and boiling wort for	J. and M. Stark	Buffalo, N. Y	Jan. 7, 1868	73, 056
Beer, Making	H. L. Bowker	Boston, Mass	Sept. 12, 1871	118, 901
Beer, Making	J. S. Bressler	Milwaukee, Wis	Mar. 6, 1866	53, 078
Beer, Making	E. J. Krause	Lancaster, Pa	June 27, 1865	48, 413
Beer, Making spruce	J. Williams	Cambridge, Mass	Feb. 17, 1820	....
Beer, Manufacture of	C. C. Haley	Troy, N. Y	Oct. 29, 1872	132, 574
Beer, Manufacture of lager	J. Schneider	Williamsburgh, N. Y	July 17, 1866	56, 453
Beer, Manufacture of small	O. F. Green and J. E. Clark	Saint Louis, Mo	Sept. 22, 1868	82, 401
Beer, Manufacturing spruce	G. Jones	Boston, Mass	Jan. 27, 1832	....
Beer-material, Concentrated	F. G. Rietsch	Rudoletz, Austria	Feb. 3, 1852	8, 708
Beer-mug	W. C. King	Pittsburgh, Pa	Dec. 30, 1873	146, 078
Beer on draught, Apparatus for preserving	J. W. Moore	Bellefonte, Pa	Aug. 19, 1873	141, 944
Beer or ale by re-brewing, Restoring sour or musty	M. Granger	Lowville, N. Y	Jan. 11, 1832	....
Beer, Pine-apple	G. Rivera	Cambridgeport, Mass	Feb. 11, 1868	74, 429
Beer-powder	J. McKellar	Thomaston, Me	May 15, 1860	28, 289
Beer-preserver and cooler	J. Lorenz	Hamilton, Ohio	May 23, 1871	115, 072
Beer, Preserving	A. Adametz	New York, N. Y	Sept. 2, 1873	142, 428
Beer-preserving apparatus	W. Dietrichsen	Newark, N. J	May 25, 1869	90, 349
Beer-preserving apparatus	A. F. W. Neynaber	Philadelphia, Pa	May 29, 1866	55, 151
Beer-preserving cellar	R. Schmid	Chicago, Ill	Mar. 16, 1869	87, 878
Beer, Root or tonic	B. Bates	Baltimore, Md	Apr. 4, 1871	113, 617
Beer, Vegetable	S. Whiton	Hartford, Conn	Mar. 21, 1846	4, 430
Beer, &c., vessel	W. Compton	New York, N. Y	Oct. 30, 1866	59, 188
Beer, water, and other liquids, Apparatus for cooling and purifying.	J. P. Gruber	New York, N. Y	Feb. 22, 1870	100, 027
Beer while on draught, Preserving	D. Wernz	New York, N. Y	Nov. 13, 1866	59, 690
Beer with cold air, Apparatus for preserving and forcing.	F. Blucher	Mascoutah, Ill	Dec. 5, 1871	121, 577
Beer-wort, &c., Method of cooling	O. P. Lewis	Cincinnati, Ohio	Oct. 25, 1870	108, 606
Belaying-cleat	J. Bangs	South Dennis, Mass	Feb. 25, 1868	74, 789
Belaying cleat	C. S. H. Foster	Deer Isle, Me	Dec. 15, 1868	84, 873
Belaying-cleat	H. Ryder	Somerville, Mass	Oct. 8, 1872	131, 974
Belaying-cleat for boat	W. W. Andrews	Warrensville, Ohio	Oct. 21, 1862	36, 696
Bell	A. G. Bevin	East Hampton, Conn	June 1, 1869	90, 811
Bell	W. Fletcher	New York, N. Y	June 1, 1869	90, 654
Bell	W. Hoyt	Brookville, Ind	Mar. 3, 1825	....
Bell	J. Kintz	West Meriden, Conn	Apr. 19, 1870	102, 016
Bell	J. Regester	Baltimore, Md	Mar. 14, 1871	112, 631
Bell	J. S. Tibbets	Evansville, Ind	Aug. 4, 1863	39, 433
Bell	H. Tyler	Utica, N. Y	Oct. 20, 1826	....

Index of patents issued from the United States Patent Office from 1790 *to* 1873, *inclusive*—Continued.

Invention.	Inventor.	Residence.	Date.	No.
Bell, Æolian chiming	H. Herrmann	New York, N. Y	Feb. 7, 1871	111, 537
Bell, Alarm	S. C. Bond	Hainesville, Ill	May 9, 1871	114, 521
Bell, Alarm	L. F. Bruce	Bridgeport, Conn	Aug. 3, 1869	93, 168
Bell, Alarm	L. Holmes	Paterson, N. J	May 18, 1869	90, 173
Bell, Alarm	T. and G. A. Pemberton	Birmingham, England	Apr. 19, 1870	102, 035
Bell, Alarm	C. Penfield	New Britain, Conn	Jan. 12, 1869	85, 760
Bell, Alarm	W. M Preston	Roxbury, N. Y	Apr. 2, 1872	125, 328
Bell, Alarm	C. A. Slack	Frenchtown, N. J	Feb. 9, 1869	86, 876
Bell, Alarm	C. Wiley	Hannibal Centre, N. Y	Mar. 23, 1869	88, 106
Bell alarm, Door	C. C. Gerhardt and G. S. Lander.	Wyandotte, Kans	Apr. 16, 1872	125, 731
Bell and bell wheel, Rotating	J. Regester	Baltimore, Md	July 13, 1869	92, 470
Bell and burglar-alarm, Door	D. L. Collins	Antwerp, N. Y	Nov. 20, 1866	59, 820
Bell and burglar-alarm, Door	E. H. Crane	Burr Oak, Mich	June 26, 1866	55, 823
Bell and burglar-alarm, Door	M. A. Genung	Granville, Ohio	June 17, 1862	35, 603
Bell and caster-stand, Call	F. A. Blatterlien	West Meriden, Conn	Nov. 8, 1870	109, 101
Bell and gong striking apparatus	C. and G. M. Stevens	Boston, Mass	May 21, 1872	126, 992
Bell and slop-bowl, Call	N. Lawrence	Taunton, Mass	June 28, 1870	104, 859
Bell and table-caster, Combined	H. A. Dierkes	New York, N. Y	Dec. 7, 1869	97, 617
Bell and vessel, Call	N. Lawrence	Taunton, Mass	Aug. 30, 1870	106, 840
Bell attachment	A. E. Taylor	Ogdensburgh, N. Y	Oct. 23, 1860	30, 509
Bell attachment, Alarm	A. Sherwood	Auburn, N. Y	Aug. 26, 1873	142, 287
Bell, Call	H. H. Abbe	East Hampton, Conn	Jan. 2, 1866	51, 784
Bell, Call	E. C. Barton	East Hampton, Conn	Aug. 20, 1872	130, 690
Bell, Call	N. L. Bradley	West Meriden, Conn	Aug. 25, 1863	39, 697
Bell, Call	E. G. Cone	East Hampton, Conn	Oct. 27, 1868	83, 468
Bell, Call	H. A. Dierkes	New York, N. Y	Aug. 8, 1871	117, 871
Bell, Call	H. A. Dierkes	New York, N. Y	Mar. 26, 1872	125, 029
Bell, Call	H. A. Dierkes and J. Fretts	New York, N. Y	Mar. 21, 1871	112, 907
Bell, Call	W. H. Nichols	East Hampton, Conn	July 16, 1867	66, 874
Bell, Call	W. H. Nichols	East Hampton, Conn	Aug. 27, 1872	130, 818
Bell, Call	E. Parker	West Meriden, Conn	Mar. 29, 1864	42, 107
Bell, Call	D. W. Sexton	East Hampton, Conn	May 3, 1864	42, 604
Bell, Call	H. Stratton	West Meriden, Conn	Feb. 11, 1873	135, 860
Bell, Call	H. A. Thompson	Hartford, Conn	June 5, 1866	55, 392
Bell, Call	C. Volger	Wilmington, Del	June 20, 1871	116, 241
Bell, Car	A. Borrowman	New York, N. Y	Apr. 16, 1867	63, 843
Bell, Car	J. Sweeney	New York, N. Y	Nov. 27, 1866	60, 083
Bell, Cast-iron	E. G. Cone	East Hampton, Conn	Aug. 6, 1867	67, 500
Bell, Ceremonial	J. H. Smith	Keokuk, Iowa	Jan. 10, 1871	110, 875
Bell, Chime and alarm	G. R. Meneely	West Troy, N. Y	Jan. 2, 1872	122, 397
Bell, Cow or sheep	S. Newton		Dec. 22, 1804	....
Bell-crank	M. C. Ames	Hartford, Conn	May 7, 1872	126, 369
Bell, Door	W. Allport	New Britain, Conn	Dec. 29, 1868	85, 352
Bell, Door	W. Allport	New Britain, Conn	Apr. 12, 1870	101, 699
Bell, Door	H. D. Blake	New Britain, Conn	Apr. 9, 1872	125, 530
Bell, Door	F. Blakemore	Philadelphia, Pa	July 2, 1872	128, 458
Bell, Door	A. T. Brooks	New Britain, Conn	Dec. 31, 1867	72, 791
Bell, Door	A. T. Brooks	New Britain, Conn	June 30, 1868	79, 308
Bell, Door	N. F. Cone	La Crosse, Wis	Sept. 8, 1863	39, 796
Bell, Door	N. F. Cone	La Crosse, Wis	Feb. 23, 1864	41, 683
Bell, Door	J. P. Connell	Kensington, Conn	Mar. 21, 1871	112, 903
Bell, Door	J. P. Connell	Kensington, Conn	June 18, 1872	127, 959
Bell, Door	J. P. Connell	Kensington, Conn	Oct. 15, 1872	132, 199
Bell, Door	M. L. Delavan and J. Dyson	New Britain, Conn	July 23, 1867	67, 030
Bell, Door	M. A. Genung	Granville, Ohio	Mar. 10, 1863	37, 863
Bell, Door	E. H. Goldman and D. W. Hisey.	Kansas, Ill	May 24, 1870	103, 446
Bell, Door	T. Lyons	Hartford, Conn	Aug. 11, 1868	80, 983
Bell, Door	W. T. Munger	New Britain, Conn	June 20, 1871	116, 082
Bell, Door	W. H. Nichols	Chatham, Conn	Jan. 30, 1866	52, 365
Bell, Door	W. H. Nichols	East Hampton, Conn	Sept. 22, 1868	82, 434
Bell, Door	C. S. Nickelson	Canton, N. Y	Oct. 24, 1865	50, 616
Bell, Door	O. A. North	New Britain, Conn	Apr. 27, 1869	89, 331
Bell, Door	O. B. Oakley and H. Rosekrans	San Francisco, Cal	Sept. 1, 1868	81, 672
Bell, Door	M. B. Ogden	Fond du Lac, Wis	Aug. 23, 1864	43, 924
Bell, Door	C. Penfield	New Britain, Conn	July 30, 1867	67, 212
Bell, Door	C. Penfield	New Britain, Conn	Jan. 12, 1869	85, 759
Bell, Door	C. W. Saladee	Circleville, Ohio	Mar. 16, 1869	87, 794
Bell, Door	E. B. Sims	Antwerp, N. Y	Sept. 16, 1873	142, 953
Bell, Door	A. A. Stuart	Cedar Rapids, Iowa	May 20, 1873	139, 206
Bell, Door	A. L. Swan	Cherry Valley, N. Y	Nov. 19, 1872	133, 269
Bell, Door	A. Turnbull	New Britain, Conn	June 13, 1865	48, 242
Bell, Door	W. H. Watrous	Hartford, Conn	Mar. 20, 1866	53, 366
Bell, Door and alarm	G. O. Lackey	Akron, Ohio	Oct. 25, 1870	108, 708
Bell, Diving	J. A. Weisse	New York, N. Y	Apr. 27, 1869	89, 453
Bell, Electro-magnetic alarm	M. G. Farmer	Salem, Mass	May 4, 1852	8, 920
Bell, Electro-magnetic alarm	C. Williams, jr., and J. Redding	Somerville and Charlestown, Mass.	Feb. 7, 1871	111, 707
Bell, Electro-magnetic fog	A. Barbarin and B. F. Simms	New Orleans, La	July 15, 1856	15, 323
Bell, Enameled	F. Raymond	Woodhaven, N. Y	June 29, 1869	92, 095
Bell, Fog	A. C. Rand and R. R. Johnson	Buffalo, N. Y	Apr. 13, 1858	19, 949
Bell for doors, Alarm	E. Barton	East Hampton, Conn	Sept. 9, 1862	36, 436
Bell for fire-engines, Alarm	J. P. Parke	Philadelphia, Pa	Dec. 19, 1808	....
Bell for horses	J. Barton	Cairo, N. Y	Mar. 18, 1834	....
Bell for horses, Alarm chime	C. Kirchkof	Newark, N. J	Sept. 29, 1863	40, 108
Bell, Gong	I. A. Bevin	Chatham, Conn	Oct. 9, 1866	58, 580
Bell, Gong	L. S. Carpenter	East Hampton, Conn	Apr. 20, 1869	89, 024
Bell, Gong	H. A. Foss	New Britain, Conn	May 28, 1872	127, 333
Bell-head, Rotary	S. M. and W. M. Fulton	Pittsburgh, Pa	Nov. 30, 1869	97, 383
Bell-holder, Stationary	S. Croll	Philadelphia, Pa	Feb. 14, 1871	111, 729
Bell, Horse	J. Barton	Middle Haddam, Conn	Feb. 14, 1854	10, 532
Bell, House	H. Barton	East Hampton, Conn	Oct. 9, 1866	58, 576
Bell, House	J. Barton	East Hampton, Conn	June 15, 1858	20, 538
Bell, House	A. L. Swan	Cherry Valley, N. Y	Dec. 26, 1871	122, 200
Bell, House	A. Turnbull	New Britain, Conn	June 12, 1866	55, 558
Bell joint	B. P. Walker	Wolverhampton, England	May 4, 1869	89, 820
Bell-lever box	B. W. Hopper	Astoria, N. Y	Jan. 8, 1869	91, 019

Index of patents issued from the United States Patent Office from 1790 *to* 1873, *inclusive*—Continued.

Invention.	Inventor.	Residence.	Date.	No.
Bell-levers, Attaching caps to bases of	E. W. Brettel	Elizabeth, N. J	Aug. 2, 1870	105, 900
Bell, Locomotive alarm	B. Briscoe	Detroit, Mich	Mar. 2, 1869	87, 462
Bell-machinery for hotels, &c	T. D. Jackson and A. Judson	Rochester, N. Y	Oct. 17, 1846	4, 816
Bell, Magnetic alarm	A. Eckert	Trenton, Ohio	Oct. 10, 1854	11, 780
Bell or gong, Door	H. H. Abbe	Chatham, Conn	July 11, 1865	48, 637
Bell or gong, Door	A. G. Dexter	San Francisco, Cal	June 13, 1865	48, 234
Bell, Portable house	A. W. Hale	New Britain, Conn	Aug. 31, 1858	21, 335
Bell, Pressure	J. Barton	Middle Haddam, Conn	Apr. 8, 1856	14, 593
Bell-pull	S. Bonsall and L. Hillebrand	Philadelphia, Pa	Mar. 3, 1868	75, 118
Bell-pull	S Bonsall and L. Hillebrand	Philadelphia, Pa	Dec. 15, 1868	84, 998
Bell-pull	C. J. Bradbury	Boston, Mass	Feb. 28, 1860	27, 330
Bell-pull	J. Garvey and M. H. Kimball	San Francisco, Cal	Nov. 3, 1868	83, 622
Bell-pull	J. Garvey and M. H. Kimball	San Francisco, Cal	Nov. 3, 1868	83, 623
Bell-pull	H. Homer	New York, N. Y	Sept. 5, 1865	49, 757
Bell-pull	W. T. Munger	New Britain, Conn	June 10, 1873	139, 687
Bell, Pull	W. M. Preston	Roxbury, N. Y	Aug. 20, 1872	130, 742
Bell-pull	J. J. C. Smith	Somerville, Mass	Jan. 18, 1870	99, 020
Bell-pull	A. L. Swan	Cherry Valley, N. Y	Nov. 7, 1871	120, 682
Bell-pull	A. L. Swan	Cherry Valley, N. Y	Aug. 19, 1873	142, 057
Bell-pull, Noiseless	J. F. Cory	New York, N. Y	Nov. 3, 1868	83, 606
Bell pulls and trips, Construction of	S. L. Covell, jr	Troy, N. Y	July 16, 1867	66, 801
Bell-ringer, Automatic	E. N. Scherr	Philadelphia, Pa	Mar. 1, 1859	23, 118
Bell-ringer, Steam	W. H. Beach	Chicago, Ill	Apr. 19, 1864	42, 428
Bell-ringer, Steam	G. B. Snow	Buffalo, N. Y	June 11, 1872	127, 933
Bell-ringer, Steam	J. West and O. M. Parker	Quincy, Ill	June 25, 1872	128, 441
Bell-ringer, Steam	J. West and O. M. Parker	Quincy, Ill	Jan. 7, 1873	134, 719
Bell-ringing apparatus	J. R. Baird	Vincennes, Ind	Nov. 17, 1857	18, 623
Bell-ringing apparatus	J. Harrison	Troy, N. Y	Aug. 28, 1860	29, 784
Bell-ringing apparatus	R. Kinsley	Springfield, Mass	June 11, 1861	32, 520
Bell-ringing apparatus for locomotive, Automatic	J. S. Lamar	Augusta, Ga	July 9, 1872	128, 801
Bell-rope supporter	W. C. Marshall	Hartford, Conn	Aug. 10, 1869	93, 461
Bell-ropes, Eye for railway-car	W. M. Walton	Newark, N. J	Dec. 21, 1869	98, 130
Bell, Self-adjusting fog	H. L. De Zeng	Geneva, N. Y	Aug. 26, 1856	15, 605
Bell, Sheet-steel	J. E. Tencate	Pittsburgh, Pa	Jan. 9, 1872	122, 679
Bell, Signal	G. F. and D. H. Benckert	Philadelphia, Pa	Oct. 11, 1859	25, 714
Bell, Signal	J. A. Woodward	Philadelphia, Pa	Nov. 15, 1859	26, 137
Bell, Signal or alarm	G. H. Hoagland	Port Jervis, N. Y	July 21, 1857	17, 836
Bell strap, Cow	J. H. and A. Hughs	Wautoma, Wis	Dec. 19, 1871	122, 020
Bell, Table or call	A. W. Turner	New York, N. Y	Jan. 30, 1866	52, 344
Bell to straps, Attaching	D. W. Welch	Middle Haddam, Conn	Aug. 10, 1869	93, 652
Bell-yoke attachment	C. H. Meneely	Troy, N. Y	May 9, 1871	114, 585
Bells and adjustable clappers for ringing fog	D. Jones, jr	Saint John, New Brunswick.	Nov. 27, 1849	6, 915
Bells, &c., Apparatus for sounding house	J. Corduan	Brooklyn, N. Y	Mar. 29, 1859	23, 353
Bells by steam, Ringing	G. B. Snow	Buffalo, N. Y	July 11, 1854	11, 307
Bells, Casting	A. Jusberg	Galva, Ill	Aug. 27, 1867	68, 206
Bells, Constructing and hanging	E. Dewey	New York, N. Y	Apr. 10, 1839	1, 114
Bells, Construction and tone of	R. Leslie		Feb. 2, 1793	....
Bells, Device for ringing street-car	C. Carr	Boston, Mass	Apr. 5, 1870	101, 580
Bells, Die for making	A. Patterson	Birmingham, Pa	Mar. 5, 1867	62, 678
Bells, Electro-magnetic machine for ringing	C. Robinson	New York, N. Y	Dec. 6, 1864	45, 347
Bells for cows, &c., Brazing and bronzing	S. Booth	Berlin, Conn	June 1, 1832	
Bells, Hanging	G. E. Baker	Waukegan, Ill	July 10, 1866	56, 160
Bells, Hanging	H. Belfied	Philadelphia, Pa	Apr. 12, 1859	23, 639
Bells, Hanging	T. H. Bell	Washington, D. C	Apr. 17, 1860	27, 880
Bells, Hanging	J. Currier	Boston, Mass	Jan. 23, 1834	....
Bells, Hanging	N. G. Du Bois	Brooklyn, N. Y	Jan. 12, 1858	19, 082
Bells, Hanging	G. Equillon	Paris, France	Oct. 26, 1869	96, 092
Bells, Hanging	S. Fuller	Boston, Mass	Dec. 26, 1833	....
Bells, Hanging	G. W. Hildreth	Lockport, N. Y	June 19, 1855	13, 089
Bells, Hanging	M. R. Jones	Troy, N. Y	Mar. 19, 1872	124, 685
Bells, Hanging	A. Laroÿe	Sas Slykens, near Ostend, Belgium.	Mar. 21, 1871	112, 932
Bells, Hanging	A. Lynar	New York, N. Y	Dec. 24, 1819	....
Bells, Hanging	G. R. Meneely	West Troy, N. Y	Sept. 7, 1858	21, 422
Bells, Hanging	G. R. Meneely	West Troy, N. Y	Oct. 9, 1860	30, 338
Bells, Hanging	J. Russell	New York, N. Y	Oct. 10, 1829	....
Bells, Hanging	E. W. Vanduzen	Cincinnati, Ohio	Oct. 23, 1866	59, 098
Bells, Hanging	E. W. Vanduzen	Cincinnati, Ohio	Oct. 23, 1866	59, 099
Bells, Hanging	J. B. Young	Harper's Ferry, Va	July 4, 1854	11, 236
Bells, Hanging door	J. O. Harris	Ottawa, Ill	June 21, 1864	43, 233
Bells, Hanging house	E. Stetson	New Bedford, Mass	Aug. 17, 1843	3, 226
Bells, Machine for making wire	R. W. Norton	New Haven, Conn	Aug. 30, 1870	106, 951
Bells, Machinery for operating fog	J. Haynes	Pembroke, Me	Apr. 2, 1861	31, 886
Bells, Machinery for ringing fog	A. Morse	Portland, Me	Aug. 27, 1861	33, 156
Bells, Manufacture of corrugated	G. S. Saxton	Saint Louis, Mo	Dec. 17, 1867	72, 422
Bells, Manufacture of cow	G. C. Albaugh	Louisville, Ky	Feb. 26, 1861	31, 521
Bells, Manufacture of cow	W. T. and L. B. Tibbals	Cobalt, Conn	Aug. 31, 1869	94, 453
Bells, Method of making smoke	J. S. and T. B. Atterbury	Pittsburgh, Pa	Aug. 14, 1866	57, 063
Bells, Mold for casting	E. Jones	Troy, N. Y	Dec. 18, 1855	13, 948
Bells, Molding	W. H. Davis	Brooklyn, N. Y	Sept. 15, 1868	82, 094
Bells, Molding and casting	B. Hanks	Albany, N. Y	Nov. 4, 1816	....
Bells on steamers, Mode of operating pilot	J. R. Hopkins	Lincoln, Me	Nov. 17, 1857	18, 641
Bells or other hollow castings, Machine for turning or plaining the inside of.	O. Jones	Troy, N. Y	Oct. 4, 1870	107, 918
Bells, Ringing	J. Harrison	New York, N. Y	Mar. 8, 1864	41, 843
Bells, Ringing	J. Harrison	New York, N. Y	Aug. 6, 1867	67, 537
Bells, Ringing	J. Harrison	East Hampton, Conn	May 20, 1873	139, 147
Bells, Ringing	B. Kitt	Cincinnati, Ohio	June 5, 1860	28, 586
Bells, Ringing	T. V. Stran	New Albany, Ind	June 29, 1852	9, 081
Bells, Ringing	P. L. Weimer	Lebanon, Pa	July 23, 1867	67, 006
Bells, Ringing fixed	A. Carson	New York, N. Y	Dec. 6, 1853	10, 300
Bells to straps, Mode of attaching horse	J. Barton	Middle Haddam, Conn	Feb. 21, 1854	10, 539
Bells to their yokes, Attaching	G. R. Meneely	West Troy, N. Y	July 28, 1868	80, 422
Bells, &c., Tuning	A. Patterson	Birmingham, Pa	Nov. 26, 1867	71, 322
Bellows	J. Arndt	Wheeling, Va	Mar. 2, 1858	19, 475
Bellows	W. T. Barnes	Buffalo, N. Y	Apr. 24, 1849	6, 400

Index of patents issued from the United States Patent Office from 1790 *to* 1873, *inclusive*—**Continued.**

Invention.	Inventor.	Residence.	Date.	No.
Bellows	L. Bishop	Readsborough, Vt	Dec. 22, 1831	....
Bellows	R. Boeklen and L. Planer	Brooklyn and New York, N. Y.	June 16, 1863	38, 880
Bellows	J. Bowden	Mitcham, England	July 10, 1869	56, 335
Bellows	J. and W. Bowden	Bushwick, N. Y	Feb. 25, 1868	74, 884
Bellows	G. Bushnell	Schodack, N. Y	Oct. 1, 1867	69, 401
Bellows	G. W. Dalbey	Wheeling, W. Va	Nov. 14, 1865	50, 912
Bellows	G. L. Dimpfel	New York, N. Y	May 23, 1834	....
Bellows	J. Dixon	Pittsborough, N. C	June 11, 1827	....
Bellows	J. Drake		May 19, 1794	....
Bellows	C. L. English	Cincinnati, Ohio	Nov. 19, 1867	71, 060
Bellows	C. D. Everett	Cleveland, Ohio	June 25, 1872	128, 378
Bellows	E. Field	Cincinnati, Ohio	July 10, 1866	56, 199
Bellows	A. F. Jones	New York, N. Y	Aug. 8, 1871	117, 783
Bellows	J. B. and J. A. Maxwell	Allegheny, Pa	Feb. 7, 1860	27, 062
Bellows	A. Miller and U. Faris	Red Rock, Iowa	Apr. 9, 1872	125, 605
Bellows	W. Pilcher	Chicago, Ill	Aug. 19, 1873	141, 888
Bellows	J. Robe	Morgantown, Va	Apr. 17, 1834	....
Bellows	J. Rumsay		Aug. 26, 1791	....
Bellows	J. H. Snyder	Rockford, Ill	Dec. 5, 1871	121, 678
Bellows	L. U. Stuart	Brooklyn, N. Y	Feb. 11, 1862	34, 380
Bellows, Application of steam to work	J. Stevens, jr		Aug. 26, 1791	....
Bellows, Blacksmith's	J. F. Cory and H. C. Webb	Brooklyn, E. D., N. Y	May 31, 1870	103, 720
Bellows by steam, Working	M. Bell	Antis Township, Pa	Apr. 24, 1838	705
Bellows, Double	W. Lillie	Edwards, N. Y	June 2, 1845	4, 066
Bellows, Elastic piston	H. Dotterer	Philadelphia, Pa	May 1, 1810	....
Bellows, Flexible sides for	A. F. Jones	New York, N. Y	Apr. 30, 1872	126, 214
Bellows, Foot	H. Neumeyer	Macungie, Pa	Oct. 25, 1864	44, 814
Bellows for blasting foundries, furnaces, forges, &c., Triangular.	W. S. Jacks	New York, N. Y	Apr. 4, 1829	....
Bellows for blow-pipes	H. Cassell and W. F. Semple	Fredericktown, Ohio	Feb. 25, 1862	34, 479
Bellows for furnaces and forges, Pump	I. Jennings	New York, N. Y	Sept. 20, 1808	....
Bellows for furnaces, &c., Construction of	H. Dotterer	Philadelphia, Pa	Apr. 14, 1825	....
Bellows for furnaces, Mode of operating	H. Crumlish	Buffalo, N. Y	Oct. 7, 1873	143, 439
Bellows for furnaces or forges, Working	J. W. Godfrey		May 25, 1803	....
Bellows for musical instruments	M. Smith	New Haven, Conn	Feb. 25, 1851	7, 947
Bellows for pumping ships, Pendulum	B. Wynkoop		June 26, 1797	....
Bellows for reed-instruments	I. T. Packard	Campello, Mass	Sept. 28, 1852	9, 290
Bellows for smiths' and furnace fires	J. R. Morrison	Springfield, Ohio	Dec. 23, 1834	....
Bellows, Forge	H. Dotterer	Philadelphia, Pa	Feb. 16, 1809	....
Bellows, Forge	W. Thompson	Detroit, Mich	Apr. 17, 1860	27, 936
Bellows, Hand	J. Grennell	Springfield, Mass	Aug. 3, 1838	872
Bellows, Making	J. J. Eddy	Providence, R. I	Sept. 9, 1825	....
Bellows, Making japanned	R. B. Richardson	Philadelphia, Pa	Aug. 22, 1817	....
Bellows-nozzle from burning, Preserving	J. Dodge, jr	Saint Albans, Vt	Jan. 20, 1824	....
Bellows or blowing-machine	E. Brady	Mount Pleasant, N. Y	Apr. 3, 1828	....
Bellows-pipe, Welding	A. Pearsall	Nashville, Tenn	Aug. 31, 1858	21, 359
Bellows, Reciprocating	J. S. Butler and A. J. Stucker	Siver City, Idaho	July 13, 1869	92, 580
Bellows, Rotary	J. Darling	Adrian, Mich	May 23, 1846	4, 533
Bellows, Rotary	H. Phinney	Kingston, N. Y	Jan. 29, 1867	61, 684
Bellows, Rotary	A. Savage and H Killam	Scottsville, N. Y	Nov. 24, 1843	3, 350
Bellows, Sand	J. W. Hendley	Washington, D. C	July 24, 1866	56, 558
Bellows, Smith's	J. P. Hemmingsen	Marshalltown, Iowa	Mar. 2, 1869	87, 412
Bellows, Smith's	A. Holmes	Pomfret, N. Y	Mar. 23, 1818	....
Bellows, Smith's	M. Loomis	Worcester, N. Y	May 2, 1848	5, 544
Bellows, Smith's	G. H. Peek	East Hamburgh, N. Y	May 3, 1870	102, 704
Bellows, Steam	J. Eastman	Bath, N. H	Apr. 8, 1831	....
Bellows-tub	A. H. McPharin	Huntingdon, Pa	June 11, 1829	....
Bellows, Windworm for blacksmith's	C. F. Conrad	Adrian, Mich	Dec. 31, 1867	72, 721
Bellows, Wooden	B. and J. Tyler	Claremont, N. H	Feb. 1, 1817	....
Belly-band fastener	C. H. Horne	Astoria, Oreg	Nov. 30, 1869	97, 401
Belt and band fastening	G. D. Young	Plymouth, Mass	Jan. 29, 1856	14, 175
Belt and band of rubber and metal, Driving	L. Sterne	London, England	Aug. 3, 1869	93, 364
Belt attachment for machinery, Slack	J. W. Howard	Greenville, Ala	Oct. 19, 1869	95, 903
Belt, Cartridge	W. B. Hayden	Columbus, Ohio	Jan. 31, 1871	111, 450
Belt, Chain	W. M. Julian and M. Kidnocker	Tarlton, Ohio	Jan. 12, 1869	85, 830
Belt-clamp	E. Ainsworth	Wilmington, Del	Oct. 14, 1873	143, 604
Belt-clasp	A. D. Ansell	Hartford, Conn	July 18, 1865	48, 784
Belt-clasp	J. Chenoweth and J. McLain	Auglaize County, Ohio	Oct. 8, 1867	69, 628
Belt-clasp	G. Churchill	Hartford, Conn	Jan. 18, 1859	22, 624
Belt-clasp	G. Cuppers	New York, N. Y	Apr. 5, 1864	42, 170
Belt-clasp	P. Harlow	Hudson, Mass	Feb. 12, 1867	61, 935
Belt-clasp	E. Hatch	Charlestown, Mass	Feb. 11, 1868	74, 354
Belt-clasp	J. H. Hawes and G. H. Bliss	Boston and West Stockbridge, Mass.	Apr. 28, 1868	77, 282
Belt-clasp	O. Kromer and C. Ohlemacher	Sandusky, Ohio	Nov. 20, 1866	59, 914
Belt-clasp	I. N. Plotts	New York, N. Y	Nov. 7, 1865	50, 841
Belt-clasp	T. W. Porter	Boston, Mass	Mar. 24, 1868	75, 970
Belt-clasp	J., S. A., G. E., and F. F. Reading.	Birmingham, England	Mar. 24, 1868	75, 976
Belt-clasp	J. T. Senn	Troy, Ala	Nov. 12, 1872	133, 061
Belt-clasp	A. M. Smith	Rochester, N. Y	June 29, 1852	9, 080
Belt clasp	C. Towns	Cleveland, Ohio	Feb. 16, 1869	87, 081
Belt-clasp, &c	G. F. White and H. Chamberlain.	Hornsey and London, England.	Oct. 30, 1866	59, 332
Belt-clasp for machinery	H. G. Ellsworth	Auburn, N. Y	Mar. 28, 1854	10, 692
Belt-clasp, Leather	L. Sanders	Brooklyn, N. Y	Mar. 4, 1873	136, 549
Belt coupling	T. H Corbett	Brooklyn, N. Y	Nov. 27, 1855	13, 843
Belt-coupling	C. Fairfax, jr	Cincinnati, Ohio	Aug. 21, 1860	29, 684
Belt-coupling	S. Green	Grand Rapids, Mich	Mar. 23, 1858	19, 735
Belt-coupling	R J. Jordon	Elkhart, Ind	July 30, 1867	67, 197
Belt-coupling	C. W. T. Krausch	Philadelphia, Pa	Apr. 28, 1868	77, 295
Belt-coupling	W. Leas	Kokomo, Ind	Jan. 23, 1866	52, 178
Belt-coupling	W. Leas	Kokomo, Ind	July 31, 1866	56, 767
Belt-coupling	T. S. Livermore	Leicester, Mass	June 23, 1868	79, 239
Belt-coupling	J. Mattix	Kokomo, Ind	Aug. 7, 1866	56, 966
Belt-coupling	T. McMullen	Osgood, Ind	Aug. 25, 1868	81, 391
Belt-coupling	S. Metzler	Naperville, Ill	June 25, 1861	32, 637

Index of patents issued from the United States Patent Office from 1790 *to* 1873, *inclusive*—Continued.

Invention.	Inventor.	Residence.	Date.	No.
Belt-coupling	A. C. G. Rathburn and A. M. Comstock.	Lyme, Conn	Feb. 7, 1865	46,264
Belt-coupling	T. Rüdiger	Oberle's Corners, Minn	June 1, 1869	90,876
Belt-coupling	E. Thayer	New York, N. Y	Jan. 8, 1867	61,118
Belt-coupling	J. L. Thomas	Newburgh, Ohio	Feb. 11, 1868	74,449
Belt-coupling	H. Underwood	New York, N. Y	Oct. 28, 1862	36,814
Belt coupling, Round	W. S. Jarboe	New York, N. Y	Oct. 1, 1872	131,879
Belt-cutter	M. G. Goodale	Lowell, Mass	Nov. 12, 1867	70,711
Belt-cutting machine	H. D. Smith	New York, N. Y	Oct. 13, 1863	40,287
Belt-cutting machine, Round	J. C. Foster	New London, Conn	June 13, 1871	115,949
Belt, Driving	W. Clissold	Dudbridge, England	Aug. 27, 1861	33,133
Belt, Driving	W. R. Colton	Syracuse, N. Y	Feb. 4, 1873	135,409
Belt, Driving	M. J. Haines	Bristol, England	Apr. 16, 1867	63,883
Belt, Driving	A. Schpakowsky	St. Petersburg, Russia	Dec. 23, 1873	145,907
Belt, Driving	T. R. White and W. G. Bedford	Philadelphia, Pa	June 25, 1867	66,111
Belt, Endless	H. Richards and J. A. Traut	New Britain, Conn	Sept. 29, 1868	82,640
Belt, Endless driving	J. F. Reigart	Washington, D. C	Dec. 9, 1873	145,447
Belt-fastener	G. M. Beardsley	Fenton, Mich	Aug. 25, 1868	81,331
Belt-fastener	S. S. Bolton	Big Rapids, Mich	Dec. 6, 1870	109,798
Belt-fastener	A. Hyde	Springfield, Mass	July 1, 1873	140,419
Belt-fastener	W. C. James	Fishersville, N. H	July 18, 1871	117,175
Belt-fastener	C. Liebrich and L. Uitting	Philadelphia, Pa	Jan. 20, 1863	37,453
Belt-fastener	M. Olmsted	Alum Creek, Tex	Nov. 28, 1872	121,405
Belt-fastener	C. O. Pike	North Leverett, Mass	Dec. 17, 1867	72,325
Belt-fastener	D. Wigger	New York, N. Y	Nov. 12, 1867	70,930
Belt-fastener	F. G. Wilson	Nashua, N. H	Dec. 15, 1868	84,925
Belt-fastening	G. W. Blake	Pepperell, Mass	Apr. 24, 1860	28,033
Belt-fastening	G. W. Blake	East Pepperell, Mass	Mar. 26, 1861	31,859
Belt-fastening	G. A. Brown	Reading, Mich	Sept. 12, 1871	118,785
Belt-fastening	M. W. Costolo	Boston, Mass	Dec. 19, 1865	51,561
Belt-fastening	A. Fickett	Rochester, N. Y	Aug. 23, 1859	25,187
Belt-fastening	W. Frazier	Hartford, Conn	Dec. 20, 1859	26,488
Belt-fastening	J. W. Hicks	Laurel, Md	Dec. 12, 1871	121,781
Belt-fastening	T. Kennedy	Mount Carmel, Conn	Sept. 15, 1868	82,227
Belt-fastening	G. Koeb and L. Houcke	Springfield, Ohio	Mar. 30, 1869	88,307
Belt-fastening	J. C. Merritt	West Point, N. Y	Aug. 27, 1872	130,816
Belt-fastening	H. Norfolk	Vicksburgh, Miss	Apr. 2, 1872	125,213
Belt-fastening	J. E. Richard	Columbia, S. C	Apr. 22, 1873	138,194
Belt-fastening	T. Ruediger	Carver County, Minn	Feb. 4, 1873	135,492
Belt-fastening	E. Schmeltz	New York, N. Y	Mar. 17, 1868	75,704
Belt-fastening	G. V. Sheffield and B. Whitcomb.	Worcester, Mass	Dec. 24, 1867	72,690
Belt-fastening	S. J. Sherman	Brooklyn, N. Y	Dec. 18, 1866	60,579
Belt-fastening	P. Subit	Boston, Mass	Oct. 7, 1873	143,474
Belt-fastening	A. Whiteley	Springfield, Ohio	Feb. 5, 1856	14,213
Belt for polishing, Endless	J. A. Traut	New Britain, Conn	Mar. 26, 1867	63,341
Belt for the body, Electric	J. E. Bazault	Paris, France	Mar. 26, 1872	125,006
Belt-gearing	J. H. Butler	Hampden, Me	Jan. 17, 1871	110,955
Belt-gearing	G. B. Hamlin	Willimantic, Conn	Oct. 11, 1870	108,259
Belt-gearing	G. B. Hamlin	Willimantic, Conn	Sept. 10, 1872	131,268
Belt-gearing	G. C. Howard	Philadelphia, Pa	Feb. 28, 1871	112,145
Belt-gearing, Idler for	T. Bell and J. F. Hillerich	Louisville, Ky	Aug. 5, 1873	141,417
Belt guide	L. R. Jenkins	Philadelphia, Pa	Sept. 11, 1866	57,917
Belt-guide	C. P. Leavitt	New York, N. Y	Aug. 9, 1870	106,179
Belt-hole cover	T. P. Rodgers	Taunton, Mass	Oct. 28, 1873	144,144
Belt-hole scupper	B. L. Wood	Taunton, Mass	Nov. 18, 1873	144,816
Belt-hook	F. J. Jones	Detroit, Mich	Mar. 31, 1868	76,082
Belt-hook	F. E. Oliver	New York, N. Y	Oct. 6, 1863	40,182
Belt-hook	H. L. Peirce	Taunton, Mass	Apr. 27, 1869	89,499
Belt-hook	C. G. Sargent	Graniteville, Mass	Dec. 15, 1868	84,968
Belt-hook	E. F. Sherman	Chicopee, Mass	July 21, 1868	80,227
Belt hook, pliers, and punch	N. E. Hale	Nashua, N. H	Sept. 27, 1859	25,567
Belt-joining clasp	B. D. Randleman	Port Louisa, Iowa	July 21, 1868	80,219
Belt joining or splicing	J. Ashworth	Lewiston, Me	Jan. 26, 1869	86,123
Belt-knife	H. Blake	East Pepperell, Mass	Aug. 11, 1868	80,899
Belt-knives, Manufacture of	B. F. Radford	Hyde Park, Mass	Dec. 19, 1871	121,963
Belt-lacer	J. K. Priest and W. Earl, jr	Nashua, N. H	Aug. 25, 1868	81,537
Belt-lacing	H. A. Alden	Matteawan, N. Y	July 31, 1860	29,348
Belt-lacing	H. C. Babcock	Cincinnati, Ohio	Nov. 12, 1867	70,775
Belt-lacing device	D. P. Davis	New York, N. Y	Dec. 3, 1867	71,713
Belt-lacing gage	H. Gould	Minneapolis, Minn	May 6, 1873	138,639
Belt-lacing, Machine for cutting	B. F. Sanborn	Boston, Mass	Sept. 11, 1866	57,976
Belt-lacing tool	J. M. Stamp	Grass Valley, Cal	Mar. 8, 1870	100,686
Belt, Lady's safety	M. G. Porter	Charlestown, Mass	Feb. 1, 1870	99,347
Belt, Lady's waist	J. M. Flagg	Providence, R. I	Feb. 6, 1872	123,471
Belt-lap cutter	C. S. Robinson	Concord, N. H	Feb. 12, 1867	62,073
Belt, Machine	T. F. Snover	Oconto, Wis	May 12, 1868	77,847
Belt on pulley, Apparatus for shifting	P. J. Zier	Pittsburgh, Pa	Feb. 6, 1872	123,543
Belt or girdle, Waist	M. I. Findley	New York, N. Y	May 26, 1868	78,367
Belt, Pulley	L. R. Faught	Philadelphia, Pa	Oct. 15, 1872	132,205
Belt-replacer	W. C. Bridges	Michigan City, Ind	Apr. 19, 1870	101,978
Belt-rivet	C. Frank	Cincinnati, Ohio	Apr. 2, 1867	63,497
Belt-riveting tool	M. D. Lawrence	Flintville, Wis	Aug. 20, 1872	130,646
Belt, Safety	N. Downing	Brooklyn, N. Y	Oct. 29, 1861	33,579
Belt-shifter	S. Forsythe	San Francisco, Cal	July 23, 1872	129,806
Belt-shifter	W. E. Leighton	Pembroke, Me	Apr. 12, 1870	101,745
Belt-shifter	J. E. Mutchler	Grand Rapids, Mich	Oct. 18, 1870	108,379
Belt-shifter	T. P. Rogers	Taunton, Mass	Aug. 27, 1872	130,870
Belt-shifter	W. H. H. Sisum	Newark, N. J	May 21, 1872	127,109
Belt-shifter	W. H. H. Sisum	Newark, N. J	Apr. 18, 1871	113,939
Belt shifter and tightener	B. O'Bryan	Marietta, Pa	June 24, 1873	140,299
Belt-shifter for machinery	L. J. Knowles	Warren, Mass	July 7, 1857	17,743
Belt-shifting apparatus	O. H. Wade	East Bridgewater, Mass	Oct. 14, 1873	143,736
Belt-shifting device	A. M. Freeland	New York, N. Y	Aug. 27, 1867	68,061
Belt-shifting device	W. Sellers	Philadelphia, Pa	Dec. 16, 1862	37,180
Belt-shipper	E. Buck	Vincennes, Ind	June 30, 1868	79,443
Belt-shipper	T. W. Frost	Dorchester, Mass	Apr. 14, 1868	76,739
Belt-shipper	J. C. Goar	Monterey, Cal	Aug. 14, 1860	29,587

Index of patents issued from the United States Patent Office from 1790 *to* 1873, *inclusive*—Continued.

Invention.	Inventor.	Residence.	Date.	No.
Belt-shipper	J. C. Goar	Jamaica Plain, Mass	Oct. 21, 1862	36, 707
Belt-shipper	J. S. Otis	Leeds, Mass	Jan. 3, 1871	110, 674
Belt-shipper	T. P. Rodgers	Taunton, Mass	Sept. 15, 1863	39, 956
Belt-shipper for mules	W. H. Saltmarsh	Waltham, Mass	Dec. 24, 1867	72, 686
Belt-splice	G. E. Burt	Harvard, Mass	July 30, 1872	129, 928
Belt-splice point finisher	J. C. McLaren	Montreal, Canada	Dec. 13, 1870	110, 060
Belt-stop for machinery, Traveling	T. E. Baden	Washington, D. C	July 26, 1870	105, 765
Belt-stretcher	S. Rogers	Pittsburgh, Pa	Jan. 9, 1866	51, 970
Belt-stud	D. M. Weston	Boston, Mass	Apr. 14, 1868	76, 861
Belt-tension apparatus	S. E. Jewett	Haverhill, Mass	July 22, 1873	141, 004
Belt-tightener	J. Albertson and D. W. Marmon	Richmond, Ind	Jan. 2, 1872	122, 345
Belt-tightener	H. C. Allyn	Falls Village, Conn	Apr. 16, 1872	125, 780
Belt-tightener	J. W. Batcheller	Oregon, Mo	Jan. 23, 1872	122, 932
Belt-tightener	M. C. Chambelin and A. Clawson.	Plain View, Minn	Feb. 8, 1870	99, 634
Belt-tightener	L. Funke	Champion Mills, N. Mex	Jan. 23, 1872	122, 943
Belt-tightener	J. M. Hawley	Holton, Ind	Sept. 10, 1867	68, 625
Belt-tightener	J. A. and H. A. House	Brooklyn, N. Y	June 23, 1863	39, 015
Belt-tightener	J. M. King	Quincy, Minn	Dec. 15, 1868	85, 012
Belt-tightener	R. N. Meriam	Worcester, Mass	Nov. 9, 1869	96, 718
Belt-tightener	J. Nichols	Fond du Lac, Wis	Dec. 20, 1864	45, 5 4
Belt-tightener	S. Patton	Chatsworth, Ill	Aug. 18, 1868	81, 285
Belt-tightener	M. Peatt	Dexter, Mich	Nov. 19, 1867	71, 052
Belt-tightener	J. Pinnell	Pittston, Pa	Dec. 24, 1872	134, 309
Belt-tightener	G. W. Runk	Franklin, La	Jan. 24, 1871	111, 257
Belt-tightener	W. Sellers	Haverhill, Mass	Dec. 23, 1873	145, 908
Belt-tightener	T. G. Stansberry	Medora, Ill	Sept. 3, 1867	68, 465
Belt-tightener	A. N. Woodard	Fenton, Mich	Apr. 19, 1870	102, 674
Belt-tool	D. A. J. Lamson	Cherry Valley, Mass	Sept. 15, 1857	18, 206
Belt, Waist	A. R. Boylson	Chicago, Ill	Nov. 17, 1868	84, 082
Belt, Waist	J. H. Vogt and G. Dietzel	New York, N. Y	Mar. 11, 1873	136, 685
Belt, Woven endless.	S. W. Baker	Providence, R. I	Oct. 22, 1861	33, 510
Belts, Composition to prevent the slipping of machine.	I. F. Eaton	Boston, Mass	May 17, 1870	103, 034
Belts, Connecting and disconnecting machinery by means of.	T. A. Madison	Terre Haute, Ind	May 31, 1859	24, 220
Belts, Fastening driving	J. C. Desumeur, C. and E. Dudin, and L. Delacourt.	Guise, France	Dec. 5, 1871	121, 452
Belts, Manufacture of India-rubber	A. H. Hook	New York, N. Y	Apr. 26, 1870	102, 268
Belts, Shifting	M. Wells	Brooklyn, N. Y	Feb. 2, 1858	19, 272
Belts, Tightening and guiding	C. K. Myers	Pekin, Ill	Dec. 14, 1869	97, 955
Belts together, Machine for drawing	E. F. Miller and B. Gardner	Williamsburgh and Hampshire County, Mass.	Oct. 24, 1865	50, 609
Belts, Tool for mending	G. W. Miller	West Meriden, Conn	Aug. 4, 1868	80, 756
Belts, Tool for use in lacing	F. W. Du Bois	New York, N. Y	May 13, 1873	138, 740
Belting	C. J. Fay	Philadelphia, Pa	June 18, 1867	65, 895
Belting	H. Lemaistre	Brussels, Belgium	Mar. 17, 1868	75, 554
Belting	W. P. Powers	North La Crosse, Wis	Aug. 26, 1873	142, 272
Belting	T. Standring	Fort Richmond, N. Y	Sept. 1, 1868	81, 823
Belting	H. Underwood	New York, N. Y	Apr. 10, 1860	27, 846
Belting, Apparatus for forming rubber	T. J. Mayall	Roxbury, Mass	Dec. 27, 1859	26, 603
Belting, Artificial leather	S. M. Allen	Woburn, Mass	Jan. 21, 1868	73, 427
Belting, Artificial leather	S. M. Allen	Woburn, Mass	July 21, 1868	80, 048
Belting, Composition for stuffing	J. Haseltine	Warren, N. H	Aug. 13, 1867	67, 760
Belting-conveyer	J. Campbell	Saint Louis, Mo	Sept. 1, 1863	39, 716
Belting, Elastic	D. C. Gately	Newtown, Conn	Mar. 13, 1860	27, 440
Belting, Elastic	J. and S. Peatfield	Ipswich, Mass	Mar. 13, 1860	27, 471
Belting, Fastening for machine	L. Smith	Buffalo, N. Y	Nov. 17, 1857	18, 659
Belting, Fastening leather	E. Osgood	Bangor, Me	May 10, 1853	9, 710
Belting, Finishing vulcanized India-rubber	C. McBurney	Boston, Mass	June 13, 1871	115, 880
Belting for driving machines	M. A. Strouvelle	Saint Louis, Mo	Sept. 17, 1867	69, 042
Belting for machinery	S. M. Allen	Woburn, Mass	Mar. 13, 1866	53, 095
Belting for machinery	J. H. Clifton	Newcastle, Pa	Aug. 16, 1859	25, 096
Belting, hose, &c., Manufacture of	W. A. Torrey	Mont Clare, N. J	July 10, 1866	56, 292
Belting, India-rubber	B. F. Lee	New York, N. Y	Dec. 20, 1859	26, 549
Belting, Lap-joint for	H. Underwood	New York, N. Y	Feb. 9, 1858	19, 318
Belting, Leather	W. Strevell, S. B. Wells, and G. B. Kerper.	Jersey City, N. J., and New York, N. Y.	Jan. 28, 1868	73, 848
Belting, Machine	J. H. Cheever	New York, N. Y	Sept. 28, 1858	21, 596
Belting, Machine	V. Fountain, jr	Castleton, N. Y	Dec. 17, 1867	72, 181
Belting, Machine	W. H. Gates	Louisville, Ky	Jan. 1, 1867	60, 713
Belting, Machine	T. J. Mayall	Boston, Mass	Feb. 7, 1860	27, 060
Belting, Machine	J. Montgomery	New York, N. Y	June 22, 1869	91, 656
Belting, Machine	H. Taylor	Trenton, N. J	Jan. 26, 1864	41, 405
Belting, Machine	J. G. Street	Brooklyn, N. Y	Apr. 26, 1870	102, 447
Belting, Machine for cutting solid leather	T. J. Dickerson and O. B. Warren.	Auburn, N. Y	Nov. 14, 1871	120, 948
Belting, Machine for making India-rubber	R. Hale	Roxbury, Mass	May 5, 1857	17, 216
Belting, Machine for making rubber	T. J. Mayall	Roxbury, Mass	Jan. 31, 1860	27, 001
Belting, Machine for making rubber	T. J. Mayall	Roxbury, Mass	Apr. 3, 1860	27, 730
Belting, Machine for removing rivets from	H. F. Crafts and W. Updegraff	Williamsport, Pa	Sept. 10, 1872	131, 152
Belting, Making rubber	D. C. Gately	Newtown, Conn	Nov. 22, 1859	26, 178
Belting, Manufacture of	J. B. Crane	Dalton, Mass	June 18, 1867	65, 879
Belting, Manufacture of	C. E. Smith	Philadelphia, Pa	May 31, 1859	24, 243
Belting, Manufacture of caoutchouc	D. C. Gately	Newtown, Conn	Dec. 20, 1859	26, 489
Belting, Manufacture of elastic	S. T. Parmelee	Edinburgh, Scotland	Apr. 26, 1859	23, 779
Belting, Manufacture of India-rubber	D. C. Gately	Newtown, Conn	Dec. 27, 1859	26, 580
Belting, Manufacture of India-rubber	D. C. Gately	Newtown, Conn	Jan. 3, 1860	26, 669
Belting, Manufacture of leather	G. O. Clark and F. G. Slemmer	Philadelphia, Pa	Dec. 10, 1872	133, 698
Belting, Manufacture of machine	J. H. Clifton	Newcastle, Pa	Aug. 16, 1859	25, 095
Belting, Manufacture of machine	G. and D. Hurn	London, England	Feb. 9, 1869	86, 755
Belting, Manufacture of paper	J. B. Crane	Dalton, Mass	Nov. 24, 1868	84, 263
Belting, Manufacture of round	M. Jewell	Hartford, Conn	June 15, 1858	20, 564
Belting, Manufacture of rubber	A. O. Bourn and I. F. Williams.	Providence and Bristol, R. I	Apr. 4, 1871	113, 486
Belting, Manufacture of rubber	G. P. Dodge	London, England	Jan. 15, 1867	61, 173
Belting, Manufacture of rubber	D. C. Gately	Newtown, Conn	Nov. 29, 1859	26, 264
Belting, Manufacture of rubber	D. C. Gately	Newtown, Conn	Nov. 29, 1859	26, 265
Belting, Manufacture of vulcanized rubber	D. C. Gately and J. B. Forsyth	Newtown, Conn., and Boston, Mass.	Aug. 5, 1873	141, 453

Index of patents issued from the United States Patent Office from 1790 *to* 1873, *inclusive*—Continued.

Invention.	Inventor.	Residence.	Date.	No.
Belting, Manufacturing rubber	D. C. Gately	Newtown, Conn	Jan. 31, 1860	26, 982
Belting, Metal	W. P. Powers	North La Crosse, Wis	Nov. 30, 1869	97, 312
Belting, Mode of	B. Chester	New York, N. Y	Dec. 22, 1857	18, 941
Belting, Mode of forming joints in India-rubber	J. McDougall	Masonville, Mich	Oct. 11, 1859	25, 749
Belting, Mold for making cushioned	S. W. Tyler	Troy, N. Y	May 2, 1871	114, 371
Belting or banding, Machine for	T. J. Mayall	Roxbury, Mass	Nov. 24, 1863	40, 699
Belting or banding, Manufacture of India-rubber	J. H. Cheever	Boston, Mass	Mar. 11, 1856	14, 389
Belting, Round	A. Holbrook, jr	Providence, R. I	June 18, 1872	128, 040
Belting, Rubber	D. C. Gately	Newtown, Conn	Nov. 22, 1859	26, 177
Belting, Rubber-coated rubber	P. Jewell, jr	Hartford, Conn	July 16, 1867	66, 848
Belting, Splicing	J. W. Lyder and H. Sheve	Alliance, Ohio	June 23, 1868	79, 136
Belting, Splicing	N. E. Smith	East Cleveland, Ohio	July 2, 1867	66, 261
Belting, Vulcanized India-rubber	D. C. Gately	Newtown, Conn	Oct. 27, 1868	83, 374
Bench:				
See Carpenter's bench.				
Clothes-wringer bench.				
Corn-husking bench.				
Currier's bench.				
Cutter-bench.				
Folding bench.				
Folding wash-bench.				
Laundry-bench.				
Revolving bench.				
Saw-bench.				
Shoemaker's bench.				
Sleigh-bench.				
Wash-bench.				
Work-bench.				
Bench, Carpenter's	J. W. Mahan	Lexington, Ill	Mar. 25, 1856	14, 511
Bench-clamp	C. W. Clapp	Wappinger's Falls, N. Y	Mar. 11, 1856	14, 390
Bench-clamp	G. Cooper	Hartford, Conn	Mar. 20, 1860	27, 592
Bench-clamp	O. L. Fenner	Rochester, N. Y	Nov. 30, 1869	97, 376
Bench-dog	A. K. and H. P. Hood	Lowell, Mass	Mar. 12, 1867	62, 748
Bench-dog	C. P. Whitman	Charlemont, Mass	Aug. 29, 1871	118, 662
Bench-drill	C. G. Miller	Brattleborough, Vt	May 26, 1868	78, 309
Bench-hook	D. Boyce	North Providence, R. I	Sept. 23, 1873	143, 054
Bench-hook	R. Frisbie	Middletown, Conn	July 10, 1860	29, 070
Bench-hook	A. Hotchkin	Schenevus, N. Y	Oct. 16, 1855	13, 678
Bench-hook	J. Humphreys	Chicopee Falls, Mass	Apr. 27, 1869	89, 483
Bench-hook	W. B. Kean	Worcester, Mass	Aug. 13, 1850	7, 565
Bench-hook	J. W. McGill	New York, N. Y	May 14, 1872	126, 644
Bench-hook	H. M. Putnam	Fitchburgh, Mass	June 20, 1871	116, 229
Bench-hook	H. J. Skinner	Dunkirk, N. Y	June 7, 1870	104, 069
Bench-hook	C. H. Weston	Lowell, Mass	Oct. 2, 1866	58, 519
Bench-hook	E. B. White	Nashua, N. H	Mar. 9, 1858	19, 595
Bench hook and clamp	E. P. Woods and A. E. Blood	Lowell and Lynn, Mass	June 27, 1865	48, 498
Bench-hook, Carpenter's	L. F. Noe	New York, N. Y	Aug. 14, 1866	57, 176
Bench-hook, Joiner's	P. P. Cowles	Cabotville, Mass	June 6, 1848	5, 619
Bench hook or dog	C. T. Crane	Lowell, Mass	May 28, 1867	65, 060
Bench-hook, Socket for	J. Sawyer	South Royalston, Mass	Feb. 28, 1854	10, 579
Bench or table	L. D. Hubbard	Worcester, Mass	Apr. 7, 1868	76, 459
Bench-pin	H. Gabelmann	Fort Scott, Kans	May 10, 1870	102, 802
Bench-pin	T. J. W. Porter	Grand Rapids, Mich	Jan. 12, 1869	85, 850
Bench-plane, Bit-fastening for cast-iron	W. S. Loughborough	Victor, N. Y	Apr. 4, 1854	10, 748
Bench-plane bit, Manner of securing	C. W. Seely and B. F. Locke	Wellington, Ohio	June 7, 1859	24, 335
Bench-rest	J. D. Spiller	Concord, N. H	June 5, 1855	13, 016
Bench-shears	J. Hill	Charlotte, Mich	Aug. 17, 1869	93, 715
Bending and folding sheet-metal, Machine for	J. B. and L. C. Clark	Plantsville, Conn	Dec. 14, 1869	97, 762
Bending and hardening springs, Machine for	C. A. Coggeshall	Bridgeport, Conn	July 6, 1869	92, 268
Bending-brake for sheet-metal	F. Yaesché	Springfield, Mass	Aug. 15, 1871	118, 090
Bending circles, Machine for	W. Boyd	Hartford, N. Y	June 2, 1868	78, 418
Bending corrugated plates of metal, Apparatus for	R. Montgomery	New York, N. Y	June 3, 1862	35, 496
Bending-device	W. Malick	Tidioute, Pa	Dec. 31, 1867	72, 874
Bending-device for wood	E. Lacey	Chicago, Ill	Jan. 24, 1871	111, 218
Bending-machine	L. H. Beckwith	Port Jervis, N. Y	Sept. 3, 1867	68, 340
Bending-machine	J. Forbes	Halifax, Canada	Sept. 12, 1871	118, 920
Bending-machine	E. D. and W. K. Gird	Cedar Lake, N. Y	Mar. 19, 1867	63, 034
Bending-machine	H. S. Golightly and C. S. Twitchell.	New Haven, Conn	Aug. 10, 1869	93, 612
Bending-machine	G. Huntington	Norwichville, Canada	Oct. 15, 1867	69, 916
Bending-machine	S. W. Kimble	Springfield, Ill	Oct. 28, 1873	144, 109
Bending-machine	D. Pierce	Almont, Mich	Nov. 23, 1869	97, 113
Bending-machine	L. Raymond	New York, N. Y	Aug. 25, 1857	18, 058
Bending-machine	W. Richardson and L. B. Müller	Baltimore, Md	Jan. 28, 1868	73, 837
Bending-machine	J. J. H. Sercombe	New London, Wis	May 10, 1870	102, 975
Bending-machine	H. Winter	Honesdale, Pa	May 3, 1870	102, 641
Bending-machine	J. N. Woodward	Aurora, Ill	Sept. 17, 1867	69, 060
Bending-machine chain, Timber	L. Heywood	Gardner, Mass	Mar. 13, 1860	27, 447
Bending-machine clamp, Wood	J. B. Van Horn	Trenton, N. J	June 30, 1868	79, 530
Bending-machine for corrugated metal plates	A. Johnson	Brooklyn, N. Y	July 26, 1870	105, 805
Bending machine, Metal	W. W. Cumberland	Newark, N. J	Dec. 12, 1854	12, 053
Bending machine, Metal	C. H. Delamater	New York, N. Y	Aug. 4, 1863	39, 382
Bending machine, Metal	W. C. Eiseman	Pittsburgh, Pa	Feb. 27, 1872	124, 123
Bending machine, Metal	S. Hall	New York, N. Y	Feb. 26, 1867	62, 327
Bending machine, Metal	J. Noland	Philadelphia, Pa	Dec. 24, 1867	72, 669
Bending machine, Metal	R. Potts	Chatham, N. Y	Oct. 15, 1867	69, 936
Bending machine, Metallic-plate	E. L. Gaylord	Terrysville, Conn	Oct. 3, 1857	18, 542
Bending machine, Metallic-plate	E. L. Gaylord	Terrysville, Conn	Aug. 18, 1857	18, 011
Bending machine, Metallic-plate	D. Howell	Louisville, Ky	Jan. 12, 1858	19, 090
Bending machine, Metallic-plate	J. Perry	Plymouth Hollow, Conn	Aug. 18, 1857	18, 019
Bending machine, Sheet-metal	R. Brady	New York, N. Y	Jan. 8, 1856	14, 049
Bending machine, Sheet-metal	G. W. Burling	Trenton, N. J	Oct. 28, 1856	15, 964
Bending machine, Sheet-metal	J. Eppley	York, Pa	May 16, 1848	5, 571
Bending machine, Sheet-metal	J. M. Grove and H. Hedrick	Anderson, Ind	July 28, 1868	80, 348
Bending machine, Sheet-metal	C. C. Hare	Kansas City, Mo	Apr. 19, 1870	102, 001
Bending machine, Sheet-metal	R. B. McConnell	Evansville, Ind	Apr. 2, 1872	125, 317
Bending machine, Sheet-metal	W. J. McLea	Leroy, N. Y	Jan. 5, 1869	85, 531
Bending machine, Sheet-metal	A. Shepard	New Britain, Conn	Dec. 8, 1868	84, 714

Index of patents issued from the United States Patent Office from 1790 *to* 1873, *inclusive*—Continued.

Invention.	Inventor.	Residence.	Date.	No.
Bending machine, Sheet-metal	W. Webster	Morrisania, N. Y	Mar 28, 1854	10, 713
Bending machine, Sheet-metal	A. W. and P. A. Whitney	Woodstock, Vt	Oct. 20, 1868	83, 347
Bending machine, Sheet-metal	J. Wright	Plantsville, Conn	Jan. 20, 1857	16, 456
Bending machine, Timber	J. D. Sarven	Maury County, Tenn	Jan. 20, 1857	16, 453
Bending machine, Volute-spring	R. Miller	Pittsburgh, Pa	Aug. 23, 1870	106, 603
Bending machine, Wire-screw eye	N. C. Perry	Chester, Conn	Aug. 23, 1870	106, 720
Bending machine, Wood	A. Bailey	Gardiner, Me	Oct. 11, 1859	25, 711
Bending machine, Wood	C. F. Beverly	Lancaster, Ohio	Oct. 13, 1857	18, 379
Bending machine, Wood	T. Blanchard	Boston, Mass	May 4, 1858	20, 137
Bending machine, Wood	I. W. Bowers	Boston, Mass	Jan. 21, 1868	73, 493
Bending machine, Wood	D. F. Breed	Valparaiso, Ind	Jan. 21, 1868	73, 496
Bending machine, Wood	D. Catchpole and J. Havens	Geneva and Auburn, N. Y	Oct. 1, 1867	69, 313
Bending machine, Wood	A. Chubb	Painesville, Ohio	Feb. 22, 1859	23, 012
Bending machine, Wood	E. A. Conner	Metropolis, Ill	Mar. 14, 1871	112, 553
Bending machine, Wood	J. Conner	Richmond, Ind	Feb. 18, 1868	74, 668
Bending machine, Wood	M. F. Connett	Evansville, Ind	July 11, 1865	48, 661
Bending machine, Wood	J. A. Dann	New Haven, Conn	Apr. 23, 1867	63, 997
Bending machine, Wood	V. De Lyon and V. Werner	Canton, Miss	Apr. 27, 1869	89, 387
Bending machine, Wood	O. Evans	Alliance, Ohio	June 5, 1866	55, 264
Bending machine, Wood	S. C. and E. O. Fink	Indianapolis, Ind	Sept. 10, 1867	67, 621
Bending machine, Wood	J. Fipps	New Albany, Ind	Nov. 8, 1870	108, 992
Bending machine, Wood	J. Fishbaugh	Tiffin, Ohio	May 1, 1866	54, 316
Bending machine, Wood	R. Fitts, jr	Fitchburgh, Mass	Mar. 2, 1869	87, 404
Bending machine, Wood	J. W. Griffiths	Brooklyn, N. Y	Jan. 2, 1866	51, 826
Bending machine, Wood	O. W. Griffiths	Charlestown, Mass	Apr. 2, 1872	125, 130
Bending machine, Wood	G. Gustafson	Chicago, Ill	Oct. 10, 1871	119, 702
Bending machine, Wood	J. Hale	Somerville, Mass	Apr. 21, 1868	77, 036
Bending machine, Wood	J. S. Hall	Jeffersonville, Ind	Aug. 25, 1868	81, 365
Bending machine, Wood	H. Hanna	Columbus, Ohio	Dec. 9, 1873	145, 416
Bending machine, Wood	E. C. Harris	New York, N. Y	Feb. 27, 1866	52, 944
Bending machine, Wood	E. P. Haskell	Harlan, Ind	Dec. 9, 1862	37, 096
Bending machine, Wood	L. Heywood	Gardner, Mass	Dec. 17, 1867	72, 292
Bending machine, Wood	L. Heywood	Gardner, Mass	Dec. 17, 1867	72, 293
Bending machine, Wood	L. Heywood	Gardner, Mass	Aug. 4, 1868	80, 627
Bending machine, Wood	M. V. B. Howe	Gardner, Mass	Jan. 28, 1868	73, 720
Bending machine, Wood	P. Hurm	Hamilton, Ohio	Aug. 1, 1865	49, 109
Bending machine, Wood	S. Keeler	Lancaster, Pa	June 13, 1865	48, 181
Bending machine, Wood	S. U. King	Windsor, Vt	Oct. 24, 1865	50, 601
Bending machine, Wood	S. Kingsland	Lyndon, Wis	July 22, 1862	35, 983
Bending machine, Wood	J. Klahr	Bernville, Pa	Dec. 10, 1867	71, 886
Bending machine, Wood	G. Kriebel	Hosensack, Pa	Oct. 24, 1865	50, 602
Bending machine, Wood	E. Lacey	Chicago, Ill	Sept. 12, 1871	118, 864
Bending machine, Wood	J. P. Selzelter	Lancaster, Pa	Feb. 21, 1865	46, 479
Bending machine, Wood	C. Leseberg	Cincinnati, Ohio	Mar. 19, 1872	124, 753
Bending machine, Wood	W. P. Letchworth	Buffalo, N. Y	Aug. 18, 1868	81, 095
Bending machine, Wood	J. C. Lucas	Newark, N. J	June 16, 1868	78, 983
Bending machine, Wood	A. Luckhaupt	Columbus, Ohio	Oct. 1, 1861	33, 395
Bending machine, Wood	O. Marland	Boston, Mass	Feb. 14, 1871	111, 855
Bending machine, Wood	A. F. Marshall	Black River, N. Y	Sept. 5, 1871	118, 736
Bending machine, Wood	J. W. Martin	Philadelphia, Pa	Feb. 15, 1870	99, 925
Bending machine, Wood	E. G. Matthews	Oakham, Mass	Apr. 25, 1871	114, 171
Bending machine, Wood	H. McDonald	Union Springs, N. Y	Jan. 22, 1861	31, 182
Bending machine, Wood	H. McDonald	Shortsville, N. Y	Oct. 24, 1871	120, 305
Bending machine, Wood	H. McDonald	Shortsville, N. Y	Oct. 24, 1871	120, 306
Bending machine, Wood	J. McMichael	Philadelphia, Pa	Dec. 20, 1870	110, 262
Bending machine, Wood	J. F. Melcher	Bloomington, Ill	Sept. 15, 1868	82, 237
Bending machine, Wood	J. Merrill	Boston, Mass	Oct. 13, 1868	83, 078
Bending machine, Wood	E. Mets	Rochester, N. Y	June 30, 1868	79, 489
Bending machine, Wood	L. E. Minott	Sheboygan, Wis	Aug. 1, 1871	117, 663
Bending machine, Wood	C. Moyer, jr	Coopersburgh, Pa	Oct. 16, 1866	58, 873
Bending machine, Wood	C. L. Nelson and O. Bostwick	Burlington, Vt	Mar. 27, 1860	27, 672
Bending machine, Wood	J. Newman	Falmouth, Me	July 31, 1866	56, 788
Bending machine, Wood	H. H. Nichols	Keeseville, N. Y	Sept. 21, 1869	95, 037
Bending machine, Wood	H. Ocorr	Sheboygan, Wis	Sept. 26, 1871	119, 393
Bending machine, Wood	J. Phillips	Chicago, Ill	Oct. 24, 1871	120, 321
Bending machine, Wood	J. H. Preston	Jefferson City, Mo	Sept. 21, 1869	95, 042
Bending machine, Wood	C. Prudden	Philadelphia, Pa	Apr. 17, 1866	54, 012
Bending machine, Wood	N. Purdy	Milwaukee, Wis	May 4, 1869	89, 790
Bending machine, Wood	J. N. Ray	Indianapolis, Ind	Dec. 13, 1864	45, 457
Bending machine, Wood	J. N. Ray	Indianapolis, Ind	May 1, 1866	54, 405
Bending machine, Wood	T. D. Roberts	Middletown, N. Y	Dec. 5, 1865	51, 353
Bending machine, Wood	E. Robinson	Raynham, Mass	Oct. 8, 1861	33, 467
Bending machine, Wood	A. Rogers	Painesville, Ohio	Mar. 20, 1860	27, 568
Bending machine, Wood	W. H. Ryan	Louisville, Ky	Aug. 2, 1864	43, 745
Bending machine, Wood	J. P. Schmucker	Ashland, Ohio	Oct. 2, 1866	58, 488
Bending machine, Wood	F. Seidle and S. Eberly	Mechanicsburgh, Pa	July 10, 1860	29, 107
Bending machine, Wood	F. Smith and P. Swope	Tiffin, Ohio	Feb. 28, 1865	46, 591
Bending machine, Wood	A. G. Snyder	Ashtabula, Ohio	Mar. 16, 1869	87, 981
Bending machine, Wood	J. K. B. Solomon	Rigglesville, Pa	Aug. 30, 1870	106, 880
Bending machine, Wood	A. P. Spaulding and E. Pierce	Westminster, Mass	Sept. 11, 1860	30, 004
Bending machine, Wood	H. E. Stager	Milwaukee, Wis	Jan. 9, 1866	51, 978
Bending machine, Wood	O. W. Stearns	Lebanon, N. H	July 2, 1867	66, 409
Bending machine, Wood	E. J. Updegraff	York, Pa	Apr. 8, 1856	14, 636
Bending machine, Wood	D. B. Wedden	Newark, N. J	Oct. 16, 1860	30, 404
Bending machine, Wood	J. L. Whipple and E. Trowbridge.	Detroit, Mich	May 28, 1872	127, 396
Bending machine, Wood	W. D. Williams	Raleigh, N. C	May 3, 1870	102, 637
Bending machinery, Wood	R. Fitts, jr	Fitchburgh, Mass	Aug. 5, 1862	36, 082
Bending machinery, Wood	J. Hann	Frenchtown, N. J	July 6, 1869	92, 187
Bending-mandrel	C. French	Seymour, Conn	Apr. 7, 1868	76, 432
Bending metal	C. H. Schubens	Newark, N. J	May 7, 1867	64, 585
Bending metals edgewise, Machine for	G. W. Brown	Galesburgh, Ill	May 25, 1869	90, 492
Bending mold-boards, Adjustable die for	J. H. Franklin	Avoca, Wis	Sept. 6, 1870	107, 032
Bending several pieces of wood of unequal lengths at once, Method of.	H. A. Barnard	Moline, Ill	Mar. 9, 1858	19, 538
Bending sheet-metal	J. Wright	Harmar, Ohio	Feb. 26, 1856	14, 332
Bending sheet-metal, Plier for	M. Rothschild	Harrisburgh, Pa	Feb, 18, 1868	74, 601
Bending sheet-metal, Roller for	D. Newton	Southington, Conn	Mar. 10, 1857	16, 804

Index of patents issued from the United States Patent Office from 1790 *to* 1873, *inclusive*—Continued.

Invention.	Inventor.	Residence.	Date.	No.
Bending timber, Method of	E. Green and M. Blackslee	Litchfield, Conn	Mar. 20, 1822	....
Bending timber, Retaining-strap for	L. E. Minott	Sheboygan, Wis	Aug. 1, 1871	117,664
Bending wood	T. Blanchard	Boston, Mass	Dec. 18, 1849	6,951
Bending wood, Method of	T. Blanchard	Boston, Mass	Oct. 21, 1856	15,944
Bending wood, Method of	R. Fitts	New Ipswich, N. H	Jan. 4, 1859	22,529
Bending wood, Method of	E. A. and C. Kilburn	Burlington, Vt	Oct. 7, 1856	15,851
Bending wood, Method of	J. C. Morris	Cincinnati, Ohio	Mar. 11, 1856	14,405
Bending wooden hoops, Machine for	J. Dobbins	Waterloo, Mich	Nov. 6, 1866	59,369
Benzine from hydrocarbon, Apparatus for removing.	H. J. Berg	Butler, Pa	Aug. 24, 1869	93,952
Benzine, Purifying	G. Lupton	Indianapolis, Ind	Dec. 13, 1870	110,054
Benzole-light	H. M. Paine	Worcester, Mass	July 13, 1852	9,119
Benzole-vapor apparatus	C. Cunningham	Nashua, N. H	Mar. 13, 1855	12,535
Benzole-vapor apparatus	O. P. Drake	Boston, Mass	Aug. 30, 1853	9,967
Berry-box	T. Mabbett, jr	Vineland, N. J	Jan. 14, 1868	73,352
Berry-box	C. Reese	Baltimore, Md	Apr. 23, 1872	126,087
Berry-boxes, Machine for cutting	C. Colby	South Pass, Ill	Aug. 27, 1867	68,171
Berth for railway-cars, Sleeping	A. Judson	Newark, N. J	July 21, 1868	80,184
Berth for railway-cars, Sleeping	D. M. Lawrence	Cincinnati, Ohio	Nov. 9, 1858	22,025
Berth for railway-cars, Sleeping	D. L. Long	Dayton, Ohio	Aug. 23, 1859	25,205
Berth for ships, Oscillating	I. A. Chomel	New York, N. Y	June 17, 1873	140,014
Berth for vessels, Oscillating	B. Weisker	New York, N. Y	Aug. 5, 1873	141,614
Berth-knee former	D. Taylor	East Boston, Mass	June 6, 1854	11,005
Berth lock, Sleeping-car	T. A. Bissell	Detroit, Mich	Dec. 23, 1873	145,835
Berth, Movable	N. W. Wheeler	Brooklyn, N. Y	Apr. 25, 1865	47,481
Berth, Oscillating	L. D. Newell	New York, N. Y	Sept. 10, 1872	131,223
Berth, Oscillating ship's	T. S. Brown	Philadelphia, Pa	Aug. 28, 1860	29,763
Berth, Ship's	I. A. Chomel	New York, N. Y	Aug. 27, 1872	130,792
Berth, Ship's	J. Evans	Denver, Colo. Ter.	Apr. 16, 1872	125,729
Berth, Ship's	H. Getty	Brooklyn, N. Y	Aug. 25, 1857	18,044
Berth, Ship's	J. Monroe	New York, N. Y	July 6, 1869	92,204
Berth, Sleeping-car	B. F. Manier	Green Island, N. Y	Apr. 4, 1871	113,539
Berth, Sleeping-car	W. B. Snow	Chicago, Ill	Feb. 25, 1868	74,773
Berth, Sleeping-car	W. B. Snow	Chicago, Ill	May 17, 1870	103,097
Berth, Step for	S. E. Siegel	New York, N. Y	July 9, 1872	128,761
Berth, Vessel	B. Weisker	New York, N. Y	June 3, 1873	139,638
Berths for vessels, Method of constructing	H. King	New York, N. Y	Sept. 4, 1841	2,240
Berths, Step-attachment for	E. Hoyt and E. P. Whitney	Stamford, Conn	Mar. 12, 1867	62,850
Bessemer-converter bottom	A. L. Rothman	Joliet, Ill	Sept. 3, 1872	131,120
Bessemer or Kelly converter, Compound for lining bottom of.	J. E. Atwood	Pittsburgh, Pa	May 13, 1873	138,780
Bessemer plant	A. L. Rothman	Joliet, Ill	Nov. 19, 1872	133,258
Beton, Hydraulic	J. Drevet	Paris, France	May 21, 1872	126,940
Bevel	L. D. Howard	Saint Johnsbury, Vt	Nov. 5, 1867	70,570
Bevel, Adjustable	E. A. Bell	Meriden, Conn	Apr. 25, 1871	114,096
Bevel, Adjustable	W. C. Ellis	Springfield, Mass	Mar. 21, 1871	112,793
Bevel, Adjustable	L. D. Howard	Saint Johnsbury, Vt	Feb. 14, 1871	111,746
Bevel, Adjustable	J. A. Traut	New Britain, Conn	May 9, 1871	114,626
Bevel and tapering gage	D. H. Tierney	New York, N. Y	June 16, 1868	79,032
Bevel and try square	J. Graham	Ludlow, Vt	Nov. 5, 1867	70,547
Bevel, Carpenter's	L. Bailey and S. D. Sargent	New Britain, Conn	Mar. 19, 1872	124,779
Bevel, Carpenter's	H. Fairbanks and I. J. Robinson.	Saint Johnsbury, Vt	Mar. 11, 1873	136,714
Bevel, Carpenter's	W. H. Laughlin	Quincy, Ill	July 23, 1872	129,737
Bevel, Carpenter's	S. D. Sargent	Hartford, Conn	July 22, 1873	141,081
Bevel, Carpenter's adjustable	I. J. Robinson and H. Fairbanks.	Saint Johnsbury, Vt	Apr. 9, 1872	125,617
Bevel, Protracting	W. W. Branch, jr	Madison, Ohio	Mar. 27, 1866	53,397
Bevel-rest	J. E. Seavey	Kennebunkport, Me	Dec. 30, 1873	146,027
Bevel, square, rule, &c., Combined	J. D. Otstot	Springfield, Ohio	July 5, 1870	105,121
Bevel-turning gage	J. W. Moore	Worcester, Mass	Sept. 21, 1869	94,968
Beverage	G. Beck	Rochester, N. Y	July 3, 1866	55,983
Beverage	A. Cochard	Port Richmond, N. Y	Jan. 31, 1865	46,081
Beverage	W. Davis	Portland, Me	June 4, 1867	65,475
Beverage	W. H. Goss	Boston, Mass	June 22, 1869	91,621
Beverage	E. G. Holland	Union Springs, N. Y	May 29, 1866	55,106
Beverage	A. C. Howell	Vienna, N. J	June 27, 1865	48,405
Beverage	J. H. Kenyon	Sempronius, N. Y	Apr. 10, 1866	53,834
Beverage	C. Nessi	San Francisco, Cal	Nov. 16, 1869	96,928
Beverage	S. B. Shaw	West Randolph, Mass	July 6, 1869	92,376
Beverage	H. Smith and H. F. Snow	Dover, N. H	July 17, 1866	56,458
Beverage	C. Sussegger	New York, N. Y	June 19, 1866	55,737
Beverage and sirup from mustang grape	J. C. Wood	Larissa, Tex	June 11, 1872	127,944
Beverage cooling and dispensing apparatus	J. Matthews	New York, N. Y	May 6, 1873	138,671
Beverage from sour or bitter oranges	P. G. Pearson	Jacksonville, Fla	Apr. 27, 1869	89,430
Beverage, Medicated	W. W. Timmons	Rahway, N. J	Feb. 2, 1869	86,471
Beverage or champagne mead	A. S. Taylor	San Francisco, Cal	Aug. 16, 1870	106,428
Beverage, Tonic	J. Mayer	Cleveland, Ohio	Jan. 26, 1869	86,172
Beverages, Composition for	R. F. Brown	Provincetown, Mass	Oct. 19, 1869	95,980
Beverages for medicinal and other purposes, Manufacture of.	E. N. Horsford	Cambridge, Mass	June 14, 1870	104,311
Bias-cutter	J. H. Goodfellow	Troy, N. Y	Sept. 13, 1870	107,248
Bias-cutting device	G. Moore	Scotch Plains, N. J	Jan. 14, 1873	134,918
Bias-marking device	G. Moore	Scotch Plains, N. J	Jan. 14, 1873	134,919
Bib, Child's	A. L. Thomson	Hudson City, N. J	May 2, 1871	114,366
Bib, Compression	W. Young	Easton, Pa	July 18, 1871	117,233
Bidet or hip-bath	D. Gibbons	Rochester, N. Y	Nov. 14, 1871	120,961
Bier	D. Fitzgerald	New York, N. Y	May 19, 1863	38,575
Bier	W. Scarlett	Aurora, Ill	Oct. 2, 1860	30,251
Bilge and keel of vessels, Block for supporting	F. Grice	Washington, D. C	Feb. 20, 1849	6,120
Bilge and leakage water indicator for ship or other vessel.	R. Shaler	Madison, Conn	Aug. 26, 1856	15,624
Bilge-block for docking ships	P. Burgess	Brooklyn, N. Y	Apr. 19, 1864	42,345
Bilge-boat, Solid	D. Brader	Beach Haven, Pa	Apr. 9, 1872	125,435
Bilge-water and for the ventilation of vessel, Apparatus for the discharge of.	J. Briggs	Louisville, Ky	Oct. 27, 1863	40,394
Bilge-water &c., Apparatus for elevating and discharging.	N. Hodge	North Adams, Mass	Oct. 19, 1852	9,344

Index of patents issued from the United States Patent Office from 1790 *to* 1873, *inclusive*—Continued.

Invention.	Inventor.	Residence.	Date.	No.
Bilge-water, Device for discharging	J. A. Miller	Paducah, Ky	Feb. 11, 1873	135, 656
Bilge-water gage	J. R. Barry	Philadelphia, Pa	Sept. 27, 1864	44, 385
Bilge-water gage	W. P. Kirkland	San Francisco, Cal	July 31, 1866	56, 763
Bilge-water gage, Vessel	C. McCloskey	East Cambridge, Mass	June 4, 1872	127, 422
Bilge-ways	S. Thomas	New York, N. Y	Oct. 10, 1829	....
Bill and currency holder	G. B. Isham	Burlington, Vt	July 7, 1863	39, 148
Bill and note holder	G. K. Snow	Watertown, Mass	Oct. 21, 1862	36, 734
Bill-folding gage	A. J. Marshall	Buffalo, N. Y	Oct. 15, 1872	132, 166
Bill-holder	A. Buswell	San Francisco, Cal	Mar. 24, 1863	37, 949
Bill-holder	J. A. M. Collins	Keokuk, Iowa	Dec. 29, 1868	85, 367
Bill-holder	E. F. French	Franklin, Vt	Mar. 17, 1857	16, 829
Bill-holder	G. Moulton	Cambridge, Mass	Aug. 27, 1867	68, 305
Bill-holder	G. W. Palmer	Boston, Mass	Mar. 27, 1855	12, 604
Bill-holder	C. T. Wakeley	Madison, Wis	Feb. 7, 1865	46, 282
Bill of credit	J. Golder	Philadelphia, Pa	Sept. 26, 1835	....
Bill of exchange	C. C. Wright	New York, N. Y	June 19, 1835	....
Bill of fare	W. A. Bury	Grosse Isle, Mich	May 8, 1860	28, 152
Bill of fare, Apparatus for displaying	J. Faye	Philadelphia, Pa	Aug. 23, 1864	43, 904
Bill-of-fare frame	C. Gloyd	Wynant, Ohio	Oct. 9, 1860	30, 311
Bill-register	J. N. Ayres	Stamford, Conn	Aug. 24, 1852	9, 211
Billiard and bagatelle table	W. A. Clark	Boston, Mass	Sept. 21, 1869	95, 084
Billiard and dining table	H. W. Collender	New York, N. Y	Apr. 21, 1868	77, 001
Billiard and dining table	F. E. Held	Chicago, Ill	Mar. 1, 1870	100, 401
Billiard and other game tables, Mode of constructing.	W. E. Bond	Cleveland, Ohio	Aug. 3, 1869	93, 272
Billiard and other tables, Device for leveling	B. W. Bull	New York, N. Y	Aug. 16, 1870	106, 317
Billiard and other tables, and for other purposes, Fabric for bed of.	L. A. Grill	New York, N. Y	July 11, 1871	116, 951
Billiard-ball	J. W. Hyatt, jr	Albany, N. Y	Oct. 10, 1865	50, 359
Billiard-ball, Vulcanite	W. H. Lippincott	Pittsburgh, Pa	May 12, 1868	77, 823
Billiard-balls and other articles, Compositions for	J. W. Hyatt, jr	Albany, N. Y	Apr. 14, 1868	76, 765
Billiard-balls and other composition articles, Manufacture of.	L. D. Benjamin, P. Wayne, and A. Stone.	Albany, N. Y	Apr. 4, 1871	113, 272
Billiard-balls, knife-handles, and other articles, Process of coating.	J. W. Hyatt, jr	Albany, N. Y	May 16, 1871	114, 945
Billiard-balls, &c., Method of coating	J. W. Hyatt, jr	Albany, N. Y	Apr. 6, 1869	88, 634
Billiard-board, Portable	L. C. Prindle	Chicago, Ill	Aug. 13, 1872	130, 389
Billiard-counter	H. Ball	Walpole, N. H	July 6, 1869	92, 247
Billiard-cue	G. Bevitt	Madison, Wis	Jan. 23, 1866	52, 128
Billiard-cue	C. Coan and A. R. Burdick	Chicago, Ill., and Racine, Wis.	Oct. 8, 1872	132, 054
Billiard-cue	C. H. Gould	New York, N. Y	June 25, 1867	66, 147
Billiard-cue	H. C. Griffin	Franklin, N. H	Mar. 24, 1868	75, 751
Billiard-cue	M. V. Ingersoll	Norwalk Bridge, Conn	Aug. 9, 1870	106, 166
Billiard-cue	C. Leicht	New York, N. Y	May 27, 1856	14, 964
Billiard-cue	J. N. McIntire	New York, N. Y	Oct. 27, 1868	83, 397
Billiard-cue	L. Nemetz	New York, N. Y	Mar. 4, 1873	136, 535
Billiard-cue	H. Platts	Ithaca, N. Y	June 3, 1873	139, 519
Billiard-cue	O. C. Wilbur, jr	Coventry, R. I	Apr. 9, 1872	125, 643
Billiard-cue, Apparatus for chalking	D. F. Shaw	Elkhart, Ind	Aug. 3, 1869	93, 356
Billiard-cue chalk-holder	J. Jenkinson	Brooklyn, N. Y	Dec. 3, 1867	71, 621
Billiard-cue clamp	O. D. Benjamin	Toledo, Ohio	Nov. 25, 1873	144, 824
Billiard-cue, Clamp for gluing the tip on	S. Gissinger	Lawrenceville, Pa	Dec. 18, 1866	60, 503
Billiard-cue, Clamp for leathering	E. Brunswick	Chicago, Ill	Apr. 2, 1867	63, 465
Billiard-cue, Composition tip for	A. Wetherbee	Waltham, Mass	Nov. 10, 1868	84, 036
Billiard-cue cutter	F. R. Gardner	Cleveland, Ohio	Aug. 27, 1872	130, 912
Billiard-cue, Fastening for the tip of	O. C. Wilbur, jr	Providence, R. I	Jan. 10, 1871	110, 883
Billiard-cue, Press for attaching leathers to	W. L. Aldrich	Atlanta, Ga	Feb. 14, 1860	27, 093
Billiard-cue rack	E. Brunswick	Chicago, Ill	Jan. 7, 1868	72, 969
Billiard-cue rack	V. A. Buschmann	Baltimore, Md	May 5, 1868	77, 452
Billiard-cue tip	H. W. Collender	New York, N. Y	Jan. 4, 1859	22, 491
Billiard-cue tip	G. W. Dickinson	Saratoga Springs, N. Y	Nov. 4, 1873	144, 324
Billiard-cue tip	J. H. Green	Christiansburgh, Iowa	Aug. 30, 1859	25, 257
Billiard-cue tip	J. A. Veazie	Boston, Mass	Dec. 10, 1867	71, 925
Billiard-cue tip and fastener	E. B. Stocking	Binghamton, N. Y	Apr. 2, 1867	63, 575
Billiard-cue trimmer	H. Pernot	New York, N. Y	Mar. 3, 1868	75, 049
Billiard-cues true, Machine for cutting ends of	J. Glynn and M. Borowsky	Placerville, Cal	July 5, 1859	24, 625
Billiard-cushion	H. A. Alden	Matteawan, N. Y	Apr. 29, 1873	138, 356
Billiard-cushion	A. Bassford	New York, N. Y	Jan. 22, 1867	61, 308
Billiard-cushion	J. Berlien	Chicago, Ill	Feb. 1, 1870	99, 393
Billiard-cushion	J. Berlien	Chicago, Ill	Jan. 21, 1870	104, 542
Billiard-cushion	H. W. Collender	New York, N. Y	Nov. 26, 1867	71, 282
Billiard-cushion	M. Delaney	New York, N. Y	Aug. 15, 1871	117, 995
Billiard-cushion	J. G. H. Meyer	San Francisco, Cal	Oct. 1, 1872	131, 890
Billiard-cushion	W. St. Martin	Cincinnati, Ohio	May 27, 1873	139, 434
Billiard-cushion	J. Syrcher	Buffalo, N. Y	Nov. 10, 1863	40, 580
Billiard-cushion	J. Wagner	Detroit, Mich	Apr. 23, 1872	126, 113
Billiard-cushion	W. K. Winant	Brooklyn, N. Y	Aug. 10, 1858	21, 159
Billiard-cushion, India-rubber	C. L. Richards	New York, N. Y	Mar. 22, 1870	101, 163
Billiard-cushion, Lining	L. Wacker	Buffalo, N. Y	July 8, 1862	35, 851
Billiard-cushion of rubber and gutta-percha	J. Murphy	New York, N. Y	Sept. 26, 1871	119, 391
Billiard-game keeper	G. Miller and J. Reichert	New York, N. Y	Aug. 21, 1866	57, 443
Billiard-game register	E. Holmes and H. C. Roome	New York, N. Y	Mar. 9, 1869	87, 674
Billiard-game register	W. A. Hough	Saint Johnsville, N. Y	Apr. 27, 1869	89, 408
Billiard-game register, Electro-magnetic	G. A. Webster	Elgin, Ill	Aug. 15, 1871	118, 082
Billiard-indicator	M. Gootman	New York, N. Y	Jan. 16, 1866	52, 041
Billiard-indicator	B. Schmitz	Brooklyn, N. Y	Jan. 10, 1865	45, 868
Billiard-light	W. St. Martin	Cincinnati, Ohio	Feb. 25, 1873	136, 275
Billiard-marker	W. H. Newell	Hudson City, N. J	Aug. 15, 1871	118, 044
Billiard-register	H. I. Behrens	New York, N. Y	Feb. 8, 1859	22, 849
Billiard-register	G. W. Earl and J. H. Hawley	Kalamazoo, Mich	Apr. 23, 1867	64, 000
Billiard-register	L. Godefroy	Pulaski, Tenn	Dec. 26, 1871	122, 167
Billiard-register	R. H. Ingersoll	Washington, D. C	Feb. 6, 1866	52, 417
Billiard-register	O. W. Kellogg and H. R. Hill	Ripon, Wis	Oct. 2, 1866	58, 429
Billiard-register	I. Kling	Seymour, Ind	June 23, 1868	79, 233
Billiard-register	J. Lee	Galesburgh, Ill	July 23, 1872	129, 670
Billiard-register	M. McCay	Topeka, Kans	Dec. 24, 1872	134, 211
Billiard-register	G. Simpson and R. M. Taylor	Waterbury, Vt	May 14, 1867	64, 718

Index of patents issued from the United States Patent Office from 1790 *to* 1873, *inclusive*—Continued.

Invention.	Inventor.	Residence.	Date.	No.
Billiard-register	F. M. Sofge	Macon, Ga	Nov. 15, 1859	26, 131
Billiard-register	A. Wirsching and W. Zocke	Williamsburgh, N. Y	May 15, 1866	54, 806
Billiard-register	F. M. Wood and G. Geer	Galesburgh, Ill	Apr. 2, 1872	125, 242
Billiard-room, Cue-rack for	H. W. Collender	New York, N. Y	July 9, 1872	128, 855
Billiard-table	A. W. Bailey	Boston, Mass	July 2, 1872	128, 580
Billiard-table	J. D. Bradley	Washington, D. C	Aug. 16, 1864	43, 829
Billiard-table	A. and F. Braun	Buffalo, N. Y	Aug. 26, 1873	142, 202
Billiard-table	E. Brunswick	Chicago, Ill	May 6, 1873	138, 609
Billiard-table	G. Buntin	Boston, Mass	May 13, 1873	138, 850
Billiard-table	G. Buntin	Boston, Mass	Aug. 5, 1873	141, 539
Billiard-table	H. W. Collender	New York, N. Y	Nov. 16, 1858	22, 064
Billiard-table	H. W. Collender	New York, N. Y	July 30, 1872	130, 020
Billiard-table	H. W. Collender	New York, N. Y	Dec. 23, 1873	145, 787
Billiard-table	F. Fedderke	New York, N. Y	Apr. 5, 1859	23, 458
Billiard-table	W. H. Griffith	New York, N. Y	Nov. 26, 1872	133, 365
Billiard-table	L. A. Grill	New York, N. Y	June 6, 1871	115, 601
Billiard-table	L. A. Grill	New York, N. Y	July 11, 1871	116, 950
Billiard-table	K. Gudenoge	San Francisco, Cal	Oct. 20, 1868	83, 153
Billiard-table	J. Hunt	New York, N. Y	Oct. 8, 1872	132, 084
Billiard-table	J. G. Kappner	New York, N. Y	Nov. 15, 1859	26, 112
Billiard-table	R. M. Knauss	Philadelphia, Pa	July 16, 1872	128, 965
Billiard-table	W. Lochhead	Brooklyn, N. Y	Jan. 10, 1871	110, 925
Billiard-table	L. C. Prindle	Chicago, Ill	June 24, 1873	140, 162
Billiard-table	C. B., J., and W. C. Rogers	Deep River, Conn	Sept. 7, 1858	21, 444
Billiard-table	P. Schouller	Boston, Mass	June 19, 1866	55, 721
Billiard-table	C. Schulenburg	Detroit, Mich	Dec. 16, 1873	145, 533
Billiard-table	G. Smith	Detroit, Mich	Aug. 22, 1871	118, 288
Billiard-table	W. H. Stroup	Pittsburgh, Pa	July 16, 1872	128, 985
Billiard-table	I. N. Swasey	Yonkers, N. Y	Dec. 20, 1870	110, 306
Billiard-table	S. H. Waldie	Belmont, Tex	Nov. 18, 1873	144, 716
Billiard-table	D. D. Winant	New York, N. Y	Nov. 2, 1858	22, 001
Billiard-table, Attachment for the legs of	W. B. Soumeillian	Philadelphia, Pa	Sept. 25, 1866	58, 313
Billiard-table bolt-head cap	J. Plunkett	New York, N. Y	July 11, 1871	116, 990
Billiard-table bridge	T. Dolan	Albany, N. Y	Mar. 12, 1867	62, 827
Billiard-table chalk	T. Harrison, W. Hollman, and W. Chart.	Detroit and Three Rivers, Mich.	Oct. 19, 1869	96, 003
Billiard-table, Chalk and sand-paper holder for	C. F. Ritchel	Chicago, Ill	Sept. 8, 1868	81, 947
Billiard-table, Chalk-holder for	H. W. Collender	New York, N. Y	Oct. 21, 1873	143, 885
Billiard-table, Chalk-holder for	J. P. Ellicott	Washington, D. C	May 22, 1860	28, 435
Billiard-table, Chalk-holder for	A. M. Leonard and B. Perkins	Ann Arbor, Mich	Aug. 25, 1868	81, 515
Billiard-table, Chalk-holder for	G. W. Morris	Minonk, Ill	Mar. 5, 1872	124, 371
Billiard-table, Chalk-holder for	H. M. Wall	Miles, Mich	Dec. 12, 1865	51, 499
Billiard-table cloth	D. Skidmore	Seneca Falls, N. Y	Jan. 21, 1868	73, 660
Billiard-table cushion	A. Bassford	New York, N. Y	July 20, 1831	....
Billiard-table cushion	A. Bassford	New York, N. Y	Dec. 5, 1848	5, 952
Billiard-table cushion	A. Bassford	New York, N. Y	Mar. 29, 1859	23, 340
Billiard-table cushion	J. M. Brunswick	Cincinnati, Ohio	Dec. 8, 1857	18, 799
Billiard-table cushion	S. Brunswick and H. Borlinghaus, sr.	Saint Louis, Mo	Oct. 28, 1873	143, 961
Billiard-table cushion	J. E. Came	Boston, Mass	Dec. 7, 1858	22, 263
Billiard-table cushion	J. E. Came	Boston, Mass	Sept. 2, 1873	142, 435
Billiard-table cushion	W. B. Carpenter	Brooklyn, N. Y	Oct. 28, 1856	15, 994
Billiard-table cushion	H. W. Collender	New York, N. Y	Dec. 8, 1857	18, 805
Billiard-table cushion	H. W. Collender	New York, N. Y	Jan. 12, 1858	19, 074
Billiard-table cushion	H. W. Collender	New York, N. Y	Mar. 29, 1859	23, 350
Billiard-table cushion	H. W. Collender	New York, N. Y	Sept. 25, 1860	30, 177
Billiard-table cushion	H. W. Collender	New York, N. Y	May 2, 1871	114, 410
Billiard-table cushion	L. Decker	Bergen, N. J	Dec. 15, 1857	18, 841
Billiard-table cushion	L. Decker	Bergen, N. J	Nov. 9, 1858	22, 020
Billiard-table cushion	L. Decker	New York, N. Y	Dec. 18, 1866	60, 657
Billiard-table cushion	M. Delany	Virginia City, Nev	Nov. 23, 1869	97, 057
Billiard-table cushion	G. W. Holman	New York, N. Y	May 4, 1858	20, 156
Billiard-table cushion	J. Natus	Detroit, Mich	Feb. 13, 1872	123, 722
Billiard-table cushion	M. Phelan	New York, N. Y	Feb. 19, 1856	14, 290
Billiard-table cushion	M. Phelan	New York, N. Y	Jan. 12, 1858	19, 101
Billiard-table cushion	G. E. Phelan	New York, N. Y	Feb. 25, 1873	136, 178
Billiard-table cushion	G. D. Sharp	New York, N. Y	Nov. 15, 1859	26, 128
Billiard-table cushion	C. E. Wilcox	Milwaukee, Wis	Sept. 3, 1867	68, 586
Billiard-table, Cushion-rail for	A. Bassford	New York, N. Y	Mar. 29, 1859	23, 341
Billiard-table, Folding	C. Croley	Cincinnati, Ohio	Mar. 30, 1858	19, 755
Billiard-table leveler	D. H. Hill	South Chester, Pa	Oct. 28, 1873	144, 028
Billiard-table, Leveling-jack for	C. Paxton and W. H. Hull	Loveland, Ohio, and Coffeyville, Kans.	Feb. 20, 1872	123, 791
Billiard-table, Light-shade for	D. Conlan	New York, N. Y	June 7, 1859	24, 285
Billiard-table, Manufacture of rubber cushion for	M. Delaney	New York, N. Y	July 11, 1871	116, 819
Billiard-table, Mode of fastening pockets to	L. Petersen	Baltimore, Md	Jan. 31, 1865	46, 136
Billiard-table, Mode of making the bed of	W. B. Billings	Chicago, Ill	Nov. 2, 1869	96, 300
Billiard-table-pocket handle	J. M. Brunswick	Cincinnati, Ohio	June 7, 1859	24, 279
Billiard-table-pocket iron	J. P. Ellicott	Washington, D. C	May 1, 1860	28, 128
Billiard-table-pocket supporter	J. E. Came and S. Havens	Boston, Mass	June 15, 1858	20, 548
Billiard-table, Portable	E. Hunt	Philadelphia, Pa	June 11, 1872	127, 886
Billiard-table, Revolving	S. Brunswick	Saint Louis, Mo	Oct. 12, 1869	95, 765
Billiard-table, Substitute for	A. G. Buzby	Philadelphia, Pa	Nov. 3, 1868	83, 759
Billiard-table top or bed	C. Croley	Cincinnati, Ohio	Mar. 9, 1858	19, 546
Billiard-wire	E. O. Dow	Chicago, Ill	Mar. 8, 1870	100, 509
Billiards, Chalk-holder for	R. H. Thomas	Stoke-on-Trent, England	Dec. 3, 1872	133, 603
Billiards, Portable	H. Meyer	Philadelphia, Pa	July 2, 1872	128, 557
Bin:				
See Ash-bin.				
Corn-bin.				
Fruit-bin.				
Grain-bin.				
Lime-bin.				
Potato-bin.				
Preserving-bin.				
Revolving bin.				
Warehouse-bin.				
Bin-cover	A. S. Maxwell	Dixon, Ill	Jan. 31, 1871	111, 363

Index of patents issued from the United States Patent Office from 1790 *to* 1873, *inclusive*—Continued.

Invention.	Inventor.	Residence.	Date.	No.
Binder for papers, Temporary	M. A. Shepard	Lebanon, Ill	Sept. 10, 1872	131, 229
Binder, Temporary	J. F. Adams	Irvington, N. Y	Sept. 10, 1872	131, 203
Binder, Temporary	W. H. Barnard	Washington, D. C	June 17, 1873	139, 998
Binder, Temporary	D. Dunton	Carpentersville, Ill	Mar. 25, 1873	137, 131
Binder, Temporary	G. W. Emerson	Chicago, Ill	July 7, 1868	79, 560
Binder, Temporary	T. J. Gaffney and A. P. Williams	Cincinnati, Ohio	Dec. 3, 1872	133, 580
Binder, Temporary	N. M. Shaffer	New York, N. Y	Jan. 14, 1868	73, 261
Binder's board, Machine for cutting	J. A. Elder	Westbrook, Me	Oct. 11, 1853	10, 122
Binding-folder	J. B. Nichols	Lynn, Mass	Aug. 29, 1854	11, 615
Binding-folder	E. P. West	Jersey City, N. J	May 13, 1873	138, 772
Binding-guide	J. S. McCurdy	New York, N. Y	Feb. 26, 1856	14, 322
Binding-guide	G. W. White	New York, N. Y	Sept. 24, 1872	131, 583
Binding, Metallic	T. B. De Forest	Birmingham, Conn	Nov. 20, 1866	59, 895
Binnacle	E. S. Ritchie	Brookline, Mass	June 27, 1865	48, 443
Bird-bath	T. W. Van Tassel	Washington, D. C	Aug. 10, 1869	93, 503
Bird-cage	H. C. W. Battermann	New York, N. Y	July 23, 1861	32, 855
Bird-cage	G. J. Bolz, M. Grebner, and J. M. Jagel.	New York, N. Y	Oct. 31, 1871	120, 483
Bird-cage	T. H. Bradley	Saint Louis, Mo	July 12, 1870	105, 164
Bird-cage	G. F. J. Colburn	Newark, N. J	Jan. 4, 1870	98, 562
Bird-cage	G. F. J. Colburn	Newark, N. J	Jan. 25, 1870	99, 164
Bird-cage	G. Günther	New York, N. Y	Apr. 29, 1873	138, 396
Bird-cage	G. Günther	New York, N. Y	Jan. 5, 1869	85, 523
Bird-cage	G. Günther	New York, N. Y	Mar. 7, 1871	112, 449
Bird-cage	E. Hutchinson	New York, N. Y	Nov. 25, 1873	144, 907
Bird-cage	E. B. Jewett	Buffalo, N. Y	Nov. 26, 1872	133, 372
Bird-cage	J. C. Jewett and J. Vogt	Buffalo, N. Y	Sept. 21, 1869	95, 116
Bird-cage	O. Lindemann	New York, N. Y	Feb. 4, 1873	135, 567
Bird-cage	J. Maxheimer	New York, N. Y	June 3, 1862	35, 459
Bird-cage	J. Maxheimer	New York, N. Y	Apr. 6, 1869	88, 654
Bird-cage	J. Maxheimer	New York, N. Y	Dec. 23, 1873	145, 886
Bird-cage	G. J. Munschauer	Buffalo, N. Y	Mar. 11, 1873	136, 611
Bird-cage	C. L. Osborn	Brooklyn, N. Y	Apr. 14, 1863	38, 177
Bird-cage	C. L. Osborn	Brooklyn, N. Y	Feb. 28, 1865	46, 579
Bird-cage	C. L. Osborn	New York, N. Y	Aug. 4, 1868	80, 561
Bird-cage	G. R. Osborn	Bridgeport, Conn	Sept. 27, 1870	107, 802
Bird-cage	G. R. Osborn and B. A. Drayton	Morrisania, N. Y	July 30, 1872	129, 980
Bird-cage	G. R. Osborn and B. A. Drayton	Morrisania, N. Y	Dec. 17, 1872	134, 094
Bird-cage	W. Staehlen	Brooklyn, N. Y	Sept. 3, 1861	33, 216
Bird-cage	J. H. Williams	New York, N. Y	May 25, 1869	90, 619
Bird-cage, Bath-tub for	G. T. and G. E. Peters	Jersey City, N. J	May 6, 1873	138, 692
Bird-cage bottom	O. Lindemann	New York, N. Y	Nov. 7, 1871	120, 755
Bird-cage bracket	A. D. Judd	New Haven, Conn	May 16, 1871	114, 948
Bird-cage, Construction of	G. Günther	New York, N. Y	Apr. 2, 1872	125, 288
Bird-cage, Construction of	G. R. Osborn and B. A. Drayton	New York, N. Y	May 9, 1871	114, 593
Bird-cage, Feed-cup for	G. Günther	New York, N. Y	Nov. 26, 1872	133, 442
Bird-cage hook	J. Bradbury	Berlin, Conn	Nov. 7, 1871	120, 569
Bird-cage hook	D. McArthur	New Haven, Conn	Nov. 5, 1872	132, 846
Bird-cage hook	J. M. Spring	New Britain, Conn	Apr. 18, 1871	113, 808
Bird-cage, Manufacture of	G. Günther	New York, N. Y	June 4, 1872	127, 598
Bird-cage, Manufacture of	J. Hepp	New York, N. Y	Mar. 12, 1872	124, 576
Bird-cage mat	O. Lindemann	New York, N. Y	Jan. 2, 1872	122, 472
Bird-cage mat	C. S. Schenck	New York, N. Y	May 2, 1871	114, 345
Bird-cage mat	J. A. Singer	New York, N. Y	Apr. 4, 1871	113, 461
Bird-cage musical attachment	G. Günther	New York, N. Y	July 17, 1866	56, 407
Bird-cage, Paper for bottom of	J. H. Singer	New York, N. Y	Feb. 7, 1871	111, 579
Bird-cage, Rod for construction of	J. H. Williams	New York, N. Y	Jan. 1, 1869	90, 906
Bird-cage, Spring-perch for	E. Aldom	New York, N. Y	Dec. 12, 1871	121, 742
Bird, Decoy	N. Wales	Boston, Mass	Feb. 11, 1868	74, 458
Bird-house	H. Miller	Johnston, R. I	May 3, 1870	102, 573
Bird-house, Metallic	H. Miller	Cranston, R. I	Apr. 14, 1868	76, 648
Bird-house, Metallic	J. Murdock	Jersey City, N. J	Mar. 16, 1869	88, 799
Bird-trap	S. M. Brooks	Memphis, Tenn	May 10, 1870	102, 761
Bird-trap	R. Rex	Charles City, Iowa	Sept. 12, 1865	49, 923
Bird-trap	J. S. Stone and G. W. Chamberlin.	Fitchburgh, Mass	Oct. 20, 1868	83, 220
Biscuit and crackers, Machine for making and baking.	B. F. Mason	Kennebunkport, Me	July 6, 1839	1, 221
Biscuit and sugar-bread, Making	E. Treadwell	New York, N. Y	May 18, 1826	....
Biscuit, Apparatus for making scroll	A. Exton	Trenton, N. J	Jan. 16, 1866	52, 036
Biscuit-board	I. R. Shank	Buffalo, Va	July 26, 1859	24, 890
Biscuit-board and flour-chest	D. E. Bryer	Logansport, Ind	Aug. 30, 1864	44, 040
Biscuit-cutter	S. E. Clapp	Cambridge, Mass	May 5, 1868	77, 583
Biscuit-cutting machine	J. and S. Turner	Toronto, Canada	Mar. 25, 1873	137, 114
Biscuit-machine	A. Exton	Trenton, N. J	Aug. 27, 1861	33, 136
Biscuit-machine, Scroll	A. Exton	Trenton, N. J	Jan. 30, 1866	52, 278
Biscuit-making machine	W. Liddle	New Orleans, La	Sept. 21, 1815	
Biscuit, Manufacture of	C. D. and C. D. Boss, jr	New London, Conn	Oct. 11, 1870	108, 097
Biscuit, Manufacture of sweet	H. Even	Pekin, Ill	Apr. 15, 1873	137, 766
Biscuit-pan	J. C. Milligan	Brooklyn, N. Y	Nov. 9, 1869	96, 605
Biscuit-pan	J. C. Milligan	Brooklyn, N. Y	Apr. 5, 1870	101, 490
Bistouries	C. C. Brown	Washington, D. C	Oct. 10, 1865	50, 330
Bisulphites, Apparatus for the manufacture of	W. J. Turner	Bradford, England	Feb. 20, 1872	123, 799

Bit:

See **Adjustable bit.**
Auger-bit.
Ax-bit.
Boring-bit.
Brace-bit.
Bridle-bit.
Carving-machine bit.
Center-bit.
Countersink-bit.
Curb-bit.
Doweling-bit.
Drill-bit.
Driving-bit.
Expansive bit.

Index of patents issued from the United States Patent Office from 1790 to 1873, inclusive—Continued.

Invention.	Inventor.	Residence.	Date.	No.
Bit—Continued.				
See **Extension-bit.**				
Horse-training bit.				
Joint-bit.				
Ox-bit.				
Plane-bit.				
Safety-bit.				
Screw-cutting bit.				
Spring-bit.				
Upper-jaw bit.				
Weaning-bit.				
Bit	H. S. Shepardson	Shelburne, Mass	Feb. 27, 1872	124, 089
Bit and auger	C. Whitehouse	Bridgtown, England	Apr. 6, 1869	88, 760
Bit and auger die	J. Swan	Seymour, Conn	Nov. 16, 1869	96, 847
Bit and drill brace	G. G. Parker and W. P. Dodge	Prospect, N. Y	Mar. 21, 1871	112, 842
Bit and other tools, Device for holding	W. H. Barber	Greenfield, Mass	May 24, 1864	42, 827
Bit and tap, Combined	S. A. Smith	New Haven, Conn	Oct. 12, 1869	95, 846
Bit-brace	C. H. Amidon	Greenfield, Mass	May 21, 1867	64, 931
Bit-brace	Q. S. Backus	Miller's Falls, Mass	Nov. 5, 1872	132, 791
Bit-brace	C. L. Butler	Greenfield, Mass	Nov. 16, 1869	96, 881
Bit-brace	W. Cleveland	Lawrence, Mass	Mar. 7, 1871	112, 419
Bit-brace	J. W. Craig	Knoxville, Ill	June 9, 1868	78, 652
Bit-brace	C. M. Daboll	New London, Conn	Mar. 24, 1868	75, 870
Bit-brace	W. P. Dolin	Charlottesville, Va	Jan. 17, 1871	110, 960
Bit-brace	S. F. Fenn	Middletown, Conn	Dec. 3, 1867	71, 727
Bit-brace	J. S. Fray	Bridgeport, Conn	Aug. 20, 1872	130, 709
Bit-brace	R. French	Seymour, Conn	Mar. 21, 1871	112, 799
Bit-brace	S. B. Hill	Chicopee, Mass	Apr. 14, 1868	76, 632
Bit-brace	W. A. Ives	New Haven, Conn	Mar. 16, 1869	87, 940
Bit-brace	W. A. Ives	New Haven, Conn	Feb. 7, 1871	111, 649
Bit-brace	J. T. Lyman	Jeffersonville, Ind	Apr. 4, 1871	113, 680
Bit-brace	C. Morrill	New York, N. Y	Dec. 10, 1861	33, 899
Bit-brace	O. Peck	Southington, Conn	May 8, 1866	54, 650
Bit-brace	W. E. Bercau	Oakland, Cal	Dec. 16, 1873	145, 524
Bit-brace	F. P. Pfleghar	New Haven, Conn	Mar. 17, 1868	75, 699
Bit-brace	H. W. Porter	Rothsville, Pa	July 7, 1857	17, 769
Bit-brace	J. Rice	Prairie Creek, Ind	Jan. 9, 1872	122, 652
Bit-brace	C. O. Richey	Aurora, Ind	June 5, 1866	55, 364
Bit-brace	S. C. Rundlett	Portland, Me	Apr. 28, 1868	77, 219
Bit-brace	W. F. Seavey	Portland, Me	July 23, 1867	67, 139
Bit-brace	H. S. Shepardson	Shelburne Falls, Mass	Mar. 1, 1870	100, 458
Bit-brace	G. Stackpole	Portland, Me	Sept. 23, 1862	36, 538
Bit-brace	G. Stackpole	New York, N. Y	Apr. 2, 1867	63, 438
Bit-brace	C. H. Stockbridge	Whateley, Mass	Feb. 19, 1867	62, 232
Bit-brace	I. C. Tate	New London, Conn	May 19, 1868	78, 157
Bit-brace	C. Whitus	Philadelphia, Pa	Apr. 9, 1872	125, 642
Bit-brace	F. A. Wood	Jersey City, N. J	Dec. 1, 1868	84, 523
Bit-brace	W. H. Woods	San Francisco, Cal	Apr. 9, 1867	63, 684
Bit to brace, Fastener of	E. Smith	Norwich, N. Y	Aug. 17, 1852	9, 209
Bit brace for boring obliquely to the stock	C. C. Plaisted	Chicopee, Mass	July 28, 1857	17, 891
Bit-brace wrench	D. Wilcox	Ansonia, Conn	Mar. 2, 1869	87, 531
Bit-fastening	H. Lettington	Norwich, N. Y	May 13, 1856	14, 876
Bit for cutting out cylindrical plugs of wood	C. W. Saladee	Columbus, Ohio	May 26, 1857	17, 395
Bit for cutting washers	H. Pennie	Buffalo, N. Y	Sept. 6, 1859	25, 348
Bit grinding and polishing machine	J. Swan	Seymour, Conn	Feb. 15, 1870	99, 970
Bit-holder	B. B. Hill	Chicopee, Mass	Sept. 29, 1857	18, 282
Bit-holder	B. B. Hill and S. W. Adams	Chicopee, Mass	Jan. 5, 1858	19, 028
Bit-holder	C. H. Stockbridge	Northampton, Mass	Nov. 10, 1868	84, 015
Bit holder	D. H. Whittemore	Worcester, Mass	Apr. 20, 1858	20, 010
Bit-holder	J. Winkelhouse	New York, N. Y	Sept. 21, 1869	95, 060
Bit-holder for bit-stock	L. C. Rodier	Springfield, Mass	Aug. 30, 1864	44, 047
Bit in brace, Device for holding	G. Benjamine	Avoca, N. Y	June 9, 1857	17, 479
Bit in brace, Securing	O. Peck	Windsor, Vt	Feb. 23, 1864	41, 718
Bit in brace, Securing	W. Wimmer	Elizabethport, N. J	May 22, 1866	54, 990
Bit in socket, Device for holding	C. F. Hunter	Adrian, Mich	Oct. 27, 1863	40, 412
Bit in stock, Method of securing	D. Kelly	Grand Rapids, Mich	Dec. 16, 1862	37, 196
Bit in stock, Method of securing	W. Tucker	Blackstone, Mass	Mar. 22, 1859	23, 326
Bit in stock, Mode of securing	A. C. Moore	Wilmington, Vt	Mar. 31, 1857	16, 931
Bit or auger in its stock or handle, Method of fastening.	L. Winslow	Rochester, N. Y	Nov. 25, 1862	37, 022
Bit or boring tool	B. Sanderson	Derby Line, Vt	May 19, 1863	38, 633
Bit or drill holder	A. J. Smith	Lynn, Mass	July 7, 1857	17, 770
Bit or drill stock	D. H. Chamberlain	Boston, Mass	Oct. 26, 1852	9, 353
Bit or drill stock	D. H. Chamberlain	Boston, Mass	Feb. 7, 1854	10, 499
Bit-stock	A. S. Alden	Chicopee, Mass	Mar. 16, 1869	87, 816
Bit-stock	C. H. Amidon	Greenfield, Mass	Oct. 3, 1865	50, 214
Bit-stock	C. H. Amidon	Greenfield, Mass	Jan. 14, 1868	73, 279
Bit-stock	M. Andrews	Bridgewater, Mass	Apr. 24, 1847	5, 087
Bit-stock	Q. S. Backus	Miller's Falls, Mass	Nov. 5, 1872	132, 790
Bit-stock	H. S. Bartholomew	Bristol, Conn	May 24, 1870	103, 281
Bit-stock	C. H. Clark	Pittsfield, Me	Aug. 2, 1870	105, 906
Bit-stock	J. F. Cory	New York, N. Y	June 25, 1861	32, 618
Bit-stock	A. P. Daboll	New London, Conn	June 17, 1873	139, 876
Bit-stock	B. Darling	Bridgewater, Mass	Oct. 20, 1868	83, 261
Bit-stock	S. W. Davis	Wilmington, Del	Aug. 18, 1868	81, 260
Bit-stock	O. Ellsworth	Boston, Mass	Sept. 25, 1866	58, 234
Bit-stock	L. Feely	Rochester, N. Y	Dec. 30, 1873	146, 056
Bit-stock	D. P. Foster	Waltham, Mass	May 4, 1869	89, 755
Bit-stock	D. P. Foster	Waltham, Mass	Mar. 29, 1870	101, 251
Bit-stock	A. D. Goodell	Florence, (Northampton,) Mass.	July 14, 1868	79, 825
Bit-stock	A. D. Goodell	Miller's Falls, Mass	June 10, 1873	139, 667
Bit-stock	A. D. Goodell	Miller's Falls, Mass	July 29, 1873	141, 345
Bit-stock	D. W. Goodell	Northampton, Mass	Nov. 28, 1865	51, 173
Bit-stock	R. S. Hildreth	South Adams, Mass	Feb. 1, 1870	99, 317
Bit-stock	I. Holliday	South Brooklyn, N. Y	May 24, 1870	103, 462
Bit-stock	S. U. King	Windsor, Vt	July 7, 1863	39, 150
Bit-stock	C. E. Lombard	Springfield, Mass	Oct. 17, 1865	50, 481

Index of patents issued from the United States Patent Office from 1790 *to* 1873, *inclusive*—**Continued.**

Invention.	Inventor.	Residence.	Date.	No.
Bit-stock	W. Lyon	Deep River, Conn	Oct. 23, 1860	30, 487
Bit-stock	C. Manson	Boston, Mass	May 2, 1871	114, 314
Bit-stock	W. H. McCoy	Erving, Mass	Aug. 15, 1871	118, 039
Bit-stock	D. A. Newton	New York, N. Y	Oct. 28, 1873	144, 123
Bit-stock	M. V. Nobles	Rochester, N. Y	June 20, 1865	48, 346
Bit-stock	W. W. S. Orbeton	Haverhill, Mass	Oct. 30, 1866	59, 254
Bit-stock	H. S. Pratt	Brooklyn, N. Y	Oct. 1, 1872	131, 829
Bit-stock	G. Richards	Richmond Centre, Wis	Oct. 27, 1868	83, 410
Bit-stock	C. B. Rose	Sunderland, Mass	Apr. 16, 1867	63, 944
Bit-stock	C. B. Rose	Sunderland, Mass	Sept. 15, 1868	82, 251
Bit-stock	S. Sawyer	Erving, Mass	Aug. 15, 1871	118, 058
Bit-stock	W. H. Sible	Harrisburgh, Pa	Mar. 9, 1869	87, 715
Bit-stock	A. W. Smith	Manchester, N. H	Jan. 2, 1866	51, 874
Bit stock	R. D. O. Smith	Washington, D. C	Sept. 8, 1868	82, 041
Bit-stock	N. Spofford	Haverhill, Mass	Nov. 1, 1859	25, 984
Bit-stock	A. Stanley	New Britain, Conn	Nov. 30, 1869	97, 455
Bit-stock	O. G. Stratton	Greenfield, Mass	Aug. 11, 1868	81, 031
Bit-stock	A. W. Streeter	Shelburne Falls, Mass	Jan. 8, 1867	61, 113
Bit-stock	J. Taylor	Hebron, Conn	June 30, 1836	....
Bit-stock	O. H. Taylor	Brooklyn, N. Y	Sept. 8, 1868	82, 044
Bit-stock	O. H. Taylor	Brooklyn, N. Y	Nov. 2, 1869	96, 505
Bit-stock	F. M. and J. W. Thompson	Greenfield, Mass	Feb. 23, 1869	87, 309
Bit-stock	J. W. and F. M. Thompson	Greenfield, Mass	Sept. 15, 1868	82, 179
Bit-stock	C. P. Whitman	Charlemont, Mass	Aug. 23, 1870	106, 754
Bit-stock and wrench	S. S. Hout and B. B. Beers	New Fairfield, Conn	July 19, 1859	24, 807
Bit stock, Composition	A. Phelps	Marlborough, Conn	Oct. 10, 1829	....
Bit stock, Drill	H. B. Williams	Cincinnati, Ohio	May 13, 1873	138, 973
Bit-stock fastener	C. P. Whitman	Charlemont, Mass	Nov. 16, 1869	97, 010
Bit-stock, Method of constructing	A. W. Streeter	Shelburne Falls, Mass	Mar. 31, 1857	16, 938
Bit to brace, Attaching	O. Peck	Windsor, Vt	Aug. 4, 1863	39, 419
Bit to brace, Attaching	O. Peck	Windsor, Vt	Oct. 13, 1863	40, 275
Bit to brace, Device for attaching	S. U. King	Windsor, Vt	June 8, 1858	20, 495
Bitting attachment	S. V. R. York	Antwerp, N. Y	Apr. 28, 1868	77, 235
Bitters and sauce bottle tube, Manufacture of	J. B. Lyon	Pittsburgh, Pa	Dec. 17, 1872	133, 947
Bitters, Cider	T. P. Devor	Millerstown, Pa	July 30, 1872	129, 937
Bitters, Stomach	M. Becker	Philadelphia, Pa	Sept. 7, 1869	94, 466
Bitumen and clay, Combining	P. Lea	Goliad, Tex	Sept. 24, 1872	131, 690
Bitumen, ore, &c., Compound of	P. Lea	Goliad, Tex	Nov. 11, 1873	144, 396
Blackboard	H. L. Andrews	Chicago, Ill	Sept. 15, 1868	82, 062
Blackboard	M. F. Cowdery	Sandusky, Ohio	Feb. 6, 1872	123, 456
Blackboard	M. F. Cowdery	Sandusky, Ohio	Feb. 6, 1872	123, 457
Blackboard	W. H. Joeckel	New York, N. Y	July 13, 1869	92, 533
Blackboard	F. G. Johnson	New York, N. Y	Dec. 3, 1872	133, 583
Blackboard	J. Reber	Nebraska, Ohio	Nov. 4, 1873	144, 289
Blackboard	T. J. Thorp	Buffalo, N. Y	Jan. 21, 1873	135, 019
Blackboard, Adjustable	P. W. Moeller	New York, N. Y	Nov. 25, 1873	144, 917
Blackboard and map-case	W. C. Herider	Miami Town, Ohio	Sept. 22, 1863	40, 035
Blackboard, Composition	R. W. Young	Rising Sun, Ind	Apr. 25, 1871	114, 246
Blackboard, &c., Composition for covering	N. V. Evans	Winslow, Ind	Jan. 16, 1872	122, 763
Blackboard, Composition for covering	J. B. Rowell	Lynn, Mass	Aug. 7, 1860	29, 552
Blackboard for school	W. Arronquier	Worcester, Mass	Dec. 10, 1867	71, 946
Blackboard, &c., Liquid slating for forming	W. W. Dudley	Centreville, Ind	Aug. 30, 1870	106, 923
Blackboard-rubber	J. F. Bigger and W. A. Pugh	Rushville, Ind	Jan. 9, 1872	122, 511
Blackboard-rubber	P. B. Horton	San Francisco, Cal	Nov. 18, 1873	144, 674
Blackboard, slate, &c., Composition for forming	W. A. Reason	Chicago, Ill	Sept. 27, 1864	44, 456
Blackboard-wiper	P. Perry	Waterford, N. Y	Sept. 30, 1862	36, 589
Blacking	J. McCrellish	Philadelphia, Pa	Nov. 20, 1866	59, 851
Blacking	W. F. Quinby	Wilmington, Del	Aug. 28, 1866	57, 567
Blacking	L. R. Rockwood	Worcester, Mass	Feb. 22, 1859	23, 065
Blacking	A. Tomlinson	Cincinnati, Ohio	Oct. 31, 1865	50, 780
Blacking and brush holder	E. H. Sweetser	Salem, Mass	Mar. 26, 1872	124, 984
Blacking and brush holder	E. H. Sweetser	Salem, Mass	Aug. 13, 1872	130, 450
Blacking and polish, Boot and shoe	H. A. Reams	Durham, N. C	Mar. 19, 1872	124, 760
Blacking-apparatus and boot-rack combined	J. M. McMaster	Rochester, N. Y	Sept. 12, 1871	118, 809
Blacking apparatus, Boot	H. H. Dodge	Georgetown, D. C	Mar. 6, 1866	52, 977
Blacking apparatus, Boot	J. Vandercar	Brooklyn, N. Y	Feb. 18, 1868	74, 637
Blacking block, Boot and shoe	F. G. Harding	Boston, Mass	June 2, 1857	17, 431
Blacking, Boot and shoe	J. H. Patterson	Glen's Falls, N. Y	Sept. 26, 1871	119, 239
Blacking-box	J. S. Brooks	Rochester, N. Y	Aug. 8, 1865	49, 224
Blacking-box	R. Cadle	Shawneetown, Ill	Mar. 23, 1869	88, 128
Blacking-box	J. H. Doughty	New York, N. Y	Mar. 14, 1865	46, 783
Blacking-box	C. H. Gatchell	Oldtown, Me	Nov. 23, 1869	97, 072
Blacking-box	E. M. Gates	Watertown, N. Y	Sept. 12, 1871	118, 924
Blacking-box	W. L Gilroy	Philadelphia, Pa	Dec. 22, 1868	85, 084
Blacking-box	S. R. Sinclaire	New York, N. Y	Nov. 14, 1871	120, 907
Blacking-box	H. Smith	Haddonfield, N. J	Jan. 30, 1872	123, 204
Blacking-box	T. H. Spencer	Providence, R. I	Aug. 21, 1866	57, 401
Blacking-box	A. M. Utley	Watertown, N. Y	Nov. 21, 1871	121, 069
Blacking-box	J. Van Santvoord	Mount Vernon, N. Y	Oct. 31, 1871	120, 554
Blacking-box and brush	C. E. Yager	Hudson, N. Y	Aug. 30, 1870	106, 909
Blacking-box and holder	G. H. Monroe	Cincinnati, Ohio	June 21, 1864	43, 221
Blacking-box, Construction of	T. H. Spencer	Providence, R. I	May 12, 1868	77, 778
Blacking-box cover	J. B. Shaler	Brooklyn, N. Y	Apr. 22, 1873	138, 102
Blacking-box, Foot-rest for	J. H. Porter	New York, N. Y	Oct. 11, 1864	44, 656
Blacking-box handle-attachment	T. S. Robinson	New York, N. Y	June 25, 1867	66, 173
Blacking-box holder	C. R. Bacon and G. D. Clark	Newark, N. J	Feb. 11, 1868	74, 276
Blacking-box holder	W. A. Field	Boston, Mass	Apr. 16, 1867	63, 876
Blacking-box holder	R. R. Forrest	Washington, Pa	Aug. 13, 1872	130, 419
Blacking-box holder	C. E. S. Jelliffe	Williamsburgh, N. Y	Aug. 1, 1865	49, 115
Blacking-box holder	H. B. Hammon	Bristolville, Ohio	Aug. 3, 1869	93, 196
Blacking-box holder	J. W. Lewis	Providence, R. I	Jan. 1, 1861	31, 065
Blacking-box holder	T. K. Payson	New York, N. Y	Nov. 12, 1867	70, 892
Blacking-box holder	J. H. Rector and A. Sweetland	Syracuse, N. Y	July 26, 1864	43, 645
Blacking-box holder	N. H. Talbot	North Easton, Mass	July 25, 1871	117, 481
Blacking-box holder	G. W. Taylor	Springfield, Vt	Jan. 1, 1867	60, 962
Blacking-box holder	A. Wilder	Calais, Me	Jan. 8, 1867	61, 128
Blacking-box holder	S. T. Wyman	Lynn, Mass	Dec. 16, 1873	145, 544
Blacking-box, Implement for holding	W. and J. Cairns	Jersey City, N. J	Jan. 13, 1857	16, 374

Index of patents issued from the United States Patent Office from 1790 *to* 1873, *inclusive*—Continued.

Invention.	Inventor.	Residence.	Date.	No.
Blacking-box, Manufacture of	G. W. Bentley	New York, N. Y	May 9, 1865	47, 676
Blacking box, Shoe	G. H. Wetjen	New York, N. Y	Sept. 12, 1871	118, 995
Blacking-case	C. E. Brown	Northampton, Mass	Apr. 9, 1867	63, 700
Blacking-case	W. P. Huges	Rochester, N. Y	July 16, 1872	129, 137
Blacking-case	G. H. Rice	Norwalk, Conn	Apr. 8, 1873	137, 722
Blacking case and closet	J. McFerran	Philadelphia, Pa	Feb. 18, 1868	74, 705
Blacking-case and night-chair, Combined	J. H. Doughty	New York, N. Y	July 24, 1866	56, 535
Blacking case, Boot	E. S. Carter	Keokuk, Iowa	Mar. 13, 1866	53, 111
Blacking-case, Portable	A. W. Overbaugh	New York, N. Y	May 21, 1867	64, 899
Blacking-dish and knife	E. L. Bolster	Waterbury, Conn	June 11, 1867	65, 637
Blacking for leather	S. Adams	Cleveland, Ohio	May 30, 1842	2, 660
Blacking for leather, &c	I. B. Merrill	Dexter, Me	Apr. 21, 1868	77, 069
Blacking for leather, Oil	S. S. Allen	Richmond, Ind	Nov. 10, 1868	83, 817
Blacking for leather, Oil	O. K. Tripp	Rochester, N. Y	Oct. 24, 1871	120, 348
Blacking for shoes, &c	J. L. Sneed	Frankfort, Ky	Nov. 18, 1873	144, 801
Blacking-fount	C. Künstler	Brooklyn, N. Y	Sept. 28, 1869	95, 357
Blacking, Japan-paste	H. Lake	San Francisco, Cal	Jan. 28, 1868	73, 730
Blacking, Leather	A. D. Strong	Ashtabula, Ohio	Aug. 12, 1873	141, 829
Blacking leather, Composition for	J. Engelhardt	Carbondale, Pa	Apr. 21, 1868	76, 897
Blacking machine, Boot and shoe	M. A. Myer	Decatur, Ill	Feb. 5, 1867	61, 755
Blacking machine, Mold	B. S. Benson	Baltimore, Md	Oct. 27, 1868	83, 448
Blacking, Manufacture of	R. Bartholow	Cincinnati, Ohio	Apr. 4, 1865	47, 082
Blacking, Manufacture of	G. W. Corey	Port Jervis, N. Y	Feb. 6, 1866	52, 391
Blacking, Manufacture of	J. L. Lucas	Saratoga Springs, N. Y	Oct. 11, 1870	108, 271
Blacking, Metal-founder's	J. C. Sellars	Birkenhead, England	June 30, 1868	79, 504
Blacking, Metal-founder's	J. C. Sellars	Birkenhead, England	May 16, 1871	114, 978
Blacking, Packing shoe	C. Herold	Pittsburgh, Pa	Aug. 27, 1872	130, 917
Blacking, Shoe and stove	J. Breinig	Allentown, Pa	Sept. 10, 1872	131, 245
Blacking-spreader	H. S. Kerr	Philadelphia, Pa	Apr. 23, 1872	125, 894
Blacking-spreader	H. S. Kerr	Philadelphia, Pa	May 7, 1872	126, 399
Blacking, Water-proof	D. L. Pickard	Rochester, N. Y	Oct. 10, 1865	50, 383
Blacksmith's butteris	E. Bacher	Findlay, Ohio	Oct. 8, 1872	132, 044
Blacksmith, Foot-lifter for	T. C. Williams	Warrenton, Mo	Apr. 4, 1871	113, 379
Blacksmith's furnace for burning anthracite coal	E. Nichols and J. Augur	Hamden and New Haven, Conn.	Aug. 21, 1839	1, 293
Blacksmith's hardy	C. F. Moore	Snedikerville, Pa	Aug. 1, 1871	117, 665
Blacksmith's hearth	J. Cavender	Milton, Ohio	Aug. 27, 1861	33, 132
Blacksmiths, Mode of blowing and striking for	S. Hoyt and E. Pierce	Pultney, N. Y	Mar. 3, 1827	....
Blacksmith's striker	G. Bell	Martinsburgh, W. Va	June 25, 1867	66, 118
Blacksmith's striker	D. S. Blue	Fort Seneca, Ohio	Nov. 27, 1855	13, 841
Blacksmith's striker	H. Kendall	East Dorset, Vt	May 12, 1857	17, 284
Blacksmith's tool	S. J. Forbes	Marshalltown, Iowa	Oct. 10, 1871	119, 837
Blacksmith's tool	T. B. and L. W. Kelly	Hammondsburgh, Pa., and Brunswick, Ohio.	Nov. 25, 1862	37, 000
Blacksmith's tool	J. F. Kernon	Saugerties, N. Y	Mar. 5, 1872	124, 362
Black-washing mold	G. Ross	Newport, Ky	May 3, 1864	42, 600
Bladder, Machine for crumbling stone in	J. L. Hannah	New York, N. Y	Aug. 19, 1834	....
Blade-case and attaching-pin, Combined	W. L. Bowser	Moncton, New Brunswick	Feb. 9, 1869	86, 637
Blank clip for swingletree, Method of forming	M. Loughran	Pittsburgh, Pa	July 11, 1865	48, 699
Blanks, Machine for nicking screw	J. C. Rhodes	East Bridgewater, Mass	Nov. 10, 1863	40, 588
Blanket	M. V. Nobles	Elmira, N. Y	Feb. 27, 1872	124, 007
Blanket-fastener	L. C. Chase	Boston, Mass	Oct. 15, 1867	69, 768
Blanket, Horse	S. W. Baker	Providence, R. I	Feb. 25, 1868	74, 742
Blankets and overcoats, Sling for carrying	J. Short	Salem, Mass	Apr. 15, 1862	35, 002
Blankets, &c., Machine for making	J. Getzendannar	Frederick County, Md	Sept. 30, 1814	....
Blast and cupola furnace	D. W. Hendrickson	Red Bank, N. J	June 17, 1873	139, 891
Blast and cupola furnace	F. Lawrence	Philadelphia, Pa	Sept. 2, 1873	142, 482
Blast and cupola furnace for melting iron	J. Pattarson	Warwick, N. Y	June 13, 1831	....
Blast and draft of furnace, forge, &c., Mode of managing.	A. Collins	Ulster, N. Y	Dec. 31, 1838	1, 054
Blast and puddling furnace for making wrought-iron direct from the ore, Combined rotary.	J. Neville	Jersey City, N. J	Apr. 29, 1873	138, 429
Blast-apparatus	B. F. Sturtevant	West Roxbury, Mass	July 13, 1869	92, 490
Blast-apparatus for furnace	R. A. Smith	Newburyport, Mass	Nov. 15, 1870	109, 356
Blast apparatus for puddling and other furnaces, Hot.	P. Hoop, jr., and R. Hoop	Berlin Cross Roads, Ohio	Oct. 27, 1868	83, 382
Blast apparatus, Hot	J. J. Pierce	Sharpsville, Pa	June 7, 1870	103, 923
Blast apparatus, Steam	G. A. Jasper	Charlestown, Mass	Oct. 4, 1864	44, 533
Blast for furnace used in smelting iron, &c., Apparatus for heating the.	J. Player	Norton, England	June 11, 1867	65, 600
Blast for iron and other furnaces	F. A. T. de Beauregard	Paris, France	Jan. 22, 1867	61, 323
Blast for smelting iron-ore, Furnace	T. Gregg	Connellsville, Pa	May 22, 1832	....
Blast-furnace	E. B. Andrews	Lancaster, Ohio	Oct. 7, 1873	143, 487
Blast-furnace	G. Atkins	Sharon, Pa	July 13, 1869	92, 505
Blast-furnace	J. G. Blunt	Leavenworth, Kans	Apr. 15, 1873	137, 884
Blast-furnace	F. Büttgenbach	Düsseldorf, Germany	Nov. 25, 1873	144, 829
Blast-furnace	I. B. B. Case	Toledo, Ohio	Sept. 12, 1871	118, 905
Blast-furnace	H. Davies	Newport, Ky	Dec. 12, 1871	121, 762
Blast-furnace	J. F. Harris	Fort Edward, N. Y	June 25, 1872	128, 304
Blast-furnace	D. W. Hendrickson	New York, N. Y	Apr. 23, 1867	64, 012
Blast-furnace	D. W. Hendrickson	Red Bank, N. J	Apr. 1, 1873	137, 366
Blast-furnace	W. Kelly	Eddyville, Ky	Jan. 20, 1857	16, 444
Blast-furnace	B. Kugler	Philadelphia, Pa	June 2, 1836	....
Blast-furnace	F. W. Lurmann	Osnabruck, Prussia	Nov. 5, 1867	70, 447
Blast-furnace	J. H. McKernan	Indianapolis, Ind	Jan. 23, 1872	123, 035
Blast-furnace	P. P. Parrott	Greenwood Iron-Works, N. Y.	Aug. 20, 1872	130, 590
Blast-furnace	J. Pattison	New York, N. Y	Dec. 10, 1872	133, 718
Blast-furnace	W. Raschette	St. Petersburg, Russia	June 21, 1864	43, 274
Blast-furnace	S. C. Salisbury	New York, N. Y	Apr. 24, 1866	54, 215
Blast-furnace	J. Steece	Lawrence County, Ohio	May 16, 1833	....
Blast-furnace	I. Tyson, jr	Baltimore, Md	Apr. 18, 1834	....
Blast-furnace	S. Wilkes	Hammondsville, Ohio	Sept. 8, 1857	18, 167
Blast-furnace	H. Wissenborn	New York, N. Y	Feb. 3, 1857	16, 560
Blast-furnace	J. V. Woodhouse	Mine La Motte, Mo	June 3, 1873	139, 489
Blast-furnace	W. Wright and G. Brown	Newcastle-upon-Tyne, England.	Aug. 12, 1856	15, 541
Blast-furnace, Air-heating pipe for	S. and J. Thomas	Catasauqua, Pa	Nov. 22, 1859	26, 212

Index of patents issued from the United States Patent Office from 1790 *to* 1873, *inclusive*—Continued.

Invention.	Inventor.	Residence.	Date.	No.
Blast-furnace alarm-apparatus	E. Davis	Millerton, N. Y	May 27, 1873	139, 300
Blast-furnace, Apparatus for charging	W. A. Miles	Salisbury, Conn	Nov. 12, 1872	132, 974
Blast-furnace, Apparatus for feeding	L. S. Goodrich	Waverly, Tenn	May 30, 1871	115, 305
Blast-furnace, Apparatus for feeding	L. S. Goodrich	Waverly, Tenn	July 23, 1872	129, 813
Blast-furnace, &c., Apparatus for heating air for	H. Davies	Portsmouth, Ohio	Oct. 22, 1861	33, 521
Blast-furnace, Apparatus for opening and closing the top of.	J. Thomas	Parryville, Pa	May 24, 1870	103, 391
Blast-furnace, Appendage to blast-pipe of	J. L. Agnew	Negaunee, Mich	July 27, 1869	93, 032
Blast furnace, Arrangement of pipes for hot	J. Young	Franklin Furnace, Ohio	Aug. 2, 1853	9, 908
Blast-furnace breast-plate	R. A. Fisher	San Francisco, Cal	July 22, 1873	141, 043
Blast-furnace charging-apparatus	J. H. Collins	Edgehill, Pa	Sept. 24, 1872	131, 667
Blast-furnace charging-apparatus	W. A. Miles	Salisbury, Conn	Aug. 20, 1872	130, 652
Blast-furnace cinder-plate	S. W. Harris	Albany, N. Y	Sept. 2, 1873	142, 463
Blast-furnace, Combination of a double traveling hearth with.	L. Sibert	Wookstock, Va	Nov. 20, 1849	6, 894
Blast-furnace, Construction and mode of working	H. Bessemer	London, England	July 25, 1871	117, 246
Blast-furnace, Construction of	S. Chubbuck and J. Briggs	Wareham, Mass	Jan. 9, 1841	1, 926
Blast-furnace, Fluxing	C. Shunk	Armstrong County, Pa	Feb. 12, 1856	14, 257
Blast-furnace for making iron	D. W. Hendrickson and J. P. McLean.	New York, N. Y	Sept. 3, 1867	68, 565
Blast-furnace for smelting iron, &c	S. W. Harris	Albany, N. Y	Sept. 2, 1873	142, 464
Blast-furnace for smelting metal	J. Barker	Baltimore, Md	Apr. 20, 1837	172
Blast-furnace hearth	G. Poe	Elk Ridge Landing, Md	June 23, 1838	804
Blast-furnace hopper	D. Bauman	Parryville, Pa	May 28, 1872	127, 214
Blast-furnace, Hot-air	R. Long	Chillicothe, Ohio	Dec. 10, 1867	72, 056
Blast furnace, Hot	A. Burtenshaw	Hope Furnace, Ohio	Mar. 29, 1870	101, 223
Blast-furnace, Method of heating	T. W. Bakewell	Cincinnati, Ohio	Nov. 6, 1855	13, 743
Blast-furnace, Oven or furnace for heating the blast of.	T. Whitwell	Stockton-on-Tees, England	July 9, 1867	66, 543
Blast-furnace, Oven or stove for heating the blast of.	J. M. Hartman	Philadelphia, Pa	July 26, 1870	105, 578
Blast-furnaces, Collecting and using the waste gases from.	J. E. A. B. De Langlade	Bordeaux, France	Mar. 19, 1872	124, 796
Blast-furnace to steam-boiler, Mode of applying the waste heat of.	M. Bell	Antis, Pa	June 10, 1840	1, 630
Blast-generator	C. C. Lloyd	Philadelphia, Pa	Apr. 17, 1849	6, 341
Blast-generator	W. H. Place	New York, N. Y	Feb. 4, 1862	34, 331
Blast-generator, Hydraulic	A. F. W. Partz	New York, N. Y	June 2, 1857	17, 447
Blast-heating furnace	H. Hamilton	Youngstown, Ohio	July 13, 1869	92, 526
Blast-heating furnace or oven	G. H. Lagorce	Newcastle, Pa	June 13, 1871	115, 870
Blast in furnace, Mode of heating	C. C. Alger	Stockbridge, Mass	June 30, 1838	819
Blast in furnace, Mode of heating air for hot	J. Jones	York, Pa	Dec. 10, 1838	1, 033
Blast, Method of generating air	I. J. Hendryx	New York, N. Y	Feb. 17, 1857	16, 645
Blast of furnace, Application of steam to	C. Foss	Madison, Ohio	Aug. 2, 1826	....
Blast oven for iron-furnace, Hot	S. and J. Thomas	Hokendauqua, Pa	June 1, 1869	90, 796
Blast oven for metallurgic furnace, Hot	T. Whitwell	Stockton-on-Tees, England	Aug. 27, 1872	130, 885
Blast oven, Hot	R. E. Brown	Saint Louis, Mo	Feb. 2, 1864	41, 421
Blast oven, Hot	D. and J. Campbell and S. Raymond.	Middletown, Pa	Sept. 14, 1869	94, 709
Blast oven, Hot	J. Fraggett	Youngstown, Ohio	May 31, 1870	103, 597
Blast oven, Hot	J. Player	Philadelphia, Pa	May 31, 1870	103, 651
Blast oven, Hot	J. Young	Portsmouth, Ohio	Sept. 3, 1872	131, 143
Blast oven or furnace, Hot	E. M. Furguson	Brady's Bend, Pa	Apr. 5, 1870	101, 449
Blast-oven pipe, Hot	J. K. McLanahan	Hollidaysburgh, Pa	July 23, 1872	129, 744
Blast-pipe for conveying heated air and gas to furnace.	R. Cook	Saratoga Springs, N. Y	Mar. 26, 1850	7, 211
Blast-producing chair, Rotary	L. R. Breisach	New York, N. Y	Feb. 16, 1858	19, 343
Blast-regulator	A. Kipp, jr	Sing Sing, N. Y	Aug. 16, 1864	43, 856
Blast, smelting, and cupola furnace	S. C. Salisbury	New York, N. Y	Mar. 23, 1869	88, 083
Blasts, &c., Machine for blowing	P. W. Mackenzie	Jersey City, N. J	Jan. 2, 1855	12, 165
Blastic compound	L. Buchholtz	Richmond, Va	Aug. 19, 1856	15, 551
Blasting by electrical currents	C. Stowell	Concord, Mass	Oct. 28, 1862	36, 824
Blasting by electricity	T. P. Shaffner	Louisville, Ky	Dec. 19, 1865	51, 671
Blasting by electricity	T. P. Shaffner	Louisville, Ky	Dec. 19, 1865	51, 674
Blasting-charge	G. Werlich	Watertown, Wis	Aug. 3, 1869	93, 379
Blasting-compound	M. Nowak	Williamsburgh, N. Y	May 24, 1864	42, 869
Blasting logs	M. Cooke	Sacramento City, Cal	May 3, 1870	102, 503
Blasting, Method of	N. Cadwallader	Birchville, Cal	May 29, 1866	55, 055
Blasting, Nitro-glycerine compound for	J. Horsley	Cheltenham, England	Dec. 28, 1869	98, 382
Blasting, Package of powder-charges for	H. M. Boies	Scranton, Pa	Nov. 11, 1873	144. 434
Blasting-plug	G. H. Felt	New York, N. Y	Feb. 27, 1866	52, 836
Blasting-plug	J. H. Holsey	Butler, Ga	Oct. 31, 1871	120, 438
Blasting-plug	J. H. Holsey	Butler, Ga	July 8, 1873	140 628
Blasting-plug	C. Monson	New Haven, Conn	Oct. 3, 1865	50, 263
Blasting-plug	D. Shoemaker	Kittaning Township, Pa	Sept. 16, 1873	142, 948
Blasting-powder	H. Biebuyck	Brussels, Belgium	Oct. 7, 1862	36, 599
Blasting-powder	W. and E. Fehleisen	Cilli City, Austria	Jan. 29, 1867	61, 659
Blasting-powder	W. H. Jackson	Salem, Mass	Mar. 23, 1869	88, 171
Blasting-powder	F. A. Jaeckel	Buckan-Magdeburg, Prussia.	May 24, 1864	42, 913
Blasting-powder	A. Murtineddu	Marseilles, France	May 12, 1857	17, 291
Blasting-powder	P. A. Oliver	New York, N. Y	July 14, 1868	80, 004
Blasting-powder	W. Silver, jr	Pittston. Pa	Nov. 22, 1853	10, 260
Blasting-powder	W. Silver, jr	Wapwallopen, Pa	July 1, 1856	15, 257
Blasting-powder	W. R Thomas and M. Emanuel, jr.	Catasauqua, Pa	Mar. 11, 1862	34, 654
Blasting-powder, Composition for	W. R. Thomas and M. Emanuel, jr.	Catasauqua, Pa	Apr. 9, 1861	32, 016
Blasting-powder, Composition for	W. R. Thomas and M. Emanuel, jr.	Catasauqua, Pa	Dec. 9, 1862	37, 117
Blasting-powder, Compound for	A. T. Rand	New York, N. Y	Oct. 29, 1867	70, 359
Blasting-powder, Machine for making	W. Silver	Bloomsburgh, Pa	Apr. 6, 1869	88, 674
Blasting-purposes, Tamping-device for	J. Shoemaker	Putneyville, Pa	Oct. 18, 1870	108, 357
Blasting rock	G. C. Bunsen	Belleville, Ill	May 30, 1865	47, 925
Blasting rock	M. Shaw	New York, N. Y	June 3, 183	....
Blasting rock, &c	C. Monson	New Haven, Conn	Apr. 1, 1851	8, 011
Blasting rock, Chemical composition for	C. Seidel	New York, N. Y	Feb. 12, 1867	61, 957
Blasting rock, Explosive composition for	E. Callow	London, England	Feb. 17, 1852	8, 734
Blasting rock, Implement for	C. F. Brown	Warren, R. I	July 11, 1854	11, 250

Index of patents issued from the United States Patent Office from 1790 *to* 1873, *inclusive*—Continued.

Invention.	Inventor.	Residence.	Date.	No.
Blasting rock, Loading the boring for	E. Gilbert	Rochester, N. Y	July 22, 1823	......
Blasting rock, Making cartridges for	J. B. Ives	Lebanon, Pa	July 17, 1826	......
Blasting rock, Method of	J. Brodie and S. H. Wheeler	San Francisco, Cal	Oct. 8, 1872	131, 995
Blasting rock, Method of	I. Whitcomb	Hingham, Mass	July 29, 1873	141, 248
Blasting rock, Mode of	J. Gilleland	Athens, Ga	Dec. 4, 1860	30, 809
Blasting rock, Plug for	J. D. Buckley and S. F. Mosher	Schagticoke, N. Y	May 17, 1859	24, 006
Blasting rock, timber, &c	J. Norton	Cork, Ireland	Aug. 8, 1854	11, 508
Blasting rock under water, Apparatus for	J. B. Eads	Saint Louis, Mo	Jan. 27, 1857	16, 473
Bl sting rock under water	B. Maillefert	New York, N. Y	Mar. 2, 1852	8, 776
Blasting-squib	S. H. Daddon	Saint Clair, Pa	Sept. 30, 1873	143, 396
Blasting, Tamping-plug for	E. Gomez	New York, N. Y	Apr. 26, 1870	102. 250
Blasting, &c., Treating explosive compounds to render them safe for.	E. A. L. Roberts	Titusville, Pa	May 20, 1873	139, 192
Bleached fabric, Process for dechlorinating	I. A. Roth	Philadelphia, Pa	Oct. 4, 1853	10, 095
Bleaching	J. B. Fuller	Norwich, Conn	June 25, 1867	66, 013
Bleaching	J. G. Kientzle	Montgomery County, Pa	Oct. 8, 1818	......
Bleaching and cleaning cloth	C. H. Farnham	Norwich, Conn	Aug. 15, 1835	......
Bleaching and cleaning vegetable fiber	E. T. Rice	New York, N. Y	Feb. 22, 1870	100, 071
Bleaching and cleansing textile fabric, Apparatus for.	S. Barlow	Middleton, England	May 27, 1862	35, 357
Bleaching and disinfecting	P. Marcelin	New York, N. Y	Apr. 9, 1872	125, 469
Bleaching and dyeing yarn, cloth, and other textile fabrics, Method of.	I. C. Colton and A. M. Hastings	Buffalo and Rochester, N.Y.	Mar. 5, 1867	62, 612
Bleaching and removing stains, Liquid for	M. E. Tompkins	Brooklyn, N. Y	Nov. 26, 1867	71, 426
Bleaching animal and vegetable fiber	C. M. E. du Motay and C. R. Maréchal.	Metz, France	Aug. 28, 1866	57, 649
Bleaching-apparatus	C. T. Appleton	Roxbury, Mass	Apr. 17, 1855	12, 709
Bleaching-apparatus	E. A. Combs	Monroe, Wis	July 14, 1868	79, 955
Bleaching-apparatus	J. Meyer	Bay Ridge, N. Y	May 17, 1864	42, 782
Bleaching-apparatus	M. Pierce	Norwich, Conn	Nov. 8, 1845	4, 257
Bleaching-apparatus	I. A. Roth and J. Lea	Philadelphia County, Pa	May 9, 1854	10, 894
Bleaching cane-juice, Apparatus for	W. A. Jordan	New Orleans, La	Apr. 2, 1867	63, 527
Bleaching cloth, yarn, &c., Process of	W. Luther	Niagara Falls, N. Y	Jan. 28, 1868	73, 733
Bleaching cotton and linen yarn, &c	J. Kendan	Leominster, Mass	Aug. 22, 1817	......
Bleaching cotton and woolen fabric	J. Jeanning	Plainville, Conn	Dec. 28, 1869	98, 387
Bleaching cotton goods by steam	A. Milne and D. S. Deane	Clinton, N. Y	July 14, 1813	......
Bleaching cotton, linen, &c	J. B. Green	Portsmouth, N. H	May 29, 1832	......
Bleaching dark soaps and "foots"	O. Loew	New York, N. Y	Mar. 29, 1870	101, 283
Bleaching fabrics, Apparatus for	L. Mendelson	New York, N. Y	Aug. 8, 1871	117, 796
Bleaching fatty substances	F. F. Mayer	New York, N. Y	Aug. 6, 1861	32, 005
Bleaching fibrous materials, Process of	H. M. Baker	Rochester, N. Y	Oct. 16, 1866	58, 935
Bleaching fibrous substances, Process for	J. Short	New York, N. Y	Jan. 23, 1866	52, 250
Bleaching ivory, bone, &c	D. K. Tuttle	New York, N. Y	Jan. 12, 1869	85, 875
Bleaching linen, cotton, &c., Apparatus for	L. W. Wright	United States	Mar. 3, 1838	622
Bleaching-liquors, Manufacture of	H. Deacon	Widnes, England	May 27, 1873	139, 239
Bleaching oil, paraffine, wax, &c., Apparatus for	C. Adams	Philadelphia, Pa	Feb. 13, 1866	52, 509
Bleaching palm-leaf, straw, &c	F. Perrin	Cambridge, Mass	July 31, 1866	56, 860
Bleaching paper-pulp and drying paper, Apparatus for.	L. Dodge	Waterford, N. Y	July 31, 1866	56, 732
Bleaching-powder, Manufacture of	H. Deacon	Widnes, England	Aug. 22, 1871	118, 210
Bleaching-powder, Manufacture of	T. Gray	Union Road, England	Mar. 19, 1867	63, 036
Bleaching-powder, sulphate, &c., Manufacture of	H. Deacon	Widnes, England	Dec. 5, 1871	121, 595
Bleaching-powder, Treating mixed gases containing chlorine for the production of.	H. Deacon	Widnes, England	Aug. 22, 1871	118, 212
Bleaching, Preparation of cloth and vegetable fiber for.	G. W. Billings	New York, N. Y	Mar. 14, 1865	46, 774
Bleaching-process	A. R. Arrott	Saint Helen's, England	Sept. 13, 1864	44, 250
Bleaching-process	J. A. Roth	Philadelphia, Pa	Nov. 18, 1856	16, 100
Bleaching-process	B. Schmidt	New York, N. Y	Jan. 12, 1869	85, 860
Bleaching rags, Apparatus for	H. Loring	Boston, Mass	June 5, 1855	13, 008
Bleaching resin	A. K. Lee	Galveston, Tex	May 27, 1873	139, 402
Bleaching-safe	C. W. Shiveley	Peru, Ind	Jan. 14, 1873	134, 941
Bleaching straw goods	A. M. Rosbugh	Panora, Iowa	May 10, 1870	102, 868
Bleaching thread, yarn, &c., Apparatus for	J. E. Clarner	Pawtucket, R. I	Nov. 15, 1870	109, 177
Bleaching with sulphur-fumes, Apparatus for	A. J. Crosby and O. W. Crow	Bluffton, Ind	Feb. 25, 1868	74, 898
Bleaching wool	J. Watteau	Antwerp, Belgium	Dec. 5, 1871	121, 564
Bleaching wool, yarn, &c	E. H. Haserick	Lake Village, N. H	July 23, 1872	129, 819
Blind, Apparatus for setting and copying music for the.	E. Marquis	Bloomington, Ind	Dec. 6, 1859	26, 361
Blind, Map-type for the	S. P. Ruggles	Boston, Mass	Oct. 22, 1872	132, 371
Blind, Musical notation for the	C. Mahony	New York, N. Y	Oct. 4, 1859	25, 657
Blind, Printing-apparatus for the	D. A. Johnston	Memphis, Tenn	Feb. 19, 1867	62, 206
Blind, Printing-instrument for the	E. A. Beach	Stratford, Conn	June 24, 1856	15, 164
Blind, Stitching-gage for the	W. H. Richardson	Fort Smith, Ark	July 8, 1873	140, 645
Blind, Type for the	S. P. Ruggles	Boston, Mass	Oct 22, 1872	132, 370
Blind, Type to print for the	H. Robyn	Saint Louis, Mo	Feb. 19, 1867	62, 156
Blind, Writing-apparatus for the	J. R. Cole	Paducah, Ky	Mar. 26, 1872	125, 024
Blind, Writing-apparatus for the	J. Synnott	San Francisco, Cal	Nov. 19, 1867	71, 084
Blind, Writing-apparatus for the	A. Von Briesen	New York, N. Y	Nov. 14, 1871	121, 026
Blind	W. E. Brock	New York, N. Y	Mar. 8, 1870	100, 593
Blind, Adjuster for window	A. A. Starr	New York, N. Y	June 26, 1855	13, 140
Blind and awning, Window and door	G. M. McMahan	Mount Sterling, Ky	Feb. 11, 1868	74, 399
Blind and curtain, Apparatus for raising and lowering window.	G. Allix	Saint Helier's, Island of Jersey.	Nov. 19, 1867	71, 114
Blind and curtain fixture, Window	J. H. Kinsman	United States Army	Apr. 5, 1864	42, 198
Blind and shutter fastener	S. Barker	New York, N. Y	May 11, 1852	8, 933
Blind and shutter fastener	W. Crighton	Fall River, Mass	May 31, 1864	42, 929
Blind and shutter fastener	C. H. Eddy	Auburn, N. Y	Mar. 19, 1867	62, 947
Blind and shutter fastener, Window	E. Jaquith	Brattleborough, Vt	Dec. 26, 1837	537
Blind and shutter fastener, Window	J. P. McKean	Washington, D. C	Apr. 24, 1841	2, 065
Blind and shutter fastening	W. B. Barnard	Waterbury, Conn	Oct. 7, 1862	36, 597
Blind and shutter fastening	F. Chase	South Sutton, N. H	Mar. 25, 1862	34, 739
Blind and shutter fastening	W. J. Decker	Nyack, N. Y	Mar. 9, 1869	87, 647
Blind and shutter fastening	J. Frick	Philadelphia, Pa	Jan. 5, 1864	41, 064
Blind and shutter fastening	D. E. True	Leominster, Mass	July 12, 1864	43, 533
Blind and shutter mover and fastener	W. Maguire	Cincinnati, Ohio	Jan. 1, 1850	6, 990
Blind and shutter opener and fastener	J. F. Lawrence and L. A. Farnsworth.	Claremont, N. H	Apr. 2, 1850	7, 246

Index of patents issued from the United States Patent Office from 1790 *to* 1873, *inclusive*—Continued.

Invention.	Inventor.	Residence.	Date.	No.
Blind and shutter operator	J. R. Creighton	Cincinnati, Ohio	Jan. 13, 1852	8, 647
Blind and shutter operator	R. V. Jones	Birmingham, Pa	Nov. 16, 1852	9, 405
Blind and shutter operator	N. W. Speers	Cincinnati, Ohio	Dec. 16, 1851	8, 597
Blind and shutter supporter	T. C. Ball	Springfield, Vt	May 20, 1862	35, 347
Blind and window fastening	G. L. Reynolds	Newburgh, N. Y	Mar. 20, 1866	53, 339
Blind-boring machine	J. H. Gibbs	Grand Rapids, Mich	Feb. 6, 1866	52, 409
Blind, Butt for	O. S. Garretson	Buffalo, N. Y	June 19, 1866	55, 643
Blind-butt, Self-locking	W. R. Goodrich	Utica, N. Y	Mar. 5, 1872	124, 263
Blind-catch	J. Currier	Portland, Me	Dec. 24, 1867	72, 609
Blind-catch	W. F. Seavey	Portland, Me	Apr. 13, 1869	88, 814
Blind, Construction of	G. W. Day	Haverhill, Mass	Oct. 15, 1872	132, 142
Blind, Device for opening and closing window	E. Bascom	New York, N. Y	Feb. 9, 1869	86, 630
Blind, Device for opening and closing window	S. W. Woodward	Buffalo, N. Y	Apr. 17, 1866	54, 077
Blind, door, &c., Window	L. Stevens and S. B. Ellithorp	Elmira, N. Y	June 26, 1855	13, 141
Blind, Enameled window	T. J. Olsaver and W. P. Elliott	Aurora, Ill	Dec. 5, 1865	51, 345
Blind-fastener	O. M. Andrews	Hecla Works, N. Y	Nov. 29, 1859	26, 319
Blind-fastener	F. and F. Babcock	Middletown, Conn	June 1, 1869	90, 625
Blind-fastener	F. and F. Babcock	Middletown, Conn	Sept. 13, 1870	107, 210
Blind-fastener	J. M. Barnaby	West Harwich, Mass	Feb. 9, 1869	86, 628
Blind-fastener	A. G. Batchelder	Lowell, Mass	June 27, 1871	116, 257
Blind-fastener	A. Bingham	Newtonville, Mass	June 25, 1867	65, 993
Blind-fastener	E. R. Chandler	Cambridgeport, Mass	Sept. 10, 1861	33, 236
Blind-fastener	H. M. Clark	Meriden, Conn	June 12, 1866	55, 577
Blind-fastener	W. H. Davis	Taunton, Mass	Oct. 4, 1859	25, 634
Blind-fastener	G. K. Dearborn	South Boston, Mass	Oct. 17, 1865	50, 532
Blind-fastener	G. K. Dearborn	Smithfield, R. I	Sept. 26, 1871	119, 266
Blind-fastener	J. Johnson and S. Ingersoll	Brooklyn, N. Y	Feb. 2, 1869	86, 552
Blind-fastener	W. S. Kirkham	Brantford, Conn	Nov. 6, 1860	38, 581
Blind-fastener	J. Murphy	Boston, Mass	Dec. 28, 1858	22, 446
Blind-fastener	W. Phelps, jr	Salem, Mass	Nov. 3, 1868	83, 789
Blind-fastener	D. B. Randall	Augusta, Me	Oct. 31, 1865	50, 775
Blind-fastener	G. T. Smith and W. E. Sparks	New Haven, Conn	Dec. 18, 1866	60, 584
Blind-fastener	S. F. Stanton	Manchester, N. H	Jan. 12, 1869	85, 867
Blind-fastener	C. E. Struck	New York, N. Y	May 31, 1870	103, 795
Blind-fastener	D. E. True	Lake Village, N. H	Dec. 4, 1855	13, 890
Blind-fastener	F. Veazie	Worcester, Mass	Dec. 1, 1868	84, 661
Blind fastener and operator	J. Aborn	Providence, R. I	May 22, 1866	54, 996
Blind fastener, Gravitation window	J. A. Carver	Taunton, Mass	Feb. 27, 1830	....
Blind fastener, Window	W. Avery	Salisbury, N. Y	Oct. 29, 1861	33, 563
Blind fastener, Window	J. R. Baker	Jersey City, N. J	July 23, 1867	67, 015
Blind fastener, Window	J. L Bassett	Bridgeport, Conn	Sept. 25, 1847	5, 305
Blind fastener, Window	C. P. Bell	Nashua, N. H	May 14, 1867	64, 740
Blind fastener, Window	J. M. Evarts	New Haven, Conn	Aug. 21, 1847	5, 252
Blind fastener, Window	S. Fansher	Southbury, Conn	Apr. 10, 1841	2, 042
Blind fastener, Window	A. T. Finch	Meriden, Conn	Oct. 16, 1860	30, 3[illegible]7
Blind fastener, Window	C. B. Francis	Newark, N. J	Jan. 1, 1867	60, 872
Blind fastener, Window	E. A. Holbrook, J. E. Dodge, and G. H. Marshall.	Watertown, N. Y	Nov. 19, 1867	71, 609
Blind fastener, Window	M. M. Isbel	New Haven, Conn	June 29, 1839	1, 210
Blind fastener, Window	D. B. and C. C. Mosher	Seneca Falls, N. Y	Apr. 14, 1868	76, 799
Blind fastener, Window	G. Welsh	Washington, D. C	Feb. 29, 1848	5, 463
Blind-fastening	E. B. Beecher, J. G. Davis, H. S. Frost, and A. G. Davis.	Westville and Watertown, Conn.	Apr. 16, 1867	63, 835
Blind-fastening	E. A. Chavantre	Newark, N. J	Nov. 19, 1867	94, 719
Blind-fastening	J. E. Cryer	Greenpoint, N. Y	Sept. 14, 1869	71, 135
Blind-fastening	S. Hall	New York, N. Y	May 23, 1865	47, 816
Blind-fastening	J. J. Henry	North White Creek, N. Y	Nov. 27, 1860	30, 730
Blind-fastening	W. C. Hicks	Boston, Mass	May 21, 1861	32, 365
Blind-fastening	W. Humiston	Meriden, Conn	Feb. 26, 1867	62, 337
Blind-fastening	S. W. Huntington	Augusta, Me	May 28, 1867	65, 086
Blind-fastening	A. S. Lesner	Milltown, Me	Aug. 1, 1871	117, 649
Blind-fastening	G. Lightfoot	Elgin, Ill	Nov. 12, 1867	70, 726
Blind-fastening	W. C. Marshall	New York, N. Y	Mar. 5, 1867	62, 656
Blind-fastening	B. Mayo	Chatham, Mass	Nov. 19, 1867	71, 031
Blind-fastening	C. A. Palmer	Newburgh, N. Y	Feb. 6, 1866	52, 442
Blind-fastening	R. Porter	Washington, D. C	Feb. 22, 1859	23, 043
Blind-fastening	S. V. Quimby and W. G. Marston	Boston, Mass., and West Fairlee, Vt.	Oct. 31, 1865	50, 733
Blind-fastening	H. Vansands	Middletown, Conn	May 5, 1857	17, 243
Blind-fastening, &c	W. Toepfer and H. Rugee	Milwaukee, Wis	Feb. 3, 1863	37, 597
Blind fastening, Window	I. Amos	Bel Air, Md	Nov. 28, 1871	121, 309
Blind fastening, Window	A. C. Arnold	Norwalk, Conn	Oct. 24, 1865	50, 546
Blind fastening, Window	J. Bacon	Bedford, Mass	Apr. 28, 1836	....
Blind fastening, Window	W. B. Barnard	Waterbury, Conn	Oct. 7, 1862	36, 596
Blind fastening, Window	S. Lichtenthaeler	Litiz, Pa	July 14, 1846	4, 633
Blind fastening, Window	J. Luther and A. Marsh	Worcester, Mass	Dec. 4, 1866	60, 210
Blind fastening, Window	J. Luther and A. Marsh	Worcester, Mass	Aug. 20, 1867	67, 891
Blind fastening, Window	F. A. Makepeace	Worcester, Mass	Oct. 2, 1866	58, 441
Blind fastening, Window	N. F. Mathewson	Barrington, R. I	July 9, 1867	66, 510
Blind fastening, Window	G. B. Melcher	Salem, Mass	Feb. 4, 1868	73, 987
Blind fastening, Window	D. Mills	Paterson, N. J	Jan. 6, 1863	37, 300
Blind fastening, Window	H. Munroe	Fall River, Mass	Oct. 2, 1866	58, 461
Blind fastening, Window	T. J. Sloan	New York, N. Y	Mar. 26, 1867	63, 317
Blind fastening, Window	J. Stewart	Utica, N. Y	Apr. 24, 1847	5, 078
Blind fastening, Window	T. Stover	Cambridgeport, Mass	July 23, 1867	67, 146
Blind fastening, Window	L. M. Townsley	Sedalia, Mo	Apr. 30, 1867	64, 264
Blind fastening, Window	A. Warner	Brooklyn, N. Y	Apr. 9, 1867	63, 768
Blind fastening, Window	A. J. Warner	Brooklyn, N. Y	Feb. 5, 1867	61, 779
Blind fastening, Window	J. W. Whittier	Cambridge, Mass	Nov. 26, 1861	33, 806
Blind fixture, Window	A. G. Batchelder	Lowell, Mass	Jan. 19, 1858	19, 170
Blind, Folding	M. Blake	Sutton, N. H	Mar. 14, 1854	10, 648
Blind, Folding window	S. S. Clark	Manchester, N. H	Apr. 7, 1857	16, 966
Blind, Folding window	F. R. Osgood	Grand Rapids, Mich	Nov. 1, 1870	108, 819
Blind for doors and windows, Metallic	W. E. Worthen	New York, N. Y	July 17, 1855	13, 279
Blind for houses, Inside	J. Wright and T. Thompson	Elizabeth, N. J	May 21, 1872	127, 006
Blind holder, Window	E. W. Bullard	Hardwick, Mass	Aug. 29, 1854	11, 597
Blind-hook blanks, Cutting and forming	G. Orr	Needham, Mass	May 12, 1868	77, 905
Blind, Inside	D. Y. Kilgore	Philadelphia, Pa	June 17, 1873	139, 899

Index of patents issued from the United States Patent Office from 1790 *to* 1873, *inclusive*—Continued.

Invention.	Inventor.	Residence.	Date.	No.
Blind, Inside	E. Metcalf	Rome, N. Y	July 29, 1873	141, 286
Blind, Inside	W. Potter, jr	Andover, N. Y	May 13, 1873	138, 761
Blind, Inside window	S. W. Shorey	Galesburgh, Ill	Dec. 17, 1867	72, 332
Blind, Iron window	H. Blakely	New York, N. Y	Jan. 23, 1855	12, 292
Blind machinery, Window	D. H. Thompson	Springfield, Mass	May 4, 1852	8, 931
Blind-making machine	J. Milne	Philadelphia, Pa	June 18, 1872	128, 057
Blind, Metal-clad wooden	W. E. Brock	New York, N. Y	June 11, 1872	127, 736
Blind, Metallic	J. M. Jomain	Paris, France	July 23, 1867	66, 967
Blind, Metallic window	G. A. Lathrop	East Saginaw, Mich	Dec. 13, 1859	26, 438
Blind, Metallic window	C. Neer	Troy, N. Y	June 15, 1858	20, 576
Blind, Metallic window, &c	C. Neer	Albany, N. Y	Jan. 14, 1862	34, 162
Blind-mortising machine	M. Buck	Lebanon, N. H	Nov. 23, 1869	97, 040
Blind-mortising machine	B. T. Norris	Lynn, Mass	Apr. 10, 1855	12, 691
Blind, Movable window	N. Poulson	Washington, D. C	July 21, 1868	80, 215
Blind opener and shutter, Curvilinear	R. B. Rollf	Cincinnati, Ohio	Apr. 17, 1849	6, 347
Blind, Opening and closing outside	J. E. Clokey	Washington, D. C	Mar. 30, 1858	19, 751
Blind, Opening and shutting window	E. Keith	Bridgewater, Mass	Jan. 25, 1833	....
Blind opener and slat regulator, Combined	A. Ball	Milford, Mass	Jan. 21, 1873	135, 063
Blind opener and slat regulator, Combined	L. Gathmann	Chicago, Ill	Feb. 27, 1872	124, 050
Blind, Operating and fastening window	W. Allen, jr	Meriden, Conn	Apr. 4, 1848	5, 497
Blind-operating device	O. Williams	Governeur, N. Y	Apr. 26, 1864	42, 523
Blind, Operating window	T. Christian	New York, N. Y	Mar. 2, 1858	19, 488
Blind, Operating window	T. Christian	New York, N. Y	Sept. 7, 1858	21, 408
Blind, Operating window	A. Ferber	Elizabeth City, N. J	July 27, 1858	20, 996
Blind-operator	J. A. Dorman and J. E. Stearns	Worcester, Mass	Sept. 28, 1858	21, 638
Blind-operator	L. N. Fay and W. Mason	West Warren, Mass	Nov. 30, 1858	22, 172
Blind-operator	L. N. Fay and W. Mason	Warren, Mass	Feb. 14, 1860	27, 119
Blind-operator	J. Johnson	Geneseo, N. Y	Feb. 28, 1860	27, 295
Blind operator and fastener	J. R. Creighton	Cincinnati, Ohio	Aug. 31, 1852	9, 231
Blind operator, Window	G. Jennison and M. F. Otis	Westborough, Mass	Apr. 4, 1871	113, 301
Blind or shutter fastener	W. Race	Seneca Falls, N. Y	Sept. 23, 1851	8, 387
Blind or shutter, Opening and closing iron	W. H. Brown	Worcester, Mass	Feb. 23, 1864	41, 676
Blind-rack	C. E. Smith	Goffstown, N. H	Nov. 16, 1869	96, 980
Blind, Railway-car	D. M. Hall	Bridgeport, Conn	Dec. 6, 1859	26, 351
Blind, "Retaining" window	N. Palmer	New York, N. Y	June 13, 1831	....
Blind rods and slats, Machine for wiring	G. Meyer	Cleveland, Ohio	Dec. 7, 1869	97, 541
Blind-rods, Hand-machine for wiring	B. C. Davis	Binghamton, N. Y	Oct. 11, 1859	25, 726
Blind-rods, Machine for trimming	B. C. Davis	Binghamton, N. Y	June 7, 1870	103, 848
Blind-rods, Machine for wiring	B. Boardman	Norwich, Conn	Sept. 1, 1857	18, 080
Blind-rods, Machine for wiring	T. R. Crosby	Newark, N. J	Dec. 13, 1859	26, 417
Blind-rods, Machine for wiring	F. H. Moore	Ithaca, N. Y	June 15, 1852	9, 032
Blind-rods, Machine for wiring	T. F. St. John	Le Roy, N. Y	Dec. 16, 1856	16, 252
Blind rods, Machine for wiring window	D. Kelly	Grand Rapids, Mich	Sept. 12, 1865	49, 888
Blind rods, Window	J. G. Baker	Washington, D. C	Oct. 18, 1859	25, 864
Blind-rods, Wiring	J. Coover	Chambersburgh, Pa	Aug. 7, 1860	29, 467
Blind-rod-wiring machine	J. Holzberger	Newark, N. J	Oct. 12, 1869	95, 800
Blind, Rolling window	S. W. Bidwell	Hartford, Conn	Oct. 5, 1858	21, 648
Blind-screen for shutter	B. J. Williams	Philadelphia, Pa	May 10, 1870	102, 900
Blind, Securing window	W. H. Mackrell	Bushwick, N. Y	June 13, 1831	....
Blind, Self-adjusting spring for window	H. Seely	Unadilla, N. Y	July 14, 1828	....
Blind, Sheet-iron	W. E. Ward	Port Chester, N. Y	Jan. 2, 1855	12, 145
Blind, Shifting	B. C. Howell, A. B. Thompson, and E. D. Snyder.	Ithaca and Owego, N. Y	Dec. 20, 1864	45, 499
Blind-shutter fastening	O. Paddock	Watertown, N. Y	Oct. 13, 1868	82, 980
Blind-slat adjuster	O. L. Houghton	Holden, Mo	Oct. 29, 1872	132, 581
Blind-slat adjuster	W. B. Sloan	Hamburgh, Iowa	May 28, 1872	127, 376
Blind-slat adjuster	W. B. Sloan	Hamburgh, Iowa	Jan. 21, 1873	135, 166
Blind-slat adjuster, Window	R. C. Beach and G. A. Berry	Tidioute, Pa	Dec. 16, 1873	145, 481
Blind-slat boring and mortising machine	T. Flesher	Dunkirk, N. Y	Nov. 21, 1871	121, 093
Blind-slat clasp and pivot, Metallic	G. R. Clarke	New York, N. Y	Mar. 5, 1867	62, 608
Blind-slat-crimping machine	M. Buck	Lebanon, N. H	Aug. 15, 1871	118, 102
Blind-slat-cutting machine	J. W. Helder	Shannon, Ill	Aug. 23, 1870	106, 694
Blind-slat fastener	G. W. Brooks	Bloomington, Ill	Sept. 9, 1873	142, 674
Blind-slat fastener	A. F. Champlin	Westerly, R. I	Oct. 28, 1873	144, 062
Blind-slat fastener	W. Sellers	New York, N. Y	Apr. 27, 1869	89, 439
Blind-slat fastener	T. G. Springer	New York, N. Y	Sept. 2, 1873	142, 526
Blind-slat fastening	J. D. Burdick	Ashway, R. I	Sept. 6, 1864	44, 072
Blind-slat fastening	J. M. Peirce	Mokena, Ill	May 19, 1868	78, 004
Blind-slat fastening	W. F. Redding	Utica, N. Y	June 16, 1863	38, 913
Blind-slat fastening	T. F. Rockwell	Yorkville, N. Y	Apr. 28, 1868	77, 218
Blind-slat fastening	T. R. Smith	Bennington, Vt	May 7, 1867	64, 588
Blind-slat holder, Venetian	J. Hampson	New Orleans, La	Aug. 21, 1841	2, 223
Blind-slat holder, Window	J. Boyd	Mamaroneck, N. Y	Dec. 1, 1868	84, 530
Blind-slat holder, Window	W. S. Mayo	New York, N. Y	Aug. 4, 1857	17, 937
Blind-slat machine	S. Jackson	Ossipee, N. H	June 19, 1860	28, 809
Blind-slat machine	W. F. Johnson and J. Doyle	Wetumpka, Ala	July 31, 1860	29, 380
Blind-slat machine	K. Kendal and L. H. Stark	Goffstown, N. H	Aug. 15, 1871	118, 023
Blind-slat machine	F. Leclere	Watertown, N. Y	Feb. 1, 1870	99, 327
Blind-slat machine	H. B. Smith	Lowell, Mass	Oct. 16, 1860	30, 429
Blind-slat machine	G. F. Woolston	Washington, D. C	Mar. 14, 1871	112, 666
Blind-slat machine, Window	E. R. Benson	Warsaw, N. Y	Oct. 4, 1853	10, 063
Blind-slat machine, Window	H. W. Farmer	Poultney, Vt	Apr. 24, 1860	27, 976
Blind-slat-mortising machine	C. W. Strout	Calais, Me	July 2, 1861	32, 727
Blind-slat operator	H. B. Lum	Sandusky, Ohio	Oct. 3, 1871	119, 625
Blind-slat operator	O. Paddock	Watertown, N. Y	Nov. 16, 1869	96, 950
Blind-slat operator	L. W. Swafford, E. Butler, and J. R. Hess.	Muscatine, Iowa	Oct. 27, 1868	83, 421
Blind-slat operator	J. B. Smith, S. A. Greely, and A. Campaigne.	Chicago, Ill	Sept. 20, 1870	107, 557
Blind-slat operator	D. R. Williams	Prospect, Conn	Mar. 19, 1850	7, 203
Blind-slat operater	Q. M. Youngs	Utica, N. Y	Aug. 29, 1871	118, 573
Blind-slat operator, Window	E. C. Byam	Fort Dodge, Iowa	Dec. 23, 1873	145, 842
Blind-slat-planing machine	C. Carlisle and L. Worcester	Woodstock, Vt	Mar. 16, 1858	19, 619
Blind-slat-planing machinery	G. Bonwill	Kent County, Del	Mar. 5, 1850	7, 133
Blind-slat-planing machinery	A. Woodworth, 3d	Worcester, Mass	Aug. 26, 1846	4, 721
Blind-slat regulator	J. T. O'Donohue and A. Leavitt.	New York, N. Y	Feb. 14, 1871	111, 770
Blind-slat staple	B. Boardman	Norwich, Conn	Mar. 30, 1858	19, 747

Index of patents issued from the United States Patent Office from 1790 *to* 1873, *inclusive.*—**Continued.**

Invention.	Inventor.	Residence.	Date.	No.
Blind-slat staple	C. H. Palmer	New York, N. Y	June 3, 1873	139, 516
Blind-slat stiles, Machine for mortising	D. T. Drake	Leominster, Mass	Apr. 28, 1857	17, 141
Blind-slat tenon	W. McFarland	New York, N. Y	Apr. 7, 1868	76, 491
Blind-slat-tenoning machine	H. Bickford	Cincinnati, Ohio	Apr. 27, 1869	89, 277
Blind-slat-tenoning machine	T. J. Bowdle, S. R. Lawder, and F. E. Johnston.	Piqua, Ohio	Oct. 6, 1868	82, 792
Blind-slat-tenoning machine	J. J. and T. Clark	Elgin, Ill	July 16, 1867	66, 678
Blind-slat-tenoning machine	J. J. Clark and T. Clark	Elgin, Ill	Oct. 6, 1868	82, 801
Blind-slat-tenoning machine	M. W. Clark	Worcester, Mass	July 28, 1868	80, 392
Blind-slat-tenoning machine	F. Douglas	Norwich, Conn	Sept. 27, 1870	107, 671
Blind-slat-tenoning machine	J. Kindleberger and W. A. Arnold.	San Francisco, Cal	Oct. 12, 1869	95, 810
Blind-slat-tenoning machine	J. M. Seymour	Newark, N. J	Apr. 24, 1866	54, 222
Blind-slat-tenoning machine	T. G. Stagg	Jersey City, N. J	Mar. 28, 1854	10, 710
Blind-slat-tenoning machine	L. F. Stevens	Elmira, N. Y	Aug. 9, 1859	25, 056
Blind slat, Window	A. Kohler	Williamsburgh, N. Y	May 7, 1872	126, 464
Blind-slat-wiring machine	P. Barry	Newark, N. J	May 21, 1867	64, 935
Blind-slat-wiring machine	B. C. Davis	Binghamton, N. Y	Oct. 15, 1872	132, 141
Blind-slat-wiring machine	E. F. Dunaway	Indianapolis, Ind	Jan. 1, 1867	60, 839
Blind-slat-wiring machine	G. V. Orton and W. H. Doane	Cincinnati, Ohio	Aug. 18, 1868	81, 103
Blind-slat-wiring machine	G. Pancake	Harrisburgh, Pa	May 12, 1868	77, 908
Blind-slats	M. W. Clark	Worcester, Mass	Feb. 22, 1870	100, 118
Blind-slats	D. Kelly	Muskegon, Mich	May 14, 1872	126, 713
Blind-slats and boring the stiles, Machine for tenoning.	H. Smith and T. J. Lumis	Norwich, Conn	Jan. 1, 1867	60, 800
Blind slats, Apparatus for operating window	S. Avery	Phœnix, N. Y	Apr. 15, 1851	8, 045
Blind-slats, Clamp for entering	J. Church	Saint Louis, Mo	July 25, 1871	117, 383
Blind-slats, Cutter-head for tenoning	M. W. Clark	Worcester, Mass	Mar. 31, 1868	76, 052
Blind-slats, Device for fastening	W. Palmer	New York, N. Y	Dec. 18, 1866	60, 545
Blind-slats, Device for forming round tenons on	T. C. Ball	Keene, N. H	Nov. 10, 1857	18, 569
Blind-slats, Device for holding	J. A. McCreary	Brooklyn, N. Y	Nov. 18, 1862	36, 960
Blind-slats, Device for moving	L. Gathmann	Chicago, Ill	Jan. 10, 1872	122, 595
Blind-slats, Machine for compressing the ends of	L. T. Smart	Manchester, N. H	May 19, 1857	17, 341
Blind-slats, Machine for cutting shoulder on	P. Clark	Nashville, Tenn	Feb. 13, 1872	123, 677
Blind-slats, Machine for cutting tenons on	S. C. Ellis	Albany, N. Y	Feb. 3, 1857	16, 534
Blind-slats, Machine for cutting tenons on	E. W. Roff	Newark, N. J	Nov. 7, 1854	11, 907
Blind-slats, Machine for entering	J. Church	Saint Louis, Mo	May 20, 1873	139, 043
Blind-slats, Machine for making	P. Schumacher	San Francisco, Cal	July 25, 1871	117, 471
Blind-slats, Machine for making	F. W. White	Norwich, Conn	Oct. 29, 1867	70, 382
Blind slats, Machine for making window	I. W. Gere	South Granby, N. Y	Nov. 30, 1858	22, 177
Blind-slats, Machine for setting staples in	J. Wyman	Schaghticoke, N. Y	Aug. 24, 1858	21, 292
Blind-slats, Machine for tenoning	L. Stevens	Elmira, N. Y	Apr. 28, 1857	17, 175
Blind-slats, &c., Machine for wiring	P. Barry	Newark, N. J	Nov. 2, 1869	96, 381
Blind-slats, Machine for wiring	E. F. Dunaway	Indianapolis, Ind	Jan. 30, 1866	52, 276
Blind-slats, Machine for wiring	J. M. Seymour and D. Whitlock	Newark, N. J	June 1, 1869	90, 789
Blind slats, Mode of adjusting window	B. C. English	Hartford, Conn	Sept. 23, 1856	15, 758
Blind-slats, Mold for making glass	E. B. Hungerford	Corning, N. Y	Mar. 18, 1873	136, 835
Blind slats, Operating window	J. D. Burdick	New Berne, N. C	Mar. 20, 1860	27, 520
Blind-slats, Spring for fastening	G. H. Dimond	Bridgeport, Conn	Sept. 17, 1867	68, 967
Blind-slats to receive the staples, Device for piercing.	J. Carpenter	Stonington, Conn	Aug. 25, 1857	18, 037
Blind slats, Window	F. Little	Saint Louis, Mo	Feb. 26, 1867	62, 348
Blind-splint machine	J. A. Welsh	Xenia, Ohio	Feb. 6, 1866	52, 473
Blind, Spring-holder for slat	W. L. Gallaudet	New York, N. Y	Nov. 11, 1856	16, 053
Blind-staple	J. B. Sargent	New Haven, Conn	Dec. 1, 1868	84, 585
Blind-staples, Machine for making and setting	J. Keith	Charlton, Mass	Mar. 14, 1871	112, 719
Blind-staples, Method of boxing	B. C. Davis	Binghamton, N. Y	May 23, 1871	115, 175
Blind-staples, Process of making	J. Keith	Providence, R. I	Aug. 27, 1872	130, 809
Blind-stile-boring machine	A. B. Carlin	Camden, N. J	Aug. 5, 1873	141, 545
Blind-stile-boring machine	S. E. Ellis	Jersey City, N. J	Oct. 29, 1872	132, 646
Blind-stile-boring machine	L. G. Kirkham	Derby, Conn	Apr. 15, 1873	137, 932
Blind-stile-boring machine	E. H. Smith	Whitestown, N. Y	Oct. 31, 1871	120, 465
Blind-stile-boring machine	J. F. Tudor	Camden, N. J	July 16, 1872	129, 439
Blind-stile-boring machine	A. M. Winn, J. Kindleberger, and W. A. Arnold.	San Francisco, Cal	Dec. 28, 1869	98, 455
Blind-stile boring and mortising machine	E. R. Longhead	Cincinnati, Ohio	May 19, 1863	38, 590
Blind-stile-mortising machine	L. M. Collins	Lebanon, N. H	May 16, 1871	114, 765
Blind-stiles, Machine for boring or mortising	L. Worcester	Lebanon, N. H	July 5, 1859	24, 688
Blind-stiles, Machine for mortising	M. W. Collins	Enfield, N. H	Aug. 6, 1872	130, 194
Blind-stiles, Machine for mortising	C. A. and E. P. Fenn and I. Cook	Saint Louis, Mo	Aug. 3, 1869	93, 291
Blind-stiles, Machine for spacing and boring	D. Dunham	Pawtucket, R. I	June 14, 1859	24, 427
Blind stop	P. A. Burgess	Butler, Mo	Dec. 26, 1871	122, 154
Blind-stop	W. A. Caswell	Providence, R. I	May 4, 1869	89, 557
Blind-stop	J. H. Cranston	Norwich, Conn	Oct. 28, 1873	144, 019
Blind-stop	J. Weber	New York, N. Y	Sept. 23, 1873	143, 207
Blind, Sun	J. Jeffreys	Upper Norwood, England	Apr. 19, 1864	42, 438
Blind, Suspending Venetian	J. Bohrer	Philadelphia, Pa	June 4, 1850	7, 408
Blind, Venetian	H. H. Andresen and H. Asbahr	Davenport, Iowa	May 19, 1868	78, 040
Blind, Venetian	G. F. Smith	Philadelphia, Pa	July 23, 1867	66, 995
Blind, Venetian	G. H. Woodworth	Brooklyn, N. Y	July 26, 1859	24, 900
Blind, Venetian window	C. D. Blinn	Port Huron, Mich	Nov. 28, 1865	51, 131
Blind, Venetian window	C. Rose	Allentown, Pa	Apr. 10, 1855	12, 695
Blind, Ventilating roller window	A. Cooper	Twickenham, England	May 2, 1871	114, 412
Blind, Ventilating window	H. Berdan and J. Bantly	Wayne, Mich	Nov. 30, 1869	97, 269
Blind, Window	W. Bellairs and H. Demott	Atkinson, Ill	May 26, 1868	78, 359
Blind, Window	W. H. Bixler	Easton, Pa	July 17, 1855	13, 251
Blind, Window	J. P. Boyd	La Porte, Ind	Oct. 25, 1870	108, 557
Blind, Window	W. E. Brock	New York, N. Y	July 4, 1871	116, 679
Blind, Window	E. Cate	Boston, Mass	May 1, 1847	5, 097
Blind, Window	E. Cate	Boston, Mass	May 8, 1847	5, 108
Blind, Window	F. Chase	South Sutton, N. H	July 24, 1855	13, 300
Blind, Window	L. W. Chase	Galion, Ohio	Mar. 27, 1866	53, 412
Blind, Window	A. M. Cochran	New York, N. Y	Feb. 3, 1857	16, 527
Blind, Window	T. Donato	New York, N. Y	June 7, 1870	103, 854
Blind, Window	D. D. Douglass	Springfield, Mass	June 11, 1861	32, 512
Blind, Window	S. Eich	East Toledo, Ohio	June 28, 1870	104, 717
Blind, Window	G. M. Fowler	Seymour, Conn	Apr. 23, 1872	125, 887
Blind, Window	A. S. Grenville and T. J. Lewis	Cambridge and Boston, Mass.	Aug. 9, 1839	1, 279

Index of patents issued from the United States Patent Office from 1790 *to* 1873, *inclusive*—Continued.

Invention.	Inventor.	Residence.	Date.	No.
Blind, Window	J. Harding	Galesburgh, Ill	Jan. 12, 1869	85, 818
Blind, Window	S. Hebron	Buffalo, N. Y	Sept. 7, 1869	94, 599
Blind, Window	A. Herder	New York, N. Y	Sept. 7, 1858	21, 417
Blind, Window	I. H. Hobbs	Philadelphia, Pa	Aug. 29, 1871	118, 454
Blind, Window, &c	H. Hoffman	New York, N. Y	Dec. 1, 1863	40, 795
Blind, Window	W. P. Hoffman	San Francisco, Cal	Aug. 27, 1867	68, 076
Blind, Window	E. B. Hungerford	Elmira, N. Y	Feb. 27, 1866	52, 854
Blind, Window	A. A. Jaqua	New York, N. Y	Dec. 7, 1869	97, 514
Blind, Window	W. Johnston	Cincinnati, Ohio	Oct. 20, 1868	83, 286
Blind, Window	D. Kelly and W. Livingston	Grand Rapids, Mich	Feb. 10, 1857	16, 632
Blind, Window	C. K. Marshall	Vicksburgh, Miss	Sept. 24, 1867	69, 231
Blind, Window	C. G. Matchett	Greenville, Ohio	Mar. 12, 1867	62, 765
Blind, Window	S. W. Merrill	Assabet, Mass	May 1, 1866	54, 385
Blind, Window	G. Munger	Cambridge, Md	Mar. 20, 1817	....
Blind, Window	L. Park	Brady, Pa	Dec. 28, 1869	98, 403
Blind, Window	L. Park	Brady, Pa	Mar. 18, 1873	136, 858
Blind, Window	J. Parkerson	Boston, Mass	Sept. 11, 1828	....
Blind, Window	A. and M. Pirz	New York, N. Y	Oct. 5, 1869	95, 511
Blind, Window	S. Pocock	Woodstock, (Ontario,) Canada.	Oct. 11, 1870	108, 182
Blind, Window	W. F. Redding	Saratoga Springs, N. Y	July 10, 1866	56, 267
Blind, Window	J. C. Reed	Stamford, Conn	Nov. 28, 1865	51, 268
Blind, Window	S. M. Sherman	Fort Dodge, Iowa	Oct. 12, 1869	95, 737
Blind, Window	S. M. Sherman	Fort Dodge, Iowa	July 11, 1871	117, 001
Blind, Window	A. P. Smith	Sterling, Ill	July 10, 1866	56, 282
Blind, Window	C. W. Smith	Evans, N. Y	June 26, 1860	28, 908
Blind, Window	H. Smith and T. J. Lumis	Norwich, Conn	July 30, 1867	67, 363
Blind, Window and door	S. Steere	Gloucester, R. I	Mar. 25, 1835	....
Blind, Window	T. J. Stratton	Cleveland, Ohio	July 16, 1872	129, 435
Blind, Window	A. Van Wagenen	Boston, Mass	June 6, 1865	48, 116
Blind, Window	C. H. Warner	Pittsfield, Mass	Apr. 14, 1868	76, 855
Blind, Window	M. C. Weld	Closter, N. J	Aug. 12, 1873	141, 841
Blind, Window	G. Wilkinson	White Creek, N. Y	Jan. 8, 1842	2, 414
Blind, Window	L. H. Wooden	Hampstead, Md	July 9, 1872	128, 937
Blind-wiring machine, Window	B. C. Davis	Binghamton, N. Y	Mar. 22, 1870	101, 107
Blind-wiring machine	W. F. Dodge	Newark, N. J	Mar. 23, 1869	88, 145
Blind-wiring machine	E. F. Dunaway	Cincinnati, Ohio	Mar. 21, 1871	112, 791
Blind-wiring machine	J. Kindleberger and W. A. Arnold.	San Francisco, Cal	Sept. 28, 1869	95, 355
Blind-wiring machine	C. McGill	New Haven, Conn	Oct. 7, 1873	143, 521
Blind-wiring machine	J. H. Nelson	Little Falls, N. Y	Oct. 31, 1871	120, 454
Blinds	R. Hutton	Williamsburgh, N. Y	Jan. 1, 1867	60, 736
Blinds and slats, Apparatus for operating window	J. Jones	Clyde, N. Y	Dec. 24, 1850	7, 856
Blinds, Apparatus for opening and closing	C. Reed and E. Howe, jr	Cambridgeport, Mass	Sept. 25, 1849	6, 748
Blinds, &c., Apparatus for painting window	S. T. Field	Worcester, Mass	Apr. 18, 1854	10, 789
Blinds, Cast and wrought metal	R. White	Washington, D. C	Jan. 20, 1852	8, 679
Blinds, Corner-support for	G. W. Day	Haverhill, Mass	June 24, 1873	140, 120
Blinds, Device for operating slats of window	L. N. Fay and W. Mason	Warren, Mass	Aug. 4, 1857	17, 923
Blinds, Device for operating window	J. McMackin	New York, N. Y	Aug. 4, 1857	17, 938
Blinds, Hanging Venetian	H. W. Safford	Philadelphia, Pa	June 30, 1863	39, 069
Blinds, Horizontal spring-catch for window	W. Phelps	Salem, Mass	July 7, 1830	....
Blinds, Locking and stopping window	J. Wetzel	Mott Haven, N. Y	May 30, 1865	48, 003
Blinds, Machine for boring window	L. S. Colburn	Oberlin, Ohio	Oct. 6, 1868	82, 803
Blinds, &c., Machine for cleaning	G. W. La Baw	Jersey City, N. J	Apr. 18, 1854	10, 792
Blinds, Machine for cutting rolls of window	L. H. Dwelley	Dorchester, Mass	Mar. 12, 1867	62, 829
Blinds, Machine for cutting slats for window	G. H. Denison	Suspension Bridge, N. Y	Jan. 26, 1864	41, 367
Blinds, Machine for making metallic slats for	J. S. Sanson and W. P. Farrand	Philadelphia, Pa	Dec. 16, 1856	16, 250
Blinds, Machine for making window	M. C. Stiles and T. S. Lewis	Hollis, Me	Jan. 24, 1854	10, 462
Blinds, Machine for mortising window	J. A. Peabody	Lowell, Mass	July 17, 1855	13, 271
Blinds, Machine for tenoning window	J. H. Palmer	Elmira, N. Y	Feb. 19, 1856	14, 289
Blinds, Machinery for boring window	J. Wiley	New Orleans, La	Dec. 25, 1849	6, 979
Blinds, Machinery for mortising frames of window	D. M. Cummings	North Enfield, N. H	Jan. 10, 1854	10, 403
Blinds, Means for holding window	H. A. Frost	Worcester, Mass	Jan. 23, 1855	12, 275
Blinds, Metallic frame for window	C. Neer	Albany, N. Y	May 17, 1859	24, 048
Blinds, Method of adjusting window	W. H. Babcock	Homer, N. Y	Oct. 12, 1858	21, 752
Blinds, Method of moving and fastening window	C. Reed	Cambridge, Mass	May 15, 1849	6, 455
Blinds, Method of opening, shutting, and fastening	W. Chase	Buffalo, N. Y	May 22, 1849	6, 466
Blinds, Mode of constructing slats for	W. E. Worthen and J. J. Althouse.	New York, N.Y	Jan. 11, 1859	22, 600
Blinds, Mode of manufacturing iron slats for window	W. E. Ward	Port Chester, N. Y	July 11, 1854	11, 311
Blinds, Mode of opening and closing window	G. Butterfield	Hopkinton, N. H	Sept. 4, 1840	1, 766
Blinds, Mode of opening and closing window	L. N. Fay and W. Mason	Warren, Mass	Aug. 4, 1857	17, 922
Blinds, Mode of operating window	J. Clark	Williamsburgh, N. Y	Jan. 18, 1859	22, 625
Blinds, Mode of raising and lowering Venetian	J. Weir	New York, N. Y	Oct. 8, 1840	1, 807
Blinds, Mortising and boring stiles for door and window.	D. Hathaway	Troy, N. Y	Feb. 5, 1833	....
Blinds, Opening and closing device for window	W. Wright	Cincinnati, Ohio	Apr. 10, 1847	5, 062
Blinds, Operating window	C. G. Bloomer	North Kingston, R. I	Oct. 18, 1859	25, 801
Blinds to windows, &c., Mode of adjusting	C. E. Parker and J. Sanger	Boston & Watertown, Mass	Aug. 7, 1855	13, 398
Blinds, Turn-buckle for window	J. L. Chapman	Philadelphia, Pa	Dec. 28, 1858	22, 470
Block:				
See Alphabet-block.				
Anvil-block.				
Asphalt-block.				
Basket-braiding block.				
Bolster block.				
Boot-crimping block.				
Brake-block.				
Bridge-block.				
Brush-block.				
Building-block.				
Butchers' block.				
Cam and spring block.				
Carriage-step block.				
Carriage top-prop block.				
Color-printing block.				
Composition-block.				
Concrete-block.				

Index of patents issued from the United States Patent Office from 1790 *to* 1873, *inclusive*—Continued.

Invention.	Inventor.	Residence.	Date.	No.
Block—Continued.				
See Conveyer-block.				
Cutting-block.				
Detaching-block.				
Flue-block.				
Gaff-block.				
Griping-block.				
Harness-pad block.				
Hat-block.				
Head-block.				
Hoisting-block.				
Horse-collar block.				
Keel-block.				
Last-block.				
Lead-block.				
Lettering-block.				
Meat-block.				
Muff-block.				
Muff-forming block.				
Music-note block.				
Oyster-block.				
Pavement-block.				
Paving-block.				
Pounding-block.				
Presser-block.				
Pulley-block.				
Pulling-block.				
Punch-block.				
Railway-brake block.				
Ribbon-block.				
Rubber-block.				
Sash-guide block.				
Saw-mill head-block.				
Shackle-block.				
Sheave-block.				
Ship's block.				
Snatch-block.				
Spelling-block.				
Stereotype-block.				
Stone-block.				
Strapping-block.				
Tackle-block.				
Toy-block.				
Toy building-block.				
T-rail block.				
Type-block.				
Waist-block.				
Wood-block.				
Block and tackle check	D. C. Guttridge and W. F. Rogers.	Canton, Ohio	Aug. 21, 1866	57, 316
Block-disengaging hook	S. Brown	San Francisco, Cal	Feb. 13, 1866	52, 636
Block-fitting machine	E. H. Woodsum	South Boston, Mass	Nov. 11, 1873	144, 427
Block press, Furnace	A. Hall	Perth Amboy, N. J	Mar. 18, 1873	136, 913
Block-sheave	C. Alger	Boston, Mass	July 1, 1836	....
Block-sheave	L. Aspinwall	Albany, N. Y	Apr. 22, 1835	....
Block-sheave	C. Curtis and T. C. Smith	Boston, Mass	Apr. 20, 1831	....
Block-sheave	N. H. Lewis	New York, N. Y	Aug. 26, 1816	....
Block-sheave, Bush for	J. Rudder	Norfolk, Va	Feb. 6, 1819	....
Blocks, hubs of wheels, &c., Bushing for	T. and D. Curtis	Washington, D. C	Jan. 23, 1826	....
Blocks, Manner of strapping	A. E. Wolf	Mystic River, Conn	Nov. 6, 1866	59, 496
Blocks, tackle, and pulleys of metal	J. Barron	Philadelphia, Pa	Oct. 20, 1834	....
Blood, Equalizing	S. R. Terrell	Burton, Miss	Feb. 5, 1836	....
Bloom-rolling machine	G. W. Billings	Chicago, Ill	May 27, 1873	139, 358
Blooming-furnace	B. B. Howell	Philadelphia, Pa	Nov. 6, 1828	....
Blotter	R. G. Allerton	New York, N. Y	June 9, 1857	17, 477
Blotter	S. H. Merrill	Washington, D. C	July 2, 1872	128, 556
Blotter	C. C. Moore	New York, N. Y	Dec. 5, 1865	51, 337
Blotter	C. C. Moore	New York, N. Y	Apr. 2, 1867	63, 417
Blotter	D. Walker	Newark, N. J	May 12, 1868	77, 783
Blotter, Automatic	J. E. Billings	Belmont, Mass	Oct. 8, 1867	69, 616
Blotter-holder	N. M. Shafer	New York, N. Y	Sept. 24, 1867	69, 131
Blotter, paper-weight, rule, cutter, and square, Combined.	A. H. Trego	Trenton, N. J	Oct. 10, 1865	50, 404
Blotter, Roll	P. B. Sheldon	Prattsburgh, N. Y	Feb. 19, 1861	31, 483
Blotter, Rotary	F. W. Frost and S. L. Hayward	Somerville, Mass	Apr. 12, 1870	101, 852
Blotting-pad	R. Boeklen	Brooklyn, N. Y	May 14, 1867	64, 623
Blotting-pad	A. Q. Collins	Cambridge, Mass	Mar. 21, 1871	112, 787
Blotting-pad	J. F. French	Boston, Mass	July 19, 1870	105, 563
Blotting-pad	C. A. Gale	Demopolis, Ala	Dec. 14, 1869	97, 900
Blotting-pad	P. Gorsline	Elizabeth, N. J	Apr. 20, 1869	89, 142
Blotting-pad	G. C. Hinman	Boston, Mass	Aug. 23, 1870	106, 696
Blotting-pad	J. M. Keep	New York, N. Y	Apr. 5, 1870	101, 627
Blotting-pad	A. B. Kellogg	Buffalo, N. Y	Dec. 20, 1870	110, 245
Blotting-pad	D. Walker	Newark, N. J	Nov. 20, 1866	59, 933
Blower	W. H. Bailey	Mahanoy City, Pa	Nov. 14, 1871	120, 929
Blower	G. W. Bigelow	New Haven, Conn	Feb. 12, 1867	61, 919
Blower	H. B. Bigelow and G. Murray	New Haven, Conn., and Cambridgeport, Mass.	July 10, 1866	56, 165
Blower	E. Carleton	Cape Elizabeth, Me	June 27, 1871	116, 265
Blower	W. S. Colwell	Pittsburgh, Pa	Sept. 27, 1870	107, 665
Blower	F. Engel	Camden, N. J	Mar. 5, 1867	62, 620
Blower	J. A. Evarts	Meriden, Conn	Dec. 11, 1866	60, 350
Blower	J. N. Gilchrist	Connersville, Ind	May 31, 1870	103, 733
Blower	D. D. Hardy and E. E. Wood	Cincinnati, Ohio	June 21, 1870	104, 585
Blower	S. H. Hartman	Allegheny, Pa	Aug. 8, 1871	117, 776
Blower	B. Hotchkiss	New Haven, Conn	Nov. 3, 1863	40, 482
Blower	R. Lapham	New York, N. Y	June 19, 1860	28, 758
Blower	W. A. Parmele	New Haven, Conn	July 30, 1867	67, 345

Index of patents issued from the United States Patent Office from 1790 *to* 1873, *inclusive*—Continued.

Invention.	Inventor.	Residence.	Date.	No.
Blower	T. C. Richards	Milwaukee, Wis	Jan. 17, 1860	26, 863
Blower	T. Rogers	Fredericktown, Ohio	Oct. 24, 1871	120, 328
Blower	P. H. Roots	Connersville, Ind	Sept. 25, 1860	30, 157
Blower	B. F. Sturtevant	Boston, Mass	Feb. 2, 1869	86, 469
Blower	B. F. Sturtevant	West Roxbury, Mass	Apr. 19, 1870	102, 062
Blower	E. Suckon and E. Habel	Oldham, England	Jan. 6, 1863	37, 325
Blower	J. A. Svedberg	Washington, D. C	Jan. 30, 1872	123, 304
Blower	H. Thirion	Miricourt, France	Oct. 29, 1861	33, 613
Blower	W. W. Webb	Indianapolis, Ind	Mar. 28, 1871	113, 229
Blower	J. M. Williams	Connersville, Ind	Jan. 29, 1867	61, 592
Blower	T. H. Willson	Harrisburgh, Pa	Mar. 26, 1861	31, 844
Blower, Air	J. W. Newcomb	New York, N. Y	Jan. 23, 1872	122, 959
Blower and exhauster, Steam	J. Howarth	Salem, Mass	Mar. 1, 1870	100, 294
Blower and rotary-engine combined	H. P. Tenant	East Germantown, Ind	May 2, 1871	114, 489
Blower, Blast	J. Braugh	Aurora, Ill	June 30, 1857	17, 664
Blower-case	P. H. Roots	Connersville, Ind	Nov. 1, 1864	44, 892
Blower-case	B. F. Sturtevant	Boston, Mass	Mar. 2, 1869	87, 523
Blower, Coal-grate	R. Fuller and T. Thomas	New York, N. Y	May 22, 1827	....
Blower, Compound	A. K. Rider	New York, N. Y	Apr. 5, 1870	101, 510
Blower, Compound	T. Shaw	Philadelphia, Pa	July 8, 1873	140, 597
Blower cross-head	P. H. and F. M. Roots	Connersville, Ind	July 24, 1866	56, 614
Blower, Fan	H. B. Adams	New York, N. Y	Jan. 31, 1860	26, 962
Blower, Fan	S. Barnhart	Chillicothe, Ohio	July 31, 1855	13, 346
Blower, Fan	J. B. Charles	Ashland, Ohio,	Feb. 14, 1860	27, 108
Blower, Fan	P. Clark	Rahway, N. J	Oct. 23, 1866	58, 985
Blower, Fan	P. Clark	Rahway, N. J	Dec. 22, 1868	85, 213
Blower, Fan	R. Cook	Saratoga Springs, N. Y	Nov. 12, 1861	33, 694
Blower, Fan	F. P. Dimpfel	New York, N. Y	Dec. 28, 1839	1, 448
Blower, Fan	J. Ericsson	New York, N. Y	Aug. 16, 1870	106, 348
Blower, Fan	W. Farmer	New York, N. Y	Oct. 18, 1864	44, 715
Blower, Fan	J. C. Gartley and J. Fox	Philadelphia, Pa	July 17, 1855	13, 260
Blower, Fan	W. Kendrick	New York, N. Y	Dec. 5, 1865	51, 321
Blower, Fan	J. F. Lietch	Greene, N. Y	Dec. 24, 1861	34, 028
Blower, Fan	C. P. Marshall	Worcester County, Mass	Feb. 3, 1857	16, 547
Blower, Fan	W. P. Miller	San Francisco, Cal	Nov. 12, 1867	70, 736
Blower, Fan	M. V. Nobles	Saint Anthony, Minn	Dec. 15, 1863	40, 982
Blower, Fan	F. Ortlieb	Williamsburgh, N. Y	Mar. 13, 1866	53, 174
Blower, Fan	N. Parrish	Kalamazoo, Mich	Dec. 17, 1867	72, 416
Blower, Fan	D. and J. R. Pollock	Lancaster, Pa	June 19, 1855	13, 100
Blower, Fan	R. Porter	Malden, Mass	Jan. 3, 1865	45, 749
Blower, Fan	T. T. Prosser	Chicago, Ill	Dec. 23, 1873	145, 753
Blower, Fan	S. W. Ruggles	Fitchburgh, Mass	Aug. 8, 1854	11, 511
Blower, Fan	C. G. Sargent	Graniteville, Mass	July 4, 1865	48, 593
Blower, Fan	I. P. Smith	Orangetown, N. Y	Mar. 14, 1846	4, 424
Blower, Fan	B. F. Sturtevant	West Roxbury, Mass	Apr. 19, 1870	102, 063
Blower, Fan	H. Sweetapple	Napa, Cal	July 31, 1860	29, 417
Blower, Fan	L. D. Wheeler	Fitchburgh, Mass	Nov. 1, 1870	108, 949
Blower, Fan	W. Winter	Plainfield, N. J	Dec. 15, 1863	40, 974
Blower, Fan-wheel	A. Westcott	Syracuse, N. Y	Mar. 26, 1867	63, 343
Blower, Fire	A. J. Redway	Cincinnati, Ohio	Jan. 19, 1869	86, 635
Blower for blacksmith's forge	J. G. Tscheulin	Baltimore, Md	May 3, 1870	102, 626
Blower for furnace, Hydraulic	R. Cook	Saratoga Springs, N. Y	Apr. 16, 1850	7, 283
Blower for furnace of steam-boiler, &c., Hydro-carbon.	H. Gerner	New York, N. Y	July 18, 1865	48, 806
Blower for furnace, Steam	J. A. Bassett	Salem, Mass	Sept. 5, 1865	49, 823
Blower for smith's forge, Fan	L. Duskin and B. Sledge	Thomasville, N. C	Sept. 6, 1870	107, 167
Blower for steam-generator	D. M. Nichols	New York, N. Y	Oct. 3, 1865	50, 265
Blower, Forge	C. West and B. K. Price	Pittsburgh, Pa	Oct. 22, 1867	70, 140
Blower, Franklin-stove	D. Stuart	Philadelphia, Pa	Aug. 27, 1850	7, 608
Blower, Furnace	M. Alden	Ralston, Pa	Apr. 18, 1848	5, 513
Blower, Furnace	B. L. Johnson	Cassawago, Pa	Apr. 4, 1848	5, 498
Blower, Gas-engine	J. H. Bean	Cincinnati, Ohio	July 1, 1873	140, 397
Blower-holder	J. B. F. Davidge	New York, N. Y	Feb. 18, 1868	74, 514
Blower, Hopper	T. S. Barnum	Sharon, Conn	Jan. 18, 1810	....
Blower, Hot-air	B. F. Sturtevant	Jamaica Plain, Mass	Feb. 22, 1870	100, 238
Blower, Hydraulic	J. Darling	Cincinnati, Ohio	Dec. 24, 1850	7, 853
Blower, Locomotive steam-engine	R. Winans	Baltimore, Md	July 29, 1837	307
Blower, Pressure	W. C. Grimes	Philadelphia, Pa	Nov. 11, 1868	83, 846
Blower, Pressure	J. Perrault	Troy, N. Y	May 10, 1870	102, 964
Blower, Pressure	B. F. Sturtevant	Boston, Mass	July 13, 1869	92, 489
Blower, Pressure	B. F. Sturtevant	Jamaica Plain, Mass	Feb. 22, 1870	100, 239
Blower, Rotary	W. B. Barnard and E. Jordan	Waterbury, Conn	Sept. 27, 1859	25, 554
Blower, Rotary	F. Hainsworth	Chicago, Ill	Apr. 3, 1866	53, 609
Blower, Rotary	W. G. Hyndman	Cincinnati, Ohio	Aug. 9, 1870	106, 165
Blower, Rotary	W. G. Hyndman	Cincinnati, Ohio	Dec. 26, 1871	122, 118
Blower, Rotary	C. W. Isbell	New York, N. Y	Mar. 17, 1868	75, 635
Blower, Rotary	R. F. Knox	San Francisco, Cal	July 1, 1873	140, 509
Blower, Rotary	T. Leffel	Springfield, Ohio	May 14, 1872	126, 639
Blower, Rotary	H. C. McIlwain and A. Brumfiel	Connersville, Ind	May 24, 1870	103, 482
Blower, Rotary	H. C. McIlwain and A. Brumfiel	Connersville, Ind	Jan. 10, 1871	110, 929
Blower, Rotary	J. Mitchell	Philadelphia, Pa	Oct. 6, 1868	82, 736
Blower, Rotary	P. H. Roots	Connersville, Ind	Oct. 9, 1866	58, 745
Blower, Rotary	P. H. and F. M. Roots	Connersville, Ind	Jan. 21, 1868	73, 654
Blower, Rotary	P. H. and F. M. Roots	Connersville, Ind	Aug. 11, 1868	81, 009
Blower, Rotary	A. Smith	New York, N. Y	Feb. 4, 1862	34, 318
Blower, Rotary	B. F. Sturtevant	Boston, Mass	Feb. 2, 1869	86, 470
Blower, Spiral cone	J. A. Svedberg	Washington, D. C	Apr. 29, 1873	138, 448
Blower-stand	B. Brundred	Oldham, N. J	Sept. 26, 1835	....
Blower, Steam	P. Bradford	New Haven, Conn	Oct. 3, 1871	119, 560
Blower, Steam	J. A. Bassett and O. C. Smith	Salem, Mass	Dec. 3, 1867	71, 569
Blower, Steam	G. W. Bright	Philadelphia, Pa	Jan. 22, 1867	61, 391
Blower, Steam	M. Foreman and J. R. Mathewson.	Philadelphia, Pa	May 7, 1867	64, 414
Blower, Steam	J. Hainsworth	Chicago, Ill	Mar. 3, 1868	75, 011
Blower, Steam	J. T. Hancock	Jamaica Plain, Mass	Aug. 3, 1869	93, 197
Blower, Steam	J. H. Johnson	Paducah, Ky	Sept. 25, 1866	58, 264
Blower, Steam	J. Simmonds	Brooklyn, N. Y	Mar. 6, 1866	53, 087
Blower, Steam	L. F. Smith	Philadelphia, Pa	Dec. 17, 1867	72, 426

Index of patents issued from the United States Patent Office from 1790 *to* 1873, *inclusive*—Continued.

Invention.	Inventor.	Residence.	Date.	No.
Blower, Steam	S. R. Wilmot	Bridgeport, Conn	Apr. 2, 1867	63,595
Blower, Steam-generator	W. H. Squires	New York, N. Y	Sept. 24, 1867	69,268
Blowers to furnaces of locomotives, Application of.	F. B. Blanchard	Brooklyn, N. Y	June 25, 1861	32,614
Blowers to furnaces of locomotives, Application of.	F. B. Blanchard	New York, N. Y	Sept. 22, 1863	40,014
Blower, Turbine fan	M. Smith	New Haven, Conn	July 25, 1865	48,986
Blower-wheel	B. F. Sturtevant	West Roxbury, Mass	Oct. 29, 1867	70,286
Blower-wheel, Pressure	B. F. Sturtevant	Jamaica Plain, Mass	Feb. 22, 1870	100,236
Blower-wheel, Pressure	B. F. Sturtevant	Jamaica Plain, Mass	Feb. 22, 1870	100,237
Blowing-apparatus	J. M. Bailey	Pittsburgh, Pa	Nov. 12, 1872	132,891
Blowing-apparatus	D. Cumming	Sorrel Horse, Pa	Apr. 27, 1858	20,045
Blowing-apparatus	R. H. Gilbert	Washington, D. C	Feb. 15, 1870	99,876
Blowing-apparatus	P. W. McKenzie	Jersey City, N. J	Oct. 25, 1864	44,809
Blowing-engine	J. Fritz and J. Moore	Bethlehem and Philadelphia, Pa.	Aug. 2, 1864	43,678
Blowing-engine	A. C. Kirk	Glasgow, Great Britain	Nov. 24, 1868	84,428
Blowing-engine	L. J. Knowles	Worcester, Mass	Sept. 12, 1871	118,949
Blowing-engine, Means for repairing	T. Critchlow	Swatara Township, Pa	Jan. 5, 1869	85,510
Blowing-machine	J. G. Baker	Philadelphia, Pa	Dec. 9, 1873	145,382
Blowing-machine	J. Dougherty	Philadelphia, Pa	July 12, 1870	105,318
Blowing-machine	M. Reichenbach and S. Golay	Paris, France	Jan. 26, 1869	86,320
Blowing-machine	J. M. Williams	Connersville, Ind	May 23, 1871	115,259
Blow-off for steam-boiler, Surface	J. S. Griffith	Saint Louis, Mo	May 9, 1871	114,551
Blow-off pipe for boiler, Surface	J. Perkins	Baltimore, Md	Apr. 4, 1871	113,716
Blow-off pipe for steam-boiler	J. S. Griffith	Saint Louis, Mo	June 20, 1871	116,179
Blow-pipe	J. Cook	New York, N. Y	May 28, 1872	127,151
Blow-pipe	W. T. Gillinder	Philadelphia, Pa	Dec. 5, 1865	51,386
Blow-pipe	J. Hollely	New York, N. Y	Mar. 16, 1858	19,636
Blow-pipe	J. S. Hull	Cincinnati, Ohio	Aug. 4, 1863	39,398
Blow-pipe	J. B. Hyde	New York, N. Y	Sept. 13, 1870	107,263
Blow-pipe	J. Kendy	San Francisco, Cal	May 16, 1865	47,721
Blow-pipe	N. W. Kingsley	New York, N. Y	Apr. 3, 1866	53,629
Blow-pipe	M. H. Knapp	Fulton, N. Y	July 16, 1872	129,569
Blow-pipe	O. L. Lawson	Crestline, Ohio	May 13, 1856	14,875
Blow-pipe	J. E. McClure and D. H. Ainsworth.	San Francisco and Salinas, Cal.	Nov. 4, 1873	144,345
Blow-pipe	J. McFarland	Clinton, Ill	Sept. 18, 1866	58,118
Blow-pipe	O. A. Moses	Charleston, S C	Aug. 21, 1866	57,362
Blow-pipe	S. B. Palmer	Tully, N. Y	Sept. 23, 1856	15,774
Blow-pipe	F. Shaller	Hudson, N. Y	Nov. 27, 1866	60,070
Blow-pipe	J. R. Street	Washington, D. C	June 7, 1870	104,078
Blow-pipe	L. B. Wilson	Cambridge, Ohio	Apr. 5, 1870	101,692
Blow-pipe, Alcohol	E. Conway	Dayton, Ohio	July 7, 1857	17,727
Blow-pipe, Construction of compound	J. B. Hyde	Newark, N. J	Oct. 18, 1859	25,832
Blow-pipe for blast-furnace	J. Barker	Baltimore, Md	Mar. 3, 1837	134
Blow-pipe for enlarging blasting cavities	A. Stickney	Norwich, Vt	Sept. 20, 1853	10,039
Blow-pipe for enlarging blasting cavities, Compound.	A. Stickney	Norwich, Vt	Sept. 20, 1853	10,040
Blow-pipe, Furnace	J. Barker	Baltimore, Md	Feb. 12, 1836	......
Blow-pipe, Hydro-oxygen	R. Hare	Philadelphia, Pa	July 25, 1845	4,104
Blow-pipe mechanisms	W. S. Burgess	Norristown, Pa	July 2, 1872	128,587
Blow-pipe operations, Composition for supports in	J. B. Hyde	Newark, N. J	Nov. 20, 1860	30,680
Blow-pipe, Self-acting	M. Pinner	New York, N. Y	Mar. 13, 1866	53,181
Blow-pipe, Spirit	H. N. Macomber	Lynn, Mass	Mar. 11, 1856	14,403
Blow-pipe, Steam	J. S. Anderson	Flintville, Wis	Jan. 21, 1873	135,058
Blow-pipe, Table	R. Somerby	Louisville, Ky	Nov. 15, 1843	3,336
Blubber, Knife for cutting up	W. Ball	New York, N. Y	Apr. 25, 1822	......
Blubber-mincing machine	G. and J. J. Kilburn	Fall River, Mass	Nov. 16, 1841	2,368
Blubber mincing-machine, Whale	B. Taber	Fairhaven, Mass	Aug. 11, 1821	......
Blue for use in laundry and bleaching, Lump	J. H. Dilks	New York, N. Y	Dec. 31, 1867	72,818
Blue, Manufacture of Prussian	J. M. Merrymon	Indianapolis, Ind	Jan. 10, 1865	45,846
Bluing, Article of	E. L. Molineux	New York, N. Y	May 26, 1868	78,313
Bluing-box	G. A. Moss	New York, N. Y	May 21, 1867	64,897
Bluing-box	A. F. Pickens	New York, N. Y	Aug. 16, 1870	106,402
Bluing cotton fiber	F. Wilkinson	Manchester, England	June 25, 1872	128,267
Bluing for use in laundry and bleaching, Process of making soluble.	J. H. Dilks	New York, N. Y	Dec. 31, 1867	72,817
Bluing, Method of preparing laundry	E. L. Molineux	New York, N. Y	Apr. 21, 1868	76,935
Bluing-paper for laundry purposes	T. Dreidel	Cincinnati, Ohio	Nov. 12, 1867	70,703
Bluing-paste	R. G. Vassar	Poughkeepsie, N. Y	Jan. 31, 1865	46,160

Board:

See Alley-board.
Arm or grain board.
Backgammon-board.
Billiard-board.
Biscuit-board.
Black-board.
Bread-board.
Bulletin-board.
Card and cribbage board.
Carriage dash-board.
Center-board.
Cheese-board.
Chess-board.
Chess and checker board.
Clapboard.
Comber-board.
Croquet-board.
Cutting-board.
Cutting or lap board.
Dash-board.
Dough-board.
Drawing-board.
End-board.
Game-board.
Guide-board.
Ironing-board.
Key-board.
Kneading-board.

Index of patents issued from the United States Patent Office from 1790 *to* 1873, *inclusive*—Continued.

Invention.	Inventor.	Residence.	Date.	No.
Board—Continued:				
See Knife-cleaning board.				
Knife-scouring board.				
Lap-board.				
Lea-board.				
Leather-board.				
Letter-board.				
Mold-board.				
Molders' match-board.				
Molding-board.				
Music black-board.				
Musical demonstrating-board.				
Organ-reed board.				
Paper board.				
Pasteboard.				
Pitch-board.				
Plow mold-board.				
Portable billiard-board.				
Press-board.				
Shearing-board.				
Shooting-board.				
Sign-board.				
Skirt-board.				
Sounding-board.				
Stove-board.				
Straw-board.				
Tally-board.				
Telegraph-switch board.				
Transposing-board.				
Tray or show board.				
Wash-board.				
Weather-board.				
Board edging and slitting machine	C. I. Hays and M. Newman	Unadilla, N. Y	Aug. 12, 1862	36, 150
Board-edging machine	J. K. Sanborn	Sandy Hill, N. Y	Oct. 24, 1871	120, 170
Board, &c., for use, Seasoning, drying, and fitting	A. Plumb	Buffalo, N. Y	Apr. 6, 1832	....
Board jointing and matching machine	D. Gleason and H. Frisbee	Bethany, N. Y	Apr. 8, 1826	....
Board-jointing apparatus	D. Foster	Whitestown, N. Y	May 28, 1850	7, 395
Board-jointing machine	E. B. Clark	Damascus, Pa	Jan. 31, 1827	....
Board raising and lowering mechanism	C. and A. Kilburn	Philadelphia, Pa	Dec. 21, 1869	98, 167
Board-slitting machine	A. T. Nichols	Norfolk, Va	Nov. 5, 1872	132, 854
Board-slitting machinery	J. R. Remington and R. Beal	Lowndes County, Ala., and Washington, D. C.	Jan. 15, 1847	4, 929
Boards and planks for making boxes, &c., Machine for mitering and dovetailing.	R. Urann	Boston, Mass	Oct. 26, 1839	1, 380
Boards, Device for tonguing and grooving tapering	J. Absterdam and W. B. Merrill.	Boston, Mass	Mar. 28, 1854	10, 703
Boards, &c., Machine for mitering and dovetailing the ends of.	A. Davis	Princeton, Mass	Aug. 21, 1839	1, 297
Boards, Machine for planing and reducing	J. Cumberland	Mobile, Ala	Mar. 6, 1847	4, 997
Boards, Machine for planing, matching, and grooving.	C. Taylor	Angola, N. Y	Feb. 7, 1829	....
Boards, Machine for squaring ends of	E. H. Rollins	Bangor, Me	June 18, 1872	127, 990
Boards, Machine for thinning	S. Case	Lumpkin, Ga	Sept. 4, 1860	29, 861
Boards, Machine for tonguing, grooving, and beading.	F. Fredly	Sugar Valley, Pa	Sept. 15, 1838	921
Boards, planks, &c., Machine for slitting	E. Olds	Brookfield, Mass	June 29, 1808	....
Boards, Shearing	S. Willard	Cincinnati, Ohio	May 9, 1832	....
Boards, Side-cutter head for jointing, tonguing, and grooving.	W. M. Hutton	Troy, N. Y	Oct. 26, 1838	988
Boards to uniform thickness, Machine for reducing and smoothing.	T. D. Knight	Charleston, Tenn	Mar. 3, 1857	16, 777
Boarding-machine	J. E. Wiggin	Stoneham, Mass	Feb. 26, 1867	62, 514
Boat:				
See Army-boat.				
Bed-boat.				
Bilge-boat.				
Buggy-boat.				
Canal-boat.				
Chain-boat.				
Collapsible boat.				
Cylinder boat.				
Dredging-boat.				
Duck-shooting boat.				
Ferry-boat.				
Folding boat.				
Gutta-percha boat.				
Ice-boat.				
Ice-breaking boat.				
Insubmersible boat.				
Life-boat.				
Log-boat.				
Metallic boat.				
Mud-boat.				
Night-soil boat.				
Oyster-boat.				
Paper boat.				
Portable boat.				
Rapid-boat.				
River-boat.				
Row-boat.				
Sectional boat.				
Shallow boat.				
Ship's boat.				
Submarine boat.				
Surf and life boat.				
Torpedo-boat.				
Tow-boat.				
Toy-boat.				

Index of patents issued from the United States Patent Office from 1790 to 1873, inclusive—Continued.

Invention.	Inventor.	Residence.	Date.	No.
Boat—Continued. *See* Train-boat. Twin-boat. Whale-boat. Wharf-boat.				
Boat	S. Elwell, jr	Gloucester, Mass	July 9, 1872	128,717
Boat	E. Fitzpatrick	Mount Morris, N. Y	Feb. 12, 1836	....
Boat	J. Hutchinson	Matteawan, N. Y	July 2, 1861	32,695
Boat	D. Keller	....	Nov. 5, 1799	....
Boat	A. Morrison	Johnstown, Pa	May 15, 1837	198
Boat	R. Porter	Billerica, Mass	May 26, 1834	....
Boat	E. and G. A. Waters	Troy, N. Y	June 30, 1868	79,421
Boat and ark	T. Symington	Baltimore, Md	July 22, 1833	....
Boat and davit tackle	E. Wakeman	Brooklyn, Cal	Apr. 2, 1867	63,585
Boat and ponton	E. L. Perry	New York, N. Y	Feb. 23, 1864	41,719
Boat and ship building	W. Phoebus	New York, N. Y	June 27, 1810	....
Boat and trunk	J. A. Olmstead	New York, N. Y	July 30, 1867	67,342
Boat, bridge, and tent, Convertible	A. O. Crane	Hoboken, N. J	Mar. 18, 1862	34,672
Boat-builders' platform	J. Granger	Zanesville, Ohio	Apr. 30, 1867	64,309
Boat-chair	R. H. Cutter	Cleveland, Ohio	July 11, 1871	116,934
Boat convertible into a land-carriage	P. Davis	Providence, R. I	Nov. 13, 1860	30,618
Boat-detaching apparatus	N. B. Allen	Newport, R. I	Jan. 1, 1867	60,661
Boat-detaching apparatus	H. Ashford	Philadelphia, Pa	Sept. 16, 1873	142,834
Boat-detaching apparatus	J. Atwood, jr., and T. M. Hatch	Provincetown, Mass	Feb. 2, 1869	86,490
Boat-detaching apparatus	D. S. Brown	Astoria, Oreg	Oct. 19, 1869	95,877
Boat-detaching apparatus	S. Brown and L. Level	New York, N. Y	Dec. 4, 1866	60,133
Boat-detaching apparatus	D. L. Cohen	Pensacola, Fla	Apr. 25, 1871	113,981
Boat-detaching apparatus	J. C. Cottingham	Philadelphia, Pa	Oct. 18, 1870	108,454
Boat-detaching apparatus	J. C. Cottingham	Philadelphia, Pa	Dec. 13, 1870	110,116
Boat-detaching apparatus	R. Creuzbaur	New York, N. Y	Dec. 17, 1867	72,172
Boat-detaching apparatus	T. L. Cuthbert	Charleston County, S. C	Oct. 20, 1868	83,136
Boat-detaching apparatus	W. A. Devon	Port Richmond, N. Y	June 25, 1867	66,096
Boat-detaching apparatus	C. A. Enell	Philadelphia, Pa	Apr. 29, 1873	138,386
Boat-detaching apparatus	H. E. Foster	Decatur, Ill	June 13, 1871	115,838
Boat-detaching apparatus	J. Foster, jr	Camden, N. J	June 1, 1869	90,738
Boat-detaching apparatus	J. Foster, jr	Camden, N. J	Aug. 2, 1870	105,931
Boat-detaching apparatus	E. J. Hill	Pimlico, England	Oct. 8, 1872	132,007
Boat-detaching apparatus	I. S. Hill and A. Burnham	Boston and Chelsea, Mass	Jan. 22, 1867	61,336
Boat-detaching apparatus	W. H. Houston	New York, N. Y	July 2, 1867	66,345
Boat-detaching apparatus	J. M. Kilner	Chester, Great Britain	Apr. 22, 1873	138,165
Boat-detaching apparatus	M. A. Lanagan	Brooklyn, N. Y	Dec. 31, 1867	72,866
Boat-detaching apparatus	S. L. Lord	Surry, Me	Feb. 15, 1870	99,922
Boat-detaching apparatus	G. W. Mallery	Mystic Bridge, Conn	Sept. 17, 1872	131,361
Boat-detaching apparatus	D. McFarland	New York, N. Y	June 18, 1872	128,054
Boat-detaching apparatus	T. H. Mortimer	Charleston, S. C	Jan. 5, 1869	85,685
Boat-detaching apparatus	F. M. Munger	United States Revenue Service.	Apr. 22, 1873	138,041
Boat-detaching apparatus	M. V. Nobles	Elmira, N. Y	Oct. 18, 1870	108,507
Boat-detaching apparatus	C. H. Nye	Vineland, N. J	Sept. 14, 1869	94,837
Boat-detaching apparatus	C. Quaritius	Canarsie, N. Y	Mar. 12, 1872	124,511
Boat-detaching apparatus	N. M. Ray	Ellsworth, Me	Sept. 29, 1868	82,550
Boat-detaching apparatus	N. M. Ray	Surry, Me	Aug. 23, 1870	106,619
Boat-detaching apparatus	N. C. Reynolds	Ellsworth, Me	Mar. 29, 1870	101,312
Boat-detaching apparatus	I. A. Richards	Middletown, Conn	May 30, 1871	115,360
Boat-detaching apparatus	J. Ross	San Francisco, Cal	Feb. 22, 1870	100,193
Boat-detaching apparatus	W. S. Ryerson	New York, N. Y	June 15, 1869	91,488
Boat-detaching apparatus	W. S. Ryerson and G. Stancliff	New York, N. Y	Apr. 5, 1870	101,665
Boat-detaching apparatus	W. S. Ryerson, G. Stancliff, and O. T. McIntosh.	New York, N. Y	Aug. 2, 1870	105,983
Boat-detaching apparatus	T. S. Seabury	Saint James, N. Y	July 23, 1872	129,755
Boat-detaching apparatus	S. Sneden	Brooklyn, N. Y	Aug. 5, 1873	141,470
Boat-detaching apparatus	G. Stancliff	New York, N. Y	Jan. 31, 1871	111,398
Boat-detaching apparatus	W. M. Van Wagenen	Newark, N. J	Apr. 16, 1861	32,107
Boat-detaching apparatus	E. Wakeman	Brooklyn, N. Y	Mar. 2, 1869	87,449
Boat-detaching apparatus	W. M. Wood	Owing's Mills, Md	June 22, 1869	91,584
Boat-detaching apparatus	W. M. Wood, jr	United States Navy	Dec. 3, 1872	133,692
Boat-detaching apparatus, Automatic	J. H. Kingsland	New York, N. Y	Sept. 14, 1869	94,758
Boat detaching block	N. B. Adams	San Francisco, Cal	Mar. 10, 1868	75,234
Boat-detaching device	W. F. Morgan	New York, N. Y	Sept. 2, 1873	142,496
Boat-detaching hook	W. W. Andrews	Warrensville, Ohio	Mar. 3, 1863	37,799
Boat-detaching method	L. H. Watson	Pittsburgh, Pa	Mar. 26, 1872	124,923
Boat-detaching tackle	J. M. Brooke	Lexington, Va	Sept. 24, 1867	69,069
Boat-detaching tackle	S. Brown	New York, N. Y	Jan. 22, 1867	61,312
Boat-detaching tackle	D. L. Cohen	Pensacola, Fla	Dec. 10, 1867	71,984
Boat-detaching tackle	W. A. Devon	Port Richmond, N. Y	Jan. 22, 1867	61,406
Boat-detaching tackle	W. C. Dodge	Washington, D. C	Apr. 9, 1867	63,620
Boat-detaching tackle	R. H. Griffith	Baltimore, Md	May 28, 1867	65,213
Boat-detaching tackle	T. C. Hargrave	Boston, Mass	July 23, 1867	67,047
Boat-detaching tackle	L. Hover	Chicago, Ill	Aug. 6, 1867	67,547
Boat-detaching tackle	E. C. Hurlbut and E. H. Snow	Middle Haddam and Hartford, Conn.	Aug. 20, 1867	67,983
Boat-detaching tackle	P. H. Jackson	New York, N. Y	Apr. 23, 1867	64,110
Boat-detaching tackle	G. H. Kempton	Hudson, N. J	Aug. 20, 1867	67,989
Boat-detaching tackle	J. A. Libbertz	Hamburg, Germany	Mar. 26, 1867	63,268
Boat-detaching tackle	G. B. Massey	New York, N. Y	Feb. 19, 1867	62,279
Boat-detaching tackle	G. B. Massey	New York, N. Y	Apr. 2, 1867	63,544
Boat-detaching tackle	M. V. Nobles	Elmira, N. Y	Mar. 19, 1867	63,081
Boat-detaching tackle	M. V. Nobles	Elmira, N. Y	Mar. 19, 1867	63,082
Boat-detaching tackle	M. V. Nobles	Elmira, N. Y	June 18, 1867	65,825
Boat-detaching tackle	C. Peterson and C. Gunner	San Francisco, Cal	Oct. 29, 1867	70,354
Boat-detaching tackle	C. H. Ramsten	Carlskrona, Sweden	June 25, 1867	66,105
Boat-detaching tackle	T. M. and T. J. Raser	Philadelphia, Pa	July 23, 1867	67,137
Boat-detaching tackle	J. R. Taylor	New York, N. Y	Jan. 1, 1867	60,963
Boat-detaching tackle	J. R. Taylor	New York, N. Y	Jan. 1, 1867	60,964
Boat-detaching tackle	J. R. Taylor	New York, N. Y	Jan. 1, 1867	60,965
Boat-detaching tackle	J. R. Taylor	New York, N. Y	Jan. 1, 1867	60,966
Boat-detaching tackle	J. R. Taylor	New York, N. Y	Jan. 15, 1867	61,279
Boat-detaching tackle	J. R. Taylor	New York, N. Y	Jan. 15, 1867	61,280

Index of patents issued from the United States Patent Office from 1790 *to* 1873, *inclusive*—Continued.

Invention.	Inventor.	Residence.	Date.	No.
Boat-detaching tackle	J. R. Taylor	New York, N. Y	Jan. 15, 1867	61, 281
Boat-framer	J. Beetle	New Bedford, Mass	May 20, 1856	14, 903
Boat from its tackle, Detaching	S. F. Blunt	Baltimore, Md	Mar. 25, 1856	14, 489
Boat-hoisting apparatus	A. F. Crosman	Steamer Ossipee, United States Navy.	Nov. 12, 1867	70, 813
Boat hoisting, lowering, and detaching apparatus	J. Humphries	Washington, D. C	Oct. 17, 1871	119, 984
Boat-hook, Detaching	W. H. Mills	Philadelphia, Pa	Apr. 23, 1872	125, 900
Boat-hook handle	F. F. Bibber	Boston, Mass	May 17, 1870	103, 007
Boat knee, Dory	E. J. Matthews	Booth Bay, Me	Apr. 7, 1868	76, 487
Boat lowering and detaching apparatus	M. Joyce	Washington, D. C	Sept. 24, 1867	69, 220
Boat-lowering apparatus	R. Creuzbaur	New York, N. Y	Dec. 17, 1867	72, 170
Boat-lowering apparatus	R. Creuzbaur	New York, N. Y	Dec. 17, 1867	72, 171
Boat-lowering apparatus	A. F. Crosman	Steamer Ossipee, United States Navy.	Dec. 17, 1867	72, 268
Boat-lowering apparatus	W. Flowers and Z. S. Patton	Bangor, Me	July 24, 1860	29, 264
Boat-lowering apparatus	G. W. Mallory	Mystic Bridge, Conn	Mar. 21, 1871	112, 939
Boat-lowering device	H. Goulding	Dedham, Mass	Oct. 16, 1866	58, 810
Boat or scow	A. R. Tewksbury	Boston, Mass	Aug. 16, 1853	9, 949
Boat-raising machine	E. Turner	Rochester, N. Y	Dec. 31, 1825	....
Boat row-lock	C. C. Burrows	Mystic River, Conn	Oct. 15, 1867	69, 899
Boat-tackle hook, Self-relieving	H. G. Guyon	Brooklyn, N. Y	Jan. 19, 1864	41, 339
Boat-timbers, Mold for	J. Wallace	Rochester, N. Y	Aug. 9, 1831	....
Boat to be worked by animal power	D. Porter	Washington, D. C	Feb. 8, 1816	....
Boat-towing mechanism	G. S. Olin	Deer Lodge, Mont	Dec. 23, 1873	145, 813
Boats against injury from snags and sawyers, Fender to protect.	I. W. Kirk	Philadelphia, Pa	Feb. 24, 1843	2, 972
Boats and other vessels, Constructing	A. Dorman	Norfolk, Va	Oct. 25, 1832	....
Boats and other vessels of sheet-iron or other metal, Making.	J. Francis	New York, N. Y	Mar. 26, 1845	3, 974
Boats and rafts, Passing	B. Sanborn	Lyman, N. H	Apr. 11, 1836	....
Boats and sea-vessels, Mode of building	W. Annesley	New York, N. Y	Sept. 12, 1816	....
Boats and vessels, Constructing	F. Rotch	New Bedford, Mass	Nov. 29, 1811	....
Boats and vessels, Machine for raising	A. Warner	Rochester, N. Y	Nov. 16, 1826	....
Boats, Arrangement of device for lowering and detaching.	H. D. Veuve	Galveston, Tex	Mar. 16, 1858	19, 666
Boats, Attachments of trains of	W. Ingalls	Sanbornton, N. H	Jan. 3, 1865	45, 720
Boats, Building	G. W. Eddy	Waterford, N. Y	Feb. 28, 1834	....
Boats, Building	W. Hopkins		May 13, 1803	....
Boats, Collapsible float for	J. MacDonough	New York, N. Y	Mar. 27, 1866	53, 461
Boats, Construction of	P. Davis	Providence, R. I	Jan. 3, 1860	26, 657
Boats, Construction of	T. Thorpe	Washington, D. C	Aug. 31, 1827	....
Boats, Detachable hook for suspending	C. L. Williams	Quincy, Fla	June 5, 1860	28, 623
Boats, Device for pitching	W. H. Richardson	Stillwater, Minn	Feb. 4, 1873	135, 490
Boats, Engine for unloading and removing weights from.	T. Potter, sr	Bowling Green, Ky	Dec. 17, 1834	....
Boats, Flexible vertically-extensible attachment for	J. R. Adams	Oakland, Cal	Sept. 23, 1873	143, 052
Boats, Float for	A. T. Boon	Galesburgh, Ill	Apr. 10, 1866	53, 775
Boats from canals, &c., Machine for raising	E. Turner	Rochester, N. Y	Nov. 18, 1825	....
Boats from davits, Apparatus for disengaging	E. Oliver and G. Myers	Rotherham, Great Britain	Mar. 27, 1866	53, 550
Boats from davits, Means for detaching	T. Huntington	New Rochelle, N. Y	June 5, 1866	55, 303
Boats from one level to another, Taking	R. P. Bell	New York, N. Y	July 13, 1827	....
Boats from their davits, Hook for attaching and detaching.	H. Davidson	United States Army	Apr. 9, 1861	32, 036
Boats, &c., Manner of constructing and propelling	B. D. Beecher	Prospect, Conn	Dec. 31, 1839	1, 459
Boats, Manufacture of	W. H. Ferguson	Forest, Ill	Oct. 14, 1873	143, 687
Boats, Means for chocking and securing	F. M. Howes	Somerville, Mass	Dec. 23, 1873	145, 875
Boats, Means for detaching	T. Shaw	Philadelphia, Pa	Apr. 1, 1873	137, 325
Boats, &c., Method of building	J. Francis	New York, N. Y	Oct. 11, 1841	2, 293
Boats, Mode of bracing and staying	F. Mertens	Cumberland, Md	May 28, 1867	65, 255
Boats, Molding-frame for the construction of	N. Thompson, jr	Brooklyn, E. D., N. Y	May 18, 1858	20, 308
Boats over dams, Passing	S. Underwood	Bath, N. H	Mar. 19, 1836	....
Boats, Protector for stems of	W. P. Davis and S. Elwell, jr	Gloucester, Mass	Sept. 21, 1869	94, 948
Boats, &c., Rowing and propelling	W. Shaw	Otsego County, N. Y	Feb. 19, 1812	....
Boats, Suspending	C. H. Hasker	Portsmouth, Va	May 29, 1860	28, 478
Boats to and from tackle, Means for attaching and detaching.	J. M. Brooke	United States Army	Aug. 5, 1856	15, 473
Boats to facilitate discharge of cargo, Fittings for	W. H. Bryan	Georgetown, D. C	July 29, 1851	8, 266
Boats, vessels, &c., Constructing	F. Rotch	Philadelphia, Pa	Aug. 3, 1812	....
Boats, Weigh-lock for	J. Brainerd	Rome, N. Y	July 17, 1839	1, 249
Bobbin	M. Bliss	Ionia, Mich	May 17, 1870	103, 009
Bobbin	H. Clarke	Dedham, Mass	June 28, 1853	9, 814
Bobbin	A. I. Earle	Valley Falls, R. I	June 3, 1873	139, 558
Bobbin	C. H. Fisk	Lowell, Mass	Nov. 26, 1867	71, 371
Bobbin	A. P. Holmes	Great Falls, N. H	Jan. 7, 1868	73, 185
Bobbin	H. J. Hubbard	Chicopee, Mass	Nov. 3, 1868	83, 635
Bobbin	D. Hussey	Nashua, N. H	Oct. 28, 1862	36, 782
Bobbin	T. G. McLaughlin	Glen Riddle, Pa	Aug. 5, 1873	141, 575
Bobbin	W. H. Ramsdell	Lowell, Mass	July 7, 1868	79, 687
Bobbin	C. H. Reynolds	North Kingston, R. I	Aug. 30, 1864	44, 016
Bobbin	G. Richardson	Lowell, Mass	July 23, 1872	129, 682
Bobbin	C. A. Shaw	Biddeford, Me	Dec. 15, 1863	40, 9[illegible]2
Bobbin	J. N. Stearns	New York, N. Y	June 20, 1871	116, 108
Bobbin, Cotton-spool	N. Rider	Millbury, Mass	Dec. 31, 1833	....
Bobbin for roving and slubbing	I. Hayden	Lawrence, Mass	Aug. 4, 1857	17, 929
Bobbin for spinning, &c	L. Ferguson	Lowell, Mass	May 1, 1866	54, 465
Bobbin for spinning, &c	B. Wilbour	Scituate, R. I	Jan. 2, 1866	51, 887
Bobbin-holder	J. Salisbury	Scituate, R. I	Jan. 25, 1870	99, 111
Bobbin-holder for spinning	J. Goulding	Worcester, Mass	Oct. 3, 1865	50, 240
Bobbin-holder for spinning	J. Goulding	Worcester, Mass	Oct. 3, 1865	50, 241
Bobbin-holder for spinning	E. Wright	Worcester, Mass	Oct. 3, 1865	50, 311
Bobbin, Metallic	W. B. Rice	Utica, N. Y	Apr. 16, 1867	63, 813
Bobbin, Metallic	C. T. Smith	Utica, N. Y	Apr. 23, 1867	64, 158
Bobbin or spool	B. Saunders	Nashua, N. H	Aug. 25, 1863	39, 681
Bobbin or spool for factories, Trussel	T. Van Riper	Paterson, N. J	Aug. 16, 1826	....
Bobbin, Spinning	J. Davis and S. W. Foster	Lowell, Mass	Mar. 26, 1867	63, 225
Bobbin, Weft	W. E. Bass	Lawrence, Mass	Jan. 29, 1867	61, 598
Bobbin, Weft	O. Pearl	Lawrence, Mass	Apr. 3, 1866	53, 665
Bobbin-winder	J. S. Demarest	Elmira, N. Y	May 28, 1872	127, 155

Index of patents issued from the United States Patent Office from 1790 *to* 1873, *inclusive*—Continued.

Invention.	Inventor.	Residence.	Date.	No.
Bobbin-winder	H. A. House	Bridgeport, Conn	Mar. 7, 1871	112, 345
Bobbin-winder	D. F. Mellen	Manchester, N. H	Sept. 15, 1863	39, 997
Bobbin-winder	E. Wilder	Springfield, Mass	Apr. 16, 1872	125, 869
Bobbin-winder, Socket	J. Thorp and S. Shepard	Taunton, Mass	Oct. 14, 1816	....
Bobbin-winders, Thread guiding and cutting device for.	T. M. Day and H. Shook	Washington, Ohio	Mar. 19, 1872	124, 667
Bobbin-winding machine	J. Adams and W. A. Tolman	Richmond, Ind	May 30, 1871	115, 264
Bobbin-winding machine	W. Atkinson	Philadelphia, Pa	Aug. 26, 1873	142, 188
Bobbin-winding machine	W. V. Gee	Philadelphia, Pa	Apr. 9, 1872	125, 564
Bobbin-winding machine	J. Goodyear	Yonkers, N. Y	Aug. 5, 1873	141, 434
Bobbin-winding machine	F. H. Morrill	Philadelphia, Pa	Sept. 21, 1869	95, 034
Bobbin-winding machine	F. H. Morrill	Philadelphia, Pa	Apr. 4, 1871	113, 325
Bobbin-winding machine, Cup of	J. W. Vaughn	New York, N. Y	May 30, 1871	115, 393
Bobbin-winding machine, Thread guide and holder for.	W. W. Altemus	Philadelphia, Pa	Dec. 27, 1870	110, 535
Bobbin-winding, Reverse motion for	G. Richardson	Lowell, Mass	Dec. 8, 1868	84, 764
Bobbins, &c., Cutting out cylinders for	L. Brown	Epsom, N. H	Apr. 24, 1849	6, 391
Bobbins, Driving	A. M. Eastman	Boston, Mass	Apr. 17, 1849	6, 344
Bobbins, Friction-brake for	T. Estes	North Bennington, Vt	Nov. 27, 1860	30, 725
Bobbins in machinery for spinning fibrous substances, Method of operating.	F. McCulley, jr	Paterson, N. J	Oct. 19, 1844	3, 801
Bobbins in shuttles for weaving, Securing	D. Leavitt	Cabotville, Mass	Aug. 18, 1842	2, 755
Bobbins in shuttles, Mode of securing	S. C. Mendenhall	Richmond, Ind	Aug. 29, 1865	49, 643
Bobbins, Machine for automatically winding	D. M. Church	Holyoke, Mass	July 16, 1872	129, 394
Bobbins, Machine for boring	V. D. Beach	Battle Creek, Mich	Dec. 3, 1867	71, 683
Bobbins, Machine for pointing the journals of	D. M. Church	Holyoke, Mass	May 13, 1873	138, 789
Bobbins, Machine for turning heads of	C. M. Spencer	Hartford, Conn	Apr. 8, 1873	137, 631
Bobbins, Machine for winding conical	G. S. Bradford	Bennington, Vt	Aug. 22, 1865	49, 582
Bobbins, Machinery for boring	C. E. Norris	Peacham, Vt	Apr. 24, 1849	6, 374
Bobbins, Machinery for winding conical	H. Marcellus and S. Ward	Amsterdam, N. Y	May 24, 1864	42, 861
Bobbins, Machinery for winding conical	C. Tompkins and J. Johnson	Troy, N. Y., and Roxbury, Mass.	Apr. 21, 1857	17, 120
Bobbins or spools, Method of strengthening	A. Hallowell	Lowell, Mass	Mar. 15, 1870	100, 754
Bobbins, Skewer for speeder	C. Harris	River Point, R. I	July 6, 1869	92, 189
Bobbins upon spindles, Driving	O. Pearl	Essex County, Mass	Jan. 29, 1850	7, 058
Bobbins, Winding yarn on weavers' or shuttle	J. Thorp	Providence, R. I	Mar. 20, 1829	....
Bodkin for inserting grooved ring into cloth	J. S. Turner	Middletown, Conn	Apr. 25, 1837	181
Bodkin for passing enlarged shank of button through cloth.	P. W. Gengembre	Boston, Mass	Sept. 27, 1864	44, 416
Body-brace	E. P. Banning	New York, N. Y	Nov. 1, 1864	44, 914
Body-brace	G. S. Browne	Hartford, Conn	Nov. 22, 1853	10, 248
Body-brace	H. Mellish	Walpole, N. H	Jan. 9, 1849	6, 023
Body brace and supporter, Child's	L. Spigelmyer	Hartleton, Pa	Apr. 12, 1870	101, 934
Body-conformator	S. O. Brigham	San Francisco, Cal	Aug. 7, 1866	56, 892
Body-protector	J. W. Torrey	Milford, Mass	Mar. 17, 1868	75, 710
Bog and bush, Machine for cutting	T. Spring	Granby, Conn	Apr. 27, 1833	....
Bog-cutter	R. Cummings	Lima, Ind	Nov. 20, 1847	5, 376
Bog-cutter	J. D. Filkins	Lima, Ind	Jan. 9, 1849	6, 018
Bog-cutter	C. E. Steller	Chicago, Ill	July 31, 1866	56, 822
Bog-cutter and drag	J. W. Newton	Geneva, Wis	Nov. 3, 1868	83, 784
Bog-cutter and drag	J. W. Newton	Geneva, Wis	Mar. 16, 1869	88, 801
Bog-cutting machine	A. Follet	Windham, Conn	Oct. 16, 1849	6, 802
Bog-cutting machine	C. Hoisington	Seward, Ill	June 30, 1863	39, 048
Bogging-machine	S. Collins	Hillsdale, N. Y	Feb. 22, 1827	....
Bogging-machine	J. A. Post	Orange County, N. Y	Apr. 23, 1828	....
Boiler:				
See Agricultural boiler.				
Back-log boiler.				
Base-burning boiler.				
Base-burning steam-boiler.				
Box boiler.				
Circular boiler.				
Clothes-boiler.				
Coffee-boiler.				
Collender-boiler.				
Cooking and washing boiler.				
Culinary boiler.				
Domestic boiler.				
Egg-boiler.				
Farmers' boiler.				
Feeder-boiler.				
Fire-extinguisher boiler.				
Heater-boiler.				
Heating-boiler.				
Hot-water boiler.				
Iron boiler.				
Kitchen-boiler.				
Lard-boiler.				
Locomotive-boiler.				
Marine steam-boiler.				
Milk-boiler.				
Paper-pulp boiler.				
Portable boiler.				
Portable-furnace boiler.				
Range-boiler.				
Revolving boiler.				
Roofing-boiler.				
Sheet-metal boiler.				
Steam-boiler.				
Steam-generating boiler.				
Steam-heater boiler.				
Stone-boiler.				
Stove-boiler.				
Sugar-boiler.				
Tubular boiler.				
Wash-boiler.				
Water-heating boiler.				
Water-tube steam-boiler.				
Boiler	H. Adler	Yellow Springs, Ohio	Oct. 15, 1867	69, 896

Index of patents issued from the United States Patent Office from 1790 *to* 1873, *inclusive*—Continued.

Invention.	Inventor.	Residence.	Date.	No.
Boiler	J. Baumeister	Detroit, Mich	Aug. 20, 1867	67, 939
Boiler	G. C. Bidwell	Philadelphia, Pa	Sept. 23, 1862	36, 501
Boiler	C. Britain	Saint Joseph, Mich	Oct. 6, 1863	40, 157
Boiler	O. S. Camp	Fairfield, Iowa	June 21, 1859	24, 442
Boiler	C. Godfrey	Brooklyn, N. Y	Dec. 11, 1860	30, 917
Boiler	M. Gore	Batavia, Ill	June 7, 1870	103, 870
Boiler	C. O. Green and R. Ham	Troy, N. Y	Apr. 21, 1868	76, 910
Boiler	J. Hurd, jr	Boston, Mass	Nov. 10, 1829	....
Boiler	C. Kieffer	Lancaster, Pa	Sept. 20, 1859	25, 511
Boiler	T. S. Lambert	Peekskill, N. Y	Oct. 6, 1863	40, 172
Boiler	A. J. Le Grand	Boonton, N. J	Aug. 17, 1869	93, 725
Boiler	J. Marshall	Hartland, Mich	Mar. 24, 1868	75, 942
Boiler	G. N. Philips	Goshen, N. Y	June 28, 1814	....
Boiler	J. R. Purdy and D. C. Barger	Peekskill, N. Y	Aug. 6, 1867	67, 452
Boiler	S. C. Riley	New York, N. Y	Oct. 11, 1841	2, 318
Boiler	W. P. Skiffington	New York, N. Y	June 20, 1871	116, 104
Boiler	D. Sullivan	Bangor, Me	Apr. 25, 1871	114, 222
Boiler alarm-gage	T. C. Banks	Wallingford, Conn	Mar. 24, 1868	75, 833
Boiler, also working stills with the same	W. Thornton		Oct. 28, 1803	....
Boiler and condenser	B. T. Babbitt	New York, N. Y	Sept. 9, 1873	142, 663
Boiler and condenser	B. T. Babbitt	New York, N. Y	Sept. 9, 1873	142, 664
Boiler and feed-water apparatus	W. M. Fuller	New York, N. Y	Oct. 15, 1872	132, 274
Boiler and furnace united	P. M. Hackley	Herkimer, N. Y	Jan. 23, 1815	....
Boiler and hot-air register, Combined	B. B. Perkins	Chestertown, Md	Feb 25, 1868	74, 936
Boiler and other furnace, Steam	C. Schemioth	London, England	Nov. 4, 1873	144, 241
Boiler and steamer	D. R. Prindle	Bethany, N. Y	Sept. 13, 1859	25, 442
Boiler and steamer, Combined	J. P. Bradford	Calamus, Iowa	Oct. 17, 1871	119, 961
Boiler and still	W. Eaton	Baltimore, Md	Oct. 29, 1810	....
Boiler and stone furnace	S. T. Savage	Albany, N. Y	June 22, 1858	20, 667
Boiler and tea-kettle	T. S. Lambert	Peekskill, N. Y	June 3, 1862	35, 455
Boiler and washer	E. C. Brooks	Norwich, N. Y	June 21, 1870	104, 550
Boiler and washing-machine	D. H. Hull	Plantsville, Conn	Apr. 5, 1870	101, 622
Boiler-bottom	G. T. Palmer	Brooklyn, N. Y	June 16, 1868	78, 993
Boiler-cleaner	W. P. Slensby	Chicago, Ill	July 30, 1867	67, 361
Boiler-cleaner	G. W. Wiswell	Pilot Knob, Mo	Nov. 19, 1867	71, 109
Boiler cupola and grate, Combined	L. Wood	New York, N. Y	Nov. 26, 1850	7, 804
Boiler explosion and burning, Preventing	W. A. Turner	Washington County, N. C	Apr. 6, 1831	....
Boiler explosion by applying a fusible alloy, Preventing.	C. Evans	Philadelphia, Pa	May 8, 1834	....
Boiler explosion, Device to prevent	A. Arnold	Tenafly, N. J	June 26, 1866	55, 801
Boiler explosion, Means of preventing	A. W. Jones	New York, N. Y	Aug. 22, 1854	11, 567
Boiler explosion, Preventing	T. Newhall, jr	Lynn, Mass	Apr. 12, 1834	....
Boiler explosion, Preventing	W. A. Turner	Washington County, N. C	Mar. 15, 1831	....
Boiler-feed regulator	W. McCormick	Philadelphia, Pa	Dec. 17, 1867	72, 413
Boiler-feed regulator	W. Webster	Morrisania, N. Y	Nov. 3, 1863	40, 523
Boiler-feed regulator, Automatic	B. Brett	Geneva, Ohio	June 10, 1873	139, 762
Boiler feed, Steam	J. S. Barden	Providence, R. I	July 16, 1872	129, 514
Boiler feed-water regulator	J. S. Barden	Providence, R. I	Sept. 21, 1869	94, 937
Boiler-feeder	J. W. Bishop	New Haven, Conn	Apr. 11, 1865	47, 180
Boiler-feeder	W. Bowman	Greenock, Scotland	Apr. 17, 1866	54, 078
Boiler-feeder	E. Brockway	Haverstraw, N. Y	Feb. 18, 1868	74, 659
Boiler-feeder	H. I. Brunner	Nazareth, Pa	June 22, 1869	91, 515
Boiler-feeder	T. W. Burger	Jersey City, N. J	Feb. 17, 1863	37, 673
Boiler-feeder	H. Burt	Newark, N. J	Aug. 31, 1869	94, 395
Boiler-feeder	J. Darling	Cincinnati, Ohio	July 2, 1867	66, 223
Boiler-feeder	H. O. Demarest	New York, N. Y	Sept. 17, 1867	68, 964
Boiler-feeder	J. W. Doughty and B. F. Olmsted.	Newburgh, N. Y	Feb. 19, 1867	62, 258
Boiler-feeder	L. Foote	Fairfield, Ohio	Apr. 19, 1870	101, 992
Boiler-feeder	C. H. Griffin	Chelsea, Mass	Nov. 20, 1866	59, 909
Boiler-feeder	A. Hammond	Jacksonville, Ill	Feb. 9, 1864	41, 504
Boiler-feeder	H. McGan	Cleveland, Ohio	May 26, 1868	78, 224
Boiler-feeder	A. F. W. Neynaber	Philadelphia, Pa	June 4, 1867	65, 418
Boiler-feeder	S. J. Parker	Williamsport, Pa	Apr. 12, 1870	101, 762
Boiler-feeder	G. T. Parry	Philadelphia, Pa	Apr. 23, 1867	64, 026
Boiler-feeder	W. C. Pickersgill	Providence, R. I	Sept. 1, 1868	81, 679
Boiler-feeder	N. Puckett	Deming, Ind	Dec. 6, 1859	26, 374
Boiler-feeder	R. Rafael	Havana, Cuba	Sept. 16, 1873	142, 866
Boiler-feeder	J. Shirk and I. W. Martin	East Earl Township, Pa	Apr. 20, 1869	89, 087
Boiler-feeder	J. M. Stiven	New York, N. Y	Aug. 26, 1862	36, 324
Boiler-feeder	A. Walker, jr	Johnsbury, Vt	Aug. 24, 1852	9, 225
Boiler-feeder	G. I. Washburn	Worcester, Mass	Jan. 26, 1864	41, 408
Boiler-feeder	W. S. and S. B. Wells	New York and Middleburgh, N. Y.	May 15, 1866	54, 798
Boiler-feeder	S. H. Wheeler	Dowagiac, Mich	Nov. 1, 1864	44, 922
Boiler-feeder	A. R. Young	Red Wing, Minn	Oct. 18, 1870	108, 423
Boiler-feeder alarm-device	J. W. Ebert and E. C. McCloy	Zanesville, Ohio	Oct. 5, 1869	95, 444
Boiler feeder and filter	J. G. Fulton	Toledo, Ohio	Apr. 1, 1873	137, 357
Boiler feeder and low-water alarm	V. D. Anderson	Kewanee, Ill	May 31, 1870	103, 539
Boiler feeder and low-water detector	J. L. Starkey	Worcester, Mass	Nov. 8, 1870	109, 149
Boiler-feeder, Automatic	H. B. Adams	Brooklyn, N. Y	Jan. 4, 1859	22, 477
Boiler-feeder, Automatic	H. Beadle	Washington, D. C	Mar. 19, 1867	62, 996
Boiler-feeder, Automatic	R. Berryman and R. N. Pratt	Hartford, Conn., and Philadelphia, Pa.	May 2, 1871	114, 255
Boiler-feeder, Automatic	J. W. Bishop	New Haven, Conn	Apr. 11, 1865	47, 181
Boiler-feeder, Automatic	J. Boden	Columbus, Pa	Mar. 23, 1869	88, 005
Boiler-feeder, Automatic	J. N. B. Bond	New York, N. Y	May 23, 1865	47, 791
Boiler-feeder, Automatic	J. E. Burdge	Cincinnati, Ohio	Oct. 19, 1869	95, 983
Boiler-feeder, Automatic	B. Chalfant	Williamsport, Pa	July 31, 1866	56, 706
Boiler-feeder, Automatic	D. L. F. Chase	Boston, Mass	Apr. 13, 1869	88, 774
Boiler-feeder, Automatic	S. Cook	Magnolia, Iowa	Nov. 30, 1869	97, 361
Boiler-feeder, Automatic	J. A. Davis	Watertown, N. Y	Sept. 29, 1868	82, 498
Boiler-feeder, Automatic	H. O. Demarest	New York, N. Y	Mar. 26, 1867	63, 227
Boiler-feeder, Automatic	S. Driver	Philadelphia, Pa	Sept. 12, 1865	49, 949
Boiler-feeder, Automatic	S. Driver	Philadelphia, Pa	Aug. 4, 1868	80, 536
Boiler-feeder, Automatic	G. W. B. Gedney and W. J. Brassington.	New York, N. Y	July 19, 1864	43, 579
Boiler feeder, Automatic	E. H. Goldman	Clayton, Ind	Mar. 15, 1870	100, 883

Index of patents issued from the United States Patent Office from 1790 *to* 1873, *inclusive*—**Continued.**

Invention.	Inventor.	Residence.	Date.	No.
Boiler-feeder, Automatic	J. Holtz	Baltimore, Md	Mar. 23, 1869	88, 168
Boiler-feeder, Automatic	J. Hoover	Lewisburgh, Ohio	Aug. 6, 1861	32, 995
Boiler-feeder, Automatic	S. B Hunt	New York, N. Y	May 13, 1862	35, 237
Boiler-feeder, Automatic	T. Lovelidge and J. Grindrod	Philadelphia, Pa	Aug. 17, 1869	93, 826
Boiler-feeder, Automatic	P. N. J. Macabies	Paris, France	May 11, 1869	90, 047
Boiler-feeder, Automatic	H. McGann	Cleveland, Ohio	Dec. 1, 1868	84, 642
Boiler-feeder, Automatic	W. Moore	Kokoma, Ind	June 5, 1866	55, 338
Boiler-feeder, Automatic	S. J. Parker	Williamsport, Pa	Apr. 30, 1867	64, 356
Boiler-feeder, Automatic	J. N. Poage	Cincinnati, Ohio	Nov. 12, 1872	133, 052
Boiler-feeder, Automatic	E. Quinn	Washington, D. C	Jan. 30, 1866	52, 376
Boiler-feeder, Automatic	E. Quinn	Washington, D. C	Mar. 6, 1866	53, 042
Boiler-feeder, Automatic	R. Rafael	New York, N. Y	Sept. 12, 1865	49, 919
Boiler-feeder, Automatic	G. A. Riedel	Philadelphia, Pa	Sept. 19, 1865	50, 034
Boiler-feeder, Automatic	G. A. Riedel	Philadelphia, Pa	Sept. 19, 1865	50, 035
Boiler-feeder, Automatic	G. A. Riedel	Philadelphia, Pa	July 3, 1866	56, 096
Boiler feeder, Automatic	G. A. Riedel	Philadelphia, Pa	July 23, 1867	66, 993
Boiler-feeder, Automatic	E. Sheppard	Philadelphia, Pa	Sept. 22, 1868	82, 357
Boiler-feeder, Automatic	T. J. Sloan	New York, N. Y	Feb. 27, 1866	52, 898
Boiler-feeder, Automatic	H. E. Stager	Milwaukee, Wis	Aug. 10, 1869	93, 564
Boiler-feeder, Automatic	D. Vaughan	Cincinnati, Ohio	Mar. 26, 1872	124, 921
Boiler-feeder, Automatic	T. B. Webster	Sekaukus, N. J	Mar. 23, 1869	88, 246
Boiler-feeder, Automatic	J. R. Widgeon	Bucyrus, Ohio	Mar. 13, 1866	53, 238
Boiler-feeder, Automatic	J. R. Widgeon and F. E. Frey	Bucyrus, Ohio	Oct. 13, 1868	802, 13
Boiler feeder, Automatic steam	J. Wheelock	Worcester, Mass	Mar. 26, 1872	125, 106
Boiler-feeder, Balance	W. D. Allen	Durhamville, N. Y	Dec. 10, 1850	7, 817
Boiler-feeder for steam-engines	E. A. Woods	Utica, N. Y	July 3, 1860	29, 028
Boiler-feeder, Self-acting	F. D. Boyle	Evansville, Ind	Aug. 4, 1863	39, 378
Boiler feeder, Steam	S. J. Parker	Rochester, N. Y	May 28, 1872	127, 362
Boiler feeder, Steam	J. M. Case	Worthington, Ohio	Mar. 29, 1870	101, 225
Boiler feeder, Steam	W. J. Sanderson and S. Stanton	Syracuse, N. Y	May 15, 1860	28, 307
Boiler-flanges, Machine for turning	E. Paye	New York, N. Y	May 29, 1866	55, 209
Boiler-float	H. McGann	Cleveland, Ohio	Dec. 1, 1868	84, 502
Boiler-flue cleaner	J. Armbruster	Petroleum Centre, Pa	Oct. 28, 1873	144, 046
Boiler-flue cleaner	W. C. Baker	New York, N. Y	Nov. 17, 1868	84, 079
Boiler-flue cleaner	G. R. Dobbins	Lowell, Mass	Aug. 25, 1868	81, 351
Boiler-flue cleaner	J. Fairclough	Saint Joseph, Mo	Nov. 30, 1869	97, 372
Boiler-flue cleaner	M. and C. H. Morse	Franklin, Mass	Mar. 30, 1869	88, 405
Boiler-flue cleaner	B. Schaefer	Chicago, Ill	Jan. 1, 1867	60, 795
Boiler-flue cleaner	G. V. Sloat	Morrisania, N. Y	Dec. 1, 1868	84, 587
Boiler-flue cleaner	J. M. Spiegle	Philadelphia, Pa	Dec. 5, 1865	51, 362
Boiler-flue-cleaning apparatus	T. H. Donohue	Washington, D. C	Sept. 16, 1873	142, 775
Boiler-flue plug	W. M. Sinclair	Baltimore, Md	July 28, 1868	80, 313
Boiler-flues by steam jet, Cleaning	C. Dasenbrock	Cincinnati, Ohio	Oct. 3, 1871	119, 581
Boiler-flues, Machine for repairing	J. Hughes	Bloomington, Ill	Aug. 23, 1870	106, 589
Boiler-flues, Method of increasing the effective length and cleansing.	A. Chapman	Fairfax, Vt	July 17, 1849	6, 595
Boiler-flues, Steam-jet head for cleaning	J. M. Wheeler	Oxford, Conn	Mar. 31, 1868	76, 125
Boiler-flues, Thimble for protecting	J. C. Farmer	Providence, R. I	May 20, 1873	138, 998
Boiler for heating-purposes	J. Bradley	Baltimore, Md	May 5, 1868	77, 576
Boiler for using anthracite coal	J. Barken	Baltimore, Md	Feb. 7, 1827	......
Boiler for vaporizing volatile fluids	J. F. Haskins	Fitchburgh, Mass	Dec. 2, 1873	145, 175
Boiler-form	J. S. Jennings	Medina, N. Y	Mar. 12, 1867	62, 852
Boiler-forming apparatus	W. W. Hornberger	Chicago, Ill	Jan. 8, 1867	61, 068
Boiler-furnace	H. F. Baker	Boston, Mass	May 30, 1846	4, 546
Boiler-furnace	T. H. Clark	Saint Louis, Mo	May 9, 1865	47, 619
Boiler-furnace	J. Cole	Cincinnati, Ohio	Aug. 22, 1848	5, 720
Boiler-furnace	T. B. Davis	Boston, Mass	Sept. 20, 1864	44, 366
Boiler-furnace	A. de Pindray	Paris, France	Mar. 24, 1868	75, 739
Boiler-furnace	W. Ennis	Philadelphia, Pa	Mar. 16, 1869	87, 917
Boiler-furnace	S. A. Ford	Chicago, Ill	Aug. 19, 1873	141, 870
Boiler-furnace	A. Friedmann and F. E. d'Erlanger.	Paris, France	Aug. 25, 1863	39, 640
Boiler-furnace	M. L. Horton	Windsor, Vt	Nov. 30, 1869	97, 296
Boiler-furnace	S. Hungerford	Slaterville, N. Y	Aug. 15, 1848	5, 712
Boiler-furnace	H. R. Ives	New Haven, Conn	May 21, 1872	126, 965
Boiler-furnace	E. L. Miller	Brooklyn, N. Y	July 7, 1846	4, 626
Boiler-furnace	J. A. Miller	New York, N. Y	Apr. 4, 1865	47, 118
Boiler-furnace	G. H. Nott	Boston, Mass	Oct. 5, 1869	95, 601
Boiler-furnace	J. A. F. Overend	San Francisco, Cal	May 3, 1870	102, 581
Boiler-furnace	J. B. Root	New York, N. Y	Aug. 5, 1873	141, 594
Boiler-furnace	E. Skelly	Plaquemine, La	Dec. 21, 1858	22, 382
Boiler-furnace	L. Skinner	Chittenango, N. Y	Feb. 7, 1871	111, 580
Boiler-furnace	A. B. Smith	Geneva, N. Y	June 3, 1873	139, 522
Boiler-furnace	B. F. Smith	New Orleans, La	Apr. 2, 1872	125, 148
Boiler-furnace	C. D. Smith	Chicago, Ill	Sept. 30, 1873	143, 305
Boiler-furnace	P. Smith	New York, N. Y	Aug. 29, 1865	49, 662
Boiler-furnace	W. C. Smith	New Brunswick, N. J	Feb. 6, 1872	123, 427
Boiler-furnace	J. A. Straight	Albion, N. Y	Aug. 17, 1869	93, 922
Boiler-furnace	E. C. Strange and G. R. Huntley	Taunton, Mass	Mar. 7, 1865	46, 725
Boiler-furnace	U. B. Stribling	Madison, Ind	Mar. 28, 1871	113, 110
Boiler-furnace	T. Vicars, sr., T. Vicars, jr., and J. Smith.	Liverpool, England	Dec. 20, 1870	110, 313
Boiler-furnace	C. D. Williams	Saint Paul, Minn	Feb. 11, 1873	135, 869
Boiler-furnace	T. B. Wilson and W. R. Shaw	Meadville, Pa	June 6, 1865	48, 125
Boiler furnace and flue	W. H. Burns	Unionville, Mo	Feb. 15, 1870	99, 8[illegible]9
Boiler-furnace caloric-regulator	J. G. and E. W. H. Cooper	Hartford, Conn	Mar. 12, 1867	62, 821
Boiler-furnace crown, Steam	L. H. Waugh, J. A. Hanglin, and J Graham.	Wyandotte, Kans	July 16, 1872	129, 634
Boiler-furnace fire-plate	E. Boileau	Saint Louis, Mo	Oct. 3, 1871	119, 562
Boiler-furnace, Oil-burning	F. Hungerford	Rochester, N. Y	Sept. 2, 1873	142, 468
Boiler-furnace, Portable	D. R. Prindle	East Bethany, N. Y	Apr. 16, 1867	63, 811
Boiler-furnace, Return-flue	C. J. Harris	Bloomington, Ill	May 2, 1871	114, 437
Boiler furnace, Steam	H. W. Adams	Philadelphia, Pa	Oct. 15, 1872	132, 192
Boiler furnace, Steam	S. P. Bartley	Columbus, Ohio	Apr. 14, 1868	76, 587
Boiler furnace, Steam	A. J. Bell and J. B. Friend	Ashland, Ky	Dec. 24, 1872	134, 243
Boiler furnace, Steam	A. W. Cram	Litchfield, Ill	Dec. 31, 1872	134, 465
Boiler furnace, Steam	G. H. Diehl	Chicago, Ill	Dec. 19, 1871	121, 993
Boiler furnace, Steam	R. S. Dillon	Detroit, Mich	Dec. 28, 1869	98, 238

Index of patents issued from the United States Patent Office from 1790 *to* 1873, *inclusive*—Continued.

Invention.	Inventor.	Residence.	Date.	No.
Boiler furnace, Steam	P. Estes	Leavenworth, Kans	Sept. 17, 1872	131, 431
Boiler furnace, Steam	M. A. Foster	Saint Louis, Mo	May 21, 1872	126, 946
Boiler furnace, Steam	M. A. Foster	Saint Louis, Mo	Oct. 15, 1872	132, 151
Boiler furnace, Steam	M. A. Foster	Saint Louis, Mo	Mar. 4, 1873	136, 497
Boiler furnace, Steam	M. A. Foster	Saint Louis, Mo	May 13, 1873	138, 875
Boiler furnace, Steam	C. H. Fox	New Orleans, La	Oct. 1, 1872	131, 871
Boiler furnace, Steam	C. H. Fox	New Orleans, La	Jan. 21, 1873	135, 105
Boiler furnace, Steam	C. H. Fox	New Orleans, La	Jan. 21, 1873	135, 106
Boiler furnace, Steam	J. B. Gardner and C. H. Swain	New York, N. Y	Aug. 11, 1868	80, 821
Boiler furnace, Steam	E. F. Griffin	Chicago, Ill	Dec. 24, 1872	134, 270
Boiler furnace, Steam	C. J. Hazstreem	Chicago, Ill	May 24, 1870	103, 324
Boiler furnace, Steam	J. Howes	Worcester, Mass	Oct. 22, 1872	132, 466
Boiler furnace, Steam	F. A. Huntington	San Francisco, Cal	Oct. 15, 1872	132, 209
Boiler furnace, Steam	S. Keyes	Bennington, Vt	Nov. 11, 1873	144, 393
Boiler furnace, Steam	T. F. Morrison	Bryan, Ohio	Apr. 25, 1871	114, 0[illegible]4
Boiler furnace, Steam	A. C. Rand	Aurora, Ill	May 6, 1873	138, 694
Boiler furnace, Steam	C. D. Smith	Chicago, Ill	Aug. 6, 1872	130, 249
Boiler furnace, Steam	C. D. Smith	Chicago, Ill	May 6, 1873	138, 588
Boiler furnace, Steam	A. Steinway	New York, N. Y	Aug. 6, 1872	130, 326
Boiler furnace, Steam	U. B. Stribling	Madison, Ind	Nov. 4, 1873	144, 296
Boiler furnace, Steam	M. K. Taylor	United States Army	Jan. 24, 1871	111, 154
Boiler furnace, Steam	R. L. Walker	Globe Village, Mass	Mar. 24, 1868	75, 817
Boiler furnace, &c., Steam	J. Westerman	Sharon, Pa	Mar. 18, 1873	136, 948
Boiler furnace, Steam	J. D. Whelpley and J. J. Storer	Boston, Mass	Mar. 31, 1868	76, 280
Boiler furnace, Tubular steam	H. W. Adams	Philadelphia, Pa	Aug. 6, 1872	132, 268
Boiler-furnaces, Apparatus for consuming smoke and gas and increasing draft in.	D. E. Somes	Washington, D. C	Apr. 28, 1868	77, 226
Boiler furnaces, Blast for steam	R. Gigodot	Lyons, France	Nov. 18, 1873	144, 757
Boiler-furnaces, Bridge-wall of	W. G. Hamilton	New York, N. Y	Jan. 4, 1859	22, 530
Boiler furnaces, Mouth-piece for steam	C. Stewart	Worcester, Mass	Oct. 28, 1873	144, 002
Boiler-gage	C. L. Frink	Rockville, Conn	June 9, 1866	55, 638
Boiler guard-plate	A. O'Neill	Portsmouth, Ohio	Feb. 20, 1866	52, 806
Boiler-head bolt-fastening	E. Clark	New York, N. Y	Apr. 24, 1866	54, 113
Boiler-head-flanging forge and furnace	J. Nixon	Altoona, Pa	June 30, 1868	79, 382
Boiler-head-flanging machine	S. Lowen	Temperanceville, Pa	July 18, 1871	117, 091
Boiler-heads, Bending flanges upon	W. W. Martin	Allegheny City, Pa	June 19, 1866	55, 687
Boiler-heads, Machine for bending flanges on	D. Howell	Louisville, Ky	Oct. 20, 1857	18, 452
Boiler-heads, Machine for bending the flanges of	J. T. G. Piedboeuf	Liege, Belgium	Oct. 22, 1872	132, 989
Boiler-heads, Machine for cutting oval holes in	J. McBride	Allegheny City, Pa	Aug. 28, 1866	57, 533
Boiler-heads, Machine for cutting oval holes in	J. McBride	Allegheny City, Pa	Aug. 28, 1866	57, 534
Boiler-heads, Machinery for forming flanges upon	J. F. Keeler	Pittsburgh, Pa	May 22, 1866	54, 920
Boiler-heads, Method of flanging	W. B. Scaife	Pittsburgh, Pa	Jan. 31, 1871	111, 478
Boiler-heads, Securing removable	J. Booth and W. Heiser	Buffalo, N. Y	June 25, 1872	128, 207
Boiler-heater	P. Colvin	Pecatonica, Ill	Oct. 8, 1872	132, 056
Boiler heater and feeder	G. C. Williams	Catskill, N. Y	Nov. 25, 1873	144, 937
Boiler heater and filter	J. J. Doughty	Lake City, Minn	Sept. 29, 1868	82, 609
Boiler heater, Steam	D. Coughlin	Fall River, Mass	Dec. 2, 1873	145, 095
Boiler or steam generator	L. M. Richards	Cazenovia, N. Y	Mar. 30, 1815	......
Boiler or steam generator	L. M. Richards	Chagrin River, Ohio	Aug. 21, 1828	......
Boiler or stew-pan	W. B. Watkins	Jersey City, N. J	Sept. 7, 1869	94, 679
Boiler-plates, Machine for punching	R. Griffin	Syracuse, N. Y	Dec. 24, 1861	33, 989
Boiler-plates, Method of riveting	J. B. Henry	New York, N. Y	Jan. 17, 1860	26, 846
Boiler-plates, Mode of preparing and screwing in	J. Goulding	Boston, Mass	June 19, 1835	......
Boiler-rivet machine	C. B. Allen	Philadelphia, Pa	July 18, 1871	117, 137
Boiler-riveting machine	S. Bennett	New Orleans, La	June 30, 1857	17, 661
Boiler safety-gage	R. H. Jackson	Sandusky, Ohio	July 9, 1867	66, 593
Boiler safety-gage	J. Marshall	Greenwich, England	Dec. 22, 1868	85, 179
Boiler-scraper	E. L. Pratt	Beverly, Mass	May 11, 1869	89, 941
Boiler-stand	J. L. Mott	Mott Haven, N. Y	Jan. 25, 1870	99, 222
Boiler stand and heater	C. F. Hitchings	New York, N. Y	Apr. 1, 1873	137, 448
Boiler-supply chamber	J. L. Sullivan	Boston, Mass	Feb. 28, 1818	......
Boiler-thimble	G. W. Duvall	Norfolk, Va	May 15, 1866	54, 703
Boiler-tube	B. Fitts	Newark, N. J	May 1, 1866	54, 318
Boiler-tube cleaner	J. B. Christoffel	New York, N. Y	Mar. 12, 1867	62, 816
Boiler-tube cleaner	P. H. Coyle	Newark, N. J	Nov. 9, 1869	96, 552
Boiler-tube cleaner	P. H. Coyle	Newark, N. J	Apr. 26, 1870	102, 374
Boiler-tube cleaner	J. Green	Boston, Mass	July 12, 1870	105, 194
Boiler-tube cleaner	W. P. Heffron	Chicago, Ill	July 23, 1867	67, 112
Boiler tube cleaner	J. E. Regan	Chicago, Ill	Nov. 9, 1869	96, 617
Boiler-tube cleaner	H. Waterman	Brooklyn, N. Y	Mar. 13, 1866	53, 203
Boiler-tube cleaner, steam-jet	R. C. Hussey	Milford, Mass	Dec. 13, 1870	110, 044
Boiler-tube cutter	W. B. Smith	Charleston, S. C	Dec. 31, 1867	72, 927
Boiler-tube-cutting tool	F. Ramsey and J. Miller	New York, N. Y	Oct. 16, 1866	58, 886
Boiler-tube ferrule	J. U. Adams	Richfield, Mich	Feb. 4, 1868	74, 028
Boiler-tube ferrule	R. A. Copeland	Brooklyn, N. Y	Apr. 6, 1869	88, 611
Boiler-tube plug	I. La Blanc	Philadelphia, Pa	Feb. 7, 1871	111, 552
Boiler-tube plug	R. L. Neill	Paterson, N. J	June 28, 1870	104, 759
Boiler-tube-scaling tool	J. Werner, jr	Prairie du Lac, Wis	Aug. 8, 1865	49, 327
Boiler-tube scraper	J. B. Christoffel	Williamsburgh, N. Y	Apr. 15, 1873	137, 826
Boiler-tube-scraping device	E. Green	Wakefield, England	Aug. 17, 1869	93, 874
Boiler-tubes, Apparatus for expanding and fastening.	R. Dudgeon	New York, N. Y	Feb. 5, 1867	61, 815
Boiler-tubes, Cleaning	J. J. Illingworth	Brooklyn, N. Y	July 25, 1865	48, 948
Boiler-tubes, Cleaning	D. McDowell	Kingston, West Indies	Mar. 6, 1866	53, 091
Boiler-tubes, Device for leaky	R. A. Copeland	Baltimore, Md	Jan. 21, 1868	73, 439
Boiler-tubes, Device for repairing	E. Clark	New York, N. Y	Sept. 1, 1863	39, 717
Boiler-tubes, Device for stopping leaks in	R. Lavery and S. Stuart	South Boston, Mass	Dec. 27, 1864	45, 615
Boiler-tubes, Drawing	S. J. Perry	Columbia, S. C	Aug. 21, 1860	29, 715
Boiler-tubes, Ferrule for	G. W. Durall	Norfolk, Va	Sept. 19, 1865	49, 987
Boiler-tubes, Ferrule for stopping leaks in	P. Quinn	South New Market, N. H	June 11, 1867	65, 603
Boiler-tubes in sea-going vessels, Mode of preventing corrosion of.	G. Hauxhurst	Somerville, Cal	Dec. 1, 1868	84, 489
Boiler tubes in tube-sheets, Securing steam	W. Moorhouse	Philadelphia, Pa	Apr. 2, 1872	125, 321
Boiler-tubes, Method of fastening	J. Bowden	New York, N. Y	Jan. 1, 1867	60, 853
Boiler-tubes, Plug for leaky	J. M. Spiegle	Philadelphia, Pa	Apr. 16, 1872	125, 764
Boiler-tubes Ring for stopping leaks in	R. Lavery and S. Stuart	South Boston, Mass	Dec. 27, 1864	45, 61[illegible]
Boiler-tubes, Spring-expanding swage for	J. McCarty	Reading, Pa	July 29, 1851	8, 256
Boiler-tubes, Steam-jet for cleaning	J. M. Wheeler	Oxford, Conn	Dec. 17, 1867	72, 348

Index of patents issued from the United States Patent Office from 1790 *to* 1873, *inclusive*—Continued.

Invention.	Inventor.	Residence.	Date.	No.
Boiler-tubes, Swaging-tool for	I. S. Hamilton	Hamilton, Ohio	Dec. 5, 1871	121, 512
Boiler-tubes, Tool for attaching	T. Prosser	New York, N. Y	Apr. 17, 1849	6, 360
Boiler-tubes, Tool for cleaning	S. Van Auken	Binghamton, N. Y	Dec. 18, 1866	60, 653
Boiler-tubes, Tool for cutting off	R. H. Burke	Greenpoint, N. Y	Jan. 22, 1867	61, 392
Boiler-tubes, Tool for cutting off	D. A. Dacey	New York, N. Y	June 6, 1865	48, 054
Boiler-tubes, Tool for cutting off	P. Hoffman	Jersey City, N. J	Mar. 26, 1867	63, 250
Boiler-tubes, Tool for cutting off	R. Lavery	South Boston, Mass	Mar. 12, 1867	62, 859
Boiler-tubes, Tool for cutting off	D. M. Nichols	New York, N. Y	Mar. 20, 1866	53, 325
Boiler-tubes, Tool for cutting off	W. P. Slensby	Chicago, Ill	May 7, 1867	64, 454
Boiler-tubes, Tool for cutting off	N. Thomas	Chicago, Ill	June 5, 1866	55, 391
Boiler-tubes, Tool for cutting off	N. Wright	Jersey City, N. J	Jan. 22, 1867	61, 499
Boiler-tubes, Tool for expanding	R. McKenzie	New Orleans, La	Oct. 26, 1869	96, 133
Boiler-tubes, Tool for expanding and cutting off	I. S. Hamilton	Hamilton, Ohio	Dec. 5, 1871	121, 513
Boiler-tubes, Tool for fastening	A. Van Guysling	North Greenbush, N. Y	Dec. 15, 1863	40, 967
Boiler-tubes, Tool for plugging leaky	J. Ritchie	Detroit, Mich	Sept. 12, 1871	118, 818
Boiler-tubes, Tool for removing scale in	P. C. Rowe	Boston, Mass	May 22, 1866	54, 962
Boiler-tubes, Tool for scaling	J. Daley and J. H. Marvill	Philadelphia, Pa	Feb. 14, 1865	46, 339
Boiler, Tubular	C. Hawthorn	Pittsburgh, Pa	Dec. 20, 1864	45, 494
Boilers and other vessels, Mode of attaching handles to.	J. H. Brown	New York, N. Y	Dec. 11, 1866	60, 337
Boilers, Apparatus for packing straw into	H. B. Meech	Fort Edward, N. Y	Sept. 11, 1866	57, 947
Boilers, Arrangement of steam-cylinders in	J. S. Shapter	New York, N. Y	Aug. 19, 1856	15, 579
Boilers, Blow-off for	E. F. Husted	Harrisburgh, Tex	Sept. 10, 1872	131, 219
Boilers, Composition for preventing and removing incrustation from.	A. Temple	Bridgeport, Conn	Aug. 22, 1865	49, 569
Boilers, Composition for preventing incrustation in	G. R. Spannagel	Saint Louis, Mo	Sept. 12, 1865	49, 931
Boilers, Composition for removing incrustation from.	J. G. Gansz and J. J. Lavo	Saint Louis, Mo	Oct. 3, 1865	50, 237
Boilers, Composition for removing incrustation from.	N. S. Thomas	Painted Post, N. Y	Aug. 14, 1866	57, 219
Boilers, Composition for removing scale from	J. Buzby	Philadelphia, Pa	Apr. 11, 1865	47, 188
Boilers, Compound for removing scale from	A. B. Auer	Babcock's Grove, Ill	May 30, 1865	47, 914
Boilers, Device for forming	G. S. Pierce	Wilkesbarre, Pa	Dec. 7, 1869	97, 549
Boilers, Device for preventing incrustations in	T. Barfield	Athens, Ill	May 4, 1869	89, 725
Boilers, Device for removing incrustations of	W. E. Everett and M. M. Thompson	New York, N. Y	Feb. 12, 1856	14, 230
Boilers, Electrical protection for	A. T. Hay	Burlington, Iowa	June 24, 1873	140, 196
Boilers, Feeding	Hunsicker & Krauss	Northampton, Pa	Aug. 20, 1835	
Boilers, Fusible plug for	J. Smith	Wentworth Road, Great Britain.	Mar. 28, 1865	47, 076
Boilers, Hand-hole joint for	G. White	Harlem, N. Y	Jan. 21, 1868	73, 557
Boilers, ice-houses, &c., to impede the passage of heat, Non-conducting composition for covering.	J. A. Jones	Middlesborough-on-Tees, Great Britain.	Mar. 26, 1867	63, 255
Boilers, Machine for forming	E. S. Sackett	Monroe, Wis	May 7, 1867	64, 577
Boilers, Man-hole lid for	R. Tippett	Harrisburgh, Pa	Apr. 15, 1873	137, 806
Boilers, Method of applying fusible metal to	E. H. Ashcroft	Boston, Mass	Mar. 19, 1850	7, 179
Boilers, Method of bracing the water-spaces of	B. O'Neill	Reading, Pa	Mar. 4, 1851	7, 962
Boilers, Method of removing incrustation from	D. Embree	Dayton, Ohio	Jan. 17, 1865	45, 914
Boilers, &c., Mode of riveting metallic plates for	R. Smith	Great Britain	Sept. 3, 1839	1, 312
Boilers, Mode of setting	E. Jenks	Portland, Me	June 1, 1805	
Boilers, Safety-plug for	T. G. Eiswald	Providence, R. I	Jan. 14, 1868	73, 311
Boilers, Setting	C. Neames	New Orleans, La	Nov. 8, 1870	109, 039
Boilers, Smoke-extinguisher for	W. H. Nobles	Saint Paul, Minn	Dec. 31, 1867	72, 886
Boilers, steam-cylinders, fittings, &c., Manufacture of.	G. W. Bollman	Pittsburgh, Pa	Oct. 24, 1871	120, 235
	F. Y. Arnold	Philadelphia, Pa	Aug. 9, 1870	106, 105
Boilers, steam-pipes, &c., Non-conducting casing for.	W. Harris	Philadelphia, Pa	Jan. 18, 1870	98, 865
Boilers, steam-pipes, &c., Non-conducting covering for.	R. McConnel	Jacksonville, Ill	Aug. 8, 1865	49, 287
Boilers, Tool for fastening tubes in				
Boilers, Water-feed for	H. C. Bristol	East China, Mich	July 1, 1873	140, 462
Boilers with water, Mode of supplying	J. Fox	Lowell, Mass	Mar. 14, 1832	
Boilers with water, Self-acting apparatus for supplying.	V. F Clenet	Paris, France	Aug. 4, 1863	39, 379
Boiling	C. Shepard	Cincinnatus, N. Y	Oct. 4, 1817	
Boiling and condensing apparatus	G. S. G. Spence	Boston, Mass	July 31, 1860	29, 414
Boiling and distilling apparatus	W. Hoffmire	San Francisco, Cal	July 31, 1860	29, 376
Boiling and puddling furnace	J. I. Williams	Etna, Pa	May 12, 1868	77, 790
Boiling and puddling furnace	J. Zimmer	Pittsburgh, Pa	May 19, 1868	78, 172
Boiling and puddling furnaces, Chill-bosh for	T. Coates	Ironton, Ohio	June 7, 1870	103, 983
Boiling and puddling furnaces, Door and bit for	W. Sims	Pittsburgh, Pa	June 11, 1867	65, 773
Boiling and puddling furnaces, Fix for	J. D. Williams	Allegheny, Pa	Jan. 4, 1870	98, 534
Boiling-apparatus	J. McCormick	Madison, Ind	June 25, 1861	32, 636
Boiling liquids	R. M. Beach	Franklin, N. Y	Apr. 7, 1831	
Bolster, Adjustable	G. Couch	Saint Louis, Mo	Oct. 31, 1871	120, 498
Bolster and pillow	T. S. Sperry	Chicago, Ill	Jan. 24, 1871	111, 270
Bolster for mattress, Adjustable	P. Kraher	Cincinnati, Ohio	Aug. 6, 1867	67, 436
Bolster, pillow, &c.	C. P. Cooper	New York, N. Y	Oct. 7, 1873	143, 502
Bolster-plate	T. Morgan	Marquette, Mich	June 21, 1870	104, 628
Bolster-plates, Manufacture of	W. J. Lewis	Pittsburgh, Pa	July 29, 1873	141, 227
Bolster-plate, Manufacturing	W. J. Lewis and H. W. Oliver, jr	Pittsburgh, Pa	June 23, 1868	79, 237
Bolster, Spindle	I. P. Richards	Whitinsville, Mass	July 13, 1869	92, 647

Bolt:

See Barn-door bolt.
Barrel-bolt.
Car-bolt.
Carriage-bolt.
Catch-bolt.
Clover-bolt.
Copper bolt.
Cupboard-bolt.
Door-bolt.
Door and gate bolt.
Draw-bolt.
Expansion-bolt.
Eye-bolt.
Flour-bolt.

Index of patents issued from the United States Patent Office from 1790 to 1873, inclusive—Continued.

Invention.	Inventor.	Residence.	Date.	No.
Bolt—Continued.				
See Flush-bolt.				
Guard-bolt.				
Insulating-bolt.				
Interlocked conical bolt.				
Key-bolt.				
King-bolt.				
Knob-bolt.				
Lathe-bilt.				
Lock-bolt.				
Locking-bolt.				
Middlings-bolt.				
Mill-bolt.				
Prison-door bolt.				
Rack-bolt.				
Rail-bolt.				
Safe-bolt.				
Safe-door bolt.				
Sash-bolt.				
Saw-frame bolt.				
Screw-bolt.				
Seal-bolt.				
Seal-locking bolt.				
Self-heading bolt.				
Self-locking bolt.				
Shingle-bolt.				
Ship-bolt.				
Shutter-bolt.				
Shutter-lock bolt.				
Slide-bolt.				
Spring-bolt.				
Spring-catch bolt.				
Square-necked bolt.				
Strap-bolt.				
T-bolt.				
Transient bolt.				
Tray bolt.				
Trunk-tray bolt.				
Wagon-box strap-bolt.				
Bolt	J. F. Carlin	Alexandria, Va	June 10, 1873	139, 767
Bolt	W. J. Clark	Southington, Conn	Oct. 20, 1863	40, 327
Bolt	W. C. Coles	Williamsburgh, N. Y	Apr. 4, 1871	113, 264
Bolt	T. A. Davies	New York, N. Y	Dec. 24, 1872	134, 262
Bolt	M. Harbster	Reading, Pa	July 26, 1870	105, 675
Bolt	G. Havell	Newark, N. J	Feb. 13, 1866	52, 570
Bolt	C. H. Hopkins	Lyndonville, Vt	Apr. 13, 1869	88, 959
Bolt	O. D. Hunter	Terrysville, Conn	Aug. 6, 1867	67, 549
Bolt	E. M. Judd	New Haven, Conn	Feb. 4, 1873	135, 560
Bolt	M. D. Kinkade	Brooklyn, N. Y	Mar. 12, 1872	124, 442
Bolt	P. Lecloux	Dijon, France	May 7, 1872	126, 404
Bolt	B. F. Lotridge	New York, N. Y	July 28, 1868	80, 295
Bolt	F. G. McClelland	Attica, Ohio	Nov. 29, 1870	109, 640
Bolt	J. H. Meissner	Jersey City, N. J	July 29, 1862	36, 014
Bolt	W. R. Miller	Baltimore, Md	Dec. 31, 1872	134, 391
Bolt	E. L. Roberts	Brooklyn, N. Y	July 27, 1869	93, 006
Bolt and bolt-head	J. Crompton	Little Falls, N. J	Apr. 9, 1867	63, 706
Bolt and latch	H. Baker	Lancaster, Pa	Jan. 21, 1868	73, 491
Bolt and nut machine, Combined	J. N. Smith	New Haven, Conn	Jan. 2, 1872	122, 411
Bolt and nut threading machine	F. S. Allen and C. F. Ritchel	New York, N. Y	Dec. 12, 1871	121, 743
Bolt and nut threading machine	J. Killefer	West Richfield, Ohio	June 15, 1869	91, 348
Bolt and rivet cutter	D. D. Cone and P. Polder	Panama, N. Y	July 1, 1873	140, 405
Bolt and rivet cutter	J. S. Henry and A. H. Reist	Manheim, Pa	Feb. 4, 1868	74, 087
Bolt and rivet cutter	C. B. Shaw	Victory, Wis	July 16, 1872	129, 370
Bolt and rivet heading machine	J. Van Brocklin	Middleport, N. Y	Aug. 20, 1850	7, 586
Bolt and rivet machine	N. Beauregard	San Francisco, Cal	Aug. 26, 1873	142, 193
Bolt and rivet machine	J. Morgan, jr	Wheeling, W. Va	Apr. 16, 1867	63, 927
Bolt and rivet machine	J. Morgan, jr	Wheeling, W. Va	May 21, 1867	64, 896
Bolt and rivet machine	J. Morgan, jr	Wheeling, W. Va	Apr. 26, 1870	102, 418
Bolt and rivet machine	J. C. Reed	Cincinnati, Ohio	Mar. 5, 1867	62, 565
Bolt and rivet machine	A. Reese	Pittsburgh, Pa	May 21, 1861	32, 381
Bolt and rivet machine	J. Wakefield	Birmingham, England	Dec. 10, 1867	72, 135
Bolt and rivet machine	W. E. Ward	Port Chester, N. Y	July 30, 1850	7, 538
Bolt and rivet machine	G. Worstenholm	Newark, N. J	Apr. 14, 1868	76, 873
Bolt and rivet making machine	J. R. Bassett	Cincinnati, Ohio	June 30, 1857	17, 660
Bolt and rivet making machine	J. Griffiths	Wheeling, W. Va	Feb. 6, 1872	123, 341
Bolt and rivet making machine	J. Paton, T. Campbell, and R. Paton.	Newburgh, Ohio	Nov. 29, 1864	45, 269
Bolt and rivet trimmer	M. D. Budd	Roscoe, Ill	June 4, 1867	65, 338
Bolt and rivet trimmer	M. Theisen	Waukon, Iowa	June 21, 1870	104, 509
Bolt and rod cutter	S. H. Preston	Worth, Pa	Apr. 15, 1873	137, 956
Bolt and rod cutter	H. Schmidt	Chicago, Ill	Sept. 2, 1873	142, 354
Bolt and rod cutter	L. H. Smith	Stryker, Ohio	Oct. 14, 1873	143, 645
Bolt and screw threading machine	R. Boeklen	New York, N. Y	June 17, 1873	139, 861
Bolt and spike drawing machine	J. C. Chapman	Charlestown, Mass	Jan. 6, 1863	37, 332
Bolt and spike drawing machine	H. E. Towle	Exeter, N. H	Sept. 3, 1861	33, 218
Bolt and spike heading machine	R. Boeklen	Brooklyn, N. Y	Mar. 7, 1871	112, 410
Bolt and spike making machine	J. S. Hall	Pittsburgh, Pa	July 28, 1868	80, 478
Bolt-blank	J. B Clark	Plantsville, Conn	Nov. 2, 1869	96, 308
Bolt-blank-heating furnace	A. Alexander	Pittsburgh, Pa	Mar. 3, 1868	74, 970
Bolt-blanks, Apparatus for heating	J. B. Clark	Plantsville, Conn	June 20, 1871	116, 158
Bolt-blanks, Automatic feeder for	W. F. Parker	Meriden, Conn	June 4, 1867	65, 423
Bolt-blanks, Machine for upsetting	J. B. Clark	Plantsville, Conn	Jan. 16, 1872	122, 706
Bolt-clamp	C. E. Phillips	Deerfield, Mass	Dec. 21, 1869	98, 105
Bolt-cutter	A. S. Bailey	Knoxville, Ill	June 25, 1867	65, 989
Bolt-cutter	W. Britton	Abingdon, Ill	May 28, 1867	65, 161
Bolt-cutter	J. R. Brown	Cambridgeport, Mass	Sept. 20, 1870	107, 438
Bolt-cutter	J. R. Brown	Cambridgeport, Mass	July 4, 1871	116, 547
Bolt-cutter	J. R. Brown	Cambridgeport, Mass	Oct. 24, 1871	120, 149

Index of patents issued from the United States Patent Office from 1790 *to* 1873, *inclusive*—Continued.

Invention.	Inventor.	Residence.	Date.	No.
Bolt-cutter	J. R. Brown	Cambridgeport, Mass	Sept. 17, 1872	131, 423
Bolt-cutter	O. E. Butler and S. P. Dunham	Marshalltown, Iowa	Dec. 14, 1869	97, 874
Bolt-cutter	A. Carbnow	Potsdam, N. Y	May 19, 1868	78, 055
Bolt-cutter	L. J. Chapman	Vallejo, Cal	Feb. 18, 1873	135, 887
Bolt-cutter	C. Dreher	Detroit, Mich	Dec. 5, 1865	51, 300
Bolt-cutter	E. W. and R. Fawcett and W. E. Sefton.	Salem, Ohio	July 29, 1873	141, 265
Bolt-cutter	H. M. Handy	Niles, Mich	Apr. 9, 1867	63, 721
Bolt-cutter	W. W. Hill	Fulton, N. Y	Mar. 25, 1873	137, 076
Bolt-cutter	H. L. Howard	Mendon, Mich	Apr. 19, 1870	102, 006
Bolt-cutter	H. Howe	Oneonta, N. Y	July 28, 1868	80, 351
Bolt-cutter	J. Johnson	West Fallowfield Township, Pa.	June 4, 1867	65, 391
Bolt-cutter	J. Johnson	West Fallowfield Township, Pa.	Mar. 31, 1868	76, 197
Bolt-cutter	J. Johnson	Cochransville, Pa	Sept. 26, 1871	119, 231
Bolt-cutter	W. Mendham	Philadelphia, Pa	Apr. 19, 1870	102, 026
Bolt-cutter	F. Miller	Indianola, Ill	May 5, 1868	77, 637
Bolt-cutter	T. W. Moore	Richmond, Ind	Feb. 18, 1868	74, 575
Bolt-cutter	T. W. Moore	Richmond, Ind	Oct. 27, 1868	83, 525
Bolt-cutter	L. J. Newlan	Barton, N. Y	July 23, 1867	66, 986
Bolt-cutter	A. W. Owen and M. E. Lilley	East Canton, Pa	June 27, 1871	116, 478
Bolt-cutter	G. A. Ries	Belvidere, N. J	Oct. 17, 1865	50, 497
Bolt-cutter	J. S. Ryder	Sing Sing, N. Y	May 13, 1873	138, 940
Bolt-cutter	E. Schlenker	Buffalo, N. Y	July 4, 1865	48, 594
Bolt-cutter	R. Sischo	Tuscola, Ill	Oct. 10, 1865	50, 395
Bolt-cutter	E. A. Sloat	Theresa, N. Y	Jan. 7, 1868	73, 205
Bolt-cutter	O. J. Smith	Wauwatosa, Wis	Nov. 13, 1866	59, 673
Bolt-cutter	J. Strayer and T. Hazelhurst	South Bend, Ind	June 25, 1867	66, 107
Bolt-cutter	W. F. Strong	Charleston, S. C	Aug. 27, 1872	130, 880
Bolt-cutter	J. K. Taylor	Bridgeport, Conn	Dec. 11, 1860	30, 895
Bolt-cutter	J. Tomlinson	Onslow, Iowa	Feb. 4, 1873	135, 497
Bolt-cutter	J. Whittaker	Sandusky, N. Y	July 25, 1871	117, 495
Bolt-cutter	W. W. Worden	Waukesha, Wis	Dec. 31, 1867	72, 954
Bolt-cutter	W. W. Wright and J. Boody	Ellsworth, N. Y	June 4, 1867	65, 523
Bolt-cutter	W. E. Yeager	Lawrence, Kans	May 2, 1871	114, 383
Bolt cutting and shearing, Compound tool for	T. Smith	California, Mo	Sept. 1, 1868	81, 697
Bolt-cutting device	G. Adair and J. F. Pool	Monroe, Wis	Apr. 27, 1869	89, 267
Bolt-cutting device	L. Daniels	Oak Hill, N. Y	May 17, 1870	103, 152
Bolt-cutting machine	M. D. Budd	Roscoe, Ill	Dec. 24, 1861	33, 982
Bolt-cutting machine	J. A. Merriman	Chicago, Ill	Oct. 18, 1864	44, 773
Bolt-cutting shears	J. Guthrie	Wilmington, Del	Sept. 4, 1866	57, 705
Bolt-cutting shears	D. H. Hickok	Rockford, Ill	Feb. 20, 1866	52, 714
Bolt-cutting shears	G. W. Hyatt	Auburn, N. Y	May 23, 1865	47, 832
Bolt-cutting shears	S. W. Wright	Ellsworth, N. Y	Feb. 26, 1867	62, 462
Bolt cutting tool	J. E. Heath	Niles, Mich	Nov. 5, 1867	70, 563
Bolt die, Square-head	W. Shields	Philadelphia, Pa	Feb. 18, 1868	74, 616
Bolt die, Square-head	W. Shields	Philadelphia, Pa	Aug. 25, 1868	81, 419
Bolt die, T-head	W. J. Lewis	Pittsburgh, Pa	May 1, 1866	54, 377
Bolt-drawing machine	C. L. Stevenson	Charlestown, Mass	Oct. 26, 1858	21, 910
Bolt-drawing machine	S. Wilmarth	Charlestown, Mass	Nov. 29, 1859	26, 315
Bolt-fastening	J. M. Hopkins	New York, N. Y	May 1, 1866	54, 348
Bolt-fastening	V. Lapham	El Paso, Ill	Nov. 12, 1867	70, 866
Bolt-fastening	P. Peterson	Abingdon, Ill	Sept. 6, 1870	107, 096
Bolt-fastening	E. E. Stubbs	West Elkton, Ohio	Mar. 10, 1868	75, 488
Bolt-feeder	J. Cornwell	Kalamazoo, Mich	Mar. 23, 1869	88, 137
Bolt-feeder	O. Van Tassell	Naperville, Ill	Sept. 28, 1869	95, 289
Bolt feeder and cooler	W. Pickens	Chicago, Ill	Feb. 4, 1868	74, 125
Bolt-feeder for mills	B. A. Wing	Galesburgh, Mich	Feb. 22, 1870	100, 094
Bolt-forging machine, Die for	J. T. Willmarth	Northbridge, Mass	Apr. 17, 1855	12, 735
Bolt-head	F. S. Miles	Philadelphia, Pa	Oct. 13, 1868	82, 973
Bolt-head die, Square	J. Gribben	Allegheny, Pa	July 31, 1866	56, 745
Bolt-head-swaging die	A. Eames	Bridgeport, Conn	Aug. 9, 1859	25, 001
Bolt-head-trimming machine	M. Rugg	Southington, Conn	Aug. 31, 1842	2, 766
Bolt-head-trimming tool	A. P. Plant	Plantsville, Conn	June 8, 1869	91, 040
Bolt-head-turning machine	W. F. Swathel	Mount Carmel, Conn	Sept. 20, 1870	107, 561
Bolt-heading device	W. M. Thoroughgood	Salisbury, Md	May 13, 1873	138, 958
Bolt-heading device	J. M. Woods	Washington, Mo	Mar. 9, 1869	87, 747
Bolt-heading die	F. Bruso	Buffalo, N. Y	Sept. 6, 1870	107, 160
Bolt-heading die	J. B. Clark	Plantsville, Conn	Jan. 30, 1872	123, 084
Bolt-heading die	P. Eley	New York, N. Y	Apr. 4, 1871	113, 280
Bolt-heading die	C. Kane	Pittsburgh, Pa	May 15, 1866	54, 738
Bolt-heading die	C. Kane	Pittsburgh, Pa	Jan. 28, 1868	73, 728
Bolt-heading die	J. Kaylor	Pittsburgh, Pa	Mar. 16, 1869	87, 780
Bolt-heading die	W. J. Lewis	Pittsburgh, Pa	Mar. 31, 1868	76, 210
Bolt-heading machine	J. R. Abbe	Providence, R. I	Jan. 19, 1869	85, 983
Bolt-heading machine	J. R. Abbe	Providence, R. I	June 6, 1871	115, 555
Bolt-heading machine	H. E. Anthony	Providence, R. I	June 19, 1866	55, 597
Bolt-heading machine	O. C. Burdict	New Haven, Conn	Sept. 3, 1867	68, 555
Bolt-heading machine	O. C. Burdict	Providence, R. I	Nov. 30, 1869	97, 351
Bolt-heading machine	G. Chapman	Rockford, Ill	June 6, 1871	115, 702
Bolt-heading machine	H. M. Clark	New Britain, Conn	Jan. 15, 1856	14, 086
Bolt-heading machine	L. L. Davis	Laconia, N. H	Feb. 6, 1866	52, 393
Bolt-heading machine	P. Eley	New York, N. Y	Aug. 1, 1871	117, 527
Bolt-heading machine	A. B. Glover	Derby, Conn	June 19, 1866	55, 646
Bolt-heading machine	A. B. Glover	Birmingham, Conn	Mar. 17, 1868	75, 675
Bolt-heading machine	S. W. Goodyear and W. F. Parker.	Meriden, Conn	June 11, 1867	65, 561
Bolt-heading machine	R. Gracey	Pittsburgh, Pa	Dec. 17, 1867	72, 288
Bolt-heading machine	R. Gracey	Pittsburgh, Pa	Oct. 27, 1868	83, 376
Bolt-heading machine	W. Grant	Boston, Mass	Dec. 26, 1848	5, 986
Bolt-heading machine	J. Greenwood	Fitchburgh, Mass	Feb. 20, 1872	123, 822
Bolt-heading machine	C. Hall	New York, N. Y	Aug. 12, 1873	141, 711
Bolt-heading machine	C. Hall and E. Hubner	New York, N. Y	Sept. 18, 1866	58, 096
Bolt-heading machine	T. Hull and N. Thomas	Chicago, Ill	Feb. 28, 1871	112, 248
Bolt-heading machine	C. E. Hunter	Hinsdale, N. H	Jan. 2, 1872	122, 385
Bolt-heading machine	J. O. Jones	Brooklyn, N. Y	July 22, 1873	141, 059
Bolt-heading machine	C. and J. Kane	Pittsburgh, Pa	Apr. 10, 1866	53, 831

Index of patents issued from the United States Patent Office from 1790 *to* 1873, *inclusive*—Continued.

Invention.	Inventor.	Residence.	Date.	No.
Bolt-heading machine	E. Kaylor	Pittsburgh, Pa	Oct. 24, 1865	50,599
Bolt-heading machine	E. Kaylor	Pittsburgh, Pa	June 12, 1866	55,501
Bolt-heading machine	L. Kirkup	Brooklyn, N. Y	Feb. 6, 1866	52,494
Bolt-heading machine	W. J. Lewis	Pittsburgh, Pa	Mar. 15, 1864	41,929
Bolt-heading machine	M. D. Marcy	Worcester, Mass	Feb. 20, 1866	52,727
Bolt-heading machine	J. W. McDermott	New York, N. Y	Feb. 6, 1866	52,433
Bolt-heading machine	W. Melvell	Paterson, N. J	July 21, 1868	80,198
Bolt-heading machine	J. Minter	Worcester, Mass	July 12, 1864	43,521
Bolt-heading machine	J. Minter	Lowell, Mass	July 10, 1866	56,247
Bolt-heading machine	D. G. Morrill	Catasauqua, Pa	Aug. 24, 1869	94,021
Bolt-heading machine	F. B. Prindle	Southington, Conn	Dec. 31, 1867	72,899
Bolt heading machine	F. B. Prindle	Southington, Conn	Jan. 11, 1870	98,798
Bolt-heading machine	T. T. Prosser	Chicago, Ill	Apr. 13, 1869	88,981
Bolt-heading machine	J. Root	New Haven, Conn	Oct. 29, 1867	70,363
Bolt-heading machine	J. Root	New Haven, Conn	Mar. 3, 1868	75,198
Bolt-heading machine	J. Root	New Haven, Conn	Mar. 23, 1869	88,214
Bolt-heading machine	J. Root	New Haven, Conn	Mar. 15, 1870	100,928
Bolt-heading machine	F. Schweizer	New York, N. Y	Nov. 7, 1865	50,848
Bolt-heading machine	F. Schweizer	New York, N. Y	Mar. 20, 1866	53,351
Bolt-heading machine	F. Schweizer	Green Point, N. Y	June 8, 1869	91,171
Bolt-heading machine	N. Starks	Albany, N. Y	Dec. 23, 1851	8,616
Bolt-heading machine	W. Swathel	Southington, Conn	Nov. 19, 1867	71,243
Bolt-heading machine	H. Thompson	Worcester, Mass	Mar. 3, 1868	75,221
Bolt-heading machine	P. P. Trayser	Baltimore, Md	Feb. 5, 1867	61,896
Bolt-heading machine	A. Tucker	Gilford, N. H	Feb. 27, 1866	52,912
Bolt-heading machine	B. C. Vanduzen	Cincinnati, Ohio	Mar. 22, 1859	23,328
Bolt-heading machine	B. Walker	Birmingham, Conn	Oct. 23, 1866	59,100
Bolt-heading machine	F. Watkins	Birmingham, England	Feb. 14, 1865	46,434
Bolt-heading machine	C. D. Wiley and M. S. Norton	Junction, Minn	June 18, 1872	128,083
Bolt-heading-machine die	J. Gribben	Allegheny, Pa	Aug. 21, 1866	57,313
Bolt-heading-machine die	J. W. Tibbet	Cincinnati, Ohio	Sept. 4, 1866	57,781
Bolt heads, Machine for making	C. H. and J. F. Emerson	New York, N. Y	Apr. 20, 1869	89,211
Bolt-heads, Machine for making	M. D. Marcy	Worcester, Mass	Aug. 24, 1869	94,122
Bolt-holder	S. R. Butterfield	Rome, N. Y	June 14, 1870	104,264
Bolt-holder	B. Perry and A. Cornish	Lee, N. Y	May 21, 1867	64,904
Bolt-holder	G. W. Phelps	Conneaut, Ohio	Aug. 11, 1868	80,998
Bolt-keeper	J. M. Mitchell	Cincinnati, Ohio	May 16, 1871	114,841
Bolt-keeper	G. M. Wood	Decatur, Ill	Oct. 16, 1866	58,929
Bolt-keys, Machine for making	N. Adams	Altoona, Pa	Oct. 15, 1867	69,740
Bolt-locking device	A. McKenny	Maumee City, Ohio	Aug. 1, 1871	117,657
Bolt-machine	J. R. Abbe	Providence, R. I	Mar. 8, 1870	100,581
Bolt-machine	A. Alexander	Pittsburgh, Pa	Aug. 15, 1865	49,484
Bolt-machine	A. R. Bailey and W. W. Knowles.	New Haven and Plantsville, Conn.	Nov. 19, 1867	70,939
Bolt-machine	H. Carter	Pittsburgh, Pa	Mar. 2, 1858	19,485
Bolt-machine	D. S. Coe	Pine Meadow, Conn	June 27, 1871	116,270
Bolt-machine	J. L. Hall	Abington, Mass	June 1, 1869	90,661
Bolt-machine	M. Hardaway	Saint Louis, Mo	Apr. 13, 1869	88,783
Bolt-machine	W. W. Hubbard	Philadelphia, Pa	Nov. 19, 1867	71,012
Bolt-machine	J. O. Jones	Brooklyn, N. Y	Apr. 8, 1873	137,688
Bolt-machine	C. and J. Kane	Pittsburgh, Pa	Apr. 10, 1866	53,832
Bolt-machine	E. Kaylor	Pittsburgh, Pa	Aug. 22, 1865	49,532
Bolt-machine	A. Marcellus	Rockford, Ill	Aug. 29, 1865	49,638
Bolt-machine	H. M. Phillips	Indianapolis, Ind	Sept. 19, 1865	50,028
Bolt-machine	E. Simkins	Allegheny, Pa	Aug. 24, 1858	21,279
Bolt-machine	T. F. Taft	Fitchburgh, Mass	Feb. 12, 1856	14,258
Bolt-machine	A. A. Wilder	Detroit, Mich	Nov. 19, 1872	133,135
Bolt-machine	S. B. Wilkins and A. Marcellus	Rockford, Ill	June 17, 1873	140,103
Bolt-machine, Method of constructing and operating the header in.	D. L. Weatherhead	Providence, R. I	May 8, 1849	6,438
Bolt-making	D. E. Adams	Allegheny City, Pa	Sept. 18, 1866	58,040
Bolt-making die	J. T. Willmarth	Northbridge, Mass	Jan. 16, 1855	12,262
Bolt making machine	A. Alexander	Pittsburgh, Pa	Dec. 17, 1867	72,152
Bolt-making machine	A. Alexander	Pittsburgh, Pa	Dec. 17, 1867	72,153
Bolt-making machine	A. Charbono	Worcester, Mass	July 9, 1872	128,853
Bolt-making machine	R. H. Cole	Saint Louis, Mo	Nov. 3, 1857	18,534
Bolt-making machine	R. Crichton	Buchanan, Pa	May 8, 1855	12,815
Bolt-making machine	C. Hall and E. Hubner	New York, N. Y	Mar. 27, 1866	53,531
Bolt-making machine	M. Hardaway	Saint Louis, Mo	Oct. 11, 1864	44,623
Bolt-making machine	W. Klemm and R. Dittrich	Pittsburgh, Pa	Feb. 25, 1868	74,834
Bolt-making machine	E. B. Locke	Exeter, N. H	Mar. 30, 1869	88,493
Bolt-making machine	C. Lusted	New York, N. Y	Sept. 12, 1865	49,902
Bolt-making machine	W. E. Ward	Port Chester, N. Y	July 10, 1855	13,241
Bolt-making machine	W. E. Ward	Port Chester, N. Y	May 12, 1863	38,518
Bolt-making machine	F. Watkins	Birmingham, England	Dec. 8, 1868	84,782
Bolt-making-machine die	R. Gracey	Pittsburgh, Pa	Oct. 27, 1868	83,377
Bolt-reel	B. C. White	Des Moines, Iowa	Jan. 31, 1871	111,501
Bolt-screwing and nut-tapping machine	W. W. Hubbard	Philadelphia, Pa	Feb. 6, 1866	52,415
Bolt-screwing machine	A. Babbett	Auburn, N. Y	Dec. 19, 1865	51,538
Bolt-straightening machine	C. E. Hunter	Hinsdale, N. H	Feb. 21, 1871	111,941
Bolt-tapping machine	F. Watkins	Birmingham, England	Dec. 13, 1864	45,452
Bolt-thread-cutting machine	J. J. Grant	Northampton, Mass	Nov. 16, 1869	96,910
Bolt-threader	G. W. Mingus	Pomeroy, Ohio	May 31, 1870	103,642
Bolt-threading die	A. Belaieff	St. Petersburg, Russia	July 13, 1869	92,567
Bolt-threading die	J. E. Weaver	Temperanceville, Pa	June 14, 1870	104,234
Bolt-threading machine	B. D. Beecher	Plantsville, Conn	May 12, 1868	77,710
Bolt-threading machine	W. B. Bement	Philadelphia, Pa	Mar. 10, 1868	75,239
Bolt-threading machine	O. P. and L. W. Briggs	Chicago, Ill	May 16, 1871	115,609
Bolt-threading machine	O. C. Burton	Unionville, Conn	Jan. 16, 1872	122,702
Bolt-threading machine	C. E. DeValin	Baltimore, Md	Nov. 2, 1869	96,403
Bolt-threading machine	G. Emig	Cincinnati, Ohio	Dec. 3, 1872	133,574
Bolt-threading machine	G. W. Frosst	Richmond, Va	May 14, 1872	126,689
Bolt-threading machine	J. J. Grant	Greenfield, Mass	May 14, 1872	126,802
Bolt-threading machine	M. Hamlin	Catawissa, Pa	July 19, 1870	105,568
Bolt-threading machine	M. Hamlin	Catawissa, Pa	July 19, 1870	105,569
Bolt-threading machine	M. Hine	New Haven, Conn	Mar. 29, 1870	101,366
Bolt-threading machine	M. Hine	New Haven, Conn	Sept. 23, 1873	143,154
Bolt-threading machine	W. Johnson	Lambertville, N. J	Aug. 24, 1869	93,998

Index of patents issued from the United States Patent Office from 1790 *to* 1873, *inclusive*—Continued.

Invention.	Inventor.	Residence.	Date.	No.
Bolt-threading machine	M. Love	Corry, Pa	Nov. 22, 1870	109, 433
Bolt-threading machine	H. Martin	Louisville, Ky	Mar. 14, 1871	112, 612
Bolt-threading machine	H. Martin	Louisville, Ky	Mar. 28, 1871	113, 072
Bolt-threading machine	S. W. Putnam, jr	Fitchburgh, Mass	Jan. 24, 1871	111, 248
Bolt-threading machine	J. F. Rodgers	South Bend, Ind	Dec. 5, 1865	51, 355
Bolt-threading machine	C. Schneider	Newark, N. J	Nov. 14, 1871	120, 903
Bolt-threading machine	J. Schuessler and J. Kennedy	La Fayette, Ind	Sept. 15, 1868	82, 252
Bolt-threading machine	F. Schweizer	New York, N. Y	Mar. 5, 1867	62, 693
Bolt-threading machine	W. Sellers	Philadelphia, Pa	Dec. 1, 1857	18, 775
Bolt-threading machine	N. Thomas	Chicago, Ill	Jan. 25, 1870	99, 259
Bolt-threading machine	J. B. Wiggenhorn	Saint Louis, Mo	Nov. 2, 1869	96, 372
Bolt-threading machine	A. Wood	Worcester, Mass	July 7, 1868	79, 714
Bolt-threading machine	S. H. Wright	Lowell, Mass	Oct. 3, 1871	119, 681
Bolt-threading machine	S. H. Wright	Lowell. Mass	Apr. 30, 1872	126, 170
Bolt-threading, Lubricating-device for	W. Armstrong	Kent, Ohio	Oct. 25, 1870	108, 673
Bolt-trimmer	J. Blackinton	Roscoe, Ill	Oct. 22, 1867	69, 962
Bolt-trimmer	W. Dunlop	Fullerton, Canada	Dec. 17, 1872	133, 972
Bolt-trimmer	H. Howe	Oneonta, N. Y	June 18, 1867	65, 913
Bolt-trimmer	G. W. Lewis	Dansville, N. Y	Dec. 8, 1868	84, 701
Bolt-trimmer	A. B. White	Mendon, Mich	Apr. 14, 1868	76, 863
Bolt-trimming machine	A. Ford	Harmonsburgh, Pa	Sept. 10, 1872	131, 159
Bolt-trimming machine	W. Hannah	Middlefield, N. Y	Dec. 23, 1856	16, 301
Bolt-turning tool	L. Burns	Port Chester, Pa	Feb. 26, 1867	62, 390
Bolt-upsetting machine	B. and J. A. Briscoe	Detroit, Mich	Mar. 21, 1871	112, 775
Bolt-washer	D. Cumming, jr	Brooklyn, N. Y	Nov. 5, 1872	132, 807
Bolt-washer	G. G. Hickman	Coatesville, Pa	Aug. 14, 1866	57, 132
Bolt-washer	G. G. Hickman	Coatesville, Pa	Sept. 10, 1867	68, 742
Bolts and nuts, Machine for making	G. R. Postlethwaite	Birmingham, Great Britain.	Jan. 17, 1871	111, 082
Bolts and rivets, Apparatus for clipping	Z. B. Cotant	Greenwich Station, Ohio	May 19, 1863	38, 564
Bolts and rivets, Machine for making	J. Howden	Glasgow, Scotland	May 10, 1864	42, 731
Bolts and rivets, Making	J. R. Bassett	Cincinnati, Ohio	Mar. 1, 1859	23, 138
Bolts and rivets, Manufacture of	W. Livingstone	Brooklyn, N. Y	Mar. 18, 1873	137, 009
Bolts and spikes, Drawing	R. Haynes	Portsmouth, Va	July 2, 1836	....
Bolts by hydraulic pressure, Machine for drawing	S. Wilmarth	Boston, Mass	Jan. 3, 1865	45, 785
Bolts, Cutter-head for pointing the ends of	O. C. Burdict	Buffalo, N. Y	Feb. 4, 1873	135, 520
Bolts, Device for cutting off	T. J. Emery	Charlestown, Mass	May 7, 1872	126, 384
Bolts, Device for rounding the ends of	J. I. Schermerhorn	Schenectady, N. Y	Nov. 11, 1873	144, 569
Bolts, Die for making	W. J. Clark	Southington, Conn	Oct. 18, 1864	44, 706
Bolts, Heading	E. and P. Coleman	Philadelphia, Pa	Sept. 16, 1856	15, 729
Bolts, Heading	J. Weathers	Greensburgh, Ind	June 11, 1861	32, 545
Bolts, Heading	G. Woodward	Brunswick, Me	Mar. 4, 1856	14, 381
Bolts, Machine for cutting off	H. M. Powers	Lancaster, Pa	Jan. 30, 1872	123, 124
Bolts, Machine for cutting off the ends of	J. Rewey	Berkshire, N. Y	May 8, 1811	....
Bolts, Machine for cutting threads on	W. W. Hubbard	Philadelphia, Pa	Aug. 22, 1865	49, 524
Bolts, &c., Machinery for heading	E. Page	Albany, N. Y	Dec. 28, 1852	9, 506
Bolts, &c., Machinery for securing	J. Caswell	Syracuse, N. Y	Nov. 30, 1852	9, 428
Bolts, Manufacture of	W. J. Lewis	Pittsburgh, Pa	Apr. 4, 1865	47, 113
Bolts, Manufacture of	A. Nailer	Buffalo, N. Y	May 3, 1870	102, 566
Bolts, Method of making	W. J. Clark	Southington, Conn	Aug. 2, 1864	43, 669
Bolts, nuts, &c., Mode of squaring and finishing the heads of.	J. Bellemere	Philadelphia, Pa	Sept. 25, 1839	1, 342
Bolts or rivets, Machine for pointing	J. Stackler	West Winsted, Conn	May 25, 1869	90, 602
Bolts, rivets, and nails, Mode of attaching the heads of.	G. B. Brayton	Providence, R. I	May 26, 1863	38, 652
Bolts, rods, &c., Tool for cutting off	H. Peters	Davenport, Iowa	June 7, 1870	104, 061
Bolts, Socket for	H. W. Collender	New York, N. Y	Sept. 15, 1857	18, 193
Bolts, Threading	H. Abbott	Wakeman, Ohio	Apr. 17, 1860	27, 943
Bolts, Trimming or cutting	H. Beckwith	Grass Lake, Mich	Mar. 24, 1863	37, 943
Bolting-chest	J. Bell	Belleville, Ill	Apr. 9, 1861	31, 947
Bolting-chest	M. C. Cogswell and J. McKiernan.	Buffalo, N. Y	June 12, 1860	28, 654
Bolting-cloth to reel, Mode of attaching	J. Woodville	Chillicothe, Ohio	Apr. 21, 1857	17, 126
Bolting-cloth to reel, Securing	G. M. Elliott	Hagerstown, Md	Mar. 19, 1834	....
Bolting, dusting, and separating the ground material, Machinery for.	J. Woodward	Philadelphia, Pa	Apr. 6, 1858	19, 892
Bolting-machine	H. Cabanes	Bordeaux, France	Nov. 18, 1873	144, 739
Bolting-machine	J. H. Jones	Dayton, Ohio	Sept. 10, 1872	131, 278
Bolting-mill	E. H. Vittecoq	Beaumontel, France	Jan. 3, 1865	45, 802
Bolting-reel	J. T. Agner	Lexington, Va	May 11, 1869	89, 963
Bolting-reel	S. Corsett and O. D. Lowe	Middleville, Mich	July 4, 1871	116, 685
Bolting-reel	F. B. Lewis	Tiffin, Ohio	Sept. 20, 1870	107, 512
Bomb for killing whales	N. Schofield and W. W. Wright	Norwich, Conn	Mar. 10, 1857	16, 819
Bomb-lance	P. B. Comins	San Francisco, Cal	May 3, 1859	23, 827
Bomb-lance for killing whales	S. Barker	Hartford, Conn	Feb. 21, 1865	46, 437
Bomb or mine, Subterrene	J. MacGregor, jr	Wilton, N. Y	Mar. 23, 1842	2, 502
Bomb, shell, or grenade	E. T. Miller	Boston, Mass	Nov. 7, 1854	11, 901
Bombs, lances, and other projectiles, Guide for	I. Goodspeed and C. Crawley	Norwich, Conn	July 16, 1861	32, 830
Bonds, &c., Filing and recording	F. Munson	Chicago, Ill	Apr. 2, 1867	63, 419
Bonds, Mode of preventing the stealing of	J. Myers, jr	Brooklyn, E. D., N. Y	Mar. 15, 1870	100, 788
Bonds, &c., Preventing alteration in numbers on	G. W. Casilear	Washington, D. C	Nov. 24, 1868	84, 341
Bone and plaster mill	S. W. Powell	Brookville, Md	Dec. 17, 1867	72, 418
Bone-black. *See* Charcoal, Animal.				
Bone-black, Apparatus and process for revivifying.	A. Lonsky	Hoboken, N. J	Jan. 7, 1873	134, 686
Bone-black, Apparatus for cooling and purifying	D. H. Turner	New York, N. Y	May 12, 1868	77, 935
Bone-black, Apparatus for revivifying	C. N. Brock	Philadelphia, Pa	July 2, 1861	32, 679
Bone-black, Apparatus for revivifying	G. Finken	New York, N. Y	Aug. 25, 1863	39, 637
Bone-black, Artificial	H. Endemann	New York, N. Y	Mar. 9, 1869	87, 653
Bone-black cleaning and purifying machine	R. Ficken and F. L. Williams	Philadelphia, Pa	Mar. 5, 1867	62, 537
Bone-black-cooling apparatus	C. H. Senff	New York, N. Y	June 27, 1871	116, 361
Bone-black drier	P. Farley	New York, N. Y	Apr. 18, 1871	113, 754
Bone-black equalizer	D. Brasill and D. A. Mullane	New Orleans, La	Aug. 17, 1869	93, 668
Bone-black for filtering, Artificial	W. H. Kelsey	Cleveland, Ohio	Jan. 9, 1872	122, 526
Bone-black furnace, Apparatus for emptying the cooling-tubes of.	E. Langen	Cologne, Prussia	Aug. 3, 1869	93, 208
Bone-black, Furnace for reburning	G. Thulemeyer	New York, N. Y	Oct. 20, 1863	40, 371
Bone-black kiln	J. O. Donner	Jersey City, N. J	Nov. 16, 1869	96, 899

Index of patents issued from the United States Patent Office from 1790 to 1873, inclusive—Continued.

Invention.	Inventor.	Residence.	Date.	No.
Bone-black kiln	E. P. Eastwick	Baltimore, Md	Dec. 18, 1866	60, 492
Bone-black kiln	E. P. Eastwick	Baltimore, Md	Apr. 6, 1869	88, 700
Bone-black kiln	E. P. Eastwick	Baltimore, Md	Apr. 6, 1869	88, 701
Bone-black kiln	E. P. Eastwick	Baltimore, Md	Apr. 6, 1869	88, 702
Bone-black, Kiln for reburning	E. P. Eastwick	Baltimore, Md	May 16, 1871	114, 780
Bone-black kiln, Reburning and purifying	A. Weber	New York, N. Y	June 4, 1867	65, 457
Bone-black, Manufacture of	C. Y. Beach	Brooklyn, N. Y	Sept. 19, 1871	119, 005
Bone-black, Oven for reburning	W. Moller	New York, N. Y	Aug. 19, 1862	36, 230
Bone-black, Purifying	T. H. Quick	New York, N. Y	May 15, 1866	54, 771
Bone-black, Purifying	T. H. Quick	New York, N. Y	Nov. 26, 1867	71, 536
Bone-black purifying, screening, and cooling apparatus, Reburnt.	M. Hanford	Boston, Mass	May 18, 1869	90, 097
Bone-black reburning and ore-reducing furnace	A. Weber	New York, N. Y	July 26, 1870	105, 747
Bone-black, Revivification of	H. Kattenhorn	New York, N. Y	Jan. 25, 1859	22, 734
Bone-black revivifying	W. Bellows	Cincinnati, Ohio	Dec. 13, 1859	26, 457
Bone-black, Revivifying	J. Forrest	New York, N. Y	Aug. 25, 1863	39, 638
Bone-black-revivifying apparatus	W. Mitchell	New York, N. Y	Mar. 13, 1860	27, 462
Bone-black, Treating and revivifying	J. Rogers and L. Reid	Brooklyn, N. Y	July 20, 1869	92, 751
Bone-black-washing apparatus	C. Kinzler	New York, N. Y	July 10, 1860	29, 128
Bone burning or charring pot, Construction of	C. W. Pierce	Albany, N. Y	June 11, 1867	65, 597
Bone-charcoal, Retort for revivifying	G. A. Jasper	Charlestown, Mass	June 4, 1867	65, 389
Bone handle for canes, &c	J. Harvey	Philadelphia, Pa	July 9, 1867	66, 586
Bone-mill	E. P. Baugh	Philadelphia, Pa	Mar. 7, 1865	46, 627
Bone, &c., Mill for pulverizing	A. Newell	Red Wing, Minn	Sept. 8, 1868	81, 933
Bone-setting apparatus	J. H. Willard	Brown Elm, Ohio	June 11, 1836	....
Bones for manure, &c., Mode of grinding	J. Wister	Greencastle, Pa	May 8, 1866	54, 635
Bones, refuse meat, &c., Apparatus for boiling	R. W. Thing	Winchester, Mass	June 18, 1872	127, 995
Bonnet	S. A. Blake	New York, N. Y	Dec. 24, 1861	33, 978
Bonnet	A. Henri	Louisville, Ky	May 1, 1860	28, 082
Bonnet	F. Howard	Somerville, Mass	June 7, 1864	43, 028
Bonnet	S. H. Lyon and W. E. Doubleday	Brooklyn, N. Y	Mar. 19, 1861	31, 724
Bonnet and bonnet-frame pressing machine	W. Osborn	Louisville, Ky	Aug. 19, 1856	15, 570
Bonnet and hat pressing machine	E. Copleston	Wrentham, Mass	Aug. 15, 1865	49, 384
Bonnet and hat pressing machine	C. Merritt	Baltimore, Md	Mar. 13, 1844	3, 477
Bonnet-binding, Making	J. L. Weaver	Orange, Mass	Jan. 31, 1865	46, 163
Bonnet-box	C. A. Taylor	Chicago, Ill	June 26, 1860	28, 916
Bonnet-forming machine	C. O. Parmenter	Amherst, Mass	Mar. 25, 1862	34, 771
Bonnet-frame	G. O. Cooper and C. Southworth	Stoughton, Mass	Mar. 20, 1866	53, 274
Bonnet-frame	G. H. Hawkins	New York, N. Y	July 22, 1862	35, 934
Bonnet-frame	W. E. Kidd	New York, N. Y	Apr. 13, 1858	19, 932
Bonnet-frame clamp	H. A. Reynolds	New York, N. Y	May 1, 1860	28, 131
Bonnet-frames, Machine for forming	S. H. Bowker	Worcester, Mass	Feb. 2, 1858	19, 231
Bonnet-frames, Machine for stiffening netting for	P. C. Ritchie	New York, N. Y	Feb. 11, 1873	135, 730
Bonnet-frames, Manufacture of	R. T. Wilde	New York, N. Y	Apr. 24, 1860	28, 028
Bonnet-frames, Mode of constructing wire	H. Weed	Philadelphia, Pa	June 27, 1854	11, 186
Bonnet-frames, Mold for pressing	N. Spence	New York, N. Y	Nov. 28, 1854	12, 004
Bonnet-front former	G. A. Cox	Brooklyn, N. Y	Jan. 17, 1860	26, 836
Bonnet-fronts, Mold for pressing	W. E. Kidd	New York, N. Y	Nov. 28, 1854	11, 996
Bonnet, Portable	T. Hammond	New York, N. Y	Oct. 30, 1844	3, 809
Bonnet-pressing machine	G. M. Richardson	Barre, Mass	Oct. 2, 1866	58, 555
Bonnet-pressing machine	C. W. Russell	Philadelphia, Pa	Aug. 16, 1859	25, 219
Bonnet-pressing machine	S. and H. Squire	Monson, Mass	Nov. 12, 1867	70, 912
Bonnet-pressing machine	H. E. West	Attleborough, Mass	Jan. 29, 1861	31, 274
Bonnet-pressing machinery	R. Murdoch	Baltimore, Md	Oct. 12, 1842	2, 817
Bonnet-stands, Branch-holder for	J. R. Palmenbery	New York, N. Y	Aug. 2, 1859	24, 950
Bonnet-tips, Apparatus for pressing	T. Kendall	New York, N. Y	Sept. 3, 1844	3, 728
Bonnets and cap-fronts, &c, Machine for manufacturing.	W. H. Morrison	Nottingham, England	Oct. 18, 1859	25, 847
Bonnets and hats and other manufactures of straw, Manner of pressing and finishing.	W. Chaplin	New York, N. Y	Sept. 10, 1840	1, 785
Bonnets and hats, Manufacturing grass	G. and S. Wells	Wethersfield, Conn	Dec. 29, 1821	....
Bonnets and hats, Pressing straw	O. Plimpton	Foxborough, Mass	Dec. 28, 1832	....
Bonnets and other articles of varying thickness, Machine for pressing straw.	H. E. West	Norton, Mass	July 6, 1858	20, 837
Bonnets, Machine for trimming straw-plait for	E. Smith	Franklin, Mass	Mar. 30, 1808	....
Bonnets, Manufacturing	T. Bracher	New York, N. Y	Dec. 3, 1861	33, 872
Bonnets, Method of shaping	C. W. Russell	Philadelphia, Pa	Aug. 16, 1859	25, 218
Bonnets, Pressing	C. C. Dow	Windham, Conn	July 10, 1849	6, 585
Bonnets, &c., Pressing straw	S. Pittee	Foxborough, Mass	Sept. 14, 1825	....
Book and manuscript holder	J. E. D. Comstock	New York, N. Y	July 10, 1866	56, 184
Book and method of binding, Copy	W. Davison	Baltimore, Md	Oct. 9, 1841	2, 286
Book and paper clamp	J. P. McLean	Brooklyn, N. Y	Sept. 23, 1873	143, 088
Book and paper trimming machine	G. Utley	Chapel Hill, N. C	Aug. 7, 1860	29, 532
Book and slate, Combined	F. Shepherd	New Haven, Conn	Nov. 2, 1858	21, 984
Book-backs, Machine for molding	J. K. Max	Springfield, Mass	Mar. 12, 1867	62, 767
Book-backs, Machine for shaping and finishing	G. H. Sanborn and J. E. Coffin	Boston, Mass., and Portland, Me.	Sept. 20, 1859	25, 548
Book-binder, Schoolboy's	T. Goodrum	Providence, R. I	Oct. 16, 1866	58, 940
Book-binders and others, Material for paste for	I. L. Plumer	Chelsea, Mass	Aug. 23, 1870	106, 617
Book-binder's beveling-machine	W. P. Chase	Boston Mass	May 19, 1868	77, 958
Book-binder's beveling-machine	H. L. Tumy	Cincinnati, Ohio	Dec. 17, 1867	72, 243
Book-binder's board, Drying	J. H. Longbotham	Brooklyn, N. Y	Jan. 3, 1854	10, 375
Book-binder's boards, Machinery for cutting	G. S. Schofield	Philadelphia, Pa	June 16, 1846	4, 575
Book-binder's clamp	G. F. Holland	Boston, Mass	June 11, 1872	127, 882
Book-binder's cloth-cutting table	C. Lemon	Washington, D. C	Dec. 27, 1864	45, 620
Book-binder's cutting-press	C. Stimpson, jr	Boston, Mass	Oct. 17, 1831	....
Book-binder's cutting press	J. J. Wells	Hartford, Conn	Mar. 23, 1824	....
Book-binder's head-band	J. Schlichting	New York, N. Y	Apr. 6, 1869	88, 587
Book-binders, Ornamenting rolls and stamps for	D. H. Mason	Philadelphia, Pa	Jan. 26, 1826	....
Book-binder's prepared paste	J. Woodward	Springfield, Mass	Feb. 20, 1866	52, 779
Book-binder's roll	J. Feely	New York, N. Y	Mar. 31, 1868	76, 065
Book-binder's sewing-table	M. T. Lincoln	Washington, D. C	Oct. 30, 1866	59, 240
Book-binder's standing press	M. K. Pelletreau	New York, N. Y	Oct. 2, 1860	30, 243
Book-binders, Supplemental table for	G. P. Goff	Washington, D. C	July 5, 1864	43, 456
Book-binder's use, Manufacture of gold-leaf for	R. E. Hastings	Philadelphia, Pa	Dec. 30, 1873	146, 001
Book-binding	W. I. Blackman	Columbus, Miss	May 4, 1869	89, 622
Book-binding	W. Daniels	Brooklyn, N. Y	Dec. 17, 1867	72, 373
Book-binding	R. De Silver	Philadelphia, Pa	Sept. 14, 1812	....

Index of patents issued from the United States Patent Office from 1790 *to* 1873, *inclusive*—Continued.

Invention.	Inventor.	Residence.	Date.	No.
Book-binding	W. Hancock	Great Britain	Oct. 28, 1837	444
Book-binding	A. Holbrook, jr	Lynn, Mass	June 23, 1868	79, 123
Book-binding	A. H. Jocelyn	New York, N. Y	Dec. 29, 1868	85, 450
Book-binding	J. S. Lever	Philadelphia, Pa	Nov. 10, 1868	83, 980
Book-binding	L. G. Matthews	New Albany, Ind.	Dec. 1, 1868	84, 564
Book-binding	J. Meer	Philadelphia, Pa	Aug. 26, 1818	....
Book-binding	G. H. Reynolds	New York, N. Y	Feb. 22, 1870	100, 069
Book-binding	I. Reynolds	Dayton, Ohio	Dec. 21, 1869	98, 191
Book-binding	I. Reynolds	Dayton, Ohio	Jan. 23, 1872	122, 965
Book-binding	J. L. Rile	New York, N. Y	June 13, 1871	115, 896
Book-binding	A. H. Rowand	Allegheny, Pa	May 12, 1857	17, 299
Book-binding	D. Shive	Philadelphia, Pa	Dec. 21, 1869	98, 200
Book-binding	H. M. Shute	Waukegan, Ill	Feb. 4, 1868	74, 014
Book-binding	G. Smith	Bridgeport, Conn	Feb. 20, 1872	123, 947
Book-binding	W. Smith	New York, N. Y	June 23, 1868	79, 269
Book-binding	D. M. Smyth	Orange, N. J	Oct. 10, 1871	119, 894
Book-binding	J. T. Sullivan	Philadelphia, Pa	July 27, 1812	....
Book-binding	J. C. Terry	Springfield, Conn	Aug. 9, 1870	106, 234
Book-binding	J. Torrey, jr	Germantown, Pa	Aug. 8, 1829	....
Book-binding	T. Towndrow	New York, N. Y	May 29, 1860	28, 521
Book-binding apparatus	I. Reynolds	Dayton, Ohio	July 4, 1871	116, 757
Book-binding machine	J. Glass	Green Point, N. Y	May 2, 1871	114, 286
Book-binding machine	R. G. Lowrey	Brooklyn, N. Y	June 6, 1871	115, 621
Book, Blank	J. C. Bonnell	Burlington, Iowa	Feb. 11, 1873	135, 760
Book, Blank	H. Fischer	Chicago, Ill	Dec. 15, 1868	84, 869
Book, Blank account	C. Hopkins	New York, N. Y	Nov. 27, 1849	6, 905
Book-brace	W. Ives	Buffalo, N. Y	Jan. 23, 1855	12, 279
Book-case	A. Lamb	Cambridge, Mass	Apr. 16, 1861	32, 069
Book-case, Folding	E. Haskell	Dover, N. H	Apr. 22, 1873	138, 022
Book-case for schools, &c	P. T. Vannice	Kewanee, Ill	Jan. 10, 1871	110, 941
Book-case, Portable	E. Haskell	Canton, Mass	Feb. 24, 1863	37, 752
Book-case, Revolving	L. B. Pert	Tecumseh, Mich	Apr. 29, 1873	138, 279
Book-clamp	J. W. Jones	Harrisburgh, Pa	Oct. 24, 1865	50, 597
Book-clamp	W. A. Miller	Paterson, N. J	Dec. 16, 1873	145, 582
Book-clamp, Portable	C. W. Holbrook and E. F. Butler	Windsor Locks, Conn	Nov. 15, 1870	109, 213
Book-clamp, Portable	C. W. Sherwood	Chicago, Ill	Mar. 17, 1868	75, 588
Book-clamp, Portable	W. C. Watson	Paterson, N. J	Mar. 17, 1868	75, 660
Book clamp, School	W. C. Watson	Paterson, N. J	Sept. 24, 1867	69, 285
Book-clasp	C. Folsom	Cambridge, Mass	July 3, 1855	13, 166
Book clip, Check	A. B. Auer	Chicago, Ill	June 15, 1869	91, 198
Book, Copy	N. P. Beers	New York, N. Y	June 7, 1870	103, 965
Book, Copy	B. G. Howes	Worcester, Mass	Dec. 10, 1867	71, 882
Book, Copy	P. F. Van Everen	Brooklyn, N. Y	May 3, 1870	102, 732
Book, Copy	J. D. Williams	New York, N. Y	Apr. 19, 1870	102, 189
Book-cover	G. F. Barden	South Adams, Mass	Aug. 8, 1865	49, 213
Book-cover	W. L. Burlock	Philadelphia, Pa	Dec. 24, 1872	134, 249
Book-cover	A. H. Jocelyn	New York, N. Y	June 24, 1873	140, 275
Book-cover	W. C. Wendell	Albany, N. Y	Apr. 6, 1869	88, 684
Book-cover and stand	W. Milliken	Cambridge, Mass	Mar. 3, 1868	75, 040
Book-cover protector	C. L. Alexander and V. A. Osborn.	Washington, D. C	Sept. 24, 1867	69, 062
Book-cover protector	J. N. Larned	Buffalo, N. Y	Sept. 17, 1872	131, 356
Book-cover protector	A. L. Sewell	Chicago, Ill	June 22, 1869	91, 570
Book-cover protector	M. Thompson	New York, N. Y	Aug. 1, 1871	117, 697
Book-cover protector	P. F. Van Everen	Brooklyn, N. Y	May 3, 1870	102, 733
Book-cover protector	A. Van Patten	Weyauwega, Wis	Nov. 24, 1868	84, 397
Book-covers, Machine for making	L. Danforth	Buffalo, N. Y	May 23, 1854	10, 961
Book-covers, Machine for making	J. W. Harrison	Washington, D. C	Feb. 14, 1865	46, 353
Book-covers, Machine for making	M. and R. H. Rumrell	Brooklyn, N. Y	Apr. 23, 1867	64, 038
Book-covers, Ornamenting	W. McAdams	Albany, N. Y	Oct. 9, 1847	5, 327
Book-curb	M. A. H. Saurman	Philadelphia, Pa	May 11, 1869	90, 052
Book-cutting machine	F. J. Austin	New York, N. Y	Jan. 19, 1864	41, 337
Book for book-keeping	J. H. Gleim	Saint Louis, Mo	Dec. 24, 1867	72, 629
Book hand-rest	W. F. West	Haverstraw, N. Y	Dec. 17, 1872	134, 116
Book, Herbal and scrap	A. M. Safford	Springfield, Mass	June 14, 1864	43, 135
Book-holder	D. F. Dimon and G. H. Carsnell	Fishkill Landing, N. Y	July 19, 1864	43, 575
Book-holder	F. Emmons	Danvers, Mass	Mar. 23, 1869	88, 022
Book-holder	W. D. Gridley	New Britain, Conn	Apr. 27, 1869	89, 307
Book-holder	G. P. Johnson	Webster's Grove, Mo	Nov. 9, 1869	96, 702
Book holder	D. Moritz and R. White	Carmansville and Mott Haven, N. Y.	Oct. 15, 1872	132, 311
Book-holder	O. B. Parker	Woodville, Mass	June 26, 1866	55, 897
Book-holder	C. Phelps	Clayton, N. Y	May 21, 1872	126, 901
Book-holder	T. E. Platt	New Haven, Conn	Mar. 12, 1867	62, 774
Book-holder	W. S. Poulson	Cadiz, Ohio	June 1, 1869	90, 870
Book-holder	E. Propst and J. F. Avis	Tipton, Ind	Aug. 9, 1870	106, 285
Book-holder	H. Sherman	Waverly, Pa	Dec. 7, 1869	97, 710
Book-holder	C. W. Simpson	Bangor, Me	July 13, 1869	92, 482
Book-holder	D. I. Stagg	New York, N. Y	June 26, 1866	55, 924
Book-holder	W. J. Thorn	Westbrook, Me	Sept. 13, 1870	107, 310
Book-holder and leaf-turner	A. McFall	Yonkers, N. Y	Aug. 1, 1871	117, 654
Book-holder for pews	N. A. Wright	Prairie du Chien, Wis	Apr. 9, 1867	63, 776
Book-holder, Portable	C. C. Moore	New York, N. Y	Feb. 28, 1871	112, 165
Book-holder, Portable	H. A. Oesterle	Philadelphia, Pa	Nov. 21, 1871	121, 191
Book-holder, Portable	C. W. Sherwood	Chicago, Ill	Feb. 7, 1871	111, 577
Book-holding press	L. Heitkamp	New York, N. Y	Oct. 7, 1873	143, 410
Book-holding stand	J. S. Brown	Washington, D. C	Feb. 2, 1864	41, 420
Book, Hymn and tune	E. Curtice	Yonkers, N. Y	Feb. 12, 1867	61, 927
Book, Ivory-covered	J. H. Pratt	New York, N. Y	Apr. 26, 1864	42, 507
Book-leaf turner	C. C. Clapp	Hartford, Conn	Jan. 2, 1866	51, 804
Book-leaf-turning device	C. C. Clapp	Hartford, Conn	Oct. 15, 1867	69, 770
Book-leaves, Machine for turning	J. H. Schomacker and M. Kuemerle.	Philadelphia, Pa	Sept. 4, 1849	6, 697
Book, Manifold writing	A. S. Solomons	Washington, D. C	Dec. 6, 1864	45, 354
Book-mark	H. M. Ward and G. E. Dolton	Monee, Ill	June 17, 1873	139, 936
Book-mark holder	P. Tomppert, jr	Louisville, Ky	Dec. 4, 1866	60, 285
Book-marker	E. Cottle	Randolph, Mass	Aug. 30, 1870	106, 785
Book, Memorandum	J. H. Guest and E. Fawcett	New Albany, Ind	May 11, 1869	89, 928

Index of patents issued from the United States Patent Office from 1790 to 1873, inclusive—Continued.

Invention.	Inventor.	Residence.	Date.	No.
Book, Memorandum	H. M. Hinsdill	Grand Rapids, Mich	Jan. 7, 1873	134, 546
Book, Memorandum	I. N. Swasey	Yonkers, N. Y	July 11, 1871	116, 884
Book or music leaf turning apparatus	J. Grant	Hampstead, England	Apr. 20, 1869	89, 214
Book-paging apparatus	G. Leverich	Wellsburgh, N. Y	May 29, 1855	12, 954
Book-paging machine	W. C. Demain	Boston, Mass	Aug. 21, 1855	13, 477
Book-paging machine	R. L. Hawes	Worcester, Mass	Dec. 9, 1856	16, 207
Book-paging machine	H. Hochstrasser	Philadelphia, Pa	Sept. 26, 1854	11, 728
Book-paging machine	J. McAdams	Brooklyn, N. Y	Mar. 9, 1869	87, 693
Book-paging machine	S. E. Parrish	New York, N. Y	Mar. 23, 1852	8, 830
Book-paging machine	H. S. Taylor	Springfield, Mass	May 16, 1854	10, 933
Book-paging machine	E. Town	Jersey City, N. J	Jan. 31, 1854	10, 485
Book paging machinery, Blank	J. L. Burdick	New York, N. Y	Feb. 14, 1854	10, 533
Book-rack	D. Lewis	Boston, Mass	Dec. 9, 1862	37, 102
Book-rack	J. K. Otis	East Cambridge, Mass	July 25, 1871	117, 321
Book-rack	J. P. Tibbits	New York, N. Y	Nov. 16, 1869	97, 003
Book rack and stand	G. Gardener	Clarksville, N. J	Apr. 26, 1870	102, 248
Book, Record	H. Arden	Brooklyn, N. Y	July 25, 1871	117, 240
Book-repository	R. L. Dodge	Gallatin, Mo	June 7, 1870	103, 853
Book-rest	C. Sulivan	Boston, Mass	May 31, 1870	103, 682
Book-rest	L. Tarring	Washington, D. C	May 27, 1873	139, 341
Book, Scrap	S. L. Clemens	Hartford, Conn	June 24, 1873	140, 245
Book, Slated	H. B. Barnes	New York, N. Y	Aug. 6, 1872	130, 179
Book-stand	J. B. Annin	Rochester, N. Y	June 24, 1873	140, 107
Book-stand	H. M. Sweeney	Boston, Mass	Apr. 9, 1872	125, 590
Book stand or holder	W. W. Marston	New York, N. Y	Aug. 29, 1865	49, 639
Book-support	J. Densmore	Holley, N. Y	Mar. 12, 1872	124, 423
Book-support	R. B. Hindle	Saint Louis, Mo	July 1, 1873	140, 369
Book-support	J., A., and W. J. A. McCausland.	Providence, R. I	Feb. 14, 1871	111, 858
Book-support	D. E. Morgan	Cincinnati, Ohio	Jan. 23, 1872	122, 957
Book-support	C. J. North	Auburn, N. Y	June 13, 1871	115, 886
Book-support	S. M. Thomson	Providence, R. I	Mar. 21, 1871	112, 984
Book-supporter	L. Willson	Indianapolis, Ind	July 9, 1872	128, 835
Book, Tablet index	J. B. Lake, jr	Baltimore, Md	July 25, 1871	117, 433
Book-trimming machine	M. Riehl	Cincinnati, Ohio	Aug. 28, 1855	13, 501
Book trimming machinery	L. F. Markham	Cambridgeport, Mass	Apr. 18, 1848	5, 523
Books, Adhesive compound for binding	D. Felt	New York, N. Y	Aug. 1, 1838	867
Books and music, Apparatus to hold and turn the leaves of.	C. B. Thayer	Boston, Mass	June 21, 1859	24, 504
Books, Apparatus for turning leaves of	H. C. Bridgham and J. M. Stewart.	New London and Norwich, Conn.	Sept. 12, 1854	11, 660
Books, Apparatus for turning leaves of	C. Desbeaux	Paris, France	Jan. 3, 1854	10, 381
Books, Binding invoice	J. L. Rile	New York, N. Y	Oct. 3, 1871	4, 577
Books, Binding mathematical	F. B. Wells	Fishkill, on the Hudson, N. Y.	Oct. 20, 1868	83, 345
Books, Binding mercantile	J. H. Gleim	Saint Louis, Mo	Oct. 20, 1868	83, 276
Books, Case for record	W. H. Somers	Urbana, Ill	Aug. 11, 1863	39, 503
Books, Clamp-strap for school	J. W. McKee	Brooklyn, N. Y	Sept. 24, 1867	69, 234
Books, Clamp-strap for school	L. F. Van De Weile	Brooklyn, N. Y	Nov. 26, 1867	71, 554
Books, &c., Clasp-lock for	L. Lall	New York, N. Y	Nov. 27, 1866	60, 017
Books, Composition for making copy-writing and drawing.	R. Wall	Philadelphia, Pa	Apr. 28, 1817	....
Books, Curving the backs of	J. A. Elder	Westbrook, Me	July 26, 1853	9, 886
Books, Embossed cover for	W. T. Anderson	Brooklyn, N. Y	July 5, 1864	43, 450
Books, Guide and blotter for writing	P. Martin	Medford, Mass	Sept. 3, 1872	131, 106
Books, Indexing	J. S. Hicks	Roslyn, N. Y	Sept. 23, 1873	143, 075
Books in the round, Machine for cutting	G. Trinks and L. Heitkamp	New York and Brooklyn, N. Y.	June 17, 1862	35, 639
Books, Machine for backing	W. Laighton	Portsmouth, N. H	Jan. 10, 1845	3, 880
Books, Machine for curving the backs of	J. E. Coffin	Portland, Me	June 12, 1866	55, 468
Books, Machine for cutting	A. O. Douglass	Philadelphia, Pa	Jan. 6, 1832	....
Books, Machine for cutting index to blank	G. Hodgkinson and T. F. Randolph.	Cincinnati, Ohio	Apr. 28, 1857	17, 149
Books, Machine for cutting the front of	I. Jones	Camden, N. J	Oct. 30, 1866	59, 229
Books, Machine for finishing the backs of	C. Starr	New York, N. Y	June 24, 1851	8, 179
Books, Machine for numbering the pages of	E. and C. E. Town	Jersey City, N. J	Oct. 5, 1858	21, 708
Books, Machine for numbering the pages of account.	J. McAdams	Boston, Mass	Aug. 12, 1851	8, 291
Books, Machine for rounding and backing	T. Bergner	Philadelphia, Pa	June 9, 1857	17, 480
Books, Machine for rounding and backing	J. E. Coffin	Westbrook, Me	July 8, 1856	15, 282
Books, Machine for rounding and backing	R. F. Foster, T. S. Baylie, and J. Harbin.	Washington, D. C	Mar. 25, 1873	137, 192
Books, Machine for rounding backs of	L. F. Markham	Cambridgeport, Mass	Dec. 5, 1854	12, 026
Books, Machine for shaping the backs of	J. E. Coffin	Portland, Me	June 14, 1859	24, 425
Books, Machine for trimming	A. C. Semple	New York, N. Y	Mar. 16, 1858	19, 654
Books, Machine for turning the leaves of	F. Suter	Brooklyn, N. Y	Dec. 22, 1857	18, 935
Books, Mode of making spring-backs for	L. Francis	New York, N. Y	Apr. 9, 1867	63, 626
Books, Paging	R. M. Leslie	Philadelphia, Pa	Feb. 15, 1853	9, 586
Books, Paging bound	J. McAdams	Boston, Mass	Mar. 29, 1853	9, 638
Books, Protecting bound	J. C. G. Kennedy	Washington, D. C	May 17, 1864	42, 779
Books, Register for account	A. F. Jones	Douglass, Mass	Sept. 29, 1863	40, 106
Books, Tool for embossing backs of	C. Starr	New York, N. Y	Jan. 21, 1851	7, 911
Boom and gaff joint	J. W. Norcross	Middletown, Conn	Dec. 27, 1864	45, 625
Boom-connection for masts	B. McGrath	Gloucester, Mass	June 12, 1866	55, 515
Boom for vessels	E. G. Gaillac	Cutler, Me	Feb. 15, 1870	99, 874
Boom-gear	C. R. Webb	Philadelphia, Pa	Apr. 14, 1868	76, 857
Boom-gudgeon	C. Sayward	Gloucester, Mass	July 14, 1868	79, 864
Boom jack or elevator	R. B. Condon	Belfast, Me	Oct. 1, 1872	131, 742
Boom-jaws, Elastic lining for	O. P. Rowland	Jamesport, N. Y	June 5, 1860	28, 608
Boom-jaws, Guide for	M. Gray	Newton, Mass	May 22, 1866	55, 001
Boom, Jib	W. E. Beman	Portland, Me	Apr. 28, 1868	77, 348
Boom-saddle and chafing-band for mast	W. T. Griffenberg	Wilmington, Del	Nov. 26, 1872	133, 439
Boom saddle and jaw	J. Davis, jr	Gloucester, Mass	Dec. 31, 1845	4, 345
Boom sling, Spritsail	G. Dowling	Fair Haven, Conn	Apr. 7, 1868	76, 309
Booms and gaffs, Band for	D. Dryburgh	Philadelphia, Pa	Oct. 5, 1869	95, 442
Booms and gaffs of vessels, Jaw for	A. Manning	Fair Haven, Conn	May 1, 1866	54, 379
Booms, Device for sheering	L. W. Pond	Eau Claire, Wis	Aug. 4, 1868	80, 663
Booms, Elastic anti-friction attachment for vessel's	H. Gregory, jr	Rockland, Me	Nov. 25, 1873	144, 975

Index of patents issued from the United States Patent Office from 1790 *to* 1873, *inclusive*—Continued.

Invention.	Inventor.	Residence.	Date.	No.
Booms, Outhaul-band for	G. Nevenger	Philadelphia, Pa	Dec. 12, 1871	121, 890
Booms, Outhaul for	G. W. Leighton and C. O. Cole	Portland, Me	Sept. 15, 1868	82, 133
Booms, Outhaul for	T. O'Neill	New York, N. Y	July 13, 1869	92, 539
Booms to masts, Connection of	A. Gill	Holmes' Hole, Mass	Apr. 25, 1871	114, 128
Booms to masts, Device for attaching	E. Rogers	Waterford, Conn	May 23, 1871	115, 241
Booms to masts, Means of attaching	M. McClain	Pemaquid, Me	Oct. 13, 1863	40, 273
Booms to masts, Truss for connecting	J. E. Tibbetts	Trenton, N. J	Aug. 17, 1869	93, 770
Boot	J. H. Agnew	Dobb's Ferry, N. Y	Apr. 28, 1863	38, 272
Boot	M. F. Chandler	Boston, Mass	Mar. 23, 1869	88, 131
Boot, &c	F. Closs	New Haven, Conn	Nov. 22, 1864	45, 138
Boot	E. D. Coffee	Holliston, Mass	Sept. 5, 1865	49, 723
Boot	L. Duvall	Big Spring, Ky	Sept. 6, 1859	25, 325
Boot	L. H. Farnsworth	Hudson, Mass	Sept. 23, 1873	143, 001
Boot	A. A. Felton and F. C. Floyd	Marblehead and Boston, Mass.	Dec. 10, 1872	133, 767
Boot	C. H. Fitch	Worcester, Mass	Nov. 8, 1870	109, 123
Boot	I. Hall	Boston, Mass	May 27, 1873	139, 385
Boot	J. Holland	Conshohocken, Pa	Jan. 7, 1868	73, 010
Boot	J. P. Jamison	New York, N. Y	Sept. 22, 1868	82, 417
Boot	P. Keffer	Reading, Pa	Dec. 13, 1859	26, 436
Boot	G. S. Lee	Worcester, Mass	Apr. 15, 1873	137, 781
Boot	J. S Lewis	Baltimore, Md	Feb. 21, 1871	112, 056
Boot	J. H. Livingston	Providence, R. I	Mar. 23, 1869	88, 050
Boot	L. O. Makepeace	Lynn, Mass	Jan. 5, 1869	85, 681
Boot	G. C. Parker	Worcester, Mass	Dec. 16, 1873	145, 589
Boot	J. H. Walker	Worcester, Mass	Nov. 1, 1870	108, 948
Boot	P. West	Worcester, Mass	Jan. 30, 1872	123, 316
Boot	P. West and G. S. Lee	Worcester, Mass	Sept. 2, 1873	142, 424
Boot	W. W. Whitcomb	Boston, Mass	Sept. 3, 1872	131, 041
Boot	W. H. Willard	Cleveland, Ohio	Sept. 2, 1862	36, 381
Boot and gaiter	J. P. Herron	Washington, D. C	Aug. 18, 1863	39, 614
Boot and gaiter strap	W. J. Turner	Utica, N. Y	May 21, 1867	65, 027
Boot and pantaloon jack	I. S. Clough and S. S. Day	Brooklyn and New York, N. Y.	Apr. 30, 1861	32, 181
Boot and shoe	J. C. Adams	New London, N. H	Mar. 5, 1867	62, 587
Boot and shoe	J. M. Allen	Fredericktown, Ohio	Sept. 18, 1860	30, 034
Boot and shoe	R. Andrews	Milwaukee, Wis	Aug. 25, 1868	81, 457
Boot and shoe	D. M. Ayer	Lewiston, Me	Jan. 8, 1867	60, 987
Boot and shoe	S. Babbitt	Brazil, Ind	Feb. 1, 1870	99, 281
Boot and shoe	C. W. Bailey	Boston, Mass	Jan. 29, 1867	61, 502
Boot and shoe	F. D. Ballou	Abington, Mass	July 4, 1865	48, 614
Boot and shoe	L. R. Blake	Abington, Mass	Aug. 14, 1860	29, 562
Boot and shoe	L. R. Blake	Quincy, Mass	Sept. 27, 1864	44, 388
Boot and shoe	P. Boisset	Paris, France	Apr. 22, 1862	35, 057
Boot and shoe	A. O. Bourn	Providence, R. I	May 30, 1871	115, 422
Boot and shoe	J. S. Bowler	Lynn, Mass	Nov. 21, 1842	2, 859
Boot and shoe	M. Bray	Newton Centre, Mass	Dec. 19, 1871	121, 927
Boot and shoe	E. Brown	South Reading, Mass	Jan. 7, 1862	34, 086
Boot and shoe	E. Brown	Burlington, Vt	Oct. 26, 1869	96, 195
Boot and shoe	T. W. Brown	Malden, Mass	Apr. 16, 1872	125, 717
Boot and shoe	F. J. Burcham	Racine, Wis	Jan. 17, 1871	111, 039
Boot and shoe	A. Burke	New York, N. Y	Aug. 2, 1870	106, 030
Boot and shoe	D. H. Campbell and E. Woodward.	Sunderland, Scotland, and Charlestown, Mass.	June 6, 1871	115, 566
Boot and shoe	S. Chamberlain	Boston, Mass	Nov. 21, 1808	....
Boot and shoe	E. Chesterman	Roxbury, Mass	June 27, 1865	48, 368
Boot and shoe	E. Chesterman	Boston, Mass	May 12, 1868	77, 717
Boot and shoe	G. P. Clark	Brooklyn, N. Y	May 2, 1865	47, 521
Boot and shoe	D. N. P. Coffin, jr	Woburn, Mass	May 5, 1863	38, 373
Boot and shoe	N. Colver	Detroit, Mich	Sept. 12, 1854	11, 662
Boot and shoe	A. O. Crane	Hoboken, N. J	Apr. 23, 1861	32, 122
Boot and shoe	G. W. Day	Charlestown, Mass	June 28, 1864	43, 295
Boot and shoe	S. F. Dexter	Paris, N. Y	Oct. 16, 1860	30, 391
Boot and shoe	J. Dick	Philadelphia, Pa	Nov. 4, 1842	2, 840
Boot and shoe	T. H. Dodge	Worcester, Mass	July 30, 1872	130, 026
Boot and shoe	C. S. Dunbrack	Swampscott, Mass	Sept. 28, 1869	95, 210
Boot and shoe	T. G. Eiswald	St. Louis, Mo	Aug. 13, 1861	33, 032
Boot and shoe	H. Eldridge	Lynn, Mass	Mar. 10, 1868	75, 398
Boot and shoe	A. B. Ely	Newton, Mass	Dec. 31, 1867	72, 727
Boot and shoe	M. R. Ethridge	Locke's Mills, Me	July 17, 1866	56, 391
Boot and shoe	M. R. Ethridge	Locke's Mills, Me	Jan. 21, 1868	73, 589
Boot and shoe	M. Evans	Russiaville, Ind	Jan. 1, 1867	60, 707
Boot and shoe	A. T. Evory and A. Heston	La Porte, Ind	Nov. 6, 1866	59, 375
Boot and shoe	J. B. Field	Boston, Mass	Jan. 9, 1872	122, 519
Boot and shoe	S. T. Fowler	Brooklyn, N. Y	Aug. 7, 1866	56, 921
Boot and shoe	W. Francis	Boston, Mass	June 23, 1812	....
Boot and shoe	B. D. Godfrey	Milford, Mass	Mar. 28, 1865	47, 010
Boot and shoe	H. V. Gould	Dover, N. H	July 18, 1871	117, 168
Boot and shoe	T. Grason	Manchester, England	Sept. 6, 1864	44, 146
Boot and shoe	T. T. Hartford	Boston, Mass	Mar. 26, 1872	125, 047
Boot and shoe	D. E. Haywood	Melrose, Mass	Oct. 27, 1863	40, 408
Boot and shoe	E. Heaton and J. L. Joyce	New Haven, Conn	Oct. 8, 1861	32, 433
Boot and shoe	J. M. Hunter	New York, N. Y	Jan. 23, 1872	122, 948
Boot and shoe	W. F. Jobbins	New York, N. Y	May 25, 1869	90, 551
Boot and shoe	C. G. Johnson	London, England	July 23, 1861	32, 875
Boot and shoe	I. T. Jones	Sandwich, Mass	Aug. 1, 1865	49, 118
Boot and shoe	J. Kimball	Boston, Mass	June 19, 1866	55, 671
Boot and shoe	O. Lafreniere	New York, N. Y	Dec. 23, 1862	37, 233
Boot and shoe	O. Lafreniere	New York, N. Y	Jan. 19, 1864	41, 305
Boot and shoe	O. Lafreniere	New York, N. Y	Nov. 28, 1865	51, 195
Boot and shoe	W. Leathe	Woburn, Mass	June 22, 1869	91, 547
Boot and shoe	D. Lenain	New York, N. Y	May 14, 1861	32, 300
Boot and shoe	N. C. Lewis, jr	Boston, Mass	July 24, 1860	29, 344
Boot and shoe	G. W. Ludlow	Elizabeth, N. J	July 7, 1863	39, 156
Boot and shoe	M. Lumsden	San Francisco, Cal	May 26, 1868	78, 385
Boot and shoe	J. Macintosh and W. Boggett	London, Great Britain	July 27, 1869	93, 103
Boot and shoe	J. C. Mack	Philadelphia, Pa	Nov. 27, 1866	60, 024
Boot and shoe	G. A. Mansfield	Boston, Mass	Oct. 1, 1861	33, 396

Index of patents issued from the United States Patent Office from 1790 *to* 1873, *inclusive*—Continued.

Invention.	Inventor.	Residence.	Date.	No.
Boot and shoe	H. B. Marshall	Waldoborough, Me	Apr. 21, 1868	77, 063
Boot and shoe	C. Mayer	New York, N. Y	Aug. 9, 1864	43, 784
Boot and shoe	E. Mayer	Philadelphia, Pa	May 14, 1867	64, 780
Boot and shoe	G. McKay	Boston, Mass	Apr. 29, 1862	35, 105
Boot and shoe	C. Meyer	New Brunswick, N. J	June 18, 1861	32, 575
Boot and shoe	C. Meyer	New Brunswick, N. J	Mar. 18, 1862	34, 689
Boot and shoe	R. A. Miller	Boston, Mass	Apr. 26, 1870	102, 297
Boot and shoe	C. Mole	London, England	Dec. 10, 1867	71, 898
Boot and shoe	J. A. E. Moroney	Pontiac, Mich	Jan. 9, 1872	122, 632
Boot and shoe	W. Murray	Chicago, Ill	July 13, 1869	92, 634
Boot and shoe	S. Neiswender and G. King	Mexico, Ind	July 16, 1872	129, 580
Boot and shoe	J. L. Newton	Boston, Mass	Oct. 17, 1865	50, 487
Boot and shoe	O. Pacalin	New York, N. Y	June 19, 1866	55, 698
Boot and shoe	M. M. Pettes	West Concord, Vt	Oct. 22, 1861	33, 540
Boot and shoe	J. C. Plumer	Portland, Me	June 4, 1861	32, 487
Boot and shoe	J. C. Plumer	Boston, Mass	Jan. 9, 1866	51, 968
Boot and shoe	H. Port	New York, N. Y	Aug. 13, 1861	33, 051
Boot and shoe	H. Port and E Surgi	New Orleans, La	Oct. 16, 1860	30, 419
Boot and shoe	W. F. Prusha and E. L. Wales	Marlborough, Mass	May 2, 1871	114, 340
Boot and shoe	T. K. Reed	North Bridgewater, Mass.	May 23, 1865	47, 859
Boot and shoe	E. P. Richardson	Lawrence, Mass	May 28, 1867	65, 120
Boot and shoe	E. P. Richardson	Lawrence, Mass	Sept. 7, 1869	94, 648
Boot and shoe	E. T. Rogers	San Francisco, Cal	Feb. 18, 1868	74, 599
Boot and shoe	H. Rogers and L. Felton	Sudbury and Marlborough, Mass.	Feb. 4, 1873	135, 591
Boot and shoe	S. Rosenheimer	New York, N. Y	June 17, 1862	35, 629
Boot and shoe	G. F. Seaver	Ashland, Mass	July 30, 1872	129, 989
Boot and shoe	F. Senn	Forestville, N. Y	July 11, 1871	116, 872
Boot and shoe	P. Shaw	Boston, Mass	June 2, 1863	38, 766
Boot and shoe	G. and G. Smith	New York, N. Y	Feb. 25, 1868	74, 858
Boot and shoe	T. B. Smith	Boston, Mass	Apr. 2, 1867	63, 568
Boot and shoe	T. B. Smith	Boston, Mass	Apr. 2, 1867	63, 569
Boot and shoe	W. Smith	Bridesburgh, Pa	July 28, 1868	80, 366
Boot and shoe	R. Sommerville	Sandusky, Ohio	Dec. 24, 1872	134, 225
Boot and shoe	V. H. Spear	Lynn, Mass	July 16, 1872	128, 984
Boot and shoe	H. P. Stewart	Oak's Corners, N. Y	Dec. 18, 1866	60, 587
Boot and shoe	O. Stoddard	Jackson, Mich	Oct. 30, 1866	59, 288
Boot and shoe	B. F. Sturtevant	Boston, Mass	June 21, 1864	43, 235
Boot and shoe	G. W. Tash	Dover, N. H	Jan. 16, 1872	122, 788
Boot and shoe	E. Thayer	New York, N. Y	Jan. 8, 1867	61, 119
Boot and shoe	G. W. Tolhurst	New York, N. Y	Mar. 12, 1867	62, 907
Boot and shoe	W. H. Towers	Boston, Mass	Apr. 13, 1869	88, 824
Boot and shoe	C. H. Trask and J. B. Johnson	Lynn, Mass	Feb. 20, 1866	52, 798
Boot and shoe	R. Vollschwitz	New York, N. Y	Jan. 22, 1867	61, 487
Boot and shoe	G. Wagner	Washington, D. C	Jan. 15, 1867	61, 287
Boot and shoe	G. W. Walker	Lowell, Mass	Aug. 17, 1869	93, 845
Boot and shoe	J. H. Walker	Worcester, Mass	Dec. 3, 1872	133, 087
Boot and shoe	P. Ware, jr	Boston, Mass	Dec. 19, 1871	122, 085
Boot and shoe	W. P. Ware	New York, N. Y	June 7, 1864	43, 052
Boot and shoe	L. H. Whitney	Washington, D. C	June 7, 1870	103, 950
Boot and shoe	W. Wickersham	Boston, Mass	Aug. 2, 1870	106, 013
Boot and shoe	H. Wight	East Cambridge, Mass	May 29, 1866	55, 188
Boot and shoe	B. C. Young	Boston, Mass	Oct. 1, 1867	69, 528
Boot and shoe and clog for the feet	G. W. Martin	Boston, Mass	Aug. 4, 1868	80, 555
Boot and shoe blacking apparatus	I. Bickhart	Harlan, Ind	Jan. 27, 1863	37, 487
Boot and shoe blacking machine	J. M. and J. Connel	Newark, Ohio	Jan. 12, 1858	19, 075
Boot and shoe blacking machine	N. Eisenmann	New York, N. Y	Nov. 28, 1871	121, 343
Boot and shoe bottomings	C. Hay	Stoneham, Mass	Oct. 13, 1868	82, 946
Boot and shoe bottoms, Machine for molding	S. W. Baldwin	Yonkers, N. Y	May 27, 1873	139, 354
Boot and shoe buffing machine	G. C. Hawkins	Boston, Mass	Dec. 16, 1873	145, 501
Boot and shoe burnisher	L. I. Bumpus	Auburn, Me	Nov. 26, 1872	133, 301
Boot and shoe burnisher	R. Chauncey	Syracuse, N. Y	Mar. 14, 1871	112, 548
Boot and shoe burnishing machine	J. H. Howard	Stoneham, Mass	Feb. 13, 1872	123, 566
Boot and shoe burnishing machine	J. H. Howard	Stoneham, Mass	June 11, 1872	127, 772
Boot and shoe buttons, Machine for facing	N. B. Jenett and E. Everson	Haverhill, Mass	June 19, 1866	55, 664
Boot and shoe calk	J. L. Wager and A. L. Scudder	Sanford, N. Y	Aug. 23, 1870	106, 747
Boot and shoe channeling instrument	J. Cole	East Bloomfield, N. Y	Nov. 26, 1830	....
Boot and shoe channeling machine	A. E. Chickering and H. B. Tuttle	Albany, N. Y	Aug. 12, 1873	141, 693
Boot and shoe channeling machine	C. S. Dunbrack	Swampscott, Mass	Oct. 14, 1873	143, 561
Boot and shoe channeling machine	B. H. Hadley	Brooklyn, N. Y	Sept. 30, 1873	143, 237
Boot and shoe channeling machine	S. D. Tripp	Lynn, Mass	Nov. 8, 1870	109, 077
Boot and shoe channeling machine	H. S. Vrooman	Boston, Mass	June 13, 1871	115, 994
Boot and shoe cleaner	J. Chesley	Concord, N. H	June 19, 1860	28, 736
Boot and shoe cleaner	J. Malarkey	New York, N. Y	Sept. 24, 1872	131, 620
Boot and shoe cleaner	L. M. Newbury	Sparta, Wis	July 10, 1866	56, 252
Boot and shoe cleaner	W. H. Smith	Newport, R. I	Dec. 7, 1869	97, 716
Boot and shoe cleaning and polishing apparatus	S. Horsley and E. H. Jones	Liverpool, England	Sept. 9, 1862	36, 408
Boot and shoe cleaning machine	T. C. Andrews	Leverington, Pa	July 18, 1865	48, 864
Boot and shoe cleaning machine	P. O. Ferheun and J. I. Ackerman	Hohokus, N. J	Aug. 27, 1872	130, 829
Boot and shoe coasting-guard	J. Fenning	Danbury, Conn	Aug. 5, 1862	36, 123
Boot and shoe conformator	L. Brooks	New York, N. Y	Apr. 28, 1868	77, 166
Boot and shoe counter	J. Brackett	Lynn, Mass	May 14, 1872	126, 778
Boot and shoe counter	G. W. Day	Haverhill, Mass	Nov. 5, 1872	132, 813
Boot and shoe counter and shank	M. E. Savoy	Corinth, N. Y	Sept. 1, 1868	81, 690
Boot and shoe counters, Cutting	S. C. Phinney	Stoughton, Mass	Nov. 14, 1871	120, 998
Boot and shoe counters, Machine for skiving	L. Hill	Stoneham, Mass	June 5, 1855	13, 003
Boot and shoe counters, Machine for skiving	H. S. Vrooman	Boston, Mass	Feb. 27, 1872	124, 098
Boot and shoe crimper	C. Glantz	Baltimore, Md	Apr. 8, 1873	137, 543
Boot and shoe crimping device	J. W. Maxfield	Potsdam, N. Y	July 16, 1867	66, 864
Boot and shoe crimping machine	L. O. Makepeace	Worcester, Mass	Sept. 10, 1872	131, 288
Boot and shoe, Elastic	P. Azam	Baltimore, Md	Sept. 24, 1816	....
Boot and shoe fastening	F. D. Ford	New Bedford, Mass	Feb. 7, 1871	111, 625
Boot and shoe fastening	G. H. Ward	Middletown, Conn	Feb. 18, 1873	136, 111
Boot and shoe feather-edging and channeling machine	L. R. Blake	Fort Wayne, Ind	June 27, 1871	116, 402
Boot and shoe fender and sheath	J. Dillingham	Turner, Me	Feb. 18, 1862	34, 461
Boot and shoe gores, Elastic material for	C. Winslow	Boston, Mass	Feb. 18, 1873	136, 119
Boot and shoe hand-burnisher	J. G. Ross	Philadelphia, Pa	June 27, 1871	116, 491

Index of patents issued from the United States Patent Office from 1790 *to* 1873, *inclusive*—**Continued.**

Invention.	Inventor.	Residence.	Date.	No.
Boot and shoe holder	J. Ellison	Boston, Mass	July 18, 1865	48,801
Boot and shoe, India-rubber	E. M. Chaffee	Roxbury, Mass	May 17, 1834	....
Boot and shoe, India-rubber	L. Elliott, jr	New Haven, Conn	Nov. 12, 1872	133,020
Boot and shoe, India-rubber	B. D. Godfrey	Milford, Mass	Mar. 18, 1862	34,682
Boot and shoe instep stretcher	M. Belding	Hartford, Conn	July 5, 1870	104,923
Boot and shoe iron	J. Knox	Auburn, N. Y	Dec. 4, 1866	60,198
Boot and shoe lacing	P. S. Foster	Richmond, Me	Mar. 9, 1869	87,655
Boot and shoe lacing	C. Goodyear, jr	New York, N. Y	June 26, 1866	55,852
Boot and shoe lacing	C. Goodyear, jr	New York, N. Y	June 26, 1866	55,853
Boot and shoe lacing	L. A. Sprague	New York, N. Y	June 26, 1866	55,923
Boot and shoe lining	T. Lucey and J. E. Murphy	Salem, Mass	Mar. 31, 1868	76,216
Boot and shoe machine	J. E. Wiggin	Stoneham, Mass	Sept. 14, 1869	94,801
Boot and shoe metal-tip	M. Pettengill	Le Roy, Minn	May 31, 1870	103,772
Boot and shoe, Miner's	G. Latham and J. Burton	Jeddo, Pa	Mar. 12, 1872	124,498
Boot and shoe, Moccasin	F. A. and F. G. Bishop	Bangor, Me	June 18, 1872	128,101
Boot and shoe, Moccasin	T. Hersey	Bangor, Me	Sept. 6, 1870	107,045
Boot and shoe, Moccasin	P. Kelleher and J. C. Randlett	Bangor, Me	Dec. 19, 1871	122,030
Boot and shoe nail	B. D. Godfrey	Milford, Mass	Feb. 9, 1869	86,832
Boot and shoe nail	B. Morahan	Brooklyn, N. Y	Sept. 13, 1864	44,212
Boot and shoe nailing machine	L. R. Blake and A. S. Libbey	Fort Wayne, Ind., and Lawrence, Mass.	Jan. 23, 1872	122,985
Boot and shoe nailing machine	W. N. Linnell	Cambridge, Mass	June 11, 1872	127,779
Boot and shoe nailing machines, Device for preparing patterns for.	A. A. Reed	North Bridgewater, Mass.	Oct. 22, 1872	132,492
Boot and shoe nailing machines, Driver for	A. S. Libbey	Lawrence, Mass	Oct. 14, 1873	143,699
Boot and shoe patterns, Graduating	N. Silvester	Boston, Mass	Sept. 19, 1865	50,043
Boot and shoe peg-attachment	D. H. Campbell and E. Woodward.	Sunderland, Scotland, and Charlestown, Mass.	June 6, 1871	115,567
Boot and shoe pegging machine	J. H. Brown	Boston, Mass	Oct. 13, 1863	40,306
Boot and shoe piping	C. Stone	Stoughton, Mass	Apr. 3, 1866	53,699
Boot and shoe rivet	P. West	Worcester, Mass	Jan. 21, 1873	135,051
Boot and shoe seam	S. A. Brackett and W. H. Whitcomb.	Boston, Mass	Dec. 10, 1872	133,826
Boot and shoe seam	L. H. Farnsworth	Hudson, Mass	May 7, 1872	126,537
Boot and shoe seam	C. F. Langford	Brooklyn, N. Y	July 30, 1872	129,970
Boot and shoe shank	A. Bertram	New Albany, Ind	July 14, 1868	79,941
Boot and shoe shank	G. Goodyear	New York, N. Y	Sept. 12, 1871	118,851
Boot and shoe shank	G. Goodyear	New York, N. Y	Mar. 18, 1873	136,991
Boot and shoe shank	T. H. Gorden and E. N. Dow	Hudson, N. Y	Nov. 5, 1872	132,719
Boot and shoe shank	A. J. and A. J. Rice	Salem, Mass	Sept. 10, 1867	68,652
Boot and shoe shank and toe laster	C. L. Graves	Osage, Iowa	Aug. 29, 1871	118,527
Boot and shoe shank-bending machine	H. Barnes	Brookfield, Mass	May 7, 1872	126,434
Boot and shoe shank-piece	J. M. Watson	Sharon, Mass	Jan. 17, 1871	111,022
Boot and shoe shanks, Machines for burnishing	A. C. Carey	Malden, Mass	Sept. 23, 1873	142,990
Boot and shoe shave	A. B. Clark	North Oxford, Mass	Mar. 15, 1870	100,724
Boot and shoe shave	L. H. Farnsworth	Hudson, Mass	Dec. 22, 1868	85,077
Boot and shoe shave	D. Harrington	Southbridge, Mass	Apr. 29, 1873	138,328
Boot and shoe shave	A. E. Johnson	Oxford, Mass	July 30, 1867	67,313
Boot and shoe shave	S. Packard	Grafton, Mass	Feb. 1, 1870	99,466
Boot and shoe shave	B. A. Stockwell	Sutton, Mass	Mar. 12, 1872	124,638
Boot and shoe shave	W. Wilber	New Salem, Mass	Feb. 16, 1869	86,958
Boot and shoe shield	J. P. Bradley	Lawrence, Mass	Oct. 29, 1867	70,157
Boot and shoe shield	O. Collier	Sacramento, Cal	Aug. 22, 1871	118,201
Boot and shoe shield, Metallic	J. Platt and M. D. Brooks	Akron, Ohio	May 24, 1859	24,150
Boot and shoe spring-shank	W. J. Neill	Lawrence, Kans	Dec. 30, 1873	145,963
Boot and shoe stay or brace	T. K. Reed	East Bridgewater, Mass	June 18, 1867	65,839
Boot and shoe stiffener	N. J. Simonds	Woburn, Mass	July 19, 1870	105,501
Boot and shoe stiffening, Machine for crimping	N. J. Simonds	Woburn, Mass	Nov. 19, 1872	133,262
Boot and shoe straightener	G. W. Wilford	Dayton, Ohio	Feb. 23, 1869	87,129
Boot and shoe stretcher	E. R. Bardin	Newburgh, N. Y	June 13, 1871	115,808
Boot and shoe stretcher	J. L. Devol	Parkersburgh, W. Va	June 29, 1869	91,920
Boot and shoe stretcher	W. Frederick	Ashland, Pa	June 19, 1866	55,637
Boot and shoe stretcher	W. Frederick	Pottsville, Pa	Apr. 7, 1868	76,431
Boot and shoe stretcher	W. Holden	Philadelphia, Pa	May 1, 1855	12,793
Boot and shoe stretcher	J. Lyons	Chippewa, Canada	Sept. 9, 1873	142,710
Boot and shoe stretcher	T. C. Maris	Marietta, Ohio	July 9, 1872	128,740
Boot and shoe stretcher	P. Veitch	San Francisco, Cal	Feb. 2, 1869	86,475
Boot and shoe stretcher	J. G. Young, jr	Auburn, Me	Oct. 28, 1862	36,818
Boot and shoe stretching instrument	G. Munro	Philadelphia, Pa	July 14, 1868	79,850
Boot and shoe stud	H. S. Walcott	Boston, Mass	Apr. 28, 1868	77,421
Boot and shoe tacking machine	D. M. Smyth	Lynn, Mass	Sept. 30, 1873	143,388
Boot and shoe tap	J. C. Hancock and J. C. and E. P. Richardson.	Charlestown and Somerville, Mass.	June 27, 1871	116,439
Boot and shoe tip	A. B. Ely	Newton, Mass	Sept. 24, 1867	69,082
Boot and shoe tip	T. J. Mayall	Roxbury, Mass	Oct. 8, 1861	33,440
Boot and shoe tip	G. A. Mitchell	Turner, Me	Mar. 12, 1861	31,673
Boot and shoe tip	F. W. Rhinelander	New York, N. Y	Dec. 16, 1873	145,590
Boot and shoe tip	N. Silverthorn	Prescott, Wis	Nov. 29, 1859	26,329
Boot and shoe tip, Detachable	M. R. Hanley	Providence, R. I	May 20, 1873	139,060
Boot and shoe toe plate	G. Beatty	Cleveland, Ohio	Aug. 31, 1869	94,382
Boot and shoe top	J. Honecker	Columbus, Ohio	June 1, 1869	90,753
Boot and shoe tree	T. R. Evans	Philadelphia, Pa	Sept. 19, 1871	119,131
Boot and shoe tree and stretcher	T. R. Evans	Philadelphia, Pa	Aug. 5, 1873	141,550
Boot and shoe trimming jack	J. Webb, jr	Portland, Me	Aug. 1, 1871	117,582
Boot and shoe uppers, Corrugation of	W. Lee	New Haven, Conn	Mar. 8, 1870	100,535
Boot and shoe uppers, Machine for cutting	P. Jackson	Saugus, Mass	Oct. 10, 1865	50,360
Boot and shoe uppers, Machine for cutting	J. P. Molliere	Lyons, France	Nov. 13, 1855	13,796
Boot and shoe uppers, Machinery for crimping	A. Knowlton	Boston, Mass	Sept. 2, 1873	142,402
Boot and shoe uppers, soles, &c., from sheets of India-rubber, Machine for cutting.	J. and E. Arthur	New Brunswick, N. J	July 24, 1855	13,296
Boot and shoe ventilator	A. Stocker	Watertown, N. Y	Nov. 12, 1867	70,915
Boot and shoe, Water-proof	D. Clarkson	New York, N. Y	Dec. 2, 1835	....
Boot and shoe, Water-proof	A. Dana	Boston, Mass	Dec. 31, 1821	....
Boot and shoe, Water-proof	S. Eells, 2d	Middletown, Conn	Apr. 23, 1830	....
Boot and shoe, Water-proof	E. L. Simpson	Bridgeport, Conn	Sept. 8, 1863	39,865
Boot and shoe, Water-proof	L. C. Tower	Rochester, N. Y	Feb. 23, 1869	87,310
Boot and shoe welt and rand	H. F. Packard	North Bridgewater, Mass	July 4, 1871	116,744
Boot and shoe welt trimmer	J. H. Allen	Wadsworth, Nev	July 18, 1871	117,138

Index of patents issued from the United States Patent Office from 1790 to 1873, inclusive—Continued.

Invention.	Inventor.	Residence.	Date.	No.
Boot and shoe welts, Machine for cutting	S. D. Tripp	Stoneham, Mass	Jan. 14, 1862	34, 172
Boot and shoe wiper	J. K. Staman	Mifflin, Ohio	June 5, 1860	28, 614
Boot and shoe, Wooden-soled	J. Fulton	Zanesville, Ohio	Oct. 3, 1865	50, 236
Boot and shoe work trimmer	A. V. Hill	Limestone, N. Y	Feb. 25, 1868	74, 757
Boot, Balmoral	H. Hayward	Fitchburgh, Mass	Jan. 9, 1866	51, 944
Boot, Balmoral or laced	L. R. Stockbridge	Haverhill, Mass	May 9, 1871	114, 724
Boot-blacking	A. Boudrou	Philadelphia, Pa	June 11, 1867	65, 535
Boot blacking and polishing machine	S. W. Huntington	Augusta, Me	Nov. 5, 1867	70, 571
Boot-blacking apparatus	H. Churchman	Horsham, England	June 29, 1869	91, 825
Boot-blacking apparatus	J. M. Connel	Newark, Ohio	Jan. 18, 1859	22, 628
Boot-blacking case	F. G. Harding	Boston, Mass	July 25, 1865	48, 931
Boot-blacking kit	S. Van Gilder	Knoxville, Tenn	July 1, 1873	140, 447
Boot-blacking machine	M. Burnell	Arundel, England	May 4, 1869	89, 552
Boot-blacking machine	N. Eisenmann	New York, N. Y	Jan. 30, 1872	123, 160
Boot-blacking machine	A. E. White	Rockford, Ill	May 10, 1864	42, 707
Boot-blacking ottoman	D. B. Boynton	Boston, Mass	Feb. 6, 1866	52, 381
Boot-blacking stand	D. Keyser	Philadelphia, Pa	May 13, 1862	35, 240
Boot, bootee, &c., Iron-bound	E. Twombly	Norway, Mass	Aug. 28, 1812	....
Boot-brushing apparatus	M. Beck and C. Schneidt	Philadelphia, Pa	July 29, 1873	141, 196
Boot, Button	J. L. Joyce	New Haven, Conn	Aug. 25, 1868	81, 378
Boot, Button	E. P. Taylor	New Bedford, Mass	July 28, 1868	80, 430
Boot-calk	H. Myer	Derby, Conn	Feb. 9, 1869	86, 857
Boot-clamp for base-ball players	E. S. Ellis	Trenton, N. J	Jan. 9, 1872	122, 587
Boot-clamp, Horizontal	E. G. Pomeroy	Newark, Ohio	Oct. 4, 1836	38
Boot-cleaning machine, Cylindrical	I. Pinkney	New York, N. Y	Apr. 16, 1817	....
Boot, Cloth	M. Crane	East Somerville, Mass	Oct. 22, 1872	132, 387
Boot-conformator	S. W. Shorey	Galesburgh, Ill	Dec. 7, 1869	97, 558
Boot, Congress	H. S. Holmes	Lynn, Mass	June 19, 1860	28, 754
Boot, Cork-sole	W. L. McCauley	Baltimore, Md	June 5, 1844	3, 615
Boot-counter machine	J. Brooks and C. F. Sylvester	North Bridgewater, Mass	July 11, 1865	48, 651
Boot-counters, Machine for cutting and skiving	V. Snell	North Bridgewater, Mass	Apr. 25, 1854	10, 835
Boot-counters, Machine for skiving	W. Butterfield and B. Stetson	Boston and Uxbridge, Mass	June 16, 1857	17, 603
Boot-counters, Machine for skiving	S. J. and C. H. Trofatter	Salem, Mass	Nov. 29, 1853	10, 288
Boot-cramp	H. L. Pierce	Saint Johnsbury, Vt	June 25, 1836	....
Boot-crimp	L. Barrett	Gainesville, N. Y	Oct. 26, 1852	9, 352
Boot-crimp	A. and G. W. Caywood	Ithaca, Ohio	Nov. 14, 1865	50, 904
Boot-crimp	J. Copeland	Weymouth, Mass	Jan. 20, 1844	3, 410
Boot-crimp	T. Daugherty	Erie, Pa	May 8, 1855	12, 816
Boot-crimp	H. S. Davis	Westernville, N. Y	June 27, 1846	4, 603
Boot-crimp	N. Dawes and H. Harrison	Little York, N. J	May 27, 1851	8, 127
Boot-crimp	N. Day	Ithaca, Ohio	Feb. 6, 1866	52, 394
Boot-crimp	E. P. Drake	Troy, N. Y	Nov. 15, 1843	3, 340
Boot-crimp	R. and S. Fairchild	Trumbull, Conn	May 10, 1845	4, 038
Boot-crimp	H. Ferre	Springfield, Mass	Sept. 25, 1839	1, 338
Boot-crimp	G. Fetter	Philadelphia, Pa	Mar. 4, 1856	14, 353
Boot-crimp	B. C. Finfrock	Stephenson's Depot, Va	Mar. 2, 1869	87, 333
Boot-crimp	E. K. Horner and W. Holland	Fayetteville, Pa	Oct. 2, 1849	6, 770
Boot-crimp	F. Kali and S. Andrews	Rochester, N. Y	Nov. 5, 1867	70, 440
Boot-crimp	F. L. Kathan and E. D. Rummer	Roscoe, Ill	June 30, 1868	79, 357
Boot-crimp	B. Livermore	Hartland, Vt	Oct. 16, 1849	6, 794
Boot-crimp	C. Lomax	Paoli, Ind	June 22, 1869	91, 756
Boot-crimp	G. W. Loy	Winchester, Va	Sept. 24, 1872	131, 552
Boot-crimp	E. Mann and A. J. F. Howard	Milford, Mass	Aug. 18, 1868	81, 096
Boot-crimp	A. Overholt	Gardenville, Pa	May 11, 1869	90, 016
Boot-crimp	S. Pasco and E. Perry	Cato, N. Y	Apr. 3, 1849	6, 260
Boot-crimp	B. Rowe	Maryland, N. Y	Dec. 17, 1842	2, 885
Boot-crimp	P. Sanders	Middletown, Pa	Nov. 26, 1872	133, 339
Boot-crimp	J. Sanderson	Cincinnati, Ohio	June 7, 1839	1, 167
Boot-crimp	C. Snow and T. N. Sadler	Spencer, Mass	May 21, 1845	4, 057
Boot-crimp	H. Stanley	Wilmington, Vt	Aug. 19, 1851	8, 305
Boot-crimp	P. Stevens, jr	Canton, Mass	July 15, 1844	3, 666
Boot-crimp	W. Taylor	Berlin, N. Y	Oct. 3, 1846	4, 793
Boot-crimp	J. Tipton and J. Carl	Malaga, Ohio	Jan. 21, 1868	73, 554
Boot-crimp	J. E. Tucker	Boston, Mass	Aug. 22, 1848	5, 723
Boot-crimp	O. J. Warren	Chicago, Ill	Mar. 15, 1864	41, 952
Boot-crimp	C White	Galway, N. Y	Feb. 12, 1845	3, 909
Boot-crimp	C. White	Galway, N. Y	Jan. 12, 1848	5, 412
Boot-crimp	J. G. Whittier	Attica, Ind	Apr. 2, 1861	31, 922
Boot-crimp	W. W. Willmott	Boston, Mass	Aug. 25, 1857	18, 074
Boot-crimp	H. Wright	Saco, Me	Jan. 21, 1868	73, 483
Boot-crimp	J. Young	West Galway, N. Y	Aug. 26, 1845	4, 165
Boot-crimper	O. M. Adams	Milford, Mass	Nov. 24, 1868	84, 402
Boot-crimper	J. B. Aikin	Somerton, Ohio	May 25, 1869	90, 477
Boot-crimper	J. Brewster	Worthington, Mass	Mar. 23, 1830	....
Boot-crimper	A. Dunbar	Sharon, Mass	Feb. 19, 1830	....
Boot-crimper	W. B. Gleason	Conneautville, Pa	Dec. 8, 1868	84, 743
Boot-crimper	H. Henley	Shoals, Ind	Apr. 4, 1871	113, 430
Boot-crimper	A. J. F. Howard	Milford, Mass	Mar. 31, 1868	76, 079
Boot-crimper	A. J. F. Howard	Milford, Mass	Aug. 18, 1868	81, 087
Boot-crimper	F. C. Jackson	Peru, Ind	Dec. 1, 1868	84, 629
Boot-crimper	S. W. Jamison	Newark, N. J	June 21, 1870	104, 596
Boot-crimper	F. P. Marcy	Keokuk, Iowa	Dec. 14, 1869	97, 944
Boot-crimper	D. C. Mowrey	Milford, Mass	June 25, 1867	66, 035
Boot-crimper	W. Polsgrove	Saint Thomas, Pa	Jan. 24, 1871	111, 245
Boot-crimper	E. Powell	New London, Ind	July 4, 1871	116, 750
Boot-crimper	E. Powell	New London, Ind	Feb. 27, 1872	124, 157
Boot-crimper	J. M. Read	Boston, Mass	July 7, 1868	79, 777
Boot-crimper	J. M. Read	Boston, Mass	June 8, 1869	91, 162
Boot crimper	P. Richmond and A. McFarland	Aberdeen and Allensville, Ind	May 19, 1868	78, 131
Boot-crimper's rub-stick	A. J. F. Howard	Milford, Mass	July 25, 1871	117, 288
Boot-crimping	O. J. Warren	New York, N. Y	Mar. 27, 1866	53, 508
Boot-crimping apparatus	M. R. Marcell	Dansville, N. Y	Nov. 25, 1873	144, 995
Boot-crimping block	H. Bordner and D. Sullivan	Virginia City, Nev	Sept. 2, 1873	142, 369
Boot-crimping device	T. Madgett	Buffalo, N. Y	Nov. 17, 1863	40, 634
Boot-crimping device	W. C. Shipherd	Saratoga Springs, N. Y	June 10, 1862	35, 547
Boot-crimping machine	J. G. Baker, jr	Philadelphia, Pa	Feb. 17, 1857	16, 670
Boot-crimping machine	J. D. Batchelor	Upton, Mass	Nov. 7, 1865	50, 788

Index of patents issued from the United States Patent Office from 1790 *to* 1873, *inclusive*—Continued.

Invention.	Inventor.	Residence.	Date.	No.
Boot-crimping machine	J. D. Black	Boston, Mass	May 10, 1859	23, 977
Boot-crimping machine	H. Conklin	Kirkwood, N. Y	July 9, 1867	66, 566
Boot crimping machine	R. H. Dorn	Port Henry, N. Y	June 2, 1868	78, 438
Boot-crimping machine	R. H. Dorn	Port Henry, N. Y	Jan. 30, 1872	123, 158
Boot-crimping machine	W. R. Dunn	Alton, Ind	May 4, 1869	89, 640
Boot-crimping machine	J. A. Eldridge	Milford, Mass	June 1, 1869	90, 828
Boot-crimping machine	B. Esch	Sandusky, Ohio	June 1, 1869	90, 648
Boot-crimping machine	H. Faus	Hayesville, Ohio	Mar. 5, 1867	62, 621
Boot crimping machine	L. W. Hayden	Wilkesbarre, Pa	Mar. 12, 1861	31, 664
Boot-crimping machine	H. B. Horton	Northville, Mich	May 1, 1855	12, 794
Boot-crimping machine	T. Howe	Worcester, Mass	June 11, 1829	....
Boot crimping machine	S. W. Jamison	New York, N. Y	May 7, 1867	64, 538
Boot-crimping machine	J. S. Landes	Lancaster, Pa	Feb. 14, 1865	46, 427
Boot-crimping machine	C. Lomax and H. Lindley	Paoli, Ind	Feb. 6, 1872	123, 492
Boot-crimping machine	S. Morehouse	Eastport, Me	June 19, 1827	....
Boot-crimping machine	J. A. Nesbit	Charlottesville, Ind	Mar. 3, 1868	75, 186
Boot-crimping machine	J. S. Nolen and C. C. Hinchman	Paulsborough, N. J	Mar. 20, 1860	27, 561
Boot-crimping machine	C. Pratt	Pratt's Hollow, N. Y	Dec. 23, 1862	37, 242
Boot-crimping machine	J. Rausch	Huntington, Ind	Dec. 20, 1870	110, 285
Boot-crimping machine	C. H. Rice	Port Henry, N. Y	Feb. 25, 1868	74, 848
Boot-crimping machine	P. Shaw	Abington, Mass	Oct. 23, 1860	30, 502
Boot-crimping machine	G. Utley	Chapel Hill, N. C	Dec. 19, 1854	12, 109
Boot-crimping machine	R. Warren	Jefferson, Ohio	Mar. 20, 1860	27, 586
Boot-crimping machine	H. Wing	Buffalo, N. Y	Apr. 12, 1864	42, 324
Boot-crimping machine	S. Witherby	Essex, Vt	May 12, 1832	....
Boot-crimping machine	G. W. Zeigler	Tiffin, Ohio	June 5, 1855	13, 023
Boot-crimping screw	B. S. Norris	Ripley, Ohio	Mar. 24, 1868	75, 784
Boot-fastening	A. A. Abbot	Boston, Mass	July 14, 1868	79, 800
Boot fastening, Gaiter	H. M. Whitmarsh	Abington, Mass	Sept. 16, 1862	36, 497
Boot-form	J. Chilcott and R. Snell	Brooklyn, N. Y	Apr. 10, 1855	12, 670
Boot-form	J. Howe	Milford, Mass	May 5, 1868	77, 489
Boot-form	L. N. Leland	Grafton, Mass	July 21, 1863	39, 293
Boot-forms, Machine for beveling	E. Holmes	Stoughton, Mass	Aug. 12, 1840	1, 716
Boot-forms, Machine for dressing	J. Burgess	Leicester, Mass	July 22, 1851	8, 246
Boot-fronts, Crimping	J. H. Punchard	Boston, Mass	May 8, 1830	....
Boot-fronts, Method of cutting	J. Dick	New York, N. Y	July 27, 1858	20, 992
Boot, Gaiter	S. Babbit	Kokomo, Ind	Mar. 7, 1865	46, 622
Boot, Gaiter	W. H. H. Babbitt	New Corner, Ind	Oct. 13, 1868	83, 017
Boot, Gaiter	C. K. Bradford	Lynn, Mass	Oct. 2, 1860	30, 197
Boot, Gaiter	J. C. Breed and C. K. Bradford	Lynn, Mass	Aug. 18, 1863	39, 544
Boot, Gaiter	C. Hersome	Boston, Mass	Nov. 19, 1872	133, 225
Boot, Gaiter	C. Hersome	Boston, Mass	Dec. 23, 1873	145, 801
Boot, Gaiter	R. B. Jacobs	Quincy, Ill	Nov. 6, 1866	59, 399
Boot, Gaiter	E. Mongaret	Newark, N J	Mar. 16, 1869	87, 869
Boot, Gaiter	T. Powell	Richland, Ind	Sept. 19, 1865	50, 030
Boot, Gaiter	J. Schroeder	Cincinnati, Ohio	July 16, 1872	129, 178
Boot, Half	L. U. Williams	La Fayette, Ind	Aug. 26, 18 3	142, 312
Boot-jack	F. Ahl	West Meriden, Conn	Dec. 28, 1858	22, 404
Boot-jack	T. Anthony	Providence, R. I	Feb. 20, 1872	123, 759
Boot-jack	H. Arnot	Barclay, Pa	June 24, 1873	140, 179
Boot-jack	S. P. Babcock	Adrian, Mich	Dec. 28, 1869	98, 219
Boot-jack	H. D. Boss	Williamsburgh, N. Y	Apr. 16, 1867	63, 844
Boot-jack	W. Bourn	Geneva, N. Y	Nov. 4, 1862	36, 831
Boot-jack	C. Brown	Charlottesville, Va	Feb. 1, 1870	99, 517
Boot-jack	E. L. Brown	Chicago, Ill	May 18, 1869	90, 076
Boot-jack	W. W. Cansler	Baltimore, Md	Jan. 25, 1859	22, 700
Boot-jack	W. Case	Russia, N. Y	Dec. 7, 1869	97, 478
Boot-jack	C. Clay	Scranton, Pa	June 11, 1867	65, 543
Boot-jack	E. Coleman	San Francisco, Cal	Feb. 7, 1871	111, 613
Boot-jack	J. Crabtree	Cincinnati, Ohio	Jan. 3, 1871	110, 743
Boot-jack	H. Crocker	Brighton, Mass	May 21, 1872	126, 876
Boot-jack	P. Cullen	Bridgeport, Conn	May 4, 1869	89, 633
Boot-jack	J. Darden	Washington, D. C	Dec. 15, 1868	84, 862
Boot-jack	H. N. Degraw	Green Island, N. Y	Feb. 22, 1859	23, 015
Boot-jack	H. N. Degraw	Green Island, N. Y	Feb. 12, 1861	31, 374
Boot-jack	H. N. Degraw	Newburgh, N. Y	Nov. 13, 1866	59, 712
Boot-jack	F. Dorsett, sr	Chicago, Ill	Oct. 7, 1873	143, 408
Boot-jack	J. Durham	Cherry Grove, Ohio	Dec. 18, 1860	30, 913
Boot-jack	F. J. French	Whitingham, Vt	June 5, 1855	13, 002
Boot-jack	J. A. McKinstry	Springfield, Mass	Dec. 17, 1872	134, 084
Boot-jack	J. G. Moulton	Boston, Mass	Oct. 29, 1867	70, 246
Boot-jack	W. S. Peffer	Carlisle, Pa	Mar. 26, 1867	63, 292
Boot-jack	A. Pierpont	New Haven, Conn	May 22, 1866	54, 948
Boot-jack	M. A. Richardson	Sherman, N. Y	July 19, 1864	43, 626
Boot-jack	M. A. Richardson	Sh rman, N. Y	May 16, 1871	114, 970
Boot-jack	M. T. Ridout	Milwaukee, Wis	Oct. 22, 1861	33, 543
Boot-jack	J. Roward	San Francisco, Cal	Mar. 14, 1871	112, 638
Boot-jack	A. P. Seymour	Hecla Works, N. Y	Oct. 29, 1867	70, 275
Boot-jack	O. S. Sikes	Saffield, Conn	May 18, 1858	20, 307
Boot-jack	S. B. Sumner	Grantville, Mass	Sept. 13, 1853	10, 016
Boot-jack	S. Thomson	Hartsville, Mass	Apr. 6, 1852	8, 865
Boot-jack	W. H. Towers	New York, N. Y	Jan. 4, 1861	32, 490
Boot-jack	J. Walker	Cincinnati, Ohio	Mar. 19, 1867	63, 121
Boot-jack	G. Wheeler	New York, N. Y	June 26, 1868	28, 927
Boot-jack	J. Wheeler	Athol, Mass	June 6, 1865	48, 122
Boot-jack	L. J. Wicks	Racine, Wis	Feb. 8, 1859	22, 923
Boot-jack	W. D. Young	Baltimore, Md	Jan. 25, 1859	22, 766
Boot-jack, Adjustable	G. Couch	Saint Louis, Mo	July 30, 1872	129, 936
Boot-jack and blacking-brush	H. J. Miller	Nashua, N. H	Sept. 24, 1867	69, 112
Boot-jack and blacking-brush	H. J. Miller	Nashua, N. H	Mar. 3, 1868	75, 181
Boot-jack and blacking-cabinet	L. P. Keach	Baltimore, Md	Jan. 26, 1869	86, 165
Boot-jack and blacking-case	B. Douglas	New York, N. Y	Sept. 25, 1866	58, 230
Boot-jack and brush	S. Kennedy	Rochester, Pa	Apr. 19, 1870	102, 014
Boot-jack and burglar-alarm combined	F. C. Goffin	Newark, N. J	Apr. 6, 1858	19, 844
Boot-jack and door-buffer	W. S. Chatham	Williamsport, Pa	Dec. 27, 1870	110, 433
Boot-jack and spur	T. Gunn	Hamilton County, Ind	July 26, 1870	105, 671
Boot-jack, nut-cracker, tack-hammer, and tack-puller.	G. R. Osborn and J. Crandell	Ilion, N. Y	Apr. 17, 1866	54, 008

Index of patents issued from the United States Patent Office from 1790 *to* 1873, *inclusive*—Continued.

Invention.	Inventor.	Residence.	Date.	No.
Boot-jack, wrench, and nail-pull	O. Shepard	Alton, Ill	July 9, 1867	66,643
Boot-lacing	W. Banister	Boston, Mass	Feb. 2, 1869	86,494
Boot-lacing device	A. McKenney and S. Carpenter.	Portland and Patten, Me	June 25, 1867	66,099
Boot-lacing, Mode of	R. Newton	Philadelphia, Pa	July 7, 1863	39,163
Boot, Lady's	E. F. Doty	Ravenna, Ohio	Sept. 13, 1870	107,231
Boot, Lady's	H. Libby	Evansville, Wis	June 13, 1865	48,188
Boot, Lady's	W. Love	McConnellsville, Ohio	Dec. 28, 1869	98,391
Boot, Lady's	D. H. Murphy	Lynn, Mass	Apr. 16, 1872	125,834
Boot-leg	D. A. Haviland	Fort Dodge, Iowa	Sept. 3, 1861	33,195
Boot-leg	C. H. Leffingwell	Providence, R. I	May 14, 1861	32,335
Boot-leg	A. P. Nash	Weymouth, Mass	May 16, 1865	47,743
Boot-leg stiffener	N. Gear	Indianapolis, Ind	Apr. 26, 1864	42,544
Boot-leg straps, Method of securing	L. J. Worden	Utica, N. Y	Jan. 26, 1858	19,227
Boot-leg straps, Mode of attaching	J. A. Pickering	Milford, Mass	Apr. 5, 1859	23,532
Boot-legs, Cramping Suwarrow	E. Lundy	New York, N. Y	Aug. 14, 1810	......
Boot-legs, Cramping Suwarrow	J. Sibley	New York, N. Y	Apr. 25, 1811	......
Boot-legs, Crimping	W. Gerrish	Poland, Me	Feb. 10, 1836	......
Boot-legs, Cutting and manufacturing	J. Vernon	Baltimore, Md	Oct. 21, 1811	......
Boot-legs, Machine for cramping	N. L. Moore	Smithfield, N. Y	Feb. 4, 1812	......
Boot-legs, Machine for crimping leather for	J. Van Winkle	Rochester, N. Y	Apr. 17, 1837	164
Boot-legs, Machine for turning	D. Bissell	Detroit, Mich	Dec. 3, 1861	33,818
Boot-legs, Machine for turning	M. C. Chamberlin and W. Filkins.	Sheldon, N. Y	Nov. 25, 1856	16,123
Boot-legs, Machine for turning	C. Collins	Warren, Ind	June 22, 1869	91,606
Boot-legs, Machine for turning	A. Ransom	Manheim, N. Y	Jan. 29, 1861	31,281
Boot-legs, Machine for turning	T. N. Saddler	Spencer, Mass	Nov. 8, 1836	73
Boot-legs, Machine for turning	J. Shearman	Fayetteville, Pa	Aug. 18, 1868	81,301
Boot-legs, Mode of closing up	P. Ware and C. R. Tilton	Newton and Tisbury, Mass	Jan. 7, 1868	73,062
Boot legs, Mode of siding	G. C. Parker	Worcester, Mass	July 8, 1873	147,726
Boot-legs, Turning	J. H. Sanford	Hopewell, N. Y	June 22, 1842	2,686
Boot-lining	H. White	La Fayette, Ind	Aug. 26, 1873	142,307
Boot, Mathematical	W. G. Crease and J. Morgan	Philadelphia, Pa	July 20, 1812	......
Boot, Moccasin	E. A. Buck	Bangor, Me	May 27, 1873	139,234
Boot, Open-backed spring-box	W. Atchison	New York, N. Y	May 23, 1815	......
Boot or bootee	T. P. Mitchell	Haverhill, Mass	Dec. 2, 1873	145,117
Boot or harness clamp	A. J. Curtis	Winterport, Me	Oct. 31, 1865	50,766
Boot or shoe, Nailed	L. R. Blake	Boston, Mass	Aug. 8, 1865	49,219
Boot or shoe, Pegged	S. Preston	Danvers, Mass	Mar. 8, 1833	......
Boot-packs, Method of cutting	W. G. Slater	Hart, Mich	Nov. 8, 1870	109,065
Boot-pattern	J. F. Daniels	Greencastle, Ind	Sept. 10, 1872	131,254
Boot-pattern	A. Forrist and C. A. Wheeler	Mount Vernon, Iowa	Sept. 24, 1861	33,342
Boot-pattern	J. Rider	Wooster, Ohio	June 2, 1845	4,068
Boot-pattern	S. C. Shive	Bloom Township, Pa	Aug. 14, 1847	5,244
Boot-pattern	E. Shopbell	Ashland, Ohio	Feb. 8, 1870	99,604
Boot-pattern	E. Shopbell	Ashland Ohio	Oct. 3, 1871	119,538
Boot, Plow	W. Swarts	Pent Water, Mich	Oct. 19, 1869	95,951
Boot, Plowman's	C. H. Ricker	Dover, N. H	Apr. 22, 1873	138,196
Boot-protector	A. G. Walker and S. Brown	Worcester, Mass	Aug. 19, 1873	142,061
Boot protector, Gaiter	J. U. Johnson	Springfield, Mass	Aug. 4, 1868	80,549
Boot, Riding	F. N. Blodgett	Boston, Mass	Dec. 15, 1863	40,901
Boot, Rubber	B. C. Young	Boston, Mass	July 23, 1872	129,776
Boot, Rubber	L. Elliott, jr	New Haven, Conn	Feb. 9, 1869	86,827
Boot, Rubber	C. S. Foster and O. Taylor	Ashland and Philadelphia, Pa.	Aug. 29, 1871	118,444
Boot, Rubber	C. E. Lins	Ashland, Pa	Aug. 15, 1871	118,158
Boot-shank, Elastic	I. Gale	Natchez, Miss	July 11, 1844	3,657
Boot-shank machine	L. Schije	Chicago, Ill	Dec. 22, 1868	85,138
Boot-shank, Steel	H. and E. Briner	Manhattanville, N. Y	July 7, 1868	79,631
Boot, shoe, or sandal, Metallic	G. W. Griswold	Abington, Pa	Apr. 5, 1864	42,187
Boot, Skating	J. W. Pearson and H. O. Peabody.	Winchester, Mass	Jan. 1, 1861	31,057
Boot-strap	F. H. Moore	Boston, Mass	Jan. 5, 1864	41,087
Boot-strap fastener	S. Walker	Boston, Mass	Dec. 20, 1859	26,552
Boot-strap guard	J. E. Curtis	Marlborough, Mass	Apr. 2, 1872	125,274
Boot-strap, Metallic	D. Forrest	Eastport, Me	Mar. 30, 1869	88,380
Boot-straps, Machine for making and wrapping webbing.	J. W. Richardson	South Braintree, Mass	Oct. 5, 1869	95,516
Boot-stretcher	D. Harris	Litchfield, Ill	June 10, 1873	139,671
Boot-stretcher	J. Harwood	Albany, N. Y	Jan. 25, 1870	99,084
Boot-stretcher	J. Hoffman	Belvidere, N. J	Aug. 8, 1871	117,777
Boot-stretcher	I. W. Myers	San Francisco, Cal	Jan. 24, 1871	111,137
Boot-tops, &c., Revolving support for printing	A. P. Nash	Weymouth, Mass	June 11, 1872	127,702
Boot-tops, Turning the edge of circular	P. C. Clapp	Stoughton, Mass	Apr. 13, 1858	19,912
Boot-tree	J. Ayars	Brookfield, N. Y	Nov. 27, 1828	......
Boot-tree	W. H. Betts and I. H. Parker	Kokomo, Ind	July 19, 1859	24,787
Boot-tree	T. Branigan	Beloit, Wis	Nov. 4, 1873	144,312
Boot-tree	W. C. Clark	Baltimore, Md	June 30, 1863	39,034
Boot tree	P. De Vries	Adrian, Mich	Nov. 19, 1867	70,975
Boot-tree	C. T. Eames	Milford, Mass	May 27, 1856	14,951
Boot-tree	O. V. Elliott	Mansfield, Pa	July 1, 1873	140,490
Boot-tree	W. B. Fay and R. W. Collier	Upton, Mass	Apr. 5, 1859	23,457
Boot-tree	D. Hastings	Deerfield, Mass	Dec. 15, 1837	514
Boot-tree	D. R. Hendrix	Pottstown, Pa	Oct. 28, 1851	8,467
Boot-tree	J. Howe	Worcester, Mass	Oct. 24, 1848	5,876
Boot-tree	W. and A. G. Kelsey	Delavan, Wis	Nov. 20, 1866	59,842
Boot-tree	R. L. Lewis	Milford, Mass	Mar. 2, 1858	19,508
Boot-tree	R. L. Lewis	Worcester, Mass	Mar. 5, 1867	62,649
Boot-tree	M. Matthews	Wayne County, Ohio	Feb. 25, 1836	......
Boot-tree	H. B. Miller	Mayville, N. Y	Aug. 25, 1828	......
Boot-tree	O. P. Richardson	Worcester, Mass	June 23, 1863	38,986
Boot-tree	J. Russell	Philadelphia, Pa	July 7, 1846	4,623
Boot-tree	D. Sadleir	McWilliamstown, Pa	Nov. 23, 1852	9,425
Boot-tree	J. H. Sampson	Grafton, Mass	Dec. 4, 1855	13,886
Boot-tree	W. Upfield	Lancaster, Ohio	June 25, 1850	7,466
Boot-tree	W. Upfield	Lancaster, Ohio	Aug. 4, 1857	17,947
Boot-tree	W. W. Willmott	Boston, Mass	Apr. 14, 1857	17,067
Boot-tree	W. W. Willmott	Boston, Mass	May 4, 1858	20,185

Index of patents issued from the United States Patent Office from 1790 *to* 1873, *inclusive*—Continued.

Invention.	Inventor.	Residence.	Date.	No.
Boot-tree	F. S. Wilt	Allentown, Pa	Dec. 17, 1867	72, 351
Boot-tree	A. J. Wisner	Homer, N. Y	July 13, 1858	20, 914
Boot-tree	H. Wright	Newcastle, Me	Jan. 16, 1849	6, 031
Boot-tree and stretcher	J. Bechtel	Roxbury, Pa	Aug. 25, 1868	81, 463
Boot-tree, Screw	G. Nichol	Madison, Va	Sept. 8, 1826	....
Boot-treeing machine	E. Hall, jr	Spencer, Mass	May 29, 1841	2, 110
Boot-trimming tool	L. C. Rogers	Danvers, Mass	Oct. 16, 1860	30, 427
Boot-tucking machine	S. C. Blodget	Rowley, Mass	June 11, 1836	....
Boot-vamps, Mode of cutting	E. L. Vertrees	Howe's Valley, Ky	Feb. 14, 1860	27, 173
Boot, Ventilated rubber	H. C. Cottrell	Norwich, Conn	Apr. 26, 1870	102, 228
Boot, Ventilated rubber	R. W. Perkins	Colchester, Conn	Apr. 26, 1870	102, 314
Boot, Ventilating	S Perry	New York, N. Y	Aug. 21, 1866	57, 371
Boot, Water-proof	A. Cushman	Auburn, Me	Mar. 5, 1872	124, 257
Boots, &c., Adjusting pattern for	F. E. Augustine	Dubuque, Iowa	July 26, 1864	43, 631
Boots and gaiters, Machine for fitting fronts of	T. Lincoln	Saint Louis, Mo	Nov. 19, 1872	133, 106
Boots and other leather stock, Punching	C. H. Helms	Poughkeepsie, N. Y	Jan. 5, 1869	85, 661
Boots and shoes, Apparatus for crimping the stiffening of.	J. W. Hatch	Rochester, N. Y	Aug. 1, 1871	117, 627
Boots and shoes, Apparatus for holding and supporting.	E. Lemercier	Paris, France	Mar. 10, 1863	37, 886
Boots and shoes, Apparatus for molding box-toes for.	D. H. Packard	North Bridgewater, Mass.	Oct. 21, 1873	143, 926
Boots and shoes, belts for ladies' dresses, &c., Clasp for.	E. W. Prussia	Marlborough, Mass	Aug. 11, 1868	80, 834
Boots and shoes, Black-ball for	G. A. Knowlton	Natick, Mass	Feb. 15, 1870	99, 915
Boots and shoes, Box-toe for	H. W. George	Danvers, Mass	Oct. 19, 1869	95, 894
Boots and shoes by rule, Cutting	S. Hart	Monkton, Vt	Mar. 3, 1821	....
Boots and shoes by substituting a metallic shank on outsole, Manner of manufacturing.	A. Thayer	Braintree, Mass	Apr. 24, 1841	2, 062
Boots and shoes, Cap for	B. F. Sage	Beverly, N. J	Jan. 23, 1872	122, 969
Boots and shoes, Channel for	R. Ashe	Boston, Mass	June 21, 1870	104, 534
Boots and shoes, Composition for printing designs on.	D. McKellar	Lowell, Mass	Mar. 2, 1869	87, 350
Boots and shoes, Compound rubber-enameled cloth for.	E. M. Stevens	Chelsea, Mass	Mar. 31, 1868	76, 265
Boots and shoes, Constructing	L. R. Blake	Boston, Mass	May 16, 1865	47, 696
Boots and shoes, Construction of	M. M. Wheeler	Galena, Ill	Feb. 25, 1873	136, 293
Boots and shoes, Cover for	W. E. Prall	Washington, D. C	Sept. 20, 1870	107, 538
Boots and shoes, Crimping	E G. Pomeroy	Lancaster, Ohio	July 27, 1824	....
Boots and shoes, Cutting	J. Chilcott and R. Snell	Brooklyn, N. Y	Sept. 13, 1853	10, 099
Boots and shoes, Cutting and forming	J. Hall	Frederick Town, Md	Dec. 15, 1815	....
Boots and shoes, Device for holding	H. T. Dillon	Big Lick, Va	Jan. 14, 1868	73, 304
Boots and shoes, Device for tapping	G. Conkling	Conklinsville, N. Y	Aug. 9, 1864	43, 814
Boots and shoes, Die for making metallic shank for.	H. F. Whidden	South Abington, Mass	Aug. 29, 1871	118, 660
Boots and shoes, Edge-protecting welt for	J. Green	Brooklyn, N. Y	Oct. 14, 1873	143, 623
Boots and shoes, Elastic calk for	G. H. Clemens	New York, N. Y	Nov. 10, 1868	83, 930
Boots and shoes, Elastic goring for	J. Haskins	Boston, Mass	Oct. 13, 1868	82, 944
Boots and shoes, Elastic goring for	C. Winslow	Boston, Mass	Mar. 12, 1872	124, 527
Boots and shoes, Fastening	B. F. Allen and J. R. Ryerson	Saint Albans, Me	July 21, 1868	80, 109
Boots and shoes, Graduating pattern for	M. Meade	Boston, Mass	May 18, 1869	90, 115
Boots and shoes, Hand nail-driver for	E. Hale	Boston, Mass	July 28, 1868	80, 477
Boots and shoes, Hand-nailer for lasting	W. E. Fischer	Boston, Mass	June 1, 1869	90, 650
Boots and shoes, Head-block for holding	I. E. Allen	Windham, Me	Feb. 20, 1866	52, 657
Boots and shoes heels made of vulcanized wood, Method of attaching to the soles of.	F. H. Morgan	Beverly, Mass	May 11, 1869	90, 049
Boots and shoes, &c., Instrument for trimming welts of.	L. Clark	South Royalston, Mass	June 13, 1854	11, 078
Boots and shoes, Jack for nailing, &c	J. G. Ross	Philadelphia, Pa	Oct. 21, 1873	143, 786
Boots and shoes, Jack for the manufacture of	C. M. Gustin	Saccarappa, Me	Dec. 18, 1866	60, 507
Boots and shoes, Jointed tip for	G. A. Mitchell	Turner, Me	May 22, 1860	28, 395
Boots and shoes, Lasting	W. Wickersham	Boston, Mass	Aug. 22, 1871	118, 319
Boots and shoes lengthwise, Instrument for stretching.	W. Jones	Norfolk, Va	Nov. 20, 1866	59, 912
Boots and shoes, Lining for	J. Adams	Kokomo, Ind	Sept. 12, 1865	49, 946
Boots and shoes, Machine for burnishing shanks of	J. K. Blanchard, F. S. Hunt, and A. C. Carey.	Lynn and Malden, Mass	June 11, 1872	127, 838
Boots and shoes, Machine for crimping leather for	S. W. Jamison	Newark, N. J	June 10, 1873	139, 717
Boots and shoes, Machine for cutting out the uppers and soles of.	W. G. Greeley	Hingham, Mass	Feb. 14, 1860	27, 123
Boots and shoes, Machine for cutting pegs out of	T. Corey	Marlborough, Mass	May 15, 1866	54, 692
Boots and shoes, Machine for cutting rands for	J. H. Walker	Worcester, Mass	Dec. 3, 1872	133, 686
Boots and shoes, Machine for distributing nails for the manufacture of.	H. P. Fairfield	Boston, Mass	June 18, 1872	128, 134
Boots and shoes, Machine for feeding and inserting rivets in.	G. V. Sheffield	Boston, Mass	Apr. 30, 1872	126, 159
Boots and shoes, Machine for holding the uppers of	E. M. Dickinson	Fitchburgh, Mass	Apr. 11, 1865	47, 191
Boots and shoes, Machine for inserting screw-pegs in.	A. Cavalli	San Francisco, Cal	Apr. 1, 1873	137, 288
Boots and shoes, Machine for lasting the uppers of	M. R. Ethridge	Bethel, Me	Mar. 17, 1863	37, 907
Boots and shoes, Machine for making	E. Barnum	Massachusetts	Jan. 9, 1808	....
Boots and shoes, Machine for making	C. H. Southall and R. Heap	Staleybridge, England	Feb. 20, 1866	52, 802
Boots and shoes, Machine for making	E. Thayer	Leicester, Mass	Mar. 14, 1812	....
Boots and shoes, Machine for making	E. Thayer	Leicester, Mass	June 9, 1826	....
Boots and shoes, Machine for making shank-pieces for.	C. H., D. D., and F. M. Blake	Worcester, Mass	Jan. 7, 1873	134, 584
Boots and shoes, Machine for nailing and pegging	J. F. Sargent	Melrose, Mass	Sept. 7, 1869	94, 654
Boots and shoes, Machine for paring the edges of	H. Hodges	Lynn, Mass	Oct. 11, 1864	44, 688
Boots and shoes, Machine for pressing seams and cutting welts of.	E. Reed	Kingston, Mass	Feb. 23, 1869	87, 200
Boots and shoes, Machine for pressing seams and cutting welts for.	E. Reed	Kingston, Mass	July 26, 1870	105, 846
Boots and shoes, Machine for skiving counters for	W. A. Bacon	Campello, Mass	May 29, 1860	28, 531
Boots and shoes, Machine for trimming and burnishing the edges of.	C. J. Addy	Boston, Mass	Sept. 16, 1873	142, 756
Boots and shoes, Machine for turning channel-flaps for.	W. E. Fischer	Boston, Mass	July 2, 1872	128, 610
Boots and shoes, Machine for wrinkling the instep of.	C. Heisterman	Brownsville, Pa	Oct. 30, 1866	59, 219

Index of patents issued from the United States Patent Office from 1790 *to* 1873, *inclusive*—Continued.

Invention.	Inventor.	Residence.	Date.	No.
Boots and shoes, Machinery for burnishing the shanks of.	J. K. Blanchard, F. S. Hunt, and A. C. Carey.	Lynn and Malden, Mass	Mar. 12, 1872	124, 479
Boots and shoes, Machinery for crimping uppers for.	C. Glantz	Baltimore, Md	Dec. 30, 1873	145, 940
Boots and shoes, Machinery for lasting	A. C. Carey	Malden, Mass	Dec. 30, 1873	146, 043
Boots and shoes, Machinery for making wood shanks for.	W. N. Sprague	Keene, N. H	Nov. 25, 1873	145, 024
Boots and shoes, Making	J. Bedford	Philadelphia, Pa	July 16, 1806	....
Boots and shoes, Making	J. Frask	Hatfield, Mass	Apr. 5, 1826	....
Boots and shoes, Making	S. B. Hitchcock and J. Bement.	Homer, N. Y	July 30, 1811	....
Boots and shoes, Making	E. Martin	Salem, N. J	Jan. 12, 1824	....
Boots and shoes, Making	W. Montgomery	Philadelphia, Pa	Feb. 11, 1807	....
Boots and shoes, &c., Making furred	S. Eells	Connecticut	Feb. 24, 1816	....
Boots and shoes, Manufacture of	E. Alexander	Providence, R. I	Aug. 4, 1863	39, 370
Boots and shoes, Manufacture of	F. D. Ballou	Abington, Mass	Jan. 10, 1860	26, 808
Boots and shoes, Manufacture of	L. R. Blake	Abington, Mass	Aug. 14, 1860	29, 561
Boots and shoes, Manufacture of	L. R. Blake	Boston, Mass	May 18, 1869	90, 225
Boots and shoes, Manufacture of	L. R. Blake	Boston, Mass	Apr. 26, 1870	102, 362
Boots and shoes, Manufacture of	J. Boyle	New York, N. Y	July 29, 1873	141, 257
Boots and shoes, Manufacture of	W. N. Brookhouse	West Danvers, Mass	Oct. 4, 1870	107, 864
Boots and shoes, Manufacture of	M. F. Chandler and P. Ware	Boston, Mass	Dec. 24, 1872	134, 252
Boots and shoes, Manufacture of	W. Duchemin and A. Jeffries	Lynn, Mass	Aug. 27, 1867	68, 330
Boots and shoes, Manufacture of	C. Hawkes	Lynn, Mass	June 26, 1866	55, 858
Boots and shoes, Manufacture of	G. Hawkes	Lynn, Mass	June 29, 1869	91, 844
Boots and shoes, Manufacture of	L. and S. B. Holden	Woburn, Mass	Apr. 2, 1861	31, 929
Boots and shoes, Manufacture of	D. Howard		Apr. 10, 1800	....
Boots and shoes, Manufacture of	J. L. Joyce	New Haven, Conn	July 29, 1873	141, 357
Boots and shoes, Manufacture of	W. and J. Keats	Street, England	Dec. 10, 1867	72, 048
Boots and shoes, Manufacture of	W. and J. Keats	Leek, England	Feb. 11, 1868	74, 229
Boots and shoes, Manufacture of	C. A. Keith	Danvers, Mass	Aug. 12, 1873	141, 797
Boots and shoes, Manufacture of	D. Peck	New York, N. Y	Aug. 14, 1821	....
Boots and shoes, Manufacture of	H. Port	New York, N. Y	Aug. 4, 1863	39, 421
Boots and shoes, Manufacture of	D. Read	New York, N. Y	Sept. 26, 1865	50, 163
Boots and shoes, Manufacture of	E. P. Richardson	Lawrence, Mass	Apr. 27, 1869	89, 504
Boots and shoes, Manufacture of	S. J. Shaw	Marlborough, Mass	Feb. 13, 1872	123, 736
Boots and shoes, Manufacture of	E. W. Southworth	Ashland, Mass	Aug. 6, 1872	130, 162
Boots and shoes, Manufacture of	H. P. Stewart	Oak's Corners, N. Y	Dec. 18, 1866	60, 588
Boots and shoes, Manufacture of	H. G. Tyer and J. Helm	New Brunswick, N. J	Mar. 27, 1855	12, 607
Boots and shoes, Manufacture of	P. Ware, jr	Boston, Mass	Sept. 2, 1873	142, 538
Boots and shoes, Manufacture of	S. H. Wherf	Roxbury, Mass	Jan. 8, 1856	14, 080
Boots and shoes, Manufacture of	W. Wickersham	Boston, Mass	Aug. 2, 1870	106, 012
Boots and shoes, Manufacture of India-rubber	S. C. Smith	New York, N. Y	Dec. 7, 1837	503
Boots and shoes, Manufacture of vulcanized rubber	J. A. Greene	Brooklyn, N. Y	Apr. 6, 1869	88, 706
Boots and shoes, Manufacturing	A. Destony	New York, N. Y	Nov. 17, 1868	84, 245
Boots and shoes, Manufacturing	C. A. Ore	Philadelphia, Pa	May 5, 1831	....
Boots and shoes, Manufacturing	M. Pennock	East Marlborough, Pa	Dec. 14, 1830	....
Boots and shoes, Manufacturing counters for	S. C. Phinney	Stoughton, Mass	Feb. 9, 1869	86, 779
Boots and shoes, Mechanism for making counters for.	E. Andrews	Welchville, Me	July 8, 1873	140, 569
Boots and shoes, Metal tips for toes of	G. A. Mitchell	Turner, Me	Jan. 5, 1858	19, 040
Boots and shoes, Metallic counters for	A. R. Spaulding	Clinton, Me	Feb. 27, 1872	124, 094
Boots and shoes, Metallic shank for	E. Heaton	New Haven, Conn	Feb. 33, 1864	41, 701
Boots and shoes, Method of constructing	J. McComber	Rockford, Ill	Nov. 22, 1870	109, 4.8
Boots and shoes, Method of stretching	G. W. Griswold	Carbondale, Pa	Aug. 31, 1858	21, 334
Boots and shoes, Mock buttons and button-holes for the ornamentation of.	W. Butterfield	Boston, Mass	July 23, 1872	129, 785
Boots and shoes, Mode of cutting	J. B. Keen	Bridgeton, N. J	June 21, 1839	1, 178
Boots and shoes, Mode of cutting channels in	L. R. Blake	Boston, Mass	Jan. 21, 1868	73, 570
Boots and shoes, Mode of cutting out	S. Marshall	Philadelphia, Pa	June 23, 1829	....
Boots and shoes, Mode of lacing	W. Banister and A. H. Rowell	Lawrence, Mass	Mar. 31, 1868	76, 038
Boots and shoes, Mode of lining	B. S. Bryant	Hanson, Mass	Feb. 13, 1872	123, 674
Boots and shoes, Mode of securing taps on	T. H. Lindley	Providence, R. I	Oct. 22, 1872	132, 474
Boots and shoes, Mode of securing tips to	J. H. Jennings	New Bedford, Mass	Mar. 13, 1866	53, 150
Boots and shoes, Nail-presenting mechanism for	A. Knowlton and H. P. Fairfield.	Boston, Mass	July 23, 1872	129, 834
Boots and shoes, Nailing	L. R. Blake and A. S. Libby	Brooklyn, N. Y., and Lawrence, Mass.	July 1, 1873	140, 400
Boots and shoes, Pattern for cutting out the uppers of.	W. W. Merriam	Oswego, N. Y	Nov. 10, 1857	18, 594
Boots and shoes, Polished water-proof	J. Rynex, J. Haskins, and S. Knower.	Boston, Mass	Sept. 23, 1829	....
Boots and shoes, Process for preparing wood for	R. T. Havens	Wilmington, Ohio	May 1, 1866	54, 339
Boots and shoes, Process of ornamenting	E. J. Scott	Glasgow, Great Britain	May 18, 1869	90, 312
Boots and shoes, Rotary edge-key for polishing	S. H. Hodges	Lynn, Mass	Feb. 6, 1872	123, 477
Boots and shoes, Scale for cutting	S. F. Burdett and H. Still	Keokuk, Iowa, and Leavenworth City, Kans.	Nov. 1, 1859	25, 947
Boots and shoes, Screw-fastening for	J. Chilcott and R. Snell	Brooklyn, N. Y	Sept. 13, 1853	10, 021
Boots and shoes, Shank-piece for	J. V. Sheffield	Northbridge, Mass	May 3, 1870	102, 721
Boots and shoes, Shank-spring for	E. Heaton	New Haven, Conn	Nov. 19, 1867	71, 003
Boots and shoes, Shipping and show-case for	M. Buhler	Lamoille, Ill	June 17, 1873	140, 007
Boots and shoes, Siding	H. T. Daggett	South Braintree, Mass	July 7, 1868	79, 733
Boots and shoes, Spring bottom for	C. S. Hale and O. C. Hubbell	Cleveland, Ohio	Mar. 24, 1868	75, 900
Boots and shoes, Spring-shanks for	J. McGinley	Philadelphia, Pa	Mar. 13, 1849	6, 173
Boots and shoes, Ventilating	E. Thomas	Philadelphia, Pa	July 28, 1868	80, 518
Boots and shoes, Ventilating	A. Stocker	Watertown, N. Y	Sept. 27, 1864	44, 467
Boots and shoes, Ventilating rubber	B. H. Webb	North Cambridge, N. Y	Mar. 8, 1864	41, 879
Boots and shoes water-proof, Composition for making.	P. G. Nagle	Philadelphia, Pa	Feb. 10, 1837	125
Boots and shoes, Water-proof-composition shanks for.	W. H. Towers	Boston, Mass	Jan. 23, 1872	122, 925
Boots and shoes water-proof, Making	A. Buffum	Smithfield, R. I	Dec. 28, 1822	....
Boots and shoes water-proof, Making	P. G. Nagle	Philadelphia, Pa	Dec. 14, 1825	....
Boots and shoes, Wear-plate for	S. Minges	Rochester, N. Y	June 30, 1868	79, 492
Boots and shoes while applying the soles, Mechanism for protecting the uppers of	J. Jenkins	Lynn, Mass	Aug. 23, 1859	25, 202
Boots and shoes whilst pegging, Mode of holding	S. Nourse	Danvers, Mass	Dec. 8, 1827	....
Boots and shoes, Wire-driving machine for	W. Wickersham	Boston, Mass	Aug. 2, 1870	106, 011
Boots and shoes with nails and pegs, Manufacturing	N. Mills	New York, N. Y	Mar. 21, 1831	....

Index of patents issued from the United States Patent Office from 1790 *to* 1873, *inclusive*—Continued.

Invention.	Inventor.	Residence.	Date.	No.
Boots and shoes with wooden pegs, screws, notches, or creases, Manufacturing.	R. U. Richards	Norfolk, Conn	May 23, 1812	
Boots and shoes without seams, Manufacture of the uppers of.	S. Middleton	England	Sept. 29, 1857	18,291
Boots, Apparatus for drawing on	J. Russell	Saint Louis, Mo	July 16, 1872	129,175
Boots, bootees, &c., Cutting uppers of	J. T. Buck	New Canaan, Conn	Oct. 27, 1835	
Boots by hinges or boot-cramps, Manufacturing	N. Ayer	Saint Johnsbury, Vt	June 19, 1835	
Boots, &c., Channel-opener for	F. Guscetti	New York, N. Y	May 13, 1873	138,883
Boots, Clamp for crimping	E. G. Pomeroy	Newark, Ohio	Oct. 4, 1836	
Boots, Clamp for crimping leather for	J. M. Read	Boston, Mass	Mar. 16, 1841	2,006
Boots, Construction of	L. R. Blake	Boston, Mass	Jan. 19, 1869	85,989
Boots, Cramping	L. Lucas	Barre, Vt	July 25, 1832	
Boots, Crimping	D. B. Barnum	New Fairfield, Conn	Aug. 5, 1833	
Boots, Cutting	J. Adams	Kokomo, Ind	June 26, 1866	55,799
Boots, Cutting	G. A. Brown	Milford, Mass	Mar. 15, 1864	41,904
Boots, Cutting	T. Cranage	Warren, Ohio	Mar. 13, 1844	3,481
Boots, Cutting	S. S. Drew	Dixon, Ill	Oct. 13, 1863	40,308
Boots, Cutting	A. F. and J. A. Jones	Lexington, Ky	Mar. 14, 1848	5,467
Boots, Cutting	D. Lynahon	Buffalo, N. Y	Oct. 18, 1853	10,134
Boots, Cutting	A. J. Moore, S. Bleistein, and S. S. Shirk.	Lebanon, Pa	July 7, 1868	79,676
Boots, Cutting gaiter	A. D. Drew	Dixon, Ill	Jan. 30, 1866	52,274
Boots, Cutting leather for	J. Morgan	Philadelphia, Pa	Apr. 10, 1811	
Boots, Cutting uppers of	J. S. Lewis	Athol, Mass	Nov. 27, 1855	13,852
Boots, Cutting uppers of	O. Thibaudeau	Montreal, Canada	Aug. 20, 1872	130,602
Boots, Cutting uppers of balmoral	H. Hayward	Fitchburgh, Mass	Jan. 9, 1866	51,943
Boots, Device for pulling on	F. H. Moore	Boston, Mass	May 23, 1865	47,847
Boots, Fastening for gaiter	P. S. Boothby	Biddeford, Me	Sept. 2, 1862	36,383
Boots, Fastening straps to	P. H. Baker	Virginia City, Nev	Nov. 17, 1868	84,157
Boots, Foxing and soling	A. G. Smith	Marathon, N. Y	May 12, 1868	77,927
Boots, Lacing	F. Borchardt	Washington, D. C	Mar. 6, 1866	52,954
Boots, Machine for blacking and cleaning	J. Folsom	Hallowell, Me	Apr. 14, 1835	
Boots, Machine for crimping and forming the front of.	C. Day	Mineral Point, Wis	Dec. 14, 1869	97,765
Boots, Machine for crimping leather for	M. S. Woodward	Marshallton, Pa	July 31, 1837	338
Boots, Machine for cutting gaiter	W. Snell	Easton, Pa	Apr. 10, 1849	6,300
Boots, Machine for cutting out	B. Millett	Woburn, Mass	June 4, 1872	127,630
Boots, Machine for drawing	L. H. Dole and N. Pickard	Rowley, Mass	July 12, 1834	
Boots, Machine for manufacture of screwed	G. Galli	San Francisco, Cal	Sept. 1, 1868	81,620
Boots, &c., Machine for piercing side-seams of	W. May	Binghamton, N. Y	May 30, 1871	115,493
Boots, Machine for rubbing and rolling seams of	A. Thompson	Boston, Mass	May 6, 1873	138,713
Boots, Machine for turning	H. Blodget and H. Boynton	Rowley, Mass	Sept. 26, 1835	
Boots, Making	H. Thayer	Monson, Mass	May 17, 1870	103,104
Boots, Making	J. Tumbull	Philadelphia, Pa	June 24, 1843	3,143
Boots, Manufacture of	J. J. Christian	Yonkers, N. Y	Aug. 21, 1866	57,289
Boots, Manufacture of	J. Scrimgeour	Brooklyn, N. Y	June 9, 1857	17,524
Boots, Manufacturing	A. Burdick	Carroll, N. Y	Oct. 31, 1829	
Boots, Manufacturing	P. Gordon		Aug. 10, 1791	
Boots or shoes from skins with the hair on, Manufacturing.	S. H. and E. Brown and H. Whitman.	Norfolk, Conn	Apr. 16, 1833	
Boots or shoes, Manufacturing	S. Gerhard and J. Vanneman	Camden, N. J	July 26, 1833	
Boots, Pattern for cutting	A. D. Drew	Dixon, Ill	May 9, 1865	47,623
Boots, Pattern for cutting	E. Shopbell	Ashland, Ohio	Jan. 27, 1863	37,526
Boots, &c., Shank for	F. P. Buren	Trenton Mo	May 13, 1873	138,851
Boots, shoes, &c., of gutta-percha combined with other fabrics, Making.	C. Keene	Sussex Place, England	May 23, 1848	5,593
Boots, shoes, socks, &c., Water-proof	W. Atkinson	Tewkesbury, Mass	Mar. 7, 1834	
Boots, Tap for rubber	F. Flynn	Smithfield, R. I	May 31, 1870	103,594
Boots, Turning	P. Stevens	Stoughton, Mass	Mar. 4, 1836	
Boots with gum-elastic gores, Gaiter	J. H. Dupont and T. Hyatt	New York, N. Y	Oct. 30, 1840	1,841
Bootees and other articles, Eyes for lacing	L. Hoffmeister	Philadelphia, Pa	Jan. 5, 1864	41,069
Bootees, Manufacture of	A. H. Silvester	Boston, Mass	Jan. 28, 1862	34,273
Borax from native borate of lime, Manufacture of	T. Bell and H. Scholefield	South Shields, England	Oct. 9, 1855	13,633
Borax from solutions used in treating wood, Process of recovering and reworking.	S. Beer	New York, N. Y	Dec. 1, 1868	84,471
Borer, Compound slide and groove	Z. Holt	Chelmsford, Mass	May 27, 1833	
Borer, Post-hole	A. F. Summers	Peoria, Ill	Aug. 25, 1868	81,555
Boring and drilling machine	J. Goodin	Centralia, Ill	July 28, 1868	80,475
Boring and drilling machine	J. J. Sheridan	New York, N. Y	Sept. 2, 1873	142,416
Boring and drilling tool	L. Dow	Piermont, N. Y	Jan. 23, 1866	52,148
Boring and grinding apparatus	S. Cary	Centreville, La	July 31, 1866	56,705
Boring and marking machine	G. M. Nickason	Ellenville, N. Y	Sept. 6, 1870	107,091
Boring and mortising machine	H. Allen	Norwich, Conn	Apr. 11, 1854	10,768
Boring and mortising machine	S. W. Bidwell	Hartford, Conn	Jan. 11, 1861	32,509
Boring and mortising machine	C. Carter	Manchester, Mich	May 22, 1849	6,458
Boring and mortising machine	J. J. Earley	Fairfield, Ohio	Sept. 15, 1868	82,096
Boring and mortising machine	J. Humphrey	Ravenna, Ohio	Dec. 28, 1869	98,265
Boring and mortising machine	A. O. Neal	Hyde Park, Mass	Jan. 26, 1869	86,176
Boring and mortising machine	R. L. Nelson	Orange Court House, Va	Apr. 25, 1871	114,183
Boring and mortising machine	J. Peters	Lebanon, Pa	Jan. 31, 1871	111,472
Boring and mortising machine	C. G. Pollock	Cincinnati, Ohio	Sept. 20, 1859	25,523
Boring and mortising machine	J. H. Power	Norwalk, Ohio	Jan. 31, 1837	115
Boring and mortising machine	R. D. Roys and N. French	Detroit, Mich	Jan. 23, 1845	3,891
Boring and mortising machine	G. H. Stevens	Lowell, Wis	Nov. 18, 1856	16,101
Boring and mortising machine	A. Swingle	Galveston, Tex	July 11, 1848	5,661
Boring and mortising machine	S. W. Wright	Athens, Ohio	July 26, 1870	105,760
Boring and mortising machine, Feed-motion for	J. M. Kendall	South Hardwick, Vt	Feb. 14, 1860	27,134
Boring and mortising machinery	J. Wilkinson	Pennfield, N. Y	May 28, 1825	
Boring and sawing machine	B. F. Goodspeed and D. H. Wiswell.	New York, N. Y	Apr. 16, 1834	
Boring and tenoning machine	T. Place	Alfred Centre, N. Y	Mar. 12, 1867	62,883
Boring and tenoning machine	A. Warren	Saugerties, N. Y	Apr. 17, 1826	
Boring and tenoning machine, Wheelwright's	C. Cowdry and O. and C. C. Tolls.	Ithaca, N. Y	May 1, 1855	12,776
Boring and tenoning machinery	D. Sperry	Colchester, Conn	Apr. 18, 1827	
Boring and turning machine, Wood	A. and F. Brown	New York, N. Y	July 3, 1855	13,158
Boring and turning machine, Wood	F. and A. Brown	New York, N. Y	Feb. 12, 1856	14,223
Boring angular holes, Machine for	B. Merritt, jr	Newton, Mass	May 24, 1864	42,863

Index of patents issued from the United States Patent Office from 1790 *to* 1873, *inclusive*—Continued.

Invention.	Inventor.	Residence.	Date.	No.
Boring-apparatus	S. Cary	Centreville, La	Aug. 7, 1866	57, 058
Boring-apparatus	S. R. and W. S. Hunter	Cortland, N. Y	May 16, 1865	47, 727
Boring-apparatus	J. Wheelock	Worcester, Mass	Oct. 11, 1870	108, 310
Boring-bar for boring and screw-cutting	E. S. Chapell	Milton, Mass	Sept. 3, 1867	68, 488
Boring-bit	S. Hipkins, jr	South Wheeling, W. Va	July 1, 1873	140, 370
Boring-bit	M. Jincks	Danville, N. Y	Apr. 2, 1867	63, 392
Boring-bit	D. Kelly	Muskegon, Mich	May 10, 1870	102, 829
Boring-bit	D. Kelly	Muskegon, Mich	July 11, 1871	116, 837
Boring-bit brace	H. S. Bartholomew	Bristol, Conn	May 28, 1867	65, 046
Boring-bit brace	J. B. Fellows	Concord, N. H	Sept. 24, 1867	69, 086
Boring-bit, Extension-shank for	W. S. Pattin	Portsmouth, Ohio	June 20, 1871	116, 090
Boring bit for boring gun-barrels, Nut	J. Green	Harper's Ferry, Va	Oct. 3, 1817	....
Boring bit, Hole	C. F. Kimball and A. Parsons	Portland, Me	Jan. 23, 1866	52, 174
Boring bit, Mineral	D. Shoemaker	Kittaning Township, Pa	Sept. 16, 1873	142, 949
Boring bit, Variable	W. Tucker	Glocester, R. I	Aug. 10, 1858	21, 160
Boring bit, Wood	C. Boernicke	Philadelphia, Pa	Apr. 16, 1867	63, 842
Boring bit, Wood	R. Cook	Saratoga Springs, N. Y	Aug. 18, 1868	81, 069
Boring-brace	J. A. and H. A. House	Bridgeport, Conn	Sept. 5, 1865	49, 758
Boring-brace attachment	J. S. Fray	Bridgeport, Conn	May 11, 1869	89, 925
Boring-drill	D. R. Erdmann	Philadelphia, Pa	Mar. 28, 1865	47, 001
Boring drill, Polygonal-hole	J. C. Broadley	Franklin, N. J	Dec. 14, 1869	97, 868
Boring-engine, Portable	T. Goodrum	Providence, R. I	Oct. 9, 1855	13, 646
Boring framing-timber, Machine for	J. Badger	Brooklyn, Conn	Jan. 20, 1838	572
Boring holes in posts, Machine for	T. T. Strode	Coatesville, Pa	Dec. 2, 1851	8, 569
Boring holes, Tool for	A. Eliaers	Boston, Mass	Dec. 20, 1864	45, 483
Boring-implements, Making	R. Cooke	Shelburne Falls, Mass	Nov. 13, 1855	13, 780
Boring-instrument, Guide-attachment for	A. Amory	New York, N. Y	July 27, 1869	92, 927
Boring-jar	J. Slusser	Cincinnati, Ohio	May 30, 1865	47, 993
Boring-machine	J. H. Aldrich	Portsmouth, N. H	Nov. 27, 1849	6, 909
Boring-machine	F. S. Allen and C. F. Ritchel	New York, N. Y	Nov. 21, 1871	121, 074
Boring-machine	R. Allison	Port Carbon, Pa	Oct. 9, 1866	58, 568
Boring-machine	C. N. Baldwin	Willington, Conn	Oct. 25, 1870	108, 552
Boring-machine	E. C. Barton	Bloomsburgh, Pa	Dec. 21, 1869	98, 014
Boring-machine	J. Bosenbury	Cherryville, N. J	Apr. 14, 1857	17, 017
Boring-machine	W. W. Carey and G. W. Harris	Lowell, Mass	Aug. 9, 1870	106, 120
Boring-machine	G. F. Cluff	Petersburgh, Ill	Nov. 7, 1871	120, 716
Boring-machine	J. B. Coffin	Mohicanville, Ohio	June 20, 1845	4, 085
Boring-machine	E. C. Cole	Pawling, N. Y	Aug. 6, 1872	130, 193
Boring-machine	A. M. Connett	Madison, Ind	July 21, 1868	80, 059
Boring-machine	G. S. Corwin	Riverhead, N. Y	Nov. 3, 1863	40, 459
Boring-machine	I. Cory	Dalton, Ind	Feb. 11, 1873	135, 693
Boring-machine	W. H. Deatrick	Heidlersburgh, Pa	Nov. 1, 1870	108, 890
Boring-machine	E. C. Dodge	Edgecomb, Me	Nov. 30, 1869	97, 280
Boring-machine	L. A. Dole	Salem, Ohio	July 6, 1858	20, 779
Boring-machine	A. Dool and P. B. White	Dowagiac, Mich	Mar. 31, 1868	76, 171
Boring-machine	E. P. Drake	Greenbush, N. Y	Apr. 23, 1861	32, 125
Boring-machine	J. Edgecomb	Worcester, Mass	Jan. 14, 1862	34, 137
Boring-machine	J. Edgecomb	Worcester, Mass	June 28, 1864	43, 364
Boring-machine	S. L. Fitts	Ashburnham, Mass	June 11, 1861	32, 550
Boring-machine	J. W. Frazee	Peoria, Ill	July 12, 1870	105, 188
Boring-machine	W. C. Freeman	Louisiana, Mo	Oct. 22, 1872	132, 393
Boring-machine	J. Gardner	Bigler, Pa	Apr. 16, 1872	125, 802
Boring-machine	F. M. Gibson	Chelsea, Mass	Dec. 7, 1869	97, 499
Boring-machine	N. Gillet and J. Smith	Cazenovia, N. Y	Apr. 8, 1812	....
Boring-machine	E. A. Goodes	Philadelphia, Pa	Jan. 25, 1859	22, 724
Boring machine	J. Hampson	Newburgh, N. Y	Dec. 7, 1869	97, 506
Boring-machine	H. Hays	New York, N. Y	July 5, 1859	24, 633
Boring-machine	J. Isenberg	McConnellstown, Pa	Feb. 26, 1867	62, 491
Boring-machine	J. Johnson	Mifflinsburgh, Pa	Mar. 13, 1847	5, 011
Boring-machine	J. Jones	Camden, N. J	Sept. 26, 1848	5, 809
Boring-machine	F. L. King	Worcester, Mass	Nov. 23, 1869	97, 202
Boring-machine	S. U. King	Windsor, Vt	May 24, 1864	42, 914
Boring-machine	S. Klahr	Reamstown, Pa	Dec. 16, 1856	16, 241
Boring-machine	C. B. Knapp	Waterloo, Wis	Apr. 2, 1867	63, 532
Boring-machine	A. Leonard	Sullivan, Ohio	Sept. 17, 1861	33, 324
Boring-machine	R. Little	Freeport, Ill	May 16, 1871	114, 953
Boring-machine	C. R. Long	Louisville, Ky	Sept. 1, 1868	81, 655
Boring-machine	H. Longwell	Mansfield, Pa	June 17, 1873	139, 901
Boring-machine	J. Longyear	Grass Lake, Mich	Feb. 18, 1862	34, 436
Boring-machine	N. R. Merchant	Guilford, N. Y	May 28, 1861	32, 449
Boring-machine	J. L. Metcalfe	Quincy, Pa	Apr. 16, 1872	125, 828
Boring-machine	J. Meyer	Brooklyn, N. Y	June 16, 1863	38, 908
Boring-machine	R. O. Moorhouse and S. L. Wiegand.	Philadelphia, Pa	Aug. 12, 1873	141, 657
Boring-machine	W. Morstatt	New York, N. Y	June 14, 1870	104, 337
Boring-machine	J. H. Pardieck	Acton, Ind	Mar. 7, 1871	112, 372
Boring-machine	Z. C. Phillips	Allegheny City, Pa	Jan. 2, 1872	122, 402
Boring-machine	C. E. Pierce	Westminster, Md	May 21, 1872	126, 902
Boring-machine	H. Pitcher	Fond du Lac, Wis	May 13, 1873	138, 927
Boring-machine	W. P. Powers	North La Crosse, Wis	Aug. 1, 1871	117, 567
Boring-machine	E. Quinn	Trenton, N. J	Aug. 25, 1857	18, 057
Boring-machine	E. Quinn	Washington, D. C	Mar. 27, 1866	53, 485
Boring-machine	B. F. Rea, M. Pyke, and D. M. Rennoe.	South Bend, Ind	Oct. 28, 1873	144, 139
Boring-machine	J. Ribon	New York, N. Y	Jan. 4, 1870	98, 633
Boring-machine	G. F. Rice	Worcester, Mass	Dec. 21, 1858	22, 379
Boring-machine	W. Samuels and G. L. Stansbury.	Jackson Township, Ind	June 3, 1856	15, 022
Boring-machine	S. T. Sanford	Fall River, Mass	July 12, 1853	9, 845
Boring-machine	J. W. Shaw	Wenona, Mich	Dec. 17, 1872	134, 059
Boring-machine	J. Simpson	Meadville, Pa	Feb. 27, 1872	124, 091
Boring-machine	D. Stanley and G. Johnson	Cincinnati, Ohio	Sept. 12, 1865	49, 933
Boring-machine	L. Stevens	Elmira, N. Y	Dec. 15, 1857	18, 872
Boring-machine	J. Strickler	Columbiana, Ohio	Aug. 26, 1873	142, 296
Boring-machine	M. Sweet	Troy, N. Y	Apr. 13, 1869	88, 821
Boring-machine	B. F. Taft	South Groton, Mass	Dec. 12, 1854	12, 078
Boring-machine	O. F. Teed	Guilford, N. Y	Oct. 15, 1872	132, 187
Boring-machine	J. Temple	Birmingham, Pa	Feb. 26, 1856	14, 327

Index of patents issued from the United States Patent Office from 1790 *to* 1873, *inclusive*—Continued.

Invention.	Inventor.	Residence.	Date.	No.
Boring-machine	A. J. Truxell	Lynchburgh, Va	Nov. 8, 1870	109,078
Boring-machine	W. Tucker	Fiskedale, Mass	Jan. 28, 1870	104,797
Boring-machine	J. Van Dyke	Grimsby, Canada	July 25, 1865	49,060
Boring-machine	J. Waugh	Elmira, N. Y	May 31, 1859	24,252
Boring-machine	J. Webster	Great Barrington, Mass	May 1, 1810	....
Boring-machine	A. Weikart	Green Village, Ohio	Nov. 1, 1845	4,251
Boring-machine	A. Weikart	Greenford, Ohio	Jan. 15, 1850	7,031
Boring-machine	W. H. Willcox	Tarrytown, N. Y	May 29, 1849	6,484
Boring-machine	E. J. Worcester	Worcester, Mass	Dec. 27, 1870	110,530
Boring-machine	A. Wyckoff and E. R. Morrison	Elmira, N. Y	Sept. 25, 1855	13,606
Boring-machine	B. F. Young	Charlestown, Mass	June 26, 1866	55,945
Boring machine, Cylinder	M. S. Otis	Rochester, N. Y	June 19, 1855	13,117
Boring machine, Perpendicular	R. Selden		July 7, 1806	....
Boring machine, Rock	G. C. Doherty	Cumberland County, Ky	Aug. 29, 1838	897
Boring machine, Timber	R. Hubbell	Hudson, Ohio	Oct. 26, 1839	1,382
Boring machine, Timber	R. Smith	Towanda, Pa	Feb. 10, 1843	2,948
Boring machine, Wood	G. Dryden	Worcester, Mass	July 7, 1868	79,640
Boring machine, Wood	H. Fleming	Vienna, N. J	Mar. 24, 1868	75,891
Boring machine, Wood	B. F. Mohr	Mifflinburgh, Pa	Apr. 20, 1869	89,163
Boring machine, Wood	A. Roff	Southport, Conn	Mar. 31, 1868	76,103
Boring machine, Wood	L. Stevens	Elmira, N. Y	Oct. 6, 1857	18,370
Boring machine, Wood	J. B. Thorp	Warrensville, Ohio	Mar. 24, 1868	75,813
Boring-machines, Adjustment in	I. W. Ward	Birmingham, Pa	Mar. 18, 1856	14,479
Boring-machines, Device for feeding and limiting the depth of hole in.	L. B Lloyd	Warwick Township, Pa	June 23, 1857	17,630
Boring machines, Device for feeding the cutters intermittently in mortise.	H. E. Paine	Troy, N. Y	July 14, 1857	17,799
Boring-machines, Method of adjusting cylinder in	W. B. Emery	Albany, N. Y	Feb. 27, 1855	12,437
Boring-machines to objects, Method of securing	P. Baylor	Salem, Ohio	July 1, 1844	3,645
Boring-machines, Variable crank for	G. C. Taft	Worcester, Mass	Dec. 10, 1867	72,114
Boring-mill	F. E. Hahn	Philadelphia, Pa	Sept. 9, 1873	142,694
Boring-mill	W. Sellers	Philadelphia, Pa	June 23, 1857	17,641
Boring-mill	J. Wheelock	Worcester, Mass	Mar. 12, 1872	124,647
Boring-mill for metal	W. Sellers	Philadelphia, Pa	Aug. 5, 1862	36,112
Boring, planing, and slotting metals, Machine for	N. Aylsworth	Rochester, N. Y	Jan. 2, 1855	12,117
Boring post-holes in the earth, Machine for	J. S. Wertz	Middletown, Iowa	July 5, 1859	24,682
Boring, reaming out, &c., Machine for	E. Johnson	Lexington Heights, N. Y	Nov. 7, 1826	....
Boring timber for mortising	B. Knight	Cranston, R. I	Oct. 23, 1834	....
Boring-tool	A. Allan	New York, N. Y	Oct. 12, 1869	95,627
Boring-tool	F. S. Allen	New York, N. Y	May 28, 1872	127,209
Boring-tool	W. Brodhead	Rondout, N. Y	Aug. 8, 1865	49,336
Boring-tool	J. C. Chapman	Cambridgeport, Mass	Apr. 23, 1867	63,992
Boring-tool	A. G. Hotchkiss	Wolcottville, Conn	July 5, 1870	104,958
Boring-tool	W. J. Johnson and G. Tainter	Newton and Watertown, Mass.	Oct. 5, 1869	95,483
Boring-tool	M. Joy	West Greenville, Pa	Aug. 8, 1865	49,277
Boring-tool	W. P. Lathrop	West Winsted, Conn	Oct. 18, 1870	108,367
Boring-tool	J. Miller	New York, N. Y	Jan. 29, 1867	61,629
Boring-tool	G. W. Moore	Harrisburgh, Pa	Feb. 20, 1872	123,786
Boring-tool	A. Steinbok	New York, N. Y	July 28, 1868	80,515
Boring-tool	T. E. Stanley	Hazelhurst, Miss	July 12, 1870	105,271
Boring-tool	C. C. Strong	Defiance, Ohio	July 6, 1869	92,395
Boring-tool	N. Thompson	Saint John's Wood, England	Jan. 8, 1867	61,121
Boring-tool	G. I. Washburn	Worcester, Mass	July 26, 1859	24,894
Boring-tool	A. F. Whitin	Whitinsville, Mass	Apr. 23, 1872	126,120
Boring-tool	E. F. Whitney and R. Jones	Union Mills, Pa	May 17, 1870	103,265
Boring-tool	A. Wippo	Chicago, Ill	Dec. 17, 1867	72,436
Boring-tool, Expanding	J. K. Kacerovsky	Bridgeport, Conn	Mar. 11, 1873	136,736
Boring-tools, Coupling for shafts of	J. Esler	Brooklyn, N. Y	July 11, 1865	48,667
Boring-tools, Coupling for shafts of	J. B. Stockton	Oil City, Pa	Oct. 3, 1865	50,287
Boring-tools, Coupling for shafts of	J. Watson	Philadelphia, Pa	Sept. 26, 1865	50,190
Boring-tools, Coupling shafts of	J. N. Bolles	Baltimore, Md	May 9, 1865	47,613
Boring-tools, Coupling shafts of	D. G. Coppin	Cincinnati, Ohio	July 18, 1865	48,793
Boring-tools, Coupling shafts of	S. Fawcett	Rochester, N. Y	Mar. 13, 1866	53,219
Boring-tools, Coupling shafts of	E. Kaylor	Pittsburgh, Pa	Sept. 26, 1865	50,138
Boring tools, Coupling shafts of	R. H. Lecky	Allegheny City, Pa	May 2, 1865	47,554
Boring-tools, Coupling shafts of	R. H. Lecky	Allegheny City, Pa	May 2, 1865	47,556
Boring-tools, Coupling shafts of	A. A. Wilson	Green Point, N. Y	May 23, 1865	47,907
Boring wood	J. B. Pell	New York, N. Y	Mar. 2, 1836	....
Bosom and collar, Combined over	I. T. Dyer	Canton, Mo	Nov. 4, 1873	144,267
Bosom, collar, and similar articles of apparel	A. Johnson	Brooklyn, N. Y	Feb. 25, 1873	136,247
Bosom-expander	D. M. Church and C. A. Ellsworth.	Birmingham, Conn	May 8, 1860	28,155
Bosom-holder, Elastic	G. W. Wright	Salem, Ohio	July 16, 1872	128,994
Bosom-pad	B. Carter	Middletown, Conn	Jan. 8, 1869	90,923
Bosom-pad	J. C. Cook	Buffalo, N. Y	Apr. 21, 1868	76,894
Bosom-pad	E. W. Glover	Medford, Mass	Oct. 22, 1867	69,988
Bosom-pad	A. H. Johnson	Hartford, Conn	May 5, 1868	77,493
Bosom-pad	H. W. Libbey	Cleveland, Ohio	Mar. 3, 1868	75,030
Bosom-pad	H. W. Libbey	Cleveland, Ohio	Mar. 3, 1868	75,031
Bosom-pad	H. M. Millar	Louisville, Ky	Oct. 21, 1873	143,921
Bosom-pad	C. A. Phillips	Chicago, Ill	Feb. 18, 1873	135,936
Bosom-pad	J. Waterman	New York, N. Y	Mar. 1, 1870	100,472
Bosom-pin	J. P. Derby	South Reading, Mass	Feb. 17, 1857	16,640
Bottle	J. N. Bodine	Bridgeton, N. J	Aug. 14, 1860	29,563
Bottle	R. Boeklen and H. T. Brown	Jersey City, N. J., and Brooklyn, N. Y.	July 25, 1854	11,353
Bottle	C. W. Cahoon	Portland, Me	Nov. 28, 1865	51,141
Bottle	C. W. Cahoon	Portland, Me	Nov. 28, 1865	51,142
Bottle	H. S. Carley	Cambridgeport, Mass	Sept. 24, 1867	69,177
Bottle	W. Clark	New York, N. Y	July 25, 1854	11,357
Bottle	H. Codd	Camberwell, England	July 23, 1872	129,652
Bottle	G. G. Hickman	Coatesville, Pa	Mar. 3, 1868	75,014
Bottle	V. H. Lyon	Indianapolis, Ind	June 17, 1873	139,905
Bottle	J. Maurer	New York, N. Y	Aug. 14, 1860	29,611
Bottle	B. C. Odell	Kingston, N. Y	Oct. 8, 1872	132,020
Bottle and bottle-stopper	H. Naylor	Oil City, Pa	Mar. 4, 1873	136,534
Bottle and box opener	G. J. Hill	Buffalo, N. Y	Aug. 29, 1871	118,453
Bottle and case, Combined	G. W. Banker and G. W. Peck	New York, N. Y	July 16, 1872	128,940

Index of patents issued from the United States Patent Office from 1790 *to* 1873, *inclusive*—Continued.

Invention.	Inventor.	Residence.	Date.	No.
Bottle and glass case	J. Mathieu	Paris, France	Sept. 6, 1870	107, 189
Bottle and jug stopper	J. B. Alexander	Washington, D. C	June 18, 1867	65, 857
Bottle and transportation case	C. Burnham and I. G. Taite	Philadelphia, Pa	June 18, 1872	128, 109
Bottle, barrel, &c., stopper	S. Marsh	Detroit, Mich	Mar. 5, 1872	124, 277
Bottle box, Porter	G. W. Righter	Philadelphia, Pa	June 5, 1860	28, 606
Bottle-cap	M. V. Olry	Philadelphia, Pa	Sept. 25, 1866	58, 281
Bottle cap and spoon, Mustard	H. J. White	Boston, Mass	Aug. 12, 1873	141, 844
Bottle cap or cover	J. G. Chillingworth	London, England	July 1, 1873	140, 346
Bottle cap or top	W. Burnet	New York, N. Y	July 2, 1867	66, 212
Bottle-capping machine	A. C. Jordan	New York, N. Y	May 14, 1872	126, 711
Bottle-capping machine	W. H. Sperling	Washington, N. J	June 20, 1871	116, 107
Bottle, Capsuling	J. Paterson	Edinburgh, North Britain	Nov. 28, 1871	121, 407
Bottle-casing	J. Dugan	New York, N. Y	Sept 9, 1873	142, 685
Bottle, Cast-iron mercury	M. Wrangle	New York, N. Y	Dec. 28, 1858	22, 476
Bottle, Caster	B. Beach	West Meriden, Conn	July 17, 1866	56, 486
Bottle, Caster	C. Casper	West Meriden, Conn	June 11, 1872	127, 740
Bottle, Caster	C. P. Crossman	West Warren, Mass	Sept. 15, 1863	39, 889
Bottle, Caster	A. Weber	Newark, N. J	July 1, 1873	140, 564
Bottle, Caster	A. Weber	Newark, N. J	Dec. 2, 1873	145, 138
Bottle-closing device	C. Herman	Baltimore, Md	June 25, 1867	66, 023
Bottle cork	I. B. Sampson	Albany, N. Y	May 13, 1873	138, 820
Bottle-cork fastener	D. Miller	Marietta, Ohio	May 14, 1861	32, 306
Bottle-cork fastening	J. Conner	Brooklyn, N. Y	June 11, 1872	127, 851
Bottle-corks, Labeling	F. W. Copcutt	New York, N. Y	Dec. 24, 1867	72, 606
Bottle-corks, Machine for making	J. Barron	Hampton, Va	Jan. 12, 1819	....
Bottle-corking	C. H. Porter	Albany, N. Y	July 30, 1867	67, 346
Bottle-corking apparatus	T. W. Gillett	New Haven, Conn	July 11, 1854	11, 281
Bottle-corking apparatus	A. C. Jordan	New York, N. Y	Aug. 26, 1873	142, 240
Bottle-corking apparatus	G. Migliavacca	Napa, Cal	Mar. 30, 1869	88, 404
Bottle-corking apparatus	H. Redlich	Chicago, Ill	May 20, 1862	35, 325
Bottle-corking apparatus	H. Unger	Logansport, Ind	Mar. 9, 1869	87, 732
Bottle-corking machine	J. Armstrong	San Francisco, Cal	Oct. 14, 1873	143, 606
Bottle-corking machine	L. S. Chichester	New York, N. Y	Feb. 22, 1859	23, 058
Bottle-corking machine	H. N. De Graw	Piermont, N. Y	Mar. 18, 1856	14, 446
Bottle-corking machine	D. Mueller	New York, N. Y	Apr. 4, 1871	113, 326
Bottle-corking machine	J. Wolff and W. N. Numsen	New York, N. Y., and Baltimore, Md.	Sept. 20, 1870	107, 582
Bottle, Cosmetic	M. H. Huntington	Watertown, N. Y	Oct. 14, 1873	143, 627
Bottle-cover	W. Bourguignon	Providence, R. I	Dec. 16, 1873	145, 484
Bottle-envelope	J. Seithen	Coblenz, Prussia	Feb. 12, 1856	14, 255
Bottle-fastener	G. Otto and G. W. Bauer	Washington, D. C	July 4, 1871	116, 742
Bottle-fastener	I. Winslow	Philadelphia, Pa	Dec. 18, 1849	6, 963
Bottle-fasteners, Key and cork-screw for	H. W. Putnam	Cleveland, Ohio	Sept. 23, 1862	36, 528
Bottle-fastening	J. Allender	New London, Conn	July 24, 1855	13, 338
Bottle-fastening	B. Bates	Baltimore, Md	Dec. 27, 1870	110, 421
Bottle-fastening	W. S. M. Beal	Baltimore, Md	Oct. 31, 1871	120, 363
Bottle-fastening	J. Jeannotat	Paterson, N. J	July 17, 1855	13, 266
Bottle-fastening	J. Spratt	Cincinnati, Ohio	Sept. 6, 1853	9, 995
Bottle-fastening	E. D. Weatherbee	Worcester, Mass	Apr. 4, 1871	113, 603
Bottle-fastening	H. and J. Wilson	Paterson, N. J., and New York, N. Y.	July 24, 1866	56, 653
Bottle-fastening, Cork-saving	C. L. Knecht	Clair Township, Pa	May 31, 1870	103, 624
Bottle-fastening, Register	J. Smylie	Philadelphia, Pa	Oct. 9, 1855	13, 659
Bottle-fastenings, Band for securing	E. D. Weatherbee	Shrewsbury, Mass	Apr. 16, 1872	125, 866
Bottle-fastenings, Ring and gudgeon for	J. C. Day	Hackettstown, N. J	Nov. 13, 1855	13, 782
Bottle-filler	T. Cochen	Williamsburgh, N. Y	June 5, 1866	55, 242
Bottle-filler	E. Jeanjaquet	New York, N. Y	May 4, 1869	89, 664
Bottle-filler	J. H. Parkhurst	Milwaukee, Wis	Sept. 17, 1872	131, 457
Bottle-filler	F. Schlich and A. Feyh	New York, N. Y	May 6, 1873	138, 534
Bottle filler and corker	T. W. Cowey	Cannonsburgh, Pa	Feb. 9, 1869	86, 816
Bottle-filling apparatus	A. Albertson	New York, N. Y	Oct. 11, 1864	44, 685
Bottle-filling apparatus	G. B. Bachmann	Brooklyn, N. Y	Nov. 10, 1868	83, 904
Bottle-filling apparatus	C. A. Gregory	Poughkeepsie, N. Y	Aug. 28, 1860	29, 781
Bottle-filling apparatus	C. A. Gregory	Poughkeepsie, N. Y	July 12, 1870	105, 328
Bottle-filling apparatus	C. A. Gregory	Quebec, Canada	Nov. 19, 1872	133, 094
Bottle-filling apparatus	J. Matthews, jr	New York, N. Y	Nov. 7, 1865	50, 832
Bottle-filling apparatus	J. Matthews, jr	New York, N. Y	Dec. 8, 1868	84, 836
Bottle-filling apparatus	J. Matthews	New York, N. Y	Apr. 29, 1873	138, 421
Bottle-filling apparatus	H. W. Putnam	Cleveland, Ohio	Oct. 29, 1861	33, 602
Bottle-filling apparatus	P. M. Sherwood	New York, N. Y	Nov. 24, 1868	84, 384
Bottle-filling apparatus	T. Simmons and D. H. Lowe	Brooklyn, N. Y	Dec. 27, 1870	110, 504
Bottle-filling machine	J. Alcorn	Charlestown, Mass	Feb. 9, 1869	86, 721
Bottle-filling machine	W. F. Davidson	Cincinnati, Ohio	Feb. 16, 1864	41, 608
Bottle filling machine	C. A. Gregory	Quebec, Canada	Aug. 6, 1872	130, 126
Bottle-filling machine	W. M. Tate	Zanesville, Ohio	Nov. 3, 1863	40, 516
Bottle, Filtering and pouring	V. M. Griswold	Peekskill, N. Y	Dec. 22, 1868	85, 088
Bottle for aërated liquids	J. D. Lynde	Philadelphia, Pa	Apr. 8, 1862	34, 894
Bottle for aërated liquids	T. Warker	New York, N. Y	Mar. 18, 1862	34, 709
Bottle for containing mercury	I. G. Johnson	Spuyten Duyvil, N. Y	Oct. 19, 1858	21, 835
Bottle for gaseous liquids	J. F. Kubly and C. F. Crailsheim.	New York, N. Y., and Paris, France.	Apr. 12, 1870	101, 889
Bottle for holding hydrofluoric acid	D. P. Webster	New York, N. Y	Apr. 14, 1868	76, 678
Bottle for infants and invalids, Cap for feeding	J. Thompson and J. G. Ingram.	London, England	May 25, 1869	90, 606
Bottle for oil	C. W. Cahoon	Portland, Me	Feb. 7, 1865	46, 214
Bottle for tooth-powder, Toilet	W. A. Spalding	Waterbury, Conn	Aug. 15, 1871	118, 163
Bottle, Glass	J. Borden and J. L. Trimble	Medford, N. J	Nov. 25, 1873	144, 885
Bottle, Glass	J. McGann	Kensington, Pa	Nov. 26, 1830	....
Bottle, Graduated	G. W. Stoeckel	Pittsburgh, Pa	Feb. 6, 1866	52, 461
Bottle-handle	G. Ireland	Birmingham, England	Jan. 30, 1866	52, 373
Bottle-holder	W. O. Pond	Mobile, Ala	May 28, 1872	127, 268
Bottle-holder and ice-box, Combined	S. R. Briggs	Owego, N. Y	July 16, 1872	129, 094
Bottle holder, Caster	A. E. Young	Dorchester, Mass	Sept. 5, 1865	49, 840
Bottle holder, Caster	A. E. Young	Dorchester, Mass	Sept. 5, 1865	49, 841
Bottle holding and transporting box	J. Matthews, jr	New York, N. Y	Mar. 5, 1867	62, 658
Bottle holding and transporting box	J. Matthews	New York, N. Y	June 10, 1873	139, 722
Bottle-lock	J. Dugan	New York, N. Y	Aug. 17, 1869	93, 866
Bottle-lock	L. Lehman	Harrisburgh, Pa	June 16, 1868	78, 976
Bottle lock	W. A. Ludden	Brooklyn, N. Y	Dec. 22, 1868	85, 111

Index of patents issued from the United States Patent Office from 1790 *to* 1873, *inclusive*—Continued.

Invention.	Inventor.	Residence.	Date.	No.
Bottle-making tool	J. Wilson	New York, N. Y	June 14, 1870	104, 390
Bottle, Medicine	W. H. Flinn	Nashua, N. H	Mar. 24, 1868	75, 744
Bottle, Medicine	W. H. Flinn	Nashua, N. H	Mar. 24, 1868	75, 745
Bottle-mold	J. L. Mason	New York, N. Y	Nov. 23, 1858	22, 129
Bottle-mold	L. Thomas	Philadelphia, Pa	June 23, 1868	79, 162
Bottle mold, Glass	J. J. Christie	Baltimore, Md	Dec. 17, 1867	72, 368
Bottle mold, Glass	J. J. Christie	Camden, N. J	Nov. 12, 1872	132, 897
Bottle mold, Glass	C. D. Fox	Philadelphia, Pa	Apr. 29, 1873	138, 323
Bottle mold, Glass	S. S. Shinn	Lancaster, N. Y	Nov. 16, 1858	22, 091
Bottle, Nursing	M. S. Burr	Boston, Mass	Aug. 27, 1867	68, 285
Bottle, Nursing	W. Hobson	Sandwich, Mass	June 17, 1873	139, 958
Bottle, Nursing	A. M. Knapp	Racine, Wis	Feb. 9, 1869	86, 762
Bottle, Nursing	F. J. La Forme	Boston, Mass	Nov. 29, 1859	26, 327
Bottle, Nursing	H. W. Libbey	Cleveland, Ohio	June 16, 1868	78, 881
Bottle, Nursing	E. J. Mallet, jr., and W. S. Ward	New York, N. Y	Apr. 26, 1870	102, 289
Bottle, Nursing	J. L. Mason	New York, N. Y	Apr. 26, 1870	102, 417
Bottle, Nursing	T. J. Mayall	Boston, Mass	July 1, 1873	140, 518
Bottle, Nursing	P. J. McElroy	East Cambridge, Mass	June 16, 1868	78, 987
Bottle, Nursing	W. B. Potter	Boston, Mass	Jan. 11, 1859	22, 579
Bottle, Nursing	E. Pratt	New York, N. Y	Aug. 9, 1845	4, 138
Bottle, Nursing	G. R. White	Boston, Mass	Apr. 22, 1873	138, 219
Bottle, Nursing	S. Zeno	New York, N. Y	Apr. 19, 1864	42, 427
Bottle-opener	C. R. Trimble	New York, N. Y	Oct. 10, 1871	119, 802
Bottle-opening instrument	J. Woolaver	Suisun, Cal	Nov. 7, 1865	50, 868
Bottle or caster, Salt	G. B. Richardson	Boston, Mass	Dec. 3, 1867	71, 643
Bottle-packer	R. T. Penick	Saint Joseph, Mo	Nov. 19, 1872	133, 173
Bottle-packing case	R. Carpenter	Roxbury, Mass	Aug. 4, 1863	39, 377
Bottle, Pepper	J. W. Gay	Brooklyn, N. Y	Jan. 26, 1864	41, 372
Bottle, Pepper caster	E. Wattis, sr	Philadelphia, Pa	Feb. 25, 1868	74, 869
Bottle-protector	O. Fitzgerald	Dexter, Me	Oct. 28, 1873	144, 082
Bottle-rack	W. Burrow	Great Malvern, Great Britain.	Dec. 22, 1868	85, 208
Bottle-rinser	J. Rone	Saint John's, Canada	Oct. 29, 1872	132, 603
Bottle, Sample	S. H. Gilman	New Orleans, La	Oct. 21, 1873	143, 753
Bottle, Screw-neck	J. L. Mason	New York, N. Y	Nov. 30, 1858	22, 186
Bottle, Show	H. Whitney	East Cambridge, Mass	Mar. 23, 1869	88, 105
Bottle, Siphon	J. B. Alexander	Washington, D. C	May 18, 1869	90, 215
Bottle, Siphon	C. J. Converse	Boston, Mass	Apr. 6, 1869	88, 610
Bottle, Siphon	G. W. Doty	Ravenna, Ohio	May 16, 1865	47, 705
Bottle, Siphon	C. Glover	New York, N. Y	Sept. 19, 1871	119, 028
Bottle, Siphon	J. E. Milliman	Jersey City, N. J	July 23, 1872	129, 852
Bottle, Siphon	T. Pimer	New London, Conn	Jan. 2, 1872	122, 403
Bottle, Spice	J. L. Likins	Vallejo, Cal	Mar. 19, 1872	124, 836
Bottle stand, Wire	G. D. Dudley	Lowell, Mass	Dec. 9, 1873	145, 284
Bottle-stopper	A. Albertson	New York, N. Y	Aug. 26, 1862	36, 266
Bottle-stopper	A. Albertson	New York, N. Y	Nov. 1, 1864	44, 912
Bottle-stopper	J. B. Alexander	Washington, D. C	July 7, 1868	79, 536
Bottle-stopper	N. Ames	Saugus Centre, Mass	Jan. 21, 1862	34, 227
Bottle-stopper	A. Barbarin	New Orleans, La	July 24, 1866	56, 516
Bottle-stopper	J. B. Basaloux	Saint Louis, Mo	Dec. 17, 1872	133, 915
Bottle-stopper	C. F. Baxter	Boston, Mass	July 14, 1863	39, 208
Bottle-stopper	J. Beard and M. Fairbanks	Boston, Mass	July 17, 1866	56, 349
Bottle-stopper	R. F. Bocemsdes	Wallingford, Conn	July 2, 1867	66, 290
Bottle-stopper	H. S. Carley	Cambridgeport, Mass	May 21, 1867	64, 838
Bottle-stopper	H. S. Carley	Cambridgeport, Mass	Sept. 3, 1867	68, 484
Bottle-stopper	H. S. Carley	Cambridgeport, Mass	Sept. 3, 1867	68, 485
Bottle-stopper	S. and S. C. Cary	New York, N. Y	Apr. 21, 1868	76, 994
Bottle-stopper	F. Catlin	New York, N. Y	Apr. 17, 1866	53, 948
Bottle-stopper	R. S. Connelly	Johnsonville, N. Y	Jan. 30, 1866	52, 269
Bottle-stopper	C. J. Converse	Boston, Mass	Dec. 22, 1868	85, 214
Bottle-stopper	J. T. Cree	Worcester, Mass	Aug. 23, 1870	106, 557
Bottle-stopper	M. C. Cronk	Auburn, N. Y	July 6, 1858	20, 778
Bottle-stopper	M. C. Cronk	Auburn, N. Y	Mar. 19, 1861	31, 703
Bottle-stopper	D. M. Cumings	Newburyport, Mass	Jan. 28, 1873	135, 267
Bottle-stopper	A. De Mestre	Bordeaux, France	Feb. 18, 1873	136, 045
Bottle-stopper	L. Dovell	Newark, N. J	Apr. 18, 1871	113, 751
Bottle-stopper	W. Ely	New York, N. Y	May 3, 1870	102, 669
Bottle-stopper	J. Ewing	New York, N. Y	July 6, 1858	20, 843
Bottle-stopper	H. B. Fox	Oxton, England	June 5, 1866	55, 438
Bottle-stopper	H. B. Fox and J. T. Hall	Oxton and Liverpool, England.	Nov. 19, 1867	70, 986
Bottle-stopper	H. Frank	Pittsburgh, Pa	Aug. 6, 1872	130, 208
Bottle-stopper	W. H. Gibbs	Cincinnati, Ohio	July 13, 1869	92, 601
Bottle-stopper	W. H. Gibbs	Cincinnati, Ohio	Aug. 3, 1869	93, 296
Bottle-stopper	W. H. Gibbs	Cincinnati, Ohio	July 26, 1870	105, 797
Bottle-stopper	C. Glover	New York, N. Y	June 27, 1871	116, 433
Bottle-stopper	C. Goldthwait	South Weymouth, Mass	Sept. 19, 1865	49, 996
Bottle-stopper	J. H. Gould	Newburyport, Mass	July 30, 1867	67, 292
Bottle-stopper	W. Graff	Philadelphia, Pa	Oct. 27, 1868	83, 485
Bottle-stopper	W. H. Hall	New York, N. Y	Sept. 13, 1864	44, 184
Bottle-stopper	J. T. Haviland	San Francisco, Cal	Jan. 5, 1869	85, 588
Bottle-stopper	P. R. Higley	Oshawa, Canada	Nov. 8, 1864	45, 005
Bottle-stopper	P. R. Higley	Oshawa, Canada	Feb. 20, 1872	123, 898
Bottle-stopper	W. L. Hoefer	Jeffersonville, N. Y	Nov. 15, 1870	109, 212
Bottle-stopper	H. Holl	Philadelphia, Pa	Feb. 26, 1867	62, 333
Bottle-stopper	W. Hunt	New York, N. Y	Jan. 4, 1853	9, 527
Bottle-stopper	T. Kendall	San Francisco, Cal	Nov. 7, 1854	11, 895
Bottle-stopper	E. and D. Kinsey	Cincinnati, Ohio	Nov. 16, 1852	9, 407
Bottle-stopper	J. Klee	Dayton, Ohio	Sept. 15, 1868	82, 230
Bottle-stopper	W. Kloenne	New York, N. Y	June 20, 1865	48, 341
Bottle-stopper	F. Kutscher	New Haven, Conn	June 15, 1869	91, 349
Bottle-stopper	F. Kutscher	New Haven, Conn	Feb. 25, 1873	136, 326
Bottle-stopper	G. W. Ladd and F. W. Copcutt	New York, N. Y	May 3, 1870	102, 685
Bottle-stopper	H. S. Lesher	New York, N. Y	Nov. 18, 1873	144, 777
Bottle-stopper	T. Lewis	Malden, Mass	Dec. 21, 1858	22, 370
Bottle-stopper	G. C. Lowe	New York, N. Y	Nov. 14, 1871	120, 821
Bottle-stopper	J. D. Lynde	Philadelphia, Pa	Mar. 25, 1862	34, 759

Index of patents issued from the United States Patent Office from 1790 *to* 1873, *inclusive*—Continued.

Invention.	Inventor.	Residence.	Date.	No.
Bottle-stopper	P. E. Malmström and P. E. Dummer.	New York, N. Y	Feb. 20, 1872	123, 920
Bottle-stopper	J. Mathews, jr	New York, N. Y	Aug. 13, 1867	67, 781
Bottle-stopper	T. J. Mayall	Boston, Mass	July 1, 1873	140, 517
Bottle-stopper	F. Miller	Newark, N. J	Aug. 21, 1866	57, 356
Bottle-stopper	T. W. Mirick	Boston, Mass	July 3, 1866	56, 079
Bottle-stopper	W. Morgenstern	New York, N. Y	Oct. 18, 1870	108, 506
Bottle-stopper	W. Morgenstern	New York, N. Y	Sept. 3, 1872	131, 111
Bottle-stopper	E. Morris, jr	New Haven, Conn	Feb. 20, 1866	52, 791
Bottle-stopper	E. D. Moyer	Philadelphia, Pa	June 20, 1865	48, 300
Bottle-stopper	J. Mulchahey	Springfield, Mass	Sept. 18, 1866	58, 174
Bottle-stopper	L. B. Myers	Elmore, Ohio	Apr. 3, 1866	53, 655
Bottle-stopper	J. Nathan	Washington, D. C	May 28, 1867	65, 259
Bottle-stopper	R. T. Osgood	Orland, Me	Aug. 1, 1865	49, 144
Bottle-stopper	J. Park	Philadelphia, Pa	June 27, 1871	116, 347
Bottle-stopper	J. H. Parkhurst	Saint Louis, Mo	Dec. 10, 1872	133, 883
Bottle-stopper	R. G. Pike	Middletown, Conn	Apr. 24, 1866	54, 201
Bottle-stopper	T. Pimer	New London, Conn	June 12, 1866	55, 589
Bottle-stopper	H. C. Pratt	Boston, Mass	June 28, 1870	104, 766
Bottle-stopper	J. A. Preston	Boston, Mass	Jan. 25, 1861	32, 647
Bottle-stopper	J. A. Preston	Boston, Mass	Aug. 5, 1862	36, 126
Bottle-stopper	H. F. Reiner	Blairsville, Pa	Feb. 18, 1873	136, 095
Bottle-stopper	L. Rhoades	Providence, R. I	Aug. 19, 1873	141, 950
Bottle-stopper	D. T. Robinson	Boston, Mass	Dec. 11, 1866	60, 424
Bottle-stopper	R. Robinson	Brooklyn, N. Y	Apr. 16, 1867	63, 942
Bottle-stopper	G. W. Rogers	New York, N. Y	Aug. 21, 1866	57, 382
Bottle-stopper	A. Scharlett	Newark, N. J	Feb. 20, 1866	52, 755
Bottle-stopper	F. Schlich	New York, N. Y	Sept. 5, 1865	49, 793
Bottle-stopper	F. Schlich	New York, N. Y	Feb. 28, 1871	112, 185
Bottle-stopper	W. A. Shaw	Boston, Mass	Dec. 17, 1861	33, 964
Bottle-stopper	N. Thompson	London, England	Dec. 4, 1866	60, 283
Bottle-stopper	N. Thompson	Brooklyn, N. Y	June 28, 1870	104, 794
Bottle-stopper	S. H. Simmons	La Fayette, Ind	Sept. 18, 1866	58, 153
Bottle-stopper	W. Vom Hofe	New York, N. Y	Nov. 5, 1867	70, 654
Bottle-ctopper	T. B. Way	Bennington, Vt	Nov. 7, 1865	50, 864
Bottle-stopper	T. A. Weber	New York, N. Y	July 2, 1867	66, 270
Bottle-stopper	E. R. Wilbur	New York, N. Y	Apr. 25, 1865	47, 483
Bottle-stopper	J. B. Williams	New York, N. Y	Feb. 9, 1858	19, 323
Bottle-stopper	G. R. Willmot	Minden, Conn	Aug. 29, 1865	49, 671
Bottle-stopper	W. and D. Wilson	New York, N. Y	Apr. 26, 1870	102, 349
Bottle-stopper	I. A. and G. E. Woodbury	East Cambridge, Mass	Sept. 19, 1865	50, 063
Bottle-stopper	J. Woolaver	Suisun, Cal	Nov. 7, 1865	50, 867
Bottle-stopper	W. Wright	Phœnicia, N. Y	Nov. 14, 1871	120, 923
Bottle-stopper	I. Zamboni	Saint Louis, Mo	Mar. 3, 1868	75, 233
Bottle-stopper and coupling	W. D. Brown	Milwaukee, Wis	June 11, 1867	65, 537
Bottle-stopper and medicine-gage	S. H. Timmons	La Fayette, Ind	Nov. 20, 1866	59, 875
Bottle-stopper device	J. Hanley	New York, N. Y	Jan. 9, 1855	12, 202
Bottle-stopper fastener	D. T. Robinson	Boston, Mass	Mar. 20, 1866	53, 342
Bottle-stopper fastening	T. A. Ashburner	Philadelphia, Pa	Mar. 13, 1855	12, 501
Bottle-stopper fastening	H. T. Dewey	Brooklyn, N. Y	Jan. 9, 1872	122, 579
Bottle-stopper fastening	H. W. Putnam	Cleveland, Ohio	Mar. 15, 1859	23, 263
Bottle-stopper fastening	R. S. Stubbs	Dover, N. H	Dec. 31, 1867	72, 761
Bottle-stopper fastening, Machine for forming wire	H. W. Putnam	Bennington, Vt	May 30, 1871	115, 354
Bottle stopper, Glass	S. A. Whitney	Glassborough, N. Y	Jan. 1, 1861	31, 046
Bottle stopper, Ink	M. Ames	New Ipswich, N. H	Mar. 27, 1866	53, 390
Bottle stopper or cap	J. Quinn and G. W. Putnam	Boston, Mass	Feb. 2, 1869	86, 446
Bottle-stopper, Self-acting	D. A. Draper	East Cambridge, Mass	Oct. 13, 1863	40, 251
Bottle stopper, Smelling	E. F. Marble and R. W. Glidden	Attleborough, Mass	Apr. 15, 1873	137, 938
Bottle-stopper, Undetachable swinging	A. H. Forbes	New York, N. Y	Apr. 10, 1849	6, 297
Bottle-stopper, Valvular	C. Patterson	Philadelphia, Pa	Dec. 30, 1873	146, 018
Bottle stopper, Volatile salts	A. Hebbard	New York, N. Y	Jan. 7, 1873	134, 600
Bottle-stoppers, caster-wheels, syringe-pistons, &c., Manufacture of.	R. B. Hugunin	Cleveland, Ohio	Mar. 6, 1866	53, 002
Bottle-stoppers, Device for cutting wire or cord from	J. A. Traut	New Britain, Conn	Sept. 6, 1870	107, 125
Bottle, &c., stoppers, Device for securing	W. H. Richards	Auburndale, Mass	Jan. 7, 1868	73, 043
Bottle-stoppers, Machine for bending wire-fastening for.	H. W. Putnam	Bennington, Vt	May 23, 1871	115, 102
Bottle, Tooth-powder	J. B. Da Camara, jr	Newark, N. J	Nov. 19, 1867	71, 142
Bottle-uncorking device	A. Manuel	Reims, France	Jan. 29, 1867	61, 624
Bottle-washer	H. B. Davison	San Francisco, Cal	Nov 19, 1867	70, 973
Bottle-washer	W. Dick	New York, N. Y	Mar. 4, 1873	136, 423
Bottle-washing apparatus	F. Schlick and A. Feyh	New York, N. Y	Apr. 22, 1873	138, 046
Bottle-washing machine	H. N. Degraw	Watervliet, N. Y	Sept. 1, 1857	18, 086
Bottle-washing machine	C. Euler	Evansville, Ind	Sept. 12, 1871	118, 917
Bottle-washing machine	J. Matthews, jr	New York, N. Y	Aug. 22, 1865	49, 538
Bottle-washing machine	A. H. Rauch	Bethlehem, Pa	June 6, 1854	11, 000
Bottle-washing machine	N. H. Sirrelle	Baltimore, Md	June 29, 1869	92, 112
Bottle-washing machine	C. L. Werk and N. Verdin	Cincinnati, Ohio, and Yorkville, Ind.	Aug. 13, 1872	130, 552
Bottle-washing machine	W. B. White and J. A. Whiteford.	Saratoga Springs, N. Y	Apr. 27, 1859	20, 113
Bottle-wrapper	A. E. Francis	Ravenna, Ohio	May 27, 1873	139, 308
Bottle-wrapper	W. A. Hinman	New York, N. Y	Dec. 12, 1871	121, 718
Bottles, &c., air-tight, Closing	R. Arthur	Washington, D. C	Aug. 15, 1854	11, 513
Bottles and jars, Revolving plug for manufacturing	J. F. Bodine	Williamstown, N. J	Apr. 12, 1859	23, 640
Bottles and jars, Tool for forming lugs in the mouths of.	A. Stone	Philadelphia, Pa	Apr. 12, 1859	23, 623
Bottles and other vessels, Device for closing	J. Mathews	New York, N. Y	Apr. 15, 1873	137, 941
Bottles and other vessels, Machine for cutting cork stoppers for.	J. Power and A. J. Bailey	Boston and Charlestown, Mass.	May 20, 1862	35, 351
Bottles and pipes, Device for closing	D. L. and R. Bollerman	New York, N. Y	July 20, 1869	92, 782
Bottles and similar articles, Forming screw-threads, &c., in the necks of glass.	A. Stone	Philadelphia, Pa	Aug. 7, 1855	13, 402
Bottles, Apparatus for changing siphon	W. P. Clark	Medford, Mass	July 16, 1872	129, 002
Bottles, Apparatus for cutting the strings that secure corks in.	G. Blanchard	New York, N. Y	June 10, 1856	15, 098
Bottles, Apparatus for filling siphon	W. Gee	New York, N. Y	Jan. 21, 1868	73, 594
Bottles, Apparatus for inserting stoppers in	A. Stone	Philadelphia, Pa	June 18, 1861	32, 590

Index of patents issued from the United States Patent Office from 1790 to 1873, inclusive—Continued.

Invention.	Inventor.	Residence.	Date.	No.
Bottles, Attaching cork-retaining bails to	L. B. Flanders	Philadelphia, Pa	June 17, 1873	139, 886
Bottles, Attaching stoppers to	J. J. Eshleman	Philadelphia, Pa	Mar. 5, 1867	62, 536
Bottles, cans, &c., Device for sealing	M. B. Espy	Philadelphia, Pa	July 14, 1857	17, 783
Bottles, Case for indelible-ink	E. Daniels	Northampton, Mass	Aug. 7, 1860	29, 469
Bottles, Cleansing	M. C. Cronk	Auburn, N. Y	Aug. 7, 1849	6, 631
Bottles, Closing	T. S. Bowman	Saint Louis, Mo	Aug. 27, 1867	68, 162
Bottles, Closing	E. Hamilton	Chicago, Ill	Jan. 5, 1864	41, 067
Bottles, Closing	J. Matthews, jr	New York, N. Y	July 18, 1865	48, 822
Bottles, Closing	N. Prescott	Dorchester, Mass	Sept. 3, 1867	68, 456
Bottles, Closing	R. Robinson	New York, N. Y	Mar. 14, 1865	46, 864
Bottles, &c., Closing	N. Thompson	Saint John's Wood, England.	Oct. 13, 1863	40, 293
Bottles, Closing or stopping	E. Hamilton	Chicago, Ill	Apr. 5, 1864	42, 188
Bottles, Construction of glass	F. N. Bullard	Worcester, Mass	Oct. 30, 1866	59, 333
Bottles, Construction of nursing	E. Dupuy	New York, N. Y	Oct. 7, 1844	3, 780
Bottles, Cork-receptacle for	A. Honrath	New York, N. Y	July 30, 1867	67, 303
Bottles, Device for packing	G. C. Furber	Yreka, Cal	Aug. 9, 1870	106, 149
Bottles, Device for securing corks in	S. L. Gouverneur	Frederick City, Md	Feb. 7, 1871	111, 657
Bottles, Device for stopping	A. Andre	Chicago, Ill	May 5, 1863	38, 367
Bottles, &c., Device for stopping	N. Thompson	Saint John's Wood, England.	Oct. 13, 1863	40, 292
Bottles, Drip-cup for	W. R. Miller	Baltimore, Md	Dec. 24, 1872	134, 302
Bottles, Ejecting-apparatus for	W. S. Ward	New York, N. Y	Jan. 14, 1873	134, 953
Bottles, Fastening metallic collars to	E. Wattis, jr	Philadelphia, Pa	Jan. 21, 1868	73, 680
Bottles, &c., for transportation, Packing	T. M. Perot	Philadelphia, Pa	Sept. 15, 1863	39, 950
Bottles, Frame for making straw envelopes for	W. F. Tillinghast	Burlington, N. J	Mar. 4, 1873	136, 394
Bottles, Frame for securing liquor and other	F. J. Miller	Brooklyn, N. Y	Mar. 27, 1866	53, 540
Bottles, Hermetically sealing	M. B. Espy	Philadelphia, Pa	Sept. 30, 1856	15, 802
Bottles, Implement for grooving the mouths of glass.	E. Bennett	Philadelphia, Pa	July 3, 1866	55, 988
Bottles, Instrument for opening	J. Matthews, jr	New York, N. Y	June 27, 1865	48, 422
Bottles, Instrument for removing effervescing fluids from.	W. W. Meglone	Nashville, Tenn	Aug. 14, 1866	57, 256
Bottles, Instrument for removing twine and wire from.	J. T. Haviland	San Francisco, Cal	June 21, 1870	104, 453
Bottles, Instrument for removing wire from	J. S. Hazard	Newport, R. I	Oct. 16, 1866	58, 820
Bottles, jars, &c., Fastening for covers or stoppers of.	F. Klee	Brooklyn, N. Y	Jan. 19, 1869	86, 082
Bottles, jars, &c., Metallic cap for	W. J. Stevenson	New York, N. Y	June 8, 1858	20, 520
Bottles, jars, &c., Mode of closing the mouths of	N. Thompson	Saint John's Wood, England.	Sept. 24, 1867	69, 143
Bottles, Locking-cap for	L. M. Ballard	New York, N. Y	May 18, 1869	90, 221
Bottles, Locking-cap for	J. S. Hough	New York, N. Y	Sept. 27, 1870	107, 778
Bottles, Machine for coloring capsules for	H. Stiassny	New York, N. Y	Apr. 23, 1872	125, 912
Bottles, Manufacture of cruet	W. Pountney	Brooklyn, N. Y	July 10, 1866	56, 263
Bottles, Manufacture of stopples for	L. Bishop	Paris, France	June 11, 1867	65, 533
Bottles, Means of holding glass	J. Wood	Brooklyn, N. Y	May 22, 1866	54, 991
Bottles, Metallic capsule for	W. Betts	Wharf Road, England	Nov. 19, 1867	71, 124
Bottles, Method of drawing fluids from	I. W. Fox	Durhamville, N. Y	May 27, 1856	14, 982
Bottles, Nipple for nursing	H. D. Lockwood	Charlestown, Mass	May 28, 1872	127, 357
Bottles of clay, Apparatus for making	E. H. and H. E. Merrill	Akron, Ohio	June 9, 1868	78, 676
Bottles of clay, Making	E. H. and C. J. Merrill	Akron, Ohio	July 31, 1847	5, 206
Bottles or other vessels made from plastic substances, Tool for forming groove around the orifice of.	A. Stone	Philadelphia, Pa	Sept. 23, 1856	15, 788
Bottles, steam-pipes, &c., Covering for	J. B Crane	Dalton, Mass	Feb. 26, 1867	62, 318
Bottles, Stopping	A. Albertson	New York, N. Y	Oct. 11, 1864	44, 684
Bottles, Stopping	A. Albertson	New York, N. Y	Nov. 15, 1864	45, 112
Bottles, Stopping	A. Wiegand	Philadelphia, Pa	Sept. 29, 1863	40, 136
Bottles, Stopping mineral-water	A. Quantin	Philadelphia, Pa	July 25, 1854	11, 385
Bottles, Tool for forming lips on necks of	T. Barrett	Charlestown, Mass	July 27, 1869	93, 039
Bottles, &c., Tool for forming mouths of	H. Frank	Pittsburgh, Pa	Aug. 6, 1872	130, 207
Bottles, Tool for forming necks and orifices of glass	R. MacLardy	Pittsburgh, Pa	Feb. 28, 1860	27, 333
Bottles, Tool for forming necks of	L. F. Smith	Stonington, Conn	Dec. 9, 1862	37, 142
Bottles, Tool for forming screw-threads in the necks of glass.	G. M. Keefer	East Birmingham, Pa	Apr. 16, 1872	125, 739
Bottles, Tool for holding glass	E. McArdle	Cambridge, Mass	May 8, 1866	54, 572
Bottles, Tool for holding glass	T. B. Hewitt	Williamstown, N. J	Dec. 30, 1873	145, 945
Bottles, Valve-gage for	A. Quantin	Philadelphia, Pa	Oct. 25, 1853	10, 158
Bottles, Washing, filling, and corking	S. A. Bille	New York, N. Y	Nov. 8, 1828	
Bottling-apparatus	J. Buser	New York, N. Y	Feb. 11, 1862	34, 347
Bottling-apparatus	H. B. Goodyear	New Haven, Conn	Mar. 7, 1865	46, 658
Bottling-apparatus	A. S. Taylor	San Francisco, Cal	Nov. 12, 1872	133, 068
Bottling fluids under gaseous pressure, Method of	A. Quantin	Philadelphia, Pa	Mar. 4, 1856	14, 368
Bottling gas-charged liquids, Machine for	E. D. Taylor	Hornellsville, N. Y	Aug. 15, 1871	118, 069
Bottling liquids	J. Klee	Dayton, Ohio	July 29, 1873	141, 360
Bottling liquids, Apparatus for	J. Matthews, jr	New York, N. Y	Mar. 6, 1866	53, 019
Bottling liquids under pressure	J. Matthews, jr	New York, N. Y	Jan. 5, 1864	41, 082
Bottling-machine	J. Armstrong and S. Marks	San Francisco, Cal	Nov. 12, 1872	133, 003
Bottling-machine	H. Carse	Pittsburgh, Pa	Dec. 8, 1868	84, 795
Bottling-machine	J. Matthews, jr	New York, N. Y	Feb. 2, 1864	41, 440
Bottling-machine	J. Matthews, jr	New York, N. Y	Jan. 29, 1867	61, 627
Bottling-machine for liquids	J. V. Mathivet	Cleveland, Ohio	Aug. 15, 1871	118, 036
Bottling-machines, Siruping-device for	W. Tollast	New York, N. Y	Apr. 23, 1872	126, 103
Bottling mineral-water, Method of	C. H. Thomas	Philadelphia, Pa	June 18, 1867	65, 842
Bottling soda, &c., Sirup-gage for	J. Schrink	Toledo, Ohio	June 28, 1864	43, 344
Bottling still liquids	J. Beard and M. Fairbanks	Boston, Mass	Dec. 6, 1864	45, 373
Bougie for stricture	R. J. Dodd	Philadelphia, Pa	Oct. 6, 1843	3, 297
Bouquet-holder	E. Dithridge	Pittsburgh, Pa	Mar. 20, 1866	53, 279
Bouquet-holder	A. D. Frye, jr	New York, N. Y	May 28, 1867	65, 204
Bouquet-holder	A. F. Ransom	Glen's Falls, N. Y	June 3, 1873	139, 477
Bouquet-holder	J. C. Reed	Boston, Mass	Aug. 29, 1871	118, 486
Bouquet-holder	E. Wilder	Springfield, Mass	July 16, 1872	129, 444
Bow-drill stock	D. F. Hartford	Boston, Mass	Aug. 27, 1867	68, 071
Bow-drill stock, Stringing	D. F. Hartford	Boston, Mass	Dec. 24, 1867	72, 632
Bow for archery	E. S. Morton	Plymouth, Mass	May 14, 1872	126, 734
Bowl, Water	B. Brower	New York, N. Y	Aug. 5, 1873	141, 421
Bowl, Water	S. Hawson and T. Sweeney	Jersey City, N. J., and Brooklyn, N. Y.	May 7, 1872	126, 457

Index of patents issued from the United States Patent Office from 1790 *to* 1873, *inclusive*—Continued.

Invention.	Inventor.	Residence.	Date.	No.
Bowls, &c., Detachable nose for	G. Raymond	Fitchburgh, Mass	May 12, 1868	77, 914
Bowls, Machine for making wooden	R. Simonds and G. N. Goodspeed.	Ludlow, Vt., and Winchenden, Mass.	July 31, 1860	29, 411
Bowls, Machinery for turning	P. Hutchins, jr	Worthington, Mass	Sept. 25, 1847	5, 306
Bowls, Method of making footed glass	E. G. Cate	Wheeling, W. Va	Oct. 28, 1873	144, 061
Bowling-alley	A. T. Peirce	Fair Haven, Conn	July 13, 1869	92, 467
Bowling-alley pin, Substitute for	J. M. Currier	Newburyport, Mass	Nov. 12, 1861	33, 695
Box:				
See Adjustable box.				
Advertising-box.				
Alkali-box.				
Ammunition-box.				
Annealing-box.				
Anti-friction box.				
Ash-box.				
Axle-box.				
Bacon and ham box.				
Ballot-box.				
Berry-box.				
Blacking-box.				
Bluing-box.				
Camera-box.				
Candle-mold box.				
Cap-box.				
Car-spring box.				
Carriage-wheel box.				
Cartouch-box.				
Cartridge box.				
Centering-box.				
Chalk-line box.				
Change-box.				
Check-box.				
Cheese-box.				
Cigar-box.				
Cigar-fuse box.				
Coal-box.				
Collar-box.				
Collection-box.				
Conductor's change-box.				
Core-box.				
Coupling-box.				
Cracker-box.				
Cylindrical box.				
Dice-box.				
Die-box.				
Dredge-box.				
Elevator-box.				
Fare-box.				
Feed-box.				
File-box.				
Filler-box.				
Fire-box.				
Fire-alarm box.				
Free-box.				
Freezing-box.				
Fruit-box.				
Fur-box.				
Game-box.				
Garbage-box.				
Grape-box.				
Hat-box.				
Honey-box.				
Hop-box.				
Hub-box.				
Jewel-box.				
Jewelry-box.				
Journal-box.				
Laundry-box.				
Letter-box.				
Lubricating-box.				
Lunch-box.				
Mail-box.				
Market-box.				
Match-box.				
Metallic box.				
Miter-box.				
Money-box.				
Moth-proof box.				
Muff-box.				
Music-box.				
Musical box.				
Nut-box.				
Oil and paint box.				
Oil-stone box.				
Packing-box.				
Paper box.				
Pasteboard box.				
Passing-box.				
Peach-box.				
Peg-box.				
Pepper-box.				
Photographic-camera box.				
Photographic-paper box.				
Pill-box.				
Pitman-box.				
Polishing-box.				
Porter's box.				
Post-office box.				

Index of patents issued from the United States Patent Office from 1790 *to* 1873, *inclusive*—Continued.

Invention.	Inventor.	Residence.	Date.	No.
Box—Continued.				
See Powder-box.				
Preserving-box.				
Preserving and packing box.				
Propagating-box.				
Pump-box.				
Resin-box.				
Retaining-box.				
Round box.				
Safe-deposit box.				
Sample box.				
Sand-box.				
Scouring-box.				
Seidlitz-powder box.				
Setting-box.				
Settling-box.				
Sheep-tagging box.				
Sheet-metal box.				
Sheet-metal-can box.				
Shingle-gaging box.				
Shoe-box.				
Shovel-drop box.				
Show-box.				
Shuttle-box.				
Shuttle-drop box.				
Signal-box.				
Silver-ware box.				
Sluice-box.				
Spice-box.				
Spindle-step box.				
Spool-box.				
Step-box.				
Stop-cock box.				
Stove-blacking box.				
Strawberry-box.				
Stuffing-box.				
Tape-box.				
Telegraph-signal box.				
Thimble-box.				
Ticket-box.				
Tin box.				
Tobacco-box.				
Tobacco-cutting box.				
Tool-box.				
Toy money-box.				
Tree-box.				
Trunk-box.				
Tumbling-box.				
Turpentine-box.				
Twine-box.				
Wagon-box.				
Wagon and sleigh box.				
Water-closet box.				
Wheel-box.				
Wood-box.				
Wooden box.				
Wool-box.				
Work-box.				
Box	J. Cohn	New York, N. Y	Nov. 21, 1871	121, 156
Box	R. B. Davis	New York, N. Y	Mar. 26, 1872	125, 028
Box	R. B. Davis	New York, N. Y	Apr. 30, 1872	126, 272
Box	W. B. Guernsey	New York, N. Y	June 23, 1868	79, 113
Box	H. R. Heyl	Philadelphia, Pa	Oct. 8, 1872	132, 074
Box	H. R. Heyl	Philadelphia, Pa	Oct. 8, 1872	132, 075
Box	J. S. Lash	Philadelphia, Pa	Dec. 16, 1873	145, 663
Box	J. J. Leighton	Boston, Mass	Nov. 17, 1868	84, 200
Box	H. Manneck	New York, N. Y	Jan. 31, 1871	111, 463
Box	J. Nelson	Rockford, Ill	Aug. 16, 1870	106, 393
Box	E. C. Patterson	Rochester, N. Y	Apr. 1, 1873	137, 385
Box	W. M. Pierce	Sandusky, Ohio	June 18, 1872	128, 170
Box	H. S. Shepardson	Shelburne Falls, Mass	Jan. 23, 1872	122, 919
Box	C. J. Siercks	Chicago, Ill	Aug. 9, 1870	106, 221
Box	J. W. Wilcox	New York, N. Y	Nov. 3, 1868	83, 812
Box and bottle cap	W. L. Troxell	Brooklyn, N. Y	July 30, 1872	130, 086
Box and hub and hanging coach-body	D. Watson	Fayette, Mass	May 29, 1832	....
Box, axle, journal, &c	D. Cumming	Sorrel Horse, Pa	Nov. 25, 1856	16, 113
Box-boiler	J. Morris	New Haven, Conn	Oct. 3, 1812	....
Box-fastening	J. J. Greenough	Syracuse, N. Y	Oct. 24, 1871	120, 192
Box for case-hardening	J. Greene	Providence, R. I	Nov. 3, 1863	40, 477
Box-former	G. W. Swan	San Francisco, Cal	Sept. 23, 1873	143, 103
Box for transporting small fruit and berries	E. Secor	Lawrence, Mich	Jan. 1, 1867	60, 796
Box-heads, &c., Removable	S. Macferren	Philadelphia, Pa	Nov. 24, 1868	84, 366
Box-hook	C. Bruso	Worcester, Mass	Apr. 19, 1870	102, 087
Box-joints, Machine for making	J. Stimpson	Baldwinsville, Mass	Nov. 8, 1859	26, 060
Box lifter	B. H. Smith	Ipswich, Mass	May 20, 1873	139, 029
Box-machine	E. G. Alden	Cambridge, Mass	Apr. 4, 1871	113, 238
Box-machine	A. Davis	Washington, D. C	Apr. 18, 1854	10, 788
Box-machine	T. Kingsford	Oswego, N. Y	Sept. 3, 1872	131, 009
Box-machine	G. P. Roberts	Saint Louis, Mo	Mar. 3, 1863	37, 827
Box-making machine	B. E. Dexter	Gowanda, N. Y	Dec. 17, 1872	134, 039
Box-making machine	S. L. Hill	Williamsburgh, N. Y	Nov. 3, 1863	40, 480
Box-making machine	E. James	Tyngsborough, Mass	Nov. 24, 1868	84, 422
Box-making machine	T. C. Luther	Lee, Mass	Oct. 11, 1864	44, 690
Box-making machine	G. W. Swan	San Francisco, Cal	Aug. 26, 1873	142, 297
Box-making machine	H. Thayer	Brooklyn, N. Y	June 21, 1864	43, 239
Box-making machinery	W. M. Davis	Gardiner, Me	Jan. 16, 1849	6, 038
Box-nailing machine	C. Bauer and W. C. Munder	Newark, N. J	Nov. 8, 1864	44, 929
Box-nailing machine	E. Beard	Ottawa, Ill	July 9, 1872	128, 698

Index of patents issued from the United States Patent Office from 1790 *to* 1873, *inclusive*—Continued.

Invention.	Inventor.	Residence.	Date.	No.
Box-nailing machine	G. Wicke	New York, N. Y	June 16, 1863	38, 924
Box-opener	A. Barbarin	New Orleans, La	Oct. 13, 1868	83, 022
Box-opener	R. Blake	Scranton, Pa	Apr. 20, 1869	89, 013
Box-opener	R. H. Chinn, G. Hall, and J. J. Fitch.	Washington, D. C., and Morgantown, W. Va.	Dec. 26, 1871	122, 106
Box-opener	J. C. Gaston	Cincinnati, Ohio	Feb. 27, 1872	124, 049
Box-opener	W. Gould	Minneapolis, Minn	Feb. 27, 1872	124, 129
Box-opener	H. H. Hall	Tioga, Pa	Aug. 25, 1868	81, 497
Box-opener	J. W. Hankenson	Minneapolis, Minn	May 28, 1872	127, 342
Box-opener	T. B. Henkle	Knightstown, Ind	Jan. 24, 1871	111, 207
Box-opener	A. Heusser	Ellington, Conn	May 24, 1870	103, 330
Box-opener	M. J. Hinden	Detroit, Mich	Nov. 11, 1873	144, 454
Box-opener	L. Holtzscheiter	Philadelphia, Pa	Oct. 27, 1868	83, 499
Box-opener	A. F. Jackson	Minneapolis, Minn	July 16, 1872	129, 412
Box-opener	M. D. Lawrence	Springfield, Mass	Oct. 3, 1865	50, 30[illegible]
Box-opener	S. Mills	Jersey City, N. J	Apr. 22, 1873	138, 177
Box-opener	N. Purdey	Providence, Pa	Oct. 11, 1870	108, 293
Box opener	P. Stone	Charlestown, Mass	Feb. 4, 1868	74, 167
Box-opener	G. C. Taft	Worcester, Mass	Oct. 21, 1851	8, 457
Box-opener	H. C. Van Gieson	Paterson, N. J	Oct. 19, 1869	95, 957
Box-opener	C. P. S. Wardwell	Lake Village, N. H	Apr. 29, 1856	14, 783
Box-opener	J. Willard	Norwich, Conn	Sept. 1, 1868	81, 855
Box-opening apparatus	C. M. O'Hara	New York, N. Y	Aug. 3, 1869	93, 335
Box-opening instrument	G. C. Taft	Worcester, Mass	Apr. 13, 1852	8, 878
Box-opening tool	E. Hambujer	New York, N. Y	Nov. 7, 1865	50, 820
Box-opening tool	L. D. Howard	Saint Johnsbury, Vt	Nov. 9, 1869	96, 700
Box-opening tool	W. M. Keague	Brooklyn, N. Y	Mar. 15, 1864	41, 925
Box-opening tool	E. C. C. Kellogg	Hartford, Conn	May 23, 1865	47, 899
Box-stamping die, Wooden	W. Weeks	Albany, N. Y	Sept. 1, 1868	81, 852
Box-stuff, Machine for cutting	C. W. Royse	Rindge, N. H	May 6, 1873	138, 584
Box-trap for animals	B. B. and J. R. Hill	Worcester, Mass	Feb. 14, 1865	46, 356
Box-trap for animals	J. Stroop and J. Esteale	Allegheny, Pa	Nov. 6, 1843	3, 326
Box with anti-friction roller	W. Edwards	Baltimore, Md	Sept. 11, 1817	....
Boxes and axles, Packing	W. H. Hovey	Hartford, Conn	Aug. 13, 1850	7, 563
Boxes and cases, Method of making	H. Thayer	Brooklyn, N. Y	June 21, 1864	43, 240
Boxes and packages, Implement for handling	C. Hoffmann	New York, N. Y	Dec. 12, 1871	121, 873
Boxes, cartridge-cases, &c., Manufacture of	C. S. Wells	Springfield, Mass	Oct. 18, 1870	108, 542
Boxes, cases, &c., End or head for	H. Thayer	Warsaw, N. Y	Dec. 20, 1864	45, 538
Boxes, Casting	W. Butler and E. B. Rice	Worcester, Mass	May 31, 1833	....
Boxes, Device for joining	W. Duryea	New York, N. Y	Sept. 11, 1860	29, 957
Boxes, &c., Eccentric fastening for	W. Youngblood	Brooklyn, N. Y	Feb. 11, 1873	135, 873
Boxes from paper-pulp, Manufacture of	S. Wheeler and E. Jerome	Albany, N. Y	Feb. 12, 1867	61, 969
Boxes, Implement for straining band about	S. Marden	Newton, Mass	Oct. 15, 1867	69, 826
Boxes, Making	H. Thayer	Brooklyn, N. Y	June 21, 1864	43, 241
Boxes, Making	J. C. Walker	Concord, N. H	Aug. 26, 1843	3, 231
Boxes, Manufacture of	A. N. Allen	Chicago, Ill	Mar. 26, 1872	124. 997
Boxes, Manufacture of	F. W. Fliedner	New York, N. Y	Jan. 31, 1865	46, 097
Boxes, Manufacture of	W. T. Slocum	Philadelphia, Pa	May 30, 1865	48, 031
Boxes, Manufacture of	C. Storer	Montreal, Canada	Nov. 26, 1872	133, 496
Boxes, Manufacture of	J. Trottier	Hennebon, France	June 11, 1872	127, 811
Boxes, Manufacturing	J. Sperry	New York, N. Y	Nov. 3, 1863	40, 507
Boxes, &c., Material for making	W. Painter	Baltimore, Md	Jan. 17, 1865	45, 950
Boxes, Method of bending wood for	S. Patterson	Berlin Heights, Ohio	Sept. 26, 1871	119, 340
Boxes, &c., Method of joining	J. Stimson	Baldwinsville, Mass	Feb. 3, 1857	16, 557
Boxes, &c., Method of joining corners of	J. Bell	Harlem, Mass	Jan. 25, 1853	9, 552
Boxes, Metallic corner-shield for	G. F. Sargent	Manchester, N. H	June 17, 1873	140, 080
Boxes, Mode of fitting heads in	C. Williams	Fallsburgh, Va	July 25, 1854	11, 397
Boxes, packages, &c., Manufacture of	H. Everett	Philadelphia, Pa	May 2, 1865	47, 532
Boxes, picture-frames, &c., Material for the manufacture of.	J. M. Merrick, jr	Boston, Mass	Dec. 15, 1868	85, 018
Boxes, Prop for holding up the lids of	A. Davis and T. Parsons	Boston, Mass	Aug. 29, 1871	118, 588
Boxes, &c., Sealing	R. M. Bartlett	Cincinnati, Ohio	Mar. 27, 1866	53, 517
Brace:				
See Ankle-brace.				
Ball-brace.				
Bit-brace.				
Body-brace.				
Boring-bit brace.				
Bracket-brace.				
Buggy-brace.				
Carriage-spring brace.				
Carriage-top brace.				
Chair-brace.				
Counter-brace.				
Crank-brace.				
Garment-brace.				
Gum-elastic brace.				
Knee-brace.				
Ratchet-brace.				
Sanitary-brace.				
Shoulder-brace.				
Spinal brace.				
Spring-brace.				
Spring and body brace.				
Suspending-brace.				
Thorough-brace.				
Trunk and piano-cover brace.				
Vehicle-spring brace.				
Wagon-brace.				
Wagon-spring brace.				
Brace	D. N. Baird	Warrenville, Ohio	Aug. 26, 1856	15, 632
Brace	A. W. Streeter	Shelburne Falls, Mass	Nov. 17, 1863	40, 652
Brace and lacing-device	S. J. Shaw and W. E. C. Worcester.	Marlborough, Mass	Oct. 16, 1866	58, 955
Brace and skirt-supporter combined	R. A. N. A. Ward	Philadelphia, Pa	Oct. 6, 1868	82, 901
Brace and suspender combined	E. L. Demorest and W. G. Cook	New York, N. Y	Mar. 9, 1869	87, 648
Brace, Attaching bit to	J. H. Nellis	Richmondville, N. Y	Feb. 12, 1867	62, 057
Brace, Attaching bit to	L. J. Parsons	New Haven, Conn	May 7, 1867	64, 444

Index of patents issued from the United States Patent Office from 1790 *to* 1873, *inclusive*—Continued.

Invention.	Inventor.	Residence.	Date.	No.
Brace, Bit	W. A. Ives	New Haven, Conn	May 7, 1872	126, 395
Brace-bit holder	J. P. Gordon	West Garland, Me	Jan. 16, 1866	52, 042
Brace-bit holder	M. V. Nobles	Rochester, N. Y	Dec. 19, 1865	51, 660
Brace-bit in socket, Mode of securing	E. W. Nichols	Worcester, Mass	May 15, 1855	12, 868
Brace-bit, Mode of securing	C. B. Rose	Sunderland, Mass	Oct. 25, 1864	44, 822
Brace-bit, Mode of securing	C. B. Rose	Sunderland, Mass	Oct. 25, 1864	44, 823
Brace-bit, Securing	J. Mix	West Cheshire, Conn	June 17, 1862	35, 618
Brace-bit stock	J. Comstock	New London, Conn	Dec. 13, 1853	10, 307
Brace-coupling, Adjustable	C. M. Hall	Yates City, Ill	Apr. 8, 1873	137, 669
Brace-drill	G. Gibbs	Canton, Ohio	Mar. 18, 1873	136, 830
Brace, &c., fastener	E. L. Parker	Birmingham, England	Jan. 3, 1871	110, 783
Brace or bit stock	D. P. Foster	Shelburne Falls, Mass	June 28, 1864	43, 302
Brace-stock, Catch for holding the bit in	C. M. Daboll	New London, Conn	May 16, 1854	10, 924
Braces or suspenders	M. Harris	New York, N. Y	Nov. 7, 1871	120, 583
Bracelet	A. O. Baker	Providence, R. I	Nov. 26, 1872	133, 291
Bracelet	W. H. Ball and T. Barnard	Newark, N. J	Sept. 13, 1870	107, 325
Bracelet	J. Barclay	Bergen, N. J	Sept. 20, 1870	107, 588
Bracelet	J. Barclay	Bergen, N. J	Apr. 9, 1872	125, 516
Bracelet	G. Burch	Newark, N. J	June 8, 1869	91, 081
Bracelet	H. Carlisle, jr	Philadelphia, Pa	Oct. 25, 1870	108, 683
Bracelet	D. D. Codding	North Attleborough, Mass.	Mar. 8, 1870	100, 504
Bracelet	T. L. Cornell	Derby, Conn	Jan. 14, 1873	134, 735
Bracelet	W. Edge	Newark, N. J	Nov. 19, 1872	133, 146
Bracelet	A. Forbes	Attleborough, Mass	Dec. 2, 1873	145, 164
Bracelet	J. W. Grant and C. H. Cook	Providence, R. I	May 27, 1873	139, 311
Bracelet	S. W. Grant	Providence, R. I	June 17, 1873	139, 888
Bracelet	C. M. Kinsel	Columbus, Ohio	Apr. 7, 1868	76, 470
Bracelet	H. Kipling	New York, N. Y	Apr. 2, 1861	31, 931
Bracelet	W. W. Knapp	Mansfield, Mass	Dec. 17, 1872	133, 943
Bracelet	M. H. Mason	Attleborough, Mass	Apr. 9, 1872	125, 402
Bracelet	J. H. Sprague	Providence, R. I	Oct. 3, 1871	119, 540
Bracelet	G. D. Stevens	New York, N. Y	Feb. 13, 1872	123, 740
Bracelet	F. M. Sweet	Syracuse, N. Y	Aug. 31, 1858	21, 381
Bracelet	J. Wilkinson	West Mount Vernon, N. Y.	July 30, 1872	130, 096
Bracelet	H. A. Williams	Mansfield, Mass	Sept. 23, 1873	143, 109
Bracelet-clasp	J. Mansure	Philadelphia, Pa	May 23, 1854	10, 973
Bracelet, &c., Constructing	J. S. Palmer	Providence, R. I	Mar. 19, 1861	31, 735
Bracelet, Constructing flat chains for	J. Christle	Newark, N. J	Jan. 5, 1864	41, 175
Bracelet-fastening	J. Bissinger	New York, N. Y	Feb. 28, 1860	27, 264
Bracelet-fastening	S. Cottle	New York, N. Y	Dec. 23, 1873	145, 788
Bracelet-fastening	R. S. Hamilton	Providence, R. I	Mar. 19, 1872	124, 815
Bracelet-fastening	F. Kursh	Philadelphia, Pa	Jan. 7, 1873	134, 681
Bracelet-fastening	B. Ranger	Brattleborough, Vt	Oct. 31, 1871	120, 535
Bracelet-fastening	B. Ranger and J. B. Smiley	Brattleborough, Vt	Sept. 30, 1873	143, 382
Bracelet-fastening	G. H. Soule	Jersey City, N. J	Aug. 4, 1868	80, 679
Bracelet-fastening	G. H. Soule	Jersey City, N. J	July 16, 1872	128, 983
Bracelet-fastening	G. O. Yeiser	Lexington, Ky	June 25, 1872	128, 447
Bracelet guard-chain	G. D. Stevens	New York, N. Y	Jan. 31, 1871	111, 489
Bracelet-joints, Manufacture of	J. B. Black	Providence, R. I	May 27, 1873	139, 361
Bracelet manufacture	G. Sanford	Providence, R. I	Oct. 2, 1860	30, 250
Bracelet, Manufacture of enameled	A. Codding, jr	North Attleborough, Mass.	June 22, 1869	91, 604
Bracelet, Manufacture of plated metal	J. Barclay	Attleborough, Mass	Aug. 24, 1869	94, 064
Bracelet, &c., Safety-clasp for	I. Hermann	New York, N. Y	Sept. 22, 1857	18, 245
Bracelet safety-latchet	G. H. Soule	Jersey City, N. J	Apr. 9, 1872	125, 495
Bracelet-snap	G. S. Grant	Providence, R. I	Apr. 20, 1869	89, 037
Bracelet stock or blank	I. B. Staples	Attleborough, Mass	Aug. 23, 1870	106, 740
Bracelet, Voltaic	T. Hall	Boston, Mass	Apr. 23, 1867	64, 098
Bracing for cylindrical structures	G. H. Johnson	Buffalo, N. Y	Mar. 9, 1869	87, 569
Bracket:				
See Adjustable bracket.				
Bird-cage bracket.				
Bucket-bracket.				
Bureau-bracket.				
Cornice-bracket.				
Creel-bracket.				
Desk-bracket.				
Eave-trough bracket.				
Extension-bracket.				
Fence-bar bracket.				
Flower-pot bracket.				
Gas-bracket.				
Hanging bracket.				
Hat and cap bracket.				
Horse-bracket.				
Iron-bracket.				
Lamp-bracket.				
Leader-bracket.				
Looking-glass bracket.				
Metallic bracket.				
Pointed bracket.				
Pump-bracket.				
Pump-handle bracket.				
Roofing-bracket.				
Scaffold-bracket.				
Scaffolding-bracket.				
Shelf-bracket.				
Shelving-bracket.				
Shingling-bracket.				
Shingling and slating bracket.				
Slating-bracket.				
Stair-rail bracket.				
Stove-door bracket.				
Toilet-bracket.				
Tongue-bracket.				
Wagon-tongue bracket.				
Warp-creel bracket.				
Wire-bracket.				
Bracket	L. Benecke	Brunswick, Mo	Apr. 29, 1873	138, 367

Index of patents issued from the United States Patent Office from 1790 *to* 1873, *inclusive*—Continued.

Invention.	Inventor.	Residence.	Date.	No.
Bracket	J. E. Chesley	Boston, Mass	Aug. 29, 1871	118, 433
Bracket	M. H. Matsinger	Philadelphia, Pa	June 27, 1865	48, 420
Bracket	G. W. Peirce	Boston, Mass	July 11, 1871	116, 862
Bracket	J. M. Whiting	Providence, R. I	May 30, 1871	115, 550
Bracket and rack	W. A. Middleton	Harrisburgh, Pa	Oct. 20, 1868	83, 195
Bracket and towel-rack combined	G. W. Peirce	Boston, Mass	Nov. 7, 1871	120, 665
Bracket-brace for stair-rails	E. Betts	Boston, Mass	Sept. 15, 1863	39, 875
Bracket-clamp	G. W. Spaulding and G. R. Smith.	Syracuse, N. Y	June 22, 1869	91, 683
Bracket-hanger	C. E. Bliven	Toledo, Ohio	Mar. 29, 1870	101, 219
Bracket-seat	J. J. Wilson	New York, N. Y	Apr. 4, 1871	113, 607
Bracket-sheave	C. H. Straffin	Boston, Mass	Apr. 9, 1872	125, 419
Bracket-shelf	I. H. Frost	Bristol, Conn	Sept. 13, 1870	107, 245
Bracket shelf and drawer	L. Gibbs	Canton, Ohio	Apr. 7, 1868	76, 434
Bracket, shelf, &c , Construction of	N. G. Burleigh	Boston, Mass	Sept. 19, 1871	119, 116
Bracket to glass, Means of attaching	L. J. Lavater	Paris, France	Dec. 31, 1867	72, 867
Brad and nail cutting machine	J. Berry and O. P. Rand	New Market, N. H	Mar. 30, 1837	155
Brad and sprig, Machine for cutting	G. Jenkins	Plymouth County, Mass	May 16, 1817	....
Brad-cutter, Revolving	A. B. Woods and E. Talbot, jr	Windsor, Conn	Feb. 13, 1835	....
Brad-cutting machine	S. Boyden, jr	Boston, Mass	June 27, 1815	....
Brad-cutting machine	W. J. Miller	Cold Spring, N. Y	June 27, 1854	11, 172
Brad-cutting machine	H. Waterman	New York, N. Y	Jan. 10, 1840	1, 465
Brad-setter	M. D. Converse	London, Ohio	Mar. 8, 1870	100, 665
Brad-setter	C. E. Smith	Columbus, Ohio	Sept. 7, 1869	94, 520
Brad-setter for glaziers' use	M. Kleeman	Columbus, Ohio	June 1, 1869	90, 668
Brad, sprig, and nail machine, Cylindrical	R. Turner	Boston, Mass	Feb 21, 1814	....
Brads, Finishing	H. Stanton	Syracuse, N. Y	Nov. 5, 1867	70, 642
Brads from iron plates, Cutting and manufacturing	E. Gammond	Gorham, Me	Oct. 20, 1830	....
Brads, Machine for cutting	J. Ellis		Feb. 10, 1807	....
Brads, nails, and spikes, Cutting	S. Belknap	Washington, D. C	Apr. 8, 1814	....
Braid	J. H. Smith and W. and A. Shedlock.	Norwich, Conn., and New York, N. Y.	Sept. 1, 1868	81, 834
Braid and embroidery die, Machine for making	J. J. and E. D. Parker	Marietta, Ohio	Oct. 2, 1866	58, 470
Braid-holder	A. F. Jennings	Fredonia, N. Y	Nov. 23, 1869	97, 198
Braid, Lapping	W. R. Arnold	Providence, R. I	June 18, 1867	65, 859
Braid-making	J. Fletcher	Providence, R. I	Aug. 11, 1868	80, 818
Braid-making machine	G. Rehfuss	Philadelphia, Pa	Jan. 28, 1868	73, 930
Braid, Manufacture of	J. W. Bowers	Newton, Mass	July 27, 1869	92, 933
Braid, Manufacture of	A. H. Boyd	Medway, Mass	Jan. 27, 1863	37, 541
Braid-pressing machinery, Trimmed	H. H. Robbins	Middleborough, Mass	Apr. 10, 1841	2, 044
Braid sizing and polishing machine	J. S. Fenner	Warren, R. I	Sept. 8, 1868	81, 886
Braid, trimming, &c., Manufacture of imitation	H. Loewenburg	New York, N. Y	Aug. 2, 1870	106, 068
Braid, trimming, &c., Process of making imitation	H. Loewenburg	New York, N. Y	Aug. 13, 1872	130, 510
Braid-winding machine	W. and A. Shedlock	New York, N. Y	Feb. 23, 1869	87, 117
Braiding-machine	J. A. Bazin	Canton, Mass	Aug. 8, 1854	11, 512
Braiding-machine	J. D. Butler	Lancaster, Mass	Feb. 9, 1869	86, 642
Braiding-machine	J. D. Butler	Lancaster, Mass	July 20, 1869	92, 938
Braiding-machine	J. D. Butler	Lancaster, Mass	Mar. 26, 1872	124, 882
Braiding-machine	H. W. Cady, J. M. Carpenter, and G. K. Winchester.	Pawtucket and Providence, R. I.	Oct. 29, 1861	33, 569
Braiding-machine	A. B. Clemons	Derby, Conn	Nov. 16, 1858	22, 100
Braiding-machine	W. Darker	West Philadelphia, Pa	Aug. 21, 1866	57, 435
Braiding-machine	E. B. Day	Boston, Mass	Feb. 7, 1860	27, 039
Braiding-machine	H. Fletcher	Providence, R. I	July 7, 1863	39, 137
Braiding-machine	J. Fletcher	Providence, R. I	July 7, 1863	39, 138
Braiding-machine	L. Hull	Charlestown, Mass	Aug. 7, 1855	13, 391
Braiding-machine	L. Hull	Charlestown, Mass	Oct. 16, 1866	58, 830
Braiding-machine	O. M. Inman	Providence, R. I	Feb. 13, 1866	52, 573
Braiding-machine	I. W. Lamb	West Novi, Mich	June 28, 1859	24, 565
Braiding-machine	G. Mason	Providence, R. I	Apr. 6, 1869	88, 576
Braiding-machine	J. McCahey and S. B. Salisbury	Providence, R. I	Mar. 21, 1871	112, 946
Braiding-machine	D. D. Sacket	Westfield, Mass	Jan. 22, 1840	1, 476
Braiding-machine	S. B. Salisbury and J. McCahey	Providence, R. I	Aug. 6, 1872	130, 319
Braiding-machine	S. Scholfield	Providence, R. I	June 29, 1869	91, 970
Braiding-machine	E. T. and E. Sizer and A. Halladay.	Westfield, Mass	Mar. 28, 1854	10, 718
Braiding-machine	T. J. Sloan	New York, N. Y	Aug. 14, 1866	57, 201
Braiding-machine	J. Thorp	Providence, R. I	Aug. 10, 1821	....
Braiding-machine	J. Thorp	Providence, R. I	July 10, 1826	....
Braiding-machine	W. Tunstill	Paterson, N. J	Aug. 19, 1862	36, 249
Braiding-machine	W. Tunstill	Paterson, N. J	May 28, 1867	65, 309
Braiding-machine	W. Tunstill	Paterson, N. J	Aug. 11, 1868	81, 038
Braiding-machine	F. L. Verkamp and C. F. Leopold.	Philadelphia, Pa	July 24, 1866	56, 643
Braiding-machine	G. R. Winchester	Providence, R. I	Jan. 1, 1861	31, 061
Braiding-machine	J. B. Wood	Providence, R. I	Dec. 22, 1863	41, 045
Braiding-machine carrier	D. Avery	Westfield, Mass	June 2, 1868	78, 411
Braiding-machine carrier	O. E. Drown	Pawtucket, R. I	June 12, 1866	55, 580
Braiding-machine carrier	S. Scholfield	Providence, R. I	Oct. 29, 1867	70, 273
Braiding-machine for covering skirt and other wire	O. R. Burnham	New York, N. Y	Apr. 25, 1865	47, 487
Braiding-machine for making three-strand braid	J. Taggart	Roxbury, Mass	Feb. 27, 1866	52, 936
Braiding-machine stop-motion	E. W. Dean	Norwich, Conn	Aug. 23, 1870	106, 672
Braiding machine stop-motion	H. B. Howe and W. J. Mackrell	New York, N. Y	Aug. 21, 1866	57, 326
Braiding-machines, Expansion-pulley for	J. Fewkes	Newton, Mass	Jan. 7, 1868	72, 991
Braiding-machines, Let-off mechanism for	J. M. Carpenter	Providence, R. I	Aug. 14, 1866	57, 084
Braiding-machines, Let-off mechanism for carrier for.	T. Greenhalgh	Raritan, N. J	Apr. 13, 1869	88, 956
Braiding-machines, Let-off and tension-device for spool of.	J. D. Butler	Lancaster, Mass	Mar. 8, 1870	100, 596
Braiding-machines, Spool let-off and tension-device.	W. J. Horstmann	Philadelphia, Pa	Sept. 21, 1869	95, 112
Braiding-machines, Take-up for	G. K. Winchester	Providence, R. I	Mar. 26, 1867	63, 194
Braiding-machines, &c., Thread-tension and delivery mechanism for.	L. Hull	Charlestown, Mass	May 10, 1864	42, 663
Braiding-machines, Thread-tension and thread-delivery in.	L. Hull	Charlestown, Mass	June 14, 1864	43, 115
Braiding-machines, Yarn-delivery apparatus for	G. F. Chambers and S. Robinson	Worcester, Mass	June 5, 1866	55, 239

Index of patents issued from the United States Patent Office from 1790 *to* 1873, *inclusive*—Continued.

Invention.	Inventor.	Residence.	Date.	No.
Braiding-machinery	L. Hull	Charlestown, Mass	Oct. 30, 1855	13,719
Brake: *See* Air-brake. Automatic brake. Car-brake. Carriage-brake. Cart-brake. Collision-brake. Cotton-lapper brake. Crimping-brake. Elevator-brake. Flax-brake. Flax-boll brake. Friction-brake. Frictional brake. Hemp-brake. Horse-power brake. Inclined-plane brake. Inclined-plane and hoist brake. Land-conveyance brake. Machinery-brake. Omnibus-brake. Pneumatic brake. Pulley-brake. Railway-brake. Railway-train brake. Self-adjusting brake. Sewing-machine brake. Shoe-brake. Sled-brake. Sleigh-brake. Spinning-mule brake. Steam-brake. Steam and air brake. Steam-power air-brake. Vehicle-brake. Wagon-brake.				
Brake and starter combined	J. A. Cody	New York, N. Y	Apr. 2, 1872	125,119
Brake-block holder	A. W. Dorr	Lake Valley, Cal	Jan. 19, 1869	86,007
Brake by signal-cord, Operating	W. G. Creamer	New Haven County, Conn	Dec. 20, 1853	10,321
Brake coupling, Steam-power	G. Westinghouse, jr	Pittsburgh, Pa	Mar. 4, 1873	136,396
Brake coupling, Steam-power	G. Westinghouse, jr	Pittsburgh, Pa	Mar. 11, 1873	136,631
Brake-mechanism	O. Grüninger	New York, N. Y	July 8, 1873	140,700
Brake-mechanism for light machinery	J. M. Cayce	Franklin, Tenn	June 25, 1872	128,283
Brake reaction movement	H. B. Martin	San Francisco, Cal	June 17, 1873	140,057
Brake-rubbers, Mode of hanging	T. C. Ball	Keene, N. H	Aug. 2, 1859	24,918
Brake-shoe	J. Christy	Philadelphia, Pa	Sept. 12, 1865	49,948
Brake-shoe	H. R. McAlister	Harrisburgh, Pa	Aug. 20, 1872	130,734
Brake-shoe	J. W. Moffitt	Harrisburgh, Pa	May 6, 1873	138,514
Brake-shoe	J. S. Whitworth	Norfolk, Va	Feb. 13, 1872	123,655
Brake shoe, Car	J. Brahn	Jersey City, N. J	Feb. 25, 1868	74,792
Brake-shoe for railway-car, Composition-metal	W. McConway	Pittsburgh, Pa	Feb. 27, 1872	124,190
Brake-shoe, Malleable-iron	J. J. Torley	Pittsburgh, Pa	June 18, 1872	127,996
Bran-duster	E. R. Benton	Milwaukee, Wis	May 7, 1850	7,336
Bran-duster	E. R. Benton	Cleveland, Ohio	July 26, 1853	9,881
Bran-duster	J. M. Carr and J Hughes	Cambridge City, Ind	Apr. 1, 1851	8,014
Bran-duster	G. Clark and P. T. Elting	Sandusky, Ohio	Feb. 12, 1861	31,371
Bran-duster	J. Cook	Saint Louis, Mo	Mar. 12, 1867	62,733
Bran-duster	J. Damp	Ashland, Ohio	Aug. 15, 1871	117,990
Bran-duster	R. M. Dempsey	Indianapolis, Ind	Dec. 18, 1849	6,952
Bran-duster	P. T. Elting	Buffalo, N. Y	May 19, 1868	77,965
Bran-duster	L. Fagin	Cincinnati, Ohio	Feb. 17, 1852	8,738
Bran-duster	W. Hall	Saint Louis, Mo	Nov. 1, 1859	25,965
Bran-duster	A. Hildreth	Newark, Ohio	Feb. 17, 1852	8,739
Bran-duster	J. W. Houghtelin	Du Quoin, Ill	June 26, 1860	28,863
Bran-duster	S. Hughes	Hamilton, Ohio	Apr. 13, 1869	88,875
Bran duster	W. W. Huntley and A. Babcock.	Silver Creek, N. Y	May 12, 1863	38,484
Bran-duster	J. Johnston	Wilmington, Del	Oct. 24, 1854	11,832
Bran-duster and flour-bolt	J. H. Jones	Yellow Springs, Ohio	Oct. 24, 1871	120,198
Bran-duster	S. W. Kirk	Coatesville, Pa	July 22, 1851	8,248
Bran-duster	S. B. Manning	Allegheny, Pa	July 27, 1858	21,009
Bran-duster	W. A. McFarlan and T. C. Carpenter.	Wilmington, Del	May 27, 1851	8,124
Bran-duster	L. S. Reynolds	Indianapolis, Ind	June 28, 1853	9,819
Bran-duster	J. T. McNally	Brooklyn, N. Y	Mar. 25, 1873	137,228
Bran-duster	G. H. Reynolds	Peoria, Ill	Nov. 20, 1866	59,862
Bran-duster	J. Richmond	Lockport, N. Y	Feb. 27, 1866	52,886
Bran-duster	M. Smith	Rochester, N. Y	June 23, 1863	39,021
Bran-duster, Feeding-screen for	G. S. Cooper	Baraboo, Wis	Jan. 21, 1873	135,085
Branch pipe through the ground without excavating, Inserting.	J. Ridgway	New York, N. Y	July 16, 1842	2,722
Brand for barrel	S. Huse	Newburyport, Mass	June 28, 1832	....
Brand for branding flour-barrel, &c	W. Ross	New Orleans, La	Dec. 2, 1825	....
Brand for marking animals	F. H. Richards	Troy, N. Y	Nov. 19, 1867	71,217
Branding and stamping iron	C. T. Holloway	Baltimore, Md	Feb. 4, 1862	34,301
Branding-apparatus	J. W. Dodge	Malden, Mass	May 4, 1869	89,829
Branding-instrument	N. J. and J. P. Wemmer	Philadelphia, Pa	Feb. 11, 1868	74,263
Branding-iron	F. L. Penny	Boston, Mass	Mar. 9, 1869	87,588
Branding-iron	C Rundquist	Mankato, Minn	Nov. 19, 1867	71,068
Branding-iron	L. Stark	Chelsea, Mass	Mar. 9, 1869	87,720
Branding-stamp	W. C. Garretson and E. Draper	Oskaloosa, Iowa	Oct. 6, 1868	82,823
Branding, stamping, marking, &c., Apparatus for	E. Barns, G. Hill, and J. B. Hawkins.	Ashtabula, Ohio	Feb. 25, 1835	....
Branding-tool	L. Stark	Chicopee, Mass	Jan. 1, 1850	6,993
Branding-tool	J. P. Worrall	Philadelphia, Pa	Jan. 23, 1866	52,238
Brandy and whisky by redistillation and filtration, Mode of purifying and rectifying.	R. Wauck	Gallipolis, Ohio	Dec. 7, 1829	....

Index of patents issued from the United States Patent Office from 1790 *to* 1873, *inclusive*—Continued.

Invention.	Inventor.	Residence.	Date.	No.
Brandy from grain or fruit, Making	C. J. Hütter		Feb. 11, 1803	......
Brandy manufacture	D. J. Browne and S. T. Bacon	Cambridge and Boston, Mass.	Jan. 22, 1867	61, 313
Brandy of domestic materials, Imitating	H. Moore	Aurora, N. Y	July 2, 1814	......
Brass, copper, and composition nails, Manufacturing.	G. W. Robinson	Attleborough, Mass	Mar. 17, 1813	......
Brass, &c., Furnace for melting	I. D. Bush	Detroit, Mich	Feb. 18, 1873	135, 967
Brass-polishing composition	W. H. Trissler	Cleveland, Ohio	Sept. 8, 1863	39, 848
Brazier	R. B. Wakefield	Springfield, Mass	Sept. 28, 1869	95, 291
Bread and cake receptacle	C. C. Savery	Philadelphia, Pa	Nov. 2, 1869	96, 487
Bread and candy cutter	F. Quinn	Keokuk, Iowa	Oct. 1, 1872	131, 781
Bread and cracker machine	W. R. Nevins and J. J. Yates	New York, N. Y	Sept. 28, 1858	21, 619
Bread and meat cutter	W. Budd and J. L. Husband	Philadelphia, Pa	Aug. 8, 1865	49, 225
Bread and meat cutter	S. Kepner	Pottstown, Pa	Apr. 10, 1866	53, 835
Bread and pastry board	J. McNamee	Easton, Pa	Jan. 1, 1861	31, 066
Bread and vegetable cutter	H. A. Titus	Gloversville, N. Y	Feb. 25, 1868	74, 864
Bread, Apparatus for making aërated	R. L. Howard and J. Danglish	London and Reading, England.	Jan. 23, 1866	52, 252
Bread, Apparatus for making aërated	A. Scratchard	Mile End, England	Apr. 26, 1870	102, 325
Bread-baking device	A. I. Quackenbush and G. Hawn.	Fort Plain, N. Y	Jan. 25, 1870	99, 109
Bread-board	W. H. Lewis	New York, N. Y	Jan. 7, 1868	73, 104
Bread-board and dough-tray	B. Fulton	Pulaski, Iowa	Oct. 17, 1871	119, 926
Bread-cutter	H. Berdan	New York, N. Y	Apr. 12, 1859	23, 548
Bread-cutter	J. Buckett	New York, N. Y	June 6, 1865	48, 127
Bread-cutter	I. S. Bunnell	Carbondale, Pa	Oct. 3, 1871	119, 569
Bread-cutter	M. Chapman	Greenfield, Mass	Feb. 2, 1858	19, 238
Bread-cutter	A. M. Dexter	Philadelphia, Pa	July 23, 1867	67, 032
Bread-cutter	G. D. Goodsell and N. E. Babcock.	Rockford, Ill	May 31, 1870	103, 601
Bread-cutter	W. S. Gray	Worcester, Mass	July 30, 1867	67, 191
Bread-cutter	C. Hansen and P. Andersen	Racine, Wis	May 14, 1872	126, 807
Bread-cutter	F. Hüllhorst	Freeport, Ill	Jan. 24, 1865	46, 603
Bread-cutter	G. L. Jones	Chicopee, Mass	Apr. 28, 1868	77, 383
Bread-cutter	A. E. and D. Lazell	Chicopee Falls, Mass	July 1, 1851	8, 193
Bread-cutter	C. Lemke	Cincinnati, Ohio	Oct 10, 1871	119, 861
Bread-cutter	J. Madden	Cleveland, Ohio	Apr. 2, 1867	63, 406
Bread-cutter	J. Madden	Cleveland, Ohio	July 30, 1867	67, 325
Bread-cutter	J. Madden and J. G. Haserot	Cleveland, Ohio	Jan. 7, 1868	73, 017
Bread-cutter	S. H. Martin and J. S. Williams	Mount Vernon, N. Y	May 7, 1872	126, 468
Bread-cutter	J. Oxley	Sheffield, England	May 10, 1864	42, 715
Bread-cutter	J. F. Plass	New York, N. Y	Apr. 25, 1865	47, 450
Bread-cutter	S. D. Simmons	San Francisco, Cal	July 25, 1865	48, 985
Bread-cutter	S. D. Simmons	Brooklyn, E. D., N. Y	June 4, 1867	65, 441
Bread-cutter	J. D. Soles	Lynn Township, Ill	Aug. 25, 1868	81, 427
Bread-cutter	G. D. Williams	Chicopee, Mass	Dec. 24, 1867	72, 709
Bread-cutting machine	R. P. March	Morristown, Pa	July 21, 1863	39, 298
Bread-cutting machine	J. Naughten	Cincinnati, Ohio	Nov. 10, 1857	18, 597
Bread, Device for raising and kneading	A. G. Good	Reading, Pa	June 15, 1869	91, 432
Bread-dough	N. Wyeth	Hingham, Mass	May 12, 1834	......
Bread-finisher	E. Treadwell	Philadelphia, Pa	Jan. 10, 1820	......
Bread-knife	J. Carrier	Marlborough, Conn	May 24, 1859	24, 099
Bread-knife	J. Frisch	Albany, N. Y	June 2, 1868	71, 586
Bread-knife	F. Roys	Berlin, Conn	Oct. 9, 1844	3, 783
Bread-machine	J. E. Hawkins	Lansingburgh, N. Y	Nov. 16, 1869	96, 804
Bread-machine	J. E. Hawkins	Lansingburgh, N. Y	Apr. 5, 1870	101, 615
Bread-machine	L. P. Jenks	Boston, Mass	May 10, 1870	102, 943
Bread, Machine for gaging size of loaf of	E. L. Moeschler	Rochester, N. Y	Feb. 26, 1867	62, 354
Bread, Machine for making ship and other	J. Starr	Middletown, Conn	Jan. 8, 1813	......
Bread machine, Loaf	J. D'Arcy	San Francisco, Cal	Nov. 19, 1867	70, 971
Bread, Machinery for manufacturing aërated	J. Danglish	Reading, England	July 4, 1865	48, 534
Bread, Making	W. Banks		Dec. 14, 1798	......
Bread-making	J. Perry	Brooklyn, N. Y	Nov. 24, 1863	40, 707
Bread-making apparatus	J. A. H. Le Gantvoort	Waupun, Wis	June 21, 1870	104, 662
Bread-making chest and table	B. Gonzales	Goodland, Ind	Oct. 28, 1862	36, 777
Bread-making machine	W. Carr	Bath, Me	July 22, 1843	3, 192
Bread-making machine	W. O. Drew	Georgetown, D. C	Mar. 31, 1863	38, 037
Bread-making machine	M. A. Jones	Frankfort, Ky	Mar. 24, 1868	75, 922
Bread-making machine	G. Richards	Manlius, N. Y	Feb. 8, 1813	......
Bread-making table	W. K. Wyckoff	Ripon, Wis	Mar. 15, 1859	23, 283
Bread, Manufacture of	B. Garvey	New York, N. Y	July 17, 1860	29, 158
Bread, Manufacture of	J. G. Moxey	Philadelphia, Pa	Dec. 19, 1865	51, 659
Bread, Manufacture of aërated	S. F. Ambler	Brooklyn, N. Y	Aug. 12, 1862	36, 134
Bread, Manufacture of light	W. J. Walker	Baltimore, Md	Apr. 9, 1867	63, 676
Bread-manufacture, Treating grain for	A. Sézille	Paris, France	Dec. 31, 1872	134, 443
Bread, meat, and vegetable cutter	G. Stackhouse	Mount Washington, Pa	May 19, 1868	78, 149
Bread, meat, &c., cutter for	D. Campbell	Elizabeth, N. J	May 30, 1865	48, 014
Bread-pan	W. A. Daggett	South Vineland, N. J	Nov. 22, 1870	109, 497
Bread-pan	W. A. Daggett	South Vineland, N. J	Dec. 6, 1870	109, 871
Bread-pan, Machine for forming	M. L. Best	Canton, Ohio	June 30, 1868	79, 438
Bread, Preparation of portable soup	G. Borden, jr	Galveston, Tex	Feb. 5, 1850	7, 066
Bread, Process for making	C. Crum	Hudson, N. Y	Mar. 6, 1855	12, 479
Bread, Raising	A. Conant	Lowell, Mass	Oct. 12, 1842	2, 816
Bread-raising apparatus	E. Buss	Springfield, Ohio	July 3, 1866	56, 002
Bread-raising apparatus	J. Sutton	New York, N. Y	Apr. 15, 1873	137, 974
Bread-raising preparation	R. P. Leonard	Keene, N. H	Apr. 27, 1869	89, 320
Bread-slicer	J. W. Currier	Springfield, Mass	May 16, 1871	114, 930
Bread-slicing machine	A. Iske and J. O. Steinheiser	Lancaster, Pa	May 28, 1872	127, 242
Bread, soap, and black lead, Machine for cutting	W. B. Vincent	Boston, Mass	Feb. 13, 1866	52, 627
Bread-toaster	H. P. Brooks	Waterbury, Conn	Aug. 16, 1870	106, 462
Bread-toaster	A. Brown	Rochester, N. Y	Aug. 23, 1870	106, 654
Bread-toaster	D. Miller	Marietta, Ohio	Mar. 12, 1872	124, 605
Bread-toaster, Revolving	J. C. Paine	Dubuque, Iowa	Feb. 19, 1867	62, 218
Bread-worker	J. H. Balderston	Colora, Md	Feb. 18, 1873	135, 959
Breaker and leveler, Combined	D. Witt	Hubbardston, Mass	Feb. 9, 1869	86, 893
Breaker-roller	E. Dowden	Dykens, Pa	Mar. 1, 1870	100, 273
Breakfast and dining table	P. Barker	Worthington, Ohio	Aug. 6, 1828	......

Index of patents issued from the United States Patent Office from 1790 *to* 1873, *inclusive*—Continued.

Invention.	Inventor.	Residence.	Date.	No.
Breaking and grinding mill	W. Beal, jr	Lowell, Mass	Nov. 14, 1848	5, 916
Breaking-machine	J. D. Whelpley	Boston, Mass	June 28, 1870	104, 910
Breakwater	D. H. Armour	Columbia, Tex	June 7, 1859	24, 271
Breakwater	L. Gutekunst	Philadelphia, Pa	Jan. 18, 1870	98, 953
Breakwater, Construction of	D. Cunningham	Dundee, Scotland	Apr. 8, 1873	137, 659
Breast-cup	A. M. Knapp	Lowell, Mass	July 22, 1873	141, 005
Breast-cup	E. Waters	Troy, N. Y	Aug. 29, 1854	11, 623
Breast-developer, Pneumatic	J. M. Clark	Lancaster, Pa	Nov. 26, 1867	71, 459
Breast-pad	R. Collins	Chicopee, Mass	Dec. 31, 1867	72, 802
Breast-pad	A. S. and J. P. McLean	Brooklyn, N. Y	Jan. 1, 1867	60, 767
Breast-pad	D. K. Wertman	Centralia, Pa	Mar. 31, 1868	76, 124
Breast-pad and perspiration shield	H. S. Lesher	Brooklyn, N. Y	May 17, 1859	24, 033
Breast-pad, Lady's	J. A. Mason	Brooklyn, N. Y	Jan. 10, 1865	45, 843
Breast-pin	T. C. Eiswald	Providence, R. I	June 28, 1870	104, 840
Breast-pin	C. A. Pease	Boston, Mass	July 30, 1872	129, 902
Breast-pin	L. Sauter	Jersey City, N. J	Oct. 16, 1866	58, 894
Breast-pin	A. Weiller	New York, N. Y	Nov. 28, 1871	121, 440
Breast-pin, ear-ring, &c., Catch-fastening for	J. N. B. Jaquith	Boston, Mass	Nov. 15, 1864	45, 044
Breast-pin fastening	S. Ayres	Danville, Ky	Aug. 31, 1869	94, 270
Breast-pin fastening	G. Buhler	Newark, N. J	Feb. 23, 1864	41, 678
Breast-pin fastening	C. G. Cahoone and B. E. Brown	Providence, R. I	Mar. 30, 1869	88, 273
Breast-pin fastening	B. F. Grinnell	New York, N. Y	Jan. 18, 1859	22, 645
Breast-pin fastening	C. F. Kolb	Philadelphia, Pa	July 28, 1857	17, 881
Breast-pin fastening	J. A. Lehman	Philadelphia, Pa	Oct. 10, 1871	119, 860
Breast-pin fastening	T. W. F. Smitten	Brooklyn, N. Y	Sept. 1, 1868	81, 835
Breast-pin fastening	A. K. P. Walter	Richmond, Me	Sept. 17, 1867	69, 052
Breast-pin guard	J. M. Ross	Springfield, Mass	Mar. 10, 1857	16, 808
Breast-pin joint	D. O. Stanley	South Attleborough, Mass.	Nov. 7, 1871	120, 601
Breast-pin safety-attachment	F. Catlin	New York, N. Y	July 6, 1869	92, 163
Breast-pin safety-attachment	C. F. Pierce	Providence, R. I	June 15, 1869	91, 261
Breast-pin, Shield to protect	J. H. Phillips	Washington, D. C	June 24, 1856	15, 212
Breast-pin, Spiral catch for	J. F. Mascher	Philadelphia, Pa	July 14, 1857	17, 800
Breast-pin, &c., Tongue for	S. Davidson	New York, N. Y	July 23, 1867	67, 029
Breast-pin, Tongue for	L. H. Sondheim	New York, N. Y	Mar. 25, 1873	137, 154
Breast-pins, &c., Safe-catch for	E. C. Benyaurd	Philadelphia, Pa	Mar. 6, 1855	12, 477
Breast-pipe	T. Lewis	Malden, Mass	Nov. 16, 1858	22, 080
Breast-protector	J. Stadermann and H. Sauerbier	New York, N. Y., and Newark, N. J.	Dec. 11, 1866	60, 436
Breast-protector	J. Stadermann and H. Sauerbier	New York, N. Y	Mar. 26, 1867	63, 323
Breast-protector, Mammiform	E. M. Marshall	Hillsdale, N. Y	June 28, 1864	43, 321
Breast-strap	A. L. Hill	Decatur, Ill	July 14, 1868	79, 977
Breast-strap for horses	C. P. Holmes	Gouverneur, N. Y	Dec. 24, 1872	134, 283
Breast-strap hook	H. Foulkes	Utica, N. Y	Aug. 28, 1866	57, 629
Breast-strap, Roller-attachment for	F. Jones	Prescott, Wis	Sept. 22, 1863	40, 045
Breast-strap shield	H. F. Willson	Elyria, Ohio	Apr. 25, 1865	47, 485
Breast-strap shield	H. F. Willson	Fort Wayne, Ind	Aug. 20, 1867	67, 934
Breast-strap slide	A. Buckham	Newark, N. J	May 3, 1864	42, 554
Breast-strap slide	J. H. Martin	Columbus, Ohio	Mar. 23, 1869	88, 192
Breast-strap slide	O. B. North	New Haven, Conn	Dec. 10, 1867	71, 901
Breast-strap slide	I. Roraback	South Bend, Ind	Aug. 13, 1867	67, 857
Breast-strap slide	S. Selleck	Rosendale, N. Y	Feb. 2, 1864	41, 450
Breast-strap slide	C. H. Stevens and W. Garrison	Syracuse, N. Y	Aug. 25, 1868	81, 553
Breasts, Lacteal or artificial method of constructing	C. M. Windship	Roxbury, Mass	Feb. 18, 1841	1, 985
Breathing-tube	J. S. Rose	Philadelphia, Pa	May 8, 1843	3, 079
Breech-loader	B. Burton	Brooklyn, N. Y	June 29, 1869	92, 013
Breech-loader	L. T. Delassize	New Orleans, La	July 20, 1869	92, 799
Breech-loader	Z. R. Von Wessely	New York, N. Y	July 13, 1869	92, 673
Breech-loader	W. G. Ward	New York, N. Y	June 29, 1869	92, 129
Breech-pin and self-capping gun, Sliding	C. Sharps	Cincinnati, Ohio	Sept. 12, 1848	5, 763
Breeching-hook	J. H. Littlefield	Cambridge Mass	June 27, 1865	48, 415
Breeching-hook	G. W. Miller	Springfield, Mass	Nov. 27, 1866	60, 033
Breeching-hook for vehicle	E. Brown	Leominster, Mass	July 4, 1865	48, 513
Breeching-loop	S. Ward	Princeton, Ind	July 9, 1872	128, 932
Breeching-strap fastening	A. W. Olds	Green Oak, Mich	Aug. 1, 1865	49, 143
Brewers and distillers, Apparatus for preparing mash for.	A. S. Swartz	Buffalo, N. Y	Mar. 26, 1867	63, 331
Brewers and distillers, Testing-instrument for	M. Augenstein	Brooklyn, N. Y	Nov. 15, 1870	109, 168
Brewer's boiler, Apparatus for heating	D. S. Blair	Albany, N. Y	Dec. 22, 1863	40, 990
Brewer's boiler, Coil for	J. Frageser and J. G. Schreiber	New York, N. Y	June 19, 1866	55, 747
Brewer's cooler	A. Wood	Pittsburgh, Pa	Sept. 15, 1857	18, 220
Brewer's mash-tun	B. G. Martin	Philadelphia, Pa	Dec. 12, 1865	51, 518
Brewer's mash-tun	J. Walker	Cincinnati, Ohio,	Feb. 12, 1867	62, 101
Breweries and bar, Collecting waste spirits from	A. Maginnis and W. McCormick.	Philadelphia, Pa	Mar. 1, 1870	100, 428
Breweries, Apparatus for mashing and cooling in	G. S. Yingling and S. F. Poorman.	Tiffin, Ohio	May 14, 1867	64, 824
Breweries, cellars, and beer-vaults, Apparatus for cooling.	L. Schulze	Philadelphia, Pa	June 4, 1872	127, 647
Breweries, Cooler for	A. Hammer	Reading, Pa	Sept. 15, 1857	18, 201
Breweries, Cooler for	A. Hammer	New York, N. Y	Apr. 18, 1865	47, 298
Breweries, Cooler for	C. R. M. Wall	Brooklyn, N. Y	Aug. 22, 1865	49, 573
Breweries, &c., Machine for washing	L. Beatty		May 17, 1816	....
Breweries, Machine for washing shavings in	F. Hinckel	Albany, N. Y	Jan. 5, 1869	85, 526
Brewing	E. Beanes	Cordwallis, Great Britain	Mar. 23, 1869	88, 001
Brewing	W. H. Elliot	New York, N. Y	Jan. 7, 1868	72, 989
Brewing and distilling cooler	L. Reich	New York, N. Y	Apr. 3, 1866	53, 677
Brewing and distilling, Cooling and condensing apparatus for.	D. E. Somes	Washington, D. C	Feb. 28, 1865	46, 594
Brewing and preserving alcoholic drinks	J. Hopkins	Brownsville, Pa	May 8, 1849	6, 430
Brewing ale, beer, &c	J. McCormick	Boston, Mass	Nov. 16, 1869	96, 939
Brewing ale, porter, &c	F. M. Ruschaupt and G. Burbenne.	New York and Williamsburgh, N. Y.	Aug. 18, 1868	81, 214
Brewing-apparatus	R. L. Lloyd and A. Partenheimer.	Philadelphia, Pa	Jan. 27, 1824	....
Brewing beer and ale	T. Behan	Norwich, N. Y	Aug. 1, 1838	868
Brewing beer and ale	L. Pasteur	Paris, France	Jan. 28, 1873	135, 245
Brewing beer and other malt-liquors	W. H. Elliot	New York, N. Y	Mar. 31, 1868	76, 177

Index of patents issued from the United States Patent Office from 1790 *to* 1873, *inclusive*—Continued.

Invention.	Inventor.	Residence.	Date.	No.
Brewing beer, &c., Process of using unmashed Indian corn in.	N. Baumann	Kalamazoo, Mich	May 18, 1869	90,066
Brewing by steam	H. Hackley	New York, N. Y	Apr. 21, 1820	......
Brewing, malting, &c., Apparatus for	A. B. Walker	Warrington, England	Apr. 14, 1868	76,653
Brewing malt-liquor, Apparatus for	P. Conday	Philadelphia, Pa	Sept. 15, 1868	82,208
Brewing malt-liquor, Process of	C. R. M. Wall	Brooklyn, N. Y	Dec. 20, 1864	45,543
Brewing, Preparing mash for	S. C. Granger	Chicago, Ill	Nov. 26, 1861	33,779
Brewing, Process for	A. Hammer	New York, N. Y	Mar. 28, 1865	47,014
Brewing with Indian corn	A. Anderson	......	Jan. 26, 1831	......
Brewing with Indian corn	L. Haecker	Altenburg, Hungary	July 1, 1862	35,752
Brewing with maize	L. Haecker	Altenburg, Hungary	Dec. 8, 1863	40,836
Brick	D. L. Bartlett and G. H. Johnson.	Baltimore, Md	Aug. 28, 1866	57,461
Brick	J. Dennen	Chicago, Ill	Aug. 26, 1873	142,087
Brick	J. Grimm	Cuyahoga Falls, Ohio	Feb. 9, 1869	86,664
Brick	S. McLaughlin	Philadelphia, Pa	May 21, 1867	64,994
Brick	L. Till	Sandusky, Ohio	June 19, 1855	13,110
Brick	N. Towson	Washington, D. C	May 16, 1848	5,570
Brick and artificial stone, Machine for making	J. Fensom	Toronto, Canada	Jan. 16, 1872	122,714
Brick and concrete press	G. A. Trear	Chicago, Ill	Dec. 1, 1868	84,485
Brick and draining pipe	J. C. Bryant	Philadelphia, Pa	May 31, 1864	42,926
Brick and lime kiln, Controlling draft in	J. Leeds	Philadelphia, Pa	July 4, 1854	11,230
Brick and mortar elevator	W. Boswell	Pontiac, Mich	Sept. 12, 1871	118,783
Brick and mortar elevator	J. McNamara	Buffalo, N. Y	Oct. 4, 1864	44,544
Brick and mortar elevator	A. Pohl	Detroit, Mich	Feb. 18, 1868	74,590
Brick and mortar hod	E. B. Black, J. Hinkle, jr., and T. S. White.	Columbia, Pa	Mar. 30, 1869	88,361
Brick and mortar lift, Alternate	B. Dearborn	Boston, Mass	June 20, 1816	......
Brick and mortar machine	A. Parker	Sweden, Me	Mar. 5, 1830	......
Brick and other molds	H. Martin	Brooklyn, N. Y	Apr. 4, 1871	113,540
Brick and pantiles, Manufacturing	S. Brouwer	......	Aug. 17, 1793	......
Brick and pottery-ware kiln	J. N. Stanley	Brooklyn, N. Y	May 9, 1865	47,665
Brick and sand drier	S. D. Rader	Williamsport, Pa	June 8, 1869	91,041
Brick and stove-lining, Fire	D. Perry	West Mansfield, Mass	Jan. 23, 1872	122,908
Brick and tempering clay, Molding	J. C. Porter	Powhatan County, Va	Dec. 29, 1830	......
Brick and tile	B. P. Coston	Stirling, Pa	Oct. 22, 1835	......
Brick and tile after molding, Machine for pressing and finishing.	J. B. Wilson and A. B. Crossman.	Malden, Mass., and Huntington, N. Y.	Sept. 30, 1841	2,271
Brick and tile kiln	H. Aiken, H. McAllister, jr., and H. G. Morris.	Pittsburgh and Philadelphia, Pa.	Mar. 2, 1869	87,455
Brick and tile kiln	S. H. Clapp	Malden, Mass	June 8, 1869	90,929
Brick and tile kiln	J. Duff	Washington, D. C	Oct. 31, 1825	......
Brick and tile kiln	A. Morand	Brooklyn, N. Y	Apr. 16, 1872	125,832
Brick and tile machine	G. Baldwin	Monroe, Mich	Mar. 4, 1873	136,481
Brick and tile machine	I. C. Bryant	Washington, D. C	Apr. 18, 1871	113,847
Brick and tile machine	H. Clayton	Dorset Square, England	June 26, 1855	13,123
Brick and tile machine	E. Fisk and B. Hinkley	Fayette, Me	Sept. 8, 1827	......
Brick and tile machine	D. French	New York, N. Y	Aug. 22, 1810	......
Brick and tile machine	G. Hadfield	......	May 15, 1800	......
Brick and tile machine	J. Hotchkiss and E. Buss	Springfield, Ohio	Sept. 17, 1867	68,986
Brick and tile machine	W. Hutchinson	Salford, Great Britain	Apr. 4, 1871	113,300
Brick and tile machine	N. and P. W. Miller	......	Jan. 5, 1804	......
Brick and tile machine	G. Scott	Philadelphia, Pa	Apr. 5, 1870	101,525
Brick and tile machine	B. F. St. John and H. Horst	Shelbyville, Ind	Oct. 17, 1865	50,5[illegible]8
Brick and tile machine	Z. Street	Salem, Ohio	July 8, 1862	35,843
Brick and tile machine, Die for	C. Chambers, jr	Philadelphia, Pa	Sept. 15, 1863	39,884
Brick and tile machine, Die for	J. Hotchkiss and E. Buss	Springfield, Ohio	Apr. 30, 1867	64,223
Brick and tile making	W. L. Drake	Evanston, Ill	July 16, 1872	129,113
Brick and tile making	J. E. Stubbs and J. Bonner	Cincinnati, Ohio	May 10, 1822	......
Brick and tile making	J. Wood	Rockland County, N. Y	Dec. 2, 1836	97
Brick and tile press	B. Porter	Salem, Mass	May 18, 1814	......
Brick and tile pressing	S. Lane	Hallowell, Me	Sept. 30, 1825	......
Brick and tile pressing machine	J. Parker	Gardiner, Me	Nov. 29, 1828	......
Brick and tile, Pressing unburnt	E. Mayo	Hallowell, Me	Dec. 9, 1828	......
Brick and tiles, Machine for heating untempered clay for.	I. Gregg	Pittsburgh, Pa	May 19, 1863	38,580
Brick-burner	W. H. Towers	Boston, Mass	Feb. 12, 1867	62,089
Brick-burner, Perpetual	Z. Shaw	Ypsilanti, Mich	Dec. 28, 1869	98,306
Brick, Burning	N. Adams and A. Noyes	Cornwall and Newburgh, N. Y.	Dec. 28, 1829	......
Brick, Burning	J. W. Andrews	Norristown, Pa	Mar. 21, 1843	3,016
Brick, Burning	T. Billesbach	Kearney Junction, Nebr	Nov. 18, 1873	144,652
Brick, Burning	D. and G. M. Blocher	Cumberland, Md	June 5, 1855	12,991
Brick, Burning	R. Crutchfield	Botetourt County, Va	Feb. 7, 1824	......
Brick, Burning	A. J. Mullen and R. Hall	Greensborough, Ala	Feb. 9, 1858	19,309
Brick by pressing, Making	S. Briggs	Salem, Mass	Jan. 11, 1823	......
Brick by pressure, Making	D. Flagg, jr	Gardiner, Me	Dec. 17, 1824	......
Brick-clay, Machine for making	D. Phillips and J. Drummond	Whitestown, N. Y	May 15, 1832	......
Brick-clay, Machine for mixing	J. Chapman	Mad River Township, Ohio.	June 2, 1832	......
Brick-cleaning machine	T. M. Schleier	Knoxville, Tenn	Dec. 10, 1872	133,893
Brick, Coloring	W. Boies	Troy, N. Y	June 4, 1872	127,551
Brick, Coloring	C. B Doty	Cortland, N. Y	Feb. 6, 1849	6,102
Brick, Composition for fire	D. Cannon and H. S. Lucas	Chester, Mass	July 24, 1847	5,198
Brick, Composition for fire	C. W. Fenton	Bennington, Vt	Sept. 22, 1837	393
Brick, Composition for fire	E. F. Rogers	Chelsea, Mass	Oct. 10, 1871	119,723
Brick, Composition for making	N. J. Wyeth	Cambridge, Mass	Mar. 28, 1844	3,517
Brick, &c., Compressing and smoothing unburnt	T. Norcross and G. Pollard	Hallowell, Me	Oct. 1, 1825	......
Brick-drier	J. K. Caldwell	Allegheny City, Pa	Jan. 21, 1868	73,433
Brick-drier	I. C. Hatch	Camden, N. J	June 9, 1868	78,802
Brick-drier	J. J. Johnston	Allegheny City, Pa	Sept. 24, 1867	69,217
Brick-drier	W. O. Leslie	Philadelphia, Pa	Aug. 18, 1868	81,094
Brick-drier	J. McDonald	Saratoga Springs, N. Y	Aug. 13, 1867	67,783
Brick-drier	J. M. Moyer	Pittsburgh, Pa	June 23, 1868	79,247
Brick-drier	S. M. Parish	Baldwinsville, N. Y	Aug. 16, 1864	43,861
Brick-drying	J. R Bowers	Concord, N. H	July 2, 1867	66,206
Brick-drying	I. Gregg	Philadelphia, Pa	July 9, 1867	66,486
Brick-drying house	G. C. Bovey	Cincinnati, Ohio	Mar. 26, 1872	125,013

Index of patents issued from the United States Patent Office from 1790 *to* 1873, *inclusive*—Continued.

Invention.	Inventor.	Residence.	Date.	No.
Brick-drying kiln	E. W. Bingham	Williamsport, Pa	June 23, 1868	79, 192
Brick-drying kiln	E. C. Dean, H. Hamilton, G. P. Tenney, and A. T. Putnam.	Detroit, Mich	Aug. 31, 1869	94, 188
Brick-drying press	J. T. Shryock	Zanesville, Ohio	June 5, 1866	55, 376
Brick-elevator	T. Mann	San Francisco, Cal	Feb. 16, 1869	86, 932
Brick-elevator	J. H. Sears and B. Merritt	Boston and Chelsea, Mass.	Mar. 19, 1861	31, 725
Brick, &c., Enamel composition for	D. W. Clark	Bennington, Vt	June 7, 1859	24, 282
Brick-finishing machine	W. S. Mayo	New York, N. Y	June 7, 1859	24, 320
Brick, Fire	E. Seavey	Boston, Mass	June 12, 1860	28, 692
Brick, Fire	R. W. Smith	Baltimore, Md	Sept. 27, 1825	....
Brick, Fire	S. K. Wellman	Nashua, N. H	Feb. 12, 1867	61, 966
Brick for building-purposes	W. Smith	Saint Louis, Mo	Apr. 7, 1868	76, 540
Brick for ceilings	M. Abord	Paris, France	Aug. 21, 1866	57, 450
Brick for constructing metallurgic furnaces	H. Frank	Pittsburgh, Pa	Dec. 31, 1872	134, 374
Brick for curved masonry-work	E. Hennessey	Washington County, D. C	Dec. 31, 1867	72, 734
Brick for pavement, &c	R. Wright	Philadelphia, Pa	Dec. 17, 1867	72, 352
Brick, Forming	C. H. Preston	New York, N. Y	June 19, 1847	5, 170
Brick from dry clay, Machine for pressing	B. H. Brown	Washington, D. C	Dec. 8, 1838	1, 026
Brick-hacking machine	D. Woodbury	Minneapolis, Minn	Nov. 5, 1867	70, 669
Brick handling or piling machine	J. Sangster	Buffalo, N. Y	Jan. 7, 1868	73, 123
Brick-hoisting machine	T. F. Christman	Wilson, N. C	June 28, 1859	24, 540
Brick, Hollow	G. H. Johnson	New York, N. Y	Jan. 3, 1871	110, 764
Brick, Implements for picking up and carrying	R. Bryden	Medford, Mass	Nov. 26, 1872	133, 299
Brick-kiln	H. W. Adams	Philadelphia, Pa	July 21, 1868	80, 046
Brick-kiln	H. W. Adams	Philadelphia, Pa	July 20, 1869	92, 770
Brick-kiln	J. Blaisius	Washington, D. C	Dec. 24, 1872	134, 124
Brick-kiln	O. Bennett	Boston, Mass	Apr. 5, 1870	101, 416
Brick-kiln	D. Blocher	Baltimore, Md	July 1, 1873	140, 343
Brick-kiln	F. F. Boudrye	San Francisco, Cal	Mar. 12, 1872	124, 538
Brick-kiln	G. C. Bovey	Cincinnati, Ohio	Jan. 15, 1867	61, 149
Brick-kiln	S. C. Brewer	Water Valley, Miss	June 20, 1871	116, 016
Brick-kiln	W. V. Cecil	Monmouth, Ill	July 20, 1869	92, 702
Brick-kiln	P. Clark	Brooklyn, N. Y	Feb. 2, 1869	86, 363
Brick-kiln	S. H. Clapp	Malden, Mass	Mar. 24, 1868	75, 730
Brick-kiln	C. B. Corey and C. M. Turner	Cleveland, Ohio	July 9, 1867	66, 464
Brick-kiln	R. D. Cox	Philadelphia, Pa	Feb. 15, 1870	99, 852
Brick-kiln	J. W. Crary	New Orleans, La	May 4, 1858	20, 146
Brick-kiln	J. Eisele	Ann Arbor, Mich	Apr. 23, 1872	126, 040
Brick-kiln	W. Ennis	Philadelphia, Pa	Jan. 25, 1870	99, 174
Brick-kiln	W. Gilbert	Detroit, Mich	Sept. 29, 1868	82, 511
Brick-kiln	J. H. Good	Pequa Township, Pa	Mar. 16, 1869	87, 838
Brick-kiln	W. S. Hall	New York, N. Y	Apr. 12, 1870	101, 870
Brick-kiln	W. Hallam	Cheltenham, Mo	Sept. 28, 1869	95, 339
Brick-kiln	S. M. Hamilton	Baltimore, Md	Feb. 15, 1870	99, 886
Brick-kiln	S. M. Hamilton	Baltimore, Md	Feb. 15, 1870	99, 887
Brick-kiln	S. M. Hamilton	Baltimore, Md	Mar. 22, 1870	101, 124
Brick-kiln	S. M. Hamilton	Baltimore, Md	May 21, 1872	127, 050
Brick-kiln	E. Haueisen, A. Wagner, and A. Nulsen.	Cincinnati, Ohio	Oct. 16, 1866	58, 941
Brick-kiln	B. R. Hawley	Normal, Ill	Aug 22, 1871	118, 364
Brick-kiln	B. R. Hawley	Normal, Ill	Apr. 2, 1872	125, 292
Brick-kiln	B. R. Hawley	Normal, Ill	July 16, 1872	129, 558
Brick-kiln	F. E. Hoffmann	Berlin, Prussia	June 6, 1871	115, 734
Brick-kiln	G. Howland	Brunswick, N. Y	July 30, 1872	129, 959
Brick-kiln	T. Lindsley	New York, N. Y	Oct. 24, 1871	120, 202
Brick-kiln	T. Lindsley	New York, N. Y	Nov. 21, 1871	121, 179
Brick-kiln	T. Lindsley	New York, N. Y	Jan. 9, 1872	122, 528
Brick-kiln	W. Linton	Baltimore, Md	Jan. 20, 1852	8, 678
Brick-kiln	A. W. M. Maass	Chicago, Ill	May 13, 1873	138, 905
Brick-kiln	W. T. Mathews	Negaunee, Mich	Apr. 19, 1870	102, 141
Brick-kiln	A. S. McBride	Saint Louis, Mo	July 9, 1867	66, 604
Brick-kiln	J. M. McCarthy	Canal Dover, Ohio	Mar. 30, 1869	88, 316
Brick-kiln	J. M. McCarthy	Canal Dover, Ohio	Mar. 28, 1871	113, 185
Brick-kiln	J. McDonald	Saratoga Springs, N. Y	Nov. 27, 1866	60, 029
Brick-kiln	J. McDonald	New York, N. Y	Oct. 27, 1868	83, 394
Brick-kiln	L. R. Norman and W. F. Dieterichs, jr.	Saint Louis, Mo	Jan. 22, 1867	61, 448
Brick-kiln	L. R. Norman and W. F. Dieterichs, jr.	Saint Louis, Mo	Jan. 22, 1867	61, 449
Brick-kiln	J. Ogle	Baltimore, Md	May 9, 1848	5, 551
Brick-kiln	C. D. Page	Grand Rapids, Mich	Mar. 29, 1864	42, 106
Brick-kiln	C. A. Parker	New Orleans, La	July 12, 1870	105, 241
Brick-kiln	N. F. Potter	Providence, R. I	May 24, 1870	103, 500
Brick-kiln	F. Power	Saint Louis, Mo	July 7, 1868	79, 595
Brick-kiln	S. D. Rader	Williamsport, Pa	June 18, 1867	65, 943
Brick-kiln	S. D. Rader	Williamsport, Pa	Sept. 15, 1868	82, 248
Brick-kiln	H. Read	Kensington, Pa	June 17, 1840	1, 638
Brick-kiln	J. V. B. Remsen	New York, N. Y	Sept. 7, 1869	94, 513
Brick-kiln	R. Robson	Buffalo, N. Y	Apr. 21, 1868	76, 946
Brick-kiln	J. Russell	Elkton, Ky	Mar. 13, 1855	12, 521
Brick-kiln	R. E. Schroeder	Rochester, N. Y	Sept. 28, 1852	9, 285
Brick-kiln	W. A. Shepard	New York, N. Y	June 11, 1867	65, 772
Brick-kiln	F. H. Smith	Baltimore, Md	Sept. 18, 1866	58, 146
Brick-kiln	R. A. Smith	Newburyport, Mass	Nov. 19, 1872	133, 127
Brick-kiln	J. S. Speights	Baltimore, Md	July 4, 1854	11, 234
Brick-kiln	E. C. Sterling	Saint Louis, Mo	Nov. 8, 1870	109, 170
Brick-kiln	F. Strayer	Clinton, Iowa	Aug. 6, 1872	130, 254
Brick-kiln	H. Tugby	Woodville, England	July 29, 1873	141, 404
Brick-kiln	B. Wallis	Baltimore, Md	June 6, 1871	115, 793
Brick-kiln	P. J. Walsh and W. J. Taylor	Oil City, Pa	Mar. 4, 1873	136, 568
Brick-kiln, &c	H. D. Whittemore	New York, N. Y	Mar. 3, 1868	75, 096
Brick-kiln	E. V. Wingard	Williamsport, Pa	Dec. 21, 1869	98, 134
Brick-kiln, Circular	A. J. Works	Fair Haven, Conn	July 20, 1869	92, 923
Brick-kiln, Covering or platting	F. E. Hoffmann	Berlin, Prussia	June 13, 1865	48, 244
Brick-kiln, Perpetual	S. R. Bakewell	Wellsburgh, Va	May 10, 1826	....
Brick knife	W. Gilbert	Detroit, Mich	Dec. 6, 1870	109, 815
Brick-lifter	J. Garity	Brewer, Me	Aug. 22, 1871	118, 230
	K. J. Rugg	Cincinnati, Ohio	Apr. 18, 1871	113, 799

Index of patents issued from the United States Patent Office from 1790 *to* 1873, *inclusive*—Continued.

Invention.	Inventor.	Residence.	Date.	No.
Brick, lime, &c., kiln	W. P. Hall	Piqua, Ohio	Nov. 14, 1871	120, 965
Brick-machine	N. Adams	Cornwall, N. Y	June 22, 1830	......
Brick-machine	N. Adams	Cornwall, N. Y	Oct. 14, 1834	......
Brick-machine	N. Adams	Newburgh, N. Y	Apr. 13, 1836	......
Brick-machine	H. Aiken	Pittsburgh, Pa	Dec. 22, 1868	85, 197
Brick-machine	F. Allen	Boston, Mass	June 22, 1858	20, 612
Brick-machine	F. Alsip	North McGregor, Iowa	Aug. 29, 1871	118, 504
Brick-machine	A. Anderson	Peekskill, N. Y	Sept. 12, 1871	118 777
Brick-machine	E. F. Andrews	Glasgow, Mo	Jan. 28, 1868	73, 863
Brick-machine	J. Armstrong	Saint Louis, Mo	Feb. 4, 1868	74, 030
Brick-machine	J. M. Austin	Georgetown, Mo	Dec. 13, 1870	109, 997
Brick-machine	C. B. Baker	Troy, N. Y	Sept. 25, 1866	58, 200
Brick-machine	J. H. Ballard and E. P. Bond	New Antioch, Ohio	Dec. 8, 1868	84, 724
Brick-machine	G. Bancker	New York, N. Y	June 15, 1858	20, 536
Brick-machine	G. Bancker	New York, N. Y	Jan. 25, 1859	22, 694
Brick-machine	H. C. Bankerd	Morgantown, W. Va	Apr. 2, 1872	125, 255
Brick-machine	J. M. Bannister	Phelps, N. Y	June 19, 1835	......
Brick-machine	T. Barker	Mexico, N. Y	Aug. 11, 1868	80, 850
Brick-machine	W. Barker and G. Martin	Schenectady, N. Y	Aug. 23, 1864	43, 893
Brick-machine	R. D. Bartlett	Bangor, Me	Aug. 8, 1854	11, 465
Brick-machine	W. C. Bartol	Huntingdon, Pa	Dec. 11, 1866	60, 323
Brick-machine	R. T. Barton	New Haven, Conn	Mar. 25, 1873	137, 122
Brick-machine	W. Beach and E. Lukens	Baltimore, Md	May 22, 1841	2, 101
Brick-machine	F. Beaujen	Parish of New Orleans, La.	Dec. 26, 1871	122, 214
Brick-machine	H. T. Beggs and J. Allen	Liberty and Lynchburgh, Va.	May 10, 1859	23, 892
Brick-machine	S. Belknap	Newburgh, N. Y	May 17, 1821	......
Brick-machine	S. W. Bennett, jr	Monroe, La	July 20, 1869	92, 778
Brick-machine	T. Bishop and D. Agnew	Vincennes, Ind	Jan. 4, 1870	98, 547
Brick-machine	E. S. Bitner	Lock Haven, Pa	Dec. 8, 1868	84, 727
Brick-machine	P. E. Bland	Saint Louis, Mo	Oct. 1, 1867	69, 308
Brick-machine	T. E. Bonner	Elkhart, Ind	Nov. 22, 1870	109, 376
Brick-machine	J. Booth	Mobile, Ala	Sept. 21, 1858	21, 545
Brick-machine	G. C. Bovey	Cincinnati, Ohio	Feb. 19, 1867	62, 249
Brick-machine	G. C. Bovey	Cincinnati, Ohio	Oct. 20, 1868	83, 244
Brick-machine	G. C. Bovey	Cincinnati, Ohio	Mar. 8, 1870	100, 590
Brick-machine	G. C. Bovey	Cincinnati, Ohio	Mar. 26, 1872	125, 012
Brick-machine	G. C. Bovey	Chillicothe, Ohio	May 13, 1873	138, 731
Brick-machine	H. Brad	Greencastle, Ind	Sept. 30, 1856	15, 798
Brick-machine	E. Braman and R. Peterson	Greencastle, Ind	July 8, 1856	15, 276
Brick-machine	J. Bretz, W. Sangster, and J. F. Bretz	Springfield, Ill	Dec. 22, 1868	85, 061
Brick-machine	A. H. Brown	Washington, D. C	Nov. 6, 1855	13, 747
Brick-machine	C. G. Brown	Caldwell N. Y	Oct. 11, 1841	2, 312
Brick-machine	L. Brown	Canandaigua, N. Y	Aug. 5, 1851	8, 269
Brick-machine	S. B. Brusstar	Kensington, Pa	Apr. 14, 1838	693
Brick-machine	M. and J. H. Bruck and F. A. Cushman.	Lebanon, N. H	May 27, 1836	14, 947
Brick-machine	A. H. and J. H. Buckwalter	Kimberton, Pa	June 17, 1862	35, 585
Brick-machine	H. C. Bull	Louisville, Ky	Feb. 20, 1866	52, 674
Brick-machine	H. Bulmer and C. Sheppard	Montreal, Canada	July 9, 1872	128, 787
Brick-machine	J. W. Burns	Henry, Ill	Apr. 23, 1867	64, 069
Brick-machine	J. D. Bush	Elyton, Ala	Oct. 21, 1873	143, 806
Brick-machine	R. and A. P. Campbell	Hillsdale, Mich	July 7, 1868	79, 633
Brick-machine	E. P. H. Capron	Springfield, Ohio	Aug. 15, 1865	49, 376
Brick-machine	E. P. H. Capron	Springfield, Ohio	May 8, 1866	54, 500
Brick-machine	E. P. H. Capron	Springfield, Ohio	July 31, 1866	56, 703
Brick-machine	E. P. H. Capron and J. F. Winchell.	Springfield, Ohio	July 31, 1866	56, 847
Brick-machine	E. P. H. Capron	Springfield, Ohio	Apr. 21, 1868	76, 888
Brick-machine	J. T. Carman	Springfield, Ill	Jan. 8, 1867	61, 048
Brick-machine	C. Carnell	Philadelphia, Pa	Feb. 2, 1858	19, 236
Brick-machine	F. L. Carnell	Philadelphia, Pa	July 16, 1872	129, 097
Brick-machine	D. Carpenter	Cortland, N. Y	July 26, 1839	1, 263
Brick-machine	J. Caswell	Syracuse, N. Y	Apr. 16, 1861	32, 046
Brick-machine	J. Caswell	Syracuse, N. Y	Apr. 16, 1861	32, 047
Brick-machine	C. Chambers, jr	Philadelphia, Pa	Oct. 6, 1863	40, 221
Brick-machine	C. Chambers, jr	Philadelphia, Pa	Nov. 1, 1870	108, 880
Brick-machine	C. Chambers, jr	Philadelphia, Pa	Dec. 6, 1870	109, 803
Brick-machine	J. L. Chapman	Philadelphia, Pa	June 14, 1870	104, 268
Brick-machine	C. Clark	Connellsville, Pa	Apr. 2, 1872	125, 272
Brick-machine	F. L. Clarke	Oakland, Ill	Apr. 5, 1870	101, 432
Brick-machine	P. Clark	Brooklyn, N. Y	Sept. 29, 1868	82, 492
Brick-machine	P. Clark	Brooklyn, N. Y	Feb. 9, 1869	86, 732
Brick-machine	W. Cole	Milan, Tenn	Aug. 19, 1873	141, 855
Brick-machine	J. B. Collen	Philadelphia, Pa	Nov. 17, 1857	18, 629
Brick-machine	A. E. Cooke	Philadelphia, Pa	July 6, 1869	92, 270
Brick-machine	J. Cooke	Muncy, Pa	July 6, 1869	92, 169
Brick-machine	J. Cöoke	Muncy, Pa	Dec. 5, 1871	121, 450
Brick-machine	F. F. Cornell, jr	New York, N. Y	July 16, 1867	66, 800
Brick-machine	J. W. Crary	New Orleans, La	Aug. 17, 1858	21, 186
Brick-machine	J. W. Crary	Pensacola, Fla	Aug. 13, 1867	67, 728
Brick-machine	W. Crighton and H. Roesler	Fort Wayne, Ind	Apr. 7, 1868	76, 405
Brick-machine	L. B. Crittenden	Pittsburgh, Pa	July 9, 1867	66, 468
Brick-machine	J. C. Culver	Saint Louis, Mo	May 7, 1872	126, 618
Brick-machine	J. B. Curtis	Hillsdale, Mich	June 4, 1867	65, 351
Brick-machine	R. Cutler and G. C. Swallow	Saint Louis and Columbia, Mo.	Mar. 17, 1868	75, 668
Brick-machine	J. Dane, D. Healy, and G. Cummings.	Derby, Vt.	Aug. 5, 1851	8, 271
Brick-machine	J. C. Dean	Chicago, Ill	Dec. 15, 1868	84, 863
Brick-machine	P. K. Dederick	Albany, N. Y	Sept. 2, 1873	142, 375
Brick-machine	W. H. Degges	Washington, D. C	June 5, 1855	12, 998
Brick-machine	L. T. Delassize	New Orleans, La	Jan. 29, 1856	14, 155
Brick-machine	P. S. Devlan	Reading, Pa	Oct. 3, 1854	11, 745
Brick-machine	P. S. Devlan	Reading, Pa	Apr. 22, 1856	14, 713
Brick-machine	P. S. Devlan	Reading, Pa	Aug. 25, 1857	18, 040
Brick-machine	L. Dieterich	Sandwich, Ill	Apr. 7, 1868	76, 415

Index of patents issued from the United States Patent Office from 1790 *to* 1873, *inclusive*—Continued.

Invention.	Inventor.	Residence.	Date.	No.
Brick-machine	T. Dixcee	Woodfield Road, England	May 7, 1867	64,504
Brick-machine	D. P. Dobbins, J. S. Richards, and J. Sangster.	Erie, Pa., and Buffalo, N. Y	Feb. 11, 1868	74,207
Brick-machine	D. P. Dobbins and J. Sangster	Buffalo, N. Y	Oct. 4, 1870	108,012
Brick-machine	J. A. Dorr, I. Hersey, and E. J. Oldfield.	New York, N. Y	Mar. 31, 1857	16,907
Brick-machine	R. A. Douglas	Chicago, Ill	Sept. 4, 1866	57,689
Brick-machine	H. Dueberg	New York, N. Y	Nov. 26, 1867	71,466
Brick-machine	F. Durand	Paris, France	May 26, 1868	78,268
Brick-machine	J. S. Elliott	East Boston, Mass	June 26, 1866	55,841
Brick-machine	J. S. Elliott	Boston, Mass	Sept. 6, 1870	107,017
Brick-machine	A. Elliot and J. O. Woodward	Taunton, Mass	Oct. 28, 1873	143,972
Brick-machine	T. and W. A. Ellis	Philadelphia, Pa	Feb. 8, 1870	99,658
Brick-machine	T. S. Emery	Philadelphia, Pa	May 10, 1870	102,929
Brick-machine	J. A. Falconer and R. Graham	Jersey City, N. J	Sept. 22, 1868	82,396
Brick-machine	E. Faron	New York, N. Y	Nov. 3, 1868	83,769
Brick-machine	A. Ferguson, D. Ralston, and G. Hildreth.	Troy, N.Y	July 11, 1871	116,942
Brick-machine	C. W. Ferguson	Jackson, Tenn	May 10, 1870	102,794
Brick-machine	H. J. Ferguson	New York, N. Y	Aug. 28, 1866	57,628
Brick-machine	G. B. Fisher	Chicago, Ill	Apr. 14, 1868	76,617
Brick-machine	T. Forbes	Kansas City, Mo	Oct. 26, 1858	21,876
Brick machine	F. E. Frey	Bucyrus, Ohio	Aug. 11, 1868	80,862
Brick-machine	W. C. Gaither	La Fayette, Ind	Nov. 26, 1872	133,429
Brick-machine	R. M. Gano	Allegheny City, Pa	Jan. 21, 1868	73,522
Brick-machine	R. M. Gano and R. S. Miller	Pittsburgh, Pa	July 28, 1868	80,470
Brick-machine	A. Gans	Lincoln, Ill	July 16, 1867	66,700
Brick-machine	B. M. Gard	Champaign County, Ohio	June 5, 1866	55,274
Brick-machine	B. M. and E. R. Gard	Urbana, Ohio, and Chicago, Ill.	Feb. 22, 1870	100,136
Brick-machine	B. M. and E. R. Gard	Urbana, Ohio, and Chicago, Ill.	Feb. 7, 1871	111,629
Brick-machine	E. R. Gard	Chicago, Ill	Jan. 23, 1866	52,156
Brick-machine	E. R. Gard	Chicago, Ill	Nov. 20, 1866	59,831
Brick-machine	E. R. Gard	Chicago, Ill	May 28, 1867	65,069
Brick-machine	I. M. Gattman	Cincinnati, Ohio	July 3, 1860	29,036
Brick-machine	E. Geary	Harrisburgh, Pa	Dec. 8, 1868	84,816
Brick-machine	G. W. B. Gedney	New York, N. Y	Sept. 18, 1855	13,572
Brick-machine	J. George	Jackson, Mich	Mar. 14, 1865	46,845
Brick-machine	J. George and H. Hague	Jackson, Mich	Apr. 3, 1866	53,600
Brick-machine	G. W. Gilbert	Pittsburgh, Pa	Aug. 15, 1835	....
Brick-machine	T. M. Gisborne	Lymington, England	Feb. 20, 1866	52,801
Brick-machine	S. Gissinger	Allegheny City, Pa	Oct. 21, 1862	36,706
Brick-machine	S. Gissinger	Allegheny City, Pa	Dec. 10, 1867	72,015
Brick-machine	D. W. Glendinning	Detroit, Mich	Oct. 14, 1873	143,569
Brick-machine	D. Goldthorp and T. Hield	Hartford, Conn	Dec. 2, 1873	145,169
Brick-machine	D. W. Gould	Independence, Iowa	Mar. 31, 1863	38,041
Brick-machine	J. F. Gould		Mar. 1, 1806	....
Brick-machine	G. Graezzle	Hamilton, Ohio	Dec. 4, 1866	60,175
Brick-machine	W. A. Graham	Carlisle, Pa	Mar. 17, 1868	75,542
Brick-machine	J. Grant	Providence, R. I	Jan. 23, 1866	52,245
Brick-machine	H. H. Gray	Haverstraw, N. Y	Dec. 1, 1868	84,624
Brick-machine	E. R. Green and H. D. Phillips, jr.	Trenton, N. J	Feb. 4, 1868	74,079
Brick-machine	I. Gregg	Pittsburgh, Pa	Aug. 5, 1851	8,284
Brick-machine	I. Gregg	Philadelphia, Pa	Sept. 19, 1865	50,070
Brick-machine	I. Gregg	Philadelphia, Pa	Oct. 30, 1866	59,212
Brick-machine	I. Gregg and C. Green	Philadelphia, Pa	June 12, 1866	55,583
Brick-machine	I. Gregg and H. Moser	Philadelphia and Pittsburgh, Pa.	May 19, 1863	38,629
Brick-machine	M. Gregg	Philadelphia, Pa	June 17, 1851	8,158
Brick-machine	J. B. Gridley	Albany, N. Y	Oct. 23, 1866	59,006
Brick-machine	F. L. Hall	Oneida, N. Y	July 4, 1871	116,587
Brick-machine	F. L. Hall	Oneida, N. Y	Apr. 30, 1872	126,291
Brick-machine	E. Hallett	Hillsdale, Mich	Aug. 14, 1866	57,128
Brick-machine	E. Hallett	Hillsdale, Mich	Apr. 4, 1871	113,514
Brick-machine	G. Halloway	Montgomery County, Pa	Apr. 21, 1836	....
Brick-machine	J. A. Hamer	Reading, Pa	July 8, 1856	15,293
Brick-machine	J. A. Hamer	Reading, Pa	Sept. 13, 1859	25,467
Brick-machine	J. S. Hanna	Mercer, Pa	Nov. 23, 1829	....
Brick-machine	G. W. Harlam	Cincinnati, Ohio	Apr. 19, 1870	102,115
Brick-machine	I. Harman	Tamaqua, Pa	Aug. 12, 1856	15,546
Brick-machine	G. V. Harper	Franklinville, N. Y	Oct. 23, 1837	439
Brick-machine	P. Hayden	Pittsburgh, Pa	Aug. 6, 1867	67,541
Brick-machine	P. Hayden	Pittsburgh, Pa	Sept. 8, 1868	81,901
Brick-machine	B. S. Heath and T. W. Gardner	La Cygne, Kans., and Huntington, Ind.	July 16, 1872	129,225
Brick-machine	C. V. Hemenway	New London, Ohio	Sept. 21, 1869	95,108
Brick machine	C. V. Hemenway	New London, Ohio	Apr. 5, 1870	101,616
Brick-machine	D. Hess	Des Moines, Iowa	July 30, 1872	130,046
Brick-machine	D. Hess	Blandville, Ky	Sept. 15, 1868	82,116
Brick-machine	D. Hess	Des Moines, Iowa	Dec. 12, 1871	121,871
Brick-machine	J. A. Hill	Greencastle, Ind	Sept. 30, 1856	15,808
Brick-machine	J. S. Hobbs and L. R. Elder	West Falmouth, Me	Jan. 4, 1870	98,495
Brick-machine	J. Hockman	Mexico, Ind	Jan. 13, 1857	16,385
Brick-machine	J. Hodge	Fair Play, S. C	Mar. 6, 1840	6,509
Brick-machine	J. Hodges	Fair Play, S. C	June 3, 1837	220
Brick-machine	W. A. Horrall	Washington, Ind	Jan. 31, 1865	46,188
Brick-machine	J. Hotchkiss	Yellow Springs, Ohio	July 17, 1860	29,171
Brick-machine	J. Hotchkiss	Springfield, Ohio	Mar. 13, 1866	53,223
Brick-machine	J. Hotchkiss and E. Buss	Springfield, Ohio	Nov. 21, 1865	51,050
Brick-machine	J. Hotchkiss and E. Buss	Springfield, Ohio	Apr. 17, 1866	53,980
Brick-machine	J. Hotchkiss and E. Buss	Springfield, Ohio	June 4, 1867	65,383
Brick-machine	J. Hotchkiss and W. H. Schofield.	Yellow Springs, Ohio	May 5, 1857	17,248
Brick-machine	J. Hotchkiss and P. W. Woliston.	Springfield, Ohio	Apr. 29, 1873	138,251
Brick-machine	G. O. Houck and H. Gore	Springfield, Ohio	Feb. 16, 1858	19,366

Index of patents issued from the United States Patent Office from 1790 *to* 1873, *inclusive*—Continued.

Invention.	Inventor.	Residence.	Date.	No.
Brick-machine	G. O. Houck and C. E. Sawyer	Springfield, Ohio	June 17, 1873	139,894
Brick-machine	A. V. Hough	Greencastle, Ind	June 26, 1855	13,129
Brick-machine	A. V. Hough and R. W. Jones	Greencastle, Ind	Mar. 17, 1857	16,839
Brick-machine	G. H. Houghton	Marlin, Tex	July 16, 1872	129,563
Brick-machine	W. H. Hovey	Springfield, Mass	Feb. 27, 1866	52,852
Brick-machine	W. H. Hovey	Springfield, Mass	Aug. 6, 1867	67,548
Brick-machine	W. H. Hovey	Springfield, Mass	Aug. 25, 1868	81,506
Brick-machine	W. H. Hovey	Springfield, Mass	July 16, 1872	129,411
Brick-machine	E. C. Hubbard	Green Bay, Wis	Jan. 24, 1871	111,210
Brick-machine	E. R. Hubbard	New York, N. Y	Jan. 7, 1873	134,672
Brick-machine	H. Hughes	Pekin, Ill	May 6, 1873	138,651
Brick-machine	D. J. Hunter	Exeter, N. H	Nov. 16, 1869	96,920
Brick-machine	D. J. Hunter	Somerville, Mass	Apr. 25, 1871	114,011
Brick-machine	D. J. Hunter	Boston, Mass	Aug. 12, 1873	141,646
Brick-machine	W. G. Hyndman	Cincinnati, Ohio	Aug. 20, 1867	67,879
Brick-machine	G. W. Ives	North Haven, Conn	Sept. 17, 1867	68,882
Brick-machine	D. J. Irwin	Noble, Ill	Mar. 4, 1873	136,520
Brick-machine	D. J. Irwin and H. S. Langdon	Noble, Ill	Aug. 5, 1873	141,560
Brick-machine	F. Jacobie	Albany, N. Y	July 12, 1864	43,505
Brick-machine	J. W. Jayne	Sandusky, Ohio	May 5, 1857	17,220
Brick-machine	J. J. Johnston	Allegheny City, Pa	Feb. 19, 1867	62,207
Brick-machine	H. Jones	Fort Madison, Iowa	Feb. 6, 1872	123,400
Brick-machine	J. Jones	Baltimore, Md	Sept. 11, 1866	57,919
Brick-machine	R. W. Jones	Greencastle, Ind	Jan. 15, 1856	14,100
Brick-machine	W. A. Jordan	Thibodeaux, La	Sept. 23, 1856	15,766
Brick-machine	J. Keller	Paducah, Ky	May 25, 1869	90,555
Brick-machine	J. Keller	Paducah, Ky	Aug. 16, 1870	106,490
Brick-machine	P. H. Kells	Adrian, Mich	Mar. 19, 1867	63,057
Brick-machine	P. H. Kells	Adrian, Mich	July 2, 1867	66,355
Brick-machine	P. H. Kells	Adrian, Mich	Aug. 25, 1868	81,511
Brick-machine	P. H. Kells	Adrian, Mich	Mar. 12, 1872	124,590
Brick-machine	E. Kingsland	New York, N. Y	May 13, 1856	14,873
Brick-machine	L. Kirk	Reading, Pa	Jan. 31, 1860	26,997
Brick-machine	L. Kirk	Reading, Pa	Mar. 4, 1862	34,580
Brick-machine	J. Klinkhardt	Saint Louis, Mo	June 29, 1869	92,059
Brick-machine	J. Klinkhardt and W. Kiburz	Saint Louis, Mo	Nov. 24, 1868	84,284
Brick-machine	J. L. Kucker	Philadelphia, Pa	May 23, 1871	115,218
Brick-machine	J. Kutts	Philadelphia, Pa	Oct. 26, 1858	21,888
Brick-machine	J. A. Lafler	Albion, N. Y	Jan. 6, 1863	37,348
Brick-machine	J. A. Lafler	Albion, N. Y	Aug. 11, 1868	80,976
Brick-machine	A. La Tourette and S. H. Smith	Venice and Waterloo, N. Y	Oct. 8, 1867	69,683
Brick-machine	J. G. Lehr and H. D. Thorp	Harlan, Ind	Feb. 2, 1869	86,423
Brick-machine	W. O. Leslie	Philadelphia, Pa	Aug. 6, 1867	67,439
Brick-machine	W. O. Leslie	Philadelphia, Pa	May 12, 1868	77,894
Brick-machine	W. O. Leslie	Philadelphia, Pa	June 30, 1868	79,362
Brick-machine	W. O. Leslie	Philadelphia, Pa	Feb. 2, 1869	86,557
Brick-machine	S. Lewis	Louisiana	Feb. 11, 1824	......
Brick-machine	D. Locke	Lexington, Mo	Apr. 19, 1859	23,691
Brick-machine	R. Lockwood and C. C. Schmitt	New York, N. Y	Mar. 2, 1869	87,416
Brick-machine	D. Lombard	Boston, Mass	Feb. 23, 1858	19,470
Brick-machine	R. Long	Columbus, Ohio	Aug. 19, 1851	8,301
Brick-machine	S. W. Long	Louisville, Ky	Nov. 5, 1867	70,585
Brick-machine	W. H. Lotz and F. Baumann	Chicago, Ill	July 23, 1867	66,979
Brick-machine	Z. Ludington	Uniontown, Pa	July 9, 1872	128,738
Brick-machine	W. H. and H. P. L. Machen, jr	Toledo, Ohio	Aug. 1, 1871	117,551
Brick-machine	J. F. Mallinckrodt	Saint Louis, Mo	July 13, 1869	92,625
Brick-machine	H. Manthé	New York, N. Y	June 14, 1870	104,330
Brick-machine	J. Marshall	Fond du Lac, Wis	May 14, 1867	64,778
Brick-machine	G. Martin	Milwaukee, Wis	July 16, 1867	66,723
Brick-machine	H. Martin	Springfield, Mass	June 27, 1865	48,419
Brick-machine	H. Martin	Keyport, N. J	Oct. 20, 1868	83,297
Brick-machine	H. Martin	Cincinnati, Ohio	Sept. 17, 1872	131,363
Brick-machine	H. Martin	Chicopee, Mass	May 13, 1873	138,754
Brick-machine	I. Martin	Newark, N. J	Apr. 7, 1868	76,337
Brick-machine	J. Martin	Jersey City, N. J	Oct. 20, 1868	83,191
Brick-machine	S. Martin	Knobnoster, Mo	Apr. 1, 1873	137,460
Brick-machine	P. Marvin	Warsaw, Ind	Mar. 12, 1867	62,764
Brick-machine	B. J. McAfee	Delphi, Ind	May 26, 1868	78,221
Brick-machine	G. W. McCann	Springfield, Ohio	Feb. 2, 1869	86,567
Brick-machine	C. C. and J. McDermid	Cambria Mills, Mich	July 9, 1867	66,605
Brick-machine	J. McDonald	New York, N. Y	Apr. 1, 1812	......
Brick-machine	J. McDonald	New York, N. Y	Aug. 6, 1867	67,561
Brick-machine	J. C. McKenzie	Adrian, Mich	Mar. 10, 1868	75,287
Brick-machine	J. N. McLean and J. Bennor	Philadelphia, Pa	May 14, 1872	126,728
Brick-machine	J. N. McLean and J. Bennor	Philadelphia, Pa	Aug. 26, 1873	142,114
Brick-machine	J. McManus	Pittsburgh, Pa	Feb. 9, 1869	86,853
Brick-machine	J. McMurtry	Lexington, Ky	Aug. 8, 1854	11,487
Brick-machine	A. R. McNair	New York, N. Y	May 15, 1866	54,752
Brick-machine	J. Melling	Bolton, England	Oct. 22, 1861	33,535
Brick-machine	W. Mendham and C. Chambers, jr.	Philadelphia, Pa	Nov. 8, 1870	109,034
Brick-machine	J. Menefee	Saint Landry Parish, La	Aug. 7, 1823	......
Brick-machine	G. Metcalf	Leland, Ill	Jan. 28, 1868	73,741
Brick-machine	J. W. Metz	Stout's Post-Office, Ohio	Jan. 28, 1873	135,353
Brick-machine	E. Miller	......	July 17, 1802	......
Brick-machine	J. C. Miller	Bushnell, Ill	Sept. 8, 1868	81,928
Brick-machine	J. Mills	Des Moines, Iowa	June 18, 1867	65,931
Brick-machine	A. F. Mitchell	Valparaiso, Ind	July 18, 1871	117,097
Brick-machine	J. M. Mitchell	Dunlap, Iowa	Dec. 9, 1873	145,358
Brick-machine	J. Moffat	Buffalo, N. Y	Mar. 8, 1836	......
Brick-machine	L. Montgomery	Tunnel, Md	July 8, 1843	3,167
Brick-machine	A. Morand	Leeds, England	June 27, 1871	116,341
Brick-machine	A. Morand	Brooklyn, N. Y	Sept. 3, 1872	131,021
Brick-machine	A. Morgan	Cedar Bayou, Tex	May 4, 1869	89,680
Brick-machine	I. Morley	Allegheny City, Pa	Aug. 15, 1865	49,430
Brick-machine	I. Morley	Pittsburgh, Pa	May 29, 1866	55,146
Brick-machine	I. Morley	Pittsburgh, Pa	Feb. 19, 1867	62,149
Brick-machine	J. M. Moyer	Pittsburgh, Pa	Sept. 17, 1867	68,894

Index of patents issued from the United States Patent Office from 1790 *to* 1873, *inclusive*—Continued.

Invention.	Inventor.	Residence.	Date.	No.
Brick-machine	C. H. Murray	Southwark, England	June 24, 1873	140, 296
Brick-machine	D. Murtha	Philadelphia, Pa	May 10, 1864	42, 680
Brick-machine	C. Murtha	Philadelphia, Pa	May 29, 1866	55, 148
Brick-machine	B. F. Nave	Roanoke, Ind	Jan. 20, 1857	16, 449
Brick-machine	J. N. Newell	Des Moines, Iowa	June 30, 1863	39, 061
Brick-machine	A. N. Newton	Richmond, Ind	Dec. 4, 1866	68, 231
Brick-machine	C. Nolan	Niles, Mich	Apr. 7, 1868	76, 504
Brick-machine	J. North	New York, N. Y	Dec. 17, 1867	72, 220
Brick-machine	A. Nulson, E. Haneisen, and A. Wagner.	Cincinnati, Ohio	Jan. 15, 1867	61, 238
Brick-machine	A. Nulsen, E. Haneisen, and A. Wagner.	Cincinnati, Ohio	Aug. 18, 1868	81, 284
Brick-machine	M. Nye and A. J. Kniseley	Chicago, Ill	Feb. 25, 1868	74, 765
Brick-machine	W. Olds	Albany, Ill	July 13, 1869	92, 636
Brick-machine	W. F. O'Reilly	Starkville, Miss	Aug. 6, 1872	130, 236
Brick-machine	J. Ormerod	Peekskill, N. Y	Jan. 9, 1872	122, 642
Brick-machine	J. W. Osgood	Columbus, Ohio	Apr. 7, 1868	76, 341
Brick-machine	D. Packard	Saint Joseph, Mo	Jan. 19, 1869	86, 032
Brick-machine	W. H. Paige	Springfield, Mass	Dec. 6, 1864	45, 377
Brick-machine	S. J. Parker	Ithaca, N. Y	Jan. 14, 1868	73, 258
Brick-machine	S. Parks	San Francisco, Cal	May 26, 1857	17, 390
Brick-machine	J. Parsons	Cleveland, Ohio	Oct. 30, 1860	30, 541
Brick-machine	L. Patterson	Parker City, Pa	Dec. 9, 1873	145, 444
Brick-machine	Z. M. Paul	Alexandria, La	Oct. 3, 1854	11, 760
Brick-machine	J. W. Pease	Belmont, N. Y	Nov. 27, 1866	60, 050
Brick-machine	J. W. Pease	Belmont, N. Y	Dec. 15, 1868	84, 901
Brick-machine	J. W. Penfield	Willoughby, Ohio	Jan. 16, 1872	122, 851
Brick-machine	H. D. Phillips, jr., and C. H. Williams.	Trenton, N. J., and Matteawan, N. Y.	Mar. 26, 1872	126, 075
Brick-machine	Z. Phillips	Osseo, Mich	Oct. 18, 1870	108, 511
Brick-machine	J. F. M. Pollock	Manchester, England	Aug. 31, 1869	94, 337
Brick-machine	J. F. M. Pollock	Leeds, Great Britain	June 27, 1871	116, 484
Brick-machine	H. B. Ramsay	Indianapolis, Ind	Aug. 26, 1856	15, 618
Brick-machine	H. B. Ramsey	Rockville, Ind	Nov. 19, 1872	133, 175
Brick-machine	J. L. Ransom	Charleston, S. C	Mar. 30, 1858	19, 792
Brick-machine	L. E. Ransom	Trenton, Mich	Feb. 4, 1868	74, 000
Brick-machine	J. Reeder	Cincinnati, Ohio	Feb. 15, 1838	601
Brick-machine	D. B. Rich and C. D. Hubbard	Cambridge, Mass	Mar. 19, 1872	124, 761
Brick-machine	J. J. Riddle	Covington, Ky	July 22, 1851	8, 231
Brick-machine	S. Rigby, 3d	Newcastle, Pa	June 14, 1870	104, 354
Brick-machine	S. C. Ripley	New York, N. Y	Mar. 7, 1854	10, 612
Brick-machine	D. Rising	Atchester, Vt	Mar. 21, 1827	....
Brick-machine	D. Rising	Colchester, N. Y	Jan. 29, 1828	....
Brick-machine	T. H. Rodgers	Chicago, Ill	Apr. 10, 1866	53, 879
Brick-machine	A. Sabbaton	Reading, Pa	May 16, 1848	5, 586
Brick-machine	S. C. Salisbury	Milwaukee, Wis	July 27, 1858	21, 025
Brick-machine	A. H. Sampson	New Orleans, La	June 14, 1853	9, 785
Brick-machine	J. Samuels	Allentown, Pa	Apr. 20, 1852	8, 892
Brick-machine	H. Sands and G. Cummings	Cambridge, Mass., and West Derby, Vt.	Sept. 6, 1853	10, 005
Brick-machine	J. Sangster	Buffalo, N. Y	Oct. 23, 1866	59, 080
Brick-machine	J. Sangster	Buffalo, N. Y	Mar. 12, 1867	62, 781
Brick-machine	J. Sangster, D. P. Dobbins, and J. S. Richards.	Buffalo, N. Y., and Erie, Pa.	Jan. 7, 1868	73, 201
Brick-machine	W. Sangster	Joliet, Ill	Dec. 17, 1867	72, 421
Brick-machine	J. F. Schuffenecker	Keokuk, Iowa	Mar. 22, 1859	23, 315
Brick-machine	J. F. Schuffenecker	Saint Louis, Mo	Feb. 7, 1865	46, 273
Brick-machine	T. E. Seay	Columbia, Va	Apr. 11, 1854	10, 759
Brick-machine	D. W. Seeley and F. Jacobie	Albany, N. Y	May 8, 1866	54, 610
Brick-machine	H. C. Sergeant	Columbus, Ohio	June 11, 1867	65, 612
Brick-machine	H. C. Sergeant	Columbus, Ohio	June 11, 1867	65, 613
Brick-machine	H. C. Sergeant	Columbus, Ohio	Sept. 17, 1867	68, 907
Brick-machine	W. F. Shanks	Louisville, Ky	Dec. 17, 1867	72, 330
Brick-machine	W. A. Shepard	New York, N. Y	Jan. 8, 1867	61, 168
Brick-machine	S. Shreffler	Joliet, Ill	Aug. 22, 1865	49, 562
Brick-machine	S. Shreffler	Joliet, Ill	Oct. 8, 1867	69, 592
Brick-machine	J. Simpson	Decatur, Ga	Apr. 25, 1846	4, 481
Brick-machine	J. Simpson	Saint Louis, Mo	Jan. 28, 1868	73, 843
Brick-machine	D. Smith	Newburyport, Mass	June 9, 1868	78, 765
Brick-machine	F. H. Smith	Baltimore, Md	Oct. 3, 1854	11, 753
Brick-machine	F. H. Smith	Baltimore, Md	July 21, 1868	80, 231
Brick-machine	J. H. Smith and C. H. Florence	Richview, Ill	Apr. 21, 1868	77, 118
Brick-machine	J. N. Smith	Jersey City, N. J	Mar. 12, 1867	62, 782
Brick-machine	J. T. Smith	Springfield, Ill	Feb. 14, 1865	46, 399
Brick-machine	P. J. Smith	Philadelphia, Pa	Apr. 28, 1868	77, 224
Brick-machine	P. J. Smith	Philadelphia, Pa	Nov. 16, 1869	96, 985
Brick-machine	R. A. Smith	Newburyport, Mass	July 9, 1872	128, 822
Brick-machine	G. L. Smull	Meadville, Pa	June 15, 1858	20, 594
Brick-machine	T. Smurfit	Davisville, Mich	Dec. 21, 1869	98, 115
Brick-machine	S. L. Speissegger	Savannah, Ga	Mar. 2, 1852	8, 780
Brick-machine	A. J. Sprague	Toledo, Ohio	July 2, 1867	66, 408
Brick-machine	E. Sprague	Allegheny City, Pa	Dec. 22, 1868	85, 142
Brick-machine	A. Stewart	Baltimore, Md	Apr. 8, 1819	....
Brick-machine	H. W. Stillman	Port Washington, Wis	Aug. 2, 1859	24, 962
Brick-machine	T. Stone	Plainfield, Ind	Dec. 3, 1867	71, 813
Brick-machine	H. H. Strawbridge and D. Tyson.	New Orleans and Covington, La.	Nov. 9, 1852	9, 394
Brick-machine	S. Strong	Washington, D. C	May 26, 1863	38, 716
Brick-machine	R. Stuckwisch	Terre Haute, Ind	Nov. 16, 1869	96, 995
Brick-machine	P. L. Sword	Adrian, Mich	Mar. 26, 1867	63, 333
Brick-machine	P. L. Sword and G. S. Tiffany	Adrian, Mich	June 14, 1864	43, 162
Brick-machine	L. Sylvester	Philadelphia, Pa	Sept. 3, 1867	68, 396
Brick-machine	J. Taggart	Roxbury, Mass	Mar. 17, 1868	75, 596
Brick-machine	S. Talbott	Richmond, Va	Apr. 8, 1840	1, 539
Brick-machine	S. H. Taylor	Jacksonville, Ill	Nov. 29, 1870	109, 777
Brick-machine	J. S. Thomas	Saint Louis, Mo	July 8, 1873	140, 742
Brick-machine	H. D. Thorp and J. G. Lehr	Harlan, Ind	Feb. 1, 1870	99, 371
Brick-machine	S. Till	Sandusky, Ohio	Aug. 28, 1855	13, 502

Index of patents issued from the United States Patent Office from 1790 *to* 1873, *inclusive*—Continued.

Invention.	Inventor.	Residence.	Date.	No.
Brick-machine	J. V. Tompkins	Canandaigua, Mich	May 16, 1871	114, 884
Brick-machine	J. Treadway	Haverstraw, N. Y	Dec. 12, 1871	121, 825
Brick-machine	S. Ustick	Philadelphia, Pa	Apr. 18, 1854	10, 777
Brick-machine	S. Ustick	Philadelphia, Pa	July 7, 1857	17, 759
Brick-machine	S. Ustick	Philadelphia, Pa	Sept. 8, 1857	18, 166
Brick-machine	S. Ustick	Philadelphia, Pa	Apr. 27, 1859	20, 107
Brick-machine	J. Van Riswick	Washington, D. C	Mar. 8, 1859	23, 204
Brick-machine	L. M. Van Sickle	Woodbridge, N. J	May 26, 1868	78, 343
Brick-machine	B. Van Vranken	Schenectady, N. Y	Feb. 25, 1868	74, 956
Brick-machine	R. A. Ver Valen	Haverstraw, N. Y	June 29, 1852	9, 082
Brick-machine	J. A. Victor	Montgomery County, Ky	Sept. 4, 1855	13, 533
Brick-machine	I. Z. A. Wagner	Philadelphia, Pa	Apr. 27, 1859	20, 109
Brick-machine	R. L. Walker	Globe Village, Mass	Aug. 21, 1866	57, 415
Brick-machine	W. S. Wallace	Americus, Ga	Aug. 28, 1860	29, 834
Brick-machine	P. and J. Walbrath	Chittenango, N. Y	Oct. 13, 1868	83, 010
Brick-machine	T. and J. Walsh and D. Evans	Brownsville, Pa	May 14, 1867	64, 726
Brick-machine	J. Ward	Boston, Mass	Oct. 14, 1862	36, 690
Brick-machine	J. Ward	Boston, Mass	Jan. 24, 1865	46, 055
Brick-machine	G. I. Washburn	Worcester, Mass	Sept. 15, 1857	18, 226
Brick-machine	G. I. Washburn and E. H. Bellows.	Worcester, Mass	Oct. 7, 1856	15, 863
Brick-machine	G. I. Washburn and E. H. Bellows.	Worcester, Mass	Apr. 21, 1857	17, 131
Brick-machine	C. Waterman	Bath, Me	July 2, 1836	....
Brick-machine	J. Watson	Buffalo, N. Y	June 5, 1866	55, 432
Brick-machine	J. Watson	Buffalo, N. Y	Mar. 12, 1867	62, 790
Brick-machine	W. S. Watson	Madison, Ind	June 7, 1859	24, 343
Brick-machine	W. S. Watson	Madison, Ind	Apr. 2, 1861	31, 918
Brick-machine	J. A. Welch	Xenia, Ohio	July 23, 1867	67, 007
Brick-machine	D. Wellington	Boston, Mass	Aug. 18, 1868	81, 813
Brick-machine	P. V. Westfall	Kalamazoo, Mich	Dec. 10, 1867	72, 137
Brick-machine	F. Whitcomb	Chicago, Ill	May 31, 1870	103, 805
Brick-machine	H. White	Cleveland, Ohio	Sept. 7, 1858	21, 458
Brick-machine	W. G. White	Bedford, Ohio	Aug. 8, 1871	117, 842
Brick-machine	J. Whiteford	Pond City, Kans	Oct. 19, 1869	95, 962
Brick-machine	R. Wildman	Danbury, Conn	Apr. 12, 1859	23, 634
Brick-machine	J. Willerd	Baltimore, Md	Jan. 29, 1840	1, 484
Brick-machine	C. H. Williams	Rhinebeck, N. Y	Apr. 5, 1870	101, 690
Brick-machine	J. R. Williams	Taunton, Mass	Nov. 8, 1870	109, 089
Brick-machine	M. D. Williams	Dakota, Iowa	Nov. 1, 1870	108, 864
Brick-machine	R. Wilson	Houston, Tex	May 30, 1848	5, 6[illegible]5
Brick-machine	C. A. Winn	Lockhaven, Pa	June 2, 1868	78, 632
Brick-machine	R. Wolff	New York, N. Y	Jan. 15, 1867	61, 298
Brick-machine	R. Wolff	New York, N. Y	May 10, 1870	102, 9[illegible]3
Brick-machine	P. N. Woliston	Springfield, Ohio	June 5, 1866	55, 400
Brick-machine	P. N. Woliston	Springfield, Ohio	Aug. 11, 1868	80, 890
Brick-machine	J. S. Wood	Hartford, Conn	May 26, 1868	78, 348
Brick-machine	W. Wood	Hartford, Conn	Feb. 3, 1857	16, 561
Brick-machine	W. Wood	Hartford, Conn	Mar. 22, 1859	23, 331
Brick-machine	W. Wood	Hartford, Conn	Aug. 14, 1860	29, 645
Brick-machine	A. Woodworth, 3d, and S. Mower.	Boston, Mass	Aug. 31, 1852	9, 238
Brick-machine	C. D. Wrightington	Boston, Mass	Apr. 5, 1870	101, 696
Brick-machine	C. D. Wrightington and B. P. Rider.	Fairhaven and Chelsea, Mass.	June 2, 1868	78, 634
Brick machine, Concrete	B. Bisbee	Ames, Iowa	Mar. 2, 1869	87, 326
Brick machine, Concrete	W. Emmons	Sandwich, Ill	Apr. 14, 1868	76, 726
Brick machine, Concrete	M. S. Harsha	Batavia, Ill	Nov. 19, 1867	70, 999
Brick machine, Concrete	F. Hawkins	Chicago, Ill	May 26, 1868	78, 284
Brick machine, Concrete	I. Pardee	Vineland, N. J	Feb. 25, 1868	74, 769
Brick machine, Concrete	J. Stewart and D. Windsor	Sandwich, Ill	May 21, 1867	65, 024
Brick-machine, Cut-off for	C. Chambers, jr	Philadelphia, Pa	Nov. 30, 1869	97, 356
Brick-machine duster	C. Chambers, jr	Philadelphia, Pa	Jan. 24, 1865	45, 974
Brick, Machine for cleaning	J. Lyon	Norfolk, Va	Apr. 20, 1869	89, 263
Brick, Machine for dressing	J. McNamara	Buffalo, N. Y	May 26, 1868	78, 388
Brick, Machine for mixing clay, mortar, &c., for making.	A. Teel	Waterloo, N. Y	Mar. 12, 1839	1, 098
Brick, Machine for mixing earth for	B. K. Hill	Richmond County, Ga	Feb. 17, 1827	....
Brick, Machine for molding and drying	C. Vassar	Poughkeepsie, N. Y	Nov. 1, 1830	....
Brick, Machine for preparing clay for making	H. Whipple	Port Richmond, N. Y	Mar. 25, 1851	8, 001
Brick, Machine for preparing mortar for the manufacture of.	O. W. Seely	Sodus, N. Y	Mar. 5, 1839	1, 095
Brick, Machine for pressing and delivering	U. Ward	Washington, D. C	Dec. 15, 1835	....
Brick, Machine for striking unburnt	M. P. Crapo	Humboldt County, Cal	July 15, 1856	15, 329
Brick, Machine for tempering clay for	R. Stillman and J. Taylor	Kensington and Spring Garden, Pa.	Apr. 21, 1842	2, 571
Brick, Machine for turning or edging	C. O. Farrington	Brewer, Me	Feb. 1, 1859	22, 790
Brick-machine, Off-bearing apparatus for	C. Chambers, jr	Philadelphia, Pa	June 28, 1870	104, 705
Brick-machine, Preparing clay for	C. M. Ferris and N. Swan	New Milford, Conn., and Patterson, N. Y.	Apr. 16, 1850	7, 286
Brick machine, Press	W. C. Grimes	York, Pa	Dec. 2, 1834	....
Brick-machine, Rotary	G. Crangle	Philadelphia, Pa	June 3, 1856	15, 005
Brick-machine, Rotary	G. Crangle	Philadelphia, Pa	Jan. 27, 1857	16, 468
Brick-machine, Rotary	G. Crangle	Philadelphia, Pa	May 5, 1857	17, 203
Brick-machine, Rotary	M. C. Motch	Covington, Ky	Dec. 31, 1867	72, 883
Brick-machine, Soak-pit of	W. H. Degges	Washington, D. C	June 5, 1855	12, 997
Brick-machine, Die or mouth for	C. H. Murray	Southwark, England	June 24, 1873	140, 297
Brick, Machinery for the manufacture of	B. D. Berry	Edwardsville, Ill	Feb. 9, 1869	86, 633
Brick, Making	A. Frost	Cazenovia, N. Y	Sept. 21, 1808	....
Brick, Making	I. H. Garretson	Richland, Iowa	July 31, 1866	56, 742
Brick, Making	A. Kinsley		Feb. 1, 1793	....
Brick, Making	F. Lambert	Los Angeles, Cal	Jan. 17, 1871	111, 066
Brick making	S. Lowery	Philadelphia, Pa	Apr. 18, 1846	4, 457
Brick, Making	W. Mitchell	Henrico, Va	June 24, 1816	....
Brick-making	T. C. Prosser	Bay City, Mich	July 24, 1866	56, 608
Brick, Making	D. Ridgeway		Mar. 7, 1792	....
Brick-making	J. C. Salomon, jr	Washington, D. C	Apr. 25, 1854	10, 831
Brick-making and clay-grinding machine	W. Mitchell	Richmond, Va	Mar. 26, 1817	....

Index of patents issued from the United States Patent Office from 1790 *to* 1873, *inclusive*—Continued.

Invention.	Inventor.	Residence.	Date.	No.
Brick-making, Machine for preparing clay for	W. N. Graves	Saint Louis, Mo	June 29, 1869	92, 040
Brick, Making fire	E. Jenks	Canaan, Conn	Apr. 17, 1807	....
Brick, Making fire-proof	T. Caldwell	Philadelphia, Pa	July 21, 1823	....
Brick, Manner of forming and pressing	J. Miller	Olean, N. Y	Mar. 2, 1869	87, 354
Brick manufacture	S. D. Arnold	Pittsfield, Mass	Aug. 24, 1869	93, 944
Brick manufacture	S. H. Bowman	Halfmoon Bay, Cal	Apr. 28, 1868	77, 163
Brick manufacture	G. S. Hewitt	Boston, Mass	Mar. 15, 1870	100, 892
Brick manufacture	T. James	Canton, Md	June 1, 1858	20, 433
Brick manufacture	F. Jones	Boston, Mass	June 19, 1866	55, 667
Brick manufacture	T. C. Kier	Pittsburgh, Pa	June 4. 1872	127, 419
Brick manufacture	N. Parmeter	Gardner, Mass	May 17, 1859	24, 050
Brick manufacture	L. E. Ransom	Havana, Ohio	Mar. 20, 1855	12, 558
Brick manufacture	S. B. Spaulding	Brandon, Vt	July 2, 1867	66, 407
Brick, Manufacture of fire	B. Kreischer	New York, N. Y	May 2, 1871	114, 449
Brick, Manufacture of fire	G. Nimmo	Jersey City, N. J.	Apr. 30, 1867	64, 352
Brick, Manufacture of fire	J. Ostrander and J. S. Heartt	Troy, N. Y	Oct. 12, 1858	21, 774
Brick, Manufacture of fire	W. Peters	Baltimore, Md	Aug. 12, 1862	36, 195
Brick, Manufacture of fire	S. K. Wellman	Nashua, N. H	Apr. 30, 1867	64, 389
Brick, Manufacture of machine	S. W. Wood	Washington, D. C	May 6, 1856	14, 846
Brick, Manufacturing	C. Colles		Jan. 26, 1793	....
Brick, Manufacturing	E. W. Crittenden	Pittsburgh, Pa	Nov. 19, 1867	70, 969
Brick, Manufacturing	I. Gregg	Philadelphia, Pa	Oct. 22, 1872	132, 463
Brick, Manufacturing	J. Gregg	New Hampshire, Ohio	May 7, 1811	....
Brick, Manufacturing	R. Mansfield		Oct. 24, 1800	....
Brick, Manufacturing	R. Robberts	Berlin Township, Ohio	Mar. 2, 1833	....
Brick-mold	J. W. Andrews	Norristown, Pa	Mar. 1, 1859	23, 067
Brick-mold	J. A. Buckwater	Kimberton, Pa	Oct. 4, 1859	25, 626
Brick-mold	G. Carnell	Philadelphia, Pa	Dec. 5, 1871	121, 449
Brick-mold	M. Elder	Lansing, Mich	Sept, 11, 1860	29, 958
Brick-mold	T. and W. A. Ellis	Philadelphia, Pa	Mar. 16, 1869	87, 8[illegible]0
Brick-mold	J. Evans	Davenport, Iowa	Sept. 17, 1867	68, 862
Brick-mold	J. M. Ferrell	Philadelphia, Pa	Aug. 7, 1866	57, 041
Brick-mold	F. M. Franklin, O. K. McIntire, and W. Whitely.	Springfield, Ohio	Oct. 15, 1867	69, 909
Brick-mold	M. T. Glinsdal	Rockford, Ill	Apr. 5, 1870	101, 608
Brick-mold	G. Graessle	Hamilton, Ohio	Nov. 15, 1870	109, 313
Brick-mold	J. A. Hamer	Reading, Pa	Nov. 23, 1858	22, 119
Brick-mold	J. A. Hamer	Reading, Pa	Aug. 2, 1859	24, 972
Brick-mold	J. A. Hamer	Reading, Pa	Jan. 3, 1860	26, 736
Brick-mold	J. A. Hamer	West Vincent, Pa	Jan. 28, 1862	34, 282
Brick-mold	J. A. Hamer and T. Lippincott	Philadelphia, Pa	Aug. 1, 1865	49, 105
Brick-mold	S. Inman	Rockford, Ill	Nov. 30, 1869	97, 297
Brick-mold	S. Inman	Rockford, Ill	Dec. 7, 1869	97, 642
Brick-mold	T. James	Baltimore, Md	May 15, 1866	54, 734
Brick-mold	W. Langster and J. Bretz	Springfield, Ill	Sept. 8, 1868	82, 037
Brick-mold	A. McAlpin	Savannah, Ga	Apr. 12, 1870	101, 896
Brick-mold	M. Newlove	Burlington, Iowa	Dec. 21, 1869	98, 181
Brick-mold	J. F. Schuffenecker	Keokuk, Iowa	Mar. 27, 1860	27, 652
Brick-mold	J. F. Schuffenecker	Saint Louis, Mo	Feb. 7, 1865	46, 272
Brick-mold	S. Shreffer	Joliet, Ill	Mar. 13, 1866	53, 191
Brick-mold	R. Stuckwisch	Terre Haute, Ind	Oct. 3, 1871	119, 667
Brick-mold	S. H. Taylor	Jacksonville, Ill	Nov. 29, 1870	109, 778
Brick-mold and bottom-board	A. T. Putnam	Detroit, Mich	May 10, 1870	102, 862
Brick-mold for brick-pressing machine	B. H. Brown	Alexandria, D. C	Oct. 23, 1837	436
Brick-mold, Hand	C. Alvord	Syracuse, N. Y	Nov. 21, 1854	11, 959
Brick-mold piston	L. Sylvester	Philadelphia, Pa	Aug. 13, 1867	67, 684
Brick-mold safety-guard	B. Porter	Jackson, Mich	Aug. 25, 1868	81, 405
Brick-mold-sanding machine	R. Lent	Glasco, N. Y	Apr. 23, 1872	126, 063
Brick, Molding	D. Fox	Ashtabula, Ohio	Aug. 15, 1831	....
Brick, Molding	N. Johnson	Noblesville, Ind	June 20, 1854	11, 127
Brick, Molding	O. W. Seely	Syracuse, N. Y	May 19, 1863	38, 607
Brick, Molding	I. Town	New Haven, Conn	Sept. 27, 1844	3, 762
Brick-molding	T. Tufts	Somerville, Mass	May 14, 1867	64, 812
Brick molding and making machine	J. Lee and J. Ring	Ogden, N. Y	June 11, 1825	....
Brick molding and pressing	D. Flagg, jr	Gardiner, Me	Mar. 19, 1828	....
Brick molding and pressing	J. McDonald and R. McQueen	New York, N. Y	Nov. 19, 1828	....
Brick molding and pressing	J. Treadway	Haverstraw, N. Y	Mar. 21, 1865	46, 956
Brick molding and pressing machine	N. Adams	Cornwall, N. Y	Sept. 8, 1837	376
Brick molding and pressing machine	J. J. Alvord	Tecumseh, Mich	May 20, 1862	35, 287
Brick molding and pressing machine	D. Barnford	Deerfield, N. Y	Apr. 28, 1832	....
Brick molding and pressing machine	C. Chaney	Prospect, Me	Sept. 11, 1834	....
Brick molding and pressing machine	D. Philips	Natchez, Miss	June 23, 1832	....
Brick molding and pressing machine	H. Waterman	Bath, Me	Nov. 4, 1837	456
Brick molding and pressing machine	S. Waterman and C. Learned	Charleston, S. C	Apr. 14, 1838	694
Brick pressing and molding machinery	J. McDonald	New York, N. Y	Apr. 24, 1827	....
Brick-molding machine	J. Bolton	Saratoga Springs, N. Y	Aug. 23, 1838	891
Brick-molding machine	J. Booth and W. H. Stevenson	Columbus, Miss	Jan. 6, 1844	3, 399
Brick-molding machine	J. Butler	Buffalo, N. Y	Dec. 13, 1853	10, 319
Brick-molding machine	J. Coppuck	Louisville, Ky	July 1, 1836	....
Brick-molding machine	J. Drummond	Whitestown, N. Y	Mar. 6, 1832	....
Brick-molding machine	E. Fish	Fayette, Me	June 9, 1826	....
Brick-molding machine	J. W. Frost	Croton, N. Y	Nov. 20, 1849	6, 882
Brick-molding machine	W. C. Grimes	York, Pa	Dec. 2, 1834	....
Brick-molding machine	J. J. Hutchison and J. D. Brandbery.	Cape Girardeau, Mo	Oct. 16, 1860	30, 406
Brick-molding machine	T. P. Kneeland	Ogden, N. Y	June 21, 1825	....
Brick-molding machine	H. Mane	Point Coupee, La	Nov. 23, 1822	....
Brick-molding machine	I. P. Owen	Upper Alton, Ill	Apr. 10, 1843	3, 041
Brick-molding machine	R. Rankin	Frankfort, Me	Dec, 24, 1834	....
Brick-molding machine	L. E. Ransom	Mill Port, N. Y	Jan. 9, 1838	548
Brick-molding machine	S. Ustick	Philadelphia, Pa	July 10, 1847	5, 187
Brick-molding machine	J. Z. A. Wagner	Philadelphia, Pa	Aug. 11, 1857	17, 999
Brick-molding machine	D. Watson	Fayetteville, Me	May 6, 1826	....
Brick-molding machine	D. Wolfe	Mount Vernon, Ohio	Dec. [illegible], 1834	....
Brick, mortar, &c., Extension-machine for raising	J. Cox	Ebensburgh, Pa	June 5, 1849	6, 491
Brick-mortar, Machine for mixing and hoisting	J. Rinehart	Danville, Ill	June 6, 1835	....
Brick or artificial stone without burning, Making	J. B. Harris	Philadelphia, Pa	Aug. 21, 1810	....
Brick or building-block, Machine for pressing hollow.	M. and J. H. Buck and F. A. Cushman.	Lebanon, N. H	Dec. 9, 1856	16, 174

Index of patents issued from the United States Patent Office from 1790 to 1873, inclusive—Continued.

Invention.	Inventor.	Residence.	Date.	No.
Brick or building-blocks, Manufacture of	H. W. Angell	Waukesha, Wis	Dec. 11, 1866	60, 320
Brick or hollow block, Building	G. H. Johnson and G. Milson	Buffalo, N. Y	Feb. 16, 1869	86, 929
Brick or tiles, Steam-boiler	W. Dillon	Wheeling, W. Va	June 4, 1872	127, 467
Brick, Paving	D. Moffet and J. Thomson	Philadelphia, Pa	Apr. 19, 1870	102, 144
Brick, Paving	J. Overmyer	Indianapolis, Ind	Aug. 20, 1872	130, 589
Brick-pottery kiln	J. B. De Palm	New York, N. Y	June 6, 1854	11, 002
Brick-press	N. Adams	Canterbury, N. Y	Apr. 17, 1849	6, 361
Brick-press	J. J. Alvord	Tecumseh, Mich	Sept. 26, 1865	50, 088
Brick-press	C. B. Baker	Troy, N. Y	June 6, 1848	5, 617
Brick-press	C. B. Baker	Troy, N. Y	Mar. 26, 1850	7, 206
Brick-press	C. B. Baker	Troy, N. Y	Aug. 1, 1854	11, 408
Brick-press	C. B. Baker and E. Gifford	Troy, N. Y	Sept. 7, 1844	3, 731
Brick-press	P. Ball	Mount Vernon, Ohio	Mar. 2, 1836	....
Brick-press	W. Barker and G. Martin	Schenectady, N. Y	May 1, 1866	54, 460
Brick-press	C. Becker	Flint, Mich	June 11, 1867	65, 635
Brick-press	E. H. Bellows	Worcester, Mass	Oct. 6, 1857	18, 318
Brick-press	J. Booth	Mobile, Ala	June 20, 1848	5, 640
Brick-press	J. Boynton	East Hartford, Conn	Aug. 5, 1856	15, 471
Brick-press	A. H. Brown	Georgetown, D. C	Sept. 5, 1854	11, 658
Brick-press	B. H. Brown	Philadelphia, Pa	Oct. 3, 1844	3, 770
Brick-press	J. T. Brown and M. Fuller	Midville, Ga	Dec. 11, 1849	6, 933
Brick-press	R. Brunson	Chicago, Ill	Apr. 23, 1867	63, 990
Brick-press	T. J. Burke	Sandwich, Ill	May 25, 1869	90, 339
Brick-press	J. Butler	Buffalo, N. Y	Jan. 29, 1850	7, 048
Brick-press	A. Carbonel	Philadelphia, Pa	May 12, 1842	2, 619
Brick-press	C. Carnell	Kensington, Pa	Jan. 15, 1850	7, 017
Brick-press	J. Chase, jr	Pegnonock, Conn	May 22, 1855	12, 898
Brick-press	J. F. Clark	Morenci, Mich	Dec. 13, 1870	110, 113
Brick-press	J. B. Collen	Reading, Pa	Jan. 1, 1856	14, 012
Brick-press	T. Conklin	Woodbridge, Miss	Jan. 23, 1841	1, 947
Brick-press	D. M. Crellis	Sandwich, N. H	Aug. 11, 1832	....
Brick-press	T. Culbertson	Cincinnati, Ohio	May 16, 1846	4, 521
Brick-press	T. Culbertson and G. Scott	Philadelphia, Pa	June 25, 1850	7, 453
Brick-press	J. Dane	West Derby, Vt	Oct. 24, 1848	5, 878
Brick-press	W. L. Drake	Sturgis, Mich	Sept. 3, 1867	68, 494
Brick-press	J. S. Elliott	West Needham, Mass	Feb. 14, 1860	27, 117
Brick-press	J. M. Enos	Saint Joseph, Mo	Apr. 14, 1868	76, 613
Brick-press	A. K. Fahnestock	Harrisburgh, Pa	Apr. 16, 1842	2, 560
Brick-press	J. Finegan	Haverstraw, N. Y	Sept. 15, 1863	39, 906
Brick-press	J. Grant	Providence, R. I	May 13, 1851	8, 093
Brick-press	I. Gregg	Pittsburgh, Pa	June 6, 1848	5, 618
Brick-press	I. Gregg	Pittsburgh, Pa	Aug. 1, 1854	11, 422
Brick-press	A. Hall	Coxsackie, N. Y	May 22, 1847	5, 128
Brick-press	R. R. Harbour	Oskaloosa, Iowa	June 9, 1857	17, 546
Brick-press	T. Hartley	Pittsburgh, Pa	Apr. 11, 1846	4, 451
Brick-press	J. W. Hope	New York, N. Y	June 4, 1850	7, 414
Brick-press	H. J. Hughes	Davenport, Iowa	Feb. 5, 1856	14, 195
Brick-press	J. Johnson	Fort Smith, Ark	Aug. 8, 1854	11, 482
Brick-press	L. Kirk	Reading, Pa	June 17, 1856	15, 135
Brick-press	A. J. Knisely	Chicago, Ill	Sept. 17, 1872	131, 353
Brick-press	W. O. Leslie	Philadelphia, Pa	Aug. 11, 1868	80, 977
Brick-press	S. Lillie, jr	Fort Wayne, Ind	Feb. 17, 1857	16, 649
Brick-press	G. A. Metcalf	Malden, Mass	May 18, 1869	90, 285
Brick-press	J. A. Millholland	Mount Savage Md	Jan. 17, 1871	111, 076
Brick-press	J. P. Owen	Cincinnati, Ohio	July 26, 1845	4, 123
Brick-press	C. D. Page	Rochester, N. Y	Sept. 17, 1867	68, 896
Brick-press	J. L. Ransom	Mill Port, N. Y	Apr. 4, 1846	4, 448
Brick-press	N. Read	Belfast, Me	Aug. 20, 1835	....
Brick-press	J. J. Riddle	Covington, Ky	Apr. 1, 1851	8, 021
Brick-press	V. Roth	Evansville, Ind	Feb. 20, 1849	6, 127
Brick-press	W. Sanford	Cambridge, Mass	Sept. 9, 1845	4, 183
Brick-press	N. Sawyer	Baltimore, Md	Sept. 27, 1844	3, 768
Brick-press	N. Sawyer	Baltimore, Md	Dec. 4, 1847	5, 386
Brick-press	N. Sawyer	Baltimore, Md	Apr. 23, 1850	7, 309
Brick-press	J. Scheitlin	Louisville, Ky	Jan. 21, 1851	7, 908
Brick-press	D. M. Seeley	Albany, N. Y	May 22, 1866	54, 967
Brick-press	J. Smedley	Columbia, Pa	Aug. 28, 1844	3, 720
Brick-press	J. N. Smith	Jersey City, N. J	May 24, 1870	103, 381
Brick-press	T. W. Smith	Alexandria, D. C	Jan. 30, 1841	1, 959
Brick-press	A. A. Torpey	Chicago, Ill	Aug. 25, 1868	81, 560
Brick-press	M. Twitchell	Gray, Me	Sept. 24, 1825	....
Brick-press	M. Twitchell	Gray, Me	June 10, 1844	3, 622
Brick-press	S. Ustick	Philadelphia, Pa	Dec. 28, 1838	1, 045
Brick-press	S. Ustick	Philadelphia, Pa	July 10, 1855	13, 239
Brick-press	W. Wadsworth	Hartford, Conn	June 26, 1835	....
Brick-press	J. Z. A. Wagner	Philadelphia, Pa	Apr. 8, 1851	8, 024
Brick-press	J. Waite	Leicester, Mass	Mar. 12, 1845	3, 942
Brick-press	W. Wakely	Taunton, Mass	Apr. 9, 1872	125, 636
Brick-press	W. B. Waldran and G. Hargitt	Shelby County, Tenn	July 10, 1849	6, 582
Brick-press	J. W. Ward	Cambridge, Mass	May 30, 1848	5, 610
Brick-press	H. Waterman	Bath, Me	Nov. 20, 1837	469
Brick-press	S. Whitman	New Albany, Ind	Apr. 23, 1850	7, 313
Brick-press	E. D. Williams and T. Tyrell	Wilmington, Del., and York, Pa.	Oct. 17, 1854	11, 817
Brick-press	A. Witmer	Henry, Ill	Sept. 6, 1864	44, 133
Brick-press	J. Woodson	Rockbridge County, Va	Sept. 10, 1829	....
Brick-press	A. Woodworth	Worcester, Mass	May 8, 1843	3, 074
Brick-press	A. Woodworth, 3d, and S. Mower.	Worcester, Mass., and Philadelphia, Pa.	Nov. 27, 1849	6, 899
Brick-press	F. Zisemann	Saint Louis, Mo	Nov. 13, 1849	6, 876
Brick-press bridle	J. G. Talcott	Glastenbury, Conn	Oct. 1, 1830	....
Brick-press bridle	J. Willard	Baltimore, Md	June 10, 1830	....
Brick press, Concrete	T. J. Burke	Chicago, Ill	Jan. 21, 1868	73, 576
Brick press, Concrete	J. H. Wirt	Delphi, Ind	Apr. 21, 1868	77, 153
Brick-press, Construction of	A. Hall	Cleveland, Ohio	Sept. 3, 1842	2, 768
Brick-press, Finishing	L. B. Crittenden	Pittsburgh, Pa	June 2, 1868	78, 434
Brick-press for pressing brick from dry clay	N. Sawyer and T. W. Smith	Washington, D. C	Aug. 13, 1838	881
Brick-press, Hydraulic	E. Rogers	Cleveland, Ohio	Sept. 23, 1856	15, 778
Brick-press mold or box	J. McKenna	Cambria, Pa	Aug. 19, 1873	141, 942

Index of patents issued from the United States Patent Office from 1790 *to* 1873, *inclusive*—Continued.

Invention.	Inventor.	Residence.	Date.	No.
Brick-press, Self-feeding	J. Winslow	Portland, Me	Dec. 8, 1826	....
Brick-press, Spiral-molding	J. Parker	Gardiner, Me	Mar. 28, 1826	....
Brick press, Unburnt	E. Mayo	Hallowell, Me	June 24, 1826	....
Brick, Pressing	H. Crofoot	Oak Park, Ill	May 14, 1867	64, 637
Brick, Pressing	J. Moore	Gardiner, Me	Aug. 24, 1825	....
Brick, Pressing	O. W. Seely	Buffalo, N. Y	Aug. 27, 1867	68, 244
Brick, Pressing	W. G. White	Bedford, Ohio	Mar. 30, 1869	88, 353
Brick, Pressing and finishing	C. Hinkle	Allentown, Pa	June 1, 1833	....
Brick pressing and molding machine	S. Costill	Philadelphia, Pa	Sept. 4, 1840	1, 769
Brick-pressing machine	C. Berrian	Clinton City, Iowa	Sept. 1, 1868	81, 738
Brick-pressing machine	M. Chittenden	Danbury, Conn	Oct. 10, 1865	50, 333
Brick-pressing machine	W. H. Combs	Fort Wayne, Ind	Jan. 7, 1868	73, 079
Brick-pressing machine	G. V. Harper	Batavia, N. Y	Apr. 25, 1838	711
Brick-pressing machine	J. Hotchkiss	Springfield, Ohio	Dec. 20, 1864	45, 563
Brick-pressing machine	J. Howe	Alna, Me	May 18, 1827	....
Brick pressing machine	T. A. Irick	Harrisonburgh, Va	July 27, 1831	....
Brick-pressing machine	J. K. Lemon	Allegheny City, Pa	Apr. 11, 1865	47, 209
Brick-pressing machine	A. McClung	Fairfield, Va	Mar. 26, 1831	....
Brick-pressing machine	A. F. Mervine	Saint Louis, Mo	Sept. 12, 1837	382
Brick-pressing machine	B. R. Reed	Danvers, Mass	May 29, 1823	....
Brick-pressing machine	B. Rolfe	Deering, N. H	Apr. 30, 1822	....
Brick-pressing machine	N. Sawyer	Mount Vernon, Ohio	Apr. 8, 1835	....
Brick-pressing machine	J. Sites	Harrisonburgh, Va	Mar. 29, 1831	....
Brick-pressing machine	W. H. Walrath	Chittenango, N. Y	Oct. 9, 1866	58, 700
Brick-pressing machine	J. Willerd	Baltimore, Md	May 2, 1826	....
Brick-pressing machinery	A. B. Crossman	Huntington, N. Y	Feb. 9, 1827	....
Brick-pressing machinery	C. Heath and J. Wilson	Malden and Boston, Mass	June 17, 1862	35, 650
Brick, Pressing unburnt	T. Norcross	Hallowell, Me	Dec. 31, 1825	....
Brick-re-pressing machine	J. K. Caldwell	Philadelphia, Pa	May 20, 1873	139, 113
Brick-re-pressing machine	H. Maurer	New York, N. Y	June 29, 1869	91, 856
Brick-re-pressing machine	A. J. Sprague	Toledo, Ohio	July 12, 1870	105, 270
Brick, Reversible curb	E. Linson	Shoemakertown, Pa	Jan. 21, 1873	134, 994
Brick, Scouring	J. Valentine	Woodbridge, N. J	Nov. 7, 1865	50, 862
Brick, stock, tile, &c	D. Flagg, jr., and A. Parker	Sweden, N. Y	Dec. 28, 1832	....
Brick, stone, and mortar, Elevating	C. G. Fairman	Lewiston, N. Y	Sept. 16, 1833	....
Brick, stone, &c., Method of moving	S. Brown	New York, N. Y	July 31, 1827	....
Brick, stove-linings, &c., Composition for fire	E. H. Richter	Taunton, Mass	Dec. 9, 1873	145, 448
Brick-striker	P. Sweet	Ashtabula, Ohio	Feb. 20, 1835	....
Brick, Striking	A. W. Duty	Sangerfield, N. Y	Mar. 7, 1833	....
Brick-striking machine	E. and D. W. Duty	Painesville, Ohio	June 7, 1828	....
Brick, tiles, and clay-ware, Making	J. Wood	Rockland County, N. Y	Mar. 3, 1829	....
Brick, tiles, &c., Forming	J. Hawkins	....	July 11, 1797	....
Brick, tiles, &c., from slag, Making	F. Fabre	Marseilles, France	May 31, 1870	103, 591
Brick, tiles, &c., Making	J. F. Gould	Newburyport, Mass	Oct. 17, 1809	....
Brick, tiles, &c., Making	E. G. Oldfield	Bordentown, N. J	Jan. 15, 1861	31, 126
Brick, tiles, &c., Making	J. M. Reid	Pittsburgh, Pa	Apr. 2, 1872	125, 144
Brick, tiles, &c., Making	D. Thackara	Woodbury, N. J	Dec. 14, 1869	97, 992
Brick, Treadle-machine for	B. R. Reed	Danvers, Mass	May 29, 1823	....
Brick, Treating clay and drying	J. Fisher	Reading, Pa	Aug. 10, 1869	93, 526
Brick wall, Composition for cleaning and renovating.	W. B. Walters	Lock Haven, Pa	Aug. 18, 1868	81, 233
Brick, Water and fire proof	T. H. Sherman	Scriba, N. Y	Dec. 30, 1833	....
Brick-work, Joint groover for	A. M. Zabriskie	Bergen Point, N. J	Feb. 26, 1867	62, 463
Brick-yard shed	R. O. Smith	Newburyport, Mass	Aug. 5, 1873	141, 520
Bricks, Arching	W. F. Quinby	Wilmington, Del	Aug. 9, 1870	106, 203
Bricks, Cutting clay into	J. Plumbe	San Francisco, Cal	June 12, 1855	13, 042
Bricks, Form of building	E. Conkling	Cincinnati, Ohio	Jan. 8, 1856	14, 052
Bricks, retorts, muffles, crucibles, &c., Manufacture of.	E. L. Seymour	New York, N. Y	Nov. 19, 1867	71, 229
Bricklayer's pointer	J. S. Merriken	Baltimore, Md	May 13, 1873	138, 911
Bridge	J. J. Alises	New York, N. Y	Oct. 4, 1864	44, 498
Bridge	C. Baker	Saint Joseph, Mo	Aug. 31, 1869	94, 272
Bridge	J. Boles, jr	Boston, Mass	May 20, 1865	47, 920
Bridge	J. Boles, jr	Boston, Mass	May 30, 1865	48, 013
Bridge	T. C. Boutel	Paris, France	May 2, 1871	114, 401
Bridge	J. Bragg	Montreal, Canada	Jan. 4, 1820	....
Bridge	J. C. Briggs	Concord, N. H	May 26, 1863	38, 653
Bridge	G. H. Bruce	Lancaster, N. Y	Jan. 7, 1862	34, 102
Bridge	T. Burr	....	Feb. 14, 1806	....
Bridge	A. M. Campbell	Newark, N. J	Dec. 27, 1870	110, 546
Bridge	S. and T. Champion	Washington, D. C	July 18, 1854	11, 322
Bridge	L. K. Cole and H. Soule, jr	Syracuse, N. Y	Feb. 13, 1866	52, 506
Bridge	G. W. Corey	Port Jervis, N. Y	July 16, 1867	66, 799
Bridge	A. Cottrell	Newport, R. I	Jan. 19, 1864	43, 099
Bridge	B. F. Davis	Hearne, Tex	Sept. 23, 1873	143, 125
Bridge	E. Denmead and W. Bollman	Marietta, Ga., and Baltimore, Md.	May 19, 1868	78, 073
Bridge	F. Dieckman	Cincinnati, Ohio	Mar. 28, 1871	113, 030
Bridge	J. H. Diedrichs	Richmond, Va	Apr. 2, 1872	125, 182
Bridge	T. Durden	Montgomery, Ala	June 1, 1858	20, 414
Bridge	J. R. Durfee	Oshkosh, Wis	Sept. 16, 1873	142, 776
Bridge	J. B. Eads	Saint Louis, Mo	Nov. 10, 1868	83, 942
Bridge	J. B. Eads	Saint Louis, Mo	May 4, 1869	89, 745
Bridge	J. B. Eads	Saint Louis, Mo	Sept. 2, 1873	142, 378
Bridge	J. B. Eads	Saint Louis, Mo	Nov. 11, 1873	144, 519
Bridge	C. H. Earle	Green Bay, Wis	Sept. 15, 1857	18, 196
Bridge	S. Ensign	New Franklin, Ohio	Nov. 9, 1869	96, 569
Bridge	A. Fink	Baltimore, Md	May 9, 1854	10, 887
Bridge	A. Fink	Louisville, Ky	Apr. 9, 1867	63, 714
Bridge	H. Flad	Saint Louis, Mo	Oct. 15, 1872	132, 271
Bridge	J. Fowler	....	Feb. 24, 1797	....
Bridge	S. F. Gassaway	Marietta, Ga	Sept. 18, 1847	5, 297
Bridge	J. Gates	Saint Louis, Mo	Feb. 11, 1873	135, 705
Bridge	L. Gay	Chili, N. Y	Nov. 4, 1846	4, 837
Bridge	J. H. Gilbert	Roxbury, Mass	Sept. 18, 1866	58, 094
Bridge	J. Glass, G. P. Schneider, and W. B. Rezner.	Cleveland, Ohio	Dec. 10, 1867	71, 868
Bridge	F. Good	New London, Pa	Nov. 4, 1837	450

Index of patents issued from the United States Patent Office from 1790 *to* 1873, *inclusive*—Continued.

Invention.	Inventor.	Residence.	Date.	No.
Bridge	J. M. Goodwin	New York, N. Y	Sept. 16, 1873	142, 785
Bridge	T. B. Gregory	Champaign, Ill	July 12, 1870	105, 195
Bridge	R. Grotz	Chicago, Ill	Apr. 2, 1867	63, 507
Bridge	E. Hamilton	Chicago, Ill	May 26, 1868	78, 202
Bridge	D. Hammond	Canton, Ohio	July 3, 1866	56, 043
Bridge	D. Hammond, M. Adler, and J. Abbott.	Canton, Ohio	Feb. 11, 1873	135, 802
Bridge	F. Harbach	Pittsfield, Mass	Aug. 12, 1846	4, 694
Bridge	G. E. Harding	New York, N. Y	Oct. 22, 1872	132, 398
Bridge	S. P. Hastings	Tonawanda, N. Y	Oct. 15, 1872	132, 284
Bridge	P. Hendricks	Floris, Iowa	Nov. 26, 1867	71, 483
Bridge	J. G. Henszey	Philadelphia, Pa	June 22, 1869	91, 745
Bridge	G. P. Herthel, jr	Saint Louis, Mo	Jan. 18, 1870	98, 866
Bridge	H. L. Hervey and R. E. Osborne.	Quincy, Ill., and Springfield, Ohio.	Aug. 21, 1855	13, 461
Bridge	G. W. O. Hüygens	Saint Louis, Mo	Apr. 1, 1856	14, 584
Bridge	W. James	Richmond, Va	Apr. 16, 1867	63, 901
Bridge	J. Johnson	Mott Haven, N. Y	July 16, 1872	129, 479
Bridge	J. E. Kanser	New York, N. Y	Nov. 7, 1865	50, 827
Bridge	J. J. Kelley	Slippery Rock, Pa	Mar. 23, 1869	88, 181
Bridge	Z. King	Cleveland, Ohio	Sept. 25, 1866	58, 266
Bridge	L. Kirkup	Brooklyn, N. Y	Oct. 24, 1871	120, 282
Bridge	M. Kremser	Cleveland, Ohio	Feb. 27, 1866	52, 860
Bridge	J. Laird	Canton, Ohio	Aug. 31, 1869	94, 321
Bridge	J. and G. F. Laird	Canton, Ohio	Aug. 31, 1869	94, 322
Bridge	G. T. Lape	Summit, N. Y	May 12, 1868	77, 741
Bridge	O. G. Leopold	Cincinnati, Ohio	Dec. 4, 1866	60, 205
Bridge	J. H. Linville	Philadelphia, Pa	Nov. 24, 1868	84, 288
Bridge	L. Liscom	Boston, Mass	Mar. 31, 1868	76, 212
Bridge	R. Lockwood	Brooklyn, N. Y	Dec. 5, 1865	51, 328
Bridge	S. H. Long	United States Engineers	Nov. 13, 1847	5, 366
Bridge	S. H. Long	United States Army	Aug. 17, 1858	21, 203
Bridge	D. C. McCallum	Owego, N. Y	Jan. 20, 1857	16, 446
Bridge	D. McCurdy	Ottawa, Ohio	June 28, 1870	104, 867
Bridge	D. McCurdy	Ottawa, Ohio	June 28, 1870	104, 868
Bridge	D. McCurdy	Ottawa, Ohio	June 28, 1870	104, 869
Bridge	H. S. McDowell	Columbus, Tex	Jan. 11, 1870	98, 699
Bridge	J. A. McKay	Auburn, Ind	June 1, 1869	90, 767
Bridge	R. S. Merrill	Boston, Mass	May 19, 1868	78, 000
Bridge	D. A. Mitchell	Chicago, Ill	July 31, 1866	56, 777
Bridge	F. Monroe	Bruce, Mich	Apr. 17, 1866	54, 004
Bridge	R. and M. J. Montgomery	New York, N. Y	Sept. 1, 1868	81, 666
Bridge	G. C. Morgan	Chicago, Ill	Aug. 22, 1871	118, 258
Bridge	T. W. H. Moseley	Covington, Ky	Feb. 3, 1857	16, 572
Bridge	T. W. H. Moseley	Boston, Mass	Oct. 23, 1866	59, 054
Bridge	T. W. H. Moseley	Boston, Mass	Aug. 30, 1870	106, 855
Bridge	A. Oudry	Paris, France	May 13, 1863	35, 251
Bridge	T. Palmer	....	Dec. 17, 1797	....
Bridge	C. H. Parker	Boston, Mass	Aug. 10, 1869	93, 638
Bridge	C. H. Parker	Boston, Mass	Jan. 4, 1870	98, 620
Bridge	C. H. Parker	Boston, Mass	Feb. 22, 1870	100, 185
Bridge	C. H. Parker	Boston, Mass	May 17, 1870	103, 233
Bridge	R. L. Partridge	Marysville, Ohio	June 11, 1872	127, 791
Bridge	C. W. Peale	....	Jan. 21, 1797	....
Bridge	O. H. Perry and W. H. Allen	Beloit, Wis	Oct. 24, 1871	120, 319
Bridge	H. Pettit	Philadelphia, Pa	Feb. 25, 1873	136, 177
Bridge	C. Pfeifer	Saint Louis, Mo	Dec. 12, 1871	121, 894
Bridge	A. J. Post	Hudson City, N. J	Aug. 25, 1868	81, 406
Bridge	A. J. Post	Hudson City, N. J	Sept. 1, 1868	81, 817
Bridge	A. Reiling	Bellevue, Iowa	Dec. 16, 1873	145, 685
Bridge	N. Rider	Worcester, Mass	Nov. 26, 1845	4, 287
Bridge	I. Rogers	Cincinnati, Ohio	Sept. 30, 1856	15, 823
Bridge	I. Rogers	Washington, D. C	Feb. 10, 1863	37, 642
Bridge	R. W. Rogers	Pittsburgh, Pa	Dec. 29, 1868	85, 332
Bridge	H. G. Russell	Lincoln, Ill	Nov. 5, 1872	132, 860
Bridge	H. A. Rust and L. Hermann	Chicago, Ill	Sept. 8, 1868	81, 950
Bridge	J. Sanderson	Fredericksburgh, Ohio	Apr. 21, 1868	77, 103
Bridge	J. Seebold	Kantz, Pa	July 19, 1870	105, 497
Bridge	J. Seebold	Kantz, Pa	Sept. 6, 1870	107, 106
Bridge	J. Seebold	Kantz, Pa	May 2, 1871	114, 497
Bridge	C. S. Smith, C. H. Latrobe, and F. H. Smith.	Baltimore, Md	Jan. 18, 1870	99, 017
Bridge	F. H. Smith	Baltimore, Md	Dec. 11, 1866	60, 434
Bridge	F. H. Smith	Baltimore, Md	Apr. 27, 1869	89, 442
Bridge	F. H. Smith	Baltimore, Md	Mar. 10, 1868	75, 477
Bridge	F. H. Smith	Baltimore, Md	May 11, 1869	89, 948
Bridge	F. H. Smith	Baltimore, Md	Oct. 26, 1869	96, 278
Bridge	F. H. Smith	Baltimore, Md	June 25, 1872	128, 449
Bridge	R. W. Smith	Tippecanoe, Ohio	July 16, 1867	66, 900
Bridge	R. W. Smith	Toledo, Ohio	Dec. 7, 1869	97, 714
Bridge	C. B. Sreeves	Atchison, Kans	May 2, 1871	114, 36[illegible]
Bridge	J. D. Steele	Pottstown, Pa	Apr. 9, 1867	63, 666
Bridge	J. Stickney	....	June 3, 1797	....
Bridge	J. Tellweger	Louisville, Ky	Apr. 2, 1872	125, 244
Bridge	J. Templeman	Georgetown, D. C	Mar. 6, 1810	....
Bridge	W. E. Thomas	Queenstown, Md	July 16, 1872	129, 374
Bridge	J. K. Thompson	Chicago, Ill	Sept. 8, 1868	81, 960
Bridge	I. Town	Fayetteville, N. C	Jan. 28, 1820	....
Bridge	I. Town	New York, N. Y	Apr. 3, 1835	....
Bridge	J. B. Tracy	Lincoln, Del	Dec. 5, 1871	121, 556
Bridge	W. P. Trowbridge	Newtown, N. Y	Sept. 7, 1869	94, 529
Bridge	L. E. Truesdell	Warren, Mass	Aug. 31, 1858	21, 388
Bridge	L. E. Truesdell	Chicago, Ill	Apr. 7, 1868	76, 5[illegible]8
Bridge	E. Trumbull	Little Falls, N. Y	July 10, 1841	2, 164
Bridge	M. Turley	Council Bluffs, Iowa	Dec. 13, 1870	110, 173
Bridge	J. Valleley	Canton, Ohio	Dec. 30, 1873	145, 034
Bridge	J. B. Valliquette	Chicago, Ill	Apr. 7, 1868	76, 560
Bridge	D. H. Van Duzer	Sugar Loaf, N. Y	Sept. 20, 1859	25, 537

Index of patents issued from the United States Patent Office from 1790 *to* 1873, *inclusive*—Continued.

Invention.	Inventor.	Residence.	Date.	No.
Bridge	L. Wernwag	Jefferson County, Va	Dec. 22, 1829	....
Bridge	I. H. Wheeler	Sciotoville, Ohio	Sept. 20, 1870	107, 576
Bridge	A. White	Boston, Mass	Feb. 3, 1852	8, 713
Bridge	T. B. White	New Brighton, Pa	Mar. 10, 1868	75, 502
Bridge, &c	R. T. L. Witty	Lowell, Mass	Oct. 14, 1835	....
Bridge	W. Woodmansee	Kingston, N. Y	Mar. 6, 1827	....
Bridge	J. S. Yerk and G. H. Heming	Tiffin, Ohio	Dec. 24, 1861	34, 023
Bridge	E. W. Young	Spring Gardens, Great Britain.	Sept. 28, 1869	95, 402
Bridge, Anchoring suspension-chain for	J. A. Roebling	Pittsburgh, Pa	Aug. 26, 1846	4, 710
Bridge and ferry stairs	D. Dunham	New York, N. Y	June 28, 1814	....
Bridge and other structure	A. Bonzano	Phœnixville, Pa	May 21, 1872	127, 017
Bridge and roof trusses, Bolster-block and pier or abutment-plate for.	J. M. Wilson	Philadelphia, Pa	Mar. 21, 1871	112, 878
Bridge and switch telegraph, Railway draw	W. C. McRea	Philadelphia, Pa	Oct. 3, 1854	11, 763
Bridge-arch	D. Forargue	Cleveland, Ohio	Apr. 2, 1872	125, 128
Bridge, Arched	J. Davenport	Massillon, Ohio	Sept. 22, 1868	82, 388
Bridge, Arched	J. B. Eads and A. Flad	Saint Louis, Mo	Oct. 12, 1869	95, 784
Bridge, Arched	W. B. Rezner	Cleveland, Ohio	July 2, 1872	128, 509
Bridge, Arched	J. Zellweger	Louisville, Ky	June 25, 1872	128, 349
Bridge, Arched trussed	H. L. Hervey	Quincy, Ill	Feb. 26, 1856	14, 314
Bridge, Automatic canal	D. Berry	Huntington, Ind	Nov. 22, 1859	26, 156
Bridge, Balance	J. Jessop	York Town, Pa	July 1, 1811	....
Bridge-block	J. R. Smith	Springfield, Mass	July 14, 1868	79, 869
Bridge, Boat and ferry	W. W. Virdin	Baltimore, Md	July 16, 1861	32, 846
Bridge-braces, Mode of connecting iron	A. D. Briggs	Springfield, Mass	June 18, 1861	32, 555
Bridge, Canal	B. G. Anderson	Chillicothe, Ohio	Oct. 14, 1856	15, 873
Bridge, Chain	J. Finley	Fayette County, Pa	June 17, 1808	....
Bridge, Chain	J. Templeman	Allegany County, Md	Aug. 16, 1808	....
Bridge, Chain suspension	E. W. and E. W. Serrell, jr	New York, N. Y	Oct. 21, 1873	143, 788
Bridge-clamp	N. Chapin	Chicago, Ill	Sept. 27, 1864	44, 397
Bridge-clamp	E. L. Truesdell	Chicago, Ill	July 26, 1870	105, 868
Bridge-column	J. Zellweger	Louisville, Ky	Dec. 16, 1873	145, 545
Bridge connections	T. C. Clarke and A. Bonzano	Philadelphia, Pa	Aug. 13, 1872	130, 479
Bridge, Corrugated-iron	R. Montgomery	New York, N. Y	Aug. 23, 1859	25, 210
Bridge counter-braces, Means for adjusting the effective length of.	D. C. McCallum	Owego, N. Y	July 15, 1851	8, 224
Bridge, Draw	G. W. R. Bayley and T. W. Nelson.	Brashear, La	Jan. 29, 1861	31, 222
Bridge, Draw	H. W. Cass	Lodi, Wis	Nov. 23, 1869	97, 044
Bridge, Draw	C. K. Marshall	New Orleans, La	July 7, 1868	79, 768
Bridge, Draw	J. L. Piper	Pittsburgh, Pa	Oct. 22, 1872	132, 410
Bridge, Draw	T. W. Pratt	Boston, Mass	Feb. 22, 1870	100, 065
Bridge, Draw	T. L. Speakman	Camden, N. J	Apr. 7, 1868	76, 542
Bridge, Draw	H. Whitney	East Cambridge, Mass	July 16, 1872	129, 638
Bridge, Drop	J. D. Woodruff and J. H. Butterworth.	Newark and Dover, N. J	Feb. 14, 1854	10, 527
Bridge, Ferry	W. J. Alsop	Camden, N. J	June 4, 1872	127, 542
Bridge, Ferry	C. J. Atkins	Louisiana, Mo	Aug. 19, 1873	141, 911
Bridge, Ferry	J. S. Bradford	New York, N. Y	Aug. 11, 1868	80, 902
Bridge, Ferry	J. A. Clarke	New York, N. Y	July 28, 1868	80, 391
Bridge, Ferry	A. C. Willson	Green Point, N. Y	Apr. 14, 1868	76, 867
Bridge, Floating	T. Schofield	Grass Valley, Cal	Mar. 20, 1860	27, 571
Bridge, Floating draw	N. B. Proctor	Burlington, Vt	May 20, 1856	14, 928
Bridge, Floating swing	J. N. Vrooman	Danube, N. Y	Apr. 15, 1840	1, 551
Bridge, Flood	T. A. Bryan	Baltimore, Md	Sept. 6, 1870	106, 995
Bridge for canal, Automatic draw	G. C. Bovey	Chillicothe, Ohio	May 8, 1860	28, 148
Bridge for playing pool	O. A. Hill	Westbrook, Me	Mar. 9, 1869	87, 566
Bridge for railway-car, Safety	A. J. Elder	Kansas City, Mo	Dec. 8, 1868	84, 808
Bridge foundation, Iron	J. B. Eads	Saint Louis, Mo	Sept. 2, 1873	142, 382
Bridge, Frame	G. W. Long	Fort Jackson, La	Mar. 10, 1830	....
Bridge, Frame	G. Tabb	Martinsburgh, Va	Feb. 23, 1816	....
Bridge-gate	E. R. Coyne	Chicago, Ill	May 2, 1871	114, 414
Bridge-gate	M. Kirsch	Chicago, Ill	Nov. 22, 1870	109, 520
Bridge-gate	J. D. Sturges	Chicago, Ill	Sept. 6, 1870	107, 119
Bridge-gate	A. Weide	Chicago, Ill	Sept. 7, 1869	94, 534
Bridge-gate	A. Wermerskirchen	Chicago, Ill	June 7, 1870	104, 086
Bridge-gate	J. Wilcke and M. Ellenbogen	Chicago, Ill	Mar. 1, 1870	100, 480
Bridge gate, Draw	G. A. May	Chicago, Ill	Oct. 12, 1869	95, 821
Bridge gate, Draw	F. Pairan	Dayton, Ohio	Apr. 18, 1871	113, 916
Bridge gate, Ferry	L. P. Decker	Brooklyn, N. Y	June 19, 1866	55, 628
Bridge gate, Swing	A. D. Northway	Kenosha, Wis	Dec. 12, 1871	121, 892
Bridge gates, Turning	R. Laderwig and A. Rosenberg	Berlin, Prussia	Feb. 27, 1872	124, 072
Bridge-girder	J. Davenport	Massillon, Ohio	Dec. 24, 1867	72, 611
Bridge-girder	J. W. Evans	New York, N. Y	Nov. 18, 1873	144, 751
Bridge-girder	P. C. Guion	Cincinnati, Ohio	Feb. 26, 1856	14, 313
Bridge-girder	D. Hammond and W. R. Reeves	Canton, Ohio	June 21, 1864	43, 202
Bridge-girder	D. Hammond and W. R. Reeves	Canton, Ohio	Feb. 2, 1869	86, 538
Bridge-girders, Compressive chords for	M. Adler	Canton, Ohio	June 25, 1872	128, 350
Bridge-guard, Ferry	J. A. Clarke	New York, N. Y	Mar. 17, 1868	75, 526
Bridge guard or barrier	J. Lehmann	Crown Point, Ind	Aug. 25, 1868	81, 383
Bridge, Iron	J. J. Beard	Columbus, Ohio	Feb. 16, 1864	41, 594
Bridge, Iron	C. Bender, C. H. Latrobe, and C. S. Smith.	New York, N. Y., Baltimore, Md., and Saint Louis, Mo.	July 29, 1873	141, 310
Bridge, Iron	H. C. Brundage	Buffalo, N. Y	Mar. 1, 1870	100, 254
Bridge, Iron	F. E. Canda	Chicago, Ill	Mar. 30, 1869	88, 446
Bridge, Iron	W. B. Cooper	Albany, N. Y	Feb. 18, 1873	135, 970
Bridge, Iron	J. B. Eads	Saint Louis, Mo	Sept. 2, 1873	142, 379
Bridge, Iron	J. B. Eads	Saint Louis, Mo	Sept. 2, 1873	142, 380
Bridge, Iron	J. B. Eads	Saint Louis, Mo	Sept. 2, 1873	142, 381
Bridge, Iron	L. Eikenberry	Easton, Pa	Jan. 25, 1859	22, 715
Bridge, Iron	L. Eikenberry	Easton, Pa	Jan. 22, 1861	31, 157
Bridge, Iron	J. P. Fisher	Rochester, N. Y	June 26, 1860	28, 845
Bridge, Iron	P. Johnson	Springfield, Mass	Nov. 18, 1873	144, 766
Bridge, Iron	M. Miller	Cleveland, Ohio	June 7, 1870	103, 911
Bridge, Iron	D. H. Morrison	Dayton, Ohio	Oct. 29, 1867	70, 245
Bridge Iron	J. D. Pierce	Columbia, Mo	Aug. 5, 1873	141, 458
Bridge, Iron	S. S. Post	Jersey City, N. J	June 16, 1863	38, 910

Index of patents issued from the United States Patent Office from 1790 *to* 1873, *inclusive*—Continued.

Invention.	Inventor.	Residence.	Date.	No.
Bridge, Iron	W. Sellers	Philadelphia, Pa	Mar. 4, 1873	136, 389
Bridge, Iron	L. E. Truesdell	Warren, Mass	June 28, 1870	104, 902
Bridge, Iron	T. B. White	New Brighton, Pa	July 2, 1867	66, 433
Bridge, Iron	T. B. White	New Brighton, Pa	Mar. 9, 1869	87, 741
Bridge, Iron	J. Yandell and J. H. Johnson	Saint Louis, Mo	Oct. 17, 1854	11, 818
Bridge, Iron-arch	L. Kittenger	Massillon, Ohio	Oct. 3, 1871	119, 466
Bridge, Iron-truss	T. C. Clarke	Philadelphia, Pa	Nov. 5, 1872	132, 803
Bridge, Iron-truss	G. Halstead	Buffalo, N. Y	June 8, 1869	91, 124
Bridge, Iron-truss	J. H. Linville	Altoona, Pa	Jan. 14, 1862	34, 183
Bridge, Lattice	H. Haupt	York, Pa	Dec. 27, 1839	1, 445
Bridge, Lattice	L. E. Truesdell	Warren, Mass	June 3, 1856	15, 048
Bridge, Lattice and truss	G. B. Manley	Cogan Station, Pa	Apr. 30, 1867	64, 340
Bridge, Lift-draw	S. Whipple	Albany, N. Y	Dec. 24, 1872	134, 338
Bridge, Lifting	S. Swartz	Buffalo, N. Y	Feb. 25, 1873	136, 278
Bridge-pier	C. H. Lilienthal	Yonkers, N. Y	Nov. 29, 1870	109, 637
Bridge-pier	E. W. Smith	New York, N. Y	June 20, 1865	48, 317
Bridge, Pier for suspension	J. Gray	Cincinnati, Ohio	Dec. 24, 1872	134, 269
Bridge-pier, Iron	E. M. Grant	Macon, Ga	Dec. 22, 1868	85, 171
Bridge, Pivot	T. C. Clarke, A. Bonzano, and J. Griffen.	Philadelphia and Phœnixville, Pa.	June 18, 1872	128, 115
Bridge, Pivot	T. C. Clarke, A. Bonzano, and J. Griffen.	Philadelphia and Phœnixville, Pa.	Feb. 11, 1873	135, 776
Bridge-railing, Baluster for iron	S. Sellers and M. M. Manly	Philadelphia, Pa	Apr. 22, 1873	138, 049
Bridge, Rolling draw	R. Crosbie	Newark, N. J	Feb. 17, 1812	....
Bridge, Safety ferry	H. Lawrence	New York, N. Y	Mar. 13, 1855	12, 523
Bridge, Safety railway draw	J. K. and W. P. Gamble	Philadelphia, Pa	July 17, 1855	13, 258
Bridge safety-switch and signal, Draw	E. H. Tobey	Saint Louis, Mo	June 24, 1873	140, 319
Bridge, Self-acting draw	A. Koch	Rocktown, Pa	July 1, 1862	35, 765
Bridge, Self-acting draw	L. Schneider and J. A. Montgomery.	Williamsport, Pa	Sept. 4, 1860	29, 917
Bridge, Self-adjusting counter-brace of truss	J. Gray	Nashville, Tenn	Dec. 27, 1859	26, 583
Bridge, Self-adjusting platform for ferry	G. Sickels	Brooklyn, N. Y	June 7, 1853	9, 772
Bridge, Self-opening canal	J. Selser	Williamsport, Pa	Oct. 29, 1861	33, 606
Bridge signal-apparatus, Railway	T. S. Hall	New Haven, Conn	Apr. 4, 1871	113, 425
Bridge signal, Draw	L. B. Edgecomb	Troy, N. Y	July 17, 1866	56, 491
Bridge signal, Railway	A. Burnham	Taunton, Mass	July 6, 1858	20, 841
Bridge signal, Railway draw	A. F. Smith	Norwich, Conn	Oct. 2, 1866	58, 492
Bridge, Spiral-braced cylinder	I. Rogers	New York, N. Y	Nov. 10, 1841	2, 347
Bridge, Suspension	C. Bender	New York, N. Y	Dec. 10, 1867	71, 955
Bridge, Suspension	E. M. Carpenter	Middletown, N. Y	May 12, 1868	77, 800
Bridge, Suspension	A. S. Hallidie	San Francisco, Cal	July 2, 1867	66, 327
Bridge, Suspension	J. Royal	White Rock, Ill	Feb. 8, 1870	99, 597
Bridge, Suspension	J. Tomlinson	Putnam, Iowa	May 28, 1861	32, 440
Bridge, Swing	J. N. King	Murray, N. Y	May 29, 1855	12, 952
Bridge, Swing	A. R. Ring	Parma, N. Y	Nov. 9, 1838	1, 004
Bridge, Swing	J. Ross	Ipswich, Mass	Jan. 2, 1849	5, 997
Bridge, Swing	G. R. Winkler	Williamsport, Pa	Jan. 19, 1869	86, 117
Bridge, Traveling	F. Field	Adrian, Mich	Apr. 25, 1854	10, 817
Bridge, Trestle	A. Derrom	Paterson, N. J	July 4, 1865	48, 530
Bridge, Truss	J. S. Adams	Elgin, Ill	Aug. 30, 1870	106, 760
Bridge, Truss	J. Anderson	Gouverneur, N. Y	Mar. 18, 1873	136, 951
Bridge, Truss	J. P. Avery	Norwich, Conn	Nov. 5, 1861	33, 629
Bridge, Truss	A. Bannister	Alameda, Cal	July 22, 1873	141, 026
Bridge, Truss	W. Batchelder	Newburyport, Mass	July 11, 1865	48, 643
Bridge, Truss	G. O. Bishop	Hannibal, Mo	Apr. 24, 1860	27, 963
Bridge, Truss	J. Bole, 2d	Boston, Mass	May 19, 1863	38, 552
Bridge, Truss	W. F. Bonnell	Portland, Me	Aug. 20, 1872	130, 561
Bridge, Truss	D. C. Bower	Troy, Ohio	June 24, 1873	140, 181
Bridge, Truss	J. C. Briggs	Concord, N. H	Nov. 23, 1858	22, 106
Bridge, Truss	J. Brown, jr	Buffalo, N. Y	July 7, 1857	17, 722
Bridge, Truss	M. S. and H. B. Cartter	Saint Louis, Mo	June 14, 1870	104, 110
Bridge, Truss	M. S. and H. B. Cartter	Saint Louis, Mo	June 4, 1872	127, 564
Bridge, Truss	H. Childs	Henniker, N. H	Aug. 12, 1846	4, 693
Bridge, Truss	T. C. Clarke, A. Bonzano, and J. Griffen.	Philadelphia and Phœnixville, Pa.	July 1, 1873	140, 471
Bridge, Truss	A. Fink	Parkersburgh, Va	Mar. 3, 1857	16, 728
Bridge, Truss	A. Fink	Louisville, Ky	July 4, 1871	116, 787
Bridge, Truss	J. Foreman	Pottstown, Pa	June 9, 1868	78, 797
Bridge, Truss	J. Foreman	Pottstown, Pa	June 14, 1870	104, 295
Bridge, Truss	T. Hassard	New York, N. Y	Jan. 15, 1846	4, 359
Bridge, Truss	G. P. Herthel, jr	Saint Louis, Mo	Nov. 20, 1866	59, 769
Bridge, Truss	G. P. Herthel, jr	Saint Louis, Mo	Nov. 26, 1867	71, 484
Bridge, Truss	G. P. Herthel, jr	Saint Louis, Mo	May 18, 1869	90, 263
Bridge, Truss	W. Howe	Springfield, Mass	Aug. 28, 1846	4, 726
Bridge, Truss	W. Johnston	Lambertville, N. J	Nov. 29, 1870	109, 628
Bridge, Truss	J. L. Jones	Saint Louis, Mo	Nov. 6, 1860	30, 577
Bridge, Truss	J. L. Jones	Saint Louis, Mo	Aug. 4, 1863	39, 447
Bridge, Truss	J. H. Junkins	Upper Sandusky, Ohio	June 4, 1861	32, 480
Bridge, Truss	C. Kellogg	Athens, Pa	June 7, 1870	104, 036
Bridge, Truss	A. McGuffie	Rochester, N. Y	Mar. 25, 1862	34, 765
Bridge, Truss	A. McGuffie	Rochester, N. Y	May 27, 1862	35, 381
Bridge, Truss	J. A. McKay	Auburn, Ind	Feb. 7, 1871	111, 662
Bridge, Truss	T. W. Pratt	Boston, Mass	Apr. 25, 1871	114, 039
Bridge, Truss	F. Schwatka	United States Army	July 29, 1873	141, 293
Bridge, Truss	E. J. Story	Getzville, N. Y	Feb. 12, 1861	31, 415
Bridge, Truss	A. S. Swartz	Buffalo, N. Y	Sept. 22, 1857	18, 253
Bridge, Truss-frame	F. C. Lowthorp	Trenton, N. J	Feb. 19, 1867	62, 278
Bridge-trusses, Arrangement of arches in	C. M. Pennington	Rome, Ga	Jan. 7, 1851	7, 890
Bridge-trusses, Bearing-block for	J. L. Piper	Altoona, Pa	Oct. 22, 1861	33, 542
Bridge, Tubular	E. A. Baldwin	Elmira, N. Y	Aug. 8, 1854	11, 467
Bridge, Tubular	G. H. White	New York, N. Y	Oct. 25, 1870	108, 663
Bridge, Tubular arched	W. S. Levake	Cleveland, Ohio	July 5, 1870	104, 969
Bridge, Tubular arch	T. W. H. Moseley	Boston, Mass	May 31, 1870	103, 765
Bridge, Turn	J. Ingersoll	Grafton, Ohio	June 23, 1863	38, 966
Bridge, Turn	G. R. Winkler	Williamsport, Pa	Oct. 8, 1872	132, 038
Bridge, Wire-truss	C. Bender	Hesse-Darmstadt, Germany	Mar. 31, 1868	76, 041
Bridge, Wooden	S. K. Long	United States Engineers	Jan. 23, 1836	....

Index of patents issued from the United States Patent Office from 1790 *to* 1873, *inclusive*—Continued.

Invention.	Inventor.	Residence.	Date.	No.
Bridge, Wooden	G. W. Thayer	Springfield, Mass	Apr. 22, 1845	4, 004
Bridge, Wooden-framed brace	S. H. Long	United States Army	Nov. 7, 1839	1, 398
Bridge, Wooden-framed suspension	S. H. Long	United States Army	Nov. 7, 1839	1, 397
Bridge, Wooden or frame	S. H. Long	Baltimore, Md	Mar. 6, 1830	....
Bridge, Wooden truss	J. Burke	Saginaw, Mich	July 18, 1871	117, 042
Bridge, Wooden truss	T. W. Pratt	Boston, Mass	Apr. 1, 1873	137, 482
Bridge, Wrought-iron	A. P. Boller	Orange, N. J	Apr. 2, 1872	125, 117
Bridge, Wrought-iron	G. Heath	Little Falls, N. Y	May 27, 1862	35, 374
Bridge, Wrought-iron	J. H. Linville and J. L. Piper	Pittsburgh and Altoona, Pa	Oct. 31, 1865	50, 723
Bridges, Arched truss for	H. Lanergan	Boston, Mass	Apr. 23, 1850	7, 305
Bridges, Arched truss for	P. L. Weimer	Lebanon, Pa	Aug. 29, 1871	118, 566
Bridges across rivers, &c., Apparatus for passing suspension-wires for.	J. A. Roebling	Pittsburgh, Pa	Jan. 26, 1847	4, 945
Bridges, Bearing-block for truss	A. D. Briggs	Springfield, Mass	July 27, 1858	20, 987
Bridges, Building	T. Burr	Burr Haven, Pa	Apr. 3, 1817	....
Bridges, Building	J. Snyder	Union County, Pa	Feb. 3, 1834	....
Bridges, buildings, &c., Composition for covering wooden.	J. Heckel and M. Eichinger	Decatur, Ill	Apr. 7, 1868	76, 443
Bridges, buildings, &c., Method of preventing decay in the timbers of.	A. Allen	Cass County, Mich	Aug. 23, 1870	106, 647
Bridges, Cable-shackle for	T. G. Hulett	Niagara, N. Y	Dec. 8, 1868	84, 827
Bridges, &c., Cast-iron arch for	G. F. Lape	Summit, N. Y	Dec. 4, 1866	60, 199
Bridges, Cast-iron chord-connection for	C. Kellogg	Philadelphia, Pa	Feb. 23, 1869	87, 174
Bridges, Chord for	C. S. Smith	Baltimore, Md	Aug. 17, 1869	93, 917
Bridges, Connecting iron girders of	J. T. Ham	Covington, Ky	June 21, 1859	24, 460
Bridges, Connecting together the braces of truss	L. E. Truesdell	Warren, Mass	May 17, 1859	24, 068
Bridges, Connection for iron and steel	T. C. Clarke and A. Bonzano	Philadelphia, Pa	July 18, 1871	117, 047
Bridges, Connection for iron and steel	T. C. Clarke and A. Bonzano	Philadelphia, Pa	July 18, 1871	117, 048
Bridges, Connection for iron and steel	T. C. Clarke and A. Bonzano	Philadelphia, Pa	July 18, 1871	117, 049
Bridges, Connection for iron and steel	T. C. Clarke and A. Bonzano	Philadelphia, Pa	July 18, 1871	117, 050
Bridges, Connection for iron and steel	T. C. Clarke and A. Bonzano	Philadelphia, Pa	Dec. 12, 1871	121, 848
Bridges, Constructing	A. Canfield	Paterson, N. J	June 29, 1833	....
Bridges, Constructing	T. C. Clarke	Philadelphia, Pa	July 15, 1873	140, 888
Bridges, Constructing	G. Law	Easton, Pa	June 12, 1835	....
Bridges, &c., Constructing frame of	W. McKibbin	San Francisco, Cal	Mar. 9, 1858	19, 573
Bridges, Construction of	J. P. Bakewell	Pittsburgh, Pa	May 15, 1827	....
Bridges, Construction of	W. Bollman	Baltimore, Md	Jan. 6, 1852	8, 624
Bridges, Construction of	A. Bradway and E. Valentine	Monson, Mass	July 6, 1852	9, 090
Bridges, Construction of	B. Conner	Portsmouth, N. H	Apr. 23, 1812	....
Bridges, Construction of	J. J. Dyster	Philadelphia, Pa	Feb. 23, 1810	....
Bridges, Construction of	J. B. Gridley	Brooklyn, N. Y	July 6, 1852	9, 093
Bridges, Construction of	Z. King	Cleveland, Ohio	Nov. 15, 1864	45, 051
Bridges, Construction of	A. McGuffie	Rochester, N. Y	Dec. 17, 1861	33, 954
Bridges, Construction of	S. B. B. Nowlan	New York, N. Y	Mar. 18, 1873	136, 935
Bridges, Construction of	T. Pope	New York, N. Y	Apr. 18, 1807	....
Bridges, Construction of	C. S. Smith, C. H. Latrobe, and F. H. Smith.	Baltimore, Md	Dec. 14, 1869	97, 975
Bridges, Construction of	E. Stanley	Bennington, N. Y	Sept. 2, 1851	8, 337
Bridges, Construction of	C. W. Warner	Essex, Vt	Aug. 27, 1872	130, 959
Bridges, Construction of	H. Wilton	Wrightsville, Pa	June 24, 1839	1, 192
Bridges, Construction of iron-truss	S. Whipple	Utica, N. Y	Apr. 24, 1841	2, 064
Bridges, Driving piles for	J. Stone	....	Mar. 10, 1791	....
Bridges, Elliptical or oval truss-frame for	J. Barnes	Springfield, Mass	Mar. 27, 1849	6, 230
Bridges, &c., Fastening and combining the truss-frames of.	J. Price and J. T. Phillips	Golden, Md	Feb. 23, 1841	1, 994
Bridges, Gate of swinging	L. Anderson	Chicago, Ill	Dec. 14, 1869	97, 855
Bridges, Girder and chord for iron	T. B. Mills and B. M. Smith	Iola, Kan	Nov. 12, 1872	132, 975
Bridges, Hydraulic spindle and turning-apparatus for draw.	P. L. Fox and G. P. Herthel, jr.	Saint Louis, Mo	June 26, 1866	55, 844
Bridges, Iron abutment for	J. S. Goshorn	Fort Wayne, Ind	Feb. 7, 1871	111, 636
Bridges, Iron truss-frame for	F. C. Lowthorp	Trenton, N. J	June 30, 1857	17, 684
Bridges, &c., Iron truss-frame for	F. C. Lowthorp	Trenton, N. J	Oct. 3, 1857	18, 548
Bridges, Key or packing block for	N. Chapin	Chicago, Ill	Sept. 27, 1864	44, 398
Bridges, Lock bar for	L. E. Truesdell	Chicago, Ill	May 26, 1868	78, 403
Bridges, Manner of crossing rivers, &c., by moving platforms suspended to.	H. Leach	Philadelphia, Pa	Mar. 4, 1842	2, 478
Bridges, Manufacture of tension eye-bar for	F. H. Smith	Baltimore, Md	June 18, 1872	128, 184
Bridges, Metallic abutment for	A. Wheelock	Fort Wayne, Ind	Feb. 15, 1870	99, 989
Bridges, Metallic beam and girder for	J. Gill	Cincinnati, Ohio	Apr. 27, 1869	89, 400
Bridges, Metallic column for	A. B. Ives	Bloomington, Ill	Dec. 31, 1867	72, 859
Bridges, Metallic shoe for truss	D. H. Morrison	Dayton, Ohio	Apr. 27, 1858	20, 082
Bridges, Method of attaching the arch to the truss-frame in.	I. D. Steele	Pottstown, Pa	Feb. 20, 1849	6, 126
Bridges, Method of building	A. Cottrell	Newport, R. I	Nov. 10, 1841	2, 334
Bridges, &c., Method of increasing strength of beams or rafters of.	J. R. Remington	Lowndes County, Ala	May 19, 1843	3, 095
Bridges, Method of turning	W. N. Berkeley	Cedar Rapids, Iowa	June 5, 1866	55, 230
Bridges, Method of turning	W. N. Berkeley	Cedar Rapids, Iowa	June 5, 1866	55, 231
Bridges, Mode of adjustment of truss-frames of	J. W. Murphy	Philadelphia, Pa	Apr. 30, 1861	32, 199
Bridges, Mode of constructing, setting, and removing.	J. Du Bois	Williamsport, Pa	Oct. 7, 1862	36, 606
Bridges, Plate for securing chord, brace, &c., of truss.	F. C. Lowthorp	Trenton, N. J	Mar. 13, 1860	27, 457
Bridges, &c., Segmental truss for	G. S. Avery	Lewisborough, N. Y	July 28, 1857	17, 864
Bridges, Shoe for truss	M. Lassig	Chicago, Ill	Nov. 19, 1872	133, 252
Bridges, Timber-splice for	J. J. Hooker	Owego, N. Y	Oct. 15, 1872	132, 287
Bridges, Transporting	T. and S. Champion	Washington, D. C	Nov. 22, 1853	10, 250
Bridges, &c., Truss-beam for	P. M. Frees and Z. King	Cincinnati and Milan, Ohio	Oct. 1, 1861	33, 384
Bridges, Truss-frame for	W. Howe	Warren, Mass	July 10, 1840	1, 685
Bridges, Truss-frame for	W. Howe	Warren, Mass	Aug. 3, 1840	1, 711
Bridges, Truss-frame for	E. Jacobs	Cincinnati, Ohio	Jan. 3, 1860	26, 680
Bridges, Truss-frame for	J. H. Linville	Philadelphia, Pa	Dec. 2, 1873	145, 114
Bridges, Truss-frame for	T. W. and C. Pratt	Norwich, Conn., and Boston, Mass.	Apr. 4, 1844	3, 523
Bridges, Truss-girder for	S. D. Kendall	Brooklyn, N. Y	Jan. 21, 1862	34, 209
Bridges, Truss-girder for	A. McGuffie	Rochester, N. Y	Feb. 4, 1862	34, 311
Bridges, Tubular connection for	J. W. Sprague	Rochester, N. Y	Oct. 18, 1859	25, 852
Bridges, Turn-table for pivot	A. Bonzano	Phoenixville, Pa	May 21, 1872	127, 018

Index of patents issued from the United States Patent Office from 1790 *to* 1873, *inclusive*—Continued.

Invention.	Inventor.	Residence.	Date.	No.
Bridges, Turn-table for swing	T. C. Clarke and A. Bonzano	Philadelphia and Phœnixville, Pa.	Oct. 15, 1872	132, 254
Bridges, Turn-table for swing	G. Walter	Phœnixville, Pa	Mar. 5, 1872	124, 400
Bridges with draws, Method of building wooden	G. Wilkinson	White Creek, N. Y	May 15, 1827	....
Bridges, Wrought-iron pier for	T. B. Mills	Iola, Kans	Oct. 15, 1872	132, 307
Bridging navigable streams	B. F. Lee	New York, N. Y	Mar. 2, 1852	8, 781
Bridle	G. A. Albright and W. R. Burns	Lancaster, Pa	Nov. 27, 1866	59, 937
Bridle	A. Bauerschmitt	Rochester, N. Y	July 7, 1868	79, 628
Bridle	W. R. Beans	Brownsburgh, Pa	Oct. 15, 1867	69, 893
Bridle	M. H. Buchanan	Washington County, Va	Feb. 15, 1870	99, 834
Bridle	W. F. Clark	Hagaman's Mills, N. Y	Nov. 17, 1868	84, 170
Bridle	M. Haberbush and E. Kreckel	Lancaster, Pa	Nov. 27, 1866	59, 956
Bridle	J. C. Haines	Lewistown, Pa	July 10, 1866	56, 213
Bridle	J. Harris	Kansas, Ill	Apr. 16, 1867	63, 886
Bridle	S. B. Hartman	Millersville, Pa	June 28, 1864	43, 308
Bridle	S. B. Hartman	Millersville, Pa	Nov. 7, 1865	50, 822
Bridle	S. B. Hartman	Millersville, Pa	Nov. 13, 1866	59, 596
Bridle	P. Laporte	Richmond, Va	Aug. 15, 1822	....
Bridle	A. H. Langholz	Chicago, Ill	Feb. 3, 1863	37, 583
Bridle	J. F. Mason	Benton's Port, Iowa	May 28, 1867	65, 101
Bridle	J. McKibben	Lima, Ohio	Dec. 8, 1868	84, 838
Bridle	R. B. Norvell	Huntsville, Ala	May 10, 1859	23, 939
Bridle	J. A. Putt	Marlborough, Ohio	Nov. 26, 1867	71, 324
Bridle	A. H. Rockwell	Harpersville, N. Y	Dec. 4, 1866	60, 254
Bridle	E. R. Terry	New Haven, Conn	June 30, 1868	79, 334
Bridle	H. Seitz	Marietta, Pa	Sept. 26, 1848	5, 804
Bridle	R. Smether	Orange County, Va	Dec. 10, 1812	....
Bridle	J. H. Wilson	Brentwood, Tenn	Dec. 10, 1872	133, 817
Bridle and bit connection	C. B. Hogg	Boston, Mass	Mar. 8, 1864	41, 845
Bridle and bit, Horse	F. L. Haskell	Leominster, Mass	Sept. 3, 1872	130, 994
Bridle and halter, Combined	J. McKibben	Lima, Ohio	Mar. 5, 1867	62, 662
Bridle-attachment, Safety	R. Finck	Lancaster, Pa	Dec. 31, 1867	72, 729
Bridle-bit	S. D. Arnold	New Britain, Conn	Aug. 7, 1866	56, 872
Bridle-bit	J. B. Baker	Syracuse, N. Y	June 7, 1859	24, 275
Bridle-bit	D. S. Balch	Bradford, Vt	Aug. 23, 1839	1, 299
Bridle-bit	A. P. Baldwin	Newark, N. J	Dec. 1, 1868	84, 469
Bridle-bit	G. W. Barnes	Mount Vernon, N. Y	Mar. 19, 1872	124, 789
Bridle-bit	S. C. Boughton	Waterford, N. Y	Sept. 12, 1871	118, 900
Bridle-bit	J. Burns	East Topham, Vt	July 2, 1872	128, 532
Bridle-bit	D. H. Carpenter	Hartford, Conn	June 17, 1873	140, 010
Bridle-bit	H. Crane	New York, N. Y	June 5, 1860	28, 563
Bridle-bit	J. M. Crawford	Philadelphia, Pa	July 2, 1867	66, 221
Bridle-bit	O. Crook	Dayton, Ohio	Aug. 6, 1867	67, 509
Bridle-bit	B. I. Day	Gibson County, Ind	May 13, 1856	14, 856
Bridle-bit	E. Day	West Springfield, Mass	Nov. 28, 1865	51, 151
Bridle-bit	G. W. Eddy	Waterford, N. Y	Oct. 15, 1872	132, 149
Bridle-bit	A. B. Ely	Newton, Mass	Mar. 23, 1869	88, 021
Bridle-bit	J. M. Fabré	Bordeaux, France	June 17, 1873	140, 023
Bridle-bit	M. J. Firey	Mansfield, Ohio	June 14, 1870	104, 133
Bridle-bit	M. J. Firey	Mansfield, Ohio	June 14, 1870	104, 134
Bridle-bit	W. S. Ford	Clinton, Ill	May 12, 1868	77, 810
Bridle-bit	M. Foreacre	New Harrisburgh, Ohio	Aug. 2, 1870	106, 044
Bridle-bit	F. N. Frost	New Britain, Conn	Aug. 7, 1866	56, 923
Bridle-bit	J. P. Gates	Lincoln, Ill	Oct. 15, 1867	69, 910
Bridle-bit	A. Gilliam	Pittsburgh, Pa	July 11, 1871	116, 945
Bridle-bit	C. L. Ginkinger	Sterling, Ill	Apr. 15, 1873	137, 913
Bridle-bit	W. B. Hayden	Columbus, Ohio	Mar. 26, 1867	63, 156
Bridle-bit	N. Hayward	Roanoke, Ind	Jan. 2, 1872	122, 383
Bridle-bit	C. M. Huckins	East Topsham, Vt	Oct. 12, 1869	95, 801
Bridle-bit	W. F. and W. R. Johnson	Wetumpka, Ala	July 24, 1860	29, 284
Bridle-bit	F. B. Kalkbrenner	Clinton, Mo	Nov. 1, 1870	108, 795
Bridle-bit	S. M. King	Lancaster, Ohio	July 31, 1866	56, 762
Bridle-bit	A. H. Langholz	Chicago, Ill	Aug. 30, 1864	44, 002
Bridle-bit	J. Letchworth	Buffalo, N. Y	June 1, 1869	90, 857
Bridle-bit	J. Letchworth	Buffalo, N. Y	Jan. 7, 1873	134, 684
Bridle-bit	W. P. Letchworth	Buffalo, N. Y	May 10, 1870	102, 838
Bridle-bit	J. Lowbridge	Pittsburgh, Pa	Sept. 24, 1872	131, 691
Bridle-bit	A. P. Mason	Gowanda, N. Y	June 15, 1869	91, 463
Bridle-bit	P. S. McGuiness	New York, N. Y	June 2, 1868	78, 466
Bridle-bit	C. H. Miller	Buffalo, N. Y	Aug. 25, 1868	81, 395
Bridle-bit	J. H. Minnich	Tuscarawas, Ohio	Mar. 16, 1869	87, 864
Bridle-bit	A. Niel	Brooklyn, N. Y	Jan. 11, 1859	22, 571
Bridle-bit	D. M. Nixon	Danville, Ill	Feb. 23, 1869	87, 281
Bridle-bit	J. K. Norton	Flushing, Ohio	Feb. 4, 1868	74, 122
Bridle-bit	J. H. J. O'Neill	New Haven, Conn	July 7, 1863	39, 165
Bridle-bit	H. Pierce	Sharon, Conn	Mar. 15, 1834	....
Bridle-bit	E. N. Price	Salem, Mass	June 13, 1854	11, 083
Bridle-bit	J. C. Price	New Philadelphia, Ohio	Mar. 26, 1867	63, 297
Bridle-bit	W. S. Robbins	New Bedford, Mass	Dec. 8, 1868	84, 843
Bridle-bit	J. M. Roberds	Washington, D. C	Feb. 26, 1861	31, 557
Bridle-bit	A. H. Rockwell	Harpersville, N. Y	Nov. 12, 1867	70, 745
Bridle-bit	H. T. Romertze	Philadelphia, Pa	May 12, 1863	38, 541
Bridle-bit	B. L. Rowley	New Britain, Conn	Mar. 11, 1873	136, 767
Bridle-bit	P. Schoonmaker	New Britain, Conn	July 28, 1868	80, 510
Bridle-bit	J. Spoonhour and S. R. Boyd	Green Township and Chambersburgh, Pa.	Dec. 3, 1867	71, 806
Bridle-bit	C. E. Stockder	West Meriden, Conn	Sept. 8, 1863	39, 843
Bridle-bit	J. A. Swan	North Anson, Me	May 17, 1870	103, 103
Bridle-bit	H. C. Thompson	Mount Sterling, Ky	June 28, 1870	104, 793
Bridle-bit	W. D. Titus and R. W. Fenwick	Brooklyn, N. Y	Feb. 13, 1855	12, 397
Bridle-bit	A. Vanauken	Ludlowville, N. Y	Nov. 14, 1871	120, 913
Bridle-bit	Vicenzo, Count de Tergolina	London, England	July 29, 1873	141, 334
Bridle-bit	A. Viridet	Glasgow, Ky	May 25, 1869	90, 612
Bridle-bit	G. Webb	Lewiston, Me	Jan. 28, 1868	73, 853
Bridle-bit	A. S. Weymouth	Boston, Mass	July 30, 1861	32, 933
Bridle-bit	B. F. Wheeler	C[illegible]is, Vt	July 12, 1870	105, 282
Bridle-bit	R. P. Whelan	Leavenworth, Kans	Dec. 31, 1867	72, 950

Index of patents issued from the United States Patent Office from 1790 *to* 1873, *inclusive*—Continued.

Invention.	Inventor.	Residence.	Date.	No.
Bridle-bit	W. F. M. Williams	Augusta, Ga	Jan. 10, 1860	26, 804
Bridle-bit	L. D. Woodmansee	Mad River Township, Ohio	Jan. 1, 1867	60, 980
Bridle-bit attachment	C. M. Alexander	Washington, D. C	Feb. 11, 1862	34, 343
Bridle-bit attachment	J. D. Tracy	Springfield, Mass	Oct. 23, 1860	30, 510
Bridle-bit, Check-piece for	W. P. Wolfington	Louisville, Ky	Feb. 27, 1872	124, 187
Bridle-bit, Combination-lever	H. M. Cornell	Brighton, Ill	Nov. 14, 1871	120, 860
Bridle-bit, Flexible	B. L. Rowley	New Brighton, Conn	Apr. 14, 1868	76, 821
Bridle-bit, Method of connecting the check-piece to the mouth-piece of.	K. Frazer	Syracuse, N. Y	May 12, 1857	17, 268
Bridle-bit mouth-piece	A. P. Baldwin	Newark, N. J	Nov. 23, 1869	97, 022
Bridle-bit, Safety	S. S. Petersheim	Upper Leacock Township, Pa.	Apr. 30, 1872	126, 231
Bridle-bit slide-bar	A. P. Mason	Franklinville, N. Y	June 6, 1871	115, 753
Bridle-blind	J. Patterson, jr	Bangor, Me	Dec. 2, 1873	145, 233
Bridle blind, Harness	J. L. Brown	Connellsville, Pa	Oct. 18, 1870	108, 440
Bridle blind, Horse	J. B. Low	Homerville, Ohio	Dec. 5, 1871	121, 528
Bridle blind, Horse	J. G. Tibbets	New York, N. Y	Oct. 9, 1841	2, 285
Bridle-blind, Operating	A. Simis	Brooklyn, N. Y	Dec. 29, 1868	85, 406
Bridle, Breaking and training	W. O. Bartlett	New York, N. Y	June 5, 1866	55, 226
Bridle brow-band	F. Meinberg	New York, N. Y	Dec. 24, 1872	134, 213
Bridle check and brace	W. F. Pendleton	King and Queen County, Va	Feb. 20, 1833	......
Bridle, Driving	A. Rice and L. Leach	Fresno, Cal	Jan. 7, 1868	73, 042
Bridle for preventing horses from kicking	D. V. Grace and J. S. Elliott	Coshocton, Ohio	Feb. 9, 1869	86, 662
Bridle-front	W. T. Merserean	Orange, N. Y	Aug. 6, 1872	130, 144
Bridle-front, Machine for forming	I. Manning	Philadelphia, Pa	Dec. 18, 1866	60, 531
Bridle, Gag-runner connection for	J. C. Covert	Farmer Village, N. Y	Dec. 10, 1872	133, 834
Bridle-halter	G. W. Griswold	Logansport, Ind	July 22, 1862	35, 932
Bridle, Harness	G. Horter	New Orleans, La	July 5, 1870	105, 081
Bridle, Horse	J. Muller	Philadelphia, Pa	Dec. 23, 1873	145, 892
Bridle, Horse	M. A. Penn	Sumter, S. C	Apr. 30, 1872	126, 326
Bridle, Horse	J. C. Smith	Brookhaven, N. Y	Mar. 23, 1842	2, 510
Bridle, Nerving	J. V. Reardon	Elkton, Md	Mar. 23, 1869	88, 210
Bridle pulley-bit	W. Brower	Baltimore, Md	July 23, 1867	66, 941
Bridle-rein spring-snap	M. X. Tschus	Bloomington, Ill	July 5, 1859	24, 677
Bridle, Riding	J. C. Smith	Brookhaven, N. Y	Sept. 17, 1842	2, 780
Bridle-rosette	W. F. Niles	Leominster, Mass	Oct. 29, 1867	70, 251
Bridle-rosette	J. O'Brien	Geneseo, Ill	May 3, 1870	102, 579
Bridle-rosette	J. O'Brien	Cedar Rapids, Iowa	Oct. 25, 1870	108, 725
Bridle, Safety	G. W. Barnes	Mount Vernon, N. Y	Aug. 11, 1868	80, 897
Bridle, Safety	A. B. Christ and H. H. Stehman	Manor Township, Pa	Jan. 23, 1866	52, 139
Bridle, Safety	D. M. Donehoo	Beaver, Pa	Jan. 29, 1867	61, 522
Bridle, Safety	D. M. Donehoo	Beaver, Pa	July 2, 1867	66, 312
Bridle, Safety	B. R. Du Val	Portsmouth, Va	Aug. 27, 1872	130, 794
Bridle, Safety	J. M. Elder and P. A. Bishop	Elyria, Ohio	May 13, 1873	138, 872
Bridle, Safety	E. R. Ferry	New Haven, Conn	Oct. 13, 1868	83, 055
Bridle, Safety	F. A. Hannaford	New York, N. Y	Oct. 22, 1867	70, 089
Bridle, Safety	S. B. Hartman	Millersville, Pa	May 28, 1867	65, 216
Bridle, Safety	E. W. Lindeman	Manor Township, Pa	Sept. 24, 1867	69, 106
Bridle, Safety	J. McKillop	Brooklyn, N. Y	June 2, 1868	78, 533
Bridle, Safety	J. Weatherhead	San José, Cal	Mar. 7, 1871	112, 398
Bridle, Safety	S. V. R. York	Antwerp, N. Y	Dec. 22, 1868	85, 158
Bridle safety-strap	H. H. Rockwell	New London, Conn	Dec. 3, 1867	71, 648
Bridle to prevent horses from kicking or running away.	J. M. Lanier	Eufaula, Ala	Oct. 26, 1858	21, 889
Bridle-winker	W. and W. F. Boyd	Watertown, Mass	June 26, 1855	13, 119
Bridle winker, Harness	W. Boyd	Mansfield, Mass	July 19, 1870	105, 544
Bridle-winker, Metallic	M. O. Stanley	South Danvers, Mass	Sept. 6, 1864	44, 121
Bridle winkers, Manufacture of harness	E. Ward	Newark, N. J	May 24, 1870	103, 530
Bridles, Machine for punching, splitting, and creasing.	J. B Gathright	Louisville, Ky	Dec. 9, 1873	145, 292
Bridles, Manufacture of	D. A. Read	New York, N. Y	Oct. 24, 1826	......
Bridles, Metallic check-piece for	J. C. Baxter	Washington, D. C	Oct. 29, 1867	70, 315
Brine and other liquid evaporator	L. R. Cornell	Syracuse, N. Y	July 18, 1871	117, 052
Brine and other liquid evaporator	S. Platt	Goderich, Canada	Feb. 2, 1869	86, 443
Brine and other liquids, Process and apparatus for purifying.	G. Clark	Buffalo, N. Y	June 6, 1871	115, 573
Brine and salted meat to remove the salt, Treating	A. Whitelaw	Glasgow, Scotland	May 31, 1864	42, 988
Brine-evaporator	C. W. Atkeson	Henderson, Ky	June 16, 1857	17, 548
Brine-evaporator	D. Brigham	New York, N. Y	Apr. 27, 1858	20, 034
Brine-evaporator	J. Buchanan	Detroit, Mich	Dec. 6, 1870	109, 802
Brine-evaporator	S. D. Gilson	Syracuse, N. Y	Sept. 19, 1871	119, 136
Brine-evaporator	M. P. Hayes	Seaforth, Canada	Sept. 5, 1871	118, 718
Brine-evaporator	J. Heim	New York, N. Y	Mar. 8, 1870	100, 528
Brine-evaporator	R. G. Leckie	Acton Vale, Canada	Jan. 30, 1872	123, 182
Brine-evaporator	R. G. Leckie	Montreal, Canada	Jan. 30, 1872	123, 181
Brine-evaporator	J. McGrew	Ravenswood, W. Va	Nov. 19, 1872	133, 239
Brine, &c., evaporator and concentrator	R. G. Leckie	Acton Vale, Canada	May 16, 1871	114, 831
Brine evaporator, Salt	C. E. Tripler	Philadelphia, Pa	Sept. 24, 1872	131, 721
Bristle-assorting machinery	N. H. Spafford	Providence, R. I	Mar. 17, 1863	37, 927
Bristle-bundling machine	N. H. Spafford	Baltimore, Md	Apr. 17, 1866	54, 033
Bristle cleaning and assorting machine	G. E. Bart	Westford, Mass	Feb. 7, 1854	10, 498
Bristle-cleaning apparatus	H. W. Mosher and J. A. Conboie	New York, N. Y	Mar. 29, 1859	23, 385
Bristle-combing machine	L. F. Lannay and W. F. Parks	Indianapolis, Ind., and Baltimore, Md.	May 8, 1866	54, 564
Bristle-combing machine	N. H. Spafford	Baltimore, Md	Apr. 17, 1866	54, 034
Bristle-separator	A. Randel	New York, N. Y	Aug. 19, 1856	15, 573
Bristle washing and combing machine combined	L. F. Lannay and W. F. Parks	Indianapolis, Ind., and Baltimore, Md.	Dec. 1, 1868	84, 561
Bristles, felt, fur, wool, &c., Cutter and trimmer for	S. F. Harlow	Boston, Mass	June 30, 1868	79, 468
Bristles, Machine for assorting	N. H. Spafford	Baltimore, Md	Oct. 30, 1866	59, 286
Bristles, Machine for combing and assorting	N. H. Spafford	Baltimore, Md	Jan. 22, 1867	61, 480
Bristles, Machine for separating	N. H. Spafford	Baltimore, Md	Feb. 20, 1866	52, 763
Bristles, &c., Machine for washing	L. F. Lannay	Indianapolis, Ind	May 19, 1868	78, 102
Broaching-machine	A. P. Stephens	Brooklyn, N. Y	July 22, 1873	141, 091
Brogan	C. E. Tyler	Georgetown, Mass	Apr. 9, 1872	125, 505
Broiler	S. Bowers	Penn Yan, N. Y	July 21, 1868	80, 125
Broiler	W. F. Browne	New York, N. Y	June 29, 1869	92, 008
Broiler	G. Chilson	Boston, Mass	May 27, 1862	35, 361

Index of patents issued from the United States Patent Office from 1790 to 1873, inclusive—Continued.

Invention.	Inventor.	Residence.	Date.	No.
Broiler	L. H. Colborne and D. H. Lowe	New York, N. Y	Feb. 7, 1871	111, 516
Broiler	D. W. Denman and W. K. Tillotson.	Detroit, Mich	Mar. 29, 1870	101, 236
Broiler	A. Dick	Buffalo, N. Y	Nov. 1, 1870	108, 768
Broiler	J. M. Dick	Buffalo, N. Y	Mar. 24, 1863	38, 018
Broiler	B. G. Fitzhugh	Frederick, Md	Oct. 17, 1871	119, 974
Broiler	A. G. Gibson	Cincinnati, Ohio	Mar. 17, 1868	75, 673
Broiler	A. L. H. Graham	Chester, Pa	Mar. 29, 1870	101, 361
Broiler	J. Grossius and D. J. Mullaney.	Cincinnati, Ohio	Sept. 3, 1872	131, 057
Broiler	W. T. Howard	Baltimore, Md	Apr. 18, 1871	113, 886
Broiler	A. C. Hull	Saint Louis, Mo	Mar. 1, 1870	100, 410
Broiler	M. T. Hynes	Boston, Mass	Jan. 3, 1871	110, 763
Broiler	H. H. Johnson	New Haven, Conn	Aug. 6, 1867	67, 435
Broiler	W. J. Johnson and H. A. Hildreth.	Newton and Lowell, Mass	Oct. 19, 1869	96, 909
Broiler	G. H. Link and C. D. Curtis	Syracuse, N. Y	Nov. 16, 1869	96, 930
Broiler	J. Mallory	Penn Yan, N. Y	Nov. 3, 1868	83, 649
Broiler	B. Marshall	Marietta, Ohio	June 21, 1870	104, 611
Broiler	B. and S. N. Marshall	Marietta, Ohio	Oct. 15, 1872	132, 167
Broiler	J. C. Nobles	Ilion, N. Y	Nov. 7, 1871	120, 768
Broiler	J. T. Page	Rochester, N. Y	Aug. 9, 1870	106, 280
Broiler	J. T. Page	Rochester, N. Y	July 16, 1872	129, 362
Broiler	E. B. Phelps and J. P. McLean.	Brookyln, N. Y	Mar. 21, 1871	112, 846
Broiler	D. E. Roe	Elmira, N. Y	June 13, 1871	115, 984
Broiler	D. E. Roe	Elmira, N. Y	Aug. 9, 1870	106, 210
Broiler	G. S. Saxton	Saint Louis, Mo	Nov. 16, 1869	96, 973
Broiler	O. J. Smith	Wauwatosa, Wis	Oct. 10, 1871	119, 892
Broiler	O. J. Smith	Wauwatosa, Wis	Oct. 1, 1872	131, 910
Broiler	R. P. Smith	New York, N. Y	Mar. 5, 1872	124, 296
Broiler	W. S. Stiver and J. S. Williams.	Brooklyn, N. Y	June 7, 1870	104, 075
Broiler	C. Walsh	Newark, N. J	Dec. 12, 1871	121, 917
Broiler	J. Willging	Bubuque, Iowa	Dec. 31, 1872	134, 410
Broiler and furnace, Combined	J. S. Runyan	Columbus, Ohio	Aug. 9, 1870	106, 288
Broiler and toaster	W. F. Collier and J. H. Bigelow	Worcester, Mass	Mar. 24, 1868	76, 032
Broiler and toaster	T. C. Law	Green Island, N. Y	Nov. 21, 1865	51, 064
Broiler and toaster	J. M. Read and J. M. Smith	Boston, Mass	Dec. 27, 1870	110, 589
Broiler and toaster	M. H. Wiley	Boston, Mass	Mar. 5, 1872	124, 306
Broiler and toaster, Wire	H. A. Hildreth and W. J. Johnson.	Lowell and Newton, Mass.	Apr. 18, 1865	47, 302
Broiler and toaster, Wire	C. L. Prouty	Worcester, Mass	June 22, 1869	91, 664
Broiler, Meat	L. Holms	Keene, N. H	Mar. 24, 1868	75, 914
Broiler, Meat	G. T. Teel	Hoboken, N. J	Aug. 30, 1864	44, 026
Broiler, Meat	H. Willard	Grand Rapids, Mich	Apr. 26, 1870	102, 348
Broiler, Reversible	S. Smith	Brooklyn, N. Y	May 7, 1872	126, 585
Broiler, Steak	D. C. Teller	Terre Haute, Ind	July 16, 1867	66, 911
Broiler, Steak	G. W. Walker	Boston, Mass	Feb. 4, 1862	34, 339
Broiling-apparatus	M. E. A. W. Evard	Leesburgh, Va	Apr. 7, 1868	76, 314
Broiling-apparatus	O. F. Morrill	Boston, Mass	Dec. 6, 1859	26, 368
Broiling-furnace and cooking-range, Combined	W. Resor	Cincinnati, Ohio	Aug. 3, 1858	21, 085
Broiling, toasting, &c., Apparatus for	H. W. Harkness and W. A. Terry.	Bristol, Conn	Aug. 24, 1858	21, 297
Bromine and iodine, Apparatus for the manufacture of.	D Alter	Freeport, Pa	Feb. 26, 1867	62, 464
Bromine, Apparatus for making	H. Lerner	Pomeroy, Ohio	July 27, 1869	93, 099
Bromine, Apparatus for the manufacture of	H. Lerner and E. C. Harpold	Mason City and Hartford City, W. Va.	Oct. 15, 1872	132, 296
Bromine, Apparatus for the manufacture of	D. C. Turner	Clifton, W. Va	Apr. 1, 1873	137, 512
Bromine, Apparatus for the production of	H. Stieren and W. A. Nisbet	Mason, W. Va., and Natrona, Pa.	May 17, 1870	103, 253
Bromine from bittern, &c., Apparatus for producing	G. Leyer and J. A. Winter	Pomeroy, Ohio	Mar. 25, 1873	137, 222
Bromine from bittern, Manufacture of	G. A. Hagemann	Natrona, Pa	Sept. 22, 1868	82, 309
Bromine, Manufacture of	D. Alter and E. Gillespie	Freeport, Pa	July 5, 1848	5, 658
Bromine, Manufacture of	J. J. Jühler	Natrona, Pa	Jan. 3, 1871	110, 662
Brooch, ear-ring, &c	H. Oliver	Philadelphia, Pa	Feb. 22, 1859	23, 042
Brooch-fastening	W. Sackermann	New York, N. Y	Oct. 18, 1870	108, 520
Bronze and other matters, Process of removing earthy matters from.	M. Smith	Somerville, Mass	Dec. 13, 1870	110, 081
Bronze or metallic powder, Surfacing fabrics with	J. B. Batchelder	Boston, Mass	Apr. 27, 1869	89, 274
Bronze-powder manufacture	L. Brandeis	Brooklyn, N. Y	Apr. 3, 1866	53, 563
Bronze-powder, Process of making	L. Brandeis	New York, N. Y	Sept. 16, 1851	8, 365
Bronzing and gilding	J. L. Duffee	Washington, D. C	Apr. 19, 1870	102, 198
Bronzing columns, pillars, monuments, &c	L. Brandeis	Brooklyn, N. Y	July 16, 1872	129, 207
Bronzing-compound	A. Towne	Philadelphia, Pa	Nov. 12, 1872	132, 993
Bronzing-liquid	H. Hoffman	New York, N. Y	Oct. 6, 1857	18, 338
Bronzing-machine	G. H. Babcock	Westerly, R. I	Oct. 25, 1859	25, 874
Bronzing-machine	E. F. Benton	Buffalo, N. Y	July 11, 1871	116, 918
Bronzing-machine	E. F. Benton	Buffalo, N. Y	Oct. 3, 1871	119, 498
Bronzing-machine	E. F. Benton	Buffalo, N. Y	Dec. 26, 1871	122, 216
Bronzing-machine	L. G. Chaput	New York, N. Y	Aug. 13, 1872	130, 475
Bronzing-machine	L. G. Chaput and W. and J. Braidwood.	New York, N. Y	Nov. 18, 1873	144, 741
Bronzing-machine	S. Crump	Brooklyn, N. Y	Aug. 23, 1870	106, 667
Bronzing-machine	J. K. Lowe	Cleveland, Ohio	Aug. 28, 1866	57, 527
Bronzing-machine	J. K. Lowe	Cleveland, Ohio	Feb. 12, 1867	61, 044
Bronzing-machine	G. L. Mayes	Buffalo, N. Y	Jan. 29, 1867	61, 550
Bronzing machine	J. H. Nevins	Williamsburgh, N. Y	Oct. 3, 1871	119, 637
Bronzing-machine	I. L. G. Rice	Cambridge, Mass	Feb. 21, 1871	111, 973
Bronzing-machine	C. Shoop	Buffalo, N. Y	Apr. 23, 1872	126, 097
Bronzing-machine	H. Skidmore	Mount Vernon, N. Y	Feb. 9, 1869	86, 875
Bronzing-machine	J. F. Tapley	Springfield, Mass	Dec. 22, 1863	41, 029
Bronzing-machine	J. F. Tapley	Springfield, Mass	Nov. 19, 1867	71, 085
Bronzing-machine, Bronze-collecting attachment to	S. Crump	New York, N. Y	Apr. 12, 1870	101, 830
Bronzing or coloring iron, Process of	H. Tucker	Newton, Mass	Dec. 15, 1863	40, 964
Bronzing-pad	L. G. Chaput	New York, N. Y	Oct. 21, 1873	143, 880
Bronzing printed work	B. G. George	London, England	May 25, 1869	90, 521
Bronzing wall-paper, Machine for	W. G. Mackay	New York, N. Y	June 25, 1861	32, 667
Broom	E. A. Anderson	Danville, Tex	Aug. 15, 1871	117, 959
Broom, &c	J. H. Anderson	Terre Haute, Ind	Aug. 2, 1870	106, 021

Index of patents issued from the United States Patent Office from 1790 *to* 1873, *inclusive*—Continued.

Invention.	Inventor.	Residence.	Date.	No.
Broom	J. and G. Bacon	Medina, Wis	Nov. 28, 1865	51, 127
Broom	N. E. Badgley	New York, N. Y	Jan. 28, 1868	73, 769
Broom	W. N. Bates	Cedar Rapids, Iowa	Sept. 15, 1863	39, 874
Broom	J. D. Bell	Wattsborough, Va	June 24, 1873	140, 180
Broom	S. Bennett	Hartford, Conn	Jan. 2, 1866	51, 791
Broom	J. D. Blood	Amsterdam, N. Y	July 16, 1872	129, 089
Broom	C. Boeckh	Toronto, Canada	Mar. 11, 1873	136, 694
Broom	T. E. C. Brinley	Louisville, Ky	Apr. 18, 1871	113, 845
Broom	E. D. Bronson and M. W. Dillingham.	Amsterdam, N. Y	July 30, 1872	129, 926
Broom	J. Buercky	Overpeck's Station, Ohio	Apr. 14, 1868	76, 710
Broom	M. Burnett	Boston, Mass	June 29, 1869	91, 907
Broom	S. Carpenter	Cedar Rapids, Iowa	Mar. 22, 1864	41, 973
Broom	J. M. Clark	Dayton, Ohio	May 23, 1865	47, 799
Broom	E. P. Cooley	New York, N. Y	July 31, 1866	56, 717
Broom	R. E. Copson	Hamburgh, Iowa	Aug. 15, 1871	117, 986
Broom	E. M. Crandal	Marshalltown, Iowa	June 28, 1870	104, 834
Broom	E. M. Crandal	Chicago, Ill	Oct. 1, 1872	131, 854
Broom	R. F. Dobson	Darlington, Wis	Feb. 27, 1872	124, 040
Broom	R. F. Dobson	Goderich, Canada	Aug. 4, 1868	80, 613
Broom	R. W. English	Buffalo, N. Y	July 2, 1872	128, 609
Broom	T. R. Evans	Blacksburgh, Va	Dec. 5, 1871	121, 501
Broom	D. P. Farnham	Rock County, Wis	Apr. 24, 1866	54, 249
Broom	W. S. Hancock	Chicago, Ill	May 10, 1870	102, 936
Broom	A. C. Hoag	Clinton, Ill	Apr. 22, 1862	35, 023
Broom	H. C. Ingersoll	Bangor, Me	July 1, 1862	35, 759
Broom	D. Kaufman	Boiling Spring, Pa	Feb. 19, 1861	31, 463
Broom	O. W. Kellogg	Ripon, Wis	Mar. 7, 1865	46, 678
Broom	T. Langdon and H. C. Kellogg	Quasqueton, Iowa	May 14, 1861	32, 299
Broom	T. Langdon and C. Weitman	Hazleton, Iowa	July 17, 1860	29, 179
Broom	H. Lumbard	Chicago, Ill	Jan. 7, 1868	73, 107
Broom	W. A. Middleton	Harrisburgh, Pa	July 18, 1871	117, 192
Broom	C. T. Moore	Concord, N. H	July 27, 1852	9, 160
Broom	H. E. Newton	Manchester, N. H	Jan. 22, 1867	61, 446
Broom	D. J. Owen	Springville Lynn, Pa	May 31, 1859	24, 231
Broom	J. H. Parsons	Jonesville, Mich	Apr. 26, 1870	102, 310
Broom	F. Reese	Calera, Ala	Apr. 8, 1873	137, 721
Broom	C. L. Reid	Louisville, Ky	Mar. 21, 1871	112, 960
Broom	S. M. Sherman	Fort Dodge, Iowa	Aug. 6, 1861	33, 010
Broom	A. Sinclair	West Waterville, Me	July 13, 1869	92, 483
Broom	W. C. Spellman	Hartford, Conn	Mar. 1, 1870	100, 336
Broom	W. C. Spellman	Hartford, Conn	Nov. 7, 1871	120, 791
Broom	G. Stackpole	Elizabeth, N. J	Apr. 5, 1870	101, 675
Broom	S. Standish	Eureka, Nev	Feb. 25, 1873	136, 273
Broom	W. H. Towers	New York, N. Y	Sept. 4, 1860	29, 926
Broom	W. H. Towers	New York, N. Y	Nov. 5, 1861	33, 685
Broom	T. Walter	Philadelphia, Pa	Mar. 1, 1870	100, 343
Broom	C. Weitman	Independence, Iowa	May 14, 1861	32, 322
Broom	W. E. S. Whitmore	Augusta, Me	June 20, 1871	116, 126
Broom	A. Willis	Chicago, Ill	Aug. 8, 1871	117, 844
Broom	J. Wisner	Aurora, N. Y	June 5, 1866	55, 409
Broom	T. Wright	New York, N. Y	Nov. 13, 1866	59, 733
Broom	D. Young	Burlington, Ill	Nov. 28, 1865	51, 253
Broom and brush	R. Kinsley	Cabotville, Mass	May 17, 1836	......
Broom and brush	T. H. Powers	Wyocena, Wis	May 13, 1856	14, 883
Broom and brush handle fastening	E. F. Dunaway	Cincinnati, Ohio	May 23, 1871	115, 040
Broom and brush handle holder	F. S. Pinkham	Boston, Mass	Jan. 19, 1869	85, 956
Broom and brush head	G. T. Reed	Philadelphia, Pa	Aug. 21, 1866	57, 378
Broom and brush head	G. T. Reed	Philadelphia, Pa	Mar. 5, 1867	62, 686
Broom and brush holder	G. B. Cunningham	Northampton, Mass	June 11, 1872	127, 743
Broom and brush holder	B. D. Wallace	Boston, Mass	Nov. 2, 1869	96, 516
Broom and brush holding clamp	A. B. Thompson	Owego, N. Y	Mar. 22, 1870	101, 062
Broom and brush, Tying	J. H. Anderson	Terre Haute, Ind	Mar. 21, 1871	113, 060
Broom and mop head	E. J. Green	Valparaiso, Ind	Oct. 10, 1865	50, 352
Broom and mop holder	H. L. Franklin and E. Clark	Nashua, N. H	Jan. 11, 1870	98, 756
Broom and mop holder	G. B. Isham	Burlington, Vt	Nov. 14, 1871	120, 818
Broom and scraper	D. Elliot	New York, N. Y	Sept. 14, 1869	94, 728
Broom and whisk	E. P. Cooley	New York, N. Y	June 14, 1870	104, 118
Broom-bag	E. D. Bronson	Amsterdam, N. Y	Feb. 18, 1873	136, 026
Broom-band	F. G. Harding	Boston, Mass	Sept. 10, 1872	131, 270
Broom-brace	F. O. Willey	Wilmette, Ill	Feb. 22, 1870	100, 238
Broom-bridle	J. F. Blondel	Thomaston, Me	Dec. 31, 1872	134, 459
Broom-bridle	J. H. Subers and J. H. Troup	Philadelphia, Pa	Oct. 17, 1871	120, 122
Broom-clamp	W. L. Babcock	Erie, Pa	June 25, 1872	128, 204
Broom-clamp	T. F. Boyer	Harrisburgh, Pa	Apr. 10, 1866	53, 776
Broom-clamp	J. Day	Murray, N. Y	Nov. 27, 1866	59, 977
Broom-clamp	J. Day	Holley, N. Y	Oct. 15, 1867	69, 780
Broom-clamp	A. L. Harwood	Malone, N. Y	May 22, 1866	54, 897
Broom-clamp	H. B. Lyon and G. M. Hopkins	Albion, N. Y	Mar. 5, 1867	62, 548
Broom-clamp	S. Mason	Indian Springs, Md	Mar. 24, 1857	16, 877
Broom-clamp	W. D. Merick	Rochester, N. Y	Apr. 9, 1872	125, 633
Broom-clasp	N. Homes	Laona, N. Y	May 14, 1861	32, 292
Broom-clasp	P. H. Miles	Boston, Mass	June 5, 1866	55, 345
Broom-clasp	S. Parr	Boston, Mass	Dec. 19, 1865	51, 614
Broom-clasp	P. B. Sheldon	Prattsburgh, N. Y	June 28, 1859	24, 582
Broom-clasp	H. Tilden	Boston, Mass	June 5, 1866	55, 394
Broom, Corn	C. L. W. Baker	Hartford, Conn	June 11, 1861	32, 504
Broom-corn and sorghum stripper	K. W. Dodena	Millwood Township, Ohio	Mar. 13, 1866	53, 126
Broom corn-breaking machine	C. Campbell	Yellow Head, Ill	May 13, 1862	35, 214
Broom-corn cutting and assorting machine	M. D. Bradley	Ballston Spa, N. Y	July 28, 1868	80, 443
Broom-corn duster	J. L. Stranahan	Chicago, Ill	Apr. 15, 1873	137, 895
Broom-corn, Machine for assorting	L. D. Grosvenor	Harvard, Mass	Jan. 1, 1851	7, 872
Broom-corn, Machine for combing seed off	G. E. Burt	Harvard, Mass	Apr. 8, 1856	14, 596
Broom-corn, Machine for cutting and separating	A. Walrath and J. Snell	Fort Plain, N. Y	Sept. 3, 1872	131, 178
Broom-corn, Machine for removing the seed from	J. D. Brown	Preble County, Ohio	Sept. 3, 1867	68, 346
Broom-corn, Machine for sizing	O. M. Truair	Mount Morris, N. Y	Dec. 17, 1861	33, 968
Broom-corn, Machine for stripping seed from	L. D. Grosvenor	South Groton, Mass	Sept. 23, 1851	8, 375
Broom-corn-scraping machine	B. J. and J. S. Billings	Gorham, N. Y	Oct. 15, 1831	......

Index of patents issued from the United States Patent Office from 1790 *to* 1873, *inclusive*—Continued.

Invention.	Inventor.	Residence.	Date.	No.
Broom-corn-seed stripper	G. E. Burt and E. A. Hildreth	Harvard, Mass	Sept. 20, 1870	107, 445
Broom-corn-stripping machine	J. B. Stine	Rohrersville, Md	Sept. 24, 1872	131, 718
Broom-corn to stripper, Machine for feeding	S. C. Kenaga	Kankakee, Ill	Oct. 15, 1872	132, 158
Broom-crotch supporter	H. Hopkins	Wellfleet, Mass	Aug. 30, 1870	106, 822
Broom, Detachable	G. H. Pierce	Richmond, Canada	Aug. 20, 1872	130, 594
Broom-envelope	E. M. Crandal	Chicago, Ill	Oct. 1, 1872	131, 855
Broom-handle	J. Strong	College Hill, Ohio	Sept. 20, 1870	107, 559
Broom-handle	W. H. Towers	New York, N. Y	Oct. 15, 1861	33, 502
Broom-handle socket	H. Achew	Cincinnati, Ohio	Dec. 20, 1870	110, 325
Broom-handles, Clamp for splicing	C. L. W. Baker	Camden, N. J	Dec. 30, 1873	145, 041
Broom-handles, Machine for sawing sticks for	T. J. Alexander	Westerville, Ohio	Sept. 20, 1853	10, 023
Broom-handles, Machine for turning	G. M. Morrow	Clarksville, Ohio	Oct. 13, 1868	83, 082
Broom-handles, moldings, &c., Apparatus for painting or coating.	S. A. Shepard	Pittsburgh, Pa	May 13, 1873	138, 767
Broom-hanger	M. A. Clifford	Boston, Mass	Sept. 23, 1873	143, 063
Broom-hanger	C. F. Lewis	Wakeman, Ohio	Oct. 8, 1872	132, 089
Broom-hanger	J. L. Patterson	Wheeler Station, Ind	Jan. 1, 1867	60, 779
Broom-head	A. Alden	East Cambridge, Mass	Apr. 14, 1868	76, 688
Broom-head	E. A. Alexander and H. C. Kellogg.	Buchanan County, Iowa	Oct. 1, 1867	69, 300
Broom-head	L. Allen	Berkeley Springs, W. Va	July 9, 1867	66, 441
Broom-head	J. M. Allison	Salina, Pa	July 27, 1869	93, 033
Broom-head	J. D. M. Armbrust	Apolloborough, Pa	June 26, 1866	55, 948
Broom-head	F. G. Bakes	Vevay, Ind	July 10, 1866	56, 162
Broom-head	H. Barker	Aurora, Ind	Apr. 23, 1867	63, 983
Broom-head	E. M. Bayne	Uniontown, Pa	Nov. 6, 1866	59, 343
Broom-head	C. Blom, jr., and J. Aling	Holland, Mich	Dec. 5, 1871	121, 578
Broom-head	F. C. Blender	Lima, Ohio	June 13, 1865	48, 232
Broom-head	W. P. Brooks	Fairmont, Minn	Dec. 18, 1866	60, 615
Broom-head	D. J. Brougher	Harrisburgh, Pa	Dec. 26, 1865	51, 689
Broom-head	J. D. Browne	Cincinnati, Ohio	Oct. 24, 1865	50, 554
Broom-head	J. Buchanan	Aurora, Ind	Nov. 21, 1865	51, 013
Broom-head	J. Buchanan	Aurora, Ind	Apr. 24, 1866	54, 104
Broom-head	H. Buck	Harrisburgh, Pa	Jan. 30, 1866	52, 263
Broom-head	H. Buck	Harrisburgh, Pa	May 8, 1866	54, 498
Broom-head	J. T. Carpenter	Harrisburgh, Pa	Feb. 26, 1867	62, 471
Broom-head	F. B. Carroll	Noblesville, Ind	Jan. 1, 1867	60, 688
Broom-head	O. L. Castle	Upper Alton, Ill	Mar. 26, 1867	63, 211
Broom-head	J. O. Clay	Hudson, Wis	Apr. 3, 1866	53, 576
Broom-head	B. C. Cook	Richmond, Va	Feb. 4, 1873	135, 525
Broom-head	N. B. Cooper	Gratis Ohio	Feb. 27, 1866	52, 823
Broom-head	J. Danner	Canton, Ohio	Apr. 10, 1866	53, 791
Broom-head	B. F. Early	Palmyra, Pa	Nov. 6, 1866	59, 372
Broom-head	J. S. Elkins	Marquette, Wis	Apr. 17, 1866	53, 962
Broom-head	S. S. Evans	El Paso, Ill	June 26, 1866	55, 843
Broom-head	D. P. Farnham	Janesville, Wis	July 30, 1867	67, 180
Broom-head	J. U. Fiester	Winchester, Ohio	Jan. 2, 1866	51, 819
Broom-head	C. Fiscus	Washington, D. C	Apr. 25, 1871	114, 120
Broom-head	G. W. Golay	Vevay, Ind	June 19, 1866	55, 769
Broom-head	S. B. Guernsey	Chicago, Ill	May 1, 1866	54, 332
Broom-head	N. Hall	New Market, Md	Mar. 20, 1866	53, 296
Broom-head	C. C. Hand	Cincinnati, Ohio	June 27, 1865	48, 392
Broom-head	M. Haneline	Clear Creek, Ind	June 26, 1866	55, 856
Broom-head	J. Harris	Marquette, Wis	May 1, 1866	54, 338
Broom-head	J. Harris	Marquette, Wis	Oct. 23, 1866	59, 011
Broom-head	A. Harroun, jr	Onondaga, N. Y	Aug. 21, 1866	57, 320
Broom-head	G. W. Hoffman	Harrisburgh, Pa	Dec. 5, 1865	51, 318
Broom-head	J. M. Hotaling	Waterport, N. Y	Jan. 29, 1867	61, 621
Broom-head	J. S. Hugg	Philadelphia, Pa	Feb. 19, 1867	62, 202
Broom-head	W. G. Hughes	Hebron, Ind	May 21, 1867	64, 979
Broom-head	B. Hunt	Farmland, Ind	Apr. 16, 1867	63, 899
Broom-head	J. E. Hunter	Mechanicsburgh, Ohio	Apr. 3, 1866	53, 624
Broom-head	J. W. Keene and W. E. Snediker	Utica, N. Y	Aug. 7, 1866	56, 953
Broom-head	E. Kelley	Locust Grove, Ohio	July 16, 1867	66, 715
Broom-head	J. Kiefer	Hamilton, Ohio	May 22, 1866	54, 921
Broom-head	M. C. Kilgore	Washington, Iowa	May 8, 1866	54, 559
Broom-head	A. B. King	Seven Mile, Ohio	Mar. 27, 1866	53, 452
Broom-head	I. Kohn	Edgerton, Ohio	Aug. 27, 1867	68, 209
Broom-head	J. Liget	Posey Township, Ind	June 5, 1866	55, 316
Broom-head	J. H. Lightner	Shirleysburgh, Pa	July 3, 1866	56, 067
Broom-head	J. C. McLelland and J. Graham	Pittsburgh, Pa	Nov. 27, 1866	60, 030
Broom-head	C. Messenger	Chicago, Ill	June 12, 1866	55, 519
Broom-head	C. Messenger	Chicago, Ill	Dec. 4, 1866	60, 221
Broom-head	W. A. Middleton	Harrisburgh, Pa	July 31, 1866	56, 859
Broom-head	W. A. Middleton	Harrisburgh, Pa	Apr. 12, 1870	101, 901
Broom-head	C. E. Miller	Amelia, Ohio	Oct. 18, 1864	44, 736
Broom-head	C. E. Miller	Cincinnati, Ohio	June 13, 1865	48, 195
Broom-head	C. E. Miller	Cincinnati, Ohio	Dec. 19, 1865	51, 658
Broom-head	H. B. Miller and M. P. Weston	Grand Rapids, Mich	Mar. 12, 1867	62, 768
Broom-head	J. A. Miller	Somerville, Ohio	Oct. 24, 1865	50, 610
Broom-head	T. Moore	Bloomington, Ill	Nov. 27, 1866	60, 038
Broom-head	W. B. Moore	Philadelphia, Pa	July 17, 1866	56, 434
Broom-head	J. H. Mumma	Harrisburgh, Pa	Apr. 3, 1866	53, 654
Broom-head	D. P. Myers	Salem, Ill	Mar. 20, 1866	53, 323
Broom-head	G. R. Nebinger	Lewisberry, Pa	May 1, 1866	54, 391
Broom-head	M. V. Nobles and J. Holcomb	Elmira, N. Y., and Towanda, Pa.	Aug. 13, 1867	67, 793
Broom-head	D. J. Owen and J. G. Brydges	Springville, Pa	Feb. 11, 1873	135, 661
Broom-head	T. G. Packer	Mexico, N. Y	Aug. 6, 1867	67, 573
Broom-head	W. Paine and R. E. Caviness	Fairfield, Iowa	July 17, 1866	56, 461
Broom-head	G. W. Parsons	Harrisburgh, Pa	Apr. 24, 1866	54, 200
Broom-head	G. W. Parsons	Harrisburgh, Pa	Jan. 9, 1867	51, 964
Broom-head	D. Peters	Eaton, Ohio	May 15, 1866	54, 765
Broom-head	T. H. Powers	Milwaukee, Wis	June 27, 1865	48, 434
Broom-head	M. Quinby and J. E. Sturdevant.	Skinner's Eddy, Pa	July 17, 1866	56, 446
Broom-head	J. W. Sanford	Bath, N. Y	Aug. 20, 1867	67, 916
Broom-head	S. S. and J. G. Sherman	McHenry, Ill	Apr. 10, 1866	53, 893

Index of patents issued from the United States Patent Office from 1790 *to* 1873, *inclusive*—Continued.

Invention.	Inventor.	Residence.	Date.	No.
Broom-head	J. E. Short	New Richmond, Ohio	Jan. 30, 1866	52, 330
Broom-head	A. Silvers	Collinsville, Ohio	Dec. 19, 1865	51, 663
Broom-head	J. A. Sinclair	Woodsfield, Ohio	July 23, 1867	67, 075
Broom-head	G. Smith	Woodstown, N. J	Aug. 29, 1871	118, 493
Broom-head	L. S. Smith	Gorsuch's Mills, Md	Nov. 19, 1867	71, 238
Broom-head	G. T. Spaulding	Brodhead, Wis	Apr. 17, 1866	54, 035
Broom-head	W. C. Spellman	Baltimore, Md	Apr. 13, 1869	88, 922
Broom-head	W. C. Spellman	Providence, R. I	May 31, 1870	103, 791
Broom-head	W. D. Stroud	Oshkosh, Wis	July 3, 1866	56, 119
Broom-head	W. D. Stroud	Oshkosh, Wis	Feb. 5, 1867	61, 774
Broom-head	J. Taylor and R. M. Lafferty	Three Rivers, Mich	Sept. 11, 1866	57, 998
Broom-head	S. D. Thurston	Somerville, Ohio	Dec. 19, 1865	51, 631
Broom-head	H. Trumbull	Central College, Ohio	July 24, 1866	56, 640
Broom-head	H. H. B. Vincent	Oshkosh, Wis	Dec. 4, 1866	60, 291
Broom-head	J. Wasson	Fairwater, Wis	June 26, 1866	55, 936
Broom-head	J. Wisner and T. Rose	Aurora and Cortlandville, N. Y.	Oct. 15, 1867	69, 884
Broom head and clamp	M. W. Owens	Waterford, Pa	Aug. 21, 1866	57, 369
Broom-head, Metallic	O. L. Castle	Upper Alton, Ill	Oct. 4, 1864	44, 515
Broom-head stock	A. D. Forbes	Rockford, Ill	June 5, 1866	55, 269
Broom-holder	F. B. Batchelder	Prairie du Chien, Wis	May 26, 1868	78, 252
Broom-holder	C. T. Beardsley	Hamden, Conn	June 20, 1871	116, 140
Broom-holder	W. H. Bixler	Easton, Pa	Nov. 22, 1870	109, 375
Broom-holder	Z. Howe	Lowell, Mich	Jan. 10, 1871	110, 917
Broom-holder	H. B. McCool	Pottsville, Pa	Mar. 12, 1872	124, 445
Broom-holder	J. G. Powell and W. A. Morse	Philadelphia, Pa	Oct. 8, 1867	69, 586
Broom-holder	C. P. Snow	Freeport, Ill	June 28, 1870	104, 787
Broom-holder	L. W. Turner	Yalesville, Conn	May 25, 1869	90, 411
Broom-holder	H. W. Warner	Watertown, Conn	July 2, 1867	66, 431
Broom-jaw	I. Webster	Bucksport, Me	June 15, 1869	91, 290
Broom machine, Splint	J. Crum and A. Larwill	Ramapo, N. Y	Mar. 27, 1849	6, 233
Broom-making	T. Floyd	Chambersburgh, Va	Jan. 26, 1858	19, 190
Broom-needle	C. M. Cowardin	Gardiner Station, Tenn	Oct. 10, 1871	119, 745
Broom or brush	M. L. Byrn	New York, N. Y	Dec. 12, 1865	51, 423
Broom or brush	J. H. Power	Middletown, Iowa	July 10, 1860	29, 099
Broom or brush head	J. E. Phillips	Philadelphia, Pa	July 4, 1865	48, 587
Broom or brush holder	A. G. Davis	Watertown, Conn	Jan. 7, 1868	73, 169
Broom or mop clamp	E. Chapman	Lacon, Ill	Sept. 13, 1870	107, 336
Broom or mop clamp	E. Chapman	Salisbury, Mo	Jan. 30, 1872	123, 083
Broom or mop holder	H. L. Franklin and E. Clark	Nashua, N. H	Sept. 14, 1869	94, 734
Broom or whisk, Clothes	B. Morahan	Brooklyn, N. Y	July 9, 1867	66, 611
Broom-protector	G. Hunt	New York, N. Y	Apr. 15, 1873	137, 774
Broom, Shear	C. Randall, J. F. Moore, and J. Randall.	West Eau Claire, Wis	Feb. 2, 1869	86, 588
Broom, Splint	J. W. Wheeler	Cleveland, Ohio	May 24, 1859	24, 181
Broom, Stable	E. J. Richmond and T. Wright	New York, N. Y	Apr. 16, 1861	32, 083
Broom, Stable, &c	T. Wright	New York, N. Y	Feb. 1, 1870	99, 512
Broom-stand	E. A. Harris	Chicago, Ill	Aug. 29, 1871	118, 451
Broom-straw, Coloring and toughening	S. Greger	Cleveland, Ohio	Nov. 12, 1872	132, 961
Broom-tying machine	J. W. Congdon	Marietta, Ohio	Sept. 12, 1871	118, 845
Broom, Whisk	E. P. Cooley	New York, N. Y	May 9, 1871	114, 646
Broom, Whisk	H. A. Lee	New York, N. Y	Mar. 5, 1872	124, 365
Broom, Whisk	R. Y. Martin	New York, N. Y	May 9, 1871	114, 697
Broom, Whisk	G. Stackpole	Elizabeth, N. J	Mar. 21, 1871	112, 976
Broom, Whisk	G. Stackpole	Elizabeth, N. J	Apr. 4, 1871	113, 358
Brooms, Attachment for	J. H. Mumma	Harrisburgh, Pa	Nov. 21, 1865	51, 074
Brooms, billiard-cues, &c., Rack for	E. Richmond	Brooklyn, N. Y	Sept. 17, 1867	69, 026
Brooms, brushes, mops, &c., Machine for making	J. M. Spooner	Belchertown, Mass	Apr. 28, 1838	718
Brooms, Construction of	A. Mitchell	Eaton, Pa	Jan. 5, 1858	19, 039
Brooms, Device for the construction of	S. M. Sherman	Fort Dodge, Iowa	Apr. 28, 1863	38, 341
Brooms, Flanged collar for	H. A. Lee	New York, N. Y	Oct. 10, 1871	119, 775
Brooms, Gudgeon for	N. Robbins, jr	Rockport, Mass	July 9, 1867	66, 633
Brooms, &c., Machine for cutting splints for manufacture of.	L. Gleason	Le Roy, N. Y	Oct. 9, 1841	2, 289
Brooms, Machine for making	J. H. Hinton	Lancaster, Pa	Mar. 13, 1844	3, 483
Brooms, Machine for making	S. Rowe	Baltimore, Md	Dec. 1, 1857	18, 770
Brooms, Machine for making	J. Thomas	West Chester, Pa	Sept. 18, 1849	6, 717
Brooms, Machine for making	S. H. Weed	Poughkeepsie, N. Y	Feb. 3, 1819	....
Brooms, Machine for manufacturing splints for	J. W. Wheeler	Cleveland, Ohio	Apr. 13, 1858	19, 971
Brooms, Machine for winding wire or twine on	W. Beaman	Salem, N. J	Aug. 17, 1843	3, 219
Brooms, Machinery for making	H. Warner	Hamilton County, Ohio	Feb. 15, 1848	5, 444
Brooms, Manufacture of	E. P. Marble	Sutton Centre, Mass	Jan. 31, 1871	111, 464
Brooms, Manufacture of	W. B. Walker	Bennington, N. H	Aug. 1, 1854	11, 451
Brooms, Metallic clasp or head for	I. Cheney	Leyden, Mass	June 25, 1839	1, 198
Brooms, Metallic head for	H. G. Smith	Muscatine, Iowa	Apr. 3, 1860	27, 746
Brooms, Metallic tip for suspending	M. C. Hawkins	Edenborough, Pa	Oct. 8, 1867	69, 664
Brooms or brushes, Manufacturing	D. Wooster	China, N. Y	Feb. 16, 1829	....
Brooms, Preparing wisps for	J. E. Phillips	Philadelphia, Pa	May 8, 1866	54, 591
Brooms, Screw-ferrule for	M. W. Dillingham	Amsterdam, N. Y	Nov. 5, 1872	132, 817
Brooms, &c., Shape and formation of hickory	A. Sclater	Oxford Township, Pa	Feb. 15, 1826	....
Brooms, Sprinkling-attachment for	P. Louis	New York, N. Y	Dec. 18, 1866	60, 637
Brush	L. Abbott	Lewiston, Me	Nov. 26, 1872	133, 347
Brush	A. Alden	New York, N. Y	May 23, 1865	47, 781
Brush	J. B. Alden	Worcester, Mass	Mar. 19, 1867	62, 987
Brush	J. Ames	Lansingburgh, N. Y	Mar. 7, 1871	112, 404
Brush	J. Ames, jr	Lansingburgh, N. Y	Aug. 13, 1872	130, 402
Brush	C. L. W. Baker	Camden, N. J	Nov. 11, 1873	144, 496
Brush	J. A. Bell	Tyrone, Pa	July 16, 1872	129, 205
Brush	C. Boeckh	Buffalo, N. Y	Sept. 21, 1869	94, 940
Brush	J. Boggs	Harrisburgh, Pa	May 1, 1866	54, 284
Brush	S. Brillinger	Clarence Centre, N. Y	Sept. 21, 1869	94, 941
Brush	A. W. Brown	Lansingburgh, N. Y	Nov. 17, 1868	84, 166
Brush	J. Buercky	Overpeck's Station, Ohio	Oct. 29, 1867	70, 161
Brush	W. W. Clark	New York, N. Y	Feb. 4, 1868	74, 048
Brush	R. J. Combs	Bergen, N. J	Oct. 11, 1870	108, 108
Brush	I. Cross	New London, Ohio	July 5, 1853	9, 827
Brush	J. F. W. Dorman	Baltimore, Md	June 9, 1868	78, 727
Brush	A. C. Estabrook	Florence, Mass	June 19, 1866	55, 764

Index of patents issued from the United States Patent Office from 1790 *to* 1873, *inclusive*—Continued.

Invention.	Inventor.	Residence.	Date.	No.
Brush	J. A. Fanshawe and J. A. Jaques.	Tottenham, England	Jan. 7, 1862	34,109
Brush	L. P. Faught	Foxborough, Mass	May 1, 1866	54,464
Brush	D. Fleming	Brooklyn, N. Y	Mar. 5, 1861	31,593
Brush	W. H. Forker	Meadville, Pa	Aug. 13, 1867	67,745
Brush	J. A. Grant	Ottawa, Canada	Nov. 14, 1871	120,964
Brush	H. A. Harvey	Orange, N. J	Nov. 21, 1871	121,050
Brush	J. Hayman	Poole, England	Dec. 24, 1872	134,278
Brush	D. L. Keller	Norwalk, Ohio	Mar. 14, 1871	112,718
Brush	L. Kissling	East Cambridge, Mass	July 16, 1872	129,143
Brush	D. W. Lapham	Baltimore, Md	Oct. 31, 1871	120,523
Brush	J. C. Lawrence	Brooklyn, N. Y	Apr. 24, 1866	54,179
Brush	T. J. Mayall	Roxbury, Mass	Mar. 5, 1861	31,613
Brush	D. A. McDonel and P. Z. Klock	Ashland, Pa	Oct. 22, 1872	132,479
Brush	F. McLaughlin	Boston, Mass	Sept. 17, 1867	69,008
Brush	F. McLaughlin	Boston, Mass	Jan. 11, 1870	98,787
Brush	J. Michales, jr., and R. J. Combs	Bergen, N. J	Dec. 26, 1871	122,186
Brush	J. Minetree	Petersburgh, Va	Sept. 12, 1871	118,871
Brush	S. Morris	Charlestown, Mass	Aug. 4, 1863	39,417
Brush	F. Murrow	Williamsburgh, N. Y	Apr. 27, 1852	8,911
Brush	W. M. Newton	Baltimore, Md	Aug. 18, 1868	81,101
Brush	C. W. Palmer	Cleveland, Ohio	June 3, 1873	139,474
Brush	J. and F. Pirrung	Chicago, Ill	Dec. 12, 1871	121,811
Brush	C. F. Russet	Brooklyn, N. Y	July 5, 1870	105,130
Brush	K. Shaler	Madison, Conn	Dec. 21, 1858	22,381
Brush	J. H. Tatum	New York, N. Y	May 11, 1858	20,226
Brush	W. Tusch	Brooklyn, N. Y	May 22, 1860	28,4[illegible]
Brush	C. Twyford	Red Bank, N. J	Mar. 29, 1864	42,131
Brush	A. Van Dusen	Chicago, Ill	Aug. 15, 1865	49,457
Brush	P. Wagner	New York, N. Y	Apr. 17, 1873	138,213
Brush	A. M. White	Port Chester, N. Y	Apr. 4, 1865	47,157
Brush	A. M. White	New York, N. Y	May 21, 1867	65,029
Brush	J. S. White	Portland, Me	Jan. 21, 1873	135,053
Brush	J. L. Whiting	Portland, Me	Aug. 4, 1863	39,439
Brush	J. L. Whiting	Boston, Mass	Sept. 8, 1868	81,966
Brush	J. L. Whiting	Boston, Mass	Mar. 29, 1870	101,332
Brush	J. L. Whiting	Boston, Mass	Aug. 30, 1870	106,902
Brush	J. L. Whiting	Boston, Mass	Sept. 27, 1870	107,742
Brush	J. L. Whiting	Boston, Mass	Nov. 8, 1870	109,084
Brush	J. L. Whiting	Boston, Mass	June 27, 1871	116,382
Brush	J. L. Whiting	Boston, Mass	Aug. 29, 1871	118,661
Brush	O. D. Woodbury	New York, N. Y	Apr. 26, 1870	102,460
Brush	E. J. Worcester	Worcester, Mass	June 18, 1867	65,982
Brush	A. A. Young	Boston, Mass	July 16, 1867	66,762
Brush and boot-jack, Blacking	S. Gissinger	Allegheny City, Pa	Feb. 20, 1866	52,787
Brush and box, Combined blacking	F. Hatch	La Crosse, Wis	Dec. 18, 1866	60,511
Brush and box, Combined blacking	H. S. Kerr	Philadelphia, Pa	Sept. 12, 1871	118,805
Brush and box, Combined blacking	W. H. H. Saunders	Troy, N. Y	Nov. 6, 1866	59,461
Brush and box, Portable shoe	E. A. Pierce	New York, N. Y	Apr. 26, 1870	102,430
Brush and broom, Dust	A. Dietz	New York, N. Y	July 3, 1866	56,018
Brush and broom handle	L. F. Cavanaugh	Newfield, N. Y	July 1, 1851	8,200
Brush and broom handle	E. Turk	Harrisburgh, Pa	May 18, 1832	....
Brush and broom holder	S. G. Groff	Vogansville, Pa	Dec. 5, 1871	121,458
Brush and card handle	D. Witt and A. Wait	Hubbardston, Mass	Nov. 16, 1869	97,012
Brush and case, Blacking	J. Schwab	Bridgeport, Conn	June 14, 1870	104,361
Brush and case combined, Shoe	C. Hamilton	New York, N. Y	Jan. 1, 1867	60,886
Brush and comb combined	A. A. Young	Boston, Mass	July 7, 1868	79,621
Brush and fan, Automatic fly	E. J. Clark	Kalamazoo, Mich	Apr. 21, 1868	76,996
Brush and fan, Automatic fly	C. C. Short	Osgood, Ind	Jan. 15, 1867	61,269
Brush and handle attachment, Whitewash	W. B. Burtnett and J. P. McIntosh.	New York and Brooklyn, N. Y.	May 30, 1865	47,927
Brush and holder, Blacking	D. Bowker	Boston, Mass	Jan. 26, 1864	41,359
Brush and mixer for stove-blacking	G. S. Goodspeed	Providence, R. I	Jan. 30, 1872	123,099
Brush and mop head	T. T. Prosser	Chicago, Ill	Dec. 24, 1867	72,677
Brush and mop holder	P. O'Brian and C. L. W. Baker	New York, N. Y., and Hartford, Conn.	Aug. 27, 1872	130,859
Brush and mop holder	I. Smith	Rochester, N. Y	May 13, 1873	138,946
Brush and mop, Scrubbing	J. F. Mason	Brooklyn, N. Y	Nov. 17, 1868	84,129
Brush and mop-wringer, Scrubbing	E. Rees	Stoddartsville, Pa	May 10, 1870	102,865
Brush and rubber, Combination for	A. S. Miles	Brooklyn, N. Y	Aug. 13, 1867	67,787
Brush and scraper, Boot and shoe	W. A. Morrison	Morrisania, N. Y	Oct. 11, 1859	25,755
Brush and scraper for sink	J. H. Osgood	Boston, Mass	Oct 17, 1871	120,090
Brush attachment	S. Pearson	Cincinnati, Ohio	Mar. 29, 1870	101,307
Brush, Automatic fan and fly	W. Carpenter	Metamora, Ill	July 30, 1861	32,970
Brush, Automatic fly	H. C. Chandler	Peru, Ind	Dec. 12, 1871	121,750
Brush, Automatic fly	W. H. Chipley	Libertytown, Md	Nov. 28, 1871	121,335
Brush, Automatic fly	B. F. Day	East Freedom, Pa	Aug. 25, 1868	81,349
Brush-back	W. U. Dudley	Port Richmond, N. Y	June 15, 1869	91,314
Brush, Barber's	J. A. Wiedersheim	Philadelphia, Pa	Apr. 2, 1867	63,448
Brush, Bathing	J. Marshall	New York, N. Y	June 1, 1869	90,763
Brush, Bathing	J. Marshall	Brooklyn, N. Y	Feb. 21, 1871	111,955
Brush, Blacking	R. Adams	Cincinnati, Ohio	Oct. 8, 1867	69,603
Brush, Blacking	R. Adams	Cincinnati, Ohio	Mar. 9, 1869	87,614
Brush, Blacking	B. F. Averill	Dunkirk, N. Y	July 11, 1871	116,794
Brush, Blacking	C. L. W. Baker	Hartford, Conn	Apr. 27, 1869	89,459
Brush, Blacking	C. Britzinghoffer	Philadelphia, Pa	Oct. 10, 1871	119,815
Brush, Blacking	G. R. Burdon	Waltham, Mass	Nov. 5, 1867	70,515
Brush, Blacking	C. D. Day	Chatham, Conn	Aug. 25, 1868	81,481
Brush, Blacking	J. H. Doughty	New York, N. Y	Feb. 18, 1868	74,674
Brush, Blacking	N. Eisenmann	New York, N. Y	Feb, 20, 1872	123,815
Brush, Blacking	W. S. Fickett	Rochester, N. Y	Aug. 12, 1873	141,638
Brush, Blacking	W. A. Green	Troy, N. Y	May 10, 1864	42,655
Brush, Blacking	J. R. Howard	Boston, Mass	Aug. 15, 1871	118,015
Brush, Blacking	H. S. Kerr	Philadelphia, Pa	Apr. 2, 1872	125,200
Brush, Blacking	G. W. Lishawa	Cincinnati, Ohio	Jan. 25, 1870	99,095
Brush, Blacking	C. W. Maguire	Brooklyn, N. Y	Apr. 1, 1873	137,376
Brush, Blacking	W. C. McGill	Cincinnati, Ohio	May 5, 1868	77,506
Brush, Blacking	C. Mulchahey	Springfield, Mass	Dec. 4, 1866	60,226

Index of patents issued from the United States Patent Office from 1790 *to* 1873, *inclusive*—Continued.

Invention.	Inventor.	Residence.	Date.	No.
Brush, Blacking	H. E. Newton	Manchester, N. H	Jan. 28, 1868	73, 746
Brush, Blacking	G. R. Owen	Utica, N. Y	Nov. 7, 1871	120, 663
Brush, Blacking	C. A. Paret	Nashville, Tenn	Nov. 12, 1867	70, 889
Brush, Blacking	A. N. Thompson	Holyoke, Mass	Oct. 26, 1869	96, 165
Brush, Blacking	F. W. C. Von Kessler and A. F. Medenwald.	Chicago, Ill	Apr. 29, 1873	138, 353
Brush, Blacking	N. G. Whitmore	Mansfield, Mass	May 1, 1866	54, 472
Brush-block	W. B. Burtnett	New York, N. Y	Apr. 30, 1867	64, 195
Brush block, Whitewash	C. Williams	Philadelphia, Pa	Feb. 23, 1858	19, 459
Brush-blocks, Machine for boring	T. Mitchell	Lansingburgh, N. Y	Jan. 10, 1860	26, 778
Brush-blocks, Machine for boring	T. Mitchell	Lansingburgh, N. Y	Feb. 19, 1861	31, 471
Brush-blocks, Machine for making	S. Costill	Philadelphia, Pa	Feb. 25, 1824	....
Brush blocks, Machine for shaping hair	A. G. Mitchell	Lansingburgh, N. Y	Jan. 31, 1860	27, 005
Brush-blocks, Machinery for boring	S. Taylor and A. R. Davis	East Cambridge, Mass	Dec. 19, 1848	5, 974
Brush-blocks with bristles, Machine for filling	S. Taylor and A. R. Davis	Cambridge and Boston, Mass.	May 1, 1845	4, 027
Brush, Boiler-flue	C. H. Bush	Fall River, Mass	June 26, 1866	55, 815
Brush, Boiler-flue	P. H. Coyle	Newark, N. J	Nov. 3, 1868	83, 607
Brush, Boiler-flue	R. King	Brooklyn, N. Y	Nov. 27, 1866	60, 013
Brush, Boiler-flue	G. P. Leonard and W. Lanster	Fall River, Mass	Mar. 29, 1864	42, 092
Brush, Boiler-flue	P. H. Ryan	Cincinnati, Ohio	July 6, 1869	92, 211
Brush, Boiler-flue	B. M. Spencer	Newark, N. J	Mar. 14, 1871	112, 645
Brush, box, and holder, Combined blacking	E. W. Woodruff	Washington, D. C	Aug. 13, 1872	130, 460
Brush-bridle	G. Hergesheimer	Philadelphia, Pa	Feb. 11, 1868	74, 224
Brush, Broom	A. S. Goodman	Duval County, Fla	May 8, 1849	6, 423
Brush, broom, and mop holder	C. L. W. Baker	Hartford, Conn	Apr. 4, 1871	113, 244
Brush, broom, &c., handles, Attachment for, (canceled.)	A. Eddowes	Frankford, Pa	Jan. 29, 1867	61, 658
Brush case, Shoe	C. D. Thum	Philadelphia, Pa	Aug. 3, 1858	21, 092
Brush, Circular	M. Robbins and C. Heery	Cincinnati, Ohio	May 3, 1864	42, 625
Brush-clamp	A. Huston	Bristol, Me	Oct. 8, 1867	69, 674
Brush clamp, Scrubbing	C. B. Clark	Buffalo, N. Y	Mar. 10, 1868	75, 369
Brush clamp, Paint	G. R. Gardiner	Westerly, R. I	July 30, 1867	67, 286
Brush clasp, Paint	G. W. Hupp	Elwood, Ind	July 12, 1870	105, 212
Brush, Cloth and hat	J. Marshall	New York, N. Y	Sept. 28, 1869	95, 244
Brush, Cloth or hair	H. Aiken	Dracut, Mass	May 24, 1830	....
Brush, Clothes	A. J. Walker	Flatbush, N. Y	Feb. 27, 1872	124, 023
Brush, Coiled-wire	J. B. Christoffel	Williamsburgh, N. Y	Apr. 29, 1873	138, 317
Brush, &c., Combined scrubbing	E. K. Wood	De Witt, Iowa	Aug. 23, 1870	106, 644
Brush, Damping	J. K. Park	Marlborough, N. Y	Jan. 1, 1867	60, 931
Brush, Device for attaching handle to whitewash	W. P. Burtnett and J. P. McIntosh.	New York and Highland Mills, N. Y.	Oct. 17, 1865	50, 449
Brush, Device for trimming cylindrical	A. G. Risley	Utica, N. Y	Dec. 13, 1870	110, 073
Brush, Dust	W. M. Conger	Newark, N. J	May 14, 1872	126, 677
Brush, Dust	G. G. Finn	Cleveland, Ohio	May 21, 1872	126, 945
Brush, Dust	A. Schelling	Erie, Pa	Mar. 12, 1872	124, 634
Brush, Dust	S. Taylor	Boston, Mass	Aug. 6, 1867	67, 461
Brush, Dust	E. Thayer	Worcester, Mass	Aug. 27, 1867	68, 259
Brush, Dusting	R. H. Aldrich	Northampton, Mass	Mar. 12, 1867	62, 724
Brush, Dusting	A. S. Hadley	Boston, Mass	Mar. 12, 1867	62, 742
Brush, Dusting	C. M. Moody	Greenfield, Mass	Apr. 18, 1865	47, 318
Brush, Dusting	T. G. Porter	Louisville, Ky	July 25, 1871	117, 456
Brush, Dusting	L. R. Witherell and A. B. Crandall.	Galesburgh, Ill	June 7, 1870	104, 092
Brush, Ear	G. W. Wood	New York, N. Y	Apr. 26, 1870	102, 351
Brush, Flexible	R. N. Eagle	Washington, D. C	Feb. 27, 1866	52, 833
Brush, Flexible-back	J. J. Adams	New York, N. Y	June 18, 1861	32, 552
Brush, Flue	J. D. Kunkel	Cincinnati, Ohio	Aug. 17, 1869	93, 723
Brush, Fly	B. F. Brown	Cotlin, Ind	Mar. 5, 1872	124, 324
Brush, Fly	W. M. Cline	Sillard's Mills, Tenn	Apr. 23, 1872	126, 022
Brush, Fly	J. E. Darnall	Washington, D. C	Nov. 29, 1870	109, 720
Brush, Fly	H. Fisher	Canton, Ohio	Nov. 14, 1865	50, 926
Brush, Fly	J. A. Lyle	Bridgeport, Ohio	Nov. 18, 1873	144, 779
Brush, Fly	H. R. Robbins	Baltimore, Md	Sept. 21, 1869	95, 141
Brush, Fly	J. M. Runyon	Pemberton, Ohio	June 25, 1872	128, 253
Brush, Fly	D. Shankland and E. B. Hopkinson.	Nevada City, Cal	Apr. 4, 1871	113, 581
Brush, Fly	S. R. Wilmot	La Fayette, Ind	Nov. 5, 1850	7, 763
Brush, Folding	P. Henrichs	Erie, Pa	June 20, 1871	116, 184
Brush, Folding hair	C. Crossman	Tompkinsville, N. Y	May 15, 1866	54, 695
Brush, Foot	J. W. Osborne	Brooklyn, N. Y	May 16, 1871	114, 847
Brush for cannon	P. Birchmeyer	Syracuse, N. Y	Feb. 21, 1865	46, 441
Brush for cleaning metallic plates	E. A. Harvey	Wilmington, Del	Sept. 5, 1865	49, 755
Brush for cleaning teeth, Finger	G. M. Allerton	Dover Plains, N. Y	June 18, 1872	128, 093
Brush for cleaning travelers	H. S. Houghton	Blackstone, Mass	June 17, 1856	15, 131
Brush for cleaning tumblers	A. Fischer	New York, N. Y	Aug. 16, 1870	106, 476
Brush for cleaning tumblers	A. Fischer	New York, N. Y	Aug. 1, 1871	117, 614
Brush for dressing warps	S. Taylor	Lowell, Mass	May 28, 1842	2, 651
Brush for dressing warps	S. Taylor	Cambridge, Mass	July 8, 1856	15, 398
Brush for finger-nails	W. Thomson	Buffalo, N. Y	Dec. 27, 1859	26, 629
Brush for foundry use, Steel-wire	P. Farley	Philadelphia, Pa	Oct. 29, 1872	132, 528
Brush for mucilage-bottle	W. Burnet	New York, N. Y	Feb. 2, 1864	41, 422
Brush for mucilage, painting, gluing, &c	J. W. McGill	Washington, D. C	July 9, 1867	66, 666
Brush for ordnance, Cleaning	J. T. Greenfield	Dover, England	Oct. 5, 1869	95, 468
Brush for washing windows	P. C. Rowe	Boston, Mass	June 21, 1859	24, 491
Brush, Fountain	D. H. Chamberlain and J. Hartshorn.	Boston, Mass	Jan. 30, 1855	12, 309
Brush, Fountain	J. Davis	McLean, N. Y	Oct. 15, 1867	69, 901
Brush, Fountain	L. B. Hoit	New York, N. Y	Feb. 22, 1859	23, 028
Brush, Fountain	R. Lapham	New York, N. Y	Dec. 9, 1873	145, 301
Brush, Fountain	Z. Poitras	Chicago, Ill	Apr. 18, 1871	113, 923
Brush, Fountain	J. B. Warren	South Danvers, Mass	Nov. 19, 1867	71, 094
Brush, Fountain blacking	J. R. Howard	Worcester, Mass	May 20, 1862	35, 316
Brush, Fountain blacking	A. D. Pentz	New York, N. Y	Feb. 7, 1871	111, 566
Brush, Fountain paint	D. J. Kellogg	Toledo, Ohio	Jan. 17, 1871	110, 978
Brush, Fountain paint	D. J. Kellogg	Toledo, Ohio	Apr. 4, 1871	113, 305
Brush from cotton-fields, Machine for cutting	E. Peck	Canton, Ill	Feb. 9, 1858	19, 311
Brush, Galvanic metal friction	L. A. Hoffman	Berlin, Prussia	June 25, 1861	32, 625

Index of patents issued from the United States Patent Office from 1790 *to* 1873, *inclusive*—Continued.

Invention.	Inventor.	Residence.	Date.	No.
Brush, Hair	J. B. Burgess	New York, N. Y	Apr. 14, 1835	....
Brush, Hair	S. M. Firey	Clear Spring, Md	Aug. 23, 1870	106, 680
Brush, Hair	J. N. George and J. R. Sanborn	Boston and Waltham, Mass	Oct. 29, 1867	70, 194
Brush, Hair	J. R. Ingersoll	New York, N. Y	May 22, 1860	28, 375
Brush, Hair	J. Mayer	Philadelphia, Pa	June 20, 1865	48, 295
Brush, Hair	J. F. McClure	Boston, Mass	Mar. 19, 1861	31, 727
Brush, Hair-dyeing	W. B. Coates	Philadelphia, Pa	June 8, 1869	91, 089
Brush-handle	A. Alden	New York, N. Y	Sept. 4, 1866	57, 656
Brush-handle	C. Alvord	Westford, Wis	Mar. 12, 1867	62, 801
Brush-handle	C. Brintzinghoffer and G. Eckert.	Philadelphia, Pa	Oct. 31, 1871	120, 466
Brush-handle	R. S. Craig	Cincinnati, Ohio	Jan. 21, 1873	135, 090
Brush-handle	S. D. Foster	Portland, Me	Apr. 2, 1872	125, 279
Brush-handle	C. L. Larder	Brooklyn, N. Y	Mar. 9, 1869	87, 684
Brush-handle	F. Rudolph and W. Kasefang	Jersey City, N. J	Nov. 29, 1864	45, 276
Brush-handle attachment	R. Wyatt	Brooklyn, N. Y	Mar. 29, 1870	101, 400
Brush handle, Lather	W. H. Miles, jr	Brooklyn, N. Y	Nov. 11, 1873	144, 553
Brush handle, Mucilage	T. N. Hickcox	Brooklyn, N. Y	Apr. 20, 1869	89, 222
Brush handle, Tooth and nail	G. A. Scott	New York, N. Y	Dec. 3, 1872	133, 600
Brush handles, Machine for driving	J. Ames, jr	Lansingburgh, N. Y	Aug. 26, 1873	142, 137
Brush handles, Machine for driving	P. Peartree	Lansingburgh, N. Y	Apr. 29, 1873	138, 344
Brush handles, Machine for driving	P. Peartree	Lansingburgh, N. Y	July 23, 1872	129, 748
Brush handles, Machine for finishing	T. Mitchell	Lansingburgh, N. Y	June 23, 1857	17, 631
Brush handles, Machine for finishing hair	T. Mitchell	Lansingburgh, N. Y	June 28, 1859	24, 571
Brush-handles, Screw-socket for	W. H. Johnson	Philadelphia, Pa	Oct. 20, 1868	83, 173
Brush, Hat	C. F. Phelps	Boston, Mass	May 19, 1863	38, 632
Brush, Hat	R. D. Radcliffe	Palmyra, N. Y	Mar. 7, 1871	112, 492
Brush, Hat	W. B. Shedd	East Boston, Mass	Aug. 19, 1862	36, 239
Brush-head	C. J. Moore and L. S. Gambold	Coatesville, Ind	June 22, 1869	91, 556
Brush, Helical-wire	F. F. Field	Stapleton, N. Y	Oct. 10, 1871	119, 752
Brush-holder	C. L. W. Baker	Camden, N. J	Dec. 16, 1873	145, 479
Brush-holder	C. B. Clark	Buffalo, N. Y	July 28, 1868	80, 453
Brush-holder	J. Messinger	Springfield, Vt	Nov. 19, 1867	71, 037
Brush-holder	B. Morahan	Brooklyn, N. Y	July 9, 1867	66, 612
Brush-holder	A. H. Trego	Philadelphia, Pa	Mar. 19, 1872	124, 865
Brush-holder and mop-head	H. P. Gregg	Cincinnati, Ohio	Sept. 1, 1868	81, 626
Brush-holder, Dust-pan	W. M. Couger	Newark, N. J	May 21, 1872	126, 931
Brush, Holder for hat	J. M. Osgood	Chelsea, Mass	Apr. 21, 1863	38, 240
Brush holder, Fountain	W. A. Shepard	Chicago, Ill	July 5, 1870	105, 135
Brush holder, Painter's	J. Lemoine	Winchendon, Mass	Apr. 30, 1872	126, 219
Brush holder, Whitewash	A. Isko	Lancaster, Pa	Sept. 24, 1867	69, 097
Brush, Horse	R. Dickson	Lansingburgh, N. Y	Apr. 8, 1873	137, 536
Brush, Horse	J. Hayworth	Manchester, England	Jan. 16, 1866	52, 118
Brush, Hydraulic	T. Wellham	Washington, D. C	Jan. 31, 1865	46, 167
Brush, India-rubber tooth	S. W. Francis	New York, N. Y	July 6, 1869	92, 298
Brush, Lather	W. H. Miles, jr	New York, N. Y	Oct. 31, 1871	120, 390
Brush, Leather-backed horse	O. Jones	South Englewood, N. J	Sept. 3, 1867	68, 513
Brush-machine	E. F. Bradley	New Haven, Conn	Apr. 1, 1873	137, 411
Brush-machine	M. Dillon	Albany, N. Y	Feb. 11, 1873	135, 634
Brush-machine	E. A. Hill	Cambridge, Mass	Sept. 16, 1873	142, 913
Brush-machine	J. Ruegg	Saint Louis, Mo	Oct. 9, 1860	30, 353
Brush-machine	E. D. Van Horn and C. N. Farr	Philadelphia, Pa	May 6, 1873	138, 545
Brush-making	W. Steel	New York, N. Y	Aug. 17, 1835	....
Brush-making	W. Steel	New York, N. Y	Aug. 17, 1835	
Brush-making machine	E. F. Bradley	London, England	Dec. 10, 1872	133, 827
Brush-making machine	J. Pickering	Philadelphia, Pa	Oct. 24, 1871	120, 164
Brush-making machine	J. Ruegg	Saint Louis, Mo	Nov. 20, 1868	30, 693
Brush-making machine	L. A. Tripp	New York, N. Y	Oct. 27, 1857	18, 528
Brush-making machine	A. M. White	Thompsonville, Conn	Dec. 22, 1868	85, 193
Brush-making machine	A. M. White	Thompsonville, Conn	June 1, 1869	90, 903
Brush-making machinery	C. D. Rogers, M. P. Wilkins, and H. A. Hawey.	Utica, N. Y., Jersey City and Orange, N. J.	June 25, 1872	128, 251
Brush manufacture	R. Ashworth	Fall River, Mass	Aug. 6, 1872	130, 270
Brush manufacture	S. Barnes	New Haven, Conn	Sept. 7, 1858	21, 464
Brush manufacture	A. R. Davis	East Cambridge, Mass	Aug. 19, 1851	8, 299
Brush manufacture	A. R. Davis	East Cambridge, Mass	May 11, 1852	8, 938
Brush manufacture	A. C. Estabrook	Florence, Mass	Feb. 4, 1873	135, 415
Brush manufacture	A. H. Kidney and R. J. Simpson.	New York, N. Y	Feb. 27, 1872	124, 141
Brush manufacture	C. F. Ruset	Brooklyn, N. Y	July 12, 1870	105, 373
Brush manufacture	M. Stewart	Philadelphia, Pa	Dec. 5, 1854	12, 037
Brush manufacture	W. H. Van Kleeck	Lansingburgh, N. Y	Jan. 23, 1872	123, 064
Brush manufacture	M. P. Wilkins and C. D. Rogers	Jersey City, N. J	Feb. 26, 1867	62, 460
Brush-manufacturing device	C. F. Russet	Brooklyn, N. Y	July 5, 1870	105, 131
Brush, Marking	E. P. Clark	Northampton, Mass	Sept. 29, 1863	40, 089
Brush, Marking	T. Fowler	Richmond Valley, N. Y	Dec. 24, 1861	33, 986
Brush, Marking	F. W. Wentworth	Boston, Mass	June 24, 1873	140, 228
Brush, Means of attaching handle to whitewash	W. B. Burtnett	New York, N. Y	May 7, 1867	64, 406
Brush, Mechanical fly	H. E. Aughinbaugh	Harrisburgh, Pa	June 6, 1871	115, 560
Brush, mop, and wringer, Combined scrubbing	J. D. Smith	Sacramento, Cal	July 8, 1873	140, 655
Brush, mop, and wringer, Scrubbing	L. Frey and J. Hahn	Chicago, Ill	Oct. 31, 1865	50, 701
Brush, Mucilage	W. W. Beach	New York, N. Y	July 9, 1867	66, 446
Brush, Mucilage	D. Bly	Macon, Ga	Mar. 30, 1869	88, 265
Brush, Mucilage	M. W. House	Cleveland, Ohio	Mar. 16, 1869	87, 776
Brush, Mucilage and marking	W. R. Anderson	New York, N. Y	Feb. 19, 1867	62, 108
Brush, Mucilage and marking	E. H. Boswell	Philadelphia, Pa	Aug. 1, 1865	49, 072
Brush, Nail	W. Thomson	Buffalo, N. Y	July 31, 1860	29, 419
Brush of saw cotton-gin	E. Keith	Bridgewater, Mass	Apr. 28, 1857	17, 155
Brush or card, Curry	J. Voak	Penn Yan, N. Y	Feb. 21, 1865	46, 510
Brush or comb, Hair	T. Divine	Charleston, S. C	June 14, 1870	104, 125
Brush or fan, Fly	M. Laveen	Moorefield, Va	Apr. 10, 1860	27, 844
Brush, Paint	J. Ames, jr	Lansingburgh, N. Y	Apr. 25, 1871	114, 086
Brush, Paint	C. S. Benjamin and E. Story	Washington, Ill	Feb. 22, 1870	100, 108
Brush, Paint	J. D. Bonney	Pembroke, Mass	July 2, 1867	66, 291
Brush, Paint	W. B. Burtnett	New York, N. Y	Aug. 23, 1870	106, 544
Brush, Paint	W. B. Burtnett	New York, N. Y	Jan. 3, 1871	110, 628
Brush, Paint	G. H. Clinnock	Brooklyn, N. Y	May 13, 1873	138, 787
Brush, Paint	W. Cover	Jenner's Cross Roads, Pa	Jan. 8, 1867	61, 053

Index of patents issued from the United States Patent Office from 1790 *to* 1873, *inclusive*—Continued.

Invention.	Inventor.	Residence.	Date.	No.
Brush, Paint	A. Cutter	Chelsea, Mass	Feb. 25, 1868	74, 752
Brush, Paint	M. Dillon	Albany, N. Y	Feb. 11, 1873	135, 635
Brush, Paint	B. A. Drayton	Utica, N. Y	July 9, 1867	66, 475
Brush, Paint	J. M. Estabrook	Milford, Mass	Oct. 29, 1867	70, 182
Brush, Paint	L. P. Faught	Foxborough, Mass	Sept. 19, 1865	50, 067
Brush, Paint	W. H. Forker	Meadville, Pa	Nov. 5, 1867	70, 542
Brush, Paint	W. H. Forker	Meadville, Pa	May 5, 1868	77, 475
Brush, Paint	F. P. Furnald, jr., R. W. Champion, and I. N. Davies.	New York and Brooklyn, N. Y., and Bergen City, N. J.	May 4, 1869	89, 649
Brush, Paint	F. P. Furnald, jr., R. W. Champion, and I. N. Davies.	New York, N. Y	Dec. 21, 1869	98, 158
Brush, Paint	H. B. Gillman and H. S. Beamish.	Milford, Mass	Aug. 27, 1867	68, 066
Brush, Paint	J. J. Gorman	Cincinnati, Ohio	Jan. 14, 1873	134, 745
Brush, Paint	J. J. Gorman	Cincinnati, Ohio	June 17, 1873	139, 887
Brush, Paint	W. Hicks	Steubenville, Ohio	Jan. 2, 1855	12, 182
Brush, Paint	E. Hiss	Delaware, Ohio	July 31, 1866	56, 750
Brush, Paint	L. A. Lightenhome	Chicago, Ill	Dec. 17, 1872	133, 996
Brush, Paint	J. Marchbank	Lansingburgh, N. Y	Nov. 20, 1866	59, 850
Brush, Paint	J. S. Martin	Boston, Mass	June 20, 1854	11, 129
Brush, Paint	A. McDonald	Washington, Ill	Nov. 1, 1870	108, 923
Brush, Paint	A. McDonald and J. M. White.	Washington, Ill	Aug. 16, 1870	106, 384
Brush, Paint	W. H. Miles, jr	Brooklyn, N. Y	Nov. 18, 1862	36, 962
Brush, Paint	J. W. Moore	Cambridgeport, Mass	Mar. 31, 1868	76, 239
Brush, Paint	G. G. Morris	Boston, Mass	Feb. 1, 1870	99, 338
Brush, Paint	J. Rayls	Washington, Ill	Jan. 25, 1870	99, 276
Brush, Paint	H. Rosenthal	New York, N. Y	Apr. 30, 1867	64, 369
Brush, Paint	H. Rosenthal	New York, N. Y	Aug. 6, 1867	67, 588
Brush, Paint	F. S. Shearer	Washington, Ill	Mar. 7, 1871	112, 387
Brush, Paint	G. L. Shuttleworth	Sharon, Vt	Sept. 6, 1870	107, 200
Brush, Paint	S. Standish	Eureka, Nev	June 24, 1873	140, 314
Brush, Paint	A. S. Thompson	Little Falls, Minn	Aug. 5, 1873	141, 522
Brush, Paint	G. K. Thompson	Greene, N. Y	Jan. 25, 1870	99, 260
Brush, Paint	P. Wagner	New York, N. Y	Feb. 25, 1873	136, 288
Brush, Paint	D. White	Portland, Me	Nov. 24, 1868	84, 327
Brush, Paint	J. S. White	Boston, Mass	May 7, 1867	64, 607
Brush, Paint	J. L. Whiting	Boston, Mass	Oct. 18, 1870	108, 420
Brush, Paint	J. N. Woodward	Aurora, Ill	May 14, 1867	64, 731
Brush, Paint and varnish	J. McKittrick	Brooklyn, N. Y	Feb. 18, 1868	74, 566
Brush, Paint and varnish	A. Randal	Pittsburgh, Pa	Jan. 30, 1872	123, 203
Brush, Paint and varnish	E. Thayer	Worcester, Mass	May 7, 1867	64, 596
Brush, Paint and varnish	G. A. White	Boston, Mass	Jan. 15, 1867	61, 292
Brush, Painters' stripping	P. Feeck	Seneca Falls, N. Y	Nov. 19, 1872	133, 088
Brush, Paper-hangers'	J. M McComb	Lancaster, Ohio	Nov. 25, 1873	144, 999
Brush, &c., Pocket	P. Henrichs	Erie, Pa	June 20, 1871	116, 185
Brush, Pocket	T. F. Stumpf	New York, N. Y	Dec. 27, 1870	110, 603
Brush-rack	E. F. Ames	Lansingburgh, N. Y	Aug. 13, 1872	130, 401
Brush-rack	J. Ames and N. H. Horton	Lansingburgh and New York, N. Y.	July 30, 1867	67, 247
Brush, Reservoir damping	W. Shriver	New York, N. Y	May 28, 1867	65, 290
Brush, Revolving hair	J. Bayston	Chicago, Ill	Feb. 25, 1868	74, 743
Brush, Rotary	G. Carlisle	Attleborough, Mass	Oct. 14, 1873	143, 666
Brush, Rotary	S. Stuart	Sterling, Mass	May 29, 1866	55, 176
Brush, Rotary hair	E. G. Camp	Bristol, England	Mar. 8, 1864	41, 885
Brush, Rotary horse	W. W. McKay	Ossian, Iowa	Dec. 8, 1868	84, 837
Brush, Sash	J. B. Wood	Lansingburgh, N. Y	Aug. 20, 1867	68, 022
Brush, Scrubbing	R. M. Bicknell and C. J. Abel.	Philadelphia, Pa	Sept. 4, 1847	5, 267
Brush, Scrubbing	A. S. Brinser	Falmouth, Pa	Oct. 8, 1872	132, 049
Brush, Scrubbing	R. M. Brooks	Griffin, Ga	July 21, 1868	80, 128
Brush, Scrubbing	G. W. Brown	Providence, R. I	July 6, 1869	92, 154
Brush, Scrubbing	G. Carver	Chambersburgh, Pa	Aug. 1, 1844	3, 689
Brush, Scrubbing	W. C. Cleveland	Cambridge, Mass	May 8, 1866	54, 505
Brush, Scrubbing	A. E. Colman	New York, N. Y	Nov. 10, 1868	83, 932
Brush, Scrubbing	W. Devines	Williamsburgh, N. Y	June 20, 1871	116, 030
Brush, Scrubbing	A. J. Doolittle	Hamden, Conn	July 16, 1867	66, 812
Brush, Scrubbing	B. Figer	Cleveland, Ohio	July 3, 1866	56, 025
Brush, Scrubbing	B. G. Fitzhugh	Frederick, Md	Sept. 24, 1872	131, 674
Brush, Scrubbing	S. Gibson	Safe Harbor, Pa	Oct. 6, 1868	82, 705
Brush, Scrubbing	S. Gibson	Safe Harbor, Pa	Oct. 26, 1869	96, 219
Brush, Scrubbing	S. A. Gibson and J. B. De Haas	Clearfield, Pa	July 18, 1871	117, 167
Brush, Scrubbing	C. F. Gillett	Sparta, Wis	July 30, 1872	129, 944
Brush, Scrubbing	R. P. Gillett	Sparta, Wis	July 3, 1866	56, 035
Brush, Scrubbing	W. L. Haller	Philadelphia, Pa	Apr. 30, 1861	32, 224
Brush, Scrubbing	D. E. Hayward	Malden, Mass	Mar. 10, 1868	75, 421
Brush, Scrubbing	C. Herald	Pittsburgh, Pa	Oct. 14, 1873	143, 695
Brush, Scrubbing	B. F. Kaller	Shrewsbury, Pa	Apr. 5, 1870	101, 630
Brush, Scrubbing	D. Minderle	Saint Louis, Mo	Sept. 13, 1870	107, 397
Brush, Scrubbing	P. O'Brian	New York, N. Y	June 27, 1871	116, 346
Brush, Scrubbing	J. Odell	Petroleum Centre, Pa	Aug. 9, 1870	106, 196
Brush, Scrubbing	R. Rockwell and Z. B. Custer.	Petroleum Centre, Pa	Dec. 6, 1870	109, 946
Brush, Scrubbing	S. C. Rundlett and R. Dodge.	Portland, Me	Mar. 26, 1867	63, 181
Brush, Scrubbing	A. Schweizer and G. Jansen	Cleveland, Ohio	Aug. 26, 1862	36, 309
Brush, Scrubbing	J. See	Philadelphia, Pa	Mar. 15, 1870	100, 932
Brush, Scrubbing	J. N. Valley and J. A. Stetson, jr.	North East, Pa	Oct. 24, 1871	120, 350
Brush, Scrubbing	R. Wyatt	Brooklyn, N. Y	Feb. 5, 1867	61, 910
Brush, Self-feeding blacking	G. Hamilton and J. Post	New York, N. Y	Feb. 23, 1864	41, 700
Brush, Self-supplying mucilage	C. Hamilton	New York, N. Y	Aug. 13, 1867	67, 648
Brush, Shaving	W. S. Jewett	New York, N. Y	Apr. 10, 1849	6, 293
Brush, Shaving	W. T. Wylie	Newcastle, Pa	May 14, 1867	64, 732
Brush-shears	C. Brombacher	New York, N. Y	Mar. 26, 1867	63, 205
Brush, Shoe	J. J. Adams	Boston, Mass	Feb. 10, 1852	8, 714
Brush, Shoe	C. L. W. Baker and L. S. Hills	Hartford, Conn	May 14, 1867	64, 736
Brush, Shoe	F. D. Holland	Lewiston, Me	May 3, 1870	102, 540
Brush, Shoe	C. F. J. Moller and C. Sholes	Milwaukee, Wis	Aug. 14, 1866	57, 168
Brush, Shoe	J. E. Nolin	Chicago, Ill	June 9, 1868	78, 683

Index of patents issued from the United States Patent Office from 1790 *to* 1873, *inclusive*—Continued.

Invention.	Inventor.	Residence.	Date.	No.
Brush, Shoe	G. R. Owen	Rome, N. Y	Aug. 29, 1871	118, 638
Brush, Shoe	E. A. G. Roulstone	Roxbury, Mass	Nov. 27, 1866	60, 063
Brush, Shoe	H. E. Scotchmer	Chicago, Ill	Apr. 3, 1866	53, 689
Brush, Shoe	G. Wale	Hoboken, N. J	Feb. 25, 1873	136, 289
Brush, Shoe	R. M. Whipple	Cambridge, Mass	Jan. 6, 1863	37, 315
Brush, Shoe and stove	F. M. Carnes	New York, N. Y	Nov. 13, 1866	59, 709
Brush-sleeve	J. S. Tilton	Philadelphia, Pa	Feb. 11, 1868	74, 258
Brush-socket	P. Wagner	New York, N. Y	Apr. 1, 1873	137, 515
Brush, Steam-generator-flue	P. H. Coyle	Newark, N. J	June 22, 1869	91, 523
Brush, Stencil	W. R. Allen	Saint Louis, Mo	May 17, 1859	23, 997
Brush, Stencil	J. S. Costello	Saint Louis, Mo	Aug. 13, 1867	67, 726
Brush, Stencil	D. K. Herr	Locust Valley, Pa	Apr. 10, 1866	53, 923
Brush stock and handle	A. A. Young	Boston, Mass	Feb. 4, 1868	74, 026
Brush stock and handle	A. A. Young	Boston, Mass	Feb. 4, 1868	74, 027
Brush-stock, Method of holding bristles in	C. D. Rogers and M. P. Wilkins	Utica, N. Y., and Jersey City, N. J.	Oct. 29, 1867	70, 270
Brush, Stove	W. Small	Brooklyn, N. Y	July 9, 1872	128, 917
Brush, Stove-polishing	L. C. Crowell	West Dennis, Mass	Sept. 7, 1869	94, 477
Brush to hat, Device for attaching	A. A. Veer	Delaware, Ohio	Apr. 9, 1872	125, 422
Brush, Toilet	S. Wilder	Hingham, Mass	Mar. 5, 1872	124, 409
Brush, Tooth	G. M. Allerton	Dover Plains, N. Y	July 9, 1872	128, 840
Brush, Tooth	P. Blake	New Haven, Conn	Mar. 8, 1830	....
Brush, Tooth	E. F. Burrows	Mystic River, Conn	May 18, 1869	90, 235
Brush, Tooth	R. K. Chandler	Ruther Glen, Va	Jan. 25, 1870	99, 161
Brush, Tooth	T. Maitland	Williamsport, Pa	June 2, 1868	78, 599
Brush, Tooth	T. F. Maury	Washington, D. C	Feb. 18, 1868	74, 560
Brush, Tooth	R. Nelson	Albany, N. Y	July 19, 1864	43, 597
Brush, Tooth	J. D. and W. O'Donoghue	New York, N. Y	July 1, 1873	140, 429
Brush, Tooth	C. Roberts	Newark, N. J	Feb. 16, 1869	87, 070
Brush, Tooth	C. Rosefield	Washington, D. C	May 8, 1866	54, 604
Brush, Tooth	T. H. Spencer	Providence, R. I	May 7, 1867	64, 455
Brush, Tooth	H. N. Wadsworth	Washington, D. C	Nov. 17, 1857	18, 653
Brush, Tooth	H. N. Wadsworth	Washington, D. C	June 19, 1860	28, 794
Brush, Tooth	W. B. Watkins	Jersey City, N. J	Apr. 13, 1869	88, 825
Brush, Tooth and nail	G. A. Scott	New York, N. Y	Dec. 3, 1872	133, 599
Brush trimmer, Tooth	J. Stone	Waterford, N. Y	Oct. 8, 1872	132, 031
Brush trimmer, Tooth	J. Stone	Waterford, N. Y	Oct. 8, 1872	132, 032
Brush, Tumbler	H. Lumbard	Chicago, Ill	Mar. 24, 1868	75, 940
Brush, Tumbler	G. Wiesler	Chicago, Ill	Sept. 1, 1868	81, 718
Brush, Warp	S. Taylor	East Cambridge, Mass	May 12, 1863	38, 516
Brush, wash-board, and roller	I. Hussey	Ironton, Ohio	Sept. 16, 1873	142, 853
Brush-wheel	P. Fecker	Providence, R. I	May 10, 1870	102, 932
Brush, Whisk	H. S. Blunt	New York, N. Y	Mar. 26, 1872	125, 010
Brush, Whisk	M. L. Dickinson	West Troy, N. Y	Sept. 30, 1862	36, 560
Brush, Whitewash	W. B. Burtnett and J. P. McIntosh	New York and Brooklyn, N. Y.	Jan. 17, 1865	45, 907
Brush, Whitewash	E. S. Curtis	Charlestown, Mass	June 2, 1820	....
Brush, Whitewash	A. Foss	Wayne County, Ohio	July 16, 1867	66, 822
Brush, Whitewash	J. Loughridge	Pittsburgh, Pa	Mar. 31, 1863	38, 051
Brush, Whitewash	D. W. Shaw and W. A. Megraw	Baltimore, Md	June 1, 1858	20, 447
Brush, Whitewash and paste	J. S. White	Portland, Me	Jan. 2, 1872	122, 421
Brush, Window	J. Stofer	Cleveland, Ohio	Nov. 20, 1866	59, 932
Brush, Window	J. S. White	Boston, Mass	Apr. 30, 1867	64, 266
Brushes, Binder for paint	J. F. Canning	Boston, Mass	Dec. 17, 1867	72, 364
Brushes, brooms, &c., Grasses for making	S. H. Weed	New York, N. Y	Feb. 28, 1825	....
Brushes, Construction of	J. Marchbank	Lansingburgh, N. Y	Aug. 30, 1870	106, 845
Brushes, Drip-guard for	W. Buck	Portsmouth, Mich	Jan. 16, 1872	122, 701
Brushes, Ferrule for paint	W. B. Burtnett	New York, N. Y	Aug. 23, 1870	106, 545
Brushes for blending colors, Manufacturing	G. M. Morris	Philadelphia, Pa	Mar. 6, 1835	....
Brushes for whitewashing, &c., Manner of attaching bristles to the stock of.	R. B. Lewis	Hallowell, Me	Feb. 23, 1841	1, 992
Brushes from broom-corn, Making	B. Croasdale	Byberry, Pa	Dec. 21, 1819	....
Brushes, Girdle for paint	A. M. Piper and O. Allen	Springfield, Mass	Feb. 2, 1869	86, 586
Brushes, Guard for paint	E. C. Haserick	Lake Village, N. H	Mar. 1, 1864	41, 772
Brushes, Machine for making	O. D. Woodbury	New York, N. Y	Dec. 27, 1870	110, 529
Brushes, Machine for manufacturing	W. A. Foskett and H. B. Tyler	New Haven, Conn	July 6, 1869	92, 181
Brushes, Machine for manufacturing the wooden part of.	T. Mitchell	Lansingburgh, N. Y	Sept. 9, 1856	15, 702
Brushes, Machine for trimming bristles of	S. Taylor	East Cambridge, Mass	May 17, 1844	3, 593
Brushes, Machinery for manufacturing	G. Willett	Saint Albans, Vt	Sept. 26, 1871	119, 439
Brushes, Machinery for manufacturing	G. Willett	Saint Albans, Vt	Sept. 26, 1871	119, 440
Brushes, Machinery for trimming	B. Babbit	New York, N. Y	Oct. 14, 1846	4, 814
Brushes, Manufacture of paint	J. T. Steer	New York, N. Y	Mar. 31, 1857	16, 950
Brushes, Manufacturing	J. J. Adams	Boston, Mass	Mar. 12, 1842	2, 490
Brushes, Preparation of bristles for	C. Williams	Philadelphia, Pa	July 19, 1853	9, 869
Brushes, Socket for paint and other	J. W. Davis	Washington, D. C	Jan. 5, 1864	41, 060
Brush-wood, Machine for shaping	J. Ames, jr	Lansingburgh, N. Y	Dec. 16, 1873	145, 546
Brushing and polishing apparatus, Hand	W. H. Willson	New York, N. Y	Sept. 22, 1868	82, 465
Buck-board spring	T. H. Brown	Chicago, Ill	May 10, 1870	102, 762
Bucket	R. T. Brown	Newport, Ky	Oct. 28, 1873	143, 959
Bucket	J. V. Deuell	Brooklyn, N. Y	Feb. 25, 1873	136, 140
Bucket	J. H. Tomlinson	Chicago, Ill	June 1, 1869	90, 798
Bucket and measure	J. Hageman	Williamsburgh, Ohio	Oct. 7, 1862	36, 613
Bucket and strainer, Milk	D. N. West	Smithsburgh, Md	Sept. 3, 1867	68, 400
Bucket, Army feed and water	R. B. Fitts	Philadelphia, Pa	Nov. 19, 1861	33, 737
Bucket bracket, Sap	J. J. Pollett	Oconomowoc, Wis	Aug. 27, 1872	130, 863
Bucket, Butter	J. H. Stimpson	Baltimore, Md	Aug. 17, 1858	21, 220
Bucket, Chamber	W. W. Knight	Philadelphia, Pa	Apr. 3, 1866	53, 632
Bucket, Chamber	C. G. Schneider	Washington, D. C	July 8, 1862	35, 841
Bucket, Coal	L. Hayner	Clifton Park, N. Y	June 22, 1869	91, 743
Bucket, Collapsible	C. W. Curtiss	New Haven, Conn	Oct. 23, 1860	30, 465
Bucket cover, Sap	R. Marshall	Hobart, N. Y	Oct. 10, 1871	119, 779
Bucket cover, Sap	D. and E. Smith	Gilsum and Keene, N. H	Oct. 17, 1871	119, 952
Bucket, Covered	F. D. Kellogg and G. N. Ives	New Haven, Conn	Nov. 28, 1871	121, 288
Bucket, Dinner	J. Haines	West Middleburgh, Ohio	Mar. 12, 1867	62, 842
Bucket, Dinner	F. E. Heinig	Louisville, Ky	Sept. 30, 1873	143, 241
Bucket, Dumping	F. B. Colton	Philadelphia, Pa	Dec. 26, 1871	122, 227
Bucket-ear	H. Callahan	Dayton, Ohio	May 7, 1867	64, 484

Index of patents issued from the United States Patent Office from 1790 *to* 1873, *inclusive*—Continued.

Invention.	Inventor.	Residence.	Date.	No.
Bucket-ear	L. Rowland	Philadelphia, Pa	May 2, 1865	47, 573
Bucket, Elevator	O. W. Clark	Appleton, Wis	Aug. 4, 1868	80, 602
Bucket, Feed	R. C. Sturges	Barnstable, Mass	Dec. 31, 1867	72, 935
Bucket, Field	G. Wood	Strasburgh, Pa	May 14, 1861	32, 328
Bucket, Folding	H. W. Wilcox	Columbus, Pa	Jan. 31, 1865	46, 168
Bucket for preserving butter, fruit, &c	E. G. Hofman	Saint Louis, Mo	Oct. 11, 1864	44, 626
Bucket for raising water, Endless-chain	J. Dutton	Aston Township, Delaware County, Pa.	Oct. 9, 1841	2, 290
Bucket for removing coal, &c	J. S. Lloyd	Salem, N. J	Apr. 24, 1860	27, 993
Bucket for the manufacture of maple-sugar, Sap	J. H. Fairchild	Highgate, Vt	June 10, 1862	35, 561
Bucket for wells, Self-tilting	A. J. Cook	North Branford, Conn	May 29, 1866	55, 062
Bucket, Hanging well	S. F. Dexter	Paris, N Y	Mar. 15, 1859	23, 233
Bucket, Hoisting	E. J. Hulbert and A. N. N. Aubin.	Portland, Conn	Oct. 31, 1871	120, 522
Bucket holder, Sap	A. Franklin	Galena, Ohio	Sept. 19, 1871	119, 024
Bucket hoop, Sap	J. Bullock	Readsborough, Vt	June 19, 1866	55, 611
Bucket, Sap	J. W. Currier	Newbury, Vt	Mar. 31, 1868	76, 168
Bucket, Sap	C. C. Post	Hinesburgh, Vt	Sept. 13, 1870	107, 407
Bucket, Sap	J. Purdy	Cherburne, N. Y	Oct. 4, 1817	....
Bucket, Self-dumping coal	J. Wüst	Philadelphia, Pa	July 13, 1858	20, 912
Bucket, Sheet-metal	W. Austin	Philadelphia, Pa	Aug. 12, 1873	141, 748
Bucket, Water	C. Ballinger	Buchanan, Pa	Apr. 30, 1872	126, 249
Bucket, Weighing	D. D. Stelle	New Brunswick, N. J	Jan. 10, 1865	45, 873
Bucket, Well	A. J. Clemmons	Aberdeen, Miss	Aug. 16, 1870	106, 326
Bucket, Well	D. P. Farnham	Johnstown Centre, Wis	Apr. 3, 1860	27, 708
Bucket, Well	C. F. Stites	Cincinnati, Ohio	July 1, 1873	140, 439
Bucket, Wood	S. S. Vail	Keokuk, Iowa	Sept. 20, 1870	107, 568
Buckets and tubs, Tool for finishing	J. W. Bartlett and A. Morris	Harmar and Marietta, Ohio	July 28, 1863	39, 334
Buckets, Apparatus for raising and tilting water	H. W. Sabin and L. B. Benton	Reed's Corner and Penn Yan, N. Y.	Mar. 13, 1849	6, 190
Buckets, Brake and ratchet for elevating and lowering.	I. T. Smith	Richmond, Va	Feb. 22, 1870	100, 201
Buckets, Construction of sheet-metal	J. Fallows	Philadelphia, Pa	Nov. 5, 1867	70, 427
Buckets, Device for holding sap	J. Bevins	Unadilla Forks, N. Y	Nov. 5, 1867	70, 394
Buckets, Ear of wooden	L. A. Fleming	New York, N. Y	Nov. 9, 1869	96, 571
Buckets in raising water from wells, Self-operating device for tilting.	D. P. Farnham	Milton, Wis	Feb. 3, 1857	16, 535
Buckets, Metallic guard for water	J. B. Hyzer	Janesville, Wis	Dec. 13, 1864	45, 412
Buckets, Mode of emptying well	E. Willard	Egremont, Mass	June 15, 1826	....
Buckets, pails, tubs, &c., Machine for making	N. Rider	Sturbridge, Mass	Apr. 1, 1829	....
Buckets, Raising tilting	I. I. Vedder and H. Vine	Schenectady, N. Y	Feb. 1, 1848	5, 425
Buckets, Stud for attaching bail to	W. Brown and C. Stevens	Saint Louis, Mo	July 2, 1872	128, 585
Buckets, Wire ear for metal	J. M. Shank	Dayton, Ohio	Aug. 3, 1869	93, 355
Buckle	J. J. Adair	Portland, Ind	Apr. 19, 1870	102, 076
Buckle	M. Adams	Chilmark, Mass	Mar. 22, 1870	100, 963
Buckle	J. K. Andrews	Antrim, Ohio	Nov. 19, 1867	71, 116
Buckle	H. Aschenbach	Washington, D. C	Oct. 31, 1865	50, 673
Buckle	W. Babin	New York, N. Y	Mar. 24, 1868	75, 722
Buckle	G. L. Bailey	Portland, Me	Apr. 8, 1862	34, 868
Buckle	T. Bailey and H. H. Young	Fairfax County, Va., and Washington, D. C.	Aug. 17, 1869	93, 792
Buckle	T. G. Bailey	Wassaic, N. Y	July 4, 1865	48, 503
Buckle	I. Banister	Newark, N. J	Mar. 15, 1864	41, 900
Buckle	I. Banister	Newark, N. J	Nov. 8, 1864	44, 927
Buckle	I. Banister	Newark, N. J	Feb. 14, 1865	46, 324
Buckle	I. Banister	Newark, N. J	July 7, 1868	79, 626
Buckle	J. Barron	Mobile, Ala	Apr. 30, 1867	64, 273
Buckle	J. Bavier	Newark, N. J	Feb. 14, 1865	46, 326
Buckle	H. Bernheimer and H. Newman	New York, N. Y	Aug. 17, 1869	93, 665
Buckle	E. C. Blakeslee	Waterbury, Conn	Nov. 19, 1867	70, 944
Buckle	J. M. Borchardt	New York, N. Y	Aug. 26, 1873	142, 076
Buckle	G. A. Brown	Kalamazoo, Mich	Feb. 2, 1869	86, 501
Buckle	J. Buche	Apple River, Ill	Nov. 28, 1871	121, 325
Buckle	I. W. Burch	Fayette, Miss	Dec. 14, 1869	97, 891
Buckle	S. P. Burdick	New York, N. Y	May 19, 1868	77, 955
Buckle	F. Busch	New York, N. Y	Oct. 8, 1872	132, 051
Buckle	W. C. Bussey	San Francisco, Cal	Aug. 20, 1872	130, 565
Buckle	D. S. Butler	Otterville, Mo	Oct. 26, 1868	96, 083
Buckle	C. A. Button	Pontiac, Mich	Dec. 4, 1866	60, 135
Buckle	C. S. Cissna	Burlington, Iowa	May 8, 1866	54, 504
Buckle	J. G. Clifton	Northfield, Ind	July 28, 1868	80, 454
Buckle	A. H. Cole	Sylvania, Ohio	Oct. 10, 1865	50, 337
Buckle	E. Cole	Fairfield, Mich	Sept. 24, 1867	69, 181
Buckle	E. Cole	Fairfield, Mich	Sept. 24, 1867	69, 182
Buckle	E. P. Corwin	Washington, Ill	Dec. 13, 1870	110, 115
Buckle	J. Cory	Wayne, Mich	Sept. 15, 1868	82, 091
Buckle	L. D. Cowles	Romeo, Mich	Aug. 4, 1868	80, 605
Buckle	L. D. Cowles	Romeo, Mich	Dec. 6, 1870	109, 871
Buckle	S P. Crafts	New Haven, Conn	Jan. 16, 1866	52, 112
Buckle	J. Cumberland and J. R. McClintock.	Mobile, Ala., and New York, N. Y.	Dec. 21, 1858	22, 352
Buckle	T. B. DeForest	Birmingham, Conn	Nov. 20, 1866	59, 896
Buckle	T. B. DeForest	Birmingham, Conn	Mar. 5, 1867	62, 533
Buckle	N. Dico	Xenia, Ind	Jan. 4, 1870	98, 572
Buckle	S. H. Doughty	Nyack, N. Y	May 6, 1873	138, 487
Buckle	F. Douglas	Norwich, Conn	July 14, 1863	39, 217
Buckle	E. F. Driggs	Brooklyn, N. Y	Oct. 24, 1871	120, 187
Buckle	T. Duncan	Brookville, Md	June 21, 1870	104, 564
Buckle	E. F. Eaton	Northford, Conn	Sept. 28, 1869	95, 332
Buckle	A. R. Egbert	Philadelphia, Pa	May 1, 1866	54, 313
Buckle	L Elsberg	New York, N. Y	Jan. 7, 1868	73, 176
Buckle	H. Fautz	Newark, N. J	July 30, 1872	130, 032
Buckle	O. A. Field	Minneapolis, Minn	Aug. 6, 1872	130, 119
Buckle	L. Fogg	Boston, Mass	June 2, 1863	38, 793
Buckle	M. Fowler	Wolcottville, Conn	Mar. 12, 1867	62, 832
Buckle	M. Fowler	Wolcottville, Conn	Oct. 6, 1868	82, 821
Buckle	N. D Fowler	Valley Ford, Cal	Jan. 3, 1872	122, 315
Buckle	R. E. Frye	Manchester, N. H	Jan. 23, 1866	52, 154

Index of patents issued from the United States Patent Office from 1790 *to* 1873, *inclusive*—Continued.

Invention.	Inventor.	Residence.	Date.	No.
Buckle	J. N. Gaston	Lyons City, Iowa	Oct. 15, 1866	63, 793
Buckle	G. L. Gerard	New Haven, Conn	Aug. 6, 1867	67, 428
Buckle	G. L. Gerard	New Haven, Conn	Jan. 14, 1868	73, 242
Buckle	F. F. Greenwood	Hornsey, England	Dec. 14, 1869	97, 909
Buckle	H. C. Griggs	Waterbury, Conn	Aug. 28, 1866	57, 632
Buckle	H. C. Griggs	Waterbury, Conn	July 2, 1867	66, 324
Buckle	H. C. Griggs	Waterbury, Conn	Apr. 21, 1868	77, 033
Buckle	J. F. J. Gunning and I. T. Myer	New York, N. Y	June 26, 1866	55, 854
Buckle	A. Hagny	Keokuk, Iowa	May 29, 1866	55, 088
Buckle	J. C. Hall	Fayette, Miss	Apr. 3, 1860	27, 714
Buckle	W. S. Hall	Quincy, Mass	Aug. 8, 1865	49, 261
Buckle	E. Hambujer	Detroit, Mich	Mar. 19, 1867	63, 040
Buckle	M. Haneline	Huntington, Ind	Mar. 9, 1869	87, 563
Buckle	J. B. R. Hardeman	Tehuacana, Tex	June 18, 1872	128, 037
Buckle	S. S. Hartshorn	Orange, Conn	July 10, 1855	13, 218
Buckle	S. S. Hartshorn	Orange, Conn	Dec. 11, 1855	13, 907
Buckle	S. S. Hartshorn	New Haven, Conn	July 24, 1860	29, 270
Buckle	S. S. Hartshorn	New Haven, Conn	Feb. 13, 1866	52, 568
Buckle	S. S. Hartshorn	New Haven, Conn	Feb. 26, 1867	62, 485
Buckle	J. W. Hatch	Manlius, N. Y	Feb. 20, 1844	3, 440
Buckle	C. B. Hatfield	Boston, Mass	June 13, 1865	48, 236
Buckle	C. B. Hatfield	Boston, Mass	Oct. 31, 1865	50, 706
Buckle	G. Havell	Newark, N. J	Aug. 23, 1864	43, 945
Buckle	G. Havell	Newark, N. J	Aug. 8, 1871	117, 882
Buckle	J. B. Hawley	New Haven, Conn	Jan. 15, 1867	61, 192
Buckle	P. P. R. Hayden	New York, N. Y	Jan. 11, 1853	9, 534
Buckle	H. Herbert	Jersey City, N. J	Oct. 27, 1868	83, 381
Buckle	J. W. Hinman	Omro, Wis	Dec. 3, 1861	33, 840
Buckle	J. P. Hisley	Syracuse, N. Y	Aug. 10, 1869	93, 440
Buckle	A. H. Hopkins	Goshen, Ind	Oct. 8, 1867	69, 669
Buckle	O. L. Hopson and H. P. Brooks	Waterbury, Conn	May 12, 1863	38, 482
Buckle	O. L. Hopson and H. P. Brooks	Waterbury and Wolcottville, Conn.	Oct. 9, 1866	58, 725
Buckle	G. C. Huntress	Elkhorn, Wis	June 11, 1867	65, 749
Buckle	A. Hurd	Yonkers, N. Y	Feb. 11, 1868	74, 367
Buckle	J. C. Hyde	West Haven, Conn	Nov. 15, 1870	109, 321
Buckle	J. H. James	Warren, R. I	May 12, 1868	77, 885
Buckle	W. B. Johnson	Bowling Green, Ky	Aug. 27, 1867	68, 204
Buckle	W. V. Kay	Chicago, Ill	June 1, 1869	90, 849
Buckle	G. R. Kelsey	Middletown, Conn	Jan. 15, 1850	7, 025
Buckle	G. R. Kelsey	West Haven, Conn	Aug. 30, 1864	43, 997
Buckle	G. R. Kelsey	West Haven, Conn	May 29, 1866	55, 113
Buckle	G. R. Kelsey	West Haven, Conn	June 5, 1866	55, 311
Buckle	H. Laurence	Manlius, N. Y	July 13, 1844	3, 662
Buckle	B. S. Lawson	New York, N. Y	July 4, 1865	48, 569
Buckle	H. Lowenberg	New York, N. Y	May 22, 1866	54, 928
Buckle	W. Maclean and J. H. Harris	Vermont, Ill	June 1, 1869	90, 859
Buckle	F. W. Maes	Ise lohn, Prussia	Jan. 28, 1868	73, 734
Buckle	C. A. Mallory	Cambridge, Mass	Mar. 6, 1866	53, 017
Buckle	T. P. Marshall	Trenton, N. J	July 5, 1859	24, 649
Buckle	C W. Martin	Mount Pleasant, Iowa	Jan. 21, 1868	73, 620
Buckle	J. F. Martin	Harrisburgh, Oreg	May 9, 1871	114, 580
Buckle	J. H. Martin	Columbus, Ohio	Apr. 18, 1871	113, 902
Buckle	J. A. Mashmeyer	Beardstown, Ill	Apr. 21, 1868	77, 064
Buckle	C. Maxwell and J. G. Lindner	Syracuse, N. Y	Apr. 2, 1872	125, 315
Buckle	J. McLellen	Chambersburgh, Pa	July 24, 1866	56, 590
Buckle	D. McMillan and A. Rowan	Webster City, Iowa	Mar. 21, 1871	112, 947
Buckle	G. O. Monroe	New York, N. Y	Dec. 26, 1865	51, 740
Buckle	C. F. Moore	Waveland, Ind	Oct. 22, 1872	132, 482
Buckle	O. W. Morley	Ellisburgh, N. Y	Dec. 9, 1862	37, 105
Buckle	J. H. Morris	Paxton, Ill	May 2, 1871	114, 461
Buckle	M. Murphy	Washington, D. C	July 31, 1866	56, 785
Buckle	H. Neumann	New York, N. Y	Feb. 12, 1867	62, 059
Buckle	H. Neumann	New York, N. Y	Jan. 21, 1868	73, 633
Buckle	T. L. Ogier	West Chester, Pa	May 14, 1867	64, 693
Buckle	W. Parsons	Palmyra, N. Y	Sept. 9, 1873	142, 721
Buckle	J. Peabody	Dixmont Centre, Me	Nov. 28, 1865	51, 210
Buckle	J. Peckham	New Haven, Conn	Sept. 4, 1866	57, 758
Buckle	J. Peckham	New Haven, Conn	May 30, 1871	115, 513
Buckle	L. W. Perkins	Beloit, Wis	Apr. 10, 1866	53, 866
Buckle	E. A. Pierce	Brighton, Mass	Apr. 22, 1862	35, 038
Buckle	I. N. Plotts	New York, N. Y	Feb. 13, 1866	52, 597
Buckle	I. N. Plotts	New York, N. Y	June 12, 1866	55, 533
Buckle	N. Post	East Cleveland, Ohio	Oct. 10, 1865	50, 386
Buckle	T. O. Potter and J. W. Smith	Boston, Mass	Jan. 9, 1872	122, 533
Buckle	F. Puetz	Buffalo, N. Y	Dec. 6, 1870	109, 941
Buckle	L. Rhoades	Newport, R. I	Oct. 22, 1867	70, 019
Buckle	F. C. Richer	Gilmer, Tex	May 4, 1869	89, 689
Buckle	E. G. Rockwood	Hillsdale, Mich	May 10, 1864	42, 689
Buckle	W. Roemer	Newark, N. J	Sept. 3, 1872	131, 028
Buckle	A. Roesler	Warsaw, Ill	May 24, 1859	24, 145
Buckle	G. W. Roland	Salem, Oregon	May 11, 1869	90, 026
Buckle	I. Roraback	South Bend, Ind	Aug. 13, 1867	67, 808
Buckle	F. A. Ross and S. De La Mater	Terre Haute, Ind	June 13, 1871	115, 985
Buckle	L. W. Russell	Galesburgh, Mich	June 19, 1866	55 717
Buckle	R. F. Russell	Hazelton, Pa	Mar. 12, 1872	124, 514
Buckle	C. W. Saladee	Putnam, Ohio	Aug. 8, 1865	49, 309
Buckle	H. Sanders	Utica, N. Y	Apr. 2, 1872	125, 335
Buckle	A. V. Sargent	Syracuse, N. Y	June 1, 1869	90, 878
Buckle	C. Sears and F. Townsend	Monmouth County, N. J., and Kings County, N. Y.	Mar. 21, 1865	46, 948
Buckle	A. B. Shaw	Medford, Mass	Aug. 12, 1873	141, 729
Buckle	J. Shepard	New Britain, Conn	Nov. 15, 1864	45, 088
Buckle	W. Shove	Elizabethport, N. J	June 13, 1854	11, 090
Buckle	D. L. Smith	Waterbury, Conn	June 13, 1865	48, 241
Buckle	E. A. Smith	Waterbury, Conn	Jan. 22, 1867	61, 477
Buckle	E. A. and D. L. Smith	Waterbury, Conn	July 31, 1866	56, 811

Index of patents issued from the United States Patent Office from 1790 *to* 1873, *inclusive*—Continued.

Invention.	Inventor.	Residence.	Date.	No.
Buckle	E. A. and D. L. Smith	Waterbury, Conn	Apr. 25, 1871	114, 214
Buckle	E. A. and D. L. Smith	Waterbury, Conn	May 30, 1871	115, 535
Buckle	J. E. Smith	Waterbury, Conn	Apr. 23, 1861	32, 152
Buckle	J. E. Smith	Waterbury, Conn	June 6, 1865	48, 135
Buckle	P. W. Smith	Abington, Ill	Oct. 13, 1868	83, 102
Buckle	W. W. Smith	Marshall, Mich	July 18, 1854	11, 336
Buckle	L. A. Sprague	Brooklyn, N. Y	May 27, 1862	35, 401
Buckle	J. Stanbrough	Newark, N. J	Nov. 7, 1865	50, 852
Buckle	C. S. Stearns and T. Corey	Marlborough, Mass	Nov. 14, 1865	50, 992
Buckle	G. E. Stedman	Boston, Mass	Dec. 6, 1870	109, 849
Buckle	G. F. Stephens	Portland, Oreg	June 6, 1871	115, 654
Buckle	J. Stevens	New York, N. Y	Jan. 12, 1864	41, 267
Buckle	S. G. Sturges	Newark, N. J	July 16, 1872	129, 257
Buckle	H. R. Swan	Norwalk, Conn	Nov. 23, 1869	97, 133
Buckle	S. A. Tenny	Muskego, Wis	Oct. 6, 1868	82, 768
Buckle	D. S. Thompson	West Haven, Conn	June 26, 1866	55, 933
Buckle	W. McK. Thornton	Clinton, Wis	Sept. 3, 1867	68, 468
Buckle	W. McK. Thornton	Clinton, Wis	May 4, 1869	89, 836
Buckle	J. Tiebout	Brooklyn, N. Y	July 24, 1860	29, 332
Buckle	I. B. Verplank	Mentz, N. Y	Mar. 9, 1844	3, 471
Buckle	J. Waterman	New York, N. Y	Feb. 13, 1866	52, 652
Buckle	D. Webber	Houlton, Me	June 20, 1871	116, 124
Buckle	W. Welsh	McHenry, Ill	Dec. 4, 1866	60, 302
Buckle	P. White	Dixmont Centre, Me	July 4, 1865	48, 610
Buckle	H. S. Wilkin	Brooklyn, N. Y	May 9, 1871	114, 631
Buckle	J. J. Wilkins	Virdey, Ill	May 15, 1866	54, 803
Buckle	H. C. Wissel	Indiana, Pa	Aug. 18, 1868	81, 318
Buckle	C. E. Woodman	Boston, Mass	Apr. 20, 1869	89, 110
Buckle	C. E. Woodman and C. B. Hatfield.	Boston, Mass	Apr. 4, 1865	47, 159
Buckle	G. P. Woodruff	Watertown, Conn	Jan. 6, 1857	16, 347
Buckle	H. S. Woodruff	Janesville, Wis	Feb. 12, 1867	61, 979
Buckle	H. S. Woodruff	Janesville, Wis	Jan. 9, 1872	122, 545
Buckle	E. L. Woods	Alliance, Ohio	Dec. 17, 1867	72, 255
Buckle	H. B. Worcester	Chicago, Ill	Apr. 23, 1872	126, 003
Buckle	W. Yost	Butterville, Ohio	July 18, 1871	117, 232
Buckle and clasp	W. Church	Birmingham, England	Feb. 18, 1840	1, 490
Buckle and loop	S. C. Talcott	Ashtabula, Ohio	July 25, 1871	117, 347
Buckle and loop, Attaching	L. C. Chase	Boston, Mass	June 14, 1864	43, 093
Buckle and ring	R. C. Dunham	New Britain, Conn	Oct. 9, 1866	58, 619
Buckle and ring, Halter	L. E. Jones and T. M. Barber	Syracuse, N. Y	Feb. 21, 1871	112, 048
Buckle and snap, Combined	S. G. and W. E. Sturges	Newark, N. J	June 22, 1869	91, 685
Buckle-attachment	W. E. Barton	East Hampton, Conn	June 6, 1865	48, 039
Buckle-attachment	R. T. Moss	Cambridge, Ohio	Oct. 22, 1867	70, 105
Buckle, Belt	I. Alexander	San Francisco, Cal	June 28, 1870	104, 683
Buckle, Belt	F. Clausen	San Francisco, Cal	Sept. 22, 1868	82, 291
Buckle, Belt	F. Clausen	San Francisco, Cal	May 25, 1869	90, 501
Buckle, Belt	W. Cummings	Sacramento, Cal	Aug. 11, 1868	80, 924
Buckle, Belt	P. M. Haas	Youngstown, Ohio	May 21, 1872	127, 048
Buckle, Belt	G. R. Kelsey	West Haven, Conn	Oct. 2, 1866	58, 430
Buckle, Belt	F. and C. Lemme	San Francisco, Cal	May 2, 1865	47, 556
Buckle, Breeching-loop	J. H. McGarrah	Princeton, Ind	Jan. 28, 1873	135, 238
Buckle, Bridle	D. S. Butler	Brownsville, Mo	Feb. 20, 1872	123, 866
Buckle-chape and strap-loop	G. Crouch	Westport, Conn	July 9, 1872	128, 714
Buckle-clasp	E. D. Ballou	Abington, Mass	May 29, 1866	55, 197
Buckle-closing die	M. Fowler	Wolcottville, Conn	Feb. 11, 1868	74, 335
Buckle, Double	E. W. Clark	Tallahassee, Fla	July 21, 1868	80, 139
Buckle-fastening	F. H. Gladding	Providence, R. I	Feb. 25, 1873	136, 131
Buckle-fastening	C. Goodyear, jr., and L. A. Sprague.	New York, N. Y	May 10, 1864	42, 726
Buckle-fastening	R. Meyer	Buffalo, N. Y	Jan. 29, 1867	61, 628
Buckle-fastening	W. Wiley, jr	Kokomo, Ind	Sept. 26, 1865	50, 195
Buckle for clothing	E. A. Smith	Waterbury, Conn	Nov. 16, 1869	96, 981
Buckle for grain-bands, bag-ties, &c	L. Kniffen	Rochester, Pa	Mar. 2, 1869	87, 500
Buckle for gaging reciprocating saws	H. W. Brown	Warren, Pa	May 27, 1873	139, 363
Buckle for harness	K. Thrashure	Manlius, N. Y	Jan. 16, 1845	3, 883
Buckle for harness	J. W. Ropp and R. C. Kibby	Elyria, Ohio	Mar. 11, 1873	136, 766
Buckle for harness, Trace	W. R. Knowles	Columbiana, Ohio	Feb. 15, 1870	99, 914
Buckle for harness-tug	R. Tattershall	Beloit, Wis	Oct. 11, 1864	44, 693
Buckle for harness, Wedge	K. Frasier	Syracuse, N. Y	Sept. 8, 1868	81, 993
Buckle for skirt-hoop	J. Stevens and J. Hanley	New York, N. Y	Dec. 21, 1858	22, 385
Buckle for waist-belt	W. U. Bohm	San Francisco, Cal	Jan. 27, 1863	37, 489
Buckle for wearing-apparel	E. Parker	Plymouth, Conn	Sept. 2, 1856	15, 666
Buckle for wearing-apparel	W. Slade	Gum Creek, Ga	Sept. 2, 1856	15, 671
Buckle gag-runner	J. McMartin	Janesville, Wis	Nov. 17, 1863	40, 636
Buckle, Girth	L. C. Chase	Boston, Mass	Nov. 8, 1859	26, 013
Buckle, Halter	F. Ditton	Auburn, N. Y	June 2, 1868	78, 516
Buckle, Halter	A. McGaffey	Burlington, Vt	Dec. 3, 1872	133, 657
Buckle, Halter	P. A. Snyder	Butterville, Ohio	July 18, 1871	117, 214
Buckle, Hames-tug	L. Wetzell	Washington, D. C	Jan. 7, 1868	73, 148
Buckle, Harness	J. Allbee	Squaw Grove, Ill	Dec. 9, 1873	145, 270
Buckle, Harness	J. Ammidon	Springfield, Mass	Jan. 7, 1862	34, 085
Buckle, Harness	A. E. Bailey and H. Nichols	Middleville, N. Y	Nov. 19, 1867	70, 940
Buckle, Harness	A. Bedford	Coldwater, Mich	June 2, 1868	78, 568
Buckle, Harness	A. Bedford	Coldwater, Mich	May 17, 1870	103, 128
Buckle, Harness	S. Bingham, jr	Poestenkill, N. Y	Dec. 10, 1850	7, 818
Buckle, Harness	O. Brown	Albia, Iowa	Oct. 31, 1871	120, 413
Buckle, Harness	A. B. Buell	Westmoreland, N. Y	June 19, 1847	5, 166
Buckle, Harness	O. S. Burges	Columbus, Ohio	Oct. 10, 1848	5, 838
Buckle, Harness	G. S. Caldwell	Syracuse, N. Y	Apr. 16, 1867	63, 782
Buckle, Harness	W. H. Cocks	Richmond, Ind	Sept. 17, 1867	68, 846
Buckle, Harness	J. W. Covel	Bangor, Me	July 31, 1860	29, 361
Buckle, Harness	L. D. Cowles	Armada, Mich	Oct. 25, 1864	44, 786
Buckle, &c., Harness	L. D Cowles	Armada, Mich	Oct. 25, 1864	44, 787
Buckle, Harness	T. Crakes	Mishawaka, Ind	June 20, 1871	116, 028
Buckle, Harness	P. S. Crawford	Rockford, Ill	May 26, 1868	78, 364
Buckle, Harness	E. S. Dawson	Syracuse, N. Y	Jan. 12, 1869	85, 798
Buckle, Harness	F. W. Dean	Tremont, Ill	May 25, 1869	90, 430

Index of patents issued from the United States Patent Office from 1790 to 1873, inclusive—Continued.

Invention.	Inventor.	Residence.	Date.	No.
Buckle, Harness	J. S. Embich	Green Village, Pa	May 21, 1850	7, 377
Buckle, Harness	K. Frazer	Fayetteville, N. Y	Mar. 26, 1845	3, 970
Buckle, Harness	K. Frazer	Syracuse, N. Y	Sept. 16, 1873	142, 907
Buckle, Harness	W. D. Hillis	Cuyahoga, Ohio	Jan. 26, 1847	4, 943
Buckle, Harness	R. A. Helmes	New York, N. Y	Sept. 19, 1848	5, 790
Buckle, Harness	J. Howey	New Pittsburgh, Ohio	Mar. 8, 1848	5, 465
Buckle, Harness	J. Ives	Mount Carmel, Conn	July 27, 1869	92, 973
Buckle, Harness	C. C. Lee	Falls Church, Va	May 27, 1873	139, 253
Buckle, Harness	A. C. Luther	Canton, Ohio	Apr. 20, 1869	89, 057
Buckle, Harness	J. H. Morris	Normal, Ill	Jan. 30, 1872	123, 281
Buckle, Harness	D. M. Nixon	Danville, Ill	May 17, 1864	42, 787
Buckle, Harness	O. B. North	New Haven, Conn	Sept. 23, 1873	143, 180
Buckle, Harness	C. B. Payne	Bloomington, Ill	May 19, 1868	78, 124
Buckle, Harness	M. W. Pond	Elyria, Ohio	Aug. 19, 1862	36, 235
Buckle, Harness	N. Post	East Cleveland, Ohio	Mar. 18, 1856	14, 469
Buckle, Harness	J. Prendergast	Boston, Mass	Nov. 10, 1857	18, 601
Buckle, Harness	J. Purplee	Wellington, Ohio	Oct. 4, 1864	44, 554
Buckle, Harness	G. Reyer	Indianapolis, Ind	Apr. 19, 1870	102, 040
Buckle, Harness	C. W. Saladee	Putnam, Ohio	May 2, 1865	47, 574
Buckle, Harness	S. S. Sargeant	Newark, N. J	May 14, 1872	126, 649
Buckle, Harness	J. L. Selleck	Baldwinsville, N. Y	Apr. 29, 1873	138, 291
Buckle, Harness	O. B. Smith	Monticello, N. Y	Dec. 21, 1858	22, 383
Buckle, Harness	M. M. Stewart	Mayfield, Ky	July 9, 1872	128, 824
Buckle, Harness	S. P. Taylor	Oxford, Ohio	June 30, 1868	79, 519
Buckle, Harness	W. H. Taylor	Baldwinsville, N. Y	July 14, 1868	79, 874
Buckle, Harness	H. Todd	Columbus, Ohio	May 22, 1846	6, 477
Buckle, Harness	S. E. Tyler and R. Tattershall.	Beloit, Wis	Dec. 5, 1865	51, 367
Buckle, Harness	D. Vogt	Trenton, Mich	May 10, 1870	102, 889.
Buckle, Harness	D. Walker	New Haven, Conn	Dec. 10, 1872	133, 904
Buckle, Harness	H. Hise	Ottawa, Ill	Aug. 25, 1863	39, 651
Buckle, Harness and trace	P. Burns	Syracuse, N. Y	Sept. 24, 1872	131, 658
Buckle, Harness-trace	H. H. Hartzell	Holden, Mo	Sept. 30, 1873	143, 284
Buckle, Harness-trace	C. S. Abeel	Marshalltown, Iowa	Aug. 22, 1871	118, 326
Buckle, Harness-tug	G. H. Buckins	Canton, Ohio	June 29, 1869	92, 011
Buckle, Harness-tug	J. H. Feraw	Hinsdale, N. Y	July 27, 1858	20, 997
Buckle, Harness-tug	F. Seifert	Knoxville, Iowa	Feb. 4, 1873	135, 601
Buckle, Harness-tug	W. Straw and R. H. Armstrong	Hudson, Mich	May 11, 1858	20, 222
Buckle, Harness-tug	J. N. Burton	Senoria, Ga	Mar. 10, 1868	75, 359
Buckle, Hat	P. J. Stoll	Marshallville, Ohio	Dec. 22, 1868	85, 186
Buckle, Hitching-strap	L. A. Sprague	New York, N. Y	Mar. 7, 1865	46, 719
Buckle, Lever	H. W. Warner	Greenfield, Mass	June 6, 1865	48, 121
Buckle, Lever	M. Fowler	Wolcottville, Conn	Aug. 20, 1867	67, 865
Buckle-making machine	J. Robbins	Auburn, N. Y	Dec. 7, 1869	97, 700
Buckle-making machine	R. Durning	Lawrenceville, Pa	Feb. 16, 1864	41, 610
Buckle-making tool	S. S. Hartshorn	Orange, Conn	Nov. 7, 1854	11, 892
Buckle manufacture	J. Green	New York, N. Y	May 25, 1807	....
Buckle, Mechanical spring	R. G. Pine	Sing Sing, N. Y	Apr. 5, 1856	14, 633
Buckle-polishing machine	K. Frazer	Syracuse, N. Y	May 6, 1862	35, 150
Buckle, ring, &c	J. Chestnut, jr	Hustontown, Pa	Dec. 21, 1869	98, 147
Buckle, Safety harness	W. W. Spencer	Pittsburgh, Pa	June 2, 1868	78, 491
Buckle, Self-fastening	M. Marvin	Salem, Oreg	Apr. 5, 1870	101, 482
Buckle, Self-tightening	J. Elleman	Providence, R. I	Oct. 10, 1865	50, 345
Buckle, Shoe	J. F. Mersell	Stamford, Conn	Nov. 19, 1867	71, 041
Buckle, Spring	C. Benedict	Waterbury, Conn	Oct. 22, 1850	7, 726
Buckle, Suspender	J. Bingham	Philadelphia, Pa	Feb. 4, 1843	2, 940
Buckle, Suspender	P. Blake	New Haven, Conn	Mar. 29, 1864	42, 067
Buckle, Suspender	H. Dubosq	Philadelphia, Pa	Apr. 25, 1844	3, 559
Buckle, Suspender	S. Hamm	Philadelphia, Pa	Sept. 23, 1873	143, 144
Buckle, Suspender	S. S. Hartzhorn	Naugatuck, Conn	Sept. 25, 1849	6, 736
Buckle, Suspender	A. Pototsky	New York, N. Y	July 14, 1868	79, 918
Buckle, Suspender	W. Scarlett	Newark, N. J	May 14, 1850	7, 370
Buckle, Suspender	A. Shenfield	New York, N. Y	Mar. 21, 1871	112, 857
Buckle, Suspender	E. Steele	Waterbury, Conn	Dec. 3, 1850	7, 814
Buckle, Suspender	O. Webb	New York, N. Y	July 5, 1817	....
Buckle, Suspender	E. L. Parker	Birmingham, England	Feb. 13, 1872	123, 640
Buckle tab-plate	A. Worster	Hannibal, N. Y	Dec. 4, 1849	6, 931
Buckle-tongue, Detachable	A. Bedford	Coldwater, Mich	Aug. 25, 1868	81, 464
Buckle-tongues, Tool for cutting holes in cloth or leather for.	R. J. Algeo	Kalamazoo, Mich	July 31, 1866	56, 689
Buckle, Trace	R. J. Baker	Madison, Wis	Aug. 27, 1867	68, 147
Buckle, Trace	E. K. Brayton and J. Baugh	Eau Claire, Wis	July 22, 1873	141, 108
Buckle, Trace	L. Brooks	Mount Pleasant, Iowa	Nov. 22, 1870	109, 378
Buckle, Trace	T. Brownlee	Fairbury, Ill	July 1, 1873	140, 463
Buckle, Trace	J. H. H. Buell	Oriskany, N. Y	Aug. 24, 1869	94, 070
Buckle, Trace	W. D. Bunker	Portage, Wis	Sept. 29, 1868	82, 489
Buckle, Trace	I. Clark	Dryden, Mich	Aug. 8, 1871	117, 742
Buckle, Trace	A. H. Cole	Adrian, Mich	May 9, 1871	114, 530
Buckle, Trace	E. A. Crownhart	Bridgeport, Conn	June 1, 1869	90, 641
Buckle, Trace	E. S. Dawson	Syracuse, N. Y	Feb. 9, 1869	86, 822
Buckle, Trace	M. H. Densmore	Shickshinny, Pa	Apr. 4, 1871	113, 406
Buckle, Trace	M. B. De Silva and G. T. Benjamin.	Cleveland, Ohio	Jan. 21, 1873	135, 093
Buckle, Trace	J. W. Denton	Paris, Ill	Aug. 14, 1866	57, 098
Buckle, Trace	C. Fillmore	Romeo, Mich	Dec. 31, 1867	72, 832
Buckle, Trace	O. Finch	Owego, N. Y	Sept. 17, 1867	68, 864
Buckle, Trace	K. Frazer	Syracuse, N. Y	Apr. 13, 1869	88, 952
Buckle, Trace	M. Gayhart	Young America, Wis	June 23, 1868	79, 064
Buckle, Trace	W. W. Gordon and D. Pettengill.	Delhi, N. Y	Oct. 6, 1868	82, 709
Buckle, Trace	W. B. Hayden	Columbus, Ohio	Oct. 8, 1867	69, 562
Buckle, Trace	H. Hise	Chicago, Ill	July 14, 1868	79, 978
Buckle, Trace	A. H. Hopson	Taylorville, Ill	Aug. 2, 1870	105, 941
Buckle, Trace	J. Kennedy	Osage Mission, Kans	Aug. 5, 1873	141, 565
Buckle, Trace	E. M. Kinne	Cuba, N. Y	Mar. 30, 1869	88, 489
Buckle, Trace	W. W. Kittleman	Bloomfield, Iowa	Dec. 27, 1870	110, 572
Buckle, Trace	W. W. Kittleman	Bloomfield, Iowa	Sept. 5, 1871	118, 724
Buckle, Trace	E. F. Lacy	Danville, Ill	Sept. 3, 1867	68, 756
Buckle, Trace	J. Lindsey	Litchfield, Ill	Mar. 26, 1872	125, 060

Index of patents issued from the United States Patent Office from 1790 *to* 1873, *inclusive*—**Continued.**

Invention.	Inventor.	Residence.	Date.	No.
Buckle, Trace	J. Lindsey	Litchfield, Ill	Nov. 26, 1872	133, 458
Buckle, Trace	D. F. Maine	Mansfield, Ohio	Sept. 19, 1865	50, 017
Buckle, Trace	J. H. Martin	Columbus, Ohio	Aug. 31, 1869	93, 229
Buckle, Trace	T. C. Martin and J. Offineer	Perrysville, Ohio	May 10, 1870	102, 843
Buckle, Trace	C. H. Miller	Buffalo, N. Y	Aug. 25, 1868	81, 394
Buckle, Trace	G. Oldham, jr	Cuba, N. Y	Apr. 25, 1871	114, 184
Buckle, Trace	C. B. Payne	Bloomington, Ill	Aug. 6, 1867	67, 448
Buckle, Trace	D. K. Peacock	Little Valley, N. Y	Aug. 8, 1871	117, 923
Buckle, Trace	W. G. Riley	Sullivan, Ind	Apr. 25, 1871	114, 201
Buckle, Trace	W. A. Robinson	Grand Rapids, Mich	Dec. 7, 1869	97, 704
Buckle, Trace	S. S. Sargeant	Newark, N. J	Nov. 26, 1872	133, 340
Buckle, Trace	C. Schwaner	Keokuk, Iowa	Sept. 10, 1867	68, 796
Buckle, Trace	C. H. Stevens	Fayetteville, N. Y	July 19, 1870	105, 509
Buckle, Trace	W. S. Thayer	Smithsborough, N. Y	Apr. 9, 1872	125, 631
Buckle, Trace	J. Thornton and C. F. Demmen	Wellsville, N. Y	Apr. 26, 1870	102, 333
Buckle, Trace	W. M. Thornton	Clinton, Wis	July 24, 1866	56, 638
Buckle, Trace	J. Welker	Attica, N. Y	June 15, 1869	91, 390
Buckle, Trace	H. S. Woodruff	Janesville, Wis	Nov. 14, 1865	50, 979
Buckle, Trace	A. Worster	Hannibal, N. Y	June 22, 1869	91, 807
Buckle, Trace	A. Worster	Syracuse, N. Y	June 22, 1869	91, 808
Buckle, Trace	A. Worster	Syracuse, N. Y	Oct. 31, 1871	120, 561
Buckle, Trace	W. Wyckoff	Chagrin Falls, Ohio	Aug. 22, 1865	49, 580
Buckle, Tug	C. W. Adams	Wheeling, W. Va	Feb. 11, 1873	135, 744
Buckle, Tug	J. C. Barrows	Centreville, Iowa	May 21, 1872	126, 920
Buckle, Tug	D. S. Butler	Otterville, Mo	June 1, 1869	90, 725
Buckle, Tug	G. P. Cole	Hudson, Mich	Mar. 31, 1868	76, 162
Buckle, Tug	H. Hise	Ottawa, Ill	Aug. 28, 1866	57, 507
Buckle, Tug	N. H. Howard	Beloit, Wis	Sept. 1, 1863	39, 734
Buckle, Tug	C. D. W. Ries	Edwards, N. Y	May 2, 1865	47, 570
Buckle, Turn	T. Gaillard	Brooklyn, N. Y	Mar. 18, 1873	136, 910
Buckle, Turn	M. White	Brooklyn, N. Y	Apr. 23, 1872	125, 999
Buckle, Wedge	M. W. Pond, jr., and A. T. Ballantine.	Titusville, Pa	Feb. 11, 1868	74, 243
Buckle with button-hole attachment	B. J. Greely	Boston, Mass	Aug. 16, 1870	106, 356
Buckles and straps, Metallic fastening for	H. H. Mansfield	South Canton, Mass	Oct. 9, 1866	58, 657
Buckles, Attachment for	C. H. Pratt	Boston, Mass	May 24, 1870	103, 395
Buckles, Clip for attaching	S. G. and W. E. Sturges	Newark, N. J	Aug. 24, 1869	94, 042
Buckles, Cutting and finishing harness	J. Carman	Hudson, N. Y	Oct. 12, 1814	....
Buckles, Machine for making suspender	C. A. Lent	Newark, N. J	Jan. 30, 1849	6, 074
Buckles, Machine for making suspender	W. Scarlett	Newark, N. J	Apr. 10, 1849	6, 311
Buckles, Machinery for making	L. P. Mitchell	Waterbury, Conn	Dec. 19, 1865	51, 606
Buckles, Machinery for making four-sided	A. and O. B. North and S. Frink.	New Britain, Conn	May 28, 1850	7, 398
Buckles, Making	G. B. McDonald	New Albany, Ind	May 14, 1867	64, 782
Buckles, Manufacture of	G. R. Kelsey	West Haven, Conn	Feb. 18, 1862	34, 429
Buckles, &c., Method of attaching loops to	L. C. Chase	Boston, Mass	June 13, 1865	48, 153
Buckles, Mode of fastening on	L. Hilbright	Newark, N. J	Oct. 4, 1864	44, 530
Buckles to straps, Mode of attaching	N. H. Furness	New York, N. Y	July 16, 1872	129, 022
Buckles to suspenders, &c., Attaching	J. Abernethy	Woodbury, Conn	July 3, 1849	6, 575
Buckwheat and barley huller	T. Register	West Township, Ohio	Apr. 5, 1832	....
Buckwheat-cleaning machine	D. Pease, jr	Floyd, N. Y	July 14, 1846	4, 628
Buckwheat-cleaning machine	J. H. Reed	Penn Township, Pa	Sept. 4, 1860	29, 940
Buckwheat-huller	C. B. Horton	Elmira, N. Y	May 22, 1855	12, 908
Buckwheat-huller	C. B. Horton	Elmira, N. Y	Aug. 13, 1861	33, 040
Buckwheat-huller	T. Nelson	Troy, N. Y	Mar. 15, 1870	100, 790
Buckwheat-huller, Cellular	A. Crawford	Wilkesbarre, Pa	Feb. 23, 1869	87, 093
Buckwheat, Hulling	W. Ager	Rohrsburgh, Pa	June 29, 1852	9, 064
Buckwheat-hulling machine	J. Baysore	Freeport, Ill	Jan. 7, 1868	73, 156
Buckwheat, Machine for scouring and hulling	J. N. Treadwell	Redding, Conn	Sept. 27, 1859	25, 593
Buckwheat-refiner	D. D. Brewster	West Laurens, N. Y	Apr. 25, 1871	114, 103
Buckwheat-size, Manufacture of	W. A. Comstock	Providence, R. I	Nov. 14, 1848	5, 915
Budding-knife	E. D. and R. Gird	Cedar Lake, N. Y., and Healdsburgh, Cal.	Feb. 4, 1862	34, 328
Buds from blossoming, Delaying	R. Moore	Rowan County, N. C	Mar. 16, 1822	....
Buffet-table	W. H. Tufts	Lynn, Mass	Nov. 11, 1873	144, 424
Buffing-knives, Implement for turning the edge of shoemakers'.	B. B. Webster	East Haverhill, Mass	May 10, 1870	102, 984
Buffing-machine	W. W. Messer, jr	Boston, Mass	Apr. 1, 1873	137, 465
Buffing wheel	E. and W. Dixon	Newark, N. J	Dec. 16, 1873	145, 492
Bug-destroying machine, Steam	J. Howlet	Greenborough, N. C	July 10, 1834	....
Bug or roach trap	J. M. Keep	New York, N. Y	May 23, 1871	115, 215
Bugs, perfuming houses, and cleaning furniture, Destroying.	J. H. Clark	Connersville, Ind	May 31, 1834	....
Buggy	J. H. Bye	Sterling, Ill	Dec. 8, 1863	40, 817
Buggy	J. R. Gilman	South Bend, Ind	Apr. 25, 1871	114, 000
Buggy	J. S. McClelland	Jefferson, Ind	Aug. 8, 1854	11, 486
Buggy	C. T. Stoddard	Colebrook, N. H	Apr. 8, 1873	137, 577
Buggy and wagon top	G. Fuchs	Milwaukee, Wis	Apr. 13, 1869	88, 954
Buggy-boats, Wheel of	P. Davis	Providence, R. I	July 12, 1859	24, 725
Buggy-body	C. P. Kimball	Portland, Me	Apr. 16, 1872	125, 817
Buggy-brake	I. L. Myers	Ferguson Township, Pa	Mar. 28, 1871	113, 193
Buggy-brake, Automatic	L. T. Conant	New Lisbon, Ohio	Aug. 23, 1870	106, 555
Buggy-coupling	L. G. Peel	Preston, Ga	Apr. 27, 1869	89, 336
Buggy-gear, Elastic	I. N. Topliff	Elyria, Ohio	June 4, 1872	127, 441
Buggy-gearing	J. B. Augur	Poughkeepsie, N. Y	May 10, 1870	102, 905
Buggy hanging-top	C. Smith	Stockbridge, Mich	Oct. 22, 1867	70, 128
Buggy, Joint-bodied	E. J. Green	Cedarville, N. Y	June 6, 1854	11, 042
Buggy, Joint-bodied	E. J. Green	Valparaiso, Ind	Apr. 12, 1859	23, 567
Buggy, Joint-bodied	E. J. Green and M. H. Wheeler	Cedarville, N. Y	Apr. 29, 1856	14, 766
Buggy pole and shaft, Combined	G. Steiner	Deedsville, Ind	July 16, 1872	129, 066
Buggy-reach	J. C. Hilsabeck	Montovallo, Mo	Nov. 7, 1871	120, 644
Buggy-reach	J. W. Reeder	West Manchester, Ohio	Oct. 15, 1872	132, 320
Buggy running-gear	E. P. Carter	Arcade, N. Y	Oct. 10, 1871	119, 819
Buggy-seat	C. Haas	Chicago, Ill	June 21, 1870	104, 583
Buggy-seat	E. T. Mithoff and J. W. Dann	Columbus, Ohio	Jan. 24, 1871	111, 232
Buggy-seat, Adjustable	D. and N. Kroninger	Eagle Point, Pa	Oct. 31, 1871	120, 444
Buggy-seat, Folding	T. H. Wood	New York, N. Y	July 26, 1870	105, 758
Buggy-seat rail	J. Carlisle	Mount Gilead, Ohio	Sept. 3, 1867	68, 486

Index of patents issued from the United States Patent Office from 1790 to 1873, inclusive—Continued.

Invention.	Inventor.	Residence.	Date.	No.
Buggy-seat shifting-rail	C. Disser	West Union, Ohio	Sept. 10, 1867	68, 717
Buggy-seats, Machine for boring	G. W. Lemley	Pavilion, N. Y	Apr. 29, 1862	35, 099
Buggy-shifting rail	H. French and R. Meyer	Buffalo, N. Y	Dec. 21, 1869	98, 048
Buggy-spring	H. H. Hill	Pontiac, Ill	July 18, 1871	117, 171
Buggy-spring	W. Humphreys	Brooklyn, N. Y	Oct. 29, 1867	70, 219
Buggy-spring	G. W. Kenan	Upper Sandusky, Ohio	Sept. 13, 1870	107, 386
Buggy-spring and attachment	G. A. Brown	Reading, Mich	Oct. 3, 1871	119, 564
Buggy-spring clip	R. A. Clark	Unadilla, N. Y	Mar. 14, 1871	112, 549
Buggy-tap	W. S. Johnson	Henderson, Ky	Jan. 21, 1868	73, 530
Buggy-top	R. Bower	Lima, Ohio	Feb. 13, 1872	123, 611
Buggy-top	A. M. Cory	New Providence, N. J	Dec. 7, 1869	97, 479
Buggy-top	L. H. Gano	Ripon, Wis	June 18, 1861	32, 558
Buggy-top	H. Hibbard	Henrietta, N. Y	July 15, 1851	8, 222
Buggy-top	J. C. McCully	Atchison, Kans	July 9, 1872	128, 895
Buggy-top	W. Tanner and W. Shimanaur	Findley, Ohio	June 11, 1872	127, 715
Buggy-top	J. S. Wayne	Quincy, Ill	Aug. 24, 1869	94, 049
Buggy-top bed	A. M. Corey	New Providence, N. J	June 7, 1870	103, 846
Buggy-top-bow setter	O. Smith	Bloomington, Ill	June 8, 1869	90, 965
Buggy-top, Detachable	A. M. Plimpton	Hornellsville, N. Y	Sept. 3, 1867	68, 455
Buggy-top fastening	D. S. Early	Hummelstown, Pa	Oct. 20, 1868	83, 269
Buggy-top, Fastening	H. F. Holt	Fredonia, N. Y	Nov. 26, 1867	71, 304
Buggy-top frame	J. H. Havens	Troy, Ohio	June 4, 1861	32, 477
Buggy-top joint and fastening	H. M. Curtis	Ypsilanti, Mich	Aug. 6, 1867	67, 418
Buggy-top prop	A. Huff and A. J. Mitchell	Louisville, Ky	Jan. 16, 1872	122, 833
Buggy-top roller	J. Palmer	Mechanicsburgh, Pa	Feb. 11, 1868	74, 414
Buggy-top, Shifting	B. L. Benson	Fairview, Ind	July 4, 1871	116, 540
Buggy-top, Shifting	R. C. Diehl	Scott Township, Columbia County, Pa.	July 1, 1873	140, 351
Buggy-top, Shifting	D. Eldridge	Salem, Ohio	Feb. 15, 1870	99, 865
Buggy-top, Shifting	T. Lodge	New Lisbon, Ohio	Oct. 20, 1868	83, 183
Buggy-top, Shifting	S. S. Meily	Lebanon, Pa	June 19, 1866	55, 690
Buggy-top, Shifting	G. Stricker	Catawissa, Pa	Aug. 9, 1870	106, 292
Buggy-tops, Device for shifting	W. B. Slutter	Warsaw, Ind	Aug. 17, 1869	93, 759
Buggy-tops, Machine for setting	J. B. Weller	Bellbrook, Ohio	Oct. 10, 1871	119, 728
Buggy-tops, Shifting-rail for	J. F. Regan	Chicago, Ill	Oct. 17, 1871	119, 951
Buggy-tops, Supporting top for folding	A. Dorm	Mount Healthy, Ohio	Nov. 8, 1870	109, 116
Buggies, Shifting-rail for	I. Bronson	Lockport, N. Y	Apr. 19, 1870	101, 980
Buggies, Shifting top-rail for	J. H. Harter	Nevada, Ohio	June 25, 1872	128, 305
Bugle, horn, cornet, &c	H. G. Lehnert	Boston, Mass	Feb. 13, 1866	52, 580
Bugle, Kent	R. Willis	West Point, N. Y	Nov. 10, 1827	
Bugles, Manufacture of keyed	G. W. Shaw	Thompson, Conn	Aug. 4, 1845	4, 132
Buhr, Metallic	A. T. Boon	Galesburgh, Ill	Nov. 10, 1863	40, 549
Buhr-mill	S. G. Rollins	Boston, Mass	Dec. 6, 1870	109, 841
Buhr-stone mill	G. Sanford	Poughkeepsie, N. Y	Feb. 1, 1859	22, 829
Builder's jack	J. F. Darley	Nebraska City, Nebr	May 28, 1872	127, 314
Building	W. Damerel	Brooklyn, N. Y	Dec. 3, 1867	71, 587
Building	W. J. Fryer, jr	New York, N. Y	Apr. 13, 1869	88, 953
Building	J. Johnson and E. D. Davis	Brooklyn, N. Y	Sept. 17, 1867	68, 884
Building	J. Potts	Harrisburgh, Pa	Aug. 3, 1869	93, 227
Building	J. Touaillon	San Francisco, Cal	Feb. 15, 1870	99, 973
Building and roof, Iron	C. H. Parker	Boston, Mass	Apr. 30, 1872	126, 323
Building-block	N. Boch	New York, N. Y	June 6, 1871	115, 685
Building-block	J. L. Boone	San Francisco, Cal	June 4, 1872	127, 554
Building-block	F. W. Colby	Streator, Ill	Oct. 21, 1873	143, 809
Building-block	G. DeWitt and J. Fairman	Belleville, N. J., and New York, N. Y.	June 24, 1873	140, 122
Building-block	A. Foster and J. A. Messinger	Milwaukee, Wis	Jan. 16, 1855	12, 264
Building-block	S. T. Fowler	Brooklyn, N. Y	June 3, 1873	139, 462
Building-block	F. J. Huber	West Cleveland, Ohio	Aug. 12, 1862	36, 155
Building-block	C. S. Hutchinson	Burlington, N. J	Mar. 6, 1866	53, 004
Building-block	T. Hyatt	New York, N. Y	Sept. 2, 1873	142, 475
Building-block	R. M. Seldis	New York, N. Y	Oct. 28, 1873	144, 149
Building-block	J. S. Stewart	Homer, N. Y	Dec. 24, 1867	72, 557
Building-block and artificial stone	J. Leffler	Rochester, N. Y	Dec. 22, 1868	85, 231
Building-block and method of forming same	A. Derrom	Paterson, N. J	Jan. 30, 1872	123, 156
Building-block, Artificial	G. E. Van Derburgh	New York, N. Y	July 11, 1865	48, 744
Building-block, Child's	C. M. Crandall	Montrose, Pa	Feb. 5, 1867	61, 721
Building-block, Compound	A. Derrom	Paterson, N. J	Feb. 6, 1872	123, 385
Building-block, Hollow	M. R. Pierce	New York, N. Y	Nov. 21, 1871	121, 193
Building-block machinery	A. L. Finch	Sing Sing, N. Y	June 4, 1867	65, 365
Building-block mould	T. J. Lowry	Conneautville, Pa	July 28, 1868	80, 358
Building-block, Silicated	G. E. Van Derburgh	New York, N. Y	July 11, 1865	48, 745
Building-blocks, Casting	J. D. Wise	German Township, Ohio	Feb. 16, 1869	86, 961
Building-blocks, Compound for forming	G. Heim	Naperville, Ill	Apr. 27, 1869	89, 311
Building-blocks, Fire-proof	W. T. Van Zandt and L. A. Tartière.	New York, N. Y	Mar. 11, 1873	136, 684
Building-blocks from clay, Machine for molding and pressing.	A. Foster and G. M. Foster	New York, N. Y., and Fairhaven, Conn.	July 22, 1856	15, 374
Building-blocks from slag, Manufacture of	C. Diebold	Lebanon, Pa.	Apr. 12, 1870	101, 835
Building-blocks, Mold for forming	A. Derrom	Paterson, N. J.	May 20, 1873	139, 050
Building-blocks, pavements, &c., Composition for forming.	S. E. Carr	Danville, Pa	Oct. 13, 1868	83, 037
Building, Corrugated metallic	R. Montgomery	New York, N. Y	Apr. 27, 1869	89, 540
Building-elevator	J. S. Baldwin	Newark, N. J	Oct. 20, 1868	83, 237
Building-elevator	D. H. Chamberlain	West Roxbury, Mass	Oct. 27, 1868	83, 461
Building-elevator	E. Keech	Poughkeepsie, N. Y	June 17, 1873	139, 896
Building-elevator	G. Müllar	New York, N. Y	Nov. 4, 1873	144, 350
Building-elevator	I. S. Schuyler and S. I. Thursby	Brooklyn, N. Y	June 17, 1873	140, 082
Building-elevator	D. B. Thompson	Brooklyn, N. Y	Mar. 3, 1868	75, 220
Building-facing	M. B. Dyott	Philadelphia, Pa	Aug. 16, 1853	9, 937
Building-facing	T. A. Hunter and J. Blewitt	New York, N. Y	Nov. 16, 1869	96, 921
Building, Fire-proof	J. J. Bartlett	New York, N. Y	Apr. 2, 1872	125, 163
Building, Fire-proof	R. M. Hoe	New York, N. Y	May 25, 1869	90, 361
Building, Fire-proof	A. K. Holte	Chicago, Ill	Mar. 4, 1873	136, 515
Building, Fire-proof	G. H. Johnson	New York, N. Y	Sept. 30, 1873	143, 351
Building, Fire-proof	J. H. Johnson and E. R. Hall	Chicago, Ill	Nov. 26, 1872	133, 448
Building, Fire-proof	E. May	Indianapolis, Ind	Jan. 26, 1869	86, 312
Building, Fire-proof	E. May	Indianapolis, Ind	May 4, 1869	89, 782

Index of patents issued from the United States Patent Office from 1790 *to* 1873, *inclusive*—Continued.

Index of patents issued from the United States Patent Office from 1790 *to* 1873, *inclusive*—Continued.

Invention.	Inventor.	Residence.	Date.	No.
Buildings, Iron front for	P. H. Jackson	New York, N. Y	Nov. 16, 1869	96, 922
Buildings, Machine for moving	S. Wells	Elmore, Ohio	May 22, 1860	28, 445
Buildings, Machine for removal of	E. Bowen	Ledyard, N. Y	Apr. 9, 1829	......
Buildings, &c., Machine for removing	L. Pullman	Portland, N. Y	Aug. 21, 1841	2, 225
Buildings, Means of ventilating	S. M. Stone	New Haven, Conn	Dec. 18, 1860	30, 938
Buildings, Method of moving	J. H. Moore	Binghamton, N. Y	May 7, 1867	64, 439
Buildings, Method of moving	J. H. Moore	Binghamton, N. Y	Apr. 21, 1868	77, 074
Buildings, Method of raising bents in	J. Van Gaasbeek	Mount Vernon, N. Y	Apr. 2, 1867	63, 583
Buildings, Method of removing	T. W. Prather	Iowa City, Iowa	Apr. 23, 1867	64, 032
Buildings, Mode of constructing	C. M. Amsden	Wooster, Ohio	May 28, 1872	127, 289
Buildings, Mode of constructing	F. O. Rogers	Niles, Mich	Dec. 11, 1866	60, 425
Buildings, Mode of constructing	A. Tanner	Hoboken, N. J	Nov. 13, 1866	59, 730
Buildings, Mode of constructing cast-iron	W. D. Terry	Boston, Mass	Feb. 5, 1856	14, 208
Buildings, Mode of constructing iron	A. J. Saxton	New York, N. Y	Aug. 8, 1854	11, 509
Buildings, Mode of constructing iron	S. J. Seely	Brooklyn, N. Y	May 7, 1861	32, 256
Buildings, Mode of covering	H. Knowles	Colchester, Conn	Oct. 11, 1828	......
Buildings, Mode of facing the walls of	T. S. Lambert	Peekskill, N. Y	Dec. 8, 1863	40, 844
Buildings, Mode of strengthening and improving sheet-iron.	S. J. Seely	Brooklyn, N. Y	Mar. 1, 1864	41, 787
Buildings of concrete, &c., Sectional mold for putting up.	J. R. Richards	Mount Joy, Pa	June 18, 1867	65, 945
Buildings, Safety-hatch for	G. N. Creamer	Trenton, N. J	Mar. 1, 1870	100, 264
Buildings, Seat for public	A. H. Allen	Boston, Mass	Dec. 5, 1854	12, 017
Buildings, Seat for public	A. Eliaers	Boston, Mass	Nov. 28, 1854	11, 991
Buildings, Service-pipe for	E. Hagan	New York, N. Y	Sept. 8, 1868	81, 895
Buildings, Sheathing-board for	W. E. Hale	Chicago, Ill	Feb. 1, 1870	99, 432
Buildings, Siding for	C. Avery	San Francisco, Cal	Sept. 13, 1870	107, 323
Buildings, Staging for	W. Arronquier	Worcester, Mass	Apr. 11, 1865	47, 179
Buildings, Stationary fire-pipe for	H. Palmieri	Philadelphia, Pa	Mar. 4, 1873	136, 537
Buildings, Steam device for washing	C. Nivert	Paris, France	Nov. 17, 1868	84, 132
Buildings, Underpinning for	I. S. Hill and A. Burnham	Boston and North Chelsea, Mass.	May 11, 1869	89, 995
Buildings, Ventilating	R. Mayo and R. Mills	Washington, D. C	Oct. 24, 1836	......
Buildings, Ventilating	L. B. Valk	Brooklyn, N. Y	Sept. 9, 1873	142, 598
Buildings, Wall for	J. Weathers	Greensburgh, Ind	Aug. 23, 1870	106, 749
Buildings, &c., Warming and ventilating	H. Ruttan	Coburg, Canada	Dec. 5, 1848	5, 958
Buildings water-proof, Mode of making	J. W. Kingman	North Bridgewater, Mass	June 10, 1862	35, 526
Bullet	J. Reim and J. Stock	New York, N. Y	Oct. 7, 1873	143, 419
Bullet-casting apparatus	E. May	Indianapolis, Ind	May 20, 1862	35, 320
Bullet-casting machine	R. Chadwick and N. Allen	Hartford, Conn	May 13, 1862	35, 275
Bullet-casting machine	J. P. Driver	Marengo, Iowa	Jan. 13, 1863	37, 389
Bullet, Elongated	E. D. Williams	Philadelphia, Pa	Dec. 9, 1862	37, 145
Bullet, Expanding	E. D. Williams	New York, N. Y	July 19, 1864	43, 615
Bullet, Explosive	S. H. Mead, jr	New York, N. Y	Dec. 10, 1872	133, 714
Bullet for fire-arms	G. W. Billings	New York, N. Y	Oct. 6, 1863	40, 153
Bullet for fire-arms	E. D. Williams	New York, N. Y	Sept. 27, 1864	44, 492
Bullet for rifled fire-arms	W. Rosée	New York, N. Y	Sept. 5, 1865	49, 792
Bullet for small-arms	E. G. Allen	Boston, Mass	Dec. 6, 1864	45, 306
Bullet for small-arms	J. G. DeCoursey	Philadelphia, Pa	Apr. 3, 1866	53, 582
Bullet for small-arms, Compound	I. W. Shaler and R. Shaler	Brooklyn, N. Y., and Mattison, Conn.	Aug. 12, 1862	36, 197
Bullet-machine	J. D. Custer	Norristown, Pa	June 14, 1864	43, 102
Bullet-machine	J. A. Knight	Saint Louis, Mo	Sept. 14, 1858	21, 505
Bullet-machine	C. H. Remington	Dubuque, Iowa	Oct. 1, 1867	69, 481
Bullet-machine	W. Spillman	Marion Station, Miss	Nov. 19, 1867	71, 075
Bullet-machine	W. H. Ward	Auburn, N. Y	Nov. 10, 1857	18, 616
Bullet machine	C. Young	Auburn, N. Y	Sept. 7, 1858	21, 463
Bullet-making machine	R. Gornall	Baltimore, Md	Apr. 30, 1861	30, 232
Bullet-making machine	J. E. Granniss	New York, N. Y	Aug. 10, 1869	93, 613
Bullet-making machine	E. Nugent	Brooklyn, N. Y	Feb. 15, 1859	22, 974
Bullet-making machine	G. R. Stetson	New Haven, Conn	Nov. 12, 1872	133, 066
Bullet-making machinery	J. Drummond	New York, N. Y	May 9, 1848	5, 563
Bullet-mold	W. Ashton	Middletown, Conn	May 1, 1855	12, 774
Bullet-mold	T. Campbell and H. B. Poorman	Saint Louis, Mo	Jan. 6, 1857	16, 327
Bullet-mold	W. A. Clark	Bethany, Conn	Apr. 15, 1862	34, 944
Bullet-mold	H. L. DeZeng	Geneva, N. Y	Mar. 31, 1857	16, 910
Bullet-mold	J. S. Keith and J. Brooks	Canton, Mass	Feb. 20, 1855	12, 411
Bullet-mold	W. M. Storm	New York, N. Y	Apr. 25, 1854	10, 834
Bullet-mold, Revolving	J. Bradford	Chicago, Ill	Jan. 28, 1873	135, 197
Bullet-patch	A. C. Hobbs	Bridgeport, Conn	Apr. 4, 1871	113, 431
Bullet-pressing die	E. Brown	Brooklyn, N. Y	Apr. 9, 1872	125, 376
Bullets, Machine for lubricating	A. Ball	Worcester, Mass	May 23, 1865	47, 784
Bullets, Machine for making elongated	E. C. Hussey	Brooklyn, N. Y	June 17, 1862	35, 651
Bullets, Machine for making hollow	R. Gornall	Baltimore, Md	Dec. 14, 1858	22, 286
Bullets, Machine for molding	L. Hobard and W. S. Brown	Lexington, Cal	Mar. 6, 1866	52, 996
Bullets, Machine for patching	S. W. Wood	Cornwall, N. Y	Feb. 25, 1873	136, 352
Bullets, &c., Manufacture of	G. W. Campbell	Belleville, N. J	Nov. 20, 1847	5, 371
Bullets, Mode of patching	F. D. Newbury	Albany, N. Y	Feb. 10, 1857	16, 629
Bullets or pills, Machine for spherifying	J. F. Ostrander	New York, N. Y	Apr. 3, 1849	6, 265
Bullets, Patching Minie	O. D. Lull	Watkins, N. Y	Dec. 1, 1863	40, 761
Bulletin-board	A. M. Ernsberger	Danville, Ill	Nov. 28, 1871	121, 347
Bullion, Process for refining	J. Reynolds	San Francisco, Cal	Mar. 20, 1866	53, 340
Bullion, Process of reducing gold	R. S. McCulloh	Princeton, N. J	Sept. 24, 1850	7, 670
Bumper-spring	L. J. Frazee	Louisville, Ky	Dec. 31, 1872	134, 471
Bumper-spring	R. Levington	Monroe, Mich	Jan. 17, 1865	45, 927
Bundling-machine	E. J. Reddy	Bayville, N. Y	Aug. 4, 1868	80, 665
Bundling-press	C. R. Taylor	Ionia, Mich	Sept. 12, 1871	118, 829
Bung	J. F. Applegate	New Albany, Ind	June 14, 1870	104, 244
Bung	J. F. Applegate	New Albany, Ind	Aug. 2, 1870	106, 023
Bung	J. F. Applegate and C. Feiock	New Albany, Ind	Mar. 22, 1870	101, 078
Bung	T. Burke	New York, N. Y	June 18, 1872	128, 014
Bung	O. R. Burnham	New York, N. Y	Dec. 12, 1865	51, 422
Bung	W. Calhoun	West Troy, N. Y	Jan. 19, 1869	85, 903
Bung	N. L. Chappell and C. H. Pettit	New York, N. Y., and Jersey City, N. J.	Mar. 10, 1868	75, 367
Bung	W. F. Class	Cleveland, Ohio	Mar. 4, 1873	136, 415
Bung	D. M. Cumings	Newburyport, Mass	July 5, 1870	104, 935
Bung	V. Fountain, jr	West New Brighton, N. Y	Sept. 20, 1870	107, 473

Index of patents issued from the United States Patent Office from 1790 *to* 1873, *inclusive*—Continued.

Invention.	Inventor.	Residence.	Date.	No.
Bung	V. Fountain, jr	West New Brighton, N. Y	Dec. 5, 1871	121, 607
Bung	H. K. Hazlett	Saint Louis, Mo	Oct. 14, 1873	143, 573
Bung	V. A. Houdaille	Paris, France	Oct. 26, 1869	96, 116
Bung	H. Hufendick and E. Spangenberg.	Saint Louis, Mo	Nov. 27, 1866	60, 011
Bung	J. Kirby	Cincinnati, Ohio	Jan. 31, 1871	111, 352
Bung	J. B. Lefferson	New York, N. Y	Feb. 18, 1868	74, 702
Bung	D. Lightenstadt and R. Pentlarge.	Brooklyn and New York, N. Y.	Apr. 26, 1870	102, 285
Bung	J. G. Marriott and J. Ruegg	Saint Louis, Mo	July 9, 1872	128, 742
Bung	D. B. Rickey	San Francisco, Cal	Feb. 6, 1872	123, 421
Bung	C. C. Stremme	Austin, Tex	Apr. 14, 1868	76, 670
Bung	A. Warth	Stapleton, N. Y	July 19, 1870	105, 613
Bung	A. Warth	Stapleton, N. Y	Aug. 2, 1870	106, 100
Bung	E. White and W. Shiloock	New York, N. Y	Jan. 28, 1868	73, 856
Bung and bung-inserter	D. B. Rickey	San Francisco, Cal	Sept. 19, 1871	119, 094
Bung and bush	J. Ruegg	Saint Louis, Mo	Oct. 22, 1867	70, 024
Bung and bushing	A. D. Behan	Syracuse, N. Y	Dec. 24, 1872	134, 242
Bung-attachment to barrels	A. Wieners	Williamsburgh, N. Y	Dec. 12, 1871	121, 834
Bung, Barrel	M. S. Drake	Newark, N. J	Jan. 22, 1867	61, 328
Bung, Beer-barrel	E. Kraff	Tyrone, Pa	Dec. 23, 1873	145, 878
Bung, Beer-cask	M. Hey	Philadelphia, Pa	May 22, 1866	54, 904
Bung-borer	J. G. Baker and H. Asbury	Philadelphia, Pa	Sept. 22, 1868	82, 374
Bung-borer	W. A. Ives	New Haven, Conn	Aug. 9, 1870	106, 168
Bung-borer	W. A. Ives	New Haven, Conn	May 23, 1871	115, 211
Bung-bush	O. P. and L. W. Briggs	Chicago, Ill	Mar. 11, 1873	136, 696
Bung-bush	W. Kenyon and A. Menzies	New York, N. Y	June 9, 1863	38, 830
Bung-bush inserter	L. Littlejohn	New York, N. Y	May 6, 1873	138, 568
Bung-bush inserter	L. Littlejohn	New York, N. Y	Dec. 23, 1873	145, 867
Bung-bushes, Apparatus for molding	J. More	New York, N. Y	Dec. 3, 1872	133, 592
Bung-bushes, Tool for seating	L. Littlejohn	New York, N. Y	Aug. 5, 1873	141, 511
Bung-cutter	J. Christiansen	Milwaukee, Wis	July 28, 1863	39, 339
Bung-cutter	B. and F. Geyler	Cincinnati, Ohio	Mar. 23, 1869	88, 030
Bung-cutter	A. J. Gibson	Cincinnati, Ohio	June 15, 1869	91, 225
Bung-cutter	A. Goetzinger	Cincinnati, Ohio	July 13, 1869	92, 439
Bung-cutter	C. W. Harris	Pittsburgh, Pa	July 21, 1863	39, 288
Bung-cutter	J. Kirby	Cincinnati, Ohio	June 7, 1859	24, 310
Bung-cutter	J. Kirby	Cincinnati, Ohio	Dec. 24, 1867	72, 505
Bung-cutter	J. I. Munroe	Woburn, Mass	Apr. 12, 1870	101, 903
Bung-cutter	J. I. Munroe	Woburn, Mass	July 9, 1872	128, 745
Bung-cutter	A. W. C. Sternberg	Davenport, Iowa	May 23, 1871	115, 253
Bung-cutter	N. J. Templeton	Cincinnati, Ohio	July 19, 1870	105, 606
Bung-cutter	J. Tiebout	Brooklyn, N. Y	June 5, 1860	28, 640
Bung-cutter	C. Van Derzee	Albany, N. Y	June 24, 1862	35, 741
Bung-cutting machine	A. Donaldson	Salem, Mass	Jan. 15, 1818	......
Bung-cutting machine	A. J. Gibson	Cincinnati, Ohio	Jan. 5, 1869	85, 579
Bung-cutting machine	J. Kirby	Cincinnati, Ohio	Sept. 12, 1848	5, 759
Bung-cutting machine	J. Kirby	Cincinnati, Ohio	Sept. 29, 1857	18, 287
Bung-cutting machine	J. Lyon and G. H. Brady	New York, N. Y	Nov. 16, 1858	22, 101
Bung-cutting machine	J. H. Murrill	Baltimore, Md	Apr. 21, 1868	76, 936
Bung-cutting machine	J. G. Schmidt	Rochester, N. Y	May 7, 1867	64, 581
Bung-cutting machine	J. E. Smith	Buffalo, N. Y	Dec. 4, 1866	60, 272
Bung-cutting machine	W. L. Standish	Pittsburgh, Pa	July 23, 1867	67, 079
Bung-cutting machine	F. Zünkeler	Cincinnati, Ohio	Sept. 13, 1864	44, 267
Bung-cutting machinery	J. P. Gaume	Cincinnati, Ohio	June 6, 1848	5, 613
Bung-elevator	D. F. Fetter	New York, N. Y	Dec. 13, 1870	110, 127
Bung-extractor	J. Kirby	Cincinnati, Ohio	Jan. 31, 1871	111, 351
Bung-extractor	F. Miller and H. Pernot	New York, N. Y	Nov. 26, 1867	71, 402
Bung, Fermentation	W. W. Woodruff	New Britain, Conn	Apr. 21, 1868	77, 154
Bung for barrels	C. A. Neuhaus	New York, N. Y	June 27, 1865	48, 427
Bung for barrels and other vessels	A. G. Day	Seymour, Conn	June 27, 1865	48, 377
Bung for beer-barrels	J. E. McBeth	New Orleans, La	Jan. 22, 1867	61, 348
Bung for casks	M. Hickey	Boston, Mass	Aug. 15, 1865	49, 473
Bung for casks	A. A. Stimson	Boston, Mass	June 30, 1868	79, 407
Bung for casks, barrels, &c	J. Miller	Buffalo, N. Y	Mar. 19, 1867	63, 076
Bung for casks or barrels	M. C. Cronk	Auburn, N. Y	Sept. 3, 1867	68, 417
Bung for coal-oil barrels, &c	J. S. Loomis and A. Thompson	Brooklyn, N. Y	Feb. 17, 1863	37, 722
Bung for oil-casks	A. Walton	Philadelphia, Pa	Sept. 16, 1862	36, 490
Bung-hole borer and reamer	J. Kirby	Cincinnati, Ohio	Sept. 20, 1859	25, 512
Bung-holes, Tool for facing	T. T. Prosser and G. W. Gillette	Chicago, Ill	Feb. 11, 1873	135, 843
Bung-inserter	J. Gillies	Glasgow, Great Britain	Dec. 19, 1871	122, 006
Bung-lock	H. Fise	Lanesville, Ind	Nov. 1, 1870	108, 897
Bung-making	W. Long and J. Garand	West Troy and Troy, N. Y	Aug. 10, 1869	93, 457
Bung-making machine	J. Batchelder	Canaan, N. H	Sept. 13, 1864	44, 251
Bung, Metallic	W. Boynton	Auburn, N. Y	Feb. 12, 1867	61, 993
Bung, Metallic	T. B. Smith	Marietta, Ohio	July 19, 1859	24, 827
Bung of cask	J. Keane	New York, N. Y	Sept. 27, 1859	25, 606
Bung, Self-venting	C. F. Spencer	Cleveland, Ohio	Aug. 19, 1873	142, 051
Bung socket and plug for barrels	S. J. Geoghegan and W. Ulmer	New York, N. Y	Feb. 24, 1863	37, 747
Bung-spout	J. Marvin and W. T. Hulse	Port Jefferson, N. Y	Aug. 29, 1871	118, 545
Bung-stave for barrels	F. Acker	San Francisco, Cal	Dec. 20, 1864	45, 553
Bung-tap for casks	G. W. Banker	Saint Louis, Mo	Feb. 16, 1869	87, 021
Bung, Valve	F. Dahis	Williamsburgh, N. Y	Feb. 8, 1859	22, 857
Bung, Vent	B. R. Cole	Buffalo, N. Y	July 11, 1871	116, 809
Bung, Vent	C. H. Miller and W. Ascough	Buffalo, N. Y	Sept. 12, 1871	118, 811
Bung with automatic vent, Metallic	A. Ruoff	Detroit, Mich	Aug. 10, 1869	93, 481
Bungs, Combined bush and wrench for	J. Lacey and G. B. Cornell	Chicago, Ill	Aug. 29, 1871	118, 617
Bungs for barrels, Machine for cutting	J. G. Schmidt	Rochester, N. Y	Dec. 11, 1866	60, 450
Bungs for casks, Machines for making	W. Donaldson	Cincinnati, Ohio	June 16, 1868	78, 938
Bungs for casks, Making	R. Pentlarge	Brooklyn, N. Y	Dec. 23, 1873	145, 897
Bungs for casks, &c., Ventilating	J. B. Melvin	Lowell, Mass	Apr. 16, 1867	63, 920
Bungs from barrels, Extracting	H. Meyers and A. Webb	Hyde Park and Scranton, Pa.	July 17, 1866	56, 431
Bungs impervious to liquids and gases, Mode of rendering wooden.	P. Geier	Cincinnati, Ohio	Feb. 23, 1869	87, 163
Bungs, Machine for cutting	R. Barlow	Philadelphia, Pa	June 29, 1833	......
Bungs, Machine for cutting	G. D. Yates	Hartford, Conn	May 2, 1835	......
Bungs, Machine for making	C. Abel	Morrisania, N. Y	Nov. 12, 1872	132, 943
Bungs, Machine for making	R. Boeklen	Brooklyn, N. Y	Jan. 28, 1873	135, 196

Index of patents issued from the United States Patent Office from 1790 to 1873, inclusive—Continued.

Invention.	Inventor.	Residence.	Date.	No.
Bungs, Manufacture of	L. Gray	Pittsburgh, Pa	Aug. 25, 1863	39, 644
Bungs. Manufacture of	B. D. Sanders	Wellsburgh, W. Va	June 1, 1869	90, 690
Bungs, Manufacturing	R. Barlow	Philadelphia, Pa	Dec. 1, 1832	....
Bungs or corks, Machine for cutting	P. Hayden	Pittsburgh, Pa	Feb. 9, 1864	41, 506
Bungs out of barrels, Instrument for wrenching	A. Herde	Baltimore, Md	Oct. 8, 1867	69, 563
Bungs, plugs, &c., Mold for casting	A. H. Rowand	Allegheny, Pa	Mar. 1, 1864	41, 786
Bunter-shape swaging and forming die	A. E. Barnard	Akron, Ohio	Dec. 3, 1867	71, 567
Buoy	W. M. Ellis	Washington, D. C	Oct. 7, 1856	15, 845
Buoy, Channel	W. W. Goff	Eagle Harbor, Mich	July 7, 1868	79, 750
Buoy-light	L. Stevens	Washington, D. C	Mar. 26, 1872	125, 096
Buoy, Tidal-alarm	J. Taggart	Roxbury, Mass	June 24, 1856	15, 200
Buoys and other floating bodies, Method of obtaining foundations and of mooring ships to.	A. Mitchell	Belfast, Ireland	Apr. 1, 1845	3, 986
Buoying vessels, Apparatus for	E. Goulard	New York, N. Y	Apr. 2, 1861	31, 885
Bureau	J. H. Better	New York, N. Y	Jan. 24, 1860	26, 881
Bureau	G. S. Graf	Pittsburgh, Pa	May 10, 1870	102, 809
Bureau	C. Kilburn	Philadelphia, Pa	Aug. 30, 1870	106, 941
Bureau	H. Rocke	New York, N. Y	Feb. 6, 1872	123, 422
Bureau and bath-tub, Combined dressing	J. E. Gilman	Hartford, Conn	Sept. 20, 1870	107, 479
Bureau and bedstead	J. Stock	El Paso, Ill	Aug. 13, 1867	67, 816
Bureau and clothes-drier, Combined	W. Hathaway	Northbridge, Mass	June 18, 1872	12[illegible], 038
Bureau and commode	G. W. Koch	New York, N. Y	Sept. 5, 1865	49, 766
Bureau and ironing-table combined	W. W. Adams	Rockford, Ill	May 20, 1873	139, 101
Bureau and trunk, Combination of	A. V. Ryder	New York, N. Y	Aug. 25, 1863	39, 679
Bureau and wardrobe, Combined	H. H. Staugaard	Chicago, Ill	Dec. 9, 1873	145, 312
Bureau and washstand, Combined	J. Schneemann	New York, N. Y	Apr. 23, 1872	125, 991
Bureau boot-blacking attachment	W. E. Phelps	Elmwood, Ill	May 24, 1870	103, 365
Bureau, Bracket	O. Andrews	Lanesville, Mass	July 23, 1872	129, 704
Bureau cabinet. Dressing	W. E. Beames	New York, N. Y	June 15, 1869	91, 409
Bureau-drawer fastening	A. J. Grant	Williamstown, N. Y	July 22, 1873	141, 001
Bureau-drawer fastening	G. Wode	Elizabethport, N. J	July 30, 1850	7, 541
Bureau dressing-table	N. Perry	Boston, Mass	Oct. 13, 1809	....
Bureau, Folding	J. E. Lawrence and J. S. Young.	Philadelphia, Pa	Dec. 26, 1871	122, 179
Bureau or wardrobe bedstead	C. May	Cincinnati, Ohio	Mar. 4, 1873	136, 443
Bureau, Portable	L. Heywood, J. L. Ross, and J. K. Otis.	Boston, Mass	Aug. 22, 1854	11, 551
Bureau, Sectional	E. Gill	New York, N. Y	Jan. 12, 1869	85, 811
Bureau toilet-attachment	H. W. Eastman	Baltimore, Md	Apr. 28, 1868	77, 263
Bureau, Wardrobe	J. H. F. Lehmann	New York, N. Y	June 11, 1872	127, 778
Bureau, wash-stand, and commode, Combined	L. Aling	Zeeland, Mich	Mar. 19, 1872	124, 653
Bureau, wash-stand, and wardrobe, Combined	T. W. Moore	New York, N. Y	Apr. 1, 1873	137, 380
Bureaus and wash-stands, Construction of	J. D. Burton	Boston, Mass	Jan. 12, 1858	19, 069
Bureaus, &c., Device for holding together the different parts of.	D. A. Mullane and J. O. L. Murray.	New Orleans, La	Feb. 1, 1870	99, 460
Bureaus, &c., Device for holding together the different parts of.	J. O. L. Murray and D. A. Mullane.	New Orleans, La	Oct. 5, 1869	95, 504
Burglar-alarm	G. S. Acker	Kalamazoo. Mich	Oct. 15, 1867	69, 888
Burglar-alarm	H. P. Beardsley and G. Wilcox.	Corunna, Mich	Sept. 22, 1868	82, 275
Burglar-alarm	H. Behn	New York, N. Y	Mar. 19, 1867	62, 998
Burglar-alarm	H. Bergstein	San Francisco, Cal	June 1, 1869	90, 809
Burglar-alarm	G. W. Bigelow	New Haven, Conn	Apr. 10, 1860	27, 772
Burglar-alarm	A. Bingham	Boston, Mass	Aug. 21, 1855	13, 478
Burglar-alarm	C. A. Blake	Philadelphia, Pa	Nov. 27, 1866	59, 949
Burglar-alarm	F. W. Blakemore	Philadelphia, Pa	Dec. 23, 1873	145, 836
Burglar-alarm	L. W. Blakeslee and A. D. Smith.	Cincinnati, Ohio	Dec. 31, 1867	72, 786
Burglar-alarm	J. G. Bolen	New York, N. Y	Oct. 21, 1851	8, 439
Burglar-alarm	O. M. Brooks and R. W. Soper.	Janesville, Wis	July 9, 1867	66, 457
Burglar-alarm	H. L. Brower	New York, N. Y	Feb. 25, 1873	136, 210
Burglar-alarm	E. Brown	Rindge, N. H	Oct. 4, 1853	10, 077
Burglar-alarm	E. Brown	Lowell, Mass	Oct. 31, 1854	11, 856
Burglar-alarm	E. Brown	Lowell, Mass	July 3, 1855	13, 157
Burglar-alarm	H. L. Brown	Middletown, Conn	Dec. 9, 1873	145, 332
Burglar-alarm	G. Bruce, sr	Sing Sing, N. Y	May 21, 1861	32, 349
Burglar-alarm	I. N. Buck	Elgin, Ill	Dec. 22, 1868	85, 066
Burglar-alarm	I. N. Buck	Elgin, Ill	Apr. 27, 1869	89, 286
Burglar-alarm	R. Bunker	Hudson, Wis	Mar. 5, 1867	62, 604
Burglar-alarm	R. M. Campbell	East Cambridge, Mass	Mar. 8, 1859	23, 215
Burglar-alarm	H. D. Chance	Allentown, Pa	June 1, 1869	90, 636
Burglar-alarm	J. G. Clark	Augusta, Ga	May 24, 1859	24, 177
Burglar-alarm	E. F. Clegg	North Harpersfield, N. Y	Nov. 29, 1870	109, 588
Burglar-alarm	G. A. Colton	Adrian, Mich	Jan. 1, 1867	60, 695
Burglar-alarm	D. Coon	Ithaca, N. Y	May 26, 1857	17, 406
Burglar-alarm	S. Coon	Ithaca, N. Y	Sept. 22, 1857	18, 236
Burglar-alarm	J. F. Coppock	West Newton, Ind	Apr. 8, 1873	137, 532
Burglar-alarm	T. P. Coulston	Philadelphia, Pa	Aug. 22, 1871	118, 345
Burglar-alarm	C. J. Crum and W. Irwin	Circleville, Ohio	Dec. 31, 1867	72, 723
Burglar-alarm	B. F. and J. F. Cunningham	Flora, Ill	Dec. 24, 1867	72, 459
Burglar-alarm	A. W. Decrow	Bangor, Me	Sept. 14, 1858	2[illegible], 555
Burglar-alarm	A. W. Decrow	Bangor, Me	Dec. 12, 1865	51, [illegible]11
Burglar-alarm	L. E. Denison	Saybrook, Conn	Oct. 22, 1840	1, 835
Burglar-alarm	C. A. Eaton	Minneapolis, Minn	Sept. 17, 1867	68, 857
Burglar-alarm	D. E. Eaton	Boston, Mass	Oct. 30, 1855	13, 738
Burglar-alarm	W. H. Emmons and F. Kissam.	New York, N. Y., and Newark, N. J.	Oct. 26, 1869	96, 094
Burglar-alarm	A. H. Enholm	Saint Louis, Mo	July 3, 1860	28, 977
Burglar-alarm	W. Farnham	Janesville, Wis	Dec. 4, 1866	60, 162
Burglar-alarm	J. Foster	Green Point, N. Y	Oct. 31, 1854	11, 859
Burglar-alarm	R. G. Fowler	Olney, Ill	Jan. 1, 1867	60, 711
Burglar-alarm	W. F. Gardiner	Bethany, Canada	Dec. 12, 1871	121, 862
Burglar-alarm	M. A. Genung	Granville, Ohio	Nov. 3, 1863	40, 469
Burglar-alarm	H. Green	Norwalk, Conn	Jan. 4, 1870	98, 489
Burglar-alarm	D. Haldeman	Morgantown, Va	Feb. 6, 1855	12, 351
Burglar-alarm	S. Hamilton, jr	Tolland, Mass	Dec. 4, 1855	13, 874
Burglar-alarm	A. F. Hammond	Houston, Ohio	May 8, 1866	54, 531
Burglar-alarm	G. R. Harding	Manchester, Va	Feb. 23, 1869	87, 256
Burglar-alarm	E. H. Hendrickson	Brooklyn, N. Y	June 24, 1873	140, 134
Burglar-alarm	M. C. Hepinstall	Enfield, N. C	Mar. 13, 1866	53, 144

Index of patents issued from the United States Patent Office from 1790 *to* 1873, *inclusive*—Continued.

Invention.	Inventor.	Residence.	Date.	No.
Burglar-alarm	H. Hersch, B. Bauman, and H. C. Locher.	Lancaster, Pa	Apr. 13, 1858	19, 973
Burglar-alarm	H. L. Hervey	Quincy, Ill	Dec. 4, 1855	13, 876
Burglar-alarm	I. and A. Herzberg	Philadelphia, Pa	Apr. 2, 1872	125, 293
Burglar-alarm	I. and A. Herzberg	Philadelphia, Pa	July 16, 1872	129, 344
Burglar-alarm	W. O. Hills	Nottingham, N. Y	Mar. 6, 1866	52, 999
Burglar-alarm	H. Holcroft	Media, Pa	Mar. 5, 1872	124, 356
Burglar-alarm	S. B. Holden	Woburn, Mass	Oct. 11, 1859	25, 788
Burglar-alarm	M. A. Holland	Passaic, N. J	Sept. 12, 1871	118, 939
Burglar-alarm	H. P. Hood	Indianapolis, Ind	Sept. 9, 1873	142, 630
Burglar-alarm	E. Hoole	New York, N. Y	May 1, 1866	54, 353
Burglar-alarm	T. N. Howell	Circleville, Ohio	Apr. 2, 1872	125, 296
Burglar-alarm	M. J. Hunt	Rising Sun, Md	Dec. 13, 1870	110, 043
Burglar-alarm	J. C. Huntly	Philadelphia, Pa	June 12, 1860	28, 671
Burglar-alarm	A. G. Hutchinson	Stoneycroft, near Liverpool, England.	July 13, 1869	92, 610
Burglar-alarm	J. J. Jackson	Curwinsville, Pa	Sept. 7, 1869	94, 604
Burglar-alarm	N. Jensen	Washington, D. C	Nov. 9, 1858	22, 024
Burglar-alarm	J. J. Kane	Brooklyn, N. Y	Dec. 30, 1873	146, 006
Burglar-alarm	P. Kane and W. Floyd	South Perry, Ohio	Dec. 3, 1861	33, 842
Burglar-alarm	A. Kazenmayer and L. Valois	Newark, N. J	Feb. 11, 1868	74, 375
Burglar-alarm	E. C. C. Kellogg	Hartford, Conn	Nov. 19, 1867	71, 180
Burglar-alarm	C. Knisely	Chicago, Ill	Mar. 24, 1868	75, 771
Burglar-alarm	N. P. Larsen	Chicago, Ill	Sept. 8, 1868	81, 913
Burglar-alarm	R. Lee	Cincinnati, Ohio	Mar. 7, 1871	112, 359
Burglar-alarm	G. A. Lilliendahl	New York, N. Y	Feb. 15, 1859	22, 959
Burglar-alarm	A. J. Loomis	Madrid, N. Y	May 23, 1865	47, 841
Burglar-alarm	G. W. Lore	Owasso, Mich	June 4, 1872	127, 421
Burglar-alarm	M. Lunt	Cambridgeport, Mass	Dec. 21, 1869	98, 078
Burglar-alarm	E. F. Mallory	West Springfield, Pa	Sept. 3, 1867	68, 520
Burglar-alarm	D. E. McDougall	Springfield, Mass	June 20, 1854	11, 131
Burglar-alarm	J. McDowell	Washington, Pa	Feb. 21, 1871	111, 959
Burglar-alarm	W. McLachlan	New York, N. Y	Nov. 4, 1856	16, 017
Burglar-alarm	W. H. McPherson	Nashville, Tenn	Mar. 18, 1873	137, 016
Burglar-alarm	E. M. and I. E. Mix	Ithaca, N. Y	Sept. 29, 1857	18, 292
Burglar-alarm	J. Morgan, jr	Wheeling, W. Va	July 25, 1871	117, 444
Burglar-alarm	M. P. Murphy	New York, N. Y	Mar. 6, 1866	53, 030
Burglar-alarm	J. T. Mygatt and J. Downing	Binghamton, N. Y	Mar. 28, 1871	113, 082
Burglar-alarm	R. W. Newbery	New York, N. Y	Oct. 24, 1871	120, 313
Burglar-alarm	M. R. Perkins	Portsmouth, N. H	July 29, 1873	141, 291
Burglar-alarm	C. E. Pierce	New York, N. Y	June 25, 1867	66, 040
Burglar-alarm	M. Pierson and M. D. Manville	Adams, N. Y	Jan. 12, 1869	85, 848
Burglar-alarm	H. H. Potter	Carthage, N. Y	July 10, 1866	56, 262
Burglar-alarm	M. L. Powell	Newcastle, Ind	July 23, 1861	32, 891
Burglar-alarm	W. Reynolds	Manchester, N. H	July 12, 1870	105, 369
Burglar-alarm	W. Reynolds	Manchester, N. H	Jan. 17, 1871	111, 003
Burglar-alarm	H. R. Robbins	Baltimore, Md	Oct. 19, 1858	21, 849
Burglar-alarm	H. R. Robbins	Baltimore, Md	June 18, 1867	65, 949
Burglar-alarm	A. J. Ross	Rochester, N. Y	May 29, 1866	55, 165
Burglar-alarm	A. Q. Ross	Cincinnati, Ohio	Sept. 27, 1859	25, 586
Burglar-alarm	A. Q. Ross	Cincinnati, Ohio	Oct. 31, 1871	120, 399
Burglar-alarm	T. Royer	Lancaster, Pa	Nov. 26, 1867	71, 537
Burglar-alarm	R. Shaler	Madison, Conn	Mar. 15, 1870	100, 933
Burglar-alarm	I. Silvernail	Byron, Mich	Oct. 26, 1869	96, 277
Burglar-alarm	G. Simpson	Waterbury, Vt	June 5, 1866	55, 378
Burglar-alarm	D. B. Skelly	Lockport, N. Y	Nov. 19, 1867	71, 073
Burglar-alarm	S. Stewart	Philadelphia, Pa	Aug. 9, 1859	25, 057
Burglar-alarm	B. L. Stone	San Francisco, Cal	Feb. 14, 1865	46, 402
Burglar-alarm	B. L. Stone	New York, N. Y	Dec. 12, 1865	51, 489
Burglar-alarm	C. H. Stone and C. W. Livingston.	South Groton, Mass	Apr. 7, 1863	38, 136
Burglar-alarm	J. H. Thorp	New York, N. Y	Dec. 6, 1870	109, 970
Burglar-alarm	J. G. Trout	Philadelphia, Pa	Apr. 2, 1867	63, 582
Burglar-alarm	C. Waterman	New York, N. Y	Mar. 19, 1867	63, 124
Burglar-alarm	R. M. Webb	New York, N. Y	July 17, 1866	56, 471
Burglar-alarm	D. Wells	Philadelphia, Pa	Jan. 2, 1855	12, 177
Burglar-alarm	I. M. Wells	Jeffersonville, Ohio	June 11, 1867	65, 710
Burglar-alarm	I. M. Wells	Jeffersonville, Ohio	Feb. 25, 1868	74, 779
Burglar-alarm	J. N. Wells	Brooklyn, N. Y	Sept. 14, 1869	94, 929
Burglar-alarm	S. Whitaker	Macon, Ill	Aug. 6, 1867	67, 619
Burglar-alarm	H. Wickham, jr	Chicago, Ill	June 19, 1866	55, 754
Burglar-alarm	J. P. Wilson and J. F. Thomas	Frankfort and Ilion, N. Y	Feb. 8, 1859	22, 911
Burglar-alarm	H. J. Wolters	Chester, Mass	Nov. 17, 1868	84, 153
Burglar-alarm	L. J. Worden and E. H. Space	Clinton, N. Y	Jan. 27, 1852	8, 699
Burglar-alarm	W. D. Wright	Baltimore, Md	Mar. 2, 1858	19, 527
Burglar-alarm	H. Yerty	Sidney, Ohio	Oct. 30, 1866	59, 307
Burglar-alarm and lock	E. Warne	Broadway, N. J	Sept. 12, 1865	49, 940
Burglar-alarm and animal-trap, Combined	G. Smith	New York, N. Y	June 11, 1861	32, 539
Burglar-alarm and door-fastener	F. Oakley	London, England	Aug. 28, 1866	57, 654
Burglar-alarm and door-fastener, Combined	W. W. Marston	New York, N. Y	May 7, 1872	126, 558
Burglar-alarm and hotel-room indicator	L. C. Gosson	Trenton, N. J	Apr. 35, 1871	114, 001
Burglar-alarm and lock-apparatus	C. E. Pierce	New York, N. Y	Dec. 3, 1867	71, 637
Burglar alarm and signal, Circuit-closer for electrical.	E. Holmes and H. C. Roome	Brooklyn, N. Y., and Jersey City, N. J.	Nov. 7, 1871	120, 744
Burglar-alarm and table-bell	A. Iske	Lancaster, Pa	Dec. 8, 1868	84, 696
Burglar-alarm apparatus for portable safes, &c	S. J. Hoffman	Mobile, Ala	May 28, 1872	127, 165
Burglar-alarm drawer	J. Murphy	Boston, Mass	Mar. 14, 1871	112, 619
Burglar-alarm, Electrical	H. E. Walter	Richfield Springs, N. Y	May 13, 1873	138, 965
Burglar-alarm, Electro-magnetic	J. Haller	Ann Arbor, Mich	Sept. 25, 1860	30, 141
Burglar-alarm, Electro-magnetic	W. J. Biggar, J. C. Blood, and D. M. Griswold.	Conneaut, Ohio	July 14, 1868	79, 895
Burglar-alarm, Electro-magnetic	J. G. Butler	Glen's Falls, N. Y	May 2, 1871	114, 406
Burglar-alarm, Electro-magnetic	E. C. Clay	Malden, Mass	Nov. 1, 1859	25, 950
Burglar-alarm, Electro-magnetic	G. E. Cock and J. H. Guest	New York, N. Y	Aug. 22, 1871	118, 199
Burglar-alarm, Electro-magnetic	J. M. Dille	Cooperstown, Pa	Nov. 29, 1870	109, 723
Burglar-alarm, Electro-magnetic	W. B Guernsey	Jersey City, N. J	Oct. 11, 1870	108, 257
Burglar-alarm, Electro-magnetic	W. B. Guernsey	Jersey City, N. J	Mar. 14, 1871	112, 704
Burglar-alarm, Electro-magnetic	W. B. Guernsey	Jersey City, N. J	Mar. 14, 1871	112, 705

Index of patents issued from the United States Patent Office from 1790 *to* 1873, *inclusive*—Continued.

Invention.	Inventor.	Residence.	Date.	No.
Burglar-alarm, Electro-magnetic	W. B. Guernsey	Jersey City, N. J	Mar. 14, 1871	112, 706
Burglar-alarm, Electro-magnetic	W. B. Guernsey	Jersey City, N. J	Apr. 30, 1872	126, 288
Burglar-alarm, Electro-magnetic	G. F. Milliken	Somerville, Mass	Oct. 11, 1859	25, 753
Burglar-alarm, Electro-magnetic	J. P. Snyder	Brooklyn, N. Y	Apr. 4, 1871	113, 589
Burglar-alarm, Electro-magnetic	W. E. Tracer	Philadelphia, Pa	Mar. 28, 1871	113, 036
Burglar-alarm for door-spring	G. W. R. Pollock	Boston, Mass	May 25, 1869	90, 575
Burglar-alarm for window	M. A. Holland	Passaic, N. J	Sept. 12, 1871	118, 940
Burglar-alarm for window	W. H. Winans	Coxsackie, N. Y	Mar. 13, 1866	53, 239
Burglar-alarm lock	A. Corey	Casstown, Ohio	May 25, 1858	20, 333
Burglar-alarm lock	A. L. Gennerat	Paris, France	May 28, 1867	65, 207
Burglar-alarm, Portable	J. Pennepacker	Philadelphia, Pa	Sept. 30, 1873	143, 379
Burglar-alarm, Portable	J. H. Thorp	Saint Louis, Mo	May 27, 1873	139, 440
Burglar-alarm, Portable	J. N. Webster and D. Fey	Peoria, Ill	July 23, 1872	129, 698
Burglar and fire alarm	J. Holmes and C. W. Nickerson.	Pittsburgh, Pa	Feb. 12, 1867	61, 033
Burglar and fire alarm combined	L. Giebrich	Ottumwa, Iowa	Oct. 24, 1871	120, 261
Burglar and fire alarm, Electro-magnetic	E. Blunt, jr	Bay Ridge, N. Y	June 30, 1868	79, 440
Burglar and fire alarm, Electro-magnetic	J. H. Guest	Brooklyn, N. Y	July 14, 1868	79, 973
Burglar and safety alarm, Electro-magnetic	A. O. Willcox	Philadelphia, Pa	July 2, 1872	128, 691
Burglar-trap	W. Carr	Yellow Springs, Ohio	May 5, 1868	77, 582
Burial-apparatus	W. H. McGavran	Connotton, Ohio	Nov. 7, 1871	120, 657
Burial-case	A. C. Barstow	Providence, R. I	Apr. 19, 1859	23, 652
Burial-case	E. H. Borton	Saint Louis, Mo	June 9, 1863	38, 857
Burial-case	J. B. Cox	San Francisco, Cal	Oct. 7, 1873	143, 438
Burial-case	A. Crosby	Westfield, N. Y	June 22, 1869	91, 611
Burial-case	J. A. Dandridge	Buffalo, N. Y	Sept. 28, 1869	95, 202
Burial-case	P. H. Griffin	Albany, N. Y	Jan. 24, 1871	111, 117
Burial-case	J. Hackett	Louisville, Ky	Dec. 5, 1871	121, 616
Burial-case	J. R. Hathaway	Westfield, N. Y	Feb. 18, 1868	74, 531
Burial-case	J. R. Hathaway	Westfield, N. Y	Feb. 2, 1869	86, 396
Burial-case	R. F. Hill	Philadelphia, Pa	Dec. 17, 1867	72, 295
Burial-case	R. J. Howdon	Cincinnati, Ohio	Mar. 19, 1872	124, 747
Burial-case	R. Hunt	Milford, N. J	June 28, 1870	104, 737
Burial-case	I. Lorde	Moline, Ill	Jan. 17, 1871	110, 984
Burial-case	J. L. Lovett, J. Wippich, and J. Wood.	Salem, Mass	Feb. 27, 1866	52, 865
Burial-case	W. K. Miller	Canton, Ohio	Aug. 28, 1866	57, 545
Burial-case	C. C. W. Morgan	Holly Springs, Miss	Nov. 19, 1872	133, 166
Burial-case	G. W. Nash	Columbus, Ohio	Aug. 6, 1872	130, 235
Burial-case	A. S. Patterson	Westfield, N. Y	Aug. 17, 1869	93, 832
Burial-case	J. Peak and F. T. Pinkham	Boston, Mass	Mar. 22, 1864	42, 018
Burial-case	G. W. Scollay	Saint Louis, Mo	Mar. 18, 1862	34, 700
Burial-case	G. W. Scollay	Washington, D. C	Feb. 20, 1866	52, 756
Burial-case	F. B. Shearer	Columbus, Ohio	Jan. 24, 1871	111, 263
Burial-case	G. Shilling	Baltimore, Md	Sept. 20, 1870	107, 552
Burial-case	I. C. Shuler	Amsterdam, N. Y	May 23, 1871	115, 116
Burial-case	E. A. Skeele	Saint Louis, Mo	Sept. 1, 1863	39, 758
Burial-case	S. Stein	Rochester, N. Y	Mar. 19, 1872	124, 769
Burial-case	J. H. Van Houten	New York, N. Y	Oct. 8, 1867	69, 596
Burial-case	F. Vester	Newark, N. J	Aug. 25, 1868	81, 437
Burial-case	J. H. Weaver	Baltimore, Md	May 7, 1861	32, 261
Burial-case	J. Weidenmann	Hartford, Conn	Feb. 13, 1872	123, 599
Burial-case, Air-tight	C. Timmerman	Amsterdam, N. Y	June 12, 1866	55, 555
Burial-case, Composition	F. W. Brown	Albany, N. Y	Mar. 28, 1871	113, 138
Burial-case, Metallic	A. C. Barstow	Providence, R. I	Jan. 27, 1863	37, 484
Burial-case, Metallic	M. H. Crane	Cincinnati, Ohio	May 5, 1863	38, 443
Burial case, Metallic	M. H. Crane	Cincinnati, Ohio	May 7, 1867	64, 496
Burial-case, Mold for cement	P. B. Viele	Rochester, N. Y	Dec. 3, 1872	133, 684
Burial-case, Self-sealing	P. F. Lawshe	Rochester, Minn	May 5, 1868	77, 627
Burial-case, Wooden	J. Gawler	Washington, D. C	May 11, 1869	89, 982
Burial-cases, Composition for the manufacture of	J. R. Hathaway	Westfield, N. Y	Oct. 27, 1868	83, 495
Burial-cases, Device for making ends of	E. T. Smith and J. S. Winston	New York, N. Y	May 30, 1871	115, 536
Burial-cases, Vent for	W. W. Woodward	Cincinnati, Ohio	Sept. 3, 1872	131, 073
Burial-casket	J. A. Fogg	Salem, Mass	May 1, 1866	54, 321
Burial-casket	H. Henika and M. F. Carder	Kalamazoo, Mich	June 3, 1873	139, 465
Burial-casket	C. S. Hurlbut	Springfield, Mass	Apr. 20, 1869	89, 148
Burial-casket	J. O. Moore	Albany, N. Y	Aug. 8, 1871	117, 800
Burial-casket	J. Scott	Philadelphia, Pa	Oct. 24, 1871	120, 171
Burial-casket	E. T. Smith and J. S. Winston	New York, N. Y	Jan. 31, 1871	111, 393
Burial-casket	S. Stein	Rochester, N. Y	Oct. 29, 1872	132, 605
Burial-casket	S. Stein	Rochester, N. Y	Jan. 7, 1873	134, 570
Burial-casket, Composition	A. N. Atwood	Philadelphia, Pa	Mar. 4, 1873	136, 404
Burial-casket fastening	E. S Earley	Philadelphia, Pa	Nov. 26, 1867	71, 468
Burial-casket handle	A. G. Chapman	East Hampton, Conn	Apr. 1, 1873	137, 289
Burial-casket handle	C. Strong	Winsted, Conn	Jan. 24, 1871	111, 271
Burial-caskets, Composition for	W. H. Ross	Brooklyn, N. Y	July 29, 1873	141, 238
Burial-caskets, Lid-fastening for	W. Hamilton	Allegheny City, Pa	Sept. 13, 1870	107, 253
Burner	S. C. Pruden	Harmony, Ohio	Jan. 28, 1868	73, 834
Burner, Aëro-gas	W. Jones	Chelsea, Mass	Jan. 5, 1869	85, 671
Burner, Aëro-gas	W. Jones and M. H. Collins	Chelsea, Mass	Aug. 7, 1866	56, 949
Burner, Aëro-vapor	W. Bryent	Boston, Mass	Nov. 21, 1865	51, 011
Burner, Aëro-vapor	O. F. Morrill	Chelsea, Mass	Oct. 20, 1863	40, 353
Burner, Air and vapor	O. F. Morrill	Boston, Mass	Oct. 20, 1857	18, 465
Burner and lamp-post, Combined vapor	B. D. Evans	Columbus, Ohio	June 1, 1869	90, 830
Burner and reflector, Gas	H. Berg and A. Blessing	Springfield, Mass	Jan. 1, 1867	60, 674
Burner and stop-cock, Gas	E. Jones	Boston, Mass	Oct. 3, 1865	50, 302
Burner and wick-tube of vapor-lamp	M. Safford	Boston, Mass	July 20, 1858	20, 977
Burner, Argand	C. E. Ball	Philadelphia, Pa	June 24, 1873	140, 108
Burner, Argand	E. P. Gleason	New York, N. Y	Dec. 17, 1867	72, 187
Burner, Argand	E. R. Walker	New York, N. Y	Dec. 17, 1867	72, 244
Burner, Argand gas	W. W. Batchelder	New York, N. Y	June 15, 1858	20, 604
Burner, Argand gas	T. Clough	New York, N. Y	May 13, 1873	138, 858
Burner, Argand gas	G. Darricott and J. Nason	Boston, Mass	July 26, 1839	1, 261
Burner, Argand gas	J. B. Fuller	Norwich, Conn	July 26, 1870	105, 795
Burner, Argand gas	G. W. Hatch	Brooklyn, N. Y	June 25, 1872	128, 307
Burner, Argand gas	H. W. Hayden	Waterbury, Conn	Jan. 2, 1872	122, 382
Burner, Argand gas	C. H. Johnson	Boston, Mass	Nov. 13, 1855	13, 739
Burner, Argand gas	C. H. Johnson	Boston, Mass	Apr. 21, 1857	17, 152

Index of patents issued from the United States Patent Office from 1790 *to* 1873, *inclusive*—Continued.

Invention.	Inventor.	Residence.	Date.	No.
Burner, Argand gas	C. H. Johnson	Boston, Mass	June 20, 1865	48, 340
Burner, Argand gas	H. Monier	Paris, France	Aug. 23, 1859	25, 269
Burner, Argand gas	R. Murray and G. Oakes	Boston, Mass	Apr. 24, 1866	54, 193
Burner, Argand gas	J. E. Stanwood	Malden, Mass	Aug. 3, 1858	21, 090
Burner, Argand gas	J. G. Webb	Williamsburgh, N. Y	Oct. 14, 1851	8, 437
Burner, Argand lamp	J. B. Fuller	Norwich, Conn	Sept. 10, 1872	131, 265
Burner, Argand lamp	H. W. Hayden	Waterbury, Conn	June 13, 1871	115, 955
Burner, Argand lamp	T. H. Mott	New York, N. Y	July 11, 1871	116, 984
Burner, Argand lamp	J. Ravoux	New York, N. Y	June 25, 1872	128, 325
Burner, Argand oil	G. K. Osborn	New York, N. Y	June 21, 1870	104, 635
Burner attachment, Gas	J. Kopp	Hoboken, N. J	Sept. 15, 1868	82, 231
Burner attachment, Gas	J. Scholl	Soho, England	Aug. 4, 1868	80, 568
Burner attachment, Gas	F. Shaller	Hudson, N. Y	Nov. 10, 1868	83, 886
Burner, Base	W. B. Treadwell	Albany, N. Y	Aug. 3, 1869	93, 248
Burner, Candle	J. A. Pease	Catskill, N. Y	Oct. 10, 1871	119, 785
Burner, Chandelier gas	R. Nutting	Randolph, Vt	Feb. 28, 1871	112, 172
Burner check, Gas	J. H. Jennings	New Bedford, Mass	Mar. 8, 1870	100, 634
Burner chimney, Gas	J. Stratton	Brooklyn, N. Y	July 25, 1865	49, 010
Burner, Coal-oil	W. W. Batchelder	New York, N. Y	Apr. 25, 1865	47, 381
Burner, Coal-oil	M. B. Dyott	Philadelphia, Pa	Jan. 6, 1863	37, 281
Burner, Coal-oil	S. Fredrick	New York, N. Y	Apr. 21, 1863	38, 223
Burner, Coal-oil	W. Fulton	Cranberry, N. J	May 27, 1862	35, 370
Burner, Coal-oil	J. G. Leffingwell	Newark, N. J	Dec. 6, 1864	45, 375
Burner, Coal-oil	W. O. B. Merrill	Philadelphia, Pa	June 10, 1862	35, 533
Burner, Coal-oil	E. Trittin	Philadelphia, Pa	Mar. 31, 1863	38, 070
Burner, Coal-oil	T. Wood	Westport, Conn	July 24, 1866	56, 655
Burner, Conical arch charcoal	E. B. Gilbert	Euphrata, N. Y	Nov. 7, 1835	....
Burner, Fluid	Y. Bailey	West Chester, Pa	Feb. 20, 1855	12, 409
Burner for Argand lamps	A. Olcott	Rochester, N. Y	Mar. 16, 1852	8, 805
Burner for carbureted air	J. A. Bassett	Salem, Mass	Mar. 14, 1865	46, 770
Burner for coal-oil lamps	J. S. Bradford	Baltimore, Md	July 1, 1862	35, 801
Burner for coal-oil lamps	J. Donning	Paterson, N. J	July 22, 1862	35, 925
Burner for coal-oil lamps	A. J. Gibson	Worcester, Mass	Apr. 1, 1862	34, 831
Burner for coal-oil lamps	A. Taplin	Providence, R. I	June 10, 1862	35, 552
Burner for coal-oil lamps	A. C. Wilhelm	Philadelphia, Pa	Sept. 8, 1863	39, 856
Burner for cooking and heating, Hydrocarbon	H. W. Dopp	Buffalo, N. Y	June 27, 1865	48, 379
Burner for cooking or heating, Vapor	H. S. Saroni	Baltimore, Md	Mar. 15, 1870	100, 807
Burner for cooking, &c., Petroleum	J. P. Hayes	Philadelphia, Pa	Sept. 26, 1865	50, 121
Burner for cooking-purposes, Gas	A. Geiss	Buffalo, N. Y	Nov. 28, 1865	51, 170
Burner for cooking-purposes, Petroleum	W. H. Smith	New York, N. Y	Jan. 2, 1866	51, 876
Burner for gas-lamps, Argand	J. G. Webb	Williamsburgh, N. Y	Aug. 7, 1849	6, 625
Burner for gas-stoves	T. H. Brymer	New York, N. Y	Mar. 27, 1866	53, 404
Burner for gas-stoves	I. R. Clark and S. T. Savage	New York and Albany, N. Y	Apr. 24, 1866	54, 114
Burner for gas-stoves	T. J. Kelly	New York, N. Y	Nov. 14, 1865	50, 938
Burner for gas-stoves	E. Osmond	Cincinnati, Ohio	Mar. 14, 1865	46, 861
Burner for heating and cooking, Wick	H. W. Dopp	Buffalo, N. Y	Sept. 11, 1866	57, 880
Burner for heating &c., Gas	E. P. Gleason	New York, N. Y	Dec. 17, 1867	72, 188
Burner for heating, Gas	J. Van	Cincinnati, Ohio	Sept. 2, 1873	142, 360
Burner for heating-purposes, Gas	H. Y. Lazear	New York, N. Y	May 7, 1867	64, 544
Burner for heating-purposes, Hydrocarbon	J. Youmans and J. Reed	Davenport, Iowa	Dec. 11, 1866	60, 455
Burner for heating-purposes, Petroleum-gas	D. L. Holden	New Orleans, La	June 25, 1867	66, 025
Burner for heating, Vapor	S. Child, jr., and R. A. Copeland	Baltimore, Md	Aug. 27, 1867	68, 045
Burner for heating, Vapor	C. W. Duncan	Baltimore, Md	July 30, 1867	67, 277
Burner for heating, &c., Vapor	R. R. Lewis	New York, N. Y	Feb. 19, 1861	31, 466
Burner for hot-air, steam, and hydrocarbon fluid, Annular petroleum.	G. L. Moody	New York, N. Y	Jan. 29, 1867	61, 632
Burner for hydrocarbon fluid	R. S. Merrill	Boston, Mass	Dec. 17, 1867	72, 414
Burner for kerosene-lamp	T. Raymond	Brooklyn, N. Y	May 12, 1863	38, 537
Burner for lamps, Aëro-vapor	O. F. Morrill	Boston, Mass	May 18, 1858	20, 289
Burner for lamps and lanterns	J. H. Irwin	Chicago, Ill	Feb. 14, 1865	46, 363
Burner for lamps, Coal-oil	J. Dodin	Brooklyn, N. Y	Dec. 23, 1862	37, 220
Burner for lamps, Coal-oil	J. T. Van Kirk	Philadelphia, Pa	Mar. 17, 1863	37, 930
Burner for locomotive head-lights	A. C. Vaughan	Philadelphia, Pa	Nov. 26, 1867	71, 450
Burner for locomotive head-lights, Lamp	D. T. Briggs	Albany, N. Y	Feb. 16, 1869	87, 024
Burner for oil-lamp	J. G. Leffingwell	Newark, N. J	Apr. 5, 1864	42, 204
Burner for pine-knots, &c	J. Price	Nashville, Tenn	Dec. 18, 1839	1, 434
Burner for purifying gas	J. Danks	Troy, N. Y	Apr. 23, 1861	32, 124
Burner for spirit-gas lamps	R. W. Sargent	Philadelphia, Pa	Oct. 12, 1852	9, 334
Burner for stills, engines, &c., Hydrocarbon	I. Helme	Philadelphia, Pa	Nov. 6, 1866	59, 393
Burner for stoves, Vapor-generating	D. E. Ryan	Saint Louis, Mo	Sept. 27, 1870	107, 816
Burner for vapor-lamps	C. F. Allen	Indianapolis, Ind	Feb. 1, 1859	22, 774
Burner for vapor-lamps	J. L. Butler, W. S. Hosford, and D. W. Smith.	Brooklyn, N. Y	Mar. 6, 1860	27, 346
Burner for vapor-lamps	T. Connelly	Philadelphia, Pa	Jan. 24, 1860	26, 888
Burner for vapor-lamps	H. W. Dopp and W. S. Mead	Buffalo, N. Y	Feb. 21, 1860	27, 211
Burner for vapor-lamps	A. Geiger	Dayton, Ohio	Jan. 31, 1860	26, 983
Burner for vapor-lamps	F. Heidrick	Philadelphia, Pa	Aug. 10, 1858	21, 166
Burner for vapor-lamps	J. S. Hull	Cincinnati, Ohio	Feb. 2, 1864	41, 434
Burner for vapor-lamps	J. S. Hull	Cincinnati, Ohio	Sept. 25, 1866	58, 261
Burner for vapor-lamps	H. Johnson	Washington, D. C	Apr. 12, 1859	23, 583
Burner for vapor-lamps	C. B. Loveless	Syracuse, N. Y	June 8, 1858	20, 498
Burner for vapor-lamps	J. K. O'Neil	Kingston, N. Y	May 18, 1858	20, 296
Burner for vapor-lamps	W. H. Racey	New York, N. Y	Sept. 20, 1859	25, 525
Burner for vapor-lamps	R. Ramsey	Philadelphia, Pa	May 24, 1859	24, 147
Burner for vapor-lamps	G. W. Randall	Boston, Mass	July 27, 1858	21, 053
Burner for vapor-lamps	E. D. Rosencrantz and W. H. Smith.	New York, N. Y	Jan. 4, 1859	22, 536
Burner for vapor-lamps	E. Trittin	Philadelphia, Pa	May 15, 1860	28, 333
Burner for vapor-lamps	T. Varney	San Francisco, Cal	May 11, 1858	20, 232
Burner for vapor-lamps	S. Wales	Boston, Mass	Dec. 28, 1858	22, 465
Burner for vapor-lamps	E. M. Williams	Philadelphia, Pa	Dec. 7, 1858	22, 270
Burner for vapor-stoves	R. L. Howell	Baltimore, Md	Jan. 22, 1867	61, 432
Burner, Gas	B. Andreas	Frankfort-on-the-Main, Germany.	May 27, 1873	139, 352
Burner, Gas	A. Arnoux	New York, N. Y	June 4, 1838	764
Burner, Gas	J. W. Averill	Newmarket, N. H	Aug. 5, 1873	141, 415
Burner, Gas	Y. Baily	Philadelphia, Pa	Oct. 12, 1858	21, 733
Burner, Gas	A. Barbarin	New Orleans, La	Sept. 1, 1868	81, 733

Index of patents issued from the United States Patent Office from 1790 to 1873, inclusive—Continued.

Invention.	Inventor.	Residence.	Date.	No.
Burner, Gas	J. F. Barker	Springfield, Mass	July 26, 1870	105,768
Burner, Gas	J. A. Bassett	Salem, Mass	May 23, 1865	47,786
Burner, Gas	W. W. Batchelder	New York, N. Y	Sept. 22, 1857	18,230
Burner, Gas	J. Battin	Philadelphia, Pa	June 6, 1846	4,563
Burner, Gas	H. Berg	Union Hill, N. J	May 23, 1865	47,787
Burner, Gas	H. Berg and A. Blessing	Springfield, Mass	Oct. 16, 1866	58,764
Burner, Gas	W. Blake	Boston, Mass	Aug. 9, 1845	4,141
Burner, Gas	W. Blake	Boston, Mass	Apr. 5, 1859	23,441
Burner, Gas	W. J. Brassington	Brooklyn, N. Y	Aug. 18, 1868	81,063
Burner, Gas	S. R. Brick	Philadelphia, Pa	May 24, 1853	9,742
Burner, Gas	J. S. Bridgman and E. G. Wellman.	Brockport, N. Y	July 7, 1868	79,545
Burner, Gas	J. Brönner	Frankfort-on-the-Main, Prussia.	Dec. 8, 1868	84,675
Burner, Gas	A. Buckham	Delhi, N. Y	Feb. 6, 1872	123,327
Burner, Gas	T. Clough	Dobbs's Ferry, N. Y	June 14, 1870	104,271
Burner, Gas	T. Clough	New York, N. Y	Nov. 7, 1871	120,715
Burner, Gas	S. L. Cole	Burlington, Vt	Nov. 29, 1859	26,255
Burner, Gas	S. L. Cole	Brooklyn, N. Y	June 2, 1868	78,578
Burner, Gas	J. B. Coolidge	Boston, Mass	Oct. 9, 1866	58,606
Burner, Gas	R. Cornelius	Philadelphia, Pa	June 22, 1858	20,626
Burner, Gas	J. Cornell and B. McDougall	Brooklyn and New York, N. Y.	Dec. 9, 1856	16,176
Burner, Gas	J. W. Cremin	New York, N. Y	Sept. 7, 1869	94,476
Burner, Gas	J. W. Cremin	New York, N. Y	Jan. 11, 1870	98,744
Burner, Gas	C. A. Cummings and C. Douglass.	New London, Conn	Jan. 15, 1856	14,091
Burner, Gas	H. H. Dodge	Georgetown, D. C	Oct. 23, 1860	30,470
Burner, Gas	V. Dubourg	Paris, France	Oct. 10, 1865	50,429
Burner, Gas	A. E. Dupas	New Orleans, La	Sept. 13, 1870	107,234
Burner, Gas	A. E. Dupas	New Orleans, La	Feb. 14, 1871	111,825
Burner, Gas	W. S. Dyer	Portland, Me	Mar. 22, 1864	41,981
Burner, Gas	E. M. Echalleus	Switzerland	Dec. 28, 1846	4,913
Burner, Gas	J. S. Fancher	Newark, N. J	Sept. 1, 1863	39,725
Burner, Gas	F. A. Fisher	Crawford, N. J	May 28, 1872	127,329
Burner, Gas	I. R. Fisher	Reading, Pa	Aug. 3, 1869	93,189
Burner, Gas	T. B. Fogarty	New York, N. Y	Feb. 18, 1873	135,978
Burner, Gas	T. B. Fogarty	New York, N. Y	Feb. 18, 1873	135,979
Burner, Gas	E. Foote	Yonkers, N. Y	Apr. 15, 1873	137,905
Burner, Gas	C. S. Ford	Philadelphia, Pa	Mar. 31, 1868	76,069
Burner, Gas	C. S. Ford	Philadelphia, Pa	June 28, 1870	104,723
Burner, Gas	C. S. Ford	Philadelphia, Pa	Oct. 25, 1870	108,579
Burner, Gas	J. S. Ford	Philadelphia, Pa	Oct. 8, 1867	69,556
Burner, Gas	C. Franz	Cincinnati, Ohio	Mar. 16, 1869	87,922
Burner, Gas	A. Fulton	Albany, N. Y	Nov. 22, 1870	109,404
Burner, Gas	S. Gardner, jr	New York, N. Y	June 16, 1868	78,871
Burner, Gas	S. Gardner	New York, N. Y	Mar. 26, 1872	125,042
Burner, Gas	A. D. Gates	Binghamton, N. Y	June 30, 1857	17,674
Burner, Gas	J. Gilfillan	Hartford, Conn	Aug. 2, 1859	24,932
Burner, Gas	R. Gill	New York, N. Y	Aug. 24, 1869	94,103
Burner, Gas	E. P. Gleason	Providence, R. I	Apr. 14, 1857	17,035
Burner, Gas	A. G. Hamaker	Peoria, Ill	Aug. 7, 1860	29,486
Burner, Gas	G. W. Hatch	Brooklyn, N. Y	Apr. 1, 1873	137,303
Burner, Gas	L. E. Hicks	New York, N. Y	Sept. 14, 1858	21,497
Burner, Gas	C. H. Johnson	Boston, Mass	June 26, 1855	13,130
Burner, Gas	W. L. Jukes	New York, N. Y	Aug. 3, 1869	93,310
Burner, Gas	W. L. Jukes	New York, N. Y	Nov. 2, 1869	96,441
Burner, Gas	W. L. Jukes	New York, N. Y	Dec. 7, 1869	97,517
Burner, Gas	F. C. Krause	New York, N. Y	Aug. 3, 1858	21,076
Burner, Gas	F. Küp	Frankfort-on-the-Main, Germany.	Oct. 10, 1865	50,431
Burner, Gas	A. M. Loeveison	Quincy, Ill	Dec. 7, 1869	97,652
Burner, Gas	J. G. Leffingwell and G. W. Thompson.	Newark, N. J., and New York, N. Y.	Apr. 2, 1861	31,893
Burner, Gas	W. H. Lindsay	Brooklyn, N. Y	Sept. 29, 1857	18,289
Burner, Gas	R. B. Locke	New York, N. Y	Apr. 7, 1868	76,333
Burner, Gas	R. B. Locke and W. B. Ulrich	New Orleans and Concordia Parish, La.	Apr. 16, 1867	63,802
Burner, Gas	F. Lunkenheimer	Cincinnati, Ohio	Sept. 8, 1863	39,820
Burner, Gas	W. Mallerd	Brooklyn, N. Y	June 27, 1854	11,168
Burner, Gas	H. L. McAvoy	Baltimore, Md	Sept. 19, 1865	50,075
Burner, Gas	J. McGlensey	Philadelphia, Pa	June 19, 1860	28,761
Burner, Gas	C. P. Miller	Philadelphia, Pa	Sept. 10, 1861	33,258
Burner, Gas	H. Monier	Paris, France	May 12, 1868	77,749
Burner, Gas	C. Monson and S. Moore	New Haven, Conn	Feb. 19, 1861	31,472
Burner, Gas	G. Mooney	Providence, R. I	Dec. 17, 1867	72,415
Burner, Gas	G. Mooney	Providence, R. I	Aug. 25, 1868	81,524
Burner, Gas	H. B. Myer	Philadelphia, Pa	Nov. 10, 1868	83,876
Burner, Gas	H. B. Myer	Philadelphia, Pa	May 30, 1871	115,506
Burner, Gas	J. Neal	Boston, Mass	May 6, 1856	14,822
Burner, Gas	A. Ostrander	New York, N. Y	Jan. 11, 1859	22,574
Burner, Gas	S. W. Pingree	Lawrence, Mass	Apr. 2, 1872	125,327
Burner, Gas	J. Poole	New York, N. Y	July 2, 1861	32,738
Burner, Gas	A. C. Rand	New York, N. Y	June 23, 1868	79,081
Burner, Gas	A. H. Ray	Boston, Mass	June 15, 1858	20,584
Burner, Gas	P. Reitmeier	Chicago, Ill	Jan. 23, 1872	123,047
Burner, Gas	J. and T. D. Richardson	New York, N. Y	June 24, 1873	140,164
Burner, Gas	J. Rigby	Cincinnati, Ohio	June 17, 1873	139,919
Burner, Gas	C. Ritter	Reading, Pa	Oct. 24, 1865	50,628
Burner, Gas	W. H. Rodgers	Brooklyn, N. Y	May 12, 1868	77,917
Burner, Gas	M. Rosenwax	New York, N. Y	Mar. 14, 1871	112,637
Burner, Gas	G. C. Roundey	New York, N. Y	May 5, 1863	38,417
Burner, Gas	E. P. Russell	Manlius, N. Y	Dec. 8, 1868	84,768
Burner, Gas	D. E. Ryan	Cincinnati, Ohio	June 3, 1873	139,481
Burner, Gas	J. Scholl	London, England	Dec. 24, 1867	72,545
Burner, Gas	I. W. Shaler	Brooklyn, N. Y	Apr. 15, 1873	137,964
Burner, Gas	W. F. Shaw	Boston, Mass	Apr. 22, 1856	14,737
Burner, Gas	I. Simmons	Baltimore, Md	June 10, 1873	139,828

Index of patents issued from the United States Patent Office from 1790 *to* 1873, *inclusive*—Continued.

Invention.	Inventor.	Residence.	Date.	No.
Burner, Gas	J. L. Smith	Louisville, Ky	May 22, 1866	54, 971
Burner, Gas	D. H. Solliday	Philadelphia, Pa	Mar. 20, 1849	6, 212
Burner, Gas	D. H. Solliday	Philadelphia, Pa	Sept. 6, 1859	25, 372
Burner, Gas	R. N. Stewart	Philadelphia, Pa	Feb. 18, 1862	34, 450
Burner, Gas	W. B. Stofer	Memphis, Tenn	Oct. 3, 1871	119, 480
Burner, Gas	J. Stratton	Philadelphia, Pa	Nov. 21, 1865	51, 121
Burner, Gas	D. Sullivan and M. McIntyre	Cincinnati, Ohio	Mar. 23, 1858	19, 720
Burner, Gas	W. Tallman	Cincinnati, Ohio	Apr. 13, 1858	19, 959
Burner, Gas	G. W. Thompson	New York, N. Y	May 1, 1860	28, 113
Burner, Gas	J. F. Tozer	Binghamton, N. Y	Oct. 5, 1858	21, 728
Burner, Gas	J. and J. Wadsworth	Marple and Salford, England.	Apr. 19, 1864	42, 421
Burner, Gas	J. C. Walsh	Lockport, N. Y	June 9, 1857	17, 530
Burner, Gas	T. Ward	Columbus, Ohio	Mar. 19, 1872	124, 866
Burner, Gas	J. Webster and O. Spencer	Cleveland, Ohio	Feb. 14, 1854	10, 529
Burner, Gas	A. H. Wood	Boston, Mass	Nov. 9, 1852	9, 396
Burner, Gas	A. H. Wood	Boston, Mass	Sept. 21, 1858	21, 586
Burner, Gas	S. E. Woodworth and J. S. Wethered.	San Francisco, Cal	June 12, 1860	28, 708
Burner, Gas	W. Wright	Saint Louis, Mo	Aug. 17, 1858	21, 229
Burner, Gas	W. Wright	New York, N. Y	Jan. 11, 1859	22, 609
Burner, Gas	W. Wright	New York, N. Y	June 12, 1860	28, 709
Burner, Gas and lamp	J. Horton	New York, N. Y	June 8, 1869	91, 020
Burner, Gas or vapor	G. P. Ganster	New York, N. Y	Aug. 20, 1872	130, 632
Burner, Globe-holder and cooler combined for vapor	J. J. and F. G. Palmer	Pittsburgh, Pa	Mar. 14, 1871	112, 624
Burner, Hollow wick lamp	J. W. Emerson	Millbury, Mass	Oct. 25, 1870	108, 694
Burner, Hydrocarbon	M. L. Callender	New York, N. Y	Nov. 20, 1860	30, 703
Burner, Hydrocarbon	M. L. Callender	New York, N. Y	Feb. 18, 1862	34, 402
Burner, Hydrocarbon	M. L. Callender	New York, N. Y	Mar. 31, 1863	38, 031
Burner, Hydrocarbon	C. Carpenter, jr	Astoria, N. Y	Feb. 11, 1868	74, 196
Burner, Hydrocarbon	F. Cook	New York, N. Y	Jan. 21, 1868	73, 506
Burner, Hydrocarbon	A. W. Cook and R. Dempster	Buffalo, N. Y	June 9, 1868	78, 788
Burner, Hydrocarbon	S. E. Crow	Stratford, England	Oct. 13, 1868	82, 925
Burner, Hydrocarbon	A. De Landsee	Paris, France	Mar. 1, 1870	100, 268
Burner, Hydrocarbon	H. Everett	New York, N. Y	July 7, 1868	79, 563
Burner, Hydrocarbon	J. Gray	San Francisco, Cal	Nov. 17, 1868	84, 110
Burner, Hydrocarbon	A. J. Griffin	Lowell, Mass	Apr. 16, 1867	63, 881
Burner, Hydrocarbon	M. Herzog and D. Cohn	Vienna, Austria, and London, England.	July 3, 1866	56, 155
Burner, Hydrocarbon	S. A. Hill	Oil City, Pa	Jan. 5, 1869	85, 663
Burner, Hydrocarbon	J. S. Hull	Cincinnati, Ohio	Mar. 3, 1868	75, 165
Burner, Hydrocarbon	W. L. Imlay	Philadelphia, Pa	Jan. 14, 1868	73, 335
Burner, Hydrocarbon	S. P. Legg	Springfield, Mass	July 28, 1868	80, 413
Burner, Hydrocarbon	D. H. Lowe	Boston, Mass	Aug. 11, 1868	80, 980
Burner, Hydrocarbon	N. May and H. F. Stivers	San Francisco, Cal	Jan. 19, 1869	86, 024
Burner, Hydrocarbon	E. P. McCarthy	San Francisco, Cal	July 27, 1869	93, 105
Burner, Hydrocarbon	H. A. V. Post and J. Garrard	Cincinnati, Ohio	Oct. 22, 1867	70, 117
Burner, Hydrocarbon	C. Sauvage	Paris, France	Feb. 9, 1869	86, 871
Burner, Hydrocarbon	J. Stratton	Philadelphia, Pa	Mar. 20, 1866	53, 383
Burner, Hydrocarbon	J. Stratton	Philadelphia, Pa	Mar. 31, 1868	76, 115
Burner, Hydrocarbon	E. W. Taylor	Norristown, Pa	Nov. 17, 1868	84, 148
Burner, Hydrocarbon	F. W. Thayer	Boston, Mass	Apr. 7, 1868	76, 552
Burner, Hydrocarbon	L. Verstraet	Paris, France	Nov. 17, 1868	84, 234
Burner, Hydrocarbon	L. Verstraet	Paris, France	Jan. 12, 1869	85, 877
Burner, Hydrocarbon	J. H. White	San Francisco, Cal	Jan. 21, 1868	73, 682
Burner, Hydrocarbon	H. W. Yerington	Jersey City, N. J	Oct. 20, 1868	83, 352
Burner, Hydrocarbon-vapor	H. W. Dopp and M. J. Stark	Buffalo, N. Y	Mar. 12, 1872	124, 425
Burner, Hydrocarbon-vapor	A. M. Mace	Springfield, Mass	Oct. 26, 1858	21, 893
Burner, Illuminating incense	E. W. Hastings	Boston, Mass	Aug. 15, 1871	118, 126
Burner, Incense	E. W. Hastings	Boston, Mass	July 25, 1871	117, 281
Burner, Kerosene	G. Neilson	Boston Mass	Sept. 27, 1864	44, 441
Burner, Kerosene	L. C. White	Waterbury, Conn	July 3, 1866	56, 131
Burner, Kerosene	S. R. Wilmot	Brooklyn, N. Y	July 18, 1865	48, 800
Burner, Kerosene-lamp	H. F. Adams and W. Berry	Syracuse, N. Y	June 24, 1862	35, 660
Burner, Kerosene-lamp	A. B. Hendryx	Derby, Conn	Aug. 29, 1865	49, 680
Burner, Kerosene-oil	P. D. Cummings	Portland, Me	Apr. 8, 1862	34, 880
Burner, Kerosene-oil	H. and J. Sangster	Buffalo, N. Y	June 27, 1865	48, 450
Burner, Lamp	J. R. Ackerman, E. B. Campbell, and N. O. Golden.	Dobb's Ferry, N. Y	Apr. 21, 1868	76, 966
Burner, Lamp	B. F. Adams	Boston, Mass	Aug. 16, 1870	106, 302
Burner, Lamp	B. F. Adams	Boston, Mass	Aug. 16, 1870	106, 303
Burner, Lamp	T. Adams	Hudson City, N. J	Sept. 15, 1868	82, 059
Burner, Lamp	T. Adams	Hudson City, N. J	July 20, 1869	92, 771
Burner, Lamp	B. R. Alden	New York, N. Y	Oct. 21, 1862	36, 694
Burner, Lamp	J. B. Alexander	Washington, D. C	Apr. 23, 1867	64, 056
Burner, Lamp	J. B. Alexander	Washington, D. C	Mar. 16, 1869	87, 898
Burner, Lamp	J. B. Alexander	Washington, D. C	Dec. 14, 1869	97, 854
Burner, Lamp	J. Allen	New York, N. Y	Aug. 25, 1868	81, 321
Burner, Lamp	J. K. Andrews	Antrim, Ohio	Nov. 13, 1866	59, 704
Burner, Lamp	N. L. Archer and C. Deavs	New York, N. Y	Feb. 19, 1867	62, 244
Burner, Lamp	L. J. Atwood	Waterbury, Conn	June 9, 1868	78, 637
Burner, Lamp	L. J. Atwood	Waterbury, Conn	Jan. 18, 1870	98, 836
Burner, Lamp	L. J. Atwood	Waterbury, Conn	Apr. 18, 1871	113, 833
Burner, Lamp	L. J. Atwood	Waterbury, Conn	Nov. 26, 1872	133, 397
Burner, Lamp	L. J. Atwood	Waterbury, Conn	May 6, 1873	138, 601
Burner, Lamp	J. R. Baker	Kendallville, Ind	Jan. 5, 1864	41, 048
Burner, Lamp	P. Baker	Chicago, Ill	July 2, 1867	66, 203
Burner, Lamp	P. Baker	Chicago, Ill	Aug. 18, 1868	81, 128
Burner, Lamp	P. Baker	Chicago, Ill	Apr. 26, 1870	102, 355
Burner, Lamp	P. Baker	Ansonia, Conn	Feb. 6, 1872	123, 376
Burner, Lamp	G. E. Baldwin	West Meriden, Conn	Dec. 10, 1867	71, 949
Burner, Lamp	G. A. Beidler	Philadelphia, Pa	Apr. 21, 1868	76, 979
Burner, Lamp	H. M. Beidler	Philadelphia, Pa	Oct. 19, 1869	95, 870
Burner, Lamp	G. Berkstresser	Bedford, Pa	June 7, 1870	103, 967
Burner, Lamp	E. Blackman	Norwalk, Conn	July 20, 1869	92, 695
Burner, Lamp	E. C. Blakeslee	Waterbury, Conn	Sept. 29, 1868	82, 480
Burner, Lamp	A. Bliss	New York, N. Y	Nov. 24, 1868	84, 406
Burner, Lamp	J. O. Blythe	Philadelphia, Pa	Mar. 24, 1863	37, 945

Index of patents issued from the United States Patent Office from 1790 *to* 1873, *inclusive*—Continued.

Invention.	Inventor.	Residence.	Date.	No.
Burner, Lamp	J. W. Brimblecom	Lynn, Mass	Apr. 14, 1868	76, 592
Burner, Lamp	C. Britain	Saint Joseph, Mich	Sept. 27, 1864	44, 393
Burner, Lamp	C. P. Brockett	New Haven, Conn	Apr. 22, 1862	35, 010
Burner, Lamp	H. Brown	New York, N. Y	Feb. 10, 1863	37, 606
Burner, Lamp	A. W. Browne	Brooklyn, N. Y	Feb. 25, 1868	74, 793
Burner, Lamp	A. W. Browne	Brooklyn, N. Y	June 15, 1869	91, 208
Burner, Lamp	G. E. Brush	Danbury, Conn	Aug. 22, 1871	118, 193
Burner, Lamp	C. W. Cahoon	Portland, Me	Dec. 15, 1868	84, 993
Burner, Lamp	G. J. Capewell	West Cheshire, Conn	Sept. 25, 1866	58, 213
Burner, Lamp	G. J. Capewell	West Cheshire, Conn	Nov. 13, 1866	59, 554
Burner, Lamp	G. J. Capewell	West Cheshire, Conn	Mar. 10, 1868	75, 363
Burner, Lamp	W. Carleton	Boston, Mass	Aug. 11, 1868	80, 909
Burner, Lamp	B. F. Chappell	Hartford, Conn	Mar. 11, 1873	136, 586
Burner, Lamp	G. Chinnock	New York, N. Y	Apr. 28, 1868	77, 254
Burner, Lamp	G. H. Chinnock	New York, N. Y	Oct. 26, 1869	96, 200
Burner, Lamp	P. J. Clark	West Meriden, Conn	June 16, 1863	38, 926
Burner, Lamp	J. R. Cole	Demopolis, Ala	Sept. 14, 1869	94, 811
Burner, Lamp	M. H. Collins	Chelsea, Mass	Mar. 1, 1870	100, 260
Burner, Lamp	H. Coulter	Philadelphia, Pa	Apr. 14, 1868	76, 716
Burner, Lamp	W. R. Cranna	San Francisco, Cal	Sept. 29, 1868	82, 495
Burner, Lamp	P. Crans, jr	Philadelphia, Pa	Apr. 17, 1866	53, 953
Burner, Lamp	R. R. Crosby	Boston, Mass	June 28, 1870	104, 709
Burner, Lamp	J. D. Custer	Norristown, Pa	Mar. 25, 1862	34, 742
Burner, Lamp	J. Dodin	Brooklyn, N. Y	Aug. 11, 1863	39, 524
Burner, Lamp	J. H. Doolittle	Ansonia, Conn	July 8, 1873	140, 687
Burner, Lamp	M. B. Dyott	Philadelphia, Pa	Dec. 7, 1858	22, 230
Burner, Lamp	R. N. Eagle	Washington, D. C	Mar. 2, 1869	87, 480
Burner, Lamp	J. P. Egan	Zanesville, Ohio	May 9, 1865	47, 625
Burner, Lamp	L. Erving	Brooklyn, N. Y	Apr. 7, 1868	76, 426
Burner, Lamp	H Fairbanks	Boston, Mass	Apr. 12, 1864	42, 281
Burner, Lamp	S. W. Fowler	Brooklyn, N. Y	May 3, 1870	102, 674
Burner, Lamp	S. W. Fowler	Brooklyn, N. Y	June 27, 1871	116, 292
Burner, Lamp	S. W. Fowler	Brooklyn, N. Y	Apr. 16, 1872	125, 672
Burner, Lamp	J. A. Frey	Washington, D. C	Dec. 18, 1866	60, 499
Burner, Lamp	J. A. Frey	New York, N. Y	May 21, 1867	64, 965
Burner, Lamp	F. H. Fuller	South Boston, Mass	July 14, 1868	79, 969
Burner, Lamp	G. P. Fuller	Humphrey, N. Y	Jan. 10, 1871	110, 846
Burner, Lamp	J. B. Fuller	Norwich, Conn	Nov. 23, 1869	97, 069
Burner, Lamp	J. B. Fuller	Norwich, Conn	May 31, 1870	103, 598
Burner, Lamp	E. A. Galbraith	Boston, Mass	June 23, 1868	79, 221
Burner, Lamp	T. W. Gardner	Terre Haute, Ind	Jan. 7, 1873	134, 657
Burner, Lamp	E. L. Gilman	Somerville, Mass	Oct. 5, 1869	95, 460
Burner, Lamp	R. Gorsline	Rochester, N. Y	July 28, 1868	80, 281
Burner, Lamp	J. Gracie and R. H. Boyd	Pittsburgh, Pa	May 5, 1868	77, 481
Burner, Lamp	J. B. Gray	Hudson, Wis	June 23, 1863	38, 960
Burner, Lamp	W. H. Gray	Saint Louis, Mo	Apr. 18, 1871	113, 876
Burner, Lamp	W. H. Gray and S. Ross	Saint Louis, Mo., and Washington, D. C.	June 13, 1871	115, 953
Burner, Lamp	E. J. Hale	Foxcroft, Me	May 16, 1871	114, 805
Burner, Lamp	T. Hall	Bergen, N. J	Sept. 18, 1866	58, 097
Burner, Lamp	T. F. Halley and J. F. Livingston.	Washington, D. C	July 25, 1871	117, 412
Burner, Lamp	J. O. Harris	Reading, Pa	May 17, 1864	42, 765
Burner, Lamp	J. O. Harris	Reading, Pa	Dec. 20, 1864	45, 493
Burner, Lamp	J. O. Harris	Reading, Pa	Mar. 21, 1865	46, 301
Burner, Lamp	H. W. Hayden	Waterbury, Conn	Aug. 16, 1870	106, 363
Burner, Lamp	H. W. Hayden	Waterbury, Conn	May 30, 1871	115, 315
Burner, Lamp	H. W. Hayden	Waterbury, Conn	May 30, 1871	115, 466
Burner, Lamp	E. K. Haynes	Boston, Mass	Jan. 26, 1869	86, 301
Burner, Lamp	E. K. Haynes	Boston, Mass	June 6, 1871	115, 608
Burner, Lamp	A. N. Henderson	Buffalo, N. Y	June 2, 1863	38, 742
Burner, Lamp	G. Hillegass	Philadelphia, Pa	Apr. 10, 1866	53, 824
Burner, Lamp	G. Hillegass	Philadelphia, Pa	July 27, 1869	93, 087
Burner, Lamp	T. Hipwell	Camden, N. J	Oct. 31, 1871	120, 383
Burner, Lamp	R. Hoadley and H. A. Shipman	Ansonia, Conn	Apr. 7, 1868	76, 450
Burner, Lamp	E. Hobbis and A. McNair	Newark and Bloomfield, N. J.	Mar. 1, 1864	41, 774
Burner, Lamp	W. W. Horton	Providence, R. I	May 5, 1868	77, 614
Burner, Lamp	J. Hubbard	Dayton, Ohio	Aug. 10, 1869	93, 442
Burner, Lamp	J. S. Hull	Cincinnati, Ohio	Aug. 4, 1863	39, 399
Burner, Lamp	J. S. Hull	Cincinnati, Ohio	Aug. 4, 1863	39, 400
Burner, Lamp	H. C. Hunt	Ottumwa, Iowa	Feb. 3, 1863	37, 600
Burner, Lamp	J. G. Hunt	Cincinnati, Ohio	Apr. 3, 1866	53, 623
Burner, Lamp	H. C. Hutchinson	Cayuga, N. Y	May 6, 1862	35, 157
Burner, Lamp	H. C. Hutchinson	Cayuga, N. Y	Mar. 8, 1864	41, 846
Burner, Lamp	J. H. Irwin	Philadelphia, Pa	Feb. 1, 1870	99, 443
Burner, Lamp	W. P. Jenney and G. W. Taylor	Fairhaven, Mass	Dec. 20, 1870	110, 241
Burner, Lamp	M. Jincks	Dansville, N. Y	May 5, 1868	77, 620
Burner, Lamp	M. Jincks	Wallace, N. Y	Nov. 30, 1869	97, 298
Burner, Lamp	J. Johnson and F. Bailey	New York, N. Y	Dec. 7, 1858	22, 253
Burner, Lamp	J. H. Johnson	Newark, N. J	May 3, 1870	102, 552
Burner, Lamp	A. Judson	Brooklyn, N. Y	Nov. 20, 1866	59, 772
Burner, Lamp	A. C. Ketcham	New York, N. Y	Oct. 21, 1862	36, 717
Burner, Lamp	A. H. Knapp	Medford, Mass	Apr. 7, 1857	16, 981
Burner, Lamp and lantern	C. W. T. Krausch	Chicago, Ill	Mar. 10, 1863	37, 868
Burner, Lamp	C. H. Kupfer	Hoboken, N. J	Sept. 22, 1863	40, 050
Burner, Lamp	G. Lavere	Bridgeport, Conn	May 12, 1868	77, 821
Burner, Lamp	J. C. Love	Philadelphia, Pa	Sept. 17, 1867	68, 889
Burner, Lamp	J. C. Love	Philadelphia, Pa	Dec. 17, 1867	72, 209
Burner, Lamp	J. C. Love	Philadelphia, Pa	June 8, 1869	91, 141
Burner, Lamp	J. C. Love	Philadelphia, Pa	Aug. 31, 1869	94, 421
Burner, Lamp	W. D. Ludlow	New York, N. Y	May 17, 1870	103, 213
Burner, Lamp	G. Lumpton	Indianapolis, Ind	Dec. 13, 1870	110, 055
Burner, Lamp	G. R. Lyon	Waterbury Conn	July 5, 1870	104, 971
Burner, Lamp	J. Magee	Chelsea, Mass	Mar. 31, 1868	76, 089
Burner, Lamp	J. Magoun	East Cambridge, Mass	Jan. 5, 1864	41, 179
Burner, Lamp	C. B. Mann	Baltimore, Md	Mar. 28, 1871	113, 183
Burner, Lamp	J. J. Marcy	Meriden, Conn	Dec. 2, 1862	37, 047
Burner, Lamp	J. J. Marcy	Meriden, Conn	July 21, 1863	39, 320

Index of patents issued from the United States Patent Office from 1790 *to* 1873, *inclusive*—Continued.

Invention.	Inventor.	Residence.	Date.	No.
Burner, Lamp	J. J. Marcy	West Meriden, Conn	Dec. 4, 1866	60, 215
Burner, Lamp	J. J. Marcy	West Meriden, Conn	Apr. 7, 1868	76, 482
Burner, Lamp	L. J. Marcy	Newport, R. I	Aug. 11, 1868	80, 984
Burner, Lamp	G. A. Mason	Chelsea, Mass	June 30, 1868	79, 369
Burner, Lamp	C. B. Matthews	Oquawka, Ill	Sept. 9, 1862	36, 417
Burner, Lamp	W. McCaine	Groton, Mass	Dec. 24, 1867	72, 659
Burner, Lamp	J. P. McGee	Trenton, Tenn	Sept. 22, 1868	82, 334
Burner, Lamp	W. C. McGill	Cincinnati, Ohio	Apr. 14, 1868	76, 645
Burner, Lamp	W. C. McGill	Cincinnati, Ohio	Feb. 21, 1871	111, 960
Burner, Lamp	J. McHenry	Cincinnati, Ohio	Aug. 4, 1863	39, 413
Burner, Lamp	R. S. Merrill	Boston, Mass	July 18, 1865	48, 824
Burner, Lamp	R. S. Merrill	Boston, Mass	July 10, 1866	56, 245
Burner, Lamp	R. S. Merrill	Boston, Mass	June 30, 1868	79, 488
Burner, Lamp	R. S. Merrill and W. Carlton	Boston, Mass	June 1, 1869	90, 863
Burner, Lamp	R. S. Merrill	Cambridge, Mass	Mar. 8, 1870	100, 623
Burner, Lamp	R. S. Merrill	Hyde Park, Mass	June 21, 1870	104, 624
Burner, Lamp	R. S. Merrill	Hyde Park, Mass	June 21, 1870	104, 625
Burner, Lamp	R. S. Merrill	Boston, Mass	Jan. 17, 1871	111, 074
Burner, Lamp	R. S. Merrill	Boston, Mass	Nov. 28, 1871	121, 394
Burner, Lamp	R. S. Merrill	Hyde Park, Mass	Dec. 9, 1873	145, 438
Burner, Lamp	C. M. Mills and G. W. Remsen	Lincoln, Del	May 5, 1868	77, 514
Burner, Lamp	J. Minifie	Baltimore, Md	July 28, 1868	80, 494
Burner, Lamp	H. K. Needham	Titusville, Pa	Mar. 24, 1868	75, 959
Burner, Lamp	G. Neilson	Boston, Mass	June 18, 1867	65, 933
Burner, Lamp	G. Neilson	Boston, Mass	Mar. 17, 1868	75, 565
Burner, Lamp	G. Neilson	Boston, Mass	Mar. 17, 1868	75, 566
Burner, Lamp	G. Neilson	Boston, Mass	Mar. 24, 1868	75, 960
Burner, Lamp	G. Neilson	Boston, Mass	Dec. 8, 1868	84, 752
Burner, Lamp	G. K. Osborn	Brooklyn, N. Y	Oct. 29, 1867	70, 252
Burner, Lamp	G. K. Osborn	Brooklyn, N. Y	Apr. 14, 1868	76, 653
Burner, Lamp	W. Painter	Fallston, Md	June 30, 1863	39, 102
Burner, Lamp	R. W. Park	Philadelphia, Pa	Mar. 17, 1868	75, 697
Burner, Lamp	A. Parsons	Portland, Me	Feb. 26, 1867	62, 438
Burner, Lamp	J. M. Perkins	Cleveland, Ohio	July 27, 1869	93, 117
Burner, Lamp	A. H. Platt	Yellow Springs, Ohio	Jan. 31, 1865	46, 138
Burner, Lamp	A. H. Platt	Philadelphia, Pa	June 2, 1868	78, 607
Burner, Lamp	S. B. Platt	Derby, Conn	June 13, 1871	115, 890
Burner, Lamp	P. Prettymann	Paradise Spring Farm, Oreg	Aug. 17, 1869	93, 744
Burner, Lamp	H. Read	Providence, R. I	Dec. 10, 1867	71, 913
Burner, Lamp	F. J. Rebbeck and E. M. Davies	Pittsburgh, Pa	Sept. 2, 1862	36, 367
Burner, Lamp	E. B. Requa	Jersey City, N. J	July 15, 1862	35, 893
Burner, Lamp	E. B Requa	South Bergen, N. J	Mar. 24, 1868	75, 978
Burner, Lamp	W. F. Rippon and G. A. Johnson.	Providence, R. I	Sept. 14, 1869	94, 776
Burner, Lamp	F. S. Robinson	Boston, Mass	Apr. 21, 1868	76, 945
Burner, Lamp	F. S. Robinson	Boston, Mass	Dec. 5, 1871	121, 550
Burner, Lamp	W. Robinson	Funkville, Pa	July 9, 1867	66, 635
Burner, Lamp	W. Robinson	Brooklyn, N. Y	June 2, 1868	78, 539
Burner, Lamp	S. Ross	Washington, D. C	Oct. 24, 1871	120, 214
Burner, Lamp	W. J. Ross	Worcester, Mass	May 12, 1868	77, 844
Burner, Lamp	T. Rowatt, jr	London, England	Jan. 29, 1867	61, 634
Burner, Lamp	D. E. Rugg	Sing Sing, N. Y	Feb. 9, 1869	86, 694
Burner, Lamp	S. Russell	Waterbury, Conn	July 16, 1872	129, 598
Burner, Lamp	J. F. Sanford	Keokuk, Iowa	Oct. 16, 1866	58, 893
Burner, Lamp	J. F. Sanford	Keokuk, Iowa	Dec. 4, 1866	60, 263
Burner, Lamp	J. F. Sanford	Keokuk, Iowa	Jan. 5, 1869	85, 697
Burner, Lamp	J. F. Sanford	Keokuk, Iowa	Feb. 9, 1869	86, 869
Burner, Lamp	O. I. Savage and G. P. Hawley	Ithaca, N. Y	May 5, 1863	38, 422
Burner, Lamp	J. W. Schreiber	New York, N. Y	June 16, 1868	79, 014
Burner, Lamp	F. Shaller	Hudson, N. Y	June 22, 1869	91, 778
Burner, Lamp	A. J. Simpson and J. B. Currier	Lowell, Mass	Sept. 2, 1862	36, 374
Burner, Lamp	A. G. Smith	Jersey City, N. J	Dec. 8, 1868	84, 775
Burner, Lamp	G. Smith	Providence, R. I	May 14, 1867	64, 804
Burner, Lamp	G. Smith	Providence, R. I	Dec. 31, 1867	72, 925
Burner, Lamp	G. L. Smith	Bridgeport, Conn	June 14, 1870	104, 366
Burner, Lamp	G. L. Smith	Bridgeport, Conn	Mar. 28, 1871	113, 104
Burner, Lamp	R. L. Smith	West Meriden, N. Y	Mar. 15, 1864	41, 965
Burner, Lamp	W. H. Smith	New York, N. Y	Mar. 29, 1864	42, 122
Burner, Lamp	W. H. Smith	New York, N. Y	June 12, 1866	55, 546
Burner, Lamp	C. St. John	Boston, Mass	Dec. 13, 1864	45, 443
Burner, Lamp	C. St. John	Charlestown, Mass	Mar. 10, 1868	75, 483
Burner, Lamp	F. A. Taber	New Bedford, Mass	Jan. 31, 1871	111, 400
Burner, Lamp	A. Taplin	Somerville, Mass	Mar. 16, 1869	87, 988
Burner, Lamp	A. Taplin	Somerville, Mass	Apr. 26, 1870	102, 449
Burner, Lamp	A. Taplin	Forrestville, Conn	May 3, 1870	102, 728
Burner, Lamp	A. Taplin	Forrestville, Conn	Jan. 24, 1871	111, 274
Burner, Lamp	A. Taplin	Forrestville, Conn	Sept. 19, 1871	119, 061
Burner, Lamp	A. Taplin and H. D. Bradley	Forrestville, Conn	Aug. 31, 1869	94, 451
Burner, Lamp	A. Thirault	Brooklyn, N. Y	Mar. 9, 1869	87, 601
Burner, Lamp	E. J. Toof	Madison, Iowa	Aug. 11, 1868	80, 843
Burner, Lamp	J. Trent	Millerton, N. Y	Apr. 4, 1871	113, 472
Burner, Lamp	E. Trittin	Philadelphia, Pa	Aug. 19, 1862	36, 262
Burner, Lamp	A. C. Vaughn	Philadelphia, Pa	Nov. 6, 1866	59, 518
Burner, Lamp	S. W. Warren	Boston, Mass	Sept. 14, 1869	94, 796
Burner, Lamp	C. C. Warwick	Philadelphia, Pa	Oct. 27, 1863	40, 439
Burner, Lamp	J. H. Weeden	Waterbury, Conn	Sept. 7, 1869	94, 533
Burner, Lamp	W. N. Weeden	Waterbury, Conn	Feb. 18, 1873	136, 113
Burner, Lamp	M. J. Wellman	New York, N. Y	Jan. 7, 1873	134, 718
Burner, Lamp	W. Westlake	Chicago, Ill	Nov. 9, 1869	96, 645
Burner, Lamp	W. Westlake	Chicago, Ill	Mar. 22, 1870	101, 066
Burner, Lamp	J. S. Wetherby	New York, N. Y	Apr. 26, 1870	102, 342
Burner, Lamp	H. J. White	Boston, Mass	May 10, 1870	102, 897
Burner, Lamp	A. Whitlock	Danbury, Conn	June 4, 1867	65, 459
Burner, Lamp	S. W. Wilcox	Mendon, Mass	Oct. 2, 1866	58, 524
Burner, Lamp	N. W. Williams	Frankford, Pa	July 8, 1862	35, 860
Burner, Lamp	J. D. Willoughby	Shippensburgh, Pa	Apr. 26, 1870	102, 457
Burner, Lamp	S. R. Wilmot	Bridgeport, Conn	Oct. 4, 1870	108, 078
Burner, Lamp	S. R. Wilmot	Bridgeport, Conn	Oct. 25, 1870	108, 667
Burner, Lamp	J. Wolstenholme	Providence, R. I	Feb. 3, 1863	37, 599

Index of patents issued from the United States Patent Office from 1790 to 1873, inclusive—Continued.

Invention.	Inventor.	Residence.	Date.	No.
Burner, Lamp	J. Wood	Nottingham, England	June 6, 1865	48, 142
Burner, Lamp	J. B. Wortham	Huntsville, Ala	May 28, 1867	65, 322
Burner, Lamp	D. P. Wright and C. Butler	Birmingham, England	Mar. 22, 1870	101, 073
Burner, Lamp	H. Wright	Pittsburgh, Pa	Jan. 27, 1863	37, 540
Burner, Lamp	M. B. Wright	West Meriden, Conn	June 23, 1863	38, 999
Burner, Lamp	M. B. Wright	West Meriden, Conn	Jan. 5, 1864	41, 172
Burner, Lamp	M. B. Wright	Meriden, Conn	June 18, 1867	65, 851
Burner, Lamp	H. M. Wyatt	Somerville, Mass	June 16, 1868	78, 978
Burner, Lamp	J. N. Wyatt	Baltimore, Md	Mar. 25, 1873	137, 276
Burner, Lamp	J. H. Wygant and C. W. Vanderbeck.	Hackensack, N. Y	Aug. 16, 1864	43, 879
Burner, Lamp gas	R. Nutting	Randolph, Vt	June 7, 1870	103, 917
Burner, Light-house lamp	I. Dunham	Bristol, Me	June 20, 1836	......
Burner, Naphtha	L. A. Gouch	Yonkers, N. Y	Dec. 17, 1867	72, 191
Burner, Night	W. P. Newhall	Peekskill, N. Y	Dec. 20, 1864	45, 513
Burner of burning-fluid lamp	C. A. Greene	Philadelphia, Pa	Apr. 21, 1857	17, 086
Burner of burning-fluid lamp	R. W. Sargent	Philadelphia, Pa	Mar. 17, 1857	16, 852
Burner, Oxyhydrocarbon-gas	J. D. Averell	New York, N. Y	Oct. 14, 1873	143, 555
Burner, Oxyhydrogen	A. W. Wilkinson	New York, N. Y	Feb. 6, 1872	123, 536
Burner, Oxyhydrogen	A. W. Wilkinson	New York, N. Y	Feb. 6, 1872	123, 537
Burner, Oxyhydrogen-gas	A. W. Wilkinson	New York, N. Y	Mar. 19, 1872	124, 869
Burner, Petroleum-gas	G. A. Hyver	New Orleans, La	Dec. 10, 1867	71, 883
Burner, Petroleum-stove	W. E. Jervey	New Orleans, La	Feb. 26, 1867	62, 338
Burner, Petroleum-stove	W. E. Jervey	New Orleans, La	Feb. 26, 1867	62, 339
Burner, Porcelain gas	T. G. Arnold and B. Irving	New York, N. Y	Mar. 12, 1867	62, 725
Burner, Reflecting gas	T. Grist	Philadelphia, Pa	June 2, 1868	78, 451
Burner, Reflecting lamp	C. H. Tessy	Utica, N. Y	June 4, 1872	127, 660
Burner, Refuse	W. Glue	Muskegon, Mich	Dec. 23, 1873	145, 861
Burner regulator, Gas	J. S. Bromhead	Peckham, England	June 29, 1869	92, 007
Burner regulator, Gas	W. Clark	Boston, Mass	Aug. 14, 1866	57, 087
Burner regulator, Gas	R. Gill	New York, N. Y	Dec. 7, 1860	97, 627
Burner regulator, Gas	W. Jones	Chelsea, Mass	Nov. 14, 1871	120, 882
Burner regulator, Gas	E. Lawler	Hartford, Conn	July 25, 1871	117, 305
Burner regulator, Gas	W. Mallerd	Brooklyn, N. Y	June 27, 1854	11, 167
Burner regulator, Gas	W. Mallerd	Bridgeport, Conn	Dec. 18, 1860	30, 977
Burner regulator, Gas	A. Mayer	Philadelphia, Pa	Sept. 12, 1854	11, 674
Burner regulator, Gas	H. Shultz	Milwaukee, Wis	Apr. 12, 1870	101, 771
Burner regulator, Gas	A. H. Wood	Boston, Mass	Feb. 19, 1861	31, 506
Burner, Revolving gas	F. McLewel	New York, N. Y	Apr. 18, 1871	113, 906
Burner, Self-closing gas	G. E. Smith	New York, N. Y	Oct. 3, 1871	119, 478
Burner, Self-generating vapor	J. S. Gray	New York, N. Y	Sept. 9, 1862	36, 404
Burner, Self-lighting gas	A. Barbarin	New Orleans, La	Sept. 1, 1868	81, 734
Burner, Self-regulating gas	G. W. Thompson	New York, N. Y	May 27, 1862	35, 402
Burner, Smoke and steam	J. W. Kingman and A. Eurgens	Laramie City, Wyo	July 29, 1873	141, 277
Burner socket, Gas	J. Todd	Madison, Ind	May 3, 1864	42, 609
Burner, Spirit-gas	C. Miller	Saint Louis, Mo	June 7, 1859	24, 355
Burner, Tar or petroleum	J. Law	London, Canada	Feb. 11, 1873	135, 649
Burner tips made from soapstone, &c., Mode of hardening gas.	H. J. Smith	Boston, Mass	Sept. 22, 1868	82, 445
Burner to burn spirits of turpentine, Argand oil	C. Carr	Philadelphia, Pa	June 20, 1840	1, 644
Burner, Tubular-wick	H. W. Dopp	Buffalo, N. Y	Nov. 20, 1866	59, 898
Burner, Vapor	J. E. Ambrose	Lombard, Ill	Aug. 17, 1869	93, 661
Burner, Vapor	E. H. Anderson	Easton, Md	Apr. 3, 1860	27, 676
Burner, Vapor	W. Aurich	Chicago, Ill	Aug. 31, 1869	94, 269
Burner, Vapor	W. Aurich	Chicago, Ill	Mar. 22, 1870	101, 079
Burner, Vapor	J. W. Baker	Columbus, Ohio	May 10, 1870	102, 749
Burner, Vapor	S. D. Baldwin	Chicago, Ill	Apr. 27, 1869	89, 271
Burner, Vapor	S. D. Baldwin	Chicago, Ill	Apr. 5, 1870	101, 413
Burner, Vapor	S. D. Baldwin	Chicago, Ill	Nov. 7, 1871	120, 613
Burner, Vapor	W. E. Bartlett	Newburgh, N. Y	May 23, 1871	115, 015
Burner, Vapor	W. W. Batchelder	New York, N. Y	Aug. 21, 1866	57, 276
Burner, Vapor	J. Benson and A. H. Watkins	Boston, Mass	Feb. 18, 1873	136, 021
Burner, Vapor	J. Benson and A. H. Watkins	Boston, Mass	Feb. 18, 1873	136, 022
Burner, Vapor	D. Berkey	Huntington, Ind	Mar. 29, 1870	101, 215
Burner, Vapor	W. Bliesner	Saint Louis, Mo	Jan. 18, 1870	98, 842
Burner, Vapor	C. B. Brown	Placerville, Cal	Aug. 22, 1871	118, 192
Burner, Vapor	W. N. Brown	Camden, N. J	May 15, 1860	28, 250
Burner, Vapor	J. Burns	Keokuk, Iowa	July 9, 1872	128, 703
Burner, Vapor	M. L. Callender	New York, N. Y	Sept. 16, 1862	36, 452
Burner, Vapor	D. H. Carpenter	Wallingford, Conn	Aug. 4, 1857	17, 916
Burner, Vapor	L. Chandor	St. Petersburg, Russia	Aug. 17, 1869	93, 674
Burner, Vapor	L. Chandor	St. Petersburg, Russia	Dec. 28, 1869	98, 228
Burner, Vapor	J. Cook	Saint Louis, Mo	Sept. 14, 1869	94, 716
Burner, Vapor	J. Cook	New York, N. Y	Nov. 7, 1871	120, 627
Burner, Vapor	W. E. Darrah	Baltimore, Md	Aug. 11, 1868	80, 926
Burner, Vapor	J. R. De Mahy	New Orleans, La	Sept. 14, 1869	94, 722
Burner, Vapor	J. R. De Mahy and J. P. Cross	New Orleans, La	Apr. 13, 1869	88, 854
Burner, Vapor	H. C. De Witt	Waukegan, Ill	Mar. 1, 1870	100, 270
Burner, Vapor	T. P. Doane	New York, N. Y	Dec. 5, 1875	121, 497
Burner, Vapor	M. B. Dyott	Philadelphia, Pa	July 4, 1871	116, 573
Burner, Vapor	M. B. Dyott	Philadelphia, Pa	Sept. 3, 1872	131, 086
Burner, Vapor	H. Fairbanks	South Brookfield, Mass	June 23, 1857	17, 622
Burner, Vapor	H. Fayette	Brooklyn, N. Y	Feb. 20, 1872	123, 879
Burner, Vapor	J. E. Findley	Memphis, Tenn	Jan. 11, 1870	98, 680
Burner, Vapor	D. R. Fletcher	Covington, Ky	June 7, 1870	104, 009
Burner, Vapor	L. Fischer	Brooklyn, N. Y	Mar. 30, 1869	88, 287
Burner, Vapor	L. Fischer	Brooklyn, N. Y	June 22, 1869	91, 533
Burner, Vapor	M. F. Gale	New York, N. Y	June 17, 1873	140, 029
Burner, Vapor	M. F. Gale	New York, N. Y	June 17, 1873	140, 030
Burner, Vapor	T. S. Gates and A. H. Fritchey	Franklin County, Ohio	June 1, 1869	90, 659
Burner, Vapor	T. S. Gates and A. H. Fritchey	Columbus, Ohio	Aug. 17, 1869	93, 698
Burner, Vapor	T. S. Gates and A. H. Fritchey	Columbus, Ohio	May 23, 1871	115, 189
Burner, Vapor	T. S. Gates and A. H. Fritchey	Columbus, Ohio	Oct. 10, 1871	119, 839
Burner, Vapor	E. Gillert	Saint Louis, Mo	Aug. 23, 1870	106, 574
Burner, Vapor	L. A. Gouch	Yonkers, N. Y	July 26, 1870	105, 669
Burner, Vapor	F. Gould	Paterson, N. J	Apr. 26, 1870	102, 251
Burner, Vapor	J. S. Gray	New York, N. Y	Dec. 27, 1864	45, 600
Burner, Vapor	B. D. Greene	Sturgis, Mich	Feb. 15, 1870	99, 878

Index of patents issued from the United States Patent Office from 1790 *to* 1873, *inclusive*—Continued.

Invention.	Inventor.	Residence.	Date.	No.
Burner, Vapor	B. D. Greene	Sturgis, Mich	Aug. 23, 1870	106, 689
Burner, Vapor	S. Holmes	High Holborn, England	Nov. 10, 1868	83, 854
Burner, Vapor	S. B. Hopkins and E. H. Anderson.	Easton, Md	Apr. 3, 1860	27, 718
Burner, Vapor	S. B. Hopkins and E. H. Anderson.	Easton, Md	May 5, 1868	77, 612
Burner, Vapor	W. W. Jacobs	Hagerstown, Md	Oct. 20, 1868	83, 171
Burner, Vapor	H. Johnson	Washington, D. C	Nov. 6, 1860	30, 578
Burner, Vapor	J. Kidd	New York, N. Y	Nov. 9, 1869	87, 680
Burner, Vapor	I. Kling	Seymour, Ind	Mar. 15, 1870	100, 774
Burner, Vapor	W. H. Lawrence	Baltimore, Md	Mar. 28, 1871	113, 178
Burner, Vapor	D. Leonard	Chicago, Ill	Oct. 25, 1870	108, 711
Burner, Vapor	J. C. Love	Philadelphia, Pa	Feb. 8, 1870	99, 688
Burner, Vapor	J. C. Love	Philadelphia, Pa	Feb. 8, 1870	99, 689
Burner, Vapor	J. C. Love	Philadelphia, Pa	Oct. 21, 1873	143, 915
Burner, Vapor	C. B. Loveless	Syracuse, N. Y	June 30, 1868	79, 485
Burner, Vapor	D. H. Lowe	Boston, Mass	Nov. 10, 1868	83, 865
Burner, Vapor	J. F. Marsh	Dubuque, Iowa	Sept. 6, 1870	107, 072
Burner, Vapor	J. H. Melick	New Germantown, N. J	July 13, 1869	92, 537
Burner, Vapor	T. Moore	Bloomington, Ill	July 19, 1870	105, 589
Burner, Vapor	T. Moore and J. O'Donald	Bloomington and Clinton, Ill.	Feb. 7, 1871	111, 666
Burner, Vapor	O. F. Morrill	Boston, Mass	Apr. 3, 1860	27, 732
Burner, Vapor	S. G. Munn	Chicago, Ill	Jan. 9, 1872	122, 637
Burner, Vapor	A. F. Noble	Indianapolis, Ind	May 23, 1871	115, 089
Burner, Vapor	F. G. Palmer	Pittsburgh, Pa	May 13, 1873	138, 923
Burner, Vapor	J. J. Palmer	Pittsburgh, Pa	Dec. 20, 1870	110, 276
Burner, Vapor	R. W. Park	Philadelphia, Pa	Aug. 9, 1870	106, 199
Burner, Vapor	R W. Park	Philadelphia, Pa	Aug. 9, 1870	106, 2[illegible]0
Burner, Vapor	R. W. Park	Philadelphia, Pa	Jan. 31, 1871	111, 471
Burner, Vapor	R. W. Park	Philadelphia, Pa	Mar. 7, 1871	112, 373
Burner, Vapor	R. W. Park	Philadelphia, Pa	Feb. 6, 1872	123, 506
Burner, Vapor	R. W. Park	Philadelphia, Pa	Feb. 27, 1872	124, 151
Burner, Vapor	G. T. Parry	Philadelphia, Pa	May 30, 1871	115, 350
Burner, Vapor	J. H. Pattee	Monmouth, Ill	Nov. 30, 1869	97, 311
Burner, Vapor	O. N. Perkins	Meriden, Conn	Nov. 12, 1872	133, 051
Burner, Vapor	A. W. Porter and J. H. Brown	New York, N. Y	May 21, 1867	65, 005
Burner, Vapor	H. M. Richmond	Buffalo, N. Y	May 21, 1867	65, 011
Burner, Vapor	C. Royle	Brooklyn, N. Y	July 16, 1872	129, 060
Burner, Vapor	W. H. Randolph	Saint Louis, Mo	Dec. 21, 1869	98, 192
Burner, Vapor	S. Rust, jr	Cincinnati, Ohio	June 11, 1872	127, 711
Burner, Vapor	H. S. Laroni	Marietta, Ohio	Aug. 28, 1866	57, 578
Burner, Vapor	H. S. Laroni	Baltimore, Md	Jan. 5, 1869	85, 698
Burner, Vapor	H. S. Laroni	Chicago, Ill	Feb. 9, 1869	86, 870
Burner, Vapor	E. Savage	Chicago, Ill	Sept. 14, 1869	94, 778
Burner, Vapor	H. Schminke	Baltimore, Md	Mar. 15, 1870	100, 930
Burner, Vapor	J. Shay	Cincinnati, Ohio	July 5, 1870	105, 133
Burner, Vapor	J. Shay	Cincinnati, Ohio	Feb. 28, 1871	112, 291
Burner, Vapor	T. Simmons	Halifax, Canada	Aug. 12, 1873	141, 823
Burner, Vapor	C. E. Smith	Columbus, Ohio	Oct. 18, 1870	108, 528
Burner, Vapor	C. E. Smith and H. J. Rice	Columbus, Ohio	Apr. 19, 1870	102, 169
Burner, Vapor	W. H. Smith	New York, N. Y	June 5, 1866	55, 383
Burner, Vapor	W. H. Smith	New York, N. Y	June 30, 1868	79, 404
Burner, Vapor	W. H. Smith	New York, N. Y	Aug. 3, 1869	93, 360
Burner, Vapor	W. H. Smith	New York, N. Y	Aug. 23, 1870	106, 736
Burner, Vapor	W. H. Smith	Brooklyn, N. Y	May 7, 1872	126, 419
Burner, Vapor	W. H. Smith and L. Fischer	Brooklyn, N. Y	May 7, 1872	126, 420
Burner, Vapor	J. Stratton	Philadelphia, Pa	Sept. 12, 1865	49, 955
Burner, Vapor	D. Symonds	Lowell, Mass	Oct. 22, 1867	70, 043
Burner, Vapor	W. Thompson	Cleveland, Ohio	Sept. 27, 1870	107, 737
Burner, Vapor	T. Tully	Saint Joseph, Mo	Apr. 26, 1870	102, 335
Burner, Vapor	W. W. Tyson	Allegheny City, Pa	Jan. 11, 1870	98, 817
Burner, Vapor	T. Ward	Columbus, Ohio	Mar. 10, 1868	75, 496
Burner, Vapor	T. Ward	Columbus, Ohio	Aug. 16, 1870	106, 522
Burner, Vapor	T. Ward	Franklin County, Ohio	Apr. 26, 1870	102, 339
Burner, Vapor	T. Ward and H. C. Hunt	Columbus, Ohio, and Chicago, Ill.	May 30, 1871	115, 398
Burner, Vapor	H. Wellington	Chicago, Ill	Jan. 11, 1870	98, 824
Burner, Vapor	H. Wellington	Chicago, Ill	Mar. 8, 1870	100, 572
Burner, Vapor	H. Wellington	Chicago, Ill	Sept. 27, 1870	107, 741
Burner, Vapor	J. White	Philadelphia, Pa	Aug. 23, 1870	106, 753
Burner, Vapor	I. Whitehouse	New York, N. Y	Aug. 22, 1871	118, 314
Burner, Vapor	G. H. Wilson	Mansfield, Ohio	June 27, 1871	116, 385
Burner, Vapor	C. Wintergerst	Mobile, Ala	Sept. 15, 1868	82, 262
Burner, Vapor	J. S. Wood	Philadelphia, Pa	Feb. 14, 1871	111, 893
Burner, Vapor	A. F. Ziegler	Columbus, Ohio	Mar. 22, 1870	101, 202
Burner, Vapor and gas	J. C. Love	Philadelphia, Pa	July 26, 1870	105, 819
Burner, Vapor-gas	J. J. Riddle	Cincinnati, Ohio	Nov. 7, 1865	50, 845
Burner, Vapor-lamp	C. M. Alexander	New Albany, Ind	June 28, 1859	24, 529
Burner, Vapor-lamp	S. Andrews	Perth Amboy, N. J	Aug. 24, 1858	21, 239
Burner, Vapor-lamp	D. H. Carpenter	New York, N. Y	May 25, 1858	20, 324
Burner, Vapor-lamp	C. A. Greene	Boston, Mass	May 4, 1858	20, 153
Burner, Vapor-lamp	W. W. Jacobs	Hagerstown, Md	Nov. 5, 1867	70, 573
Burner, Vapor-lamp	A. M. Mace	Springfield, Mass	June 22, 1858	20, 648
Burner, Water	S. Morey	Oxford, N. H	Dec. 11, 1817	......
Burner, Water	S. Morey	Oxford, N. H	Dec. 19, 1818	......
Burner, Water	J. S. Sullivan	Boston, Mass	Dec. 11, 1818	......
Burners and lamps, Heating-shade for gas	A. H. Lochman	York, Pa	Mar. 27, 1866	53, 456
Burners, Apparatus for automatically regulating the flame of gas.	I. and H. Herzberg	Philadelphia, Pa	Apr. 16, 1872	125, 679
Burners, Apparatus for cooking by coal-oil	O. F. Morrill	Chelsea, Mass	May 22, 1866	55, 033
Burners, Apparatus for supplying air to hydrocarbon.	J. Stratton	Philadelphia, Pa	May 11, 1869	89, 952
Burners, Apparatus for suspending glass bowls under gas.	J. C. Conner	New York, N. Y	Feb. 20, 1872	123 870
Burners, Attachment for gas	J. C. Love	Philadelphia, Pa	July 27, 1869	93, 101
Burners, Attachment for gas	R. N. Stewart	Philadelphia, Pa	Sept. 14, 1869	94, 922
Burners, Attachment for lamp	R. Hutton	Williamsburgh, N. Y	Sept. 21, 1869	95, 115

Index of patents issued from the United States Patent Office from 1790 *to* 1873, *inclusive*—Continued.

Invention.	Inventor.	Residence.	Date.	No.
Burners, Automatic attachment to gas	A. R. Marshall	Stratford, Conn	Aug. 26, 1856	15, 614
Burners, Automatic cut-off for gas	J. B. Smith	Pittston, Pa	Feb. 11, 1868	74, 440
Burners, Automatic lighting-attachment for lamps and taper.	W. H. Weeks	New York, N. Y	Jan. 25, 1870	99, 123
Burners, Carbureting-attachment for gas	J. H. Steiner	Reading, Pa	Nov. 1, 1870	108, 844
Burners, Chimney-holder for gas	E. P. Gleason	New York, N. Y	Nov. 13, 1866	59, 588
Burners, Chimney-holder for gas	G. Mooney	Providence, R. I	July 16, 1872	129, 158
Burners, Clock-attachment to gas	P. C. Bensel	New York, N. Y	May 14, 1872	126, 620
Burners, Clutch for supporting water-bowls under gas.	S. Gardner	New York, N. Y	June 4, 1872	127, 590
Burners, Construction of gas	J. McHenry	Cincinnati, Ohio	Mar. 17, 1857	16, 848
Burners, Construction of gas	J. Stevens and J. Johnson	New York, N. Y	Sept. 27, 1859	25, 591
Burners, Construction of gas	H. K. Symmes	Newton, Mass	Oct. 4, 1859	25, 686
Burners, Construction of lamp	W. N. Weeden	Waterbury, Conn	Oct. 1, 1872	131, 917
Burners, Construction of vapor	W. W. Batchelder	New York, N. Y	Oct. 4, 1859	25, 621
Burners, Construction of vapor	W. H. Hunt	Brooklyn, N. Y	Oct. 18, 1859	25, 830
Burners, Construction of vapor-lamp	W. W. Batchelder	New York, N. Y	Oct. 18, 1859	25, 799
Burners, Device for drawing wick through	F. A. Blaetterlein	West Meriden, Conn	Dec. 31, 1867	72, 785
Burners, Device for regulating by electricity the issue of gas from.	C. W. Smith	Evans, N. Y	May 18, 1858	20, 305
Burners, Electric lighting-attachment for gas	H. C. Appleby	Conneaut, Ohio	June 25, 1872	128, 201
Burners for holding vessels or shades over the flame, Attachment to lamp or gas.	W. L. Fish	Newark, N. J	Nov. 17, 1863	40, 617
Burners for lamps, Mode of suspending	A. Moore and J. A. Cole	Northville, N. Y	Mar. 7, 1865	46, 691
Burners for vapor-lamps, Construction of	R. Steel	Philadelphia, Pa	Nov. 1, 1859	25, 986
Burners, Globe and holder for gas	G. M. Ramsay	New York, N. Y	June 17, 1873	140, 075
Burners, Lamp for lighting gas	R. Thompson	Lowell, Mass	Dec. 17, 1850	7, 848
Burners, Light-intensifying attachment for gas	S. T. Bacon	Boston, Mass	June 11, 1872	127, 829
Burners, Manufacture of Argand	E. S. Archer	New York, N. Y	Jan. 31, 1865	46, 063
Burners, Manufacture of Argand gas	G. Mooney	Providence, R. I	Jan. 14, 1868	73, 367
Burners or lamps, Shade-holder for gas	G. Wedekind	Philadelphia, Pa	May 20, 1862	35, 341
Burners, Perforation in lamp	W. H. Smith	New York, N. Y	Mar. 3, 1863	37, 845
Burners, Protector for gas or vapor	R. Nutting	Randolph, Vt	May 2, 1871	114, 329
Burners, Regulating-attachment for gas	W. Mallerd	Bridgeport, Conn	May 8, 1866	54, 569
Burners, Regulating gas	H. B. Myer	Philadelphia, Pa	Nov. 10, 1868	83, 990
Burners, Regulator for gas	W. Mallerd	Bridgeport, Conn	Apr. 18, 1865	47, 362
Burners, Rest for supporting glass bowls underneath horizontal gas.	S. Gardner	New York, N. Y	Apr. 30, 1872	126, 140
Burners, Revolving attachment for gas	J. O. Belknap	Mobile, Ala	Apr. 1, 1873	137, 406
Burners, Self-lighting attachment for gas	S. Gardiner, jr	New York, N. Y	Feb. 16, 1869	87, 039
Burners, Shade for lamp and gas	M. H. Collins	Chelsea, Mass	Nov. 8, 1870	108, 972
Burners, Shade-ring for lamp	H. W. Hayden	Waterbury, Conn	Dec. 14, 1869	97, 772
Burners, Shield for gas	I. W. Shaler	Brooklyn, N. Y	Sept. 27, 1870	107, 824
Burners, Shield for gas	H. J. Wattles	Rockford, Ill	June 21, 1870	104, 516
Burners, Shield for naphtha-gas	H. H. Edgerton	Fort Wayne, Ind	July 25, 1871	117, 396
Burners, Stop-cock and regulator for gas	A. E. Deloueste	Buffalo, N. Y	Feb. 18, 1873	135, 895
Burners to lamps, Attaching	W. P. Miller	San Francisco, Cal	July 31, 1866	56, 775
Burners to lamps, Attaching	H. Weston	Towanda, Pa	May 21, 1867	64, 925
Burners to lamps, Device for attaching	J. B. Alexander	Washington, D. C	Nov. 26, 1867	71, 261
Burners, Tip for gas	R. Lanstrom	Cincinnati, Ohio	Dec. 31, 1867	72, 741
Burners, Valve for gas	A. Mayer	Philadelphia, Pa	Mar. 13, 1855	12, 515
Burners, Ventilating-fan for gas	H. Seher	Saint Louis, Mo	Dec. 18, 1866	60, 566
Burners, Wick-adjuster for Argand lamp	W. B. Curtiss	Bridgeport, Conn	June 7, 1870	103, 993
Burners, Wick-tube for lamp	C. L. Daboll	New London, Conn	May 19, 1863	38, 567
Burning-fluid	B. H. Chadbourn	Saint Louis, Mo	Oct. 30, 1866	59, 182
Burning-fluid	J. P. Cross	Watertown, N. Y	Aug. 14, 1866	57, 095
Burning-fluid	H. H. Etter	Washington, D. C	Mar. 19, 1867	63, 027
Burning-fluid	G. L. Fattie	Buffalo, N. Y	Nov. 20, 1866	59, 797
Burning-fluid	G. W. Flowers and J. C. and D. W. Happersett.	Urbana, Ohio	Feb. 25, 1868	74, 756
Burning-fluid	A. Gesner	Williamsburgh, N. Y	May 23, 1855	12, 936
Burning-fluid	J. Griffin	Stanford, N. Y	Mar. 8, 1859	23, 167
Burning-fluid	J. G. Hester	Raleigh, N. C	Sept. 24, 1867	69, 210
Burning-fluid	W. W. Jacobs	Hagerstown, Md	Mar. 19, 1867	63, 054
Burning-fluid	J. Jann	New Windsor, Md	Feb. 13, 1866	52, 574
Burning-fluid	J. Jann	New Windsor, Md	Sept. 4, 1866	57 727
Burning-fluid	J. J. Kamm	Fort Wayne, Ind	Apr. 9, 1867	63, 727
Burning-fluid	J. T. Leete	New York, N. Y	Apr. 30, 1867	64, 232
Burning-fluid	W. R. Loomis, N. Wells, H. Hitchcock, and S. G. Stryker.	Elmira, N. Y	July 2, 1867	66, 364
Burning-fluid	D. Mansfield	Oshkosh, Wis	June 26, 1866	55, 880
Burning-fluid	T. Martin and J. G. Evans	Muscatine, Iowa	Dec. 3, 1867	71, 629
Burning-fluid	G. McLean	Brooklyn, N. Y	Oct. 29, 1867	70, 346
Burning-fluid	G. H. Mellen and J. C. Hazleton.	Washington, D. C	Sept. 4, 1866	57, 749
Burning-fluid	C. W. Pinkham	Fond du Lac, Wis	Mar. 25, 1862	34, 772
Burning-fluid	W. B. Rogers	Chicago, Ill	Dec. 18, 1866	60, 559
Burning-fluid	J. B. Scott	Hyattsville, Md	Sept. 18, 1866	58, 180
Burning-fluid	E. D. Seely	Brookline, Mass	Aug. 21, 1866	57, 390
Burning-fluid	G. W. Spangle	Clifton Springs, N. Y	Oct. 16, 1866	58, 905
Burning-fluid	M. L. Stoddard	Corning, N. Y	July 30, 1867	67, 373
Burning-fluid	A. C. Vaughan	Rainsburgh, Pa	Apr. 3, 1866	53, 709
Burning-fluid	I. B. Wiggin	Washington, D. C	Apr. 9, 1867	63, 777
Burning-fluid	W. Wilber	New York, N. Y	Mar. 8, 1859	23, 210
Burning-fluid composition	B. F. Hebard	Neponset, Mass	Feb. 19, 1861	31, 457
Burning-fluid compound	A. Gesner	Williamsburgh, N. Y	May 29, 1855	12, 987
Burning-fluid for illumination, Compound	H. B. Brace and W. T. Swart	Canandaigua, N. Y	May 8, 1866	54, 495
Burning-fluid, Illuminating	L. D. Ferguson	Corning, N. Y	Apr. 2, 1867	63, 376
Burning-fluids, Composition for	C. N. Tyler	Buffalo, N. Y	Mar. 24, 1863	38, 015
Burning fluids, Composition for treating	H. B. Wellman	Indianapolis, Ind	Oct. 22, 1867	70, 054
Burning-fluids, Manufacture of	W. Corfield	Philadelphia, Pa	Apr. 17, 1866	54, 060
Burning-fluids, Manufacture of	W. Corfield	Philadelphia, Pa	Apr. 17, 1866	54, 061
Burning-fluids, Manufacture of	L. L. Hill	Greenport, N. Y	June 15, 1858	20, 558
Burning-fluids, Mode of preparing	R. E. Campbell	New York, N. Y	Oct. 30, 1866	59, 177
Burning fluids, Testing	P. Millspaugh	Kent, Conn	May 28, 1872	127, 259
Burning hydrocarbons and bagasse	E. S. Roman	Cantrelle, La	Apr. 23, 1872	125, 989
Burning-kiln	J. Green	Saint Louis, Mo	Jan. 12, 1869	85, 813
Burning-kiln	B. Kreischer	New York, N. Y	Sept. 1, 1868	81, 793

Index of patents issued from the United States Patent Office from 1790 *to* 1873, *inclusive*—Continued.

Invention.	Inventor.	Residence.	Date.	No.
Burning-kiln	B. Kreischer	New York, N. Y	Aug. 17, 1869	93, 891
Burning-kiln	G. A. Wedekind and H. Dueberg.	New York, N. Y	Dec. 15, 1868	85, 042
Burning liquid-fuel and generating steam, Apparatus for.	W. T. Scheide	Tidioute, Pa	June 24, 1873	140, 220
Burning liquids lighter than water, Apparatus for.	A. T. Schmidt	Pittsburgh, Pa	Nov. 11, 1862	36, 917
Burning ornamental figures upon wood	H. Wood	New York, N. Y	Feb. 19, 1850	7, 111
Burnisher	C. Frampton	Brooklyn, N. Y	Apr. 20, 1858	19, 988
Burnishing-machine	C. H. Helms	Poughkeepsie, N. Y	Nov. 28, 1865	51, 180
Burnishing-machine	D. F. Randall	Hartford, Conn	Oct. 22, 1861	33, 558
Burnishing-machine	N. C. Stow	Stoneham, Mass	Jan. 30, 1866	52, 336
Burnishing-machine	A. M. White	Hartford, Conn	Nov. 20, 1860	30, 707
Burnishing-machine	L. S. White	Hartford, Conn	Aug. 24, 1858	21, 304
Burnishing-machine	L. S. White	Waterbury, Conn	Apr. 5, 1859	23, 520
Burnishing-machine	L. S. White	Waterbury, Conn	Oct. 11, 1859	25, 783
Burnishing machine, Metal	J. Stever	Bristol, Conn	May 1, 1855	12, 799
Burnishing metal, Process for	E. Satterlee	Albany, N. Y	Mar. 20, 1849	6, 200
Burr-cylinder	J. Bidwell	New York, N. Y	Nov. 15, 1859	26, 081
Burr, Locking	J. J. Steward	Big Prairie, Ohio	June 1, 1869	90, 699
Burring-cylinder	C. G. Sargent	Lowell, Mass	Oct. 9, 1849	6, 778
Burring-machine	A. Crane	Lowell, Mass	Oct. 11, 1845	4, 231
Burring-machine	Z. Parkhurst	New York, N. Y	Sept. 11, 1847	5, 282
Burring-machine	O. W. Stow	Southington, Conn	Dec. 28, 1858	22, 459
Burring-machine	T. S. Washburn	Lowell, Mass	Oct. 11, 1845	4, 230
Burring-machine burr-box	Z. Parkhurst	Milford, Mass	July 10, 1866	56, 257
Burring-machines, Guard-cylinder for	F. A. Calvert	Lowell, Mass	Aug. 14, 1847	5, 240
Burring-machines, Guard or stripper for	A. Wright	Lowell, Mass	Jan. 23, 1849	6, 057
Burring-machines, &c., Stripping and feed roll for.	J. K. Proctor	Philadelphia, Pa	Nov. 26, 1872	133, 482
Bush, Lubricating	S. H. Everett	Milton, Ohio	Sept. 4, 1866	57, 695
Bush-spindle	S. Merchant	Ohio County, Va	July 22, 1834	....
Bushing, Anti-frictional	F. S. Burr	Brooklyn, N. Y	Apr. 1, 1873	137, 415
Bushing for wheels	T. Blake	Stockton, Cal	Oct. 6, 1868	82, 790
Bustle	J. I. Barnum	Brooklyn, N. Y	June 11, 1872	127, 832
Bustle	M. H. Beckworth	Camden, N. Y	Aug. 24, 1869	93, 950
Bustle	I. W. Birdseye	Birmingham, Conn	Nov. 7, 1871	120, 616
Bustle	M. K. Bortree	Jackson, Mich	Feb. 25, 1873	136, 127
Bustle	M. Boupart	Fordham, N. Y	Sept. 16, 1873	142, 764
Bustle	J. Broughton	Brooklyn, N. Y	Nov. 1, 1870	108, 872
Bustle	J. Broughton	Brooklyn, N. Y	Nov. 1, 1870	108, 873
Bustle	G. P. Bryant	Boston, Mass	July 30, 1872	129, 883
Bustle	B. Davis	Brooklyn, N. Y	Oct. 18, 1859	25, 865
Bustle	A. Douglas	New York, N. Y	Apr. 21, 1857	17, 082
Bustle	T. A. Earl	North Attleborough, Mass.	Oct. 11, 1859	25, 786
Bustle	T. F. Fessenden	Providence, R. I	May 28, 1872	127, 327
Bustle	I. W., jr., and A. H. Hakes	Norwich, Conn	Apr. 19, 1850	23, 681
Bustle	J. H. Hall	Ansonia, Conn	Feb. 11, 1873	135, 801
Bustle	J. B. Loomis	Chelsea, Mass	Aug. 18, 1868	81, 281
Bustle	H. H. May	Birmingham, Conn	June 25, 1872	128, 412
Bustle	D. McInroy	Brooklyn, E. D., N. Y	Oct. 4, 1870	108, 039
Bustle	C. B. Pollock	Norwich, Conn	July 9, 1872	128, 813
Bustle	C. A. Postley	Jersey City, N. J	Dec. 7, 1858	22, 242
Bustle	D. G. Rollin	New York, N. Y	Sept. 14, 1869	94, 915
Bustle	D. Smith	Litchfield Corner, Me	Nov. 12, 1872	133, 062
Bustle	S. Smith	Boston, Mass	Dec. 17, 1872	134, 010
Bustle	S. and D. L. Smith	Skowhegan, Me	Aug. 13, 1872	130, 446
Bustle	T. S. Sperry	New York, N. Y	Jan. 28, 1873	135, 379
Bustle	T. Stockman	New York, N. Y	Feb. 11, 1873	135, 858
Bustle	G. R. Taber	Boston, Mass	Apr. 30, 1872	126, 164
Bustle	A. W. Thomas	Philadelphia, Pa	June 25, 1872	128, 337
Bustle	A. W. Thomas	Philadelphia, Pa	July 15, 1873	140, 966
Bustle	A. J. Thompson	Malden, Mass	Oct. 18, 1859	25, 854
Bustle	J. Wehl	New York, N. Y	Aug. 6, 1872	130, 345
Bustle	G. W. Yerby	New York, N. Y	Oct. 18, 1859	25, 861
Bustle and skirt	H. N. Daggett	Attleborough, Mass	July 13, 1858	20, 865
Bustle, Extension	G. V. Pierce	Jersey City, N. J	May 23, 1871	115, 235
Bustle for lady's dress	G. V. and E. A. Pierce	New York, N. Y	Nov. 23, 1858	22, 133
Bustle hoops, Clasp for the ends of	F. S. Otis	Brooklyn, N. Y	Jan. 17, 1860	26, 876
Bustle, Lady's	J. W. Bradley	New York, N. Y	Nov. 22, 1859	26, 159
Bustle, Lady's	B. F. Moore	New York, N. Y	Aug. 23, 1859	25, 211
Bustles, Fastening hoop-ends in tabs of	L. De Forest	Birmingham, Conn	Sept. 23, 1862	36, 510
Butcher's cleaver	C. Hart	Unionville, Conn	Oct. 1, 1872	131, 754
Butcher's knife	P. Houseman and C. C. Campbell.	Rural Retreat, Va	Oct. 11, 1870	108, 141
Butcher's spreader	F. B. Guthrie	Crawfordsville, Ind	Sept. 3, 1872	131, 096
Butt	T. W. Brown	Reading, Pa	Jan. 19, 1869	85, 902
Butt	J. A. Frary	New Britain, Conn	Feb. 2, 1869	86, 524
Butt-blanks, Machine for trimming	C. Kenney	West Troy, N. Y	July 26, 1844	3, 682
Butt-centering machine	E. Parker	New Britain, Conn	July 19, 1870	105, 428
Butt-machine	T. Tracy	New Britain, Conn	Mar. 19, 1867	62, 981
Butt, Post	J. Holland	Conshohocken, Pa	May 15, 1860	28, 275
Butt, Reversible	G. W. Field	Lowell, Mass	Nov. 29, 1870	109, 604
Butt, Reversible	J. Kindleberger and W. A. Arnold	San Francisco, Cal	Sept. 14, 1869	94, 756
Butt, Spring	T. Fredericks	Newark, N. J	Feb. 16, 1869	86, 978
Butt, Spring	L. Preston	Elizabethport, N. J	June 1, 1869	90, 686
Butt, Spring	W. Wells	Ashtabula, Ohio	Nov. 17, 1868	84, 240
Butts or hinges for hanging and fastening window blinds and shutters, Method of constructing.	A. C. Palmer	Utica, N. Y	Jan. 27, 1843	2, 936
Butts, Process of finishing sheet metal	E. C. Blakeslee	Waterbury, Conn	Dec. 17, 1872	134, 029
Butter and cheese, Preparing milk for	C. Ainsworth	Littleton, N. Y	Nov. 19, 1833	....
Butter and cutting curd, Machine for working	J. Buck	Chatham, Conn	Dec. 5, 1825	....
Butter and lard package	J. L. Cone	Waterloo, N. Y	July 4, 1871	116, 684
Butter and lard tub, Covered	S. E. Southland	Jamestown, N. Y	Sept. 11, 1866	57, 989
Butter and other articles, Vessel for packing	W. Pratt	New York, N. Y	June 14, 1870	104 198
Butter, Apparatus for making	A. Denison	Woodville, Ohio	Sept. 21, 1869	95, 096
Butter-box	W. B Guernsey	Norwich, N. Y	Feb. 26, 1867	62, 410
Butter, &c., Box and bag for packing	J. D. Smedley	Chicago, Ill	Mar. 31, 1868	76, 110
Butter, Box or tub for putting up	J. Parce	Fairport, N. Y	Oct. 4, 1864	44, 551

Index of patents issued from the United States Patent Office from 1790 *to* 1873, *inclusive*—Continued.

Invention.	Inventor.	Residence.	Date.	No.
Butter-bucket	J. T. Dumont	Kansas City, Mo	Mar. 11, 1873	136, 713
Butter-bucket	J. Gilberds and M. Harris	Jamestown, N. Y	Sept. 2, 1873	142, 456
Butter bucket or package	N. C. Burnap	Argusville, N. Y	Dec. 3, 1872	133, 628
Butter by machinery, Mode of making	E. Waterhouse	Gardiner, Me	Sept. 29, 1825	....
Butter-carrier	J. B. Crozier	Cumberland, Ohio	Jan. 16, 1872	122, 758
Butter-carrier	B. Yaw	New Concord, Ohio	Nov. 12, 1872	132, 998
Butter, Case for packing	S. H. Matthews	Nottingham, Pa	Sept. 3, 1872	131, 015
Butter-coloring compound	G. Bogart, W. H. Cramer, and W. L. Lewis.	Laporte City, Iowa	Sept. 16, 1873	142, 891
Butter-cooler	E. Kaufman and A. Weber	Philadelphia, Pa	Sept. 8, 1868	82, 005
Butter-cooler	G. W. Smith	Hartford, Conn	July 5, 1859	24, 636
Butter-cooler	J. H. Stimpson	Baltimore, Md	May 15, 1855	12, 876
Butter-cooler	J. H. Stimpson	Baltimore, Md	July 13, 1858	20, 902
Butter, Curing and preserving	J. F. Saiger	Shelby, Ohio	Dec. 28, 1869	98, 421
Butter-cutter	A. N. Merritt	Gardner, Mass	Oct. 8, 1867	69, 690
Butter cutter and lifter, Gaged	T. E. Colbrunn	Cleveland, Ohio	Aug. 26, 1873	142, 207
Butter cutter and stamp, Combined	N. Clough	Lowell, Mass	Nov. 23, 1869	97, 049
Butter, Device for purifying	G. W. Putnam	Smithfield, N. Y	Mar. 4, 1862	34, 592
Butter-dish	W. E. Hawkins	Wallingford, Conn	Feb. 11, 1868	74, 221
Butter-dish	N. Lawrence	Taunton, Mass	Dec. 10, 1867	71, 889
Butter-dish	W. W. Lyman	West Meriden, Conn	July 28, 1868	80, 297
Butter-dish	I. A. Richards	Brookfield, Mass	Feb. 18, 1868	74, 722
Butter-dish drainer	F. C. Booth	Meriden, Conn	Apr. 23, 1872	125, 876
Butter-excavator	N. J. Eaton	Montana, Iowa	Jan. 18, 1870	98, 938
Butter-extractor	T. Curtis	Holly, Mich	June 14, 1870	104, 120
Butter-firkin clamp	D. M. Lockridge	Otto, N. Y	Nov. 2, 1869	96, 451
Butter from cheese, whey, &c., Making	H. C. Markham	West Turin, N. Y	Dec. 11, 1866	60, 399
Butter from cream, Apparatus for reducing	S. D. Warner and O. W. Thomas	Richmond, Ill	Feb. 22, 1870	100, 224
Butter from firkins, Implement for cutting	N. Woodbury	Salem, Mass	May 11, 1852	8, 947
Butter from milk, Producing	N. Orcutt	Binghamton, N. Y	May 17, 1864	42, 821
Butter from whey, Manufacture of	J. Suggett	Cortlandville, N. Y	Dec. 18, 1866	60, 656
Butter from whey, Manufacturing	I. Page	Adams, N. Y	June 23, 1868	79, 078
Butter-jar	J. Smith	East Liverpool, Ohio	Sept. 21, 1869	95, 156
Butter-kettle	J. Liming	Philadelphia, Pa	July 11, 1871	116, 844
Butter-knife holder	H. Benton	Guilford, Conn	Apr. 19, 1864	42, 339
Butter-knife rest	B. D. Reed	Westfield, Mass	July 13, 1869	92, 468
Butter-knife rest	B. D. Reed	Westfield, Mass	Jan. 21, 1873	135, 156
Butter-machine	D. Rogers	Mount Gilead, Ohio	Aug. 24, 1869	94, 033
Butter, Machine for manufacturing	A. Curtiss	Paris, N. Y	Apr. 20, 1830	....
Butter, Machine for weighing and printing	W. S. Reinert	Spring Garden, Pa	May 9, 1854	10, 896
Butter, Making	B. G. and G. Burtis	New York, N. Y	Feb. 27, 1807	....
Butter, Making	D. H. McGregory	Detroit, Mich	Sept. 10, 1867	68, 639
Butter-making	A. Mot	Washington, D. C	July 19, 1870	105, 479
Butter-making apparatus	F. P. Deuel	Tecumseh, Mich	Mar. 23, 1869	88, 016
Butter-making mold	G. Matthewman and A. Leininger.	Brooklyn, N. Y	June 6, 1865	48, 082
Butter, Manufacture of	J. P. Crampton	Madison County, Ind	Nov. 5, 1867	70, 417
Butter, Manufacture of	Z. Phinney	Greene County, N. Y	Mar. 4, 1814	....
Butter, Method of preparing coloring-matter for	D. W. Dake	Brooklyn, N. Y	Nov. 27, 1866	59, 975
Butter, milk, and other articles, Cooler for preserving.	W. Garrard	Fallston, Pa	Nov. 27, 1866	59, 993
Butter, Mode of churning	J. Strobeck	Parishville, N. Y	Nov. 19, 1867	71, 082
Butter, Mode of purifying rancid	C. Peck	Marshall, Ill	Nov. 2, 1869	96, 477
Butter, Mode of putting up and preserving	J. Wilcox	Springfield, Mass	July 30, 1867	67, 391
Butter, Mode of restoring rancid	J. W. Prentiss	Pultney, N. Y	Oct. 4, 1859	25, 672
Butter, Mode of working and pressing	T. D. Gail	Eden, N. Y	Oct. 10, 1840	1, 815
Butter-mold	O. Allen	San Francisco, Cal	Feb. 14, 1865	46, 319
Butter-mold	J. S. Bullar	Chagrin Falls, Ohio	Apr. 17, 1866	54, 054
Butter-mold	A. J. Derrick	Sheridan, Nev	Dec. 5, 1871	121, 496
Butter-mold	H. W. Hopkins	Milford, N. H	Jan. 1, 1867	60, 894
Butter-mold	M. T. Ridout	Milwaukee, Wis	May 27, 1862	35, 393
Butter-mold	D. G. Williams	Quincy, Mich	Apr. 9, 1872	125, 644
Butter mold and print	J. S. Corya	Dupont, Ind	Nov. 23, 1869	97, 653
Butter-molds, Tool for boring	J. S. Bullard	Chagrin Falls, Ohio	June 14, 1864	43, 092
Butter molding and working machine	J. C. Hervey	Cincinnati, Ohio	Apr. 27, 1869	89, 480
Butter-molding machine	A. Nudd	Wampum, Wis	June 6, 1865	48, 088
Butter-package	H. P. Adams	Cleveland, Ohio	Oct. 29, 1872	132, 555
Butter-package	J. Andrews	Cleveland, Ohio	Apr. 15, 1873	137, 752
Butter-package	T. W. Ryding	Tully, N. Y	Oct. 19, 1869	95, 937
Butter-package	J. Thayer	Palmyra, Wis	Nov. 2, 1869	96, 506
Butter, Package for preserving and transporting	E. H. Benners	Brooklyn, N. Y	Oct. 8, 1872	131, 930
Butter, Packing and preserving	W. B. Guernsey	Norwich, N. Y	Feb. 26, 1867	62, 411
Butter-paddle	D. L. Grover	Groton, N. Y	May 13, 1873	138, 800
Butter-pail	S. D. King	Middletown, N. Y	June 4, 1872	127, 612
Butter-pail	A. Robertson	East Rupert, Vt	May 9, 1871	114, 714
Butter-print press	L. Coates and J. T. Criswell	Collamer, Pa	Sept. 16, 1873	142, 838
Butter-printer	W. C. Stern and J. W. Robinson.	London Grove, Pa	Nov. 8, 1870	109, 070
Butter-printer	B. Yaw	New Concord, Ohio	Nov. 12, 1872	132, 999
Butter-printing machine	D. E. Dutrow	Washington, D. C	Jan 28, 1873	135, 323
Butter-printing machine	A. F. Hines	Washington, D. C	July 8, 1873	140, 704
Butter-printing machine	J. S. Miller and S. L. Wiegand	Philadelphia, Pa	May 22, 1860	28, 394
Butter, Process of purifying	L. S. Robbins	New York, N. Y	July 28, 1868	80, 506
Butter, Refining	G. W. Kirchhöffer	Chicago, Ill	June 10, 1873	139, 796
Butter-rolls, Forming and gaging	C. H. Fancher	San Francisco, Cal	Mar. 25, 1873	137, 064
Butter-safe	T. Pyle	Oxford, Pa	Feb. 27, 1872	124, 158
Butter-stamp	T. Mabbett, jr	Vineland, N. J	May 14, 1872	126, 641
Butter-tray, Portable	L. Beemer	Libertyville, N. J	May 9, 1871	114, 516
Butter-trier	W. H. Sloan	Saint Louis, Mo	Oct. 1, 1867	69, 498
Butter-tub	A. R. Bailey	Elmore, Vt	Aug. 25, 1868	81, 459
Butter-tub	G. S. Batcheller	Saratoga Springs, N. Y	Mar. 2, 1869	87, 323
Butter-tub	A. J. Drake	Middletown, N. Y	Sept. 16, 1873	142, 843
Butter-tub	J. Gilberds	Jamestown, N. Y	Dec. 26, 1871	122, 166
Butter-tub	D. A. Gilbert	Morristown, Vt	Apr. 21, 1868	77, 030
Butter-tub	D. A. Gilbert	Morristown, Vt	June 16, 1868	78, 955
Butter-tub	E. Guyer	Wolcott, Vt	Oct. 13, 1868	82, 939
Butter-tub	J. M. Hale	Georgia Plains, Vt	Nov. 9, 1869	96, 584
Butter-tub	A. B. Harris	Morrisville, Vt	June 16, 1868	78, 963

Index of patents issued from the United States Patent Office from 1790 *to* 1873, *inclusive*—Continued.

Invention.	Inventor.	Residence.	Date.	No.
Butter-tub	H. Holt	Brooklyn, N. Y	Nov. 19, 1872	133, 226
Butter-tub	D. Lown	Poughkeepsie, N. Y	July 20, 1869	92, 734
Butter-tub, &c	W. Marsh	Waterford, Pa	Feb. 14, 1871	111, 758
Butter-tub	A. T. Peck	Scott, N. Y	Sept. 2, 1862	36, 361
Butter-tub	A. C. Sawyer	Canton, N. Y	Sept. 14, 1869	94, 917
Butter-tub	G. Smith	Castile, N. Y	Oct. 22, 1872	132, 495
Butter-tub	C. H. White	Emmett Township, Mich	Mar. 28, 1871	113, 122
Butter tub and cooler	A. J. Connelly and T. Benjamin	Philadelphia, Pa	June 11, 1872	127, 742
Butter-tub, &c., Hasp for	H. C. Carter	New York, N. Y	Nov. 25, 1873	144, 889
Butter-worker	J. P. Adams	Whitney's Point, N. Y	Sept. 12, 1865	51, 408
Butter-worker	J. P. Adams	Whitney's Point, N. Y	May 1, 1866	54, 272
Butter-worker	J. P. Adams and J. P. Corbin	Whitney's Point, N. Y	Aug. 9, 1870	106, 104
Butter-worker	J. A. Allen	Deerfield, Mass	Feb. 12, 1861	31, 364
Butter-worker	N. W. Bancroft	Burlington, Vt	Aug. 3, 1858	21, 106
Butter-worker	J. H. Bennett	Bennington, Vt	Mar. 25, 1856	14, 530
Butter-worker	N. Bennett	Sherman, N. Y	Sept. 17, 1861	33, 290
Butter-worker	C. A. Boynton	Hyde Park, Vt	July 17, 1860	29, 135
Butter-worker	J. M. Brush	Eyota, Minn	Nov. 22, 1870	109, 381
Butter-worker	S. H. Bush	Boistfort, Wash	Dec. 16, 1873	145, 624
Butter-worker	E. Butler and G. M. Peck	Pompey, N. Y., and Abington, Pa.	Oct. 27, 1857	18, 497
Butter-worker	J. L. Colburn	West Burke, Vt	Apr. 7, 1868	76, 304
Butter-worker	G. S. Coleman	Alexandria, Va	Oct. 3, 1871	119, 575
Butter-worker	W. Cunningham	Oxford, N. Y	Jan. 21, 1868	73, 582
Butter-worker	D. W. Dake	Chicago, Ill	Jan. 14, 1873	134, 788
Butter-worker	D. W. Dake	Beloit, Wis	Nov. 18, 1873	144, 662
Butter-worker	D. W. Dake	Beloit, Wis	Dec. 9, 1873	145, 338
Butter-worker	E. J. Dickey	Hopewell Cotton Works, Pa.	July 12, 1853	9, 838
Butter-worker	O. L. Dow	Hancock, N. H	Apr. 5, 1870	101, 444
Butter-worker	R. R. Eastman	Granby, Mass	Nov. 27, 1866	59, 985
Butter-worker	E. Farnum	Blackstone, Mass	Feb. 5, 1867	61, 822
Butter-worker	C. W. Gage	Homer, N. Y	June 16, 1857	17, 564
Butter-worker	H. Garrett	Richmond, Mo	Feb. 4, 1868	74, 076
Butter-worker	C. L. Gilpatrick	Lewiston, Me	Sept. 8, 1863	39, 808
Butter-worker	W. N. Golden	Coldwater, Mich	June 3, 1873	139, 571
Butter-worker	E. Gore	Bennington, Vt	July 25, 1854	11, 370
Butter-worker	W. H. Hart	Medfield, Mass	June 26, 1866	55, 857
Butter-worker	L. S. Ingraham	Grafton, Ohio	Aug. 7, 1860	29, 492
Butter-worker	N. Johnson	Guilford, N. Y	Jan. 12, 1864	41, 223
Butter-worker	J. Jones	Philadelphia, Pa	July 5, 1859	24, 698
Butter-worker	S. Keen	East Bridgewater, Mass	Feb. 18, 1868	74, 546
Butter-worker	L. L. Kellogg	Leon, N. Y	July 25, 1871	117, 428
Butter-worker	R. Lapham	New York, N. Y	May 14, 1872	126, 638
Butter-worker	J. B. Lyons	Milton, Conn	Nov. 24, 1863	40, 697
Butter-worker	J. McBride	Ithaca, N. Y	June 10, 1873	139, 723
Butter-worker	P. P. Meredith	Stevensville, Mont	Sept. 26, 1871	119, 385
Butter-worker	S. E. Morse	Montgomery, Vt	Mar. 8, 1864	41, 853
Butter-worker	W. C. Moser	East Nantmeal Township, Pa.	Nov. 26, 1867	71, 318
Butter-worker	B. D. Pease	Madison, Pa	Nov. 26, 1861	33, 794
Butter-worker	G. M. Rhoades	Hamilton, N. Y	July 2, 1861	32, 717
Butter-worker	E. Ring	Mecklenburgh, N. Y	July 25, 1854	11, 387
Butter-worker	J. Robinson	Calais, Vt	June 19, 1866	55, 714
Butter-worker	J. Romans	Homeville, Pa	May 28, 1872	127, 271
Butter-worker	G. Ruston	Freeport, Ill	June 4, 1872	127, 520
Butter-worker	E. E. Scott	Beloit, Wis	Sept. 30, 1873	143, 260
Butter-worker	W. S. and E. H. Shoemaker	Towsontown, Md., and Columbus, Ohio.	Apr. 19, 1870	102, 055
Butter-worker	W. E. Skinner	Milford, Mich	May 14, 1867	64, 803
Butter-worker	W. E. Skinner	Milford, Mich	Aug. 6, 1872	130, 159
Butter-worker	C. F. Smith	Litchfield, Conn	Mar. 13, 1866	53, 192
Butter-worker	I. L. Smith and C. C. Colburn	Burlington, Vt., and Massena, N. Y.	Nov. 17, 1857	18, 649
Butter-worker	J. M. Smith	Lyme, Conn	Jan. 12, 1858	19, 103
Butter-worker	L. A. Smith	Pineville, Pa	Aug. 23, 1853	9, 959
Butter-worker	O. Snow	Burlington, Vt	Sept. 21, 1869	95, 158
Butter-worker	H. Soggs	Columbus, Pa	Aug. 16, 1859	25, 145
Butter-worker	P. M. Stackhouse	Pineville, Pa	July 8, 1873	140, 739
Butter-worker	L. S. Starrett	Newburyport, Mass	Mar. 21, 1865	46, 953
Butter-worker	L. W. Stites	Brooklyn, Ohio	May 15, 1860	28, 312
Butter-worker	M. Sweet	Sidney, N. Y	Feb. 10, 1863	37, 664
Butter-worker	A. Swift	East Elmore, Vt	Apr. 5, 1864	42, 238
Butter-worker	S. H. Wade	Montgomery Centre, Vt	May 28, 1867	65, 315
Butter-worker	E. L. Walker	Twin Grove, Wis	Dec. 5, 1871	121, 559
Butter-worker	E. P. Walker	Belchertown, Mass	Apr. 21, 1868	77, 135
Butter-worker	W. Weaver	Phœnixville, Pa	July 11, 1871	117, 022
Butter-worker, &c	J. T. Whipple	Chicago, Ill	Jan. 5, 1869	85, 551
Butter-worker	W. H. Wiley	Lockport, N. Y	June 12, 1860	28, 706
Butter-worker	A. Willard	Vergennes, Vt	June 2, 1868	78, 502
Butter-worker	I. M. Williams	Blanchester, Ohio	Feb. 6, 1855	12, 366
Butter-worker	Z. Williams	Ithaca, N. Y	Sept. 7, 1858	21, 460
Butter-worker	P. G. Woodard	Waterford, Pa	Mar. 19, 1861	31, 758
Butter-worker	P. G. Woodard	Waterford, Pa	May 23, 1871	115, 144
Butter worker and mold	T. D. Smith	Independence, Ohio	July 3, 1866	56, 112
Butter worker and packer	C. F. Barager	Candor, N. Y	Oct. 23, 1866	58, 971
Butter, Working	J. Seymour	Coventry, N. Y	Dec. 20, 1859	26, 527
Butter worker and printing machine	A. S. and S. B. McDowell	Philadelphia and Montgomery County, Pa.	Oct. 15, 1867	69, 923
Butter-working machine	S. Bishop	Redding, Conn	Jan. 8, 1842	2, 419
Butter-working machine	G. Hotchkiss	Windsor, N. Y	June 7, 1859	24, 305
Butter-working machine	E. H. Merryman	Springfield, Ill	Nov. 27, 1849	6, 898
Butter-working machine	M. A. Richardson	Sherman, N. Y	June 10, 1862	35, 543
Butter-working machine	E. Wilson and M. Lee	Redding, Conn	Feb. 7, 1842	2, 447
Butteris	S. Davis	New Trenton, Ind	July 4, 1871	116, 692
Butteris	D. I. Foust	Crestline, Ohio	Mar. 19, 1872	124, 804
Butteris, Blacksmiths'	R. Kilmer and J. W. Williams	Newtown, Pa	Apr. 21, 1857	17, 128
Butteris, Blacksmiths'	J. H. Rhamy	Findley, Ohio	June 11, 1872	127, 708

Index of patents issued from the United States Patent Office from 1790 *to* 1873, *inclusive*—Continued.

Invention.	Inventor.	Residence.	Date.	No.
Button	J. N. Allen	Providence, R. I	June 17, 1873	139,991
Button	H. Ansley	Washington, D. C	Aug. 4, 1868	80,699
Button	J. F. Bapterosses	Paris, France	Jan. 19, 1858	19,120
Button	E. T. Barnum	Waterbury, Conn	Sept. 12, 1865	49,846
Button	C. Beeker and M. Wise	New York, N. Y	Dec. 28, 1869	98,338
Button	P. H. Benedict	Syracuse, N. Y	Mar. 14, 1865	46,772
Button	E. Bredt	New York, N. Y	May 5, 1868	77,443
Button	A. W. Browne	Brooklyn, N. Y	Sept. 25, 1866	58,332
Button	A. W. Browne	Brooklyn, N. Y	May 14, 1867	64,742
Button	F. M. Bugbee	Kingsville, Ohio	Jan. 28, 1868	73,871
Button	L. L. Burdon	Providence, R. I	June 29, 1869	91,906
Button	G. J. Capewell	Cheshire, Conn	July 16, 1872	128,948
Button	G. J. Capewell	Cheshire, Conn	July 16, 1872	128,949
Button	H. Carlos	New York, N. Y	Jan. 8, 1867	60,997
Button	M. P. Carpenter	New York, N. Y	Apr. 15, 1873	137,824
Button	V. Charlet	Hoboken, N. J	July 30, 1867	67,267
Button	E. A. Cobb	Philadelphia, Pa	July 18, 1871	117,150
Button	I. Cole	New York, N. Y	Mar. 30, 1869	88,451
Button	J. F. Collins and A. Neill	New York, N. Y	May 31, 1870	103,571
Button	G. Cooke	Winchester, Mass	Dec. 10, 1867	71,988
Button	A. P. Critchlow	Northampton, Mass	Nov. 30, 1869	97,363
Button	P. Davey	Ironton, Ohio	Nov. 29, 1859	26,258
Button	F. Douglas	Norwich, Conn	Apr. 10, 1866	53,796
Button	J. Durand	New York, N. Y	Sept. 9, 1873	142,686
Button	R. N. Eagle and C. D. Smith	Washington, D. C	Aug. 14, 1866	57,101
Button	A. A. Feldtrappe and R. Dutfoy	Paris, France	Oct. 13, 1863	40,254
Button	C. A. French, 2d	Boston, Mass	May 1, 1866	54,324
Button	P. W. Gengembre	Boston, Mass	Jan. 19, 1864	41,292
Button	P. W. Gengembre	Boston, Mass	Dec. 13, 1864	45,400
Button	P. W. Gengembre	Boston, Mass	Dec. 13, 1864	45,401
Button	P. W. Gengembre	Boston, Mass	Dec. 13, 1864	45,402
Button	H. Gerner	New York, N. Y	Nov. 14, 1865	50,927
Button	R. H. Guilford	West Cheshire, Conn	Sept. 3, 1867	68,562
Button	R. H. Guilford	West Cheshire, Conn	May 24, 1870	103,451
Button	J. Hatch	Attleborough, Mass	Feb. 20, 1845	3,915
Button	C. L. Horack	Willimantic, Conn	Nov. 16, 1869	96,808
Button	D. Howarth	Portland, Me	May 8, 1866	54,549
Button	L. A. Jefferson	Bridgeport, Conn	Feb. 23, 1869	87,262
Button	J. Jenk	Washington, D. C	Apr. 9, 1867	63,642
Button	J. M. Johnson	Mamaroneck, N. Y	Apr. 17, 1866	53,986
Button	J. M. Johnson	New York, N. Y	May 15, 1866	54,736
Button	J. Kenmuir	Leavenworth, Kans	Dec. 28, 1869	98,271
Button	M. R. Kenyon	Providence, R. I	Mar. 16, 1869	87,854
Button	J. Koberle	Saint Louis, Mo	Dec. 1, 1868	84,555
Button	S. Krom	New York, N. Y	June 5, 1860	28,587
Button	S. B. Lane	Waterbury, Conn	Nov. 19, 1872	133,160
Button	H. Link	Little Falls, N. Y	Aug. 17, 1869	93,726
Button	C. M. Loomis	Hartford, Conn	June 6, 1865	48,079
Button	F. Loos	Germantown, Pa	Jan. 29, 1867	61,547
Button	P. F. Mauras	New York, N. Y	Dec. 4, 1866	60,217
Button	E. Maynard	Washington, D. C	May 23, 1865	47,843
Button	G. W. McGill	Washington, D. C	May 28, 1867	65,250
Button	G. A. Meacham	New York, N. Y	Jan. 10, 1860	26,776
Button	G. A. Meacham	New York, N. Y	July 2, 1861	32,703
Button	G. A. Meacham	New York, N. Y	Aug. 13, 1861	33,046
Button	G. A. Meacham	New York, N. Y	Apr. 8, 1862	34,897
Button	R. J. Monks	Boston, Mass	May 28, 1872	127,183
Button	M. D. Moore	Brooklyn, N. Y	May 19, 1868	78,116
Button	C. Mudler	Cleveland, Ohio	June 2, 1868	78,535
Button	C. N. Nickerson	Gloucester, Mass	Oct. 27, 1868	83,403
Button	O. Paddock	Watertown, N. Y	May 18, 1869	90,119
Button	F. I. Palmer	Springfield, Mass	Mar. 27, 1866	53,471
Button	F. J. Peabody	Medford, Mass	Aug. 4, 1868	80,563
Button	F. J. Perkins	Salem, Mass	Apr. 28, 1868	77,313
Button	G. W. Phillips	Fresh Pond, N. Y	Aug. 1, 1871	117,564
Button	C. M. Platt	Waterbury, Conn	July 10, 1866	56,261
Button	C. M. Platt	Waterbury, Conn	Nov. 14, 1871	121,000
Button	H. Prouhet	Saint Louis, Mo	Dec. 3, 1867	71,641
Button	J. N. Prugger	New York, N. Y	June 4, 1867	65,503
Button	W. H. Reed	Philadelphia, Pa	June 6, 1865	48,095
Button	W. H. Reed	New York, N. Y	Feb. 19, 1867	62,153
Button	J. L. Remlinger	Providence, R. I	Aug. 17, 1869	93,748
Button	P. E. Richardière	Paris, France	Feb. 24, 1863	37,774
Button	C. E. Richards	Attleborough, Mass	Feb. 13, 1866	52,608
Button	E. A. Robinson	Waterbury, Conn	Aug. 14, 1866	57,260
Button	C. Robitaille	Brooklyn, N. Y	Apr. 17, 1866	54,071
Button	A. H. Savage	Ashtabula, Ohio	Sept. 3, 1872	131,067
Button	G. A. Schultz	Louisville, Ky	Apr. 4, 1871	113,577
Button	A. Selkirk	Albany, N. Y	Mar. 6, 1866	53,086
Button	A. M. Smith	Brooklyn, N. Y	Sept. 25, 1866	58,310
Button	D. M. Somers	Brooklyn, N. Y	Oct. 22, 1867	70,039
Button	D. M. Somers and W. S. Atwood	Brooklyn, N. Y., and Newark, N. J.	Dec. 31, 1867	72,929
Button	C. F. Spencer	Rochester, N. Y	Oct. 23, 1866	59,090
Button	J. A. Spooner and J. Ellerby	Brooklyn, N. Y	Feb. 1, 1870	99,365
Button	J. R. Spooner	Lowell, Ohio	June 25, 1867	66,053
Button	A. E. Thurber	Providence, R. I	Mar. 18, 1873	136,882
Button	W. W. Wade	Medford, Mass	Mar. 23, 1869	88,099
Button	F. Washbourne	Brooklyn, N. Y	Apr. 22, 1873	138,107
Button	B. B. Whaley	Brooklyn, E. D., N. Y	Oct. 18, 1864	44,759
Button	E. S. Wheeler	Westport, Conn	Jan. 3, 1865	45,781
Button	E. S. Wheeler	Westport, Conn	Mar. 24, 1868	76,016
Button	W. B. White	North Attleborough, Mass.	Aug. 22, 1865	49,595
Button	J. F. Wild	New York, N. Y	Mar. 14, 1865	46,840
Button	F. Wittram	San Francisco, Cal	Oct. 6, 1868	82,778
Button and lacing hooks, Machine for setting	H. C. Bradford	Providence, R. I	Feb. 11, 1873	135,763
Button and locket, Combined sleeve	W. C. Almy	Providence, R. I	May 7, 1872	126,368
Button and stud attachment	H. Freeman	Petroleum Centre, Pa	May 13, 1873	138,745

Index of patents issued from the United States Patent Office from 1790 *to* 1873, *inclusive*—Continued.

Invention.	Inventor.	Residence.	Date.	No.
Button and stud, Shirt	J. T. Thornton	Providence, R. I	May 6, 1873	138,595
Button and stud, Sleeve	B. Clayton	Philadelphia, Pa	Feb. 12, 1867	62,008
Button-attachment for apparel	I. Lewine	New York, N. Y	June 19, 1866	55,798
Button-backs, Machine for forming	J. C. Cooke	Waterbury, Conn	July 27, 1852	9,146
Button blanks, Machine for grinding pearl	J. D. Hall	Philadelphia, Pa	Jan. 25, 1870	99,183
Button, Clasp	A. Flatley	Brooklyn, N. Y	Mar. 11, 1873	136,648
Button, Covered	P. Kirkham	Birmingham, England	Aug. 14, 1849	6,651
Button, Covered	N. C. Newell	Springfield, Mass	Aug. 28, 1866	57,555
Button, Cuff	F. McCarthy	Chicago, Ill	June 14, 1873	140,211
Button, Cuff and collar	H. W. Little	Muncie, Ind	Jan. 23, 1872	122,899
Button, Dead-eyed wood	D. G. Fowler	New Branford, Conn	May 8, 1833	....
Button, Detachable metal	T. Kirk	Waterbury, Conn	Nov. 8, 1864	44,958
Button-dies, Machine for grinding steel	W. Robinson	Attleborough, Mass	Mar. 3, 1829	....
Button-fastener	J. C. Gaston	Cincinnati, Ohio	May 24, 1870	103,443
Button-fastener	R. B. Griffin, jr	Baltimore, Md	Nov. 26, 1867	71,382
Button-fastener	H. W. Hewett	New York, N. Y	Dec. 10, 1840	1,889
Button-fastener	J. Lafvendahl	East Boston, Mass	Apr. 16, 1861	32,105
Button-fastener	G. A. Meacham	New York, N. Y	Apr. 8, 1862	34,895
Button-fastener	A. M. Richmond	New York, N. Y	Apr. 25, 1871	114,200
Button-fastener	E. Ripley	Troy, N. Y	July 21, 1868	80,224
Button-fastener	R. Weaver	Washington, D. C	June 16, 1868	78,903
Button-fastening	J. F. Blood	Providence, R. I	June 1, 1869	90,631
Button-fastening	E. Brady	Philadelphia, Pa	Nov. 9, 1869	96,544
Button-fastening	A. Brookmann	Chicago, Ill	May 6, 1873	138,608
Button-fastening	D. Cannata	Jersey City, N. J	June 17, 1873	140,008
Button-fastening	G. J. Capewell	West Cheshire, Conn	Feb. 12, 1867	62,001
Button-fastening	G. D. Clark	Plainville, Conn	Aug. 4, 1868	80,710
Button-fastening	F. Dahis	Brooklyn, N. Y	Nov. 14, 1865	50,988
Button-fastening	I. F. Eaton	Brooklyn, N. Y	Oct. 21, 1873	143,892
Button-fastening	B. Geiger and H. Wocher	Philadelphia, Pa	Nov. 9, 1869	96,689
Button-fastening	L. Goodwin	New York, N. Y	June 29, 1858	20,707
Button-fastening	W. Holloway	Buffalo, N. Y	July 8, 1873	140,627
Button-fastening	H. Humphrey	Adrian, Mich	Oct. 29, 1867	70,218
Button-fastening	J. Johnson	New York, N. Y	May 8, 1866	54,556
Button-fastening	J. M. Johnson	New York, N. Y	Dec. 4, 1866	60,194
Button-fastening	W. Kuhlenschmidt	New York, N. Y	Apr. 23, 1861	32,136
Button-fastening	F. M. La Boiteaux	Cottage Hill, Ohio	Nov. 8, 1870	109,024
Button-fastening	E. F. Lee	New York, N. Y	Nov. 8, 1870	109,134
Button-fastening	H. S. Magrane	Hoboken, N. J	Jan. 28, 1868	73,735
Button-fastening	M. Ormsbee	Brooklyn, N. Y	June 10, 1873	139,808
Button-fastening	C. H. Reid	Danbury, Conn	Aug. 19, 1873	142,043
Button-fastening	A. Rix	San Francisco, Cal	Sept. 22, 1868	82,440
Button fastening	M. Rosenthal	Philadelphia, Pa	Mar. 17, 1868	75,650
Button-fastening	J. B. Rumsey	Tiffin, Ohio	Apr. 17, 1866	54,022
Button-fastening	W. B. Smith	New York, N. Y	Mar. 4, 1873	136,556
Button-fastening	E. J. F. Stolpe	Washington, D. C	May 6, 1873	138,708
Button-fastening	J. R. Underhill	Brooklyn, N. Y	June 11, 1867	65,708
Button-fastening	E. S. Wheeler	Westport, Conn	Aug. 12, 1873	141,743
Button-fastening	M. Wise	New York, N. Y	Apr. 19, 1870	102,190
Button for fastening cupboard and other doors, Plate turn.	P., E. W., and J. A. Blake	New Haven, Conn	Mar. 21, 1843	3,017
Button-forming maching	S. G. Pitts	Leominster, Mass	Sept. 15, 1868	82,243
Button, Garment	A. C. Wilhelm	Philadelphia, Pa	Feb. 11, 1873	135,739
Button, Glass	A. W. Demuth	New York, N. Y	May 31, 1864	42,931
Button, Glass	A. M. Smith	Brooklyn, N. Y	Jan. 28, 1873	135,375
Button, Glass	A. W. Welton	Cheshire, Conn	Oct. 12, 1852	9,337
Button grinding and polishing machine	G. J. Capewell	West Cheshire, Conn	Aug. 28, 1866	57,471
Button-holder	M. J. Cooper	New York, N. Y	Dec. 16, 1873	145,556
Button-hole	E. F. Dieterichs	Philadelphia, Pa	July 10, 1866	56,190
Button-hole	H. B. Fairman	New York, N. Y	June 27, 1865	48,484
Button-hole	C. D. Smith	Washington, D. C	Aug. 14, 1866	57,202
Button-hole and buckle	L. A. Kettle	Philadelphia, Pa	Aug. 24, 1869	94,002
Button-hole clamp and guide	L. Reding	Norwalk, Ohio	Mar. 28, 1871	113,093
Button-hole cutter	P. Bauer	Newark, N. J	July 16, 1872	129,309
Button-hole cutter	C. E. Bean	Oswego, N. Y	June 11, 1872	127,835
Button-hole cutter	T. W. Brown	Boston, Mass	May 23, 1854	10,972
Button-hole cutter	T. C. Bush	New Haven, Conn	Sept. 13, 1870	107,332
Button-hole cutter	D. G. Chase	Boston, Mass	Dec. 28, 1869	98,349
Button-hole cutter	W. Chicken	Boston, Mass	July 7, 1857	17,725
Button-hole cutter	J. S. Crane	Lake Village, N. H	Feb. 11, 1868	74,315
Button-hole cutter	D. H. Cunningham	Waltham, Mass	Sept. 6, 1870	107,009
Button-hole cutter	C. N. Cutter	Worcester, Mass	Jan. 1, 1867	60,700
Button hole cutter	A. Day	Detroit, Mich	Feb. 18, 1873	135,893
Button hole cutter	T. B. Doolittle	Bridgeport, Conn	Sept. 3, 1872	131,085
Button-hole cutter	W. W. Egnew	Detroit, Mich	Jan. 30, 1872	123,250
Button-hole cutter	J. V. D. Eldredge	Detroit, Mich	Dec. 31, 1872	134,366
Button-hole cutter	I. J. Fearing	South Weymouth, Mass	May 14, 1861	32,281
Button-hole cutter	W. Fitzgerald	Boston, Mass	Apr. 30, 1867	64,300
Button-hole cutter	H. Hempel	New York, N. Y	June 11, 1867	65,671
Button-hole cutter	J. A. and H. A. House	Bridgeport, Conn	Apr. 3, 1866	53,619
Button-hole cutter	A. Humann	New York, N. Y	Feb. 25, 1868	74,827
Button-hole cutter	C. M. Johnson	Hartford, Conn	Dec. 10, 1872	133,708
Button-hole cutter	F. C. Leypoldt	Philadelphia, Pa	Dec. 11, 1860	30,925
Button-hole cutter	F. C. Leypoldt	Philadelphia, Pa	July 25, 1865	48,962
Button-hole cutter	W. Lieber	New York, N. Y	Aug. 21, 1866	57,342
Button-hole cutter	A. J. Lytle	West Union, Ohio	Oct. 20, 1868	83,293
Button-hole cutter	W. C. McGill	Cincinnati, Ohio	June 3, 1873	139,472
Button-hole cutter	W. A. Morse	Philadelphia, Pa	July 18, 1871	117,100
Button-hole cutter	H. H. Pinkham	New Market, N. H	July 9, 1872	128,902
Button-hole cutter	W. S. Porter	Boston, Mass	Dec. 15, 1868	84,902
Button hole cutter	J. G. Powell	Philadelphia, Pa	May 9, 1871	114,598
Button-hole cutter	F. G. Sanborn	Boston, Mass	Jan. 24, 1865	46,028
Button-hole cutter	G. H. Seymour and W. B. Barnard.	Waterbury, Conn	Aug. 28, 1866	57,584
Button-hole cutter	M. M. Shellaberger	Rudolph County, Ind	Feb. 23, 1864	41,725
Button-hole cutter	E. Stern and J. S. Newell	Dorchester and Newton, Mass.	Dec. 9, 1862	37,115
Button-hole cutter	C. F. Tebbets and G. Nutting	Fitchburgh, Mass	May 27, 1873	139,437

Index of patents issued from the United States Patent Office from 1790 to 1873, inclusive—Continued.

Invention.	Inventor.	Residence.	Date.	No.
Button-hole cutter	C. Van Hoosen	New York, N. Y	June 17, 1873	139, 983
Button-hole cutter	F. H. Walker	Boston, Mass	June 30, 1868	79, 418
Button-hole cutter	F. H. Walker	Boston, Mass	Dec. 14, 1869	97, 996
Button-hole cutter and punch	A. L. Whitney	Fitchburgh, Mass	Apr. 8, 1873	137, 745
Button-hole cutter, eyelet-puncher, and scissors-sharpener, Combined.	W. C. McGill	Cincinnati, Ohio	June 11, 1872	127, 783
Button-hole-cutter gage	G. Stackpole	Elizabeth, N. J	Apr. 16, 1872	125, 702
Button-hole-cutting machine	G. Rehfuss	Philadelphia, Pa	Aug. 2, 1864	43, 707
Button-hole guide-plate	A. W. Webster	Ansonia, Conn	June 27, 1871	116, 378
Button-hole clasp, Metallic	J. R. Little	Jamaica Plains, N. Y	Sept. 21, 1869	94, 965
Button-hole stitch	D. W. G. Humphrey	Chelsea, Mass	Oct. 7, 1862	36, 616
Button-hole stitch	W. Weitling	New York, N. Y	Feb. 18, 1862	34, 454
Button-hole-stitching machine	D. W. G. Humphrey	Chelsea, Mass	June 13, 1871	115, 857
Button-hole-stitching machine	D. W. G. Humphrey	Chelsea, Mass	Feb. 6, 1872	123, 348
Button-hole-stitching machine	R. H. Peabody	Chelsea, Mass	June 9, 1868	78, 821
Button-hole twist, Preparing	R. Morrison	Yonkers, N. Y	Mar. 8, 1870	100, 548
Button-holes, Apparatus for making	F. Lindner	Dayton, Ohio	Oct. 9, 1866	58, 654
Button-holes, Apparatus for piercing cloth for	D. W. Whitney	New York, N. Y	Sept. 30, 1862	36, 594
Button-holes, Cutting	C. Currier	Providence, R. I	June 22, 1858	20, 632
Button-holes, Edge-finish for	D. W. G. Humphrey	Chelsea, Mass	Apr. 9, 1872	125, 393
Button-holes, Guide for working	O. Avery	Bethany, Pa	Oct. 14, 1856	15, 872
Button-holes in fabrics, Mode of weaving	L. J. Knowles	Warren, Mass	Mar. 7, 1865	46, 679
Button-holes, Machine for cutting and re-enforcing	J. T. Bruen and G. M. Jacobs	New York, N. Y	Nov. 27, 1866	59, 957
Button-holes, Mode of making	H. Loewenberg	Boston, Mass	May 6, 1862	35, 163
Button-holes, Mode of securing	J. Walters	New York, N. Y	Dec. 3, 1872	133, 554
Button-hook and handkerchief-holder combined	G. D. Stevens	New York, N. Y	Oct. 4, 1870	108, 063
Button-hook, Lacing	D. Heaton	Providence, R. I	Nov. 26, 1872	133, 312
Button-hooks, Device for setting	J. S. Palmer	Providence, R. I	May 25, 1869	90, 382
Button-hooks, Instrument for setting	J. S. Palmer	Providence, R. I	Sept. 28, 1869	95, 257
Button-hooks, Machine for making	J. S. Palmer	Providence, R. I	June 1, 1869	90, 681
Button-hooks, Machine for making	L. Towne	Providence, R. I	Feb. 2, 1869	86, 473
Button-hooks, Machine for setting	P. Esser and F. A. Steere	North Providence, R. I	May 19, 1868	78, 077
Button, Imitation	S. G. Jones	North Bridgewater, Mass	Feb. 27, 1872	124, 139
Button, Imitation	W. H. Shurtleff	Providence, R. I	Jan. 30, 1872	123, 299
Button-key	J. Lawrence	New Haven, Conn	Oct. 18, 1864	44, 771
Button key	H. St. John	New Haven, Conn	Nov. 24, 1863	40, 714
Button, Lacing	D. Heaton	Providence, R. I	July 16, 1872	129, 474
Button, Lacing	D. Heaton	Providence, R. I	Nov. 19, 1872	133, 223
Button, Lacing	D. Heaton	Providence, R. I	Dec. 10, 1872	133, 857
Button-lacing hook	W. H. Shurtleff	Providence, R. I	June 11, 1867	65, 615
Button-machine	C. Volkert	New York, N. Y	Sept. 20, 1870	107, 570
Button-machinery	L. E. Hicks	Middletown, Conn	Apr. 24, 1847	5, 079
Button-making die	A. C. Sweetland	North Attleborough, Mass.	Sept. 26, 1865	50, 185
Button-making machine	J. F. Bapterosses	Paris, France	Aug. 12, 1873	141, 749
Button-making machine	H. Bock	New York, N. Y	Feb. 14, 1865	46, 330
Button-making machine	W. Kraemer	Cincinnati, Ohio	Oct. 30, 1866	59, 234
Button-making machine	L. D. Phillips	New York, N. Y	Nov. 27, 1866	60, 053
Button-manufacture	L. Moses	New York, N. Y	Feb. 25, 1868	74, 764
Button-manufacture	T. Prosser	Paterson, N. J	July 29, 1841	2, 199
Button-manufacture, Machine for forming collets, washers, &c., for.	M. Ferre	Williamsburgh, Mass	July 28, 1842	2, 740
Button mold for casting wire-eyed buttons, Old single-jointed pewter.	J. B. Collins	Meriden, Conn	Apr. 12, 1815	....
Button-molds, Manufacture of	J. Hayden and R. Hyde	Williamsburgh and Chesterfield, Mass.	Aug. 7, 1849	6, 635
Button-molds, Manufacturing and polishing wooden	J. Norton	Southington, Conn	June 13, 1816	....
Button-molds of wood, Manufacturing	A. Mathews	Southington, Conn	Apr. 26, 1815	....
Button-molds, Tool for making	J. T. Hawkins	Salisbury, Vt	Sept. 23, 1873	143, 073
Button-molding die	W. M. Welling	New York, N. Y	Oct. 25, 1870	108, 660
Button needle and fastener	J. A. Ostburg	Boston, Mass	June 20, 1871	116, 087
Button or fastener, Sleeve	S. J. Shaw	Marlborough, Mass	Dec. 27, 1864	45, 646
Button or garment-fastener	R. Bocklen and W. Stoehlen	Brooklyn, N. Y	Aug. 13, 1861	33, 025
Button or lacing hooks, Machine for setting	H. C. Bradford	Providence, R. I	June 14, 1870	104, 257
Button or lacing hooks, Machine for setting	H. C. Bradford	Providence, R. I	Feb. 27, 1872	124, 029
Button or stud	J. B. Carter	Hartsville, Ind	Aug. 31, 1869	94, 183
Button or stud	C. A. Newton	Providence, R. I	Aug. 8, 1871	117, 916
Button or stud	D. R. Rae and T. F. Lewis	Providence, R. I	Feb. 27, 1872	124, 010
Button or stud, Collar	J. I. Waddell	Annapolis, Md	Apr. 15, 1873	137, 807
Button or stud, Shirt	G. R. Burdon	Providence, R. I	July 16, 1872	128, 946
Button or stud, Shirt	L. W. Fifield	Worcester, Mass	Jan. 17, 1871	111, 049
Button or stud, Shirt	T. J. Holmes	Malden, Mass	Dec. 30, 1873	145, 948
Button-ornamenting machine	J. Tunnicliff and P. Cahill	Northampton, Mass	Nov. 26, 1867	71, 429
Button, Picture	R. E. Hitchcock	Waterbury, Conn	Aug. 13, 1861	33, 038
Button-planing machine	J. G. Vallentine	Naugatuck, Conn	Mar. 7, 1865	46, 763
Button-rings, Machine for making	S. B. Lane	Waterbury, Conn	Sept. 3, 1867	68, 516
Button-riveting machine	J. J. Mervesp	New York, N. Y	Jan. 4, 1870	98, 512
Button-shank gage	J. F. Russell	Washington, D. C	Apr. 26, 1870	102, 436
Button shanks, Method of forming sleeve	W. H. Blake	Waterbury, Conn	Nov. 23, 1869	97, 156
Button, Silk-covered	H. Heinemann	New York, N. Y	July 8, 1851	8, 205
Button, Sleeve	S. Bogart	New York, N. Y	Jan. 12, 1864	41, 191
Button, Sleeve	J. Caldwell and H. B. Winslow	Providence, R. I	July 30, 1872	130, 015
Button, Sleeve	J. P. Derby	Boston, Mass	Mar. 31, 1857	16, 909
Button, Sleeve	J. C. Harrington and D. S. Cooke.	Providence, R. I	Nov. 5, 1872	132, 761
Button, Sleeve	W. H. Wilson	Providence, R. I	Feb. 24, 1863	37, 784
Button, Spring-catch	T. K. Reed and H. F. Packard	East Bridgewater and North Bridgewater, Mass.	Aug. 23, 1864	43, 928
Button, stud, &c	D. Rait	New York, N. Y	Apr. 6, 1852	8, 860
Button, stud, &c	T. S. Sedgwick	Onargo, Ill	Aug. 18, 1868	81, 300
Button, Suspender	E. Smith	Naugatuck, Conn	Apr. 16, 1861	32, 087
Button, Swivel	E. H. Porter	New Britain, Conn	Mar. 20, 1866	53, 333
Button, Window	E. K. Breckenridge	West Meriden, Conn	Nov. 30, 1869	97, 348
Buttons, American wire-eyed	F. Hayden	Waterbury, Conn	Oct. 1, 1830	....
Buttons, Apparatus for centering shanks or eyes on	W. H. Blake	Waterbury, Conn	Feb. 1, 1870	99, 394
Buttons, Attaching the shanks to mineral and composition.	C. L. Potter	Providence, R. I	Dec. 13, 1870	110, 070
Buttons, &c., Attachment of	N. Ames	Saugus Centre, Mass	Mar. 22, 1864	41, 969

Index of patents issued from the United States Patent Office from 1790 *to* 1873, *inclusive*—Continued.

Invention.	Inventor.	Residence.	Date.	No.
Buttons cast in mold, Construction of an eye for metal.	I. Ives	Bristol, Conn	Aug. 7, 1815	....
Buttons, Composition for	E. S. Wheeler	Westport, Conn	Nov. 14, 1865	50, 974
Buttons, Composition for making wire-eyed	S. H. Woodruff and A. Matthews.	Connecticut	Mar. 1, 1816	....
Buttons, Detachable covering for	F. H. Gould	Newark, N. J	Apr. 14, 1868	76, 746
Buttons, Die for forming sleeve	H. Ansley	Washington, D. C	Apr. 5, 1870	101, 565
Buttons, Fastener for sleeve	H. Cogswell	Providence, R. I	May 11, 1858	20, 194
Buttons, Finishing wired-neck metal	H. Mathews	Southington, Conn	Sept. 12, 1815	....
Buttons from straw boards, Manufacture of	E. M. Pomeroy	Wallingford, Conn	Aug. 21, 1849	6, 654
Buttons, handles for knives, and for other purposes, Material for the manufacture of.	L. E. Chittenden	Washington, D. C	Jan. 24, 1865	45, 977
Buttons, Holding shanks of molds for glass	G. Matthewman	Williamsburgh, N. Y	Oct. 10, 1865	50, 374
Buttons in the mold, Setting the eyes of metal	I. Ives	Bristol, Conn	Aug. 7, 1815	....
Buttons, Machine for attaching	W. S. Platt and G. J. Capewell	Cheshire, Conn	Sept. 5, 1871	118, 743
Buttons, Machine for making covered	W. W. Wade	Medford, Mass	Apr. 27, 1869	89, 450
Buttons, Machine for making wire-necked metal	H. Mathews	Southington, Conn	Setp. 12, 1815	....
Buttons, Machine for molding glass	C. Vigneron	Providence, R. I	Apr. 19, 1870	102, 070
Buttons, Machine for pressing glass	T. Guilford	Cheshire, Conn	Apr. 10, 1866	53, 816
Buttons, Making paper	E. M. Pomeroy	Wallingford, Conn	Sept. 23, 1843	3, 281
Buttons, Manufacture of	P. Kirkham	Waterbury, Conn	Dec. 18, 1849	6, 953
Buttons, Manufacture of	C. M. Platt	Waterbury, Conn	May 6, 1873	138, 525
Buttons, Manufacture of wooden	L. L. and A. L. Platt	Newtown, Conn	June 27, 1854	11, 176
Buttons, Manufacturing	C. Goodyear	Philadelphia, Pa	Jan. 12, 1831	....
Buttons, Manufacturing coat and waistcoat	G. W. Robinson		Mar. 24, 1804	....
Buttons, Manufacturing cord	N. Perkins	Wawarsing, N. Y	Aug. 17, 1852	9, 210
Buttons, Manufacturing dead-eyed wood	D. Fowler	Branford, Conn	June 13, 1831	....
Buttons, Manufacturing metallic	J. O. M. Ingersoll	Ithaca, N. Y	Aug. 18, 1857	18, 014
Buttons, Manufacturing metallic flexible-shank	J. Hayden, jr	Williamsburgh, Mass	Feb. 17, 1831	....
Buttons, Means for fastening	I. F. Eaton	Brooklyn, N. Y	Oct. 21, 1873	143, 891
Buttons, Mode of fastening	W. A. and F. V. Johnson and H. S. Worth.	Oxford, Pa	Oct. 7, 1873	143, 512
Buttons, Mode of finishing glass	S. Richards	Attleborough, Mass	Aug. 19, 1822	....
Buttons, Mold for casting metallic	G. W. Robinson	New Haven, Conn	May 14, 1812	....
Buttons, Mold for fancy	F. Maass	Newark, N. J	Oct. 21, 1873	143, 916
Buttons on cloth, Machine for riveting	W. J. Gordon and E. D. Gilbert	Philadelphia, Pa	Mar. 7, 1865	46, 660
Buttons on heads, &c., of whips, Forming worked	R. S. Brown	Philadelphia, Pa	Feb. 21, 1842	2, 462
Buttons, Ornamenting metallic	H. W. Hayden	Waterbury, Conn	July 11, 1854	11, 263
Buttons, Preparing wood for the manufacture of	R. H. Isbell	New Milford, Conn	Dec. 24, 1872	134, 286
Buttons, Process of varnishing	E. M. Pomeroy	Wallingford, Conn	Feb. 19, 1850	7, 102
Buttons, Punching-machine for manufacture of covered.	A. C. Arnold	Norwalk, Conn	Apr. 23, 1842	2, 580
Buttons, Smoothing and rounding the edges of metallic.	I. Ives	Bristol, Conn	Aug. 7, 1815	....
Buttons to cards, Securing	E. S. Boynton	New York, N. Y	Aug. 29, 1865	49, 674
Buttons to cloth, Method of attaching	H. S. Poole	Boston, Mass	Aug. 11, 1841	2, 219
Buttons to cloth, Method of making and attaching metallic.	F. Hayden	Waterbury, Conn	July 10, 1840	1, 682
Buttons to fabrics, Apparatus for securing	G. Rehfuss	Philadelphia, Pa	Feb. 27, 1866	52, 885
Buttons to fabrics, Apparatus for securing	G. Rehfuss	Philadelphia, Pa	Feb. 27, 1866	52, 884
Buttons to fabrics, Device for securing	D. M. Somers	Brooklyn, N. Y	Oct. 12, 1869	95, 847
Buttons to fabrics, Fastening	W. H. Reed	Philadelphia, Pa	Nov. 15, 1864	45, 077
Buttons to fabrics, Instrument for attaching	H. Mauch	Providence, R. I	Oct. 20, 1868	83, 193
Buttons to fabrics, Instrument for attaching	D. M. Somers and W. S. Atwood	Brooklyn, N. Y	Nov. 3, 1868	83, 669
Buttons to fabrics, Machine for attaching	G. J. Capewell	West Cheshire, Conn	Apr. 14, 1868	76, 600
Buttons to fabrics, Machine for riveting	W. J. Gordon	Philadelphia, Pa	Jan. 22, 1867	61, 334
Buttons to fabrics, Machine for riveting	J. J. Mervesp	Brooklyn, N. Y	Oct. 29, 1872	132, 537
Buttons, &c., to fabrics, Mode of attaching	H. B. Walbridge	Brooklyn, N. Y	Nov. 19, 1872	133, 273
Buttons to fabrics, Securing	G. J. Capewell	West Cheshire, Conn	June 8, 1869	91, 083
Buttons to garments, Device for attaching	G. J. Capewell	Cheshire, Conn	July 16, 1872	128, 950
Buttons to garments, Securing	J. W. Roberts	New Monmouth, N. J	July 24, 1866	56, 612
Buttons to garments, Securing	C. F. Spencer	Rochester, N. Y	Mar. 12, 1867	62, 784
Buttons, Tool for setting hook	H. C. Bradford	Providence, R. I	Nov. 7, 1871	120, 703
Buttons, &c., Treating wood for the manufacture of sleeve.	O. Knipfer	Bridgeport, Conn	May 14, 1872	126, 816
Buttons, Turning	D. French	Bridgeport, Pa	Apr. 23, 1816	....
Buttons while polishing, Hold-fast for	I. Ives	New Haven, Conn	June 18, 1814	....
C.				
Cab	J. Pol	New York, N. Y	Jan. 29, 1867	61, 675
Cab and Cradle	G. H. Henkel	Hartford City, Ind	Aug. 10, 1869	93, 437
Cab, Brakeman's	H. C. Glasgow	Chicago, Ill	Apr. 1, 1862	34, 829
Cabbage-cutter	H. Aeuer	Muscatine, Iowa	Jan. 22, 1867	61, 304
Cabbage-cutter	H. Carver	Edinburgh, Va	Aug. 26, 1851	8, 310
Cabbage-cutter	A. Fischer	Dayton, Ohio	Apr. 27, 1858	20, 054
Cabbage-cutter	J. D. Schaub	Birmingham, Pa	Apr. 5, 1870	101, 520
Cabbage-cutter	F. S. Vogel and J. R. Albright	Lancaster, Pa	Mar. 30, 1869	88, 349
Cabbage-cutter	J. C. Wilkins	Fox Chase, Pa	Jan. 3, 1860	26, 725
Cabbage-cutting machine	F. Berkeymeyer and J. Dengler	Greenwich, Pa	Nov. 14, 1826	....
Cabbage-cutting machine	G. G. Elias	Lancaster, Pa	Nov. 22, 1859	26, 173
Cabin-chair	W. Thomas	Hingham, Mass	Apr. 10, 1855	12, 703
Cabin, Floating safety	M. I. Butler	Nashville, Tenn	Oct. 25, 1859	25, 881
Cabin for steam and other vessels, Floating	W. R. Jackson	Baltimore, Md	June 5, 1855	13, 006
Cabin-table	H. J. Nichols	Searsport, Me	Aug. 23, 1870	106, 608
Cabinet	R. Bacon	Boston, Mass	Feb. 15, 1870	99, 746
Cabinet and bed	R. F. Clayton	Boston, Mass	Jan. 30, 1872	123, 085
Cabinet, Blacking	L. P. Keach	Baltimore, Md	July 20, 1869	92, 838
Cabinet-chair	S. Pearson	Boston, Mass	Oct. 18, 1859	25, 849
Cabinet-chair and secretary combined	C. S. Trevitt	Washington, D. C	Apr. 8, 1873	137, 738
Cabinet, College	W. W. Lovering	San Francisco, Cal	Dec. 18, 1866	60, 529
Cabinet for ladies	A. J. Forbes	San Francisco, Cal	Nov. 16, 1869	96, 788
Cabinet-hook	J. B. Sargent	New Haven, Conn	June 15, 1869	91, 369
Cabinet-maker's tool	R. W. Tanner	Albany, N. Y	Dec. 14, 1869	97, 833
Cabinet-purposes, Combination of wood and paper for.	A. C. Spencer	Bridgeport, Conn	June 30, 1868	79, 511
Cabinet-ventilator	S. Harnden	Reading, Mass	Apr. 20, 1869	89, 219
Cabinet-work, Machine for polishing	J. B. Morris	Ansonia, Conn	July 24, 1860	29, 293

Index of patents issued from the United States Patent Office from 1790 *to* 1873, *inclusive*—Continued.

Invention.	Inventor.	Residence.	Date.	No.
Cable-chain	I. Jennings	New York, N. Y	Oct. 29, 1821	....
Cable, Electric	J. L. Arman	Bordeaux, France	Jan. 21, 1868	73, 487
Cable for railway-guide, Propelling	C. T. Harvey	Tarrytown, N. Y	July 2, 1867	66, 330
Cable hook, Chain	E. Applegate	Wilmington, Del	Apr. 29, 1856	14, 758
Cable-link-bending machine	R. M. Green	Baltimore, Md	Mar. 12, 1867	62, 839
Cable meter, Drill	L. C. Bristol	Victor, N. Y	Apr. 3, 1866	53, 564
Cable-stopper	W. H. Bridge	Boston, Mass	Feb. 23, 1858	19, 403
Cable-stopper	E. R. Cheney and J. J. Emery	Boston, Mass., and Owl's Head, Me.	June 21, 1870	104, 425
Cable-stopper	E. R. Cheney and J. J. Emery	Boston, Mass., and Owl's Head, Me.	Aug. 2, 1870	106, 031
Cable-stopper	P. H. Jackson	New York, N. Y	Aug. 10, 1858	21, 135
Cable-stopper	J. M. Kilner	Chester, England	June 24, 1873	140, 202
Cable-stopper	P. Murray	Wilwaukee, Wis	June 5, 1866	55, 343
Cable-stopper	G. S. Perkins	Essex, Mass	Jan. 12, 1864	41, 235
Cable-stopper	J. Reed	Marshfield, Mass	Jan. 30, 1855	12, 325
Cable-stopper	J. Robinson	Barnegat, N. J	Dec. 9, 1862	37, 109
Cable-stopper	J. Stitt	San Francisco, Cal	July 19, 1870	105, 510
Cable stopper, Chain	E. R. Cheney	Boston, Mass	Jan. 30, 1872	123, 152
Cable stopper, Chain	J. E. Crane	Lowell, Mass	Apr. 5, 1853	9, 645
Cable stopper, Chain	J. E. Crane	Lowell, Mass	Jan. 19, 1858	19, 131
Cable stopper, Chain	M. P. Mix	New York, N. Y	Nov. 3, 1838	998
Cable stopper, Chain	O. Nichols	Lowell, Mass	Apr. 4, 1854	10, 724
Cable stopper, Chain	O. Nichols	Lowell, Mass	Apr. 4, 1854	10, 725
Cable stopper, Chain	C. Perley	New York, N. Y	June 13, 1854	11, 080
Cable stopper, Chain	C. Perley	New York, N. Y	Nov. 29, 1859	26, 292
Cable stopper, Chain	J. Tucker	Washington, D. C	Jan. 3, 1860	26, 722
Cable, Submarine	P. S. Devlan	Jersey City, N. J	Sept. 15, 1863	39, 896
Cable surge-reliever	J. J. Emery	Owl's Head, Me	Sept. 21, 1869	94, 950
Cables and spin ropes of any size, Machine to lay	W. Schultz	Baltimore, Md	June 24, 1809	....
Cables, Apparatus for improving the quality of	N. Cutting	Washington, D. C	Aug. 14, 1806	....
Cables, Apparatus for working and stoppering chain.	W. H. Harfield	London, England	Aug. 5, 1862	36, 085
Cables, Apparatus for working and stoppering chain.	W. H. Harfield	London, England	Aug. 5, 1862	36, 086
Cables, Arrangement of means for working and stoppering chain.	T. Brown	London, England	July 25, 1854	11, 404
Cables, Elastic chain or surge-spring for ships'	A. H. Wright	Philadelphia, Pa	Mar. 13, 1860	27, 505
Cables from sudden strain, Apparatus for relieving ships'.	L. Bissell	Brooklyn, N. Y	Feb. 28, 1845	3, 932
Cables, Link-shackle of chain	W. J. Hotchkiss	Derby, Conn	Jan. 1, 1861	31, 021
Cables, Insulating submarine	G. B. Simpson	Washington, D. C	May 21, 1867	65, 019
Cables, Machine for adjusting chain	C. Hall	New York, N. Y	Aug. 14, 1866	57, 127
Cables, Machine for laying bridge	J. Gray	Cincinnati, Ohio	Dec. 26, 1871	122, 247
Cables, &c., Machine for proving	J. Judge	Washington, D. C	Nov. 29, 1828	....
Cables, Machine for wrapping wire	J. Gray	Cincinnati, Ohio	Dec. 26, 1871	122, 246
Cables, Making chain	E. Bartholomew	Boston, Mass	May 17, 1826	....
Cables, Pressure-stopper for chain	J. Emerson	Worcester, Mass	Sept. 25, 1855	131, 593
Cables, Surge-reliever for	J. Bingham	Philadelphia, Pa	July 24, 1860	29, 338
Caboose	J. Youle		May 25, 1795	....
Caboose	G. Youle	New York, N. Y	Mar. 21, 1806	....
Caboose	G. Youle	New York, N. Y	July 27, 1809	....
Caboose, Construction of iron	W. Ashbridge		June 28, 1803	....
Caboose, Ship's	J. Wilson	New York, N. Y	Aug. 17, 1843	3, 224
Cage, Eccentric	T. E. Rollins	Corning, N. Y	Dec. 27, 1864	45, 639
Cage, Wire	M. Grebner	New York, N. Y	June 10, 1873	139, 784
Caisson	A. G. Wolfgram	Apex, N. C	Nov. 25, 1873	145, 039
Caisson, Cast-iron	J. P. Duffey	Philadelphia, Pa	July 13, 1852	9, 113
Caisson for sinking pier	J. B. Eads	Saint Louis, Mo	Aug. 15, 1871	118, 118
Caissons, Dumb waiter for	H. Flad	Saint Louis, Mo	Nov. 22, 1870	109, 505
Caissons for founding piers of bridges and for laying pipes under water, Constructing	G. Daniels	Philadelphia, Pa	Feb. 28, 1834	....
Cake and bread, Box for preserving	A. R. Ledoux	Cornwall Landing, N. Y	Feb. 4, 1873	135, 433
Cake and fruit basket	C. H. Latham	Lowell, Mass	Aug. 15, 1871	118, 027
Cake baker, Griddle	J. Mull	Troy, N. Y	Apr. 3, 1866	53, 744
Cake-closet	G. A. Higgins	New York, N. Y	Apr. 5, 1870	101, 618
Cake-cutter	G. R. Peckham	Worcester, Mass	Nov. 17, 1857	18, 647
Cake-cutter	G. O. Sanderson	Boston, Mass	May 19, 1868	78, 137
Cake-cutter	J. Webster	Charlestown, Mass	May 20, 1873	139, 217
Cake-cutter and rolling-pin	I. N. Pyle	Decatur, Ind	Oct. 31, 1865	50, 732
Cake-cutting machine	D. J. Little	Albany, N. Y	Oct. 8, 1872	132, 034
Cake-machine	J. Repetti	Philadelphia, Pa	May 25, 1869	90, 577
Cake-mixer	T. Holmes	Williamsburgh, N. Y	Aug. 22, 1871	118, 243
Cake-mixer	J. Lafetra	New York, N. Y	Oct. 13, 1868	83, 069
Cake-pan	W. C. Butler	Louisville, Ky	Aug. 27, 1872	130, 843
Cake-pan	J. B. Firth	Brooklyn, N. Y	July 8, 1873	140, 619
Cake-pan	J. H. Smith	Brooklyn, N. Y	Aug. 2, 1870	105, 991
Cake-pan	J. H. Smith	Brooklyn, N. Y	Aug. 2, 1870	105, 992
Cake-stirrer	S. M. Clark	Beaver Dam, Wis	Nov. 28, 1871	121, 336
Cakes, Cutting and panning	J. H. Shrote	Baltimore, Md	Oct. 11, 1859	25, 767
Calcimines	E. B. Benedict	Susquehanna Depot, Pa	July 23, 1872	129, 645
Calcium-light, Method of producing	C. L. Coombs and J. A. Basset	Washington, D. C., and Salem, Mass.	Aug. 3, 1869	93, 178
Calcium-magnesium-light, Producing	C. A. Dresser	New York, N. Y	Dec. 10, 1867	71, 860
Calculating and registering machine	T. T. Strode	Mortonville, Pa	Feb. 4, 1868	74, 170
Calculating-apparatus	A. Johnson	Springfield, Mass	Dec. 22, 1868	85, 229
Calculating-balance	B. W. Ogburn	Whittles' Mills, Va	July 14, 1868	80, 003
Calculating device, Price	A. Sinclair	West Waterville, Me	Dec. 14, 1869	97, 974
Calculating interest, Machine for	J. Hatfield	Glen's Falls, N. Y	May 6, 1844	3, 574
Calculating-machine	J. B. Alexander	Baltimore, Md	Mar. 15, 1864	41, 898
Calculating-machine	E. D. Barbour	Boston, Mass	Aug. 13, 1872	130, 404
Calculating-machine	E. D. Barbour	Boston, Mass	Nov. 19, 1872	133, 188
Calculating-machine	J. J. Bavanowski	Poland	Sept. 5, 1848	5, 746
Calculating-machine	G. W. Edelman	Philadelphia, Pa	Dec. 22, 1846	4, 902
Calculating-machine	T. Esersky	St. Petersburg, Russia	Nov. 11, 1873	144, 523
Calculating-machine	G. B. Grant	Cambridge, Mass	July 16, 1872	129, 335
Calculating-machine	G. B. Grant	Cambridge, Mass	Apr. 29, 1873	138, 245
Calculating-machine	W. M. Haines	Rochester, N. Y	May 1, 1849	6, 403

Index of patents issued from the United States Patent Office from 1790 *to* 1873, *inclusive*—Continued.

Invention.	Inventor.	Residence.	Date.	No.
Calculating-machine	L. Keiler	Catawissa, Pa	Jan. 21, 1868	73, 449
Calculating-machine	J. A. Loomis and A. Johnson	Springfield, Mass	Jan. 28, 1868	73, 732
Calculating-machine	A. Mendenhall	Cerro Gordo, Ind	Aug. 13, 1867	67, 786
Calculating-machine	J. W. Nystrom	Philadelphia, Pa	Mar. 4, 1851	7, 961
Calculating-machine	D. D. Parmelee	New Paltz, N. Y	Feb. 5, 1850	7, 074
Calculating-machine	A. C. Pierson	Rahway, N. J	Mar. 12, 1867	62, 882
Calculating-machine	A. C. Pierson	Rahway, N. J	Feb. 4, 1868	73, 995
Calculating-machine	J. H. R. Reffelt	Hoboken, N. J	Sept. 14, 1869	94, 772
Calculating-machine	W. Robjohn	New York, N. Y	Aug. 6, 1872	130, 244
Calculating-machine	N. Rowland	Hilltown, Pa	Jan. 9, 1866	51, 972
Calculating-machine	R. Teasdale	Alberton, Ga	Dec. 5, 1871	121, 687
Calculating-machine	S. S. Young	Eaton, Ohio	July 24, 1849	6, 602
Calculator	A. Brody	Union, Mich	Apr. 26, 1870	102, 216
Calculator, Arithmetical	J. P. Miller	Boston, Mass	Sept. 1, 1863	39, 740
Calculator, Interest	B. C. Coldwell	Lockhaven, Pa	July 16, 1872	129, 531
Calculator, Tax	C. D. Crane	Fort Wayne, Ind	May 5, 1863	38, 374
Calculator, Tax	I. S. Hyatt	Rockford, Ill	May 12, 1868	77, 736
Calculi, Apparatus for removing	W. A. Dudley	Petersburgh, Va	Oct. 23, 1860	30, 471
Caldron	H. Newsham	Baltimore, Md	Feb. 12, 1856	14, 271
Caldron, Agricultural	E. E. Sill and A. H. Bennett	Rochester, N. Y	Nov. 29, 1870	109, 679
Caldron and furnace for agriculturists and others, Combined.	J. L. Mott	New York, N. Y	Dec. 1, 1840	1, 873
Caldron, steam-boiler, and furnace, Combined	L. E. Hopkins	New York, N. Y	Oct. 11, 1841	2, 306
Calendar	C. W. Bryan	Springfield, Mass	Feb. 26, 1867	62, 313
Calendar	G. L. Coburn	Hartford, Conn	Oct. 12, 1869	95, 655
Calendar	J. L. Dagg	Philadelphia, Pa	May 4, 1830	....
Calendar	Vve. A. Maurin and G. Toiray	Paris, France	May 17, 1870	103, 216
Calendar	W. McAdams	Newton, Mass	Sept. 26, 1871	119, 381
Calendar	J. M. Patton	Indianapolis, Ind	Jan. 26, 1864	41, 391
Calendar	J. F. Tannatt	Springfield, Mass	Mar. 1, 1870	100, 338
Calendar	A. S. Vose	Randolph, Vt	Mar. 17, 1868	75, 659
Calendar	W. P. Ware	New York, N. Y	July 17, 1866	56, 470
Calendar, Advertising	R. C. Ogden	New York, N. Y	May 7, 1872	126, 481
Calendar-catch	R. C. Ogden	New York, N. Y	June 24, 1873	140, 300
Calendar, Counting-house	J. Bryner	Peoria, Ill	Feb. 28, 1860	27, 267
Calendar for almanacs	J. H. Mead	New York, N. Y	Nov. 15, 1864	45, 064
Calendar, Hand-setting	C. H. Wight	Bal imore, Md	Mar. 29, 1870	101, 333
Calendar, Perpetual	S. L. Barinds	Saint Joseph, Mo	May 26, 1868	78, 356
Calendar, Perpetual	H. Harris	Newark, N. J	Aug. 7, 1866	56, 935
Calendar, Perpetual	H. W. Holly	Norwich, Conn	Jan. 3, 1865	45, 795
Calendar, Perpetual	C. T. Pooler	Deansville, N. Y	May 12, 1868	77, 837
Calendar, Pocket	D. E. Crosby	Brooklyn, N. Y	Feb. 16, 1864	41, 605
Calendar, Pocket	H. C. Foote	McGaheysville, Va	July 17, 1860	29, 156
Calendar, Portable	J. N. Dudley	Mitchell, Iowa	May 6, 1862	35, 143
Calendar, Writing-desk	S. J. Tucker	Philadelphia, Pa	June 22, 1869	91, 690
Calender-roll	G. G. Clark	Providence, R. I	Mar. 16, 1858	19, 623
Calender-rolls, Device for starting and stopping	W. T. Porter	Wilmington, Del	June 23, 1868	79, 080
Calender-rolls, Feed-regulator for	M. Lawler	Holyoke, Mass	Jan. 14, 1873	134, 810
Calender-rolls, Manufacturing	J. Worsley	Providence, R. I	Dec. 23, 1856	16, 319
Calendering-machine	H. E. Rogers	South Manchester, Conn	Oct. 5, 1869	95, 606
Calendering-machine	D. Scofield	Gloucester, N. J	Sept. 16, 1873	142, 872
Calf-weaner	T. A. R. Keech	Bladensburgh, Md	Dec. 26, 1871	122, 119
Calf-weaning device	J. J. Welling	Cedar Falls, Iowa	Dec. 10, 1861	33, 912
Calico, Machine for printing	J. Green	Lowell, Mass	Nov. 21, 1865	51, 039
Calico, Machine for rinsing	G. I. Prentiss	Tiverton, R. I	Apr. 4, 1846	4, 445
Calico, Roller for printing	J. Hope	Providence, R. I	Oct. 5, 1858	21, 723
Calipers	J. Atkins	Washington, D. C	Mar. 24, 1868	75, 827
Calipers	H. A. Boardman	New Haven, Conn	Feb. 5, 1867	61, 799
Calipers	C. E. Brown	Florence, Mass	June 20, 1865	48, 257
Calipers	J. H. Call	Springfield, Mass	May 28, 1861	32, 410
Calipers	J. H. Culver	San Francisco, Cal	Apr. 30, 1867	64, 203
Calipers	C. A. Fairfield	Springfield, Mass	July 21, 1863	39, 281
Calipers	T. Goodrum	Providence, R. I	Sept. 29, 1868	82, 514
Calipers	F. Gould	Huntington, N. Y	Apr. 12, 1859	23, 564
Calipers	W. P. Hopkins	Lawrence, Mass	Jan. 3, 1871	110, 657
Calipers	C. Jillson	Worcester, Mass	Aug. 16, 1864	43, 854
Calipers	B. G. Martin	Philadelphia, Pa	Dec. 12, 1865	51, 517
Calipers	G. L. McKnight	Worcester, Mass	May 28, 1867	65, 253
Calipers	W. H. Miner	Boston, Mass	Oct. 14, 1873	143, 584
Calipers	W. A. Morse	Boston, Mass	Jan. 26, 1864	41, 387
Calipers	T. C. Page and G. W. Hadley	Chicopee, Mass	Feb. 8, 1870	99, 698
Calipers	H. K. Porter	Boston, Mass	June 7, 1870	103, 926
Calipers	S. Sawyer	Fitchburgh, Mass	July 7, 1868	79, 603
Calipers	P. Seaver	Oxford, Mass	June 27, 1854	11, 180
Calipers	W. A. Sharpe	Syracuse, N. Y	Oct. 25, 1870	108, 640
Calipers	W. Siar	Franklin, Pa	Feb. 15, 1870	99, 961
Calipers	R. D. O. Smith	Washington, D. C	Dec. 20, 1864	45, 529
Calipers	P. Soper	Buffalo, N. Y	Aug. 28, 1866	57, 588
Calipers	C. D. Sutton	Kensico, N. Y	Jan. 3, 1860	26, 716
Calipers	D. C. Talbot	Worcester, Mass	Jan. 27, 1863	37, 531
Calipers	F. C. Washburn	Millville, Mass	Jan. 10, 1865	45, 894
Calipers	A. V. D. Westervelt	New Brunswick, N. J	Feb. 5, 1867	61, 902
Calipers	S. Whalen	Burnt Mills, N. Y	Dec. 26, 1865	51, 767
Calipers	A. E. Whitmore	Boston, Mass	July 5, 1870	105, 153
Calipers and dividers	M. G. Imbach	Hartford, Conn	Nov. 19, 1867	71, 178
Calipers and dividers	J. D. Moon	Chelsea, Mass	Sept. 7, 1858	21, 435
Calipers and dividers	L. Shelters	Manchester, N. H	July 30, 1867	67, 360
Calipers and dividers	E. Wright	Worcester, Mass	Sept. 24, 1867	69, 292
Calipers and dividers, Process of making	J. B. Gooding	Waltham, Mass	July 10, 1866	56, 323
Calipers and poising-tool, Jewelers' combined	F. E. Allen	Keene, N. H	Feb. 28, 1871	112, 203
Calipers and T-square	J. Bennor	Philadelphia, Pa	Dec. 10, 1867	71, 843
Calipers die, Spring	N. C. Smith	Berlin, Conn	Sept. 14, 1869	94, 920
Calipers for measuring irregular forms	E. T. Miller	Charlestown, Mass	May 27, 1856	14, 984
Calipers for taper surfaces	E. C. C. Kellogg	Hartford, Conn	July 16, 1872	129, 349
Calipers gage, Fixed	J. Richards	Cincinnati, Ohio	Oct. 15, 1867	69, 953
Calipers, rule, and wire-gage combined	F. A. Adams	Shelburne Falls, Mass	Jan. 5, 1869	85, 553
Calipers, Sliding	A. E. Whitmore	Boston, Mass	July 27, 1869	93, 026
Calipers, Sliding	A. E. Whitmore	Somerville, Mass	Apr. 5, 1870	101, 689
Calipers, Spring	W. W. Smith	Boston, Mass	Oct. 1, 1850	7, 689

Index of patents issued from the United States Patent Office from 1790 *to* 1873, *inclusive*—Continued.

Invention.	Inventor.	Residence.	Date.	No.
Calipers, Stem	W. C. Cross	Boston, Mass	July 16, 1872	129, 536
Calipers, Transverse	W. J. Van Ness	Baltimore, Md	Oct. 30, 1849	6, 841
Calk bar, Toe	R. B. Caswell	Springfield, Mass	Mar. 7, 1871	112, 321
Calk-sharpener	H. Kline	Marshalltown, Iowa	Dec. 1, 1868	84, 5[illegible]3
Calk-sharpening machine	G. W. Lane	Chichester, N. H	May 23, 1871	115, 068
Calks and blanks for the same, Manufacture of toe	T. Dooley	South Boston, Mass	June 2, 1868	78, 581
Calks, Blank or bar for toe	P. F. Burke	Worcester, Mass	Feb. 2, 1869	86, 503
Calking ships, &c., Composition for	H. Le Grand	Havre, France	Oct. 22, 1872	132, 365
Calligraph	C. Thurber	Worcester, Mass	June 23, 1857	17, 647
Calligraph	C. Thurber	Brooklyn, N. Y	Nov. 27, 1860	30, 777
Caloric-engine	W. T. Hallefas	New York, N. Y	Apr. 18, 1871	113, 764
Caloric-engine	A. A. Henderson	Portsmouth, N. H	Oct. 9, 1860	30, 315
Caloric-engine	H. Messer	Roxbury, Mass	Oct. 6, 1863	40, 178
Caloric-engine	N. A. Otto	Deutz, Germany	Dec. 2, 1873	145, 123
Caloric-engine	T. Schwartz	New York, N. Y	Dec. 20, 1864	45, 526
Caloric-engine	H. D. Wallen, jr	Fort Columbus, N. Y	Oct. 13, 1868	83, 114
Caloriferes	S. Whitmarsh	Northampton, Mass	Aug. 17, 1852	9, 206
Cam, Adjustable	C. F. Annan	Boston, Mass	Dec. 27, 1870	110, 537
Cam-groove-cutting machine	J. S. George and W. Houghtaling.	Bridgeport, Conn	Apr. 1, 1873	137, 358
Cam-groove, Machine for dressing	G. M. Pratt and V. A. King	Middletown, Conn	Feb. 4, 1873	135, 584
Cam-lever	S. Whitman	Londonderry, Vt	Feb. 15, 1870	99, 991
Cam-press	T. R. Hopkins	Petersburgh, Va	Aug. 23, 1859	25, 232
Cam-press	E. Thomas	Beverly, Va	Dec. 21, 1858	22, 387
Cam, Sectional adjustable slotted	F. D. Elderfield	Saint Louis, Mo	Aug. 5, 1873	141, 548
Cam, Variable	W. H. Andrews	New Haven, Conn	Nov. 26, 1861	33, 771
Cams for pressing and raising weights, &c., Combination of.	A. S. Greenville	Cambridgeport, Mass	Dec. 30, 1835	....
Caboose	J. Bouis	Baltimore, Md	Dec. 31, 1812	....
Camera:				
See Multiplying-camera.				
Photographic camera.				
Solar-camera.				
Stereoscopic camera.				
Camera and developing-box, Combined	E. C. Ratzell	Philadelphia, Pa	Mar. 7, 1871	112, 380
Camera-box support	O. Ackerman	Carthage, N. Y	Apr. 25, 1871	114, 082
Camera-boxes, Guide-slide for	O. Loehr	New York, N. Y	Apr. 2, 1872	125, 202
Camera focusing-attachment	A. St. Clair	Cedar Rapids, Iowa	Dec. 31, 1872	134, 402
Camera-obscura	G. F. Kolb	Philadelphia, Pa	Apr. 15, 1862	34, 970
Camera-screen	F. Peabody	Louisville, Ky	Feb. 1, 1870	99, 343
Camera-stand	N. S. Bowdish	Richfield Springs, N. Y	July 29, 1873	141, 256
Camera-stand	E. M. Corbett	New York, N. Y	June 24, 1862	35, 671
Camera-stand	A. R. Cribfield	Lincoln, Ill	Apr. 4, 1871	113, 267
Camera-stand	W. W. Dames	San José, Cal	Apr. 19, 1870	101, 985
Camera-stand	R. B. Douglas	Cleveland, Ohio	Apr. 12, 1864	42, 279
Camera-stand	J. W. Harper	Cleveland, Ohio	Dec. 5, 1865	51, 311
Camera-stand	B. Holler	Brooklyn, N. Y	Mar. 18, 1873	136, 997
Camera-stand	H. J. Lewis	Brooklyn, N. Y	Mar. 29, 1859	23, 428
Camera-stand	T. H. McBride and D. Baldwin	East Birmingham, Pa	Sept. 27, 1864	44, 438
Camera-stand	C. H. Snively	Millersburgh, Pa	Nov. 7, 1871	120, 790
Camera-stand	E. P. Spahn	Newark, N. J	May 7, 1872	126, 586
Camera-stand	I. H. Stoddard	Amenia, N. Y	Aug. 17, 1869	93, 763
Camera-stand	A. E. Turnbull	Upper Sandusky, Ohio	Sept. 7, 1869	94, 672
Camera-stand	J. A. Whipple	Boston, Mass	Aug. 18, 1863	39, 602
Camp-baker, Folding	F. Lehnen	Marquette, Mich	Sept. 30, 1873	143, 288
Camp-chair	P. J. Hardy	New York, N. Y	Aug. 20, 1861	33, 082
Camp-chest	A. S. Carner	Brooklyn, N. Y	Nov. 19, 1861	33, 734
Camp-chest	F. Graves	Avoca, N. Y	Oct. 4, 1864	44, 578
Camp-chest	C. W. Irwin	Saint Louis, Mo	Dec. 17, 1861	33, 946
Camp-chest	G. C. Lane	Buffalo, N. Y	Oct. 22, 1861	33, 529
Camp-chest	G. Parr	Buffalo, N. Y	June 25, 1861	32, 643
Camp chest and table combined	W. Chase	Buffalo, N. Y	Oct. 29, 1861	33, 570
Camp-kit	J. Walkey	Monroe, Wis	Dec. 20, 1864	45, 542
Camp pan and baker	C. Bush	New York, N. Y	Oct. 22, 1861	33, 514
Camp-stool	A. Ashald	Garrettsville, Ohio	Sept. 17, 1861	33, 286
Camp table or stool	J. Cram	Boston, Mass	May 12, 1863	38, 470
Camphor for preserving furs, &c., Preparing crude	W. F. Simes	Philadelphia, Pa	June 11, 1872	127, 807
Camphor, Preparing	W. F. Simes	Philadelphia, Pa	Mar. 11, 1873	136, 623
Camphor, Refining	H. Mead	New Haven, Conn	Feb. 12, 1810	....
Camphor, Refining and making	J. Green	Charlestown, Mass	Apr. 27, 1812	....
Camphor-wash mixture	C. W. Crozier	Knoxville, Tenn	July 11, 1854	11, 247
Can:				
See Alkali-can.				
Armored can.				
Butter-can.				
Confectioner's can.				
Cotton-can.				
Cotton or woolen can.				
Cotton-sliver can.				
Dairy-can.				
Dinner-can.				
Dredging-can.				
Fruit-can.				
Grocer's can.				
Hydrocarbon-liquid can.				
Lard-can.				
Lard and packing can.				
Liquid-can.				
Marking-can.				
Measuring-can.				
Metal can.				
Metallic can.				
Milk-can.				
Nitro-glycerine can.				
Oil-can.				
Oil-dripping can.				
Organ-can.				
Oyster-can.				
Oyster-packing can.				
Packing-can.				

Index of patents issued from the United States Patent Office from 1790 *to* 1873, *inclusive*—Continued.

Invention.	Inventor.	Residence.	Date.	No.
Can—Continued.				
See Packing and atomizing can.				
Paint-can.				
Preserving-can.				
Rectangular can.				
Roofing-can.				
Roving-can.				
Safety-can.				
Sealed can.				
Sealing-can.				
Seamless can.				
Sheet-metal can.				
Shipping-can.				
Sirup-can.				
Spice-can.				
Sprinkling-can.				
Tea-can.				
Tin can.				
White-lead can.				
Wood-incased can.				
Wooden-cased can.				
Can	A. D. Armstrong	Pittsburgh, Pa	Sept. 24, 1867	69, 154
Can	C. A. Murdock	Milwaukee, Wis	May 20, 1873	139, 181
Can and bottle, Sealing	J. D. Willoughby	Carlisle, Pa	Jan. 4, 1859	22, 535
Can and jar, Sealing	J. Bellerjeau	Philadelphia, Pa	Mar. 31, 1868	76, 149
Can-cap, Uncut	J. I. Livingston	Pittsburgh, Pa	Mar. 2, 1869	87, 415
Can-cover lock	H. W. Shepard	New York, N. Y	Apr. 9, 1872	125, 622
Can-cover lock	H. W. Shepard	New York, N. Y	Apr. 9, 1872	125, 623
Can-filler	R. Newton	Millville, N. J	May 6, 1873	138, 574
Can filling and soldering apparatus	L. C. Straub	Pittsburgh, Pa	June 20, 1871	116, 114
Can-handle	L. A. Sunderland	Madison, Ohio	Nov. 30, 1869	97, 326
Can-hook	G. Webber	Portland, Me	Sept. 11, 1849	6, 702
Can-making die	J. L. Gray	Baltimore, Md	Apr. 2, 1867	63, 503
Can-manufacture	J. T. Ackley and J. K. Trux	Philadelphia, Pa	Apr. 10, 1866	53, 765
Can-opener	H. C. Alexander	New York, N. Y	Nov. 16, 1869	96, 761
Can-opener	H. C. Alexander	New York, N. Y	Jan. 25, 1870	99, 046
Can-opener	R. H. Atwell	Baltimore, Md	Feb. 9, 1869	86, 626
Can-opener	A. Barker	Wyoming, Pa	May 17, 1870	103, 125
Can-opener	F. G. Beach	Hartford, Conn	Apr. 6, 1869	88, 536
Can-opener	W. M. Bleakley	Verplank, N. Y	June 29, 1869	91, 902
Can-opener	W. M. Bleakley	Verplank, N. Y	Oct. 19, 1869	95, 873
Can-opener	S. O. Church	West Meriden, Conn	Jan. 15, 1867	61, 161
Can-opener	M. C. Davis	Folsom, Cal	July 13, 1869	92, 520
Can-opener	E. F. Dewey	San Francisco, Cal	Sept. 28, 1869	95, 205
Can-opener	E. M. Dewey	San Francisco, Cal	Aug. 29, 1871	118, 593
Can-opener	W. L. Hubbell	Brooklyn, N. Y	Oct. 22, 1867	69, 996
Can-opener	G. C. Humphreys	Washington, D. C	Nov. 17, 1868	84, 122
Can-opener	G. G. Joyce	Baltimore, Md	Aug. 10, 1869	93, 541
Can-opener	J. Kaufman	New York, N. Y	Sept. 6, 1870	107, 061
Can-opener	O. J. Livermore	Worcester, Mass	June 26, 1866	55, 878
Can-opener	W. W. Lyman	Meriden, Conn	July 12, 1870	105, 346
Can-opener	W. W. Lyman	West Meriden, Conn	July 19, 1870	105, 583
Can-opener	T. A. McFarland	Meadville, Pa	May 21, 1867	64, 891
Can-opener	C. J. C. Petersen	Port Chester, N. Y	June 17, 1873	140, 072
Can-opener	A. C. Platt	Sandusky, Ohio	Aug. 23, 1870	106, 723
Can-opener	J. J. Reed	Lyons, Iowa	Mar. 25, 1873	137, 149
Can-opener	C. F. Ritchel	Chicago, Ill	May 12, 1868	77, 916
Can-opener	L. B. Smith	West Meriden, Conn	June 17, 1873	140, 088
Can-opener	N. F. Stone	Chicago, Ill	Apr. 14, 1868	76, 669
Can-opener	W. Thomas	Geneseo, Ill	Nov. 26, 1872	133, 500
Can-opener	S. E. Totten	Brooklyn, N. Y	Jan. 22, 1867	61, 484
Can-opener	E. J. Warner	Waterbury, Conn	Jan. 5, 1858	19, 063
Can-opener	J. A. Wells	Holly Springs, Miss	Aug. 10, 1869	93, 505
Can-opener	J. Wood	New York, N. Y	July 8, 1873	140, 604
Can-opener	F. S. Wyman	Chicago, Ill	July 28, 1868	80, 326
Can-opener and knife or fork, Combined	T. Kenderdine	Lisbon, Iowa	Aug. 26, 1873	142, 109
Can-opener and pipe-cutter	D. A. Barnes	Chicago, Ill	Oct. 28, 1873	144, 051
Can-opening machine	W. H. Forker	Meadville, Pa	Oct. 29, 1867	70, 188
Can-opening tool	G. A. Dickson	Woodcock Township, Pa	Dec. 24, 1867	72, 464
Can-opening tool	M. T. McCormick	Meadville, Pa	Apr. 7, 1868	76, 490
Can-opening tool	E. T. Orne	Chicago, Ill	Nov. 6, 1866	59, 513
Can or bottle stopper	J. Drenton	Philadelphia, Pa	Dec. 23, 1862	37, 221
Can or canister top	A. Bliss	Newark, N. J	Oct. 21, 1851	8, 440
Can or flask	J. Dunton	Philadelphia, Pa	Mar. 3, 1863	37, 843
Can screw-tap	L. R. Boyd	New York, N. Y	June 13, 1871	115, 927
Can-sealing apparatus	H. Y. Wildey	Philadelphia, Pa	Oct. 16, 1860	30, 410
Can-top	L. F. Betts	Chicago, Ill	Sept. 1, 1868	81, 585
Can-top	C. E. Dayton	Meriden, Conn	May 16, 1871	114, 931
Cans and bottles, Elastic cap for sealing	R. Davis	Brookhaven, N. Y	Feb. 24, 1857	16, 684
Cans and boxes, Manufacturing	J. W. Millett	Batchellerville, N. Y	Oct. 18, 1864	44, 737
Can and kettle bails, Construction of ears for	T. Evans	Newark, N. J	Apr. 14, 1868	76, 728
Cans and vessels, Apparatus for exhausting air from and hermetically sealing.	A. M. Purnell	Washington, D. C	Nov. 25, 1856	16, 126
Cans, Apparatus for sealing	J. Green	Cincinnati, Ohio	July 11, 1854	11, 297
Cans, barrels, &c., Apparatus for testing	W. D. Brooks	Baltimore, Md	Dec. 5, 1871	121, 581
Cans, Construction of	H. W. Millar	Utica, N. Y	Sept. 19, 1865	50, 022
Cans for transportation, Packing	J. F. Drummond	New York, N. Y	Feb. 25, 1862	34, 547
Cans, &c., Handle for	N. Bray	Boston, Mass	Feb. 2, 1869	86, 500
Cans, &c., Lid or cover for	J. Bryant	Brooklyn, N. Y	Nov. 10, 1863	40, 552
Cans, &c., Machine for applying screw mouth-pieces to.	H. O. Lothrop	Milford, Mass	Aug. 28, 1866	57, 524
Cans, Machine for fastening bottoms to	R. Brady	New York, N. Y	Sept. 28, 1869	95, 190
Cans, Machine for heading	E. Norton and A. H. Fancher	Chicago, Ill	Oct. 29, 1872	132, 595
Cans, Machine for holding and soldering	J. Redheffer	Kansas City, Mo	Feb. 25, 1873	136, 265
Cans, Machine for making sheet-metal screw-necks and screw-caps for.	J. H. Stone	Hamilton, Canada West	Nov. 28, 1871	121, 435
Cans, &c., Machinery for laying roving in	J. W. Strange	Taunton, Mass	Oct. 23, 1847	5, 337
Cans, Method of hermetically sealing	W. Borrman	Cincinnati, Ohio	Aug. 25, 1857	18, 035

Index of patents issued from the United States Patent Office from 1790 *to* 1873, *inclusive*—Continued.

Invention.	Inventor.	Residence.	Date.	No.
Cans, Mode of laying roving in	J. Tatham and D. Cheetham	Rochdale, England	Nov. 18, 1845	4, 277
Cans, Safety-cap for	H. C. Alexander	New York, N. Y	June 21, 1870	104, 402
Cans, Safety-cap for	H. C. Alexander	New York, N. Y	July 12, 1870	105, 157
Cans, Screw-cap for	A. N. Lapierre	New York, N. Y	Nov. 2, 1869	96, 447
Cans, Seam for	A. L. Webster	Cleveland, Ohio	July 5, 1870	105, 152
Cans, Self-locking lid for	J. W. Weimel and W. Domer	Pittsburgh, Pa	Oct. 22, 1872	132, 507
Cans while being soldered, Device for holding	T. Kerr and J. C. Kelly	Edinburgh, Ind	Apr. 7, 1868	76, 330
Cans with tomatoes, &c., Apparatus for filling	C. S. Bucklin	Red Bank, N. J	Oct. 14, 1873	143, 613
Canning and preserving meat, &c., Process and apparatus for.	N. H. Shipley	Baltimore, Md	Feb. 6, 1872	123, 366
Canal	W. H. Doane	Cincinnati, Ohio	Sept. 5, 1871	118, 699
Canal	H. Hill	Cabin Creek, W. Va	Apr. 22, 1873	138, 156
Canal and navigation thereof	H. H. Day	New York, N. Y	Mar. 12, 1867	[illegible]
Canal and other boats, Machine for ascertaining the weight of cargoes in.	A. Amsden	Rochester, N. Y	June 27, 1838	810
Canal and river boat	S. Doolittle	Utica, N. Y	Nov. 19, 1833	....
Canal and river boat	A. Plantou	Philadelphia, Pa	Mar. 18, 1835	....
Canal and river lock	J. Davies	Schuylkill Haven, Pa	Nov. 6, 1860	30, 564
Canal-boat	B. T. Babbitt	New York, N. Y	June 11, 1872	127, 826
Canal-boat	T. W. Bakewell	Cincinnati, Ohio	Aug. 25, 1829	....
Canal-boat	E. C. Bancroft	Syracuse, N. Y	Jan. 2, 1872	122, 351
Canal-boat	J. Bromwell	Cincinnati, Ohio	June 23, 1829	....
Canal-boat	J. C. Cook	New Haven, Conn	July 9, 1872	128, 713
Canal-boat	J. M. Dodge	Newark, N. J	Aug. 6, 1872	130, 110
Canal-boat	J. English	Syracuse, N. Y	Nov. 28, 1871	121, 346
Canal-boat	J. E. Gibson	Port Carbon, Pa	July 20, 1858	20, 944
Canal-boat	H. J. Hatch	Chicago, Ill	Apr. 30, 1872	126, 204
Canal-boat	C. W. Hermance	Schuylersville, N. Y	Jan. 2, 1872	122, 463
Canal-boat	D. Hinchy	Santiago, Chili	Feb. 27, 1872	124, 060
Canal-boat	R. Hooper	Baltimore, Md	July 25, 1871	117, 419
Canal-boat	J. Huges	New Berne, N. C	Feb. 27, 1872	124, 136
Canal-boat	W. A. Leggo	Montreal, Canada	May 6, 1873	138, 664
Canal-boat	J. McCausland	Kingston, N. Y	June 14, 1834	....
Canal-boat	J., J., and J. McCausland	Kingston and Esopus, N. Y	Sept. 21, 1858	21, 572
Canal-boat	N. H. Murray	Louisville, Ky	June 4, 1872	127, 424
Canal-boat	W. H. Newell	Jersey City, N. J	Jan. 23, 1872	123, 039
Canal-boat	I. M. Perry	Slate Cut, Ind	Mar. 12, 1872	124, 622
Canal-boat	B. Phillips	Philadelphia, Pa	Apr. 21, 1828	....
Canal-boat	E. Randolph	Salt Creek, Ohio	Jan. 9, 1838	560
Canal-boat	E. Reed	Pottstown, Pa	Apr. 6, 1833	....
Canal-boat	H. Wickham, jr	Chicago, Ill	Jan. 16, 1872	122, 874
Canal-boat and other vessels for the transportation of grain.	D. E. Somes	Washington, D. C	Oct. 6, 1868	82, 887
Canal-boat fender	J. Rowland	Hancock, Md	June 5, 1847	5, 147
Canal-boat for transshipment of merchandise	J. O. Conner	Philadelphia, Pa	Oct. 14, 1835	....
Canal-boat stanchion	C. Van Name	Binghamton, N. Y	June 11, 1861	32, 544
Canal-boat, Steam	D. W. Crocker	Philadelphia, Pa	May 3, 1834	....
Canal-boat, Steam	J. Elgar	Baltimore, Md	Nov. 7, 1835	....
Canal-boat, Steam	R. Hugunin	Albany, N. Y	Mar. 24, 1823	....
Canal-boat, Steam	E. Lynch	Georgetown, D. C	Oct. 15, 1872	132, 300
Canal-boat, Steam	G. Parker	Worcester, Mass	Mar. 27, 1849	6, 239
Canal-boat, Steam	J. F. Wright	Erie, Pa	Feb. 14, 1828	....
Canal-boats and other vessels, Method of unloading.	W. Loughridge	Weverton, Md	May 9, 1854	10, 891
Canal-boats, Arrangement of lever and catch for tow-line of.	J. W. Caldwell	Rochester, N. Y	Nov. 14, 1854	11, 957
Canal-boats, Attaching paddle-wheels to	R. Jane	Otego, N. Y	Nov. 24, 1857	18, 694
Canal-boats, Attachment of the tow-line of	C. W. Gage	Homer, N. Y	Apr. 21, 1863	38, 224
Canal-boats, Attachment of whiffletree to the tow-line of.	A. Clow	Port Byron, N. Y	May 20, 1862	35, 300
Canal-boats, barges, &c., Constructing and applying locomotive-engines to.	C. Bonycastle	Charlottesville, Va	Jan. 21, 1834	....
Canal-boats, Construction of	A. Judson	Unadilla, N. Y	Aug. 17, 1858	21, 201
Canal-boats, Construction of	G. B. Martin	New York, N. Y	Jan. 7, 1873	134, 555
Canal-boats, Construction of	A. Van Order	Ithaca, N. Y	Sept. 10, 1861	33, 272
Canal-boats, Construction of	B. Williams	Mendon, N. Y	Mar. 3, 1829	....
Canal-boats, Deck for	A. A. Bissell	Lockport, N. Y	Oct. 30, 1866	59, 171
Canal-boats for carrying horses by which they are towed, Construction of.	J. H. Long	Lewistown, Pa	Apr. 5, 1838	676
Canal-boats for repairing, Machine for raising	N. R. Penrose and S. F. Palmer	Beaver Meadow, Pa	May 11, 1839	1, 143
Canal-boats from water, Arrangement of means for freeing.	W. Loughridge	Weverton, Md	July 11, 1854	11, 293
Canal-boats in sections and manner of uniting them, Constructing.	R. Frazer	Waynesburgh, Pa	Sept. 10, 1840	1, 776
Canal-boats, &c., Lessening friction in	P. Freeman	Morris County, N. J	June 1, 1825	....
Canal-boats, Machine for moving	W. R. King	Washington, D. C	Feb. 1, 1870	99, 323
Canal-boats, Machinery for towing	O. Dehesnil and M. Eyth	Brussels, Belgium, and Stuttgart, Wurtemberg.	May 31, 1870	103, 724
Canal-boats, Machinery for weighing	B. Bull	New York, N. Y	June 20, 1826	....
Canal-boats, Means of snubbing	T. W. Edgar	Espy, Pa	Mar. 18, 1873	136, 982
Canal-boats, Method of discharging cargo from	A. Young	Georgetown, D. C	May 2, 1854	10, 843
Canal-boats, Mode of unloading	T. Sharp	Chicago, Ill	Aug. 25, 1863	39, 683
Canal-boats on inclined planes, Raising and lowering.	W. Knight	Morristown, N. J	Aug. 5, 1828	....
Canal-boats or sections thereof, Revolving cradle for unloading.	J. Elgar and B. Hallowell	Baltimore, Md., and Alexandria, Va.	Apr. 10, 1849	6, 303
Canal-boats out of the water for repair, Machine for lifting.	S. C. Jones	Rochester, N. Y	June 16, 1826	....
Canal-boats, Raising and lowering	W. Wiard	Avon, N. Y	May 1, 1828	....
Canal-boats, Sheet-iron twin for	L. Parmelee	Poughkeepsie, N. Y	Nov. 26, 1835	....
Canal-boats, ships, &c., Raising	W. W. Smith	Rochester, N. Y	Feb. 23, 1830	....
Canal-boats, Singletree safety-hook for towing	T. Jackson	Reading, Pa	June 25, 1839	1, 196
Canal-boats to be repaired, Machine for raising	S. Doolittle	Rochester, N. Y	Oct. 25, 1825	....
Canal-boats, Tow-line attachment for	C. Bentz	Mindenville, N. Y	Sept. 13, 1864	44, 154
Canal-boats, Tow-line for	J. Sanford	Weston, Conn	Feb. 16, 1829	....
Canal-boats, Towing	P. Cooper	New York, N. Y	Jan. 24, 1820	....
Canal-boats, Towing	C. T. Harvey	Tarrytown, N. Y	Feb. 2, 1869	86, 395
Canal-boats, Towing	J. L. Simms and W. T. Duvall	Georgetown, D. C	July 30, 1872	130, 080

Index of patents issued from the United States Patent Office from 1790 *to* 1873, *inclusive*—Continued.

Invention.	Inventor.	Residence.	Dato.	No.
Canal-boats, Towing	I. P. Tice	New York, N. Y	Feb. 25, 1873	136, 343
Canal-boats, Towing-apparatus for	W. O. Buchanan	Montreal, Canada	Mar. 5, 1872	124, 326
Canal-boats, Towing-apparatus for	A. H. Emery and G. Leverich	New York, N. Y	June 21, 1870	104, 459
Canal-boats, Towing-apparatus for	S. F. Palmer	New York, N. Y	Apr. 26, 1853	9, 687
Canal-boats, Towing-apparatus for	S. W. and J. F. Palmer	Auburn, N. Y	Nov. 8, 1870	109, 045
Canal-boats, Towing-hook for	J. and A. H. Doty	West Falls, N. Y	May 24, 1870	103, 310
Canal-digging machine	J. Humes	Richmond, Va	Nov. 10, 1821	....
Canal-gates, Mode of closing wickets in	D. N. Kownover	Danville, Pa	June 13, 1854	11, 077
Canal-lock	G. Bender	New York, N. Y	Aug. 26, 1822	....
Canal-lock	M. Bishop	Putnam, Ohio	Sept. 17, 1867	68, 833
Canal-lock	J. Burt	Detroit, Mich	May 28, 1867	65, 054
Canal-lock	H. H. Day	New York, N. Y	Oct. 8, 1867	69, 659
Canal-lock	J. W. Gentry and G. W. Barcus.	Peytona, W. Va	July 4, 1871	116, 583
Canal-lock	G. Heath	Little Falls, N. Y	June 25, 1867	66, 151
Canal-lock	G. Heath	Annapolis, Md	Dec. 6, 1870	109, 899
Canal-lock	D. Rodgers	Little Falls Town, N. Y	Jan. 4, 1828	....
Canal-lock	S. J. Seely	Brooklyn, N. Y	Dec. 23, 1862	37, 265
Canal-lock	I. Townsend	Capeville, Va	May 28, 1872	127, 389
Canal-lock	W. W. Virdin	Havre de Grace, Md	Jan. 20, 1852	8, 668
Canal-lock	J. White	Philadelphia, Pa	Oct. 19, 1819	....
Canal-lock	D. Wilkinson	Cohoes, N. Y	May 14, 1836	....
Canal-lock	C. W. Williams	Port Jervis, N. Y	Nov. 29, 1859	26, 314
Canal-lock gate	V. Brown	Clifton Park, N. Y	May 14, 1836	....
Canal-lock gate	R. English	Legro, Ind	July 1, 1841	2, 154
Canal-lock gate	F. Livingston	Waterford, N. Y	Apr. 13, 1838	687
Canal-lock gate	C. Neer	Troy, N. Y	Mar. 9, 1852	8, 789
Canal-lock gate	S. J. Seely	New York, N. Y	Nov. 3, 1857	18, 555
Canal-lock gate	S. S. Seely	Albany, N. Y	May 8, 1860	28, 201
Canal-lock gate	R. Taylor	New York, N. Y	Mar. 26, 1861	31, 836
Canal-lock gate	D. Wilkinson	Cohoes, N. Y	Aug. 17, 1835	....
Canal-lock gate	C. W. Williams	Port Jervis, N. Y	Apr. 12, 1859	23, 635
Canal-lock gate	G. W. Wood and L. O. Webster	Utica, N. Y	June 13, 1854	11, 103
Canal-lock-gate step	H. Rexford	Sandy Hill, N. Y	Apr. 28, 1868	77, 405
Canal-lock gates, Device for raising	W. Thomas	Ottawa, Ill	Apr. 4, 1865	47, 139
Canal-lock gates, Operating	K. T. Hurlburt and H. Thompson.	Port Byron, N. Y	Oct. 9, 1860	30, 323
Canal-lock gates, Valve or wicket for	G. Heath	Little Falls, N. Y	July 1, 1862	35, 756
Canal-lock gates, Wicket for	A. H. Griggs	Newark, N. J	May 17, 1864	42, 761
Canal-lock indicator	V. Brown	Clifton Park, N. Y	May 14, 1836	....
Canal-lock mechanism	G. Heath	Annapolis, Md	Dec. 27, 1870	110, 459
Canal-lock paddle-gate	J. F. King	Waterford, N. Y	Nov. 29, 1828	....
Canal-locks, Making paddle-gates for	D. Rodgers	Little Falls, N. Y	Feb. 17, 1826	....
Canal-locks, Preserving wooden	S. Guilford	Lebanon, Pa	July 10, 1826	....
Canal-locks, Saving water in	J. P. Fairlamb	Wilmington, Del	Nov. 27, 1823	....
Canal-locks, Saving water in passing boats through	J. Dewees	Pottsville, Pa	Dec. 31, 1832	....
Canal-locks, &c., Sliding valve for	W. Lake	Richmond, Va	June 7, 1839	1, 163
Canal-locks, Sluice-gate for	G. W. Hildreth	Lockport, N. Y	Mar. 19, 1840	1, 517
Canal-locks, Wicket-gate for	W. L. Potter	Clifton Park, N. Y	Aug. 31, 1839	1, 310
Canal-navigation by boats	S. D. Ingham	New Hope, Pa	Apr. 19, 1831	....
Canal stop-dam	I. Knapp	Medina, N. Y	May 13, 1862	35, 242
Canal waste-way and sluice-gate	A. J. Whitney	Harrisburgh, Pa	Dec. 6, 1870	109, 984
Canal-wicket	J. D. Marshbank	Lancaster, Pa	Feb. 2, 1864	41, 463
Canals and railways, Transportation on	J. Elgar	Baltimore, Md	Nov. 7, 1835	....
Canals and removing earth, Digging	J. Brainard	Rome, N. Y	Feb. 10, 1826	....
Canals, Balance-lock for	J. White	Philadelphia, Pa	May 17, 1839	1, 155
Canals, Constructing	J. Dougherty	Philadelphia, Pa	Feb. 24, 1843	2, 973
Canals, Discharging water from floating dry-docks for.	S. F. Palmer	New York, N. Y	Feb. 1, 1859	22, 818
Canals, Double-inclined plane for	A. Tanner	Warren, Pa	Apr. 15, 1825	....
Canals, Feed-regulator for	C. Ross	West Buddick, Ohio	May 21, 1850	7, 384
Canals, Feeding	J. G. Brewer	Lone Tree, Nebr	Nov. 26, 1872	133, 353
Canals, &c., from being closed by ice, Mode of preventing.	R. A. Chesebrough	New York, N. Y	June 1, 1869	90, 727
Canals, Machine for ditching and excavating ground for.	G. Henricks	Urbana, Ohio	Aug. 5, 1829	....
Canals, Machine for raising or lowering boats on	E. Morris	Bloomfield, Pa	Oct. 13, 1829	....
Canals, &c., Means of keeping open the navigation of	J. Mullaly	New York, N. Y	Oct. 21, 1873	143, 833
Canals, Mode of hauling earth from	O. Phelps	Lansing, N. Y	July 16, 1827	....
Canals, Mode of raising boats up and down elevations in.	R. Graves	Brooklyn, N. Y	July 26, 1827	....
Canals, Safety-gate for	J. Van Dorn	Glen, N. Y	May 14, 1827	....
Canceling-apparatus	T. S. Hudson	East Cambridge, Mass	Oct. 30, 1866	59, 225
Cancer, Method of drawing and curing	J. Ware	Philadelphia, Pa	Sept. 20, 1815	....
Candle	J. L. Field	Kensington, Great Britain.	May 12, 1868	77, 725
Candle	H. Ryder	New Bedford, Mass	Oct. 17, 1871	120, 105
Candle and lamp wick, Treating	T. S. Williams and F. A. Taber.	Boston, Mass	May 24, 1870	103, 534
Candle and ornament holder for Christmas-trees	G. Anton	Philadelphia, Pa	Apr. 27, 1869	89, 270
Candle apparatus, Mold	G. Kendall	Providence, R. I	May 3, 1853	9, 696
Candle-cap	J. H. Foote	Pittsfield, Mass	Nov. 17, 1868	84, 103
Candle-cutting apparatus	J. Jones	Brooklyn, N. Y	Apr. 15, 1856	14, 662
Candle-dipping machine	J. Aborn	Burlington County, N. J	Mar. 20, 1821	....
Candle-dipping machine	J. A. and A. F Jones	Lexington, Ky	May 16, 1848	5, 576
Candle-dipping machine	C. A. McPhetridge	Saint Louis, Mo	Dec. 9, 1856	16, 211
Candle-dipping machine	V. Squarza	New York, N. Y	Mar. 4, 1856	14, 376
Candle-guard	B. Morgan	San Francisco, Cal	July 22, 1873	141, 066
Candle-holder	E. Daire	Amiens, France	July 19, 1864	43, 628
Candle holder	C. Kirchof	Newark, N. J	Dec. 24, 1867	72, 506
Candle-holder	M. Schall	New York, N. Y	Sept. 24, 1867	69, 254
Candle-holder	J. Verch	Albany, N. Y	July 10, 1866	56, 293
Candle-holder, Economical	P. Ward and B. Allison	Philadelphia, Pa	July 20, 1812	....
Candle machine, Mold	W. C. Childs	Boston, Mass	Oct. 28, 1856	15, 968
Candle machine, Mold	B. D. Sanders	Holliday's Cove, Va	Mar. 3, 1857	16, 754
Candle machines, Construction of	M. Massey	Cleveland, Ohio	Apr. 24, 1860	27, 995
Candle-making	F. Fuller	New York, N. Y	Sept. 29, 1821	....
Candle-making	B. F. Shelabarger	Mifflintown, Pa	June 5, 1847	5, 139
Candle-making apparatus	W. Humiston	Troy, N. Y	Dec. 23, 1851	8, 607
Candle-making machine	E. Cowles	Hounslow, England	Jan. 19, 1869	86, 059

Index of patents issued from the United States Patent Office from 1790 to 1873, inclusive—Continued.

Invention.	Inventor.	Residence.	Date.	No.
Candle-making machine	J. Jones	Baltimore, Md	Oct. 26, 1858	21 882
Candle-making machine	J. Stainthorp	Buffalo, N. Y	Mar. 6, 1855	12, 492
Candle-making, Preparing fats for	M. W. Brown	Buffalo, N. Y	Oct. 13, 1857	18, 381
Candle-manufacture	J. G. Davis	Buffalo, N. Y	Aug. 13, 1850	7, 559
Candle-manufacturing machine	H. Grambo	Philadelphia, Pa	Jan. 1, 1867	60, 718
Candle, Many-wicked	B. D. Sanders	Holliday's Cove, Va	Sept. 30, 1856	15, 821
Candle, Miner's	P. R. Gottsein	Houghton, Mich	Apr. 12, 1870	101, 857
Candle, Mold	J. Dunlap	New Holland, Pa	Mar. 8, 1836	....
Candle, Mold	H. Halvorson	Cambridge, Mass	Dec. 13, 1859	26, 429
Candle, Mold and dipped	J. Zwisler	New York, N. Y	July 1, 1814	....
Candle-mold apparatus	L. C. Ashley	Troy, N. Y	Dec. 25, 1855	13, 973
Candle-mold apparatus	H. Camp	Dunkirk, N. Y	Jan. 22, 1850	7, 033
Candle-mold apparatus	W. Humiston	Troy, N. Y	Apr. 4, 1854	10, 730
Candle-mold apparatus	W. Humiston	Troy, N. Y	July 24, 1855	13, 334
Candle-mold box	G. A. Stanley	Cleveland, Ohio	May 8, 1860	28, 210
Candle-mold machine	L. C. Ashley	Troy, N. Y	Jan. 9, 1855	12, 193
Candle-mold machine	D. E. and M. Battershall	Troy, N. Y	Dec. 20, 1853	10, 331
Candle-mold machine	A. Hengstenberg	Muscatine, Iowa	Nov. 11, 1856	16, 056
Candle-mold tip	I. Cole	Brooklyn, N. Y	Apr. 29, 1873	138, 380
Candle-molds, Machine for making	M. Burlingame	Garrattsville, N. Y	Dec. 7, 1869	97, 749
Candle-molds, Tubular plunger for	H. Ryder	New Bedford, Mass	Nov. 6, 18 6	59, 460
Candle-molding	J. Drummond	New York, N. Y	Feb. 20, 1846	4, 389
Candle-molding apparatus	J. Gamble and J. S. Hill	Cincinnati, Ohio	Dec. 30, 1841	2, 405
Candle-molding apparatus	G. A. Stanley	Cleveland, Ohio	May 8, 1860	28, 209
Candle-molding apparatus	G. A. Stanley	Cleveland, Ohio	July 31, 1860	29, 416
Candle-molding machine	T. King	Troy, N. Y	Nov. 6, 1860	30, 605
Candle-molding machine	J. Robingson	New Brighton, Pa	Sept. 2, 1856	15, 668
Candle-molding machine	G. Roth	Cincinnati, Ohio	Oct. 28, 1862	36 798
Candle-molding machine	J. Wales	New York, N. Y	Mar. 28, 1871	113, 117
Candle-molding machinery	J. H. Tuck	Nantucket, Mass	July 11, 1837	265
Candle or taper holder, Wax	J. Fritsch	Carlstadt, N. J	Aug. 8, 1871	117, 763
Candle-snuffers	O. W. Stow and A. Barnes	Southington, Conn	Nov. 24, 1857	18, 713
Candle, Tubular	F. Taber	Boston, Mass	May 24, 1870	103, 388
Candle-wick	J. H. Tatum	New York, N. Y	Aug. 21, 1860	29, 729
Candle-wick	C. A. Wortendyke	Godwinville, N. J	Mar. 30, 1852	8, 849
Candle-wick	C. A. Wortendyke	Godwinville, N. J	Jan. 1, 1861	31, 049
Candle-wick, Cotton-yarn	G. Dickinson	New York, N. Y	Feb. 21, 1828	....
Candle-wick, Preparation of	S. K. Weeden	Providence, R. I	Aug. 23, 1859	25, 227
Candle-wicks, Dipping and cutting	W. Morey	Worcester, Mass	Aug. 15, 1835	....
Candles, Apparatus for making dipped	P. R. Gottstein	Houghton, Mich	Dec. 10, 1867	72, 019
Candles, Apparatus for making mold	A. L. Brown	New Haven, Conn	Oct. 2, 1849	6, 759
Candles, Apparatus for molding	H. Leonard and H. Ryder	New Bedford, Mass	Dec. 13, 1859	26, 440
Candles, Apparatus for molding	A. Meucci	Clifton, N. Y	Sept. 25, 1860	30, 180
Candles by machinery, Dipping	W. Day	Gardiner, Me	Mar. 11, 1826	....
Candles by machinery, Making	J. M. Tard	Trenton, N. J	Dec. 4, 1821	....
Candles, Casting or molding spermaceti	W. W. Swain	New Bedford, Mass	Jan. 14, 1820	....
Candles, Composition for	J. L. Klein	New York, N. Y	Nov. 22, 1864	45, 160
Candles, Composition for and mode of making	P. Farrell	New York, N. Y	Dec. 16, 1833	....
Candles, Composition for coating	C. Morfit	New York, N. Y	Jan. 10, 1860	26, 780
Candles, Drip-cup for	D. M. Ayer	Lawrence, Mass	July 15, 1873	140, 755
Candles from paraffine, Manufacture of	C. Havard	New York, N. Y	Sept. 19, 1865	50, 000
Candles, Machine for dipping	J. Aborn	Trenton, N. J	May 19, 1832	....
Candles, Machine for dipping	S. Blydenburg	....	Mar. 22, 1798	....
Candles, Machine for dipping	S. Stansbury	New York, N. Y	Jan. 6, 1809	....
Candles, Machine for making mold	A. Black	New York, N. Y	May 27, 1862	35, 359
Candles, Machine for manufacturing dip	W. Miller	New York, N. Y	Jan. 6, 1809	....
Candles, Machine for molding	W. Thomas	New York, N. Y	Apr. 3, 1860	27, 763
Candles, Machinery for molding	F. Meyrose	Saint Louis, Mo	Aug. 7, 1860	29, 506
Candles, Machinery for molding	G. A. Stanley	Cleveland, Ohio	Aug. 2, 1859	24, 960
Candles, Machinery for molding	G. A. Stanley	Cleveland, Ohio	Jan. 10, 1860	26, 797
Candles, Machinery for molding	G. A. Stanley	Cleveland, Ohio	Jan. 10, 1860	26, 798
Candles, Making	J. S. Sampson	....	June 26, 1797	....
Candles, Making mold	D. M. Randolph	Richmond, Va	Aug. 1, 1815	....
Candles, Making tallow and wax	J. Zwisler	Baltimore, Md	Apr. 2, 1812	....
Candles, Manufacture of	H. Halvorson	Cambridge, Mass	Jan. 10, 1860	26, 763
Candles, Manufacture of	C. Harvard	New York, N. Y	May 2, 1871	114, 438
Candles, Manufacture of	E. Marsh	Alton, Ill	Apr. 30, 1842	2, 600
Candles, Manufacture of	A. Meucci	Clifton, N. Y	Jan. 25, 1859	22, 739
Candles, Manufacture of	J. H. Tatum	New York, N. Y	Oct. 5, 1858	21, 706
Candles, Manufacture of	J. H. Tatum	New York, N. Y	Jan. 11, 1859	22, 592
Candles, Manufacture of	J. K. Truax	Pittsburgh, Pa	Apr. 9, 1872	125, 632
Candles, Manufacture of molded	J. L. Field	Lambeth, Great Britain	Jan. 31, 1865	46, 197
Candles, Manufacture of paraffine	J. L. Klein	New York, N. Y	Nov. 22, 1864	45, 161
Candles, Manufacture of paraffine	E. C. Leonard	New Bedford, Mass	Nov. 22, 1859	26, 193
Candles, Manufacture of paraffine	H. Ryder	New Bedford, Mass	Nov. 20, 1866	59, 749
Candles, Manufacturing	R. Robotham	....	June 2, 1794	....
Candles, Manufacturing	J. S. Sampson	....	Aug. 6, 1790	....
Candles, Manufacturing dipped	T. M. Scott	Falls Township, Pa	Apr. 1, 1830	....
Candles, Manufacturing mold	T. Hewitt, jr	Philadelphia, Pa	May 4, 1831	....
Candles, Manufacturing mold	T. Hewitt, jr	Philadelphia, Pa	June 13, 1831	....
Candles, Mode of ornamenting	A. Field and W. B. Nation	Lambeth and Surrey County, England.	Apr. 13, 1869	88, 779
Candles, Mold for making	T. J. Dyer and A. T. Richmond.	New Bedford, Mass	July 20, 1831	....
Candles, Molding	J. Mower and S. P. Bower	Strasburgh, Pa	Mar. 3, 1837	136
Candles, Molding paraffine	H. Leonard	New Bedford, Mass	Feb. 8, 1859	22, 921
Candles, Preparation of tallow for making	F. Garcin	Philadelphia, Pa	Mar. 11, 1856	14, 397
Candles, Preparing tallow for the manufacture of	J. Kirkman	New York, N. Y	Oct. 31, 1840	1, 844
Candles, Process of treating fatty bodies for the manufacture of.	H. Mége	Paris, France	May 24, 1864	42, 901
Candles, Treating fatty matter for the manufacture of.	A. R. St. Cyr	Lyons, France	Apr. 12, 1870	101, 935
Candlestick	C W. Blakeslee	Northfield, Conn	Aug. 21, 1855	13, 450
Candlestick	W. Church	England	Mar. 28, 1840	1, 528
Candlestick	F. C. Cone	San Francisco, Cal	Oct. 11, 1870	108, 109
Candlestick	B. Dearborn	....	Oct. 29, 1804	....
Candlestick	C. H. Doughty	Newburgh, N. Y	Feb. 25, 1873	136, 223
Candlestick	S. Gardiner, jr	New York, N. Y	Feb. 15, 1870	99, 875

Index of patents issued from the United States Patent Office from 1790 *to* 1873, *inclusive*—Continued.

Invention.	Inventor.	Residence.	Date.	No.
Candlestick	S. D. Hill	Downieville, Cal	Aug. 26, 1873	142, 162
Candlestick	W. H. H. Hinds	Groton, Mass	Oct. 31, 1865	50, 708
Candlestick	W. H. H. Hinds	Groton, Mass	Sept. 8, 1868	82, 000
Candlestick	W. H. H. Hinds	Ayer, Mass	Dec. 16, 1873	145, 502
Candlestick	A. E. Lyman	Williamsburgh, Mass	Mar. 25, 1862	34, 758
Candlestick	J. Manning	Middletown, Conn	Jan. 1, 1851	7, 875
Candlestick	S. L. Marsden and S. R. Burrell	Westville, Conn., and New York, N. Y.	Nov. 26, 1861	33, 788
Candlestick	H. Ogborn	Richmond, Ind	Dec. 3, 1867	71, 782
Candlestick	Z. I. Pratt	Chicago, Ill	Mar. 18, 1873	136, 864
Candlestick	A. Rix	San Francisco, Cal	Sept. 25, 1866	58, 296
Candlestick	F. A. Rockwell	Ridgefield, Conn	Dec. 16, 1851	8, 594
Candlestick	A. E. Rogers	La Grande, Oreg	Mar. 18, 1873	137, 029
Candlestick	H. E. Rogers	South Manchester, Conn	July 24, 1860	29, 314
Candlestick	A. E. Rogers and G. W. Webb	La Grande, Oreg	June 18, 1872	127, 989
Candlestick	T. Rose	Cortlandville, N. Y	June 23, 1857	17, 640
Candlestick	H. Ryder	New Bedford, Mass	July 12, 1870	105, 257
Candlestick, &c	S. Slocomb	Cambridge, Mass	Nov. 2, 1858	21, 987
Candlestick	J. Spratt	Cincinnati, Ohio	Nov. 10, 1857	18, 606
Candlestick	F. A. Taber	Baltimore, Md	May 4, 1869	89, 811
Candlestick	A. Whiteley	Springfield, Ohio	Jan. 8, 1856	14, 074
Candlestick	J. Williams	San Mateo, Cal	May 14, 1872	126, 765
Candlestick	H. Zahn	San Francisco, Cal	Dec. 14, 1869	97, 848
Candlestick	H. Zahn	San Francisco, Cal	July 26, 1870	105, 875
Candlestick, Camp	C. Alexander	Washington, D. C	Jan. 7, 1862	34, 045
Candlestick for ships, Swinging	J. J. Walton and E. A. Barrett	New York, N. Y	Feb. 25, 1873	136, 290
Candlestick for tents, shops, &c., Adjustable	A. E. Lyman	Williamsburgh, Mass	Jan. 12, 1864	41, 226
Candlestick, Miner's safety	T. A. Washburn	Gold Hill, Nev	Dec. 17, 1872	134, 113
Candle stick or holder	S. J. Rockwood	Elsah, Ill	Jan. 21, 1868	73, 652
Candlestick, Pewter	J. Witherie	Boston, Mass	Oct. 24, 1817	......
Candlestick, Portable	C. P. Gorely	Boston, Mass	Nov. 19, 1867	71, 161
Candlestick, Press-mold	S. T. Barnes	Columbus, Ohio	July 19, 1853	9, 852
Candlestick, Sheet-metal	J. W. Smith	Poultney, Vt	July 25, 1854	11, 389
Candlesticks, Construction of iron	W. P. Merriam, N. C. Harris, W. Wheeler, and E. N. Merriam.	Poultney, Vt	Apr. 26, 1853	9, 685
Candy-box	G. Ruger	La Fayette City, Ind	June 4, 1872	127, 431
Candy-cigar machine	H. Geilhausen	New York, N. Y	Aug. 22, 1865	49, 515
Candy-cutter	J. W. Kaskel	Paducah, Ky	Dec. 30, 1873	146, 076
Candy-cutter	F. Quinn	Keokuk, Iowa	Oct. 14, 1873	143, 590
Candy cutter, Corn	D. Davenport	Albany, N. Y	Aug. 20, 1867	67, 848
Candy-cutting machine	C. Wentz, A. Green, and O. P. Conner.	Trenton, N. J	Feb. 7, 1871	111, 592
Candy-jar	W. C. King	Pittsburgh, Pa	Dec. 17, 1872	134, 072
Candy-jar lid	W. C. King	Pittsburgh, Pa	May 6, 1873	138, 503
Candy-machine	J. S. Batchelder	Fort Wayne, Ind	June 7, 1870	103, 962
Candy-machine	G. K. Farrington and S. Brown, jr	Xenia, Ohio	May 11, 1858	20, 240
Candy, Machine for making sugar	B. O'Brien	Rochester, N. Y	Jan. 13, 1852	8, 657
Candy-making machine	T. and G. M. Mills	Philadelphia, Pa	Feb. 14, 1871	111, 765
Candy-manufacture	H. C. Wilkins	Albany, N. Y	Apr. 10, 1866	53, 913
Candy, Manufacture of	W. J. Burns	Georgetown, Ky	Jan. 30, 1872	123, 149
Candy, Medical compound or phosphated	C. S. Allen	New York, N. Y	Apr. 16, 1872	125, 714
Candy-molding machine	E. K. Powers	Grand Rapids, Mich	Sept. 22, 1868	82, 347
Candy-rolling machine	R. M. Marshall	Dayton, Ohio	July 23, 1861	32, 882
Candy-slicing machine	J. P. Anderson	Philadelphia, Pa	Aug. 1, 1871	117, 589
Candy, Sugar	H. and G. Garrison	Newburgh, N. Y	Apr. 10, 1844	3, 534
Candy-tongs, &c., Machine for making	T. and G. M. Mills	Philadelphia, Pa	Sept. 1, 1868	81, 665
Candy-twisting machine	J. Gardner	Philadelphia, Pa	Nov. 3, 1857	18, 541
Candy ware, Manufacturing sugar	L. Astolfi	Philadelphia, Pa	Nov. 20, 1812	......
Candying-machine	F. W. C. Spiess	New York, N. Y	Feb. 11, 1873	135, 672
Cane	F. Eckstein	New York, N. Y	Dec. 30, 1873	146, 054
Cane	M. Osborn	Albion, Mich	May 20, 1873	139, 020
Cane and billiard-cue, Combined walking	C. A. Bogert	Bay City, Mich	Dec. 26, 1871	122, 218
Cane and lamp, Combined	T. Crossley	Bridgeport, Conn	Nov. 26, 1867	71, 460
Cane and maize cutter	J. W. Cormack	Quincy, Ill	Nov. 1, 1853	10, 178
Cane and seat combined	C. H. Dascomb	Cleveland, Ohio	Aug. 20, 1861	33, 073
Cane and sorghum stripper	J. A. Campbell	Stow, Ohio	May 21, 1867	64, 836
Cane and sorghum stripper	C. P. Hale	Calhoun, Ky	Feb. 26, 1867	62, 413
Cane and stool, Convertible	J. Wade	Richmond, Ind	Jan. 7, 1862	34, 096
Cane and stubble shaver	P. G. Kleinpeter	Plaquemine, La	Aug. 10, 1869	93, 453
Cane and telescope combined	G. W. Wilson	Concord, N. H	July 28, 1868	80, 324
Cane and thermometer, Combined	J. L. Reber	Philadelphia, Pa	Dec. 24, 1867	72, 539
Cane and umbrella combined	H. Beebe	Hudson City, N. J	Nov. 5, 1867	70, 506
Cane and umbrella combined	G. Bockstaller	New York, N. Y	Jan. 7, 1868	73, 074
Cane and umbrella handle	A. Wanner	Hoboken, N. J	Oct. 11, 1870	108, 218
Cane and whip, Combined	C. L. Bushnell	Jefferson, Ohio	Jan. 11, 1870	98, 664
Cane and whip, Combined	C. L. Bushnell	Cherry Valley, Ohio	Nov. 15, 1870	109, 297
Cane and willow stripper	A. F. Ward and J. H. Bean	Marietta, Ohio	Nov. 24, 1868	84, 324
Cane-cleaner	S. Bean	Syracuse, Ohio	Mar. 3, 1868	74, 976
Cane, Compartment	L. C. Heylin	Philadelphia, Pa	Oct. 13, 1868	82, 949
Cane-coverer	J. Allison	Saint Martinsville, La	Mar. 6, 1860	27, 339
Cane-coverer	E. H. Angamer	New Orleans, La	Aug. 28, 1860	29, 842
Cane-coverer	J. P. Bryan	Princeton, Ky	June 24, 1843	3, 151
Cane-crusher, Hydraulic	C. A. Dickinson	Rosedale, La	Apr. 1, 1873	137, 349
Cane-crushing mill	H. C. Emery	Lincoln, Ohio	July 5, 1859	24, 621
Cane-cutter	J. P. Bryan	Princeton, Ky	July 8, 1843	3, 158
Cane-cutter	E. Owen	Byron, Ind	May 30, 1846	4, 551
Cane fiber for paper and other purposes, Treating	B. A. Lavender and H. Lowe	Baltimore, Md	Apr. 4, 1854	10, 722
Cane for paying omnibus-fare	S. W. Francis	New York, N. Y	Mar. 30, 1858	19, 765
Cane, Physician's	S. T. Trowbridge	Decatur, Ill	Jan. 3, 1860	26, 721
Cane-handle	J. Harvey and E. Ford	Philadelphia, Pa	Oct. 2, 1860	30, 225
Cane-handle	H. Nitzsche	Philadelphia, Pa	Aug. 21, 1866	57, 367
Cane-juice, Apparatus for applying sulphurous acid gas in purification of.	J. Communy	New Orleans, La	Dec. 11, 1860	30, 870
Cane-juice, Apparatus for bleaching and defecating	J. C. Marsh	Alexandria, La	Dec. 28, 1869	98, 285
Cane-juice, Apparatus for clarifying	W. A. Jordan	New Orleans, La	July 12, 1864	43, 509
Cane-juice, Apparatus for defecating	H. B. Bond	Terre Browne Parish, La	Oct. 6, 1868	82, 682
Cane-juice, Apparatus for defecating	C. K. Marshall	New Orleans, La	Mar. 29, 1870	101, 288

Index of patents issued from the United States Patent Office from 1790 *to* 1873, *inclusive*—Continued.

Invention.	Inventor.	Residence.	Date.	No.
Cane-juice, Apparatus for defecating	E. H. Wheeler	New Orleans, La	May 8, 1860	28,225
Cane-juice, Apparatus for extracting	D. Moffat	New Orleans, La	Sept. 23, 1873	143,089
Cane-juice, Apparatus for separating gum from	H. Gortner	Deavertown, Ohio	Sept. 18, 1866	58,095
Cane-juice, Apparatus for treating	W. L. Shaffer and J. F. Wood	Chacahoula Station, La	Oct. 22, 1872	132,417
Cane-juice, Apparatus for treating	G. R. Taylor	Thibodeaux, La	July 1, 1873	140,443
Cane-juice boxes, Construction of	L. Tregre	Parish Saint John the Baptist, La.	Sept. 6, 1859	25,354
Cane-juice by means of sulphurous-acid gas, Apparatus for clarifying.	A. Labauve	Convent, La	Aug. 24, 1869	94,008
Cane-juice, Clarification of	J. Spangenberg	Jefferson Parish, La	Mar. 27, 1849	6,219
Cane-juice, Clarifying	F. Domenech	Ponce, Island of Puerto Rico.	Dec. 27, 1859	26,573
Cane-juice, Clarifying	W. J. McIntosh	Georgia	Mar. 7, 1828	....
Cane-juice-clarifying apparatus	P. Marceline and E. Eude	New Orleans, La	May 22, 1860	28,385
Cane-juice concentrated by steam	D. Strobel, jr	Washington, D. C	May 9, 1831	....
Cane-juice, Defecating	D. F. Boyd	Mansfield, Ohio	Feb. 7, 1860	27,033
Cane-juice, Defecating	A. B. Sharp	Paincourtville, La	Jan. 7, 1868	73,054
Cane-juice, Defecating	L. Wray	London, England	June 30, 1857	17,713
Cane-juice, Defecating and bleaching	M. Landry	Iberville Parish, La	Aug. 8, 1871	117,787
Cane-juice defecating and bleaching machine	P. Paille	Saint James Parish, La	Jan. 19, 1869	85,955
Cane-juice, Defecating and clarifying	R. A. Stewart	Parish of Saint Bernard, La	Jan. 11, 1859	22,590
Cane-juice, Evaporating	S. A. Poché	Saint James Parish, La	June 21, 1870	104,491
Cane-juice evaporation	R. Wright	London, England	Oct. 16, 1860	30,401
Cane-juice evaporator	H. Bessemer	Baxter House, England	Mar. 8, 1853	9,607
Cane juice evaporator	J. S. Blymyer	Mansfield, Ohio	Dec. 20, 1870	110,192
Cane-juice evaporator	T. Boutté	New Iberia, La	Apr. 7, 1868	76,392
Cane-juice evaporator	M. S. Bringier	Ascension Parish, La	Sept. 21, 1869	94,942
Cane-juice evaporator	M. S. Bringier	Ascension Parish, La	Oct. 26, 1869	96,081
Cane-juice evaporator	W. Edgerton	Spiceland, Ind	Aug. 15, 1865	49,392
Cane-juice evaporator	F. C. Fulghum	Arba, Ind	Sept. 13, 1864	44,177
Cane-juice evaporator	F. Groves	New Oxford, Pa	Mar. 21, 1865	46,969
Cane-juice evaporator	L. S. Hereford	West Baton Rouge Parish, La.	Sept. 21, 1869	94,959
Cane-juice evaporator	S. A. Poché	Saint James Parish, La	Nov. 17, 1868	84,134
Cane-juice, &c., evaporator	E. Sperry	Chicago, Ill	Apr. 4, 1871	113,590
Cane-juice, &c., evaporator and concentrator	A. Fryer	Manchester, England	May 28, 1867	65,205
Cane-juice evaporator, Setting kettles for	F. M. Brignac	Saint James Parish, La	Oct. 3, 1871	119,501
Cane-juice, Machine for bleaching	E. Skelly	Plaquemine, La	Oct. 26, 1869	96,155
Cane juice, Machine for expressing	H. Bessemer	Middlesex County, England	June 3, 1851	8,137
Cane-juice, Mode of applying sulphurous-acid gas in the defecation of.	R. B. Brashear	Pattersonville, La	Dec. 6, 1859	26,401
Cane-juice, Mode of defecating	L. Reid and E. H. Swift	New York, N. Y., and Middlefield, Conn.	Dec. 4, 1866	60,242
Cane-juice, Mode of defecating	L. Reid and E. H. Swift	New York, N. Y., and Middlefield, Conn.	Dec. 4, 1866	60,243
Cane-juice, Mode of defecating	L. Reid and E. H. Swift	New York, N. Y., and Middlefield, Conn.	Dec. 4, 1866	60,244
Cane-juice, Mode of defecating	L. Reid and E. H. Swift	New York, N. Y., and Middlefield, Conn.	Dec. 4, 1866	60,245
Cane-juice, Mode of defecating	L. Reid and E. H. Swift	New York, N. Y., and Middlefield, Conn.	Dec. 4, 1866	60,246
Cane-juice, Mode of treating	J. E. Pattison	Houma, La	Sept. 11, 1866	57,958
Cane-juice with sulphur fumes, Apparatus for treating.	W. G. Billiu	Thibodeaux, La	Aug. 19, 1873	141,914
Cane-juice with sulphurous acid, Apparatus for treating.	J. E. Pattison	Thibodeaux, La	Jan. 21, 1868	73,640
Cane-juice, &c., with sulphurous-acid gas, Apparatus for impregnating.	E. Skelly	Plaquemine, La	Sept. 15, 1868	82,164
Cane-juice with sulphurous-acid gas, Apparatus for treating.	J. W. Austin	Plaquemine, La	Nov. 29, 1870	109,571
Cane-juice with sulphurous-acid gas, Apparatus for treating.	J. Dymond	New York, N. Y	Jan. 7, 1873	134,655
Cane-juice with sulphurous-acid gas, Apparatus for treating.	G. A. Huwald	Knoxville, Tenn	Aug. 20, 1861	33,086
Cane-juice with suphurous-acid gas, Apparatus for treating.	T. Morillon and U. Naquin	La Fourche Parish, La	July 2, 1867	66,377
Cane-juice, &c., with sulphurous acids, Bleaching	B. R. Hawley	Normal, Ill	July 16, 1872	129,341
Cane-mill	L. Biddle	Knoxville, Iowa	Jan. 21, 1870	104,544
Cane-mill	W. Briggs	Greencastle, Iowa	Nov. 20, 1866	59,889
Cane-mill	B. F. Cauffman	Millerstown, Pa	Jan. 28, 1868	73,779
Cane-mill	D. M. Cook	Mansfield, Ohio	Aug. 11, 1863	39,467
Cane-mill	J. W. Richardson, D. L. Davis, and W. C. Howell.	Sligo, Ohio	May 22, 1866	54,960
Cane mill	H. B. Stephens and D. J. Powers	Madison, Wis	Oct. 26, 1869	96,283
Cane-mill	O. E. Woodbury	Madison, Wis	Nov. 14, 1865	50,978
Cane-mill and steam-engine	J. Moore	Madison, Ind	Feb. 18, 1868	74,574
Cane, &c., Mill for grinding	I. A. Hedges	Cincinnati, Ohio	Feb. 1, 1859	22,802
Cane-press	E. Powell	Conneautville, Pa	Apr. 10, 1860	27,830
Cane, Rifle	A. Crow	Middlefield, Mass	May 8, 1860	28,160
Cane-scraper	L. A. Bringier and N. B. Trist	Ascension Parish and New Orleans, La.	July 13, 1869	92,420
Cane seat	G. W. Martin	Boston, Mass	July 28, 1868	80,300
Cane-seat chairs, Construction of	C. M. Rohr	Portland, Oreg	June 14, 1870	104,207
Cane seat for chairs	W. F. Howe	Galveston, Tex	Sept. 3, 1872	131,060
Cane seat for chairs	J. B. Sawyer	Templeton, Mass	June 11, 1861	32,535
Cane-stripper	J. H. Barley	Sedalia, Mo	Sept. 24, 1867	69,161
Cane-stripper	W. N. Barr	Richmond, Ind	Sept. 18, 1866	58,049
Cane-stripper	J. C. Brown	Crawfordsville, Ind	Oct. 1, 1867	69,310
Cane-stripper	W. Gladden and R. F. Bishop	Chrome Hill, Md	Oct. 10, 1865	50,350
Cane-stripper	J. A. Hall	Memphis, Tenn	Aug. 8, 1865	49,260
Cane-stripper	J. A. Hall	Columbus, Ohio	Apr. 10, 1866	53,818
Cane-stripper	C. A. Hege	Friedberg, N. C	Oct. 11, 1870	108,135
Cane-stripper	R. C. James	Denison, Tex	July 22, 1873	141,057
Cane-stripper	M. Mellinger	Dayton, Ohio	Jan. 1, 1867	60,768
Cane-stripper	I. E. and J. A. Overpeck	Overpeck Station, Ohio	Mar. 12, 1867	62,876
Cane-stripper	H. Rockwell	Roanoke, Ind	Dec. 12, 1865	51,482
Cane-stripper	G. E. Sellers	Seller's Landing, Ill	Jan. 24, 1865	46,031

Index of patents issued from the United States Patent Office from 1790 *to* 1873, *inclusive*—Continued.

Invention.	Inventor.	Residence.	Date.	No.
Cane-stripper	W. Todd	Barnesville, Ohio	Apr. 5, 1864	42, 241
Cane stripper and knife combined	R. Conarroe	Camden, Ohio	July 3, 1866	56, 009
Cane-stripping device	T. W. Peirce	Minneapolis, Minn	Sept. 5, 1865	49, 785
Cane-stripping machine	D. C. Flint	Bushnell, Ill	May 22, 1866	54, 880
Cane-stubble shaver	F. F. Patout	Jeanerett, La	May 23, 1871	115, 094
Cane, Sword	S. A. Hudson	Worcester, Mass	Sept. 9, 1851	8, 346
Cane, umbrella, pistol, dagger, and camp-stool combined.	D. Morrison	Portland, Me	Apr. 2, 1867	63, 552
Cane, Walking	A. Cain	Holyoke, Mass	July 12, 1859	24, 718
Cane, Walking	C. K. Peney	Worcester, Mass	Sept. 30, 1873	143, 248
Canes, crutches, &c., Mode of constructing the bottom ends of ferrules for.	J. Ball	Buffalo, N. Y	Oct. 11, 1841	2, 297
Canister: *See* Grocer's canister. Paint-canister. Powder-canister. Tea-canister.				
Canister	E. Mather	Chicago, Ill	July 23, 1872	129, 842
Canisters, Measuring-stopple for	W. M. Smith	Washington, D. C	Sept. 25, 1866	58, 311
Cannon	J. Adams	Cleveland, Ohio	Oct. 25, 1859	25, 929
Cannon	J. W. Cochran	Lowell, Mass	Mar. 23, 1836	....
Cannon	A. Krupp	Essen, Prussia	Nov. 27, 1855	13, 851
Cannon	J. A. Terrell	Bloomfield, Ky	Dec. 29, 1868	85, 491
Cannon and fire-arms, Construction of	N. A. Patterson	Kingston, Tenn	Nov. 3, 1863	40, 498
Cannon and other fire-arms, Manufacture of	C. W. Lancaster	London, England	July 5, 1853	9, 830
Cannon, Apparatus for planing the chambers of	A. Alexander	Pittsburgh, Pa	Oct. 13, 1863	40, 225
Cannon, Automatic	C. E. Barnes	Lowell, Mass	July 8, 1836	15, 315
Cannon, &c., Automatic finger for closing the vents of	A. Le Mat	New Orleans, La	June 7, 1859	24, 313
Cannon-balls, Flask for casting packings around	L. Evans	Morgantown, Va	Oct. 30, 1860	30, 533
Cannon-ball, &c., from malleable iron, lead, &c., Machine for manufacturing.	L. Grandy and T. Osgood	Troy, N. Y	Feb. 3, 1841	1, 965
Cannon-balls, &c., Mode of forming molds for casting.	A. Elliot	Philadelphia, Pa	Oct. 6, 1843	3, 295
Cannon, Bolt and disk sectional	J. Fitzgerald	New York, N. Y	Feb. 6, 1849	6, 090
Cannon, Boring	L. A. B. Walbach	United States Army	Aug. 23, 1853	9, 961
Cannon boring and rifling apparatus	P. Foy	New York, N. Y	Mar. 11, 1862	34, 661
Cannon, Breech-loading	H. J. Allen	Arkadelphia, Ark	Apr. 25, 1871	113, 963
Cannon, Breech-loading	J. W. Hollensbury	Alexandria, Va	Dec. 28, 1858	22, 427
Cannon, Breech-loading	E. Marshall	New York, N. Y	Dec. 14, 1858	22, 299
Cannon, Breech-loading	J. H. Merrill	Baltimore, Md	Mar. 22, 1859	23, 306
Cannon, Breech-loading	J. H. Murrill	Baltimore, Md	June 15, 1858	20, 608
Cannon, Breech-loading	W. E. Osborn	Milton, N. Y	Sept. 12, 1854	11, 678
Cannon, Breech-loading	J. B. Prescott	Waterford, N. Y	Jan. 28, 1862	34, 263
Cannon, Breech-loading	E. S. Wright and T. P. Gould	Buffalo, N. Y	Dec. 14, 1858	22, 325
Cannon by electricity, Firing	G. W. Beardslee	College Point, N. Y	Aug. 18, 1863	39, 543
Cannon, Casting	A. H. Emery	New York, N. Y	Oct. 4, 1864	44, 521
Cannon, Casting	J. B. Tarr	Fairhaven, Mass	Jan. 18, 1870	98, 894
Cannon, Constructing	C. Perley	New York, N. Y	May 5, 1863	38, 4[illegible]9
Cannon, Device for firing	D. Treadwell	Cambridge, Mass	Nov. 25, 1862	37, 017
Cannon, Discharging	H. Whitcomb	Adams, N. Y	Oct. 25, 1826	....
Cannon-elevating apparatus	G. M. Ransom	United States Navy	Nov. 15, 1859	26, 124
Cannon for chain-shot	E. Gordon	Hingham, Mass	Feb. 17, 1836	....
Cannon for chain-shot	A. Lemmer	Newark, N. J	Nov. 18, 1851	8, 528
Cannon for chain-shot, Double	W. M. Jeffers	Elmira, N. Y	June 21, 1859	24, 518
Cannon, Forging	A. Hitchcock	New York, N. Y	Aug. 8, 1865	49, 266
Cannon, Friction-primer for	W. Ball	Chicopee, Mass	Mar. 23, 1852	8, 820
Cannon-hoops, Manufacture of	D. Treadwell	Cambridge, Mass	May 10, 1864	42, 701
Cannon-lock	E. Hidden and S. Sawyer	New York, N. Y	Apr. 29, 1842	2, 594
Cannon-lock, Compression	J. Shaw	Philadelphia, Pa	Dec. 3, 1832	....
Cannon-lock, Portable	J. Shaw	Philadelphia, Pa	Dec. 3, 1832	....
Cannon, Machine for boring the chambers of	W. McClery	Pittsburgh, Pa	Apr. 12, 1864	42, 332
Cannon, Machine for turning rimbase of	E. Kaylor	Pittsburgh, Pa	Oct. 24, 1865	50, 598
Cannon, Machinery for welding and forming wrought-iron.	D. Treadwell	Cambridge, Mass	June 20, 1846	4, 589
Cannon, Malleable-iron	G. W. Chapman	New York, N. Y	June 16, 1836	....
Cannon, Manufacture of	T. N. Hornsby	Simpsonville, Ky	Sept. 3, 1867	68, 509
Cannon, Manufacture of	D. Treadwell	London, England	Dec. 11, 1855	13, 927
Cannon, Manufacture of wrought	H. Ames	Falls Village, Conn	Aug. 16, 1864	43, 825
Cannon, Manufacture of wrought-iron	J. Griffen	Safe Harbor, Pa	Dec. 25, 1855	13, 984
Cannon, Manufacture of wrought-iron and steel	R. Cook	Saratoga, N. Y	Feb. 1, 1842	2, 443
Cannon, Manufacturing cast-iron	C. Alger	Boston, Mass	May 30, 1837	208
Cannon, Many-barrelled	G. Natcher	Sidney, Ohio	Dec. 27, 1864	45, 623
Cannon, Many-chambered breech-loading	C. C. Terrel	Shullsburgh, Wis	Feb. 5, 1856	14, 215
Cannon, Method of making wrought-iron	S. Hunt	Torrington, Conn	Nov. 24, 1843	3, 358
Cannon, Mode of charging	J. Dodge	Dummerston, Vt	July 15, 1856	15, 357
Cannon, Mode of discharging	J. Dodge	Dummerston, Vt	Aug. 4, 1857	17, 920
Cannon, Mode of loading rifled	L. Houghton	Philadelphia, Pa	Apr. 3, 1855	12, 629
Cannon, Mounting and maneuvering	O. Hopkins	New York, N. Y	Dec. 17, 1861	3[illegible], 944
Cannon of wrought iron or wrought iron and steel, Method of making.	D. Treadwell	Cambridge, Mass	Feb. 12, 1845	3, 906
Cannon, Pendulum-sight for	R. Smith	Brooklyn, N. Y	Oct. 13, 1863	40, 288
Cannon, Percussion-lock and vent for	W. H. Bell	Norfolk, Va	Oct. 1, 1830	....
Cannon, Percussion-lock for	R. Beale	Washington, D. C	Jan. 13, 1835	....
Cannon, Percussion-lock for	E. Hidden	New York	Jan. 14, 1831	....
Cannon, Percussion-lock for	E. Hidden	New York	Dec. 16, 1831	....
Cannon, Percussion-lock for	E. Hidden	New York	Aug. 20, 1834	....
Cannon, Percussion-lock for	J. Shaw	Philadelphia, Pa	Oct. 24, 1828	....
Cannon, Percussion-primer for	L. and W. H. Bell	Fortress Monroe, Va	Dec. 8, 1829	....
Cannon, Pointing	J. Black	Northumberland, Pa	Jan. 17, 1818	....
Cannon-pointing apparatus	J. Hobday	Portsmouth, Va	May 30, 1837	207
Cannon, Primer and percussion for	J. Shaw	Philadelphia, Pa	Dec. 3, 1832	....
Cannon-rammer	M. C. Borgia	Philadelphia, Pa	Oct. 22, 1861	33, 513
Cannon, Repeating	J. H. Headrick	Wythe County, Va	Dec. 20, 1870	110, 233
Cannon, Repeating	S. Huffman	Charlestown, Ill	Jan. 23, 1855	12, 295
Cannon, Repeating	T. Tufts	Somerville, Mass	Mar. 7, 1865	46, 762
Cannon, Revolving	E. Brehm	Jersey City, N. J	Dec. 20, 1870	110, 194
Cannon, Revolving	A. Schausten	Michigan City, Ind	Feb. 4, 1868	74, 152

Index of patents issued from the United States Patent Office from **1790** *to* **1873,** *inclusive*—**Continued.**

Index of patents issued from the United States Patent Office from 1790 *to* 1873, *inclusive*—Continued.

Invention.	Inventor.	Residence.	Date.	No.
Caoutchouc to cloths, &c., Application of	E. M. Chaffee	Roxbury, Mass	Aug. 31, 1836	....
Caoutchouc to cloth, &c., Method of applying	C. Meyer	New Brunswick, N. J	Mar. 19, 1861	31, 729
Caoutchouc, Tool for manufacturing goods of	D. D. Parmelee	New York, N. Y	Oct. 5, 1858	21, 697
Caoutchouc, Treating	L. Hull	Charlestown, Mass	Dec. 23, 1862	37, 231
Caoutchouc, Treatment of	A. G. Day	Seymour, Conn	Aug. 10, 1858	21, 122
Caoutchouc, Treatment of	M. Mattson	Boston, Mass	Mar. 1, 1859	23, 103
Caoutchouc, Vulcanizing	B. W. Franklin	New York, N. Y	Oct. 8, 1861	33, 464
Caoutchouc-vulcanizing apparatus	G. E. Hayes	Buffalo, N. Y	Mar. 5, 1861	31, 599
Caoutchouc with sulphur, Mode of preparing	N. Hayward	Woburn, Mass	Feb. 24, 1839	1, 090
Cap	J. W. Bryant	Welaka, Fla	Sept. 23, 1862	36, 549
Cap	I. K. Gittens, jr	Green Point, N. Y	Oct. 1, 1861	33, 386
Cap	S. Goldstone	Philadelphia, Pa	Sept. 4, 1866	57, 702
Cap	M. N. Myers	Boston, Mass	May 6, 1873	138, 681
Cap	B. Scharl	Boston, Mass	Mar. 18, 1873	137, 032
Cap	C. and F. Stattmann	Chicago, Ill	Oct. 23, 1866	59, 092
Cap and nozzle, Metallic	E. S. Covell	Brooklyn, N. Y	June 7, 1870	103, 988
Cap, Army and Navy	W. F. Warburton	Philadelphia, Pa	Aug. 26, 1862	36, 315
Cap-box	A. F. Gove	Lincoln, Vt	May 14, 1872	126, 692
Cap, Fireman's	P. Ackerman	Bangor, Me	July 2, 1872	128, 451
Cap-fronts, Machine for cutting	G. Burgess	New York, N. Y	Feb. 26, 1850	7, 114
Cap-fronts, Mode of binding	L. Londinsky	New York, N. Y	June 26, 1860	28, 878
Cap, Military	S. Mossman	Cleveland, Ohio	Aug. 25, 1863	39, 667
Cap, Military	J. S. Smith	New York, N. Y	Aug. 21, 1860	29, 724
Cap, Military	J. F. Whipple	New York, N. Y	July 16, 1861	32, 849
Cap, Military fatigue	C. Kollinky, J. Ehrlich, and J. De Zeyk.	Washington, D. C	Nov. 5, 1861	33, 651
Cap setting and extracting device	E. S. Holmes	Grand Rapids, Mich	June 10, 1873	139, 674
Cap-shearing block	R. Cook and T. Hanford	New Hartford, N. Y	Oct. 3, 1871	119, 449
Cap, Traveling	J. F. Sargent	Tunbridge, Vt	Mar. 29, 1870	101, 383
Cap, Water-proof and elastic	L. L. Macomber	Gardiner, Me	Oct. 11, 1831	....
Caps, Former for forming seamless felt	J. L. Bridge and W. B. Lodge	Vernon, N. Y	May 8, 1860	28, 149
Caps, Machine for cutting leather visors for	J. Hawkins	Roxbury, Mass	Feb. 30, 1832	....
Caps, Mode of preventing corrosion in metallic	L. R. Boyd	New York, N. Y	Mar. 30, 1869	88, 439
Capped nails	C. Walsh	Newark, N. J	Sept. 30, 1873	143, 393
Capstan	J. S. Brown	Schenectady, N. Y	July 4, 1871	116, 548
Capstan	E. Buel	Silver Creek, N. Y	Apr. 19, 1870	102, 088
Capstan	D. N. B. Coffin, jr	Boston, Mass	July 2, 1867	66, 299
Capstan	D. N. B. Coffin, jr	Newton Centre, Mass	Mar. 15, 1870	100, 859
Capstan	Z. E. Coffin	Newton Centre, Mass	June 24, 1862	35, 730
Capstan	H. Davies	Portsmouth, Ohio	Apr. 26, 1864	42, 465
Capstan	J. Edson	Boston, Mass	Aug. 21, 1866	57, 302
Capstan	J. Edson	Boston, Mass	Nov. 12, 1867	70, 820
Capstan	J. Edson	Boston, Mass	Nov. 9, 1869	96, 568
Capstan	J. Ericsson	New York, N. Y	Jan. 18, 1870	98, 940
Capstan	J. Gardner	New York, N. Y	Aug. 3, 1869	93, 295
Capstan	W. D. Grimshaw	New York N. Y	Nov. 28, 1865	51, 174
Capstan	W. D. Grimshaw	Newark, N. J	Dec. 18, 1866	60, 505
Capstan	A. Morse	Boston, Mass	Mar. 12, 1836	....
Capstan	J. Reed	Marshfield, Mass	Apr. 5, 1859	23, 499
Capstan	J. W. Riggs	Wade, Ohio	Sept. 6, 1870	107, 104
Capstan	P. Roberts	New York, N. Y	Feb. 17, 1852	8, 746
Capstan	W. W. Vanderbilt	New York, N. Y	Dec. 1, 1868	84, 660
Capstan	C. L. Willis	Washington, D. C	Apr. 6, 1869	88 761
Capstan and crab, Portable	A. Elmen	Shabbona Grove, Ill	Aug. 9, 1859	25 079
Capstan and windlass	D. N. B. Coffin, jr	Boston, Mass	Nov. 27, 1866	59, 969
Capstan and windlass	I. G. Morgan	Ithaca, N. Y	Oct. 20, 1863	40, 352
Capstan and windlass and cable-stopper, Ship's	J. Grylls	Portsea, England	Dec. 12, 1842	2, 876
Capstan and windlass, Arrangement of ship's	J. Reed	Marshfield, Mass	June 20, 1854	11, 139
Capstan and windlass, Combined	J. S. Getchell	Machias, Me	May 14, 1861	32, 284
Capstan and windlass, Ship's	J. Emerson	Worcester, Mass	June 17, 1856	15, 123
Capstan for ships, Compound	C. Perley	New York, N. Y	Aug. 4, 1857	17, 940
Capstan for ships, Rail	W. H. Willard	Cleveland, Ohio	Aug. 11, 1863	39, 518
Capstan, Portable	E. G. Ament	Oswego, Ill	Sept. 13, 1864	44, 268
Capstan, Portable	G. Cook	Bristol Station, Ill	May 28, 1861	32, 414
Capstan, Power	D. N. B. Coffin, jr	Boston, Mass	Nov. 21, 1865	51, 107
Capstan, Reversible	J. A. H. Ellis and A. Gordon	Rochester, N. Y	June 27, 1854	11, 156
Capstan, Ship's	R. Dunbar and J. F. Robertson	Buffalo, N. Y	June 30, 1857	17, 715
Capstan, Ship's	S. Gaty	Saint Louis, Mo	May 27, 1856	14, 983
Capstan, Ship's	I. F. Holloway	Saline Mines, Ill	June 28, 1859	24, 558
Capstan, Ship's	J. B. Holmes	New York, N. Y	July 17, 1855	13, 262
Capstan, Ship's	J. B. Holmes	New York, N. Y	Mar. 3, 1857	16, 774
Capstan, Ship's	S. Huse	New York, N. Y	Aug. 11, 1857	17, 969
Capstan, Ship's	D. Knowlton	Camden, Me	Aug. 11, 1857	17, 971
Capstan, Ship's	D. Knowlton	Camden, Me	June 21, 1859	24, 467
Capstan, Ship's	C. E. Marwick	Portland, Me	Aug. 25, 1857	18, 053
Capstan, Ship's	C. Oaks	Rochester, N. Y	May 16, 1835	....
Capstan, Ship's	C. Perley	New York, N. Y	Oct. 21, 1856	15, 933
Capstan, Ship's	C. Perley	New York, N. Y	Mar. 5, 1861	31, 619
Capstan, Ship's	D. and G. Tallcot	Oswego, N. Y	Mar. 4, 1856	14, 377
Capstan, Ship's	D. and G. Tallcot	Oswego, N. Y	May 27, 1856	14, 986
Capstan, Ship's	J. R. Taylor	New York, N. Y	Apr. 27, 1858	20, 131
Capstan, Ship's	I. Wilson and F. D. Beckwith	New London, Conn	June 15, 1837	238
Capstan, Steamboat	J. Schaffer	Manchester, Pa	Oct. 21, 1856	15, 954
Capstan, Steamboat	J. Schaffer	West Manchester, Pa	Mar. 31, 1857	16, 935
Caps an, Steam-working	J. S. McMillin	Pittsburgh Pa	Feb. 20, 1866	52, 739
Capstan, Variable	J. E. Andrews	Boston, Mass	Apr. 24, 1849	6, 396
Capstan, Winch	D. Knowlton	Camden, Me	June 19, 1866	55, 673
Capstans of vessels, Application of steam-power to	J. S. McMillin	Pittsburgh, Pa	Apr. 16, 1867	63, 917
Capstans, Purchase for	J. Edgecomb	Gardiner, Me	Mar. 28, 1871	113, 032
Capsules, Machine for making plated	J. J. Baranowski	London, England	Apr. 1, 1873	137, 403
Capsules, Manufacture of	D. Dick	New York, N. Y	Aug. 8, 1865	49, 242
Capsules, Metallic	W. Betts	City Road, England	Sept. 3, 1861	33, 179
Capsules of copaiba, Making	P. Cauhaupe	New Lebanon, N. Y	Dec. 24, 1861	33, 983
Car, Aërial	W. Morrow	San Francisco, Cal	Mar. 30, 1869	88, 324
Car, Aërial steam	F. Marriott	San Francisco, Cal	Nov. 23, 1869	97, 100
Car and cable coupling	W. S. Nearing	Morris Run, Pa	Dec. 31, 1872	134, 437
Car and carriage axles and boxes, Railway	D. C. Force and F. Davis	Baltimore, Md	Nov. 6, 1834	....
Car and carriage wheels, Chilling	P. Davis	Baltimore, Md	July 29, 1834	....

Index of patents issued from the United States Patent Office from 1790 *to* 1873, *inclusive*—Continued.

Invention.	Inventor.	Residence.	Date.	No.
Car and driving-wheel, Railway	N. C. Lombard	Cambridge, Mass	Sept. 4, 1869	95, 122
Car and dwelling ventilator	J. Bradley	New York, N. Y	Apr. 25, 1871	114, 162
Car and engine wheels, Constructing railway	J. Stimpson	Baltimore, Md	Oct. 23, 1834	....
Car and locomotive axles, Mode of strengthening railway.	Z. Durkee	Philadelphia, Pa	Jan. 9, 1838	549
Car and locomotive-engine wheel	R. Winans	Baltimore, Md	Nov. 19, 1833	....
Car and mail-bag receiver, Mail	W. G. Sanford	Union City, N. Y	May 26, 1868	78, 331
Car and school seat	W. H. Joeckel	New York, N. Y	Dec. 10, 1861	33, 887
Car and switch, Railway	I. N. Stanley	Philadelphia, Pa	Mar. 12, 1840	1, 512
Car and tender loading apparatus	J. Williams	Bell's Depot, Tenn	Dec. 26, 1871	122, 207
Car and track, Coal	G. Martz	Pottsville, Pa	Nov. 23, 1869	97, 207
Car and track for elevating on inclined plane	J. W. Pearce	Suisun, Cal	Sept. 1, 1868	81, 677
Car and truck connection	J. J. Sherman	Albany, N. Y	Apr. 14, 1863	38, 182
Car and truck, Railway	W. Youmans	Lansingburgh, N. Y	Nov. 12, 1872	133, 001
Car and truck-wheel, Coal	J. Patterson	Pittsburgh, Pa	Mar. 15, 1870	100, 795
Car and vehicle spring	W. R. Nichols and C. W. Pickering.	Philadelphia, Pa	Apr. 26, 1870	102, 432
Car and window washer	P. S. Devlan	Jersey City, N. J	Feb. 25, 1873	136, 310
Car apron or duster and bridge, Railway	W. H. Ward	Auburn, N. Y	Nov. 26, 1867	71, 558
Car-axle	B. Ahrens and D. Gerhard	Reading, Pa	Sept. 16, 1873	142, 757
Car-axle	J. Anthony	Greenbush, N. Y	Dec. 24, 1867	72, 584
Car-axle	F. A. Braymer and M. W. Lyman.	Chicago, Ill	Dec. 24, 1872	134, 247
Car-axle	W. A. Brickill	New York, N. Y	July 9, 1867	66, 556
Car-axle	L. Brown and J. and J. Leland	Worcester, Mass	Apr. 24, 1860	27, 966
Car-axle	S. S. Burt	Marquette, Mich	Nov. 12, 1867	70, 799
Car-axle	J. W. Clark	Manchester, Wis	Nov. 7, 1865	50, 796
Car-axle	R. Cromelien	Washington, D. C	July 7, 1868	79, 555
Car-axle	W. B. Fahnestock	Lancaster, Pa	May 19, 1868	77, 968
Car-axle	M. P. Hadley	Bluffton, Wis	June 11, 1872	127, 868
Car-axle	J. W. Hard	Decorah, Iowa	Mar. 28, 1865	47, 015
Car-axle	J. W. Hard	Decorah, Iowa	Mar. 26, 1872	125, 046
Car-axle	B. J. La Mothe	New York, N. Y	Sept. 27, 1864	44, 434
Car-axle	G. W. Miltimore	Janesville, Wis	Dec. 10, 1872	133, 790
Car-axle	G. W. Miltimore	Janesville, Wis	Nov. 4, 1873	144, 347
Car-axle	H. Mooers	Toledo, Ohio	Feb. 16, 1869	86, 937
Car-axle	W. Phelps	Sycamore, Ill	July 3, 1860	29, 001
Car-axle	T. R. Timby	Tarrytown, N. Y	Mar. 11, 1873	136, 791
Car-axle	C. D. Tisdale	East Boston, Mass	Aug. 11, 1863	39, 533
Car-axle	E. H. Williamson	Philadelphia, Pa	June 18, 1872	128, 084
Car-axle adjustment	R. Eaton	Eaton Lodge, England	Feb. 1, 1870	9[illegible], 418
Car-axle and axle-box	G. H. Perry and C. B. Hawley	Salt Lake City, Utah	Oct. 8, 1872	132, 100
Car-axle and wheel	J. Montgomery	Croton, N. Y	June 2, 1868	78, 467
Car-axle bearing	L. Brauer	Washington, D. C	Oct. 12, 1869	95, 867
Car-axle bearing, Railway	E. D. Murfey	New York, N. Y	June 27, 1871	116, 475
Car-axle cap	W. Werts	Pana, Ill	June 9, 1868	78, 776
Car-axle, Divided	S. S. Hickok	Methuen, Mass	July 30, 1872	130, 047
Car-axle, Divided	D. B. Hunt	San Francisco, Cal	May 4, 1869	89, 767
Car-axle, Divided	J. K. Nelson	Green Point, N. Y	Nov. 10, 1868	83, 991
Car-axle, Divided	R. Vose	New York, N. Y	Apr. 22, 1856	14, 747
Car-axle, Divided railway	W. S. Loughborough	Victor, N. Y	June 22, 1852	9, 056
Car-axle, Hollow	W. A. Lewis	Chicago, Ill	Oct. 10, 1871	119, 866
Car axle, Railway	R. N. Allen	Pittsford, Vt	May 27, 1873	139, 226
Car axle, Railway	J. Armstrong	New Orleans, La	Mar. 16, 1869	87, 818
Car axle, Railway	H. D. Burghardt	Pittsfield, Mass	Aug. 17, 1869	93, 671
Car axle, Railway	C. Cole	Troy, N. Y	Mar. 15, 1870	100, 860
Car axle, Railway	D. M. Cummings	Enfield, N. H	June 15, 1869	91, 424
Car axle, Railway	J. H. Linsmore	Boston, Mass	June 11, 1872	127, 857
Car axle, Railway	E. Doty and G. W. Miltimore	Janesville, Wis	Oct. 10, 1871	119, 831
Car axle, Railway	S. H. Dubois	Buffalo, N. Y	Jan. 6, 1863	37, 280
Car axle, Railway	R. Dutch	Jersey City, N. J	July 13, 1869	92, 592
Car axle, Railway	A. E. Elmer	Greenfield, Mass	Jan. 15, 1867	61, 180
Car axle, Railway	P. G. Gardiner	New York, N. Y	Feb. 12, 1856	14, 232
Car axle, Railway	C. P. Hewett	Kingston, Wis	May 16, 1865	47, 723
Car axle, Railway	F. Hudner	New York, N. Y	Jan. 25, 1870	99, 195
Car axle, Railway	E. T. Ligon	Demopolis, Ala	Mar. 9, 1869	87, 687
Car axle, Railway	E. J. Mallett	New York, N. Y	Nov. 22, 1859	26, 194
Car axle, Railway	J. E. McConnell	Wolverton, England	Oct. 31, 1854	11, 869
Car axle, Railway	N. Miller and E. Bancroft	Franklin, Mass., and Providence, R. I.	Dec. 15, 1843	3, 377
Car axle, Railway	J. Montgomery	Baltimore, Md	Apr. 24, 1860	28, 004
Car axle, Railway	J. C. Nye	Cincinnati, Ohio	Mar. 29, 1864	42, 105
Car axle, Railway	A. E. Smith	Bronxville, N. Y	Jan. 8, 1861	31, 092
Car axle, Railway	F. Sturneyk	Saint Paul, Minn	Aug. 23, 1870	106, 635
Car axle, Railway	T. C. Theaker	New York, N. Y	Apr. 15, 1873	137, 977
Car axle, Railway	L. E. Truesdell	Chicago, Ill	Mar. 22, 1870	101, 187
Car axle, Railway	J. Van Slooten	New Orleans, La	May 18, 1869	90, 321
Car-axle-straightening apparatus	B. S. Shoter	Whistler, Ala	Jan. 18, 1870	99, 014
Car-axles and bearings, Mode of oiling	B. D. Stevens	Lawrence, Mass	June 23, 1863	39, 017
Car-axles and shafts, Construction of	J. Montgomery	Croton Landing, N. Y	Feb. 14, 1871	111, 862
Car axles, Automatic lubricator for railway	W. Baker	Utica, N. Y	Aug. 11, 1857	17, 957
Car axles, Automatic lubricator for railway	A. B. Latta	Cincinnati, Ohio	Aug. 11, 1857	17, 972
Car axles, Box and journal for railway	I. P. Wendell	Philadelphia, Pa	Mar. 2, 1858	19, 530
Car axles, Box-case and lubricator for railway	G. W. and J. C. Geisendorff	Cincinnati, Ohio	Apr. 6, 1858	19, 840
Car-axles, Machine for determining the load of	J. H. Erhardt	Dresden, Saxony	Feb. 13, 1866	52, 653
Car-axles, Machine for rolling	T. Cooper	Cincinnati, Ohio	Oct. 17, 1865	50, 458
Car-axles, Oil-box for	J. F. Sharp	Wilmington, Del	Mar. 16, 1869	87, 880
Car axles, Oil-box for	T. B Stewart	Hartford, Conn	May 31, 1870	103, 676
Car-axles or shaftings, Bearing for	C. Williams	Adrian, Mich	Feb. 9, 1864	41, 563
Car-axles, Preventing consequences of breaking	J. S. Kite	Philadelphia, Pa	July 14, 1834	....
Car-axles, Ribbed metal plates to construct hollow	W. A. Lewis	Chicago, Ill	Mar. 19, 1872	124, 834
Car-axles with boxes, Device for connecting	T. Pritchard	Philadelphia, Pa	June 24, 1873	140, 303
Car, Barrel	W. A. Plantz	Iowa Falls, Iowa	May 23, 1871	115, 097
Car-basket racks, Mode of attaching	M. Tower	Boston, Mass	July 11, 1871	116, 891
Car basket, Railway	J F. Hudson	East Cambridge, Mass	July 1, 1873	140, 534
Car basket, Railway	M. Tower	Boston, Mass	Nov. 14, 1871	121, 024
Car bodies and trucks, Connecting and disconnecting.	C. D. Tisdale	Boston, Mass	June 10, 1873	139, 835

Index of patents issued from the United States Patent Office from 1790 *to* 1873, *inclusive*—Continued.

Invention.	Inventor.	Residence.	Date.	No.
Car-bodies, Hanging	L. B. Thyng	Lowell, Mass	Nov. 18, 1845	4, 276
Car bodies, Mode of supporting railway	R. Imlay	Philadelphia, Pa	Sept. 21, 1837	389
Car-bodies with trucks, Mode of connecting	C. S. Moore	Alexandria, Va	July 3, 1860	28, 995
Car-body elevator, Railway	R. Wells	Jeffersonville, Ind	Sept. 7, 1869	94, 680
Car-body frame	S. Merrick	New Brighton, Pa	Apr. 2, 1867	63, 548
Car-body, Iron	T. E. Warren	Troy, N. Y	Oct. 18, 1853	10, 142
Car body, Railway	G. S. Hacker	Charleston, S. C	Jan. 21, 1841	1, 937
Car body, Railway	R. L. Omensetter	Philadelphia, Pa	May 28, 1872	127, 186
Car boiler and engine, Street	W. Baxter	Newark, N. J	Nov. 14, 1871	120, 930
Car bolster, Railway	J. Marquis and J. W. Kimmel	Crestline, Ohio	May 28, 1867	65, 098
Car bolt, Freight	C. E. Shanahan	South Bend, Ind	Nov. 5, 1872	132, 739
Car box case and pedestal, Railway	J. C. Geisendorff	Cincinnati, Ohio	July 13, 1858	20, 871
Car box, Railway	G. F. Lynch	Milwaukee, Wis	Jan. 23, 1866	52, 181
Car box, Railway	D. Matthew	Philadelphia, Pa	Jan. 12, 1858	19, 095
Car box, Railway	I. P. Wendell	Philadelphia, Pa	Oct. 16, 1866	58, 927
Car-brake	A. M. Allen	New York, N. Y	Jan. 11, 1870	98, 731
Car-brake	A. I. Ambler	Chicago, Ill	Sept. 22, 1863	40, 005
Car-brake	A. I. Ambler	Chicago, Ill	Aug. 30, 1864	44, 036
Car-brake	W. W. Babcock	Harmar, Ohio	Oct. 27, 1868	83, 442
Car-brake	W. Ball	Oregon, Mo	June 25, 1872	128, 277
Car-brake	J. L. Barnes	Etna Green, Ind	Oct. 15, 1867	69, 892
Car-brake	W. T. Batty	Canton, Ohio	Apr. 28, 1868	77, 159
Car-brake	J. H. Beatty	Franklin, Pa	Jan. 19, 1869	85, 987
Car-brake	C. Bemis	Mishawaka, Ind	Jan. 1, 1867	60, 850
Car-brake	W. P. Blades	Baltimore, Md	July 14, 1868	79, 805
Car-brake	V. W. Blanchard	Bridport, Vt	Mar. 13, 1866	53, 106
Car-brake	A. A. Bliven	Jersey City, N. J	Mar. 30, 1869	88, 438
Car-brake	G. W. Bridgman	Somerville, Mass	June 11, 1861	32, 549
Car-brake	J. F. Brode	Memphis, Tenn	Nov. 21, 1871	121, 150
Car-brake	J. M. Brown	Cincinnati, Ohio	July 22, 1862	35, 967
Car-brake	W. H. Browne and W. M. Pegram.	Baltimore, Md	Mar. 11, 1873	136, 581
Car-brake	J. N. Brush	Olmstead County, Minn	Apr. 22, 1873	138, 126
Car-brake	W. H. Burke	Brighton, Mass	Mar. 17, 1868	75, 518
Car-brake	G. E. Burt	Harvard, Mass	Nov. 6, 1866	59, 355
Car-brake	W. L. Burt	Boston, Mass	Aug. 1, 1865	49, 082
Car-brake	J. A. Campbell	South Boston, Mass	Feb. 4, 1868	73, 950
Car-brake	F. A. Canfield	Dover, N. J	Jan. 9, 1872	122, 564
Car-brake	F. M. Chapman	New York, N. Y	Dec. 3, 1872	133, 629
Car-brake	L. Clark	Pine Island, Minn	June 23, 1868	79, 206
Car-brake	J. Cockshoot, jr., and H. Weatherill.	Manchester, Great Britain	Oct. 13, 1868	83, 042
Car-brake	J. M. Collins	New Bedford, Mass	Mar. 7, 1865	46, 642
Car-brake	G. W. Combs	Canandaigua, N. Y	Apr. 14, 1868	76, 715
Car-brake	T. B. Comins, jr	Lowell, Mass	Sept. 10, 1867	68, 700
Car-brake	T. B. Comins, jr	Lowell, Mass	Sept. 10, 1867	68, 701
Car-brake	W. G. Creamer	Brooklyn, N. Y	Jan. 21, 1868	73, 509
Car-brake	D. S. Cross	Cincinnati, Ohio	Feb. 3, 1863	37, 568
Car-brake	J. Davis	Allegheny City, Pa	May 29, 1866	55, 067
Car-brake	J. Davis	Allegheny City, Pa	Aug. 20, 1867	67, 962
Car-brake	S. Davis	Dartmouth, Mass	Dec. 10, 1867	71, 856
Car-brake	F. Dengler	North Vernon, Ind	Jan. 7, 1868	72, 984
Car-brake	P. S. Devlan	Elizabethport, N. J	July 9, 1861	32, 758
Car-brake	J. D'homergue	New York, N. Y	Dec. 20, 1853	10, 336
Car-brake	D. Dick and O. W. Preston, jr	Corning, N. Y	Jan. 7, 1868	72, 985
Car-brake	S. R. Dimmock	Syracuse, N. Y	Aug. 8, 1865	49, 243
Car-brake	S. F. Dimock	Spencer, Ohio	May 7, 1867	64, 409
Car-brake	D. H. Dotterer	Philadelphia, Pa	Dec. 22, 1868	85, 374
Car brake	W. H. Dunham and J. Widney	Allegheny City, Pa	Dec. 3, 1867	71, 591
Car-brake	E. H. Dunn	Paris, France	Dec. 3, 1867	71, 720
Car-brake	W. Ebbitt	New York, N. Y	Jan. 23, 1872	123, 003
Car-brake	W. Ebbitt	New York, N. Y	June 24, 1873	140, 260
Car-brake	J. B. Ellinwood	Hillsborough, N. H	Sept. 26, 1848	5, 807
Car-brake	E. Farnsworth	Tipton, Pa	Apr. 22, 1873	138, 141
Car-brake	A. W. Filer and L. T. Hatfield	Danby, Ill	Aug. 29, 1871	118, 443
Car-brake	N. B Forrest	Auburn, N. Y	July 3, 1866	56, 030
Car-brake	W. S. Foster	Foster's Crossing, Ohio	Dec. 10, 1872	133 844
Car-brake	J. W. Gibbs	New Haven, Conn	Apr. 17, 1866	53, 971
Car-brake	J. C. Gibson	Sacramento, Cal	Sept. 24, 1872	131, 678
Car-brake	H. A. Goodman	Omaha, Neb	Aug. 2, 1870	106, 052
Car-brake	W. D. Goodnow	Albany, N. Y	Oct. 18, 1864	44, 718
Car-brake	M. W. Griswold	New York, N. Y	Aug. 9, 1870	106, 152
Car-brake	S. Guilbert	New York, N. Y	May 1, 1866	54, 333
Car-brake	C. H. Gustin	Worcester, Mass	July 18, 1865	48, 810
Car-brake	O. J. Hardgrove	Canton, Ohio	Feb. 27, 1866	52, 848
Car-brake	E. P. Harrington	Volusia, N. Y	Nov. 11, 1873	144, 537
Car-brake	S. E. Harrison	New Haven, Conn	May 28, 1872	127, 233
Car-brake	G. H. Henkel	Middletown, Ohio	Apr. 24, 1866	54, 152
Car-brake	A. C. Herron	West Farms, N. Y	Feb. 12, 1861	31, 387
Car-brake	A. Higley	South Bend, Ind	Dec. 18, 1866	60, 515
Car-brake	A. Higley	South Bend, Ind	Feb. 19, 1867	62, 198
Car-brake	A. Higley	South Bend, Ind	Feb. 19, 1867	62, 199
Car-brake	A. Higley	South Bend, Ind	Mar. 12, 1867	62, 847
Car-brake	J. Hirst	Jamaica, N. Y	Dec. 1, 1868	84, 490
Car-brake	J. Hough and J. Moore	Buckingham and Bart, Pa	Dec. 14, 1858	22, 291
Car brake	R. Hurd	Morrison, Ill	June 28, 1870	104, 738
Car-brake	J. W. Jacobs	Jeffersonville, Ind	Mar. 12, 1872	124, 581
Car-brake	E. F. Jewett	Plainville, Ohio	May 1, 1860	28, 087
Car-brake	G. N. Jones	Oshkosh, Wis	Sept. 22, 1868	82, 324
Car-brake	M. Karg	Albany, N. Y	Nov. 18, 1873	144, 681
Car-brake	J. Katzenberg	New York, N. Y	Dec. 31, 1867	72, 861
Car-brake	W. R. and H. E. Kay	Westerly, R. I	June 25, 1861	32, 665
Car-brake	P. Keffer	Reading, Pa	Dec. 18, 1860	30, 946
Car-brake	P. Keffer	Reading, Pa	Oct. 27, 1868	83, 508
Car-brake	T. A. Kelley	Cleveland, Ohio	Nov. 26, 1867	71, 391
Car-brake	W. H. Kilburn	Kennedy, N. Y	June 19, 1866	55, 670
Car-brake	W. Kimball	Woburn, Mass	June 18, 1872	127, 979
Car-brake	J. L. Knowlton	Sharon Hill, Pa	Mar. 25, 1873	137, 079

Index of patents issued from the United States Patent Office from 1790 *to* 1873, *inclusive*—Continued.

Invention.	Inventor.	Residence.	Date.	No.
Car-brake	A. Langellier	Concord, N. H	Sept. 23, 1873	143, 083
Car-brake	C. B. Lashar	New York, N. Y	Dec. 24, 1861	34, 000
Car-brake	J. W. Latcher and W. J. Powell	Amsterdam, N. Y	Dec. 27, 1864	45, 617
Car-brake	S. M. Lee	New London, Iowa	Apr. 16, 1867	63, 909
Car-brake	S. M. Lee	New London, Iowa	July 28, 1868	80, 290
Car-brake	A. Z. Long	Scranton, Pa	Oct. 29, 1867	70, 235
Car-brake	G. F. Lynch	Milwaukee, Wis	Jan. 16, 1866	52, 063
Car-brake	G. F. Lynch	Milwaukee, Wis	Aug. 14, 1866	57, 161
Car-brake	S. Marden	Newton, Mass	Feb. 11, 1868	74, 235
Car-brake	J. McCabe	Rondout, N. Y	Jan. 23, 1872	123, 033
Car-brake	S. McCambridge	Philadelphia, Pa	Nov. 28, 1865	51, 204
Car-brake	S. McCambridge	Philadelphia, Pa	May 8, 1866	54, 573
Car-brake	S. McCambridge	Philadelphia, Pa	Feb. 5, 1867	61, 844
Car-brake	S. McCambridge	Philadelphia, Pa	July 28, 1868	80, 419
Car-brake	S. McCambridge and E. G. Martin.	Philadelphia, Pa	Sept. 18, 1866	58, 115
Car-brake	S. McCambridge and E. G. Martin.	Philadelphia, Pa	July 28, 1868	80, 420
Car-brake	H. A. Mears	Pecatonica, Ill	May 1, 1860	28, 098
Car-brake	G. L. Miller	De Witt, N. Y	Aug. 18, 1868	81, 192
Car-brake	G. S. Miller, E. B. Peck, and W. Olmsted.	Thompsonville, Bridgeport, and Thompsonville, Conn.	Feb. 23, 1864	41, 744
Car-brake	J. Mitchell	La Porte, Ind	Dec. 10, 1867	72, 069
Car-brake	J. H. Moore and J. E. Gary	Chicago, Ill	Feb. 11, 1868	74, 240
Car-brake	J. H. Moore and J. E. Gary	Chicago, Ill	Dec. 29, 1868	85, 326
Car-brake	W. E. Moore	Crawfordsville, Ind	Mar. 15, 1859	23, 259
Car-brake	B. Morahan	New York, N. Y	June 28, 1864	43, 370
Car-brake	W. T. Morrow	Chicago, Ill	Dec. 22, 1863	41, 042
Car-brake	D. Myers	Chicago, Ill	Jan. 30, 1866	52, 310
Car-brake	D. Myers	Chicago, Ill	Jan. 30, 1866	52, 311
Car-brake	D. Myers	Chicago, Ill	Jan. 30, 1866	52, 312
Car-brake	D. Myers	Chicago, Ill	Oct. 16, 1866	58, 877
Car-brake	D. Myers	Chicago, Ill	Dec. 4, 1866	60, 228
Car-brake	D. Myers	Chicago, Ill	June 30, 1868	79, 379
Car-brake	W. Naylor	Mildmay Park, England	Nov. 4, 1873	144, 351
Car-brake	W. Nelson	Flatbush, N. Y	Dec. 10, 1872	133, 879
Car-brake	H. W. Norvill	Livingston, Ala	Aug. 21, 1860	29, 712
Car-brake	G. F. Outten	Norfolk, Va	Oct. 25, 1859	25, 911
Car-brake	P. Pardee	New Haven, Conn	Jan. 1, 1867	60, 930
Car-brake	W. T. Parsons	Thomasville, Ga	June 9, 1868	78, 759
Car-brake	J. B. Pelton	Mount Pleasant, Md	Dec. 31, 1872	134, 484
Car-brake	J. B. Pelton	Mount Pleasant, Md	Nov. 18, 1873	144, 787
Car-brake	J. Pettingill, jr	Jackson, N. H	July 16, 1861	32, 841
Car-brake	G. Quick and J. N. Wallis	Fleming, N. Y	Nov. 26, 1867	71, 325
Car-brake	J. W. Reid	New York, N. Y	Sept. 18, 1866	58, 136
Car-brake	G. H. Reynolds	Parsons, Kans	Feb. 20, 1872	123, 840
Car-brake	J. W. Rice	Springfield, Mass	Apr. 23, 1867	64, 036
Car-brake	E. O. Richard	Quebec, Canada	Sept. 23, 1873	143, 187
Car-brake	D. T. Robinson	Boston, Mass	Mar. 27, 1866	53, 544
Car-brake	F. Root	Boston, Mass	Aug. 25, 1868	81, 411
Car-brake	A. M. Rouse	Saint Louis, Mo	Jan. 14, 1873	134, 934
Car-brake	J. B. Rumsey	Tiffin, Ohio	Mar. 13, 1866	53, 235
Car-brake	L. D. Rundell	South Westerlo, N. Y	Oct. 30, 1866	59, 273
Car-brake	W. H. M. Sanger	Bay Side, N. Y	Jan. 21, 1873	135, 161
Car-brake	A. H. Sassaman	Lebanon, Pa	June 23, 1868	79, 149
Car-brake	S. W. Y. Shimonsky	Cheyenne, Dak	Aug. 25, 1868	81, 414
Car-brake	G. H. Seymour	Newark, Ohio	Apr. 5, 1870	101, 667
Car-brake	J. Shannon	Cohoes, N. Y	May 6, 1873	138, 535
Car-brake	J. Sheward and G. A. Stanbery.	Dunkirk, N. Y	Apr. 23, 1867	64, 040
Car-brake	W. S. Shotwell	Paterson, N. J	Sept. 24, 1867	69, 262
Car-brake	T. E. Sickels	Kennett's Square, Pa	July 12, 1870	105, 264
Car-brake	C. W. Singer	Anderson Store, Va	Oct. 16, 1866	58, 956
Car-brake	L. S. Sisson	West Edmeston, N. Y	Dec. 15, 1868	85, 034
Car-brake	C. A. Smith	Jersey City, N. J	May 16, 1871	114, 983
Car-brake	J. L. Smith	Tuscola, Ill	Sept. 3, 1867	68, 519
Car-brake	J. W. W. Smith	Canton, Mo	Feb. 23, 1869	87, 214
Car-brake	W. H. Smith and J. Steger	New York, N. Y	Feb. 25, 1868	74, 860
Car-brake	C. Spofford	Boston, Mass	Nov. 12, 1867	70, 911
Car-brake	J. Steger	New York, N. Y	Oct. 27, 1868	83, 418
Car-brake	J. Steger	New York, N. Y	June 28, 1870	104, 897
Car-brake	E. Stiles	New York, N. Y	Apr. 14, 1868	76, 843
Car-brake	W. Stinehart and J. Taggart	Charlestown, Mass	Feb. 27, 1849	6, 150
Car-brake	B. Tatham and J. Steger	New York, N. Y	Oct. 27, 1868	83, 422
Car-brake	J. Temple	Selin's Grove, Pa	June 26, 1866	55, 964
Car-brake	J. Temple	Bellefonte, Pa	Jan. 30, 1872	123, 133
Car-brake	J. Temple	Mooresburgh, Pa	Aug. 5, 1873	141, 608
Car-brake	E. Thayer	Chatham, N. Y	Aug. 12, 1846	4, 690
Car-brake	S. H. Timmons	La Fayette, Ind	Dec. 4, 1866	60, 284
Car-brake	W. W. Todd and J. Vandercar	Brooklyn, N. Y	Sept. 19, 1865	50, 051
Car-brake	H. H. Trenor	New York, N. Y	Oct. 17, 1865	50, 517
Car-brake	H. H. Trenor	New York, N. Y	Aug. 7, 1866	57, 015
Car-brake	L. A. Velu and E. F. and L. E. A. Fosse.	Paris, France	Nov. 27, 1866	60, 094
Car-brake	I. H. Voorhies	Ypsilanti, Mich	Mar. 25, 1873	137, 264
Car-brake	J. N. Ward	Brooklyn, N. Y	Nov. 2, 1858	21, 996
Car-brake	W. Warinner	Creelsborough, Ky	Sept. 9, 1873	142, 658
Car-brake	W. Warner	Philadelphia, Pa	Mar. 20, 1866	53, 365
Car-brake	W. Warwick and H. C. Duggan	Pittsburgh, Pa	Nov. 4, 1873	144, 240
Car-brake	A. A. Weidemeyer	Williamsburgh, N. Y	Jan. 2, 1872	122, 419
Car-brake	J. White and T. Lingle	South Amboy, N. J	Mar. 10, 1868	75, 501
Car-brake	E. Whitehead	Cincinnati, Ohio	July 10, 1866	56, 302
Car-brake	T. J. Whitney	Whitpain Township, Pa	May 19, 1868	78, 166
Car-brake	J. G. Wiggin	Moultonville, N. H	Dec. 9, 1873	145, 470
Car-brake	W. E. Wilcox	Cleveland, Ohio	Feb. 6, 1866	52, 501
Car-brake	J. W. Williams	Somerville, N. J	Mar. 12, 1867	62, 912
Car-brake	J. F. Wood, 2d	Cohocton, N. Y	Oct. 22, 1867	70, 145

Index of patents issued from the United States Patent Office from 1790 *to* 1873, *inclusive*—Continued.

Invention.	Inventor.	Residence.	Date.	No.
Car-brake	T. C. Woods	Marion County, Ky	Aug. 25, 1868	81, 568
Car-brake	J. E. Worthman	Mobile, Ala	Dec. 10, 1872	133, 913
Car-brake	W. C. Wright	New York, N. Y	Jan. 15, 1861	31, 137
Car brake and starter	E. Bonneval	Boston, Mass	Mar. 14, 1871	112, 534
Car brake and starter	J. A. Cody	Cleveland, Ohio	May 9, 1871	114, 529
Car brake and starter	J. A. Cole	Adams, N. Y	July 27, 1869	93, 059
Car brake and starter	M. Henderson	Detroit, Mich	May 26, 1868	78, 371
Car brake and starter	R. Heneage	Buffalo, N. Y	Sept. 17, 1867	68, 875
Car brake and starter	C. M. Hinckley	Boston, Mass	Mar. 26, 1872	125, 049
Car brake and starter	C. L. Irving	Indianapolis, Ind	Apr. 27, 1869	89, 410
Car brake and starter	R. R. James	Rising Sun, Ind	July 2, 1872	128, 491
Car brake and starter	W. J. Johnson	New Orleans, La	Aug. 11, 1868	80, 964
Car brake and starter	E. P. Jones	Shell Mound, Miss	Mar. 15, 1870	100, 960
Car brake and starter	J. M. McMaster	Rochester, N. Y	July 6, 1869	92, 335
Car brake and starter	J. M. McMaster	Rochester, N. Y	Jan. 11, 1870	98, 788
Car brake and starter	C. D. Moody	Saint Louis, Mo	May 12, 1868	77, 903
Car brake and starter	D. M. Moore	Windsor, Vt	July 20, 1869	92, 869
Car brake and starter	J. Paradis	Brooklyn, N. Y	Mar. 21, 1871	112, 954
Car brake and starter	I. Rider	Indianapolis, Ind	Feb. 16, 1869	87, 068
Car brake and starter	L. J. Smith and D. S. Knight	Hamilton, Ohio, and New York, N. Y.	Apr. 28, 1868	77, 223
Car brake and starter	W. M. Starr	Washington, D. C	Oct. 24, 1871	120, 341
Car brake and starter	A. B. Vandemark	Phelps, N. Y	Feb. 9, 1869	86, 885
Car brake and starter	J. Wiley, 2d	South Reading, Mass	Nov. 19, 1867	71, 102
Car brake and starter	J. W. Wilson	New York, N. Y	Feb. 1, 1870	99, 509
Car brake and starter	J. S. Wood	Lansing, Mich	Aug. 4, 1868	80, 695
Car brake and starter, Combined	C. B. Broadwell	New Orleans, La	July 23, 1872	129, 782
Car brake and starter combined, Street	C. L. Irving	Indianapolis, Ind	Oct. 21, 1873	143, 909
Car brake and starter, Railway	R. R. Carpenter	Tippecanoe, Ohio	Mar. 30, 1869	88, 447
Car brake and starting apparatus	T. W. Murray	New York, N. Y	Sept. 3, 1867	68, 452
Car-brake apparatus, Steam-power	G. Westinghouse, jr	Pittsburgh, Pa	June 6, 1871	115, 667
Car-brake, Atmospheric	W. L. Chambers	Pleasant Unity, Pa	July 6, 1869	92, 265
Car-brake, Atmospheric	S. Marsh	Littleton, N. H	Apr. 12, 1870	101, 895
Car-brake attachment	J. Kirkley	Chicago, Ill	Sept. 22, 1868	82, 329
Car-brake, Automatic	J. R. Crabill	La Crosse, Ill	July 11, 1871	116, 814
Car-brake, Automatic	J. Hartman, jr	Philadelphia, Pa	Dec. 26, 1865	51, 715
Car-brake, Automatic	M. Wilkin and J. Clark	London, England	Mar. 15, 1870	100, 958
Car brake, Automatic railway	G. En-Earl	Boston, Mass	June 3, 1873	139, 557
Car brake, Automatic railway	P. R. Higley	Oshawa, Canada	Mar. 8, 1864	41, 890
Car brake, Automatic railway	W. R. Jackson	Baltimore, Md	Sept. 8, 1857	18, 150
Car-brake beam	D. Wellington	Virginia City, Nev	Dec. 16, 1873	145, 605
Car-brake block	B. De Vout	Harrisburgh, Pa	Dec. 5, 1865	51, 298
Car brake, Bumper railway	L. Brauer	Sommerville, Tenn	May 12, 1857	17, 257
Car-brake, Coal	D. Wetsel	Morris, Pa	May 20, 1873	139, 097
Car-brake, Eccentric	E. Morris	Bloomfield, N. J	Sept. 19, 1838	928
Car-brake, Electro-magnetic	F. F. A. Achard	Paris, France	Nov. 20, 1866	59, 805
Car-brake, Electro-magnetic	H. S. Daggett	La Fayette, Ind	July 11, 1871	116, 818
Car-brake, Electro-magnetic	J. Olmsted	Galesburgh, Ill	Apr. 27, 1869	89, 495
Car-brake, Electro-magnetic	J. Olmsted	Knoxville, Ill	Jan. 18, 1870	98, 880
Car-brake for city-railways	R. W. Jenks, jr., and F. A. Steere.	Providence, R. I	Apr. 17, 1860	27, 910
Car-brake for horse-railways	J. Stephenson	New York N. Y	Feb. 23, 1869	87, 122
Car brake, Hand	J. McGinn	Norwalk, Ohio	Dec. 10, 1872	133, 872
Car brake, Horse	G. R. Barker	Germantown, Pa	Sept. 17, 1861	33, 287
Car brake, Horse	J. Stephenson	New York, N. Y	Dec. 27, 1859	26, 626
Car brake, Hydraulic	J. F. Taylor	Charleston, S. C	Nov. 11, 1873	144, 578
Car-brake, Hydraulic or air	W. M. Henderson	Philadelphia, Pa	Oct. 28, 1873	143, 980
Car brake, Hydraulic railway	W. M. Henderson	Philadelphia, Pa	Aug. 12, 1873	141, 790
Car-brake, Iron	S. Morse	Springfield, Mass	Sept. 6, 1853	10, 004
Car-brake, Magnetic	J. Olmstead	Knoxville, Ill	Jan. 8, 1867	61, 089
Car-brake pipe, Atmospheric	G. Westinghouse, jr	Pittsburgh, Pa.	Nov. 29, 1870	109, 695
Car-brake, Pneumatic	D. Myers and A. B. Pullman	Chicago, Ill	Nov. 1, 1870	108, 932
Car-brake, Propelling	R. Heneage	Buffalo, N. Y	Feb. 26, 1867	62, 419
Car brake, Railway	L. Adams	Mattoon, Ill	Oct. 28, 1873	144, 013
Car brake, Railway	A. M. Allen	New York, N. Y	June 15, 1869	91, 404
Car brake, Railway	A. I. Ambler	Milwaukee, Wis	Apr. 1, 1862	34, 855
Car brake, Railway	A. I. Ambler	Milwaukee, Wis	May 27, 1 62	35, 408
Car brake, Railway	A. I. Ambler	Milwaukee, Wis	June 17, 1862	35, 647
Car brake, Railway	A. I. Ambler	Milwaukee, Wis	June 24, 1862	35, 727
Car brake, Railway	A. I. Ambler	Chicago, Ill	Sept. 22, 1863	40, 007
Car brake, Railway	J. S. Anderson	Flintville, Wis	June 3, 1873	139, 527
Car brake, Railway	F. Armstrong	New Orleans, La	Aug. 9, 1859	24, 981
Car brake, Railway	D. Arndt and D. C. Washington	Cleveland, Ohio	Sept. 19, 1871	119, 103
Car brake, Railway	S. and W. W. Balkwill	Cleveland, Ohio	July 30, 1872	129, 880
Car brake, Railway	V. Barnes	Washington, D. C	Mar. 11, 1856	14, 385
Car brake, Railway	L. B. Batcheller	Arlington, Vt	Jan. 24, 1854	10, 440
Car brake, Railway	F. G. Bates	Springfield, Mass	Feb. 18, 1873	136, 019
Car brake, Railway	A. Bean	Dedham, Mass	Mar. 6, 1860	27, 344
Car brake, Railway	N. Berkeley	Aldie, Va	Aug. 3, 1869	93, 166
Car brake, Railway	J. T. Blois	Jonesville, Mich	Apr. 20, 1869	89, 014
Car brake, Railway	M. S. Borthwick	Montana, Iowa	Nov. 9, 1869	96, 542
Car brake, Railway	J. L., L., and D. W. Branch	Charleston, S. C	Jan. 5, 1858	19, 012
Car brake, Railway	F. E. Canda	Chicago, Ill	May 16, 1865	47, 700
Car brake, Railway	J. H Champlin	Essex, Conn	Jan. 24, 1865	45, 975
Car brake, Railway	H. E Chapman	Albany, N. Y	Dec. 7, 1858	22, 229
Car brake, Railway	S F. Closser	Salt Lake City, Utah	July 23, 1872	129, 789
Car brake, Railway	H. M. Collier	Binghamton, N. Y	June 29, 1858	20, 769
Car brake, Railway	M. P. Coons	Brooklyn, N. Y	Aug. 15, 1854	11, 517
Car brake, Railway	W. E. Cooper	Dunkirk, N. Y	Apr. 12, 1859	23, 663
Car brake, Railway	W. G. Creamer	New York, N. Y	Nov. 4, 1856	16, 004
Car brake, Railway	J. M. Crosby	Marathon, N. Y	Aug. 24, 1869	94, 083
Car brake, Railway	J. M. Crosby and W. Ballard	Caroline, N. Y	Feb. 8, 1870	99, 539
Car brake, Railway	D. S. Cross	Cincinnati, Ohio	Apr. 4 1865	47, 091
Car brake, Railway	G. W. Crowe	Cincinnati, Ohio	July 1, 1873	140, 349
Car brake, Railway	C. D. Culver	Mauch Chunk, Pa	Dec. 29, 1868	85, 369
Car brake, Railway	G. W. Cummings	Philadelphia, Pa	June 1, 1858	20, 468
Car brake, Railway	J. Darling	Whistler, Ala	Apr. 16, 18 2	125, 665
Car brake, Railway	M. F. Daugherty	Portsmouth, Va	Oct. 3, 1871	119, 582

Index of patents issued from the United States Patent Office from 1790 *to* 1873, *inclusive*—Continued.

Invention.	Inventor.	Residence.	Date.	No.
Car brake, Railway	H. Davis	Baltimore, Md	Sept. 13, 1859	25, 392
Car brake, Railway	J. Davis	Allegheny City, Pa	Oct. 20, 1863	40, 332
Car brake, Railway	J. Davis	Allegheny City, Pa	June 7, 1864	43, 010
Car brake, Railway	D. Derr	Bellefonte, Pa	Feb. 26, 1861	31, 583
Car brake, Railway	S. R. Dimock	Pittsfield, Mass	Dec. 9, 1862	37, 088
Car brake, Railway	G. Dorsch	Schenectady, N. Y	Apr. 13, 1858	19, 917
Car brake, Railway	C. B. Eaton	Grafton, Ill	June 20, 1871	116, 169
Car brake, Railway	C. H. Eisenbrandt	Baltimore, Md	May 25, 1858	20, 339
Car brake, Railway	D. M. Elder	Monmouth, Ill	Feb. 8, 1870	99, 545
Car brake, Railway	J. F. Elder	Blairsville, Pa	Jan. 28, 1873	135, 212
Car brake, Railway	R. M. Evans	Laconia, N. H	Apr. 8, 1856	14, 640
Car brake, Railway	K. Exeter, sr	Munich, Bavaria	July 30, 1872	130, 029
Car brake, Railway	K. Exeter, sr	Munich, Bavaria	July 30, 1872	130, 030
Car brake, Railway	D. H. Feger	Cincinnati, Ohio	Mar. 23, 1858	19, 734
Car brake, Railway	F. M. Finnell	Covington, Ky	Nov. 29, 1870	109, 727
Car brake, Railway	W. G. Foster	Apalachin, N. Y	Feb. 1, 1870	99, 424
Car brake, Railway	E. D. B. Freer	Lima Centre, Mich	Oct. 26, 1869	96, 098
Car brake, Railway	M. S. Frost	Detroit, Mich	June 3, 1856	15, 038
Car brake, Railway	J. C. Gove	Cleveland, Ohio	Apr. 4, 1871	113, 290
Car brake, Railway	S. Gumaer	Chicago, Ill	Jan. 26, 1858	19, 192
Car brake, Railway	J. T. Guthrie	Pittsburgh, Pa	Dec. 29, 1868	85, 302
Car brake, Railway	W. Hall	North Adams, Mass	July 20, 1852	9, 135
Car brake, Railway	D. Harrigan	Winchester, Mass	Nov. 4, 1856	16, 011
Car brake, Railway	O. J. Harrington	Manchester, Pa	Mar. 15, 1864	41, 917
Car brake, Railway	O. J. Harrington	Manchester, Pa	Apr. 5, 1864	42, 189
Car brake, Railway	A. Higley	South Bend, Ind	Aug. 14, 1866	57, 133
Car brake, Railway	A. Higley	Cleveland, Ohio	Apr. 12, 1870	101, 874
Car brake, Railway	L. Hill and S. D. Tripp	Stoneham and Lynn, Mass	Mar. 7, 1871	112, 456
Car brake, Railway	R. W. Hix	New York, N. Y	Dec. 11, 1860	30, 921
Car brake, Railway	B. Holly	Seneca Falls, N. Y	Feb. 10, 1852	8, 730
Car brake, Railway	T. Hopper	Newark, N. J	June 1, 1858	20, 429
Car brake, Railway	T. and H. Hopper and C. C. Coats.	Newark, N. J	June 7, 1870	103, 886
Car brake, Railway	J. C. Hosmer	Boston, Mass	Apr. 30, 1872	126, 208
Car brake, Railway	G. Hotchkiss	Windsor, N. Y	Mar. 20, 1855	12, 552
Car brake, Railway	J. Hough	Buckingham, Pa	Apr. 14, 1863	38, 166
Car brake, Railway	J. Houston and E. Ross	Manchester, N. H	July 20, 1852	9, 141
Car brake, Railway	E. N. Huntsman	Allegheny, Pa	Feb. 8, 1870	99, 678
Car brake, Railway	M. Ingersoll	Grafton, Ohio	Mar. 6, 1866	53, 006
Car brake, Railway	E. P. Jones	Shell Mound, Miss	Mar. 15, 1870	100, 900
Car brake, Railway	J. Jones	Utica, N. Y	July 29, 1862	36, 007
Car brake, Railway	J. Jones	West Albany, N. Y	Sept. 12, 1865	49, 885
Car brake, Railway	E. M. Judd	New Britain, Conn	Dec. 9, 1862	37, 098
Car brake, Railway	L. Kirk	Reading, Pa	May 10, 1859	23, 923
Car brake, Railway	J. J. Kiser	Middletown, Ind	Oct. 26, 1869	96, 240
Car brake, Railway	J. and W. Kitchen and S. Samuels.	Accrington, England	Nov. 26, 1867	71, 392
Car brake, Railway	B. Kraft	Reading, Pa	Apr. 20, 1852	8, 899
Car brake, Railway	J. S. Lamar	Augusta, Ga	July 30, 1872	129, 968
Car brake, Railway	J. W. Latcher	Northville, N. Y	Feb. 14, 1865	46, 366
Car brake, Railway	C. H. Lathrop	Jersey City, N. J	Dec. 26, 1871	122, 264
Car brake, Railway	D. P. Léfebre and L. P. Dorre	France	Mar. 29, 1870	101, 280
Car brake, Railway	A. H. Lighthall	Albany, N. Y	May 10, 1870	102, 840
Car brake, Railway	H. A. Lincoln and H. T. Douglass.	New Haven, Conn	Aug. 2, 1859	24, 943
Car brake, Railway	J. P. Martin	Philadelphia, Pa	July 13, 1852	9, 116
Car brake, Railway	W. S. Martin	Waukegan, Ill	Dec. 22, 1863	41, 013
Car brake, Railway	A. F. McCrone	Ellicott's Mills, Md	Apr. 25, 1865	47, 440
Car brake, Railway	J. W. McGlashan	Montreal, Canada	July 18, 1871	117, 190
Car brake, Railway	T. G. McLaughlin	Kensington, Pa	May 4, 1852	8, 924
Car brake, Railway	T. G. McLaughlin	Philadelphia, Pa	Aug. 15, 1854	11, 527
Car brake, Railway	T. J. Mead	Port Byron, N. Y	Sept. 6, 1859	25, 345
Car brake, Railway	G. H. Merriam	Portland, Me	Feb. 13, 1872	123, 718
Car brake, Railway	J. Mitchell	Osceola, Iowa	Aug. 18, 1857	18, 018
Car brake, Railway	W. Montgomery	Roxbury, Mass	July 6, 1852	9, 098
Car brake, Railway	G. W. Morris	Pittsburgh, Pa	May 6, 1873	138, 678
Car brake, Railway	J. Mulvey and C. Ohlemacher	Aurora, Ill	Sept. 3, 1861	33, 210
Car brake, Railway	D. Myers	South Bend, Ind	Dec. 23, 1862	37, 238
Car brake, Railway	D. Myers	South Bend, Ind	July 21, 1863	39, 306
Car brake, Railway	W. Naylor	Mildmay Park, England	July 23, 1872	129, 855
Car brake, Railway	M. P. Norton	Tinmouth, Vt	Nov. 7, 1854	11, 904
Car brake, Railway	S. Nunamaker	Osceola Mills, Pa	Aug. 13, 1872	130, 526
Car brake, Railway	M. Obermiller	Tiffin, Ohio	Jan. 3, 1860	26, 702
Car brake, Railway	J. Paradis	Brooklyn, N. Y	Oct. 24, 1871	120, 316
Car brake, Railway	L. Parker	Manchester Station, Conn	July 9, 1861	32, 783
Car brake, Railway	D. J. Parmele	San Francisco, Cal	Dec. 15, 1868	84, 963
Car brake, Railway	T. Payne	Detroit, Mich	Dec. 14, 1869	97, 800
Car brake, Railway	O. L. Pease and L. E. Skinner	Xenia and Cincinnati, Ohio	Jan. 4, 1870	98, 621
Car brake, Railway	W. Perkins	Plympton, Mass	July 19, 1859	24, 818
Car brake, Railway	W. D. Pope	Gadsden, Ala	Sept. 9, 1873	142, 649
Car brake, Railway	W. Portlock and J. R. Dodds	New London, Iowa	Nov. 3, 1868	83, 661
Car brake, Railway	N. Potter	Hillsdale, Mich	Feb. 2, 1858	19, 260
Car brake, Railway	I. N. Pyle	Decatur, Ind	Aug. 11, 1863	39, 495
Car brake, Railway	S. Randall	Centreville, R. I	Nov. 3, 1868	83, 728
Car brake, Railway	B. F. Reimer	Philadelphia, Pa	June 20, 1854	11, 138
Car brake, Railway	E. E. Rice	Hallowell, Me	June 26, 1855	13, 139
Car brake, Railway	I. L. G. Rice	Cambridge, Mass	Nov. 1, 1870	108, 830
Car brake, Railway	J. W. Rice	Springfield, Mass	Aug. 3, 1858	21, 086
Car brake, Railway	J. W. Rice	Springfield, Mass	Dec. 28, 1858	22, 455
Car brake, Railway	J. W. Rice	Springfield, Mass	June 29, 1869	92, 097
Car brake, Railway	H. T. Robbins	Lowell, Mass	Sept. 4, 1849	6, 689
Car brake, Railway	M. H. Rumpf	Paris, France	Oct. 20, 1868	83, 321
Car brake, Railway	L. A. Russell	Shrewsbury, Vt	Mar. 8, 1871	112, 499
Car brake, Railway	E. W. Sandford	Brooklyn, N. Y	Aug. 16, 1870	106, 412
Car brake, Railway	J. Schoenherr	Reading, Pa	Aug. 10, 1852	9, 189
Car brake, Railway	E. S. Scripture and J. H. Darrah.	Brooklyn, N. Y., and Aurora, Ill.	June 29, 1869	92, 104
Car brake, Railway	M. Shimer	Union Township, Pa	Dec. 5, 1854	12, 033

Index of patents issued from the United States Patent Office from 1790 *to* 1873, *inclusive*—Continued.

Invention.	Inventor.	Residence.	Date.	No.
Car brake, Railway	H. C. Sides	Baltimore, Md	July 12, 1843	3, 171
Car brake, Railway	E. Slater	Girard, Pa	Jan. 11, 1870	98, 716
Car brake, Railway	R. L. Smith	Philadelphia, Pa	Apr. 14, 1857	17, 058
Car brake, Railway	T. W. Smith	Alexandria, Va	Jan. 19, 1858	19, 157
Car brake, Railway	W. F. H. Smith	Milwaukee, Wis	Nov. 14, 1871	121, 013
Car brake, Railway	G. A. Somerby and C. W. Fogg.	Waltham, Mass	Aug. 14, 1855	13, 440
Car brake, Railway	W. Somerville, jr	Mitchell Station, Va	Aug. 21, 1860	29, 725
Car brake, Railway	L. Stebbins	Hinsdale, N. Y	Jan. 21, 1873	135, 017
Car brake, Railway	J. Steger	New York, N. Y	Jan. 25, 1570	99, 257
Car brake, Railway	F. A. Stevens	Burlington, Vt	Nov. 25, 1851	8, 552
Car brake, Railway	W. F. Stewart	Patuxent Forge, Md	Oct. 4, 1859	25, 708
Car brake, Railway	S. R. Stinard	Paterson, N. J	Oct. 19, 1869	96, 051
Car brake, Railway	J. W. Swales	San Francisco, Cal	Sept. 7, 1869	94, 525
Car brake, Railway	B. Tatham and J. Steger	New York, N. Y	June 8, 1869	91, 181
Car brake, Railway	A. Tatzel, sr., and F. Kinn	New York, N. Y	Mar. 26, 1872	124, 985
Car brake, Railway	B. O. Thompson	Chicago, Ill	Nov. 3, 1868	83, 674
Car brake, Railway	L. F. Thompson and A. G. Bachelder.	Charlestown and Lowell, Mass.	July 6, 1852	9, 109
Car brake, Railway	C. W. Tierney	Altoona, Pa	Feb. 7, 1871	111, 584
Car brake, Railway	A. F. Toulmir	Ellicott's Mills, Md	Nov. 29, 1859	26, 307
Car brake, Railway	F. Townsend	Albany, N. Y	July 6, 1869	92, 230
Car brake, Railway	I. Townsend	Capeville, Va	Nov. 21, 1871	121, 140
Car brake, Railway	L. W. Tracy	New York, N. Y	Mar. 21, 1871	112, 865
Car brake, Railway	L. Treadwell	New York, N. Y	Apr. 3, 1849	6, 273
Car brake, Railway	G. Trinks	Jersey City, N. J	Apr. 5, 1853	9, 655
Car brake, Railway	E. B. Turner	Providence, R. I	Mar. 13, 1860	27, 486
Car brake, Railway	A. P. Tutton	Reading, Pa	Apr. 19, 1859	23, 727
Car brake, Railway	J. B. Van Dyne	Nashville, Tenn	Dec. 7, 1869	97, 730
Car brake, Railway	E. P. Vinning	Grand Rapids, Mich	Apr. 12, 1870	101, 791
Car brake, Railway	R. M. Wade	Wadesville, Va	Apr. 7, 1857	17, 004
Car brake, Railway	W. B. Wait	Portsmouth, N. H	July 27, 1858	21, 058
Car brake, Railway	T. Walber	New York, N. Y	Mar. 16, 1852	8, 812
Car brake, Railway	I. Walker	Lynn, Mass	Sept. 7, 1869	94, 530
Car brake, Railway	J. N. Walker	Cincinnati, Ohio	Sept. 16, 1862	36, 489
Car brake, Railway	L. Watson	South Plymouth, Mich	Dec. 20, 1861	45, 545
Car brake, Railway	J. E. Weaver	Lancaster, Pa	Feb. 7, 1871	111, 591
Car brake, Railway	A. Wellshmidt	Albany, N. Y	Nov. 7, 1871	120, 603
Car brake, Railway	G. Westinghouse, jr	Pittsburgh, Pa	Sept. 9, 1873	142, 600
Car brake, Railway	A. L. Whipple	Elmira, N. Y	Nov. 30, 1858	22, 213
Car brake, Railway	S. M. Whipple	North Adams, Mass	Mar. 9, 1858	19, 599
Car brake, Railway	T. J. Whitney	Broad Axe, Pa	Aug. 4, 1863	39, 440
Car brake, Railway	A. Whittemore	Cambridgeport, Mass	Oct. 15, 1872	132, 225
Car brake, Railway	G. W. Windsor	Allegheny, Pa	Jan. 26, 1858	19, 223
Car brake, Railway	G. W. B. Yocom and E. Cowan	Arcata, Cal	Nov. 2, 1869	96, 375
Car brake, Railway	G. W. Zeigler	Tiffin, Ohio	May 11, 1858	20, 237
Car-brake, Re-action	J. Hall	Granville, Ohio	Apr. 22, 1862	35, 020
Car-brake, Safety	A. S. Martin	Washington, D. C	Aug. 30, 1870	106, 847
Car brake, Self-acting coal	J. D. Leonard	Cleveland, Ohio	Sept. 9, 1873	142, 572
Car-brake, Self-acting safety	G. S. Griggs	Roxbury, Mass	Dec. 31, 1839	1, 452
Car-brake shoe	J. Bing	Philadelphia, Pa	Oct. 6, 1863	40, 156
Car-brake shoe	J. Bing	Philadelphia, Pa	Sept. 25, 1866	58, 207
Car-brake shoe	E. L. Countiss	Philadelphia, Pa	Mar. 19, 1867	62, 940
Car-brake shoe	S. D. Danfield	Philadelphia, Pa	Oct. 24, 1865	50, 656
Car-brake shoe	G. Drake	Farmington, Me	Mar. 3, 1868	75, 001
Car-brake shoe	C. H. Sollers and J. Rhoads	Harrisburgh, Pa	Nov. 21, 1865	51, 093
Car-brake shoe	I. P. Wendell	Philadelphia, Pa	Oct. 1, 1867	69, 377
Car-brake shoe	H. Werntz	Pine Grove, Pa	Mar. 20, 1866	53, 369
Car-brake shoe	J. Wood	Red Bank, N. J	Jan. 5, 1864	41, 114
Car-brake shoe	J. Wood	Red Bank, N. J	Nov. 15, 1864	45, 106
Car-brake shoe, Railway	S. B. Gardner	Freeport, Ill	Oct. 12, 1869	95, 677
Car-brake, Steam	V. C. Buckner	Boonville, Mo	Jan. 2, 1866	51, 797
Car-brake, Steam	T. and H. Hopper	Newark, N. J	Oct. 1, 1867	69, 437
Car-brake, Steam-actuated	J. T. Bassett	Galesburgh, Ill	Oct. 17, 1871	119, 913
Car-brake, Steam and air	J. W Gardner	Cleveland, Ohio	Jan. 23, 1872	122, 884
Car-brake, Steam and air	J. Y. Smith	Pittsburgh, Pa	July 23, 1872	129, 868
Car brake, Steam railway	H. Miller	New York, N. Y	Jan. 2, 1855	12, 166
Car brake, Street	J. Stephenson	New York, N. Y	June 4, 1872	127, 525
Car-brake to prevent accidents in descending inclined planes on railways.	M. W. King	New York, N. Y	May 8, 1840	1, 587
Car-brake, Vacuum	R. J. Wilson	Pittsburgh, Pa	Nov. 14, 1871	120, 922
Car-brake wheel, Railway	C. A. Waterbury	New York, N. Y	May 13, 1873	138, 968
Car-brakes, Arrangement of bumper for self-acting.	J. J. McComb	New Orleans, La	June 12, 1855	13, 038
Car-brakes, Brake-block for	J. P. Levan	Altoona, Pa	Apr. 26, 1864	42, 493
Car-brakes, Connecting a series of	P. Moody	Camden, N. J	Sept. 26, 1854	11, 732
Car-brakes, Graduating the tension of	W. Loughridge	Weverton, Md	Apr. 10, 1855	12, 685
Car brakes, Lever of railway	L. Paige	Cavendish, Vt	Mar. 25, 1856	14, 515
Car-brakes, Machinery for operating	J. Marks	Dunkirk, N. Y	Mar. 28, 1854	10, 701
Car-brakes, Machinery for operating	J. Marks and J. Howarth	Boston and Salem, Mass	Mar. 28, 1854	10, 702
Car brakes, Machinery for operating railway	J. R. Grant	Utica, N. Y	Aug. 25, 1842	2, 760
Car brakes, Mechanism for operating railway	E. R. Roe	Bloomington, Ill	Aug. 11, 1857	17, 983
Car brakes, Method of applying railway	I. J. Webber	Salem, Mass	July 7, 1857	17, 763
Car-brakes, Method of applying steam-power to	W. Loughridge	Weverton, Md	Apr. 19, 1864	42, 385
Car-brakes, Method of engaging and disengaging self acting.	G. F. Leach	Boston, Mass	Nov. 14, 1854	11, 937
Car-brakes, Operating	N. Hodge	North Adams, Mass	Oct. 2, 1849	6, 762
Car-brakes, Operating	W. Loughridge	Weverton, Md	Apr. 12, 1864	42, 301
Car-brakes, Operating	G. W. Mitchell	Jackson, Tenn	Dec. 6, 1859	26, 365
Car-brakes, Operating	J. Y. Smith	Pittsburgh, Pa	Aug. 6, 1872	130, 323
Car-brakes, Operating	L. Stebbins	Hartford, Conn	Apr. 18, 1848	5, 510
Car brakes, Operating railway	S. N. Goodale	Saint Louis, Mo	Oct. 31, 1871	120, 429
Car brakes, Operating railway	M. McGee	Jackson, Mich	Mar. 9, 1858	19, 574
Car brakes, Operating railway	D. Mumma, jr	Harrisburgh, Pa	Nov. 15, 1859	26, 117
Car brakes, Operating railway	A. G. Safford	Boston, Mass	Dec. 23, 1862	37, 246
Car brakes, Operating railway	C. B. Turner	Buffalo, N. Y	Nov. 14, 1848	5, 918
Car-brakes, Radial arm for	T. G. McLaughlin	Philadelphia, Pa	June 20, 1854	11, 130
Car brakes, Rubber of railway	H. M. Collier	Binghamton, N. Y	Oct. 20, 1857	18, 435
Car, Brick	J. K. Caldwell	Pittsburgh, Pa	Aug. 13, 1867	67, 632

Index of patents issued from the United States Patent Office from 1790 to 1873, inclusive—Continued.

Invention.	Inventor.	Residence.	Date.	No.
Car, Brick-carrying	J. K. Caldwell	Pittsburgh, Pa	Mar. 10, 1868	75, 243
Car-bridge, Safety	E. Sturgeon	Columbiana, Ohio	Sept. 8, 1868	81, 958
Car buffer and draw-bar, Railway	E. L. Caum	Patterson, Pa	Dec. 3, 1867	71, 570
Car buffer, Railway	P. Allen and B. Valiquette	Rutland, Vt	Apr. 6, 1869	88, 600
Car-buffer spring, Railway	J. Haldeman	West Point, Va	Apr. 2, 1872	125, 191
Car-buffer spring, Railway	C. W. Saladee	Saint Catharine's, Canada	Jan. 28, 1873	135, 369
Car bumper and draw-head, Railway	A. Ross and J. Arthur	Freeport, Ill	Nov. 15, 1870	109, 347
Car-bumper and draw-head spring	C. F. Allen	Paw Paw, Mich	May 6, 1862	35, 196
Car bumper and draw-head spring, Railway	J. C. Jackson	Rochester, N. Y	Jan. 20, 1863	37, 448
Car-bumper attachment	J. P. Laird	Altoona, Pa	Feb. 7, 1865	46, 248
Car-bumper brake, Railway	F. Armstrong	New Orleans, La	Nov. 11, 1856	16, 042
Car bumper, Railway	J. Taney and J. H. Brown	Bangor, Me	Oct. 13, 1868	83, 006
Car, carriage, truck, &c., for railways	J. Harrison, jr	Philadelphia, Pa	Apr. 24, 1838	706
Car, Cattle	J. H. Aldrich	Nashua, N. H	Apr. 9, 1867	63, 686
Car, Cattle	E. Fontaine	Fort Wayne, Ind	Mar. 31, 1868	76, 067
Car, Cattle	M. T. Kehoo	Amsterdam, N. Y	Jan. 1, 1867	60, 906
Car, Cattle	Z. Street	Salem, Ohio	Nov. 2, 1869	96, 500
Car, Cattle	L. Swearingen	Valley River Falls, Va	May 29, 1860	28, 517
Car, Cattle	A. Welch	Southall, England	June 4, 1872	127, 443
Car chair, Railway	W. Palmer	New York, N. Y	May 30, 1871	115, 510
Car-chock for railway-rail	W. O. Stephenson	Marion County, Ind	Mar. 11, 1873	136, 625
Car, Circular	J. Crane	Schenectady, N. Y	Oct. 1, 1830	
Car, City-railway	W. C. Allison	Philadelphia, Pa	Jan. 3, 1860	26, 730
Car, City-railway	W. C. Allison	Philadelphia, Pa	May 29, 1860	28, 530
Car, City-railway	J. Grice and R. H. Long	New York, N. Y., and Philadelphia, Pa.	Dec. 11, 1860	30, 918
Car, City-railway	J. Harris, jr	Roxbury, Mass	Sept. 4, 1860	29, 882
Car, City-railway	J. Ruth	Philadelphia, Pa	Apr. 14, 1863	38, 181
Car, Coal	T. McCrory	Fayette City, Pa	June 14, 1870	104, 332
Car, Compartment	W. D. Mann	Mobile, Ala	Jan. 9, 1872	122, 622
Car-conductors, Check on	P. F. Milligan	Washington, D. C	Apr. 9, 1867	63, 649
Car connection, Railway	W. P. Anderson	New York, N. Y	May 18, 1869	90, 216
Car, Convertible freight	T. Fogg	Saint Mary's, Canada	Apr. 22, 1873	138, 142
Car, Convertible freight	W. Worsley	Little Falls, N. J	Apr. 15, 1873	137, 871
Car couch, Railway	W. A. Brown	Philadelphia, Pa	July 17, 1860	29, 137
Car couch, Railway	J. B. Creighton	Tiffin, Ohio	Sept. 28, 1858	21, 600
Car couch, Railway	E. C. Knight	Philadelphia, Pa	Feb. 28, 1860	27, 297
Car couch, Railway	F. R. Myers and F. H. Furniss	Cleveland, Ohio	Sept. 7, 1858	21, 436
Car couch, Railway	G. W. Williamson	Goldsborough, Pa	Apr. 6, 1869	88, 685
Car couch, Railway	J. Hartman, jr	Philadelphia, Pa	Sept. 7, 1858	21, 469
Car-coupler	S. K. Christy	Noblesville, Ind	Nov. 3, 1868	83, 761
Car-coupler	T. Ray	Pelham, Canada	May 23, 1871	115, 104
Car-coupler	C. E. Stevens	New York, N. Y	Mar. 22, 1859	23, 321
Car-coupler	W. D. Tisdale	Cortland, N. Y	Mar. 11, 1873	136, 627
Car-coupler	P. Weatherbee	Port Washington, Ohio	June 21, 1870	104, 670
Car-coupler	J. B. Atwater	Berlin, Wis	Mar. 8, 1859	23, 145
Car coupler and buffer	E. Miller	Janesville, Wis	Jan. 31, 1865	46, 126
Car coupler and buffer, Railway	P. G. Gardiner	New York, N. Y	Oct. 25, 1870	108, 698
Car-couplers, Buffer-head for	W. Rickards, jr	Franklin, Pa	July 19, 1870	105, 491
Car-coupling	G. S. Acker	Kalamazoo, Mich	Oct. 20, 1868	83, 124
Car-coupling	J. W. Adams	Spring Creek, Wis	May 26, 1868	78, 173
Car-coupling	L. Adams	Blanchester, Ohio	Dec. 13, 1859	26, 403
Car-coupling	M. K. Adams	Mountain Eagle, Pa	Dec. 3, 1867	71, 673
Car-coupling	H. Adkins	Plymouth, Ill	Sept. 18, 1866	58, 041
Car-coupling	J. H. Akin	Wyandotte City, Kans	Mar. 5, 1872	124, 310
Car-coupling	O. Albrecht	Philadelphia, Pa	June 17, 1873	139, 850
Car-coupling	H. Allen and A. H. Baldwin	Houston, Tex	Sept. 24, 1872	131, 587
Car-coupling	J. W. Allen and E. C. Boyles	Mount Pleasant, Iowa	Jan. 2, 1872	122, 304
Car-coupling	R. N. Allen	Pittsford, Vt	July 19, 1870	105, 407
Car-coupling	J. D. Alvord	Springfield, Mass	Sept. 18, 1849	6, 728
Car-coupling	A. I. Ambler	Milwaukee, Wis	Sept. 2, 1862	36, 382
Car-coupling	A. A. Ames	Minneapolis, Minn	Oct. 8, 1872	132, 042
Car-coupling	J. D. Anderson	Corry, Pa	May 28, 1867	65, 151
Car-coupling	W. S. Anderson	Shelbyville, Tenn	Aug. 4, 1868	80, 584
Car-coupling	T. Andress	Auroraville, Wis	Nov. 11, 1873	144, 428
Car-coupling	J. N. Anthoine	Biddeford, Me	Oct. 29, 1872	132, 618
Car-coupling	J. D. M. Armbrust	Apolloborough, Pa	Dec. 1, 1868	84, 465
Car-coupling	R. S. Arnall	Wright City, Mo	Mar. 12, 1867	62, 803
Car-coupling	R. S. Arnall	Wright City, Mo	Aug. 13, 1867	67, 703
Car-coupling	T. Arndt	Mount Joy, Pa	Dec. 19, 1865	51, 535
Car-coupling	T. Arndt	Mount Joy, Pa	Aug. 25, 1868	81, 323
Car-coupling	T. H. Arnold	Arlington, Ill	June 9, 1863	38, 801
Car-coupling	B. Atkinson	Indianapolis, Ind	July 8, 1873	140, 667
Car-coupling	A. A. Atwater	Smith Valley, N. Y	Nov. 11, 1873	144, 494
Car-coupling	J. B. Atwater	Berlin, Wis	Feb. 1, 1859	22, 776
Car-coupling	G. Aulick	Winchester, Va	Apr. 18, 1854	10, 787
Car-coupling	C. P. Bachelder	Franklin, N. H	Dec. 24, 1867	72, 441
Car-coupling	J. Bailor	Cannon City, Minn	Sept. 4, 1866	57, 658
Car-coupling	P. Baker	Oakland, Md	Nov. 12, 1867	70, 776
Car-coupling	W. R. Baker	Trempealeau, Wis	June 14, 1870	104, 098
Car-coupling	A. C. Baker and J. Van Dyne	Hyde Park, N. Y	Sept. 2, 1862	36, 331
Car-coupling	C. W. Baldwin	Boston, Mass	Aug. 28, 1866	57, 618
Car-coupling	H. Baldwin	Nashville, Tenn	Apr. 16, 1850	7, 280
Car-coupling	H. Bale	Petersburgh, Ill	Oct. 22, 1872	132, 429
Car-coupling	D. H. Ball	Sinnamahoning, Pa	Jan. 31, 1871	111, 419
Car-coupling	D. H. Ball	Sinnamahoning, Pa	July 4, 1871	116, 663
Car-coupling	T. A. Banta	Bloomington, Ind	Oct. 29, 1872	132, 619
Car-coupling	E. W. Barker	Portland, Me	Dec. 17, 1872	133, 963
Car-coupling	E. W. Barker	Portland, Me	July 15, 1873	140, 805
Car coupling	E. T. Barlow	San Francisco, Cal	Aug. 4, 1863	39, 373
Car-coupling	E. T. Barlow	San Francisco, Cal	Nov. 12, 1872	133, 006
Car-coupling	P. C. Barlow	Beverly, W. Va	July 25, 1871	117, 368
Car coupling	W. F. Barlow	Monmouth, Ill	Dec. 10, 1867	71, 951
Car-coupling	J. D. Barnard	Frostburgh, Md	Sept. 12, 1871	118, 836
Car-coupling	W. B. Barnes	Plymouth, Mass	July 29, 1873	141, 258
Car-coupling	H. W. Barnum	Omaha, Nebr	June 4, 1872	127, 450
Car-coupling	J. W. Barrett	Calhoun, Ga	June 27, 1871	116, 397

Index of patents issued from the United States Patent Office from 1790 *to* 1873, *inclusive*—Continued.

Invention.	Inventor.	Residence.	Date.	No.
Car-coupling	R. M. Barthelmess and C. C. Millar.	Savannah, Ga	Nov. 3, 1868	83,591
Car-coupling	J. Bassler	Galesburgh, Ill	Mar. 19, 1872	124,780
Car-coupling	J. W. Bates	Minneapolis, Minn	Jan. 7, 1873	134,629
Car-coupling	R. F. Baughn	Lexington, Miss	July 6, 1869	92,249
Car-coupling	D. S. Beals	Adrian, Mich	July 21, 1868	80,117
Car-coupling	G. M Beardsley	Fenton, Mich	Nov. 5, 1867	70,505
Car-coupling	T. Beddow and A. T. Jackson	New Albany, Ind	July 22, 1873	141,103
Car-coupling	J. B. Behrens	Pearl, Ill	June 4, 1867	65,331
Car-coupling	J. S. Bell	Hackettstown, N. J	June 21, 1864	43,175
Car-coupling	W. G. Bell	Pittsburgh, Pa	Dec. 15, 1868	84,933
Car-coupling	W. W. Bell	Chicago, Ill	Feb. 15, 1870	99,817
Car-coupling	W. E. Beman	Portland, Me	Mar. 18, 1873	136,958
Car-coupling	L. D. Bennett	Willimantic, Conn	Aug. 26, 1873	142,142
Car-coupling	C. W. Benson	Frederick City, Md	Feb. 15, 1870	99,818
Car-coupling	W. Bergmann	Philadelphia, Pa	July 3, 1866	55,989
Car-coupling	J. Bestwick, jr., and A. Alden	Dedham, Mass	Mar. 20, 1860	27,516
Car-coupling	A. and S. Beucus	Waupun, Wis	June 21, 1870	104,543
Car coupling	B. Bevelander	Boston, Mass	Mar. 9, 1869	87,622
Car-coupling	W. C. Bibb	Morgan County, Ga	Aug. 22, 1871	118,184
Car-coupling	E. Bickell	Milton, Pa	July 29, 1873	141,312
Car-coupling	E. R. Bigelow	Salem, Mass	Dec. 31, 1867	72,717
Car-coupling	C. S. Bigler	Harrisburgh, Pa	Feb. 11, 1873	135,686
Car-coupling	T. L. Birch and J. C. Noble	Washington, Pa	Feb. 4, 1862	34,289
Car-coupling	G. S. Bishop	Washington, D. C	Oct. 12, 1858	21,737
Car-coupling	G. S. Bishop	Washington, D. C	Mar. 3, 1863	37,805
Car-coupling	W. J. Blackman	Columbus, Miss	Feb. 23, 1869	87,135
Car-coupling	H. Blackmore	Pittsburgh, Pa	Mar. 22, 1864	42,041
Car-coupling	T. B. Blackstone	Chicago, Ill	Nov. 10, 1868	83,907
Car-coupling	H. I. Blakeslee	Concord, Pa	June 23, 1863	38,939
Car-coupling	H. Blanchard, jr	Boston, Mass	Dec. 6, 1870	109,797
Car-coupling	A. W. Bohaker	Granville, Nova Scotia	Oct. 7, 1873	143,491
Car-coupling	H. W. Boifeuillet	Savannah, Ga	Dec. 29, 1868	85,359
Car-coupling	C. C. Bole	Conemaugh, Pa	Dec. 10, 1872	133,824
Car-coupling	E. Bolman and G. S. W. Clarke	Halifax, Nova Scotia	May 16, 1871	114,755
Car-coupling	T. H. Bomar	Atlanta, Ga	Sept. 1, 1868	81,741
Car-coupling	A. T. Boon and D. M. Osborn	Galesburgh, Ill	Apr. 9, 1867	63,696
Car-coupling	J. Boothroyd	Michigan City, Ind	Jan. 24, 1871	111,170
Car-coupling	L. J. Bosworth	Monmouth, Ill	Apr. 18, 1871	113,730
Car-coupling	J. W. Boughton	Appleton, Wis	July 18, 1865	48,787
Car-coupling	J. W. Boughton	Appleton, Wis	Apr. 3, 1866	53,560
Car-coupling	L. Boyd and P. Kriegbaum	Springfield, Ohio	Nov. 12, 1867	70,689
Car-coupling	W. A. Boyden	Harrisburgh, Pa	Sept. 23, 1873	143,055
Car-coupling	W. L. Braddock	Boston, Mass	Mar. 30, 1869	88,440
Car-coupling	D. Bradford	Hamilton, Canada	Jan. 28, 1873	135,316
Car-coupling	A. J. Braley	Berlin, Vt	Jan. 26, 1869	86,129
Car-coupling	A. Branshaw	Fond du Lac, Wis	May 11, 1869	89,913
Car-coupling	G. Brenner and G. T. Polk	Poughkeepsie, N. Y	Mar. 14, 1871	112,536
Car-coupling	J. Broadley	Bradford, England	Aug. 13, 1872	130,408
Car-coupling	W. C. Brooks	Stoneham, Pa	Oct. 28, 1873	144,015
Car-coupling	P. Brown	Louisville, Ky	Oct. 15, 1872	132,238
Car-coupling	P. Brown	Louisville, Ky	Nov. 26, 1872	133,406
Car-coupling	H. A. Buck	Meadville, Pa	Apr. 19, 1864	42,343
Car-coupling	R. B. Buckner	Gilroy, Cal	Aug. 12, 1873	141,756
Car-coupling	J. H. Bull	Hereford, Md	July 11, 1871	116,801
Car-coupling	C. T. Burchardt	New York, N. Y	July 28, 1868	80,271
Car-coupling	C. C. Burns	Greensburgh, Ind	Mar. 6, 1866	52,962
Car-coupling	W. E. Bush	Damascus, Pa	Oct. 20, 1868	83,252
Car-coupling	W. C. Bussey	Rockgrove, Ill	July 17, 1847	5,194
Car-coupling	W. C. Bussey	Jackson, Cal	Aug. 30, 1864	43,969
Car-coupling	R. Butt	Harrisburgh, Pa	Nov. 11, 1873	144,439
Car-coupling	C. C. Cady	West Union, Iowa	Jan. 8, 1867	61,047
Car-coupling	T. Caldwell and L. C. Wilcox	Buffalo, N. Y	Feb. 23, 1869	87,142
Car-coupling	W. Callow, jr	Baltimore, Md	May 30, 1871	115,433
Car-coupling	D. C. Camerer	Martinsburgh, Pa	Aug. 19, 1873	141,991
Car-coupling	E. Campbell	Medusa, N. Y	June 13, 1871	115,819
Car-coupling	G. G. Campbell	Janesville, Wis	Oct. 24, 1865	50,559
Car-coupling	S. O. Campbell	Tipton, Mo	Dec. 14, 1869	97,876
Car-coupling	R. Campion and J. W. Thompson, jr.	Camden, N. J	Mar. 23, 1869	88,130
Car-coupling	M. H. Card and J. W. Stewart	Fulton, Ill., and Lyons, Iowa.	May 15, 1866	54,683
Car-coupling	M. H. Card and T. Tripp	Chicago, Ill	July 19, 1864	43,569
Car-coupling	S. P. Carll and A. Shute	Richmond, Ind	June 28, 1870	104,704
Car-coupling	J. W. Carrier	Springfield, Mass	Mar. 29, 1859	23,423
Car-coupling	E. Cary	Burlington, Iowa	Aug. 2, 1864	43,734
Car-coupling	E. Cary	Burlington, Iowa	Mar. 13, 1866	53,215
Car-coupling	E. W. Chadwick	Edgartown, Mass	Aug. 18, 1868	81,067
Car-coupling	J. H. Chadwick	Bristol, R. I	Nov. 17, 1868	84,049
Car-coupling	S. E. Chaney	Hillsborough, Ohio	July 11, 1865	48,656
Car-coupling	A. F. Chandler	Fall River, Mass	Nov. 25, 1873	144,954
Car-coupling	W. J. Chaplin	Kalamazoo, Mich	Feb. 11, 1873	135,775
Car-coupling	J. A. J. Chapman	Kansas City, Mo	Jan. 30, 1872	123,150
Car-coupling	A. P. Chatham	Canoga, N. Y	Nov. 1, 1853	10,176
Car-coupling	A. H. Clark	Fond du Lac, Wis	July 16, 1867	66,677
Car-coupling	A. H. Clark	Otisville, Mich	Oct. 12, 1869	95,652
Car-coupling	W. C. Clark	Portland, Me	July 4, 1865	48,617
Car-coupling	W. H. H. Clark	Burlington, Iowa	July 27, 1869	93,055
Car-coupling	G. E. Clarke	Racine, Wis	Dec. 26, 1865	51,695
Car-coupling	T. P. Clines	Louisville, Ky	Dec. 3, 1872	133,565
Car-coupling	C. Clinton	Blooming Grove, N. Y	Feb. 14, 1865	46,336
Car-coupling	C. D. Clinton	Peoria, Ill	July 17, 1866	56,371
Car-coupling	W. B. Coates	Philadelphia, Pa	Jan. 22, 1867	61,399
Car-coupling	J. T. L. Cochran	Columbia, Tenn	Nov. 25, 1873	144,956
Car-coupling	S. M. Cochran	Baltimore, Md	Jan. 1, 1851	7,866
Car-coupling	W. A. Cochran	Flat Rock, Ind	Nov. 18, 1873	144,658
Car-coupling	F. Coffin	Claremont, N. H	Oct. 27, 1868	83,464
Car-coupling	G. W. Coffin	Pittsburgh, Pa	Feb. 18, 1873	136,033

Index of patents issued from the United States Patent Office from 1790 *to* 1873, *inclusive*—Continued.

Invention.	Inventor.	Residence.	Date.	No.
Car-coupling	C. M. Colby	Corinth, Vt	Aug. 29, 1871	118, 436
Car-coupling	V. and E. Cole	Detroit, Mich	Dec. 18, 1866	60, 479
Car-coupling	J. Coleman	Lynchburgh, Va	May 3, 1870	102, 500
Car-coupling	W. T. Collins	Winchester, Ill	Feb. 18, 1873	135, 890
Car-coupling	G. Collyer	Philadelphia, Pa	Feb. 10, 1863	37, 612
Car-coupling	T. F. Conner	Odin, Ill	Oct. 22, 1867	69, 972
Car-coupling	C. C. Converse	Erie, Pa	Nov. 11, 1873	144, 386
Car-coupling	J. M. Cook	Washington, D. C	Sept. 8, 1868	81, 985
Car-coupling	W. Cook	Belvidere, Ill	Feb 26, 1867	62, 475
Car-coupling	J. G. W. Coolidge	Portland, Me	Aug. 29, 1865	49, 676
Car-coupling	W. R. Coovert	London, Pa	Nov. 4, 1873	144, 257
Car-coupling	S. A. Corser	Holyoke, Mass	Nov. 7, 1865	50, 797
Car-coupling	C. B. Cotler	Harrisburgh, Pa	June 22, 1858	20, 627
Car-coupling	J. Couch	Harrison, Me	Aug. 15, 1865	49, 385
Car-coupling	R. A. Cowell	Cleveland, Ohio	Mar. 5, 1867	62, 613
Car-coupling	S. H. Cowles	Oakville, Conn	Mar. 24, 1868	75, 866
Car-coupling	E. S. Cram	New Hampton, N. H	Apr. 20, 1869	89, 202
Car-coupling	J. Crist	Tiffin, Ohio	Aug. 26, 1873	142, 209
Car-coupling	T. H. Cross	East Boston, Mass	Sept. 23, 1873	142, 993
Car-coupling	G. D. Crosthwait	Murfreesborough, Tenn	Feb. 18, 1873	136, 039
Car-coupling	W. W. Culpepper	Augusta, Ga	June 12, 1860	28, 660
Car-coupling	G. E. Cuming	La Fayette, Ind	June 18, 1867	65, 882
Car-coupling	J. W. Currier	Springfield, Mass	June 1, 1869	90, 730
Car-coupling	S. Daggett, jr	Charleston, S. C	Mar. 20, 1860	27, 531
Car-coupling	R. Daily	Canal Township, Pa	Mar. 24, 1868	75, 738
Car-coupling	G. E. Darling and M. Heus	Marytown, Wis	July 4, 1871	116, 566
Car-coupling	A. H. S. Davis	Farmington, Me	Jan. 2, 1872	122, 368
Car-coupling	P. W. Davis	Portland, Oreg	Aug. 15, 1871	117, 992
Car-coupling	W. A. Dean	New Lexington, Ohio	Nov. 26, 1872	133, 421
Car-coupling	J. C. Dearborn	Candia, N. H	Dec. 20, 1870	110, 214
Car-coupling	P. H. Decker and N. Lee	Chicago, Ill	Aug. 12, 1873	141, 635
Car-coupling	T. W. Defrees	South Bend, Ind	June 7, 1870	103, 850
Car-coupling	C. R. Densmore and J. A. Vrooman.	Oil City, Pa	Oct. 5, 1869	95, 570
Car-coupling	J. Depeu	Peekskill, N. Y	July 9, 1867	66, 571
Car-coupling	J. Depeu and J. D. Hall	Peekskill, N. Y	July 2, 1867	66, 309
Car-coupling	J. Dinsmore	Dinsmore Station, Pa	Apr. 30, 1872	126, 136
Car-coupling	M. Disney	San Francisco, Cal	Dec. 3, 1867	71, 717
Car-coupling	M. Disney	Oakland, Cal	Nov. 19, 1872	133, 085
Car-coupling	M. Disney	Oakland, Cal	June 24, 1873	140, 124
Car-coupling	W. J. Dodge	Kasoag, N. Y	July 29, 1862	35, 994
Car-coupling	M. C. Doubleday	Sharon, Vt	Feb. 25, 1873	136, 222
Car-coupling	R. H. Dowling	Newark, Ohio	Sept. 9, 1873	142, 682
Car-coupling	R. H. Dowling and E. S. Perry	Fenton, Mich., and Clay Lick, Ohio.	Nov. 7, 1871	120, 726
Car-coupling	J. L. Duncan	Fremont, Ohio	Sept. 23, 1873	142, 998
Car-coupling	J. Dunott	Philadelphia, Pa	Dec. 7, 1869	97, 620
Car-coupling	H. W. Earl	Baltimore, Md	July 4, 1871	116, 696
Car-coupling	C. Eastin	Louisville, Ky	Jan. 9, 1872	122, 586
Car-coupling	J. Ebee	Hummelstown, Pa	Oct. 29, 1867	70, 326
Car-coupling	G. Edmonds	New Orleans, La	Oct. 14, 1873	143, 680
Car-coupling	J. Elbertson	Kirksville, Mo	Nov. 10, 1868	83, 945
Car-coupling	A. J. Elder	Kansas City, Mo	Oct. 13, 1868	83, 052
Car-coupling	J. Enos	Chicago, Ill	Oct. 21, 1873	143, 894
Car-coupling	J. M. Enos	Saint Joseph, Mich	Aug. 26, 1873	142, 219
Car-coupling	J. M. and J. Enos	Saint Joseph, Mich	Nov. 21, 1871	121, 047
Car-coupling	L. A. Evans	Chester, Pa	Oct. 11, 1870	108, 124
Car-coupling	J. M. Everhart	Pittston, Pa	July 7, 1868	79, 643
Car-coupling	R. F. Fairlie	London, England	Feb. 23, 1869	87, 158
Car-coupling	H. Fake	Chicago, Ill	Sept. 6, 1864	44, 085
Car-coupling	M. G. Farnam	Clifton, Canada	Jan. 2, 1872	122, 375
Car-coupling	M. Ferren	Freedom, N. H	Sept. 30, 1873	143, 231
Car-coupling	T. Firth and J. W. Hollis	Cincinnati, Ohio	Sept. 24, 1872	131, 673
Car-coupling	F. A. Fleming	Curwensville, Pa	Oct. 21, 1873	143, 752
Car-coupling	H. H. Fleming	Kokomo, Ind	Nov. 7, 1865	50, 811
Car-coupling	C. S. Flower and C. F. Graves	Kickapoo City, Kans., and Hickory, Iowa.	Nov. 26, 1872	133, 364
Car-coupling	E. L. Foreman	Rantoul, Ill	Apr. 25, 1871	114, 124
Car-coupling	W. H. Forker	Meadville, Pa	Sept. 27, 1864	44, 484
Car-coupling	J. J. Fort	Oshkosh, Wis	Mar. 17, 1863	37, 908
Car-coupling	J. Fortier	Fairport, N. Y	Feb. 4, 1868	73, 964
Car-coupling	C. Foster	Oshkosh, Wis	Mar. 8, 1864	41, 837
Car-coupling	R. G. Fowler	Olney, Ill	Sept. 18, 1866	58, 090
Car-coupling	F. Fox and F. A. Howarth	Providence, R. I	Feb. 4, 1873	135, 537
Car-coupling	J. J. Fraikin	Fort Howard, Wis	Aug. 5, 1873	141, 432
Car-coupling	E. J. Frazier	Peoria, Ill	June 5, 1866	55, 271
Car-coupling	A. M. Freeman and A. M. Stoner.	Springfield, Ohio	Jan. 8, 1867	61, 003
Car-coupling	J. P. Freeman	Dalton, Ga	Oct. 20, 1868	83, 272
Car-coupling	G. B. French	Dunbarton, N. H	Jan. 29, 1867	61, 616
Car-coupling	J. H. Froman	Plattsburgh, Mo	Aug. 30, 1870	106, 807
Car-coupling	J. A. Gale	West Medway, Mass	June 24, 1873	140, 131
Car-coupling	C. Gallagher	Taunton, Mass	Dec. 24, 1872	134, 199
Car-coupling	M. Gally	Marion, N. Y	Mar. 5, 1867	62, 625
Car-coupling	P. G. Gardiner	New York, N. Y	Apr. 9, 1872	125, 561
Car-coupling	P. G. Gardiner	New York, N. Y	Apr. 9, 1872	125, 562
Car-coupling	M. C. Gardner	Rochester, N. Y	Dec. 8, 1863	40, 833
Car-coupling	C. H. Gerhart	Farmer City, Ill	Mar. 25, 1873	137, 067
Car-coupling	E. M. George	Three Rivers, Mich	May 16, 1871	114, 797
Car-coupling	J. B. German	Walnut Hills, Ohio	Mar. 26, 1867	63, 239
Car-coupling	M. F. Gibbs	Livonia, N. Y	June 4, 1867	65, 373
Car-coupling	E. N. Gifford	Cleveland, Ohio	Aug. 26, 1873	142, 226
Car-coupling	F. M. Gifford	Brant, N. Y	Mar. 7, 1865	46, 656
Car-coupling	J. W. Gillam	Newton, N. J	July 18, 1871	117, 067
Car-coupling	J. W. and J. W. Gillam, jr	Newton, N. J	Feb. 18, 1873	135, 984
Car-coupling	H. C. Gilliland	Wellsville, Mo	June 27, 1871	116, 431
Car-coupling	S. Gissinger	Lawrenceville, Pa	Nov. 12, 1867	70, 8[illegible]9
Car-coupling	H. C. Glasgow	Chicago, Ill	June 9, 1863	38, 819

Index of patents issued from the United States Patent Office from 1790 *to* 1873, *inclusive*—Continued.

Invention.	Inventor.	Residence.	Date.	No.
Car-coupling	H. C. Glasgow	Chicago, Ill	Mar. 1, 1864	41, 767
Car-coupling	H. C. Glasgow	Cleveland, Ohio	Oct. 6, 1868	82, 707
Car-coupling	H. C. Glasgow	Cleveland, Ohio	Oct. 6, 1868	82, 708
Car-coupling	J. L. D. Good	Jefferson, Pa	Jan. 7, 1873	134, 648
Car-coupling	M. Goolaren	Taunton, Mass	May 13, 1873	138, 748
Car-coupling	R. Goole	Abingdon, Ill	Dec. 24, 1867	72, 482
Car-coupling	J. M. Gow	Rock Island, Ill	Feb. 9, 1864	41, 502
Car-coupling	J. M. Gow	Rock Island, Ill	July 5, 1864	43, 399
Car-coupling	W. F. Grassler	Muncy, Pa	Oct. 11, 1870	108, 254
Car-coupling	W. F. Grassler	Muncy, Pa	Nov. 26, 1872	133, 437
Car-coupling	W. F. Grassler and A. J. Cutter	Lewisburgh, Pa	Aug. 26, 1862	36, 293
Car-coupling	W. F. Grassler	Muncy, Pa	Oct. 31, 1871	120, 430
Car-coupling	R. W. Green	Bradford, Pa	Feb. 12, 1867	61, 028
Car-coupling	S. Gregory	Jonesville, Mich	Mar. 28, 1871	113, 160
Car-coupling	J. Griffith and W. F. Miller	Adamsburgh, Pa	Sept. 30, 1873	143, 281
Car-coupling	P. I. Gross	Manheim Centre, N. Y	Feb. 21, 1865	46, 465
Car-coupling	J. Growdon	New Enterprise, Pa	Feb. 25, 1873	136, 237
Car-coupling	J. Gum	Marseilles, Ill	Oct. 21, 1873	143, 756
Car-coupling	D. C. Guttridge	Canton, Ohio	Oct. 9, 1866	58, 723
Car-coupling	F. B. Hall	Hartford, Conn	Apr. 30, 1861	32, 223
Car-coupling	A. S. and H. H. Hallett	East Abington, Mass	Dec. 5, 1871	121, 617
Car-coupling	E. C. Hamlin	Pavilion, N. Y	Jan. 14, 1862	34, 143
Car-coupling	J. B. Hards and W. Hodnett	Chicago, Ill	Aug. 2, 1870	106, 057
Car-coupling	C. R. Hardy	Lexington, Ind	Aug. 4, 1868	80, 730
Car-coupling	D. Harger	Des Moines, Iowa	Jan. 28, 1868	73, 889
Car-coupling	J. B. Harper	Saint John, Mo	June 13, 1871	115, 845
Car-coupling	J. Harris	Green Lake, Wis	Feb. 6, 1866	52, 410
Car-coupling	J. H. Harris	Virginia, Ill	Apr. 2, 1867	63, 514
Car-coupling	A. Hartman	Canton, Ohio	Aug. 28, 1866	57, 504
Car-coupling	A. Hartman	Canton, Ohio	May 28, 1867	65, 079
Car-coupling	E. W. Harvey	Springfield, Mass	June 4, 1872	127, 601
Car-coupling	N. Haskins	Hillsborough County, N. H.	Oct. 29, 1850	7, 741
Car-coupling	W. D. Hatch	Worcester, Mass	Oct. 2, 1849	6, 754
Car-coupling	W. W. Haver	Schuyler, N. Y	Sept. 16, 1873	142, 849
Car-coupling	H. Hawley	Lynchburgh, Va	Mar. 12, 1872	124, 490
Car-coupling	G. W. Haynie	Olney, Ill	Mar. 10, 1868	75, 420
Car-coupling	J. C. Heaton	Fitchburgh, Mich	Feb. 18, 1868	74, 532
Car-coupling	A. Hebbard	Galesburgh, Ill	Apr. 13, 1858	19, 925
Car-coupling	R. Hemming	Boston, Mass	Nov. 6, 1847	5, 354
Car-coupling	G. H. Henfield	San Francisco, Cal	Sept. 3, 1867	68, 439
Car-coupling	T. R. Herd	Allegheny, Pa	Dec. 28, 1869	98, 377
Car-coupling	J. W. Hess	Montandon, Pa	Dec. 16, 1873	145, 652
Car-coupling	T. Hess	West Penn, Pa	Feb. 22, 1870	100, 147
Car-coupling	J. B. Heverling	Greenville, Ohio	Nov. 21, 1871	121, 051
Car-coupling	O. Hewes	Kankakee, Ill	Aug. 4, 1868	80, 735
Car-coupling	J. Hiers	West Manayunk, Pa	Aug. 26, 1873	142, 230
Car-coupling	A. Hillman	Devonshire, England	Dec. 17, 1867	72, 200
Car-coupling	J. H. Hills	Burlington, Vt	Apr. 21, 1863	38, 225
Car-coupling	W. W. Hinman and W. W. Nevill.	Jacksonville, Ill	May 29, 1866	55, 104
Car-coupling	J. E. Hitt	Bement, Ill	Feb. 4, 1873	135, 552
Car-coupling	C. M. Hoag	Nassau, N. Y	Aug. 30, 1870	106, 936
Car-coupling	A. I. Hobbs	Kokomo, Ind	Sept. 12, 1865	49, 879
Car-coupling	H. Holden	Hartford, Conn	Mar. 19, 1867	63, 049
Car-coupling	J. F. L. Holman	Hamilton, Canada	Sept. 23, 1873	143, 071
Car-coupling	C. Holtz	Chicago, Ill	Oct. 2, 1866	58, 418
Car-coupling	J. Hood	Milwaukee, Wis	Jan. 31, 1860	26, 990
Car-coupling	W. H. Hoover	North Benton, Ohio	Mar. 10, 1868	75, 270
Car-coupling	C. L. Horack	Hastings, Minn	Sept. 26, 1871	119, 359
Car-coupling	C. L. Horack	Hastings, Minn	May 14, 1872	126, 702
Car-coupling	C. F. Hornbeck	Slatersville, N. Y	Jan. 11, 1870	98, 770
Car-coupling	F. E. Howard	Geneseo, N. Y	Sept. 2, 1873	142, 395
Car-coupling	H. R. Howe	Hartwick, N. Y	July 28, 1868	80, 486
Car-coupling	H. R. Howe	Hartwick, N. Y	Feb. 9, 1869	86, 673
Car-coupling	H. R. Howe	Hartwick, N. Y	July 4, 1871	116, 595
Car-coupling	E. P. Howland	Worcester, Mass	Sept. 15, 1863	39, 927
Car-coupling	E. P. Howland	Worcester, Mass	Apr. 4, 1865	47, 108
Car-coupling	E. C. Hubbard	Green Bay, Wis	Nov. 26, 1872	133, 369
Car-coupling	H. Hughes	Sarona, N. Y	June 26, 1866	55, 870
Car-coupling	T. G. Hughes, A. Nutting, and L. Aldrich.	Horton, Iowa	Jan. 9, 1872	122, 610
Car-coupling	F. A. Hull	Belvidere, Ill	July 14, 1868	79, 982
Car-coupling	F. M. Hunt	Marshallville, Ga	Apr. 2, 1872	125, 298
Car-coupling	H. C. Hunt	Amboy, Ill	Dec. 3, 1861	33, 841
Car-coupling	O. Z. Hurd and J. W. Ardinger.	Mount Pulaski, Ill	Jan. 26, 1869	86, 161
Car-coupling	R. H. Huston	Keokuk, Iowa	May 28, 1867	65, 228
Car-coupling	F. A. Illingworth	Waltham, Mass	Dec. 26, 1871	122, 175
Car-coupling	W. M. Inge and E. P. Wheeler	Corinth, Miss	May 7, 1872	126, 460
Car-coupling	M. Ingersoll	Grafton, Ohio	May 22, 1866	54, 913
Car-coupling	J. S. Ingram	Louisville, Ky	Nov. 5, 1872	132, 835
Car-coupling	W. M. Ingstrum	Hornellsville, N. Y	Mar. 22, 1870	101, 018
Car-coupling	G. W. Irish	Memphis, N. Y	Oct. 25, 1870	108, 598
Car-coupling	O. P. Ives	Syracuse, N. Y	May 31, 1870	103, 748
Car-coupling	W. R. Jamison	Taylorstown, Pa	July 30, 1867	67, 311
Car-coupling	E. H. Janney	Alexandria, Va	Apr. 21, 1868	77, 046
Car-coupling	E. A. Janney	Alexandria, Va	Apr. 29, 1873	138, 405
Car-coupling	E. V. Jeinsen	New York, N. Y	Feb. 19, 1867	62, 203
Car-coupling	J. B. Johnson	Laurel, Ind	Jan. 28, 1868	73, 902
Car-coupling	J. T. Johnson and N. T. Smith	Grand Rapids, Mich	Mar. 15, 1864	41, 924
Car-coupling	W. J. Johnson	New Orleans, La	June 21, 1870	104, 462
Car-coupling	W. Johnston	Decatur, Mich	May 9, 1871	114, 566
Car-coupling	G. H. Jones and L. D. Boyce	Rochester, N. Y	Apr. 26, 1870	102, 271
Car-coupling	J. H. Jones	Williamsport, Pa	Aug. 13, 1867	67, 767
Car-coupling	J. W. Jones	Hereford, Md	Apr. 25, 1871	114, 015
Car-coupling	J. W. Jones	Hereford, Md	Mar. 19, 1872	124, 684
Car-coupling	J. W. Jones	Towsontown, Md	Mar. 4, 1873	136, 523
Car-coupling	A. J. Jourde	Houston, Tex	June 18, 1872	128, 047
Car-coupling	J. H. Kavanagh	Joliet, Ill	Nov. 27, 1866	60, 012

Index of patents issued from the United States Patent Office from 1790 *to* 1873, *inclusive*—**Continued.**

Invention.	Inventor.	Residence.	Date.	No.
Car-coupling	J. Keck	Dwight, Ill	Dec. 9, 1873	145, 428
Car-coupling	E. L. Keeler	Pittsburgh, Pa	Aug. 2, 1859	24, 938
Car-coupling	E. B. Keith	Galesburgh, Mich	Jan. 11, 1870	98, 775
Car-coupling	E. H. Keith	Peoria, Ill	Dec. 11, 1866	60, 384
Car-coupling	J. Kelly	Woodberry, Md	Jan. 1, 1867	60, 741
Car-coupling	C. H. Kendall	McConnellsburgh, Pa	Jan. 7, 1873	134, 676
Car-coupling	J. E. Kendall	Cairo, Ill	June 25, 1872	128, 314
Car-coupling	P. Kendrick	Trenton, N. J	Sept. 9, 1873	142, 634
Car-coupling	P. Kennedy, jr, and A. B. Diss	Watertown, N. Y	Feb. 11, 1873	135, 714
Car-coupling	J. H. Kenworthey	West Point, Ind	Oct. 24, 1871	120, 199
Car-coupling	L. Kenworthy	Adel, Iowa	Dec. 24, 1872	134, 148
Car-coupling	W. Kenyon	Crawfordsville, Ind	May 7, 1872	126, 554
Car-coupling	J. Kingsbury	Ravenna, Ohio	Sept. 4, 1866	57, 732
Car-coupling	F. S. Kirkbride	Wyandotte, Kans	June 11, 1872	127, 894
Car-coupling	G. L. Kitson	Philadelphia, Pa	Dec. 12, 1865	51, 461
Car-coupling	A. K. Kline	Readington, N. J	Mar. 5, 1872	124, 274
Car-coupling	A. K. Kline	Readington, N. J	Mar. 11, 1873	136, 655
Car-coupling	P. Knobbock	Wyandotte, Kans	June 30, 1868	79, 481
Car-coupling	C. B. Knowles	Nashville, Tenn	July 23, 1872	129, 668
Car-coupling	C. H. Knowlton	Rockland, Me	Nov. 26, 1872	133, 456
Car-coupling	C. H. Knowlton	Rockland, Me	Apr. 1, 1873	137, 455
Car-coupling	G. E. Knox	Ballston Spa, N. Y	Sept. 23, 1873	143, 165
Car-coupling	G. Koeb	Springfield, Ohio	July 29, 1873	141, 278
Car-coupling	C. Köhler	Galena, Ill	Dec. 8, 1868	84, 700
Car-coupling	X. Krapf and J. F. Boerckel	Allentown, Pa	Dec. 9, 1873	145, 353
Car-coupling	C. F. Kuhnle and A. Row	Philadelphia, Pa	Mar. 19, 1872	124, 750
Car-coupling	J. Lacey and G. Watkins	Bristol, Wis	Sept. 19, 1865	50, 010
Car-coupling	J. J. Lahaye	Reading, Pa	Oct. 29, 1872	132, 672
Car-coupling	N. E. Laman	Dayton, Ohio	May 17, 1870	103, 207
Car-coupling	T. R. Land	Grass Valley, Cal	Dec. 30, 1873	146, 0[illegible]8
Car-coupling	J. K. Landis	Palmyra, Pa	Nov. 3, 1868	83, 715
Car-coupling	E. Lane	Old Town, Me	May 7, 1872	126, 401
Car-coupling	A. Langellier	Concord, N. H	Aug. 12, 1873	141, 648
Car-coupling	A. Langellier	Concord, N. H	Nov. 18, 1873	144, 624
Car-coupling	T. Lanstone	Washington, D. C	June 13, 1871	115, 964
Car coupling	G. C. Lawton	Syracuse, N. Y	June 6, 1865	48, 076
Car-coupling	W. Layland	Mixersville, Ind	Jan. 18, 1859	22, 657
Car-coupling	C. Layton	Matawan, N. J	June 8, 1869	91, 028
Car-coupling	C. Layton	Matawan, N. J	May 28, 1872	127, 251
Car-coupling	L. J. Lececq	Argenteuil, France	July 7, 1868	79, 664
Car-coupling	S. D. Lecompte	Leavenworth County, Kans.	Feb. 4, 1868	73, 984
Car-coupling	W. Leib and G. B. Hornbeek	Winchester, Ill	Feb. 26, 1867	62, 345
Car-coupling	H. L. B. Lewis	New York, N. Y	Sept. 18, 1849	6, 730
Car-coupling	I. B. Lewis	Albion, Mich	Aug. 5, 1873	141, 445
Car-coupling	J. E. Lewis	Boston, Mass	July 30, 1872	129, 896
Car-coupling	E. Lindsley	Neenah, Wis	May 7, 1867	64, 547
Car-coupling	D. Lippy	Mansfield, Ohio	Sept. 11, 1866	57, 931
Car-coupling	R. Lloyd	Cleveland, Ohio	Oct. 21, 1873	143, 828
Car-coupling	J. B. Lochbaum	Chambersburgh, Pa	Aug. 20, 1872	130, 585
Car-coupling	S. D. Locke	Janesville, Wis	Aug. 8, 1865	49, 348
Car-coupling	S. L. Loomis	Byron, N. Y	Nov. 12, 1867	70, 729
Car-coupling	A. Z. Long	Scranton, Pa	June 8, 1869	91, 140
Car-coupling	J. Long	Mechanicsburgh, Pa	Dec. 21, 1869	98, 076
Car-coupling	J. T. Lowrey	La Fayette, Ind	Jan. 19, 1864	41, 308
Car-coupling	H. C. Lowrie	Fort Wayne, Ind	Sept. 2, 1873	142, 487
Car-coupling	G. W. Loyd	Markleysburgh, Pa	Apr. 23, 1872	125, 967
Car-coupling	A. A. Lusk and S. A. Wells	Wooster, Ohio	Apr. 3, 1866	53, 639
Car-coupling	I. V. and W. J. Lynn	Pittsburgh, Pa	Mar. 23, 1869	88, 189
Car-coupling	S. Madden	Eureka South, Cal	Oct. 13, 1868	82, 966
Car-coupling	S. Mahurin	Liberty, Ill	Dec. 6, 1870	109, 830
Car-coupling	M. B. Malott	Richmond, Ind	July 26, 1870	105, 706
Car-coupling	J. C. Maloy	Johnstown, Pa	Dec. 2, 1873	145, 067
Car-coupling	J. A. Manning	Ashtabula, Ohio	July 28, 1868	80, 493
Car-coupling	H. E. Marchand	Louisville, Ky	May 21, 1872	126, 973
Car-coupling	H. E. Marchand	Pittsburgh, Pa	Mar. 25, 1873	137, 225
Car-coupling	J. L. Mareness	Constantine, Mich	May 14, 1872	126, 724
Car-coupling	A. S. Markham	Bushnell, Ill	Dec. 5, 1865	51, 330
Car-coupling	C. Markley	New York, N. Y	July 19, 1870	105, 470
Car-coupling	B. F. Marsden	Galena, Ill	Dec. 20, 1864	45, 509
Car-coupling	L. Marshall	Mount Sterling, Ill	Feb. 6, 1872	123, 496
Car-coupling	F. W. Marston and O. Jones	Philadelphia, Pa	Mar. 18, 1873	137, 014
Car-coupling	J. A. Mason	Keokuk, Iowa	June 28, 1870	104, 750
Car-coupling	C. Maus	Danville, Pa	Nov. 23, 1869	97, 208
Car-coupling	J. Mayben	Milroy, Pa	Jan. 24, 1871	111, 224
Car-coupling	W. H. Mayo	Hillsburgh, Nova Scotia	May 21, 1867	64, 890
Car-coupling	A. McCarter	Lancaster, Pa	Mar. 26, 1867	63, 277
Car-coupling	D. McComas	Calhoun, Mo	Sept. 12, 1871	118, 957
Car-coupling	G. W. McEwen	Lock Haven, Pa	Dec. 3, 1872	133, 539
Car-coupling	C. McInturff	Greeneville, Tenn	Dec. 8, 1868	84, 747
Car-coupling	E. T. McKean	Hammonton, N. J	Feb. 22, 1870	100, 173
Car-coupling	H. McKee	Chandlerville, Ill	June 21, 1864	43, 219
Car-coupling	J. McLaughlin	Duncannon, Pa	Jan. 1, 1867	60, 920
Car-coupling	A. McLean	Boston, Mass	Apr. 20, 1869	89, 061
Car-coupling	J. S. McMurray	Toronto, Canada	Mar. 3, 1868	75, 178
Car-coupling	W. H. Meadows	McMinnville, Tenn	Dec. 27, 1870	110, 580
Car-coupling	J. H. Mears and G. Cameron	Oshkosh, Wis	Oct. 18, 1859	25, 844
Car-coupling	T. and J. W. Meikle	Louisville, Ky	Sept. 9, 1873	142, 713
Car-coupling	J. W. Melcher	Oshkosh, Wis	Apr. 7, 1868	76, 494
Car-coupling	A. Melot and J. T. Fry	Reading, Pa	Sept. 26, 1865	50, 148
Car-coupling	J. P. Mendenhall	Farmington, Ill	July 31, 1860	29, 395
Car-coupling	G. H. Merriam	Portland, Me	Aug. 20, 1872	130, 651
Car-coupling	J. A. Merrill and D. W. Kempton.	Denver, Colo	Oct. 7, 1873	143, 524
Car-coupling	F. H. Meyers	Wilmington, Del	Jan. 1, 1867	60, 923
Car-coupling	A. Middleton, jr	Philadelphia, Pa	Sept. 2, 1873	142, 492
Car-coupling	J. T. Middleton	Harveysburgh, Ohio	Nov. 23, 1869	97, 212
Car-coupling	W. J. Millar	McKeesport, Pa	Dec. 24, 1867	72, 664

Index of patents issued from the United States Patent Office from 1790 *to* 1873, *inclusive*—Continued.

Invention.	Inventor.	Residence.	Date.	No.
Car-coupling	H. Millard	Philadelphia, Pa	May 13, 1873	138, 912
Car-coupling	C. L. Miller	Cuba, N. Y	Sept. 9, 1873	142, 715
Car-coupling	D. M. Miller	South Bend, Ind	Sept. 30, 1873	143, 368
Car-coupling	D. W. Miller and M. Brestle, jr.	Middletown, Pa	Oct. 1, 1867	69, 464
Car-coupling	E. Miller	Janesville, Wis	Mar. 31, 1863	38, 057
Car-coupling	J. Miller	Cuba, N. Y	Jan. 29, 1867	61, 553
Car-coupling	J. Miller	Carrollton, Ohio	Oct. 15, 1867	69, 927
Car-coupling	J. Miller	Carrollton, Ohio	Oct. 22, 1867	70, 103
Car-coupling	S. Miller	Holley, N. Y	July 8, 1873	140, 720
Car-coupling	J. D. Mills	Alexandria, Va	Sept. 30, 1873	143, 369
Car-coupling	S. Mills	Madison, Wis	Dec. 24, 1867	72, 665
Car-coupling	S. Mills	Madison, Wis	Apr. 21, 1868	77, 071
Car-coupling	S. Milne	Perthshire, Scotland	Oct. 8, 1867	69, 574
Car-coupling	T. S. Minniss	Meadville, Pa	Nov. 20, 1866	59, 858
Car-coupling	T. S. Minniss	Meadville, Pa	May 14, 1867	64, 689
Car-coupling	I. N. Mitchell	Arcanum, Ohio	July 20, 1869	92, 866
Car-coupling	W. B. Mitchell	Chicago, Ill	Dec. 16, 1873	145, 584
Car-coupling	G. W. Moffitt	Washington, D. C	Dec. 17, 1861	33, 958
Car-coupling	B. Monroe	Bristol, R. I	June 23, 1868	79, 245
Car-coupling	L. Monzert	New York, N. Y	Aug. 25, 1868	81, 397
Car-coupling	B. D. Moody	Darksville, Mo	May 20, 1873	139, 016
Car-coupling	H. T. Moody	Newburyport, Mass	Mar. 9, 1869	87, 582
Car-coupling	L. Moody	Malden, Mass	Jan. 3, 1865	45, 733
Car-coupling	L. Moody	Malden, Mass	Sept. 26, 1865	50, 152
Car-coupling	L. Moody	Malden, Mass	Dec. 19, 1865	51, 608
Car-coupling	B. Moore	Seaford, Del	Jan. 7, 1873	134, 695
Car-coupling	F. Moore and J. A. Baker	Carrollton, Ohio	Oct. 8, 1867	69, 693
Car-coupling	G. R. Moore	Lyons, Iowa	Nov. 19, 1867	71, 202
Car-coupling	G. R. Moore	Lyons, Iowa	Nov. 19, 1867	71, 203
Car-coupling	G. R. Moore	Philadelphia, Pa	Dec. 23, 1873	145, 746
Car-coupling	S. C. Moore	Last Chance, Iowa	June 25, 1872	128, 237
Car-coupling	T. M. Moore	Newton, N. J	Apr. 28, 1868	77, 398
Car-coupling	H. H. Morgan and A. Gerry	San Francisco, Cal	Nov. 28, 1871	121, 252
Car-coupling	T. Morgan	Marquette, Mich	July 4, 1871	116, 738
Car-coupling	D. C. Morris	New Sharon, Iowa	Apr. 8, 1873	137, 617
Car-coupling	J. C. Morris	Greeneville, Tenn	Dec. 20, 1870	110, 269
Car-coupling	J. P. Morris	Bloomington, Ill	July 23, 1867	67, 064
Car-coupling	R. B. Morris	Harrisonville, Mo	Jan. 2, 1872	122, 480
Car-coupling	W. C. Morse	Boston, Mass	Aug. 8, 1865	49, 292
Car-coupling	J. Morton	Cincinnati, Ohio	Dec. 31, 1872	134, 393
Car-coupling	J. Moulton	Moultonville, N. H	Mar. 11, 1873	136, 610
Car-coupling	J. Mount	Monroe Township, N. J	Jan. 3, 1871	110, 770
Car-coupling	H. Mulhollen	Fostoria, Pa	Apr. 1, 1873	137, 470
Car-coupling	J. F. Mullowny, J. Cope, and S. R. Brooks.	Pittsburgh, Connellsville, and Pittsburgh, Pa.	Dec. 2, 1873	145, 228
Car-coupling	W. Musgrove and J. S. Sharp	Studholm Parish and Sussex Parish, Canada.	Aug. 27, 1872	130, 934
Car-coupling	J. W. Musser and R. Carkhuff.	Lewisburgh, Pa	Feb. 14, 1871	111, 767
Car-coupling	J. F. Nagel	Ligonier, Pa	May 18, 1869	90, 290
Car-coupling	F. Nalley	Battle Ground, Ind	Dec. 26, 1871	122, 189
Car-coupling	D. S. Neal	Lynn, Mass	Sept. 3, 1850	7, 620
Car-coupling	R. Neisch and C. G. Hirner	Allentown, Pa	Nov. 5, 1872	132, 852
Car-coupling	W. L. Newell and J. S. Simmerman.	Millville, N. J	Feb. 20, 1866	52, 736
Car-coupling	W. Nichols	Centralia, Ill	May 30, 1871	115, 507
Car-coupling	J. A. Niman	Mansfield, Ohio	July 3, 1866	56, 085
Car-coupling	A. T. Norgan	Pottsville, Pa	Aug. 2, 1870	105, 970
Car-coupling	N. W. Northrup	Greene, N. Y	Oct. 21, 1862	36, 758
Car-coupling	S. G. Northrop	Wilmington, N. C	June 11, 1872	127, 790
Car-coupling	G. W. Noyes	Norwich, Conn	Aug. 27, 1867	68, 105
Car-coupling	S. O'Blenis	Greensburgh, Pa	July 30, 1867	67, 209
Car-coupling	H. L. Ogden	Atkinson, Ill	Nov. 20, 1866	59, 860
Car-coupling	J. H. Oliver	Baltimore, Md	Aug. 29, 1871	118, 476
Car-coupling	J. H. J. O'Neill	New Haven, Conn	May 21, 1867	64, 898
Car-coupling	A. Oot	Minetto, N. Y	Oct. 23, 1860	30, 496
Car-coupling	A. V. B. Orr	Steeleville, Pa	July 16, 1867	66, 733
Car-coupling	J. H. Osgood, jr	Boston, Mass	Apr. 14, 1863	38, 178
Car-coupling	J. Osman and J. F. Potter	Linden Hall, Pa	Nov. 17, 1868	84, 066
Car-coupling	S. A. Otis	Boston, Mass	Oct. 7, 1873	143, 459
Car-coupling	C. Ottinger	Philadelphia, Pa	July 9, 1872	128, 900
Car-coupling	E. E. Packer, jr., and J. Daley	Philadelphia, Pa	Dec. 18, 1866	60, 544
Car-coupling	S. K. Paden	Pulaski, Pa	July 16, 1872	129, 163
Car-coupling	A. G. Page	Augusta, Me	Nov. 28, 1865	51, 209
Car-coupling	J. R. Palmer	Mariposa, Cal	Dec. 6, 1870	109, 932
Car-coupling	J. H. Parsons	Quincy, Mich	Feb. 13, 1866	52, 592
Car-coupling	J. H. Parsons	Quincy, Mich	Mar. 12, 1867	62, 772
Car-coupling	W. B. Parsons	Short Tract, N. Y	May 25, 1869	90, 573
Car-coupling	W. B. Parsons	Short Tract, N. Y	Apr. 19, 1870	102, 149
Car-coupling	W. R. Patton	Des Moines, Iowa	Aug. 16, 1870	106, 399
Car-coupling	D. H. Payne and G. Boxley	Troy, N. Y	Mar. 6, 1866	53, 037
Car-coupling	J. H. Payne	Roserville, Minn	Oct. 14, 1873	143, 636
Car-coupling	H. C. Payson	Haydenville, Mass	Sept. 24, 1867	69, 123
Car-coupling	G. L. Peabody	Buxton, Me	Aug. 6, 1872	130, 238
Car-coupling	J. Pearson	Sacramento, Cal	Oct. 1, 1872	131, 826
Car-coupling	E. W. Peloubet	Newark, N. J	July 29, 1873	141, 290
Car-coupling	D. O. Pender	Palmyra, Mich	Apr. 15, 1873	137, 860
Car-coupling	T. N. Pengelly	Detroit, Mich	Feb. 25, 1873	136, 176
Car-coupling	A. Perrin	Cleveland, Ohio	June 23, 1868	79, 254
Car-coupling	W. V. Perry	Beaver Dam, Wis	July 25, 1871	117, 453
Car-coupling	J. Pettengill, jr	Lisbon, N. H	July 2, 1867	66, 385
Car-coupling	J. M. Phelps	Cogan Station, Pa	Mar. 11, 1873	136, 758
Car-coupling	A. Pierce	Olmsted, Ohio	Sept. 11, 1866	57, 962
Car-coupling	H. W. Pike	Middleport, N. Y	Apr. 23, 1872	125, 985
Car-coupling	W. W. Pitman	Freehold, N. J	May 28, 1872	127, 188
Car-coupling	E. F. Plant	Burr Oak, Mich	Mar. 11, 1873	136, 759
Car-coupling	H. Plumb	Philadelphia, Pa	May 23, 1871	115, 236
Car-coupling	W. D. Pope	Gadsden, Ala	Oct. 14, 1873	143, 638

Index of patents issued from the United States Patent Office from 1790 *to* 1873, *inclusive*—Continued.

Invention.	Inventor.	Residence.	Date.	No.
Car-coupling	A. Porter	Irving, Ill	May 30, 1871	115, 517
Car-coupling	C. Porter	Nunda, N. Y	Mar. 4, 1873	136, 540
Car-coupling	H. H. Potter	Sterlingville, N. Y	Dec. 9, 1873	145, 446
Car-coupling	R. S. Potter	Chicago, Ill	Aug. 28, 1860	29, 815
Car-coupling	T. K. Power	Arsenal Post-office, Pa	Mar. 2, 1869	87, 366
Car-coupling	T. D. Powers	Rochester, Wis	Oct. 16, 1866	58, 883
Car-coupling	D. R. Pratt	Worcester, Mass	Dec. 12, 1848	5, 961
Car-coupling	A. J. Prescott	Catawissa, Pa	Sept. 6, 1870	107, 196
Car-coupling	A. Pritz	Dayton, Ohio	Mar. 25, 1873	137, 238
Car-coupling	S. Puffer	Oxford, N. Y	June 11, 1867	65, 602
Car-coupling	H. Purlier, J. Harlan, and E. C. Cheek.	Cincinnati, Ohio	Apr. 19, 1859	23, 705
Car-coupling	A. Pursell	Williamsport, Pa	Apr. 8, 1873	137, 719
Car-coupling	G. W. Putnam	South Glen's Falls, N. Y	Dec. 5, 1871	121, 660
Car-coupling	G. W. Putnam	Moreau, N. Y	June 4, 1872	127, 644
Car-coupling	G. W. Putnam	South Glen's Falls, N. Y	Oct. 14, 1873	143, 716
Car-coupling	S. S. Quest	Wellsburgh, W. Va	Aug. 28, 1866	57, 566
Car-coupling	G. Quick and J. N. Wallis	Fleming, N. Y	May 22, 1866	54, 955
Car-coupling	C. M. Radcliff	Piermont, N. Y	Apr. 19, 1864	42, 400
Car-coupling	J. W. Radebaugh	Lancaster, Ohio	Nov. 19, 1872	133, 252
Car-coupling	J. C. Ransier	Lyons, N. Y	Apr. 5, 1859	23, 498
Car-coupling	R. Rawson	Quincy, Mich	Apr. 3, 1860	27, 737
Car-coupling	A. Ray	Granville, Mo	Apr. 6, 1869	88, 736
Car-coupling	J. H. Reed	New Haven, Conn	Apr. 25, 1865	47, 455
Car-coupling	J. H. Reed	New Haven, Conn	Sept. 4, 1866	57, 767
Car-coupling	J. V. Reed	New Albany, Ind	Sept. 16, 1873	142, 943
Car-coupling	S. Reed	Liberty, Pa	Nov. 11, 1873	144, 476
Car-coupling	S. H. and G. A. Reed	Fredericktown, Ohio	Feb. 21, 1871	111, 972
Car-coupling	N. Reeves	Du Quoin, Ill	Apr. 16, 1867	63, 940
Car-coupling	S. B. Replogle	Roaring Springs, Pa	Mar. 4, 1873	136, 547
Car-coupling	S. B. Replogle	Roaring Springs, Pa	Sept. 16, 1873	142, 815
Car-coupling	H. Resley	Cumberland, Md	Jan. 14, 1868	73, 385
Car-coupling	W. A. and J. M. Rex, jr	Butler, Ind	Aug. 9, 1870	106, 204
Car-coupling	C. M. Reynolds	Champaign, Ill	Apr. 9, 1867	63, 750
Car-coupling	W. Richards, jr	Franklin, Pa	May 30, 1871	115, 522
Car-coupling	R. Rickkon	Rochester, N. Y	May 24, 1859	24, 146
Car-coupling	L. P. Rider	Pittsburgh, Pa	Nov. 11, 1873	144, 412
Car-coupling	J. Ridings and J. O. Roberts	New Castle, Del	Aug. 13, 1867	67, 677
Car-coupling	J. D. Riggs	Buckley, Ill	Feb. 6, 1872	123, 510
Car-coupling	M. Rinehart	Monroe, Mich	Jan. 17, 1865	45, 939
Car-coupling	A. G. Ritz	Elizabethtown, Ind	Jan. 12, 1869	85, 767
Car-coupling	H. R. Robbins	Baltimore, Md	Sept. 26, 1871	119, 469
Car-coupling	J. N. Robbins	Goshen, Ohio	Nov. 9, 1869	96, 732
Car-coupling	N. Robbins, jr	Rockport, Mass	June 12, 1866	55, 537
Car-coupling	C. F. Rodrick	Lynn, Mass	May 7, 1867	64, 574
Car-coupling	S. Rogers	London, England	July 1, 1873	140, 545
Car-coupling	S. O. Rogers, jr	Stamfordville, N. Y	Mar. 3, 1868	75, 055
Car-coupling	W. B. Rogerson and H. Beyea	Paterson, N. J	Apr. 18, 1871	113, 932
Car-coupling	H. S. Root	Muncy, Pa	Feb. 22, 1870	100, 192
Car-coupling	B. R. Rose	Kansas City, Mo	Sept. 27, 1870	107, 814
Car-coupling	J. Royal, M. A. Lentz, and J. M. Deibert.	La Fayette, Ind	May 23, 1871	115, 242
Car-coupling	J. Royal, M. A. Lentz, and J. M. Deibert.	La Fayette, Ind	Feb. 11, 1873	135, 849
Car-coupling	L. Ruel	Saint Johnsbury, Vt	Oct. 28, 1873	144, 034
Car-coupling	H. M. Rulon	Monmouth, Ill	May 30, 1865	47, 985
Car-coupling	J. B. Rumsey	Washington, D. C	Aug. 21, 1866	57, 385
Car-coupling	J. C. Rupp and S. Ott	Newark, Del	June 6, 1871	115, 643
Car-coupling	E. Russell	Waynesburgh, Pa	Dec. 1, 1868	84, 648
Car-coupling	H. G. Russell	Lincoln, Ill	Nov. 11, 1873	144, 479
Car-coupling	T. W. and T. D. Ryan	Maumee City, Ohio	Oct. 21, 1873	143, 932
Car-coupling	J. S. Sammons	New York, N. Y	June 5, 1860	28, 609
Car-coupling	A. Sanders	Penn Yan, N. Y	Sept. 1, 1868	81, 689
Car-coupling	S. S. Sartwell	Camden, N. Y	May 2, 1871	114, 343
Car-coupling	A. Shorg and B. Van Valkenburgh.	Cobleskill, N. Y	Sept. 23, 1873	143, 099
Car-coupling	L. O. Schultz	Mattoon, Ill	Nov. 5, 1867	70, 625
Car-coupling	P. H. Schuyler	Lyme, Ohio	Aug. 27, 1867	68, 120
Car-coupling	E. R. Scott	Noblesville, Ind	Oct. 14, 1873	143, 720
Car-coupling	L. Scott and P. Trimner	Burgettstown, Pa	July 28, 1868	80, 426
Car-coupling	J. Seislove	Tiffin, Ohio	Oct. 14, 1873	143, 593
Car-coupling	W. F. Senior	Ripley, Ohio	Sept. 16, 1873	142, 873
Car-coupling	C. S. Servoss	Wilmington, N. C	May 7, 1872	126, 494
Car-coupling	F. J. Seybold	Chicago, Ill	Aug. 19, 1873	141, 895
Car-coupling	J. B. Shelly	Richland Township, Pa	Apr. 16, 1872	125, 698
Car-coupling	H. J. Shepardson	Shelburne Falls, Mass	Nov. 7, 1865	50, 882
Car-coupling	J. Shepler	Lambertville, N. J	May 9, 1871	114, 612
Car-coupling	G. W. Shingleton	Auburn, N. Y	June 4, 1867	65, 513
Car-coupling	L. H. Shuler	Crawfordsville, Ind	Aug. 7, 1860	29, 526
Car-coupling	W. P. Siddens	Danville, Ill	June 18, 1872	128, 182
Car-coupling	W. P. Siddens	State Line City, Ind	July 15, 1873	140, 957
Car-coupling	J. E. Simpson	Malden Bridge, N. Y	July 13, 1869	92, 547
Car-coupling	J. P. Sinclair	Little Prairie Ronde, Mich	Aug. 19, 1862	36, 241
Car-coupling	W. H. Skidmore	Secor, Ill	Dec. 19, 1871	122, 070
Car-coupling	S. Slack	Oxford, Pa	May 10, 1864	42, 696
Car-coupling	D. V. B. Smart	Troy, N. Y	Apr. 14, 1868	76, 833
Car-coupling	H. A. Smead and C. H. Huntly	Pavilion, N. Y	May 27, 1862	35, 399
Car-coupling	E. D. Smith	Sutton Township, Canada	Oct. 14, 1873	143, 728
Car-coupling	J. Smith and J. F. Irvin	La Porte, Ind	May 21, 1867	64, 918
Car-coupling	J. E. Smith	Jackson, Pa	Sept. 16, 1873	142, 818
Car-coupling	J. L. Smith	Penn Township, Pa	Oct. 29, 1867	70, 279
Car-coupling	J. W. and J. P. Smith	Elder's Ridge, Pa	Jan. 28, 1868	73, 935
Car-coupling	L. J. Smith	New York, N. Y	Oct. 4, 1870	107, 971
Car-coupling	O. L. Smith and J. F. Utton	Providence, R. I	Sept. 12, 1871	118, 824
Car-coupling	S. H. Smith	Venice Centre, N. Y	Sept. 24, 1867	69, 266
Car-coupling	T. Smith	Waterford, N. Y	July 22, 1873	141, 177
Car-coupling	T. B. Smith and A. Hinchman	Pleasant Hill, Mo	Mar. 16, 1869	87, 882
Car-coupling	W. B. Snedaker	Phœnix, N. Y	Oct. 18, 1870	108, 404

Index of patents issued from the United States Patent Office from 1790 *to* 1873, *inclusive*—Continued.

Invention.	Inventor.	Residence.	Date.	No.
Car-coupling	W. B. Snedaker	Syracuse, N. Y	Nov. 4, 1873	144, 293
Car-coupling	P. H. Snelling	Wartrace, Tenn	July 30, 1867	67, 367
Car-coupling	D. B. Snyder	Millville, N. J	Aug. 21, 1866	57, 397
Car-coupling	H. Soggs	Columbus, Pa	Feb. 5, 1867	61, 771
Car-coupling	H. Soggs	Columbus, Pa	Mar. 30, 1869	88, 342
Car-coupling	J. P. Spangle	Hopewell Centre, N. Y	Aug. 14, 1866	57, 206
Car-coupling	J. P. Spangle	Hopewell Centre, N. Y	Apr. 9, 1867	63, 662
Car-coupling	T. W. Sparks	Rochester, N. Y	July 25, 1871	117, 475
Car-coupling	J. F. Spaulding	Rutland, Vt	Apr. 20, 1869	89, 089
Car-coupling	C. Spear	Chapinsville, N. Y	Nov. 6, 1866	59, 517
Car-coupling	G. D. Spooner and J. F. F. Hale	Rutland, Vt	Aug. 14, 1866	57, 208
Car-coupling	G. C. Sprangler	Allegheny City, Pa	Apr. 4, 1871	113, 462
Car-coupling	E. Staples and W. W. Gould	Skowhegan, Me	Mar. 5, 1867	62, 700
Car-coupling	L. S. Starrett	Athol, Mass	Feb. 7, 1871	111, 695
Car-coupling	A. B. Steele	Winsted, Conn	Aug. 25, 1868	81, 551
Car-coupling	T. Steinhauer	Syracuse, N. Y	Mar. 31, 1868	76, 112
Car-coupling	D. M. Steward	Dayton, Ohio	July 14, 1868	80, 026
Car-coupling	O. S. St. John	Willoughby, Ohio	Dec. 8, 1868	84, 716
Car-coupling	O. S. St. John	Willoughby, Ohio	Dec. 29, 1868	85, 489
Car-coupling	J. T. Stoakes	New Church Parish, England.	Dec. 31, 1867	72, 933
Car-coupling	G. F. Stone	Baltimore, Md	Apr. 30, 1872	126, 344
Car-coupling	T. B. Stout	Keyport, N. J	Aug. 8, 1854	11, 498
Car-coupling	W. A. Stowell	Mooretown, Vt	May 21, 1867	64, 921
Car-coupling	X. St. Pierre	Ophir City, Utah	Oct. 28, 1873	144, 038
Car-coupling	A. Strain	Greenfield, Ohio	Dec. 9, 1873	145, 371
Car-coupling	A. F. Street	Zanesville, Ohio	Dec. 13, 1870	110, 088
Car-coupling	A. Stroh	Port Jervis, N. Y	June 11, 1861	32, 540
Car-coupling	I. W. Stuart, jr	Charlottesville, Ind	Mar. 19, 1867	62, 978
Car-coupling	F. D. Sturges and W. M. Young	Mount Vernon, Ohio	Apr. 27, 1869	89, 355
Car-coupling	J. R. Succop	Birmingham, Pa	Feb. 18, 1873	136, 011
Car-coupling	A. W. Sullenberger	Laurel, Ind	Mar. 10, 1868	75, 489
Car-coupling	D. Sutherland	Milo, Ill	Nov. 19, 1872	133, 268
Car-coupling	J. M. Sutton	Sutton's Corners, Pa	Mar. 18, 1873	136, 879
Car-coupling	H. C. Swan	Washington, D. C	Jan. 30, 1872	123, 207
Car-coupling	J. Swan	Baltimore, Md	Oct. 22, 1867	70, 131
Car-coupling	B. F. Sweet	Fond du Lac, Wis	Oct. 29, 1861	33, 612
Car-coupling	N. Swigart	West Richfield, Ohio	July 16, 1872	129, 300
Car-coupling	P. Swineford	Middleburgh, Pa	Oct. 21, 1873	143, 791
Car-coupling	A. B. and F. S. Taft	Montreal, Canada	Apr. 3, 1866	53, 764
Car-coupling	F. F. Taylor and G. Withington	San Francisco and Ione City, Cal.	Sept. 16, 1873	142, 962
Car-coupling	O. Taylor	Grand Rapids, Wis	Oct. 14, 1873	143, 646
Car-coupling	J. Temple	Bellefonte, Pa	Apr. 5, 1870	101, 543
Car-coupling	J. Temple	Bellefonte, Pa	Sept. 19, 1871	119, 196
Car-coupling	J. Temple	Bellefonte, Pa	Jan. 30, 1872	123, 134
Car-coupling	J. Temple	Danville, Pa	Feb. 18, 1873	135, 948
Car-coupling	J. Temple	Mooresburgh, Pa	Aug. 19, 1873	142, 059
Car-coupling	G. B. Terry and W. G. Hawley	Pittsford and Gorham, N. Y	Sept. 27, 1870	107, 837
Car-coupling	E. D. Thompson	Lawrence, Kans	July 12, 1870	105, 278
Car-coupling	J. A. Thompson	Readington, N. J	Dec. 24, 1872	134, 329
Car-coupling	P. M. Thompson	Toronto, Canada	Sept. 16, 1873	142, 965
Car-coupling	W. M. Thompson	Carlisle, Pa	Oct. 22, 1872	132, 501
Car-coupling	S. Thorp	Turner, Me	Aug. 19, 1862	36, 247
Car-coupling	S. and W. Thorpe	Turner, Me	Feb. 18, 1868	74, 631
Car-coupling	A. C. Tichnor	Council Bluffs, Iowa	Sept. 8, 1868	81, 961
Car-coupling	W. E. Tickler and E. T. and D. M. Marshall.	Pierceton, Ind	July 31, 1866	56, 866
Car-coupling	J. Timms and W. P. Brown	Malta, Ohio	Sept. 19, 1871	119, 097
Car-coupling	C. D. Tisdale	East Boston, Mass	June 2, 1863	38, 792
Car-coupling	D. J. Tittle	Albany, N. Y	Nov. 27, 1866	60, 092
Car-coupling	D. J. Tittle	Albany, N. Y	July 2, 1867	66, 418
Car-coupling	W. Todd	Portland, Me	Jan. 28, 1873	135, 387
Car-coupling	J. B. Tracy	Lincoln, Del	Dec. 12, 1871	121, 823
Car-coupling	J. B. Tracy	Lincoln, Del	Oct. 7, 1873	143, 429
Car-coupling	H. Trefry	Winfield, Mich	June 20, 1871	116, 238
Car-coupling	A. H. Trego	Lambertville, N. J	June 11, 1861	32, 543
Car-coupling	T. B. Tremper	Rockland Lake, N. Y	Dec. 5, 1871	121, 690
Car-coupling	H. H. Trenor	New York, N. Y	Aug. 7, 1866	57, 014
Car-coupling	J. Trent	Millertown, N. Y	Oct. 1, 1867	69, 511
Car-coupling	T. Tripp	Chicago, Ill	Mar. 21, 1865	46, 958
Car-coupling	A. Truxell and F. J. Williams	Toledo, Ohio	Feb. 11, 1873	135, 866
Car-coupling	N. A. Tucker	Burlington, Vt	Feb. 10, 1863	37, 649
Car-coupling	E. A. Turner	New York, N. Y	May 28, 1867	65, 310
Car-coupling	R. D. Turner	New York, N. Y	July 8, 1862	35, 849
Car-coupling	W. E. Twining	Morrison, Ill	Mar. 24, 1868	76, 006
Car-coupling	J. N. and S. T. Vandergrit, S. D. French, and E. S. Stone.	Wabash, Ind	Jan. 7, 1868	73, 140
Car-coupling	J. B. Van Deusen	New York, N. Y	Mar. 4, 1862	34, 604
Car-coupling	J. Van Dyne	Crum Elbow, N. Y	Dec. 15, 1863	40, 966
Car-coupling	I. W. Van Houten	Philadelphia, Pa	July 3, 1860	29, 022
Car-coupling	I. L. Vansant	Glasgow, Del	Mar. 30, 1869	88, 424
Car-coupling	M. Van Slyke and D. W. Wood	Rome, N. Y	Nov. 2, 1869	96, 514
Car-coupling	J. Vansteenberg	Chicago, Ill	May 17, 1870	103, 108
Car-coupling	W. Van Valkinburgh	Smithville, N. Y	July 31, 1866	56, 837
Car-coupling	J. J. Vaughan	Alexandria, Va	May 22, 1860	28, 444
Car-coupling	J. B. Vedder	Gloversville, N. Y	May 30, 1871	115, 394
Car-coupling	H. B. Verrie and D. G. Wightman.	North Kingston, R. I	Nov. 23, 1869	97, 250
Car-coupling	L. and W. H. Waddell	Staunton, Va	June 5, 1860	28, 620
Car-coupling	W. H. and L. Waddell and J. A. Lutz.	Churchville, Va	Sept. 16, 1873	142, 970
Car-coupling	J. Waite	Yarmouth, Me	Sept. 9, 1873	142, 750
Car-coupling	W. Walker	Woodside, Cal	Sept. 20, 1870	107, 571
Car-coupling	S. J. Wallace	Carthage, Ill	Oct. 23, 1866	59, 101
Car-coupling	W. K. Wallace and E. C. Rutledge.	Crawfordsville and Shannondale, Ind.	May 9, 1871	114, 752
Car-coupling	W. V. Wallace	New York, N. Y	Apr. 27, 1869	89, 525

Index of patents issued from the United States Patent Office from 1790 *to* 1873, *inclusive*—Continued.

Invention.	Inventor.	Residence.	Date.	No.
Car-coupling	D. Walter	Evansport, Ohio	Nov. 19, 1872	133, 181
Car-coupling	W. E. Warner and M. J. Palmer	Newark and Syracuse, N. Y	Aug. 6, 1867	67, 468
Car-coupling	W. Y. Warner	Wilmington, Del	Mar. 19, 1867	62, 984
Car-coupling	W. Y. Warner	Wilmington, Del	Nov. 26, 1867	71, 559
Car-coupling	M. Waterbury	Polo, Ill	Feb. 11, 1862	34, 384
Car-coupling	C. Weaver	Easton, Pa	Dec. 4, 1866	60, 299
Car-coupling	G. C. E. Weber	Cleveland, Ohio	Jan. 9, 1872	122, 690
Car-coupling	H. Webster	Elgin, Ill	Aug. 4, 1863	39, 436
Car-coupling	H. Webster	Elgin, Ill	Jan. 16, 1866	52, 097
Car-coupling	G. H. Weeks	Allegan, Mich	Jan. 10, 1871	110, 880
Car-coupling	W. Weiler	Washington, N. J	Jan. 21, 1868	73, 681
Car-coupling	E. F. Wells	New York, N. Y	Nov. 29, 1864	45, 288
Car-coupling	J. M. Wells	Painesville, Ohio	Nov. 4, 1873	144, 298
Car-coupling	C. O. Wheeler	Matteson, Ill	Apr. 24, 1866	54, 237
Car-coupling	E. P. Wheeler	Corinth, Miss	Aug. 6, 1872	130, 262
Car-coupling	J. M. Wheeler	Afton, Iowa	Sept. 17, 1872	131, 416
Car-coupling	J. M. Wheeler and C. W. Chase	Batavia, Iowa	Oct. 19, 1869	96, 065
Car-coupling	J. P. Whipple	Whitewater, Wis	Apr. 8, 1873	137, 642
Car-coupling	J. White	Harrison, Ohio	Aug. 4, 1868	80, 688
Car-coupling	N. S. White	Port Chester, N. Y	Oct. 4, 1864	44, 571
Car-coupling	J. Whiteford	Pond City, Kans	Feb. 22, 1870	100, 092
Car-coupling	D. G. Whitmore	Bridgewater, Va	Nov. 9, 1869	96, 749
Car-coupling	E. R. Whitney	Plattsburgh, N. Y	Apr. 7, 1868	76, 368
Car-coupling	E. R. Whitney	Plattsburgh, N. Y	June 23, 1868	79, 283
Car-coupling	O. J. Whitney	Clifton Springs, N. Y	July 16, 1867	66, 922
Car-coupling	N. Whitten	Etna, Me	May 5, 1868	77, 558
Car-coupling	C. Whitus	Philadelphia, Pa	Apr. 23, 1872	126, 000
Car-coupling	W. F. Wickersham and E. Roush.	Springfield, Ill	Apr. 21, 1868	77, 150
Car-coupling	J. Widney	Allegheny City, Pa	July 5, 1864	43, 447
Car-coupling	J. Widney	Carlisle, Pa	Apr. 24, 1866	54, 238
Car-coupling	J. Wilcox	Canton, Pa	Aug. 12, 1873	141, 846
Car-coupling	E. Wiley	Philadelphia, Pa	May 20, 1873	139, 221
Car-coupling	W. Williams	Harrisonburgh, Va	Jan. 2, 1872	122, 424
Car-coupling	C. C. Wilson	Kewanee, Ill	June 30, 1863	39, 087
Car-coupling	E. W. Wilson and J. E. Erwin	Springfield, Mass	Nov. 14, 1865	50, 994
Car-coupling	G. W. Wilson	Abingdon, Ill	Sept. 4, 1866	57, 807
Car-coupling	J. F. Wilson	Pittsburgh, Pa	Jan. 19, 1869	85, 979
Car-coupling	L. B. Wilson	Caldwell, Ohio	Nov. 26, 1872	133, 512
Car-coupling	L. B. Wilson	Caldwell, Ohio	Feb. 11, 1873	135, 871
Car-coupling	J. T. Wilson and T. J. Louis	East Liberty and Port Perry, Pa.	Jan. 15, 1867	61, 296
Car-coupling	W. Winer	Union City, Ind	Apr. 5, 1870	101, 694
Car-coupling	W. M. Wiswell	Portland, Me	Oct. 28, 1873	144, 008
Car-coupling	J. B. Witherle	Upton, Mass	July 15, 1856	15, 355
Car-coupling	G. Withington and F. F. Taylor.	Ione City and San Francisco, Cal.	Mar. 18, 1873	136, 889
Car-coupling	H. E. Wolcott	Elbridge, N. Y	Mar. 12, 1872	124, 648
Car-coupling	R. T. Wolcott	Claremont, N. H	June 21, 1870	104, 678
Car-coupling	G. E. Wood	Providence, R. I	May 10, 1864	42, 710
Car-coupling	M. R. Wood	Chicago, Ill	Nov. 11, 1873	144, 490
Car-coupling	B. A. Worden	Scranton, Pa	Apr. 23, 1872	126, 122
Car-coupling	G. Worden	Pittston, Pa	Jan. 2, 1872	122, 509
Car-coupling	G. Worden	Pittston, Pa	Dec. 31, 1872	134, 503
Car-coupling	J. F. Wotring	Willey, W. Va	June 25, 1867	66, 061
Car-coupling	J. C. Wrenshall	Baltimore, Md	June 21, 1870	104, 679
Car-coupling	D. D. Wright	Oakville, Canada	Apr. 27, 1869	89, 544
Car-coupling	D. D. Wright	Oakville, Canada	Aug. 20, 1872	130, 779
Car-coupling	E. M. Wright	Wyandotte, Kans	May 24, 1864	42, 897
Car-coupling	G. Yates	West Dresden, N. Y	June 14, 1859	24, 422
Car-coupling	J. C. Young	Bloomington, Ind	June 8, 1869	91, 194
Car-coupling	J. J. Zabriskie, jr., and C. M. Powers.	Ridgewood, N. J	Dec. 31, 1872	134, 414
Car-coupling	A. Zeigler	Cumberland, Md	Sept. 23, 1873	143, 215
Car-coupling	S. H. Zink	Knobnoster, Mo	Jan. 28, 1873	135, 255
Car-coupling	E. Zorger	Greensburgh, Ind	Jan. 1, 1867	60, 815
Car coupling and buffer	A. Stevens	Portland, Me	May 14, 1872	126, 849
Car-coupling apparatus	R. F. Randolph, jr	East Palestine, Ohio	Sept. 20, 1870	107, 623
Car-coupling attachment	W. G. Brown and J. W. Jedkins.	Monmouth, Me	Jan. 28, 1873	135, 262
Car-coupling, Automatic	W. Bragg	Connersville, Ind	Oct. 27, 1868	83, 358
Car-coupling, Automatic	D. P. Cory	Cranford, N. J	Jan. 3, 1871	110, 741
Car-coupling, Automatic	W. W. Jeffery and C. Snyder	Greenview and Middleton, Ill.	Nov. 10, 1868	83, 857
Car-coupling, Automatic	P. Laflin	Warren, Mass	Oct. 27, 1868	83, 512
Car-coupling, Automatic	J. McLain and J. Kelsey	Saint Mary's, Ohio	May 25, 1869	90, 569
Car-coupling, Automatic	H. T. Romertye	Philadelphia, Pa	Jan. 28, 1862	34, 284
Car-coupling, Automatic	H. Tarbox	Warwick, R. I	Nov. 3, 1868	83, 673
Car coupling, City-railway	G. Collyer and A. H Patterson	Philadelphia, Pa	July 17, 1860	29, 142
Car-coupling, Device for	L. L. Bond	Chicago, Ill	Mar. 11, 1873	136, 640
Car coupling-hook, Coal	F. Bush	Boonton, N. J	Sept. 19, 1871	119, 117
Car coupling, Horse-railway	B. E. Sampson	Boston, Mass	July 27, 1858	21, 026
Car-coupling-link, Railway	G. Ryer	Bridgeport, Conn	July 16, 1872	129, 368
Car-coupling-link supporter	J. U. Fiester	Winchester, Ohio	Sept. 3, 1872	130, 986
Car-coupling pins, Die for making	C. H. Williams	Cleveland, Ohio	July 15, 1873	140, 980
Car-coupling pins, Die for manufacturing	C. H. Williams	Cleveland, Ohio	July 15, 1873	140, 981
Car-coupling pins, Machine for rolling	F. W. Davidson	Cleveland, Ohio	Nov. 16, 1869	96, 782
Car-coupling pins, Machine for rolling	W. Davis and J. While	Cleveland, Ohio	Feb. 22, 1870	100, 125
Car coupling, Railway	S. J. Anderson	Cazenovia, N. Y	Mar. 15, 1870	100, 837
Car coupling, Railway	R. K. Andrews	South Valley, N. Y	Aug. 17, 1869	93, 850
Car coupling, Railway	A. A. Atwater	Trumansburgh, N. Y	Nov. 7, 1871	120, 610
Car coupling, Railway	H. A. Barnes	Milwaukee, Wis	Nov. 1, 1859	25, 938
Car coupling, Railway	R. W. Baylor	Norfolk, Va	Sept. 7, 1869	94, 547
Car coupling, Railway	G. S Bishop	Washington, D. C	July 27, 1858	20, 983
Car coupling, Railway	J. Blakeney	Springfield, Ohio	May 17, 1870	103, 130
Car coupling, Railway	E. A Bohne	Brookhaven, Wis	Aug. 31, 1869	94, 386
Car coupling, Railway	H. F. Breneman	Rapho Township, Pa	Aug. 10, 1869	93, 407
Car coupling, Railway	W. H. Burridge and N. L. Post	Cleveland, Ohio	May 4, 1858	20, 139

Index of patents issued from the United States Patent Office from 1790 *to* 1873, *inclusive*—Continued.

Invention.	Inventor.	Residence.	Date.	No.
Car coupling, Railway	L. F. Buschmann	New York, N. Y	Feb. 8, 1870	99, 632
Car coupling, Railway	J. Campbell, V. B. Lighthizer, and P. Shannon.	Steubenville, Ohio	Aug. 24, 1858	21, 244
Car coupling, Railway	P. Campbell	Carrolltown, Pa	May 16, 1871	114, 758
Car coupling, Railway	S. O. Campbell	Centretown, Mo	Aug. 9, 1870	106, 119
Car coupling, Railway	R. Carkhuff	Lewisburgh, Pa	May 24, 1864	42, 835
Car coupling, Railway	R. D. Chatterton	Bath, Great Britain	June 28, 1864	43, 375
Car coupling, Railway	T. J. Christy	Noblesville, Ind	Apr. 3, 1866	53, 574
Car coupling, Railway	W. C. Clark	Portland, Me	Dec. 8, 1863	40, 820
Car coupling, Railway	J. M. Connel	Newark, Ohio	Jan. 5, 1858	19, 021
Car coupling, Railway	M. Connelly	Baltimore, Md	Oct. 5, 1869	95, 431
Car coupling, Railway	J. W. Corey	Crawfordsville, Ind	Sept. 14, 1858	21, 486
Car coupling, Railway	W. Cottrell	Bordentown, N. J	Oct. 5, 1869	95, 434
Car coupling, Railway	R. A. Cowell	Cleveland, Ohio	Oct. 6, 1868	82, 807
Car coupling, Railway	R. A. Cowell	Cleveland, Ohio	Sept. 6, 1870	107, 008
Car coupling, Railway	J. A. Cutting and G. Butterfield	Boston, Mass	Oct. 7, 1846	4, 802
Car coupling, Railway	J. Davis	Allegheny City, Pa	Oct. 20, 1863	40, 331
Car coupling, Railway	T. J. Delany	Loudon County, Va	May 17, 1870	103, 156
Car coupling, Railway	J. L. Devol and A. L. Peadro	Parkersburgh, W. Va	Aug. 16, 1870	106, 471
Car coupling, Railway	L. M. Doddridge	Portland, Ind	Mar. 9, 1869	87, 649
Car coupling, Railway	G. W. Doolittle	Richfield Springs, N. Y	Jan. 26, 1858	19, 186
Car coupling, Railway	T. B. Dora	Mattoon, Ill	Sept. 7, 1869	94, 579
Car coupling, Railway	J. W. H. Doubler	Chicago, Ill	Mar. 1, 1870	100, 272
Car coupling, Railway	H. Dübs and S. G. Goodall-Copestake.	Glasgow, Great Britain	Mar. 7, 1871	112, 431
Car coupling, Railway	C. H. Eisenbrandt	Baltimore, Md	May 24, 1859	24, 109
Car coupling, Railway	J. Elbertson	Kirksville, Mo	July 6, 1869	92, 293
Car coupling, Railway	J. T. England	Baltimore, Md	Dec. 4, 1855	13, 869
Car coupling, Railway	W. J. Evans	Homer, Iowa	Mar. 8, 1870	100, 515
Car coupling, Railway	A. C. Fell	Caledonia, Ohio	Mar. 12, 1872	124, 430
Car coupling, Railway	H. L. Fidler	Congress, Ohio	Dec. 24, 1872	134, 135
Car coupling, Railway	J. C. Fischer and W. E. Kittredge.	Milwaukee, Wis	July 26, 1870	105, 791
Car coupling, Railway	C. Flanders	Charlestown, Mass	Dec. 16, 1856	16, 230
Car coupling, Railway	A. Geiss	Buffalo, N. Y	Mar. 22, 1870	100, 999
Car coupling, Railway	F. E. Gleason	Columbus, Ohio	May 18, 1858	20, 264
Car coupling, Railway	J. G. Goshon	Shippensburgh, Pa	Dec. 27, 1859	26, 568
Car coupling, Railway	W. F. Grassler	Muncy, Pa	May 17, 1870	103, 172
Car coupling, Railway	R. Green	Boonton, N. J	Feb. 8, 1870	99, 562
Car coupling, Railway	W. H. Hall	Malone, N. Y	Jan. 19, 1869	86, 011
Car coupling, Railway	G. Harris	Ipswich, Mass	Apr. 20, 1869	89, 145
Car coupling, Railway	A. S. Hart	San Francisco, Cal	Dec. 21, 1869	98, 058
Car coupling, Railway	D. Hart	Romulus, N. Y	June 7, 1870	104, 024
Car coupling, Railway	A. Hebert	Malone, N. Y	June 14, 1870	104, 306
Car coupling, Railway	A. G. Heckrotte	New York, N. Y	July 17, 1855	13, 264
Car coupling, Railway	S. R. Herd	Allegheny City, Pa	Feb. 22, 1870	100, 146
Car coupling, Railway	J. W. Hess	Montandon, Pa	Aug. 9, 1870	106, 159
Car coupling, Railway	J. R. Hill	Millville, N. J	Nov. 7, 1865	50, 824
Car coupling, Railway	N. Hill	Leavenworth City, Kans	Dec. 14, 1869	97, 918
Car coupling, Railway	O. Hood, jr	Turner, Me	July 17, 1860	29, 219
Car coupling, Railway	D. A. Hopkins	Elmira, N. Y	Aug. 1, 1854	11, 428
Car coupling, Railway	D. D. Howe	Beaver Dam, Wis	Nov. 10, 1868	83, 966
Car coupling, Railway	L. and J. K. Huddle	Tiffin, Ohio	Jan. 24, 1871	111, 123
Car coupling, Railway	G. C. Hugg	Berlin, N. J	May 31, 1870	103, 618
Car coupling, Railway	R. M. Hughes	Pleasant Grove, Pa	Mar. 5, 1861	31, 603
Car coupling, Railway	L. Ibeck	Kickapoo, Ill	Nov. 7, 1871	120, 746
Car coupling, Railway	J. H. Johnson	Dresden, Mo	Aug. 16, 1870	106, 367
Car coupling, Railway	W. C. Johnson	Fort Madison, Iowa	Nov. 30, 1869	97, 410
Car coupling, Railway	J. H. Jones	Scio, N. Y	Apr. 10, 1855	12, 680
Car coupling, Railway	O. Kelsey	Commerce, Mich	July 20, 1869	92, 839
Car coupling, Railway	C. P. Kenyon	Wilson, N. C	Sept. 14, 1858	21, 502
Car coupling, Railway	J. D. Kerrison	New York, N. Y	Nov. 16, 1869	96, 813
Car coupling, Railway	L. Laley	Goshen, Ind	Nov. 16, 1869	96, 816
Car coupling, Railway	J. Larimore and J. M. Williams	Connersville, Ind	Sept. 28, 1869	95, 360
Car coupling, Railway	G. Lavally, jr	Champlain, N. Y	June 26, 1860	28, 872
Car coupling, Railway	A. Lhapham and D. H. Burns	Brooklyn, N. Y	May 25, 1858	20, 392
Car coupling, Railway	L. D. Livermore	Hartland, Vt	Nov. 11, 1851	8, 514
Car coupling, Railway	H. E. Loane	Baltimore, Md	Mar. 23, 1858	19, 705
Car coupling, Railway	H. L. Lockwood	Denmark, Iowa	Nov. 30, 1869	97, 303
Car coupling, Railway	F. B. Lord	Cincinnati, Ohio	Nov. 16, 1869	96, 932
Car coupling, Railway	D. Lynahon and C. J. Wing	Buffalo, N. Y	Oct. 28, 1856	15, 997
Car coupling, Railway	J. Marston	Saratoga Springs, N. Y	Oct. 26, 1869	96, 130
Car coupling, Railway	G. W. McEuen and C. Eves	Millville, Pa	Oct. 31, 1871	120, 526
Car coupling, Railway	J. McKinney	Lansing, Mich	Sept. 4, 1860	29, 900
Car coupling, Railway	E. D. Meier	Saint Louis, Mo	June 6, 1871	115, 627
Car coupling, Railway	G. H. Merriam	Portland, Me	Aug. 8, 1871	117, 909
Car coupling, Railway	J. S. Merriken	Baltimore, Md	Oct. 19, 1869	96, 070
Car coupling, Railway	J. Miller	Olean, N. Y	Nov. 14, 1854	11, 940
Car coupling, Railway	W. H. H. Miller	Williamsport, Pa	Aug. 28, 1860	29, 806
Car coupling, Railway	R. L. Mills and P. Carpenter	Lancaster, Ohio	July 5, 1859	24, 651
Car coupling, Railway	J. A. Morrison	Brady's Bend, Pa	Oct. 12, 1869	95, 711
Car coupling, Railway	C. Norpel	Newark, Ohio	Dec. 27, 1859	26, 608
Car coupling, Railway	N. Norris	Buchanan, Mich	Oct. 4, 1870	107, 950
Car coupling, Railway	J. M. Osgood, jr., and F. B. Shaw	Boston, Mass	Jan. 15, 1861	31, 127
Car coupling, Railway	G. W. Parshall	Middlefield, N. Y	Apr. 19, 1859	23, 698
Car coupling, Railway	J. Pearson	Sterling, Iowa	Jan. 26, 1858	19, 204
Car coupling, Railway	P. Perry	Troy, N. Y	Oct. 26, 1858	21, 901
Car coupling, Railway	S. D. Pratt	Penn Yan, N. Y	Oct. 3, 1871	119, 475
Car coupling, Railway	W. Prosser	Kendall, N. Y	July 21, 1857	17, 845
Car coupling, Railway	W. V. Pulliam	Kansas City, Mo	Oct. 12, 1869	95, 835
Car coupling, Railway	J. H. Quackenbush	Owasso, Mich	July 6, 1858	20, 817
Car coupling, Railway	M. Quinn	Wataga, Ill	Oct. 26, 1869	96, 146
Car coupling, Railway	J. F. Rague	Dubuque, Iowa	Dec. 29, 1857	18, 990
Car coupling, Railway	E. Rice	Canandaigua, N. Y	Mar. 6, 1855	12, 490
Car coupling, Railway	J. W. Rice	Springfield, Mass	Mar. 30, 1858	19, 794
Car coupling, Railway	J. Riley	Chicago, Ill	July 6, 1869	92, 365
Car coupling, Railway	A. Roll	South Amboy, N. J	Nov. 28, 1865	51, 223
Car coupling, Railway	A. H. Roward	Allegheny, Pa	Jan. 1, 1861	31, 036
Car coupling, Railway	J. Ryan	Wilmington, Del	July 17, 1855	13, 274

Index of patents issued from the United States Patent Office from 1790 *to* 1873, *inclusive*—Continued.

Invention.	Inventor.	Residence.	Date.	No.
Car coupling, Railway	J. Schneider	Chicago, Ill	Jan. 5, 1858	19, 049
Car coupling, Railway	G. Shatswell	Waukegan, Ill	Nov. 16, 1869	96, 976
Car coupling, Railway	G. C. Sherman	Chicago, Ill	Sept. 26, 1871	119, 416
Car coupling, Railway	J. C. Smith	Stoughstown, Pa	July 6, 1869	92, 383
Car coupling, Railway	M. B. Smith	Mattoon, Ill	Feb. 21, 1871	111, 985
Car coupling, Railway	O. L. Smith	Providence, R. I	Mar. 30, 1869	88, 337
Car coupling, Railway	E. F. Sneider	West Penn, Pa	Feb. 22, 1870	100, 204
Car coupling, Railway	D. G. W. Snyder	Williamsport, Md	July 6, 1869	92, 386
Car coupling, Railway	A. L. Spear	Flint, Mich	Nov. 23, 1869	97, 242
Car coupling, Railway	W. T. Spies	Baltimore, Md	May 20, 1862	35, 333
Car coupling, Railway	J. R. Swift	New Orleans, La	Nov. 29, 1859	26, 303
Car coupling, Railway	A. B. Thompson	Oswego, N. Y	Dec. 22, 1868	85, 147
Car coupling, Railway	S. Thorp and W. D. Shurtleff	Turner, Me	July 17, 1860	29, 205
Car coupling, Railway	W. C. Tilton	Spring Place, Ga	July 20, 1869	92, 902
Car coupling, Railway	J. B. Tracy	Lincoln, Del	Oct. 10, 1871	119, 897
Car coupling, Railway	J. Turnbull	Edinburgh, Scotland	June 29, 1869	92, 126
Car coupling, Railway	J. Turner	East Nassau, N. Y	July 20, 1852	9, 144
Car coupling, Railway	E. M. Van Hoesen and N. H. Brown.	Syracuse, N. Y	Sept. 26, 1871	119, 431
Car coupling, Railway	J. B. Vedder	Gloversville, N. Y	Jan. 11, 1870	98, 723
Car coupling, Railway	W. V. Wallace	New York, N. Y	Aug. 3, 1869	93, 375
Car coupling, Railway	W. V. Wallace	New York, N. Y	Oct. 19, 1869	96, 059
Car coupling, Railway	J. C. Ward	Charleston, S. C	Oct. 14, 1856	15, 909
Car coupling, Railway	D. Warren	Gettysburgh, Pa	Jan. 14, 1859	24, 420
Car coupling, Railway	R. Wells	Bloomington, Ill	Aug. 3, 1869	93, 378
Car coupling, Railway	A. Wellshmidt	Albany, N. Y	Jan. 23, 1872	122, 928
Car coupling, Railway	N. H. Wentworth and M. L. Ames.	Somersworth, N. H	Mar. 15, 1859	23, 282
Car coupling, Railway	G. W. Wheat	Philipsburgh, Pa	Oct. 4, 1870	107, 988
Car coupling, Railway	O. M. Whitman	North Haverhill, N. H	Aug. 3, 1869	93, 381
Car coupling, Railway	J. C. Wilson	Appleton, Wis	Oct. 5, 1869	95, 546
Car coupling, Railway	G. Winters	Portsmouth, Pa	Sept. 16, 1851	8, 359
Car coupling, Railway	S. W. Wood	Washington, D. C	Feb. 12, 1856	14, 265
Car coupling, Railway	O. D. Woodruff	Southington, Conn	June 8, 1869	90, 977
Car coupling, Reversible railway	J. Boothroyd	Michigan City, Ind	Sept. 8, 1857	18, 132
Car-coupling, Self-acting	A. G. Safford	Boston, Mass	Dec. 11, 1849	6, 947
Car coupling, Self-disengaging railway	J. C. Price	New Philadelphia, Ohio	Feb. 17, 1857	16, 654
Car coupling, Spring	F. Steinhart	Dansville, N. Y	Apr. 26, 1859	23, 793
Car coupling, Street	J. Stephenson	New York, N. Y	Jan. 7, 1873	134, 616
Car-couplings, Automatic attachment for	G. H. Merriam	Portland, Me	Dec. 16, 1873	145, 673
Car-couplings, Link-guide for	G. Stillson	Binghamton, N. Y	Dec. 24, 1872	134, 173
Car-couplings, Link-guide for	W. Warinner and W. L. D. Johnson.	Creelsborough, Ky	Nov. 4, 1873	144, 297
Car crank, Hand	P. Greel	Meadville, Pa	Feb. 9, 1864	41, 574
Car-door	J. H. Robertson	New York, N. Y	Apr. 18, 1871	113, 931
Car door, Baggage	H. L. Clarke	La Porte, Ind	Sept. 5, 1854	11, 656
Car-door fastening	J. K. Smith	Port Clinton, Pa	Feb. 12, 1836	....
Car-door fastening, Railway	H. D. Bird	Petersburgh, Va	Jan. 29, 1867	61, 652
Car door, Grain	J. Bassler	Galesburgh, Ill	Mar. 12, 1872	124, 535
Car door, Grain	T. A. Bissell and R. Miller	Aurora, Ill	Oct. 15, 1872	132, 237
Car door, Grain	S. E. Bright	Elkhorn, Wis	Sept. 10, 1867	68, 694
Car door, Grain	S. E. Knott	Chicago, Ill	May 25, 1869	90, 560
Car-door holder	G. Crandell	Washington, D. C	Apr. 19, 1864	42, 452
Car-door, &c., lock	G. B. F. Cooper	New Albany, Ind	May 14, 1867	64, 635
Car-door lock	A. V. Hartwell	Chicago, Ill	Oct. 26, 1869	96, 230
Car-door lock	J. L. Howard	Hartford, Conn	Jan. 18, 1870	98, 967
Car door, Railway	T. R. Leighton	Cameron, Mo	Mar. 9, 1869	87, 685
Car door, Railway	B. Martin	Prairie du Chien, Wis	Apr. 30, 1867	64, 236
Car door, Railway	C. Barker	Knox County, Ill	May 17, 1870	103, 126
Car door, Railway freight	H. L. Clark	Rahway, N. J	Aug. 29, 1871	118, 514
Car door, Street	D. R. Hart	Saint Louis, Mo	Aug. 16, 1870	106, 362
Car doors, Lock for freight	W. S. Brewster	Chicago, Ill	Feb. 4, 1873	135, 402
Car doors, Opening railway	T. Castor	Philadelphia, Pa	July 2, 1861	32, 681
Car-doors, Operating	J. G. and G. M. Brill	Philadelphia, Pa	Dec. 23, 1873	145, 839
Car-doors, Operating	J. A. O'Haire	West Troy, N. Y	Sept. 16, 1873	142, 810
Car-doors, &c., Seal for	D. W. Long	Baltimore, Md	July 14, 1857	17, 796
Car, Dumping	O. M. Avery	Pensacola, Fla	Apr. 29, 1873	138, 361
Car, Dumping	C. P. Bailey	Zanesville, Ohio	Jan. 24, 1854	10, 441
Car, Dumping	A. J. Ballard	Cohoes, N. Y	Oct. 18, 1870	108, 313
Car, Dumping	J. W. Bancroft	Philadelphia, Pa	Jan. 8, 1867	61, 038
Car, Dumping	C. Barrett	Boston, Mass	Dec. 19, 1871	121, 981
Car, Dumping	C. Barrett	Boston, Mass	Feb. 20, 1872	123, 762
Car, Dumping	C. Barrett	Boston, Mass	July 16, 1872	129, 515
Car, Dumping	T. Bootsmann	Tompkinsville, N. Y	June 14, 1870	104, 254
Car, Dumping	T. Bootsmann	New Brighton, N. Y	May 27, 1873	139, 362
Car, Dumping	L. C. Brady	Chambersville, Pa	Nov. 18, 1873	144, 654
Car, Dumping	J. Braidwood	Wilmington, Ill	July 24, 1866	56, 518
Car, Dumping	D. J. Brimmer	North Petersburgh, N. Y	Mar. 28, 1871	113, 013
Car, Dumping	N. W., C. W., and N. H. Camp.	Saratoga Springs, N. Y., and Wilkesbarre, Pa.	Aug. 8, 1871	117, 739
Car, Dumping	W. Chisholm	Cleveland, Ohio	June 30, 1868	79, 445
Car, Dumping	F. B. Colton	Philadelphia, Pa	Dec. 26, 1871	122, 226
Car, Dumping	J. G. Conlon	New Orleans, La	July 13, 1869	92, 426
Car, Dumping	D. S. Cook	Wrightsville, Pa	Apr. 23, 1872	126, 029
Car, Dumping	P. Daniels	Jackson, Mich	July 28, 1868	80, 339
Car, Dumping	P. K. Dederick	Albany, N. Y	Nov. 15, 1870	109, 188
Car, Dumping	J. Disterdick	Kelley's Mills, Ohio	Apr. 23, 1872	126, 036
Car, Dumping	J. R. Dubois	Virginia City, Nev	Nov. 5, 1872	132, 818
Car, Dumping	J. Elliott, J. P. Harrington, and W. R. Davenport.	Erie, Pa	Nov. 25, 1873	144, 966
Car, Dumping	C. Gates	Burlington, Ind	Aug. 17, 1869	93, 872
Car, Dumping	D. Glover	Cass Township, Pa	July 12, 1859	24, 732
Car, Dumping	G. B. Goodwin and S. McCord	Milwaukee, Wis	Dec. 22, 1868	85, 086
Car, Dumping	J. M. Goodwin	Cleveland, Ohio	Aug. 13, 1872	130, 424
Car, Dumping	J. M. Hanford and J. Wood	Middletown, N. Y	June 3, 1873	139, 575
Car, Dumping	W. H. Harding and G. F. Morse	Portland, Me	Apr. 4, 1871	113, 655
Car, Dumping	G. W. Hart	Aurora, Ind	Dec. 1, 1857	18, 743
Car, Dumping	E. C. Hegeler	La Salle, Ill	Sept. 27, 1870	107, 683

Index of patents issued from the United States Patent Office from 1790 *to* 1873, *inclusive*—**Continued.**

Invention.	Inventor.	Residence.	Date.	No.
Car, Dumping	J. Hughes	New Berne, N. C	July 15, 1873	140, 779
Car, Dumping	J. Hughes	New Berne, N. C	Aug. 19, 1873	142, 021
Car, Dumping	E. H. Jackson	Boston, Mass	July 31, 1866	56, 756
Car, Dumping	I Keith	West Sandwich, Mass	July 13, 1869	92, 457
Car, Dumping	J. Kimbel	Zanesville, Ohio	Oct. 3, 1854	11, 750
Car, Dumping	S. D. King	Middletown, N. Y	Nov. 8, 1870	109, 020
Car, Dumping	S. D. King	Middletown, N. Y	Feb. 21, 1871	112, 052
Car, Dumping	S. D. King	Middletown, N. Y	Oct. 10, 1871	119, 856
Car, Dumping	S. D. King	Middletown, N. Y	Apr. 2, 1872	125, 307
Car, Dumping	W. R. Maffit	Wilkesbarre, Pa	Oct. 4, 1864	44, 539
Car, Dumping	C. W. McLean and W. M. Elton	New Berne, N. C., and Brooklyn, N. Y.	July 9, 1872	128, 744
Car, Dumping	T. McVay	Braddock's Field, Pa	Sept. 12, 1871	118, 959
Car, Dumping	W. Merrington and N. Kirkwood.	McKeesport, Pa	Nov. 1, 1870	108, 810
Car, Dumping	A. Nettleton	Springfield, Mass	Jan. 30, 1849	6, 065
Car, Dumping	J. G. Payson	Foxborough, Mass	July 1, 1873	140, 534
Car, Dumping	A. Peteler	New Brighton, N. Y	Mar. 16, 1869	87, 966
Car, Dumping	A. Peteler	New Brighton, N. Y	May 30, 1871	115, 514
Car, Dumping	A. H. Petsch	Charleston, S. C	Aug. 1, 1854	11, 441
Car, Dumping	W. Riley, jr	Terre Haute, Ind	Oct. 25, 1870	108, 632
Car, Dumping	W. Robinson	Bellfontaine, Ohio	Dec. 31, 1872	134, 488
Car, Dumping	S. Rousculp and I. B. Shambaugh.	Thornville and Scio, Ohio	Aug. 29, 1871	118, 557
Car, Dumping	C. H. Sage	Fulton, N. Y	Aug. 31, 1869	94, 445
Car, Dumping	W. A. Sharp	Tama City, Iowa	Apr. 18, 1871	113, 804
Car, Dumping	E. C. Smeed	Lawrence, Kans	Oct. 18, 1870	108, 400
Car, Dumping	O. L. Smith	Providence, R. I	Mar. 22, 1870	101, 057
Car, Dumping	E. Thompson	Hokah, Minn	Nov. 26, 1867	71, 552
Car, Dumping	J. M. Thompson	Quincy, Cal	Aug. 6, 1872	130, 334
Car, Dumping	J. M. Thompson	Quincy, Cal	Aug. 6, 1872	130, 335
Car, Dumping	J. C. and F. A. Wiswell	Lennoxville, Canada, and Beebe Plain, Vt.	June 20, 1871	116, 127
Car, Dumping	J. C. and F. A. Wiswell	Lennoxville, Canada, and Beebe Plain, Vt.	May 14, 1872	126, 859
Car-dumping apparatus	E. Cockill	Llewellyn, Pa	Sept. 19, 1871	119, 122
Car, Dumping railway	A. B. Cooley	Philadelphia, Pa	July 3, 1860	28, 969
Car duster and ventilator, Railway	J. P. Curry	New York, N. Y	Aug. 30, 1870	106, 786
Car duster, Railway	L. M. Platt	Chicago, Ill	June 15, 1869	91, 366
Car draw-bar, Street	J. C. Gove	Cleveland, Ohio	Mar. 2, 1869	87, 486
Car draw-head, Railway	H. C. Lowrie	Fort Wayne, Ind	Dec. 23, 1873	145, 881
Car, Drying	J. K. Caldwell	Allegheny City, Pa	Nov. 16, 1869	96, 882
Car, Earth	R. H. Emerson	Chicago, Ill	Apr. 11, 1854	10, 775
Car elevating and transferring apparatus, Railway	W. T. Beekman	Petersburgh, Ill	Dec. 16, 1873	145, 611
Car-elevator	H. Arden	Cincinnati Ohio	May 5, 1868	77, 706
Car elevator, Coal	P. H. Lamey	Wiconisco, Pa	Feb. 4, 1873	135, 565
Car elevator, Coal	P. H. Lamey	Wiconisco, Pa	July 22, 1873	141, 148
Car, Extension railway	J. S. Brown	Lowell, Mass	Jan. 15, 1856	14, 083
Car, Fire-proof railway	G. C. Bestor	Peoria, Ill	Aug. 30, 1870	106, 771
Car-floor slat-matting	S. Lewis	Williamsburgh, N. Y	Sept. 12, 1871	118, 952
Car, Foot	N. Hodge	North Adams, Mass	June 22, 1852	9, 054
Car for carrying petroleum	J. Clark	Canandaigua, N. Y	Nov. 3, 1863	40, 458
Car for carrying petroleum	J. Clark	Pontiac, Mich	Mar. 18, 1873	136, 813
Car for carrying petroleum, &c	J. Scott	Lawrenceville, Pa	Jan. 20, 1863	37, 461
Car for carrying petroleum, &c	S. J. Seely	Brooklyn, N. Y	June 2, 1863	38, 765
Car for coal-mines, Dumping	M. G. Smith and W. P. Stephens	Kingston, Pa	Oct. 9, 1866	58, 736
Car for day and night service, Railway	J. B. Creighton	Tiffin, Ohio	May 18, 1858	20, 254
Car for dumping earth, &c	M. Berney	Syracuse, N. Y	Sept. 11, 1849	6, 712
Car for elevated railways	J. B. Newbrough	New York, N. Y	Dec. 5, 1871	121, 539
Car for grading, Dumping	W. Price	Cincinnati, Ohio	Nov. 2, 1869	96, 481
Car for inclined planes, Safety	J. Tittle	Johnstown, Pa	Feb. 10, 1846	4, 381
Car for mines, Ore and timber	G. Williams	Sterling, Colo	July 17, 1866	56, 480
Car for plank-roads, wooden rails, &c	G. Morgan	Calhoun, Tenn	June 11, 1850	7, 427
Car for preventing accidents, Railway	F. W. Jenkins	Brooklyn, N. Y	June 19, 1866	55, 663
Car for rail and other roads	J. A. Whitford	Saratoga Springs, N. Y	Jan. 29, 1841	1, 931
Car for railways and locomotive-engine wheel	M. W. Baldwin	Philadelphia, Pa	June 29, 1833	....
Car for railways, Hand	A. Welsch	Chicago, Ill	Dec. 13, 1859	26, 453
Car for railways, Metallic	B. J. La Mothe	New York, N. Y	Sept. 24, 1861	33, 350
Car for railways, Sleeping	C. L. Harrington	Buffalo, N. Y	Aug. 3, 1858	21, 070
Car for railways, Sleeping	E. Wheeler	Elmira, N. Y	Aug. 3, 1858	21, 099
Car for removing gravel and earth on railways	G. Palmer	Greenbush, N. Y	Jan. 16, 1843	2, 909
Car for transportation of coal	L. Myers	Philadelphia, Pa	June 24, 1851	8, 177
Car for transportation of coal, &c	R. Winans	Baltimore, Md	June 26, 1847	5, 175
Car for transporting and drying peat	D. E. Teal	Norwich, N. Y	Feb. 11, 1868	74, 255
Car for transporting petroleum	J. and A. Densemore	Meadville, Pa	Apr. 10, 1866	53, 794
Car for turning curves, Railway	J. H. Quail	Philadelphia, Pa	Feb. 28, 1844	3, 457
Car-frame	R. M. C. Parker	Memphis, Tenn	Jan. 7, 1873	134, 701
Car frame, Railway	H. Bachman	Lancaster, Pa	May 2, 1835	....
Car frames with metallic joints, Constructing railway.	T. L. Nickols	Alexandria, Va	Dec. 4, 1860	30, 833
Car, Freight	J. D. Billings	Rutland, Vt.	May 31, 1859	24, 189
Car, Freight	S. W. Downey	Piedmont, W. Va	Jan. 14, 1868	73, 308
Car, Freight	C. R. Foote and J. Orton	Williamstown, Mass	Jan. 12, 1864	41, 212
Car, Freight	L. Myers	Philadelphia, Pa	July 1[illegible], 1865	48, 830
Car, Freight	T. R. Timby	Tarrytown, N. Y	Oct. 31, 1871	120, 552
Car, Grain	A. E. Gordon	New Brunswick, N. J	Dec. 5, 1871	121, 612
Car, Grain	W. H. Long and J. E. Lavey	Plymouth, Ind	May 26, 1868	78, 383
Car, Grain	S. W. Wood	Cornwall, N. Y	Dec. 24, 1872	134, 181
Car, Grain-drying	C. S. Dole	Chicago, Ill	May 9, 1871	114, 534
Car, Grain-dumping	L. Bammerlin	Massillon, Ohio	Jan. 3, 1871	110, 724
Car, Gravel	T. C. Hendry	Conyers, Ga	Dec. 13, 1859	26, 460
Car, Hand	W. T. Beekman	Petersburgh, Ill	Feb. 21, 1871	112, 008
Car, Hand	L. J. Cathell	Salisbury, Md	Jan. 30, 1872	123, 322
Car, Hand	J. Collins	Fairview, Ohio	Jan. 16, 1872	122, 814
Car, Hand	M. E. Hastings	Salisbury, Md	June 27, 1871	116, 527
Car, Hand	J. D. Hinckley	Adrian, Mich	June 17, 1873	140, 039
Car, Hand	D. M. Hunt	Southampton Mills, Pa	Mar. 11, 1873	136, 652
Car, Hand	D. Johnston	Eddyville, Iowa	June 20, 1871	116, 062

Index of patents issued from the United States Patent Office from 1790 *to* 1873, *inclusive*—Continued.

Invention.	Inventor.	Residence.	Date.	No.
Car, Hand	L. H. Kenyon	East Greenwich, R. I	Oct. 7, 1873	143, 513
Car hanger, Railway	T. S. Hudson	Cambridge, Mass	Apr. 30, 1872	126, 299
Car-heater	J. C. Eckert	Dayton, Ohio	Oct. 20, 1868	83, 142
Car-heater	A. H. Lighthall and C. F. North	Cohoes, N. Y	Mar. 3, 1868	75, 172
Car-heater	W. S. McNeil and O. S. Cadwell, jr.	Springfield, Mass	Aug. 25, 1868	81, 392
Car-heater	T. Smith and J. O. Reilly	Baltimore, Md	Oct. 6, 1868	82, 985
Car heater and ventilator	E. Himrod	Dunmore, Pa	Mar. 16, 1869	87, 846
Car heater and ventilator	W. S. McNeil	Springfield, Mass	June 8, 1869	91, 148
Car heater and ventilator, Railway	S. Darling	Bangor, Me	June 30, 1868	79, 449
Car heater and ventilator, Railway	S. W. Francis	Newport, R. I	Apr. 28, 1868	77, 270
Car heater and ventilator, Railway	T. H. Mott	New York, N. Y	July 29, 1873	141, 372
Car heater and ventilator, Railway	E. L. Roberts	New York, N. Y	Aug. 18, 1868	81, 211
Car heater and ventilator, Railway	T. F. Strong	Fond du Lac, Wis	July 2, 1861	32, 726
Car heater and ventilator, Railway	A. Weeks	Minneapolis, Minn	Apr. 27, 1869	89, 529
Car heater and ventilator, Railway	C. F. Whorf	Saint Louis, Mo	Apr. 22, 1873	138, 062
Car heater, Passenger	G. K. Dearborn	Abington, Mass	Jan. 28, 1862	34, 280
Car heater, Railway	J. G. Allen	Philadelphia, Pa	Sept. 6, 1870	107, 149
Car heater, Railway	J. G. Allen	Philadelphia, Pa	July 11, 1871	116, 914
Car heater, Railway	I. R. Amsdam	Buffalo, N. Y	Aug. 25, 1868	81, 322
Car heater, Railway	T. B. Atterbury	Pittsburgh, Pa	June 18, 1872	127, 945
Car heater, Railway	W. C Baker	New York, N. Y	Mar. 10, 1868	75, 345
Car heater, Railway	W. H. Beal	Philadelphia, Pa	Mar. 9, 1869	87, 532
Car heater, Railway	B. Berghausen and A. L. Kiesling.	Cologne, Germany	Nov. 4, 1873	144, 183
Car heater, Railway	E. Bradley and J. G. Fulghum	Nashville, Tenn	June 18, 1872	128, 103
Car heater, Railway	H. M. Britton	Cincinnati, Ohio	Sept. 8, 1868	81, 873
Car heater, Railway	C. Carpenter	Buffalo, N. Y	Aug. 18, 1868	81, 137
Car heater, Railway	D. C. Chipman	Noblesville, Ind	June 29, 1869	91, 824
Car heater, Railway	A. C. Crary	Utica, N. Y	Mar. 2, 1869	87, 472
Car heater, Railway	A. Davidsohn	Saint Louis, Mo	Dec. 8, 1868	84, 735
Car heater, Railway	I. Dripps	Fort Wayne, Ind	Oct. 6, 1868	82, 810
Car heater, Railway	W. B. Farwell	New York, N. Y	Sept. 15, 1868	82, 210
Car heater, Railway	J. Gibson, jr	Albany, N. Y	Dec. 28, 1869	98, 248
Car heater, Railway	P. Grandjean	Paris, France	Oct. 25, 1870	108, 702
Car heater, Railway	W. G. Kendrick	Wilmington, Del	Aug. 13, 1867	67, 770
Car heater, Railway	J. E. Kline	Wheeling, W. Va	Sept. 21, 1869	95, 026
Car heater, Railway	G. S. Koontz, E. Potts, and M. D. Lewis.	Washington, D. C., and Baltimore, Md.	Mar. 23, 1869	88, 183
Car heater, Railway	F. McManus	Ellenburgh Centre, N. Y	Mar. 23, 1869	88, 155
Car heater, Railway	F. M. Meddock	Mainsville, Ohio	Oct. 13, 1868	83, 076
Car heater, Railway	W. Meller	McKeesport, Pa	Mar. 8, 1870	100, 543
Car heater, Railway	J. P. Michaels and L. W. Broadwell.	Vienna, Austria	Jan. 28, 1873	135, 239
Car heater, Railway	A. C. Miner	Philadelphia, Pa	June 16, 1868	78, 990
Car heater, Railway	W. Pauli	Alexandria, Va	Nov. 27, 1860	30, 753
Car heater, Railway	C. F. Pike	Providence, R. I	Mar. 26, 1872	124, 973
Car heater, Railway	J. Pino	New York, N. Y	Oct. 9, 1860	30, 347
Car heater, Railway	J. Rice	Bloomington, Ind	July 28, 1868	80, 505
Car heater, Railway	H. R. Robbins	Baltimore, Md	May 2, 1871	114, 349
Car heater, Railway	A. M. Rodgers	Brooklyn, N. Y	Apr. 4, 1871	113, 348
Car heater, Railway	C. Sanborn	Chichester, N. H	July 20, 1869	92, 753
Car heater, Railway	E. S. Scripture and W. Stockman.	Brooklyn, N. Y	Sept. 9, 1873	142, 737
Car heater, Railway	J. Shackleton	Rahway, N. J	May 30, 1871	115, 533
Car heater, Railway	F. Shaller	Hudson, N. Y	Mar. 1, 1870	100, 333
Car heater, Railway	C. Smith	Hermon, Me	Jan. 26, 1869	86, 255
Car heater, Railway	E. Spencer	Elizabeth, N. J	Dec. 15, 1868	84, 969
Car heater, Railway	F. J. Steinhauser and H. M. Shreiner.	Lancaster, Pa	Nov. 2, 1869	96, 498
Car heater, Railway	G. F. Stone	Baltimore, Md	Apr. 30, 1872	126, 343
Car heater, Railway	B. D. Thompson	New York, N. Y	Sept. 26, 1871	119, 249
Car heater, Railway	A. Weeks	Minneapolis, Minn	Jan. 5, 1869	85, 712
Car heater, Railway	J. H. Weibel	Geneva, Switzerland	Nov. 11, 1873	144, 425
Car heater, Railway	W. Westlake	Milwaukee, Wis	July 29, 1862	36, 044
Car heater, Steam	G. B. Riggins	New York, N. Y	June 18, 1872	128, 174
Car heater, Steam railway	A. C. Crary	Utica, N. Y	June 22, 1869	91, 524
Car heater, Street	J. Gibson, jr	Albany, N. Y.	Jan. 1, 1867	60, 714
Car heaters, Coupling for steam	W. N. McDuffey and B. F. Jacques.	Petersburgh, Va	Jan. 21, 1873	135, 137
Car heaters, Pipe connection in railway	M. S. Bolt	Elmira, N. Y	Jan. 5, 1869	85, 559
Car-hooks, Machine for bending	D. G. Morris	Catasauqua, Pa	Aug. 10, 1869	93, 467
Car-hooks, Machine for bending and punching	D. G. Morris	Catasauqua, Pa	Aug. 9, 1870	106, 188
Car, Horse-railway	D. T. Robinson	Boston, Mass	Jan. 8, 1867	61, 021
Car, Ice	F. L. Kidder	Brooklyn, N. Y	Mar. 18, 1862	34, 718
Car indicator, Railway	C. H. Coffin	Boston, Mass	June 10, 1873	139, 771
Car indicator, Street-railway	J. Knight	Philadelphia, Pa	June 30, 1868	79, 480
Car, Iron	R. Montgomery	New York, N. Y	Aug. 7, 1860	29, 510
Car, Iron railway	J. Davenport	Massillon, Ohio	Mar. 22, 1859	23, 333
Car, Iron railway	H. Merrill	New York, N. Y	Mar. 31, 1863	38, 082
Car, Iron railway	J. Miner and S. Merrick	New Brighton, Pa	Jan. 10, 1860	26, 777
Car jack, Railway	E. C. Patterson	Rochester, N. Y	June 16, 1868	78, 994
Car-jumper	C. C. Dow	Philadelphia, Pa	Mar. 2, 1869	87, 399
Car-light	R. Cathcart	Baltimore, Md	Oct. 9, 1860	30, 297
Car-links, Manufacture of	G. H. Sellers	Phœnixville, Pa	Mar. 29, 1864	42, 147
Car-loading machine	S. J. Bingham	Garlandsville, Miss	Jan. 30, 1872	123, 146
Car-lock	H. Ritchie	Newark, N. J	Apr. 7, 1857	17, 013
Car lock, Freight	O. F. Burton	New York, N. Y	July 16, 1872	129, 524
Car lock, Freight	H. C. Jones	Newark, N. J	Dec. 11, 1855	13, 911
Car lock, Freight	J. F. Keeler	Cleveland, Ohio	Oct. 2, 1860	30, 229
Car lock, Freight	T. Slaight	Newark, N. J	Sept. 23, 1856	15, 783
Car lock, Railway	C. T. Gibson	Baltimore, Md	Aug. 8, 1871	117, 764
Car, Locomotive	W. Romans	Columbus, Ohio	Jan. 28, 1862	34, 270
Car, Locomotive	H. F. Shaw	West Roxbury, Mass	Oct. 3, 1865	50, 282
Car, Locomotive and railway	J. P. Fairlamb and L. C. Judson.	Philadelphia, Pa	Jan. 9, 1838	558
Car, Marine	A. Bloomquist and C. Crook	New York, and Yonkers, N. Y.	July 17, 1866	56, 351
Car, Marine	G. H. Young	Charlestown, Mass	July 24, 1866	56, 660

Index of patents issued from the United States Patent Office from 1790 *to* 1873, *inclusive*—Continued.

Invention.	Inventor.	Residence.	Date.	No.
Car-mat	J. O'Neill	Brooklyn, N. Y	Dec. 16, 1873	145, 558
Car, Metallic railway	B. J. La Mothe	New York, N. Y	July 26, 1870	105, 699
Car, Metallic railway	J. A. Roebling	Trenton, N. J	Oct. 16, 1860	30, 426
Car-mover	F. L. Bailey	Freeport, Ind	Dec. 8, 1868	84, 789
Car-mover	S. Becker and P. Loucks	York, Pa	Apr. 26, 1870	102, 357
Car-mover	R. A. Cowell	Cleveland, Ohio	July 29, 1873	141, 327
Car-mover	J. Douglass	McConnellstown, Pa	Apr. 7, 1868	76, 421
Car-mover	H. B. Morrison	Le Roy, N. Y	Mar. 10, 1868	75, 447
Car-mover	E. Springer	Davis, Ill	Apr. 12, 1870	101, 780
Car mover, Railway	E. Spinger	Davis, Ill	Nov. 10, 1868	84, 012
Car-moving device	J. Foreman	Pottstown, Pa	Apr. 26, 1870	102, 388
Car, Oil-tank	G. W. Ilgenfritz and M. Schall	York, Pa	July 7, 1868	79, 573
Car or carriage spring, Railway	G. Douglass	Scranton, Pa	Dec. 29, 1857	18, 961
Car, Passenger-railway	T. Castor	Philadelphia, Pa	Feb. 21, 1860	27, 203
Car, Passenger-railway	E. Y. Robbins	Cincinnati, Ohio	Nov. 3, 1868	83, 731
Car, Peat	T. J. Wells	Saint Anthony, Minn	Jan. 22, 1867	61, 493
Car, Pendulous railway	S. T. Jones	Philadelphia, Pa	Feb. 22, 1830	......
Car, Petroleum	J. Densmore and G. W. N. Yost.	Meadville and Corry, Pa	June 26, 1866	55, 830
Car, Petroleum	J. Densmore and G. W. N. Yost.	Meadville and Corry, Pa	June 26, 1866	55, 831
Car, Petroleum	J. and A. Densmore and W. N. Yost.	Meadville and Corry, Pa	June 26, 1866	55, 832
Car-platform	H. S. Wilcox	West Meriden, Conn	Aug. 27, 1867	68, 327
Car-platform bridge	B. G. Fitzhugh	Ellicott City, Md	July 7, 1868	79, 646
Car-platform bridge	A. Rank	Salem, Ohio	Apr. 28, 1868	77, 321
Car platform, Railway	R. A. Cowell	Cleveland, Ohio	Jan. 10, 1871	110, 901
Car platform, Railway	J. Gilmer	Monticello, Fla	Oct. 5, 1869	95, 579
Car platform, Railway	H. Holcroft and C. S. Smith	Chester Valley, Pa	June 2, 1863	38, 743
Car platform, Safety	C. R. Abbot	Elmira, N. Y	Sept. 3, 1867	68, 543
Car-platform stakeholder	C. H. Bryan	Racine, Wis	July 25, 1865	48, 899
Car, Pneumatic	L. Ransom	Lansingburgh, N. Y	Dec. 10, 1867	72, 082
Car, Pneumatic railway	H. G. Yates	Kalamazoo, Mich	Sept. 9, 1873	142, 605
Car, Pneumatic street	D. Myers	Chicago, Ill	Oct. 18, 1870	108, 380
Car, Pneumatic street	C. W. Wailey	New Orleans, La	Nov. 24, 1868	84, 447
Car propelled by horse-power, Railway	D. W. Crocker	Philadelphia, Pa	Oct. 30, 1834	......
Car-propeller	J. Day	Holley, N. Y	Sept. 2, 1873	142, 443
Car-propeller	G. W. Earl	New York, N. Y	Oct. 28, 1873	144, 076
Car-propeller	P. W. Phillips	Bristol, England	Dec. 13, 1864	45, 450
Car-propeller	R. Steel and S. Austin	Philadelphia, Pa	Nov. 4, 1873	144, 235
Car-propelling apparatus	C. T. Harvey	Tarrytown, N. Y	Apr. 16, 1867	63, 887
Car-propelling apparatus	J. Roy	New Orleans, La	Dec. 27, 1870	110, 501
Car propulsion	R. De Leidi	Brooklyn, N. Y	Feb. 18, 1873	135, 972
Car-pusher	A. S. Bailey	Paxton, Ill	June 3, 1873	139, 529
Car-pusher	R. Lane	Freeport, Ill	Oct. 4, 1870	107, 926
Car-pushing device	R. Odenath	Philadelphia, Pa	June 18, 1872	128, 165
Car pushing-jack, Railway	R. A. Cowell	Cleveland, Ohio	Dec. 7, 1869	97, 480
Car pushing-jack, Railway	E. Hutson	Brockport, N. Y	Jan. 7, 1868	73, 101
Car-rack	A. Bridges	New York, N. Y	May 3, 1870	102, 482
Car-rack	T. S. Hudson	Cambridge, Mass	Apr. 30, 1872	126, 298
Car, Railway	E. H. Ashcroft	Lynn, Mass	May 15, 1866	54, 662
Car, Railway	A. E. Beach	Stratford, Conn	Sept. 5, 1865	49, 695
Car, Railway	W. Bell	Perth Amboy, N. J	Feb. 13, 1872	123, 610
Car, Railway	G. and G. T. Benjamin and H. S. Weston.	Millersburgh, Ohio	Dec. 18, 1866	60, 465
Car, Railway	H. Buck	Polo, Ill	Mar. 4, 1873	136, 359
Car, Railway	I. Bullock	New York, N. Y	Oct. 11, 1841	2, 316
Car, Railway	J. Busser	Troy, Ohio	Dec. 28, 1869	98, 227
Car, Railway	T. Castor	Philadelphia, Pa	Dec. 6, 1864	45, 316
Car, Railway	G. W. Cook	Rock Island, Ill	Dec. 26, 1865	51, 698
Car, Railway	E. Crane	Dorchester, Mass	Nov. 22, 1859	26, 166
Car, Railway	W. D. Crane	New York, N. Y	Dec. 7, 1869	97, 608
Car, Railway	L. B. Crittenden	Pittsburgh, Pa	July 9, 1867	66, 467
Car, Railway	M. M. Crooker	Rutland, Vt	May 26, 1868	78, 188
Car, Railway	J. Davis and W. Ashdown	Baltimore, Md	Mar. 4, 1836	......
Car, Railway	E. H. Derby and T. West	Boston and Roxbury, Mass	Aug. 22, 1865	49, 508
Car, Railway	G. W. Eddy	Waterford, N. Y	Feb. 11, 1868	74, 325
Car, Railway	J. Elgar	Baltimore, Md	Oct. 1, 1830	......
Car, Railway	J. S. Fairfax	Wheeling, W. Va	Jan. 1, 1867	60, 868
Car, Railway	J. P. Fairlamb	Philadelphia, Pa	Jan. 19, 1833	......
Car, Railway	D. Fitzgerald	New York, N. Y	May 11, 1869	90, 060
Car, Railway	J. Foreman	Pottstown, Pa	Sept. 10, 1867	67, 619
Car, Railway	W. C. Fuller	England	Oct. 23, 1846	4, 827
Car, Railway	E. D. Gird	Syracuse, N. Y	Mar. 19, 1872	124, 734
Car, Railway	A. Gregg	Forest City, Cal	Feb. 19, 1867	62, 127
Car, Railway	J. W. Griffiths	East Boston, Mass	May 9, 1871	114, 675
Car, Railway	G. S. Hacker	Charleston, S. C	Mar. 5, 1850	7, 140
Car, Railway	W. S. Hall	Quincy, Mass	July 4, 1865	48, 553
Car, Railway	C. T. Harvey	Tarrytown, N. Y	July 7, 1868	79, 756
Car, Railway	S. C. Hawkins	Patchogue, N. Y	Oct. 17, 1865	50, 472
Car, Railway	J. H. Kaufman	Lisburn, Pa	Mar. 20, 1860	27, 547
Car, Railway	I. Knight	Baltimore, Md	Apr. 28, 1836	......
Car, Railway	B. J. La Mothe	New York, N. Y	Apr. 4, 1854	10, 721
Car, Railway	M. C. Lawless	Montana, Iowa	Oct. 12, 1869	95, 697
Car, Railway	E. T. Ligon	Demopolis, Ala	Nov. 3, 1868	83, 717
Car, Railway	I. G. Macfarlane	Wilkinsburgh, Pa	Apr. 23, 1872	125, 969
Car, Railway	I. G. Macfarlane	Wilkinsburgh, Pa	Sept. 17, 1872	131, 360
Car, Railway	I. G. Macfarlane	Wilkinsburgh, Pa	June 10, 1873	139, 680
Car, Railway	F. Marin	Geneva, Switzerland	Sept. 30, 1873	143, 363
Car, Railway	C. McWayne	Sacramento, Cal	Oct. 22, 1861	33, 534
Car, Railway	E. Miller	Brooklyn, N. Y	July 24, 1866	56, 594
Car, Railway	W. Miller and H. E. Towle	Boston, Mass., and New York, N. Y.	May 3, 1864	42, 590
Car, Railway	J. Miner and S. Merrick	New Brighton, Pa	Nov. 29, 1859	26, 282
Car, Railway	W. Partridge	Philadelphia, Pa	Aug. 29, 1865	49, 648
Car, Railway	S. Pennock	Kennett Square, Pa	Dec. 31, 1867	72, 891
Car, Railway	A. Planton	Philadelphia, Pa	June 3, 1837	218
Car, Railway	B. P. Power and J. Coyne	Baltimore, Md	Feb. 16, 1869	87, 002

Index of patents issued from the United States Patent Office from 1790 *to* 1873, *inclusive*—Continued.

Invention.	Inventor.	Residence.	Date.	No.
Car, Railway	J. R. Reader	New York, N. Y	Nov. 13, 1866	59, 652
Car, Railway	A. V. Rider	New York, N. Y	May 14, 1872	126, 837
Car, Railway	D. T. Robinson	Boston, Mass	Oct. 29, 1867	70, 269
Car, Railway	J. F. Rodgers	Troy, N. Y	Sept. 26, 1848	5, 815
Car, Railway	C. W. Saladee	Pittsburgh, Pa	Dec. 9, 1873	145, 365
Car, Railway	C. W. Saladee	Pittsburgh, Pa	Dec. 9, 1873	145, 366
Car, Railway	S. J. Seely	Buffalo, N. Y	Apr. 24, 1860	28, 041
Car, Railway	S. Skillman	Jersey City, N. J	Jan. 10, 1865	45, 871
Car, Railway	S. Snyder	Lancaster, Pa	Nov. 10, 1832	....
Car, Railway	S. Vallo	Philadelphia, Pa	Apr. 4, 1865	47, 156
Car, Railway	J. C. Wands	Nashville, Tenn	Aug. 19, 1873	142, 062
Car, Railway	C. Waterbury	Bridgeport, Conn	June 29, 1852	9, 084
Car, Railway	P. H. Watson	Ashtabula, Ohio	Jan. 5, 1869	85, 720
Car, Railway	R. T. M. Wells	Franklin Centre, Vt	Jan. 3, 1865	45, 778
Car, Railway	R. T. M. Wells	Roxbury, Mass	July 10, 1866	56, 301
Car, Railway	R. Winans	Baltimore, Md	Oct. 1, 1834	....
Car, Railway	J. Withers	Bart Township, Pa	Mar. 25, 1835	....
Car, Railway	J. Withycombe and C. Reiblein.	Cleveland, Ohio	Jan. 26, 1864	41, 409
Car, Railway	J. Wright	Columbia, Pa	Sept. 10, 1829	....
Car, Railway and dumping	A. Welsch	Chicago, Ill	Sept. 29, 1863	40, 134
Car, Railway bathing	L. Bagger	Washington, D. C	Nov. 26, 1872	133, 398
Car, Railway cattle	E. Payne and J. D. Cleghorn	Chicago, Ill	Nov. 9, 1869	96, 722
Car, Railway cattle	C. F. Pike	Providence, R. I	Feb. 1, 1870	99, 471
Car, Railway cattle	J. B. Shafer	Grafton, Va	July 31, 1860	29, 409
Car, Railway cattle	J. W. Street	Marshalltown, Iowa	Nov. 2, 1869	96, 362
Car, Railway dining	G. M. Pullman	Chicago, Ill	Apr. 27, 1869	89, 537
Car, Railway dumping	T. A. McFarland	Meadville, Pa	May 24, 1864	42, 862
Car, Railway freight	J. H. Aldrich	Nashua, N. H	Nov. 5, 1867	70, 384
Car, Railway freight	R. Eaton	Montreal, Canada	Jan. 1, 1867	60, 865
Car, Railway freight	W. A. Goodwin	Newton, Mass	Feb. 23, 1869	87, 098
Car, Railway freight	R. H. Gordon, sr	Cleveland, Ohio	May 27, 1873	139, 243
Car, Railway freight	B. P. Lamason	Milton, Pa	Feb. 11, 1873	135, 717
Car, Railway freight	L. Savage	Ashtabula, Ohio	Mar. 17, 1868	75, 581
Car, Railway freight	J. P. Woodbury	Boston, Mass	Aug. 31, 1869	94, 461
Car, Railway hand	H. L. Brown	Adrian, Mich	Sept. 7, 1869	94, 469
Car, Railway hand	H. Fisher	Alliance, Ohio	Nov. 29, 1859	26, 263
Car, Railway hand	J. C. Hearne	Pleasant Hill, Mich	June 6, 1871	115, 609
Car, Railway hotel	G. M. Pullman	Chicago, Ill	Apr. 27, 1869	89, 538
Car, Railway passenger	J. Stephenson	New York, N. Y	Apr. 22, 1833	....
Car, Railway passenger	J. T. Worley	Cleveland, Ohio	Dec. 26, 1871	122, 208
Car, Railway sleeping	J. S. Du Bois	Saint Louis, Mo	Nov. 8, 1870	109, 119
Car, Railway sleeping	G. W. Hunt	Hopkinton, Mass	Dec. 10, 1867	72, 045
Car, Railway sleeping	T. Luce and J. H. Morrison	Detroit, Mich	Jan. 4, 1859	22, 506
Car, Railway stock	G. R. Blanchard	Baltimore, Md	Aug. 25, 1868	81, 468
Car, Railway stock	J. S. Kendall	Northfield, Minn	May 25, 1869	90, 557
Car, Railway stock	J. S. Kendall	Northfield, Minn	June 1, 1869	90, 851
Car, Railway stock	H. Lee	Beloit, Wis	June 1, 1869	90, 856
Car, Railway stock	W. M. Lyon	Salem, Ohio	May 23, 1871	115, 074
Car, Railway stock	A. Rank	Salem, Ohio	Mar. 12, 1872	124, 625
Car, Railway stock	A. Rank	Salem, Ohio	Mar. 12, 1872	124, 626
Car, Railway stock	A. Rank, H. King, and J. Sharp	Salem, Ohio	Feb. 14, 1871	111, 873
Car, Railway stock	S. W. Remer	Taunton, Mass	Mar. 5, 1872	124, 387
Car, Railway stock	W. Stark, J. G. Fisher, and S. Fitch.	White Pigeon, Mich., and Toledo, Ohio.	Nov. 21, 1871	121, 212
Car, Railway stock	W. Stark, J. G. Fisher, and S. Fitch.	White Pigeon, Mich., and Toledo, Ohio.	Nov. 21, 1871	121, 213
Car, Railway street	S. H. Lettle	Rosemond, Ill	June 17, 1873	140, 052
Car, Refrigerating	A. Booth	Chicago, Ill	Sept. 13, 1870	107, 217
Car, Refrigerator	L. Kniffen	Worcester, Mass	Sept. 24, 1867	69, 223
Car, Refrigerator railway	J. D. Potts and B. P. Lamason	Philadelphia and Milton, Pa.	June 28, 1870	104, 765
Car-register	L. H. French	Philadelphia, Pa	Aug. 28, 1860	29, 778
Car-register	J. and C. Gschwind	New York, N. Y	Nov. 6, 1866	59, 387
Car register, Railway	S. F. Covington	Indianapolis, Ind	Dec. 27, 1859	26, 565
Car register, Railway	P. S. Gerhardt	Philadelphia, Pa	June 30, 1868	79, 339
Car-replacer	C. C. Ash	Manor Hill, Pa	May 16, 1871	114, 745
Car-replacer	F. L. Bailey	Freeport, Ind	Aug. 14, 1855	13, 417
Car-replacer	D. H. Ball and J. Brooks	Sinnamahoning, Pa	Jan. 3, 1871	110, 723
Car-replacer	J. T. Baxter	Finksburgh, Md	Sept. 17, 1872	131, 384
Car-replacer	J. T. Baxter	Finksburgh, Md	Apr. 29, 1873	138, 311
Car-replacer	A. G. Black	Wooster, Ohio	Oct. 22, 1867	70, 069
Car-replacer	J. G. Burkhardt	Brooklyn, N. Y	Nov. 4, 1873	144, 188
Car-replacer	P. Cartwright	New York, N. Y	Mar. 5, 1872	124, 251
Car-replacer	G. Chambers	Ithaca, N. Y	Dec. 4, 1866	60, 137
Car-replacer	R. Davis	Utica, N. Y	June 30, 1868	79, 325
Car-replacer	N. H. Edgerton	Pottsville, Pa	June 18, 1867	65, 799
Car-replacer	D. Edwards	New York, N. Y	Apr. 7, 1868	76, 424
Car-replacer	B. W. Felton	Roxbury, Mass	Dec. 3, 1867	71, 596
Car-replacer	D. Fisher and W. Cumming	Oil City, Pa	Jan. 28, 1868	73, 789
Car-replacer	G. Fowler	Macon, Ill	Dec. 20, 1864	45, 485
Car-replacer	C. Hurst	New York, N. Y	Mar. 3, 1868	75, 020
Car-replacer	S. S. Jamison, jr	Saltsburgh, Pa	Apr. 14, 1868	76, 768
Car-replacer	C. King	Morristown, N. J	July 2, 1867	66, 359
Car-replacer	B. S. Lawson	New York, N. Y	Nov. 23, 1869	97, 204
Car-replacer	J. Mabie	New York, N. Y	Nov. 29, 1864	45, 254
Car-replacer	N. Pullman	New Oregon, Iowa	Apr. 7, 1868	76, 520
Car-replacer	H. Schreiner	Philadelphia, Pa	Dec. 22, 1868	85, 137
Car-replacer	L. Straus	Louisville, Ky	Oct. 29, 1867	70, 375
Car-replacer	A. N. Towne	Chicago, Ill	Jan. 28, 1868	73, 940
Car-replacer	H. Voth	Saint Louis, Mo	Mar. 25, 1873	137, 265
Car-replacer	A. Whittemore	Cambridgeport, Mass	Aug. 26, 1873	142, 135
Car-replacer or guide-rail	J. P. Lipps	Newark, N. J	July 21, 1868	80, 191
Car replacer, Railway	R. Harper	Chelsea, Mass	Dec. 15, 1863	40, 928
Car replacer, Railway	H. H. Holmes	Elmira, N. Y	Feb. 1, 1870	99, 318
Car replacer, Railway	E. Snyder	Providence, Pa	July 6, 1869	92, 387
Car-replacing jack	D. Moritz	Buffalo, N. Y	Sept. 12, 1871	118, 965
Car, Reversible street	W. T. Jenks	Toledo, Ohio	May 20, 1873	139, 063

Index of patents issued from the United States Patent Office from 1790 *to* 1873, *inclusive*—Continued.

Invention.	Inventor.	Residence.	Date.	No.
Car, Revolving	D. Lott	Lottsville, Pa	Aug. 1, 1865	49, 126
Car-roof	J. L. Burnham	Nashville, Tenn	Apr. 2, 1872	125, 266
Car-roof	N. C. Day	Nashua, N. H	July 16, 1872	129, 215
Car-roof	J. Garry	Cleveland, Ohio	Oct. 11, 1870	108, 130
Car-roof	J. B. Slichter	Chicago, Ill	Aug. 26, 1873	142, 129
Car-roof	J. C. Wands	Nashville, Tenn	Mar. 4, 1873	136, 569
Car roof, Railway	C. Dummeldinger	Cleveland, Ohio	June 18, 1867	65, 888
Car roof, Railway	W. H. Myers	Norwich, Conn	Aug. 14, 1866	57, 173
Car roof, Railway	J. Palmer	Cleveland, Ohio	June 21, 1864	43, 223
Car roof, Railway	J. Palmer	Cleveland, Ohio	July 2, 1867	66, 379
Car roof, Railway	B. F. Pickett	Nashville, Tenn	Apr. 26, 1870	102, 316
Car roof, Railway	O. V. Scaife	Pittsburgh, Pa	Apr. 5, 1864	42, 229
Car roof, Railway	J. Stephenson	New York, N. Y	Jan. 22, 1867	61, 482
Car roof, Railway	A. P. Winslow	Cleveland, Ohio	Aug. 9, 1859	25, 071
Car, Rotating dumping	W. A. Hawkes	Corinth, N. Y	June 7, 1859	24, 301
Car safety-apparatus, Railway	K. E. Holmes	Cambridgeport, Mass	Oct. 10, 1871	119, 766
Car, Safety railway	W. Kinkead	Elkton, Md	Dec. 29, 1837	535
Car safety-shoe, Railway	S. W. Emery and E. P. Doyen	Portland, Me	Mar. 12, 1872	124, 560
Car-seal	J. Dewe	Toronto, Canada	Nov. 22, 1870	109, 394
Car-seat	W. L. Bass	Cambridge, Mass	July 20, 1852	9, 128
Car-seat	S. G. Blackman	Waterbury, Conn	Aug. 4, 1868	80, 703
Car-seat	A. C. Blondyn	Saint Joseph, Mo	Oct. 26, 1858	21, 870
Car-seat	E. Booth and E. Ripley	Troy, N. Y	Nov. 11, 1851	8, 508
Car-seat	J. Briggs	Boston, Mass	July 6, 1852	9, 091
Car-seat	J. Casseday	Philadelphia, Pa	Feb. 22, 1870	100, 116
Car-seat	J. R. Chiles	Richmond, Va	July 30, 1867	67, 165
Car-seat	W. G. Creamer	Brooklyn, N. Y	Jan. 21, 1868	73, 510
Car-seat	J. S. Diack	Aurora, Ill	May 11, 1869	89, 976
Car-seat	A. S. Dotter	Philadelphia, Pa	July 2, 1867	66, 228
Car-seat	F. Farrel	Cincinnati, Ohio	June 3, 1873	139, 561
Car-seat	W. M. Henderson	Baltimore, Md	Aug. 16, 1859	25, 116
Car-seat	S. B. Holden	Sedalia, Mo	Apr. 7, 1868	76, 325
Car-seat	J. F. C. Hollings	Detroit, Mich	Nov. 19, 1872	133, 102
Car-seat	P. P. Joseff	Philadelphia, Pa	Nov. 2, 1858	21, 967
Car-seat	C. P. Kimball	Portland, Me	July 14, 1868	79, 837
Car-seat	E. Lockwood and G. W. Pitman	Bordentown, N. J	June 13, 1865	48, 191
Car-seat	E. Lockwood and G. W. Pitman	Bordentown, N. J	Apr. 10, 1866	53, 842
Car-seat	S. McGregor	Logansport, Ind	Mar. 27, 1860	27, 645
Car-seat	P. W. Nolan	New York, N. Y	Dec. 2, 1873	145, 229
Car-seat	W. Painter	Wilmington, Del	Aug. 31, 1858	21, 356
Car-seat	F. I. Palmer	Knoxville, Tenn	Mar. 13, 1860	27, 469
Car-seat	W. G. Philips and N. Coleman	Newport, Del	May 7, 1872	126, 485
Car-seat	T. Rainey	New York, N. Y	July 23, 1861	32, 892
Car-seat	E. Ripley and E. L. Brundage	Troy, N. Y	Dec. 9, 1851	8, 583
Car-seat	C. Rowland	Washington, D. C	Mar. 24, 1868	75, 982
Car-seat	J. W. Sibbet	Cincinnati, Ohio	Nov. 2, 1858	21, 985
Car-seat	R. Stilwell and E. L. Brundage	New York and Troy, N. Y	Apr. 22, 1851	8, 059
Car-seat	A. D. Tate	Peekskill, N. Y	Nov. 2, 1869	96, 504
Car-seat	F. F. Wagner	Harrisburgh, Pa	Sept. 1, 1868	81, 712
Car-seat	W. Wells	Salem, Mass	Aug. 9, 1870	106, 297
Car-seat	J. S. Wheat	South Wheeling, W. Va	May 28, 1867	65, 318
Car-seat	N. S. Whipple	Detroit, Mich	Apr. 2, 1872	125, 364
Car-seat, Adjustable	C. A. Buck and J. Lovett	Saint Louis, Mo	Aug. 1, 1871	117, 599
Car-seat, Adjustable	C. Stevenbank and J. Quinn	Wilmington, Del	May 30, 1871	115, 539
Car seat, Adjustable railway	W. N. Bragg	Richmond, Va	Oct. 9, 1866	58, 713
Car seat and berth, Railway	S. C. Case	Detroit, Mich	June 22, 1858	20, 622
Car seat and berth, Railway	Z. Cobb	Chicago, Ill	July 6, 1858	20, 777
Car seat and chair	W. N. Bragg	Richmond, Va	June 2, 1868	78, 570
Car seat and couch	H. L. Arnold	Elk Horn, Wis	Dec. 21, 1858	22, 338
Car seat and couch	W. M. Baker	Walpole, Ind	Jan. 25, 1859	22, 693
Car seat and couch	G. W. Fairchild	Holyoke, Mass	Dec. 14, 1858	22, 283
Car seat and couch	I. N. Forrester	Fairfax Court-House, Va	Aug. 24, 1858	21, 251
Car seat and couch	R. E. Fowler	Clayton, N. Y	Sept. 7, 1858	21, 412
Car seat and couch	K. Freeman	Fond du Lac, Wis	Aug. 31, 1858	21, 331
Car seat and couch	A. M. Holmes	Eaton, N. Y	Sept. 14, 1858	21, 536
Car seat and couch	I. W. Lamb	Salem, Mich	Sept. 3, 1867	68, 368
Car seat and couch	D. Pennoyer	North East, N. Y	Aug. 28, 1860	29, 813
Car seat and couch	T. Sharp	Carlisle, Pa	Aug. 14, 1866	57, 198
Car seat and couch	T. C. Theaker	Bridgeport, Ohio	Apr. 27, 1869	89, 542
Car seat and couch, Railway	H. Allen	New York, N. Y	June 12, 1866	55, 448
Car seat and couch, Railway	E. Burke	Philadelphia, Pa	Oct. 23, 1860	30, 517
Car seat and couch, Railway	J. H. Fisher	Placerville, Cal	Oct. 23, 1860	30, 472
Car seat and couch, Railway	J. Good	Philadelphia, Pa	Feb. 22, 1859	23, 061
Car seat and couch, Railway	J. L. Hamilton	Saint Joseph, Mo	Aug. 6, 1872	130, 219
Car seat and couch, Railway	W. R. Jackson	Baltimore, Md	Apr. 12, 1859	23, 581
Car seat and couch, Railway	T. E. McNeill	Philadelphia, Pa	May 31, 1859	24, 225
Car seat and couch, Railway	E. D. Sargent	Indianapolis, Ind	Aug. 9, 1859	25, 049
Car seat and couch, Railway	N. Thompson, jr	Brooklyn, N. Y	Dec. 28, 1858	22, 462
Car seat and couch, Railway	T. T. Woodruff	Alton, Ill	Dec. 2, 1856	16, 159
Car seat and couch, Railway	T. T. Woodruff	Alton, Ill	Dec. 2, 1856	16, 160
Car seat and couch, Railway	T. T. Woodruff	Philadelphia, Pa	May 31, 1859	24, 257
Car seat and couch, Railway	T. T. Woodruff	Philadelphia, Pa	Jan. 24, 1860	26, 942
Car seat and couch, Sleeping	T. E. McNeill	Philadelphia, Pa	May 24, 1859	24, 136
Car seat and desk combined, Railway	W. C. Huffman	Toledo, Ohio	Feb. 4, 1873	135, 557
Car seat and sleeping-couch, Railway	P. G. Green	Chicago, Ill	Dec. 21, 1858	22, 364
Car-seat arm	A. L. Babcock	New Haven, Conn	Feb. 12, 1867	61, 914
Car-seat arm, Railway	A. Prier	Milwaukee, Wis	Nov. 28, 1871	121, 255
Car-seat back	T. E. Warren	Troy, N. Y	July 30, 1850	7, 539
Car-seat back, Removable	P. F. Duchemin	Somerville, Mass	Jan. 24, 1871	111, 184
Car-seat backs, Support for	G. Higginson	Newark, N. J	Aug. 4, 1868	80, 736
Car-seat bolt	C. A. McEvoy	Richmond, Va	May 8, 1860	28, 190
Car-seat cushions, Redyeing	T. Brown	Albany, N. Y	July 31, 1866	56, 697
Car-seat frame	G. Buntin	Boston, Mass	Dec. 5, 1871	121, 486
Car-seat frames, Arm-rest for	T. S. Hudson	Cambridge, Mass	June 18, 1872	128, 044
Car seat, Horse	M. T. Glynn and J. L. Goodman.	Boston, Mass	Sept. 26, 1871	119, 344
Car-seat indicator	F. H. Carney	Boston, Mass	Oct. 16, 1866	58, 771

Index of patents issued from the United States Patent Office from 1790 to 1873, inclusive—Continued.

Invention.	Inventor.	Residence.	Date.	No.
Car-seat link-joint	R. Hitchcock	Springfield, Mass	Feb. 11, 1868	74, 359
Car-seat lock	G. W. R. Bayley and J. McCluskey.	Algiers, La	Jan. 7, 1868	73, 155
Car-seat lock	S. B. Bowles	Brooklyn, N. Y	Feb. 26, 1867	62, 311
Car-seat lock	A. Duncan and J. M. Ziegler	Aurora, Ill	Feb. 21, 1865	46, 456
Car-seat lock	M. P. Ford	Columbus, Ohio	July 9, 1867	66, 479
Car-seat lock	E. Hambuger	Detroit, Mich	Sept. 3, 1867	68, 436
Car-seat lock	A. Loeffelholz and A. Prier	Milwaukee, Wis	Dec. 20, 1870	110, 254
Car-seat lock	G. McGregor	Cincinnati, Ohio	Sept. 29, 1863	40, 113
Car-seat lock	E. Miller	New York, N. Y	Aug. 2, 1870	105, 965
Car-seat lock	R. B. More	Cincinnati, Ohio	July 21, 1863	39, 303
Car seat, Railway	J. M. Allen	Washington, D. C	Nov. 21, 1871	121, 145
Car seat, Railway	A. S. Babbit	Keesville, N. Y	Sept. 27, 1864	44, 379
Car seat, Railway	C. P. Bailey	Zanesville, Ohio	Aug. 3, 1852	9, 161
Car seat, Railway	C. P. Bailey	Muskingum, Ohio	July 12, 1853	9, 851
Car seat, Railway	C. P. Bailey	Zanesville, Ohio	Oct. 10, 1854	11, 772
Car seat, Railway	C. P. Bailey	Zanesville, Ohio	Oct. 13, 1857	18, 375
Car seat, Railway	J. M. Baird	Wheeling, Va	Aug. 17, 1858	21, 178
Car seat, Railway	A. Barbarin	New Orleans, La	July 13, 1869	92, 418
Car seat, Railway	A. Barney	Wilmington, Del	Jan. 7, 1873	134, 627
Car seat, Railway	H. S. Blood	Jefferson, La	Sept. 29, 1868	82, 482
Car seat, Railway	J. Briggs	Boston, Mass	Feb. 15, 1853	9, 583
Car seat, Railway	J. A. Brown	Bath, Me	Sept. 3, 1867	68, 347
Car seat, Railway	A. B. Buell	Westmoreland, N. Y	May 11, 1852	8, 935
Car seat, Railway	D. Buzzell	Charlestown, Mass	Apr. 13, 1858	19, 910
Car seat, Railway	D. H. Chamberlain	West Roxbury, Mass	June 20, 1871	116, 154
Car seat, Railway	W. L. Childs	Piermont, N. Y	Feb. 1, 1859	22, 782
Car seat, Railway	J. H. Cocke	Bremo, Va	June 19, 1855	13, 079
Car seat, Railway	W. Crandell	Westfield, N. Y	Jan. 3, 1871	110, 746
Car seat, Railway	J. S. Denman	Brooklyn, N. Y	Jan. 12, 1858	19, 079
Car seat, Railway	J. C. De Witt	West Bloomfield, N. J	Aug. 31, 1858	21, 326
Car seat, Railway	A. B. Dinsmore	Springfield, Mass	Sept. 26, 1871	119, 331
Car seat, Railway	G. L. Dulaney	Mount Jackson, Va	Dec. 28, 1858	22, 471
Car seat, Railway	I. Fay	Cambridgeport, Mass	Sept. 20, 1853	10, 029
Car seat, Railway	W. Graham and L. McLaren	Philadelphia, Pa	Sept. 26, 1854	11, 724
Car seat, Railway	J. T. Hammitt	Philadelphia, Pa	Dec. 5, 1854	12, 023
Car seat, Railway	S. Hickok	Buffalo, N. Y	Aug. 9, 1853	9, 919
Car seat, Railway	S. B. Holden	Meadville, Pa	Aug. 25, 1863	39, 652
Car seat, Railway	E. Jeffers	Dorchester, Mass	Aug. 21, 1855	13, 464
Car seat, Railway	B. J. La Mothe	New York, N. Y	July 14, 1857	17, 794
Car seat, Railway	B. J. La Mothe	Detroit, Mich	May 13, 1873	138, 899
Car seat, Railway	C. B. Lasher	New York, N. Y	Oct. 4, 1864	44, 536
Car seat, Railway	C. M. Mann	Detroit, Mich	Aug. 31, 1858	21, 352
Car seat, Railway	F. Martin	Aurora, Ind	Feb. 12, 1867	62, 047
Car seat, Railway	M. M. Martin	Cochran, Ind	May 26, 1868	78, 304
Car seat, Railway	M. M. Martin	Cochran, Ind	Aug. 1, 1871	117, 553
Car seat, Railway	M. M. and F. Martin	Aurora, Ind	Aug. 21, 1866	57, 351
Car seat, Railway	G. T. McLauthlin	Boston, Mass	Mar. 4, 1856	14, 364
Car seat, Railway	J. McMurtry	Fayette County, Ky	July 27, 1858	21, 052
Car seat, Railway	T. E. McNeill	Philadelphia, Pa	Jan. 11, 1859	22, 568
Car seat, Railway	J. Millar	Paterson, N. J	June 22, 1858	20, 654
Car seat, Railway	E. Miller	New York, N. Y	Apr. 12, 1870	101, 755
Car seat, Railway	W. E. Milligan	New York, N. Y	Apr. 25, 1854	10, 818
Car seat, Railway	E. H. Olmstead	Savannah, Ga	Aug. 6, 1867	67, 572
Car seat, Railway	J. M. O'Neill	Clinton, Iowa	Aug. 30, 1870	106, 953
Car seat, Railway	J. I. Pease	Stockbridge, Mass	Aug. 23, 1870	106, 615
Car seat, Railway	G. W. Perry and J. D. Billings	Wilmington, Del	Dec. 8, 1868	84, 755
Car seat, Railway	G. W. Perry and J. D. Billings	Wilmington, Del	May 18, 1869	90, 297
Car seat, Railway	S. M. Perry	New York, N. Y	July 27, 1852	9, 155
Car seat, Railway	S. H. Rhoades and W. Carroll	Clyde, Ohio	Oct. 15, 1867	69, 938
Car seat, Railway	G. W. Sayre	Pisgah, Ohio	Apr. 18, 1865	47, 342
Car seat, Railway	P. I. Schopp	Louisville, Ky	June 25, 1872	128, 428
Car seat, Railway	E. F. Shornberger	Philadelphia, Pa	Jan. 3, 1865	45, 759
Car seat, Railway	A. D. Smith	Meredith, N. Y	Apr. 3, 1855	12, 644
Car seat, Railway	A. M. Smith	Rochester, N. Y	Aug. 21, 1855	13, 471
Car seat, Railway	C. A. Smith	Piermont, N. Y	Sept. 13, 1859	25, 448
Car seat, Railway	A. W. Snow	Norwich, Conn	June 26, 1849	6, 552
Car seat, Railway	W. H. Soper	Baltimore, Md	Aug. 23, 1870	106, 739
Car seat, Railway	D. Stone	Milwaukee, Wis	Oct. 5, 1858	21, 727
Car seat, Railway	J. B. Sutherland	Detroit, Mich	Sept. 28, 1869	95, 283
Car seat, Railway	J. H. Swan	New York, N. Y	Sept. 22, 1857	18, 252
Car seat, Railway	T. C. Theaker	Bridgeport, Ohio	Aug. 10, 1869	93, 660
Car seat, Railway	W. B. Thomas and S. Hickok	Buffalo, N. Y	Apr. 25, 1854	10, 837
Car seat, Railway	F. F. Wagner	Harrisburgh, Pa	Oct. 6, 1868	82, 899
Car seat, Railway	F. F. Wagner and P. P. Dickinson.	Harrisburgh, Pa	June 21, 1859	24, 511
Car seat, Railway	W. M. Warren	Watertown, Conn	July 26, 1853	9, 880
Car seat, Railway	W. M. Warren	Watertown, Conn	Aug. 23, 1853	9, 960
Car seat, Railway	E. Wheeler	Elmira, N. Y	Sept. 20, 1859	25, 499
Car seat, Railway	A. M. White	Thompsonville, Conn	July 14, 1868	80, 041
Car seat, Railway	G. Willard	Boston, Mass	May 6, 1856	14, 841
Car seat, Railway	D. H. Wiswell	Buffalo, N. Y	Nov. 16, 1852	9, 413
Car seat, Reversible	T. J. Close	Philadelphia, Pa	May 3, 1870	102, 498
Car seat, Reversible	P. F. Duchemin	Somerville, Mass	Mar. 18, 1873	136, 822
Car-seat, Revolving and reclining	A. Rapp	Elmira, N. Y	Dec. 10, 1872	133, 721
Car seat, Sleeping	R. Dirks	Philadelphia, Pa	Aug. 9, 1859	24, 998
Car seat, Street	C. H. Foster	San Francisco, Cal	Mar. 2, 1869	87, 335
Car-seats, Head-rest for	J. C. Giffing	New York, N. Y	Dec. 26, 1871	122, 165
Car-seats, Head-rest for	E. M. Judd	New Haven, Conn	Dec. 6, 1870	109, 823
Car-seats, Head-rest for	H. W. Safford	New York, N. Y	Dec. 28, 1869	98, 420
Car-seats, Head-rest for	G. Sanford	New York, N. Y	Apr. 29, 1862	35, 112
Car seats, Head-rest for railway	N. Gates	Middletown, Ohio	June 20, 1865	48, 270
Car seats, Head-rest for railway	W. M. McCauley	Washington, D. C	Sept. 1, 1857	18, 122
Car seats, Head-rest for railway	W. R. Phelps	New York, N. Y	July 4, 1865	48, 586
Car-seats, Hinged joint for	G. W. Perry	Wilmington, Del	Feb. 6, 1872	123, 417
Car seats, Lock for railway	A. H. Churchill	Boston, Mass	Sept. 17, 1872	131, 331
Car-seats, Metallic band for trimming	D. F. Randall	Chicopee, Mass	Jan. 28, 1868	73, 927
Car-seats, Mode of locking	D. H. Baker	Jersey City, N. J	Aug. 9, 1864	43, 753
Car seats, Pocket for railway	S. Simonson	Bridgeport, Conn	Aug. 31, 1869	94, 428

Index of patents issued from the United States Patent Office from 1790 *to* 1873, *inclusive*—Continued.

Index of patents issued from the United States Patent Office from 1790 to 1873, inclusive—Continued.

Invention.	Inventor.	Residence.	Date.	No.
Car-spring	E. J. Horner	Wilmington, Del	May 11, 1869	89, 999
Car spring	C. R. Hurlburt	Seymour, Conn	Dec. 14, 1858	22, 292
Car-spring	C. F. Jeffries	Philadelphia, Pa	Dec. 7, 1869	97, 645
Car-spring	A. H. King	Rahway, N. J	Feb. 27, 1872	123, 999
Car-spring	A. H. King	Rahway, N. J	July 9, 1872	128, 733
Car-spring	W. Kingsley	New York, N. Y	May 8, 1860	28, 237
Car-spring	J. Leland	Springfield, Mass	Jan. 24, 1871	111, 127
Car-spring	G. H. Lewis	Providence, R. I	June 28, 1864	43, 316
Car-spring	W. J. F. Liddell	Erie, Pa	Mar. 6, 1860	27, 411
Car-spring	W. J. F. Liddell	Erie, Pa	Mar. 20, 1866	53, 376
Car-spring	W. Marshall	New York, N. Y	June 14, 1864	43, 121
Car-spring	J. R. Mathews	New London, Conn	Oct. 5, 1869	95, 497
Car-spring	J. J. McCormick and J. E. Jerrold.	Paterson, N. J	July 17, 1860	29, 222
Car-spring	G. W. McMinn	Cincinnati, Ohio	Nov. 26, 1867	71, 317
Car-spring	A. Middleton, jr	Philadelphia, Pa	Apr. 23, 1872	126, 073
Car-spring	A. Middleton, jr	Philadelphia, Pa	May 6, 1873	138, 675
Car-spring	J. S. Miller	Springfield, Mass	Aug. 20, 1867	67, 897
Car-spring	A. W. Moses	Philadelphia, Pa	Sept. 23, 1873	143, 175
Car-spring	J. Murray	New York, N. Y	June 6, 1865	48, 084
Car-spring	D. Myers	Chicago, Ill	Dec. 4, 1866	60, 227
Car-spring	E. Myers	Jersey City, N. J	May 13, 1873	138, 920
Car-spring	W. Newbauer	Philadelphia, Pa	Aug. 1, 1865	49, 137
Car-spring	W. R. Nichols	Philadelphia, Pa	July 3, 1866	56, 084
Car-spring	W. R. Nichols	Philadelphia, Pa	May 19, 1868	78, 120
Car-spring	F. E. Oliver	New York, N. Y	May 10, 1864	42, 682
Car-spring	J. G. Pugsley	New York, N. Y	Aug. 4, 1863	39, 422
Car-spring	J. B. Quirk	Philadelphia, Pa	Dec. 17, 1872	133, 952
Car-spring	J. B. Quirk	Philadelphia, Pa	Apr. 29, 1873	138, 433
Car-spring	W. F. Ray	Fort Wayne, Ind	Dec. 10, 1867	71, 912
Car-spring	F. W. Rhinelander	New York, N. Y	May 31, 1870	103, 778
Car-spring	G. A. Riedel	Philadelphia, Pa	Apr. 19, 1864	42, 402
Car-spring	G. A. Riedel	Philadelphia, Pa	Aug. 2, 1864	43, 708
Car-spring	G. A. Riedel	Philadelphia, Pa	Apr. 18, 1865	47, 373
Car-spring	H. E. Roeder	New York, N. Y	Sept. 24, 1861	33, 366
Car-spring	A. H. Rowand	Allegheny, Pa	Oct. 9, 1866	58, 677
Car-spring	T. Shaw	Philadelphia, Pa	Aug. 22, 1865	49, 560
Car-spring	J. J. C. Smith	Philadelphia, Pa	Nov. 7, 1865	50, 849
Car-spring	J. J. C. Smith	Philadelphia, Pa	Nov. 7, 1865	50, 850
Car-spring	J. B. Smyth	Philadelphia, Pa	Apr. 24, 1866	54, 227
Car-spring	W. Toshach	New York, N. Y	Jan. 1, 1867	60, 805
Car-spring	G. L. Turner	New York, N. Y	June 18, 1861	32, 591
Car-spring	U. P. Vidal	Philadelphia, Pa	Apr. 26, 1864	42, 519
Car-spring	R. Vose	New York, N. Y	Jan. 3, 1860	26, 723
Car-spring	R. Vose	New York, N. Y	June 5, 1860	28, 619
Car-spring	R. Vose	New York, N. Y	June 12, 1860	28, 703
Car-spring	R. Vose	New York, N. Y	Mar. 11, 1862	34, 657
Car-spring	R. Vose	New York, N. Y	Oct. 28, 1862	36, 813
Car-spring	R. Vose	New York, N. Y	May 19, 1863	38, 642
Car-spring	R. Vose	New York, N. Y	Oct. 6, 1863	40, 218
Car-spring	R. Vose	New York, N. Y	Dec. 5, 1865	51, 368
Car-spring	R. Vose	New York, N. Y	Aug. 27, 1867	68, 133
Car-spring	R. Vose	New York, N. Y	Aug. 27, 1867	68, 134
Car-spring	R. Vose	New York, N. Y	Dec. 10, 1867	71, 926
Car-spring	R. Vose	New York, N. Y	Dec. 31, 1867	72, 941
Car-spring	R. Vose	New York, N. Y	July 19, 1870	105, 514
Car-spring	R. Vose and C. D. Gibson	New York, N. Y	June 2, 1863	38, 777
Car-spring	L. Vote	Pittsburgh, Pa	Nov. 5, 1872	132, 879
Car-spring	C. Wheeler, jr	Auburn, N. Y	Dec. 14, 1869	97, 842
Car-spring box	J. W. Evans	New York, N. Y	Jan. 5, 1869	85, 518
Car spring, Conical plate railway	J. J. Speed, jr., and J. A. Bailey	Detroit, Mich	Sept. 11, 1855	13, 552
Car-spring, Corrugated spiral	C. French	Seymour, Conn	Apr. 15, 1862	34, 952
Car-spring fastener	J. W. Wood	Philadelphia, Pa	Aug. 30, 1864	44, 033
Car-spring, India-rubber	S. Peatfield	Ipswich, Mass	Sept. 28, 1858	21, 624
Car-spring, India-rubber	R. Vose	New York, N. Y	Aug. 28, 1866	57, 600
Car spring, India-rubber railway	T. F. Allen	Dyersville, Iowa	Oct. 23, 1860	30, 450
Car-spring, Metallic	A. Egan	Pesth, Hungary	Sept. 10, 1872	131, 154
Car-spring, Metallic	D. Johnson	Chicago, Ill	Sept. 9, 1856	15, 698
Car-spring, Nest-spiral	R. Vose	New York, N. Y	Sept. 26, 1871	119, 253
Car-spring, Nest-spiral	R. Vose	New York, N. Y	Sept. 26, 1871	119, 254
Car-spring, Pneumatic	J. Merlett	Bound Brook, N. J	Jan. 9, 1867	51, 956
Car spring, Railway	T. F. Allen	Dyersville, Iowa	Jan. 15, 1861	31, 102
Car spring, Railway	T. F. Allyn	Nyack, N. Y	Jan. 11, 1870	98, 732
Car spring, Railway	T. F. Allyn	Nyack, N. Y	July 16, 1872	129, 448
Car spring, Railway	T. F. Allyn	Nyack, N. Y	Nov. 19, 1872	133, 074
Car spring, Railway	J. F. Babcock	Boston, Mass	Mar. 22, 1870	101, 080
Car spring, Railway	W. Barry and G. Franklin	Philadelphia, Pa	Dec. 21, 1869	98, 140
Car spring, Railway	J. C. Blair	Pittsburgh, Pa	Dec. 29, 1857	18, 950
Car spring, Railway	A. Bridges	New York, N. Y	July 29, 1862	36, 050
Car spring, Railway	E. Cliff	Oswego, N. Y	May 9, 1871	114, 645
Car spring, Railway	J. W. Cochran	New York, N. Y	July 6, 1869	92, 267
Car spring, Railway	J. W. Cochran	New York, N. Y	Oct. 10, 1871	119, 823
Car spring, Railway	W. F. Converse	Harrison, Ohio	Nov. 8, 1859	26, 019
Car spring, Railway	W. F. Converse	Harrison, Ohio	Aug. 1, 1865	49, 086
Car spring, Railway	J. W. Culmer	Pittsburgh, Pa	Jan. 7, 1873	134, 645
Car spring, Railway	D. G. Daniels	Chicago, Ill	Jan. 11, 1870	98, 672
Car spring, Railway	A. B. Davis	Philadelphia, Pa	Feb. 15, 1859	22, 941
Car spring, Railway	A. M. De Hart	Reading, Pa	May 4, 1858	20, 148
Car spring, Railway	R. Dudley and B. Hershey	Erie, Pa	Feb. 28, 1871	112, 229
Car spring, Railway	G. Elliot	Saint Louis, Mo	Sept. 19, 1871	119, 129
Car spring, Railway	J. W. Evans	New York, N. Y	Mar. 26, 1861	31, 792
Car spring, Railway	J. W. Evans	New York, N. Y	Jan. 6, 1863	37, 283
Car spring, Railway	J. W. Evans	New York, N. Y	June 25, 1872	128, 292
Car spring, Railway	J. J. Fields	Brooklyn, N. Y	July 27, 1858	20, 998
Car spring, Railway	H. Gardiner	New York, N. Y	June 1, 1858	20, 418
Car spring, Railway	H. Gardiner	New York, N. Y	Apr. 7, 1863	38, 105
Car spring, Railway	H. Gardiner	New York, N. Y	Nov. 15, 1870	109, 312
Car spring, Railway	P. G. Gardiner	New York, N. Y	Dec. 11, 1855	13, 905

Index of patents issued from the United States Patent Office from 1790 *to* 1873, *inclusive*—Continued.

Invention.	Inventor.	Residence.	Date.	No.
Car spring, Railway	P. G. Gardiner	New York, N. Y	Mar. 10, 1863	37, 862
Car spring, Railway	P. G. Gardiner	New York, N. Y	July 6, 1869	92, 182
Car spring, Railway	P. G. Gardiner	New York, N. Y	July 6, 1869	92, 183
Car spring, Railway	P. G. Gardiner	New York, N. Y	Aug. 24, 1869	93, 983
Car spring, Railway	P. G. Gardiner	New York, N. Y	Aug. 24, 1869	93, 984
Car spring, Railway	P. G. Gardiner	New York, N. Y	Aug. 24, 1869	94, 101
Car spring, Railway	P. G. Gardiner	New York, N. Y	Oct. 29, 1872	132, 570
Car spring, Railway	G. F. Godley	New York, N. Y	June 24, 1873	140, 192
Car spring, Railway	A. Hay	Philadelphia, Pa	Aug. 30, 1859	25, 264
Car spring, Railway	H. Jeffrey and H. Fisher	Aurora Station, Ind	Apr. 23, 1872	125, 959
Car spring, Railway	D. Johnson	Chicago, Ill	Jan. 24, 1860	26, 905
Car spring, Railway	A. H. King	Rahway, N. J	Nov. 5, 1872	132, 765
Car spring, Railway	A. Middleton, jr	Philadelphia, Pa	July 16, 1872	129, 155
Car spring, Railway	J. Millholland	Baltimore, Md	Sept. 23, 1843	3, 278
Car spring, Railway	J. Mitchell	Sheffield, England	July 19, 1870	105, 478
Car spring, Railway	S. Morse	Springfield, Mass	Feb. 23, 1858	19, 435
Car spring, Railway	J. Murphy	New York, N. Y	Aug. 21, 1866	57, 365
Car spring, Railway	H. M. Paine	Worcester, Mass	Oct. 27, 1857	18, 515
Car spring, Railway	A. Potts	Philadelphia, Pa	Dec. 27, 1870	110, 497
Car spring, Railway	F. M. Ray	Catskill, N. Y	Nov. 3, 1838	1, 000
Car spring, Railway	F. W. Rhinelander	New York, N. Y	Feb. 1, 1870	99, 475
Car spring, Railway	P. Riley	Shamokin, Pa	June 21, 1839	1, 180
Car spring, Railway	D. B. Rogers	Pittsburgh, Pa	Feb. 23, 1858	19, 448
Car spring, Railway	D. B. Rogers and J. A. Wood	Pittsburgh, Pa	July 26, 1859	24, 888
Car spring, Railway	C. Russell	Massillon, Ohio	July 21, 1863	39, 310
Car spring, Railway	W. Slicer	Baltimore, Md	June 17, 1873	140, 084
Car spring, Railway	W. Tosbach	New York, N. Y	Aug. 30, 1864	44, 027
Car spring, Railway	G. L. Turner	New York, N. Y	Aug. 21, 1860	29, 732
Car spring, Railway	R. Vose	New York, N. Y	July 21, 1863	39, 314
Car spring, Railway	R. Vose	New York, N. Y	June 28, 1870	104, 907
Car spring, Railway	H. Waterman	Hudson, N. Y	Jan. 26, 1858	19, 219
Car spring, Railway	A. T. Watson	Castleton, N. Y	Mar. 20, 1860	27, 587
Car spring, Railway	A. T. Watson	New York, N. Y	Apr. 24, 1866	54, 234
Car spring, Railway	R. Winans	Baltimore, Md	June 14, 1834	....
Car spring, Railway	J. E. Wootten	Philadelphia, Pa	Apr. 21, 1863	38, 255
Car spring, Railway	S. Yerkes, jr	Philadelphia, Pa	May 9, 1871	114, 741
Car-spring, Spiral	F. W. Rhinelander	New York, N. Y	Mar. 15, 1870	100, 926
Car-spring, Vulcanized India-rubber	H. A. Alden	Matteawan, N. Y	Jan. 4, 1870	98, 540
Car-spring, Vulcanized rubber	H. W. Beins	New York, N. Y	July 5, 1859	24, 693
Car-springs, End-fastener for	G. Elliot	Saint Louis, Mo	June 28, 1870	104, 718
Car-springs from mandrels, Machine for disengaging	P. G. Gardiner	New York, N. Y	Mar. 31, 1857	16, 916
Car springs, Hanging or arranging railway	W. E. Cooper	Dunkirk, N. Y	Nov. 8, 1859	26, 068
Car springs, Machine for creasing plates for railway	P. G. Gardiner	New York, N. Y	Mar. 31, 1857	16, 915
Car-springs, Machine for making	A. Hebbard	Springfield, Mass	Mar. 1, 1870	100, 400
Car-springs, Machine for testing and measuring the strength of.	P. G. Gardiner	New York, N. Y	Mar. 30, 1858	19, 767
Car-springs, Manufacture of	N. and A. Middleton, jr	Philadelphia, Pa	Feb. 18, 1873	135, 929
Car springs, Packer for railway	P. G. Gardiner	New York, N. Y	Nov. 30, 1869	97, 385
Car-springs, Tempering steel	P. G. Gardiner	New York, N. Y	Sept. 28, 1858	21, 603
Car-springs, Wrought-iron mold for vulcanizing rubber.	C. H. Franklin	Jersey City, N. J	Sept. 14, 1869	94, 815
Car-stake holder	G. A. Brown	Kalamazoo, Mich	Apr. 14, 1868	76, 596
Car-stake holder	O. R. Parmele	Aurora, Ill	Dec. 17, 1867	72, 226
Car-stake holder	W. P. Wentworth	Detroit, Mich	May 26, 1868	78, 405
Car-stake holder, Platform	W. J. Willits	Detroit, Mich	May 12, 1868	77, 943
Car-stake holder, Railway	A. R. Burdick	Racine, Wis	Jan. 10, 1865	45, 891
Car-stake supporter	W. H. Masterman and W. T. Hooper.	Stockton, Cal	Feb. 4, 1873	135, 436
Car-standard	R. Clarke	Mount Vernon, Ohio	June 30, 1868	79, 315
Car-standard, Folding	S. Little	Loretto, Pa	June 28, 1864	43, 317
Car standard, Railway	S. S. Beeman	Saint Albans, Vt	July 18, 1871	117, 146
Car-starter	P. Bates, jr	Pittsburgh, Pa	Jan. 28, 1873	135, 193
Car-starter	L. D. Benner	Boston, Mass	Jan. 14, 1873	134, 840
Car-starter	J. Bevan	Hudson, N. Y	Sept. 13, 1870	107, 330
Car-starter	I. N. Bevans	Thomaston, Conn	June 30, 1868	79, 529
Car-starter	R. Bogardus	New York, N. Y	June 21, 1870	104, 411
Car-starter	R. Bogardus	New York, N. Y	July 2, 1872	128, 527
Car-starter	J. Boswell and J. W. Brindle	Wilmington, Ohio	Feb. 23, 1869	87, 136
Car-starter	S. R. and A. C. Bradley	Brooklyn, N. Y	May 13, 1873	138, 848
Car-starter	C. B. Broadwell	New Orleans, La	Mar. 19, 1872	124, 662
Car-starter	W. H. Butler and R. G. Hatfield.	New York, N. Y	June 12, 1866	55, 464
Car-starter	T. Cooper	Philadelphia, Pa	Apr. 29, 1873	138, 233
Car-starter	J. Corbeil	Lind, Wis	July 29, 1873	141, 326
Car-starter	A. H. Crozier	New York, N. Y	Nov. 4, 1873	144, 260
Car-starter	A. B. Davis	Catahoula Parish, La	Aug. 3, 1869	93, 283
Car-starter	G. W. Davis and A. E. Smith	Providence, R. I	May 4, 1869	89, 636
Car-starter	D. A. Dickinson	Baltimore, Md	Oct. 18, 1870	108, 460
Car-starter	G. Fetter	Philadelphia, Pa	May 20, 1873	139, 134
Car-starter	G. W. Field and J. Steger	New York, N. Y	June 3, 1873	139, 563
Car-starter	D. W. Garst	Washington, D. C	Nov. 19, 1872	133, 218
Car-starter	J. W. Houghtelin	Detroit, Mich	Apr. 30, 1867	64, 224
Car-starter	T. W. Johnston	Chicago, Ill	July 15, 1873	140, 781
Car-starter	J. B. Jones	Philadelphia Pa	Feb. 18, 1873	135, 920
Car-starter	W. A. Jordan	New Orleans, La	Nov. 11, 1873	144, 546
Car-starter	W. Keeler	Towanda, Pa	Sept. 24, 1872	131, 689
Car-starter	S. H. Kenny	Dwight, Ill	Nov. 19, 1872	133, 157
Car-starter	G. B. Kirkham	New York, N. Y	Jan. 31, 1871	111, 353
Car-starter	W. W. Lamberton and D. R. Lemman.	New Orleans, La	Feb. 4, 1873	135, 564
Car-starter	C. P. Leavitt	New York, N. Y	Dec. 5, 1871	121, 635
Car-starter	G. Lowden	Brooklyn, N. Y	June 10, 1873	139, 721
Car-starter	C. J. Moore	Joliet, Ill	July 22, 1873	141, 159
Car-starter	W. H. Newton	Newport, R. I	Jan. 3, 1871	110, 760
Car-starter	J. North	New York, N. Y	Dec. 5, 1871	121, 540
Car-starter	B. F. Oakes	Milford, Me	Oct. 21, 1873	143, 837
Car-starter	L. Rodenhausen	Philadelphia, Pa	Mar. 24, 1868	75, 981
Car-starter	L. C. Rodier and F. G. Bates	Springfield, Mass	Apr. 29, 1873	138, 440

Index of patents issued from the United States Patent Office from 1790 *to* 1873, *inclusive*—**Continued.**

Invention.	Inventor.	Residence.	Date.	No.
Car-starter	W. E. Snediker	Charlotteburgh, N. J	Nov. 19, 1872	133, 265
Car-starter	J. Steger	New York, N. Y	Dec. 29, 1868	85, 407
Car-starter	J. F. Stokes	Philadelphia, Pa	Apr. 6, 1869	88, 675
Car-starter	J. F. Stokes	Philadelphia, Pa	June 27, 1871	116, 508
Car-starter	W. M. and W. E. Stratton	West Troy, N. Y	May 21, 1872	126, 993
Car-starter	R. R. Taylor	Reading, Pa	Mar. 2, 1869	87, 524
Car-starter	J. G. Thompson	Stockton, N. Y	Dec. 16, 1873	145, 600
Car-starter	I. C. Wallace and J. W. Andrews.	Detroit, Mich	Dec. 30, 1873	146, 112
Car-starter	M. N. Ward	Bangor, Me	Oct. 26, 1869	96, 172
Car-starter	W. M. Watson	Tonica, Ill	June 4, 1872	127, 531
Car-starter	J. J. Wheeler	Grinnell, Iowa	Jan. 21, 1873	135, 187
Car-starter	A. Whittimore	Cambridgeport, Mass	June 24, 1873	140, 229
Car-starter	A. Whittimore	Cambridgeport, Mass	Oct. 21, 1873	143, 948
Car-starter	A. Whittimore	Cambridgeport, Mass	Dec. 9, 1873	145, 468
Car-starter	G. A. Wilbur	Skowhegan, Me	July 13, 1869	92, 499
Car-starter	N. J. Wilkinson	Kalamazoo, Mich	Mar. 4, 1873	136, 398
Car-starter	V. Wirick	Chicago, Ill	Dec. 2, 1873	145, 076
Car-starter	F. J. Wright and L. W. Wandell.	New York, N. Y	Jan. 24, 1871	111, 294
Car starter and brake	T. R. Sinclaire	New York, N. Y	July 31, 1866	56, 810
Car starter and brake	J. Wiley, 2d	South Reading, Mass	Mar. 12, 1867	62, 911
Car-starter, Electrical	J. and W. H. Clark	Philadelphia, Pa	Nov. 26, 1867	71, 458
Car-starter for street-railways	M. Toulmin	New Orleans, La	Aug. 5, 1873	141, 474
Car starter, Horse	H. Schreiner	Philadelphia, Pa	Mar. 14, 1871	112, 640
Car starter, Railway	A. Amory	New York, N. Y	May 30, 1871	115, 266
Car starter, Railway	C. B. Broadwell	New Orleans, La	July 26, 1870	105, 635
Car starter, Railway	C. B. Broadwell	New Orleans, La	Aug. 6, 1872	130, 275
Car starter, Railway	D. Cumming	Sorrel Horse, Pa	Apr. 26, 1859	23, 757
Car starter, Railway	A. A. Wilder	Detroit, Mich	Feb. 19, 1867	62, 302
Car starter, Street	A. B. Davis	Catahoula Parish, La	Aug. 3, 1869	93, 282
Car starter, Street	T. S. E. Dixon	Janesville, Wis	Nov. 9, 1869	96, 557
Car starter, Street	G. P. Frick	Baltimore, Md	Oct. 11, 1870	108, 127
Car starter, Street	B. Lepper	Saint Louis, Mo	June 7, 1870	103, 897
Car starter, Street	H. Welch	Schuylersville, N. Y	June 21, 1870	104, 673
Car starter, Street-railway	J. D. S. Newell	New Orleans, La	Oct. 1, 1872	131, 772
Car starting and stopping apparatus	J. B. Waring	New York, N. Y	Mar. 24, 1868	75, 818
Car-starting apparatus	J. Steger	New York, N. Y	July 9, 1867	66, 648
Car-starting apparatus	J. Steger	New York, N. Y	Jan. 28, 1868	73, 764
Car starting-apparatus, City-railway	S. N. Sanford	Cleveland, Ohio	Apr. 3, 1860	27, 742
Car, Steam	H. F. Knapp	New York, N. Y	July 23, 1872	129, 833
Car, Steam street	T. C. Robinson and G. P. Clark	Boston, Mass	Dec. 29, 1868	85, 331
Car-step, Adjustable	W. Neumann	Saint Louis, Mo	Aug. 18, 1868	81, 194
Car, Stock	G. A. Egan	Covington, Ky	Feb. 4, 1873	135, 532
Car, Stock	G. W. Fox	Laramie, Wyo	May 16, 1871	114, 795
Car, Stock	T. E. Knauss	Zaleski, Ohio	Dec. 26, 1871	122, 120
Car, Stock	H. Lee	Beloit, Wis	Oct. 8, 1872	132, 088
Car, Stock	T. J. McCarty	Salem, Ohio	Oct. 7, 1873	143, 414
Car, Stock	A. Rank	Salem, Ohio	Feb. 14, 1871	111, 872
Car, Stock	A. Rank	Salem, Ohio	July 11, 1871	116, 866
Car, Stock	O. Severance	Middlebury, Vt	Nov. 18, 1873	144, 798
Car, Stock	W. Stark	Bronson, Mich	Aug. 2, 1864	43, 720
Car, Stock	Z. Street	Salem, Ohio	Aug. 30, 1870	106, 887
Car, Stock	Z. Street	Salem, Ohio	Aug. 30, 1870	106, 888
Car, Stock	Z. Street	Salem, Ohio	Mar. 26, 1872	125, 097
Car, Stock	A. Welch	Southall, England	Dec. 24, 1872	134, 335
Car, Stock and freight	J. B. Calkins	Pacific, Mo	July 11, 1871	116, 804
Car, Stock and freight	J. B. Calkins	Pacific, Mo	Dec. 26, 1871	122, 104
Car-strap	H. S. Vrooman	Boston, Mass	Jan. 2, 1872	122, 417
Car strap, Street	M. Warne	Philadelphia, Pa	Apr. 1, 1873	137, 397
Car, Street	C. B. Broadwell	New Orleans, La	Feb. 28, 1871	112, 215
Car, Street	M. De Graff	Chicago, Ill	Aug. 24, 1869	94, 086
Car, Street	Z. Eastman	Chicago, Ill	June 8, 1869	91, 105
Car, Street	J. I. Herrick	Milwaukee, Wis	May 5, 1868	77, 485
Car, Street	W. T. Jenks	Toledo, Ohio	Jan. 28, 1873	135, 277
Car, Street	J. F. Madison and H. McLaughlin.	Saint Louis, Mo	Aug. 3, 1869	93, 322
Car, Street	J. A. Miller	New York, N. Y	Jan. 26, 1864	41, 386
Car, Street	J. E. Ridgeway	Philadelphia, Pa	Apr. 26, 1870	102, 435
Car, Street	P. Shaw	Boston, Mass	Apr. 20, 1869	89, 086
Car, Street	J. Stephenson	New York, N. Y	Jan. 22, 1867	61, 481
Car, Street	J. Stephenson	New York, N. Y	Feb. 23, 1869	87, 120
Car, Street	J. Stephenson	New York, N. Y	May 14, 1872	126, 756
Car, Street	J. Stephenson	New York, N. Y	May 14, 1872	126, 757
Car, Street	J. Stephenson	New York, N. Y	June 4, 1872	127, 526
Car, Street	M. P. Turner	Des Moines, Iowa	Feb. 8, 1870	99, 732
Car, Street-railway	E. L. Dorsey	Washington, D. C	May 7, 1872	126, 449
Car, Street-railway	J. J. Gutierrez	Jefferson Parish, La	June 7, 1870	103, 874
Car, Street-railway	W. H. T. Hughes	Brooklyn, N. Y	Dec. 17, 1872	134, 060
Car, Street-railway	S. T. Jones	Philadelphia, Pa	Feb. 22, 1830	....
Car, Street-railway	J. Stephenson	New York, N. Y	Nov. 29, 1864	45, 281
Car, Street-railway	J. Stephenson	New York, N. Y	Feb. 23, 1869	87, 121
Car, Street steam-railway	J. P. Woodbury	Boston, Mass	Jan. 24, 1865	46, 043
Car, Street steam-railway	J. P. Woodbury	Boston, Mass	May 30, 1865	48, 008
Car-tank cover	L. C. Cattell	Cleveland, Ohio	Dec. 14, 1869	97, 879
Car, track, and operating apparatus for mines	W. B. Frue	Houghton, Mich	Mar. 22, 1870	101, 115
Car, Track-laying	J. S. Lake	Smith's Landing, N. J	Jan. 22, 1867	61, 344
Car unloader, Grain	M. W. Bosworth	Binghamton, N. Y	Dec. 23, 1873	145, 780
Car-ventilator	B. T. Babbitt	New York, N. Y	Jan. 30, 1855	12, 306
Car-ventilator	J. B. Bausman	Rochester, Pa	Apr. 9, 1861	31, 943
Car-ventilator	W. C. Betts	Brooklyn, N. Y	Jan. 17, 1871	110, 950
Car-ventilator	J. A. Caldwell	Springfield, Mass	Aug. 28, 1866	57, 470
Car-ventilator	J. J. Crowley	Whistler, Ala	Feb. 18, 1873	135, 971
Car-ventilator	J. M. Dexter	Elmira, N. Y	Jan. 10, 1871	110, 906
Car-ventilator	C. C. Gerhardt	Wyandotte, Kans	Mar. 4, 1873	136, 502
Car-ventilator	G. Hardy	Lawrence, Mass	May 29, 1866	55, 094
Car-ventilator	M. T. Hitchcock	Springfield, Mass	Feb. 18, 1868	74, 534
Car-ventilator	M. T. Hitchcock	Springfield, Mass	Dec. 23, 1873	145, 871

Index of patents issued from the United States Patent Office from 1790 *to* 1873, *inclusive*—Continued.

Invention.	Inventor.	Residence.	Date.	No.
Car-ventilator	R. Hitchcock	Springfield, Mass	Aug. 20, 1867	67, 877
Car-ventilator	R. Hitchcock	Springfield, Mass	Mar. 24, 1868	75, 910
Car-ventilator	R. Hitchcock	Springfield, Mass	Sept. 7, 1869	94, 494
Car-ventilator	W. H. Hunt	Bolton, Conn	Aug. 30, 1870	106, 940
Car-ventilator	S. C. Maine	Boston, Mass	Sept. 16, 1873	142, 799
Car-ventilator	A. J. Morrison	Troy, N. Y	Sept. 16, 1873	142, 807
Car-ventilator	D. C. Richardson	Lawrence, Mass	Feb. 25, 1868	74, 849
Car-ventilator	W. M. Russell and D. E. Holmes.	Cincinnati, Ohio	Aug. 18, 1868	81, 297
Car-ventilator	A. B. Spencer	Rochester, N. Y	Apr. 5, 1864	42, 234
Car-ventilator	C. Wadsworth	New York, N. Y	May 6, 1862	35, 190
Car-ventilator and dust-screen, Pocket	J. B. Timberlake	Salem, Ohio	May 27, 1873	139, 278
Car-ventilator and refrigerator	W. E. Phelps	Elmwood, Ill	Sept. 27, 1870	107, 716
Car-ventilator, Automatic	I. Buckingham	Seymour, Conn	May 12, 1868	77, 867
Car-ventilator, Eduction	M. T. Hitchcock	Springfield, Mass	Sept. 28, 1869	95, 346
Car ventilator, Railway	M. C. Andrews	Lawrence, Mass	May 26, 1868	78, 174
Car ventilator, Railway	G. B. Armstrong and G. F. McLellan.	Chicago, Ill., and Washington, D. C.	Dec. 27, 1870	110, 416
Car ventilator, Railway	G. W. R. Bayley and J. McCluskey.	Algiers, La	July 7, 1868	79, 541
Car ventilator, Railway	L. C. Beardsley	Cleveland, Ohio	Oct. 29, 1861	33, 565
Car ventilator, Railway	J. Bevan	Jersey City, N. J	June 13, 1854	11, 101
Car ventilator, Railway	I. Bonnell, jr	Chicago, Ill	Mar. 8, 1870	100, 589
Car ventilator, Railway	W. A. Brown	Philadelphia, Pa	Dec. 24, 1861	33, 979
Car ventilator, Railway	J. H. Bruce	Wyandotte, Kans	June 4, 1872	127, 559
Car ventilator, Railway	W. Conard	Burlington, N. J	July 5, 1870	105, 046
Car ventilator, Railway	W. G. Creamer	Brooklyn, N. Y	Aug. 20, 1861	33, 071
Car ventilator, Railway	W. G. Creamer	Brooklyn, N. Y	Nov. 12, 1867	70, 812
Car ventilator, Railway	W. G. Creamer	Brooklyn, N. Y	May 11, 1869	89, 974
Car ventilator, Railway	W. G. Creamer	Brooklyn, N. Y	Feb. 20, 1872	123, 770
Car ventilator, Railway	H. A. Custis	Richmond, Va	May 31, 1870	103, 722
Car ventilator, Railway	S. Darling	Providence, R. I	Mar. 15, 1870	100, 870
Car ventilator, Railway	I. Dripps	Fort Wayne, Ind	Mar. 24, 1868	75, 878
Car ventilator, Railway	W. B. Dunning	Geneva, N. Y	Dec. 17, 1867	72, 274
Car ventilator, Railway	E. B. Forbush	Buffalo, N. Y	Feb. 21, 1860	27, 212
Car ventilator, Railway	D. H. Fox and J. Fink	Reading, Pa	May 8, 1855	12, 818
Car ventilator, Railway	T. W. Freeborn	Newport, R. I	May 14, 1872	126, 797
Car ventilator, Railway	F. H. Furniss	Cleveland, Ohio	Sept. 17, 1861	33, 302
Car ventilator, Railway	R. C. Graves	Barnesville, Ohio	Apr. 16, 1867	63, 880
Car ventilator, Railway	G. B. Hall and J. Shaffer	Kansas City, Mo	Sept. 26, 1871	119, 349
Car ventilator, Railway	A. Hapgood	Worcester, Mass	May 22, 1860	28, 365
Car ventilator, Railway	C. H. Haskins	Saint Louis, Mo	Nov. 26, 1867	71, 385
Car ventilator, Railway	R. Heneage and F. W. Breed	Buffalo, N. Y	May 26, 1868	78, 206
Car ventilator, Railway	G. A. Hines	Brattleborough, Vt	Oct. 17, 1871	119, 982
Car ventilator, Railway	M. T. Hitchcock	Springfield, Mass	July 28, 1868	80, 411
Car ventilator, Railway	M. T. Hitchcock	Springfield, Mass	Jan. 18, 1870	98, 965
Car ventilator, Railway	R. Hitchcock	Springfield, Mass	Oct. 26, 1869	96, 114
Car ventilator, Railway	J. L. Howard	Hartford, Conn	Sept. 15, 1863	39, 925
Car ventilator, Railway	J. L. Howard	Hartford, Conn	Oct. 29, 1867	70, 216
Car ventilator, Railway	J. L. Howard	Hartford, Conn	July 18, 1871	117, 076
Car ventilator, Railway	J. L. Howard	Hartford, Conn	July 25, 1871	117, 289
Car ventilator, Railway	M. G. Imbach	Hartford, Conn	Dec. 17, 1867	72, 203
Car ventilator, Railway	N. Jones	Buffalo, N. Y	June 27, 1871	116, 319
Car ventilator, Railway	D. C. Justison	Wilmington, Del	Nov. 14, 1871	120, 980
Car ventilator, Railway	S. E. Kirkpatrick	Saint Albans, Vt	Apr. 9, 1872	125, 461
Car ventilator, Railway	J. Lesperance	Saint Louis, Mo	Feb. 9, 1869	86, 766
Car ventilator, Railway	S. C. Maine	Boston, Mass	Apr. 9, 1872	125, 401
Car ventilator, Railway	T. J. Mell	Macon, Ga	May 18, 1869	90, 181
Car ventilator, Railway	S. Merrick	New Brighton, Pa	May 26, 1863	38, 691
Car ventilator, Railway	D. Merrill	Worcester, Mass	Apr. 8, 1862	34, 898
Car ventilator, Railway	J. Miller and W. Ketting	Jersey City, N. J	Sept. 10, 1861	33, 259
Car ventilator, Railway	J. H. Moore	Chicago, Ill	Jan. 29, 1867	61, 555
Car ventilator, Railway	J. H. Moore	Chicago, Ill	Oct. 1, 1867	69, 357
Car ventilator, Railway	C. Newcomb	New York, N. Y	Oct. 9, 1860	30, 340
Car ventilator, Railway	E. Norton	Brooklyn, N. Y	Dec. 21, 1869	98, 095
Car ventilator, Railway	D. W. Noyes	Bennington, Vt	Mar. 11, 1873	136, 756
Car ventilator, Railway	R. L. Omensetter	Philadelphia, Pa	Mar. 12, 1872	124, 613
Car ventilator, Railway	R. Reniff and W. W. Buttolph	Bloomington, Ill	Feb. 19, 1867	62, 225
Car ventilator, Railway	W. M. Russell and D. E. Holmes.	Cincinnati, Ohio	Feb. 9, 1869	86, 780
Car ventilator, Railway	H. J. Ruttan	Coburg, Canada	Feb. 6, 1872	123, 513
Car ventilator, Railway	A. G. Safford	Boston, Mass	Oct. 19, 1869	95, 938
Car ventilator, Railway	T. H. B. Sanders	Pittsburgh, Pa	Nov. 14, 1865	50, 960
Car ventilator, Railway	W. Schairath	Bielefeld, Prussia	Nov. 10, 1868	84, 004
Car ventilator, Railway	J. D. Simmons	Quincy, Ill	Dec. 23, 1873	145, 760
Car ventilator, Railway	O. Slagle	London, Ohio	Dec. 21, 1869	98, 201
Car ventilator, Railway	A. B. Spencer	Rochester, N. Y	Oct. 6, 1863	40, 199
Car ventilator, Railway	G. Spencer	Utica, N. Y	Nov. 8, 1853	10, 216
Car ventilator, Railway	W. C. Stickney	Steubenville, Ohio	Apr. 16, 1872	125, 855
Car ventilator, Railway	O. J. Styner and J. Egan	La Fayette, Ind	Nov. 2, 1869	96, 501
Car ventilator, Railway	A. B. Sweetland	Fitchburgh, Mass	July 4, 1871	116, 774
Car ventilator, Railway	A. B. Sweetland	Fitchburgh, Mass	Jan. 23, 1872	123, 059
Car ventilator, Railway	O. W. Swift	New Haven, Conn	Aug. 8, 1871	117, 941
Car ventilator, Railway	J. B. Talmadge	Winsted, Conn	Mar. 14, 1865	46, 831
Car ventilator, Railway	J. A. Thompson	Cayuga, N. Y	Oct. 29, 1861	33, 614
Car ventilator, Railway	C. P. Tillinghast	Providence, R. I	Jan. 3, 1871	110, 803
Car ventilator, Railway	J. G. Treadwell	Albany, N. Y	Oct. 11, 1859	25, 776
Car ventilator, Railway	A. P. Vining	Scranton, Pa	Oct. 17, 1865	50, 522
Car ventilator, Railway	W. Westlake	Milwaukee, Wis	July 29, 1862	36, 063
Car ventilator, Railway	J. S. Williamson and B. J. Bicknell.	Memphis, Tenn	Aug. 27, 1872	130, 965
Car ventilator, Railway	E. H. Winchell	New York, N. Y	Oct. 10, 1871	119, 809
Car ventilator, Railway	J. Wright	Richmond, Va	July 16, 1872	129, 201
Car ventilator, Railway	R. B. Wright	Norfolk, Va	Oct. 23, 1860	30, 515
Car-ventilators, Device for operating	G. W. Perry	Wilmington, Del	Apr. 30, 1872	126, 327
Car ventilators, Device to open railway	W. C. Stickney and J. McGee	Steubenville, Ohio	Dec. 22, 1868	85, 143
Car ventilating and warming apparatus, Railway	J. J. Watson, W. Hardiker, and T. Toye.	Buffalo, N. Y	Dec. 18, 1860	30, 996

Index of patents issued from the United States Patent Office from 1790 *to* 1873, *inclusive*—Continued.

Invention.	Inventor.	Residence.	Date.	No.
Car ventilating-attachment, Railway	J. Shaw	Bridgeport, Conn	Oct. 15, 1867	69,847
Car, wagon, and other vehicles	T. Stone	Plainfield, Ind	June 30, 1868	79,410
Car-washer	E. W. Leavens	Boston, Mass	Dec. 24, 1872	134,295
Car, Weighing	J. H. Zinn	Kingston, Tenn	June 36, 1866	55,946
Car-wheel	R. N. Allen and L. W. Kimball	Pittsford, Vt	May 11, 1869	89,908
Car-wheel	A. Atwood	New York, N. Y	July 28, 1868	80,380
Car-wheel	A. Atwood	Brooklyn, N. Y	July 1, 1873	140,339
Car-wheel	J. E. Atwood	Pittsburgh, Pa	Nov. 11, 1873	144,495
Car-wheel	J. Baker	Boston, Mass	Oct. 4, 1853	10,062
Car-wheel	J. Baldwin	Lynn, Mass	Nov. 1, 1870	108,825
Car-wheel	G. S. Bosworth	Troy, N. Y	Oct. 23, 1860	30,458
Car-wheel	S. B. Chapman	New York, N. Y	Mar. 17, 1868	75,521
Car-wheel	J. L. Constable	New York, N. Y	Feb. 20, 1866	52,683
Car-wheel	J. M. Cook	Taunton, Mass	Dec. 5, 1848	5,953
Car-wheel	A. F. Cooper	Cambridge, Mass	Oct. 14, 1873	143,560
Car-wheel	W. Dickinson	Brooklyn, N. Y	Oct. 16, 1866	58,786
Car-wheel	J. V. Dinsmore	Milford, Mass	May 14, 1867	64,642
Car-wheel	G. Dock	Wiconisco, Pa	Mar. 3, 1868	74,999
Car-wheel	G. W. Eddy	Waterford, N. Y	Dec. 26, 1845	4,330
Car-wheel	G. W. Eddy	Waterford, N. Y	Jan. 8, 1850	6,999
Car-wheel	G. Elmslie	Toronto, Canada	Nov. 4, 1873	144,195
Car-wheel	W. B. Fahnestock	Lancaster, Pa	Mar. 30, 1858	19,763
Car-wheel	D. P. Fales	West Poultney, Vt	Sept. 27, 1853	10,057
Car-wheel	J. Farnsworth, jr	Madison, Ind	Nov. 1, 1853	10,184
Car-wheel	H. B. Fernald	Dedham, Mass	June 29, 1869	91,926
Car-wheel	A. C. Fletcher	New York, N. Y	Apr. 30, 1872	126,280
Car-wheel	A. C. Fletcher	New York, N. Y	Dec. 23, 1873	145,856
Car-wheel	F. A. Tonts	Bloomington, Ill	May 13, 1873	138,798
Car-wheel	D. Forrest	Eastport, Me	July 23, 1867	67,036
Car-wheel	T. Glasco	Wilmington, Del	Apr. 4, 1846	4,447
Car-wheel	W. Goodman	Boston, Mass	Dec. 27, 1870	110,563
Car-wheel	F. Harbach	Pittsfield, Mass	Nov. 6, 1847	5,360
Car-wheel	F. Harbach	Pittsfield, Mass	Nov. 6, 1847	5,361
Car-wheel	J. Harris	Marquette, Wis	Oct. 24, 1865	50,579
Car-wheel	J. Harris	Marquette, Wis	Nov. 5, 1867	70,560
Car-wheel	C. Hart	Bridgeport, Conn	Dec. 13, 1853	10,309
Car-wheel	C. T. Harvey	Tarrytown, N. Y	July 7, 1868	79,757
Car-wheel	R. B. Hugunin	Cleveland, Ohio	Mar. 2, 1869	87,496
Car-wheel	L. B. Hunt	New York, N. Y	July 11, 1871	116,961
Car-wheel	G. K. Kane	York, Pa	Mar. 6, 1847	5,002
Car-wheel	J. W. Kern	Portland, Oreg	Feb. 18, 1873	136,068
Car-wheel	J. S. Kingsland	Stryker, Ohio	Feb. 27, 1866	52,859
Car-wheel	C. E. Kleinschmidt	Cleveland, Ohio	Nov. 19, 1867	71,021
Car-wheel	W. A. Lewis	Chicago, Ill	Oct. 10, 1871	119,867
Car-wheel	H. C. Lockwood	Baltimore, Md	Nov. 26, 1872	133,461
Car-wheel	Z. H. Mann	Newport, Ky	Oct. 4, 1853	10,073
Car-wheel	R. C. Mansell	Ashford, England	Oct. 9, 1866	58,741
Car-wheel	W. V. Many	Albany, N. Y	Feb. 1, 1848	5,424
Car-wheel	G. B. Massey	New York, N. Y	June 18, 1867	65,925
Car-wheel	H. Merrill	Brooklyn, N. Y	Apr. 22, 1873	138,036
Car-wheel	H. W. Moore	Jersey City, N. J	June 22, 1869	91,762
Car-wheel	H. W. Moore	Jersey City, N. J	June 7, 1870	104,053
Car-wheel	G. Natorp	New York, N. Y	Nov. 3, 1868	83,722
Car-wheel	D. P. Nickerson	Cleveland, Ohio	May 28, 1867	65,109
Car-wheel	W. H. Nobles	Saint Paul, Minn	Oct. 13, 1868	82,978
Car-wheel	J. Pugh	Franklin, Tenn	June 15, 1858	20,583
Car-wheel	P. Putnam	Laconia, N. H	June 28, 1864	43,337
Car-wheel	J. Raddin	Lynn, Mass	May 14, 1867	64,796
Car-wheel	J. Raddin	Lynn, Mass	Feb. 25, 1868	74,959
Car-wheel	J. Raddin	Lynn, Mass	Apr. 14, 1868	76,811
Car-wheel	R. Ramsay	Irwin's Station, Pa	Jan. 14, 1873	134,929
Car-wheel	F. M. Ray	New York, N. Y	Dec. 17, 1846	4,894
Car-wheel	E. S. Robinson	New York, N. Y	Sept. 25, 1866	58,298
Car-wheel	J. C. Rupp and S. Ott	Newark, Del	Jan. 9, 1872	122,661
Car-wheel	E. Sampson	Lansingburgh, N. Y	May 7, 1867	64,579
Car-wheel	J. K. Sax and G. W. Kear	Pittston and Kingston, Pa.	Sept. 9, 1873	142,587
Car-wheel	S. J. Seely	New York, N. Y	Jan. 31, 1865	46,145
Car-wheel	T. Sharp	Chicago, Ill	Sept. 29, 1863	40,123
Car-wheel	G. W. and H. Sizer	Springfield, Mass	Apr. 17, 1847	5,076
Car-wheel	A. Small	York, Pa	Dec. 10, 1846	4,882
Car-wheel	J. Y. Smith	Pittsburgh, Pa	Feb. 9, 1869	86,878
Car-wheel	S. P. Smith	Troy, N. Y	May 29, 1860	28,512
Car-wheel	W. W. Snow	Jersey City, N. J	Feb. 19, 1861	31,488
Car-wheel	G. W. and L. W. Snyder	Port Carbon and Lykens, Pa.	Mar. 4, 1873	136,392
Car-wheel	W. R. Thomas	Catasauqua, Pa	June 16, 1868	79,029
Car-wheel	W. R. Thomas	Catasauqua, Pa	Aug. 17, 1869	93,769
Car-wheel	J. and N. Tiranoff and A. Kousakoff	St. Petersburg, Russia	Sept. 16, 1873	142,968
Car-wheel	F. W. Townrow	United States Army	Jan. 2, 1872	122,340
Car-wheel	S. A. Traugh	Cincinnati, Ohio	Nov. 19, 1867	71,247
Car-wheel	N. K. Wade and J. Kaye	Pittsburgh, Pa.	Sept. 4, 1860	29,927
Car-wheel	A. B. Wakefield and B. A. Berryman	Saint Louis, Mo	Mar. 25, 1873	137,266
Car-wheel	P. H. Watson	Ashtabula, Ohio	May 9, 1871	114,628
Car-wheel	J. W. Weston	New York, N. Y	Apr. 5, 1870	101,686
Car-wheel	W. Willoughby	Markwell, Miss	June 15, 1858	20,610
Car-wheel	W. Wilmington	Toledo, Ohio	July 28, 1868	80,323
Car-wheel	W. Wilmington	Toledo, Ohio	Sept. 22, 1868	82,466
Car-wheel	W. Wilmington	Toledo, Ohio	Feb. 11, 1873	135,870
Car-wheel	C. Wilson and J. H. McNall	Clinton, Pa	Feb. 5, 1867	61,786
Car-wheel, Adjustable	G. G. Sobdell	Wilmington, Del	Feb. 20, 1872	123,919
Car-wheel, Adjustable-gage	W. B. Snow	Brooklyn, N. Y	Aug. 15, 1871	118,065
Car wheel and axle	D. Babson, 2d	Rockport, Mass	Mar. 6, 1866	52,949
Car wheel and axle	F. Hudner	New York, N. Y	Aug. 31, 1869	94,414
Car wheel and axle	J. S. Upton	Battle Creek, Mich	June 4, 1872	127,661
Car wheel and axle	J. A. Woodbury	Boston, Mass	Mar. 9, 1869	87,746
Car wheel and axle, Changeable-gage	P. Putnam	Laconia, N H	Aug. 29, 1871	118,642

Index of patents issued from the United States Patent Office from 1790 *to* 1873, *inclusive*—Continued.

Invention.	Inventor.	Residence.	Date.	No.
Car wheel and axle, Changeable-gage	C. D. Tisdale	Boston, Mass	Aug. 15, 1871	118, 170
Car wheel and axle, Railway	T. C. Hargrave	Boston, Mass	Aug. 18, 1868	81, 083
Car wheel and axle, Railway	W. Hudgin	Athens, Ga	Feb. 7, 1871	111, 543
Car wheel and axle, Railway	P. Putnam	Laconia, N. H	Dec. 7, 1869	97, 552
Car wheel and axle, Railway	C. D. Tisdale	Boston, Mass	May 25, 1869	90, 410
Car-wheel and car-axle, Combined	N. W. Northrup	Greene, N. Y	Oct. 28, 1862	36, 790
Car wheel and frog	W. H. Childe	Gainesville, Ala	Sept. 15, 1868	82, 086
Car-wheel, Anti-friction	G. C. Beecher	Livonia, N. Y	Jan. 6, 1863	37, 269
Car-wheel brake	W. McCammon	Albany, N. Y	Dec. 12, 1846	4, 886
Car-wheel, Cast-iron	A. Atwood	Troy, N. Y	May 15, 1847	5, 112
Car-wheel, Cast-iron	E. B. Baker	Charleston District, S. C	Jan. 9, 1849	6, 024
Car-wheel, Cast-iron	J. Boon	Lancaster, Pa	Apr. 16, 1850	7, 282
Car-wheel, Cast-iron	T. S. Bourshett	Little Falls, N. Y	Nov. 13, 1849	6, 863
Car-wheel, Cast-iron	A. G. Bristol and J. C. Jackson	Rochester, N. Y	Feb. 24, 1852	8, 751
Car-wheel, Cast-iron	A. T. Converse and W. S. Cooley.	Norwich, Conn	Jan. 9, 1849	6, 026
Car-wheel, Cast-iron	J. M. Cook	Taunton, Mass	Jan. 9, 1849	6, 022
Car-wheel, Cast-iron	L. Dean and A. Higham	Utica, N. Y	Jan. 9, 1849	6, 019
Car-wheel, Cast-iron	P. Dorsch	Schenectady, N. Y	June 15, 1852	9, 017
Car-wheel, Cast-iron	T. J. Eddy	Waterford, N. Y	July 29, 1851	8, 255
Car-wheel, Cast-iron	A. Fuller	Boston, Mass	Aug. 27, 1850	7, 599
Car-wheel, Cast-iron	P. G. Gardiner	New York, N. Y	Mar. 11, 1851	7, 967
Car-wheel, Cast-iron	G. W. Glass	Allegheny City, Pa	Apr. 25, 1854	10, 816
Car-wheel, Cast-iron	C. Hart	New York, N. Y	Dec. 25, 1849	6, 971
Car-wheel, Cast-iron	C. Hart and N. Washburn	Rochester, N. Y	Apr. 3, 1849	6, 269
Car-wheel, Cast-iron	A. Hebbard	Worcester, Mass	May 20, 1851	8, 106
Car-wheel, Cast-iron	A. Hebbard	Worcester, Mass	June 29, 1852	9, 071
Car-wheel, Cast-iron	J. Henry	Lynchburgh, Va	June 20, 1854	11, 124
Car-wheel, Cast-iron	L. Kinsley	Norfolk County, Mass	Mar. 12, 1850	7, 169
Car-wheel, Cast-iron	L. Kinsley	Norfolk County, Mass	Mar. 12, 1850	7, 170
Car-wheel, Cast-iron	G. G. Lobdell	Wilmington, Del	July 14, 1868	79, 842
Car-wheel, Cast-iron	G. G. Lobdell	Wilmington, Del	Jan. 12, 1869	85, 745
Car-wheel, Cast-iron	W. V. Many	Albany, N. Y	May 1, 1847	5, 095
Car-wheel, Cast-iron	G. R. McFarlane	Hollidaysburgh, Pa	Jan. 14, 1851	7, 903
Car-wheel, Cast-iron	H. W. Moore	Bridgeport, Conn	Mar. 2, 1852	8, 777
Car-wheel, Cast-iron	H. W. Moore	Jersey City, N. J	Sept. 20, 1858	21, 614
Car-wheel, Cast-iron	O. Moulton	Blackstone, Mass	Mar. 2, 1852	8, 772
Car-wheel, Cast-iron	A. L. Mowry	Cincinnati, Ohio	June 2, 1857	17, 442
Car-wheel, Cast-iron	G. Peacock	Selma, Ala	Nov. 26, 1867	71, 408
Car-wheel, Cast-iron	T. Perkins and W. McMahon	Baltimore, Md	Apr. 10, 1843	3, 037
Car-wheel, Cast-iron	D. Prew	Taunton, Mass	June 13, 1854	11, 056
Car-wheel, Cast-iron	D. R. Rall	Rochester, N. Y	June 1, 1852	8, 985
Car-wheel, Cast-iron	H. H. Scoville	Chicago, Ill	July 13, 1852	9, 121
Car-wheel, Cast-iron	B. Severson	Schenectady, N. Y	Oct. 21, 1851	8, 455
Car-wheel, Cast-iron	B. Severson	Philadelphia, Pa	July 4, 1854	11, 233
Car-wheel, Cast-iron	J. B. Tarr	Fairhaven, Mass	July 7, 1868	79, 611
Car-wheel, Cast-iron	S. Thurston	Scranton, Pa	May 25, 1852	8, 975
Car-wheel, Cast-iron	W. B. Treadwell	Albany, N. Y	Jan. 9, 1849	6, 021
Car-wheel, Cast-iron	S. Truscott	Columbia, Pa	Jan. 16, 1849	6, 030
Car-wheel, Cast-iron	I. Van Kuran	Rochester, N. Y	May 1, 1849	6, 415
Car-wheel, Cast-iron	I. Van Kuran	Boston, Mass	May 20, 1851	8, 108
Car-wheel, Cast-iron	N. Washburn	Worcester, Mass	Oct. 8, 1850	7, 710
Car-wheel, Cast-iron	A. Whitney	Philadelphia, Pa	May 22, 1847	5, 126
Car-wheel, Cast-iron	A. Whitney	Philadelphia, Pa	May 22, 1847	5, 127
Car-wheel, Cast-iron	A. Whitney	Philadelphia, Pa	Mar. 19, 1850	7, 202
Car-wheel, Cast-iron	W. Wilmington	Toledo, Ohio	Aug. 9, 1870	106, 243
Car-wheel, Cast-iron	H. H. Wiser	Rochester, N. Y	Dec. 4, 1849	6, 939
Car-wheel, Cast-iron	H. W. Woodruff	Watertown, N. Y	Jan. 6, 1852	8, 644
Car-wheel, Cast-iron plate	H. Felton, P. D. Cummings, and H. Hinckley.	Portland, Me	Jan. 23, 1849	6, 041
Car wheel, Cast-iron railway	W. W. Bergstresser	Harrisburgh, Pa	Jan. 11, 1840	1, 470
Car wheel, Cast-iron railway	J. M. Ross	Springfield, Mass	Mar. 17, 1857	16, 851
Car wheel, Cast-iron railway	B. Severson	Schenectady, N. Y	Sept. 17, 1850	7, 658
Car wheel, Cast-iron railway	J. M. Sigourney	Watertown, N. Y	Oct. 21, 1856	15, 935
Car-wheel, Cast-metal	G. G. Lobdell	Wilmington, Del	Apr. 15, 1862	34, 972
Car-wheel, Cast-steel	C. W. Stafford	Saybrook, Conn	Sept. 6, 1864	44, 120
Car-wheel, Cast-steel	J. B. Tarr	Chicago, Ill	July 30, 1867	67, 227
Car-wheel, Changeable-gage	J. Hamilton and G. F. Morse	Portland, Me	Oct. 18, 1870	108, 475
Car-wheel, Elastic	J. J. Sherman	Albany, N. Y	Apr. 23, 1872	126, 096
Car-wheel, Elastic	J. M. Whiting	Providence, R. I	Nov. 14, 1871	121, 032
Car-wheel, Elastic	J. A. Woodbury	Boston, Mass	July 25, 1871	117, 498
Car wheel for curved roads, Railway	A. Look	Waterford, N. Y	Dec. 31, 1833	......
Car-wheel for lubricating purposes	J. W. Latcher	Albany, N. Y	May 22, 1866	54, 925
Car-wheel gears, Self-adjusting	G. Sewell	Brooklyn, N. Y	May 16, 1871	114, 979
Car wheel, Locomotive	J. B. Tarr	Fairhaven, Mass	Jan. 18, 1870	98, 895
Car-wheel lubricator	W. A. Bullard	Kalamazoo, Mich	Dec. 9, 1873	145, 395
Car-wheel lubricator	W. W. Crane	Auburn, N. Y	Nov. 26, 1872	133, 418
Car-wheel lubricator	W. W. Martin	Allegheny City, Pa	Jan. 21, 1868	73, 621
Car-wheel lubricator	J. H. Murray	Hollidaysburgh, Pa	Sept. 30, 1873	143, 370
Car-wheel lubricator	J. M. Porter	Frostburgh, Md	May 2, 1871	114, 473
Car-wheel lubricator	C. Smith	Irwin Station, Pa	Oct. 24, 1871	120, 337
Car-wheel mold	J. K. Sax	Pittston, Pa	June 28, 1870	104, 777
Car-wheel mold	W. E. Worth	San Francisco, Cal	Jan. 3, 1871	110, 717
Car-wheel-molding flask	J. O'Brian	Louisville, Ky	Sept. 3, 1872	131, 115
Car wheel, Railway	R. N. Allen	Pittsford, Vt	July 16, 1872	128, 939
Car wheel, Railway	J. Baker	Lancaster, Pa	Mar. 20, 1835	......
Car wheel, Railway	T. C. Ball	Keene, N. H	Nov. 9, 1858	22, 049
Car wheel, Railway	A. G. Barrett	Barrett, Kans	Nov. 21, 1871	121, 075
Car wheel, Railway	C. K. Bradford	Lynnfield, Mass	May 10, 1870	102, 908
Car wheel, Railway	G. B. Bryant	Pottsville, Pa	Oct. 31, 1871	120, 489
Car wheel, Railway	H. C. Bulkley	Springfield, Mass	Mar. 30, 1858	19, 810
Car wheel, Railway	D. Cockley	Lancaster, Pa	Oct. 21, 1839	1, 377
Car wheel, Railway	E. Crane	Dorchester, Mass	Nov. 8, 1859	26, 022
Car wheel, Railway	T. C. Craven	Albany, N. Y	July 23, 1872	129, 655
Car wheel, Railway	T. Curtis	New Hudson, Mich	Sept. 6, 1864	44, 078
Car wheel, Railway	C. Delafield and F. G. Johnson	Castleton and Northfield, N. Y.	Aug. 31, 1869	94, 293
Car wheel, Railway	T. J. Eddy	Waterford, N. Y	Aug. 2, 1853	9, 897

Index of patents issued from the United States Patent Office from 1790 *to* 1873, *inclusive*—Continued.

Invention.	Inventor.	Residence.	Date.	No.
Car wheel, Railway	J. N. Farrar	Pepperell, Mass	Feb. 8, 1870	99, 547
Car wheel, Railway	J. N. Farrar	Pepperell, Mass	Nov. 15, 1870	109, 310
Car wheel, Railway	A. C. Fletcher	New York, N. Y	Mar. 23, 1869	88, 154
Car wheel, Railway	A. C. Fletcher	New York, N. Y	May 4, 1869	89, 752
Car wheel, Railway	W. J. Fryer, jr., and W. Freeborn.	New York, N. Y	Mar. 6, 1866	52, 986
Car wheel, Railway	A. Fuller	Providence, R. I	Oct. 14, 1840	1, 824
Car wheel, Railway	P. G. Gardiner	New York, N. Y	May 1, 1847	5, 098
Car wheel, Railway	J. D. Green	Troy, N. Y	Mar. 29, 1870	101, 259
Car wheel, Railway	J. B. Handyside	Glasgow, Great Britain	Oct. 31, 1871	120, 432
Car wheel, Railway	N. Hodge	Adams, Mass	Nov. 18, 1851	8, 526
Car wheel, Railway	N. Hodge	North Adams, Mass	June 1, 1852	8, 979
Car wheel, Railway	W. W. Hubbell and R. H. Hubbell.	Philadelphia and Delaware County, Pa.	Mar. 30, 1858	19, 776
Car wheel, Railway	L. B. Hunt	Leverett, Mass	Nov. 30, 1869	97, 407
Car wheel, Railway	L. B. Hunt	New York, N. Y	Nov. 15, 1870	109, 320
Car wheel, Railway	L. B. Hunt	New York, N. Y	Jan. 24, 1871	111, 211
Car wheel, Railway	A. C. Ketchum	New York, N. Y	July 15, 1856	15, 339
Car wheel, Railway	A. Komp	New York, N. Y	Jan. 25, 1870	99, 210
Car wheel, Railway	A. B. Latta	Cincinnati, Ohio	Dec. 1, 1857	18, 752
Car wheel, Railway	E. A. Lester	Boston, Mass	Aug. 9, 1859	25, 020
Car wheel, Railway	G. G. Lobdell	Wilmington, Del	Feb. 19, 1861	31, 467
Car wheel, Railway	H. C. Lockwood	Baltimore, Md	Nov. 14, 1871	120, 985
Car wheel, Railway	W. H. Mason	Boston, Mass	Feb. 23, 1869	87, 180
Car wheel, Railway	J. L. Mott	New York, N. Y	Mar. 21, 1854	10, 672
Car wheel, Railway	J. T. Owen	Philadelphia, Pa	Nov. 30, 1869	97, 433
Car wheel, Railway	J. T. Owen	Philadelphia, Pa	Dec. 21, 1869	98, 099
Car wheel, Railway	W. H. Paige	Springfield, Mass	July 22, 1873	141, 011
Car wheel, Railway	S. E. Parish	Nashville, Tenn	Feb. 16, 1858	19, 380
Car wheel, Railway	M. Phelan	Beaver County, Pa	Dec. 1, 1857	18, 767
Car wheel, Railway	D. Prew	Providence, R. I	Nov. 26, 1872	133, 336
Car wheel, Railway	J. Raddin	Lynn, Mass	Jan. 5, 1869	85, 693
Car wheel, Railway	F. M. Ray, sr	New York, N. Y	Apr. 23, 1872	126, 085
Car wheel, Railway	J. Rogers	Cincinnati, Ohio	July 27, 1869	93, 123
Car wheel, Railway	S. Rogers	Pittsburgh, Pa	Feb. 23, 1858	19, 445
Car wheel, Railway	J. K. Sax and G. W. Kear	Kingston, Pa	Apr. 6, 1869	88, 743
Car wheel, Railway	W. Smith	Pittsburgh, Pa	Jan. 10, 1860	26, 821
Car wheel, Railway	W. W. Spafford	Peterborough, N. H	Feb. 14, 1860	27, 163
Car wheel, Railway	J. H. Steiner	Philadelphia, Pa	June 19, 1860	28, 816
Car wheel, Railway	E. Stiles	New York, N. Y	Apr. 14, 1868	76, 844
Car wheel, Railway	A. W. Straub	Philadelphia, Pa	Jan. 16, 1866	52, 089
Car wheel, Railway	T. R. Timby	Saratoga, N. Y	Mar. 2, 1869	87, 445
Car wheel, Railway	F. H. Trevithick, jr	New York, N. Y	Aug. 2, 1870	106, 001
Car wheel, Railway	O. O. Van Orman	Harrisville, Ohio	Sept. 10, 1861	33, 273
Car wheel, Railway	Z. Washburn	Hopedale, Mass	Nov. 30, 1869	97, 329
Car wheel, Railway	R. A. Wilder	Schuylkill Haven, Pa	Mar. 28, 1854	10, 714
Car wheel, Railway	J. A. Woodbury	Boston, Mass	Apr. 9, 1872	125, 649
Car wheel, Railway	G. W. N. Yost	Corry, Pa	Aug. 3, 1869	93, 386
Car-wheel, Self-oiling	W. Owen	Hubbard, Ohio	June 29, 1869	91, 863
Car-wheel, Steel	J. B. Tarr	Fairhaven, Mass	June 28, 1870	104, 792
Car wheel to prevent friction, Railway	S. Krauser	Reading, Pa	Nov. 2, 1831	....
Car-wheel with independent flange and tread	W. S. McLean	Pittsburgh, Pa	Aug. 1, 1854	11, 438
Car-wheel, Wrought-iron	H. Aiken	Franklin, N. H	Oct. 1, 1850	7, 676
Car wheel, Wrought-iron-plate railway	G. W. Alden	New York, N. Y	May 12, 1857	17, 250
Car wheels and axles by incasing, Protecting	A. L. Finch	New Britain, Conn	Apr. 20, 1852	8, 886
Car-wheels, Annealing and cooling cast-iron	A. Whitney	Philadelphia, Pa	Apr. 25, 1848	5, 531
Car-wheels, Annealing-pit for annealing	W. J. Cochran	Baltimore, Md	Nov. 3, 1868	83, 605
Car-wheels, Apparatus for casting	S. H. Whitaker	Covington, Ky	Nov. 5, 1867	70, 486
Car-wheels, Apparatus for regulating the contraction of.	S. Truscott	Columbia, Pa	July 16, 1850	7, 515
Car-wheels, Casting	L. H. Allen	Tamaqua, Pa	June 10, 1856	15, 109
Car-wheels, Casting	J. W. Ells	Pittsburgh, Pa	July 25, 1871	117, 397
Car-wheels, Casting	D. Finley	Champlain, N. Y	May 4, 1858	20, 151
Car-wheels, Casting	J. K. and N. B. Morange	Pittsburgh, Pa	May 16, 1871	114, 958
Car-wheels, Casting	A. W. Moses and J. H. Springer	Philadelphia, Pa	Nov. 8, 1859	26, 045
Car-wheels, Casting	A. A. Needham	Rockford, Ill	Dec. 22, 1857	18, 924
Car-wheels, Casting	C. Needham	Worcester, Mass	Jan. 3, 1871	110, 779
Car-wheels, Casting	R. Poole	Baltimore, Md	Apr. 20, 1858	20, 022
Car-wheels, Casting	J. K. Sax	Pittston, Pa	Jan. 9, 1872	122, 538
Car-wheels, Casting	J. Segmüller	Pittsburgh, Pa	Mar. 5, 1872	124, 223
Car-wheels, Casting	N. Washburn	Worcester, Mass	Apr. 23, 1872	125, 917
Car wheels, Casting railway	N. Aylsworth	Rochester, N. Y	Mar. 24, 1857	16, 868
Car-wheels, Center-chill for	J. W. Latcher	Albany, N. Y	Aug. 7, 1866	56, 958
Car-wheels, Chill for casting	T. Sharp	Chicago, Ill	Oct. 6, 1863	40, 190
Car-wheels, Chill for casting	W. Wilmington	Toledo, Ohio	Dec. 15, 1868	85, 046
Car-wheels, Clamp to be used in the manufacture of wrought-iron.	H. Aiken	Franklin, N. H	Apr. 2, 1850	7, 229
Car-wheels, Constructing	W. M. Arnold	New York, N. Y	June 4, 1872	127, 544
Car-wheels, Constructing	J. K. and N. B. Morange	Pittsburgh, Pa	Sept. 6, 1870	107, 083
Car wheels, Constructing railway	W. R. Thomson	Cleveland, Ohio	June 10, 1856	15, 092
Car wheels, Constructing railway	A. Tiers	Kensington, Pa	Dec. 2, 1835	......
Car-wheels, Cooling	R. Poole	Baltimore, Md	July 13, 1858	20, 924
Car-wheels, Cooling cast-iron	J. M. Sigourney	Watertown, N. Y	Sept. 18, 1855	13, 585
Car-wheels, Cope for casting	A. Alling	Chicago, Ill	Dec. 27, 1870	110, 534
Car-wheels, Core-box for	T. Maher	Cleveland, Ohio	Oct. 19, 1869	95, 919
Car-wheels, Device for annealing	H. W. Moore	Bridgeport, Conn	Dec. 5, 1865	51, 338
Car-wheels, Drilling and boring	D. Walker	Gunpowder, Md	Dec. 28, 1832	....
Car-wheels, Fastening the disk and rim of	A. Breaer	Saugatuck, Conn	May 30, 1854	10, 983
Car-wheels, Flask for casting	W. R. Thomas	Catasauqua, Pa	Oct. 29, 1867	70, 290
Car-wheels, Machine for forging	W. M. Lee	New York, N. Y	Jan. 2, 1855	12, 135
Car-wheels, Machine for making cast-steel	J. B. Tarr	Chicago, Ill	July 23, 1867	67, 000
Car-wheels, Machine for making wrought-iron	N. Starks	Albany, N. Y	Oct. 29, 1850	7, 746
Car-wheels, Machine for making wrought-iron	J. C. Vaughn	Greenbush, N. Y	June 24, 1851	8, 173
Car-wheels, Making	S. Vanstone	Providence, R. I	June 4, 1867	65, 454
Car-wheels, Making locomotive-engine and railway	H. R. Dunham	New York, N. Y	Feb. 15, 1838	611
Car-wheels, Making wrought-iron	G. B. Hartson	New York, N. Y	July 11, 1854	11, 243
Car-wheels, Manner of constructing railway	W. Creed	Boston, Mass	Feb. 24, 1843	2, 979
Car-wheels, Manufacture of	E. Finch	Liverpool, England	Aug. 21, 1849	6, 657

Index of patents issued from the United States Patent Office from 1790 *to* 1873, *inclusive*—Continued.

Invention.	Inventor.	Residence.	Date.	No.
Car-wheels, Manufacture of	J. Scoville	Buffalo, N. Y	June 26, 1866	55, 916
Car wheels, Manufacture of railway	J. D. Murphy	Pottsville, Pa	May 16, 1865	47, 720
Car-wheels, Manufacturing cast-iron	G. S. Bosworth	Troy, N. Y	Oct. 19, 1858	21, 863
Car-wheels, Metal for and mode of manufacturing	W. G. Hamilton	New York, N. Y	Dec. 22, 1868	85, 089
Car-wheels, Method of attaching and lubricating	W. R. Reece	Tremont, Pa	Feb. 13, 1866	52, 603
Car-wheels, Method of constructing	C. Kingsland	McKeesport, Pa	Dec. 17, 1867	72, 405
Car wheels, Method of making cast-iron railway	E. A. Lester	Boston, Mass	Aug. 10, 1844	3, 700
Car-wheels, Method of regulating the contraction of.	J. Murphy	Kensington, Pa	Aug. 7, 1849	6, 633
Car-wheels, Mode of cutting the seats, slots, or grooves for the reception of keys in the hubs and axles of.	T. J. Butler	Johnstown, Pa	Dec. 31, 1839	1, 458
Car-wheels, Mode of lubricating	W. Youmans	Lansingburgh, N. Y	June 13, 1865	48, 230
Car-wheels, Mode of making cast-iron	J. Bonny, C. Bush, and G. C. Lobdell.	Wilmington, Del	Mar. 17, 1838	637
Car-wheels, Mode of making cast-iron	H. Mooers	Beaver Meadows, Pa	Mar. 10, 1838	634
Car-wheels, Mode of making cast-iron	S. Truscott, G. Wolf, and J. Dougherty.	Columbia, Pa	Mar. 17, 1838	640
Car wheels, Mode of making cast-iron railway	W. W. Pennell	Lancaster, Pa	Feb. 8, 1839	1, 076
Car-wheels, Mold for casting	J. B. Tarr	Chicago, Ill	Nov. 5, 1867	70, 482
Car-wheels, Mold for casting	J. B. Tarr	Fairhaven, Mass	Sept. 16, 1873	142, 821
Car-wheels, Mold for casting	N. Washburn	Worcester, Mass	June 20, 1871	116, 246
Car-wheels, Mounting and securing	W. A. Lewis	Chicago, Ill	Mar. 19, 1872	124, 835
Car-wheels, Tempering the tread of	J. A. Woodbury	Boston, Mass	Apr. 9, 1872	125, 651
Car-wheels to axles, Applying	H. Gartside	Baltimore, Md	Aug. 6, 1872	130, 124
Car-wheels to axles, Mechanism for adjusting and securing.	G. G. Lobdell	Wilmington, Del	May 2, 1871	114, 452
Car-wheels to axles, Mode of attaching	G. M. Palmer	Clinton, Mass	Jan. 7, 1862	34, 070
Car-wheels to axles, Mode of attaching	G. Sewell	Brooklyn, N. Y	Jan. 30, 1866	52, 328
Car-wheels to axles, Mode of attaching	R. S. Tarrey	Bangor, Me	July 5, 1864	43, 440
Car-wheels to axles, Securing	A. M. Leod and G. G. Lobdell	Wilmington, Del	May 24, 1870	103, 484
Car-wheels to different gages, Means for adjusting	W. B. Snow	New York, N. Y	Aug. 23, 1870	106, 737
Car-wheels, Treating cast iron for the manufacture of.	H. M. Woodward	Saint Louis, Mo	Nov. 10, 1868	84, 041
Car-wheels upon axles, Adjusting	J. Kaiser	Wilmington, Del	Feb. 20, 1872	123, 913
Car-wheels upon axles, Mode of adjusting	H. Helm	Allegheny City, Pa	Nov. 7, 1865	50, 823
Car-wheels upon axles, Securing	J. L. Mott	New York, N. Y	June 6, 1854	11, 041
Car, Wheelbarrow dumping	H. J. Peters	Quebec, Canada	Apr. 15, 1873	137, 862
Car-window	J. Beetle	New Bedford, Mass	May 27, 1856	14, 945
Car-window	G. S. Knapp	Winona, Minn	Jan. 18, 1870	98, 981
Car-window	W. McCaull	Philadelphia, Pa	Feb. 20, 1872	123, 833
Car-window	G. W. Perry	Wilmington, Del	Sept. 21, 1869	95, 134
Car-window deflector	W. Conard	Burlington, N. J	Sept. 6, 1870	107, 163
Car window fixture, Railway	J. D. Hall	Philadelphia, Pa	Oct. 18, 1864	44, 770
Car-window protector	H. G. Carr	Lewistown, Pa	June 4, 1867	65, 342
Car window, Railway	J. K. Andrews	Antrim, Ohio	Oct. 9, 1866	58, 710
Car window, Railway	T. W. Emery	Buffalo, N. Y	Mar. 15, 1864	41, 911
Car window, Railway	G. Mann, jr	Ottawa, Ill	Jan. 5, 1864	41, 079
Car window, Railway	E. A. Schoeller	Connellsville, Pa	Mar. 14, 1871	112, 639
Car window, Railway	G. Spencer	Utica, N. Y	Aug. 15, 1854	11, 535
Car-window screen, Railway	J. E. Earle	New Haven, Conn	Dec. 30, 1873	146, 053
Car-window screen, Railway	F. U. Stokes	Cincinnati, Ohio	Jan. 22, 1867	61, 367
Car-window shield	H. S. Hale	Philadelphia, Pa	Dec. 30, 1873	146, 066
Car-window stop, Adjustable	C. Page	Meriden, Conn	July 22, 1873	141, 165
Car-window, Ventilating	C. B. Knevals	New York, N. Y	Dec. 23, 1873	145, 805
Car-window, Ventilating	G. Mann, jr	Ottawa, Ill	Nov. 7, 1865	50, 831
Car-window ventilator and cinder and dust guard	O. C. Rife	North Manchester, Ind	July 1, 1873	140, 541
Car-windows, Dust-guard for	O. C. Rife	North Manchester, Ind	Jan. 28, 1873	135, 365
Car windows, Opening and shutting device for railway.	J. F. Lash	Toronto, Canada	Nov. 15, 1870	109, 326
Car-windows, Portable deflector for	J C. Stoddard	Worcester, Mass	May 1, 1866	54, 436
Car-windows, Spring-attachment to	G. Cornwall	Brooklyn, N. Y	Jan. 21, 1873	135, 086
Car, Wrecking	G. Herrick	Nashville, Tenn	Aug. 21, 1866	57, 323
Cars, Adjustable canopy for railway	I. E. Jones	Cincinnati, Ohio	May 17, 1859	24, 030
Cars, Adjustable draw-head for railway	D. P. Corey and J. Crane, jr	Crawford, N. J	May 30, 1871	115, 439
Cars, Adjusting coupling-links of railway	T. Andrews	North Easton, Mass	Feb. 5, 1861	31, 289
Cars, Air-spring for railway	G. M. Alsop	Philadelphia, Pa	May 31, 1859	24, 184
Cars and applying brakes, Disconnecting railway	J. Buck	Fitchburgh, Mass	Sept. 30, 1856	15, 833
Cars and canal-boats up and down inclined planes, Conveying railway.	G. Brown	New York, N. Y	Oct. 6, 1837	299
Cars and carriages by horse-power, Propelling	H. G. Vanderwerken	Greenbush, N. Y	Dec. 1, 1857	18, 781
Cars and carriages, Spring for railway	P. G. Gardiner	New York, N. Y	Apr. 26, 1859	23, 766
Cars and locomotives, Mode of preventing jarring and jolting of railway.	R. A. Riley	Greenfield, Ind	Feb. 25, 1862	34, 525
Cars and locomotives, Self-separating link for connecting railway.	C. H. Hunt and W. Browne	Fredericksburgh, Va	Dec. 26, 1837	538
Cars and locomotives when without steam, Windlass for moving.	C. Page	West Meriden, Conn	Oct. 5, 1858	21, 696
Cars and other vehicles, Apparatus for heating	C. C. Converse	Brooklyn, N. Y	Feb. 19, 1867	62, 254
Cars, Apparatus for bending and punching frames of draw-heads for railway.	P. L. Weimer	Lebanon, Pa	Oct. 31, 1865	50, 755
Cars, Apparatus for city-railway	S. Green	Lambertville, N. J	July 12, 1859	24, 776
Cars, Apparatus for feeding and watering stock in	W. Stark, J. G. Fisher, and S. Fitch.	White Pigeon, Mich., and Toledo, Ohio.	Nov. 21, 1871	121, 214
Cars, Apparatus for forming bumper-carriers for railway.	W. C. Allison	Philadelphia, Pa	June 25, 1867	65, 987
Cars, Apparatus for receiving and delivering mail on railway.	A. Jordan	Washington, D. C	Oct. 10, 1865	50, 435
Cars, Apparatus for supplying pneumatic springs to railway.	S. G. Randall	New York, N. Y	Nov. 8, 1864	44, 972
Cars, Applying fly-wheels to hand	C. T. Kipp and J. Lawrenson	New York, N. Y	July 14, 1857	17, 793
Cars, Applying steam-power to street	L. W. Langdon	Northampton, Mass	Nov. 30, 1869	97, 300
Cars, Arranging couches in railway	E. C. Knight	Philadelphia, Pa	June 28, 1859	24, 563
Cars, Attaching railway	L. Reckering and J. Lightner	Boston, Mass	June 28, 1836	......
Cars, Automatic stop for mining	J. Tamblin	Virginia City, Nev	Dec. 8, 1868	84, 778
Cars, Basket-back for railway	W. G. Creamer	Brooklyn, N. Y	July 6, 1869	92, 174
Cars, Bearing for railway	D. R. Pratt	Worcester, Mass	Oct. 30, 1860	30, 547
Cars, Boat for transporting railway	J. Wheelock	Lancaster, N. Y	Jan. 25, 1859	22, 759

Index of patents issued from the United States Patent Office from 1790 *to* 1873, *inclusive*—Continued.

Invention.	Inventor.	Residence.	Date.	No.
Cars, boats, &c., by condensed air, Propelling	A. McGrew	Cincinnati, Ohio	Oct. 27, 1835	....
Cars, Body-bolster for railway	A. Ward	Altoona, Pa	June 21, 1859	24, 512
Cars, Bolster for railway	J. Christy	Philadelphia, Pa	Feb. 5, 1867	61, 713
Cars, Bow-spring for railway	T. F. Allyn	Nyack, N. Y	Sept. 8, 1868	81, 969
Cars, Box for railway	R. Levington	Monroe, Mich	Apr. 17, 1849	6, 348
Cars, Brake-block for railway	L. Paige	Cavendish, Vt	Jan. 16, 1855	12, 245
Cars, Brake-block for railway	W. Wilmore	Scranton, Pa	Feb. 2, 1869	86, 613
Cars, Brake for checking and starting	R. Grant	New York, N. Y	Oct. 10, 1854	11, 785
Cars, Brake for stopping and starting	G. E. Burt and E. A. Hildreth	Harvard, Mass	Sept. 10, 1867	68, 601
Cars, Brake-head for railway	N. P. Stevens	Keene, N. H	Apr. 19, 1859	23, 722
Cars, Breakage-signal for railway	S. Jackson	Philadelphia, Pa	Jan. 2, 1872	122, 386
Cars, Brick-drier for	C. Chambers, jr	Philadelphia, Pa	Mar. 30, 1869	88, 274
Cars, Buffer for railway	R. D. Chatterton	Bath, England	Feb. 9, 1864	41, 483
Cars, Buffer-head and draw-bar for	J. T. Wilson	Pittsburgh, Pa	Apr. 4, 1871	113, 608
Cars, Buffer-spring for railway	A. H. Rowand	Allegheny City, Pa	Sept. 29, 1863	40, 122
Cars, Bumper arrangement for uncoupling railway	W. O. George	Richmond, Va	Oct. 7, 1856	15, 839
Cars by compressed air, Apparatus for propelling street.	W. H. H. Bowers	Franklin, Ky	Sept. 2, 1873	142, 433
Cars by "safety-box," Strengthening axles of railway.	P. and W. C. Allison	Philadelphia, Pa	Nov. 3, 1841	2, 322
Cars by stationary power, Propelling	A. R. Crihfield	Lincoln, Ill	Mar. 12, 1872	124, 557
Cars, Cattle feeding and watering device for railway	W. Reid	Granton, North Britain	Mar. 17, 1868	75, 701
Cars, Center-plate for railway	G. W. Bennett	White Haven, Pa	Mar. 3, 1868	75, 115
Cars, Change-gate for railway	J. B. Slawson	New York, N. Y	Jan. 4, 1868	73, 396
Cars, Coiled spring for railway	C. French	Seymour, Conn	Oct. 7, 1856	15, 869
Cars, Collision-apparatus for railway	J. Kulinski	Charleston, S. C	Sept. 9, 1856	15, 699
Cars, Combined brake and propelling mechanism for	J. W. Hill	Jefferson, Iowa	July 29, 1873	141, 272
Cars, Connecting-link for railway	R. Hemming	Boston, Mass	Oct. 11, 1845	4, 233
Cars, Connection for railway	J. R. Crabill	La Crosse, Ill	Apr. 4, 1871	113, 633
Cars, Connection of brakes with	J. Kimball and H. Rice	Concord, N. H	Jan. 1, 1859	6, 988
Cars, Constructing	S. J. Seely	Brooklyn, N. Y	May 12, 1863	38, 507
Cars, Constructing and applying bumpers and draft-springs on railway.	F. M. Ray	Catskill, N. Y	July 29, 1841	2, 201
Cars, Constructing and connecting railway	R. Grant	Philadelphia, Pa	July 22, 1837	294
Cars, Constructing railway	L. I. Germain	Catskill, N. Y	May 7, 1839	1, 145
Cars, Construction of	J. E. Leeper	Godfrey, Ill	June 20, 1871	116, 069
Cars, Construction of iron railway	S. Merrick	New Brighton, Pa	Oct. 2, 1866	58, 448
Cars, Construction of peat	J. Bundy	West Liberty, Iowa	May 19, 1868	78, 051
Cars, Construction of railway	C. M. Atkins	Pottsville, Pa	Oct. 28, 1862	36, 762
Cars, Construction of railway	W. A. Davis	Baltimore, Md	Oct. 5, 1838	963
Cars, Construction of railway	D. H. Dotterer	Philadelphia, Pa	Feb. 14, 1865	46, 341
Cars, Construction of railway	K. Thompson and S. D. Beerbrower.	Oxford, Pa	Dec. 2, 1873	145, 255
Cars, Construction of railway	W. G. Von Straden	Strathroy, Canada	Sept. 24, 1872	131, 725
Cars, Contrivance for protecting passengers in railway.	S. F. Holbrook	Boston, Mass	Feb. 21, 1854	10, 547
Cars, Coupling and uncoupling	W. A. Herrick	Leeds, Me	June 19, 1860	28, 751
Cars, Coupling-head for railway	R. M. Hughes	Oxford, Pa	Feb. 24, 1863	37, 787
Cars, Deck-sash or ventilator-opener for railway	J. E. Cross	Adrian, Mich	July 22, 1873	141, 037
Cars, Deflector for railway	J. A. Rockwood	Normal, Ill	Jan. 9, 1872	122, 655
Cars, Detective register for doors of railway	W. C. Smith	Georgetown, D. C	Apr. 26, 1859	23, 791
Cars, Device for applying steam-power to street-railway.	W. W. Crane	Auburn, N. Y	May 13, 1873	138, 793
Cars, Device for changing the gage of railway	D. Todd	Detroit, Mich	Jan. 30, 1872	123, 308
Cars, Device for discharging grain from railway	S. W. Hawes	Jersey City, N. J	Oct. 17, 1871	120, 065
Cars, Device for partially excluding dust from railway.	V. P. Corbett	New York, N. Y	Nov. 13, 1855	13, 779
Cars, Device for propelling street	S. Jones and B. Terfloth	New Orleans, La	July 5, 1870	105, 088
Cars, Device for turning street	B. I. Day	Vanderburgh County, Ind	Jan. 12, 1869	85, 732
Cars, Door-catch for railway	C. Graham	Kingston, Pa	Apr. 16, 1872	125, 808
Cars, Double and single tree attachment for street	J. F. Lowe	Louisville, Ky	Nov. 8, 1870	109, 029
Cars, Draft-bar for horse	J. Trent	Milleıton, N. Y	Feb. 8, 1870	99, 730
Cars, Draft-bar for railway	H. J. Lombaert	Philadelphia, Pa	Mar. 12, 1861	31, 670
Cars, Draft-pole for horse	S. A. Otis	Boston, Mass	Dec. 5, 1871	121, 656
Cars, Draw-bar iron for buffers of railway	J. T. Wilson	Pittsburgh, Pa	June 11, 1872	127, 819
Cars, Draw-head and bumper for railway	D. Shaaber	Reading, Pa	June 13, 1871	115, 986
Cars, Draw-head for railway	W. S. Shotwell	Paterson, N. J	July 30, 1867	67, 368
Cars, Draw-spring for railway	R. A. Wilder	Cressona, Pa	May 19, 1863	38, 618
Cars, Driving-gear for hand	J. D. Hinckley	Adrian, Mich	Apr. 15, 1873	137, 922
Cars, Dumping-platform for railway	J. W. Harrison	Niles, Mich	Oct. 21, 1873	143, 758
Cars, Dust-arrester for railway	W. M. K. Thornton	Rolla, Mo	Apr. 12, 1870	101, 787
Cars, Dust-deflector for railway	J. P. O. Lownsdale	Portland, Oreg	June 29, 1869	91, 947
Cars, Dust-deflector for windows of railway	J. M. Cook	Taunton, Mass	Oct. 16, 1855	13, 676
Cars, Dust-shield for railway	W. M. K. Thornton	Saint Louis, Mo	Oct. 3, 1871	119, 545
Cars, Easing shocks in stopping railway	C. Davenport	Cambridge, Mass	Sept. 9, 1835	....
Cars, Electro-motor for	S. Jones	New Orleans, La	Jan. 31, 1871	111, 348
Cars, Elevator for loading	R. D. Chatterton	Bath, England	Feb. 9, 1864	41, 482
Cars, Elliptic cushion for railway	S. R. Jones	York, Pa	Apr. 27, 1858	20, 070
Cars, Equalizing beam and lever in railway	J. H. Dennis	Louisville, Ky	Nov. 26, 1861	33, 776
Cars, Excluding dust from railway	J. M. Cook	Taunton, Mass	Apr. 19, 1853	9, 671
Cars, Excluding dust from railway	E. Hamilton	Bridgeport, Conn	May 27, 1851	8, 121
Cars, Excluding dust from railway	O. Newton and J. A. Crever	Pittsburgh, Pa	Mar. 14, 1854	10, 626
Cars, Excluding dust from railway	E. C. Salisbury	New York, N. Y	July 31, 1855	13, 364
Cars, Excluding dust from railway	J. Wood	Jersey City, N. J	Feb. 26, 1856	14, 331
Cars, Extension holding-strap for street	T. Cogswell	Boston, Mass	Mar. 13, 1866	53, 116
Cars, Folding-seat for street-railway	K. Egan	New York, N. Y	Apr. 29, 1873	138, 321
Cars for preserving and transporting meat, fish, and vegetables, Construction of railway.	C. F. Pike	Providence, R. I	Dec. 31, 1867	72, 895
Cars, Framing railway	T. M. Mullen	Philadelphia, Pa	Apr. 3, 1860	27, 734
Cars from being thrown from the track, Machine for preventing engines and railway.	A. L. Johnson	Baltimore, Md	Nov. 1, 1859	25, 970
Cars from one track to another, Method of transferring.	J. Denhard	Reading, Pa	June 25, 1867	66, 132
Cars from one track to another, Mode of switching off railway.	M. Semple	Philadelphia, Pa	June 7, 1859	24, 336
Cars from one track to another, Transferring	W. Wharton, jr	Philadelphia, Pa	Sept. 18, 1860	30, 100
Cars from one track to another, Transferring railway.	J. Ashenfelder	Philadelphia, Pa	Oct. 9, 1860	30, 284

Index of patents issued from the United States Patent Office from 1790 *to* 1873, *inclusive*—Continued.

Invention.	Inventor.	Residence.	Date.	No.
Cars from running off the track, Preventing	L. Ball	Auburn, N. Y	June 22, 1858	20, 614
Cars from the track, Device for removing	D. R. Erdmann	Philadelphia, Pa	Sept. 5, 1865	49, 737
Cars, Furnace for railway	H. M. Hutchinson	Baltimore, Md	Mar. 13, 1860	27, 449
Cars, Grain-door for	L. F. Frazee	Jersey City, N. J	May 9, 1871	114, 661
Cars, Grain-door for	H. Stahlnecker	Allentown, Pa	Jan. 23, 1872	123, 055
Cars, Grain-door for	H. Stahlnecker	Allentown, Pa	Jan. 23, 1872	123, 056
Cars, Grain-door for railway	M. H. Card and L. Safford	Freeport and Chicago, Ill	Aug. 22, 1871	118, 339
Cars, Grain-door for railway	G. B. Rich	La Fayette, Ind	Mar. 19, 1867	62, 972
Cars, Guard for railway	F. M. Daunoy	New Orleans, La	Mar. 14, 1867	64, 639
Cars, Guide-wheel for railway	J. B. Wickersham	New York, N. Y	Mar. 3, 1857	16, 762
Cars, Head-rest for railway	R. Hamilton	Franklin, Ind	Feb. 26, 1867	62, 484
Cars, Head-rest for railway	P. A. La France and J. D. Dinsmore.	Elmira, N. Y	Sept. 11, 1866	57, 926
Cars, Head-rest for railway	W. B. Slaughter	Chicago, Ill	Aug. 19, 1856	15, 581
Cars, Head-supporter for railway	J. N. Williams	Dubuque, Iowa	Feb. 13, 1855	12, 400
Cars, Heating and ventilating apparatus for railway.	C. F. Allen and L. W. Campbell.	Aurora, Ill	Nov. 5, 1867	70, 495
Cars, Heating and ventilating railway	D. H. Dotterer	Philadelphia, Pa	Feb. 25, 1868	74, 899
Cars, Heating and ventilating railway	S. Lloyd	Washington, D. C	Feb. 11, 1868	74, 386
Cars, Heating and ventilating railway	T. H. B. Sanders	Philadelphia, Pa	May 29, 1866	55, 168
Cars, Heating railway	R. A. Chesebrough	New York, N. Y	Jan. 28, 1868	73, 780
Cars, Heating railway	A. C. Crary	Utica, N. Y	May 12, 1868	77, 871
Cars, Hoisting and dumping coal	G. Martz	Pottsville, Pa	Apr. 3, 1855	12, 633
Cars, &c., Implement for sealing railway	F. W. A. Krause	Baltimore, Md	Oct. 13, 1857	18, 400
Cars, &c., India-rubber spring for	T. B. De Forest	New York, N. Y	Aug. 16, 1859	25, 100
Cars, Instrument for coupling railway	A. McOmber and M. Ward	Schenectady, N. Y	Mar. 8, 1870	100, 542
Cars, Jaw for dump and coal	J. M. Foss	Saint Albans, N. Y	June 25, 1872	128, 219
Cars, Lavatory attachment for railway	J. G. Murdock	Cincinnati, Ohio	Oct. 26, 1869	96, 259
Cars, Lever for railway	J. Noble	Rochester, N. Y	Nov. 19, 1867	71, 046
Cars, Lifting-jack for moving rail	N. B. Carpenter and J. Powers	New York, N. Y	Feb. 6, 1855	12, 345
Cars, Lighting railway	A. Longstreet	Chicago, Ill	Apr. 27, 1869	89, 539
Cars, Link for coupling railway	G. W. Putnam	South Glen's Falls, N. Y	Oct. 1, 1872	131, 905
Cars, Link-guide for coupling	N. L. Post	East Cleveland, Ohio	Feb. 20, 1872	123, 930
Cars, Loading and unloading baggage and freight from.	O. C. Brown	Iberia, Ohio	May 6, 1873	138, 557
Cars, Loading cattle upon	L. O. Cottle	Cedar Rapids, Iowa	June 18, 1872	127, 961
Cars, Loading railway flat	G. W. Penniston	North Vernon, Ind	May 6, 1873	138, 577
Cars, Lock for doors of baggage	T. Slaight	Newark, N. J	Sept. 11, 1866	57, 986
Cars, Lock for railway	T. Slaight	Newark, N. J	Feb. 5, 1861	31, 332
Cars, Locking-door for railway	E. W. Morse	Chicago, Ill	Nov. 22, 1864	45, 165
Cars, locomotives, &c., Spring for railway	W. Duff	Baltimore, Md	Jan. 9, 1841	1, 928
Cars, Machine for bending truck-sides for railway	J. H. Helm	Pittsburgh, Pa	Sept. 23, 1873	143, 010
Cars, Machine for moving railway	C. W. T. Krausch	Chicago, Ill	Mar. 3, 1863	37, 818
Cars, Machine for replacing railway	H. N. Degraw	Piermont, N. Y	Jan. 29, 1856	14, 154
Cars, Machine for unloading railway	J. Dable	Chicago, Ill	July 28, 1868	80, 276
Cars, Machine for propelling	G. I. Beauregard	New Orleans, La	Nov. 30, 1869	97, 343
Cars, Machinery for propelling railway	J. Busser and J. B. Harmer	Philadelphia, Pa	May 8, 1860	28, 153
Cars, Machine for unloading flat	G. S. Caldwell	Auburn, N. Y	May 6, 1873	138, 476
Cars, Machinery for unloading railway	J. Dable	Chicago, Ill	Aug. 7, 1866	56, 906
Cars, Metallic safety-seat for railway	H. Martin	Chicago, Ill	Jan. 15, 1867	61, 222
Cars, Method of preventing dust, &c., from entering the windows of railway.	P. M. Pyfer	Baltimore, Md	Mar. 10, 1857	16, 806
Cars, Method of self-ventilation for railway	V. P. Corbett	Corbettsville, N. Y	Mar. 20, 1855	12, 541
Cars, Mode of applying steam as a motor to city-railway.	W. Darker, jr	Philadelphia, Pa	Nov. 22, 1859	26, 230
Cars, Mode of arranging couches in railway	E. C. Knight	Philadelphia, Pa	Sept. 27, 1859	25, 570
Cars, Mode of attaching stakes to railway	J. Fortun	Cienfuegos, Island of Cuba	Nov. 28, 1865	51, 278
Cars, Mode of collecting letters on street-railway	J. B. Murray	New York, N. Y	May 20, 1862	35, 323
Cars, Mode of confining the driver's seat in city-railway.	W. C. Allison	Philadelphia, Pa	Nov. 8, 1859	26, 066
Cars, Mode of constructing the heating and lighting apparatus on railway.	A. J. Gibson	Cincinnati, Ohio	Dec. 8, 1868	84, 740
Cars, Mode of coupling and uncoupling	S. H. Hamilton	Bushnell, Ill	Sept. 5, 1865	49, 754
Cars, Mode of dumping railway	W. Pearce and J. Lowrie	Piedmont, Va	Apr. 21, 1857	17, 114
Cars, Mode of fastening and unfastening drop-doors in coal.	W. Burt	Marquette, Mich	Dec. 11, 1866	60, 340
Cars, Mode of loading dirt	C. Phillips	Detroit, Mich	Jan. 8, 1856	14, 066
Cars, Mode of operating and dumping earth	R. Ray	Louisport, Ky	Sept. 4, 1855	13, 528
Cars, Mode of operating railway	C. E. Willis	New York, N. Y	Oct. 4, 1864	44, 574
Cars, Mode of switching street	J. S. Reid	Muncie, Ind	July 25, 1865	48, 978
Cars, Mode of watering cattle on railway	W. Robinson	Rochester, N. Y	Oct. 29, 1861	33, 605
Cars off the track, Rail for switching	J. C. Mather	New York, N. Y	Aug. 24, 1858	21, 266
Cars, Oil-box for railway	E. F. Hurlbut and R. S. Potter	Chicago, Ill	May 10, 1864	42, 732
Cars on rails, Apparatus for retaining	W. Payne	New York, N. Y	Mar. 5, 1850	7, 146
Cars on railways, Machinery for moving	A. Foster and H. Brown	New York, N. Y	Jan. 18, 1859	22, 638
Cars on railways, Mode of propelling	C. T. Harvey	Tarrytown, N. Y	May 8, 1866	54, 537
Cars on railways, Propelling	J. A. Bennet	Kings County, N. Y	Oct. 9, 1860	30, 287
Cars on the track, Adjustable rail for replacing	C. Perley	New York, N. Y	June 13, 1854	11, 079
Cars on the track, Adjustable rail for replacing	J. A. Stephan	La Fayette, Ind	Oct. 11, 1859	25, 771
Cars on the track, Device for moving	J. W. Pettengill	Rockford, Ill	Sept. 25, 1866	58, 285
Cars on track, Machine for replacing railway	J. White	Boston, Mass	June 22, 1858	20, 675
Cars, Operating dumping	A. C. Johnson	Meadville, Pa	Apr. 18, 1854	10, 805
Cars, Operating dumping	R. Ray	Louisport, Ky	Sept. 4, 1855	13, 529
Cars over elevations, Taking	S. Cram	New York, N. Y	June 11, 1836	......
Cars over obstructions, Conveying city-railway	P. I. Biderman	Philadelphia, Pa	Oct. 9, 1860	30, 289
Cars, Passenger-support for city	J. Hanley	New York, N. Y	Dec. 13, 1864	45, 497
Cars, Pedestal for railway	D. H. Feger	New York, N. Y	Aug. 4, 1857	17, 953
Cars, Pedestal for railway	I. P. Wendell and S. Ustick	Philadelphia, Pa	Oct. 10, 1865	50, 427
Cars, Pedestal for street railway	D. B. Hart	Saint Louis, Mo	Apr. 23, 1872	126, 054
Cars, Pipe-coupling for heating	H. R. Robbins	Baltimore, Md	Jan. 18, 1870	99, 003
Cars, Pipe-coupling for heating and ventilating railway.	F. R. Hunt	Leavenworth, Kans	Mar. 15, 1870	100, 896
Cars, Platform between railway	J. Newman	Baltimore, Md	July 26, 1859	24, 885
Cars, Plow for unloading	J. McMullin	Casey, Iowa	Jan. 28, 1873	135, 351
Cars, Pole-adjuster for horse	A. G. Safford	Boston, Mass	Apr. 25, 1871	114, 205
Cars, Pole for horse	T. Pendergast	Saint Louis, Mo	Dec. 29, 1868	85, 473
Cars. Pole for horse-railway	D. F. Robinson	Boston, Mass	Dec. 4, 1866	60, 250
Cars. Portable seat for drivers upon	J. F. Campbell and C. Tinney	Williamsburgh, N. Y	May 21, 1867	64, 837

Index of patents issued from the United States Patent Office from 1790 to 1873, inclusive—Continued.

Invention.	Inventor.	Residence.	Date.	No.
Cars, Preventing dust from entering railway	D. S. Darling	Brooklyn, N. Y	Jan. 10, 1854	10, 402
Cars, Propelling	I. Avery	Tunkhannock, Pa	Sept. 25, 1847	5, 308
Cars, Propelling	C. T. Harvey	Tarrytown, N. Y	Apr. 16, 1867	63, 888
Cars, Propelling	C. T. Harvey	Tarrytown, N. Y	May 21, 1867	64, 862
Cars, Propelling	W. J. Sage	Steubenville, Ohio	Jan. 28, 1862	34, 269
Cars, Propelling railway	S. G. Randall	New York, N. Y	May 23, 1865	47, 858
Cars, Propelling street	H. C. Bull and B. Bloomfield	New Orleans, La	Mar. 5, 1872	124, 327
Cars, Propelling street	J. W. Conway	Madison, Ind	Aug. 28, 1866	57, 623
Cars, Propelling street	G. S. Petry	Troy Grove, Ill	Apr. 9, 1867	63, 652
Cars, Propelling street	J. L. Simms and W. T. Duvall	Georgetown, D. C	Apr. 2, 1872	125, 224
Cars, Propulsion of street	C. M. Mann	Detroit, Mich	June 6, 1865	48, 081
Cars, Pull-iron for horse	A. S. Jimmerson	New York, N. Y	Jan. 5, 1869	85, 670
Cars, Register for horse	C. B. Angell	Coventry, R. I	May 12, 1863	38, 456
Cars, Register for street	J. B. Greenhut	Chicago, Ill	Dec. 12, 1865	51, 453
Cars, Retracker for railway	J. D. Potts	Pittsburgh, Pa	Nov. 26, 1861	33, 795
Cars, Rolling shoes for replacing	W. L. Gilroy	Philadelphia, Pa	May 26, 1863	38, 671
Cars, Running-gear for city-railway	J. Grice and R. H. Long	New York, N. Y., and Philadelphia, Pa.	Feb. 21, 1860	27, 217
Cars, Running-gear for railway	S. B. Childs	Syracuse, N. Y	Sept. 25, 1866	58, 216
Cars, Running-gear for railway	T. A. Davies	New York, N. Y	Dec. 9, 1851	8, 576
Cars, Running-gear for railway	J. H. Dennis	Louisville, Ky	Jan. 14, 1862	34, 134
Cars, Running-gear for railway	S. B. Driggs	New York, N. Y	Mar. 14, 1865	46, 784
Cars, Running-gear for railway	R. Hornbrook	Cincinnati, Ohio	Apr. 24, 1860	27, 982
Cars, Running-gear for railway	S. Howard	Elyria, Ohio	May 28, 1861	32, 423
Cars, Running-gear for railway	J. Ingersoll	Grafton, Ohio	Aug. 24, 1858	21, 259
Cars, Running-gear for railway	I. Knight	Baltimore, Md	Mar. 28, 1848	5, 493
Cars, Running-gear for railway	J. S. Lester	Atlanta, Ga	Oct. 7, 1873	143, 516
Cars, Running-gear for railway	W. Nebinger	Sharpsburgh, Md	Oct. 21, 1851	8, 451
Cars, Running-gear for railway	J. R., D. W., and J. Perry	Wilkesbarre, Pa	Feb. 9, 1869	86, 685
Cars, Running-gear for railway	D. G. Smith	Carbondale, Pa	Dec. 5, 1854	12, 036
Cars, Running-gear for railway	J. Stephenson	New York, N. Y	July 25, 1865	49, 004
Cars, Running-gear for railway	J. Stimpson	Baltimore, Md	Oct. 23, 1834	....
Cars, Running-gear for railway	H. D. Taylor	Newark, N. J	Feb. 3, 1852	8, 710
Cars, Running-gear for railway	H. H. Trenor	New York, N. Y	Oct. 17, 1865	50, 518
Cars, Running-gear for railway	T. Wilson	Silver Creek, Ill	Feb. 25, 1862	34, 544
Cars, Running-gear for street	Z. Eastman	Chicago, Ill	May 25, 1869	90, 513
Cars, Running-gear for street-railway	A. McNair	Newark, N. J	May 13, 1862	35, 244
Cars, Safety-attachment for railway	H. A. Newhall	Newton, Mass	May 18, 1858	20, 293
Cars, Safety-attachment for street-railway	J. Fogarty	Brooklyn, N. Y	Sept. 7, 1869	94, 587
Cars, Safety-attachment for railway	G. W. Brady	New York, N. Y	Mar. 3, 1868	74, 980
Cars, Safety-attachment for railway	C. Mahon	Washington, D. C	Oct. 23, 1855	13, 705
Cars, Safety-brake for railway	F. Wolf	Philadelphia, Pa	Sept. 20, 1864	44, 361
Cars, Safety bridge and gate for railway	A. H. Allen	Hartford, Conn	Aug. 13, 1867	67, 701
Cars, Safety-bridge for railway	L. Hempstead and L. S. Hills	Hartford, Conn	Aug. 21, 1866	57, 453
Cars, Safety-bridge for railway	E. Sturgeon	Columbiana, Ohio	Aug. 15, 1871	118, 167
Cars, Safety-bridge for railway	L. Traxler	Butler, Ohio	July 2, 1867	66, 419
Cars, Safety-coupling for railway	E. H. Anderson	Milford, Del	Feb. 3, 1857	16, 520
Cars, Safety-guard for railway	J. G. Crocker	Utica, N. Y	Jan. 1, 1856	14, 014
Cars, Safety-guard for railway	T. Gillen	Philadelphia, Pa	June 23, 1863	39, 011
Cars, Safety-guard for railway	S. Males	Cincinnati, Ohio	June 11, 1867	65, 582
Cars, Safety-guard for railway	W. Siefert	New York, N. Y	Jan. 29, 1867	61, 637
Cars, Safety-guard for railway	T. Walter	Philadelphia, Pa	Feb. 23, 1869	87, 126
Cars, Safety-guard for railway	J. A. Wilkinson	Wilson, N. Y	July 26, 1870	105, 753
Cars, Safety-hatch for railway	R. Liston	Albany, N. Y	Feb. 18, 1873	135, 923
Cars, Safety-padlock for railway	G. W. Stevens	Albany, N. Y	Jan. 7, 1868	73, 133
Cars, Safety-platform between railway	J., jr., and S. V. Kline	Chicago, Ill	Apr. 15, 1856	14, 665
Cars, Safety-platform for	H. E. Marchand	Pittsburgh, Pa	May 6, 1873	138, 669
Cars, Safety-platform for railway	R. Strode	Coatesville, Pa	Nov. 18, 1873	144, 804
Cars, Safety-step for railway	B. F. Beckwith, J. H. Rynerson, and A. N. Clark.	Clayton, Ind	Mar. 5, 1872	124, 318
Cars, Safety-top for railway	A. Potts	Philadelphia, Pa	Aug. 11, 1857	17, 982
Cars, Sand-distributer for railway	C. M. Bromwick	South Boston, Mass	Dec. 19, 1865	51, 646
Cars, Scraper-attachment to	E. B. Wells	Northampton, Mass	Jan. 7, 1868	73, 212
Cars, Seal-bolt for railway	P. H. Mann and G. P. Terry	Albany, N. Y	Dec. 15, 1868	84, 891
Cars, Seal-bolt for railway	J. P. Moore	Boston, Mass	Dec. 1, 1868	84, 503
Cars, &c., Seal for railway freight	H. D. Mears and W. Houlton, jr	Baltimore, Md	July 14, 1857	17, 801
Cars, &c., Seal for railway freight	H. D. Mears and W. Houlton, jr	Baltimore, Md	July 14, 1857	17, 802
Cars, Seat for driver and conductor of street	T. G. Hambly	San Francisco, Cal	Nov. 7, 1865	50, 819
Cars, Settee for railway-passenger	J. B. Easland	Bridgeport, Conn	Sept. 2, 1862	36, 343
Cars, Shackle for railway	H. A. Barns	Milwaukee, Wis	May 13, 1862	35, 257
Cars, Singletree for horse	J. Wills	Rock Island, Ill	Apr. 15, 1873	137, 990
Cars, Sleeping-chair for railway	J. Danner	Canton, Ohio	Dec. 27, 1859	26, 571
Cars, Sliding door for railway box	J. Bassler	Galesburgh, Ill	Aug. 2, 1870	105, 888
Cars, Sliding door for railway	A. G. Safford	Boston, Mass	Aug. 8, 1865	49, 308
Cars, Spiral spring for railway	J. Murray	New York, N. Y	Oct. 15, 1872	132, 213
Cars, Spiral spring for railway	F. M. Ray	New York, N. Y	Apr. 18, 1854	10, 784
Cars, Spring craft and bumper for railway	P. Alberson	New Haven, Conn	Sept. 8, 1838	908
Cars, Spring-platform for railway	C. H. Lewis	Malden, Mass	Mar. 25, 1856	14, 508
Cars, Spring-seat for railway	T. J. Gifford	Salem, Mass	Nov. 1, 1870	108, 777
Cars, Sprinkling-attachment for railway	H. Mitchell	Cincinnati, Ohio	June 19, 1860	28, 764
Cars, Stake-holder for platform	E. A. Eddy	Racine, Wis	July 18, 1865	48, 799
Cars, Stake-holder for platform	E. A. Eddy	Racine, Wis	July 18, 1865	48, 800
Cars, Stake-holder for railway	T. A. Slack	Peoria County, Ill	Dec. 10, 1867	72, 097
Cars, Standard for lumber	H. Jacob	Loretto, Pa	Apr. 5, 1864	42, 194
Cars, Starting	H. H. Covert	Detroit, Mich	Aug. 28, 1866	57, 624
Cars, Starting	A. F. French	Franklin, Vt	Aug. 1, 1865	49, 195
Cars, Starting	J. McMurtry	Lexington, Ky	Feb. 25, 1868	74, 927
Cars, Starting and stopping	A. G. Crossman	Huntington, N. Y	Nov. 12, 1867	70, 814
Cars, Starting and stopping	J. Phillips, D. W. Southwick, and D. A. Arnold.	Pawtucket, R. I	Oct. 22, 1867	70, 114
Cars, Starting and stopping	P. Rhoads	Carlisle, Pa	Mar. 8, 1871	112, 494
Cars, Starting and stopping city-railway	J. H. Wygant	Hackensack, N. J	Aug. 7, 1860	29, 536
Cars, Starting and stopping street	E. M. Scott	Auburn, N. Y	Apr. 30, 1867	64, 373
Cars, Starting-apparatus for horse	A. Cary	Worcester, Mass	Jan. 21 1862	34, 199
Cars, Starting-apparatus for railway	E. Woodward and J. S. Millett	Charlestown, Mass	Sept. 29, 1868	82, 671
Cars, Starting-apparatus for street	T. B. Jordan	Gloucester, N. J	Apr. 9, 1867	63, 726
Cars, Starting city-railway	G. P. Frick	Baltimore, Md	June 7, 1859	24, 293
Cars, Starting city-railway	G. Hamel	Abington, Pa	Nov. 15, 1859	26, 161
Cars, Starting horse-railway	W. Henderson and J. W. Fowle	Boston, Mass	May 16, 1865	47, 722

Index of patents issued from the United States Patent Office from 1790 *to* 1873, *inclusive*—Continued.

Invention.	Inventor.	Residence.	Date.	No.
Cars, Starting railway	W. Lawton	New York, N. Y	Sept. 5, 1865	49, 770
Cars Starting railway	W. Palmer	New York, N. Y	Oct. 3, 1854	11, 759
Cars, Starting railway	T. R. Sinclaire	New York, N. Y	Dec. 19, 1865	51, 626
Cars, Starting street	J. Adams	Newark, Del	Feb. 5, 1867	61, 699
Cars, Starting street	A. S. Armstrong	Saint Bernard Parish, La	Feb. 26, 1867	62, 304
Cars, Starting street	J. S. Briggs	South Bend, Ind	Feb. 25, 1862	34, 477
Cars, Starting street	T. F. Kums and W. W Burson	Rockford, Ill	Sept. 11, 1866	57, 925
Cars, State-room for railway	W. Brown	Duncannon, Pa	Feb. 4, 1868	73, 949
Cars, Steam-brake for railway	S. N. Goodale	Cleveland, Ohio	May 30, 1865	47, 943
Cars, Steam-brake for railway	N. Nilson	Minneapolis, Minn	Nov. 11, 1873	144, 410
Cars, Steam-coupling for railway	S. A. Appold	Baltimore, Md	Aug. 23, 1870	106, 648
Cars, Steam-heater for railway	C. A. Mabie, T. H. Whitcomb, and J. T. Brown.	Albion, N. Y	Jan. 2, 1872	122, 326
Cars, Steam-pipe coupling for railway	J. Conner	Philadelphia, Pa	Dec. 13, 1870	110, 114
Cars, steamers, &c., Apparatus for illuminating railway.	W. Foster, jr., and G. P. Ganster.	New York, N. Y	Dec. 8, 1868	84, 814
Cars, Step for street	A. A. Duly	New York, N. Y	Jan. 11, 1870	98, 750
Cars, Stopping and starting	T. S. Bigelow	Lake Mills, Wis	June 30, 1863	39, 091
Cars, Stopping and starting	E. T. Colburn	Boston, Mass	Nov. 5, 1867	70, 528
Cars, Stopping and starting	J. Higgin	Manchester, England	Jan. 8, 1861	31, 080
Cars, Stopping and starting	C. S. Hunt	Terre Bonne Parish, La	Mar. 10, 1868	75, 426
Cars, Stopping and starting	J. E. Kelsey	Providence, R. I	May 1, 1866	54, 363
Cars, Stopping and starting city-railway	E. S. Ritchie	Brookline, Mass	July 3, 1860	29, 007
Cars, Stopping and starting railway	J. A. Emerick	Philadelphia, Pa	Apr. 23, 1861	32, 127
Cars, Stopping and starting railway	F. C. Kutt	Hackensack, N. J	Aug. 21, 1860	29, 700
Cars, Stopping and starting railway	P. Louis	New York, N. Y	Jan. 15, 1861	31, 140
Cars, Stopping and starting railway	B. Morohan	Brooklyn, N. Y	Jan. 15, 1861	31, 124
Cars, Stopping and starting railway	J. Steger	New York, N. Y	Sept. 15, 1863	39, 999
Cars, Stopping and starting street	J. M. Starr	Richmond, Ind	Nov. 11, 1873	144, 484
Cars, Street-indicator for city	W. Brown	Duncannon, Pa	Sept. 20, 1870	107, 440
Cars, Support and advertising medium for street	M. Warne	Philadelphia, Pa	Oct. 31, 1871	120, 471
Cars, Support for passengers in	H. S. Vrooman	Boston, Mass	Apr. 28, 1868	77, 338
Cars, Suspending upper berths in sleeping	E. H. Paine	Louisville, Ky	Jan. 11, 1870	98, 792
Cars, Swing-jack for railway	J. H. Clark	Westbrook, Me	Dec. 26, 1865	51, 696
Cars, Switch for changing gage of	T. Fogg	St. Mary's, Canada	Apr. 25, 1871	114, 123
Cars, Switching-apparatus for street-railway	P. S. Dusouchet	New Orleans, La	Aug. 11, 1868	80, 932
Cars through tunnel, Method of transporting	J. H. Crane	Charlestown, Mass	May 14, 1867	64, 636
Cars to railway-tracks of different gages, Mode of applying.	C. D. Tisdale	East Boston, Mass	Mar. 10, 1863	37, 889
Cars to turn short curves, Railway track or rail for	H. M. Naglee	Philadelphia, Pa	Aug. 22, 1840	1, 726
Cars, Tongue-support for railway street	J. M. Tiernan	Pittsburgh, Pa	Dec. 17, 1867	72, 430
Cars, Trussed-spring bolster for railway	J. Anderson	Wilmington, N. C	Aug. 12, 1873	141, 618
Cars, Turning railway	A. Shermer	Philadelphia, Pa	Sept. 9, 1835	....
Cars, Unloading	J. Dahle	Chicago, Ill	Nov. 16, 1869	96, 781
Cars, Unloading coal and other	A. Patrick	Alleghany County, Md	Aug. 15, 1854	11, 530
Cars, Unloading grain	D. J. Whittemore	Milwaukee, Wis	Oct. 16, 1866	58, 928
Cars, Unloading grain from	E. M. Clark	Detroit, Mich	Nov. 8, 1864	44, 937
Cars, Unloading railway	F. Haase and W. Rost	Oak Park, Ill	June 11, 1867	65, 743
Cars upon railway-track, Replacing	L. B. Flanders	Dunkirk, N. Y	Dec. 6, 1853	10, 301
Cars upon springs, Mode of suspending	A. Bridges	Newton, Mass	June 12, 1866	55, 457
Cars upon the track, Apparatus for replacing railway.	S. P. Coon	Milwaukee, Wis	June 19, 1855	13, 080
Cars upon the track, Manner of replacing railway	S. H. Bean	Philadelphia, Pa	Oct. 11, 1845	4, 232
Cars upon the track, Mechanism for retaining	G. P. Ketcham	Bedford, Ind	Feb. 6, 1855	12, 371
Cars upon the track, Spiral wheel for replacing railway.	R. F. R. Lewis	Annapolis, Md	Apr. 10, 1855	12, 684
Cars upon the track, Switch for replacing	N. Pullman	New Oregon, Iowa	May 1, 1866	54, 478
Cars, Ventilating and excluding dust from railway	R. Cook	Saratoga Springs, N. Y	Aug. 19, 1851	8, 298
Cars, Ventilating and excluding dust from railway	A. B. Spencer	Rochester, N. Y	May 4, 1858	20, 176
Cars, Ventilating and warming railway	A. J. Marshall	Warrenton, Va	Aug. 20, 1867	67, 894
Cars, Ventilating grain	W. S. Sampson	New York, N. Y	Oct. 17, 1871	120, 109
Cars, Ventilating railway	J. G. Allen	Philadelphia, Pa	Aug. 12, 1873	141, 617
Cars, Ventilating railway	C. Atwood	Birmingham, Conn	July 10, 1855	13, 204
Cars, Ventilating railway	N. S. Barnum and L. Whitney	New Haven, Conn	Jan. 6, 1852	8, 622
Cars, Ventilating railway	H. Bradford and E. Morris	New York, N. Y	June 4, 1850	7, 409
Cars, Ventilating railway	S. A. Clemens	Springfield, Mass	Nov. 22, 1853	10, 251
Cars, Ventilating railway	J. Cunningham	Reading, Pa	Mar. 12, 1850	7, 158
Cars, Ventilating railway	G. F. Foote	Buffalo, N. Y	July 11, 1854	11, 268
Cars, Ventilating railway	C. D. Gibson	New York, N. Y	July 7, 1863	39, 140
Cars, Ventilating railway	T. S. Lambert	Peekskill, N. Y	Dec. 8, 1863	40, 843
Cars, Ventilating railway	H. L. B. Lewis	Philadelphia, Pa	Oct. 30, 1855	13, 725
Cars, Ventilating railway	W. H. Medcalfe	Baltimore, Md	Jan. 22, 1856	14, 139
Cars, Ventilating railway	G. Neilson	Boston, Mass	Jan. 24, 1854	10, 449
Cars, Ventilating railway	W. Pauli	Alexandria, Va	May 8, 1855	12, 827
Cars, Ventilating railway	C. Pepper	Albany, N. Y	Apr. 20, 1858	20, 021
Cars, Ventilating railway	C. Reed and B. K. Mould	Chicago, Ill	Aug. 8, 1854	11, 494
Cars, Ventilating railway	H. J. Ruttan	Coburg, Canada West	Jan. 9, 1866	52, 009
Cars, Ventilating railway	A. F. Smith and W. Wagner	Norwich, Conn., and Palatine Bridge, N. Y.	Sept. 23, 1862	36, 536
Cars, Ventilating railway	J. K. Taylor	Binghamton, N. Y	Oct. 30, 1855	13, 732
Cars, Ventilating railway	R. R. Taylor	Reading, Pa	Feb. 22, 1859	23, 049
Cars, Ventilating-window for railway	S. Darling	Providence, R. I	Aug. 23, 1870	106, 558
Cars, Ventilating-window for railway	R. Monroe, E. Stone, and E. St. John.	Elmira, N. Y	Dec. 19, 1865	51, 607
Cars, Ventilating-window for railway	M. C. Murphy	Boston, Mass	June 21, 1870	104, 486
Cars, Ventilating-window for railway	G. Neilson	Boston, Mass	May 30, 1854	10, 9?0
Cars, Ventilating-window for railway	H. M. Paine	Worcester, Mass	Jan. 6, 1852	8, 645
Cars, Warming passenger	J. B. Johnson	Lynn, Mass	June 17, 1862	35, 612
Cars, Warming railway	A. C. Crary	Utica, N. Y	Apr. 13, 1869	88, 8 2
Cars, Warming railway	G. W. Eddy	Waterford, N. Y	Mar. 31, 1868	76, 175
Cars, Warming railway	T. N. Morse	Fairhaven, Mass	Nov. 16, 1869	96, 945
Cars, Warming railway	T. Townsend	Albany, N. Y	July 24, 1846	4, 654
Cars, Water-can for railway	L. T. Richer	New York, N. Y	Apr. 3, 1866	53, 748
Cars, Water-tank for railway stock	J. B. Calkins	Pacific, Mo	Feb. 25, 1873	136, 132
Cars with steam, Apparatus for heating railway	G. W. Samson	Washington, D. C	July 16, 1861	32, 843
Cars, Wooden mat for	W. Groat	Green Island, N. Y	June 12, 1866	55, 489
Caramels, Manufacture of dry	T. Hyatt	Philadelphia, Pa	Mar. 13, 1866	53, 249
Carbine-socket	J. S. P. Taylor	Oxford, Ohio	June 27, 1865	48, 462

Index of patents issued from the United States Patent Office from 1790 *to* 1873, *inclusive*—Continued.

Invention.	Inventor.	Residence.	Date.	No.
Carbon and other oils, Vessel for transporting	R. C. Glyde	Pittsburgh, Pa	Feb. 11, 1862	34,359
Carbon, Apparatus for manufacturing sulphuret of	E. Deiss	Paris, France	Apr. 27, 1858	20,047
Carbon-battery connection	A. S. Ogden	Newark, N. J	June 26, 1866	55,892
Carbon-battery connector	C. T. Chester	Hackensack Township, N. J	Nov. 28, 1865	51,144
Carbon, Black pigment for mineral	P. O'Reilly	Hartford, Conn	June 8, 1869	91,038
Carbon for use in the arts, Preparation for mineral	J. Dickinson	Bay Ridge, N. Y	June 1, 1869	90,824
Carbon, Mode of baking articles composed of	D. G. B. Fowler	New York, N. Y	Nov. 23, 1858	22,115
Carbon-tool	H. and J. L. Young	Middletown, Conn., and New York, N. Y.	Oct. 12, 1869	95,866
Carbonating and dispensing artificial mineral-waters, Mode of.	D. and T. Kolb	Washington, D. C	Feb. 23, 1869	87,177
Carbonic-acid engine	N. H. Barbour	Auburn, N. Y	Mar. 14, 1865	46,769
Carbonic-acid-gas engine	J. C. Salomon, jr	Cincinnati, Ohio	Dec. 9, 1851	8,577
Carbonic-acid-gas generator	O. Zwietusch	Milwaukee, Wis	Dec. 23, 1873	145,774
Carbonic oxide for treating metals, Generating and applying.	T. Shaw	Philadelphia, Pa	Nov. 17, 1868	84,220
Carboys and bottles, Machine for forming rings on	T. Barrett	Charlestown, Mass	June 30, 1868	79,434
Carboys, Machine for finishing	L. Hyde	Ellenville, N. Y	Feb. 22, 1859	23,063
Carboys, Tilting-stand for	A. W. Caverly	New York, N. Y	Nov. 11, 1873	144,442
Carbureter	H. C. Appleby	Conneaut, Ohio	June 23, 1868	79,048
Carbureter	J. D. Averell	New York, N. Y	July 30, 1872	130,004
Carbureter	J. D. Averell	New York, N. Y	Dec. 24, 1872	134,240
Carbureter	W. W. Bierce	Cleveland, Ohio	Jan. 7, 1868	73,073
Carbureter	A. F. H. Braun	San Francisco, Cal	June 25, 1872	128,356
Carbureter	M. P. Coons	Brooklyn, N. Y	Aug. 11, 1868	80,918
Carbureter	W. H. Covel	New York, N. Y	June 15, 1869	91,213
Carbureter	J. R. Cross	New York, N. Y	Apr. 30, 1872	126,189
Carbureter	O. P. Drake	Boston, Mass	Oct. 1, 1872	131,814
Carbureter	C. F. Dunderdale	New York, N. Y	May 25, 1869	90,436
Carbureter	C. F. Dunderdale	New York, N. Y	Mar. 1, 1870	100,274
Carbureter	G. Edmonds	New Orleans, La	May 23, 1871	115,182
Carbureter	W. H. Elston	Marshall, Mich	Feb. 25, 1873	136,228
Carbureter	J. B. Fish	Providence, Pa	May 21, 1872	127,039
Carbureter	F. A. Fisher	Crawford, N. J	June 4, 1872	127,409
Carbureter	F. A. Fisher	Elizabeth, N. J	July 22, 1873	140,998
Carbureter	T. B. Fogarty	New York, N. Y	Feb. 18, 1873	135,981
Carbureter	T. B. Fogarty	Warren, Mass	Feb. 18, 1873	135,982
Carbureter	E. J. Fraser	Erie, Pa	Aug. 20, 1867	67,971
Carbureter	C. N. Gilbert, J. F. Barker, and E. N. Ives.	Springfield, Mass	Jan. 8, 1867	61,004
Carbureter	T. Holmes	Brooklyn, N. Y	Feb. 11, 1873	135,806
Carbureter	H. Holton	Saint Louis, Mo	Apr. 2, 1872	125,194
Carbureter	E. S. Hutchinson and H. L. McAvoy.	Baltimore, Md	Jan. 9, 1866	51,946
Carbureter	J. B. Hyde	New York, N. Y	July 16, 1872	129,566
Carbureter	M. W. Kidder	Lowell, Mass	Sept. 13, 1870	107,268
Carbureter	J. F. G. Kromschroeder	London, England	Apr. 1, 1873	137,307
Carbureter	J. F. and G. E. Lockwood	Taylorsville, Ill	Sept. 2, 1873	142,545
Carbureter	J. B. Lyman	Rockford, Ill	Dec. 30, 1873	146,082
Carbureter	W. T. McMillen	Cincinnati, Ohio	Oct. 7, 1873	143,523
Carbureter	J. Millward	Fayetteville, N. Y	Aug. 16, 1870	106,389
Carbureter	E. L. Mix	Rochester, N. Y	May 11, 1869	90,012
Carbureter	J. Musgrave	Cincinnati, Ohio	Nov. 25, 1873	144,858
Carbureter	T. W. Ofeldt	Newark, N. J	Sept. 17, 1872	131,369
Carbureter	G. H. Peacock	Fairport, N. Y	Aug. 27, 1867	68,231
Carbureter	F. S. Pease	Buffalo, N. Y	Nov. 6, 1866	59,446
Carbureter	C. H. Pierson	Plainfield, N. J	May 28, 1872	127,366
Carbureter	F. Ransom	Buffalo, N. Y	July 30, 1867	67,216
Carbureter	J. F. M. Rigod	Paris, France	Oct. 8, 1872	132,025
Carbureter	I. W. Shaler	Brooklyn, N. Y	Oct. 7, 1873	143,534
Carbureter	I. Simmons	Baltimore, Md	Dec. 2, 1873	145,248
Carbureter	H. Slatter	Covington, Ky	Sept. 22, 1868	82,359
Carbureter	J. B. Terry	Brooklyn, N. Y	Dec. 17, 1872	133,957
Carbureter	W. Thompson and J. E. Hall	Cleveland, Ohio	Dec. 3, 1867	71,665
Carbureter	A. W. Wilkinson	New York, N. Y	Feb. 6, 1872	123,539
Carbureter	A. W. Wilkinson	New York, N. Y	June 17, 1873	140,105
Carbureter	C. B. Willoughby	Uhricksville, Ohio	June 23, 1868	79,290
Carbureter	H. Woodward	London, England	Aug. 18, 1868	81,238
Carbureter, Air	E. S. Archer	New York, N. Y	Sept. 6, 1864	44,060
Carbureter, Air	J. F. Barker	Springfield, Mass	Apr. 21, 1868	76,880
Carbureter, Air	J. F. Barker	Springfield, Mass	June 6, 1871	115,562
Carbureter, Air	J. F. Barker and C. N. Gilbert	Springfield, Mass., and New York, N. Y.	Aug. 3, 1869	93,267
Carbureter, Air	J. F. Barker and C. N. Gilbert	Springfield, Mass., and New York, N. Y.	Aug. 3, 1869	93,268
Carbureter, Air	J. F. Barker and C. N. Gilbert	Springfield, Mass., and New York, N. Y.	Aug. 17, 1869	93,795
Carbureter, Air	J. A. Bassett	Salem, Mass	Mar. 14, 1865	46,771
Carbureter, Air	J. A. Bassett	Salem, Mass	Apr. 18, 1865	47,272
Carbureter, Air	J. A. Bassett	Salem, Mass	Oct. 31, 1865	50,675
Carbureter, Air	C. C. Beers	Boston, Mass	Apr. 25, 1871	113,968
Carbureter, Air	A. D. Bell	San Francisco, Cal	Feb. 28, 1871	112,111
Carbureter, Air	D. Bickford	Boston, Mass	Nov. 28, 1865	51,128
Carbureter, Air	J. F. Birchard	Milwaukee, Wis	July 11, 1865	48,706
Carbureter, Air	F. H. Brown	Chicago, Ill	June 26, 1866	55,949
Carbureter, Air	F. H. Brown	Chicago, Ill	June 26, 1866	55,950
Carbureter, Air	W. H. Burkland	London, England	Feb. 14, 1865	46,432
Carbureter, Air	H. A. Chapin	New York, N. Y	Jan. 24, 1871	111,175
Carbureter, Air	J. Chase	Windsor Locks, Conn	Nov. 14, 1865	50,905
Carbureter, Air	J. Chase	Windsor Locks, Conn	Nov. 14, 1865	50,987
Carbureter, Air	W. H. Clarke	Saint Anthony's Falls, Minn.	Mar. 26, 1867	63,215
Carbureter, Air	W. F. Cozzens and J. H. Jones	Saint Louis, Mo	Nov. 12, 1867	70,809
Carbureter, Air	H. G. Dayton	Maysville, Ky	May 21, 1872	127,031
Carbureter, Air	S. R. Divine	New York, N. Y	Sept. 4, 1866	57,686
Carbureter, Air	B. Douglas and W. H. Walton	New York and Brooklyn, N. Y.	Jan. 29, 1867	61,656

Index of patents issued from the United States Patent Office from 1790 *to* 1873, *inclusive*—Continued.

Invention.	Inventor.	Residence.	Date.	No.
Carbureter, Air	O. P. Drake	Boston, Mass	May 6, 1862	35, 144
Carbureter, Air	O. P. Drake	Boston, Mass	Apr. 24, 1866	54, 132
Carbureter, Air	C. M. Drennan	Boston, Mass	Sept. 26, 1865	50, 103
Carbureter, Air	C. F. Dunderdale	New York, N. Y	Nov. 30, 1869	97, 283
Carbureter, Air	E. Dunscomb	Boston, Mass	May 9, 1865	47, 679
Carbureter, Air	H. F. Eberts and J. Fanning	Detroit, Mich	Nov. 30, 1869	97, 285
Carbureter, Air	H. Fairbanks	Boston, Mass	Apr. 10, 1866	53, 798
Carbureter, Air	R. B. Fitts	Philadelphia, Pa	May 16, 1871	114, 787
Carbureter, Air	T. F. Frank	Buffalo, N. Y	Nov. 27, 1866	59, 991
Carbureter, Air	F. Hainsworth	Chicago, Ill	June 27, 1865	48, 391
Carbureter, Air	D. Hurd	Chicago, Ill	May 2, 1865	47, 550
Carbureter, Air	E. S. Hutchinson and H. L. McAvoy.	Baltimore, Md	Feb. 5, 1867	61, 739
Carbureter, Air	J. H. Irwin	Chicago, Ill	Apr. 11, 1865	47, 256
Carbureter, Air	J. H. Irwin	Chicago, Ill	Aug. 22, 1865	49, 526
Carbureter, Air	J. H. Irwin	Chicago, Ill	Oct. 3, 1865	50, 250
Carbureter, Air	J. H. Irwin	Chicago, Ill	Oct. 3, 1865	50, 251
Carbureter, Air	J. H. Irwin	Chicago, Ill	June 25, 1867	66, 153
Carbureter, Air	J. H. Irwin and I. Simmons	Chicago, Ill	Apr. 11, 1865	47, 258
Carbureter, Air	T. Judd and E. Doty	Janesville, Wis	Apr. 29, 1873	138, 409
Carbureter, Air	C. B. Loveless	Syracuse, N. Y	Jan. 2, 1866	51, 841
Carbureter, Air	C. B. Loveless	Syracuse, N. Y	Apr. 10, 1866	53, 843
Carbureter, Air	W. M. Marshall	Philadelphia, Pa	July 7, 1868	79, 667
Carbureter, Air	H. L. McAvoy	Baltimore, Md	Aug. 23, 1864	43, 948
Carbureter, Air	H. L. McAvoy	Baltimore, Md	Nov. 22, 1864	45, 206
Carbureter, Air	H. L. McAvoy	Baltimore, Md	Dec. 13, 1864	45, 456
Carbureter, Air	H. L. McAvoy	Baltimore, Md	Feb. 7, 1865	46, 302
Carbureter, Air	H. L. McAvoy	Baltimore, Md	Sept. 19, 1865	50, 076
Carbureter, Air	H. L. McAvoy	Baltimore, Md	Aug. 14, 1866	57, 164
Carbureter, Air	H. L. McAvoy	Baltimore, Md	Sept. 11, 1866	57, 940
Carbureter, Air	D. McDonald	Albany, N. Y	Aug. 21, 1866	57, 442
Carbureter, Air	S. T. McDougall	New York, N. Y	June 5, 1866	55, 324
Carbureter, Air	J. McGeary	Salem, Mass	Oct. 16, 1866	58, 861
Carbureter, Air	P. Mihan	Boston, Mass	July 11, 1865	48, 772
Carbureter, Air	P. Mihan	Boston, Mass	Aug. 28, 1866	57, 543
Carbureter, Air	M. J. A. Mille	Paris, France	Aug. 22, 1865	49, 596
Carbureter, Air	G. Odiorne	Boston, Mass	Nov. 1, 1864	44, 883
Carbureter, Air	H. Oertel	Memphis, Tenn	Sept. 13, 1870	107, 403
Carbureter, Air	A. Patterson	Birmingham, Pa	Oct. 2, 1866	58, 471
Carbureter, Air	F. S. Pease	Buffalo, N. Y	Aug. 6, 1867	67, 576
Carbureter, Air	R. H. Plass	New York, N. Y	June 21, 1870	104, 642
Carbureter, Air	J. T. and R. H. Plass	New York, N. Y	Sept. 15, 1868	82, 244
Carbureter, Air	E. A. Pond	Rutland, Vt	Oct. 17, 1865	50, 491
Carbureter, Air	E. A. Pond and M. S. Richardson.	Rutland, Vt	Sept. 19, 1865	50, 029
Carbureter, Air	E. A. Pond and M. S. Richardson.	Rutland, Vt	Mar. 27, 1866	53, 481
Carbureter, Air	A. W. Porter	New York, N. Y	Sept. 3, 1872	131, 025
Carbureter, Air	A. C. Rand	Union Mills, Pa	Feb. 26, 1867	62, 364
Carbureter, Air	G. Rex	Philadelphia, Pa	May 9, 1871	114, 709
Carbureter, Air	G. Reznor	Mercer, Pa	Mar. 26, 1872	125, 085
Carbureter, Air	J. Richard	New York, N. Y	Nov. 3, 1868	83, 730
Carbureter, Air	J. F. Rowley, W. M. Sloane, and J. E. Woodruff.	Buffalo, N. Y	Aug. 28, 1866	57, 639
Carbureter, Air	J. Sangster	Buffalo, N. Y	Mar. 10, 1868	75, 468
Carbureter, Air	J. Sangster	Buffalo, N. Y	Mar. 10, 1868	75, 469
Carbureter, Air	I. W. Shaler	Brooklyn, N. Y	Nov. 16, 1869	96, 942
Carbureter, Air	W. A. Simonds	Boston, Mass	Mar. 21, 1865	46, 976
Carbureter, Air	W. A. Simonds	Boston, Mass	Aug. 15, 1865	49, 448
Carbureter, Air	B. Sloper	Saint Louis, Mo	June 12, 1871	115, 988
Carbureter, Air	J. F. Spence	Brooklyn, N. Y	Jan. 16, 1866	52, 087
Carbureter, Air	J. F. Spence	Williamsburgh, N. Y	Sept. 4, 1866	57, 788
Carbureter, Air	J. F. Spence	Brooklyn, N. Y	Aug. 2, 1870	105, 994
Carbureter, Air	J. F. Spence and L. D. Tousley	Newark, N. J	Jan. 25, 1870	99, 274
Carbureter, Air	J. H. Springer and J. C. McDonald.	Philadelphia, Pa	July 16, 1867	66, 749
Carbureter, Air	L. Stevens	Fitchburgh, Mass	Oct. 2, 1866	58, 559
Carbureter, Air	L. Stevens	Fitchburgh, Mass	Nov. 6, 1866	59, 473
Carbureter, Air	L. Stevens	Fitchburgh, Mass	Nov. 6, 1866	59, 474
Carbureter, Air	L. Stevens	Fitchburgh, Mass	May 28, 1867	65, 296
Carbureter, Air	J. Stratton	Philadelphia, Pa	Mar. 31, 1868	76, 114
Carbureter, Air	J. B. Terry	Auburndale, Mass	Feb. 7, 1865	46, 280
Carbureter, Air	J. B. Terry	Auburndale, Mass	Sept. 12, 1865	49, 934
Carbureter, Air	J. B. Terry	Auburndale, Mass	June 19, 1866	55, 741
Carbureter, Air	W. Thompson	Cleveland, Ohio	Mar. 27, 1866	53, 504
Carbureter, Air	E. F. Van Houten	Newark, N. J	Aug. 19, 1873	141, 968
Carbureter, Air	J. H. Van Houten	Newark, N. J	Nov. 22, 1870	109, 562
Carbureter, Air	J. S. Wood	Philadelphia, Pa	Nov. 24, 1868	84, 332
Carbureter, Air and gas	A. Barbarin	New Orleans, La	Oct. 5, 1869	95, 412
Carbureter, Air and gas	J. F. Barker and C. N. Gilbert	Springfield, Mass	July 16, 1867	66, 777
Carbureter, Air and gas	A. Bartholf	New York, N. Y	Oct. 18, 1870	108, 432
Carbureter, Air and gas	H. A. Chapin	New York, N. Y	Oct. 4, 1870	108, 005
Carbureter, Air and gas	H. G. Dayton	Maysville, Ky	Oct. 8, 1872	131, 943
Carbureter, Air and gas	O. P. Drake	Boston, Mass	Oct. 1, 1872	131, 815
Carbureter, Air and gas	J. Jonson	Toledo, Ohio	May 4, 1869	89, 665
Carbureter, Air and gas	F. H. Lutkewitte	Saint Louis, Mo	Apr. 4, 1871	113, 317
Carbureter, Air and gas	L. Marks	San Francisco, Cal	May 2, 1871	114, 316
Carbureter, Air and gas	W. J. Nichols	Buffalo, N. Y	July 13, 1869	92, 635
Carbureter, Air and gas	O. Tirrill	Brooklyn, N. Y	Aug. 22, 1871	118, 302
Carbureter, Air and gas	S. Whitney	Flushing, N. Y	June 6, 1871	115, 798
Carbureter, Air, gas, &c	A. C. Messenger	Syracuse, N. Y	June 19, 1866	55, 778
Carbureter and generator, Gas	F. A. Fisher	Westfield Township, N. J	Feb. 21, 1871	112, 026
Carbureter and regulator, Gas	J. A. Bassett	Salem, Mass	June 25, 1867	66, 067
Carbureter-burner	I. W. Shaler	Brooklyn, N. Y	Sept. 30, 1873	143, 335
Carbureter, Coal-gas	W. A. Earseman and R. W. Gray	Pittsburgh, Pa	Dec. 31, 1867	72, 825
Carbureter-feeder	E. M. Smith	New York, N. Y	Oct. 23, 1866	59, 142
Carbureter for locomotive head-lights	F. S. Pease	Buffalo, N. Y	Oct. 22, 1867	70, 014
Carbureter, Gas	J. L. Bartlett	Stockton, Cal	Oct. 4, 1870	107, 853

Index of patents issued from the United States Patent Office from 1790 *to* 1873, *inclusive*—Continued.

Invention.	Inventor.	Residence.	Date.	No.
Carbureter, Gas	J. A. Bassett	Salem, Mass	Mar. 4, 1862	34, 557
Carbureter, Gas	J. A. Bassett	Salem, Mass	July 29, 1862	35, 984
Carbureter, Gas	J. A. Bassett	Salem, Mass	Aug. 18, 1863	39, 541
Carbureter, Gas	J. A. Bassett	Salem, Mass	June 25, 1867	66, 066
Carbureter, Gas	W. W. Bierce	Cleveland, Ohio	Feb. 12, 1867	61, 918
Carbureter, Gas	W. W. Bierce	Cleveland, Ohio	July 23, 1867	66, 937
Carbureter, Gas	J. F. Boynton	Syracuse, N. Y	Sept. 25, 1866	58, 209
Carbureter, Gas	G. W. Coleman	Kalamazoo, Mich	Apr. 23, 1872	126, 024
Carbureter, Gas	G. B. Dyer	New York, N. Y	Aug. 3, 1869	93, 288
Carbureter, Gas	S. Gwynn	New York, N. Y	Apr. 28, 1863	38, 357
Carbureter, Gas	P. Hogan	Albany, N. Y	Apr. 17, 1866	53, 979
Carbureter, Gas	A. K. Johnston	New York, N. Y	Sept. 4, 1866	57, 729
Carbureter, Gas	J. Kidd	New York, N. Y	Oct. 20, 1868	83, 289
Carbureter, Gas	J. Kidd	New York, N. Y	Mar. 9, 1869	87, 682
Carbureter, Gas	W. H. Laubach	Philadelphia, Pa	May 14, 1867	64, 776
Carbureter, Gas	S. T. McDougall	New York, N. Y	Jan. 3, 1865	45, 729
Carbureter, Gas	A. W. Porter	New York, N. Y	Nov. 25, 1873	144, 863
Carbureter, Gas	W. A. Simonds	Boston, Mass	Apr. 23, 1867	64, 156
Carbureter, Gas	T. D. Worrall	New York, N. Y	Sept. 4, 1866	57, 812
Carbureter, Gas	W. M. Wright	Baltimore, Md	July 17, 1866	56, 503
Carbureter, Gas and air	A. T. Boon and A. D. Perry	Galesburgh, Ill	July 28, 1868	80, 268
Carbureter, Gas and air	J. F. Boynton	Syracuse, N. Y	Jan. 22, 1867	61, 309
Carbureter, Gas and air	M. P. Coons	Brooklyn, N. Y	July 4, 1871	116, 563
Carbureter, Gas and air	D. Hall	New York, N. Y	Apr. 2, 1867	63, 511
Carbureter, Gas and air	J. Kidd	London, England	Mar. 12, 1867	62, 855
Carbureter, Gas and air	G. W. Porter	Boston, Mass	Apr. 30, 1867	64, 361
Carbureter, Gas and air	J. S. Stephenson	Cleveland, Ohio	Mar. 26, 1867	63, 326
Carbureter, Hydrogen-gas	J. Ambuhl	Morristown, N. J	May 16, 1871	114, 744
Carbureter, Hydrogen-gas	R. V. De Guinon	Jersey City, N. J	Dec. 3, 1872	133, 569
Carbureters, Air-supplying machine for	P. Kelly	Dayton, Ohio	Nov. 9, 1869	96, 704
Carbureters, Automatic feed for	E. S. Hutchinson	Baltimore, Md	Oct. 2, 1866	58, 422
Carbureters, Blast-apparatus for	F. S. Pease	Buffalo, N. Y	June 11, 1867	65, 594
Carbureters, Blast-apparatus for	F. S. Pease	Buffalo, N. Y	June 11, 1867	65, 595
Carbureters, Capillary material for filling gas and air.	J. A. Bassett	Salem, Mass	Jan. 1, 1867	60, 670
Carbureters in railway-cars, Apparatus for forcing air into.	C. F. Dunderdale	New York, N. Y	Mar. 15, 1870	100, 737
Carbureters in railway-cars, &c., Automatic blast for.	E. S. Hutchinson	Baltimore, Md	Oct. 30, 1866	59, 226
Carbureters, &c., Machine for producing blast in gas.	E. A. Pond and M. S. Richardson.	Rutland, Vt	June 18, 1867	65, 939
Carbureters, Rotary air-wheel for gas	J. F. Spence and L. D. Towsley.	Brooklyn, N. Y., and Newark, N. J.	Sept. 20, 1870	107, 635
Carbureters, Siphon-can for	R. S. Osborn	Belleville, N. J	Feb. 20, 1872	123, 926
Carbureting air	C. Lawrence	Cincinnati, Ohio	Mar. 8, 1870	100, 534
Carbureting air	J. C. Pedrick	Washington, D. C	July 9, 1867	66, 622
Carbureting air and applying the same, Apparatus for.	A. Brin	Paris, France	Sept. 1, 1868	81, 590
Carbureting air and gas, Apparatus for	J. B. Hyde	New York, N. Y	Sept. 13, 1870	107, 262
Carbureting air and gas, Apparatus for	O. Terrill	Boston, Mass	Aug. 31, 1869	94, 360
Carbureting air and regulating its flow, Apparatus for	J. S. Wood	Philadelphia, Pa	July 9, 1867	66, 545
Carbureting air, Apparatus for	A. Barbarin and A. E. Dupas	New Orleans, La., and Paris, France.	Oct. 26, 1869	96, 074
Carbureting air, Apparatus for	A. E. Dupas and A. Barbarin	Paris, France, and New Orleans, La.	June 28, 1870	104, 716
Carbureting air, Apparatus for	J. F. Lafroque	Paris, France	Sept. 14, 1869	94, 898
Carbureting air, Apparatus for	J. D. Spang	Dayton, Ohio	Nov. 8, 1870	109, 148
Carbureting air, Combination apparatus for	L. Stevens	Fitchburgh, Mass	Sept. 10, 1867	68, 666
Carbureting air, Process of	J. H. Irwin	Chicago, Ill	Apr. 11, 1865	47, 257
Carbureting air to produce inflammable gas, Machine for.	L. Stevens	Fitchburgh, Mass	Apr. 9, 1867	63, 667
Carbureting and applying air for lighting and heating, Apparatus for.	J. Tiffany	Albany, N. Y	Nov. 2, 1869	96, 364
Carbureting air and enriching gas and for illuminating and heating, Liquids for.	J. F. Boynton	Syracuse, N. Y	Sept. 18, 1866	58, 054
Carbureting-apparatus	T. Judd and C. H. Pierson	Philadelphia, Pa	July 8, 1873	140, 711
Carbureting-apparatus	W. A. Simonds	Boston, Mass	July 12, 1870	105, 378
Carbureting apparatus, Gas	J. Gair	New York, N. Y	Feb. 15, 1870	99, 769
Carbureting-attachment for gas-burner	J. H. Butler, jr	Park Ridge, Ill	Sept. 10, 1872	131, 210
Carbureting gas	J. A. Bassett	Salem, Mass	June 25, 1867	66, 068
Carbureting gas and air	J. F. Boynton	Syracuse, N. Y	Nov. 5, 1867	70, 512
Carbureting gas and air, Liquid for	L. E. Holden	Cleveland, Ohio	Jan. 29, 1867	61, 662
Carbureting gas and oil for the same	J. Kidd	New York, N. Y	Mar. 9, 1869	87, 681
Carbureting gas for heating and illuminating, Mode of.	J. A. Bassett	Salem, Mass	June 25, 1867	66, 069
Carbureting gas from steam and hydrocarbons	W. H. Gwynne	White Plains, N. Y	July 28, 1863	39, 342
Carbureting gas, Hydrocarbon-fluid for	J. A. Bassett	Salem, Mass	June 18, 1867	65, 866
Carbureting gas-lamp	C. E. Ball	New York, N. Y	Oct. 15, 1872	132, 132
Carbureting gas, Liquid for	J. A. Bassett	Salem, Mass	May 21, 1867	64, 831
Carbureting gas, Method of	C. M. Williams	New York, N. Y	Jan. 8, 1867	61, 033
Carbureting gas, Process and material for	J. A. Bassett	Salem, Mass	Sept. 1, 1868	81, 736
Carbureting hydrogen-gas	C. B. Loveless	Syracuse, N. Y	June 13, 1871	115, 873
Carbureting illuminating-gas	W. H. Burridge	Cleveland, Ohio	Jan. 29, 1867	61, 606
Carbureting illuminating-gas	T. M. Fell	Brooklyn, N. Y	Sept. 10, 1872	131, 157
Carbureting-lamp	A. W. Porter and J. S. Gray	New York, N. Y	Feb. 20, 1872	123, 929
Carbureting-lamp	J. W. Post	New York, N. Y	Nov. 19, 1872	133, 118
Carbureting-machine	A. E. Dupas and A. Barbarin	New Orleans, La	Mar. 28, 1870	113, 147
Carbureting machine, Air	A. Barbarin	New Orleans, La	Oct. 26, 1869	96, 073
Carbureting machine, Air	T. F. Frank	Buffalo, N. Y	Oct. 20, 1868	83, 147
Carbureting machine, Air	J. H. Van Houten	Newark, N. J	Jan. 21, 1873	135, 020
Carbureting-machine, Apparatus for forcing air into.	J. F. Birchard	Milwaukee, Wis	May 8, 1866	54, 491
Carbureting wheel, Air	J. H. Van Houten	Newark, N. J	Nov. 22, 1870	109, 561
Card and brush, Combined	W. H. Prouty	Hanson, Mass	July 7, 1868	79, 596
Card and cribbage board	R. S. Jennings	Philadelphia, Pa	Jan. 14, 1868	73, 337
Card and other cylinders, Device for traversing	F. Herboth	Newark, N. J	June 3, 1873	139, 466
Card and other wire, Drawing	R. Prouty	Spencer, Mass	Jan. 19, 1822	......

Index of patents issued from the United States Patent Office from 1790 *to* 1873, *inclusive*—Continued.

Invention.	Inventor.	Residence.	Date.	No.
Card and pin-cushion, Business	D. C. Beamer	Philadelphia, Pa	Sept. 3, 1867	68, 339
Card-board and scab-board, Machine for planing	W. Kimball and S. Willard, jr	Mount Vernon, N. H	Oct. 23, 1813	....
Card-board cutter	L. Knickerbocker	Philadelphia, Pa	June 2, 1863	38, 750
Card-board drier	E. F. Bailey	Holderness, N. H	Aug. 27, 1867	68, 146
Card-board drier	E. F. Bailey	Ashland, N. H	July 2, 1872	128, 455
Card board, Playing	R. S. Jennings	Philadelphia, Pa	Aug. 27, 1867	68, 302
Card-board, Smoothing	D. Stearns	Brattleborough, Vt	Mar. 10, 1814	....
Card-boards for the handles and finishing them, Preparing.	H. Whittemore	Newton, N. Y	Apr. 27, 1832	....
Card-cabinet	A. T. Woodward	New York, N. Y	June 14, 1870	104, 391
Card-case	G. H. and J. James	London, England	July 20, 1869	92, 832
Card-case	F. A. Lamontague	Montreal, Canada	Sept. 25, 1866	58, 357
Card-case	G. V. Metzel	Baltimore, Md	July 27, 1869	92, 986
Card-case	H. H. Miller and T. Bennett	Saint Louis, Mo	June 3, 1873	139, 513
Card-clothing	A. F. Bishop, J. H. Aiken, J. M. Pendleton, and A. W. Gates.	Norwalk, Conn., and New York, N. Y.	July 6, 1869	92, 149
Card-clothing	A. W. Gates	New York, N. Y	June 21, 1870	104, 445
Card-clothing	R. Kitson	Lowell, Mass	Oct. 5, 1858	21, 685
Card-clothing clamp	J. O. Lewis	Worcester, Mass	Jan. 4, 1870	98, 507
Card clothing, Manufacture of	E. S. Lawrence	Worcester, Mass	Oct. 20, 1868	83, 179
Card-clothing, Trimming	E. B. Howe	Lowell, Mass	Dec. 23, 1856	16, 275
Card clothing upon carding-cylinders, Clamps for stretching.	J. O. Lewis	Worcester, Mass	Apr. 11, 1865	47, 210
Card cutting and assorting machine, Playing	V. E. Mauger	New York, N. Y	Nov. 21, 1871	121, 117
Card-cutting machine	E. J. Hunt	Concord, N. H	Feb. 4, 1868	74, 095
Card-cutting machinery	E. Cowles	Cleveland, Ohio	June 8, 1869	90, 931
Card-cylinder	J. M. Stone	North Andover, Mass	July 14, 1868	79, 872
Card-cylinder cleaner	J. H. Smith and E. Squires	Raritan, N. J	Sept. 10, 1872	131, 185
Card-cylinders, Grinding and cleaning	W. B. Ingram	Manchester, Conn	Apr. 6, 1869	88, 635
Card-cylinders, Machinery for grinding	J. Parker	Biddeford, Me	Mar. 3, 1857	16, 744
Card-exhibiting apparatus	I. M. Miller	Huntsville, Ala	Oct. 24, 1871	120, 309
Card-exhibitor	W. Duryea	New York, N. Y	Apr. 10, 1855	12, 674
Card, Flexible hand	G. F. Ells	Douglas, N. Y	Aug. 8, 1871	117, 755
Card for carding fibrous substances, Self stripping	H. Barbour and J. Gleason	Lowell, Mass	Dec. 4, 1844	3, 843
Card for carding wool, cotton, &c	E. Faber	Pittsburgh, Pa	Jan. 24, 1832	....
Card for hooks and eyes, Paper	D. Fowler, jr	Northford, Conn	Apr. 21, 1863	38, 259
Card for liquids, Sample	H. Mustedt	New York, N. Y	Nov. 16, 1869	96, 930
Card for social games	P. West and G. S. Lee	Worcester, Mass	Sept. 2, 1873	142, 423
Card-globe, Folding	D. Townsend	Fiddletown, Cal	Feb. 16, 1869	87, 082
Card-globe, Folding	D. Townsend	Felchville, Vt	Mar. 22, 1870	101, 185
Card-grinder	R. Kitson	Lowell, Mass	Nov. 11, 1851	8, 512
Card-grinder	B. S. Roy	Lowell, Mass	Aug. 18, 1868	81, 213
Card-grinding cylinder	J. O. Lewis	Worcester, Mass	Dec. 24, 1867	72, 509
Card-grinding machine	T. C. Kirkham	Ancaster, Canada	July 1, 1873	140, 376
Card-grinding machine	S. G. Ladd and G. W. Crown	Lowell, Mass	Mar. 26, 1867	63, 165
Card-grinding machine	J. Robb	Lawrence, Mass	Feb. 2, 1869	86, 452
Card, Hand	G. F. Ells	Troy, N. Y	Apr. 23, 1867	64, 004
Card, Hand	E. S. Ells and G. F. Ells	Fair Haven, Vt., and Troy, N. Y.	Oct. 25, 1864	44, 791
Card, Hand	R. H. Waite	Hubbardston, Mass	Feb. 11, 1868	74, 261
Card-handles, Machine for finishing	I. S. Waite	Hubbardston, Mass	Mar. 3, 1868	75, 082
Card-holder	W. I. Adams	New York, N. Y	Mar. 17, 1868	75, 507
Card-holder	S. E. Adamson	Philadelphia, Pa	Dec. 28, 1869	98, 327
Card-holder	B. Brower	New York, N. Y	Apr. 30, 1872	126, 259
Card-holder	J. Dean	Worcester, Mass	Aug. 26, 1862	36, 283
Card-holder	C. R. Doane	Brooklyn, N. Y	Dec. 21, 1869	98, 154
Card-holder	P. B. Groat	Hannibal, Mo	Dec. 17, 1872	134, 051
Card-holder	S. L. Hill	Brooklyn, N. Y	Dec. 10, 1867	71, 878
Card-holder	A. A. Marks	New York, N. Y	Sept. 25, 1866	58, 270
Card-holder	W. R. Oatley	Rochester, N. Y	Aug. 24, 1869	94, 127
Card-holder	H. H. Pember	New York, N. Y	May 21, 1867	64, 902
Card-holder	L. A. Roberts	Boston, Mass	Apr. 5, 1864	42, 260
Card-holder	W. H. Ryninger	Baltimore, Md	Oct. 15, 1872	132, 325
Card-holder	E. Stewart	Fort Madison, Iowa	July 18, 1871	117, 120
Card-holder	J. H. Caterson	Philadelphia, Pa	Oct. 21, 1873	143, 878
Card-mount	E. F. Stephens	Towanda, Pa	May 28, 1867	65, 131
Card or label holder	C. C. Blakemore	Washington, D. C	June 23, 1868	79, 194
Card or ticket case	W. I. Lindlon	Cleveland, Ohio	Dec. 19, 1871	122, 041
Card, Postal	J. R. Smith	Rockville, Canada	Oct. 15, 1872	132, 330
Card-press, Cameo	E. D. Averell	New York, N. Y	Nov. 5, 1867	70, 500
Card-rack	J. F. Curtis	Chicago, Ill	Nov. 1, 1870	108, 887
Card-rack	D. A. Danforth and F. M. Winchester.	Elkhart, Ind	May 27, 1873	139, 298
Card-rack	J. J. Gray	Boston, Mass	Jan. 28, 1873	135, 272
Card-rack	J. M. Keep	New York, N. Y	Oct. 2, 1866	58, 428
Card-rack	M. Lepp	Albany, N. Y	July 28, 1868	80, 415
Card-rack	E. R. McKean	Washington, D. C	Nov. 26, 1872	133, 378
Card-rack	G. A. Nelson	Chicago, Ill	Oct. 22, 1867	70, 107
Card-rack	L. H. Olmstead	Brooklyn, N. Y	Mar. 8, 1870	100, 656
Card-rack	L. C. Rindle	Chicago, Ill	Oct. 17, 1871	119, 995
Card-rack	E. Safford	Boston, Mass	Sept. 4, 1866	57, 776
Card-rack	H. R. Van Eps	Peoria, Ill	Apr. 29, 1873	138, 454
Card-rack	C. A. Wall	Grand Rapids, Mich	Oct. 17, 1871	120, 134
Card-rack	W. Wendell	Milwaukee, Wis	Sept. 20, 1870	107, 574
Card-rack	C. F. Wilson	Brooklyn, N. Y	Jan. 11, 1870	98, 831
Card, Reading	H. F. Bond	Waltham, Mass	Feb. 28, 1860	27, 266
Card, Reading	B. Snyder	Trenton, N. J	Aug. 19, 1862	36, 242
Card-safe	C. S. Crane	Selma, Alabama	Feb. 18, 1868	74, 509
Card-setting machine	J. Russell	Springfield, Mass	Oct. 8, 1867	69, 709
Card-setting machine	J. Russell	Springfield, Mass	Feb. 25, 1868	74, 853
Card-setting-machine cutter	D. McFarland	Worcester, Mass	Dec. 7, 1869	97, 667
Card-setting machinery	T. A. Dickinson	Worcester, Mass	Mar. 10, 1868	75, 388
Card strapper and grinder	W. H. Chandler	North Scituate, R. I	Mar. 17, 1868	75, 624
Card-stripping machinery	J. F. Foss	Lowell, Mass	May 7, 1872	126, 387
Card-stripping machines, Cam for	S. L. Crockett and B. T. Mills	Lowell, Mass	Apr. 21, 1868	76, 895
Card-stripping mechanism	G. E. Taft	Northbridge, Mass	Apr. 9, 1872	125, 501
Card-stripping mechanism, Traversing cams for	G. E. Taft	Whitinsville, Mass	July 8, 1873	140, 740

Index of patents issued from the United States Patent Office from 1790 *to* 1873, *inclusive*—Continued.

Invention.	Inventor.	Residence.	Date.	No.
Card-teeth, Bracing and supporting	C. Speer	New York, N. Y	Nov. 16, 1852	9,411
Card-teeth-cutting machine	J. Lamb	Leicester, Mass	Aug. 1, 1827	....
Card-teeth for machine-case	J. L. Tuttle	New York, N. Y	Oct. 14, 1856	15,905
Card-teeth, Machine	W. Montgomery	Roxbury, Mass	June 13, 1854	11,070
Card-teeth, Machine for grinding	G. Emerson	North Providence, R. I	Dec. 5, 1848	5,948
Card-teeth, Machine for setting	D. McFarland	Worcester, Mass	June 25, 1867	66,098
Card-teeth, Machine for sticking	G. W. Coats and J. Russell	Springfield, Mass	Aug. 1, 1854	11,434
Card-teeth, Process of making and using elastic steel.	E. Jenks	Colebrook, Conn	Nov. 13, 1813	....
Cards and drawing can-hoops, Bending iron	J. Butterworth	Philadelphia County, Pa	July 20, 1831	....
Cards, Apparatus for shuffling and dealing	H. E. Piquet	Sartrouville, France	Oct. 27, 1868	83,540
Cards, Box for supplying business	W. and W. H. Lewis	New York, N. Y	Nov. 22, 1853	10,255
Cards, Cleaning machine	G. Wellman	Lowell, Mass	Dec. 6, 1853	10,298
Cards, Composition for preparing cotton and wool machine.	J. M. Gates	Warren, Mass	Sept. 14, 1815	....
Cards, &c., Cylinder for carrying and supporting	S. R. Parkhurst	West Bloomfield, N. J	Jan. 23, 1849	6,043
Cards, filleting or cut, prick, and set machine	R. Meriam	Leicester, Mass	May 2, 1831	....
Cards for carding cotton, wool, &c., Manufacture of	G. Taber	Canton, Ohio	Aug. 1, 1838	863
Cards, labels, &c., Method of manufacturing show	A. J. Connell	New York, N. Y	Nov. 3, 1868	83,763
Cards, Machine for cleaning machine	H. Woodman	Biddeford, Me	Mar. 13, 1860	27,495
Cards, Machine for grinding and facing	G. Drake	Windsor, Conn	Feb. 28, 1817	....
Cards, Machine for grinding cotton	N. Smith and A. Crandall	North Kingston, R. I	Mar. 14, 1854	10,643
Cards, Machine for removing wire teeth from	J. A. Baham, R. C. Wilson, and S. French.	Auburn, N. Y	Aug. 4, 1868	80,586
Cards, Making band and other	E. L. Sprague	Leicester, Mass	Mar. 30, 1868	76,111
Cards, Manufacturing machine and animal	W. Wheeler	West Poultney, Vt	July 5, 1859	24,684
Cards, Manufacturing wool	A. Whittemore	....	June 5, 1797	....
Cards, Mode of ornamenting show	F. B. Scott	Buffalo, N. Y	Aug. 15, 1865	49,444
Cards, Playing	H. Billings	Boston, Mass	Aug. 26, 1873	142,075
Cards, Playing	J. J. Levy	New York, N. Y	Sept. 15, 1868	82,134
Cards, Playing	J. Stevens	Mount Vernon, N. Y	Sept. 21, 1869	95,160
Cards, Rub-rolls for condenser	W. Ferguson	Amsterdam, N. Y	Dec. 3, 1872	133,638
Cards, Scientific playing	C. Goodwin	Chicago, Ill	Feb. 8, 1870	99,561
Cards, Sliding slate for computation	G. N. Jackson	Chicago, Ill	Aug. 7, 1866	56,943
Cards, Straining and grinding	J. Boynton	Windham County, Conn	Mar. 30, 1811	....
Cards, Writing and drawing	H. A. Clark and H. J. Griswold	Boston, Mass., and Norwich, Conn.	Aug. 7, 1866	56,903
Carding and drawing engines, Wool	C. Jackson and J. Movi	Cazenovia, N. Y	Feb. 5, 1850	7,072
Carding and mixing wool and cotton	S. H. Adams and J. A. Wood	Cohoes, N. Y	June 11, 1850	7,421
Carding and other machines, Apparatus for feeding wool to.	J. S. Bolette	Gaffontaine Cornesse, Belgium.	Aug. 23, 1864	43,959
Carding and other machines, Construction of cylinders for.	J. M. Stone	North Andover, Mass	Feb. 21, 1871	111,988
Carding and other preparing machines, Feeding device for.	E. Pettitt	Manchester, England	Feb. 7, 1871	111,674
Carding and picking machines, Roller feed for	J. Dempster	Naugatuck, Conn	June 26, 1866	55,828
Carding and spinning machine, Yarn	M. Chase	Baltimore, Md	Mar. 23, 1842	2,511
Carding and spinning wool	P. C. Curtis	Paris, N. Y	Oct. 15, 1823	....
Carding and spinning wool by one continued operation, Machine for.	A. Holmes	Pomfret, N. Y	Jan. 8, 1810	....
Carding by which variegated slivers are produced	J. Holmes and E. French	Lee, Mass	May 18, 1852	8,948
Carding, Composition to be applied before	Z. Parkhurst	Grafton, Mass	Aug. 19, 1826	....
Carding-cylinders, Clothing for	C. G. Sargent and F. A. Calvert	Lowell, Mass	Mar. 9, 1858	19,585
Carding-cylinders, Machine for cleaning	A. A. Hawley	Methuen, Mass	Aug. 29, 1865	49,679
Carding-cylinders, Machinery for grinding card-teeth of.	C. Hardy	Biddeford, Me	Feb. 5, 1861	31,315
Carding-cylinders, Stripper for	F. M. Abbott and E. F. Fields	Boston, Mass., and Lewiston, Me.	Feb. 25, 1868	74,874
Carding-engine	S. R. and G. W. Ballard	Coldwater, Mich	Mar. 19, 1867	62,921
Carding-engine	J. Boyd	Philadelphia, Pa	May 24, 1859	24,092
Carding-engine	J. Boynton	South Coventry, Conn	July 12, 1843	3,174
Carding-engine	G. Bradley	Paterson, N. J	Nov. 19, 1861	33,762
Carding-engine	J. and J. Butterworth	Trenton, N. J	Nov. 9, 1869	96,671
Carding-engine	F. A. Calvert	Lowell, Mass	Sept. 22, 1863	40,018
Carding-engine	E. C. Cleveland and J. M. Bassett.	Worcester, Mass	Jan. 26, 1869	86,211
Carding-engine	J. Davis	East Wilton, N. H	Dec. 17, 1861	33,972
Carding-engine	J. Dyson	Fulton, S. C	Feb. 20, 1849	6,135
Carding-engine	J. Dyson	Fulton, S. C	Dec. 6, 1859	26,347
Carding-engine	J. Dyson	Philadelphia, Pa	Sept. 15, 1863	39,902
Carding-engine	O. F. Fitch	Morristown, Ind	Mar. 12, 1867	62,830
Carding-engine	J. Fitton	Cavendish, Vt	Dec. 11, 1860	30,875
Carding-engine	J. Fitton	Cavendish, Vt	Jan. 14, 1862	34,138
Carding-engine	D. T. Gage	Philadelphia, Pa	Feb. 2, 1864	41,462
Carding-engine	H. N. Gambril and S. F. Burgee	Woodbury, Md	Sept. 1, 1857	18,124
Carding-engine	B. H. Jenks	Bridesburgh, Pa	Mar. 2, 1869	87,342
Carding-engine	D. S. Kimball	Lowell, Mass	Nov. 19, 1861	33,744
Carding-engine	E. Leigh	Manchester, England	Sept. 26, 1865	50,211
Carding-engine	H. Marsden and T. H. Blamires	Huddersfield, England	Dec. 5, 1865	51,333
Carding-engine	T. McQuirk and O. Cole	Millville, Mass	Nov. 20, 1866	59,919
Carding-engine	H. L. Moulton	Camden, N. J	Oct. 10, 1865	50,377
Carding-engine	W. K. Platt	Philadelphia, Pa	Sept. 1, 1863	39,776
Carding-engine	R. Plews	Smithfield, R. I	July 28, 1863	39,355
Carding-engine	C. Pooley	Charlton-upon-Medlock, England.	Aug. 22, 1846	4,706
Carding-engine	A. D. Shattuck	Grafton, Mass	Sept. 23, 1856	15,781
Carding-engine	A. D. Shattuck	Grafton, Mass	Sept. 23, 1856	15,784
Carding-engine	H. Storms	Ann Arbor, Mich	Nov. 11, 1862	36,923
Carding-engine	W. H. Walton and G. H. Phinney.	Brooklyn and New York, N. Y.	Sept. 22, 1857	18,257
Carding-engine	S. Wethered	Baltimore, Md	Aug. 16, 1859	25,153
Carding-engine	J. C. Whitin	Strothbridge, Mass	July 7, 1863	39,187
Carding-engine, Condensing	I. Stead	Philadelphia, Pa	Apr. 29, 1862	35,114
Carding engine, Cotton	W. B. Leonard	Fishkill, N. Y	Sept. 16, 1833	....
Carding-engine doffer	B. D. Wheat	Mount Carmel, Ill	Jan. 19, 1869	85,975
Carding-engine doffer	R. Lord and L. Hutton	Rittenhouse, Pa	Jan. 31, 1865	46,120
Carding-engine feeding-attachment	J. Lawton	Glenham, N. Y	Sept. 28, 1869	95,237

Index of patents issued from the United States Patent Office from 1790 *to* 1873, *inclusive*—Continued.

Invention.	Inventor.	Residence.	Date.	No.
Carding-engine feeding-device	G. Bruce	Corydon, Ind	Dec. 11, 1866	60, 338
Carding-engine feeding-mechanism	A. A. Dow	Glenham, N. Y	Sept. 28, 1869	95, 208
Carding-engine feeding-mechanism	T. Sampson	Wanskuck, R. I	Mar. 24, 1868	75, 798
Carding-engine guide	J. Bachelder	Norwich, Conn	Jan. 28, 1868	73, 866
Carding-engine rack	L. Monroe	Lowell, Mass	Mar. 10, 1868	75, 290
Carding-engine sliver-guide	H. Kent	Lewiston, Me	Feb. 25, 1862	34, 508
Carding-engine winding-frame	W. and J. Leach	New Harmony, Ind	Aug. 25, 1868	81, 514
Carding-engines, Adjusting feed-roller for	L. Holmes	Newton, Mass	Apr. 13, 1869	88, 874
Carding-engines, Apparatus for spinning direct from doffers of.	T. Welham	Philadelphia, Pa	Oct. 23, 1866	59, 106
Carding-engines, Arrangement of the feeding or delivering roller of.	H. Barbour	Lowell, Mass	May 19, 1843	3, 094
Carding-engines, Attaching card-clothing to cylinders of.	D. H. Rowe	Pana, Ill	Dec. 22, 1868	85, 133
Carding-engines, Cleaning cards of	S. Greene	Woonsocket, R. I	Nov. 21, 1854	11, 964
Carding-engines, Cleaning top-flats of	W. H. Walton	New York, N. Y	Dec. 9, 1856	16, 196
Carding-engines, Lifting-flats in self-stripping	B. Dobson and W. Slater	Bolton, England	Aug. 24, 1869	94, 089
Carding-engines, Lubricating the traversing-guide in machines for feeding.	B. S. Roy	Olneyville, R. I	Mar. 8, 1870	100, 671
Carding-engines, Machine for oiling wool for	J. H. Aiken	Norwalk, Ohio	Oct. 30, 1866	59, 308
Carding-engines, Machine for preparing cotton for	R. Pilson	Laurel, Md	Oct. 30, 1866	59, 259
Carding-engines, Machine for spooling wool from the breaker of.	Z. Allen	Providence, R. I	Sept. 10, 1839	1, 316
Carding-engines, Machine for stripping top-flat of	S. L. Crockett and B. T. Mills	Lowell, Mass	Oct. 23, 1866	58, 989
Carding-engines, Machinery for cleaning top-flats of	H. Woodman	Biddeford, Me	July 8, 1856	15, 313
Carding-engines, Machinery for feeding	S. R. Parkhurst	New York, N. Y	Jan. 6, 1863	37, 304
Carding-engines, Machinery for stripping top-flats of.	G. Wellman	Lowell, Mass	Jan. 27, 1857	16, 504
Carding-engines, Means of operating doffer-combs of	A. A. Bennett	Norwalk, Conn	May 29, 1866	55, 046
Carding-engines, Mode of treating wood for manufacture of.	B. H. Jenks	Bridesburgh, Pa	May 29, 1866	55, 111
Carding-engines or preparing cotton, wool, &c., for spinning, Machinery or apparatus for making laps for feeding.	J. Mason	Rochdale, Great Britain	Feb. 10, 1843	2, 947
Carding-engines, Stop-motion for feeding-mechanism of.	W. E. Ainsworth and A. D. Wright.	Lowell, Mass	Mar. 19, 1867	62, 917
Carding-engines, Waste-saver for	C. F. Morrison	Rifton Glen, N. Y	Oct. 27, 1868	83, 526
Carding-engines, Waste-saving attachment to	A. A. Bennett	Norwalk, Conn	Apr. 18, 1865	47, 274
Carding-engines, Wool-oiling apparatus for	P. C. Kirk and M. Pendergast	Lawrence, Mass	Dec. 11, 1866	60, 386
Carding-engines, &c., Wool-oiling machinery for	W. H. Salisbury	Providence, R. I	Oct. 2, 1866	58, 516
Carding fibrous substances, Machine for	H. Wightman	Pittsburgh, Pa	Feb. 12, 1845	3, 904
Carding hair of neat-cattle	W. Shotwell and A. Kinder	New York, N. Y	Nov. 4, 1813	....
Carding-machine	E. Atkinson	Brookline, Mass	June 25, 1872	128, 202
Carding-machine	R. Bartlett	Ripley, Ohio	Oct. 25, 1832	....
Carding-machine	A. A. Bennett and G. Vine	Norwalk, Conn	Jan. 5, 1869	85, 556
Carding-machine	C. Bishop	Newtown, Conn	Sept. 26, 1846	4, 778
Carding-machine	T. G. Boone	Brooklyn, N. Y	Mar. 20, 1849	6, 197
Carding-machine	F. T. Chase and J. H. Platt	Dudley, Mass	June 4, 1872	127, 566
Carding-machine	L. Colburn	Fairfield, Vt	Nov. 20, 1832	....
Carding-machine	J. Davis	East Milton, N. H	Oct. 13, 1857	18, 423
Carding-machine	J. Davis	East Milton, N. H	Feb. 12, 1861	31, 425
Carding-machine	J. Dempster and H. Holcroft	Media, Pa	Nov. 9, 1869	96, 556
Carding-machine	E. French	North Adams, Mass	Apr. 9, 1872	125, 447
Carding-machine	H. N. Gambrill and S. F. Burgee	Woodbury Mills, Md	Feb. 27, 1855	12, 469
Carding-machine	C. S. Goodwin	Springfield, Mass	Mar. 5, 1872	124, 351
Carding-machine	P. S. Haines	Newburgh, N. Y	June 6, 1865	48, 059
Carding-machine	D. W. Hayden	Windham, Conn	Oct. 2, 1849	6, 758
Carding-machine	J. Haythorn and J. Martin	Thompsonville, Conn	Mar. 24, 1868	75, 906
Carding-machine	H. Houghton	Somers, Conn	Apr. 21, 1857	17, 094
Carding-machine	G. Jaquith	Concord, Mass	June 7, 1870	103, 889
Carding-machine	C. Jones	Yonkers, N. Y	May 14, 1872	126, 813
Carding-machine	P. L. King and E. Blasdell	Sparta and Lawrenceburgh Township, Ind.	Oct. 14, 1830	....
Carding-machine	J. McCarty	Somerset, Pa	June 19, 1849	6, 539
Carding-machine	J. McDonald	Detroit, Mich	June 26, 1866	55, 881
Carding-machine	M. W. Obenchain	Springfield, Ohio	July 20, 1846	4, 642
Carding-machine	L. O'Brien	Indianapolis, Ind	Oct. 6, 1863	40, 181
Carding-machine	H. Palmerler	Shoreham, Vt	May 25, 1811	....
Carding-machine	S. H. Parkhurst	Providence, R. I	Oct. 10, 1835	....
Carding-machine	S. R. Parkhurst	West Bloomfield, N. J	June 20, 1848	5, 643
Carding-machine	S. R. Parkhurst	New York, N. Y	May 21, 1861	32, 380
Carding-machine	D. Pheteplace	Lewiston, Me	June 18, 1872	127, 985
Carding-machine	C. E. Price and J. Haythorn	Thompsonville, Conn	Aug. 31, 1858	21, 364
Carding-machine	L. St. George	North Bellingham, Mass	May 28, 1872	127, 199
Carding-machine	D. Tainter	Worcester, Mass	Apr. 18, 1865	47, 345
Carding-machine	D. Tainter	Worcester, Mass	May 9, 1865	47, 668
Carding-machine	E. Thompson	Lowell, Mass	June 9, 1826	....
Carding-machine	G. Thresh	Oxford, Me	Sept. 6, 1870	107, 204
Carding-machine	S. Tillon	Newtown, Conn	Sept. 8, 1827	....
Carding-machine	U. G. Warner	Ripley, Ohio	Oct. 1, 1830	....
Carding-machine	J. Woodard	Portage County, Ohio	May 2, 1821	....
Carding-machine alarm	J. Haythorn and C. E. Price	Thompsonville, Conn	Aug. 18, 1868	81, 084
Carding-machine by gleaning-savers	J. Hugus	Greenbush, Pa	Sept. 15, 1815	....
Carding-machine card	W. Turner, S. Shore, and W. Halliwell.	Rochdale, England	Apr. 24, 1866	54, 270
Carding-machine card	E. Waite	Franklin City, Mass	May 1, 1866	54, 450
Carding-machine cleaner and grinder	L. W. Boynton	Hartford, Conn	Dec. 13, 1870	110, 004
Carding machine, Cloth	J. Jessup	Orange County, N. Y	Feb. 6, 1813	....
Carding-machine condenser, Woolen	L. L. Gowdy	Montgomery, N. Y	Oct. 11, 1841	2, 311
Carding-machine, Condensing	J. C. Shaw	Manayunk, Pa	Jan. 28, 1868	73, 760
Carding machine, Cotton or wool	E. and A. Crane	Lowell, Mass	Jan. 30, 1841	1, 962
Carding machine, Cotton or wool	J. Munroe	Palmer, Mass	Oct. 11, 1841	2, 314
Carding-machine cylinders, Means for securing the clothing on.	H. Bennett	Waterbury, Conn	Nov. 7, 1871	120, 615
Carding-machine doffer	C. Atwood	Middletown, Conn	Nov. 1, 1830	....
Carding-machine doffer-combs, Mechanism for operating,	J. K. Proctor	Philadelphia, Pa	June 24, 1873	140, 304
Carding-machine doffer-stripper	A. M. Comstock	Holden, Mass	Dec. 10, 1872	133, 832

Index of patents issued from the United States Patent Office from 1790 *to* 1873, *inclusive*—Continued.

Invention.	Inventor.	Residence.	Date.	No.
Carding-machine feed-mechanism	W. A. Lawton	Providence, R. I	Nov. 29, 1870	109, 634
Carding-machine feed-roller	H. G. Elsworth	Enfield, Conn	Sept. 4, 1847	5, 272
Carding-machine feeder	G. S. Harwood	Boston, Mass	Nov. 19, 1867	71, 601
Carding-machine feeder, Weighing-attachment for	P. C. Evans and H. J. H. King.	Brimscombe, England, and Glasgow, Scotland.	Oct. 4, 1870	107, 890
Carding-machine feeding-apparatus	J. W. Barbour	Winooski Falls, Vt	July 9, 1872	128, 697
Carding-machine feeding-apparatus	D. Tainter	Worcester, Mass	July 30, 1867	67, 226
Carding-machine feeding-device	J. W. Barbour	Winooski Falls, Vt	Sept. 27, 1870	107, 751
Carding-machine feeding-mechanism	W. Clissold	Stroud, England	Apr. 23, 1867	64, 077
Carding-machine feeding-mechanism	B. W. Taugee	Woodville, R. I	Feb. 16, 1869	86, 953
Carding-machine feeding-mechanism	B. W. Taugee	Dorrville, R. I	Feb. 16, 1869	86, 954
Carding-machine for preparing bats for felting	S. G. Blackman	Norwalk, Conn	Feb. 5, 1850	7, 065
Carding-machine grinding-attachment	A. J. Burke	Mansfield, Conn	June 14, 1870	104, 262
Carding-machine guide	F. W. and E. Albertine	Hanover, Conn	June 25, 1867	66, 112
Carding-machine roll	E. S. Worcester	Worcester, Mass	Feb. 20, 1872	123, 968
Carding-machine stripper	L. M. Capron	Worcester, Mass	Nov. 16, 1869	96, 884
Carding-machine teeth	C. Hindle, J. C. Milton, and O. Arnold.	Worcester, Mass	July 23, 1872	129, 662
Carding-machine teeth	W. H Whiting	Wellington, Conn	Sept. 27, 1870	107, 842
Carding-machine to form bats for wool hats	A. Guild	Dedham, Mass	Mar. 31, 1806	....
Carding machine, Wool	J. H. Arnold	Belmont, Ohio	Sept. 25, 1826	....
Carding machine, Wool	C. Atwood	Middletown, Conn	Apr. 18, 1829	....
Carding machine, Wool	D. M. Bacon	Huntingdon, Pa	Sept. 10, 1828	....
Carding machine, Wool	G. Booth	Poughkeepsie, N. Y	Oct. 13, 1812	....
Carding machine, Wool	J. Davis	Wilton, N. H	Aug. 4, 1863	39, 381
Carding machine, Wool	N. Freeman	Lowell, Mass	Dec. 31, 1839	1, 462
Carding machine, Wool	E. Hale, jr., and S. and J. Hale.	Haverhill, Mass	Feb. 18, 1825	....
Carding machine, Wool	F. Nowell	Lowell, Mass	June 3, 1856	15, 016
Carding machine, Wool	S. C. Philbrick	Rockville, Conn	Dec. 3, 1867	71, 784
Carding machine, Wool	J. W. Shankland	Summerfield, Ohio	May 5, 1829	....
Carding machine, Wool	H. A. Shannon	Columbia County, N. Y	Apr. 5, 1828	....
Carding machine, Wool	D. Smith	Philadelphia, Pa	June 14, 1809	....
Carding machine, Wool	D. Tainter	Worcester, Mass	Sept. 29, 1863	40, 131
Carding machines, Apparatus for stripping top-flats of.	E. C. Pfaff	Chemnitz, Saxony	Apr. 4, 1871	113, 339
Carding-machines, Cleaning top-cards of	H. Woodman	Biddeford, Me	Aug. 1, 1854	11, 448
Carding-machines, Composition coating for drawing-rollers for.	B. Borden, G. Townsend, and E. A. Green.	Pawtucket and Lincoln, R. I.	Aug. 20, 1872	130, 694
Carding-machines, Condensing-tube for	W. Germain	Rockbottom, Mass	Sept. 24, 1867	69, 205
Carding-machines, &c., Condensing-tube for	A. W. Skinner	Providence, R. I	Jan. 21, 1873	135, 015
Carding-machines, Construction of	N. Rider	Sturbridge, Mass	Mar. 12, 1825	....
Carding-machines, Dividing-apparatus for condensing.	E. Bede	Verviers, Belgium	May 7, 1872	126, 439
Carding-machines, Machine for drawing and spinning wool, &c., from.	J. Goulding	Worcester, Mass	Oct. 5, 1869	95, 580
Carding-machines, &c., Machine for feeding wool, &c., to.	W. Clissold	Stroud, England	July 20, 1869	92, 705
Carding-machines, Machine for grinding top-cards and the workers, strippers, and licker-in cylinders of.	C. Hardy	Biddeford, Me	Apr. 16, 1867	63, 884
Carding-machines, Machinery for cleaning top-cards of.	H. Woodman	Biddeford, Me	Dec. 1, 1857	18, 787
Carding-machines, Machinery for oiling wool in	J. Eccles	Philadelphia, Pa	May 16, 1865	47, 767
Carding-machines, Machinery for oiling wool in	G. S. Harwood	Boston, Mass	Jan. 31, 1865	46, 104
Carding-machines, Machinery for oiling wool in	J. W. Hussey	Boston, Mass	Jan. 31, 1865	46, 189
Carding-machines, Machinery for oiling wool in	C. Jones	Boston, Mass	July 31, 1866	56, 758
Carding-machines, Machinery for oiling wool in	J. Shinn	Philadelphia, Pa	Jan. 31, 1865	46, 194
Carding-machines, Mechanism for adjusting roller of.	E. Lord	Todmorden, England	Aug. 15, 1871	118, 029
Carding-machines, Mechanism for stripping top-flats of.	F. Morf	Wetzikon, Switzerland	May 10, 1870	102, 851
Carding-machines, Operating condensing-roller in.	E. Stafford	Philadelphia, Pa	Sept. 3, 1867	68, 464
Carding-machines, Self-feed for	R. W. Lewis	Beacon Falls, Conn	Jan. 15, 1867	61, 219
Carding-machines, Stop-mechanism for	C. W. Anderson	Grosvenor Dale, Conn	Mar. 1, 1870	100, 247
Carding-machines, Stripper-carrying frame for	J. F. Foss	Lowell, Mass	May 23, 1871	115, 185
Carding-machines, Stripping top-flats of	G. Wellman	Lowell, Mass	Mar. 18, 1856	14, 481
Carding-machines, Waste-removing device for	G. W. Craner	Darby, Pa	July 15, 1873	140, 814
Carding, Machinery for cutting and preparing flax, &c., for.	J. and C. Beach, T. Beach, and W. G. R. Mowry.	Connecticut, Massachusetts, and Rhode Island.	Aug. 29, 1848	5, 734
Carding, spinning, and roping machine	T. Norton and G. Biddis	Milford, Pa	Apr. 15, 1813	....
Carding wool	J. Boynton	South Coventry, Conn	Apr. 11, 1829	....
Carding wool	J. Boynton	South Coventry, Conn	Mar. 1, 1833	....
Carding wool	L. French	New Haven, Conn	May 18, 1814	....
Carding wool and other fibrous material	C. Wing	Gardiner, Me	Oct. 21, 1830	....
Carding wool, cotton, &c	D. Adams	Richmond, Va	July 27, 1833	....
Cargoes of vessels, Apparatus for removing	J. Manderson and S. Favinger.	Philadelphia, Pa	Jan. 9, 1866	52, 001
Carmine, Method of manufacturing	G. A. Siegle	Brooklyn, N. Y	July 16, 1867	66, 897
Carpentering	J. Bailey	Philadelphia, Pa	Apr. 7, 1827	....
Carpenter's bench	J. Jones	Newark, N. J	Jan. 28, 1873	135, 342
Carpenter's bench	R. C. Love	Augusta, Me	Feb. 7, 1871	111, 553
Carpenter's bench	J. W. Mahan	Lexington, Ill	Sept. 16, 1856	15, 739
Carpenter's bench	A. B. Marshall	Medfield, Mass	Apr. 9, 1867	63, 645
Carpenter's bench	R. McConnell	Lawrenceville, Pa	Oct. 16, 1866	58, 858
Carpenter's bench	F. Starke	Dayton, Ohio	Sept. 12, 1871	118, 980
Carpenter's bench-clamp	J. E. A. Gibbs	Mill Point, Va	Feb. 17, 1857	16, 642
Carpenter's bench-dog	E. B. McCoy	Winsted, Conn	May 19, 1868	77, 998
Carpenter's bench-gage	F. A. Traut	New Britain, Conn	Nov. 18, 1862	36, 973
Carpenter's bench-hook	S. Swan	New York, N. Y	June 2, 1868	78, 620
Carpenter's bench-hook	C. H. Weston	Lowell, Mass	Feb. 4, 1868	74, 182
Carpenter's bench, Stop for	L. Augur and J. L. Lord	Chester, Conn	May 29, 1849	6, 485
Carpenter's brace and bit fastener	H. Perkins	North Bridgewater, Mass	Nov. 1, 1853	10, 189
Carpenter's bracket, &c., Holding-bolt for	J. W. Kennedy	Plainfield, Conn	Mar. 23, 1858	19, 700
Carpenter's clamp	J. Cadwell	Cincinnati, Ohio	Apr. 24, 1860	27, 968
Carpenter's clamp	B. H. Green	Princeton, N. J	Aug. 9, 1853	9, 917
Carpenter's clamp	C. L. Jones	Richmond, Va	Nov. 27, 1860	30, 735
Carpenter's gage	T. E. Barrow	Mansfield, Ohio	Apr. 13, 1869	88, 833
Carpenter's gage	A. H. Blaisdell	Newton Corners, Mass	June 23, 1868	79, 052
Carpenter's gage	W. Brodhead	Meadville, Pa	Apr. 21, 1868	76, 884

Index of patents issued from the United States Patent Office from 1790 to 1873, inclusive—Continued.

Invention.	Inventor.	Residence.	Date.	No.
Carpenter's gage	O. Brown and T. F. Berry	Capron, Ill	July 7, 1868	79, 632
Carpenter's gage	J. Bryant	Brooklyn, N. Y	Aug. 19, 1856	15, 556
Carpenter's gage	F. W. Coy	Boston, Mass	May 19, 1868	78, 065
Carpenter's gage	B. T. Currier	Boston, Mass	July 11, 1865	48, 663
Carpenter's gage	M. Horton	Brooklyn, N. Y	May 23, 1865	47, 824
Carpenter's gage	G. T. Lape	New York, N. Y	Feb. 5, 1867	61, 840
Carpenter's gage	P. Lawyer	Richmondville, N. Y	Mar. 19, 1867	63, 062
Carpenter's gage	J. McCrum	Locust Grove, Ohio	July 11, 1865	48, 703
Carpenter's gage	G. Miller	Washington, D. C	Jan. 17, 1865	45, 932
Carpenter's gage	G. Miller	Washington, D. C	Oct. 17, 1865	50, 484
Carpenter's gage	R. Phillips	Gardiner, Me	Jan. 15, 1867	61, 248
Carpenter's gage	M. C. Robichau	Marblehead, Mass	July 22, 1873	141, 014
Carpenter's gage	E. Sahm	Greenville, Pa	Nov. 4, 1873	144, 359
Carpenter's gage	D. W. Simmons	Lynn, Mass	May 30, 1871	115, 368
Carpenter's gage	H. P. Sisson	Rutland, Vt	Dec. 27, 1870	110, 596
Carpenter's gage	H. Skinner and W. Greenhalgh	West Farms, N. Y	Aug. 22, 1854	11, 572
Carpenter's gage	J. A. Traut	New Britain, Conn	Oct. 22, 1872	132, 421
Carpenter's gage	J. A. Traut	New Britain, Conn	Aug. 5, 1873	141, 475
Carpenter's implement	E. Gray	Oldtown, Me	Sept. 7, 1869	94, 688
Carpenter's shooting-board	J. Jones	Newark, N. J	Jan. 3, 1871	110, 765
Carpet	J. Cochran, jr	Malden, Mass	Mar. 1, 1870	100, 377
Carpet	G. Crompton	Worcester, Mass	Jan. 31, 1871	111, [illegible]25
Carpet	G. Crompton	Worcester, Mass	June 18, 1872	128, 023
Carpet	T. Crossley	Roxbury, Mass	Mar. 16, 1852	8, 798
Carpet	T. Crossley	Bridgeport, Conn	Oct. 12, 1869	95, 775
Carpet	J. Dornan	Philadelphia, Pa	Sept. 9, 1873	142, 681
Carpet	D. Hirschberg	Baltimore, Md	Nov. 16, 1869	96, 915
Carpet	G. Iskiyan	New York, N. Y	Dec. 23, 1873	145, 803
Carpet	R. Scott	Philadelphia, Pa	Sept. 16, 1873	142, 947
Carpet-bag	C. F. Blakslee	Brooklyn, N. Y	Mar. 19, 1867	63, 002
Carpet-bag	N. Groel	Newark, N. J	Nov. 7, 1865	50, 815
Carpet-bag	J. M. Matthews	New York, N. Y	Apr. 12, 1859	23, 596
Carpet-bag	A. J. Robrecht	Newark, N. J	Sept. 22, 1868	82, 352
Carpet-bag	E. A. G. Roulstone	Roxbury, Mass	Oct. 30, 1866	59, 270
Carpet-bag	E. A. G. Roulstone	Roxbury, Mass	Mar. 26, 1867	63, 178
Carpet-bag	E. A. G. Roulstone	Roxbury, Mass	Mar. 26, 1867	63, 179
Carpet-bag	A. Sonnekalb and J. W. Lieb	Newark, N. J	Feb. 27, 1866	52, 902
Carpet-bag	F. J. Thring	New York, N. Y	Jan. 31, 1854	10, 484
Carpet-bag	J. Zepfel	New York, N. Y	June 9, 1857	17, 545
Carpet-bag button and pin	W. Roemer	Newark, N. J	Dec. 28, 1869	98, 300
Carpet-bag catch	G. Crouch	New York, N. Y	Mar. 9, 1869	87, 549
Carpet-bag clasp	J. Sellers and A. L. Pennock	Philadelphia, Pa	May 25, 1840	1, 620
Carpet-bag frame	P. Born	New York, N. Y	July 10, 1866	56, 168
Carpet-bag frame	G. Havell	Newark, N. J	June 5, 1866	55, 290
Carpet-bag frame	F. Herman	Newark, N. J	Feb. 2, 1869	86, 542
Carpet-bag frame	S. Lagowitz	Newark, N. J	July 7, 1863	39, 152
Carpet-bag frame	S. Lagowitz	Newark, N. J	Sept. 13, 1864	44, 202
Carpet-bag frame	A. Sonnekalb and J. W. Lieb	Newark, N. J	Aug. 21, 1866	57, 399
Carpet-bag frame	C. and Z. Walsh	Newark, N. J	Aug. 14, 1866	57, 262
Carpet-bag-frame-bending machine	H. Havell	Newark, N. J	Nov. 10, 1868	83, 852
Carpet-bag frames, &c., Machinery for bending	E. L. Gaylord	Newark, N. J	Dec. 7, 1852	9, 445
Carpet-bag lock	Z. Walsh	Newark, N. J	Jan. 24, 1860	26, 959
Carpet-beater	W. H. Hankinson	New York, N. Y	Oct. 25, 1870	108, 589
Carpet-beater	W. H. Hankinson	New York, N. Y	May 20, 1873	139, 145
Carpet-beater	J. Hothersall	New York, N. Y	Apr. 16, 1872	125, 815
Carpet-beater	J. Hothersall	New York, N. Y	Dec. 17, 1872	134, 059
Carpet-beater	J. Hothersall	New York, N. Y	Sept. 23, 1873	143, 155
Carpet-beater	J. Hothersall and J. Banks	New York, N. Y	July 11, 1871	116, 958
Carpet-beater	S. Jordan	Brooklyn, N. Y	Aug. 17, 1869	93, 718
Carpet-beater	W. G. Mowry and C. Pulis	New York, N. Y	July 1, 1873	140, 525
Carpet-beater	J. W. and W. A. Wheeler	Cleveland, Ohio	Dec. 26, 1871	122, 297
Carpet beater and cleaner	A. Cutler and E. S. Wright	New York, N. Y	Apr. 10, 1860	27, 780
Carpet beater and cleaner	W. H. Hankinson	New York, N. Y	Feb. 12, 1867	61, 029
Carpet beater and cleaner	G. P. Mitchell	Philadelphia, Pa	Sept. 13, 1864	44, 210
Carpet beater and cleaner	A. Stevenson	New York, N. Y	Dec. 14, 1869	97, 824
Carpet beating and brushing machine	W. Peters	Charlestown, Mass	July 31, 1849	6, 618
Carpet-beating machine	J. Harris, jr., and D. Holmes	Roxbury and Chelsea, Mass	Feb. 23, 1858	19, 465
Carpet-beating machine	T. and W. H. Jordan	Brooklyn and New York, N. Y.	Aug. 1, 1871	117, 543
Carpet-beating machine	L. Miner	San Francisco, Cal	Nov. 2, 1869	96, 461
Carpet-binding, Metal	R. P. Johnston	Steubenville, Ohio	Aug. 15, 1871	118, 019
Carpet-cleaner	H. W. Bates	Allston, Mass	July 22, 1873	140, 990
Carpet-cleaner	A. M. George and J. W. Carter	Nashua, N. H., and Brooklyn, N. Y.	June 4, 1861	32, 473
Carpet-cleaner	H. H. Lindhorst	Saint Louis, Mo	Jan. 31, 1871	111, 355
Carpet-cleaner	T. Lüke	Saint Louis, Mo	June 11, 1867	65, 683
Carpet-cleaner	H. L. Nichols	New York, N. Y	May 22, 1860	28, 398
Carpet-cleaner	A. W. Noney	Bridgeport, Conn	Aug. 17, 1858	21, 211
Carpet-cleaner	J. Spaulding	San Francisco, Cal	July 29, 1873	141, 243
Carpet-cleaner	R. Terry and F. W. Hafkemeyer	Chicago, Ill	Oct. 25, 1870	108, 651
Carpet-cleaner	G. W. Young	San Francisco, Cal	Sept. 24, 1867	69, 152
Carpet cleaning and finishing machine	J. Wilkinson, jr	Leeds, England	Apr. 2, 1872	125, 240
Carpet-cleaning machine	G. W. Bishop	Stamford, Conn	May 26, 1868	78, 178
Carpet-cleaning machine	J. C. Craft	Baltimore, Md	Mar. 7, 1871	112, 421
Carpet-cleaning machine	D. A. Drew	Philadelphia, Pa	Jan. 25, 1870	99, 073
Carpet-cleaning machine	W. H. Jordan	New York, N. Y	Apr. 5, 1864	42, 254
Carpet-cleaning machine	H. H. Lindhorst	Saint Louis, Mo	June 14, 1870	104, 171
Carpet-cleaning machine	T. Lüke	Saint Louis, Mo	Mar. 23, 1869	88, 188
Carpet-cleaning machine	W. McArthur	Philadelphia, Pa	Aug. 28, 1866	57, 532
Carpet-cleaning machine	W. McArthur	Philadelphia, Pa	Mar. 16, 1869	87, 952
Carpet-cleaning machine	E. S Poucher	New York, N. Y	May 25, 1869	90, 390
Carpet-cleaning machine	G. C. Smith and C. T. Story	Boston, Mass	Nov. 12, 1872	132, 927
Carpet-cleaning machine	J. Wentworth	Palatine, N. Y	Nov. 6, 1849	6, 855
Carpet, Device for slitting and looping rags for	L. C. Palmer	Howard, Pa	Nov. 21, 1871	121, 124
Carpet-duster	T. Ferry	Wilmington, Del	Aug. 23, 1870	106, 679
Carpet-duster	J. Rolls	New York, N. Y	Nov. 6, 1860	30, 590
Carpet-fabric, &c	T. Crossley	Bridgeport, Conn	Feb. 27, 1866	52, 825
Carpet-fabric	S. Hunter and W. Kerr	Philadelphia, Pa	July 23, 1872	129, 829

Index of patents issued from the United States Patent Office from 1790 *to* 1873, *inclusive*—Continued.

Invention.	Inventor.	Residence.	Date.	No.
Carpet-fabric	W. Wallace	Philadelphia, Pa	June 14, 1870	104, 232
Carpet fabric, Felted	T. Crossley	Bridgeport, Conn	Oct. 12, 1869	95, 776
Carpet-fastener	W. B. Blaisdell and J. E. Atwood.	Lynn, Mass	Sept. 12, 1865	49, 850
Carpet-fastener	C. B. Browne	Geneseo, Ill	Feb. 18, 1873	135, 884
Carpet-fastener	F. O. Clark	Des Moines, Iowa	Aug. 13, 1872	130, 360
Carpet-fastener	J. C. Craft	Baltimore, Md	Dec. 3, 1872	133, 568
Carpet-fastener	S. Culver	Newark, N. J	Nov. 17, 1857	18, 631
Carpet-fastener	R. De Charms	Philadelphia, Pa	Dec. 21, 1858	22, 354
Carpet-fastener	M. Dewey and I. Phillips	Clarendon, N. Y	Aug. 31, 1858	21, 325
Carpet-fastener	W. Filkins	Lancaster, N. Y	May 25, 1858	20, 341
Carpet-fastener	S. N. French	Fitchburgh, Mass	June 8, 1869	91, 111
Carpet-fastener	J. H. Friederich	Rochester, N. Y	Mar. 6, 1866	52, 985
Carpet-fastener	R. A. Gawler	Concord, N. H	Oct. 23, 1866	59, 002
Carpet-fastener	A. Givanaan	Washington, D. C	Apr. 18, 1871	113, 874
Carpet-fastener	M. Gramiss	Waterbury, Conn	Apr. 12, 1859	23, 566
Carpet-fastener	C. Gullmann	Poughkeepsie, N. Y	Sept. 11, 1866	57, 895
Carpet-fastener	C. Gullmann	Poughkeepsie, N. Y	Apr. 9, 1867	63, 718
Carpet-fastener	J. W. Hart and O. Norton	New Britain, Conn	Feb. 11, 1868	74, 352
Carpet-fastener	C. Harting	Washington, D. C	May 7, 1872	126, 456
Carpet-fastener	C. Harting	Washington, D. C	Sept. 30, 1873	143, 238
Carpet-fastener	L. S. Hicks	Omro, Wis	July 9, 1867	66, 588
Carpet-fastener	J. O. Jones	Boston, Mass	June 27, 1865	48, 410
Carpet-fastener	B. D. Kested	Glen, N. Y	Oct. 21, 1873	143, 784
Carpet-fastener	G. G. Noyes	Worcester, Mass	Sept. 20, 1859	25, 519
Carpet-fastener	J. Reynolds	New Britain, Conn	Aug. 31, 1858	21, 365
Carpet-fastener	J. W. Saumenig	Baltimore, Md	Apr. 30, 1872	126, 334
Carpet-fastener	A. M. Smith	New York, N. Y	May 31, 1859	24, 242
Carpet-fastener	J.V.C. Smith	New York, N. Y	June 22, 1869	91, 573
Carpet-fastener	M. D. and S. A. Snyder	Clarendon, N. Y	June 28, 1859	24, 586
Carpet-fastener	M. D. and S. A. Snyder	Clarendon, N. Y	Apr. 16, 1861	32, 089
Carpet-fastener	C. F. Spencer	Rochester, N. Y	Mar. 22, 1859	23, 319
Carpet fastener	J. H. Stanton	Franklin, Ohio	Oct. 11, 1870	108, 200
Carpet-fastener	J. A. Taylor	Cowlesville, N. Y	Apr. 26, 1859	23, 797
Carpet-fastener	T. H. Tyndale	Springfield, Ill	Mar. 16, 1869	87, 991
Carpet-fastener	C. A. Wakefield	Dalton, Mass	Jan. 19, 1858	19, 164
Carpet-fastener	R. W. Walker	Washington, D. C	Apr. 19, 1870	102, 185
Carpet-fastener	W. Weaver	Salem, Ohio	Oct. 30, 1866	59, 298
Carpet-fastener	G. E. West and W. R. Cunningham.	La Fayette, Ind	Oct. 22, 1867	70, 055
Carpet-fastener	A. J. Williams	Seneca Falls, N. Y	Mar. 25, 1873	137, 118
Carpet fastener and protector, Stair	J. Conner, jr	Chicago, Ill	May 26, 1868	78, 263
Carpet fastener, Stair	C. Rice	New York, N. Y	Dec. 27, 1859	26, 617
Carpet-fastening	G.W. Andrews and J. P. Burnham.	Chicago, Ill	Aug. 29, 1865	49, 600
Carpet-fastening	W. Barnan	Deerfield, Mass	Sept. 27, 1864	44, 384
Carpet-fastening	D. N. B. Coffin, jr	Newton, Mass	June 9, 1857	17, 488
Carpet-fastening	S. R. C. Denison	Rochester, N. Y	Aug. 5, 1856	15, 478
Carpet-fastening	G. Gregory	New Haven, Conn	Oct. 15, 1872	132, 282
Carpet-fastening	E. Jackman and E. G. Dunham	Portland, Conn	July 10, 1855	13, 220
Carpet-fastening	W. S. Loughborough	Rochester, N. Y	June 5, 1855	13, 009
Carpet-fastening	J. J. Märki	Chicago, Ill	June 28, 1870	104, 864
Carpet-fastening	J. Matson	Bridgeport, Conn	July 23, 1867	66, 981
Carpet-fastening	F. Miller	New York, N. Y	June 19, 1855	13, 094
Carpet-fastening	W. H. Penrose	Philadelphia, Pa	July 28, 1857	17, 890
Carpet-fastening	J. A. Robbins	Medford, Mass	Mar. 14, 1871	112, 633
Carpet-fastening	R. E. Schroeder	Rochester, N. Y	July 28, 1857	17, 897
Carpet-holder	E. Shopbell	Ashland, Ohio	May 29, 1866	55, 170
Carpet-holder	F. Smith	Alexandersville, Ohio	Feb. 4, 1868	74, 160
Carpet-holder	H. Thayer	Warsaw, N. Y	Apr. 6, 1858	19, 882
Carpet-lining	G. W. Chipman	Melrose, Mass	Dec. 18, 1866	60, 476
Carpet-lining	G. W. Chipman	Boston, Mass	Aug. 11, 1868	80, 913
Carpet-lining	J. F. Fales	Walpole, Mass	Feb. 27, 1866	52, 835
Carpet-lining	W. Fuzzard	Chelsea, Mass	Apr. 23, 1867	64, 092
Carpet-lining	J. R. Harrington	New York, N. Y	Apr. 23, 1861	32, 134
Carpet-lining	J. R. Harrington	Brooklyn, N. Y	Sept. 27, 1870	107, 682
Carpet-lining	J. R. Harrington	Brooklyn, N. Y	Apr. 18, 1871	113, 765
Carpet-lining machine	J. R. Harrington	Brooklyn, N. Y	July 29, 1873	141, 270
Carpet lining	M. A. Johnson	Lowell, Mass	June 18, 1867	65, 814
Carpet-lining	J. C. Mayall	Boston, Mass	Jan. 31, 1871	111, 465
Carpet-lining	M. Mayall	Roxbury, Mass	Sept. 15, 1868	82, 236
Carpet-linings, &c., Fabric for	J. L. Kendall and R. H. Trested	Foxborough, Mass., and New York, N. Y.	May 17, 1870	103, 198
Carpet-linings, Machine for folding	J. C. Mayall	Boston, Mass	June 27, 1871	116, 467
Carpet-linings, Machine for making	J. Foster, jr., and F. J. Dill	Camden, N. J., and Foxborough, Mass.	Feb. 9, 1869	86, 830
Carpet-linings, Machine for making	J. R. Harrington	Dayton, Ohio	Apr. 1, 1856	14, 585
Carpet-linings, Machine for making	J. R. Harrington	New York, N. Y	Sept. 11, 1860	29, 970
Carpet-linings, Machine for making	J. R. Harrington	Brooklyn, N. Y	Jan. 2, 1872	122, 381
Carpet-linings, Machine for manufacture of paper	C. A Pease	Astoria, N. Y	Jan. 17, 1871	111, 081
Carpet-linings, Machine for manufacturing	E. H. Bailey	Brooklyn, N. Y	Sept. 23, 1873	143, 053
Carpet-linings, Manufacture of	S. M. Allen	Boston, Mass	Feb. 24, 1863	37, 728
Carpet-linings, Manufacture of	J. R. Harrington	Brooklyn, N. Y	July 23, 1867	67, 048
Carpet-linings, Manufacture of	W. A. Mauran	Providence, R. I	Aug. 2, 1870	105, 960
Carpet matting or lining	P. Sweeny	New York, N. Y	Oct. 15, 1872	131, 333
Carpet nails or tacks	W. P. Patton	Harrisburgh, Pa	May 19, 1863	38, 598
Carpet pad, Stair	G. W. Chipman	Melrose, Mass	Apr. 3, 1866	53, 573
Carpet, Printed	T. Crossley	Roxbury, Mass	Aug. 16, 1853	9, 935
Carpet, Printed piled	T. Crossley	Bridgeport, Conn	June 10, 1873	139, 706
Carpet-printing block	T. Crossley	Bridgeport, Conn	Aug. 16, 1870	106, 468
Carpet-printing machine	T. Crossley	Bridgeport, Conn	Oct. 12, 1869	95, 777
Carpet-protector	H. W. Eskildson	Boston, Mass	July 11, 1871	116, 822
Carpet-rag cutter	W. Eberhard	Akron, Ohio	Sept. 13, 1870	107, 236
Carpet-rag cutter and looper, Combined	S. P. Converse	Greenville, Ohio	July 9, 1872	128, 857
Carpet-rag looper	G. W. Ansley	Marengo Township, Mich	July 9, 1872	128, 774
Carpet-rag looper	W. Bollinger	Millerstown, Pa	Sept. 17, 1872	131, 420
Carpet-rag looper	W. Clayton, jr	Bristol, Conn	July 15, 1873	140, 889
Carpet-rag looper	J. F. Cooper	Frankton, Ind	Dec. 2, 1873	145, 094

Index of patents issued from the United States Patent Office from 1790 *to* 1873, *inclusive*—Continued.

Invention.	Inventor.	Residence.	Date.	No.
Carpet-rag looper	F. J. Egger	Springfield, Ohio	Sept. 16, 1873	142, 778
Carpet-rag looper	G. L. Price	Clifton Springs, N. Y	Feb. 2, 1869	86, 445
Carpet-rag looper and button-hole cutter, Combined.	G. W. Morris and W. Lenhart.	Corry, Pa	Nov. 11, 1873	144, 472
Carpet-rag looper, button-hole cutter, and press-board combined.	S. M. Whitten	Algansee, Mich	Mar. 11, 1873	136, 633
Carpet-rod	H. Thry	New York, N. Y	Feb. 9, 1869	86, 714
Carpet-sack	A. Rose	Bath, N. Y	Jan. 22, 1867	61, 467
Carpet, Stamped paper	F. Guy	Baltimore, Md	Feb. 23, 1819	....
Carpet-stretcher	H. Blan	Washington, D. C	Feb. 2, 1858	19, 230
Carpet-stretcher	W. A. Boles	Shelbyville, Ind	Aug. 5, 1873	141, 418
Carpet-stretcher	J. H. Bosworth	Bath, Me	Dec. 18, 1866	60, 469
Carpet-stretcher	J. Boyd	Lowell, Mass	Dec. 19, 1865	51, 545
Carpet-stretcher	I. W. Bragg	Cincinnati, Ohio	Feb 15, 1859	22, 930
Carpet-stretcher	W. Brown	New York, N. Y	Feb. 9, 1869	86, 728
Carpet-stretcher	E. W. Bullard	Barre, Mass	May 28, 1872	127, 303
Carpet-stretcher	W. P. D. Claybrook	Palmyra, Mo	Jan. 30, 1872	123, 239
Carpet-stretcher	F. J. Collier	Cannonsburgh, Pa	Mar. 21, 1865	46, 882
Carpet-stretcher	W. C. Conant	New York, N. Y	Oct. 5, 1858	21, 654
Carpet-stretcher	S. G. Dare	New York, N. Y	Aug. 11, 1868	80, 925
Carpet-stretcher	J. H. De Poe	Boonton, N. J	Aug. 22, 1871	118, 214
Carpet-stretcher	A. L. Dunbar	Sheldon, Ill	Jan. 14, 1868	73, 310
Carpet-stretcher	G. O. Dunlap	Chicopee, Mass	Feb. 12, 1867	61, 928
Carpet-stretcher	S. Elliott	Sonora, Cal	Aug. 1, 1871	117, 528
Carpet-stretcher	A. C. Ellis	Birmingham, Mich	July 23, 1872	129, 802
Carpet-stretcher	L. Frankhauser	Columbus, Ohio	Jan. 14, 1873	134, 873
Carpet-stretcher	C. E. Gale	Aurelius, N. Y	Jan. 31, 1871	111, 337
Carpet-stretcher	J. B. Greenalgh	Providence, R. I	Jan. 19, 1869	85, 925
Carpet-stretcher	J. S. Green and T. D. Bradt	Watertown, N. Y	May 6, 1873	138, 560
Carpet-stretcher	G. S. Greenleaf and C. Buckland.	Springfield, Mass	Feb. 12, 1861	31, 383
Carpet-stretcher	H. M. Hartshorn	Malden, Mass	June 26, 1860	28, 858
Carpet-stretcher	N. Hill	Caton, N. Y	Feb. 14, 1865	46, 357
Carpet-stretcher	H. Hungerford	New York, N. Y	Oct. 16, 1866	58, 831
Carpet-stretcher	H. Hungerford	New York, N. Y	Jan. 29, 1867	61, 663
Carpet-stretcher	W. J. Johnson	Newton, Mass	Feb. 19, 1867	62, 138
Carpet-stretcher	J. Kaily	Canton, Ohio	Oct. 23, 1866	59, 129
Carpet-stretcher	P. Kelly	New York, N. Y	Apr. 29, 1873	138, 333
Carpet-stretcher	C. A. Lindner	Cincinnati, Ohio	Oct. 9, 1866	58, 653
Carpet-stretcher	J. Lindsay	New York, N. Y	May 14, 1872	126, 818
Carpet-stretcher	J. Lindsay	New York, N. Y	Aug. 5, 1873	141, 446
Carpet-stretcher	J. Luther	Buchanan, Mich	Nov. 11, 1873	144, 464
Carpet-stretcher	J. B. Martindale	Newcastle, Ind	Dec. 4, 1866	60, 216
Carpet-stretcher	C. S. McRobert	Plymouth, Mich	Feb. 4, 1868	73, 986
Carpet-stretcher	H. A. Mead	Cuba, N. Y	July 25, 1865	48, 968
Carpet-stretcher	A. F. Michael and H. N. Duncan.	Lockport, N. Y	July 1, 1873	140, 522
Carpet-stretcher	W. Minster	Washington, D. C	July 28, 1868	80, 495
Carpet-stretcher	G. Mosman	Chicopee, Mass	Jan. 2, 1866	51, 853
Carpet-stretcher	G. Mosman	Chicopee, Mass	May 14, 1867	64, 788
Carpet-stretcher	G. G. Mudge	Pittsburgh, Ind	July 18, 1865	48, 829
Carpet-stretcher	S. Pennock	Geneva, Ill	Jan. 5, 1869	85, 690
Carpet-stretcher	H. Powelson	New Brunswick, N. J	Sept. 23, 1873	143, 032
Carpet-stretcher	D. Pray	Boston, Mass	June 4, 1872	127, 515
Carpet-stretcher	A. S. Richards	Montgomery County, Md	Nov. 1, 1870	108, 938
Carpet-stretcher	H. Ridley	Hartford, Conn	Aug. 24, 1858	21, 303
Carpet-stretcher	W. A. Robinson	Grand Rapids, Mich	Oct. 22, 1867	70, 023
Carpet-stretcher	C. Rückert	New York, N. Y	Jan. 26, 1869	86, 253
Carpet-stretcher	E. P. Shaffer	Rochester, N. Y	Apr. 25, 1871	114, 053
Carpet-stretcher	N. K. Shelton	Wabash, Ind	Feb. 25, 1873	136, 339
Carpet-stretcher	L. B. Southworth	Deep River, Conn	Apr. 7, 1868	76, 355
Carpet-stretcher	S. Stevenson	Dansville, N. Y	Nov. 3, 1868	83, 741
Carpet-stretcher	G. W. Story	Kansas City, Mo	Apr. 5, 1870	101, 677
Carpet-stretcher	T. B. Stout	Keyport, N. J	June 8, 1869	91, 179
Carpet-stretcher	W. W. Taylor	Newark, N. J	July 9, 1867	66, 649
Carpet-stretcher	C. Terry	Saint Louis, Mo	Apr. 29, 1873	138, 470
Carpet-stretcher	H. N. Tucker	Stoughton, Mass	Dec. 3, 1872	133, 552
Carpet-stretcher	E. W. Twing	Springfield, Mass	June 26, 1866	55, 935
Carpet-stretcher	O. Vanorman	Fon du Lac, Wis	Sept. 10, 1872	131, 196
Carpet-stretcher	O. Vanorman	Fon du Lac, Wis	Apr. 8, 1873	137, 739
Carpet-stretcher	I. P. Warner	Marengo, Ill	Aug. 13, 1867	67, 690
Carpet-stretcher	J. Warner	New Britain, Conn	Mar. 9, 1858	19, 596
Carpet-stretcher	J. W. Weatherby	Kingsville, Ohio	Oct. 18, 1853	10, 143
Carpet-stretcher	W. Weaver	Salem, Ohio	Apr. 30, 1867	64, 387
Carpet-stretcher	W. Wheeler	West Poultney, Vt	Oct. 18, 1859	25, 858
Carpet-stretcher	D. White	Normal, Ill	Apr. 1, 1873	137, 522
Carpet-stretcher	T. Wilson and J. W. Appleyard	Chicago, Ill	Apr. 20, 1869	89, 107
Carpet-stretcher	E. Wood	Monson, Mass	Aug. 12, 1862	36, 187
Carpet stretcher and hammer	F. M. Osborn	Dover Plains, N. Y	May 29, 1866	55, 154
Carpet stretcher and holder	J. S. Munger	Olean, N. Y	June 11, 1867	65, 686
Carpet stretcher and nailer	E. S. Wheeler	Westport, Conn	June 15, 1869	91, 392
Carpet-stretcher and tack-driver	W. Brown	Springfield, Mass	Feb. 5, 1867	61, 803
Carpet-stretcher and tack-driver	W. Brown	New York, N. Y	July 30, 1867	67, 159
Carpet-stretcher and tack-hammer, Combined	W. Brown	New York, N. Y	Aug. 8, 1871	117, 735
Carpet-stretcher and tack-holder	F. Ashley	New York, N. Y	Oct. 16, 1866	58, 751
Carpet-stretcher and tack-holder	F. W. Judd	New Britain, Conn	Dec. 14, 1869	97, 929
Carpet-stretcher and tack-holder	R. M. Mansur	Augusta, Me	June 23, 1868	79, 242
Carpet stretcher and tacker	J. K. Bancroft and R. C. Bache	Philadelphia, Pa	May 24, 1870	103, 410
Carpet stretcher and tacker	Z. A. Ward	Pittsfield, Mass	Dec. 16, 1873	145, 604
Carpet-stretcher, tack driver and puller combined.	L. Colby and J. D. Gilman	Morrisville, Vt., and Boston, Mass.	Jan. 19, 1869	85, 908
Carpet-stretcher, tack-holder, and nail-puller, Combined.	H. C. Velie	Poughkeepsie, N. Y	Aug. 20, 1872	130, 774
Carpet-sweeper	J. B. Baker	Syracuse, N. Y	July 27, 1869	92, 929
Carpet-sweeper	W. G. Budlong	Hartford, Conn	May 24, 1859	24, 176
Carpet-sweeper	A. C. Carey	Ipswich, Mass	Oct. 19, 1858	21, 815
Carpet-sweeper	J. H. Crane	Charlestown, Mass	Aug. 16, 1859	25, 099
Carpet-sweeper	H. Davis	Bethlehem, Conn	May 24, 1859	24, 103

Index of patents issued from the United States Patent Office from 1790 *to* 1873, *inclusive*—Continued.

Invention.	Inventor.	Residence.	Date.	No.
Carpet-sweeper	J. Edson	Boston, Mass	Oct. 5, 1858	21,660
Carpet-sweeper	J. Edson	Boston, Mass	Apr. 5, 1859	23,526
Carpet-sweeper	J. Edson	Boston, Mass	Aug. 16, 1859	25,104
Carpet-sweeper	D. C. Hall	Hannibal, Mo	July 11, 1871	116,833
Carpet-sweeper	D. Harris	Boston, Mass	Oct. 5, 1858	21,673
Carpet-sweeper	H. H. Herrick	East Boston, Mass	Aug. 17, 1858	21,233
Carpet-sweeper	H. H. Herrick	East Boston, Mass	Sept. 6, 1859	25,332
Carpet-sweeper	D. Hess	West Union, Iowa	July 10, 1860	29,077
Carpet-sweeper	R. C. Higgins	Boston, Mass	May 9, 1871	114,680
Carpet-sweeper	R. C. Higgins and A. Fuller	Boston, Mass	Nov. 16, 1869	96,807
Carpet-sweeper	W. Miller	Boston, Mass	June 24, 1873	140,293
Carpet-sweeper	H. A. Palmer and A. H. Spencer.	Hingham and Boston, Mass	Dec. 30, 1873	146,016
Carpet-sweeper	N. B. Pratt	Deep River, Conn	Feb. 8, 1859	22,890
Carpet-sweeper	N. B. Pratt	Deep River, Conn	Feb. 15, 1859	22,975
Carpet-sweeper	S. F. Pratt	Roxbury, Mass	Jan. 18, 1859	22,671
Carpet-sweeper	G. Purrington, jr., and J. H. Purrington.	New York, N. Y., and Mattapoisett, Mass.	Jan. 1, 1867	60,785
Carpet-sweeper	S. P. Rowell	Reading, Mass	Oct. 5, 1858	21,701
Carpet-sweeper	R. Shaler	Madison, Conn	Sept. 7, 1858	21,451
Carpet-sweeper	G. F. Taylor	New York, N. Y	July 6, 1869	92,226
Carpet-sweeper	O. H. Weed	Boston, Mass	May 18, 1869	90,211
Carpet-sweeper	O. H. Weed	Boston, Mass	May 30, 1871	115,399
Carpet-sweeper	F. O. Willey and D. B. McEnery.	La Fayette, Ind	Apr. 2, 1872	125,369
Carpet-sweeper, Fan	A. C. Carey	Lynn, Mass	Jan. 4, 1859	22,488
Carpet-sweeper guard	A. J. Knight	New York, N. Y	May 24, 1870	103,344
Carpet-sweepers, Driving-gear wheel for	A. J. Knight	Brooklyn, N. Y	July 12, 1870	105,223
Carpet-sweeping machines, Guard for	G. F. Taylor	New York, N. Y	Sept. 29, 1868	82,566
Carpet-sweeping machines, Gear for	G. F. Taylor	New York, N. Y	June 7, 1870	103,939
Carpet-sweeping machines, Gear for	G. F. Taylor	New York, N. Y	Aug. 16, 1870	106,429
Carpet-tacks	T. A. Mitchell	Washington, D. C	Mar. 29, 1870	101,374
Carpet-tacking machine	F. H. Staffer	Philadelphia, Pa	Apr. 26, 1870	102,331
Carpet to floor, Securing	E. Jackman	Portland, Conn	Dec. 12, 1854	12,061
Carpet-underlie	N. Edwards	Jericho, Vt	July 26, 1870	105,661
Carpet-wadding	T. H. Dunham	Boston, Mass	May 1, 1866	54,312
Carpet-wadding	T. H. Dunham	Boston, Mass	Apr. 23, 1867	64,084
Carpet-wadding	E. Waite	Franklin, Mass	July 9, 1867	66,539
Carpet, washing, scouring, and fulling cloth and working hat-bodies, Machine for making wool or fur.	H. Stayton	Lockport, N. Y	Nov. 25, 1829	....
Carpet, Woolen	W. Harrington	Harrison, N. Y	June 11, 1829	....
Carpets and other fabrics from jute, flax, &c., Manufacture of.	T. Crossley	Bridgeport, Conn	Jan. 21, 1868	73,512
Carpets and rugs, Manufacture of double-pile	J. Goulding	Worcester, Mass	Jan. 20, 1857	16,437
Carpets and rugs, Manufacturing	J. G. McNair	West Farms, N. Y	Oct. 24, 1854	11,834
Carpets, chests, drawers, wardrobes, &c., Aromatic lining for.	J. M. Perkins	Plainfield, N. J	Apr. 7, 1868	76,342
Carpets, Compound for cleaning	L. Marks	Cincinnati, Ohio	Dec. 27, 1870	110,481
Carpets, coverlids, diapers, &c., Weaving	E. Meily, jr., and J. and S. Mellinger.	Lebanon, Pa	Mar. 1, 1834	....
Carpets, Fabric for underlaying	W. S. Pratt	Brooklyn, N. Y	Nov. 14, 1856	16,036
Carpets, Finishing	S. Fay	Lowell, Mass	Oct. 23, 1855	13,702
Carpets, Flooring or dust-rack for	G. J. Colby	Waterbury, Vt	July 25, 1865	48,907
Carpets, Implement for slitting and joining rags for	J. Beal	Port Gibson, N. Y	Aug. 10, 1869	93,514
Carpets, Implement for slitting and joining rags for	M. Green	Coldwater, Mich	Apr. 25, 1871	114,004
Carpets, Implement for slitting and looping rags for	D. A. Russell	Windham, Ohio	Feb. 21, 1871	111,977
Carpets, Machine for sizing backs of	H. G. Thompson	New York, N. Y	Oct. 21, 1862	36,738
Carpets, Making ingrain	J. and N. Haight	New York, N. Y	Aug. 12, 1820	....
Carpets, Making ingrain	D. B. Kerr	New York, N. Y	Sept. 23, 1856	15,767
Carpets, Making ingrain	W. A. Prince	New York, N. Y	Feb. 12, 1823	....
Carpets, Making woolen	W. Harrington	Harrison, N. Y	Mar. 3, 1829	....
Carpets, Manufacture of	H. A. Clark	Boston, Mass	Sept. 2, 1873	142,439
Carpets, Manufacturing	J. G. McNair	West Farms, N. Y	Aug. 7, 1855	13,395
Carpets, Manufacturing two-ply	T. Crossley	Boston, Mass	Aug. 22, 1854	11,564
Carpets, Manufacturing two and three ply	A. Smith	West Farms, N. Y	Dec. 10, 1850	7,825
Carpets, Slitting and joining rags for	B. F. Cady	Chittenango, N. Y	Jan. 25, 1870	99,154
Carpets, Taking figures of ingrain	W. Sherwood	Somersworth, N. H	Apr. 20, 1830	....
Carpets, Tool for laying and stretching	F. Cist, W. Kossak, and W. H. Godfrey.	Saint Louis, Mo	June 12, 1860	28,650
Carpets, Weaving Brussels and other pile	J. B. Buzzell	Clinton, Mass	Aug. 22, 1871	118,338
Carpeting and embossed matting, India-rubber	J. H. Cheever	New York, N. Y	Apr. 12, 1870	101,823
Carpeting and rugs, Manufacture of	J. Humphries	New York, N. Y	Dec. 10, 1838	1,028
Carpeting, Felt for	A. Byington	Herkimer, N. Y	May 6, 1829	....
Carriage	B. F. Adams	Bangor, Me	Aug. 17, 1869	93,791
Carriage	R. W. Benedict	Brant, N. Y	June 10, 1856	15,097
Carriage	C. Brown	Adrian, Mich	Nov. 10, 1868	83,917
Carriage	S. C. Brown and L. J. Hicks	Macedon, N. Y	July 17, 1837	279
Carriage	C. T. Bush	Rensselaerville, N. Y	Apr. 3, 1866	53,563
Carriage	J. S. Cisco	Xenia, Ohio	Aug. 28, 1855	13,487
Carriage	I. A. Clippinger	Newton, Iowa	Nov. 16, 1869	96,887
Carriage	L. W. Coe	Auburn, Me	Feb. 20, 1872	123,809
Carriage	G. R. Comstock	Manheim, N. Y	Jan. 23, 1855	12,272
Carriage	C. Conderman	Hornellsville, N. Y	June 18, 1867	65,876
Carriage	L. D. Cowles	Armada, Mich	Jan. 21, 1862	34,197
Carriage	J. W. Crannell	Olivet, Mich	Jan. 6, 1857	16,353
Carriage	J. Curtis	Cincinnati, Ohio	Mar. 26, 1867	63,223
Carriage	J. Curtis	Cincinnati, Ohio	Oct. 3, 1871	119,579
Carriage	H. E. Delessert	Paris, France	Aug. 12, 1873	141,704
Carriage	C. N. Dennett	Amesbury, Mass	Nov. 15, 1870	109,189
Carriage	J. L. Dolson	Charlotte, Mich	May 25, 1869	90,351
Carriage	G. B. Durkee	Alden, N. Y	May 27, 1851	8,118
Carriage	G. W. Efner and S. A. Sperry	Ann Arbor, Mich	Sept. 4, 1860	29,866
Carriage	E. and C. Everett, jr	Washington, D. C	Dec. 17, 1850	7,835
Carriage	T. A. and A. F. Fisher	Beardstown, Ill	Apr. 30, 1867	64,299
Carriage	A. W. Forwood	Scott County, Ky	Mar. 27, 1847	5,034

Index of patents issued from the United States Patent Office from 1790 *to* 1873, *inclusive*—Continued.

Invention.	Inventor.	Residence.	Date.	No.
Carriage	J. Fox	Manchester, N. J	Aug. 10, 1852	9,183
Carriage	E. Frauer	Cobleskill, N. Y	Mar. 1, 1864	41,765
Carriage	D. Freeman	Burford, Canada	Oct. 21, 1856	15,923
Carriage	J. Gale, M. B. Ames, and F. Blaisdale.	Lawrence, Mass	Oct. 8, 1867	69,559
Carriage	L. Glesenkamp	Pittsburgh, Pa	Apr. 15, 1873	137,914
Carriage	S. P. Graham	Columbus, Ohio	Sept. 13, 1870	107,360
Carriage	A. S. Grant	Waupun, Wis	May 17, 1864	42,760
Carriage	C. H. Guard	Toronto, Canada	Feb. 21, 1871	111,928
Carriage	B. Hale	Newburyport, Mass	Oct. 4, 1817	....
Carriage	G. W. Ham	Parsonfield, Me	Sept. 13, 1870	107,252
Carriage	J. C. Ham	New York, N. Y	July 27, 1869	92,964
Carriage	N. L. Hatch	Cape Elizabeth, Me	Dec. 22, 1868	85,093
Carriage	J. Hatfield	Cleveland, Ohio	June 18, 1867	65,910
Carriage	G. L. Haussknecht	New Haven, Conn	Dec. 16, 1851	8,588
Carriage	J. R. Hiller	Woodland, Cal	Aug. 30, 1870	106,817
Carriage	E. Hitt	Katonah, N. Y	Feb. 18, 1868	74,689
Carriage	M. G. Hubbard	New York, N. Y	July 24, 1855	13,311
Carriage	S. Jackson	Newark, N. J	Apr. 6, 1869	88,637
Carriage	J. Jones	Clyde, N. Y	Jan. 14, 1851	7,906
Carriage	C. P. Kimball	Portland, Me	Nov. 15, 1864	45,050
Carriage	J. T. Kimball	Kennebunk, Me	July 22, 1845	4,121
Carriage	L. King	Madison, N. Y	Dec. 23, 1851	8,609
Carriage	T. Knox	Snickersville, Va	Mar. 15, 1828	....
Carriage	L. Kutscher	New York, N. Y	Nov. 27, 1866	60,016
Carriage	L. W. Mason	Shelburne Falls, Mass	Apr. 11, 1865	47,211
Carriage	J. R. McAlister	Heuvelton, N. Y	July 13, 1869	92,536
Carriage	H. Moon	Red Creek, N. H	Mar. 3, 1868	75,183
Carriage	J. H. Moore	Warren, Mass	Apr. 9, 1867	63,739
Carriage	G. H. and E. Morgan	Edgware Road, England	Nov. 6, 1866	59,525
Carriage	G. H. and E. Morgan	Edgware Road, England	Nov. 6, 1866	59,526
Carriage	J. J. Morris	New Bedford, N. J	Aug. 28, 1866	57,548
Carriage	E. C. Newton	Batavia, Ill	Nov. 28, 1871	121,295
Carriage	W. Ottmann	New York, N. Y	Nov. 26, 1872	133,475
Carriage	T. B. Patten	West Amesbury, Mass	Jan. 28, 1873	135,287
Carriage	J. Patterson	Franklinsville, N. Y	Apr. 16, 1850	7,290
Carriage	B. Peck	Philadelphia, Pa	July 5, 1814	....
Carriage	T. W. and H. K. Porter	Boston, Mass	Mar. 24, 1868	75,973
Carriage	U. Reynolds	New York, N. Y	May 1, 1866	54,410
Carriage	E. Robbins	Worcester, Mass	Apr. 18, 1865	47,334
Carriage	E. Robbins	Worcester, Mass	June 20, 1865	48,310
Carriage	J. Rock	Hastings, England	Sept. 17, 1867	68,904
Carriage	W. Ross	New York, N. Y	Jan. 14, 1809	....
Carriage	J. L. Rowley	Steuben County, Ind	June 13, 1854	11,085
Carriage	J. S. Royce	Leicester, N. Y	May 8, 1847	5,107
Carriage	C. W. Saladee	Saint Catharine's, Canada	Oct. 17, 1871	120,106
Carriage	B. E. Sampson	Boston, Mass	Oct. 25, 1864	44,826
Carriage	A. Searls	San Francisco, Cal	Dec. 10, 1867	72,091
Carriage	I. M. Singer	New York, N. Y	Oct. 25, 1859	25,920
Carriage	E. Soper	New York, N. Y	Dec. 3, 1867	71,805
Carriage	J. C. Spencer	Phelps, N. Y	May 27, 1851	8,120
Carriage	L. Theobald	Bremen, Ind	Dec. 28, 1869	98,443
Carriage	C. Thomas	Boston, Mass	Dec. 27, 1870	110,606
Carriage	C. Thomas	Boston, Mass	Apr. 4, 1871	113,705
Carriage	L. H. Thomas	Waterbury, Vt	May 10, 1864	42,700
Carriage	G. K. Tichenor	Wayland, N. Y	Sept. 2, 1873	142,531
Carriage	S. Titcomb	Amesbury, Mass	Aug. 18, 1868	81,227
Carriage	J. D. Van Hoevenbergh	Kingston, N. Y	Nov. 27, 1866	60,093
Carriage	J. D. Van Hoevenbergh	Kingston, N. Y	May 26, 1868	78,340
Carriage	M. S. Watkins	Somerville, Tenn	June 25, 1850	7,467
Carriage	F. M. Watson	Columbia, Tenn	July 22, 1873	141,190
Carriage	J. Whitehead	Ames Station, Iowa	July 28, 1868	80,377
Carriage	E. S. Wicklin and J. D. Weaver	Carlinville, Ill	Sept. 4, 1860	29,931
Carriage	J. B. Withey	Chicago, Ill	Sept. 24, 1872	131,728
Carriage	C. B. Wood	New York, N. Y	May 14, 1861	32,327
Carriage, Ambulance	T. Wilkins	Greenville, Ill	Nov. 22, 1864	45,209
Carriage and buggy top bow-irons	G. and A. Woeber	Davenport, Iowa	Oct. 15, 1867	69,885
Carriage and car axle	S. Coleman	Mount Pleasant, Va	Mar. 2, 1836	....
Carriage and car axles, Oiling	W. Kenworthy and J. H. Pollitt	Buchanan, Pa	Dec. 7, 1869	97,651
Carriage and car wheel	A. F. Cooper	San Francisco, Cal	May 31, 1870	103 573
Carriage and car wheel and axle-box	J. A. Maynard	Newtonville, Mass	July 12, 1870	105,228
Carriage and cart boxes, Casting	C. Darby	Worcester, Mass	Jan. 22, 1825	....
Carriage and cradle, Child's	C. W. Higgins	Waukesha, Wis	Nov. 5, 1867	70,566
Carriage and gig springs, Connecting	A. Davis	Easton, Md	Mar. 18, 1835	....
Carriage and harness, Wheel	G. Barnard	United States	May 25, 1838	748
Carriage and machinery when wheel is used	Williams and Wing	Hartford, Conn	May 16, 1835	....
Carriage and other vehicle wheel	A. F. Cooper	San Francisco, Cal	Feb. 15, 1870	99,851
Carriage and perambulator handle, Child's	W. E. Crandall	New York, N. Y	Aug. 2, 1870	106,033
Carriage and railway-car spring	H. Gardiner	New York, N. Y	Jan. 3, 1860	26,668
Carriage and sleigh combined	J. S. Drake	New York, N. Y	Aug. 3, 1869	93,184
Carriage and sleigh top	D. K. Hickok	Morrisville, Wis	Mar. 24, 1868	75,908
Carriage and velocipede, Baby	J. Scheen	New Haven, Conn	Apr. 2, 1867	63,433
Carriage and velocipede combined, Child's	J. C. Cline	Philadelphia, Pa	Nov. 2, 1869	96,396
Carriage and wagon brake	J. G. Bicknell	Cambridge, Mass	Mar. 1, 1870	100,357
Carriage and wagon jack	W. N. Rowe	Sharpsburgh, Md	Aug. 16, 1859	25,141
Carriage and wagon seat	C. K. Lehmann	Savannah, Ohio	Oct. 7, 1873	143,515
Carriage and wagon spring	P. Hill	Veteran, N. Y	Nov. 25, 1837	489
Carriage and wagon spring	G. W. La Baw	Jersey City N. J	Feb. 25, 1862	34,549
Carriage and wheel, Railway	R. Grant	Baltimore, Md	Nov. 3, 1838	999
Carriage attachment	G. J. Capewell	West Cheshire, Conn	Sept. 17, 1867	68,951
Carriage-axle	S. Adams	New York, N. Y	Feb. 1, 1848	5,426
Carriage-axle	S. Barker	Hartford, Conn	Oct. 30, 1866	59,162
Carriage-axle	W. D. Bollinger	Cedar Rapids, Iowa	Sept. 29, 1868	82,483
Carriage-axle	W. A. Clark	Woodbridge, Conn	Mar. 22, 1870	101,097
Carriage-axle	S. S. Cook	Woonsocket, R. I	Dec. 9, 1873	145,280
Carriage axle	A. B. Crandall	Sumner, Ill	May 28, 1872	127,222
Carriage axle	D. Dalzell	South Egremont, Mass	Jan. 4, 1870	98,473
Carriage-axle	T. H. Elder	Chicago, Ill	Mar. 30, 1869	88,376

Index of patents issued from the United States Patent Office from 1790 *to* 1873, *inclusive*—Continued.

Invention.	Inventor.	Residence.	Date.	No.
Carriage-axle	T. Falvey	Racine, Wis	Feb. 5, 1867	61, 821
Carriage-axle	J. J. Flack	Joliet, Ill	Sept. 4, 1849	6, 688
Carriage-axle	S. Forrester	Allegheny, Pa	Dec. 7, 1869	97, 622
Carriage-axle	S. Forrester	Allegheny, Pa	Dec. 7, 1869	97, 623
Carriage-axle	M. R. Freeman	Macon, Ga	July 26, 1870	105, 793
Carriage-axle	K. Goddard	Philadelphia, Pa	June 15, 1852	9, 023
Carriage-axle	J. Grabach	Clyde, Ohio	Aug. 17, 1869	93, 707
Carriage-axle	E. H. Green	Baltimore Md	June 13, 1854	11, 063
Carriage-axle	W. N. Hall	Springfield, Tex	July 26, 1870	105, 801
Carriage-axle	W. T. Harrington	Roxbury, Mass	May 24, 1864	42, 911
Carriage-axle	E. W. Ives	Hamden, Conn	Apr. 25, 1871	114, 147
Carriage-axle	H. Killam	New Haven, Conn	Sept. 27, 1870	107, 783
Carriage-axle	H. Killam	New Haven, Conn	Jan. 3, 1871	110, 766
Carriage-axle	W. A. Lewis	Joliet, Ill	Mar. 8, 1870	100, 646
Carriage-axle	L. Littlejohn	New York, N. Y	June 7, 1870	103, 900
Carriage-axle	G. A. Lloyd	San Francisco, Cal	Mar. 29, 1870	101, 370
Carriage-axle	J. H. Moffett	Reading, Mich	June 29, 1869	92, 081
Carriage-axle	D. W. Phillips and W. Maher	Covington, N. Y	Apr. 23, 1828	....
Carriage-axle	J. A. C. Ruffner	Hillsdale, Pa	Feb. 9, 1869	86, 693
Carriage-axle	W. H. Saunders	Hastings, N. Y	June 27, 1854	11, 179
Carriage axle	F. Scherb	New York, N. Y	Apr. 5, 1870	101, 521
Carriage-axle	I. Slack	Avondale, Pa	Oct. 10, 1846	4, 809
Carriage-axle	A. E. Smith	Bronxville, N. Y	Oct. 19, 1869	96, 043
Carriage-axle	A. E. Smith	Bronxville, N. Y	Dec. 28, 1869	98, 436
Carriage-axle	A. E. Smith	Bronxville, N. Y	July 12, 1870	105, 266
Carriage-axle	A. E. Smith	Bronxville, N. Y	Sept. 6, 1870	107, 112
Carriage-axle	A. E. Smith	Bronxville, N. Y	Nov. 29, 1870	109, 681
Carriage-axle	T. Spurrier	Sharon, Pa	June 8, 1869	90, 968
Carriage-axle	G. H. Thomas	New York, N. Y	June 6, 1865	48, 111
Carriage-axle	O. Tracy	Fitchburgh, Mass	Dec. 12, 1846	4, 889
Carriage-axle	F. Volkmann and A. Miller	Philadelphia, Pa	Mar. 23, 1869	88, 240
Carriage-axle	E. F. Wagner	Houston, Tex	June 13, 1871	115, 913
Carriage-axle	E. Wells	New Haven, Conn	Mar. 14, 1871	112, 757
Carriage-axle	J. F. Wilbur and T. Tuttle	Pownal, Me	Feb. 16, 1869	86, 959
Carriage-axle	J. B. Wilson	Philadelphia, Pa	Sept. 14, 1869	94, 932
Carriage-axle	J. T. Wilson	Pittsburgh, Pa	Mar. 9, 1869	87, 609
Carriage-axle adjuster	J. Childs	West Troy, N. Y	Mar. 5, 1867	62, 607
Carriage-axle and axle box	B. A. Berryman	Saint Louis, Mo	Jan. 21, 1873	135, 068
Carriage axle and box	C. Beach	Newark, N. J	June 26, 1807	....
Carriage axle and box	F. Davis	Baltimore, Md	Nov. 24, 1834	....
Carriage axle and boxing	J. W. and A. W. Beer	Rural Valley, Pa	Aug. 8, 1871	117, 730
Carriage-axle and carriage-yoke	S. Rowell	Amesbury, Mass	Dec. 6, 1870	109, 842
Carriage axle and hub	L. Blair	Painesville, Ohio	Sept. 10, 1867	68, 596
Carriage axle and journal	L. Dorman	Worcester, Mass	Aug. 14, 1866	57, 100
Carriage-axle, Anti-friction	A. Allcott	Haverhill, Mass	Mar. 29, 1870	101, 205
Carriage-axle, Anti-friction	E. C. Otis	Voluntown, Conn	July 24, 1866	56, 601
Carriage-axle arm	A. J. Bell	Greenupsburgh, Ky	Mar. 20, 1860	27, 517
Carriage-axle connection	J. W. Kingsbury	New Bedford, Mass	Oct. 5, 1869	95, 488
Carriage-axle die	W. W. Simmons	Birmingham, Conn	Dec. 8, 1868	84, 845
Carriage-axle gage	H. Harper	Berlin, Wis	Nov. 18, 1862	36, 951
Carriage-axle gage	M. Merk	Rochester, N. Y	Sept. 14, 1869	94, 834
Carriage-axle iron	S. J. Dennis	Boston, Mass	May 13, 1873	138, 795
Carriage-axle lubricator	N. B. Brown	Antwerp, N. Y	Apr. 24, 1866	54, 246
Carriage-axle lubricator	J. E. Emerson	San Francisco, Cal	Apr. 10, 1860	27, 785
Carriage-axle lubricator	J. Killefer	West Richfield, Ohio	Apr. 26, 1870	102, 408
Carriage-axle lubricator	E. Thomas	Kickapoo, Ill	Nov. 5, 1867	70, 650
Carriage-axle lubricator	A. A. Vedder	Lysander, N. Y	Oct. 6, 1857	18, 364
Carriage axle, Reversible	J. R. Renkin	Hillsdale, Pa	Oct. 12, 1869	95, 726
Carriage axle shield	B. F. Robbins	Harwich, Mass	Sept. 26, 1871	119, 281
Carriage-axles, Device for oiling	J. Vanderpool	Hackensack, N. J	Mar. 1, 1870	100, 342
Carriage-axles, Gage for setting	L. Taylor	Marlborough, Mass	May 21, 1872	126, 995
Carriage-axles, Machine for making	B. W. Foster	Auburn, N. Y	Mar. 24, 1868	75, 893
Carriage-axles, Machine for making	C. Young	Auburn, N. Y	May 8, 1866	54, 638
Carriage-axles, Machine for making	C. Young	Auburn, N. Y	May 8, 1866	54, 639
Carriage-axles, Machine for rolling	T. H. Miller	Lancaster, Pa	Aug. 27, 1861	33, 155
Carriage-axles, Machine for rolling	J. C. Richardson	New Haven, Conn	Mar. 11, 1873	136, 620
Carriage-axles, Machine for rolling	J. C. Richardson	Philadelphia, Pa	Dec. 16, 1873	145, 526
Carriage-axles, Machine for turning	J. G. Aram	Cordova, Ill	May 28, 1872	127, 211
Carriage-axles, Machine for turning	H. E. Forrest	Boston, Mass	Apr. 15, 1873	137, 769
Carriage-axles, Machine for turning	W. Jones	Brooklyn, N. Y	Nov. 25, 1873	144, 908
Carriage-axles, Machine for upsetting	Z. Doolittle	Perry, Ga	July 6, 1858	20, 780
Carriage-axles, Manufacture of	J. C. and J. Lones, C. Vernon, and E. Holden.	Southwick, England	Dec. 16, 1873	145, 510
Carriage-axles, Manufacture of	A. and J. Schirck and J. Smith	Rochester, N. Y	Mar. 4, 1873	136, 460
Carriage-axles, Manufacture of	A. E. Smith	Bronxville, N. Y	Aug. 2, 1870	106, 086
Carriage-axles, Method of forming	J. Le Ferre	Charlestown, Mass	May 26, 1868	78, 213
Carriage-axles, Oil-cup for	L. Gregory	Battle Creek, Mich	June 21, 1864	43, 197
Carriage-axles, Pipe box or skein for	J. Jones	Galway, N. Y	Apr. 1, 1845	3, 984
Carriage-axles, Sand-box for	J. S. Steele	Rockingham, Vt	Feb. 19, 1867	62, 231
Carriage-axles, Securing nuts to	K. Goddard	Philadelphia, Pa	May 27, 1856	14, 953
Carriage-axles, Wrench for nuts of	C. N. Morgan	Granby, Mass	Aug. 7, 1866	56, 980
Carriage-axle-tree	A. Johnson	Newark, N. J	Feb. 9, 1833	....
Carriage axle-tree, Railway and other wheeled	R. Winans	Baltimore, Md	July 20, 1831	....
Carriage-axle-trees, Gage for setting	L. D. Cross	Phelps, N. Y	Mar. 18, 1873	136, 819
Carriage-axle-trees, Lessening the friction in	R. Kilborn	Great Barrington, Mass	Apr. 12, 1808	....
Carriage-axle-trees, Method of turning	J. Hennon	Brighton, Pa	July 1, 1856	15, 233
Carriage-axle-trees, Mode of greasing	F. Biemler	New York	Aug. 22, 1816	....
Carriage, Baby	C. H. Amidon	Miller's Falls, Mass	Jan. 7, 1873	134, 623
Carriage, Baby	J. Higgins	Salem Crossing, Ind	June 7, 1870	103, 882
Carriage, Baby	S. McIntire	Brooklyn, N. Y	Mar. 11, 1873	136, 607
Carriage Baby	S. S. Woodruff	Brooklyn, N. Y	May 14, 1867	64, 822
Carriage, Ball	A. D. Bishop	New York, N. Y	Mar. 29, 1864	42, 066
Carriage-bodies, &c., Compound or rough stuff to be applied to.	E. Scharnikow	Margarettville, N. Y	Aug. 30, 1870	106, 875
Carriage-bodies, Construction of	S. P. Graham	Richland Centre, Ind	Sept. 17, 1867	68, 980
Carriage-bodies, Device for adjusting and hanging	J. D. Cole	Phelps, N. Y	Apr. 13, 1869	88, 944
Carriage-bodies, Hanging	D. B. Barnum	New Fairfield, Conn	Oct. 16, 1839	1, 369
Carriage-bodies, Hanging	G. A. Blair and A. S. Gladding	Johnsonburgh, N. Y	Jan. 21, 1868	73, 492

Index of patents issued from the United States Patent Office from 1790 to 1873, inclusive—Continued.

Invention.	Inventor.	Residence.	Date.	No.
Carriage-bodies, Hanging	I. W. Brittan	Medina, Ohio	Oct. 26, 1839	1,381
Carriage-bodies, Hanging	B. F. Brown	Dorchester, Mass	Dec. 12, 1854	12,051
Carriage-bodies, Hanging	J. M. De Ruyter	New York, N. Y	Mar. 6, 1847	4,998
Carriage-bodies, Hanging	W. Doulin	Youngstown, Ohio	Nov. 1, 1859	25,958
Carriage-bodies, Hanging	S. Fairchild	Trumbull, Conn	Jan. 18, 1848	5,413
Carriage-bodies, Hanging	A. Hosmer	Bath, Ohio	June 2, 1845	4,065
Carriage-bodies, Hanging	M. G. Hubbard	Hume, N. Y	Feb. 12, 1850	7,085
Carriage-bodies, Hanging	M. G. Hubbard	Geneva, N. Y	Nov. 26, 1850	7,796
Carriage-bodies, Hanging	I. Jackson	West Grove, Pa	Mar. 20, 1849	6,213
Carriage-bodies, Hanging	J. Jones	Clyde, N. Y	July 22, 1851	8,234
Carriage-bodies, Hanging	J. Jones	Clyde, N. Y	July 22, 1851	8,236
Carriage-bodies, Hanging	E. Lane	Philadelphia, Pa	June 2, 1863	38,751
Carriage-bodies, Hanging	L. C. Miner	Hartford, Conn	June 14, 1859	24,398
Carriage-bodies, Hanging	A. W. Porter	Little Falls, N. Y	Nov. 13, 1855	13,797
Carriage-bodies, Hanging	J. Reynolds	Newberry, Pa	July 9, 1844	3,653
Carriage-bodies, Hanging	H. Scharch	New York, N. Y	Oct. 23, 1866	59,081
Carriage-bodies, Hanging	J. Speed	Crawfordsville, Ind	Feb. 27, 1847	4,989
Carriage-bodies, Hanging	C. C. Stringfellow and D. W. Surles.	Lumpkin, Ga	Jan. 15, 1861	31,134
Carriage-bodies, Hanging	A. Vallerchamp	McDowellsville, Pa	Oct. 12, 1839	1,362
Carriage-bodies, Hanging	C. J. Woolson	Cleveland, Ohio	Sept. 11, 1847	5,281
Carriage bodies, Making	J. Reeder	Lebanon, Ohio	Mar. 4, 1828	
Carriage-bodies, Mode of suspending	H. W. Libbey	Cleveland, Ohio	Jan. 28, 1868	73,821
Carriage-body	H. Dunlap	Georgetown, D. C	Aug. 27, 1812	......
Carriage-body	W. C. and J. Dunn	New York, N. Y	June 11, 1861	32,513
Carriage-body	E. Kirsch	New Haven, Conn	July 31, 1860	29,386
Carriage-body	C. Thomas	Boston, Mass	Jan. 31, 1871	111,491
Carriage-body	E. M. Williams	Columbus, Ohio	Jan. 30, 1872	123,213
Carriage-body brace	J. H. Ormsby	Dixon, Ill	Oct. 12, 1869	95,828
Carriage-body brace	J. Taylor	Beloit, Ohio	Oct. 12, 1869	95,851
Carriage-body loop	F. A. Cowles	Plantsville, Conn	Mar. 15, 1870	100,864
Carriage-body loop	C. H. Guard	Troy, N. Y	Aug. 25, 1863	39,699
Carriage-body loop	W. W. Knowles	Plantsville, Conn	July 18, 1871	117,179
Carriage body, Standing-roof landau	E. Gibson	New York	May 10, 1811	
Carriage-body-supporting irons, Elastic bearing for	J. D. Sarven	Columbia, Tenn	Feb. 4, 1868	74,011
Carriage-body-supporting irons, Elastic bearing for	J. D. Sarven	Columbia, Tenn	Feb. 4, 1868	74,012
Carriage-body top	U. Reynolds	New York, N. Y	Sept. 25, 1866	58,292
Carriage-bolt	O. C. Burdict	New Haven, Conn	Sept. 3, 1867	68,554
Carriage-bolt	O. C. Burdict	New Haven, Conn	Sept. 26, 1871	119,309
Carriage-bolt	I. Newton	Cleveland, Ohio	Aug. 12, 1873	141,661
Carriage-bolt	A. Pond	Hamden, Conn	Apr. 11, 1865	47,245
Carriage-bolt	J. Strawn	Plymouth, Mich	Apr. 7, 1868	76,546
Carriage-bolts, Die for making	E. H. Plant	Plantsville, Conn	Mar. 18, 1873	137,023
Carriage-bolts, Machine for making	W. Koplin	Newcastle, Pa	Mar. 19, 1867	62,963
Carriage-bolts, Machine for making	J. T. Wood and E. C. Smith	Allegheny County, Pa	Dec. 6, 1864	45,367
Carriage-bolts, Method of Making	S. Frisbie and A. S. Upson	Farmington, Conn	Oct. 1, 1867	69,333
Carriage-boot	P. T. Gates	Plattsburgh, N. Y	Jan. 15, 1867	61,188
Carriage-bow	I. N. Topliff	Adrian, Mich	Dec. 27, 1870	110,513
Carriage-bow cover and slat-iron, Combined	I. N. Topliff	Elyria, Ohio	May 16, 1871	114,885
Carriage-bow iron	H. M. Bidwell	New Haven, Conn	Jan. 16, 1866	52,016
Carriage-bow iron	A. W. Clark and G. W. Marble	Charlestown, N. H	Nov. 3, 1868	83,693
Carriage-bow iron	G. W. Slater	New Haven, Conn	Apr. 9, 1867	63,660
Carriage-bow setter	H. C. Hoover	Greencastle, Pa	Mar. 24, 1868	75,916
Carriage-bows, Attaching props to	L. Sawyer	South Amesbury, Mass	May 15, 1866	54,782
Carriage-bows, Attaching the props of	D. B. Wright and L. Sawyer	South Amesbury, Mass	Aug. 31, 1858	21,391
Carriage-box	T. B. Whitlock		Feb. 23, 1802	......
Carriage box and axle	J. A. Smith	New Haven, Conn	Mar. 23, 1838	651
Carriage box, Wheel	J. Nicholson	Herkimer, N. Y	Apr. 28, 1810	......
Carriage boxes and seats, Corner-support for	W. Fisher	Marathon, N. Y	Sept. 6, 1870	107,025
Carriage boxes and seats, Metallic corner-support for.	O. H. Smith and W. Fisher	Marathon, N. Y	Mar. 22, 1870	101,172
Carriage-brace	J. Howe	Mount Pleasant, Iowa	Oct. 6, 1868	82,839
Carriage-brace	J. B. Pelton	Sandusky, N. Y	Jan. 15, 1867	61,246
Carriage-brace	M. Powo	Belvidere, N. J	Feb. 4, 1868	74,139
Carriage-brace joint	F. B. Morse	New Haven, Conn	June 11, 1867	65,761
Carriage-brake	W. Ballard	Lapeer, N. Y	June 23, 1868	79,186
Carriage-brake	E. Behr	New York, N. Y	Aug. 7, 1860	29,453
Carriage-brake	A. S. Boyer	Bernville, Pa	Dec. 21, 1869	98,018
Carriage-brake	J. E. Briggs	Watertown, N. Y	Apr. 2, 1861	31,867
Carriage-brake	L. R. Carpenter	Lancaster, Ohio	Dec. 3, 1861	33,865
Carriage-brake	E. Chapman	Akron, Ohio	Sept. 26, 1846	4,775
Carriage-brake	W. Clayton	Marshallton, Pa	Aug. 18, 1846	4,701
Carriage-brake	H. B. S. Davis	Farmington, Me	Sept. 20, 1870	107,462
Carriage-brake	G. L. Dickson	Carbondale, Pa	June 1, 1858	20,412
Carriage-brake	W. Ellmaker and C. Hurst	Earl Township, Pa	Apr. 22, 1862	35,016
Carriage-brake	D. D. Gibson and W. Cobbs	Damascoville, Ohio	Feb. 12, 1845	3,907
Carriage-brake	G. Griest	Adams County, Pa	June 5, 1849	6,489
Carriage-brake	G. Hauck	Mechanicsburgh, Pa	June 9, 1857	17,506
Carriage-brake	S. D. Kimble	Allegheny City, Pa	Oct. 20, 1868	83,174
Carriage-brake	B. F. Leet	Dayton, Nev	Dec. 28, 1869	98,277
Carriage-brake	M. K. Lewis	Iowa City, Iowa	Apr. 8, 1862	34,892
Carriage-brake	M. K. Lewis	Iowa City, Iowa	June 2, 1863	38,752
Carriage-brake	M. S. Marshall and J. G. Bicknell.	Somerville and Cambridge, Mass.	Feb. 9, 1869	86,770
Carriage brake	A. B. McFarlan	Downingtown, Pa	Apr. 17, 1849	6,355
Carriage-brake	W. D. Sheldon	Wolcott, N. Y	June 7, 1864	43,044
Carriage-brake	J. Sollenberger	Higginsport, Ohio	Mar. 7, 1854	10,604
Carriage-brake	W. T. Welch, jr	Churchville, Md	Mar. 12, 1850	7,177
Carriage-brake	A. Wnuck	Cincinnati, Ohio	Mar. 14, 1871	112,665
Carriage-brake	M. S. Woodward	Marshallton, Pa	Dec. 4, 1843	3,361
Carriage-brake, Adjustable	J. A. Letts	Trumansburgh, N. Y	Jan. 29, 1861	31,249
Carriage-brake mechanism	L. Wilber	Putney, Vt	Jan. 20, 1863	37,468
Carriage brake, Railway	P. J. O. Conway	Philadelphia, Pa	Sept. 10, 1840	1,777
Carriage-brake, Self-acting	S. Cope	Salem, Ohio	Dec. 22, 1846	4,900
Carriage-brake, Self-acting	J. Du Bois, jr	Cascade, Pa	June 13, 1846	4,571
Carriage-brake, Self-acting	W. Ellmaker	New Holland, Pa	Oct. 22, 1861	33,522
Carriage brake, Self-acting	A. Larrowe	Cohocton, N. Y	Mar. 27, 1860	27,639
Carriage-brake, Self-acting	Z. C. Robbins	Washington, D. C	June 13, 1846	4,569

Index of patents issued from the United States Patent Office from 1790 *to* 1873, *inclusive*—Continued.

Invention.	Inventor.	Residence.	Date.	No.
Carriage-brakes, Apparatus for operating	C. Stoner	Gettysburgh, Pa	Aug. 28, 1847	5, 264
Carriage-brakes, Means of operating	W. Gourley and I. Krebs	Winchester, Va	Jan. 4, 1859	22, 498
Carriage-brakes, Means of operating	I. Krebs	Winchester, Va	Jan. 4, 1859	22, 504
Carriage, &c., Buoyant	H. Stanton	Washington, D. C	Feb. 27, 1847	4, 984
Carriage-button	W. P. Bateman	Barrington, R. I	Oct. 1, 1867	69, 506
Carriage-button	C. H. Field	Providence, R. I	Dec. 27, 1870	110, 358
Carriage-button, Elastic	T. E. King	Painesville, Ohio	May 28, 1867	65, 691
Carriage, chair, and cradle, Combined	G. W. Hawk	Chicago, Ill	June 21, 1864	43, 206
Carriage check brace	I. D. Johnson	Kennett Square, Pa	Apr. 28, 1868	77, 196
Carriage, Child's	A. F. R. Arndt	Detroit, Mich	Oct. 21, 1873	143, 803
Carriage, Child's	C. Askam	Philadelphia, Pa	May 14, 1861	32, 271
Carriage, Child's	G. L. Atwater	New Haven, Conn	June 4, 1872	127, 404
Carriage, Child's	J. A. Bechler	Philadelphia, Pa	Jan. 7, 1873	134, 630
Carriage, Child's	J. Bein and W. Ulrich	Newark, N. J	Feb. 11, 1868	74, 284
Carriage, Child's	F. Boylston	New York, N. Y	Oct. 6, 1868	82, 683
Carriage, Child's	F. Boylston	New York, N. Y	Apr. 9, 1872	125, 533
Carriage, Child's	R. G. Britton	Springfield, Vt	Jan. 24, 1871	111, 103
Carriage, Child's	R. G. Britton	Springfield, Vt	Feb. 11, 1873	135, 764
Carriage, Child's	A. Christian	New York, N. Y	Apr. 24, 1866	54, 111
Carriage, Child's	I. Cole	Newark, N. J	Jan. 30, 1872	123, 241
Carriage, Child's	B. P. Crandall	New York, N. Y	Dec. 28, 1869	98, 351
Carriage, Child's	J. A. Crandall	Brooklyn, N. Y	Jan. 11, 1870	98, 670
Carriage, Child's	J. A. Crandall	Brooklyn, N. Y	Jan. 14, 1873	134, 856
Carriage, Child's	J. C. Crandall	New York, N. Y	Jan. 3, 1871	110, 744
Carriage, Child's	W. E. Crandall	New York, N. Y	Feb. 22, 1870	100, 121
Carriage, Child's	W. E. Crandall	New York, N. Y	Jan. 3, 1871	110, 745
Carriage, Child's	R. P. Crandall and J. A. Conover.	New York, N. Y	Apr. 2, 1867	31, 872
Carriage, Child's	C. W. F. Dare	New York, N. Y	Jan. 23, 1872	122, 998
Carriage, Child's	A. Dick	Buffalo, N. Y	Apr. 23, 1867	64, 088
Carriage, Child's	R. G. Elder	New York, N. Y	May 10, 1870	102, 927
Carriage, Child's	R. G. Elder	New York, N. Y	Mar. 14, 1871	112, 696
Carriage, Child's	J. A. H. Ellis	Springfield, Vt	June 22, 1869	91, 615
Carriage, Child's	I. N. Forrester	Bridgeport, Conn	Dec. 27, 1870	110, 561
Carriage, Child's	I. N. Forrester	Bridgeport, Conn	Nov. 8, 1870	108, 993
Carriage, Child's	A. D. Fowler	Newark, N. J	Dec. 31, 1867	72, 833
Carriage, Child's	T. Galt	Rock Island, Ill	May 13, 1873	138, 746
Carriage, Child's	T. Galt	Rock Island, Ill	Oct. 21, 1873	143, 818
Carriage, Child's	L. Havasy	Hoboken, N. J	Oct. 14, 1873	143, 624
Carriage, Child's	J. N. Hazelip	Baltimore, Md	Dec. 9, 1873	145, 298
Carriage, Child's	C. Holt	Jersey City, N. J	Nov. 7, 1871	120, 645
Carriage, Child's	J. G. Kamphaus	Pittsburgh, Pa	Oct. 14, 1873	143, 628
Carriage, Child's	J. G. Krieger	Washington, D. C	Oct. 3, 1871	119, 523
Carriage, Child's	H. Lutz	New York, N. Y	Sept. 19, 1871	119, 162
Carriage, Child's	C. G. Macht	New York, N. Y	Feb. 6, 1872	123, 408
Carriage, Child's	T. A. Madison	Terre Haute, Ind	Apr. 2, 1867	63, 541
Carriage, Child's	G. Martienssen	Brooklyn, N. Y	June 6, 1871	115, 623
Carriage, Child's	O. Mather	Newport, Ky	July 18, 1865	48, 821
Carriage, Child's	G. Maynard	Greenfield, Mass	Dec. 14, 1858	22, 300
Carriage, Child's	W. P. McKinstry	New York, N. Y	Aug. 31, 1858	21, 353
Carriage, Child's	G. H. Mellen	New York, N. Y	Mar. 29, 1870	101, 295
Carriage, Child's	H. H. and J. W. Prindle	Sandusky, Ohio	Nov. 19, 1867	71, 061
Carriage, Child's	H. M. Richardson	Boston, Mass	Oct. 7, 1873	143, 421
Carriage, Child's	G. F. C. Rosenthal	Philadelphia, Pa	Aug. 9, 1870	106, 286
Carriage, Child's	G. L. Shepard	Albany, N. Y	July 8, 1873	140, 598
Carriage, Child's	L. Snow and J. Haseltine	Boston, Mass	Dec. 16, 1873	145, 692
Carriage, Child's	F. Snyder	New York, N. Y	June 17, 1873	139, 978
Carriage, Child's	C. Spring and A. Spring	Dorchester and Weston, Mass.	Mar. 19, 1867	62, 976
Carriage, Child's	E. Talkingham	San Francisco, Cal	Dec. 5, 1871	121, 453
Carriage, Child's	W. H. Towers	Boston, Mass	Oct. 28, 1873	144, 166
Carriage, Child's	D. Troxell	Newark, N. J	July 16, 1872	129, 072
Carriage, Child's	H. W. Warner	Greenfield, Mass	Aug. 29, 1871	118, 656
Carriage, Child's	F. H. Way	Brooklyn, N. Y	Nov. 19, 1872	133, 130
Carriage, Child's	H. Weed	New Haven, Conn	Sept. 10, 1872	131, 317
Carriage, Child's	W. B. Whitney	Leominster, Mass	Mar. 14, 1871	112, 662
Carriage, Child's	J. B. Wightman	Brooklyn, N. Y	June 13, 1871	115, 920
Carriage, Child's	J. B. Wightman	Brooklyn, N. Y	Jan. 7, 1873	134, 576
Carriage, Child's folding	T. G. Stagg	East New York, N. Y	July 19, 1870	105, 505
Carriage, Child's folding	H. Lutz	New York, N. Y	Nov. 7, 1871	120, 757
Carriage, Child's three-wheeled	J. H. Gould	New York, N. Y	June 10, 1856	15, 071
Carriage-circle	J. R. McGuire	Warren, Ohio	Sept. 24, 1867	69, 233
Carriage-circle	E. Wilson	Plattsburgh, N. Y	June 11, 1872	127, 818
Carriage-circle-bending machine	S. S. Daniels	Kendallville, Ind	Mar. 31, 1868	76, 059
Carriage-circle coupling	G. G. Larkin	West Amesbury, Mass	Apr. 5, 1864	42, 199
Carriage-clip	L. J. M. Baker	Enon, Ohio	Dec. 31, 1867	72, 779
Carriage-clip	A. Bixby	Lansing, Mich	Oct. 12, 1869	95, 761
Carriage-clip	W. Bound	Hackensack, N. J	Jan. 29, 1867	61, 601
Carriage-clip	E. Hoxie	Montezuma, N. Y	Mar. 26, 1867	63, 252
Carriage-clip	L. A. Johnson	Candor, N. Y	Apr. 5, 1870	101, 467
Carriage-clip	T. McCreary	Matteawan, N. Y	July 28, 1868	80, 360
Carriage-clip	F. B. Morse	Milwaukee, Wis	Feb. 6, 1866	52, 440
Carriage-clip	F. B. Morse	Plantsville, Conn	Oct. 26, 1869	96, 252
Carriage-clip	F. B. Morse	Plantsville, Conn	Oct. 25, 1870	108, 720
Carriage-clip	F. B. Morse	Plantsville, Conn	Dec. 27, 1870	110, 581
Carriage-clip	G. B. Salmon	Saint Paul, Minn	May 14, 1867	64, 711
Carriage-clip	M. Seward	New Haven, Conn	Mar. 14, 1871	112, 743
Carriage-clip bar	F. B. Morse	Plantsville, Conn	Dec. 20, 1870	110, 386
Carriage-clip blank and die for making	H. M. Beecher	Plantsville, Conn	Sept. 12, 1871	118, 780
Carriage-clip die	W. Noble	Derby, Conn	June 27, 1871	116, 345
Carriage-clip-forging die	M. Seward	New Haven, Conn	Apr. 25, 1871	114, 051
Carriage-clip-forming die	W. W. Knowles	Plantsville, Conn	June 27, 1871	116, 455
Carriage-clip-forming die	W. W. Knowles and L. S. White.	Plantsville, Conn	June 27, 1871	116, 456
Carriage-clip-forming machine	W. S. Ward	Plantsville, Conn	Apr. 1, 1873	137, 517
Carriage-clip-swaging die	F. B. Morse	Plantsville, Conn	Oct. 11, 1870	108, 283
Carriage clips, Apparatus for compressing and inserting rubber blocks into.	T. H. Brown and C. E. Gilman	Chicago, Ill	Apr. 4, 1871	113, 624

Index of patents issued from the United States Patent Office from 1790 *to* 1873, *inclusive*—Continued.

Invention.	Inventor.	Residence.	Date.	No.
Carriage-clips, Die for bending	M. Seward	New Haven, Conn	Nov. 26, 1872	133, 492
Carriage-clips, Die for forging	F. B. Morse	Plantsville, Conn	Nov. 15, 1870	109, 336
Carriage-clips, Die for forging French	F. B. Morse	Plantsville, Conn	Sept. 7, 1869	94, 630
Carriage-clips, Die for forming	E. Meeker	Bridgeport, Conn	Sept, 16, 1873	142, 804
Carriage-clips, Die for forming blanks for	F. B. Morse	Plantsville, Conn	May 14, 1872	126, 646
Carriage-clips, Die for swaging	R. R. Miller	Plantsville, Conn	May 27, 1873	139, 413
Carriage-clips, Machine for forming	W. S. Ward	Plantsville, Conn	Dec. 10, 1872	133, 906
Carriage-clips, Machine for making	D. Wilcox and R. McChesney	Derby, Conn	Dec. 14, 1869	98, 002
Carriage-clips, Manufacture of	E. Meeker	Bridgeport, Conn	Oct. 15, 1867	69, 827
Carriage-corner body-irons	S. Z. Leslie and T. W. Porter	Hartland, Me., and Boston, Mass.	Oct. 22, 1867	70, 006
Carriage-coupling	A. Booth	Springfield, Ill	Aug. 10, 1869	93, 404
Carriage-coupling	J. Bundy	Irondequoit, Mo	July 11, 1865	48, 653
Carriage-coupling	J. H. Burrell, jr	Charlestown, Mass	Aug. 13, 1867	67, 631
Carriage-coupling	T. Chope	Detroit, Mich	Mar. 18, 1856	14, 443
Carriage coupling	H. Conderman	Haskinville, N. Y	Feb. 19, 1867	62, 253
Carriage-coupling	W. T. Dole	Peabody, Mass	Dec. 26, 1871	122, 235
Carriage-coupling	A. J. Gibson	Clinton, Mass	Sept. 12, 1854	11, 668
Carriage-coupling	W. Greenleaf	Greenfield, Ohio	Apr. 8, 1856	14, 607
Carriage-coupling	L. Grim	Fort Branch, Ind	Jan. 26, 1869	86, 150
Carriage-coupling	C. H. Guard	Troy, N. Y	July 26, 1864	43, 640
Carriage-coupling	G. W. Hatch	Parkman, Ohio	June 13, 1844	3, 627
Carriage-coupling	J. Hollinger	Millersburgh, Ohio	Apr. 18, 1871	113, 884
Carriage-coupling	A. S. Johnson	Waupun, Wis	Sept. 1, 1868	81, 642
Carriage-coupling	G. P. Kimball	San Francisco, Cal	Sept. 1, 1863	39, 785
Carriage-coupling	N. B. Livingston	Portland, Ind	July 25, 1854	11, 381
Carriage-coupling	H. J. and W. Pringle	Columbus, Ohio	Apr. 13, 1869	88, 904
Carriage-coupling	I. and J. S. Van Pelt	Petersburgh, Va	Apr. 21, 1868	77, 132
Carriage-cover	E. H. Elliott	Charlestown, Mass	Nov. 4, 1873	144, 193
Carriage cover	J. I. Rankin	Astoria, N. Y	May 19, 1863	38, 601
Carriage, cradle, swing, and baby-walker, Combined	W. H. Flanigan	Philadelphia, Pa	Mar. 15, 1870	100, 882
Carriage-cramp	S. T. Sanford	Fall River, Mass	Jan. 9, 1855	12, 218
Carriage, Cross-cramp for four-wheel	S. Glover	Newton, Conn	May 4, 1805	....
Carriage cross strap	H. Killam	New Haven, Conn	Feb. 11, 1868	74, 380
Carriage-curtain	H. Donnell	Philadelphia, Pa	Jan. 14, 1873	134, 862
Carriage-curtain	J. H. Spencer	Philadelphia, Pa	Feb. 9, 1869	86, 706
Carriage-curtain button	S. P. Campbell	Quincy, Mass	Jan. 28, 1868	73, 778
Carriage-curtain button	A. Comstock and J. H. Nicholson.	Chicopee, Mass	June 7, 1870	103, 985
Carriage-curtain button	E. Howell	Ashtabula, Ohio	Aug. 27, 1867	68, 199
Carriage-curtain button	N. F. Mathewson	Barrington, R. I	Jan. 19, 1869	85, 946
Carriage-curtain button-hole	E. L. Yancy	Utica, N. Y	July 16, 1867	66, 760
Carriage-curtain button-hole	E. L. Yancy	Utica, N. Y	July 16, 1867	66, 761
Carriage-curtain button-hole shield	W. G. Queal	Otego, N. Y	Aug. 25, 1868	81, 538
Carriage-curtain eyelet	C. W. Acker	Watertown, N. Y	Dec. 18, 1866	60, 608
Carriage-curtain fastener	H. M. Bidwell	New Haven, Conn	Nov. 21, 1871	121, 076
Carriage-curtain fastener	E. Crawford and J. H. Birch	Burlington, N. J	Mar. 27, 1866	53, 419
Carriage-curtain fastener	H. Drake	Attleborough, Mass	June 10, 1873	139, 663
Carriage-curtain fastener	A. Hunt and S. Mero, jr	Camden, Me	Sept. 17, 1867	68, 880
Carriage-curtain fastener	J. Huston, jr., and O. W. Stanford.	Sharonville, Ohio	Mar. 31, 1868	76, 080
Carriage-curtain fastener	T. D. Marsh	Jersey, Ohio	June 11, 1872	127, 780
Carriage-curtain fastener	W. H. Morse	West Newbury, Mass	Jan. 10, 1871	110, 863
Carriage-curtain fastener	H. E. Pond	Franklin, Mass	Aug. 4, 1868	80, 662
Carriage-curtain fastener	W. C. Shipherd	Cleveland, Ohio	Oct. 1, 1872	131, 791
Carriage-curtain fastener	S. Shute, jr	Richmond, Ind	Mar. 28, 1871	113, 215
Carriage-curtain fastener	J. Taynton	Port Jervis, N. Y	Jan. 30, 1872	123, 132
Carriage-curtain fastener	J. Taynton	Port Jervis, N. Y	Apr. 9, 1872	125, 502
Carriage-curtain fastener	N. M. Terry	Lyme, Conn	May 27, 1873	139, 438
Carriage-curtain fastener	E. P. Whitney	Stamford, Conn	Feb. 18, 1868	74, 649
Carriage-curtain fastener	A. Weeber	Davenport, Iowa	Sept. 17, 1867	68, 925
Carriage-curtain fastening	J. Adams	Newark, Del	Jan. 4, 1870	98, 538
Carriage-curtain fastening	A. M. Bardwell	Amherst, Mass	Dec. 10, 1872	133, 693
Carriage-curtain fastening	F. Baumgartner	Brooklyn, N. Y	Mar. 1, 1870	100, 355
Carriage-curtain fastening	F. Baumgartner	Brooklyn, N. Y	Dec. 9, 1873	145, 328
Carriage-curtain fastening	W. Z. W. Chapman	New York, N. Y	May 3, 1859	23, 821
Carriage-curtain fastening	W. Z. W. and J. W. Chapman	New York, N. Y., and Hyannis, Mass.	Oct. 22, 1861	33, 515
Carriage-curtain fastening	A. B. Clark	Richmond, Ind	Oct. 31, 1871	120, 493
Carriage-curtain fastening	J. C. Fish	Barnstable, Mass	Oct. 13, 1868	83, 056
Carriage-curtain fastening	J. H. Gleason	Troy, N. Y	May 13, 1873	138, 799
Carriage-curtain fastening	T. McPherson	Burlington, N. J	Feb. 19, 1867	62, 145
Carriage-curtain fastening	T. A. Mitchell	Washington, D. C	Dec. 24, 1867	72, 666
Carriage-curtain fastening	E. Shepard	New York, N. Y	Oct. 27, 1868	83, 555
Carriage-curtain fastening	J. Zimmerman	Powhatan, Md	Aug. 29, 1871	118, 665
Carriage-curtain fixture	A. C. Babcock and J. Duffy	New Haven, Conn	Apr. 30, 1867	64, 186
Carriage-curtain fixture	C. E. Thompson	New Haven, Conn	Mar. 31, 1868	76, 273
Carriage-curtain knob	W. B. Douglass	Newark, N. J	Feb. 7, 1871	111, 522
Carriage-curtain knob	C. M. Munson	Greenfield, Mass	July 8, 1873	147, 723
Carriage-curtain knob, Elastic	W. H. Hawkins	Cleveland, Ohio	Dec. 3, 1867	71, 613
Carriage-curtain lock	G. Cook	New Haven, Conn	Jan. 6, 1852	8, 626
Carriage-curtain lock	S. Marshall	Wilmington, Del	Feb. 22, 1859	23, 036
Carriage-curtain strap	H. Cornell and A. H. Marshall	Wilmington, Del	Oct. 3, 1871	119, 450
Carriage-curtains, Button for fastening	S. Bidwell	Bordentown, N. J	Apr. 16, 1867	63, 837
Carriage-curtains, Button-hole lining for	J. Barclay	Attleborough, Mass	Apr. 14, 1868	76, 695
Carriage-curtains, Elastic button-hole for	S. C. Talcott	Ashtabula, Ohio	Nov. 13, 1866	59, 679
Carriage-curtains, Frame for the glasses of	W. F. Beaton	Philadelphia, Pa	Aug. 27, 1867	68, 154
Carriage-curtains, Frame for the glasses of	W. H. Rhodes	Lancaster, Pa	Oct. 10, 1871	119, 722
Carriage-curtains, Knob-hole for	C. W. Holland	Fredonia, N. Y	Nov. 13, 1866	59, 718
Carriage-cushion, Spring	T. S. Burnett	Syracuse, N. Y	Feb. 21, 1871	111, 903
Carriage dash-board	J. Bland	Thomaston, Ga	Jan. 25, 1870	99, 140
Carriage dash-board	J. Curtis	Cincinnati, Ohio	Jan. 23, 1872	122, 996
Carriage dash-board	W. Franklin and T. M. Smith	East Randolph, N. Y	Mar. 24, 1868	75, 746
Carriage dash-board	A. B. Hurd	Painted Post, N. Y	Nov. 7, 1871	120, 647
Carriage dash-board	L. Lupton	Winchester, Va	Oct. 1, 1850	7, 685
Carriage-dash fall, Mode of fastening apron-hooks to	J. Curtis	Cincinnati, Ohio	Nov. 5, 1872	132, 712
Carriage dash-boards, Metallic binding for	L. J. Parsons	New Bedford, Mass	Oct. 20, 1868	83, 205

Index of patents issued from the United States Patent Office from 1790 *to* 1873, *inclusive*—Continued.

Invention.	Inventor.	Residence.	Date.	No.
Carriage-dash rail	B. B. Noyes and O. G. Stratton	Greenfield, Mass	July 12, 1870	105, 362
Carriage-dash rail	B. B. Noyes and O. G. Stratton	Greenfield, Mass	Feb. 11, 1873	135, 834
Carriage, Dash-rail for child's	M. Hine	New Haven, Conn	Apr. 1, 1873	137, 446
Carriage-door	J. Carson	Buffalo, N. Y	Nov. 11, 1873	144, 440
Carriage-door	J. Cunningham	Rochester, N. Y	Apr. 22, 1873	138, 008
Carriage-door	P. Devilliard and A. Postweiler	Paris France	Oct. 8, 1867	69, 641
Carriage-door	P. Leek	Hartford, Conn	Dec. 22, 1868	85, 108
Carriage-door	A. Wright	Wilmington, Del	Dec. 12, 1871	121, 922
Carriage-door and window-sash support	J. C. Ham	New York, N. Y	May 2, 1871	114, 436
Carriage-door handle	E. Wells	New Haven, Conn	Aug. 15, 1871	118, 083
Carriage door, Landau	H. Killam	New Haven, Conn	May 14, 1872	126, 715
Carriage door, Landau	E. Wells	New Haven, Conn	Mar. 15, 1870	100, 956
Carriage door, Landau	F. Wood	Bridgeport, Conn	Mar. 6, 1866	53, 075
Carriage-door lock	J. Ives	Hampden, Conn	July 20, 1846	4, 646
Carriage draft-poles, Device for sustaining	S. B. Whitney	Coxsackie, N. Y	Feb. 4, 1868	74, 184
Carriage, Dumping	J. R. True	Richmond, Me	Aug. 15, 1871	118, 072
Carriage-evener	C. L. Ames	Bangor, Me	Nov. 26, 1867	71, 262
Carriage, Extension	J. A. Naylor	Rahway, N. J	Sept. 4, 1860	29, 902
Carriage extension-seat	F. J. Flowers	Rahway, N. J	Dec. 27, 1859	26, 578
Carriage, Folding	T. G. Stagg	East New York, N. Y	Dec. 24, 1872	134, 226
Carriage folding-step	G. Gregory	New Haven, Conn	May 9, 1871	114, 673
Carriage folding-step	G. H. Vollhardt	New Haven, Conn	June 6, 1871	115, 664
Carriage folding-step	E. Wells	New Haven, Conn	June 6, 1871	115, 665
Carriage folding-top	W. G. Foglesong	Xenia, Ohio	Apr. 17, 1855	12, 751
Carriage foot-rail	T. W. and H. K. Porter	Boston, Mass	Sept. 23, 1873	143, 031
Carriage for common roads, Steam	F. Alger	Boston, Mass	June 13, 1871	115, 802
Carriage for railways, Steam	J. H. Moore and W. P. Parrott	Boston, Mass	Dec. 2, 1851	8, 561
Carriage for tram-ways, Steam	J. Grantham	London, England	Sept. 2, 1873	142, 459
Carriage forward-gear	J. J. D. Meincke	Milwaukee, Wis	Apr. 28, 1868	77, 395
Carriage, Four-wheel	T. Brooks and D. W. Eames	Rutland, N. Y	Dec. 6, 1827	....
Carriage, Four-wheel	J. B. Swan	Albany, N. Y	July 24, 1823	....
Carriage front, Landaulet	E. Wells	New Haven, Conn	Dec. 10, 1872	133, 910
Carriage-front, Removable	C. F. Albright and L. Burkhard.	Pottsville, Pa	Sept. 24, 1861	33, 326
Carriage-gate	J. Ellis	Detroit, Mich	Jan. 28, 1862	34, 247
Carriage-gearing	C. Custer	Norristown, Pa	Sept. 27, 1870	107, 668
Carriage gearing	C. Custer	Norristown, Pa	Mar. 21, 1871	112, 905
Carriage gearing	J. R. McAlister	Richville, N. Y	Sept. 4, 1866	57, 745
Carriage-gearing	C. W. Saladee	Circleville, Ohio	Feb. 9, 1869	86, 866
Carriage-guard	F. B. Shaw	Boston, Mass	Mar. 5, 1867	62, 569
Carriage-guard	I. M. Singer	Paris, France	Feb. 5, 1867	61, 883
Carriage, Hand	W. Allen and J. W. Bond	Saint Paul, Minn	June 6, 1871	115, 674
Carriage, Hand	D. S Shanabrook	Greencastle, Pa	June 29, 1869	92, 106
Carriage, Hand-motive	J. Allgaier	Philadelphia, Pa	Feb. 22, 1870	100, 097
Carriage, Hand-propelling	R. Kind	New York, N. Y	Feb. 13, 1872	123, 707
Carriage handle, Child's	R. G. Elder	New York, N. Y	Feb. 1, 1870	99, 420
Carriage handle, Child's	J. H. White	Newark, N. J	Feb. 20, 1872	123, 960
Carriage head-block	J. C. Bates	Warrensburgh, Mo	Aug. 3, 1869	93, 163
Carriage head-block	T. M. Cluxton	Rising Sun, Ind	Oct. 20, 1868	83, 256
Carriage head-block	D. Hinman	Southington, Conn	May 9, 1871	114, 681
Carriage head-block	F. Van Patten	Auburn, N. Y	July 23, 1872	129, 768
Carriage head-block-plate-forming die	W. Terrell	Ansonia, Conn	Sept. 26, 1871	119, 286
Carriage hold-back	W. R. Baker	Wellington Square, Canada	May 16, 1871	114, 746
Carriage hold-back	J. Davis	Elmira, N. Y	Oct. 22, 1861	33, 555
Carriage hold-back	T. F. Griffiths	Dansville, N. Y	Sept. 2, 1862	36, 347
Carriage hold-back	H. A. Harris	Battle Creek, Mich	Dec. 23, 1862	37, 227
Carriage hold-back	J. Hunlock	Wyoming Valley, Pa	Jan. 25, 1870	99, 196
Carriage hold-back	T. F. Kiff	Havana, N. Y	Sept. 21, 1869	94, 962
Carriage hold-back	J. A. Lannert	Cleveland, Ohio	Sept. 26, 1871	119, 370
Carriage hold-back	J. A. Mackinnon	Cleveland, Ohio	Oct. 20, 1868	83, 294
Carriage hold-back	R. Nutting	Randolph, Vt	May 19, 1863	38, 597
Carriage hold-back	R. Rolph	Coventry, N. Y	Sept. 15, 1863	39, 958
Carriage hold-back	D. Terry	Wakeman, Ohio	Sept. 23, 1873	143, 199
Carriage hold-back	A. Webster	Montpelier, Vt	Sept. 25, 1855	13, 604
Carriage hold-back	E. Wilson	Northbridge, Mass	Nov. 6, 1866	59, 494
Carriage hook, Landau	O. S. Osborn	New Haven, Conn	June 3, 1873	139, 515
Carriage, Hose	I. S. Schuyler and L. A. Rockwell.	New York, N. Y	Sept. 7, 1858	21, 449
Carriage-hound	J. Maddock	Dubuque, Iowa	Sept. 24, 1861	33, 355
Carriage-iron	H. K. Porter	Boston, Mass	Dec. 30, 1873	146, 093
Carriage-jack	M. Andriot	Mount Washington, Ohio	July 24, 1866	56, 510
Carriage-jack	M. Andriot	Mount Washington, Ohio	Sept. 25, 1866	58, 198
Carriage-jack	D. Bull	Amboy, Ill	Nov. 16, 1869	96, 880
Carriage-jack	J. O. Burch	Buffalo, N. Y	July 6, 1869	92, 157
Carriage-jack	J. Card	Cleveland, Ohio	Sept. 25, 1860	30, 120
Carriage-jack	O. Churchill	Canton, Pa	Feb. 25, 1868	74, 894
Carriage-jack	T. G. Clinton and G. H. and E. H. Knight.	Cincinnati, Ohio	Mar. 5, 1850	7, 135
Carriage-jack	G. L. Cummings	New York, N. Y	Mar. 17, 1863	37, 902
Carriage-jack	G. L. Cummings	New York, N. Y	Aug. 1, 1865	49, 090
Carriage-jack	W. S. Douglass	Richmond, Vt	Dec. 7, 1869	97, 618
Carriage-jack	D. Elliott and E. Seely	New York, N. Y	Aug. 3, 1869	93, 289
Carriage-jack	J. F. Emmert	Quincy, Pa	Sept. 17, 1867	68, 970
Carriage-jack	A. W. Field	Vergennes, Vt	Apr. 25, 1865	47, 408
Carriage-jack	R. Fink	Batavia, Ill	May 9, 1865	47, 628
Carriage-jack	A. R. Giles	Adams, N. Y	Aug. 2, 1870	106, 050
Carriage-jack	T. L. Goble	Orange, N. Y	Apr. 30, 1867	64, 302
Carriage-jack	J. H. Hadley	Boston, Mass	Mar. 16, 1869	87, 927
Carriage-jack	H. L. Hammond	Providence, R. I	Nov. 3, 1868	83, 773
Carriage-jack	J. F. Hammond	Providence, R. I	Oct. 23, 1866	59, 008
Carriage-jack	D. Hiestand	Pughtown, Pa	May 16, 1871	114, 816
Carriage-jack	D. Hiestand	Pughtown, Pa	Mar. 25, 1873	137, 202
Carriage-jack	A. Higley	South Bend, Ind	Oct. 31, 1865	50, 707
Carriage-jack	F. Hovey	New York, N. Y	Dec. 28, 1869	98, 379
Carriage-jack	T. W. Johnston	Richmond, Me	Jan. 25, 1870	99, 090
Carriage-jack	A. W. Keeler and J. Eckert	La Fayette, N. Y	Apr. 13, 1869	88, 789
Carriage-jack	D. Marshall	Northville, Mich	Aug. 16, 1870	106, 379
Carriage-jack	J. P. Moore	Boston, Mass	Sept. 3, 1867	68, 451
Carriage-jack	A. Myers	Van Wert, Ohio	Feb. 18, 1868	74, 576

Index of patents issued from the United States Patent Office from 1790 *to* 1873, *inclusive*—Continued.

Invention.	Inventor.	Residence.	Date.	No.
Carriage-jack	C. H. Paine	Providence, R. I	Sept. 30, 1862	36, 587
Carriage-jack	C. H. Paine	Providence, R. I	Mar. 22, 1864	42, 050
Carriage-jack	C. H. Paine	Providence, R. I	Feb. 4, 1868	73, 992
Carriage-jack	C. H. Paine	Providence, R. I	Mar. 2, 1869	87, 358
Carriage-jack	C. J. Philleo	Kenyonville, N. Y	Aug. 3, 1869	93, 343
Carriage-jack	J. J. Pike	Chelsea, Mass	Oct. 9, 1860	30, 346
Carriage-jack	J. C. Plumer	Boston, Mass	Jan. 2, 1866	51, 861
Carriage-jack	P. C. Porter	Augusta, Me	Sept. 3, 1867	68, 572
Carriage-jack	O. T. Potter	Scott, N. Y	Jan. 22, 1867	61, 455
Carriage-jack	H. L. Riouff	Pleasant Ridge, Ohio	July 18, 1871	117, 114
Carriage-jack	J. F. Seaman	Cortlandville, N. Y	Aug. 10, 1869	93, 484
Carriage-jack	H. S. Shepardson	Shelburne Falls, Mass	Jan. 1, 1867	60, 945
Carriage-jack	J. B. Small	Boston, Mass	Nov. 12, 1867	70, 908
Carriage-jack	J. Stener	Albany, N. Y	July 20, 1869	92, 899
Carriage-jack	O. B. Sutton	Kensico, N. Y	Sept. 3, 1867	68, 395
Carriage-jack	J. N. Thatcher	Martinsburgh, W. Va	May 25, 1869	90, 473
Carriage-jack	I. Varney	Kennebunk, Me	Nov. 12, 1867	70, 763
Carriage-jack	A. H. Wellbrock	Boston, Mass	May 23, 1871	115, 140
Carriage-jack	J. B. White	Detroit, Mich	Nov. 1, 1868	84, 665
Carriage-jack	D. R. Wight	Southbridge, Mass	Dec. 23, 1873	145, 922
Carriage-jack	J. E. Woll	Allegheny City, Pa	Aug. 17, 1869	93, 786
Carriage-jack, bag-holder, and weighing-scale combined.	N. Berkeley	Aldie, Va	Sept. 5, 1871	118, 678
Carriage-joint	R. Nickson	Akron, Ohio	Sept. 10, 1867	68, 780
Carriage, Joint-bodied	A. H. Niles	Georgetown, N. Y	May 15, 1855	12, 869
Carriage jump-seat	C. P. Kimball	Portland, Me	Sept. 24, 1867	69, 102
Carriage-knob	R. P. Cowles	New Haven, Conn	June 27, 1865	48, 373
Carriage-knob	E. S. Wheeler	Westport, Conn	Jan. 19, 1869	85, 976
Carriage-clamp support	T. P. White	Bridgeport, Conn	June 27, 1871	116. 381
Carriage, Land	R. C. Parvin	Philadelphia, Pa	Jan. 16, 1872	122, 850
Carriage, Land and steam	S. Fairman	Nassau, N. Y	Mar. 27, 1830	......
Carriage, Landau	H. Killam	New Haven, Conn	Dec. 10, 1872	133, 862
Carriage, Landau	E. Wells	New Haven, Conn	Mar. 14, 1871	112, 758
Carriage, Landaulet	E. Wells	New Haven, Conn	Dec. 6, 1870	109, 981
Carriage-lock	J. A. Bower	Middlefield, Ohio	Apr. 28, 1868	77, 162
Carriage lock or brake, Spring	W. Clayton	Marshallton, Pa	June 18, 1842	2, 676
Carriage-lock-plate blanks, Machine for cutting	T. Insull and J. F. Dovey	Philadelphia, Pa	Sept. 16, 1873	142, 854
Carriage, Locomotive	R. Winans	Baltimore, Md	July 28, 1846	4, 665
Carriage, Locomotive	R. Winans	Baltimore, Md	Oct. 14, 1846	4, 812
Carriage, Locomotive and other railway	A. C. Jones	Philadelphia, Pa	Oct. 10, 1840	1, 812
Carriage loop and billet cover	N. Jenny, jr	Pittsburgh, Pa	Dec. 8, 1868	84, 829
Carriage-makers' tools	G. Atkinson	San Francisco, Cal	Mar. 7, 1871	112, 311
Carriage, Mechanism for connecting a horse with a.	A. Colburn and E. G. Stanley.	Lynn and Fitchburgh, Mass.	Dec. 17, 1867	72, 167
Carriage-motor	D. Morrison	Portland, Me	Dec. 31, 1867	72, 881
Carriage moved by hand	D. Nine, jr	Reading, Pa	May 4, 1813	......
Carriage neck-yoke leathers	H. Sanders	Utica, N. Y	Dec. 13, 1870	110, 074
Carriage-nut	J. P. Skinner	Plantsville, Conn	Oct. 17, 1871	120, 004
Carriage or buggy seat, Shifting	S. W. Beach	South Bend, Ind	June 14, 1870	104, 248
Carriage or perambulator, Child's	B. P. Crandall, jr	Williamsburgh, N. Y	Apr. 4, 1871	113, 402
Carriage panels and veneers off circular logs, Machine for cutting.	J. White and P. Quimby	Belfast, Me	Sept. 12, 1829	......
Carriage-perch	I. P. Bacon	Bedford, Mass	Mar. 14, 1871	112, 528
Carriage-perch	W. H. Cooper and G. Gregory.	New Haven, Conn	Jan. 21, 1868	73, 438
Carriage-perch	L. E. Stilwill	Franklinville, N. Y	Dec. 2, 1851	8, 568
Carriage perch connection	D. A. Mathews	Geneva, N. Y	Mar. 17, 1868	75, 688
Carriage-perch coupling	W. S. Lord	Pulaski, Tenn	Oct. 28, 1856	15, 979
Carriage-pole	B. Foltz	Rockford, Ill	Jan. 5, 1869	85, 519
Carriage-pole	C. K. Mellinger	Millersville, Pa	Oct. 1, 1867	69, 356
Carriage-pole	V. N. Mitchell	Concord, N. H	Apr. 7, 1868	76, 496
Carriage-pole	H. W. Painter	New Haven, Conn	Aug. 14, 1866	57, 179
Carriage-pole	G. N. Shaw	Muir, Mich	Feb. 2, 1869	86, 459
Carriage-pole, Adjustable	M. A. Keon	Catskill, N. Y	Sept. 15, 1868	82, 127
Carriage-pole, Adjustable	L. C. Miner	Hartford, Conn	Aug. 4, 1863	39, 415
Carriage pole and shaft combined	A. Moore and J. Aylwerd	Mission San José, Cal	Aug. 10, 1869	93, 466
Carriage pole and thill	J. G. Perry	Kingston, R. I	Mar. 27, 1866	53, 478
Carriage-pole attachment	H. F. Edwards	Worcester, Mass	Mar. 23, 1869	88, 149
Carriage-pole, Convertible	F. Fowler	West Haven, Conn	Apr. 21, 1868	76, 906
Carriage-pole coupling	C. A. Moore	Akron, Ohio	Sept. 6, 1870	107, 082
Carriage-pole support	H. B. Hurd	Aurora, Ill	Apr. 28, 1868	77, 382
Carriage-pole supporter	S. E. Bolles	Mattapoisett, Mass	Nov. 26, 1861	33, 811
Carriage-pole tip	A. Benedict	Albany, N. Y	May 12, 1868	77, 796
Carriage-pole tip	S. Mason	Hamden, Conn	Dec. 23, 1873	145, 885
Carriage-poles, Hook for supporting	S. S. Bent	Portchester, N. Y	Apr. 20, 1869	89, 118
Carriage-prop	W. Finn	Poughkeepsie, N. Y	Aug. 20, 1867	67, 968
Carriage-prop	L. Sawyer	South Amesbury, Mass	Apr. 5, 1870	101, 518
Carriage-prop	C. Thomas	West Newbury, Mass	Sept. 22, 1857	18, 254
Carriage-prop block	C. H. Davis	Syracuse, N. Y	Dec. 10, 1872	133, 764
Carriage-prop joint	A. Searls	Newark, N. J	Sept. 24, 1872	131, 568
Carriage-prop joint	E. W. Waite	New Haven, Conn	Aug. 11, 1868	80, 845
Carriage-prop joint, Adjustable	A. Searls	San Francisco, Cal	Nov. 5, 1867	70, 628
Carriage, Propeller	G. T. Palmer	Brooklyn, N. Y	May 21, 1872	126, 900
Carriage-protector	G. B. Brown	Newburgh, N. Y	May 27, 1873	139, 364
Carriage-rail, Shifting	W. H. Keeney	Waupun, Wis	July 16, 1872	129, 140
Carriage-raising machine	E. Huson	Ithaca, N. Y	Feb. 18, 1862	34, 427
Carriage, Riding	W. and J. Jessup	Guilford, N. C	June 1, 1827	......
Carriage-rocker	A. S. Wells	Hopkinton, R. I	Apr. 23, 1872	125, 998
Carriage rocker, Child's	A. D. Juelson	Wooster, Ohio	Mar. 10, 1868	75, 278
Carriage, Rocking	A. Armando	New York, N. Y	Sept. 21, 1869	94, 990
Carriage, Rocking	A. C. Griswold and W. R. Griswold.	Hartford and Durham, Conn.	June 7, 1859	24, 299
Carriage, Rolling lever	H. Chapman	Corinth, N. Y	Mar. 24, 1831	......
Carriage-runner attachment	J. W. Moore	Watseka, Ill	Jan. 18, 1870	98, 990
Carriage running-gear	N. Adams	Cornwall, N. Y	Feb. 11, 1862	34, 342
Carriage running-gear	J. C. Anderson and A. W. Benson.	Saginaw, Mich	Aug. 15, 1871	118, 093
Carriage running-gear	J. Calef	Buffalo, N. Y	Aug. 16, 1859	25, 091
Carriage running-gear	F. Crick	Beamesville, Ohio	Aug. 14, 1866	57, 094

Index of patents issued from the United States Patent Office from 1790 *to* 1873, *inclusive*—Continued.

Invention.	Inventor.	Residence.	Date.	No.
Carriage running-gear	G. L. Haussknecht	New Haven, Conn	Jan. 13, 1852	8,648
Carriage running-gear	W. Hemme	Michigan Valley, Kans	June 11, 1872	127,767
Carriage running-gear	W. Hemme	Michigan Valley, Kans	July 30, 1872	129,954
Carriage running-gear	R. Kline and R. M. Jack	Pottstown, Pa	May 20, 1873	139,164
Carriage running-gear	W. A. Lewis	Chicago, Ill	July 23, 1872	129,838
Carriage running-gear	R. Murdoch	Baltimore, Md	June 24, 1856	15,189
Carriage running-gear	R. Murdoch	Baltimore, Md	May 19, 1857	17,337
Carriage running-gear	T. O'Brien	Quincy, Mich	June 21, 1870	104,634
Carriage running-gear	J. Pine	New York, N. Y	Nov. 5, 1850	7,758
Carriage running-gear	J. Rancevan	Carthage, Ohio	Mar. 28, 1871	113,092
Carriage running-gear	C. F. Verleger	Baltimore, Md	Feb. 3, 1852	8,711
Carriage running-gear	J. L. Ware	Mantorville, Minn	Mar. 4, 1873	136,570
Carriage running-gear	E. Wigle	Bay City, Mich	Apr. 25, 1871	114,072
Carriage running-gear	E. Wigle	Bay City, Mich	Feb. 27, 1872	124,026
Carriage running-gear	J. B. Withey	Lexington, Mich	Jan. 28, 1868	73,941
Carriage running-gear	J. B. Withey	Detroit, Mich	Dec. 27, 1870	110,528
Carriage running-gear, Land	J. Blocher	Buffalo, N. Y	Apr. 2, 1867	63,358
Carriage, Safety	R. Beale	Washington, D. C	May 12, 1832	....
Carriage safety-attachment	C. Ducrux	New York, N. Y	Jan. 1, 1867	60,705
Carriage-sash fastener and supporter	W. Stewart	Hartford, Conn	May 4, 1869	89,604
Carriage sash-strap roller	J. H. Bloodgood	Bridgeport, Conn	May 17, 1870	103,131
Carriage-seat	J. H. Adams	Portland, Me	June 2, 1868	78,563
Carriage-seat	D. Aspinwall	South Bend, Ind	Feb. 13, 1872	123,547
Carriage-seat	E. Chamberlin	Lansingburgh, N. Y	Dec. 24, 1867	72,602
Carriage-seat	G. R. Comstock	Manheim, N. Y	Jan. 23, 1855	12,273
Carriage-seat	C. D. Flynt	Collinsville, Ill	May 28, 1867	65,201
Carriage seat	H. H. Forbes and H. C. Sears	New Bedford, Mass	Apr. 21, 1868	77,023
Carriage-seat	J. Gale and M. B. Ames	Lawrence, Mass	June 29, 1869	92,036
Carriage-seat	S. P. Graham	Columbus, Ohio	Oct. 5, 1869	95,466
Carriage-seat	S. P. Graham	London, Canada	June 13, 1871	115,842
Carriage seat	S. P. Graham	London, Canada	Mar. 12, 1872	124,433
Carriage-seat	E. H. Harris	Palmetto, Ga	Nov. 8, 1859	26,032
Carriage-seat	G. L. Hudson	Romeo, Mich	Nov. 2, 1869	96,434
Carriage-seat	I. Kinney	Woodstock, Canada	Aug. 17, 1869	93,825
Carriage-seat	C. K. Mellinger	Lebanon, Pa	Jan. 7, 1873	134,692
Carriage-seat	J. N. Miller	Bellefontaine, Ohio	July 22, 1873	141,065
Carriage-seat	F. B. Morse	Milwaukee, Wis	May 15, 1866	54,755
Carriage-seat	A. J. Ritter	Rahway, N. J	Aug. 25, 1863	39,677
Carriage-seat	G. Feichert	New Haven, Conn	Sept. 15, 1863	39,971
Carriage-seat	H. Timkin	Saint Louis, Mo	Aug. 13, 1872	130,546
Carriage-seat	S. Toomey	Wilmot, Ohio	Jan. 11, 1870	98,721
Carriage-seat	R. Walker	Batavia, N. Y	Jan. 23, 1866	52,229
Carriage-seat	L. J. Woodruff	Mohawk, N. Y	Sept. 11, 1866	58,016
Carriage-seat, Adjustable	D. Aspinwall	South Bend, Ind	Dec. 31, 1872	134,452
Carriage-seat, Adjustable	S. W. Beach	Ypsilanti, Mich	Apr. 4, 1871	113,387
Carriage-seat, Adjustable	S. W. Beach	Ypsilanti, Mich	May 23, 1871	115,152
Carriage-seat, Adjustable	G. and D. Cook	New Haven, Conn	Feb. 3, 1857	16,528
Carriage-seat, Adjustable	D. N. Flanders	South Royalton, Vt	Feb. 12, 1856	14,231
Carriage-seat, Adjustable	J. Fleming	Philadelphia, Pa	Apr. 7, 1868	76,428
Carriage-seat, Adjustable	W. H. Gregg and W. Bowe	Wilmington, Del	Aug. 24, 1869	93,987
Carriage-seat, Adjustable	J. A. Naylor	Rahway, N. J	June 26, 1860	28,892
Carriage-seat, Adjustable	H. H. Potter	Carthage, N. Y	Dec. 14, 1858	22,304
Carriage-seat, Adjustable	I. L. Vansant	Red Lion, Del	June 19, 1860	28,792
Carriage-seat, Adjustable	A. Wright	Wilmington, Del	Apr. 12, 1870	101,959
Carriage seat and top	L. Z. Dodds	South Bend, Ind	Apr. 4, 1871	113,504
Carriage seat and top	E. L. Yancey	Batavia, N. Y	Sept. 24, 1872	131,585
Carriage seat and top, Combined	F. Jackson	Sparta, Ohio	Jan. 1, 1867	60,737
Carriage-seat back	J. Burt	Sturgis, Mich	Nov. 5, 1867	70,517
Carriage-seat back, Adjustable	E. A. Rice	Wilmington, Vt	June 18, 1872	128,171
Carriage-seat backs, Manufacture of	B. Hurlburt	Fort Wayne, Ind	June 27, 1871	116,449
Carriage-seat clamp	H. J. Northrup	Lansingburgh, N. Y	Jan. 24, 1871	111,140
Carriage-seat, Elastic or spring	J. Nichols	Providence, R. I	Oct. 2, 1822	....
Carriage-seat fastener	L. D. Belnap	Jonesville, Mich	Aug. 27, 1872	130,785
Carriage-seat fastener	M. Brockway, jr	Angola, Ind	July 8, 1873	140,572
Carriage seat fastening	W. Beers	Milan, Ohio	Apr. 27, 1869	89,276
Carriage-seat fastening	J. H. Fellows	Alba, Pa	July 26, 1870	105,789
Carriage-seat fastening	M. C. Remington	Auburn, N. Y	May 28, 1867	65,118
Carriage-seat joint	J. A. Hanna	Bel Air, Md	Nov. 28, 1871	121,362
Carriage-seat lock	W. Conway	Rushville, N. Y	Oct. 20, 1868	83,258
Carriage-seat, Self-adjusting	J. C. Kimball	New Haven, Conn	Jan. 1, 1861	31,023
Carriage-seat shifting-rail	F. Baumgarten	Brooklyn, N. Y	Apr. 23, 1867	64,008
Carriage-seat shifting-rail	C. Disser	West Union, Ohio	Apr. 19, 1870	101,986
Carriage-seat shifting-rail	H. Dressel	Springfield, Vt	July 29, 1873	141,335
Carriage-seat shifting-rail	J. Fellows	Chicago, Ill	Oct. 2, 1866	58,542
Carriage-seat shifting-rail	S. Toomey	Wilmot, Ohio	Apr. 21, 1868	76,959
Carriage-seat shifting-rail	J. Zahn	Fredonia, N. Y	Mar. 5, 1867	62,719
Carriage-seat, Sliding	S. N. Beecher	Milford, Conn	Mar. 24, 1868	75,841
Carriage-seat, Sliding	W. A. Bird	Newark, N. J	Apr. 17, 1860	27,881
Carriage-seat, Sliding	R. T. Briggs	Amesbury, Mass	Nov. 22, 1870	109,377
Carriage-seat spring	W. Scott	Plymouth, Mich	Nov. 5, 1867	70,475
Carriage-seat, Spring-back	N. Cowles and A. Hulbert	Edgefield, S. C	Oct. 11, 1859	25,725
Carriage seats, Attaching the rails of	C. Schofield	Trumbull, Conn	Apr. 19, 1859	23,712
Carriage-seats, Clamp for movable	S. W. Beach	Ypsilanti, Mich	Nov. 8, 1870	109,102
Carriage-seats, Construction of	B. Hurlburt	Milford, Conn	Nov. 6, 1866	59,508
Carriage-seats. Corner-post for	S. S. Stanley	Sandusky, Ohio	Jan. 30, 1872	123,302
Carriage seats, Head-rest for railway	E. Hambujer	Detroit, Mich	Mar. 29, 1864	42,084
Carriage-seats, Mode of attaching	C. Krebs	West Springfield, Mass	Dec. 7, 1869	97,524
Carriage-shackle	J. Brennan	New Haven, Conn	May 14, 1867	64,626
Carriage-shackle	W. F. Gilbert	Derby, Conn	Sept. 1, 1868	81,768
Carriage-shackle	W. W. Knowles	Southington, Conn	Apr. 24, 1866	54,253
Carriage-shackle	G. G. Larkin	West Amesbury, Mass	Sept. 22, 1868	82,420
Carriage-shackle	L. M. Lawless	Geneseo, Ill	Sept. 5, 1871	118,729
Carriage-shackle	J. Low	New Britain, Conn	Feb. 9, 1869	86,847
Carriage-shackle	F. B. Morse	New Haven, Conn	June 30, 1868	79,494
Carriage-shackle	F. B. Morse	Plantsville, Conn	Oct. 26, 1869	96,256
Carriage-shackle	J. H. J. O'Neill	New Haven, Conn	Oct. 2, 1866	58,465
Carriage-shackle	G. T. Pearsall	Apalachin, N. Y	Apr. 31, 1867	64,249
Carriage-shackle	J. E. Prudden	Birmingham, Conn	Oct. 29, 1876	70,358

Index of patents issued from the United States Patent Office from 1790 to 1873, inclusive—Continued.

Invention.	Inventor.	Residence.	Date.	No.
Carriage-shackle	T. S. Smith	New Haven, Conn	Nov. 13, 1866	59, 728
Carriage-shackle	F. M. Weller	Evanston, Ill	May 28, 1867	65, 317
Carriage-shackle	L. Wilkinson	New Haven, Conn	Sept. 17, 1867	68, 924
Carriage-shackle-blank die	W. B. Smith	Plantsville, Conn	Aug. 9, 1870	106, 225
Carriage-shackle die	R. R. Miller	Plantsville, Conn	Jan. 4, 1870	98, 610
Carriage-shackle die	L. S. White	Plantsville, Conn	Apr. 12, 1870	101, 952
Carriage-shackle-eyes, Die for forging	H. M. Beecher	Plantsville, Conn	Apr. 30, 1872	126, 252
Carriage-shackle-forming die	H. M. Beecher	Plantsville, Conn	June 20, 1871	116, 012
Carriage-shackles, Machine for drilling	J. B. Clark	Plantsville, Conn	Oct. 4, 1870	107, 875
Carriage-shackles, Process of forging	F. B. Morse	Plantsville, Conn	Aug. 9, 1870	106, 190
Carriage-shaft connection	C. Tholl	Boston, Mass	Oct. 29, 1867	70, 289
Carriage-shaft-holding attachment	A. C. Maxfield	Biddeford, Me	Oct. 14, 1873	143, 582
Carriage-shaft iron	W. W. Knowles	Plantsville, Conn	May 13, 1873	138, 753
Carriage shaft or pole coupling	C. G. Dudley and J. Gulden	Key Port, N. J	May 3, 1870	102, 515
Carriage-shaft-shackle die	H. M. Beecher	Plantsville, Conn	Aug. 24, 1869	93, 951
Carriage shifting-seat	C. R. Mellinger	Manor Township, Pa	Nov. 20, 1866	59, 920
Carriage shifting-top	A. Soursin	Saint Louis, Mo	July 16, 1867	66, 746
Carriage slat-iron	F. B. Morse	Plantsville, Conn	Oct. 26, 1869	96, 253
Carriage-spring	D. S. Abbott	Ischua, N. Y	Dec. 13, 1870	199, 993
Carriage-spring	J. B. Ashley	New Bedford, Mass	Nov. 11, 1867	71, 118
Carriage-spring	J. Bacon	Bedford, Mass	Mar. 9, 1842	2, 484
Carriage-spring	J. Balbach	San José, Cal	Feb. 2, 1869	86, 348
Carriage-spring, &c	M. L. Ballard	Canton, Ohio	Apr. 3, 1866	53, 556
Carriage-spring	B. M. Beckwith	Plattsburgh, N. Y	July 3, 1866	55, 985
Carriage-spring	L. Bissell	New York, N. Y	May 20, 1851	8, 105
Carriage-spring	L. Bissell	New York, N. Y	Nov. 4, 1851	8, 498
Carriage-spring	P. E. Bomboy	Espy, Pa	Dec. 3, 1867	71, 574
Carriage-spring	M. H. Bonwill	Canterbury, Del	Nov. 17, 1818	....
Carriage-spring	P. Broadbooks	Batavia, N. Y	July 25, 1871	117, 253
Carriage-spring	R. B. Brown	Essex, Vt	Dec. 14, 1841	2, 385
Carriage-spring	J. Bullock	Windsor, N. Y	Dec. 9, 1873	145, 333
Carriage spring	C. H. Butterfield	Sturbridge, Mass	Jan. 1, 1867	60, 831
Carriage-spring	A. Buzzell	West Fairlee, Vt	Sept. 15 1868	82, 082
Carriage-spring	H. S. Clark	Wyalusing, Pa	Aug. 30, 1859	25, 246
Carriage-spring	J. Cone	Reading, Conn	Aug. 26, 1808	....
Carriage-spring	J. Curtis	Cincinnati, Ohio	Nov. 11, 1873	144, 514
Carriage-spring	T. De Witt	Detroit, Mich	Sept. 17 1867	68, 966
Carriage-spring	G. Douglass	Bridgeport, Conn	June 2, 1868	78, 439
Carriage-spring	W. H. English	Macon, Ga	Sept. 24, 1867	69, 197
Carriage-spring	J. U. Fiester	Winchester, Ohio	June 17, 1856	15, 126
Carriage-spring	J. M. Forrest	Norfolk, Va	Jan. 31, 1860	26, 981
Carriage-spring	J. S. Foster	Salem, Mass	June 14, 1870	104, 135
Carriage-spring	C. B. Galentine	Brooklyn Centre, Ohio	July 14, 1863	39, 224
Carriage-spring	P. G. Gardiner	New York, N. Y	Apr. 2, 1861	31, 881
Carriage-spring	E. L. Gaylord	Terryville, Conn	Sept. 29, 1868	82, 510
Carriage-spring	J. W. Gilmer and W. H. De Valin.	Sacramento, Cal	Aug. 17, 1869	93, 703
Carriage-spring	D. R. Gould and W. S. Wickham.	Chestertown, N. Y	Sept. 6, 1870	107, 173
Carriage-spring	A. B. Greenwalt	Baltimore, Md	Aug. 21, 1866	57, 312
Carriage-spring	C. S. S. Griffing	Salem, Ohio	Mar. 25, 1873	137, 197
Carriage-spring	G. R. Groot	Cincinnati, Ohio	July 13, 1869	92, 525
Carriage-spring	C. H. Guard	Brownville, N. Y	June 10, 1851	8, 152
Carriage-spring	C. F. Hall	Rockford, Ill	Apr. 29, 1873	138, 247
Carriage-spring	E. Hall	Oxford, N. Y	July 4, 1871	116, 705
Carriage-spring	G. W. Harlan	Cincinnati, Ohio	Sept. 9, 1873	142, 627
Carriage-spring	G. L. Haussknecht	New Haven, Conn	July 15, 1851	8, 221
Carriage-spring	B. T. Henry	New Haven, Conn	May 21, 1867	64, 866
Carriage-spring	B. T. Henry	New Haven, Conn	Mar. 8, 1870	100, 626
Carriage-spring	W. B. Higgins	San Francisco, Cal	May 12, 1868	77, 880
Carriage-spring	E. C. Hodge and D. H. Mann	Oneonta and Delhi, N. Y	June 22, 1869	91, 538
Carriage-spring	B. R. Hood	Clinton, N. C	Oct. 20, 1857	18, 451
Carriage-spring	G. Hopson	Bridgeport, Conn	Oct. 14, 1873	143, 576
Carriage-spring	M. C. Hubbard	Rochester, N. Y	July 22, 1851	8, 244
Carriage-spring	M. G. Hubbard	New York, N. Y	May 15, 1855	12, 890
Carriage-spring	M. G. Hubbard	New York, N. Y	May 22, 1855	12, 910
Carriage-spring	R. and I. C. Humphries	Cantwell's Bridge, Del	Apr. 22, 1814	....
Carriage-spring	N. Hungerford	Ithaca, N. Y	Mar. 31, 1836	....
Carriage-spring	F. A. Huntington	San Francisco, Cal	Dec. 15, 1868	85, 008
Carriage-spring	H. T. Hyde	Troy, N. Y	Apr. 3, 1849	6, 276
Carriage-spring	J. M. and E. Ingold	Allegheny, Pa	Mar. 5, 1867	62, 543
Carriage-spring	E. Ives and J. Hill	New Haven, Conn	June 3, 1808	....
Carriage-spring	J. Jackson	Owego, N. Y	Oct. 20, 1868	83, 170
Carriage-spring	G. W. La Baw and P. F. Campbell.	Jersey City, N. J	Aug. 19, 1862	36, 220
Carriage-spring	D. M. Lane	West Philadelphia, Pa	June 8, 1858	20, 497
Carriage-spring	G. H. Laub	Newark, Mo	Aug. 14, 1860	29, 605
Carriage-spring	A. A. Livingston	Wilmington, Ill	Dec. 3, 1872	133, 651
Carriage-spring	J. R. Locke	San Francisco, Cal	Oct. 6, 1868	82, 726
Carriage-spring	J. R. Locke	San Francisco, Cal	Oct. 6, 1868	82, 727
Carriage-spring	G. S. Manning	Springfield, Ill	June 5, 1866	55, 319
Carriage-spring	E. Maynard	Brooklyn, N. Y	Mar. 1, 1859	23, 102
Carriage-spring	E. Maynard	Brooklyn, N. Y	Apr. 3, 1860	27, 731
Carriage-spring	A. R. Miller	Attica, N. Y	Nov. 26, 1861	33, 789
Carriage-spring	J. M. Miller	Cincinnati, Ohio	Dec. 29, 1868	85, 323
Carriage-spring	R. Montgomery	New York, N. Y	Feb. 5, 1856	14, 197
Carriage-spring	F. B. Morse	Plantsville, Conn	Nov. 1, 1870	108, 929
Carriage-spring	T. Murgatroyd, jr	Smithville, Canada	Apr. 24, 1855	12, 764
Carriage-spring	T. Murgatroyd	Hiawatha, Kans	May 27, 1873	139, 325
Carriage-spring	H. H. Olds	New Haven, Conn	Aug. 22, 1865	49, 545
Carriage-spring	H. Pace, sr	Cincinnati, Ohio	Oct. 14, 1835	....
Carriage-spring	R. W. Parker	Woburn, Mass	Nov. 29, 1864	45, 265
Carriage-spring	L. H. Parsons	Stockbridge, Mass	June 23, 1824	....
Carriage-spring	W. Patton	Towanda, Pa	Jan. 9, 1838	569
Carriage-spring	S. B. Peet	New York, N. Y	Sept. 13, 1859	25, 439
Carriage-spring	C. P. Phillips	Syracuse, N. Y	Nov. 22, 1864	45, 173
Carriage-spring	J. W. Post	Lansing, Mich	Jan. 7, 1873	134, 703

Index of patents issued from the United States Patent Office from 1790 to 1873, inclusive—Continued.

Invention.	Inventor.	Residence.	Date.	No.
Carriage-spring	B. F. Power	Morgan County, Ohio	June 29, 1869	92, 093
Carriage-spring	D. R. Pratt	Worcester, Mass	Mar. 20, 1849	6, 215
Carriage-spring	J. B. Pressey and D. Sheets	Suisun City, Cal	Oct. 16, 1860	30, 422
Carriage-spring	T. H. Prushaw	Fredonia, N. Y	Nov. 22, 1870	109, 544
Carriage-spring	J. G. Reiff	Farmersville, Pa	Oct. 8, 1867	69, 568
Carriage-spring	G. N. Reynolds	Charleston, S. C	Apr. 30, 1819	....
Carriage-spring	J. D. Richardson	Houston, Tex	Apr. 29, 1873	138, 438
Carriage-spring	A. J. Ritter	Rahway, N. J	June 27, 1865	48, 444
Carriage-spring	B. H. Roberts	Fall River, Mass	Nov. 24, 1868	84, 378
Carriage-spring	J. D. Sarven	Columbia, Tenn	Feb. 4, 1868	74, 009
Carriage-spring	J. D. Sarven	Columbia, Tenn	Feb. 4, 1868	74, 010
Carriage-spring	A. Selkirk	Albany, N. Y	Sept. 26, 1865	50, 209
Carriage-spring	W. Sharp	Burdett, N. Y	Apr. 7, 1838	682
Carriage-spring	T. J. Shears	Ypsilanti, Mich	Oct. 20, 1868	83, 217
Carriage-spring	C. Sheldon	Middleburgh, N. Y	Jan. 25, 1870	99, 249
Carriage-spring	D. Shockey	Waynesborough, Pa	Dec. 28, 1869	98, 433
Carriage spring	A. E. Smith	Bronxville, N. Y	Sept. 21, 1869	94, 981
Carriage-spring	W. Smith	Shrewsbury, Pa	Oct. 11, 1864	44, 668
Carriage-spring	E. T. Sprout	Springville, Pa	July 18, 1848	5, 674
Carriage-spring	S. Steward	Trenton, N. J	Oct. 25, 1870	108, 649
Carriage-spring	A. C. Stowe	San José, Cal	Nov. 24, 1868	84, 316
Carriage-spring	A. C. Stowe	San José, Cal	Dec. 22, 1868	85, 144
Carriage-spring	W. W. Sutliff	Town Line, Pa	Mar. 24, 1868	75, 999
Carriage-spring	W. Taylor	East Zora, Canada	Nov. 7, 1865	50, 886
Carriage-spring	G. W. Tew	Kansas City, Mo	Dec. 21, 1869	98, 207
Carriage-spring	W. S. Thomas	Norwich, N. Y	Oct. 30, 1849	6, 839
Carriage-spring	W. F. Vernier	Philadelphia, Pa	Nov. 24, 1868	84, 446
Carriage-spring	R. Walker	Harrisville, Ohio	Dec. 9, 1873	145, 374
Carriage-spring	R. E. Walker	Dresden, Mo	Aug. 2, 1870	106, 005
Carriage-spring	W. Wharton	Birmingham, England	Feb. 10, 1863	37, 653
Carriage-spring	J. B. Whitcomb	Beloit, Wis	Feb. 18, 1873	136, 116
Carriage-spring	W. F. Whitney	Milton, N. Y	Nov. 3, 1868	83, 810
Carriage-spring	D. D. Wisell	Zanesville, Ind	Oct. 10, 1871	119, 907
Carriage-spring	G. Woeber	Davenport, Iowa	Dec. 1, 1863	40, 787
Carriage-spring	A. E. Wolcott	Chicago, Ill	May 18, 1869	90, 144
Carriage-spring	A. E. Wolcott	Chicago, Ill	May 18, 1869	90, 145
Carriage-spring	A. E. Wolcott	Chicago, Ill	May 18, 1869	90, 146
Carriage-spring	T. H. Wood	New York, N. Y	May 27, 1873	139, 348
Carriage-spring	E. G. Woodside	Augusta, Mo	May 4, 1838	725
Carriage-spring	E. M. Wright	Wyandotte, Kans	Nov. 1, 1864	44, 908
Carriage-spring, Adjustable	I. Carter	Champlain, N. Y	Feb. 7, 1860	27, 036
Carriage-spring, Adjustable	R. S. Morse	Dixfield, Me	Nov. 1, 1853	10, 188
Carriage-spring, Adjustable	G. Palmer	Littlestown, Pa	July 24, 1860	29, 306
Carriage-spring, and attaching carriage-bodies thereto.	J. Ives and J. Walters	Brooklyn, N. Y	Oct. 6, 1830	....
Carriage spring and coupling	F. De Witt	Detroit, Mich	May 7, 1867	64, 501
Carriage-spring and mode of attachment	C. W. Saladee	St. Catharine's, Canada	Feb. 20, 1872	123, 937
Carriage-spring attachment	O. E. Bennett	Cannonsville, N. Y	Oct. 3, 1871	119, 560
Carriage-spring bars, Metallic scroll-ends for	A. L. Warburton and H. Bendir	Fort Wayne, Ind	Feb. 4, 1868	74, 178
Carriage, Spring-body	A. Moffit	Brownsville, Pa	Aug. 8, 1854	11, 490
Carriage-spring brace	J. H. Chadwick	Wheaton, Ill	Feb. 18, 1868	74, 499
Carriage-spring brace	T. Dutton	Washington, D. C	Oct. 14, 1856	15, 914
Carriage-spring brace	W. Evans	Eureka, Wis	Feb. 15, 1870	99, 867
Carriage-spring brace	C. C. Gleason	Wauconda, Ill	Sept. 26, 1865	50, 114
Carriage-spring brace	L. C. Miller	Humphrey, N. Y	Apr. 28, 1868	77, 306
Carriage-spring brace	L. C. Miller	Humphrey, N. Y	Jan. 24, 1871	111, 229
Carriage-spring brace	D. W. Norris	Paxton, Ill	Apr. 18, 1871	113, 915
Carriage-spring brace	E. T. Sprouth	Dimock, Pa	Sept. 7, 1844	3, 729
Carriage-spring brace and clip	E. J. Green	Valparaiso, Ind	June 7, 1864	43, 019
Carriage-spring clip	D. W. Baird	Geneva, N. Y	June 10, 1873	139, 648
Carriage-spring clip and brace	J. H. Deal	Hornellsville, N. Y	July 23, 1867	66, 953
Carriage-spring clip and plate	H. M. Beecher	Plantsville, Conn	Jan. 2, 1872	122, 428
Carriage-spring-clip plate	N. C. Dean	Union Springs, N. Y	Jan. 14, 1873	134, 859
Carriage-spring clips, Manufacture of	H. M. Beecher	Plantsville, Conn	Jan. 30, 1872	123, 079
Carriage-spring coupling	U. S. Hall	Watkins, N. Y	Dec. 10, 1872	133, 705
Carriage-spring, Elastic	T. Long	Vandalia, Ill	Nov. 5, 1867	70, 446
Carriage-spring, Elliptic	H. R. Hawkins	Akron, Ohio	Dec. 10, 1867	72, 035
Carriage-spring for light vehicles	R. Burns, jr	New York, N. Y	July 18, 1854	11, 345
Carriage-spring guard	O. H. Wheeler	Hamlin, Mich	June 14, 1870	104, 239
Carriage-spring guard	T. Winans	Baltimore, Md	Feb. 16, 1858	19, 396
Carriage-spring-head ears, Die for forging	J. Evans	New Haven, Conn	June 14, 1870	104, 292
Carriage-spring-head ears, Die for forging	J. Evans	New Haven, Conn	June 21, 1870	104, 440
Carriage-spring-head-forging die	J. H. Mason	New Haven, Conn	June 20, 1871	116, 206
Carriage-spring heads, Machine for filing	C. B. and G. Gilbert	Philadelphia, Pa	Nov. 26, 1872	133, 432
Carriage-spring heads, Machine for heading	J. Evans	New Haven, Conn	Oct. 29, 1861	33, 580
Carriage-spring heads, Machine for milling	W. Evans	New Haven, Conn	Oct. 4, 1870	108, 013
Carriage-spring heads, Manufacturing	W. R. Petrie	Westville, Conn	June 14, 1870	104, 194
Carriage-spring jack	R. R. Miller	Plantsville, Conn	Feb. 22, 1870	100, 176
Carriage-spring perch	E. R. Ferry	New Haven, Conn	Oct. 26, 1869	96, 216
Carriage spring-seat	F. M. Hubbard	Ripon, Wis	July 31, 1866	56, 753
Carriage-spring supporter	D. H. Brown	Utica, Mo	Feb. 14, 1871	111, 809
Carriage-spring supporter	N. A. Newton	Schoolcraft, Mich	May 21, 1872	127, 096
Carriage-spring, Thorough-brace	J. Mix	New Haven, Conn	June 17, 1808	....
Carriage spring. Wheel	W. Wing and H. Salisbury	Hartford, Conn	Aug. 29, 1805	....
Carriage-spring, Wooden	L. Birckhead	Philadelphia, Pa	Dec. 30, 1814	....
Carriage-spring, Wooden	J. Greenwood	New Castle, Del	Apr. 28, 1818	....
Carriage-spring, Wooden	L. H. Thrall	Troy, N. Y	Nov. 10, 1813	....
Carriage-springs, Adjustable shackle for	J. Bullard	North Hyde Park, Vt	Dec. 13, 1870	110, 006
Carriage-springs, Adjustable shackle for tightening	F. P. Hutchinson	Manchester, N. H	June 25, 1872	128, 399
Carriage-springs, Arrangement of	R. P. March	Jeffersonville, Pa	Apr. 7, 1857	16, 986
Carriage-springs, Arrangement of	C. A. McElroy	Delaware, Ohio	Mar. 10, 1857	16, 802
Carriage-springs, Arrangement of	T. Phillips	Ann Arbor, Mich	Mar. 26, 1861	31, 823
Carriage-springs, Arrangement of	E. Roughton	Frostburgh, Md	Feb. 26, 1861	31, 560
Carriage-springs, Attaching	U. S. Hall	Chemung, N. Y	May 9, 1871	114, 676
Carriage-springs, Attaching	D. A. Morton	Groton, N. Y	Jan. 9, 1838	551
Carriage-springs, Attaching	L. O. Rice	Berlin, Canada	July 6, 1858	20, 820
Carriage-springs, Connecting	J. A. Topliff and G. H. Ely	Elyria, Ohio	Dec. 19, 1871	122, 079
Carriage-springs, Device for holding	R. Rowell and F. H. Briggs	Boston, Mass	Feb. 27, 1866	52, 889

Index of patents issued from the United States Patent Office from 1790 to 1873, inclusive—Continued.

Invention.	Inventor.	Residence.	Date.	No.
Carriage-springs, Equalizing	D. G. Rollin	New York, N. Y	Jan. 12, 1858	19, 102
Carriage-springs, Forming heads of	S. H. Hartman	Pittsburgh, Pa	May 18, 1858	20, 268
Carriage-springs, Goose-neck for hanging	S. M. Wier	New Haven, Conn	May 10, 1870	102, 999
Carriage-springs, Machine for forming	W. Harty	Bridgeport, Conn	Mar. 28, 1871	113, 164
Carriage-springs, Machine for grinding	H. M. Wentworth	Gardiner, Me	Mar. 21, 1871	112, 873
Carriage-springs, Machine for shaping light	L. H. Richardson	Westfield, Mass	June 18, 1872	128, 173
Carriage-springs, Manner of applying	J. S. Tough	Baltimore, Md	June 9, 1843	3, 126
Carriage-springs, Manufacturing steel	L. Gleason	Stafford, N. Y	Dec. 24, 1834	....
Carriage-springs, Method of forming heads of	B. T. Henry	New Haven, Conn	Dec. 28, 1869	98, 376
Carriage-springs, Method of manufacturing leaves of steel.	M. Seabury	Waterville, N. Y	Apr. 29, 1839	1, 137
Carriage-springs, Mill for rolling steel	J. Stondinger	Newark, N. J	July 17, 1830	....
Carriage-springs, Mode of adjusting	M. G. Hubbard	Penn Yan, N. Y	July 1, 1856	15, 234
Carriage-springs, Mode of applying	I. Jones, A. M. Eels, and H. Griswold.	Delaware County, N. Y	July 16, 1838	842
Carriage-springs, Mode of equalizing the action of.	C. W. Saladee	Saint Catharine's, Canada.	Apr. 9, 1872	125, 413
Carriage-stay	G. G. Larkin	Portland, Me	Aug. 23, 1870	106, 593
Carriage, Steam	G. W. Barnett	Urbana, Ohio	Nov. 22, 1864	45, 130
Carriage, Steam	L. Bigelow	Petersham, Mass	Aug. 29, 1823	....
Carriage, Steam	Z. P. Dederick and I. Grass	Newark, N. J	Mar. 24, 1868	75, 874
Carriage, Steam	M. Fletcher	Louisville, Ky	Sept. 4, 1866	57, 696
Carriage, Steam	J. S. Hall	Pittsburgh, Pa	Mar. 31, 1857	16, 919
Carriage, Steam	J. V. Merrick	Philadelphia, Pa	Apr. 17, 1860	27, 920
Carriage, Steam land	C. H. Baker	Red Wing, Minn	July 3, 1860	28, 955
Carriage, Steam land	J. B. McKinley	Spencer County, Ky	Feb. 14, 1871	111, 761
Carriage-step	F. Baker	New York, N. Y	Apr. 2, 1867	63, 458
Carriage-step	H. T. Betts	Springfield, Mass	June 4, 1861	32, 460
Carriage-step	C. H. Gould	Boston, Mass	Apr. 8, 1873	137, 547
Carriage-step	G. A. Keene	Lynn, Mass	Dec. 24, 1872	134, 291
Carriage-step	G. A. Keene	Lynn, Mass	Feb. 11, 1873	135, 815
Carriage-step	W. H. Knowles	Plantsville, Conn	July 26, 1870	105, 876
Carriage-step	D. and E. Z. Little	Gettysburgh, Pa	Oct. 7, 1846	4, 795
Carriage-step	F. B. Morse	Plantsville, Conn	Mar. 22, 1870	101, 150
Carriage-step	F. B. Morse	Plantsville, Conn	Apr. 4, 1871	113, 551
Carriage-step	F. B. Morse	Plantsville, Conn	Aug. 29, 1871	118, 632
Carriage-step	G. Panchot	Hastings, Minn	Oct. 20, 1868	83, 305
Carriage-step	C. Parker and W. Volger	Canterbury, N. H	Oct. 1, 1867	69, 364
Carriage-step	J. Pendergast	New Haven, Conn	May 19, 1872	124, 697
Carriage-step	G. M. Plympton	New York, N. Y	Feb. 18, 1868	74, 589
Carriage-step	E. Wells	New Haven, Conn	Apr. 2, 1872	125, 235
Carriage-step	J. H. Yager	Trenton, Ohio	Apr. 2, 1867	63, 601
Carriage step and wheel fender	J. W. Gosling	Cincinnati, Ohio	May 4, 1869	89, 759
Carriage-step block	P. F. Hulbert	Chatham, N. Y	Mar. 20, 1866	53, 302
Carriage step cover and wheel fender	J. Curtis	Cincinnati, Ohio	Dec. 8, 1868	84, 802
Carriage step cover and wheel fender	J. Roberts	Cincinnati, Ohio	May 25, 1869	90, 584
Carriage-step die	F. B. Morse	Plantsville, Conn	Dec. 12, 1871	121, 803
Carriage-step-forming die	L. Burns	Port Chester, N. Y	May 24, 1870	103, 296
Carriage-step guard	R. H. Goodwin, G. L. Gamage, and R. J. P. Goodwin.	Boston and Lynn, Mass., and Manchester, N. H.	Jan. 4, 1870	98, 488
Carriage-step screen	A. L. Ross	Cleves, Ohio	May 5, 1868	77, 531
Carriage-step shield	C. H. Gould	Boston, Mass	Sept. 20, 1870	107, 482
Carriage-steps, India-rubber tread for	G. A. Keene	Newburyport, Mass	June 11, 1867	65, 754
Carriage-steps, Method of forming	W. N. Knowles	Plantsville, Conn	June 7, 1870	104, 040
Carriage-steps, Method of making	F. B. Morse	Plantsville, Conn	Apr. 25, 1871	114, 025
Carriage, Street	G. S. McHenry	Kansas City, Mo	Aug. 22, 1871	118, 255
Carriage, Street-railway	W. L. Burt	Cambridge, Mass	Aug. 12, 1862	36, 137
Carriage stump-joint	F. B. Morse	New Haven, Conn	July 21, 1868	80, 202
Carriage stump-joint	F. B. Morse	New Haven, Conn	Aug. 18, 1868	81, 193
Carriage stump-joint	F. B. Morse	Plantsville, Conn	Oct. 19, 1869	96, 024
Carriage stump-joint	F. B. Morse	Plantsville, Conn	Oct. 19, 1869	96, 025
Carriage, Swing	W. Hewett	Trenton, N. J	Apr. 14, 1868	76, 629
Carriage, Three-wheeled	C. H. Barrows	Willimantic, Conn	June 21, 1870	104, 409
Carriage, Three-wheeled	J. Gehr	Mercersburgh, Pa	Oct. 1, 1867	69, 335
Carriage, Three-wheeled pleasure	C. W. Saladee	Columbus, Ohio	July 15, 1856	15, 345
Carriage-thill	J. P. Tyson	Philadelphia, Pa	Mar. 17, 1868	75, 711
Carriage-thill hold-back	P. S. Van Wagner	Saltfleet Township, Canada	Nov. 16, 1869	96, 851
Carriage to be moved by manual labor, Four-wheeled.	L. Cook	Whitestown, N. Y	Mar. 28, 1811	....
Carriage to be propelled by mechanical powers	J. J. Staples		Apr. 25, 1794	....
Carriage to overcome obstructions in roads	H. Knowles	Colchester, Conn	Aug. 29, 1833	....
Carriage-tongue, Child's	E. A. Morse	Rutland, Vt	Feb. 7, 1871	111, 558
Carriage-tongue shackle	F. R. Pollard	Canaan, N. H	July 24, 1866	56, 606
Carriage-top	W. Bauder	Circleville, Ohio	May 16, 1871	114, 751
Carriage-top	N. Benedict	Aurelius, N. Y	Jan. 12, 1858	19, 065
Carriage-top	G. Bockstaller	Chicago, Ill	June 3, 1873	139, 450
Carriage-top	P. Boyden	Sandy Creek, N. Y	Mar. 29, 1859	23, 347
Carriage-top	N. G. Burr	Homer, N. Y	Aug. 18, 1868	81, 135
Carriage-top	H. M. Colton	Southbridge, Mass	Apr. 30, 1872	126, 130
Carriage-top	C. A. Dearborn	New Bedford, Mass	Oct. 28, 1873	144, 069
Carriage top	L. Z Dodds and R. Walsh	Three Rivers, Mich	Aug. 29, 1865	49, 611
Carriage-top	D. B. Dorsey	Chillicothe, Mo	Dec. 23, 1873	145, 730
Carriage-top	S. Emmons and E. S. Simpson	Geneva, N. Y	Nov. 19, 1867	71, 151
Carriage-top	J. Enders	Boston, Mass	June 27, 1865	48, 386
Carriage-top	J. M. Freeman	Belleville, N. Y	Feb. 14, 1860	27, 121
Carriage-top	J. N. Gill	Oshkosh, Wis	Nov. 18, 1873	144, 612
Carriage-top	A. J. Hall and R. Patten	Morristown, Vt	Nov. 22, 1859	26, 183
Carriage-top	H. Hayes	Quincy, Ill	Mar. 11, 1856	14, 400
Carriage-top	S. T. Huntington	Syracuse, N. Y	June 20, 1854	11, 126
Carriage-top	K. T. Hurlburt	Lyons, N. Y	July 11, 1865	48, 688
Carriage top	M. T. Jackson	Montrose, Pa	Mar. 1, 1870	100, 295
Carriage-top	R. S. Jennings	Waterbury, Conn	Mar. 31, 1857	16, 925
Carriage-top	R. S. Jennings	Waterbury, Conn	June 30, 1857	17, 680
Carriage-top	A. McKenzie	Westminster, England	Sept. 19, 1871	119, 166
Carriage-top	R. Miller	Middlefield, Ohio	Apr. 4, 1854	10, 723
Carriage-top	P. Owens	Chicago, Ill	Sept. 24, 1867	69, 119
Carriage-top	W. I. Peck	Carthage, N. Y	July 29, 1873	141, 377
Carriage-top	L. Righter	Salem, Ohio	Jan. 12, 1869	85, 853
Carriage-top	W. F. Rundell	Genoa, N. Y	Nov. 27, 1866	60, 066

Index of patents issued from the United States Patent Office from 1790 *to* 1873, *inclusive*—Continued.

Invention.	Inventor.	Residence.	Date.	No.
Carriage-top	J. F. Sargent	North Tunbridge, Vt	Aug. 4, 1868	80, 670
Carriage-top	H. Sayler	Saint Paris, Ohio	Nov. 18, 1873	144, 704
Carriage-top	F. C. Schaffer	Brooklyn, N. Y	Sept. 13, 1859	25, 445
Carriage-top	W. Schoch	Plumsteadville, Pa	Feb. 8, 1870	99, 601
Carriage-top	E. S. Scripture	Brooklyn, N. Y	June 9, 1868	78, 836
Carriage-top	W. Smith and E. M. Pike	McDonough, N. Y	Jan. 28, 1868	73, 763
Carriage-top	J. H. Snyder	Rockford, Ill	Aug. 19, 1873	142, 050
Carriage-top	R. H. Wright	New Bloomfield, Pa	Apr. 6, 1869	88, 597
Carriage-top	G. H. Young	New York, N. Y	May 20, 1873	139, 225
Carriage top and back, Shifting	R. M. Stivers and G. W. V. Smith.	New York, N. Y	Sept. 23, 1862	36, 539
Carriage-top bow	J. F. Fowler	Alliance, Ohio	June 28, 1870	104, 724
Carriage-top bows, Machine for adjusting	A. Soursin	Saint Louis, Mo	July 16, 1867	66, 747
Carriage-top brace	J. L. Allen	New Haven, Conn	Oct. 15, 1850	7, 715
Carriage-top brace, Concealed jointed	D. W. Baird	Geneva, N. Y	Sept. 30, 1873	143, 316
Carriage-top-brace joint	G. Gregory and F. B. Morse	New Haven, Conn	Apr. 16, 1867	63, 792
Carriage-top-brace joint	F. B. Morse	New Haven, Conn	Apr. 23, 1867	64, 024
Carriage-top brace, Stump-joint for	F. Van Patten and E. D. Clapp	Auburn, N. Y	Nov. 29, 1870	109, 782
Carriage-top button-hole	S. A. Budd	Cleveland, Ohio	Nov. 12, 1867	70, 797
Carriage top, Calash	G. and D. Cook	New Haven, Conn	Jan. 27, 1857	16, 467
Carriage-top fixture	N. B. Richardson	Somerville, N. J	Nov. 1, 1870	108, 831
Carriage-top, Folding	T. H. Wood	New York, N. Y	July 12, 1870	105, 287
Carriage-top frame	J. H. Flagg	Perkinsville, Vt	Feb. 25, 1868	74, 809
Carriage-top joint	W. H. Busser	Sidney, Ohio	Jan. 26, 1869	86, 278
Carriage-top joint	J. H. Combs	Stamford, Conn	Oct. 14, 1873	143, 669
Carriage-top joint	A. C. Shelton and B. Tuttle	Plymouth, Conn	Dec. 29, 1857	18, 992
Carriage-top joint	W. B. C. Stirling and J. W. Pohlman.	Batavia, Ohio	Dec. 2, 1873	145, 135
Carriage-top joint	R. W. Stone	Solsville, N. Y	Nov. 3, 1857	18, 558
Carriage-top joint	J. A. Topliff and G. H. Ely	Elyria, Ohio	Oct. 15, 1872	132, 188
Carriage-top, Molded	G. L. Swett and J. P. Lockey	Leominster, Mass	Nov. 15, 1870	109, 272
Carriage-top, Moving	O. E. Mallory	Batavia, N. Y	June 28, 1870	104, 863
Carriage-top prop	C. R. Abbot	Elmira, N. Y	May 28, 1867	65, 150
Carriage-top prop	J. Bauer	Cincinnati, Ohio	Dec. 3, 1872	133, 619
Carriage-top prop	L. W. Blessing	Wilmington, Del	Sept. 16, 1873	142, 889
Carriage-top prop	F. A. Bradley	New Haven, Conn	Sept. 15, 1868	82, 203
Carriage-top prop	F. A. Bradley	New Haven, Conn	May 16, 1871	114, 909
Carriage-top prop	G. Cook and H. I. Kimball	New Haven, Conn	Dec, 27, 1859	26, 564
Carriage-top prop	W. P. Elam	Petersburgh, Ill	Dec. 30, 1873	145, 935
Carriage-top prop	R. S. Grummon	Newark, N. J	Apr. 23, 1867	64, 010
Carriage-top prop	F. B. Morse	Plantsville, Conn	Oct. 26, 1869	96, 254
Carriage-top prop	J. F. Mullen	New York, N. Y	Jan. 21, 1868	73, 632
Carriage-top prop	W. B. Pardee	New Haven, Conn	Dec. 1, 1868	84, 573
Carriage-top prop	A. Searls	San Francisco, Cal	Nov. 19, 1867	71, 228
Carriage-top prop	A. Searls	Newark, N. J	Aug. 2, 1870	105, 985
Carriage-top prop	A. Searls	Newark, N. J	Mar. 5, 1872	124, 291
Carriage-top prop	A. Searls	Newark, N. J	Sept. 24, 1872	131, 569
Carriage-top prop	D. M. Valentine	New York, N. Y	Apr. 16, 1872	125, 772
Carriage-top-prop block	W. N. Barnett	Urbana, Ohio	Jan. 23, 1866	52, 126
Carriage-top-prop block	W. H Stickel	Knightstown, Ind	July 30, 1867	67, 371
Carriage-top-prop-block washer	A. D. Westbrook	Astoria, N. Y	Dec. 26, 1871	122, 205
Carriage-top-prop joint	C. W. Saladee	Saint Catharine's, Canada	Oct. 17, 1871	120, 107
Carriage-top-prop joint	E. W. Waite	New Haven, Conn	Sept. 22, 1868	82, 456
Carriage-top-prop rest	H. W. Libbey	Cleveland, Ohio	June 4, 1867	65, 403
Carriage-top protector	R. Nickson	Akron, Ohio	Oct. 23, 1866	59, 056
Carriage-top rest	C. C. Lawrence and J. Lewis	Marengo, Mich	Nov. 30, 1869	97, 301
Carriage-top rest	J Lewis	Marengo, Mich	July 5, 1870	105, 097
Carriage-top rest	W. O. Snyder	Philadelphia, Pa	Jan. 19, 1869	86, 107
Carriage-top setter	J. Poffenbarger	Farmersville, Ohio	Sept. 23, 1873	143, 183
Carriage-top, Shifting	A. V. Heyden	Milwaukee, Wis	Oct. 23, 1866	59, 014
Carriage-top, Shifting	G. R. Lucas	Mannsville, N. Y	Jan. 5, 1869	85, 680
Carriage-top, Shifting	O. E. Mallory	Batavia, N. Y	Aug. 30, 1870	106, 947
Carriage-top, Shifting	O. E. Mallory	Batavia, N. Y	Nov. 22, 1870	109, 528
Carriage-top, Shifting	O. E. Mallory	Batavia, N. Y	Apr. 30, 1872	126, 313
Carriage-top support	H. J. Lingenfelter	Glen, N. Y	Nov. 5, 1872	132, 766
Carriage-tops, Apparatus for raising and lowering	A. Quigley	Sheldrake, N. Y	Apr. 22, 1856	14, 735
Carriage-tops, Apparatus for setting bows for	E. Hayes and M. Hayes	Wheeling, Va., and Washington, Pa.	July 3, 1855	13, 171
Carriage-tops, Attaching movable	J. S. Belcher	Albany, N. Y	July 10, 1860	29, 049
Carriage-tops, Box-loop for	C. H. Davis	Syracuse, N. Y	Sept. 21, 1869	95, 004
Carriage-tops, Elevating and lowering	J. L. Allen	Syracuse, N. Y	Feb. 19, 1850	7, 097
Carriage-tops, Lowering, raising, and fastening	Z. S. Ogden	Glen's Falls, N. Y	Jan. 3, 1854	10, 373
Carriage-tops, Machine for making braces for	B. F. Hooper	Birmingham, Conn	Apr. 9, 1861	32, 028
Carriage-tops, Mode of adjusting	C. W. Saladee	Columbus, Ohio	Sept. 9, 1856	15, 725
Carriage-tops, Mode of adjusting	C. W. Saladee	Columbus, Ohio	Sept. 1, 1857	18, 106
Carriage-tops, Mode of attaching	W. Harrocks	Poughkeepsie, N. Y	July 21, 1868	80, 180
Carriage-tops, Raising	J. L. Allen	New Haven, Conn	Jan. 14, 1851	7, 897
Carriage-tops, Raising and lowering	S. Goddard and H. Warfield	Truxton, N. Y	Feb. 19, 1850	7, 100
Carriage-tops, Raising and lowering	G. Stover	Centre Hill, Pa	Oct. 30, 1866	59, 289
Carriage-tops, Raising and lowering	J. R. Winchester	Medina, N. Y	June 20, 1854	11, 144
Carriage-tops, Shifting-rail for	P. G. Clancy	Augusta, Me	Sept. 17, 1867	68, 843
Carriage-tops, Shifting-rail for	S. Johns	Waupun, Wis	Apr. 24, 1866	54, 169
Carriage-tops, Shifting-rail for	W. Reynolds	New York, N. Y	May 28, 1867	65, 119
Carriage-tops, Shifting-rail for	J. J. Waldron	East Durham, N. Y	Mar. 2, 1869	87, 527
Carriage-tops, Slat-iron for	J. W. Sheppard	Plantsville, Conn	Feb. 20, 1872	123, 942
Carriage-tops, Slat-iron for	G. W. Traphagen	Glen's Falls, N. Y	Oct. 30, 1866	59, 322
Carriage transom or king-bolt and rocker-plate	J. A. Judd	Newton, Mass	Oct. 3, 1871	119, 463
Carriage, Traveling	G. H. Richards	New London, Conn	July 8, 1821	....
Carriage-trimming	C. Bried	Newark, N. J	May 21, 1867	64, 942
Carriage-trimmings, Manufacture of rubber-coated	J. W. Munson	Bridgeport, Conn	June 7, 1870	104, 056
Carriage, Two-wheeled	J. Page	New York	Sept. 14, 1840	1, 789
Carriage, wagon, &c	E. Lane	Philadelphia, Pa	Nov. 28, 1865	51, 196
Carriage-washer, Wooden	T. M. Hart	New Bedford, Mass	Dec. 1, 1868	84, 488
Carriage-washing machine	O. P. Weston	Shattuckville, Mass	Aug. 15, 1871	118, 085
Carriage, Water-proof mail	B. B. Pleasants	Brookville, Md	Dec. 7, 1837	5, 9
Carriage-wheel	G. L. Ackerman	Troy, N. Y	Apr. 16, 1845	4, 001
Carriage wheel	L. Adams	Amherst, Mass	Mar. 31, 1868	76, 033
Carriage-wheel	C. Anderegg	Lawrenceburgh, Ind	July 16, 1892	129, 449

Index of patents issued from the United States Patent Office from 1790 *to* 1873, *inclusive*—Continued.

Invention.	Inventor.	Residence.	Date.	No.
Carriage-wheel	C. H. Appel	Allentown, Pa	Feb. 6, 1872	123, 375
Carriage-wheel	S. Atha	West Liberty, Ohio	Nov. 9, 1869	96, 656
Carriage-wheel	C. C. Ayer	Chelsea, Mass	Feb. 5, 1867	61, 793
Carriage-wheel	C. C. Ayer	Chelsea, Mass	Nov. 12, 1867	70, 681
Carriage-wheel	J. R. Baird	Vincennes, Ind	Mar. 8, 1870	100, 584
Carriage-wheel	J. C. Baker	Adams Centre, N. Y	July 9, 1861	32, 742
Carriage-wheel	A. Beale	Stamford, Conn	Sept. 10, 1867	68, 594
Carriage-wheel	N. S. Bean	Manchester, N. H	Feb. 5, 1867	61, 703
Carriage-wheel	H. Beebe	Haverstraw, N. Y	Mar. 12, 1834	......
Carriage-wheel	I. E. Bower	Bainbridge, Ga	Nov. 14, 1871	120, 847
Carriage-wheel	R. Brooks, jr	Rockport, Mass	Oct. 15, 1867	69, 757
Carriage-wheel	T. Brownfield	Georges Township, Pa	Aug. 19, 1856	15, 554
Carriage-wheel	S. R. Bryant	Waterford, Pa	Aug. 20, 1872	130, 621
Carriage-wheel	S. B. Buck	Elyria, Ohio	July 19, 1870	105, 546
Carriage-wheel	G. G. W. Burnham	Baltimore County, Md	Apr. 18, 1871	113, 851
Carriage-wheel	J. G. Buzzell	Lynn, Mass	Oct. 1, 1867	69, 403
Carriage-wheel	J. G. Buzzell	Lynn, Mass	Oct. 6, 1868	82, 690
Carriage-wheel	J. G. Buzzell	Lynn, Mass	Jan. 18, 1870	98, 920
Carriage-wheel	C. F. Carman	Hamburgh, Iowa	June 8, 1869	90, 993
Carriage-wheel	D. W. Clark and S. H. Gray	Bridgeport, Conn	May 22, 1855	12, 897
Carriage-wheel	C. Clarke	Coral, Ill	Aug. 18, 1868	81, 141
Carriage-wheel	J. Coney	South Boston, Mass	Mar. 16, 1869	87, 823
Carriage-wheel	S. D. Cook	West Liberty, Ohio	June 25, 1872	128, 212
Carriage-wheel	J. M. Coombs	Boston, Mass	Apr. 9, 1867	63, 705
Carriage-wheel	W. Cooper	Columbia, S. C	Mar. 3, 1825	......
Carriage-wheel	S. D. Craft	New York, N. Y	Nov. 7, 1871	120, 720
Carriage-wheel	C. Cummings	Providence, R. I	Mar. 16, 1869	87, 913
Carriage-wheel	A. J. Curtis	Winterport, Me	Jan. 2, 1866	51, 808
Carriage-wheel	D. P. Davis	New York, N. Y	July 27, 1869	93, 067
Carriage-wheel	W. H. De Valin	Sacramento, Cal	Sept. 1, 1868	81, 758
Carriage-wheel	H. E. Dodson	West Liberty, Ohio	Oct. 3, 1871	119, 511
Carriage-wheel	F. N. Draper and A. Danison	West Liberty, Ohio	Sept. 19, 1871	119, 082
Carriage-wheel	H. C. and J. W. Drew	Waterloo, Mich	Feb. 27, 1866	52, 832
Carriage-wheel	J. Eastman and G. C. Rix	Bath, N. H	Dec. 23, 1830	......
Carriage-wheel	W. P. Elam	Petersburgh, Ill	Sept. 21, 1869	95, 098
Carriage-wheel	C. W. Fillmore	Marengo, Ill	Feb. 6, 1872	123, 468
Carriage-wheel	J. P. Fisher	Rochester, N. Y	Dec. 18, 1860	30, 915
Carriage-wheel	H. K. Flinchbaugh	Conestoga Centre, Pa	Aug. 18, 1863	39, 561
Carriage-wheel	W. K. Foster	Bangor, Me	Nov. 22, 1864	45, 150
Carriage-wheel	W. K. Foster	Bangor, Me	Oct. 8, 1867	69, 557
Carriage-wheel	C. M. Foulke	Philadelphia, Pa	Apr. 6, 1869	88, 705
Carriage-wheel	G. W. Gilbert	Radnor, Pa	May 26, 1863	38, 669
Carriage-wheel	J. Goodman	Blackfriars Road, England	Feb. 21, 1865	46, 524
Carriage-wheel	J. S. Graves	Lima, N. Y	Oct. 10, 1871	119, 841
Carriage-wheel	D. Grim	Pittsburgh, Pa	Mar. 16, 1869	87, 926
Carriage-wheel	J. W. Guider	Saint Joseph, Mo	Apr. 12, 1870	101, 860
Carriage-wheel	H. Gwynn	Baltimore, Md	Dec. 16, 1873	145, 646
Carriage-wheel	G. G. Hickman	Coatesville, Pa	Apr. 24, 1866	54, 155
Carriage-wheel	S. B. Hindman	Richmond, Ind	Oct. 1, 1872	131, 757
Carriage-wheel	I. Holmes	Moscow, N. Y	Jan. 1, 1847	4, 918
Carriage-wheel	D. A. Johnson and F. M. Gibson	Chelsea, Mass	Oct. 23, 1860	30, 479
Carriage-wheel	M. Johnson	Three Rivers, Mich	Mar. 26, 1872	125, 053
Carriage-wheel	J. A. Johnston	Topeka, Kans	May 21, 1872	127, 063
Carriage-wheel	J. A. Johnston	Topeka, Kans	May 21, 1872	127, 064
Carriage-wheel	J. B. Jones	Sparta, Ill	Sept. 29, 1868	82, 530
Carriage-wheel	T. M. Jones and C. W. Fillmore	Chicago, Ill	Nov. 1, 1870	108, 794
Carriage-wheel	G. Kenny	Nashua, N. H	Aug. 18, 1868	81, 175
Carriage-wheel	H. Keyes	Terre Haute, Ind	Dec. 28, 1869	98, 273
Carriage-wheel	A. H. Leach	Marathon, N. Y	Apr. 26, 1870	102, 280
Carriage-wheel	C. Leavitt	Cleveland, Ohio	Jan. 14, 1862	34, 149
Carriage-wheel	O. Look	Bridgeport, Conn	Aug. 31, 1869	94, 325
Carriage-wheel	J. H. Loughridge	Key West, Fla	Aug. 8, 1871	117, 902
Carriage-wheel	W. H. Marshall	Sutton, N. H	Nov. 19, 1867	71, 193
Carriage-wheel	W. S. Mayo	New York, N. Y	Feb. 9, 1869	86, 772
Carriage-wheel	R. W. McClelland	Springfield, Ill	June 29, 1869	91, 859
Carriage-wheel	R. W. McClelland	Springfield, Ill	Feb. 8, 1870	99, 691
Carriage-wheel	J. McCreery	Springfield, Ill	May 30, 1871	115, 497
Carriage-wheel	C. A. Miller	Urbana, Ohio	Apr. 2, 1872	125, 206
Carriage-wheel	W. F. Moody	Auburn, N. Y	Sept. 28, 1869	95, 370
Carriage-wheel	W. F. Morton	New Haven, Conn	May 28, 1867	65, 257
Carriage-wheel	W. F. Morton	New Haven, Conn	Feb. 4, 1868	74, 117
Carriage-wheel	J. and T. Nevison, jr	Morgan, Ohio	June 30, 1868	79, 495
Carriage-wheel	T., jr., and J. Nevison	Morgan, Ohio	Feb. 18, 1868	74, 579
Carriage-wheel	A. M. Ocobock	Toledo, Ohio	Feb. 20, 1872	123, 836
Carriage-wheel	J. O'Connor	Jackson, Mo	Apr. 16, 1872	125, 840
Carriage-wheel	C. Palmer	Hamilton, Nev	Jan. 9, 1872	122, 646
Carriage-wheel	A. Prentiss	Prentiss Vale, Pa	May 14, 1867	64, 794
Carriage-wheel	A. Prentiss	Otto, Pa	May 14, 1867	64, 795
Carriage-wheel	V. Price	New York, N. Y	May 4, 1869	89, 835
Carriage-wheel	W. Race and B. Holly	Seneca Falls, N. Y	Jan. 30, 1855	12, 317
Carriage-wheel	J. Raddin	Lynn, Mass	June 13, 1865	48, 207
Carriage-wheel	J. Raddin	Lynn, Mass	Dec. 24, 1867	72, 538
Carriage-wheel	J. A. Reed	New Market, N. J	Sept. 14, 1869	94, 839
Carriage-wheel	J. Ridge	Wayne County, Ind	Sept. 20, 1870	107, 543
Carriage-wheel	J. Ridge	Richmond, Ind	July 30, 1872	130, 074
Carriage-wheel	J. Rowe	Cincinnati, Ohio	Jan. 26, 1847	4, 946
Carriage-wheel	J. D. Sarven	Columbia, Tenn	June 9, 1857	17, 520
Carriage-wheel	J. Scott	Ocala, Fla	Dec. 26, 1865	51, 756
Carriage-wheel	G. W. Seymour	Whitney's Point, N. Y	Sept. 1, 1868	81, 828
Carriage-wheel	S. S. Sherman and S. D. Piper	West Eau Claire, Wis	Aug. 3, 1869	93, 357
Carriage-wheel	J. T. Shimer	Easton, Pa	Aug. 25, 1868	81, 545
Carriage-wheel	S. E. Shute and W. C. Starr	Richmond, Ind	July 23, 1872	129, 686
Carriage-wheel	J. Y. Sitton	Due West, S. C	Apr. 25, 1871	114, 058
Carriage-wheel	J. Skelley	Brooklyn, N. Y	Jan. 30, 1855	12, 330
Carriage-wheel	A. B. Smith	Plattsmouth, Nebr	May 19, 1868	78, 143
Carriage-wheel	T. R. Smith	San Francisco, Cal	Sept. 7, 1869	94, 521
Carriage-wheel	H. G. Spafford	Albany, N. Y	Nov. 25, 1814	......
Carriage-wheel	J. C. Sparks	Philadelphia, Pa	Mar. 3, 1868	75, 213

Index of patents issued from the United States Patent Office from 1790 *to* 1873, *inclusive*—Continued.

Invention.	Inventor.	Residence.	Date.	No.
Carriage-wheel	C. S. Stearns	Marlborough, Mass	Feb. 9, 1869	86, 709
Carriage-wheel	C. S. Stearns	Marlborough, Mass	Apr. 6, 1869	88, 748
Carriage-wheel	R. W. Thomson	England	May 8, 1847	5, 104
Carriage-wheel	E. Tolles	Hartford, Conn	Dec. 27, 1839	1, 440
Carriage-wheel	I. Van Gorder	Warren, Ohio	Jan. 7, 1833	....
Carriage-wheel	S. Vreeland	Cuba, N. Y	July 20, 1869	92, 915
Carriage-wheel	R. Walker	Batavia, N. Y	Sept. 11, 1866	58, 030
Carriage-wheel	I. B. Ward	Camden, N. J	Dec. 25, 1849	6, 977
Carriage-wheel	S. M. H. Ward	New York, N. Y	Dec. 22, 1868	85, 259
Carriage-wheel	H. G. Weibling	Denver City, Colo	Jan. 20, 1863	37, 480
Carriage-wheel	E. S. Winchester	Boston, Mass	Nov. 28, 1865	51, 251
Carriage-wheel	M. Wonser	Norwalk, Ohio	May 17, 1870	103, 114
Carriage-wheel	G. B. Woodard and A. B. Woodard.	Bolivar and Alfred Centre, N. Y.	July 26, 1864	43, 651
Carriage-wheel	J. Woodburn	Saint Louis, Mo	July 24, 1866	56, 681
Carriage-wheel	J. Woodburn	Saint Louis, Mo	Mar. 12, 1872	124, 468
Carriage-wheel	E. G. Woodside	San Francisco, Cal	June 1, 1869	90, 714
Carriage, Wheel	S. W. Dana	Rutland, Vt	Aug. 30, 1808	....
Carriage, Wheel	D. F. Goodhue and E. H. Carey.	Cincinnati, Ohio	May 14, 1861	32, 285
Carriage, Wheel	W. H. Lewis	Greenwood, Mass	Aug. 23, 1864	43, 919
Carriage, Wheel	B. J. Malfeson	Burlington, N. J	Apr. 6, 1824	....
Carriage, Wheel	A. K. Stone	Oronoco, Minn	Apr. 9, 1867	63, 668
Carriage, Wheel	A. K. Stone	Oronoco, Minn	Apr. 9, 1867	63, 669
Carriage wheel and axle	J. H. Lewis	Duxbury, Mass	Oct. 22, 1867	70, 098
Carriage wheel and axle, Railway	C. D. Tisdale	Boston, Mass	Dec. 14, 1869	97, 994
Carriage-wheel and dress-guard	W. R. Bush	Albany, N. Y	Mar. 13, 1860	27, 427
Carriage-wheel axle	G. W. Ray	Rankin's Depot, Tenn	May 14, 1872	126, 742
Carriage-wheel axle and box	A. R. Reynolds	Skaneateles, N. Y	Dec. 5, 1840	1, 883
Carriage-wheel axle and hub	H. F. Phillips	Skaneateles, N. Y	Sept. 18, 1841	2, 260
Carriage-wheel bearings, Lubricator for	W. P. White	Orland, Me	Nov. 19, 1867	71, 101
Carriage-wheel box	L. Brown	Berkshire, Mass	May 30, 1806	....
Carriage-wheel box	W. A. Clark	Westville, Conn	Dec. 12, 1871	121, 706
Carriage-wheel box	J. Gridley	Boston, Mass	May 23, 1806	....
Carriage-wheel box	C. H. Holdredge	Westerly, R. I	Dec. 1, 1868	84, 548
Carriage-wheel box	R. W. McClelland	Pekin, Ill	Oct. 12, 1858	21, 766
Carriage-wheel box	W. Sharp	Millport, N. Y	Feb. 28, 1860	27, 314
Carriage-wheel box	A. Woodruff	Hartford, Conn	Aug. 6, 1872	130, 172
Carriage-wheel box, Self-fastening	T. Massey	New London, Conn	June 11, 1829	....
Carriage-wheel brake	C. Walker	Lewisburgh, Va	July 8, 1839	1, 222
Carriage-wheel brakes, Mode of operating	W. Dunning	Dunningsville, Pa	Nov. 1, 1845	4, 254
Carriage-wheel dress-protector	P. G. Hubert and J. W. Pitney.	New York, N. Y	Apr. 27, 1869	89, 315
Carriage-wheel, Elastic	E. S. Pratt and J. B. Thompson.	Boston, Mass	Mar. 15, 1864	41, 910
Carriage-wheel, Elastic	S. D. Woodbury	Lynn, Mass	Nov. 26, 1861	33, 809
Carriage-wheel fender	S. R. Ramsdell	Providence, R. I	Sept. 4, 1866	57, 765
Carriage-wheel for preventing horses from running away.	R. Jarvis	Boston, Mass	Mar. 29, 1834	....
Carriage-wheel gear	A. Baxter	Howard, N. Y	Apr. 19 1870	102, 079
Carriage-wheel, Iron	R. Crane, jr	Woodbury, Conn	May 4, 1805	....
Carriage-wheel, Iron	J. D. Murphey	Baltimore, Md	Apr. 19, 1859	23, 695
Carriage-wheel jack	H. Hooton and J. G. Bicknell	Boston and Cambridge, Mass.	May 31, 1859	24, 213
Carriage-wheel lubricator	S. S. Vollum and W. H. Green	New York, N. Y	Dec. 6, 1870	109, 977
Carriage-wheel, Metallic	W. Beach	Baltimore, Md	Mar. 2, 1858	19, 478
Carriage-wheel, Metallic	A. P. Ware	Camden County, N. J	Feb. 12, 1867	61, 962
Carriage-wheel-painting machine	S. B. Fuller	Worthington, Mass	Sept. 23, 1856	15, 759
Carriage wheel, Railway	J. P. Allen	Manchester, Mass	Apr. 20, 1869	89, 003
Carriage wheel, Railway	J. Elgar	Philadelphia, Pa	Nov. 19, 1833	....
Carriage wheel, Railway	A. R. Morrill	Northfield, Vt	Feb. 15, 1859	22, 972
Carriage-wheel, Self-moving	R. W. Thomson	Edinburgh, Great Britain	Nov. 9, 1869	96, 635
Carriage-wheel-spoke socket	A. J. Carleton	Springfield, Mass	Apr. 12, 1870	101, 821
Carriage-wheel-spoking pit	J. Collins	Crawfordsville, Ind	Aug. 8, 1871	117, 745
Carriage-wheel, Spring	J. Lamb and C. H. Root	McDonough, N. Y	Jan. 1, 1851	7, 874
Carriage-wheel, Spring	B. Ruggles	Warsaw, N. Y	Sept. 8, 1834	....
Carriage-wheel, Steam	R. W. Thomson	Edinburgh, Great Britain	Dec. 21, 1869	98, 126
Carriage-wheel-tenon-cutting machine	J. Reynolds	Dutchess, N. Y	June 24, 1815	....
Carriage-wheel-washing device	W. T. Sweet	Fayette, N. Y	Apr. 16, 1867	63, 960
Carriage-wheel wrench	L. B. Fisk	Lockport, N. Y	Apr. 21, 1868	76, 900
Carriage-wheel, Wrought-iron	J. McCollum	Wilsonville, Ala	May 10, 1844	3, 578
Carriage-wheels after tiring, Machine for immersing	J. Morris	Bloomfield, N. J	Sept. 23, 1843	3, 276
Carriage-wheels, Anti-friction box for	J. B. Blanchard	Sacramento, Cal	Sept. 17, 1861	33, 292
Carriage-wheels, Box-setter for	W. Witbeck	Troy, N. Y	Nov. 6, 1866	59, 495
Carriage-wheels, Confining	C. Force	Baltimore, Md	Apr. 28, 1836	....
Carriage-wheels, Facilitating the making of	B. L. Greenough	Lebanon, N. H	Sept. 12, 1815	....
Carriage-wheels, Fastening	B. Berndt and F. Barsch	Williamsport, Pa	Dec. 12, 1871	121, 841
Carriage-wheels, Fastening	W. Elder	Mill Hall, Pa	Dec. 27, 1870	110, 557
Carriage-wheels, Hooping or tiring	R. Hunt, 2d	Litchfield County, Conn	Mar. 20, 1821	....
Carriage-wheels, Lubricating axle-boxes of	W. Diller	Lancaster, Pa	Mar. 9, 1858	19, 551
Carriage-wheels, Machine for compressing	G. Cook	New Haven, Conn	July 21, 1863	39, 275
Carriage-wheels, Machine for compressing	H. Killam	New Haven, Conn	Oct. 15, 1867	69, 818
Carriage-wheels, Machine for finishing the exterior of rims of.	R. Fretz	Montville, Ohio	July 5, 1859	24, 623
Carriage-wheels, Machine for making	C. H. Guard	Brownville, N. Y	Sept. 7, 1852	9, 242
Carriage-wheels, Machine for making	C. H. Guard	Troy, N. Y	Feb. 4, 1862	34, 300
Carriage-wheels, Machine for making	G. W. Hatch	Parkman, Ohio	Dec. 22, 1863	41, 004
Carriage-wheels, Making	J. Whitaker	Dingman's Ferry, Pa	June 24, 1822	....
Carriage-wheels, Method of attaching	J. H. Riemkasten	Franklin Grove, Ill	Oct. 27, 1863	40, 431
Carriage-wheels, Method of boxing	C. Schmidt	Union, Me	Mar. 11, 1856	14, 434
Carriage-wheels, Method of constructing	E. S. Scripture	Syracuse, N. Y	Apr. 1, 1845	3, 979
Carriage-wheels, Method of constructing	S. Toomey	Wilmot, Ohio	May 5, 1868	77, 680
Carriage-wheels, Mode of applying anti-friction rollers and balls to the axles and boxes of.	J. G. Tibbets	New York, N. Y	July 22, 1839	1, 254
Carriage-wheels, Mold for casting boxes for	L. Farwell	Lancaster, Mass	Mar. 5, 1810	....
Carriage-wheels, Mold for casting boxes for	S. Smith, 2d	Amherst, Mass	July 6, 1810	....
Carriage-wheels, Removable runner for	G. A. Keene	Newburyport, Mass	Sept. 6, 1864	44, 141
Carriage-wheels to axles, Securing	I. Bicknell	Cincinnati, Ohio	July 12, 1870	105, 298
Carriage-wheels to axles, Securing	H. S. Fischer and D. Bruhn	Xenia, Ohio, and Eckenforde, Prussia.	Jan. 28, 1873	135, 327
Carriage-wheels to axles, Securing	E. P. Hoyt	New York, N. Y	Sept. 10, 1861	33, 251

Index of patents issued from the United States Patent Office from 1790 *to* 1873, *inclusive*—Continued.

Invention.	Inventor.	Residence.	Date.	No.
Carriage-whip holder	E. W. Scott	Wauregan, Conn	Dec. 27, 1870	110, 503
Carriage-window	J. F. Dohan	Binghamton, N. Y	Dec. 27, 1870	110, 444
Carriage-window	J. Henogen and E. Kruger	Saint Louis, Mo	Aug. 26, 1873	142, 229
Carriage-window	C. M. Oblenis	New York, N. Y	May 9, 1871	114, 592
Carriage-window	J. Ogden and D. R. Hart	Troy, N. Y	Mar. 21, 1846	4, 433
Carriage-window	J. T. Ogden	Boston, Mass	Feb. 20, 1855	12, 430
Carriage-window frame	F. Baker	New York, N. Y	May 21, 1867	64, 829
Carriage-window holder	J. H. Moore and W. Johnston	New Haven, Conn	Apr. 28, 1868	77, 397
Carriage window, Railway	E. T. Colburn	Boston, Mass	Jan. 1, 1867	60, 694
Carriage-window-sash holder	S. E. Totten, jr	New York, N. Y	Mar. 11, 1873	136, 793
Carriage with calash top	G. Holloway	Baltimore, Md	Sept. 9, 1825	....
Carriage with shifting seat	G. Simon	Reading, Pa	Dec. 20, 1853	10, 342
Carriage-work, Collar for ornamental	M. Seward	New Haven, Conn	Feb. 5, 1861	31, 330
Carriage-wrench	E. T. Ford	Stillwater, N. Y	July 13, 1869	92, 438
Carriages, Adjustable pole for	S. H. Bishop	Orange, Conn	May 5, 1857	17, 191
Carriages, Adjustable pole for	J. E. Prudden	Birmingham, Conn	Apr. 2, 1867	63, 423
Carriages and placing steps by the sides, Hanging	J. I. Wells	Hartford, Conn	Jan. 24, 1815	....
Carriages, Anti-rattler for	D. C. Gately	Newtown, Conn	Feb. 2, 1869	86, 387
Carriages, Apparatus for propelling	T. A. Háres	New York, N. Y	July 6, 1869	92, 306
Carriages, Apparatus for propelling	D. Swank	Newton, Iowa	July 13, 1869	92, 668
Carriages, Axle and thorough-box for	W. Slicer	Baltimore, Md	July 5, 1837	255
Carriages, Block for top-prop joint for	J. J. Wilson	New York, N. Y	Apr. 4, 1871	113, 477
Carriages, boats, &c., Machine for propelling	B. Langdon	Whitehall, N. Y	Feb. 22, 1817	....
Carriages, boats, mills, &c., by weights, Machine for propelling.	M. F. Colburn	Baltimore, Md	Apr. 7, 1829	....
Carriages, Body-loop for	H. A. Luttgens	Paterson, N. J	Oct. 17, 1871	120, 079
Carriages, Body-loop-head for	D. Wilcox	Ansonia, Conn	May 18, 1869	90, 326
Carriages, Boot-attachment to	M. M. Follett	Westborough Post-Office, Mass.	Feb. 4, 1868	74, 071
Carriages, &c., Box for receiving money in	J. Rodgers	New York, N. Y	Feb. 23, 1858	19, 446
Carriages, Button-hole for	S. C. Talcott	Ashtabula, Ohio	July 2, 1867	66, 413
Carriages by steam, Propelling	E. Clarke	New York, N. Y	Sept. 9, 1825	....
Carriages by steam, Propelling	C. Reynolds	East Windsor, Conn	Aug. 21, 1811	....
Carriages, Calash or folding top for	I. Cogswell, jr	Eastville, Ill	Feb. 24, 1863	37, 734
Carriages, Central-draft joint of	L. O. Rice	Caistorville, Canada	June 23, 1857	17, 638
Carriages, Changing the speed of steam	J. Griffin	Louisville, Ky	Feb. 19, 1861	31, 455
Carriages, Combined safety attachment and brake for.	C. Ducreux	New York, N. Y	Jan. 12, 1869	85, 883
Carriages, Combined step-cover and wheel-fender for.	J. W. Gosling	Cincinnati, Ohio	Feb. 26, 1867	62, 406
Carriages, Constructing	R. Nutting	Randolph, Vt	July 14, 1857	17, 805
Carriages, Constructing	E. W. Seymour	Lisle, N. Y	May 6, 1862	35, 204
Carriages, Constructing	J. C. Stroebel		Sept. 19, 1803	....
Carriages, Constructing	J. B. Swan	Albany, N. Y	Sept. 29, 1818	....
Carriages, &c., Construction and application of wheels to.	G. D. Bridgeman	New Haven, Conn	July 16, 1829	....
Carriages, Construction of	F. Biemler	New York, N. Y	Jan. 21, 1812	....
Carriages, Construction of	D. Neal	Milford, Del	Feb. 26, 1812	....
Carriages, Coupling four-wheeled	A. Osgood	Charlestown, Ohio	Jan. 27, 1843	2, 932
Carriages, Device for propelling	T. U. and H. H. Hamilton	Panama, N. Y	July 13, 1869	92, 607
Carriages, Die for forging offsets or stay-ends for	D. Wilcox	Birmingham, Conn	Nov. 4, 1873	144, 375
Carriages, Die for forging, trimming, and shaping spring-clips for.	F. B. Morse	Plantsville, Conn	Feb. 7, 1871	111, 668
Carriages, Double perch and fifth-wheel for	D. C. Doran	Mount Healthy, Ohio	Aug. 15, 1871	118, 115
Carriages, Draft-clip tie for	P. Blake	New Haven, Conn	May 26, 1863	38, 650
Carriages, Dress-guard for	G. W. Raitt	Cincinnati, Ohio	Oct. 16, 1866	58, 885
Carriages, Elastic running-gear for	G. E. Garretson	Russellville, Ky	Nov. 29, 1870	109, 608
Carriages, Elliptical spring for	F. Hatch and J. W. Terry	South Cortland, N. Y	July 10, 1839	1, 230
Carriages, Elliptical spring for	S King	Suffield, Conn	Oct. 31, 1839	1, 387
Carriages, Extension-pole and hold-back for	W. W. Rexford	Lock Sheldrake, N. Y	Nov. 24, 1868	84, 376
Carriages, Extension-reach for	E. Wilson	Prattsburgh, N. Y	Oct. 16, 1855	13, 690
Carriages, Extension-seat for	J. A. Naylor	Rahway, N. J	Aug. 21, 1860	29, 709
Carriages, Fan-attachment to child's	J. Reimsch	New York, N. Y	Sept. 24, 1867	69, 127
Carriages, Flap-fastener for the dasher of	R. Clingen	Boston, Mass	May 17, 1870	103, 142
Carriages, Folding double-step for	J. Pendergast	New Haven, Conn	June 11, 1872	127, 707
Carriages for children, Hanging	G. Maynard	Greenfield, Mass	July 27, 1858	21, 012
Carriages, Front gear for	C. L. Leonard	Wells' Bridge, N. Y	May 14, 1872	126, 720
Carriages, Front platform for	J. Heiden	New York, N. Y	Feb. 2, 1869	86, 397
Carriages, Hanging	S. Fairchild	Trumbull, Conn	Mar. 14, 1846	4, 414
Carriages, Hanging bodies of two-wheeled	I. Woodcock	Worcester, Mass	Apr. 18, 1848	5, 514
Carriages, &c., Heating	A. McWilliams	Washington, D. C	Nov. 29, 1832	....
Carriages, Horizontal wooden spring for	L. Sommer	Downington, Pa	July 22, 1811	....
Carriages, Horse-holder attachment to	G. W. Goodwyn	Petersburgh, Va	Feb. 21, 1871	111, 924
Carriages into sleighs, Converting wheel	O. B. Hale	Chicopee, Mass	Jan. 1, 1867	60, 884
Carriages, Leading-bar of four-horse	L. W. Stockton	Uniontown, Pa	Sept. 29, 1825	....
Carriages, Machinery for propelling steam	E. Ware	Bayonne, N. J	Aug. 20, 1867	68, 015
Carriages, Main bolt or goose-neck stay on	A. B. Shaeffer	Ephratah, Pa	May 26, 1868	78, 240
Carriages, Main spring for	J. Mix	New Haven, Conn	Apr. 18, 1807	....
Carriages, Manufacture of rub-irons for	H. W. Oliver, jr	Pittsburgh, Pa	Mar. 18, 1873	136, 857
Carriages, Measuring distances traveled by	Fuller and Richardson	Brunswick, Me	May 29, 1835	....
Carriages, Metal-plated shoe for	J. Du Bois	Williamsport, Pa	May 19, 1863	38, 571
Carriages, Method of attaching top to seat of	L. Jacobs and E. C. Landon	Castile, N. Y	Nov. 20, 1855	13, 821
Carriages, Method of forming body-loops for	R. R. Miller	Plantsville, Conn	Jan. 3, 1871	110, 773
Carriages, Mode of attaching animals to	H. B. Hale and T. Flagler	Grass Lake, Mich	Mar. 10, 1868	75, 262
Carriages, Mode of attaching sleigh-runners to wheel.	H. G. Guyon	New York, N. Y	Sept. 13, 1838	916
Carriages, Mode of connecting and disconnecting railway.	J. Stimpson	Baltimore, Md	Dec. 10, 1840	1, 885
Carriages, Mode of connecting springs to children's	J. H. Flagg	New York, N. Y	Oct. 15, 1872	132, 150
Carriages, Mode of making the skeins of axle-arms for.	G. Schreyer	Columbus, Ohio	May 7, 1861	32, 255
Carriages, Mode of regulating the speed of	T. Blanchard	Springfield, Mass	Dec. 28, 1825	....
Carriages, Mode of suspending or hanging pleasure	J. Henderson	Guilford, Conn	Dec. 26, 1808	....
Carriages, Motive-mechanism for	G. Kilner and F. H. Simmons	Sullivan, Ill	June 7, 1870	103, 895
Carriages, Motive-power for	S. L. Langdon	New Orleans, La	Mar. 8, 1870	100, 643
Carriages, Movable top for	J. C. Kimball	New Haven, Conn	Sept. 13, 1859	25, 420
Carriages on inclined plane, Transporting	M. Robinson	Henrico County, Va	Apr. 9, 1828	....
Carriages, Pole-iron socket for	C. Reynolds	New York, N. Y	Aug. 21, 1866	57, 379

Index of patents issued from the United States Patent Office from 1790 *to* 1873, *inclusive*—Continued.

Invention.	Inventor.	Residence.	Date.	No.
Carriages, Preventing rattling in	W. S. Chapman	Cincinnati, Ohio	Aug. 8, 1854	11, 471
Carriages, Propelling	C. K. Bradford	Lynn, Mass	Mar. 13, 1866	53, 214
Carriages, Propelling	W. Faris		Apr. 29, 1797	....
Carriages, Propelling wheeled	D. S. Fisher	Cedar Spring, Ind	May 28, 1867	65, 199
Carriages, Rub-iron for	F. B. Morse	Milwaukee, Wis	June 19, 1866	55, 692
Carriages, Runner-attachment for	C. F. Brigham	Worcester, Mass	Dec. 22, 1868	85, 2[illegible]4
Carriages, Saddle-clip for	F. Seward	New Haven, Conn	May 28, 1872	127, 275
Carriages, Sand-cap for	A. O. Coleburn and H. T. Stanard.	Wayne, Mich	Apr. 14, 1868	76, 605
Carriages, &c., Securing the wheels of	A. Bruns	Davenport, Iowa	Mar. 23, 1858	19, 676
Carriages, Shifting-rail for	A. E. Bailey	Middleville, N. Y	Apr. 2, 1867	63, 457
Carriages, Shifting-rail for	H. F. Holt	Fredonia, N. Y	July 2, 1867	66, 343
Carriages, Shifting-top rail for	C. H. Ayres	Hightstown, N. J	Oct. 8, 1872	131, 928
Carriages so as to ease the lateral motion of the bodies thereof, Constructing railway.	A. Bridges and C. Davenport	Cambridgeport, Mass	May 4, 1841	2, 071
Carriages, Spiral spring for	W. Croasdale	Hartsville, Pa	Feb. 3, 1837	116
Carriages, Spring and oscillating axle for children's	J. Loughridge	Pittsburgh, Pa	Apr. 23, 1872	126, 066
Carriages, Spring-perch for	S. Tomlinson	Bridgeport, Conn	May 16, 1842	2, 624
Carriages, Spring-power for propelling	W. K. Chase	Charlestown, Mass	Apr. 3, 1866	53, 572
Carriages, Steering-apparatus for steam	E. C. Jones	Pittsburgh, Pa	Aug. 11, 1857	17, 970
Carriages, Sustaining weights and applying power to wheeled.	S. Chapman, jr	Windsor, Mass	Mar. 2, 1835	....
Carriages, Third seat for	A. E. Thayer	Plymouth, Conn	Sept. 24, 1867	69, 270
Carriages, Thorough-brace for	F. A. Jewett	Abington, Mass	Aug. 19, 1856	15, 567
Carriages, Top iron and prop for	C. W. Saladee and W. Bauder	Circleville, Ohio	Feb. 9, 1869	86, 696
Carriages, Top iron and prop for	C. W. Saladee and W. Bauder	Circleville, Ohio	Feb. 9, 1869	86, 697
Carriages, Top-prop nut for	J. Ives	Mount Carmel, Conn	July 9, 1867	66, 496
Carriages, Turning-circle for	G. Kenny	Milford, N. H	Nov. 25, 1856	16, 122
Carriages, Turning-plate for	T. A. Edmison	Lake Port, Mich	Sept. 24, 1867	69, 194
Carriages, Umbrella-support for	A. Clarke	Sheboygan Falls, Wis	Feb. 6, 1872	123, 380
Carriages, Wear-iron for	I. G. Lefler	Philadelphia, Pa	Sept. 8, 1857	18, 153
Carriages when descending hills, Impeding	G. Bunker	Truxton, N. Y	July 5, 1833	....
Carriages when descending hills, Mode of retarding	E. Slifer	Boonsborough, Md	Jan. 16, 1826	....
Carriages with the perch by means of springs, Connecting the body of.	G. Nichols	Trumbull, Conn	Apr. 10, 1844	3, 527
Carriages with the reach and head-block or rocker, Coupling the forward axle-tree of four-wheeled.	A. H. Hart	Chagrin Falls, Ohio	Sept. 23, 1843	3, 274
Carriages, Wooden C-spring for	W. Robinson	Wilmington, Del	Nov. 27, 1810	....
Carrier: *See* Butter-carrier. Egg-carrier. Grain-carrier. Hay-carrier. Hod-carrier. Ice-carrier. Lathe-carrier. Sod-carrier. Spur-carrier. Straw-carrier. Trace-carrier. Water-carrier.				
Carronade, Elevating screw-box and cap of	E. Hidden	New York, N. Y	Apr. 26, 1826	....
Carrousel	W. Schneider	Davenport, Iowa	July 25, 1871	117, 336
Carryall	F. Shelton	Jackson, N. C	Dec. 2, 1834	....
Cart	N. R. Baldwin	Afton, N. Y	Feb. 5, 1861	31, 293
Cart	Z. Butt	Ocala, Fla	Mar. 6, 1866	52, 963
Cart	J. W. Cahoon	Philadelphia, Pa	Dec. 31, 1867	72, 795
Cart	H. Holcroft and C. S. Smith	Media, Pa	Aug. 30, 1864	43, 991
Cart	B. Ricketson and A. B. Smith	New Bedford, Mass., and Clinton, Pa.	July 12, 1864	43, 549
Cart and brake, Self-adjusting	I. Showalter	Chester County, Pa	July 18, 1871	117, 213
Cart and carriage wheel	W. Woodbridge	Kennebec, Me	May 17, 1836	....
Cart and railway for leveling hills, &c., Chain	J. H. Morison	Boscawen, N. H	Sept. 28, 1812	....
Cart and wagon, Dumping	I. B. Conklin	Pemberton, N. J	Feb. 16, 1864	41, 604
Cart and wagon, Dumping	W. W. Rogers	Hampden Corner, Me	Aug. 4, 1868	80, 669
Cart, Ash	R. A. Smith	Philadelphia, Pa	Jan. 3, 1865	45, 761
Cart-bodies, Mode of operating	T. Mussey	New London, Conn	Nov. 1, 1845	4, 250
Cart-body	W. Bradley	Vienna, Me	June 2, 1868	78, 419
Cart-body catch	C. F. Chew	Swedesborough, N. J	Apr. 23, 1872	125, 935
Cart-body fastening	E. Spalding	Plainfield, N. H	July 4, 1871	116, 641
Cart-body fastening, Tip	J. E. Seavey	Kennebunkport, Me	May 21, 1867	64, 913
Cart-boxes, Mold for running	S. Smith, 2d	Amherst, Mass	Oct. 17, 1809	....
Cart-brake	H. Holcroft and C. S. Smith	Media, Pa	Oct. 3, 1865	50, 248
Cart-brake	S. Y. Ives	Meriden, Conn	Feb. 5, 1867	61, 740
Cart, Brick	J. Evans	Philadelphia, Pa	May 31, 1870	103, 589
Cart, Coal	R. Heckscher	New York, N. Y	Feb. 7, 1860	27, 049
Cart, Dumping	J. H. C. Applegate	Bridgeton, N. J	Dec. 29, 1868	85, 353
Cart, Dumping	O. Benson and J. G. Falk	Chicago, Ill	Oct. 3, 1871	119, 444
Cart, Dumping	T. Blodgett	Belchertown, Mass	Sept. 6, 1864	44, 068
Cart, Dumping	C. A. Carpenter	Glen Cove, N. Y	May 17, 1864	42, 746
Cart, Dumping	G. L. Collins	Trenton, N. J	Nov. 21, 1871	121, 026
Cart, Dumping	F. Dengler	North Vernon, Ind	July 27, 1869	93, 069
Cart, Dumping	S. Doubleday	Baltimore, Md	June 4, 1872	127, 407
Cart, Dumping	B. G. Fitzhugh	Frederick, Md	Oct. 17, 1871	119, 976
Cart, Dumping	B. G. Fitzhugh	Frederick, Md	Apr. 22, 1873	138, 013
Cart, Dumping	B. G. Fitzhugh	Frederick, Md	June 17, 1873	140, 025
Cart, Dumping	N. W. Godfrey	Locust Valley, N. Y	Sept. 17, 1867	68, 977
Cart, Dumping	R. Ham and J. Durgin	Bangor, Me	Aug. 11, 1868	80, 823
Cart, Dumping	W. Hand	Plainfield, N. J	Apr. 4, 1871	113, 293
Cart, Dumping	A. A. Jennings	Webster, N. Y	Jan. 28, 1868	73, 725
Cart, Dumping	J. S. Lash	Carlisle, Pa	Sept. 13, 1859	25, 424
Cart, Dumping	G. E. Newell	Pawtucket, R. I	Jan. 28, 1868	73, 918
Cart, Dumping	R. A. Smith	Philadelphia, Pa	Nov. 8, 1864	44, 983
Cart, Dumping	M. F. Wickersham	Springfield, Ill	Sept. 7, 1869	94, 681
Cart, Dumping	M. F. Wickersham	Springfield, Ill	Sept. 21, 1869	95, 176
Cart, Dumping	F. H. Williams	Washington, D. C	Sept. 21, 1869	94, 988
Cart, Dumping	G. Zoanny	Napa, Cal	Dec. 3, 1872	133, 615
Cart, Excavating	W. Harles	Washington, Mo	Feb. 4, 1873	135, 546

Index of patents issued from the United States Patent Office from 1790 *to* 1873, *inclusive*—Continued.

Invention.	Inventor.	Residence.	Date.	No.
Cart for removing earth	J. Price	Lockport, N. Y	May 18, 1827	....
Cart, Hand	W. B. Glover	Boston, Mass	Sept. 3, 1872	131, 095
Cart, Hand-dumping	W. Farmer	New York, N. Y	Mar. 16, 1869	87, 831
Cart, Hay-rack for	H. R. Hawkins	Akron, Ohio	July 10, 1860	29, 075
Cart, Hemp	Z. Feagan	Palmyra, Mo	Mar. 26, 1861	31, 795
Cart-loading apparatus	B. G. Fitzhugh	Frederick, Md	Apr. 15, 1873	137, 768
Cart, Log	A. Kirkwood	Jackson County, Miss	Aug. 29, 1871	118, 460
Cart, Log	J. Stitt	Saint John's, Mich	Nov. 17, 1868	84, 145
Cart, Manure	T. L. Cotten	Madison County, Miss	Aug. 9, 1870	106, 125
Cart, Manure	J. K. Holland	Beaufort County, N. C	July 9, 1850	7, 491
Cart, Manure	D. Reid	Washington, N. C	May 3, 1853	9, 697
Cart, Manure	B. Teague	Moscow, Tenn	Aug. 22, 1871	118, 298
Cart, Milk	J. Harris	New York, N. Y	June 6, 1871	115, 606
Cart, Night	J. H. Lynch	Baltimore, Md	Aug. 20, 1867	67, 892
Cart or wagon, Weighing	N. E. Doane	Hannibal, Mo	Jan. 22, 1861	31, 155
Cart, Ox	M. S. Woodbury	Bethel, Vt	Feb. 2, 1869	86, 619
Cart, Self-balanced hand	G. Coolidge	Watertown, Mass	June 13, 1831	....
Cart, Self-loading	L. A. Beardsley	Fredericksburgh, Va	Dec. 22, 1868	85, 200
Cart, Self-loading	J. S. Brown	Washington, D. C	June 9, 1857	17, 486
Cart, Self-loading	Z. Butt	Lincolntown, N. C	Mar. 13, 1855	12, 504
Cart, Self-loading	R. B. Cantrell	Tenafly, N. J	May 13, 1873	138, 854
Cart, Self-loading	J. A. Sprague and B. O'Connor	Dayton, Ohio	Mar. 20, 1855	12, 562
Cart, Self-loading	G. W. Whitson	Asneville, N. C	July 21, 1868	80, 259
Cart, Self loading and dumping	W. H. Herbert	New Albany, Ind	Feb. 13, 1872	123, 565
Cart, Self loading and dumping	J. B. Hulbert	Hermon, N. Y	May 3[illegible], 1871	115, 321
Cart, Self loading and dumping	C. C. McKinley	Champaign, Ill	July 16, 1872	129, 154
Cart, Self loading and dumping	S. Stone	Kirkersville, Ohio	Aug. 1, 1854	11, 442
Cart, Self loading and dumping	B. T. Stowell	Waddam's Grove, Ill	Apr. 27, 1852	8, 913
Cart, Self loading and unloading	J. Wilkinson	Hopewell Cotton-Works Post-Office, Pa.	May 1, 1855	12, 804
Cart, Self-loading hay	E. Holt	Wheaton, Ill	Jan. 3, 1865	45, 715
Cart, S lf-loading hay	I. M. Williams	Blanchester, Ohio	Nov. 3, 1868	83, 749
Cart, Self-weighing	E. Blackman	Danbury, Conn	Oct. 7, 1862	36, 631
Cart, Street-sprinkling	L. Rodenhausen	Philadelphia, Pa	Aug. 13, 1867	67, 805
Cart, Water	H. Austin	East Liberty, Ohio	Nov. 27, 1860	30, 710
Cart, wagon, &c., Box and axle-tree for	W. Batchelor	New York, N. Y	May 2, 1816	....
Cart, Weighing	G. M. Barth	Philadelphia, Pa	Feb. 14, 1860	27, 184
Cart, Weighing	N. B. Livingston	Portland, Ind	Sept. 30, 1851	8, 391
Cart, Weighing	J. W. Martin	Burlington, N. J	May 20, 1856	14, 939
Cart, Weighing	J. W. Martin	Philadelphia, Pa	Nov. 2, 1858	21, 999
Cart, wheelbarrow, and turnip-drill, Combined	A. M. Newland	Olivet, Mich	July 5, 1870	10[illegible], 983
Carts, &c., Apparatus for unloading	C. Downer	Philadelphia, Pa	July 24, 1849	6, 609
Carts, &c., Combination of springs with the back-chains of.	T. S. Speakman	Philadelphia, Pa	July 18, 1848	5, 673
Carte de visite exhibitor	P. G. Hubert	New York, N. Y	Mar. 19, 1867	62, 959
Cartouch-box	R. Dingee	New York, N. Y	Aug. 15, 1835	....
Cartouch-box	J. Mix	New York, N. Y	June 3, 1813	....
Cartouch-box, brush, &c., Machine for boring blocks for.	V. Bogenreiff	Georgetown, D. C	July 10, 1810	....
Cartridge	J. Abraham and T. R. Bayliss	Birmingham, England	Oct. 27, 1868	83, 434
Cartridge	C. R. Alsop	Middletown, Conn	Jan. 27, 1863	37, 481
Cartridge	C. F. Brown	Warren, R. I	May 29, 1855	12, 942
Cartridge	S. Crispin	New York, N. Y	Apr. 12, 1864	42, 329
Cartridge	G. H. Daw	London, England	May 4, 1869	89, 563
Cartridge	J. H. Ferguson	Baltimore, Md	June 28, 1859	24, 548
Cartridge	E. Gonez	New York, N. Y	Jan. 25, 1870	99, 078
Cartridge	E. Gonez	New York, N. Y	Jan. 25, 1870	99, 079
Cartridge	B. B. Hotchkiss	Sharon, Conn	July 10, 1860	29, 080
Cartridge	W. W. Hubbell	Philadelphia, Pa	Apr. 23, 1872	126, 058
Cartridge	S. Jackson	Philadelphia, Pa	Jan. 10, 1865	45, 830
Cartridge	E. Lindner	New York, N. Y	May 12, 1857	17, 287
Cartridge	J. Logan and G. E. Hart	Newark, N. J	Sept. 16, 1873	142, 924
Cartridge	J. C. Mayberry	White Rock, Ill	June 24, 1862	35, 699
Cartridge	E. Maynard	Washington, D. C	June 17, 1856	15, 141
Cartridge	E. Maynard	Washington, D. C	Sept. 29, 1863	40, 111
Cartridge	I. M. Milbank	Greenfield Hill, Conn	May 30, 1871	115, 498
Cartridge	R. R. Moffatt	Brooklyn, N. Y	Dec. 20, 1870	110, 264
Cartridge	G. W. Morse	Baton Rouge, La	Oct. 28, 1856	15, 996
Cartridge	G. W. Morse	Baton Rouge, La	June 29, 1858	20, 727
Cartridge	A. N. Newton	Richmond, Ind	Mar. 20, 1855	12, 556
Cartridge	W. F. Parker	Meriden, Conn	Mar. 23, 1869	88, 202
Cartridge	W. R. Pomery	Millersburgh, Ohio	Aug. 5, 1862	36, 108
Cartridge	E. O. Potter	New York, N. Y	July 22, 1862	35, 949
Cartridge	J. Riedel	Pleasant Hill, Ky	Sept. 9, 1856	15, 707
Cartridge	T. P. Shaffner	Louisville, Ky	Dec. 19, 1865	51, 672
Cartridge	G. Smith	Buttermilk Falls, N. Y	June 30, 1857	17, 702
Cartridge	W. H. Smith	Charlestown, Mass	Feb. 8, 1870	99, 721
Cartridge	D. Smith and J. W. Storrs	Springfield, Mass	Apr. 20, 1869	89, 088
Cartridge	H. Smith and D. B. Wesson	Norwich, Conn	Aug. 8, 1854	11, 496
Cartridge	J. R. Van Vechten	New York, N. Y	Mar. 9, 1869	87, 735
Cartridge	C. S. Wells	Springfield, Mass	May 30, 1871	115, 548
Cartridge	C. S. Wells	Springfield, Mass	Jan. 2, 1872	122, 504
Cartridge	R. White	Davenport, Iowa	Nov. 26, 1861	33, 805
Cartridge	R. White	Lowell, Mass	Feb. 28, 1871	112, 305
Cartridge	N. G. Whitmore	Mansfield, Mass	June 28, 1870	104, 912
Cartridge	N. G. Whitmore	Mansfield, Mass	Aug. 20, 1872	130, 679
Cartridge	D. E. Williams	Davenport, Iowa	Oct. 18, 1870	108, 543
Cartridge	F. Wohlgemuth	New York, N. Y	Nov. 2, 1869	96, 373
Cartridge and percussion-cap box, Combined	J. T. Warren and R. A. Chesebrough.	Stafford and New York, N. Y.	Nov. 8, 1864	44, 999
Cartridge-belt	A. Mills	United States Army	Aug. 20, 1867	67, 898
Cartridge, Blasting	H. S. Lucas	Chester, Mass	Oct. 9, 1866	58, 656
Cartridge, Blasting	T. P. Shaffner	Louisville, Ky	Dec. 19, 1865	51, 673
Cartridge-box	A. A. Bennett	Cincinnati, Ohio	Jan. 27, 1863	37, 485
Cartridge-box	E. Blakeslee	Plymouth, Conn	Dec. 20, 1864	45, 469
Cartridge-box	T. F. Brabson	Brooklyn, N. Y	May 13, 1873	138, 847
Cartridge-box	R. L. Bryan and J. A. Bigelow	Franklin, Mich	Jan. 30, 1866	52, 357

Index of patents issued from the United States Patent Office from 1790 to 1873, inclusive—Continued.

Invention.	Inventor.	Residence.	Date.	No.
Cartridge-box	F. Bush	Boston, Mass	Dec. 23, 1862	37,216
Cartridge-box	F. Bush	Boston, Mass	Feb. 10, 1863	37,607
Cartridge-box	F. Chillingworth	Springfield, Mass	June 15, 1869	91,419
Cartridge-box	S. Crispin	New York, N. Y	Dec. 1, 1868	84,616
Cartridge-box	S. A. Day	Bowling Green, Ohio	Jan. 11, 1870	98,748
Cartridge-box	A. Domis	New York, N. Y	Apr. 1, 1862	34,820
Cartridge-box	J. Elbertson	Kirksville, Mo	Apr. 21, 1868	77,016
Cartridge-box	W. H. Elliott	Ilion, N. Y	June 17, 1873	139,885
Cartridge-box	J. W. Frazier	Newark, N. J	Feb. 20, 1872	123,883
Cartridge-box	J. W. Frazier	Newark, N. J	Feb. 20, 1872	123,884
Cartridge-box	W. Freeborn	Tivoli, N. Y	Dec. 10, 1867	72,011
Cartridge-box	C. Howlett	New York, N. Y	May 28, 1867	65,225
Cartridge-box	G. Jassath	New York, N. Y	Sept. 17, 1861	33,306
Cartridge-box	J. R. King	Washington, D. C	Nov. 15, 1870	109,365
Cartridge-box	A. D. Laidley	Philadelphia, Pa	Aug. 15, 1865	49,420
Cartridge-box	M. C. Leonard	Washington, D. C	Jan. 22, 1867	61,345
Cartridge-box	J. C. Ludlam	Brooklyn, N. Y	Dec. 20, 1870	110,379
Cartridge-box	J. C. McGinness	Washington, D. C	June 15, 1869	91,249
Cartridge-box	S. McKeever	United States Army	June 10, 1873	139,846
Cartridge-box	S. McKeever	United States Army	June 10, 1873	139,847
Cartridge-box	J. Miller	United States Army	Mar. 26, 1872	125,069
Cartridge-box	W. H. Morris	Cold Spring, N. Y	July 23, 1867	67,065
Cartridge-box	A. C. Newcomb and B. Lyon	Springfield, Mass	Dec. 3, 1867	71,633
Cartridge-box	H. E. Paine	Troy, N. Y	Dec. 22, 1868	85,241
Cartridge-box	J. Pease	Boston, Mass	Oct. 31, 1865	50,730
Cartridge-box	J. I. Pittman	New York, N. Y	Jan. 12, 1869	85,849
Cartridge-box	W. Rossiter	Newark, N. J	Aug. 8, 1865	49,304
Cartridge-box	P. F. Schneider	Hartford, Conn	Jan. 21, 1868	73,549
Cartridge-box	J. S. Smith	Brooklyn, N. Y	Mar. 12, 1861	31,680
Cartridge-box	P. S. and F. M. Thomson	Hudson City, N. J	Nov. 12, 1872	132,992
Cartridge-box	J. T. Warren	Stafford, N. Y	June 28, 1864	43,373
Cartridge-box	H. S. Weston	Akron, Ohio	July 12, 1864	43,539
Cartridge-box	M. V. B. White	Troy, N. Y	Feb. 14, 1865	46,411
Cartridge-box fastening	W. Z. W. Chapman	New York, N. Y	Mar. 4, 1862	34,563
Cartridge box fastening	J. M. Willson	Washington, D. C	Dec. 6, 1859	26,4[illegible]2
Cartridge-box, Revolving	C. Howlett	Manchester, Conn	Aug. 22, 1865	49,523
Cartridge-box, Revolving	P. F. Schneider	Hartford, Conn	July 31, 1866	56,804
Cartridge-bullet	W. H. Dibble	Middletown, Conn	Sept. 29, 1863	40,092
Cartridge-cap extractor	J. Logan and D. W. Eldridge	Boston, Mass	Dec. 13, 1870	110,052
Cartridge-cap extractor	W. C. Pickersgill	Providence, R. I	Dec. 14, 1869	97,805
Cartridge capping and uncapping device	G. S. Green, jr	Mott Haven, N. Y	Mar. 25, 1873	137,071
Cartridge capping and uncapping device	G. V. McGraw	Mankato, Minn	July 8, 1873	140,718
Cartridge capping and uncapping device	J. F. Nettleton	Branford, Conn	Apr. 22, 1873	138,180
Cartridge capping device	E. A. Kelsey	West Meriden, Conn	Aug. 20, 1872	130,723
Cartridge-capping machine	E. A. Kelsey	West Meriden, Conn	Aug. 20, 1872	130,724
Cartridge-capping tool	J. L. Raub	Meriden, Conn	Apr. 8, 1873	137,720
Cartridge-case	S. A. Day	Bowling Green, Ohio	Apr. 26, 1870	102,375
Cartridge-case	B. C. English	Hartford, Conn	Oct. 8, 1861	32,429
Cartridge-case	G. P. Foster	Providence, R. I	Apr. 10, 1860	27,791
Cartridge-case	J. P. Lindsay	New York, N. Y	July 24, 1860	29,287
Cartridge-case	G. W. Merse	Baton Rouge, La	May 11, 1858	20,214
Cartridge-case	T. J. Powers	New York, N. Y	May 17, 1870	103,079
Cartridge case	O. Shevenell	Marion, Ala	Apr. 19, 1870	102,051
Cartridge-case	S. W. Wood	Cornwall, N. Y	Oct. 15, 1872	132,227
Cartridge-case cleaner	F. H. Aiken	Franklin, N. H	June 8, 1869	91,058
Cartridge-case for revolving fire-arms	J. H. Vickers	Worcester, Mass	Sept. 8, 1863	39,869
Cartridge-case holder, Metallic	S. C. Greene	Vicksburgh, Miss	Mar. 22, 1870	101,121
Cartridge-case, Metallic	W. C. Dodge	Washington, D. C	July 4, 1865	48,536
Cartridge-case, Metallic	E. C. Dunning	Bridgeport, Conn	Mar. 18, 1862	34,713
Cartridge-case, Metallic	E. Maynard	Washington, D. C	Jan. 11, 1859	22,565
Cartridge-case, Metallic	W. Tibbals	Hartford, Conn	Feb. 23, 1869	87,125
Cartridge-case, Metallic	S. W. Wood	Cornwall, N. Y	Oct. 28, 1873	144,012
Cartridge-case, Applying percussion priming to	A. K. Johnston and L. Dow	New York, N. Y	May 10, 1864	42,666
Cartridge-cases, Forming	C. Sharps	Philadelphia, Pa	Nov. 13, 1860	30,647
Cartridge-cases, Machine for heading	T. I. Powers	New York, N. Y	June 18, 1867	65,940
Cartridge-cases, Machine for loading metallic	C. H. Davis	Philadelphia, Pa	Aug. 9, 1864	43,780
Cartridge-cases, Machine for making	T. J. Powers	New York, N. Y	Apr. 24, 1866	54,255
Cartridge-cases, Machine for making percussion	E. Allen	Worcester, Mass	Feb. 14, 1860	27,094
Cartridge-cases, Machine for necking	B. Payne	South Coventry, Conn	Oct. 17, 1865	50,489
Cartridge-cases, Manufacture of metallic	D. Williamson	New York, N. Y	June 29, 1869	92,136
Cartridge-cases, Material for	A. B. Ely	Newton, Mass	Apr. 13, 1869	88,948
Cartridge-charger	A. C. Hobbs	Bridgeport, Conn	June 15, 1869	91,442
Cartridge-charger	P. Powell	Cincinnati, Ohio	Dec. 22, 1868	85,189
Cartridge-cutter, Fire-arm-muzzle	C. H. Bradley	West Chester, Pa	Jan. 28, 1862	34,235
Cartridge-ejector	J. M. Marlin	Hartford, Conn	Apr. 5, 1870	101,637
Cartridge-ejector for breech-loading fire-arms	C. H. Ballard	Worcester, Mass	Apr. 9, 1867	63,605
Cartridge-ejector for breech-loading fire-arms	H. Hammond	Hartford, Conn	Dec. 31, 1867	72,849
Cartridge-ejector for breech-loading fire-arms	W. H. and G. W. Miller	West Meriden, Conn	Aug. 27, 1867	68,099
Cartridge-ejector for revolving fire-arms	J. T. and G. M. Smith and J. J. Sweeney.	Springfield, Mass	Mar. 18, 1873	136,871
Cartridge-ejector for revolving fire-arms	C. S. Wells	New Haven, Conn	Dec. 10, 1872	133,732
Cartridge, Electrical gun	T. P. Mott and S. Gardner, jr	New York, N. Y	May 6, 1873	138,679
Cartridge-envelope	A. K. Johnston and L. Dow	Middletown, Conn., and Topeka, Kans.	Oct. 1, 1861	33,393
Cartridge, Explosive	W. Schmitz	Philadelphia, Pa	Dec. 29, 1868	85,482
Cartridge-extractor	H. Lord	Hartford, Conn	Jan. 14, 1868	73,351
Cartridge-extractor for fire-arms	H. Reynolds	Springfield, Mass	Nov. 22, 1864	45,176
Cartridge extractor, Ordnance	N. Thompson	Brooklyn, N. Y	Nov. 19, 1872	133,271
Cartridge-feeder for gun-hammers	T. Shaw	Philadelphia, Pa	Dec. 21, 1869	98,198
Cartridge-filling machine	W. C. Dodge and R. D. O. Smith	Washington, D. C	July 17, 1866	56,489
Cartridge-filling machine	T. J. Powers	New York, N. Y	Jan. 22, 1867	61,456
Cartridge-filling machine	J. G. Stowe and E. F. Allen	Providence, R. I	July 12, 1864	43,550
Cartridge, Fixed	G. Buckel and E. Dorsch	Monroe, Mich	July 22, 1856	15,369
Cartridge for artillery and blasting	T. P. Shaffner	Louisville, Ky	Sept 14, 1869	94,847
Cartridge for atmospheric fire-arms	P. Giffard	Paris, France	Dec. 17, 1872	134,048
Cartridge for breech-loading fire-arms	J. D. Greene	Cambridge, Mass	Sept. 8, 1859	18,143
Cartridge for breech-loading fire-arms	B. King	Providence, R. I	Mar. 4, 1862	34,579
Cartridge for breech-loading fire-arms	I. M. Millbank	Greenfield Hill, Conn	Feb. 6, 1872	123,352

Index of patents issued from the United States Patent Office from 1790 *to* 1873, *inclusive*—Continued.

Invention.	Inventor.	Residence.	Date.	No.
Cartridge for breech-loading fire-arms	D. Moore	Williamsburgh, N. Y	Oct. 31, 1854	11, 870
Cartridge for breech-loading fire-arms, Nippled	T. E. Sturtevant	Boston, Mass	June 12, 1866	55, 552
Cartridge for breech-loading fire-arms, Shot	C. E. Sneider	Baltimore, Md	July 4, 1871	116, 640
Cartridge for breech-loading guns	W. W. Marston and F. Goodell	New York, N. Y	May 18, 1852	8, 956
Cartridge for breech-loading rifled fire-arms	H. Berdan	New York, N. Y	Feb. 7, 1865	46, 292
Cartridge for cannon, Metal	W. E. Moore	Crawfordsville, Ind	Apr. 28, 1863	38, 322
Cartridge for fire-arms	J. W. Cochran	New York, N. Y	Mar. 13, 1860	27, 428
Cartridge for fire-arms	E. Gomez and W. Mills	New York, N. Y	Aug. 24, 1858	21, 253
Cartridge for fire-arms	T. T. S. Laidley	Watertown, Mass	June 24, 1873	140, 144
Cartridge for fire-arms	I. M. Milbank	Greenfield Hill, Conn	Sept. 3, 1872	131, 016
Cartridge for fire-arms	I. M. Milbank	Greenfield Hill, Conn	Sept. 3, 1872	131, 017
Cartridge for fire-arms	I. M. Milbank	Greenfield Hill, Conn	Sept. 3, 1872	131, 018
Cartridge for fire-arms	J. Rupertus	Philadelphia, Pa	Oct. 8, 1867	69, 707
Cartridge for fire-arms	A. Shannon	New York, N. Y	Mar. 4, 1862	34, 615
Cartridge for fire-arms	J. W. Smith	Iowa Point, Kans	Jan. 30, 1866	52, 370
Cartridge for fire-arms	S. W. Wood	Cornwall, N. Y	Oct. 28, 1873	144, 011
Cartridge for fire-arms, Envelope of	A. K. Johnston and L. Dow	Middletown, Conn., and Topeka, Kans.	Jan. 7, 1862	34, 061
Cartridge for loose ammunition, Metallic	W. H. Risley	Berlin, Conn	Mar. 27, 1866	53, 490
Cartridge for ordnance	T. Yates	Milwaukee, Wis	Dec. 27, 1864	45, 666
Cartridge for revolving fire-arms	J. M. Cooper	Pittsburgh, Pa	Dec. 6, 1864	45, 319
Cartridge for revolving fire-arms	D. Williamson	Brooklyn, N. Y	Jan. 5, 1864	41, 183
Cartridge for small-arms	T. Cullen	San Francisco, Cal	Jan. 7, 1868	72, 982
Cartridge for small-arms	D. M. Mefford	Cincinnati, Ohio	Apr. 18, 1865	47, 317
Cartridge heads, Method of re-enforcing metallic	S. W. Wood	Cornwall, N. Y	May 7, 1872	126, 613
Cartridge-holder	L. J. Gaines	West Meriden, Conn	Mar. 18, 1873	136, 987
Cartridge-holder	J. S. Smith	Brooklyn, N. Y	Dec. 1, 1868	84, 651
Cartridge-loader	G. H. Ferriss	Utica, N. Y	Sept. 12, 1871	118, 849
Cartridge-loader	E. Maynard	Washington, D. C	Apr. 2, 1861	31, 898
Cartridge-loader	J. M. Taylor	Lexington, Ky	July 13, 1869	92, 669
Cartridge-loader	J. S. Warner	Ogdensburgh, N. Y	Dec. 3, 1872	133, 607
Cartridge loader, Pocket	G. F. Dutch	New York, N. Y	Aug. 13, 1872	130, 414
Cartridge-loading apparatus	F. A. Thuer	East Hartford, Conn	Jan. 4, 1870	98, 529
Cartridge-loading device	T. L. Sturtevant	Framingham, Mass	July 22, 1873	141, 185
Cartridge-loading machine	M. V. B. Hill	Bridgeport, Conn	Sept. 28, 1869	95, 345
Cartridge-loading machine	A. C. Hobbs	Bridgeport, Conn	Feb. 18, 1873	135, 913
Cartridge-loading machine	T. L. Sturtevant	Framingham, Mass	Apr. 29, 1873	138, 294
Cartridge-machine	J. H. Gill	Philadelphia, Pa	Dec. 14, 1869	97, 904
Cartridge-machine	D. Smith	Springfield, Mass	Nov. 27, 1866	60, 074
Cartridge-magazine	W. H. Elliot	New York, N. Y	Sept. 12, 1871	118, 916
Cartridge-making machine	W. H. Horstman and H. J. Behrens.	New York, N. Y	Sept. 15, 1863	39, 923
Cartridge-making machine	B. S. Roberts	United States Army	Mar. 9, 1869	87, 593
Cartridge, Metallic	E. Allen	Worcester, Mass	Sept. 25, 1860	30, 109
Cartridge, Metallic	E. Allen	Worcester, Mass	May 16, 1865	47, 688
Cartridge, Metallic	E. G. Allen	Boston, Mass	Feb. 16, 1864	41, 590
Cartridge, Metallic	W. Bakewell	Pittsburgh, Pa	July 7, 1863	39, 109
Cartridge, Metallic	H. Berdan	New York, N. Y	Feb. 27, 1866	52, 818
Cartridge, Metallic	H. Berdan	New York, N. Y	Sept. 29, 1868	82, 587
Cartridge, Metallic	A. J. Bergen	Brooklyn, N. Y	Feb. 26, 1867	62, 466
Cartridge, Metallic	C. J. Bergen	Brooklyn, N. Y	May 17, 1864	42, 815
Cartridge, Metallic	A. S. Blake	Waterbury, Conn	June 5, 1866	55, 233
Cartridge, Metallic	E. M. Boxer	Woolwich, England	June 29, 1869	91, 818
Cartridge, Metallic	F. E. Boyd	Boston, Mass	Feb. 8, 1870	99, 528
Cartridge, Metallic	B. Burton	Brooklyn, N. Y	Feb. 25, 1873	136, 130
Cartridge, Metallic	J. J. Chaudun	Paris, France	July 20, 1869	92, 795
Cartridge, Metallic	J. W. Cochran	New York, N. Y	Nov. 7, 1871	120, 625
Cartridge, Metallic	J. W. Cochran	New York, N. Y	May 28, 1872	127, 308
Cartridge, Metallic	J. S. Crary	Salem, N. Y	July 25, 1871	117, 388
Cartridge, Metallic	S. Crispin	New York, N. Y	Apr. 4, 1871	113, 634
Cartridge, Metallic	C. F. and J. E. De Dartein	Strasburg, France	Nov. 7, 1871	120, 630
Cartridge, Metallic	A. C. Depen and J. Slatcher	Bridgeport, Conn	Dec. 7, 1869	97, 615
Cartridge, Metallic	D. C. Farrington	Lowell, Mass	Dec. 31, 1872	134, 368
Cartridge, Metallic	S. Foreband and H. C. Wadsworth.	Worcester, Mass	Dec. 5, 1871	121, 606
Cartridge, Metallic	A. N. C. Garard	Paris, France	Apr. 19, 1870	102, 109
Cartridge, Metallic	R. J. Gatling	Hartford, Conn	May 3, 1870	102, 675
Cartridge, Metallic	I. P. Gillespie	New Albany, Ind	Oct. 15, 1861	33, 481
Cartridge, Metallic	E. Gomez	New York, N. Y	Feb. 8, 1870	99, 666
Cartridge, Metallic	A. Hall	Danville, Iowa	Sept. 15, 1863	39, 915
Cartridge, Metallic	A. C. Hobbs	Bridgeport, Conn	July 18, 1871	117, 173
Cartridge, Metallic	B. B. Hotchkiss	New York, N. Y	Aug. 31, 1869	94, 210
Cartridge, Metallic	J. C. Howe	Worcester, Mass	Aug. 16, 1864	43, 851
Cartridge, Metallic	H. Kellogg	New Haven, Conn	July 15, 1862	35, 878
Cartridge, Metallic	C. D. Leet and B. B. Hotchkiss	Vienna, Austria, and New York, N. Y.	Dec. 28, 1869	98, 278
Cartridge, Metallic	J. Logan and D. W. Eldredge	Boston, Mass	Dec. 7, 1869	97, 537
Cartridge, Metallic	E. Martin	Springfield, Mass	Mar. 23, 1869	88, 191
Cartridge, Metallic	E. Martin	Springfield, Mass	July 12, 1870	105, 348
Cartridge, Metallic	E. Martin	Springfield, Mass	Feb. 14, 1871	111, 856
Cartridge, Metallic	E. Maynard	Washington, D. C	Sept. 8, 1863	39, 823
Cartridge, Metallic	E. Maynard	Washington, D. C	Sept. 29, 1863	40, 112
Cartridge, Metallic	J. V. Meigs	Washington, D. C	Mar. 2, 1869	87, 352
Cartridge, Metallic	J. V. Meigs	Washington, D. C	June 8, 1869	90, 951
Cartridge, Metallic	H. Metcalfe	New York, N. Y	Nov. 14, 1871	120, 980
Cartridge, Metallic	I. M. Milbank	Greenfield Hill, Conn	Aug. 10, 1869	93, 545
Cartridge, Metallic	I. M. Milbank	Greenfield Hill, Conn	Aug. 10, 1869	93, 546
Cartridge, Metallic	I. M. Milbank	Greenfield Hill, Conn	Jan. 2, 1872	122, 599
Cartridge, Metallic	I. M. Milbank	Greenfield Hill, Conn	Feb. 6, 1872	124, 351
Cartridge, Metallic	I. M. Milbank	Greenfield Hill, Conn	Apr. 16, 1872	125, 830
Cartridge, Metallic	I. M. Milbank	Greenfield Hill, Conn	Feb. 25, 1873	136, 168
Cartridge, Metallic	R. R. Moffatt	Brooklyn, N. Y	Dec. 20, 1870	110, 265
Cartridge, Metallic	S. Newhouse	Oneida, N. Y	Jan. 18, 1870	98, 995
Cartridge, Metallic	W. I. Page	East Boston, Mass	Dec. 6, 1870	109, 931
Cartridge, Metallic	A. Payne	Bridgeport, Conn	Oct. 31, 1871	120, 529
Cartridge, Metallic	A. Payne	Bridgeport, Conn	Dec. 12, 1871	121, 808
Cartridge, Metallic	G. R. Pierce	Grand Rapids, Mich	Oct. 24, 1871	120, 323
Cartridge, Metallic	T. J. Powers	New York, N. Y	Oct. 17, 1865	50, 536

Index of patents issued from the United States Patent Office from 1790 to 1873, inclusive—Continued.

Invention.	Inventor.	Residence.	Date.	No.
Cartridge, Metallic	T. J. Powers	New York, N. Y	June 13, 1871	115,892
Cartridge, Metallic	T. J. Powers	New York, N. Y	June 20, 1871	116,094
Cartridge, Metallic	B. S. Roberts	United States Army	Feb. 23, 1869	87,297
Cartridge, Metallic	T. J. Rodman and S. Crispin	Watertown, Mass., and New York, N. Y.	Dec. 15, 1863	40,[illegible]88
Cartridge, Metallic	C. Sharps	Philadelphia, Pa	Apr. 15, 1862	34,987
Cartridge, Metallic	D. Smith	Springfield, Mass	June 15, 1869	91,278
Cartridge, Metallic	W. H. Smith	Charlestown, Mass	Dec. 28, 1869	98,439
Cartridge, Metallic	W. S. Smoot	Ilion, N. Y	June 20, 1871	116,105
Cartridge, Metallic	W. S. Smoot	Ilion, N. Y	Oct. 24, 1871	120,338
Cartridge, Metallic	H. C. Spaulding	Brooklyn, N. Y	Jan. 24, 1865	46,034
Cartridge, Metallic	G. R. Stetson	New Haven, Conn	Oct. 31, 1871	120,403
Cartridge, Metallic	W. Tibbals	South Coventry, Conn	May 25, 1869	90,607
Cartridge, Metallic	R. White	Lowell, Mass	Dec. 14, 1869	97,843
Cartridge, Metallic	R. White	Lowell, Mass	Jan. 10, 1871	110,881
Cartridge, Metallic	W. H. Wills	Boston, Mass	Nov. 29, 1864	45,292
Cartridge, Metallic	O. F. Winchester	New Haven, Conn	Jan. 1, 1867	60,814
Cartridge-opener	J. S. Butterfield and S. Marshall.	Philadelphia, Pa	May 13, 1856	14,850
Cartridge packing-case	J. W. Frazier	New York, N. Y	July 16, 1872	129,545
Cartridge, Patched	W. H. Elliott	Plattsburgh, N. Y	July 15, 1862	35,878
Cartridge percussion-primer	R. R. Moffatt	Brooklyn, N. Y	Dec. 20, 1870	110,266
Cartridge-pouch	H. Hammond	Hartford, Conn	Feb. 26, 1867	62,415
Cartridge, Primed metallic	S. Crispin	New York, N. Y	Dec. 15, 1863	40,978
Cartridge, Primed metallic	W. W. Marston	New York, N. Y	Nov. 3, 1863	40,490
Cartridge, Primed metallic	E. K. Root	Hartford, Conn	Oct. 11, 1864	44,660
Cartridge-primer	I. C. Farrington	Lowell, Mass	Dec. 17, 1872	133,929
Cartridge-primer	A. C. Hobbs and J. Orcutt	Bridgeport, Conn	Oct. 24, 1871	121,196
Cartridge-primer	B. B. Hotchkiss	New York N. Y	Feb. 15, 1870	99,899
Cartridge-primer	I. M. Milbank	Greenfield Hill, Conn	May 31, 1870	103,641
Cartridge primer, Fire-arm	H. Smith and D. B. Wesson	Norwich and New Haven, Conn.	Jan. 22, 1856	14,147
Cartridge-retractor for breech-loading fire-arms	E. Allen	Worcester, Mass	Mar. 7, 1865	46,617
Cartridge-retractor for breech-loading fire-arms	A. Ball	Windsor, Vt	Jan. 1, 1867	60,664
Cartridge-retractor for breech-loading fire-arms	F. Beals	New Haven, Conn	Feb. 7, 1865	46,207
Cartridge-retracter for breech-loading fire-arms	J. W. Cochran	New York, N. Y	Feb. 20, 1866	52,679
Cartridge-retractor for breech-loading fire-arms	G. P. and G. F. Foster	Mohawk, N. Y	Sept. 19, 1865	49,994
Cartridge-retractor for breech-loading fire-arms	J. Gray	Medford, Mass	June 20, 1865	48,337
Cartridge-retractor for breech-loading fire-arms	E. Maynard	Washington, D. C	July 25, 1865	48,968
Cartridge-retractor for breech-loading fire-arms	E. S. Piper	Springfield, Mass	Dec. 5, 1865	51,391
Cartridge-retracter for breech-loading fire-arms	W. S. Smoot	Washington, D. C	Aug. 27, 1867	68,250
Cartridge-retracter for breech-loading fire-arms	C. M. Spencer	South Manchester, Conn	July 29, 1862	36,062
Cartridge-retractor for breech-loading fire-arms	T. L. Sturtevant	Boston, Mass	Nov. 7, 1865	50,854
Cartridge-retractor for breech-loading fire-arms	F. Truelender	Salem, N. J	May 10, 1864	42,702
Cartridge-retractor for breech-loading fire-arms	H. H. Wolcott	Yonkers, N. Y	June 13, 1865	48,227
Cartridge-retractor for magazine fire-arms	E. Stabler	Montgomery County, Md	Dec. 6, 1864	45,356
Cartridge-retractor for many-chambered fire-arms	W. C. Dodge	Washington, D. C.	Jan. 17, 1865	45,912
Cartridge-retractor for revolving fire-arms	W. Clews	Ilion, N. Y	Feb. 25, 1873	136,134
Cartridge-retractor for revolving fire-arms	H. Hammond	Bridgeport, Conn	Jan. 23, 1866	52,165
Cartridge-retractor for revolving fire-arms	R. D. O. Smith	Washington, D. C	July 12, 1864	43,529
Cartridge-retractor for revolving fire-arms	S. W. Wood	Cornwall, N. Y	Jan. 16, 1866	52,105
Cartridge-shell	C. H. Todd	Montgomery, Ala	Mar. 16, 1869	87,990
Cartridge-shell ejector for revolving fire-arms	G. H. Harrington	Worcester, Mass	Feb 7, 1871	111,534
Cartridge-shell extractor	J. M. Martin	Hartford, Conn	Feb. 8, 1870	99,690
Cartridge-shell for drill-purposes	A. C. Hobbs	Bridgeport, Conn	Sept. 26, 1871	119,357
Cartridge-shell for magazine fire-arms	G. H. Dupee	New Haven, Conn	Feb. 13, 1872	123,622
Cartridge-shell-forming machine	A. C. Hobbs and T. V. Boyden.	Bridgeport, Conn	Sept. 14, 1869	94,745
Cartridge-shells, Construction of	J. F. Cranston	Springfield, Mass	Aug. 25, 1868	81,478
Cartridge-shells, Device for manufacturing metallic.	E. A. Worthen	Springfield, Mass	Nov. 29, 1870	109,791
Cartridge-shells, Die for making	J. Gardner	New Haven, Conn	May 21, 1872	126,949
Cartridge-shells, Machine far drawing	A. S. Warner	Springfield, Mass	July 10, 1866	56,338
Cartridge-shells, Machine for forming external recesses in the heads of.	A. C. Hobbs	Bridgeport, Conn	Sept. 14, 1869	94,744
Cartridge-shells, Machine for heading	T. V. Boyden	Bridgeport, Conn	May 9, 1871	114,639
Cartridge-shells, Machine for making	W. A. McIntire	Springfield, Mass	Apr. 16, 1867	63,915
Cartridge-shells, Machine for re-enforcing	C. S. Wells	Bridgeport, Conn	Apr. 9, 1872	125,508
Cartridge-shells, Machine for tapering	S. W. Wood	Cornwall, N. Y	May 7, 1872	126,609
Cartridge-shells, Machine for tapering	S. W. Wood	Cornwall, N. Y	May 7, 1872	126,610
Cartridge-shells, Machine for tapering	S. W. Wood	Cornwall, N. Y	May 7, 1872	126,611
Cartridge-shells, Machine for tapering	S. W. Wood	Cornwall, N. Y	May 7, 1872	126,612
Cartridge-shells, Manufacture of	D. Smith	Springfield, Mass	Dec. 7, 1869	97,561
Cartridge-shells with bullets, Machine for interlocking.	A. C. Hobbs	Bridgeport, Conn	Oct. 28, 1873	143,981
Cartridge, Shot	L. B. Bruen	Brooklyn, N. Y	Jan. 27, 1863	37,491
Cartridge, Shot	B. L. Budd	New York, N. Y	Mar. 25, 1862	34,806
Cartridge, Shot	A. R. Davis	East Cambridge, N. Y	Mar. 20, 1855	12,545
Cartridge, Shot	W. O. Howard	New York, N. Y	Mar. 3, 1868	75,019
Cartridge, Shot	W. B. Johns	United States Army	July 14, 1857	17,792
Cartridge, Shot	C. W. Lancaster	London, England	Dec. 7, 1869	97,653
Cartridge, Shot	A. D. Laws	Bridgeport, Conn	Sept. 3, 1872	131,104
Cartridge, Shot	C. W. Lovett, jr.	Boston, Mass	Apr. 4, 1871	113,677
Cartridge, Shot	R. R. Moffatt	Brooklyn, N. Y	Dec. 20, 1870	110,383
Cartridge, Shot	S. W. Paine	Williamsport, Pa	Jan. 31, 1871	111,377
Cartridge, Shot	S. W. Paine	Rochester, N. Y	Feb. 25, 1873	136,336
Cartridge, Shot	C. E. Sneider	Baltimore, Md	May 10, 1870	102,984
Cartridge, Shot	C. E. Sneider	Baltimore, Md	Sept. 10, 1872	131,189
Cartridge, Shot metallic	E. K. Root	Hartford, Conn	May 5, 1863	38,414
Cartridge, Skin	J. Hotchkiss	Middletown, Conn	Feb. 11, 1862	34,367
Cartridge, Skin	W. M. Storm	New York, N. Y	Oct. 29, 1861	33,611
Cartridge, Solid	R. Bartholow	United States Army	Aug. 5, 1862	36,066
Cartridge-tearer for muskets	D. Kelly	Grand Rapids, Mich	Dec. 16, 1862	37,171
Cartridge tube and conveyer forming a fire-arm	C. W. Büchel	New York, N. Y	Feb. 20, 1849	6,136
Cartridge-tube, &c., for fire-arms, Detached metallic	D. Minesinger	Beaver, Pa	Feb. 27, 1849	6,139
Cartridge, Water-proof	R. Bartholow	United States Army	May 21, 1861	32,345
Cartridges and setting and ejecting caps, Implement for loading.	H. B. Hooker	Rochester, N. Y	May 21, 1872	126,962

Index of patents issued from the United States Patent Office from 1790 *to* 1873, *inclusive*—Continued.

Invention.	Inventor.	Residence.	Date.	No.
Cartridges, Apparatus for setting bullets in	W. C. Pickersgill	Providence, R. I	Dec. 14, 1869	97, 806
Cartridges, Apparatus for setting caps in metallic	W. C. Pickersgill	Providence, R. I	Dec. 14, 1869	97, 804
Cartridges, Applying fire-extinguishing	W. M. Storm	New York, N. Y	Oct. 9, 1855	13, 660
Cartridges, Applying percussion priming to	A. K. Johnston and L. Dow	New York, N. Y	May 10, 1864	42, 667
Cartridges around bullets, Implement for compressing.	J. S. Adams	Taunton, Mass	May 30, 1865	48, 010
Cartridges, Ball	R. O. Doremus and B. L. Budd.	New York, N. Y	Mar. 18, 1862	34, 725
Cartridges, Ball	L. Wells	Astoria, N. Y	Sept. 15, 1857	18, 217
Cartridges, caps, &c., Forming metallic cases for	S. W. Wood	Cornwall, N. Y	Oct. 28, 1873	144, 010
Cartridges, Device for crimping	C. E. Sneider	Baltimore, Md	Sept. 10, 1872	131, 188
Cartridges, Fabric for envelope for	A. K. Johnston and L. Dow	New York, N. Y	May 10, 1864	42, 668
Cartridges, Filling metallic	H. Smith and D. B. Wesson	Springfield, Mass	Apr. 17, 1860	27, 933
Cartridges for ordnance, &c., Graduating accelerating.	J. M. Crockett	Newbern, Va	Sept. 10, 1867	68, 609
Cartridges from breech-loading fire-arms, Nipple for discharging or withdrawing.	W. C. Hicks	New Haven, Conn	Mar. 10, 1857	16, 797
Cartridges, Implement for capping	H. M. Bronson	Sandusky, Ohio	Nov. 11, 1873	144, 435
Cartridges, Implement for uncapping, capping, and charging.	W. M. Fowler	Brooklyn, N. Y	Aug. 26, 1873	142, 157
Cartridges, Instrument for extracting caps from	W. Clews	Ilion, N. Y	Oct. 3, 1871	119, 506
Cartridges, Machine for attaching balls to	D. Ellis and G. R. Stetson	New Haven, Conn	June 6, 1865	48, 056
Cartridges, Machine for cupping metallic	T. J. Powers	New York, N. Y	Apr. 11, 1865	47, 246
Cartridges, Machine for manufacture of metallic	E. Allen	Worcester, Mass	Mar. 19, 1861	31, 695
Cartridges, Machine for priming metallic	T. J. Powers	New York, N. Y	Aug. 14, 1866	57, 258
Cartridges, Machine for tapering metallic	S. W. Wood	Cornwall, N. Y	May 7, 1872	126, 608
Cartridges made of fusible metals, &c	W. H. Tooth	Broxton, England	Mar. 4, 1873	136, 468
Cartridges, Means for uncapping	C. A. King	Springfield, Mass	Oct. 17, 1871	120, 075
Cartridges, Metallic wad for	E. Maynard	Washington, D. C	Apr. 19, 1864	42, 388
Cartridges, Method of attaching balls to wooden	W. Hunt	New York, N. Y	Aug. 10, 1848	5, 699
Cartridges, Packing	S. Colt	Hartford, Conn	Mar. 15, 1859	23, 230
Cartridges, Packing	E. K. Root	Hartford, Conn	Jan. 18, 1859	22, 675
Cartridges, Packing	C. Sharps	Philadelphia, Pa	July 10, 1860	29, 108
Cartridges, Priming	G. A. Fitch	Kalamazoo, Mich	Oct. 16, 1866	58, 800
Cartridges, Priming	A. Moffatt	Washington, D. C	Mar. 13, 1865	53, 168
Cartridges, Priming	C. E. Sneider	Baltimore, Md	Oct. 11, 1864	44, 692
Cartridges, Priming	T. L. Sturtevant	Boston, Mass	Mar. 27, 1866	53, 501
Cartridges, Priming metallic	H. Berdan	New York, N. Y	Mar. 20, 1866	53, 388
Cartridges, Priming metallic	A. J. Bergen	Brooklyn, N. Y	Feb. 26, 1867	62, 467
Cartridges, Priming metallic	B. Burton	Brooklyn, N. Y	Aug. 11, 1868	81, 058
Cartridges, Priming metallic	J. F. Cranston	Springfield, Mass	Sept. 17, 1867	68, 960
Cartridges, Priming metallic	J. F. Cranston	Springfield, Mass	Jan. 28, 1868	73, 877
Cartridges, Priming metallic	S. Crispin	New York, N. Y	Aug. 8, 1865	49, 237
Cartridges, Priming metallic	R. J. Gathing	Indianapolis, Ind	June 16, 1868	78, 953
Cartridges, Priming metallic	C. Jackson and J. G. Pusey	Providence, R. I	Oct. 24, 1865	50, 592
Cartridges, Priming metallic	T. T. S. Laidley	Springfield, Mass	Dec. 5, 1865	51, 324
Cartridges, Priming metallic	T. T. S. Laidley	Springfield, Mass	June 19, 1866	55, 676
Cartridges, Priming metallic	E. Martin	Springfield, Mass	July 18, 1865	48, 820
Cartridges, Priming metallic	E. Maynard	Washington, D. C	Dec. 13, 1864	45, 420
Cartridges, Priming metallic	E. Maynard	Tarrytown, N. Y	Oct. 23, 1866	59, 044
Cartridges, Priming metallic	E. Maynard	Tarrytown, N. Y	Jan. 15, 1867	61, 225
Cartridges, Priming metallic	H. Meigs, jr	Bergen Point, N. J	Jan. 28, 1868	73, 739
Cartridges, Priming metallic	I. M. Milbank	Greenfield Hill, Conn	Feb. 19, 1867	62, 283
Cartridges, Priming metallic	T. J. Powers	New York, N. Y	Apr. 24, 1866	54, 254
Cartridges, Priming metallic	T. J. Powers	New York, N. Y	June 1, 1869	90, 871
Cartridges, Priming metallic	J. Rider	Newark, Ohio	Nov. 5, 1867	70, 612
Cartridges, Priming metallic	E. K. Root	Hartford, Conn	Nov. 15, 1864	45, 079
Cartridges, Priming metallic	C. Sharps	Philadelphia, Pa	Dec. 1, 1863	40, 772
Cartridges, Priming metallic	D. Smith	Springfield, Mass	June 11, 1867	65, 774
Cartridges, Priming metallic	C. E. Sneider	Baltimore, Md	Nov. 22, 1864	45, 210
Cartridges, Priming metallic	T. L. Sturtevant	Boston, Mass	Apr. 17, 1866	54, 038
Cartridges, Priming metallic	W. Tibbals	South Coventry, Conn	May 26, 1868	78, 337
Cartridges, Scraping disk or wad for ordnance	J. M. Connel	Newark, Ohio	Nov. 29, 1864	45, 227
Cartridges water-proof, Rendering	A. K. Johnston and L. Dow	New York, N. Y	June 24, 1862	35, 687
Cartridges, Water-proof	R. O. Doremus and B. L. Budd.	New York, N. Y	Mar. 25, 1862	34, 744
Cartridges, Winged metallic	A. D. Perry	New York, N. Y	Mar. 5, 1850	7, 147
Carving and engraving, Hand-tool for	G. B. Soley	Philadelphia, Pa	Mar. 21, 1871	112, 860
Carving and molding machine, Wood	M. T. Boult	Battle Creek, Mich	July 4, 1871	116, 543
Carving and ornamenting machine, Wood	M. T. Boult	Battle Creek, Mich	June 15, 1869	91, 410
Carving device, Wood	N. Ruger	West Farms, N. Y	July 29, 1856	15, 441
Carving-fork	H. Garbanati	Brooklyn, N. Y	Dec. 27, 1859	26, 579
Carving-fork and knife-sharpener combined	E. S. Scofield	Rochester, N. Y	July 21, 1868	80, 226
Carving in wood, Imitation of open	W. H. May	Bridgeport, Conn	Jan. 29, 1867	61, 549
Carving-knife	C. W. Sykes	New York, N. Y	Nov. 8, 1859	26, 061
Carving-knife	O. W. Taft	New York, N. Y	May 3, 1870	102, 617
Carving knife and fork holder	A. T. Foster	Albany, N. Y	Dec. 10, 1867	72, 009
Carving-machine	H. Augur	New Haven, Conn	Dec. 23, 1846	4, 906
Carving-machine	H. Augur	New Haven, Conn	Jan. 23, 1849	6, 058
Carving-machine	C. E. Bacon	Buffalo, N. Y	Sept. 21, 1852	9, 269
Carving-machine	V. W. Blanchard	Bridport, Vt	May 11, 1869	89, 911
Carving-machine	J. W. Campbell	New York, N. Y	Aug. 15, 1871	117, 978
Carving-machine	L. S. Chichester	Williamsburgh, N. Y	June 3, 1851	8, 141
Carving-machine	H. Cottrell	Newark, N. J	Nov. 18, 1873	144, 745
Carving-machine	H. Grubenbecher	New York, N. Y	Apr. 15, 1873	137, 837
Carving-machine	A. Henkel	New York, N. Y	Apr. 13, 1869	88, 869
Carving-machine	A. McCreight	Tranquility, Ohio	Mar. 31, 1868	76, 093
Carving-machine	G. Merrill	Newburyport, Mass	Sept. 15, 1868	82, 145
Carving-machine	G. Merrill	Newburyport, Mass	Sept. 22, 1868	82, 430
Carving-machine	B. J. Tayman	Philadelphia, Pa	May 21, 1872	126, 994
Carving-machine	H. Thomas	Brooklyn, N. Y	June 17, 1873	139, 980
Carving-machine bit	C. F. Bauersfeld	Cincinnati, Ohio	Aug. 1, 1854	11, 407
Carving machine, Wood	H. H. Adams	Newburyport, Mass	Sept. 1, 1868	81, 722
Carving machine, Wood	A. Basse	Quincy, Ill	Dec. 3, 1867	71, 568
Carving machine, Wood	I. Lindsley	Providence, R. I	Dec. 22, 1857	18, 917
Carving machine, Wood	I. M. Singer	New York, N. Y	Dec. 11, 1855	13, 921
Carving machine, Wood	J. Westworth	Chicago, Ill	June 6, 1871	115, 669
Carving, polishing, &c., Machine for	R. T. Smith	Nashua, N. H	Nov. 12, 1872	132, 928
Carving-table	R. E. Deane	New York, N. Y	Oct. 2, 1866	58, 387
Carving-table	W. Howard	Watertown, N. Y	July 16, 1872	129, 565
Carving-tables, Heating	S. Patrick	Galesburgh, Ill	Feb. 21, 1865	46, 491
Carving-tool	M. T. Boult	Battle Creek, Mich	Mar. 14, 1871	112, 680

Index of patents issued from the United States Patent Office from 1790 *to* 1873, *inclusive*—Continued.

Invention.	Inventor.	Residence.	Date.	No.
Carving wood	M. T. Boult	Battle Creek, Mich	Mar. 15, 1870	100, 848
Carving wood or metal, Machine for	I. M. Singer	Pittsburgh, Pa	Apr. 10, 1849	6, 310
Case:				
See Album-case.				
Bacon and ham case.				
Blacking-case.				
Bolt-case.				
Book-case.				
Bottle-case.				
Bottle and glass case.				
Bottle and transportation case.				
Burial-case.				
Card-case.				
Cartridge-case.				
Cast-metal case.				
Clock-case.				
Condenser-case.				
Corn-sheller case.				
Corpse preserving case.				
Daguerreotype-case.				
Document-case.				
Drawing-case.				
Dressing-case.				
Fan-case.				
Filing-case.				
Fire-kindler case.				
Flower-case.				
Folding case.				
Glass case.				
Governor-case.				
Gridiron-case.				
Jewel-case.				
Jewelry and silver-ware case.				
Kneeling-case.				
Lace and ribbon case.				
Lamp-case.				
Lamp-ratchet-wheel case.				
Latch-case.				
Lock-case.				
Match-case.				
Medicine-case.				
Melodeon-case.				
Metal case.				
Miniature-case.				
Moth-proof case.				
Needle-case.				
Odometer-case.				
Oil-case.				
Organ-case.				
Packing-case.				
Pen and pencil case.				
Pencil-case.				
Photographic-card case.				
Piano-forte case.				
Picture-case.				
Pie-case.				
Pin-case.				
Pin-package case.				
Post-office letter-case.				
Postage-stamp case.				
Postal-card case.				
Preserving-case.				
Printer's type-case.				
Pulley-case.				
Pump-case.				
Razor-case.				
Reading-case.				
Record-book case.				
Ribbon-case.				
Ribbon and velvet case.				
Salinometer-case.				
Sample-case.				
Scissors-case.				
Sewing-case.				
Sewing-machine case.				
Sheet-metal case.				
Shipping-case.				
Show-case.				
Skirt-case.				
Spectacle-case.				
Spool-case.				
Spool-exhibiting case.				
Spool-silk case.				
Spool-thread case.				
Steam-gage case.				
Stereoscope-case.				
Tape-measure case.				
Ticket-case.				
Tobacco-case.				
Tobacco-pipe case.				
Toilet-case.				
Transportation-case.				
Turbine-case.				
Turbine water-wheel case.				
Type-case.				
Valve case.				
Watch-case.				
Water-wheel case.				
Wood-case.				
Work-case.				

Index of patents issued from the United States Patent Office from 1790 *to* 1873, *inclusive*—Continued.

Invention.	Inventor.	Residence.	Date.	No.
Case-stand	A. F. De Puy	New York, N. Y	Aug. 30, 1870	106, 921
Cash-box, Alarm	J. D. Perkins	New York, N. Y	May 18, 1869	90, 187
Casing for reverberatory and other furnaces	W. F. Durfee	Bridgeport, Conn	Feb. 23, 1869	87, 152
Cask	J. Connolly	Boston, Mass	July 21, 1863	39, 274
Cask and barrel	M. Hickey	Boston, Mass	Oct. 16, 1866	58, 821
Cask and barrel	J. Marshall	Brooklyn, N. Y	Mar. 12, 1872	124, 601
Cask and barrel for oil	A. Thompson	Brooklyn, N. Y	Jan. 20, 1863	37, 477
Cask and barrel headings	H. Andrews	Canaan, Conn	Mar. 2, 1836	....
Cask and box identifying mark	E. A. Locke	Boston, Mass	Oct. 16, 1866	58, 847
Cask, barrel, &c	J. Merrill	Boston, Mass	Oct. 2, 1866	58, 449
Cask, barrel, &c	J. Merrill	Boston, Mass	Oct. 2, 1866	58, 450
Cask, barrel, &c	J. Merrill	Boston, Mass	Nov. 13, 1866	59, 622
Cask, barrel, and keg	J. Merrill	Boston, Mass	Oct. 2, 1866	58, 451
Cask, Beer	M. Seitz	Williamsburgh, N. Y	May 27, 1873	139, 334
Cask, Brewers'	J. Wiley	Allegheny City, Pa	June 18, 1872	128, 003
Cask, Brewers' stocking	J. Jakel	Detroit, Mich	July 29, 1873	141, 354
Cask-filler	D. Cope	Liverpool, England	June 7, 1870	103, 845
Cask for containing fermentable beverages	J. Hamilton and R. Paterson	Glasgow, Great Britain	Feb. 22, 1870	100, 029
Cask for fermenting wine	J. Glenert	Washington, Mo	July 16, 1867	66, 701
Cask for holding oil, quicksilver, &c	A. D. Campbell and E. Perce	New York, N. Y	Jan. 7, 1862	34, 051
Cask for lager-beer	J. J. Schillinger	New York, N. Y	May 3, 1870	102, 716
Cask for paste	H. Braunhold	New York, N. Y	Apr. 4, 1871	113, 622
Cask for preserving beer, &c	J. Haege	Shiloh, Ill	Jan. 24, 1865	45, 994
Cask-handle	J. L. Jones	Utica, N. Y	Feb. 9, 1869	86, 759
Cask-heads, Machine for holding	J. D. Leach	Penobscot, Me	Aug. 28, 1866	57, 522
Cask-heater	S. Burgess	Wayne, Pa	Apr. 21, 1857	17, 076
Cask-hoops, Machine for shaving	J. G. Morgan	Colton, N. Y	Sept. 4, 1866	57, 753
Cask-hoops, Machine for shaving	S. Wagner	Monroe, Wis	Oct. 4, 1864	44, 567
Cask-machine	S. King	Suffield, Conn	Mar. 23, 1836	....
Cask, Metallic	W. B. Scaife	Pittsburgh, Pa	Apr. 24, 1860	28, 014
Cask, Oil-proof	G. E. Van Derburgh	Mamaroneck, N. Y	Mar. 18, 1862	34, 707
Cask or barrel	J. Marshall	Brooklyn, N. Y	July 18, 1871	117, 185
Cask, Sheet-metal	S. J. Seely	Brooklyn, N. Y	Aug. 12, 1862	36, 175
Cask-stopper faucet-attachment	C. Raggio	Memphis, Tenn	Nov. 9, 1869	96, 726
Cask-washing machine	W. Johnson	Milwaukee, Wis	May 25, 1869	90, 550
Cask-washing machine	W. Robinson	Bridgewater, England	Dec. 1, 1863	40, 797
Cask, Water-tight	D. C. Rand and M. Wadhams	Perinton, N. Y	Aug. 12, 1862	36, 168
Casks and barrels, Device for raising	R. Smith	Brooklyn, N. Y	Jan. 14, 1868	73, 398
Casks and barrrels, Machine for pitching	A. Warth	Stapleton, N. Y	May 16, 1871	114, 999
Casks and barrels, Machinery for making	J. Hale	Oakum, Mass	June 21, 1828	....
Casks and barrels, Method of repairing	C. A. Baldwin	Boston, Mass	Oct. 11, 1864	44, 591
Casks, Apparatus for applying liquids to	J. O. Woodruff	Albany, N. Y	July 17, 1866	56, 483
Casks, Apparatus for pitching	F. Brenner	Cincinnati, Ohio	Mar. 25, 1873	137, 173
Casks, Apparatus for steaming lard and oil	C. J. Yergason	Brooklyn, N. Y	May 17, 1870	103, 116
Casks, Apparatus for transferring liquids from	D. Sexton	San Gabriel, Cal	Nov. 22, 1864	45, 183
Casks, barrels, and kegs, Manufacture of	J. Merrill	Boston, Mass	May 29, 1866	55, 137
Casks, barrels, &c., Apparatus for pitching	G. Meyer and J. Dick	Quincy, Ill	Aug. 5, 1873	141, 449
Casks, barrels, hogsheads, &c., Application of hot water or steam in making.	J. Thompson	Philadelphia, Pa	Dec. 30, 1829	....
Casks, &c., Composition for coating ale	C. P. Matthews	Grantham, England	Sept. 5, 1871	118, 738
Casks, Composition for lining ale and beer	J. Werner	Mannheim, Baden	Jan. 25, 1870	99, 124
Casks, Construction of metallic	J. I. Bard	New Orleans, La	Apr. 8, 1873	137, 590
Casks containing liquids, Ventilation of	L. Whilhelm	Buffalo, N. Y	June 19, 1860	28, 799
Casks, Device for closing the mouths of	J. F. C. Rider	South New Market, N. H	Apr. 2, 1872	125, 220
Casks from saturated timber, Machine for cutting headings for.	L. Wells	Hartsville, N. Y	Jan. 27, 1838	582
Casks, &c., from solid pieces, Method of turning	J. P. Osborn	Staunton, N. J	June 20, 1854	11, 136
Casks, Gage for	J. W. Cochran	New York, N. Y	Apr. 28, 1857	17, 185
Casks, &c., Gage for contents of	J. K. Barney	Warren, R. I	Oct. 19, 1858	21, 809
Casks, Gaging and ullaging	W. W. Cooper	Washington, D. C	Oct. 31, 1865	50, 689
Casks, Hoop-lock for	T. Hanvey	Elma, N. Y	July 12, 1864	43, 494
Casks, Hooping	D. and E. Perry	Pawtucket, R. I	Mar. 19, 1867	63, 088
Casks, Instrument for gaging	J. K. Barney	Warren, R. I	June 21, 1859	24, 438
Casks, Instrument for gaging	E. R. McKean	Nashville, Tenn	July 12, 1870	105, 352
Casks, Instrument for measuring liquid in	W. C. McCarthy	Pittsburgh, Pa	Nov. 21, 1865	51, 069
Casks, Joint of metallic	M. Wappich	Sacramento, Cal	July 9, 1867	66, 540
Casks, Lining for coal-oil	S. G. Morrison	Williamsport, Pa	Sept. 16, 1862	36, 474
Casks, Machine for making heads of	E. P. Spaulding	Saint Louis, Mo	Mar. 7, 1865	46, 717
Casks, Machine for pitching lager-beer and other	H. A. Ratterman	Cincinnati, Ohio	June 7, 1870	104, 062
Casks, Machine for chamfering, crozing, and howeling.	P. Estes	Adrian, Mich	Dec. 3, 1846	4, 869
Casks, Machinery for making	J. Hamilton	New York, N. Y	Apr. 13, 1852	8, 873
Casks, Making	U. Emmons	Green County, N. Y	Jan. 8, 1823	....
Casks, Making	J. Little	Hillsborough, N. H	July 25, 1818	....
Casks, Manufacturing	W. Reid	West Hebron, N. Y	Apr. 19, 1864	42, 445
Casks, Metal lock for wooden hoops for	H. W. Catlin	Burlington, Vt	July 19, 1864	43, 621
Casks, &c., Method of cooling and drawing fluid from.	F. Espenschade	Williamsport, Pa	June 24, 1856	15, 177
Casks, Method of securing heads in seamless	D. Phillips and W. Reid	Shaftsbury and West Arlington, Vt.	Apr. 16, 1867	63, 810
Casks, Mode of cleaning musty beer and other	D. Willard, jr	Pittsford, N. Y	Aug. 31, 1869	94, 460
Casks oil-proof, Rendering	S. Gardner	New York, N. Y	Oct. 3, 1865	50, 238
Casks or barrels, Machine for driving hoops on	J. A. Loomis	Fond du Lac, Wis	Aug. 15, 1865	49, 425
Casks or barrels, Rolling	E. L. Collins	Wellfleet, Mass	Jan. 5, 1864	41, 139
Casket-handle	D. Leonard	Winsted, Conn	Jan. 4, 1870	98, 506
Casket-handle	A. McGuire	Winchester, Conn	Aug. 16, 1870	106, 496
Casket-handle	W. M. Smith	West Meriden, Conn	Nov. 12, 1867	70, 759
Casket-handle	C. Strong	Winsted, Conn	May 7, 1872	126, 500
Casket-handle	H. C. Wilcox	West Meriden, Conn	Aug. 10, 1869	93, 508
Casket, Jewel	J. Mathiew	Paris, France	Jan. 11, 1870	98, 785
Caskets, Manufacture of	T. Hegarty	Saint Louis, Mo	Jan. 15, 1861	31, 119
Cast-iron chair	J. L. Mott	New York, N. Y	Oct. 2, 1847	5, 317
Cast-metal case for spring-balances	J. Chatillon	New York, N. Y	Dec. 10, 1867	71, 980
Cast-metal dies, Manufacture of	L. Chapman	Collinsville, Conn	Nov. 23, 1869	97, 046
Cast metal, Pressing	J. B. Tarr	Fairhaven, Mass	Oct. 31, 1871	120, 467
Cast metal, Surfacing articles of	H. Tucker	Newton, Mass	June 1, 1869	90, 892
Casts and fancy articles, Composition for forming	M. Schall	New York, N. Y	Mar. 3, 1868	75, 058
Casts, Composition for making anatomical	J. Hurford	Salem, Ohio	Mar. 13, 1866	53, 225

Index of patents issued from the United States Patent Office from 1790 to 1873, inclusive—Continued.

Invention.	Inventor.	Residence.	Date.	No.
Casts, fancy articles, toys, &c., Composition for	M. Schall	New York, N. Y	June 30, 1863	39, 070
Casts for fancy articles, Compound for making	H. Hirsch	New York, N. Y	Mar. 5, 1872	124, 355
Casts for pads, Apparatus for taking	F. Kesmodel	San Francisco, Cal	July 9, 1861	32, 773
Casts from the faces of living persons, Mode of taking.	C. Mills	Washington, D. C	Apr. 4, 1865	47, 121
Casts, Mode of preparing plaster	T. Taylor	Washington, D. C	Oct. 6, 1868	82, 891
Castanets	G. F. Fessenden	Arlington, Mass	June 1, 1869	90, 832
Caster	J. T. Barnes	Hudson City, N. J	June 25, 1867	66, 116
Caster	T. Beach	Freeport, Pa	Aug. 7, 1866	57, 036
Caster	J. Bradbury	Berlin, Conn	Nov. 7, 1871	120, 568
Caster	M. S. Brownell	Adrian, Mich	May 4, 1869	89, 628
Caster	S. Curtis	New York, N. Y	Sept. 30, 1873	143, 334
Caster	J. N. Dinsmore	Kendall Mills, Me	June 18, 1872	128, 129
Caster	W. P. Elliott	Cincinnati, Ohio	Apr. 6, 1869	88, 558
Caster	F. G. Ford	New York, N. Y	Aug. 3, 1869	93, 292
Caster	F. A. Gardner	Danbury, Conn	Jan. 24, 1871	111, 193
Caster	E. G. Gory	Cincinnati, Ohio	June 11, 1872	127, 862
Caster	G. E. Grosse	Massillon, Ohio	Apr. 26, 1870	102, 253
Caster	J. Kintz	West Meriden, Conn	Sept. 20, 1870	107, 507
Caster	J. Kintz	West Meriden, Conn	June 6, 1871	115, 619
Caster	C. Knisely	Chicago, Ill	Apr. 19, 1864	42, 381
Caster	C. Knisely	Chicago, Ill	Apr. 19, 1864	42, 382
Caster	C. Lewando	Boston, Mass	Nov. 17, 1868	84, 126
Caster	W. C. McGill	Cincinnati, Ohio	Apr. 27, 1869	89, 418
Caster	J. Miller	Olean, N. Y	Feb. 16, 1869	86, 997
Caster	G. B. Munn	New Brunswick, N. J	July 16, 1872	128, 972
Caster	J. W. Pugh	Grand Rapids, Mich	June 9, 1868	78, 690
Caster	J. M. Riley	Newark, N. J	Sept. 11, 1866	57, 971
Caster	T. L. Rivers	Saint Louis, Mo	June 29, 1869	91, 873
Caster	J. Toler	Newark, N. J	May 20, 1873	139, 212
Caster	A. C. Twining	New Haven, Conn	Feb. 4, 1868	74, 019
Caster	G. W. Waitt	Philadelphia, Pa	Dec. 23, 1873	145, 769
Caster	J. White	Providence, R. I	Dec. 24, 1867	72, 575
Caster Adjustable	W. C. Dodge	Washington, D. C	May 17, 1864	42, 754
Caster and cake-basket	J. W. Larimore	Chicago, Ill	Feb. 16, 1869	87, 054
Caster and fan, Automatic	E. A. H. Nordyke	Richmond, Ind	Oct. 20, 1857	18, 466
Caster and fruit and cake dish	J. W. Larimore	Chicago, Ill	Feb. 2, 1869	86, 422
Caster and spoon-holder, Combined	L. Evans	Pittsburgh, Pa	Nov. 29, 1870	109, 602
Caster attachment for furniture	J. H. Travis	Charlestown, Mass	June 17, 1873	140, 097
Caster, Ball	A. F. Ahrens	Philadelphia, Pa	May 8, 1847	5, 102
Caster, Ball	S. S. Hickok and D. B. Clement	Boston, Mass	Apr. 14, 1868	76, 630
Caster, Ball	F. S. Smith	New Haven, Conn	Nov. 26, 1867	71, 546
Caster, Ball	L. Wilkinson	Boston, Mass	June 9, 1868	78, 850
Caster, Ball furniture	J. Holmes	Boston, Mass	Apr. 1, 1862	34, 833
Caster, Ball furniture	J. C. Pedrick	Washington, D. C	Aug. 16, 1859	25, 138
Caster, Ball furniture	B. A. Russell	Deep River, Conn	Jan. 25, 1859	22, 751
Caster, Bottle	E. Gleason	Dorchester, Mass	Oct. 21, 1856	15, 946
Caster, Chair	T. Fry	Brooklyn, N. Y	Apr. 10, 1860	27, 792
Caster for bedsteads, Trap	D. Henderson	Boston, Mass	Mar. 15, 1864	41, 919
Caster for billiard-tables and furniture	W. Reagan	New York, N. Y	June 14, 1870	104, 201
Caster for trunks and furniture, Ball	J. Knight	Newark, N. J	Aug. 26, 1856	15, 611
Caster-frame	F. J. Miller	Brooklyn, N. Y	Feb. 12, 1867	61, 947
Caster, Furniture	A. T. Adams	Indianapolis, Ind	Oct. 6, 1868	82, 783
Caster, Furniture	A. C. Arnold and O. G. Hanschildt.	Norwalk, Conn	May 7, 1872	126, 433
Caster, Furniture	M. L. Babb	Cape Elizabeth, Me	June 13, 1865	48, 146
Caster, Furniture	G. L. Bailey	Portland, Me	Mar. 20, 1855	12, 578
Caster, Furniture	D. S. Barnes	New York, N. Y	July 5, 1859	24, 607
Caster, Furniture	J. T. Barnes	Hudson City, N. J	Oct. 30, 1866	59, 163
Caster, Furniture	R. J. Beardsley	Brooklyn, N. Y	Feb. 14, 1865	46, 334
Caster, Furniture	L. Bertsche	Allegheny, Pa	Jan. 18, 1870	98, 912
Caster, Furniture	W. I. Blackman	Columbus, Miss	Jan. 30, 1872	123, 147
Caster, Furniture	H. D. Blake	New Hartford Centre, Conn	Apr. 27, 1858	20, 031
Caster, Furniture	J. H. Bloomfield	Chicago, Ill	Nov. 11, 1862	36, 890
Caster, Furniture	S. A. Bracket	Boston, Mass	Dec. 31, 1872	134, 350
Caster, Furniture	J. D. Bradley	Washington, D. C	Aug. 16, 1864	43, 830
Caster, Furniture	J. Brown	Utica, N. Y	Apr. 21, 1868	76, 885
Caster, Furniture	J. D. Browne	Madisonville, Ohio	June 18, 1872	128, 107
Caster, Furniture	T. P. Burger	Oyster Bay, N. Y	Dec. 15, 1857	18, 839
Caster, Furniture	L. F. Cerf	New York, N. Y	Nov. 26, 1867	71, 363
Caster, Furniture	S. Chandler	New York, N. Y	June 8, 1869	91, 086
Caster, Furniture	S. Chandler	New York, N. Y	Nov. 2, 1869	96, 393
Caster, Furniture	F. G. Ford	Washington, D. C	Mar. 28, 1865	47, 007
Caster, Furniture	F. G. Ford	New York, N. Y	Oct. 23, 1866	58, 999
Caster, Furniture	F. G. Ford	New York, N. Y	May 17, 1870	103, 165
Caster, Furniture	F. G. Ford	Bridgeton, N. J	Oct. 8, 1872	132, 066
Caster, Furniture	E. E. Furney	Chicopee, Mass	June 9, 1868	78, 798
Caster, Furniture	G. H. Glad	Boston Highlands, Mass	Nov. 11, 1873	144, 530
Caster, Furniture	C. R. Gorgas and W. H. Smith	Wooster, Ohio	July 9, 1861	32, 766
Caster, Furniture	E. Hambujer	Detroit, Mich	Mar. 19, 1867	62, 954
Caster, Furniture	J. W. Hewitt and G. R. Lynch	Allegheny City, Pa	July 7, 1868	79, 571
Caster, Furniture	S. S. Hickock	Methuen, Mass	Dec. 19, 1871	122, 016
Caster, Furniture	P. B. Holmes	New York, N. Y	Oct. 3, 1865	50, 249
Caster, Furniture	J. S. and J. W. Hyatt	Albany, N. Y	July 9, 1872	128, 884
Caster, Furniture	J. Johnson	Brooklyn, N. Y	Mar. 4, 1873	136, 522
Caster, Furniture	W. Johnson	Milwaukee, Wis	June 19, 1866	55, 665
Caster, Furniture	T. M. Kane and C. Brown	Goshen, N. Y	Feb. 6, 1866	52, 420
Caster, Furniture	J. Kinzer	Pittsburgh, Pa	Feb. 16, 1858	19, 369
Caster, Furniture	A. H. Lightall	Albany, N. Y	Dec. 3, 1872	133, 650
Caster, Furniture	E. Lindner	New York, N. Y	Feb. 26, 1861	31, 548
Caster, Furniture	S. N. Long	South Chatham, Mass	May 12, 1868	77, 743
Caster, Furniture	W. T. Mersereau	Newark, N. J	Apr. 4, 1865	47, 117
Caster, Furniture	H. Munroe	Fall River, Mass	Mar. 17, 1868	75, 564
Caster, Furniture	H. Munroe	Fall River, Mass	Aug. 11, 1868	80, 990
Caster, Furniture	H. Naylor	Pekin, Ill	Oct. 16, 1866	58, 879
Caster, Furniture	P. B. O'Brien and W. E. Sparks	New Haven, Conn	Dec. 10, 1867	71, 904
Caster, Furniture	J. Parry	Manchester, Great Britain	Jan. 28, 1873	135, 286
Caster, Furniture	J. C. Plumer	Boston, Mass	June 5, 1866	55, 358

Index of patents issued from the United States Patent Office from 1790 *to* 1873, *inclusive*—Continued.

Invention.	Inventor.	Residence.	Date.	No.
Caster, Furniture	S. C. Pratt	Boston, Mass	Aug. 11, 1868	81, 001
Caster, Furniture	J. M. Riley	Newark, N. J	Nov. 25, 1862	37, 008
Caster, Furniture	J. B. Sargent	New Haven, Conn	July 12, 1870	105, 374
Caster, Furniture	C. B. Sheldon	New York, N. Y	Dec. 24, 1872	134, 223
Caster, Furniture	C. B. Sheldon	New York, N. Y	Mar. 25, 1873	137, 249
Caster, Furniture	C. B. Sheldon	New York, N. Y	Mar. 25, 1873	137, 250
Caster, Furniture	C. B. Sheldon	New York, N. Y	Mar. 25, 1873	137, 251
Caster, Furniture	C. B. Sheldon	New York, N. Y	Apr. 8, 1873	137, 728
Caster, Furniture	C. B. Sheldon	New York, N. Y	June 3, 1873	139, 619
Caster, Furniture	F. Smith	Boston, Mass	Jan. 1, 1867	60, 799
Caster, Furniture	I. A. Stafford	Essex, N. Y	Oct. 23, 1860	30, 505
Caster, Furniture	A. G. Stevens	Manchester, N. H	Apr. 25, 1871	114, 061
Caster, Furniture	S. B. Ulmann	New York, N. Y	Feb. 12, 1850	7, 093
Caster, Furniture	W. W. Wade	Springfield, Mass	May 17, 1853	9, 733
Caster, Furniture	A. Walker	New Haven, Conn	Jan. 3, 1865	45, 773
Caster, Furniture	L. S. White	Chicopee, Mass	Jan. 31, 1854	10, 488
Caster, Furniture	L. S. White	Chicopee, Mass	Aug. 8, 1854	11, 506
Caster, Furniture	J. A. Wiedersheim	Philadelphia, Pa	Oct. 2, 1866	58, 523
Caster, Furniture	J. H. Wilhelm and F. G. Ensign	Chicago, Ill	Dec. 26, 1865	51, 770
Caster, Furniture	C. G. Wilson	Brooklyn, N. Y	Sept. 7, 1869	94, 682
Caster, Glass table	A. E. Young	Dorchester, Mass	July 1, 1862	35, 799
Caster, Globe	W. Hunt	New York, N. Y	Apr. 22, 1833	....
Caster or bottle-holder	C. H. Latham	Lowell, Mass	Oct. 1, 1872	131, 822
Caster, Pepper	L. Soehlmann	Jersey City, N. J	Sept. 25, 1866	58, 312
Caster, Pepper	F. Wolf	South Boston, Mass	Aug. 17, 1869	93, 785
Caster, Pickle	H. C. Wilcox	West Meriden, Conn	Feb. 20, 1872	123, 961
Caster, Revolving	C. H. Latham and J. S. Lugg	Lowell, Mass	July 4, 1871	116, 722
Caster, Revolving bottle	E. Gleason	Dorchester, Mass	Dec. 1, 1857	18, 740
Caster, Salt and pepper	J. Bird	Philadelphia, Pa	Sept. 9, 1873	142, 670
Caster, Sewing-machine	L. O. Allen	Gardiner, Me	Aug. 25, 1868	81, 454
Caster, Sewing-machine	T. T. Bishop and H. J. Merret	Evansville, Ind	Mar. 28, 1871	113, 135
Caster, Sewing-machine	D. J. Clark	Evansville, Ind	June 4, 1872	127, 571
Caster, Sewing-machine	E. R. Clark	Marshfield, Ind	Sept. 9, 1873	142, 615
Caster, Sewing-machine	F. H. Duncan	Evansville, Ind	Aug. 15, 1871	118, 117
Caster, Sewing-machine	W. P. Elliot	New Haven, Conn	Apr. 12, 1870	101, 843
Caster, Sewing-machine	W. P. Elliot	New Haven, Conn	Apr. 12, 1870	101, 844
Caster, Sewing-machine	W. D. Hatch	Antrim, N. H	Oct. 3, 1871	119, 606
Caster, Sewing-machine	R. Hathaway	Chicopee, Mass	Mar. 24, 1868	75, 755
Caster, Sewing-machine	H. Jones	Richmond, Ind	May 23, 1871	115, 060
Caster, Sewing-machine	J. B. Lincoln	Providence, R. I	Mar. 25, 1873	137, 141
Caster, Sewing-machine	S. McAfferty	North Shenango, Pa	July 16, 1872	129, 354
Caster, Sewing-machine	J. H. Plank	Bloomfield, Iowa	June 3, 1873	139, 606
Caster, Sewing-machine	G. K. Proctor	Salem, Mass	Oct. 17, 1871	120, 098
Caster, Sewing-machine	G. K. Proctor	Salem, Mass	June 3, 1873	139, 608
Caster, Sewing-machine	J. K. Proctor	Malden, Mass	Nov. 25, 1873	145, 011
Caster, Sewing-machine	J. Robertson	Northampton, Mass	July 29, 1873	141, 236
Caster, Sewing-machine	B. F. Ryder	New York, N. Y	Apr. 12, 1870	101, 924
Caster, Sewing-machine	B. F. Ryder	New York, N. Y	Mar. 14, 1871	112, 740
Caster, Sewing-machine	J. B. Sargent	New Haven, Conn	May 31, 1870	103, 782
Caster, Sewing-machine	H. A. Skinner	Worcester, Mass	Nov. 7, 1871	120, 783
Caster, Sewing-machine	J. E. Smith	Sharpsburgh, Ky	Sept. 30, 1873	143, 387
Caster, Sewing-machine	C. F. Stafford and L. Stansberry	Evansville, Ind	June 6, 1871	115, 779
Caster, Sewing-machine	N. D. Stoops	Newark, N. J	Oct. 10, 1865	50, 402
Caster, Sewing-machine	J. M. Veasey	Denver, Colo	Mar. 29, 1870	101, 328
Caster, Sewing-machine	J. M. Veasey	Denver, Colo	July 16, 1872	129, 629
Caster, Sewing-machine	J. N. Wilkins	Chicago, Ill	June 12, 1866	55, 567
Caster, Sewing-machine	W. W. Wright	Lynn, Mass	Feb. 27, 1872	124, 106
Caster-socket	A. C. Twining	New Haven, Conn	Mar. 3, 1868	75, 080
Caster, Spring	E. J. Hall	Highgate, Vt	Dec. 17, 1861	33, 974
Caster, Spring	E. J. Hall	Highgate, Vt	Feb. 18, 1862	34, 463
Caster-stand	G. D. Dudley	Lowell, Mass	Jan. 14, 1873	134, 738
Caster-stand	C. Reistle	Brooklyn, N. Y	Dec. 31, 1867	72, 758
Caster, Stove-leg	E. H. Harding	Philadelphia, Pa	July 16, 1872	129, 550
Caster, Stove-leg, &c	H. A. Humphrey	Milwaukee, Wis	May 30, 1871	115, 322
Caster, Sirup	E. Bigelow	Springfield, Mass	Apr. 6, 1858	19, 824
Caster, Table	H. A. Dirkes	New York, N. Y	Oct. 5, 1869	95, 572
Caster, Table	H. Fachs	New York, N. Y	Sept. 16, 1873	142, 905
Caster, Table	J. Gibson, jr	Albany, N. Y	Aug. 8, 1871	117, 765
Caster, Table	R. Gleason, jr	Dorchester, Mass	Mar. 8, 1859	23, 218
Caster, Table	H. A. Hiestand	Hellam, Pa	Oct. 15, 1872	132, 285
Caster, Table	C. H. Latham	Lowell, Mass	June 28, 1870	104, 743
Caster, Table	E. B. Manning	Middletown, Conn	Oct. 15, 1872	132, 301
Caster, Table	A. B. Searles	Providence, R. I	Apr. 13, 1869	88, 911
Caster, Table	A. B. Searles	Providence, R. I	Apr. 13, 1869	88, 912
Caster, Table	T. Shaw	Providence, R. I	Mar. 25, 1873	137, 248
Caster, Table	C. B. Sheldon	New York, N. Y	July 29, 1873	141, 390
Caster, Table	D. Sherwood	Lowell, Mass	Feb. 25, 1868	74, 945
Caster, Table	D. Sherwood	Lowell, Mass	Feb. 2, 1869	86, 463
Caster, Table	D. Sherwood	Lowell, Mass	Dec. 28, 1869	98, 432
Caster, Table	D. Sherwood	Lowell, Mass	Mar. 22, 1870	101, 169
Caster, Table	D. Sherwood	Lowell, Mass	Nov. 1, 1870	108, 838
Caster, Table	D. Sherwood and G. D. Dudley	Lowell, Mass	July 26, 1870	105, 732
Caster, Table	D. Sherwood, E. P. Woods, and G. D. Dudley.	Lowell, Mass	July 26, 1870	105, 731
Caster, Table	W. M. Smith	Augusta, Ga	Feb. 15, 1870	99, 965
Caster, Table	G. D. Woodworth	Cincinnati, Ohio	July 16, 1872	129, 507
Caster, Table-car or	E. L. Crandall	Williamstown, N. Y	Apr. 27, 1869	89, 294
Caster, Trunk	J. W. C. and J. E. Haskell	Chicago, Ill	Sept. 22, 1868	82, 310
Caster, Trunk	G. Havell	Newark, N. J	Dec. 10, 1872	133, 856
Caster, Trunk	W. O. Headley	Newark, N. J	July 25, 1865	48, 937
Caster, Trunk	L. Horton and J. E. McGaw	Manchester, N. H	Mar. 3, 1868	75, 017
Caster, Trunk	H. T. Lee	Jersey City, N. J	Oct. 11, 1864	44, 689
Caster, Trunk	T. L. Rivers	Newark, N. J	Nov. 7, 1871	120, 670
Caster, Trunk	T. L. Rivers	Newark, N. J	July 2, 1872	128, 659
Caster, Trunk	M. Schwerin	Newark, N. J	Oct. 11, 1870	108, 193
Caster, Trunk	A. J. Sessions	Bristol, Conn	Oct. 11, 1870	108, 300
Caster, Trunk	W. H. and L. Young	Boston, Mass	Dec. 31, 1867	72, 956
Caster, Trunk	W. H. and L. Young	Boston, Mass	July 7, 1868	79, 798

Index of patents issued from the United States Patent Office from 1790 *to* 1873, *inclusive*—Continued.

Invention.	Inventor.	Residence.	Date.	No.
Caster, Wall or bracket	A. J. Forbes and W. G. Fletcher	Boston, Mass	June 29, 1869	91, 835
Caster-wheel	T. Beach	New York, N. Y	Apr. 19, 1864	43, 338
Caster wheel, Furniture	C. B. Sheldon	New York, N. Y	June 24, 1873	140, 310
Caster wheel, Furniture	C. B. Sheldon	New York, N. Y	June 24, 1873	140, 311
Caster-wheel, Glass	J. B. Capewell	Gloucester, N. J	Nov. 7, 1865	50, 795
Casters and applying them to bedsteads, Mode of constructing.	P., E. W., and J. A. Blake	New Haven, Conn	June 30, 1838	821
Casters, Construction of cruet	W. E. Hawkins	New York, N. Y	Oct. 29, 1867	70, 209
Casters, Detachable fixture for	A. C. Twining	New Haven, Conn	May 18, 1869	90, 137
Casters, Device for supporting furniture	H. E. Richards	Newark, N. J	Dec. 7, 1858	22, 243
Casters, Finishing furniture	P. B. Tyler	Springfield, Mass	Oct. 14, 1856	15, 902
Casters for stove-legs, Adjustable stop for	W. Coughlin	Clarksville, Ohio	Sept. 27, 1870	107, 762
Casters from vulcanizable compounds, Manufacture of.	C. Goodyear, jr	New York, N. Y	May 13, 1862	35, 231
Casters, Manufacture of ball	R. Hinton	Roxbury, Mass	Dec. 14, 1852	9, 464
Casters, Roller of furniture	P. B. Tyler and B. Lathrop	Springfield, Mass	Aug. 8, 1854	11, 502
Casters to furniture, Mode of attaching	W. B. Bartram	Norwalk, Conn	Jan. 30, 1866	52, 257
Casters to trunks, Attaching	I. H. Giffing	New York, N. Y	Apr. 12, 1859	23, 672
Casters to trunks, Attaching	L. S. Maring	Fall River, Mass	Nov. 27, 1855	13, 853
Casters, Tool for boring recesses for	B. F. Graves	Boston, Mass	Aug. 22, 1854	11, 587
Casting and annealing articles made of scoria	W. H. Smith	Philadelphia, Pa	Mar. 22, 1859	23, 317
Casting-apparatus	J. P. Broadmeadow	Bridgeport, Conn	Apr. 4, 1871	113, 249
Casting, Apparatus for attaching pieces of metal to each other by.	H. B. Osgood	Thompsonville, Conn	Jan. 13, 1852	8, 658
Casting chilled plates	R. Poole	Baltimore, Md	Aug. 2, 1859	24, 976
Casting chilled rolls, Method of giving rotary motion to metal in.	J. C. Parry	Pittsburgh, Pa	Feb. 26, 1850	7, 125
Casting-composition	G. Nimmo	Jersey City, N. J	July 25, 1865	48, 973
Casting copper cylinders	F. Adams	Somerville, Mass	Apr. 30, 1861	32, 169
Casting-flask	G. P. Sisson	Florence, Mass	Oct. 1, 1867	69, 372
Casting for steam-pipes, boilers, &c	W. T. Kosinski	Philadelphia, Pa	Sept. 13, 1870	107, 387
Casting hollow articles	J. Brunner	New York, N. Y	Oct. 12, 1869	95, 645
Casting hollow ware	J. Ziegler	Dayton, Ohio	Nov. 3, 1868	83, 751
Casting hollow ware, &c., Method of making patterns for.	E. Ripley	Troy, N. Y	Aug. 31, 1844	3, 724
Casting lugs and dovetails, Device for	G. W. Herrick	Stuyvesant, N. Y	Oct. 6, 1868	82, 715
Casting metal	E. Ripley	Troy, N. Y	Jan. 15, 1856	14, 115
Casting metal	R. Ross	Middlebury, Vt	Mar. 22, 1870	101, 046
Casting metal, Apparatus for forming molds for	A. H. Lowell	Manchester, N. H	June 18, 1867	65, 922
Casting metal, Method of	J. Rives	Paris, France	Sept. 29, 1868	82, 642
Casting, Metal or second patterns for	F. N. Still	New York, N. Y	Jan. 28, 1851	7, 915
Casting metals under pressure, Method of	J. J. C. Smith	Philadelphia, Pa	May 18, 1869	90, 318
Casting, Method of supporting chills in	W. H. Wiley	Fredonia, N. Y	July 28, 1868	80, 321
Casting-mold	S. A. Corser	Northampton, Mass	July 10, 1860	29, 123
Casting-mold	J. R. Davies	Racine, Wis	Sept. 10, 1872	131, 214
Casting-mold	J. R. Davis	Racine, Wis	Mar. 10, 1863	37, 857
Casting-mold	W. Murdock	Jersey City, N. J	May 13, 1863	35, 248
Casting-mold	M. Nelson	New York, N. Y	Apr. 21, 1857	17, 109
Casting-mold	W. B. Robinson	Detroit, Mich	Dec. 10, 1872	133, 801
Casting-mold	N. Thompson	Brooklyn, N. Y	Dec. 12, 1871	121, 736
Casting mold, Metal	E. N. Cleaves	Boston, Mass	Feb. 23, 1869	87, 071
Casting mold, Metal	J. Farrar and W. Groves	Providence, R. I	July 30, 1867	67, 181
Casting molds and cores, Composition for metal	C. Grauer	Philadelphia, Pa	May 18, 1869	90, 257
Casting-molds, Device for forming	W. T. Horrobin	Biddeford, Me	Feb. 28, 1865	46, 564
Casting-molds for metal trundle-heads, Method of making.	E. Lester	Little Falls, N. Y	Feb. 12, 1807	....
Casting molds, Lining for metal	J. Farrar and L. F. Whiting	Providence, R. I., and Boston, Mass.	Jan. 26, 1869	86, 145
Casting-molds, Making	R. Jobson	Wordsley, England	Mar. 29, 1859	23, 375
Casting-molds, Making	F. N. Still	Buffalo, N. Y	Oct. 7, 1846	4, 803
Casting-molds, Manufacture of dry	J. P. Townsend	New York, N. Y	Aug. 16, 1864	43, 885
Casting-molds, Method of constructing metallic	W. Hainsworth	Sharpsville, Pa	Mar. 9, 1869	87, 663
Casting-molds, Method of making	J. G. Holt	Chicago, Ill	June 5, 1866	55, 294
Casting or smut machine	D. Tomlinson	Brookfield, Conn	May 20, 1812	....
Casting ornamental borders	J. L. Lewis	....	Oct. 8, 1824	....
Casting-pattern	A. L. Finch	Sing Sing, N. Y	Apr. 26, 1870	102, 384
Casting-pattern	J. L. Jackson	New York, N. Y	June 7, 1870	103, 888
Casting-pattern	L. H. Mans	Danville, Pa	Feb. 5, 1836	....
Casting pipe	A. Brady	New York, N. Y	Sept. 18, 1860	30, 044
Casting pipe	T. J. Lovegrove	Philadelphia, Pa	May 23, 1865	47, 901
Casting pipe	G. Rogers	Philadelphia, Pa	Mar. 5, 1867	62, 688
Casting pot-hole covers, Pattern for	W. J. Fryer	West Troy, N. Y	July 2, 1867	66, 317
Casting-press for metal	J. B. Tarr	Fairhaven, Mass	Aug. 31, 1869	94, 253
Casting refractory metals, Apparatus for	M. Smith	Philadelphia, Pa	Oct. 22, 1867	70, 038
Casting rolls, Method of giving rotary motion to fluid iron in.	J. C. Parry	Pittsburgh, Pa	May 21, 1850	7, 383
Casting under pressure, Mold for	J. B. Tarr	Fairhaven, Mass	Dec. 20, 1870	110, 307
Castings and preparing cast-iron patterns, Process of reducing iron.	C. R. Ely	Sheldon, Vt	Apr. 2, 1861	31, 880
Castings, Apparatus for making molds for metal	A. Weaber	Philadelphia, Pa	July 13, 1869	92, 677
Castings, Apparatus for molding	P. W. Lamb	Albany, N. Y	June 12, 1866	55, 504
Castings, Apparatus for producing refined iron and steel.	J. W. Middleton	Philadelphia, Pa	July 30, 1872	129, 899
Castings, Apparatus for treating malleable-iron	A. F. Andrews	New Haven, Conn	Jan. 30, 1872	123, 073
Castings, Blackwashing mold for	W. and D. Ferguson	New York, N. Y	Mar. 13, 1860	27, 438
Castings, Cleaning	A. Ralston	West Middletown, Pa	Mar. 8, 1859	23, 193
Castings, Composition for making chilled	J. Reichenbach	Allegheny, Pa	Apr. 20, 1869	89, 076
Castings direct from the blast-furnace, Producing	J. W. Middleton	Philadelphia, Pa	Oct. 18, 1870	108, 375
Castings, Furnace for tempering	W. M. Watson	Tonica, Ill	Oct. 7, 1873	143, 548
Castings, Grinder for cleaning	G. Miller	Providence, R. I	July 26, 1870	105, 711
Castings, Grinder or rattler for cleaning	G. Miller	Johnston, R. I	Feb. 22, 1870	100, 175
Castings, Machine for making cast-steel	J. B. Tarr	Chicago, Ill	Aug. 28, 1866	57, 644
Castings, Machine for molding for metal	D. Brown	Baltimore, Md	June 27, 1854	11, 191
Castings, Machine for smoothing oxide and sand on	B. Seymour	Utica, N. Y	Dec. 2, 1835	....
Castings, Machine for trimming	A. S. Gear	Boston, Mass	Apr. 23, 1872	125, 948
Castings, Making chilled	G. W. Bolman and W. Neemes	Pittsburgh, Pa	Oct. 31, 1865	50, 680
Castings, Making chilled	W. Butler	Little Falls, N. Y	Mar. 18, 1856	14, 442

Index of patents issued from the United States Patent Office from 1790 *to* 1873, *inclusive*—Continued.

Invention.	Inventor.	Residence.	Date.	No.
Castings, Malleable-iron	A. F. Andrews	New Haven, Conn	Jan. 30, 1872	123, 072
Castings, Metal for	G. Whitney	Philadelphia, Pa	July 16, 1872	129, 261
Castings, Metallic pin for cooling the interior of	W. H. Saunders	Greensburgh, N. Y	Nov. 26, 1835	....
Castings, Mill for cleaning	H. R. Remsen	Albany, N. Y	Apr. 7, 1857	17, 012
Castings, Mode of making cast-steel	T. J. Chubb	Williamsburgh, N. Y	June 8, 1869	90, 925
Castings, Mold for making	S. A. Carser	Northampton, Mass	Apr. 5, 1864	42, 251
Castings, Mold for steel	P. G. Gardiner	New York, N. Y	Apr. 12, 1859	23, 670
Castings, Molding for metal	J. P. Broadmeadow	Bridgeport, Conn	Nov. 29, 1859	26, 321
Castings, Preparing metallic patterns for	T. G. Bucklin	West Troy, N. Y	May 8, 1849	6, 440
Castings, Process of making thin iron	H. and W. E. Bleecker and S. D. Vose	Albany, N. Y	Dec. 25, 1849	6, 969
Castings, Process of chilling	J. L. Mott	New York, N. Y	June 13, 1848	5, 636
Castings, Split	J. Yocom, jr	Philadelphia, Pa	June 20, 1871	116, 134
Castings, Tempering steel	C. Parking and S. Trethewey	Pittsburgh, Pa	July 20, 1869	92, 747
Castings, Tumbling-barrel for cleaning	F. N. Bixby	West Meriden, Conn	Apr. 1, 1873	137, 409
Cat-ball	T. H. Joyce	New York, N. Y	Nov. 7, 1871	120, 650
Cat-head and shank-painter stopper	C. Perley	New York, N. Y	Apr. 2, 1850	7, 249
Cat hook and stopper	D. H. Cousins	Surry, Me	Aug. 6, 1872	130, 198
Catamenial and urinal bandage and receptacle	M. Vedder	New York, N. Y	May 20, 1862	35, 338
Catamenial bandage	C. E. Clark	Boston, Mass	Feb. 22, 1859	23, 059
Catamenial bandage	F. Dahis and F. Doermer	Brooklyn, N. Y	July 31, 1860	29, 362
Catamenial bandage	D. F. Robertson	Middletown, N. Y	Dec. 20, 1864	45, 523
Catamenial guard and supporter	J. A. Belvin, jr	Baltimore, Md	Sept. 10, 1867	68, 688
Catamenial sack	A. F. Baum	New York, N. Y	June 2, 1868	78, 414
Catamenial sack	J. C. Bewzinger	Catonsville, Md	Sept. 4, 1866	57, 665
Catamenial sack	G. E. Brinckerhoff	Brooklyn, N. Y	Nov. 17, 1868	84, 083
Catamenial sack	W. A. Dinsmore	South Boston, Mass	July 26, 1870	105, 785
Catamenial sack	H. W. Libbey	Cleveland, Ohio	Mar. 10, 1868	75, 434
Catamenial sack	C. Manheim	New York, N. Y	Mar. 3, 1868	75, 036
Catamenial sack	E. L. Perry	New York, N. Y	Sept. 12, 1865	49, 915
Catamenial supporter	A. A. Starr	New York, N. Y	Aug. 22, 1854	11, 574
Cataract-instrument	T. R. Williams	Philadelphia, Pa	June 14, 1826	....
Cataract by tubes and cannular points, Removing	F. B. Shaw	Philadelphia, Pa	July 27, 1815	....
Catch-bolt	W. Salisbury	Wheeling, Va	Nov. 29, 1859	26, 296
Catch for clasps	J. Bingham	Philadelphia, Pa	Sept. 11, 1847	5, 285
Catch for hanging drapery	A. and C. A. Warner	Bristol, Conn	Jan. 4, 1859	22, 521
Catch, Hand	J. P. R. James	Pepin, Minn	Aug. 27, 1867	68, 082
Caterpillar-destroying compound	G. F. Whisenant	Chapel Hill, Tex	Jan. 14, 1873	134, 959
Caterpillar-nests, Instrument for destroying	J. S. Needham	South Danvers, Mass	Dec. 13, 1864	45, 428
Caterpillars, Instrument for destroying embryo	A. Casebeer	Sipesville, Pa	Dec. 4, 1866	60, 136
Catheter and syringe, Combined	N. B. Sornborger	Northampton, Mass	Jan. 14, 1868	73, 402
Catheter, Metallic flexible	I. Balch and M. Carter	Salisbury, Mass	Oct. 26, 1816	....
Cattle, Apparatus for breachy	J. P. Leddy and W. Rogers	Mount Carroll, Ill	May 15, 1860	28, 286
Cattle, Apparatus for watering	S. W. Wood	Rochester, N. Y	Oct. 28, 1851	8, 479
Cattle-blinder	M. Backmayer	Lyons, N. Y	Apr. 29, 1873	138, 224
Cattle-card	W. M. Warren	Watertown, Conn	Mar. 26, 1872	124, 991
Cattle, Compound for destroying vermin in	F. Kalteyer	San Antonio, Tex	Aug. 10, 1869	93, 622
Cattle, Device for removing obstacles from the throats of.	C. Schule	North East, Pa	Apr. 2, 1872	125, 337
Cattle, Device for watering	O. Wheedon	Medina, N. Y	Mar. 13, 1866	53, 207
Cattle during transportation, Device for feeding	W. Reid	Granton, Scotland	Mar. 16, 1869	87, 970
Cattle-fastening	K. Gibbs	Berwick, Me	Nov. 26, 1861	33, 780
Cattle-fastening device	S. E. Southland	Jamestown, N. Y	May 20, 1862	35, 331
Cattle from jumping, Device for preventing	C. Bettinger	South Danville, N. Y	Sept. 21, 1869	95, 072
Cattle-gag	W. Kegg	Lasellsville, N. Y	Oct. 23, 1866	59, 035
Cattle-gate, Railway	J. Bowman and J. B. Overmyer	Reading, Ohio	Oct. 26, 1869	96, 193
Cattle-gate, Railway	M. Hall, jr	Osborn, Ohio	July 12, 1859	24, 778
Cattle-guard gate	A. Freeman	Lowell, Mich	Nov. 5, 1867	70, 429
Cattle-guard gate, Railway	J. H. Mallory	La Porte, Ind	Mar. 7, 1871	112, 473
Cattle-guard, Railway	C. Brenneman	Orrville, Ohio	Nov. 26, 1867	71, 272
Cattle-guard, Railway	C. Caton	Boyd's Mills, Ohio	Aug. 1, 1871	117, 600
Cattle-guard, Railway	J. Forrest	Manchester, N. Y	Aug. 13, 1861	33, 035
Cattle-guard, Railway	P. Mougey	Marshallsville, Ohio	May 5, 1868	77, 642
Cattle-guard, Railway	J. L. Rowley	Angola, Ind	Mar. 29, 1859	23, 398
Cattle-guard, Railway	S. Strack	Dover Township, Pa	Oct. 24, 1871	120, 343
Cattle-guard, Railway	H. Tagart	Jacksontown, Ohio	May 5, 1868	77, 677
Cattle-guard, Railway	R. M. Yorks	Schoolcraft, Mich	Apr. 29, 1873	138, 308
Cattle, Herding and securing	J. Wilkinson	Urbana, Ill	June 2, 1868	78, 631
Cattle-hitch	C. H. Sawyer	Hollis, Me	Aug. 7, 1866	57, 049
Cattle-horn tip	J. C. Thompson	Charlestown, Mass	Feb. 9, 1869	86, 882
Cattle, Instrument for relieving choke and bloat in	N. Q. Munger and G. B. Pomeroy.	Casco and South Haven, Mich.	June 29, 1869	92, 085
Cattle-leading clasp	J. Welton	Waterbury, Conn	Oct. 9, 1855	13, 672
Cattle, Machine for watering	M. Smith	....	Mar. 16, 1804	....
Cattle-mill	La Paype, sr	Baltimore, Md	Nov. 4, 1805	....
Cattle, Mode of watering	A. Allen	Troy, Pa	July 10, 1824	....
Cattle-pricker	R. A. Carson and W. T. Peter	Briensburgh, Ky	Feb. 25, 1868	74, 795
Cattle, Remedy for murrain in	H. Jacobs	Fayetteville, Tenn	Aug. 17, 1869	93, 717
Cattle, Salt-trough for	T. Gladding	East Berlin, Conn	Jan. 19, 1869	86, 066
Cattle, Shears for marking	S. D. Baldwin	Chicago, Ill	May 23, 1865	47, 785
Cattle, Skinning	H. S. Lewis	Communipaw, N. J	Apr. 16, 1867	63, 910
Cattle-stanchion	E. S. Alvord	Harmony, N. Y	May 10, 1870	102, 904
Cattle-stanchion	N. W. Boody	Westbrook, Me	Jan. 4, 1870	98, 467
Cattle-stanchion	J. B. Crowell	Newport, N. H	Oct. 30, 1866	59, 180
Cattle-stanchion	W. C. Gifford	Jamestown, N. Y	May 10, 1870	102, 804
Cattle-stanchion	G. W. Hatch	Parkman, Ohio	July 30, 1850	7, 524
Cattle-stanchion	G. A. Keene	Newburyport, Mass	Mar. 15, 1864	41, 926
Cattle-stanchion	J. Manley	Hope, Me	Apr. 23, 1867	64, 017
Cattle-stanchion	C. H. Mann	Fairlee, Vt	Sept. 3, 1867	68, 372
Cattle-stanchion	H. Maycock	Verona, N. Y	Jan. 3, 1865	45, 728
Cattle-stanchion	J. A. Rosback	Hermon, N. Y	Jan. 24, 1871	111, 255
Cattle-stanchion	L. S. Safford	Hope, Me	Oct. 13, 1868	82, 995
Cattle-stanchion	C. W. Sawdey	Poolville, N. Y	Nov. 4, 1873	144, 360
Cattle-stanchion	J. D. Scott	Alviso, Cal	Aug. 30, 1870	106, 876
Cattle-tie	W. Allport	New Britain, Conn	Jan. 26, 1869	86, 119
Cattle-tie	E. P. Banks	Portland, Me	May 14, 1867	64, 621
Cattle-tie	G. Hull	Port Crane, N. Y	Aug. 28, 1860	29, 790
Cattle-tie	S. T. Hutchins	North Anson, Me	Mar. 21, 1871	112, 922

Index of patents issued from the United States Patent Office from 1790 *to* 1873, *inclusive*—Continued.

Invention.	Inventor.	Residence.	Date.	No.
Cattle-tie	J. Ives	Mount Carmel, Conn	Aug. 29, 1865	49, 630
Cattle-tie	A. F. Migeon	Wolcottville, Conn	Mar. 30, 1869	88, 403
Cattle-tie	C. S. Rundlett	Portland, Me	Mar. 12, 1867	62, 780
Cattle-tie	H. C. Small	Portland, Me	June 19, 1866	55, 727
Cattle-tie	J. Wiard	New Britain, Conn	June 9, 1868	78, 849
Cattle-tie for stalls	C. M. Baker	Bingham, Me	Jan. 15, 1867	61, 138
Cattle-tie, Safety	C. P. Winslow	Westborough, Me	Mar. 2, 1869	87, 454
Cavil	J. A. Wood	Pittsburgh, Pa	Jan. 9, 1872	122, 692
Caviar, Manufacture of	R. G. Westacott	Worcester, Mass	Jan. 7, 1851	7, 895
Ceiling and floor, Fire-proof	J. W. Bassett	New York, N. Y	Apr. 22, 1873	138, 118
Ceiling and roof, Fire-proof	J. Gilbert	Philadelphia, Pa	May 14, 1867	64, 659
Ceiling and wall for building	C. N. Poole	Sandwich, Ill	Apr. 18, 1871	113, 793
Ceiling and wall ventilator	O. S. Trexler	Naperville, Ill	Sept. 10, 1867	68, 807
Ceiling, Fire-proof	F. Baumann and G. F. Letz	Chicago, Ill	June 7, 1870	103, 963
Ceiling, Fire-proof	J. B. Cornell	New York, N. Y	Mar. 23, 1858	19, 682
Ceiling, Fire-proof	B. J. Harris	Richmond, Va	May 17, 1814	....
Ceiling, Fire-proof	S. P. Snead	Louisville, Ky	June 15, 1869	91, 375
Ceiling-ornament	E. T. Bussell	Indianapolis, Ind	Dec. 12, 1871	121, 845
Celery-collar	J. Simpson	York, England	July 23, 1872	129, 758
Cell-lock	E. Kershaw	Boston, Mass	Jan. 29, 1856	14, 178
Cellar, Artificial	V. Haeffner	Dobb's Ferry, N. Y	Dec. 9, 1862	37, 095
Cellar, vault, &c	H. Hallark	New York, N. Y	Oct. 24, 1826	....
Cellar, Water-proof	T. New	Brooklyn, N. Y	Apr. 4, 1871	113, 328
Cellars, Apparatus for draining	A. F. Erich	Baltimore, Md	Apr. 3, 1866	53, 594
Cellars, Mode of constructing walls and floors of	A. R. Moen	New York, N. Y	Feb. 26, 1856	14, 323
Cellars, vaults, magazines, &c., Construction of	E. Ludlow	New York, N. Y	Dec. 14, 1824	....
Cellars, water-closets, &c., Apparatus for emptying	R. Boeklen	Brooklyn, N. Y	Aug. 26, 1873	142, 200
Cement	D. Arndt	Cleveland, Ohio	Aug. 30, 1870	106, 911
Cement	H. Billings	Beardstown, Ill	May 1, 1860	28, 055
Cement	C. Fricke	Mobile, Ala	May 1, 1860	28, 070
Cement	C. Fricke	Mobile, Ala	May 29, 1860	28, 464
Cement	A. S. Jourdan	Nashville, Tenn	Nov. 20, 1866	59, 771
Cement	P. A. Letourneur	New Orleans, La	Sept. 11, 1860	29, 979
Cement	E. V. Machette, jr., and E. M. Crary.	Philadelphia, Pa	Dec. 1, 1868	84, 638
Cement	W. McRay	Ottawa, Canada	Nov. 12, 1872	132, 973
Cement	N. Parmeter	Gardner, Mass	May 3, 1859	23, 856
Cement	C. Saffray	New York, N. Y	May 5, 1868	77, 537
Cement	J Stansfield	Brooklyn, N. Y	Sept. 26, 1865	50, 185
Cement	D. B. Tooly	Albion, N. Y	Sept. 10, 1867	68, 671
Cement	G. W. Upham	Amherst, N. H	Sept. 22, 1868	82, 454
Cement	B. S. Welch	Brooklyn, N. Y	May 18, 1852	8, 962
Cement	L. Woolworth	Albion, Wis	Apr. 2, 1867	63, 449
Cement and appliances for preparing and using the same.	J. McKenzie and J. M. Stebbins.	Philadelphia, Pa	Nov. 21, 1871	121, 118
Cement and artificial stone, Manufacture of	G. S. Dean	San Francisco, Cal	Jan. 23, 1872	122, 880
Cement and as a plastic material for molding various articles, Composition to be used as a.	S. Sorel	Paris, France	Mar. 6, 1866	53, 092
Cement and metallic pipe, Combination	H. Knight	Jersey City, N. J	Aug. 27, 1861	33, 152
Cement and metallic pipe, Combination	H. Knight	Brooklyn, N. Y	Apr. 7, 1863	38, 112
Cement, Apparatus for spreading	J. H. Pulte	Cincinnati, Ohio	Sept. 4, 1866	57, 764
Cement, Artificial stone or marble	O. Parker	New York, N. Y	Sept. 9, 1835	....
Cement, Asphaltic	C. G. Reinhold	Milton, Pa	Mar. 21, 1865	46, 975
Cement, Asphaltic	A. Straub	Milton, Pa	Nov. 17, 1863	40, 649
Cement, Bituminous compound for	C. Poullalier	New York, N. Y	Mar. 3, 1838	620
Cement, Chemical water-proof	C. Fletcher	Boston, Mass	Nov. 23, 1831	....
Cement-compound	J. Fairchild	Eagleville, Ohio	Sept. 10, 1867	68, 720
Cement-compound	J. R. Remington	Macon County, Ala	July 4, 1854	11, 237
Cement, Compound hydraulic	A. Pfund	New York, N. Y	Feb. 4, 1873	135, 582
Cement-felt for covering steam boilers, pipes, &c	W. T. Kosinski	Brooklyn, N. Y	Jan. 29, 1867	61, 648
Cement, Fire and water	J. Coburn	Middlesex County, Mass	Sept. 3, 1828	....
Cement, Fire and water proof	S. R. Scharf	Baltimore, Md	Apr. 6, 1869	88, 745
Cement, Fire and water proof	W. Yates and D. Dolan	Manchester, England	Mar. 14, 1846	4, 420
Cement for architectural purposes, Artificial-stone	O. Parker	New York, N. Y	Sept. 9, 1835	....
Cement for artificial stone	J. A. Henshaw	Cambridge, Mass	Apr. 9, 1872	125, 390
Cement for blocks, pillars, &c	C. Clinton	Minisink, N. Y	Oct. 28, 1835	....
Cement for calking ships, &c	E. Heylin	Rochester, N. Y	June 22, 1869	91, 537
Cement for cisterns	Carson and Roberts	York, N. Y	Oct. 17, 1835	....
Cement for coating and protecting wood	H. M. Westman	East Boston, Mass	Oct. 25, 1870	108, 661
Cement for coating wood, &c	A. Pelletier	Washington, D. C	Apr. 14, 1868	76, 806
Cement for covering buildings, &c	J. Bump	Kirtland, Ohio	Sept. 3, 1840	1, 765
Cement for fastening door-knobs, &c	N. B. Hall and H. Jones	Branford, Conn	June 2, 1868	78, 592
Cement for fixing door-knobs, &c	P. Kennedy	New York, N. Y	June 11, 1867	65, 680
Cement for forming artificial stone	J. D. Greenwood and R. W. Keen.	England	Oct. 9, 1839	1, 361
Cement for grinding cylinders	J. Stephan	Boston, Mass	Oct. 28, 1851	8, 474
Cement for leather	A. Leach	Lynn, Mass	Nov. 20, 1866	59, 774
Cement for leather and other substances	S. F. Hilton	Providence, R. I	Aug. 13, 1861	33, 065
Cement for leather, &c., Water-proof	R. Hinshelwood and C. A. A. During.	New York, N. Y	May 12, 1863	38, 480
Cement for lining oil-barrels	S. H. Jones	Sandy Spring, Md	Aug. 20, 1867	67, 880
Cement for making water-tight joints in coping, roofing, &c.	J. Judge	New York, N. Y	Jan. 19, 1869	86, 018
Cement for manufacture of cement-pipes, Preparation of.	J. W. Stockwell	Portland, Me	July 19, 1870	105, 511
Cement for mending china, glass, &c	R. W. Patten	New York, N. Y	Mar. 12, 1872	124, 617
Cement for pavements, &c., Composition	A. M. Shaw	Lebanon, N. H	Mar. 10, 1868	75, 473
Cement for pavements, drain-pipes, &c., Asphalt	E. J. De Smedt	New York, N. Y	Feb. 7, 1871	111, 520
Cement for pavements, walks, roofs, &c	G. H. Smith	New York, N. Y	May 17, 1870	103, 095
Cement for paving and building	J. E. Hover	Philadelphia, Pa	Oct. 4, 1870	107, 910
Cement for preparing masts, spars, &c	A. Jeffery	Great Britain	Apr. 6, 1843	3, 027
Cement for roofs or walls of buildings	C. Clinton	New York, N. Y	July 13, 1827	....
Cement for sealing preserve-cans	J. B. Wilson	Fisherville, N. J	Mar. 1, 1864	41, 823
Cement for setting slates, making gutters, &c., Composition for.	J. Fullagar and M. Byrne	New York, N. Y	June 15, 1869	91, 223
Cement for steam-joints	J. G. Kilgour	Brooklyn, N. Y	Nov. 2, 1865	51, 060
Cement for stone	H. Schneider	Cleveland, Ohio	Jan. 2, 1866	51, 867

Index of patents issued from the United States Patent Office from 1790 *to* 1873, *inclusive*—Continued.

Invention.	Inventor.	Residence.	Date.	No.
Cement for walks, floors, pavements, &c	J. S. Baldwin, W. H. Jones, and E. N. Gibbs.	Elmira, Rochester, and Elmira, N. Y.	Feb. 12, 1867	61,984
Cement for walls, &c	A. Scott	New Berne, N. C	Aug. 15, 1814	....
Cement for wood, brick, stone, or iron work	R. Walsh	Boston, Mass	Jan. 5, 1832	....
Cement from basanite, Hydraulic	E. C. Warner	Albany, N. Y	Oct. 6, 1837	371
Cement from slag, Preparing	J. J. Bodmer	Newport, Great Britain	Nov. 5, 1867	70,510
Cement, Gum-elastic	C. Goodyear	New Haven, Conn	Sept. 9, 1835	....
Cement, Hydraulic	L. Kidder	New York, N. Y	Jan. 15, 1836	....
Cement, Hydraulic	O. Parker	New York, N. Y	Sept. 9, 1835	....
Cement, Hydraulic	Parker, Clowes, and Garfield	New York, N. Y	Aug. 27, 1835	....
Cement, Hydraulic	J. White	Syracuse, N. Y	Jan. 23, 1836	....
Cement impervious to moisture, Hardening and rendering.	S. Goodwin	New York, N. Y	Apr. 16, 1841	2,048
Cement in making cistern-reservoir, &c., Mode of applying.	T. Coyle	Baltimore, Md	Aug. 16, 1837	358
Cement-kiln	J. E. Park	Rutherford County, Tenn	Dec. 3, 1872	133,664
Cement-lined pipe	M. Stephens	Brooklyn, N. Y	Apr. 15, 1873	137,970
Cement-lined pipe	M. Stephens	Brooklyn, N. Y	Sept. 2, 1873	142,528
Cement, Making	W. H. Smith	Georgetown, D. C	June 10, 1844	3,621
Cement, Manufacture of	J. E. Park	La Vergne, Tenn	May 13, 1873	138,924
Cement, Manufacture of	D. O. Saylor	Allentown, Pa	Sept. 26, 1871	119,413
Cement, Manufacture of hydraulic	D. O. Saylor	Allentown, Pa	May 21, 1872	126,989
Cement, Marble	G. G. Gariboldi	Buffalo, N. Y	Aug. 20, 1867	67,868
Cement, Mastic	W. H. Chase	United States Engineers	Apr. 16, 1845	3,999
Cement, mastic, and japan from grahamite, Manufacture of.	H. Wurtz	New York, N. Y	Aug. 13, 1867	67,696
Cement, Mode of incorporating bituminous liquids with wet earths for.	W. H. Johnson	Springfield, Ill	Dec. 9, 1856	16,208
Cement of boiled coal-tar and earth	H. P. Gengembre	Allegheny City, Pa	July 11, 1854	11,269
Cement or earthen tubes, Mold for	B. S. and C. M. Pierce	New Bedford, Mass	Aug. 1, 1854	11,440
Cement or hydraulic lime, Water-proof	R. Leckie	Washington, D. C	Mar. 31, 1829	....
Cement or mortar, Composition	W. Bleser	New York, N. Y	Nov. 22, 1859	26,158
Cement or mortar, Water-proof	S. Gilford	Washington, D. C	Jan. 16, 1827	....
Cement or pigments, Water-proof	E. Deutsch	Havre, France	May 25, 1844	3,598
Cement, paint, &c., Flexible water and air proof	J. T. Howe	New York, N. Y	Jan. 31, 1829	....
Cement pipe	H. Knight	Brooklyn, N. Y	Nov. 27, 1866	60,014
Cement pipe, Branch	E. Lockhart, F. Roberts, and H. Knight.	Louisville, Ky., and Brooklyn, N. Y.	July 28, 1868	80,357
Cement-pipe joint	M. Stephens	Brooklyn, N. Y	Aug. 24, 1869	94,146
Cement-pipe machine	E. T. Jewett	Saint Albans Bay, Vt	June 28, 1864	43,310
Cement-pipe mold	D. Copeland, jr	Rochester, N. Y	May 23, 1871	115,029
Cement-pipe mold	H. Knight	Jersey City, N. J	Apr. 8, 1862	34,890
Cement-pipe mold	H. Knight	Jersey City, N. J	May 13, 1862	35,243
Cement-pipe mold	F. I. Sage	Springfield, Mass	July 29, 1873	141,388
Cement pipes, Apparatus for coating	E. Dayton	Meriden, Conn	Nov. 2, 1869	96,313
Cement pipes, Branch coupling for	M. Stephens	Brooklyn, N. Y	Apr. 2, 1872	125,345
Cement pipes, Composition for	J. A. Middleton	New York, N. Y	Dec. 10, 1872	133,874
Cement pipes, Device for lining	M. Stephens	Brooklyn, N. Y	Apr. 23, 1872	126,100
Cement pipes, Forming hydraulic	J. B. and W. F. Poague	Fancy Hill, Va	Nov. 29, 1853	10,286
Cement pipes, Machine for making	J. E. Earle	New Haven, Conn	Sept. 5, 1865	49,827
Cement pipes, Machine for making	H. Holden	New Haven, Conn	Sept. 20, 1864	44,369
Cement pipes, Machine for making	J. Howarth	Salem, Mass	July 14, 1863	39,233
Cement pipes, Machine for molding	H. Knight	Jersey City, N. J	June 24, 1862	35,692
Cement pipes, Machinery for making	W. Goodwin	New Haven, Conn	Sept. 5, 1865	49,828
Cement pipes, Manufacture of	J. A. Middleton	New York, N. Y	Dec. 10, 1872	133,875
Cement pipes, Manufacture of	A. P. Stephens	Brooklyn, N. Y	June 4, 1872	127,438
Cement pipes, Manufacture of hydraulic	H. Knight	Jersey City, N. J	May 14, 1861	32,298
Cement pipes, Manufacture of water-proof	A. F. Jaloureau	Paris, France	May 24, 1859	24,125
Cement pipes, &c., Means for making	D. S. Ogden	New York, N. Y	May 26, 1863	38,695
Cement pipes, Mold for molding	H. Knight	Jersey City, N. J	May 8, 1860	28,184
Cement, Plaster called the threefold	E. Godfrey	New York, N. Y	Sept. 9, 1825	....
Cement, Plastic	G. E. Hopkins	Harwich, Mass	Sept. 7, 1869	94,495
Cement, Rubber	H. J. Ball	Oswego, N. Y	Nov. 29, 1870	109,570
Cement, Silicated	G. E. Vanderburgh	Mamaroneck, N. Y	May 29, 1860	28,541
Cement tiles, pipes, pavements, &c., Composition for forming.	I. Marsh, jr., and G. Marsh	Milton and Lewisburgh, Pa.	June 9, 1863	38,833
Cement to be used in sewers and drains and for constructing flues and other parts of buildings.	J. Kleckner	Mottville, Mich	Mar. 8, 1870	100,706
Cement to be used under water	J. Grant	Saratoga, N. Y	Nov. 11, 1817	....
Cement, Treating argillaceous limestone to obtain hydraulic.	F. Coignet	Paris, France	Feb. 8, 1870	99,637
Cement walk, pavement, and road	C. Burgess	Rochester, N. Y	Sept. 27, 1870	107,756
Cement, Water	S. Guilford	Lebanon, Pa	Jan. 11, 1826	....
Cement, Water	T. F. Purcell and A. B. McFarlin.	Williamsport, Pa	Dec. 30, 1833	....
Cement, Water	D. M. Randolph	Richmond, Va	Sept. 26, 1821	....
Cement, Water-proof	A. Brower	New York, N. Y	July 27, 1858	20,985
Cement, Water-proof	G. W. Caton	Canandaigua, N. Y	Feb. 26, 1867	62,393
Cement, Water-proof	T. S. Clark	Charlestown, Mass	July 26, 1870	105,648
Cement, Water-proof	W. Y. Singleton	Springfield, Ill	Dec. 26, 1845	4,333
Cementation of rails, axles, &c., Method of effecting the.	T. W. Dodds	Rotherham, England	July 2, 1867	66,310
Center and lathe dog, Combined	A. J. Truxell	Lynchburgh, Va	Nov. 8, 1870	109,079
Center-bit	W. H. McMillar and S. Devoe	New York, N. Y	Aug. 13, 1867	67,785
Center-bit	W. H. Richards	Deerfield Corners, N. Y	Apr. 16, 1872	125,759
Center-bit, Fastening	A. W. Streeter	Shelburne Falls, Mass	Jan. 23, 1855	12,289
Center-board	A. G. Crossman	Huntington, N. Y	May 23, 1871	115,173
Center-board	F. Dominy	Peconic, N. Y	Nov. 19, 1867	70,977
Center-board	G. M. Fay	Eureka, Cal	Aug. 22, 1865	49,514
Center-board	B. Kennon	New Orleans, La	May 11, 1869	90,046
Center-board	J. G. Saunders	Narragansett, R. I	Oct. 1, 1867	69,492
Center board and box for vessels	D. P. Nickerson	Cleveland, Ohio	Jan. 22, 1867	61,351
Center-board, Folding	J. M. Hoffman	Buffalo, N. Y	Apr. 10, 1849	6,299
Center-board for sailing-vessels	W. F. Davis	Boston, Mass	Mar. 5, 1867	62,614
Center-board for vessels	J. Call	Richmond, Me	Sept. 9, 1873	142,677
Center-board for vessels	R. Chambers	Detroit, Mich	Jan. 8, 1867	61,049
Center-board for vessels	A. G. Crossman	Huntington, N. Y	Apr. 25, 1871	113,984
Center-board for vessels	J. Dean	Detroit, Mich	Oct. 22, 1872	132,390

Index of patents issued from the United States Patent Office from 1790 *to* 1873, *inclusive*—Continued.

Invention.	Inventor.	Residence.	Date.	No.
Center-board for vessels	J. Dillon	New York, N. Y	July 27, 1869	93, 070
Center-board for vessels	J. Espalla	Mobile, Ala	Nov. 25, 1873	144, 968
Center-board for vessels	D. G Gerard	Patchogue, N. Y	July 3, 1860	28, 979
Center-board for vessels	F. J. Macfarlan	San Francisco, Cal	Dec. 10, 1867	72, 058
Center-board for vessels	J. J. Moule	Huntington, N. Y	June 20, 1871	116, 079
Center-board for vessels	J. F. Potts	Apalachicola, Fla	Nov. 16, 1858	22, 088
Center-board for vessels	N. Pratt	Nicholson, Pa	Mar. 1, 1859	23, 114
Center-board for vessels	D. Robertson	Detroit, Mich	Feb. 4, 1868	74, 003
Center-board, Jointed	T. Maskell	Franklin, La	Oct. 9, 1849	6, 774
Center-board of navigable vessels	B. Joline	Westfield, N. Y	Apr. 20, 1858	19, 996
Center-board vessels, Construction of	C. E. Ketchum and W. L. Hunt	Port Jefferson, N. Y	July 10, 1860	29, 085
Center-board winch	H. V. Corbett	Buffalo, N. Y	Nov. 26, 1867	71, 283
Center-board winch	E. C. Hammond	Oswego, N. Y	Nov. 24, 1868	84, 275
Center-boards and rudders for shoal-water vessels, Arrangement of.	G. Chase	Prudence Island, R. I	Feb. 8, 1853	9, 574
Center-boards, Device for operating	J. N. Buell	Middletown, Conn	Nov. 15, 1864	45, 018
Center-boards for vessels, Attachment for	G. Storer and G. W. Storer	New Britain and Portland, Conn.	Jan. 8, 1867	61, 029
Center-boards for vessels, Method of hanging	W. W. Bates	Chicago, Ill	May 14, 1867	64, 739
Center-boards of vessels, Hanging	J. F. Hall	Westerly, R. I	Feb. 20, 1866	52, 708
Center-boards to vessels, Attaching	G. S. Burrows	Mystic River, Conn	Nov. 18, 1856	16, 084
Center-table	E. Lampman	Catskill, N. Y	Dec. 18, 1866	60, 525
Center-turning tool	H. D. Richardson	East Hampton, Mass	May 4, 1869	89, 687
Centering and squaring device	C. O. Gardiner	Springfield, Ohio	Dec. 4, 1866	60, 169
Centering bar of iron	N. F. Newell	Northbridge, Mass	Jan. 29, 1861	31, 257
Centering box or bearing, Self	M. S. Bullock	Pottsville, Pa	Apr. 18, 1871	113, 739
Centering-device	E. E. Lazell	Philadelphia, Pa	Sept. 15, 1868	82, 131
Centering-device	N. L. Revere	Worcester, Mass	Mar. 2, 1869	87, 431
Centering-implement	F. Deluce	Boston, Mass	Feb. 18, 1862	34, 412
Centering-machine	E. McNiel	Groton, N. Y	May 30, 1871	115, 337
Centering-machine	E. F. Whiton	West Stafford, Conn	July 14, 1857	17, 814
Centering-tool	W. I. Alvord	Bridgeport, Conn	Nov. 9, 1869	96, 653
Centering-tool	R. Hayworth	South New Market, N. H	Sept. 3, 1867	68, 504
Centrifugal battery	A. Potts	Philadelphia, Pa	May 19, 1857	17, 339
Centrifugal crushing-mill	C. H. Griffin	Chelsea, Mass	Nov. 22, 1864	45, 213
Centrifugal disk revolving in air and water	T. W. Rammell	London, England	May 10, 1864	42, 716
Centrifugal drawing-machine	H. N. Glass and H. W. Bartol	Philadelphia, Pa	May 16, 1865	47, 714
Centrifugal machine	R. J. Barr	Philadelphia, Pa	Aug. 4, 1868	80, 702
Centrifugal machine	E. J. M. Becker	New York, N. Y	May 20, 1873	139, 108
Centrifugal machine	J. F. Brinjes	White Chapel, England	Nov. 2, 1869	96, 304
Centrifugal machine	J. D. Browne	Cincinnati, Ohio	June 12, 1866	55, 461
Centrifugal machine	C. T. Burchardt	New York, N. Y	Mar. 4, 1873	136, 485
Centrifugal machine	A. Chenu	Paris, France	Sept. 3, 1872	131, 081
Centrifugal machine	J. Cottle	Boston, Mass	July 16, 1872	129, 322
Centrifugal machine	G. B. Hartson and E. J. Woolsey	New York, N. Y	Feb. 13, 1866	52, 569
Centrifugal machine	S. S. Hepworth	Boston, Mass	Sept. 22, 1868	82, 314
Centrifugal machine	S. S. Hepworth	Cold Spring, N. Y	June 14, 1870	104, 149
Centrifugal machine	H. Hunt	Oskaloosa, Iowa	Mar. 27, 1866	53, 448
Centrifugal machine	J. O. Joyce	Dayton, Ohio	Aug. 28, 1866	57, 517
Centrifugal machine	A. Mackey and E. Müller	New York and Williamsburgh, N. Y.	Mar. 27, 1866	53, 463
Centrifugal machine	J. Sparrow	Portland, Me	May 6, 1873	138, 705
Centrifugal machine for filtering liquids	H. N. Fryatt	Belleville, N. J	June 3, 1862	35, 441
Centrifugal machine or hydro-extractor	E. C. Cleveland	Worcester, Mass	Aug. 20, 1867	67, 953
Centrifugal machine, Self-balancing	D. M. Weston	Boston, Mass	Sept. 8, 1868	82, 049
Centrifugal machines, Apparatus for liquoring sugar in.	F. Seiberlich	Charlestown, Mass	Aug. 8, 1865	49, 310
Centrifugal machines, Apparatus for supplying liquor to.	W. R. Meins	Boston, Mass	Oct. 16, 1866	58, 864
Centrifugal machines, Discharger for	H. Merrill	Brooklyn, N. Y	Apr. 23, 1867	64, 182
Centrifugal machines, Feeder for	H. Merrill	Brooklyn, N. Y	Apr. 23, 1867	64, 183
Centrifugal machines, Mode of liquoring sugar in	R. T. Sprague	Boston, Mass	Aug. 20, 1867	67, 924
Centrifugal power-regulator, Pneumatic	C. A. Sullivan	Starkville, Miss	Dec. 28, 1869	98, 314
Centrifugal sugar-machine	H. W. Bartol	Philadelphia, Pa	Oct. 26, 1869	96, 186
Centrifugal sugar-machine	J. R. Brown	Winchester, Ind	Nov. 27, 1866	59, 956
Centrifugal sugar-machine	J. Corby	Gravesend, England	Mar. 12, 1872	124, 484
Centrifugal sugar-machine	J. Cottle	Boston, Mass	Apr. 16, 1872	125, 724
Centrifugal sugar-machine	P. Cramer	Jersey City, N. J	Nov. 4, 1873	144, 319
Centrifugal sugar-machine	W. Elmenhorst and F. O. Mathiessen.	Jersey City, N. J	Apr. 21, 1868	77, 017
Centrifugal sugar-machine	G. E. Evans	Boston, Mass	Sept. 3, 1867	68, 356
Centrifugal sugar-machine	G. E. Evans	Boston, Mass	Oct. 1, 1867	69, 328
Centrifugal sugar-machine	A. Fesca	Berlin, Prussia	Mar. 26, 1872	125, 036
Centrifugal sugar-machine	C. Fischer	Gloucester, N. J	Nov. 18, 1873	144, 752
Centrifugal sugar-machine	B. Hasket and W. B. Cox	West Milton, Ohio	Nov. 13, 1866	59, 598
Centrifugal sugar-machine	J. B. Hill	Allegheny City, Pa	Dec. 10, 1867	71, 877
Centrifugal sugar-machine	D. King	Brooklyn, N. Y	Nov. 25, 1851	8, 545
Centrifugal sugar-machine	H. W. and R. Lafferty	Gloucester City, N. J	Mar. 23, 1869	88, 184
Centrifugal sugar-machine	H. W. and R. Lafferty	Gloucester City, N. J	Mar. 23, 1869	88, 185
Centrifugal sugar-machine	H. W. and R. Lafferty	Gloucester City, N. J	Aug. 16, 1870	106, 491
Centrifugal sugar-machine	A. Mackey	New York, N. Y	June 18, 1867	65, 923
Centrifugal sugar-machine	A. Mackey	New York, N. Y	Mar. 17, 1868	75, 687
Centrifugal sugar-machine	H. Priew	Magdeburg, Prussia	Feb. 11, 1873	135, 727
Centrifugal sugar-machine	P. Tully	San Francisco, Cal	July 19, 1870	105, 520
Centrifugal sugar-machine	W. Van Anden	Poughkeepsie, N. Y	June 10, 1851	8, 156
Centrifugal sugar-machine	L. Weinrich	Berlin, Prussia	Apr. 2, 1872	125, 358
Centrifugal sugar-machine	C. O. West and J. Carey	Martinsville, Ohio	May 1, 1866	54, 456
Centrifugal sugar-machine	D. M. Weston	Boston, Mass	Apr. 9, 1867	63, 770
Centrifugal sugar-machine	H. A. Wilder	Millville, N. Y	Nov. 9, 1869	96, 752
Centrifugal sugar-machines, Device for feeding	A. Kusenberg	Philadelphia, Pa	Jan. 5, 1869	85, 528
Centrifugal ventilator	A. P. Blake	Milton, Mass	Jan. 31, 1865	46, 067
Centripetal press	J. E. Serrell and D. Smith	New York, N. Y	July 3, 1849	6, 563
Centrolinead	W. J. von Kammerhueber	Washington, D. C	July 15, 1856	15, 359
Cerasine or wax, Treating bituminous materials for the manufacture of.	H. Ujhely and C. Beurle	Vienna, Austria	Sept. 3 1872	131, 137
Cereals, Apparatus for decorticating and cleaning	W. W. Gibson	Edinburgh, North Britain	July 14, 1868	79, 970
Cerotypography, Feed-motion for	J. McElhenan	Brooklyn, N Y	Aug. 17, 1858	21, 208
Cess-pool	G. T. Bohen	San Francisco, Cal	Dec. 27, 1864	45, 577

Index of patents issued from the United States Patent Office from 1790 *to* 1873, *inclusive*—Continued.

Invention.	Inventor.	Residence.	Date.	No.
Cess-pools, &c., Apparatus for cleaning	W. C. McCarthy	Pittsburgh, Pa	Apr. 22, 1873	138, 034
Cess-pools, sinks, &c., Apparatus for cleaning	J. P. F. Datichy	Brooklyn, N. Y	Sept. 16, 1873	142, 840
Cess-pools, vaults, privies, &c., Apparatus for cleaning.	W. C. McCarthy	Pittsburgh, Pa	June 28, 1870	104, 866
Chafe-iron	E. Hackett	Concord, N. H	July 21, 1868	80, 166
Chafe-iron for wheeled-vehicle	E. P. Roche	Bath, Me	June 4, 1872	127, 645
Chaffing mill, Grain	G. Wilcox	Neenah, Wis	June 16, 1868	78, 907
Chain	J. Blocher	Williamsville, N. Y	Dec. 4, 1860	30, 797
Chain	J. F. Brewer	Plantsville, Conn	Apr. 27, 1869	89, 282
Chain	P. D. Cummings	Portland, Me	June 5, 1860	28, 630
Chain	J. Good	Brooklyn, N. Y	Oct. 18, 1870	108, 473
Chain	J. G. Jung	Newark, N. J	Apr. 6, 1869	88, 643
Chain	W. Wallace	Ansonia, Conn	Feb. 5, 1867	61, 777
Chain, Adjustable runner to be attached to	A. Adamson	Washington, D. C	Mar. 5, 1867	62, 589
Chain, Adjusting-slide for	R. J. Pond	Morrisania, N. Y	Apr. 5, 1870	101, 503
Chain and rope on canal and railway, Using	J. M. Palisse and S. S. Durfee	Hudson, N. Y	Feb. 25, 1835	......
Chain-blocking device	P. Kendrick	Trenton, N. J	Sept. 15, 1868	82, 124
Chain-boats for inland navigation	G. F. Saltonstall	Society Hill, S. C	July 12, 1834	......
Chain, Box-slide for ornamental	O. J. Valentine	Newark, N. J	Mar. 18, 1873	136, 946
Chain, Braced	J. H. Edwards	Lanark, Ill	Feb. 27, 1872	124, 042
Chain-bucket wheel, Construction of the buckets in the.	S. Dutton, jr	Aston, Pa	Feb. 26, 1840	1, 499
Chain-cable, Machine for manufacturing	W. D. Grimshaw	Newark, N. J	Aug. 13, 1867	67, 751
Chain, Casting	C. C. Van Alstine	New Haven, Conn	Aug. 25, 1868	81, 563
Chain, Chain-pump	J. M. Connel	Newark, Ohio	June 21, 1864	43, 256
Chain, Cog	T. Shaw	Philadelphia, Pa	Nov. 12, 1867	70, 752
Chain-connecting link	S. Vanstone	Providence, R. I	Oct. 29, 1867	70, 380
Chain-coupling	M. Osborn	Cleveland, Ohio	Jan. 9, 1872	122, 644
Chain-coupling link	S. T. Lamb	New Albany, Ind	Nov. 21, 1871	121, 225
Chain, Elastic	C. E. Richards	North Attleborough, Mass	May 1, 1866	54, 411
Chain elevator and bucket	J. A. Ball	Grass Valley, Cal	Nov. 16, 1869	96, 866
Chain, Flat	C. King and S. P. Johnson	Cleveland, Ohio	Sept. 14, 1869	94, 757
Chain for hauling down ships, Safety	J. Fales and T. D. Brown	Bedford, Mass	Mar. 9, 1833	......
Chain for ornament	L. Towne	Providence, R. I	Nov. 1, 1864	44, 899
Chain for power-press	N. Chapman	Mystic River, Conn	Jan. 1, 1856	14, 009
Chain for water-elevator	P. Anderson	East Avon, N. Y	May 2, 1865	47, 595
Chain, Frame	D. Leslie	New York, N. Y	Nov. 19, 1827	......
Chain from sheet-metal, Making	L. Towne	Providence, R. I	Jan. 10, 1860	26, 801
Chain, &c., from slipping, Wheel to prevent	J. Cooper	Augusta County, Va	June 17, 1823	......
Chain-holder	S. Gladding	Providence, R. I	July 11, 1865	48, 675
Chain-hook	M. Colgan	Port Jervis, N. Y	Apr. 11, 1865	47, 241
Chain-hook	G. H. Draper	North Attleborough, Mass.	May 12, 1863	38, 532
Chain-hook	W. Straw	Dalton, N. H	July 11, 1854	11, 306
Chain-hook	J. R. Thorne	Waldoborough, Me	June 27, 1871	116, 511
Chain-hook	J. Weed	Lyons, Iowa	Oct. 15, 1872	132, 224
Chain, Hook for vest	A. Wallach	New York, N. Y	June 7, 1859	24, 362
Chain-hook or cable-stopper	E. Davidson	Providence, R. I	Feb. 6, 1866	52, 392
Chain hook, Ox	F. G. Johnson	Brooklyn, N. Y	May 12, 1863	38, 490
Chain, Horse	H. W. Knowlton	Saratoga Springs, N. Y	July 25, 1865	48, 957
Chain inclinometer	H. Schüssler	San Francisco, Cal	Jan. 14, 1868	73, 391
Chain lanyard	R. Rogers	Dover, N. H	Feb. 6, 1823	......
Chain-link	J. P. Kirk	Austin, Tex	Nov. 13, 1860	30, 632
Chain-link	J. Packer	New York, N. Y	July 14, 1863	39, 239
Chain-link	W. C. Short	Providence, R. I	July 5, 1870	105, 004
Chain-link	S. Vanstone	Providence, R. I	Apr. 7, 1868	76, 562
Chain-link	G. W. N. Yost	Yellow Springs, Ohio	Sept. 11, 1860	30, 019
Chain link and bar	D. A. Peloubet	Hudson City, N. J	Oct. 25, 1870	108, 621
Chain-link-bending machine	H. Boyd	East Bridgewater, Mass	Oct. 24, 1871	120, 148
Chain-link, Elastic	E. Myers	Creagerstown, Md	July 11, 1871	116, 986
Chain-link flattening and bending machine	P. Kendrick	Trenton, N. J	July 28, 1868	80, 354
Chain-link rods, Machine for preparing	G. Homfray	Halesowen Parish, England	Feb. 26, 1867	62, 335
Chain-link-trimming machine	F. Van Patten	Ilion, N. Y	Nov. 21, 1865	51, 102
Chain-links, Apparatus for bending	P. L. Weimer	Lebanon, Pa	Oct. 31, 1865	50, 758
Chain-links, Apparatus for welding	H. Reynolds	Aurora, N. Y	Aug. 31, 1869	94, 441
Chain-links, Apparatus for welding	E. Tangye	Brussels, Belgium	May 14, 1867	64, 722
Chain-links, Device for swaging	V. Draper	North Attleborough, Mass.	Sept. 26, 1865	50, 200
Chain-links, Machine for attaching ornaments to	V. Draper	Attleborough, Mass	July 5, 1870	105, 054
Chain-links, Machine for bending	E. L. Keeler	Allegheny, Pa	Aug. 7, 1866	57, 0 3
Chain-links, Machine for bending	C. B. Long	Worcester, Mass	Dec. 16, 1873	145, 511
Chain-links, Machine for bending	D. Roy	Mill Creek Township, Pa	Dec. 2, 1873	145, 072
Chain-links, Machine for cutting coiled bars for	G. Homfray	Halesowen Parish, England	Feb. 26, 1867	62, 488
Chain-links, Machine for making	A. M. George	Nashua, N. H	Feb. 27, 1855	12, 439
Chain-links, Machine for making	H. Reynolds	Aurora, N. Y	May 3, 1870	102, 711
Chain-links, Machine for making	J. H. Snyder	Troy, N. Y	Aug. 16, 1870	106, 419
Chain links, Machine for making ornamental	E. I. Levavasseur	Paris, France	Apr. 29, 1873	138, 413
Chain-links, Machine for stamping	P. S. Bishop	Attleborough, Mass	Apr. 24, 1866	54, 101
Chain-links, Machine for welding	W. B. Hayden	Columbus, Ohio	Sept. 20, 1870	107, 613
Chain links, Ornamental	J. J. Freeman	Attleborough, Mass	Mar. 11, 1873	136, 594
Chain-lock	L. F. Cahn	New York, N. Y	Nov. 7, 1871	120, 619
Chain-locker pipe	C. Perley	New York, N. Y	Nov. 6, 1855	13, 760
Chain, Loop for bearing	J. Bird	New York, N. Y	Aug. 6, 1867	67, 486
Chain-machine	G. W. Binnix	Sunbury, Pa	Apr. 1, 1873	137, 408
Chain-machine	A. J. Clemmons	Aberdeen, Mass	Dec. 12, 1871	121, 849
Chain-machine	J. Daykin and W. B. Case	Cleveland, Ohio	Oct. 28, 1873	143, 958
Chain-machine	L. Fitzpatrick and J. Schinneller.	Temperanceville, Pa	Jan. 21, 1868	73, 518
Chain-machine	B. Hershey	Erie, Pa	May 20, 1873	139, 151
Chain-machine	F. Leonard	Cleveland, Ohio	July 8, 1873	140, 583
Chain-machine	F. Leonard	Cleveland, Ohio	Dec. 2, 1873	145, 063
Chain-machine	L. Towne	Providence, R. I	Oct. 20, 1857	18, 490
Chain machinery, Arrangement of jack	H. Marshall and S. S. Cook	Stamford, Conn	Mar. 16, 1852	8, 815
Chain-making	C. Sleppy	Newport, Pa	Sept. 27, 1853	10, 050
Chain-making machine	J. Copley, jr	Allegheny City, Pa	Sept. 10, 1867	68, 709
Chain-making machine	W. Dennison	Cambridge, Mass	Nov. 23, 1869	97, 173
Chain-making machine	L. King	East Cleveland, Ohio	Aug. 9, 1864	43, 778
Chain-making machine	W. J. Lewis	Pittsburgh, Pa	Jan. 12, 1858	19, 094
Chain-making machine	E. H. Perry	Providence, R. I	Aug. 31, 1858	21, 362
Chain-making machine	L. Souther	Springfield, Ill	Aug. 12, 1873	141, 734

Index of patents issued from the United States Patent Office from 1790 *to* 1873, *inclusive*—Continued.

Invention.	Inventor.	Residence.	Date.	No.
Chain-making machine	E. Weissenborn	New York, N. Y	Dec. 11, 1855	13,929
Chain, Neck	G. D. Stevens	New York, N. Y	Aug. 22, 1871	118,294
Chain or band	J. H. Doerr	Philadelphia, Pa	Feb. 1, 1870	99,298
Chain or belt, Driving	W. Clissold	Dudbridge, England	Dec. 15, 1863	40,910
Chain, Ornamental	W. F. Davis	North Attleborough, Mass	Sept. 16, 1873	142,841
Chain, Ornamental, &c	E. S. Richards	Attleborough, Mass	June 9, 1863	38,842
Chain-plate attachment	T. W. Porter, J. D. Leach, and S. Hutchings.	Boston, Mass., and Penobscot, Me.	Apr. 27, 1869	89,339
Chain power, Endless	W. McCreery	Pittsburgh, Pa	Jan. 7, 1868	73,191
Chain, Runner for	W. H. Stroup	Pittsburgh, Pa	Mar. 26, 1867	63,329
Chain-shackle	E. Bangs	Provincetown, Mass	Feb. 8, 1870	99,623
Chain-shackle	J. Snelling	East Boston, Mass	Apr. 13, 1858	19,955
Chain, Sheet-metal	J. Lancelott	Cranston, R. I	May 4, 1858	20,183
Chain shot battery	C. B. Thayer	Boston, Mass	July 24, 1860	29,331
Chain, Spring	H. Turner	Boston, Mass	Mar. 24, 1868	75,814
Chain-stopper	W. D. Baker	East Abington, Mass	Aug. 17, 1869	93,793
Chain-stopper	W. H. Gray	Dover, N. H	Aug. 24, 1858	21,296
Chain-stretching machine	C. Hall	New York, N. Y	Aug. 30, 1864	43,987
Chain, Surveyor's	J. M. Grumman	Brooklyn, N. Y	Apr. 19, 1859	23,680
Chain toggle, Rafting	E. Evans	North Tonawanda, N. Y	Aug. 13, 1872	130,363
Chain-wheel	J. Hines and E. D. McCord	Sandy Hill, N. Y	Nov. 20, 1829	......
Chain-wheel and paddle for boat	R. R. Livingston	Clermont, N. Y	June 13, 1812	......
Chain-wheels, Right and left	R. Bulkley	New York, N. Y	Oct. 8, 1821	......
Chains, bracelets, &c., Method of forming the parts, links, &c., of.	P. M. C. Béziel	Paris, France	Feb 10, 1867	62,178
Chains, Construction of ornamental	I. Lindsley	Pawtucket, R. I	Mar. 19, 1872	124,837
Chains, Machine for making	J. C. Brown	Providence, R. I	Nov. 29, 1859	26,247
Chains, Machine for making	W. Malick	Erie, Pa	Aug. 3, 1869	93,213
Chains, Machine for making	E. H. Perry	Providence, R. I	July 20, 1858	20,955
Chains, Machine for making ball	D. T. Munger	Waterbury, Conn	Nov. 19, 1867	71,042
Chains, Machine for making jack	C. Atwood and G. Kellogg	Birmingham, Conn	Nov. 12, 1850	7,768
Chains, Machine for making links of cable and other	A Homfray	Witley Lodge, England	Nov. 12, 1867	70,845
Chains, Machine for making links of jack	A. Wyckoff	Wellsburgh, N. Y	Feb. 14, 1854	10,528
Chains, Machine for making ornamental	W. Bancroft and J. Wood	Birmingham, England	Nov. 12, 1872	132,947
Chains, Machine for making ornamental	J. A. Field	Providence, R. I	Nov. 26, 1872	133,427
Chains, Machine for twisting curb	L. Towne	Providence, R. I	Aug. 18, 1857	18,027
Chains, Machine used in the manufacture of	W. C. Edge	Newark, N. J	Apr. 23, 1872	125,940
Chains, Machinery for making	J. M. Crawford	Newcastle, Pa	Feb. 17, 1852	8,737
Chains, Making gold	I. Lindsley	Providence, R. I	Aug. 7, 1860	29,503
Chains, Making ornamental	J. Lancelott	Cranston, R. I	Mar. 22, 1859	23,303
Chains, Making ornamental	J. Lancelott	South Providence, R. I	Oct. 18, 1859	25,837
Chains, Manufacture of	W. B. Wadsworth	Cleveland, Ohio	Mar. 23, 1869	88,100
Chains, Manufacture of flat ornamental chains	J. J. Freeman	Attleborough, Mass	Jan. 16, 1872	122,824
Chains, Manufacture of ornamental	E. N. Foote	Northampton, Mass	June 25, 1872	128,381
Chains, Method of constructing	J. G. Jung	Newark, N. J	Nov. 24, 1868	84,359
Chains, Method of fastening blocks in triple	P. Kendrick	Trenton, N. J	Aug. 26, 1873	142,110
Chains, Method of making iron	E. Weissenborn	New York, N. Y	Aug. 6, 1861	33,014
Chains, Safety-attachment for watch	M. Pollak	New York, N. Y	Aug. 3, 1869	93,344
Chains, Tool for making jack	W. Todd	Stamford, Conn	Apr. 8, 1851	8,029
Chair: *See* Accoucheur's chair. Adjustable chair. Arm-chair. Baby's chair. Barber's chair. Boat-chair. Cabinet-chair. Cane-seat chair. Canopied chair. Card-board chair. Child's chair. Cottage-chair. Cradle-chair. Crib-chair. Dentist's chair. Dining-chair. Eagle-chair. Easy-chair. Electric chair. Enema-chair. Excursion-chair. Exercising-chair. Extension-chair. Fan-chair. Folding-chair. Garden-chair. Harness-maker's chair. High-chair. Invalid-chair. Knock-down chair. Lounging-chair. Metallic chair. Night-chair. Nursery-chair. Obstetric chair. Obstetrical chair. Opera-chair. Operating-chair. Oscillating chair. Photograph-chair. Photographer's posing-chair. Posing-chair. Rail-chair. Railway-chair. Railway-coupling chair Reception-chair.				

Index of patents issued from the United States Patent Office from 1790 *to* 1873, *inclusive*—Continued.

Invention.	Inventor.	Residence.	Date.	No.
Chair—Continued.				
See Reclining-chair.				
Recumbent chair.				
Revolving chair.				
Rocking-chair.				
School-chair.				
Sea sitting-chair.				
Sewing-machine chair.				
Sheep-shearing chair.				
Spittoon-chair.				
Split-bottom chair.				
Spring-chair.				
Spring-back chair.				
Step-chair.				
Step-ladder chair.				
Swinging chair.				
Switch-chair.				
Swiveled chair.				
Tilting chair.				
Tip-chair.				
Tipping chair.				
Toilet-chair.				
Wall-wainscot chair.				
Chair	D. L. Akers	Evansville, Ind	Dec. 3, 1872	133, 555
Chair, &c	L. H. Baker	Washington, D. C	June 25, 1872	128, 272
Chair	L. H. Baker	Washington, D. C	Jan. 14, 1873	134, 834
Chair	J. Baughman and B. R. Chalk	Mount Washington, Md	Mar. 18, 1873	136, 956
Chair	E. F. Benjamin	Utica, N. Y	Aug. 17, 1835	......
Chair	M. Brennan	Cambridgeport, Mass	July 2, 1872	128, 459
Chair	P. Buckley	Vienna, N. J	July 13, 1869	92, 576
Chair	J. Chase	Orange, Mass	Nov. 9, 1869	96, 548
Chair	F. J. Coates	Cincinnati, Ohio	Aug. 20, 1867	67, 956
Chair	D. C. Colby	Washington, D. C	Sept. 29, 1868	82, 602
Chair	H. Cole	Cincinnati, Ohio	Nov. 12, 1867	70, 698
Chair	M. C. Cronk	Auburn, N. Y	May 22, 1866	54, 869
Chair	J. Daley	Baltimore, Md	Feb. 9, 1827	......
Chair	J. Defoe	Detroit, Mich	Aug. 27, 1872	130, 903
Chair	G. O. Donnell	New Lebanon, N. Y	Mar. 2, 1852	8, 771
Chair	W. T. Doremus	New York, N. Y	June 17, 1873	139, 948
Chair	G. Feldkamp	Cincinnati, Ohio	Oct. 28, 1873	144, 080
Chair	J. Fernald	Boston, Mass	July 22, 1856	15, 405
Chair	L. W. Ferris	Owego, N. Y	Mar. 27, 1855	12, 587
Chair	G. Gardner	Glen Gardner, N. J	Apr. 4, 1871	113, 287
Chair	W. Gardner	Glen Gardner, N. J	June 3, 1873	139, 568
Chair	J. Habermehl	Wheeling, W. Va	June 20, 1865	48, 272
Chair	J. Habermehl	Wheeling, W. Va	July 24, 1866	56, 554
Chair	H. S. Hale	Philadelphia, Pa	Dec. 30, 1873	146, 065
Chair	J. T. Hammitt	Philadelphia, Pa	Dec. 7, 1852	9, 449
Chair	W. W. Haupt	Mountain City, Tex	Oct. 17, 1871	120, 064
Chair	W. W. Haupt	Mountain City, Tex	Nov. 5, 1872	132, 762
Chair	D. B. Hedden	Newark, N. J	Nov. 17, 1868	84, 118
Chair	L. Heywood	Gardner, Mass	Jan. 7, 1873	134, 545
Chair	G. H. Hoagland	Port Jervis, N. Y	Oct. 1, 1867	69, 435
Chair	J. A. and H. A. House	Bridgeport, Conn	June 13, 1865	48, 177
Chair	C. B. Howard	Atlanta, Ga	May 14, 1872	126, 703
Chair	G. Hunzinger	New York, N. Y	Mar. 30, 1869	88, 297
Chair	W. H. Jockel	New York, N. Y	Jan. 18, 1870	98, 973
Chair	M. W. King	New York, N. Y	Aug. 25, 1840	1, 737
Chair	A. Lapham	Paterson, N. J	Oct. 5, 1869	95, 491
Chair	J. Lee	New York, N. Y	July 27, 1869	92, 977
Chair	J. Lewis	New York, N. Y	Nov. 19, 1872	133, 233
Chair	L. Lindquest	Pittsburgh, Pa	July 13, 1869	92, 620
Chair	C. R. Long	Louisville, Ky	Jan. 11, 1870	98, 783
Chair	C. R. Long	Louisville, Ky	Dec. 13, 1870	110, 053
Chair	C. O. Lundberg	Chicago, Ill	Oct. 6, 1863	40, 175
Chair	R. S. Mains	New York, N. Y	July 9, 1872	128, 804
Chair	J. Morrison	Indianapolis, Ind	Aug. 8, 1871	117, 915
Chair	J. Morrison and H. M. Hutchinson.	Philadelphia, Pa	Nov. 4, 1873	144, 349
Chair	G. W. Morstatt	New York, N. Y	Mar. 5, 1872	124, 281
Chair	C. M. O'Hara	Hillsborough, Ohio	Apr. 30, 1872	126, 322
Chair	C. M. O'Hara	Hillsborough Ohio	Oct. 1, 1872	131, 892
Chair	V. P. Parkhurst	Templeton, Mass	May 15, 1866	54, 762
Chair	J. K. Rowley	Chicago, Ill	May 13, 1873	138, 939
Chair	J. F. Sargent	South Strafford, Vt	Apr. 24, 1866	54, 217
Chair	C. C. Schmitt	New York, N. Y	Oct. 9, 1866	58, 682
Chair	P. M., O., and A. S. Snell	Williamsburgh, Ohio	Nov. 14, 1871	120, 833
Chair	A. Taylor	Brooklyn, N. Y	Jan. 30, 1872	123, 131
Chair	D. E. Teal	Norwich, N. Y	May 5, 1863	38, 428
Chair	J. H. Travis	Charlestown, Mass	June 6, 1871	115, 787
Chair	C. Tucker	Terre Haute, Ind	Jan. 31, 1871	111, 494
Chair	J. Ungerer	Brooklyn, N. Y	Oct. 16, 1866	58, 917
Chair	H. Willard	Grand Rapids, Mich	Apr. 12, 1870	101, 957
Chair	G. C. Winchester	Ashburnham, Mass	July 18, 1871	117, 132
Chair	G. C. Winchester and M. V. B. Howe.	Ashburnham, Mass	Jan. 1, 1867	60, 978
Chair	J. C. Zimmerman	Thomaston, Ga	Dec. 20, 1870	110, 413
Chair, Adjustable	G. Wilson	Chicago, Ill	Sept. 20, 1870	107, 581
Chair and bed, Combined	J. H. Green	Louisville, Ky	Nov. 5, 1872	132, 826
Chair and bed, Combined	G. D. Heatwole	Bridgewater, Va	Apr. 4, 1871	113, 662
Chair and bed, Combined	J. E. Pitcher	Louisville, Ky	Aug. 1, 1871	117, 565
Chair and bedstead	J. H. Greenleaf	New Haven, Conn	May 28, 1867	65, 072
Chair and cane, Combined	D. O. Parker	Liverpool, Nova Scotia	Aug. 8, 1871	117, 810
Chair and car seat	B. M. Darling	Woonsocket, R. I	Jan. 26, 1869	86, 141
Chair and car spring seat	W. T. Doremus	New York, N. Y	Dec. 17, 1872	133, 970
Chair and cradle	G. W. Hawk	Chicago, Ill	May 2, 1865	47, 542
Chair and cradle, Combined	A. Abrahams	Syracuse, N. Y	Sept. 19, 1871	119, 071
Chair and cradle, Combined	E. Stoney	Walkerton, Canada	Apr. 18, 1871	113, 812

Index of patents issued from the United States Patent Office from 1790 *to* 1873, *inclusive*—Continued.

Invention.	Inventor.	Residence.	Date.	No.
Chair and cradle rocker	C. Wetterhan	Fond du Lac, Wis	June 8, 1869	91, 187
Chair and crib, Combined	S. Ray and M. R. Shalters	Alliance, Ohio	Feb. 7, 1860	27, 069
Chair and crib for children, Combined	W. B. Carpenter	New York, N. Y	Feb. 6, 1855	12, 357
Chair and couch	J. E. Jonett	New York, N. Y	Aug. 6, 1867	67, 556
Chair and desk	C. B. Sherman	Troy, N. Y	Sept. 10, 1867	68, 659
Chair and desk, Combined	W. A. Ehlman	Milwaukee, Wis	July 24, 1866	56, 540
Chair and desk, Combined	E. J. Smith	Washington, D. C	June 25, 1872	128, 432
Chair and fan	A. Dyson	Saint Louis, Mo	Nov. 23, 1869	97, 063
Chair and fan, Combined	F. Handschuh	Allentown, Pa	July 10, 1866	56, 214
Chair and furniture tip	F. H. Holton	Brooklyn, N. Y	Sept. 27, 1870	107, 685
Chair and ladder	W. Bergmann	Philadelphia, Pa	Aug. 15, 1871	118, 096
Chair and lounge	E. Hagan	New York, N. Y	Nov. 15, 1870	109, 202
Chair and lounge	P. J. Hardy	New York, N. Y	Aug. 13, 1867	67, 758
Chair and lounge	M. P. Harley	Philadelphia, Pa	July 26, 1870	105, 677
Chair and lounge	G. Knell	Moorestown, N. J	Mar. 31, 1868	76, 292
Chair and lounge	T. Winter	Boston, Mass	May 5, 1868	77, 697
Chair and lounge, Combined	G. W. Cropcup	Philadelphia, Pa	July 16, 1872	129, 537
Chair and lounge, Combined	F. J. Gardner	Washington, N. C	Jan. 25, 1859	22, 722
Chair and lounge combined, Invalid	G. G. Boardman	Albia, Iowa	Mar. 18, 1873	136, 962
Chair and other seat, Spring-bottom for	P. Gallagher	Pleasant Unity, Pa	Dec. 28, 1858	22, 419
Chair and secretary, Combined	G. C. Taylor	Thibodeaux, La	Sept. 2, 1873	142, 418
Chairs and settees into their frames, Machine for pressing the seats and backs of.	G. A. Watkins	Cavendish, Vt	July 13, 1869	92, 496
Chair and sofa	A. H. Hoffmeier	Lancaster, Pa	Oct. 1, 1872	131, 758
Chair and sofa bottom	J. Sutter	New York, N. Y	May 3, 1864	42, 602
Chair and sofa seat	F. Wittram	San Francisco, Cal	Dec. 7, 1869	97, 579
Chair and step-ladder	A. Madson	Paterson, N. J	Jan 12, 1869	85, 835
Chair and step-ladder combined	B. F. Green	Syracuse, N. Y	July 1, 1873	140, 363
Chair and step-ladder combined	I. Jörgenson and R. Olson	Racine, Wis	Apr. 4, 1871	113, 304
Chair and step-ladder, Combined	A. Liesche	Syracuse, N. Y	Aug. 2, 1870	106, 067
Chair and step-ladder combined	A. Liesche	Syracuse, N. Y	Jan. 30, 1872	123, 183
Chair and step-ladder combined	A. Liesche	Syracuse, N. Y	June 18, 1872	127, 980
Chair and step-ladder combined	C. B. Rawson	Worcester, Mass	Dec. 3, 1872	133, 594
Chair and stool irons	H. Ocorr	Sheboygan, Wis	June 13, 1871	115, 979
Chair and stool seat	F. W. Dickerman	New York, N. Y	Nov. 7, 1871	120, 632
Chair-back-scraping machine	E. S. French	Westminster, Mass	Oct. 4, 1870	108, 015
Chair-backs and cradle-ends	T. W. Moore	New York, N. Y	May 21, 1872	126, 978
Chair-backs, Bows for	C. Kilburn	Philadelphia, Pa	Jan. 14, 1873	134, 896
Chair-backs, Machine for cutting	J. A. Dyer	Newburgh, Ohio	May 21, 1861	32, 398
Chair-backs, Machine for cutting	E. Edwards and J. Cowee, jr	Keene, N. H	Mar. 18, 1862	34, 675
Chair-backs, Machine for dressing	S. L. Fitts	Ashburnham, Mass	July 14, 1863	39, 220
Chair-backs, Machine for manufacturing	S. E. Foster	Fitchburgh, Mass	July 13, 1858	20, 918
Chair-backs, Machine for molding	J. Lemman	Cincinnati, Ohio	Aug. 17, 1869	93, 893
Chair-backs, Machine for molding	J. Lemman	Cincinnati, Ohio	Apr. 18, 1871	113, 898
Chair-backs, Machinery for sawing arabesque	J. H. Better	New York, N. Y	July 31, 1847	5, 208
Chair bedstead and crib	W. McGregor	Chicago, Ill	May 17, 1870	103, 066
Chair-bottom	Z. B. Bellows	Cortlandville, N. Y	Mar. 15, 1859	23, 225
Chair-bottom	J. Elliot, jr	Philadelphia, Pa	May 26, 1812	....
Chair bottom	A. Ellis	Dedham, Mass	June 12, 1811	
Chair-bottom	P. C. Ingersoll	Green Point, N. Y	Feb. 1, 1870	99, 441
Chair-bottom	C. W. Royse	Peterborough, N. H	Nov. 19, 1867	71, 067
Chair-bottom	C. Russell	Wilmington, Ohio	Sept. 26, 1865	50, 166
Chair bottom and back	W. Bramhill	New York, N. Y	Nov. 28, 1865	51, 132
Chair-bottoms, Machine for molding	J. Lemman	Cincinnati, Ohio	Jan. 17, 1871	110, 981
Chair-brace	G. F. Ells	Deposit, N. Y	Mar. 18, 1873	136, 824
Chair-cane	R. Porter	Washington, D. C	July 11, 1854	11, 271
Chair, Cane-seat	J. R. Cannon	New Albany, Ind	Sept. 7, 1858	21, 409
Chair, couch, &c	C. Adams	Boston, Mass	Apr. 18, 1831	....
Chair, couch, and stretcher	J. E. Jouett	Brooklyn, N. Y	Nov. 6, 1866	59, 403
Chair-coupling clamp	J. P. Brooks	Newark, N. J	Feb. 13, 1872	123, 549
Chair, cradle, cot, &c	J. T. Wightman	Charleston, S. C	Oct. 5, 1869	95, 544
Chair, Desk or table attachment for	S. L. Bligh	Pit Hole City, Pa	Oct. 24, 1871	120, 234
Chair fan-attachment	B. L. Jordan	Mercersburgh, Pa	June 17, 1873	140, 047
Chair foot-rest	G. L. Badlam and C. W. Lang	Brandon, Vt	May 4, 1869	89, 827
Chair foot-rest	E. Collins	New York, N. Y	Jan. 14, 1873	134, 733
Chair-frame	G. Gardner	Glen Gardner, N. J	Sept. 9, 1873	142, 625
Chair-frame	M. E. Halsey	New York, N. Y	Sept. 26, 1854	11, 725
Chair frame	L. Heywood and E. E. Horton	Gardner, Mass	Jan. 7, 1873	134, 544
Chair-frame	W. M. Smith	Columbus, Ohio	Dec. 28, 1869	98, 440
Chair-frame	W. M. Smith	Columbus, Ohio	Dec. 27, 1870	110, 598
Chair head-rest	C. A. Mills	Dubuque, Iowa	Oct. 21, 1856	15, 931
Chair-legs, Tip for	E. Coogan and H. Miller	Washington, D. C	Nov. 10, 1868	83, 933
Chair-legs, Tip for	E. S. Winchester	Boston, Mass	Nov. 5, 1867	70, 490
Chair, lounge, and step-ladder	J. Gerdom, jr	West Albany, N. Y	Sept. 17, 1867	68, 973
Chair-manufacturing machine	I. Lewis	Irville, Ohio	May 25, 1830	....
Chair or stool support	H. Wadsworth	Duxbury, Mass	May 7, 1872	126, 429
Chair-posts, Manufacturing	E. Farrington	New York, N. Y	Oct. 26, 1816	....
Chair rail, Continuous	C. A. Stancliff and J. Mingis	Williamsport, Pa	July 6, 1858	20, 828
Chair-rims, &c., Method of bending	S. M. Barrett	Cincinnati, Ohio	July 6, 1869	92, 145
Chair-rocker	M. Wilcox and C. Depledge	Cassadaga, N. Y	Nov. 25, 1873	145, 036
Chair-rocker	L. B. Yale	Bainbridge, N. Y	Mar. 17, 1868	75, 617
Chair-rocker	W. J. Zakrzewska	Berlin, Prussia	Apr. 12, 1870	101, 802
Chair-rocker and lounge	W. H. Whiterow	New Albany, Ind	Mar. 5, 1872	124, 238
Chair-rocker and lounge combined	H. Haidt	New York, N. Y	Oct. 22, 1872	132, 397
Chair-rocker, Detachable	W. Pinkerman	Bridgeport, Conn	Feb. 20, 1866	52, 741
Chair-rocker, Elastic	J. Barron	Cincinnati, Ohio	Nov. 26, 1867	71, 357
Chair-seat	L. B. Batcheller	Rochester, N. Y	Sept. 11, 1860	30, 022
Chair-seat	A. Bingham	Sherry, N. H	June 19, 1866	55, 605
Chair-seat	A. Bingham	Surry, N. H	Aug. 6, 1867	67, 485
Chair-seat	O. A. Bingham	Gardner, Mass	Sept. 7, 1869	94, 553
Chair-seat	O. A. Bingham	Cavendish, Vt	Mar. 22, 1870	101, 090
Chair-seat	H. Bjorkman and C. J. Lagergren.	Glen Gardner, N. J	Oct. 22, 1872	132, 431
Chair-seat	H. Bjorkman and C. J. Lagergren.	Glen Gardner, N. J	Dec. 3, 1872	133, 622
Chair-seat	H. Buchter	Louisville, Ky	Oct. 20, 1868	83, 250
Chair-seat	G. Buckle	Detroit, Mich	Dec. 29, 1868	85, 363
Chair-seat	E. L. Buckingham	Jefferson, Wis	Sept. 15, 1868	82, 081

Index of patents issued from the United States Patent Office from 1790 to 1873, inclusive—Continued.

Invention.	Inventor.	Residence.	Date.	No.
Chair-seat	P. Buckley	Binghamton, N. Y	Feb. 13, 1872	123, 550
Chair-seat	W. G. Bulgin	Vienna, N. J	June 7, 1870	103, 973
Chair, &c., seat	W. T. Doremus	New York, N. Y	Feb. 18, 1873	135, 973
Chair-seat	J. W. Drummond	Skaneateles, N. Y	Dec. 16, 1851	8, 585
Chair-seat	L. W. Field	Camden, Ill	Sept. 14, 1869	94, 731
Chair-seat	G. Gardner	Glen Gardner, N. J	May 21, 1872	127, 044
Chair-seat	G. Gardner	Glen Gardner, N. J	May 21, 1872	127, 045
Chair-seat	G. Heesen	Tecumseh, Mich	Oct. 1, 1867	69, 430
Chair-seat	H. C. Knowlton	Gardner, Mass	Nov. 3, 1868	83, 713
Chair-seat	C. Platt	New Haven, Conn	Oct. 27, 1868	83, 541
Chair-seat	J. P. Sinclair	Elbridge, N. Y	July 29, 1873	141, 295
Chair-seat	C. D. Smith	Templeton, Mass	Apr. 5, 1870	101, 671
Chair-seat	L. A. Smith	Cincinnati, Ohio	July 23, 1867	67, 077
Chair-seat	G. A. Watkins	Proctorsville, Vt	Apr. 7, 1868	76, 569
Chair-seat	G. A. Watkins	Proctorsville, Vt	Apr. 28 1868	77, 228
Chair-seat	G. A. Watkins	Cavendish, Vt	Sept. 6, 1870	107, 136
Chair-seat	G. A. Watkins	Cavendish, Vt	Sept. 6, 1870	107, 137
Chair-seat	G. A. Watkins	Proctorsville Post-Office, Vt	Aug. 15, 1871	118, 079
Chair-seat	R. Wood	West Troy, N. Y	Mar. 29, 1864	42, 150
Chair-seat	B. F. Wright	Charlestown, Mass	June 14, 1870	104, 243
Chair seat and back	A. B. Clark	Gardner, Mass	Nov. 12, 1872	132, 951
Chair seat and back	W. T. Doremus	New York, N. Y	Jan. 7, 1873	134, 593
Chair seat and back	W. S. Farmer	New York, N. Y	May 1, 1866	54, 314
Chair seat and back	C. D. Flynt	New York, N. Y	Sept. 24, 1872	131, 677
Chair seat and back	H. T. Pratt	Fitchburgh, Mass	July 29, 1862	36, 026
Chair seat and back and bed-bottom, Elastic	L. Daft	New York, N. Y	Mar. 9, 1869	87, 644
Chair-seat and sofa-bottom	J. W. Kimball and J. Mahady	Cambridge, Mass	May 24, 1864	42, 858
Chair-seat-dressing machine	I Baker	Warwick, Mass	Sept. 12, 1848	5, 760
Chair-seat frame	H. Buchter	Louisville, Ky	Nov. 21, 1871	121, 080
Chair-seat frame	H. Buchter	Louisville, Ky	June 4, 1872	127, 560
Chair-seat frame	G. A. Watkins	Proctorsville, Vt	May 14, 1867	64, 727
Chair-seat frames, Machine for grooving	G. A. Watkins	Cavendish, Vt	Sept. 6, 1870	107, 135
Chair-seat material	A. N. Elliott	Barre, Mass	Mar. 29, 1870	101, 241
Chair-seat, Metallic	S. A. Snyder	Clarendon, N. Y	Aug. 2, 1870	105, 993
Chair seat or back	G. Gardner	Glen Gardner, N. J	Sept. 9, 1873	142, 624
Chair seat or back	R. W. Myers	New York, N. Y	July 29, 1873	141, 374
Chair-seat-planing machine	E. Q. Smith	Cincinnati, Ohio	May 12, 1857	17, 300
Chair-seat supporter	G. Butterfield and A. G. Treadwell.	Boston, Mass	Jan. 21, 1868	73, 499
Chair-seat, Tilting	M. V. B. Howe	Gardner, Mass	Jan. 28, 1868	73, 721
Chair-seat, Wooden	L. Heywood	Gardner, Mass	Dec. 24, 1867	72, 634
Chair-seat, Wooden	G. C. Winchester	Ashburnham, Mass	Jan. 12, 1869	85, 883
Chair-seats, Device for manufacturing	O. A. Bingham	Cavendish, Vt	Apr. 26, 1870	102, 212
Chair-seats, Device for weaving	G. A. Watkins	Proctorsville, Vt	Mar. 10, 1868	75, 500
Chair-seats, &c., Fabric for	J. W. Owen	New Haven, Conn	Apr. 27, 1869	89, 496
Chair-seats, Frame for weaving	W. Aldrich and E. T. Watkins	Gardner, Mass	Dec. 3, 1872	133, 616
Chair-seats into their frames, Device for pressing	W. Aldrich and A. F. Spaulding	Proctorsville and Northfield, Vt.	Oct. 10, 1871	119, 732
Chair-seats, &c., Machine for boring and mortising	W. L. Harley	Chagrin Falls, Ohio	Aug. 14, 1840	1, 723
Chair-seats, Machine for dressing	H. Meyer	Grafton, Wis	Mar. 31, 1868	76, 226
Chair-seats, Machine for making	S. W. Cross	Waupun, Wis	Oct. 22, 1872	132, 447
Chair-seats, Machine for making	E. S. Jackson	Bethel, Vt	Oct. 11, 1870	108, 147
Chair-seats, Machine for making	J. C. McCormick	Smicksburgh, Pa	Nov. 2, 1869	96, 458
Chair-seats, Machine for making	E. Ransom	Flint, Mich	Nov. 14, 1865	50, 955
Chair-seats, Preparing raw-hide for use in	B. F. Wright and J. Rowe, jr	Charlestown and Wilmington, Mass.	July 26, 1870	105, 759
Chair-seating	M. Kohn	Hartford, Conn	Nov. 31, 1871	121, 112
Chair, sofa, and car-seat	J. W. Cochran	New York, N. Y	May 22, 1866	54, 863
Chair-splints, Machine for cutting	T. N. Davey	Jeffersonville, Ind	Jan. 2, 1865	45, 790
Chair-spring	A. M. Blake	Canton, Ohio	Apr. 28, 1868	77, 161
Chair-spring, &c	W. T. Doremus	New York, N. Y	Oct. 28, 1873	144, 020
Chair-spring	W. T. Doremus	New York, N. Y	Nov. 18, 1873	144, 664
Chair-spring	T. C. Purington	Lincoln, Cal	Nov. 30, 1869	97, 314
Chair-spring	T. E. Warren	Troy, N. Y	Sept. 25, 1849	6, 740
Chair, table, and bureau combined	H. Martindale	Newark, Ind	Jan. 7, 1873	134, 691
Chair, table, and lounge combined	J. Croghan	Chicago, Ill	Mar. 25, 1873	137, 062
Chair, table, and stand combined	A. W. Viles	Elkhorn, Wis	Aug. 17, 1869	93, 773
Chair-tip	E. Coogan	Washington, D. C	Jan. 18, 1870	98, 926
Chair-tops, Turning	H. Cooke	New York, N. Y	Nov. 30, 1832	......
Chair and basket splints, Machine for making	J. B. Poe	Worthington, Ind	Aug. 19, 1873	141, 948
Chairs, Attachable rocker for	P. J. and J. H. Welsh	Philadelphia, Pa	May 13, 1873	138, 826
Chairs, Construction of	A. and O. A. Bingham	Keene, N. H., and Gardner, Mass.	Apr. 17, 1866	54, 053
Chairs, Construction of	H. H. Evarts	Chicago, Ill	Feb. 22, 1870	100, 132
Chairs, Construction of	R. Fitts, jr	Fitchburgh, Mass	Mar. 16, 1869	87, 768
Chairs, Construction of	H. I. Seymour	Troy, N. Y	May 31, 1870	103, 786
Chairs, Construction of	D. E. Teal	New Lisbon, N. Y	July 19, 1870	105, 515
Chairs, Construction of	D. E. Teal	New Lisbon, N. Y	July 19, 1870	105, 516
Chairs, Construction of	D. E. Teal	New Lisbon, N. Y	July 19, 1870	105, 517
Chairs, &c., Elastic bottom for	L. Spooner	Boston, Mass	June 3, 1856	15, 021
Chairs, &c., Elastic tip for legs of	S. Van Patten	Albany, N. Y	May 3, 1870	102, 627
Chairs, &c., Heating-device for	C. S. Hunt	Terre Bonne Parish, La	June 9, 1868	78, 742
Chairs, Manufacturing	E. Q. Smith	Cincinnati, Ohio	Aug. 12, 1856	15, 530
Chairs, Needle for weaving cane seats for	D. C. Mosher	Washington, D. C	Sept. 16, 1873	142, 808
Chairs, Seat and back for	F. Carre	Paris, France	May 15, 1866	54, 828
Chairs, Table-attachment for	W. Zimmerman	Bloomington, Ill	Mar. 23, 1869	88, 255
Chairs, sofas, &c., Spring-seat for	C. Robinson	Cambridgeport, Mass	Mar. 9, 1858	19, 582
Chairs, sofas, &c., Mode of securing seats and backs of.	L. Baker	Templeton, Mass	June 28, 1870	104, 815
Chaise-body, Grecian	P. Dow	Boston, Mass	Aug. 15, 1810	......
Chalk-holder	L. S. Woodbury	Ogdensburgh, N. Y	Aug. 13, 1872	130, 389
Chalk-line box	S. Beyl	Osborn, Ohio	Dec. 22, 1868	85, 202
Chalk-line box	E. Braley	Carmel, Me	July 25, 1871	117, 375
Chalk-line holder	N. Robbins	Rockport, Mass	July 16, 1872	129, 250
Chalk-line marker	J. W. Carter	New York, N. Y	May 28, 1867	65, 057
Chalk-line reel	J. Bathgate	Cincinnati, Ohio	Nov. 12, 1867	70, 684
Chalk-line reel	M. V. B. Knowles	Wakefield, R. I	Dec. 1, 1868	84, 554
Chalk-line reel	G. Sickles	Middletown, Conn	Dec. 31, 1839	1, 451

Index of patents issued from the United States Patent Office from 1790 *to* 1873, *inclusive*—Continued.

Invention.	Inventor.	Residence.	Date.	No.
Chalk-line winder	J. H. Rose	Mount Sterling, Ill	Oct. 30, 1866	59, 268
Chalking lines, Method of	S. B. Knight	North Providence, R. I	Mar. 27, 1855	12, 599
Chamber and other vessels, Cover for	J. S. Davidson and N. Lorton	Cranberry, N. J	Aug. 18, 1868	81, 261
Chamber-closet, Portable	W. J. Lyman	East Hampton, Mass	Feb. 18, 1868	74, 558
Chamber-closet, Portable	W. J. Lyman	East Hampton, Mass	July 14, 1868	79, 994
Chamber-pot	F. Imhorst	New York, N. Y	Mar. 19, 1872	124, 824
Chamber-pot	A. Rankin	New York, N. Y	Jan. 30, 1866	52, 320
Chamber-vessel	C. H. Berry	East Somerville, Mass	Apr. 5, 1870	101, 419
Chamber-vessel	A. Rankin	New York, N. Y	June 5, 1866	55, 361
Chamber-vessel	V. Rhodes	Memphis, Tenn	Dec. 20, 1870	110, 286
Chamber-vessel	C. Robinson	Boston, Mass	Apr. 14, 1868	76, 817
Chamber-vessel cover	W. Stockton	New York, N. Y	May 10, 1870	102, 877
Chamber-vessel seat	I. Freed	Harrisburgh, Pa	Oct. 29, 1867	70, 191
Chamber-utensil	J. C. Stoddard	Worcester, Mass	Sept. 27, 1859	25, 592
Chameleon whirligig	L. O. Franke	Baltimore, Md	July 26, 1870	105, 792
Chameleotrope	S. W. Anderson	New York, N. Y	Mar. 1, 1870	100, 248
Chamfering and crozing machine	H. Wilde	Newark, N. J	Aug. 12, 1862	36, 186
Chamfering-machine	B. F. Mattox and S. Corson	West Ridge, Ill	Mar. 23, 1869	88, 153
Chamfering-machine	J. Stufflebeen	Milwaukee, Wis	May 24, 1864	42, 886
Chamfering-tool	W. Johnson, 2d	Hampstead, N. H	June 21, 1859	24, 465
Chandelier	W. L. Bradley	West Meriden, Conn	Jan. 7, 1873	134, 585
Chandelier	T. Buckley	New York, N. Y	Jan. 4, 1870	98, 469
Chandelier	T. Buckley	New York, N. Y	Nov. 7, 1871	120, 707
Chandelier	J. A. Evarts	West Meriden, Conn	May 26, 1868	78, 365
Chandelier	I. P. Frink	New York, N. Y	Apr. 9, 1872	125, 560
Chandelier	C. F. Jacobsen	New York, N. Y	June 8, 1869	91, 135
Chandelier	C. F. Jacobsen	New York, N. Y	Mar. 7, 1871	112, 351
Chandelier	J. Kintz	West Meriden, Conn	May 20, 1873	139, 066
Chandelier	P. Loth	New York, N. Y	Apr. 18, 1865	47, 312
Chandelier	R. J. Skinner	Chicago, Ill	May 14, 1867	64, 719
Chandelier	R. J. Skinner	Chicago, Ill	Mar. 3, 1868	75, 064
Chandelier	E. M. Smith	New York, N. Y	Oct. 17, 1871	120, 114
Chandelier	J. F. Travis	New York, N. Y	Dec. 15, 1868	84, 975
Chandelier and lamp, Apparatus for raising and lowering.	H. S. Hall	Boston, Mass	Dec. 28, 1869	98, 255
Chandelier attachment, Gas	A. Bliss	New York, N. Y	Nov. 5, 1867	70, 398
Chandelier center	J. Kintz	West Meriden, Conn	Oct. 10, 1871	119, 773
Chandelier center	J. Kintz	West Meriden, Conn	Aug. 26, 1873	142, 165
Chandelier center	J. Meah	Meriden, Conn	Oct. 10, 1871	119, 780
Chandelier center-light fixture	R. F. White	Hoboken, N. J	July 18, 1871	117, 131
Chandelier drop-light	C. E. Cornelius	Philadelphia, Pa	Dec. 2, 1873	145, 053
Chandelier, Drop-light	W. C. Vosburg	Brooklyn, N. Y	Apr. 7, 1868	76, 360
Chandelier, Extension	L. Hull	Charlestown, Mass	Aug. 26, 1873	142, 107
Chandelier, Extension	H. Tucker	Newton, Mass	Arg 17, 1869	93, 927
Chandelier extension-fixtures	J. A. Evarts	West Meriden, Conn	Apr. 21, 1868	77, 019
Chandelier extension-tube	E. Russell	Waterbury, Conn	Apr. 9, 1872	125, 619
Chandelier-hook	W. Hubbard	West Meriden, Conn	Sept. 3, 1872	131, 100
Chandelier or hanging lamp	W. Lawrence	Meriden, Conn	Mar. 23, 1831	....
Chandelier or hanging lamp	W. Lawrence	Meriden, Conn	Mar. 10, 1834	....
Chandelier, Reflecting	C. F. Jacobsen	New York, N. Y	Mar. 12, 1872	124, 582
Chandelier, Sliding	G. F. Blaisse	Philadelphia, Pa	June 17, 1873	140, 003
Chandelier, Sliding or extension	J. Brannen	Philadelphia, Pa	July 1, 1873	140, 459
Chandelier, Vapor	C. E. Smith and H. J. Rice	Columbus, Ohio	Aug. 23, 1870	106, 629
Chandeliers from the crystallization of salts, Manufacturing.	F. Ransom	Buffalo, N. Y	Nov. 3, 1829	....
Chandeliers, lamps, &c., in imitation of glass, Manufacture from the crystallization of salts.	F. B. Merrill	Buffalo, N. Y	June 13, 1829	....
Change-box	J. B. Eustis	New Orleans, La	June 11, 1867	65, 555
Change-box, Railway-conductor's	J. B. Slawson	New York, N. Y	Aug. 20, 1872	130, 756
Change-gate	A. W. Wood	Saint Louis, Mo	July 15, 1873	140, 986
Channeling and beveling machine	I. Manning	Philadelphia, Pa	Aug. 27, 1867	68, 094
Channeling and edging tools	L. Bauer	San Francisco, Cal	Apr. 1, 1873	137, 404
Channeling-machine	G. McKay	Boston, Mass	Apr. 5, 1864	42, 211
Channeling-machine	G. McKay and L. B. Blake	Boston and Quincy, Mass	Aug. 23, 1864	43, 923
Channeling-machine	C. Stoddard	North Brookfield, Mass	Apr. 5, 1864	42, 237
Channeling-machine	H. S. Vrooman	Hoboken, N. J	Jan. 1, 1867	60, 807
Channeling-tool	A. Bottum	Bridgeport, Conn	Dec. 8, 1863	40, 814
Channeling-tool	F. E. Droll	Saint Charles, Mo	Feb. 25, 1868	74, 806
Channeling-tool	G. D. Edmands	Saugus, Mass	Mar. 30, 1869	88, 283
Channeling-tool	J. B. Johnson	Lynn, Mass	July 28, 1863	39, 366
Channeling-tool	G. W. Pruyne	Mexico, N. Y	May 5, 1868	77, 653
Channeling-tool for harness-maker	E. D. Gould	Lockport, N. Y	Feb. 18, 1862	34, 421
Charcoal, Animal: *See* Bone-black.				
Charcoal, Apparatus for cooling and saving	G. A. Jasper	Charlestown, Mass	Mar. 22, 1870	101, 019
Charcoal, Apparatus for the manufacture of	L. S. Goodrich	Waverly, Tenn	July 23, 1873	129, 814
Charcoal, Burning	A. Grimes	Lancaster, N. Y	Dec. 23, 1856	16, 304
Charcoal, Charring wood for	I. Doolittle	Bennington, Vt	Dec. 14, 1829	....
Charcoal, Cleansing and revivifying	G. A. Jasper	Charlestown, Mass	Apr. 18, 1865	47, 308
Charcoal, Cleansing animal	G. A. Jasper	Charlestown, Mass	Mar. 27, 1866	53, 534
Charcoal, Cooling and purifying animal	D. H. Turner	New York, N. Y	Sept. 17, 1867	68, 915
Charcoal for rectifying spirits, Retort for preparing	J. McCann	Saint Louis, Mo	Apr. 27, 1869	89, 492
Charcoal for refining sugar, Revivifying animal	A. H. Leplay and J. F. J. Cruisinier.	Paris, France	May 6, 1862	35, 160
Charcoal from peats, Making	C. Fales	Worcester, Mass	Feb. 11, 1807	....
Charcoal-furnace	C. W. Briggs	Springfield, Mass	Sept. 1, 1868	81, 742
Charcoal-furnace	W. T. Downs	Saint Louis, Mo	Dec. 28, 1869	98, 239
Charcoal-furnace	J. McNeill	New York, N. Y	Mar. 27, 1855	12, 602
Charcoal-furnace	A. J. Redway	Cincinnati, Ohio	July 16, 1867	66, 738
Charcoal-furnace	W. S. Wright	Saint Louis, Mo	Oct. 6, 1863	40, 207
Charcoal-kiln	C. T. Harvey	Chicago, Ill	Aug. 5, 1862	36, 090
Charcoal-kiln, Mode of forming	M. Carroll	Pellico Plains, Tenn	Apr. 28, 1838	720
Charcoal-making apparatus	K. S. Chaffee	Cambridge, Mass	July 24, 1866	56, 528
Charcoal manufacture	W. P. McConnell	Washington, D. C	Nov. 4, 1851	8, 492
Charcoal, Manufacturing	A. Kurtz	Philadelphia, Pa	June 28, 1816	....
Charcoal, Mode of treating animal	E. Beanes	London, England	Aug. 2, 1864	43, 748
Charcoal, Process and apparatus for the manufacture of.	L. S. Goodrich	Waverly, Tenn	July 23, 1872	129, 815

Index of patents issued from the United States Patent Office from 1790 *to* 1873, *inclusive*—Continued.

Invention.	Inventor.	Residence.	Date.	No.
Charcoal, Process of screening	J. S. Evans	Irondale, Mo	Dec. 15, 1868	84, 866
Charcoal, Purification of animal	C. F. L. Wandel	Bernburg, Germany	May 25, 1869	90, 617
Charcoal, Purifying animal	H. Eissfeldt and C. Thumb	Sollingen, Duchy of Brunswick, and Magdeburg, Prussia.	Apr. 4, 1871	113, 279
Charcoal, Treating animal	E. Beanes	London, England	Apr. 5, 1864	42, 156
Charcoal used in rectifying spirits, Revivifying	C. L. Fleischmann	Washington, D. C	Feb. 25, 1873	136, 147
Charcoal used in rectifying spirits, Revivifying	A. Ernst	Baltimore, Md	Feb. 25, 1873	136, 229
Charring wood	S. S. Perry	Charles City County, Va	Apr. 8, 1856	14, 619
Chart-holder	D. Shryock	Hannibal, Mo	Sept. 19, 1871	119, 054
Chart-roller	E. L. Hagar	Empire City, Colo	May 12, 1868	77, 813
Charts, landscapes, &c., Machine for copying	C. Thompson	Bristol, R. I	Feb. 5, 1806	....
Charts, Method of finding courses and bearings on marine.	E. R. Knorr	Washington, D. C	May 1, 1860	28, 089
Chartæ lusoriæ, Fair dealer of the	R. Bailey	Washington, D. C	May 15, 1812	....
Chasing-mill	W. M. Force	Newark, N. J	Sept. 20, 1864	44, 298
Chasing-mill	C. Moore	Trenton, N. J	May 18, 1858	20, 290
Check and driving line	C. M. Alexander	Washington, D. C	July 14, 1868	79, 932
Check, Baggage	G. Bailey	Buffalo, N. Y	June 3, 1862	35, 431
Check, Baggage	E. Hoole	Mount Vernon, N. Y	Oct. 5, 1858	21, 677
Check, Baggage	J. Murdock, jr., and W. W. Spencer.	Cincinnati, Ohio	Aug. 22, 1865	49, 543
Check-box, Conductors'	J. F. Gaunt	Newark, N. J	June 10, 1873	139, 780
Check-box, Conductors'	T. W. Knox	New York, N. Y	May 9, 1865	47, 646
Check-box, Railway-conductors'	E. Keith	Buffalo, N. Y	June 4, 1872	127, 487
Check-box, Railway-conductors' registering	J. S. Stridiron	Detroit, Mich	May 28, 1872	127, 200
Check-holder	A. A. Baker	Camden, N. J	Mar. 28, 1871	113, 131
Check-hook	C. A. Ball	New York, N. Y	Apr. 3, 1866	53, 726
Check-hook	A. Bedford	Cold Water, Mich	Nov. 3, 1868	83, 594
Check-hook	H. W. Burress	New Brighton, Pa	Jan. 21, 1873	135, 075
Check-hook	H. A. Collins	Springfield, Mass	Aug. 16, 1864	43, 834
Check-hook	E. Deming	Middletown, Conn	Sept. 3, 1867	68, 559
Check-hook	N. Dieterich	Sandwich, Ill	July 9, 1867	66, 473
Check-hook	J. H. Jones	Williamsport, Pa	Sept. 17, 1867	68, 990
Check-hook	C. B. Payne	Clinton, Ill	July 20, 1869	92, 879
Check-hook	A. V. M. Sprague	Rochester, N. Y	July 1, 1873	140, 555
Check-hook	G. Theobalt	Springfield, Mass	Dec. 7, 1869	97, 569
Check hook and terret fastening	A. L. Hill	Decatur, Ill	June 23, 1868	79, 227
Check hook and terret fastening	A. L. Hill	Decatur, Ill	Oct. 20, 1868	83, 157
Check-protector	J. Adair	Pittsburgh, Pa	Mar. 5, 1872	124, 309
Check-rein	T. Heaton	Cornwall, N. Y	Apr. 9, 1867	63, 635
Check-rein holder	D. Reynolds	Rockford, Ill	Nov. 30, 1869	97, 443
Check-runner	J. Haggerty	Corry, Pa	Aug. 26, 1873	142, 160
Checks, drafts, &c., Fluid for writing on	C. L. Lawrence	New York, N. Y	June 6, 1871	115, 744
Checks, drafts, &c., from fraud, Mode of securing	J. Hester	Knoxville, Ill	Mar. 24, 1868	75, 758
Checks, drafts, &c., Manufacturing blanks for	J. Atwater	New Haven, Conn	Jan. 21, 1829	....
Checks, drafts, &c., Manufacturing blanks for	J. Atwater	New Haven, Conn	Aug. 18, 1829	....
Checks, drafts, &c., Manufacturing blanks for	J. Atwater and N. and S. S. Jocelyn.	New Haven, Conn	June 11, 1829	....
Checks for preventing forgeries	J. D. Pope	Baltimore, Md	Mar. 8, 1834	....
Checker-men	J. W. Hyatt, jr	Albany, N. Y	June 15, 1869	91, 233
Checkers, Construction of	J. V. H. Nott	New York, N. Y	Dec. 24, 1867	72, 670
Cheese and butter cutting machine	J. H. Thomas	Lynn, Mass	Feb. 14, 1871	111, 788
Cheese and cider press	L. Caswell	Harrison, Me	Dec. 18, 1826	....
Cheese, Apparatus for manufacturing	P. W. Strong	Evans' Mills, N. Y	Nov. 23, 1869	97, 131
Cheese-bandage	T. Hamer and G. W. Robinson	Boylston and Sandy Creek, N. Y.	June 4, 1872	127, 600
Cheese-bandage	H. N. Kimball	Watertown, N. Y	Aug. 6, 1867	67, 557
Cheese, Bandaging and boxing	J. Blood	Watertown, N. Y	May 10, 1870	102, 755
Cheese-box	F. Blecka	Elgin, Ill	Nov. 2, 1869	96, 302
Cheese-box	D. M. Cole	Elgin, Ill	Feb. 25, 1868	74, 797
Cheese-box	J. J. Hecox	Lyons, N. Y	Nov. 7, 1871	120, 741
Cheese-box	V. P. Kimball	Watertown, N. Y	May 18, 1869	90, 107
Cheese-box	N. and F. Lewis	Adams, N. Y	Mar. 26, 1867	63, 267
Cheese-box	C. Stoll	Mokena, Ill	May 1, 1866	54, 437
Cheese-box and butter-tub	F. H. Wilson	Weathersfield, N. Y	Jan. 7, 1868	73, 151
Cheese-box detector	J. P. Conkling	Batchellerville, N. Y	May 22, 1866	54, 866
Cheese-box, Machine for making	T. Hanvey	Elma, N. Y	Mar. 12, 1867	62, 844
Cheese, Boxing, bandaging, and preparing	A. M. Utley, H. N. Kimball, and W. Reynolds.	Watertown, N. Y	Oct. 6, 1868	82, 895
Cheese, butter, and bread cutter	B. F. Adams	Bangor, Me	Nov. 11, 1851	8, 501
Cheese, cider, &c., Machine to press	I. Nichols	Otsego County, N. Y	Oct. 12, 1809	....
Cheese, cider, &c., Pressing	D. Holmes	Paris, Me	Feb. 6, 1832	....
Cheese, Compound for preserving	F. Raymond and A. Miller	Cleveland, Ohio	Oct. 27, 1868	83, 408
Cheese-cover	E. L. Pratt	Philadelphia, Pa	June 21, 1859	2, 452
Cheese-curd cutter	W. A. Bemis	Spencer, Mass	Aug. 27, 1867	68, 279
Cheese-curd cutter	H. A. Blakeman	Cuyler, N. Y	Jan. 19, 1864	41, 271
Cheese-curd cutter	I. Hunter, jr	New Braintree, Mass	Dec. 23, 1834	....
Cheese-curd cutter	H. Keeney	Potter Centre, N. Y	Aug. 29, 1865	49, 631
Cheese-curd from whey, Mode of separating	S. Greene	Rome, N. Y	May 28, 1867	65, 211
Cheese-curd, Machine for cutting	A. Morris	Sackett's Harbor, N. Y	May 2, 1829	....
Cheese-curd rake	E. A. Palmer	Clayville, N. Y	June 23, 1868	79, 250
Cheese-curd sink	H. C. Markham	Collinsville, N. Y	Mar. 7, 1871	112, 364
Cheese, Curing	A. Holdredge	West Burlington, N. Y	Dec. 28, 1869	98, 262
Cheese-cutter	S. R. Bailey	Bath, Me	Mar. 30, 1869	88, 357
Cheese-cutter	J. G. Barker	Watertown, Mass	Aug. 14, 1860	29, 559
Cheese-cutter	G. W. Cushman	Aiken, S. C	Jan. 25, 1870	99, 168
Cheese-cutter	J. G. Dreher	Pine Grove, Pa	Dec. 1, 1868	84, 539
Cheese-cutter	J. Haines	Middleburgh, Ohio	July 10, 1866	56, 212
Cheese-cutter	G. C. and P. B. Jones	Alna, Me	Feb. 16, 1869	87, 049
Cheese-cutter	J. Locke	Lewisburgh, Pa	May 13, 1873	138, 903
Cheese-cutter	W. A. McDonald	Alna, Me	Mar. 30, 1869	88, 500
Cheese-cutter	M. Morse and P. W. Sawyer	Gray, Me	May 26, 1868	78, 228
Cheese-cutter	T. H. Pollock and D. Bliven	Greenville, Conn	Apr. 5, 1859	23, 492
Cheese-cutter	J. W. Row	Lewisburgh, Pa	Oct. 1, 1872	131, 784
Cheese-cutter	D. Stevens	Newark, N. J	June 7, 1859	24, 339
Cheese cutter and box	S. S. Brown	Woonsocket, R. I	Oct. 6, 1868	82, 795
Cheese-cutting machine	J. G. Baker	Philadelphia, Pa	Nov. 28, 1871	121, 314

Index of patents issued from the United States Patent Office from 1790 *to* 1873, *inclusive*—Continued.

Invention.	Inventor.	Residence.	Date.	No.
Cheese-cutting machine	W. K. Foster	Bangor, Me	Aug. 24, 1852	9, 214
Cheese-cutting machine	W. Rhoads, jr., and T. Gerhard.	Reading, Pa	Oct. 22, 1867	70, 020
Cheese for market, Preparing	L. J. Randoll	Chardon, Ohio	Mar. 6, 1866	53, 043
Cheese-frame	A. C. Ainger and S. W. Webster	Stockholm, N. Y	Dec. 9, 1862	37, 140
Cheese-frame	T. M. Brintnall	Medina, Ohio	Aug. 15, 1871	117, 974
Cheese, fruit, &c., Press for	B. G. Martin	New York, N. Y	Jan. 14, 1873	134, 910
Cheese, Gear-press for	H. Erwin	Boston, Mass	Dec. 28, 1808	....
Cheese-hoop	J. Beach	De Ruyter, N. Y	May 23, 1854	10, 956
Cheese-hoop	J. Beach	De Ruyter, N. Y	Oct. 18, 1859	25, 800
Cheese-hoop	A. F. Bent	Antwerp, N. Y	Dec. 22, 1868	85, 201
Cheese-hoop	G. B. Boomer	Syracuse, N. Y	Feb. 4, 1873	135, 514
Cheese-hoop	L. Chapin	Antwerp, N. Y	June 9, 1868	78, 643
Cheese-hoop	M. B. Fraser	Rome, N. Y	Jan. 9, 1872	122, 520
Cheese-hoop	E. Hodgkins	Carthage, N. Y	Jan. 2, 1872	122, 464
Cheese-hoop	A. Holdredge and B. F. and H. H. Harrington.	West Burlington and New Berlin, N. Y.	Mar. 14, 1871	112, 597
Cheese-hoop	O. A. King	Bedford, Ohio	Aug. 13, 1867	67, 658
Cheese-hoop	H. W. Millar	Utica, N. Y	May 29, 1866	55, 140
Cheese-hoop	H. W. Millar	Utica, N. Y	Nov. 16, 1869	96, 942
Cheese-hoop	E. A. Palmer	Clayville, N. Y	Nov. 13, 1866	59, 636
Cheese-hoop	H. A. Roe	Madison, Ohio	Aug. 28, 1860	29, 824
Cheese-hoop	J. L. Sprague	Hermon, N. Y	Sept. 7, 1869	94, 661
Cheese-hoop	W. Sternberg	Bridgeport, N. Y	Mar. 21, 1871	112, 977
Cheese-hoop	W. Sternberg	Bridgeport, N. Y	Apr. 18, 1871	113, 809
Cheese-hoop	S. Purdy	Whitestown, N. Y	Apr. 7, 1868	76, 345
Cheese-hoop	C. P. S. Wardwell	Lake Village, N. H	June 9, 1857	17, 536
Cheese-hoop	S. Wilson	Watertown, N. Y	Sept. 13, 1870	107, 315
Cheese-hoop, Adjustable	W. P. Thomson	Watertown, N. Y	Mar. 22, 1870	101, 184
Cheese-hoop flange	A. Chandler	Rome, N. Y	May 2, 1871	114, 263
Cheese-hoop follower	H. Cooper	Watertown, N. Y	May 4, 1869	89, 560
Cheese, Hoop for curing and packing	W. B. Nickelson	Lowville, N. Y	Nov. 13, 1866	59, 630
Cheese-hoop making	H. M. Viets	Carlisle, Ohio	June 23, 1868	79, 164
Cheese-hoops, Casting metallic	T. Brown	Georgetown, N. Y	Mar. 23, 1858	19, 677
Cheese-hoops, Expanding	R. Van Horn	Northfield, Ohio	Aug. 28, 1866	57, 599
Cheese-hoops, &c., Machine for cutting and slitting	P. Bryant	Chesterfield, Mass	May 15, 1849	6, 447
Cheese, Machine for cutting the curd of	S. Thomson	Monterey, Mass	Mar. 7, 1865	46, 735
Cheese, Machine for turning and curing	H. Webber	East Richfield, N. Y	Apr. 22, 1835	....
Cheese, Machine for turning and curing	H. Webber	East Richfield, N. Y	Apr. 22, 1835	....
Cheese-making apparatus	E. L. Yancy and C. E. Dorman	Batavia and Pembroke, N. Y.	Nov. 1, 1870	108, 866
Cheese-making machine	B. Armstrong	Huntsburgh, Ohio	Jan. 14, 1868	73, 281
Cheese manufacture	J. W. Andrews and N. J. Ogden	Dryden, N. Y	Apr. 13, 1869	88, 830
Cheese manufacture	J. Budlong	Chardon, Ohio	Apr. 10, 1866	53, 777
Cheese manufacture	W. S. Cornell	New York, N. Y	Mar. 31, 1868	76, 167
Cheese manufacture	W. McAllister	Gerry, N. Y	Nov. 13, 1860	30, 638
Cheese manufacture	F. A. Redington and G. McCluer.	Fredonia, N. Y	Feb. 8, 1859	22, 891
Cheese, Mode of applying nets to pine-apple	L. M. Norton	Litchfield, Conn	June 4, 1827	....
Cheese, Mode of covering	U. Bushnell	Gustavus, Ohio	Jan. 6, 1852	8, 625
Cheese, Mode of preparing rennet for use in making	L. B. Arnold	Lansing, N. Y	June 15, 1869	91, 298
Cheese, Netting for forming pine-apple	W. Starr	Norway, N. Y	Dec. 27, 1831	....
Cheese-press	J. Aiken	Warner, N. H	Jan. 6, 1863	37, 327
Cheese-press	P. Allen	Burlington, N. Y	May 13, 1809	....
Cheese-press	R. Allen	Cleveland, Ohio	Dec. 13, 1870	110, 105
Cheese-press	J. Arnold	Pawlings, N. Y	Oct. 26, 1807	....
Cheese-press	J. Arnold	Harmony, N. Y	Apr. 2, 1841	2, 036
Cheese-press	B. Atwell	Burlington, N. Y	July 18, 1809	....
Cheese-press	C. Auborn	Watertown, N. Y	Apr. 9, 1861	31, 940
Cheese-press	A. G. Bagg	Holland Patent, N. Y	Jan. 25, 1870	99, 132
Cheese-press	E. Barnes	North Brookfield, Mass	Sept. 5, 1834	....
Cheese-press	S. Bartlett	Hocking County, Ohio	Feb. 28, 1842	2, 469
Cheese-press	J. Bigelow	Montpelier, Vt	Jan. 25, 1816	....
Cheese-press	G. B. Boomer, T. G. Morse, and R. E. Boschert.	Phœnix, N. Y	Nov. 1, 1870	108, 753
Cheese-press	J. Card	Fairport, Ohio	Oct. 22, 1850	7, 728
Cheese-press	D. A. Church	Friendship, N. Y	May 4, 1841	2, 072
Cheese-press	C. A. Codding	Augusta, Mich	July 16, 1861	32, 820
Cheese press	C. A. Codding	Augusta, Mich	Jan. 28, 1862	34, 241
Cheese-press	G. R. Comstock	Manheim, N. Y	July 24, 1855	13, 337
Cheese-press	S. Cope	Enterprise, Ill	Mar. 29, 1859	23, 352
Cheese-press	E. J. Crane	La Porte, Ind	Nov. 12, 1867	70, 810
Cheese-press	H. Dickerman	Great Barrington, Mass	Dec. 27, 1808	....
Cheese-press	A. H. Emery	Mexico, N. Y	June 21, 1859	24, 449
Cheese-press	J. Erdle	South Bristol, N. Y	Sept. 1, 1868	81, 614
Cheese-press	J. A. Fletcher	Irasburgh, Vt	June 20, 1840	1, 650
Cheese-press	M. B. Fraser	Steuben, N. Y	July 21, 1868	80, 158
Cheese-press	A. Gaunt	Springfield, N. J	July 6, 1839	1, 218
Cheese-press	J. Goulding	Worcester, Mass	May 27, 1814	....
Cheese-press	W. H. Guy	Jonesville, Mich	Sept. 10, 1861	33, 248
Cheese-press	M. A. Hackley	Belleville, N. Y	Mar. 22, 1853	9, 625
Cheese-press	C. L. Haines	North Newburgh, Me	Apr. 18, 1871	113, 763
Cheese-press	L. Hale	Hollis, N. H	June 30, 1838	818
Cheese-press	B. F. Harriman	Warner, N. H	July 8, 1862	35, 820
Cheese-press	J. Hibbard	Weathersfield, N. Y	May 25, 1858	20, 346
Cheese-press	B. Hinkley	Fayette, Me	Oct. 15, 1832	....
Cheese-press	S. Y. Ives	Meriden, Conn	Aug. 25, 1868	81, 509
Cheese-press	L. Kellogg	Ravenna, Ohio	Feb. 6, 1849	6, 101
Cheese-press	H. Kendall	East Dorset, Vt	Oct. 26, 1858	21, 883
Cheese-press	S. Kibbe	Schoharie, N. Y	July 18, 1834	....
Cheese-press	L. Martin	Granville, N. Y	Dec. 19, 1808	....
Cheese-press	S. B. McCullough and J. R. West.	Rock Springs, Md., and Lancaster County, Pa	July 6, 1869	92, 201
Cheese-press	C. C. Musselman	Somerset, Pa	May 21, 1867	64, 998
Cheese-press	N. Norcross	Livermore, Me	Apr. 19, 1864	42, 393
Cheese-press	M. Owen	Potsdam, N. Y	Jan. 5, 1864	41, 169
Cheese-press	W. C. Pancost	Geneva Township, Ohio	Jan. 8, 1856	14, 065
Cheese-press	J. Paterson	Indianapolis, Ind	Apr. 23, 1861	32, 142
Cheese-press	D. Phelps	Bangor, Me	June 12, 1835	....

Index of patents issued from the United States Patent Office from 1790 *to* 1873, *inclusive*—Continued.

Invention.	Inventor.	Residence.	Date.	No.
Cheese press	R. Porter	Billerica, Mass	Apr. 15, 1840	1, 554
Cheese-press	J. Pride	Potsdam, N. Y	Dec. 9, 1824	....
Cheese-press	W. H. Ragan	Fillmore, Ind	June 4, 1867	65, 431
Cheese-press	E. Raymond	Sherburn, N. Y	Feb. 21, 1810	
Cheese-press	C. H. Robertson	Middleport, N. Y	Dec. 15, 1857	18, 862
Cheese-press	A. N. Severance	Cherry Valley, Ohio	Mar. 26, 1850	7, 225
Cheese-press	W. H. Stevens	Winona, Minn	Nov. 6, 1866	59, 475
Cheese-press	J. D. Stratton and T. Wilson	Mackinaw, Ill	Sept. 15, 1868	82, 174
Cheese-press	M. C. Taft	Potsdam, N. Y	Apr. 10, 1860	27, 845
Cheese-press	L. C. Tanney	Olmstead, Ohio	Dec. 13, 1864	45, 444
Cheese-press	C. Taylor	Little Falls, N. Y	Mar. 22, 1859	23, 323
Cheese-press	W. Thomas and W. Rhoades	Mukwonago, Wis	Dec. 11, 1866	60, 438
Cheese-press	W. W. Townsend	Addison County, Vt	June 13, 1812	....
Cheese-press	W. W. Townsend	Shoreham, Vt	Dec. 18, 1839	1, 433
Cheese-press	J. L. Treat	New York, N. Y	Sept. 29, 1863	40, 150
Cheese-press	A. Tyrrill	Fowler, Ohio	Dec. 31, 1833	....
Cheese-press	E. Warner	Waterbury, Conn	Apr. 12, 1808	....
Cheese-press	R. Webb and J. Cox	Madison, Conn	Feb. 27, 1830	....
Cheese-press	S. White	Bridgewater, Vt	Aug. 15, 1837	355
Cheese-press	P. Wilbor	Milan, Ohio	Aug. 22, 1854	11, 580
Cheese-press	P. Wilson	Newport, Me	June 23, 1868	79, 286
Cheese-press	L. C. Winslow	Canton, N. Y	Jan. 5, 1864	41, 113
Cheese press and hoop	D. P. Nickerson	Cleveland, Ohio	Aug. 27, 1861	33, 157
Cheese, &c., press, Combined	C. C. Musselman	Somerset, Pa	Jan. 11, 1870	98, 702
Cheese-press, "Lever and eccentric-wheel"	J. S. and E. Pulsifer	Ipswich, Mass	Feb. 14, 1831	....
Cheese-press, Self acting	I. Carter	Plattsburgh, N. Y	Aug. 14, 1849	6, 641
Cheese-press, Self-acting	B. Gillett and L. Allis	Hartford, Conn	Aug. 26, 1851	8, 311
Cheese-press, Self-acting	E. Hall	Persia, N. Y	Dec. 4, 1847	5, 381
Cheese-press, Self-acting	D. Hitchcock and C. Stone	New York	July 24, 1828	
Cheese-press, Self-acting	C. B. and J. Kingsbury	Utica, N. Y	Dec. 18, 1847	5, 397
Cheese-press, Self-acting	W. Leach	Clarkson, N. Y	Feb. 8, 1859	22, 920
Cheese-press, Self-acting	S. Mann	Alstead, N. H	Sept. 25, 1849	6, 742
Cheese-press, Self-acting	J. Martin, jr.	Azatalan, Wister	Nov. 26, 1844	3, 840
Cheese-press, Self-acting	B. H. Otis	Cleveland, Ohio	Mar. 27, 1849	6, 225
Cheese-press, Self-acting	S. W. Ruggles	Fitchburgh, Mass	July 4, 1854	11, 241
Cheese-press, Self-acting	C. Stone	Rootstown, Ohio	Aug. 7, 1847	5, 220
Cheese-press, Self-acting	C. Stone and F. K. and G. S. Collins.	Rootstown and Ravenna, Ohio.	July 28, 1842	2, 739
Cheese-press, Self acting	C. Stone and F. K. and G. S. Collins.	Rootstown and Ravenna, Ohio.	Aug. 6, 1842	2, 747
Cheese-press, Self-acting	C. Stone and G. S. Collins	Rootstown and Ravenna, Ohio.	May 29, 1845	4, 063
Cheese-press, Self-acting	J. Underwood	Montpelier, Vt	June 25, 1850	7, 465
Cheese-press, Self-adjusting	R. Porter	Billerica, Mass	Feb. 6, 1835	....
Cheese-press, Self-propelling hanging-lever	J. S. Cram	Hanover, N. H	Dec. 27, 1831	....
Cheese-presses, Construction of	H. Kendall	Danby, Vt	July 15, 1843	3, 178
Cheese, Pressing	W. McAllister	Gerry, N. Y	Sept. 4, 1860	29, 899
Cheese, Pressing	H. A. Stone	Battle Creek, Mich	Mar. 26, 1861	31, 834
Cheese-pressing machine	L. Caswell	Harrison, Me	Dec. 9, 1825	....
Cheese-pressing machine	M. Norton	Goshen, Conn	Oct. 13, 1830	....
Cheese, Process for making	H. O. Freeman	Sherburne, N. Y	Feb. 18, 1873	136, 051
Cheese, Process for manufacturing	M. A. Sheaffer	Elizabethtown, Pa	Dec. 7, 1869	97, 709
Cheese, Process of manufacturing German hand	F. C. and T. F. Mende	Philadelphia, Pa	Oct. 5, 1869	95, 500
Cheese safe, gage, and cutter	E. G. Bulgin	Vienna, N. J	July 7, 1868	79, 546
Cheese-shelf	A. N. Severance	Cherry Valley, Ohio	May 1, 1847	5, 093
Cheese, Shelving for curing and storing	J. H. Phillips	Colebrook, Ohio	Nov. 24, 1857	18, 702
Cheese-skipper extractor	C. Green	Osseo, Mich	Feb. 4, 1873	135, 542
Cheese-table	E. L. Yancy	Batavia, N. Y	Dec. 15, 1868	84, 980
Cheese-table, shelving, &c	A. M. Utley	Watertown, N. Y	May 4, 1869	89, 814
Cheese, Turning	R. Scott	Watertown, N. Y	Mar. 6, 1866	53, 048
Cheese-turning apparatus	J. Q. Black	Richland Centre, Wis	Feb. 28, 1871	112, 112
Cheese-turning apparatus	A. Mears	Brashear, N. Y	Oct. 6, 1863	40, 214
Cheese-turning cover	Q. C. Culley	Ashtabula, Ohio	Apr. 21, 1868	77, 004
Cheese-vat	A. B. Armstrong	Dorset, Vt	Aug. 29, 1871	118, 507
Cheese-vat	G. Austin	Denmark, N. Y	Mar. 19, 1867	62, 991
Cheese-vat	A. G. Bagg	Holland Patent, N. Y	Nov. 27, 1866	59, 941
Cheese-vat	D. F. Barclay	Elgin, Ill	July 8, 1873	140, 670
Cheese-vat	H. H. Bent	Antwerp, N. Y	Mar. 17, 1868	75, 515
Cheese-vat	J. A. Carlisle and G. A. Bowers	Elgin, Ill	July 10, 1866	56, 176
Cheese-vat	P. Calvin	Pecatonica, Ill	July 28, 1868	80, 335
Cheese-vat	J. H. Crumb and L. Sears	De Ruyter, N. Y	June 18, 1867	65, 881
Cheese-vat	F. E. Day	Kennedy, N. Y	Dec. 26, 1871	122, 232
Cheese-vat	L. C. Hains	Bedford, Ohio	June 19, 1866	55, 649
Cheese-vat	L. C. Hains	Bedford, Ohio	Apr. 9, 1867	63, 630
Cheese-vat	W. Howard	Watertown, N. Y	May 12, 1868	77, 734
Cheese-vat	J. Jones, C. D. Faulkner, F. L. Jones, and H. K. Faulkner.	Utica, N. Y	Jan. 30, 1872	123, 107
Cheese-vat	R. M. Livingston	Manteno, Ill	Apr. 20, 1869	89, 157
Cheese-vat	F. X. Manahan	Utica, N.Y	Feb. 25, 1862	34, 511
Cheese-vat	D. W. Maples	Homer, N. Y	Apr. 30, 1861	32, 198
Cheese-vat	H. W. Millar	Brooklyn, N. Y	Nov. 8, 1870	109, 139
Cheese-vat	W. H. Obitts	Elyria, Ohio	Nov. 22, 1870	109, 541
Cheese-vat	W. H. Obitts	Elyria, Ohio	Apr. 30, 1872	126, 321
Cheese-vat	W. Ralph	Holland Patent, N. Y	Sept. 25, 1860	30, 156
Cheese-vat	W. Ralph	Utica, N. Y	Nov. 15, 1864	45, 076
Cheese-vat	H. A. Roe	West Andover, Ohio	Dec. 12, 1854	12, 070
Cheese-vat	H. A. Roe	West Andover, Ohio	June 22, 1858	20, 663
Cheese-vat	H. A. Roe	Madison, Ohio	Jan. 27, 1863	37, 524
Cheese-vat	O. Sage	Wellington, Ohio	Oct. 4, 1859	25, 576
Cheese-vat	O. Sage	Wellington, Ohio	May 20, 1862	35, 327
Cheese-vat	O. Sage	Wellington, Ohio	Mar. 5, 1867	62, 690
Cheese-vat	E. G. Seeger, jr	Ilion, N. Y	Nov. 5, 1872	132, 864
Cheese-vat	A. Slaughter	Middletown, N. Y	Feb. 12, 1867	61, 959
Cheese-vat	E. H. and W. A. Stuart	Cedarville, N. Y	Feb. 21, 1871	112, 090
Cheese-vat	A. Westcott	Syracuse, N. Y	Jan. 22, 1867	61, 372
Cheese-vat	C. M. Wilkins	Madison, Ohio	Nov. 22, 1859	26, 222
Cheese vat, Apparatus for agitation of milk in	J. C. House	Lowville, N. Y	Aug. 27, 1867	68, 198
Cheese-vat fastener	S. P. Halleck	Oriskany, N. Y	July 25, 1871	117, 411

Index of patents issued from the United States Patent Office from 1790 *to* 1873, *inclusive*—Continued.

Invention.	Inventor.	Residence.	Date.	No.
Cheese-vat for forming pine-apple cheese	L. M. Norton	Goshen, Conn	Apr. 17, 1810	....
Cheese-vat heater	H. Cooper	Watertown, N. Y	Feb. 12, 1861	31, 373
Cheese-vat heater	H. W. Miller	Utica, N. Y	June 18, 1867	65, 823
Cheese-vat heater	H. H. Roe	Madison, Ohio	Sept. 10, 1872	131, 302
Cheese-vat operator	C. M. Wilkins	West Andover, Ohio	Apr. 2, 1861	31, 925
Cheese, Vat or press-box for	A. N. Severance	Cherry Valley, Ohio	Apr. 30, 1850	7, 328
Cheese-vat, Steam	J. A. Davis	Watertown, N. Y	Oct. 12, 1869	95, 662
Chemical catholicon	I. W. Smith	Lockport, N. Y	May 28, 1830	....
Chemise, Lady's	R. L. Jones	Sacramento, Cal	Aug. 3, 1869	93, 390
Chemise, Nursing	H. Wolf	New York, N. Y	Oct. 29, 1872	132, 613
Chemistry, Method of teaching the rudiments of	S. M. Gaines	Glasgow, Ky	Dec. 29, 1868	85, 299
Chenille	W. Canter	New York, N. Y	Oct. 29, 1867	70, 163
Chenille, Machine for making	J. Thomas	New York, N. Y	Aug. 30 1864	44, 050
Chenille, Machine for manufacturing	W. Canter	New York, N. Y	July 21, 1868	80, 134
Chenille, Machine for manufacturing	G. Comings and L. Mensing	New York, N. Y	Jan. 13, 1863	37, 385
Chenille, Machinery for manufacturing	W. Canter	New York, N. Y	July 22, 1862	35, 969
Chenille, Machinery for manufacturing	W. Canter	New York, N. Y	Jan. 13, 1863	37, 415
Cherry-pitter	W. C. Barr	Macon City, Mo	Feb. 2, 1869	86, 349
Cherry-seeder	L. P. Evans	Springville, Pa	May 14, 1867	64, 651
Cherry-stoner	E. Buck and E. W. Kirk	Cincinnati, Ohio	Nov. 25, 1873	144, 887
Cherry-stoner	H. Buckwalter	Kimberton, Pa	Nov. 17, 1863	40, 604
Cherry-stoner	A. M. Comstock	Galesburgh, Ill	Dec. 5, 1871	121, 592
Cherry-stoner	F. G. and E. A. Floyd	Macomb, Ill	Jan. 22, 1867	61, 331
Cherry-stoner	G. Geer	Galesburgh, Ill	Apr. 9, 1867	63, 716
Cherry-stoner	G. Geer	Galesburgh, Ill	July 7, 1868	79, 748
Cherry-stoner	J. S. Lash	Philadelphia, Pa	Aug. 12, 1873	141, 649
Cherry stoner	J. Marchant	Farmington, Ill	Apr. 19, 1870	102, 021
Cherry-stoner	A. Rakestraw	Peoria, Ill	June 23, 1868	79, 257
Cherry-stoner	O. L. Robinson	Owasso, Mich	July 16, 1867	66, 889
Cherry-stoner	E. Smith	Farmington, Ill	Aug. 6, 1867	67, 600
Cherry-stoner	J. W. Thompson	Salem, Ohio	Oct. 30, 1866	59, 321
Cherry-stoner	T. Van Kannel	Chester, Ill	May 5, 1863	38, 434
Cherry-stoner	D. E. Warner	Chicago, Ill	Jan. 21, 1873	135, 050
Cherry stoner	W. Weaver	Phœnixville, Pa	May 15, 1866	54, 797
Cherry stoner	J. N. Webster	Peoria, Ill	May 31, 1870	103, 802
Cherry-stoner	C. E. Wright	Auburn, N. Y	Oct. 2, 1866	58, 531
Cherry-stoner	R. Wright	Brooklyn, N. Y	Feb. 4, 1868	74, 187
Cherry-stoner	R. Wright	Brooklyn, N. Y	May 5, 1868	77, 561
Cherry-stoning machine	J. Baker	Washington, D. C	Mar. 2, 1858	19, 476
Cherry-stoning machine	E. C. Custer	Evansburgh, Pa	July 26, 1859	24, 856
Cherry-stoning machine	T. Van Kannel	Chester, Ill	Oct. 14, 1862	36, 683
Cherry-stoning machine	T. Van Kannel	Cincinnati, Ohio	June 6, 1865	48, 137
Chess and checker board	S. L. Fleishman	Pittsburgh, Pa	Nov. 12, 1872	132, 905
Chess-board	F. C. Schaefer	Dubuque, Iowa	Feb. 27, 1866	52, 891
Chest:				
See Bolting-chest.				
Camp-chest.				
Fire-proof chest.				
Flour-chest.				
Meal-chest.				
Medicine-chest.				
Steam-chest.				
Chest and box fastening	J. H. Marvil	Laurel, Del	May 14, 1872	126, 819
Chest and table	G. W. Ziegler	Saint Louis, Mo	Sept. 5, 1865	49, 821
Chest, Butter	F. S. Sears	Charlestown, Mass	June 22, 1869	91, 776
Chest-expander	S. M. Barnett	New Orleans, La	Feb. 5, 1867	61, 702
Chest-expander	S. M. Barnett	New York, N. Y	Apr. 4, 1871	113, 384
Chest of drawers	C. H. Walker	Warren, Mass	May 20, 1862	35, 340
Chest-protector	P. Lear	Boston, Mass	Feb. 28, 1871	112, 258
Chest-protector	E. F. Wilder	West Roxbury, Mass	Mar. 25, 1873	137, 275
Chest-protector and shirt-bosom support, Combined	J. A. Aston and C. C. McMurphy	Leavenworth, Kans	Sept. 2, 1873	142, 366
Chevron	F. S. Johnston	Philadelphia, Pa	Jan. 21, 1873	135, 124
Chignon	M. M. Cohen	Boston, Mass	Aug. 2, 1870	105, 914
Chignon	E. Ulman	New York, N. Y	Mar. 5, 1872	124, 299
Child's chair	J. H. Apel	Boston, Mass	Dec. 22, 1868	85, 050
Child's chair	R. Ardrey	Philadelphia, Pa	Dec. 23, 1873	145, 829
Child's chair	J. F. Downing	Erie, Pa	May 6, 1873	138, 620
Child's chair	J. F. Downing	Erie, Pa	July 15, 1873	140, 900
Child's chair	J. F. Harris and E. D. Childs	Charlestown, Mass	Apr. 8, 1873	137, 610
Child's chair	A. H. Wohser	San Francisco, Cal	June 21, 1870	104, 517
Child's chair and carriage	J. Lee	New York, N. Y	June 15, 1869	91, 350
Child's chair and wagon	J. Knowles	Lowell, Mass	July 28, 1846	4, 666
Child's chair and walker	C. Holtz	Chicago, Ill	May 8, 1866	54, 546
Chill for casting chilled rolls	R. Tolmie	Wilmington, Del	Feb. 20, 1872	123, 953
Chill for mold-boards and other castings	W. Rall	South Bend, Ind	Dec. 31, 1872	134, 439
Chill-molds, Method of taking air off	G. R. H. Leffler	Baltimore, Md	Apr. 17, 1866	53, 993
Chill plate and flask	D. Long and S. A. Miller	Louisville, Ky	Aug. 27, 1872	130, 926
Chilled rollers, Mold for casting	A. Hammond	Jacksonville, Ill	Oct. 6, 1863	40, 167
Chilled rolls, Casting	G. G. Lobdell	Wilmington, Del	June 10, 1873	139, 798
Chilled rolls, Casting	J. R. Pond	New Hartford, Conn	Aug. 6, 1872	130, 151
Chilled wheels, Manufacture of cast	G. Whiting	Philadelphia, Pa	July 16, 1872	129, 382
China, glass, and other articles, Mode of protecting	A. C. Rand	Union Mills, Pa	Apr. 9, 1867	63, 653
China, glass, &c., Composition for mending	H. Bascom	Weston, Mass	June 10, 1811	....
China, glass, &c., Solution for gilding	L. P. Angenard	New York, N. Y	June 14, 1864	43, 180
Chimney	R. M. Bassett and G. Mallory	Birmingham and Watertown, Conn.	Dec. 1, 1863	40, 730
Chimney	J. Browell	San Francisco, Cal	Apr. 1, 1873	137, 412
Chimney	C. H. Brown	Atlantic, Iowa	Oct. 17, 1871	120, 031
Chimney	D. Cooley, jr., and G. N. Philips	....	Aug. 25, 1800	....
Chimney	A. B. Ewing	Lewisburgh, Tenn	Mar. 28, 1870	113, 156
Chimney	G. Gridley	Newburyport, Mass	Mar. 13, 1808	....
Chimney	J. Heed	Temperanceville, Ohio	July 2, 1867	66, 339
Chimney	S. Hoke	Mount Pleasant, Ind	Oct. 20, 1868	83, 282
Chimney	S. Hopkins	Napoleon, Ohio	June 13, 1871	115, 961
Chimney	L. Lecesne	Philadelphia, Pa	June 26, 1809	....
Chimney	J. Leeds	Philadelphia, Pa	Mar. 8, 1859	23, 179
Chimney	B. F. Mann	Oakland, Cal	Oct. 12, 1869	95, 702

Index of patents issued from the United States Patent Office from 1790 *to* 1873, *inclusive*—Continued.

Invention.	Inventor.	Residence.	Date.	No.
Chimney	S. M. McCord	Springfield, Ohio	Aug. 1, 1871	117, 555
Chimney	W. Musbach and C. R. Smith	Middletown, N. Y	July 6, 1869	92, 341
Chimney	S. Oakman	Boston, Mass	July 20, 1871	117, 320
Chimney	C. C. Phelps	Janesville, Wis	Apr. 25, 1865	47, 448
Chimney	E. S. Phelps, jr	Wyanet, Ill	Dec. 10, 1867	72, 151
Chimney	P. Portois	San Francisco, Cal	Sept. 21, 1869	94, 972
Chimney	W. G. Reed	Chelsea, Mass	Apr. 20, 1869	89, 075
Chimney	F. Richardson	Hebron, Ill	Apr. 19, 1870	102, 042
Chimney	C. W. Saladee and T. R. Eddy	Newark, Ohio	Oct. 16, 1866	58, 892
Chimney	C. F. Smith and J. Speth	Aurora, Ill	Apr. 2, 1867	63, 436
Chimney	A. J. Sprague	Springfield, Mo	Sept. 13, 1870	107, 421
Chimney	R. Stuart		Feb. 24, 1797	
Chimney	B. W. Taber	Quaker Street, N. Y	July 24, 1860	29, 330
Chimney	A. G. D. Tuthill	Utica, N. Y	June 8, 1826	
Chimney	A. S. Whittemore	Willimantic, Conn	Nov. 12, 1867	70, 927
Chimney	A. Wilhelms	St. Petersburg, Russia	Dec. 1, 1868	84, 600
Chimney, Air-back for	J. Root	Southington, Conn	Feb. 13, 1812	
Chimney and fire-place	J. Briggs	Toronto, Canada	July 9, 1872	128, 847
Chimney and fire-place	J. Briggs	Toronto, Canada	May 20, 1873	139, 111
Chimney and fire-place	M. Perin	Connersville, Ind	Sept. 26, 1835	
Chimney and fire-place	A. Sphon	Comrio, Pa	Sept. 22, 1814	
Chimney-attachment	A. H. Lanphear	Atchison, Kans	Sept. 27, 1870	107, 694
Chimney-attachment to increase the draft and prevent smoking.	J. Hurd, jr	Stoneham, Mass	Apr. 24, 1841	2, 063
Chimney-cap	M. Anderson	Brooklyn, N. Y	Aug. 6, 1867	67, 477
Chimney-cap	T. D. Bailey	Lowell, Mass	Mar. 14, 1871	112, 529
Chimney-cap	L. F. Betts	Albion, Mich	Jan. 6, 1863	37, 270
Chimney-cap	T. Boyd	Cambridgeport, Mass	July 22, 1873	141, 107
Chimney-cap	W. Brownell	Newport, R. I	Sept. 16, 1856	15, 726
Chimney-cap	W. S. Burch	Washington, D. C	July 14, 1846	4, 631
Chimney-cap	W. Chappell	Buffalo, N. Y	Oct. 6, 1868	82, 692
Chimney-cap	M. Chase	Boston, Mass	Mar. 7, 1846	4, 392
Chimney-cap	J. Clark	Washington, D. C	Sept. 5, 1854	11, 640
Chimney-cap	M. H. Collins	Boston, Mass	Apr. 2, 1850	7, 232
Chimney-cap	W. Creutzfeldt	Washington, D. C	Nov. 6, 1847	5, 356
Chimney-cap	J. W. Davies	Richmond, Va	Nov. 6, 1855	13, 772
Chimney-cap	G. W. Demond	Boston, Mass	May 2, 1865	47, 525
Chimney-cap	C. Douglas	Cleveland, Ohio	Nov. 23, 1858	22, 112
Chimney-cap	C. Douglas	Hebron, Conn	May 24, 1859	24, 107
Chimney-cap	T. Ewbank	New York, N. Y	Dec. 17, 1842	2, 888
Chimney-cap	F. A. Finn	New York, N. Y	Dec. 12, 1848	5, 962
Chimney-cap	T. J. Fitzpatrick	New Orleans, La	July 24, 1860	29, 262
Chimney-cap	A. Hamann	Washington, D. C	July 22, 1845	4, 117
Chimney-cap	J. Hammond	Adams Centre, N. Y	Oct. 8, 1867	69, 660
Chimney-cap	J. P. Hayes	Boston, Mass	Oct. 10, 1848	5, 839
Chimney-cap	E. Hinkley and G. W. Crowell	Cleveland, Ohio	Sept. 19, 1865	50, 086
Chimney-cap	W. H. Horton	Jersey City, N. J	Sept. 19, 1865	50, 002
Chimney-cap	W. H. Horton	Jersey City, N. J	Mar. 20, 1866	53, 301
Chimney-cap	A. B. Johnson	Washington, Ind	Dec. 19, 1871	122, 022
Chimney-cap	B. Kihlholz	Saint Louis, Mo	Oct. 26, 1858	21, 884
Chimney-cap	J. H. Kirkwood	Cleveland, Ohio	Sept. 12, 1865	49, 890
Chimney-cap	T. A. Mann	West Greenville, Pa	Apr. 26, 1864	42, 496
Chimney-cap	I. Mayhew	Albion, Mich	Sept. 1, 1857	18, 100
Chimney-cap	B. F. and J. G. Miller	New York, N. Y	Sept. 13, 1870	107, 282
Chimney-cap	I. L. Mott	New York, N. Y	Dec. 17, 1842	2, 887
Chimney-cap	E. Myrick	Harvard, Mass	May 25, 1869	90, 380
Chimney-cap	W. J. Pettingell	Lowell, Mass	July 11, 1871	116, 864
Chimney-cap	H. Reynolds	New York, N. Y	Jan. 30, 1866	52, 322
Chimney-cap	A. M. Rice	Boston, Mass	Apr. 9, 1850	7, 275
Chimney-cap	C. W. Russell	Washington, D. C	Dec. 16, 1851	8, 595
Chimney-cap	C. K. Scudder	Brooklyn, N. Y	May 15, 1849	6, 449
Chimney-cap	J. H. Shedd and B. Worcester	Waltham, Mass	July 3, 1866	56, 109
Chimney-cap	A. L. Sweet	Norwich, Conn	Feb. 16, 1864	41, 649
Chimney-cap	J. Tomlinson	Racine, Wis	Sept. 8, 1863	39, 846
Chimney-cap	H. J. Weed	Cazenovia, N. Y	Apr. 14, 1868	76, 859
Chimney-cap	E. Whiteley	Boston, Mass	Mar. 5, 1850	7, 152
Chimney-cap, Adjustable	P. H. Carlin	Brooklyn, N. Y	Oct. 15, 1872	132, 249
Chimney-cap, Cast-metal	W. Green	Snyder Township, Pa	Oct. 12, 1869	95, 795
Chimney-cap for increasing draft in furnace, &c	J. P. Espy	Philadelphia, Pa	June 29, 1833	
Chimney-cap for preventing the emission of sparks	E. A. G. Young	New Castle, Del	July 22, 1833	
Chimney-cap or conical ventilator	P. Lear	Medford, Mass	Jan. 23, 1866	52, 177
Chimney cap or cowl	C. G. Miller	Cincinnati, Ohio	Nov. 12, 1867	70, 734
Chimney-cap plates, Fastening for	W. Griffiths	Philadelphia, Pa	Sept. 13, 1864	44, 182
Chimney-caps, Mode of constructing	B. Kenney	Northfield, Mass	July 22, 1839	1, 257
Chimney, Cast-iron	D. June	Fremont, Ohio	Jan. 22, 1867	61, 342
Chimney, Cast-iron	A. W. McMillen	Chicago, Ill	Mar. 31, 1868	76, 094
Chimney-clasp	C. F. Espick	Plymouth, Ind	June 2, 1868	78, 584
Chimney-cleaner	T. H. Donohue	Washington, D. C	May 5, 1868	77, 466
Chimney-cleaner	G. S. Knapp	Winona, Minn	Jan. 18, 1870	98, 980
Chimney-cleaner and fire-safe, Combined	C. F. Eckstein and L. Jurgens	Prescott, Wis	June 25, 1872	128, 376
Chimney-cleaning machine	J. Bruff	Baltimore, Md	Dec. 16, 1814	
Chimney-cleaning machine	S. Dow	Elizabeth Borough, N. J	June 11, 1829	
Chimney-collar	J. H. Congdon	Coventry, R. I	May 23, 1871	115, 168
Chimney-collar	J. Stokely	Hiram, Ohio	Jan. 29, 1867	61, 639
Chimney-cowl	W. N. Abbott	New York, N. Y	Apr. 25, 1871	113, 961
Chimney-cowl	S. M. Allen	Boston, Mass	Apr. 21, 1842	2, 568
Chimney-cowl	C. W. Bache	Philadelphia, Pa	July 26, 1870	105, 624
Chimney-cowl	C. B. Barlow	Portsmouth, N. H	Apr. 26, 1870	102, 208
Chimney-cowl	D. C. Battey and C. L. Svenson.	Topeka, Kan	Aug. 10, 1869	93, 400
Chimney-cowl	A. Bedlow	Newport, R. I	June 7, 1859	24, 276
Chimney-cowl	J. Burgy	Powhatan Point, Ohio	Aug. 5, 1873	141, 541
Chimney-cowl	E. P. H. Capron	Springfield, Ohio	Mar. 1, 1870	100, 372
Chimney-cowl	L. N. Chapin	New Lisbon Township, N.Y	Jan. 25, 1870	99, 162
Chimney-cowl	T. W. Chatfield	Utica, N. Y	Oct. 14, 1856	15, 916
Chimney-cowl	A. E. Clement	Wapakonetta, Ohio	June 30, 1863	79, 457
Chimney-cowl	J. J. Currier	Gloucester, Mass	Sept. 29, 1868	82, 497
Chimney-cowl	H. S. Decker	New York, N. Y	Sept. 1, 1868	81, 609

Index of patents issued from the United States Patent Office from 1790 to 1873, inclusive—Continued.

Invention.	Inventor.	Residence.	Date.	No.
Chimney-cowl	N. Douglas	Goshen, Ind	Apr. 5, 1870	101, 443
Chimney-cowl	W. C. Frailey	Ironton, Ohio	Aug. 4, 1868	80, 618
Chimney-cowl	G. H. Garderwine	Nokomis, Ill	Sept. 16, 1873	142, 909
Chimney-cowl	B. J. Goodsell	Pent Water, Mich	Nov. 19, 1867	70, 993
Chimney-cowl	D. Hahn	Reserve, Ind	Mar. 22, 1870	101, 123
Chimney-cowl	M. H. Hale and S. Horton	Newburyport, Mass	Feb. 17, 1857	16, 644
Chimney-cowl	E. Hewett	Saint Leonard's, England	Nov. 2, 1869	96, 430
Chimney-cowl	B. Irrgang	Philadelphia, Pa	Oct. 20, 1868	83, 166
Chimney-cowl	D. F. Jauss	Harrisburgh, Pa	Apr. 21, 1868	77, 047
Chimney-cowl	M. Lockhart	Douglas, Isle of Man	Mar. 1, 1870	100, 422
Chimney-cowl	P. Mihan	Boston, Mass	Dec. 16, 1856	16, 246
Chimney-cowl	G. Millard	Waterbury, Conn	May 15, 1860	28, 291
Chimney-cowl	F. W. Mulvany	New York, N. Y	July 16, 1872	129, 160
Chimney-cowl	W. H. Myers	Philadelphia, Pa	Apr. 8, 1873	137, 561
Chimney-cowl	E. T. Noualhier	Paris, France	Sept. 29, 1868	82, 632
Chimney-cowl	J. F. Pond	Cleveland, Ohio	June 1, 1869	90, 684
Chimney-cowl	J. H. Richardson	Philadelphia, Pa	May 20, 1873	139, 080
Chimney-cowl	J. G. Roth	New York, N. Y	Oct. 10, 1871	119, 885
Chimney-cowl	T. S. Speakman	Camden, N. J	May 25, 1869	90, 600
Chimney-cowl	P. Sumner	New York, N. Y	Feb. 20, 1843	2, 964
Chimney-cowl	G. W. Thatcher	Philadelphia, Pa	Aug. 12, 1856	15, 536
Chimney-cowl	C. F. Thomas	Taunton, Mass	Jan. 29, 1856	14, 171
Chimney-cowl	G. G. Thomas	Saint Louis, Mo	July 11, 1871	116, 887
Chimney-cowl	C. Turner	Southampton, England	July 2, 1872	128, 514
Chimney-cowl	F. Villard	Mount Eaton, Ohio	Feb. 15, 1870	99, 983
Chimney-cowl	J. A. Wagoner	Kilgore, Ohio	June 18, 1867	65, 969
Chimney-cowl	J. Walker	Boston, Mass	July 18, 1871	117, 129
Chimney-cowl	J. W. Whiting and C. E. Whittemore	Bay City, Mich	Feb. 9, 1869	86, 891
Chimney-cowl, Self-clearing	C. H. Watkins	New York, N. Y	Aug. 19, 1856	15, 584
Chimney-fastener	R. W. Hawkins	Pittsburgh, Pa	May 5, 1863	38, 388
Chimney, fire-place, air-flue, and ventilator	J. C. Brush	Washington, D. C	Jan. 28, 1814	....
Chimney-flue	W. J. Fryer, jr	Albany, N. Y	Feb. 9, 1864	41, 498
Chimney-flue	C. T. Harvey	Chicago, Ill	Mar. 22, 1864	41, 993
Chimney-flue base	C. F. Rothe	Baltimore, Md	Aug. 13, 1872	130, 539
Chimney-furnace	L. White	Waterbury, Conn	Oct. 14, 1873	143, 739
Chimney-heater	T. M. Aspinall and S. H. Whitlock	Piqua, Ohio	Apr. 10, 1866	53, 770
Chimney holder and fastening	R. H. Dewey	Pittsfield, Mass	Feb. 6, 1866	52, 399
Chimney-jack, Ship	J. Kingsborough	Cleveland, Ohio	Apr. 30, 1867	64, 229
Chimney joint	M. H. Kelsey	Red Bank, N. J	May 7, 1867	64, 428
Chimney-lining	C. Schivers and S. Ustick	Philadelphia, Pa	Oct. 15, 1861	33, 498
Chimney or flue attachment	J. E. Starns	Worthington, Ind	Oct. 7, 1873	143, 540
Chimney or globe holder	R. Wright and J. H. Race	Brooklyn, N. Y	Aug. 5, 1873	141, 478
Chimney-safe	G. B. Clarke	Leonardsville, N. Y	Feb. 20, 1855	12, 404
Chimney scraping and sweeping machine	J. Hunt	Baltimore, Md	July 12, 1817	....
Chimney-scraping machine	M. Benson	Cincinnati, Ohio	May 15, 1823	....
Chimney, Smoke preventer	L. Silliman	Albany, N. Y	July 20, 1831	....
Chimney smoking, Apparatus to prevent	U. B. A. Lange	Philadelphia, Pa	Oct. 14, 1834	....
Chimney-stack	B. F. Miller	New York, N. Y	Oct. 2, 1855	13, 620
Chimney-stack blower	N. L. Blanchard	Spuyten Duyvil, N. Y	July 25, 1871	117, 252
Chimney stack or cap	M. M. Camp	New Haven, Conn	Sept. 4, 1855	13, 520
Chimney, Steamboat	W. J. Hamilton	Cairo, Ill	Feb. 20, 1872	123, 825
Chimney-stop	C. H. Earle	De Pere, Wis	Aug. 29, 1871	118, 521
Chimney, Stove	C. Varle	Fredericktown, Md	July 1, 1814	....
Chimney-sweeping	W. Hall	New York, N. Y	Dec. 19, 1818	....
Chimney-sweeping machine	S. S. Edmonston	New York, N. Y	Nov. 8, 1820	....
Chimney-sweeping machine	A. Lane	New York, N. Y	Sept. 22, 1818	....
Chimney-sweeping machine	J. W. Moore	Washington, D. C	Aug. 8, 1821	....
Chimney-top	T. Boyd	Cambridge, Mass	Sept. 14, 1869	94, 864
Chimney-top	J. C. H. Brown	Jersey City, N. J	Dec. 16, 1873	145, 621
Chimney-top	G. Elbreg	Cincinnati, Ohio	Feb. 17, 1863	37, 683
Chimney-top	G. Elbreg	Cincinnati, Ohio	Feb. 2, 1864	41, 427
Chimney-top	H. English	Wilmington, Del	Sept. 15, 1868	82, 098
Chimney-top	A. L. Geserick	Saint Louis, Mo	Dec. 13, 1864	45, 403
Chimney-top	J. Gorton	New York, N. Y	Feb 25, 1873	136, 234
Chimney-top	R. W. Griffith	Georgetown, D. C	Feb. 14, 1871	111, 839
Chimney-top	N. Hackett	Albany, N. Y	Jan. 1, 1861	31, 015
Chimney-top	B. A. Henriksen	San Francisco, Cal	Oct. 10, 1865	50, 356
Chimney-top	M. S. Kavanagh	Detroit, Mich	Apr. 4, 1871	113, 527
Chimney-top	J. D. Kennard	Barnesville, Ohio	July 2, 1872	128, 633
Chimney-top	T. Ketchen	New York, N. Y	Dec. 5, 1871	121, 633
Chimney-top	H. Markthaler	Elizabeth, N. J	June 29, 1869	91, 950
Chimney-top	M. E. Mead	Darien, Conn	Nov. 29, 1870	109, 642
Chimney-top	H. Palmer	Barnesville, Ohio	Jan. 7, 1873	134, 700
Chimney-top	C. M. Reynolds	Mifflin, Wis	Jan. 11, 1870	98, 800
Chimney-top	W. Richards	London, England	May 2, 1871	114, 342
Chimney-top	D. S. Robinson	Pittsburgh, Pa	Mar. 29, 1870	101, 313
Chimney-top	J. Snively	Williamsburgh, Pa	Aug. 27, 1867	68, 126
Chimney-top	G. Wingate	Boston, Mass	Nov. 11, 1873	144, 586
Chimney-top, Iron	J. Pettingell	Lowell, Mass	Oct. 9, 1860	30, 345
Chimney-top, Revolving	C. L. Garfield	Albany, N. Y	Aug. 26, 1873	142, 225
Chimney-top, Revolving	F. Villard	Mount Eaton, Ohio	Mar. 5, 1867	62, 578
Chimney-tops, Self-regulating draft for	J. A. Royce	Lee, Mass	Sept. 23, 1856	15, 779
Chimney, Ventiduct-topping	J. Ennis	Brooklyn, N. Y	Oct. 1, 1830	....
Chimney-ventilator	M. Collins	Boston, Mass	Apr. 25, 1846	4, 487
Chimney-ventilator	J. J. Pemberton	Oakland, Ill	July 27, 1869	92, 999
Chimneys and fire-places, Construction of	L. Kingsbury	Livonia, N. Y	Feb. 9, 1825	....
Chimneys and flues, Construction of	J. Kleckner	Mottville, Mich	Mar. 8, 1870	100, 705
Chimneys and railway-car, Cowl for	S. Lutz	Philadelphia, Pa	Jan. 9, 1872	122, 619
Chimneys, Apparatus for increasing draft of	A. Niel	Brooklyn, N. Y	Jan. 11, 1859	22, 570
Chimneys, Building	J. Gilbert	Stark County, Ohio	Nov. 13, 1844	3, 826
Chimneys by means of steam, Method of creating draft in	W. H. Martin	Brooklyn, N. Y	Mar. 31, 1868	76, 219
Chimneys, Cap for regulating the draft of	J. Hurd	Stoneham, Mass	Dec. 12, 1844	3, 854
Chimneys, Carbon-arrester for	H. Chase	Lynn, Mass	Sept. 2, 1856	15, 645
Chimneys, Cleaning	T. Hinkley	Hallowell, Me	Oct. 7, 1831	....
Chimneys, Constructing	H. Antes	Harrisburgh, Pa	Oct. 24, 1831	....
Chimneys, Construction of	J. B. Kelsey	Newburyport, Mass	Sept. 11, 1847	5, 287

Index of patents issued from the United States Patent Office from 1790 *to* 1873, *inclusive*—Continued.

Invention.	Inventor.	Residence.	Date.	No.
Chimneys, Construction of factory	S. Rodman	New Bedford, Mass	July 11, 1848	5,664
Chimneys, Curing smoky	D. Bain	Baltimore, Md	Feb. 5, 1833	....
Chimneys, Curing smoky	S. Chase	Williamsport, Pa	July 26, 1824	....
Chimneys, Curing smoky	J. Plant	Washington, D. C	Nov. 21, 1845	4,279
Chimneys, Curing smoky back and hearth of	H. Pollock	Baltimore, Md	Mar. 11, 1834	....
Chimneys, Device for lowering and raising steamboat.	C. and R. Hawthorn	Allegheny, Pa	Feb. 20, 1872	123,896
Chimneys, Device for lowering and raising steamboat.	W. Weaver	New Albany, Ind	Jan. 30, 1872	123,314
Chimneys, Device for preventing leakage about	A. Lang	Buffalo, N. Y	Dec. 21, 1869	98,074
Chimneys, Draft in	E. Skinner	Sandwich, N. H	Apr. 3, 1834	....
Chimneys, Fire-frame for	J. Correja	Brooklyn, N. Y	Dec. 3, 1867	71,708
Chimneys, fire-guards &c., Cast-iron copings for	C. Neer	Waterford, N. Y	Sept. 11, 1829	....
Chimneys, fire-places, &c., Construction of	E. Rand	Hudson, N. Y	Mar. 27, 1816	....
Chimneys, Guard for preventing leakage around	J. F. Schuyler	Fostoria, Ohio	Dec. 2, 1873	145,245
Chimneys, Machine for cleaning	J. Slaughter	Strasburgh, Pa	Nov. 4, 1846	4,835
Chimneys, Mode of constructing the draft of	G. H. Crosley	New Haven, Ohio	Jan. 28, 1840	1,479
Chimneys, Preventing smoke and extinguishing fire in.	J. Sullivan	Washington, D. C	Apr. 27, 1832	....
Chimneys, Preventive for smoking	F. Crey	Baltimore, Md	Oct. 8, 1821	....
Chimneys, Rain-excluder for	J. J. Thomas		Oct. 2, 1804	....
Chimneys, Regulating draft in steamboat and other	J. B. Campbell	Cincinnati, Ohio	Mar. 19, 1867	62,937
Chimneys, stoves, and fire-places, Construction of	L. D. Detroismonts	New York, N. Y	Apr. 5, 1817	....
Chimneys to prevent smoking, Building	A. H. Read	Montrose, Pa	Apr. 3, 1829	....
Chimneys to prevent smoking, Constructing	O. Richardson	Medway, Mass	Apr. 18, 1810	....
Chimneys, Wind-guard for	F. M. Lutler	New York, N. Y	Aug. 10, 1858	21,115
Chisel	R. and S. Fairchild	Trumbull, Conn	Feb. 27, 1847	4,990
Chisel	C. E. L. Jelliffe	Bayonne, N. J	Oct. 7, 1873	143,413
Chisel, Cutting and clearing	G. Page	Keene, N. H	Aug. 14, 1833	....
Chisel, Dovetailing	N. H. Robinson	Owasso, Mich	May 31, 1870	103,659
Chisel for cutting gains	H. Bigelow	Skowhegan, Me	Aug. 16, 1870	106,310
Chisel for opening boxes	D. N. Dunzack	Salem, Mass	June 26, 1860	28,848
Chisel for pointing pickets	A. Damitio	Detroit, Mich	Jan. 30, 1872	123,243
Chisel, Grooving	H. G. Terwilliger	Scranton, Pa	Mar. 8, 1870	100,690
Chisel-handle	R. V. Hilton, J. G. Webster, and H. E. Wheeler.	Lowell, Mass	May 12, 1868	77,881
Chisel-holder	J. Russell	Sing Sing, N. Y	Apr. 10, 1866	53,884
Chisel-holder for file-cutting machine	W. J. Birdsall	Newark, N. J	Nov. 23, 1869	97,030
Chisel in mortising-machine, Device for reversing	C. B. Rogers	Norwich, Conn	Oct. 27, 1857	18,521
Chisel in mortising-machine, Method of reversing	D. M. Cummings and P. C. Cambridge, jr.	North Enfield, N. H	Nov. 3, 1857	18,535
Chisel. Mortising	O. Adams and J. Hatch	San Francisco, Cal	Sept. 15, 1868	82,058
Chisel, Mortising	M. Feigel	New Utrecht, N. Y	June 22, 1869	91,531
Chisel, Mortising	J. B. Fisher	Beaver Dam, Wis	Feb. 8, 1859	22,863
Chisel, Mortising	C. J. Heistand	Rapho, Pa	July 28, 1857	17,878
Chisel, Mortising	C. Hinz	San Francisco, Cal	Aug. 10, 1869	93,534
Chisel, Mortising	G. P. Ketcham	Bedford, Ind	May 12, 1857	17,285
Chisel, Mortising	I. W. McGaffey	Philadelphia, Pa	Jan. 10, 1854	10,421
Chisel, Mortising	G. Page	Keene, N. H	July 7, 1835	....
Chisel, Mortising	C. Rhinehart	Marietta, Pa	Apr. 8, 1835	....
Chisel, Mortising	J. A. Scroggs	Burlington, Vt	Sept. 8, 1857	18,161
Chisel, Mortising	L. S. Shuler and J. Carpenter	Jeffersonville, Ind	Aug. 27, 1872	130,874
Chisel, Mortising	J. M. Smith	Warren, N. J	July 28, 1868	80,428
Chisel, Mortising-machine	E. McConnell	Sharon Pa	Mar. 28, 1871	113,075
Chisel, Stone-cutter's	W. C. Peckham	Troy, Ohio	Oct. 29, 1872	132,541
Chisel, Turning	R. P. Buttles and S. Sweet	Mansfield, Pa	Apr. 19, 1870	101,981
Chisels, Socket-handle for	L. F. Richardson	Worcester, Mass	June 19, 1855	13,101
Chisels to mandrels, Joint for uniting mortising	J. R. Perry	Port Clinton, Pa	Nov. 4, 1856	16,020
Chloride of calcium engine	E. Lamm	New Orleans, La	Mar. 12, 1872	124,594
Chlorine, Apparatus for manufacture of	H. Deacon	Widnes, England	Aug. 22 1871	118,209
Chlorine, Apparatus for producing	H. Deacon	Widnes, England	Aug. 22, 1871	118,211
Chlorine, bleaching-powder, carbonate of soda, and other products, Process of preparing.	T. Macfarlane	Acton Vale, Canada	Aug. 22, 1865	49,597
Chlorine, Compound for neutralizing	E. N. Horsford	Cambridge, Mass	Oct. 10, 1854	11,786
Chlorine, Manufacture of	H. Deacon	Appleton, England	Dec. 29, 1868	85,370
Chlorine, Manufacture of	H. Deacon	Lancaster County, England	July 29, 1873	141,333
Chlorine, Producing	L. E. Aubertin	Paris, France	Dec. 24, 1872	134,190
Chocolate-drops and other confections, Apparatus for manufacture of.	W. C. Murdock and E. K. Haynes.	Boston, Mass	Jan. 4, 1870	98,615
Chocolate-drops and other confections, Apparatus for manufacture of.	S. A. Parker	Cambridge, Mass	Oct. 17, 1871	119,946
Chocolate, Grinding	G. W. Waite	Baltimore, Md	June 25, 1836	....
Chocolate ingredients, Heating	G. W. Waite	Baltimore, Md	June 25, 1836	....
Chocolate, Manufacture of	J. M. O. Tamin	New York, N. Y	Aug. 12, 1873	141,832
Chocolate, Molding	G. W. Waite	Baltimore, Md	June 25, 1836	....
Chocolate-paste	L. F. Leger	New York, N. Y	Oct. 5, 1869	95,492
Chocolate, Soluble	J. Corell	New York, N. Y	Apr. 22, 1873	138,007
Cholera, Method of treating	A. C. Campbell	Mount Morris, N. Y	Mar. 27, 1866	53,407
Chopper: *See* Beefsteak-chopper. Cornstalk-chopper. Cotton-chopper. Grain-chopper. Meat-chopper. Sausage-chopper. Stalk-chopper. Vegetable-chopper.				
Chopping-knife, Hand	F. M. Untiedt	East Orange, N. J	June 22, 1869	91,795
Chopping-machine	J. A. Eberly	Reamstown, Pa	Jan. 23, 1872	123,004
Chopping-machine	E. Newcomer	Columbia, Pa	May 9, 1871	114,589
Chord-bar die	W. Forshaw	Chicago, Ill	Nov. 22, 1870	109,506
Chord-bar-head die	F. J. Smith	Chicago, Ill	Apr. 5, 1870	101,529
Chromates, Process for obtaining	J. C. Booth	Philadelphia, Pa	July 19, 1853	9,853
Chrome compound	A. K. Eaton	New York, N. Y	Apr. 28, 1863	38,297
Chrome vermillion, Process of preparing	J. Huber	New York, N. Y	Sept. 20, 1864	44,370
Chromic yellow, Making	R. A. Tilghman	Philadelphia, Pa	Jan. 16, 1843	2,910
Chromium, Manufacture of the salts of	B. Marqulies	Trieste, Austria	Aug. 29, 1865	49,682
Chronograph, Electro-ballistic	P. Le Boulengé	Antwerp, Belgium	Jan. 2, 1866	51,905

Index of patents issued from the United States Patent Office from 1790 to 1873, inclusive—Continued.

Invention.	Inventor.	Residence.	Date.	No.
Chronometer	P. Bantel	New York, N. Y	July 21, 1868	80, 051
Chronometer escapement	R. Barclay	Buffalo, N. Y	Oct. 14, 1862	36, 637
Chronometer escapement	C. Fasoldt	Albany, N. Y	Mar. 7, 1865	46, 652
Chronometer escapement	J. Fulton	Louisville, Ky	Mar. 3, 1857	16, 727
Chronometer escapement	V. Giroud	New York, N. Y	Dec. 12, 1848	5, 963
Chronometer escapement	W. H. Hammond	New York, N. Y	Aug. 30, 1859	25, 261
Chronometer escapement	P. Humbert	Boston, Mass	Feb. 26, 1861	31, 543
Chronometer escapement	J. Karr	Washington, D. C	Sept. 20, 1864	44, 317
Chronometer escapement	J. Karr	Washington, D. C	Nov. 28, 1865	51, 191
Chronometer escapement	T. Morison	New York, N. Y	Oct. 19, 1858	21, 865
Chronometer escapement	T. Morrison	Kingston, N. Y	June 12, 1860	28, 719
Chronometer escapement	A. H. Potter	Williamsburgh, N. Y	Jan. 21, 1868	73, 646
Chronometer escapement	G. P. Reed	Boston, Mass	Apr. 7, 1868	76, 346
Chronometer escapement	G. P. Reed	Boston, Mass	Mar. 9, 1869	87, 707
Chronometer escapement	H. Rothfelder	New York, N. Y	July 11, 1865	48, 726
Chronometer escapement	O. W. Waste	Pittsford, N. Y	Sept. 24, 1844	3, 759
Chronometer for longitude	J. Sheldon	Millville, N. Y	Nov. 20, 1849	6, 877
Chronometer, Solar	L. Mifflin	Germantown, Pa	May 21, 1867	64, 892
Chronometric lock	W. L. Bass	Boston Mass	Dec. 23, 1851	8, 603
Chronometric lock	A. Holbrook	Milford, Mass	Apr. 13, 1858	19, 927
Chronometric lock	A. Holbrook and H. D. Fish	Milford and Hardwick, Mass.	Apr. 28, 1857	17, 150
Chronometric lock	J. Y. Savage	New York, N. Y	Oct. 9, 1847	5, 321
Chuck	T. R. Almond	Fitchburgh, Mass	Aug. 19, 1873	141, 978
Chuck	T. G. Arnold	New York, N. Y	Dec. 4, 1866	60, 116
Chuck	W. F. Bacon	Skowhegan, Me	Aug. 31, 1869	94, 376
Chuck	S. H. Bellows	Middletown, Conn	Aug. 6, 1872	130, 103
Chuck	G. E. Brettell	Rochester, N. Y	May 25, 1869	90, 338
Chuck	J. C. Chapman	Cambridgeport, Mass	Dec. 4, 1866	60, 138
Chuck	W. T. Cole	New York, N. Y	Sept. 24, 1867	69, 184
Chuck	S. G. Dare	New York, N. Y	Dec. 22, 1868	85, 073
Chuck	G. B. Fairman	Rochester, N. Y	Mar. 30, 1869	88, 377
Chuck	C. F. Hadley	Chicopee, Mass	Nov. 17, 1868	84, 057
Chuck	G. W. Harris and W. H. Haight	New York, N. Y	June 25, 1867	66, 081
Chuck	B. Haviland	Hudson, N. Y	Aug. 25, 1868	81, 500
Chuck	E. Horton	Windsor Locks, Conn	Aug. 26, 1873	142, 163
Chuck	M. Love	Corry, Pa	Nov. 22, 1870	109, 434
Chuck	J. W. Martin and E. Parry	Northern Liberties, Pa	Aug. 28, 1849	6, 681
Chuck	R. W. Morse	East Berlin, Conn	Aug. 7, 1866	56, 981
Chuck	L. H. Olmsted	Stamford, Conn	May 15, 1866	54, 760
Chuck	L. Parmelee	New Haven, Conn	Oct. 28, 1873	144, 128
Chuck	B. W. Peirce	New Bedford, Mass	Oct. 22, 1867	70, 112
Chuck	R. and T. Ross	Middlebury, Vt	Aug. 21, 1866	57, 383
Chuck	I. Smith	New York, N. Y	July 10, 1866	56, 329
Chuck	I. Smith	New York, N. Y	July 10, 1866	56, 330
Chuck	J. M. Smith	Seymour, Conn	Sept. 3, 1867	68, 537
Chuck	W. Staub	Philadelphia, Pa	Oct. 31, 1865	50, 747
Chuck	S. P. M. Tasker	Philadelphia, Pa	July 26, 1870	105, 860
Chuck	J. F. Thomas	Ilion, N. Y	Oct. 18, 1870	108, 535
Chuck	R. H. Thorn	Syracuse, N. Y	Sept. 18, 1866	58, 152
Chuck	A. B. Underhill	Meadville, Pa	Dec. 6, 1864	45, 364
Chuck	J. H. Westcott	Oneida, N. Y	Feb. 25, 1873	136, 349
Chuck	D. E. Whitton	West Stafford, Conn	Oct. 9, 1866	58, 704
Chuck	E. S. Williams	Cambridge, Mass	Sept. 28, 1869	95, 297
Chuck and countersink, Combined	D. Argerbright	Gratis, Ohio	Aug. 21, 1860	29, 657
Chuck, Arrangement for lathe	L. A. Dole	Salem, Ohio	July 25, 1854	11, 364
Chuck, Automatic friction dog-lathe	L. Coes	Worcester, Mass	May 17, 1870	103, 015
Chuck, Auxiliary jaw for planer	C. H. Riggs	Windsor Locks, Conn	June 1, 1869	90, 783
Chuck, Centering	G. H. Miller	Binghamton, N. Y	Sept. 9, 1873	142, 642
Chuck, Centering and squaring	G. R. Parker	Worcester, Mass	June 4, 1867	65, 422
Chuck, Dredge	D. Keller	Baltimore, Md	Jan. 19, 1869	85, 936
Chuck, Drill	E. H. Babcock	Canandaigua, N. Y	June 16, 1868	78, 916
Chuck, Drill	Q. S. Backus	Winchendon, Mass	Sept. 29, 1868	82, 583
Chuck, Drill	J. Fox	Philadelphia, Pa	Feb. 7, 1860	27, 044
Chuck, Drill	F. G. Johnson	Brooklyn, N. Y	July 3, 1866	56, 059
Chuck, Drill	G. W. Miller	Woonsocket, R. I	July 27, 1869	93, 108
Chuck, Drill	L. Parmelee	New Haven, Conn	Sept. 30, 1873	143, 377
Chuck, Drill	P. Philippi	Beardstown, Ill	Sept. 19, 1871	119, 173
Chuck, Drill	C. S. Wells and L. H. Mayot	Springfield, Mass	Mar. 20, 1866	53, 368
Chuck, Expanding	W. Webb	Waterbury, Conn	Mar. 23, 1869	88, 245
Chuck for boring and mortising machine	E. K. Wisell	Warren, Ohio	Jan. 22, 1850	7, 045
Chuck for boring fire-arm cylinders	C. H. Alsop	Middletown, Conn	July 2, 1861	32, 675
Chuck for centering, &c	D. N. Smith	Boston, Mass	Aug. 3, 1858	21, 088
Chuck for cutting barrel-heads	F. Fruit	Jefferson City, Mo	Nov. 15, 1853	10, 228
Chuck for cutting disks of paper, &c	M. Kaefer	New York, N. Y	Apr. 10, 1860	27, 810
Chuck for grinding crystal	J. S. Warner	Ogdensburgh, N. Y	Jan. 11, 1870	98, 820
Chuck for holding bodies of different shapes, Planer	R. Henderson	Thomaston, Conn	June 22, 1869	91, 625
Chuck for holding buttons	E. Russell	Naugatuck, Conn	Dec. 4, 1866	60, 258
Chuck for holding drills	A. F. Cushman	Hartford, Conn	Dec. 24, 1872	134, 259
Chuck for holding drills	W. X. Stevens	East Brookfield, Mass	May 21, 1872	126, 910
Chuck for holding pipe-fittings	M. Walty	Buffalo, N. Y	Aug. 5, 1873	141, 612
Chuck for holding valve-cock to be dressed	J. L. Hayden	Haydensville, Mass	Oct. 3, 1871	119, 516
Chuck for iron-planing	M. C. Gardner	Rochester, N. Y	June 25, 1867	66, 014
Chuck for lathe, Centering	S. Keagy	Mineral Point, Pa	Mar. 20, 1860	27, 598
Chuck for lathe, Self-centering	G. O. Buckley	New Bedford, Mass	Feb. 28, 1871	112, 119
Chuck for lathe, Self-centering	G. O. Buckley	New Bedford, Mass	Feb. 20, 1872	123, 863
Chuck for lathe, Slide	C. F. Stackpole	Woburn, Mass	Aug. 16, 1870	106, 420
Chuck for lathes and planes, Magnetic	J. A. Jaques, J. T. Oakley, and L. Sterne.	Tottenham, Bermondsey, and Westminster, England.	Apr. 15, 1873	137, 875
Chuck for metal-turning lathe	A. F. Cushman	Hartford, Conn	Dec. 5, 1871	121, 494
Chuck for planing-machine	J. S. Hoar	West Acton, Mass	Feb. 11, 1868	74, 360
Chuck for screw-cutting	C. Deavs	New York, N. Y	Nov. 28, 1871	121, 340
Chuck for screw-cutting	R. Nuttall and J. Kirkpatrick	Allegheny, Pa	May 4, 1858	20, 168
Chuck for screw-cutting	R. Nuttall and J. Kirkpatrick	Allegheny, Pa	May 24, 1859	24, 152
Chuck for screw-cutting lathe	E. C. Plimpton and S. Taylor	Bridgeport, Conn	Jan. 30, 1872	123, 197
Chuck for sleigh-bells, Lathe	W. E. Barton	East Hampton, Conn	Dec. 9, 1873	145, 383
Chuck for turning eccentrics	J. W. Russell	Springfield, Mass	June 19, 1855	13, 104
Chuck for turning elliptical cylinders	P. S. Cahoon and S. F. Ross	La Grange, Mo	Jan. 2, 1855	12, 180

Index of patents issued from the United States Patent Office from 1790 *to* 1873, *inclusive*—Continued.

Invention.	Inventor.	Residence.	Date.	No.
Chuck for turning staves	F. Robbins	Acton, Mass	July 7, 1863	39, 173
Chuck, Gripe	D. Peeler	Boston, Mass	Mar. 31, 1836	....
Chuck, Jeweler's	L. S. Hill	Grand Rapids, Mich	Jan. 25, 1870	99, 191
Chuck, Lathe	C. E. Albro	Fulton, N. Y	Oct. 17, 1871	120, 015
Chuck, Lathe	C. Archer	Nelsonville, Ohio	July 23, 1872	129, 705
Chuck, Lathe	W. Bellows	Cincinnati, Ohio	May 25, 1869	90, 487
Chuck, Lathe	J. A. Bunce	East Berlin, Conn	Mar. 27, 1866	53, 405
Chuck, Lathe	S. B. Burritt	New York, N. Y	June 20, 1865	48, 259
Chuck, Lathe	J. Christman and W. Gilfillan	Syracuse, N. Y	Apr. 14, 1863	38, 148
Chuck, Lathe	S. J. Cone	Middletown, Conn	July 4, 1865	48, 521
Chuck, Lathe	A. F. Cushman	Hartford, Conn	Jan. 10, 1871	110, 903
Chuck, Lathe	H. M. Darling	Bridgeport, Conn	Oct. 26, 1869	96, 207
Chuck, Lathe	F. Davison	Richmond, Va	Mar. 24, 1868	75, 872
Chuck, Lathe	J. S. Detrick	San Francisco, Cal	Aug. 11, 1868	80, 928
Chuck, Lathe	S. Goodfellow	Troy, N. Y	Aug. 17, 1858	21, 232
Chuck, Lathe	S. C. Goodsell	Westville, Conn	Apr. 8, 1873	137, 674
Chuck, Lathe	W. Grant	Boston, Mass	Jan. 23, 1849	6, 046
Chuck, Lathe	E. Horton	Windsor Locks, Conn	Nov. 13, 1855	13, 787
Chuck, Lathe	J. Hyde	Troy, N. Y	Dec. 9, 1851	8, 580
Chuck, Lathe	W. Johnson	Lambertville, N. J	Apr. 5, 1859	23, 472
Chuck, Lathe	W. Johnson	Lambertville, N. J	July 29, 1873	141, 276
Chuck, Lathe	A. Judson	Brooklyn, N. Y	Dec. 8, 1868	84, 698
Chuck, Lathe	W. H. King	Newark, N. J	Nov. 22, 1870	109, 423
Chuck, Lathe	J. L. Mason	New York, N. Y	Mar. 30, 1858	19, 786
Chuck, Lathe	J. S. Moody	Saco, Me	Aug. 25, 1868	81, 523
Chuck, Lathe	M. Neckermann	Pittsburgh, Pa	Apr. 8, 1856	14, 632
Chuck, Lathe	J. O'Connor	Buffalo, N. Y	Oct. 8, 1867	69, 577
Chuck, Lathe	C. H. Reid	Danbury, Conn	Aug. 12, 1873	141, 817
Chuck, Lathe	W. A. Reilly	Cincinnati, Ohio	May 30, 1865	47, 981
Chuck, Lathe	J. Rich	Painesville, Ohio	Aug. 31, 1869	93, 239
Chuck, Lathe	J. Rich	Painesville, Ohio	Nov. 9, 1869	96, 730
Chuck, Lathe	J. Rich	Painesville, Ohio	June 7, 1870	104, 063
Chuck, Lathe	J. Rich	Painesville, Ohio	Sept. 13, 1870	107, 291
Chuck, Lathe	J. Rich	Painesville, Ohio	Mar. 28, 1871	113, 095
Chuck, Lathe	E. A. L. Roberts	New York, N. Y	Oct. 4, 1859	25, 574
Chuck, Lathe	R. Rothwell	Dover, N. H	Apr. 3, 1866	53, 683
Chuck, Lathe	T. Shrewsbury	Camden, N. J	Oct. 8, 1872	132, 109
Chuck, Lathe	J. Sutter	New York, N. Y	Dec. 6, 1864	45, 359
Chuck, Lathe	F. Tully	Philadelphia, Pa	Nov. 19, 1872	133, 272
Chuck, Lathe	S. G. Twambly	Saco, Me	Aug. 13, 1861	33, 056
Chuck, Lathe	L. Von Gunten	Cincinnati, Ohio	Dec. 4, 1866	60, 292
Chuck, Lathe	A. H. Wagner	Prairie City, Ill	July 4, 1871	116, 649
Chuck, Lathe	J. R. Washburn	West Stafford, Conn	Sept. 29, 1868	82, 571
Chuck, Lathe	D. E. Whiton	West Stafford, Conn	Oct. 20, 1868	83, 349
Chuck, Machinists' lathe	A. F. Cushman	Hartford, Conn	Nov. 14, 1871	120, 863
Chuck, Nut-squaring	H. F. Wheeler	Boston, Mass	Aug. 4, 1868	80, 792
Chuck or holder, Self-centering	T. H. Worrall	Lawrence, Mass	July 4, 1865	48, 612
Chuck, Planer	R. N. Bruce	Springfield, Mass	Mar. 3, 1868	74, 982
Chuck, Planer	C. H. Riggs	Windsor Locks, Conn	May 19, 1868	78, 132
Chuck, Planer	W. H. Warren	Worcester, Mass	Mar. 10, 1868	75, 498
Chuck, Planing	A. Newell and W. Pim	Chicago, Ill	May 3, 1870	102, 577
Chuck, Screw-cutting	F. H. Higgins	Bordentown, N. J	Jan. 23, 1866	52, 247
Chuck, Screw-cutting	T. Kennedy	Mount Carmel, Conn	Jan. 30, 1866	52, 294
Chuck, Scroll	A. F. Cushman	Hartford, Conn	Dec. 5, 1865	51, 384
Chuck, Self-centering	E. B. Beach	West Meriden, Conn	Sept. 6, 1864	44, 067
Chuck, Self-centering	T. Brooks	Middletown, Conn	Oct. 10, 1865	50, 329
Chuck, Self-centering	S. T. Jackson	Sheboygan Falls, Wis	Oct. 24, 1865	50, 593
Chuck, Self-centering	T. H. Worrall	Manchester, N. H	Mar. 7, 1865	46, 747
Chuck, Self-centering	J. W. Bartlett and A. Morris	Harmar and Marietta, Ohio	Sept. 8, 1863	39, 789
Chuck, Universal	M. Alden	Philadelphia, Pa	Aug. 11, 1863	39, 457
Chuck, Universal	S. S. Hogle	Rockville, N. Y	Nov. 16, 1841	2, 366
Chuck, Universal	W. H. King	Troy, N. Y	Apr. 25, 1865	47, 428
Chuck, Warping	J. T. Haskins	Rockport, Mass	June 16, 1868	78, 964
Chuck, Watch-makers'	H. H. Heskett	Le Roy, Ill	July 4, 1871	116, 591
Chuck, Watch-makers'	J. Mansir	Richmond, Me	Nov. 14, 1871	120, 988
Chuck, Watch-makers'	A. K. P. Walker	Richmond, Me	Aug. 1, 1871	117, 703
Chuck, Watch-makers' lathe	W. Kerr	Boston, Mass	May 30, 1871	115, 324
Chuck, Watch-makers' lathe	S. S. Lavey	Plymouth, Ind	Sept. 17, 1867	68, 998
Chuck, Watch-makers' lathe	J. Stark	Waltham, Mass	Aug. 12, 1873	141, 824
Chuck, Watch-makers' lathe	W. Stephens	Richmond, Ind	Mar. 10, 1857	16, 811
Chuck, Watch-makers' lathe	G. H. Waldin	Burlington, Iowa	Feb. 22, 1859	23, 050
Chuck, Wood-lathe	R. H. Dowling	Fenton, Mich	May 23, 1871	115, 179
Chucks, Device for holding and carrying pipe-fittings	A. J. Peavey and F. B. Cotton	Boston, Mass	Dec. 23, 1873	145, 896
Chucks, Jaw for lathe	M. C. Johnson	New Bedford, Mass	Aug. 26, 1873	142, 239
Church and school seat	C. Perley	New York, N. Y	May 24, 1859	24, 151
Church-pew head rest	J. H. Weeden	Waterbury, Conn	June 15, 1869	91, 389
Church-seat	C. O. Mealia	Brooklyn, N. Y	Jan. 16, 1872	122, 730
Churn	H. Abbott	North Huron, N. Y	June 19, 1860	28, 723
Churn	H. W. Adams	Milton, Pa	Jan. 7, 1868	72, 959
Churn	L. Adams	Redding, Conn	Apr. 16, 1842	2, 559
Churn	J. Aiken	Warner, N. H	Mar. 7, 1865	46, 749
Churn	D. C. Aldrich	Anamosa, Iowa	Feb. 7, 1865	46, 202
Churn	J. Alexander	Gallipolis, Ohio	Feb. 2, 1869	86, 343
Churn	W. Alexander	Union Valley, N. Y	May 14, 1867	64, 618
Churn	H. Allen, jr	Wallingford, Conn	Mar. 14, 1865	46, 766
Churn	L. O. Allen	Gardiner, Me	Apr. 10, 1866	53, 766
Churn	S. S. Allen	Belvidere, N. Y	May 4, 1869	89, 616
Churn	C. T. Anderson	Hyattstown, Md	June 18, 1861	32, 553
Churn	C. T. Anderson	Clarksburgh, Md	May 2, 1865	47, 507
Churn	C. T. Anderson	Clarksburgh, Md	Apr. 26, 1870	102, 354
Churn	J. Andrews	Woburn, Mass	Apr. 23, 1850	7, 298
Churn	C. Angerine	New York, N. Y	Aug. 17, 1835	....
Churn	O. A. Anthony	Mayfield, N. Y	Sept. 3, 1872	131, 076
Churn	J. S. Appleton	White River Junction, Vt	Aug. 17, 1858	21, 176
Churn	W. F. Armstrong and M. Payne	Cardington, Ohio	July 16, 1861	32, 814
Churn	R. S. Arnall	Wright City, Mo	July 23, 1867	67, 095

Index of patents issued from the United States Patent Office from 1790 *to* 1873, *inclusive*—Continued.

Invention.	Inventor.	Residence.	Date.	No.
Churn	E. Archer	Davisville, W. Va	Aug. 5, 1873	141, 480
Churn	M. B. Atkinson	Georgetown, D. C	Apr. 12, 1870	101, 810
Churn	A. Austin	Altona, Ill	Sept. 27, 1859	25, 553
Churn	S. S. Ayers	Plainfield, N. J	Mar. 12, 1867	62, 804
Churn	L. Bacon	Charlotte, Mich	July 26, 1864	43, 632
Churn	S. L. Bagley	Hillsdale, N. Y	Mar. 24, 1827	....
Churn	C. Baker	Charlton, N. Y	June 13, 1831	....
Churn	D. D. Baker	West Alexandria, Ohio	Apr. 21, 1868	76, 879
Churn	H. D. Baker	Pittstown Corners, N. Y	Feb. 16, 1858	19, 334
Churn	H. F. Baker	Centreville, Ind	Apr. 10, 1849	6, 298
Churn	T. Baker	Saint John's, Mich	Mar. 15, 1870	100, 711
Churn	A. Bailey	Poultney, Vt	Oct. 6, 1831	....
Churn	T. K. Bailey	Lockport, N. Y	Apr. 24, 1866	54, 092
Churn	A. Baird	Schoharie, N. Y	Mar. 25, 1825	....
Churn	S. Ballard, sr	Sullivan, Ind	May 7, 1867	64, 471
Churn	J. A. Balzart	Piqua, Ohio	Sept. 18, 1866	58, 047
Churn	H. B. Barber	Scott, N. Y	Dec. 24, 1867	72, 588
Churn	H. Barr	Independence, Iowa	Nov. 18, 1862	36, 935
Churn	W. E. Barr	Ripley's, W. Va	Aug. 27, 1872	130, 836
Churn	J. C. Barrett	Collins, N. Y	Oct. 19, 1829	....
Churn	D. Bartholomew and D. C. Dinsmore.	Kirkville, Iowa	May 19, 1868	78, 042
Churn	N. S. Barton	Mannsville, N. Y	Aug. 31, 1869	94, 381
Churn	M. L Bauder	Elyria, Ohio	Jan. 19, 1858	19, 117
Churn	C. F. Baylor	Clinton, N. J	Apr. 25, 1865	47, 383
Churn	J. Beach	Northfield, Mass	Jan. 24, 1811	....
Churn	S. and M. Bear	Versailles, Ohio	Sept. 24, 1867	69, 065
Churn	W. Beaton	Grinnell, Iowa	Oct. 4, 1864	44, 505
Churn	J. Beckley	Stiles, Iowa	May 6, 1873	138, 604
Churn	E. H. Beckwith	Westernville, N. Y	Nov. 20, 1866	59, 811
Churn	L. W. Beecher	Avon, N. Y	Nov. 24, 1857	18, 671
Churn	C. H. and E. G. Beeman	North Fairfax, Vt	July 13, 1869	92, 566
Churn	B. Beers	New Fairfield, Conn	Dec. 8, 1857	18, 797
Churn	G. Beisner	Chicago, Ill	Nov. 3, 1868	83, 595
Churn	H. C. Bell	Heyworth, Ill	Mar. 24, 1868	75, 842
Churn	H. C. Bell	Edina, Mo	July 30, 1872	129, 918
Churn	E. O. Bennett	Mount Pleasant, Iowa	Oct. 1, 1867	69, 391
Churn	J. Bennett	Brutus, N. Y	July 25, 1832	....
Churn	H. Bentley	Ballston, N. Y	Apr. 20, 1844	3, 556
Churn	E. L. Bergstresser	Berrysburgh, Pa	Feb. 21, 1865	46, 440
Churn	G. Berkstresser	Bedford, Pa	Aug. 31, 1869	94, 176
Churn	S. Besser	Dorchester, Ill	July 7, 1868	79, 542
Churn	W. T. Best	Scranton, Pa	Apr. 28, 1868	77, 160
Churn	H. H. Bigard, E. H. Kellogg, and N. A. Prentiss.	Fowler, N. Y	Feb. 15, 1870	99, 819
Churn	A. Bills	Baltimore, Md	July 29, 1816	....
Churn	A. G. Binns	Goshen Township, Ohio	Aug. 26, 1862	36, 269
Churn	T. Bisbing	Buckstown, Pa	Dec. 10, 1867	71, 958
Churn	C. G. Bishop	Poughkeepsie, N. Y	July 18, 1865	48, 786
Churn	L. Bissell	North Bergen, N. Y	July 10, 1860	29, 052
Churn	D. O. Blair	Abingdon, Ill	Feb. 19, 1867	62, 179
Churn	F. Blecka	Elgin, Ill	June 25, 1867	66, 122
Churn	J. C. Bodine	Camden, N. J	Nov. 20, 1866	59, 759
Churn	T. H. Body	White Creek, Wis	Jan. 26, 1869	86, 274
Churn	T. Bogan	Lacon, Ill	Aug. 27, 1867	68, 034
Churn	L. Bogner	Rochester, Ind	Aug. 6, 1872	130, 104
Churn	J. Bolton	Philadelphia, Pa	Aug. 2, 1808	....
Churn	C. A. Boone	Shickshinny, Pa	May 3, 1870	102, 481
Churn	J. Borst	East Cobbleskill, N. Y	Jan. 22, 1867	61, 389
Churn	J. P. Bortle	Claverack, N. Y	Apr. 19, 1825	....
Churn	J. Borton and W. H. Hartly	Quaker City, Ohio	Feb. 11, 1873	135, 761
Churn	C. L. Bottum	Dansville, N. Y	Feb. 24, 1863	37, 730
Churn	C. Boyer	Marshall, Mich	Oct. 8, 1867	69, 620
Churn	J. Boyers	Granville, Va	Aug. 14, 1855	13, 421
Churn	J. Boyers	Orrville, Ohio	July 3, 1866	55, 996
Churn	W. Boynton, jr	Auburn, N. Y	Apr. 15, 1862	34, 936
Churn	W. Boynton	Auburn, N. Y	Feb. 20, 1866	52, 664
Churn	C. H. Bradley	Coatesville, Pa	Apr. 28, 1868	77, 252
Churn	R. Bradley	Williston, Vt	July 21, 1835	....
Churn	J. W. Bradly and G. H. Jordon.	Rocheport, Mo	June 23, 1868	79, 197
Churn	A. H. Brainerd	Rome, N. Y	Apr. 23, 1867	63, 989
Churn	V. M. R. Branch	Richmond, Va	Sept. 1, 1868	81, 589
Churn	M. Bratt	Maysville, Ky	Oct. 23, 1866	58, 976
Churn	C. W. Brewer	Racine, Wis	Dec. 29, 1868	85, 272
Churn	I. Brewster	Blenheim, N. Y	May 28, 1830	....
Churn	I. Brewster	Blenheim, N. Y	Dec. 31, 1833	....
Churn	I. Brewster	Schoharie County, N. Y	Jan. 7, 1835	....
Churn	J. C. Bridgman	London, Ohio	May 22, 1866	54, 846
Churn	W. W. Brigg	Home, Tenn	July 20, 1869	92, 788
Churn	G. N Brigham	Montpelier, Vt	July 24, 1866	56, 520
Churn	J. Brinkerhoff	Auburn, N. Y	Dec. 23, 1862	37, 214
Churn	J. L. and T. R. Britt	Raleigh, N. C	Oct. 14, 1873	143, 612
Churn	A. Broadway	Monson, Mass	Nov. 25, 1816	....
Churn	A. H. Brown	Springfield, Vt	Oct. 8, 1867	69, 623
Churn	D. C. Brown	New York, N. Y	Jan. 24, 1860	26, 880
Churn	G. C. Brown	Atlanta, Ga	Mar. 22, 1870	101, 091
Churn	H. Brown	New York, N. Y	Apr. 6, 1858	19, 828
Churn	J. Brown	Hallowell, Me	June 18, 1834	....
Churn	J. F. Brown	New London, Conn	July 2, 1867	66, 295
Churn	M. M. Brown	Pimento, Ind	Oct. 8, 1867	69, 624
Churn	W. Brown	Duncannon, Pa	May 11, 1858	20, 189
Churn	L. A. Brown and H. Bigelow	Hartford, Conn	Oct. 5, 1852	9, 311
Churn	E. Brown and A. Keith	Winfield, N. Y	June 19, 1822	....
Churn	E. T. Brownfield	Smithfield, Pa	Mar. 17, 1868	75, 621
Churn	G. N. Broughton	Bellefontaine, Ohio	Aug. 13, 1872	130, 473
Churn	S. W. Bruce	Laguardo, Tenn	Jan. 18, 1870	98, 918
Churn	J. H. Bruner	McKean, Ohio	Feb. 9, 1832	....

Index of patents issued from the United States Patent Office from 1790 *to* 1873, *inclusive*—Continued.

Invention.	Inventor.	Residence.	Date.	No.
Churn	A. P. Bryson	Prospect, Pa	Dec. 1, 1868	84, 532
Churn	G. G. Buchanan	Cotton Plant, Miss	Nov. 11, 1873	144, 437
Churn	R. Buck	Baltimore, Md	June 23, 1807	....
Churn	T. W. Buck	Fawn River, Mich	Oct. 8, 1867	69, 538
Churn	W. E. Budd	Chatham, N. J	Apr. 4, 1871	113, 251
Churn	E. Buel	Sheridan, N. Y	Oct. 11, 1864	44, 595
Churn	J. M. Buell	Zanesville, Ohio	Aug. 7, 1860	29, 460
Churn	J. H. Bump	Morris, N. Y	Oct. 26, 1858	21, 871
Churn	F. and L. Burdick	South East, N. Y., and Lockhaven, Pa.	Oct. 5, 1869	95, 560
Churn	T. I. Burhyte	Fond du Lac, Wis	Feb. 26, 1867	62, 470
Churn	T. J. Burke and S. B. Gassette	Chicago, Ill	Oct. 31, 1865	50, 684
Churn	W. H. Burnham and B. Hibbard.	Cortland Village, N. Y	July 29, 1856	15, 412
Churn	W. and D. C. Burson	Salineville, Ohio	Sept. 18, 1866	58, 059
Churn	E. Burt	Princeton, N. J	Nov. 20, 1809	....
Churn	S. Bushnell, 2d	Saybrook, Conn	Dec. 20, 1830	....
Churn	M. Byard	Milan, Ind	Oct. 13, 1857	18, 384
Churn	P. Byrns and G. Stannard	Mindora, Wis	Nov. 5, 1867	70, 406
Churn	M. P. Callender	Astoria, Oreg	Sept. 3, 1872	130, 976
Churn	M. Calvert	Marshall, Ohio	June 23, 1868	79, 103
Churn	J. D. Campbell and J. Elliott	Jamestown, Ind	Sept. 24, 1872	131, 660
Churn	S. N. Campbell	Elgin, Ill	Oct. 18, 1859	25, 804
Churn	W. Campbell	Waterloo, Pa	Sept. 13, 1859	25, 383
Churn	A. Carbnow	Potsdam, N. Y	Dec. 18, 1866	60, 473
Churn	W. L. Card	Gardner, Ill	Feb. 12, 1867	61, 9[illegible]5
Churn	J. Carlton	Walla Walla, Wash	Dec. 15, 1868	84, 858
Churn	R. Carmack	Marengo, Ill	Apr. 16, 1872	125, 660
Churn	D. H. Carpenter and H. L. Slaghs.	Hector and Lodi, N. Y	Feb. 18, 1868	74, 498
Churn	A. C. Carrell	Granville, Ohio	Oct. 25, 1832	....
Churn	J. S. Carson	Brookhaven, Miss	Aug. 25, 1868	81, 341
Churn	C. H. Carver	Taunton, Mass	Feb. 25, 1868	74, 796
Churn	S. S. Case	Marion, N. Y	Nov. 3, 1868	83, 815
Churn	S. S. Case	Marion, N. Y	May 16, 1871	114, 922
Churn	J. E. Casey	Cortland Village, N. Y	Mar. 27, 1866	53, 521
Churn	H. Caslow	York, Pa	June 14, 1870	104, 267
Churn	C. J. Chalfant	Unionville, Pa	July 30, 1867	67, 266
Churn	W. C. Chamberlain	Dubuque, Iowa	Jan. 29, 1867	61, 515
Churn	W. C. Chamberlain	Dubuque, Iowa	Aug. 3, 1869	93, 174
Churn	N. P. Chaney	Potsdam, N. Y	Sept. 1, 1868	81, 599
Churn	F. B. Chapman	Salisbury, Mo	Feb. 21, 1871	111, 907
Churn	J. W. Chapman	Madison, Ind	Sept. 5, 1871	118, 687
Churn	N. Chapman	Milford, Mass	Jan. 30, 1866	52, 268
Churn	N. Chapman	Milford, Mass	Apr. 13, 1869	88, 847
Churn	J. Chapple	Jasper, N. Y	Jan. 18, 1870	98, 924
Churn	J. I. Cheatham	Athens, Ga	July 6, 1869	92, 164
Churn	E. Chipman	New York, N. Y	Jan. 12, 1864	41, 193
Churn	E. Chipman	New York, N. Y	Mar. 21, 1865	46, 880
Churn	J. B. Christian and J. P. Beach	Hamburgh, Iowa	Aug. 13, 1872	130, 477
Churn	J. M. Chritton	Joliet, Ill	Jan. 15, 1867	61, 160
Churn	N. B. Clabaugh	Frederick, Md	July 2, 1867	66, 219
Churn	J. Clark	Brush Valley, Pa	Jan. 10, 1871	110, 897
Churn	S. Clark	Morris County, N. J	Oct. 4, 1822	....
Churn	G. B. Clarke	Leonardsville, N. Y	Sept. 23, 1851	8, 373
Churn	U. L. Clark	Manor Township, Pa	Dec. 31, 1833	....
Churn	G. Clayton and C. B. Allen	Willoughby, Ohio	Jan. 21, 1868	73, 503
Churn	E. B. Clement	Barnet, Vt	Jan. 20, 1852	8, 671
Churn	E. B. Clement	Barnet, Vt	Feb. 13, 1855	12, 377
Churn	E. B. Clement	Barnet, Vt	Feb. 21, 1860	27, 205
Churn	D. Cleveland	Sheffield, Mass	July 7, 1808	....
Churn	D. Cleveland	Sheffield, Mass	July 11, 1808	....
Churn	L. Cleveland	Holliston, Mass	Feb. 7, 1811	....
Churn	A. Clift, jr	Mystic River, Conn	May 22, 1866	54, 861
Churn	H. Clift	Mystic River, Conn	Sept. 25, 1866	58, 337
Churn	M. Clifton	Peoria, Ill	Feb. 16, 1869	87, 027
Churn	J. Closs	Decatur, Ind	May 31, 1859	24, 197
Churn	D. Clough	Auburn, N. Y	Feb. 5, 1867	61, 716
Churn	P. L. Clow	Cohoes, N. Y	Nov. 15, 1859	26, 092
Churn	J. C. Clymer	Galion, Ohio	Mar. 26, 1867	63, 216
Churn	J. Cochran, jr	Auburn, Mo	Dec. 12, 1871	121, 755
Churn	J. T. Coe	Chambersburgh, Pa	Feb. 18, 1868	74, 505
Churn	T. Coffield and B. Egli	Natrona, Pa	Nov. 8, 1870	108, 969
Churn	J. B Coffman	New Richmond, Ohio	Jan. 19, 1869	86, 002
Churn	L. S. Colburn	Oberlin, Ohio	June 30, 1863	39, 035
Churn	C. Colby	San Francisco, Cal	Nov. 19, 1867	70, 959
Churn	E. Coleman	Woburn, Mass	Mar. 9, 1869	87, 639
Churn	T. Conely	Grafton, Ill	June 25, 1872	128, 364
Churn	A. J. Conner	Louisville, Ky	Apr. 14, 1868	76, 606
Churn	A. L. Converse	Springfield, Ill	May 5, 1868	77, 460
Churn	G. W. Cook	Saint Louis, Mo	Feb. 28, 1844	3, 460
Churn	L. M. Cook	Owatonna, Minn	July 23, 1867	66, 949
Churn	W. M. Cook	Lyons, Iowa	Nov. 13, 1866	59, 560
Churn	D. C. Cooley	Wilkesbarre, Pa	Feb. 23, 1869	87, 147
Churn	J. Cooper	Dublin, Ind	July 10, 1866	56, 340
Churn	N. B. Cooper	Gratis, Ohio	July 3, 1860	28, 970
Churn	G. W. Corbit, J. M. Orput, and G. M. Case.	Malta, Ill	Apr. 28, 1868	77, 172
Churn	A. B. Corby	Binghamton, N. Y	May 5, 1868	77, 462
Churn	A. and H. K. Cornell	Cambridge, N. Y	Jan. 15, 1816	....
Churn	J. Copeland and G. P. Martin	Quasqueton, Iowa	June 17, 1862	35, 588
Churn	G. W. Cottingham and J. H. Binkley.	Columbus, Tex	Oct. 8, 1872	131, 939
Churn	F. Cotton	New York, N. Y	Jan. 9, 1835	....
Churn	E. P. and J. A. Cowles	Oakfield, N. Y	Mar. 3, 1857	16, 717
Churn	A. J. Cox	Indianapolis, Ind	Mar. 26, 1872	124, 883
Churn	J. A. Cozad	Mercer, Pa	July 27, 1869	93, 061
Churn	J. Crail	Warren, Ohio	May 5, 1831	....

Index of patents issued from the United States Patent Office from 1790 to 1873, inclusive—Continued.

Invention.	Inventor.	Residence.	Date.	No.
Churn	J. Cram	Chicago, Ill	July 16, 1867	66, 681
Churn	J. Cram	Chicago, Ill	Jan. 10, 1871	110, 902
Churn	G. Cramton	Marshall, Mich	Apr. 12, 1864	42, 275
Churn	A. G. Crane	Ottumwa, Iowa	June 11, 1872	127, 852
Churn	F. J. Crissey	Leesburgh, Va	June 26, 1866	55, 825
Churn	M. C. Crank	Auburn, N. Y	Aug. 13, 1861	33, 030
Churn	J. E. Cryer	Green Point, N. Y	July 23, 1867	67, 105
Churn	A. W. Cunningham	West Middletown, Pa	Feb. 28, 1860	27, 276
Churn	W. H. Curtin and W. Lammers	Clement and Breeze, Ill	July 26, 1870	105, 652
Churn	S. T. Curtis	El Paso, Ill	June 9, 1868	78, 791
Churn	G. A. Dabney	San José, Cal	Feb. 2, 1869	86, 371
Churn	C. H. Dana	West Lebanon, N. H	July 14, 1857	17, 781
Churn	R. Daniels	Woodstock, Vt	May 21, 1872	127, 030
Churn	F. Danzenbaker	Bridgeton, N. J	Sept. 18, 1866	58, 228
Churn	J. Darrow	Warren, Ohio	June 2, 1832	....
Churn	J. J. Davelin	Philadelphia, Pa	Mar. 19, 1867	63, 023
Churn	J. B. Davidson	Oberlin, Ohio	Apr. 19, 1864	42, 354
Churn	W. C. T. Davidson and W. H. Durrett.	Hannibal, Mo	June 4, 1872	127, 406
Churn	J. Davies	Mazomania, Wis	July 9, 1867	66, 469
Churn	J. Davis, 2d	Lake Village, N. H	Mar. 13, 1866	53, 123
Churn	J. Davis, jr	Lake Village, N. H	Nov. 20, 1866	59, 790
Churn	R. W. Davis	Rodgersville, N. Y	Apr. 2, 1850	7, 233
Churn	R. W. Davis	Rodgersville, N. Y	Feb. 28, 1854	10, 566
Churn	W. Davis	Arrow Rock, Mo	Oct. 1, 1867	69, 324
Churn	J. Davison	Plymouth, Mich	July 2, 1867	66, 306
Churn	J. T. Dawson	Frostburgh, Md	Apr. 9, 1867	63, 708
Churn	L. Day	Buffalo, N. Y	Jan. 31, 1860	26, 970
Churn	A. B. Dean	Louisville, Ky	Aug. 3, 1869	93, 285
Churn	H. Decker	Lebanon, Ohio	Feb. 26, 1867	62, 329
Churn	G. Deckman	Malvern, Ohio	Apr. 2, 1867	63, 485
Churn	L. Dederick	New York, N. Y	Sept. 20, 1870	107, 464
Churn	D. Deshon, 2d	Somerset, Pa	Apr. 3, 1860	27, 701
Churn	P. S. Devlan	Reading, Pa	June 14, 1859	24, 382
Churn	J. L. Devol	Parkersburgh, W. Va	May 3, 1870	102, 509
Churn	H. Dickerman	Great Barrington, Mass	May 9, 1809	....
Churn	S. Dickerman and S. M. Parsons.	Meriden, Conn	Oct. 1, 1830	....
Churn	I. L. Dickinson	Richmond, Ind	Jan. 31, 1854	10, 487
Churn	J. Dodder	Washington, Iowa	Jan. 30, 1866	52, 272
Churn	H. Doolittle	Alton, Ill	June 23, 1863	38, 951
Churn	H. Doolittle	Aurora, Ind	Oct. 11, 1864	44, 614
Churn	E. L. Dorsey	Greenwood, Ind	Apr. 12, 1859	23, 556
Churn	Z. S. Doty	Groton, N. Y	Mar. 17, 1829	....
Churn	Z. S. Doty	Massillon, Ohio	Aug. 31, 1831	....
Churn	J. H. Doughty	Adamsville, Ohio	Oct. 8, 1861	33, 469
Churn	J. H. Doughty	Adamsville, Ohio	Feb. 18, 1862	34, 416
Churn	W. C. Douthett	Rochelle, Ill	May 12, 1868	77, 873
Churn	W. C. Douthett	Rochelle, Ill	Oct. 13, 1868	83, 050
Churn	G. H. Dow	Freeport, Ill	July 7, 1868	79, 739
Churn	N. Drew	Howell, Mich	July 16, 1867	66, 687
Churn	S. G. Dugdale	Richmond, Ind	July 15, 1851	8, 219
Churn	J. G. Dungan	Steubenville, Ohio	Oct. 9, 1855	13, 638
Churn	S. P. Dunham and A. Hipple	Killbourne, Ohio	Apr. 17, 1860	27, 894
Churn	D. Dunton	Brooklyn, N. Y	Apr. 3, 1866	53, 591
Churn	P. Dunwald	Corning, N. Y	Feb. 19, 1861	31, 447
Churn	D. Dutcher	Springfield, N. Y	June 10, 1851	8, 143
Churn	W. T. Eastes	Madison County, Ind	Nov. 26, 1867	71, 469
Churn	B. R. Eaton	Clifton, Wis	Feb. 19, 1867	62, 121
Churn	A. G. Eddy	Ashfield, Mass	Dec. 9, 1862	37, 090
Churn	S. D. Edgar	Dayton, Ohio	Mar. 23, 1869	88, 148
Churn	J. B. Edgell, E. A. Alexander, and H. C. Kellogg.	Quasqueton, Iowa	Jan. 20, 1863	37, 437
Churn	W. Edmister and S. Johnson	Mount Vernon, Ohio	July 11, 1865	48, 666
Churn	J. P. Edmonds	Rochelle, Ill	July 9, 1867	66, 478
Churn	J. J. Edwards	Columbus, Wis	Sept. 27, 1864	44, 410
Churn	C. L. Eggert	Lawrence, Kans	Aug. 20, 1867	67, 964
Churn	R. and W. J. Elarton	Hillsborough, Iowa	Sept. 28, 1869	95, 212
Churn	S. S. Elder	Springfield, Ill	Dec. 1, 1868	84, 619
Churn	A. H. Elliott	Albion, Mich	May 3, 1870	102, 518
Churn	C. H. Elliott	York, Pa	Apr. 28, 1868	77, 431
Churn	A. N. Elzy	Placerville, Cal	Nov. 26, 1867	71, 289
Churn	P. Embree	West Chester, Pa	Mar. 20, 1860	27, 536
Churn	E. R. Embry	Richmond, Ky	Nov. 23, 1869	97, 179
Churn	G. W. Emerson	Peru, Ill	Apr. 30, 1867	64, 210
Churn	S. F. Emerson	Seville, Ohio	Dec. 22, 1863	40, 997
Churn	S. F. Emerson	Seville, Ohio	June 16, 1863	38, 891
Churn	F. M. English	Evansville, Ind	July 11, 1871	116, 940
Churn	W. Estabrook	Salem, Ohio	Nov. 19, 1833	....
Churn	J. W. Evans	Forsyth, Ga	May 29, 1860	28, 462
Churn	T. R. Evans	Blacksburgh, Va	Apr. 5, 1870	101, 599
Churn	J. J. Everst	Cumberland, Md	Nov. 26, 1867	71, 472
Churn	J. Ewing	Worcester, Mass	Jan. 29, 1830	....
Churn	B. Fairbanks	Worcester, Mass	Mar. 5, 1810	....
Churn	F. T. Fairchild	Sheboygan, Wis	May 3, 1870	102, 523
Churn	G. K. Farrington	Xenia, Ohio	Sept. 28, 1858	21, 637
Churn	J. Fassauer	Wheeling, Iowa	Sept. 1, 1868	81, 615
Churn	W. S. Ferrier	Indiana, Pa	Aug. 7, 1866	56, 918
Churn	T. T. Ffirth	Camden, N. J	Dec. 23, 1862	37, 254
Churn	J. U. Fiester	Winchester, Ohio	Feb 26, 1856	14, 309
Churn	J. U. Fiester	Winchester, Ohio	Feb. 15, 1859	22, 945
Churn	J. E. Finley	Memphis, Tenn	Mar. 12, 1867	62, 740
Churn	J. E. Finley	Memphis, Tenn	Sept. 3, 1867	68, 560
Churn	J. E. Finley	Memphis, Tenn	May 5, 1868	77, 601
Churn	J. E. Finley	Memphis, Tenn	May 4, 1869	89, 645
Churn	J. E. Finley	Memphis, Tenn	May 16, 1871	114, 785
Churn	D. Fisher	College Corner, Ohio	Jan. 1, 1851	7, 870
Churn	M. Fisk	Adrian, Mich	Dec. 5, 1871	121, 604

Index of patents issued from the United States Patent Office from 1790 *to* 1873, *inclusive*—Continued.

Invention.	Inventor.	Residence.	Date.	No.
Churn	D. A. Fiske	Delavan, Wis	Mar. 27, 1866	53, 432
Churn	D. A. Fiske	Delavan, Wis	Aug. 4, 1868	80, 616
Churn	G. C. Fitch	Randolph, N. Y	Mar. 17, 1868	75, 539
Churn	J. P. Fitch	New York, N. Y	Jan. 17, 1860	26, 841
Churn	J. H. Fleming	Groton, Ohio	May 26, 1868	78, 199
Churn	B. S. Fletcher	Cornish, N. H	May 22, 1866	54, 878
Churn	O. V. Flora and J. S. Bogle	Madison, Ind., and Springfield, Ohio.	Apr. 2, 1867	63, 493
Churn	N. C. Folger	New Orleans, La	Oct. 13, 1868	83, 057
Churn	J. W. Forsyth	Leesburgh, Va	Oct. 9, 1866	58, 628
Churn	C. Foss	Perry, Ohio	June 11, 1829	......
Churn	A. W. Foster	Millbridge, Me	Apr. 5, 1870	101, 453
Churn	G. W. Fowler	Jenner's Cross Roads, Pa	Aug. 27, 1867	68, 060
Churn	J. Fowler and W. L. Walter	Homer, N. Y	July 30, 1861	32, 939
Churn	D. Frankfoder	Wakarusa, Ind	Mar. 21, 1871	112, 796
Churn	S. D. Frazier	Tekonsha, Mich	Feb. 23, 1864	41, 693
Churn	T. Freeman	Westfield, Ill	Mar. 7, 1871	112, 332
Churn	D. Frey and D. Sanders	Homer, N. Y	Aug. 16, 1864	43, 844
Churn	J. P. Friest	Chillicothe, Mo	Nov. 18, 1873	144, 755
Churn	W. Fritts	Flanders, N. J	Dec. 19, 1871	122, 004
Churn	I. N. Frost	Peoria, Ill	May 26, 1868	78, 368
Churn	A. Fuqua	Sandford, Ind	Aug. 12, 1873	141, 709
Churn	W. Furbish	Hallowell, Me	Mar. 6, 1833	......
Churn	N. Gabel	Gratis, Ohio	Aug. 20, 1861	33, 080
Churn	C. W. Gage	Homer, N. Y	Nov. 15, 1864	45, 033
Churn	A. S. Galliher	Bristol, Tenn	Mar. 9, 1869	87, 657
Churn	A. O. Gallup and E. A. Hewitt	Salem and New London, Conn.	Apr. 18, 1865	47, 292
Churn	A. O. Gallup and E. A. Hewitt	Salem and Groton, Conn	May 18, 1869	90, 255
Churn	M. Gascon	Malvern, Ohio	Aug. 30, 1870	106, 929
Churn	H. Gardiner	New York, N. Y	Dec. 23, 1862	37, 224
Churn	H. Gardiner	New York, N. Y	Feb. 22, 1870	100, 025
Churn	J. C. Gaston	Cincinnati, Ohio	Apr. 16, 1867	63, 791
Churn	J. C. Gaston	Cincinnati, Ohio	June 25, 1867	66, 015
Churn	J. C. Gaston	Cincinnati, Ohio	Dec. 24, 1867	72, 475
Churn	D. Gates	Captina Post-Office, Ohio	May 14, 1867	64, 758
Churn	J. W. Gault	Pleasant Township, Ohio	Sept. 24, 1867	69, 088
Churn	J. Geiger	Peoria County, Ill	Mar. 30, 1869	88, 289
Churn	C. George	Ligonier, Pa	June 18, 1867	65, 899
Churn	J. B. Ghormley	Bellefontaine, Ohio	Oct. 10, 1865	50, 340
Churn	J. Glattner	Suspension Bridge, N. Y	Mar. 9, 1869	87, 659
Churn	A. J. Gibson	Worcester, Mass	Sept. 13, 1864	44, 179
Churn	A. and Z. A. Gifford	Somerset, N. Y	Jan. 17, 1871	110, 968
Churn	G. W. Gilbert	Bettsville, Ohio	Jan. 7, 1862	34, 110
Churn	J. C. Gilbert	Galesburgh, Ill	July 14, 1868	79, 821
Churn	G. E. Gill and J. B. Tillinghast	Chillicothe, Ohio	June 19, 1849	6, 538
Churn	A. E. Gillilan	Marion, Iowa	June 2, 1868	78, 589
Churn	T. Gillmor	Saint Louis, Mo	July 3, 1866	56, 036
Churn	E. L. Gilman	Somerville, Mass	Oct. 8, 1867	69, 653
Churn	C. L. Gilpatrick	Saco, Me	Sept. 13, 1859	25, 405
Churn	S. Gissinger	Allegheny, Pa	Sept. 20, 1859	25, 502
Churn	S. Gissinger	Allegheny City, Pa	July 14, 1863	39, 225
Churn	S. Gissinger	Manchester, Pa	Feb. 2, 1864	41, 429
Churn	J. Gire	Louden City, Ill	Oct. 10, 1871	119, 840
Churn	D. H. Gobin	Springfield, Ill	July 16, 1872	129, 471
Churn	G. Goewey	New York, N. Y	Dec. 17, 1861	33, 973
Churn	S. M. Golden	Marcelline, Ill	Jan. 22, 1867	61, 415
Churn	J. L. Good	Elizabethtown, Pa	Mar. 9, 1869	87, 660
Churn	R. Goodwin	North Bergen, N. Y	May 4, 1842	2, 603
Churn	W. F. Goodwin	Metuchen, N. J	Apr. 12, 1870	101, 856
Churn	G. W. Goodwyn	Petersburgh, Va	May 19, 1868	78, 081
Churn	W. B. Gordnier	Cowdersport, Pa	May 8, 1860	28, 170
Churn	M. C. Gordon	Knightstown, Ind	Oct. 15, 1867	69, 795
Churn	W. L. Gordon	Dalton, Ga	Aug. 17, 1869	93, 818
Churn	E. Gore	Bennington, Vt	Jan. 2, 1855	12, 127
Churn	C. R. Gorgas	Wooster, Ohio	May 31, 1864	42, 939
Churn	B. Graham	Lyons, Iowa	June 5, 1866	55, 281
Churn	M. Granger	Syracuse, N. Y	July 14, 1830	......
Churn	A. W. Gray	Bennington, Ohio	May 21, 1867	64, 971
Churn	S. Gray	Genesee, N. Y	Aug. 17, 1810	......
Churn	G. H. Gregory	North Wilton, Conn	Dec. 21, 1869	98, 161
Churn	T. F. Griffiths	Dansville, N. Y	Oct. 27, 1863	40, 404
Churn	E. Groat	Napa, Cal	Sept. 5, 1871	118, 712
Churn	E. Groat	Napa, Cal	Sept. 5, 1871	118, 713
Churn	G. Groom	Brockville, Canada	July 18, 1871	117, 070
Churn	H. H. Grover	Auburn, N. Y	Nov. 30, 1829	......
Churn	D. L. Grover and L. S. Wright	Groton, N. Y	Aug. 12, 1862	36, 147
Churn	J. J. Gruver and A. D. Wiggins	New Market, Ohio	Nov. 5, 1867	70, 431
Churn	H. S. Gurney and H. Merrill	Memphis, Mich	Mar. 21, 1871	112, 806
Churn	J. Gustine	New Petersburgh, Ohio	May 14, 1834	......
Churn	A. Guthrie	Chicago, Ill	July 26, 1859	24, 907
Churn	B. Haden, jr	Trenton, N. J	June 15, 1822	......
Churn	H. Hagans	Brandonville, Va	July 23, 1861	32, 868
Churn	T. Haigh	Harrisburgh, Pa	Sept. 8, 1868	81, 896
Churn	J. Hale	Hillsborough, N. H	Dec. 15, 1819	......
Churn	A. Hall	Loysville, Ohio	Oct. 9, 1849	6, 777
Churn	A. W. Hall	New York, N. Y	Jan. 31, 1865	46, 185
Churn	A. W. Hall	New York, N. Y	June 5, 1866	55, 287
Churn	A. W. Hall	New York, N. Y	June 12, 1866	55, 491
Churn	S. Z. Hall	Camden, N. J	Aug. 2, 1864	43, 686
Churn	S. Z. Hall	Camden, N. J	Nov. 22, 1864	45, 155
Churn	W. S. Hall	Quincy, Mass	Aug. 2, 1859	24, 934
Churn	J. A. Ham and W. Carpenter, jr	Barry, Mo	Aug. 31, 1869	94, 311
Churn	F. Hamblin	Madrid Springs, N. Y	Feb. 1, 1870	99, 312
Churn	W. Hamilton	Saint Catharine, Mo	Mar. 19, 1861	31, 715
Churn	A. Hamlin	Schoharie, N. Y	Mar. 13, 1866	53, 141
Churn	B. Handforth	Chicago, Ill	Feb. 19, 1867	62, 129
Churn	J. A. Hanger	Staunton, Va	Aug. 31, 1869	94, 312

Index of patents issued from the United States Patent Office from 1790 *to* 1873, *inclusive*—Continued.

Invention.	Inventor.	Residence.	Date.	No.
Churn	J. O. and G. C. Hanger	Churchville, Va	Jan. 18, 1870	98, 958
Churn	A. Hanson	Windham, Me	Mar. 2, 1836	....
Churn	J. W. Hardie	New York, N. Y	Mar. 13, 1860	27, 446
Churn	E. T. Harlan	Star City, Ind	Mar. 31, 1868	76, 185
Churn	J. Harper	Hillsborough, Iowa	Mar. 30, 1869	88, 474
Churn	S. Harper	Lawrence, Kans	Aug. 17, 1869	93, 879
Churn	T. B. Harper	Xenia, Ohio	Apr. 27, 1858	20, 062
Churn	B. N. Harris	Talbotton, Ga	July 6, 1869	92, 308
Churn	F. M. Harris	Winnamac, Ind	June 14, 1870	104, 304
Churn	J. D. Harrison	Middletown, Ohio	Apr. 23, 1872	126, 052
Churn	R. H. Harrison	Washington, D. C	Mar. 28, 1854	10, 690
Churn	M. S. Harsha	Sycamore, Ill	Dec. 13, 1859	26, 459
Churn	A. H. Hart	Stockbridge, Wis	Aug. 25, 1863	39, 650
Churn	G. Hart	Atwater, Ohio	June 20, 1865	48, 277
Churn	G. B. Hart	Buffalo Township, Iowa	Oct. 1, 1872	131, 877
Churn	C. Harvey	Cairo, N. Y	June 25, 1872	128, 306
Churn	M. B. Hassler	Columbia City, Ind	Sept. 13, 1859	25, 410
Churn	J. Hatfield and H. M. Goldsmith	Falmouth, Ind., and Burlington, Iowa.	July 13, 1858	20, 878
Churn	J. Hathaway	Canandaigua, N. Y	Aug. 22, 1828	....
Churn	G. W Hawk	Chicago, Ill	Apr. 2, 1867	63, 515
Churn	O. Hawley and J. W. Ward	Wheeling, W. Va	July 2, 1867	66, 235
Churn	S. Hays	Poultney, N. Y	Mar. 8, 1808	....
Churn	N. S. Hazen	La Fayette, Ind	May 31, 1870	103, 612
Churn	A. J. Heavner	Time, Ill	Jan. 14, 1868	73, 327
Churn	W. H. Henderson	West Point, Ill	July 21, 1868	80, 174
Churn	J. Henry	Steubenville, Ohio	Oct. 2, 1866	58, 415
Churn	H. Hensel	Moon Township, Pa	Aug. 1, 1871	117, 629
Churn	W. A. Herrick	Greene, Me	July 21, 1835	....
Churn	H. L. Hervey	Philadelphia, Pa	Aug. 8, 1865	49, 345
Churn	C. Hess	Lyons City, Iowa	May 26, 1868	78, 286
Churn	C. Hess	Lyons, Iowa	Apr. 6, 1869	88, 568
Churn	A. E. Hewett	Homer, N. Y	Apr. 26, 1864	42, 481
Churn	S. Hewit	Seneca Falls, N. Y	July 14, 1857	17, 790
Churn	S. Hewit	Seneca Falls, N. Y	May 1, 1866	54, 343
Churn	S. Hewit	Seneca Falls, N. Y	Nov. 10, 1868	83, 853
Churn	J. Hewitt	Carmichael, Pa	Aug. 12, 1862	36, 151
Churn	P. Hill	Millport, N. Y	Feb. 4, 1868	74, 088
Churn	T. M. Hill	Eaton, Ohio	Dec. 5, 1865	51, 320
Churn	H. A. Hincheer	Hustonville, Ky	Apr. 1, 1873	137, 369
Churn	F. M. Hindman and N. Hiatt	Sidney, Iowa	Aug. 23, 1870	106, 695
Churn	E. Hinman	Syracuse, N. Y	Aug. 11, 1863	39, 529
Churn	E. Hitchcock	Sturbridge, Mass	June 8, 1869	91, 016
Churn	D. F. Hitt	Galena, Ill	Oct. 7, 1842	2, 800
Churn	A. D. Hoffman	Minneapolis, Minn	May 19, 1868	78, 091
Churn	W. H. Holdam	Crab Orchard, Ky	Apr. 22, 1873	138, 089
Churn	S. B. Holden	Woburn, Mass	Apr. 20, 1869	89, 046
Churn	E. F. Holloway	Knightstown, Ind	June 20, 1865	48, 278
Churn	R. G. Holmes	Worcester, Mass	May 8, 1860	28, 177
Churn	C. P. Holmes and A. L. Howell	Gouverneur and Mohawk, N. Y.	Mar. 1, 1870	100, 292
Churn	G. P. Hopkins	Cabot, Vt	Mar. 29, 1859	23, 374
Churn	W. B. Hopkins	Oakfield, N. Y	Mar. 26, 1861	31, 806
Churn	W. Hosier	Washington Township, Ind	May 30, 1865	47, 952
Churn	N. Hospers	Pelia, Iowa	Aug. 13, 1872	130, 428
Churn	M. B. Hough	Dover, Ohio	Jan. 21, 1839	1, 072
Churn	J. W. F. How	Douglas County, Oreg	Dec. 17, 1872	133, 940
Churn	S. B. Howd	Arcadia, N. Y	Dec. 12, 1846	4, 887
Churn	N. Howe	Windham, Conn	Apr. 25, 1808	....
Churn	H. W. Howland	Calhoun, Ill	Sept. 24, 1867	69, 213
Churn	J. Houston	Lake Village, N. H	Nov. 17, 1863	40, 623
Churn	E. Hoyt	Stamford, Conn	July 3, 1866	56, 050
Churn	S. Huff	New Vienna, Ohio	Apr. 24, 1849	6, 389
Churn	J. S. Huffman	Brownsburgh, Va	May 25, 1869	90, 544
Churn	A. Humphrey	Gray, Me	May 9, 1825	....
Churn	W. Humphrey	Ohio	Jan. 5, 1814	....
Churn	G. W. Hurst	Chestertown, Md	Apr. 9, 1867	63, 639
Churn	S. Hutchings and J. D. Leach	Penobscot, Me	Aug. 14, 1860	29, 597
Churn	C. Hutchins	Baldwin City, Kans	Oct. 1, 1872	131, 821
Churn	P. Hutchinson	Boston, Mass	Nov. 26, 1867	71, 489
Churn	T. H. Hutchinson	Gorham, N. H	Jan. 19, 1869	85, 933
Churn	H. Hutchison	Three Rivers, Mich	July 18, 1865	48, 815
Churn	A. Huyck	Our Town, Wis	June 18, 1867	65, 915
Churn	D. Hyde	Bridgeport, N. Y	Oct. 22, 1867	69, 998
Churn	B. Illingworth	Freeport, Ill	Sept. 11, 1866	57, 913
Churn	B. Illingworth	Freeport, Ill	Jan. 28, 1868	73, 895
Churn	W. L. Imlay	Philadelphia, Pa	Sept. 26, 1865	50, 129
Churn	J. Irving	Boston, Mass	Apr. 30, 1816	
Churn	J. Jackson	Coopersville, Mich	June 20, 1871	116, 194
Churn	W. Jackson and J. Clarke	Syracuse, N. Y	May 7, 1861	32, 250
Churn	J. N. Jacobs	Crittenden, Ky	Mar. 1, 1870	100, 412
Churn	H. E. James	West Alexandria, Pa	May 11, 1869	90, 002
Churn	J. C. Jay and J. Younce	Wabash, Ind	June 4, 1867	65, 488
Churn	T. A. Jebb	Buffalo, N. Y	Nov. 15, 1859	26, 109
Churn	T. A. Jebb	Buffalo, N. Y	July 2, 1861	32, 697
Churn	F. A. Jewett	Shrewsbury, Mass	June 16, 1868	78, 877
Churn	D. Johnson	New York, N. Y	Sept. 14, 1858	21, 501
Churn	J. C. Johnson	Uniontown, Pa	July 9, 1872	128, 730
Churn	J. B. Johnston	Saint Matthews, Ky	Jan. 25, 1870	99, 206
Churn	G. Jones	Grand Rapids, Mich	June 1, 1869	90, 665
Churn	L. Jones	Canton, Conn	Dec. 10, 1808	....
Churn	J. A. Jordan	Shelbyville, Tenn	Mar. 30, 1858	19, 782
Churn	J. A. Jordan	Shelbyville, Tenn	Dec. 19, 1871	122, 026
Churn	J. W. Jordan	Lexington, Va	Dec. 13, 1870	110, 047
Churn	R. Justis	Dublin, Ind	Jan. 21, 1862	34, 208
Churn	R. Keese	Cardington, Ohio	June 13, 1865	48, 182
Churn	R. Keese	Bennington, Ohio	Apr. 9, 1867	63, 728
Churn	J. W. Kelberg	Pittsburgh, Pa	Oct. 16, 1860	30, 409

Index of patents issued from the United States Patent Office from 1790 *to* 1873, *inclusive*—Continued.

Invention.	Inventor.	Residence.	Date.	No.
Churn	J. D. Kellogg, jr	Northampton, Mass	Sept. 4, 1866	57, 730
Churn	W. Kelly	Hastings, Mich	May 24, 1859	24, 127
Churn	W. C. Kemp	Palmyra, Mo	Sept. 14, 1869	94, 753
Churn	E. Kenney	Livermore, Me	Sept. 19, 1865	50, 009
Churn	J. Kepler	Crawfordsville, Ind	June 2, 1868	78, 526
Churn	S. A. Kerr	Arbor hill, Va	June 19, 1860	28, 757
Churn	J. J. Kimball	Naperville, Ill	May 2, 1871	114, 306
Churn	M. W. Kilgore	Baltimore, Md	Nov. 19, 1867	71, 182
Churn	P. Killin	Mount Healthy, Ohio	July 12, 1864	43, 513
Churn	P. Killin	Mount Healthy, Ohio	Aug. 30, 1864	43, 999
Churn	A. Kindermann	Cleveland, Ohio	July 13, 1869	92, 458
Churn	W. E. Kinert	Bluffton, Ind	Dec. 29, 1868	85, 452
Churn	J. King	Suckasunny, N. J	Jan. 5, 1869	85, 594
Churn	J. King	Suckasunny, N. J	Sept. 14, 1869	94, 830
Churn	S. P. Kingsley	Springfield, Wis	May 28, 1867	65, 234
Churn	N. S. Kinyon	Chenango Forks, N. Y	Jan. 22, 1867	61, 436
Churn	L. Kittinger and L. Ruch	Canal Fulton, Ohio	July 2, 1872	128, 549
Churn	J. Klingensmith	Warren, Ohio	Nov. 30, 1869	97, 413
Churn	C. B. Knowles	Fayette, Me	Aug. 2, 1832	....
Churn	A. Köhler and M. W. Wilson	Noblesville, Ind	Jan. 19, 1869	86, 083
Churn	H. Kuhlmann	Cincinnati, Ohio	Aug. 30, 1870	106, 836
Churn	A. Ladd	Saint Lawrence, N. Y	Oct. 4, 1870	108, 035
Churn	H. R. Ladd	Orwell, Ohio	May 19, 1863	38, 587
Churn	L. Lake	Middlebury, Pa	July 19, 1859	24, 845
Churn	L. Lamb	Berlin, Conn	Sept. 2, 1856	15, 661
Churn	W. Lamb	Rochelle, Ill	Mar. 24, 1868	75, 772
Churn	J. Lane	Homer, N. Y	May 15, 1815	....
Churn	S. S. Langdon	Cleveland, Ohio	Aug. 16, 1859	25, 126
Churn	W. W. Lapham	Decatur, Ill	Oct. 18, 1864	44, 732
Churn	R. Lapham and R. P. Wilson	New York, N. Y	Feb. 8, 1859	22, 879
Churn	B. U Lapish	Durham, N. H	Aug. 15, 1809	....
Churn	P. C. Lamb	Allentown, Pa	Mar. 31, 1868	76, 206
Churn	S. Leach	Wilbraham, Mass	June 1, 1869	90, 758
Churn	L. Leavenworth	Trumansburgh, N. Y	Mar. 18, 1856	14, 458
Churn	H. Leber	Bellfair Mills, Va	Sept. 15, 1868	82, 235
Churn	C. H. Lee	Oskaloosa, Iowa	Nov. 12, 1867	70, 725
Churn	J. Lees	Racine, Wis	Oct. 28, 1862	36, 787
Churn	S. F. Lefler	Racine, Wis	Jan. 5, 1858	19, 034
Churn	J. J. Lehaye	Reading, Pa	Aug. 16, 1859	25, 162
Churn	G. H. Lenher	Richmond, Va	Dec. 11, 1866	60, 391
Churn	E. C. Leonard	Binghamton, N. Y	June 12, 1866	55, 508
Churn	J. S. Lewis	Elkport, Iowa	June 7, 1870	103, 898
Churn	P. F. Lewis	Columbus, Pa	Apr. 20, 1869	89, 155
Churn	J. Liebhaber	Bless, Bavaria	Apr. 21, 1868	76, 928
Churn	G. Lightfoot	Elgin, Ill	Dec. 9, 1856	16, 210
Churn	F. A. Lindal	Stockton, N. Y	Dec. 14, 1869	97, 937
Churn	N. H. Lindley	Redding, Conn	Aug. 25, 1842	2, 758
Churn	N. H. Lindley and W. Perry	Redding and Bridgeport, Conn.	Mar. 10, 1843	2, 993
Churn	W. H. Link	Shanesville, Ohio	Oct. 10, 1871	119, 870
Churn	T. Ling	Portland. Me	Aug. 21, 1844	3, 714
Churn	N. B. Livingston	Portland, Ind	July 6, 1852	9, 097
Churn	A. Lloyd	Millersburgh, Ill	Sept. 7, 1869	94, 619
Churn	J. A. Lloyd	Saint Paul, Minn	June 30, 1863	39, 055
Churn	M. Lockwood	Cuba, N. Y	Nov. 26, 1872	133, 462
Churn	T. E. Lockwood	Cincinnati, Ohio	July 17, 1866	56, 425
Churn	M. C. Longacre	Cleveland, Ohio	Feb. 5, 1861	31, 355
Churn	W. Loomis	Ashford, Conn	June 7, 1832	....
Churn	A. D. Lord	Bethany, N. Y	Aug. 20, 1872	130, 730
Churn	P. S. Lowell	Farmington, Me	May 29, 1835	....
Churn	D. Lown	Poughkeepsie, N. Y	Nov. 3, 1868	83, 779
Churn	S. D. Lucas	Winterpock, Va	Sept. 21, 1869	95, 029
Churn	J. Luccock and J. M. Gowdy	Peoria, Ill	Apr. 29, 1862	35, 101
Churn	E. Lynch	Buffalo, N. Y	Apr. 10, 1860	27, 864
Churn	H. N. Mackey	Morgantown, Va	Apr. 28, 1857	17, 159
Churn	H. H. Macklin	New Springfield, Ohio	Oct. 22, 1867	70, 009
Churn	J. O. Maclaskey	Perth Amboy, N. J	Dec. 27, 1870	110, 616
Churn	J. Macnish	Berlin, Wis	Apr. 20, 1858	20, 025
Churn	J. Macnish	Berlin, Wis	July 6, 1858	20, 803
Churn	J. Macnish	Berlin, Wis	July 6, 1858	20, 804
Churn	R. Mahr	New York, N. Y	Nov. 19, 1867	71, 029
Churn	D. L. Main	Brooklyn, Mich	Aug. 25, 1868	81, 518
Churn	C. A. Maltby	Roland, Ill	July 5, 1870	104, 972
Churn	P. J. Manning	Troy, Ill	Sept. 21, 1869	95, 124
Churn	G. W. Manson	Buxton, Me	Feb. 12, 1867	62, 046
Churn	J. A. Marden	Boston, Mass	Oct. 29, 1872	132, 591
Churn	M. R. Marcell	Dansville, N. Y	July 27, 1858	21, 010
Churn	J. A. Marine	Mooresville, Ind	Mar. 19, 1872	124, 841
Churn	T. R. Markillie	Winchester, Ill	Aug. 27, 1861	33, 154
Churn	J. L Marsh	Centreville, Ind	Mar. 9, 1869	87, 691
Churn	D. Marshall	Genoa, N. Y	June 5, 1866	55, 320
Churn	J. Masten	Lee, Ohio	Oct. 21, 1873	143, 919
Churn	C. Matheney	Greensburgh, Ind	Dec. 6, 1870	109, 919
Churn	W. D. Matthews	Columbia, Tenn	Nov. 6, 1866	59, 428
Churn	J. Mayes	Oxford, N. Y	May 14, 1872	126, 725
Churn	J. Mayhew	West Tisbury, Mass	July 19, 1870	105, 471
Churn	J. Maxey	Kewanna, Ind	Dec. 17, 1867	72, 310
Churn	R. Maxwell	Lewis County, Va	Aug. 24, 1852	9, 218
Churn	M. A. McAfee	Talbotton, Ga	Mar. 30, 1869	88, 399
Churn	J. McBride	Ithaca, N. Y	Feb. 7, 1871	111, 555
Churn	W. H. McClintock	Frankfort, Ohio	May 31, 1859	24, 223
Churn	J. McCoord	Washington, N. Y	June 13, 1809	....
Churn	R. B. McCormick	Bloomington, Ill	Apr. 9, 1872	125, 601
Churn	R. T. McCormick	Greencastle Junction, Ind	Aug. 23, 1870	106, 599
Churn	W. R. McCutcheon	Washington, Iowa	Sept. 11, 1866	57, 942
Churn	W. R. McCutcheon	Washington, Iowa	June 4, 1867	65, 496
Churn	J. McElroy	Allegheny City, Pa	Aug. 10, 1869	93, 462
Churn	W. D. McFadden	Senatobia, Miss	Feb. 23, 1869	87, 185

Index of patents issued from the United States Patent Office from 1790 to 1873, inclusive—Continued.

Invention.	Inventor.	Residence.	Date.	No.
Churn	H. B. McFall	Mount Solon, Va	June 7, 1870	103, 908
Churn	G. H. McGlothlen	Chariton, Iowa	Aug. 30, 1870	106, 848
Churn	W. McKeever	Staunton, Va	May 7, 1872	126, 560
Churn	J. McKenzie	Portland, Me	Aug. 27, 1867	68, 097
Churn	J. J. McLane	Sagetown, Ill	June 9, 1868	78, 809
Churn	J. McLaughlin	Goshen, Ohio	July 13, 1852	9, 117
Churn	W. S. McManus and R. S. Merryman.	Brunswick, Me	June 1, 1869	90, 674
Churn	E. McMillan, jr	Wilmington, Ohio	Nov. 20, 1866	59, 802
Churn	D. C. McNeil	De Witt, Iowa	Sept. 3, 1867	68, 523
Churn	A. H. McWaine	Shickshinny, Pa	Dec. 28, 1869	98, 288
Churn	G. McWilliams	Fostoria, Ohio	Jan. 7, 1868	73, 023
Churn	J. Megown	New London, Mo	Mar. 19, 1867	63, 074
Churn	J. H. Mendenhall	Cerro Gordo, Ind	Sept. 4, 1866	57, 750
Churn	J. P. Meranda	Springfield, Ohio	July 19, 1870	105, 475
Churn	S. H. Merriditb	Oxford, Ohio	Feb. 9, 1864	41, 525
Churn	D. C. Merrill	South Paris, Me	Mar. 26, 1867	63, 282
Churn	J. O. Merrill	Chichester, N. H	May 10, 1859	23, 986
Churn	C. Merriman, jr	Middletown, Conn	Apr. 21, 1836	....
Churn	C. Messenger	Cleveland, Ohio	Mar. 23, 1869	88, 156
Churn	W. P. and H. T. Messick	Clarkesville, Tex	Sept. 24, 1872	131, 623
Churn	J. Mewhinney	Liberty Township, Ohio	July 30, 1831	....
Churn	J. R. Mickey	Waterford, Pa	Feb. 26, 1861	31, 554
Churn	J. R. Mickey	Chicago Ill	Sept. 18, 1866	58, 192
Churn	J. L. Middleton	Zanesville, Ohio	Sept. 1, 1868	81, 810
Churn	C. J. Miller, jr	Richmond, Ky	Nov. 9, 1869	96, 604
Churn	F. Miller	Frostburgh, Md	Nov. 23, 1869	97, 103
Churn	M. B. Miller	Peoria County, Ill	Mar. 30, 1869	88, 320
Churn	I. Millington	Warren, N. Y	Mar. 2, 1809	....
Churn	A. J. Mills	Scott, N. Y	Feb. 5, 1867	61, 850
Churn	M. B. Mills	East Mendon, N. Y	Aug. 27, 1867	68, 100
Churn	S. Mills	Clinton, Ill	June 9, 1868	78, 812
Churn	J. Mitchell	Gasport, N. Y	Mar. 8, 1859	23, 186
Churn	J. Mitchell	Aleppo Township, Pa	Apr. 17, 1860	27, 927
Churn	J. E. Mitchell	Paris, Canada	Dec. 10, 1872	133, 717
Churn	J. H. Monce	Hopkinsville, Ohio	June 23, 1868	79, 244
Churn	O. F. Montfort	Dearborn, Mich	July 11, 1871	116, 855
Churn	H. H. Montgomery	Greensburgh, Ind	Sept. 26, 1871	119, 387
Churn	A. Moe	Plainfield, N. J	Aug. 17, 1869	93, 827
Churn	S. Moon, jr	Downingtown. Pa	Oct. 24, 1808	....
Churn	L. Mooney	Allegheny City, Pa	Nov. 27, 1866	60, 035
Churn	M. Moses	Malone, N. Y	Dec. 10, 1872	133, 878
Churn	J. Moran	Millport, Pa	Oct. 8, 1872	132, 096
Churn	C. R. Morehouse	Cardington, Ohio	Nov. 21, 1865	51, 073
Churn	E. Morgan	French Creek, N. Y	Dec. 7, 1869	97, 674
Churn	W. Morgan	Middlebrook, Va	Mar. 20, 1860	27, 559
Churn	T. and L. Morrison	Groton, N. Y	Oct. 16, 1829	....
Churn	M. Morse and P. W. Sawyer	Gray, Me	Apr. 14, 1868	76, 649
Churn	G. W. Moster and E. C. Packer.	Alliance, Ohio	Jan. 19, 1869	86, 094
Churn	H. Mott	Bellevue, Mich	May 8, 1866	54, 583
Churn	J. Moyers	Hillsborough, Ohio	Sept. 13, 1870	107, 284
Churn	A. P. Moyers, I. Searles, and G. W. Spencer.	Prattsville, N. Y	June 17, 1862	35, 620
Churn	S. W. Mudge	Rome, N. Y	Sept. 10, 1861	33, 262
Churn	S. W. Mudge	Rome, N. Y	Nov. 5, 1861	36, 659
Churn	H. Mumford	Ulster County, N. Y	Mar. 28, 1811	....
Churn	C. Murdock	Baltimore, Md	Feb. 20, 1849	6, 133
Churn	R. Murphy	Jasper, N. Y	July 5, 1864	43, 423
Churn	R. Murphy	Jasper, N. Y	Feb. 16, 1869	86, 999
Churn	J. W. Myers	Lyons, Iowa	Apr. 30, 1867	64, 348
Churn	J. Neal	Sheboygan, Wis	June 4, 1867	65, 417
Churn	M. Neal	Kalamazoo, Mich	Oct. 25, 1864	44, 836
Churn	M. Neal	Kalamazoo, Mich	Jan. 30, 1866	52, 313
Churn	J. L. Nelson	Lewisburgh, W. Va	Aug. 10, 1869	93, 550
Churn	W. Newberry	Clarksville, Mo	Oct. 29, 1867	70, 249
Churn	A. Newbrough	Madisonville, Ky	Aug. 1, 1865	49, 138
Churn	D. Newbrough	Clarksburgh, Ind	Mar. 6, 1860	27, 377
Churn	W. Newbrough	Mohican, Ohio	Apr. 15, 1856	14, 677
Churn	A. A. Newman	Sparta, Ill	Aug. 29, 1865	49, 646
Churn	E. P. Newman	New Albany, Ind	May 12, 1868	77, 753
Churn	J. L. Nettleton	West Cheshire, Conn	June 15, 1869	91, 360
Churn	H. A. Novers and C. Ross	Claremont, N. H	Nov. 20, 1860	30, 686
Churn	J. P. Nichols	New Richmond, Ohio	Jan. 11, 1870	98, 703
Churn	H. C. Nicholson	Mount Washington, Ohio	June 2, 1857	17, 444
Churn	T. Nicholson	Nelson County, Ky	Aug. 10, 1809	....
Churn	T. Nicholson	New Market, Va	Mar. 31, 1836	....
Churn	A. S. Norcross	Hallowell, Me	Sept. 19, 1834	....
Churn	T. S. Nutter	Harrisburgh, Ohio	Aug. 3, 1869	93, 219
Churn	S. J. O'Brien	New York	Dec. 22, 1829	....
Churn	J. O'Donald	Clinton, Ill	Jan. 16, 1866	52, 066
Churn	J. Oothoudt	Lebanon, N. Y	Nov. 10, 1829	....
Churn	J. Oothoudt	Minneapolis, Minn	Aug. 4, 1868	80, 658
Churn	J. H. Ormsby and R. S. Harton	Holden, Mo	June 14, 1870	104, 191
Churn	D. Osborn	Berkshire County, Mass	Jan. 8, 1810	....
Churn	D. Osgood, jr	Blue Hill, Me	July 9, 1838	831
Churn	J. E. Overaker	Redwood, N. Y	Apr. 6, 1869	88, 662
Churn	S. R. Owen	Stewartsville, Mo	Dec. 15, 1868	85, 026
Churn	R. D. Ozburn	Lena, Ill	Sept. 7, 1869	94, 506
Churn	J. Paff	Eden, N. Y	July 17, 1860	29, 189
Churn	E. Page	Streetsborough, Ohio	Feb. 9, 1858	19, 310
Churn	G. N. Palmer	Greene, N. Y	Aug. 31, 1869	94, 435
Churn	G. N. Palmer	Greene, N. Y	Apr. 26, 1870	102, 425
Churn	J. T. Palmer	Greencastle, Ind	Sept. 13, 1864	44, 216
Churn	S. M. Palmer	Greene, N. Y	Jan. 30, 1872	123, 289
Churn	V. Palmer	Castalia, Ohio	Dec. 11, 1866	60, 414
Churn	J. Park	Joliet, Ill	July 31, 1860	29, 397
Churn	T. B. Parke	Near Downieville, Cal	June 27, 1871	116, 348
Churn	J. Parker	Biddeford, Me	June 10, 1862	35, 540

Index of patents issued from the United States Patent Office from 1790 *to* 1873, *inclusive*—Continued.

Invention.	Inventor.	Residence.	Date.	No.
Churn	J. R. Parker	Sing Sing, N. Y	May 31, 1859	24, 233
Churn	R. Parker	North Cohocton, N. Y	Sept. 3, 1861	33, 213
Churn	W. Parks	Meadville, Pa	Feb. 7, 1871	111, 562
Churn	J. D. Parrot	Morristown, N. J	Nov. 13, 1866	59, 637
Churn	A. Patterson	Birmingham, Pa	Mar. 1, 1859	23, 111
Churn	M. Payne	Cardington, Ohio	Dec. 9, 1862	37, 166
Churn	T. Payne	Grand Rapids, Mich	Dec. 24, 1867	72, 535
Churn	A. Pease	Weston, Vt	Sept. 16, 1856	15, 741
Churn	J. N. Pease	Panama, N. Y	Mar. 7, 1865	46, 695
Churn	L. P. Pease	Mount Carmel, Ill	Oct. 16, 1855	13, 686
Churn	R. H. Peck	Wolcott, Vt	Nov. 26, 1861	33, 815
Churn	W. C. Peck	Wheeling, W. Va	Oct. 1, 1867	69, 475
Churn	J. Pelsor	Brooklyn, Ill	Dec. 24, 1867	72, 673
Churn	P. Penington	Union City, Ind	July 7, 1868	79, 773
Churn	W. H. Pennock	Mermaid, Del	May 24, 1870	103, 364
Churn	G. S. Perfater	Camp Point, Ill	Mar. 28, 1871	113, 089
Churn	H. Phelps	Williston, Vt	July 21, 1835	....
Churn	E. J. Phillips	Prescott, Wis	Apr. 11, 1865	47, 219
Churn	E. H. Philo	Half Moon, N. Y	June 9, 1863	38, 865
Churn	H. C. Pierce	Homer, N. Y	Apr. 29, 1862	35, 110
Churn	T. Pierce	Hartwick, N. Y	Nov. 10, 1841	2, 343
Churn	D. Pierpont	New Haven, Conn	Apr. 22, 1808	....
Churn	J. Pike	Syracuse, N. Y	Dec. 18, 1860	30, 930
Churn	J. G. Pike	Lisbon, N. Y	Apr. 28, 1838	719
Churn	A. J. Pope	Strongsville, Ohio	Dec. 4, 1866	60, 240
Churn	E. Porter	Tallmadge, Ohio	June 7, 1864	43, 040
Churn	E. Porter	Tallmadge, Ohio	Oct. 15, 1867	69, 934
Churn	R. Porter	Bellerica, Mass	May 10, 1838	735
Churn	S. T. W. Potter	Scott, N. Y	June 5, 1866	55, 360
Churn	A. M. Powell	Collinsville, Ill	June 22, 1869	91, 562
Churn	S. Power	Lawrenceville, Pa	June 23, 1832	....
Churn	E. L. Pratt	Philadelphia, Pa	Aug. 30, 1859	25, 306
Churn	D. K. Price	Ossian, N. Y	Jan. 19, 1864	41, 322
Churn	I. T. Price	Leesville, Ohio	Sept. 25, 1866	58, 288
Churn	J. F. Quimby	Stetson, Me	Apr. 17, 1866	54, 013
Churn	A. Ralston	West Middletown, Pa	Sept. 21, 1858	21, 575
Churn	B. Randall	North Pownal, Me	Mar. 11, 1835	....
Churn	D. B. Randall	West Glover, Vt	Mar. 27, 1866	53, 487
Churn	J. Rankin	New York, N. Y	Apr. 4, 1865	47, 129
Churn	J. Rankin	New York, N. Y	Jan. 16, 1866	52, 075
Churn	J. Rankin	New York, N. Y	Mar. 20, 1866	53, 335
Churn	J. Rankin and J. N. McIntire	New York, N. Y	June 21, 1864	43, 227
Churn	F. Ransom	Buffalo, N. Y	Aug. 30, 1864	44, 015
Churn	G. S. Rarey	Columbus, Ohio	Apr. 27, 1858	20, 089
Churn	J. B. Raynor	Mazo Manie, Wis	Feb. 25, 1868	74, 941
Churn	A. Reed	Columbia County, N. Y	Apr. 25, 1808	....
Churn	J. H. Reed	La Fayette, Ind	Jan. 3, 1871	110, 678
Churn	C. C. Reese	Attica, Ind	June 16, 1868	79, 002
Churn	W. W. Reid	Rochester, N. Y	Oct. 16, 1860	30, 424
Churn	J. K. Reiner	Line Lexington, Pa	Nov. 27, 1866	60, 058
Churn	J. Rengel	Lancaster, N. Y	Dec. 17, 1872	134, 100
Churn	J. L. Ressler	Ramsburgh, Pa	Sept. 7, 1869	94, 645
Churn	T. Rettew	West Vincent Township, Pa.	Nov. 5, 1867	70, 470
Churn	E. Reynolds	Omro, Wis	Sept. 7, 1869	94, 646
Churn	A. Rhoades	Pontiac, Mich	June 6, 1865	48, 096
Churn	C. Rhodes	Morrow, Ohio	May 18, 1852	8, 959
Churn	H. Rice	Youngstown, Ohio	Sept. 13, 1864	44, 219
Churn	T. Rich	Kingston, N. Y	Apr. 14, 1868	76, 815
Churn	I. S. Richardson	Boston, Mass	June 10, 1851	8, 146
Churn	J. C. Richardson and L. Taylor	Prairie du Chien and Jordan, Wis.	Oct. 25, 1870	108, 631
Churn	J. Ricker	Bangor, Me	Aug. 9, 1834	....
Churn	J. Rickey	Centre Township, Ohio	Oct. 25, 1832	....
Churn	J. Risher	Delaware, Ohio	Dec. 24, 1867	72, 682
Churn	S. Risler	Locktown, N. J	Dec. 14, 1869	97, 815
Churn	S. Risler	Locktown, N. J	Nov. 29, 1870	109, 759
Churn	J. L. Riter and R. C. Swann	Brownsville, Ind	Apr. 30, 1867	64, 367
Churn	F. W. Robbins	Solon, Ohio	Dec. 3, 1867	71, 645
Churn	Z. C. Robbins	Saint Louis, Mo	June 26, 1849	6, 556
Churn	Z. C. Robbins	Saint Louis, Mo	Jan. 8, 1850	7, 066
Churn	C. H. L. Roberts	Morrison, Ill	Apr. 23, 1867	64, 149
Churn	F. L. Roberts	Jacksonville, Ill	Nov. 16, 1869	96, 967
Churn	H. Roberts	Seneca Falls, N. Y	Feb. 5, 1836	....
Churn	H. and R. V. Robie	Eaton, N. Y	Nov. 16, 1858	22, 090
Churn	D. Robinson	Franklin, N. Y	May 3, 1811	....
Churn	F. J. Robinson	Laconia, N. H	Apr. 10, 1866	53, 876
Churn	J. J. Robinson	Clinton, Ill	Dec. 5, 1865	51, 354
Churn	W. Robinson	Bellefontaine, Ohio	Dec. 23, 1862	37, 244
Churn	W. Robinson	Bellefontaine, Ohio	Oct. 24, 1865	50, 631
Churn	W. C. Robinson	Saltsburgh, Pa	Feb. 25, 1868	74, 943
Churn	H. Rohrer	Strasburgh Township, Lancaster County, Pa.	Dec. 6, 1859	26, 377
Churn	A. Rose	Penn Yan, N. Y	June 29, 1858	20, 730
Churn	A. A. Rose	Binghamton, N. Y	Feb. 26, 1867	62, 502
Churn	T. Rose	Cortlandville, N. Y	June 17, 1862	35, 628
Churn	T. Rose	Cortlandville, N. Y	Apr. 7, 1868	76, 350
Churn	L. Rosenkrans	Urbana, N. Y	May 19, 1827	....
Churn	A. A. Ross	Horicon, N. Y	July 28, 1868	80, 507
Churn	N. D. Ross	Braintrim, Pa	June 5, 1860	28, 607
Churn	R. F. Ross	Marshall County, Miss	Feb. 22, 1870	100, 076
Churn	T. Ross	Pickaway, Ohio	Jan. 26, 1831	....
Churn	J. H. Roundey	Old Town, Me	June 25, 1867	66, 042
Churn	J. H. Rowe	Fort Wayne, Ind	June 16, 1868	79, 008
Churn	J. H. Rowley	Vanceburgh, Ky	Dec. 10, 1867	72, 087
Churn	M. V. B. Rowley	Worcester, N. Y	Sept. 17, 1867	69, 029
Churn	W. M. Rumrill	Roanoke, Ind	Mar. 23, 1869	88, 078
Churn	H. D. Rumsey	Homer, N. Y	Sept. 10, 1867	68, 655
Churn	J. B. Rumsey	Port Huron, Mich	Jan. 28, 1868	73, 932

Index of patents issued from the United States Patent Office from 1790 *to* 1873, *inclusive*—Continued.

Invention.	Inventor.	Residence.	Date.	No.
Churn	L. Runyon	Newark, N. Y	Oct. 17, 1871	120, 103
Churn	E. P. Russell	Manlius, N. Y	Nov. 3, 1868	83, 664
Churn	E. W. Russell	Ashley, Mo	Oct. 15, 1867	69, 843
Churn	J. L. Rust	Oquaka Junction, Ill	July 3, 1866	56, 103
Churn	J. L. Rust	Keithburgh, Ill	July 11, 1871	116, 871
Churn	I. F. Ryerson	Pierceton, Ind	Feb. 27, 1866	52, 890
Churn	H. Saggan	Newark, N. J	Oct. 11, 1870	108, 297
Churn	C. W. Saladee	Newark, Ohio	Jan. 19, 1869	85, 967
Churn	G. H. Sanborn	Boston, Mass	Aug. 28, 1866	57, 573
Churn	J. F. Sanborn	Hardwick, Vt	June 27, 1865	48, 449
Churn	W. W. Sanborn	Lyons City, Iowa	May 21, 1867	65, 014
Churn	J. Sangster	Buffalo, N. Y	Mar. 20, 1860	27, 570
Churn	J. R. Sapp	Danville, Ohio	Aug. 2, 1870	106, 079
Churn	S. E. Saul	New York, N. Y	Apr. 4, 1865	47, 134
Churn	H. Saxton	Paris, N. Y	Nov. 6, 1829	....
Churn	G. W. Sayre	Pisgah, Ohio	Aug. 9, 1864	43, 797
Churn	J. B. Schermerhorn	New York, N. Y	June 5, 1844	3, 616
Churn	C. Schifferly	Bourbon, Ind	Jan. 7, 1868	73, 125
Churn	A. Schlingman, D. Glander, and J. Campbell.	West Alexandria, Ohio	Nov. 3, 1868	83, 732
Churn	J. B. Schuette	Lockington, Ohio	Apr. 5, 1870	101, 666
Churn	H. Schültdrees	Brookville, Ind	Apr. 25, 1871	114, 050
Churn	L. Scott	Burgettstown, Pa	May 19, 1868	78, 139
Churn	J. C. Scribner	Holderness, N. H	Mar. 20, 1866	53, 352
Churn	J. F. Seaman	Cortlandville, N. Y	May 29, 1866	55, 169
Churn	A. G. Searls	Cleveland, Ohio	Mar. 25, 1862	34, 784
Churn	D. W. Seeley	Albany, N. Y	Aug. 27, 1861	33, 164
Churn	O. Seely	Syracuse, N. Y	Nov. 22, 1864	45, 182
Churn	O. W. Seely	Buffalo, N. Y	Aug. 28, 1866	57, 582
Churn	J. W. Sefton	Gettysburgh, Pa	Mar. 20, 1866	53, 380
Churn	J. S. Severance	Homer, N. Y	Feb. 20, 1866	52, 767
Churn	A. M. Seymour, jr	Madison, Wis	Dec. 3, 1867	71, 652
Churn	L. W. Shaeffer	West Milton, Ohio	Sept. 10, 1867	68, 798
Churn	A. Shaffer	Vandalia, Mich	Apr. 4, 1871	113, 580
Churn	S. G. Shanks	Richmond, Ky	July 12, 1870	105, 262
Churn	Z. B. Shannon	Port Washington, Ohio	Oct. 30, 1866	59, 282
Churn	J. Shappell	Lynnville, Pa	Dec. 20, 1870	110, 296
Churn	J. Shaver	Lawrence, Ill	Aug. 8, 1871	117, 821
Churn	B. W. Shaw and G. A. Simmons	Morristown, Vt	Mar. 3, 1868	75, 205
Churn	C. A. Shaw	Biddeford, Me	Dec. 9, 1856	16, 193
Churn	E. F. Shaw	Wyoming, Mich	July 14, 1868	80, 021
Churn	J. and W. A. Shaw	Hinkley, Ohio	Sept. 22, 1868	82, 356
Churn	T. D. Shaw	Westfield, Ohio	Jan. 15, 1867	61, 268
Churn	A. Shebie	Philadelphia, Pa	Nov. 27, 1811	....
Churn	D. Shelden	Pultney, N. Y	Sept. 13, 1827	....
Churn	N. Shelton	Odessa, N. Y	Feb. 15, 1870	99, 960
Churn	J. H. Shepard	Canaan, Mass	Oct. 1, 1811	....
Churn	N. H. Sherburne	Campton, Ill	July 13, 1858	20, 898
Churn	D. Sherman and R. W. Fenwick.	Uniontown, Md., and Washington, D. C.	July 31, 1860	29, 445
Churn	H. Shively and R. S. McEwen	Fredericksburgh, Ohio	Mar. 6, 1830	....
Churn	R. W. Shriner	Woodland, Mich	Jan. 1, 1867	60, 946
Churn	A. Shuffert and G. Cooper	Wyandotte, Mich	May 26, 1868	78, 398
Churn	F. Shurr and J. Palmer	Milwaukee, Wis	Jan. 2, 1866	51, 873
Churn	R. L. Shute	Philadelphia, Pa	Nov. 14, 1865	50, 963
Churn	H. Sidle	Dillsburgh, Pa	Nov. 26, 1861	33, 802
Churn	H. Sidle	Minneapolis, Minn	June 25, 1867	66, 178
Churn	I. J. Siler	Arcanum, Ohio	Jan. 21, 1868	73, 659
Churn	J. Simpson	Libertyville, Ill	June 1, 1869	90, 882
Churn	H. Skinner	Attica, N. Y	Dec. 16, 1851	8, 596
Churn	J. C. Slaughter	Crumpton, Md	May 12, 1868	77, 926
Churn	W. Slaughter	Westerville, Ohio	Jan. 19, 1864	41, 328
Churn	A. Smith	Tyngsborough, Mass	Apr. 11, 1810	....
Churn	B. J. Smith	Osage Mission, Kans	Nov. 22, 1870	109, 556
Churn	D. Smith	Emmetsburgh, Md	Apr. 21, 1831	....
Churn	H. E. Smith	New York, N. Y	Aug. 6, 1872	130, 322
Churn	J. W. X. Smith	Independence, Iowa	Feb. 8, 1870	99, 720
Churn	S. Smith	Yohoghany, Pa	Dec. 14, 1869	97, 820
Churn	S. Smith	Clarksburgh, N. Y	Dec. 14, 1869	97, 976
Churn	T. K. Smith	Oskaloosa, Iowa	Nov. 27, 1866	60, 077
Churn	W. C. Smith	Yantic, Conn	July 9, 1867	66, 646
Churn	W. C. Smith	Yantic, Conn	Sept. 22, 1868	82, 446
Churn	J. F. Smith and W. Brown	Galena and Rose, N. Y	Aug. 31, 1858	21, 374
Churn	T. C. Smith and N. L. Francis	Oquawka, Ill	Sept. 19, 1871	119, 193
Churn	H. Soggs	Columbus, Pa	Nov. 29, 1864	45, 277
Churn	H. Soggs	Columbus, Pa.	July 25, 1865	48, 987
Churn	G. Sowle and P. Brewer	Blenheim, N. Y	Mar. 22, 1830	....
Churn	E. Spain	Mount Holly, N. J	Apr. 23, 1828	....
Churn	E. Spain	Philadelphia, Pa	May 16, 1848	5, 585
Churn	E. Spain	Philadelphia, Pa	Sept. 29, 1868	82, 652
Churn	M. H. Spaulding	Morrisville, Vt	July 16, 1867	66, 901
Churn	G. Spayd	Alma, Mich	Sept. 12, 1871	118, 826
Churn	I. H. Spelman	Bazetta, Ohio	Oct. 15, 1867	69, 855
Churn	A. L. Sperry	Auburn, Ind	Aug. 30, 1859	25, 285
Churn	G. Sperry	Potter's Corners, Pa	Jan. 1, 1867	60, 952
Churn	J. Springer	Clinton, Wis	July 2, 1872	128, 512
Churn	E. N. Sprinkle	Marion, Va	Nov. 1, 1859	25, 985
Churn	S. Stackhouse	Cazenovia, N. Y	Mar. 11, 1808	....
Churn	J. Stadler and G. M. Streng	Detroit and Plymouth, Mich.	Aug. 11, 1868	80, 882
Churn	C. W. Stafford	Burlington, Iowa	Nov. 16, 1858	22, 093
Churn	O. W. Stanford	Cincinnati, Ohio	June 12, 1860	28, 696
Churn	O. W. Stanford	Lebanon, Ohio	July 3, 1866	56, 115
Churn	R. Stanley	Chariton, Ohio	Jan. 29, 1867	61, 691
Churn	C. B. Stanton	Scott, N. Y	Mar. 12, 1867	62, 899
Churn	C. C. Stearns	Bucksport, Me	Aug. 17, 1835	....
Churn	B. H. Steele	Barnesville, Ohio	May 16, 1871	114, 873
Churn	M. V. B. Stinemetz	Annville, Pa	Feb. 5, 1867	61, 7[illegible]2

Index of patents issued from the United States Patent Office from 1790 *to* 187 3, *inclusive*—Continued.

Invention.	Inventor.	Residence.	Date.	No.
Churn	A. Stephens	Washington, Iowa	Nov. 29, 1864	45, 280
Churn	J. H. Stephens	Orange Court-House, Va	Mar. 23, 1869	88, 095
Churn	T. B. Stephens	Washington, Iowa	Apr. 12, 1870	101, 936
Churn	D. H. Stevens	Enosburgh Falls, Vt	Sept. 10, 1872	131, 231
Churn	G. R. Stevens	Chicago, Ill	July 7, 1868	79, 702
Churn	I. V. Stevens	Pomeroy, Ohio	Feb. 5, 1861	31, 338
Churn	J. Stilwell	Griffin, Ga	May 21, 1861	32, 388
Churn	W. W. Stilwell	Oxford, Wis	Jan. 31, 1871	111, 490
Churn	V. Stirewalt	Albany, Ga	Feb. 12, 1861	31, 413
Churn	O. Stoddard	Busti, N. Y	Feb. 15, 1859	22, 989
Churn	H. C. Stoll	Mokena, Ill	Sept. 4, 1866	57, 794
Churn	S. S. Stokes	Westborough, Ohio	June 30, 1868	79, 409
Churn	W. H. Stonaker	Cooperstown, Pa	Aug. 15, 1865	49, 450
Churn	J. Stubbs	Dublin, Ind	July 5, 1859	24, 673
Churn	T. Stumm	Ada, Ohio	June 18, 1872	128, 080
Churn	A. Stump	Bodega, Cal	June 23, 1868	79, 156
Churn	C. Sweeney	East Bloomfield, N. Y	Dec. 22, 1868	85, 187
Churn	S. Sweet, jr	Readfield, Mass	June 13, 1831	......
Churn	J. B. Sweetland	Pontiac, Mich	Oct. 18, 1864	44, 755
Churn	J. B. Sweetland	Pontiac, Mich	Feb. 5, 1867	61, 892
Churn	J. G. Talbot	Sloansville, N. Y	Sept. 13, 1870	107, 306
Churn	W. H. Tambling	Mazo Manie, Wis	Mar. 12, 1867	62, 785
Churn	W. H. Tambling	Berlin, Wis	June 29, 1858	20, 740
Churn	J. Taylor	Rushville, Ill	Sept. 13, 1859	25, 453
Churn	J. G. Taylor	East Bethlehem, Pa	Sept. 10, 1867	68, 805
Churn	D. E. Teal	Norwich, N. Y	Feb. 6, 1866	52, 467
Churn	C. M. Terrell and N. W. Hussey	Oskaloosa, Iowa	Dec. 17, 1867	72, 338
Churn	J. E. Thomas	Winchester, Ohio	Feb. 12, 1836	......
Churn	J. Thompson	Vevay, Ind	Oct. 28, 1862	36, 811
Churn	J. Thompson	Vevay, Ind	Mar. 22, 1864	42, 029
Churn	J. Thompson	Vevay, Ind	Oct. 24, 1865	50, 645
Churn	J. W. Thompson	Bureau Junction, Ill	Sept. 1, 1868	81, 842
Churn	J. S. Thomson	Wyalusing, Pa	July 27, 1839	1, 265
Churn	L. Thomson	Sturbridge, Mass	June 24, 1809	......
Churn	A. E. Thorn	Fletcher, Ohio	Apr. 2, 1867	63, 580
Churn	T. J. Thorn	Skaneateles, N. Y	Jan. 30, 1866	52, 340
Churn	T. Thorne	Clinton, N. Y	Apr. 12, 1833	......
Churn	F. Thorpe	Shelbyville, Ill	Sept. 23, 1856	15, 787
Churn	W. Tibbits	La Fayette, N. Y	May 24, 1864	42, 887
Churn	J. A. Tice	Windsor, Ill	Mar. 30, 1869	88, 529
Churn	N. C. Tiffany and E. Robison	Caroline, N. Y	Dec. 14, 1830	......
Churn	W. H. Tillou and S. Shumway	Le Roy, N. Y	June 26, 1866	55, 934
Churn	J. Tingley	Waterford, N. J	Aug. 11, 1863	39, 507
Churn	J. Tingley	Philadelphia, Pa	Feb. 25, 1868	74, 954
Churn	J. Toll	Locust Grove, Ohio	Feb. 25, 1868	74, 955
Churn	E. Townsend	Canaan, N. Y	July 12, 1808	......
Churn	W. W. Townsend	Sandy Hill, N. Y	Mar. 22, 1808	......
Churn	A. Traver and P. Nichols	Troy, N. Y	Oct. 10, 1871	119, 898
Churn	D. E. True	Lake Village, N. H	Sept. 8, 1857	18, 165
Churn	W. H. Truesdell	Elgin, Ill	Dec. 22, 1857	18, 936
Churn	W. B. Tucker	Hillsborough, Ohio	Mar. 12, 1867	62, 786
Churn	W. B. Tucker	Columbus, Ohio	Dec. 1, 1868	84, 658
Churn	J. Turner	Poland, Me	Aug. 15, 1835	......
Churn	C. R. Tuttle	New Brighton, Pa	Nov. 25, 1862	37, 016
Churn	W. Twining	Philadelphia, Pa	Feb. 27, 1872	124, 176
Churn	S. Tyler	New Gloucester, Me	June 25, 1836	......
Churn	S. S. Ulrey	North Manchester, Ind	Sept. 21, 1869	95, 167
Churn	C. M. Vail	Susquehanna Depot, Pa	Feb. 16, 1858	19, 389
Churn	C. D. Van Allen	Syracuse, N. Y	July 23, 1861	32, 907
Churn	C. D. Van Allen	Syracuse, N. Y	June 17, 1862	35, 638
Churn	J. Van Aukin, 2d	Knox, N. Y	Sept. 28, 1831	......
Churn	J. Vanatter	Stratford, Canada	Jan. 16, 1872	122, 870
Churn	A. J. Vannatta	Vanatta, Ohio	Dec. 4, 1866	60, 289
Churn	J. Vandolah and E. Curry	Dillsborough, Ind	Dec. 22, 1857	18, 937
Churn	I. Van Gorder	Warren, Ohio	June 27, 1832	......
Churn	C. A. Van Horn	Chenango, N. Y	Aug. 6, 1867	67, 615
Churn	S. Van Meter	Henderson, Ill	June 28, 1870	104, 800
Churn	G. H. Van Vleck	Buffalo, N. Y	May 8, 1860	28, 219
Churn	D. Variel	Minot, Me	July 1, 1836	......
Churn	W. A. Vertrees	Winchester, Mo	Dec. 9, 1856	16, 203
Churn	J. Vincent and S. Leslie	Quasqueton, Iowa	Feb. 11, 1862	34, 383
Churn	C. Vogt and X. Krapf	Allentown, Pa	July 23, 1867	67, 085
Churn	A. Votaw	New Garden, Ohio	June 11, 1867	65, 620
Churn	G. Votee	Berlin, Conn	May 10, 1809	......
Churn	W. H. and L. Waddell	Churchville, Va	June 16, 1868	79, 035
Churn	A. C. Wade	Paris, Ill	Aug. 7, 1866	57, 051
Churn	I. M. Wade	Clinton, Mich	June 26, 1855	13, 133
Churn	A. Walker	Claremont, N. H	June 23, 1863	39, 023
Churn	M. D. Wallace	White Creek, N. Y	May 7, 1867	64, 463
Churn	G. B. Waller	Franklin, Ill	Nov. 27, 1866	60, 097
Churn, &c	M. Walton	Marlborough, Ohio	Oct. 8, 1867	69, 598
Churn	D. T. Ward	Mansfield, Ohio	Dec. 18, 1860	30, 941
Churn	D. T. Ward	Cardington, Ohio	Nov. 28, 1865	51, 246
Churn	D. T. Ward	Cardington, Ohio	Jan. 21, 1868	73, 479
Churn	E. Ward	New York, N. Y	Feb. 14, 1860	27, 175
Churn	J. S. Ward	Plattsburgh, Mo	Dec. 5, 1871	121, 560
Churn	C. Warner	Louisville, Ky	June 12, 1849	6, 527
Churn	C. D. Warner	Syracuse, N. Y	June 11, 1867	65, 623
Churn	G. W. Warren	Alma, N. Y	Apr. 20, 1869	89, 098
Churn	T. A. Warner	Gettysburgh, Pa	Apr. 16, 1867	63, 969
Churn	C. H. Warren and A. C. Baldwin	Tiffin, Ohio	Dec. 20, 1864	45, 544
Churn	P. H. Watson	Rockford, Ill	May 16, 1848	5, 573
Churn	J. Watts	Brazil, Ind	Sept. 1, 1868	81, 716
Churn	J. H. Weaver and D. M. Mefford	Chillicothe, Ohio, and Jeffersonville, Ind	May 19, 1863	38, 635
Churn	B. Webb	Tolland, Conn	Aug. 22, 1816	......
Churn	C. Webb	Wallingford, Conn	Nov. 26, 1840	1, 860

Index of patents issued from the United States Patent Office from 1790 *to* 1873, *inclusive*—Continued.

Invention.	Inventor.	Residence.	Date.	No.
Churn	E. Webber	Gardiner, Me	Dec. 5, 1854	12, 043
Churn	H. Webster	Ogdensburgh, N. Y	Feb. 6, 1855	12, 370
Churn	W. Weddington	Winterset, Iowa	July 23, 1867	67, 090
Churn	F. Wegner and C. Schleeter	West Troy, N. Y	Feb. 15, 1870	99, 985
Churn	M. D. Wells	Morgantown, Va	July 17, 1855	13, 281
Churn	R. P. Wells	Dayton, Ohio	Feb. 8, 1870	99, 612
Churn	I. M. West	Wilmington, Ohio	June 27, 1865	48, 469
Churn	I. M. West	Wilmington, Del	Aug. 28, 1866	57, 606
Churn	L. Westbrook	New York, N. Y	May 10, 1859	23, 968
Churn	A. Westcott	Syracuse, N. Y	Jan. 7, 1862	34, 082
Churn	A. Westcott	Syracuse, N. Y	Apr. 29, 1862	35, 120
Churn	A. Westcott	Syracuse, N. Y	Mar. 15, 1864	41, 954
Churn	A. Westcott	Syracuse, N. Y	Mar. 7, 1865	46, 739
Churn	A. Westcott	Syracuse, N. Y	Feb. 27, 1866	52, 919
Churn	A. Westcott	Syracuse, N. Y	June 12, 1866	55, 566
Churn	A. Westcott	Syracuse, N. Y	Apr. 21, 1868	77, 144
Churn	A. Westcott	Syracuse, N. Y	Dec. 8, 1868	84, 784
Churn	H. P. Westcott	Seneca Falls, N. Y	Apr. 4, 1865	47, 145
Churn	H. P. Westcott	Seneca Falls, N. Y	Nov. 13, 1866	59, 691
Churn	I. E. Weston and A. Streeter	Winchendon, Mass	May 31, 1870	103, 803
Churn	A. Wharff	New Gloucester, Me	June 16, 1836	....
Churn	S. A. Wheelock	Charlton, Mass	Sept. 2, 1862	36, 379
Churn	S. J. Whipple	Orville, Cal	Jan. 21, 1868	73, 556
Churn	H. Whisler	New Market, Ohio	June 5, 1866	55, 407
Churn	M. Whisler	New Market, Ohio	Jan. 29, 1867	61, 645
Churn	E. P. Whitcomb	Coldwater, Mich	Apr. 10, 1866	53, 912
Churn	C. N. White	Batesville, Miss	Aug. 25, 1868	81, 445
Churn	H. W. White	Olney, Ill	May 8, 1866	54, 632
Churn	J. White	Cleveland, Ohio	Mar. 27, 1866	53, 512
Churn	N. Whitney	Augusta, Me	June 7, 1827	....
Churn	R. W. Whitney	South Berwick, Me	July 7, 1863	39, 205
Churn	L. Whittier	Vienna, Me	Feb. 10, 1836	....
Churn	F. Whitton	South Carrollton, Ky	Dec. 1, 1868	84, 668
Churn	L. Wiard and W. H. Nelson	Spring Township, Pa	Dec. 24, 1867	72, 576
Churn	L. J. Wicks	Racine, Wis	Mar. 8, 1859	23, 208
Churn	L. J. Wicks	Racine, Wis	July 5, 1859	24, 686
Churn	D. Wedmayer	Lansing, Mich	July 6, 1869	92, 411
Churn	A. Wieting	Fort Plain, N. Y	Mar. 5, 1872	124, 304
Churn	M. J. Wikoff	Stout's Post-Office, Ohio	July 26, 1870	105, 751
Churn	E. Wilcox	Hamburgh, Iowa	Aug. 22, 1871	118, 320
Churn	J. H. Wildasin and J. A. Peek	Saint Charles, Iowa	Sept. 14, 1869	94, 802
Churn	A. G. Wilkins, G. N. Grodle, and F. L. Niner.	Cooperstown, Pa	May 25, 1869	90, 414
Churn	S. B. Willard	Bloomfield, N. Y	May 18, 1808	....
Churn	D. A. Willbanks	Harmony Grove, Ga	Feb. 13, 1872	123, 656
Churn	E. Willemin	Salem Township, Ohio	Oct. 25, 1832	....
Churn	C. B. Williams	Bourbon, Ind	Feb. 11, 1868	74, 467
Churn	J. B. Williams	Glastenbury, Conn	May 30, 1871	115, 552
Churn	R. Williams	Yonkers, N. Y	Oct. 8, 1872	132, 037
Churn	A. Willson	Colden, N. Y	Feb. 14, 1860	27, 181
Churn	G. W. Wilson	Freeport, Ill	Apr. 2, 1867	63, 596
Churn	R. Wilson	Rees Corners, Md	Aug. 3, 1869	93, 264
Churn	S. C. Wilson	Olney, Ill	Aug. 27, 1867	68, 273
Churn	T. J. Wilson	New Lisbon, Ind	Jan. 23, 1872	122, 978
Churn	W. E. Wilson	Central Station, W. Va	May 8, 1866	54, 634
Churn	A. Wing	Washington, N. Y	May 26, 1806	....
Churn	B. L. Winner	Belvidere, Ill	Aug. 15, 1865	49, 464
Churn	D. H. Wiswell	Buffalo, N. Y	Mar. 13, 1860	27, 504
Churn	M. Witmer	Cedar Rapids, Iowa	Mar. 9, 1869	87, 743
Churn	J. Witmore	Litchfield, Conn	Dec. 8, 1809	....
Churn	G. L. Witsil	Philadelphia, Pa	Oct. 27, 1863	40, 450
Churn	G. L. Witsil	Philadelphia, Pa	Sept. 27, 1864	44, 480
Churn	G. L. Witsil	Philadelphia, Pa	July 3, 1866	56, 153
Churn	G. L. Witsil	Philadelphia, Pa	Apr. 30, 1867	64, 393
Churn	G. Wolf	Williamsport, Md	Jan. 16, 1866	52, 103
Churn	I. Wood	Fayette County, Ind	June 26, 1835	....
Churn	M. Wood	Babe's Corners, Mich	June 9, 1868	78, 706
Churn	J. K. Wood and D. R. Speer	Allegheny City and Pittsburgh, Pa.	Dec. 1, 1868	84, 524
Churn	D. M. Woodin	Brandon, Wis	May 29, 1860	28, 528
Churn	C. Wright and W. Phelps	Sycamore, Ill	June 19, 1860	28, 802
Churn	O. Wyman	Dedham, Mass	July 17, 1835	....
Churn	J. M. Yeager	Lanesville, Ind	Aug. 29, 1871	118, 664
Churn	J. Young	Adrian, Ohio	July 24, 1866	56, 661
Churn	C. S. Young, H. Wissinger, and T. T. Williams.	Minta, Pa	Sept. 19, 1871	119, 068
Churn	C. E Zimmerman	Cincinnati, Ohio	Aug. 11, 1868	81, 056
Churn and butter-worker	O. R. Fyler	Brattleborough, Vt	July 27, 1852	9, 148
Churn and butter-worker	S. L. Hall	West Salem, Mass	Aug. 4, 1868	80, 624
Churn and butter-worker	A. Hawver and W. C. Hanyen	Galva, Ill	Jan. 14, 1862	34, 144
Churn and butter-worker	A. Willard	Boston, Mass	Sept. 23, 1851	8, 382
Churn and butter-worker combined	J. Randall	Grand Rapids, Mich	Aug. 22, 1865	49, 550
Churn and butter-worker, Combined	S. H. Scribner	Stowe, Vt	Aug. 6, 1867	67, 594
Churn and butter-worker combined	C. H. Smith	Boston, Mass	Jan. 19, 1869	85, 970
Churn and egg-beater combined	G. C. Westover	Paducah, Ky	Jan. 22, 1867	61, 494
Churn and ice-cream freezer	C. Higley	Port Byron, N. Y	June 30, 1868	79, 349
Churn and ice-cream freezer	T. W. Pomeroy	East Hampton, Mass	June 26, 1866	55, 960
Churn and ice-cream freezer combined	G. C. Westover	Paducah, Ky	Jan. 28, 1868	73, 766
Churn and pump power	F. Danzenbaker	Bridgeton, N. J	May 21, 1867	64, 950
Churn and washing-machine	J. Devotie	Vernon, N. Y	Dec. 5, 1808	....
Churn and washing-machine	H. Kingsberry	Hebron, Conn	June 26, 1830	....
Churn and washing-machine	T. Ling	Winthrop, Me	Sept. 9, 1835	....
Churn and washing machine	C. Otis	Finksburgh, Md	June 12, 1835	....
Churn and washing-machine	I. Parke	Delhi, N. Y	Jan. 16, 1835	....
Churn and washing-machine	J. M. Peirce	Mokena, Ill	Nov. 6, 1866	59, 447
Churn and washing-machine	D. Read	Slaterville, N. Y	June 16, 1828	....
Churn, Apparatus for operating	P. L. Jordan	Lexington, Miss	May 4, 1871	113, 435
Churn, Atmospheric	S. Ballard	Meig's Creek, Ohio	May 16, 1848	5, 567

Index of patents issued from the United States Patent Office from 1790 *to* 1873, *inclusive*—Continued.

Invention.	Inventor.	Residence.	Date.	No.
Churn, Atmospheric	N. Chapin	Cortlandville, N. Y	May 9, 1848	5, 560
Churn, Atmospheric	J. C. Coult and A. B. Davis	Philadelphia, Pa	May 15, 1849	6, 457
Churn, Atmospheric	P. F. Ellicott	Philadelphia, Pa	June 25, 1850	7, 454
Churn, Atmospheric	S. F. Emerson	Canaan, Ohio	Apr. 2, 1850	7, 236
Churn, Atmospheric	S. P. Francisco	Reading, Pa	June 19, 1849	6, 544
Churn, Atmospheric	D. C. Hall	Hannibal, Mo	Apr. 14, 1868	76, 753
Churn, Atmospheric	W. H. Johnson and P. Lewis	Springfield, Ill	May 9, 1848	5, 561
Churn, Atmospheric	R. McCutcheon	Towanda, Pa	June 2, 1857	17, 439
Churn, Atmospheric	J. O'Neil	Xenia, Ohio	July 30, 1850	7, 531
Churn, Atmospheric	W. D. Prindle and C. M. York	Tiffin, Ohio	Jan. 29, 1867	61, 567
Churn, Atmospheric	J. Young	West Galway, N. Y	Jan. 22, 1850	7, 046
Churn, Atmospheric-air	E. V. Coe	Warwick, N. Y	Apr. 7, 1831	....
Churn-attachment	W. M. Pyle	Greensburgh, Ind	Mar. 27, 1866	53, 484
Churn, Balanced lever	C. Angerine	New York, N. Y	Apr. 19, 1832	....
Churn, Barrel	L. H. Muzzey	Springfield, Ohio	Mar. 22, 1864	42, 013
Churn, Barrel	S. H. Swasey	Morristown, Vt	Jan. 7, 1868	73, 134
Churn, Barrel	P. Thompson	Springfield, Ohio	Mar. 22, 1864	42, 030
Churn, beer-cooler, &c	S. S. Elder	Springfield, Ill	Dec. 18, 1866	60, 493
Churn-body	J. C. Hills	Willoughby, Ohio	May 24, 1864	42, 851
Churn, Cradle	J. Barber	Caroline, N. Y	Jan. 28, 1830	....
Churn-dasher	E. J. Ashton	Utica, Ind	Nov. 14, 1871	120, 928
Churn-dasher	A. A. Avery	Cardiff, N. Y	Mar. 19, 1867	62, 992
Churn-dasher	C. J. Baldwin	Norwalk, Ohio	Jan. 26, 1869	86, 269
Churn-dasher	A. E. Banks	Detroit, Mich	Dec. 17, 1867	72, 154
Churn-dasher	G. W. Barker	Saint John's, Mich	Dec. 9, 1873	145, 272
Churn-dasher	H. F. Bartlett	La Grange, Mo	Aug. 16, 1870	106, 453
Churn-dasher	J. W. Barton	Clifton Springs, N. Y	Feb. 18, 1868	74, 485
Churn-dasher	S. E. Bauder	Birmingham, Ohio	Oct. 11, 1870	108, 232
Churn-dasher	S. E. Bauder	Birmingnam, Ohio	Mar. 12, 1872	124, 536
Churn-dasher	A. Belt	Newton, Iowa	Dec. 14, 1869	97, 859
Churn-dasher	A. T. Bleyley	Conception, Mo	Sept. 1, 1868	81, 586
Churn-dasher	F. Bosom	Saint Louis, Mich	Dec. 28, 1869	98, 221
Churn-dasher	T. Bosom	Jonesville, Mich	May 10, 1870	102, 758
Churn-dasher	D. Boyd	Vevay, Ind	Dec. 30, 1873	145, 985
Churn-dasher	A. H. Brainerd	Rome, N. Y	June 2, 1868	78, 511
Churn-dasher	I. Brewbaker	Fincastle, Va	Apr. 19, 1870	102, 084
Churn-dasher	J. A. Brooks	Madison County, Tenn	Dec. 3, 1872	133, 626
Churn-dasher	A. C. Brown	Sycamore, Ill	Apr. 14, 1863	38, 145
Churn-dasher	R. Brown	Columbus, Miss	Dec. 27, 1870	110, 429
Churn-dasher	W. C. Broyhill and W. D. Sperry	Tremont, Ill	Nov. 14, 1871	120, 849
Churn-dasher	J. M. Buchanan	Lawrenceville, Ill	May 4, 1869	89, 629
Churn-dasher	I. N. Buck	Elgin, Ill	Oct. 13, 1857	18, 382
Churn-dasher	G. Bunting	Liberty, Ind	Nov. 22, 1859	26, 226
Churn-dasher	J. Carl	Grenada, Miss	Apr. 20, 1869	89, 125
Churn-dasher	J. S. Carson	Brookhaven, Miss	Aug. 25, 1868	81, 340
Churn-dasher	J. W. Cheney and B. Ingalls	Shelbyville, Ill	June 22, 1869	91, 715
Churn-dasher	C. L. Cole	Bushnell, Ill	Mar. 9, 1869	87, 638
Churn-dasher	W. M. Combs	Auburn, N. Y	Feb. 18, 1873	136, 035
Churn dasher	I. B. Compton	Seward, Nebr	Oct. 14, 1873	143, 559
Churn-dasher	E. G. Connelly	Jasper, Ind	July 17, 1866	56, 374
Churn-dasher	N. B. Cooper	Gratis, Ohio	Nov. 8, 1859	26, 020
Churn-dasher	A. W. Cramer	Bethany, Pa	Nov. 15, 1864	45, 026
Churn-dasher	H. A. Crance	Lewisburgh, Pa	Aug. 11, 1868	80, 921
Churn dasher	R. Crawford	Mercer, Pa	Mar. 3, 1868	74, 995
Churn-dasher	T. Crutcher	Edgefield, Tenn	Dec. 21, 1869	98, 152
Churn-dasher	J. J. Cumming	Independence, Mo	Aug. 27, 1867	68, 049
Churn-dasher	M. O. Davis	Warrensburgh, N. Y	Jan. 21, 1868	73, 583
Churn-dasher	G. Deckman	Malvern, Ohio	July 31, 1866	56, 728
Churn-dasher	S. F. Dolloff	Bloomington, Ill	May 23, 1871	115, 177
Churn-dasher	G. H. Dow	Freeport, Ill	Sept. 3, 1867	68, 354
Churn-dasher	P. Edgerton	Rutland, Vt	June 14, 1870	104, 127
Churn-dasher	M. Fish	Adrian, Mich	Sept. 7, 1869	94, 586
Churn-dasher	J. M. Fletcher	Sidney, Ohio	Nov. 16, 1869	96, 906
Churn-dasher	G. R. Forsyth	Pemberton, Ohio	Feb. 19, 1867	62, 124
Churn-dasher	D. K. France	Congress, Ohio	Sept. 13, 1859	25, 494
Churn-dasher	E. H. Funk	Newark, Ohio	Sept. 15, 1868	82, 216
Churn-dasher	J. D. Garlick	Lyons, N. Y	Feb. 26, 1850	7, 118
Churn-dasher	J. Gelston	Cincinnati, Ohio	May 22, 1866	54, 891
Churn-dasher	A. J. Gibson	Cincinnati, Ohio	June 12, 1866	55, 484
Churn-dasher	W. H. H. Gorham and B. H. Williams.	Greenwich, Ohio	Aug. 23, 1870	106, 575
Churn-dasher	H. Grass	Olney, Ill	Aug. 27, 1867	68, 067
Churn-dasher	J. A. Gridley	Southampton, Mass	May 1, 1849	6, 411
Churn-dasher	D. L. Grover	Groton, N. Y	Oct. 17, 1871	119, 980
Churn-dasher	W. J. Hale	Ashley, Ill	Apr. 13, 1869	88, 957
Churn-dasher	A. W. Hall	New York, N. Y	Dec. 11, 1866	60, 363
Churn-dasher	M. A. Hamilton	Detroit, Mich	Nov. 9, 1869	96, 693
Churn-dasher	J. Harris	Palestine, Ill	Aug. 30, 1870	106, 935
Churn-dasher	M. R. Heliker	Norwalk, Ohio	Dec. 29, 1868	85, 445
Churn-dasher	G. Heliker and A. Burlen	Greenwich, Ohio	July 5, 1870	105, 074
Churn-dasher	M. Heliker and O. A. White	Norwalk, Ohio	Oct. 13, 1868	82, 947
Churn-dasher	J. M. Hill	Fairfield, Ill	July 12, 1870	105, 332
Churn-dasher	J. Hobbs	North Sandford, N. Y	Sept. 26, 1865	50, 122
Churn-dasher	A. J. Hudson	Camden, Tenn	Dec. 2, 1873	145, 107
Churn-dasher	M. B. Hudson	Canandaigua, N. Y	Apr. 9, 1867	63, 723
Churn-dasher	A. B. Hurd	Watkins, N. Y	July 23, 1867	67, 118
Churn-dasher	A. B. Hutchins	Patchogue, N. Y	Mar. 5, 1867	62, 633
Churn-dasher	S. Jackson	Jay, Me	June 11, 1836	....
Churn-dasher	M. Jincks	Steuben County, N. Y	June 30, 1863	39, 050
Churn-dasher	D. Johnson	Chicago, Ill	Apr. 11, 1865	47, 259
Churn-dasher	C. Johnston	Clarksville, Mo	Mar. 28, 1871	113, 172
Churn-dasher	W. F. Jones	Easton, Kans	Apr. 25, 1871	114, 016
Churn-dasher	W. F. Jones	Easton, Kans	May 30, 1871	115, 323
Churn-dasher	W. Kegg	Lasellsville, N. Y	Aug. 3, 1869	93, 207
Churn-dasher	J. D. Kellogg, jr	Northampton, Mass	June 9, 1868	78, 749
Churn-dasher	C. King	Scipio, N. Y	Apr. 4, 1846	4, 449
Churn-dasher	A. Kirlin	Rock Island, Ill	June 5, 1866	55, 313
Churn-dasher	G. Lange	East Saginaw, Mich	June 15, 1869	91, 454

Index of patents issued from the United States Patent Office from 1790 to 1873, inclusive—Continued.

Invention.	Inventor.	Residence.	Date.	No.
Churn-dasher	J. Leaken	Clinton, Ill	Oct. 29, 1867	70, 229
Churn-dasher	E. H. Lord and W. Thomson	Homer, N. Y	Dec. 15, 1868	84, 953
Churn-dasher	M. E. J. Marr	Jefferson, La	Apr. 7, 1868	76, 484
Churn-dasher	J. Marsh	Petersburgh, Ill	July 30, 1850	7, 527
Churn-dasher	J. L. Marsh	Richmond, Ind	Feb. 11, 1868	74, 396
Churn-dasher	D. W. Maurice	Springfield, Ohio	May 22, 1866	54, 930
Churn dasher	A. W. McClure	Richmond, Ind	Nov. 5, 1872	132, 767
Churn-dasher	J. W. McClure	Jefferson City, Mo	May 17, 1870	103, 219
Churn-dasher	J. W. McClure	Saint Louis, Mo	May 30, 1871	115, 496
Churn-dasher	G. W. McCormick	Saint Francis County, Ark	May 27, 1873	139, 406
Churn-dasher	D. McCurdy	Ottawa, Ohio	Mar. 14, 1865	46, 810
Churn-dasher	H. McDonough	New York, N. Y	Mar. 10, 1868	75, 286
Churn-dasher	S. McGuffin	Rising Sun, Ind	Apr. 30, 1867	64, 238
Churn-dasher	F. McTarnahan	Santa Clara County, Cal	July 30, 1867	67, 205
Churn-dasher	J. B. Mellor	New Hope, Mo	Dec. 27, 1870	110, 487
Churn-dasher	D. S. Miller	West Alexandria, Ohio	June 30, 1868	79, 491
Churn-dasher	H. Mulholland	Union Mills, Pa	July 25, 1871	117, 446
Churn-dasher	L. T. Newell	Geneva, Ohio	June 14, 1870	104, 189
Churn-dasher	M. A. Ober	Chazy, N. Y	Oct. 1, 1867	69, 363
Churn-dasher	F. Ogden	Fisherville, Ky	Nov. 29, 1870	109, 654
Churn-dasher	D. K. Overhiser	Williamsport, Pa	Aug. 10, 1869	93, 471
Churn dasher	W. S. Owen	Council Bluffs, Iowa	Apr. 27, 1869	89, 497
Churn-dasher	M. J. Palmer	Homer, N. Y	Mar. 25, 1862	34, 770
Churn-dasher	M. J. Palmer and H. R. Ingalls	Homer and Groton, N. Y	Jan. 6, 1863	37, 357
Churn-dasher	B. M. Parks	Saint Louis, Mo	Mar. 24, 1868	75, 963
Churn-dasher	J. W. Pettengill	Rockford, Ill	Aug. 20, 1867	68, 000
Churn-dasher	G. H. Pool	New York, N. Y	May 15, 1866	54, 769
Churn-dasher	H. S. Potter	Hawley, Pa	Oct. 17, 1871	120, 097
Churn-dasher	G. Radbruch	Hoboken, N. J	Oct. 5, 1869	95, 514
Churn-dasher	F. Ransom	Buffalo, N. Y	Sept. 24, 1867	69, 248
Churn-dasher	G. Ridler	Rickardsville, Iowa	Nov. 4, 1873	144, 290
Churn-dasher	J. T. Rittenhouse	Urbana, Ill	Nov. 20, 1866	59, 781
Churn-dasher	H. B. Robinson	Birmingham, Ala	Dec. 16, 1873	145, 592
Churn-dasher	R. W. Robinson	Clinton, Ill	Nov. 19, 1867	71, 065
Churn-dasher	A. Rose	Penn Yan, N. Y	Oct. 25, 1859	25, 914
Churn-dasher	N. Routzahn	Middletown, Md	Mar. 19, 1850	7, 199
Churn-dasher	L. Scott	Burgettstown, Pa	Nov. 8, 1870	109, 061
Churn-dasher	J. M. See	Griffin, Ga	May 2, 1871	114, 478
Churn-dasher	R. S. Sherman	Napanock, N. Y	Aug. 6, 1850	7, 550
Churn-dasher	J. Smith	Kansas, Ill	Oct. 8, 1867	69, 718
Churn-dasher	M. J. Smith	Dansville, N. Y	Nov. 7, 1865	50, 851
Churn-dasher	N. H. Spencer	Canandaigua, N. Y	Oct. 23, 1866	59, 031
Churn-dasher	A. Sperry	Tremont, Ill	Dec. 21, 1869	98, 117
Churn-dasher	O. Sperry and J. W. Hopson	Hartfield, N. Y	Mar. 23, 1869	88, 226
Churn-dasher	H. Stanton	Richfield, N. Y	Dec. 18, 1849	6, 961
Churn-dasher	N. Starbuck	Wilmington, Ohio	Mar. 21, 1865	46, 952
Churn-dasher	H. N. Stearns	Chardon, Ohio	Mar. 11, 1862	34, 652
Churn-dasher	A. T. Still	Baldwin City, Kans	Dec. 19, 1871	122, 075
Churn-dasher	B. F. Stoler and I. W. Warner	Ladoga and Crawfordsville, Ind.	July 25, 1871	117, 479
Churn-dasher	S. Stout	Tremont, Ill	Apr. 4, 1871	113, 361
Churn-dasher	H. Tilden	Philadelphia, Pa	Nov. 22, 1864	45, 194
Churn-dasher	J. Thomas	Huntingdon, Pa	May 7, 1867	64, 460
Churn-dasher	M. H. Thomas	Dansville, N. Y	Oct. 6, 1868	82, 892
Churn-dasher	T. W. Tyler	Corry, Pa	Aug. 4, 1868	80, 685
Churn-dasher	P. Van Antwerp	Coeymans, N. Y	Feb. 16, 1843	2, 957
Churn-dasher	A. C. Vaughn	Johnstown, Pa	Mar. 6, 1860	27, 396
Churn-dasher	R. A. Vick	Holly Springs, Miss	Aug. 13, 1872	130, 395
Churn-dasher	W. and M. C. Walker	Lancaster, Pa	Aug. 6, 1850	7, 554
Churn-dasher	D. F. Wallace and D. T. Cockerill.	Ripley, Ohio	Aug. 6, 1867	67, 617
Churn-dasher	J. J. Watson	Buffalo, N. Y	Oct. 16, 1860	30, 438
Churn-dasher	F. H. Weaver	Marietta, Ga	Sept. 6, 1870	107, 138
Churn-dasher	H. L. Wells	Chillicothe, Mo	May 9, 1871	114, 736
Churn-dasher	M. D. Wells	Morgantown, Va	Mar. 19, 1861	31, 756
Churn-dasher	E. B. West	Saint Anthony, Minn	May 12, 1868	77, 939
Churn-dasher	H. P. Westcott	Seneca Falls, N. Y	Aug. 18, 1863	39, 600
Churn-dasher	O. A. White	Norwalk, Ohio	May 25, 1869	90, 413
Churn-dasher	J. E. Williams and M. Lemon	Binghamton, N. Y	May 12, 1868	77, 789
Churn-dasher	P. Wineman	Loyds ille, Ohio	July 26, 1859	24, 898
Churn-dasher	T. H. Withers and J. Dolfinger	Louisville, Ky	Apr. 6, 1869	88, 688
Churn-dasher	B. Wright	Cardiff, N. Y	May 31, 1864	42, 984
Churn-dasher	E. M. Wright	Wilmington, Ohio	May 16, 1865	47, 763
Churn-dasher	S. Yates	Marshall, Mo	Oct. 13, 1868	83, 015
Churn dasher	T. P. York	Terre Haute, Ind	Sept. 28, 1869	95, 401
Churn-dasher, Adjustable	T. G. Clinton and G. H. and E. H. Knight.	Cincinnati, Ohio	Oct. 2, 1849	6, 764
Churn dasher and lid	S. P. Hopkins	Port Deposit, Md	Aug. 25, 1868	81, 372
Churn dasher, Atmospheric	S. Case and A. W. Pratt	Pultneyville, N. Y	June 19, 1866	55, 622
Churn dasher, Atmospheric	J. C. Gaston	Cincinnati, Ohio	May 21, 1867	64, 969
Churn dasher, Atmospheric	W. M. Wright	Pittsburgh, Pa	Sept. 11, 1849	6, 711
Churn-dasher-crank fixture	C. Blush and M. Harbster	Lucas, Ohio, and Reading, Pa.	Mar. 22, 1870	100, 969
Churn-dasher head	B. F. and A. H. Stover	Ladoga, Ind	May 19, 1868	78, 025
Churn-dasher, Metallic	O. Updike	Grass Lake, Mich	May 10, 1870	102, 887
Churn-dasher, Reciprocating	T. Earle	Valley Falls, R. I	Aug. 19, 1873	142, 007
Churn-dasher rod	M. G. Decrow	Newark, Ohio,	Oct. 17, 1871	120, 046
Churn-dasher, Rotary	L. W. Colver	Saint Louis, Mo	Sept. 18, 1849	6, 726
Churn-dasher, Rotary	D. N. Egbert	Hudson, Ohio	Sept. 18, 1849	6, 727
Churn-dasher, Rotary	H. S. Wells	Oquawka, Ill	Jan. 2, 1872	122, 505
Churn-dasher, Spiral	C. R. and J. Hight	Geneva, Ill	July 9, 1850	7, 490
Churn-dasher staff	D. L. Grover	Groton, N. Y	Apr. 9, 1872	125, 453
Churn-dashers, Device for moving	A. W. Hall	New York, N. Y	Mar. 21, 1865	46, 970
Churn-dashers, Fastening blade of	S. T. Lamb	New Washington, Ind	Oct. 16, 1860	30, 411
Churn-dashers, Operating	W. Kegg	Lasellsville, N. Y	Sept. 21, 1869	95, 023
Churn-dashers, Working rotary and vertical	W. R. Nash	Bridgeport, Conn	May 7, 1850	7, 348
Churn, Dog	F. Trax'er	Scottsburgh, N. Y	June 12, 1866	55, 556
Churn, Double-dasher	E. Mitchell	Pittston, Me	May 22, 1841	2, 102

Index of patents issued from the United States Patent Office from 1790 *to* 1873, *inclusive*—Continued.

Invention.	Inventor.	Residence.	Date.	No.
Churn, Double revolving	L. Barney and A. A. Beach	Groton, N. Y	Feb. 5, 1830	....
Churn, Floating-wheel	E. Dewey	Butternuts, N. Y	Oct. 1, 1830	....
Churn-gearing	A. C. Mason	Springfield, Vt	July 31, 1866	56, 771
Churn, Geneva double-dasher	W. Sutton	Geneva, N. Y	Oct. 1, 1830	....
Churn, Horizontal	J. Davidson	Cheshire, N. H	Apr. 3, 1812	....
Churn, Horizontal-vibrating	T. Pratt	Cheshire County, N. H	Sept. 12, 1811	....
Churn, Labor-saving	A. Bristol	Hillsdale, N. Y	Apr. 28, 1830	....
Churn, Lever	W. Cook	Southport, N. Y	Dec. 7, 1829	....
Churn, Lever	P. H. Kemball	Salem, Mass	June 13, 1831	....
Churn, Lever	J. Ladd	Holderness, N. H	June 22, 1832	....
Churn-lid screen	E. Reynolds	Winneconne, Wis	Apr. 13, 1869	88, 983
Churn-machine	P. Cornell	Brutus, N. Y	Oct. 1, 1830	....
Churn-motion	D. Morris	Bartlett, Ohio	July 14, 1868	80, 001
Churn-motor	J. B. Sweetland	Pontiac, Mich	May 20, 1873	139, 208
Churn-operating apparatus	G. A. Dabney	San José, Cal	Mar. 4, 1862	34, 567
Churn-operating apparatus	D. W. Ketcham	Owasso, Mich	Mar. 26, 1872	124, 900
Churn-operating apparatus	H. J. Klingenberg and J. J. Mau.	Davenport, Iowa	June 29, 1869	91, 942
Churn-operating device	E. R. Hall and W. H. Town	Syracuse, N. Y	Mar. 15, 1870	100, 753
Churn-operating device	A., J., and C. Lamb	Jeffersonville, N. Y	Jan. 27, 1863	37, 511
Churn-operating device	O. W. Seely	Syracuse, N. Y	Mar. 24, 1863	38, 013
Churn-operating machinery	C. S. Williamson	Covert, N. Y	Dec. 28, 1869	98, 454
Churn, Oscillating	E. W. Bullard	Barre, Mass	Nov. 7, 1871	120, 708
Churn, Oscillating	N. E. Wilson	Wilsonburgh, W. Va	July 19, 1870	105, 533
Churn, Oval	J. Oothoudt	Lebanon, N. Y	Mar. 10, 1830	....
Churn, Pendulum	C. Augevine	New York, N. Y	Feb. 28, 1831	....
Churn, Pendulum	S. Hitchcock	Vernon, N. Y	Aug. 7, 1813	....
Churn, Pneumatic	A. P. Smith	Litchfield, Conn	May 2, 1865	47, 607
Churn-power	J. Budd	Pittsford, N. Y	June 4, 1867	65, 337
Churn-power	J. Christley	Slippery Rock, Pa	May 7, 1867	64, 490
Churn-power	B. F. Frampton	Punxatawney, Pa	July 18, 1871	117, 166
Churn-power	A. Fuller and W. H. Howe	Amsterdam and Port Jackson, N. Y.	Oct. 4, 1864	44, 523
Churn-power	I. S. Goolman	Monrovia, Ind	Apr. 6, 1869	88, 629
Churn-power	L. A. Haight	Cairo, N. Y	Aug. 29, 1871	118, 529
Churn-power	J. V. Hartman	Marathon, Ohio	Apr. 9, 1867	63, 632
Churn-power	E. Hoag	Rensselaerville, N. Y	Dec. 1, 1863	40, 751
Churn-power	D. J. Knapp	Fallsburgh, N. Y	May 29, 1866	55, 120
Churn-power	G. Lewis	Panama, N. Y	June 7, 1864	43, 076
Churn-power	W. A. Lewis	Springfield, Vt	Mar. 18, 1873	136, 925
Churn-power	D. McCurdy	Ottawa, Ohio	June 15, 1869	91, 467
Churn-power	G. T. Montague	Keokuk, Iowa	June 21, 1870	104, 627
Churn-power	J. D. Parrot	Morristown, N. J	Sept. 8, 1863	39, 827
Churn-power	R. N. Smith	New York, N. Y	Apr. 19, 1870	102, 171
Churn-power	H. Swarthout	Altay, N. Y	Mar. 12, 1867	62, 905
Churn-power	J. B. Sweetland	Fostoria, Ohio	Jan. 5, 1864	41, 107
Churn-power, Endless platform for	A. B. Smith	Rochester, Pa	Sept. 1, 1868	81, 833
Churn propelled by weights	A. Bacon	Windsor, N. Y	Oct. 10, 1835	....
Churn, Reciprocating	E. Barrett and G. W. Fenimore.	Franklin, Tenn	July 15, 1873	140, 871
Churn, Reciprocating	L. I. Bodenhamer	Kernersville, N. C	July 15, 1873	140, 880
Churn, Reciprocating	G. P. Coan	Wyandotte, Mich	June 10, 1873	139, 658
Churn, Reciprocating	H. Hatch	Jefferson, Me	Aug. 26, 1873	142, 102
Churn, Reciprocating	G. M. Heim	Brownsville, Ind	May 13, 1873	138, 888
Churn, Reciprocating	A. D. Huntley	Smyrna Mills, Me	Oct. 21, 1873	143, 906
Churn, Reciprocating	L. B. Keeler	Westerville, Ohio	June 24, 1873	140, 280
Churn, Reciprocating	M. Litty	Williamsville, Nebr	Apr. 22, 1873	138, 168
Churn, Reciprocating	M. W. Staples	Richmond, Va	July 15, 1873	140, 796
Churn, Reciprocating	W. M. Thompson and J. L. Mahurin.	Rockfield, Ind	July 1, 1873	140, 445
Churn, Reciprocating	E. T. Wheeler	Cannelton, Ind	Oct. 21, 1873	143, 800
Churn, Revolving	N. King	Washington, D. C	Mar. 10, 1808	....
Churn, Revolving-dasher	H. Branch	New York, N. Y	Apr. 18, 1832	....
Churn, Rocking	W. Keeler and J. Waring	....	June 23, 1803	....
Churn, Rocking	D. Parmele	Dutchess County, N. Y	Sept. 8, 1808	....
Churn, Rocking	J. G. Philip	Kinderhook, N. Y	Feb. 15, 1827	....
Churn, Rocking	A. C. Stiles	South Bloomfield, Ohio	Sept. 27, 1844	3, 764
Churn, Rocking	E. Thomas	Harrisonburgh, Va	May 30, 1838	753
Churn, Rotary	J. W. Arnold	Fairport, N. Y	Oct. 8, 1872	131, 990
Churn, Rotary	W. Brand and G. Puder	Evansville, Ind	June 24, 1873	140, 112
Churn, Rotary	W. H. Bunch	Windsor, N. C	Aug. 12, 1873	141, 692
Churn, Rotary	F. E. Clarkson	Highland, Kans	Apr. 29, 1873	138, 229
Churn, Rotary	A. L. Cornell	New York, N. Y	Nov. 15, 1859	26, 093
Churn, Rotary	B. Eason	Easonville, Ala	Feb. 25, 1873	136, 226
Churn, Rotary	H. H. Grover	North Cohocton, N. Y	Nov. 15, 1853	10, 245
Churn, Rotary	J. Harwood	Cobleskill, N. Y	Sept. 24, 1867	69, 093
Churn, Rotary	E. A. Hewitt	Groton, Conn	July 29, 1873	141, 348
Churn, Rotary	O. B. Loomis	Windsor, Conn	Apr. 2, 1850	7, 247
Churn, Rotary	W. McClure	Sinking Spring, Ohio	Jan. 28, 1873	135, 281
Churn, Rotary	G. Walker	Whitley's Point, Ill	May 27, 1873	139, 345
Churn, Spiral-spring	L. Hinkson	Hallowell, Me	Jan. 7, 1835	....
Churn, Swinging	W. F. and N. Davis	Castleton, Vt	Nov. 23, 1852	9, 416
Churn, Thermometer	J. K. Pilkay	Carlisle, Pa	Feb. 16, 1869	87, 064
Churn-thermometer	J. H. Smiley	Caroline, N. Y	Oct. 31, 1871	120, 464
Churn, Tin	T. E. Warner	Harford County, Md	Feb. 11, 1834	....
Churn-top	E. Cox	Point Pleasant, Ohio	Apr. 17, 1866	53, 952
Churn, Valve-piston	B. Cushwa	Clear Spring, Md	Jan. 24, 1829	....
Churn, Ventilating portable	E. P. Williams	Yorkville, S. C	May 26, 1868	78, 249
Churns and washing-machines, Operating	J. T. Bever	Springville, Ill	Jan. 6, 1863	37, 330
Churns, Construction of	A. and W. A. Crowell	Salisbury, Conn	June 20, 1840	1, 648
Churns, Cutting floats of	R. Ryerson	Jay, Me	July 17, 1835	....
Churns, Device for operating	H. E. Addis	Springfield, Ill	July 7, 1863	39, 107
Churns, Device for operating	O. Edson	Franklinville, N. Y	May 26, 1863	38, 663
Churns, Device for operating	G. C. Ferris	Sharon, Wis	Apr. 14, 1863	38, 158
Churns, Device for operating	H. Soggs	Columbus, Pa	Dec. 1, 1863	40, 775
Churns, Machine attached to the top of	L. Stevens	....	May 8, 1804	....
Churns, &c., Machine for operating	I. Morse	West Franklin, Pa	Apr. 21, 1868	77, 077
Churns, Machinery for operating	J. J. Taylor	Attica, Ind	Aug. 11, 1863	39, 532

Index of patents issued from the United States Patent Office from 1790 *to* 1873, *inclusive*—Continued.

Invention.	Inventor.	Residence.	Date.	No.
Churns, Machinery for working	P. B. Pratt	Kent, Conn	July 22, 1833	....
Churns, Manner of working	J. H. Monce	Hopkinsville, Ohio	Feb. 4, 1868	74, 114
Churns, &c., Mechanical movement for operating	J. P. Nichols	New Richmond, Ohio	Dec. 4, 1866	60, 232
Churns, Mechanism for operating	G. True	Funchal Island, Madeira	Mar. 29, 1864	42, 129
Churns, Mode of constructing and operating	S. P. W. Douglass	Lansingburgh, N. Y	Jan. 9, 1838	554
Churns, Mode of operating	J. Redding	New Castle, Ind	June 6, 1865	48, 094
Churns, Mode of operating	I. M. Williams	Blanchester, Ohio	Oct. 1, 1861	33, 422
Churns, Operating	A. G. Brush	Great Bend, Pa	June 15, 1858	20, 545
Churns, Operating	J. Forsyth	Wheeling, Va	Nov. 9, 1858	22, (22
Churns, &c., Operating	M. Swan	Potter Hill, N. Y	Aug. 17, 1858	21, 221
Churning and working butter	W. D. Baughn	Milford, Mich	July 9, 1867	66, 445
Churning-apparatus	J. P. Curtis	Wytheville, Va	June 28, 1870	104, 710
Churning-apparatus	J. Letort	Wytheville, Va	July 5, 1870	104, 968
Churning-apparatus	E. J. Moore	Westfield, N. Y	Apr. 7, 1868	76, 497
Churning-apparatus	E. J. Moore	Westfield, N. Y	Oct. 13, 1868	83, 080
Churning-apparatus	J. Thompson	Hartville, Ohio	July 2, 1867	66, 416
Churning, Apparatus for tempering cream preparatory to.	E. T. Brownfield	Smithfield, Pa	Mar. 17, 1868	75, 622
Churning, Art of	I. F. Waring	Columbia Township, Ohio	Apr. 10, 1830	....
Churning butter, Mode of	I. Herrick	Merrimack, N. H	Apr. 21, 1868	76, 913
Churning, Inclined wheel applied to	G. F. Reeve and J. Ketcham	Orange County, N. Y	Aug. 10, 1821	....
Churning-machine	B. Atwell and I. Nichols	Otsego County, N. Y	Oct. 14, 1809	....
Churning-machine	E. Bacon	Hancock County, Mass	July 26, 1810	....
Churning-machine	I. Baker		Feb. 20, 1802	....
Churning-machine	S. H. Baker	Wells Township, Bradford County, Pa.	Jan. 10, 1828	....
Churning-machine	S. Clarke	Parkman, Me	Apr. 22, 1835	....
Churning-machine	G. Crandell	Rhinebeck, N. Y	Aug. 20, 1867	67, 960
Churning-machine	B. Cutler	Washington, N. Y	Oct. 3, 1807	....
Churning-machine	N. Harrington	Whitestown, N. Y	Oct. 11, 1809	....
Churning-machine	E. G. Jones	Otsego, N. Y	Apr. 21, 1809	....
Churning-machine	M. Knight	Pownal, Me	May 9, 1835	....
Churning-machine	A. Murray	Athens, N. Y	June 27, 1829	....
Churning-machine	J. Pierce		Apr. 10, 1802	....
Churning-machine	W. A. Rhoades	Lincolnville, Pa	July 6, 1869	92, 360
Churning-machine	G. Sanford	Ellenville, N. Y	May 4, 1852	8, 927
Churning-machine	J. Scripture	Alford, Mass	Dec. 15, 1807	....
Churning-machine	D. G. Taylor	Campbellsville, Ky	Dec. 7, 1869	97, 565
Churning, Machine for saving cream while	M. Love	Corry, Pa	July 21, 1868	80, 078
Churning, Power for	A. A. Drake	Flanders, N. J	July 8, 1862	35, 813
Chute and fish-way	J. D. Brewer	Muncy, Pa	Apr. 30, 1872	126, 257
Chute and water-wheel	H. Van Dewater	Philadelphia, Pa	Sept. 19, 1848	5, 785
Chute for dams	A. Addis	Washington Township, Pa.	Oct. 4, 1834	....
Chute for delivering timber	J. W. Haines	Genoa, Nev	Sept. 20, 1870	107, 611
Chute for horizontal water-wheels	I. Mallery	Etna, N. Y	Aug. 30, 1859	25, 270
Chute for river or canal navigation	A. Livermore	Tremont, Pa	Nov. 27, 1860	30, 739
Chute for water-wheels	C. B. Whitney	Ithaca, N. Y	Jan. 12, 1858	19, 115
Chute-gate	J. M. Thompson	Quincy, Cal	Sept. 17, 1872	131, 477
Chute, Lumber	W. Van Name and J. A. Wakefield.	Chippewa Falls, Wis	Jan. 28, 1873	135, 302
Cider, ale, &c., Compound for refining	W. M. Davis	Cleveland, Ohio	May 28, 1867	65, 063
Cider and bark mill	I. Quintard	Stanfield, Conn	Apr. 5, 1806	....
Cider and cheese press	E. Benedict	New Marlborough, Mass	Mar. 21, 1808	....
Cider and cheese press	E. Benedict	New Marlborough, Mass	June 18, 1808	....
Cider and vinegar from the offals of a distillery, Making.	W. Richardson	Kingston, N. Y	Sept. 26, 1816	....
Cider and wine mill	T. F. Deniston, G. C. Prosser, and J. J. Deniston.	Torrey, N. Y	Aug. 17, 1869	93, 861
Cider and wine mill	W. B. Farrar	Greensborough, N. C	Apr. 4, 1871	113, 283
Cider and wine mill	J. R. Gates	Louisville, Ky	Oct. 29, 1861	33, 581
Cider and wine mill	A. McCreight and R. W. Glasgow.	Tranquility, Ohio	Mar. 18, 1873	136, 847
Cider and wine mill	J. Walton	Sunfish, Ohio	June 2, 1868	78, 558
Cider and wine press	C. Beach	Penn Yan, N. Y	Aug. 21, 1866	57, 277
Cider and wine press	O. Kromer	Sandusky, Ohio	July 23, 1872	129, 836
Cider, &c., Concentrating and preserving	G. Borden, jr.	Amenia, N. Y	July 22, 1862	35, 919
Cider, Furnace for clarifying	C. Cory	Lima, Ind	Sept. 13, 1864	44, 162
Cider into wine. Converting	H. Seabolt	Scipio, N. Y	Nov. 11, 1812	....
Cider, Machine for making	M. B. Bliss	Pittston, Me	Apr. 21, 1829	....
Cider, Machine for preparing apples for the manufacture of.	G. B. Hamlin	Willimantic, Conn	Aug. 16, 1870	106, 359
Cider-making	G. B. Hamlin	Willimantic, Conn	Feb. 21, 1871	111, 930
Cider-making machine	W. Lucas	Rushville, Ill	Dec. 31, 1867	72, 872
Cider-mill	H. Abbott	Huron, N. Y	Dec. 23, 1856	16, 261
Cider-mill	W. Aikin and W. W. Drummond.	Louisville, Ky	Apr. 30, 1872	126, 244
Cider-mill	T. Appelget	Princeton, N. J	Nov. 20, 1866	59, 807
Cider-mill	R. C. Archibald	La Fayette, Ind	Aug. 27, 1867	68, 145
Cider-mill	W. Barr	Ypsilanti, Mich	July 11, 1871	116, 798
Cider-mill	J. Bauman	Shepherdstown, Pa	June 27, 1854	11, 151
Cider-mill	P. J. Berlin	Blairsville, Pa	Dec. 16, 1862	37, 147
Cider-mill	N. Booth	Cheshire, Conn	July 12, 1834	....
Cider-mill	J. Bowen	Yellow Bud, Ohio	Oct. 13, 1863	40, 237
Cider-mill	J. Bowen and A. T. Foster	Clarksburgh, Ohio	Apr. 18, 1871	113, 731
Cider-mill	E. W. Branch	East Henrietta, N. Y	Jan. 8, 1867	60, 992
Cider-mill	E. W. Branch	East Henrietta, N. Y	July 2, 1867	66, 207
Cider-mill	S. and R. W. Caldwell	Chillicothe, Ohio	Apr. 16, 1861	32, 044
Cider-mill	C. L. Carter	Union City, Ind	June 15, 1869	91, 210
Cider-mill	J. Case	Selma, Ohio	Aug. 10, 1852	9, 178
Cider-mill	N. Chapin	Cortlandville, N. Y	Nov. 21, 1848	5, 936
Cider-mill	N. Chapin	Syracuse, N. Y	Sept. 2, 1851	8, 339
Cider-mill	O. Clarke	Rockford, Ill	June 14, 1864	43, (95
Cider-mill	W. and L. Clayton	West Philadelphia, Pa	July 11, 1865	48, 657
Cider-mill	G. R. Cluxton	Ripley, Ohio	Aug. 6, 1872	130, 192
Cider-mill	J. A. Crever and F. H. Keeney	Cincinnati, Ohio	Aug. 1, 1865	49, 088
Cider-mill	G. W. D. Culp	Allenville, Ind	Nov. 14, 1846	4, 852
Cider-mill	R. M. Curtice	North Adams, Mich	Jan. 10, 1860	26, 752
Cider-mill	P. Dickson	Erie, Pa	Sept. 27, 1864	44, 408

Index of patents issued from the United States Patent Office from 1790 *to* 1873, *inclusive*—Continued.

Invention.	Inventor.	Residence.	Date.	No.
Cider-mill	J. Diehl and W. Wilson	Brooklyn, N. Y	Aug. 6, 1861	32, 988
Cider-mill	N. Eaton	Woburn, Mass	Dec. 10, 1872	133, 840
Cider-mill	J. Eiberneiser	Cincinnati, Ohio	Oct. 26, 1858	21, 874
Cider-mill	O. S. Garretson	Buffalo, N. Y	Jan. 11, 1870	98, 757
Cider-mill	L. Gebhart	Orangeville, Ohio	Jan. 21, 1868	73, 523
Cider-mill	S. A. Hebard	North Stamford, Conn	Sept. 1, 1863	39, 731
Cider-mill	M. W. Hilton	Bloomington, Ind	Feb. 23, 1858	19, 421
Cider-mill	A. D. Hoffman	Belleville, Mich	Nov. 8, 1859	26, 037
Cider-mill	S. J. Homan	Walden, N. Y	June 5, 1866	55, 297
Cider-mill	H. Hurd	Spring Hill, Ill	May 24, 1864	42, 912
Cider-mill	S. G. Hurlbut	Cleveland, Ohio	Sept. 6, 1864	44, 096
Cider-mill	S. Jackson	Hamilton, Ohio	Jan. 15, 1850	7, 022
Cider-mill	E. Jenkins	Harmony, Pa	May 17, 1836	....
Cider-mill	T. J. Kindleberger	Springfield, Ohio	May 29, 1855	12, 953
Cider-mill	T. J. Kindleberger	Springfield, Ohio	Feb. 9, 1864	41, 514
Cider-mill	J. Krauser	Reading, Pa	Aug. 30, 1853	9, 972
Cider-mill	H. Lightner	Neffs Mills, Pa	Aug. 8, 1871	117, 898
Cider-mill	J. Lytch	Laurinburgh, N. C	Nov. 22, 1870	109, 527
Cider-mill	B. Mackerley	New Petersburgh, Ohio	Nov. 14, 1856	16, 040
Cider-mill	S. Males	Cincinnati, Ohio	Sept. 6, 1870	107, 187
Cider-mill	S. Males	Cincinnati, Ohio	Feb. 13, 1866	52, 583
Cider-mill	W. M. McDowell and C. E. Baechtel.	Hagerstown, Md	Aug. 29, 1865	49, 641
Cider-mill	J. McGrew	Ravenswood, W. Va	Sept. 17, 1872	131, 453
Cider-mill	N. E. Miller	Canaan Centre, Ohio	Aug. 5, 1873	141, 582
Cider-mill	W. S. Oborn	Marion, Ohio	July 7, 1868	79, 771
Cider-mill	H. E. Paine and S. H. Russel	Leroy, Ohio	Mar. 5, 1827	....
Cider-mill	H. K. Parsons	Harrisburgh, Pa	Oct. 11, 1864	44, 651
Cider-mill	N. A. Patterson	Knoxville, Tenn	Oct. 3, 1871	119, 472
Cider-mill	D. F. Phillips	Republic, Ohio	Nov. 25, 1851	8, 549
Cider-mill	E. H. Philo	Half Moon, N. Y	Mar. 27, 1860	27, 650
Cider-mill	C. Pool and M. Eddy	Blissfield, Mich	Sept. 26, 1865	50, 158
Cider-mill	E. S. Purdy	Croton, N. Y	Mar. 12, 1867	62, 884
Cider-mill	J. Redlein	Brooklyn, N. Y	Jan. 23, 1866	52, 203
Cider-mill	C. Rice	Barre, Mass	Nov. 11, 1830	....
Cider-mill	W. Rice	Fitchburgh, Mass	June 3, 1873	139, 615
Cider-mill	L. O. Rockwood	Ottawa, Ill	June 17, 1873	139, 975
Cider-mill	J. Rosenkrans	Avoca, N. Y	May 3, 1859	23, 862
Cider-mill	G. R. Ruland and W. W. Green, jr.	Byron Centre, N. Y	June 21, 1864	43, 229
Cider-mill	J. E. Saugi	Washington, D. C	Jan. 28, 1873	135, 370
Cider-mill	M. P. Schenck	Fulton, N. Y	May 6, 1873	138, 533
Cider-mill	D. Scully	Baltimore, Md	July 5, 1864	43, 434
Cider-mill	G. Seger	Humberstone, Canada	May 23, 1871	115, 112
Cider-mill	H. Sells	Vienna, Canada	Sept. 4, 1866	57, 833
Cider-mill	T. Sharp	Louisville, Ky	Nov. 24, 1863	40, 711
Cider-mill	W. Shaw	New Gordon, Ohio	Nov. 20, 1866	59, 864
Cider-mill	C. Sheaffer	Lebanon, Pa	Mar. 19, 1836	....
Cider-mill	H. O. Sheidley	Republic, Ohio	Dec. 29, 1857	18, 993
Cider-mill	T. A. Shinn	Baden, Pa	Aug. 27, 1867	68, 246
Cider-mill	T. Showerman	Covington, N. Y	July 20, 1825	....
Cider-mill	J. K. P. Smith	Jeffersonville, Ind	Oct. 25, 1870	108, 645
Cider-mill	A. Spencer	Grampian Hills, Pa	July 1, 1862	35, 787
Cider-mill	M. Stevens	Lucas, Ohio	May 11, 1858	20, 220
Cider-mill	L. R. Taylor	Clark Township, Ind	Sept. 13, 1870	107, 424
Cider-mill	J. H. Thomas and P. P. Mast	Springfield, Ohio	Nov. 21, 1865	51, 101
Cider-mill	T. Van Kannel	Cincinnati, Ohio	July 2, 1867	66, 427
Cider-mill	H. T. Watkins	Anderson, Ind	Feb. 5, 1861	31, 344
Cider-mill	J. I. White	Zanesville, Ohio	Dec. 3, 1872	133, 609
Cider-mill	W. N. Whitely, J. Fassler, and O. S. Kelly.	Springfield, Ohio	Jan. 14, 1868	73, 417
Cider-mill	H. L. Whitman	Saint Louis, Mo	Nov. 19, 1872	133, 133
Cider-mill	J. H. Williams	Sandusky, Ohio	Nov. 13, 1866	59, 700
Cider-mill	J. D. Willoughby	Shippensburgh, Pa	Mar. 3, 1868	75, 101
Cider-mill	D. T. Willson	Harrisburgh, Pa	Jan. 2, 1872	122, 508
Cider-mill	C. Wilson	Clinton, Pa	Sept. 15, 1868	82, 261
Cider-mill	L. Wilson	Springfield, Ohio	Mar. 5, 1867	62, 585
Cider-mill	L. Wilson	Springfield, Ohio	Dec. 7, 1869	97, 743
Cider-mill	D. Zeigler	Lewistown, Pa	July 25, 1854	11, 401
Cider mill and fruit-press	L. A. Warner	Freeport, Ill	Oct. 2, 1866	58, 514
Cider mill and press	U. Emmons	New York, N. Y	Jan. 31, 1829	....
Cider mill and press	F. and G. F. Hovey	New York, N. Y	June 1, 1869	90, 843
Cider mill and press	M. Jones	East Saginaw, Mich	Oct. 29, 1872	132, 536
Cider mill and press	H. and H. S. Paul	Kittery, Me., and Cambridge, Mass.	Nov. 26, 1872	133, 476
Cider mill and press	O. Pettibone	Hartford, Conn	Mar. 1, 1815	....
Cider mill and press	E. Wangaman	Blairsville, Pa	July 21, 1868	80, 253
Cider mill and press	M. Winger	Ephratah, Pa	Dec. 10, 1867	71, 935
Cider mill and press, Combined	O. M. Brock	Monroeton, Pa	July 18, 1871	117, 041
Cider mill and press combined	D. H. Krauser	Pottsville, Pa	Apr. 18, 1871	113, 778
Cider mill and press, Portable	F. Fredley	Sugar Valley, Pa	Aug. 3, 1838	875
Cider-mill, Cast-iron	P. Pryer	Genesee County, N. Y	May 29, 1835	....
Cider-mill, Convertible	S. Males	Cincinnati, Ohio	Mar. 17, 1857	16, 847
Cider-mill, corn-sheller, and fodder-cutter, Mode of combining.	J. P. Adams	Chester, Ill	Aug. 25, 1863	39, 620
Cider-mill press	M. Stevens	Smithville, Ohio	July 31, 1866	56, 824
Cider-mill, Rotary-cylinder	S. Freeman, jr	New Marlborough, Mass	Mar. 14, 1828	....
Cider or wine mill	S. Krauser	Reading, Pa	Oct. 30, 1855	13, 741
Cider-press	A. Brooks	Tolland, Conn	Oct. 5, 1869	95, 558
Cider-press	O. Clark	Rockford, Ill	Apr. 26, 1864	42, 528
Cider-press	J. H. Harper	Pittsburgh, Pa	Nov. 19, 1867	71, 165
Cider-press	J. Holbrook	Sherburne, Mass	July 18, 1871	117, 075
Cider-press	F. Hovey	New York, N. Y	Apr. 26, 1870	102, 402
Cider press	J. Maerhoffer	Boonville, Mo	June 1, 1869	90, 672
Cider-press	D. Pride	Potsdam, N. Y	June 11, 1829	....
Cider-press	E. Primm	Petersburgh, Ill	Nov. 26, 1867	71, 533
Cider-press	D. Reed	Easton, Mass	Mar. 3, 1829	....
Cider-press	C. Ritter	Reading, Pa	June 21, 1859	24, 490

Index of patents issued from the United States Patent Office from 1790 to 1873, inclusive—Continued.

Invention.	Inventor.	Residence.	Date.	No.
Cider-press	J. Schaffer and E. Stoner	Westminster, Md	Oct. 5, 1869	95, 608
Cider-press	A. W. Shidler	South Bend, Ind	Mar. 17, 1868	75, 589
Cider-press	A. D. Strong	Ashtabula, Ohio	July 2, 1867	66, 262
Cider-press	C. H. Thomas	Milton, N. Y	Oct. 24, 1865	50, 665
Cider-press, Lever	H. Peters	Penn Township, Pa	Apr. 29, 1873	138, 280
Cider-press, Screw	S. Sanderson	Shelby, N. Y	Oct. 1, 1872	131, 785
Cider, Putting up and preserving	M. Halsted and J. L. Wheeler	Brooklyn, N. Y	Mar. 18, 1873	136, 914
Cider, wine, &c., with sulphurous acid, Apparatus for treating.	D. M. Mefford	Huron County, Ohio	Dec. 31, 1872	134, 389
Cigar	T. Blanchard	Boston, Mass	Mar. 30, 1858	19, 746
Cigar	S. Davis	Montreal, Canada	Oct. 11, 1864	44, 695
Cigar	D. Davison	New York, N. Y	July 4, 1865	48, 527
Cigar	F. L. Hilbright	Newark, N. J	July 28, 1868	80, 284
Cigar	W. C. Kneeland	Brooklyn, N. Y	Dec. 10, 1861	33, 889
Cigar	T. W. Kreitz	Quincy, Ill	May 28, 1872	127, 172
Cigar	I. Lindsley	Providence, R. I	July 31, 1860	29, 436
Cigar	E. B. Mead	Pittsfield, Mass	June 24, 1873	140, 153
Cigar	C. Quartley	Baltimore, Md	Feb. 11, 1868	74, 246
Cigar	W. P. Surgey	Hackney, Great Britain	Dec. 9, 1856	16, 200
Cigar	M. Turley and J. M. Innes	Council Bluffs, Iowa	June 11, 1872	127, 939
Cigar	C. Van Dyeck	Nashville, Tenn	Nov. 6, 1866	59, 485
Cigar	C. Walton	Washington, D. C	June 6, 1865	48, 119
Cigar and cigarette	T. Griffin	Roxbury, Mass	Mar. 23, 1869	88, 033
Cigar and cigarette making machine	J. and A. Marengo	Burlington, Vt	June 16, 1868	78, 985
Cigar and cigarette manufacture	L. Morgenthan	Manheim, Baden	Mar. 14, 1865	46, 855
Cigar and match case	P. J. Clark	West Meriden, Conn	Apr. 3, 1860	27, 693
Cigar and pipe mouth-piece	B. S. Stokes	Manchester, N. H	June 3, 1862	35, 491
Cigar-ash holder, &c	A. Philipp	Manchester, Great Britain	Mar. 30, 1869	88, 408
Cigar binders, fillers, and wrappers, Forming	R. W. Heywood	Baltimore, Md	May 28, 1872	127, 235
Cigar-box	H. Fowler	Detroit, Mich	Dec. 9, 1873	145, 290
Cigar-box	P. Greenwald	Syracuse, N. Y	June 4, 1872	127, 597
Cigar-box	W. B. Hale, jr	Milford, Mass	Aug. 20, 1872	130, 713
Cigar-box	E. Henkle	North Scituate, R. I	July 26, 1870	105, 681
Cigar-box	C. Jerome	New Haven, Conn	Mar. 1, 1870	100, 296
Cigar-box	E. C. Patterson	Rochester, N. Y	Oct. 28, 1873	144, 129
Cigar-box	W. R Rhoades	Auburn, N. Y	Jan. 16, 1872	122, 854
Cigar-box	M. Richardson	Brooklyn, N. Y	Mar. 29, 1870	101, 381
Cigar-box	H. Schmeer	Syracuse, N. Y	Apr. 23, 1872	125, 910
Cigar-box	T. A. Wiley	Lancaster, Pa	July 4, 1871	116, 782
Cigar-box, Perforated	R. S. Jennings	New York, N. Y	Jan. 25, 1870	99, 204
Cigar-box, Water-proof	J. L. Thomson	Syracuse, N. Y	June 27, 1871	116, 373
Cigar-boxes from redwood, Manufacture of	C. A. Hooper	San Francisco, Cal	Dec. 23, 1873	145, 735
Cigar-bunch pressing and molding machine	W. D. Brewer	Charlestown, Mass	Jan. 21, 1873	135, 030
Cigar-bunch-shaping mold	A. Pearl	Brooklyn, N. Y	July 18, 1871	117, 200
Cigar-bunches, Machine for making	S. Scholfield	Providence, R. I	Dec. 10, 1872	133, 725
Cigar-bunches, Machine for making	S. Scholfield	Providence, R. I	July 15, 1873	140, 848
Cigar-bunches, Machine for making	W. Weis	Saint Paul, Minn	Nov. 15, 1870	109, 280
Cigar-bunches, Machine for making	D. A. Wightman	East Greenwich, R. I	July 15, 1873	140, 860
Cigar-bunches, Machine for making	D. A. Wightman	East Greenwich, R. I	July 15, 1873	140, 861
Cigar-bunches, Machine for pressing	G. Pierce	Boston, Mass	Jan. 14, 1873	134, 768
Cigar-bunches, Machine for pressing and molding	W. D. Brewer	Charlestown, Mass	July 25, 1871	117, 377
Cigar-bunches, Manufacture of	A. Pearl	New York, N. Y	May 17, 1870	103, 236
Cigar-bunches, Manufacture of	S. Scholfield	Providence, R. I	July 16, 1872	129, 599
Cigar-bunches, Mold for shaping and drying	A. Pearl	New York, N. Y	Oct. 11, 1870	108, 290
Cigar-bunches, Retainer for holding pressed and molded.	W. D. Brewer	Charlestown, Mass	Aug. 29, 1871	118, 428
Cigar-bunches, Retainer for pressed and molded	W. D. Brewer	Charlestown, Mass	Jan. 21, 1873	135, 031
Cigar-bunching machine	H. B. Bunster	New York, N. Y	Mar. 4, 1873	136, 411
Cigar-bunching machine	S. T. Hennaman	Baltimore, Md	Aug. 13, 1872	130, 496
Cigar-bundling apparatus	C. A. Siecke	Philadelphia, Pa	May 19, 1868	78, 142
Cigar-bundling machine	F. A. Henckell	New York, N. Y	Mar. 2, 1869	87, 493
Cigar-case	C. A. Perry	Elkhorn, Wis	Nov. 21, 1865	51, 080
Cigar-case	S. N. Risley	Brooklyn, N. Y	Dec. 15, 1868	84, 906
Cigar-clamp	N. A. Buhl	New York, N. Y	Oct. 24, 1871	120, 240
Cigar-cutter	F. Funke	Detroit, Mich	July 2, 1872	128, 481
Cigar-cutter	R. A. Stendell	New York, N. Y	July 16, 1872	129, 182
Cigar-cutter	L. S. Stimson	Lowell, Mass	Nov. 8, 1870	109, 153
Cigar cutter and holder combined	J. Brady	Brooklyn, N. Y	Dec. 17, 1872	134, 030
Cigar cutting and perforating machine	J., T., and A. Levy	New York, N. Y	Apr. 15, 1873	137, 850
Cigar-drier	O. Vallandingham	Saint Louis, Mo	Mar. 23, 1869	88, 237
Cigar-drying mold	N. W. Palmer	New York, N. Y	June 3, 1873	139, 517
Cigar-ends, Machine for cutting off	F. W. Hoffmann	Morrisania, N. Y	Sept. 3, 1867	68, 441
Cigar-fillers, Apparatus for arranging	S. Scholfield	Providence, R. I	July 15, 1873	140, 850
Cigar-fillers, Machine for bunching and binding	H. Schmidt	New York, N. Y	Sept. 24, 1872	131, 709
Cigar-fillers, Machine for making	G. A. Reniger	Stuttgart, Wurtemberg	May 1, 1866	54, 476
Cigar-fillers, Machine for preparing	J. Maffet	Lancaster, Pa	May 16, 1832	....
Cigar-fillers, Machine for pressing and shaping	C. H. Palmer	New York, N. Y	Jan. 28, 1873	135, 243
Cigar-filling	S. Scholfield	Providence, R. I	Jan. 17, 1871	111, 089
Cigar-fillings, Mold for drying	S. B. Jerome	New Haven, Conn	Sept. 13, 1870	107, 266
Cigar-fuse box	H. Reimann	Brooklyn, N. Y	Apr. 3, 1866	53, 678
Cigar-gage and butt-cutter	H. Peterson	Chicago, Ill	Sept. 10, 1867	68, 785
Cigar-header	G. Moebs	Detroit, Mich	May 26, 1858	78, 312
Cigar-heading socket	T. Thorp	New York, N. Y	Mar. 13, 1860	27, 484
Cigar-holder	C. Appel	Hoboken, N. J	May 21, 1867	64, 826
Cigar-holder	L. Auguste	New York, N. Y	Nov. 29, 1864	45, 215
Cigar-holder	J. K. Chase	New York, N. Y	Mar. 12, 1872	124, 545
Cigar-holder	M. Johnston	South Boston, Mass	Mar. 26, 1861	31, 808
Cigar-holder	T. J. Lewis	Chicago, Ill	Apr. 22, 1873	138, 032
Cigar-holder	J. D. W. Olney	Providence, R. I	Oct. 29, 1872	132, 685
Cigar-holder	W. K. Vanderslice, jr	San Francisco, Cal	Jan. 21, 1868	73, 478
Cigar-holder	M. V. B. Young	New York, N. Y	July 28, 1868	80, 527
Cigar-holder and hat-hook	C. Gschwind and J. Grether	Union Hill, N. J	Aug. 14, 1866	57, 126
Cigar holder and perforator	R. J. Sheehy	Boston, Mass	June 4, 1872	127, 649
Cigar-holding case	S. R. Wilmot	Bridgeport, Conn	Feb. 23, 1869	87, 314
Cigar-knife	J. H. Abbott	Frederick City, Md	Aug. 16, 1864	43, 823
Cigar labeling device	R. H. Andrews	Washington, D. C	Feb. 14, 1871	111, 804
Cigar-light	W. Belcher	New Haven, Conn	Oct. 2, 1866	58, 537
Cigar-lighter	M. F. Gale	New York, N. Y	Nov. 21, 1871	121, 049

Index of patents issued from the United States Patent Office from 1790 *to* 1873, *inclusive*—Continued.

Invention.	Inventor.	Residence.	Date.	No.
Cigar-lighter	E. P. Gleason	New York, N. Y	Nov. 13, 1866	59, 587
Cigar-lighter	J. B. Miller	Rondout, N. Y	Oct. 31, 1871	120, 453
Cigar-lighter, Gas	J. W. Tracy	Saint Louis, Mo	Nov. 26, 1867	71, 428
Cigar-lighter, Gas jet	G. B. Snow and T. G. Lewis	Buffalo, N. Y	Feb. 6, 1866	52, 457
Cigar-lighter, Suspended gas	D. F. Brandon	Chicago, Ill	Jan. 5, 1869	85, 636
Cigar-lighters, Machine for making	H. Reimann	Hartford, Conn	Nov. 24, 1857	18, 706
Cigar-lighters, Manufacture of friction-match	W. Porter	Saint Stephen, Canada	Sept. 27, 1870	107, 810
Cigar-lighters, Match for	W. Porter	Saint Stephens' Parish, Canada.	Nov. 29, 1870	109, 660
Cigar-lighters, Plane for cutting	W. H. Drips	Cincinnati, Ohio	Aug. 26, 1873	142, 216
Cigar-lighting cinder	H. Reimann	Hartford, Conn	Mar. 23, 1858	19, 717
Cigar-lighting cinders, Apparatus for containing and igniting.	H. Reimann	Hartford, Conn	Mar. 9, 1858	19, 580
Cigar-machine	J. Allen and J. Fanning	Brooklyn, N. Y	July 26, 1870	105, 620
Cigar-machine	W. G. Ayres and S. L. Cole	Brooklyn, N. Y	Feb. 15, 1870	99, 745
Cigar-machine	I. S. Barber	New York, N. Y	June 24, 1862	35, 728
Cigar-machine	R. A. Bright, jr	Providence, R. I	Oct. 20, 1868	83, 247
Cigar-machine	C. G. A. Brinckmann	New York, N. Y	Sept. 6, 1864	44, 070
Cigar-machine	G. Buckle	Monroe, Mich	Sept. 5, 1865	49, 709
Cigar-machine	G. B. Clarke	New York, N. Y	Apr. 20, 1869	89, 200
Cigar-machine	R. M. Cole	Burlington, Vt	Apr. 13, 1869	88, 850
Cigar-machine	S. L. Cole	Brooklyn, N. Y	June 20, 1871	116, 026
Cigar-machine	W. Dawson	Huntington, Conn	Apr. 29, 1856	14, 763
Cigar-machine	J. de Bary	Offenbach, Germany	May 28, 1861	32, 405
Cigar-machine	C. J. Delbridge	Rochester, N. Y	July 15, 1873	140, 896
Cigar-machine	H. Erfurth	Leipsic, Germany	June 18, 1872	128, 133
Cigar-machine	T. Ernst	Fort Madison, Iowa	July 15, 1873	140, 906
Cigar-machine	E. J. Fisk	South Byron, N. Y	Dec. 19, 1865	51, 578
Cigar-machine	I. A. Heald	Carlisle, Mass	May 31, 1864	42, 946
Cigar-machine	I. A. Heald	Washington, D. C	Feb. 2, 1869	86, 541
Cigar-machine	I. A. Heald	Lowell, Mass	Apr. 18, 1871	113, 880
Cigar-machine	J. T. Hennaman	Baltimore, Md	July 1, 1873	140, 501
Cigar-machine	R. W. Heywood	Baltimore, Md	May 21, 1872	126, 960
Cigar-machine	J. C. Hintz	Cincinnati, Ohio	Oct. 5, 1869	95, 478
Cigar-machine	W. W. Huse	Brooklyn, N. Y	Feb. 12, 1861	31, 390
Cigar-machine	J. Lauritzen	Newark, N. J	Oct. 4, 1870	107, 927
Cigar-machine	F. Meyer and H. Schild	New York, N. Y	Jan. 18, 1870	98, 877
Cigar-machine	F. C. Miller	Cincinnati, Ohio	Jan. 23, 1872	122, 955
Cigar-machine	C. Müller	Albany, N. Y	July 21, 1868	80, 204
Cigar-machine	C. Müller	Albany, N. Y	Oct. 26, 1869	96, 257
Cigar-machine	H. Müller and C. Major	New York, N. Y	Feb. 19, 1861	31, 518
Cigar-machine	R. Neisch	Allentown, Pa	Dec. 20, 1870	110, 270
Cigar-machine	R. Neisch	Allentown, Pa	July 16, 1872	129, 489
Cigar-machine	J. Ochs	Litiz, Pa	Sept. 10, 1872	131, 177
Cigar-machine	J. O'Reilly	Baltimore, Md	Oct. 25, 1870	108, 629
Cigar-machine	W. H. Pease	Fulton, Wis	July 2, 1872	128, 560
Cigar-machine	J. Prentice	New York, N. Y	Sept. 6, 1864	44, 114
Cigar-machine	J. Prentice and W. F. Wuterich.	New York, N. Y	Nov. 26, 1867	71, 532
Cigar-machine	G. A. Reiniger	Stuttgart, Wurtemberg	May 29, 1866	55, 217
Cigar-machine	R. B. Robbins	Adrian, Mich	Mar. 12, 1872	124, 629
Cigar-machine	H. Shild and G. J. Prentice	New York, N. Y	Jan. 18, 1870	98, 890
Cigar-machine	S. Scholfield	Providence, R. I	Aug. 9, 1870	106, 214
Cigar-machine	S. Scholfield	Providence, R. I	Nov. 28, 1871	121, 298
Cigar-machine	S. Scholfield	Providence, R. I	Sept. 24, 1872	131, 567
Cigar-machine	S. Scholfield	Providence, R. I	July 15, 1873	140, 845
Cigar-machine	S. Scholfield	Providence, R. I	July 15, 1873	140, 846
Cigar-machine	G. W. Tanner	Providence, R. I	July 25, 1871	117, 482
Cigar-machine	G. W. Tanner	Providence, R. I	Sept. 17, 1872	131, 474
Cigar-machine	G. W. Tanner and F. D. Bliss	Providence, R. I	July 26, 1870	105, 859
Cigar-machine	I. Ten Eyck and J. O. Reilley	New York, N. Y	Sept. 13, 1870	107, 307
Cigar-machine	J. Thompson	New York, N. Y	Feb. 14, 1865	46, 404
Cigar-machine	H. E. Tylander	Keokuk, Iowa	May 27, 1873	139, 280
Cigar-machine	J. L. Weatherhead	Philadelphia, Pa	Apr. 16, 1872	125, 709
Cigar-machine	J. Wettstein	Baltimore, Md	Nov. 23, 1869	97, 255
Cigar-machine roll	D. A. Wightman	East Greenwich, R. I	July 15, 1873	140, 859
Cigar-machinery	W. Kramer and J. Wise	Milwaukee, Wis., and New York, N. Y.	Aug. 20, 1867	67, 887
Cigar-maker's filler-box	G. Barry	Chicago, Ill	Dec. 26, 1871	122, 211
Cigar-maker's mold	J. Prentice	New York, N. Y	Jan. 12, 1869	85, 764
Cigar-making apparatus	J. F. Shepard	Hampton Falls, N. H	May 12, 1868	77, 846
Cigar-making machine	L. Beauché	Paris, France	Feb. 16, 1858	19, 341
Cigar-making machine	J. Hafer and J. A. Henderson	Bedford, Pa	June 18, 1867	65, 805
Cigar-making machine	H. J. Hall	Brookline, Mass	Jan. 23, 1872	123, 014
Cigar-making machine	I. A. Heald	Carlisle, Mass	May 22, 1866	54, 900
Cigar-making machine	J. A. Heald	Springfield, Mass	Aug. 6, 1861	32, 993
Cigar-making machine	W. W. Huse	Brooklyn, N. Y	Dec. 6, 1864	45, 326
Cigar-making machine	J. McKee and T. W. Fletcher	New York, N. Y	Mar. 16, 1869	87, 955
Cigar-making machine	M. V. McKinney	Louisville, Ky	Sept. 19, 1871	119, 040
Cigar-making machine	H. Pierce	Winchester, N. H	Dec. 10, 1846	4, 885
Cigar-making machine	T. Thorp	New York, N. Y	Dec. 6, 1859	26, 382
Cigar-making machine	T. Thorp	New York, N. Y	May 1, 1860	28, 115
Cigar-making machine	A. Weeks	Syracuse, N. Y	June 30, 1868	79, 423
Cigar-making machine	J. Wettstein	Baltimore, Md	May 16, 1871	115, 002
Cigar-making machine	F. Wuterich	New York, N. Y	Dec. 18, 1860	31, 002
Cigar-making machine	F. Wuterich	New York, N. Y	Nov. 5, 1861	33, 687
Cigar-manufacture	W. C. Kneeland	Brooklyn, N. Y	July 28, 1868	80, 287
Cigar-mold	G. Barry	Chicago, Ill	Feb. 6, 1872	123, 323
Cigar-mold	J. Baxter	Detroit, Mich	Mar. 26, 1872	125, 005
Cigar-mold	N. H. Borgfeldt	New York, N. Y	Nov. 22, 1870	109, 490
Cigar-mold	N. H. Borgfeldt	New York, N. Y	Jan. 3, 1871	110, 732
Cigar-mold	N. H. Borgfeldt	New York, N. Y	Oct. 29, 1872	132, 622
Cigar-mold	N. Dubrul	Joliet, Ill	May 9, 1871	114, 655
Cigar-mold	N. Dubrul	Joliet, Ill	May 16, 1871	114, 932
Cigar-mold	N. Du Brul	Chicago, Ill	Sept. 9, 1873	142, 683
Cigar-mold	I. Guthman	Morrison, Ill	June 25, 1872	128, 386
Cigar-mold	B. Hawkins	Trenton, N. J	June 11, 1872	127, 764
Cigar-mold	J. J. Lahaye and J. M. Lyons	Reading, Pa	Dec. 19, 1871	122, 034
Cigar-mold	T. D. McGuire	Brooklyn, N. Y	May 6, 1873	138, 672

Index of patents issued from the United States Patent Office from 1790 *to* 1873, *inclusive*—Continued.

Invention.	Inventor.	Residence.	Date.	No.
Cigar-mold	F. C. Miller	Cincinnati, Ohio	Feb. 4, 1873	135, 572
Cigar-mold	C. H. Palmer	New York, N. Y	Mar. 11, 1873	136, 615
Cigar-mold	J. Prentice	New York, N. Y	May 23, 1871	115, 101
Cigar-mold	S. G. Rice	Albany, N. Y	July 2, 1872	128, 510
Cigar-mold	S. G. Rice	Albany, N. Y	Dec. 31, 1872	134, 398
Cigar mold and press	J. Ryan	Detroit, Mich	Oct. 3, 1871	119, 653
Cigar-mold clamp	S. Wilmot	Detroit, Mich	Mar. 4, 1873	136, 399
Cigar-mold press	G. Barry	Chicago, Ill	Feb. 27, 1872	123, 972
Cigar-mold press	F. Brown	New York, N. Y	Oct. 22, 1872	132, 351
Cigar-mold press	F. Brown	New York, N. Y	Oct. 22, 1872	132, 352
Cigar-molds, Machine for filling	F. C. Miller	Cincinnati, Ohio	July 23, 1872	129, 674
Cigar-molds, Machine for making	M. H. Eeimerdinger	New York, N. Y	May 21, 1872	126, 957
Cigar-molds, Tool for boring	H. F. Moeller and H. P. Brandt	Davenport, Iowa	Nov. 21, 1871	121, 059
Cigar-molding apparatus	J. Charter	Sterling, Ill	Apr. 23, 1872	126, 018
Cigar-molding device	T. J. Winship	Montreal, Canada	Apr. 8, 1873	137, 749
Cigar, money, &c., box	N. Thompson	Brooklyn, E. D., N. Y	Dec. 1, 1868	84, 655
Cigar mouth-piece	J. Ball	Elmira, N. Y	Mar. 28, 1865	46, 986
Cigar mouth-piece	I. S. Barber	New York, N. Y	Nov. 13, 1866	59, 544
Cigar, Medicated	J. Barrett	Chicago, Ill	Mar. 2, 1869	87, 320
Cigar, Medicated	L. Walther	New York, N. Y	Oct. 12, 1869	95, 858
Cigar perforator	E. A. Konter	Brooklyn, N. Y	Dec. 10, 1872	133, 864
Cigar-piercer	H. N. Foster	East Greenwich, R. I	Mar. 15, 1870	100, 744
Cigar-pipe	H. E. Doster	Bethlehem, Pa	Sept. 1, 1868	81, 610
Cigar-pipe	A. A. Pathi	Paris, France	Oct. 6, 1868	82, 743
Cigar-pipe	E. Schlichting	New York, N. Y	Dec. 31, 1867	72, 912
Cigar-point splitter	A. Sickenberger	Chicago, Ill	Nov. 11, 1873	144, 570
Cigar-press	J. Campbell	Lancaster, Pa	Oct. 2, 1866	58, 538
Cigar-press	G. Heiss	Lancaster, Pa	Nov. 20, 1866	59, 911
Cigar-press	A. Iske	Lancaster, Pa	July 3, 1866	56, 055
Cigar-press	M. Leippe	Lancaster, Pa	July 24, 1866	56, 578
Cigar-press	G. Studer	Richmond, Ind	Dec. 29, 1868	85, 344
Cigar-pressing machine	J. Campbell	Lancaster, Pa	June 19, 1866	55, 619
Cigar-punch	S. J. Bestor	Hartford, Conn	June 14, 1870	104, 250
Cigar-rack for hats	F. I. Miller	Brooklyn, N. Y	June 19, 1860	28, 763
Cigar-sample box	J. C. W. Frishmuth	Philadelphia, Pa	May 23, 1871	115, 188
Cigar-shaped pipe	S. N. Buynitzky	Washington, D. C	Mar. 4, 1873	136, 487
Cigar-shaped smoker	D. D. Foley	Washington, D. C	Aug. 23, 1864	43, 906
Cigar-shield	G. E. Brinkerhoff	New York, N. Y	Mar. 15, 1870	100, 850
Cigar-tip	J. Clark	Lowell, Mass	July 13, 1869	92, 423
Cigar-tip	W. M. Herron	Allegheny, Pa	Mar. 4, 1873	136, 514
Cigar-tip cutter	F. Timm and T. Fantini	Stapleton, N. Y	Feb. 25, 1873	136, 285
Cigar-tip machine	W. W. Huse	Brooklyn, N. Y	Dec. 6, 1864	45, 325
Cigar-tips, Composition for covering	J. H. Harris	Pittsfield, Mass	May 14, 1872	126, 698
Cigar-tips, Forming	S. L. Cole	Brooklyn, N. Y	Oct. 18, 1870	108, 331
Cigar-tips, Instrument for cutting	C. Froehlich	Philadelphia, Pa	Dec. 3, 1867	71, 600
Cigar-trimmer	J. H. Christman	Syracuse, N. Y	July 16, 1872	129, 101
Cigar-trimmer	A. Cramer	Detroit, Mich	June 24, 1873	140, 117
Cigar-trimmer	A. Hettinger	Philadelphia, Pa	May 20, 1873	139, 152
Cigar-trimmer	S. Scholfield	Providence, R. I	July 15, 1873	140, 847
Cigar-tuck cutter	S. Scholfield	Providence, R. I	Dec. 24, 1872	134, 222
Cigar-wrapper	H Durell	Morrisania, N. Y	Sept. 14, 1858	21, 558
Cigar-wrapper	C. E. Raffee	Barrington, R. I	Sept. 19, 1865	50, 038
Cigar-wrapper	J. S. Suter and G. M. Palmer	Baltimore, Md	Oct. 5, 1858	21, 704
Cigar-wrappers, Apparatus for cutting	R. Appleby	Beverly, N. J	Apr. 8, 1873	137, 524
Cigar-wrappers, Cutting	S. Scholfield	Providence, R. I	Oct. 15, 1872	132, 183
Cigar-wrappers, Manufacture of	F. Dixon	Lynn, Mass	Sept. 27, 1859	25, 604
Cigar-wrapping machine	S. Scholfield	Providence, R. I	July 15, 1873	140, 849
Cigar-wrapping machine	G. W. Tanner and F. D. Bliss	Providence, R. I	July 5, 1870	105, 140
Cigars and plug-tobacco, Machine for forming	W. Hall and E. J. Bennett	Boston, Mass	Nov. 30, 1869	97, 292
Cigars, Apparatus for cutting the ends of	M. J. Hinden	Detroit, Mich	Mar. 21, 1871	112, 808
Cigars, Bunch-mold machine for	J. Wettstein	Highland, Ill	Dec. 31, 1872	134, 502
Cigars, &c., Compound for lighting	C. J. M. Sohet and H. C. T. Molvant.	New York, N. Y	Nov. 20, 1866	59, 868
Cigars, Device for perforating	J. Houghton and G. Wingfield	New York and Brooklyn, N. Y.	May 21, 1867	64, 978
Cigars, Forming bunches for mold	C. M. Mann	Detroit, Mich	Sept. 17, 1872	131, 362
Cigars for tying, Clamp for holding	M. Ali	Philadelphia, Pa	May 15, 1866	54, 811
Cigars, &c., Friction-match for lighting	H. Reiman	Brooklyn, N. Y	Nov. 7, 1865	50, 843
Cigars, Implement for cutting, perforating, and lighting.	C. F. Smith	Springfield, Mass	Apr. 7, 1868	76, 537
Cigars, Instrument for perforating	O. Guinand	Vicksburgh, Miss	July 2, 1867	66, 325
Cigars, Lighting	W. B Doudes	Canton, Ohio	July 30, 1867	67, 276
Cigars, Machine for compressing and cutting the filling for.	W. W. Huse	Brooklyn, N. Y	July 16, 1867	66, 845
Cigars, Machine for cutting off	C. and F. W. Hoffman	Morrisania, N. Y	June 13, 1865	48, 174
Cigars, Machine for cutting off	F. W. Hoffman	Morrisania, N. Y	Nov. 13, 1866	59, 606
Cigars, Machine for cutting off	P. Zern and W. Warwick	Pittsburgh, Pa	May 28, 1867	65, 325
Cigars, Machine for cutting off the ends of	J. G. Maier and C. W. Schaeffer	Baltimore, Md	Nov. 9, 1869	96, 714
Cigars, Machine for examining and replacing	J. Levy	Wolcottville, Conn	Aug. 29, 1871	118, 541
Cigars, Machine for making	W. Dawson	Huntington, Conn	June 15, 1852	9, 016
Cigars, Machine for making the bodies of	G. A. Reiniger	Stuttgart, Wurtemburg	Oct. 29, 1861	33, 603
Cigars, Machine for manufacturing	J. Ball	Elmira, N. Y	Jan. 10, 1865	45, 808
Cigars, Machine for putting on the wrappers of	G. A. Reiniger	Stuttgart, Wurtemburg	Oct. 29, 1861	33, 604
Cigars, Machine for trimming the ends of	P. A. La France	Elmira, N. Y	Feb. 9, 1869	86, 764
Cigars, Machinery for making	G. Barker	New York, N. Y	July 9, 1861	32, 743
Cigars, Machinery for making	S. E. Hartwell and W. M. and D. Fowler.	New York, N. Y., and Waterbury, Conn.	May 8, 1847	5, 110
Cigars, Making	J. Ball	Buffalo, N. Y	Oct. 12, 1842	2, 809
Cigars, Method and apparatus for making	E. A. Metz	Cincinnati, Ohio	Sept. 5, 1871	118, 739
Cigars, Molding	J. Charter	Sterling, Ill	July 11, 1871	116, 930
Cigars, &c., Mouth-piece for	W. Thompson	Dublin, Ireland	Mar. 31, 1868	76, 119
Cigars, &c., Powder for lighting	C. W. Roseling	Cleveland, Ohio	Apr. 18, 1865	47, 335
Cigars, Process of treating and drying	J. Cuddy	Pittsburgh, Pa	Mar. 4, 1873	136, 490
Cigars, Self-igniting	J. Marck	New York, N. Y	Apr. 16, 1834	....
Cigars, snuff, &c., Process of preparing plants to be used in.	P. V. Ramel	Paris, France	Dec. 14, 1869	97, 962

Index of patents issued from the United States Patent Office from 1790 *to* 1873, *inclusive*—Continued.

Index of patents issued from the United States Patent Office from 1790 *to* 1873, *inclusive*—Continued.

Invention.	Inventor.	Residence.	Date.	No.
Clamp—Continued.				
See Flask-clamp.				
Floor-clamp.				
Flooring-clamp.				
Fruit-jar clamp.				
Furnacemen's clamp.				
Furnace-mold clamp.				
Gas-fitter's clamp.				
Glass-blower's clamp.				
Hand-clamp.				
Harness-clamp.				
Harness-pad clamp.				
Hawser-clamp.				
Hitching-clamp.				
Hog-holding clamp.				
Hub-clamp.				
Iron-structure clamp.				
Joiner's clamp.				
Line-clamp.				
Molder's clamp.				
Mop-clamp.				
Music-stand clamp.				
Newspaper-clamp.				
Paper-clamp.				
Picket-holding clamp.				
Picture-frame clamp.				
Pin-clamp.				
Pipe and faucet clamp.				
Planking-clamp.				
Pole-clamp.				
Portable clamp.				
Rail-clamp.				
Regulating-clamp.				
Rigging-clamp.				
Rigging-setting clamp.				
Rope-clamp.				
Rope or wire clamp.				
Saddler's clamp.				
Safety-clamp.				
Sash and door clamp.				
Saw-clamp.				
Saw-filing clamp.				
Scaffold-pole clamp.				
Screw-clamp.				
Sewing-clamp.				
Sheet-piling clamp.				
Ship-planking clamp.				
Shoemaker's clamp.				
Slate-frame clamp.				
Stage-clamp.				
Stay-log clamp.				
Stitching-clamp.				
Strap-clamp.				
Suspending-clamp.				
Tassel-clamp.				
Thill-coupling clamp.				
Timber-holding clamp.				
Trunk-clamp.				
Tube and rod clamp.				
Tubing-clamp.				
Watchmaker's chain-clamp.				
Whip-making clamp.				
Whip-socket clamp.				
Clamp	G. F. Almy	Toledo, Ohio	Mar. 19, 1872	124, 776
Clamp	A. Anderson	Madison, Wis	Aug. 11, 1868	80, 894
Clamp	C. S. Bonney	Philadelphia, Pa	Feb. 18, 1873	135, 882
Clamp	S. C. Bradley	New Haven, Conn	Feb. 15, 1870	99, 826
Clamp	M. V. Brigham	Mannsville, N. Y	Sept. 7, 1869	94, 557
Clamp	M. G. Burkhardt	Cincinnati, Ohio	Dec. 14, 1869	97, 872
Clamp	C. B. Canfield	Oriskany, N. Y	May 3, 1870	102, 487
Clamp	G. H. Coe and G. H. Snow	New Haven, Conn	Dec. 10, 1867	71, 983
Clamp	T. O. Cornish	Woonsocket, R. I	Dec. 12, 1871	121, 851
Clamp	J. J. Dominic	Gallupville, N. Y	Jan. 17, 1871	110, 961
Clamp	L. Feely	Rochester, N. Y	Aug. 19, 1873	142, 010
Clamp	F. Glasser	Mystic Bridge, Conn	Sept. 13, 1870	107, 247
Clamp	G. W. Goulden	Waverly, N. Y	Sept. 28, 1869	95, 337
Clamp	E. B. Hayes	Vergennes, Vt	Mar. 22, 1870	101, 126
Clamp	F. M. Holmes	Boston, Mass	Sept. 9, 1873	142, 697
Clamp	N. J. Holt and J. Leach	Howell and Detroit, Mich	June 24, 1873	140, 136
Clamp	P. F. Hulbert	Chatham, N. Y	June 5, 1866	55, 302
Clamp	J. S. Ladow	Mechanicsville, N. Y	Jan. 25, 1870	99, 212
Clamp	G. D. Lambert	New Haven, Conn	Mar. 8, 1870	100, 642
Clamp	J. J. Lebeau	Cincinnati, Ohio	June 21, 1870	104, 606
Clamp	J. F. Metz	Baltimore, Md	May 21, 1872	126, 895
Clamp	S. A. Morse	New Bedford, Mass	Apr. 16, 1867	63, 929
Clamp	W. F. Otis	New London, Ohio	Oct. 29, 1872	132, 538
Clamp	O. L. Payne	Batavia, Ill	July 13, 1869	92, 642
Clamp	W. H. Payne and H. Imhof	Lockport, N. Y	Sept. 16, 1873	142, 938
Clamp	J. H. Phillips	Troy, N. Y	June 18, 1872	128, 664
Clamp	E. K. Purdy	Schoolcraft, Mich	Oct. 13, 1868	82, 985
Clamp	W. Sailer	Philadelphia, Pa	Aug. 18, 1868	81, 298
Clamp	W. Sailer	Philadelphia, Pa	Nov. 16, 1869	96, 971
Clamp	J. E. Sinclair	Worcester, Mass	July 15, 1873	140, 852
Clamp	W. Strevell	Jersey City, N. J	Jan. 1, 1867	60, 959
Clamp	J. A. Traut	New Britain, Conn	Sept. 10, 1872	131, 195
Clamp	L. Wharton	Salem, Ohio	Aug. 30, 1870	106, 900
Clamp and wrench, Combined	E. Maguire	Princeton, Ill	Nov. 26, 1872	133, 325
Clamp-mill for turning metals	W. H. Brainard	Branford, Conn	July 16, 1872	129, 092

Index of patents issued from the United States Patent Office from 1790 *to* 1873, *inclusive*—Continued.

Invention.	Inventor.	Residence.	Date.	No.
Clamp-milling machine	A. B. Lawther	Stonington, Conn	Apr. 5, 1864	42, 201
Clamping-device	E. King	Fredonia, N. Y	Sept. 17, 1867	68, 886
Clamping-device	E. Simmons	South Providence, R. I	Sept. 26, 1865	50, 175
Clamping-device	C. Stout	Waverly Heights, Pa	June 17, 1873	139, 931
Clamping-machine	J. H. Humes	East Saginaw, Mich	June 14, 1870	104, 315
Clamping-machine	E. Wheeler	Marlborough, Mass	Jan. 27, 1857	16, 506
Clamping-machine, Carpenter's	W. R. Axe	Beloit, Wis	Jan. 22, 1861	31, 145
Clamping-pontil	F. H. James and N. B. Gatchell	Lancaster, N. Y	Nov. 21, 1865	51, 058
Clapboard	H. Osgood	Hartland, Me	Nov. 7, 1865	50, 859
Clapboard	A. A. Wilder	Detroit, Mich	Jan. 10, 1860	26, 803
Clapboard and lath cutting machine	E. Day	Grand Detour, Ill	June 18, 1842	2, 674
Clapboard and shingle sawing machine	J. Jaquith	Sherburn, Mass	June 26, 1817	....
Clapboard-gage	W. E Babcock	East Pembroke, N. Y	June 27, 1871	116, 256
Clapboard-gage	J. C. Biddlecom	Macedon, N. Y	May 3, 1870	102, 649
Clapboard-gage	A. Carson	Memphis, Tenn	July 2, 1867	66, 216
Clapboard-gage	A. Deys	Rockford, Ill	Oct. 4, 1870	108, 010
Clapboard-gage	G. Hall	Middletown, Ohio	June 4, 1867	65, 378
Clapboard-gage	E. Horton	Dundee, N. Y	July 12, 1870	105, 210
Clapboard-gage	S Inman	Rockford, Ill	Sept. 3, 1861	33, 199
Clapboard-gage	G. Smith	Omaha, Nebr	Nov. 30, 1869	97, 452
Clapboard-gage	H. Van Deusen	Phelps, N. Y	Dec. 6, 1859	26, 386
Clapboard-joint	W. Baker	Utica, N. Y	May 16, 1854	10, 903
Clapboard-machine	B. Corser	Mount Morris, N. Y	Jan. 16, 1849	6, 033
Clapboard-machine	C. S Davis	Orono, Me	Dec. 7, 1869	97, 484
Clapboard-machine	O. R. and L. C. Kendall	Groton, N. H	Mar. 30, 1869	88, 306
Clapboard-machine	D. Newton	Dalton, N. H	Jan. 31, 1833	....
Clapboard-machine	A. A. Wilder	Detroit, Mich	Oct. 30, 1855	13, 734
Clapboard or siding gage	J. W. Arnold	Fairport, N. Y	July 9, 1872	128, 696
Clapboard sawing and planing machine	E. Parker	Rock Island, Ill	Dec. 20, 1853	10, 341
Clapboard-sawing machine	D. Bartholomew	Wolcott, Vt	July 8, 1843	3, 165
Clapboard-sawing machine	S. Goss	Milford, N. H	June 3, 1837	226
Clapboard-sawing machine	A. A. Holmes	Epsom, N. H	Sept. 2, 1873	142, 339
Clapboard-sawing machine	D. F. Mellen	Wentworth, N. H	Aug. 22, 1854	11, 568
Clapboard-sawing machine	O. and D. Parker	Hubbardstown, Mass	Dec. 24, 1834	....
Clapboard-sawing machine	C. Tyler	Milford, N. H	Aug. 30, 1838	900
Clapboard-sawing machinery	E. Pray, E. Benjamin, and J. S. Stone.	Maine	Dec. 28, 1824	....
Clapboard, tipping, &c., Sawing	E. Carleton and S. Whiting, jr	Bath, N. H	Mar. 25, 1830	....
Clapboards from round logs, Making	H. Whittemore	Worcester, Mass	Aug. 26, 1831	....
Clapboards, Gage for holding	W. H. Cummings and I. Babcock.	Boonsborough, Iowa	May 28, 1867	65, 178
Clapboards, Machine for beveling the ends of	S. W. Curtis	Stoughton, Mass	May 26, 1843	3, 104
Clapboards, Machine for planing	J. Atkins	Augusta, Me	Feb. 25, 1873	136, 201
Clapboards, Machine for sapping logs for	E Webber	Gardiner, Me	June 27, 1871	116, 377
Clapboarding	F. Buscher	Dunkirk, N. Y	Jan. 9, 1872	122, 562
Clapboarding-clamp	W. P. Wentworth	Detroit, Mich	Mar. 12, 1867	62, 792
Clapboarding-gage	H. D. Vandercook	Marshall, Mich	July 17, 1860	29, 206
Clarifying and condensing the juice of fruit	C. Cory	Lima, Ind	May 24, 1864	42, 841
Clarinet	A. Fritsche	New York, N. Y	Jan. 10, 1871	110, 845
Clarinet	J. Rebhun	New York, N. Y	July 27, 1869	93, 005
Clasp:				
See Bag-clasp.				
Band-clasp.				
Bank-bill clasp.				
Bed-clothes clasp.				
Belt-clasp.				
Broom-clasp.				
Buckle-clasp.				
Cattle-leading clasp.				
Colter-clasp.				
Compression-clasp.				
Corset-clasp.				
Corset-busk clasp.				
Curtain-clasp.				
Dress-looping clasp.				
Elastic-tube clasp.				
Glove-clasp.				
Halter-clasp.				
Harness-clasp.				
Harness-maker's clasp.				
Harness-tug clasp.				
Hoop-skirt clasp.				
Lamp-shade clasp.				
Leather-strap clasp.				
Lever-clasp.				
Lightning-rod clasp.				
Limb-clasp.				
Mail-bag clasp.				
Metallic clasp.				
Necktie clasp.				
Paint-brush clasp.				
Pocket-book clasp.				
Regulating-clasp.				
Rope clasp.				
Safety-clasp.				
Scarf-clasp.				
Seat-clasp.				
Shoe-clasp.				
Side-arm clasp.				
Skirt-hoop clasp.				
Spring-clasp.				
Stay-log clasp.				
Stocking-supporter clasp.				
Suspender-clasp.				
Traveling-bag clasp.				
Trunk-clasp.				
Tug-clasp.				
Wearing-apparel clasp.				
Whip-socket clasp.				

Index of patents issued from the United States Patent Office from 1790 *to* 1873, *inclusive*—Continued.

Invention.	Inventor.	Residence.	Date.	No.
Clasp	J. H. Eastman	Boston, Mass	Dec. 2, 1873	145, 097
Clasp	C. T. P. Ware	New York, N. Y	Jan. 17, 1854	10, 436
Clasp-hook	D. Hayes	Cambridge, Mass	June 30, 1868	79, 347
Clasp-ring	L. A. Sanford	Wolcott, Conn	Sept. 29, 1868	82, 643
Clasps, Machine for making	J. Cutler	Chicopee, Mass	Dec. 13, 1859	26, 419
Clavicle-adjuster	A. M. Day	Bennington, Vt	July 5, 1853	9, 828
Claw-bar	G. H. Beard	Cincinnati, Ohio	June 12, 1860	28, 645
Claw-bar	G. Brownell	Mitchell, Ind	Nov. 14, 1865	50, 899
Claw-bar	D. Christie	Chillicothe, Ohio	July 18, 1871	117, 046
Claw-bar	M. Cornelius	Cincinnati, Ohio	Oct. 4, 1864	44, 576
Claw-bar	M. Hennasy	Cranford, N. J	July 21, 1868	80, 072
Claw-bar	H. Jeffrey	Vincennes, Ind	Oct. 10, 1865	50, 361
Claw-bar	I. Lamplugh	Peoria, Ill	Dec. 16, 1862	37, 172
Claw-bar	W. S. Terhune	Ridgewood, N. J	Apr. 1, 1873	137, 333
Claw-bar	C. Winter	Chillicothe, Ohio	Mar. 7, 1871	112, 401
Clay and cement pipe mold	P. McIntyre	Hartford, Conn	Oct. 15, 1872	132, 168
Clay and cement pipe mold	J. Sharpe	Paterson, N. J	July 8, 1873	140, 734
Clay and peat press	B. Van Vrauken	Schenectady, N. Y	June 19, 1866	55, 748
Clay, Apparatus for treating	I. Gregg	Philadelphia, Pa	July 9, 1867	66, 488
Clay articles, Mold for forming	C. A. Fischer	Baltimore, Md	July 15, 1873	140, 910
Clay-cleaning apparatus	L. P. Norton	Bennington, Vt	Jan. 28, 1868	73, 921
Clay-crusher	A. Hall	Perth Amboy, N. J	Jan. 11, 1870	98, 764
Clay for potters' use, Machine for condensing pap or slops of.	J. Muir	New York, N. Y	Jan. 3, 1865	45, 736
Clay for potters' use, Process of preparing	J. Muir	New York, N. Y	Jan. 3, 1865	45, 737
Clay grinding and separating mill, Combined	F. B. Norton and F. Hancock	Worcester, Mass	Apr. 7, 1868	76, 505
Clay, &c., grinding machine	A. Alexander	Pittsburgh, Pa	Apr. 2, 1872	125, 159
Clay-grinding machine	B. Porter	Jackson, Mich	Apr. 10, 1866	53, 868
Clay-grinding machine	W. H. Thomas	Chicago, Ill	July 9, 1867	66, 536
Clay-heating apparatus	I. Gregg	Philadelphia, Pa	July 9, 1867	66, 487
Clay-kneading machine	H. H. Thayer	Sandwich, Mass	June 5, 1855	13, 017
Clay, Machine for breaking	N. Boynton	Danville, Vt	Feb. 27, 1833	......
Clay, Machine for pulverizing	J. O'Neil	Kingston, N. Y	July 10, 1855	13, 224
Clay, Machine for tempering and preparing	N. Adams	Cornwall, N. Y	Dec. 16, 1833	......
Clay, Machine for working	H. Leguay	Saint Louis, Mo	Sept. 14, 1858	21, 506
Clay-machinery	G. D. Goodrich	Chicago, Ill	Jan. 26, 1869	86, 148
Clay-mill	L. Moore	Baraboo, Wis	Aug. 4, 1868	80, 652
Clay-mill	N. F. Potter	Providence, R. I	July 14, 1868	80, 008
Clay-mill	J. A. Vaughn, H. B. Camp, H. E. Merrill and C. J. Merrill.	Cuyahoga Falls and Akron, Ohio, and Alton, Ill.	Dec. 9, 1873	145, 373
Clay-mill	C. Webster, W. Camp, and S. L. Stall.	Akron, Ohio	Mar. 22, 1864	42, 055
Clay-mixer	A. J. Knisely	Chicago, Ill	Oct. 29, 1872	132, 670
Clay-mixing machine	A. Parkhurst	Scriba, N. Y	Apr. 11, 1831	......
Clay, Mode of dressing damp	J. Steele	Buffalo, N. Y	May 9, 1865	47, 684
Clay mold for casting metal	J. J. C. Smith	Philadelphia, Pa	Dec. 29, 1868	85, 340
Clay-molding machine	S. Hambleton and G. P. Herthel, jr.	Saint Louis, Mo	Feb. 16, 1869	87, 043
Clay-molding machine	R. Hill	Saint Louis, Mo	Mar. 23, 1869	88, 166
Clay-molding machine	T. Hoadley	Cleveland, Ohio	Sept. 7, 1858	21, 419
Clay pipe	T. Wickersham	Newbury, Pa	May 13, 1828	......
Clay-pipe die	G. D. and H. A. Goodrich	Joliet, Ill	Oct. 16, 1866	58, 851
Clay pipe for conduits	S. Bartlett	Hartford, Conn	Oct. 1, 1805	......
Clay-pipe machine	G. D. Goodrich	Joliet, Ill	Feb. 5, 1867	61, 732
Clay-pipe machine	C. Stotz and G. Smith	Perth Amboy, N. J	Dec. 29, 1868	85, 490
Clay-pipe machine	S. Ustick	Philadelphia, Pa	Mar. 20, 1860	27, 602
Clay-pipe manufacture	G. D. and H. A. Goodrich	Joliet, Ill	July 2, 1867	66, 319
Clay-pipe manufacture	H. A. Goodrich	Joliet, Ill	July 2, 1867	66, 320
Clay-pipe manufacture	J. Putnam	Salem, Mass	Sept. 30, 1851	8, 403
Clay-pipe-pressing machine	T. Shaw	Philadelphia, Pa	Dec. 28, 1869	98, 430
Clay pipes, Apparatus for turning heavy	W. Wassall	Wellsville, Ohio	Aug. 31, 1869	94, 364
Clay pipes, Machine for forming	C. P. Wardwell	Lake Village, N. H	Apr. 21, 1857	17, 125
Clay pipes, Machine for making	H. Augood and S. Ustick	Mansfield Township, N. J., and Philadelphia, Pa.	May 24, 1859	24, 174
Clay pipes, Machine for making	J. Jones	Baltimore, Md	Jan. 25, 1859	22, 730
Clay pipes, Machine for making	W. Linton	Baltimore, Md	Aug. 23, 1859	25, 233
Clay pipes, Making	J. Bower	East Bethlehem, Pa	Dec. 1, 1826	......
Clay pipes, Manufacture of	G. D. Goodrich	Joliet, Ill	Jan. 7, 1868	72, 997
Clay pipes, Manufacture of	H. A. Goodrich and J. Amos	Joliet, Ill	Feb. 4, 1868	74, 078
Clay-pulverizer and stone-separator	F. H. Smith	Baltimore, Md	Mar. 9, 1869	87, 716
Clay pulverizing and cleaning machine	G. F. Blake	Medford, Mass	Nov. 26, 1861	33, 810
Clay-pulverizing machine	G. C. Bovey	Chillicothe, Ohio	May 20, 1873	139, 110
Clay retorts, Decarbonizing	G. W. Edge	Jersey City, N. J	Sept. 19, 1865	49, 989
Clay, Tempering	S. Miller and G. Roller	Manchester Post-Office, Md	July 31, 1846	4, 676
Clay, Tempering	J. W. Ward	Cambridge, Mass	June 6, 1848	5, 615
Clay-tempering apparatus	L. E. Ransom	Trenton, Mich	May 4, 1869	89, 685
Clay-tempering machine	S. Barnes	Rochester, Pa	June 25, 1872	128, 278
Clay-tempering machine	J. D. Custer	Norristown, Pa	June 7, 1859	24, 350
Clay-tempering machine	G. E. Noyes	Washington, D. C	Sept. 28, 1869	95, 374
Clay-tempering mill	G. Carnell, S. Williams, and W. Ellis.	Philadelphia, Pa	Oct. 27, 1868	83, 361
Clay-tempering wheel	F. L. Carnell	Philadelphia, Pa	May 16, 1871	114, 919
Clay tubes, Machine for making	J. H. Rowell and H. Wise	Fredericktown, Pa	May 10, 1827	......
Clay-washing and stone-separating machine	E. Wilzinski	Chicago, Ill	Feb. 25, 1868	74, 967
Clay-working machine	D. and G. Duchemin	Cincinnati, Ohio	Dec. 2, 1851	8, 565
Cleaner:				
See Boiler-flue cleaner.				
Boot and shoe cleaner.				
Cane-cleaner.				
Card-cylinder cleaner.				
Carpet-cleaner.				
Cartridge-case cleaner.				
Chimney-cleaner.				
Clothes-cleaner.				
Coffee-cleaner.				
Coffee and grain cleaner.				
Comb-cleaner.				

Index of patents issued from the United States Patent Office from 1790 *to* 1873, *inclusive*—Continued.

Invention.	Inventor.	Residence.	Date.	No.
Cleaner—Continued.				
See Cotton-cleaner.				
Cotton-seed cleaner.				
Dish-cleaner.				
Flax and hemp cleaner.				
Floor-cleaner.				
Flue-cleaner.				
Fork-cleaner.				
Garden-walk cleaner.				
Grain-cleaner.				
Grain and malt cleaner.				
Gun-cleaner.				
Harvester-track cleaner				
Hop-pole cleaner.				
Knife-cleaner.				
Knife and fork cleaner.				
Lamp-chimney cleaner.				
Ore-cleaner.				
Plow-cleaner.				
Rag-cleaner.				
Railway-track cleaner.				
Rice-cleaner.				
Sadiron-cleaner.				
Sewer-cleaner.				
Shoe-cleaner.				
Silk-cleaner.				
Slate cleaner.				
Stable-cleaner.				
Steam-boiler-flue cleaner.				
Steam-jet boiler-tube cleaner.				
Stove-pipe cleaner.				
Street-cleaner.				
Track-cleaner.				
Tube-cleaner.				
Watch-cleaner.				
Well-cleaner.				
Wheat-cleaner.				
Window-cleaner.				
Cleaning and boring device	J. B. Jordan	Aurora, Wis	June 30, 1868	79, 356
Cleaning and polishing attachment to sheet-metal rolls.	J. B. Hastings	Ironton, Ohio	July 6, 1869	92, 309
Cleansing-liquid	E. J. Balcear	Martinez, Cal	May 18, 1869	90, 064
Cleat and capstan, Combined	D. Snedeker	Lockport, N. Y	June 29, 1869	92, 116
Cleat and chock, Iron	F. B. Stevens and W. Brown	Weehawken and Hoboken, N. J.	Apr. 27, 1869	89, 514
Cleat chock	A. Lake	Smith's Landing, N. J	Nov. 5, 1867	70, 580
Cleat for sail-boats, Automatic	L. Hill	Alexandria, Va	Dec. 3, 1867	71, 615
Cleat or ring-bolt, Shifting	J. E. Murray	Provincetown, Mass	June 8, 1869	90, 952
Cleaver	C. Hammond	Philadelphia, Pa	Aug. 10, 1869	93, 614
Clevis	W. W. Atteberry	Chesterfield, Ill	Aug. 10, 1869	93, 396
Clevis	D. M. Castle	Constantine, Mich	Dec. 20, 1870	110, 200
Clevis	E. Evans	Montgomery, Ala	June 1, 1869	90, 736
Clevis	I. Evans	Lebanon, Ohio	May 16, 1848	5, 581
Clevis	G. W. Holton	Berlin, Ky	Apr. 27, 1869	89, 407
Clevis	E. R. Kagarice	New Enterprise, Pa	Nov. 5, 1872	132, 724
Clevis	E. A. Palmer	Clayville, N. Y	Sept. 16, 1856	15, 743
Clevis	S. W. Pope	Louisville, Ky	Aug. 15, 1871	118, 050
Clevis	E. M. Potter	Kalamazoo, Mich	Jan. 7, 1868	73, 197
Clevis	J. H. Shaw	Inlet, Ill	May 23, 1871	115, 113
Clevis	F. Sigrist	Napa County, Cal	Nov. 27, 1860	30, 766
Clevis	Z. B. Sims	Bonham, Tex	Aug. 31, 1869	94, 348
Clevis	R. C. Whitehouse	Booth Bay, Me	Apr. 28, 1868	77, 426
Clevis and plow-shares, Apparatus for rolling	F. Murray	Pittsburgh, Pa	Apr. 23, 1867	64, 131
Clevis and stirrup bending machine	W. C. Kaiser	Louisville, Ky	Aug. 9, 1870	106, 172
Clevis bar and hook	R. Gibbs	Spring Hill, Mo	May 27, 1873	139, 381
Clevis-blank die	J. Kritch	Cleveland, Ohio	Aug. 25, 1868	81, 380
Clevis-blanks, Apparatus for bending	T. Meikle	Louisville, Ky	July 27, 1869	93, 106
Clevis-blank, Machine for rolling	M. Loughran	Pittsburgh, Pa	Dec. 24, 1867	72, 512
Clevis, Clamp for making	V. M. Chafee	Xenia, Ill	Jan. 31, 1860	26, 968
Clevis, Elastic link for	L. Fliedner	Cleveland, Ohio	Apr. 6, 1869	88, 619
Clevis-iron	T. P. Warren	Norfolk, Va	June 9, 1868	78, 775
Clevis-making machine	J. S. Hall	Pittsburgh, Pa	Oct. 4, 1870	107, 901
Clevis, Substitute for	J. and W. D. Howell and J. Sipe.	Clark County, Ohio	Jan. 15, 1850	7, 021
Clevis, Three-horse	J. Fowler	Allegan, Mich	June 30, 1868	79, 463
Clevis, Three-horse	S. H. Frederick	Matteson, Mich	July 27, 1869	92, 953
Clew-blocks to clews, Means of attaching	E. A. Sawyer	Portland, Me	Jan. 5, 1864	41, 096
Clew-thimble	T. Carroll	Middletown, Conn	May 17, 1864	42, 817
Clew-thimble	W. W. Wilcox	Middletown, Conn	Apr. 14, 1863	38, 192
Climbing poles, Machine for	H. D. Chapman	Baltimore, Md	Mar. 11, 1851	7, 966
Climozonator	W. Elmer	New York, N. Y	Feb. 23, 1869	87, 155
Clinch-ring die	S. Vanstone	Providence, R. I	Mar. 15, 1870	100, 823
Clinch-rings, Machine for making	J. A. Coleman	Providence, R. I	June 13, 1865	48, 157
Clinching and nipping tool	E. Warren	Marshall, Mich	Nov. 3, 1863	40, 522
Clinching and nipping tool	D. A. Wilson	Cambridge, Vt	Oct. 17, 1865	50, 539
Clinching-iron	D. H. Williams	Antwerp, N. Y	May 29, 1866	55, 189
Clinometer and level	J. L. L. Knox	Pittsburgh, Pa	Dec. 31, 1867	72, 740
Clip:				
See Album-clip.				
Axle-clip.				
Carriage-clip.				
Carriage and saddle clip.				
Carriage-saddle clip.				
Carriage-spring clip.				
Crinoline-clip.				
Curtain-tassel clip.				
Harness-clip.				
Letter-clip.				

Index of patents issued from the United States Patent Office from 1790 *to* 1873, *inclusive*—Continued.

Invention.	Inventor.	Residence.	Date.	No.
Clip—Continued.				
See Paper-clip.				
Ribbon-roll clip.				
Shaft-clip.				
Spring-clip.				
Suspension-clip.				
Tassel-clip.				
Ticket-clip.				
Trace-clip.				
Tug-clip.				
Wagon-clip.				
Wagon-axle-skein clip.				
Whiffletree-clip.				
Clip-circle	I. N. Topliff	Adrian, Mich	Oct. 15, 1867	69, 866
Clipping apparatus, Animal	A. D. Kenshaw	London, England	Apr. 21, 1868	77, 093
Clipping-machine	J. W. Moyer	Cooperstown, N. Y	Apr. 2, 1872	125, 209
Clipping machine, Horse	P. Adie	Strand, England	June 30, 1868	79, 293
Clipping-shears	G. F. Evans	Norway, Me	Sept. 19, 1871	119, 019
Clipping-shears	C. F. Harlow	Boston, Mass	Aug. 8, 1871	117, 774
Clipping-shears	G. H. Pratt	Boston, Mass	June 20, 1871	116, 216
Clipping-shears	J. K. Priest and R. T. Smith	Nashua, N. H	Jan. 16, 1872	122, 852
Clipping-shears	R. T. Smith and J. K. Priest	Nashua, N. H	Apr. 23, 1872	125, 911
Clipping-shears	J. C. Wilson, A. Walker, and J. Foster.	New York, N. Y	Dec. 15, 1868	84, 926
Clipping-shears	J. C. Wilson, A. Walker, and J. Foster.	New York, N. Y	Sept. 14, 1869	94, 803
Clipping-shears	R. Wyatt	New York, N. Y	Aug. 22, 1871	118, 417
Clipping shears, Horse	S. H. Folsom	East Cambridge, Mass	May 19, 1863	38, 576
Cloak and coat suspender	J. D. Leach and E. S. Wardwell	Penobscot and Bucksport, Me.	July 7, 1868	79, 580
Cloak and tent, Convertible	W. B. Johns	Georgetown, D. C	Oct. 22, 1861	33, 528
Cloak, bed, tent, &c., Combined	T. C. Brecht and S. B. Sigismond.	United States Navy and Washington, D. C.	Oct. 14, 1862	36, 685
Cloak, Military	F. W. Weiss	Mount Vernon, N. Y	Dec. 10, 1861	33, 910
Cloak stand, Lady's	J. R. Palmenberg	New York, N. Y	July 25, 1865	48, 989
Cloaks, Cutting	A. S. Thompson	Albion, Pa	Dec. 11, 1855	13, 926
Clock	R. T. Andrews	Plymouth Hollow, Conn	May 19, 1863	38, 543
Clock	W. B. Barnes	Forestville, Conn	Nov. 6, 1860	30, 558
Clock	S. Blydenburgh and W. Beebe.	New York, N. Y	Apr. 26, 1833	....
Clock	C. Boardman and J. A. Wells	Hartford, Conn	Jan. 1, 1847	4, 914
Clock	J. Bogardus	New York, N. Y	Mar. 2, 1830	....
Clock	L. F. and W. W. Carter	Bristol, Conn	Feb. 19, 1867	62, 112
Clock	A. D. Crane	Caldwell, N. J	Mar. 18, 1829	....
Clock	A. D. Crane	Newark, N. J	Feb. 10, 1841	1, 973
Clock	R. P. Cunningham	Eastford, Conn	Dec. 22, 1857	18, 890
Clock	J. D. Custer	Norristown, Pa	Nov. 24, 1830	....
Clock	T. A. Davies	New York, N. Y	Jan. 15, 1846	4, 334
Clock	T. A. Davies	New York, N. Y	Aug. 12, 1846	4, 687
Clock	H. G. Dyar	New York, N. Y	Nov. 6, 1827	....
Clock	A. Gould	Henrietta, N. Y	Oct. 1, 1830	....
Clock	J. Henry	Maysville, Ky	Sept. 13, 1820	....
Clock	J. Ives	Bristol, Conn	July 1, 1836	....
Clock	C. Kirk	Bristol, Conn	Aug. 26, 1843	3, 233
Clock	C. Kirk	Bristol, Conn	Apr. 3, 1847	5, 045
Clock	W. Lindon	New Haven, Conn	Nov. 26, 1867	71, 519
Clock	J. B. Mayer	Niagara Falls, N. Y	Sept. 15, 1868	82, 266
Clock	J. M. Schrock and J. G. Fisher.	Millersburgh, Ohio, and Quincy, Ill.	Oct. 3, 1846	4, 791
Clock	C. Schwippl	New York, N. Y	June 21, 1864	43, 230
Clock	J. S. Seger	New York, N. Y	May 22, 1833	....
Clock	C. and M. Stevens	Boston, Mass	Mar. 4, 1862	34, 599
Clock	E. Terry	Plymouth, Conn	Mar. 4, 1826	....
Clock	S. B. Terry	Waterbury, Conn	Mar. 31, 1868	76, 117
Clock	S. B. Terry	Waterbury, Conn	Dec. 1, 1868	84, 517
Clock, &c	T. Thompson	Washington, D. C	Jan. 25, 1870	99, 261
Clock	H. Twiss	Meriden, Conn	May 13, 1834	....
Clock	S. Willard	Boston, Mass	Dec. 8, 1819	....
Clock, Alarm	R. Woolworth	New Haven, Conn	Apr. 7, 1868	76, 370
Clock, Alarm	J. E. Buerk	Boston, Mass	Sept. 18, 1866	58, 058
Clock-alarm	D. M. Charters	Xenia, Ohio	Oct. 21, 1873	143, 881
Clock, Alarm	J. Decker	Sparta, N. J	Mar. 5, 1867	62, 531
Clock, Alarm	J. F. Mascher	Philadelphia, Pa	Feb. 8, 1859	22, 883
Clock-alarm	W. A. Terry	Bristol, Conn	Oct. 29, 1867	70, 376
Clock, Alarm	J. S. Turner	New Haven, Conn	July 13, 1852	9, 123
Clock-alarm attachment	C. H. Warner	New Haven, Conn	Apr. 7, 1868	76, 567
Clock-alarm, Self-acting	J. H. Davis	Chillicothe, Mo	Dec. 13, 1870	110, 016
Clock and advertiser combined	O. N. Angell	Johnston, R. I	June 20, 1834	....
Clock and burgular alarm, Combined	G. M. Levette	Indianapolis, Ind	Feb. 18, 1868	74, 704
Clock and fly-trap, Combined	G. K. Proctor	Beverly, Mass	Aug. 7, 1860	29, 520
Clock and watch escapement	C. Kallmann	Newbury, N. Y	Apr. 6, 1869	88, 718
Clock and watch key	D. J. Mozart, L. Beach, and L. Hubbell.	New York, N. Y., and Farmington and Bristol, Conn.	Dec. 8, 1863	40, 851
Clock and watch movement	P. J. Hoffliger	Philadelphia, Pa	Sept. 30, 1873	143, 285
Clock, Astronomical	D. J. Mozart	New York, N. Y	Aug. 5, 1862	36, 103
Clock-bell	H. Miller	East Hanover, Pa	May 5, 1825	....
Clock, Burglar-alarm	L. C. Butch	Lancaster, Ohio	May 31, 1870	103, 559
Clock, Burglar-alarm	J. Mathewman	New Haven, Conn	July 6, 1858	20, 810
Clock, Calendar	G. D. Sargent	Boston, Mass	July 6, 1858	20, 852
Clock, Calendar	W. A. Atkins and J. C. Burritt	Ithaca, N. Y	Sept. 19, 1854	11, 711
Clock, Calendar	W. H. Akins and J. C. Burritt	Berkshire and Ithaca, N. Y	Nov. 17, 1857	18, 665
Clock, Calendar	E. Allen	Glastenbury, Conn	Apr. 15, 1856	14, 645
Clock-calendar	E. Allen	Glastenbury, Conn	Sept. 2, 1865	15, 637
Clock, Calendar	A. Boardman	Forestville, Conn	July 2, 1867	66, 289
Clock, Calendar	W. W. Carter	Bristol, Conn	Sept. 15, 1863	39, 883
Clock, Calendar	W. K. Chase	Charlestown, Mass	Oct. 19, 1865	51, 556
Clock, Calendar	C. M. Clinton and L. Mood	Ithaca, N. Y	June 25, 1867	66, 003
Clock, Calendar	C. M. Clinton and L. Mood	Ithaca, N. Y	July 30, 1867	67, 166
	C. M. Clinton and L. Mood	Ithaca, N. Y	July 9, 1872	128, 854

Index of patents issued from the United States Patent Office from 1790 to 1873, inclusive—Continued.

Invention.	Inventor.	Residence.	Date.	No.
Clock, Calendar	C. M. Clinton and L. Mood	Ithaca, N. Y	Nov. 11, 1873	144, 384
Clock-calendar	A. Frankfeid	New York, N. Y	Oct. 14, 1873	143, 618
Clock, Calendar	D. J. Gale	Sheboygan Falls, Wis	Nov. 16, 1869	96, 792
Clock, Calendar	J. H. H. Hawes	Ithaca, N. Y	May 17, 1853	9, 727
Clock, Calendar	H. B. Horton	Ithaca, N. Y	Apr. 18, 1865	47, 306
Clock, Calendar	H. B. Horton	Ithaca, N. Y	Aug. 28, 1866	57, 510
Clock, Calendar	H. B. Horton and M. L. Wood	Ithaca, N. Y	June 11, 1867	65, 748
Clock, Calendar	S. P. La Due	Rockford, Iowa	Sept. 13, 1859	25, 468
Clock, Calendar	B. B. Lewis	Bristol, Conn	Feb. 4, 1862	34, 341
Clock, Calendar	B. B. Lewis	Hartford, Conn	June 21, 1864	41, 214
Clock, Calendar	B. B. Lewis	Bristol, Conn	Dec. 29, 1868	85, 456
Clock, Calendar	G. Maranville	Hampton Corners, N. Y	Mar. 5, 1861	31, 612
Clock, Calendar	E. M. and J. E. Mix	Ithaca, N. Y	Jan. 31, 1860	27, 023
Clock, Calendar	E. M. and J. E. Mix	Ithaca, N. Y	Mar. 4, 1862	34, 613
Clock, Calendar	E. P. Monroe	Albany, N. Y	Feb. 10, 1857	16, 628
Clock, Calendar	D. J. Mozart	New York, N. Y	Feb. 28, 1865	46, 577
Clock, Calendar	D. J. Mozart, L. Beach, and L. Hubbell.	New York, N. Y., and Farmington and Bristol, Conn.	Jan. 5, 1864	41, 122
Clock, Calendar	G. B. Owen	New York, N. Y	Apr. 24, 1866	54, 198
Clock, Calendar	E. Prichard	Waterbury, Conn	June 5, 1860	28, 603
Clock-calendar	J. K. Seem	Canton, Pa	Jan. 7, 1868	73, 127
Clock-calendar	J. K. Seem	Macomb, Ill	Dec. 24, 1872	134, 319
Clock, Calendar	H. Skinner	Huron, Ohio	Mar. 2, 1858	19, 519
Clock, Calendar	T. T. Strode	Mortonville, Pa	Sept. 25, 1860	30, 166
Clock, Calander	T. T. Strode	Mortonville, Pa	Aug. 1, 1865	49, 169
Clock, Calendar	W. A. Terry	Bristol, Conn	June 16, 1868	79, 026
Clock-calendar	W. A. Terry	Bristol, Conn	Jan. 25, 1870	99, 258
Clock, Calendar	M. J. Whitmore	Potsdam, N. Y	Jan. 13, 1857	16, 418
Clock, Calendar	J. Williams	Hartford, Conn	Sept. 19, 1854	11, 713
Clock, Calendar	J. Williams	Hartford, Conn	July 24, 1855	13, 341
Clock-calendar	N. T. Worthley	Brunswick, Me	Sept. 16, 1873	142, 975
Clock-calendar movement	G. R. Williams	Ithaca, N. Y	Sept. 16, 1873	142, 829
Clock-case	H. B. Horton	Ithaca, N. Y	Aug. 28, 1866	57, 511
Clock.case	D. Monnin	Paris, France	Dec. 3, 1867	71, 631
Clock.case	G. B. Owen	New York, N. Y	Aug. 5, 1862	36, 105
Clock.case	C. W. Roberts	Chicago, Ill	Dec. 24, 1872	134, 168
Clock.case, Composition	S. B. H. Vance and E. M. Smith	New York, N. Y	June 30, 1868	79, 521
Clock.case, Glass	C. A. Moore	Westbrook, Conn	Oct. 17, 1871	119, 990
Clock case making	J. Ives	Bristol, Vt	Mar. 21, 1822	....
Clock cases, Adjustable foot for	O. R. Luther	Waterbury, Conn	Apr. 26, 1870	102, 413
Clock collet die	H. C. Thompson	Bristol, Conn	Apr. 21, 1868	76, 958
Clock-, Cuckoo	C. K. Giles	Chicago, Ill	Oct. 8, 1867	69, 561
Clock-dial	S. E. Root	Bristol, Conn	May 10, 1859	23, 950
Clock-dial, Composition	S. Barnes	New Haven, Conn	Oct. 6, 1868	82, 788
Clock-dial sash	G. Hills	Plainville, Conn	Dec. 20, 1864	45, 497
Clock-dials, Enamel for	T.G. Leibnaw and A. G. Heaney	Plainville, Conn	Apr. 12, 1870	101, 891
Clock-dials, Setting	G. Hills	Plainville, Conn	July 5, 1870	105, 077
Clock, Electric	L. Bradley	Jersey City, N. J	Sept. 28, 1869	95, 316
Clock, Electric	W. M. Davis	Cincinnati, Ohio	Oct. 24, 1871	120, 185
Clock, Electric	A. Hall	Loydsville, Ohio	Sept. 26, 1854	11, 723
Clock, Electric	V. Himmer	New York, N. Y	Jan. 4, 1870	98, 593
Clock, Electric	V. Himmer	New York, N. Y	Jan. 4, 1870	98, 594
Clock, Electric	V. Himmer	New York, N. Y	June 4, 1872	127, 483
Clock, Electric	M. Hipp	Neufchatel, Switzerland	June 1, 1869	90, 841
Clock, Electric	S. A. Kennedy	Attleborough, Pa	Feb. 1, 1870	99, 321
Clock, Electric	S. A. Kennedy, S. W. Holt, and J. Gerlach.	Attleborough and Philadelphia, Pa.	Dec. 3, 1867	71, 624
Clock, Electric	F. J. Ritchie	Edinburgh, Scotland	Oct. 21, 1873	143, 847
Clock, Electric	E. Wilson	New York, N. Y	Feb. 1, 1870	99, 386
Clock, Electric	E. Wilson	Elizabeth, N. J	Feb. 27, 1872	124, 104
Clock, Electric-circuit-breaking	E. Holmes	New York, N. Y	Mar. 26, 1867	63, 158
Clock, Electro-magnetic watch	J. M. Batchelder	Cambridge, Mass	Sept. 3, 1872	130, 971
Clock-escapement	C. J. Addy	Roxbury, Mass	Nov. 22, 1859	26, 150
Clock-escapement	B. Bacon	Morrison, Ill	May 1, 1866	54, 277
Clock-escapement	C. Fasoldt	Albany, N. Y	Apr. 8, 1873	137, 603
Clock-escapement	E Groux	Rome, N. Y	Aug. 21, 1866	57, 315
Clock-escapement	W. Hart	Mayville, Wis	Dec. 24, 1861	33, 990
Clock-escapement	W. Hart	Mayville, Wis	Aug. 4, 1863	39, 395
Clock-escapement	H. T. Hewitt	Scotch Plains, N. J	Nov. 21, 1865	51, 044
Clock-escapement	W. C. Kellum	San Francisco, Cal	Nov. 3, 1868	83, 775
Clock-escapement	W. C. Kellum	San Francisco, Cal	Nov. 3, 1868	83, 776
Clock-escapement	B. B. Lewis	Bristol, Conn	Aug. 30, 1870	106, 843
Clock-escapement	J. C. Pitel	West Meriden, Conn	Apr. 23, 1861	32, 144
Clock-escapement	A. Platt	New York, N. Y	Nov. 4, 1873	144, 287
Clock-escapement	C. Reinhart	New Haven, Conn	June 16, 1868	79, 003
Clock-escapement	E. K. Reynolds	New York, N. Y	Oct. 2, 1855	13, 623
Clock-escapement	M. Tromly	Cincinnati, Ohio	Aug. 24, 1869	94, 148
Clock-escapement	W. J. Weaver and J. M. Sandifer.	Somerset, Ky	Aug. 8, 1865	49, 323
Clock for steam-boilers, Detective	H. Pieper	New York, N. Y	June 4, 1872	127, 641
Clock-fronts, Hinging	A. Allen	New Haven, Conn	Nov. 26, 1867	71, 438
Clock, Galvanic	M. G. Farmer	Salem, Mass	Sept. 21, 1852	9, 279
Clock, Globe	S. E. G. Rawson	Saratoga Springs, N. Y	Jan. 22, 1867	61, 459
Clock, Globe	T. R. Timby	Saratoga Springs, N. Y	May 2, 1865	47, 585
Clock, Hydraulic	O. Abbruzzo	St. Margherita, Italy	Nov. 26, 1867	71, 437
Clock, Illuminated	J. Glenn	New York, N. Y	Dec. 6, 1853	10, 302
Clock, Illuminated	E. Wade	Elizabeth, N. J	June 10, 1873	139, 839
Clock-key, Safety	P. Laflin	Warren, Mass	Feb. 9, 1869	86, 845
Clock, Machine	G. Richardson	Lowell, Mass	Mar. 3, 1868	75, 196
Clock-machinery	W. Dean	Pleasant Valley, N. Y	Dec. 31, 1821	....
Clock, Marine	V. Giroud	New York, N. Y	Nov. 3, 1863	40, 474
Clock, Marine	L. Hubbell	Bristol, Conn	Nov. 28, 1865	51, 184
Clock-movement	B. B. Lewis	Bristol, Conn	Apr. 18, 1871	113, 781
Clock-movement	M. Tromly	Mount Vernon, Ill	Nov. 3, 1868	83, 802
Clock-movements, Equalizing spring for	H. Smith	Boston, Mass	Nov. 13, 1866	59, 668
Clock-movements, Machine for turning pillars for.	W. H. Nettleton, C. Raymond, and A. Hatch.	Bristol, Conn	Nov. 17, 1857	18, 661

Index of patents issued from the United States Patent Office from 1790 *to* 1873, *inclusive*—Continued.

Invention.	Inventor.	Residence.	Date.	No.
Clock, Musical	T. A. Kohn	New York, N. Y	Apr. 24, 1866	54, 175
Clock or time-piece	W. Pardee	Albany, N. Y	May 22, 1835	
Clock-pendulum	F. Kesselmeier	Wooster, Ohio	Apr. 10, 1844	3, 531
Clock-pendulum	R. Leslie		Jan. 30, 1793	
Clock, Pendulum	G. M. Phelps	Troy, N. Y	July 10, 1860	29, 097
Clock-pendulum	W. D. Whalen	Northville, Mich	Dec. 4, 1866	60, 303
Clock-pendulum rod	R. W. Norton	New Haven, Conn	Feb. 7, 1871	111, 672
Clock-pillar	W. H. Nettleton	Bristol, Conn	Mar. 29, 1864	42, 104
Clock-plates, Mode of boring	J. Curtis	Cairo, N. Y	Aug. 22, 1814	
Clock, Programme	S. F. Estell	Richmond, Ind	Jan. 11, 1870	98, 678
Clock, Programme	S. F. Estell	Chicago, Ill	Oct. 10, 1871	119, 833
Clock, Programme	S. F. Estell	Chicago, Ill	Sept. 30, 1873	143, 230
Clock, Public	A. D. Crane	Boston, Mass	Feb. 16, 1858	19, 351
Clock-registering attachment	S. Foumier	New Orleans, La	July 6, 1858	20, 786
Clock regulator, Marine	A. I. Goodrich	Waterbury, Conn	Dec. 3, 1867	71, 738
Clock, Repeating	C. W. Roberts	Austin, Ill	Dec. 8, 1868	84, 709
Clock, Self-winding	E. Casselberry	Saint Louis, Mo	Aug. 2, 1842	2, 745
Clock, Self-winding	R. Hitchcock	Watertown, N. Y	Sept. 10, 1861	33, 250
Clock-spring	J. S. Ives	Bristol, Conn	May 23, 1836	
Clock-springs by currents of air, Means of winding	C. B. Hoard	Watertown, N. Y	Apr. 3, 1860	27, 721
Clock-springs, Method of tempering	W. Barnes	Bristol, Conn	Nov. 27, 1866	59, 943
Clock-springs, Mode of applying	J. S. Ives	New York, N. Y	May 4, 1838	723
Clock, Striking	L. Myers	Philadelphia, Pa	Sept. 10, 1872	131, 175
Clock-striking attachment	N. E. Mulford	Madison, N. J	Apr. 21, 1868	77, 080
Clock-striking mechanism	G. A. Jones and H. H. Warner	New York, N. Y., and Bristol, Conn.	May 30, 1871	115, 481
Clock striking mechanism	J. B. Mayer	Niagara Falls, N. Y	Sept. 15, 1868	82, 267
Clock-striking mechanism	C. W. Roberts	Chicago, Ill	Dec. 12, 1871	121, 780
Clock-striking movement	J. H. La Bau	Brooklyn, N. Y	June 22, 1869	91, 753
Clock, Swinging	G. Herrmann	Newport, R. I	July 25, 1871	117, 284
Clock, Telegraph	J. Chandler	Syracuse, N. Y	Nov. 5, 1867	70, 410
Clock, Thirty-hour brass and wooden	E. Terry	Litchfield County, Conn	June 12, 1816	
Clock, Thirty-hour wooden	E. Terry	Plymouth, Conn	Mar. 4, 1826	
Clock, Thirty-hour wooden-wheeled	E. Terry	Plymouth, Conn	May 18, 1825	
Clock, Thirty-hour wooden-wheeled	E. Terry	Plymouth, Conn	July 5, 1826	
Clock, Thirty-hour wooden-wheeled	E. Terry	Plymouth, Conn	Sept. 9, 1825	
Clock, Time and striking parts of a	I. Ives	Bristol, Conn	June 24, 1809	
Clock, time-keeper, and watch	E. Terry		Nov. 17, 1797	
Clock, Tower	M. G. Crane	Roxbury, Mass	Oct. 8, 1861	33, 462
Clock, Turret	C. F. Johnson	Owego, N. Y	July 28, 1846	4, 662
Clock, Watch	J. Hamblet, jr	Boston, Mass	July 1, 1862	35, 755
Clock, Watchman's	W. Winter	Plainfield, N. J	July 19, 1864	43, 616
Clock, Watchman's time-checking	I. G. Blake	Worcester, Mass	Apr. 25, 1871	113, 970
Clock-weights, Construction of	R. F. Bond	Cambridge, Mass	Apr. 3, 1860	27, 684
Clock-wheels, Casting	L. J. Kilborn	Pennsylvania	Oct. 13, 1809	
Clock, Winding	R. Hitchcock	Watertown, N. Y	Jan. 29, 1861	31, 242
Clock, Wooden-wheeled	E. Terry	Plymouth, Conn	May 26, 1823	
Clock-work-turning gage	H. F. Henderson and J. E. Ladd	Bristol, Conn	Dec. 27, 1870	110, 460
Clocks and dial-indicators, Escapement for electrical.	B. F. Edmands and J. Hamblet, jr.	Boston and Charlestown, Mass.	Nov. 26 1867	71, 470
Clocks and other time-pieces, Manner of applying alarms to.	B. Knight	Slatersville, R. I	Sept. 10, 1840	1, 780
Clocks, Attachment for alarm	E. T. Quimby	New Ipswich, N. H	June 14, 1859	24, 433
Clocks, Attachment for pendulum	O. P. McDonald	Carbondale, Ill	May 5, 1868	77, 504
Clocks, Balance-pendulum for	R. R. Ramsdell and G. A. Whitcomb.	Marlborough Depot, N. H.	Oct. 18, 1870	108, 390
Clocks by currents of air, Winding	R. Hitchcock	Watertown, N. Y	Jan. 13, 1863	37, 397
Clocks, Compensating-pendulum for	W. L. Coffinberry	Grand Rapids, Mich	Dec. 28, 1858	22, 413
Clocks, Dead-beat verge for	N. Pomeroy	Hartford, Conn	July 13, 1869	92, 644
Clocks, Engine for cutting and pointing the teeth of wheels and pinions for.	J. Curtis and D. Bradley	Cairo, N. Y., and Connecticut.	Aug. 22, 1814	
Clocks, Escapement of	O. R. Tyler	Bradford, Vt	Sept. 6, 1833	
Clocks, Glass wheel for	J. P. Bakewell	Pittsburgh, Pa	Oct. 1, 1830	
Clocks, Illuminating public	T. I. Bailey	Nashville, Tenn	Sept. 12, 1865	51, 411
Clocks, Lighting-attachment for alarm	H. X. Wright	Memphis, Tenn	Mar. 28, 1871	113, 236
Clocks, &c., Machine for cutting wheels for wooden	A. Hopkins	Litchfield, Conn	Aug. 22, 1814	
Clocks, &c., Machine for pointing wire for	A. Sperry	Waterbury, Conn	Aug. 22, 1814	
Clocks, &c., Machine for winding up	J. B. Powell	Philadelphia, Pa	Aug. 2, 1859	24, 977
Clocks, Manufacturing wooden	O. R. Tyler	Chelsea, Vt	June 13, 1831	
Clocks, Mode of applying the vibratory springs of balance.	S. B. Terry	Plymouth, Conn	Nov. 29, 1853	10, 277
Clocks, Multiple time-dial for	J. F. Niehaus	Saint Louis, Mo	Mar. 26, 1872	125, 073
Clocks or time-pieces by atmospheric condensed air, Machine for propelling.	A. Morse, jr	Bloomfield, Me	Sept. 18, 1835	
Clocks or time-pieces, Escapement of	J. Fulton	Shelby County, Ky	Dec. 30, 1835	
Clocks or watches, Alarm-bell to be fixed to	B. F. Freymuth	Philadelphia, Pa	Dec. 22, 1814	
Clocks, Pinion of	I. Ives	Bristol, Conn	Feb. 24, 1812	
Clocks, Pinion-rolling and pinion-wheel for	J. Ives	Hartford County, Conn	Apr. 12, 1833	
Clocks, Propelling-power for	J. Ives	Bristol, Conn	Feb. 24, 1845	3, 928
Clocks, Spring, combined spiral, applied to	S. B. Terry	Plymouth, Conn	Nov. 9, 1830	
Clocks, Striking-attachment to	A. Johnson	West Newton, Ind	Apr. 30, 1867	64, 328
Clocks, Striking-movement of	F. Kienast	Ansonia, Conn	Oct. 18, 1870	108, 362
Clocks, Striking-part of	J. Bogardus	New York	May 18, 1832	
Clocks, Striking-part of	J. Ives	Hartford County, Conn	Apr. 12, 1833	
Clocks, Striking-part of	N. Jerome	Bristol, Conn	June 27, 1839	1, 200
Clocks, Striking-part of	L. J. Kilborn	Pennsylvania	Oct. 12, 1809	
Clocks, Striking-part of	G. Parker	Utica, N. Y	Apr. 7, 1832	
Clocks, Striking-part of	R. Porter	Billerica, Mass	June 22, 1832	
Clocks, Striking-part of	J. S. Seger	New York	Nov. 27, 1832	
Clocks, Striking-part of steeple	G. Denble	Canton, Ohio	July 25, 1854	11, 362
Clocks, Striking-works for	B. Bacon	Morrison, Ill	Sept. 11, 1866	57, 843
Clocks, Striking-works of	G. H. Leeds and C. N. Thorpe	Philadelphia, Pa	May 20, 1873	139, 166
Clocks, Suspending pendulum of	G. Heninger	Lena, Ill	Feb. 26, 1867	62, 330
Clocks, Suspending the balance-wheel of	E. Terry	Plymouth, Conn	Aug. 9, 1845	4, 1[illegible]9
Clocks, Time-part of common wooden	J. Harrison	Boston, Mass	Aug. 22, 1814	
Clocks to indicate the comparative time in all longitudes, Panoramic attachment for.	G. M. Stone	Fredericksburgh, Va	June 21, 1859	24, 501
Clocks, Tubular shaft for	G. H. Blakesley	Bristol, Conn	Apr. 12, 1870	101, 815

Index of patents issued from the United States Patent Office from 1790 *to* 1873, *inclusive*—Continued.

Invention.	Inventor.	Residence.	Date.	No.
Clocks, Turning and slitting pinions for wooden	H. Bronson and J. Curtis	Waterbury, Conn., and Cairo, N. Y.	Aug. 22, 1814	....
Clocks vertically, Device for adjusting	J. F. Keeler	Cleveland, Ohio	May 1, 1860	28, 088
Clocks, Winding-machine for	R. Ward	Waterbury, Conn	Nov. 5, 1829	....
Clocks, Winding-stop for weight	O. H. Woodworth	Columbia City, Ind	May 19, 1868	78, 170
Clocks, Wire bell for	H. C. Thompson	Bristol, Conn	Nov. 5, 1867	70, 484
Clod and corn-stalk fender	G. H. Jackson	College Corner, Ind	Feb. 15, 1870	99, 905
Clod breaker and pulverizer	J. B. Turner	Jacksonville, Ill	July 6, 1869	92, 233
Clod breaking and pulverizing machine	H. H. Hull	Bergen, N. Y	Sept. 26, 1871	119, 361
Clod-crusher	T. H. Ashton	Defiance, Ohio	Oct. 6, 1868	82, 784
Clod-crusher	J. G. Ernst	York, Pa	Aug. 4, 1863	39, 389
Clod-crusher	W. Fenstermacher	Shippensburgh, Pa	Jan. 3, 1865	45, 705
Clod-crusher	S. Hewit	Seneca Falls, N. Y	Apr. 7, 1863	38, 109
Clod-crusher	C. Mahan, sr	Jamestown, Ohio	Oct. 29, 1861	33, 624
Clod-crusher	J. B. Okey	Indianapolis, Ind	Feb. 21, 1871	112, 071
Clod-crusher	J. Schlosser	Piqua, Ohio	Sept. 5, 1871	118, 749
Clod-crusher	E. B. Way	Jerseyville, Ill	Mar. 29, 1859	23, 415
Clod-crusher and harrow	G. W. Dubuisson	Jerusalem, N. Y	July 7, 1863	39, 132
Clod-crusher and harrow combined	W. Young and J. Wortham	Shelbyville, Tenn	Sept. 9, 1873	142, 755
Clod crusher and pulverizer	J. W. Pollock	Bryan, Ohio	Aug. 1, 1871	117, 680
Clod crusher and pulverizer	C. R. Ramsey	Farmers' Centre, Ohio	June 14, 1870	104, 353
Clod-fender	D. Applegate	Noblesville, Ind	Aug. 31, 1869	94, 268
Clod-fender	F. L. Bailey	Freeport, Ind	Mar. 7, 1871	112, 405
Clod-fender	J. C. and W. F. Curryer	Thorntown, Ind	Apr. 27, 1869	89, 470
Clod-fender	A. L. Dirst	Seward, Ill	Feb. 18, 1873	135, 896
Clod-fender	W. L. Dearth and G. P. Bondebush.	Jefferson, Ind	Dec. 21, 1869	98, 041
Clod-fender	F. M. Gardner	Brown Township, Ohio	Oct. 12, 1869	95, 676
Clod-fender	R. T. Gillespie	Millport, Ohio	Feb. 7, 1871	111, 631
Clod-fender	R. T. Gillespie	Millport, Ohio	Apr. 16, 1872	125, 732
Clod-fender	R. Harpster	West Cairo, Ohio	Nov. 22, 1870	109, 511
Clod-fender	J. W. Loveless	Clark's Hill, Ind	July 20, 1869	92, 852
Clod-fender	F. M. and J. D. Lowden	Lawrence, Ind	Dec. 14, 1869	97, 785
Clod-fender	J. Lowe	Lebanon, Ind	Oct. 1, 1867	69, 352
Clod-fender	D. O. Moore and F. Reid	Everton, Ind	Jan. 18, 1870	98, 989
Clod-fender	G. L. Perry	Berlin, Wis	Aug. 22, 1871	118, 267
Clod-fender	G. Seeger, J. W. Loveless, and J. W. Thorp.	Clark's Hill, Ind	June 22, 1869	91, 777
Clod-fender	J. W. Tull	Zionsville, Ind	Dec. 21, 1869	98, 128
Clod-fender and cultivator combined	B. F. Neely	Daleville, Ind	May 10, 1870	102, 959
Clod-fender for growing plants	L. H. Shular	Crawfordsville, Ind	Sept. 14, 1869	94, 784
Clod-fender, Rotary	L. M. Doddridge	New Mount Pleasant, Ind	Sept. 14, 1869	94, 875
Clod-fender, Rotary	J. F. Woolley	Pleasant Ridge, Ohio	Sept. 7, 1869	94, 684
Clods in the field, Implement for dividing	J. N. Davis	Martinsville, Ohio	Sept. 27, 1864	44, 404
Clog	J. A. Davis	Watertown, N. Y	Nov. 24, 1868	84, 267
Clog or patten	C. W. Stearns	Springfield, Mass	Apr. 22, 1851	8, 053
Closet: *See* Cake-closet. Chamber-closet. Earth-closet. Hot-closet. Water-closet.				
Closet and bed	W. Kelley	Bath, Me	Apr. 19, 1870	102, 012
Closet for milk	E. H. Nash	Westford, Conn	July 27, 1858	21, 016
Cloth and blankets from sheep's wool, Making	A. Barnes, S. Gray, and J. Clark	Windham, Conn	Feb. 3, 1813	....
Cloth and clothes pin	R. Hathaway	Chicopee, Mass	June 2, 1868	78, 523
Cloth and fabric, Manufacture of rubber and other coated.	J. W. Cobb and E. A. Hill	Melrose and Quincy, Mass	Aug. 11, 1868	80, 809
Cloth and hat-bodies, Machine for shrinking bodies in the manufacture of felt.	H. A. Wells	New York, N. Y	Sept. 11, 1841	2, 245
Cloth and other fabrics, Compound for cleaning	F. T. Huntoon	Fulton City, Ill	May 31, 1870	103, 747
Cloth and paper fabrics, Manufacture of combined	J. H. Newton	Holyoke, Mass	Mar. 1, 1870	100, 436
Cloth and similar fabrics and slabs for pavements, Manufacture of floor.	F. Walton	Linoleum Works, Staines, England.	Feb. 23, 1869	87, 227
Cloth, Apparatus for coating and flocking	E. M. Chaffee	Providence, R. I	Sept. 20, 1864	44, 281
Cloth, Apparatus for felting	J. Andrews	Belleville, N. J	Jan. 31, 1845	3, 900
Cloth, Apparatus for frizzling	S. M. Moschcowitz	New York, N. Y	Feb. 15, 1870	99, 782
Cloth, Apparatus for singeing cotton	J. A. Miller	New York, N. Y	Jan. 14, 1868	73, 362
Cloth, Apparatus for sponging	E. Utley	Appleton, Wis	Mar. 12, 1872	124, 461
Cloth, Bolting	R. Dawson	....	May 12, 1796	....
Cloth brushing and winding machine	R. C. Varnel	West Somers, N. Y	Mar. 13, 1844	3, 487
Cloth-bushes in holes, Mode of securing	C. Bollermann	New York, N. Y	May 17, 1864	42, 740
Cloth by felting, Machine for making	O. Barrett, jr	Troy, N. Y	Dec. 8, 1812	....
Cloth by machinery, Mode of making	H. Raymond	New York, N. Y	June 27, 1829	....
Cloth by thrashing, Raising nap on woolen	T. Hurd and J. Fox	Lowell, Mass	June 23, 1830	....
Cloth by water, Machine for shearing woolen or other.	F. B. Kellogg	Marlborough, Mass	Nov. 22, 1805	....
Cloth, Calendering	Z. Bliss	Johnston, R. I	Oct. 17, 1835	....
Cloth, Carding and raising a nap on woolen	J. A. Christie	Elizabethtown, N. J	Jan. 24, 1816	....
Cloth, Cement for edges of	R. Brackett	Boston, Mass	Mar. 14, 1833	....
Cloth, Circular tenter-bar for drying	S. R. Parkhurst	Meriden, R. I	Dec. 2, 1834	....
Cloth, Composition for coating marine	J. G. Colcord	Boston, Mass	Aug. 6, 1872	130, 279
Cloth, Composition for stiffening felt	P. O'Rork	Norwalk, Conn	May 29, 1866	55, 153
Cloth, card, and paper cutter combined	I. B. Millner	Watkins, N. Y	Apr. 2, 1872	125, 319
Cloth-creaser	W. J. Crane	Carbondale, Pa	Mar. 10, 1868	75, 378
Cloth-creasing frame	A. W. Todd	Chicago, Ill	Feb. 18, 1868	74, 636
Cloth, Cross-cut eclipse shearing-machine for shearing woolen.	R. Daniels	Woodstock, Vt	Apr. 7, 1838	681
Cloth, &c., Crossing the fibers in forming the bats for felt.	A. C. Arnold	Norwalk, Conn	June 10, 1851	8, 147
Cloth-cutter	B. Hansell, J. McCann, and S. McCambridge.	Philadelphia, Pa	Sept. 8, 1863	39, 862
Cloth-cutting machine	V. H. Buschman	Baltimore, Md	May 15, 1866	54, 681
Cloth-cutting machine	S. M. Eisman	New York, N. Y	July 30, 1872	129, 939
Cloth-cutting machine	I. Fenno	Boston, Mass	Jan. 14, 1873	134, 793
Cloth-cutting machine	I. Fenno and P. Howe	Boston, Mass	July 16, 1872	129, 327
Cloth-cutting machine	I. Fenno and P. Howe	Boston, Mass	Aug. 27, 1872	130, 910
Cloth-cutting machine	N. C. Fluck	Gloucester, England	Oct. 28, 1873	144, 023

Index of patents issued from the United States Patent Office from 1790 *to* 1873, *inclusive*—Continued.

Invention.	Inventor.	Residence.	Date.	No.
Cloth-cutting machine	A. Heller	New York, N. Y	Apr. 30, 1872	126, 207
Cloth-cutting machine	J. Kent	New York, N. Y	Mar. 11, 1873	136, 602
Cloth-cutting machine	F. Koch and R. Brass	Williamsburgh, N. Y	July 16, 1872	129, 285
Cloth-cutting machine	W. Raenchle	Philadelphia, Pa	Apr. 18, 1871	113, 927
Cloth-cutting machine	A. Warth	Stapleton, N. Y	Apr. 9, 1872	125, 638
Cloth-cutting machine	A. Warth	Stapleton, N. Y	Aug. 6, 1872	130, 343
Cloth-cutting machine	A. Warth	Stapleton, N. Y	Aug. 6, 1872	130, 344
Cloth-cutting machine	E. B. Wells	New York, N. Y	July 25, 1871	117, 352
Cloth-cutting machine	G. Westerhauser	Brooklyn, N. Y	June 18, 1872	128, 194
Cloth-damping machine	W. Hebdon	New York, N. Y	Mar. 16, 1869	87, 930
Cloth-deseaming machine	T. E. Chase	Boston, Mass	Apr. 30, 1872	126, 264
Cloth, Device for turning down the edges of elastic	G. H. Chesbro	Stafford, Conn	Apr. 27, 1858	20, 037
Cloth, &c., Die for taking impressions from	J. J. C. Smith	Somerville, Mass	Apr. 4, 1871	113, 699
Cloth-dotter	S. S. Gould	Worcester, Mass	May 30, 1865	47, 944
Cloth doubling and folding machine	J. W. Farwell	Lewiston, Me	Feb. 27, 1872	124, 044
Cloth, Dressing	C. W. Cook	Lowell, Mass	Jan. 23, 1834	....
Cloth-dressing apparatus	J. Johnston and J. D. Snyder	Saltsburgh, Pa	Mar. 13, 1849	6, 188
Cloth-dressing machine	I. Sanford		Mar. 27, 1799	....
Cloth-dressing machinery	J. C. Carlisle	Armagh, Pa	Oct. 21, 1846	4, 821
Cloth, Dressing woolen	S. R. Parkhurst	Mendon, Mass	Dec. 23, 1833	....
Cloth, Dressing woolen and cotton	C. W. Cook	Lowell, Mass	May 30, 1833	....
Cloth, Dressing woolen or mixed	C. W. Cook	Lowell, Mass	Apr. 23, 1831	....
Cloth, Drying	D. W. Kennedy	Staunton, Va	June 27, 1854	11, 163
Cloth, Drying	R. Preston	North Pownal, Vt	Feb. 14, 1854	10, 522
Cloth-drying, &c., Circular revolving tenter-bar for	S. R. Parkhurst	Worcester, Mass	Oct. 28, 1835	....
Cloth-drying machine	N. P. Akin	Philmont, N. Y	Sept. 24, 1872	131, 586
Cloth-drying machine	C. F. Bennett	Philadelphia, Pa	Sept. 12, 1865	49, 947
Cloth-drying machine	L. W. Boynton	Hartford, Conn	Apr. 2, 1867	63, 461
Cloth-drying machine	A. Chambers	Providence, R. I	Nov. 3, 1868	83, 690
Cloth-drying machine	J. Hurd, jr	Boston, Mass	Jan. 23, 1830	....
Cloth-drying machinery	C. F. Bennett	Warehouse Point, Conn	Nov. 15, 1859	26, 141
Cloth-drying machinery	B. Sexton	East Windsor, Conn	May 8, 1860	28, 202
Cloth, Elastic	H. H. Day	New York, N. Y	Aug. 30, 1859	25, 249
Cloth, Elastic gore	C. Winslow	Lynn, Mass	Aug. 4, 1857	17, 950
Cloth, Evaporating and condensing machinery for	J. Goulding and R. Bracket	Lynn, Mass	Aug. 15, 1835	....
Cloth, Feathered	A. A. Condit	Muncie, Ind	Jan. 8, 1867	61, 000
Cloth, felt, &c., Water-proof coating for	J. Ditto	New York, N. Y	Apr. 15, 1862	34, 947
Cloth, Felting	J. Weight	Lawrence, Mass	Jan. 6, 1852	8, 636
Cloth-felting machinery	G. G. Bishop	Norwalk, Conn	Mar. 23, 1852	8, 821
Cloth-finishing machine	J. Earnshaw	East Greenwich, R. I	Dec. 1, 1868	84, 483
Cloth-finishing machine	G. C. Howard	Philadelphia, Pa	Aug. 18, 1868	81, 170
Cloth-finishing machine	D. Hussey	Nashua, N. H	Apr. 20, 1869	89, 048
Cloth-finishing machines, Tray for	A. Woolson	Springfield, Vt	July 22, 1873	141, 192
Cloth-finishing machinery	H. James	Norwalk, Conn	Feb. 9, 1864	41, 581
Cloth-finishing press	C. Heubach	Chicago, Ill	May 23, 1871	115, 203
Cloth-finishing, Steam-cylinder for	A. Webster	Seneca Falls, N. Y	Nov. 5, 1867	70, 658
Cloth, Finishing woolen	Z. Allen	Providence, R. I	Feb. 23, 1830	....
Cloth, Folding and measuring	W. C. Wright	Boston, Mass	June 20, 1854	11, 145
Cloth folding and measuring machine	A. M. Cheney	Charlotte, Mich	Mar. 18, 1873	136, 967
Cloth folding and measuring machine	J. Spalding	Morristown, Vt	Aug. 28, 1841	2, 230
Cloth folding and measuring machinery	J. Baxendale	Providence, R. I	Dec. 11, 1855	13, 900
Cloth folding and measuring machinery	S. C. Durgin	North Chelmsford, Mass	Mar. 9, 1844	3, 469
Cloth, Folding and uniting guide for the edges of two pieces of.	I. M. Rose	New York, N. Y	Feb. 20, 1866	52, 749
Cloth-folding instrument	H. Moschcowitz	New York, N. Y	Mar. 26, 1872	124, 968
Cloth-folding machine	D. R. Ambrose and O. L. Reynolds.	Portsmouth and Dover, N. H.	July 22, 1851	8, 240
Cloth-folding machine	H. Dunphy	New York, N. Y	Nov. 12, 1867	70, 706
Cloth-folding machine	J. D. Elliot	Grafton, Mass	July 9, 1861	32, 761
Cloth-folding machine	J. D. Elliot	Grafton, Mass	Apr. 30, 1867	64, 208
Cloth-folding machine	W. F. Heywood	Cumberland, R. I	July 3, 1866	56, 048
Cloth-folding machine	A. Simpson	Woonsocket, R. I	Dec. 28, 1869	98, 308
Cloth-folding machine	W. Wheeler, jr	North Providence, R. I	May 7, 1867	64, 466
Cloth-folding machinery	A. C. Cary and D. C. Bagley	Amesbury, Mass	Feb. 12, 1850	7, 079
Cloth for cutting, Gage for marking	E. E. Emery	Boston, Mass	Oct. 14, 1873	143, 681
Cloth for felting hat-bodies, &c	W. Fuzzard	Charlestown, Mass	Apr. 1, 1856	14, 559
Cloth, Forming bats for felt	M. D. Whipple	Charlestown, Mass	Oct. 26, 1858	21, 930
Cloth, Forming, heating, &c., metal plates in pressing woolen.	V. Vedder	Amsterdam, N. Y	Feb. 6, 1811	....
Cloth, Forming nap on woolen	Z. Allen	Providence, R. I	Feb. 2, 1830	....
Cloth from wool and silk, Manufacture of	R. Daniels	Woodstock, Vt	Oct. 8, 1840	1, 809
Cloth from wool and silk, Manufacture of	R. Daniels	Woodstock, Vt	Oct. 10, 1840	1, 813
Cloth, Fulling	M. W. Northrup and R. R. Dillon.	Newton, N. J	Dec. 4, 1823	....
Cloth, Fulling	R. Winans	Warwick, N. Y	June 26, 1821	....
Cloth fulling and milling machine, Felt	H. A. Wells	New York	Sept. 18, 1841	2, 255
Cloth-fulling machine	R. Hunt	Orange, Mass	July 5, 1870	105, 082
Cloth-fulling machinery	J. H. Jennings and T. Brierly	Clayville, N. Y	Jan. 24, 1854	10, 453
Cloth, Gage for cutting bias pieces of	S. T. Taylor	New York, N. Y	Aug. 23, 1870	106, 741
Cloth, &c., Gas-proof	W. B. S. Taylor	New York, N. Y	May 14, 1872	126, 851
Cloth, Gig-mill for dressing	A. Woolson	Springfield, Vt	Apr. 19, 1853	9, 678
Cloth, Gig-mill for napping	E. Gessner	Awe, Saxony	Feb. 24, 1857	16, 685
Cloth-guiding apparatus	E. O. Potter	North Providence, R. I	Apr. 16, 1867	63, 939
Cloth-hardening machine, Felt	H. A. Wells	New York	Sept. 18, 1841	2, 256
Cloth-holder for washing crockery, &c	C. F. Greeley	East Kingston, N. H	Aug. 2, 1859	24, 933
Cloth-holder in needle-work	N. Daniels	Milford, Mass	Dec. 27, 1859	26, 570
Cloth, Hurd's patent metallic napper for	J. M. Pratt	Dudley, Mass	Oct. 3, 1838	957
Cloth, Implement for cutting	G. W. Griswold	Carbondale, Pa	Oct. 18, 1853	10, 132
Cloth, Instrument for ripping sutures in	T. B. Converse	New York, N. Y	July 11, 1865	48, 662
Cloth in the piece, Machinery for fulling	M. D. Whipple	Charlestown, Mass	Oct. 26, 1858	21, 931
Cloth in the process of fulling, Machine for stretching.	B. D. Whitney and G. W. Lawton.	Winchendon, Mass	Mar. 25, 1840	1, 522
Cloth, Machine called "The wheel of knives," for shearing and raising the nap on.	S. G. Dorr		Oct. 20, 1792	....
Cloth-machine, Cross-napping	J. Taylor and J. Smith	New Lebanon, N. Y	May 4, 1842	2, 605
Cloth, Machine for brushing and finishing	S. Hart	Hempstead, N. Y	July 27, 1812	....
Cloth, Machine for cutting out	J. Harraday	New York, N. Y	May 30, 1854	10, 986
Cloth, Machine for dressing and drying woolen	D. Henderson	Merrimack, N. H	July 8, 1862	35, 823

Index of patents issued from the United States Patent Office from 1790 *to* 1873, *inclusive*—Continued.

Index of patents issued from the United States Patent Office from 1790 to 1873, inclusive—Continued.

Invention.	Inventor.	Residence.	Date.	No.
Cloth-plaiting machine	J. A. Sawer	Worcester, Mass	Feb. 6, 1872	123, 425
Cloth-press	C. H. Weston and J. Dennis	Lowell, Mass	Jan. 26, 1869	86, 334
Cloth-press	C. H. Weston and J. Dennis	Lowell, Mass	May 17, 1870	103, 262
Cloth-press	C. H. Weston and J. Dennis	Lowell, Mass	May 17, 1870	103, 263
Cloth-press, Steam	J. J. Crawford	Glasgow, Scotland	Nov. 16, 1869	96, 893
Cloth-press, Steam	J. Dennis and C. H. Weston	Lowell, Mass	Aug. 19, 1873	142, 006
Cloth-pressing machine	P. Howe	Boston, Mass	June 20, 1871	116, 057
Cloth-pressing machine	P. Howe	Boston, Mass	Oct. 10, 1871	119, 708
Cloth-pressing machine	P. Howe	Boston, Mass	Dec. 12, 1871	121, 720
Cloth-pressing, Steam device for	C. H. Weston	Lowell, Mass	Mar. 16, 1869	87, 994
Cloth, Process of manufacturing	A. Ruzé	Gaillon, France	June 27, 1871	116, 497
Cloth, Process of milling, fulling, &c	M. Lee	Otsego, N. Y	Nov. 6, 1817	......
Cloth-rack	H. C. Smith, D. A. Kelly, and J. E. Murdock, jr.	Clarksville, Ohio	Feb. 11, 1868	74, 439
Cloth-rack	A. W. Voegtly	Hannibal, Ohio	Sept. 10, 1872	131, 235
Cloth, Raising nap on	W. Burt		July 23, 1797	......
Cloth, Raising nap on	W. Duncan and J. Davidson	Acworth, N. H	May 29, 1815	......
Cloth, Raising nap on	J. Mathes	Barre, Vt	May 27, 1813	......
Cloth, Raising nap on	J. Wilkinson	New York, N. Y	May 12, 1825	......
Cloth scouring and fulling machine	H. Hickox	Rutland, N. Y	Mar. 26, 1825	......
Cloth, Shears for shearing	W. Stillman	Westerly, R. I	Apr. 28, 1815	......
Cloth, Shears for shearing woolen	W. Stillman	Westerly, R. I	June 11, 1811	......
Cloth-shears, Revolving	J. Collins	Anson, Me	June 10, 1817	......
Cloth, Shearing	G. Booth	Poughkeepsie, N. Y	Oct. 13, 1812	......
Cloth, Shearing	B. Cummings	Palmer, Mass	Mar. 2, 1811	......
Cloth, Shearing	R. Daniels	Woodstock, Vt	May 13, 1834	......
Cloth, Shearing	R. Daniels	Woodstock, Vt	Aug. 22, 1834	......
Cloth, Shearing	D. Dewey	Poultney, Vt	June 27, 1809	......
Cloth, Shearing	W. Kennedy	Frankford, Ky	June 12, 1812	......
Cloth, Shearing	H. Mathews	Delaware	May 15, 1810	......
Cloth, Shearing	J. Molleneux	Hempstead, N. Y	Apr. 30, 1811	......
Cloth, Shearing	H. Osborn and W. L. Fraser		Apr. 25, 1816	......
Cloth, Shearing	J. D. Smith	Fredericktown, Md	May 24, 1816	......
Cloth, Shearing	E. Sprague	Danbury, Conn	May 9, 1810	......
Cloth, Shearing	S. Stewart, E. Hovey, and J. Henderson.	Pittstown, N. Y	June 21, 1808	......
Cloth, Shearing	B. Swift	New York, N. Y	May 24, 1810	......
Cloth, Shearing	E. Willmarth	Rumney, Mass	Feb. 28, 1811	......
Cloth, Shearing and laying the nap of	L. Dickerman	Schaghticoke, N. Y	July 18, 1812	......
Cloth, Shearing broad	J. Davidson	Springfield, Vt	May 29, 1834	......
Cloth-shearing machine	G. Bostwick	Brookfield, Conn	Sept. 28, 1812	......
Cloth-shearing machine	E. Burt	Princeton, N. J	June 23, 1807	......
Cloth-shearing machine	Z. Cary	Oxford County, Me	Feb. 20, 1821	......
Cloth-shearing machine	J. Collins	Anson, Me	Mar. 6, 1827	......
Cloth-shearing machine	M. Craven	Dedham, Mass	Dec. 12, 1871	121, 760
Cloth-shearing machine	D. Dewy	Rutland County, Va	Mar. 16, 1818	......
Cloth-shearing machine	R. Dorr	Kenderbook, N. Y	Mar. 8, 1807	......
Cloth-shearing machine	E. Heald	Norridgewock, Me	Dec. 4, 1822	......
Cloth-shearing machine	W. Hovey	Worcester, Mass	Dec. 17, 1824	......
Cloth-shearing machine	S. Hills	Hudson, N. Y	Oct. 22, 1813	......
Cloth-shearing machine	E. Hotchkiss and A. Jaque	Brattleborough, Vt	Dec. 15, 1819	......
Cloth-shearing machine	G. C. Kellogg	New Hartford, Conn	Mar. 4, 1811	......
Cloth-shearing machine	I. and G. C. Kellogg	New Hartford, Conn	Apr. 7, 1828	......
Cloth-shearing machine	C. H. Orth and F. Strohn	Steubenville, Ohio	Oct. 11, 1814	......
Cloth-shearing machine	S. Parsons	Hoosick, N. Y	Mar. 2, 1819	......
Cloth-shearing machine	E. Remington	Frankfort, N. Y	Aug. 13, 1817	......
Cloth-shearing machine	I. Sanford	Providence, R. I	Mar. 26, 1814	......
Cloth-shearing machine	W. Stillman	Westerly, R. I	Jan. 17, 1818	......
Cloth-shearing machine	B. Swift	Washington, N. Y	July 1, 1814	......
Cloth-shearing machine	B. Swift	Washington, N. Y	Feb. 7, 1824	......
Cloth-shearing machine	J. A. Thurston	Providence, R. I	Oct. 24, 1871	120, 219
Cloth-shearing machine	M. D. Whipple	Charlestown, Mass	Dec. 1, 1857	18, 796
Cloth-shearing machine	A. Woolson	Springfield, Vt	May 28, 1850	7, 407
Cloth-shearing machine	A. Woolson	Springfield, Vt	Aug. 16, 1864	43, 878
Cloth-shearing-machine rest	A. Woolson	Springfield, Vt	Oct. 29, 1872	132, 706
Cloth-shearing machine, Shears for	S. A. Britt	Cazenovia, N. Y	Aug. 10, 1829	......
Cloth-shearing machine, Woolen	E. Stowell	Worcester, Mass	Apr. 26, 1808	......
Cloth-shearing machines, Guiding-apparatus for	L. M. Collins	Lebanon, N. H	Apr. 16, 1872	125, 722
Cloth, Shearing woolen and other	S. Kellogg		Jan. 31, 1795	......
Cloth, Shuttle for weaving	J. Baldwin	Nashua, N. H	Jan. 31, 1840	1, 485
Cloth, silk, paper, &c., Compound to remove grease from.	M. B. Kimm	Grand Rapids, Mich	Feb. 22, 1870	100, 043
Cloth sizing and dressing machine	P. Moody	Waltham, Mass	Jan. 17, 1818	......
Cloth-smoothing device	C. L. Frink	Rockville, Conn	Dec. 3, 1867	71, 599
Cloth-sponging apparatus	I. A. Davis	Philadelphia, Pa	July 12, 1870	105, 182
Cloth-sponging apparatus	J. R. Paul	Philadelphia, Pa	Nov. 26, 1867	71, 529
Cloth-sponging apparatus	L. Rothschild	New Haven, Conn	Apr. 19, 1864	42, 406
Cloth-sponging machine	G. Dayspring and E. Fitzki	Washington, D. C	Sept. 14, 1869	94, 720
Cloth, Spreading and drying India-rubber upon	W. Atkinson	New York	Aug. 15, 1835	......
Cloth, Spreading-roller for stretching	J. I. Hillard	Fall River, Mass	Apr. 15, 1856	14, 659
Cloth, Steam cylinder for finishing	A. Brown	Springfield, Vt	Mar. 12, 1872	124, 481
Cloth steaming and shrinking machine	W. Hebdon	New York, N. Y	Oct. 5, 1869	95, 584
Cloth-steaming apparatus	L. M. Heery	Hinsdale, Mass	May 23, 1871	115, 200
Cloth-steaming machine	H. S. Green	Sherbrooke, Canada	Feb. 4, 1873	135, 543
Cloth-steaming machine	C. H. Weston	Lowell, Mass	Jan. 14, 1873	134, 956
Cloth stretcher and drier	E. B. Bigelow	Boston, Mass	Sept. 5, 1848	5, 754
Cloth stretcher and smoother	J. Butcher	Lowell, Mass	Nov. 5, 1850	7, 764
Cloth, Stretching and drying	P. Hild	New York, N. Y	June 11, 1872	127, 691
Cloth, Stretching and drying	D. and H. Stearns	Pittsfield, Mass	Feb. 21, 1854	10, 545
Cloth stretching and drying machine	W. Bailey	Brooklyn, N. Y	May 20, 1873	139, 038
Cloth stretching and drying machine	T. Barrows	Dedham, Mass	Dec. 2, 1851	8, 563
Cloth stretching and drying machinery	B. J. Tayman	Philadelphia, Pa	Nov. 14, 1854	11, 952
Cloth-stretching machine	S. H. Austin	Providence, R. I	Apr. 12, 1870	101, 811
Cloth-stretching machine	E. C. Cleveland	Worcester, Mass	Dec. 4, 1860	30, 801
Cloth-stretching machine	A. C. Corpe	Stafford, Conn	Dec. 10, 1867	71, 920
Cloth-stretching machine	H. P. Roche	Utica, N. Y	Dec. 3, 1867	71, 647
Cloth-stretching roller	N. Simmons	Providence, R. I	Dec. 4, 1855	13, 888
Cloth-strip bristles	A. P. Peyroux	New Orleans, La	Oct. 24, 1871	120, 320

Index of patents issued from the United States Patent Office from 1790 to 1873, inclusive—Continued.

Invention.	Inventor.	Residence.	Date.	No.
Cloth, Substitute for teasel and card for raising the nap of.	I. Sanford	Providence, R. I	Mar. 26, 1814	
Cloth-tearing engine	J. Biddis		May 6, 1800	
Cloth, Teazling	J. Beckwith	Lyme, Conn	May 16, 1817	
Cloth, Teazling	J. Olney	Westmoreland, N. Y	Mar. 29, 1817	
Cloth teazling and napping machine	A. Foster	Whitestown, N. Y	Aug. 24, 1821	
Cloth teazling or napping machine	B. Swasey	Mount Vernon, Me	Aug. 8, 1837	350
Cloth, Tenter-bar for shaping articles of	S. L. Fiske	Philadelphia, Pa	Mar. 3, 1868	75, 143
Cloth, Tentering	W. Shaw and P. G. Green	Wales, Mass	June 20, 1854	11, 141
Cloth tentering and drying machine	A. Avery	Worcester, Mass	Aug. 24, 1869	94, 166
Cloth tentering and drying machine	J. S. Winsor	Providence, R. I	Apr. 9, 1861	32, 021
Cloth-tentering machine	G. S. Rogers	Thetford, Vt	Dec. 26, 1871	122, 283
Cloth, the weft of which is made of hair, grass, &c.	J. Downie	Paterson, N. J	Dec. 12, 1865	51, 436
Cloth upon loom, Apparatus for measuring and marking.	F. W. Howe	North Chelmsford, Mass	June 6, 1846	4, 556
Cloth used by bank-note engravers, Cleaning	H. M. Baker	New York, N. Y	Feb. 25, 1868	74, 878
Cloth, Varnish for enameling	W. H. Haines	Newark, N. J	Jan. 9, 1866	51, 998
Cloth-varnishing machine	S. B. B. Nowlan	New York, N. Y	Sept. 11, 1866	57, 955
Cloth, Vibrating machine for napping	S. Duncan	Northampton, N. Y	Jan. 21, 1828	
Cloth washing and fulling machine	S. Arnold	Philmont, N. Y	Oct. 17, 1871	119, 958
Cloth washing, rinsing, and squeezing machine	J. Lee, jr	Charlestown, Mass	Dec. 24, 1867	72, 647
Cloth washing, scouring, fulling, and cleaning machine.	D. S. Dean		Feb. 29, 1804	
Cloth, Weaving double	S. Fay	Lowell, Mass	June 20, 1854	11, 120
Cloth while fulling, &c., Machine for overhauling or taking wrinkles out of.	J. Tillon	New Haven, Conn	Dec. 14, 1840	1, 902
Cloth-whipping machine	E. Eaton	Norwalk, Conn	Mar. 2, 1869	87, 330
Cloth winding and folding machine	T. P. Forsyth	Dalton, Ind	Apr. 18, 1854	10, 782
Cloth-winding machine	G. F. Hargis	Decatur, Ill	Mar. 23, 1869	88, 257
Cloth with India rubber, &c., Machine for coating	G. S. Dwight	New York, N. Y	Feb. 16, 1869	87, 087
Cloth without spinning and weaving, Machine for manufacturing felt.	T. R. Williams	United States	Dec. 14, 1840	1, 897
Cloth without spinning or weaving, Mode of manufacturing.	N. Peck and D. Taylor	Fairfield County, Conn	June 29, 1829	
Cloth, Woven	A. Ruzé	Gaillon, France	June 27, 1871	116, 498
Clothes and flower sprinkler	D. Bickford	Boston, Mass	Jan. 15, 1867	61, 146
Clothes and hat hook	G. B. Fowler	New York, N. Y	Dec. 15, 1863	40, 923
Clothes and hat hook	R. W. Randall	Biddeford, Me	Feb. 14, 1871	111, 777
Clothes and hat rack	C. Bradfield	Newark, N. J	Mar. 7, 1865	46, 751
Clothes and hat rack	H. E. Fickett	Brooklyn, N. Y	Aug. 23, 1864	43, 905
Clothes and hat rack combined	O. F. Burgess	Decorah, Iowa	May 21, 1872	127, 022
Clothes and mop wringer	C. E. Gage	Fond du Lac, Wis	Jan. 1, 1867	60, 761
Clothes and mop wringer	L. Hannum	Cortland, N. Y	May 12, 1867	62, 843
Clothes and picture hanger	F. M. Everingham	Collingwood, N. Y	Sept. 10, 1867	68, 616
Clothes, Apparatus for ironing	C. R. Hoyt	New York, N. Y	Dec. 3, 1867	71, 758
Clothes-bar	H. Willard	Vergennes, Vt	Feb. 18, 1862	34, 456
Clothes-beater	A. Jameson	Boston, Mass	Apr. 16, 1872	125, 738
Clothes-boiler	D. Kellogg	Ypsilanti, Mich	Dec. 8, 1868	84, 832
Clothes-boiler	J. H. Rickett	West Dover, Vt	June 29, 1869	91, 967
Clothes-boiler	L. H. Whitney	Washington, D. C	Sept. 14, 1869	94, 931
Clothes boiler, drainer, &c., Combined	F. Browning	Watchet, England	Dec. 19, 1865	51, 670
Clothes-basket	J. Stimpson	Baldwinsville, Mass	July 25, 1865	49, 007
Clothes-brushing machine	W. M. Blume	New York, N. Y	May 28, 1872	127, 301
Clothes-clamp	L. H. Cushman	Monmouth, Me	Nov. 17, 1857	18, 6 2
Clothes-clamp	E. Hedge and T. H. Fleming	Liverpool and Canton, Ill	June 27, 1871	116, 307
Clothes-clamp	C. L. Poorman	Bellaire, Ohio	May 23, 1871	115, 099
Clothes-clamp	J. Sadgebury	Philadelphia, Pa	Nov. 6, 1855	13, 765
Clothes-clamp	W. H. Towers	Philadelphia, Pa	Nov. 21, 1854	11, 977
Clothes-clamp	W. H. Towers	Philadelphia, Pa	Feb. 12, 1856	14, 262
Clothes-clamp	W. Tunstill	Brooklyn, N. Y	July 8, 1873	140, 659
Clothes-clamp	C. Warner	New York, N. Y	Oct. 4, 1859	25, 695
Clothes-cleaner	J. Braun	Rochester, N. Y	Dec. 22, 1868	85, 060
Clothes-cleansing pestle	E. Pollard	Albany, N. Y	Dec. 28, 1858	22, 451
Clothes, Compound for extracting oil, paint, grease, &c., from.	C. B. Skiff	Jersey City, N. J	June 22, 1869	91, 680
Clothes, Compound for imparting a gloss to	W. D. Beaumont	Baltimore, Md	Oct. 8, 1850	7, 695
Clothes, Device to aid in ironing	C. S. Whipple	Waterford, Conn	Oct. 15, 1872	131, 340
Clothes, Dry-scouring	T. L. Jennings	New York	Mar. 3, 1821	
Clothes-drier	W. H. Acker	Tarrytown, N. Y	Dec. 15, 1868	84, 928
Clothes-drier	L. J. Adams	Hoosick Falls, N. Y	Oct. 26, 1869	96, 181
Clothes-drier	G. W. Ainsworth	Waterbury, Vt	Apr. 16, 1872	125, 713
Clothes-drier	O. P. Allen	Rindge, N. H	Sept. 11, 1860	29, 943
Clothes-drier	C. R. Anderson	Saint Louis, Mo	Dec. 6, 1870	109, 860
Clothes-drier	I. B. Arnold	Providence, R. I	Oct. 1, 1867	69, 302
Clothes-drier	W. Arronquier	Worcester, Mass	May 26, 1868	78, 351
Clothes-drier	C. Bange	Saint Louis, Mo	May 5, 1868	77, 440
Clothes-drier	J. Barnett	Dayton, Ohio	Mar. 20, 1866	53, 257
Clothes-drier	J. R. Bassett	Cincinnati, Ohio	June 18, 1872	127, 948
Clothes-drier	C. B. Bennett	Amboy, Ill	Dec. 31, 1867	72, 782
Clothes-drier	H. S. Black	Buchanan, Mich	May 9, 1871	114, 519
Clothes-drier	J. M. Blake	Buffalo, N. Y	Sept. 14, 1869	94, 703
Clothes-drier	H. S. Blood	New Orleans, La	Apr. 5, 1864	42, 160
Clothes-drier	J. B. Blood	Lynn, Mass	Sept. 22, 1868	82, 280
Clothes-drier	H. C. Boardman	Morrisville, Vt	Dec. 11, 1860	30, 868
Clothes drier	A. Bosworth	Champlain, N. Y	May 17, 1870	103, 133
Clothes-drier	R. H. Boughner	East Germantown, Ind	Mar. 23, 1869	88, 123
Clothes-drier	J. Briggs	Peterborough, N. H	May 31, 1870	103, 712
Clothes-drier	J. L. Brigham	Saint Paul, Minn	Jan. 17, 1871	110, 952
Clothes-drier	B. Britten	Galena, Ill	Mar. 5, 1867	62, 600
Clothes-drier	B S. Brown	Chicago, Ill	Feb. 22, 1870	100, 114
Clothes-drier	B. S. Brown	New York, N. Y	Mar. 5, 1872	124, 248
Clothes-drier	H. Brown	New York, N. Y	Aug. 20, 1867	67, 959
Clothes-drier	O. C. Brown	Iberia, Ohio	July 17, 1866	56, 360
Clothes-drier	W. L. Brown	Shortsville, N. Y	May 24, 1870	103, 424
Clothes-drier	D. M. Buckley	Athens, Ohio	Apr. 21, 1868	76, 990
Clothes-drier	E. Bucklin, jr	Pawtucket, R. I	May 16, 1865	47, 698
Clothes-drier	W. H. Buell	Union City, Mich	June 21, 1870	104, 652
Clothes-drier	J. Burch and B. Webb	Warren, N. Y	Nov. 25, 1873	144, 951

Index of patents issued from the United States Patent Office from 1790 to 1873, inclusive—Continued.

Invention.	Inventor.	Residence.	Date.	No.
Clothes-drier	M. C. Burr	Owatanna, Minn	Aug. 19, 1862	36, 204
Clothes-drier	E. Buss	Yellow Springs, Ohio	Apr. 2, 1861	31, 868
Clothes-drier	F. R. Butler	Rocky Hill, Conn	Aug. 31, 1869	94, 182
Clothes-drier	J. M. Butters	Lovell, Me	Aug. 22, 1865	49, 498
Clothes-drier	J. M. Butters	North Fryeburgh, Me	June 2, 1868	78, 575
Clothes-drier	J. Caffrey	Covington, Ky	Mar. 12, 1872	124, 543
Clothes-drier	J. Caffrey	Covington, Ky	Oct. 15, 1872	132, 196
Clothes-drier	L. S. Calkins	El Paso, Ill	July 23, 1867	67, 102
Clothes-drier	M. T. Campbell	Lima, Pa	Aug. 18, 1868	81, 249
Clothes-drier	C. Carrier	Oswego, N. Y	July 20, 1869	92, 794
Clothes-drier	E. Carter	Rensselaer, N. Y	Sept. 7, 1869	94, 562
Clothes-drier	R. D. Chandler	Fair Haven, N. J	Oct. 29, 1867	70, 165
Clothes-drier	A. J. Chase	Boston, Mass	Jan. 5, 1869	85, 562
Clothes-drier	T. H. Chubb and W. G. Marston	Post Mills, Vt	May 9, 1871	114, 527
Clothes-drier	E. P. Clark	Millbury, Mass	July 20, 1869	92, 797
Clothes-drier	H. H. Clark	Cincinnati, Ohio	Mar. 5, 1872	124, [illegible]
Clothes-drier	J. V. Clark	Camden, N. J	Apr. 12, 1870	101, 712
Clothes-drier	S. Cole	Pawtucket, R. I	July 25, 1865	48, 910
Clothes-drier	T. C. Collins	Little Hocking, Ohio	Sept. 7, 1869	94, 567
Clothes-drier	I. C. Conner	Dover, N. H	Sept. 4, 1866	57, 681
Clothes-drier	I. A. Coons	Middletown, Ohio	July 30, 1867	67, 270
Clothes-drier	E. B. Corby	Bloomfield, N. J	Dec. 27, 1870	110, 551
Clothes-drier	G. F. Couty	Paris, France	Nov. 7, 1871	120, 719
Clothes-drier	F. Crandall	Erie, Pa	July 9, 1867	66, 466
Clothes-drier	M. C. Cronk	Auburn, N. Y	Nov. 8, 1859	26, 023
Clothes-drier	D. E. Crosby and S. E. Strickland.	South Vineland, N. J	Aug. 24, 1869	93, 970
Clothes-drier	H. P. Crouse	Hartland, Mich	Apr. 7, 1868	76, 305
Clothes-drier	E. Culver	Shelburne, Mass	Feb. 21, 1865	46, 452
Clothes-drier	L. Cutting	San Francisco, Cal	Sept. 27, 1870	107, 763
Clothes-drier	W. A. Daggett	Landis Township, N. J	Oct. 4, 1870	108, 008
Clothes-drier	J. D. Davenport	North Providence, R. I	May 19, 1868	78, 069
Clothes-drier	A. Day	Skowhegan, Me	Feb. 7, 1871	111, 519
Clothes-drier	I. N. Deal	Brooklyn, N. Y	Feb. 11, 1868	74, 318
Clothes-drier	J. K. Derby	Jamestown, N. Y	June 13, 1871	115, 829
Clothes-drier	D. M. Devoe	New York, N. Y	Feb. 24, 1863	37, 739
Clothes-drier	E. Dickerman	Richmond, Vt	Oct. 30, 1860	30, 530
Clothes-drier	O. R. Dinsmoor	Auburn, N. H	July 13, 1858	20, 868
Clothes drier	J. P. Dorman	Galesburgh, Ill	July 4, 1865	48, 537
Clothes-drier	J. H. Doughty	New York, N. Y	Mar. 28, 1865	46, 999
Clothes-drier	J. H. Doughty	New York, N. Y	Aug. 27, 1867	68, 054
Clothes-drier	H. Du Bois	Marlborough, N. Y	Mar. 10, 1868	75, 251
Clothes-drier	J. H. Durand	Niles, Mich	Jan. 1, 1861	31, 013
Clothes-drier	J. T. Elliott	Grand Rapids, Mich	Feb. 5, 1867	61, 817
Clothes-drier	J. T. Elliott	Grand Rapids, Mich	Dec. 20, 1870	110, 347
Clothes-drier	J. Emmert	Dunleith, Ill	Oct. 25, 1870	108, 575
Clothes-drier	J. W. Epperson and T. W. Farrell.	Woodhull, Ill	Apr. 10, 1866	53, 797
Clothes-drier	J. B. Fellows	Augusta, Me	Nov. 28, 1871	121, 349
Clothes-drier	H. E. Fickett	Glen's Falls, N. Y	Sept. 4, 1860	29, 870
Clothes-drier	M. Flanders	Kendall, N. Y	May 16, 1871	114, 934
Clothes-drier	E. S. French	Westminster, Mass	Jan. 21, 1873	136, 107
Clothes-drier	W. Galaway	Sheboygan Falls, Wis	May 22, 1866	54, 886
Clothes-drier	C. A. Gale	Boston, Mass	Nov. 1, 1859	26, 004
Clothes-drier	S. L. George	Decatur, Mich	Aug. 7, 1866	56, 926
Clothes-drier	E. G. Gibson	Owego, N. Y	Sept. 28, 1858	21, 639
Clothes-drier	R. Gilbert	Morrisville, Vt	Aug. 28, 1860	29, 780
Clothes-drier	E. B. Gildersleeve	Wading River, N. Y	Mar. 18, 1873	136, 831
Clothes-drier	J. and W. Gnau	Louisville, Ky	Nov. 15, 1870	109, 200
Clothes-drier	C. Goldthwait	South Weymouth, Mass	Feb. 25, 1862	34, 487
Clothes-drier	C. Goldthwait	South Weymouth, Mass	Feb. 17, 1863	37, 686
Clothes-drier	F. W. Goodale and J. J. Brennan.	Danbury, Conn	May 30, 1871	115, 461
Clothes-drier	H. Gransden	Dubuque, Iowa	July 9, 1867	66, 582
Clothes-drier	A. Graves	Roscoe, Ill	June 24, 1873	140, 193
Clothes-drier	W. S. Graves and A. S. Capron	Kansas City, Mo., and Grass Lake, Mich.	Aug. 17, 1869	93, 708
Clothes-drier	J. S. Gray	Boston, Mass	Jan. 2, 1872	122, 378
Clothes drier	J. Greenhalgh, jr	Glendale, R. I	July 10, 1867	68, 735
Clothes-drier	A. W. Griffith	Boston, Mass	June 9, 1868	78, 661
Clothes-drier	A. M. Gurley	Waterville, N. Y	Feb. 9, 1869	86, 665
Clothes-drier	W. J. Hadden	Clinton, Ill	Dec. 3, 1872	133, 530
Clothes-drier	J. C. Haines	West Philadelphia, Pa	Apr. 19, 1870	101, 999
Clothes-drier	E. R. Hall	Utica, N. Y	Apr. 7, 1868	76, 441
Clothes-drier	R. Hamblin	Minshawaka, Ind	June 18, 1867	65, 905
Clothes-drier	J. J. Hamilton	Newcastle, Ind	Mar. 30, 1858	19, 772
Clothes-drier	G. H. Hammond	Davenport, N. Y	Aug. 4, 1868	80, 729
Clothes-drier	H. J. Hancock	New York, N. Y	Apr. 14, 1868	76, 756
Clothes-drier	A. Harbison	Newcastle, Pa	Sept. 7, 1869	94, 595
Clothes-drier	O. H. Harding and D. G. Williams.	Quincy, Mich	July 25, 1871	117, 414
Clothes-drier	O. H. Harding and J. Willis	Quincy, Mich	June 4, 1872	127, 481
Clothes-drier	T. G. Harold	Brooklyn, N. Y	Jan. 14, 1864	43, 113
Clothes-drier	E. Hause	Tecumseh, Mich	Apr. 6, 1869	88, 566
Clothes-drier	A. P. Hawse and G. R. Shippy	Wolcott, Vt	Apr. 6, 1869	88, 631
Clothes-drier	B. B. Hawse	Morrisville, Vt	Jan. 3, 1860	26, 674
Clothes-drier	M. R. Heliker	Norwalk, Ohio	Sept. 14, 1869	94, 822
Clothes-drier	D. K. Hickok	Morrisville, Vt	Dec. 20, 1859	26, 496
Clothes-drier	D. K. Hickok	Morrisville, Vt	Aug. 21, 1860	29, 694
Clothes-drier	A. H. Hill	Saint Johnsbury, Vt	Aug. 29, 1871	118, 612
Clothes-drier	J. O. Hill	Carbondale, Ill	June 11, 1872	127, 881
Clothes-drier	I. Hogeland	Indianapolis, Ind	Dec. 13, 1870	110, 137
Clothes-drier	A. S. Hopson	Plain View, Minn	Sept. 22, 1868	82, 408
Clothes-drier	L. Horn	Wolfborough, N. H	Dec. 15, 1863	40, 936
Clothes-drier	M. D. Hotchkiss	Sheboygan Falls, Wis	May 15, 1866	54, 727
Clothes-drier	H. A. Houghton	Lyme, N. H	May 27, 1862	35, 375
Clothes-drier	L. D. Howard	Saint Johnsbury, Vt	Feb. 14, 1871	111, 747
Clothes-drier	D. Hull	Newburgh, N. Y	Apr. 9, 1861	31, 980

Index of patents issued from the United States Patent Office from 1790 to 1873, inclusive—Continued.

Invention.	Inventor.	Residence.	Date.	No.
Clothes-drier	C. A. Hunt	Urbana, Ill	June 24, 1862	35, 684
Clothes drier	C. R. Hurlbut	Yorkshire, N. Y	May 24, 1859	24, 123
Clothes-drier	L. T. Ilgen	Cedarville, Ill	Aug. 14, 1866	57, 144
Clothes-drier	C. H. Jackson	Saint Louis, Mo	Nov. 21, 1865	51, 057
Clothes-drier	D. Johnson	Chicago, Ill	Aug. 2, 1859	24, 974
Clothes-drier	J. Johnson	New York, N. Y	Aug. 14, 1860	29, 599
Clothes-drier	J. Johnson	New York, N. Y	Oct. 30, 1860	30, 551
Clothes-drier	J. Johnson	Perry, Ill	June 13, 1871	115, 866
Clothes-drier	L. N. Johnson and B. Selloway	Montpelier, Vt	Aug. 24, 1869	93, 999
Clothes-drier	W. Johnston	Appleton, Wis	June 30, 1868	79, 474
Clothes-drier	J. Kaspar	Pomeroy, Ohio	Nov. 15, 1870	109, 219
Clothes-drier	G. A. Keene	Lynn, Mass	Dec. 12, 1871	121, 722
Clothes-drier	D. J. Kellogg	Rochester, N. Y	June 13, 1865	48, 183
Clothes-drier	H. H. Kendrick	Fulton, N. Y	May 10, 1870	102, 831
Clothes-drier	J. W. Kenning	Quincy, Mich	Dec. 13, 1870	110, 145
Clothes-drier	G. King	Eminence, Ky	May 11, 1869	89, 870
Clothes-drier	H. N. King and A. Z. Mason	Adrian, Mich	Oct. 1, 1867	69, 349
Clothes-drier	H. Knight	Westminster, Mass	Oct. 31, 1871	120, 443
Clothes-drier	A. Krotzer	Woodville, Ohio	Mar. 11, 1862	34, 637
Clothes-drier	C. H. L'Amoureux	New York, N. Y	Apr. 23, 1872	126, 062
Clothes-drier	J. Lee	Galesburgh, Ill	Sept. 24, 1861	33, 353
Clothes-drier	L. Ling	Pulaski, N. Y	Nov. 8, 1864	44, 964
Clothes-drier	C. F. Linscott	Chicago, Ill	July 4, 1871	116, 606
Clothes-drier	H. J. Lockwood	Wayne, Pa	Nov. 7, 1871	120, 756
Clothes-drier	J. C. Longshore	Mansfield, Ohio	Oct. 19, 1869	95, 917
Clothes-drier	I. Lynde	Marathon, N. Y	Mar. 3, 1863	37, 819
Clothes-drier	H. G. Mack	Oswego, N. Y	July 27, 1869	92, 982
Clothes-drier	N. F. Mathewson	Bloomington, Ill	Sept. 24, 1872	131, 696
Clothes-drier	W. C. Maynard	Marathon, N. Y	Dec. 19, 1871	122, 043
Clothes-drier	G. McNeil	Chestnut Hill, Pa	Feb. 11, 1862	34, 372
Clothes-drier	R. Merrill	Elmira, N. Y	Feb. 21, 1860	27, 231
Clothes-drier	A. S. Miller	Nottingham, Ind	Sept. 9, 1873	142, 714
Clothes-drier	C. G. Miller	Brattleborough, Vt	Nov. 3, 1863	40, 493
Clothes-drier	D. Miller	Marietta, Ohio	Oct. 31, 1871	120, 452
Clothes-drier	R. M. Miller	Port Andrew, Wis	Dec. 28, 1869	98, 398
Clothes-drier	A. J. Mills and E. M. Hewett	Scott, N. Y	Apr. 23, 1867	64, 129
Clothes-drier	W. N. and A. K. Moore	Neenah, Wis	Nov. 29, 1870	109, 647
Clothes-drier	R. M. Morriell	Plymouth, Ind	Sept. 3, 1867	68, 526
Clothes-drier	S. Morrill	Andover, N. H	Nov. 11, 1856	16, 065
Clothes drier	D. Morris	Cutler, Ohio	May 24, 1870	103, 486
Clothes-drier	J. F. Mullowny	Pittsburgh, Pa	June 20, 1871	116, 081
Clothes-drier	J. Mumma	Middletown, Ohio	Jan. 19, 1869	86, 095
Clothes-drier	F. S. Nettleton and E. J. Fuller	Leominster, Mass	Mar. 22, 1870	101, 153
Clothes-drier	G. W. Newell	Lawrence, Mass	Feb. 17, 1863	37, 724
Clothes-drier	J. J. Newman	Middletown, Ohio	Aug. 6, 1867	67, 567
Clothes-drier	R. L. Normando	Higginsville, N. Y	Nov. 28, 1872	121, 404
Clothes-drier	W. B. Noyes	Dorchester, N. H	Feb. 18, 1868	74, 715
Clothes-drier	T. Oakley	Booneville, N. Y	Aug. 10, 1869	93, 551
Clothes-drier	R. H. Oates	Toronto, Canada	Nov. 6, 1866	59, 527
Clothes-drier	A. O'Dell	Napanee, Canada	Nov. 1, 1870	108, 818
Clothes-drier	G. Oldham	Westfield, N. Y	Aug. 24, 1869	94, 023
Clothes-drier	W. W. S. Orberton	Haverhill, Mass	May 29, 1866	55, 208
Clothes-drier	J. P. Packer	Flemington, Pa	Dec. 10, 1872	133, 882
Clothes-drier	G. W. Page	Henrietta, Ohio	Nov. 12, 1872	132, 920
Clothes-drier	E. L. Parker	Painesville, Ohio	Feb. 21, 1871	112, 072
Clothes-drier	S. Parker	Forest Grove, N. J	Mar. 27, 1866	53, 541
Clothes-drier	A. H. Patch	Hamilton, Mass	June 28, 1870	104, 761
Clothes-drier	A. H. Patch	Hamilton, Mass	Feb. 7, 1871	111, 563
Clothes-drier	W. P. Patton	Harrisburgh, Pa	May 19, 1863	38, 631
Clothes-drier	J. A. Pease	New York, N. Y	Oct. 21, 1862	36, 726
Clothes-drier	A. W. Phillips	Fairfield, N. Y	June 25, 1872	128, 323
Clothes-drier	R. Phillips	Boston, Mass	Mar. 8, 1870	100, 662
Clothes-drier	E. N. Porter and I. J. Currier	Wolcott, Vt	Mar. 28, 1871	113, 204
Clothes-drier	E. T. Porter	Washington, D. C	Sept. 28, 1858	21, 626
Clothes-drier	Z. B. Putnam and H. H. McDonald.	Belfast, Me	Dec. 26, 1871	122, 280
Clothes-drier	G. Race	Norwich, N. Y	June 28, 1859	24, 575
Clothes-drier	D. B. Randall and A. A. Williams.	Glover, Vt	Sept. 17, 1867	69, 023
Clothes-drier	W. F. Redding	Saratoga Springs, N. Y	July 9, 1867	66, 520
Clothes-drier	J. W. Reed and H. A. Jones	Pittsburgh, Pa	Aug. 22, 1871	118, 272
Clothes-drier	G. W. Richardson	Taunton, Mass	Sept. 6, 1870	107, 199
Clothes-drier	C. Robinson	Cambridgeport, Mass	Oct. 2, 1860	30, 249
Clothes-drier	C. Robinson	Cambridgeport, Mass	Apr. 23, 1861	32, 147
Clothes-drier	C. B. Rogers	Plainfield, N. J	Feb. 26, 1867	62, 443
Clothes-drier	E. D. Sanford	Baltimore, Md	June 25, 1867	66, 045
Clothes-drier	C. G. Sargent	Chelsea, Mass	Jan. 29, 1861	31, 260
Clothes-drier	A. F. Saunders	Chelsea, Mass	Aug. 25, 1863	39, 706
Clothes-drier	A. Scheff	Raymondville, N. Y	Sept. 13, 1870	107, 295
Clothes-drier	H. Schryver	Kingston, N. Y	May 3, 1870	102, 717
Clothes-drier	J. Seeman and S. P. Catrow	Middletown, Ohio	Jan. 22, 1867	61, 473
Clothes-drier	C. C. Sheldon	Randolph, N. Y	Jan. 23, 1872	123, 051
Clothes-drier	E. Sims	Antwerp, N. Y	Jan. 31, 1865	46, 151
Clothes-drier	G. P. Sisson	Florence, Mass	Aug. 27, 1867	68, 317
Clothes-drier	W. A. Sleppy	Berwick, Pa	Dec 19, 1871	122, 071
Clothes-drier	D. C. Smart	Cambridgeport, Mass	July 4, 1871	116, 639
Clothes-drier	C. F. Smith and A. Heather	Chicago, Ill	Sept. 19, 1871	119, 191
Clothes-drier	G. C. Smith	Hamilton, Ohio	Oct. 3, 1871	119, 539
Clothes-drier	G. C. Smith and C. H. Dietrich	Middleville, Mich	Nov. 7, 1871	120, 7[illegible]5
Clothes-drier	O. S. Smith and C. R. Hopkins	Middletown, Conn	Nov. 21, 1871	121, 208
Clothes-drier	T. F. Snover	Menasha, Wis	Aug. 14, 1866	57, 204
Clothes-drier	L. A. Stare	Oconomowoc, Wis	Apr. 19, 1870	102, 059
Clothes-drier	A. Stechschult	Glendorf, Ohio	Mar. 28, 1871	113, 106
Clothes-drier	E. E. Stedman	Randolph, Ohio	Aug. 1, 1871	117, 695
Clothes-drier	H. M. Stevenson	South Peacham, Vt	Jan. 3, 1871	110, 691
Clothes-drier	R. B. Stillman	Almond, N. Y	June 16, 1868	78, 902
Clothes-drier	A. C. Stowe	San José, Cal	Apr. 2, 1872	125, 227
Clothes-drier	A. J. Stowell	Dunlap, Iowa	June 4, 1872	127, 527

Index of patents issued from the United States Patent Office from 1790 *to* 1873, *inclusive*—Continued.

Invention.	Inventor.	Residence.	Date.	No.
Clothes-drier	L. M. Streetor	Oshkosh, Wis	June 14, 1870	104, 227
Clothes-drier	H. D. Struse	Brooklyn, N. Y	Sept. 29, 1868	82, 657
Clothes-drier	T. A. Summers	Rochester, N. Y	Sept. 27, 1864	44, 469
Clothes-drier	H. Swan	Woodstock, Vt	Jan. 17, 1871	111, 011
Clothes-drier	A. L. Taylor	Springfield, Vt	Sept. 7, 1869	94, 527
Clothes-drier	E. B. Taylor	Natick, Mass	Mar. 7, 1865	46, 731
Clothes-drier	S. H. Tift	Morrisville, Vt	July 20, 1858	20, 964
Clothes-drier	S. H. Tift	Morrisville, Vt	July 27, 1858	21, 035
Clothes-drier	G. F. Tilton	Salem, Mass	Nov. 14, 1865	50, 968
Clothes-drier	S. H. Titus	Pennington, N. J	May 28, 1867	65, 305
Clothes-drier	L. A. Towne	La Crosse, Wis	July 20, 1869	92, 905
Clothes-drier	S. I. Trask	Guilford, N. Y	Apr. 3, 1866	53, 705
Clothes-drier	W. Tredenick	Providence, R. I	Dec. 27, 1870	110, 607
Clothes-drier	J. N. Valley	Detroit, Mich	May 17, 1870	103, 257
Clothes-drier	H. J. and D. D. Van Valkenburgh.	Rabbit River, Mich	Sept. 17, 1872	131, 483
Clothes-drier	J. K. Wagner	Potsdam, N. Y	Feb. 14 1871	111, 888
Clothes-drier	J. Walker	Kansas City, Mo	Dec. 17, 1867	72, 341
Clothes-drier	L. B. Waterman	Chicago, Ill	July 28, 1868	80, 316
Clothes-drier	O. H. Waters	Baltimore, Md	Sept. 13, 1859	25, 474
Clothes-drier	J. R. Watkins	Maine Prairie, Minn	June 30, 1868	79, 524
Clothes-drier	D. R. Watson	Moawequa, Ill	Jan. 14, 1873	134, 828
Clothes-drier	S. Way	La Porte, Ind	Feb. 23, 1869	87, 312
Clothes-drier	M. B. Wheaton	New York, N. Y	Apr. 14, 1868	76, 862
Clothes-drier	J. B. White	Detroit, Mich	Dec. 8, 1868	84, 849
Clothes-drier	L. M. Whitman	Sterling, Ill	May 9, 1871	114, 630
Clothes-drier	H. Willard	Vergennes, Vt	June 18, 1867	65, 979
Clothes-drier	E. B. Winship	Racine, Wis	July 16, 1872	129, 385
Clothes-drier	L. Winterhalder and D. Wilson	New York, N. Y	Apr. 13, 1869	88, 937
Clothes-drier	D. Witt	Hubbardston, Mass	Dec. 29, 1868	85, 499
Clothes-drier	M. H. Wood	Carlisle, Ohio	Aug. 8, 1871	117, 847
Clothes-drier	G. L. Woods	Manchester, N. H	May 28, 1872	127, 206
Clothes-drier	S. Woodward	Sutton, N. H	Nov. 21, 1854	11, 985
Clothes-drier	L. Woodworth	Morrison, Ill	Jan. 22, 1867	61, 498
Clothes-drier	L. C. Wright	Lockport, N. Y	Nov. 20, 1866	59, 935
Clothes-drier, Adjustable	H. S. Boynton	Cortland, Vt	Apr. 14, 1868	76, 589
Clothes-drier and awning combined	C. E. Hyde	Oswego, N. Y	Nov. 28, 1871	121, 371
Clothes-drier and ironing-board	M. Power	Chicago, Ill	Nov. 21, 1871	121, 127
Clothes-drier and ironing-table	W. Reichenback	Chicago, Ill	Apr. 6, 1869	88, 739
Clothes-drier and pendant seat combined	J. Hirons	Buchanan, Mich	Nov. 7, 1871	120, 585
Clothes-drier and stand	W. H. Earnest	Parkersburgh, W. Va	Sept. 3, 1867	68, 355
Clothes-drier and stand combined	G. Favinger	Pittsford, Mich	June 11, 1867	65, 735
Clothes-drier and stove-pipe shelf	G. E. Hoyt	Hebron, N. H	Dec. 7, 1869	97, 511
Clothes-drier attachment for stove-pipes	A. P. Anthony	Morrison, Ill	May 28, 1867	65, 152
Clothes-drier, Detachable stove-pipe	H. Whitney	Chicago, Ill	June 22, 1869	91, 803
Clothes, Drier for scoured	J. Braun	Bridgewater, Pa	Sept. 24, 1867	69, 171
Clothes-drier for stove-pipes	J. B. Waterman	Chicago, Ill	Dec. 7, 1869	97, 736
Clothes-drier, Revolving	K. H. Elliott and J. Brown	Morrisville, Vt	Nov. 12, 1861	33, 697
Clothes-drier, Revolving	J. McCaskey, jr	Orville, Ohio	June 1, 1869	90, 764
Clothes-drier, Rotating	W. Boyers	Mount Carroll, Ill	Mar. 25, 1862	34, 802
Clothes-drier, Stove-pipe	C. M. Herreman	Mankato, Minn	Aug. 22, 1871	118, 241
Clothes-drier, Wardrobe	B. S Brown	New York, N. Y	Mar. 5, 1872	124, 247
Clothes-drier, Window	W. Powers	Preston, N. Y	Jan. 6, 1863	37, 359
Clothes-drier, Window	J. R. and J. E. Shepard	Waukegan, Ill	July 15, 1862	35, 898
Clothes-drying frame	C. Stone	Ravenna, Ohio	June 22, 1858	20, 669
Clothes-drying machine	J. O. Luther and P. Staab	Grafton, Wis	June 27, 1871	116, 333
Clothes-drying machine	W. Price	Cincinnati, Ohio	Dec. 20, 1864	45, 521
Clothes-fastener	P. Frost	Springfield, Vt	Mar. 8, 1864	41, 838
Clothes-fastener	L. Leavenworth	Trumansburgh, N. Y	Mar. 8, 1859	23, 178
Clothes-frame	E. R. Bigelow	Salem, Mass	Sept. 10, 1861	33, 232
Clothes-frame	C. A. Boynton	Hyde Park, Vt	July 3, 1860	28, 961
Clothes-frame	H. Buell	Mount Morris, N. Y	June 7, 1864	43, 069
Clothes-frame	J. Burr	Baltimore, Md	July 26, 1859	24, 904
Clothes-frame	P. Cody	Hamilton, N. Y	Sept. 25, 1860	30, 124
Clothes-frame	D. C. Colby	Keene, N. H	Jan. 25, 1859	22, 767
Clothes-frame	W. A. Daggett	South Vineland, N. J	Apr. 20, 1869	89, 131
Clothes-frame	J. Danner	Canton, Ohio	July 28, 1863	39, 340
Clothes-frame	J. Danner	Canton, Ohio	Jan. 26, 1869	86, 140
Clothes-frame	E. Dickerman	Middlefield, Conn	Aug. 20 1861	33, 075
Clothes-frame	J. E. Earle	New Haven, Conn	Nov. 25, 1862	36, 994
Clothes-frame	C. J. Ferguson	New York, N. Y	July 17, 1860	29, 151
Clothes-frame	H. T. Field and D. Harrington	Worcester, Mass	Nov. 5, 1872	132, 717
Clothes-frame	H. M. Fletcher	Newport, N. H	Apr. 5, 1859	23, 460
Clothes frame	J. Fraser	Rochester, N. Y	May 15, 1860	28, 266
Clothes-frame	L. F. Frazee	Tottenville, N. Y	Feb. 26, 1861	31, 538
Clothes-frame	J. Gasser	Toledo, Ohio	July 12, 1859	24, 731
Clothes-frame	J. Greek	Evansville, Ind	Feb. 23, 1864	41, 698
Clothes-frame	A. Greenwood	Toulon, Ill	July 17, 1860	29, 163
Clothes-frame	W. Hathaway	Worcester, Mass	Dec. 21, 1858	22, 398
Clothes-frame	W. Hathaway	Providence, R. I	July 3, 1860	28, 983
Clothes-frame	D. Henshaw	Fitchburgh, Mass	Feb. 1, 1859	22, 804
Clothes-frame	D. E. Holmes	Halifax, Mass	Oct. 11, 1859	25, 737
Clothes-frame	H. J. Holmes	Warren, Mass	July 12, 1859	24, 739
Clothes-frame	H. Hempton	Fairhaven, Mass	Feb. 26, 1850	7, 120
Clothes-frame	L. L. Knight	Barre, Mass	Apr. 16, 1861	32, 104
Clothes-frame	M. J. Knox	Knox Corners, N. Y	June 18, 1861	32, 569
Clothes-frame	M. J. Knox	Knox Corners, N. Y	Apr. 22, 1862	35, 027
Clothes-frame	S. T. Lamb	New Washington, Ind	Sept. 18, 1860	30, 069
Clothes-frame	H. Littlejohn	Troy, N. Y	Jan. 7, 1862	34, 113
Clothes-frame	J. P. Mayhugh	Leitersburgh, Md	Feb. 24, 1863	37, 762
Clothes-frame	J. C. Miller	Rockport, Ind	Dec. 30, 1873	146, 087
Clothes-frame	H. A. Nutting	South Amherst, Ind	May 10, 1859	23, 941
Clothes-frame	E. Page	Streetsborough, Ohio	June 15, 1858	20, 579
Clothes-frame	S. W. and J. F. Palmer	Auburn, N. Y	Oct. 18, 1859	25, 868
Clothes-frame	H. Parkhurst	De Kalb, Ill	Aug. 9, 1859	25, 039
Clothes-frame	R. Ramsey	New Wilmington, Pa	Apr. 12, 1859	23, 607
Clothes-frame	D. Read	Hamilton, N. Y	Mar. 29, 1859	23, 392

Index of patents issued from the United States Patent Office from 1790 *to* 1873, *inclusive*—Continued.

Invention.	Inventor.	Residence.	Date.	No.
Clothes-frame	T. S. Scoville	New York, N. Y	Jan. 3, 1860	26, 710
Clothes-frame	L. B Waterman	Chicago, Ill	Feb. 15, 1859	22, 993
Clothes, Hand-rubber for washing	G. F. J. Colburn	Newark, N. J	Mar. 8, 1870	100, 603
Clothes-hanger	J. O. Montignani	Albany, N. Y	Oct. 14, 1862	36, 666
Clothes hanger	A. J. Walker	New York, N. Y	June 14, 1864	43, 144
Clothes-holder	W. Z. Brown	Decatur, Ill	Nov. 7, 1871	120, 570
Clothes-holder for clothes-lines	M. V. Bulla	South Bend, Ind	Nov. 22, 1870	109, 382
Clothes-hook	G. M. Hubbard	New Haven, Conn	Feb. 13, 1872	123, 699
Clothes-hook	O. A. North	New Britain, Conn	Jan. 12, 1869	85, 756
Clothes-hook	S. S. Putnam	Dorchester, Mass	Sept. 17, 1867	69, 021
Clothes-hook	S. S. Putnam	Dorchester, Mass	Oct. 8, 1867	69, 587
Clothes-hook	H. M. Whitmarsh and S. S. Putnam.	Abington and Dorchester, Mass.	Jan. 29, 1867	61, 590
Clothes hook and line holder combined	T. Weaver	Harrisburgh, Pa	Aug. 18, 1868	81, 312
Clothes-horse	E. Culver, jr	Shelburne, Falls, Mass	Aug. 17, 1858	21, 231
Clothes-horse	T. S. Lewis	Kendall's Mills, Me	Dec. 28, 1858	22, 435
Clothes-horse, &c., Combined	H. L. Stillson	Plattsburgh, N. Y	Sept. 15, 1868	82, 256
Clothes-horse frames, Connecting	H. Luther	Providence, R. I	May 19, 1841	2, 098
Clothes-ironing apparatus	C. Alden	Cassadaga, N. Y	Nov. 8, 1859	26, 005
Clothes-ironing machine	J. Shaefer	Lancaster, Pa	Sept. 7, 1858	21, 450
Clothes, Knuckle-protector for washing	G. W. Doty	Wooster, Ohio	Dec. 5, 1871	121, 498
Clothes-line	J. W. Bliss	Hartford, Conn	Apr. 23, 1867	64, 063
Clothes-line	P. C. Johnson	Central City, Colo	Nov. 30, 1869	97, 409
Clothes-line adjuster	R. White	Newport, Pa	Oct. 6, 1868	82, 911
Clothes line and clamp	A. D. Rust	Vernon, Mich	Feb. 5, 1867	61, 875
Clothes-line clamp	C. C. Fellows	Centre Sandwich, N. H	Oct. 22, 1867	69, 977
Clothes-line clamp	C. W. Howard	Philadelphia, Pa	Oct. 2, 1866	58, 419
Clothes-line clamp	G. R. Nebinger	Lewisberry, Pa	Feb. 18, 1868	74, 713
Clothes-line clamp	T. W. Owens	Granville, Ohio	July 19, 1870	105, 483
Clothes-line clamp	W. S. and E. H. Shoemaker	Towsontown, Md., and Lancaster, Ohio.	May 11, 1869	90, 055
Clothes-line clamp	A. T. Thayer and J. B. Le Maire.	New York, N.Y., and Hudson City, N. J.	Oct. 1, 1872	131, 912
Clothes-line clamp	G. W. Wilbar	Taunton, Mass	Feb. 1, 1870	99, 385
Clothes-line clamp	W. Winter	Philadelphia, Pa	Aug. 14, 1866	57, 263
Clothes-line clip	J. S. Rowley	Chateaugay, N. Y	July 16, 1867	66, 892
Clothes-line conductor	D. Reed and A. Shaeffer	Medway, Ohio	Apr. 4, 1871	113, 563
Clothes-line fastener	J. G. and P. A. Ames	Baltimore, Md	June 24, 1873	140, 234
Clothes-line fastener	F. Clymer	Galion, Ohio	Mar. 9, 1869	87, 636
Clothes-line fastener	E. K. Elliott and N. J. Baxter	Cuyahoga Falls, Ohio, and Washington, D. C.	Aug. 2, 1870	105, 927
Clothes-line fastener	H. J. Hendey	Wolcottsville, Conn	Feb. 14, 1871	111, 843
Clothes-line fastener	B. B. Herrick	Decatur, Mich	Nov. 29, 1870	109, 733
Clothes-line fastener	J. Johnson, S. J. Smith, and S. Ingersoll.	Brooklyn, N. Y	Mar. 23, 1869	88, 175
Clothes-line fastener	M. H. Lineback	Greenfield, Ind	Sept. 21, 1869	95, 120
Clothes-line fastener	M. McManus	East Saginaw, Mich	Jan. 2, 1872	122, 476
Clothes-line fastener	H. Ogborn	Richmond, Ind	Mar. 1, 1870	100, 438
Clothes-line fastening	S. A. Barr	Pittsburgh, Pa	July 30, 1867	67, 156
Clothes-line holder	H Browne	Des Moines, Iowa	May 31, 1870	103, 714
Clothes-line holder	D. Bull	Amboy, Ill	Sept. 28, 1869	95, 316
Clothes-line holder	A. Comey	Cuyahoga Falls, Ohio	Feb. 25, 1868	74, 798
Clothes-line holder	W. Cooke	New York, N. Y	Aug. 19, 1873	141, 921
Clothes-line holder	A. Cooper	Harrisburgh, Pa	Oct 19, 1869	95, 881
Clothes-line holder	J. Davis	Harrisburgh, Pa	May 17, 1870	103, 155
Clothes-line holder	E. Dingman	Liverpool, N. Y	June 20, 1871	116, 034
Clothes-line holder	J. W. Gladding	Normal, Ill	Apr. 28, 1868	77, 274
Clothes-line holder	J. B. Habecker	Newport, Pa	Nov. 21, 1871	121, 167
Clothes-line holder	A. L. Hurtt	Monticello, Ind	May 3, 1870	102, 548
Clothes-line holder	C. A. Kalck	Philadelphia, Pa	June 21, 1870	104, 597
Clothes-line holder	H. Morgan	Springfield, Mass	May 12, 1868	77, 750
Clothes-line holder	H. Ogborn	Richmond, Ind	July 6, 1869	92, 206
Clothes-line holder	O. S. Perkins and J. R. Richards	Mount Joy, Pa	July 7, 1868	79, 593
Clothes-line holder	J. C. Rankin	Mount Vernon, N. Y	Mar. 15, 1870	100, 803
Clothes-line holder	P. Riordan	Washington, D. C	Dec. 31, 1867	72, 908
Clothes-line holder	J. Robbins	Centralia, Ill	Sept. 14, 1869	94, 914
Clothes-line holder	A. J. Simpson	Washington, D. C	July 2, 1867	66, 403
Clothes-line holder	E. W. Talbott	Napoleon, Ohio	Jan. 3, 1871	110, 802
Clothes-line holder	C. L. Topliff	Brooklyn, N. Y	May 9, 1871	114, 728
Clothes-line holder	J. L. Wilcox	Preble, N. Y	Nov. 5, 1867	70, 665
Clothes-line holder and stretcher	N. E. Buffington	North Providence, R. I	Jan. 21, 1870	104, 419
Clothes-line hook	J. Garvey and M. H. Kimball	San Francisco, Cal	Nov. 1, 1870	108, 776
Clothes-line hook	E. H. Gray	Winchester, Ill	May 5, 1868	77, 605
Clothes-line hook	J. L. Howard	New York, N. Y	July 30, 1867	67, 196
Clothes-line-hook block	J. W. Norcross	East Boston, Mass	July 30, 1867	67, 341
Clothes-line housing	A. Trumbull	New Britain, Conn	May 30, 1871	115, 544
Clothes-line operator	C. Barron	La Fayette, Ind	Oct. 18, 1870	108, 431
Clothes-line pole	C. B. Brown	New Rochelle, N. Y	Mar. 11, 1873	136, 641
Clothes-line post	B. Chesnut	Philadelphia, Pa	Oct. 19, 1858	21, 818
Clothes-line protector and stretcher	W. J. Ripley	Cincinnati, Ohio	May 2, 1871	114, 345
Clothes-line reel	J. Brizee	Alvarado, Cal	May 13, 1873	138, 785
Clothes-line reel	G. F. Corliss	Mansfield, Ohio	Nov. 11, 1873	144, 3[illegible]7
Clothes-line reel	S. and J. Drake	Canton, Ill	July 2, 1872	128, 539
Clothes-line reel	G. B. Griffin	Madison, Wis	Aug. 20, 1867	67, 872
Clothes-line reel	G. Holman	Waterville, N. Y	Nov. 15, 1870	109, 318
Clothes-line reel	I. E. Loughborough	Pittsford, N. Y	Jan. 29, 1867	61, 672
Clothes-line reel	H. W. Prouty and A. Thompson	Boston, Mass	Mar. 19, 1872	124, 698
Clothes-line reel	D. D. Pugh	Brooklyn, N. Y	June 21, 1870	104, 495
Clothes-line reel	C. Rosenthal	Philadelphia, Pa	July 22, 1873	141, 015
Clothes-line reel	J. D. Starritt	Chicago, Ill	Aug. 6, 1867	67, 459
Clothes-line reel	C. H. Straffin	Boston, Mass	Nov. 21, 1871	121, 135
Clothes-line reel	J. Valentine and H. B. Stevens	Buffalo, N. Y	Dec. 15, 1868	84, 919
Clothes line reel and house	M. H. Card and A. Sallee	Fulton, Ill	Feb. 12, 1867	61, 924
Clothes-line reel, Automatic	W. Farrah	Des Moines, Iowa	Aug. 9, 1870	106, 146
Clothes-line sliding-support	M. Bubser	Jersey City, N. J	Jan. 14, 1873	134, 849
Clothes-line spring-clamp	E. S. Haskins	Boston, Mass	Mar. 14, 1854	1[illegible], 635
Clothes-line stretcher	S. Crowell	Philadelphia, Pa	Jan. 26, 1869	86, 139
Clothes-line support	J. E. Roache	New York, N. Y	Aug. 26, 1873	142, 278

Index of patents issued from the United States Patent Office from 1790 *to* 1873, *inclusive*—Continued.

Invention.	Inventor.	Residence.	Date.	No.
Clothes-line supporter	J. Andrews	New Bedford, Mass	Oct. 26, 1869	96, 183
Clothes-line supporter	T. Riley	Williamsburgh, N. Y	July 4, 1871	116, 632
Clothes-line supporter	F. W. Tilton and M. C. Swift	New Bedford, Mass	Aug. 18, 1868	81, 310
Clothes-lines, &c., Device for fastening	A. E. Schartz	New York, N. Y	Mar. 11, 1873	136, 768
Clothes-lines, Machinery for straining and preserving.	E. Allyn and C. B. Hildreth	Boston, Mass	Nov. 26, 1840	1, 863
Clothes-lines, Spring-clamp for	F. S. Hotchkiss and C. W. Blakeslee.	Northfield, Conn	Dec. 13, 1853	10, 311
Clothes-lines, Spring-clamp for	D. M. Smith	Springfield, Vt	Oct. 25, 1853	10, 163
Clothes, Machine for washing and pressing	J. Read	New York, N. Y	Apr. 10, 1839	1, 116
Clothes-mangle	D. Arndt	Toledo, Ohio	Mar. 2, 1869	87, 457
Clothes-mangle	F. A. Desloge	Saint Louis, Mo	June 8, 1869	91, 095
Clothes-mangle	S. Short	Cincinnati, Ohio	Dec. 23, 1873	145, 910
Clothes-mangle	H. E. Smith	New York, N. Y	Nov. 2, 1869	96, 359
Clothes-mangle	H. E. Smith	New York, N. Y	Nov. 2, 1869	96, 360
Clothes-mangle	E. D. Taylor and D. Cohn	Hornellsville, N. Y	Nov. 23, 1869	97, 245
Clothes, Mode of marking and cutting out	A. Ward	Huntsville, Ala	June 16, 1821	....
Clothes, Mode of scouring and washing	J. Goulding	Dedham, Mass	Aug. 24, 1827	....
Clothes or towel rack	F. A. Balch	Hingham, Wis	July 9, 1867	66, 442
Clothes-pin	S. Aldrich	Springfield, Vt	Sept. 14, 1852	9, 257
Clothes-pin	T. C. Ball	Bellows Falls, Vt	Apr. 2, 1872	125, 253
Clothes-pin	O. A. Bishop	Chicago, Ill	Mar. 7, 1871	112, 315
Clothes-pin	H. T. Boutell	Springfield, Vt	Aug. 27, 1867	68, 161
Clothes-pin	G. Bradley and N A. Walker	Rockford, Ill	Nov. 29, 1870	109, 580
Clothes-pin	A. W. Brinkerhoff	Upper Sandusky, Ohio	July 23, 1867	67, 160
Clothes-pin	R. G. Britton	Springfield, Vt	Oct. 1, 1867	69, 398
Clothes-pin	L. Bullock	Bellows Falls, Vt	Oct. 29, 1872	132, 627
Clothes-pin	B. Burling	Whitehall, N. Y	Sept. 26, 1871	119, 311
Clothes-pin	W. E. Burlingame	Willet, N. Y	Aug. 17, 1869	93, 672
Clothes-pin	E. A. Cone	Milford, Mich	Sept. 9, 1862	36, 394
Clothes-pin	N. B. Cooper	Liberty, Ind	Nov. 26, 1867	71, 457
Clothes-pin	J. O. Couch	Middlefield, Conn	June 9, 1868	78, 651
Clothes-pin	W. M. Doty	New York, N. Y	Mar. 26, 1867	63, 231
Clothes-pin	W. M. Doty	New York, N. Y	Dec. 31, 1867	72, 823
Clothes-pin	W. M. Doty	New York, N. Y	May 5, 1868	77, 598
Clothes-pin	R. Emerson	Rockford, Ill	June 11, 1872	127, 751
Clothes-pin	G. K. Farrington	Alcatraz Island, Cal	June 28, 1870	104, 720
Clothes-pin	P. Finley	Memphis, Tenn	Apr. 28, 1868	77, 268
Clothes-pin	T. L. Goble	Orange, N. Y	Dec. 18, 1866	60, 627
Clothes-pin	K. H. Goss	Cedar Springs, Mich	Apr. 4, 1871	113, 289
Clothes-pin	J. Greenwood	Fitchburgh, Mass	Nov. 15, 1864	45, 119
Clothes-pin	J. Haigney and F. M. Hedman	East Boston, Mass	Sept. 15, 1868	82, 114
Clothes-pin	H. D. Haraden	Hartford, Vt	Apr. 21, 1868	77, 037
Clothes-pin	D. C. Harlow	Hannibal, Mo	July 30, 1872	129, 950
Clothes-pin	G. A. Harris	Buchanan, Mich	Jan. 3, 1871	110, 649
Clothes-pin	D. K. Hickok	Morrisville, Vt	Mar. 10, 1868	75, 422
Clothes-pin	R. W. Huston	Washington, D. C	Mar. 26, 1872	124, 955
Clothes-pin	J. P. R. James	Read's Landing, Minn	Sept. 3, 1867	68, 510
Clothes-pin	C C. Johnson	Springfield, Vt	Apr. 2, 1867	63, 393
Clothes-pin	P. Johnson	Wauconda, Ill	Mar. 30, 1869	88, 390
Clothes-pin	D. D. Jones	Scranton, Pa	Jan. 21, 1868	73, 609
Clothes-pin	J. E. Lines	Bryan, Ohio	Aug. 10, 1869	93, 456
Clothes-pin	A. B. Lipsey	West Hoboken, N. J	Apr. 4, 1871	113, 313
Clothes-pin	A. B. Lipsey	West Hoboken, N. J	Feb. 27, 1872	124, 071
Clothes-pin	S. B. Lucas	Hinsdale, N. Y	Apr. 2, 1872	125, 311
Clothes-pin	L. Matthews	Antrim, Ohio	Sept. 29, 1868	82, 624
Clothes-pin	H. Mellish	Walpole, N. H	Oct. 17, 1871	119, 938
Clothes-pin	H. Mellish	Walpole, N. H	Sept. 23, 1873	143, 024
Clothes-pin	W. Miller	Boston, Mass	Dec. 20, 1870	110, 263
Clothes-pin	W. Patterson	Lowell, Mass	May 14, 1867	64, 698
Clothes-pin	D. Pierce	Sunapee, N. H	May 25, 1858	20, 364
Clothes-pin	S. Prior	Salem, N. J	Mar. 22, 1832	....
Clothes-pin	J. A. Rand	Morrisville, Vt	Sept. 8, 1868	81, 945
Clothes-pin	J. G. Roth	New York, N. Y	June 30, 1868	79, 502
Clothes-pin	R. S. Sanborn	Ripon, Wis	Sept. 11, 1866	57, 977
Clothes-pin	H. W. Sargeant, jr	Boston, Mass	Dec. 13, 1864	45, 458
Clothes-pin	H. W. Sargeant, jr	Boston, Mass	Apr. 11, 1865	47, 223
Clothes-pin	I. W. Searles	Tiffin, Ohio	June 21, 1870	104, 654
Clothes-pin	E. Seaver	Boston, Mass	July 9, 1867	66, 523
Clothes-pin	E. Seaver	Boston, Mass	May 5, 1868	77, 540
Clothes-pin	L. T. Simon	New York, N. Y	June 1, 1869	90, 881
Clothes-pin	D. M. Smith	Springfield, Vt	Apr. 9, 1867	63, 759
Clothes-pin	L. and C. H Sprague	Henderson, N. Y	Dec. 26, 1871	122, 199
Clothes-pin	D. M. Strain, jr	Des Moines, Iowa	July 13, 1869	92, 488
Clothes-pin	A. L. Taylor	Springfield, Vt	Apr. 7, 1868	76, 547
Clothes-pin	W. H. Towers	New York, N. Y	June 21, 1859	24, 503
Clothes-pin	C. N. Tyler	New York, N. Y	Mar. 23, 1869	88, 097
Clothes-pin	V. D. Urso and B. Charles	Evansville, Ind	Aug. 12, 1873	141, 740
Clothes-pin	F. Walker	New Orleans, La	Feb. 9, 1869	86, 888
Clothes-pin	W. G. Ward	Savona, N. Y	Sept. 11, 1866	58, 007
Clothes-pin	W. G. Ward	Savona, N. Y	Feb. 19, 1867	62, 240
Clothes-pin	H. J. Wattles	Rockford, Ill	Feb. 28, 1871	112, 303
Clothes-pin	H. J. Wattles	Rockford, Ill	Nov. 28, 1871	121, 303
Clothes-pin	W. Wellington	Rockford, Ill	Mar. 1, 1870	100, 345
Clothes-pin	W. Wellington	Rockford, Ill	Feb. 4, 1873	135, 501
Clothes-pin	H. Wells	Cambridge, Mass	July 9, 1872	128, 934
Clothes-pin	W. H. Wells and J. Hawse	Newport Centre, Vt	Sept. 24, 1867	69, 228
Clothes pin holder	V. Reifsnider	Chicago, Ill	Aug. 17, 1869	93, 905
Clothes-pin joints, Machine for wiring	A. C. Mason	Springfield, Vt	Sept. 6, 1859	25, 344
Clothes-pin machine	H. and M. Blake	Hartland, Vt	May 1, 1855	12, 775
Clothes-pin machine	S. Inman	Rockford, Ill	Feb. 27, 1872	124, 137
Clothes-pin-making machine	A. J. Ockington	Stratford Hollow, N. H	Sept. 14, 1869	94, 838
Clothes pin or clamp	G. F. Barden	Dover, N. H	Nov. 13, 1866	59, 545
Clothes pin or clamp	P. Gardner	Gloucestershire, England	July 20, 1869	92, 715
Clothes pin or clasp	A. L. Keeports and W. Yount	Littlestown, Pa	Mar. 7, 1871	112, 466
Clothes-pin-slitting machine	O. P. Allen	Rindge, N. H	May 14, 1850	7, 355
Clothes-pins, Arbor for turning	B. B. and A. J. Ockington	Stratford, N. H	July 5, 1870	104, 988
Clothes-pins, Machine for cutting slots in	J. Humphrey	Keene, N. H	Mar. 31, 1857	16, 923

Index of patents issued from the United States Patent Office from 1790 *to* 1873, *inclusive*—Continued.

Invention.	Inventor.	Residence.	Date.	No.
Clothes-pins, Machine for making	B. B. and A. J. Ockington	Stratford Hollow, N. H	Dec. 16, 1873	145, 587
Clothes-pins, Machine for making	E. Parker	Burlington, Iowa	Jan. 15, 1856	14, 110
Clothes-pins, Machine for making	G. W. Parker	Fitzwilliam, N. H	Mar. 18, 1856	14, 466
Clothes-pins, Machine for making	J. B. Smith	Sunapee, N. H	Dec. 29, 1868	85, 338
Clothes-pins, Machine for slitting	J. B. Smith	Sunapee, N. H	Feb. 21, 1854	10, 553
Clothes-pins, Machinery for making	R. Emerson	Rockford, Ill	July 16, 1872	129, 118
Clothes-pins, Machinery for making	J. B. Smith	Sunapee, N. H	Oct. 28, 1873	144, 152
Clothes-pins, Machinery for turning	A. Greenwood	Marlborough, N. H	Dec. 11, 1849	6, 935
Clothes-pins, Tool for slotting	J. Humphrey	Keene, N. H	Dec. 28, 1858	22, 430
Clothes-pinchers	H. Kewley	Port Hudson, Mich	May 31, 1864	42, 951
Clothes-pole	W. W. Armington	New Haven County, Conn	Oct. 2, 1866	58, 535
Clothes-pole	F. W. Tilton	New Bedford, Mass	June 19, 1866	55, 744
Clothes-pole, Extension	J. and W. Denton	Amsterdam, N. Y	Sept. 5, 1871	118, 698
Clothes-post	C. P. Jadwin	Carbondale, Pa	July 30, 1867	67, 310
Clothes-post, Extension	G. Dittenhaver	Napoleon, Ohio	Jan. 7, 1868	73, 171
Clothes-pounder	A. W. Bunnell	Linesville Station, Pa	July 2, 1872	128, 531
Clothes-pounder	A. W. Hall	New York, N. Y	Oct. 1, 1872	131, 752
Clothes-pounder	S. F. Hawley	Sandy Hill, N. Y	Oct. 21, 1873	143, 901
Clothes-pounder	S. Hinkel	Goodville, Pa	Apr. 22, 1873	138, 024
Clothes-pounder	S. T. McDougal	Brooklyn, N. Y	July 7, 1868	79, 671
Clothes-pounder	B. S. Morgan	Delhi, Iowa	Mar. 17, 1868	75, 562
Clothes-pounder	M. K. Morris	Louisville, Ky	Aug. 13, 1872	130, 523
Clothes-pounder	J. W. Norton	Pioneer, Pa	Nov. 15, 1870	109, 240
Clothes-pounder	D. W. Rawson	Croydon, N. H	Apr. 22, 1873	138, 044
Clothes-pounder	E. S. Saxton	Greenfield, Ill	July 22, 1873	141, 082
Clothes-pounder	O. J. Stickles	Canton, N. Y	Sept. 21, 1869	94, 983
Clothes-pounder	S. Thomson	West Otis, Mass	May 5, 1857	17, 240
Clothes-press	J. S. Nicholson	Anamosa, Iowa	Sept. 22, 1868	82, 341
Clothes-presser and water-expeller	T. Fowler	Seymour, Conn	Jan. 26, 1869	86, 146
Clothes-pressing machine	J. Sage	Berlin, Conn	Apr. 14, 1810	......
Clothes-pressing machine	H. E. Smith	Fitchburgh, Mass	May 20, 1873	139, 201
Clothes, Pressing water from	C. Angerine	New York, N. Y	Feb. 25, 1836	......
Clothes-rack	J. Alcorn	Charlestown, Mass	June 8, 1869	90, 979
Clothes-rack	H. M. Andrews and J. H. Cleveland.	North Bay, N. Y	Mar. 28, 1871	113, 129
Clothes-rack	H. Bauman and U. Mueller	Canton, Ohio	Oct. 4, 1870	107, 996
Clothes-rack	T. D. Berry	Lowell, Mass	Aug. 16, 1859	25, 086
Clothes-rack	C. K. Breneman	Newport, Pa	Sept. 22, 1868	82, 378
Clothes-rack	C. B. Crosby	Cortlandville, N. Y	Nov. 28, 1865	51, 149
Clothes-rack	W. S. Foster	Marilla, N. Y	Feb. 22, 1859	23, 020
Clothes-rack	G. W. and E. J. Godfrey	Leslie, Mich	May 28, 1872	127, 232
Clothes-rack	O. C. Green	Dublin, Ind	Oct. 18, 1859	25, 820
Clothes-rack	A. A. Harris	Ravenna, Ohio	Aug. 10, 1858	21, 131
Clothes-rack	J. Hatfield	Sparta, Wis	Oct. 26, 1869	96, 107
Clothes-rack	V. M. Heath	Morristown, Vt	Sept. 7, 1869	94, 598
Clothes-rack	I. Hogeland	Indianapolis, Ind	Dec. 8, 1868	84, 694
Clothes-rack	J. Johnson	New York, N. Y	July 5, 1870	105, 086
Clothes-rack	E. F. Lyman	Indianapolis, Ind	Dec. 10, 1867	72, 057
Clothes-rack	D. Miller	Marietta, Ohio	Dec. 19, 1871	121, 953
Clothes-rack	J. O. Montignani	Albany, N. Y	Sept. 5, 1865	49, 779
Clothes-rack	J. O. Montignani	Albany, N. Y	Oct. 23, 1866	59, 133
Clothes-rack	H. W. Ross	Brooklyn, N. Y	Nov. 12, 1872	133, 058
Clothes-rack	S. I. Russell	Chicago, Ill	Dec. 22, 1857	18, 930
Clothes-rack	C. J. and G. W. Schaefer	Yonkers, N. Y	Sept. 3, 1872	131, 068
Clothes-rack	A. G. Schmidt	New York, N. Y	Nov. 7, 1871	120, 671
Clothes-rack	C. Stone	Ravenna, Ohio	Jan. 12, 1858	19, 107
Clothes-rack	H. Thomas and R. Wallace	New York, N. Y	July 6, 1869	92, 400
Clothes-rack	E. Werden	Pittsfield, Mass	July 27, 1869	93, 025
Clothes-rack	C. H. Wolcott	Jamestown, N. Y	Aug. 2, 1870	106, 102
Clothes-rack	C. H. Wolcott	Jamestown, N. Y	Aug. 22, 1871	118, 416
Clothes-rack	G. Young, jr	Saratoga Springs, N. Y	July 20, 1858	20, 974
Clothes-rack, Adjustable	J. Brainerd and W. H. Burridge.	Cleveland, Ohio	Jan. 5, 1864	41, 134
Clothes rack and drier	A. Hornor	Ross, Ind	Aug. 10, 1869	93, 620
Clothes rack and stand	W. B. Kimball	Peterborough, N. H	Aug. 10, 1869	93, 626
Clothes-receptacle, ironing-board, and drier, Combined.	S. T. McDougall	Brooklyn, N. Y	Mar. 14, 1871	112, 614
Clothes-reel	D. L. Huff	Bay City, Mich	Sept. 9, 1873	142, 631
Clothes-reel	J. McMahon	Ann Arbor, Mich	July 22, 1873	141, 064
Clothes, Roller for expressing water from	J. Allender	New London, Conn	Jan. 11, 1859	22, 539
Clothes, &c., Rotary pounder for washing	C. Aumock	Elbridge, N. Y	Mar. 10, 1838	626
Clothes-rubber	S. Foster	Lansing, Mich	Feb. 28, 1871	112, 235
Clothes, Rubber for washing	J. Charlton	Allegheny City, Pa	Apr. 14, 1868	76, 603
Clothes-scourer	D. Dickinson	Chatham, Conn	Oct. 1, 1830	......
Clothes-sprinkler	F. Ashley	New York, N. Y	Oct. 30, 1866	59, 158
Clothes-sprinkler	E. T. Colburn	Boston, Mass	Dec. 12, 1865	51, 509
Clothes-sprinkler	S. N. Davies	Muskegon, Mich	Nov. 6, 1860	30, 565
Clothes-sprinkler	S. G. Dugdale	Richmond, Ind	Nov. 30, 1869	97, 369
Clothes-sprinkler	W. P. Gannett	Boston, Mass	May 22, 1866	55, 008
Clothes-sprinkler	J. L. Kendall	New York, N. Y	June 26, 1866	55, 956
Clothes-sprinkler	E. H. Kirkham	Boston, Mass	Mar. 3, 1868	75, 168
Clothes-sprinkler	M. Moriarty	Bangor, Me	May 15, 1866	54, 819
Clothes-sprinkler	H. A. Morse	Canton, Mass	July 3, 1866	56, 082
Clothes-sprinkler	T. Payne	Ridgefield, Conn	Jan. 18, 1859	22, 669
Clothes-sprinkler	E. Schnurr	Monroe, Mich	Sept. 5, 1865	49, 794
Clothes-sprinkler	I. W. Shaler	Brooklyn, N. Y	Mar. 19, 1867	63, 107
Clothes-sprinkler	T. R. Smith and J. Mitchell	Jacksonville, Ill	Jan. 18, 1870	99, 021
Clothes-sprinkler	W. E. Staniford	Bucksport, Me	Feb. 27, 1866	52, 935
Clothes-sprinkler	E. B. Taylor	South Sudbury, Mass	May 28, 1867	65, 299
Clothes-sprinkler	J. W. Walters	Tiffin, Ohio	June 30, 1868	79, 531
Clothes-sprinkler	W. V. Wilson	Philadelphia, Pa	Apr. 14, 1868	76, 870
Clothes-squeezer	F. Arnold	Middle Haddam, Conn	Oct. 23, 1860	30, 451
Clothes-squeezer	G. D. Trumpore	Newark, N. J	Nov. 6, 1860	30, 596
Clothes-stick	C. Kuder	Rochester, N. Y	Apr. 2, 1867	63, 397
Clothes-washer	W. R. Brooks	Phelps, N. Y	June 3, 1873	139, 494
Clothes-washer	H. D. Crooker	Chester, Pa	Feb. 18, 1873	136, 038
Clothes-washer	J. K. Dugdale	White Water, Ind	Jan. 18, 1870	98, 855
Clothes-washer	J. K. Dugdale	White Water, Ind	Jan. 18, 1870	98, 856

Index of patents issued from the United States Patent Office from 1790 *to* 1873, *inclusive*—**Continued.**

Invention.	Inventor.	Residence.	Date.	No.
Clothes-washer	F. Ernst	San Francisco, Cal	Aug. 20, 1867	67, 858
Clothes-washer	W. I. Miller and B. O. Irons	Linesville, Pa	Mar. 19, 1872	124, 844
Clothes-washer	W. H. Nice	Wadsworth, Ohio	Dec. 17, 1872	134, 000
Clothes-washer	H. J. Noyes	Ashtabula, Ohio	July 2, 1872	128, 501
Clothes-washer	H. W. Pell	Rome, N. Y	Sept. 26, 1871	119, 399
Clothes-washer	P. Read	Hinesburgh, Vt	Nov. 26, 1872	133, 485
Clothes-washer	D. P. Sulouff	Milton, Pa	Nov. 21, 1871	121, 216
Clothes-washer	D. P. Sulouff	Milton, Pa	Nov. 21, 1871	121, 217
Clothes-washer	G. J. Terrel	North Blanford, Mass	June 17, 1873	139, 979
Clothes-washer	T. Weaver	Harrisburgh, Pa	Aug. 6, 1872	130, 168
Clothes washer and drier	E. Rees	Stoddartsville, Pa	Feb. 8, 1870	99, 594
Clothes washer and wringer	A. Clark	Plymouth, Ohio	Nov. 22, 1870	109, 385
Clothes washer and wringer	J. Cram	Boston, Mass	Sept. 15, 1863	39, 990
Clothes washer and wringer	R. G. Holmes	Worcester, Mass	Nov. 26, 1861	33, 812
Clothes-washer, Hand	P. Falardo and G. H. Snow	Newark, N. J., and New Haven, Conn.	Feb. 15, 1870	99, 870
Clothes-washer, Steam	M. S. Lamb	Davenport, Iowa	Feb. 22, 1870	100, 156
Clothes-washer, Vacuum	M. Johnstone	Spartanburgh, S. C	July 30, 1872	129, 961
Clothes, Washing	N. Briggs		Mar. 28, 1797	
Clothes, Washing	E. Weld		June 26, 1799	
Clothes washing and cleansing machine	W. Price	Cincinnati, Ohio	Mar. 7, 1865	46, 701
Clothes-washing and cloth-fulling machine	O. D. Wade	China, N. Y	Sept. 9, 1835	
Clothes-washing and corn-shelling machine	B. Rice	Denmark, N. Y	Nov. 23, 1827	
Clothes washing and drying machine	T. Pearce	Hartwick, N. Y	Dec. 29, 1832	
Clothes washing and wringing machine, Lavater	E. Weld		Sept. 17, 1800	
Clothes-washing machine	W. E. Arnold	Haddam, Conn	Aug. 3, 1829	
Clothes-washing machine	S. Benson	Waterville, Me	May 25, 1840	1, 616
Clothes-washing machine	W. H. Blood	San Francisco, Cal	Jan. 12, 1864	41, 190
Clothes-washing machine	R. Brainard	Haddam, Conn	Nov. 23, 1829	
Clothes-washing machine	S. Hinds	Montrose, Pa	Apr. 25, 1829	
Clothes-washing machine	I. Leavitt, A. Gilmore, and W. Sturtevant.	Turner, Me	Aug. 9, 1839	1, 282
Clothes-washing machine	M. Mellinger	Dayton, Ohio	Aug. 12, 1862	36, 162
Clothes-washing machine	R. W. Oliphant	North Granville, N. Y	June 7, 1838	769
Clothes-washing machine	L. Procter	New York, N. Y	Nov. 16, 1841	2, 369
Clothes-washing machine	S. Swett, jr	Portsmouth, N. H	July 12, 1839	1, 236
Clothes-washing machine	H. N. Walter	Norwich, N. Y	June 22, 1841	2, 137
Clothes-washing machine	G. Waterman	Johnston, R. I	May 11, 1841	2, 084
Clothes, Washing, pressing, ironing, crimping, and drying.	J. Newhall	Dayton, Ohio	June 13, 1831	
Clothes washing, rinsing, and wringing machine	W. J. Fulsom and J. Hayden	Hallowell, D. M	Dec. 17, 1805	
Clothes-washing rubber	H. Burk	Mineral Point, Ohio	July 24, 1866	56, 523
Clothes-whisk and hat-brush combined	A. M. Richmond	New York, N. Y	Apr. 23, 1872	126, 090
Clothes-wringer	A. Albrecht	Philadelphia, Pa	Mar. 14, 1871	112, 669
Clothes-wringer	C. H. Amidon	Greenfield, Mass	May 21, 1867	64, 932
Clothes-wringer	A. M. Bailey	Middlefield, Conn	Mar. 21, 1871	112, 884
Clothes-wringer	A. M. Bailey and J. O. Couch	Middlefield, Conn	Jan. 14, 1862	34, 178
Clothes-wringer	S. A. Bailey	Waterford, Mass	Mar. 13, 1866	53, 098
Clothes-wringer	C. E. Bancroft	Waterbury, Vt	May 20, 1862	35, 290
Clothes-wringer	C. H. Bangs	Farmington, Me	June 11, 1872	127, 831
Clothes-wringer	O. D. Barrett	Fulton, N. Y	Oct. 1, 1861	33, 411
Clothes-wringer	E. G. W. Bartlett	Providence, R. I	May 24, 1870	103, 282
Clothes-wringer	E. G. W. Bartlett	Providence, R. I	May 23, 1871	115, 150
Clothes-wringer	E. G. W. Bartlett	Providence, R. I	July 4, 1871	116, 665
Clothes-wringer	E. G. W. Bartlett	Providence, R. I	Apr. 30, 1872	126, 173
Clothes-wringer	C. W. Bassett	Newton, Mass	May 14, 1872	126, 774
Clothes-wringer	E. Blakeman and J. R. Gill	Charleston, Ill	Oct. 25, 1864	44, 779
Clothes-wringer	T. O. Bogert	Cincinnati, Ohio	Jan. 7, 1873	134, 633
Clothes-wringer	J. Brinkerhoff	Auburn, N. Y	Dec. 26, 1871	122, 220
Clothes-wringer	W. P. Brooks and J. D. Hartzell	Ozawkie, Kans	July 1, 1837	140, 344
Clothes-wringer	H. J. Burr	Bloomfield, Conn	Sept. 30, 1873	143, 220
Clothes-wringer	E. P. H. Capron	Hudson, N. Y	Aug. 29, 1871	118, 431
Clothes-wringer	C. E. Carter	Martinsville, Ohio	May 23, 1871	115, 161
Clothes-wringer	R. S. Cathcart	Cincinnati, Ohio	Sept. 17, 1872	131, 392
Clothes-wringer	R. S. Cathcart	Cincinnati, Ohio	Dec. 31, 1872	134, 461
Clothes-wringer	R. S. Cathcart	Cincinnati, Ohio	Jan. 7, 1873	134, 641
Clothes-wringer	D. B. Clement	Milton, Mass	Apr. 8, 1862	34, 920
Clothes-wringer	D. B. Clement	Milton, Mass	Jan. 20, 1863	37, 472
Clothes-wringer	J. D. Cochran	Milford, N. H	June 24, 1862	35, 670
Clothes-wringer	G. J. Colby	Waterbury, Vt	Dec. 4, 1860	30, 802
Clothes-wringer	D. M. Cole	Elgin, Ill	June 18, 1872	128, 119
Clothes-wringer	D. M. Cole	Elgin, Ill	June 3, 1873	139, 545
Clothes-wringer	W. Cooper	Independence, Iowa	Jan. 4, 1870	98, 566
Clothes-wringer	W. Cooper	Ypsilanti, Mich	June 4, 1872	127, 577
Clothes-wringer	E. A. Corbin and H. Albrecht	Philadelphia, Pa	June 24, 1873	140, 249
Clothes-wringer	E. H. Covel	New York, N. Y	Nov. 12, 1867	70, 808
Clothes-wringer	E. H. Covel	New York, N. Y	Feb. 11, 1868	74, 201
Clothes-wringer	P. Cramer	Providence, R. I	July 28, 1868	80, 336
Clothes-wringer	P. Cramer	Barrington, R. I	Aug. 24, 1869	94, 082
Clothes-wringer	R. R. Crosby and J. Harris	Boston, Mass	Oct. 29, 1861	33, 575
Clothes-wringer	C. A. Cummings and F. M. Swallow.	Worcester, Mass	Feb. 11, 1862	34, 395
Clothes-wringer	J. E. Daniels and G. S. Kendall	Boston, Mass	July 8, 1862	35, 810
Clothes-wringer	C. H. De Knight	Pittsburgh, Pa	Aug. 9, 1870	106, 136
Clothes-wringer	E. Dickerman	Richmond, Vt	Apr. 10, 1860	27, 781
Clothes wringer	J. C. Dickey	Saratoga Springs, N. Y	Aug. 28, 1866	57, 485
Clothes-wringer	W. S. Douglas	Richmond, Vt	Jan. 21, 1868	73, 584
Clothes-wringer	S. F. Emerson	Seville, Ohio	May 5, 1863	38, 380
Clothes-wringer	T. B. Emerson	Seville, Ohio	Oct. 29, 1867	70, 181
Clothes-wringer	P. Falardo	Danbury, Conn	Aug. 3, 1869	93, 188
Clothes-wringer	R. E. Ferguson	Chicago, Ill	Oct. 4, 1870	107, 893
Clothes-wringer	R. E. Ferguson	Chicago, Ill	Mar. 26, 1872	124, 942
Clothes-wringer	T. J. U. Fisk	Macon City, Mo	June 25, 1872	128, 294
Clothes-wringer	H. G. Folger	Wadsworth, Ohio	July 4, 1865	48, 545
Clothes-wringer	M. M. Follet	Lake City, Minn	Dec. 14, 1869	97, 899
Clothes-wringer	J. Fox	Farmersville, Iowa	Dec. 26, 1871	122, 164
Clothes-wringer	A. C. Gallahue	New York, N. Y	Oct. 30, 1866	59, 208
Clothes-wringer	G. E. Gault and A. W. Winall	Cincinnati, Ohio	Jan. 7, 1873	134, 658

Index of patents issued from the United States Patent Office from 1790 *to* 1873, *inclusive*—Continued.

Invention.	Inventor.	Residence.	Date.	No.
Clothes-wringer	W. Gibbs and R. J. Bell	Carlisle, Pa	Apr. 1, 1862	34, 827
Clothes-wringer	R. Gipson	Shelby, Ohio	Oct. 25, 1864	44, 796
Clothes-wringer	R. Gipson	Shelby, Ohio	June 12, 1866	55, 486
Clothes-wringer	D. D. Gitt	Arendtsville, Pa	May 24, 1864	42, 908
Clothes-wringer	F. A. Gleason	Brooklyn, N. Y	Sept. 10, 1867	68, 732
Clothes-wringer	A. H. Goss	Auburn, N. Y	Apr. 2, 1872	125, 129
Clothes-wringer	C. Graham	Cincinnati, Ohio	Dec. 31, 1872	134, 472
Clothes-wringer	G. B. Griffin	Madison, Wis	June 4, 1861	32, 475
Clothes-wringer	L. Hale	Hollis, N. H	Oct. 31, 1871	120, 512
Clothes-wringer	J. W. Hampton	Mount Pleasant, Iowa	Nov. 14, 1871	120, 968
Clothes-wringer	D. M. Harris and S. S. Burnet	Salem, Mass	May 13, 1862	35, 234
Clothes-wringer	J. Harrison and G. W. Harris	Brooklyn and New York, N. Y.	July 16, 1867	66, 835
Clothes-wringer	J. Harrison and G. W. Harris	Brooklyn and New York, N. Y.	July 16, 1867	66, 836
Clothes-wringer	C. E. Haynes	Boston, Mass	May 13, 1873	138, 803
Clothes-wringer	S. S. Hemenway	Boston, Mass	May 30, 1865	47, 951
Clothes-wringer	C. W. Higgins and A. S. Willard.	Somerville and Charlestown, Mass.	June 19, 1860	28, 752
Clothes-wringer	R. G. Holmes	Worcester, Mass	Sept. 8, 1863	39, 813
Clothes-wringer	C. H. Hudson	New York, N. Y	July 25, 1871	117, 291
Clothes-wringer	R. B. Hugunin	Cleveland, Ohio	Mar. 10, 1868	75, 425
Clothes-wringer	J. Johnson	New York, N. Y	Jan. 28, 1862	34, 281
Clothes-wringer	J. Johnson	New York, N. Y	June 28, 1864	43, 368
Clothes-wringer	G. H. Kidney	Cleveland, Ohio	July 22, 1862	35, 940
Clothes-wringer	A. King	Philadelphia, Pa	Mar. 21, 1871	112, 815
Clothes-wringer	E. King	Dunkirk, N. Y	Jan. 9, 1872	122, 527
Clothes-wringer	C. H. Knox	Mount Pleasant, Iowa	July 2, 1867	66, 362
Clothes-wringer	J. H. Kooser	Pekin, Ill	Oct. 15, 1872	132, 294
Clothes-wringer	A. Lapham and F. E. Pratt	New York and Mott Haven, N. Y.	Apr. 28, 1868	77, 386
Clothes-wringer	J. S. Lash	Philadelphia, Pa	Oct. 23, 1866	59, 041
Clothes-wringer	J. Lee	Galesburgh, Ill	Feb. 6, 1866	52, 423
Clothes-wringer	D. Lyman	Middlefield, Conn	Oct. 17, 1865	50, 482
Clothes-wringer	D. Lyman	Middlefield, Conn	Feb. 27, 1866	52, 867
Clothes-wringer	D. Lyman	Middlefield, Conn	Mar. 26, 1867	63, 272
Clothes-wringer	D. Lyman	Middlefield, Conn	Mar. 26, 1867	63, 273
Clothes-wringer	D. Lyman	Middlefield, Conn	Dec. 17, 1867	72, 407
Clothes-wringer	A. Magowan	Trenton, N. J	Apr. 12, 1870	101, 749
Clothes-wringer	J. Makechney	Trenton, N. J	Feb 6, 1872	123, 495
Clothes-wringer	M. Mallon	Rahway, N. J	July 16, 1872	129, 043
Clothes-wringer	J. S. Maughlin and W. C. Marr.	Onawa, Ind	Feb. 27, 1872	124, 073
Clothes-wringer	T. E. McDonald	Trenton, N. J	Mar. 5, 1872	124, 367
Clothes-wringer	T. F. McDonald	Trenton, N. J	Jan. 28, 1873	135, 237
Clothes-wringer	T. E McDonald	New Brunswick, N. J	Nov. 18, 1873	144, 627
Clothes-wringer	J. McLaughlin	Steubenville, Ohio	Oct. 16, 1855	13, 684
Clothes-wringer	J. McLaughlin	Steubenville, Ohio	Sept. 20, 1870	107, 518
Clothes-wringer	J. M. McMaster	Rochester, N. Y	Sept. 17, 1867	68, 893
Clothes-wringer	W. T. McMillen and E. P. Conrick.	Cincinnati, Ohio, and Delavan, Wis.	July 24, 1866	56, 591
Clothes-wringer	C. V. Mead	Trenton, N. J	Sept. 26, 1871	119, 384
Clothes-wringer	R. O. Meldrum and A. B. Paxson.	Griffin's Mills and East Hamburgh, N. Y.	Jan. 24, 1860	26, 914
Clothes-wringer	A. B Mercier	Providence, R. I	July 31, 1866	56, 773
Clothes-wringer	C. Messenger	Warren, Ohio	Mar. 11, 1862	34, 639
Clothes-wringer	B. D. Morrell	Lisbon, N. H	Aug. 25, 1863	39, 666
Clothes-wringer	H. Nash	Cincinnati, Ohio	Oct. 9, 1866	58, 669
Clothes-wringer	A. O'Dell	Napanee, Canada	Oct. 25, 1870	108, 620
Clothes-wringer	J. O'Donald	Clinton, Ill	Dec. 26, 1865	51, 743
Clothes-wringer	C. H. Packard	North Bridgewater, Mass	Feb. 18, 1862	34, 445
Clothes-wringer	C. H. Packard	North Bridgewater, Mass	Sept. 8, 1863	39, 864
Clothes-wringer	A. H. Page	South Boston, Mass	Jan. 29, 1867	61, 680
Clothes-wringer	G. Palmer	Littlestown, Pa	Jan. 22, 1867	61, 353
Clothes-wringer	S. W. and J. F. Palmer	Auburn, N. Y	May 18, 1869	90, 185
Clothes-wringer	S. W. and J. F. Palmer	Auburn, N. Y	Dec. 5, 1871	121, 657
Clothes-wringer	J. A. Park	Lansing, Mich	Oct. 29, 1872	132, 687
Clothes-wringer	J. N. Pease	Panama, N. Y	May 16, 1865	47, 774
Clothes-wringer	J. N. Pease	Panama, N. Y	May 15, 1866	54, 764
Clothes-wringer	J. N. Pease and G. Lewis	Panama, N. Y	July 24, 1866	56, 676
Clothes-wringer	M. Pierce	Winona, Minn	Apr. 28, 1868	77, 315
Clothes-wringer	J. N. Poage	Cincinnati, Ohio	Dec. 31, 1872	134, 485
Clothes-wringer	J. F. Pond	Cleveland, Ohio	Jan. 2, 1866	51, 862
Clothes-wringer	T. Pool	Brunswick, Ohio	June 24, 1862	35, 705
Clothes-wringer	G. S. Prindle	Washington, D. C	Feb. 4, 1873	135, 585
Clothes-wringer	H. W. Putnam	Cleveland, Ohio	Dec. 17, 1861	33, 960
Clothes-wringer	H. W. Putnam	Cleveland, Ohio	Sept. 16, 1862	36, 480
Clothes-wringer	B. Reed	Allegheny, Pa	Apr. 14, 1868	76, 659
Clothes-wringer	O. Reeves	Greenport, N. Y	July 17, 1866	56, 449
Clothes-wringer	N. A. Rhoads	Waterbury, Vt	Mar. 11, 1862	34, 646
Clothes-wringer	N. A. Rhoads	Waterbury, Vt	Jan. 27, 1863	37, 521
Clothes-wringer	W. B. Rhoads	South Dedham, Mass	Jan. 25, 1861	32, 659
Clothes-wringer	H. Robbins	Cincinnati, Ohio	July 31, 1866	56, 865
Clothes-wringer	C. Robinson	Boston, Mass	July 8, 1873	140, 592
Clothes-wringer	J. G. Roth	New York, N. Y	Apr. 12, 1870	101, 922
Clothes-wringer	S. P. Rowell	Melrose, Mass	July 15, 1862	35, 910
Clothes-wringer	W. Rowell	New York, N. Y	Sept. 3, 1867	68, 576
Clothes-wringer	E. P. Russell	Manlius, N. Y	Dec. 14, 1869	97, 965
Clothes-wringer	E. P. Russell	Manlius, N. Y	Feb. 20, 1872	123, 794
Clothes-wringer	A. F. Saunders	Chelsea, Mass	June 17, 1862	35, 631
Clothes-wringer	J. Seaman	Groton, N. Y	Nov. 18, 1873	144, 705
Clothes-wringer	I. A. Sergeant	Springfield, Ohio	July 27, 1858	21, 029
Clothes-wringer	L. Shmetzer	Chicago, Ill	July 12, 1870	105, 375
Clothes-wringer	M. E. Smilie and W. Cooley	Waterbury, Vt	May 7, 1872	126, 584
Clothes-wringer	G. Smith	Highland Falls, N. Y	Aug. 10, 1869	93, 488
Clothes-wringer	H. E. Smith	New York, N. Y	Mar. 3, 1868	75, 066
Clothes-wringer	H. E. Smith	New York, N. Y	May 30, 1870	115, 372
Clothes-wringer	H. E. Smith	New York, N. Y	June 4, 1872	127, 521
Clothes-wringer	H. E. Smith	New York, N. Y	Sept. 24, 1872	131, 714

Index of patents issued from the United States Patent Office from 1790 to 1873, inclusive—Continued.

Invention.	Inventor.	Residence.	Date.	No.
Clothes-wringer	H. E. Smith	New York, N. Y	Sept. 24, 1872	131,715
Clothes-wringer	H. E. Smith	Fitchburgh, Mass	May 20, 1873	139,200
Clothes-wringer	R. Smith	Towanda, Pa	Dec. 1, 1857	18,777
Clothes-wringer	C. F. Spaulding	Saint Johnsbury, Vt	May 3, 1864	42,607
Clothes-wringer	E. Spaulding	Morrisville, Vt	Apr. 16, 1861	32,090
Clothes-wringer	A. H. Spencer	Providence, R. I	Feb. 1, 1870	99,364
Clothes-wringer	J. H. Stedman	Randolph, Vt	Sept. 10, 1861	33,271
Clothes-wringer	E. M. Stevens	Boston, Mass	Apr. 29, 1862	35,115
Clothes-wringer	W. M. Storm	New York, N. Y	Mar. 17, 1868	75,707
Clothes-wringer	L. H. Thomas	Waterbury, Vt	Mar. 25, 1862	34,795
Clothes-wringer	A. Tower	New York, N. Y	Nov. 2, 1869	96,511
Clothes-wringer	W. H. Towers	Boston, Mass	Apr. 25, 1871	114,231
Clothes-wringer	G. P. Towle	Boston, Mass	July 15, 1862	35,911
Clothes-wringer	N. Van Auken	Amsterdam, N. Y	May 20, 1862	35,335
Clothes-wringer	R. Vose and J. W. Evans	New York, N. Y	Nov. 5, 1872	132,782
Clothes-wringer	S. Walker	Boston, Mass	June 17, 1862	35,643
Clothes-wringer	J. Webb	Spartanburgh, Pa	Sept. 15, 1868	82,259
Clothes-wringer	H. Weble	Hoboken, N. J	May 8, 1866	54,631
Clothes-wringer	N. B. White	South Dedham, Mass	Mar. 4, 1862	34,618
Clothes-wringer	W. Whitney	Baldwinsville, Mass	Dec. 3, 1861	33,861
Clothes-wringer	D. F. Williams	Cumberland, R. I	Jan. 2, 1866	51,888
Clothes-wringer	R. P. Wilson	New York, N. Y	June 24, 1862	35,726
Clothes-wringer	G. L. Witsil	Philadelphia, Pa	May 28, 1867	65,148
Clothes-wringer	G. L. Witsil	Philadelphia, Pa	Mar. 3, 1868	75,231
Clothes-wringer	J. Young	Amsterdam, N. Y	Apr. 23, 1867	64,180
Clothes-wringer	J. Young	Amsterdam, N. Y	Mar. 23, 1869	88,110
Clothes wringer and mangle	S. W. and J. F. Palmer	Auburn, N. Y	Nov. 15, 1864	45,071
Clothes wringer and mangle, Combined	B. O. Thompson	Chicago, Ill	June 22, 1869	91,794
Clothes wringer and mangle combined	J. D. Tift	Cuyahoga Falls, Ohio	June 24, 1862	35,716
Clothes wringer and washer roller	R. B. Hugunin	New York, N. Y	July 17, 1866	56,418
Clothes-wringer attachment	C. L. Carter	Union City, Ind	Aug. 27, 1867	68,043
Clothes-wringer bench	S. A. Bailey	Waterford, Mass	Mar. 13, 1866	53,099
Clothes-wringer, Centrifugal	H. Rosamyer, jr	Rochester, Pa	Aug. 19, 1873	142,045
Clothes-wringer clamp	J. D. Burdick	Ashway, R. I	June 28, 1864	43,286
Clothes-wringer roll	H. E. Smith	Cincinnati, Ohio	Feb. 13, 1866	52,615
Clothes-wringer rolls, Spindle for	T. E. McDonald	Trenton, N. J	Sept. 26, 1871	119,235
Clothes-wringer roller	E. P. H. Capron	Hudson, N. Y	Aug. 5, 1873	141,424
Clothes-wringer roller	J. Critcherson	Boston, Mass	Feb. 11, 1862	34,394
Clothes-wringer roller	J. B. Forsyth	Roxbury, Mass	Nov. 20, 1866	59,798
Clothes-wringer roller	E. King	Dunkirk, N. Y	Aug. 20, 1872	130,582
Clothes-wringer roller	J. Murphy	New York, N. Y	Oct. 16, 1866	58,874
Clothes-wringers, Roller for	J. F. Holt	Providence, R. I	July 25, 1865	49,030
Clothes-wringer roller and fastening	R. B. Hugunin	Cleveland, Ohio	Mar. 6, 1866	53,001
Clothes-wringers, Composition-roll for	W. H. Towers	Boston, Mass	June 21, 1870	104,514
Clothes-wringers, Gearing for	G. C. Wright	Leroy, Ohio	May 17, 1870	103,269
Clothes-wringers, Preparing wood for the bearings in	J. Brinkerhoff	Auburn, N. Y	Oct. 15, 1872	132,194
Clothes-wringers to tubs, Mode of attaching	D. Lyman	Middlefield, Conn	Apr. 2, 1861	31,895
Clothes, Wringing	J. H. Clark	Westbrook, Me	May 8, 1860	28,156
Clothes-wringing	J. R. Gill, W. E. Palmer, and W. W. Webb	Alton, Ill	Mar. 4, 1862	34,610
Clothes-wringing hook	J. H. Pratt	Lynn, Mass	Feb. 21, 1871	112,074
Clothes-wringing machine	S. A. Bailey	New London, Conn	Apr. 5, 1859	23,436
Clothes-wringing machine	R. P. Bradley	Cuyahoga Falls, Ohio	Aug. 12, 1856	15,543
Clothes-wringing machine	E. Dickerman	Middlefield, Conn	Feb. 18, 1862	34,459
Clothes-wringing machine	K. H. Elliott	Eden, Vt	Apr. 8, 1862	34,885
Clothes-wringing machine	S. F. Emerson	Seville, Ohio	Sept. 16, 1862	36,458
Clothes-wringing machine	E. L. Hagar	Frankfort, N. Y	June 1, 1858	20,470
Clothes-wringing machine	R. G. Holmes	Worcester, Mass	Sept. 8, 1863	39,814
Clothes-wringing machine	R. B. Hugunin	Cleveland, Ohio	Sept. 20, 1864	44,313
Clothes-wringing machine	W. Joslin	Cleveland, Ohio	Aug. 19, 1862	36,218
Clothes-wringing machine	D. Lyman	Middlefield, Conn	Feb. 17, 1863	37,699
Clothes-wringing machine	J. P. Martin	Philadelphia, Pa	Oct. 5, 1852	9,302
Clothes-wringing machine	T. H. Peavey	Montville, Me	Apr. 12, 1859	23,646
Clothes-wringing machine	N. W. Peebles	Brunswick, Ohio	Sept. 2, 1862	36,362
Clothes-wringing machine	H. W. Putnam	Cleveland, Ohio	Oct. 21, 1862	36,728
Clothes-wringing machine	I. D. Robinson	Waterbury, Vt	Oct. 28, 1862	36,860
Clothes-wringing machine	S. Squires	Boston, Mass	Nov. 18, 1862	36,983
Clothes-wringing machine	D. W. Swift	West Falmouth, Mass	Jan. 28, 1862	34,276
Clothes-wringing machine	J. W. Wheeler and H. S. Bishop	Cleveland, Ohio	Feb. 10, 1863	37,655
Clothes-wringing press	C. Robinson	Boston, Mass	Apr. 4, 1871	113,347
Clothing-clasp	C. Seaver, jr	Boston, Mass	Apr. 4, 1865	47,135
Clothing, Dummy for displaying	W. E. Brock	New York, N. Y	Apr. 7, 1868	76,394
Clothing, Mode of fitting	R. Bisbee	Boston, Mass	Mar. 3, 1868	74,978
Clothing, Stand for displaying	W. E. Brock	New York, N. Y	July 23, 1867	67,021
Clothing, Submarine	C. S. Merriman	Afton, Iowa	Aug. 10, 1869	93,544
Clover and other grass seeds, Machine for hulling	A. Keagy	Morrison's Cove, Pa	June 24, 1839	1,190
Clover and other seeds, Cleaning	J. Roberts, jr		Feb. 13, 1796	....
Clover and other seeds, Hulling	W. Jones and A. Miller	Hagerstown, Md	Sept. 27, 1864	44,431
Clover and rice hulling machine	J. W. Matthews and M. S. Kahle	Lexington, Va	Jan. 17, 1834	....
Clover and rice machine	Brayley and Walker	Phillips, Me	June 6, 1835	....
Clover-bolt	E. K. Collins	Chili, N. Y	Aug. 16, 1859	25,097
Clover chaff from straw, Bolt for separating	J. Bolton	Harrisburgh, Pa	July 28, 1831	....
Clover, Cleaning	J. Goodyear, jr	South Middletown, Pa	June 2, 1836	....
Clover-huller	J. Allen	Frease's Store, Ohio	Feb. 13, 1855	12,372
Clover-huller	J. Birdsell	Hamorton, Pa	May 23, 1848	5,599
Clover-huller	A. B. Crawford	Wooster, Ohio	Nov. 28, 1854	11,990
Clover-huller	J. Hibbs	Tullytown, Pa	Jan. 30, 1855	12,314
Clover-huller	M. H. Mansfield	Ashland, Ohio	June 21, 1864	43,217
Clover-huller	J. R. Mitchell	Danville, Pa	July 20, 1843	3,182
Clover-huller	A. Overocker	McHenry, Ill	July 5, 1859	24,657
Clover-huller	C. Reif	Hartleton, Lewis Township, Pa.	May 31, 1859	24,237
Clover-huller	C. Reif	Lewis Township, Pa	Apr. 21, 1863	38,242
Clover-huller	R. Stadden	Milton, Pa	Sept. 10, 1850	7,637
Clover huller and cleaner	T. Church	Lewisburgh, Pa	Jan. 9, 1872	122,568
Clover-huller, Concave of	T. Carpenter	Manlius, N. Y	Mar. 21, 1854	10,667
Clover hulling and thrashing machine	J. C. Birdsell	Rush, N. Y	May 18, 1858	20,249
Clover-hulling cylinders, Fastening the teeth to	S. Karns	Bloody Run, Pa	Nov. 8, 1853	10,208

Index of patents issued from the United States Patent Office from 1790 to 1873, inclusive—Continued.

Invention.	Inventor.	Residence.	Date.	No.
Clover-hulling machine	A. B. Crawford	Wooster, Ohio	Dec. 31, 1844	3,867
Clover-hulling machine	N. Eames	Hanover, Pa	June 21, 1859	24,447
Clover-hulling machine	J. D. Forrey	Lewistown, Pa	Dec. 18, 1860	30,962
Clover-hulling machine	H. Hozer	Wooster, Ohio	June 14, 1845	4,080
Clover-hulling machine	M. H. Mansfield	Mifflintown, Pa	June 6, 1846	4,558
Clover-hulling machine	S. W. Powell	Turbet, Pa	Aug. 16, 1845	4,146
Clover-machine	W. Loomis	Ashford, Conn	July 2, 1836	
Clover-machine	M. H. Mansfield	Ashland, Ohio	Mar. 25, 1862	34,764
Clover-machine	J. Ross	Boundbrook, N. J	Feb. 6, 1835	
Clover-machine	D. Whiting	Ashland, Ohio	May 31, 1870	103,806
Clover, Machine for rubbing out	N. Price	Frederick County, Md	Mar. 2, 1811	
Clover-picker	W. T. Mills	Galesburgh, Mich	Feb. 1, 1859	22,817
Clover-seed and cleaning grain, Screen for hulling	M. H. Mansfield	Ashland, Ohio	Aug. 1, 1854	11,459
Clover-seed and corn sheller, &c	R. Rittenhouse	Amwell Township, Pa	Mar. 19, 1833	
Clover-seed bolting and cleaning machine	J. C. Birdsell	West Henrietta, N. Y	Dec. 13, 1859	26,409
Clover-seed cleaner	G. Faber	Chambersburgh, Pa	Apr. 5, 1832	
Clover-seed cleaner, &c	H. Hoth	Wild, Me	Sept. 26, 1835	
Clover-seed, Cleaning	J. Manning	Lambertsville, N. Y	Jan. 6, 1836	
Clover-seed, Cleaning	W. Sampson and C. Curtis	New Haven, Vt	Jan. 6, 1809	
Clover-seed-cleaning machine	J. Botton	Warren, N. Y	Jan. 28, 1822	
Clover-seed-cleaning machine	J. Cottle		Mar. 21, 1803	
Clover-seed-cleaning machine	D. Rankin	Augusta County, Va	Mar. 18, 1834	
Clover-seed-cleaning machine	A. Spicer		Apr. 22, 1802	
Clover-seed-cleaning machine	H. Spickard	Fincastle, Va	Aug. 16, 1814	
Clover-seed from the "pod," Machine for separating	T. Kirk		July 28, 1803	
Clover-seed, Gathering	J. Smith	Fredericksburgh, Va	Apr. 28, 1815	
Clover-seed huller	J. Bolton	Warren, N. Y	Jan. 9, 1823	
Clover-seed huller	G. and E. Faber	Pittsburgh, Pa	Oct. 13, 1832	
Clover-seed huller	J. Gans	Fayette County, Pa	Mar. 12, 1833	
Clover-seed huller	T. Gollogher	Fairfield County, Ohio	Feb. 28, 1833	
Clover-seed huller	G. Monohon	Augusta Township, Ohio	Apr. 27, 1832	
Clover-seed huller	P. Reading	Trenton, N. J	July 31, 1832	
Clover-seed huller	C. Seabold, jr	New Berlin, Pa	Mar. 6, 1832	
Clover-seed huller	S. West	Harford County, Md	Jan. 16, 1835	
Clover-seed huller and cleaner	W. C. Grimes	York, Pa	Mar. 3, 1841	1,996
Clover-seed huller and grain-chopper	G. Wales	Centre Township, Pa	July 24, 1832	
Clover-seed, Hulling	S. M. Baily	Attleborough, Pa	Aug. 23, 1834	
Clover-seed, Hulling	C. B. Baldwin	Faircastle, Va	June 16, 1836	
Clovcr seed, Hulling	J. Hopper and A. Doughty	Moresborough, Pa	June 30, 1836	
Clover-seed, Hulling	E. Horn	Rochester County, Va	Jan. 30, 1834	
Clover-seed, Hulling	G. B. and W. F. Poague	Lexington, Va	Feb. 17, 1836	
Clover-seed, Hulling	J. Pollock	Richmond, Ind	Feb. 19, 1850	7,104
Clover-seed, Hulling	H. Robbins	Kennebec, Me	June 20, 1836	
Clover-seed, Hulling	W. Rowe	Phillips, Me	Apr. 25, 1834	
Clover-seed, Hulling and cleaning	M. H. Mansfield	Ashland, Ohio	Feb. 27, 1855	12,462
Clover-seed, Hulling and cleaning	B. Wood	Doylestown, Pa	Aug. 11, 1834	
Clover-seed hulling and cleaning machine	I. V. Blackwell	Ovid, N. Y	Mar. 30, 1858	19,745
Clover-seed hulling and cleaning machine	F. E. Cook	Guilford, Ohio	June 4, 1861	32,470
Clover-seed hulling and cleaning machine	J. Kuhn	Centreville, Pa	June 25, 1861	32,630
Clover-seed hulling and cleaning machine	A. Matthews	Island Creek, Ohio	Dec. 13, 1831	
Clover-seed hulling and cleaning machine	P. Reading	Trenton, N. J	Oct. 25, 1831	
Clover-seed hulling and cleaning machine	D. S. Wagener	Penn Yan, N. Y	Feb. 12, 1861	31,420
Clover-seed hulling and cleaning machine	S. White	Chambersburgh, Pa	Aug. 10, 1828	
Clover-seed hulling and cleaning machine	W. Williams	Buckingham Township, Pa	Aug. 12, 1831	
Clover-seed hulling and thrashing machine	H. Bangs	New York, N. Y	Apr. 29, 1834	
Clover-seed-hulling machine	W. M. Barton	Cheeks X Roads, Tenn	Dec. 20, 1837	525
Clover-seed-hulling machine	J. Flook	Middletown, Md	May 10, 1838	733
Clover-seed-hulling machine	G. Gardner	York Springs, Pa	May 16, 1848	5,580
Clover-seed-hulling machine	W. C. Grimes	York, Pa	Nov. 8, 1834	
Clover-seed-hulling machine	W. C. Grimes	York, Pa	Nov. 8, 1834	
Clover-seed-hulling machine	D. Hunsicker	Hartley, Pa	May 8, 1838	728
Clover-seed-hulling machine	W. E. Lukens	Cadiz, Ohio	Mar. 26, 1834	
Clover-seed-hulling machine	M. F. Noraconk and D. Hoats	Milton, Pa	Dec. 18, 1860	30,982
Clover-seed-hulling machine	J. Worrell		Feb. 27, 1804	
Clover-seed machine	I. Correll	New Philadelphia, Ohio	Dec. 31, 1833	
Clover-seed machine	L. Keck	Tyrone Township, Pa	Dec. 16, 1833	
Clover-seed machine	W. Kuhn, G. Hassinger, and W. Stiles.	Union County, Pa	Dec. 24, 1833	
Clover-seed machine	C. Landes	Mount Sidney, Va	Dec. 24, 1833	
Clover-seed machine	E. Rider	Lovingston, Va	Dec. 28, 1833	
Clover-seed machine	H. Robbins	Wilton, Me	Sept. 6, 1833	
Clover-seed shelling and hulling machine	M. Zorger	York Post-Office, Pa	May 10, 1805	
Clover-seed-shelling machine	J. Gorgas	Fredericksburgh, Pa	Jan. 15, 1830	
Clover-seed-shelling machine	M. Withers		Apr. 30, 1804	
Clover-seeds, &c., Machine for cleaning	S. Arnold	Botetourt, Va	Jan. 25, 1815	
Clover-seeds, Machine for cleaning	I. H. Haskins	Columbia, N. Y	May 6, 1824	
Clover-seeds, Machine for cleaning	M. Miller		July 19, 1802	
Clover-seeds, Machine for gathering	D. K. Wertman and W. H. Reinbold.	Centralia, Pa	Oct. 2, 1866	58,517
Clover-seeds, Machine for getting out	S. Condon	New Hartford, Conn	Mar. 3, 1809	
Clover-seeds, Machine for getting out	W. Loomis	Ashford, Conn	Apr. 27, 1822	
Clover-seeds, Machine for hulling and cleaning	B. Sutton	Romulus, N. Y	Feb. 10, 1825	
Clover-seeds, Machine for saving	M. S. Kahle	Lexington, Va	May 6, 1856	14,816
Clover-seeds, &c., Machine for separating	H. Hunsicker	Lewisburgh, Pa	Feb. 12, 1861	31,389
Clover-seeds, Machine far washing, churning, and getting out.	R. Patton	Columbia, Pa	June 13, 1808	
Clover-seeds, rice, and barley, Machine for hulling	W. Manning	Westfield Township, Pa	Nov. 24, 1830	
Clover-seeds, Separating chaff from	R. Ranger	Wilton, Me	Aug. 21, 1834	
Clover-seeds, Thrashing and cleaning	S. Raub, jr	Wilkesbarre, Pa	Mar. 10, 1834	
Clover-seeds, wheat, &c., from the husks, Separating.	D. Buckman		Dec. 21, 1803	
Clover-separator	C. L. Allen	Flat Rock, Mich	Mar. 18, 1873	136,805
Clover-separator	C. Reif	Hartleton, Pa	Aug. 8, 1854	11,493
Clover separator and huller	D. M. Heikes	Franklin Township, Pa	May 12, 1868	77,731
Clover-stripper and hay-rake, Combined	R. H. Blair and A. W. Beatty	Saltsburgh, Pa	Oct. 22, 1861	33,512
Club-feet and crooked legs, Apparatus for straightening.	H. R. Allen	Charleston, Ill	Jan. 28, 1868	73,768
Club-feet apparatus	E. H. Chamberlin	Cincinnati, Ohio	Dec. 11, 1847	5,393

Index of patents issued from the United States Patent Office from 1790 *to* 1873, *inclusive*—Continued.

Invention.	Inventor.	Residence.	Date.	No.
Club-feet, Apparatus for the cure of	Z. Hussey	Chillicothe, Ohio	Dec. 14, 1852	9, 472
Club-feet, Apparatus for treating	G. B. Wood	Chicago, Ill	Aug. 30, 1870	106, 907
Club-feet, Apparatus for treatment of	J. B. Brown	Boston, Mass	Apr. 30, 1840	1, 572
Club-feet, Surgical instrument for treatment of	E. A. Grant	Louisville, Ky	May 9, 1871	114, 669
Clutch	J. Hartman, jr	Philadelphia, Pa	Apr. 26, 1864	42, 4[illegible]3
Clutch	A. Heth and G. Hall	Adams, N. Y	May 28, 1867	65, 220
Clutch	S. Ingersoll	Stamford, Conn	May 12, 1863	38, 486
Clutch	W. Knight and N. W. Lewis	Bridgeport, Conn	June 24, 1873	140, 142
Clutch	W. I. Risedorph	Albany, N. Y	Sept. 7, 1869	94, 649
Clutch	H. F. Shaw	West Roxbury, Mass	Aug. 10, 1869	93, 486
Clutch	J. W. Tuttle	Newton, Mass	Oct. 11, 1870	108, 303
Clutch, Anchor	H. Fleming	Chicago, Ill	Oct. 11, 1864	44, 621
Clutch and brake, Friction	D. Banks, jr	New York, N. Y	June 8, 1869	90, 982
Clutch and brake, Friction	D. Banks, jr	Middletown, Conn	Sept. 10, 1872	131, 207
Clutch applied to rudder-head, Spring	E. O. P. Andrews	Boston, Mass	Aug. 10, 1848	5, 700
Clutch, Centrifugal friction	R. Reynolds	Stockport, N. Y	Mar. 3, 1857	16, 748
Clutch, Chain	H. Pitcher	Fond du Lac, Wis	Dec. 13, 1870	110, 069
Clutch, Combined friction and ratchet	A. W. Hall	Northfield, Vt	Dec. 19, 1871	122, 009
Clutch, Elevating	C. C. Blodgett	Watertown, N. Y	Apr. 14, 1868	76, 588
Clutch for belt-pulley, Friction	J. H. Blessing	Albany, N. Y	Apr. 18, 1871	113, 841
Clutch for connecting and disconnecting machinery.	D. M. Smyth	New York, N. Y	Dec. 19, 1865	51, 665
Clutch for engaging and disengaging gearing, Friction.	G. D. Emerson	Calumet, Mich	Apr. 12, 1870	101, 846
Clutch for flour-packer	J. T. Noye	Buffalo, N. Y	June 17, 1856	15, 145
Clutch for hydraulic-wheel press	A. B. Couch	Worcester, Mass	Sept. 14, 1869	94, 717
Clutch for machinery, Spiral friction	J. Hanley	New York, N. Y	Dec. 18, 1866	60, 509
Clutch for power-press	N. C. Stiles	Meriden, Conn	Jan. 30, 1866	52, 335
Clutch for suspending hay-fork	W. Dixon and L. Heath	Adams, N. Y	Oct. 1, 1867	69, 326
Clutch for water-wheel, Self-releasing	G. W. Wesley	Meadville, Pa	Apr. 18, 1871	113, 826
Clutch, Friction	H. Aiken	Philadelphia, Pa	Mar. 12, 1872	124, 529
Clutch, Friction	S. B. Alger	Oswego, N. Y	Apr. 1, 1873	137, 335
Clutch, Friction	E. Allen	Norwich, Conn	May 20, 1873	138, 980
Clutch, Friction	D. Banks	New York, N. Y	May 25, 1869	90, 481
Clutch, Friction	N. Barlow	Saint Louis, Mo	Aug. 13, 1850	7, 557
Clutch, Friction	F. G. Bates	Springfield, Mass	Oct. 15, 1872	132, 234
Clutch, Friction	E. B. Bigelow	Boston, Mass	Aug. 17, 1869	93, 798
Clutch, Friction	J. S. Brown	Pawtucket, R. I	Apr. 26, 1864	42, 525
Clutch, Friction	W. H. Brown	Worcester, Mass	Nov. 28, 1865	51, 134
Clutch, Friction	W. C. Burch and G. D. Oatley	Gloucester, N. J	June 13, 1871	115, 818
Clutch, Friction	C. E. Burwell	Springfield, Mass	July 9, 1872	128, 705
Clutch, Friction	C. W. Cardot	Jamestown, N. Y	June 8, 1869	90, 992
Clutch, Friction	G. Clisbee	Marlborough, Mass	July 18, 1871	117, 234
Clutch, Friction	T. Coldwell	Newburgh, N. Y	Nov. 15, 1870	109, 299
Clutch, Friction	G. D. Emerson	Calumet, Mich	Aug. 10, 1869	93, 604
Clutch, Friction	P. Ferguson and F. G. Bates	New Haven, Conn., and Springfield, Mass.	Jan. 16, 1872	122, 822
Clutch, Friction	E. T. Ford	Stillwater, N. Y	Sept. 1, 1868	81, 617
Clutch, Friction	J. J. Grant	Greenfield, Mass	Aug. 26, 1873	142, 227
Clutch, Friction	J. J. Grant	Greenfield, Mass	Sept. 9, 1873	142, 564
Clutch, Friction	A. W. Hall	East Lebanon, N. H	Feb. 4, 1868	74, 080
Clutch, Friction	M. Hawkins	Birmingham, Conn	Nov. 16, 1869	96, 913
Clutch, Friction	G. W. Hedges	San Francisco, Cal	Oct. 31, 1871	120, 434
Clutch, Friction	S. Ingersoll	Brooklyn, N. Y	Jan. 25, 1870	99, 199
Clutch, Friction	W. W. Jerome, S. B. Alger, and C. H. Sage.	Norwich, N. Y	Nov. 29, 1870	109, 740
Clutch, Friction	J. B. Johnson and W. H. Birch	San Francisco, Cal	Apr. 7, 1868	76, 328
Clutch, Friction	O. Lull	Rochester, N. Y	May 31, 1870	103, 634
Clutch, Friction	L. H. Olmsted	Newark, N. J	Sept. 27, 1864	44, 448
Clutch, Friction	F. A. Pratt	Hartford, Conn	Feb. 22, 1870	100, 064
Clutch, Friction	G. S. Reynolds	Lebanon, N. H	Apr. 4, 1871	113, 566
Clutch, Friction	R. Reynolds	Stockport, N. Y	Dec. 24, 1861	34, 012
Clutch, Friction	G. Sickels	Brooklyn, N. Y	Mar. 2, 1852	8, 782
Clutch, Friction	H. K. Smith	Norwich, Conn	Dec. 5, 1865	51, 360
Clutch, Friction	T. Stebins	San Francisco, Cal	Oct. 1, 1872	131, 795
Clutch, Friction	T. A. Weston	Ridgewood, N. J	July 30, 1872	129, 914
Clutch, Friction	S. C. Wright	Fitchburgh, Mass	Apr. 16, 1867	63, 977
Clutch, Friction	W. Wright	New York, N. Y	June 15, 1852	9, 043
Clutch-gearing	L. S. Fithian	Brooklyn, N. Y	Aug. 8, 1871	117, 758
Clutch-gearing	E. Nicholson	Cleveland, Ohio	Nov. 1, 1870	108, 817
Clutch, Harvester	F. Bramer	Little Falls, N. Y	Aug. 22, 1871	118, 333
Clutch in machine for packing flour	S. Taggart	Indianapolis, Ind	Dec. 19, 1854	12, 107
Clutch, Lathe	R. Allen	Jersey City, N. J	Dec. 4, 1866	60, 113
Clutch, Machinery	F. G. Bates	Springfield, Mass	Mar. 12, 1872	124, 476
Clutch, Machinery	W. B. Duckworth	Hamden, Conn	Apr. 22, 1873	138, 139
Clutch, Machinery	T. F. Hammer	Branford, Conn	July 11, 1865	48, 769
Clutch, Machinery	W. E. Hawkins	Wallingford, Conn	Feb. 27, 1872	124, 058
Clutch, Machinery	J. E. Plummer and J. P. Noyes	Binghamton, N. Y	Nov. 22, 1870	109, 450
Clutch mechanism	D. M. Weston	Boston, Mass	Oct. 3, 1871	119, 677
Clutch mechanism, Friction	J. J. Coburn	Philadelphia, Pa	Apr. 22, 1873	138, 075
Clutch mechanism, Friction	M. G. Crane and A. M. Polsey	Newton and Boston, Mass.	Feb. 15, 1870	99, 761
Clutch mechanism, Friction pawl or	W. R. Close	Bangor, Me	June 27, 1871	116, 413
Clutch or pulley, Friction	H. S. Shepardson	Shelburne Falls, Mass	June 14, 1864	43, 138
Clutch or rope-holder	C. A. Emery	Springfield, Mass	Sept. 19, 1865	49, 990
Clutch, Sail	E. A. Sawyer	Portland, Me	June 6, 1865	48, 103
Clutch-shipper	F. J. Plummer	Worcester, Mass	Aug. 27, 1867	68, 108
Clutch, Slaughtering	R. Savage	Chicago, Ill	Mar. 23, 1869	88, 217
Clutch, Tubing	J. N. Angier	Titusville, Pa	Nov. 16, 1869	96, 766
Coach, &c	B. Hale	Newburyport, Mass	Jan. 17, 1818	....
Coach adapted to the prairies, Steam	J. Semple	Alton, Ill	May 1, 1845	4, 029
Coach-alarm	W. Hunt	New York, N. Y	July 30, 1827	....
Coach and other carriage	F. Bremler	New York, N. Y	Apr. 24, 1810	....
Coach, barouche, &c	R. Gedney	New York, N. Y	Apr. 7, 1832	....
Coach-door	J. Penfield and D. F. Woolsey	Bridgeport, Conn	Oct. 12, 1869	95, 832
Coach-doors, Stop for	H. Killam	New Haven, Conn	Dec. 10, 1861	33, 892
Coach-lace, Making	W. G. Barnet	Newark, N. J	Nov. 18, 1825	....
Coach, &c., Mail	S. Draper	Brookfield, Mass	Aug. 26, 1819	....
Coach or carriage	J. Allgaier	Philadelphia, Pa	June 10, 1873	139, 753

Index of patents issued from the United States Patent Office from 1790 *to* 1873, *inclusive*—Continued.

Invention.	Inventor.	Residence.	Date.	No.
Coach or carriage	J. Allgaier	Philadelphia, Pa	July 8, 1873	140, 666
Coach-pad	B. F. Hooper	Newark, N. J	Apr. 14, 1868	76, 762
Coach-panel	E. A. Lester	Boston, Mass	Mar. 18, 1835	....
Coach, pleasure-carriage, railway-car, &c	T. Shriver	Cumberland, Md	Nov. 7, 1839	1, 399
Coach-step	G. Carter	Newark, N. J	Nov. 26, 1831	......
Coach-step	C. M. King	Newark, N. J	Dec. 31, 1833	....
Coach-tongues, Mode of supporting	Z. B. Wakeman	Beloit, Wis	Mar. 10, 1857	16, 813
Coaches, &c., Hanging curtains or blinds in	P. Harvey	Philadelphia, Pa	May 9, 1807	....
Coaches, Making post and other	R. Ward	Schenectady, N. Y	June 2, 1825	....
Coaches, Securing baggage on post	J. Stratton	Greenfield, Mass	June 29, 1830	....
Coal and generating gas, Cooking	W. G. Valentin	London, England	Jan. 13, 1863	37, 412
Coal and grain boat elevator	S. K. Hoxsie	Philadelphia, Pa	May 18, 1869	90, 103
Coal and ore separator	H. Bradford	Reading, Pa	Sept. 30, 1873	143, 219
Coal and ore separators, Feeding-device for	H. Bradford	Reading, Pa	Sept. 30, 1873	143, 323
Coal and ores, Desulphurizing	J. I. Storer	Philadelphia, Pa	Apr. 9, 1861	32, 012
Coal and ores, Purification of	J. Burroughs	Ridgway, Pa	Apr. 3, 1860	27, 689
Coal and ores, Purification of	B. F. Penimen	New York, N. Y	Oct. 25, 1864	44, 817
Coal and other minerals, Machinery for moving	J. E. Wooten	Reading, Pa	June 7, 1870	104, 093
Coal and other minerals, Mode of cutting	T. Harrison	Tudhoe, England	May 3, 1864	42, 626
Coal and other substances, Machine for breaking	R. A. Wilder	Cressona, Pa	Mar. 18, 1873	136, 887
Coal and rock cutting machine	H. F. Brown	Indianapolis, Ind	Sept. 16, 1873	142, 896
Coal and slate separator	H. Bradford	Reading, Pa	Oct. 29, 1872	132, 623
Coal and wood box	J. Mallin	Chicago, Ill	May 2, 1871	114, 312
Coal, &c., Apparatus for applying oil and water to assist the combustion of.	S. C. Salisbury	New York, N. Y	Apr. 24, 1866	54, 216
Coal, &c., Apparatus for burning fine	E. Thayer	Worcester, Mass	Sept. 13, 1864	44, 263
Coal, Apparatus for raising and dumping	E. Morris	Bergen, N. J	July 8, 1856	15, [illegible]00
Coal, Artificial lump	R. Covert	Brooklyn, N. Y	Nov. 8, 1864	44, 940
Coal, Artificial lump	R. Covert	Brooklyn, N. Y	June 27, 1865	48, 372
Coal-barge	L. F. Frazee	Jersey City, N. J	Dec. 13, 1870	110, 130
Coal-barge	R. Hartley	Pittsburgh, Pa	May 31, 1870	103, 607
Coal-barge	J. A. Preston	New Haven, Conn	Sept. 24, 1867	69, 245
Coal-barge	J. A. Preston	New Haven, Conn	May 24, 1870	103, 502
Coal, Bit for boring	S. Gissinger	Manchester, Pa	Dec. 17, 1867	72, 389
Coal, &c., Bit for boring	S. Hipkins, jr	Wheeling, W. Va	Jan. 14, 1873	134, 805
Coal-boring machine	S. Hipkins, jr	Wheeling, W. Va	Jan. 14, 1873	134, 804
Coal-boring machine	D. Shoemaker	Kittaning Township, Pa	Sept. 16, 1873	142, 950
Coal-box	T. L. Bozart, jr	Indianapolis, Ind	Jan. 24, 1871	111, 171
Coal-box	C. W. Coffin	Pittsburgh, Pa	Oct. 31, 1871	120, 494
Coal box	B. R. Deacon	Montreal, Canada	Apr. 12, 1870	101, 833
Coal-box	C. Hoefinghoff	Cincinnati, Ohio	July 13, 1869	92, 447
Coal-box	S. A. Simison	Earlville, Ill	Jan. 14, 1868	73, 394
Coal-box and fire-iron holder, Combined	A. F. Tripp	Buffalo, N. Y	Dec. 24, 1872	134, 230
Coal-box and fire-iron standard	E. B. Jewett	Buffalo, N. Y	Dec. 10, 1872	133, 780
Coal-breaker	H. Bradford	Reading, Pa	Oct. 21, 1873	143, 745
Coal-breaker	W. De Haven and P. Umholtz	Minersville, Pa	Feb. 8, 1848	5, 440
Coal-breaker	J. A. Dickson	Scranton, Pa	July 4, 1865	48, 532
Coal-breaker	J. Fox	Philadelphia, Pa	Apr. 25, 1865	47, 409
Coal-breaker	B. Haywood	Pottsville, Pa	May 21, 1845	4, 058
Coal-breaker	T. Petherick	Pottsville, Pa	Apr. 22, 1856	14, 732
Coal-breaker	W. Richardson	Philadelphia, Pa	Sept. 2, 1845	4, 172
Coal-breaker	P. Umholtz	Tremont, Pa	Mar. 20, 1860	27, 581
Coal-breaker	P. Umholtz	Tremont, Pa	June 13, 1865	48, 224
Coal-breaker	R. A. Wilder	Cressona, Pa	Sept. 2, 1873	142, 426
Coal breaker and separator	L. P. Garner	Ashland, Pa	May 25, 1869	90, 439
Coal breaking and screening machine	J. Battin	Philadelphia, Pa	Oct. 6, 1843	3, 292
Coal-breaking machine	J. Battin	Philadelphia, Pa	Feb. 12, 1844	3, 438
Coal-breaking machine	A. Bolton	Port Carbon, Pa	Mar. 2, 1858	19, 481
Coal-breaking machine	J. R. Deihm and J. Snell	Pottsville, Pa	Oct. 27, 1857	18, 501
Coal-breaking machine	L. P. Garner	Ashland, Pa	Feb. 19, 1861	31, 451
Coal-breaking machine	J. Jones and S. P. Bidder, jr	Stratford and Mitcham, England.	Aug. 31, 1869	94, 216
Coal-breaking machine	C. W. Kennedy and R. T. Brown	Williamsburgh, N. Y	May 10, 1859	23, 922
Coal-breaking machine	R. A. Wilder	Cressona, Pa	Feb. 19, 1861	31, 502
Coal-breaking machine	R. A. Wilder	Cressona, Pa	Dec. 24, 1872	134, 233
Coal-breaking machines, Conveyer and separator for.	R. A. Wilder	Cressona, Pa	Dec. 24, 1872	134, 234
Coal-breaking machinery	G. E. Hoyt	Brooklyn, N. Y	Feb. 21, 1860	27, 220
Coal-breaking machinery	I. P. Lykens	Pottsville, Pa	Feb. 14, 1860	27, 191
Coal-breaking roll	W. R. Reece	Tremont, Pa	Sept. 8, 1863	39, 835
Coal burner, Anthracite	J. N. Olney	Providence, R. I	Dec. 26, 1833	....
Coal-chute	E. Maguire	Kewanee, Ill	Aug. 20, 1867	67, 995
Coal-chute	H. Merriman	Bloomington, Ill	Dec. 8, 1868	84, 749
Coal-chute	W. E. Phelps	Elmwood, Ill	Nov. 28, 1871	121, 413
Coal-chute	L. D. and C. C. Roberts	Cleveland, Ohio	May 4, 1869	89, 793
Coal-chute	J. Rhodes	Dunkirk, N. Y	June 6, 1871	115, 640
Coal-chute	C. W. Williams	Port Jervis, N. Y	Nov. 5, 1872	132, 884
Coal-chute, Extension	J. Heatherington	Bellaire, Ohio	Feb. 18, 1868	74, 686
Coal, Composition for desulphurizing	J. H. Connelly	Wheeling, W. Va	Aug. 9, 1864	43, 760
Coal conveyer and and elevator	J. T. Hoole	Buffalo, N. Y	Sept. 24, 1872	131, 683
Coal conveying and dumping apparatus	H. C. Clark and R. B. Little	Providence, R. I	June 2, 1866	78, 430
Coal, Conveying, cleaning, and assorting	J. G. Brant	Cumberland, Md	Mar. 14, 1846	4, 415
Coal, Cooling animal	W. Moller	New York, N. Y	Feb. 5, 1867	61, 851
Coal-cracker	T. Poore	Carbondale, Pa	May 12, 1857	17, 294
Coal-crusher	H. C. Rogers	Scranton, Pa	June 14, 1864	43, 133
Coal-cutting machine	G. E. Donisthorpe	Leeds, England	Sept. 22, 1868	82, 391
Coal-cutting machine	G. E. Donisthorpe	Leeds, England	Sept. 29, 1868	82, 608
Coal-cutting machine	S. Firth	Leeds, England	Apr. 8, 1873	137, 669
Coal delivering sack	W. S. Shackleton	Cleveland, Ohio	Apr. 4, 1871	113, 353
Coal delivering sack	W. S. Shackleton	Cleveland, Ohio	Sept. 9, 1873	142, 741
Coal, Device for boring and excavating	A. Buchanan	Brooklyn, N. Y	June 20, 1865	48, 258
Coal-digger	W. Ward	Pittsburgh, Pa	Apr. 12, 1870	101, 793
Coal-discharging apparatus	E. F. Flood	Chicago, Ill	Mar. 30, 1869	88, 379
Coal-drilling machine	J. Grimm	Darlington Township, Pa	Nov. 23, 1869	97, 080
Coal, Drier for preparing waste	T. M. Mitchell	Philadelphia, Pa	July 20, 1869	92, 737
Coal-dumping apparatus	R. Jenkins and T. Woods	Allegheny County, Pa	June 6, 1871	115, 738
Coal-dumping apparatus	E. R. Kerr	Kewanee, Ill	Oct. 2, 1866	58, 549
Coal dust and cinders, Mode of utilizing	A. L. Fleury	Pittsburgh, Pa	Feb. 13, 1866	52, 554

Index of patents issued from the United States Patent Office from 1790 to 1873, inclusive—Continued.

Invention.	Inventor.	Residence.	Date.	No.
Coal-dust as fuel in furnaces, Process of using anthracite.	A. Berney	Jersey City, N. J	Apr. 15, 1873	137, 820
Coal-dust for fuel, Mode of preparing	A. D. Ditmars	Lancaster, Pa	Aug. 18, 1868	81, 149
Coal-dust, Furnace for burning	G. B. Deppen and E. Levengood.	Myerstown, Pa	Nov. 16, 1858	22, 067
Coal-dust, peat, &c., Method of consolidating	W. J. Cheyney and E. T. Dieterichs.	Wallingford and Philadelphia, Pa.	Mar. 14, 1865	46, 777
Coal-elevator	A. S. Bailey	Chicago, Ill	Feb. 22, 1870	100, 103
Coal elevator	L. S. Chichester	New York, N. Y	Aug. 17, 1869	93, 676
Coal-elevator	J. C. Clifford	Yonkers, N. Y	Oct. 3, 1871	119, 574
Coal elevator	J. P. Tucker	South Reading, Mass	June 12, 1866	55, 557
Coal elevator and distributor	H. C. Clark and R. B. Little	Providence, R. I	Oct. 1, 1867	69, 407
Coal elevator and weigher	L. S. Chichester	Brooklyn, N. Y	Aug. 17, 1869	93, 677
Coal elevator bucket	A. B. Nimbs	Buffalo, N. Y	Sept. 6, 1864	44, 108
Coal for metallurgic operations, Preparing	W. J. Lynd	Golden City, Colo	Jan. 4, 1870	98, 607
Coal for smelting ores, Preparing	W. J. Lynd	Golden City, Colo	Jan. 4, 1870	98, 606
Coal for welding iron, Method of desulphurizing	J. H. Elward	Polo, Ill	Feb. 20, 1866	52, 699
Coal from slate, Apparatus for separating	B. F. Day	Hazleton, Pa	July 9, 1872	128, 791
Coal furnace, Anthracite	M. B. Bulkley	Pottsville, Pa	Jan. 16, 1835	....
Coal furnace, Anthracite	W. Church	Boston, Mass	Nov. 20, 1826	....
Coal furnace, Anthracite	D. Olmsted	New Haven, Conn	Nov. 5, 1834	....
Coal furnace, Anthracite	J. L. Sullivan	New York, N. Y	Nov. 26, 1825	....
Coal furnace, Anthracite	O. Woodruff	New York, N. Y	Nov. 7, 1825	....
Coal, Furnace for burning fossil	D. Stansbury	Belleville, N. J	Apr. 15, 1825	....
Coal-hod	J. H. Brown	Lynn, Mass	Aug. 5, 1873	141, 485
Coal-hod	E. R. Hall	Buffalo, N. Y	July 17, 1866	56, 408
Coal-hod	L. W. Hemp	Saint Louis, Mo	Dec. 31, 1872	134, 379
Coal-hod	R. W. Huston	Calais, Me	Nov. 22, 1859	26, 188
Coal-hod	E. A. Jeffery	New Haven, Conn	Aug. 23, 1864	43, 916
Coal-hod	E. A. Jeffery	New Haven, Conn	Feb. 27, 1866	52, 856
Coal-hod	C. F. Kneeland	Buffalo, N. Y	Aug. 12, 1856	15, 519
Coal-hod	J. A. Lawson	Troy, N. Y	Aug. 27, 1867	68, 213
Coal-hod	W. N. Martin	Bristol, R. I	Nov. 7, 1854	11, 898
Coal-hod	J. R. Miller	Cincinnati, Ohio	May 3, 1864	42, 623
Coal-hod	F. G. and W. F. Needringhaus.	Saint Louis, Mo	Oct. 20, 1868	83, 301
Coal-hod	J. Pfitzinger	Buffalo, N. Y	Oct. 22, 1867	70, 015
Coal-hod	J. Pilbeam	Seneca Falls, N. Y	Sept. 22, 1863	40, 061
Coal-hod	J. H. Pocock	Chicago, Ill	Nov. 20, 1866	59, 925
Coal-hod	H. B. Safford	Lancaster, N. Y	Sept. 9, 1873	142, 734
Coal-hod	C. Smith	Brooklyn, N. Y	Oct. 24, 1871	120, 336
Coal-hod	J. G. Somes	Charlestown, Mass	Apr. 27, 1869	89, 444
Coal-hod	S. Thomson, G. Smith, and J. S. Jennings.	Brooklyn, N. Y	Mar. 20, 1866	53, 384
Coal-hod	W. Wilson	Boston, Mass	May 3, 1870	102, 639
Coal-hod	A. A. Yeatman and J. M. Mason.	Washington, D. C	Jan. 15, 1867	61, 301
Coal hod and screen	A. Porter	Charlestown, Mass	Feb. 25, 1868	74, 770
Coal hod and sifter, Combined	R. Bogle	Milton, Pa	Dec. 26, 1871	122, 147
Coal hod and sifter, Combined	B. F. Brown	Portland, Me	Nov. 27, 1866	59, 954
Coal-hod, ash-sifter, and slop-pail combined	C. Jones	Brooklyn, N. Y	May 30, 1867	48, 020
Coal-hod bottom	E. W. Kimball	Hudson, N. Y	Jan. 21, 1868	73, 535
Coal-hod, Screen	M. S. Nichols and R. Weaver	Central Village, Conn	Mar. 15, 1870	100, 791
Coal-hod stand	W. M. Conger	Newark, N. J	May 14, 1872	126, 676
Coal-hods, Construction of sheet-metal	M. G. Fagan	Troy, N. Y	July 7, 1868	79, 564
Coal-hods, Construction of sheet-metal	J. McCoy	Philadelphia, Pa	Nov. 19, 1867	71, 032
Coal hoister and conveyer	J. Green	New York, N. Y	Aug. 1, 1871	117, 534
Coal, Hoisting and dumping	G. Martz	Pottsville, Pa	Oct. 30, 1866	59, 243
Coal hoisting and dumping apparatus	W. B. Culver	Scranton, Pa	Jan. 4, 1859	22, 492
Coal hoisting and dumping machine	G. Martz	Pottsville, Pa	Apr. 13, 1858	19, 939
Coal-hole cover	J. Gault	Boston, Mass	July 23, 1861	32, 865
Coal-hole cover	L. A. Kimberly	New Haven, Conn	Nov. 22, 1870	109, 422
Coal-hole cover	F. H. Moore	Boston, Mass	Apr. 3, 1855	12, 637
Coal-hole cover, Safety	F. H. Moore	Boston, Mass	Jan. 15, 1856	14, 107
Coal-hole guard	J. C. Jenks	Boston, Mass	Mar. 4, 1865	14, 359
Coal-holes, Composition-cover for	J. F. Wood	Everett, Mass	Nov. 26, 1872	133, 513
Coal-holes, Safety-port for	S. W. Frost	Boston, Mass	Apr. 3, 1855	12, 623
Coal-hopper and platform-scales, Combined	F. E. and L. I. Howe	New York, N. Y., and Boston, Mass.	May 5, 1868	77, 488
Coal in cellars, Machine for depositing	W. Bell	Boston, Mass	Feb. 26, 1856	14, 301
Coal into blocks for fuel, Press for compacting the waste particles of.	T. M. Mitchell	Philadelphia, Pa	Nov. 30, 1869	97, 309
Coal into solid lumps, Method of converting fine	W. Easby	Washington, D. C	Aug. 29, 1848	5, 739
Coal. Kiln for charring	C. Kownslar	Mill Creek, Va	June 20, 1825	....
Coal-lifter	J. B. Creemer	Philadelphia, Pa	Oct. 23, 1855	13, 699
Coal-loading device	D. Risher, jr	Dravosburgh, Pa	Sept. 8, 1868	81, 946
Coal, Machine for breaking and screening	H. Thomas	Beaver Meadow, Pa	Dec. 5, 1843	3, 368
Coal, Machine for cracking anthracite	J. S. Hubbell	New York, N. Y	Aug. 17, 1835	....
Coal, Machine for cutting	J. Gillott and P. Copley	Barnsley, England	Nov. 11, 1873	144, 529
Coal, Machine for desulphurizing bituminous	D. Morgan	Hammondsville, Ohio	Jan. 1, 1869	90, 772
Coal, Machine for loading	C. Bushor	Philadelphia, Pa	July 2, 1861	32, 680
Coal, &c., Machine for loading	C. Bushor	Philadelphia, Pa	May 20, 1862	35, 297
Coal, Machine for separating slate and other foreign substances from.	E. Borda and D. Glover	Woodside, Pa	Oct. 13, 1857	18, 380
Coal, Machine for slating	T. Garretson	Pottsville, Pa	Mar. 30, 1858	19, 768
Coal, Machine for slating	J. Gass	Trevorton, Pa	Nov. 24, 1857	18, 687
Coal, Machine for undermining	G. D. Whitcomb	Chicago, Ill	Aug. 29, 1871	118, 501
Coal, Method of burning waste	J. D. Whelpley and J. J. Storer	Boston, Mass	Mar. 13, 1866	53, 208
Coal, minerals, &c., Method of mining	W. Locke, J. Warrington, W. E. Carrett, W. E. Marshall, and J. Telford.	Kippax, England	May 15, 1866	54, 833
Coal-mining apparatus	G. E. Donisthorpe	Leeds, England	Sept. 1, 1868	81, 759
Coal-mining apparatus	G. E. Donisthorpe	Leeds, England	Sept. 1, 1868	81, 760
Coal-mining machine	E. K. and J. M. Bruce	Liberty, Pa	Sept. 19, 1865	49, 972
Coal-mining machine	W. Case	Troy, Ill	June 3, 1873	139, 541
Coal-mining machine	C. A. Chamberlin	Allegheny City, Pa	Mar. 9, 1858	19, 543
Coal-mining machine	G. E. Donisthorpe	Leeds, England	Apr. 7, 1868	76, 417
Coal-mining machine	G. E. Donisthorpe	Leeds, England	Apr. 7, 1868	76, 418
Coal-mining machine	G. E. Donisthorpe	Leeds, England	Apr. 7, 1868	76, 419

Index of patents issued from the United States Patent Office from 1790 *to* 1873, *inclusive*—Continued.

Invention.	Inventor.	Residence.	Date.	No.
Coal-mining machine	G. E. Donisthorpe	Leeds, England	Sept. 22, 1868	82, 390
Coal-mining machine	C. L. Driesslein	Chicago, Ill	Mar. 11, 1873	136, 712
Coal-mining machine	G. Elbreg	Brazil, Ind	Jan. 14, 1873	134, 870
Coal-mining machine	J. S. Fisk and J. Westerman	Meadville and Sharon, Pa	Jan. 17, 1865	45, 917
Coal-mining machine	W. W. Grier and R. H. Boyd	Hulton, Pa	July 12, 1864	43, 493
Coal-mining machine	W. W. Grier and R. H. Boyd	Hulton, Pa	Oct. 24, 1865	50, 577
Coal-mining machine	F. M. W. Price	Danville, Ill	Sept. 9, 1873	142, 582
Coal-mining machine	E. Simkins	Allegheny, Pa	Oct. 26, 1858	21, 908
Coal-mining machine	H. Spear	Portland, Me	Dec. 10, 1872	133, 900
Coal-mining-machine drill-carriage	S. Gissinger	Allegheny City, Pa	Dec. 17, 1867	72, 390
Coal-mining machinery	J. Alexander	Gartsherrie, Scotland	Feb. 18, 1873	135, 874
Coal-mining machinery	J. G. Jones	Monmouthshire, Wales	Dec. 5, 1865	51, 405
Coal-mining tool	I. Lamplugh	Peoria, Ill	Dec. 19, 1871	122, 035
Coal, Mixing apparatus for preparing waste	T. M. Mitchell	Philadelphia, Pa	July 20, 1869	92, 738
Coal, &c., Mode of pulverizing and preparing for use.	J. E. Lundgren	Stockholm, Sweden	Apr. 5, 1864	42, 257
Coal-oil for heating purposes, Apparatus for burning.	H. W. Dopp	Buffalo, N. Y	Jan. 20, 1863	37, 436
Coal-oil furnace	T. Shaw	Philadelphia, Pa	May 29, 1860	28, 539
Coal-oil heater	H. W. Dopp	Buffalo, N. Y	July 7, 1863	39, 129
Coal or charcoal, Furnace for burning stone	J. Eslin	Philadelphia, Pa	June 19, 1830	....
Coal, ore, &c., washer	M. Evrard	St. Etienne, France	Aug. 26, 1873	142, 221
Coal, Preparing	H., jr., and E. Mayhew	Lancaster, Pa	Mar. 6, 1860	27, 373
Coal, Process and apparatus for utilizing waste	T. M. Mitchell	Philadelphia, Pa	July 20, 1869	92, 736
Coal, Producing combustion by applying high steam and hydrogen gas to anthracite.	P. Chase	New York, N. Y	Oct. 7, 1831	....
Coal products to obtain benzole, Treating	J. Rowley	Camberwell, England	Mar. 22, 1870	101, 048
Coal raising and dumping apparatus	J. Delaney	Ashland, Pa	June 8, 1869	91, 093
Coal raising and moving	A. Lawton	Philadelphia, Pa	Nov. 16, 1869	96, 927
Coal-raising apparatus	A. Lawton	Philadelphia, Pa	July 13, 1869	92, 461
Coal-screen	E. G. Belknap	Philadelphia, Pa	May 16, 1871	114, 753
Coal-screen	H. L. Cake	Pottsville, Pa	Jan. 24, 1860	26, 890
Coal-screen	J. R. Deihm and J. Snell	Pottsville, Pa	June 21, 1864	43, 185
Coal-screen	R. B. Douty	Shamokin, Pa	May 19, 1863	38, 569
Coal-screen	T. Farron	Lincoln Township, Pa	Jan. 10, 1872	122, 589
Coal-screen	J. P. Fennell	Philadelphia, Pa	June 19, 1855	13, 083
Coal-screen	B. L. Fetherolf	Tamaqua, Pa	Jan. 23, 1872	122, 882
Coal-screen	L. P. Garner	Ashland, Pa	May 11, 1869	89, 927
Coal-screen	G. E. Hoyt and F. Nishwitz	Brooklyn, N. Y	Jan. 19, 1858	19, 175
Coal-screen	J. Laubenstein	Minersville, Pa	Feb. 4, 1862	34, 306
Coal-screen	G. Martz	Pottsville, Pa	Mar. 6, 1855	12, 486
Coal-screen	A. M. Olds	Chicago, Ill	Jan. 24, 1865	46, 019
Coal-screen	J. A. Robinson	Pittston, Pa	Mar. 28, 1865	47, 041
Coal-screen	J. B. Smith	Dunmore, Pa	Aug. 7, 1866	57, 002
Coal-screen	J. Snell and J. R. Deihm	Pottsville, Pa	May 24, 1859	24, 155
Coal-screen	J. Snell and J. R. Deihm	Pottsville, Pa	Sept. 10, 1861	33, 268
Coal-screen	W. Sparks	New York, N. Y	Sept. 12, 1865	49, 932
Coal-screen	W. Sparks	New York, N. Y	Nov. 16, 1869	96, 991
Coal-screen	E. Thomas	Shickshinny, Pa	Dec. 6, 1870	109, 969
Coal-screen	G. Whittle	New York, N. Y	Oct. 29, 1867	70, 305
Coal-screen	S. W. Woodward	Buffalo, N. Y	Apr. 29, 1873	138, 355
Coal screen and chute	M. R. Roberts	San Francisco, Cal	June 17, 1873	140, 077
Coal-screen, Corrugated-iron revolving	E. W. Weston	Providence, Pa	Nov. 19, 1867	71, 099
Coal-screens, Wire cloth for	C. P. Leitzinger	Scranton, Pa	June 18, 1872	128, 177
Coal-screens, Wire-cloth for	J. W. Brock	Scranton, Pa	Nov. 12, 1872	132, 949
Coal screening or sifting machine	B. S. Hort	Kensington, Pa	Feb. 7, 1842	2, 451
Coal-scuttle	S. J. Anderson	Cazenovia, N. Y	Apr. 26, 1870	102, 202
Coal-scuttle	A. Bardell and S. Smith	New York, N. Y	Mar. 12, 1867	62, 807
Coal-scuttle	F. L. Blair	Allegheny, Pa	Jan. 11, 1870	98, 661
Coal-scuttle	J. A. Bragaw	Kingston, N. Y	Oct. 10, 1871	119, 738
Coal-scuttle	M. L. Byrn	New York, N. Y	Nov. 28, 1865	51, 139
Coal-scuttle	G. Chambers	Ithaca, N. Y	Dec. 13, 1864	45, 388
Coal-scuttle	H. C. Clark	Williamsburgh, N. Y	Dec. 24, 1872	134, 253
Coal-scuttle	J. W. Coombs	Mount Vernon, N. Y	June 3, 1862	35, 436
Coal-scuttle	B. F. Cowan	New York, N. Y	Jan. 8, 1867	61, 054
Coal-scuttle	J. L. Ellithorp and P. Sloan	Canajoharie, N. Y	May 25, 1869	90, 437
Coal-scuttle	E. Eltinge	Kingston, N. Y	Dec. 11, 1866	60, 349
Coal-scuttle	E. Eltinge	Kingston, N. Y	June 20, 1871	116, 036
Coal-scuttle	B. Fielding	Cincinnati, Ohio	July 4, 1871	116, 699
Coal-scuttle	G. H. and D. W. Hazelton	Philadelphia, Pa	Apr. 4, 1871	113, 429
Coal-scuttle	C. F. Henis	Cincinnati, Ohio	Nov. 27, 1866	60, 003
Coal-scuttle	C. Hodgetts	Williamsburgh, N. Y	May 31, 1870	103, 615
Coal-scuttle	W. Hoffmire	New York, N. Y	Aug. 13, 1872	130, 498
Coal-scuttle	J. Holden	Philadelphia, Pa	Aug. 14, 1866	57, 137
Coal-scuttle	G. Howland	Brunswick, N. Y	Feb. 4, 1873	135, 554
Coal-scuttle	J. W. Jarboe	Brooklyn, E. D., N. Y	May 4, 1869	89, 771
Coal-scuttle	T. T. Markland, jr	Philadelphia, Pa	Nov. 1, 1864	44, 875
Coal-scuttle	E. Mather	Chicago, Ill	Sept. 12, 1871	118, 807
Coal-scuttle	W. Miller	New York, N. Y	Mar. 10, 1863	37, 871
Coal-scuttle	O. Morse	Rochester, N. Y	July 21, 1868	80, 203
Coal-scuttle	J. Myers, jr	New York, N. Y	July 29, 1856	15, 434
Coal-scuttle	J. T. Page	Rochester, N. Y	Dec. 22, 1868	85, 240
Coal-scuttle	J. C. Parrish	Petersburgh, Va	May 2, 1871	114, 335
Coal-scuttle	J. Pfeifer	Philadelphia, Pa	Jan. 22, 1867	61, 356
Coal-scuttle	J. Pfeifer	Philadelphia, Pa	June 13, 1871	115, 888
Coal-scuttle	H. S. Pratt	Hartford, Conn	Nov. 14, 1865	50, 953
Coal-scuttle	H. S. Reynolds	New York, N. Y	June 28, 1870	104, 885
Coal-scuttle	M. Saulson	Troy, N. Y	July 13, 1869	92, 546
Coal-scuttle	T. and J. M. Scantlin	Evansville, Ind	July 7, 1868	79, 692
Coal-scuttle	S. B. Sexton	Baltimore, Md	Jan. 31, 1865	46, 148
Coal-scuttle	D. Smith	Albany, N. Y	Dec. 10, 1872	133, 806
Coal-scuttle	G. Smith	Brooklyn, N. Y	July 31, 1866	56, 812
Coal-scuttle	G. Smith	Williamsburgh, N. Y	Feb. 27, 1872	124, 093
Coal-scuttle	G. D. Smith	Washington, D. C	Apr. 16, 1867	63, 953
Coal-scuttle	T. Smith	Cincinnati, Ohio	Oct. 31, 1865	50, 744
Coal-scuttle	A. S. Thompson	Little Falls, Minn	Sept. 30, 1873	143, 311
Coal-scuttle	J. G. Treadwell	Albany, N. Y	Apr. 23, 1861	32, 158

Index of patents issued from the United States Patent Office from 1790 *to* 1873, *inclusive*—Continued.

Invention.	Inventor.	Residence.	Date.	No.
Coal-scuttle	W. B. Treadwell	Albany, N. Y	Apr. 14, 1863	38, 188
Coal-scuttle	W. B. Treadwell	Albany, N. Y	Jan. 12, 1864	41, 244
Coal-scuttle	J. J. White	Norfolk, Va	Sept. 23, 1873	143, 208
Coal-scuttle	D. Wight	New London, Conn	Jan. 15, 1867	61, 295
Coal-scuttle	D. Wright and W. A. Kirby	Auburn, N. Y	Mar. 12, 1867	62, 915
Coal-scuttle and ash-screen, Combined	A. T. Carling and L. Rockwell	Ellenville, N. Y	Oct. 31, 1865	50, 685
Coal-scuttle and ash-sifter combined	C. McCleaver	Newark, N. J	Dec. 17, 1872	134, 081
Coal-scuttle and ash-sifter, Combined	A. McNeill	Washington, D. C	Apr. 20, 1858	20, 000
Coal-scuttle and ash-sifter, Combined	T. Parker	Philadelphia, Pa	Sept. 20, 1864	44, 336
Coal scuttle and sifter	C. L. W. Baker	Hartford, Conn	Aug. 7, 1866	56, 874
Coal scuttle and sifter	W. Hazlet and W. H. Flanigan	Philadelphia, Pa	May 3, 1870	102, 679
Coal-scuttle cover	I. Chase, jr	Boston, Mass	Jan. 8, 1856	14, 050
Coal-scuttles, Construction of	J. and J. Scott and W. Miller	Brooklyn, N. Y	Nov. 19, 1867	71, 226
Coal-separator	J. B. Wilford	Reading, Pa	Sept. 30, 1873	143, 395
Coal-shafts, Safety-cage for	D. Glover	Cass Township, Schuylkill County, Pa.	June 7, 1859	24, 296
Coal-shovel	J. F. Brewer	Plantsville, Conn	Feb. 22, 1870	100, 008
Coal-shovel	M. G. Fagan	Troy, N. Y	Aug. 24, 1869	94, 094
Coal-shovel	J. H. Farmer	Detroit, Mich	Aug. 8, 1871	117, 756
Coal-shovel handle	J. Pfeifer	Philadelphia, Pa	Sept. 4, 1866	57, 759
Coal-sifter	S. Langmaid	Lawrence, Mass	July 16, 1867	66, 718
Coal-sifter	H. Wells	Hartford, Conn	Dec. 31, 1839	1, 450
Coal-splitting machine	J. H. Lyon	Baltimore, Md	Feb. 23, 1858	19, 429
Coal-tar, Composition made from	F. M. Hillstream	Lawrence, Kans	May 31, 1870	103, 614
Coal-tar in iron-furnaces, Use of	I. G. Johnson	Spuyten Duyvil, N. Y	July 14, 1857	17, 791
Coal-tar to manufacture roofing, Process of treating	A. H. Perkins	Janesville, Wis	Oct. 7, 1862	36, 632
Coal-tar, Utilization of	R. S. Child	Philadelphia, Pa	July 9, 1861	32, 753
Coal-trap, Safety	T. W. Pratt	Boston, Mass	Oct. 10, 1865	50, 387
Coal, Treating acid tars from	G. Chevrier	Paris, France	Aug. 30, 1870	106, 915
Coal, Tub for handling	P. K. Dederick	Albany, N. Y	Aug. 20, 1872	130, 571
Coal washer and separator	A. Komp	New York, N. Y	June 18, 1867	65, 818
Coal-washing machine	J. P. Evans	Borough of Hazleton, Pa	Sept. 14, 1858	21, 559
Coal-washing machine	G. Lander	New York, N. Y	May 30, 1871	115, 486
Coat	P. Cohen	Saint Joseph, Mo	Sept. 24, 1867	69, 180
Coat	H. Kuhlman	Boston, Mass	May 14, 1872	126, 718
Coat and hat hook	G. F. J. Colburn	Newark, N. J	Nov. 1, 1864	44, 853
Coat and hat hook	J. Danner	Canton, Ohio	Mar. 6, 1866	52, 974
Coat and hat hook	N. and H. L. Judd	New Haven, Conn., and Brooklyn, N. Y.	May 10, 1870	102, 945
Coat and hat hook	J. B. Sargent	New Haven, Conn	Nov. 12, 1867	70, 749
Coat and hat hook	W. Schmitt	New York, N. Y	Sept. 17, 1867	69, 032
Coat and hat rack	J. M. Keep	New York, N. Y	Nov. 8, 1870	109, 017
Coat and hat rack	S. Macferran and S. Ustick	Philadelphia, Pa	Feb. 21, 1865	46, 528
Coat and hat rack, Folding	J. H. Monce	New York, N. Y	Sept. 6, 1870	107, 080
Coat drying and pressing apparatus	J. Braun	Rochester, N. Y	Dec. 22, 1868	85, 058
Coat-form	W. B. Olds	Meriden, Conn	Apr. 6, 1852	8, 858
Coat-hook and line-holder	W. A. Middleton	Harrisburgh, Pa	Mar. 23, 1869	88, 157
Coat-pattern, Adjustable	G. P. Sweezy	Riverhead, N. Y	Nov. 8, 1870	109, 076
Coat-rack	J. B. Lumsey	Sturgis, Mich	June 23, 1868	79, 259
Coat-sleeves, Device for closing the ends of	H. F. Herkner	New York, N. Y	May 31, 1870	103, 816
Coat sling or carrier	D. F. Drake	Somerville, Mass	Nov. 25, 1862	37, 029
Coat support	R. C. Kelly	West Meriden, Conn	Dec. 1, 1868	84, 495
Coat with inner sleeve	J. W. Moyer	Cherry Valley, N. Y	Jan. 24, 1865	46, 016
Coats and vests, Taking the form and measure of gentlemen to cut.	T. Vandoren, sr	Washington, D. C	Dec. 17, 1867	72, 432
Coats, Block for stretching	S. M. Perkins	Springfield, Pa	July 27, 1852	9, 154
Coats, Drafting out fore part of	A. Ward	Moyamensing, Pa	Sept. 28, 1837	415
Coats in one piece, Mode of cutting bodies of	W. M. Wiswell	Portland, Me	Apr. 10, 1839	1, 119
Coats, Instrument for drafting	S. Corley	Lexington, S. C	Dec. 29, 1857	18, 958
Coats, &c., Lock for	C. B. Trimble	New York, N. Y	Dec. 12, 1865	51, 524
Coating iron, &c., Composition for	F. J. Seybold	Chicago, Ill	Feb. 11, 1873	135, 668
Coating metals	J. D. Grüneberg and S. H. Gilbert.	Spring Hill, N. J	Mar. 24, 1868	75, 898
Coating or electroplating iron, brass, copper, &c., with tin.	W. E. Tilley	Middlesex County, England	June 18, 1872	128, 081
Coating with tin and other metals, Apparatus for	E. H. Davies	Pittsburgh, Pa	July 9, 1872	128, 862
Coating wood and iron, Composition for	E. F. Prentiss	Philadelphia, Pa	Aug. 26, 1873	142, 117
Coating wood and other articles to render them acid-proof, Compound for.	R. Newell	Philadelphia, Pa	July 1, 1873	140, 530
Coating wood and other materials, Composition for	C. L. Robertson	Providence, R. I	Apr. 2, 1867	63, 428
Cob and feed mill	H. Ormsby and E. R. Sumner	Beloit, Wis	July 23, 1861	32, 914
Cob and grain mill	J. R. Marston	New York, N. Y	Apr. 19, 1859	23, 693
Cob and stalk cutter	T. B. Jones	Carloville, Ala	Oct. 11, 1853	10, 110
Cock	C. M. Alburger	Philadelphia, Pa	May 21, 1867	64, 851
Cock	A. S. Davendorf	Galesburgh, Ill	July 8, 1873	140, 685
Cock	J. W. Faxon	Boston, Mass	Oct. 14, 1873	143, 683
Cock	J. P. Gallagher	Saint Louis, Mo	July 11, 1865	48, 673
Cock	W. H. Hedges and M. E. Campfield.	Newark, N. J	Jan. 24, 1871	111, 205
Cock	E. Hubball	Baltimore, Md	Oct. 4, 1820	......
Cock	N. Jenkins	Boston, Mass	Apr. 18, 1865	47, 309
Cock	H. Jones	Philadelphia, Pa	Feb. 27, 1872	123, 996
Cock	J. J. Lowry	Pittsburgh, Pa	Jan. 5, 1864	41, 077
Cock	S. Norton	Stockport, Great Britain	May 23, 1871	115, 091
Cock	M. T. F. O'Donnell	Boston, Mass	Sept. 24, 1872	131, 556
Cock	B. Pickering	Dayton, Ohio	July 24, 1860	29, 309
Cock	J. Powell	Cincinnati, Ohio	Dec. 5, 1865	51, 349
Cock	T. Prosser	Brooklyn, E. D., N. Y	Feb. 27, 1872	124, 082
Cock	J. Regester	Baltimore, Md	July 11, 1865	48, 721
Cock	J. Regester	Baltimore, Md	July 18, 1871	117, 112
Cock	H. S. Ross	Chicago, Ill	Nov. 19, 1872	133, 120
Cock	H. S. Ross	Chicago, Ill	Aug. 26, 1873	142, 122
Cock	D. H. Stickney	Cincinnati, Ohio	Oct. 18, 1859	25, 853
Cock	H. Strater, jr	Boston, Mass	July 16, 1872	129, 616
Cock	T. H. Thayer	New Haven, Conn	Aug. 2, 1870	106, 093
Cock	J. W. Trafton	Springfield, Mass	Aug. 9, 1870	106, 237
Cock, Air-cushion stop	D. A. Robinson	Washington, D. C	Aug. 6, 1872	130, 243
Cock, Ale or beer	H. Getty	Hoboken, N. J	Mar. 14, 1871	112, 701

Index of patents issued from the United States Patent Office from 1790 to 1873, inclusive—Continued.

Invention.	Inventor.	Residence.	Date.	No.
Cock and alarm-whistle, Gage	A. Miller	Cleveland, Ohio	June 29, 1858	20,726
Cock and case for pipes leading from street-mains, Stop.	J. G. Weldon	Pittsburgh, Pa	June 7, 1864	43,056
Cock and check-valve, Combined stop	A. T. Waldron	Waterford, N. Y	May 13, 1873	138,963
Cock and faucet, Valve	A. Crossley	Philadelphia, Pa	Apr. 16, 1872	125,792
Cock and filter in combination, Stop	A. and H. Johnson	New York, N. Y	Nov. 27, 1849	6,910
Cock and fixed socket, Combined removable	C. Sullivan	Boston, Mass	Feb. 7, 1871	111,699
Cock and swinging joint, Gas	C. F. Thieme	Philadelphia, Pa	Nov. 25, 1856	16,129
Cock, Angle valve	J. H. Davis	Allegheny City, Pa	Feb. 8, 1870	99,650
Cock, Anti-concussion water	M. W. Bailey	Pottstown, Pa	Feb. 13, 1872	123,606
Cock, Automatic gas	C. MacRae	New York, N. Y	July 10, 1866	56,240
Cock, Automatic lubricating water and gas	E. F. Brooks	Baltimore, Md	Nov. 18, 1873	144,735
Cock, Barreling	A., J., and T. McKenna	Pittsburgh, Pa	Jan. 8, 1867	61,015
Cock, Basin	A. Crossley	Philadelphia, Pa	Sept. 24, 1872	131,601
Cock, Basin	A. D. Davis	Chicago, Ill	Aug. 20, 1872	130,570
Cock, Basin	C. Harrison	New York, N. Y	Apr. 8, 1856	14,628
Cock, Basin	R. Leitch	Baltimore, Md	Mar. 3, 1857	16,736
Cock, Basin	C. A. Newton	Providence, R. I	Dec. 20, 1870	110,387
Cock, Basin	G. W. Randall	Boston, Mass	May 10, 1859	23,989
Cock, Basin stop	H. Eling	New York, N. Y	Aug. 21, 1855	13,457
Cock, Bib	J. H. Davis	Allegheny City, Pa	Feb. 8, 1870	99,649
Cock, Block-tin stop	J. Morris	New Haven, Conn	Mar. 7, 1816	....
Cock, Boiler gage	J. French	Newport, Ky	Aug. 20, 1867	67,866
Cock, Boiler gage	W. T. Howard	Baltimore, Md	Oct 2, 1866	58,546
Cock, Boiler gage	J. G. Raymond	Rondout, N. Y	Oct. 8, 1867	69,702
Cock, Bottle	L. A. Perrault	Natchez, Miss	Mar. 1, 1870	100,321
Cock box for water and gas mains, Stop	O. F. Woodford	Chicago, Ill	Aug. 10, 1869	93,657
Cock boxes, Top of gas and water stop	W. W. Pullis	Saint Louis, Mo	Nov. 8, 1870	109,049
Cock boxing, Stop	W. H. Graham	Saint Louis, Mo	Oct. 28, 1873	143,978
Cock case, Stop	J. G. Weldon	Pittsburgh, Pa	Nov. 1, 1864	44,904
Cock, Cast-iron valve	J. H. Davis	Allegheny City, Pa	Feb. 8, 1870	99,653
Cock, Check-valve	J. H. Davis	Allegheny City, Pa	Feb. 8, 1870	99,652
Cock, Compression	E. Allt	New York, N. Y	July 1, 1873	140,337
Cock, Compression	O. N. Ames	Haydenville, Mass	Aug. 29, 1871	118,505
Cock, Compression	H. J. Bailey and J. Porteous	Pittsburgh, Pa	Dec. 3, 1872	133,558
Cock, Compression	G. E. Boisselier	Saint Louis, Mo	Dec. 8, 1868	84,792
Cock, Compression	W. Dinnen	Detroit, Mich	Apr. 4, 1871	113,503
Cock, Compression	C. A. Howard	New Haven, Conn	June 9, 1868	78,804
Cock, Compression	J. Maclaren	Scranton, Pa	Mar. 19, 1872	124,840
Cock, Compression	A. Pearce	Cambridgeport, Mass	Dec. 17, 1872	134,095
Cock, Compression	C. Perkes	Philadelphia, Pa	Oct. 19, 1869	96,033
Cock, Confluent	J. H. G. Hawes	Newark, N. J	May 24, 1870	103,329
Cock, Conical stop	J. E. Jones	Tidioute, Pa	Mar. 26, 1872	124,959
Cock connection, Locked plug	J. B. Edson	Brooklyn, N. Y	Jan. 9, 1872	122,517
Cock, Cut-off stop	W. H. Pollard	Seneca Falls, N. Y	July 30, 1867	67,214
Cock, Cylinder	D. B. Dennison	Ottumwa, Iowa	Feb. 25, 1873	136,221
Cock, Cylinder	M. B. Mason and J. S. McCrum	Kansas City, Mo	Oct. 17, 1871	120,080
Cock, Cylinder	J. Porteous	Cincinnati, Ohio	June 27, 1871	116,485
Cock, Diaphragm stop	W. E. Banta	Springfield, Ohio	Oct. 4, 1870	107,852
Cock, Draft	W. H. Bate	East Somerville, Mass	Dec. 20, 1870	110,334
Cock, Draft	W. P. Clark	Medford, Mass	Dec. 12, 1871	121,754
Cock, Draft	A. J. Morse	Boston, Mass	June 22, 1869	91,763
Cock, Drop-handle urn	H. and E. L. Bailey	Brooklyn, N. Y., and Philadelphia, Pa.	Aug. 14, 1866	57,246
Cock, Electric gas-stop	J. A. Heyl	Boston, Mass	Oct. 16, 1866	58,943
Cock-eye	C. Bochmer, jr	Madison, Wis	Nov. 6, 1866	59,347
Cock-eye	J. Haggerty	East Springfield, Pa	Sept. 10, 1867	68,737
Cock, Filter stop	A. and H. Johnson	New York, N. Y	Nov. 14, 1848	5,926
Cock, Filtering	D. Bartlett, jr	Boston, Mass	Oct. 15, 1850	7,716
Cock, Filtering	L. P. Jenks and F. Draper	Boston and East Cambridge, Mass.	Nov. 2, 1858	21,964
Cock, Filtering	J. F. Ostrander	New York, N. Y	Apr. 4, 1846	4,440
Cock, Filtering	T. F. Wenman	New York, N. Y	Apr. 4, 1846	4,437
Cock, Filtering stop	S. H. Lewis	New York, N. Y	Jan. 26, 1847	4,939
Cock, Filtering stop	W. Read	New York, N. Y	Oct. 24, 1846	4,824
Cock for boilers, Gage	W. G. Thomas	Centralia, Pa	Jan. 21, 1868	73,672
Cock for carbureters, &c	S. Rust, jr	Cincinnati, Ohio	Nov. 29, 1870	109,671
Cock for controlling discharge through two pipes	J. and T. D. Richardson	New York, N. Y	Aug. 26, 1873	142,120
Cock for drawing beer	J. Moffet	New York, N. Y	Sept. 16, 1873	142,860
Cock for drawing liquids	E. Dickinson	Amherst, Mass	Dec. 17, 1823	....
Cock for drawing liquor	J. Bliss	New York, N. Y	Sept. 24, 1821	....
Cock for filling soda-bottles, Confluent	C. G. Ferron	New York, N. Y	July 16, 1872	129,016
Cock for filtering, Stop	R. E. House	New York, N. Y	June 27, 1846	4,593
Cock for flexible tubes, Compression	H. Fairbanks	Boston, Mass	Jan. 17, 1871	110,964
Cock for gas-burners	S. B. H. Vance	New York, N. Y	Mar. 31, 1868	76,277
Cock for gas-burners, Automatic stop	G. E. Smith	San Francisco, Cal	Dec. 15, 1868	84,914
Cock for hot water and steam, Stop	J. Sheriff	Pittsburgh, Pa	Jan. 16, 1849	6,032
Cock for hydrants, Stop	H. R. Dunham	New York, N. Y	Jan. 10, 1843	2,908
Cock for hydrants, Valve	S. Pfleger	Reading, Pa	May 31, 1870	103,773
Cock for hydrants, Valve	B. Stancliffe	Philadelphia, Pa	May 15, 1827	....
Cock for hydraulic presses, Connecting stop	C. Wilson	New York, N. Y	Oct. 28, 1843	3,319
Cock for locomotives, &c, Cylinder	C. H. Hopkins	Lyndonville, Vt	Mar. 11, 1873	136,724
Cock for mains, Branch stop	R. A. Hill	Washington, D. C	May 25, 1869	90,541
Cock for pumps, Blow-off	T. Shaw	Philadelphia, Pa	Aug. 19, 1862	36,238
Cock for racking off beer	F. Wagner	Danville, Pa	Aug. 4, 1868	80,787
Cock for soda-water and other liquids, Draft	J. D. O'Donnell	Washington, D. C	Oct. 21, 1873	143,777
Cock for soda-water and sirup	W. Gee	New York, N. Y	Apr. 22, 1873	138,143
Cock for soda-water apparatus, Draft	W. P. Clark	Boston, Mass	July 31, 1866	56,709
Cock for soda-water apparatus, Draft	L. D. Hoyt and R. Murray	Medford and Boston, Mass	June 27, 1865	48,489
Cock for steam-heaters, Air	S. J. Gold	New Haven, Conn	Feb. 26, 1856	14,312
Cock for steam-radiators, Automatic air	J. R. Nichols	Boston, Mass	July 16, 1872	129,291
Cock for water-pipes	W. Johnson	Philadelphia, Pa	Dec. 29, 1868	85,309
Cock, Gage	H. J. Bailey	Pittsburgh, Pa	Aug. 5, 1873	141,416
Cock, Gage	P. Ball	Worcester, Mass	Apr. 3, 1866	53,555
Cock, Gage	J. F. Bellemere and F. R. Fleer	Reading, Pa	July 22, 1873	141,104
Cock, Gage	J. Broughton	New York, N. Y	Sept. 12, 1865	51,418
Cock, Gage	J. B. Christoffel	Williamsburgh, N. Y	Mar. 22, 1870	100,981
Cock, Gage	W. H. Downing	Shamburgh, Pa	June 4, 1872	127,468

Index of patents issued from the United States Patent Office from 1790 *to* 1873, *inclusive*—Continued.

Invention.	Inventor.	Residence.	Date.	No.
Cock, Gage	C. L. Frink	Rockville, Conn	June 19, 1866	55,640
Cock, Gage	A. Fuller	Brooklyn, N. Y	July 9, 1867	66,579
Cock, Gage	H. Hise	Ottawa, Ill	Jan. 6, 1863	37,292
Cock, Gage	L. T. Hulbert	Painesville, Ohio	Jan. 16, 1872	122,766
Cock, Gage	B. E. Lehnan and R. Ross	Bethlehem, Pa	June 6, 1871	115,746
Cock, Gage	P. McGrath	Plattsburgh, N. Y	Aug. 8, 1871	117,907
Cock, Gage	R. L. Mills	Lancaster, Pa	Nov. 30, 1858	22,189
Cock, Gage	A. A. Murray	Baltimore, Md	Nov. 21, 1873	143,923
Cock, Gage	N. Ray	Schenectady, N. Y	Dec. 17, 1872	134,098
Cock, Gage	S. Shepherd	Nashua, N. H	Sept. 20, 1864	44,347
Cock, Gage	L. F. Smith	Philadelphia, Pa	July 19, 1870	105,503
Cock, Gage	F. Stebbins	Hinsdale, N. H	Aug. 11, 1868	81,027
Cock, Gage	E. A. Walker	Nashville, Tenn	July 25, 1865	49,016
Cock, Gage	G. L. Watson	Nesquehoning, Pa	Dec. 21, 1869	98,132
Cock, Gage	C. T. Woodman	Boston, Mass	Nov. 28, 1865	51,274
Cock, Gage	J. E. Wootten	Philadelphia, Pa	Mar. 1, 1859	23,136
Cock, Gas	P. Keller	New York, N. Y	Jan. 21, 1868	73,534
Cock, Gas	J. G. Leffingwell	Newark, N. J	Feb. 19, 1861	31,465
Cock, Gas	C. B. Littlefield	Boston, Mass	Oct. 8, 1867	69,566
Cock, Gas	E. M. Morris	Baltimore, Md	Dec. 9, 1873	145,442
Cock, Gas	M. Stratton	Philadelphia, Pa	Mar. 2, 1869	87,522
Cock, Gas	J. C. Wightman	Boston, Mass	July 9, 1861	32,804
Cock, Gas stop	J. Humphrey	Boston, Mass	July 22, 1866	15,376
Cock, Globe	J. Powell	Cincinnati, Ohio	May 2, 1865	47,565
Cock, Globe	J. Worcester	Newport, Ky	June 19, 1866	55,756
Cock, Globe-valve	J. H. Davis	Allegheny City, Pa	Feb. 8, 1870	99,651
Cock, Globe-valve	F. Lunkenheimer	Cincinnati, Ohio	Mar. 7, 1865	46,685
Cock, Grease	R. M. Wade	Wadesville, Va	June 8, 1852	9,010
Cock handle, Boiler gage	E. H. Ashcroft	Lynn, Mass	Jan. 1, 1867	60,817
Cock, Hopper	P. Regitz	Chicago, Ill	June 15, 1869	91,483
Cock, Hydrant	E. Hubball	Baltimore, Md	May 11, 1841	2,086
Cock, Hydrant stop	H. L. Frailey	Lancaster, Pa	Apr. 3, 1866	53,736
Cock, Hydrant stop	L. Magers, F. Davis, and W. Dukehart.	Baltimore, Md	June 22, 1842	2,683
Cock, Hydrant waste	G. W. Robertson	Philadelphia, Pa	Jan. 10, 1860	26,785
Cock key, Hydrant	P. H. Griffin	Albany, N. Y	Oct. 6, 1868	82,827
Cock, Lock	H. Essex	Meadville, Pa	Apr. 16, 1872	125,799
Cock, Lock stop	J. E. Jones	Tidioute, Pa	Feb. 21, 1871	111,943
Cock, Locked	H. Essex	Meadville, Pa	Mar. 21, 1871	112,910
Cock, Locking stop	E. Davis and J. M. Stiles	Camden, N. J	Mar. 14, 1871	112,693
Cock lubricator, Gas	C. H. Johnson	Boston, Mass	Mar. 10, 1857	16,820
Cock, Making stop	G. Fenn	New York, N. Y	July 12, 1823	....
Cock, Metallic screw	J. Garey	Baltimore, Md	Apr. 23, 1822	....
Cock, Oil	J. Hare, jr	Paterson, N. J	Apr. 17, 1860	27,902
Cock or faucet	C. Harrison	New York, N. Y	Sept. 8, 1868	91,900
Cock or lock for the passage of fluids	J. Witherle	Boston, Mass	Jan. 16, 1811	....
Cock or outlet-valve, Waste	S. J. Olsson	Chicago, Ill	Oct. 28, 1873	143,995
Cock, Pull	G. Leach	New York, N. Y	June 5, 1860	28,588
Cock, Pull	H. Strater, jr	Boston, Mass	July 16, 1872	129,069
Cock, Racking	A. Roos	New York, N. Y	Nov. 11, 1873	144,565
Cock, Regulating	C. E. Seal	Winchester, Va	Sept. 2, 1873	142,521
Cock, Rotating stop	D. Hurd	Chicago, Ill	May 2, 1865	47,546
Cock, Safety	J. Stowell	Charlestown, Mass	Aug. 6, 1867	67,460
Cock, Screen stem-valve	S. D. Fales	Smithfield, R. I	Apr. 4, 1865	47,097
Cock, Self-acting basin	S. C. Wentworth	Haydenville, Mass	Nov. 19, 1872	133,131
Cock, Self-boring stop	A. Weed	Boston, Mass	June 2, 1868	78,499
Cock, Self-closing	J. E. Boyle	New York, N. Y	July 16, 1872	128,998
Cock, Self-closing	M. S. Clark	New York, N. Y	Aug. 12, 1873	141,630
Cock, Self-closing	W. Dalziel	New York, N. Y	Feb. 7, 1871	111,518
Cock, Self-closing	N. Jenkins	Boston, Mass	June 27, 1865	48,407
Cock, Self-closing	J. M. Meharg	Montreal, Canada	Aug. 8, 1871	117,908
Cock, Self-closing auger stop	A. Weed	Boston, Mass	Mar. 8, 1870	100,697
Cock, Self-closing stop	J. Pigot	Brooklyn, N. Y	May 6, 1873	138,580
Cock, Self-lubricating stop	G. Parker	Boston Highlands, Mass	Sept. 3, 1872	131,065
Cock, Spring valve	M. C. Hawkins, J. W. Goodwin, and J. Cummings.	Erie, Pa	Dec. 15, 1857	18,847
Cock, Steam	A. Fuller	Cincinnati, Ohio	Aug. 31, 1858	21,332
Cock, Steam	A. Hallowell	Lowell, Mass	May 28, 1867	65,075
Cock, Steam	A. Hallowell and H. R. Barker	Lowell, Mass	Nov. 28, 1865	51,177
Cock, Steam	W. Johnson and M. Silmser	Auburn, N. Y	July 19, 1859	24,808
Cock, Steam	R. Ross and W. Holland	Philadelphia, Pa	Aug. 24, 1858	21,276
Cock, Steam	T. Sanford	Claremont, N. H	May 14, 1861	32,314
Cock, Steam	J. B. Sargent and F. W. Towne	Fitchburgh, Mass	June 27, 1865	48,452
Cock, Steam	A. Swardkins	South Boston, Mass	Apr. 30, 1861	32,212
Cock, Steam	A. Tyler and G. F. Kendall	Fitchburgh, Mass	Aug. 29, 1865	49,666
Cock, Steam	J. L. Winslow	Westbrook, Me	Aug. 17, 1858	21,237
Cock, Steam and liquid stop	J. Breeden	Birmingham, England	Apr. 19, 1870	101,977
Cock, Steam-boiler alarm gage	H. Wilkins	Brownsville, Pa	Apr. 19, 1864	42,425
Cock, Steam-boiler gage	F. W. Bacon	West Newton, Mass	July 5, 1859	24,692
Cock, Steam-boiler gage	J. Holdcraft	Philadelphia, Pa	Mar. 6, 1866	53,079
Cock, Steam-boiler gage	W. Painter	Baltimore, Md	Jan. 16, 1872	122,847
Cock, Steam-boiler gage	J. H. Vreeland and C. H. Lathrop.	Rutherford Park and Jersey City, N. J.	Oct. 1, 1872	131,799
Cock, Steam-boiler try	J. F. Cook	Baltimore, Md	Sept. 20, 1859	25,544
Cock, Steam-boiler try	J. Cumming	Boston, Mass	Oct. 18, 1859	25,810
Cock, Steam-gage	E. H. Ashcroft	Lynn, Mass	Aug. 6, 1867	67,479
Cock, Steam-gage	A. Bisbee	Chelsea, Mass	Sept. 18, 1855	13,563
Cock, Steam-gage	S. Blackman	Reading, Pa	Dec. 7, 1869	97,595
Cock, Steam-gage	J. Broughton	New York, N. Y	Feb. 13, 1866	52,521
Cock, Steam-gage	T. B Dexter	Lynn, Mass	Oct. 15, 1867	69,782
Cock, Steam-gage	V. Giroud	New York, N. Y	Aug. 29, 1865	49,623
Cock, Steam-gage	F. Henke	Scranton, Pa	May 15, 1866	54,722
Cock, Steam-gage	J. Sanders	East Boston, Mass	June 4, 1867	65,511
Cock, Steam-gage	J. C. Schaefer	Philadelphia, Pa	Feb. 6, 1866	52,452
Cock, Steam-gage	D. Williams and E. Joseph	Gallipolis, Ohio	June 7, 1870	104,090
Cock, Steam, gas, and water stop	W. H. De Valin	Sacramento, Cal	Dec. 1, 1868	84,617
Cock, Steam-generator gage	M. M Robinson	Dongola, Ill	Mar. 20, 1866	53,344
Cock, Steam-generator gage	T. Shaw	Philadelphia, Pa	May 21, 1867	65,018
Cock, Steam-generator gage	J. Waters	Minneapolis, Minn	Mar. 16, 1869	87,992

Index of patents issued from the United States Patent Office from 1790 *to* 1873, *inclusive*—Continued.

Invention.	Inventor.	Residence.	Date.	No.
Cock, Steam-generator try	J. Regester	Baltimore, Md	Mar. 20, 1866	53, 337
Cock, Steam-radiator	S. J. Gold	New Haven, Conn	Mar. 25, 1856	14, 500
Cock, Steam, water, &c	W. Thomas	New York, N. Y	Mar. 25, 1856	14, 526
Cock, Stop	H. Allen	New York, N. Y	Nov. 12, 1841	2, 351
Cock, Stop	G. C. Bailey	Pittsburgh, Pa	June 4, 1872	127, 547
Cock, Stop	J. W. Birkett	Brooklyn, N. Y	Sept. 16, 1873	142, 763
Cock, Stop	J. Breeden	Birmingham, England	Feb. 22, 1870	100, 113
Cock, Stop	H. A. Chapin	Springfield, Mass	Aug. 21, 1860	29, 744
Cock, Stop	J. C. Chapman	Cambridgeport, Mass	Sept. 7, 1869	94, 563
Cock, Stop	T. C. Clarke	Philadelphia, Pa	Mar. 14, 1846	4, 419
Cock, Stop	D. N. B. Coffin, jr	Lynn, Mass	Jan. 9, 1855	12, 195
Cock, Stop	Z. E. Coffin	Boston, Mass	Apr. 30, 1867	64, 198
Cock, Stop	W. S. Cooper	Philadelphia, Pa	June 24, 1873	140, 248
Cock, Stop	G. Cuppers	New York, N. Y	Sept. 13, 1864	44, 164
Cock, Stop	T. Daniels	Toledo, Ohio	Feb. 14, 1860	27, 112
Cock, Stop	A. Davis	Boston, Mass	Oct. 16, 1839	1, 371
Cock, Stop	W. H. De Valin	Sacramento, Cal	May 1, 1866	54, 305
Cock, Stop	J. Doering	Philadelphia, Pa	Jan. 7, 1868	72, 986
Cock, Stop	G. W. Eddy	Waterford, N. Y	June 3, 1873	139, 559
Cock, Stop	G. D. Hadley	Cincinnati, Ohio	July 2, 1867	66, 326
Cock, Stop	C. F. Johnson and J. J. Speed, jr	Oswego and Ithaca, N. Y	June 21, 1839	1, 181
Cock, Stop	M. C. Kilgore	Washington, Iowa	Apr. 3, 1866	53, 627
Cock, Stop	H. P. Kreiner	Berlin, Prussia	Jan. 19, 1869	86, 020
Cock, Stop	C. W. Laurie	Philadelphia, Pa	Sept. 23, 1873	143, 167
Cock, Stop	B. E. Lehman	Bethlehem, Pa	Oct. 16, 1866	58, 845
Cock, Stop	J. Maclaren	Scranton, Pa	Jan. 23, 1872	123, 031
Cock, Stop	P. E. Malmström and P. Dummer.	New York, N. Y	July 25, 1871	117, 438
Cock, Stop	J. P. Mern	New York, N. Y	Mar. 25, 1873	137, 144
Cock, Stop	F. J. Miller	New York, N. Y	Nov. 21, 1843	3, 344
Cock, Stop	H. C. Montgomery	Boston, Mass	Oct. 15, 1872	132, 308
Cock, Stop	J. Morris	New Haven, Conn	Mar. 1, 1815	......
Cock, Stop	J. Morris	New Haven, Conn	May 15, 1822	......
Cock, Stop	H. Müller	Vienna, Austria	Oct. 24, 1871	120, 311
Cock, Stop	R. Nicoll	New York, N. Y	June 5, 1860	28, 596
Cock, Stop	C. C. Parsons	Boston, Mass	June 19, 1866	55, 700
Cock, Stop	O. C Phelps	Boston, Mass	Mar. 14, 1854	10, 640
Cock, Stop	J. Radston	San Francisco, Cal	Nov. 7, 1871	120, 668
Cock, Stop	J. Regester	Baltimore, Md	Sept. 15, 1868	82, 161
Cock, Stop	J. Regester and W. H. Bowen	Baltimore, Md	Sept. 2, 1873	142, 350
Cock, Stop	J. Ridgway	New York, N. Y	Apr. 16, 1842	2, 563
Cock, Stop	F. Roach	Boston, Mass	June 19, 1866	55, 784
Cock, Stop	J. Robertson	New York, N. Y	May 8, 1840	1, 583
Cock, Stop	L. C. Rodier and F. G. Bates	Springfield, Mass	Nov. 18, 1873	144, 700
Cock, Stop	C. Schultz and T. Warker	New York, N. Y	Mar. 31, 1868	76, 106
Cock, Stop	J. Seeberger	Troy, N. Y	Feb. 8, 1870	99, 714
Cock, Stop	D. Sharman	Warren, N. Y	July 5, 1814	......
Cock, Stop	E. F. Spaulding	Cambridgeport, Mass	Aug. 5, 1873	141, 471
Cock, Stop	E. Stebbins	Chicopee, Mass	Apr. 19, 1859	23, 721
Cock, Stop	J. Stevens	New York, N. Y	Aug. 27, 1872	130, 876
Cock, Stop	H. Stratter, jr	Roxbury, Mass	June 26, 1866	55, 931
Cock, Stop	T. Stubblefield	Columbus, Ga	Apr. 26, 1859	23, 796
Cock, Stop	I. C. Tate	New London, Conn	June 14, 1859	24, 416
Cock, Stop	P. Walker	......	Apr. 10, 1800	......
Cock, Stop	U. West and G. Dobbs	New York, N. Y	Apr. 30, 1842	2, 596
Cock, Stop	A. W. Wilkinson	New York, N. Y	Feb. 6, 1872	123, 534
Cock, Stop	E. Wright	Boston, Mass	Oct. 4, 1853	10, 082
Cock, Stop and waste	W. Z. Hatcher	Philadelphia, Pa	Aug. 22, 1854	11, 550
Cock, Supply	W. S. Carr	New York, N. Y	Jan. 5, 1858	19, 013
Cock, Supply and waste	W. S. Cooper	Philadelphia, Pa	Aug. 1, 1871	117, 606
Cock, Swing compression basin	W. Gordon	Philadelphia, Pa	May 28, 1872	127, 340
Cock, Tap	C. Renard, M. Perret, and J. C. Voituret.	Mâcon, France	Nov. 9, 1869	96, 618
Cock, Three-way	F. S. Pease	Buffalo, N. Y	June 27, 1865	48, 430
Cock to prevent water freezing in pent-stocks, &c.	D. Lathrop, jr	Norwich, Conn	Dec. 5, 1832	......
Cock, Two-way	D. F. Dodge	Lowville, N. Y	Nov. 30, 1869	97, 366
Cock, Two-way stop	R. Leitch	Baltimore, Md	Aug. 19, 1862	36, 222
Cock, Valve	S. Adams	Boston, Mass	May 25, 1858	20, 314
Cock, Valve	R. Berryman	Boston, Mass	Dec. 6, 1870	109, 796
Cock, Valve	W. Chesley	Cincinnati, Ohio	Jan. 31, 1865	46, 077
Cock, Valve	J. Cluley and B. Stancliff	Philadelphia, Pa	Nov. 8, 1820	......
Cock, Valve	W. S. Cooper	Philadelphia, Pa	Mar. 10, 1868	75, 376
Cock, Valve	B. Eakins	Spring Garden, Pa	Apr. 4, 1854	10, 733
Cock, Valve	J. B. Gibson	Cincinnati, Ohio	July 21, 1868	80, 066
Cock, Valve	J. Griffiths	Philadelphia, Pa	Feb. 14, 1854	10, 516
Cock, Valve	N. Jenkins	Boston, Mass	Aug. 1, 1865	49, 116
Cock, Valve	J. C. Macdonald	Cincinnati, Ohio	Sept. 14, 1858	21, 510
Cock, Valve	J. C. Macdonald	Saint Louis, Mo	Sept. 27, 1870	107, 697
Cock, Valve	R. Nickerson and A. B. Cotton	Athens, Ga	May 8, 1860	28, 192
Cock, Valve	T. Ramsden and H. M. Davis	Allegheny City and Pittsburgh, Pa	July 20, 1869	92, 748
Cock, Valve	J. R. and H. S. Robinson	Clinton, Mass	Aug. 31, 1858	21, 366
Cock, Valve	R. P. Ross	Bethlehem, Pa	July 2, 1872	128, 663
Cock, Valve	P. C. Rowe	Boston, Mass	Mar. 26, 1872	125, 086
Cock, Valve	F. D. Sanno	Philadelphia, Pa	June 29, 1821	......
Cock, Valve	C. R. Vaillant	Whistler, Ala	Jan. 4, 1870	98, 646
Cock, Valve	J. Walsh	Philadelphia, Pa	Jan. 30, 1872	123, 311
Cock, Valve	J. Walsh	Philadelphia, Pa	Aug. 12, 1873	141, 839
Cock, Valve	J. Wilson	Philadelphia, Pa	Mar. 1, 1870	100, 481
Cock, Water	J. Gregg	Manchester, N. H	Apr. 20, 1869	89, 143
Cock, Water	P. A. Mayor	New York, N. Y	July 16, 1872	129, 485
Cock, Water	H. S. North and T. Thompson	Middletown, Conn	May 11, 1869	90, 015
Cock, Water	G. Youle		Aug. 25, 1803	......
Cock, Water and air	J. L. Chapman	Baltimore, Md	Oct. 14, 1841	2, 304
Cock, Water-basin	H. W. Smith	Hartford, Conn	Mar. 8, 1859	23, 198
Cock, Water-basin	H. W. Smith	Hartford, Conn	Apr. 12, 1859	23, 619
Cock, Water-closet	J. Broughton	New York, N. Y	Oct. 25, 1864	44, 783
Cock, Water-closet	H. H. Craigie	New York, N. Y	May 11, 1869	89, 856

Index of patents issued from the United States Patent Office from 1790 *to* 1873, *inclusive*—Continued.

Invention.	Inventor.	Residence.	Date.	No.
Cock, Water-closet	H. Jones	Philadelphia, Pa	Jan. 4, 1870	98, 599
Cock, Water-closet	B. P. Shear	Charlestown, Mass	Sept. 19, 1871	119, 058
Cock, Water-closet	D. Wellington	Boston, Mass	Nov. 15, 1859	26, 145
Cock, Water-closet self-closing	C. Harrison	New York, N. Y	July 6, 1869	92, 190
Cock, Way	S. P. Mervine, jr	Philadelphia, Pa	June 18, 1872	128, 160
Cock, Weighted gage	W. H. McMillan	Philadelphia, Pa	Mar. 5, 1872	124, 214
Cock, Weighted gage	S. P. M. Tasker and W. H. McMillan.	Philadelphia, Pa	Mar. 25, 1873	137, 259
Cocks, Apparatus for automatically operating gas	S. Gardiner, jr	New York, N. Y	Aug. 15, 1871	118, 808
Cocks, Box for street stop	H. A. Moore and E. O. and S. C. Frink.	Indianapolis, Ind	Feb. 18, 1868	74, 573
Cocks by electro-magnetism, Turning gas	S. Gardiner, jr	New York, N. Y	June 19, 1866	55, 642
Cocks, Case for water and gas	G. McIlwain	Philadelphia, Pa	Oct. 29, 1861	33, 592
Cocks for hydrants, Constructing	J. Martin	Baltimore, Md	Feb. 10, 1841	1, 972
Cocks in pipes under hydrostatic pressure, Machine for tapping and inserting stop.	H. Allen	New York, N. Y	Nov. 21, 1843	3, 348
Cocks, Locking-case for stop	V. T. Hall	Brooklyn, N. Y	May 11, 1869	89, 930
Cocks, Locking-device for stop	V. T. Hall	Brooklyn, N. Y	May 11, 1869	89, 929
Cocks, Machine for dressing metal	H. Essex	Meadville, Pa	Nov. 11, 1873	144, 524
Cocks, Machine for finishing plugs of stop	J. W. Lyon	Brooklyn, N. Y	Mar. 6, 1860	27, 370
Cocks, Machine for forming guard or gas	J. W. Lyon	Brooklyn, N. Y	Feb. 21, 1860	27, 228
Cocks, Manufacture of	C. A. Creasey	Philadelphia, Pa	Apr. 29, 1842	2, 589
Cocks, Manufacture of stop	J. L. Lowry	Pittsburgh, Pa	Sept. 20, 1864	44, 326
Cocks, Method of governing the action of valve	F. H. Bartholomew	New York, N. Y	June 20, 1854	11, 113
Cocks, Method of operating stop	D. N. Dunzack	Salem, Mass	Jan. 31, 1860	26, 975
Cocks for steam-cylinders, Device for operating	W. H. Woods	San Francisco, Cal	Aug. 10, 1869	93, 509
Cocks or valves of water and gas pipes, Stop-box for	J. Smith	Saint Louis, Mo	Aug. 4, 1868	80, 776
Cocks, Plug for arresting flow of stop	J. H. Rhodes	Brooklyn, N. Y	July 16, 1872	129, 168
Cocks, Tool for making plugs for gas and water	W. Tweedle	Providence, R. I	Apr. 20, 1869	89, 184
Cocks, Valvular arrangement for basin	E. G. Burnham	Springfield, Mass	June 9, 1857	17, 539
Cocks with pipes, Connecting	D. A. Webster	New York, N. Y	June 29, 1852	9, 085
Cocks with safety-plugs, Apparatus or	P. A. Chambeaux	Paris, France	Dec. 30, 1873	146, 045
Cockle and garlic separator	J. W. Neal	Big Lick, Va	July 16, 1867	66, 872
Cockle and garlic separator	J. W. Neal and A. J. Truxell	Big Lick, Va	Oct. 20, 1868	83, 198
Cockle, &c., from wheat, Machine for separating	G. H. Rich	Geneva, Ill	Aug. 26, 1873	142, 119
Cockle-separating machine	F. W. Mace	Waukaw, Wis	June 29, 1869	92, 073
Cockle-separator	S. W. Andrews and L. Godfrey	Greeneville, Tenn	July 13, 1869	92, 559
Cocoa, Preparation of	P. Pearson	Leeds, Great Britain	Mar. 16, 1869	87, 965
Cocoa-nut cutter	D. Fitzgerald	New York, N. Y	Dec. 22, 1831	....
Cocoa-nut cutter and grater	J. Gardner	Philadelphia, Pa	Aug. 6, 1867	67, 529
Cocoa-nut, Desiccated	G. B. Williams	New York, N. Y	Nov. 17, 1868	84, 241
Cocoa-nuts for pies, &c., Preparing	N. A. Classon	Windham, Conn	Dec. 23, 1873	145, 722
Cocoa-nuts, Manufacture of desiccated	G. W. Waitt	Philadelphia, Pa	July 7, 1868	79, 790
Cocoa-nuts, Machine for granulating	T. and G. M. Mills	Philadelphia, Pa	June 10, 1873	139, 804
Cocoons, Construction of lodgements in cocooneries for attachment of the.	S. M. Jenkins	Easton, Md	Sept. 28, 1839	1, 347
Cocoonery for removing filth &c., from the hurdles	R. Benson	Smyrna, Del	Jan. 10, 1840	1, 468
Coffee and grain cleaner and separator	R. Frisby	Cleveland, Ohio	Sept. 5, 1871	118, 710
Coffee and other substances, Apparatus for making extracts and decoctions from.	L. Brauer	Washington, D. C	Dec. 1, 1868	84, 609
Coffee and spice mill	A. Clark	Berlin, Conn	July 20, 1832	....
Coffee and spice mill	E. Parker and H. White	Meriden, Conn	June 22, 1832	....
Coffee and spice mill	H. Petrie	Chicago, Ill	Sept. 21, 1869	95, 135
Coffee and spice mill	C. W. Pierce	Oak Hill, N. Y	Jan. 6, 1863	37, 378
Coffee and spice mill	A. Sizer	Cheshire, Conn	Dec. 22, 1831	....
Coffee and spice mill handle	J. Kinzer	Pittsburgh, Pa	May 15, 1866	54, 740
Coffee and tea drawer	J. O. Shriner	Newcastle, Ind	Sept. 12, 1865	49, 929
Coffee and tea maker	A. Bencini	Milton, N. C	Sept. 27, 1838	952
Coffee and tea pot	G. Heiss and M. Schmidt	Houston, Tex	June 15, 1869	91, 333
Coffee and tea pot	J. M. Ingraham	New York, N. Y	Mar. 30, 1858	19, 780
Coffee and tea pot	E. A. Kelsey	Meriden, Conn	July 23, 1861	32, 876
Coffee and tea pot	S. McGuffin	Rising Sun, Ind	May 5, 1868	77, 507
Coffee and tea pot	M. Simons	Middletown, Conn	May 11, 1868	90, 030
Coffee and tea pot	T. B. and J. Stout	Keyport, N. J	June 22, 1869	91, 787
Coffee and tea pot	R. Strickland	Albany, N. Y	Nov. 10, 1868	84, 018
Coffee and tea pot handle	W. Westlake	Chicago, Ill	Mar. 16, 1869	87, 809
Coffee and tea pot strainer	N. Ames	Saugus Centre, Mass	Sept. 17, 1861	33, 316
Coffee and tea pot strainer	W. Case	Waterloo, Iowa	July 30, 1861	32, 969
Coffee and tea pot strainer	M. Simons	Middletown, Conn	Mar. 5, 1867	62, 697
Coffee and tea steamer	C. G. Murch	Chicago, Ill	Oct. 13, 1868	82, 976
Coffee and tea steeper	J. B. Wakeman and A. M. Bush	Hancock, N. Y	June 7, 1870	103, 946
Coffee and water cup for soldiers	C. L. Barritt	New York, N. Y	Sept. 22, 1863	40, 017
Coffee, Apparatus and process for roasting	O. H. Taylor	Brooklyn, N. Y	Nov. 10, 1868	83, 889
Coffee, Apparatus and process for treating	J. Ashcroft	Brooklyn, N. Y	Apr. 18, 1871	113, 832
Coffee, Apparatus for maturing raw or green	J. Ashcroft	Brooklyn, N. Y	Oct. 31, 1871	120, 410
Coffee-berries, Machine for breaking	I. Adams	Boston, Mass	Jan. 13, 1835	....
Coffee-berries, Machine for breaking outer husks of	T. Ditson	Boston, Mass	Jan. 13, 1835	....
Coffee-boiler	E. B. Manning	Middletown, Conn	July 12, 1870	105, 347
Coffee boiler	C. H. Scholle	Cincinnati, Ohio	June 8, 1869	91, 170
Coffee-boiler	W. H. Stewart and J. H. Tilley	Orion, Wis	June 29, 1869	91, 879
Coffee-boiler	N. Waterman	Boston, Mass	Sept. 8, 1863	39, 851
Coffee-boiler	E. F. Woodward	Brooklyn, N. Y	Mar. 25, 1862	34, 801
Coffee-boiler, Alarm	E. K. Sargeant	Boonton, N. J	Sept. 26, 1865	50, 169
Coffee can and crusher	D. C. Colby	Washington, D. C	June 4, 1867	65, 474
Coffee, &c., cleaner	J. Johnson	New York, N. Y	Apr. 18, 1871	113, 890
Coffee-cleaner	S. G. Taylor	Baltimore, Md	Mar. 22, 1870	101, 180
Coffee cleaner and polisher	J. W. Brady	Catonsville, Md	July 27, 1869	93, 045
Coffee cleaner and polisher	J. H. Brookmire	Saint Louis, Mo	Dec. 6, 1870	109, 865
Coffee cleaner and polisher	J. H. Brookmire	Saint Louis, Mo	Oct. 15, 1872	132, 136
Coffee cleaner and polisher	H. C. Lockwood	Baltimore, Md	Jan. 25, 1870	99, 215
Coffee cleaner and polisher	W. Newell	Philadelphia, Pa	July 13, 1858	20, 891
Coffee cleaner, drier, and polisher	W. Newell	Philadelphia, Pa	Sept. 18, 1860	30, 083
Coffee-cleaning machine	J. W. Brady	Catonsville, Md	Mar. 8, 1870	100, 492
Coffee-cleaning machine	J. W. Brady	Baltimore, Md	Mar. 8, 1870	100, 493
Coffee-cleaning machine	W. H. Elton	Baltimore, Md	May 3, 1870	102, 519
Coffee-cleaning machine	J. M. Moore	New York, N. Y	July 16, 1872	129, 359
Coffee, Compound for coating roasted	J. T. Cooke	Brooklyn, N. Y	Oct. 1, 1872	131, 852
Coffee, &c., cooler	J. Burns	New York, N. Y	Jan. 22, 1867	61, 393

Index of patents issued from the United States Patent Office from 1790 *to* 1873, *inclusive*—Continued.

Invention.	Inventor.	Residence.	Date.	No.
Coffee-cooler	D. G. Harrison and J. Reynolds	Cincinnati, Ohio	June 30, 1863	39, 098
Coffee-crushing machine	A. Schacht	Chicago, Ill	July 30, 1872	130, 078
Coffee for transportation, Mode of preparing	T. H. Berry	Lynn, Mass	Apr. 23, 1867	63, 987
Coffee-generator	C. Fobes	Whitewater, Wis	July 30, 1867	67, 185
Coffee, grain, &c., Process for cleaning	G. W. Hungerford	Chicago, Ill	Sept. 19, 1871	119, 149
Coffee-grinder	C. Krutz	Adrian, Mich	May 28, 1872	127, 249
Coffee, Grinding	T. Bruff, sr		Jan. 8, 1798	
Coffee-grinding mill	L. S. Chichester	New York, N. Y	Oct. 9, 1860	30, 298
Coffee-grinding mill	J. Fitzgerald	New York, N. Y	June 25, 1845	4, 089
Coffee-grinding mill	L. R. Livingston and C. Adams	Pittsburgh, Pa	Sept. 25, 1840	1, 795
Coffee-grinding mill	H. Twiss	Meriden, Conn	June 19, 1837	243
Coffee-grinding mill	C. W. Van Vliet	Fishkill Landing, N. Y	Nov. 20, 1855	13, 837
Coffee-huller	A. Angell	Newburgh, N. Y	Nov. 13, 1866	59, 538
Coffee-huller	D. Lombard	Boston, Mass	Jan. 28, 1868	73, 819
Coffee-huller	N. Read	Belfast, Me	Sept. 10, 1822	
Coffee-huller	T. T. Woodruff	Philadelphia, Pa	Nov. 5, 1872	132, 887
Coffee huller and polisher	C. De St. Charles	Jalapa, Mexico	Nov. 3, 1868	83, 609
Coffee hulling and scouring machine	R. P. Walker	New York, N. Y	Dec. 20, 1853	10, 328
Coffee-hulling machine	R. Abbey	New York, N. Y	Jan. 1, 1825	
Coffee-hulling machine	R. Anderson	Brooklyn, N. Y	June 26, 1866	55, 947
Coffee, Machine for cleaning	W. H. Elton	Baltimore, Md	Apr. 27, 1869	89, 298
Coffee, Machine for scouring and polishing	W. Newell	Philadelphia, Pa	July 19, 1859	24, 817
Coffee, Machine for washing and cleaning	S. Thompson	Baltimore, Md	Oct. 18, 1870	108, 407
Coffee-maker	J. Denley and T. H. Heberling	Warsaw, Ill	Aug. 3, 1858	21, 066
Coffee-maker	F. Liesche	East New York, N. Y	June 12, 1866	55, 509
Coffee-maker	J. V. Meigs	Washington, D. C	May 22, 1866	54, 933
Coffee-maker	J. Petsch and S. N. Buynitzky	Hanover, Prussia, and St. Petersburg, Russia.	Feb. 25, 1868	74, 937
Coffee-maker	J. W. Thyng	Salem, Mass	Dec. 17, 1867	72, 429
Coffee-maker	N. Waterman	Suffolk County, Mass	Feb. 26, 1850	7, 129
Coffee-maker	H. A. Zopff	Milwaukee, Wis	Oct. 10, 1865	50, 412
Coffee-making vessel	A. Berney	Jersey City, N. J	Mar. 16, 1869	87, 903
Coffee, Manufacture of roasted	E. E. Rinehart	Pittsburgh, Pa	June 29, 1869	91, 870
Coffee, Method of cleaning and polishing	W. Newell	Philadelphia, Pa	Oct. 3, 1857	18, 552
Coffee-mill	J. R. Adams	New York, N. Y	Dec. 17, 1867	72, 259
Coffee-mill	W. H. Barnes	New London, Conn	Sept. 8, 1868	81, 869
Coffee-mill	C. H. Beatty	Wheeling, Va	June 27, 1854	11, 152
Coffee-mill	T. W. Brown	Boston, Mass	July 19, 1870	105, 545
Coffee-mill	J. Carrington	Wallingford, Conn	Apr. 3, 1829	
Coffee-mill	N. Chapman	Hopedale, Mass	Oct. 16, 1866	58, 772
Coffee-mill	A. Clark	Berlin, Conn	Aug. 14, 1833	
Coffee-mill	I. F. Colby	Washington, D. C	June 9, 1868	78, 648
Coffee-mill	C. R. Edwards	Niagara City, N. Y	Mar. 13, 1866	53, 129
Coffee-mill	R. B. Fitts	Philadelphia, Pa	Mar. 22, 1859	23, 290
Coffee-mill	S. C. Jantzen	Philadelphia, Pa	June 17, 1873	139, 960
Coffee-mill	J. G. Lane	Washington, N. Y	Oct. 16, 1866	58, 946
Coffee-mill	W. J. Lane	Washington, N. Y	Jan. 14, 1868	73, 347
Coffee-mill	A. Lepage	Woodhaven, N. Y	Oct. 24, 1871	120, 287
Coffee-mill	J. Luther	Warren, R. I	Aug. 11, 1843	3, 215
Coffee mill	J. C. Milligan	Brooklyn, N. Y	Aug. 2, 1870	105, 966
Coffee mill, &c	E. Morse and C. Putnam	Knoxville, Tenn	Sept. 9, 1835	
Coffee-mill	C. and E. Parker	Meriden, Conn	May 5, 1868	77, 649
Coffee mill	E. Parker	Meriden, Conn	Oct. 29, 1861	33, 599
Coffee-mill	J. and E. Parker	Meriden, Conn	Feb. 7, 1860	27, 065
Coffee-mill	C. W. Peckham	New Haven, Conn	Apr. 13, 1836	
Coffee-mill	F. Raymond	Woodhaven, N. Y	Feb. 14, 1871	111, 778
Coffee-mill	F. C. Richer	Gilmer, Tex	Aug. 16, 1870	106, 407
Coffee-mill	J. Rittenhouse	Germantown, Pa	July 2, 1839	1, 215
Coffee-mill	L. Schulz	Cincinnati, Ohio	June 10, 1873	139, 824
Coffee-mill	B. Smith	Cincinnati, Ohio	June 9, 1818	
Coffee-mill	E. Watrous	Mystic River, Conn	July 28, 1868	80, 317
Coffee-mill	J. Watrous, jr	Mystic River, Conn	Aug. 6, 1867	76, 618
Coffee-mill and apple-parer	D. C. Warner	Chicago, Ill	Jan. 31, 1871	111, 406
Coffee-mill, Double	T. H. Witherby and J. Torrey	Millbury, Mass	July 17, 1832	
Coffee-mill fastener	J. Winkler	Hudson City, N. J	Feb. 2, 1869	86, 434
Coffee-mill, Portable	S. H. Witmer	Cincinnati, Ohio	June 28, 1864	43, 374
Coffee-mill, Steel	I. Wilson	New London, Conn	Mar. 6, 1818	
Coffee, Mode of extracting the strength from	L. Martelley	New York, N. Y	Sept. 30, 1825	
Coffee or tea pot	E. B. Manning	Middletown, Conn	July 19, 1870	105, 584
Coffee-package	H. C. Lockwood	Baltimore, Md	Aug. 26, 1873	142, 248
Coffee, pepper, &c	D. Richmond	McArthurstown, Ohio	Oct. 14, 1835	
Coffee-percolator	J. H. Nason	Franklin, Mass	Dec. 26, 1865	51, 741
Coffee-polisher	E. J. Codd	Baltimore, Md	July 12, 1870	105, 175
Coffee-polisher	J. T. Randall	Baltimore, Md	May 3, 1870	102, 591
Coffee polisher and cleaner	W. Newell	Philadelphia, Pa	Oct. 12, 1869	95, 827
Coffee, Polishing	J. T. Randall	Baltimore, Md	Mar. 14, 1871	112, 738
Coffee polishing and cleaning	W. and S. Thompson	New York, N. Y., and Baltimore, Md.	Jan. 31, 1871	111, 403
Coffee-polishing machine	J. Guardiola	Chocolá, Central America	July 30, 1872	129, 947
Coffee-polishing machine	C. C. Warren and J. B. Baldy	Toledo, Ohio	Dec. 6, 1870	109, 978
Coffee-pot	N. Agnew	Delaware, Canada	Feb. 7, 1871	111, 505
Coffee-pot	N. Barlow	New York, N. Y	Oct. 19, 1858	21, 808
Coffee-pot	J. Beaumont	Hartford, Conn	Nov. 6, 1860	30, 559
Coffee-pot	E. S. Behringer	Brooklyn, N. Y	July 30, 1872	129, 882
Coffee-pot	J. Blackie	New York, N. Y	May 7, 1867	64, 476
Coffee-pot	W. S. Blaisdell and E. Estabrook	Factory Point, Vt	Aug. 26, 1873	142, 195
Coffee-pot	E. Blunt, jr	New York, N. Y	Nov. 3, 1868	83, 686
Coffee-pot	E. Blunt, jr	New York, N. Y	Aug. 10, 1869	93, 587
Coffee pot	B. Bordman	Malden, Mass	July 14, 1868	79, 943
Coffee-pot	R. Brown and E. A. Blaisdell	Goffstown, N. H	Nov. 14, 1871	120, 936
Coffee-pot	S. I. S. Cawthon and A. F. Tatom	Troy, Ala	Nov. 14, 1871	120, 938
Coffee-pot	W. Chesterman	Centralia, Iowa	July 19, 1859	24, 795
Coffee-pot	W. Chesterman	Centralia, Iowa	Jan. 24, 1860	26, 889
Coffee-pot	H. J. Childs	New York, N. Y	Jan. 7, 1873	134, 514
Coffee-pot	H. J. Childs	New York, N. Y	Jan. 7, 1873	134, 515
Coffee-pot	H. J. Childs	New York, N. Y	Mar. 11, 1873	136, 703
Coffee-pot	E. H. Covell	New York, N. Y	Feb. 8, 1859	22, 856
Coffee-pot	J. Cragg	Baltimore, Md	Oct. 31, 1871	120, 417

Index of patents issued from the United States Patent Office from 1790 *to* 1873, *inclusive*—Continued.

Invention.	Inventor.	Residence.	Date.	No.
Coffee-pot	S. Crowell, jr	Palmyra, N. Y	Sept. 13, 1859	25, 391
Coffee-pot	J. Denley	Warsaw, Ill	July 24, 1860	29, 253
Coffee-pot	J. Dodge	New Orleans, La	Nov. 9, 1869	96, 558
Coffee-pot	J. Dodge	New Orleans, La	July 5, 1870	105, 051
Coffee-pot	J. G. Dyer	Chicago, Ill	June 16, 1868	78, 943
Coffee-pot	O. T. Eddy	Philadelphia, Pa	Sept. 20, 1859	25, 495
Coffee-pot	W. Edson	Boston, Mass	Oct. 2, 1866	58, 396
Coffee-pot	W. H. Elliot	Plattsburgh, N. Y	Jan. 25, 1859	22, 716
Coffee-pot	W. C. C. Erskine	Nether Kinnedder, Dunfermline, Scotland.	May 18, 1869	90, 159
Coffee-pot	G. K. Farrington	Xenia, Ohio	Jan. 24, 1860	26, 945
Coffee-pot	H. B. Fay	New York, N. Y	Oct. 25, 1859	25, 890
Coffee-pot	W. Funk and G. W. Fort	Warrensburgh, Mo	Aug. 23, 1870	106, 572
Coffee-pot	J. F. Fitch, B. G. Devoe, and R. W. B. McLellan.	Vandalia, Ill	Mar. 15, 1870	100, 880
Coffee-pot	D. G. Fletcher	Racine, Wis	May 31, 1859	24, 261
Coffee-pot	J. H. Treeto	Wheaton, Ill	Feb. 22, 1859	23, 021
Coffee-pot	D. S. French	Montgomery, Ala	Mar. 28, 1871	113, 156
Coffee-pot	H. P. Gatchell	Ravenna, Ohio	Nov. 22, 1859	26, 229
Coffee-pot	C. D. Goodrich	Ann Arbor, Mich	Jan. 30, 1872	123, 258
Coffee-pot	J. C. Gove	Cleveland, Ohio	Mar. 13, 1866	53, 139
Coffee-pot	B. Grigsby and T. C. Braselton	Princeton, Ind	Feb. 4, 1873	135, 544
Coffee-pot	J. E. Hall	Cleveland, Ohio	Sept. 25, 1855	13, 595
Coffee pot	N. D. Hartley and M. S. Morehouse.	Quincy, Ill	Apr. 22, 1862	35, 021
Coffee-pot	J. Heberling	Mount Pleasant, Ohio	May 30, 1871	115, 467
Coffee-pot	J. W. Hedenberg	Saint Louis, Mo	Jan. 4, 1859	22, 501
Coffee-pot	L. Hildenbrand	Michigan City, Ind	Dec. 6, 1870	109, 901
Coffee-pot	M. Hofman	San Francisco, Cal	Dec. 3, 1872	133, 582
Coffee-pot	G. Hotte	New York, N. Y	Feb. 1, 1870	99, 438
Coffee-pot	N. Hotz	Greenpoint, N. Y	Aug. 18, 1868	81, 169
Coffee-pot	E. H. Huch	Brunswick, Germany	Jan. 7, 1873	134, 603
Coffee-pot	W. N. Hutchinson	Bideford, England	Mar. 12, 1872	124, 579
Coffee-pot	P. H. Inman and C. B. Withington.	Janesville, Wis	Apr. 25, 1871	114, 012
Coffee-pot	G. Jones	Saugerties, N. Y	Apr. 16, 1867	63, 903
Coffee-pot	R. H. Kuper	Lockport, N. Y	Apr. 4, 1871	113, 674
Coffee-pot	R. H. Kuper	New York, N. Y	May 7, 1872	126, 555
Coffee-pot	J. L. Labiaux	Newark, N. J	Feb. 23, 1869	87, 179
Coffee-pot	N. Lawrence	Taunton, Mass	June 22, 1869	91, 754
Coffee-pot	J. H. Lee	Charlestown, Mass	Nov. 28, 1865	51, 198
Coffee-pot	J. H. Lee	Charlestown, Mass	Dec. 4, 1866	60, 204
Coffee-pot	J. E. Lewis	Kittery, Me	Aug. 3, 1869	93, 317
Coffee-pot	L. H. Little	Copake, N. Y	Mar. 21, 1865	46, 917
Coffee-pot	J. MacGregor, jr	Troy, N. Y	Apr. 11, 1854	10, 752
Coffee-pot	J. Magee	Lawrence, Mass	Mar. 6, 1860	27, 371
Coffee-pot	J. T. McNamee	Baltimore, Md	May 23, 1871	115, 080
Coffee-pot	S. Mento	Alliance, Ohio	Mar. 8, 1870	100, 544
Coffee-pot	C. A. Merchant and G. L. Patterson.	Frankfort, Ky	Feb. 1, 1859	22, 816
Coffee-pot	W. J. Miller	Middletown, Conn	Feb. 14, 1871	111, 861
Coffee-pot	E. Moneuse and L. Duparquet	New York, N. Y	July 27, 1869	92, 987
Coffee-pot	E. Moneuse and L. Duparquet	New York, N. Y	Oct. 5, 1869	95, 599
Coffee-pot	E. Moneuse and L. Duparquet	New York, N. Y	Oct. 12, 1869	95, 710
Coffee-pot	H. W. Mosher	Warren, Ill	Apr. 23, 1861	32, 141
Coffee-pot	E. B. Mudge	Yonkers, N. Y	Oct. 8, 1872	131, 965
Coffee-pot	J. Nason	New York, N. Y	Mar. 3, 1868	75, 045
Coffee-pot	G. Neilson	Boston, Mass	Oct. 25, 1859	25, 909
Coffee-pot	G. Neilson	Boston, Mass	July 24, 1860	29, 302
Coffee-pot	J. B. Parish	Cleveland, Ohio	Apr. 12, 1859	23, 603
Coffee-pot	J. W. Patterson	New York, N. Y	Nov. 21, 1871	121, 126
Coffee-pot	J. W. Patterson and J. S. Hill	Cincinnati, Ohio	Apr. 7, 1868	76, 514
Coffee-pot	P. Perry	Charlestown, Mass	Sept. 17, 1867	69, 017
Coffee-pot	E. Pincus and D. B. Emerick	Philadelphia, Pa	Nov. 21, 1865	51, 083
Coffee-pot	J. R. Remington	Lowndes County, Ala	Dec. 22, 1846	4, 804
Coffee-pot	D. T. Robinson	Boston, Mass	Jan. 8, 1867	61, 022
Coffee-pot	D. Rowland	Washington, D. C	Sept. 17, 1844	3, 749
Coffee-pot	F. I. Sage	Cromwell, Conn	July 18, 1871	117, 210
Coffee-pot	S. T. Savage	Greenbush, N. Y	June 15, 1869	91, 274
Coffee-pot	S. T. Savage	Brooklyn, N. Y	Apr. 5, 1870	101, 517
Coffee-pot	D. M. Skinner	Sandwich Centre, N. H	Jan. 5, 1869	85, 540
Coffee-pot	J. B. Smith	Milwaukee, Wis	Nov. 19, 1867	71, 236
Coffee-pot	J. B. Smith	Milwaukee, Wis	Nov. 19, 1867	71, 237
Coffee-pot	J. B. Smith	Milwaukee, Wis	Dec. 29, 1868	85, 339
Coffee-pot	J. B. Smith	Milwaukee, Wis	Mar. 12, 1872	124, 454
Coffee-pot	O. F. Stedman	Westfield, N. Y	Sept. 14, 1869	94, 787
Coffee-pot	D. Stewart	Annapolis, Md	Apr. 9, 1861	32, 011
Coffee-pot	J. F. Still	West Farms, N. Y	Nov. 8, 1870	109, 152
Coffee-pot	M. J. Stubbings	Youngstown, Ohio	Sept. 23, 1873	143, 102
Coffee-pot	J. H. Swing	Cleveland, Ohio	Apr. 25, 1865	47, 466
Coffee-pot	F. I. Tarlton	Baltimore, Md	Dec. 3, 1872	133, 681
Coffee-pot	H. Thal and G. Schlottmann	New Haven, Conn	Nov. 10, 1868	84, 020
Coffee-pot	H. Tilden	Boston, Mass	Jan. 8, 1867	61, 122
Coffee-pot	O. M. Tinkham	Pomfret, Vt	May 23, 1871	115, 134
Coffee-pot	H. Von Holten	Hoboken, N. J	Sept. 21, 1869	95, 056
Coffee-pot	C. B. Waite and J. W. Senner	Fredericksburgh, Va	Apr. 22, 1856	14, 748
Coffee-pot	J. C. Walker	Waco Village, Tex	Aug. 20, 1867	67, 932
Coffee-pot	A. B. Walters	Philadelphia, Pa	June 1, 1869	90, 707
Coffee-pot	S. V. Warner	Buffalo, N. Y	Oct. 19, 1869	95, 959
Coffee-pot	J. M. Webb	Somerville, Tenn	Feb. 26, 1856	14, 334
Coffee-pot	J. E. Weber and P. Knutson	La Crosse, Wis	Nov. 25, 1873	144, 936
Coffee-pot	W. Westlake	Chicago, Ill	Oct. 2, 1866	58, 518
Coffee-pot	J. P. Williams	Mobile, Ala	Mar. 29, 1870	101, 336
Coffee-pot	W. M. Williams	Saint Louis, Mo	Nov. 24, 1868	84, 456
Coffee-pot	P. B. Willoughby and H. G. Phelps.	Judd, Wis	July 27, 1869	93, 153
Coffee-pot	E. F. Woodward	Brooklyn, N. Y	June 4, 1867	65, 461
Coffee-pot	T. Yates	Dubuque, Iowa	June 19, 1860	28, 803

Index of patents issued from the United States Patent Office from 1790 *to* 1873, *inclusive*—Continued.

Invention.	Inventor.	Residence.	Date.	No.
Coffee-pot	I. Yeamans	Brooklyn, N. Y	June 22, 1869	91, 696
Coffee-pot	H. Young	New York, N. Y	Jan. 3, 1865	45, 787
Coffee-pot	J. Zimmerman	Royalton Centre, N. Y	Dec. 17, 1867	72, 353
Coffee-pot attachment	M. De Graff	Chicago, Ill	Jan. 25, 1870	99, 068
Coffee-pot, boiler, digester, oven, and lamp	J. Montgomery	New York, N. Y	June 1, 1869	90, 676
Coffee-pot digester	W. L. Gilroy	Philadelphia, Pa	Nov. 23, 1869	97, 077
Coffee-pot holder	J. T. Words and E. H. Leseman	Toledo, Ohio	Nov. 28, 1871	121, 444
Coffee-pot, pitcher, &c	J. Gibson, jr	Albany, N. Y	Dec. 28, 1869	98, 244
Coffee-pot stand	O. Ferris	Pawling, N. Y	Oct. 31, 1871	120, 425
Coffee-pot strainer	G. A. Barron	Pembroke, Me	Dec. 27, 1870	110, 420
Coffee-pot strainer	C. L. Gilpatric	Hyde Park, Mass	Feb. 22, 1870	100, 137
Coffee, Prepared	B. B. Lewis	Bristol, Conn	Mar. 2, 1869	87, 346
Coffee, Preparation for clarifying	C. L. Tucker	Chicago, Ill	May 2, 1871	114, 492
Coffee, Preparation of	T. Hyatt	Philadelphia, Pa	Mar. 13, 1866	53, 250
Coffee, Process for cleaning and polishing	C. C. Warren and J. B. Baldy	Toledo, Ohio	June 14, 1870	104, 383
Coffee, Process for polishing	G. W. Hungerford	Chicago, Ill	Sept. 19, 1871	119, 148
Coffee, Process for preparing	L. D. Gale	Washington, D. C	June 20, 1865	48, 268
Coffee, Process of cleaning	W. H. Butler	Chicago, Ill	Jan. 2, 1872	122, 306
Coffee, Process of cleaning and renovating	W. H. Butler	Chicago, Ill	June 18, 1872	127, 955
Coffee, Process of treating	J. W. Gillies	New York, N. Y	June 27, 1871	116, 299
Coffee, rice, &c., Pestle for hulling	J. Guardiola	Chocolá, Central America	Apr. 30, 1872	126, 900
Coffee, Roasted	J. Arbuckle, jr	Allegheny City, Pa	Jan. 21, 1868	73, 486
Coffee-roaster	J. Ashcroft	Brooklyn, N. Y	Oct. 17, 1871	119, 959
Coffee-roaster	B. T. Babbitt	New York, N. Y	Mar. 24, 1868	75, 829
Coffee roaster	R. L. Bate and J. Caulkins	Adrian, Mich	Nov. 1, 1859	25, 941
Coffee-roaster	F. M. Bode	Vienna, Austria	July 28, 1868	80, 329
Coffee-roaster	A. Broman	Kewanee, Ill	Dec. 3, 1872	133, 625
Coffee-roaster	J. Brown	Southampton, Ill	Sept. 24, 1872	131, 597
Coffee-roaster	R. Brown	Ashtabula, Ohio	Apr. 6, 1858	19, 827
Coffee-roaster	J. Burns	New York, N. Y	Oct. 18, 1864	44, 704
Coffee-roaster	C. C. Butt	Duck Hill, Miss	Jan. 2, 1872	122, 357
Coffee-roaster	J. W. Carter	Boston, Mass	Nov. 12, 1846	4, 849
Coffee-roaster	A. Cohn	Louisville, Ky	Sept. 14, 1869	94, 810
Coffee-roaster	L. B. Crittenden	Pittsburgh, Pa	Apr. 6, 1869	88, 612
Coffee-roaster	B. F. Dailey and E. Dougherty, jr.	Mount Pleasant, Iowa	Dec. 3, 1872	133, 633
Coffee-roaster	W. L. Dalbey	Richmond, Ind	July 9, 1872	128, 715
Coffee-roaster	N. Davis	Boston, Mass	Dec. 13, 1870	110, 017
Coffee-roaster	G. W. Dodson	Mitchell, Ind	Apr. 23, 1872	126, 037
Coffee-roaster	G. W. Dodson	Mitchell, Ind	May 27, 1873	139, 377
Coffee-roaster	J. E. Edmundson	Bartlett, Ohio	Sept. 22, 1868	82, 393
Coffee-roaster	J. Galloway	Chetopah, Kans	Sept. 21, 1869	95, 015
Coffee-roaster	J. Galloway	Webster, Ill	May 30, 1871	115, 302
Coffee-roaster	H. M. Gilbert	Ada, Ohio	Sept. 27, 1870	107, 771
Coffee-roaster	W. L. Gilroy	Philadelphia, Pa	Apr. 12, 1859	23, 563
Coffee-roaster	T. J. Hall	Bryan, Tex	Nov. 23, 1869	97, 083
Coffee-roaster	J. Hart	Kekoskee, Wis	Apr. 15, 1873	137, 839
Coffee-roaster	T. Heermans	Mitchellville, Tenn	Sept. 7, 1858	21, 416
Coffee-roaster	T. Heermans	Mitchellville, Tenn	Jan. 18, 1859	22, 649
Coffee-roaster	T. Heermans	Pleasant Hill, Mo	June 30, 1868	79, 469
Coffee-roaster	T. Heermans	Mattoon, Ill	Aug. 12, 1873	141, 789
Coffee-roaster	J. C. Holston	Derry, N. H	Apr. 8, 1862	34, 923
Coffee-roaster	L. Houcke	Springfield, Ohio	Jan. 10, 1871	110, 916
Coffee-roaster	T. Hoyt	New York, N. Y	Sept. 22, 1863	40, 037
Coffee-roaster	F. Humphrey	Philadelphia, Pa	Mar. 13, 1866	53, 148
Coffee-roaster	E. J. Hyde	Philadelphia, Pa	July 1, 1862	35, 758
Coffee-roaster	E. J. Hyde	Philadelphia, Pa	Feb. 2, 1864	41, 436
Coffee-roaster	J. Jay	Jonesborough, Ind	Dec. 7, 1869	97, 644
Coffee-roaster	T. Joyce	Brooklyn, N. Y	Nov. 18, 1862	36, 958
Coffee-roaster	C. L. Kelling	Mechanicsburgh, Pa	Apr. 15, 1862	34, 968
Coffee-roaster	R. Landstrom	Boston, Mass	Mar. 13, 1860	27, 454
Coffee-roaster	W. J. Lane	Millbrook, N. Y	Dec. 16, 1873	145, 576
Coffee-roaster	A. Larson	Galesburgh, Ill	Apr. 9, 1872	125, 579
Coffee-roaster	N. Linden	Chicago, Ill	Nov. 15, 1870	109, 228
Coffee-roaster	R. Little	Middle Branch, Ohio	Aug. 28, 1860	29, 798
Coffee-roaster	H. C. Lockwood	Baltimore, Md	Dec. 23, 1873	145, 880
Coffee-roaster	I. Long	Terre Haute, Ind	May 11, 1869	89, 934
Coffee-roaster	C. Mackh	Elgin, Ill	June 22, 1869	91, 553
Coffee-roaster	B. K. Maltby	Cincinnati, Ohio	May 26, 1868	78, 386
Coffee-roaster	B. K. Maltby	Cincinnati, Ohio	Apr. 27, 1869	89, 490
Coffee-roaster	D. D. Martin	Cincinnati, Ohio	Nov. 28, 1871	121, 292
Coffee-roaster	H. B. Masser	Sunbury, Pa	Aug. 7, 1866	56, 964
Coffee-roaster	G. W. Merrick	Adrian, Mich	May 21, 1872	127, 086
Coffee roaster	C. A. Mills	Hazel Green, Wis	Apr. 28, 1863	38, 320
Coffee-roaster	C. A. Mills	Bristol, Conn	Nov. 7, 1865	50, 878
Coffee-roaster	R. L. Mills	Springfield, Ohio	May 23, 1871	115, 228
Coffee-roaster	S. Nowlan	New York, N. Y	June 23, 1863	38, 977
Coffee-roaster	A. Obst	Cambridgeport, Mass	May 2, 1871	114, 330
Coffee-roaster	C. T. Palmer	Norwich, Conn	Oct. 19, 1869	96, 030
Coffee-roaster	J. B. Peake and J. A. Lusby	Washington, D. C	June 29, 1869	92, 090
Coffee-roaster	H. W. Persing	Chicago, Ill	May 25, 1869	90, 386
Coffee-roaster	C. J. C. Petersen	Davenport, Iowa	Oct. 19, 1858	21, 845
Coffee-roaster	S. Pierce	Troy, N. Y	Aug. 14, 1855	13, 447
Coffee-roaster	P. Plant	Washington, D. C	July 31, 1866	56, 851
Coffee-roaster	A. Ransom	Hartford, Conn	Dec. 31, 1833	....
Coffee-roaster	J. R. Remington	Baltimore, Md	Jan. 7, 1847	4, 922
Coffee-roaster	H. E. Richard	Newark, N. J	Oct. 8, 1861	33, 453
Coffee-roaster	E. Schneider and A. Kolman	New Tripoli, Pa	Nov. 24, 1857	18, 710
Coffee-roaster	J. P. Simmons	Baldwinsville, N. Y	May 3, 1859	23, 867
Coffee-roaster	A. Stillman	Poland, N. Y	Dec. 28, 1849	1, 919
Coffee-roaster	N. S. Thompson	Richmond, Ind	Aug. 29, 1871	118, 497
Coffee-roaster	S. Tower	Grand Rapids, Mich	Aug. 31, 1858	21, 387
Coffee-roaster	S. and N. Van Dyk	Saint Louis, Mo	Aug. 20, 1861	33, 113
Coffee-roaster	W. H. Welch	Bloomington, Ill	June 17, 1873	139, 984
Coffee-roaster	A. W. Weyburn	Minneapolis, Minn	Feb. 25, 1873	136, 292
Coffee-roaster	E. Whiteley	Boston, Mass	Apr. 22, 1851	8, 057
Coffee-roaster	N. L. Whitney	Effingham, Ill	Feb. 5, 1867	61, 905
Coffee-roaster	B. I. Williams	Lansing, Mich	Apr. 2, 1872	125, 241

Index of patents issued from the United States Patent Office from 1790 *to* 1873, *inclusive*—Continued.

Invention.	Inventor.	Residence.	Date.	No.
Coffee-roaster	P. Williamson	Baltimore, Md	Mar. 20, 1820	....
Coffee-roaster	T. R. Wood	Cincinnati, Ohio	Apr. 17, 1849	6, 345
Coffee-roaster	F. W. Zochert	Watertown, Wis	Aug. 3, 1869	93, 389
Coffee-roaster and grain-drier	G. D. Jones	New York, N. Y	Feb. 7, 1865	46, 301
Coffee-roaster cover	L. J. Dyke	Hamilton, N. Y	Mar. 18, 1873	136, 980
Coffee-roaster operated by steam	I. C. Smith	New York, N. Y	June 4, 1872	127, 654
Coffee, Roasting	W. Johnston and J. D. Flausburg.	Philadelphia, Pa	Nov. 17, 1868	84, 193
Coffee, Roasting	W. H. Trissler and E. Brecht.	Fairview, Pa	Apr. 30, 1850	7, 329
Coffee-roasting machine	J. D. Harrington	Rochester, N. Y	May 17, 1859	24, 024
Coffee-roasting machine	P. Williamson	Baltimore, Md	June 18, 1825	....
Coffee-roasting process	J. W. Gillies	New York, N. Y	June 27, 1871	116, 298
Coffee-settler	W. F. Rossman	Hudson, N. Y	Apr. 25, 1865	47, 460
Coffee-settling composition	G. W. Carleton	Brunswick, Me	July 31, 1866	56, 704
Coffee, spice, &c., Box for	H. W. Hutchins	Livermore Falls, Me	Jan. 4, 1870	98, 502
Coffee-steeper	J. D. Adney	Chicago, Ill	June 10, 1873	139, 645
Coffee-steeper	R. S. Sanborn	Sycamore, Ill	Apr. 2, 1861	31, 910
Coffee substitute	E. Dugdale	Griffin, Ga	Oct. 21, 1873	143, 889
Coffee substitute	E. Dugdale	Griffin, Ga	Oct. 21, 1873	143, 890
Coffee, tea, and spice can	J. M. Earl	Springfield, Mass	Dec. 13, 1870	110, 125
Coffee, Treating damaged	F. Evans	New York, N. Y	Jan. 14, 1873	134, 791
Coffee, Treating damaged	F. Evans	New York, N. Y	Jan. 14, 1873	134, 792
Coffee-urn	C. Hitchcock	Milldale, Conn	Dec. 16, 1873	145, 653
Coffee-urn	G. Jones	New Haven, Conn	Dec. 8, 1868	84, 830
Coffee-urn	E. Martin	Waterbury, Conn	Apr. 19, 1870	102, 139
Coffee-washing machine	J. Guardiola	Chocolá, Central America	July 16, 1872	129, 278
Coffer-dam	E. Bell	Saint Paul, Minn	Jan. 21, 1868	73, 566
Coffer-dam	J. Braidwood	Wilmington, Ill	June 5, 1866	55, 235
Coffer-dam	W. Easby	Washington, D. C	Nov. 19, 1833	....
Coffer-dam	A. Folsom	New York, N. Y	Jan. 31, 1865	46, 098
Coffer-dam	A. Gilmore	Buffalo, N. Y	Apr. 10, 1866	53, 919
Coffer-dam	C. H. Sanborn	Roxbury, Mass	Feb. 26, 1867	62, 444
Coffer-dam	S. S. Walley	Charlestown, Pa	Sept. 15, 1846	4, 759
Coffer-dam	J. E. Walsh	New York, N. Y	Jan. 24, 1871	111, 156
Coffer-dam	J. E. Walsh	New York, N. Y	Jan. 24, 1871	111, 157
Coffer-dam	J. E. Walsh	New York, N. Y	Apr. 4, 1871	113, 374
Coffer-dam	J. E. Walsh	New York, N. Y	Oct. 24, 1871	120, 352
Coffer-dam and boat	W. H. Applegate	Le Claire, Iowa	May 21, 1867	64, 933
Coffer-dam, Portable and convertible	S. Lewis	Williamsburgh, N. Y	July 6, 1869	92, 324
Coffer-dam, Sectional	T. Bracher	Rahway, N. J	Oct. 19, 1869	95, 976
Coffin	W. G. Algeo	Allegheny, Pa	Jan. 11, 1870	98, 730
Coffin	S. Avery	Phœnix, N. Y	Oct. 17, 1871	120, 021
Coffin	S. Avery and L. Delill	Phœnix, N. Y	Apr. 7, 1868	76, 376
Coffin	J. D. Bayliss	Alexandria, Va	Jan. 5, 1869	85, 507
Coffin	W. Broomhall	Circleville, Ohio	May 25, 1869	90, 491
Coffin	J. Burns	Providence, R. I	July 17, 1866	56, 364
Coffin	C. L. Carter and E. Jones	New York, N. Y	May 29, 1860	28, 451
Coffin	D. Clarke	Ipswich, Mass	July 10, 1866	56, 182
Coffin	E. H. Covel	New York, N. Y	May 10, 1864	42, 641
Coffin	J. E. Cox	Cincinnati, Ohio	Sept. 6, 1870	107, 164
Coffin	E. Ellingen	Mineral Point, Wis	Sept. 8, 1868	81, 885
Coffin	T. B. Estep	Cincinnati, Ohio	Oct. 1, 1867	69, 419
Coffin	T. B. Estep and W. C. Hefferman.	Cincinnati, Ohio	Oct. 1, 1867	69, 420
Coffin	J. E. Evarts	Madison, Conn	Aug. 11, 1868	80, 816
Coffin	L. Fay	Cincinnati, Ohio	May 31, 1864	42, 936
Coffin	A. D. Fisk	New York, N. Y	Nov. 14, 1848	5, 920
Coffin	J. A. Fogg	Salem, Mass	Feb. 6, 1866	52, 405
Coffin	J. A. Fogg	Stockport, England	Sept. 4, 1866	57, 829
Coffin	J. Good	Philadelphia, Pa	Dec. 12, 1854	12, 058
Coffin	T. Graham	Philadelphia, Pa	Apr. 25, 1865	47, 414
Coffin	A. W. Hendrick	Batavia, Ill	Sept. 21, 1869	95, 109
Coffin	M. M. Hersman	Delavan, Ill	Dec. 19, 1871	121, 943
Coffin	F. H. Hill	Chicago, Ill	Nov. 10, 1868	83, 964
Coffin	T. Holmes	Washington, D. C	May 10, 1864	42, 659
Coffin	D. and S. E. Hooker	West Poultney, Vt	Mar. 2, 1858	19, 503
Coffin	J. M. Lyeth	Baltimore, Md	June 27, 1854	11, 166
Coffin	W. J. Lyman and A. E. Lyman	East Hampton and Williamsburgh, Mass.	Apr. 29, 1862	35, 102
Coffin	M. R. Margerum	Trenton, N. J	Sept. 22, 1868	82, 333
Coffin	H. Marshall	Cincinnati, Ohio	Oct. 4, 1859	25, 659
Coffin	S. Merrick	New Brighton, Pa	Nov. 17, 1868	84, 130
Coffin	J. D. Nietscke	Somerset, Ohio	Sept. 24, 1867	69, 238
Coffin	P. E. Ober	Beverly, Mass	Jan. 21, 1873	135, 006
Coffin	C. Potter and E. Jones	Pawlet, Vt., and Middle Granville, N. Y.	Apr. 7, 1868	76, 344
Coffin	C. E. H. Richardson	Philadelphia, Pa	Jan. 4, 1859	22, 537
Coffin	D. Sholl	Cincinnati, Ohio	Mar. 27, 1855	12, 605
Coffin	I. C. Shuler	Amsterdam, N. Y	Mar. 29, 1859	23, 401
Coffin	I. C. Shuler	Amsterdam, N. Y	July 12, 1870	105, 377
Coffin	B. Smith	Cincinnati, Ohio	July 14, 1868	79, 921
Coffin	H. Smith, jr	Summit, N. Y	Apr. 23, 1867	64, 159
Coffin	H. D. Sprague	Portland, Me	July 10, 1866	56, 287
Coffin	T. M. Taylor	New York, N. Y	Mar. 22, 1870	101, 181
Coffin	F. Tighe	Rockland, Me	Sept. 23, 1873	143, 204
Coffin	P. Washburn, H. G. O. White, and G. A. Copeland.	Taunton, Mass	June 27, 1854	11, 185
Coffin	J. C. Williams	Newton, N. J	Jan. 28, 1868	73, 857
Coffin	S. H. Young	Saint Louis, Mo	Sept. 29, 1863	40, 137
Coffin, Artificial stone or marble	J. White	New York	July 18, 1835	....
Coffin-bier	P. Joyce	Rochester, N. Y	Mar. 9, 1869	87, 570
Coffin, Corpse-preserving	J. Y. Black	Cleveland, Ohio	Dec. 20, 1870	110, 337
Coffin, Deodorizing	A. E. Lyman	New York, N. Y	Apr. 18, 1865	47, 313
Coffin-fastening	W. S. Crane	Rome, Ga	Oct. 21, 1873	143, 748
Coffin-fastening	W. Hamilton	Allegheny City, Pa	June 28, 1870	104, 730
Coffin for use in case of doubtful death	C. H. Eisenbrandt	Baltimore, Md	Nov. 15, 1843	3, 335
Coffin from paper or other pulp	F. Keenan	Brownville, N. Y	Feb. 20, 1872	123, 781
Coffin, Glass	G. B. Field	Saint Louis, Mo	Dec. 11, 1860	30, 874

Index of patents issued from the United States Patent Office from 1790 *to* 1873, *inclusive*—Continued.

Invention.	Inventor.	Residence.	Date.	No.
Coffin, Glass	G. W. Scollay	Saint Louis, Mo	Oct. 2, 1860	30, 252
Coffin, Glass	J. N. and T. Wallis	Fleming, N. Y	May 20, 1873	139, 095
Coffin-handle	I. Almy	Farmer, N. Y	Apr. 25, 1865	47, 375
Coffin-handle	A. B. Bailey	Cobalt, Conn	Sept. 13, 1870	107, 212
Coffin-handle	A. Clark	Amsterdam, N. Y	Dec. 26, 1865	51, 693
Coffin-handle	N. Hayden	Essex, Conn	Oct. 22, 1872	132, 399
Coffin-handle	J. S. Ray	East Haddam, Conn	Mar. 1, 1870	100, 445
Coffin-handle	J. S. Ray	East Haddam, Conn	Apr. 15, 1873	137, 958
Coffin-handle	H. Rogers, jr	West Meriden, Conn	May 6, 1873	132, 696
Coffin-handle	C. Strong	Winsted, Conn	Dec. 14, 1869	97, 827
Coffin-handle	C. Strong	Winsted, Conn	May 13, 1873	138, 768
Coffin-handle	B. W. Wooster	Albany, N. Y	Feb. 11, 1862	34, 387
Coffin-handles, Casting	D. W. Sexton	East Hampton, Conn	June 13, 1865	48, 215
Coffin-head brace	J. Gawler	Washington, D. C	Nov. 9, 1869	96, 581
Coffin-lid	J. S. Merrill	Poland, Me	May 26, 1863	38, 713
Coffin-lids, Hinging	J. C. Seelye	Cambridge, Mass	July 18, 1865	48, 842
Coffin, Metallic	J. A. Gray	Richmond, Va	June 11, 1836	....
Coffin, Metallic	G. Nearstheimer	Cincinnati, Ohio	Apr. 6, 1869	88, 728
Coffin, Metallic	J. H. Renshaw	Knoxville, Tenn	Feb. 12, 1861	31, 401
Coffin, Metallic	I. C. Shuler	Amsterdam, N. Y	Apr. 12, 1859	23, 616
Coffin-nail cap	S. A. Barker	Providence, R. I	Jan. 11, 1870	98, 733
Coffin nails and screw heads, Machine for finishing	A. A. Randall	New Haven, Conn	June 19, 1866	55, 783
Coffin-plate	G. Brabrook	Taunton, Mass	July 1, 1873	140, 457
Coffin-plate	W. H. Green	Meriden, Conn	Apr. 19, 1864	42, 367
Coffin-plate	G. B. Ransom	Chester, Conn	Jan. 21, 1873	135, 155
Coffin-receptacle	J. H. Shields	Louisville, Ky	Nov. 21, 1871	121, 206
Coffin, Refrigerator	J. S. Cox	Delaware, Ohio	Mar. 10, 1868	75, 377
Coffin, Sheet-metal	J. D. Farrington, jr., and T. P. Austin.	New York, N. Y., and Portland, Conn.	Nov. 18, 1873	144, 604
Coffin, Sheet-metal	I. C. Shuler	Amsterdam, N. Y	Dec. 6, 1859	26, 379
Coffin, Sheet-metal	W. S. Wood	New York, N. Y	Dec. 24, 1872	134, 182
Coffin-sides, Apparatus for bending	J. P. Albin	Cincinnati, Ohio	Dec. 23, 1873	145, 827
Coffin-tops, Press for making	A. W. Hendrick	Batavia, Ill	Oct. 5, 1869	95, 475
Coffin, Ventilating	A. E. Lyman	Williamsburgh, Mass	Oct. 14, 1862	36, 660
Coffin, Wooden	F. Brübach	Lancaster, Pa	Oct. 9, 1860	30, 292
Coffin, Wooden	M. R. Margerum	Trenton, N. J	Sept. 26, 1865	50, 144
Coffins and other articles from asphaltic composition, Manufacture of.	D. W. Denton	Ithaca, N. Y	Jan. 19, 1864	41, 284
Coffins, Apparatus for supporting and lowering	C. A. Thompson and J. O. Coleman.	Hopkinsville, Ky	Dec. 5, 1871	121, 688
Coffins, &c., Cement for coating and sealing	I. Charles	Allegheny City, Pa	May 16, 1871	114, 761
Coffins, Compound for covering	J. W. Bower	Greencastle, Ind	June 6, 1871	115, 687
Coffins, Constructing	I. C. Shuler	Amsterdam, N. Y	Apr. 27, 1858	20, 095
Coffins, Constructing sheet-metal	I. C. Shuler	Amsterdam, N. Y	June 14, 1859	24, 409
Coffins, Construction of	F. Skiff	New York, N. Y	Apr. 20, 1847	5, 084
Coffins, Construction of glass	J. R. Cannon	New Albany, Ind	Oct. 25, 1859	25, 883
Coffins, Construction of sheet-metal	I. C. Shuler	Amsterdam, N. Y	July 5, 1859	24, 635
Coffins, Construction of sheet-metal	I. C. Shuler	Amsterdam, N. Y	Sept. 6, 1859	25, 350
Coffins, Covering	L. Snider	Indianapolis, Ind	Oct. 4, 1859	25, 682
Coffins from hydraulic cement, Constructing	Dayton, Hoyt, and White	Salina, N. Y	June 6, 1835	....
Coffins from hydraulic cement, Constructing	J. White	New York, N. Y	July 18, 1835	....
Coffins, Fastening-key for	J. Homrighouse	Royalton, Ohio	Mar. 19, 1872	124, 745
Coffins, Head-rest for	J. P. Waugh	South Scriba, N. Y	Aug. 29, 1871	118, 565
Coffins, Life-detector for	T. A. Schroeder and H. Wuest.	Hoboken, N. J	Dec. 5, 1871	121, 666
Coffins, &c., Lifting-handle for	T. M. Taylor	New York, N. Y	June 28, 1870	104, 899
Coffins, Manufacturing cement	M. Leonard	Syracuse, N. Y	July 2, 1839	1, 214
Coffins, Measuring-scale for	J. W. Hyde	Lewiston, Ill	Sept. 25, 1866	58, 262
Coffins, Mode of disinfecting	S. H. Young	Saint Louis, Mo	May 7, 1867	64, 467
Coffins, &c., Molding for	G. S. Eaton	Williamsburgh, N. Y	June 18, 1872	127, 966
Coffins, &c., Name-plate for	G. Brabrook	Taunton, Mass	Mar. 24, 1868	75, 726
Coffins, Preserving the wood of	J. M. Dufournet and L. Clemendot.	Paris, France	May 12, 1868	77, 8[illegible]6
Cog-gearing spring	J. L. Rees	Peoria, Ill	June 4, 1867	65, 506
Cog-wheel	F. A. Morley	Sodus Point, N. Y	Feb. 10, 1863	37, 634
Cog-wheel coupling, Universal	E. Smith	Adams County, Ill	Jan. 10, 1871	110, 938
Cog-wheel-cutting machinery	J. P. Gaume	Cincinnati, Ohio	June 6, 1848	5, 614
Cog-wheel for gearing	H. I. Crandall	New Bedford, Mass	Oct. 27, 1868	83, 469
Cog-wheels, Connection of	A. Skaats, jr	New Haven, Conn	May 3, 1864	42, 605
Cog-wheels, &c., Machine for making patterns for	W. Obenchain	Logansport, Ind	Apr. 5, 1859	23, 484
Cog-wheels, Machinery to supersede the use of	J. Cooper	Staunton, Va	Feb. 22, 1826	....
Coil, Electro-magnetic induction	J. S. Camacho	Havana, Cuba,	Apr. 29, 1873	138, 316
Coil, Induction	J. Kidder	New York, N. Y	Jan. 16, 1866	52, 054
Coil-spring and its attachments	T. Rose and P. S. Buell	Cortland and Windsor, N. Y	Dec. 7, 1869	97, 705
Coil-springs, Mandrel for	R. Vose and J. Anderson	New York, N. Y	Jan. 5, 1869	85, 549
Coil-springs, Manufacture of	W. H. Ward	Auburn, N. Y	Mar. 4, 1873	136, 473
Coiled springs, Machine for making	J. Harrison, jr	New York, N. Y	Jan. 27, 1857	16, 483
Coiled springs, Machine for packing	J. W. Evans	New York, N. Y	Mar. 2, 1869	87, 332
Coin assorting and computing apparatus	E. Baltzley	New Philadelphia, Ohio	Feb. 13, 1872	123, 607
Coin-assorting apparatus	J. W. Meaker	Chicago, Ill	Apr. 7, 1868	76, 492
Coin-assorting apparatus	J. W. Meaker	Chicago, Ill	Apr. 7, 1868	76, 493
Coin-counting implement	A. Whittemore	Cambridgeport, Mass	Jan. 14, 1873	134, 962
Coin-counting machine	M. F. Bonzano	New Orleans, La	June 17, 1856	15, 117
Coin-counting machine	R. Tyler	New Orleans, La	Oct. 11, 1841	2, 320
Coin-detector	H. Maranville	Clinton, Ohio	Feb. 14, 1860	27, 140
Coin detector, Counterfeit	W. Painter	Fallston, Md	July 8, 1862	35, 834
Coin detector, Counterfeit	G. B. Smith	Baltimore, Md	Sept. 6, 1853	9, 997
Coin-holder	L. Landry	San Francisco, Cal	May 29, 1866	55, 121
Coke, Desulphurizing	G. Nock	Pittsburgh, Pa	May 29, 1860	28, 543
Coke from anthracite and other coal, Manufacturing	M. Isaacs	Philadelphia, Pa	Apr. 7, 1831	....
Coke from Colorado and other coal, Process of preparing.	W. J. Lynd	Golden City, Colo	Aug. 10, 1869	93, 629
Coke from lignites, Process of making	H. Engelmann	Salt Lake City, Utah	Mar. 11, 1873	136, 592
Coke-furnace	T. Price	Steubenville, Ohio	Dec. 6, 1870	109, 940
Coke, Manufacture of	W. McPheeters and C. Pearce.	Harmony, Ind	Aug. 1, 1871	117, 714
Coking fossil-coal, Modes of	G. Lander	Irwin, Pa	Mar. 19, 1872	124, 688
Coking fossil-coal, Mode of	G. Lander	Irwin, Pa	Mar. 19, 1872	124, 689
Coking wood by waste-heat of iron-furnaces	A. H. Tait	Plattsburgh, N. Y	Aug. 22, 1848	5, 722
Colander boiler	B. F. Porter	Manchester, N. H	May 21, 1867	94, 905

Index of patents issued from the United States Patent Office from 1790 *to* 1873, *inclusive*—Continued.

Invention.	Inventor.	Residence.	Date.	No.
Collapsible boat	C. F. Lichtner	Chicago, Ill	Aug. 25, 1863	39,661
Collapsible boat	N. Thompson, jr	Williamsburgh, N. Y	Aug. 28, 1855	13,510
Collapsible boat	N. Thompson, jr	Brooklyn, N. Y	Feb. 19, 1858	19,317
Collar	J. T. Bruen	Brooklyn, N. Y	Jan. 7, 1873	134,638
Collar	H. G. Emery and M. C. Fuller	Boston, Mass	July 19, 1870	105,442
Collar	A. Flatley	Brooklyn, N. Y	Apr. 1, 1873	137,353
Collar	J. H. Parmelee	Chicopee, Ill	July 23, 1867	67,134
Collar	G. F. Rice	Worcester, Mass	Mar. 25, 1873	137,096
Collar	C. W. Saladee	Circleville, Ohio	Jan. 26, 1869	86,180
Collar	C. W. Saladee	Circleville, Ohio	Feb. 23, 1869	87,116
Collar and bosom combined	E. Hatch	Charlestown, Mass	Feb. 11, 1868	74,355
Collar and bosom, Detached shirt	C. E. Richards	North Attleborough, Mass	Dec. 20, 1864	45,566
Collar and bosom, Shirt	C. E. Richards	North Attleborough, Mass	July 25, 1865	49,039
Collar and cravat combined	F. D. James	Tamworth, N. H	June 17, 1873	139,959
Collar and cuff	J. Blakey and H. B. Fox	Liverpool and Oxton, England.	May 5, 1868	77,573
Collar and cuff	E. P. Furlong	Springfield, Mass	Nov. 18, 1873	144,611
Collar and cuff box, Combined	J. C. Bauer	New York, N. Y	Oct. 18, 1870	108,315
Collar and cuff drier	D. W. De Forest	Petersburgh, Va	Nov. 25, 1873	144,838
Collar and cuff, Lady's	W. E. Lockwood	Philadelphia, Pa	Apr. 26, 1859	23,771
Collar and cuff, Water-proof	G. W. Ray	Springfield, Mass	June 13, 1865	48,239
Collar and hame	W. O'Brien and H. Wentworth	Omaha, Nebr	Mar. 29, 1870	101,303
Collar and hame	J. L. Wooden	Greensburgh, Ind	Mar. 8, 1870	100,701
Collar and hame, Combined	F. Jones	Burlington, Iowa	July 13, 1869	92,534
Collar and hame, Harness	J. H. Ferguson	Greenville, Ind	Feb. 22, 1870	100,022
Collar and hame, Horse	O. Cann	Coldwater, Mich	Apr. 4, 1871	113,253
Collar and hame, Horse	H. Connick	Albert Lea, Minn	Aug. 24, 1869	93,968
Collar and hame, Horse	A. Dunbar	New York, N. Y	May 19, 1868	78,075
Collar and hame, Horse	A. Dunbar	Woodstock, Canada	Oct. 3, 1871	119,585
Collar and hame, Horse	M. Killacky	Philadelphia, Pa	Feb. 28, 1865	46,606
Collar and hame, Horse	S. B. Stewart	Centre Township, Pa	June 30, 1863	39,076
Collar and hame, Horse	E. Wilder	South Hingham, Mass	Aug. 3, 1869	93,262
Collar and harness, Combined horse	G. W. N. Yost	Cincinnati, Ohio	Jan. 19, 1858	19,169
Collar and muff combined	M. Wannagat	New York, N. Y	Apr. 16, 1872	125,862
Collar and neck-tie	W. Hunter	Berkeley Springs, W. Va	Apr. 24, 1866	54,168
Collar and neck-tie attachment	C. W. Powell	Milford, Conn	Nov. 13, 1866	59,646
Collar and neck-tie combined	G. F. Perkins	New York, N. Y	Mar. 3, 1868	75,047
Collar and neck-tie fastener	E. Raw	New York, N. Y	Oct. 20, 1868	83,318
Collar and spreader for double harness, Breast	J. M. Myers	Louisville, Ky	Aug. 6, 1867	67,566
Collar and tag cutting machine	S. S. Gray	Boston, Mass	Dec. 30, 1873	145,941
Collar attachment, Shirt	J. Proud	New York, N. Y	July 24, 1866	56,610
Collar, Axle	E. S. Scripture	New York, N. Y	June 4, 1861	32,499
Collar, Belvidere shirt	W. J. Cantelo and R. M. Kerrison.	Philadelphia, Pa	Nov. 19, 1833	....
Collar-blank	G. K. Snow	Watertown, Mass	Oct. 29, 1872	132,545
Collar block	J. Jacobs	Oneida, Ill	Feb. 4, 1868	73,977
Collar block, Horse	A. Benham	Belvidere, Ill	Mar. 8, 1848	5,464
Collar block, Horse	E. L. Brazenor	Birmingham, Great Britain	Apr. 12, 1870	101,819
Collar block, Horse	W. H. Bustin	Watertown, Mass	Dec. 19, 1848	5,975
Collar block, Horse	E. D. Gould	Darien, N. Y	Apr. 6, 1858	19,846
Collar block, Horse	B. W. McClure	Wyoming, Iowa	Oct. 11, 1864	44,642
Collar block, Horse	B. W. McClure and G. Marsh	Pike Hollow, N. Y	Aug. 24, 1858	21,307
Collar block, Horse	P. Moodey	Indianapolis, Ind	June 26, 1855	13,132
Collar block, Horse	O. Morgan	Henry, Ill	July 2, 1861	32,708
Collar block, Horse	T. J. Van Benschoten	Poughkeepsie, N. Y	July 3, 1855	13,189
Collar, Breast	G. W. Blaksley	Rockford, Ill	Apr. 13, 1869	88,838
Collar, Breast	S. E. Stowell	Charlestown, Mass	Jan. 16, 1866	52,088
Collar, Caoutchouc shirt	J. A. Pease	New York, N. Y	Apr. 7, 1863	38,122
Collar cap, Harness	A. P. Mason	Franklinville, N. Y	June 27, 1871	116,465
Collar cap, Horse	R. J. Algeo	Kalamazoo, Mich	Oct. 29, 1872	132,515
Collar cap, Horse	R. J. Algeo	Kalamazoo, Mich	Mar. 18, 1873	136,804
Collar cap, Horse	D. Curtis	Chicago, Ill	Aug. 2, 1870	105,917
Collar cap, Horse	D. Curtis	Sun Prairie, Wis	May 14, 1872	126,681
Collar cap, Horse	A. P. Mason	Franklinville, N. Y	Mar. 12, 1872	124,444
Collar cap, Horse	J. Sellors	Bellevue, Mich	Apr. 19, 1870	102,052
Collar cap, Horse	J. F. Walsh	Hazel Green, Wis	Oct. 4, 1870	108,071
Collar-clasp	V. Fogerty	Boston, Mass	June 21, 1864	43,190
Collar, Coach	W. Hottensteen		Apr. 10, 1800	....
Collar, Dog	P. Daly and J. Barry	Boston, Mass	Apr. 30, 1872	126,135
Collar-edge protector	C. W. Saladee	Newark, Ohio	Oct. 22, 1867	70,025
Collar-fastening	M. C. Battey	Washington, D. C	May 11, 1869	89,967
Collar-fastening	C. H. L. Roberts and W. C. Dudley.	Morrison, Ill	Oct. 6, 1868	82,879
Collar fastening, Horse	A. Bratnober	Webster County, Iowa	Oct. 25, 1870	108,559
Collar fastening, Horse	H. H. Fleming	Kokomo, Ind	Nov. 21, 1865	51,033
Collar fastening, Horse	J. P. and J. E. Force	Constantine, Mich	Dec. 8, 1868	84,687
Collar fastening, Horse	E. B. Gould	Liberty, Ind	Jan. 26, 1869	86,298
Collar fastening, Horse	B. H. Hobart and D. C. Lampman.	Troy, Pa	Oct. 6, 1868	82,836
Collar fastening, Horse	M. F. McIntyre	Girard, Pa	May 25, 1869	90,374
Collar fastening, Horse	W. A. Robinson	Grand Rapids, Mich	Apr. 27, 1869	89,346
Collar fastening, Horse	W. A. Sharp and J. A. Shannon	Tama City, Iowa	Aug. 25, 1868	81,416
Collar fastening, Horse	E. H. Sprague	Mount Vernon, Ohio	Mar. 25, 1873	137,105
Collar fastening, Horse	A. Van Fleet	Ashton, Ill	July 21, 1868	80,246
Collar fastening, Horse	E. L. Welbourn	Union City, Ind	Oct. 18, 1870	108,418
Collar fastening, Horse	D. C. Westfall	Mifflin, Pa	Oct. 1, 1867	69,520
Collar fastening, Horse	C. Wheeler	Warsaw, Ohio	Oct. 24, 1871	120,356
Collar fastening, Shirt	G. W. Brientnall	Lancaster, Pa	Aug. 5, 1873	141,484
Collar fastening, Shirt	F. Hess	Baltimore, Md	May 21, 1867	64,867
Collar fastening, Shirt	C. E. Palmer	Newburyport, Mass	July 28, 1868	80,424
Collar fastening, Shirt	G. Straszer	Manchester, Mo	Dec. 30, 1873	145,974
Collar fastening, Shirt	A. Wood	East Henrietta, N. Y	Jan. 15, 1867	61,300
Collar for attaching cow-bells	T. H. Body	Kilbourn City, Wis	May 10, 1870	102,756
Collar for dress-coats, &c	H. Clark	Brooklyn, Conn	Nov. 7, 1828	....
Collar for drill-rod	V. T. Priest	Decatur, Ill	July 24, 1866	56,607
Collar for harness	J. W. Briggs	Cleveland, Ohio	June 3, 1851	8,133
Collar for ladies and gentlemen	F. Field	Troy, N. Y	Aug. 4, 1863	39,390
Collar, Fur	W. King	New York, N. Y	Mar. 24, 1868	75,928

Index of patents issued from the United States Patent Office from 1790 *to* 1873, *inclusive*—Continued.

Invention.	Inventor.	Residence.	Date.	No.
Collar Fur	J. H. Kappelhoff and S. Rauh	New York, N. Y	July 12, 1870	105, 340
Collar, Fur	R. M. Seldis	New York, N. Y	June 15, 1869	91, 371
Collar gage, Horse	J. M. Everitt	Hackettstown, N. J.	June 17, 1873	139, 949
Collar, Horse	C. Alvord	Westford, Wis	Aug. 28, 1866	57, 457
Collar, Horse	W. M. Baker	Greenwich Station, Ohio	Dec. 7, 1869	97, 587
Collar, Horse	E. Batwell	Ypsilanti, Mich	Apr. 15, 1873	137, 879
Collar, Horse	A. Beckwith	New Orleans, La	Dec. 7, 1869	97, 592
Collar, Horse	E. W. Briding and F. G. Maxwell.	Baltimore, Md	Nov. 27, 1860	30, 715
Collar, Horse	J. Bullock	Baltimore, Md	Jan. 10, 1860	26, 748
Collar, Horse	H. C. Call	Sterling, Conn	Nov. 14, 1835	....
Collar, Horse	S. G. Cheever and J. Forgie	Boston, Mass	Oct. 8, 1867	69, 627
Collar, Horse	J. Cogan	Cambridge, Mass	Feb. 9, 1869	86, 812
Collar, Horse	C. C. Cotton	Buchanan, Mich	July 9, 1872	128, 858
Collar, Horse	A. J. Cronk	Peoria, Ill	Aug. 6, 1867	67, 508
Collar, Horse	C. K. Cuckler	Columbus, Ohio	Oct. 19, 1858	21, 821
Collar, Horse	F. C. Curtis	Columbia, S. C	Aug. 4, 1845	4, 136
Collar, Horse	A. Ducastel	New York, N. Y	Mar. 1, 1870	100, 381
Collar, Horse	C. R. Durfee	Rochester, N. Y	Oct. 9, 1866	58, 620
Collar, Horse	C. Durrant	Lyndonville, N. Y	June 16, 1863	38, 889
Collar, Horse	S. B. Edson	Kokomo, Ind	Mar. 28, 1865	47, 000
Collar, Horse	E. P. Edstrom	Somerville, Mass	Apr. 30, 1867	64, 296
Collar, Horse	E. P. Edstrom	Somerville, Mass	Oct. 22, 1867	70, 079
Collar, Horse	J. Ende	Buffalo, N. Y	Dec. 18, 1860	30, 912
Collar, Horse	J. Englaender	Sedalia, Mo	June 7, 1870	104, 000
Collar, Horse	J. Englaender	Sedalia, Mo	July 19, 1870	105, 443
Collar, Horse	C. Fisher	Waukon, Iowa	Feb. 9, 1864	41, 497
Collar, Horse	C. J. Fisher	Waukon, Iowa	July 17, 1860	29, 154
Collar, Horse	C. J. Fisher	Waukon, Iowa	Sept. 5, 1865	49, 743
Collar, Horse	M. Gordon	Brighton, Iowa	Oct. 15, 1872	132, 206
Collar, Horse	W. H. Gray	New York, N. Y	Apr. 25, 1871	114, 131
Collar, Horse	W. Guilfoyle	New York, N. Y	Apr. 8, 1873	137, 608
Collar, Horse	L. Guinniss	Danville, Ill	Mar. 28, 1871	113, 161
Collar, Horse	I. H. Hall and J. Lowrey	Wheeling, Va	Sept. 21, 1852	9, 272
Collar, Horse	T. Harvey	Baltimore, Md	Apr. 12, 1859	23, 575
Collar, Horse	T. Harvey	Baltimore, Md	Apr. 12, 1859	23, 576
Collar, Horse	J. G. Haymaker	Salem Cross Roads, Pa	Oct. 1, 1867	69, 429
Collar, Horse	H. Holton	Rochester, N. Y	May 16, 1833	....
Collar, Horse	J. Hopkins	Warren County, Ohio	Jan. 23, 1836	....
Collar, Horse	G. Horter	New Orleans, La	Apr. 5, 1870	101, 619
Collar, Horse	P. B. Horton	San Francisco, Cal	Aug. 27, 1872	130, 919
Collar, Horse	I. Houghtling	Houghton, Mich	Apr. 7, 1868	76, 456
Collar, Horse	R. Humphrey	West Troy, N. Y	Apr. 5, 1870	101, 623
Collar, Horse	W. Kays	Boston, Mass	June 7, 1870	104, 035
Collar, Horse	L. B. Kenny	Charlotte, Mich	July 19, 1870	105, 580
Collar, Horse	O. Lafreniere	New York, N. Y	Jan. 11, 1859	22, 562
Collar, Horse	O. Lafreniere	Brooklyn, N. Y	July 8, 1873	140, 714
Collar, Horse	H. B. Latham	Huntington, N. Y	Apr. 6, 1852	8, 856
Collar, Horse	W. Leonard	Boston, Mass	Sept. 3, 1867	68, 369
Collar, Horse	D. Lincoln	Johnsonsburgh, N. Y	July 30, 1867	67, 201
Collar, Horse	J. R. Lindner	New York, N. Y	Sept. 6, 1853	9, 990
Collar, Horse	J. C. Mahaffey	Little York, Ill	Mar. 30, 1869	88, 312
Collar, Horse	C. K. Marshall	New Orleans, La	Jan. 12, 1869	85, 747
Collar, Horse	C. K. Marshall	New Orleans, La	July 26, 1870	105, 821
Collar, Horse	G. F. Marshall	Cleveland, Ohio	Nov. 15, 1864	45, 060
Collar, Horse	B. W. McClure	Wyoming, Iowa	Mar. 30, 1869	88, 400
Collar, Horse	J. Meyer	Williamsburgh, N. Y	July 20, 1869	92, 864
Collar, Horse	T. Moore	New York, N. Y	Jan. 14, 1868	73, 368
Collar, Horse	J. Nack	Hudson, N. Y	Aug. 27, 1872	130, 935
Collar, Horse	G. F. Parsons	Baltimore, Md	Nov. 29, 1864	45, 267
Collar, Horse	N. Post	Madrid, N. Y	July 22, 1843	3, 194
Collar, Horse	I. A. Powell	Morrison, Ill	Nov. 13, 1866	59, 647
Collar, Horse	R. Rickey	Rutland, Ohio	Nov. 11, 1851	8, 517
Collar, Horse	D. T. Robinson	Boston, Mass	May 7, 1867	64, 448
Collar, Horse	H. Sanders	Utica, N. Y	Apr. 5, 1870	101, 515
Collar, Horse	J. N. Schmitz	Kilbourn City, Wis	June 10, 1873	139, 822
Collar, Horse	J. W. Schwaner	Egg Harbor City, N. J	Apr. 18, 1871	113, 802
Collar, Horse	S. Shattuc	Henrietta, Ohio	Dec. 18, 1855	13, 965
Collar, Horse	T. J. Shipley and W. A. Moody	Montezuma, Iowa	Oct. 8, 1867	69, 712
Collar, Horse	A. Snively	Terre Haute, Ind	Feb. 20, 1872	123, 948
Collar, Horse	J. H. Sperbeck	Warsaw, N. Y	Mar. 17, 1868	75, 590
Collar, Horse	A. A. Stannard	Ithaca, N. Y	Oct. 17, 1865	50, 505
Collar, Horse	J. A. Sutherland	Elmwood, Ill	Sept. 22, 1868	82, 361
Collar, Horse	S. P. Taylor	Oxford, Ohio	Aug. 4, 1860	80, 682
Collar, Horse	W. J. Thorn	New York, N. Y	Feb. 28, 1871	112, 196
Collar, Horse	W. M. Thornton	Bloomsburgh, Pa	June 21, 1853	9, 806
Collar, Horse	J. C. Tredway	Buffalo, N. Y	Feb. 19, 1867	62, 301
Collar, Horse	J. H. Van Sice	Buffalo, N. Y	Nov. 10, 1863	40, 582
Collar, Horse	J. L. Van Wert	Tolland, Mass	Jan. 11, 1870	98, 722
Collar, Horse	E. Webber	Portage, Mich	Apr. 14, 1868	76, 858
Collar, Horse	L. Wegmann and C. F. Diessel	Allegheny City, Pa	Dec. 31, 1867	72, 947
Collar, Horse	G. Werner and R. Robinson	Canajoharie, N. Y	June 22, 1836	....
Collar, Horse	E. Whitney	Chenango, N. Y	June 4, 1834	....
Collar, Horse	E. Whitney	Albany, N. Y	Oct. 27, 1863	40, 440
Collar, Horse	J. Whitney	Franklinville, N. Y	Feb. 27, 1872	124, 184
Collar, Horse	C. Wolf	Rantoul, Ill	Nov. 27, 1866	60, 107
Collar, Horse	C. Wolf	Danville, Ill	May 5, 1868	77, 698
Collar, Horse	J. J. Wright	Richmond, Va	Oct. 1, 1872	131, 925
Collar, Horse	W. Youngblood	New York, N. Y	May 5, 1868	77, 701
Collar, Horse breast	W. T. Albro	Ellisburgh, N. Y	July 2, 1872	128, 520
Collar, Horse breast	R. D. Kendall	Richville, N. Y	July 9, 1872	128, 890
Collar, Horse breast	W. E. Leonard	Boston, Mass	Dec. 20, 1870	110, 377
Collar, Horse breast	J. Nellis	Ypsilanti, Mich	May 30, 1871	115, 346
Collar, Horse breast	J. G. Taylor	Philadelphia, Pa	Nov. 27, 1860	30, 775
Collar, Horse breast	G. Van Wagenen	Pittsburgh, Pa	Apr. 2, 1872	125, 353
Collar, Lady's	C. O. Crosby	New Haven, Conn	Jan. 5, 1864	41, 140
Collar, Lady's imitation	C. Lang	New York, N. Y	Aug. 14, 1866	57, 154
Collar lining, Horse	D. Curtis	Sun Prairie, Wis	Apr. 29, 1873	138, 383

Index of patents issued from the United States Patent Office from 1790 to 1873, inclusive—Continued.

Invention.	Inventor.	Residence.	Date.	No.
Collar-machine	H. F. Knapp	New York, N. Y	Apr. 28, 1868	77, 294
Collar machine	J. F. Walker	Albany, N. Y	Nov. 3, 1868	83, 814
Collar machine, Horse	D. Cleavland	Owego, N. Y	Sept. 13, 1834	......
Collar-making die	G. K. Snow	Watertown, Mass	Oct. 29, 1872	132, 546
Collar, Man's	D. Mackay	United States Army	Mar. 27, 1866	53, 462
Collar, Metallic	L. Billon	Brooklyn, N. Y	Sept. 5, 1865	49, 702
Collar, Metallic horse	C. K. Marshall	New Orleans, La	June 23, 1868	79, 137
Collar, Metallic shirt	O. Ernst	New York, N. Y	Aug. 30, 1864	43, 981
Collar, Molded cloth and paper	S. S. Gray	Boston, Mass	Jan. 19, 1864	41, 294
Collar-molding apparatus	S. S. Gray	Boston, Mass	Nov. 19, 1867	70, 996
Collar-molding machine	G. H. Spaulding	Norwich, Conn	May 26, 1868	78, 242
Collar-pad	P. H. Beaver	Montandon, Pa	Mar. 25, 1873	137, 049
Collar-pad	G. P. Cole	Hudson, Mich	July 18, 1871	117, 151
Collar, Pad and lining for horse	S. W. Baker	Providence, R. I	Mar. 9, 1869	87, 619
Collar pad, Breast	R. E. Miles	Louisville, Ky	Feb. 16, 1869	86, 995
Collar pad, Horse	J. Frazer	Dowagiac, Mich	Nov. 8, 1870	108, 995
Collar pad, Horse	J. Frazer	Dowagiac, Mich	Mar. 21, 1871	112, 797
Collar pad, Horse	L. L. Hull	Oskaloosa, Iowa	Mar. 14, 1871	112, 713
Collar pad, Horse	J. S. Huston	Mechanicsburgh, Pa	Apr. 26, 1870	102, 403
Collar pad, Horse	T. Newbold	Salisbury, Mo	Nov. 5, 1872	132, 769
Collar pad, Horse	J. F. Welsh	Hazel Green, Wis	Dec. 27, 1870	110, 518
Collar-pad press, Horse	J. Frazer	Dowagiac, Mich	Mar. 29, 1870	101, 252
Collar-pads, Crimp for	J. S. Barkdull	Ballston, N. Y	July 13, 1844	3, 661
Collar, Reversible	C. W. Saladee	Circleville, Ohio	Jan. 26, 1869	86, 338
Collar, Rudder	S. Leach	Penobscot, Me	Mar. 1, 1870	100, 419
Collar-shell-forming die, Horse	J. W. Schwaner	Egg Harbor City, N. J	Jan. 3, 1871	110, 685
Collar, Shirt	S. A. Bemis	Springfield, Mass	Feb. 13, 1866	52, 516
Collar, Shirt	O. P. Dorman	New York, N. Y	Apr. 8, 1873	137, 537
Collar, Shirt	A. A. Evans	Boston, Mass	July 18, 1865	48, 802
Collar, Shirt	S. S. Gray	Boston, Mass	June 23, 1863	38, 961
Collar, Shirt	W. Hunt	New York, N. Y	July 25, 1854	11, 376
Collar, Shirt	W. Hunt	New York, N. Y	Jan. 1, 1856	14, 019
Collar, Shirt	J. A. Pease	New York, N. Y	June 16, 1863	38, 909
Collar, Shirt	V. N. Taylor	Springfield, Mass	Mar. 26, 1872	124, 920
Collar, Sleeping	L. Dederick	New York, N. Y	May 17, 1870	103, 026
Collar-stay	S. Kaufman	Fairbury, Ill	Nov. 10, 1868	83, 973
Collar, Steel shirt	L. Billon	Brooklyn, N. Y	Apr. 19 1864	42, 429
Collar, Steel shirt	R. Woodward, J. Priest, and O. Ernst.	New York, N. Y	Aug. 30, 1864	44, 051
Collar stuffer, Horse	E. B. Miller	Greeneville, Tenn	Mar. 17, 1868	75, 641
Collar-stuffing machine, Horse	S. B. McCorkle	Greeneville, Tenn	July 14, 1868	79, 997
Collar-stuffing machine, Horse	L. P. Woods	Indiana, Pa	Jan. 30, 1872	123, 139
Collar top, Horse	I. Hicks	Hartford, Wis	Oct. 4, 1870	107, 905
Collar, Turn-over shirt	N. Evans, jr	Boston, Mass	Feb. 23, 1864	41, 692
Collar, Water-proof	H. C. Gibson	Camden, N. J	Apr. 10, 1866	53, 808
Collar, Wooden horse	Z. C. Robbins	Washington, D. C	Aug. 29, 1871	118, 555
Collars and neck-ties, Stand for displaying	S. S. Gray and C. E. Staniels	Boston, Mass	May 7, 1872	126, 541
Collars, Apparatus for forming	S. S. Gray	Boston, Mass	July 5, 1864	43, 401
Collars, Apparatus for stuffing horse	W. C. Habberton	Mount Carmel, Ill	Jan. 3, 1860	26, 671
Collars, Attachment for breast	R. E. Miles	Louisville, Ky	Sept. 10, 1867	68, 774
Collars, Block for horse	L. S. Davis	New Paris, Ohio	Feb. 7, 1854	10, 502
Collars, Blocking hame	N. Post	Norfolk, N. Y	July 24, 1834	......
Collars, Clasp for securing shirt	J. Barbier	Boston, Mass	May 28, 1867	65, 155
Collars, Compound fabric for the production of shirt.	W. Hunt	New York, N. Y	May 4, 1869	89, 768
Collars, cuffs, and other articles of wearing-apparel, Coating and water-proofing.	S. W. H. Ward	New York, N. Y	Nov. 10, 1868	83, 893
Collars, cuffs, bosoms, and other articles of wearing-apparel, Fabric for.	W. E. Lockwood	Philadelphia, Pa	July 20, 1869	92, 851
Collars, cuffs, &c, Fabric for the manufacture of	J. Restein	Philadelphia, Pa	Aug. 24, 1869	94, 030
Collars, cuffs, &c., Fabric for the manufacture of enameled.	H. C. Gibson	Camden, N. J	July 12, 1864	43, 489
Collars, cuffs, &c., Manufacture of	A. Taylor	Philadelphia, Pa	Oct. 17, 1865	50, 513
Collars, Cutting top of horse	T. Deming	East Hartford, Conn	Mar. 6, 1835	......
Collars, Device for fastening shirt	P. W. Smith	Chicopee Falls, Mass	Oct. 20, 1868	83, 332
Collars, Device for forming horse	I. T. Crum	Chicago, Ill	Feb. 12, 1867	62, 014
Collars, Expanding-block for horse	R. R. Gray	Crawfordsville, Ind	June 19, 1855	13, 087
Collars, Fabric for covering horse	E. Sullivan	New York, N. Y	June 16, 1868	79, 024
Collars, Fabric for shirt	C. F. Pidgin	Boston, Mass	Nov. 16, 1869	96, 954
Collars, Fastening and strengthening horse	E. Harbaugh	Washington, Iowa	Dec. 29, 1868	85, 382
Collars from sheets of paper, &c., Method of cutting	G. K. Snow	Watertown, Mass	Oct. 29, 1872	132, 547
Collars, Lining for horse	D. Curtis	Sun Prairie, Wis	May 14, 1872	126, 682
Collars, Locking-device for dog	A. R. Scott	Albany, N. Y	June 6, 1871	115, 647
Collars, Machine for applying re-enforcing patches to button-holes of.	H. F. Cary	Boston, Mass	Nov. 10, 1868	83, 927
Collars, Machine for applying strengthening-patch to button holes of.	E. F. Bradley	Derby, Conn	Jan. 19, 1869	85, 899
Collars, Machine for cutting out	G. K. Snow	Watertown, Mass	Oct. 29, 1872	132, 544
Collars, Machine for filling horse	A. Schrick and H. Hildenbraud.	Saint Louis, Mo	Nov. 13, 1866	59, 726
Collars, Machine for filling horse	A. Schrick and H. Hildenbrand.	Saint Louis, Mo	Dec. 17, 1867	72, 423
Collars, Machine for forming horse	F. Cunningham	Chicago, Ill	Feb. 27, 1866	52, 826
Collars, Machine for forming horse	I. Davis	Mechanicsburgh, Ohio	Nov. 4, 1851	8, 482
Collars, Machine for making horse	C. Angevine	New York, N. Y	Oct. 27, 1835	......
Collars, Machine for measuring horses for	W. H. Flynn	Somerville, Mass	Feb. 8, 1870	99, 660
Collars, Machine for pressing and stuffing horse	W. Guilfoyle	New York, N. Y	June 6, 1871	115, 727
Collars, Machine for stretching horse	H. Barton	West Carlisle, Ohio	July 9, 1838	827
Collars, Machine for stretching horse	A. Schrick and H. Hildenbrand.	Saint Louis, Mo	Nov. 13, 1866	59, 727
Collars, Machine for stretching or adjusting size of horse.	J. P. Osborn	Reddington, N. J	Jan. 30, 1841	1, 961
Collars, Machine for stuffing horse	J. Albright	Greeneville, Tenn	Nov. 25, 1856	16, 107
Collars, Machine for stuffing horse	W. Fauntleroy	New Harmony, Ind	Sept. 15, 1868	82, 100
Collars, Machine for stuffing horse	W. Haworth	New York, N. Y	June 26, 1847	5, 177
Collars, Machine for stuffing horse	W. H. Haworth	Philadelphia, Pa	Mar. 17, 1857	16, 834
Collars, Machine for stuffing horse	J. W. Howell	New Paris, Ohio	Feb. 14, 1854	10, 534
Collars, Machine for stuffing horse	S. B. McCorkle	Greeneville, Tenn	Dec. 18, 1855	13, 949

Index of patents issued from the United States Patent Office from 1790 *to* 1873, *inclusive*—Continued.

Invention.	Inventor.	Residence.	Date.	No.
Collars, Machine for stuffing horse	L. Plouk	Newton, N. C	July 6, 1858	20, 816
Collars, Machine for stuffing horse	H. G. Robertson	Greeneville, Tenn	June 10, 1856	15, 107
Collars, Machine for stuffing horse	J. C. Tobias	Lincoln, Ill	July 2, 1857	17, 902
Collars, Machine for stuffing horse	W. L. Whittaker	Cumberland, Md	Mar. 6, 1865	12, 494
Collars, Machine for stuffing horse	T. Wiles	Somerset, Ohio	Mar. 21, 1845	3, 964
Collars, M chine to manufacture horse	W. Criswell	Butler, Pa	Oct. 16, 1849	6, 804
Collars, Machinery for making shirt	O. W. Edson	Troy, N. Y	Feb. 26, 1856	14, 308
Collars, Making horse	T. W. Murphey	New Egypt, N. J	Dec. 5, 1865	51, 341
Collars, Manufacture of japanned or enameled horse.	G. and H. Duxon and H. Perry	Brooklyn, N. Y, and Newark, N. J.	Dec. 12, 1871	121, 859
Collars, Mode of cutting leather for horse	T. Parkinson	Sparta, N. Y	July 17, 1841	2, 183
Collars, Mold-press for horse	M. C. Chamberlin	Sheldon, N. Y	May 6, 1856	14, 803
Collars, Spring for horse	B. J. Barton and R. J. Stanley	Washington, Iowa	Dec. 7, 1869	97, 589
Collars, Stuffing and stretching horse	W. Haworth	Dayton, Ohio	May 7, 1845	4, 037
Collars, Stuffing horse	G. W. Hobart	Silverton, Oreg	May 18, 1869	90, 170
Collars, Sweat-shield for horse	C. B. Hogg	Boston, Mass	Feb. 23, 1869	87, 257
Collarette	C. O. Crosby	New Haven, Conn	Mar. 15, 1864	41, 907
Collection-box, Portable	T. J. Homer	Saint Louis, Mo	Dec. 6, 1859	26, 356
Collet	S. A. Morse	East Bridgewater, Mass	Mar. 3, 1864	42, 592
Collice	J. Flock	Newark, N. J	May 10, 1864	42, 652
Collimator	G. N. Fairchild and T. Joyce	New York, N. Y	Dec. 18, 1860	30, 914
Collision-brake	C. B. Guy	Lybrand, Iowa	Mar. 7, 1865	46, 662
Collodion and its compounds, Machine for treating	J. A. McClelland	Louisville, Ky	June 1, 1869	90, 766
Collodion and its compounds, Mode of producing useful articles from.	J. A. McClelland	Louisville, Ky	Oct. 26, 1869	96, 132
Collodion and its compounds, Process of coating objects with.	J. A. McClelland	Louisville, Ky	July 26, 1870	105, 823
Collodion compound	J. A. McClelland	Louisville, Ky	Oct. 21, 1873	143, 772
Collodion, Method of making solid	J. W. and I. S. Hyatt	Albany, N. Y., and Rockford, Ill.	June 15, 1869	91, 341
Collodion, Preparing soluble cotton for the manufacture of.	H. T. Anthony	New York, N. Y	Oct. 21, 1873	143, 865
Collodion, Solidified	C. A. Seely	New York, N. Y	June 23, 1868	79, 261
Colors and dyes, Process of treating asphaltum to obtain.	J. Brönner and H. Gutzkow	Frankfort-on-the-Main, Prussia.	Dec. 7, 1869	97, 597
Colors and lakes, Apparatus for preparing	J. Lucas	Philadelphia, Pa	Oct. 25, 1870	108, 713
Colors and paints from ores, &c., Preparing	H. Alexander	Baltimore, Md	Jan. 14, 1813	....
Colors and painting, Mixing	L. Allwine		Feb. 24, 1797	....
Colors and pigments, Manufacture of	E. Harrsch	New York, N. Y	Mar. 30, 1869	88, 291
Colors and their application to fabrics, Manufacture of.	A. Paraf	New York, N. Y	Dec. 20, 1870	110, 277
Colors, Compressing and imprinting	L. J. B. Wells	Philadelphia, Pa	Jan. 24, 1817	....
Colors for calico, Making permanent	A. Boulu	Philadelphia, Pa	Dec. 24, 1814	....
Colors for painting, printing, &c., Making	F. Guy		Dec. 30, 1803	....
Colors for printing and dyeing, Manufacture and application of.	A. Paraf	New York, N. Y	Nov. 15, 1870	109, 341
Colors from analine, Method of preparing	X. Karcheski	Belleville, N. J	Mar. 14, 1865	46, 804
Colors in calico-printing, Material for transferring	C. A. Broquette	France	Apr. 15, 1851	8, 035
Colors, Making yellow, nankeen, or buff	J. B. Nones	Philadelphia, Pa	Apr. 28, 1825	....
Colors, Manufacture of	E. Harrsch	New York, N. Y	May 25, 1869	90, 359
Colors, Manufacture of	A. Leykauf	Nürnberg, Bavaria	Feb. 23, 1869	87, 270
Colors, Mode of applying distemper	F. G. Spilsbury, F. Corbaux, and A. S. Byrne.	England	July 10, 1840	1, 676
Colors, Mode of forming yellow	T. Bedwell		Apr. 20, 1796	....
Colors, Press for printing different	S. Brown	Syracuse, N. Y	Jan. 2, 1855	12, 178
Colors, printers' ink, &c., Grinding painters'	J. Cist		Dec. 3, 1803	....
Colors to cloth, &c., Composition for restoring	J. Warncke	Buffalo, N. Y	Mar. 3, 1863	37, 838
Colors with lime, Preparing	J. Nichols, jr	Broadalbin, N. Y	Apr. 12, 1815	....
Colored woolens, &c., Mode of producing mixed	S. Vigoureux	Rheims, France	Mar. 8, 1864	41, 878
Coloring compound or Bremen blue	C. H. Petsch	Morrisania, N. Y	Sept. 10, 1872	131, 297
Coloring fabrics, Re	J. M. Wallace	New York, N. Y	Apr. 25, 1871	114, 069
Coloring-machine, Parti	S. Smith	Acton, Mass	Feb. 21, 1854	10, 560
Coloring material for dyeing and printing	C. Lauth	Paris, France	Oct. 26, 1869	96, 242
Coloring matter for dyers, Producing	J. Eberhardt	Philadelphia, Pa	Mar. 21, 1865	46, 888
Coloring matter or paint, Preparing green	Baron A. De Carrendeffez	New York, N. Y	Oct. 12, 1807	....
Coloring matter, Vegetable	C. Seidel	New York, N. Y	Mar. 31, 1868	76, 107
Coloring matters from anthracine, Manufacture of	W. H. Perkin	Sudbury, England	June 4, 1872	127, 426
Coloring matters from vegetable substances, Producing.	L. M. F. Bretonniére	Laval, France	Sept. 16, 1873	142, 892
Coloring matters, Manufacturing	J. Holliday	Huddersfield, Great Britain	Dec. 5, 1865	51, 404
Coloring matters, Manufacturing	H. Stephens	Great Britain	Oct. 28, 1837	441
Coloring muslin, paper, &c., Process for	F. Beck	New York, N. Y	Nov. 9, 1869	96, 661
Coloring paper and other fabrics	F. Beck	New York, N. Y	Oct. 4, 1870	107, 997
Coloring the surface of metals, Process for	M. Edwards	Cambridge, Mass	Jan. 31, 1860	26, 978
Colortrope	O. Nicholson	New York, N. Y	Oct. 11, 1870	108, 170
Colter	R. Emerson	Rockford, Ill	Mar. 11, 1873	136, 647
Colter	H. M. Skinner	Rockford, Ill	Sept. 13, 1870	107, 298
Colter and gage-wheel combined, Revolving extension.	W. H. Willard	Cleveland, Ohio	Dec. 2, 1862	37, 065
Colter-clasp	C. H. Thomas	Cassville, N. Y	Aug. 12, 1873	141, 675
Colter-cleaner	A. B. Mattoon	Auburn, N. Y	June 15, 1869	91, 464
Colter-holder	J. Aughe	Dayton, Ohio	Mar. 10, 1868	75, 237
Colter, Revolving	R. L. Pitcher and R. Ellwood	Sycamore, Ill	Jan. 19, 1869	85, 957
Colter, Revolving	C. E. Steller	Chicago, Ill	Feb. 23, 1869	87, 220
Colter, Rotary	F. J. Underwood	Rock Island, Ill	Oct. 19, 1869	96, 057
Column, Building	J. L. Mott	New York, N. Y	June 11, 1836	....
Column, Fire-proof	W. A. Berkey	Grand Rapids, Mich.	Jan. 16, 1872	122, 797
Column, Fire-proof	T. Hyatt	New York, N. Y	Dec. 2, 1873	145, 180
Column, Iron	W. A. Gunn	Lexington, Ky	Jan 2, 1872	122, 379
Column, Metallic	C. Bender	Phœnixville, Pa	Nov. 28, 1871	121, 318
Column, Metallic	S. Brandeis	Philadelphia, Pa	Aug. 26, 1873	142, 201
Column, Metallic	J. B. Cornell	New York, N. Y	Feb. 11, 1868	74, 312
Column, Metallic	J. A. Kay	Saint Charles, Mo	Aug. 15, 1871	118, 021
Column, Metallic	R. Montgomery	New York, N. Y	Jan. 19, 1869	86, 028
Column, Metallic	J. L. Piper	Pittsburgh, Pa	May 13, 1873	138, 817
Column, Metallic	J. L. Piper	Pittsburgh, Pa	Aug. 12, 1873	141, 665
Column, Metallic	F. H. Smith	Baltimore, Md	Nov. 28, 1871	121, 432
Column, Metallic	F. H. Smith	Baltimore, Md	May 14, 1872	126, 841

Index of patents issued from the United States Patent Office from 1790 *to* 1873, *inclusive*—Continued.

Invention.	Inventor.	Residence.	Date.	No.
Column, Metallic	G. Walters and T. Shaffer	Phœnixville, Pa	Feb. 16, 1869	87, 016
Column or tube	F. H. Smith	Baltimore, Md	Nov. 2, 1869	96, 357
Column, Rolled iron or steel	F. H. Smith	Baltimore, Md	May 24, 1870	103, 380
Column, Wrought-iron	A. Bonzano	Phœnixville, Pa	May 21, 1872	127, 019
Column, Wrought-iron	T. C. Clarke	Philadelphia, Pa	Sept. 24, 1872	131, 502
Column, Wrought-iron	L. Herman	Detroit, Mich	Jan. 21, 1873	134, 987
Column, Wrought-iron	J. P. Kennedy	New York, N. Y	Sept. 12, 1871	118, 948
Column, Wrought-iron	J. W. Murphy	Philadelphia, Pa	Dec. 13, 1870	110, 154
Column, Wrought-iron	J. H. Linville	Philadelphia, Pa	Oct. 22, 1872	132, 475
Column, Wrought-iron	C. S. Smith	Baltimore, Md	Nov. 19, 1872	133, 263
Column, Wrought-iron	G. Walters and T. Shaffer	Phœnixville, Pa	Oct. 20, 1868	83, 226
Column, Wrought iron and steel	C. H. Parker	Boston, Mass	May 9, 1871	114, 705
Columns, Connection for compound tubular metallic.	J. A. Kay	Saint Charles, Mo	Aug. 20, 1872	130, 721
Columns, Flask for casting iron	H. Demmick	New York, N. Y	Aug. 21, 1860	29, 745
Columns from corrosion, Art of protecting metallic	J. H. Linville	Philadelphia, Pa	July 16, 1872	129, 241
Columns, Machine for facing ends of	D. Kennedy	Detroit, Mich	July 16, 1872	129, 568
Columns, &c., Machine for ornamenting	C. Durand	Springfield, N. J	Sept. 22, 1818	....
Columns, Manufacture of wrought-iron	D. Halstead	Buffalo, N. Y	June 8, 1869	91, 125
Columns, Method of constructing	G. Walters and T. Shaffer	Phœnixville, Pa	Sept. 29, 1868	82, 663
Columns, &c., Method of constructing	G. Walters and T. Shaffer	Phœnixville, Pa	Sept. 29, 1868	82, 664
Columns, Method of constructing wrought-iron	G. Walters and T. Shaffer	Phœnixville, Pa	Oct. 27, 1868	83, 425
Columns, Method of securing together the sides of cast-metal.	A. J. Bowers	Richmond, Va	Jan. 11, 1859	22, 542
Columns, shafts, braces, &c., Construction of	S. J. Reeves	Philadelphia, Pa	June 17, 1862	35, 582
Comb:				
See Band-comb.				
Carding-comb.				
Curry-comb.				
Dressing-comb.				
Enameled-metal comb.				
Fine-tooth comb.				
Fountain-comb.				
Hair-comb.				
Head-comb.				
Horn comb.				
Horn and tortoise-shell comb.				
Ivory comb.				
Lady's comb.				
Long comb.				
Metal comb.				
Metal-backed comb.				
Pocket-comb.				
Round comb.				
Shell-comb.				
Shell and horn comb.				
Side-comb.				
Toilet-comb.				
Ventilating-comb.				
Comb	J. H. Briggs	Brooklyn, N. Y	Oct. 15, 1867	69, 755
Comb	E. Brown	Wappinger's Falls, N. Y	Jan. 30, 1866	52, 262
Comb	H. Brown and S. N. Noyes	West Newbury, Mass	June 1, 1869	90, 816
Comb	T. L. Calkins	Hartford, Conn	Mar. 17, 1857	16, 859
Comb	F. A. L. Cassidey	Newnansville, Fla	Mar. 5, 1867	62, 606
Comb	G. F. J. Colburn	Newark, N. J	Jan. 17, 1865	45, 909
Comb	G. F. J. Colburn	Newark, N. J	Jan. 31, 1865	46, 082
Comb	J. S. Dickinson	Essex, Conn	Apr. 7, 1868	76, 414
Comb	J. Emerson	Lowell, Mass	Dec. 18, 1866	60, 494
Comb	C. Foster	Wappinger's Falls, N. Y	Jan. 28, 1868	73, 884
Comb	J. H. Hicks	Brooklyn, N. Y	Nov. 27, 1866	60, 004
Comb	O. Johnson	Grand Lodge, Mich	Jan. 30, 1872	123, 177
Comb	E. H. Knight	Washington, D. C	Dec. 4, 1866	60, 197
Comb	C. H. Noyes	Brooklyn, N. Y	Apr. 4, 1871	113, 330
Comb	E. M. Noyes	Newark, N. J	Apr. 24, 1860	28, 008
Comb	E. M. Noyes	Newark, N. J	June 7, 1864	43, 038
Comb	J. P. Noyes	Newark, N. J	June 5, 1866	55, 349
Comb	H. M. Paine	Newark, N. J	Apr. 6, 1869	88, 582
Comb	W. Pauly	College Point, N. Y	Dec. 17, 1867	72, 324
Comb	L. Picot	Hoboken, N. J	Nov. 13, 1866	59, 645
Comb	L. Picot	Hoboken, N. J	Aug. 11, 1868	80, 833
Comb	L. Picot	Hudson City, N. J	Apr. 4, 1871	113, 340
Comb	J. C. Reed	New York, N. Y	June 14, 1870	104, 202
Comb	I. Rice	New York, N. Y	Nov. 20, 1866	59, 926
Comb	C. L. Robertson	Providence, R. I	Apr. 13, 1869	88, 811
Comb	T. Schreiber	Wheeling, W. Va	Nov. 13, 1866	59, 664
Comb	J. Smith	Leominster, Mass	Sept. 29, 1868	82, 560
Comb	I. H. Southworth	Essex, Conn	Oct. 10, 1865	50, 400
Comb	E. Sperry	New Haven, Conn	Oct. 10, 1829	....
Comb and brush	T. Lanston	Washington, D. C	May 30, 1871	115, 488
Comb-attachment	G. Stackhouse	Mount Washington, Pa	June 29, 1869	91, 984
Comb-blanks, Device for softening	W. and H. W. Noyes	Newburyport, Mass	Apr. 30, 1872	126, 154
Comb-blanks, Machine for sizing	W. Fosket and B. S. Stedman	Meriden, Conn	Aug. 26, 1856	15, 634
Comb-cleaner	C. E. Gibbs	Boston, Mass	May 10, 1870	102, 803
Comb-cleaner	A. Sahlstrom	Chicago, Ill	June 27, 1871	116, 500
Comb-cleaner	E. J. Toof	Fort Madison, Iowa	Apr. 28, 1868	77, 416
Comb-cleaner	C. P. S. Wardwell	Lake Village, N. H	Dec. 18, 1860	30, 949
Comb-cutting machine	H. S. Cook	Leominster, Mass	June 3, 1851	8, 129
Comb-cutting machine	S. Curtis	Newtown, Conn	Nov. 18, 1851	8, 522
Comb-cutting machine	N. Toot	Philadelphia, Pa	Oct. 28, 1819	....
Comb-cutting machine	B. Haskell	Paris, France	Nov. 3, 1819	....
Comb-cutting machine	T. W. Hill	Leominster, Mass	June 10, 1851	8, 144
Comb-cutting machine	D. E. Noyes	Philadelphia, Pa	Oct. 14, 1819	....
Comb-dressing machinery	C. B. Rogers	Saybrook, Conn	Dec. 20, 1845	4, 321
Comb-dressing machinery	E. Wilson	Redding, Conn	Mar. 14, 1848	5, 472
Comb for combing wool, flax, cotton, &c	J. B. Siccardi and J. Hide	New York, N. Y	Feb. 14, 1865	46, 397
Comb-frame	W. Rasey	Remington, Ind	Sept. 30, 1873	143, 383
Comb-holder	E. E. Wheeler	South Norwalk, Conn	Nov. 4, 1873	144, 242
Comb machine, Twining	D. Carrington and I. Ives	Litchfield, Conn	Aug. 20, 1813	....

Index of patents issued from the United States Patent Office from 1790 *to* 1873, *inclusive*—Continued.

Invention.	Inventor.	Residence.	Date.	No.
Comb-making machine	J. Pitts, C. Houghton, and J. Rice, jr.	Lancaster, Mass	Oct. 1, 1830	
Comb, Metal	A. Porter	Hartford, Conn	June 14, 1834	
Comb or hair-pin, Ventilating	E. Clark	Louisville, Ky	Sept. 3, 1872	130, 978
Comb-presses, Fluting and bending	R. Munson	New York, N. Y	May 22, 1832	
Comb-sawing machine	W. Booth	College Point, N. Y	Mar. 5, 1872	124, 243
Comb-sawing machine	G. F. H. Brown	Leominster, Mass	Aug. 11, 1868	80, 855
Comb-teeth, Cutting	L. Adams	Redding, Conn	May 2, 1835	
Comb-teeth, Cutting	J. Brown	Springfield, Mass	Dec. 8, 1824	
Comb-teeth, Cutting	M. J. Littleboy	Philadelphia, Pa	May 28, 1816	
Comb-teeth, Cutting and pointing	I. Tryon		Feb. 22, 1798	
Comb-teeth, Die for cutting metallic	C. Foster	Wappinger's Falls, N. Y	June 2, 1868	78, 585
Comb-teeth, Machine for sawing	P. Pratt	New York, N. Y	Apr. 20, 1825	
Comb-teeth, Machinery for cutting	W. Noyes, jr	West Newbury, Mass	June 7, 1859	24, 358
Combs, Cutting	J. H. Derby	Leominster, Mass	Mar. 26, 1814	
Combs, Cutting horn and tortoise-shell for	S. Allen		Apr. 1, 1815	
Combs, Machine for cutting all kinds of	T. G. Thomas	Philadelphia, Pa	June 14, 1817	
Combs, Machine for cutting the teeth of	J. S. Ives	Bristol, Conn	Oct. 8, 1840	1, 811
Combs, Machine for making	P. Pratt		Apr. 12, 1799	
Combs, Machine for making	P. Pratt	Saybrook, Conn	Mar. 5, 1808	
Combs, Machine for manufacturing	D. Pettibone	Philadelphia, Pa	Aug. 11, 1818	
Combs, Machine for slitting tortoise-shell for making.	W. Redheffer	Penn Township, Pa	June 18, 1842	2, 679
Combs, Manufacturing	J. Brown	Providence, R. I	June 11, 1829	
Combs, Manufacturing	A. B. Newton	Baltimore, Md	Mar. 31, 1834	
Combs, Manufacturing	B. and W. Redheffer	Penn Township, Pa	May 26, 1834	
Combs, Manufacturing	E. Sperry	New Haven, Conn	Mar. 27, 1816	
Combs, Mode of forming and shaping	S. Lambert	Hanover County, Va	Jan. 9, 1829	
Combs, Ornamenting and making	E. Mustin	Philadelphia, Pa	June 27, 1829	
Combs, Press for molding	W. and H. W. Noyes	Newburyport, Mass	Sept. 24, 1872	131, 626
Combs, &c., Pressing and rendering transparent horse-hoofs, &c., for making.	E. M. Canviss	Southington, Conn	May 6, 1812	
Combs, Rolling backs of	N. Bishop	Danbury, Conn	Nov. 17, 1827	
Combs, Rubbing down and polishing	L. B. Prindle and D. Curtis	Newtown, Conn	Oct. 10, 1829	
Combs, Steel plate for pressing horn for	U. Bailey	West Newbury, Mass	Feb. 2, 1828	
Combs from cattle-hoofs, Mode of making	R. Gedney	New York	June 26, 1809	
Comber-boards, Mechanism for operating	J. S. Templeton	Glasgow, Great Britain	Oct. 4, 1870	107, 979
Combing-machine	I., B., and H. Smith and C. Bradley.	Bradford, England	Apr. 1, 1873	137, 326
Combing-machine	E. Tavernier	Lille, France	Jan. 9, 1872	122, 678
Combing-machine comb	C. Weiler	Landenburgh, Pa	Apr. 18, 1871	113, 825
Combing-machines, Mechanism for operating the nippers of.	H. Daniels	Pawtucket, R. I	Aug. 17, 1869	93, 811
Combination and register padlock	H. S. Shepardson	Shelburne Falls, Mass	Sept. 13, 1870	107, 414
Combination, Bench	W. Weaver	Nashua, N. H	Nov. 5, 1867	70, 657
Combination-lock	M. Adams	Chilmark, Mass	Nov. 2, 1869	96, 376
Combination-lock	F. W. Alexander	Baltimore, Md	Apr. 30, 1861	32, 171
Combination-lock	S. Bentley and C. Mee	Detroit, Mich	Mar. 15, 1870	100, 714
Combination-lock	H. Clarke	Baltimore, Md	Feb. 4, 1873	135, 523
Combination-lock	S. Colton	Philadelphia, Pa	Jan. 6, 1844	3, 395
Combination-lock	J. W. H. Doubler	Darlington, Wis	May 23, 1871	115, 037
Combination-lock	H. Gross	Tiffin, Ohio	May 11, 1869	89, 969
Combination-lock	P. W. Hall	Calvert, Tex	July 1, 1873	140, 500
Combination-lock	W. N. Hall	Springfield, Tex	Sept. 12, 1871	118, 930
Combination-lock	A. D. Hoffman	Chicago, Ill	Aug. 19, 1873	141, 876
Combination-lock	F. B. Kalkbrenner	Clinton, Mo	Oct. 18, 1870	108, 360
Combination-lock	W. A. Kerr	Williamsport, Pa	July 16, 1872	129, 482
Combination-lock	I. W. Lamb	Salem, Mich	July 20, 1869	92, 844
Combination-lock	H. S. Leland	Mount Union, Ohio	Mar. 30, 1869	88, 490
Combination-lock	S. Loyd	New York, N. Y	Apr. 4, 1871	113, 316
Combination-lock	F. P. Marsden	Galena, Ill	Dec. 26, 1871	122, 124
Combination-lock	S. Miller	Gratis, Ohio	Aug. 8, 1871	117, 912
Combination-lock	J. Moffet	New York, N. Y	Apr. 9, 1872	125, 475
Combination-lock	J. H. Morse	Peoria, Ill	Dec. 4, 1860	30, 830
Combination-lock	G. H. Peacock	Webster, N. Y	Oct. 17, 1871	119, 948
Combination-lock	G. H. Peacock	Webster, N. Y	June 11, 1872	127, 706
Combination-lock	S. Perry	Newport, N. Y	June 22, 1858	20, 658
Combination-lock	G. M. Phelps	Williamsburgh, N. Y	Dec. 23, 1862	37, 241
Combination-lock	J. Pigot	Brooklyn, N. Y	July 16, 1872	129, 052
Combination-lock	O. E. Pillard	New Britain, Conn	June 1, 1869	90, 682
Combination-lock	N. Reed	Otisville, N. Y	Aug. 71, 1869	93, 747
Combination-lock	S. K. Seelye	Hudson, Mich	Nov. 22, 1870	109, 457
Combination-lock	J. T. Taylor	Newnan, Ga	Sept. 17, 1872	131, 475
Combination-lock	A. B. Vandemark	Phelps, N. Y	Dec. 1, 1868	84, 659
Combination-lock	E. Vorbe	San Francisco, Cal	May 4, 1869	89, 817
Combination-lock	J. B. White	Detroit, Mich	Dec. 1, 1868	84, 664
Combination-lock	J. B. White	Detroit, Mich	Aug. 27, 1872	130, 964
Combination-lock	J. P. White	Savannah, Ga	Sept. 21, 1869	95, 058
Combination lock and key	S. L. Chase	Lockport, N. Y	Nov. 6, 1847	5, 359
Combination-lock for doors	S. Andrews	Perth Amboy, N. J	Sept. 30, 1841	2, 276
Combination-lock for doors	W. C. Bussey	San Francisco, Cal	Apr. 7, 1868	76, 300
Combination-lock for doors, safes, &c	R. Newell	New York, N. Y	Sept. 17, 1844	3, 747
Combination-padlock	M. P. Thatcher	Pontiac, Mich	Dec. 7, 1869	97, 568
Combination-tool	C. A. Foster	Fitchburgh, Mass	Apr. 19, 1870	102, 105
Combined supporter	S. Dike	New York, N. Y	Aug. 24, 1869	94, 087
Combining or mixing two or more substances, Apparatus for.	G. W. Wilson	Chelsea, Mass	June 6, 1871	115, 799
Combustion-power	A. Day	Bordentown, N. J	Jan. 19, 1811	
Combustion, System of supporting	J. A. Bassett and E. L. Norfolk	Salem, Mass	Mar. 7, 1865	46, 624
Comminuting various substances, Apparatus for	G. B. Canning	Brooklyn, N. Y	Mar. 11, 1873	136, 700
Commode	J. D. Averell	New York, N. Y	Oct. 7, 1873	143, 488
Commode	W. Prangley	Salisbury, England	Nov. 5, 1872	132, 775
Commode and wash-stand combined	S. P. Boone	Americus, Ga	Aug. 29, 1871	118, 510
Commode, Chamber	T. Elkins	Albany, N. Y	Jan. 9, 1872	122, 518
Commode, Chamber	E. S. Farson	Philadelphia, Pa	Sept. 15, 1868	82, 099
Commode, Disinfecting	H. J. Alvord	Detroit, Mich	July 31, 1866	56, 870
Commode, Earth	G. F. Stone	Baltimore, Md	Apr. 9, 1872	125, 627
Commode, Earth	E. Woodruff	Elizabeth, N. J	Nov. 1, 1870	108, 950
Commode, Folding	R. G. Elder	New York, N. Y	Nov. 28, 1871	121, 344

Index of patents issued from the United States Patent Office from 1790 to 1873, inclusive—Continued.

Invention.	Inventor.	Residence.	Date.	No.
Commutator, Electric	L. B. Firman	Chicago, Ill	Mar. 24, 1868	75, 743
Compass	P. Roessler	New Haven, Conn	Sept. 19, 1865	50, 037
Compass	H. Stewart	Clinton, Mass	Apr. 21, 1868	77, 123
Compass	O. Stoddard	Detroit, Mich	Aug. 27, 1872	130, 879
Compass and magnetic needle	J. Hauks	Troy, N. Y	July 22, 1833	....
Compass, Azimuth	E. S. Ritchie	Brookline, Mass	Aug. 1, 1865	49, 157
Compass, Beam	J. Lyman	Lenox, Mass	Feb. 21, 1871	111, 954
Compass, Beam	M. Toulmin	New Orleans, La	Sept. 16, 1873	142, 823
Compass, Correcting the deviation of the mariner's	L. F. A. Arson	Paris, France	May 31, 1870	103, 701
Compass for determining variations from local causes.	J. R. St. John	New York, N. Y	Mar. 2, 1852	8, 785
Compass for mining	J. Blomgren	New York, N. Y	Oct. 25, 1870	108, 555
Compass, Instrument for determining the variation of the.	E. S. Ritchie	Brookline, Mass	Jan. 3, 1865	45, 753
Compass-joint	T. Alteneder	Philadelphia, Pa	Feb. 14, 1871	111, 715
Compass joint, Measuring	T. Alteneder	Philadelphia, Pa	July 16, 1850	7, 501
Compass, Liquid	J. and G. H. Bliss	Brooklyn, N. Y	Jan. 24, 1871	111, 169
Compass, Liquid	E. Blunt	Brooklyn, N. Y	Dec. 5, 1865	51, 290
Compass, Magnetic	H. W. Hunter	New York, N. Y	May 6, 1862	35, 156
Compass, Marine	S. Custer	Salem, Va	June 26, 1866	55, 827
Compass, Marine	E. S. Ritchie	Brookline, Mass	July 19, 1870	105, 492
Compass, Mariner's	J. Ball	Buffalo, N. Y	Mar. 6, 1835	....
Compass, Mariner's	J. and G. H. Bliss	Brooklyn, N. Y	May 24, 1870	103, 286
Compass, Mariner's	J. and G. H. Bliss	Brooklyn, N. Y	May 24, 1870	103, 287
Compass, Mariner's	H. Colby	Rochester, N. Y	May 1, 1847	5, 096
Compass, Mariner's	S. Custer	Salem, Va	July 16, 1867	66, 805
Compass, Mariner's	J. S. Pender	New York, N. Y	Sept. 27, 1864	44, 451
Compass, Mariner's	E. S. Ritchie	Brookline, Mass	Sept. 9, 1862	36, 422
Compass, Mariner's	E. S. Ritchie	Brookline, Mass	Apr. 7, 1863	38, 125
Compass, Mariner's	E. S. Ritchie	Brookline, Mass	Apr. 7, 1863	38, 126
Compass, Mariner's	E. S. Ritchie	Brookline, Mass	Apr. 10, 1866	53, 875
Compass, Mariner's and surveyor's	W. C. Poole	Lancaster, Pa	July 31, 1840	1, 707
Compass, Mariner's or surveyor's	W. Russell	New Bedford, Mass	Dec. 1, 1809	....
Compass, Mariner's time	R. Reeder	Cincinnati, Ohio	Feb. 9, 1847	4, 964
Compass-needle, Magnetizing	S. Custer	Salem, Va	July 16, 1867	66, 806
Compass needle, Marine and surveying	M. Smith	New York, N. Y	June 15, 1830	....
Compass or theodolite, Surveying	J. Eames	Newry, Me	Feb. 11, 1835	....
Compass-protractor	F. H. West	San Francisco, Cal	Sept. 25, 1860	30, 187
Compass, quadrant, and protractor	F. Whiteley	Stanardsville, Va	Dec. 6, 1836	99
Compass, Self-registering ship's	R. H. Peverly	Chelsea, Mass	June 3, 1856	15, 017
Compass, Ship's	James, Earl of Caithness	Middlesex County, England	Nov. 5, 1867	70, 520
Compass, Ship's	J. Prime	Washington, N. C	Feb. 12, 1856	14, 251
Compass, Ship's	G. W. Wood	Brooklyn, N. Y	Mar. 21, 1871	112, 999
Compass, Solar	H. O. Cook	New York, N. Y	Aug. 12, 1873	141, 766
Compass, Solar	B. S. Lyman	Philadelphia, Pa	July 18, 1871	117, 184
Compass, Surveying	G. W. Dickinson, jr	Breckenridge, Va	Feb. 21, 1860	27, 210
Compass, Surveying	W. J. Young	Philadelphia, Pa	Jan. 17, 1832	....
Compass, Surveying	W. J. Young	Philadelphia, Pa	Jan. 11, 1834	....
Compass, Surveyor's	N. Bassett	Wilmington, Del	June 28, 1836	....
Compass, Surveyor's	S. Kern	Strasburgh, Va	July 28, 1846	4, 675
Compass, Surveyor's	J. Locke	Cincinnati, Ohio	July 16, 1850	7, 510
Compass, Surveyor's	H. B. Martin	Santa Rosa, Cal	June 7, 1864	43, 036
Compass, Surveyor's	S. R. Miller	Front Royal, Va	Oct. 22, 1835	....
Compass, Universal	S. Dew	Romney, Va	Apr. 13, 1822	....
Compasses and calipers	J. E. Earle	Leicester, Mass	Aug. 1, 1854	11, 420
Compasses and calipers	F. P. Pfleghar and W. Schollhorn.	New Haven, Conn	Jan. 9, 1866	51, 967
Compasses and calipers, Combined beam	W. Burrows	New York, N. Y	Dec. 29, 1868	85, 430
Compasses, Card for liquid	E. S. Ritchie	Brookline, Mass	June 2, 1863	38, 762
Compasses, Binnacle for mariner's	G. W. Richey and H. E. Bixby	Saint Louis, Mo	July 4, 1871	116, 631
Compasses, Card for mariner's liquid	E. S. Ritchie	Brookline, Mass	Jan. 24, 1871	111, 254
Compasses, Construction of	T. Hagerty	Richmond, Va	July 30, 1872	130, 036
Compasses, Construction of	T. Hagerty	Richmond, Va	July 30, 1872	130, 037
Compasses, Construction of spring	T. Hagerty	Richmond, Va	July 30, 1872	130, 038
Compasses, Electro-magnetic attachment to ships'.	A. Foucaut	Orleans, France	July 19, 1870	105, 562
Compasses insensible to local attraction, Rendering.	J. S. Gisbone and W. Simpson	Birkenhead and Liverpool, England.	Mar. 8, 1864	41, 839
Compasses, Local-attraction indicator for ships'	H. Glover	Brooklyn, N. Y	Sept. 17, 1872	131, 435
Compasses, Mode of fixing mariners'	L. Langley	Gosport, Va	June 23, 1828	....
Compasses, Pencil-attachment for	C. Schott	Nashville, Tenn	July 12, 1870	105, 376
Compasses, Pencil-attachment to	C. L. Tyler	Ithaca, N. Y	Nov. 2, 1869	96, 366
Compasses, Preventing the deviation of ships'	J. W. Girdlestone	Strand, England	Oct. 25, 1870	108, 585
Compasses used in calking seams	G. Dowling	Fair Haven, Conn	June 11, 1867	65, 654
Compensating or equilibrium spring	C. Shea	Newark, N. J	Oct. 12, 1869	95, 736
Composing-stick	L. Buschmann	Newark, N. J	Oct. 21, 1873	143, 875
Composing-stick	A. F. Cloudman and G. W. Coffin.	Brooklyn, N. Y., and Charlestown, Mass.	Oct. 18, 1870	108, 451
Composing-stick	J. M. Eaton	Charlestown, Mass	Aug. 10, 1869	93, 426
Composing-stick	F. W. Murray	Cincinnati, Ohio	Mar. 16, 1869	88, 800
Composing-stick	R. W. Thing	Boston, Mass	Jan. 26, 1869	86, 257
Composing-stick	W. T. Tillinghast	Dayton, Ohio	May 22, 1866	54, 979
Composing-stick	J. L. Wait	East Cambridge, Mass	May 19, 1868	78, 033
Composing-stick	J. Wilson	Boston, Mass	Aug. 6, 1872	130, 170
Composing-stick	R. C. Young	Middletown, Conn	Oct. 18, 1870	108, 549
Composing-stick, Printer's	S. W. Brown	Syracuse, N. Y	May 22, 1860	28, 436
Composing-stick, Printer's	A. Calhoun	Hartford, Conn	Aug. 31, 1858	21, 321
Composing-stick, Printer's	O. F. Grover	Middletown, Conn	July 15, 1856	15, 358
Composing-stick, Printer's	P. S. Hoe	New York, N. Y	Nov. 5, 1872	132, 722
Composing stick, Printer's	J. and W. Tidgewell	Middletown, Conn	June 2, 1857	17, 457
Composing-stick, Printer's	W. T. Tillinghast	Dayton, Ohio	Jan. 27, 1857	16, 500
Composing-stick, Printer's	D. Winder	Cincinnati, Ohio	Apr. 7, 1857	17, 007
Composite pipe	A. P. Stephens	Brooklyn, N. Y	May 26, 1868	78, 236
Composition-box	T. B. Gunning	New York, N. Y	Oct. 3, 1871	119, 603
Composition handle or pull	W. B. Gleason	Boston, Mass	Mar. 22, 1870	101, 119
Compositor's stand	F. Vallee	Philadelphia, Pa	Apr. 8, 1873	137, 639
Compost	E. Blanchard	Greenfield Mills, Md	Aug. 9, 1859	24, 988
Compost for soil	A. B. Martin	Baltimore, Md	June 2, 1819	....
Compound engine	A. Hartupee	Pittsburgh, Pa	Apr. 16, 1872	125, 812

Index of patents issued from the United States Patent Office from 1790 *to* 1873, *inclusive*—Continued.

Invention.	Inventor.	Residence.	Date.	No.
Compound engine	W. Wright	Newburgh, N. Y	Nov. 18, 1873	144, 817
Compound gage	A. Williams	Philadelphia, Pa	May 26, 1857	17, 403
Compound steam-engine	W. W. Crane	Auburn, N. Y	June 3, 1873	139, 549
Compound steam-engine	L. Huntoon	Natick, Mass	Jan. 28, 1873	135, 223
Compound tool	J. Dillon	New York, N. Y	Dec. 30, 1873	145, 992
Compress	C. W. Armstrong	Detroit, Mich	Nov. 30, 1869	97, 265
Compress, Double-acting	J. H. Fellows	Lewisport, Ky	Dec. 20, 1864	45, 559
Compress-table	H. A. Burr	Brooklyn, N. Y	July 21, 1868	80, 132
Compressed-air engine	E. H. Grant	Washington, D. C	Sept. 7, 1869	94, 594
Compressed-air motor	H. Bushnell	New Haven, Conn	Apr. 15, 1873	137, 889
Compressor, Spiral-curvilinear	W. Nelson	Batavia, N. Y	Nov. 13, 1828	....
Compressing and baling press	D. S. Gardner and N. A. Manning.	Greene, N. Y	Mar. 25, 1862	34, 750
Compressing and beater press	G. Ertel	Liberty, Ill	Mar. 31, 1868	76, 063
Compression-lock	J. Powell	Cincinnati, Ohio	Jan. 18, 1870	99, 001
Computing-apparatus	T. Wallworth	Newton Heath, Great Britain.	Feb. 18, 1868	74, 641
Computing-apparatus	E. Wright	Medford, Mass	Aug. 17, 1869	93, 849
Computing-machine	J. H. Chidester	Cleveland, Ohio	Dec. 18, 1866	60, 475
Computing-machine	A. W. Davies	Cleveland, Ohio	June 18, 1867	65, 883
Concave-cutting device	H. C. Rosin	Chicago, Ill	Aug. 29, 1871	118, 489
Concentrator	J. Hendy	San Francisco, Cal	Apr. 17, 1866	53, 976
Concentrator	J. Hendy	San Francisco, Cal	May 19, 1868	78, 089
Concentrator, Metallic	Z. Wheeler	San Francisco, Cal	Sept. 6, 1864	44, 130
Conception, Preventing	J. B. Beers	Rochester, N. Y	Aug. 28, 1846	4, 729
Concertina, Key-face for	L. A. Seward	New Orleans, La	Oct. 26, 1869	96, 150
Concrete, Apparatus for feeding and mixing gravel, &c., in forming.	A. D. Foote	Washington, D. C	July 16, 1872	129, 330
Concrete arches for buildings, Construction of	C. C. Dennett	Nottingham, England	Jan. 4, 1870	98, 571
Concrete block for buildings and other purposes	T. Heap	Saint Joseph, Mo	Dec. 28, 1869	98, 373
Concrete-block machine	L. Dodge and L. J. Magnusson	Chicago, Ill	Dec. 29, 1868	85, 291
Concrete-block-making machine	O. V. Evans	Ripley, Ohio	Sept. 22, 1868	82, 301
Concrete-block mold	J. P. Campbell	Dayton, Ohio	June 4, 1872	127, 562
Concrete-block press	L. S. Warner	Chicago, Ill	May 12, 1868	77, 854
Concrete building-block press	O. L. Jordan	Dowagiac, Mich	Jan. 21, 1868	73, 600
Concrete-composition for walks, flooring, &c	R. S. Lewis	Rockville, Conn	Nov. 22, 1870	109, 524
Concrete for pipes, bricks, &c	T. J. Barron	Brooklyn, N. Y	Feb. 14, 1871	111, 806
Concrete for pipes, tubes, buildings, &c	T. J. Barron	Brooklyn, E. D., N. Y	Jan. 24, 1871	111, 166
Concrete, &c., Machine for mixing	S. Putnam and T. Burt	Rockville, Conn	Jan. 17, 1871	111, 000
Concrete-pipes, Machine for making	G. L. Eagan	San Francisco, Cal	Apr. 9, 1872	125, 551
Condensation and refrigeration, Method of	N. W. Wheeler	Brooklyn, N. Y	July 3, 1866	56, 130
Condenser	E. H. Ashcroft	Lynn, Mass	Nov. 5, 1867	70, 389
Condenser	B. T. Babbitt	New York, N. Y	Sept. 1, 1863	39, 709
Condenser	A. C. Brown	Philadelphia, Pa	July 10, 1860	29, 054
Condenser	S. W. Brown	Lowell, Mass	Dec. 19, 1854	12, 089
Condenser	T. Callan	Philadelphia, Pa	Sept. 12, 1865	49, 855
Condenser	L. Cook	Whitestown, N. Y	Mar. 26, 1811	....
Condenser	J. H. Fairchild	Highgate, Vt	Mar. 27, 1866	53, 528
Condenser	A. C. Fletcher	New York, N. Y	July 11, 1865	48, 710
Condenser	A. Hartupee and H. P. Gengembre.	Pittsburgh, Pa	Dec. 13, 1864	45, 408
Condenser	J. Houpt	Springtown, Pa	Dec. 14, 1869	97, 923
Condenser	G. A. Jasper	Charlestown, Mass	July 5, 1864	43, 413
Condenser	W. Jeggle and L. A. Brooks	Chicago, Ill	Apr. 2, 1867	63, 391
Condenser	W. A. Lighthall	New York, N. Y	Feb. 7, 1865	46, 254
Condenser	W. A. Lighthall	New York, N. Y	Apr. 21, 1868	76, 929
Condenser	J. F. Llewellyn	Louisville, Ky	Sept. 11, 1866	57, 932
Condenser	P. Mihan and M. A. Lane	Boston and Charlestown, Mass.	Sept. 18, 1860	30, 078
Condenser	F. Ortlieb	Williamsburgh, N. Y	Sept. 3, 1867	68, 382
Condenser	W. Phelan	Peoria, Ill	Apr. 14, 1868	76, 655
Condenser	W. Phelan	Peoria, Ill	Oct. 20, 1868	83, 309
Condenser	F. Ransom	Buffalo, N. Y	Oct. 13, 1868	83, 092
Condenser	J. Root	Painesville, Ohio	May 26, 1834	....
Condenser	T. Shaw	Philadelphia, Pa	Sept. 28, 1869	95, 275
Condenser	J. F. Spence	Williamsburgh, N. Y	Mar. 27, 1866	53, 498
Condenser	J. M. Spiegle	Philadelphia, Pa	Jan. 17, 1865	45, 941
Condenser	G. Stump	New York, N. Y	Aug. 25, 1863	39, 688
Condenser	H. A. Towne	Hannibal, Mo	Aug. 19, 1873	142, 060
Condenser	G. R. Vanderbilt	Mount Vernon, N. Y	Mar. 15, 1864	41, 949
Condenser	G. I. Washburn	Worcester, Mass	Dec. 8, 1863	40, 872
Condenser	J. C. Wharton	Nashville, Tenn	Sept. 11, 1866	58, 0[illegible]9
Condenser	F. Wegmann	Naples, Italy	Aug. 30, 1870	106, 973
Condenser	N. W. Wheeler	Brooklyn, N. Y	July 9, 1867	66, 541
Condenser	W. L. and T. Winans	London, England, and Baltimore, Md.	Oct. 27, 1868	83, 430
Condenser, Air	O. Abruzzo	New York, N. Y	Nov. 19, 1867	70, 934
Condenser, Air-heating steam	P. J. Rice and D. A. Scott	Ashtabula and Cincinnati, Ohio.	Aug. 9, 1870	106, 206
Condenser and boiler, Steam-engine	B. Crawford	Allegheny City, Pa	Sept. 7, 1844	3, 732
Condenser and feed-water apparatus, Combined	J. S. Gibson	Martinsburgh, W. Va	Nov. 4, 1873	144, 331
Condenser and feed-water heater	W. Phelan	Peoria, Ill	May 10, 1870	102, 965
Condenser and feed-water heater, Steam-engine	J. Dilworth and J. C. Hodgins	Toronto, Canada	Feb. 21, 1871	111, 916
Condenser and feeder, Steam-boiler	F. A. Fischer	Hartzburgh, Germany	May 20, 1873	139, 136
Condenser and lime-extracting heater	J. Huntington	Cleveland, Ohio	Mar. 15, 1870	100, 766
Condenser and refiner for spirituous liquors	W. Neil	San Francisco, Cal	May 6, 1873	138, 517
Condenser and stuffing-box, Vapor-engine	P. V. Du Trembley	Paris, France	Dec. 4, 1849	6, 929
Condenser and water-heater, Steam-engine	J. S. Hooton	New Carlisle, Ind	Mar. 5, 1861	31, 600
Condenser-case	W. A. Lighthall	New York, N. Y	Feb. 7, 1865	46, 253
Condenser, Coal-gas	A. Babbitt and W. W. Binney	Auburn, N. Y	June 7, 1870	103, 824
Condenser, Coal-oil	J. F. Bennett	Pittsburgh, Pa	May 22, 1860	28, 341
Condenser, Coal-oil	W. G. W. Jaeger	Baltimore, Md	June 28, 1859	24, 561
Condenser, Compound surface	W. A. Lighthall	New York, N. Y	July 4, 1871	116, 728
Condenser for alcohol-stills	E. Smeeth	Chicago, Ill	Oct. 26, 1869	96, 157
Condenser for brewers' boilers	C. Clifford	Fulton, N. Y	Dec. 15, 1868	84, 995
Condenser for carbonic acid and for drawing off and applying the same for cooling and freezing.	T. S. C. Lowe	New York, N. Y	Apr. 2, 1867	63, 405
Condenser for carding-engines	G. R. Gardiner	Westerly, R. I	Nov. 5, 1867	70, 543

Index of patents issued from the United States Patent Office from 1790 *to* 1873, *inclusive*—Continued.

Invention.	Inventor.	Residence.	Date.	No.
Condenser for carding-machines	N. Lucas	Norwich, Conn	Sept. 23, 1873	143, 020
Condenser for coal-oil stills	C. W. Grannis	Gowanda, N. Y	Sept. 9, 1862	36, 403
Condenser for distillers, Refrigerating	F. Haeck	Brussels, Belgium	Nov. 8, 1864	45, 003
Condenser for distillation	T. O'Connor	New York	July 9, 1807	....
Condenser for fibrous material	W. H. Howard	Philadelphia, Pa	May 15, 1855	12, 863
Condenser for gas-works	P. Munzinger	Philadelphia, Pa	Nov. 8, 1870	109, 142
Condenser for heating mash in distilling	A. Anderson	....	Jan. 26, 1801	....
Condenser for illuminating-gas, Friction	W. H. St. John and S. O. Rockwell.	Brooklyn, N. Y., and Jersey City, N. J.	Apr. 4, 1871	113, 463
Condenser for lard-rendering kettles, Vapor	W. M. Bartram	Philadelphia, Pa	Sept. 15, 1868	82, 195
Condenser for list-speeders	W. Mattison	Northbridge, Mass	Sept. 1, 1857	18, 121
Condenser for making potable water	W. A. Lighthall	New York, N. Y	May 20, 1862	35, 319
Condenser for marine engines	J. Houpt	Springtown, Pa	July 19, 1870	105, 457
Condenser for marine engines, Surface	D. Carpenter	Brooklyn, N. Y	Sept. 12, 1854	11, 661
Condenser for oil-stills	J. Adair	Pittsburgh, Pa	June 10, 1862	35, 497
Condenser for oil-stills	H. W. Faucett and T. McGowan.	Titusville and Meredith, Pa.	Aug. 8, 1871	117, 873
Condenser for rectifying-apparatus	A. Feubert	Buffalo, N. Y	Aug. 29, 1871	118, 601
Condenser for spirits, steam, and other vapors	A. W. Cram	Saint Louis, Mo	Jan. 19, 1869	85, 910
Condenser for stills	T. J. Dean	Saint Louis, Mo	July 20, 1869	92, 941
Condenser for stills	J. Harrison and C. Low	Fawn Grove, Pa	May 24, 1870	103, 326
Condenser for stills	C. Morfit	Baltimore, Md	July 2, 1867	66, 243
Condenser for stills	H. C. Sherman	Buffalo, N. Y	Mar. 11, 1862	34, 648
Condenser for stills	C. E. Werner	New Castle, Ill	Dec. 6, 1853	10, 305
Condenser for stills	J. Yates and E. Deuell	Brooklyn, N. Y	Apr. 25, 1871	114, 245
Condenser for vacuum-pans, &c	C. W. Dyrant and J. Griffith	New York, N. Y	Oct. 15, 1872	132, 263
Condenser, Gas	I. N. Stanley	Brooklyn, N. Y	Apr. 18, 1871	113, 942
Condenser, Gas	E. Thompson	Madison, Wis	May 24, 1870	103, 392
Condenser, Hydro-atmospheric	J. P. F. Datichy	New York, N. Y	Aug. 30, 1864	43, 976
Condenser, Jet	J. P. F. Datichy	West Hoboken, N. J	Jan. 15, 1867	61, 167
Condenser, Marine	B. T. Babbitt	New York, N. Y	Sept. 9, 1873	142, 665
Condenser of steam-engines and apparatus for supplying boilers with water.	J. Echols	Columbus, Ga	Aug. 11, 1841	2, 212
Condenser, Oil-vapor	I. W. Wetmore	Erie, Pa	Sept. 15, 1863	39, 978
Condenser or cooler for alcoholic and other liquids	J. R. Neil	Brooklyn, N. Y	Oct. 17, 1871	119, 942
Condenser or refrigerator	W. A. Lighthall	New York, N. Y	Feb. 14, 1865	46, 368
Condenser, Potable water	W. A. Lighthall	New York, N. Y	Apr. 22, 1862	35, 029
Condenser-pump, Employment of an auxiliary engine in combination with the.	J. Ericsson	New York, N. Y	Apr. 3, 1849	6, 255
Condenser, Revolving	J. F. Spence	Williamsburgh, N. Y	May 1, 1866	54, 428
Condenser, Steam	J. L. Alberger	Buffalo, N. Y	May 20, 1873	139, 102
Condenser, Steam	J. L. Alberger	Buffalo, N. Y	June 24, 1873	140, 232
Condenser, Steam	B. C. Atkinson	Newburyport, Mass	June 22, 1869	91, 506
Condenser, Steam	A. Cail	Paris, France	Oct. 10, 1871	119, 817
Condenser, Steam	J. N. Dennisson	Newark, N. J	Feb. 8, 1859	22, 916
Condenser, Steam	J. K. Ferguson	Portland, Ky	July 31, 1866	56, 738
Condenser, Steam	J. Houpt	Springtown, Pa	Apr. 19, 1870	102, 005
Condenser, Steam	C. Hughes	Colen, Cuba	Nov. 30, 1869	97, 405
Condenser, Steam	T. L. Jones	Natchez, Miss	Apr. 26, 1870	102, 272
Condenser, Steam	A. Kennedy and J. H. Berkshire	Muscatine, Iowa	Feb. 27, 1872	124, 065
Condenser, Steam	J. T. King	New York, N. Y	Feb. 12, 1856	14, 244
Condenser, Steam	E. Korting	Vienna, Austria	Jan. 14, 1873	134, 900
Condenser, Steam	E. Korting	Vienna, Austria	July 29, 1873	141, 361
Condenser, Steam	W. A. Lighthall	New York, N. Y	Dec. 17, 1861	33, 952
Condenser, Steam	W. R. Pitts and G. K. Gluyas	San Francisco, Cal	Oct. 1, 1872	131, 779
Condenser, Steam	A. Van Orsdale	Jasper, N. Y	Sept. 8, 1868	81, 964
Condenser, Steam	W. Wright	New York, N. Y	July 11, 1871	116, 909
Condenser, Steam-engine	J. Absterdam	New York, N. Y	Oct. 2, 1866	58, 362
Condenser, Steam-engine	H. Anderson	Peoria County, Ill	Sept. 27, 1870	107, 748
Condenser, Steam-engine	E. Baldwin	Philadelphia, Pa	July 9, 1850	7, 486
Condenser, Steam-engine	L. Bollman	New York, N. Y	Apr. 3, 1855	12, 619
Condenser, Steam-engine	W. Craig and H. L. Brevoort	Brooklyn, N. Y	Mar. 28, 1871	113, 021
Condenser, Steam-engine	B. Crawford	Pittsburgh, Pa	Nov. 1, 1853	10, 179
Condenser, Steam-engine	J. T. Denniston	Lyons, N. Y	Aug. 26, 1856	15, 604
Condenser, Steam-engine	A. C. Fletcher	New York, N. Y	June 30, 1863	39, 040
Condenser, Steam-engine	S. A. Goodwin	Buffalo, N. Y	Feb. 18, 1873	135, 906
Condenser, Steam-engine	J. Houpt	Springtown, Pa	May 25, 1869	90, 542
Condenser, Steam-engine	B. F. Lemmon	New Albany, Ill	Aug. 7, 1860	29, 499
Condenser, Steam-engine	W. A. Lighthall	New York, N. Y	Dec. 17, 1861	33, 951
Condenser, Steam-engine	D. Matthew	Philadelphia, Pa	Sept. 2, 1856	15, 663
Condenser, Steam-engine	J. M. Miller	New York, N. Y	Dec. 4, 1866	60, 222
Condenser, Steam-engine	F. Ransom	Buffalo, N. Y	June 22, 1869	91, 770
Condenser, Steam-engine	J. Shirt and C. Briggs	Tamworth, Great Britain	Dec. 8, 1868	84, 773
Condenser, Steam-engine	T. E. Sickels	Kennett's Square, Pa	July 7, 1863	39, 176
Condenser, Steam-engine	T. Skidmore	New York, N. Y	Dec. 1, 1821	....
Condenser, Steam-engine	A. F. Smith	Norwich, Conn	Sept. 15, 1863	39, 965
Condenser, Steam-engine	F. G. Smith	Columbia, Tenn	July 11, 1854	11, 274
Condenser, Steam-engine	H. K. Stephens	Cincinnati, Ohio	Dec. 4, 1843	3, 366
Condenser, Steam-engine	F. B. Stevens	Weehawken, N. J	Dec. 3, 1861	33, 856
Condenser, Steam-engine	F. B. Stevens	Weehawken, N. J	Dec. 3, 1861	33, 857
Condenser, Steam-engine	F. B. Stevens	New York, N. Y	Aug. 4, 1863	39, 429
Condenser, Steam-engine	F. B. Stevens	New York, N. Y	Nov. 3, 1863	40, 510
Condenser, Steam-engine	A. C. Twining	New Haven, Conn	Aug. 12, 1862	36, 183
Condenser, Steam-engine	G. Wales	Brooklyn, N. Y	Jan. 21, 1873	135, 180
Condenser, Steam-engine	G. I. Washburn	Worcester, Mass	Oct. 6, 1863	40, 204
Condenser, Steam-engine	H. Waterman	Hudson, N. Y	Dec. 12, 1854	12, 079
Condenser, Steam-engine	W. Wright	New York, N. Y	Feb. 6, 1872	123, 541
Condenser, Steam-engine jet	G. W. Hall	Havana, N. Y	May 14, 1872	126, 696
Condenser, Steam-engine jet	J. Houpt	Springtown, Pa	Sept. 17, 1872	131, 351
Condenser, Steam-engine, &c., surface	J. Cragg and S. Archbold	Baltimore, Md., and Washington, D. C.	Feb. 19, 1861	31, 441
Condenser, Steam-engine surface	P. Hogg	New York, N. Y	Oct. 30, 1855	13, 721
Condenser, Steam-engine surface	J. M. Miller	New York, N. Y	May 20, 1856	14, 923
Condenser, Steam-engine surface	J. P. Pirsson	New York, N. Y	Apr. 2, 1850	7, 250
Condenser, Steam-engine surface	W. Sewell	New York, N. Y	Mar. 20, 1860	27, 573
Condenser, Steam-engine surface	A. Smith	Niagara Falls, N. Y	Dec. 20, 1859	26, 529
Condenser, Steam-engine surface	N. Thompson, jr	Williamsburgh, N. Y	May 20, 1856	14, 932
Condenser, Steam-pump	D. Stoddart	San Francisco, Cal	Dec. 16, 1873	145, 538
Condenser, Surface	D. Barnum	New York, N. Y	May 24, 1859	24, 087

Index of patents issued from the United States Patent Office from 1790 *to* 1873, *inclusive*—Continued.

Invention.	Inventor.	Residence.	Date.	No.
Condenser, Surface	G. H. Corliss	Providence, R. I	Aug. 26, 1862	36, 280
Condenser, Surface	A. C. Fletcher	New York, N. Y	Apr. 17, 1866	53, 967
Condenser, Surface	P. Hammond	Castleton, N. Y	Jan. 20, 1863	37, 444
Condenser, Surface	B. Holly	Lockport, N. Y	Apr. 4, 1871	113, 668
Condenser, Surface	J. Houpt	Springfield, Pa	Apr. 13, 1869	88, 960
Condenser, Surface	R. Lighthall	Brooklyn, N. Y	Apr. 12, 1870	101, 892
Condenser, Surface	J. Nason	New York, N. Y	May 29, 1866	55, 149
Condenser, Surface	F. Ortlieb	Williamsburgh, N. Y	Nov. 27, 1866	60, 042
Condenser, Surface	T. Prosser	New York, N. Y	Dec. 15, 1857	18, 861
Condenser, Surface	W. Sewell	Brooklyn, N. Y	Oct. 31, 1854	11, 883
Condenser, Surface	F. B. Stevens	New York, N. Y	Oct. 28, 1862	36, 807
Condenser, Surface	F. B. Stevens	New York, N. Y	Oct. 28, 1862	36, 809
Condenser, Surface	F. B. Stevens	New York, N. Y	Oct. 28, 1862	36, 810
Condenser, Surface	A. C. Stimers	New York, N. Y	Feb. 9, 1864	41, 545
Condenser, Surface	J. J. W. Watson and W. H. Smith.	Paris and Nantes, France.	May 10, 1864	42, 720
Condenser, Surface	N. W. Wheeler	Brooklyn, N. Y	June 25, 1867	66, 194
Condenser, Surface	I. S. Wilson and H. See	Philadelphia, Pa	Mar. 3, 1868	75, 102
Condenser-tube fastening	S. Archbold	Chester, Pa	Sept. 30, 1873	143, 314
Condenser-tube fastening	G. H. Reynolds	New York, N. Y	June 20, 1871	116, 098
Condenser, Tubular	W. A. Lighthall	New York, N. Y	Jan. 31, 1865	46, 117
Condenser, Tubular	N. W. Wheeler	Brooklyn, N. Y	Apr. 25, 1865	47, 480
Condensers and heaters, Construction of tubular	W. West and A. Mills	New York, N. Y	Nov. 25, 1856	16, 120
Condensers, Connecting the tubes with the heads of surface.	H. Allen	New York, N. Y	Jan. 14, 1868	73, 221
Condensers, Device for cleaning surface	H. C. Bucknam	Bucksport, Me	Sept. 16, 1873	142, 766
Condensers, Joint for tubes of surface	J. V. V. Booraem	Jersey City, N. J	Sept. 22, 1863	40, 013
Condensers, Means of securing tubes in surface	W. A. Lighthall	New York, N. Y	Mar. 3, 1868	75, 173
Condensers, Method of affixing tubes in steam	W. F. Brooks	New York, N. Y	Mar. 31, 1863	38, 027
Condensers, Method of setting tubes for	W. A. Lighthall	New York, N. Y	Dec. 17, 1861	33, 950
Condensers of steam-engines, Mode of discharging water and air from.	R. G. Eunson	New York, N. Y	Sept. 23, 1842	2, 786
Condensers of stills, Construction of	P. Mihan	Boston, Mass	May 15, 1860	28, 325
Condensers, Shield for surface	F. B. Stevens	New York, N. Y	Oct. 28, 1862	36, 808
Condensers, Steam-enginery for surface	W. A. Lighthall	New York, N. Y	Apr. 28, 1869	89, 156
Condensers, Tube for surface	D. H. Chamberlain	West Roxbury, Mass	Nov. 19, 1861	33, 763
Condensers, Tube-joint for	H. Allen	New York, N. Y	July 20, 1858	20, 927
Condensing and evaporating apparatus	J. J. Miller	Chicago, Ill	Dec. 23, 1862	37, 258
Condensing apparatus for forcing liquids, Air	H. Strater, jr	Roxbury, Mass	Apr. 21, 1863	38, 250
Condensing-apparatus for marine steam-engines	J. Houpt	Springtown, Pa	Nov. 4, 1873	144, 203
Condensing-apparatus for salt and gases	J. C. Salomon, jr	Baltimore, Md	June 16, 1857	17, 586
Condensing-apparatus, Steam-engine	D. Treadwell	Cambridge, Mass	Mar. 31, 1840	1, 534
Condensing coal-oil, Apparatus for	W. T. Barnes	Buffalo, N. Y	Aug. 2, 1859	24, 920
Condensing-cover	A. H. Bartlett	Spuyten Duyvil, N. Y	Apr. 5, 1859	23, 437
Condensing-engine	W. A. Lighthall	New York, N. Y	Dec. 10, 1867	72, 054
Condensing-engines, Mode of operating air pumps and condensers of.	R. F. Soper	Philadelphia, Pa	Aug. 28, 1849	6, 673
Condensing, evaporating, and cooling apparatus	J. J. Miller	Chicago, Ill	Nov. 17, 1863	40, 662
Condensing mercury in amalgamating-vessels, Mode of.	V. Ryerson	New York, N. Y	June 4, 1861	32, 488
Condensing smoke and gases	I. Bloom	East Woburn, Mass	Nov. 29, 1853	10, 271
Condensing steam and cooling the liquor	B. F. Joslin	New York, N. Y	Feb. 9, 1825	......
Condensing steam and elevating feed-water, Apparatus for.	J. Briggs	Louisville, Ky	Oct. 27, 1863	40, 392
Condensing steam and feeding steam-boilers	J. P. Pirsson, jr	New York, N. Y	Feb. 10, 1846	4, 383
Condensing steam or cooling water, Method of supplying water to steam-vessels for the purpose of.	W. A. Lighthall	New York, N. Y	Mar. 5, 1861	31, 609
Condensing-tub	H. Witmer	Lancaster, Pa	May 17, 1805	......
Condensing-tub for heating water for distillation	A. Anderson	Philadelphia, Pa	Oct. 27, 1813	......
Condensing-tube	S. Wilson	Auburn, N. Y	Jan. 27, 1863	37, 538
Condensing vapors and gases and for evaporating liquids, Apparatus for.	A. F. W. Partz	New York, N. Y	June 2, 1857	17, 448
Condensing-worm, Tinned sheet-copper	J. Taylor		June 30, 1795	......
Condiment or relish	W. F. Swasey	San Francisco, Cal	July 19, 1870	105, 514
Conductor-pipes, Construction of	J. C. Rohrman	Philadelphia, Pa	Nov. 19, 1867	71, 222
Conductor-pipes, Construction of sheet-metal	W. Austin	Philadelphia, Pa	June 2, 1868	78, 564
Conductor's bracket	P. Gantz	Albany, N. Y	Nov. 30, 1869	97, 289
Conductor's register	W. C. McGill	Cincinnati, Ohio	Apr. 14, 1868	76, 646
Conduit, Clay	S. Bartlett	Hartford, Conn	Jan. 15, 1809	......
Conduit for water around canal-locks, Self-regulating.	J. Clarke	Westmoreland County, Pa.	Jan. 25, 1831	......
Conduit-pipe	J. Clark and L. Crowell	Westminster, Va	Dec. 18, 1818	......
Conduit-pipe	E. Putnam	Albany, N. Y	Nov. 18, 1818	......
Conduit-pipe	W. S. Sampson	New York, N. Y	Mar. 3, 1863	37, 828
Conduit-pipe coupling	U. West	New York, N. Y	July 23, 1842	2, 738
Conduit-pipes, Applying ice to	H. Cunningham	Johnstown, N. Y	June 29, 1825	......
Conduit-pipes from clay, Making	R. Ramsey	Hanover, N. H	June 24, 1808	......
Conduit-pipes, Preventing the waste of water in	E. Putnam	Albany, N. Y	Dec. 31, 1816	......
Cone and chimney-elevator	W. R. P. Cross	Portland, Me	Mar. 22, 1864	42, 045
Cone springs, Making	W. H Goodale	Colton, N. Y	Mar. 23, 1869	88, 031
Confection	L. P. Eleuterius	New Orleans, La	Apr. 9, 1872	125, 446
Confection	J. Q. Hill	Worcester, Mass	June 22, 1869	91, 632
Confection	E. C. Maltby and E. Smith	Northford, Conn	June 25, 1867	66, 158
Confection	D. L. Page	Lowell, Mass	Aug. 25, 1868	81, 401
Confection from raisins	J. B. Bidwell	Grand Rapids, Mich	Nov. 23, 1869	97, 155
Confection or cake	W. Manning	Chelmsford, Mass	Jan. 7, 1868	73, 018
Confection-pan	H. Wathew	Philadelphia, Pa	Oct. 30, 1866	59, 325
Confection-pan, Steam	G. H. Cross	Montpelier, Vt	May 7, 1867	64, 497
Confectioners' and grocers' can	J. S. Batchelder	Fort Wayne, Ind	Sept. 12, 1871	118, 899
Confectioner's mold	H. S. Howe	Rutland, Vt	Mar. 14, 1871	112, 710
Confectioner's paste and other materials, Machine for making and breaking.	R. Duff	New York, N. Y	May 23, 1871	115, 039
Confectioner's pipe, Machine for making	O. R. Chase	Boston, Mass	Aug. 2, 1864	43, 735
Confectionery	L. E. Chase	Watertown, Mass	Mar. 1, 1870	100, 375
Confectionery	F. Walter	Saint Louis, Mo	Nov. 19, 1867	71, 089
Confectionery, Apparatus for making	W. H. Seidenstricker	Hagerstown, Md	May 6, 1873	138, 587
Confectionery, Apparatus for manufacturing	W. F. Goward	Boston, Mass	Feb. 7, 1871	111, 638

Index of patents issued from the United States Patent Office from 1790 *to* 1873, *inclusive*—Continued.

Invention.	Inventor.	Residence.	Date.	No.
Confectionery, Applying sirup by steam to	R. L. and A. Stuart	New York, N. Y	Mar. 7, 1833	....
Confectionery-cutter	M. Laemmel	Bay Ridge, N. Y	Jan. 30, 1872	123, 265
Confectionery for druggists	A. Seitz	Hoboken, N. J	Nov. 29, 1870	109, 677
Confectionery-jar	J. S. Batchelder	Fort Wayne, Ind	Sept. 12, 1871	118, 898
Confectionery, Machine for making	S. S. Campbell and J. Goodwin	Philadelphia, Pa	Oct. 7, 1862	36, 601
Confectionery, Machine for mixing paste for	R. Duff	New York, N. Y	May 2, 1871	114, 423
Confectionery, Machine for making	J. Gardner	Philadelphia, Pa	Dec. 22, 1868	85, 083
Confectionery, Manufacture of	L. E. Chase	Charlestown, Mass	May 1, 1866	54, 296
Confectionery, Manufacture of	G. and C. B. Miller	Philadelphia, Pa	Oct. 18, 1870	108, 505
Confectionery or cocoa-nut candy	H. N. and T. J. Harbach	Philadelphia, Pa	Mar. 28, 1871	113, 047
Conformator	E. Roseen	New York, N. Y	Nov. 5, 1867	70, 621
Conic sections and the lines of the globe, Apparatus for illustrating.	F. Shepherd	New Haven, Conn	Dec. 22, 1857	18, 931
Conical-edged knife, Machinery for grinding	J. L. Plimpton	Westfield, Mass	May 4, 1852	8, 926
Connecting-rod	E. Brown	New York, N. Y	Sept. 22, 1868	82, 379
Connecting-rod	W. G. Freeman	Richmond, Va	Jan. 10, 1871	110, 842
Connecting-rod	J. F. Haskins	Fitchburgh, Mass	Dec. 24, 1872	134, 276
Connecting-rod	E. S. Pierce	Hartford, Conn	Mar. 3, 1868	75, 051
Connecting-rod	T. T. Prosser	Chicago, Ill	Dec. 23, 1873	145, 752
Connecting-rod	W. L. Switzer	Zanesville, Ohio	Dec. 23, 1873	145, 917
Connecting-rod	S. N. Wate, jr	Danville, Pa	July 9, 1872	128, 831
Connecting-rod	S. N. Wate, jr	Danville, Pa	Nov. 19, 1872	133, 182
Connecting-rod	B. F. Wilson	Salem, Ohio	July 22, 1873	141, 191
Connecting-rod	J. W. Zinn	Morrison's Mills, Fla	Dec. 19, 1871	122, 095
Connecting-rod adjustment	R. D. O. Smith	Washington, D. C	June 16, 1868	79, 019
Connecting-rod coupling	H. S. Dodge	Lockport, N. Y	Dec. 12, 1865	51, 512
Connecting-rod for locomotives	D. Pollock	Lancaster, Pa	Dec. 10, 1861	33, 902
Connecting-rod for machinery	T. Hall	Bergen, N. J	Mar. 19, 1867	63, 039
Connecting-rod joint	T. Shaw	Philadelphia, Pa	Feb. 8, 1866	52, 455
Connecting-rod, Steam-engine, &c	L. Bissell	New York, N. Y	May 14, 1850	7, 357
Connecting-rods, Adjusting the brasses of	J. R. Sees	New York, N. Y	Mar. 18, 1856	14, 473
Connecting-rods, Apparatus for adjusting eccentrics to.	G. Fowler	Philmont, N. Y	June 1, 1869	90, 655
Connecting-rods applied to cranks	R. Boeklen	Jersey City, N. J	Oct. 11, 1859	25, 716
Connecting-rods, &c., Method of securing keys in	G. H. Coney	Boston, Mass	Sept. 11, 1855	13, 541
Convalescent chair	J. Mead	Albany, N. Y	Mar. 13, 1817	....
Converters, Lining Bessemer	A. L. Holler	Brooklyn, N. Y	June 21, 1870	104, 592
Convertible fork and hook	L. Riggs	Lansing, Mich	Jan. 14, 1868	73, 390
Conveyor block	W. De Witt	Tallmadge, Mich	Apr. 26, 1870	102, 377
Conveyor, elevator, and hopper-boy	S. Walgemore	Washington County, D. C	Feb. 22, 1812	....
Conveyor-"flight"	J. M. Lemon	Polk City, Iowa	June 15, 1869	91, 351
Cook-fork	H. Smith	Des Moines, Iowa	Apr. 14, 1868	76, 835
Cooking and air-heating furnace	O. Packard	Roxbury, Mass	May 26, 1842	2, 642
Cooking and boiling, Conjurer for	T. Passmore		Dec. 23, 1796	....
Cooking and evaporating apparatus	T. J. Newby	Richmond, Ind	Oct. 11, 1870	108, 288
Cooking and heating apparatus, Petroleum	J. S. Hull	Cincinnati, Ohio	Sept. 25, 1866	58, 260
Cooking and steam-heating apparatus	E. Whiteley	Cambridge, Mass	Oct. 13, 1863	40, 300
Cooking and warming apartments, Apparatus for	D. Stephens	Kirtland, Ohio	Oct. 18, 1837	431
Cooking and warming apparatus	D. Willis	New York, N. Y	July 25, 1854	11, 399
Cooking and washing, Boiler attachment for	I. C. Schramm	Des Moines, Iowa	Oct. 1, 1872	131, 789
Cooking-apparatus	B. Antognini	New Orleans, La	Aug. 14, 1847	5, 239
Cooking-apparatus	E. E. Bennett	Kingsbury, N. Y	Feb 15, 1828	....
Cooking-apparatus	J. Bennett	Brutus, N. Y	July 25, 1832	....
Cooking-apparatus	H. C. Berry	Wauseon, Ohio	Aug. 16, 1870	106, 457
Cooking-apparatus	S. D. and W. T. Day	Westfield, Mass	June 8, 1832	....
Cooking-apparatus	M. J. Dé Leon	Bal imore, Md	June 24, 1873	140, 253
Cooking apparatus	J. U. Fiester	Winchester, Ohio	June 10 1862	35, 512
Cooking-apparatus	W. Fulton	Elizabeth City, N. J	Feb. 4, 1862	34, 297
Cooking-apparatus	W. Fulton	Elizabeth, N. J	Dec. 9, 1862	37, 092
Cooking-apparatus	J. M. Gale and I. M. Avery	New York, N. Y	June 2, 1868	78, 588
Cooking-apparatus	J. Gallagher	Cleveland, Ohio	Sept. 20, 1870	107, 477
Cooking-apparatus	N. Gergen	Buffalo, N. Y	Nov. 21, 1871	121, 164
Cooking-apparatus	J. Grime	Beekmantown, N. Y	Dec. 28, 1840	1, 918
Cooking-apparatus	R. W. Hill	Naugatuck, Conn	Aug. 23, 1859	25, 196
Cooking-apparatus	C. Hood	Seneca Falls, N. Y	Apr. 15, 1873	137, 843
Cooking-apparatus	J. Jennings	New York, N. Y	May 14, 1830	....
Cooking-apparatus	A. C. Kasson	Milwaukee, Wis	July 10, 1866	56, 232
Cooking-apparatus	J. Morris	New Haven, Conn	Oct. 3, 1812	....
Cooking-apparatus	W. W. Parrott	Boston, Mass	Mar. 11, 1837	142
Cooking-apparatus	G. Pierce	New York, N. Y	May 6, 1856	14, 828
Cooking-apparatus	J. Singer	Chicago, Ill	Dec. 1, 1863	40, 773
Cooking-apparatus	J. Smolinski	New York, N. Y	May 25, 1852	8, 974
Cooking apparatus	E. Sperry	Chicago, Ill	July 25, 1871	117, 476
Cooking-apparatus	D. Westerfield	New York, N. Y	Mar. 24, 1827	....
Cooking-apparatus	E. Whiteley	Cambridge, Mass	Aug. 18, 1863	39, 603
Cooking-apparatus	J. Zimmerman	Royalton, N. Y	May 9, 1865	47, 675
Cooking-apparatus, Alcohol	T. G. Clinton	Washington, D. C	Nov. 25, 1856	16, 112
Cooking-apparatus and house-warmer	R. Johnson	Baltimore, Md	Nov. 19, 1833	....
Cooking-apparatus and refrigerator	J. and I. Newburg	New York, N. Y	Nov. 6, 1866	59, 441
Cooking-apparatus, Camp	D. McKenzie	Brooklyn, N. Y	Sept. 24, 1861	33, 357
Cooking-apparatus, Gas	W. L. Fish	Newark, N. J	Feb. 24, 1863	37, 745
Cooking-apparatus, Gas	D. G. Haskins	Cambridge, Mass	Oct. 29, 1867	70, 207
Cooking apparatus, Gas and steam	J. S. Gallaher, jr	Washington, D. C	Feb. 12, 1856	14, 233
Cooking-apparatus, Lamp	J. G. Covey	Saint John, Canada	May 21, 1872	126, 874
Cooking-apparatus, Lamp for	S. Cooper	Washington, D. C	June 18, 1872	128, 022
Cooking-apparatus, Portable	G. B. Isham	Burlington, Vt	Jan. 19, 1869	85, 934
Cooking-apparatus, Portable	D. F. Jauss	Harrisburgh, Pa	Nov. 19, 1872	133, 155
Cooking-apparatus, Portable	J. and W. Toothill	Wallingford, Conn	Dec. 10, 1861	33, 909
Cooking-apparatus, Portable	G. Yinger	Harrisburgh, Pa	Feb. 11, 1873	135, 682
Cooking-apparatus, Portable lamp	A. M. Silver and F. White	London, England	Feb. 18, 1873	136, 104
Cooking apparatus, Steam	J. Bouis	Baltimore, Md	Jan. 6, 1812	....
Cooking apparatus, Steam	J. O. Clay	Hudson, Wis	Nov. 23, 1869	97, 048
Cooking apparatus, Steam	J. G. Covey	Saint John, Canada	May 30, 1871	115, 437
Cooking apparatus, Steam	H. W. Horton	Wheaton, Ill	May 24, 1859	24, 178
Cooking apparatus, Steam	J. A. Little	Cartersburgh, Ind	May 28, 1872	127, 253
Cooking apparatus, Steam	B. G. Martin	New York, N. Y	Apr. 19, 1864	42, 442
Cooking apparatus, Steam	F. Millikin	Boston, Mass	Apr. 21, 1863	38, 238
Cooking apparatus, Steam	A. F. W. Neynaber	Philadelphia, Pa	Nov. 28, 1865	51, 208

Index of patents issued from the United States Patent Office from 1790 *to* 1873, *inclusive*—Continued.

Invention.	Inventor.	Residence.	Date.	No.
Cooking apparatus, Steam	E. Savage	Chicago, Ill	Sept. 29, 1868	82, 557
Cooking apparatus, Steam	C. W. Sterick	Mechanicsburgh, Pa	July 16, 1872	129, 613
Cooking apparatus, Steam	L. K. Williams	Hudson, Mich	Apr. 2, 1872	125, 155
Cooking-apparatus, Water-heater surrounding fire-pot of.	E. Whiteley	Boston, Mass	June 17, 1856	15, 156
Cooking, Applying heat by means of iron castings and grates for.	P. E. Sanborn	Troy, N. Y	June 11, 1829	....
Cooking-boiler	L. S. De Bibory	Baltimore, Md	May 11, 1852	8, 939
Cooking by steam, Vessel for	L. S. Hall	Washington, D. C	Feb. 16, 1864	41, 618
Cooking-furnace	H. Engelmann	Salt Lake City, Utah	Aug. 12, 1873	141, 778
Cooking-furnace	H. Engelmann	Salt Lake City, Utah	Aug. 12, 1873	141, 779
Cooking-furnace	H. Gleason	Boston, Mass	Dec. 15, 1837	513
Cooking furnace and oven, Portable	S. B. Badger	Dover, N. H	Jan. 30, 1829	....
Cooking furnace, Charcoal	A. Madison	Cincinnati, Ohio	Dec. 28, 1869	98, 282
Cooking-furnace, Portable	A. Cross	Cazenovia, N. Y	May 12, 1830	....
Cooking-furnace, Portable	W. Haggerty, C. Lawrence, and T. Frazer.	New York, N. Y	May 22, 1830	....
Cooking-furnace, Portable	J. D. Kellogg, jr	Northampton, Mass	Aug. 10, 1869	93, 538
Cooking-furnace, Portable	R. H. Waldron	Portsmouth, N. H	Mar. 30, 1869	88, 352
Cooking furnace, Summer	R. O. Stevenson	Baltimore, Md	Apr. 5, 1870	101, 676
Cooking-machine	F. Butler		Aug. 22, 1800	....
Cooking-machine	J. Seger	New York, N. Y	Nov. 14, 1815	....
Cooking-machine, Portable	D. Asher	Philadelphia, Pa	Feb. 18, 1825	....
Cooking or boiling apparatus	T. K. Anderson	Boston, Mass	Mar. 22, 1833	....
Cooking-table	E. D. Dodge	Worcester, Mass	Sept. 6, 1870	107, 015
Cooking-utensil	B. W. Dunklee	Boston, Mass	Feb. 14, 1860	27, 115
Cooking-utensil	T. Godfrey	New York, N. Y	Jan. 19, 1864	41, 338
Cooking-utensil	J. Mansfield	Jefferson, Wis	Feb. 4, 1873	135, 483
Cooking-utensil	J. A. Morrison	Brady's Bend, Pa	Apr. 19, 1870	102, 031
Cooking-utensil	E. L. Packard	Stoughton, Mass	Oct. 18, 1870	108, 385
Cooking-utensil	J. Smith and I. E. Hall	Logan, Ohio	Aug. 11, 1868	80, 881
Cooking-utensil	S. Spoor	Phelps, N. Y	Jan. 19, 1869	86, 108
Cooking-utensil	E. J. Sprague	Youngstown, Ohio	May 27, 1873	139, 433
Cooking-utensil	F. P. Warren	East Court Gosham, Great Britain.	Apr. 23, 1872	126, 114
Cooking-utensil for boiling and steaming	J. Stevens	Middletown, Md	Oct. 31, 1848	5, 890
Cooking-utensil, Hollow cast-iron	A. V. Van Hoevenbergh	Southside, N. Y	Apr. 20, 1858	20, 009
Cooking-utensils, Cover for	L. B. Oviatt	Brooklyn Village, Ohio	Sept. 27, 1870	107, 803
Cooking-vessel	M. V. B. Johnson	Holden, Mo	May 24, 1870	103, 468
Cooking-vessel	W. Y. Thomson	Oyster Bay, N. Y	Mar. 4, 1873	136, 467
Cooking-vessel	W. W. Tice	California, Ohio	Sept. 17, 1872	131, 479
Cooking-vessel for frying, steaming, &c	R. Broome	Central Falls, R. I	Apr. 30, 1867	64, 278
Cooking vessel or boiler	W. Y. Thomson	Oyster Bay, N. Y	Oct. 15, 1872	132, 334
Cooking, washing, &c., Apparatus for	S. S. Fitch	New York, N. Y	Nov. 28, 1865	51, 164
Cooking with quick-lime, Apparatus for	W. W. Albro	Binghamton, N. Y	Mar. 4, 1856	14, 340
Cooler:				
See Ale, beer, and water cooler.				
Beer-cooler.				
Brewer's cooler.				
Brewing and distilling cooler.				
Butter-cooler.				
Butter-tub cooler.				
Coffee-cooler.				
Corpse-cooler.				
Dairy-cooler.				
Double cooler.				
Filter-cooler.				
Flour-cooler.				
Graduated cooler.				
Lard-cooler.				
Lining for cooler.				
Liquid-cooler.				
Liquor-cooler.				
Mash-cooler.				
Milk-cooler.				
Provision-cooler.				
Refrigerator water-cooler.				
Sirup-cooler.				
Soap-cooler.				
Soda-water cooler.				
Vapor-cooler.				
Water-cooler.				
Water and beer cooler.				
Wine-cooler.				
Cooler	G. D. Blocher	Indianapolis, Ind	Aug. 21, 1866	57, 281
Cooler	W. F. Messenger and H. Rehahn.	New York, N. Y	Sept. 22, 1857	18, 263
Cooler	C. F. Pike	Providence, R. I	Jan. 25, 1870	99, 107
Cooler, Air	P. Nézeraux	Paris, France	Mar. 11, 1873	136, 753
Cooler, Air, liquid, &c	D. E. Somes	Washington, D. C	Aug. 6, 1872	130, 250
Cooler and filter	H. W. Fisher	Philadelphia, Pa	Sept. 24, 1867	69, 200
Cooler and freezer	A. C. Twining	New Haven, Conn	Dec. 24, 1861	34, 018
Cooler and freezer	A. C. Twining	New Haven, Conn	Apr. 22, 1862	35, 051
Cooler and preserver	C. F. Pike	Providence, R. I	Nov. 7, 1871	120, 667
Cooler and preserver	D. E. Somes	Washington, D. C	Mar. 8, 1870	100, 683
Cooler and refrigerator	J. Agate	Pittsford, N. Y	Aug. 31, 1869	94, 373
Cooler and refrigerator	A. Koch	Baltimore, Md	May 5, 1868	77, 624
Cooler and refrigerator	P. H. Vander Weyde	New York, N. Y	July 19, 1870	105, 609
Cooler and ventilator	J. J. Schillinger	New York, N. Y	Aug. 17, 1869	93, 840
Cooler, Room	A. M. Lesley	New York, N. Y	June 3, 1873	139, 470
Coolers and condensers, Construction of tube-sheets for.	W. A. Lighthall	New York, N. Y	June 11, 1861	32, 522
Cooling air in buildings and chambers, Mode of	D. E. Somes	Washington, D. C	June 27, 1865	48, 456
Cooling air in buildings, Mode of	D. E. Somes	Washington, D. C	Oct. 11, 1864	44, 672
Cooling air, Mode of	D. E. Somes	Washington, D. C	Nov. 28, 1865	51, 237
Cooling and conveying up meal, &c	G. F. Sal onstall		Aug. 21, 1801	....
Cooling and disinfecting apparatus	S. C. Maine	Boston, Mass	Dec. 4, 1866	60, 212

Index of patents issued from the United States Patent Office from 1790 *to* 1873, *inclusive*—Continued.

Invention.	Inventor.	Residence.	Date.	No.
Cooling and preserving meats, fruits, vegetables, &c., Apparatus for.	C. F. Pike	Providence, R. I	Dec. 11, 1866	60, 552
Cooling and preserving milk and other liquids	D. E. and F. C. Somes	Washington, D. C	May 16, 1871	114, 985
Cooling and preserving perishable articles, Process and apparatus for.	D. E. Somes	Washington, D. C	Oct. 19, 1869	96, 048
Cooling and purifying air in breweries, packing-houses, &c.	L. Schulze	Philadelphia, Pa	Aug. 19, 1873	142, 046
Cooling and ventilating dwellings and other buildings.	D. E. Somes	Washington, D. C	Feb. 28, 1865	46, 596
Cooling and ventilating rooms, &c., Method of	A. S. Lyman	New York, N. Y	Mar. 25, 1856	14, 510
Cooling and ventilating ships and other vessels	D. E. Somes	Washington, D. C	Feb. 28, 1865	46, 593
Cooling and ventilating ships and other vessels, Mode of.	D. E. Somes	Washington, D. C	June 27, 1865	48, 457
Cooling buildings and making ice, Apparatus for	W. S. Mason	Raleigh, N. C	June 21, 1870	104, 614
Cooling, freezing, and heating apparatus	D. E. Somes	Washington, D. C	Oct. 15, 1867	69, 955
Cooling, freezing, and preserving solids, liquids, and gases.	D. E. Somes	Washington, D. C	Feb. 28, 1871	112, 294
Cooling parts of the body, Apparatus for	O. M. Fuller	Catasauqua, Pa	Aug. 2, 1864	43, 679
Cooling preserving-house, packing-house, refrigerator, &c.	D. E. Somes	Washington, D. C	Feb. 28, 1865	46, 595
Cooling rooms and buildings, Apparatus for	T. D. Kingan	Indianapolis, Ind	Apr. 29, 1873	138, 411
Cooling storage-rooms, Apparatus for	J. Ring	Saint Louis, Mo	July 15, 1873	140, 792
Cooling-tub for water and beer	A. Haustetter	Philadelphia, Pa	July 12, 1864	43, 498
Coop, Chicken	J. R. Achenbach	Saddle River, N. J	Nov. 15, 1870	109, 166
Coop, Chicken	S. S. Bent	Port Chester, N. Y	May 19, 1868	78, 044
Coop, Chicken	W. J. H. Kappe	Quincy, Ill	June 24, 1873	140, 277
Coop, Chicken	J. H. Mabbett	Jersey City, N. J	May 19, 1868	78, 109
Coop, Folding	E. P. Lawrence	Worcester, Mass	Dec. 9, 1873	145, 302
Coop, Folding chicken	G. E. Cleeton	New Haven, Conn	Dec. 7, 1869	97, 605
Coop, Folding chicken	E. J. Wilcox	Joy Mills, Pa	Nov. 14, 1871	120, 921
Coop, Fowl	J. B. Smith	North Haven, Conn	Aug. 29, 1871	118, 494
Coop, Poultry	S. S. Bent	Port Chester, N. Y	Mar. 23, 1869	88, 118
Coop, Poultry	J. Golder	Fort Recovery, Ohio	Dec. 22, 1868	85, 226
Coop, Poultry	M. Potter	Girard, Pa	Apr. 2, 1872	125, 142
Coop, Poultry	W. J. Sloan	Smith's Ferry, Pa	Mar. 16, 1869	87, 798
Cooper's croze	J. F. Brodhead	Kingston, N. Y	Oct. 8, 1838	967
Cooper's croze and howel	W. R. Davis	Union City, Pa	Sept. 23, 1873	142, 995
Cooper's hoops out of split timber, Machine for making.	J. McWhorter	Brookville, Ind	Nov. 6, 1843	3, 328
Cooper's howel, Adjustable	M. Moriarty	Bangor, Me	July 16, 1872	129, 419
Cooper's tool	J. Christy	Clyde, Ohio	Dec. 21, 1869	98, 029
Cooper's tool	J. P. Heacock	Marlborough, Ohio	Nov. 27, 1855	13, 846
Cooper's tool	M. Sutton	Penfield, N. Y	Sept. 9, 1835	....
Cooper's tool	J. P. Wood	Troy, Ohio	Apr. 17, 1860	27, 955
Cooper's ware, Machinery for dressing headings for.	E. Greenlee	Summer Hill, Pa	July 21, 1863	39, 268
Cooper's ware, Sawing round bodies for	S. Newton	Hamilton County, Ohio	June 29, 1833	....
Cooper's work, Machinery for	H. Wright	Onondaga, N. Y	Oct. 23, 1822	....
Cooper's work, Manufacture of	H. Waters	Watertown, N. Y	Apr. 3, 1828	....
Coopering-machine	O. Barber	Hartford, Conn	Jan. 8, 1810	....
Cop-tube	J. Eaton	Townsend Harbor, Mass	Apr. 13, 1858	19, 918
Cop-tube	J. Marland and E. Crockett	Lawrence, Mass	Apr. 28, 1857	17, 164
Cop-tube	N. Whitmore	Somerville, Mass	Jan. 6, 1857	16, 363
Cop-tube and machine for making	J. E. Coffin	Portland, Me	Jan. 2, 1872	122, 362
Cop-tube machines, Die for	J. Eaton	Townsend Harbor, Mass	Jan. 30, 1855	12, 311
Cop-tube-making machine	H. and J. Douglas	Glasgow, Scotland	Oct. 12, 1869	95, 670
Cop-tubes, Machine for making	R. Douglas	Lowell, Mass	Apr. 30, 1872	126, 192
Cop-tubes, Machinery for making	W. Whitmore	Lynn, Mass	June 12, 1847	5, 157
Cop-waste, Machine for picking	A. A. Wood	Jersey City, N. J	June 22, 1858	20, 677
Cop-winding machine	W. Chadwick and S. Lownds	Brooklyn, N. Y	Oct. 7, 1873	143, 497
Cop-winding machine	B. H. Jenks	Bridesburgh, Pa	Aug. 21, 1866	57, 330
Cops, Composition for pasting	J. Dunkerley and J. Knight	Paterson, N. J	Jan. 19, 1864	41, 286
Copper and brass goods, Process of utilizing the waste formed in cleaning.	F. Wilcox	Waterbury, Conn	Nov. 2, 1869	96, 525
Copper and cast-steel, Combining	J. Park, jr	Pittsburgh, Pa	Jan. 14, 1868	73, 375
Copper and in separating other metals therefrom, Manufacture of.	J. B. Elkington	Birmingham, England	Feb. 22, 1870	100, 131
Copper and other ores, Process for the reduction of	H. Stull	Ione City, Cal	Dec. 30, 1873	145, 031
Copper boiler	A. C. Brownell	Brooklyn, N. Y	Sept. 15, 1863	39, 881
Copper bolt for ship-building	Berg and Ivers	New York	Mar. 15, 1824	....
Copper bolts, Uniting	G. Morgan	Onancock, Va	May 2, 1843	3, 070
Copper, Coating metals with	O. Gauduin, J. B. J. Mignon, and S. H. Rouart.	Paris, France	July 22, 1873	141, 132
Copper for boilers, &c., Mode of preparing sheet	A. O'Neill	Portsmouth, Ohio	Apr. 6, 1869	88, 660
Copper from its ores, Extracting	N. Haskell	San Francisco, Cal	Mar. 22, 1870	101, 009
Copper from its ores, Extracting	T. S. Hunt and J. Douglass, jr	Montreal and Quebec, Canada.	Feb. 9, 1869	86, 754
Copper from ores, Extracting	F. W. Dähne	Swansea, Great Britain	June 10, 1862	35, 509
Copper, Manufacture of sheet	L. Powe	Pittsburgh, Pa	Oct. 28, 1862	36, 822
Copper, Melting	J. Kintz	West Meriden, Conn	Sept. 26, 1871	119, 365
Copper, nickel, and cobalt, Separating	A. Monnier	Philadelphia, Pa	Sept. 29, 1863	40, 116
Copper-ores, Process of reducing	J. Napier	Shacklewell, England	June 5, 1847	5, 140
Copper-ores, Reduction of	J. Napier	Shacklewell Lane, England	Feb. 29, 1848	5, 461
Copper-plates, Finishing	B. F. Dudley	Boston, Mass	Oct. 18, 1870	108, 337
Copper, Process for production of sulphate and oxides of.	H. Holland	Westfield, Mass	Dec. 27, 1859	26, 589
Copper pyrites, Treating	C. M. T. Du Motay	Paris, France	Oct. 18, 1870	108, 462
Copper-roasting furnaces, Purifying the waste gases from.	A. Bigelow and J. S. Baldwin	Newark, N. J	May 18, 1869	90, 067
Copper sheathing, Machinery for cutting and punching.	W. H. Danforth	Salem, Mass	June 13, 1848	5, 625
Copper sheathing-nails, &c., Mold for casting	T. McIlwham	Princeton, N. J	Oct. 6, 1807	....
Copper tubes, Casing	J. F. Guthrie	South Bridgewater, Mass	Mar. 26, 1872	125, 044
Copper tubes, Casting	J. F. Guthrie	Somerville, Mass	Nov. 16, 1869	96, 911
Copper tubes, Machinery for making	E. Hamilton	Bridgeport, Conn	May 28, 1850	7, 397
Copper vessels, Planishing	H. Kay and T. Avery, jr	Brooklyn and Morrisania, N. Y.	July 17, 1860	29, 177
Copper, Welding	E. Renaud	Washington, D. C	Dec. 16, 1873	145, 686

Index of patents issued from the United States Patent Office from 1790 *to* 1873, *inclusive*—Continued.

Invention.	Inventor.	Residence.	Date.	No.
Copper, Welding	C. L. Schurr and W. G. Rehbein	Baltimore, Md	Nov. 14, 1871	121, 009
Copperas, Manufacture of	R. D. Birch	Philadelphia, Pa	Jan. 31, 1871	111, 305
Copperas, Method of making	I. Tyson	Baltimore, Md	Feb. 15, 1827	....
Copy-holder	J. S. Butler	Silver City, Idaho	Nov. 2, 1869	96, 391
Copy-holder	W. R. Carter	Brooklyn, N. Y	Oct. 14, 1873	143, 616
Copy-holder	L. G. Davis	Salem, Ind	Dec. 17, 1872	134, 036
Copy-holder	H. B. Denny	Washington, D. C	July 13, 1869	92, 588
Copy-holder	S. French and R. D. Chase	Orange, Mass	May 13, 1873	138, 877
Copy-holder	A. B. Manard	Rockford, Ill	Nov. 28, 1871	121, 291
Copy-holder	C. B. Moseley and L. L. Woolley.	Medford, Mass	Jan. 22, 1867	61, 444
Copy-holder	H. A. Tremper	Hammonton, N. J	Dec. 24, 1867	72, 569
Copy-holder	A. Westcott	Syracuse, N. Y	Aug. 17, 1869	93, 931
Copy-holder	H. R. Williams	Buffalo, N. Y	Mar. 14, 1871	112, 664
Copy-holder, Compositor's	P. A. La France	Elmira, N. Y	Nov. 12, 1867	70, 865
Copy-holder, Printer's	H. W. Knight	Seneca Falls, N. Y	Nov. 2, 1869	96, 445
Copy-holder, Printer's	W. H. and C. J. Young	Cambridge, Mass	Sept. 14, 1869	94, 807
Copies from manuscript, Process of multiplying	A. C. F. N. De St. Victor	Paris, France	Jan. 21, 1868	73, 514
Copies of manuscript, &c., Mode of taking	M. B. Hardin	New York, N. Y	Feb. 5, 1867	61, 733
Copying	C. Willcox	New Haven, Conn	Oct. 24, 1871	120, 223
Copying and folding press	S. W. Odell	Saint Louis, Mo	Dec. 9, 1873	145, 306
Copying-apparatus	W. J. Purcell	New York, N. Y	June 27, 1871	116, 351
Copying-apparatus, Portable	W. Van Anden	Poughkeepsie, N. Y	Sept. 7, 1858	21, 456
Copying colored and defective drawings	I. Rehn	Washington, D. C	May 16, 1871	114, 968
Copying designs, Apparatus for	T. Hall	Newton, Mass	Oct. 1, 1872	131, 819
Copying figures, &c., Machine for	F. C. Meyer	Philadelphia, Pa	Apr. 24, 1860	28, 040
Copying letters, Press for	G. Burnham	Philadelphia, Pa	Oct. 8, 1850	7, 696
Copying-machine	L. Carpenter	Oswego, N. Y	Dec. 31, 1842	2, 894
Copying manuscript	J. Jones	Clyde, N. Y	June 1, 1852	8, 980
Copying-press	C. Adams	Pittsburgh, Pa	Jan. 24, 1854	10, 438
Copying-press	J. H. Atwater	Providence, R. I	Jan. 13, 1863	37, 379
Copying-press	S. G. Cabell	Quincy, Ill	June 18, 1867	65, 873
Copying-press	C. Chambers, jr	Philadelphia, Pa	Oct. 2, 1860	30, 204
Copying-press	E. Clark	New York, N. Y	Dec. 18, 1860	30, 959
Copying-press	D. G. Coppin	Cincinnati, Ohio	July 16, 1872	129, 321
Copying-press	J. Fensom	Toronto, Canada	July 11, 1871	116, 824
Copying-press	J. Fowler	Saugatuck, Mich	June 29, 1869	92, 032
Copying-press	G. C. Gage	Waterford, N. Y	Apr. 20, 1869	89, 035
Copying-press	C. F. Grabo	Boston, Mass	Aug. 14, 1860	29, 585
Copying-press	W. H. Hawkins	Cleveland, Ohio	Mar. 29, 1870	101, 262
Copying-press	C. Y. Heckler	Philadelphia, Pa	May 21, 1861	32, 363
Copying-press	F. Hovey	New York, N. Y	Mar. 29, 1864	42, 141
Copying-press	T. P. How	Brooklyn, N. Y	Apr. 7, 1868	76, 457
Copying-press	J. M. Keep	New York, N. Y	Aug. 10, 1869	93, 536
Copying-press	C. Knauer	Pittsburgh, Pa	June 3, 1856	15, 011
Copying-press	P. Lawrence and G. Jefferys	New York, N. Y	May 30, 1867	48, 021
Copying-press	A. Le Clercq	New York, N. Y	Dec. 26, 1871	122, 265
Copying-press	E. and J. B. Platt	Clarke County, Ga	Oct. 26, 1858	21, 902
Copying-press	W. Shriver	New York, N. Y	June 13, 1865	48, 217
Copying-press	W. Shriver	New York, N. Y	Feb. 13, 1872	123, 584
Copying-press	E. H. Smith	New York, N. Y	June 14, 1853	9, 786
Copying-press	W. M. Smith	Washington, D. C	May 12, 1867	17, 307
Copying-press	W. M. Smith	Washington, D. C	Feb. 23, 1869	87, 304
Copying-press	J. D. Spaulding and D. N. B. Coffin, jr.	Boston, Mass	Jan. 16, 1866	52, 086
Copying-press	G. C. Tait	Worcester, Mass	June 21, 1864	43, 266
Copying-press	G. Tangye	Birmingham, England	Oct. 26, 1869	96, 285
Copying-press	W. W. Underhill	Boston, Mass	Aug. 10, 1869	93, 567
Copying-press	A. Whitcomb	Worcester, Mass	Jan. 4, 1859	22, 524
Copying-press	A. Whitcomb	Worcester, Mass	July 4, 1871	116, 781
Copying-press	A. A. Wilder	Detroit, Mich	Mar. 11, 1851	7, 978
Copying-press bed	W. Shriver	New York, N. Y	Aug. 6, 1872	130, 320
Copying-press damper	A. Pearl	New York, N. Y	Oct. 13, 1868	83, 088
Copying press, Letter	G. C. Taft	Worcester, Mass	Apr. 24, 1860	28, 020
Copying-press, Portable	J. H. Atwater	Providence, R. I	Apr. 2, 1861	31, 860
Copying-press, Portable	H. M. Paine	Worcester, Mass	Oct. 2, 1849	6, 752
Copying-press table	B. Brower	New York, N. Y	Aug. 5, 1873	141, 422
Copying surface by electricity	A. Bain	London, England	Dec. 5, 1848	5, 957
Copying writing, maps, &c., Process of	J. Underwood	London, England	Mar. 31, 1863	38, 086
Cord and line reel	H. W. Chamberlain	Jersey City, N. J	Oct. 2, 1866	58, 378
Cord-catch	C. C. Moore	New York, N. Y	May 3, 1870	102, 575
Cord-clamp	S. W. Meredith	Greensburgh, Ind	Mar. 12, 1872	124, 604
Cord-clamp	S. W. Meredith	Greensburgh, Ind	Aug. 6, 1872	130, 143
Cord-coupling	L. I. Ware	Warren, R. I	May 2[illegible], 1851	8, 069
Cord-covering machine	J. Bachelder	Norwich, Conn	Sept. 1, 1868	81, 577
Cord-covering machine	J. Buser	New York, N. Y	Sept. 5, 1865	49, 825
Cord-covering machine	T. N. Dale, jr., and G. Kraink	Paterson, N. J	Sept. 6, 1870	107, 610
Cord-covering machine	J. E. Gillespie and T. Kohn	Hartford, Conn	Apr. 1, 1873	137, 360
Cord-covering machine	H. Heinemann and J. Buser	New York, N. Y	Dec. 18, 1860	30, 944
Cord-covering machine	R. Lewis	New York, N. Y	Apr. 4, 1871	113, 675
Cord-covering machine	W. H. Palmer, jr	Middletown, Conn	Dec. 1, 1868	84, 506
Cord-covering machine	J. Turner	Norwich, Conn	Sept. 1, 1868	81, 845
Cord folding and measuring machine	G. P. Farmer	Philadelphia, Pa	May 10, 1870	102, 931
Cord-holder	J. P. Gruger and C. Makinson	Lancaster, Pa	Apr. 21, 1868	77, 034
Cord-holder for picture-frames, &c	H. D. Heureuse	New York, N. Y	June 8, 1869	90, 998
Cord-hook for door-springs, Adjustable	W. B. Barnard	Bristol, Conn	Mar. 5, 1850	7, 132
Cord-making machine	H. Boardman	Lancaster, Pa	Feb. 2, 1869	86, 497
Cord-making machine	H. Boardman	Lancaster Township, Pa	Aug. 29, 1871	118, 426
Cord-making machine	W. Guest	London, England	July 27, 1869	92, 960
Cord-making machine	J. McIntire	Hopewell Cotton Works, Pa	Jan. 25, 1870	99, 099
Cord-making machine	E. Otis	New Haven, Conn	Dec. 23, 1873	145, 747
Cord-making machine	G. T. Wright	New Preston, Conn	Aug. 3, 1869	93, 265
Cord-making machinery	W. E. Nichols	East Haddam, Conn	Dec. 11, 1849	6, 942
Cord plaiting and twisting	W. H. Zahn	New York, N. Y	May 8, 1855	12, 845
Cord-stretcher	R. White	Decatur, Ill	Oct. 22, 1867	70, 057
Cord twisting and plaiting machine	J. T. Williams	Newark, N. J	May 3, 1864	42, 613
Cord-winder	C. Goodwin	New Brunswick, N. J	Apr. 15, 1862	34, 954
Cords, braids, &c., Machine for starching and glazing.	D. McInroy	New York, N. Y	Dec. 12, 1865	51, 471

Index of patents issued from the United States Patent Office from 1790 *to* 1873, *inclusive*—Continued.

Invention.	Inventor.	Residence.	Date.	No.
Cords, Guide for laying	W. Taylor	Berlin, N. Y	Feb. 19, 1861	31, 494
Cords into a series of united skeins, Machine for winding.	E. Brown	South Otselic, N. Y	Feb. 6, 1866	52, 382
Cords, Machine for dressing worsted	I. E. Palmer	Hackensack, N. J	Apr. 18, 1871	113, 917
Cords, &c., Machine for drying sized or dyed	D. McInroy	New York, N. Y	July 30, 1867	67, 328
Cords, Machine for making covered	J. Turner	Norwich, Conn	June 15, 1869	91, 498
Cords, Machine for weaving a covering for	J. Danby	New York, N. Y	Mar. 20, 1866	53, 372
Cords, Machinery for manufacturing plaited	C. Feickert	New York, N. Y	Mar. 9, 1858	19, 554
Cords, Machinery for weaving shade	T. Nelson	Troy, N. Y	Dec. 16, 1856	16, 248
Cords, Method of double-stretching strands of banding.	T. Unsworth	Manchester, England	Mar. 26, 1872	124, 988
Cords or bands, Machinery for making	T. Unsworth and E. Whalley	Preston, England	Aug. 22, 1871	118, 408
Cords or covering around cords, Machine for weaving.	F. Veerkamp and F. Leopold	Philadelphia, Pa	Aug. 23, 1864	43, 937
Cords, ropes, &c., Coupling-attachment for	J. L. Howard	Hartford, Conn	Apr. 17, 1860	27, 906
Cords, ropes, &c., Machine for making	J. O. Mathieu	Paris, France	Jan. 24, 1865	46, 057
Cords, ropes, &c., Machine for manufacturing	J. A. De Maniquet	Paris, France	July 5, 1864	43, 461
Cords, Stop-motion for machine for covering	R. Lewis	New York, N. Y	Apr. 9, 1872	125, 467
Cords, wires, &c., Covering	F. Beck	New York, N. Y	Dec. 1, 1863	40, 801
Cordage and improving its quality, Machine for manufacturing.	R. Fulton and N. Cutting	New York, N. Y., and Washington, D. C.	Mar. 4, 1808	....
Cordage by machinery, Mode of making	R. Graves	Brooklyn, N. Y	July 25, 1827	....
Cordage, Flexible	H. H. Matteson	Buffalo, N. Y	Jan. 10, 1854	10, 411
Cordage-laying machine	R. Greaves	Boston, Mass	Feb. 1, 1821	....
Cordage-machine	J. P. Arnold	Louisville, Ky	Mar. 24, 1857	16, 867
Cordage-machine	J. P. Arnold	Louisville, Ky	May 19, 1857	17, 310
Cordage-machine	C. Cobb	Plymouth, Mass	July 17, 1866	56, 372
Cordage-machine	A. Hathaway	Boston, Mass	July 9, 1839	1, 228
Cordage-machine	J. W. Peer	Schenectady, N. Y	July 6, 1852	9, 101
Cordage-machine	J. Pine	Hoosick Falls, N. Y	Feb. 24, 1857	16, 694
Cordage, Machine for making	A. Hill	Walpole, N. Y	June 30, 1812	....
Cordage, Machine for making	W. Joslin	Waterford, N. Y	Mar. 23, 1852	8, 825
Cordage, Machine for making	I. E. Palmer	Hackensack, N. J	Mar. 16, 1869	87, 964
Cordage, Machine for manufacturing	S. and J. A. Bazen	Canton, Mass	Oct. 17, 1817	....
Cordage, Machine for manufacturing long	W. E. Meginnis	Philadelphia, Pa	Nov. 9, 1839	1, 404
Cordage, Machine for pressing tar out of yarn for	T. Barnitt	Philadelphia, Pa	May 20, 1822	....
Cordage-machines, Laying top for	W. Robinson	Warsaw, N. Y	Jan. 20, 1857	16, 452
Cordage-machinery	H. T. Jennings, C. S. Collier, and T. P. How.	Bethany and Buffalo, N. Y.	Nov. 16, 1852	9, 414
Cordage-machinery	H. Pearce	Cincinnati, Ohio	May 22, 1855	12, 920
Cordage-machinery	G. Stephenson	Northfield, Ind	May 31, 1859	24, 244
Cordage-machinery	P. B. Tyler	Springfield, Mass	June 27, 1854	11, 202
Cordage, Machinery for braiding	J. A. Bazin	Canton, Mass	June 29, 1858	20, 690
Cordage, Machinery for making	B. F. Adams	New Bedford, Mass	June 13, 1846	4, 574
Cordage, Machinery for making	F. Kellsey	Middletown, Conn	Feb. 8, 1825	....
Cordage, Machinery for making	R. Porter	Washington, D. C	Feb. 21, 1854	10, 559
Cordage, Machinery for making cotton	F. Slaughter and D. Perry	Fredericksburgh, Va	Jan. 1, 1850	6, 992
Cordage, Machinery for making cotton	F. Slaughter and D. Perry	Fredericksburgh, Va	Jan. 8, 1850	7, 007
Cordage, Machinery for manufacturing	W. Joslin	Waterford, N. Y	Jan. 19, 1847	4, 931
Cordage, Making	E. Bartholomew	Boston, Mass	May 17, 1826	....
Cordage, Making	J. Drummond	Brooklyn, N. Y	July 22, 1833	....
Cordage, Making and laying	C. L. Sargent	Suffolk, Mass	Oct. 4, 1817	....
Cordage-making machine	J. Hall	Munroe Township, N. Y	June 11, 1867	65, 565
Cordage-making machine	W. B. Leonard	Fishkill, N. Y	Jan. 9, 1819	....
Cordage-making machine	D. Perry	Fredericksburgh, Va	June 15, 1852	9, 040
Cordage, Manufacturing	G. Parkinson	....	June 16, 1794	....
Cordage, Manufacturing	J. Pitman	....	May 25, 1795	....
Cordage, &c., Mode of laying	R. Graves	Boston, Mass	Nov. 22, 1820	....
Cordage, Top-sled for	R. Graves	Boston, Mass	Mar. 28, 1821	....
Cordage uninflammable, Process for rendering	J. H. Johnson	New Orleans, La	Nov. 26, 1850	7, 797
Cordage, webbing, &c., Manufacturing braided	J. A. Bazin	Canton, Mass	June 29, 1858	20, 691
Corded bindings for India-rubber and other fabrics, Machine for making.	C. A. Ensign	Naugatuck, Conn	July 17, 1866	56, 393
Cording, webbing, &c., Machine for making	J. A. Bazin	Canton, Mass	July 17, 1866	56, 485
Cordial, Anti-bilious stomach	S. Lozarus	....	Dec. 21, 1802	....
Core-bar	R. T. Crane	Chicago, Ill	Aug. 4, 1868	80, 607
Core-bar for casting pipes	J. Enright	Louisville, Ky	Oct. 13, 1868	82, 931
Core-bar for casting pipes	J. Enright	Louisville, Ky	June 14, 1870	104, 397
Core-bar for castings	D. Gallagher	Cincinnati, Ohio	Sept. 23, 1873	143, 005
Core-bar for forming cores for casting pipes	G. Peacock	West Troy, N. Y	Feb. 8, 1853	9, 577
Core-bar for molds used in casting metal	F. Shickle	Saint Louis, Mo	June 29, 1869	92, 108
Core-bar for pipe-molding	J. Demarest	Mott Haven, N. Y	Apr. 8, 1856	14, 637
Core-barrel	J. E. Thomas	Pittsburgh, Pa	Oct. 4, 1864	44, 563
Core-barrel, Collapsing	R. Lye	Pittsburgh, Pa	May 28, 1872	127, 178
Core-barrel, Collapsing	W. Smith	Pittsburgh, Pa	Jan. 17, 1871	111, 091
Core-barrel for castings	J. B. Aston	Pittsburgh, Pa	Sept. 9, 1873	142, 662
Core-barrel for castings	W. Smith	Pittsburgh, Pa	Sept. 9, 1873	142, 746
Core-box	J. C. Chapman	Waltham, Mass	Feb. 6, 1872	123, 329
Core-box	J. S. Harper	Baltimore, Md	June 5, 1860	28, 574
Core-box	A. Van Horn	New York, N. Y	Jan. 27, 1857	16, 502
Core-carriage	S. Fulton	Conshohocken, Pa	Apr. 16, 1861	32, 060
Core, Collapsing	A. T. Brodie and R. R. Smith	Pittsburgh, Pa	Jan. 24, 1871	111, 173
Core, Dry-sand	W. Gage and R. B. Felthousen	Buffalo, N. Y	July 7, 1857	17, 732
Core, Dry-sand	W. Gage and R. B. Felthousen	Buffalo, N. Y	Dec. 29, 1857	18, 964
Core for casting	J. Herald	Unadilla, N. Y	Dec. 13, 1870	110, 038
Core for casting iron, glass, &c., Expansible	A. Balding	Wheeling, W. Va	Mar. 8, 1870	100, 585
Core for castings, Iron	A. C. Mott	Philadelphia, Pa	Jan. 4, 1870	98, 516
Core for dikes	W. A. Thompson	Brooklyn, N. Y	Jan. 30, 1872	123, 306
Core for molding plastic substances	J. Pilgrim	New Britain, Conn	Dec. 28, 1858	22, 450
Core-making machine	B. S. Benson	Baltimore, Md	July 7, 1868	79, 629
Core or form for making castings, Composition	A. Burt	Detroit, Mich	July 20, 1869	92, 699
Core-powder, Composition for	W. B. Lupton	Pittsburgh, Pa	Mar. 27, 1866	53, 460
Core, Sectional	W. A. Butler	New York, N. Y	May 13, 1873	138, 853
Core-spindle for casting	D. A. Webster and G. F. Burroughs.	New York, N. Y., and Lumberton, N. J.	Oct. 20, 1857	18, 481
Core-spindle joint	W. F. Perkins	Boston, Mass	Apr. 30, 1872	126, 230
Core, Supporting	E. B. Phillips	Cambridge, Mass	Feb. 14, 1871	111, 775

Index of patents issued from the United States Patent Office from 1790 *to* 1873, *inclusive*—Continued.

Invention.	Inventor.	Residence.	Date.	No.
Core, Venting	G. G. Cressey	Philadelphia, Pa	Oct. 20, 1868	83, 135
Cores, Apparatus for making solid	S. J. Peet	New York, N. Y	Mar. 1, 1870	100, 318
Cores Device for forming	H. Parker	Gananoque, Canada	Dec. 23, 1873	145, 748
Cores, Flask for forming	E. R. Austin	Norwalk, Conn	Dec. 24, 1867	72, 586
Cores for axle-skeins and hub-boxes, Making sand	J. G. Holt	Chicago, Ill	Nov. 20, 1866	59, 840
Cores for castings, Composition for making	E. Rees	Cincinnati, Ohio	Dec. 17, 1850	7, 843
Cores for castings, Machine for making and holding.	L. H. Crocker	Cincinnati, Ohio	Oct. 22, 1850	7, 729
Cores for castings, Making	I. Kellogg	New Hartford, Conn	Apr. 25, 1846	4, 483
Cores for castings, Mode of making and venting	J. Harrison, jr	Philadelphia, Pa	Aug. 8, 1865	49, 264
Cores for foundery purposes, Venting	H. Tucker	Newton, Mass	Nov. 13, 1866	59, 684
Cores for molding articles of lead and other metals, Method of making.	L. Brandeis	Brooklyn, N. Y	Dec. 22, 1868	85, 056
Cores for molding iron, Compound for forming	J. I. Vinton	Altoona, Pa	Apr. 20, 1869	89, 187
Cores for pipe-castings, Method of making	W. E. Bird	West Bridgewater, Mass	Aug. 27, 1867	68, 033
Cores for stench-traps and other metal castings, Manufacture of.	C. Lightbody, jr	Brooklyn, N. Y	Oct. 15, 1872	132, 299
Cores from hollow castings, Method of loosening metallic.	J. C. Parry	Pittsburgh, Pa	Dec. 24, 1850	7, 859
Cores in flasks, Implement for undercutting sand	W. L. McDowell	Philadelphia, Pa	Feb. 22, 1870	100, 172
Cores in molds, Anchor for securing	Z. Ellis	Philadelphia, Pa	Jan. 11, 1870	98, 676
Cores, Machine for making pipe	W. and G. Braid	New York, N. Y	Aug. 15, 1865	49, 368
Cores, Machine for producing	S. J. Peet	New York, N. Y	Mar. 1, 1870	100, 320
Cores, Manufacture of	S. Fulton	Conshohocken, Pa	June 8, 1869	91, 008
Cores, Molding	J. P. Davis	Middletown, Conn	May 30, 1865	47, 937
Cores, Molding and compressing	C. Warner	Louisville, Ky	Jan. 9, 1849	6, 013
Cores, Venting metallic	A. Shepard	New Britain, Conn	Nov. 17, 1868	84, 072
Corer and cutter, Apple	J. Bowsir	Basil, Ohio	Dec. 6, 1864	45, 312
Corer and cutter, Apple	A. Frost	Seymour, Ind	Aug. 11, 1868	80, 938
Corer and cutter, Apple	S. Mead	Fort Branch, Ind	Mar. 5, 1872	124, 368
Corer and cutter, Apple	J. M. Meschutt	Newark, N. J	July 16, 1872	129, 289
Corer and cutter, Apple	C. D. Read	Lowell, Mass	Feb. 16, 1869	87, 004
Corer and quarterer, Apple	J. M. Hassam	Mount Vernon, Me	Oct. 27, 1868	83, 494
Corer and quarterer, Apple	B. J. McFeely	Chestnut Springs, Pa	Sept. 27, 1870	107, 794
Corer and quarterer, Apple	J. S. Tripp	South Danby, N. Y	Sept. 13, 1864	44, 239
Corer and slicer, Apple	G. Custer	Norristown, Pa	July 23, 1867	66, 951
Corer and slicer, Apple	E. Dean and C. O. Potter	Jersey City, N. J	July 9, 1872	128, 865
Corer and slicer, Apple	R. J. Dodd	Orth, Ind	Apr. 12, 1870	101, 838
Corer and slicer, Apple	R. Onderdonk	New York, N. Y	Feb. 20, 1866	52, 740
Corer and slicer, Apple	D. R. Reed	Tekonsha, Mich	July 14, 1868	80, 010
Corer and slicer, Apple	I. Rogers	West Chehalem, Oreg	July 14, 1868	80, 014
Corer and slicer, Apple	S. Saucerman	Freeport, Ill	July 25, 1865	48, 981
Corer and slicer, Apple	H. W. Williams	Galesburgh, Ill	Feb. 6, 1872	123, 540
Corer and slicer, Apple	G. C. Wright	Le Roy, Ohio	Oct. 22, 1867	70, 062
Corer and slicer, Apple	M. B. Wright	West Meriden, Conn	Dec. 28, 1869	98, 457
Corer and slicer, Fruit	C. E. Thurston and J. H. Wilkinson.	South New Market, N. H	May 7, 1872	126, 426
Corer, Apple	A. N. Alcott	Gowanda, N. Y	Feb. 23, 1858	19, 402
Corer, Apple	S. C. Collins	Oregon, Mo	Nov. 14, 1871	120, 857
Corer, Apple	M. M. Hatch	Portland, Me	July 21, 1868	80, 172
Corer, Apple	M. P. Smith	Baltimore, Md	Sept. 3, 1872	131, 126
Corer, Apple	G. L. Swett and B. F. Drake	Leominster, Mass	Sept. 28, 1869	95, 286
Corer, Fruit	E. J. Marsters	Shaw's Flat, Cal	Aug. 23, 1870	106, 706
Coring and quartering apples	C. Gates	Rutland County, Vt	Dec. 15, 1810	......
Coring and quartering machine, Apple	C. Lounsberry	Nichols, N. Y	Aug. 10, 1858	21, 141
Coring and quartering machine, Apple	J. J. Vankersen	Kalamazoo, Mich	Mar. 6, 1866	53, 063
Coring and slicing machine, Apple	C. H. Gifford	Auburn, Mass	Jan. 5, 1864	41, 147
Coring, slicing, and stringing machine, Apple	N. Bennett	Sherman, N. Y	June 13, 1865	48, 149
Cork	W. H. Towers	New York, N. Y	Dec. 23, 1862	37, 249
Cork, Air-tight	J. S. Davison	Cranberry, N. J	Dec. 1, 1863	40, 742
Cork and bung cutting machinery	W. R. Crocker	Norwich, Conn	Jan. 27, 1863	37, 543
Cork, Artificial	L. Boch and A. F. Wheeler	Sheboygan, Wis	Aug. 8, 1865	49, 220
Cork, Artificial	E. L. Perry and E. D. Lazell	Brooklyn, N. Y	Sept. 12, 1865	49, 916
Cork, Bottle	W. H. Newton	Newport, R. I	Feb. 21, 1871	112, 070
Cork-composition, Water-proof	A. Stevens	New York, N. Y	Dec. 7, 1858	22, 246
Cork-cutter	E. F. Harrington	Boston, Mass	Apr. 12, 1870	101, 730
Cork-cutter	L. Hills	Boston	June 18, 1827	......
Cork-cutter	E. O. Schartau	Philadelphia, Pa	Dec. 13, 1870	110, 075
Cork-cutting machine	R. P. Abernethy	Cincinnati, Ohio	July 6, 1858	20, 770
Cork-cutting machine	R. P. Abernethy and M. M. Wombaugh.	Cincinnati, Ohio	July 6, 1858	20, 771
Cork-cutting machine	A. F. Allen	Providence, R. I	July 16 1872	129, 304
Cork-cutting machine	O. Arnold	Pawtucket, R. I	Oct. 26, 1869	96, 184
Cork-cutting machine	A. J. Bailey	Charlestown, Mass	May 29, 1866	55, 036
Cork-cutting machine	H. Boardman	Lancaster, Pa	July 15, 1862	35, 867
Cork-cutting machine	E. A. Brimson	New York, N. Y	Aug. 24, 1869	93, 958
Cork-cutting machine	E. Conroy	Boston, Mass	Nov. 2, 1858	21, 944
Cork-cutting machine	E. Conroy	Boston, Mass	Feb. 4, 1862	34, 291
Cork-cutting machine	T. S. Disston	Philadelphia, Pa	Jan. 21, 1873	134, 982
Cork-cutting machine	L. W. Felt	Keene, N. H	June 25, 1867	66, 141
Cork-cutting machine	L. W. Felt	Keene, N. H	Apr. 14, 1868	76, 615
Cork-cutting machine	I. Goodspeed	Norwich, Conn	Oct. 27, 1863	40, 444
Cork-cutting machine	G. Hammer	Philadelphia, Pa	Oct. 14, 1851	8, 422
Cork-cutting machine	G. Hammer	Philadelphia, Pa	Feb. 15, 1859	22, 949
Cork-cutting machine	G. Hammer	Philadelphia, Pa	May 22, 1866	54, 896
Cork-cutting machine	G. Hammer	Philadelphia, Pa	Nov. 11, 1868	83, 850
Cork-cutting machine	P. Hayden	Pittsburgh, Pa	Dec. 19, 1865	51, 587
Cork-cutting machine	W. King	New York, N. Y	July 8, 1851	8, 208
Cork-cutting machine	G. Lloyd	Philadelphia, Pa	Sept. 29, 1863	40, 144
Cork-cutting machine	A. Mackie	Philadelphia, Pa	Jan. 9, 1866	52, 000
Cork-cutting machine	C. R. Macy	Hyde Park, N. Y	Oct. 31, 1840	1, 842
Cork-cutting machine	A. Millar	New York, N. Y	Jan. 29, 1861	31, 253
Cork-cutting machine	A. Millar	New York, N. Y	Apr. 9, 1861	31, 990
Cork-cutting machine	G. Purves	New York, N. Y	Aug. 22, 1871	118, 269
Cork-cutting machine	G. Rawlings	Philadelphia, Pa	Oct. 30, 1827	......
Cork-cutting machine	S. Sawyer	Boston, Mass	June 10, 1840	1, 623
Cork-cutting machinery	H. F. Cox and A. Millar	Jersey City, N. J., and New York, N. Y.	Aug. 20, 1861	33, 122
Cork-cutting tool	A. Brass	Newark, N. J	Nov. 15, 1859	26, 083

Index of patents issued from the United States Patent Office from 1790 *to* 1873, *inclusive*—Continued.

Invention.	Inventor.	Residence.	Date.	No.
Cork-drawer	C. Alexander	Washington, D. C	Aug. 7, 1860	29, 539
Cork-drawer	C. Chinnock	Brooklyn, N. Y	Apr. 14, 1863	38, 147
Cork-extractor	J. Autenrith	Philadelphia, Pa	Sept. 1, 1868	81, 728
Cork-extractor	E. A. Burgess	New Haven, Conn	May 21, 1861	32, 396
Cork-extractor	E. Button	Annapolis, Md	June 2, 1868	78, 513
Cork-extractor	G. L. Gibson, jr	Concord, N. C	Apr. 27, 1869	89, 477
Cork-extractor	C. Gooch	Cincinnati, Ohio	Jan. 25, 1870	99, 080
Cork-extractor	C. F. Hunt	Brooklyn, N. Y	July 8, 1873	140, 706
Cork-extractor	J. P. Miers and J. Groendyke	Lebanon, N. J	Jan. 26, 1864	41, 385
Cork-extractor	J. L. Morrill	New York, N. Y	May 17, 1864	42, 784
Cork-extractor	J. Morton	Philadelphia, Pa	Jan. 14, 1868	73, 370
Cork-extractor	C. Rosenberry	Chicago, Ill	Oct. 16, 1866	58, 889
Cork-extractor	J. Walker	Cincinnati, Ohio	Jan. 22, 1867	61, 488
Cork-extractor	W. G. Waterman	Middletown, Conn	Dec. 17, 1867	72, 247
Cork-extractor	C. G. Wilson	Brooklyn, N. Y	July 13, 1869	92, 552
Cork-fastener	E. D. Weatherbee	Worcester, Mass	May 11, 1869	90, 037
Cork for filling beds, mattresses, &c	S. Bates	Boston, Mass	Dec. 28, 1832	......
Cork-inserting apparatus	W. Rheiner and L. H. Wolff	Detroit, Mich	Jan. 15, 1867	61, 255
Cork into strips, Machinery for cutting	C. Gregor	New York, N. Y	Mar. 18, 1862	34, 716
Cork-machine	A. Albertson	New York, N. Y	May 3, 1859	23, 880
Cork-machine	H. Boardman	Lancaster, Pa	Sept. 19, 1865	49, 971
Cork-machine	E. Conroy	South Boston, Mass	June 16, 1857	17, 557
Cork-machine	I. Goodspeed	Norwich, Conn	June 4, 1861	32, 497
Cork-machine	H. Locke	South Boston, Mass	June 21, 1859	24, 471
Cork-machine	A. Millar	New York, N. Y	Apr. 16, 1861	32, 075
Cork-machine	J. Power	Boston, Mass	July 3, 1855	13, 200
Cork-machine	E. O. Schartau	Philadelphia, Pa	Dec. 13, 1870	110, 076
Cork, Metal-capped	E. Street	East Haven, Conn	Apr. 19, 1870	102, 181
Cork or stopper holder	G. H. Plummer	New York, N. Y	Dec. 2, 1873	145, 235
Cork-polishing machine	H. F. Cox and A. Millar	Jersey City, N. J., and New York, N. Y.	July 19, 1859	24, 841
Cork-press	C. L. Lochman	Carlisle, Pa	Aug. 27, 1867	68, 093
Cork-presser	J. Ewing	New York, N. Y	June 28, 1870	104, 841
Cork-pull	F. G. Bielefield and C. C. E. Schwartz.	Berlin, Prussia, and Hamburgh.	Apr. 4, 1865	47, 161
Cork-pull	S. E. Clapp	Cambridge, Mass	Jan. 21, 1868	73, 435
Cork-pull	C. Loffler	Hoboken, N. J	Oct. 30, 1866	59, 241
Cork-pull	G. W. Schermerhorn	East Limington, Me	Sept. 22, 1868	82, 355
Cork-pull	C. T. Simpers	Philadelphia, Pa	Nov. 14, 1871	120, 830
Cork-pull	J. D. Van Zandt	Brooklyn, N. Y	July 30, 1867	67, 234
Cork-pull	D. Williamson	New York, N. Y	Feb. 25, 1868	74, 966
Cork-pull	W. C. Wyckoff	Brooklyn, N. Y	May 21, 1861	32, 394
Cork-screw	J. Adt	Wolcottville, Conn	May 8, 1866	54, 640
Cork-screw	P. Blake	New Haven, Conn	Mar. 27, 1860	27, 665
Cork-screw	J. Bussey	Cincinnati, Ohio	Feb. 18, 1868	74, 495
Cork-screw	M. L. Byrn	New York, N. Y	Mar. 27, 1860	27, 615
Cork-screw	C. Chinnock	Brooklyn, N. Y	Nov. 24, 1863	40, 674
Cork-screw	S. E. Clapp	Cambridge, Mass	Feb. 11, 1868	74, 199
Cork-screw	J. L. Clark	Chester, Conn	Nov. 22, 1864	45, 137
Cork-screw	W. Dickson	Albany, N. Y	Aug. 2, 1870	106, 036
Cork-screw	J. E. Earle	New Haven, Conn	June 23, 1868	79, 216
Cork-screw	W. Fradgley	Greenbush, N. Y	June 10, 1862	35, 514
Cork-screw	S. McCown	Stamford, Conn	Nov. 26, 1867	71, 316
Cork-screw	W. C. McGill	Cincinnati, Ohio	Jan. 8, 1867	61, 080
Cork-screw	C. L. Ridgway	Boston, Mass	Aug. 18, 1868	81, 292
Cork-screw	A. T. Russel	New York, N. Y	Jan. 21, 1862	34, 216
Cork-screw	J. A. Smith	Brooklyn, N. Y	Dec. 6, 1870	109, 958
Cork-screw	G. Twigg	Birmingham, England	Jan. 21, 1868	73, 677
Cork-screw	W. H. Van Gieson	Passaic, N. J	Jan. 22, 1867	61, 485
Cork-screw attachment	W. Dickson	Albany, N. Y	June 6, 1871	115, 585
Cork-screw, cork-puller, and can-opener combined	J. Harrigan	Fort Wadsworth, N. Y	July 25, 1871	117, 278
Cork-screws, Holding-frame for	H. M. Creamer	Brooklyn, N. Y	Apr. 14, 1863	38, 155
Cork, Sirup	A. J. Morse	Boston, Mass	June 8, 1869	91, 152
Cork-slicing machine	J. Power	Boston, Mass	Aug. 8, 1865	49, 299
Cork-trimming machine	B. T. Roath	Norwich, Conn	Aug. 16, 1864	43, 867
Cork-wood, Machine for slicing	H. Boardman	Lancaster, Pa	Dec. 6, 1864	45, 310
Corks by steam, Mode of softening	B. Potter, jr	Charlestown, Mass	Nov. 18, 1856	16, 098
Corks for stoppers, Machine for cutting	J. D. Crocker	Norwich, Conn	Mar. 25, 1862	34, 741
Corks from bottles, Instrument for extracting	J. T. Ashley	Brooklyn, N. Y	Oct. 23, 1866	58, 969
Corks in bottles, &c., Method of fastening	G. Bousigues, (dit Bley)	Rheims, France	Mar. 6, 1866	53, 090
Corks in bottles, Mode of securing	R. M. Huston	Calais, Me	May 31, 1859	24, 215
Corks, Machine for manufacturing	W. R. Crocker	Norwich, Conn	Oct. 30, 1855	13, 714
Corks, Machinery for cutting	P. C. Traver	Newburgh, N. Y	Mar. 6, 1847	5, 003
Corks of oil-cans, Method of protecting	E. A. More	Saint Louis, Mo	July 25, 1865	48, 972
Corks, rubber, &c., Compound of	L. Bauhoefer	Philadelphia, Pa	Sept. 3, 1867	68, 408
Corn and cane cutter	J. Peck	Oakland, Tenn	Aug. 28, 1844	3, 721
Corn and clover shelling and thrashing machine, combined with a grist-mill.	D. Mullier	Wooster, Ohio	Jan. 24, 1832	......
Corn and cob breaking and grinding machine	J. and W. Murray	Baltimore, Md	Feb. 12, 1842	2, 459
Corn and cob crusher	A. Glover	Powhatan Point, Ohio	June 5, 1860	28, 571
Corn and cob crusher	J. S. Griffith	Huntingdon, Pa	Feb. 6, 1855	12, 348
Corn and cob crusher	J. L. McKnight	Buchanan, Va	Dec. 15, 1843	3, 373
Corn and cob crusher	J. M. Mowrer	Millheim, Pa	Feb. 15, 1870	99, 933
Corn and cob cutter	S. B. Shinn	Philadelphia, Pa	May 3, 1859	23, 865
Corn and cob grinding mill	S. A. Bantz and W. Andrew	Frederick, Md	July 22, 1851	8, 243
Corn and cob grinding mill	S. L. Herr	Mexico, Pa	Apr. 4, 1844	3, 524
Corn and cob mill	S. A. Briggs	Philadelphia, Pa	July 3, 1860	29, 033
Corn and cob mill	J. De Frain	Philadelphia, Pa	Feb. 15, 1859	22, 997
Corn and cob mill	A. Fisher	Tuscaloosa, Ala	Aug. 28, 1847	5, 266
Corn and cob mill	R. D. Granger	Philadelphia, Pa	Sept. 18, 1855	13, 569
Corn and cob mill	H. Hall	Mansfield, Ohio	Dec. 15, 1857	18, 844
Corn and cob mill	D. S. James	New Market, Va	Aug. 21, 1855	13, 476
Corn and cob mill	J. O. Joyce	Cincinnati, Ohio	Aug. 5, 1856	15, 488
Corn and cob mill	J. H. King	Martinsburgh, W. Va	Nov. 26, 1872	133, 455
Corn and cob mill	E. A. Knowlton	Columbia, S. C	Feb. 12, 1845	3, 908
Corn and cob mill	R. F. Maynard	Baltimore, Md	Apr. 7, 1857	16, 988
Corn and cob mill	J. M. Miller	Mobile, Ala	Jan. 20, 1843	2, 923
Corn and cob mill	G. Patten	Washington, D. C	Oct. 30, 1855	13, 730

Index of patents issued from the United States Patent Office from 1790 *to* 1873, *inclusive*—Continued.

Invention.	Inventor.	Residence.	Date.	No.
Corn and cob mill	C. Roberts	Belleville, Ill	July 1, 1856	15, 255
Corn and cob mill	J. P. Ross	Lewisburgh, Pa	Nov. 1, 1845	4, 249
Corn and cob mill	W. Sailor	Philadelphia, Pa	May 17, 1859	24, 082
Corn and cob mill	T. B. Stout	Keyport, N. J	Dec. 25, 1855	14, 002
Corn and cob mill	T. Ucker and A. Hutchins	Amanda, Ohio	July 21, 1868	80, 244
Corn and cob mill	B. Winter	Buckingham Court-House, Va.	Apr. 27, 1858	20, 121
Corn and cob mills, Securing the legs of sectional	R. F. Maynard	Baltimore, Md	Apr. 7, 1857	16, 987
Corn and cob crushing machine	S. K. Gauntt	Greenville, Tenn	May 25, 1827	......
Corn and fertilizer dropper	A. Towberman and J. Keys	Washington, Ill	Nov. 2, 1869	96, 510
Corn and grain drying machine	B. Parry	Bucks County, Pa	July 7, 1810	......
Corn and grain ventilator and drier	D. A. Dickenson	Baltimore, Md	June 16, 1868	78, 935
Corn and grinding bark, Cracking	M. J. Whiton	Amsterdam, N. Y	Apr. 30, 1834	......
Corn and malt, Kiln drying	J. Hughson	Clinton County, N. Y	July 18, 1811	......
Corn and other grain, Process and apparatus for curing	H. H. Beach	Rome, N. Y	May 7, 1872	126, 511
Corn and for other purposes, Machine for shelling and cleaning.	L. Stevens	......	May 8, 1804	......
Corn and other seed planting machine	E. L. Miller	Brooklyn, N. Y	Apr. 10, 1841	2, 047
Corn and phosphate drill	J. W. Wood and G. Moore	New Leeds Corner and Fairview School-House, Md.	Jan. 31, 1871	111, 416
Corn and potato coverer	J. Swart	Hoffman's Ferry, N. Y.	Aug. 4, 1868	80, 680
Corn, Band-tightener for shocks of	J. C. Jay	Bear Creek Township, Ind.	Mar. 9, 1869	87, 678
Corn balls, Machine for molding parched	R. Arnold	Hartford, Conn	Nov. 20, 1860	30, 661
Corn-breaker	W. Furbish	Hallowell, Me	Dec. 26, 1833	......
Corn-breaking machine	W. Furbish	Hallowell, Me	Mar. 15, 1833	......
Corn-cake cutter	L. Felker	Tewksbury, Mass	Dec. 10, 1867	71, 864
Corn-cake machine	C. C. Harriman	Warner, N. H	Mar. 5, 1867	62, 628
Corn-cake machine	H. and C. Littlefield	Tewksbury, Mass	July 30, 1867	67, 202
Corn-cake machine	W. Manning	Chelmsford, Mass	July 10, 1866	56, 241
Corn-cake mold	B. Witherell	Charlestown, Mass	Dec. 3, 1867	71, 831
Corn-cakes, Machine for making	W. Manning	Chelmsford, Mass	Nov. 22, 1870	109, 529
Corn, cob, and husk mill	E. Mire	New Orleans, La	Aug. 7, 1860	29, 509
Corn-cob cutter	I. Straub	Cincinnati, Ohio	July 25, 1854	11, 390
Corn-coverer	A. J. Combs	Olney, Ill	Aug. 27, 1867	68, 329
Corn-coverer	J. D. Haynie	New Antioch, Ohio	June 9, 1868	78, 665
Corn-coverer	I. N. Monroe	Bridgeport, Ill	Oct. 25, 1870	108, 717
Corn-coverer	J. A. Moore	Salem, N. J	Nov. 25, 1873	144, 918
Corn-coverer	J. Tweedy	Vernon, Ind	Jan. 28, 1873	135, 301
Corn-coverer	D. Weygandt	Jeromesville, Ohio	June 4, 1872	127, 534
Corn-crib	C. B. Clark	Pleasant Grove, Iowa	Mar. 26, 1872	125, 021
Corn-crib	N. T. Fitch	Forsyth, Ill	Oct. 22, 1867	70, 080
Corn-crib	A. B. Furbee	Dresden, Ohio	Dec. 6, 1859	26, 348
Corn-crib	J. M. Hughes and A. J. Mapes	Independence, Mo	Apr. 29, 1873	138, 332
Corn-crib and thrashing-floor, Combined	J. R. Jordan and J. Campbell	West Alexandria, Ohio	Dec. 15, 1868	85, 011
Corn-crusher	W. Beal	Lowell, Mass	May 30, 1854	10, 979
Corn-crusher	A. Bolander	Akron, Ohio	Aug. 22, 1871	118, 188
Corn-crusher	T. B. Coursey	Frederica, Del	May 31, 1859	24, 198
Corn-cutter	H. V. Corbett	Allendale, Mich,	Dec. 13, 1870	110, 014
Corn cutter, Green	V. Baker	Otisfield, Me	May 31, 1870	103, 543
Corn, Device for marking and planting	J. P. Zeller	South Bend, Ind	Sept. 7, 1869	94, 687
Corn, Disk for shelling	J. P. Smith	Hummelstown, Pa	Nov. 25, 1856	16, 127
Corn-drill	R. F. Patton	Quincy, Ohio	Sept. 21, 1869	95, 132
Corn-drill	B. W. and N. T. Remy	Brookville, Ind	Mar. 28, 1871	113, 205
Corn-dropper	R. M. and W. H. Boman	London, Ohio	Dec. 16, 1873	145, 552
Corn-dropper	M. R. W. Caldwell	Jackson, Ohio	Nov. 29, 1870	109, 584
Corn-dropper	L. M. Doddridge, H. Reitenour, and J. B. Swhier.	New Mount Pleasant, Ind	May 4, 1869	89, 742
Corn-dropper	J. H. Gross	Niantic, Ill	Jan. 24, 1871	111, 200
Corn-dropper	J. H Junkins	Upper Sandusky, Ohio	July 30, 1872	129, 963
Corn-dropper	R. B. Killin	Canton, Ohio	June 11, 1867	65, 578
Corn-dropper	C. E. Lipe	Fort Plain, N. Y	Sept. 3, 1867	68, 760
Corn-dropper	J. Nevison	Morgan, Ohio	Dec. 1, 1868	84, 571
Corn-dropper	J. L. Smith	Pemberton, N. J	June 7, 1870	103, 936
Corn-dropper	L. Weaver	Canton, Ohio	Oct. 15, 1867	69, 876
Corn dropper and cultivator, Combined	P. P. Gardner	Stoneborough, Pa	July 6, 1869	92, 184
Corn-drier	S. Bernheisel	Tyrone Township, Pa	Feb. 5, 1856	14, 181
Corn, Edible preparation from Indian	J. W. Haskins	Charlestown, Mass	Feb. 12, 1867	61, 936
Corn-elevator	A. Erkenbrecher	Cincinnati, Ohio	Aug. 27, 1867	68, 293
Corn fodder, Cutting and grinding	B. and D. H. Harnish	Lancaster Pequea, Pa	Aug. 22, 1871	118, 234
Corn-fodder cutting and grinding machine	A. Goodhart	Newville, Pa	Aug. 20, 1867	67, 975
Corn for grinding, Process for preparing	W. Standing	Duquoin, Ill	June 10, 1873	139, 743
Corn fork, Green	W. L. Gilroy	Philadelphia, Pa	June 1, 1869	90, 836
Corn from the cob, Device for cutting	F. A. Morley	New York, N. Y	Aug. 21, 1866	57, 361
Corn from the cob, Device for cutting green	J. Burt and L. F. Dunn	Oneida, N. Y	June 19, 1866	55, 614
Corn from the cob for table use, Apparatus for stripping.	W. C. McGill	Cincinnati, Ohio	Apr. 10, 1866	53, 849
Corn from the cob, Implement for cutting green	W. L. Gilroy	Philadelphia, Pa	May 25, 1869	90, 522
Corn from the cob, Knife for cutting green	I. Winslow	Philadelphia, Pa	Dec. 5, 1865	51, 379
Corn from the cob, Knife for removing	J. W. Jones	Portland, Me	Apr. 24, 1866	54, 170
Corn from the cob, Machine for cutting green	W. B. Coates	Philadelphia, Pa	May 13, 1856	14, 855
Corn from the cob, Machine for removing green	T. S. Lewis	Portland, Me	Aug. 24, 1869	94, 013
Corn from the cob, Machine for separating green	H. Walsh	Philadelphia, Pa	Sept. 30, 1856	15, 835
Corn from the stalks, Machine for severing the ears of.	A. J. and J. A. French	Franklin, Vt	Oct. 27, 1857	18, 505
Corn, Gathering and husking	J. D Hill	Fort Scott, Kans	Sept. 3, 1867	68, 507
Corn grinder and crusher	W. D Wilson	Richmond, Ind	May 29, 1855	1[illegible], 977
Corn grinding and crushing machine	A. P. H. Jordon	Madisonville, Tenn	Oct. 28, 1835	......
Corn grinding and shelling machine	G. M. Weaver	Montgomery County, Pa	June 12, 1835	......
Corn ground, Machine for furrowing	W. H. Warwick	Dunlevy, Ohio	Oct. 16, 1866	58, 923
Corn, Hand implement for severing the butts and separating husks from ears of.	I. N. Whitaker	Pecatonica, Ill	June 2, 1857	17, 466
Corn holder, Hot	W. A. Morgan and T. B. Mosher	Brooklyn and New York, N. Y.	Aug. 31, 1869	94, 330
Corn-houses, Method of ventilating	N. Seitz	Mellmore, Ohio	June 14, 1859	24, 407
Corn huller, sheller, and feed-cutter	T. Thomasson	Calhoun, Mo	Nov. 8, 1870	109, 157
Corn, Hulling	W. Carpenter	Newcastle, Pa	Jan. 23, 18[illegible]6	......

Index of patents issued from the United States Patent Office from 1790 to 1873, inclusive—Continued.

Invention.	Inventor.	Residence.	Date.	No.
Corn-husks, Machine for hackling	W. A. Fullerton	Louisville, Ky	July 11, 1854	11, 277
Corn-husks, Machine for hackling	G. B. Stacy	Richmond, Va	Mar. 23, 1869	88, 092
Corn-husks, Machine for stemming	D. M. Mefford	Jeffersonville, Ind	Oct. 2, 1860	30, 239
Corn-husks, Slitting	A. Barrett	Baltimore, Md	Apr. 21, 1836	....
Corn-husker	L. A. Aspinwall	Albany, N. Y	Apr. 12, 1870	101, 809
Corn-husker	E. M. Bates	East Rochester, Ohio	June 18, 1867	65, 867
Corn-husker	D. Bedell	Seneca Falls, N. Y	Nov. 17, 1857	18, 625
Corn-husker	D. Bookwalter	Gardner, Ill	May 18, 1869	90, 230
Corn-husker	G. K. Brown	Moultenborough, N. H	Oct. 20, 1857	18, 433
Corn-husker	T. J. Brown	Clio, Ohio	Aug. 21, 1866	57, 283
Corn-husker	R. Bryson	Schenectady, N. Y	Oct. 13, 1857	18, 385
Corn-husker	I. S. Bunnell	Carbondale, Pa	May 19, 1868	78, 052
Corn-husker	T. J. Burgess	Rondout, N. Y	June 28, 1870	104, 693
Corn-husker	J. M. Carl sle	Sumter, S. C	Jan. 7, 1873	134, 590
Corn-husker	G. W. Carr	Philadelphia, Pa	Oct. 14, 1873	143, 667
Corn-husker	E. H. Carver and G. M. Baker	Humberstone, Canada, and Buffalo, N. Y.	Apr. 4, 1871	113, 495
Corn-husker	J. Cawthra	Rochester, N. Y	Nov. 10, 1857	18, 571
Corn-husker	J. Cawthra	Rochester, N. Y	May 18, 1858	20, 253
Corn-husker	J. C. Clapp	Seneca Falls, N. Y	June 7, 1859	24, 281
Corn-husker	H. A. Doster	Bethlehem, Pa	Nov. 17, 1857	18, 658
Corn-husker	A. R. Davis	East Cambridge, Mass	Feb. 9, 1858	19, 325
Corn-husker	A. R. Davis	East Cambridge, Mass	Jan. 25, 1859	22, 710
Corn-husker	H. A. Doster	Bethlehem, Pa	Mar. 22, 1859	23, 295
Corn-husker	J. M. Evarts	New Haven, Conn	Mar. 12, 1872	124, 486
Corn-husker	N. Evinger	Sandford, Ind	Mar. 15, 1870	100, 740
Corn-husker	J. and J. L. Fagan	San Antonio River, Tex	Mar. 9, 1858	19, 552
Corn-husker	E. Field	Geneseo, Ill	July 6, 1869	92, 295
Corn-husker	C. Ford	Forest City, Ill	Mar. 10, 1868	75, 403
Corn-husker	E. F. French	Franklin, Vt	May 12, 1857	17, 269
Corn-husker	E. F. French	New York, N. Y	May 16, 1865	47, 710
Corn-husker	J. H. Gano	Tremont, Ohio	Sept. 11, 1866	57, 888
Corn-husker	A. M. George	Nashua, N. H	Oct. 6, 1857	18, 331
Corn-husker	H. P. Gerrish	Sandoval, Ill	Oct. 6, 1857	18, 332
Corn-husker	J. W. Glass	Richland, Ind	Apr. 2, 1867	63, 501
Corn-husker	J. I. Gorton	Sing Sing, N. Y	Dec. 15, 1868	84, 876
Corn-husker	S. A. Gould	Seneca Falls, N. Y	Nov. 10, 1857	18, 584
Corn-husker	M. H. Gragg	South Boston, Mass	July 19, 1859	24, 843
Corn husker	S. N. Gragg	Shelburne Falls, Mass	July 5, 1859	24, 628
Corn-husker	A. Graham	Roxbury, Mass	Oct. 20, 1857	18, 447
Corn-husker	L. A. Grover	Roxbury, Mass	July 6, 1858	20, 849
Corn-husker	H. L. Hall	Woodbridge, Iowa	Apr. 6, 1869	88, 708
Corn-husker	J. M. Hartnett	Waukegan, Ill	Sept. 1, 1868	81, 631
Corn-husker	B. Hazen	Cincinnati, Ohio	June 22, 1858	20, 637
Corn-husker	J. D. Heaton and W. A. Clark	Dixon, Ill	Jan. 19, 1858	19, 142
Corn-husker	J. B. Heich	Cincinnati, Ohio	Oct. 13, 1857	18, 396
Corn-husker	H. N. Hill	Pontiac, Mich	July 30, 1867	67, 301
Corn-husker	J. Hindman	Olathe, Kans	Jan. 22, 1867	61, 428
Corn-husker	J. Hood	Milwaukee, Wis	Sept. 13, 1870	107, 259
Corn-husker	L. H. Johnson	Branford, Conn	Apr. 7, 1868	76, 464
Corn-husker	S. Johnston	West Shelby, N. Y	Mar. 27, 1860	27, 638
Corn-husker	W. D. Jones	Hagaman's Mills, N. Y	Jan. 5, 1869	85, 593
Corn-husker	W. D. Jones	Hagaman's Mills, N. Y	May 18, 1869	90, 175
Corn-husker	C. A. Lang	Palmyra, Me	June 3, 1873	139, 586
Corn-husker	L. Leavenworth	Trumansburgh, N. Y	June 15, 1858	20, 568
Corn-husker	C. N. Lewis	Seneca Falls, N. Y	Nov. 17, 1857	18, 644
Corn-husker	C. N. Lewis	Seneca Falls, N. Y	May 4, 1858	20, 163
Corn-husker	D. Lombard	Boston, Mass	Feb. 9, 1858	19, 326
Corn-husker	B. B. Meacham	Ridleysville, Fla	May 25, 1858	20, 360
Corn-husker	L. R. Mears	South Abington, Mass	June 22, 1858	20, 653
Corn-husker	D. M. Mefford	Perrysburgh, Ohio	Dec. 22, 1857	18, 922
Corn-husker	D. M. Mefford	Jeffersonville, Ind	June 26, 1860	28, 884
Corn-husker	A. Merwin	Saint Joseph, Mich	Jan. 3, 1871	110, 772
Corn-husker	T. Percival	Augusta, Me	May 26, 1868	78, 321
Corn-husker	O. S. Perkins and L. A. Crandall.	New Haven, Conn	May 25, 1869	90, 574
Corn-husker	C. J. C. Peterson	Davenport, Iowa	Aug. 31, 1858	21, 363
Corn-husker	I. Philbrook	Shelby County, Ill	Mar. 13, 1866	53, 180
Corn-husker	P. Philip	Stockport, N. Y	Aug. 9, 1870	106, 276
Corn-husker	P. Philip	Stockport, N. Y	Feb. 11, 1873	135, 840
Corn-husker	P. Philip	Stockport, N. Y	Feb. 11, 1873	135, 841
Corn-husker	W. Pickett and A. Hills	Naugatuck, Conn	Mar. 2, 1858	19, 512
Corn-husker	O. M. Pond	Independence, Iowa	Feb. 19, 1867	62, 151
Corn-husker	W. N. Rowe	Sharpsburgh, Md	Feb. 8, 1859	22, 894
Corn-husker	J. Russell	Brooklyn, N. Y	May 18, 1869	90, 311
Corn-husker	D. Sager	New York, N. Y	July 2, 1867	66, 255
Corn-husker	D. Sager	New York, N. Y	Oct. 3, 1871	119, 654
Corn-husker	J. Sechrist	Connellsville, Pa	Sept. 24, 1867	69, 258
Corn-husker	G. F. Shaw	Woburn, Mass	July 12, 1859	24, 765
Corn-husker	S. A. Skinner	Lawrence, Mass	Nov. 17, 1857	18, 662
Corn-husker	D. C. Smith	Tecumseh, Mich	May 11, 1858	20, 223
Corn-husker	D. C. Smith	Tecumseh, Mich	Aug. 2, 1859	24, 958
Corn-husker	T. S. Smith	Cincinnati, Ohio	Mar. 15, 1870	100, 941
Corn-husker	W. H. Smith	Newport, R. I	Oct. 6, 1857	18, 358
Corn-husker	P. P. Snell	Manheim, N. Y	Feb. 6, 1872	123, 428
Corn-husker	N. T. Spear	Boston, Mass	Sept. 14, 1858	21, 522
Corn-husker	N. T. Spear	New York, N. Y	Apr. 10, 1860	27, 841
Corn-husker	J. Stengel and C. C. Davy	Croton and Big Prairie, Mich.	July 11, 1871	117, 011
Corn-husker	G. E. Stewart	East Saginaw, Mich	July 7, 1868	79, 703
Corn-husker	J. Ure	Saginaw, Mich	Oct. 21, 1873	143, 797
Corn-husker	F. M. Walker	Greensborough, N. C	Feb. 9, 1858	19, 320
Corn-husker	G. R. Walker	Washington, D. C	Mar. 5, 1861	31, 637
Corn-husker	L. F. Ward	Marathon, N. Y	Feb. 23, 1858	19, 458
Corn-husker	R. Warriner, J. H. Baker, and G. B. Slocum.	Saratoga Springs, N. Y	June 7, 1870	104, 085
Corn-husker	D. F. Welsh	Bucyrus, Ohio	Aug. 16, 1870	106, 436
Corn-husker	S. Wesson	Worcester, Mass	Sept. 22, 1868	82, 369

Index of patents issued from the United States Patent Office from 1790 *to* 1873, *inclusive*—Continued.

Invention.	Inventor.	Residence.	Date.	No.
Corn-husker	A. Whitcomb	Worcester, Mass	Apr. 27, 1869	89, 533
Corn-husker	D. Williams	Saginaw City, Mich	Aug. 27, 1867	68, 272
Corn-husker	J. F. Winchell	Springfield, Ohio	Jan. 1, 1867	60, 977
Corn-husker	C. O. Yale	Rome, N. Y	June 16, 1868	78, 908
Corn husker and picker	S. H. Mitchell	Lacon, Ill	May 7, 1872	126, 565
Corn husker and sheller	M. Jones	Amelia County, Va	May 5, 1868	77, 494
Corn husker and sheller and grain thrasher and cleaner, Combined.	J. W. Huntoon	Washington, D. C	Sept. 13, 1870	107, 378
Corn husker and sheller, Hand	J. M. Gray	Louisville, Ky	Dec. 27, 1870	110, 565
Corn-husker and stalk-cutter	E. Briggs, sr	Fayette, Iowa	July 9, 1867	66, 455
Corn-husker, Field	J. H. Whitney and W. W. Marsh.	Rochester, Minn., and De Kalb, Ill.	Nov. 7, 1871	120, 689
Corn-husker, Hand	S. L. Bligh	Sandy Lake, Pa	Aug. 16, 1870	106, 311
Corn-husker, Hand	A. W. Brinkerhoff	Upper Sandusky, Ohio	Apr. 23, 1872	125, 931
Corn-husker, Hand	H. L. Hall	Woodbridge, Iowa	Feb. 25, 1873	136, 157
Corn-husker, Hand	J. G. Johnson	Carthage, Ill	July 25, 1871	117, 296
Corn-husker, Hand	S. H. Mitchell	Lacon, Ill	Dec. 3, 1872	133, 659
Corn-husker, Hand	C. M. O'Hara	Cincinnati, Ohio	Aug. 1, 1871	117, 670
Corn-husker roller	J. Russell	Brooklyn, N. Y	Feb. 13, 1872	123, 580
Corn husker, sheller, and cleaner	C. J. Legg	Penn Yan, N. Y	June 6, 1865	48, 077
Corn husker, sheller, and stripper	F. Hafelfinger and R. N. Eagle	Washington, D. C	Nov. 24, 1868	84, 419
Corn-husker, straw and stalk cutter	R. Warriner and J. H. Baker	Saratoga Springs, N. Y	June 23, 1868	79, 166
Corn husking and shelling machine	A. Lane	Moscow, Ky	Feb. 27, 1872	124, 068
Corn husking and shelling machine	J. Wind	Thomasville, Ga	Dec. 4, 1860	30, 860
Corn-husking bench	J. E. Draper	Northville, Mich	Feb. 28, 1871	112, 131
Corn-husking device	A. W. Brinkerhoff	Upper Sandusky, Ohio	Sept. 2, 1862	36, 333
Corn-husking glove	A. C. Meyn	Jerseyville, Ill	May 7, 1872	126, 474
Corn-husking implement	O. J. Warren	Des Moines, Iowa	Feb. 25, 1873	136, 194
Corn-husking instrument	H. Bushnell	Norwich, Conn	Nov. 14, 1826	
Corn-husking machine	L. A. Aspinwall	Albany, N. Y	Nov. 18, 1873	144, 590
Corn-husking machine	G. W. Bachman	Clifton Springs, N. Y	July 7, 1857	17, 720
Corn-husking machine	D. Bookwalter	Gardner, Ill	Oct. 13, 1868	83, 033
Corn-husking machine	R. Bryson	Schenectady, N. Y	Dec. 9, 1856	16, 204
Corn-husking machine	E. H. Carver	Preble, N. Y	May 31, 1870	103, 561
Corn-husking machine	A. W. Case	South Manchester, Conn	Dec. 13, 1864	45, 387
Corn-husking machine	J. Cutler	Putney, Vt	July 31, 1837	324
Corn-husking machine	D. A. Dickinson	Baltimore, Md	Apr. 30, 1867	64, 206
Corn-husking machine	W. Emery, jr	Chester, Ill	July 7, 1857	17, 731
Corn-husking machine	H. P. Gerrish	Boscawen, N. H	Nov. 4, 1856	16, 008
Corn-husking machine	E. S. Holmes	Lockport, N. Y	Feb. 10, 1857	16, 633
Corn-husking machine	J. M. Hubbard	West Haven, Conn	Dec. 19, 1865	51, 589
Corn-husking machine	A. R. Hurst	New Cumberland, Pa	Mar. 31, 1857	16, 924
Corn-husking machine	M. C. Jeffers	New York, N. Y	Oct. 18, 1870	108, 484
Corn-husking machine	H. W. Knowlton	Saratoga Springs, N. Y	Jan. 14, 1868	73, 251
Corn-husking machine	W. Lewis	Seneca Falls, N. Y	Mar. 3, 1857	16, 737
Corn-husking machine	J. Massey	Buffalo, N. Y	Mar. 3, 1857	16, 740
Corn-husking machine	J. Naeher	North Orange, N. J	May 17, 1859	24, 047
Corn-husking machine	J. Perkins	West Killingley, Conn	Nov. 4, 1856	16, 023
Corn-husking machine	J. Russell	Brooklyn, N. Y	Mar. 21, 1871	112, 967
Corn-husking machine	J. Russell	New York, N. Y	June 11, 1872	127, 801
Corn-husking machine	D Sager	New York, N. Y	June 16, 1868	79, 011
Corn-husking machine	W. H. Smith	Newport, R. I	Oct. 28, 1856	15, 985
Corn-husking machine	M. W. Stevens and E. G. Kingsley.	Stoughton, Mass	Oct. 20, 1857	18, 473
Corn-husking machine	O. Stoddard	Busti, N. Y	June 3, 1856	15, 047
Corn-husking machine	H. Strait	Covington, Ky	Mar. 3, 1857	16, 758
Corn-husking machine	J. Taggart and L. A. Grover	Roxbury, Mass	Dec. 9, 1856	16, 201
Corn-husking roll	P. Philip	Stockport, N. Y	Mar. 14, 1871	112, 735
Corn-husking shield	A. C. Robinson	Louisiana, Mo	Apr. 16, 1867	63, 941
Corn-husking thimble	G. W. Gash and C. L. Owens	Clinton, Ill	Apr. 15, 1873	137, 771
Corn in the cob, Cutting and grinding	J. Urney	Wilmington, Del	Oct. 25, 1845	4, 244
Corn in the cob, Grinding	W. Mayo	Henrico, Va	May 15, 1812	
Corn in the ear and other grain, Grinding	T. Briggs	East Bloomfield N. Y	Jan. 10, 1834	
Corn into meal, Machine for grinding	T. Baker	Statesburgh, S. C	Jan. 21, 1829	
Corn-knife	G. Stevens	Zionsville, Ind	Apr. 29, 1873	138, 447
Corn knife, Green	W. L. Gilroy	Philadelphia, Pa	Sept. 22, 1868	82, 306
Corn knife, Green	J. Harrington	New London, Conn	Sept. 22, 1868	82, 403
Corn, &c., Machine for covering	T. Williams	Pittsylvania, Va	Dec. 19, 1822	
Corn, Machine for cutting the stalks of standing	W. B. Coates	Philadelphia, Pa	Dec. 9, 1856	16, 177
Corn, Machine for facilitating the husking of	G. Young, jr	Saratoga Springs, N. Y	Nov. 17, 1857	18, 656
Corn, Machine for grinding cobs and	S. Stukey	Sugar Grove, Ohio	Oct. 16, 1866	58, 912
Corn, &c., Machine for grinding shelled	J. Webster	Philadelphia, Pa	July 31, 1826	
Corn, Machine for marking and covering	E. Barto	Tiffin, Ohio	Nov. 10, 1868	83, 905
Corn, &c., Machine for planting	E. Bunce	Westford, Mass	July 16, 1838	841
Corn, Machine for removing husks from	J. Young	Varick, N. Y	Dec. 24, 1861	34, 033
Corn, Machine for shelling and cleaning	J. S. Gardiner	Canandaigua, N. Y	June 11, 1829	
Corn-meal for shipment, Preparing	L. S. Chichester	Brooklyn, N. Y	Feb. 25, 1873	136, 306
Corn-mill	C. Leavitt	Cleveland, Ohio	May 11, 1858	20, 208
Corn-mill	J. W. Taylor	Philadelphia, Pa	May 31, 1859	24, 249
Corn, Mill for grinding and crushing	J. C. and C. B. Baldwin	Virginia	June 26, 1839	1, 199
Corn, Mill for shelling and grinding	H. R. Miller	Louisville, Ky	Dec. 5, 1854	12, 030
Corn picker and husker, Field	G. and C. Meader	Prairie Centre, Ill	Dec. 13, 1870	110, 061
Corn picking and husking machine	S. R. Kenyon	Greenville, R. I	Aug. 27, 1867	68, 085
Corn picking and husking machine	S. R. Kenyon	Greenville, R. I	Mar. 28, 1871	113, 174
Corn picking and husking machine	S. R. Kenyon	Greenville, R. I	Nov. 28, 1871	121, 384
Corn-picking machine	S. W. May	Galesburgh, Ill	Dec. 28, 1858	22, 440
Corn-planter	H. Blair	Glenross, Md	Oct. 14, 1834	
Corn-planter	T. V. Bush	Clark County, Ky	Dec. 17, 1834	
Corn-planter, Hand	C. L. Green	Johnson's Creek, N. Y	Jan. 2, 1866	51, 825
Corn planting and working machine	W. Ross	Middle Baxton Township, Pa.	June 24, 1826	
Corn-planting, Furrowing-device for	J. Plumb	Clarksville, N. J	Feb. 13, 1866	52, 598
Corn-planting machine	J. Blocher	Lancaster, Pa	Apr. 12, 1814	
Corn-popper	J. H. Bigelow	Worcester, Mass	Aug. 3, 1869	93, 271
Corn-popper	W. F. Collier	Worcester, Mass	July 20, 1869	92, 939
Corn-popper	W. F. Collier	Worcester, Mass	Jan. 31, 1871	111, 319
Corn-popper	A. F. Curtis and O. N. Palmer	Delaware, Ohio	May 14, 1872	126, 680
Corn-popper	D. A. Denison	Troy, Mich	Dec. 17, 1867	72, 173

Index of patents issued from the United States Patent Office from 1790 *to* 1873, *inclusive*—Continued.

Invention.	Inventor.	Residence.	Date.	No.
Corn-popper	G. D. Dudley	Lowell, Mass	Aug. 13, 1867	67, 736
Corn-popper	B. B. and J. R. Hill	Worcester, Mass	July 27, 1869	92, 967
Corn-popper	J. W. Howe and J. K. Barton	Worcester, Mass	Dec. 10, 1867	71, 881
Corn-popper	W. J. Johnson	Newton, Mass	July 4, 1871	116, 600
Corn-popper	E. G. Kinsley	Stoughton, Mass	Nov. 12, 1867	70, 723
Corn-popper	W. W. S. Orbeton	Haverhill, Mass	Oct. 30, 1866	59, 253
Corn-popper	C. H. S. Schultz	Cincinnati, Ohio	Oct. 29, 1867	70, 368
Corn-popper	M. H. Wiley	Boston, Mass	Mar. 5, 1872	124, 305
Corn-popper and coffee-roaster	L. A. Warner	Freeport, Ill	Apr. 7, 1868	76, 362
Corn-popper and coffee roaster	L. A. Warner	Freeport, Ill	July 19, 1870	105, 530
Corn, Preparation from Indian	E. F. Prentiss and C. C. Parsons.	Philadelphia, Pa., and Boston, Mass.	Mar. 26, 1867	63, 175
Corn, Preparing seed	J. M. Petit	Monroe Township, Ohio	Mar. 21, 1871	112, 845
Corn preserved green, Indian	I. Winslow	Philadelphia, Pa	Apr. 8, 1862	34, 928
Corn, Process for treating Indian	E. F. Prentiss and C. C. Parsons.	Philadelphia, Pa., and Boston, Mass.	Mar. 26, 1867	63, 174
Corn-rows, &c., Laying of	P. Moseley	Benton, Miss	Dec. 12, 1842	2, 882
Corn-separator	E. McLane	Young America, Ill	May 4, 1869	89, 589
Corn-shaving machine	E. Watts	Keyport, N. J	Aug. 29, 1871	118, 564
Corn-sheller	A. Adams	Sandwich, Ill	Sept. 28, 1858	21, 594
Corn-sheller	A. Adams	Sandwich, Ill	Aug. 6, 1861	32, 971
Corn-sheller	A. Adams	Sandwich, Ill	May 15, 1866	54, 659
Corn-sheller	A. Adams	Sandwich, Ill	Feb. 20, 1872	123, 758
Corn-sheller	A. Adams	Sandwich, Ill	Jan. 28, 1873	135, 306
Corn-sheller	C. Adams	Oak Hill, N. Y	Aug. 12, 1856	15, 502
Corn-sheller	C. Adams	Pittsburgh, Pa	Aug. 17, 1858	21, 174
Corn-sheller	H. A. Adams	Sandwich, Ill	Oct. 15, 1872	132, 128
Corn-sheller	A. B. Allen	New York, N. Y	July 15, 1862	35, 864
Corn-sheller	J. M. C. Armsby	Worcester, Mass	Jan. 7, 1851	7, 881
Corn-sheller	D. Bacon	Brewersville, Ind	Aug. 25, 1868	81, 458
Corn-sheller	C. B. Baldwin	Cincinnati, Ohio	July 16, 1842	2, 718
Corn-sheller	I. C. Baldwin	Staunton, Va	Oct. 11, 1836	....
Corn-sheller	C. Beach	Penn Yan, N. Y	Jan. 1, 1867	60, 825
Corn-sheller	W. Beach	Baltimore, Md	Apr. 18, 1846	4, 460
Corn-sheller	W. Beach	Baltimore, Md	Dec. 26, 1848	5, 982
Corn-sheller	P. Bergen	New York, N. Y	Mar. 30, 1858	19, 809
Corn-sheller	G. Bevitt	Madison, Wis	July 2, 1867	66, 205
Corn-sheller	C. Bishop	Norwalk, Ohio	Oct. 9, 1855	13, 634
Corn-sheller	T. W. Bishop	Austin, Ind	Nov. 5, 1867	70, 396
Corn-sheller	W. Black	Allegheny City, Pa	Oct. 21, 1856	15, 920
Corn-sheller	F. A. Bolles	Unadilla, N. Y	June 18, 1867	65, 868
Corn-sheller	M. Bomberger	Hummelstown, Pa	Jan. 4, 1859	22, 481
Corn-sheller	J. Bowles	Augusta, Ga	May 4, 1869	89, 550
Corn-sheller	C. H. Brady	Mount Joy, Pa	Sept. 17, 1867	68, 941
Corn-sheller	C. H. Brady	Mount Joy, Pa	Mar. 31, 1868	76, 154
Corn-sheller	B. Bridendolph	Clear Spring, Md	Dec. 20, 1859	26, 471
Corn-sheller	J. D. Briggs	Saratoga, N. Y	June 14, 1845	4, 079
Corn-sheller	J. Brinkerhoff	Auburn, N. Y	Apr. 12, 1864	42, 271
Corn-sheller	J. Brinkerhoff	Auburn, N. Y	Feb. 28, 1865	46, 540
Corn-sheller	J. Brinkerhoff	Auburn, N. Y	Aug. 13, 1867	67, 713
Corn-sheller	J. Brinkerhoff	Auburn, N. Y	Mar. 3, 1868	74, 981
Corn-sheller	F. C. Brown and J. R. Shangle	Hightstown, N. J	Apr. 22, 1873	138, 125
Corn-sheller	L. S Bundy and L. F. Edgerton	Hyde Park, Vt	May 14, 1861	32, 273
Corn-sheller	W. R. Burns	Lancaster, Pa	Apr. 3, 1866	53, 567
Corn-sheller	N. Burr	Batavia, Ill	Nov. 8, 1859	26, 010
Corn-sheller	T. D. Burrall	Geneva, N. Y	Dec. 6, 1845	4, 300
Corn-sheller	T. D. Burrall	Geneva, N. Y	Mar. 24, 1863	38, 002
Corn-sheller	J. R. Cadwell	Onondaga, Mich	Feb. 13, 1872	123, 676
Corn-sheller	M. Carpenter	Lancaster, Pa	Oct. 25, 1832	....
Corn-sheller	M. Carpenter	Lancaster, Pa	Aug. 8, 1833	....
Corn-sheller	C. W. Carter	Westville, Ind	Mar. 29, 1859	23, 424
Corn-sheller	J. A. Cauldwell	Horseheads, N. Y	Dec. 1, 1868	84, 610
Corn-sheller	R. Chalfant	West Grove, Pa	Feb. 27, 1847	4, 979
Corn-sheller	P. C. Chipron	Highland, Ill	Feb. 11, 1868	74, 306
Corn-sheller	A. and C. N. Clow	Port Byron, N. Y	June 26, 1855	13, 122
Corn-sheller	D. Codd	Ottawa, Canada	Nov. 10, 1868	83, 832
Corn-sheller	S. S. Cole	Henryville, Ind	Mar. 30, 1869	88, 367
Corn-sheller, &c	G. S. Coleman	Alexandria, Va	Sept. 27, 1870	107, 761
Corn-sheller	W. Colwell	Chillicothe, Ill	Oct. 23, 1866	58, 986
Corn-sheller	E. Converse	Cincinnati, Ohio	Jan. 7, 1846	4, 348
Corn-sheller	A. M. Cook	Milford, Mass	Oct. 6, 1857	18, 325
Corn-sheller	H. W. Cornell	Owego, N. Y	Nov. 2, 1869	96, 306
Corn-sheller	H. W. Cornell	Owego, N. Y	Aug. 1, 1871	117, 607
Corn-sheller	H. W. Cornell	Owego, N. Y	June 18, 1872	128, 124
Corn-sheller	C. S. C. Crane	Taunton, Mass	May 27, 1856	14, 990
Corn-sheller	G. Danforth	Friendsville, Ill	Aug. 21, 1860	29, 673
Corn-sheller	A. B. Davis	Philadelphia, Pa	Apr. 13, 1858	19, 915
Corn-sheller	A. B. Davis and T. Crook, jr	Philadelphia, Pa	Feb. 18, 1862	34, 410
Corn-sheller	J. Davis	Oaks, Wis	Feb. 20, 1866	52, 690
Corn-sheller	L. H. Davis	Kennett s Square, Pa	Sept. 6, 1853	10, 002
Corn-sheller	M. Day	Baltimore, Md	July 14, 1868	79, 961
Corn-sheller	M. Day	Baltimore, Md	July 14, 1868	79, 962
Corn-sheller	L. Day	Baltimore, Md	June 13, 1871	115, 939
Corn sheller	J. H. Derby	Leominster, Mass	Nov. 10, 1841	2, 337
Corn-sheller	P. Dickinson	Amherst, Mass	Sept. 6, 1853	10, 003
Corn-sheller	A. Dillman	Plainfield, Ill	Sept. 8, 1859	18, 139
Corn-sheller	O. R. Dinsmore	West Chester, N. H	Feb. 24, 1839	1, 088
Corn-sheller	J. Dodson	Guilford, N. C	June 15, 1826	....
Corn-sheller	S. E. Donnell	Greensborough County, N. C.	Apr. 5, 1833	....
Corn-sheller	E. Doud	Oshkosh, Wis	Nov. 17, 1868	84, 173
Corn-sheller	N. Drake	Newton, N J	Apr. 3, 1860	27, 705
Corn-sheller	Dunbar and Powers	Portland, Me	June 26, 1835	....
Corn-sheller	B. Edwards	Laceyville, Pa	May 16, 1848	5, 578
Corn sheller	D. Eldridge	Philadelphia, Pa	Nov. 19, 1850	7, 781
Corn-sheller	D. Eldridge	Philadelphia, Pa	June 1, 1852	8, 978
Corn-sheller	F. Elliot	Greensborough, N. C	Nov. 19, 1833	....
Corn-sheller	S. Elliott	Wayne County, Ind	Oct. 9, 1855	13, 640

Index of patents issued from the United States Patent Office from 1790 to 1873, inclusive—Continued.

Invention.	Inventor.	Residence.	Date.	No.
Corn-sheller	F. Fanning	Atchison, Kans	June 8, 1869	91, 003
Corn-sheller	P. Ferrier	Ypsilanti, Mich	Mar. 16, 1869	87, 920
Corn-sheller	D. E. Field	Leaksville, N. C	Feb. 15, 1870	99, 768
Corn-sheller	S. Field	Oakham, Mass	Jan. 18, 1870	98, 944
Corn-sheller	G. W. Fitts	South Hampton, N. H	July 11, 1865	48, 670
Corn-sheller	S. Fletcher and J. P. Pike	Bloomfield, Me	Jan. 3, 1860	26, 662
Corn-sheller	J. S. Fowler	Peoria, Ill	May 13, 1862	35, 227
Corn-sheller	C. C. French	West Stockbridge, Mass	Jan. 15, 1861	31, 115
Corn-sheller	W. Gee	Prince George's County, Va	May 10, 1832	....
Corn-sheller	B. Gilbert	Pittsburgh, Pa	Jan. 3, 1854	10, 390
Corn-sheller	W. Gilman	Ottawa, Ill	June 12, 1866	55, 485
Corn-sheller	P. A. Gladwin	Chester, Conn	Apr. 24, 1840	1, 567
Corn-sheller	L. B. Glover	Easton, Conn	July 31, 1846	4, 677
Corn-sheller	G. Goewey	Philadelphia, Pa	Feb. 10, 1863	37, 662
Corn-sheller	G. Goewey	Philadelphia, Pa	Oct. 18, 1864	44, 768
Corn-sheller	G. Goewey	Philadelphia, Pa	Oct. 29, 1867	70, 195
Corn-sheller	N. Goldsborough	Easton, Md	Feb. 12, 1841	1, 975
Corn-sheller	J. W. Gordon	Jamestown, N. C	Oct. 7, 1846	4, 806
Corn-sheller	J. Gould	Grinnell, Iowa	Dec. 10, 1867	72, 020
Corn-sheller	H. A. Graeff	Birdsborough, Pa	July 9, 1867	66, 485
Corn-sheller	S. L. Graves	Springfield, Ill	Sept. 17, 1850	7, 653
Corn-sheller	A. W. Gray	Middletown, Vt	Mar. 31, 1836	....
Corn-sheller	D. G. Greene	North Bridgewater, Mass	Mar. 9, 1858	19, 603
Corn-sheller	R. Green	Cussawago, Pa	May 18, 1858	20, 266
Corn-sheller	S. Gumaer	Aurora, Ill	Nov. 7, 1854	11, 891
Corn-sheller	D. C. Guttridge	Pittsburgh, Pa	Aug. 29, 1871	118, 605
Corn-sheller	D. Hall	Strasburgh Township, Pa	Nov. 12, 1832	....
Corn-sheller	J. R. Hamilton	Portland, Oreg	Jan. 14, 1868	73, 323
Corn-sheller	J. R. Hamilton	Kingston, Minn	Nov. 15, 1870	109, 315
Corn-sheller	D. W. Harris and E. P. Carter	Yorkshire, N. Y	Nov. 6, 1849	6, 847
Corn-sheller	J. S. Harris	Poultney, Vt	Sept. 18, 1835	....
Corn-sheller	G. W. Hathaway	Tioga, Pa	Oct. 16, 1860	30, 403
Corn-sheller	M. Hausman	Huntington, Ind	Sept. 4, 1860	29, 886
Corn-sheller	J. M. Hawley	Holton, Ind	Jan. 5, 1869	85, 590
Corn-sheller	A. Higley	Warren, Ohio	Mar. 17, 1863	37, 912
Corn-sheller	F. Hollen and A. C. Holland	Marion, Ill	Apr. 15, 1873	137, 842
Corn-sheller	D. S. Hollister	Baltimore, Md	June 13, 1846	4, 572
Corn-sheller	W. W. Holt	Dunkirk, N. Y	Apr. 9, 1872	125, 392
Corn-sheller	M. and S. Housman	Huntington, Ind	Dec. 11, 1866	60, 375
Corn-sheller	M. and S. Housman	Huntington, Ind	Feb. 11, 1868	74, 364
Corn-sheller	W. H. Hovey	Springfield, Mass	Apr. 19, 1859	23, 686
Corn-sheller	R. Hoyt	Miamisburgh, Ohio	Oct. 31, 1834	....
Corn-sheller	L. T. Hulbert and A. P. Teachout.	Painesville and Madison, Ohio.	June 20, 1871	116, 192
Corn-sheller	D. Hutchinson	Fort Ancient, Ohio	Jan. 17, 1865	45, 924
Corn-sheller	E. L. Hutchinson	Auburn, N. Y	Aug. 5, 1873	141, 559
Corn-sheller	J. Hutchison	Three Rivers, Mich	Nov. 19, 1872	133, 228
Corn-sheller	W. Janes	Ashford, Conn	May 6, 1819	....
Corn-sheller	G. F. Johnson	Marshall, Iowa	Aug. 25, 1868	81, 375
Corn-sheller	J. B. Johnson	Alma, Ill	Aug. 12, 1873	141, 795
Corn-sheller	W. Johnson	Boscawen, N. H	Dec. 17, 1834	....
Corn-sheller	J. J. Johnston	Allegheny, Pa	Dec. 11, 1855	13, 912
Corn-sheller	J. J. Johnston	Allegheny, Pa	Sept. 23, 1856	15, 765
Corn-sheller	J. J. Johnston	Allegheny, Pa	Apr. 19, 1859	23, 687
Corn-sheller	A. H. Jones	Fallsington, Pa	Nov. 5, 1861	33, 648
Corn-sheller	E. Jordan	Pickens County, Ala	Oct. 4, 1870	107, 919
Corn-sheller	J. Jordan	Red Wing, Minn	June 29, 1869	92, 056
Corn-sheller	E. Kelley	Locust Grove, Ohio	Mar. 16, 1869	88, 790
Corn-sheller	I. Kepler	Milton, Pa	Apr. 24, 1849	6, 382
Corn-sheller	C. Ketchum	Penn Yan, N. Y	Feb. 7, 1865	46, 245
Corn-sheller	N. S. Ketchum	Marshalltown, Iowa	June 4, 1872	127, 610
Corn-sheller	M. C. Kilgore	Washington, Iowa	Mar. 12, 1867	62, 857
Corn-sheller	S. H. Kisinger	Williamsport, Md	Oct. 31, 1839	1, 385
Corn-sheller	W. D. Leavitt	New Orleans, La	Aug. 10, 1869	93, 455
Corn-sheller	C. J. Legg	Penn Yan, N. Y	Sept. 22, 1863	40, 052
Corn-sheller	A. F. Severance	Concord, N. H	Mar. 29, 1864	42, 114
Corn-sheller	J. R. Lindner	Cincinnati, Ohio	Feb. 2, 1858	19, 253
Corn-sheller	W. Linsley	Waddam Township, Ill	Mar. 9, 1852	8, 788
Corn-sheller	W. J. Ludlow	Cleveland, Ohio	Feb. 2, 1869	86, 562
Corn-sheller	W. H. Main	Liverpool, Ohio	June 22, 1858	20, 650
Corn-sheller	W. Manning	South Trenton, N. J	Aug. 4, 1843	3, 208
Corn-sheller	J. Marshall	Cordova, Ill	Dec. 9, 1873	145, 356
Corn-sheller	E. Mathers	Morgantown, Va	Apr. 29, 1856	14, 771
Corn-sheller	T. J. Mayall	Boston, Mass	Apr. 23, 1872	126, 067
Corn-sheller	W. M. Mayall	Gray, Me	Nov. 16, 1869	96, 938
Corn-sheller	G. Maynard	Greenfield, Mass	Sept. 12, 1854	11, 672
Corn-sheller	W. McAll	Talladega, Ala	Apr. 13, 1844	3, 540
Corn-sheller	T. U. McFarlan and L. H. Davis.	Salem, Ohio, and West Chester, Pa.	Apr. 6, 1858	19, 862
Corn-sheller	R. M. McGrath	La Fayette, Ind	Apr. 15, 1873	137, 942
Corn-sheller	J. H. McPheeters and P. P. Gross.	Palmyra, Mo	July 7, 1868	79, 675
Corn-sheller	S. McQuiston	Morris, Ill	Oct. 8, 1861	33, 442
Corn-sheller	J. A. Merriman	Chicago, Ill	Feb. 18, 1868	74, 569
Corn-sheller	W. Miller	Bloomington, Ind	Aug. 3, 1869	93, 328
Corn-sheller	J. Miller and N. Dubrul	Joliet, Ill	May 23, 1871	115, 227
Corn-sheller	E. L. Millis	Rochester Depot, Ohio	Aug. 9, 1853	9, 924
Corn-sheller	A. C. Mills	Oaktown, Ind	July 28, 1868	80, 362
Corn-sheller	E. Morrison	Franklin, N. H	June 10, 1856	15, 105
Corn-sheller	E. Morse	Knoxville, Tenn	Sept. 9, 1835	....
Corn-sheller	J. W. Morton	Brunswick, Ohio	Nov. 24, 1857	18, 700
Corn-sheller	J. Mumma	Middletown, Pa	June 12, 1849	6, 526
Corn-sheller	J. Murray	Baltimore, Md	Oct. 7, 1846	4, 805
Corn-sheller	F. Nelson	Wyandotte, Mich	Jan. 24, 1871	111, 139
Corn-sheller	L. T. Newell	Springville, N. Y	Mar. 5, 1867	62, 671
Corn-sheller	T. J. Newland	Wolcott, Vt	July 17, 1860	29, 188
Corn-sheller	G. and H. O'Conner	Mishawaka, Ind	July 11, 1871	116, 987
Corn-sheller	I. E. Overpeck	Overpeck Station, Ohio	Nov. 27, 1866	60, 046

Index of patents issued from the United States Patent Office from 1790 to 1873, inclusive—Continued.

Invention.	Inventor.	Residence.	Date.	No.
Corn-sheller	C. Page	Sanbornton, N. H	Mar. 13, 1834	
Corn-sheller	E. Parker	Baltimore, Md	Apr. 20, 1858	20, 003
Corn-sheller	J. J. Parker	Marietta, Ohio	Sept. 29, 1857	18, 296
Corn-sheller	S. J. Parmele	Killingworth, Conn	Dec. 24, 1861	34, 008
Corn-sheller	A. H. Patch	Hamilton, Mass	July 4, 1871	116, 627
Corn-sheller	A. H. Patch	Hamilton, Mass	Apr. 2, 1872	125, 217
Corn-sheller	W. P. Patton and W. A. Middleton.	Harrisburgh, Pa	Jan. 25, 1870	99, 106
Corn-sheller	R. Pleifer	Linz, Austria	Jan. 18, 1870	98, 999
Corn-sheller	C. G. Pressey and J. B. Dudley	Andover, N. H	Mar. 30, 1843	3, 025
Corn-sheller	D. O. Prouty and E. Whitman	Philadelphia, Pa., and Baltimore, Md.	May 29, 1849	6, 421
Corn-sheller	I. G. Putnam	Tioga, Pa	May 22, 1860	28, 440
Corn-sheller	E. Rand and A. L. Norcross	Hallowell, Me	Feb. 13, 1835	
Corn-sheller	J. B. Rand and W. A. N. Long	Fisherville, N. H	Aug. 13, 1861	33, 052
Corn-sheller	J. Rankin	Binghamton, N. Y	Jan. 28, 1873	135, 362
Corn-sheller	P. Reading	Trenton, N. J	Dec. 27, 1824	
Corn-sheller	P. Reading	Batavia, Ohio	Sept. 25, 1841	2, 266
Corn-sheller	W. Reading	Washington, D. C	July 13, 1852	9, 120
Corn-sheller	D. K. Reeder	Elliottsburgh, Pa	June 5, 1866	55, 362
Corn-sheller	G. W. Reid	Evansville, Ind	May 3, 1853	9, 698
Corn-sheller	G. W. Reisinger	Harrisburgh, Pa	Sept. 13, 1870	107, 410
Corn-sheller	G. W. Reisinger	Harrisburgh, Pa	Dec. 6, 1870	109, 945
Corn-sheller	I. I. Richardson	New York, N. Y	Apr. 17, 1849	6, 320
Corn-sheller	S. Richardson	Rochester, N. Y	June 17, 1862	35, 625
Corn-sheller	J. W. Ricker	Boston, Mass	Mar. 25, 1862	34, 775
Corn-sheller	J. W. Ricker	Chelsea, Mass	May 3, 1870	102, 593
Corn-sheller	J. W. Ricker and T. S. Lewis	Chelsea, Mass	Aug. 8, 1865	49, 303
Corn-sheller	J. Robb	Lewistown, Pa	Dec. 25, 1855	13, 997
Corn-sheller	C. Robbins and R. P. Burlingame	Chicago, Ill	Apr. 1, 1862	34, 846
Corn-sheller	W. Roberts	Farnham, N. Y	June 16, 1868	79, 007
Corn-sheller	J. Ross	Roundbrook, N. J	Apr. 12, 1833	
Corn-sheller	J. M. Ross	Linden Hall, Pa	June 29, 1869	92, 103
Corn-sheller	S. P. Ross and N. Haller	Pittsburgh and Allegheny City, Pa.	Aug. 31, 1869	94, 444
Corn-sheller	S. W. Ryckman	Pontiac, Mich	Feb. 7, 1860	27, 077
Corn-sheller	N. C. and E. A. Scofield	Norwich, Conn	Jan. 20, 1843	2, 924
Corn-sheller	S. C. Scofield	Chicago, Ill	June 6, 1871	115, 646
Corn-sheller	F. H. Schroeder	Bushnell, Ill	May 30, 1865	47, 988
Corn-sheller	H. Sells	Vienna, Canada	Jan. 16, 1872	122, 740
Corn-sheller	J. H. Sharp	Wortsville, N. J	Apr. 28, 1868	77, 410
Corn-sheller	H. F. and G. F. Shaw	West Roxbury, Mass	Oct. 31, 1865	50, 740
Corn-sheller	W. E. Sheffield	New York	Apr. 15, 1823	
Corn-sheller	E. F. Sherman	Chicopee Falls, Mass	Aug. 21, 1866	57, 393
Corn-sheller	E. F. Sherman	Chicopee, Mass	Feb. 18, 1868	74, 726
Corn-sheller	J. Shipe	Upper Augusta, Pa	Aug. 2, 1864	43, 713
Corn-sheller	A. Siddall	Ransom, Mich	June 28, 1859	24, 583
Corn-sheller	I. B. Siddle	Caswell County, N. C	Mar. 19, 1867	63, 112
Corn-sheller	C. Sines	Village Green, Pa	Oct. 10, 1848	5, 848
Corn-sheller	J. Small	Bridgewater, Pa	Mar. 27, 1849	6, 228
Corn-sheller	J. P. Small	Gilmanton, N. H	July 21, 1835	
Corn-sheller	E. A. Smead	Tioga, Pa	June 5, 1860	28, 612
Corn-sheller	A. B. Smith	Wellsburgh, W. Va	Nov. 14, 1871	120, 909
Corn-sheller	A. J. Smith	Piqua, Ohio	Aug. 29, 1854	11, 620
Corn-sheller	F. N. Smith	Kinderhook, N. Y	June 1, 1843	3, 114
Corn-sheller	H. E. Smith	Philadelphia, Pa	Dec. 9, 1856	16, 191
Corn-sheller	J. P. Smith	Hummelstown, Pa	Jan. 18, 1853	9, 549
Corn-sheller	J. P. Smith	Hummelstown, Pa	Nov. 7, 1854	11, 913
Corn-sheller	J. P. Smith	Hummelstown, Pa	Feb. 20, 1855	12, 454
Corn-sheller	J. P. Smith	Hummelstown, Pa	Dec. 11, 1855	13, 922
Corn-sheller	J. P. Smith	Hummelstown, Pa	Mar. 4, 1856	14, 374
Corn-sheller	J. P. Smith	Hummelstown, Pa	Jan. 19, 1858	19, 160
Corn-sheller	J. P. Smith	Hummelstown, Pa	Feb. 8, 1859	22, 898
Corn-sheller	J. P. Smith	Hummelstown, Pa	June 19, 1860	28, 784
Corn-sheller	J. P. Smith	Hummelstown, Pa	May 6, 1862	35, 185
Corn-sheller	J. P. Smith	Hummelstown, Pa	Dec. 15, 1868	85, 035
Corn-sheller	J. P. Smith	Hummelstown, Pa	June 22, 1869	91, 572
Corn-sheller	E. E. Stedman	Randolph, Ohio	May 21, 1867	64, 919
Corn-sheller	D. C. Sterry, 1st	Worcester, Mass	Sept. 3, 1872	131, 129
Corn-sheller	A. H. Stevens	Geneva, N. Y	May 16, 1848	5, 569
Corn-sheller	A. H. Stevens	Warsaw, N. Y	Apr. 22, 1856	14, 745
Corn-sheller	E. M. Stevens	Boston, Mass	Dec. 23, 1856	16, 291
Corn-sheller	A. Stickney	Concord, N. H	Oct. 6, 1857	18, 361
Corn-sheller	A. Stickney	Concord, N. H	Oct. 29, 1861	33, 627
Corn-sheller	W. D. Stroud	Oshkosh, Wis	Apr. 9, 1867	63, 671
Corn-sheller	P. Sweeney	New York, N. Y	Jan. 31, 1865	46, 155
Corn-sheller	A. Swift	Wolcott, Vt	Apr. 8, 1862	34, 912
Corn-sheller	P. P. Taft	Taftsville, Vt	July 6, 1858	20, 831
Corn-sheller	J. H. Taylor and A. J. Cowles	Westfield, N. Y	Feb. 11, 1835	
Corn-sheller	S. Terry	Boscawen, N. H	Nov. 26, 1867	71, 551
Corn-sheller	C. Thomas	Lancaster, Pa	Sept. 3, 1832	
Corn-sheller	A. B. Thompson	Oswego, N. Y	June 7, 1870	103, 941
Corn-sheller	A. B. Thompson	Oswego, N. Y	Oct. 31, 1871	120, 549
Corn-sheller	G. W. Tolhurst	Liverpool, Ohio	Oct. 4, 1859	25, 688
Corn-sheller	M. Trimble	Princeton, Ill	Mar. 26, 1861	31, 839
Corn-sheller	G. Turner	Cambridge, Mass	Apr. 29, 1862	35, 119
Corn-sheller	J. Turner	Poland, Me	Aug. 15, 1835	
Corn-sheller	A. B. Vant and A. M. Cook	Milford, Mass	Jan. 11, 1859	22, 595
Corn-sheller	J. Warren	Lodi, Ohio	Feb. 18, 1868	74, 643
Corn-sheller	J. Warren	Lodi, Ohio	Apr. 28, 1868	77, 227
Corn-sheller	J. R. Warrington	Damascoville, Ohio	May 2, 1848	5, 548
Corn-sheller	W. F. Waters	Dunkirk, N. Y	Aug. 20, 1872	130, 777
Corn-sheller	H. P. Watts	Lynchburgh, Va	Nov. 16, 1869	97, 008
Corn-sheller	T. Weaver	Harrisburgh, Pa	July 6, 1869	92, 410
Corn-sheller	T. Weaver	Harrisburgh, Pa	Aug. 24, 1869	94, 050
Corn-sheller	D. G. Wells	Joliet, Ill	Jan. 16, 1872	122, 749
Corn-sheller	W. Wells	Boston, Mass	Jan. 4, 1859	22, 523
Corn-sheller	F. C. Whiley	Lancaster, Ohio	May 21, 1872	127, 127

Index of patents issued from the United States Patent Office from 1790 *to* 1873, *inclusive*—Continued.

Invention.	Inventor.	Residence.	Date.	No.
Corn-sheller	W. H. Whiterow	New Albany, Ind	Aug. 18, 1868	81, 121
Corn-sheller	W. H. Whiterow and W. Detrick.	New Albany, Ind	Mar. 16, 1869	87, 811
Corn-sheller	J. A. Whitford	Saratoga Springs, N Y	Jan. 23, 1841	1, 946
Corn-sheller	J. R. Wilbur	Chicopee, Mass	Feb. 25, 1868	74, 965
Corn-sheller	C. Willis	Chelsea, Mass	Jan. 27, 1841	1, 952
Corn-sheller	J. B. Wolford	Lancaster, Ohio	July 26, 1870	105, 872
Corn-sheller	T. Wright	New Village, N. J	June 12, 1838	778
Corn-sheller	G. A. Xander	Hamburgh, Pa	Jan. 3, 1854	10, 376
Corn-sheller	J. C. Zimmerman	Eberly's Mill, Pa	June 9, 1868	78, 714
Corn-sheller and apple-grinder, Combined	M. H. Ripley and W. N. Temple	Minneapolis, Minn	Sept. 15, 1868	82, 250
Corn-sheller and bean-thrasher	B. P. Pendexter	Mechanics' Falls, Me	June 26, 1866	55, 899
Corn sheller and cleaner	R. Gray	Northfield, N. H	Jan. 16, 1835	....
Corn sheller and cleaner	J. Hubler and R. M. McGrath	La Fayette, Ind	July 1, 1862	35, 757
Corn sheller and cleaner	J. C. Richards	La Fayette, Ind	Sept. 25, 1860	30, 185
Corn sheller and cleaner	N. Shock	Baltimore, Md	Sept. 8, 1868	82, 039
Corn sheller and cleaner	M. F. Williamson and J. J. Swigert.	Hyattsville, Ohio	July 23, 1861	32, 909
Corn-sheller and fanning-mill combined	J. P. Hoagland and G. E. Moser	Centralia, Pa	Sept. 10, 1867	68, 744
Corn-sheller and grain-hulling machine	J. Mercer	Harrisville, Ohio	June 24, 1839	1, 187
Corn sheller and separator	B. and D. H. Harnish	Lancaster and Pequea, Pa	May 21, 1872	127, 051
Corn sheller and separator	E. Knapp	Jamestown, N. Y	Sept. 27, 1864	44, 433
Corn sheller and separator feeder combined	J. Bernheisel, sr	Green Park, Pa	Dec. 31, 1867	72, 783
Corn-sheller and straw-cutter combined	J. P. Smith	Schuylkill Haven, Pa	Apr. 9, 1872	125, 415
Corn-sheller and vegetable-slicer combined	J. P. Smith	Hummelstown, Pa	May 3, 1870	102, 607
Corn-sheller and wheat-thrasher, &c	Z. Phinney	Cairo, N. Y	Feb. 16, 1815	....
Corn sheller case	A. Borneman	Lancaster, Ohio	Jan. 21, 1873	135, 029
Corn-sheller for table use	E. L. Tevis	Philadelphia, Pa	Jan. 16, 1866	52, 092
Corn sheller, Green	V. Barker	Otisfield, Me	Nov. 9, 1869	96, 658
Corn-sheller, Hand	G. C. Ballard	Cleveland, Ohio	Nov. 5, 1872	132, 708
Corn-sheller, Hand	O. A. Bryhn and W. T. Farre	Montreal, Canada	June 27, 1871	116, 262
Corn-sheller, Hand	C. Christian	Milwaukee, Wis	Mar. 10, 1868	75, 368
Corn-sheller, Hand	J. C. Cunyer	Thorntown, Ind	Apr. 13, 1869	89, 001
Corn-sheller, Hand	J. E. Finley	Memphis, Tenn	Oct. 11, 1870	108, 126
Corn-sheller, Hand	J. O. Fraizer	Worthington, Ind	Nov. 12, 1872	132, 958
Corn-sheller, Hand	J. M. Hawley	Alma, Ill	Feb. 11, 1873	135, 709
Corn-sheller, Hand	A. McLean and J. H. Ross	Carondelet, Mo	June 18, 1872	128, 056
Corn-sheller, Hand	W. A. Middleton	Harrisburgh, Pa	July 26, 1870	105, 825
Corn-sheller, Hand	C. M. O'Hara	Bolivar, Tenn	Nov. 23, 1869	97, 109
Corn-sheller, Hand	C. M. O'Hara	Cincinnati, Ohio	May 2, 1871	114, 331
Corn-sheller, Hand	C. M. O'Hara	Cincinnati, Ohio	July 18, 1871	117, 105
Corn-sheller, Hand	C. H. Pickering	Memphis, Tenn	May 24, 1870	103, 497
Corn-sheller, Hand	T. Weaver	Harrisburgh, Pa	Apr. 12, 1870	101, 949
Corn-sheller, Hand	T. Weaver	Harrisburgh, Pa	Nov. 26, 1872	133, 346
Corn-sheller, Hand	W. H. Wilson	Boston, Mass	Jan. 30, 1872	123, 319
Corn-sheller, Longitudinal	G. E. Waring	Poundridge, N. Y	Mar. 16, 1827	....
Corn-sheller, Phinney's	H. Gallagher	Pittsburgh, Pa	Oct. 29, 1818	....
Corn-sheller teeth	H. H. Reuter	New Hope, Mo	Jan. 23, 1872	122, 968
Corn-shellers, Cob-carrier for	H. H. Eby	Mendota, Ill	Jan. 14, 1873	134, 790
Corn-shellers, Concave for	L. J. Miller	Cincinnati, Ohio	Oct. 8, 1872	132, 017
Corn-shellers, Concave of	D. Hoats	Milton, Pa	Jan. 22, 1850	7, 036
Corn-shellers, Device for feeding corn to	H. C. Robison	Monmouth, Ill	Jan. 10, 1865	45, 862
Corn-shellers, Feeding-elevator for	P. Kaufman	Hudson, Ill	Aug. 12, 1873	141, 717
Corn-shellers separately or jointly with a fan or cutter, Arrangement of machinery for operating.	W. L. Potter	Clifton Park, N. Y	Apr. 5, 1859	23, 488
Corn, Shelling	I. A. Hedges	Elmira, N. Y	Feb. 3, 1836	....
Corn, Shelling	H. G. Neal	Poultney, Vt	Feb. 10, 1836	....
Corn, Shelling	I. Smith	Downington, Pa	Feb. 13, 1836	....
Corn shelling and cleaning machine	L. A. Beebe	Chicago, Ill	Oct. 21, 1862	36, 700
Corn shelling and cleaning machine	J. Kamp	Vanderburgh County, Ind	Jan. 31, 1871	111, 350
Corn shelling and cleaning machine	J. Reed	Marshfield, Mass	Sept. 1, 1831	....
Corn shelling and cleaning machine	J. W. Webb	Mount Morris, N. Y	July 16, 1838	840
Corn-shelling and grain and plaster grinding mill	B. M. Kemp	Fort Plain, N. Y	Jan. 31, 1829	....
Corn shelling and grinding	S. Slater and S. Noblit	Philadelphia, Pa	May 12, 1834	....
Corn shelling and grinding machine	G. Seymour	Cedar Rapids, Iowa	July 25, 1861	32, 898
Corn shelling and grinding machinery	S. Fowks	Catskill, N. Y	May 4, 1825	....
Corn shelling and husking machine	S. S. Allen	Miamisburgh, Ohio	Jan. 15, 1840	1, 472
Corn shelling and winnowing machine	B. Clough	Natick, Mass	Feb. 17, 1863	37, 674
Corn-shelling machine	A. Bailey	Poultney, Vt	Nov. 1, 1825	....
Corn-shelling machine	J. C. Baldwin	Staunton, Va	Oct. 11, 1836	45
Corn-shelling machine	J. Barnes and N. T. Loomis	Cornwall, Conn	Apr. 13, 1831	....
Corn-shelling machine	J. Brown	Providence, R. I	Mar. 18, 1828	....
Corn-shelling machine	C. Bulkeley	Manchester, Vt	May 25, 1824	....
Corn-shelling machine	E. W. Coffin	Gilead, Me	Dec. 23, 1834	....
Corn-shelling machine	B. Crooker	New York, N. Y	Dec. 2, 1824	....
Corn-shelling machine	T. J. Dean	Virgil, N. Y	Dec. 8, 1828	....
Corn-shelling machine	L. E. Denison	Saybrook, Conn	Oct. 8, 1838	972
Corn-shelling machine	L. E. Denison	Saybrook, Conn	Aug. 12, 1839	1, 283
Corn-shelling machine	J. Fuller	Rockland, Va	Mar. 19, 1816	....
Corn-shelling machine	P. Grosjean	Louisville, Ky	July 29, 1828	....
Corn-shelling machine	W. C. Hawley	New York, N. Y	June 22, 1825	....
Corn-shelling machine	J. Heavin	Fredericktown, Md	May 8, 1810	....
Corn-shelling machine	D. Hitchcock	New York, N. Y	June 13, 1831	....
Corn-shelling machine, &c	C. Hoxie	Hudson, N. Y	Dec. 8, 1825	....
Corn-shelling machine	W. Hoyt	Vernon, Ind	Apr. 29, 1828	....
Corn-shelling machine	W. Hoyt	Vernon, Ind	Dec. 17, 1830	....
Corn-shelling machine	W. Hoyt	Vernon, Ind	June 13, 1831	....
Corn-shelling machine	S, Lane	Hallowell, Me	Sept. 9, 1825	....
Corn-shelling machine	N. Lindsay	Catskill, N. Y	Mar. 28, 1821	....
Corn-shelling machine	H. Lippold	Silver Creek, N. Y	Aug. 15, 1871	118, 139
Corn-shelling machine	A. Little	Columbia, Conn	Dec 23, 1824	....
Corn-shelling machine	W. McIlroy and B. and W. Boon	Greenwich, N. J	Mar. 30, 1839	1, 112
Corn-shelling machine	J. Moon	Morrisville, Pa	June 27, 1825	....
Corn-shelling machine	T. Newman	Guilford County, N. C	Feb. 7, 1827	....
Corn-shelling machine	J. Parker and H. M. Smith	Richmond, Va	Apr. 9, 1831	....
Corn-shelling machine	W. R Parker	Milton, Del	July 9, 1839	1, 227
Corn-shelling machine	L. Peck	Brookfield, Conn	Apr. 9, 1824	....
Corn-shelling machine	P. Pilsbury	....	Oct. 25, 1803	....

Index of patents issued from the United States Patent Office from 1790 *to* 1873, *inclusive*—Continued.

Invention.	Inventor.	Residence.	Date.	No.
Corn-shelling machine	R. Porter	Billerica, Mass	Sept. 12, 1838	912
Corn-shelling machine	F. Price	New York	May 4, 1825	....
Corn-shelling machine	J. S. Rackham	Waterport, N. Y	Oct. 20, 1868	83, 314
Corn-shelling machine	H. and E. J. Rosevelt	New York, N. Y	Jan. 26, 1821	....
Corn shelling machine	E. Russell	Greensborough, N. C	Oct. 24, 1826	....
Corn-shelling machine	S. Spooner	Petersham, Mass	Feb. 27, 1823	....
Corn-shelling machine	G. W. Tolhurst	Liverpool, Ohio	Nov. 30, 1858	22, 206
Corn-shelling machine	T. Wellman, jr	Brooklin, Vt	Oct. 9, 1834	....
Corn-shelling machine	L. Whitney	New York, N. Y	Nov. 13, 1824	....
Corn-shelling machine	L. J. Wicks	Racine, Wis	Aug. 24, 1858	21, 288
Corn-shelling machine and family mill	E. Bond	Mendon, N. Y	June 13, 1831	....
Corn-shield	B. Brandreth	Sing Sing, N. Y	June 4, 1872	127, 454
Corn-shock binder	J. E. Hunter	Mechanicsburgh, Ohio	May 25, 1869	90, 451
Corn-shock binder	C. Stowe	Braceville, Ohio	Apr. 24, 1860	28, 019
Corn-shocks, Device for loading and unloading	W. M. Mason	Polo, Ind	Mar. 22, 1864	42, 008
Corn-shocker	G. E. Johnson	Smith Valley, N. Y	Dec. 9, 1873	145, 423
Corn shucking and shelling machine	S. Kingsbery	Carrollton, Ga	Oct. 6, 1857	18, 342
Corn shucking and shelling machine	J. J. Rallow	Fredericksburgh, Va	July 3, 1855	13, 182
Corn-stacking horse	L. S. Barker	Pittsford, Mich	Dec. 19, 1865	51, 540
Corn-stalk chopper	J. Hollingsworth	Chicago, Ill	Oct. 29, 1872	132, 578
Corn-stalk cutter	R. H. Avery	Galesburgh, Ill	Aug. 6, 1872	130, 100
Corn-stalk cutter	S. Bryan	Jefferson, Wis	May 14, 1867	64, 743
Corn-stalk cutter	M. Clark	Oakley, Ill	June 15, 1869	91, 420
Corn-stalk cutter	G. W. Cole	Canton, Ill	July 14, 1863	39, 214
Corn-stalk cutter	J. W. Cornell	Lawn Ridge, Ill	Aug. 8, 1871	117, 746
Corn-stalk cutter	T. A. Galt and G. S. Tracy	Sterling, Ill	Nov. 28, 1871	131, 353
Corn-stalk cutter	A. Hunt	Macomb, Ill	Sept. 10, 1867	68, 749
Corn-stalk cutter	H. Jackson	Elmira, Ill	Mar. 9, 1869	87, 676
Corn-stalk cutter	J. R. Marshall	Marine, Ill	Aug. 21, 1860	29, 705
Corn-stalk cutter	H. Martin	Round Grove, Ill	June 4, 1872	127, 623
Corn-stalk cutter	G. D. McClure	Denver Station, Ill	Apr. 10, 1866	53, 848
Corn-stalk cutter	A. J. Nebergall	Cedar County, Iowa	Sept. 6, 1870	107, 090
Corn-stalk cutter	J. B. Sherlock	Port Byron, Ill	Oct. 19, 1869	96, 042
Corn-stalk cutter	S. Walters	Dallas City, Ill	Mar. 5, 1872	124, 401
Corn-stalk cutter	D. Wilde	Washington, Iowa	July 16, 1872	129, 384
Corn stalk cutter	J. Wood	Pella, Iowa	Feb. 7, 1871	111, 709
Corn-stalk cutter and corn-husker	J. Russell	Brooklyn, N. Y	Dec. 18, 1866	60, 643
Corn-stalk cutter and cultivator combined	M. Gordon	Washington, Iowa	Apr. 21, 1868	76, 909
Corn-stalk cutter and stripper	N. Gabel	Preble County, Ohio	Jan. 30, 1866	52, 282
Corn-stalk cutter preparatory to plowing	J. D. and J. H. Wilde	Washington, Iowa	Feb. 18, 1868	74, 652
Corn-stalk cutting and grinding machine	W. G. Huyett	Williamsburgh, Pa	Dec 22, 1857	18, 905
Corn-stalk cutting and raking machine	T. M. Hill and S. D. Tuttle	Eaton, Ohio	Nov. 27, 1866	60, 098
Corn-stalk-cutting knife	G. D. Goodsell and N. E. Babcock	Rockford, Ill	Feb. 8, 1870	99, 560
Corn-stalk-cutting machine	A. Glendening	Butler County, Pa	Sept. 25, 1823	....
Corn-stalk-cutting machine	J. M. Goff	Ionia, Ill	Sept. 11, 1866	57, 893
Corn-stalk-cutting machine	H. Johnston	Collinsville, Ill	Apr. 12, 1859	23, 644
Corn-stalk knife	J. M. Brick	Marlton, N. J	Apr. 22, 1873	138, 069
Corn-stalk pith for use in the arts, Preparing	W. M. Bryant	Alexandria, Va	May 3, 1870	102, 484
Corn-stalks in the field, Machine for cutting	H. Bilharz	Seneca, Ill	July 2, 1867	66, 287
Corn-stalks in the field, Machine for cutting up	F. M. Green	Sullivan, Ill	Aug. 24, 1858	21, 254
Corn-stalks, Machine for cutting and crushing	H. and A. Hersh	Lancaster County, Pa	Mar. 9, 1858	19, 561
Corn-stalks, Machine for cutting and grinding	E. Potts	Tredyffrin Township, Pa	July 31, 1847	5, 207
Corn-stalks, Machine for cutting and pulverizing	W. M. Taylor	Newburgh, Ohio	Nov. 19, 1872	133, 270
Corn-stalks, Machine for cutting standing	F. J. Freeman	Heyworth, Ill	Apr. 2, 1861	31, 928
Corn-stalks, Machine for shocking	S. B. Lawrence	Hookstown, Pa	July 24, 1860	29, 286
Corn-stalks on the ground, Device for cutting	T. W. McDill	Perry, Ill	July 11, 1865	48, 704
Corn-stubbles preparatory to plowing, Machine for cutting.	J. Augspurger	Trenton, Ohio	Sept. 1, 1857	18, 076
Corn-weeder	S. March	Norfolk, Va	Apr. 16, 1867	63, 912
Corn with malt extract, Treating Indian	M. Cziner	New York, N. Y	Nov. 7, 1871	120, 721
Corns, bunions, &c., Composition for curing	A. J. Ferguson	Sharon, Pa	Aug. 25, 1868	81, 486
Corns, bunions, &c., Implements for removing	C. L. Roorback	Saint Clair, Pa	Dec. 13, 1870	110, 162
Corns, Composition for curing	G. Oakley	Quincy, Ill	July 27, 1869	92, 993
Corns, Eradicating and curing	W. Davis	Williamsburgh, Va	Sept. 9, 1835	....
Corns, Eradicator for	C. Wheat	Geneva, N. Y	Oct. 12, 1858	21, 790
Corns, Eradicator for	P. Williamson	New York, N. Y	May 17, 1838	742
Corns, Instrument for removing	J. C. Jacobsohn	New York, N. Y	Oct. 22, 1872	122, 468
Corns, Tincture for curing	E. Smith	New Brunswick, N. J	Jan. 20, 1826	....
Corner protector, Stair and room	H. C. Richards	Cincinnati, Ohio	July 31, 1866	56, 800
Corner-strip	G. Corbett	Pittsburgh, Pa	Sept. 24, 1872	131, 506
Cornet, &c	I. Fiske	Worcester, Mass	Oct. 30, 1866	59, 204
Cornet, &c	I. Fiske	Worcester, Mass	Feb. 11, 1868	74, 331
Cornet	L. Schreiber	New York, N. Y	Sept. 12, 1865	49, 925
Cornet, &c	L. Schreiber	New York, N. Y	May 7, 1867	64, 582
Cornet-bells, Machine for forming	L. W. Spencer	New York, N. Y	Apr. 9, 1867	63, 761
Cornets, Machine for cutting the wind-passages in the rotary valves of.	L. W. Spencer	New York, N. Y	Apr. 16, 1867	63, 819
Cornets, Machine for forming the branch-tubes of valve-cases for.	L. W. Spencer	New York, N. Y	Apr. 9, 1867	63, 760
Cornice and gutter, Metallic	J. B. Cornell	New York, N. Y	Nov. 4, 1873	144, 259
Cornice-bracket, Adjustable metallic	A. T. Perkins and N. Waterman	Toledo, Ohio	Mar. 5, 1872	124, 382
Cornice, Curtain	C. W. Hill	New York, N. Y	June 15, 1869	91, 441
Cornice, Drapery	H. R. Watson	Albany, N. Y	Nov. 12, 1872	132, 941
Cornice, Extension window	A. Pohl	Brooklyn, N. Y	May 28, 1872	127, 367
Cornice for window-curtains, Adjustable	O. L. Gardner	New York, N. Y	Apr. 20, 1869	89, 139
Cornice-mold	M. Meany, J. McGinnis, and W. Cunningham.	Brooklyn, N. Y	Mar. 29, 1870	101, 293
Cornice-moldings, Machine for making	C. L. Wood and C. A. Sheridan.	Cleveland, Ohio	Apr. 5, 1870	101, 556
Cornice, Ornamental	Y. J. McGeary	Newark, N. J	Feb. 15, 1870	99, 930
Cornice-plastering mold	C. P. Walter	Aston Township, Pa	Oct. 9, 1866	58, 701
Cornice, Running	A. Leverty	Bridgeport, Conn	May 26, 1868	78, 381
Cornice, Stucco	A. Derrom	Paterson, N. J	May 20, 1873	139, 049
Cornice-tool	J. P. Ballantine	Detroit, Mich	Dec. 10, 1872	133, 820
Cornice, Window	D. Coney and J. H. Norcross	Augusta, Me., and Melrose, Mass.	Mar. 12, 1872	124, 549
Cornice, Window	R. N. Hoffman	Chicago, Ill	Mar. 15, 1870	100, 763
Cornice, Window	A. J. Holmes	Saratoga Springs, N. Y	Apr. 16, 1867	63, 893
Cornice, Window	W. Lloyd	New York, N. Y	Sept. 24, 1872	131, 551

Index of patents issued from the United States Patent Office from 1790 to 1873, inclusive—Continued.

Invention.	Inventor.	Residence.	Date.	No.
Cornice, Window	W. Lloyd	New York, N. Y	Sept. 17, 1872	131, 358
Cornice, Window-curtain	C. W. Hill	New York, N. Y	June 22, 1869	91, 746
Cornice, Window-curtain	W. Lloyd	New York, N. Y	Dec. 19, 1871	122, 037
Cornices and moldings, Tool for forming plaster	S. Groesbeck	New York, N. Y	Sept. 17, 1850	7, 654
Cornices, Construction of metallic	J. M. Blackburn, C. L. Wood, B. K. Price, and C. A. Sheridan.	Cleveland, Ohio	Apr. 5, 1870	101, 572
Cornices, Machine for bending sheet-metal for	C. A. Buttles and D. Murphy	Milwaukee, Wis	Oct. 12, 1869	95, 650
Cornices, Machine for rolling	A. Johnson	Cairo, N. Y	Dec. 22, 1857	18, 906
Cornices of sheet-metal, Machine for forming	J. Lee	Bolivar, Ohio	June 12, 1860	28, 676
Cornices to windows, Mode of attaching	A. Peple	East Billerica, Mass	Mar. 22, 1870	101, 158
Cornish-engine	J. Storer	Peekskill, N. Y	Aug. 30, 1870	106, 886
Cornucopia, Transparent	W. Lohse	New York, N. Y	Dec. 12, 1871	121, 791
Corpse-cooler	S. H. Crump	Reading, Pa	July 29, 1873	141, 331
Corpse-cooler	E. S. Earley	Philadelphia, Pa	Jan. 28, 1873	135, 210
Corpse-cooler	J. Hoffman	Toledo, Ohio	Sept. 2, 1873	142, 393
Corpse-cooler	F. N. Troll	Baltimore, Md	Sept. 2, 1873	142, 533
Corpse-preserver	R. C. Andrus	Poughkeepsie, N. Y	Dec. 6, 1870	109, 861
Corpse-preserver	G. D. Blocher	Indianapolis, Ind	Feb. 26, 1867	62, 310
Corpse-preserver	L. D. Bunn	Morristown, N. J	Oct. 28, 1862	36, 765
Corpse-preserver	J. T. Carpenter	Downingtown, Pa	May 2, 1871	114, 407
Corpse-preserver	J. L. Clark	Providence, R. I	May 25, 1869	90, 426
Corpse-preserver	S. Cobb	Cincinnati, Ohio	Jan. 27, 1857	16, 466
Corpse-preserver	C. W. Compton	Newark, N. J	Nov. 5, 1867	70, 530
Corpse-preserver	J. E. Cox	Baltimore, Md	Aug. 22, 1871	118, 346
Corpse-preserver	H. M. Diggins	Cincinnati, Ohio	Apr. 18, 1871	113, 861
Corpse-preserver	H. V. Griffith	Altoona, Pa	Oct. 18, 1870	108, 350
Corpse-preserver	D. L. Holden	Covington, Ky	Feb. 13, 1872	123, 697
Corpse-preserver	J. Kunsman, sr	Reading, Pa	May 22, 1866	55, 013
Corpse-preserver	H. Kersten	Wheeling, W. Va	Sept. 16, 1873	142, 919
Corpse-preserver	H. Lee	Washington, D. C	May 17, 1870	103, 060
Corpse-preserver	E. F. Lenox and C. Eckhart	Trenton, N. J	June 21, 1870	104, 607
Corpse-preserver	M. R. Margerum	Trenton, N. J	May 2, 1871	114, 315
Corpse-preserver	M. E. Mott	Rouse's Point, N. Y	May 26, 1868	78, 314
Corpse-preserver	G. W. Nash	Columbus, Ohio	May 23, 1871	115, 229
Corpse-preserver	C. O. Peck	Pittsfield, Mass	Dec. 9, 1873	145, 307
Corpse-preserver	C. F. Pike	Providence, R. I	Dec. 31, 1867	72, 893
Corpse-preserver	A. G. Reed	Philadelphia, Pa	Mar. 30, 1869	88, 510
Corpse-preserver	J. J. Reicherts	Delaware, Ohio	Sept. 29, 1868	82, 552
Corpse-preserver	S. Stockton and J. Schepler	Lambertville, N. J	Sept. 27, 1870	107, 831
Corpse-preserver	P. Taltavull	Washington, D. C	Apr. 29, 1873	138, 449
Corpse-preserver	J. C. Taylor	Trenton, N. J	July 5, 1870	105, 013
Corpse-preserver	J. F. and E. G. Waters	Philadelphia, Pa	Mar. 5, 1872	124, 300
Corpse-preserver	B. Wilson	Morristown, N. J	Nov. 3, 1863	40, 529
Corpse-preserving	C. A. Seely and C. J. Eames	New York, N. Y	Feb. 18, 1868	74, 607
Corpse preserving and embalming	J. A. Mitchell	Maysville, Ky	Aug. 6, 1872	130, 232
Corpse-preserving box	P. Wendhiser	Rockville, Conn	Dec. 17, 1867	72, 346
Corpse-preserving case	J. Gravenstine	Philadelphia, Pa	July 25, 1871	117, 407
Corpse-preserving case	A. G. Reed	Philadelphia, Pa	July 12, 1870	105, 368
Corpse-preserving case	J. Searing	Morristown, N. J	Dec. 6, 1864	45, 349
Corpse-preserving case	J. S. Waterman	Roxbury, Mass	Apr. 23, 1867	64, 172
Corpse-preserving case	J. S. Waterman	Roxbury, Mass	Apr. 7, 1868	76, 568
Corpse-preserving casket	N. F. Curran	Baltimore, Md	Sept. 24, 1872	131, 668
Corpse-preserving casket	C. H. Kimball	Quincy, Mass	Jan. 28, 1868	73, 729
Corpse-preserving composition	J. M. Gallaher	Carthage, N. Y	Oct. 28, 1873	144, 085
Corpse-receptacle	T. Holmes	Washington, D. C	July 21, 1863	39, 291
Corpses, Device for handling	H. Huber	Crestline, Ohio	Nov. 26, 1872	133, 447
Corpses in place, Device for securing the features of.	A. S. Chesebrough	Hartford, Conn	May 3, 1870	102, 490
Corpses, Refrigerator for	R. Fredrick and G. A. Trump	Baltimore, Md	May 9, 1846	4, 504
Corrugated boiler	R. Montgomery	New York, N. Y	Oct. 29, 1850	7, 742
Corrugated fabrics, Manufacturing	F. Baare and J. G. Garelly	New York, N. Y	July 5, 1859	24, 691
Corrugated knives, Machine for grinding	J. B. Wilson	New York, N. Y	Nov. 9, 1869	96, 756
Corrugated metals, Rolling	R. Montgomery	New York, N. Y	Dec. 27, 1859	26, 607
Corrugated plates, Manufacture of	S. J. Seely	Brooklyn, N. Y	Sept. 9, 1862	36, 424
Corrugated steel plates as an article of manufacture, Curved.	R. Montgomery	New York, N. Y.	Mar. 19, 1867	63, 077
Corrugating and molding sheet-metals, Machine for.	A. Johnson	Brooklyn, N. Y	Feb. 8, 1870	99, 684
Corrugating circular metal plates	R. B. Hugunin	Cleveland, Ohio	Apr. 19, 1864	42, 377
Corrugating-machine	J. G. Baker	Washington, D. C	Dec. 15, 1863	40, 897
Corrugating machine, Metal	S. J. Seely	Brooklyn, N. Y	Feb. 3, 1863	37, 593
Corrugating machine, Sheet-metal	W. Mann	Newcastle, Pa	Apr. 12, 1870	101, 750
Corrugating machine, Sheet-metal	G. R. Moore	Lyons, Iowa	Feb. 18, 1868	74, 572
Corrugating machine, Sheet-metal	F. Roys	East Berlin, Conn	June 3, 1873	139, 480
Corrugating metal plates	J. Francis	New York, N. Y	June 2, 1863	38, 799
Corrugating metal plates	R. Montgomery	New York, N. Y	Feb. 21, 1854	10, 549
Corrugating metal plates, Machine for	A. W. Gray	Middletown, Vt	Dec. 10, 1872	133, 773
Corrugating metal plates, Machine for	R. Montgomery	New York, N. Y	Apr. 12, 1859	23, 599
Corrugating metals, Apparatus for	J. Montgomery	New York, N. Y	May 10, 1870	102, 849
Corrugating metals, Machine for	J. Moffet	New York, N. Y	July 16, 1872	129, 290
Corrugating metals, Machine for	R. Montgomery	New York, N. Y	Dec. 22, 1868	85, 262
Corrugating metallic sheets	R. Montgomery	New York, N. Y	June 21, 1859	24, 480
Corrugating sheet-iron, Machine for	A. and J. Reese	Pittsburgh, Pa	Feb. 25, 1873	136, 267
Corrugating sheet-metal	T. W. H. Moseley	Boston, Mass	Aug. 4, 1863	39, 418
Corrugating sheet-metal	W. E. Worthen and H. B. Renwick.	New York, N. Y	July 5, 1859	24, 689
Corrugating sheet-metals, Machine for	S. G. Booth	New York, N. Y	Apr. 4, 1854	10, 732
Corrugating sheet-metals, Machine for	R. Montgomery	New York, N. Y	Apr. 26, 1859	23, 774
Corrugating sheet-metals, Machine for	F. Roys	East Berlin, Conn	Mar. 15, 1870	100, 929
Corrugating sheet-metals, Machine for	J. Wilson, C. Green, and W. Wilson, jr.	Wilmington, Del	Apr. 19, 1859	23, 736
Corrugating sheet-metals, Roller for	S. G. Booth	New York, N. Y	Jan. 23, 1855	12, 268
Corset	M. Adler	New Haven, Conn	July 16, 1872	129, 264
Corset	M. H. Beckworth	Camden, N. Y	June 11, 1867	65, 636
Corset	J. Bowers	New York, N. Y	June 6, 1865	48, 045
Corset	M. P. Bray	Birmingham, Conn	Apr. 4, 1871	113, 392
Corset	M. P. Bray	New Haven, Conn	Apr. 23, 1872	125, 877
Corset	A. J. Brooks	Philadelphia, Pa	Aug. 26, 1862	36, 272
Corset	M. Brush	New York, N. Y	July 21, 1815	......
Corset	W. J. Cantello	New York, N. Y	Dec. 31, 1821	......

Index of patents issued from the United States Patent Office from 1790 *to* 1873, *inclusive*—Continued.

Invention.	Inventor.	Residence.	Date.	No.
Corset	M. L. Changeur	Paris, France	July 16, 1867	66, 794
Corset	L. L. Chapman	Camden, N. J	Dec. 15, 1863	40, 907
Corset	M. Cohn	New York, N. Y	Apr. 15, 1873	137, 893
Corset	S. Collins	New Haven, Conn	Apr 26, 1864	42, 460
Corset	J. S. Crotty	New York, N. Y	Sept. 3, 1872	130, 983
Corset	C. Z. Cummings	Buffalo, N. Y	Jan. 14, 1868	73, 235
Corset	S. A. Dake	Eureka, Wis	Nov. 19, 1872	133, 209
Corset	H. H. Dayton	Worcester, Mass	July 7, 1863	39, 126
Corset	R. De Baun	Chicago, Ill	Oct. 3, 1871	119, 583
Corset	T. B. De Forest	Birmingham, Conn	July 28, 1868	80, 462
Corset	S. Dixon	Nottingham, England	Jan. 21, 1873	135, 097
Corset	E. Drucker	Paris, France	May 7, 1867	64, 530
Corset	D. H. Fanning	Worcester, Mass	May 25, 1869	90, 353
Corset	D. H. Fanning	Worcester, Mass	Mar. 1, 1870	100, 278
Corset	S. B. Fisler	Newark, N. J	Feb. 23, 1869	87, 160
Corset	H. S. Flood	San Francisco, Cal	Sept. 10, 1872	131, 262
Corset	L. H. Foy	Worcester, Mass	Sept. 15, 1863	39, 908
Corset	L. H. Foy	Worcester, Mass	Sept. 15, 1863	39, 909
Corset	L. H. Foy	Newton Centre, Mass	July 7, 1868	79, 647
Corset	T. S. Gilbert	Birmingham, Conn	Aug. 10, 1869	93, 528
Corset	T. S. Gilbert	Birmingham, Conn	Feb. 8, 1870	99, 664
Corset	T. S. Gilbert	Birmingham, Conn	Apr. 4, 1871	113, 419
Corset	T. S. Gilbert	Birmingham, Conn	Oct. 1, 1872	131, 873
Corset	T. S. Gilbert	Birmingham, Conn	July 22, 1873	141, 134
Corset	C. A. Griswold	Willimantic, Conn	Feb. 5, 1867	61, 825
Corset	C. A. Griswold	Willimantic, Conn	July 4, 1871	116, 585
Corset	E. Hambujer	Detroit, Mich	June 26, 1866	55, 972
Corset	T. F. Hamilton	New Haven, Conn	June 13, 1871	115, 954
Corset	T. F. Hamilton	New Haven, Conn	Feb. 25, 1873	136, 320
Corset	C. L. Hamlin	New York, N. Y	Nov. 14, 1871	120, 967
Corset	J. W. Lane	Bristol, England	Jan. 14, 1873	134, 902
Corset	L. H. and A. F. Loomer and L. M. Smith.	Birmingham, Conn	Dec. 16, 1873	145, 665
Corset	H. A. Lyman	London, England	Nov. 30, 1869	97, 418
Corset	H. E. Marchand	Louisville, Ky	Jan. 8, 1869	90, 950
Corset	A. S. McLean	Williamsburgh, N. Y	Dec. 28, 1858	22, 443
Corset	E. J. Meriman	New York, N. Y	Sept. 8, 1868	81, 926
Corset	E. J. Meriman	New York, N. Y	Feb. 2, 1869	86, 570
Corset	W. Miller	New York, N. Y	July 30, 1867	67, 333
Corset	W. W. Netterfield	Rochester, N. Y	Sept. 15, 1868	82, 147
Corset	S. Ottenheimer	New York, N. Y	Apr. 23, 1872	126, 079
Corset	L. A. Palmer	Boston, Mass	Nov. 16, 1869	96, 951
Corset	S. M. Perry	Plainfield, N. J	July 31, 1866	56, 794
Corset	C. Preston	Detroit, Mich	May 23, 1865	47, 856
Corset	M. Russell	Union Bridge, Md	Apr. 8, 1873	137, 724
Corset	C. D. Rutherford	Brooklyn, N. Y	Dec. 11, 1866	60, 428
Corset	M. Sebillo	New York, N. Y	Sept. 15, 1863	39, 964
Corset	W. H. Stroup	Pittsburgh, Pa	Sept. 24, 1867	69, 140
Corset	J. Waterman	New York, N. Y	Oct. 7, 1873	143, 479
Corset	A. M. Weber	Oshkosh, Wis	Apr. 15, 1873	137, 985
Corset, Abdominal	J. P. McLean	New York, N. Y	Nov. 8, 1859	26, 039
Corset, Abdominal	C. Smith	Louisville, Ky	Apr. 27, 1869	89, 513
Corset, Abdominal and skirt-supporter	J. McNeven	New York, N. Y	Aug. 11, 1868	80, 988
Corset and bustle, Combined	L. H. Foy	Worcester, Mass	May 1, 1866	54, 323
Corset and skirt-supporter	E. Blakeslee	Plymouth, Conn	June 19, 1866	55, 608
Corset and skirt-supporter combined	W. Bacheller	West Newbury, Mass	Jan. 22, 1867	61, 380
Corset and skirt-supporter combined	L. Spigelmeyer	Easton, Pa	Nov. 21, 1871	121, 210
Corset and skirt-supporter, Combined	M. J. C. Vanorstrand	Pekin, Ill	Dec. 19, 1871	122, 081
Corset-busk	E. Heaton	New Haven, Conn	Feb. 16, 1869	86, 920
Corset-busk	T. Wallace, jr	Ansonia, Conn	Oct. 13, 1863	40, 298
Corset-busk clasp	W. Devines	Brooklyn, N. Y	Apr. 3, 1866	53, 734
Corset, Child's	H. G. Emery and M. C. Fuller	Boston, Mass	Mar. 26, 1872	124, 891
Corset, Child's	M. B. Solomon	Charlestown, Mass	Apr. 7, 1868	76, 354
Corset-clasp	M. E. Bulkley	Providence, R. I	May 28, 1867	65, 163
Corset-clasp	J. Burke	Boston, Mass	Oct. 7, 1873	143, 403
Corset-clasp	W. B. Cargill	New Haven, Conn	Mar. 24, 1868	75, 856
Corset-clasp	J. W. Carter	Greenville, N. J	Aug. 14, 1866	57, 085
Corset-clasp	C. O. Crosby	New Haven, Conn	Apr. 16, 1867	63, 863
Corset-clasp	L. Hill	North Brookfield, Mass	Mar. 12, 1867	62, 848
Corset-clasp	P. Lippmann	New York, N. Y	Sept. 30, 1873	143, 359
Corset-clasp	J. P. MacLean	Brooklyn, N. Y	Nov. 18, 1873	144, 689
Corset-clasp	T. A. Nelson	Birmingham, Conn	July 21, 1863	39, 322
Corset clasp and spring	P. H. Niles	Boston, Mass	Dec. 20, 1870	110, 273
Corset-clasp	J. B. Roby	Cambridge, Mass	Aug. 26, 1873	142, 279
Corset-clasp	F. Straus	New York, N. Y	Aug. 6, 1869	88, 752
Corset-clasp	R. Wallace and D. G. and H. E. Fowler.	Wallingford, Conn	Oct. 17, 1865	50, 524
Corset-fastening	J. Bowers	New York, N. Y	Mar. 19, 1867	63, 005
Corset-fastening	M. P. Bray	Ansonia, Conn	Mar. 22, 1870	100, 970
Corset-fastening	W. B. Cargill	New Haven, Conn	Jan. 28, 1868	73, 873
Corset-fastening	W. B. Cargill	New Haven, Conn	May 19, 1868	78, 056
Corset-fastening	W. B. Cargill	Waterbury, Conn	Feb. 8, 1870	99, 633
Corset-fastening	S. Chapman	New York, N. Y	June 14, 1864	43, 094
Corset-fastening	J. F. Dubber	Brooklyn, N. Y	May 7, 1867	64, 507
Corset-fastening	M. E. Frentz	New Albany, Ind	June 8, 1869	91, 112
Corset-fastening	L. Jarchow	New York, N. Y	Dec. 7, 1869	97, 643
Corset-fastening	J. P. Love	New York, N. Y	May 28, 1867	65, 246
Corset-fastening	F. W. Marston	Boston, Mass	Dec. 14, 1869	97, 788
Corset-fastening	P. H. Niles and F. W. Marston	Boston, Mass	Dec. 15, 1868	84, 899
Corset-fastening	C. Schieck	New York, N. Y	Dec. 27, 1870	110, 593
Corset-fastening	G. O. Schneller	Ansonia, Conn	May 27, 1873	139, 424
Corset-fastening	H. N. Sherman	Beloit, Wis	Jan. 26, 1869	86, 183
Corset-fastening	M. T. Smith	New York, N. Y	Aug. 10, 1869	93, 489
Corset-fastening	A. W. Webster	Ansonia, Conn	Nov. 19, 1867	71, 092
Corset-fastening, Supporting-busk for	M. E. Bulkley	Providence, R. I	Dec. 26, 1871	122, 152
Corset-pad	M. P. R. Tilton	Trenton, N. J	Dec. 20, 1870	110, 310
Corset-pad, Elastic	B. Bernstein	New York, N. Y	Aug. 6, 1872	130, 182
Corset-rings, &c., Groove in	C. Buckland	Middletown, Conn	June 2, 1836	....

Index of patents issued from the United States Patent Office from 1790 *to* 1873, *inclusive*—Continued.

Invention.	Inventor.	Residence.	Date.	No.
Corset, Skeleton	D. H. Fanning	Worcester, Mass	Feb. 15, 1870	99, 871
Corset, Skeleton	T. S. Gilbert	Birmingham, Conn	Apr. 4, 1871	113, 418
Corset skirt-supporter	J. W. Brooks	Boston, Mass	June 8, 1869	90, 920
Corset skirt-supporter	J. W. Brooks	Boston, Mass	Jan. 25, 1870	99, 146
Corset skirt-supporter	J. W. Brooks	Boston, Mass	May 31, 1870	103, 556
Corset skirt-supporter	L. H. Foy	Worcester, Mass	July 22, 1862	35, 930
Corset skirt-supporter	L. H. Foy	Worcester, Mass	Sept. 15, 1863	39, 910
Corset skirt-supporter	L. H. Foy	Worcester, Mass	Sept. 15, 1863	39, 911
Corset skirt-supporter	L. H. Foy	Worcester, Mass	Mar. 22, 1864	41, 987
Corset skirt-supporter	L. H. Foy	Worcester, Mass	Nov. 29, 1864	45, 296
Corset, Skirt-supporting	C. A. Griswold	Willimantic, Conn	July 10, 1866	56, 210
Corset, Skirt-supporting	L. S. Scofield	Boston, Mass	Aug. 23, 1864	43, 930
Corset, Skirt-supporting	S. Young	Elmira, N. Y	Nov. 7, 1871	120, 808
Corset, Spinal	A. Abbé	Boston, Mass	May 26, 1857	17, 356
Corset-spring	H. Bennett	New York, N. Y	Nov. 9, 1869	96, 665
Corset-spring	S. H. Brown and C. H. Willets	New York, N. Y	Apr. 14, 1868	76, 708
Corset-spring	T. B. De Forest and T. S. Gilbert.	Birmingham, Conn	Jan. 21, 1868	73, 440
Corset-spring	J. Hananer and N. Sartor	New York, N. Y	June 7, 1870	104, 022
Corset-spring	T. A. Hares	New York, N. Y	Aug. 30, 1870	106, 934
Corset-spring	F. W. Marston	Philadelphia, Pa	Aug. 16, 1870	106, 381
Corset-spring	G. O. Schneller	Ansonia, Conn	Feb. 20, 1872	123, 938
Corset-spring	G. O. Schneller	Ansonia, Conn	Sept. 10, 1872	131, 305
Corset-spring	G. O. Schneller	Ansonia, Conn	Dec. 31, 1872	134, 490
Corset-spring	G. O. Schneller	Ansonia, Conn	July 8, 1873	147, 733
Corset-spring, Extension	S. H. Barnes	New York, N. Y	July 17, 1866	56, 345
Corset-spring fastener	E. Kunze	Buchholz, Saxony	Apr. 30, 1872	126, 217
Corset-spring-punching machine	P. Brooks	Waterbury, Conn	Apr. 18, 1871	113, 737
Corset steel	M. Adler	New Haven, Conn	Sept. 3, 1872	130, 967
Corset steel	F. L. and S. H. Barnes	New York, N. Y	Aug. 10, 1869	93, 399
Corset steel	C. S. Chaffee	Birmingham, Conn	Jan. 21, 1873	135, 079
Corset steel	J. L. Fitzpatrick	Waterbury, Conn	Nov. 9, 1869	96, 685
Corset steels, &c., Japaning	C. H. Bassett	Derby, Conn	Apr. 26, 1864	42, 457
Corset-weaving	B. J. Goullioud	Paris, France	Jan. 27, 1863	37, 546
Corset, Woven	J. Ottenheimer	New York, N. Y	Nov. 25, 1873	144, 921
Corsets and abdominal supporters, Manufacturing	E. Adams	Boston, Mass	Jan. 21, 1841	1, 940
Corsets and bustles, Manufacturing	D. Lamoureux	New York, N. Y	Jan. 4, 1859	22, 532
Corsets, Bust or stay for	T. B. De Forest	Birmingham, Conn	Sept. 15, 1868	82, 209
Corsets, Device for attaching steels to	W. A. and H. E. Starrett	Lawrence, Kans	Oct. 4, 1870	107, 974
Corsets, Device for measuring, laying out, and forming.	M. Williams	Camden, Ohio	Mar. 29, 1870	101, 337
Corsets, Fastening for stays of	J. W. Brooks	Boston, Mass	Apr. 20, 1869	89, 018
Corsets, Machine for rounding whalebone for	J. A. Sevey	Boston, Mass	June 15, 1869	91, 370
Corsets, Skirt-supporting hook for	D. H. Fanning	Worcester, Mass	Nov. 15, 1870	109, 309
Corsets, Steam former for shaping and forming	C. Heptonstall	Providence, R. I	Aug. 2, 1870	106, 060
Corselets for medicinal and other purposes, Manufacturing.	A. Abbe	Worcester, Mass	Apr. 2, 1841	2, 035
Corundum-wheels, Mode of making	A. W. Calder	San Francisco, Cal	Aug. 10, 1869	93, 413
Corundum-wheels, Mold for making	E. H. Danforth	Jamestown, N. Y	June 13, 1865	48, 160
Cosmetic	I. M. Wilson	Sequin, Tex	Feb. 25, 1868	74, 871
Cosmetic compound	H. Z. Sill	Pittsburgh, Pa	Feb. 27, 1872	124, 018
Costumery	W. F. Munroe	Boston, Mass	July 2, 1872	128, 559
Cot and chair, Combined folding	T. C. Potter	Williston, Vt	June 25, 1872	128, 423
Cot and chest, Combined camp	T. J. Griffin	Brooklyn, N. Y	Jan. 28, 1862	34, 250
Cot and stretcher, Field-hospital	W. Dann	Cincinnati, Ohio	Dec. 20, 1864	45, 479
Cot-bath, Elastic	W. C. Palmer	New York, N. Y	Nov. 24, 1832	....
Cot, Camp	C. Mettam	New York, N. Y	July 23, 1861	32, 884
Cot, Camp and hospital	G. M. Powell, C. D. Lincoln, and G. Evans.	River Falls, Wis., Biddeford, Me., and Richburgh, N. Y.	Sept. 15, 1863	39, 954
Cot, chair, &c., Reclining valetudinary	J. Mead	Albany, N. Y	Mar. 18, 1816	....
Cot, Hammock	J. C. Day	Jersey City, N. J	Oct. 28, 1862	36, 768
Cot, lounge, and chair	G. B. Gourley and O. G. Brady	New York, N. Y	Sept. 3, 1861	33, 192
Cot or bedstead	W. Wells	Harrisburgh, Pa	July 23, 1861	32, 919
Cot or cross-bedstead sofa	G. Sickels	Middletown, Conn	Sept. 25, 1840	1, 800
Cot, Soldier's	E. W. Pierce and W. J. Clark	Southington, Conn	Jan. 7, 1862	34, 089
Cot, Spring	T. J. Griffin	Brooklyn, N. Y	Apr. 8, 1873	137, 675
Cottage-chair	C. O. Collignon	Closter, N. Y	Mar. 10, 1868	75, 373
Cotton after ginning, Machine for cleaning and moating.	J. Simons and J. McJames		May 17, 1804	......
Cotton and corn-stalk cutter	G. S. Roudebrush	Natchez, Miss	Nov. 13, 1860	30, 645
Cotton and corn-stalk puller	W. Altick	Dayton, Ohio	July 24, 1866	56, 509
Cotton and corn-stalks, Machine for extracting	J. Bishop	Austin, Tex	Feb. 19, 1861	31, 437
Cotton and corn-stalks, Machine for pulling and cutting.	H. F. Hicks	Grand View, Ind	Nov. 8, 1859	26, 034
Cotton and feeding to gin, Cleaning seed	M. B. Clarke	Newnan, Ga	Dec. 19, 1854	12, 091
Cotton and hay press	B. B. Alfred	La Grange, Ga	Nov. 19, 1867	71, 113
Cotton and hay press	B. B. Alfred	La Grange, Ga	Nov. 26, 1867	71, 439
Cotton and hay press	G. H. Aylworth	Brighton, Ill	Nov. 7, 1871	120, 612
Cotton and hay press	S. T. Baker	Portland, Me	Dec. 17, 1834	....
Cotton and hay press	L. Balles, J. Prescott, and W. A. Bickford.	Memphis, Tenn	Feb. 13, 1841	1, 979
Cotton and hay press	W. Bullock	Jersey City, N. J	Sept. 28, 1843	3, 288
Cotton and hay press	N. Chapman	Mystic River, Conn	Jan. 10, 1860	26, 750
Cotton and hay press	N. Chapman	Milford, Mass	Jan. 15, 1867	61, 159
Cotton and hay press	A. W. Clarkson	Due West, S. C	Feb. 27, 1872	124, 035
Cotton and hay press	E. S. Collins	Trenton, Tenn	Dec. 5, 1871	121, 591
Cotton and hay press	E. S. Collins	Trenton, Tenn	Dec. 16, 1873	145, 629
Cotton and hay press	J. W. Conway	Franklin, Ind	Mar. 13, 1860	27, 429
Cotton and hay press	J. W. Conway	Madison, Ind	Apr. 2, 1867	63, 477
Cotton and hay press	J. G. Cummings	Columbus, Miss	July 5, 1870	104, 936
Cotton and hay press	J. K. Davis	Monticello, S. C	Sept. 28, 1869	95, 328
Cotton and hay press	W. Deering	Louisville, Ky	Sept. 10, 1867	68, 613
Cotton and hay press	J. S. Duffy	Battle Ground, Ind	Mar. 29, 1870	101, 239
Cotton and hay press	E. Eliason, jr	Fredericksburgh, Va	Jan. 21, 1835	....
Cotton and hay press	P. G. Gardiner	New York, N. Y	Apr. 16, 1842	2, 555
Cotton and hay press	S. C. Goodsell and D. Frisbie	New Haven, Conn	Sept. 25, 1866	58, 341
Cotton and hay press	R. Greene	Greenville, N. C	Dec. 7, 1869	97, 631

Index of patents issued from the United States Patent Office from 1790 *to* 1873, *inclusive*—Continued.

Invention.	Inventor.	Residence.	Date.	No.
Cotton and hay press	J. P. Herron	Atlanta, Ga	Sept. 13, 1870	107, 371
Cotton and hay press	H. F. Hicks	Grand View, Ind	July 24, 1860	29, 341
Cotton and hay press	J. J. Hines	Evergreen, Ala	Sept. 7, 1869	94, 602
Cotton and hay press	M. Jaques and H. Freeman	Woodbridge, N. J	Oct. 2, 1811	....
Cotton and hay press	J. E. Kelsey and J. A. Potter	Poughkeepsie, N. Y., and Providence, R. I.	Sept. 25, 1840	1, 798
Cotton and hay press	D. Knowles	Philadelphia, Pa	Nov. 22, 1870	109, 424
Cotton and hay press	C. K. Marshall	New Orleans, La	Apr. 18, 1871	113, 900
Cotton and hay press	J. Massey	Thomasville, Ga	Dec. 15, 1857	18, 854
Cotton and hay press	J. C. McCurry	Wall Hill, Miss	Jan. 30, 1872	123, 275
Cotton and hay press	A. McGowen	Houston, Tex	Aug. 8, 1871	117, 794
Cotton and hay press	J. L. Nelson	Greenville, N. C	Jan. 2, 1872	122, 483
Cotton and hay press	J. D. Nix	Noble, Ill	Nov. 21, 1871	121, 121
Cotton and hay press	G. W. Penniston	North Vernon, Ind	Dec. 1, 1857	18, 766
Cotton and hay press	L. Potter	Warren, Ohio	Apr. 24, 1847	5, 088
Cotton and hay press	D. Reynolds	Prospect, Md	Oct. 14, 1873	143, 641
Cotton and hay press	J. G. Rux	Raymond, Miss	May 21, 1867	64, 909
Cotton and hay press	W. Russell	Atlanta, Ga	Feb. 18, 1868	74, 603
Cotton and hay press	W. M. Smith	Augusta, Ga	Aug. 16, 1870	106, 417
Cotton and hay press	W. M. Smith	Augusta, Ga	May 23, 1871	115, 126
Cotton and hay press	R. Stallings	Louisburgh, N. C	Apr. 4, 1871	113, 700
Cotton and hay press	E. Stockbridge	Houston, Tex	July 5, 1859	24, 672
Cotton and hay press	C. W. Stopple	Houston, Tex	Aug. 9, 1870	106, 229
Cotton and hay press	R. Triplett	Davis County, Ky	May 19, 1834	....
Cotton and hay press	C. P. Wagner	New York, N. Y	May 14, 1867	64, 814
Cotton and hay press	G. P. Webster, jr	Cross Bridge, Tenn	July 25, 1871	117, 491
Cotton and hay press	T. D. Wilson	Corydon, Ind	June 6, 1827	....
Cotton and hay press	S. Wolff	Vicksburgh, Miss	May 31, 1859	24, 256
Cotton and hay press, Portable	A. G. Murray	Annapolis, Md	Jan. 9, 1835	....
Cotton and linen waste, Treating	E. N. Horsford	Cambridge, Mass	Sept. 15, 1857	18, 203
Cotton and other fabrics, Sizing	J. W. Speyer	Hamburg, Germany	May 24, 1870	103, 384
Cotton and other fibrous materials, Combing	J. Noble	Leeds, England	June 27, 1854	11, 199
Cotton and other fibrous materials for carding, Machine for preparing.	R. Kitson	Lowell, Mass	Sept. 22, 1863	40, 049
Cotton and other fibrous materials, Machine for preparing.	C. Whipple	Providence, R. I	Dec. 6, 1864	45, 371
Cotton and other fibrous materials, Machine for treating compressed.	W. Wanklyn	Albion Mills, Bury, England.	Feb. 23, 1864	47, 731
Cotton and other fibrous materials, Pressing	H. L. Conner	Natchez, Miss	July 5, 1833	....
Cotton and other fibrous substances, Cleaning	J. C. Hurd	Medway, Mass	Apr. 18, 1854	10, 803
Cotton and other fibrous substances, Cleaning	E. Lord	Todmorden, Great Britain	Jan. 9, 1866	52, 008
Cotton and other plants, Cultivation of	P. Poullain	Greensborough, Ga	May 26, 1868	78, 233
Cotton and other presses	P. G. Gardiner	New York, N. Y	Feb. 28, 1845	3, 930
Cotton and other presses, Self-adjusting platens for	J. C. Colt	New York, N. Y	Apr. 22, 1845	4, 006
Cotton and other seeds, Cleaning	T. Rowe	New York, N. Y	Mar. 16, 1869	87, 975
Cotton and other seeds, Hulling	L. D. Cavarly	Waterford, Conn	Feb. 12, 1834	....
Cotton and sugar-cane elevator	E. Price	Waterproof, La	May 20, 1856	14, 927
Cotton and tobacco press	N. Cheek	Chapel Hill, N. C	Apr. 12, 1870	101, 822
Cotton and tobacco press	B. R. Curtis	Richmond, Va	Jan. 29, 1827	....
Cotton and wool factories, Straightening and soldering drums or cylinders for.	L. N. Perry	Worcester, Mass	Jan. 23, 1829	....
Cotton and wool ginning machine	S. R. Parkhurst	New York, N. Y	May 1, 1845	4, 023
Cotton and wool picking and cleaning machine	S. R. Parkhurst	Bloomfield, N. J	July 24, 1866	56, 675
Cotton and wool, Roping and spinning	J. Brown	Providence, R. I	Aug. 8, 1812	....
Cotton and woolen rolls, Machine for splicing	G. Barton, jr	Shaftsburgh, Vt	Apr. 22, 1826	....
Cotton and woolen waste from oil, grease, &c., Process of cleaning.	H. M. Baker	Washington, D. C	June 15, 1869	91, 407
Cotton, Apparatus for preparing	C. Lewandowski	Paris, France	Jan. 1, 1867	60, 754
Cotton-bagging, Machine for rolling and measuring.	T. H. Murphy	New Orleans, La	May 17, 1859	24, 046
Cotton-bagging, Manufacture of	T. Tebow	Lexington, Ky	May 30, 1871	115, 388
Cotton-bale fastening	M. Tildesley	Willenhall, England	Oct. 12, 1869	95, 854
Cotton-bale hoop	G. W. Penniston	North Vernon, Ind	Feb. 1, 1859	22, 823
Cotton-bale tie	J. Booth	Saint Louis, Mo	May 15, 1866	54, 675
Cotton-bales, Applying and securing metal bands on.	J. J. McComb	Liverpool, England	Oct. 23, 1866	59, 151
Cotton-bales, tobacco-hogsheads, &c., Packing-machine for.	T. Jennings	Stateburgh, S. C	Mar. 3, 1829	....
Cotton-band, Metallic	R. Lewis	Charleston, S. C	Apr. 5, 1859	23, 474
Cotton-band, Metallic	R. Lewis	Charleston, S. C	Apr. 5, 1859	23, 475
Cotton bats or laps, Method of making	J. C. Smith and G. C. Kellogg	New Hartford, Conn	Feb. 20, 1843	2, 969
Cotton-batting	H. B. and H. T. Lawton	Cohoes and Troy, N. Y	Mar. 13, 1849	6, 189
Cotton-batting	E. P. Rider	Brooklyn, N. Y	May 18, 1852	8, 958
Cotton-batting, Manufacture of	J. Essex	Bennington, Vt	Feb. 10, 1846	4, 372
Cotton-batting, Mode of glazing	D. Goff	Rehoboth, Mass	Jan. 15, 1846	4, 353
Cotton-batting, Sizing and drying	H. A. Stearns	Cincinnati, Ohio	Sept. 19, 1848	5, 773
Cotton-batting sizing and drying apparatus	E. P. Rider	New York, N. Y	July 30, 1850	7, 533
Cotton beating and picking machine	D. Harding	Lowell, Mass	June 25, 1867	66, 020
Cotton before ginning, Machine for cleaning	B. Seguine	Alexandria, La	July 3, 1843	3, 154
Cotton being carded, Mechanism for mixing soapstone with.	T. Welham	Philadelphia, Pa	Apr. 13, 1869	88, 997
Cotton, Bleaching damaged	J. B. Richards	Kilburn, England	Dec. 23, 1873	145, 816
Cotton, Bleaching stained	J. A. Meany	Brooklyn, N. Y	Apr. 3, 1866	53, 646
Cotton burring and ginning machine	J. Goulding	Worcester, Mass	May 31, 1864	42, 940
Cotton-can	R. R. Hulme	Providence, R. I	June 4, 1872	127, 486
Cotton, Carding, drawing, and roping	W. Tiffany and P. Allen	Otsego, N. Y	July 17, 1815	....
Cotton-check	S. W. Odell	Ouachita Parish, La	Aug. 31, 1869	94, 334
Cotton-chopper	C. Bailey and G. K. Bagby	Kinston, N. C	Apr. 29, 1873	138, 363
Cotton-chopper	J. Caston	Bowdon, Ga	Dec. 23, 1873	145, 849
Cotton-chopper	C. B. Douglas	Montgomery, Ala	Apr. 4, 1871	113, 640
Cotton-chopper	J. M. Garrett	Brenham, Tex	Dec. 30, 1873	146, 061
Cotton-chopper	R. C. Holt	Morehouse Parish, La	June 14, 1870	104, 154
Cotton-chopper	W. Holt	Dawson, Ga	May 3, 1870	102, 541
Cotton-chopper	J. R. Hood	Wedowee, Ala	Jan. 31, 1871	111, 346
Cotton-chopper	I. J. Kidd	Young's Settlement, Tex	Apr. 28, 1868	77, 293
Cotton-chopper	J. A. Lutz	Waynesborough, Va	Nov. 1, 1870	108, 918
Cotton-chopper	E. T. Matthews	Galveston, Tex	Oct. 8, 1872	132, 015
Cotton-chopper	A. G. Powell	Smithfield, N. C	Dec. 19, 1871	122, 057

Index of patents issued from the United States Patent Office from 1790 *to* 1873, *inclusive*—Continued.

Invention.	Inventor.	Residence.	Date.	No.
Cotton-chopper	W. Price	Mount Olive, N. C	Apr. 8, 1873	137, 715
Cotton-chopper	K. Puckett	Parish of Morehouse, La	Mar. 26, 1867	63, 176
Cotton-chopper	W. C. Tilton	Spring Place, Ga	May 6, 1873	138, 596
Cotton-chopper	E. B. Turnipseed	Columbia, S. C	May 16, 1871	114, 889
Cotton-chopper	D. F. Welsh	Nevada, Ohio	Jan. 17, 1871	111, 023
Cotton chopper and cultivator	J. D. Dunn	Griffin, Ga	Feb. 22, 1870	100, 128
Cotton chopper and cultivator	J. M. Harcrow	Marshall, Tex	Apr. 16, 1872	125, 811
Cotton chopper and cultivator	M. L. Nearn	Double Bridges, Tenn	Nov. 4, 1873	144, 281
Cotton chopper and cultivator, Combined	A. F. Roberts	Knoxville, Tenn	Oct. 28, 1873	143, 997
Cotton chopper and cultivator, Combined	J. B. Underwood	Fayetteville, N. C	July 8, 1873	140, 746
Cotton-chopper and grain-cultivator	D. Mosely	Ozark, Ark	Mar. 7, 1871	112, 481
Cotton chopper and scraper	H. B. Cage	Madison Station, Wis	Oct. 12, 1869	95, 651
Cotton chopper and scraper, Combined	W. C. Bibb	Madison, Ga	Feb. 28, 1871	112, 209
Cotton chopper and thinner	D. P. Lewis	Huntsville, Ala	Jan. 22, 1867	61, 438
Cotton chopper, cultivator, and drill	J. A. Hall	Keokuk, Iowa	Aug. 1, 1865	49, 104
Cotton chopper, scraper, and cultivator	F. A. Leonhard	Columbia, Tenn	Sept. 12, 1871	118, 951
Cotton-chopper, Self-adjusting rotary	S. A. Jefferson	Franklin, Tenn	Oct. 18, 1870	108, 358
Cotton-cleaner	S. C. Ames	Washington, Ark	Sept. 11, 1860	29, 945
Cotton-cleaner	J. L. Coker	Hartsville, S. C	Dec. 10, 1872	133, 761
Cotton-cleaner	A. S. Eastham	Wharton, Tex	Mar. 12, 1861	31, 661
Cotton-cleaner	J. Gilliam	Reedy Fork, S. C	Oct. 10, 1829	......
Cotton-cleaner	I. Hayden	Lawrence, Mass	Sept. 11, 1860	29, 971
Cotton-cleaner	E. A. Hearne	Lowndes County, Ala	Feb. 19, 1861	31, 456
Cotton-cleaner	J. Johnson	Hempstead County, Ark	Oct. 20, 1857	18, 454
Cotton-cleaner	W. H. Johnson	Richmond, Ark	Oct. 2, 1860	30, 227
Cotton-cleaner	J. H. Kinyon and J. Hollingsworth.	Chicago, Ill	Apr. 22, 1856	14, 725
Cotton-cleaner	F. McCarthy	Demopolis, Ala	Feb. 12, 1845	3, 912
Cotton-cleaner	J. Ralston	Brenham, Tex	Feb. 4, 1873	135, 586
Cotton-cleaner	J. Ryder, W. Carpenter, and H. R. Jolley.	Clinton, La	Dec. 18, 1860	30, 989
Cotton-cleaner	Z. B. Sims	Bonham, Tex	July 4, 1871	116, 764
Cotton-cleaner	C. Smith	Knoxville, Tex	Sept. 11, 1860	30, 003
Cotton-cleaner	C. S. Tarpley	Jackson, Miss	Jan. 1, 1861	31, 069
Cotton-cleaner	J. W. Thorn	Courtland, Ala	Oct. 16, 1860	30, 435
Cotton-cleaner	J. W. Thorn	Courtland, Ala	Feb. 25, 1868	74, 776
Cotton-cleaner	J. Wind	Thomasville, Ga	Nov. 13, 1847	5, 369
Cotton cleaner, Sea-island	J. Reed	Marshfield, Mass	Aug. 10, 1827	......
Cotton, Cleaning	T. Shapard	Haywood County, Tenn	Dec. 24, 1867	72, 691
Cotton cleaning and ginning machine	G. Macdonald	Aston, England	Jan. 8, 1867	61. 078
Cotton-cleaning machine	F. A. Calvert and C. G. Sargent	Lowell and Westford, Mass	Feb. 3, 1857	16, 526
Cotton-cleaning machine	L. S. Chichester	New York, N. Y	July 14, 1857	17, 819
Cotton-cleaning machine	R. J. Clay	Williamsburgh, N. Y	Jan. 1, 1867	60, 859
Cotton-cleaning machine	L. T. Clement	Smyrna, Tenn	Oct. 26, 1869	96, 202
Cotton-cleaning machine	J. Gilmore	New Orleans, La	May 22, 1860	28, 361
Cotton-cleaning machine	R. H. Hilton	New Berne, N. C	Feb. 25, 1868	74, 822
Cotton-cleaning machine	J. E. Hooper	Baltimore, Md	Jan. 30, 1866	52, 289
Cotton-cleaning machine	B. Jackman	Louisville, Ky	Dec. 11, 1860	30, 880
Cotton-cleaning machine	C. Leavitt	Quincy, Ill	Apr. 11, 1854	10, 762
Cotton-cleaning machine	T. Oliver	Yazoo City, Miss	May 18, 1858	20, 270
Cotton-cleaning machine	G. T. Sadler	Lawrence County, Ala	Mar. 10, 1823	......
Cotton-cleaning machinery	S. W. Brown	Lowell, Mass	Oct. 24, 1854	11, 826
Cotton-cleaning machinery	I. Hayden	Lawrence, Mass	Feb. 11, 1862	34, 363
Cotton-cleaning machinery	D. Hess	West Union, Iowa	Oct. 25, 1859	25, 897
Cotton, clover, and other seed huller	J. Whiteman	Philadelphia, Pa	June 26, 1835	......
Cotton-coated seeds, Process for separating the fibers from the husks in	T. Rose and R. E. Gibson	Earlestown, England	July 26, 1870	105, 729
Cotton, &c., combing machine	H. Conant	Providence County, R. I	Feb. 8, 1870	99, 641
Cotton-combing machine	I. Dimock	Florence, Mass	Sept. 12, 1865	49, 865
Cotton-combing machine	C. Whipple and R. J. Stafford	Providence and Smithfield, R. I.	Mar. 4, 1862	34, 606
Cotton-combing machine, Feeding-device for	C. F. Hadley and E. Johnson	Chicopee, Mass., and Wethersfield, Conn.	June 15, 1869	91, 229
Cotton-compress	E. L. Morse	Saint Louis, Mo	Nov. 3, 1868	83, 783
Cotton, &c., compress	E. L. Morse	Saint Louis, Mo	Apr. 16, 1867	63, 808
Cotton-compresser	H. Waterman	Bath, Me	May 10, 1838	731
Cotton, Condensing	A. Man	Smithfield, R. I	Apr. 28, 1836	......
Cotton-cultivator	C. T. Reams	Raleigh, N. C	Apr. 15, 1873	137, 959
Cotton, &c., Double-speeder for roping	P. Moody	Boston, Mass	Apr. 3, 1819	......
Cotton, Double-speeder for roving	A. Arnold	North Providence, R. I	Aug. 10, 1836	......
Cotton, &c., Drawing	C. Whipple	Providence, R. I	Feb. 16, 1858	19, 394
Cotton drawing and spinning machine	J. B. Fuller	Norwich, Conn	Aug. 25, 1868	81, 357
Cotton, &c., Drying	J. Philbrick	Cold Springs, Miss	Feb. 12, 1836	......
Cotton, Drying wet seed	G. G. Henry	Mobile, Ala	Dec. 6, 1859	26, 353
Cotton-duck, Dressing	H. N. Gambrill	Baltimore, Md	Oct. 21, 1851	8, 444
Cotton, &c., Dust-room connected with machines for picking.	R. Kitson	Lowell, Mass	Aug. 19, 1862	36, 219
Cotton, Dust-room in cleaning	J. Whitehill	Newburgh, N. Y	July 16, 1867	66, 921
Cotton-elevator	W. Potter and E. Crane	Lowell, Mass	Sept. 3, 1867	68. 386
Cotton-flannel, Manufacturing	A. S. Carleton	Clinton, Mass	Mar. 24, 1857	16, 870
Cotton, flax, &c., Can for manufacturing	H. Meses	Paterson, N. J	Mar. 12, 1833	......
Cotton for molding purposes, Preparing plastic	J. M. Legaré	Aiken, S. C	Dec. 29, 1857	18, 980
Cotton, &c., for spinning, Machine for preparing	J. Taylor	Oldham, England	Mar. 29, 1864	42, 127
Cotton from the bolls in the field, Picking	S. S. Rembert and J. Prescott	Memphis, Tenn	Sept. 10, 1850	7, 631
Cotton from the seed, Cleaning	J. Reed	Marshfield, Mass	Feb. 3, 1826	......
Cotton-gin	F. T. Ackland, H. G. Mitchell, and M. Mustapha.	Zagazig, Egypt	Sept. 18, 1866	58, 187
Cotton-gin	C. N. Andrews	Saint Louis, Mo	Sept. 19, 1871	119. 001
Cotton-gin	W. P. Baker	Boston, Mass	Nov. 20, 1838	1, 013
Cotton-gin	B. G. Beadle	Memphis, Tenn	Nov. 1, 1869	25, 943
Cotton-gin	J. Beath	Boston, Mass	July 12, 1839	1, 233
Cotton-gin	W. Bell		Nov. 24, 1803	......
Cotton-gin	W. F. Bowen	Stark, Fla	July 11, 1871	116, 924
Cotton-gin	J. B. Brackett and W. Dearborn.	Boston, Mass	Mar. 26, 1867	63, 136
Cotton-gin	J. B. Brackett and W. Dearborn.	Boston, Mass	Mar. 9, 1869	87, 535
Cotton-gin	C. Brakell	Oldham, England	Dec. 5, 1865	51, 402

Index of patents issued from the United States Patent Office from 1790 *to* 1873, *inclusive*—Continued.

Invention.	Inventor.	Residence.	Date.	No.
Cotton-gin	R. M. Brooks	Woodbury, Ga	Apr. 14, 1868	76, 594
Cotton-gin	H. W. Brown	Millville, N. J	Mar. 23, 1858	19, 679
Cotton-gin	I. F. Brown	Columbus, Ga	Apr. 23, 1861	32, 116
Cotton-gin	I. F. Brown	New London, Conn	Sept. 1, 1863	39, 767
Cotton-gin	A. H. Burdine	Chulahoma, Miss	Aug. 28, 1860	29, 765
Cotton-gin	L. Campbell	Columbus, Miss	Jan. 10, 1854	10, 401
Cotton-gin	L. Campbell	Columbus, Miss	May 22, 1855	12, 894
Cotton-gin	E. Carver, jr	Bridgewater, Mass	June 13, 1823	....
Cotton-gin	J. E. Carver	Bridgewater, Mass	Jan. 2, 1866	51, 892
Cotton-gin	J. G. Case	New London, Conn	Feb. 20, 1872	123, 868
Cotton-gin	J. A. Chanfourier	Paris, France	June 11, 1867	65, 643
Cotton-gin	L. J. Chichester	New York, N. Y	Feb. 9, 1858	19, 324
Cotton-gin	L. S. Chichester	Brooklyn, N. Y	July 25, 1854	11, 355
Cotton-gin	L. S. Chichester	New York, N. Y	Nov. 20, 1855	13, 815
Cotton-gin	L. S Chichester	New York, N. Y	Feb. 3, 1857	16, 565
Cotton-gin	L. S. Chichester	New York, N. Y	Oct. 12, 1858	21, 795
Cotton-gin	L. S. Chichester	New York, N. Y	Jan. 10, 1860	26, 815
Cotton-gin	H. Clark	New London, Conn	Feb. 25, 1836	....
Cotton-gin	H. Clark	Eufaula, Ala	Aug. 29, 1848	5, 740
Cotton-gin	H. Clark	Newport, Fla	Feb. 13, 1855	12, 376
Cotton-gin	R. J. Clay	Green Point, N. Y	Mar. 9, 1869	87, 635
Cotton-gin	J. M. Clough	Ilion, N. Y	June 7, 1870	103, 981
Cotton-gin	J. M. Clough	Ilion, N. Y	Nov. 19, 1872	133, 080
Cotton-gin	P. E. Collins	Mobile, Ala	Oct. 4, 1859	25, 630
Cotton-gin	H. Conklin	Poughkeepsie, N. Y	June 7, 1839	1, 168
Cotton-gin	S. T. Conn	New York, N. Y	June 11, 1829	....
Cotton-gin	E. A. Cooper	Westminster, Great Britain	Apr. 3, 1866	53, 753
Cotton-gin	T. C. Craven	Greenbush, N. Y	Feb. 9, 1864	41, 487
Cotton-gin	O. Crawford		June 22, 1807	....
Cotton-gin	E. Cruse	Charleston, S. C	Nov. 18, 1816	....
Cotton-gin	B. Dobson and W. Slater	Bolton, England	Mar. 30, 1869	88, 374
Cotton-gin	J. Du Bois	Greensborough, Ala	Jan. 8, 1850	6, 998
Cotton-gin	J. Du Bois	Greensborough, Ala	Apr. 27, 1858	20, 051
Cotton-gin	J. Du Bois	Greensborough, Ala	July 18, 1871	117, 058
Cotton-gin	F. Durand	Paris, France	Feb. 24, 1863	37, 743
Cotton-gin	F. Durand	Paris, France	Sept. 19, 1865	50, 080
Cotton-gin	H. L. Emery	Albany, N. Y	Sept. 4, 1860	29, 867
Cotton-gin	H. L. Emery	Albany, N. Y	Mar. 30, 1869	88, 464
Cotton-gin	W. B. Emery	Albany, N. Y	May 9, 1865	47, 626
Cotton-gin	A. Fessenden	Beaufort, S. C	July 9, 1867	66, 577
Cotton-gin	C. L. Fleischmann	New York, N. Y	Jan. 16, 1866	52, 037
Cotton-gin	C. L. Fleischmann	Washington, D. C	May 4, 1869	89, 751
Cotton-gin	H. H. Fultz	Lexington, Miss	Oct. 9, 1855	13, 641
Cotton-gin	J. E. Furguson	Micanopy, Fla	Jan. 1, 1861	31, 062
Cotton-gin	N W. Gaddy	Nichols, S. C	Nov. 18, 1873	114, 668
Cotton-gin	T. C. Garlington	La Fayette, Ala	Feb. 23, 1858	19, 415
Cotton-gin	E. Gottheil	Galveston, Tex	Apr. 19, 1859	23, 679
Cotton-gin	J. Goulding	North Wilbraham, Mass	Aug. 28, 1860	29, 780
Cotton-gin	B. D. Gullett	Aberdeen, Miss	Jan. 10, 1854	10, 406
Cotton-gin	B. D. Gullett	Aberdeen, Miss	Feb. 23, 1858	19, 417
Cotton-gin	B. D. Gullett	Amite City, La	Sept. 7, 1869	94, 488
Cotton-gin	B. D. Gullett	Amite City, La	July 1, 1873	140, 365
Cotton-gin	R. R. Gwathmey	Philadelphia, Pa	Apr. 5, 1870	101, 610
Cotton-gin	S. Z. Hall	Sing Sing, N. Y	Mar. 16, 1869	87, 771
Cotton-gin	S. Z. Hall	Ossining, N. Y	July 5, 1870	105, 070
Cotton-gin	R. Hancock, sr., and E. W. Carr	Philadelphia, Pa	Feb. 6, 1811	....
Cotton-gin	B. Hempstead	Little Rock, Ark	Sept. 23, 1873	143, 074
Cotton-gin	W. L. Henderson	Comrawattee, West India	Nov. 28, 1871	121, 365
Cotton-gin	H. Holmes		May 12, 1796	....
Cotton-gin	W. W. Howell	Columbus, Miss	Feb. 28, 1860	27, 290
Cotton-gin	B. H. Jenks and W. A. Tuttle	Philadelphia, Pa	Jan. 3, 1860	26, 681
Cotton-gin	I. Jennings	New York, N. Y	Jan. 17, 1817	....
Cotton-gin	A. Jones	New Orleans, La	Apr. 25, 1837	180
Cotton-gin	E. Jones	London, England	Sept. 24, 1872	131, 546
Cotton-gin	A. P. Keith	Bridgewater, Mass	Jan. 2, 1855	12, 132
Cotton-gin	E. Keith	Bridgewater, Mass	Dec. 22, 1846	4, 901
Cotton-gin	E. Keith	Bridgewater, Mass	Jan. 27, 1857	16, 488
Cotton-gin	J. W. Kokemuller	Bluffton, S. C	Aug. 13, 1867	67, 773
Cotton-gin	T. J. Laws	Washington, Ark	Mar. 16, 1852	8, 803
Cotton-gin	W. T. Layton	Darlington, S. C	May 22, 1849	6, 463
Cotton-gin	C. Leavitt and W. H. Burridge	Cleveland, Ohio	Nov. 24, 1868	84, 287
Cotton-gin	J. Leavitt, C. Lane, and T. Leavitt.	Suffield, Conn	Mar. 24, 1825	....
Cotton-gin	W. B. Lindsay	New Orleans, La	May 27, 1856	14, 965
Cotton-gin	W. H. Livingston	New York, N. Y	Apr. 5, 1864	42, 206
Cotton-gin	J. Lynch	Tuscaloosa, Ala	July 22, 1833	....
Cotton-gin	O. W. Massey	Macon, Ga	July 4, 1871	116, 615
Cotton-gin	M. McAuley	Thomas County, Ga	Apr. 24, 1849	6, 378
Cotton-gin	F. McCarthy	Demopolis, Ala	July 3, 1840	1, 675
Cotton-gin	F. McCarthy	Orange Springs, Fla	July 30, 1867	67, 327
Cotton-gin	J. McCreight	Winnsborough, S. C	July 2, 1836	....
Cotton-gin	W. and J. McCreight	Winnsborough, S. C	Feb. 5, 1836	....
Cotton-gin	R. A. L. McCurdy	Sabine Parish, La	June 26, 1855	13, 131
Cotton-gin	R. McKenna	White's Station, Tenn	May 31, 1870	103, 638
Cotton-gin	R. McKenna	White's Station, Tenn	Oct. 18, 1870	108, 500
Cotton-gin	R. McKenna	White's Station, Tenn	Mar. 5, 1872	124, 278
Cotton-gin	R. McKenna	White's Station, Tenn	July 9, 1872	128, 743
Cotton-gin	F. M. McMeekin	Morrison's Mills, Fla	Dec. 11, 1866	60, 402
Cotton-gin	F. M. McMeekin	Morrison's Mills, Fla	May 23, 1871	115, 226
Cotton-gin	C. A. McPhetridge	Saint Louis, Mo	Nov. 18, 1856	16, 096
Cotton-gin	W. McLendon	Greenville, Ga	Dec. 27, 1859	26, 604
Cotton-gin	J. B. Mell	Riceborough, Ga	Apr. 3, 1855	12, 635
Cotton-gin	J. B. Mell	Riceborough, Ga	Oct. 21, 1856	15, 930
Cotton-gin	J. B. Miles	Chicot, Ark	July 22, 1856	15, 381
Cotton-gin	O. Murray	New Market, N. H	Feb. 4, 1873	135, 485
Cotton-gin	D. G. Olmstead	Vicksburgh, Miss	Jan. 12, 1858	19, 097
Cotton-gin	D. G. Olmstead	Vicksburgh, Miss	Dec. 20, 1859	26, 516
Cotton-gin	J. F. Orr	Orrville, Ala	Jan. 13, 1857	16, 394

Index of patents issued from the United States Patent Office from 1790 *to* 1873, *inclusive*—Continued.

Invention.	Inventor.	Residence.	Date.	No.
Cotton-gin	J. F. Orr	Orrville, Ala	Jan. 5, 1858	19,041
Cotton-gin	E. Osgood	Boston, Mass	May 11, 1858	20,216
Cotton-gin	E. Osgood	Boston, Mass	Oct. 18, 1859	25,848
Cotton-gin	E. Osgood	New York, N. Y	Dec. 22, 1863	41,046
Cotton-gin	E. Osgood	Boston, Mass	Aug. 13, 1872	130,438
Cotton-gin	H. C. Parkhurst	New York, N. Y	Aug. 31, 1858	21,357
Cotton-gin	S. R. Harkhurst	New York, N. Y	Sept. 11, 1849	6,703
Cotton-gin	S. R. Parkhurst	New York, N. Y	Apr. 23, 1850	7,307
Cotton-gin	S. R. Parkhurst	New York, N. Y	Apr. 27, 1858	20,086
Cotton-gin	S. R. Parkhurst	Bloomfield, N. J	Jan. 9, 1867	51,962
Cotton-gin	S. R. Parkhurst	Mont Clair, N. J	Jan. 11, 1870	98,705
Cotton-gin	N. A. Patterson	Kingston, Tenn	Oct. 23, 1860	30,497
Cotton-gin	S. Pennoyer	Cross River, N. Y	July 24, 1822	....
Cotton-gin	J. Perkins	Bridgewater, Mass	June 14, 1834	....
Cotton-gin	D. Philips	Jefferson County, Miss	Apr. 3, 1829	....
Cotton-gin	J. Platt and W. Richardson	Oldham, Great Britain	Oct. 28, 1862	36,789
Cotton-gin	A. A. Porter	Griffin, Ga	Dec. 22, 1868	85,128
Cotton-gin	D. Pratt	Prattville, Ala	July 14, 1857	17,806
Cotton-gin	M. E. Pratt	Prattville, Ala	July 15, 1873	140,791
Cotton-gin	W. F. Pratt	East Bridgewater, Mass	Aug. 30, 1859	25,307
Cotton-gin	W. F. Pratt	East Bridgewater, Mass	July 6, 1869	92,208
Cotton-gin	W. A. Purdom	Jackson, Miss	Nov. 4, 1856	16,022
Cotton-gin	P. Reading	Trenton, N. J	Apr. 13, 1836	....
Cotton-gin	C. G. Sargent	Westford, Mass	Apr. 20, 1869	89,083
Cotton-gin	C. G. Sargent	Westford, Mass	May 11, 1869	89,890
Cotton-gin	C. G. Sargent and A. B. Ely	Westford and Newton, Mass.	May 11, 1869	89,891
Cotton-gin	P. C. Sawyer	Macon, Ga	June 10, 1873	139,820
Cotton-gin	H. V. Scattergood	Albany, N. Y	June 12, 1866	55,542
Cotton-gin	H. V. Scattergood	Albany, N. Y	June 25, 1867	66,202
Cotton-gin	J. Simpson	Lewisville, S. C	Aug. 14, 1855	13,441
Cotton-gin	C. Spofford and C. H. Hersey	Boston, Mass	June 19, 1866	55,787
Cotton-gin	C. Spofford and C. H. Hersey	Boston, Mass	Aug. 20, 1867	67,923
Cotton-gin	J. Stevens	New York, N. Y	May 30, 1865	48,032
Cotton-gin	R. A. Stough	Griffin, Ga	Apr. 13, 1869	88,924
Cotton-gin	L. G. Sturdevant	Delaware, Ohio	July 23, 1841	2,190
Cotton-gin	J. Tetlon	Taunton, Mass	July 13, 1858	20,904
Cotton-gin	E. Town	Washington, D. C	May 8, 1866	54,625
Cotton-gin	J. L. Tuttle	New York, N. Y	Oct. 14, 1856	15,906
Cotton-gin	J. L. Tuttle	Bridesburgh, Pa	Sept. 21, 1858	21,582
Cotton-gin	J. A. Ventress	Woodville, Miss	June 29, 1858	20,747
Cotton-gin	P. Von Schmidt	Washington, D. C	Oct. 17, 1846	4,817
Cotton-gin	J. Watrous, jr	Mystic River, Conn	Apr. 10, 1866	53,910
Cotton-gin	J. W. Webb	Cotton Valley, Ala	May 18, 1869	90,210
Cotton-gin	H. L. Weeks	Hannahatchie, Ga	Sept. 6, 1853	9,998
Cotton-gin	E. Whiting		Jan. 22, 1801	....
Cotton-gin	W. Whittemore, jr	West Cambridge, Mass	May 25, 1839	1,158
Cotton-gin	F. L. Wilkinson	Adams' Run, S. C	Mar. 9, 1858	19,598
Cotton-gin	J. Wilson	Anderson Court-House, S. C	Sept. 27, 1859	25,600
Cotton-gin	J. N. Wilson and G. W. Payne	Memphis, Tenn	Apr. 27, 1859	20,120
Cotton-gin	A. Q. Withers	Byhalia, Miss	Oct. 5, 1858	21,714
Cotton-gin	F. Wuterich and J. Koerber	New York, N. Y	Nov. 8, 1859	26,065
Cotton gin and machine cards, Manufacturing cylinders for.	J. L. Tuttle	New York, N. Y	Oct. 14, 1856	15,904
Cotton gin and picker	J. B. Brackett and W. Dearborn	Boston, Mass	Aug. 27, 1867	68,035
Cotton gin and picker	E. Osgood	Boston, Mass	Jan. 15, 1867	61,239
Cotton-gin attachment	H. P. Harrell	Roxobel, N. C	Sept. 19, 1871	119,140
Cotton-gin bush	E. Keith	Bridgewater, Mass	Sept. 19, 1845	4,196
Cotton-gin, Common foot	I. B. Barnes	Beaufort, S. C	Nov. 6, 1826	....
Cotton-gin feeder	S. Z. Hall	Seguin, Tex	Sept. 11, 1860	29,968
Cotton-gin feeder	L. Z. Hall	Sing Sing, N. Y	Sept. 10, 1872	131,163
Cotton-gin feeder	N. Hoggatt	Madison Parish, La	July 18, 1871	117,074
Cotton-gin feeder	J. Prescott	Rockford, Ill	Oct. 13, 1857	18,410
Cotton-gin feeder	J. Ralston	Brenham, Tex	Sept. 6, 1870	107,102
Cotton-gin feeding-attachment	S. Z. Hall	Camden, N. J	Aug. 6, 1867	67,533
Cotton-gin flue	J. W. Gaines	Clarksville, Tex	Aug. 13, 1872	130,423
Cotton-gin flue	J. B. Peyton	Raymond, Miss	Feb. 5, 1861	31,326
Cotton-gin, Foot	W. Gould	McIntosh County, Ga	Feb. 20, 1821	....
Cotton-gin for ginning long staple and other kinds of cotton, Roller.	E. Carver	Bridgewater, Mass	Sept. 27, 1838	949
Cotton, Gin for linting	G. W. Payne	Memphis, Tenn	Aug. 23, 1870	106,613
Cotton-gin frame, Cast-iron	E. A. Lester	Boston, Mass	Jan. 8, 1831	....
Cotton-gin grate	E. Keith	Bridgewater, Mass	Mar. 4, 1836	....
Cotton-gin knife-roller	T. H. Rushton and W. Dobson	Bolton, England	June 24, 1873	140,218
Cotton-gin, Long-stapled	C. Willey, jr	Chicago, Ill	Apr. 27, 1852	8,907
Cotton-gin, Railway	D. Philips	Georgetown, Pa	May 22, 1841	2,103
Cotton-gin rib	I. F. Brown	Columbus, Ga	July 18, 1854	11,317
Cotton-gin rib	W. J. Horton	Newburgh, Ala	July 20, 1869	92,724
Cotton-gin rib	J. Robison	New York, N. Y	Oct. 15, 1872	132,217
Cotton-gin ribs, Manufacturing	J. W. Webb	Cotton Valley, Ala	Feb. 24, 1857	16,699
Cotton-gin, Roller	W. Bell and J. S. De Montmollin		Mar. 7, 1803	....
Cotton-gin, Roller	E. Carver	Bridgewater, Mass	Jan. 17, 1842	2,429
Cotton-gin, Roller	T. Ely	New York, N. Y	Dec. 11, 1845	4,302
Cotton-gin, Roller	R. Reynolds, jr	Beaufort, S. C	Feb. 2, 1844	3,425
Cotton-gin, Roller	J. Schley	Columbus, Ga	Nov. 14, 1848	5,921
Cotton-gin roller	W. Wanklyn	Albion Mills, Bury, Great Britain.	Nov. 15, 1864	45,109
Cotton-gin roller, Metallic fluted	G. F. Saltonstall		Sept. 2, 1801	....
Cotton-gin saw	A. D. Brown	Columbus, Ga	Aug. 28, 1855	13,484
Cotton-gin saw	E. Carver	Bridgewater, Mass	Jan. 4, 1845	3,875
Cotton-gin saw	T. C. Craven	Philadelphia, Pa	May 30, 1871	115,441
Cotton-gin saw	P. Gardner	Woodville, Mass	Apr. 22, 1830	....
Cotton-gin, Saw	T. J. James	Princeton, Miss	May 7, 1842	2,608
Cotton-gin, Saw	H. H. Kelley	Port Gibson, Miss	May 19, 1843	3,091
Cotton-gin, Saw	C. A. McPhetridge	Natchez, Miss	Apr. 24, 1841	2,068
Cotton-gin, Saw	W. B. Stewart	Cincinnati, Ohio	May 19, 1843	3,097
Cotton-gin saw	W. Sutton	Washington, D. C	Nov. 24, 1868	84,392
Cotton-gin-saw teeth	T. B. Oglesby	Hempstead, Tex	May 6, 1873	138,575

Index of patents issued from the United States Patent Office from 1790 *to* 1873, *inclusive*—Continued.

Invention.	Inventor.	Residence.	Date.	No.
Cotton-gin-saw teeth, Machine for punching and swaging.	J. M. Clough	Ilion, N. Y	Jan. 2, 1872	122, 307
Cotton-gin-saw teeth, Manufacture of	T. C. Craven	Northampton, Mass	May 21, 1872	126, 875
Cotton-gin saws, Filing	C. P. Pool	New Market, Ala	Aug. 9, 1859	25, 042
Cotton-gin saws, Machine for filing	J. T. Turner	Bridgewater, Mass	June 2, 1857	17, 458
Cotton-gin, Self-feeding	J. A. Chaufourier	Paris, France	Apr. 12, 1870	101, 709
Cotton-gin sharpener	A. H. Burdine	Chulahoma, Miss	July 19, 1859	24, 790
Cotton-gins and corn-mills, Running-gear for	B. Huckaby	Huntsville, Ala	Aug. 16, 1824	....
Cotton-gins, Boxing for	W. S. Cooley	Norwich, Conn	Jan. 7, 1835	....
Cotton gins, &c., Construction of horse-power for	J. M. Albertson	New London, Conn	June 7, 1870	103, 821
Cotton-gins, Construction of rubber rollers for	W. Dearborn	Boston, Mass	Aug. 27, 1867	68, 052
Cotton-gins, Constructing saw-cylinders for	A. Jones	New York, N. Y	July 20, 1842	2, 731
Cotton-gins, Electroplating with nickel the metallic parts of.	L. Watrous	Mystic River, Conn	Nov. 22, 1870	109, 475
Cotton-gins, Feeder for roller	L. J. Mallard and W. S. Baker	Riceborough, Ga	June 17, 1856	15, 138
Cotton-gins, Grate of saw	A. Washburn	Bridgewater, Mass	June 16, 1841	2, 133
Cotton-gins, Mechanism for driving	W. L. May	Linwood, Ala	Nov. 14, 1869	97, 945
Cotton-gins, Propelling	P. Knox	Augusta, Ga	Dec. 7, 1826	....
Cotton-gins, Roller for	I. F. Brown	New London, Conn	Jan. 3, 1865	45, 695
Cotton-gins, Saw and grate for	E. Carver, jr	Bridgewater, Mass	June 12, 1823	....
Cotton-gins, Saw-cylinder for	J. Idler	Philadelphia, Pa	Dec. 1, 1837	498
Cotton-gins, Tool for turning cylinder-rings for	J. Gibbons	West Troy, N. Y	June 16, 1868	78, 954
Cotton, Ginning	J. Murry		Dec. 23, 1796	....
Cotton, Ginning	R. Watkins		Dec. 23, 1796	....
Cotton, Ginning and cleaning	S. Sawyer	Boston, Mass	Mar. 30, 1833	....
Cotton, Ginning and cleaning	C. Speer	New York, N. Y	Oct. 10, 1854	11, 795
Cotton ginning and cleaning by a combination of rolls.	S. Sawyer	Boston, Mass	Mar. 30, 1833	....
Cotton ginning, carding, and spinning machine	J. M'Bride	Nashville, Tenn	Aug. 8, 1805	....
Cotton-ginning machine	W. Bell and S. De Montmollin		Apr. 7, 1802	....
Cotton-ginning machine	R. R. Gathmey	Middletown, Ky	Dec. 31, 1867	72, 846
Cotton-ginning machine	I. Jennings	New York, N. Y	Dec. 27, 1815	....
Cotton-ginning machine	L. H. Moseley	Poughkeepsie, N. Y	Nov. 25, 1837	486
Cotton-ginning machine	J. Stevens	Poughkeepsie, N. Y	Nov. 25, 1837	485
Cotton-ginning machine	E. Whitney		Mar. 14, 1794	....
Cotton-ginning machinery	J. Percy	Albany, N. Y	Apr. 29, 1862	35, 109
Cotton-glazing machine	S. Turner	Cranston, R. I	Oct. 6, 1835	....
Cotton goods, Machine for finishing	C. S. Davis	Harrisburgh, Pa	Jan. 24, 1860	26, 895
Cotton goods, Warping, dressing, &c	J. Morgan	Baltimore, Md	Apr. 5, 1821	....
Cotton, hay, and cheese press	E. and L. L. Macomber	Gardiner, Me	Aug. 15, 1835	....
Cotton, hay, and hemp press	W. H. Morris	Troy, Tenn	May 24, 1870	103, 356
Cotton, hay, &c., Apparatus for compressing	F. Weldon	Kilmorony, Ireland	Mar. 25, 1873	137, 160
Cotton, hay, &c., Machine for binding and securing	J. Simmons	Philadelphia, Pa	Nov. 7, 1809	....
Cotton-hiller	A. W. Washburn	Yazoo City, Miss	Mar. 25, 1856	14, 539
Cotton-hopper	G. F. Saltonstall	Fayetteville, N. C	Dec. 31, 1817	....
Cotton-inspecting machine	L. Viales	New Orleans, La	Oct. 29, 1818	....
Cotton in the field, Machine for picking	J. W. Thorn	Courtland, Ala	Oct. 6, 1857	18, 363
Cotton in the seed, Machine for separating trash from.	J. Idler	Philadelphia, Pa	Dec. 31, 1838	1, 050
Cotton into yarn, Manufacturing	L. Crane	Ware, Mass	Mar. 26, 1825	....
Cotton laps, Machine for making	T. C. Craven	Hudson, N. Y	Feb. 4, 1873	135, 528
Cotton-lapper brake	D. Hussey	Nashua, N. H	Oct. 16, 1866	58, 945
Cotton-lapper brake	R. Kitson	Lowell, Mass	Sept. 22, 1863	40, 048
Cotton-lapper brake	R. Kitson	Lowell, Mass	Mar. 3, 1868	75, 169
Cotton-lapping machines, Friction-brakes for	E. Van Winkle	Paterson, N. J	Aug. 10, 1869	93, 649
Cotton, linen, and other fabrics to prevent them from burning, Solution to be applied to.	J. McGill	Boston, Mass	May 1, 1866	54, 382
Cotton, linen, &c., Printing and dyeing	A. Paraf	Mulhouse, France	Nov. 7, 1865	50, 885
Cotton-lint room, Fire-proof	R. R. Warren	Raleigh, N. C	Aug. 19, 1873	142, 065
Cotton-lint room, Revolving	W. T. Crenshaw	Burton, Tex	May 13, 1873	138, 865
Cotton, Long trunks for cleaning	I. Hayden	Lawrence, Mass	Dec. 1, 1857	18, 742
Cotton, Machine for cleaning	J. Kershaw	Paterson, N. J	Aug. 11, 1868	80, 826
Cotton, Machine for cleaning	B. J. F. Owen	Memphis, Tenn	Dec. 15, 1868	85, 024
Cotton, &c., Machine for cleaning and opening	J. E. Van Winkle	Paterson, N. J	June 2, 1863	38, 794
Cotton, Machine for cleaning unginned	R. M. Livingstone	Mobile, Ala	July 20, 1843	3, 190
Cotton, Machine for forming rolls of	C. White, J. Goulding, and S. Lowder.	Dedham, Mass	June 25, 1824	....
Cotton, Machine for ginning and cleaning	L. T. Clement	Smyrna, Tenn	July 27, 1869	93, 057
Cotton, Machine for packing	O. Stith	Lawrenceville, Va	Aug. 25, 1829	....
Cotton, &c., Machine for preparing	W. and F. W. Crighton	Manchester, England	Jan. 22, 1867	61, 322
Cotton, &c., Machine for roping and spinning	W. Carmichall	Sand Lake, N. Y	June 10, 1826	....
Cotton, Machine for roping and spinning	P. Moody	Boston, Mass	Feb. 19, 1821	....
Cotton, Machine for separating seed from	J. Eve		Dec. 6, 1803	....
Cotton, Machine for topping	A. A. Dickson	Griffin, Ga	Oct. 4, 1853	10, 065
Cotton, Machine for whipping and cleaning	S. P. Mason	Killingly, Conn	July 8, 1834	....
Cotton, Machine for winding	D. Brown, jr	Warren, R. I	Feb. 5, 1807	....
Cotton, Machinery for cleaning	S. W. Brown	Lowell, Mass	Sept. 8, 1855	13, 564
Cotton, Machinery for cleaning	R. M. Livingston	Mobile, Ala	Mar. 6, 1847	4, 999
Cotton, Machinery for cleaning	R. Needham	Memphis, Tenn	Mar. 13, 1847	5, 008
Cotton, Machinery for combing	M. D. Whipple	Charlestown, Mass	Oct. 26, 1858	21, 932
Cotton, Machinery for ginning or picking	F. A. Calvert	Lowell, Mass	Sept. 17, 1850	7, 646
Cotton, &c., Machinery for picking	R. Kitson	Lowell, Mass	Oct. 31, 1854	11, 865
Cotton-machinery, Sizing-roll for	J. M. Dunlop	Manchester, Great Britain	Nov. 1, 1864	44, 911
Cotton, Manner of constructing and affixing the ribs of saw-gins for ginning.	A. Copeland	Bridgewater, Mass	Apr. 8, 1840	1, 547
Cotton, Manufacturing	E. Congdon	Norwich, Conn	Mar. 12, 1824	....
Cotton, Manufacturing	J. T. Sharrock	Winchester, Va	June 20, 1823	....
Cotton, Mode of making machine and hand cards for	W. Barker	Smithfield, R. I	Aug. 12, 1815	....
Cotton, Mode of packing	W. C. Thomas	Richmond, Va	Feb. 15, 1827	....
Cotton-opener	R. Kitson	Lowell, Mass	Oct. 4, 1870	107, 922
Cotton-opener	R. Kitson	Lowell, Mass	Nov. 11, 1873	144, 394
Cotton-openers, &c., Feed-mechanism for	W. Noton	Oldham, England	Feb. 25, 1873	136, 173
Cotton-openers, &c., Feed-regulator for	D. Harding	Lowell, Mass	Aug. 26, 1873	142, 101
Cotton-openers, Feed-regulator for	R. Kitson	Lowell, Mass	Dec. 9, 1873	145, 300
Cotton-openers, &c., Feeding-mechanism for	W. E. Whitehead	Miles Platting, England	Dec. 20, 1870	110, 318
Cotton opening and cleaning machine	S. Fay	Lowell, Mass	Jan. 22, 1867	61, 411
Cotton opening and cleaning machine	R. Kitson	Lowell, Mass	Dec. 11, 1866	60, 387
Cotton-opening machine	W. C. Jillson and A. B. Palmer	Willimantic, Conn	Dec. 20, 1870	110, 368

Index of patents issued from the United States Patent Office from 1790 *to* 1873, *inclusive*—Continued.

Invention.	Inventor.	Residence.	Date.	No.
Cotton or hay press	J. S. Allums	Cuseta, Ga	Dec. 18, 1866	60, 457
Cotton or hay press	U. Page	Ringgold, La	July 5, 1870	104, 991
Cotton or hay press	J. Wentz	Girard, Ala	Apr. 14, 1868	76, 860
Cotton or other bale-goods, Machine for pressing	J. Idler		Sept. 24, 1802	....
Cotton or wool, Machinery for roving	W. Whitehead	Paterson, N. J	Apr. 11, 1825	....
Cotton or wool, Vibrating rolling-machine for	J. Hopkins	Strasburgh, Pa	Jan. 14, 1824	....
Cotton or woolen can	H. W. Shepard	New York, N. Y	Feb. 11, 1873	135, 853
Cotton or woolen can	H. W. Shepard	New York, N. Y	Feb. 11, 1873	135, 854
Cotton-packer	L. S. Chichester	New York, N. Y	Dec. 20, 1859	26, 546
Cotton, Packing	W. Hart	Petersburgh, Va	May 4, 1825	....
Cotton, Packing and pressing	J. Mead	Aurora, Ind	Aug. 14, 1847	5, 235
Cotton-packing machine	W. J. Cocke	Cabin Point, Va	Feb. 4, 1828	....
Cotton-packing machine	J. Cooke	Fayetteville, N. C	Oct. 12, 1821	....
Cotton-packing machine	L. Laysard	Halifax, N. C	Sept. 28, 1825	....
Cotton-picker	W. Apperly	Louisville, Ky	Aug. 2, 1870	106, 022
Cotton-picker	J. E. Carver	Bridgewater, Mass	May 14, 1867	64, 629
Cotton-picker	R. F. Cooke	Newark, N. J	May 26, 1868	78, 362
Cotton-picker	R. F. Cooke	Brooklyn, N. Y	Mar. 29, 1870	101, 231
Cotton-picker	B. I. Dreeson	Marion County, Tex	Oct. 18, 1870	108, 461
Cotton-picker	B. J. Dreeson and J. L. Buskett	Schleswig, Germany, and Saint Louis, Mo.	July 4, 1871	116, 694
Cotton-picker	W. Goodwin and S. D. G. Niles	Helena, Ark	Mar. 28, 1871	113, 159
Cotton-picker	J. Griffin	Louisville, Ky	July 3, 1860	28, 980
Cotton-picker	J. Griffin	Louisville, Ky	Jan. 22, 1861	31, 165
Cotton-picker	J. Griffin	Louisville, Ky	Mar. 5, 1861	31, 596
Cotton-picker	G. A. Howe	Brooklyn, N. Y	Mar. 13, 1866	53, 147
Cotton-picker	G. A. Howe	Brooklyn, N. Y	Mar. 27, 1866	53, 447
Cotton-picker	J. Hughes	New Berne, N. C	Mar. 28, 1871	113, 170
Cotton-picker	L. Jennings	Brooklyn, N. Y	July 31, 1860	29, 435
Cotton-picker	W. D. Ludlow	New York, N. Y	Jan. 30, 1866	52, 303
Cotton-picker	D. M. McRae	Webberville, Tex	Aug. 23, 1870	106, 601
Cotton-picker	J. G. Page	Memphis, Tenn	Dec. 4, 1866	60, 235
Cotton, &c., picker	G. F. Palmer	Rochester, N. H	Feb. 8, 1870	99, 699
Cotton-picker	C. Payne and B. Vandecar	Brandon, Vt., and Waterford, N. Y.	Feb. 11, 1868	74, 242
Cotton-picker	A. Pettingill	East Livermore, Me	Jan. 5, 1869	85, 532
Cotton-picker	N. F. Sandelin	New York, N. Y	Mar. 26, 1872	124, 978
Cotton-picker	I. S. Schuyler	New York, N. Y	July 30, 1861	32, 968
Cotton-picker	B. G. Shields	Marlin, Tex	Sept. 16, 1856	15, 746
Cotton-picker	Z. B. Sims	Bonham, Tex	Sept. 3, 1872	131; 124
Cotton-picker	H. Stevens	Memphis, Tenn	Oct. 25, 1864	44, 828
Cotton-picker	A. Tiensch	Memphis, Tenn	Apr. 10, 1866	53, 901
Cotton picker and cleaner	S. H. Gilman	Galveston, Tex	Sept. 1, 1868	81, 622
Cotton picker and cleaner	Z. B. Sims	Bonham, Tex	Aug. 31, 1869	94, 352
Cotton, Picker and lap machine for manufacturing	J. Whitehead	Godwinville, N. J	July 22, 1833	....
Cotton-picker cylinder	J. Pitts	Lancaster, Mass	Feb. 28, 1854	10, 578
Cotton-picker, Electrical	R. F. Cook	Brooklyn, N. Y	June 7, 1870	103, 986
Cotton-picker, Hand	W. B. Cargill	Waterbury, Conn	Mar. 6, 1860	27, 348
Cotton-picker, Hand	G. A. Howe	Worcester, Mass	Dec. 4, 1855	13, 877
Cotton-picker wallet	G. H. Peabody	Columbus, Ga	Nov. 29, 1859	26, 288
Cotton-pickers, Screen-cylinder for	R. Kitson	Lowell, Mass	Apr. 21, 1868	76, 926
Cotton-picking device	J. Griffin	Louisville, Ky	June 5, 1866	55, 283
Cotton-picking machine	J. Pinnell and H. Maxon	Barbourville, Va	Sept. 10, 1828	....
Cotton-picking machine	W. E. Prall	Knoxville, Tenn	Feb. 27, 1866	52, 860
Cotton-plant protector	A. L. Blanc	Louisiana, La	Mar. 22, 1870	101, 028
Cotton-plants, Apparatus for destroying worms on	M. Perl	Houston, Tex	June 15, 1869	91, 365
Cotton-plants, Destroying worms in	T. W. Mitchell	Richmond, Tex	Jan. 3, 1871	110, 774
Cotton-plants, Machine for cutting up	B. T. Currier	Bath, Me	Mar. 6, 1860	27, 352
Cotton-plants, Machine for destroying worms on	W. Ewing	Columbia, La	Oct. 19, 1869	95, 995
Cotton-plants, Machine for thinning	C. A. McCaughan	Moscow, Tenn	Feb. 5, 1867	61, 845
Cotton-plants, Machine for thinning	J. Weaver	Washington, D. C	July 5, 1837	253
Cotton-plants, Preventing rot &c., in	S. W. Pomeroy	Brighton, Mass	May 6, 1824	....
Cotton-press	E. H. Adams	Taladega, Ala	May 17, 1859	23, 995
Cotton-press	W. W. Anderson	Wartrace, Tenn	Jan. 16, 1872	122, 752
Cotton-press	J. B. Armstrong	Barnwell D., S. C	Dec. 20, 1853	10, 329
Cotton-press	T. Ashcraft	Randolph County, Ala	Mar. 6, 1849	6, 158
Cotton-press	Z. Atkinson	Richmond, Ga	Feb. 1, 1859	22, 775
Cotton-press	A. Baldwin	New York, N. Y	Oct. 5, 1869	95, 411
Cotton-press	W. C. Banks	Como Depot, Miss	June 18, 1867	65, 865
Cotton-press	W. C. Banks	Como Depot, Miss	Jan. 23, 1872	122, 983
Cotton-press	C. J. Beasley	Petersburgh, Va	Sept. 21, 1869	94, 993
Cotton-press	C. J. Beasley	Petersburgh, Va	Feb. 22, 1870	100, 105
Cotton-press	C. J. Beasley	Petersburgh, Va	Nov. 28, 1871	121, 317
Cotton-press	W. I. Blackman	Columbus, Miss	June 14, 1870	104, 103
Cotton-press	J. P. Bolin	Orangeburgh County, S. C	June 20, 1871	116, 144
Cotton-press	S. Booton	Seguin, Tex	Oct. 18, 1870	108, 320
Cotton-press	T. J. Bottoms and J. A. Bullock	Thomas County, Ga	Aug. 31, 1858	21, 317
Cotton-press	M. W. Bradford	Greenwood, La	Mar. 5, 1872	124, 245
Cotton-press	W. Bradley	West Point, Ga	June 18, 1872	128, 104
Cotton-press	J. M. Brooks, jr	Griffin, Ga	Mar. 11, 1873	136, 697
Cotton-press	R. M. Brooks	Greenville, Ga	Oct. 2, 1860	30, 200
Cotton-press	R. M. Brooks	Woodbury, Ga	Apr. 14, 1868	76, 593
Cotton-press	R. M. Brooks	U. S., Ga	Mar. 7, 1871	112, 413
Cotton-press	A. D. Brown	Clinton, Ga	Jan. 22, 1850	7, 047
Cotton-press	W. Bullock	Jersey City, N. J	Jan. 4, 1845	3, 873
Cotton-press	C. A. Caldwell	Concord, N. C	Oct. 5, 1869	95, 563
Cotton-press	J. Carson	Raleigh, N. C	Nov. 4, 1830	....
Cotton-press	F. A. L. Cassidey	Newnansville, Fla	May 14, 1867	64, 630
Cotton-press	J. Chambers	Alexandria, La	May 19, 1843	3, 087
Cotton-press	N. Chapman	Mystic River, Conn	Aug. 8, 1854	11, 472
Cotton-press	N. Chapman	Mystic River, Conn	Jan. 12, 1858	19, 071
Cotton-press	N. Cheek	Chapel Hill, N. C	Dec. 27, 1870	110, 549
Cotton-press	C. C. Conner	Ripley, Tenn	Apr. 4, 1871	113, 265
Cotton-press	C. C. Conner	Ripley, Tenn	July 23, 1872	129, 792
Cotton-press	W. P. Craig	Newport, Ky	May 7, 1861	32, 240
Cotton-press	J. G. Cummings	Columbus, Miss	Aug. 20, 1867	67, 961
Cotton-press	G. C. Davies	Dayton, Ohio	Apr. 25, 1865	47, 490
Cotton-press	H. J. Davis	Wetumpka, Ala	May 2, 1871	114, 271

Index of patents issued from the United States Patent Office from 1790 *to* 1873, *inclusive*—Continued.

Invention.	Inventor.	Residence.	Date.	No.
Cotton-press	T. F. De Bruler	Rockport, Ind	Apr. 26, 1859	23, 759
Cotton-press	P. K. Dederick	Albany, N. Y	Jan. 7, 1873	134, 592
Cotton-press	J. P. Derden	Bastrop, La	Mar. 4, 1873	136, 422
Cotton-press	A. Devall	New Orleans, La	Apr. 17, 1847	5, 067
Cotton-press	S. Dike	Columbia, S. C	Oct. 16, 1860	30, 394
Cotton-press	J. A. Disbrow and J. E. Cronk	Poughkeepsie, N. Y	Feb. 2, 1858	19, 279
Cotton-press	E. Duchamp	Saint Martinsville, La	Feb. 23, 1858	19, 413
Cotton-press	J. C. Duvoll	Sardis, Miss	June 11, 1867	65, 655
Cotton-press	E. N. Elliott	Port Gibson, Miss	Nov. 22, 1859	26, 174
Cotton-press	J. T. Elliot	Carrollton, Miss	Oct. 22, 1859	7, 731
Cotton-press	G. Faulkner	Warrenton, N. C	Apr. 30, 1872	126, 194
Cotton-press	C. I. Fay	North Lincoln, Me	Jan. 31, 1854	10, 472
Cotton-press	W. Field	Providence, R. I	Apr. 6, 1858	19, 838
Cotton-press	G. Fitz	Clinton, Miss	June 3, 1837	224
Cotton-press	S. Fry	New York, N. Y	May 7, 1842	2, 611
Cotton-press	H. H. Fultz	Lexington, Mass	Aug. 14, 1855	13, 426
Cotton-press	C. Gardiner	Richmond, Ala	Oct. 16, 1847	5, 332
Cotton-press	P. G. Gardiner	New York, N. Y	Mar. 20, 1844	3, 503
Cotton-press	P. G. Gardiner	New York, N. Y	Sept. 4, 1860	29, 874
Cotton-press	P. G. Gardiner	New York, N. Y	Oct. 23, 1860	30, 473
Cotton-press	P. G. Gardiner	New York, N. Y	Mar. 26, 1861	31, 796
Cotton-press	F. Gilbert	Natchez, Miss	Apr. 9, 1861	31, 967
Cotton-press	G. Glenn	Louisburgh, N. C	Apr. 12, 1826	....
Cotton-press	A. M. Glover	Waterborough, S. C	July 3, 1855	13, 199
Cotton-press	G. W. Grader	Memphis, Tenn	Sept. 26, 1871	119, 346
Cotton-press	J. B. Gridley	Louisville, Ky	Feb. 14, 1871	111, 739
Cotton-press	I. Griffin	Milford, Ga	Jan. 8, 1861	31, 070
Cotton-press	J. H. Groves	Greenwood, Miss	Jan. 2, 1872	22, 458
Cotton-press	J. T. Ham	Sinatobia, Miss	Jan. 10, 1860	26, 764
Cotton-press	M. B. Hand	Handsborough, Miss	Aug. 16, 1859	25, 161
Cotton-press	R. J. Harrison	Raleigh, N. C	Feb. 14, 1871	111, 840
Cotton-press	J. Hawthorn	Thomas County, Ga	May 17, 1859	24, 026
Cotton-press	W. Haynie	Memphis, Tenn	Dec. 6, 1870	109, 897
Cotton-press	J. R. Hitchcock and W. F. Serrel.	New York, N. Y	Sept. 30, 1839	1, 349
Cotton-press	C. C. Howard	Lowndesborough, Ala	Mar. 11, 1873	136, 725
Cotton-press	H. Hughes	Port Gibson, Miss	May 12, 1857	17, 278
Cotton-press	C. S. Hunt	Bridgewater, Mass	Jan. 15, 1856	14, 098
Cotton-press	T. R. Jackson	Dallas, Tex	Sept. 30, 1873	143, 349
Cotton-press	R. Jernigan	Waynesborough, N. C	May 15, 1827	....
Cotton-press	W. J. Johnson	Mobile, Ala	Mar. 13, 1849	6, 175
Cotton-press	A. Jones	New Orleans, La	Oct. 26, 1838	992
Cotton-press	M. M. Jones	Morrisville, N. Y	May 15, 1860	28, 281
Cotton-press	J. B. Knight	New Orleans, La	Sept 26, 1871	119, 367
Cotton-press	W. W. Knowles	Bastrop, Tex	Aug. 30, 1870	106, 833
Cotton-press	S. Lamb	New York, N. Y	May 12, 1843	3, 086
Cotton-press	S. Lamb	New York, N. Y	Mar. 16, 1844	3, 489
Cotton-press	W. D. Leavitt	New Orleans, La	Aug. 20, 1872	130, 728
Cotton-press	T. D. Leonard	Waco, Tex	Nov. 18, 1873	144, 775
Cotton-press	L. Lewis	Vicksburgh, Miss	Mar. 2, 1852	8, 774
Cotton-press	L. Lewis	Vicksburgh, Miss	Aug. 17, 1852	9, 197
Cotton-press	L. Lewis	Vicksburgh, Miss	Oct. 17, 1871	119, 934
Cotton-press	R. Lewis	Charleston, S. C	Mar. 18, 1873	136, 841
Cotton-press	N. I. Lilly	Selma, Ala	Mar. 8, 1859	23, 180
Cotton-press	E. W. Long and I. N. Patten	Memphis, Tenn	Mar. 21, 1871	112, 822
Cotton-press	C. N. Lovejoy	Columbia, S. C	Oct. 25, 1859	25, 904
Cotton-press	J. Loving	Moscow, Tenn	Mar. 9, 1858	19, 571
Cotton-press	M. Lynch	Horn Lake, Miss	July 30, 1872	130, 060
Cotton-press	B. G. Martin	New York, N. Y	Mar. 18, 1873	136, 846
Cotton-press	C. Martratt	Waterford, N. Y	Oct. 26, 1858	21, 894
Cotton-press	H. Mason	Lancaster, Mass	July 24, 1860	29, 290
Cotton-press	R. Mauldin	Marietta, Miss	Aug. 15, 1871	118, 142
Cotton-press	A. Z. McBride	Hannahatchee, Ga	Feb. 5, 1861	31, 323
Cotton-press	T. H. McCray	Tellico, Tex	Mar. 13, 1860	27, 459
Cotton-press	E. McDonnell	Baltimore, Md	Mar. 19, 1867	62, 965
Cotton-press	A. B. McGonnigil	Helena, Ark	July 4, 1871	116, 733
Cotton-press	J. W. McIntyre	Memphis, Tenn	Apr. 16, 1867	63, 916
Cotton-press	T. E. McNeill	Lynchburgh, Va	Nov. 20, 1866	59, 853
Cotton-press	J. W. Miller	Shannon, Miss	May 14, 1872	126, 731
Cotton-press	C. W. Millerd	Monticello, Ark	Mar. 23, 1869	88, 159
Cotton-press	G. Milliran	Byhalia, Miss	June 5, 1860	28, 592
Cotton-press	J. Mitchell	Rutherford, Tenn	Mar 12, 1836	....
Cotton-press	J. J. Morrison	Atlanta, Ga	Mar. 2, 1869	87, 505
Cotton-press	M. Murchison	Denmark, Tenn	Oct. 23, 1860	30, 493
Cotton-press	W. B. North	Jersey City, N. J	June 13, 1848	5, 632
Cotton-press	D. G. Olmstead	Vicksburgh, Miss	Jan. 26, 1858	19, 202
Cotton-press	W. T. Opie	Scarborough, Ga	Sept. 4, 1860	29, 904
Cotton-press	M. L. Parry	Galveston, Tex	Jan. 27, 1857	16, 494
Cotton-press	W. W. Patrick	Midway, S. C	June 17, 1873	139, 970
Cotton-press	I. N. Patten and E. W. Long	Shelby County, Tenn	Dec. 20, 1870	110, 278
Cotton-press	P. Payne	Claiborne County, Miss	Aug. 10, 1829	....
Cotton-press	P. Payne and J. Rundell	Port Gibson, Miss	July 20, 1831	....
Cotton-press	P. Payne and J. Rundell	Claiborne County, Miss	Jan. 25, 1833	....
Cotton-press	G. Peck	Fairfield, Conn	Mar. 9, 1844	3, 466
Cotton-press	W. Pendleton and H. M. Boardman.	Augusta, Ga	May 10, 1870	102, 963
Cotton-press	D. Philips	Natchez, Miss	Apr. 27, 1832	....
Cotton-press	P. W. Porter	Columbia, Tenn	Aug. 11, 1842	2, 752
Cotton-press	R. Powers	Charleston, Miss	Feb. 4, 1873	135, 486
Cotton-press	J. Prescott	Memphis, Tenn	Nov. 9, 1844	3, 815
Cotton-press	J. Price	Nashville, Tenn	Dec. 7, 1839	1, 426
Cotton-press	J. P. Pridgeon	Marion Court-House, S. C	July 15, 1873	140, 841
Cotton-press	W. F. Provost	Barnwell D., S. C	Sept. 14, 1844	3, 739
Cotton-press	W. F. and C. J. Provost	Barnwell D., S. C	Nov. 21, 1843	3, 347
Cotton-press	W. F. and C. J. Provost	Selma, Ala	Nov. 20, 1855	13, 826
Cotton-press	W. F. and C. J. Provost	Selma, Ala	Oct. 21, 1856	15, 936
Cotton-press	H. W. Randle	Burnsville, Ala	Feb. 16, 1858	19, 381
Cotton-press	H. W. Randle	Burnsville, Ala	May 3, 1859	23, 860

Index of patents issued from the United States Patent Office from 1790 *to* 1873, *inclusive*—Continued.

Invention.	Inventor.	Residence.	Date.	No.
Cotton-press	F. M. Ray	Catskill, N. Y	Aug. 25, 1840	1,738
Cotton-press	S. S. Rembert	Memphis, Tenn	Aug. 22, 1871	118,275
Cotton-press	J. R. Remington	Montgomery, Ala	Apr. 15, 1843	3,051
Cotton-press	W. H. Reynolds	New Orleans, La	May 6, 1873	138,583
Cotton-press	S. D. Roberts	Washington, La	July 2, 1867	66,394
Cotton-press	H. Ross	Rockport, Ind	Mar. 23, 1858	19,714
Cotton-press	I. G. Roux	Raymond, Miss	Mar. 29, 1859	23,397
Cotton-press	T. H. Scarborough	Brownsville, Tenn	Oct. 16, 1834	....
Cotton-press	M. M. Scherer	Batesville, Ark	Sept. 2, 1873	142,517
Cotton-press	J. Schley	Savannah, Ga	Aug. 15, 1871	118,060
Cotton-press	R. Scott, jr	Madison, Ind	Oct. 9, 1860	30,356
Cotton-press	J. C. Sellers	Woodville, Miss	Aug. 28, 1860	29,827
Cotton-press	W. Sewell, jr	Macon, Ga	June 15, 1844	3,631
Cotton-press	W. A. Shepard	New York, N. Y	Oct. 24, 1865	50,635
Cotton-press	H. Shrader	Burnsville, Ala	Mar. 30, 1858	19,821
Cotton press	F. Simmons	New Orleans, La	Apr. 25, 1871	114,056
Cotton-press	T. D. Simpson and W. S. Lamkin.	Marshall, Tex	Dec. 26, 1871	122,197
Cotton-press	J. Slocum	Syracuse, N. Y	Apr. 16, 1845	3,997
Cotton-press	R. Smith	Towanda, Pa	May 29, 1847	5,135
Cotton-press	R. Smith	Towanda, Pa	Dec. 29, 1857	18,995
Cotton-press	T. S. Smith	Tooner's Station, Tenn	July 29, 1873	141,393
Cotton-press	N. W. Speers	Memphis, Tenn	Dec. 20, 1870	110,304
Cotton-press	W. T. Stewart	Moscow, Tenn	July 30, 1872	129,9[illegible]8
Cotton-press	H. D. Stover and W. Hutchinson.	New York, N. Y	Oct. 8, 1867	69,721
Cotton-press	E. Stuart	Shufordsville, Miss	July 23, 1867	67,147
Cotton-press	U. T. Stuart and C. E. Stewart	Fayette County, Tenn	Mar. 8, 1859	23,201
Cotton-press	W. H. Tappey, W. C. Lumsden, and A. Steel.	Petersburgh, Va	Sept. 15, 1868	82,178
Cotton-press	J. Templeton	Florence, Ga	Sept. 5, 1871	118,756
Cotton-press	D. R. Forbet	Columbus, Ga	July 30, 1867	67,230
Cotton-press	P. B. Tyler	Philadelphia, Pa	Jan. 16, 1845	3,885
Cotton-press	G. Utley	Chapel Hill, N. C	Feb. 21, 1871	112,095
Cotton-press	W. C. Van Hoesen	Catskill, N. Y	Apr. 2, 1841	2,026
Cotton-press	W. C. Van Hoesen	Catskill, N. Y	Feb. 1, 1842	2,445
Cotton-press	I. J. Way	Memphis, Tenn	Nov. 14, 1865	50,971
Cotton, &c., press	W. and D. K. West	Kentish Town, England	July 16, 1872	129,636
Cotton-press	P. B. Wever	Scarborough, Ga	Aug. 21, 1860	29,736
Cotton-press	O. Whitney	Augusta, Me	Mar. 17, 1843	3,006
Cotton-press	D. Wilkinson	North Providence, R. I	Dec. 8, 1825	....
Cotton-press	J. T. Williams	Blakely, Ga	Dec. 17, 1872	134,021
Cotton-press	P. Williams	Lodi, Miss	Dec. 27, 1859	27,635
Cotton-press	P. Williams	Winona, Miss	Jan. 21, 1868	73,683
Cotton-press	P. and R. A. Williams	Winona, Miss	Sept. 16, 1873	142,882
Cotton-press	R. G. Williams	Hannahatchee, Ga	July 20, 1858	20,973
Cotton-press	U. T. Wilson	Independence, Miss	July 11, 1871	117,027
Cotton-press	J. G. Winger	Vicksburgh, Miss	July 13, 1852	9,124
Cotton-press	F. W. Witting	Yorktown, Tex	Feb. 16, 1858	19,399
Cotton-press	D. H. Wooldridge	Humboldt, Tenn	June 10, 1873	139,845
Cotton-press	P. M. Wright	New York, N. Y	Nov. 26, 1844	3,835
Cotton-press	Y. F. Wright	Green Hill, Ga	May 1, 1860	28,124
Cotton-press	Y. F. Wright	Green Hill, Ga	July 30, 1867	67,243
Cotton-press	R. N. Wyatt	Tehula, Miss	Mar. 5, 1872	124,307
Cotton-press and feeder	H. Zellner	Columbia, Tenn	Oct. 15, 1867	69,887
Cotton-press and horse-power combined	J. M. Albertson	New London, Conn	Oct. 10, 1871	119,731
Cotton press and tramper	C. K. Marshall	New Orleans, La	Apr. 18, 1871	113,901
Cotton press, gin, and horse-power combined	J. M. Shaw, sr	Water Valley, Miss	May 28, 1872	127,276
Cotton-press, Horizontal rack and pinion	Z. Bronson	Jasper County, Ga	July 7, 1829	....
Cotton-press, Lever	C. Williams	Petersburgh, Va	June 3, 1826	....
Cotton-press, Self-acting	J. Grout	Hocking City, Ohio	June 12, 1855	13,060
Cotton-press, Steam	A. Baldwin	New York, N. Y	May 17, 1870	103,123
Cotton-press, Steam	T. J. De Yampert	Mobile, Ala	Oct. 27, 1857	18,502
Cotton-press, Steam	C. Hurst	New Orleans, La	Aug. 22, 1848	5,730
Cotton-press, Steam	J. Roy	New Orleans, La	Dec. 15, 1857	18,864
Cotton-press, Water-power	J. Durden	Washington, Ala	June 23, 1825	....
Cotton-presses, Self-releasing door of	G. W. Penniston	North Vernon, Ind	Apr. 14, 1857	17,053
Cotton pressing and packing machine	J. H. McClelland	Fairfield District, S. C	Dec. 28, 1826	....
Cotton-pressing machine	J. Boatwright and J. Nathans	Richland, S. C	Nov. 21, 1826	....
Cotton-pressing machine	J. N. Gordon	Plymouth, N. C	Oct. 8, 1825	....
Cotton-pressing machine	F. Shaler	Georgetown, Ohio	May 5, 1824	....
Cotton-pressing machine	R. V. W. and J. Thorne	New York, N. Y	Feb. 22, 1810	....
Cotton-pressing machine	P. White	Chatham County, N. C	Feb. 19, 1827	....
Cotton, Process of treating raw	J. C. Hurd	Medway, Mass	May 12, 1857	17,279
Cotton-railways, Drawing-head for	W. B. Leonard	Fishkill, N. Y	Sept. 16, 1833	....
Cotton-renovating machine	J. B. Lyons	Milton, Conn	Dec. 13, 1870	110,149
Cotton-rolls, Process for utilizing immature	J. Hughes	New Berne, N. C	May 23, 1871	115,209
Cotton roller-gin	W. Whittemore, jr	West Cambridge, Mass	Mar. 7, 1834	....
Cotton roller-gin	W. Whittemore, jr	West Cambridge, Mass	May 29, 1835	....
Cotten, Rolling-machine for cleaning	G. F. Saltonstall	....	May 14, 1803	....
Cotton, Rolling or twisting rovings of	W. Whitehead, jr	Paterson, N. J	June 14, 1826	....
Cotton, &c., Roping	L. Brewster	Poughkeepsie, N. Y	Dec. 17, 1834	....
Cotton, Roping	W. Fowler	Fishkill, N. Y	Mar. 23, 1836	....
Cotton roping, Machine for making	P. Moody	Boston, Mass	Jan. 19, 1821	....
Cotton roping, Machine for making	S. Shepard	Taunton, Mass	Mar. 23, 1821	....
Cotton, Roping, twisting, and winding	D. Grive	Providence, R. I	Sept. 18, 1818	....
Cotton-roping, Counter-twist speeder for	J. Whitehead	Manchester, Va	May 29, 1841	2,108
Cotton roping, Machine for making	J. Whitehead	Manchester, Va	June 24, 1839	1,186
Cotton roping, slack threads, &c., Making	S. Whiting	Franklin, Mass	Nov. 17, 1824	....
Cotton, Roving	J. A. Bradshaw	Foxborough, Mass	Dec. 8, 1832	....
Cotton-roving, Drawing-can for	M. Mead	Lowell, Mass	May 6, 1862	35,169
Cotton-roving machine	A. Arnold	North Providence, R. I	Jan. 21, 1823	....
Cotton-roving machine	W. Bryant	Davidson County, Tenn	Apr. 29, 1816	....
Cotton roving, &c., Machine for condensing or preparing.	J. Graves	New Ipswich, N. H	Sept. 1, 1843	3,250
Cotton-roving machines, Speeder for	W. Mason	Taunton, Mass	May 4, 1838	724
Cotton roving, Manufacturing	G. Brewster	Poughkeepsie, N. Y	Mar. 28, 1827	....
Cotton, Roving or roping	L. H. Mosely	Lisbon, Conn	Nov. 19, 1833	....

Index of patents issued from the United States Patent Office from 1790 *to* 1873, *inclusive*—Continued.

Invention.	Inventor.	Residence.	Date.	No.
Cotton roving, Speeder for making	S. P. Mason	Leesville, Conn	Dec. 29, 1830	....
Cotton-sampling borer	E. A. Clark	Jackson, Tenn	Aug. 19, 1873	141, 998
Cotton, Saw-mill for cleaning	G. F. Saltonstall	....	Jan. 4, 1803	....
Cotton-scraper	C. H. Burbidge	Middletown, Conn	Nov. 27, 1860	30, 718
Cotton-scraper	J. M. Cobb	Jackson, Tenn	June 26, 1860	28, 835
Cotton-scraper	I. W. Collins and R. Y. Wilkinson.	Clinton, La	Nov. 27, 1860	30, 721
Cotton-scraper	M. Earnhart	Cold Water, Miss	Nov. 15, 1859	26, 096
Cotton-scraper	W. C. Finney	Fayette County, Tenn	Apr. 24, 1849	6, 379
Cotton-scraper	T. T. Fleming	Memphis, Tenn	Aug. 27, 1867	68, 296
Cotton-scraper	C. A. Gaines	Watson, Miss	Oct. 5, 1858	21, 667
Cotton-scraper	J. Henderson	Bluff Springs, Miss	Mar. 8, 1859	23, 171
Cotton-scraper	J. D. Houston	Pope's Depot, Miss	Jan. 15, 1861	31, 122
Cotton-scraper	W. Jarrell	Humboldt, Tenn	July 29, 1873	141, 355
Cotton scraper	I. J. Kidd	Young's Settlement, Tex	Apr. 28, 1868	77, 292
Cotton-scraper	W. L. Milholen	Centre, Ala	Nov. 27, 1860	30, 746
Cotton-scraper	J. H. Mitchell	Germantown, Tenn	Sept. 13, 1859	25, 434
Cotton-scraper	P. Sharkey	Brownsville, Miss	Apr. 26, 1859	23, 788
Cotton-scraper	J. S. Smith	Helena, Ark	Feb. 9, 1869	86, 704
Cotton-scraper	H. Starke	Wilkinson County, Miss	Feb. 3, 1824	....
Cotton-scraper	A. W. Washburn	Yazoo City, Miss	Mar. 25, 1856	14, 540
Cotton-scraper	J. G. Winger	Vicksburgh, Miss	Oct. 27, 1857	18, 525
Cotton scraper and chopper	N. M. Hale	Cleborne, Tex	Nov. 8, 1870	109, 124
Cotton scraper and cultivator, Combined	B. F. Bowling	Holly Springs, Miss	Dec. 28, 1869	98, 341
Cotton, Screen for machine for treating	R. Kitson	Lowell, Mass	Aug. 11, 1868	80, 975
Cotton-seed and corn grinding machine	J. Martin	Petersburgh, Va	May 16, 1835	....
Cotton-seed and guano distributer	J. T. Graves	Wilson, N. C	Oct. 18, 1870	108, 349
Cotton seed and rice hulling	J. Kellogg	New Hanover County, N. C	Mar. 30, 1835	....
Cotton-seed cleaner	T. W. Brown	Cudworth, Barnsley, England.	Aug. 4, 1868	80, 593
Cotton-seed cleaner	W. H. Delamere	Deptford, England	May 28, 1872	127, 224
Cotton-seed cleaner	A. Wells	Morgantown, W. Va	Dec. 4, 1866	60, 300
Cotton-seed-cleaning machine	A. I. Hardin	Shelby, N. C	Nov. 6, 1860	30, 573
Cotton seed-cleaning process	J. G. Page	Rockford, Ill	Dec. 26, 1865	51, 745
Cotton-seed heater	W. M. Force	Newark, N. J	Mar. 26, 1872	124, 945
Cotton-seed huller	H. C. Bradford	Providence, R. I	Jan. 17, 1871	111, 034
Cotton-seed huller	W. P. Callahan and D. R. De Bush.	Dayton, Ohio	Jan. 18, 1870	98, 921
Cotton-seed huller	W. R. Fee	Cincinnati, Ohio	Sept. 13, 1870	107, 349
Cotton-seed huller	W. R. Fee	Cincinnati, Ohio	Nov. 1, 1870	108, 896
Cotton-seed huller	J. Harrington	New London, Conn	Dec. 13, 1870	110, 134
Cotton-seed huller	L. Johnson	Madison, Ga	May 29, 1833	....
Cotton-seed huller	C. A. Lowber	Medina, N. Y	Dec. 6, 1859	26, 372
Cotton-seed huller	P. Martin	New Orleans, La	July 31, 1860	29, 393
Cotton-seed huller	G. H. Peabody	New York, N. Y	July 27, 1869	92, 997
Cotton-seed huller	J. Perkins	Providence, R. I	June 19, 1866	55, 701
Cotton-seed huller	P. Reading	Trenton, N. J	Apr. 13, 1836	....
Cotton-seed huller	E. C. Singer	New Orleans, La	July 5, 1870	105, 136
Cotton-seed huller	J. D. Stillman	Memphis, Tenn	June 29, 1869	91, 880
Cotton-seed huller	W. H. Trott, D. R. Torbett, and F. A. Pomeroy.	Columbus, Ga	Jan. 2, 1872	122, 341
Cotton-seed huller	A. J. Vandegrift	Covington, Ky	June 28, 1870	104, 799
Cotton-seed huller	F. A. Wells	Memphis, Tenn	Oct. 26, 1869	96, 177
Cotton-seed huller	W. Wilber	New Orleans, La	Sept. 11, 1855	13, 556
Cotton-seed, Hulling	H. Hubbard	Claremont, N. H	Apr. 19, 1834	....
Cotton-seed, Hulling	J. Walker	Dover, England	Mar. 27, 1855	12, 610
Cotton-seed hulling and separating machine	W. Wilber	New York, N. Y	Jan. 27, 1857	16, 509
Cotton-seed-hulling machine	W. R. Fee	Cincinnati, Ohio	Aug. 11, 1857	17, 961
Cotton-seed-hulling machine	J. Lineback	Salem, N. C	Mar. 31, 1814	....
Cotton-seed-hulling machine	Miller and Lawes	Washington County, Miss	Oct. 27, 1835	....
Cotton-seed-hulling machine	J. B. Ruperts	Jersey City, N. J	Jan. 30, 1866	52, 321
Cotton-seed-hulling machine	J. Smith	Petersburgh, Va	Mar. 15, 1845	3, 951
Cotton-seed-hulling machine	J. Tiffany	Syracuse, N. Y	Sept. 18, 1860	30, 096
Cotton-seed machine	F. A. E. G. De Massas	Hoxton, England	Jan. 16, 1866	52, 116
Cotton-seed mill	I. F. Brown	New London, Conn	Jan. 23, 1872	122, 877
Cotton-seed, rice, &c., machine	N. Mixon	Madison, Ga	Dec. 20, 1833	....
Cotton-seed separator and planter	S. W. Thompson	Otsego, Ohio	Oct. 19, 1869	96, 054
Cotton seeds and motes, Machine for linting and relinting.	G. W. Grader	Memphis, Tenn	Aug. 31, 1869	94, 304
Cotton-seeds and separating the hull from the kernel, Machine for hulling and husking.	F. Follet	Petersburgh, Va	Dec. 15, 1829	....
Cotton-seeds by roller gin, Separating	W. and W. Whittemore, jr	West Cambridge, Mass	Mar. 21, 1833	....
Cotton-seeds, Engine for separating fibers from the husks of.	T. Rose and R. E. Gibson	Earlestown, England	July 26, 1870	105, 728
Cotton-seeds for extracting oil, Preparing	O. Reichenbach	Norristown Township, Pa.	Oct. 23, 1855	13, 708
Cotton-seeds for planting, Machine for preparing	R. C. Wrenn	Covington, Ky	Dec. 4, 1855	13, 894
Cotton seeds for planting, Method of preparing	W. Blessing	Jeffersonville, Ohio	July 3, 1866	56, 140
Cotton-seeds for planting, Preparing	P. C. Ingersoll	Green Point, N. Y	June 25, 1867	66, 089
Cotton-seeds for planting, Preparing	R. M. Lafferty	Three Rivers, Mich	July 7, 1868	79, 765
Cotton-seeds, Hulling and breaking	C. Leck and W. Jenks	Columbia, S. C	Dec. 17, 1834	....
Cotton-seeds, Machine for detaching short fibers from.	L. P. Jenks	Boston, Mass	June 25, 1861	32, 626
Cotton-seeds, Machine for hulling or husking	J. Smith	Petersburgh, Va	Dec. 28, 1829	....
Cotton-seeds, Machine for husking and separating the hulls from kernels of.	F. Follet	Petersburgh, Va	Jan. 21, 1829	....
Cotton-seeds, Machine for planting	N. E. Badgley	New York, N. Y	Oct. 16, 1866	58, 752
Cotton-seeds, Machine for planting	I. W. McGaffey	Chicago, Ill	July 17, 1866	56, 430
Cotton-seeds, Machine for removing lint from	W. F. Pratt	East Bridgewater, Mass	June 29, 1869	91, 866
Cotton-seeds, Machinery for breaking	T. Rose and R. E. Gibson	Oxton and New Brighton, England.	July 27, 1869	93, 124
Cotton-seeds, Process for hulling	D. W. Messer	Boston, Mass	July 24, 1855	13, 317
Cotton-seeds, Process of cleaning	J. Kirkman	Peoria, Ill	Nov. 26, 1867	71, 499
Cotton-seeds, Process of separating fibers from	J. Duval	New Orleans, La	Aug. 2, 1870	105, 923
Cotton-seeds, Treating	J. J. Powers	Memphis, Tenn	Mar. 14, 1871	112, 628
Cotton, silk, &c., Roving, spinning, and doubling	J. Jones	Manchester, England	May 16, 1835	....
Cotton-sliver can	W. Hamer	Little Lever, near Bolton, England.	Oct. 8, 1867	69, 658
Cotton-speeder	S. Parmelee	Poughkeepsie, N. Y	Aug. 29, 1833	....

Index of patents issued from the United States Patent Office from 1799 *to* 1873, *inclusive*—Continued.

Invention.	Inventor.	Residence.	Date.	No.
Cotton-spindle	N. Rider	Dudley, Mass	Nov. 3, 1832	....
Cotton-spindles, Step and bolster for	C. H. Chapman	Shirley, Mass	Feb. 8, 1870	99, 534
Cotton spreader and picker	J. C. Whitin	Northbridge, Mass	July 20, 1832	....
Cotton-stalk cutter or pulverizer	G. Gorman	Lamar, Miss	Sept. 20, 1853	10, 043
Cotton-stalk knocker	M. M. Carruth	Helena, Ark	Apr. 15, 1873	137, 825
Cotton-stalks in the field, Machine for cutting	F. Bradshaw	Greene County, Ala	Feb. 12, 1850	7, 078
Cotton-stalks, Machine for cutting standing	J. W. Bocage	Cypress Mills, Ark	Sept. 11, 1855	13, 538
Cotton-stalks, Machine for cutting standing	S. Bowerman	Detroit, Mich	Aug. 28, 1855	13, 505
Cotton-stalks, Machine for gathering	R. Ray	Louisport, Ky	Aug. 14, 1855	13, 436
Cotton-stalks, Machine for pulling and cutting	S. Beers	Naugatuck, Conn	Jan. 17, 1860	26, 828
Cotton-sweep	T. E. C. Brinly	Louisville, Ky	Nov. 22, 1870	109, 491
Cotton, Teeth of machine for picking	B. M. Smith	Brooklyn, N. Y	May 8, 1866	54, 619
Cotton-thinner	J. Gatling	Murfreesborough, N. C	June 19, 1835	....
Cotton-thinner	E. M. Harris and J. Cleghorn	Cass County, Ga	Aug. 21, 1847	5, 257
Cotton-thinning machine	I. W. Burch	Fayette, Miss	Oct 5, 1869	95, 422
Cotton-thinning machine	E. M. Greeson	Americus, Ga	Apr. 12, 1870	101, 962
Cotton-tie	E. B. Bishop	New Orleans, La	July 23, 1867	67, 098
Cotton-tie	W. F. Buckelew	Shreveport, La	Apr. 16, 1867	63, 849
Cotton-tie	W. I. Carroll	Natchez, Miss	Jan. 23, 1866	52, 137
Cotton-tie	J. H. Fraley	New Orleans, La	Dec. 10, 1867	71, 867
Cotton-tie	J. W. Hedenberg	Chicago, Ill	Dec. 4, 1866	60, 187
Cotton-tie	J. F. Milligan	Saint Louis, Mo	Nov. 6, 1866	59, 512
Cotton-tie	M. A. Tarleton	New Orleans, La	Oct. 30, 1866	59, 293
Cotton-tie	C. W. Wailey	New Orleans, La	Oct. 9, 1866	58, 698
Cotton-tie fastening	T. McIntire	Portsmouth, Ohio	Sept. 11, 1866	57, 945
Cotton-tie-forming die	C. W. Wailey	New Orleans, La	July 30, 1867	67, 236
Cotton-tie stretcher	S. Mather	New Braunfels, Tex	Aug. 22, 1871	118, 254
Cotton to gin, Machinery for opening and feeding	M. B. Clarke	Newnan, Ga	Dec. 4, 1855	13, 898
Cotton to market, Rafting baled	G. R. Griffith	Mobile, Ala	Sept. 25, 1841	2, 267
Cotton-ties, Fastening ends of	A. Barbarin	New Orleans, La	June 25, 1867	66, 065
Cotton, Utilizing the silky down of the wild	M. H. Simpson	Boston, Mass	Mar. 7, 1871	112, 391
Cotton wadding	C. L. Fleischman	Washington, D. C	June 19, 1847	5, 161
Cotton wadding, Apparatus for drying the sizing on	O. Tenny	Dorchester, Mass	Jan. 10, 1845	3, 881
Cotton-waste and other fibers from oil, &c., Process of cleaning.	H. M. Baker	Washington, D. C	Jan. 11, 1870	98, 658
Cotton-waste and other fibrous materials, Cleaning	H. M. Baker	Brooklyn, N. Y	Nov. 7, 1871	120, 564
Cotton-waste, Cleaning	H. M. Baker	New York, N. Y	July 28, 1868	80, 381
Cotton-waste, Cleaning	G. Sagar	Chicago, Ill	June 17, 1873	139, 921
Cotton-waste for cleaning machinery, Preparing	A. Peple	Billerica, Mass	Nov. 1, 1870	108, 821
Cotton-waste for cleaning machinery, Preparing certain kinds of.	A. Peple	East Billerica, Mass	Feb. 15, 1870	99, 945
Cotton-waste, Machinery for picking and separating	D. Goff	Pawtucket, R. I	Aug. 4, 1868	80, 724
Cotton waste or rags previous to their being operated on by the cutting and dusting machinery, Machine for dressing.	E. Smith	North Sudbury, Mass	Sept. 10, 1840	1, 782
Cotton-waste, &c., Process of cleaning	G. W. Sylvester	Belleville, N. J	Jan. 3, 1871	110, 800
Cotton-waste-separating machine	F. S. Robinson	Boston, Mass	Oct. 7, 1862	36, 625
Cotton-whipper	W. Hopkins	Plainfield, Conn	Mar. 6, 1823	....
Cotton whipper and cleaner	H. G. Davis	Clarke County, Ala	May 16, 1845	4, 050
Cotton-whipper, Cylindrical grate	J. S. Simmons	Scituate, R. I	Oct. 1, 1830	....
Cotton-whipper, Oblique	L. Osgood	Abington Post-Office, Conn	Oct. 27, 1835	....
Cotton-whipper, Revolving	E. Baker	Warwick, R. I	June 4, 1830	....
Cotton-winding, Double-speeder for	W. Hines	Coventry, R. I	Feb. 6, 1819	....
Cotton, Wire-screen for cleaning	J. E. Crane	Lowell, Mass	Jan. 24, 1860	26, 887
Cotton, wool, fur, and other fibrous materials, Machinery for cleaning and separating.	I. Hayden	Lawrence, Mass	Mar. 17, 1857	16, 833
Cotton-wool, Machine for cleaning and picking	E. Raymond	Norton, Mass	Jan. 18, 1812	....
Cotton-worm destroyer	J. Helm	Hochheim, Tex	May 20, 1873	139, 062
Cotton-worm destroyer	J. W. Johnson	Columbus, Tex	Dec. 16, 1873	145, 571
Cotton-worm destroyer	J. W. Johnson	Columbus, Tex	Dec. 16, 1873	145, 572
Cotton-worms, &c., Machine for exterminating	C. Steinmann	Napoleonville, La	Feb. 4, 1868	74, 165
Couch, Accommodation	J. Mead	Albany, N. Y	Mar. 14, 1817	....
Couch, Accouchement	E. L. Moore	San Francisco, Cal	Mar. 11, 1873	136, 663
Couch, Adjustable	G. Widmer	New York, N. Y	May 12, 1868	77, 856
Coach and car truss	G. F. Chalender	Burlington, Iowa	Oct. 7, 1873	143, 498
Couch, Folding	A. J. Bell	Cincinnati, Ohio	Mar. 31, 1868	76, 148
Couch, Hospital	J. Keck	Barrington, Ill	Dec. 24, 1872	134, 147
Couch, Invalid	C. L. Taillant	New York, N. Y	Dec. 20, 1859	26, 534
Couch, Nursing	J. H. Cogshall	Lexington, Mich	Feb. 12, 1867	62, 010
Couch or cradle	A. E. and J. B. Blood and F. W. Pope.	Lynn, Mass	Aug. 11, 1868	80, 801
Couch or cradle	R. Hale	Chicago, Ill	Jan. 28, 1868	73, 802
Couch, Variety	E. Carver	Bridgewater, Mass	June 12, 1838	775
Counter and desk seat	W. R. Pomeroy	Millersburgh, Ohio	May 28, 1867	65, 115
Counter and show-case	L. F. Vienot	New York, N. Y	Apr. 12, 1870	101, 790
Counter-brace, Metallic	J. L. Cooper	Preston, Conn	Oct. 13, 1868	82, 921
Counter-knife	J. Feed	Reading, Pa	Apr. 5, 1870	101, 542
Counter-seat	D. H. Krauser	Pottsville, Pa	May 10, 1870	102, 947
Counter-shears	N. Stow	Binghamton, N. Y	Mar. 4, 1873	136, 558
Counter, Show	A. Beardsley	Mount Zion, Iowa	Oct. 26, 1869	96, 188
Counter-supporter	J. Reising	Aurora, Ill	Oct. 23, 1866	59, 075
Counterfeits, Check to detect	J. Perkins	....	Mar. 19, 1799	....
Counterpane	Z. Allen	Providence, R. I	Aug. 23, 1853	9, 962
Countersink	S. L. Abbott	Deering, Me	July 25, 1871	117, 237
Countersink	A. G. Bachelder	Lowell, Mass	May 10, 1853	9, 705
Countersink	R. P. Buttles	Mansfield, Pa	July 6, 1869	92, 261
Countersink	W. H. Dodge	Mishawaka, Ind	Jan. 30, 1872	123, 246
Countersink	S. E. Holbrook, jr	Charlestown, Mass	May 24, 1870	103, 461
Countersink	L. H. Hunt	Rockingham, Vt	Feb. 9, 1869	86, 674
Countersink	L. H. Hunt	Saxton's River, Vt	May 21, 1872	126, 963
Countersink	F. H. Palmer	Foxcroft, Me	Feb. 28, 1871	112, 176
Countersink	M. M. Petts	Worcester, Mass	Dec. 27, 1870	110, 494
Countersink	W. B. Shedd	East Boston, Mass	Nov. 1, 1870	108, 837
Countersink	H. S. Shepardson	Shelburne Falls, Mass	Mar. 8, 1864	41, 894
Countersink	D. F. Sutton	Toledo, Ohio	Nov. 8, 1870	109, 154
Countersink	A. Wheeler	Brattleborough, Vt	Apr. 12, 1870	101, 796
Countersink	P. A. Whitney	Woodstock, Vt	Oct. 20, 1868	83, 348
Countersink	A. Williams	Wellsville, Ohio	Feb. 20, 1866	52, 776

Index of patents issued from the United States Patent Office from 1790 *to* 1873, *inclusive*—Continued.

Invention.	Inventor.	Residence.	Date.	No.
Countersink and bit	W. A. Clark	Woodbridge, Conn	Aug. 17, 1869	93, 807
Countersink and plug borer	W. H. Dodge	Mishawaka, Ind	Feb. 6, 1872	123, 463
Countersink-bit	H. C. Lewis	Essex, Conn	Dec. 4, 1866	60, 207
Countersink-gage	A. Wheeler	Brattleborough, Vt	July 11, 1871	116, 901
Counting-machine	J. A. Bazin	Canton, Mass	Dec. 23, 1856	16, 323
Counting-machine	S. Comfort	Morrisville, Pa	Feb. 20, 1866	52, 681
Counting-machine	J. Dolbeer	San Francisco, Cal	Dec. 20, 1864	45, 482
Counting-machine	P. Stillman	New York, N. Y	Aug. 22, 1854	11, 577
Counting-machine	U. Turnar	Versailles, Ky	July 2, 1867	66, 269
Counting-machine	T. J. Young	Philadelphia, Pa	Mar. 11, 1862	34, 665
Counting-machine and machine for indicating motion.	W. Y. Sterling	Bridgeport, Conn	Sept. 19, 1854	11, 703
Counting-register	A. P. Atkinson	Vermont, Ill	Nov. 7, 1871	120, 609
Counting-register	H. Chandler	Buffalo, N. Y	July 30, 1872	130, 017
Counting-register	J. S. Detrick	San Francisco, Cal	Aug. 4, 1868	80, 612
Counting-register	R. P. Hinds	Chicago, Ill	Jan. 28, 1873	135, 336
Counting-register	W. H. McNary	Brooklyn, N. Y	May 17, 1870	103, 221
Counting-register	C. W. Pyle	Wilmington, Del	Apr. 18, 1871	113, 794
Counting-register	G. Sickels	Boston, Mass	Sept. 28, 1869	95, 389
Counting-register	G. H. Van Vleck	Buffalo, N. Y	Mar. 25, 1873	137, 262
Counting-register	A. M. White	Bridgeport, Conn	Apr. 30, 1872	126, 361
Counting-register, Locomotive	E. P. Curtis	Cleveland, Ohio	Apr. 2, 1872	125, 179
Coupling:				
See Aqueduct-coupling.				
Auger-shaft coupling.				
Axle-coupling.				
Barge-coupling.				
Belt-coupling.				
Brace-coupling.				
Brake-coupling.				
Car-coupling.				
Car-heater coupling.				
Carriage-felly coupling.				
Chain-coupling.				
Cord-coupling.				
Drill-rod coupling.				
Drop-light coupling.				
Elastic coupling.				
Felly-joint coupling.				
Fire-engine coupling.				
Flexible coupling.				
Friction coupling.				
Gas and water main coupling.				
Gas and water pipe coupling.				
Harvester-pitman coupling.				
Hollow-shaft coupling.				
Hose-coupling.				
Hose and pipe coupling.				
Leader-pipe coupling.				
Lightning-rod coupling.				
Locomotive-coupling.				
Melodeon-coupling.				
Mill-shaft coupling.				
Octave-coupling.				
Organ-coupling.				
Perch-coupling.				
Pipe-coupling.				
Pitman-coupling.				
Platform spring-coupling.				
Plow-coupling.				
Pulley-coupling.				
Pump-rod coupling.				
Rail-coupling.				
Railway-coupling.				
Railway-rail coupling.				
Rigging-coupling.				
Rod-coupling.				
Rope-band coupling.				
Safety-coupling.				
Seeding-machine coupling.				
Shaft-coupling.				
Shaft and pole coupling.				
Shafting-coupling.				
Spring-coupling.				
Steam-heater coupling.				
Steam or air brake coupling.				
Stove-pipe coupling.				
Strap-coupling.				
Sucker-rod coupling.				
Telegraph-wire coupling.				
Thill-coupling				
Tongue-coupling.				
Train-boat coupling.				
Tumbling-shaft coupling.				
Union coupling.				
Universal coupling.				
Universal-joint coupling.				
Vehicle-coupling.				
Vehicle-spring coupling.				
Wagon-coupling.				
Wagon-reach coupling.				
Wheel-felly coupling.				
Whiffletree-coupling.				
Wire-rigging coupling.				
Coupling	J. Henermann	Davenport, Iowa	Sept. 15, 1868	82, 117
Coupling	M. Newman, 2d. N. C. Whitcomb, and G. C. Cole.	Lanesborough, Pa., and Hartford, Conn.	June 13, 1854	11, 074
Coupling and brake	T. A. Westen	King's Norton, England	Mar. 3, 1868	75, 227

Index of patents issued from the United States Patent Office from 1790 *to* 1873, *inclusive*—Continued.

Invention.	Inventor.	Residence.	Date.	No.
Coupling and steering apparatus	J. McCreary	Middletown, Pa	July 23, 1872	129, 844
Coupling-bar for railway car, locomotive, &c	W. D. Chesnut	Wilmington, Del	Feb. 20, 1844	3, 445
Coupling-box for shafting	W. B. Dunning	Geneva, N. Y	June 22, 1858	20, 634
Coupling-iron for railway and other carriages	T. G. Owen	Baltimore, Md	July 1, 1840	1, 666
Coupling-jack	H. A. Brown and E. B. Keith	Galesburgh, Miss	Sept. 27, 1870	107, 656
Coupling-pins, Machine for making	C. H. Williams	Cleveland, Ohio	Dec. 13, 1870	110, 179
Coupling-pipe	W. Hudgin	Washington, D. C	Apr. 6, 1858	19, 852
Coupling-pipe, Wrought-iron	S. Vanstone	Providence, R. I	June 28, 1870	104, 9[illegible]5
Coupling-spring, Gum-elastic	S. M. Hoover	Carlisle, Pa	Apr. 18, 1865	47, 305
Coupon, Mode of numbering	S. M. Clark	Washington, D. C	July 9, 1867	66, 461
Court-plaster, Preparation of	L. C. Gale	Chicago, Ill	May 1, 1866	54, 326
Coverlet	C. K. Pevey	Worcester, Mass	July 15, 1873	140, 948
Coverlet	H. Wettstein	Philadelphia, Pa	Jan. 29, 1867	61, 588
Cover-lifter, hammer, &c., Combined	R. M. Lafferty	Three Rivers, Mich	June 4, 1867	65, 397
Cow and sheep rack	R. Lovett	Canton, Ohio	May 15, 1866	54, 745
Cow-catcher	C. Darling	Utica, N. Y	June 8, 1852	8, 996
Cow-catcher	T. B. Smith	Triune, Tenn	July 11, 1854	11, 273
Cow-catcher, Railway	W. J. Orr	Manorville, Pa	Dec. 31, 1867	72, 752
Cow-tail holder	N. Dutton	Janesville, Wis	Apr. 30, 1867	64, 207
Cow-tail holder	W. H. Gray	Ashfield, Mass	Mar. 14, 1871	112, 587
Cow-tail holder	J. Knight	Whitestown, N. Y	May 31, 1870	103, 753
Cow-tail holder	G. Tanner	Freetown, N. Y	June 21, 1870	104, 661
Cows and calves from sucking and for leading cattle, Device for preventing.	A. S. Haven	Barre, Mass	Feb. 4, 1868	74, 086
Cows from kicking, Device for preventing	T. Pyle	Nottingham, Pa	Aug. 27, 1872	130, 943
Cows from kicking, Device for preventing	H. J. Sadler and S. H. Spencer	Mecca, Ohio	July 8, 1873	140, 594
Cows while being milked, Device for confining	L. Brown	Evans, N. Y	Nov. 13, 1866	59, 550
Cowl: *See* Car-cowl. Chimney-cowl. Emery-wheel cowl. Ventilating-cowl.				
Cowl	C. W. Atkeson	Saint Louis, Mo	Mar. 3, 1868	75 109
Cowl	W. F. G. Beeuwkes	Holland, Mich	Jan. 22, 1867	61, 387
Crabs, Device for securing and feeding soft	C. Drexler	Washington, D. C	Dec. 17, 1867	72, 177
Cracker and biscuit cutting	J. Clark and H. Henderson	Baltimore, Md	June 13, 1831	....
Cracker and biscuit cutting	S. P. Clark	Baltimore, Md	Nov. 7, 1835	....
Cracker and biscuit cutting	J. P. Fairlamb and M. Dunot	Wilmington, Del	June 13, 1831	....
Cracker and biscuit cutting	T. and T. H. Havener	Washington, D. C	Oct. 17, 1835	....
Cracker and biscuit cutting	W. R. Nivens	New York, N. Y	Oct. 17, 1835	....
Cracker and biscuit cutting	J. S. Stiles	Baltimore, Md	Aug. 9, 1831	....
Cracker and biscuit machine	T. Bladen	Philadelphia, Pa	Mar. 16, 1830	....
Cracker, biscuit, &c	T Bladen	Philadelphia, Pa	Oct. 1, 1830	....
Cracker, biscuit, &c	D. Poole	Dauphin County, Pa	May 20, 1834	....
Cracker-box	C. F. Thurston	Cambridgeport, Mass	Mar. 12, 1872	124, 459
Cracker-crusher	A. Clarke and T. Reece	Philadelphia, Pa	Jan. 15, 1867	61, 163
Cracker cutter and stamping machine	C. Baldwin	Goffstown, N. H	Nov. 7, 1848	5, 906
Cracker-cutting machine	E. O. Brinkerhoff	New York, N. Y	Aug. 25, 1863	39, 628
Cracker-cutting machine	C. P. Forbes	Baltimore, Md	July 17, 1841	2, 180
Cracker-cutting machine	W. R. Nevins	New York, N. Y	Nov. 10, 1841	2, 344
Cracker-cutting machine	W. R. Nevins	New York, N. Y	Aug. 20, 1850	7, 580
Cracker-cutting machinery	W. Perkins	Boston, Mass	Apr. 2, 1841	2, 029
Cracker-dough, Rolling and cutting	W. R. Nevins	New York, N. Y	Mar. 2, 1836	....
Cracker-machine	W. Cairns	Jersey City, N. J	Nov. 12, 1872	133, 014
Cracker-machine	W. Cairns	Jersey City, N. J	Jan. 31, 1871	111, 315
Cracker-machine	P. Emmons	New York, N. Y	Feb. 13, 1855	12, 383
Cracker-machine	J. Fox	Lansingburgh, N. Y	Feb. 1, 1859	22, 793
Cracker-machine	J. Fox	Lansingburgh, N. Y	Oct. 14, 1873	143, 686
Cracker-machine	W. M. Garrison	New York, N. Y	June 5, 1866	55, 276
Cracker-machine	G. Y. Gray	Niles, Mich	May 11, 1869	89, 985
Cracker-machine	J. E. Hawkins	Lansingburgh, N. Y	June 25, 1867	66, 082
Cracker-machine	J. and J. C. Holyland	Rochester, N. Y	Sept. 28, 1858	21, 6 6
Cracker-machine	J. and J. C. Holyland	Rochester, N. Y	Nov. 27, 1860	30, 732
Cracker-machine	J. Johnson and O. Freeman	Boston, Mass	May 17, 1844	3, 585
Cracker-machine	G. J. Kingsbury	Rochester, N. Y	Oct. 14, 1873	143, 579
Cracker-machine	C. Marsh, 2d	Natchez, Miss	July 24, 1860	29, 295
Cracker-machine	C. Marsh, 2d	Natchez, Miss	Nov. 9, 1869	96, 715
Cracker-machine	J. McCollum	New York, N. Y	Mar. 23, 1852	8, 828
Cracker-machine	J. McCollum	New York, N. Y	Feb. 15, 1859	22, 966
Cracker-machine	C. Neer	Troy, N. Y	June 15, 1858	20, 577
Cracker-machine	L. A. Rockwell	Tremont, N. Y	June 17, 1873	139, 920
Cracker-machine	G. R. Skillman	Baltimore, Md	Apr. 4, 1871	113, 356
Cracker-machine	T. Sloat	Brooklyn, N. Y	Apr. 20, 1869	89, 250
Cracker-making machine	R. Darling	East Greenwich, R. I	Sept. 30, 1841	2, 277
Cracker-making machine	D. Steward	Philadelphia, Pa	Nov. 26, 1867	71, 420
Cracker-making machine	H. Winslow	Swansey, Mass	Dec. 14, 1840	1, 896
Cracker-molding machine	P. Emmons	New York, N. Y	Aug. 1, 1854	11, 418
Cracker or meat biscuit	E. Greenfield	New York, N. Y	Feb. 15, 1870	99, 880
Cracker, Oyster	W. P. Lyon	Portchester, N. Y	May 8, 1866	54, 566
Cracker rolling and cutting machine	D. Pool	Philadelphia, Pa	Dec. 24, 1824	....
Crackers, biscuit, &c., Cutting	J. Clarke and H. Hen erson	Philadelphia, Pa	Sept. 18, 1833	....
Crackers, biscuit, &c., Machine for cutting	J. Cooper	Philadelphia, Pa	Mar. 11, 1830	....
Crackers, biscuit, pilot-bread, &c., Machine for cutting.	J. Clark and H. Henderson	Baltimore, Md	Sept. 13, 1830	....
Crackers, Machine for cutting	G. M. Kendall	Catskill, N. Y	May 3, 1833	....
Crackers, Machine for cutting	L. Kirtland	New Haven, Conn	Sept. 4, 1834	....
Crackers, Manufacture of cream	D. M. Holmes	Brooklyn, N. Y	Oct. 9, 1866	58, 644
Crackers, Mode of making cylindrical strips of dough in the manufacture of.	F. C. Treadwell, jr., and H. McCollum.	New York, N. Y., and Windham, Conn.	Oct. 11, 1859	25, 775
Crackers, Molding	J. G. Driscoll	Auburn, N. Y	Nov. 19 1833	....
Crackers, ship-biscuit, &c	J. and C. Bruce	Kings County, N. Y	Mar. 13, 1832	....
Crackers, ship-bread, &c	N. Daskam and D. G. Wood	Geneva, N. Y	Aug. 5, 1830	....
Cracking and grinding mill	W. Beal, jr., and B. S. Hale	Norway, Me., and Lowell, Mass.	Dec. 17, 1846	4, 895
Cracklings cutting machine	A. Smith	Cincinnati, Ohio	Feb. 5, 1867	61, 884
Cracklings, &c., Cutting-machine for reducing	A. Smith	Cincinnati, Ohio	Nov. 26, 1867	71, 545
Cracklings from roasted pork, Cutting-tools for the manufacture of.	L. F. Lannay	Baltimore County, Md	Apr. 2, 1872	125, 137

Index of patents issued from the United States Patent Office from 1790 *to* 1873, *inclusive*—Continued.

Invention.	Inventor.	Residence.	Date.	No.
Cradle	W. Bedle	Keyport, N. J	June 5, 1866	55, 227
Cradle	J. B. Charlton	Kalamazoo, Mich	Apr. 4, 1871	113, 497
Cradle	F. Chichester	Milwaukee, Wis	July 22, 1873	141, 033
Cradle	P. Cooper	Hempstead, N. Y	Mar. 27, 1815	....
Cradle	D. Cox	Cincinnati, Ohio	Feb. 5, 1867	61, 720
Cradle	W. T. Doremus	New York, N. Y	Apr. 29, 1873	138, 320
Cradle	A. Dick	Buffalo, N. Y	Jan. 23, 1866	52, 146
Cradle	D. A Dunham	Pilatka, Fla	Oct. 29, 1867	70, 179
Cradle	W. H. Earnest	Clarksburgh, Va	May 20, 1862	35, 307
Cradle	W. H Earnest	Parkersburgh, W. Va	June 25, 1872	128, 375
Cradle	C. A. Fenner	Mystic River, Conn	Nov. 11, 1873	144, 391
Cradle	M. Griffin	Brooklyn, N. Y	Jan. 7, 1873	134, 661
Cradle	L. Heywood	Gardner, Mass	Jan. 21, 1873	135, 117
Cradle	L. and A. R. Hill	Alexandria, Va	Jan. 28, 1868	73, 892
Cradle	S. W. Knowles	New Haven, Conn	July 2, 1867	66, 361
Cradle	R. W. Myers	Clarksville Post-Office, N. J.	Feb. 28, 1871	112, 271
Cradle	A. H. Ordway	Haverhill, Mass	Jan. 10, 1871	110, 931
Cradle	D. O. Parker	Liverpool, Nova Scotia	Aug. 8, 1871	117, 809
Cradle	S. Pope	Windsor, Conn	June 15, 1813	....
Cradle	D. Sonder	Houston, Ohio	Apr. 15, 1873	137, 868
Cradle	L. Sperry and L. Robinson	East Windsor Hill and New Haven, Conn.	July 20, 1869	92, 896
Cradle	A. Spiegel	Indianapolis, Ind	Oct. 5, 1869	95, 530
Cradle	H. H. Wiggers	Cincinnati, Ohio	Aug. 13, 1872	130, 554
Cradle	F. R. Wolfinger	Vermont, Ill	July 5, 1864	43, 448
Cradle and baby-walker, Circular swinging	W. M. Messick	Louisville, Ky	July 30, 1867	67, 331
Cradle and chair	A. Berny	Williamsburgh, N. Y	July 3, 1866	55, 990
Cradle and chair, Combination of	A. Hicks	Flushing, N. Y	May 10, 1864	42, 730
Cradle and chair, Combined	F. E. Coomes	Berlin, Wis	July 31, 1866	56, 718
Cradle and chair rocker	S. Simmons	Baltimore, Md	Dec. 21, 1819	....
Cradle and churn, Propelling	E. Whitman, jr	Winthrop, Me	Mar. 27, 1835	....
Cradle and rocking-chair combined	E. Hambujer	Detroit, Mich	Feb. 4, 1873	135, 425
Cradle and tête-à-tête	G. H. Hazlewood	Boston, Mass	July 5, 1853	9, 829
Cradle, Automatic	D. Alger	Bryon, Mich	Sept. 6, 1870	107, 148
Cradle, Automatic	S. G. Delano	Grand Blanc, Mich	Sept. 1, 1868	81, 757
Cradle, Automatic	S. G. Delano	Argentine, Mich	Mar. 15, 1870	100, 735
Cradle, Automatic	I. S. Goodman	Winterset, Iowa	Apr. 22, 1873	138, 144
Cradle, Automatic	W. V. Van Dervoort and R. B. Walker.	New Antioch, Ohio	June 10, 1873	139, 838
Cradle-chair	J. H. Havens	Lewiston, N. Y	Aug. 5, 1862	36, 091
Cradle-chair	M. Studley	South Yarmouth, Mass	Mar. 2, 1869	87, 378
Cradle, Child's	J. L. Riter	Brownsville, Ind	Jan. 16, 1872	122, 737
Cradle, Child's	L. K. Selden	Haddam, Conn	Apr. 10, 1860	27, 835
Cradle, Child's	W. J. Stowell	Baltimore, Md	Feb. 21, 1871	111, 989
Cradle, crib, and standing-stool combined	C. E. Nurse	Chesterfield Factory, N. H	Nov. 19, 1872	133, 169
Cradle, Fleecing	J. K. Alwood	Delta, Ohio	Nov. 3, 1868	83, 754
Cradle, Folding	C. Klein	Albany, N. Y	Feb. 25, 1868	74, 917
Cradle, Folding	J. A. Latham and G. W. Tileston.	New Haven, Conn	Aug. 23, 1864	43, 947
Cradle, Folding	T. W. Moore	New York, N. Y	July 11, 1871	116, 856
Cradle, Folding	L. K. Selden	Haddam, Conn	May 24, 1859	24, 142
Cradle, Hydrostatic steam	C. Miner	Lyme, Conn	Nov. 16, 1827	....
Cradle, Infant's	T. C. Ball	Keene, N. H	Nov. 16, 1858	22, 056
Cradle, Infant's	S. F. Brooks	Weston, Mass	Apr. 24, 1860	27, 964
Cradle or crib	E. N. Cutter	Chicago, Ill	Apr. 6, 1869	88, 553
Cradle or crib	J. M. Read	New York, N. Y	Feb. 27, 1830	....
Cradle or crib, Folding	M. McNamara	Philadelphia, Pa	Jan. 21, 1873	135, 138
Cradle or crib, Portable	D. M. Reynolds	Chicago, Ill	Nov. 8, 1870	109, 050
Cradle, Portable	W. R. Evans	Philadelphia, Pa	Sept. 12, 1871	118, 918
Cradle, Portable	W. S. Harris	Trenton, N. J	Dec. 30, 1873	146, 069
Cradle, &c., rocker	R. A. Jackson	Alliance, Ohio	June 20, 1871	116, 061
Cradle-rocker	P. Power	Chicago, Ill	Oct. 8, 1867	69, 698
Cradle-rocker	H. R. Taylor	Roxbury, Mass	Jan. 16, 1843	2, 915
Cradle, Rocking	W. A. N. Long	Fisherville, N. H	Aug. 13, 1861	33, 045
Cradle, Rocking	W. D. Tewksbury	Cuylersville, N. Y	June 7, 1859	24, 340
Cradle, Self-moving	A. Buchenberger	New York, N. Y	Nov. 3, 1829	....
Cradle, Self-rocking	C. H. Helmkamp	Reading, Ohio	Feb. 9, 1864	41, 507
Cradle, Self-rocking	C. Holtz	Saint Louis, Mo	May 27, 1873	139, 247
Cradle, Self-rocking	B. Brazelle	Nashville, Ill	May 15, 1866	54, 815
Cradle, Self-rocking	D. Walker	Newark, N. J	Feb. 25, 1862	34, 535
Cradle, Self-rocking	H. G. Williams	Hamilton, Iowa	Mar. 7, 1865	46, 744
Cradle, Sofa	H. A. Axtell	Westfield, Mass	Feb. 25, 1868	74, 876
Cradle, Spring	F. Chichester	Milwaukee, Wis	July 22, 1873	141, 032
Cradle, Spring-rocking	J. B. Malbert and A. Cheviron.	Saint Louis, Mo	May 18, 1858	20, 284
Cradle, stool, and chair	J. B. Warren	Davenport, Mass	June 5, 1866	55, 401
Cradle, Swing	P. P. Carroll	Washington, D. C	Apr. 14, 1868	76, 601
Cradle, Swing	Z. Mills	Hartford, Conn	July 17, 1806	....
Cradle, Swinging	S. W. Knowles	Middletown, Conn	Oct. 28, 1851	8, 468
Cradle, Swinging	J. McKinney	West Chester, Ohio	June 13, 1831	....
Cradle, Swinging	L. F. Whitaker	Raleigh, N. C	Oct. 21, 1851	8, 461
Cradle, Wire	L. Chevallier and R. Brass	Williamsburgh, N. Y., and Waterbury, Conn.	Jan. 11, 1870	98, 742
Cradles to rocking-chairs, Coupling for connecting.	J. Reeves	Cincinnati, Ohio	Sept. 9, 1873	142, 583
Cranberry-gatherer	W. Hall	Dennis, Mass	Sept. 8, 1868	81, 897
Cranberry-gatherer	E. D. Miller	Dorchester, Mass	Nov. 20, 1866	59, 857
Cranberry-gatherer	G. Shove	Yarmouth, Mass	Mar. 7, 1865	46, 760
Cranberry-gatherer	G. Shove	Yarmouthport, Mass	Jan. 22, 1867	61, 476
Cranberry-gatherer	C. Thacher	Yarmouth, Mass	June 6, 1865	48, 136
Cranberry picker	J. P. Prickett	Medford, N. J	Mar. 22, 1870	101, 040
Cranberry-picker	J. Weston	Duxbury, Mass	Dec. 30, 1873	145, 979
Cranberry-separator	D. Perham	Tyngsborough, Mass	Jan. 19, 1858	19, 151
Cranberry-winnower	P. Flanders	Lowell, Mass	Apr. 4, 1854	10, 736
Cranberries for preservation, Putting up	A. Sampson	Providence, R. I	Dec. 2, 1862	37, 058
Cranberries, Machine for cleaning and assorting	G. L. Merrill	Chicago, Ill	July 23, 1867	67, 130
Crane	A. S. Batten	Topsham, Vt	Dec. 3, 1867	71, 682
Crane	H. Bradford	New York, N. Y	July 8, 1834	....
Crane	R. Briggs	Philadelphia, Pa	July 15, 1873	140, 882

Index of patents issued from the United States Patent Office from 1790 *to* 1873, *inclusive*—Continued.

Invention.	Inventor.	Residence.	Date.	No.
Crane	B. J. Burnett	New York, N. Y	Dec. 25, 1855	13, 976
Crane	B. J. Burnett	Mount Vernon, N. Y	Feb. 7, 1865	46, 213
Crane	F. Farrel	Ansonia, Conn	May 13, 1873	138, 876
Crane	G. Hunziker	Summit, Miss	July 16, 1872	129, 282
Crane	E. Marsh	Oswego, N. Y	Feb. 12, 1836	....
Crane	J. W. Middleton	Philadelphia, Pa	Mar. 21, 1871	112, 829
Crane	J. L. Pennock	Coatesville, Pa	Dec. 2, 1873	145, 234
Crane	F. Rumpf	Cold Spring, N. Y	Mar. 6, 1866	53, 044
Crane	G. Sherwood	Erie, Pa	June 30, 1836	....
Crane	J. F. Smith	Portsmouth, Va	July 10, 1860	29, 109
Crane	C. Williams	New York, N. Y	Aug. 18, 1868	81, 317
Crane, Balance	L. Henry	Paris, France	Nov. 9, 1844	3, 813
Crane, Blacksmith's	W. Maher	Slack, Ky	Dec. 23, 1856	16, 283
Crane, Compensating forge	T. Morris	Clarbeston, England	Nov. 5, 1867	70, 455
Crane, Dredging	G. Wood and J. King	Philadelphia, Pa	Nov. 9, 1858	22, 052
Crane, Elevating	J. T. Wright	Nashua, N. H	Aug. 22, 1865	49, 579
Crane for loading or discharging goods on board of vessels, Labor-saving.	T. Godwin	New York, N. Y	Oct. 2, 1838	953
Crane for raising brick, Portable	E. Mann and G. Hill	Rochester, N. Y	July 21, 1827	....
Crane, Hoisting	J. Y. Parce	Fairport, N. Y	July 26, 1859	24, 912
Crane, Hydraulic	D. S. Hines	Brooklyn, N. Y	July 23, 1867	67, 113
Crane, Hydraulic	S. K. Welman	Nashua, N. H	July 7, 1868	79, 710
Crane, Portable	L. A. Beardsley	South Edmeston, N. Y	June 25, 1861	32, 608
Crane, Portable	S. R. Marshall	Wilksbarre, Pa	Jan. 5, 1864	41, 161
Crane, Portable	C. Snyder and S. M. Smith	Hawley, Pa	Oct. 9, 1860	30, 358
Crane, Power	W. F. Durfee	New Bedford, Mass	Aug. 11, 1868	80, 930
Crane, Rotary	J. S. Coffinan	Greenville, Ind	Dec. 31, 1867	72, 801
Crane, Yard and tank	L. Y. Ketcham and J. Taynton	Port Jervis, N. Y	Feb. 15, 1870	99, 912
Crank	C. T. Moore	Gilmanton, N. H	Apr. 27, 1869	89, 423
Crank	A. Shedlock	New York, N. Y	July 16, 1872	128, 981
Crank, Anti-friction	B. S. Oliver	Salem, Mass	Mar. 9, 1821	....
Crank-box	G. Westinghouse	Schenectady, N. Y	Mar. 20, 1860	27, 588
Crank-brace for augers, Adjustable	J. Gourlay	Ogdensburgh, N. Y	Dec. 18, 1855	13, 943
Crank-connection	E. R. Cole	Pawtucket, R. I	Aug. 20, 1867	67, 957
Crank-engine for raising and throwing water	J. Clap	Montague, Mass	Nov. 21, 1810	....
Crank-expander	L. Swenson	North Cape, Wis	July 30, 1872	129, 996
Crank, friction-wheel, and brake, Combined	R. M. Van Sickler	New York, N. Y	Oct. 8, 1867	69, 727
Crank-gearing	T. J. McGowan	Cincinnati, Ohio	July 25, 1871	117, 312
Crank-heating apparatus	J. Miller	New York, N. Y	Apr. 15, 1873	137, 944
Crank-indicators, Arrangement of machinery for actuating the.	S. B. Hutchins	Oswegatchie, N. Y	June 3, 1851	8, 131
Crank-motion	W. Bricknell	Hartford, Me	Sept. 3, 1867	68, 482
Crank-motion	E. Quinn	Brooklyn, N. Y	June 27, 1871	116, 487
Crank-motion	T. Taylor	New Orleans, La	July 23, 1867	67, 001
Crank-movement	H. C. Bradford	Providence, R. I	Jan. 21, 1873	135, 073
Crank-pin box	T. Welch	Churchville, N. Y	Aug. 1, 1865	49, 185
Crank pins, Apparatus for turning locomotive	W. Blythe and N. Hayes	Alexandria, Va	Apr. 14, 1868	76, 702
Crank-pins, Machine for turning	H. S. Smith and W. D. Whitmore.	Bloomington, Ill	July 5, 1870	105, 007
Crank-pins, Machine for turning	M. G. Wood	Boston, Mass	Apr. 18, 1871	113, 959
Crank-pins of locomotives, Machine for returning.	N. Wright	Cleveland, Ohio	Feb. 15, 1870	99, 993
Crank, ratchet, and brake, Combined	R. B. McElrath	Newburgh, N. Y	June 13, 1871	115, 881
Crank, Serial	C. F. Ritchel	Chicago, Ill	Aug. 4, 1868	80, 769
Crank-shaft eccentric	J. F. Haskins	Fitchburgh, Mass	Apr. 1, 1873	137, 441
Crank, Substitute	A. Brooks	Crawford County, Ind	Aug. 3, 1858	21, 061
Crank-wheel motion	J. Wallace	Columbia, S. C	Sept. 25, 1824	....
Crank-wrist	C. B. Garlinghouse	Allensville, Ind	June 30, 1863	39, 034
Crank-wrist	J. M. Long	Hamilton, Ohio	Apr. 21, 1863	38, 233
Crank-wrist connection	J. Clayton	Brooklyn, N. Y	May 2, 1865	47, 522
Cranks by crank-pins, Mode of connecting	F. E. Sickels	New York, N. Y	Sept. 19, 1845	4, 199
Cranks on rods, Machine for bending	E. Lord	Todmorden, Great Britain	Mar. 13, 1866	53, 242
Cranks over dead-points, Carrying	F. Glass	Knightstown, Ind	Jan. 26, 1864	41, 373
Cranks to machinery, Attaching	A. Westcott	Syracuse, N. Y	May 23, 1865	47, 885
Cranks, Wheel used in machinery instead of	J. Steadwell, 3d	Carmel, N. Y	Nov. 25, 1823	....
Crate: *See* Folding-crate. Fruit-crate. Fruit-box crate. Peach-crate. Sewing-machine crate.				
Crate, Folding	L. A. Lindsey	Jackson, Miss	Mar. 1, 1870	100, 299
Crate, Folding	F. R. Van Dake	Jackson, Miss	Feb. 15, 1870	99, 977
Cravat	O. Ernst	New York, N. Y	Mar. 27, 1866	53, 428
Cravat-holder	J. N. Thomson	North Attleborough, Mass	June 27, 1871	116, 510
Cravat, Vulcanized rubber	W. W. Beach	New York, N. Y	July 19, 1864	43, 620
Crayon, Composition	W. Compton	New York, N. Y	June 8, 1869	91, 090
Crayon composition	I. N. Peirce	Boston, Mass	Dec. 20, 1864	45, 518
Crayon-holder	A. F. Howard	Waltham, Mass	Apr. 15, 1873	137, 844
Crayon-holder	R. Wright	New York, N. Y	July 28, 1868	80, 378
Crayon-holder, Tailors'	J. A. Gooch	Biddeford, Me	Aug. 20, 1872	130, 634
Crayon, Rubber	D. F. Pond	New Haven, Conn	Sept. 21, 1852	9, 277
Crayon-sharpener, Tailors'	R. R. Miles	Wabash, Ind	Dec. 21, 1869	98, 085
Crayons, Manufacture of nitrate of silver	S. P. Wheeler	Bridgeport, Conn	Nov. 22, 1859	26, 223
Cream-freezer	E. S. Farson	Philadelphia, Pa	Mar. 23, 1858	19, 733
Cream-freezer	E. C. Seaman	Philadelphia, Pa	Oct. 3, 1848	5, 821
Cream-freezer	S. W. Smith	Brooklyn, N. Y	Mar. 15, 1859	23, 271
Cream-freezer	E. S. Torrey	New York, N. Y	Feb. 5, 1867	61, 895
Cream from milk, Extracting	S. Davis	New York, N. Y	Oct. 1, 1830	....
Cream from whey, Method of extracting	K. Egger	South Cortland, N. Y	Sept. 25, 1866	58, 232
Cream-heater	J. C. Beddol and J. S. Coon	Branchport, N. Y	Mar. 22, 1870	101, 086
Cream-paste press and strainer	L. Bishop	Cortlandville, N. Y	Feb. 20, 1866	52, 661
Cream, Process of preparing	C. D. Birdseye	New York, N. Y	Sept. 17, 1850	7, 644
Cream-saver	O. Abell	Witoka, Minn	June 18, 1872	128, 090
Cream-saver	S. E. Mallett	Corry, Pa	Aug. 25, 1868	81, 519
Creeper	L. Witting	Philadelphia, Pa	Jan. 12, 1858	19, 109
Cresset or barrel-heater	W. B. Geddis	Rochester, N. Y	Mar. 5, 1867	62, 540
Crevasses, Method of stopping	L. A. Gossin	La Fourche Parish, La	Jan. 14, 1868	73, 318
Crib	W. B. Carpenter	West Topsham, Vt	Aug. 19, 1873	141, 918

Index of patents issued from the United States Patent Office from 1790 *to* 1873, *inclusive*—Continued.

Invention.	Inventor.	Residence.	Date.	No.
Crib and bedstead	G. T. Palmer	Brooklyn, N. Y	May 19, 1868	78, 123
Crib and bedstead, Extension	C. H. Hudson	New York, N. Y	Nov. 30, 1869	97, 403
Crib and chair	J. E. Small	Berlin, Wis	Oct. 23, 1866	59, 087
Crib and cradle	W. H. Earnest	Parkersburgh, W. Va	June 14, 1870	104, 288
Crib and cradle	H. W. Eastman	Baltimore, Md	Jan. 31, 1865	46, 091
Crib and cradle combined, Folding	J. B. Brolaski	Saint Louis, Mo	Sept. 3, 1872	131, 049
Crib and cradle, Folding	D. M. Reynolds	Chicago, Ill	Mar. 15, 1870	100, 925
Crib and walking-stool	S. S. Burr	Dedham, Mass	Jan. 1, 1867	60, 684
Crib, Child's	J. Batley	Orange, N. J	Nov. 5, 1872	132, 745
Crib, Child's	D. Cox	Cincinnati, Ohio	Mar. 26, 1872	124, 884
Crib, Child's	G. W. Ennis	Orange, N. J	Mar. 14, 1871	112, 570
Crib, Child's	W. L. Gerard	Fort Wayne, Ind	Dec. 11, 1866	60, 357
Crib, Child's	J. D. Lewis	Toronto, Canada	Dec. 2, 1873	145, 218
Crib, Child's	A. R. Swartz	Carlisle, Pa	May 11, 1869	89, 953
Crib, Extension	S. S. Burr	Dedham, Mass	Sept. 28, 1869	95, 318
Crib, Folding	A. J. Bettridge	North Bridgewater, Mass	Mar. 26, 1872	124, 877
Crib, Folding	C. Kilburn	Philadelphia, Pa	Jan. 14, 1873	134, 895
Crib or cradle	L. A. Chichester	Poughkeepsie, N. Y	Oct. 12, 1869	95, 768
Crib or cradle	W. T. Hazard	Randolph, Mass	May 2, 1871	144, 292
Crib or hammock, Swinging	D. O'Grady	Charleston, Ill	Mar. 25, 1873	137, 235
Crib, Reversible	A. E. Eaton	Portland, Me	May 23, 1871	115, 042
Crib, Traveling	S. L. Mercer	Washington, D. C	Sept. 24, 1872	131, 698
Cribs or cradles, Means of rocking	J. S. Ryan	Berlin, Wis	Jan. 1, 1867	60, 793
Cribbage-board, &c	J. Gill	New York, N. Y	Dec. 22, 1863	40, 999
Cribbing-preventer	B. J. Davis and I. S. Cramer	Sergeantsville, N. J	Mar. 10, 1868	75, 382
Cribbing-preventer	M. H. Sullivan	Providence, R. I	May 19, 1868	78, 155
Cricket and commode	J. C. Knowles	New Bedford, Mass	May 31, 1870	103, 754
Cricket-bat	M. Doherty	Boston, Mass	Feb. 22, 1859	23, 017
Cricket-wicket	W. Hanlon	Philadelphia, Pa	July 23, 1861	32, 869
Crimping and fluting machine	R. Werner	Hoboken, N. J	Jan. 7, 1873	134, 621
Crimping and fluting machine	R. Werner	Hoboken, N. J	June 24, 1873	140, 326
Crimping-brake	J. Howe	Milford, Mass	Aug. 31, 1869	94, 213
Crimping clamp	L. Hill	North Brookfield, Mass	Aug. 25, 1868	81, 503
Crimping-clamp	J. Kunstler	Collinsville, Conn	July 25, 1871	117, 304
Crimping-clamp	P. A. Schoellhorn	Buffalo, N. Y	Aug. 19, 1873	141, 956
Crimping-forms, Machine for shaping	J. H. Jellison	Milford, Mass	Sept. 19, 1865	50, 074
Crimping-machine	D. Bissell	Detroit, Mich	Aug. 28, 1860	29, 757
Crimping-machine	W. Butterfield	Boston, Mass	Mar. 8, 1870	100, 498
Crimping-machine	S. G. Cabell	Quincy, Ill	July 17, 1866	56, 365
Crimping-machine	L. Crevissier and L. Lecamp	Reims, France	Oct. 29, 1872	132, 637
Crimping-machine	J. P. Jamison	New York, N. Y	May 15, 1866	54, 735
Crimping-machine	S. W. Jamison	New York, N. Y	Feb. 15, 1870	99, 906
Crimping-machine	J. Joslyn	Canton, N. Y	Feb. 19, 1867	62, 208
Crimping-machine	M. R. Lemman and W. A. L. Kirk.	Hamilton, Ohio	Sept. 26, 1871	119, 234
Crimping-machine	L. P. Lum	Newburgh, N. Y	Dec. 9, 1873	145, 303
Crimping-machine	R. A. York	Reading, Mich	Apr. 7, 1868	76, 576
Crimping-pin	C. N. and A. C. Tyler	Buffalo, N. Y	Nov. 10, 1868	83, 892
Crinoline-clip	H. Fletcher	London, England	Apr. 29, 1862	35, 091
Crocheting and knitting needles, Machine for forming barbs of.	J. P. Tirrell	North Bridgewater, Mass	Apr. 2, 1867	63, 442
Crockery stilts	P. Pointon	Trenton, N. J	Oct. 13, 1863	40, 310
Crockery-ware, Heated metallic mold for forming.	E. Leak, H. Moore, and J. Taylor.	Trenton, N. J	May 21, 1872	127, 074
Cromamometro	P. Pettinos	Philadelphia, Pa	Mar. 3, 1829	......
Croquet, Apparatus for parlor	A. P. Eastman	Washington, D. C	Nov. 8, 1870	109, 120
Croquet, Apparatus for parlor	W. S. Messinger	Roxbury, Mass	May 29, 1866	55, 139
Croquet-arch	F. M. Clarke	Washington, D. C	Aug. 8, 1871	117, 865
Croquet-ball, Rubber	J. H. Tuttle	East Hampton, Mass	Jan. 23, 1872	122, 926
Croquet-board	J. Federhen	Boston, Mass	Nov. 6, 1866	59, 376
Croquet-boards, Method of mounting	H. R. Heyl	Philadelphia, Pa	June 14, 1870	104, 151
Croquet-mallet	S. Byrnes	Boston, Mass	Nov. 21, 1865	51, 016
Croquet-mallet	E. A. Ross	Albany, N. Y	Mar. 4, 1873	136, 388
Croquet-wicket	F. W. Smith, jr	Bridgeport, Conn	Mar. 30, 1869	88, 335
Croquet-wicket driver	E. A. Barker	Providence, R. I	Aug. 26, 1873	142, 072
Croqueterie	L. and M. Bradley	Springfield, Mass	May 22, 1866	54, 848
Croqueterie	G. L. Morse	Harrison, N. J	Feb. 26, 1867	62, 495
Cross-bar lock for doors, &c	J. E. Hanger	Staunton, Va	June 16, 1868	78, 960
Cross-bow	E. C. H. Nye	Acushnet, Mass	Mar. 26, 1867	63, 172
Cross-head	J. West	Bethlehem, Pa	May 1, 1866	54, 457
Cross-head shifter	C. R. Joyce	Alexandria, Va	Feb. 8, 1870	99, 686
Cross, Memorial	E. W. Hall	New York, N. Y	Feb. 13, 1872	123, 694
Crosses, Mode of mounting ornamental	W. B. Bennett	Providence, R. I	Aug. 17, 1869	93, 664
Croup and other diseases, Apparatus for treating	G. Déclat	Paris, France	Dec. 7, 1869	97, 613
Crout-cutter	J. G. Schwarz	Indianapolis, Ind	Aug. 31, 1869	94, 244
Crow and tamping bar	A. Wright and A. F. Tew	Westfield, N. Y	May 6, 1873	138, 599
Crow-bar, Compound	I. J. Cole	Piermont, N. Y	Jan. 23, 1855	12, 288
Crow-killer	N. J. Tilghman	Salisbury, Md	July 5, 1853	9, 835
Croze and howel, Cooper's	A. Busenger	Mount Solon, Va	Oct. 3, 1871	119, 570
Croze, Cooper's	J. F. Applegate and C. Feiock	New Albany, Ind	Apr. 25, 1871	114, 089
Croze, Cooper's	V. Applegate, M. P. Jacobs, and S. F. Roby.	Harrison County, Ind	Feb. 28, 1871	112, 205
Croze, Cooper's	J. Bailey	Springport, Ind	June 17, 1873	139, 852
Croze, Cooper's	C. O. Cook	Rockford, Ill	Aug. 11, 1868	80, 917
Croze, Cooper's	J. C. Hofer	Bellaire, Ohio	June 9, 1868	78, 740
Crozing-knives, Method of operating	G. Finn	Oswego, N. Y	Sept. 20, 1859	25, 498
Crozing-machine	V. W. Houck	Buffalo, N. Y	June 10, 1862	38, 519
Crozing-machine	W. R. Middleton	Cleveland, Ohio	June 13, 1871	115, 974
Crozing-machine	A. Wilbur	Lancaster, Pa	Mar. 21, 1854	10, 684
Crucible and metal-heating furnace	B. A. Mason	New York, N. Y	Dec. 3, 1872	133, 538
Crucible, Black-lead	G. Nimmo	Jersey City, N. J	May 31, 1864	42, 959
Crucible for melting iron and steel	J. E. Atwood	Pittsburgh, Pa	Oct. 8, 1870	108, 084
Crucible for melting iron, steel, &c	W. F. Dunbarr	Pittsburgh, Pa	Jan. 9, 1872	122, 585
Crucible for melting metals, &c	J. L. North	Ansonia, Conn	Feb. 27, 1872	124, 076
Crucible for melting metals	E. R. Playle	Great Bend, Pa	Mar. 23, 1869	88, 072
Crucible for melting metals, &c	R. Yeilding	Detroit, Mich	Nov. 7, 1871	120, 807
Crucible for metallic baths	B. S. Stokes	Manchester, N. H	July 25, 1865	49, 009

Index of patents issued from the United States Patent Office from 1790 *to* 1873, *inclusive*—Continued.

Invention.	Inventor.	Residence.	Date.	No.
Crucible-making machine	S. R. Thompson	Portsmouth, N. H	Nov. 27, 1866	60, 090
Crucible-mold	T. G. French	Jersey City, N. J	Nov. 7, 1865	50, 812
Crucible-repairing process	W. F. Sherman	Bucksport, Me	Feb. 16, 1869	87, 075
Crucible-stand for fusing metals	J. C. McManus	Providence, R. I	June 14, 1870	104, 179
Crucibles and in melting steel, Mode of repairing plumbago	N. Washburn	Worcester, Mass	July 13, 1869	92, 676
Crucibles and other articles of earthenware, Mode of forming.	J. Akrill	Williamsburgh, N. Y	Oct. 26, 1852	9, 351
Crucibles and potter's ware, Molding	A. Newkumet	Philadelphia, Pa	Jan. 26, 1869	86, 243
Crucibles, Apparatus for making plumbago	W. Smith	Pittsburgh, Pa	Nov. 3, 1863	40, 506
Crucibles, Apparatus for molding	J. L. Presby	Taunton, Mass	Apr. 24, 1866	54, 204
Crucibles, Drying and preparing	G. Nimmo	Jersey City, N. J	July 11, 1865	48, 713
Crucibles, glass-melting pots, &c., Composition for making.	W. A. Fischer	Allegheny City, Pa	Nov. 7, 1871	120, 579
Crucibles, Machine for making	J. Winkle	Pittsburgh, Pa	Jan. 12, 1869	85, 775
Crucibles, Making	J. Dalliba	Watervliet, N. Y	Sept. 26, 1823	....
Crucibles, Manufacture of	A. K. Eaton	Piermont, N. Y	Sept. 14, 1869	94, 726
Crucibles, Manufacture of	G. Nimmo	Jersey City, N. J	Aug. 1, 1865	49, 141
Crucibles, Molding	G. Nimmo	Jersey City, N. J	Aug. 1, 1865	49, 140
Cruet	T. H. Mead	Boston, Mass	Jan. 4, 1870	98, 511
Cruet or bottle for caster	J. O. Mead	Philadelphia, Pa	Feb. 15, 1859	22, 970
Cruet or bottle, Vinegar	G. W. and G. H. Simmons	Bennington, Vt	May 17, 1859	24, 083
Cruet or decanter	A. H. Newton	Worcester, Mass	Apr. 8, 1862	34, 899
Cruet-stand	J. W. Ells	Pittsburgh, Pa	Feb. 21, 1871	111, 920
Crumb-remover	A. B. Hoffman	Roxbury, Mass	Nov. 27, 1866	60, 009
Crupper	F. Howes	Boston, Mass	May 30, 1865	47, 953
Crupper, Horse	P. A. La France	Elmira, N. Y	Feb. 23, 1869	87, 269
Crupper, Spring	E. Powell	Conneautville, Pa	Jan. 1, 1867	60, 784
Crusher:				
See Bark-crusher.				
Beef-steak crusher.				
Cane-crusher.				
Clod-crusher.				
Corn and cob crusher.				
Culinary crusher.				
Drug-crusher.				
Fruit and vegetable crusher.				
Grape-crusher.				
Ice-crusher.				
Meat-crusher.				
Ore-crusher.				
Quartz-crusher.				
Rock and ore crusher.				
Stalk-crusher.				
Stone-crusher.				
Vegetable-crusher.				
Crusher and press for culinary purposes	G. B. Fowler	Brooklyn, N. Y	Mar. 4, 1873	136, 498
Crusher, cultivator, and harrow, Combined	J. B. McClean and C. A. Mayes	Franklin, Ky	Dec. 31, 1872	134, 388
Crusher, harrow, and roller combined	J. Simpson	Charleston, Ill	Nov. 24, 1868	84, 386
Crushing and baling machine	J. Price, jr	Petaluma, Cal	Nov. 28, 1865	51, 212
Crushing and grinding machine	J. W. Rutter	Saint Louis, Mo	Mar. 23, 1869	88, 216
Crushing and grinding mill	J. Weigle	Swan Station, Pa	Feb. 6, 1855	12, 356
Crushing and grinding mill	A. P. Norton	Pittsburgh, Pa	Apr. 25, 1848	5, 533
Crushing and grinding rock, phosphates, &c., Machine for.	F J. Kimball	Philadelphia, Pa	May 2, 1871	114, 303
Crushing and pressing machine	C. and O. Waste	Cameron, Ill	June 7, 1864	43, 054
Crushing-machine	J. Rowe	Athens, Ala	Apr. 24, 1840	1, 560
Crushing machine, Rock, ore, &c	W. F. Goodwin and C. R. Squire	East New York and New York, N. Y.	Oct. 8, 1867	69, 656
Crushing-mill	J. W. Smith	Columbus, Ga	Mar. 18, 1873	137, 035
Crushing mill, Apple, sugar cane, &c	J. H. Ellis	Brooklyn, Pa	Feb. 4, 1862	34, 296
Crushing-mill, Pendulum	J. Hart	Sweden Centre, N. Y	Nov. 8, 1870	109, 006
Crush-mills, Flask for casting stamp-shoes for	H. Bolthoff	Central City, Colo	Oct. 25, 1870	108, 556
Crushing-press	E. Chipman	New York, N. Y	Oct. 20, 1863	40, 325
Crushing, rolling, and kneading machine	C. Bates	Kingston, Mass	Aug. 7, 1866	56, 877
Crutch	G. T. Allamby and J. G. Bugbee	Bangor, Me	Apr. 18, 1865	47, 265
Crutch	G. T. Allamby and J. G. Bugbee	Bangor, Me	Dec. 26, 1865	51, 677
Crutch	A. Bickel	Philadelphia, Pa	Jan. 17, 1860	26, 829
Crutch	A. Bickel	Philadelphia, Pa	May 16, 1865	47, 695
Crutch	S A. Bracket	Boston, Mass	Dec. 31, 1872	134, 351
Crutch	A. E. Brown	Baltimore, Md	June 30, 1868	79, 305
Crutch	L. Crandall	Plainfield, N. J	Oct. 27, 1863	40, 443
Crutch	L. Crandall	New York, N. Y	May 21, 1872	127, 028
Crutch	H. G. Davis	New York, N. Y	Jan. 13, 1863	37, 387
Crutch	J. S. Gallagher, jr	Washington, D. C	Jan. 4, 1853	9, 518
Crutch	T. E. Gordon	Cleveland, Ohio	Nov. 29, 1864	45, 242
Crutch	T. E. Gordon	Brooklyn, Ohio	Jan. 3, 1865	45, 709
Crutch	J. A. Lobb	Independence, Mo	May 12, 1868	77, 824
Crutch	E. T. Pearl	Milwaukee, Wis	Nov. 28, 1871	121, 254
Crutch	J. C. Rhodes	Stillwater, Minn	Jan. 14, 1868	73, 386
Crutch	D. Ring	Damariscotta, Me	Sept. 1, 1863	39, 751
Crutch	R. W. Ware	Chicago, Ill	Aug. 21, 1866	57, 416
Crutch	J. D. W. Wemple	Albany, N. Y	Sept. 8, 1863	39, 854
Crutch	J. Whittemore	South Reading, Mass	Oct. 21, 1862	36, 743
Crutch	P. R. Wimer	Trenton, N. J	May 10, 1870	102, 901
Crutch-foot, Revolving	J. G. Bugbee	Bangor, Me	Aug. 15, 1865	49, 371
Crutch-handle	S. Kreger	Philadelphia, Pa	July 4, 1871	116, 603
Crutch or perambulator, Wheel	S A. Darrack	Orange, N. J	Aug. 6, 1872	130, 283
Cryptographic alphabet	E. H. Hawley	Signal Corps, Army of Potomac.	July 11, 1865	48, 681
Crystal-cutting machine	P. Pryibil	New York, N. Y	Sept. 23, 1862	36, 548
Crystal fountain	J. C. Johnson	Louisville, Ky	Apr. 28, 1868	77, 290
Crystallized metallic surfaces, Protecting and beautifying.	H. M. Johnston	New York, N. Y	July 19, 1870	105, 576
Crystallized salts upon glass, mica, &c., Preserving and beautifying.	O. S. Follet	New York, N. Y	July 19, 1870	105, 560
Cue-cutter	J. Huber	Saint Louis, Mo	May 10, 1870	102, 821
Cue-trimmer	D. Aldrich	New York, N. Y	May 5, 1868	77, 436

Index of patents issued from the United States Patent Office from 1790 *to* 1873, *inclusive*—Continued.

Invention.	Inventor.	Residence.	Date.	No.
Cuff	L. H. Foy	New Haven, Conn	Aug. 27, 1872	130, 801
Cuff	I. Levine	New York, N. Y	Mar. 5, 1867	62, 648
Cuff	E. E. Mack	Albany, N. Y	Dec. 30, 1873	145, 957
Cuff	I. M. Post	Philadelphia, Pa	Mar. 11, 1873	136, 760
Cuff-fastener	A. Douglas	English Neighborhood, N. J	Apr. 19, 1864	42, 434
Cuff-holder	G. W. James	Jersey City, N. J	Oct. 28, 1873	144, 101
Cuff, Reversible	L. H. Foy	New Haven, Conn	Apr. 15, 1873	137, 967
Cuff-supporter	T. Cogswell	Boston, Mass	Mar. 22, 1870	100, 985
Cuffs, Fabric for	H. H. Thayer and W. H. Hart, jr	Philadelphia, Pa	May 2, 1871	114, 490
Culinary apparatus	R. H. Cazier	Philadelphia, Pa	July 22, 1873	141, 114
Culinary apparatus	J. S. Field	Brooklyn, N. Y	Sept. 8, 1868	81, 887
Culinary apparatus	I. Former	Grafton, Va	Oct. 2, 1860	30, 214
Culinary apparatus	G. B. Isham	Burlington, Vt	May 12, 1868	77, 884
Culinary apparatus	T. W. Moore	New York, N. Y	Jan. 21, 1868	73, 630
Culinary apparatus	F. Morandi	Malden, Mass	Mar. 31, 1868	76, 231
Culinary apparatus	W. W. S. Orbeton	Bradford, Mass	July 28, 1868	80, 303
Culinary apparatus	V. M. Thomas	Brandon, Vt	July 28, 1868	80, 315
Culinary apparatus	J. Van	Cincinnati, Ohio	Aug. 23, 1870	106, 745
Culinary apparatus	B. Wardwell	Providence, R. I	May 4, 1869	89, 613
Culinary apparatus	H. A. Zopff	Milwaukee, Wis	May 5, 1868	77, 702
Culinary apparatus, Steam	C. I. Paine	Young America, Ill	Jan. 18, 1870	98, 881
Culinary baking apparatus	P. Wilcox	Springfield, Mass	Feb. 18, 1829	....
Culinary boiler	E. K. Ames	Chicago, Ill	Feb. 2, 1869	86, 488
Culinary boiler	A. Anderson	London, Canada	Feb. 13, 1872	123, 662
Culinary boiler	G. W. Bliss	Brooklyn, N. Y	Mar. 1, 1870	100, 361
Culinary boiler	P. Boesn and M. Bedessem	Kenosha, Wis	Mar. 28, 1871	113, 011
Culinary boiler	J. Bowlin	Charlestown, Mass	Dec. 17, 1872	133, 918
Culinary boiler	I. S. Bunnell	Carbondale, Pa	May 10, 1870	102, 765
Culinary boiler	F. M. Carnes	Rochester, N. Y	Mar. 15, 1870	100, 853
Culinary boiler	J. H. Corey	New York, N. Y	Oct. 14, 1873	143, 673
Culinary boiler	W. H. Crosby	Washington, D. C	Sept. 30, 1862	36, 558
Culinary boiler	T. Dare	Osceola, Iowa	July 30, 1861	32, 964
Culinary boiler	F. W. Dembois	East Saginaw, Mich	Mar. 27, 1866	53, 421
Culinary boiler	J. D. Durham	Dalston, England	Feb. 20, 1872	123, 876
Culinary boiler	S. S. Fitch	New York, N. Y	Apr. 8, 1873	137, 604
Culinary boiler	B. French	Rochester, N. Y	Dec. 12, 1871	121, 770
Culinary boiler	J. Gibbs	Opelousas, La	Feb. 20, 1872	123, 820
Culinary boiler	W. H. Henderson	Franklin, Ind	Sept. 3, 1867	68, 564
Culinary boiler	T. D. Ingersoll	Monroe, Mich	Mar. 26, 1861	31, 807
Culinary boiler	I. Kinney	London, Canada	Aug. 27, 1872	130, 921
Culinary boiler	R. B. Lewis	Quincy, Ill	Mar. 25, 1873	137, 140
Culinary boiler	S. P. Loomis	Philadelphia, Pa	Feb. 2, 1869	86, 428
Culinary boiler	L. McLellan	Gorham, Me	Apr. 27, 1869	89, 419
Culinary boiler	F. Meyer	New York, N. Y	May 30, 1871	115, 338
Culinary boiler	O. Poole	Detroit, Mich	Feb. 11, 1868	74, 419
Culinary boiler	T. T. Prosser	Chicago, Ill	July 24, 1866	56, 609
Culinary boiler	R. Russell	New Bedford, Mass	Jan. 2, 1872	122, 407
Culinary boiler	S. E. Saul	New York, N. Y	May 22, 1866	54, 964
Culinary boiler	J. R. Stafford	Cleveland, Ohio	June 12, 1847	5, 153
Culinary boiler	P. L. Suine	Shirleysburgh, Pa	June 30, 1863	39, 077
Culinary boiler	J. S. Totten	Lebanon, Ohio	Nov. 2, 1869	96, 509
Culinary boiler	I. H. West and T. L. Camp	Evans, N. Y	Feb. 18, 1868	74, 647
Culinary boiler	R. C. Whitehouse	Hogdon's Mills, Me	July 25, 1871	117, 356
Culinary boiler	F. Wilcox	Newark, N. J	Apr. 20, 1869	89, 189
Culinary boiler	J. Zimmerman	Royalton Centre, N. Y	Dec. 17, 1867	72, 354
Culinary-boiler cover	P. J. and J. Abbott	Dexter, Me	Oct. 11, 1870	108, 082
Culinary boiler, Portable	A. W. Spencer	Cazenovia, N. Y	June 19, 1835	....
Culinary cabinet	J. Billings	Brighton, Mass	Mar. 20, 1866	53, 260
Culinary fixtures for anthracite coal, Construction of	J. F. Walters	Philadelphia, Pa	June 8, 1827	....
Culinary fork and spoon	A. Hills	Naugatuck, Conn	Apr. 10, 1860	27, 802
Culinary furnace	J. Reed	Marshfield, Mass	Jan. 5, 1831	....
Culinary pot	J. D. Flansburgh	Philadelphia, Pa	Jan. 21, 1862	34, 202
Culinary pot	J. S. Kidd and M. Melville	Brooklyn, N. Y	July 16, 1872	129, 034
Culinary sink	Z. Hunt	Hudson, N. Y	July 10, 1866	56, 225
Culinary uses, Compound for	H. W. Bradley	Binghamton, N. Y	Jan. 3, 1871	110, 626
Culinary uses, Shortening for	H. W. Bradley	Plainfield, N. J	Oct. 17, 1871	120, 026
Culinary utensil	H. G. Dunkelberger	Shamokin, Pa	Nov. 11, 1873	144, 518
Culinary utensils for broiling and frying	S. Lee	Taunton, Mass	June 11, 1872	127, 960
Culinary vessel	J. C. Nobles	Ilion, N. Y	Nov. 7, 1871	120, 769
Culinary vessel	W. A. Barlow	Chicago, Ill	Dec. 16, 1873	145, 550
Culinary vessel	W. H. Bennett	New York, N. Y	Jan. 21, 1868	73, 568
Culinary vessel	W. H. Bloom	Tiffin, Ohio	Apr. 25, 1871	113, 971
Culinary vessel	W. J. Burnett	Cairo, Ill	Apr. 9, 1872	125, 538
Culinary vessel	J. H. Chappell	New York, N. Y	Jan. 31, 1871	111, 317
Culinary vessel	S. W. M. Chattaway	Middletown, Conn	June 6, 1871	115, 703
Culinary vessel	E. Clark	Buffalo, N. Y	July 26, 1870	105, 779
Culinary vessel	C. E. Corbett	Binghamton, N. Y	Mar. 15, 1870	100, 863
Culinary vessel	H. L. Duncklee	Boston, Mass	Apr. 7, 1868	76, 310
Culinary vessel	C. Estabrooks	Calais, Me	Mar. 29, 1870	101, 244
Culinary vessel	C. W. Fuller	Earlville, Ill	June 2, 1868	78, 587
Culinary vessel	C. W. Hermance	Schuylerville, N. Y	May 17, 1870	103, 180
Culinary vessel	L. Hermance	Lansingburgh, N. Y	Mar. 21, 1871	112, 916
Culinary vessel	B. M. Hermance	Troy, N. Y	Dec. 13, 1870	110, 039
Culinary vessel	C. Jessup	New Haven, Conn	Mar. 7, 1871	112, 352
Culinary vessel	C. A. Johnson	Des Moines, Iowa	July 7, 1868	79, 574
Culinary vessel	A. C. Kasson	Milwaukee, Wis	Nov. 17, 1868	84, 194
Culinary vessel	G. Landrine	Jersey City, N. J	Feb. 22, 1870	100, 045
Culinary vessel	S. Lee	Taunton, Mass	Aug. 29, 1871	118, 402
Culinary vessel	A. F. Marston	Clinton, La	Sept. 1, 1868	81, 658
Culinary vessel	O. M. Mitchell	Marathon, N. Y	July 20, 1869	92, 867
Culinary vessel	F. Morandi	Malden, Mass	Jan. 31, 1871	111, 369
Culinary vessel	W. H. Murch	Portland, Me	Apr. 4, 1871	113, 442
Culinary vessel	G. L. Page	Wallingford, Conn	July 16, 1872	129, 293
Culinary vessel	J. W. Patterson	Cincinnati, Ohio	Feb. 4, 1868	74, 123
Culinary vessel	H. Poole	Richmond, Ind	Aug. 25, 1868	81, 534
Culinary vessel	E. Ripley	Troy, N. Y	Jan. 5, 1864	41, 095
Culinary vessel	J. C. Smith	Rochester, N. Y	May 3, 1870	102, 608
Culinary vessel	H. Zachgo	South Brooklyn, N. Y	June 22, 1869	91, 697

Index of patents issued from the United States Patent Office from 1790 *to* 1873, *inclusive*—Continued.

Invention.	Inventor.	Residence.	Date.	No.
Culinary vessel, Nickel-lined	S. C. Moore	Boston, Mass	July 6, 1869	92, 337
Culinary vessel, Steam	C. Waterman	Providence, R. I	Nov. 2, 1869	96, 518
Culinary vessels, Pivoted lid for	J. H. McConnell	Beaver Falls, Pa	Nov. 26, 1872	133, 326
Culinary vessels, Sheet-copper plate for	A. O'Neil	Portsmouth, Ohio	Aug. 27, 1867	68, 331
Cultivating and improving earth	J. Hall		Nov. 19, 1814	
Cultivating and tilling the ground	J. D. Prescott	Chesterville, Me	May 15, 1834	
Cultivating corn and beans, Machine for	E. Spooner		Jan. 25, 1799	
Cultivating-hook	J. F. Leitch	Oxford, N. Y	June 7, 1870	104, 042
Cultivating implement, Garden	D. Mack	Barnesville, Kans	Nov. 25, 1873	144, 912
Cultivating land by steam	J. Fowler, jr	Cornhill, England	Jan. 10, 1865	45, 892
Cultivating-machine	C. G. Grabo	Greenfield, Mich	May 26, 1863	38, 676
Cultivating-machine	W. F. Quimby and G. G. Lobdell.	Stanton and Wilmington, Del.	Apr. 7, 1863	38, 124
Cultivating machine, Cotton	W. Altick	Dayton, Ohio	Apr. 23, 1867	64, 057
Cultivating machine, Cotton	W. P. Sample	Bedford County, Tenn	Sept. 3, 1831	
Cultivator	H. A. Adams	Sandwich, Ill	May 31, 1870	103, 537
Cultivator	J. Adams	Manteno, Ill	June 21, 1870	104, 399
Cultivator	W. Adams	Detroit, Mich	Aug. 3, 1858	21, 055
Cultivator	A. Agnew and W. Morrison	Chester County and Chadd's Ford, Pa.	July 10, 1860	29, 043
Cultivator	M. Alden	Auburn, N. Y	May 10, 1859	23, 886
Cultivator	J. T. D. Alexander	Maryenna, Tex	Jan. 15, 1861	31, 101
Cultivator	D. R. Allen	Cumberland, Me	Feb. 26, 1867	62, 385
Cultivator	A. H. Allison	Charlottesville, Ind	Jan. 29, 1867	61, 649
Cultivator	A. H. Allison	Charlottesville, Ind	Oct. 15, 1867	69, 743
Cultivator	A. H. Allison	Charlottesville, Ind	Sept. 15, 1868	82, 060
Cultivator	J. H. Allison	Eureka, Ill	May 7, 1867	64, 469
Cultivator	P. H. Allstott	Jeffersonville, Ind	July 18, 1865	48, 783
Cultivator	C. Alvord	Westford, Wis	Dec. 15, 1868	84, 931
Cultivator	C. Alvord	Courtland, Wis	Apr. 26, 1870	102, 201
Cultivator	W. D. Ament	Muscatine, Iowa	Jan. 10, 1865	45, 807
Cultivator	W. D. Ament	Muscatine, Iowa	July 25, 1865	48, 884
Cultivator	A. Anderson	London, Canada	Sept. 18, 1866	58, 188
Cultivator	J. H. and E. H. Anderson	Easton, Md	Nov. 27, 1860	30, 709
Cultivator	A. P. Anderson and B. Edwards	Princeton, Ill	July 10, 1866	56, 157
Cultivator	W. W. Andrew	La Porte, Ind	Mar. 5, 1872	124, 241
Cultivator	W. J. Andrews	Columbia, Tenn	Sept. 11, 1866	57, 841
Cultivator	W. J. Andrews	Columbia, Tenn	Feb. 26, 1867	62, 386
Cultivator	J. and G. W. Ansley	Marengo, Mich	Nov. 28, 1871	121, 267
Cultivator	B. Anyan	Fitchville, Ohio	Mar. 24, 1868	75, 826
Cultivator	J. Armstrong	Elmira, Ill	Dec. 26, 1865	51, 680
Cultivator	J. Armstrong, jr	Elmira, Ill	Feb. 19, 1867	62, 109
Cultivator	T. Arndt	Mount Joy, Pa	Nov. 24, 1868	84, 338
Cultivator	O. J. Arnold	Mount Ida, Wis	Feb. 26, 1867	62, 387
Cultivator	H. B. Arnoldt and J. Grimm	Saint Louis, Mo	Feb. 4, 1868	74, 031
Cultivator	W. J. Arrington	Louisville, Ga	Aug. 15, 1871	117, 961
Cultivator	J. N. Arvin	Valparaiso, Ind	July 24, 1866	56, 511
Cultivator	J. Austin	Rockford, Ill	Mar. 15, 1864	41, 899
Cultivator	I. Avery	Ottawa, Ill	July 17, 1866	56, 344
Cultivator	R. H. Avery	Galesburgh, Ill	Dec. 27, 1870	110, 419
Cultivator	R. H. Avery	Galesburgh, Ill	May 28, 1872	127, 293
Cultivator	J. Ayres and M. F. Hunt	Bushnell, Ill	July 23, 1872	129, 642
Cultivator	D. C. Baker	Fulton, N. Y	May 13, 1873	138, 838
Cultivator	J. M Baker	Fayetteville, Tex	Sept. 7, 1869	94, 542
Cultivator	N. Baker	Flowerfield, Mich	Aug. 15, 1848	5, 715
Cultivator	W. T. Baker	Lancaster, Tex	Nov. 2, 1869	96, 379
Cultivator	L. Baldwin	Stone Mills, N. Y	Apr. 8, 1873	137, 649
Cultivator	P. O. Baldwin	Spring Lake, Mich	July 18, 1871	117, 034
Cultivator	W. M. Ball	Morristown, Ind	Apr. 16, 1867	63, 830
Cultivator	W. M. Ball	Morristown, Ind	July 23, 1867	66, 934
Cultivator	T. J. Ball and J. Post	Pittsfield, Mich	Apr. 6, 1852	8, 850
Cultivator	J. T. Baltimore	Marble Rock, Iowa	Dec. 31, 1867	72, 780
Cultivator	W. Bancroft	Whiteford, Ohio	Nov. 14, 1854	11, 924
Cultivator	J. Banks	Dadeville, Ala	Mar. 30, 1858	19, 742
Cultivator	W. Bankson	Mount Pleasant, Iowa	Feb. 28, 1865	46, 537
Cultivator	I. Barber, jr	La Porte, Ind	Jan. 3, 1865	45, 687
Cultivator	I. Barber	La Porte, Ind	Dec. 3, 1867	71, 834
Cultivator	J. A. Bardel	Freeport, Ill	Sept. 12, 1865	49, 845
Cultivator	J. H. Barley	Longwood, Mo	Sept. 4, 1866	57, 660
Cultivator	J. H. Barley	Sedalia, Mo	Sept. 24, 1867	69, 162
Cultivator	H. Barnes	Burlington, Wis	May 1, 1866	54, 280
Cultivator	N. Barnes	East Hampton, N. Y	Sept. 10, 1840	1, 779
Cultivator	M. Barnett and E. Wood	Hardinsburgh, Ind	Nov. 12, 1867	70, 779
Cultivator	F. Barney	Bloomington, Ill	July 12, 1864	43, 470
Cultivator	A. S. Barnwell	Savannah, Ga	Mar. 12, 1867	62, 808
Cultivator	M. J. Barr	Centreville, Ind	Mar. 5, 1872	124, 316
Cultivator	O. Barr and F. F. Cox	Beloit, Wis	May 15, 1866	54, 669
Cultivator	F. M. Barrier	Stevenson, Ala	Feb. 4, 1868	73, 945
Cultivator	H. Barsalow	Saint Anne, Ill	Jan. 8, 1867	61, 040
Cultivator	L. B Barton	Metamora, Ill	Apr. 25, 1865	47, 380
Cultivator	L. B. Barton	Metamora, Ill	Oct. 17, 1865	50, 439
Cultivator	W. E. Bates	Elmore, Ill	May 16, 1865	47, 693
Cultivator	I. Bates, A. Wood, and D. Wells	Adams, N. Y	May 28, 1842	2, 647
Cultivator	J. W. and L. Batson	Clarksville, Md	May 24, 1859	24, 089
Cultivator	J. N. Baumann	Mascatine, Iowa	July 29, 1862	35, 985
Cultivator	W. T. Bazemore	Bibb County, Ga	Aug. 29, 1854	11, 593
Cultivator	C. Beach and T. Brown	Jacksontown, Ohio	Feb. 12, 1861	31, 367
Cultivator	T. Beale	New Milford, Ill	Oct. 9, 1866	58, 578
Cultivator	H. Bean and J. D. Tyson	Schuylkill and Lower Providence, Pa.	Oct. 15, 1867	69, 750
Cultivator	E. C. Bean and F. N. Welden	Rockford, Ill	Nov. 2, 1869	96, 382
Cultivator	G. Beeching	Augusta, N. Y	Oct. 17, 1848	5, 859
Cultivator	J. Behet	Rockford, Ill	Feb. 18, 1873	135, 964
Cultivator	C. Belden	Middlebury, Ohio	Feb. 13, 1866	52, 515
Cultivator	H. M. Belden	Farmington, Ohio	Dec. 4, 1860	30, 796
Cultivator	A. C. Belt	Goresville, Va	Nov. 12, 1867	70, 785
Cultivator	I. A. Benedict	West Springfield, Pa	June 15, 1869	91, 303
Cultivator	D. G. Benner	Holmesville, Ohio	Feb. 8, 1870	99, 625

Index of patents issued from the United States Patent Office from 1790 *to* 1873, *inclusive*—Continued.

Invention.	Inventor.	Residence.	Date.	No.
Cultivator	A. Bennett	Rockford, Ill	Jan. 14, 1868	73, 224
Cultivator	J. S. and W. Benson	Lebanon, Pa	Sept. 3, 1867	68, 549
Cultivator	L. B. Benton	Penn Yan, N. Y	Nov. 27, 1860	30, 711
Cultivator	J. Bergen	Plain Township, Ohio	Aug. 28, 1866	57, 619
Cultivator	J. Bergen	Canton, Ohio	Apr. 23, 1867	64, 061
Cultivator	T. F. Bertrand and P. Sames	Rockford, Ill	Apr. 24, 1866	54, 099
Cultivator	T. F. Bertrand and P. Sames	Rockford, Ill	Jan. 1, 1867	60, 916
Cultivator	J. T. Bever	Bethel, Ill	June 27, 1865	48, 358
Cultivator	J. T. Bever	Bethel, Ill	Dec. 5, 1865	51, 286
Cultivator	C. Billups	Norfolk, Va	May 7, 1872	126, 513
Cultivator	C. Billups	Norfolk, Va	Jan. 21, 1873	135, 070
Cultivator	C. Bird	Ackley, Iowa	Mar. 29, 1870	101, 216
Cultivator	J. C. Bird	Rising Sun, Md	June 11, 1867	65, 719
Cultivator	A. M. Black	Auburn, Ill	Nov. 11, 1862	36, 889
Cultivator	A. M. Black	Auburn, Ill	July 3, 1866	55, 993
Cultivator	T. Black	Princeville, Ill	Aug. 28, 1860	29, 758
Cultivator	W. R. Blanchard	Hertford, N. C	Jan. 26, 1869	86, 205
Cultivator	B. C. Blomsten	Waupaca, Wis	Aug. 22, 1871	118, 186
Cultivator	A. R. Blood, A. Hathaway, and V. R. Beach.	Independence, Iowa	June 2, 1868	78, 417
Cultivator	A. R. Blood, A. Hathaway, and V. R. Beach.	Independence, Iowa	June 30, 1868	79, 304
Cultivator	D. G. Blue	Winfield, Iowa	Jan. 6, 1863	37, 271
Cultivator	J. Bohan	New Hartford, Iowa	Oct. 12, 1869	95, 762
Cultivator	J. W. Booker	Fairmont, Ill	May 26, 1863	38, 651
Cultivator	J. W. Boosinger	Marine, Ill	Sept. 4, 1866	57, 668
Cultivator	E. Boughton	East Bloomfield, N. Y	Jan. 31, 1854	10, 467
Cultivator	A. Bouton	Napa, Cal	Oct. 3, 1865	50, 217
Cultivator	H. W. Bowen	Providence, R. I	June 4, 1872	127, 453
Cultivator	J. Bower	Dayton, Ohio	Apr. 19, 1870	101, 976
Cultivator	H. Bowers	New Hudson, Mich	Aug. 27, 1861	33, 129
Cultivator	D. W. Bowman	Tippecanoe City, Ohio	Dec. 5, 1871	121, 484
Cultivator	J. C. Boyd	Milroy, Ind	Jan. 14, 1868	73, 291
Cultivator	H. Boys	Rushville, Ind	May 28, 1867	65, 158
Cultivator	B. C. Bradley	Chicago, Ill	Feb. 4, 1873	135, 401
Cultivator	G. Bradley	Rockford, Ill	June 11, 1872	127, 674
Cultivator	S. Brady	Salona, Pa	Apr. 6, 1842	2, 540
Cultivator	B. F. Brate	New Scotland, N. Y	Mar. 4, 1873	136, 409
Cultivator	M. Breneman	East Donegal Township, Pa.	Jan. 29, 1867	61, 512
Cultivator	J. Brewer	Albany, Ill	May 16, 1865	47, 699
Cultivator	J. Brewer	Albany, Ill	Aug. 1, 1865	49, 075
Cultivator	E. Briggs	Medina, Ohio	July 17, 1860	29, 136
Cultivator	E. Briggs	Cleveland, Ohio	July 29, 1873	141, 199
Cultivator	T. E. C. Brinly	Louisville, Ky	Sept. 25, 1860	30, 119
Cultivator	T. E. C. Brinly	Louisville, Ky	June 20, 1871	116, 147
Cultivator	A. C. Brinser	Middletown, Pa	Aug. 24, 1869	94, 067
Cultivator	J. H. Brinton	Thornbury Township, Pa.	Dec. 15, 1868	84, 935
Cultivator	H. C. Bristol	Ravenna, Ohio	Apr. 9, 1867	63, 698
Cultivator	J. T. Brittain	Springfield, Ohio	Oct. 17, 1871	120, 028
Cultivator	S. W. Brock	Niantic, Ill	Nov. 23, 1869	97, 032
Cultivator	G. W. Bronson	Ottawa, Ill	Jan. 17, 1871	111, 037
Cultivator	J. E. Brooks	Gooding's Grove, Ill	Dec. 24, 1867	72, 598
Cultivator	R. M. Brooks	Greenville, Ga	June 26, 1860	28, 829
Cultivator	W. Brooks	Lexington, Ga	Nov. 5, 1872	132, 796
Cultivator	R. Brotten	Oskaloosa, Iowa	Sept. 10, 1861	33, 235
Cultivator	F. E. Brown	Hightstown, N. J	Mar. 25, 1862	34, 732
Cultivator	G. W. Brown	Tylersville, Ill	June 5, 1849	6, 511
Cultivator	G. W. Brown	Galesburgh, Ill	Aug. 28, 1860	29, 762
Cultivator	J. Brown	Lawn Ridge, Ill	Nov. 14, 1854	11, 929
Cultivator	W. P. Brown	Malta, Ohio	July 9, 1872	128, 701
Cultivator	M. Bruner, jr	Fremont, Ohio	Mar. 15, 1870	100, 720
Cultivator	R. C. Buckley	Peoria, Ill	Dec. 24, 1872	134, 248
Cultivator	M. H. Bucknall	Darien, Wis	Sept. 11, 1866	57, 856
Cultivator	C. J. Buckner	Paxton, Ill	Feb. 23, 1864	41, 677
Cultivator	R. Bullock	South Mills, N. C	July 10, 1866	56, 173
Cultivator	R. I. Burbank	Boston, Mass	Feb. 2, 1869	86, 502
Cultivator	J. M. Burke	Dansville, N. Y	Nov. 15, 1864	45, 019
Cultivator	J. Burnham	La Salle, Ill	Feb. 4, 1868	74, 044
Cultivator	J. Burnham and W. C. Lathrop	La Salle, Ill	Sept. 11, 1866	57, 858
Cultivator	J. Burns	Franklin, Ohio	July 28, 1863	39, 337
Cultivator	W. J. Burton	Turtle, Wis	Sept. 26, 1865	50, 093
Cultivator	N. Buttler	Otterville, Mo	Aug. 10, 1869	93, 412
Cultivator	S. S. Bushnel	Horicon, Wis	Jan. 14, 1873	134, 728
Cultivator	W. Bushnell	Easton, Pa	Mar. 27, 1860	27, 614
Cultivator	E. T. Bussell	Indianapolis, Ind	July 23, 1872	129, 712
Cultivator	H. F. Byerly	Clinton, Ill	May 22, 1866	54, 854
Cultivator	J. E. Byers	Butler, Pa	Mar. 5, 1872	124, 329
Cultivator	S. M. Cain and W. Stelfox	Austin, Tex	June 4, 1861	32, 468
Cultivator	G. Calkins	El Paso, Ill	Sept. 5, 1865	49, 715
Cultivator	J. F. Cameron	Livingston County, Mo	June 26, 1860	28, 831
Cultivator	J. Canfield	Sabula, Iowa	July 26, 1864	43, 635
Cultivator	J. Canfield and C. Hess	Washington and Lyons, Iowa.	Mar. 27, 1866	53, 408
Cultivator	T. F. Capp	Bloomington, Ill	May 23, 1871	115, 159
Cultivator	A. Carey	Rome, Ga	Feb. 14, 1860	27, 106
Cultivator	P. S. Carhart	Collamer, N. Y	June 10, 1862	35, 505
Cultivator	P. S. Carhart	Collamer, N. Y	May 9, 1865	47, 618
Cultivator	J. H. Carlow	Kidder, Mo	Dec. 27, 1870	110, 548
Cultivator	A. F. Carlson	Attica, Ind	Jan. 30, 1872	123, 235
Cultivator	D. F. Carr	East Union Township, Ohio.	June 8, 1869	91, 084
Cultivator	H. Carr	Wooster, Ohio	Mar. 8, 1870	100, 499
Cultivator	H. Carr	Wooster, Ohio	Mar. 8, 1870	100, 500
Cultivator	H. Carr	Wooster, Ohio	Mar. 8, 1870	100, 501
Cultivator	N., jr., and J. Carr	Monmouth, Ill	July 23, 1861	32, 858
Cultivator	N., jr., and J. Carr	Monmouth, Ill	Jan. 5, 1869	85, 509
Cultivator	C. H. and S. E. Carrington	Weymouth, Ohio	Oct. 12, 1858	21, 739

Index of patents issued from the United States Patent Office from 1790 to 1873, inclusive—Continued.

Invention.	Inventor.	Residence.	Date.	No.
Cultivator	N. C. Carter	Union City, Ind	Oct. 9, 1860	30, 295
Cultivator	J. Case	La Fayette, Ind	Feb. 2, 1864	41, 423
Cultivator	J. Case	Springfield, Ohio	Apr. 17, 1866	53, 947
Cultivator	J. Case	La Fayette, Ind	Jan. 31, 1871	111, 432
Cultivator	A. B. Cass	Muscatine, Iowa	June 24, 1862	35, 668
Cultivator	A. B. Cass	Chicago, Ill	Jan. 3, 1865	45, 700
Cultivator	L. J. Caswell	Scott Township, Ind	Nov. 6, 1866	59, 358
Cultivator	N. A. Cates	Thorndike, Me	Oct. 1, 1867	69, 314
Cultivator	W. W. Cato	Hicksford, Va	June 25, 1872	128, 360
Cultivator	I. N. Cauthorn	Carthage, Ill	Jan. 14, 1873	134, 732
Cultivator	R. S. Cavett	Bedias, Tex	Nov. 18, 1873	144, 740
Cultivator	J. Caylor	Fairview, Ind	July 15, 1873	140, 885
Cultivator	M. and J. Caywood	Peoria County, Ill	Feb. 16, 1869	86, 905
Cultivator	I. H. Chappell	Lawrence, Kans	Apr. 2, 1872	125, 271
Cultivator	I. H. Chappell and J. Montgomery.	Decatur, Ill	Dec. 1, 1868	84, 611
Cultivator	J. Charlton	Allegheny, Pa	June 26, 1860	28, 833
Cultivator	E. Children	Liberty, Wis	Oct. 2, 1866	58, 380
Cultivator	E. Children	Lancaster, Wis	Feb. 12, 1867	62, 004
Cultivator	J. G. Christopher	Byron, Ill	Apr. 10, 1860	27, 776
Cultivator	A. L. Chubb	Grand Rapids, Mich	Jan. 4, 1870	98, 560
Cultivator	D. Churchill	Ionia, Ill	Jan. 7, 1868	72, 975
Cultivator	D. Churchill and S. C. Brewer	Ionia, Ill	Oct. 17, 1865	50, 453
Cultivator	C. Clark	Andersonville, Ga	July 3, 1860	28, 968
Cultivator	C. A. Clark	Bloomfield, Iowa	Nov. 27, 1860	30, 720
Cultivator	C. M. Clark	Seward, Nebr	Apr. 23, 1872	125, 936
Cultivator	M. M. Clark	Industry, Ill	Dec. 15, 1863	40, 909
Cultivator	W. F. Clark	Hagaman's Mills, N. Y	Feb. 5, 1867	61, 805
Cultivator	G. D. Cleaveland	Flint, Ind	Sept. 10, 1872	131, 150
Cultivator	G. M. Clements	Kenduskeag, Me	Nov. 11, 1862	36, 894
Cultivator	A. P. Clements and J. C. Nealey	Monroe, Me	Apr. 26, 1870	102, 223
Cultivator	J. H. Clifton	Newcastle, Pa	Sept. 15, 1868	82, 089
Cultivator	L. Clifton	Barry, Ill	Nov. 10, 1868	83, 931
Cultivator	B. M. Close	West Camden, N. Y	June 15, 1869	91, 212
Cultivator	I. Cobb	Westminster, Vt	Dec. 7, 1829	....
Cultivator	B. C., T. W., and J. M. Cochran	Pana, Ill	June 18, 1867	65, 875
Cultivator	C. A. Cogswell	Maquoketa, Iowa	May 26, 1868	78, 186
Cultivator	D. C. Colby	Claremont, N. H	Nov. 11, 1862	36, 895
Cultivator	J. H. Coleman	Columbia, Mo	Jan. 19, 1869	86, 003
Cultivator	J. Collins	Farmington, Ill	Feb. 18, 1862	34, 406
Cultivator	A. B. Colver and J. Priest	Albany, Oreg	Jan. 28, 1873	135, 264
Cultivator	T. S. Cone and H. S. Potter	Oneida, Ill	May 28, 1861	32, 413
Cultivator	J. H. Conley	Moingona, Iowa	May 16, 1871	114, 767
Cultivator	A. Connelley	Milan, Ind	Mar. 23, 1869	88, 136
Cultivator	J. W. Connely	Charleston, Ill	Feb. 19, 1867	62, 185
Cultivator	J. W. Connely	Charleston, Ill	Sept. 24, 1867	69, 186
Cultivator	L. W. and R. G. Conner	Troy, Iowa	Mar. 28, 1871	113, 019
Cultivator	G. W. Conolly	Rochester, N. Y	Aug. 2, 1864	43, 671
Cultivator	I. Constant	Buffalo Heart Grove, Ill	Nov. 4, 1851	8, 483
Cultivator	A. M. Cook	Chicago, Ill	Feb. 9, 1864	41, 569
Cultivator	G. W. Cook	Macon, Ill	July 14, 1868	79, 956
Cultivator	J. Cook	Collinsville, Ohio	Oct. 2, 1866	58, 383
Cultivator	P. Conrod	Keithsburgh, Ill	June 16, 1863	38, 884
Cultivator	A. Cooper	Macomb, Ill	Dec. 23, 1873	145, 848
Cultivator	I. Cooper	Saybrook, Ohio	May 14, 1861	32, 276
Cultivator	N. B. Cooper	Gratis, Ohio	Aug. 12, 1862	36, 139
Cultivator	J. Copeland	Quasqueton, Iowa	Dec. 12, 1865	51, 428
Cultivator	T. R. Cormick	Cap Au Gris, Mo	July 21, 1863	39, 270
Cultivator	W. T. Coulter, G. F. Trabue, and W. A. Lowery.	Hardinsburgh, Ind	Nov. 24, 1868	84, 413
Cultivator	S. Cowan	Bloomfield, Iowa	Aug. 18, 1863	39, 553
Cultivator	B. S. Cox	Paulsborough, N. J	Jan. 4, 1870	98, 471
Cultivator	B. S. Cox	Paulsborough, N. J	Oct. 15, 1872	132, 201
Cultivator	J. Cox and J. A. Throp	Three Rivers, Mich	Mar. 19, 1861	31, 702
Cultivator	J. Cox and J. A. Throp	Three Rivers, Mich	Aug. 16, 1864	43, 836
Cultivator	R. D. Craft	La Porte, Ind	June 22, 1869	91, 718
Cultivator	A. Crafts and E. Weeks	Auburn, Ohio	Jan. 8, 1850	6, 997
Cultivator	R. Craig	State Line City, Ind	Mar. 6, 1860	27, 406
Cultivator	C. C. Creek	Liberty, Ind	Aug. 27, 1867	68, 290
Cultivator	S. Crossley	Rock Island, Ill	Mar. 19, 1872	124, 724
Cultivator	J. Crowther	Oxford, Mich	Mar. 3, 1868	75, 127
Cultivator	S. Crutcher	Little Mount, Ky	Nov. 4, 1873	144, 321
Cultivator	D. Culver	Kingston, Pa	July 9, 1872	128, 860
Cultivator	W. H. Cummings	Boonesborough, Iowa	Aug. 20, 1872	130, 760
Cultivator	J. Cunningham	Marshall, Mo	Jan. 18, 1859	22, 630
Cultivator	H. W. Curtis	Worcester, Mass	Apr. 24, 1866	54, 122
Cultivator	J. Custer	Sandusky, Ohio	June 26, 1866	55, 826
Cultivator	W. Custer	Shannondale, Ind	Oct. 20, 1868	83, 259
Cultivator	D. P. Daggett	Palmyra, N. Y	Dec. 15, 1857	18, 840
Cultivator	W. P. Dale	Agricultural College, Pa	Mar. 11, 1873	136, 708
Cultivator	C. H. Dana	West Lebanon, N. H	July 25, 1854	11, 361
Cultivator	C. Daniel	Lamonte, Mo	Aug. 28, 1866	57, 484
Cultivator	J. Danner	Milton, Ill	Dec. 23, 1862	37, 219
Cultivator	J. H. Davey	Rockford, Ill	June 30, 1868	79, 450
Cultivator	E. Davies, jr	Carthage, Ill	Sept. 17, 1867	68, 848
Cultivator	D. Davis	Fredericksburgh, Va	July 17, 1835	....
Cultivator	D. M. Davis	Ashbury, Ill	Aug. 23, 1864	43, 901
Cultivator	F. Davis	Lima, Ohio	Jan. 3, 1860	26, 656
Cultivator	J. Davis	Allegheny, Pa	July 5, 1864	43, 395
Cultivator	J. R. Davis	Bloomfield, Iowa	Dec. 15, 1863	40, 915
Cultivator	C. H. Dawson	Jacksonville, Ill	Mar. 8, 1859	23, 159
Cultivator	S. Day	Delavan, Ill	Nov. 10, 1868	83, 838
Cultivator	W. Day	Morristown, N. J	Feb. 23, 1869	87, 151
Cultivator	C. Debolt	Ottawa, Ill	Oct. 9, 1860	30, 302
Cultivator	I. Denham	Lanesfield, Kans	Oct. 3, 1865	50, 229
Cultivator	D. Dennett	Buxton, Me	Dec. 19, 1865	51, 567
Cultivator	E. W. Dennis	Peoria, Ill	Apr. 7, 1868	76, 411
Cultivator	O. H. Dennis	Altona, Ill	May 17, 1859	24, 013

Index of patents issued from the United States Patent Office from 1790 *to* 1873, *inclusive, &c.*—Continued.

Invention.	Inventor.	Residence.	Date.	No.
Cultivator	E. M. Dever and I. C. Pratt	Peoria, Ill	May 10, 1864	42, 646
Cultivator	G. W. Deweese	Lima, Ohio	Apr. 7, 1868	76, 412
Cultivator	C. B. Deyo	Marengo, Ill	Feb. 13, 1866	52, 542
Cultivator	E. F. Dhart	Swan Creek, Ill	Sept. 24, 1867	69, 081
Cultivator	R. D. Dodge	Adel, Iowa	June 24, 1862	35, 673
Cultivator	E. Doolittle	Pawnee, Ill	Jan. 7, 1868	73, 172
Cultivator	G. W. Doolittle	Lincoln, Ill	Oct. 2, 1866	58, 389
Cultivator	W. C. Doss	Lavaca, Tex	May 10, 1859	23, 906
Cultivator	W. D. Dorsey	Decatur, Ill	Apr. 3, 1860	27, 703
Cultivator	W. D. Dorsey	Decatur, Ill	Apr. 28, 1863	38, 295
Cultivator	J. W. Doud	Forestville, Iowa	Jan. 7, 1868	73, 173
Cultivator	J. W. Doud	Ward's Corner, Iowa	July 1, 1873	140, 484
Cultivator	L. H. Doyle	Waterloo, Iowa	July 9, 1861	32, 760
Cultivator	L. H. Doyle	Waterloo, Iowa	Dec. 9, 1862	37, 089
Cultivator	W. A. Dryden	Monmouth, Ill	Jan. 15, 1861	31, 112
Cultivator	W. A. Dryden	Monmouth, Ill	Dec. 29, 1868	85, 292
Cultivator	W. A. Dryden	Monmouth, Ill	Jan. 10, 1871	110, 836
Cultivator	W. A. and C. E. Dryden	Monmouth, Ill	July 30, 1867	67, 173
Cultivator	W. A. Dryden and J. M. Turnbull.	Monmouth, Ill	Sept. 14, 1869	94, 725
Cultivator	J. B. Duane	Schenectady, N. Y	Jan. 4, 1859	22, 494
Cultivator	W. Duffner	Petersburgh, Ind	Dec. 10, 1867	72, 004
Cultivator	J. K. Dugdale	Richmond, Ind	May 13, 1862	35, 223
Cultivator	J. Dundas	Little Rock, Ill	Feb. 8, 1859	22, 859
Cultivator	A. P. Durant	Atlanta, Ill	May 31, 1864	42, 932
Cultivator	G. M. Dwight	Oregon, Ill	Dec. 15, 1868	85, 001
Cultivator	S. Dwight	Byron, Ill	Feb. 5, 1861	31, 305
Cultivator	S. Dwight	Rockford, Ill	Apr. 3, 1866	53, 592
Cultivator	W. Dysert	Gettysburgh, Pa	Aug. 28, 1846	4, 725
Cultivater	W. Dyzert	Gettysburgh, Pa	Aug. 16, 1844	3, 709
Cultivator	D. S. Early	Hummelstown, Pa	Oct. 6, 1868	82, 814
Cultivator	D. S. Early	Hummelstown, Pa	Mar. 2, 1869	87, 400
Cultivator	N. Earlywine	Centreville, Iowa	June 4, 1872	127, 469
Cultivator	C. Eastburn	Spencer County, Ky	May 31, 1859	24, 203
Cultivator	J. S. Eastman	Baltimore, Md	June 30, 1836	....
Cultivator	D. B. Eberly	Pine Village, Ind	Oct. 10, 1871	119, 750
Cultivator	D. B. Eberly	Pine Village, Ind	Dec. 16, 1873	145, 636
Cultivator	R. N. Eby	Upper Leacock Township, Pa.	Aug. 27, 1867	68, 056
Cultivator	D. Edelman	Madison, Ind	Aug. 1, 1871	117, 524
Cultivator	D. Edwards	Marong, Australia	Mar. 28, 1870	113, 148
Cultivator	D. Edwards	Melbourne, Australia	Feb. 18, 1873	136, 047
Cultivator	I. Edwards	Paoli, Ind	Sept. 19, 1846	4, 762
Cultivator	G. Ekstrand and A. P. Cassel	Wataga, Ill	Aug. 23, 1865	49, 512
Cultivator	A. Eldred	Oppenheim, N. Y	Dec. 20, 1845	4, 320
Cultivator	T. E. Ellett	Monmouth, Ill	July 3, 1866	56, 022
Cultivator	J. L. Ellis	Concord, Ill	Nov. 18, 1862	36, 945
Cultivator	T. J. Ellis	Decatur, Ill	Aug. 15, 1871	118, 000
Cultivator	R. Ellwood and R. L. Pitcher	Sycamore, Ill	Oct. 29, 1872	132, 527
Cultivator	R. Ellwood and R. L. Pitcher	Sycamore, Ill	July 1, 1873	140, 353
Cultivator	B. M. Ely	Perry, Mo	Mar. 2, 1869	87, 401
Cultivator	C. W. Emerson	Albany, N. Y	Apr. 23, 1861	32, 128
Cultivator	E. Emmert	Franklin Grove, Ill	July 31, 1860	29, 368
Cultivator	J. Endsley and E. Fletcher	Abington, Ind	May 18, 1858	20, 260
Cultivator	G. A. Erickson	Swede Bend, Iowa	Sept. 1, 1863	39, 724
Cultivator	J. W. Ernst	Heidelburgh, Pa	Nov. 19, 1867	71, 152
Cultivator	C. Escudier	Iberia Parish, La	Oct. 3, 1871	119, 588
Cultivator	A. Eshleman	Martinsville, Pa	Dec. 6, 1870	109, 812
Cultivator	G. Essington	Plainfield, Ill	Feb. 8, 1859	22, 860
Cultivator	G. Esterly	Heart Prairie, Wis	Apr. 22, 1856	14, 715
Cultivator	G. W. Esterly and J. Van De Water.	Whitewater, Wis	Oct. 1, 1872	131, 861
Cultivator	O. Etnier	Mount Union, Pa	Apr. 23, 1867	64, 086
Cultivator	S. P. Etter	Scotland, Pa	Sept. 18, 1866	58, 082
Cultivator	D. Z. Evans	Town Point, Md	Dec. 26, 1871	122, 240
Cultivator	C. A. Ewick	Rushville, Ind	Oct. 29, 1867	70, 328
Cultivator	J. F. Eylar	Scott, Ohio	Apr. 24, 1860	27, 975
Cultivator	J. Faning and F. Legler	Burlington, Iowa	Nov. 7, 1871	120, 732
Cultivator	J. P. Farnutzer	Fond du Lac, Wis	Apr. 30, 1867	64, 262
Cultivator	C. K. Farr	Hinds County, Miss	May 9, 1854	10, 886
Cultivator	C. K. Farr	Auburn, Miss	Aug. 1, 1854	11, 454
Cultivator	G. Fawcett	Farmington, Ill	Apr. 30, 1872	126, 276
Cultivator	J. W. Fawkes	Decatur, Ill	Jan. 24, 1865	45, 987
Cultivator	J. Ferguson	Huntley Grove, Ill	Nov. 30, 1869	97, 377
Cultivator	J. Fernald	Frankfort, Ind	Nov. 21, 1865	51, 031
Cultivator	B. F. Field	Sheboygan Falls, Wis	Sept. 29, 1863	40, 095
Cultivator	B. F. Field	Sheboygan Falls, Wis	Sept. 29, 1863	40, 096
Cultivator	J. Fink	Baldwinsville, N. Y	Dec. 24, 1861	33, 985
Cultivator	R. A. Fish	Worcester, Mass	Aug. 12, 1873	141, 707
Cultivator	W. D. Fisher	Freeport, Ill	May 1, 1866	54, 317
Cultivator	O. F. Fitch	Morristown, Ind	June 26, 1860	38, 843
Cultivator	J. S. Fleming	Island Creek, Ohio	Aug. 29, 1871	118, 522
Cultivator	G. A. Forsgard	Houston, Tex	Sept. 10, 1872	131, 160
Cultivator	W. J. Forshee	Indianapolis, Ind	Oct. 6, 1857	18, 330
Cultivator	J. P. Fostevin	Racine, Wis	Sept. 19, 1865	50, 052
Cultivator	J. H. Frampton	Hopewell, Ohio	Sept. 13, 1859	25, 403
Cultivator	J. Frank	Webster City, Iowa	Aug. 6, 1867	67, 522
Cultivator	J. Frank	Webster City, Iowa	Feb. 18, 1868	74, 678
Cultivator	J. H. Frank	Millheim, Pa	Sept. 30, 1873	143, 338
Cultivator	J. T. Frankeberger	Hensly, Ill	Dec. 24, 1867	72, 622
Cultivator	D. D. Franklin	Flora, Ill	Aug. 10, 1869	93, 611
Cultivator	W. Frantz	Piqua, Ohio	Jan. 7, 1868	73, 087
Cultivator	N. W. Fraser and A. J. McLellan.	La Porte, Ind	Aug. 10, 1858	21, 128
Cultivator	P. F. Freeland	Newark, Ill	Aug. 9, 1859	25, 006
Cultivator	J. C. French	Monmouth, Ill	Nov. 6, 1866	59, 381
Cultivator	J. C. French	Monmouth, Ill	Mar. 16, 1869	87, 923
Cultivator	A. Friberg	Moline, Ill	May 14, 1867	64, 657

Index of patents issued from the United States Patent Office from 1790 *to* 1873, *inclusive*—Continued.

Invention.	Inventor.	Residence.	Date.	No.
Cultivator	A. Friberg	Moline, Ill	June 11, 1872	127, 757
Cultivator	J. Fridy	West Donegal Township, Pa.	Oct. 16, 1866	58, 805
Cultivator	E. W. Fuller	Martinsville, La.	Jan. 22, 1861	31, 163
Cultivator	B. and C. Furnas	Ononwa, Iowa	Nov. 18, 1862	36, 948
Cultivator	W. Furnas	Ononwa, Iowa	Feb. 16, 1864	41, 614
Cultivator	C. Furst	Chicago, Ill	May 30, 1871	115, 459
Cultivator	L. M. Ganong	Friar's Point, Miss	Mar. 4, 1873	136, 500
Cultivator	H. D. Ganse	Freehold, N. J	Mar. 27, 1855	12, 589
Cultivator	H. D. Ganse	Freehold, N. J	Aug. 26, 1856	15, 606
Cultivator	C. Gardner	Hoosick, N. Y	Mar. 19, 1861	31, 710
Cultivator	J. M. Garnett	Loretto, Va	Feb. 3, 1836	....
Cultivator	G. Garrett	Elkhart City, Ill	June 9, 1868	78, 799
Cultivator	W. K. Garrison	Abingdon, Ill	Jan. 29, 1867	61, 661
Cultivator	R. Garter	Grand Rapids, Mich	Dec. 31, 1867	72, 835
Cultivator	R. R. Gaskill	Wyanet, Ill	Oct. 7, 1862	36, 609
Cultivator	H. A. Gaston	Stockton, Cal	May 19, 1868	78, 080
Cultivator	I. N. Gates	Burnside, Ill	Nov. 23, 1869	97, 073
Cultivator	W. Geahr	New Holland, Pa	Sept. 4, 1866	57, 699
Cultivator	J. B. Geisinger and D. H. S. Williams.	Montville, Ohio	Dec. 18, 1860	30, 963
Cultivator	J. Gerber	Rockford, Ill	Aug. 16, 1870	106, 478
Cultivator	J. Gerber	Rockford, Ill	Oct. 18, 1870	108, 471
Cultivator	J. Gerber and H. Brown	Rockford, Ill	Apr. 4, 1871	113, 511
Cultivator	J. Gibbs	Newark, Ohio	Dec. 1, 1857	18, 739
Cultivator	H. H. Gibson	Quincy, Ill	Oct. 15, 1872	132, 277
Cultivator	G. T. Gifford	Monmouth, Ill	Sept. 4, 1866	57, 700
Cultivator	I. B. Gilbert	Lewisville, Ind	Jan. 12, 1869	85, 736
Cultivator	J. R. Gilbert and S. R. Weston	Starkville and Dawson, Ga	Sept. 4, 1860	29, 876
Cultivator	J. G. B. Gill	Chester Court-House, S. C	Jan. 12, 1869	85, 812
Cultivator	H. Gilliard	Mount Hope, Wis	Dec. 27, 1859	26, 581
Cultivator	E. S. Gillies	Albany, Wis	June 20, 1865	48, 271
Cultivator	D. C. Gilliland	Brownsville, Ohio	Dec. 24, 1861	33, 988
Cultivator	J. Gilpatric	Biddeford, Me	Feb. 26, 1867	62, 405
Cultivator	J. H. Given, H. Hutsonpiller, and C. Gilbert.	Des Moines, Iowa	Mar. 7, 1865	46, 657
Cultivator	S. M. Goff	East Addison, Vt	Feb. 12, 1861	31, 382
Cultivator	S. Goldsmith	Wataga, Ill	Aug. 14, 1860	29, 583
Cultivator	A. Gordon	Rochester, N. Y	Apr. 9, 1867	63, 717
Cultivator	M. L. Gorham	Rockford, Ill	Dec. 5, 1871	121, 613
Cultivator	M. L. Gorham	Rockford, Ill	Oct. 1, 1872	131, 818
Cultivator	O. P. Goslee	Glastenbury, Conn	Apr. 2, 1861	31, 884
Cultivator	E. M. Graham	Vernon, La	May 20, 1873	139, 001
Cultivator	H. J. Graham	Monmouth, Ill	Nov. 5, 1867	70, 430
Cultivator	J. L. Graham	Bentley Station, Ill	Aug. 16, 1870	106, 354
Cultivator	B. A. Grant	Mount Pleasant, Iowa	Dec. 5, 1865	51, 306
Cultivator	I. A. Green	Henry, Ill	Mar. 11, 1862	34, 630
Cultivator	W. W. Green	Chelsea, Ill	Mar. 6, 1860	27, 363
Cultivator	T. Green and J. Sommer	Metamora, Ill	Feb. 11, 1868	74, 344
Cultivator	E. S. Gregory	Lockport, N. Y	Aug. 10, 1869	93, 530
Cultivator	W. H. Griffith	Lockhart, Tex	July 23, 1872	129, 730
Cultivator	A. M. Griswold	Momence, Ill	July 30, 1867	67, 295
Cultivator	A. M. Griswold	Momence, Ill	Oct. 27, 1868	83, 487
Cultivator	A. Grohmann	South Saginaw, Mich	Mar. 23, 1869	88, 034
Cultivator	G. Gross, jr	Germantown, Ohio	Sept. 26, 1848	5, 890
Cultivator	J. Gross	Manilla, Iowa	Apr. 1, 1862	34, 828
Cultivator	J. Gross and J. C. Tunison	Decatur, Ill	Dec. 10, 1867	72, 027
Cultivator	A. F. Grove	James Creek, Pa	Nov. 13, 1866	59, 592
Cultivator	S. Gulick	Kline's Grove, Pa	Feb. 14, 1865	46, 349
Cultivator	J. and S. Gum	Marseilles, Ill	Dec. 18, 1860	30, 966
Cultivator	J. and S. Gum	Marseilles, Ill	June 18, 1861	32, 560
Cultivator	D. Guptail	Elgin, Ill	Oct. 15, 1867	69, 801
Cultivator	J. M. Gustin	Wilmington, Ohio	Aug. 12, 1873	141, 786
Cultivator	J. Guyer	Westport, Conn	Mar. 13, 1860	27, 445
Cultivator	C. S. Gwinnup	Milroy, Ind	Feb. 26, 1867	62, 326
Cultivator	W. G. Halburt	Columbus, Miss	Jan. 7, 1873	134, 662
Cultivator	C. M. and D. E. Hall	Uniontown, Ill	Mar. 13, 1860	27, 627
Cultivator	D. E. Hall	Abingdon, Ill	Nov. 10, 1857	18, 587
Cultivator	S. Hall	Russellville, Ky	Nov. 14, 1865	50, 930
Cultivator	T. J. Hall	Bryan, Tex	Sept. 28, 1869	95, 338
Cultivator	J. H. Hamilton	Stevenson, Ala	Sept. 20, 1870	107, 484
Cultivator	A. P. Hammon, J. H. and S. Lincoln, and T. W. Hammon.	Montfort, Wis	Jan. 1, 1867	60, 722
Cultivator	J. R. Hand	College Corner, Ohio	Feb. 4, 1868	74, 082
Cultivator	J. R. Hand	Billingsville, Ind	Apr. 13, 1869	88, 868
Cultivator	R. Haney and J. S. Estes	Peoria, Ill	Aug. 24, 1869	94, 106
Cultivator	M. E. Hanover and D. D. Bailey.	Lamoille, Ill	June 30, 1868	79, 344
Cultivator	J. Harge	Shiloh, Ill	Mar. 1, 1864	41, 769
Cultivator	W. O. Hargrave	Ripon, Wis	Jan. 7, 1868	73, 094
Cultivator	A. A. Harmon	Olney, Ill	Nov. 17, 1868	84, 184
Cultivator	J. Harris	Janesville, Wis	Feb. 23, 1869	87, 101
Cultivator	N. J. Harris	Meredosia, Ill	July 25, 1871	117, 279
Cultivator	V. P. Harris	Greensburgh, Ind	Oct. 11, 1870	108, 260
Cultivator	J. O. Harris and W. F. Slewder	Ottawa, Ill	Feb. 14, 1860	27, 125
Cultivator	A. Harrison	Blissfield, Mich	Sept. 2, 1845	4, 170
Cultivator	A. T. Harrison	Clinton, Ill	May 8, 1866	54, 534
Cultivator	N. S. Harryman	Frankfort, Ind	Feb. 11, 1862	34, 362
Cultivator	C. A. Harper	Wheeling, Ind	Jan. 7, 1868	73, 181
Cultivator	C. A. Harper	Wheeling, Ind	Dec. 8, 1868	84, 823
Cultivator	J. Harper	Hillsborough, Iowa	Mar. 28, 1865	47, 017
Cultivator	G. D. Hart	Lycoming County, Pa	Apr. 2, 1867	63, 384
Cultivator	G. D. Hart	Muncy, Pa	May 26, 1868	78, 203
Cultivator	V. Harwell	Walker County, Ga	June 5, 1860	28, 576
Cultivator	J. Haskell	Lisbon, Me	Feb. 26, 1867	62, 329
Cultivator	M. Haskins and D. B. Hart	Mentor, Ohio	July 9, 1867	66, 491
Cultivator	G. F. Hassenpflug and G. Barnhart.	Green Township, Ohio	Apr. 4, 1865	47, 105

Index of patents issued from the United States Patent Office from 1790 *to* 1873, *inclusive*—Continued.

Invention.	Inventor.	Residence.	Date.	No.
Cultivator	J. W. Hatcher	Bethesda, Tenn	Oct. 24, 1871	120, 274
Cultivator	L. D. Haughey	Atlanta, Ill	Mar. 7, 1865	46, 666
Cultivator	L. Haverstick	Manor, Pa	Nov. 25, 1873	144, 846
Cultivator	L. E. Hawkins	Sangamon, Ill	Aug. 14, 1860	29, 592
Cultivator	A. W. Hawley	Milan, Ohio	Dec. 22, 1857	18, 900
Cultivator	W. H. Haworth	Towanda, Ill	Aug. 27, 1861	33, 176
Cultivator	G. J. Hayes	Ionia, Mich	Mar. 2, 1869	87, 492
Cultivator	C. W. S. Heaton	Salem, Ill	Mar. 12, 1861	31, 665
Cultivator	C. W. S. Heaton	Belleville, Ill	Jan. 20, 1863	37, 474
Cultivator	C. W. S. Heaton	Belleville, Ill	Apr. 21, 1863	38, 261
Cultivator	C. W. S. Heaton	Salem, Ill	Aug. 11, 1863	39, 528
Cultivator	H. I. Heaton	Peoria, Ill	Dec. 16, 1862	37, 166
Cultivator	T. Heermans	Sumner, Tenn	Mar. 8, 1859	23, 170
Cultivator	C. Hefft	Tazewell County, Ill	Jan. 28, 1868	73, 891
Cultivator	A. T. Heflin	Monmouth, Ill	May 14, 1867	64, 668
Cultivator	A. T. Heflin	Monmouth, Ill	Sept. 22, 1868	82, 406
Cultivator	J. M. Heiges	York, Pa	Jan. 7, 1873	134, 665
Cultivator	S. L Heisey	West Donegal, Pa	Aug. 6, 1867	67, 430
Cultivator	J. Helm	Hockheim, Tex	Apr. 22, 1873	138, 085
Cultivator	R. H. Henry	Monmouth, Ill	Jan. 5, 1869	85, 662
Cultivator	S. Henry	Chenoa, Ill	Feb. 14, 1865	46, 355
Cultivator	I. Henton	Shelbyville, Ill	Mar. 6, 1866	52, 998
Cultivator	I. Henton	Shelbyville, Ill	July 13, 1869	92, 529
Cultivator	J. B. Herman	Mount Vernon, Iowa	Oct. 23, 1866	59, 012
Cultivator	J. T. Herndon	Bancroft, Mo	Jan. 28, 1868	73, 805
Cultivator	J. B. Herr	West Tampeter Township, Pa.	Oct. 23, 1866	59, 013
Cultivator	S. H. Herrick	Grinnell, Iowa	July 2, 1867	66, 340
Cultivator	T. Hicks	Pecatonica, Ill	Feb. 9, 1869	86, 750
Cultivator	G. W. Hildreth	Lockport, N. Y	May 21, 1861	32, 366
Cultivator	P. Hildreth	Beloit, Wis	Mar. 27, 1866	53, 445
Cultivator	J. H. Hill	Clinton, Ill	June 23, 1868	79, 119
Cultivator	J. B. Hinckley	Nacogdoches County, Tex.	Feb. 18, 1873	135, 912
Cultivator	J. Hinds and J. Gee	Conologue, Ill	Dec. 22, 1868	85, 095
Cultivator	C. Hoagland	Delavan, Ill	Mar. 24, 1868	75, 911
Cultivator	S. Hoake	Frederick, Md	Mar. 27, 1860	27, 632
Cultivator	J. C. Hoffeditz	Mercersburgh, Pa	Jan. 1, 1867	60, 892
Cultivator	S. Hoffheins	Hamilton Township, Pa	Apr. 24, 1866	54, 160
Cultivator	A. Hoffman and H. W. Limebeck.	Half Day, Ill	Nov. 19, 1861	33, 740
Cultivator	S. B. Hoisington	Galesburgh, Ill	Aug. 9, 1870	106, 161
Cultivator	L. M. Holland	Galesburgh, Ill	Apr. 14, 1868	76, 634
Cultivator	J. Hollinger	Millersburgh, Ohio	Sept. 4, 1866	57, 719
Cultivator	J. Hollinger	Millersburgh, Ohio	May 21, 1867	64, 808
Cultivator	J. Hollingsworth	Chicago, Ill	May 9, 1865	47, 641
Cultivator	C. Holman	Cameron, Ill	Mar. 27, 1866	53, 446
Cultivator	B. Holtz and W. Enoch	Springfield, Ohio	June 20, 1865	48, 281
Cultivator	L. Homrighouse	Baltimore, Ohio	Sept. 27, 1870	107, 777
Cultivator	H. T. Hooker	Skaneateles, N. Y	Dec. 15, 1863	40, 935
Cultivator	H. Hoover	Hemlo, Ill	Sept. 11, 1866	57, 906
Cultivator	W. A. Hopkins	Vicksburgh, Miss	June 29, 1858	20, 712
Cultivator	S. G. Horming	Mount Carroll, Ill	June 6, 1865	48, 066
Cultivator	J. Houck	Clinton, Ind	Feb. 2, 1858	19, 248
Cultivator	H. Howe	Darlington, Wis	June 6, 1865	48, 068
Cultivator	H. Howe	Oneonta, N. Y	Mar. 19, 1867	63, 050
Cultivator	H. Howe	Oneonta, N. Y	Dec. 17, 1867	72, 297
Cultivator	H. Howe	Oneonta, N. Y	Nov. 8, 1870	109, 129
Cultivator	W. H. Howell	Ewingsville, N. J	Dec. 26, 1865	51, 721
Cultivator	D. C. Hubbard	Okolona, Miss	July 6, 1858	20, 798
Cultivator	S. Hubbard and J. Graves	Wilmington, Vt	June 21, 1823	....
Cultivator	E. S. Huff	Zanesville, Ohio	Aug. 28, 1860	29, 789
Cultivator	J. Huff	Young America, Ill	Jan. 26, 1869	86, 160
Cultivator	M. H. Hullinger	Granville, Ill	Feb. 2, 1864	41, 435
Cultivator	M. H. Hullinger	Granville, Ill	Mar. 20, 1866	53, 303
Cultivator	J. M. Hume	Colchester, Ill	May 21, 1867	64, 878
Cultivator	A. Hunt	Macomb, Ill	Sept. 28, 1869	95, 352
Cultivator	A. Hunt	Macomb, Ill	Apr. 4, 1871	113, 520
Cultivator	H. C. Hunt	Amboy, Ill	Nov. 13, 1866	59, 720
Cultivator	G. E. Hutchinson	Cleveland, Ohio	Apr. 16, 1872	125, 681
Cultivator	G. L. Hutchinson	White Rock, Ill	Jan. 16, 1866	52, 049
Cultivator	B. S. Hyers	Pekin, Ill	Apr. 27, 1869	89, 409
Cultivator	B. S. Hyers	Pekin, Ill	Jan. 18, 1870	98, 970
Cultivator	N. Ide	Shelby, N. Y	Apr. 18, 1846	4, 459
Cultivator	J. Imel	Liberty, Ind	Jan. 2, 1855	12, 163
Cultivator	C. B. Ingersoll	Morris, Ill	Nov. 25, 1862	36, 998
Cultivator	J. W. Ingle and R. H. Wright	Livingston, Ill	Mar. 7, 1865	46, 675
Cultivator	H. Ingraham	Naples, N. Y	Dec. 27, 1864	45, 612
Cultivator	H. Ingraham	Naples, N. Y	Aug. 1, 1865	49, 111
Cultivator	H. Ingraham	Naples, N. Y	Nov. 21, 1865	51, 055
Cultivator	H. Ingraham	Naples, N. Y	July 9, 1867	66, 495
Cultivator	N. L. Isgrigg	Moore's Hill, Ind	Feb. 22, 1870	100, 040
Cultivator	J. B. Jay	Arlington, Ill	Apr. 6, 1869	88, 638
Cultivator	P. R. Jenkins	Cottonville, Iowa	Oct. 17, 1871	120, 073
Cultivator	C. M. Jenne	Young America, Ill	Oct. 18, 1864	44, 729
Cultivator	C. M. Jenne	Young America, Ill	July 18, 1865	48, 817
Cultivator	C. M. Jenne	Young America, Ill	Oct. 1, 1867	69, 442
Cultivator	J. W. Jessop	Harveysburgh, Ohio	Dec. 29, 1868	85, 387
Cultivator	G. Jessup	Shortsville, N. Y	Nov. 5, 1872	132, 764
Cultivator	F. C. Jewell	Rahway, N. J	Dec. 13, 1870	110, 046
Cultivator	T. Jobe	Clarksville, Ohio	Feb. 5, 1867	61, 836
Cultivator	C. H. Johnson	Morristown, N. J	July 20, 1869	92, 835
Cultivator	J. P. Johnson	Macon, Ill	Dec. 3, 1867	71, 623
Cultivator	M. Johnson	Three Rivers, Mich	July 19, 1870	105, 460
Cultivator	W. D. Johnson	Raleigh, N. C	Sept. 13, 1859	25, 416
Cultivator	I. B. Jones	Xenia, Ohio	Aug. 12, 1862	36, 158
Cultivator	E. P. Jones and J. L. Harrell	Hertford, N. C	Sept. 12, 1871	118, 946
Cultivator	H. P. Jordan	Victoria, Tex	Apr. 23, 1872	125, 961
Cultivator	W. H. Jordan	Roseville, Ind	Oct. 28, 1862	36, 785

Index of patents issued from the United States Patent Office from 1790 *to* 1873, *inclusive*—Continued.

Invention.	Inventor.	Residence.	Date.	No.
Cultivator	W. T. Jordan	Newnan, Ga	Nov. 28, 1871	121, 381
Cultivator	R. F. Joynes	Bristol, R. I	June 4, 1861	32, 479
Cultivator	E. Julier	Beverly, Ohio	Mar. 13, 1860	27, 450
Cultivator	A. Keck	Montgomery, Ill	Jan. 10, 1865	45, 833
Cultivator	P. H. Kecks	Morgantown, Va	May 31, 1853	9, 754
Cultivator	J. Keezer	Chillicothe, Ohio	June 18, 1861	32, 567
Cultivator	A. M. Keith	Kosciusko, Miss	Sept. 13, 1859	25, 419
Cultivator	J. H. B. Keller	Chambersburgh, Pa	Oct. 27, 1868	83, 509
Cultivator	L. W. Kelley	Brunswick, Ohio	May 11, 1858	20, 207
Cultivator	W. H. Kelly	Onondaga County, N. Y	Mar. 4, 1862	34, 578
Cultivator	A. H. Kennedy	Oberlin, Ohio	Apr. 4, 1871	113, 306
Cultivator	B. F. Kessler	Wilmington, Ill	Jan. 16, 1866	52, 053
Cultivator	J. Kessler	York County, Pa	Jan. 29, 1867	61, 622
Cultivator	W. Kiddoo	Keithsburgh, Ill	Oct. 30, 1866	59, 231
Cultivator	A. B. King	Camden, Ohio	May 7, 1867	64, 429
Cultivator	A. B. King	Camden, Ohio	Nov. 23, 1869	97, 201
Cultivator	J. R. King	Raleigh, Tenn	Dec. 27, 1859	26, 597
Cultivator	M. D. King	King's Ferry, N. Y	July 9, 1872	128, 734
Cultivator	W. H. L. King	Princeton, Iowa	Sept. 26, 1865	50, 141
Cultivator	T. and R. Kinghorn	Morgan, Ohio	June 26, 1860	28, 870
Cultivator	H. R. Kinney	Portsmouth, Ohio	Jan. 3, 1860	26, 682
Cultivator	J. Kirkman	Peoria, Ill	Dec. 27, 1864	45, 615
Cultivator	W. Kleffel	Maple Creek, Neb	May 14, 1872	126, 815
Cultivator	J. G. Knapp and S. D. Libby	Madison and Blooming Grove, Wis.	Jan. 25, 1870	99, 208
Cultivator	E. H. Knight	Unadilla, Mich	May 7, 1867	64, 542
Cultivator	J. M. Knox	Rensselaer, Ind	Apr. 15, 1873	137, 933
Cultivator	S. A. Knox	Worcester, Mass	Mar. 13, 1855	12, 536
Cultivator	J. Koehn	Canton, Ohio	June 18, 1867	65, 918
Cultivator	P. Kribs	Jefferson Furnace, Pa	July 5, 1859	24, 644
Cultivator	H. K. Krieble	Lansdale, Pa	Feb. 25, 1873	136, 249
Cultivator	G. W. Kring	Fairbury, Ill	Jan. 5, 1869	85, 675
Cultivator	C. and P. G. Krogh	Krohville, Wis	Dec. 10, 1872	133, 865
Cultivator	S. A. Kroner	New Britain, Pa	Nov. 5, 1867	70, 578
Cultivator	O. Kugler	Raritan, N. J	July 22, 1873	141, 145
Cultivator	O. Kugler	Three Bridges, N. J	July 22, 1873	141, 146
Cultivator	H. P. Kynett	Lisbon, Iowa	Aug. 1, 1871	117, 547
Cultivator	J. Lacey	Chicago, Ill	Feb. 23, 1864	41, 709
Cultivator	J. Lacey	Chicago, Ill	Feb. 23, 1864	41, 743
Cultivator	J. Lacey	Chicago, Ill	July 4, 1865	48, 627
Cultivator	J. Lacey	Chicago, Ill	Apr. 10, 1866	53, 837
Cultivator	H. Laird	Mechanicsburgh, Pa	May 24, 1870	103, 476
Cultivator	H. J. Lake	Conquest, N. Y	Nov. 27, 1860	30, 737
Cultivator	A. Lamb	Skaneateles, N. Y	Feb. 6, 1866	52, 421
Cultivator	H. Landes	Bath, Pa	Mar. 23, 1869	88, 045
Cultivator	W. B. Lane and W. Coultor	Organ Spring, Ind	Feb. 26, 1867	62, 344
Cultivator	J. B. Lang	Ithaca, N. Y	July 30, 1872	129, 894
Cul'ivator	G. Large	Rosemond, Ill	Jan. 5, 1864	41, 159
Cultivator	J. T. W. Larrabee	Newton, Ind	Mar. 19, 1872	124, 831
Cultivator	G. W. Lawbaugh and J. Williams.	Shanesville, Ohio	Aug. 13, 1861	33, 044
Cultivator	L. Leber	Springfield, Ill	Aug. 21, 1860	29, 704
Cultivator	G. Lechtenthaler	Limestoneville, Pa	July 25, 1854	11, 379
Cultivator	T. M. Lee	Broad Ford, Va	Oct. 12, 1858	21, 763
Cultivator	Z. W. and E. D. Lee	Blakeley, Ga	July 31, 1860	29, 389
Cultivator	R. A. Leeper and Z. B. Kidder	San José, Ill	Feb. 12, 1861	31, 393
Cultivator	A. B. Lefler	Canton, Ind	Jan. 22, 1861	31, 178
Cultivator	F. C. Leffler	Highland Township, Iowa	July 25, 1865	48, 959
Cultivator	A. Leigh	Clinton Station, N. J	Apr. 14, 1863	38, 171
Cultivator	A. Leonard	Newell's Run, Ohio	Apr. 15, 1873	137, 782
Cultivator	M. Lewellin	Bergen, N. Y	Mar. 25, 1873	137, 221
Cultivator	J. Lewis	Washington, D. C	July 12, 1870	105, 225
Cultivator	W. Lewis	Oxford, N. C	Sept. 23, 1873	143, 019
Cultivator	C. Lidren	Aurora, Ill	Jan. 31, 1865	46, 116
Cultivator	J. R. Little	Galesburgh, Ill	Mar. 30, 1869	88, 311
Cultivator	J. R. Little	Galesburgh, Ill	Sept. 6, 1870	107, 069
Cultivator	J. R. Little	Galesburgh, Ill	Oct. 4, 1870	108, 037
Cultivator	J. B. Livezey	Clarksborough, N. J	July 10, 1860	29, 087
Cultivator	J. H. Lockie	Humphrey, N. Y	Oct. 26, 1869	96, 126
Cultivator	I. Long	Terre Haute, Ind	Sept. 7, 1858	21, 428
Cultivator	P. Long	Penn Township, Pa	May 27, 1873	139, 320
Cultivator	I. Lord and S. Woodman	Saco, Me	Feb. 12, 1867	61, 943
Cultivator	W. C. Lostutter and S. Wolcott.	Rising Sun, Ind	July 31, 1860	29, 391
Cultivator	J. P. Lotz	Lacon, Ill	Jan. 4, 1870	98, 604
Cultivator	I. Low	East Fairfield, Ohio	Nov. 15, 1870	109, 229
Cultivator	M. F. Lowth and T. J. Howe	Owatonna, Minn	Apr. 23, 1867	64, 119
Cultivator	M. F. Lowth and T. J. Howe	Owatonna, Minn	Sept. 22, 1868	82, 423
Cultivator	J. Lueth	Kankakee, Ill	Sept. 7, 1869	94, 497
Cultivator	G. H. Lund	Mackford, Wis	Sept. 17, 1872	131, 404
Cultivator	W. Lyman, jr	Malone, N. Y	May 27, 1862	35, 422
Cultivator	E. P. Lynch	Davenport, Iowa	Sept. 14, 1869	94, 903
Cultivator	E. P. Lynch	Davenport, Iowa	Sept. 23, 1873	143, 172
Cultivator	T. Mabbett, sr	Vineland, N. J	May 10, 1870	102, 841
Cultivator	I. B. Mahon	Marion, Ohio	May 19, 1863	38, 593
Cultivator	I. B. Mahon	Dunkirk, Ohio	Jan. 29, 1867	61, 674
Cultivator	I. B. Mahon	Dunkirk, Ohio	Sept. 17, 1867	69, 003
Cultivator	I. B. Mahon	Dunkirk, Ohio	Dec. 20, 1870	110, 256
Cultivator	I. B. Mahon	Dunkirk, Ohio	Dec. 20, 1870	110, 257
Cultivator	P. Maier	Mapleton, Wis	June 13, 1871	115, 875
Cultivator	J. Mallon	Lockport, Ill	Oct. 27, 1868	83, 519
Cultivator	J. Mallon and H. Von Phul, jr.	Holly Wood, La	Dec. 12, 1871	121, 794
Cultivator	H. Mann	San Francisco, Cal	July 14, 1857	17, 797
Cultivator	H. Mann	East Attleborough, Mass	Dec. 28, 1858	22, 437
Cultivator	A. J. Manny	Freeport, Ill	Sept. 19, 1865	50, 018
Cultivator	A. M. Manny	Lena, Ill	Nov. 21, 1871	121, 114
Cultivator	A. M. Manny	Lena, Ill	Nov. 21, 1871	121, 115
Cultivator	J. P. Manny	Rockford, Ill	June 3, 1873	139, 471

Index of patents issued from the United States Patent Office from 1790 to 1873, inclusive—Continued.

Invention.	Inventor.	Residence.	Date.	No.
Cultivator	T. J. Marinus	Independence, Iowa	Mar. 16, 1869	87, 951
Cultivator	J. Markel	Monticello, Ill	Mar. 19, 1861	31, 725
Cultivator	A. S. Markham	Monmouth, Ill	Aug. 4, 1863	39, 412
Cultivator	A. S. Markham	Bushnell, Ill	Feb. 6, 1866	52, 429
Cultivator	D. and A. S. Markham and D. Eldred.	Monmouth, Ill	Mar. 8, 1859	23, 182
Cultivator	C. Marsh, 2d	Natchez, Miss	Aug. 13, 1872	130, 384
Cultivator	C. Marsh, 2d	Natchez, Miss	Nov. 26, 1872	133, 467
Cultivator	R. Marsh	Steubenville, Ohio	Apr. 25, 1865	47, 439
Cultivator	W. W. Marsh and H. McIntyre.	Sycamore, Ill	July 16, 1872	128, 969
Cultivator	T. J. Martin	Willow Hill, Ill	Jan. 26, 1869	86, 170
Cultivator	R. E. Mason	Norway, Ill	July 2, 1872	128, 497
Cultivator	W. May	Winchester, Ohio	Oct. 16, 1860	30, 447
Cultivator	L. D. McClintick	Glenwood, Iowa	Dec. 22, 1868	85, 233
Cultivator	R. McCorkell	Philadelphia, Pa	Oct. 3, 1865	50, 257
Cultivator	R. McCorkell	Philadelphia, Pa	Aug. 4, 1868	80, 556
Cultivator	W. McCormick	Muscatine, Iowa	May 15, 1866	54, 751
Cultivator	W. J. McCoy	Cartersville, Ga	Nov. 27, 1860	30 744
Cultivator	A. S. McDermott	Ogden, Iowa	Feb. 13, 1872	123, 715
Cultivator	T. W. McDill	Oquawka, Ill	Aug. 21, 1860	29, 707
Cultivator	J. C. McDonald	Nevada, Iowa	Apr. 8, 1873	137, 703
Cultivator	T. E. McDonald	New Brunswick, N. J	Nov. 24, 1868	84, 430
Cultivator	A. S. McDonell	Osgood Township, Canada	Nov. 4, 1873	144, 346
Cultivator	E. McEwen	Lisbon, Ill	Aug. 22, 1865	49, 540
Cultivator	P. McGlew	Des Moines, Iowa	Feb. 9, 1864	41, 523
Cultivator	J. McIlvain	Tennessee, Ill	May 18, 1869	90, 112
Cultivator	K. McKinnon	Pleasant Hill, Ala	Feb. 25, 1873	136, 254
Cultivator	D. McNabb	Moscow, Mich	Jan. 3, 1865	45, 730
Cultivator	D. McNeely and C. J. Cady	Spurgeon, Ind	Sept. 1, 1868	81, 807
Cultivator	T. McQuiston	Morning Sun, Ohio	Oct. 18, 1859	25, 843
Cultivator	H. S. Mead	Gloversville, N. Y	July 4, 1865	48, 575
Cultivator	C. E. Mead and G. E. Stevenson	Denmark, Iowa	Aug. 4, 1863	39, 450
Cultivator	C. F. Megquier	Eureka, Ill	Mar. 12, 1867	62, 868
Cultivator	R. M. Melton	Criglersville, Va	May 31, 1859	24, 227
Cultivator	A. Merrill	Ingersoll, Canada	June 11, 1872	127, 785
Cultivator	N. Messenger	Newark, Ill	Nov. 27, 1869	30, 745
Cultivator	W. Mettler	Frankfort, Ill	June 14, 1864	43, 123
Cultivator	I. Miers	Clay Lick, Ohio	Dec. 11, 1860	30, 926
Cultivator	J. Millard	Winslow, Ind	Dec. 29, 1868	85, 467
Cultivator	E. and B. Miller	Rising Sun, Ind	Sept. 13, 1859	25, 432
Cultivator	G. D. Miller	Lovington, Ill	Nov. 15, 1864	45, 066
Cultivator	I. Miller	Worth, Mich	July 26, 1870	105, 828
Cultivator	W. Millerton	Jacksonville, Ill	Mar. 4, 1873	136, 477
Cultivator	J. Mills	Reading, Ill	May 9, 1865	47, 655
Cultivator	J. Mills	Johnstown, Wis	Feb. 18, 1873	135, 997
Cultivator	J. R. Mills	Bloomfield, Iowa	Jan. 5, 1864	41, 086
Cultivator	S. G. Mills	Des Moines, Iowa	Jan. 23, 1866	52, 184
Cultivator	E. Minnich	McKee's Half Falls, Pa	Feb. 14, 1865	46, 376
Cultivator	J. R. Minter	Unionville, S. C	Feb. 6, 1872	123, 410
Cultivator	F. and P. A. Misner	Fox, Ill	Jan. 3, 1860	26, 695
Cultivator	S. H. Mitchell	El Paso, Ill	Dec. 1, 1863	40, 766
Cultivator	S. H. Mitchell	El Paso, Ill	June 14, 1870	104, 334
Cultivator	P. Monaghan	Camac, Ga	Dec. 27, 1859	26, 606
Cultivator	J. B. Moody	Pembroke, Ky	Apr. 27, 1869	89, 422
Cultivator	J. B. Moody	Louisville, Ky	July 12, 1870	105, 358
Cultivator	W. A. Moody	Montezuma, Iowa	Sept. 3, 1867	68, 525
Cultivator	G. Moore	Moline, Ill	Sept. 10, 1867	68, 643
Cultivator	G. Moore	Moline, Ill	July 2, 1872	128, 499
Cultivator	M. H. Moore and A. Satterwhite.	Rome, Ga	Nov. 27, 1860	30, 748
Cultivator	J. B. Moorhead and T. A. and G. G. Pool.	Bellefontaine, Ohio	Sept. 3, 1861	33, 208
Cultivator	B. S. Morgan	Delhi, Iowa	Oct. 5, 1858	21, 690
Cultivator	B. S. Morgan	Delhi, Iowa	Oct. 11, 1859	25, 754
Cultivator	A. H. Morrel	Marlin, Tex	Apr. 10, 1855	12, 690
Cultivator	J. E. Morrison	Washington, D. C	Apr. 24, 1866	54, 191
Cultivator	I. J. Morrow	Everton, Ind	Sept. 21, 1869	95, 035
Cultivator	J. M. Morse	Sandwich, Ill	Oct. 1, 1867	69, 468
Cultivator	S. Mowry	Womelsdorf, Pa	Aug. 9, 1859	25, 031
Cultivator	J. Mumford and J. W. Wilson	Clarksburgh, Ohio	May 28, 1861	32, 431
Cultivator	T. Murphey	Cincinnati, Ohio	Feb. 14, 1860	27, 144
Cultivator	J. Murphy	Albany, Ga	July 16, 1867	66, 871
Cultivator	W. H. Murrey	Brodhead, Wis	May 14, 1872	126, 828
Cultivator	B. F. Muschert	Morrisville, Pa	July 8, 1873	147, 724
Cultivator	A. H. Myers	Hermon, Ill	Aug. 8, 1871	117, 803
Cultivator	J. Neff, jr	Pultney, N. Y	May 8, 1860	28, 191
Cultivator	J. Neff, jr	Pultney, N. Y	Mar. 3, 1868	75, 185
Cultivator	J. Neff, jr	Pultney, N. Y	Sept. 20, 1870	107, 526
Cultivator	J. Neidich and E. R. Girvin	Lancaster County, Pa	Nov. 27, 1860	30, 751
Cultivator	O. S. Neisler	Indianola, Iowa	Jan. 2, 1872	122, 400
Cultivator	R. Nelson	West Point, Ind	Jan. 15, 1844	3, 406
Cultivator	J. B. Netherland	Near Louisville, Ga	Mar. 6, 1860	27, 376
Cultivator	W. Nevins	Irving, N. Y	Nov. 25, 1862	37, 005
Cultivator	E. Newlon	Monmouth, Ill	Aug. 27, 1867	68, 104
Cultivator	D. J. Noble	New Boston, Ill	Mar. 20, 1866	53, 326
Cultivator	J. S. Nolen	Paulsborough, N. J	Oct. 3, 1871	119, 639
Cultivator	D. Norman	Crawfordsville, Ind	July 16, 1872	129, 420
Cultivator	C. P. Norton	Roseville, Ill	Jan. 22, 1867	61, 450
Cultivator	C. P. Norton	Prairie City, Iowa	Oct. 18, 1870	108, 384
Cultivator	W. Notman	Deerfield, Ohio	Mar. 8, 1870	100, 655
Cultivator	A. B. C. Nusbaum	Sacramento, Cal	Feb. 22, 1870	100, 183
Cultivator	A. T. Odell	Royalton, N. Y	July 17, 1847	5, 195
Cultivator	W. H. Older	Packwaukee, Wis	Jan. 26, 1864	41, 389
Cultivator	A. W. Olds	Green Oak, Mich	Oct. 21, 1862	36, 724
Cultivator	H. Ogborn	Green's Fork, Ind	July 10, 1860	29, 094
Cultivator	H. Ogborn and G. Taylor	Green's Fork and Richmond, Ind.	July 28, 1857	17, 909

Index of patents issued from the United States Patent Office from 1790 *to* 1873, *inclusive*—Continued.

Invention.	Inventor.	Residence.	Date.	No.
Cultivator	B. F. Osborn	Nashville, Tenn	Aug. 30, 1870	106, 864
Cultivator	J. D. Osborn	Goshen, Ind	Nov. 29, 1864	45, 264
Cultivator	H. W. Ostrom	Grand Rapids, Mich	June 7, 1870	103, 918
Cultivator	G. W. Owens	Fairfield, Iowa	Aug. 29, 1871	118, 549
Cultivator	W. J. Oxer	Williamsport, Ind	Feb. 12, 1867	62, 060
Cultivator	L. Packard	Galesburgh, Ill	Aug. 9, 1859	25, 057
Cultivator	J. S. Pardon	Summerfield, Ill	July 29, 1862	36, 019
Cultivator	J. G. Page	Rockford, Ill	Aug. 22, 1865	49, 547
Cultivator	I. B. Palamountain	Tarborough, N. C	May 10, 1859	23, 942
Cultivator	I. A. Palmer	Monmouth, Ill	Feb. 14, 1865	46, 383
Cultivator	I. A. Palmer	Monmouth, Ill	Sept. 10, 1867	68, 784
Cultivator	I. A. Palmer	Monmouth, Ill	Dec. 29, 1868	85, 471
Cultivator	C. S. Pangborn and G. W. Beers.	Onarga, Ill	Apr. 3, 1866	53, 663
Cultivator	W. Y. Parker	Verona, Miss	Feb. 11, 1873	135, 726
Cultivator	R. B. and J. R. Parks	Neponset, Ill	Jan. 8, 1867	61, 091
Cultivator	R. B. and J. R. Parks	Neponset, Ill	June 22, 1869	91, 661
Cultivator	E. Parmele and G. Curkendall	Davenport, Iowa, and Moline, Ill.	July 23, 1872	129, 859
Cultivator	E. Parmele and R. N. Patterson.	Davenport, Iowa	May 15, 1866	54, 763
Cultivator	J. Paterson	Medina, N. Y	Apr. 17, 1847	5, 075
Cultivator	D. Pattee	Ypsilanti, Mich	June 20, 1848	5, 639
Cultivator	J. H. Pattee	Monmouth, Ill	Mar. 5, 1872	124, 218
Cultivator	J. H. Pattee	Monmouth, Ill	Jan. 21, 1873	135, 148
Cultivator	E. C. Patterson	Chicago, Ill	Jan. 17, 1865	45, 934
Cultivator	D. H. Paul	De Witt, Iowa	July 6, 1869	92, 350
Cultivator	J. J. Paxson	Middletown, Ind	Mar. 6, 1860	27, 378
Cultivator	R. L. Payne	Riceville, Va	Aug. 30, 1870	106, 956
Cultivator	S. G. Peabody	Champaign, Ill	Dec. 31, 1867	72, 889
Cultivator	S. G. Peabody	Champaign, Ill	Dec. 1, 1868	84, 575
Cultivator	J. C. Pearl	Mendota, Ill	Oct. 12, 1869	95, 831
Cultivator	H. N. Pease	Toledo, Ohio	Oct. 4, 1870	107, 954
Cultivator	E. Peck	Chicago, Ill	Oct. 22, 1867	70, 111
Cultivator	J. Peeler	Tallahassee, Fla	June 21, 1859	24, 486
Cultivator	L. D. Pelton and J. Barrow	Harrison, Ohio	June 18, 1867	65, 938
Cultivator	W. C. Percy	Bayou Sara, La	June 4, 1872	127, 508
Cultivator	F. Perez	Havana, Cuba	Sept. 30, 1873	143, 380
Cultivator	C. D. Perkins	Princeville, Ill	Sept. 3, 1872	131, 066
Cultivator	H. H. Perkins	Osceola, Ill	Apr. 1, 1873	137, 478
Cultivator	A. S. Perrigo	Sandwich, Ill	Oct. 5, 1869	95, 505
Cultivator	F. L. Perry	Canandaigua, N. Y	June 29, 1869	91, 963
Cultivator	F. L. Perry	Canandaigua, N. Y	Nov. 2, 1869	96, 344
Cultivator	F. L. Perry	Canandaigua, N. Y	Oct. 10, 1871	119, 786
Cultivator	G. Perry	Muscatine, Iowa	Apr. 10, 1866	53, 867
Cultivator	C. B. Pettengill	Hebron, Me	Dec. 17, 1867	72, 227
Cultivator	E. Phifer	Trenton, N. J	May 23, 1865	47, 904
Cultivator	E. Phifer	Trenton, N. J	May 7, 1867	64, 563
Cultivator	E. Phifer	Trenton, N. J	June 30, 1868	79, 387
Cultivator	W. W. Philler	Port Byron, Ill	Nov. 13, 1866	59, 643
Cultivator	J. W. Philp	Humboldt, Tenn	June 28, 1870	104, 763
Cultivator	H. L. Pigg	Knobnoster, Mo	July 2, 1872	128, 653
Cultivator	E. W. Pike	Galesburgh, Ill	Oct. 29, 1867	70, 258
Cultivator	W. H. Platt	Dayton, Ohio	Mar. 18, 1873	137, 024
Cultivator	T. Poling	Guthrie, Iowa	Aug. 13, 1872	130, 440
Cultivator	O. M. Pond	Independence, Iowa	Mar. 25, 1873	137, 094
Cultivator	T. W. Poole	Brunswick, Ohio	Sept. 28, 1858	21, 625
Cultivator	M. Porter and C. E. and G. F. Jenkins.	Terre Haute, Ill	Aug. 28, 1866	57, 562
Cultivator	J. Potts	Bridgeport, W. Va	Feb. 19, 1867	62, 223
Cultivator	T. J. Potts and P. C. Yost	Hamilton, Ill	Apr. 18, 1865	47, 332
Cultivator	J. Powell	Sullivan, Ill	Mar. 30, 1869	88, 409
Cultivator	E. Pratt	Grand Detour, Ill	Oct. 18, 1864	44, 747
Cultivator	A. Preston	Unionville, Ohio	May 17, 1859	24, 053
Cultivator	H. Preston	Oxfordville, Wis	July 28, 1868	80, 502
Cultivator	W. Rice	Goldsborough, N. C	June 6, 1854	11, 008
Cultivator	W. Rice	Mount Olive, N. C	Feb. 12, 1861	31, 400
Cultivator	G. W. Prugh and W. H. Beard	Armington, Ill	Feb. 20, 1866	52, 743
Cultivator	I. N. Pyle	Decatur, Ind	Nov. 15, 1859	26, 121
Cultivator	I. N. Pyle	Pleasant Mills, Ind	Apr. 25, 1871	114, 040
Cultivator	J. A. Quick	South Danville, N. Y	May 4, 1869	89, 791
Cultivator	W. F. Quinby	Stanton, Del	Mar. 19, 1861	31, 738
Cultivator	A. Ralston	West Middletown, Pa	Sept. 2, 1845	4, 171
Cultivator	N. A. Rand	Winslow, Ill	July 7, 1868	79, 598
Cultivator	H. C. Ravenscraft	Kingwood, Va	Nov. 27, 1860	30, 758
Cultivator	W. B. Read	Gallatin, Tenn	Jan. 24, 1871	111, 250
Cultivator	J. Rebman	Binkley's Bridge, Pa	Nov. 15, 1870	109, 247
Cultivator	J. Rebman	Manheim Township, Pa	May 7, 1872	126, 412
Cultivator	J. E. Reed	Mineville, N. Y	June 6, 1871	115, 771
Cultivator	S. Reed	Rising Sun, Md	June 16, 1868	79, 001
Cultivator	T. M. Reed	Germantown, Ohio	Oct. 18, 1870	108, 392
Cultivator	E. Reese	Eutaw, Ala	Aug. 15, 1871	118, 053
Cultivator	F. Reese	Wilsonville, Ala	Dec. 17, 1872	134, 099
Cultivator	J. Reichard	Fayetteville, Pa	Aug. 18, 1868	81, 108
Cultivator	J. K. Reiner	Line Lexington, Pa	Oct. 2, 1866	58, 476
Cultivator	L. Repp	Tiffin, Ohio	Mar. 19, 1867	62, 971
Cultivator	J. H. Reynerson	Pleasant Plain, Iowa	Apr. 2, 1867	63, 426
Cultivator	J. H. Reynerson	Pleasant Plain, Iowa	Mar. 17, 1868	75, 647
Cultivator	E. D. and O. B. Reynolds	North Bridgewater, Mass	Dec. 24, 1872	134, 166
Cultivator	P. B. and L. C. Reynolds	Prophetstown, Ill	Feb. 24, 1863	37, 789
Cultivator	W. Rhodes and M. Porter	Lovington, Ill	May 2, 1865	47, 568
Cultivator	H. Rice	Springfield, Ohio	Feb. 13, 1866	52, 606
Cultivator	R. Rice	Georgetown, Ill	Apr. 28, 1863	38, 337
Cultivator	J. C. Rich	Penfield, N. Y	Dec. 11, 1866	60, 422
Cultivator	G. R. Richardson and J. Behel.	Earlville and Rockford, Ill.	June 3, 1873	139, 478
Cultivator	J. C. Rickerd	Lewisville, Ind	Feb. 4, 1868	74, 138
Cultivator	J. J. Rider	Wilton Junction, Iowa	Nov. 22, 1864	45, 177
Cultivator	M. Rigell	Dawson, Ga	Aug. 14, 1860	29, 618

Index of patents issued from the United States Patent Office from 1790 *to* 1873, *inclusive*—Continued.

Invention.	Inventor.	Residence.	Date.	No.
Cultivator	W. S. Riggs	Hightstown, N. J	Mar. 19, 1861	31, 742
Cultivator	J. Righter	Clarksburgh, Va	Dec. 22, 1857	18, 928
Cultivator	R. B. Robbins	Adrian, Mich	Aug. 5, 1873	141, 462
Cultivator	C. Roberts	Three Rivers, Mich	Oct. 28, 1862	36, 859
Cultivator	C. Roberts	Three Rivers, Mich	Nov. 25, 1862	37, 009
Cultivator	C. Roberts	Three Rivers, Mich	Nov. 25, 1862	37, 010
Cultivator	C. Roberts	Three Rivers, Mich	Jan. 10, 1865	45, 860
Cultivator	C. Roberts	Three Rivers, Mich	Jan. 10, 1865	45, 861
Cultivator	C. Roberts	Three Rivers, Mich	June 27, 1865	48, 445
Cultivator	C. S. Roberts	Lyons, Iowa	Nov. 6, 1866	59, 454
Cultivator	E. B. Roberts	Rochester, N. Y	Mar. 17, 1868	75, 578
Cultivator	T. B. Roberts	Franklin, Ill	June 20, 1871	116, 100
Cultivator	T. A. Robertson	Friendship, Md	Oct. 27, 1857	18, 520
Cultivator	T. A. Robertson	Washington, D. C	Mar. 29, 1859	23, 395
Cultivator	H. H. Robertson and C. G. Carr	Kingston, Mo	Nov. 6, 1860	30, 589
Cultivator	L. Robinson	West Cambridge, Mass	Sept. 30, 1856	15, 819
Cultivator	W. R. Robinson	Mattoon, Ill	Aug. 27, 1872	130, 869
Cultivator	S. Rockafellow	Muscatine, Iowa	Aug. 25, 1863	39, 703
Cultivator	A. Roden	Mumford, Ala	Dec. 23, 1873	145, 901
Cultivator	W. Rodgers	Lynnville, Ind	Oct. 6, 1868	82, 753
Cultivator	D. B. Rogers	Seneca Falls, N. Y	Jan. 16, 1849	6, 037
Cultivator	D. B., S., and L. Rogers	Pittsburgh, Pa	Mar. 9, 1858	19, 584
Cultivator	M. L. Rogers	Spring, Pa	Dec. 27, 1859	26, 618
Cultivator	T. B. Rogers	Weathersfield, Conn	Apr. 2, 1867	63, 430
Cultivator	J. Root	Hartland, N. Y	Aug. 23, 1870	106, 728
Cultivator	H. M. Rose	Clinton, Ill	Jan. 24, 1871	111, 256
Cultivator	J. J. Rose	Elmwood, Ill	Aug. 17, 1869	93, 749
Cultivator	A. W. Ross	Northfield, Mass	Oct. 30, 1866	59, 269
Cultivator	H. S. Ross	Millville, Ohio	Apr. 19, 1870	102, 045
Cultivator	A. P. Routt	Liberty Mills, Va	Apr. 30, 1867	64, 370
Cultivator	A. P. Routt	Liberty Mills, Va	July 14, 1868	80, 015
Cultivator	J. S. and I. Rowell	Beaver Dam, Wis	July 3, 1866	56, 102
Cultivator	J. E. Rowland	Hagerstown, Md	May 7, 1867	64, 576
Cultivator	J. Rue	Englishtown, N. J	Aug. 9, 1859	25, 048
Cultivator	J. Rue	Englishtown, N. J	Apr. 30, 1872	126, 156
Cultivator	R. Rust	Odessa, Del	May 20, 1873	139, 193
Cultivator	S. A. Sabin	Pecatonica, Ill	Oct. 5, 1869	95, 520
Cultivator	N. F. Sandelin	New York, N. Y	Feb. 27, 1872	124, 014
Cultivator	J. R. Sanders	Davenport, Iowa	Oct. 1, 1872	131, 908
Cultivator	R. Sandiford	Joliet, Ill	Dec. 1, 1868	84, 649
Cultivator	J. W. Sanford	Byron, Ill	Dec. 3, 1867	71, 794
Cultivator	J. Sattison	Ripley Township, Ohio	Jan. 3, 1871	110, 682
Cultivator	M. Sattley	Taylorville, Ill	Feb. 14, 1871	111, 781
Cultivator	W. G. Savage	Clinton, Ill	June 6, 1865	48, 102
Cultivator	J. Savill	Monmouth, Ill	Feb. 20, 1866	52, 754
Cultivator	R. Sawyer	Wales, Me	Aug. 10, 1858	21, 170
Cultivator	A. Sawyer and H. Barnes	Burlington, Wis	Feb. 25, 1862	34, 528
Cultivator	E. H. Sawyers	Orleans, Iowa	Jan. 10, 1865	45, 866
Cultivator	E. H. Sawyers	West Grove, Iowa	Feb. 24, 1863	37, 775
Cultivator	C. H. Sayre	Utica, N. Y	Aug. 25, 1857	18, 073
Cultivator	G. H. Schanck	Libertyville, Ill	Sept. 8, 1863	39, 837
Cultivator	J. D. Scheetz and R. Adams	Robesonia, Pa	May 7, 1867	64, 580
Cultivator	H. Schreiner, jr	Berrysburgh, Pa	July 14, 1857	17, 821
Cultivator	J. Schröder	Kickapoo, Ill	Sept. 24, 1867	69, 225
Cultivator	C. Schwanger	Mount Joy Township, Pa	Jan. 12, 1869	85, 862
Cultivator	M. Schwartz	Canterbury, N. H	Jan. 30, 1872	123, 128
Cultivator	T. S. Scoville	New York, N. Y	Mar. 6, 1860	27, 388
Cultivator	W. Seely	Chillicothe, Ill	Oct. 11, 1859	25, 764
Cultivator	E. S. Seger and J. C. Ormiston	Erie, Ill	Oct. 30, 1866	59, 279
Cultivator	G. and J. Seibert	Ashley, Ill	Sept. 29, 1868	82, 646
Cultivator	J. B. Sexton	Pella, Iowa	Jan. 14, 1868	73, 392
Cultivator	J. B. Sexton	Pella, Iowa	Apr. 13, 1869	88, 915
Cultivator	J. B. Sexton	Pella, Iowa	Nov. 22, 1870	109, 554
Cultivator	W. H. and L. Seymour	Weymouth, Ohio	Sept. 11, 1860	30, 000
Cultivator	S. B. Shank	Millersville, Pa	Aug. 15, 1871	118, 159
Cultivator	J. Shannon	Dakota, Wis	Aug. 14, 1860	29, 626
Cultivator	D. W. Shares	Hamden, Conn	Aug. 1, 1854	11, 460
Cultivator	A. Shaw	Monmouth, Ill	June 29, 1869	91, 876
Cultivator	A. Shaw	Monmouth, Ill	Nov. 30, 1869	97, 319
Cultivator	S. C. Sheller	Lewisburgh, Pa	Feb. 13, 1872	123, 583
Cultivator	J. Shepard	Newport, Me	June 29, 1869	91, 973
Cultivator	J. Sherrill	Harrisburgh, Oreg	Jan. 7, 1873	134, 615
Cultivator	A. T. Sherwood	Amador, Cal	Oct. 1, 1872	131, 909
Cultivator	L. Sherwood	Marine, Ill	Jan. 3, 1865	45, 758
Cultivator	T. N. Sherwood	Dunlapsville, Ind	Jan. 23, 1866	52, 213
Cultivator	N. S. Shields	Rockford, Ill	Aug. 23, 1870	106, 733
Cultivator	J. Simpson	Cordova, Ill	Nov. 16, 1869	96, 979
Cultivator	M. P. Simpson	Rosemond, Ill	May 2, 1871	114, 483
Cultivator	M. P. Simpson and J. P. Ellacott	Pana and Shelbyville, Ill	July 1, 1873	140, 437
Cultivator	Z. B. Sims	Bonham, Tex	Jan. 21, 1873	135, 165
Cultivator	J. E. Sisson	Bastrop, La	Dec. 16, 1873	145, 535
Cultivator	J. H. Skelly	Aroma, Ill	July 21, 1868	80, 094
Cultivator	M. H. Skiff	Cornwall Bridge, Conn	Dec. 8, 1863	40, 859
Cultivator	H. M. Skinner	Rockford, Ill	Oct. 18, 1870	108, 398
Cultivator	J. B. Skinner	Rockford, Ill	Jan. 23, 1866	52, 217
Cultivator	J. B. Skinner	Rockford, Ill	Dec. 21, 1869	98, 112
Cultivator	J. B. Skinner	Rockford, Ill	Mar. 15, 1870	100, 812
Cultivator	J. B. Skinner	Rockford, Ill	July 23, 1872	129, 759
Cultivator	D. S. Slater	Poynett, Wis	Mar. 3, 1868	75, 211
Cultivator	D. Slaughter	West Hempfield Township, Pa.	Mar. 2, 1869	87, 373
Cultivator	S. Sloan	Kewanee, Ill	Apr. 5, 1864	42, 232
Cultivator	S. W. Slocum and E. Phillips	Fulton City, Ill	Apr. 26, 1864	42, 513
Cultivator	G. Slusser	Hillsborough, Ohio	Dec. 16, 1873	145, 690
Cultivator	J. Smalley	Bound Brook, N. J	Mar. 29, 1859	23, 402
Cultivator	A. Smith	Westfield, Ohio	June 14, 1859	24, 411
Cultivator	A. C. Smith	Fayetteville, N. C	Oct. 17, 1871	120, 113
Cultivator	A. C. Smith	Fayetteville, N. C	Dec. 16, 1873	145, 691

Index of patents issued from the United States Patent Office from 1790 *to* 1873, *inclusive*—Continued.

Invention.	Inventor.	Residence.	Date.	No.
Cultivator	G. Smith	Baltimore, Md	Mar. 27, 1860	27, 654
Cultivator	G. H. Smith	Des Moines, Iowa	Sept. 13, 1870	107, 416
Cultivator	H. B. Smith	Eureka, Ill	Sept. 1, 1863	39, 760
Cultivator	H. B. Smith	Eureka, Ill	June 5, 1866	55, 381
Cultivator	I. R. Smith	Elgin, Ill	Apr. 24, 1860	28, 016
Cultivator	J. A. Smith	Lacon, Ill	Oct. 26, 1869	96, 279
Cultivator	J. D. Smith	Peoria, Ill	Apr. 26, 1864	42, 514
Cultivator	N. E. Smith	Springdale, Iowa	Aug. 25, 1863	39, 686
Cultivator	N. S. Smith	Buffalo, N. Y	Aug. 31, 1858	21, 377
Cultivator	P. E. Smith	Scotland Neck, N. C	Dec. 20, 1870	110, 302
Cultivator	P. W. Smith	Abingdon, Ill	Sept. 10, 1867	68, 662
Cultivator	W. Smith	Boonville, Ind	July 27, 1869	93, 017
Cultivator	W. D. Smith	Homerville, Ga	Nov. 26, 1872	133, 387
Cultivator	W. E. Smith	Oquawka, Ill	July 23, 1867	67, 078
Cultivator	W. H. Smith	Wyanet, Ill	Jan. 15, 1861	31, 132
Cultivator	F. L. Smithson	Mecklenburgh County, Va	Apr. 3, 1855	12, 653
Cultivator	W. Smook	Nine Eagles, Iowa	June 29, 1869	91, 981
Cultivator	S. Snider	Taylorsville, Ky	Nov. 7, 1871	120, 789
Cultivator	J. Snyder	Rock Lick, W. Va	Feb. 11, 1868	74, 441
Cultivator	E. M. Sorley	Neenah, Wis	July 10, 1866	56, 286
Cultivator	J. W. Spangler	Jackson Township, Pa	July 18, 1871	117, 215
Cultivator	J. W. Spangler	York, Pa	Mar. 26, 1872	125, 093
Cultivator	A. J. Sparks	Wyanet, Ill	Dec. 15, 1863	40, 959
Cultivator	S. T. Spaulding	North Cohocton, N. Y	Dec. 21, 1869	98, 202
Cultivator	J. A. Spear, jr	Manchester, Pa	Apr. 9, 1861	32, 010
Cultivator	A. B. Spies	Sterling, Ill	Dec. 1, 1868	84, 588
Cultivator	P. Sprague	Pecatonica, Ill	May 4, 1869	89, 801
Cultivator	R. H. Springstead	Constantine, Mich	Oct. 20, 1863	40, 367
Cultivator	A. B. Springsteen	Schodack Landing, N. Y	May 21, 1872	126, 991
Cultivator	D. S. Stafford	Decatur, Ill	Jan. 15, 1861	31, 133
Cultivator	D. S. Stafford	Decatur, Ill	Jan. 6, 1863	37, 309
Cultivator	D. S. Stafford	Decatur, Ill	Sept. 17, 1867	68, 910
Cultivator	F. Stamm	Lancaster, Pa	Mar. 19, 1861	31, 753
Cultivator	J. K. Staman	Mifflin, Ohio	Jan. 10, 1860	26, 796
Cultivator	A. F. and T. M. Stansbury	Lewiston, Ill	Mar. 12, 1867	62, 898
Cultivator	A. Stark	Topeka, Kans	Sept. 18, 1866	58, 147
Cultivator	E. Starr	Royal Oak, Mich	July 10, 1866	56, 288
Cultivator	W. H. Startzman	Big Lick, Va	Oct. 27, 1868	83, 564
Cultivator	C. C. Stearns	Homer, Ill	Aug. 4, 1863	39, 428
Cultivator	C. E. Steller	Chicago, Ill	Dec. 10, 1867	72, 107
Cultivator	L. Stevens	Dover, Ky	Nov. 27, 1860	30, 771
Cultivator	T. S. Stevens	Pepperell, Mass	Oct. 19, 1858	21, 857
Cultivator	U. T. Stewart	Fayette County, Tenn	June 18, 1867	65, 961
Cultivator	W. L. Stewart	Rushville, Ind	Dec. 29, 1868	85, 488
Cultivator	G. Stiber	Cogan Station, Pa	June 4, 1872	127, 658
Cultivator	A. F. Stillwell	Fayette, Iowa	Sept. 4, 1866	57, 792
Cultivator	G. B. St. John	Kalamazoo, Mich	July 25, 1865	49, 008
Cultivator	G. B. St. John	Brooklyn, Mich	Dec. 8, 1868	84, 776
Cultivator	G. B. St. John	Kalamazoo, Mich	Apr. 5, 1870	101, 536
Cultivator	W. W. St. John	Saint Louis, Mo	Nov. 21, 1865	51, 096
Cultivator	A. D. Stocking	Dowagiac, Mich	May 29, 1866	55, 175
Cultivator	G. W. Stockton	Oquawka, Ill	May 14, 1867	64, 721
Cultivator	J. C. Stoddard	Worcester, Mass	Mar. 29, 1859	23, 407
Cultivator	J. C. Stoddard	Worcester, Mass	June 21, 1859	24, 500
Cultivator	M. Stoll	Conestoga Township, Pa	Mar. 28, 1871	113, 109
Cultivator	M. Stoll	Conestoga Township, Pa	Aug. 13, 1872	130, 393
Cultivator	O. Stone	Ionia, Mich	May 26, 1868	78, 243
Cultivator	A. F. Stoner	West Unity, Ohio	Nov. 10, 1868	84, 016
Cultivator	C. E. Storrs, W. E. Keyes, and D. W. Jones.	Grandville, Mich	Dec. 24, 1867	72, 560
Cultivator	I. Stout	Tremont, Ill	Apr. 16, 1861	32, 092
Cultivator	I. and S. Stout	Tremont, Ill	Dec. 1, 1863	40, 776
Cultivator	A. J. Stover	Sandyville, Iowa	Feb. 27, 1866	52, 905
Cultivator	D. C. Stover	Lanark, Ill	July 10, 1866	56, 289
Cultivator	D. C. Stover	Lanark, Ill	Mar. 9, 1869	87, 724
Cultivator	D. C. Stover	Lanark, Ill	Nov. 1, 1870	108, 945
Cultivator	J. G. Stowe	Bloomington, Ill	June 11, 1872	127, 935
Cultivator	J. G. Stowe	Bloomington, Ill	Nov. 12, 1872	133, 067
Cultivator	W. Strieby	Wagontown, Pa	Mar. 12, 1861	31, 683
Cultivator	J. C. Stroud	Lockhart, Tex	Sept. 29, 1868	82, 562
Cultivator	W. D. Stroud	Oshkosh, Wis	Mar. 21, 1871	112, 862
Cultivator	J. Stryker	Six-Mile Run, N. J	Apr. 17, 1855	12, 744
Cultivator	E. E. Stubbs	West Elkton, Ohio	Aug. 13, 1867	67, 817
Cultivator	J. Summers	Raleigh, Va	Nov. 24, 1857	18, 714
Cultivator	B. J. Svenson	Manor Station, Tex	Aug. 20, 1872	130, 765
Cultivator	J. Swart	Hoffman's Ferry, N. Y	Feb. 23, 1864	41, 728
Cultivator	M. Sweet	Richland, Ind	Sept. 10, 1867	68, 803
Cultivator	J. W. Swickard	Galva, Ill	Feb. 18, 1873	136, 108
Cultivator	C. W. Taliaferro	Keithsburgh, Ill	Nov. 11, 1862	36, 922
Cultivator	J. E. Tate	Columbia, Tenn	Apr. 23, 1867	64, 165
Cultivator	A. C. Taylor	North Fairfield, Ohio	Nov. 7, 1871	120, 794
Cultivator	C. F. Taylor	Vassalborough, Me	Feb. 25, 1868	74, 775
Cultivator	J. W. Taylor	Ashland, Va	Feb. 12, 1861	31, 416
Cultivator	W. Taylor	Mansfield, Mass	Feb. 11, 1873	135, 736
Cultivator	W. A. Taylor and W. W. Graves.	Fort Adams, Miss	Aug. 14, 1860	29, 637
Cultivator	H. M. Teasdale	Dansville, N. Y	June 20, 1865	48, 324
Cultivator	D. C. Teller	Terre Haute, Ind	Feb. 20, 1866	52, 770
Cultivator	E. Terrill	Cold Water, Mich	June 23, 1868	79, 161
Cultivator	R. Thayer and J. McClelland	Pittsburgh, Pa	Oct. 10, 1865	50, 403
Cultivator	C. Thede	Aledo, Ill	Apr. 29, 1873	138, 451
Cultivator	J. Thirlwell	Galesburgh, Ill	Jan. 14, 1859	24, 418
Cultivator	J. S. Thomas	Mifflintown, Pa	Mar. 28, 1871	113, 223
Cultivator	J. H. Thomas and P. P. Mast	Springfield, Ohio	Mar. 28, 1865	47, 055
Cultivator	J. H. Thomas and P. P. Mast	Springfield, Ohio	Jan. 16, 1866	52, 093
Cultivator	J. H. Thomas, P. P. Mast, and T. Harding.	Springfield, Ohio	Dec. 12, 1865	51, 494
Cultivator	A. Thompson	Ottawa, Ill	July 23, 1872	129, 693

Index of patents issued from the United States Patent Office from 1790 *to* 1873, *inclusive*—Continued.

Invention.	Inventor.	Residence.	Date.	No.
Cultivator	J. J. Thompson	Columbus, Ohio	July 4, 1871	116, 776
Cultivator	W. L. Thompson	Rockville, Ind	Sept. 24, 1872	131, 720
Cultivator	J. J. Thompson and V. F. Collier.	Richwood, Ohio	Oct. 13, 1868	83, 111
Cultivator	P. W. Thomson	Truro, Ill	Feb. 9, 1864	41, 550
Cultivator	T. Thorley	Southfield, Mich	Sept. 15, 1868	82, 180
Cultivator	J. A. Throop and J. Cox	Three Rivers, Mich	Jan. 27, 1863	37, 532
Cultivator	J. B. Tibbets	Portland, Mich	Aug. 9, 1870	106, 235
Cultivator	E. S. Tichenor	Jacksonville, N. Y	Aug. 12, 1862	36, 180
Cultivator	H. H. Tietjens	Lyons, Iowa	Sept. 3, 1867	68, 469
Cultivator	H. Tilden	Davenport, Iowa	Aug. 6, 1872	130, 337
Cultivator	B. Tinkham	Cameron, Ill	Dec. 11, 1860	30, 897
Cultivator	J. B. Tipton	Peoria, Ill	Feb. 16, 1869	87, 080
Cultivator	L. K. Tipton	Easton, Mo	Aug. 13, 1872	130, 454
Cultivator	J. C. Tobias and W. N. Bates	El Paso, Ill	Jan. 26, 1869	86, 329
Cultivator	G. W. Tolhurst	Liverpool, Ohio	Jan. 4, 1859	22, 520
Cultivator	F. W. Tolley	Coxsackie, N. Y	July 9, 1872	128, 827
Cultivator	F. W. Tolley	Coxsackie, N. Y	Dec. 31, 1872	134, 446
Cultivator	S. A. Tombs	Ashley, Mo	May 10, 1864	42, 737
Cultivator	J. H. Tomlinson	Mount Carroll, Ill	Nov. 7, 1871	120, 684
Cultivator	A. Tompkins	Paris, Ill	Aug. 29, 1871	118, 562
Cultivator	Z. Toms and L. W. McMullan	Hertford, N. C	Apr. 30, 1872	126, 165
Cultivator	J. P. Testevin	Racine, Wis	Aug. 18, 1863	39, 597
Cultivator	P. R. Totten	Adams, Ill	Oct. 27, 1868	83, 569
Cultivator	A. C. Tower	Mendota, Ill	Jan. 14, 1873	134, 824
Cultivator	J. Townsend	Head of Sassafras, Md	Dec. 5, 1865	51, 365
Cultivator	F. Trigalet	Astoria, N. Y	Oct. 22, 1872	132, 504
Cultivator	A. G. Tucker	Richview, Ill	July 5, 1864	43, 441
Cultivator	W. Tucker	Blackstone, Mass	Oct. 12, 1858	21, 787
Cultivator	J. B. Turner	Jacksonville, Ill	Mar. 12, 1861	31, 682
Cultivator	J. B. Turner	Jacksonville, Ill	Dec. 3, 1861	33, 860
Cultivator	T. Turner	Marysville, Ohio	Dec. 14, 1858	22, 316
Cultivator	L. G. Tuttle	North Haven, Conn	Feb. 19, 1867	62, 302
Cultivator	S. D. Tuttle and J. H. Gans	Eaton, Ohio	Oct. 8, 1867	69, 725
Cultivator	J. W. Tyson	Lower Providence, Pa	Aug. 14, 1866	57, 222
Cultivator	J. W. Tyson	Lower Providence, Pa	Feb. 26, 1867	62, 511
Cultivator	C. A. Uhl	Millersburgh, Ohio	June 24, 1862	35, 718
Cultivator	F. J. Underwood	Rock Island, Ill	Dec. 29, 1868	85, 412
Cultivator	P. C. Van Brocklin	Buffalo, N. Y	Mar. 24, 1863	37, 988
Cultivator	G. W. Van Brunt	Horicon, Wis	Jan. 26, 1869	86, 261
Cultivator	E. W. Vangundy	Galesburgh, Ill	Feb. 2, 1864	41, 454
Cultivator	J. Van Horn	Plainfield, Ill	July 22, 1862	35, 961
Cultivator	R. P. Van Horn	Jacksontown, Ohio	Mar. 27, 1855	12, 609
Cultivator	R. P. Van Horne	Gratiot, Ohio	May 29, 1860	28, 523
Cultivator	H Van Meter	Macomb, Ill	May 18, 1869	90, 208
Cultivator	F. Veal	Hallettsville, Tex	June 21, 1859	24, 597
Cultivator	W. F. Veber	Bowling Green, Ohio	Apr. 2, 1861	31, 917
Cultivator	W. H. Vick	Holly Springs, Miss	May 14, 1872	126, 658
Cultivator	J. Vowles	New Hudson, Mich	Feb. 14, 1860	27, 174
Cultivator	J. Vowles	New Hudson, Mich	May 21, 1861	32, 391
Cultivator	J. Vowles	Milford, Mich	Nov. 10, 1868	84, 029
Cultivator	J. Waddell	Liberty, Ind	Sept. 19, 1871	119, 206
Cultivator	H. Wadsworth	Duxbury, Mass	Aug. 3, 1869	93, 374
Cultivator	C. L. Waffle	Sharon Centre, Ohio	July 26, 1870	105, 745
Cultivator	G. Waggoner	New Kingston, Pa	Jan. 14, 1873	134, 949
Cultivator	T. Waite	Plymouth, Ohio	Nov. 17, 1868	84, 238
Cultivator	L. Walker	Victoria, Tex	May 23, 1871	115, 138
Cultivator	L. B. Walker	Chicago, Ill	Sept. 12, 1865	49, 938
Cultivator	W. T. Walker	Fontenoy Mills, Ga	May 13, 1873	138, 964
Cultivator	E. Walker and J. J. Piatt	La Porte, Ind	Aug. 31, 1869	94, 456
Cultivator	E. Walker and J. J. Piatt	La Porte, Ind	Oct. 26, 1869	96, 250
Cultivator	E. Walker and A. M. Weed	La Porte, Ind	Oct. 29, 1867	70, 381
Cultivator	W. Walton	East Palestine, Ohio	June 2, 1868	78, 626
Cultivator	B. F. Ward	Indian Springs, Ga	Jan. 4, 1870	98, 650
Cultivator	C. Warner	Monroe, Ohio	Mar. 26, 1872	124, 922
Cultivator	J. Warner	Reading, Pa	Mar. 13, 1849	6, 167
Cultivator	A. Warren	Westport, Conn	Aug. 9, 1859	25, 066
Cultivator	G. W. Warren	Macomb, Ill	July 10, 1866	56, 295
Cultivator	L. B. Waterman	Chicago, Ill	May 13, 1862	35, 282
Cultivator	L. B. Waterman	Chicago, Ill	Nov. 25, 1862	37, 019
Cultivator	L. B. Waterman	Chicago, Ill	June 21, 1864	43, 268
Cultivator	W. M. Watkins	Talcott, Va	Nov. 28, 1871	121, 439
Cultivator	C. H. Watson	Washington, D. C	July 2, 1872	128, 685
Cultivator	C. H. Watson	Washington, D. C	July 2, 1872	128, 686
Cultivator	W. T. Watson	Nottingham, Md	Dec. 11, 1866	60, 449
Cultivator	H. J. Wattles	Rockford, Ill	Apr. 6, 1869	88, 758
Cultivator	H. J. Wattles	Rockford, Ill	Aug. 10, 1869	93, 651
Cultivator	H. J. Wattles	Rockford, Ill	Dec. 7, 1869	97, 573
Cultivator	S. Way	La Porte, Ind	Nov. 24, 1868	84, 449
Cultivator	T. U. Webb	Springfield, Ill	May 29, 1866	55, 185
Cultivator	N. G. Webber	East Springfield, Pa	May 10, 1870	102, 983
Cultivator	H. H. Webster	Claremont, N. H	Aug. 29, 1865	49, 667
Cultivator	W. S. Weir, jr	Monmouth, Ill	May 28, 1861	32, 442
Cultivator	W. S. Weir, jr	Monmouth, Ill	Dec. 23, 1862	37, 251
Cultivator	W. S. Weir, jr	Monmouth, Ill	Mar. 13, 1866	53, 205
Cultivator	S. G. Welch	Athens, Ill	Feb. 9, 1864	41, 5[illegible]5
Cultivator	H. Weld	Black Walnut, Ill	Oct. 10, 1871	119, 904
Cultivator	F. N. Welden	Rockford, Ill	Mar. 21, 1871	112, 994
Cultivator	C. Wells	Sandwich, Ill	Jan. 26, 1869	86, 262
Cultivator	C. C. Wells	Lyons, Iowa	Jan. 30, 1866	52, 348
Cultivator	E. Wells	Auburn, Miss	Aug. 28, 1860	29, 836
Cultivator	H. Wells	Walnut Grove, Ill	June 21, 1859	24, 514
Cultivator	P. F. Wells	Milford, Mich	June 17, 1873	139, 985
Cultivator	W. J. Wells	Sidney, Ohio	June 22, 1869	91, 802
Cultivator	N. Werts	Magnolia, Ill	May 3 1870	102, 631
Cultivator	P. L. West	Bath, Ill	Apr. 2, 1867	63, 447
Cultivator	L. Wetherell	Worcester, Mass	Dec. 22, 1857	18, 939
Cultivator	C. L. Whaite	Waverly, Pa	Sept. 6, 1870	107, 142

Index of patents issued from the United States Patent Office from 1790 *to* 1873, *inclusive*—Continued.

Invention.	Inventor.	Residence.	Date.	No.
Cultivator	E. P. Wheeler	Nebraska City, Nebr	Apr. 10, 1866	53, 911
Cultivator	N. Whitehall	Rob Roy, Ind	Oct. 27, 1857	18, 530
Cultivator	J. Whiteside and H. F. Crabill	Fuller's Corners, Ind	Nov. 22, 1859	26, 216
Cultivator	S. M. Whiting	Galesburgh, Ill	Jan. 15, 1867	61, 294
Cultivator	J. M. Whiting	Bolton, Mass	Mar. 1, 1859	23, 135
Cultivator	E. Wiard	Louisville, Ky	Mar. 4, 1873	136, 574
Cultivator	J. Widman	Panola, Ill	Jan. 7, 1868	73, 217
Cultivator	T. Wiles and J. M. Ginnis	Muscatine, Iowa	Jan. 3, 1865	45, 782
Cultivator	J. Wilhelm	Muscatine, Iowa	Feb. 9, 1864	41, 562
Cultivator	L. H. Wilkinson	Michigan City, Ind	Jan. 1, 1869	90, 905
Cultivator	E. Wilcox	Delhi, Iowa	Dec. 15, 1863	40, 973
Cultivator	C. Willard	Newtown, Pa	Feb. 5, 1867	61, 906
Cultivator	A. Williams	Berea, Ohio	June 21, 1864	43, 249
Cultivator	J. M. Williams	Greenville, Ga	June 26, 1860	28, 929
Cultivator	W. Willmot	Wilmington, Del	Nov. 30, 1858	22, 215
Cultivator	F. O. Wilson	Mount Olive, N. C	Mar. 6, 1860	27, 402
Cultivator	J. Wilson	Somerford, Ohio	July 23, 1867	67, 093
Cultivator	J. A. Wilson	Hamburgh, Iowa	Dec. 28, 1869	98, 326
Cultivator	W. C. Wilson	Brunswick, Ill	Nov. 7, 1871	120, 692
Cultivator	W. I. Wilson	Franklin, Ind	May 24, 1859	24, 171
Cultivator	B. W. Wingo	Chillicothe, Mo	Nov. 25, 1873	145, 038
Cultivator	A. Q. Withers	Red Banks, Miss	Dec. 1, 1857	18 785
Cultivator	B. Witter	Sherrill's Mount, Iowa	Feb. 11, 1873	135, 872
Cultivator	F. Wolf	Brooklyn, N. Y	Aug. 14, 1860	29, 644
Cultivator	J. Wood	Wedowee, Ala	Apr. 16, 1872	125, 872
Cultivator	J. F. Wood	Houma, La	Sept. 25, 1860	30, 173
Cultivator	J. A. and S. S. Woodward and T. Mason.	Sandwich, Ill	Sept. 15, 1868	82, 191
Cultivator	J. A. and S. S. Woodward and T. Mason.	Sandwich, Ill	June 29, 1869	91, 894
Cultivator	R. Wottring	Prospect, Ohio	Apr. 5, 1870	101, 695
Cultivator	G. W. N. Yost	Port Gibson, Miss	Mar. 20, 1855	12, 571
Cultivator	G. W. N. Yost	Yellow Springs, Ohio	Aug. 28, 1860	29, 849
Cultivator	G. W. N. Yost	Yellow Springs, Ohio	Sept. 4, 1860	29, 934
Cultivator	A. Young	Millstadt, Ill	Nov. 6, 1866	59, 499
Cultivator	B. F. Young	Toulon, Ill	May 8, 1866	54, 637
Cultivator	B. F. Young	Toulon, Ill	Feb. 16, 1869	87, 019
Cultivator	B. F. Young	Toulon, Ill	Nov. 15, 1870	109, 363
Cultivator	J. Young	Joliet, Ill	June 7, 1859	24, 348
Cultivator	S. L. Young	Scottville, Ill	Feb. 25, 1873	136, 297
Cultivator	W. B. Young	Chicago, Ill	Sept. 10, 1867	68, 679
Cultivator	L. G. Youngs	Wilmington, Ill	July 11, 1865	48, 758
Cultivator	W. P. Zane	Woolwich, N. J	Mar. 27, 1855	12, 611
Cultivator	G. W. Zeigler	Tiffin, Ohio	Mar. 13, 1866	53, 212
Cultivator	G. W. Zeigler	Maumee City, Ohio	Feb. 9, 1869	86, 794
Cultivator	J. P. Zeller	South Bend, Ind	July 13, 1869	92, 686
Cultivator	J. Zimmerman	Oswego, Ill	July 29, 1856	15, 453
Cultivator and chopper	Z. Doolittle and A. M. Crowder.	Houston Factory, Ga	Feb. 18, 1868	74, 519
Cultivator and corn-mill, Combined	C. L. Wilcox	Wayne Township, Ohio	July 12, 1870	105, 285
Cultivator and corn-stalk cutter, Combined	S. T. Miller	Empire Prairie, Mo	Jan. 2, 1872	122, 479
Cultivator and ditcher, Combined	N. Hawkes	Appleton, Me	Oct. 16, 1866	58, 819
Cultivator and grain-drill, Combined	J. Gire	Tipton, Ill	Sept. 13, 1870	107, 359
Cultivator and grain-drill, Combined adjustable	T. Lumsdon	Waterford, Ohio	Sept. 6, 1870	107, 186
Cultivator and gang-plow	I. B. Mahon	Dunkirk, Ohio	Mar. 24, 1868	75, 941
Cultivator and gang-plow, Wheel	W. Hammond	Marshall, Mich	May 29, 1866	55, 093
Cultivator and harrow	S. Conrad	Petaluma, Cal	Apr. 7, 1868	76, 578
Cultivator and harrow	E. T. Russel	Indianapolis, Ind	Aug. 9, 1870	106, 118
Cultivator and harrow	T. Short	Fairmount, Ill	Feb. 7, 1865	46, 274
Cultivator and harrow, Combined	H. Benedict	Detroit, Mich	July 27, 1869	93, 042
Cultivator and harrow, Combined	E. Emmert	Franklin Grove, Ill	Aug. 17, 1869	93, 692
Cultivator and harrow, Combined	J. Maxton	Saybrook, Ill	Feb. 14, 1871	111, 857
Cultivator and harrow, Combined	E. D. and O. B. Reynolds	North Bridgewater, Mass	Jan. 24, 1865	46, 025
Cultivator and harrow, Combined	C. Rich and O. L. Neisler	Poughkeepsie, N. Y	Jan. 21, 1868	73, 547
Cultivator and harrow, Combined rotary	H. T. Taplin	South New Market, N. H	Sept. 28, 1869	95, 394
Cultivator and harrow teeth	R. West and H. F. Paul	Concord, N. H	Oct. 8, 1867	69, 731
Cultivator and hoe, Combined	H. P. Eckler	Catskill, N. Y	June 2, 1868	78, 442
Cultivator and horseshoe	J. W. Blake	Jefferson, Wis	Nov. 28, 1871	121, 228
Cultivator and manure-drag, Combined	P. H. Stauffer	Lehighton, Pa	Jan. 19, 1869	86, 110
Cultivator and other teeth, Friction-block for attaching.	W. Workman and J. Hitchcock	Ripon, Wis	Dec. 13, 1870	110, 103
Cultivator and planter	E. M. Wright	Wilmington, Ohio	Aug. 1, 1865	49, 204
Cultivator and planter, Combined	B. Anyan	Fitchville, Ohio	Apr. 20, 1869	89, 113
Cultivator and planter combined	P. Burress	Braidwood, Ill	Jan. 3, 1871	110, 736
Cultivator and planter, Combined	S. B. Conover	New York, N. Y	Apr. 2, 1867	63, 476
Cultivator and planter combined	C. L. Lee	Fitchville, Ohio	June 23, 1868	79, 236
Cultivator and planter combined	E. Spangler	York, Pa	Jan. 16, 1872	122, 861
Cultivator and plow	W. H. Damron, R. H. Massey, and L. F. Whitman.	Macomb, Ill	Jan. 7, 1868	73, 081
Cultivator and plow	S. F. Seely	Sylvania, Ohio	Aug. 6, 1867	67, 595
Cultivator and plow	S. F. Seely	Whiteford, Mich	Dec. 15, 1868	84, 911
Cultivator and plow	L. R. Wright	Troy, N. Y	Dec. 3, 1872	133, 614
Cultivator and plow combined	L. R. Wright	Troy, N. Y	Aug. 11, 1868	81, 054
Cultivator and plow, Combined	I. Young	Byhalia, Miss	Feb. 12, 1867	61, 980
Cultivator and plow, Cotton	J. R. Thomas	Mifflin, Pa	Nov. 22, 1870	109, 558
Cultivator and plow shovel fastener	J. Pierpont and S. S. Tuttle	La Harpe, Ill	Oct. 18, 1870	108, 512
Cultivator and plow, Wheeled	S. Fisher	Hightstown, N. J	Oct. 5, 1869	95, 453
Cultivator and potato-digger	C. P. Goss	Saint Johnsbury, Vt	May 27, 1862	35, 371
Cultivator and potato-digger	J. G. Lacy	Eureka, Wis	Mar. 26, 1872	125, 057
Cultivator and potato digger, Combined	M. and J. W. Chandler	East Corinth, Me	July 4, 1865	48, 616
Cultivator and potato-digger, Combined	M. Darling and H. Gray	Marathon, N. Y	Apr. 28, 1868	77, 176
Cultivator and potato-digger, Combined	J. D C. Outwater	Newark, N. J	Nov. 26, 1867	71, 525
Cultivator and potato-plow, Corn	J. Kurtz	Clinton Township, Pa	July 16, 1867	66, 717
Cultivator and roller, Combined	L. G. and E. E. Reed	Stark County, Ill	June 21, 1870	104, 498
Cultivator and seed-planter	D. B. Morgan	Washington, Ohio	Feb. 9, 1869	86, 776
Cultivator and seed-planter	R. H. Springsteed	Wooster, Ohio	Feb. 12, 1845	3, 901
Cultivator and seed-planter, Combined	D. E. Holt	Wilkinson County, Miss	Nov. 2, 1869	96, 322
Cultivator and seeder	L. Bishop	Talladega, Ala	Mar. 24, 1868	75, 845
Cultivator and seeder	C. Lobdell	Fort Hill, Ill	Dec. 1, 1868	84, 499

Index of patents issued from the United States Patent Office from 1790 *to* 1873, *inclusive*—Continued.

Invention.	Inventor.	Residence.	Date.	No.
Cultivator and seeder	C. E. Miller	Amelia, Ohio	Jan. 19, 1864	41, 314
Cultivator and seeder	T. L. Ray	Flora, Ill	Oct. 11, 1864	44, 658
Cultivator and seeder, Combined	M. Baer	Logansport, Ind	July 25, 1871	117, 361
Cultivator and seeder, Combined	A. Dodge	Georgetown, D. C	Apr. 17, 1866	53, 960
Cultivator and seeder, Combined	J. Lewis	Washington, D. C	Nov. 22, 1870	109, 432
Cultivator and seeder, Combined	A. S. Markham	Bushnell, Ill	Oct. 29, 1867	70, 345
Cultivator and seeder, Combined	C. Norwood	Bloomington, Ill	May 30, 1865	47, 975
Cultivator and seeder, Combined	R. J. Robeson and W. Nash	Oskaloosa, Iowa	Nov. 1, 1870	108, 823
Cultivator and seeder, Combined	S. C. Schofield	Chicago, Ill	Oct. 26, 1869	96, 271
Cultivator and seeding-machine	C. Denton	Pekin, Ill	Mar. 27, 1866	53, 525
Cultivator and seeding-machine, Combined	C. Churchill	New Hartford, Iowa	Dec. 31, 1867	72, 799
Cultivator and seeding-machine, Combined	A. P. Durant	Atlanta, Ill	May 20, 1862	35, 306
Cultivator and shovel-plow	T. F. Hamilton	Geneseo, Ill	Aug. 22, 1871	118, 362
Cultivator and shovel-plow	G. W. Parsons and W. S. Finney	Harrisburgh, Pa	May 16, 1871	114, 964
Cultivator and stalk-cutter	J. G. Johnson	Carthage, Ill	July 20, 1869	92, 834
Cultivator and stalk-cutter, Combined	E. F. Dehart	Swan Creek, Ill	Jan. 5, 1869	85, 570
Cultivator and sulky-plow	J. H. Barringer	Hillsborough, Ill	Jan. 8, 1867	61, 039
Cultivator and weeder	S. M. Marshall and J. W. Coburn	Dracut, Mass	Oct. 3, 1838	955
Cultivator and weeder, Hand	W. G. Comstock	East Hartford, Conn	June 1, 1869	90, 639
Cultivator and weeder, Root	C. Jarvis	Ellsworth, Me	Jan. 10, 1865	45, 831
Cultivator and weeding-machine	M. Johnson	Three Rivers, Mich	May 23, 1871	115, 213
Cultivator-bar elevator	J. S. and I. Rowell	Beaver Dam, Wis	June 2, 1868	78, 484
Cultivator-blades, Mode of attaching	J. L. Fountain	New Milford, Ill	May 8, 1866	54, 522
Cultivator, Bog-cutting	E. L. Freeman	Ann Arbor, Mich	June 21, 1853	9, 796
Cultivator, Comb ned corn and cotton	W. C. Gaines	Salem, Va	June 27, 1871	116, 425
Cultivator, Combined double-shovel and two-horse	S. G. Rayl	Agency City, Iowa	Nov. 16, 1869	96, 938
Cultivator, Combined harrow and shovel	A. B. Baum	Grantville, Pa	July 26, 1870	105, 770
Cultivator, Corn	W. R. Adams	Independence, Mo	Mar. 24, 1868	75, 718
Cultivator, Corn	J. B. Baker	Onondaga, N. Y	Mar. 31, 1857	16, 906
Cultivator, Corn	S. Bell and G. W. Bronson	La Salle County, Ill	May 4, 1869	89, 729
Cultivator, Corn	N. G. Blauser	Etna, Ohio	Aug. 19, 1873	141, 988
Cultivator, Corn	P. F. Brittain	Geneseo, Ill	Aug. 20, 1867	67, 949
Cultivator, Corn	G. D. Brown	Lebanon, Ill	May 30, 1871	115, 427
Cultivator, Corn	A. Campbell	Oxford, Ind	June 2, 1868	78, 424
Cultivator, Corn	A. Canfield	Lyons City, Iowa	Aug. 27, 1867	68, 165
Cultivator, Corn	J. Chapman	Linn County, Iowa	Sept. 12, 1865	49, 856
Cultivator, Corn	J. Clements	Blooming Grove, Ind	Sept. 27, 1870	107, 760
Cultivator, Corn	W. Emmons and D. A. Wells	Sandwich, Ill	Aug. 10, 1869	93, 605
Cultivator, Corn	I. Emrick	Vernon, Mich	Feb. 4, 1873	135, 414
Cultivator, Corn	J. C. Erwood	Vernon, Ind	Mar. 1, 1870	100, 277
Cultivator, Corn	C. Flory	East Donegal, Pa	Oct. 22, 1867	69, 981
Cultivator, Corn	W. Gilman	Ottawa, Ill	Sept. 13, 1870	107, 357
Cultivator, Corn	A. J. Grush	Springfield, Ill	Nov. 16, 1869	96, 801
Cultivator, Corn	J. Harper	Salem, Iowa	Mar. 28, 1865	47, 016
Cultivator, Corn	J. C. Holmes	Wyoming, Pa	Dec. 14, 1869	97, 921
Cultivator, Corn	N. S. Johnson	Maquoketa, Iowa	Feb. 4, 1868	73, 979
Cultivator, Corn	A. Kinyon	Amboy, Ill	Dec. 20, 1864	45, 503
Cultivator, Corn	A. Martin	Oquawka, Ill	Aug. 6, 1867	67, 441
Cultivator, Corn	L. B. Moore	Janesville, Wis	Sept. 4, 1866	57, 752
Cultivator, Corn	D. J. Noble	New Boston, Ill	Sept. 18, 1866	58, 125
Cultivator, Corn	W. H. Parlin	Canton, Ill	Sept. 10, 1867	68, 648
Cultivator, Corn	H. S. Potter	Fairfield, Iowa	Aug. 28, 1866	57, 563
Cultivator, Corn	E. F. Rate	Cedar County, Iowa	Oct. 13, 1868	83, 093
Cultivator, Corn	W. N. Rhinehart and H. Felker	Miami City, Ohio	Feb. 5, 1867	61, 760
Cultivator, Corn	R. B. Robbins	Adrian, Mich	Feb. 23, 1869	87, 296
Cultivator, Corn	S. B. Shank	Manor Township, Pa	Jan. 19, 1869	85, 968
Cultivator, Corn	J. B. Smith	Norfolk, Va	Apr. 15, 1839	1, 121
Cultivator, Corn	G. Sprague	Spring Hill, Kans	Apr. 30, 1867	64, 376
Cultivator, Corn	L. O. Stevens	Pekin, Ill	Oct. 9, 1866	58, 693
Cultivator, Corn	J. S. Stukey	Sugar Grove, Ohio	Feb. 6, 1866	52, 462
Cultivator, Corn	C. W. Taliaferro	Keithsburgh, Ill	July 3, 1866	56, 122
Cultivator, Corn	D. W. Travis	Enfield, N. Y	Dec. 22, 1868	85, 189
Cultivator, Corn	D. W. Travis	Enfield, N. Y	Oct. 26, 1869	96, 167
Cultivator, Corn	D. Wilde	Washington, Iowa	Dec. 20, 1864	45, 549
Cultivator, Corn	M. Wilson	Marquette, Wis	Sept. 10, 1867	68, 678
Cultivator, Corn and cotton	G. T. Bennett	Mount Olive, N. C	Apr. 24, 1860	27, 962
Cultivator, Corn and cotton	J. W. Milroy	Galveston, Ind	Feb. 4, 1868	73, 988
Cultivator, Corn and cotton	R. F. Patton	Quincy, Ohio	July 13, 1869	92, 641
Cultivator, Corn and cotton	J. H. Robinson	Selma, Ala	Sept. 21, 1869	95, 142
Cultivator, Cotton	J. Adams	Clarksville, Tex	Aug. 13, 1867	67, 700
Cultivator, Cotton	R. N. Adams	Greenfield, Ohio	Apr. 2, 1867	63, 451
Cultivator, Cotton	S. W. Akin	Maury County, Tenn	Mar. 20, 1849	6, 204
Cultivator, Cotton	W. J. Andrews	Columbia, Tenn	Mar. 22, 1870	101, 077
Cultivator, Cotton	W. Baker	Memphis, Tenn	July 16, 1872	129, 081
Cultivator, Cotton	F. L. Bates	Bremond, Tex	May 6, 1873	138, 555
Cultivator, Cotton	G. W. Beard	Canton, Miss	Aug. 9, 1859	24, 983
Cultivator, Cotton	N. Bell and H. Winfield	Pantego Township, N. C	Jan. 7, 1873	134, 632
Cultivator, Cotton	J. C. Bethea	Blakely, Ga	Feb. 5, 1867	61, 705
Cultivator, Cotton	W. W. Blair	Lebanon, Tenn	June 19, 1866	55, 607
Cultivator, Cotton	I. W. Burch	Fayette, Miss	Dec. 14, 1869	97, 870
Cultivator, Cotton	M. B. Camp	Riley Centre, Kans	Jan. 9, 1872	122, 515
Cultivator, Cotton	C. Cannaday	Indianapolis, Ind	Jan. 4, 1859	22, 487
Cultivator, Cotton	C. Casey	Goldsborough, N. C	Mar. 27, 1860	27, 616
Cultivator, Cotton	C. W. Chambers and I. Washam	Talladega, Ala	June 25, 1867	66, 128
Cultivator, Cotton	T. Dale	Russellville, Ky	Sept. 19, 1871	119, 015
Cultivator, Cotton	S. C. Darden	Connersville, Miss	June 1, 1869	90, 734
Cultivator, Cotton	T. H. Dodge	Washington, D. C	July 17, 1860	29, 147
Cultivator, Cotton	R. I. Draughon	Claiborne, Ala	Nov. 9, 1869	96, 562
Cultivator, Cotton	E. Enete	Catahoula Parish, La	Sept. 7, 1869	94, 481
Cultivator, Cotton	D. P. Forney	Jacksonville, Ala	Oct. 20, 1857	18, 442
Cultivator, Cotton	A. M. and B. H. Foster	Hallettsville, Tex	Dec. 18, 1866	60, 625
Cultivator, Cotton	L. M. Ganong	Friar's Point, Miss	July 12, 1870	105, 326
Cultivator, Cotton	R. J. Gatling	Indianapolis, Ind	July 3, 1860	28, 978
Cultivator, Cotton	C. Gibbon	Hicksford, Va	July 30, 1867	67, 289
Cultivator, Cotton	N. A. H. Goddin	Wilson, N. C	Dec. 4, 1860	30, 810
Cultivator, Cotton	E. H. and E. B. Goelet	Goldsborough, N. C	Dec. 24, 1867	72, 479

Index of patents issued from the United States Patent Office from 1790 *to* 1873, *inclusive*—Continued.

Invention.	Inventor.	Residence.	Date.	No.
Cultivator, Cotton	A. J. Going	Clinton, La	Apr. 6, 1860	88, 627
Cultivator, Cotton	W. W. Golsan	Autaugaville, Ala	Sept. 4, 1860	29, 877
Cultivator, Cotton	N. Gotten	Union Depot, Tenn	Jan. 1, 1867	60, 879
Cultivator, Cotton	J. A. Hall	Columbus, Ohio	Oct. 30, 1866	59, 215
Cultivator, Cotton	J. M. Hall	Warrenton, Ga	Apr. 21, 1857	17, 091
Cultivator, Cotton	J. M. Hall	Warrenton, Ga	Jan. 18, 1859	22, 647
Cultivator, Cotton	J. A. Hartsfield	Kinston, N. C	July 17, 1860	29, 166
Cultivator, Cotton	F. E. Heiway and J. J. Walls	Hazlehurst, Miss	June 25, 1872	128, 393
Cultivator, Cotton	L. Henderson	Manson, N. C	May 19, 1868	78, 088
Cultivator, Cotton	R. B. Henderson	Warren County, N. C	Mar. 26, 1867	63, 246
Cultivator, Cotton	J. Hinman and D. S. French	Watertown, Mass., and Marietta, Ga.	July 10, 1860	29, 127
Cultivator, Cotton	W. F. Johnson	Wetumpka, Ala	June 5, 1860	28, 583
Cultivator, Cotton	M. B. Lamar	Atlanta, Ga	Aug. 22, 1871	118, 371
Cultivator, Cotton	P. M. Leatherman	Woodville, Miss	Oct. 11, 1870	108, 159
Cultivator, Cotton	W. McCracken	Bainbridge, Ind	Mar. 5, 1867	62, 660
Cultivator, Cotton	J. L. Middlebrooks	Salem, Ga	Aug. 7, 1860	29, 507
Cultivator, Cotton	E. H. Nelson	Gainesville, Ala	Sept. 2, 1873	142, 501
Cultivator, Cotton	T. Newcomb and G. W. Byrd	Smith's Fork, Tenn	Jan. 3, 1860	26, 699
Cultivator, Cotton	R. H. Purnelle	Beulah, Miss	Oct. 3, 1871	119, 644
Cultivator, Cotton	G. W. Rice	Demopolis, Ala	Apr. 16, 1861	32, 082
Cultivator, Cotton	M. Rigell and W. D. Ivey	Dawson and Milford, Ga	July 17, 1860	29, 196
Cultivator, Cotton	W. J. Rivers	Sumter District, S. C	June 28, 1859	24, 580
Cultivator, Cotton	A. A. Roberts and B. Davis	La Grange, Ga	May 26, 1857	17, 391
Cultivator, Cotton	N. Rogers	Thomas County, Ga	Nov. 27, 1860	30, 763
Cultivator, Cotton	W. A. Rogers	Sommerville, Ala	Apr. 1, 1842	2, 526
Cultivator, Cotton	J. C. Sellers	Woodville, Miss	Sept. 25, 1860	30, 163
Cultivator, Cotton	N. F. Sandelin	Mott Haven, N. Y	Mar. 26, 1872	125, 087
Cultivator, Cotton	J. Shaw	Richland, Ga	Jan. 13, 1857	16, 401
Cultivator, Cotton	A. Smith	Ashville, Ala	July 6, 1858	20, 823
Cultivator, Cotton	M. Snow	Auburn, Miss	Mar. 27, 1860	27, 659
Cultivator, Cotton	J. Speer	Hazlehurst, Miss	July 31, 1860	29, 413
Cultivator, Cotton	H. G. Street	Liberty, Miss	Nov. 27, 1860	30, 773
Cultivator, Cotton	R. A. Vick	Byhalia, Miss	Oct. 20, 1857	18, 478
Cultivator, Cotton	J. R. Wallace and B. A. McClain.	Murfreesborough, Tenn	Apr. 9, 1867	63, 767
Cultivator, Cotton	G. W. N. Yost	Yellow Springs, Ohio	July 17, 1860	29, 211
Cultivator, Cotton	C. Zocher	Augusta, Ga	June 10, 1873	139, 750
Cultivator, Cotton and cane	T. E. Shannon	Woodville, Miss	July 21, 1857	17, 849
Cultivator, Cotton and corn	M. Reed	Little Rock, Ark	Oct. 18, 1870	108, 518
Cultivator-coupling	J. Pierpont	La Harpe, Ill	Oct 8, 1867	69, 697
Cultivator-coupling	S. M. Whitney	Galesburgh, Ill	Nov. 13, 1866	59, 698
Cultivator, Device for steering	S. Lapham	Salem, Ohio	Nov. 1, 1853	10, 197
Cultivator, Double-shovel	S. B. Forbes	New Cumberland, W. Va	May 5, 1868	77, 474
Cultivator draft-hook	A. Adams	Sandwich, Ill	June 10, 1873	139, 752
Cultivator drag-bar	C. Alvord	Courtland, Wis	Oct. 20, 1868	83, 235
Cultivator, Equalizing	M. Eichholtz	Troy, Ohio	Apr. 6, 1869	88, 557
Cultivator, Field and garden	E. L. Dorsey	Union, Ind	June 5, 1866	55, 256
Cultivator, Finishing	B. Johnston	New Iberia, La	Apr. 5, 1870	101, 468
Cultivator for dirting cotton	J. S. Smith	Helena, Ark	Jan. 26, 1869	86, 185
Cultivator-frames, Attaching teeth to	G. W. Zeigler	Tiffin, Ohio	Oct. 2, 1866	58, 533
Cultivator-frames to wagon axle-trees, Attachment of.	D. Fuller	Fullersburgh, Ill	Nov. 8, 1870	108, 996
Cultivator, Gang	C. Belden	Middlebury, Ohio	Jan. 2, 1866	51, 790
Cultivator, Garden	J. H. Chapman	Newton, Iowa	Mar. 30, 1869	88, 275
Cultivator, Garden	J. M. Culver	Gilbertsville, Iowa	May 4, 1869	89, 634
Cultivator, Garden	J. A. Hall	Columbus, Ohio	Oct. 16, 1866	58, 814
Cultivator, Garden	C. F. Ruggles	Henderson, Ky	May 30, 1871	115, 526
Cultivator, Garden	G. W. Smith	Mount Olivet, Ky	May 21, 1867	64, 917
Cultivator, Garden	I. B. Smith	Norfolk, Va	Apr. 10, 1839	1, 117
Cultivator, Garden	W. S. Spratt	Allegheny City, Pa	Feb. 9, 1869	86, 879
Cultivator, Garden or hand	M. D. Cone and A. N. Douglass	Port Gibson, N. Y	Feb. 26, 1867	62, 317
Cultivator, Garden plow	C. A. Harris	Austin, Ark	Oct. 15, 1867	69, 803
Cultivator guard-attachment	J. Armstrong, jr	Elmira, Ill	Apr. 13, 1869	88, 940
Cultivator guard-attachment	T. B. McConaughey	Newark, Del	Sept. 4, 1866	57, 747
Cultivator, Hand	L. L. Beach	Mount Upton, N. Y	Dec. 28, 1869	98, 337
Cultivator, Hand	R. Blum	Champaign, Ill	Apr. 16, 1867	63, 840
Cultivator, Hand	D. Boggs and H. Rohs	Cynthiana, Ky	Sept. 26, 1871	119, 261
Cultivator, Hand	N. B. Chase and C. W. Saunders.	Wilkinsonville, Mass	Mar. 6, 1855	12, 478
Cultivator, Hand	E. Clark	Rushville, Ind	Oct. 19, 1869	95, 986
Cultivator, Hand	P. S. Clinger	Conestoga Centre, Pa	Apr. 3, 1860	27, 694
Cultivator, Hand	S. A. Conant	Centralia, Ill	Dec. 2, 1873	145, 153
Cultivator, Hand	L. J. Dawdy	Hamburgh, Iowa	Feb. 25, 1873	136, 220
Cultivator, Hand	L. Duval	Big Spring, Ky	Feb. 8, 1870	99, 543
Cultivator, Hand	R. B. Fitts and W. Thackara	Philadelphia, Pa	Aug. 18, 1863	39, 560
Cultivator, Hand	J. H. Gill	Mount Pleasant, Ohio	July 27, 1869	93, 080
Cultivator, Hand	W. C. Goodwin	Hamden, Conn	Dec. 10, 1867	72, 018
Cultivator, Hand	J. D. and A. M. Halsted	Rye, N. Y	May 20, 1862	35, 313
Cultivator, Hand	E. E. Hawley	Middletown, Conn	Dec. 5, 1848	5, 956
Cultivator, Hand	W. G. Jones	Marshall, Tex	Feb. 27, 1872	124, 140
Cultivator, Hand	D. C. Jordon	Centre Port, N. Y	May 22, 1860	28, 376
Cultivator, Hand	E. G. Mathews	Newton, Mass	June 8, 1869	91, 144
Cultivator, Hand	J. Naugle	Mooresville, Ind	Mar. 21, 1865	46, 926
Cultivator, Hand	J. A. Robinson	Poplin, N. H	Feb. 20, 1855	12, 428
Cultivator, Hand	A. W. Ross	Northfield, Mass	Nov. 25, 1873	144, 927
Cultivator, Hand	G. W. Rue	Hamilton, Ohio	Mar. 25, 1873	137, 098
Cultivator, Hand	J. Scheibleim and J. Heitzman	Philadelphia, Pa	Dec. 15, 1868	84, 910
Cultivator, Hand	G. Smith	Omaha City, Nebr	Jan. 16, 1866	52, 083
Cultivator, Hand	B. Taylor	Forestville, Minn	Aug. 11, 1868	81, 034
Cultivator, Hand garden	P. Byrns	Mindoro, Wis	July 13, 1869	92, 581
Cultivator, Hand-hoe	W. Goodwin	Marblehead, Mass	Mar. 19, 1872	124, 810
Cultivator, Harrow	F. Mishwitz	Brooklyn, N. Y	Dec. 7, 1869	97, 680
Cultivator, Harrow	H. W. Ostrom	Grand Rapids, Mich	Sept. 24, 1867	69, 239
Cultivator-hoe	J. H. Foster	Charlottesville, Va	Aug. 17, 1869	93, 869
Cultivator, Horse-hoe	A. Webb	Bangor, Me	Aug. 8, 1865	49, 354
Cultivator-joint	W. A. Dryden	Monmouth, Ill	June 22, 1869	91, 727
Cultivator-joint and coupling	W. H. Edwards	Moline, Ill	Nov. 10, 1868	83, 944

Index of patents issued from the United States Patent Office from 1790 *to* 1873, *inclusive*—Continued.

Invention.	Inventor.	Residence.	Date.	No.
Cultivator, marker, and cover, Combined	H. J. Coyle	Buffalo, N. Y	Aug. 15, 1871	117, 989
Cultivator, Meadow	F. P. Devenport	Carthage, Ill	Feb. 28, 1871	112, 130
Cultivator, Two-winged shifting-tooth	W. Beach	Philadelphia, Pa	Oct. 25, 1832	....
Cultivator or hoe-harrow	N. J. Shull	Bensalem, Pa	Nov. 26, 1836	....
Cultivator or hoe-harrow	N. J. Shull	Bensalem Township, Bucks County, Pa.	Nov. 26, 1836	88
Cultivator or hoe-harrow, &c., Teeth for	W. Beach	Philadelphia, Pa	Nov. 8, 1825	....
Cultivator-plow	C. C. Baum	Oxford, Iowa	Dec. 26, 1865	51, 682
Cultivator-plow	G. G. Black	Crossinville, Ohio	July 14, 1857	17, 777
Cultivator-plow	V. Bolis	Saint Mary's Parish, La	Sept. 27, 1870	107, 754
Cultivator-plow	M. Crenshaw	Springfield, Tex	Mar. 25, 1856	14, 533
Cultivator-plow	D. Culver	Kingston, Pa	May 23, 1871	115, 033
Cultivator-plow	J. Doak	Keithsburgh, Ill	Nov. 29, 1864	45, 231
Cultivator-plow	J. R. Finley	Delphi, Ind	May 2, 1865	47, 534
Cultivator-plow	W. O. Gibson	Charleston, S. C	Oct. 23, 1866	59, 003
Cultivator-plow	J. S. Gilmore	Millersburgh, Ill	Jan. 23, 1866	52, 159
Cultivator-plow	W. Gowen	Bartlett, Tenn	Feb. 21, 1871	111, 925
Cultivator-plow	G. W. Hatfield	Holton, Ind	Feb. 5, 1867	61, 828
Cultivator, Plow	L. Homrighouse	Baltimore, Md	May 3, 1870	102, 543
Cultivator, Plow	E. S. Huff	Zanesville, Ohio	June 29, 1869	91, 847
Cultivator-plow	M. J. Hunt	Cincinnati, Ohio	Sept. 30, 1842	2, 792
Cultivator-plow	W. S. Hyde	Townsend, Ohio	June 21, 1853	9, 798
Cultivator-plow	A. J. Lewis	Pittsburgh, Pa	June 8, 1869	91, 031
Cultivator-plow	W. Looker	Graham, Mo	Mar. 9, 1869	87, 690
Cultivator-plow	N. McKay	Columbia, Mo	Aug. 17, 1869	93, 730
Cultivator-plow	K. McKinnon	Pleasant Hill, Ala	Dec. 2, 1873	145, 222
Cultivator-plow	J. Meyer	Linden Hall, Pa	Nov. 11, 1862	36, 909
Cultivator-plow	J. G. Miner	Nashville, Tenn	June 28, 1870	104, 754
Cultivator-plow	T. H. Miner and S. Heavenridge	Greenfield, Ind	Mar. 20, 1866	53, 322
Cultivator-plow	W. D. Nichols	Chicago, Ill	Sept. 11, 1866	57, 952
Cultivator-plow	J. Singer	Mendota, Ill	Jan. 31, 1871	111, 391
Cultivator-plow	W. H. Startzman	Big Lick, Va	Sept. 3, 1867	68, 393
Cultivator-plow	W. W. Stillman	Mount Hawley, Ill	Aug. 7, 1866	57, 007
Cultivator-plow	L. Strickland	Talleyrand, Iowa	Jan. 23, 1866	52, 222
Cultivator-plow	L. S. Tewell	Elbinsville, Pa	May 9, 1871	114, 726
Cultivator-plow	M. Tolle	Newport, Ky	June 16, 1857	17, 594
Cultivator-plow	L. M. Whitman	Weedsport, N. Y	Oct. 11, 1853	10, 123
Cultivator-plow	S. A. Wray	Greenfield, Ind	Jan. 1, 1867	60, 982
Cultivator-plow	W. E. Wyche	Brookville, N. C	Jan. 8, 1856	14, 075
Cultivator-plow	W. E. Wyche	Brookville, N. C	Feb. 26, 1856	14, 333
Cultivator-plow axle, Wheeled	G. W. C. Gillespie	Burlington, Iowa	Sept. 9, 1851	8, 348
Cultivator-plow, Double	P. Coonrod	Keithsburgh, Ill	Dec. 24, 1867	72, 456
Cultivator-plow fender	W. E. Moore	Crawfordsville, Ind	May 25, 1869	90, 376
Cultivator-plow, harrow, and roller, Combined	S. C. Thornton	Macomb, Tex	Nov. 3, 1868	83, 675
Cultivator-point	R. Creswell, jr	Chambersburgh, Pa.	Sept. 5, 1848	5, 755
Cultivator, Potato and corn	J. M. Davidson	Pulaski, Pa	Mar. 16, 1869	87, 914
Cultivator-rake	J. T. Van Wyck	Poughkeepsie, N. Y	Nov. 8, 1870	109, 082
Cultivator, Reversible	S. G. Tufts	Mainville, Ohio	June 15, 1869	91, 497
Cultivator, Revolving	W. A Estes	South China, Me	Sept. 7, 1869	94, 482
Cultivator, Revolving	A. J. Stephens	El Dorado, Wis	Mar. 9, 1869	87, 721
Cultivator, Revolving	G. Whitlock	Crown Point, N. Y	July 10, 1841	2, 163
Cultivator, Rice	G. W. Cooper	Ogeechee, Ga	June 18, 1867	65, 877
Cultivator, Rice	G. W. Cooper	Ogeechee, Ga	Aug. 4, 1868	80, 604
Cultivator, Rotary	E. T. Bussell	Shelbyville, Ind	June 15, 1858	20, 605
Cultivator, Rotary	G. Collins	Fremont, Nebr	Mar. 5, 1872	124, 332
Cultivator, Rotary	C. Comstock	Milwaukee, Wis	Feb. 26, 1861	31, 531
Cultivator, Rotary	M. Decelle	Newburgh, Ohio	June 11, 1872	127, 746
Cultivator, Rotary	B. F. Field	Sheboygan Falls, Wis	July 19, 1859	24, 799
Cultivator, Rotary	G. B. Field	Saint Louis, Mo	Mar. 14, 1854	10, 624
Cultivator Rotary	J. C. Fitzgerald	Willet, N. Y	Dec. 4, 1866	60, 164
Cultivator, Rotary	H. M. Johnson	Carlisle, Pa	June 27, 1854	11, 162
Cultivator, Rotary	G. F. Lynch	Milwaukee, Wis	Aug. 4, 1868	80, 643
Cultivator, Rotary	S. Mahurin	Clayton, Ill	Oct. 20, 1868	83, 187
Cultivator, Rotary	R. McKinley	Hyde Park, N. Y	May 20, 1873	139, 013
Cultivator, Rotary	J. W. Milroy	Haywood, Cal	Jan. 21, 1873	135, 002
Cultivator, Rotary	C. N. Poundstone	Livonia, Ill	Feb. 11, 1873	135, 664
Cultivator, Rotary	P. E. Royse	New Albany, Ind	Feb. 17, 1852	8, 747
Cultivator, Rotary	P. Shaw	Abington, Mass	Mar. 14, 1854	10, 646
Cultivator, Rotary	J. D. Starritt	Chicago, Ill	Nov. 11, 1873	144, 419
Cultivator, Rotary	A. Thompson	Ottumwa, Iowa	Sept. 25, 1866	58, 319
Cultivator, Rotary	J. Young	Joliet, Ill	June 28, 1859	24, 597
Cultivator, Rotary root-digging	S. Snow and A. Hine	Fayetteville and La Fayette, N. Y.	Oct. 11, 1853	10, 113
Cultivator, Rotating	T. Uehling	Logan, Neb	Oct. 19, 1869	95, 956
Cultivator, Rotating-tooth	J. Densmore	Holley, N. J	Dec. 3, 1867	71, 589
Cultivator, scraper, and chopper, Cotton	J. H. W. Young	Henderson, Tex	Dec. 27, 1870	110, 531
Cultivator, Seed-planting	G. Phillips	Philadelphia, Pa	Nov. 15, 1853	10, 236
Cultivator, seeder, and stalk-cutter, Combined	J. and W. Whait	Independence, Iowa	Feb. 20, 1872	123, 959
Cultivator, Seeding	A. P. Durant	Atlanta, Ill	July 16, 1861	32, 825
Cultivator, Seeding	J. B. Ewell	Baltimore, Md	Aug. 20, 1867	67, 860
Cultivator, Seeding	T. A. Galt	Sterling, Ill	Dec. 11, 1860	30, 876
Cultivator, Seeding	J. Goodman and S. Rote	Lancaster, Pa	Jan. 15, 1861	31, 118
Cultivator, Seeding	H. Hutchinson	Three Rivers, Mich	May 21, 1867	64, 981
Cultivator, Seeding	J. Reichard	Guilford Township, Pa	Dec. 16, 1833	....
Cultivator, Seeding	C. T. Settle	San José, Cal	Feb. 26, 1861	31, 564
Cultivator, Seeding	N. S. Thompson	Richmond, Ind	May 16, 1871	114, 881
Cultivator, Seeding	S. D. Tracy	Vernon, N. Y	Dec. 27, 1859	26, 630
Cultivator, Seeding	N. Whitehall	Newtown, Ind	May 17, 1859	24, 071
Cultivator-shovels, Attachment for	D. C. Stover	Dayton, Ohio	Apr. 14, 1868	76, 647
Cultivator, Steam	A. J. Stevens	San Francisco, Cal	Aug. 10, 1869	93, 494
Cultivator, Sugar-cane	A. Trouard	New Orleans, La	Jan. 21, 1873	135, 174
Cultivator, Sulky	F. G. Blauser	Etna, Ohio	July 26, 1870	105, 620
Cultivator, Sulky	E. S. Easterday	Nokomis, Ill	Feb. 25, 1868	74, 754
Cultivator, Sulky	F. Farnsworth	Frankfort, Ill	Apr. 12, 1870	101, 720
Cultivator, Sulky	O. L. Gaylord	Plainfield, Ill	Oct. 12, 1869	95, 790
Cultivator, Sulky	W. Harvey	Volga City, Iowa	July 7, 1868	79, 570
Cultivator, Sulky	P. Hewitt	Farmland, Ind	Oct. 31, 1871	120, 435

Index of patents issued from the United States Patent Office from 1790 *to* 1873, *inclusive*—Continued.

Invention.	Inventor.	Residence.	Date.	No.
Cultivator, Sulky	H. P. Jordan	Victoria, Tex	Jan. 10, 1871	110, 853
Cultivator, Sulky	R. B. Robbins	Adrian, Mich	Mar. 8, 1871	112, 496
Cultivator, Sulky	R. B. Robbins	Adrian, Mich	Nov. 9, 1869	96, 733
Cultivator, Sulky	J. Robinson	Plainfield, Ill	June 22, 1869	91, 566
Cultivator, Sulky	I. Welty	Olney, Ill	July 21, 1868	80, 102
Cultivator, Sulky	N. Whitehall	Newtown, Ind	Aug. 29, 1871	118, 567
Cultivator, Sulky	N. Wilson	Saint Louis, Mich	June 29, 1869	92, 135
Cultivator-teeth	G. C. Aiken	Nashua, N. H	Feb. 14, 1860	27, 092
Cultivator-teeth	G. C. Aiken	Nashua, N. H	Apr. 24, 1860	27, 956
Cultivator-teeth	W. Beach	Philadelphia, Pa	Apr. 13, 1826	....
Cultivator-teeth	S. Beckwith	Oshkosh, Wis	Mar. 15, 1870	100, 713
Cultivator-teeth	J. Birdsell	Hamorton, Pa	Nov. 9, 1844	3, 819
Cultivator-teeth	M. Bucklin	Grafton, N. H	Feb. 2, 1858	19, 234
Cultivator-teeth	N. Chappell	Lima, N. Y	Dec. 3, 1867	71, 581
Cultivator-teeth	J. P. Cramer	Schuylerville, N. Y	Jan. 6, 1857	16, 364
Cultivator-teeth	G. Custer	Monroe, Mich	Dec. 2, 1862	37, 071
Cultivator-teeth	L. Dayley	Minaville, N. C	Aug. 22, 1871	118, 208
Cultivator-teeth	D. Dean	Brighton, Mich	June 30, 1868	79, 451
Cultivator-teeth	J. D. De Turk	Exeter, Pa	Mar. 17, 1868	75, 669
Cultivator-teeth	H. Francisco	Lake Mills, Wis	Mar. 7, 1865	46, 654
Cultivator-teeth	E. L. Freeman	Brownville, N. Y	Sept. 8, 1857	18, 174
Cultivator-teeth	E. L. Freeman	Williamstown, N. Y	Apr. 20, 1869	89, 137
Cultivator-teeth	H. F. French	Boston, Mass	Aug. 17, 1869	93, 695
Cultivator-teeth	W. P. and T. H. Ford	Concord, N. H	Feb. 15, 1859	22, 946
Cultivator-teeth	F. R. Forsythe	Cape Vincent, N. Y	Aug. 4, 1857	17, 925
Cultivator-teeth	H. B. Hammon	Bristolville, Ohio	Apr. 10, 1860	27, 797
Cultivator-teeth	B. F. Hisert	Norton Hill, N. Y	Sept. 3, 1867	68, 440
Cultivator-teeth	J. S. Honey	Hartford, Ohio	Apr. 17, 1849	6, 336
Cultivator-teeth	W. H. Kelly	Lysander, N. Y	Oct. 27, 1863	40, 414
Cultivator-teeth	L. Lamborn	Kennett's Square, Pa	Mar. 26, 1850	7, 220
Cultivator-teeth	M. F. Lowth and T. J. Howe	Owatonna, Minn	Mar. 10, 1868	75, 436
Cultivator-teeth	D. C. Matteson and T. P. Williamson.	Stockton, Cal	Apr. 9, 1867	63, 647
Cultivator-teeth	G. Maynard	Ilion, N. Y	June 28, 1859	24, 568
Cultivator-teeth	W. Morrison	Chadd's Ford, Pa	Apr. 16, 1861	32, 077
Cultivator-teeth	E. B. Pratt	Monroe, Wis	July 20, 1869	92, 880
Cultivator-teeth	W. B. Ready	Sacramento, Cal	July 28, 1868	80, 503
Cultivator-teeth	D. B. Rogers	Pittsburgh, Pa	May 8, 1860	28, 198
Cultivator-teeth	D. B. Rogers	Stafford, N. Y	Nov. 1, 1845	4, 245
Cultivator-teeth	L. Rogers	Pittsburgh, Pa	Feb. 4, 1868	74, 004
Cultivator-teeth	H. Sanders	Utica, N. Y	Nov. 29, 1859	26, 297
Cultivator-teeth	C. H. Sayre	Utica, N. Y	Oct. 20, 1857	18, 471
Cultivator-teeth	C. H. Sayre	Utica, N. Y	June 19, 1860	28, 780
Cultivator-teeth	C. H. Sayre and G. Klinck	Utica, N. Y	Feb. 12, 1856	14, 254
Cultivator-teeth	J. Stockdale	Ypsilanti, Mich	Jan. 30, 1855	12, 332
Cultivator-teeth	G. F. Stroud	Oshkosh, Wis	July 29, 1873	141, 400
Cultivator-teeth	J. Turner and T. P. Smith	Sunapee, N. H	June 26, 1860	28, 945
Cultivator-teeth	B. Van Braeklin	Le Roy, N. Y	July 20, 1869	92, 910
Cultivator-teeth, Adjustable	W. M. Hurlburt	Winona, Minn	Jan. 26, 1869	86, 306
Cultivator-teeth, Die for cutting and forming	D. B. Rogers	Stafford, N. Y	Nov. 8, 1845	4, 265
Cultivator-teeth, Hanging	J. Fowler and F. M. Bacon	Watertown, Wis	Mar. 28, 1865	47, 008
Cultivator-teeth, Machine for forming	D. B. Rogers	Pittsburgh, Pa	Mar. 28, 1854	10, 706
Cultivator-teeth, Machine for pointing blanks for	J. Pedder and G. Abel	West Pittsburgh, Pa	Mar. 7, 1871	112, 374
Cultivator-teeth, Manufacture of	J. Behel	Rockford, Ill	June 3, 1873	139, 535
Cultivator-teeth, Mode of attaching	J. C. Hoffeditz	Mercersburgh, Pa	Jan. 1, 1867	60, 893
Cultivator-teeth, Operating	T. W. Hammon	Montfort, Wis	Dec. 27, 1864	45, 671
Cultivator-teeth, Plate for blanks of	W. H. Singer	Pittsburgh, Pa	Nov. 30, 1869	97, 451
Cultivator-teeth, Rolled steel plates for making	W. W. and D. D. Skinner	Des Moines, Iowa	Sept. 26, 1871	119, 419
Cultivator-teeth, Spring	H. Francisco	White Water, Wis	Nov. 5, 1861	33, 641
Cultivator-teeth to beams, Attaching	M. F. Lowth and O. H. Porter	Wabasha, Minn	July 18, 1871	117, 692
Cultivator, Vine, &c	J. Mason	Union, N. J	May 28, 1842	2, 650
Cultivator, Walking	A. Friberg	Moline, Ill	Feb. 7, 1871	111, 627
Cultivator, Walking	E. P. Lynch and H. R. Raff	Davenport, Iowa	Sept. 7, 1869	94, 623
Cultivator, Walking	J. Vanluvanee and H. Smith	Moline, Ill	Mar. 21, 1871	112, 990
Cultivator weed-cutter	C. Rodger	Montpelier, Vt	June 25, 1850	7, 463
Cultivator, Wheel	D. S. Billings	Conneaut, Pa	Feb. 1, 1848	5, 429
Cultivator, Wheel	G. Bradley	Rockford, Ill	Dec. 9, 1873	145, 331
Cultivator, Wheel	W. F. and G. Coulter and J. A. Lanery.	Hardinsburgh, Ind	Mar. 16, 1869	87, 824
Cultivator, Wheel	A. P. Durant and D. M. Buckley.	Atlanta, Ill	Feb. 28, 1865	46, 550
Cultivator, Wheel	N. Earlywine	Centreville, Iowa	Jan. 2, 1872	122, 373
Cultivator, Wheel	J. Eshleman	Canaan Centre, Ohio	Aug. 2, 1870	106, 039
Cultivator, Wheel	E. D. Hatch	Oconomowoc, Wis	Aug. 23, 1870	106, 579
Cultivator, Wheel	J. F. Matchet and P. W. Smith	Paris, Mo	Oct. 14, 1873	143, 631
Cultivator, Wheel	W. M. Pitts	Holden, Mo	Sept. 26, 1871	119, 404
Cultivator, Wheel	J. H. Randolph, jr	Bayou Goula, La	May 13, 1873	138, 931
Cultivator, Wheel	S. Rockafellow	Moline, Ill	Feb. 23, 1872	126, 092
Cultivator, Wheel	F. P. Root	Sweden, N. Y	June 8, 1852	9, 003
Cultivator, Wheel	T. C. Sebring	Milford, Mich	Dec. 5, 1871	121, 670
Cultivator, Wheel	W. A. Sisson	Sheffield, Ill	Aug. 27, 1867	68, 124
Cultivator, Wheel	W. H. Strong	Seneca County, Ohio	Dec. 30, 1873	146, 030
Cultivator, Wheel	J. A Viars	Sherman, Tex	Sept. 19, 1871	119, 205
Cultivator, Wheeled	S. H. Dwight and W. B. Chambers.	Decatur, Ill	Mar. 29, 1870	101, 240
Cultivators, Insect-destroying attachment for	C. T. Hurd	Victoria, Tex	Dec. 30, 1873	145, 949
Cultivators, Shovel-teeth for	T. Harding	La Fayette, Ind	July 19, 1870	105, 570
Culvert	C. McIntire	Easton, Pa	Sept. 16, 1862	36, 472
Culvert, Iron	M. G. Freeman	Wenona, Ill	Dec. 6, 1870	109, 886
Culvert, Iron	M. G. Freeman	Bloomington, Ill	May 9, 1871	114, 662

Cup:

See Breast-cup.
Dentists' impression-cup.
Drinking-cup.
Egg-cup.
Ejecting-cup.
Elevator-cup.
Eye-cup.

Index of patents issued from the United States Patent Office from 1790 *to* 1873, *inclusive*—Continued.

Invention.	Inventor.	Residence.	Date.	No.
Cup—Continued.				
See Grease-cup.				
Ink-cup.				
Lubricating-cup.				
Molasses-cup.				
Mucilage-cup.				
Nursery-cup.				
Oil-cup.				
Soap-cup.				
Sponge-cup.				
Surgical cup.				
Teat-cup.				
Tin cup.				
Cup and bell	T. Leach	Taunton, Mass	Aug. 2, 1870	106, 066
Cup and saucer, Metallic Columbian	J. Love	Baltimore, Md	June 14, 1814	....
Cup for effervescing drinks	A. Rottanzi	San Francisco, Cal	July 7, 1868	79, 689
Cups, Device for holding lids to	P. I. Schopp	Louisville, Ky	Aug. 10, 1869	93, 562
Cupboard and extension-table combined	B. F. Partridge	Syracuse, N. Y	Feb. 4, 1873	135, 440
Cupboard and sink	A. Iske	Lancaster, Pa	Aug. 7, 1860	29, 493
Cupboard and table	J. C. Mack	Bristol, Conn	Sept. 22, 1868	82, 426
Cupboard-catch	P. D. F. Goeney	Albany, N. Y	Sept. 15, 1868	82, 108
Cupboard-catch	I. Ottner	New Britain, Conn	Nov. 5, 1872	132, 771
Cupboard, Portable	J. L. Prescott	North Berwick, Me	Nov. 6, 1866	59, 516
Cupboard, Revolving	W. Wright	Bloomfield, N. J	Dec. 14, 1869	97, 980
Cupola and blast furnace	J. R. Grout	Detroit, Mich	July 31, 1866	56, 746
Cupola and blast furnace	D. W. Hendrickson	Red Bank, N. J	Aug. 13, 1872	130, 370
Cupola and blast furnace	C. Truesdale	Cincinnati, Ohio	Aug. 25, 1868	81, 561
Cupola and other furnace	A. Cowan and R. H. Starr	New Haven, Conn	July 9, 1867	66, 465
Cupola and other furnace	P. W. McKenzie and C. W. Isbell.	Jersey City, N. J., and New York, N. Y.	May 29, 1866	55, 204
Cupola and other furnace	L. Quincke	Joliet, Ill	Oct. 7, 1873	143, 463
Cupola and other melting and smelting furnace	J. Absterdam	New York, N. Y	Feb. 9, 1869	86, 795
Cupola and other metallurgic furnaces	O. Bolton, jr	Pittsburgh, Pa	Mar. 25, 1873	137, 125
Cupola, blast, and smelting furnace	H. Fayette	Portchester, N. Y	Oct. 29, 1867	70, 329
Cupola-furnace	F. C. Adams	Cincinnati, Ohio	May 12, 1868	77, 795
Cupola-furnace	R. Barckley	Philadelphia, Pa	Aug. 4, 1863	39, 372
Cupola-furnace	N. Burdick	Albany, N. Y	June 16, 1868	78, 924
Cupola-furnace	A. G. Cook	Burlington, Vt	June 13, 1871	115, 823
Cupola-furnace	J. Dougherty	Philadelphia, Pa	May 14, 1867	64, 645
Cupola-furnace	J. Dougherty	Philadelphia, Pa	May 2, 1871	114, 422
Cupola-furnace	J. H. Eddy	Taunton, Mass	Aug. 4, 1868	80, 537
Cupola-furnace	J. Howarth	Salem, Mass	Jan. 14, 1868	73, 333
Cupola-furnace	J. and T. Insull	New Haven, Conn	May 8, 1866	54, 552
Cupola-furnace	A. A. Lincoln	Norton, Mass	Oct. 31, 1865	50, 769
Cupola-furnace	P. W. Mackenzie	Jersey City, N. J	Aug. 25, 1857	18, 051
Cupola-furnace	P. W. Mackenzie	Blauveltville, N. Y	May 6, 1873	138, 510
Cupola-furnace	J. B. Pearse	Swatara Township, Pa	Apr. 22, 1873	138, 184
Cupola-furnace	A. Pevy	Lowell, Mass	Oct. 24, 1865	50, 623
Cupola-furnace	C. Truesdale	Cincinnati, Ohio	May 1, 1866	54, 470
Cupola-furnace	C. Truesdale	Cincinnati, Ohio	Dec. 11, 1866	60, 440
Cupola-furnace	H. Warral	New York	Dec. 31, 1821	....
Cupola-furnace for melting iron	I. D. Marshbank	Harrisburgh, Pa	July 16, 1872	129, 151
Cupola or blast furnace	A. G. Cook	Burlington, Vt	Feb. 20, 1866	52, 684
Cupping	L. Tillotson	Thompson, Ohio	Aug. 26, 1856	15, 626
Cupping and breast glass	W. S. Thomas	Norwich, N. Y	Mar. 16, 1852	8, 869
Cupping-apparatus	W. D. Hooper	Liberty, Va	Sept. 17, 1867	68, 985
Cupping-apparatus	M. Mattson	New York, N. Y	Aug. 13, 1867	67, 663
Cupping-instrument	R. J. Dodd	Philadelphia, Pa	Apr. 13, 1844	3, 537
Cupping-instrument	E. J. Leyburn	Lexington, Va	Sept. 10, 1867	68, 630
Cupping-instrument	S. McLean	Reynale's Basin, N. Y	Aug. 19, 1856	15, 568
Curb and gutter, Combined metallic street	W. E. Worthen	New York, N. Y	May 10, 1859	23, 973
Curb-bit	W. C. Baker	New York, N. Y	Jan. 23, 1866	52, 125
Curb-bit	T. S. Smith	Lowell, Mass	Oct. 13, 1863	40, 312
Curb, Excavating and well-boring	J. Cowperthwaite, sr		Apr. 13, 1809	....
Curb for street, road, &c	O. Faurot and W. P. Harris	Brooklyn, N. Y	Apr. 14, 1868	76, 614
Curbing	J. Miller	Pittsburgh, Pa	Sept. 3, 1867	68, 379
Curbing, Street	D. Herr	Lockhaven, Pa	Apr. 16, 1872	125, 814
Curculio-catcher	F. J. Claxton and C. D. Stevens	Saint Louis, Mo	Apr. 18, 1871	113, 855
Curculio-catcher	M. M. Hooton	Clinton County, Ill	July 12, 1870	105, 333
Curculio-trap	J. Smith	New Market, Ohio	Jan. 1, 1867	60, 951
Curd-agitator	D. C. Hall	Barnes' Corners, N. Y	June 22, 1869	91, 622
Curd-agitator	J. H. Maydole	Eaton, N. Y	Dec. 1, 1863	40, 764
Curd-agitator	B. G. Swain	Colden, N. Y	Sept. 13, 1870	107, 423
Curd-breaker	P. Crary	Lowville, N. Y	Jan. 21, 1868	73, 508
Curd-cutter	F. G. Abbey	Sandisfield, Mass	Sept. 6, 1864	44, 058
Curd-cutter	J. Crosby	Rome, N. Y	Mar. 8, 1864	41, 833
Curd-cutter	J. B. Lyons	Milton, Conn	Apr. 5, 1864	42, 209
Curd-cutter	J. H. Maydole	Eaton, N. Y	Oct. 18, 1864	44, 733
Curd-cutter	C. Wadsworth	East Livermore, Me	Oct. 3, 1865	50, 290
Curd-grinder	C. W. Terpening	Geneseo, Ill	Aug. 17, 1869	93, 768
Curd-mill	J. Macadam	Little Falls, N. Y	June 30, 1868	79, 365
Curd-mill	W. Ralph	Utica, N. Y	July 21, 1868	80, 089
Curing provisions	D. E. Somes	Biddeford, Me	Nov. 13, 1868	30, 658
Curl, Artificial	J. Graham	New York, N. Y	Mar. 11, 1873	136, 720
Curling-iron	J. O. Bentley and J. Jackson	Philadelphia, Pa	Oct. 11, 1870	108, 234
Curling-iron	H. Christian	New York, N. Y	Sept. 26, 1865	50, 095
Curling-iron	S. E. Condon	Williamsburgh, N. Y	Feb. 25, 1868	74, 800
Curling-iron	H. D. Jennings	Ilion, N. Y	July 4, 1865	48, 626
Curling-iron	J. S. Morgan	Brooklyn, N. Y	Nov. 11, 1873	144, 471
Curling-iron	L. Rosenstein and S. Feder	Memphis, Tenn	July 22, 1873	141, 079
Curling-iron	J. H. Williams	Middletown, Conn	July 16, 1872	129, 445
Curling-pin, Hair	J. Hall	Watertown, Mass	Feb. 20, 1872	123, 778
Curling-rod for paper cigar-lighters	W. D. Young	South Pittsburgh, Pa	Apr. 5, 1870	101, 561
Currency and stamp box	L. L. Tower	Cambridgeport, Mass	July 7, 1863	39, 184
Currency note, Diagram for testing the value of mutilated.	L. Fox	New York, N. Y	Aug. 1, 1865	49, 099
Current and paddle wheel	J. H. Hanchett	Beloit, Wis	Apr. 7, 1857	16, 975
Current and tide mill	H. Allen	Fayette County, Tenn	Dec. 30, 1820	....

Index of patents issued from the United States Patent Office from 1790 *to* 1873, *inclusive*—Continued.

Invention.	Inventor.	Residence.	Date.	No.
Current and tide wheel	A. Madison	Detroit, Mich	July 10, 1829	....
Current-changing apparatus	C. C. Hare	Kansas City, Mo	June 8, 1869	91, 126
Current-mill	G. H. D. Gray	Southampton, Va	Sept. 25, 1822	....
Current-mill	J. Wallace	Hartford, Ky	Sept. 11, 1829	....
Current-reversers and circuit-breakers	J. E. Smith	New York, N. Y	July 16, 1872	129, 607
Current-wheel	P. Chesley	Candia, N. H	Oct. 28, 1856	15, 970
Current-wheel	J. Clarke		Dec. 31, 1793	
Current-wheel	J. Dennison	Hillsborough, N. H	July 20, 1869	92, 800
Current-wheel	C. Morehouse	Wayland, N. Y	May 16, 1871	114, 959
Current-wheel	W. S. Smith	Cedar Rapids, Iowa	Mar. 27, 1855	12, 606
Current-wheel	W. Tuder	Moffettown, Tex	June 6, 1871	115, 660
Current-wheel	C. Weeke	Saint Charles, Mo	July 29, 1873	141, 301
Current-wheel	E. A. White	Boston, Mass	Feb. 8, 1870	99, 614
Current-wheel and condensing-engine, Combined	A. Spear	Cape Elizabeth, Me	Mar. 15, 1870	100, 946
Current-wheel, Apparatus for	J. Secor	Saint Louis, Mo	Feb. 20, 1849	6, 132
Current-wheel, Hydraulic	W. P. Wing	Greenwich, Mass	Sept. 22, 1837	403
Currents, Means for creating artificial	E. Bantz	Baltimore, Md	June 17, 1873	139, 854
Currier's arm or grain-board	A. Chase	Weare, N. H	Apr. 5, 1864	42, 165
Currier's beam and knife	J. D. Willoughby	Carlisle, Pa	Aug. 17, 1852	9, 207
Currier's beam, Constructing the face of	I. Lindsey	Charlestown, Mass	Jan. 27, 1841	1, 955
Currier's bench, Clamp for holding leather to the	J. Shimer	Bronson, Mich	May 26, 1868	78, 334
Currier's knife	T. C. Barr	Clark County, Ky	Jan. 28, 1833	
Currier's knife	G. Featherston	Au Sable Forks, N. Y	Jan. 6, 1863	37, 285
Currier's knife	L. A. Gignac	Troy, N. Y	Aug. 31, 1869	94, 197
Currier's knife	J. Glenn and J. Herd	Urbana, Ohio	Dec. 16, 1833	
Currier's knife	I. H. Harrington	Albany, Y. Y	June 26, 1828	....
Currier's knife	J. P. Hawks	Troy, N. Y	Jan. 9, 1866	51, 942
Currier's knife	W. P. Moses	Exeter, N. H	June 5, 1860	28, 594
Currier's knife	F. Palmer	Littleton, N. H	Feb. 17, 1826	....
Currier's knife	S. Parsons	Parsonsfield, Mass	Oct. 25, 1810	....
Currier's knife	D. Tomlinson	Brookfield, Conn	July 6, 1820	....
Currier's knife and double trimmer	L. Townsend	Farmington, Me	Jan. 16, 1835	....
Currier's knife for shaving leather	J. Batchelder	Lancaster, N. H	Nov. 3, 1825	....
Curriers' knives, Constructing and grinding	S. Brooks	New York	Nov. 13, 1826	....
Curriers' knives, Grinding	H. Cunningham	Albany, N. Y	Oct. 12, 1869	95, 778
Curriers' knives, Machine for sharpening	J. Bachelder	Lancaster, N. H	Nov. 9, 1824	....
Curriers' scourer	J. Hankey	North Cambridge, Mass	Feb. 6, 1866	52, 491
Curriers' slicker	G. T. Collins	North Eastham, Mass	June 6, 1871	115, 709
Curriers' slicker	D. Peters and J. W. Pauly	Keokuk, Iowa	Feb. 12, 1867	62, 064
Curriers' slicker, Whitening	D. Peters and W. D. Wilson	Keokuk, Iowa	Apr. 19, 1864	42, 397
Curriers' tools	J. T. Barnstead	Peabody, Mass	Feb. 21, 1871	111, 901
Curriers' tools	C. A. Gardner and J. A. Enos	Peabody, Mass	Apr. 8, 1873	137, 671
Curry-comb	W. F. Arnold	Winthrop, Conn	Oct. 23, 1866	59, 118
Curry-comb	J. M. Baker	Marshfield, Ohio	Sept. 21, 1869	94, 991
Curry-comb	H. L. Baldwin	Branford, Conn	June 11, 1861	32, 505
Curry-comb	J. H. Barringer, jr	Hillsborough, Ill	Mar. 3, 1868	74, 974
Curry-comb	W. Beach	Philadelphia, Pa	Mar. 13, 1849	6, 181
Curry-comb	C. B. Bristol	New Haven, Conn	Mar. 1, 1864	41, 755
Curry-comb	C. B. Bristol	New Haven, Conn	Nov. 3, 1868	83, 757
Curry-comb	J. Chaumont	Woodhaven, N. Y	June 27, 1871	116, 268
Curry-comb	E. L. Evans	Providence, R. I	Nov. 25, 1856	16, 115
Curry-comb	E. L. Evans	Providence, R. I	Jan. 26, 1858	19, 188
Curry-comb	J. Filson, jr	Racine, Ohio	Feb. 6, 1872	123, 469
Curry-comb	F. N. Frost	New Britain, Conn	Nov. 27, 1866	59, 992
Curry-comb	N. C. Harris	Poultney, Vt	July 5, 1864	43, 405
Curry-comb	N. C. Harris and A. Butler	Poultney, Vt	Oct. 6, 1857	18, 337
Curry-comb	D. E. Hayward	Malden, Mass	July 16, 1872	129, 224
Curry-comb	A. Hotchkiss	Sharon, Conn	Mar. 13, 1849	6, 178
Curry-comb	A. A. and A. Hotchkiss	Sharon, Conn	July 8, 1856	15, 294
Curry-comb	J. E. Insley	Philadelphia, Pa	June 22, 1869	91, 637
Curry-comb	L. P. Jenks	Boston, Mass	July 13, 1869	92, 454
Curry-comb	J. Jones	Bristol, Conn	Aug. 28, 1847	5, 263
Curry-comb	W. P. Kellogg	Troy, N. Y	July 2, 1872	128, 632
Curry-comb	W. P. Kellogg and M. Sweet	Lansingburgh and Troy, N. Y.	Sept. 24, 1867	69, 100
Curry-comb	L. Knapp	Woodhaven, N. Y	Mar. 1, 1870	100, 417
Curry-comb	J. W. Latcher	Albany, N. Y	July 21, 1868	80, 190
Curry-comb	W. E. Lawrence	New York, N. Y	Dec. 27, 1870	110, 576
Curry-comb	W. E. Lawrence	New York, N. Y	June 6, 1871	115, 733
Curry-comb	H. Mithoff	Columbus, Ohio	Aug. 17, 1869	93, 782
Curry-comb	S. A. Morton	Wadesborough, N. C	Dec. 23, 1873	145, 891
Curry-comb	B. F. Neal	Poultney, Vt	Oct. 6, 1863	40, 179
Curry-comb	E. M. Noyes	Binghamton, N. Y	Sept. 17, 1867	69, 016
Curry-comb	R. D. Porter	Zanesville, Ohio	Dec. 10, 1861	33, 904
Curry-comb	A. H. Reid	Brandywine Manor, Pa	Dec. 2, 1873	145, 242
Curry-comb	B. W. Remington	Providence, R. I	Feb. 11, 1868	74, 423
Curry-comb	J. W. Rockwell	Ridgefield, Conn	May 26, 1863	38, 696
Curry-comb	J. W. Rockwell	Richfield, Conn	Dec. 15, 1863	40, 946
Curry-comb	C. W. Saladee	Putnam, Ohio	Dec. 27, 1864	45, 640
Curry-comb	C. W. Saladee	Putnam, Ohio	Mar. 7, 1865	46, 710
Curry-comb	R. C. Shippey	Union, N. Y	Aug. 20, 1872	130, 755
Curry-comb	M. Sweet	Troy, N. Y	Oct. 21, 1862	36, 736
Curry-comb	M. Sweet	Troy, N. Y	June 12, 1866	55, 553
Curry-comb	M. Sweet	Troy, N. Y	Apr. 20, 1869	89, 182
Curry-comb	J. W. Trussell	Troy, N. Y	Nov. 25, 1873	145, 033
Curry-comb	J. E. Yager	Barboursville, Va	Nov. 23, 1869	97, 258
Curry-comb	S. J. Wheeler	New Britain, Conn	Jan. 22, 1861	31, 199
Curry-comb	W. Wheeler	West Poultney, Vt	Apr. 25, 1846	4, 470
Curry-comb	N. Whipple	Ledyard, Conn	Oct. 29, 1872	132, 551
Curry-comb	T. Wilkinson	Cambridge, N. Y	Aug. 16, 1844	3, 706
Curry-comb and brush	H. C. Smith	Coxsackie, N. Y	Oct. 3, 1871	119, 659
Curry-combs, Construction of	W. Wheeler	Troy, N. Y	Mar. 1, 1853	9, 605
Curry-combs, Cutting the bars and teeth of	W. Wheeler	West Poultney, Vt	Oct. 25, 1853	10, 166
Curry-combs, Method of casting the backs upon the teeth of.	J. M. Gardner	Troy, N. Y	Mar. 18, 1851	7, 986
Curry-combs, Mode of making	N. C. Sandford	Meriden, Conn	Nov. 3, 1838	997
Curry-combs, Riveting	B. B. Hotchkiss	Sharon, Conn	Jan. 1, 1861	31, 064
Currying and leather-dressing machine	R. P. Boyce	Erata, Miss	Apr. 17, 1860	27, 885

Index of patents issued from the United States Patent Office from 1790 to 1873, inclusive—Continued.

Invention.	Inventor.	Residence.	Date.	No.
Currying and tempering knife	D. S. Delano	Cornwall, Conn	Apr. 6, 1833	....
Currying cattle, Cards for	C. S. Dickerman	Lansingburgh, N. Y	Feb. 9, 1858	19, 288
Currying, Process for preparing stuffing for	C. L. Morehouse	Cleveland, Ohio	June 5, 1866	55, 426
Curtain-cord fastener	J. P. Arnold	Philadelphia, Pa	May 27, 1873	139, 228
Curtain-cord fastener	C. Gammel	Utica, N. Y	May 13, 1873	138, 878
Curtain	R. Neuninger and F. Brauer	Newark, N. J	Feb. 21, 1871	112, 067
Curtain and carpet fastener	P. Miles	New York, N. Y	Mar. 30, 1869	88, 501
Curtain and map fixture	J. W. Burns	Medway, Ohio	Jan. 25, 1870	99, 151
Curtain and shawl stretcher	J. Nicklin	Cleveland, Ohio	Dec. 21, 1869	98, 094
Curtain-brake	T. S. Baylie and G. H. Maurer	Washington, D. C	Jan. 3, 1871	110, 726
Curtain-clasp	B. F. Watson and A. Shepard	Bridgeport, Ill	Nov. 24, 1868	84, 448
Curtain-clasp	J. G. Whittier and T. M. Powell	Attica, Ind	Sept. 26, 1865	50, 193
Curtain-cord bracket	T. L. Pye	New York, N. Y	Dec. 18, 1860	30, 985
Curtain-cord clamp	G. W. Bishop	Saratoga, N. Y	May 21, 1872	127, 016
Curtain-cord fastener	H. Lull	Hoboken, N. J	May 27, 1873	139, 404
Curtain-cord rack	W. D. Ludlow	New York, N. Y	Jan. 14, 1873	134, 812
Curtain-cord retainer	H. Holcroft	Media, Pa	May 9, 1871	114, 560
Curtain-cord tightener	J. E. Baum	Philadelphia, Pa	Aug. 30, 1870	106, 770
Curtain-cord tightener	W. H. Betts	Brooklyn, N. Y	Feb. 9, 1869	86, 634
Curtain-cord tightener	T. C. Lippincott	Philadelphia, Pa	Aug. 27, 1867	68, 092
Curtain-cord tightener	J. O. Montignani	Albany, N. Y	Jan. 12, 1864	41, 231
Curtain, Electro-magnetic burglar-proof	E. Holmes and H. C. Roome	Brooklyn, N. Y., and Jersey City, N. J.	Nov. 14, 1871	120, 875
Curtain-eyelet	J. Starkey	Portland, Me	Oct. 30, 1866	59, 320
Curtain-fastener	J. F. A. Blätterlein	West Meriden, Conn	July 29, 1873	141, 313
Curtain-fastener, Reversible	H. Binder	Saint Louis, Mo	Sept. 12, 1871	118, 781
Curtain-fastening	J. R. Cook	Winsted, Conn	Aug. 15, 1865	49, 381
Curtain fastening, Window	H. Guyer	Albany, N. Y	Mar. 11, 1851	7, 971
Curtain-fixture	J. D. Ayers	East Greensborough, Vt	June 22, 1869	91, 591
Curtain-fixture	A. C. Babcock	New Haven, Conn	Jan. 11, 1860	26, 824
Curtain-fixture	J. B. Bailey	New York, N. Y	May 23, 1865	47, 891
Curtain-fixture	J. B. Bailey	New York, N. Y	June 12, 1866	55, 454
Curtain-fixture	J. B. Bailey	New York, N. Y	Feb. 23, 1869	87, 234
Curtain-fixture	H. O. Baker	New York, N. Y	July 3, 1866	55, 978
Curtain-fixture	T. C. Baldwin	Newton, Mass	Nov. 30, 1858	22, 153
Curtain-fixture	A. P. Barlow	Claremont, N. H	Dec. 3, 1867	71, 679
Curtain-fixture	J. H. Barnes and T. W. Brown	Brooklyn, N. Y	Apr. 3, 1866	53, 724
Curtain-fixture	J. E. Baum	Philadelphia, Pa	July 18, 1871	117, 037
Curtain-fixture	J. E. Baum	Philadelphia, Pa	July 18, 1871	117, 038
Curtain-fixture	J. E. Baum	Philadelphia, Pa	Sept. 26, 1871	119, 299
Curtain-fixture	R. Beal	Commerce, Mich	Nov. 1, 1870	108, 752
Curtain-fixture	G. T. Beardsley	Ithaca, N. Y	June 30, 1868	79, 435
Curtain-fixture	G. W. Beers	Bridgeport, Conn	Apr. 1, 1873	137, 405
Curtain-fixture	H. Binder	Saint Louis, Mo	Sept. 17, 1872	131, 328
Curtain-fixture	E. T. Briggs	Boston, Mass	Sept. 3, 1867	68, 552
Curtain-fixture	E. T. Briggs	Boston, Mass	Feb. 20, 1872	123, 861
Curtain-fixture	J. G. Brothwell	Wolcottville, Conn	Dec. 6, 1870	109, 800
Curtain-fixture	E. Brown	Waterbury, Conn	Apr. 3, 1860	27, 686
Curtain-fixture	S. S. Brown	Woonsocket, R. I	Oct. 6, 1868	82, 796
Curtain-fixture	W. Brown	West Cambridge, Mass	Feb. 18, 1868	74, 661
Curtain-fixture	C. M. Brown and C. L. Gates	Hartford, Conn	Feb. 14, 1871	111, 723
Curtain-fixture	C. Buckley	Meriden, Conn	July 11, 1871	116, 929
Curtain-fixture	C. Buckley	Meriden, Conn	Jan. 2, 1872	122, 356
Curtain-fixture	C. Buckley	West Meriden, Conn	Feb. 18, 1873	136, 030
Curtain-fixture	W. N. Bulkley	Brooklyn, N. Y	Oct. 3, 1871	119, 568
Curtain-fixture	T. Burgen	Rutland, Vt	Feb. 15, 1870	99, 837
Curtain-fixture	H. H. Burritt	Newark, N. J	Sept. 24, 1872	131, 496
Curtain-fixture	H. H. Burritt	Newark, N. J	Oct. 7, 1873	143, 494
Curtain-fixture	H. H. Burritt	Belleville, N. J	Dec. 23, 1873	145, 785
Curtain-fixture	N. Campbell	Rochester, N. Y	May 16, 1871	114, 917
Curtain-fixture	N. Campbell	Rochester, N. Y	May 16, 1871	114, 918
Curtain-fixture	N. Campbell	Rochester, N. Y	Oct. 10, 1871	119, 740
Curtain-fixture	N. Campbell	Rochester, N. Y	Sept. 3, 1872	130, 977
Curtain-fixture	W. Campbell	New York, N. Y	Apr. 18, 1871	113, 852
Curtain-fixture	R. Cassady	Newport, N. J	Feb. 9, 1869	86, 810
Curtain-fixture	H. W. Chace	Fall River, Mass	Oct. 8, 1861	33, 461
Curtain-fixture	J. Chace	Rochester, N. Y	Mar. 24, 1868	75, 860
Curtain-fixture	H. N. Chapman	Washington, D. C	Mar. 16, 1869	87, 757
Curtain-fixture	D. G. Chase	Boston, Mass	June 4, 1872	127, 565
Curtain-fixture	J. Chase and W. S. Loughborough.	Rochester, N. Y	Apr. 25, 1865	47, 392
Curtain-fixture	A. M. Cheney	Charlotte, Mich	Apr. 20, 1869	89, 126
Curtain-fixture	D. Clagett	Hagerstown, Md	Mar. 5, 1872	124, 330
Curtain-fixture	H. Clayton	Lexington, Ky	Sept. 26, 1871	119, 321
Curtain-fixture	W. Cleveland	New York, N. Y	July 24, 1860	29, 242
Curtain-fixture	B. F. Cloud	Philadelphia, Pa	May 11, 1869	89, 917
Curtain-fixture	W. C. Clover	Brooklyn, N. Y	Nov. 1, 1870	108, 761
Curtain-fixture	W. R. Cole	Baltimore, Md	Dec. 17, 1872	133, 921
Curtain-fixture	M. Converse	Jordan, N. Y	May 28, 1867	65, 058
Curtain-fixture	H. T. Cooper	New York, N. Y	June 15, 1869	91, 422
Curtain-fixture	J. M. Corns	Black Rock, N. Y	Feb. 18, 1868	74, 508
Curtain-fixture	G. B. Cowles	Bridgeport, Conn	Jan. 21, 1873	134, 980
Curtain-fixture	J. Crandell and P. W. Smith	Chicopee, Mass	July 13, 1869	92, 428
Curtain-fixture	J. P. Crawford	Carmichael's, Pa	Dec. 21, 1869	98, 039
Curtain-fixture	W. R. P. Cross	Portland, Me	Aug. 14, 1866	57, 096
Curtain-fixture	J. Cunningham	Philadelphia, Pa	Mar. 12, 1867	62, 824
Curtain-fixture	T. Curley	Troy, N. Y	Mar. 24, 1868	75, 869
Curtain-fixture	J. W. Currier and J. M. Thompson.	Holyoke, Mass	Nov. 3, 1857	18, 536
Curtain-fixture	J. David	New York, N. Y	Sept. 24, 1867	69, 189
Curtain-fixture	J. David	New York, N. Y	Aug. 18, 1868	81, 258
Curtain-fixture	A. S. Dickinson	Washington, D. C	Dec. 29, 1868	85, 433
Curtain-fixture	A. S. Dickinson	New York, N. Y	Oct. 19, 1869	95, 993
Curtain-fixture	L. Dodge	Syracuse, N. Y	June 29, 1869	91, 923
Curtain-fixture	E. Doen	New Britain, Conn	Apr. 9, 1867	63, 621
Curtain-fixture	J. Doyle	Hoboken, N. J	July 20, 1869	92, 710
Curtain-fixture	J. Doyle	Hoboken, N. J	Sept. 27, 1870	107, 764
Curtain-fixture	L. J. Earll	Union City, Pa	July 15, 1873	140, 901

Index of patents issued from the United States Patent Office from 1790 *to* 1873, *inclusive*—Continued.

Invention.	Inventor.	Residence.	Date.	No.
Curtain-fixture	C. Eaton	New York, N. Y	Nov. 28, 1871	121, 342
Curtain-fixture	C. Eaton	New York, N. Y	Oct. 22, 1872	132, 453
Curtain-fixture	M. R. Fenton	Washington, D. C	June 25, 1867	66, 010
Curtain-fixture	H. Fininley	New York, N. Y	Dec. 29, 1868	85, 295
Curtain-fixture	C. Fisher	Milton, Mass	Jan. 31, 1860	26, 980
Curtain-fixture	A. C. Flint	Boston, Mass	Nov. 3, 1863	40, 466
Curtain-fixture	J. W. Foard	San Francisco, Cal	July 20, 1869	92, 812
Curtain-fixture	C. H. Fowler	West Roxbury, Mass	May 28, 1867	65, 067
Curtain-fixture	C. E. Fritts	Oneonta, N. Y	May 11, 1869	89, 926
Curtain-fixture	C. E. Fritts	Oneonta, N. Y	May 6, 1873	138, 493
Curtain-fixture	G. P. Fuller	Philadelphia, Pa	Sept. 28, 1869	95, 218
Curtain-fixture	G. P. Fuller	Humphrey, N. Y	May 17, 1870	103, 167
Curtain-fixture	G. Gatty	New York, N. Y	Apr. 16, 1861	32, 062
Curtain-fixture	A. F. Gerald	Kendall's Mills, Me	Oct. 6, 1868	82, 824
Curtain-fixture	A. F. Gerald	Fairfield, Me	Sept. 17, 1872	131, 346
Curtain-fixture	H. B. Gingrich	Bradford, Ohio	Dec. 3, 1872	133, 641
Curtain-fixture	A. J. Goodrich	Wolcottville, Conn	Jan. 16, 1872	122, 826
Curtain-fixture	N. S. Graves	Boston, Mass	Aug. 11, 1857	17, 967
Curtain-fixture	J. Gray	Medford, Mass	Nov. 28, 1871	121, 242
Curtain-fixture	W. D. Gridley	New Britain, Conn	Mar. 24, 1868	75, 750
Curtain-fixture	L. B. Gusman	Philadelphia, Pa	Feb. 17, 1857	16, 673
Curtain-fixture	T. Hagerty	Richmond, Va	Feb. 27, 1872	124, 131
Curtain-fixture	J. F. Hall	Bangor, Me	Mar. 9, 1858	19, 560
Curtain-fixture	J. F. Hall	Bangor, Me	May 22, 1860	28, 363
Curtain-fixture	J. F. Hall	Bangor, Me	Sept. 4, 1860	29, 935
Curtain-fixture	A. N. Hallowell and H. S. Rue	Philadelphia, Pa	Oct. 1, 1872	131, 753
Curtain-fixture	B. Handforth	Chicago, Ill	Oct. 23, 1866	59, 009
Curtain-fixture	B. Handforth	Chicago, Ill	Nov. 10, 1868	83, 957
Curtain-fixture	O. Hanks	Cincinnati, Ohio	Aug. 27, 1867	68, 189
Curtain-fixture	T. G. Harold	Brooklyn, N. Y	Mar. 5, 1861	31, 643
Curtain-fixture	T. G. Harold and G. L. Kelty	Brooklyn and New York, N. Y.	Dec. 11, 1860	30, 878
Curtain-fixture	S. Hartshorn	New York, N. Y	Oct. 11, 1864	44, 624
Curtain-fixture	S. Hartshorn	New York, N. Y	July 9, 1872	128, 798
Curtain-fixture	S. F. Hay	New York, N. Y	July 30, 1872	129, 889
Curtain-fixture	W. B. Hazzard	Philadelphia, Pa	June 18, 1872	127, 975
Curtain-fixture	S. and M. Henry	Chenoa, Ill	Dec. 23, 1873	145, 868
Curtain-fixture	C. F. Herrick	Independence, Iowa	Aug. 20, 1867	67, 876
Curtain-fixture	J. J. Hessler	Reading, Pa	June 25, 1867	66, 084
Curtain-fixture	A. Hicks	Factoryville, N. Y	Dec. 6, 1864	45, 374
Curtain-fixture	W. C. Hicks	Summit, N. J	July 2, 1872	128, 489
Curtain-fixture	R. E. Hitchcock	Waterbury, Conn	Dec. 28, 1869	98, 261
Curtain-fixture	F. Hobart	Mount Pleasant, Iowa	Oct. 17, 1871	119, 983
Curtain-fixture	J. B. Holmes, jr	New York, N. Y	Oct. 4, 1859	25, 644
Curtain-fixture	M. G. Imbach	Bethlehem, Pa	Dec. 28, 1869	98, 384
Curtain-fixture	C. R. Jenkins	Philadelphia, Pa	Jan. 28, 1868	73, 811
Curtain-fixture	A. D. and E. M. Judd	New Haven, Conn	Apr. 1, 1873	137, 450
Curtain-fixture	E. M. Judd	New Britain, Conn	Jan. 29, 1861	31, 246
Curtain-fixture	E. M. Judd	New Britain, Conn	Oct. 20, 1863	40, 381
Curtain-fixture	E. M. Judd	New Haven, Conn	Dec. 19, 1865	51, 594
Curtain-fixture	E. M. Judd	Wolcottville, Conn	Mar. 5, 1867	62, 635
Curtain-fixture	H. L. Judd	New Britain, Conn	Dec. 13, 1864	45, 413
Curtain-fixture	J. W. King	New York, N. Y	Aug. 24, 1869	94, 006
Curtain-fixture	I. Kinman	Freeport, Ill	May 12, 1868	77, 890
Curtain-fixture	G. R. Kelsey	West Haven, Conn	Nov. 12, 1861	33, 708
Curtain-fixture	G. L. Kelty and T. G. Harold	New York and Brooklyn, N. Y.	Mar. 6, 1860	27, 410
Curtain-fixture	W. C. Kennedy	Commerce, Mich	Mar. 21, 1871	112, 814
Curtain-fixture	S. Ker	Washington, D. C	Feb. 4, 1873	135, 562
Curtain-fixture	A. H. Knapp	Newton Centre, Mass	Jan. 1, 1867	60, 908
Curtain-fixture	A. H. Knapp	Newton Centre, Mass	May 25, 1869	90, 366
Curtain-fixture	A. H. Knapp	Newton Centre, Mass	Sept. 5, 1871	118, 725
Curtain-fixture	A. H. Knapp	Newton Centre, Mass	July 30, 1872	129, 893
Curtain-fixture	A. A. Knapp and G. W. Bailey	Newton Centre and Boston, Mass.	Apr. 26, 1870	102, 409
Curtain-fixture	C. F. Knauer	Pittsburgh, Pa	Jan. 15, 1867	61, 212
Curtain-fixture	C. F. Knauer	Pittsburgh, Pa	Jan. 21, 1868	73, 612
Curtain-fixture	E. B. Lake	Bridgeport, N. J	July 23, 1867	66, 970
Curtain-fixture	W. A. Lamberson and T. O. Morton.	New York, N. Y	Jan. 30, 1866	52, 296
Curtain-fixture	B. Landon	Canton, Pa	Nov. 5, 1872	132, 726
Curtain-fixture	A. E. Lazell	West Meriden, Conn	June 10, 1873	139, 719
Curtain-fixture	C. Lee and J. Pandler	Washington, D. C	June 25, 1867	66, 094
Curtain-fixture	J. D. Legg	Long Eddy, N. Y	Aug. 18, 1868	81, 280
Curtain-fixture	J. D. Legg	Long Eddy, N. Y	Sept. 19, 1871	119, 158
Curtain-fixture	J. D. and I. W. Legg	Long Eddy, N. Y	May 5, 1868	77, 499
Curtain-fixture	M. Loeb	Chicago, Ill	Nov. 30, 1869	97, 304
Curtain-fixture	D. E. Long	Pawtucket, R. I	Aug. 4, 1868	80, 750
Curtain-fixture	J. S. Lovejoy	Washington, D. C	Jan. 4, 1870	98, 508
Curtain-fixture	T. Lovell	Cincinnati, Ohio	Apr. 23, 1872	125, 966
Curtain-fixture	A. Lovie	Philadelphia, Pa	Dec. 15, 1868	85, 016
Curtain fixture	H. and A. Lovie	Philadelphia, Pa	Feb. 5, 1867	61, 745
Curtain-fixture	H. and A. Lovie	Philadelphia, Pa., and Prussia.	Oct. 15, 1867	69, 823
Curtain-fixture	T. Lyons	Brooklyn, N. Y	Oct. 22, 1867	70, 099
Curtain-fixture	H. E. Marchand	Pittsburgh, Pa	Nov. 4, 1873	144, 342
Curtain-fixture	T. J. Marinus	Independence, Iowa	Jan. 7, 1868	73, 108
Curtain-fixture	J. Y. Marsh	New York, N. Y	Feb. 26, 1861	31, 550
Curtain-fixture	M. S. Marshall	Somerville, Mass	Feb. 16, 1869	86, 933
Curtain-fixture	W. W. Massey	Baltimore, Md	Dec. 16, 1873	145, 669
Curtain-fixture	E. Mentz	Philadelphia, Pa	May 1, 1866	54, 384
Curtain-fixture	W. T. Mersereau	Newark, N. J	Oct. 8, 1867	69, 572
Curtain-fixture	L. E. Michell	Cincinnati, Ohio	Aug. 4, 1868	80, 650
Curtain-fixture	P. Miles	Hartford, Conn	May 15, 1855	12, 866
Curtain-fixture	P. Miles	Hartford, Conn	Aug. 26, 1856	15, 615
Curtain-fixture	P. Miles	New York, N. Y	Sept. 26, 1865	50, 149
Curtain-fixture	P. Miles	New York, N. Y	Mar. 26, 1867	63, 169
Curtain-fixture	J. P. Miller	Somerset, Pa	May 26, 1868	78, 311
Curtain-fixture	J. Montgomery	Marseilles, Ill	May 24, 1870	103, 353

Index of patents issued from the United States Patent Office from 1790 to 1873, inclusive—Continued.

Invention.	Inventor.	Residence.	Date.	No.
Curtain-fixture	J. A. Morrison	Pittsburgh, Pa	Jan. 12, 1869	85, 751
Curtain-fixture	C. C. Moore	New York, N. Y	Dec. 9, 1873	145, 359
Curtain-fixture	B. Moser	Brooklyn, N. Y	Jan. 12, 1869	85, 842
Curtain-fixture	B. Moser	Philadelphia, Pa	May 25, 1869	90, 571
Curtain-fixture	B. Moser	Waltham, Mass	May 16, 1871	114, 843
Curtain-fixture	B. Moser	Waltham, Mass	June 27, 1871	116, 343
Curtain-fixture	M. H. Mosman	Waterbury, Conn	Nov. 22, 1870	109, 539
Curtain-fixture	G. W. Nell	Philadelphia, Pa	June 5, 1866	55, 344
Curtain-fixture	G. W. Nell	Philadelphia, Pa	May 11, 1869	89, 883
Curtain-fixture	P. H. Niles	Boston, Mass	Jan. 2, 1855	12, 150
Curtain-fixture	P. H. Niles	Boston, Mass	Sept. 18, 1855	13, 588
Curtain-fixture	J. Norman	Brooklyn, N. Y	July 4, 1871	116, 624
Curtain-fixture	J. L. Oliver	Boston, Mass	Nov. 15, 1870	109, 241
Curtain-fixture	P. O'Thayne	New York, N. Y	Jan. 12, 1869	85, 845
Curtain-fixture	P. O'Thayne and D. W. Canfield	New York, N. Y	Jan. 12, 1869	85, 846
Curtain-fixture	C. C. Parker	Brooklyn, N. H	Apr. 16, 1867	63, 934
Curtain-fixture	F. G. Peoble	New York, N. Y	Oct. 4, 1870	107, 956
Curtain-fixture	H. H. Phillips	Buffalo, N. Y	Feb. 27, 1872	124, 153
Curtain-fixture	H. H. Phillips	Buffalo, N. Y	Feb. 27, 1872	124, 154
Curtain-fixture	P. W. Phillips	Salem, Mass	Feb. 21, 1871	111, 967
Curtain-fixture	P. W. Phillips	Salem, Mass	Apr. 25, 1871	114, 036
Curtain-fixture	P. W. Phillips	Salem, Mass	Nov. 14, 1871	120, 997
Curtain-fixture	P. W. Phillips	Salem, Mass	Apr. 1, 1873	137, 480
Curtain-fixture	H. Polley	San Francisco, Cal	Sept. 10, 1872	131, 299
Curtain-fixture	L. C. Prindle	Chicago, Ill	May 16, 1871	114, 852
Curtain-fixture	R. B. Prindle	Norwich, N. Y	June 29, 1869	92, 094
Curtain-fixture	S. S. Putnam	Boston, Mass	Oct. 3, 1854	11, 757
Curtain-fixture	S. S. Putnam	Dorchester, Mass	Mar. 5, 1861	31, 623
Curtain-fixture	S. S. Putnam	Dorchester, Mass	July 1, 1862	35, 779
Curtain-fixture	S. S. Putnam	Dorchester, Mass	Sept. 2, 1862	36, 366
Curtain-fixture	S. S. Putnam	Dorchester, Mass	Feb. 4, 1868	74, 135
Curtain-fixture	S. S. Putnam	Dorchester, Mass	Jan. 19, 1869	86, 100
Curtain-fixture	W. Rice	Philadelphia, Pa	Jan. 24, 1860	26, 925
Curtain-fixture	T. C. Richards	Milwaukee, Wis	Mar. 4, 1862	34, 593
Curtain-fixture	C. J. Roberts	Chicago, Ill	Jan. 5, 1869	85, 695
Curtain-fixture	C. Robin	Chester, Conn	Mar. 7, 1871	112, 383
Curtain-fixture	A. Roelofs	Philadelphia, Pa	Oct. 24, 1871	120, 168
Curtain-fixture	A. Roelofs	Philadelphia, Pa	Oct. 31, 1871	120, 538
Curtain-fixture	A. Roelofs	Philadelphia, Pa	May 14, 1872	126, 835
Curtain-fixture	A. Roelofs	Philadelphia, Pa	July 16, 1872	129, 172
Curtain-fixture	F. Root	Chelsea, Mass	May 17, 1870	103, 084
Curtain-fixture	F. Root	Hartford, Conn	Aug. 9, 1870	106, 211
Curtain-fixture	F. Root	Hartford, Conn	Oct. 11, 1870	108, 1[illegible]0
Curtain-fixture	C. Rose	Allentown, Pa	Jan. 8, 1867	61, 106
Curtain-fixture	L. Ross	Springfield, Mass	Dec. 11, 1866	60, 426
Curtain-fixture	L. L. Sawyer	San Francisco, Cal	July 11, 1871	116, 996
Curtain-fixture	L. J. Schaefer	Philadelphia, Pa	Mar. 31, 1868	76, 252
Curtain-fixture	F. B. Scott	Lancaster, N. Y	Feb. 27, 1872	124, 088
Curtain-fixture	F. B. Scott	Lancaster, N. Y	May 28, 1872	127, 274
Curtain-fixture	W. E. Scranton	New Haven, Conn	Oct. 25, 1870	108, 734
Curtain-fixture	J. Shorey	Lowell, Mass	Aug. 11, 1868	81, 019
Curtain-fixture	J. Shorey	Lowell, Mass	Oct. 22, 1872	132, 376
Curtain-fixture	J. Shorey and F. H. Butler	Lowell, Mass	Dec. 28, 1869	98, 307
Curtain-fixture	D. C. Smart	Cambridgeport, Mass	Sept. 12, 1854	11, 683
Curtain-fixture	J. Smith	New York, N. Y	Apr. 3, 1860	27, 747
Curtain fixture	R. L. Smith	Wolcottville, Conn	Aug. 16, 1870	106, 416
Curtain-fixture	W. R. Smith	Seaforth, Canada	Nov. 7, 1871	120, 787
Curtain-fixture	L. Smith and S. Foster, jr	Des Moines, Iowa	Nov. 24, 1868	84, 443
Curtain-fixture	H. C. Spalding	Brooklyn, N. Y	Feb. 22, 1859	23, 047
Curtain-fixture	J. Stephens	New York, N. Y	July 1, 1856	15, 258
Curtain-fixture	J. Stephens	Fairview, Ohio	Nov. 15, 1870	109, 266
Curtain-fixture	J. Stevens and W. B. Fay	Chicopee Falls, Mass	May 1, 1866	54, 434
Curtain-fixture	T. Stewart	Philadelphia, Pa	Nov. 29, 1870	169, 683
Curtain-fixture	J. S. Stewart and S. B. Pierce	Homer, N. Y	July 24, 1866	56, 630
Curtain-fixture	C. F. Stock	Peoria, Ill	Feb. 13, 1872	123, 591
Curtain-fixture	A. Studley	Natick, Mass	Mar. 3, 1868	75, 072
Curtain-fixture	T. Symonds	Portland, Me	Apr. 22, 1873	138, 054
Curtain-fixture	W. H. Tambling	Eleroy, Ill	Jan. 10, 1871	110, 939
Curtain-fixture	J. B. Tarr	Chicago, Ill	Apr. 3, 1866	53, 703
Curtain-fixture	J. W. Taylor, jr	Dubuque, Iowa	Nov. 27, 1866	60, 087
Curtain-fixture	C. M. Thielcke	Chicago, Ill	Aug. 30, 1870	106, 890
Curtain-fixture	L. H. Thomas	Waterbury, Vt	Apr. 2, 1867	63, 440
Curtain-fixture	H. L. Traphagen	New York, N. Y	Feb. 18, 1873	135, 950
Curtain-fixture	H. H. Trenor	New York, N. Y	Aug. 22, 1865	49, 571
Curtain-fixture	A. E. Tripp	Springfield, Ohio	Aug. 23, 1870	106, 637
Curtain-fixture	L. A. Tripp	Middletown, N. Y	Feb. 19, 1867	62, 235
Curtain-fixture	L. A. Tripp and S. M. Boyd	New York, N. Y	Jan. 19, 1869	86, 044
Curtain-fixture	J. and W. Turnbull	Vancouver, Wash	Mar. 1, 1870	100, 341
Curtain-fixture	E. Turner	Wolcottsville, N. Y	May 2, 1871	114, 494
Curtain-fixture	H. Voigt	Buffalo, N. Y	July 23, 1867	67, 086
Curtain-fixture	R. Vose	New York, N. Y	Jan. 31, 1871	111, 405
Curtain-fixture	F. Walker	New Orleans, La	July 15, 1873	140, 976
Curtain-fixture	H. K. Warner and C. E. Smith	Rochester, Minn	Dec. 23, 1873	145, 821
Curtain-fixture	T. N. Webb	Baltimore, Md	Aug. 15, 1871	118, 080
Curtain-fixture	T. N. Webb	Baltimore, Md	Aug. 15, 1871	118, 081
Curtain-fixture	T. N. Webb	Baltimore, Md	Dec. 26, 1871	122, 294
Curtain-fixture	W. J. Wells	Pittsburgh, Pa	Sept. 10, 1872	131, 318
Curtain-fixture	I. B. Werner	Rossville, Ill	Sept. 24, 1872	131, 643
Curtain-fixture	W. A. Wheeldon	Chicago, Ill	Feb. 25, 1873	136, 195
Curtain-fixture	C. H. Wheeler	Boston, Mass	Apr. 14, 1857	17, 059
Curtain-fixture	G. M. White	New Haven, Conn	Aug. 7, 1866	57, 053
Curtain-fixture	G. M. White and C. S. Meeker	New Haven, Conn	June 2, 1868	78, 628
Curtain-fixture	L. White	Hartford, Conn	Jan. 15, 1856	14, 123
Curtain-fixture	L. White	Hartford, Conn	Nov. 15, 1859	26, 146
Curtain-fixture	L. Whitehead	Buffalo, N. Y	Dec. 15, 1857	18, 878
Curtain-fixture	J. S. Whitney	Lowell, Mass	Apr. 26, 1870	102, 344
Curtain-fixture	J. G. Whitwell	New York, N. Y	Sept. 25, 1860	30, 172
Curtain-fixture	F. Wieterich and K. Hagen	New York, N. Y	Sept. 2, 1856	15, 676

Index of patents issued from the United States Patent Office from 1790 to 1873, inclusive—Continued.

Invention.	Inventor.	Residence.	Date.	No.
Curtain-fixture	W. C. Wilcox	Waltham, Mass	May 30, 1871	115, 402
Curtain-fixture	J. H. Wilhelm	Chicago, Ill	July 26, 1870	105, 871
Curtain-fixture	J. H. Wilhelm	Chicago, Ill	Nov. 8, 1870	109, 087
Curtain-fixture	H. Wocher and B. Geiger	Philadelphia, Pa	Dec. 31, 1867	72, 767
Curtain-fixture	H. Wocher and B. Geiger	Philadelphia, Pa	Dec. 31, 1867	72, 768
Curtain-fixture	J. F. Wollensak	Chicago, Ill	Apr. 16, 1872	125, 712
Curtain-fixture	J. F. Wollensak	Chicago, Ill	Apr. 16, 1872	125, 870
Curtain-fixture	J. F. Wollensak	Chicago, Ill	Nov. 25, 1873	145, 040
Curtain-fixture	W. H. Woods	Philadelphia, Pa	Dec. 10, 1867	72, 150
Curtain-fixture	W. H. Woods	Philadelphia, Pa	Aug. 4, 1868	80, 793
Curtain-fixture	T. K. Work	Hartford, Conn	Apr. 20, 1858	20, 013
Curtain-fixture	J. Wyatt	Philadelphia, Pa	Jan. 4, 1870	98, 654
Curtain-fixture	W. P. Yates	Elmira, N. Y	Sept. 27, 1870	107, 747
Curtain-fixture	A. Young	Philadelphia, Pa	Mar. 24, 1868	75, 821
Curtain-fixture	J. L. Young	New York, N. Y	May 14, 1872	126, 666
Curtain-fixture	W. H. and L. Young	Boston, Mass	Nov. 3, 1868	83, 680
Curtain-fixture rack	J. Pusey	Philadelphia, Pa	Apr. 1, 1873	137, 320
Curtain-fixture, Self winding and balancing	F. C. D. McKay	Elmira, N. Y	Apr. 19, 1870	102, 142
Curtain fixture, Spring	J. Shorey	Lowell, Mass	July 26, 1870	105, 854
Curtain fixture, Window	R. Ballou, jr., and B. F. Hooper	Albany, N. Y	Apr. 7, 1857	16, 960
Curtain fixture, Window	J. Gibbs	Brooklyn, N. Y	May 19, 1863	38, 579
Curtain fixture, Window	N. H. McLean	United States Army	Oct. 2, 1860	30, 236
Curtain fixture, Window	P. Miles	Hartford, Conn	Apr. 7, 1857	16, 989
Curtain fixture, Window	P. Miles	New Britain, Conn	June 21, 1859	24, 479
Curtain fixture, Window	G. L. Miller	De Witt, N. Y	May 5, 1868	77, 638
Curtain fixture, Window	S. S. Putnam	Boston, Mass	Apr. 15, 1851	8, 041
Curtain fixture, Window	A. W. Tanner and O. P. Gorton	Paw Paw, Mich	Apr. 3, 1860	27, 749
Curtain fixture, Window	T. C. Williams	Randolph, Wis	Mar. 21, 1871	112, 876
Curtain-fixtures, Cord-tightener for	A. Carter	New York, N. Y	June 8, 1869	90, 922
Curtain-fixtures, Cord-tightener for	A. S. Dickinson	New York, N. Y	Oct. 19, 1869	95, 994
Curtain-fixtures, Cord-tightener for	E. M. Judd	New Britain, Conn	Aug. 5, 1862	36, 094
Curtain-fixtures, Cord-tightener for	H. L. Judd	Brooklyn, N. Y	Mar. 23, 1869	88, 179
Curtain-fixtures, Cord-tightener for	H. L. Judd	Brooklyn, N. Y	Feb. 4, 1873	135, 561
Curtain-fixtures, Cord-tightener for	E. Turner	Wolcottville, Conn	Dec. 21, 1869	98, 209
Curtain-fixtures, Self-holding clamp for	F. C. Payne	New York, N. Y	Sept. 15, 1863	39, 949
Curtain-holder	G. W. Gount, P. P. Reister, A. L. Keeports, and W. Yount.	Littlestown, Pa	Jan. 25, 1870	99, 271
Curtain holder, Window	J. S. and J. O. Burch	Buffalo, N. Y	Aug. 31, 1869	94, 393
Curtain-hook	H. L. Judd	Brooklyn, N. Y	June 10, 1873	139, 676
Curtain or shade fixture	W. Campbell	New York, N. Y	Oct. 11, 1864	44, 602
Curtain-rack	J. F. Calhoun	Wolcottville, Conn	May 24, 1859	24, 097
Curtain-rack	F. and W. Schmidt	Cincinnati, Ohio	Oct. 3, 1871	119, 536
Curtain-roller	C. M. Fairman and J. A. Henderson.	Pittsburgh, Pa	Oct. 1, 1872	131, 864
Curtain-roller	C. Fisher	Milton, Mass	May 26, 1857	17, 368
Curtain-roller	P. Miles	Hartford, Conn	Mar. 3, 1857	16, 741
Curtain-roller	D. H. Chamberlain	West Roxbury, Mass	Aug. 21, 1855	13, 481
Curtain-roller	D. H. Chamberlain and J. Hartshorn.	Boston, Mass	Jan. 23, 1855	12, 271
Curtain-roller	D. N. B. Coffin, jr	Newton, Mass	Oct. 6, 1857	18, 323
Curtain-roller	F. W. Urann	Saxonville, Mass	Mar. 13, 1855	12, 524
Curtain-roller, Extension	T. Van Wagoner	Newark, N. J	Jan. 4, 1870	98, 649
Curtain-roller for windows, Adjustable	E. S. Clark	Suffolk County, Mass	Feb. 19, 1850	7, 098
Curtain-roller holder	B. G. Fitzhugh	Frederick, Md	Apr. 22, 1873	138, 014
Curtain-roller, Spring	E. M. Judd	Wolcottville, Conn	Sept. 28, 1869	95, 232
Curtain-roller, Spring	B. B. Webster	Boston, Mass	May 15, 1855	12, 881
Curtain-roller stop	R. Wangeman	Chicago, Ill	Jan. 4, 1870	98, 532
Curtain roller, Window	W. Gorton	New York, N. Y	July 18, 1871	117, 068
Curtain roller, Window	W. W. Grier	Hulton, Pa	May 13, 1873	138, 749
Curtain-rollers, Fixture for	J. Hartshorn and D. H. Chamberlain.	Boston, Mass	Apr. 3, 1855	12, 661
Curtain-rollers, Fixture for	C. H. Wheeler	Boston, Mass	Apr. 7, 1857	17, 006
Curtain-rollers, Fixture for	L. White	Hartford, Conn	July 28, 1857	17, 911
Curtain-rollers; Hub or spool for	F. Pilling	Washington, D. C	Mar. 27, 1866	53, 480
Curtain screw and key, Window	H. Andrew	New York, N. Y	Mar. 6, 1822	....
Curtain slide, Window	G. Trinks	Jersey City, N. J	June 26, 1860	28, 922
Curtain-support	C. F. Herrick	Independence, Iowa	Apr. 2, 1867	63, 386
Curtain suspension, Window	G. H. Marden	Charlestown, Mass	Oct. 17, 1848	5, 868
Curtain-weight, Adjustable	W. F. Shaw	Boston, Mass	Feb. 23, 1869	87, 209
Curtain, Window	G. Butterfield and D. Bowker	Boston, Mass	Oct. 16, 1860	30, 385
Curtain, Window	H. Foster	Port Jervis, N. Y	Feb. 13, 1866	52, 555
Curtain, Window	D. G. Morgan	Jordan, N. Y	July 30, 1867	67, 337
Curtain, Window	W. H. Morrison	Indianapolis, Ind	Oct. 9, 1866	58, 668
Curtain, Window and bed	W. F. Phyfe	New York, N. Y	Feb. 19, 1830	....
Curtains, Apparatus for operating window	J. Stephens and J. Collins	Fairview, Ohio	July 19, 1870	105, 508
Curtains, blinds, and maps, Mode of hanging	J. F. Landis	Harrisburgh, Pa	Feb. 13, 1866	52, 577
Curtains, Cord-tightener for window	M. Hey	Philadelphia, Pa	Nov. 21, 1865	51, 045
Curtains, Manufacture of window and bed	S. Abbot and J. Prentis	Chester, Mass	Aug. 23, 1825	....
Curtains, Mode of hanging window	H. Aiken	Philadelphia, Pa	Oct. 19, 1869	95, 965
Curtains, Spring-roller for	J. and J. Hartshorn	Boston, Mass	May 1, 1855	12, 792
Curtains, Spring-roller for window	B. Bray	Salem, Mass	Sept. 5, 1854	11, 638
Curved pipes, Mold for casting	N. Thompson	Brooklyn, N. Y	June 25, 1872	128, 264
Curved pipes, Mold for casting	S. Williams	Foxborough, Mass	May 15, 1866	54, 826

Cushion:

See Air-cushion.
Billiard-cushion.
Billiard-table cushion.
Car-seat cushion.
Elastic cushion.
Elliptic cushion.
Hydraulic cushion.
Lubricating-cushion.
Mop-cushion.
Needle-cushion.
Pin-cushion.
Rubber cushion.
Seat-cushion.
Sleeping-cushion.

Index of patents issued from the United States Patent Office from 1790 *to* 1873, *inclusive*—Continued.

Invention.	Inventor.	Residence.	Date.	No.
Cushion—Continued:				
See Sofa-cushion.				
Spring-cushion.				
Spring carriage-cushion.				
Cushion, mattress, &c	T. H. O'Brien	Providence, R. I	Oct. 4, 1870	108, 043
Cushion or mattress	L. Banhoefer	Philadelphia, Pa	Nov. 27, 1866	59, 945
Cushion-spring	J. W. Evans	New York, N. Y	Mar. 26, 1861	31, 793
Cushions, mattresses, &c., Apparatus for diffusing vapor through.	W. Welch	Montreal, Canada	Feb. 15, 1870	99, 986
Cuspadore	E. A. Heath	New York, N. Y	Oct. 10, 1871	119, 705
Cuspadore	E. A. Heath	New York, N. Y	Oct. 10, 1871	119, 706
Cutlery	J. Baldwin	Northampton, Mass	Sept. 24, 1867	69, 157
Cutlery	C. L. Butler	Greenfield, Mass	Dec. 19, 1865	51, 648
Cutlery	W. T. Clement	Northampton, Mass	Nov. 10, 1868	83, 831
Cutlery	R. E. Curtis	Great Bend, Pa	Apr. 28, 1868	77, 175
Cutlery	E. Day	Chicago, Ill	July 21, 1868	80, 062
Cutlery	R. H. Fisher	West Meriden, Conn	June 23, 1868	79, 061
Cutlery	D. Fitzpatrick	West Winsted, Conn	Feb. 13, 1866	52, 552
Cutlery	M. W. Lyman	Chicago, Ill	Jan. 17, 1871	110, 986
Cutlery	S. Mason	Beaver Falls, Pa	June 30, 1868	79, 370
Cutlery	W. C. Mason	Beaver Falls, Pa	June 30, 1868	79, 371
Cutlery	T. McGrah	Sheffield, England	Mar. 5, 1867	62, 661
Cutlery	J. H. Nichols and W. Bower	Beaver Falls, Pa	Apr. 26, 1870	102, 306
Cutlery	H. T. Reeves	Beaver Falls, Pa	Nov. 7, 1871	120, 774
Cutlery	W. W. Robinson	Ripon, Wis	July 3, 1866	56, 099
Cutlery	M. Rubel	Chicago, Ill	May 26, 1868	78, 328
Cutlery	M. Rubel	Chicago, Ill	Jan. 26, 1869	89, 252
Cutlery	H. Sanderson	Sheffield, England	May 28, 1867	65, 123
Cutlery	W. Sanderson	New York, N. Y	June 18, 1867	65, 952
Cutlery	W. Sanderson	New York, N. Y	Aug. 10, 1869	93, 560
Cutlery	A. L. Taylor	Springfield, Vt	July 21, 1868	80, 238
Cutlery	W. D. Woods	Bennington, N. H	Oct. 6, 1868	82, 779
Cutlery and tool handle	J. T. Haviland	San Francisco, Cal	Dec. 20, 1870	110, 358
Cutlery, Attaching handles to	M. Chapman	Greenfield, Mass	Jan. 4, 1859	22, 527
Cutlery, Attaching handles to	M. Chapman	Greenfield, Mass	Dec. 20, 1859	26, 478
Cutlery, Attaching handles to	J. Flood	New Haven, Conn	June 9, 1868	78, 796
Cutlery, Attaching handles to	J. W. Gardner	Shelburne Falls, Mass	Dec. 23, 1862	37, 225
Cutlery, Attaching handles to	J. W. Gardner	Shelburne Falls, Mass	Oct. 5, 1869	95, 457
Cutlery, Attaching handles to	J. W. Gardner	Shelburne Falls, Mass	June 7, 1870	103, 867
Cutlery, Attaching handles to	S. Mason and E. Binns	Beaver Falls, Pa	Apr. 20, 1869	89, 059
Cutlery, Casting handles of table	N. S. Clement	New Britain, Conn	Apr. 6, 1869	88, 544
Cutlery cleaning and polishing machine	M. N. Armstrong	Williamsburgh, N. Y	Jan. 24, 1842	2, 435
Cutlery, Construction of table	A. E. Elmer	Windsor, Vt	Feb. 16, 1869	87, 032
Cutlery, Device for grinding	J. P. Curtiss	New Britain, Conn	Dec. 3, 1867	71, 586
Cutlery, edge-tools, &c., Substance for making	T. H. Jenkins	New York, N. Y	Dec. 26, 1865	51, 724
Cutlery, Fastening handles to	E. G. Ost	Shelburne Falls, Mass	Feb. 15, 1870	99, 785
Cutlery, Forging	H. B. Harvey	West Meriden, Conn	Nov. 26, 1867	71, 482
Cutlery-grinding apparatus	W. Fosket	Meriden, Conn	Feb. 11, 1868	74, 214
Cutlery grinding and polishing machine	J. Dodge	Waterford, N. Y	July 21, 1863	39, 317
Cutlery-grinding machine	R. S. Gladwin	Meriden, Conn	July 17, 1866	56, 403
Cutlery-handle	J. D. Frary	New Britain, Conn	Dec. 28, 1869	98, 369
Cutlery-handle	M. Chapman	Greenfield, Mass	Aug. 7, 1860	29, 465
Cutlery-handle	J. W. Gardner	Shelburne Falls, Mass	May 3, 1859	23, 837
Cutlery-handle	T. D. Lakin	Hancock, N. H	Aug. 28, 1866	57, 521
Cutlery-handle blank	I. G. Hotchkiss	Naugatuck, Conn	Feb. 16, 1869	86, 983
Cutlery, Handle for table and other	M. Chapman	Greenfield, Mass	Sept. 13, 1870	107, 223
Cutlery-handle, Metallic	L. S. White	Waterbury, Conn	Mar. 27, 1866	53, 546
Cutlery handle, Pocket	S. Barnes	New Haven, Conn	Feb. 11, 1868	74, 279
Cutlery-handle, Securing	R. H. Fisher	Beverly, N. J	Apr. 10, 1860	27, 790
Cutlery handle, Table	R. H. Fisher	Beaver Falls, Pa	Mar. 16, 1869	87, 767
Cutlery handle, Table	J. W. Gardner	Shelburne Falls, Mass	Feb. 1, 1859	22, 795
Cutlery handles, Construction of table	W. Sanderson	New York, N. Y	June 9, 1868	78, 834
Cutlery, Manufacture of	G. Parr	Buffalo, N. Y	Nov. 27, 1866	60, 048
Cutlery, Manufacture of blades for pocket	W. H. and G. W. Miller	Meriden, Conn	Aug. 2, 1870	106, 073
Cutlery, Manufacture of bolster for	J. D. Frary	New Britain, Conn	Apr. 4, 1871	113, 648
Cutlery, Manufacture of table	J. W. Gardner	Shelburne Falls, Mass	Mar. 27, 1866	53, 529
Cutlery, Manufacture of table	A. McGuire	Winsted, Conn	Aug. 9, 1864	43, 785
Cutlery, Manufacture of table	F. W. Presber and P. Shieber	Winchester, Conn	May 24, 1864	42, 872
Cutlery, Manufacture of table	L. Rice	West Winsted, Conn	Aug. 30, 1864	44, 056
Cutlery, &c., Metallic compound for coating	J. Zimmer	Cleveland, Ohio	Jan. 21, 1873	135, 028
Cutlery, Method of attaching the tangs to handles of table.	D. N. Ropes	Meriden, Conn	May 29, 1849	6, 482
Cutlery, Method of cleaning and polishing	W. Vine	New York, N. Y	Feb. 28, 1844	3, 458
Cutlery, Mode of attaching handles to table	J. W. Gardner	Shelburne Falls, Mass	Mar. 5, 1867	62, 626
Cutlery, &c., Plating scales with hard rubber for the manufacture of.	F. Beals	New Haven, Conn	Dec. 15, 1868	84, 932
Cutlery, Pocket	J. Carreer	Southington, Conn	Dec. 22, 1868	85, 211
Cutlery, Pocket	P. S. Hoe	New York, N. Y	Nov. 12, 1867	70, 918
Cutlery, Pocket	D. R. Hundley	Mountain Home, Ala	Oct. 25, 1870	108, 596
Cutlery, Pocket	W. H. and G. W. Miller	Meriden, Conn	May 24, 1870	103, 352
Cutlery, Pocket	F. B. Perry	Northampton, Mass	Nov. 19, 1867	71, 211
Cutlery, Pocket	D. E. Smith	Bronxville, N. Y	July 6, 1869	92, 217
Cutlery-sharpening implement	A. Thayer	Albany, N. Y	June 9, 1868	78, 771
Cutlery, spoons, &c., Machine for producing uniform twist in table.	E. W. Sperry	Wolcottville, Conn	Jan. 5, 1869	85, 542
Cutlery, Table	M. Chapman	Greenfield, Mass	Nov. 5, 1867	70, 525
Cutlery, Table	W. Clayton	Bristol, Conn	July 28, 1868	80, 334
Cutlery, Table	E. G. Durant	Northampton, Mass	Oct. 18, 1870	108, 463
Cutlery, Table	J. W. Gardner	Shelburne Falls, Mass	Mar. 6, 1860	27, 357
Cutlery, Table	J. W. Gardner	Shelburne Falls, Mass	Sept. 22, 1863	40, 031
Cutlery, Table	W. Hubbard	Meriden, Conn	Apr. 16, 1867	63, 795
Cutlery, Table	E. G. Lamson	Windsor, Vt	Mar. 22, 1870	101, 139
Cutlery, Table	S. Mason	Beaver Falls, Pa	Mar. 16, 1869	87, 786
Cutlery, Table	S. Mason and E. Binns	Beaver Falls, Pa	Mar. 16, 1869	87, 787
Cutlery, Table	C. McDonald	Ottawa, Ill	Jan. 21, 1873	135, 136
Cutlery, Table	N. Mies	Buckland, Mass	Feb. 7, 1865	46, 304
Cutlery, Table	J. H. Nichols and W. Bower	Beaver Falls, Pa	Jan. 25, 1870	99, 104
Cutlery, Table	C. L. Robertson	Providence, R. I	Apr. 13, 1869	88, 810

Index of patents issued from the United States Patent Office from 1790 to 1873, inclusive—Continued.

Invention.	Inventor.	Residence.	Date.	No.
Cutlery, Table	M. Smith	Meriden, Conn	Oct. 3, 1848	5,827
Cutlery, Table	J. Stott	Philadelphia, Pa	Apr. 2, 1867	63,439
Cutlery, Table	W. D. Woods	Bennington, N. H	Apr. 2, 1867	63,598
Cut-nail machine	W. Haddock	Pittsburgh, Pa	Nov. 25, 1873	144,845
Cut-nail machine	E. A. Kimball	Abington, Mass	Dec. 16, 1873	145,504
Cut-nail machine	J. Russell	Providence, R. I	Nov. 4, 1873	144,229
Cut-nail machine	J. P. Sherwood	Fort Edward, N. Y	Aug. 26, 1851	8,326
Cut nails and brads from rolled iron, Making cold	J. Elgar		Dec. 16, 1803	
Cut nails, cut points, and cut brads, Manufacturing	J. Perkins	Boston, Mass	July 17, 1810	
Cut nails from iron-hoops, &c., rendered tough	N. Kent		May 1, 1801	
Cut nails from Muntz's metal	S. L. Crocker	Taunton, Mass	Apr. 17, 1849	6,354
Cut nails, Heading	J. Davey	Fairhaven, Vt	Apr. 7, 1814	
Cut nails, Machine for heading	S. Wilmot, jr	New Haven, Conn	Apr. 21, 1809	
Cut nails, Machine for making	D. J. Farmer	Wheeling, W. Va	Aug. 31, 1869	94,194
Cut nails, Manufacturing	P. Cliff		Nov. 16, 1796	
Cut nails, Manufacturing	D. French		Dec. 23, 1796	
Cut nails while being headed, Device for holding	D. Savery	Wheeling, W. Va	Sept. 22, 1868	82,354
Cut-off	H. Allen	New York, N. Y	Aug. 29, 1848	5,745
Cut-off	F. E. Sickels	New York, N. Y	Feb. 24, 1852	8,760
Cut-off	F. B. Stevens	Weehawken, N. J	Dec. 3, 1861	33,858
Cut-off, Adjustable	S. H. Gilman	Cincinnati, Ohio	Mar. 18, 1851	7,987
Cut-off, Adjustable	J. King	Bordentown, N. J	Mar. 20, 1849	6,209
Cut-off and horse-power indicator, Adjustable	A. Stuckenrath	New York, N. Y	May 1, 1866	54,439
Cut-off and regulating cock, Gas	C. E. Seal	Winchester, Va	July 1, 1873	140,549
Cut-off and regulator valve	R. Stewart	Elmira, N. Y	May 19, 1863	38,610
Cut-off and reversing valve-gear for engines	R. M. Fryer	Nashville, Tenn	Feb. 11, 1873	135,794
Cut-off and steam-stop of rotary engines	J. W. Webb	Ledyard, N. Y	May 15, 1849	6,448
Cut-off and steam valve	A. Kendall	Cleveland, Ohio	Apr. 12, 1870	101,886
Cut-off and steam valves, Method of connecting the action of	B. H. Bartol	Cold Spring, N. Y	Sept. 20, 1844	3,755
Cut-off and working the valves of steam-engines	G. H. Corliss	Providence, R. I	Mar. 10, 1849	6,162
Cut-off apparatus, Steam-engine	R. Miller	Pittsburgh, Pa	Nov. 29, 1859	26,281
Cut-off apparatus, Steam-engine	A. S. Walbridge	Malone, N. Y	Sept. 10, 1861	33,274
Cut-off apparatus, Steam-engine	D. A. Woodbury	Rochester, N. Y	May 31, 1870	103,698
Cut-off apparatus, Variable	T. May	Brooklyn, N. Y	Nov. 8, 1870	109,137
Cut-off arrangement for steam-valves	J. Hornig	Newark, N. J	Nov. 22, 1859	26,186
Cut-off attachment to slide-valves	T. Sault	New Haven, Conn	Mar. 28, 1871	113,101
Cut-off, Automatic water	D. F. Sweet	Otsego, Mich	Feb. 8, 1870	99,727
Cut-off, Automatic water-spout	E. Stewart	Fort Madison, Iowa	Mar. 28, 1871	113,108
Cut-off, Blast-furnace	H. Davis	Newport, Ky	Apr. 18, 1871	113,745
Cut-off, Cistern	F. Fischer	Quincy, Ill	Dec. 13, 1870	110,128
Cut-off, Cistern	G. W. Howell	Covington, Ky	Sept. 9, 1873	142,569
Cut-off, Cistern	J. P. Watson	Rochester, Minn	Mar. 1, 1870	100,474
Cut-off, Cistern	C. Wuerz	Kimmswick, Mo	Sept. 16, 1873	142,832
Cut-off, Electro-magnetic-engine	L. H. McCullough	Richmond, Ind	Feb. 26, 1867	62,352
Cut-off for cistern-leaders	W. N. Hicks and F. Welker	Saint Louis, Mo	Nov. 27, 1866	60,005
Cut-off for cisterns, Rain-water	X. Amoun	Sidney, Ohio	Sept. 14, 1869	94,694
Cut-off for cisterns, Water	J. R. Manny	Chicago, Ill	Oct. 4, 1870	107,936
Cut-off for fire-plug systems	F. W. Eames	Grand Rapids, Mich	Feb. 18, 1873	135,976
Cut-off for oscillating-engines	W. Craig	Binghamton, N. Y	Oct. 15, 1861	33,477
Cut-off for pipes	J. H. Perkins	Omaha, Nebr	Nov. 2, 1869	96,478
Cut-off for rain-water pipes	C. Avery and J. W. Wetmore	Erie, Pa	July 25, 1871	117,493
Cut-off for relays and other electro-magnetic instruments.	F. M. Perry	Barton, Vt	Aug. 2, 1870	105,975
Cut-off for water-conductors	E. B. Armstrong	Columbus, Ohio	Sept. 10, 1867	68,682
Cut-off for water-conductors	J. Ash	Sterling, Ill	Sept. 19, 1865	49,960
Cut-off for water-conductors	H. Myers, jr	McLean, Ill	June 13, 1871	115,977
Cut-off for water-spouts	H. W. Mosher	Aurora, Ill	Oct. 1, 1867	69,359
Cut-off for engines of water-works, Variable	B. Holly	Lockport, N. Y	Apr. 25, 1871	114,010
Cut-off gear	G. H. Corliss	Providence, R. I	July 29, 1851	8,253
Cut-off gear for puppet-valve engines, Adjustable	H. Allen and D. G. Wells	New York, N. Y	Feb. 15, 1853	9,582
Cut-off gear, Steam-engine	E. R. Arnold	Providence, R. I	May 17, 1859	23,998
Cut-off gear, Steam-engine	J. Broughton	New York, N. Y	Dec. 21, 1858	22,344
Cut-off gear, Steam-engine	C. P. Buckingham	Mount Vernon, Ohio	Aug. 23, 1859	25,174
Cut-off gear, Steam-engine	P. W. Gates, D. R. Fraser, and T. Chalmers.	Chicago, Ill	Dec. 21, 1858	22,361
Cut-off gear, Steam-engine	P. W. Gates, D. R. Fraser, and T. Chalmers.	Chicago, Ill	July 5, 1859	24,624
Cut-off gear, Steam-engine	O. Leonard	Somerville, Mass	Mar. 18, 1856	14,486
Cut-off gear, Steam-engine	A. K. Rider	Hydeville, Vt	Dec. 24, 1861	34,013
Cut-off gear, Steam-engine	J. B. Root	New York, N. Y	Oct. 12, 1869	95,838
Cut-off gear, Steam-engine	H. Withington	Philadelphia, Pa	Sept. 20, 1859	25,541
Cut-off gear, Steam-engine variable	A. Foster and N. Sutton	New York, N. Y	July 19, 1859	24,801
Cut-off gear, Steam-engine variable	A. L. Holley	New York, N. Y	Apr. 19, 1859	23,684
Cut-off gear, Steam-engine variable	B. Hotchkiss	New Haven, Conn	Apr. 19, 1859	23,685
Cut-off gear, Steam-engine variable	J. Tremper	Buffalo, N. Y	Dec. 2, 1862	37,063
Cut-off gear, Steam-engine variable	W. W. Wade	Springfield, Mass	Dec. 11, 1855	13,934
Cut-off gear, Steam-engine variable	D. A. Woodbury	Rochester, N. Y	Apr. 19, 1859	23,737
Cut-off gear, Variable	J. E. McKay	New York, N. Y	Sept. 12, 1871	118,958
Cut-off, Governor	W. McCammon	Albany, N. Y	May 22, 1849	6,462
Cut-off, Governor variable	H. Waterman	New York, N. Y	Mar. 4, 1851	7,964
Cut-off, Hydrant	E. Bailey	Baltimore, Md	Aug. 26, 1862	36,319
Cut-off motion for puppet-valves	S. H. Gilman	Cincinnati, Ohio	Dec. 10, 1850	7,830
Cut-off, Piston-valve	G. McKay	Pittsfield, Mass	Apr. 17, 1849	6,324
Cut-off, Rain-water	J. Abercrombie and E. D. Miner.	Morrisania, N. Y	Aug. 13, 1872	130,400
Cut-off, Rain-water	L. Baltz	Saint Louis, Mo	Aug. 13, 1872	130,352
Cut-off, Rain-water	D. W. Doty	Aurora, Ill	May 29, 1866	55,072
Cut-off, Rain-water	E. Fleming	Ann Arbor, Mich	Aug. 17, 1869	93,694
Cut-off, Rain-water	R. S. Laird and W. F. Stone	Sandwich, Ill	Oct. 20, 1868	83,176
Cut-off, Rain-water	T. Lee	Cincinnati, Ohio	Apr. 16, 1872	125,742
Cut-off, Rain water	B. Rein	Marietta, Ohio	Aug. 15, 1871	118,054
Cut-off, Rain-water	C. and W. Scales	New Albany, Ind	Sept. 3, 1872	131,121
Cut-off, Rain-water	J. Spear	Carbondale, Ill	Sept. 29, 1868	82,653
Cut-off, Rain-water	J. Van Norman and W. Young	Easton, Pa	Nov. 3, 1868	83,808
Cut-off, Rain-water	G. W. Wetmore	Erie, Pa	Mar. 25, 1873	137,268
Cut-off regulator, Steam-engine	H. A. Luttgens	Paterson, N. J	Dec. 12, 1854	12,064
Cut-off, Steam-engine	J. F. Allen	New York, N. Y	Mar. 10, 1857	16,781

Index of patents issued from the United States Patent Office from 1790 *to* 1873, *inclusive*—Continued.

Invention.	Inventor.	Residence.	Date.	No.
Cut-off, Steam-engine	H. Allen and D. G. Wells	New York, N. Y	June 21, 1853	9, 792
Cut-off, Steam-engine	C. Carr	Boston, Mass	Apr. 27, 1869	89, 289
Cut-off, Steam-engine	S. Dunbar	New York, N. Y	June 1, 1869	90, 643
Cut-off, Steam-engine	T. Hansbrow	Sacramento, Cal	Jan. 12, 1869	85, 817
Cut-off, Steam-engine	H. O. Lothrop	Milford, Mass	Mar. 3, 1868	75, 834
Cut-off, Steam-engine	J. McPherson	Brooklyn, N. Y	June 30, 1868	79, 375
Cut-off, Steam-engine	J. H. Paine	Hartford, Conn	Jan. 24, 1865	46, 020
Cut-off, Steam-engine	G. W. Rawson	Cambridgeport, Mass	Nov. 12, 1867	70, 743
Cut-off, Steam-engine	G. W. Rawson	Cambridgeport, Mass	Oct. 6, 1868	82, 751
Cut-off, Steam-engine	G. J. Roberts	Dayton, Ohio	Sept. 29, 1868	82, 556
Cut-off, Steam-engine	A. P. Samuel	New York, N. Y	Jan. 19, 1858	19, 154
Cut-off, Steam-engine	J. S. Shapter	New York, N. Y	Apr. 15, 1856	14, 699
Cut-off, Steam-engine	A. Stuckenrath	New York, N. Y	Jan. 1, 1867	60, 804
Cut-off, Steam-engine	W. Watts and F. A. Phelps	Newark, N. J	Jan. 10, 1871	110, 943
Cut-off, Steam-engine	J. Widmer	New Haven, Conn	Aug. 31, 1858	21, 399
Cut-off, Steam-engine	D. A. Woodbury	Rochester, N. Y	Apr. 5, 1859	23, 521
Cut-off, Steam-engine	W. Wright	New York, N. Y	Oct. 10, 1871	119, 908
Cut-off, Steam-engine adjustable	A. Hartupee and J. Morrow	Pittsburgh, Pa	Sept. 2, 1856	15, 650
Cut-off, Steam-engine adjustable	H. J. and T. Hawkins	Mobile, Ala	June 24, 1856	15, 181
Cut-off, Steam-engine adjustable	J. R[illegible]es	Pittsburgh, Pa	May 9, 1871	114, 601
Cut-off, Steam-engine adjustable	W. Wright	Hartford, Conn	Nov. 25, 1856	16, 132
Cut-off, Steam-engine automatic	D. A. Woodbury	Rochester, N. Y	Sept. 27, 1870	107, 746
Cut-off, Steam-engine-governor	J. Felber	Saint Louis, Mo	Jan. 29, 1867	61, 528
Cut-off, Steam-engine-governor	G. Milbank	Chillicothe, Mo	Jan. 25, 1870	99, 220
Cut-off, Steam-engine variable	A. Crosby	Fredonia, N. Y	Jan. 19, 1858	19, 134
Cut-off, Steam-engine variable	W. B. Cross	Sacramento, Cal	Jan. 24, 1871	111, 180
Cut-off, Steam-engine variable	W. B. Cross	Sacramento, Cal	Aug. 8, 1871	117, 747
Cut-off, Steam-engine variable	A. Crumbie and R. D. Briggs	Brooklyn, N. Y	May 3, 1859	23, 830
Cut-off, Steam-engine variable	D. Fellenbaum	Lancaster, Pa	June 5, 1860	28, 632
Cut-off, Steam-engine variable	C. H. Reynolds	Lewiston, Me	Sept. 16, 1856	15, 745
Cut-off, Steam-engine variable	M. C. Taylor	Grass Valley, Cal	Dec. 21, 1869	98, 122
Cut-off, Steam-engine variable	W. Wright	New York, N. Y	May 20, 1873	139, 224
Cut-off valve	J. M. Albertson	New London, Conn	Oct. 10, 1865	50, 320
Cut-off valve	H. Allen	New York, N. Y	Mar. 20, 1847	5, 025
Cut-off valve	G. H. Babcock and S. Wilcox, jr	Providence, R. I	Apr. 24, 1866	54, 090
Cut-off valve	H. H. Grame	New York, N. Y	Mar. 20, 1847	5, 026
Cut-off valve	L. Griswold and G. Caul	Portland and York, Wis	July 23, 1867	67, 043
Cut-off valve	J. L. Hewes	Newark, N. J	Apr. 24, 1866	54, 153
Cut-off valve	C. H. Parshall	Detroit, Mich	June 28, 1864	43, 334
Cut-off valve	A. K. Rider	Hydeville, Vt	May 6, 1862	35, 176
Cut-off valve	I. M. Scott and W. R. Eckart	San Francisco, Cal	June 25, 1867	66, 046
Cut-off valve	R. Stewart	Elmira, N. Y	Jan. 23, 1866	52, 221
Cut-off valve	D. Stoddart	Cincinnati, Ohio	May 29, 1855	12, 966
Cut-off valve, Adjustable	G. W. Smith	New Haven, Conn	Dec. 26, 1871	122, 198
Cut-off valve, Balance	B. F. McKinley	Falmouth, Ky	June 12, 1866	55, 518
Cut-off-valve check for steam-engines	W. Wright	Hartford, Conn	June 21, 1856	15, 208
Cut-off-valve gear	A. W. Almqvist and F. W. Ofeldt.	New York, N. Y	Feb. 8, 1870	99, 518
Cut-off-valve gear	W. Dawes	Leeds, England	Nov. 2, 1869	96, 400
Cut-off-valve gear	G. W. Fisher and H. Reid	Saint Louis, Mo	Dec. 20, 1870	110, 223
Cut-off-valve gear	A. W. Foster, jr	Pittsburgh, Pa	Jan. 2, 1866	51, 821
Cut-off-valve gear	A. Kendall	Buffalo, N. Y	May 19, 1868	77, 986
Cut-off-valve gear	K. H. Loomis	Baltimore, Md	Aug. 8, 1865	49, 284
Cut-off-valve gear	W. McClintock	Wilmington, Del	Oct. 3, 1865	50, 304
Cut-off-valve gear	W. G. Pike	Philadelphia, Pa	Nov. 20, 1866	59, 777
Cut-off-valve gear	J. T. Rich	Philadelphia, Pa	Apr. 3, 1866	53, 680
Cut-off-valve gear	J. B. Schenck	Ansonia, Conn	Apr. 17, 1855	12, 729
Cut-off-valve gear	S. Stanton	Newburgh, N. Y	July 14, 1868	80, 025
Cut-off-valve gear	R. Stewart	Elmira, N. Y	May 19, 1863	38, 611
Cut-off-valve gear	R. Stewart	Elmira, N. Y	June 30, 1863	39, 075
Cut-off-valve gear	C. W. Wailey	New Orleans, La	Jan. 11, 1870	98, 724
Cut-off-valve gear	S. H. Whitmore	Decatur, Ill	Mar. 28, 1871	113, 232
Cut-off-valve gear	W. Wright	New York, N. Y	Mar. 24, 1868	76, 027
Cut-off-valve gear	J. N. Wrigley and G. Smith	Newark, N. J	Oct. 1, 1867	69, 526
Cut-off valve, Governor	K. H. Loomis	Baltimore, Md	Aug. 23, 1864	43, 920
Cut-off-valve motion	S. W. Rogers	Baltimore, Md	Dec. 21, 1852	9, 488
Cut-off-valve motion	F. B. Stevens	Weehawken, N. J	Dec. 3, 1861	33, 855
Cut-off valve, Oscillating engine	H. E. Canfield	New York, N. Y	Oct. 9, 1855	13, 637
Cut-off valve, Rotary balanced	P. Shellenback and F. Brilmayer.	Hamilton, Ohio	Mar. 11, 1873	136, 622
Cut-off valve, Sliding	S. P. Winne	Albany, N. Y	Apr. 10, 1849	6, 306
Cut-off valve, Steam	R. Sanderson	Cleveland, Ohio	July 28, 1868	80, 509
Cut-off valve, Steam	W. M. Stevenson	Sharon, Pa	Nov. 26, 1867	71, 547
Cut-off-valve gear and valve	G. E. Noyes	Washington, D. C	July 5, 1870	105, 119
Cut-off-valve gear and valve, Variable	W. Brown	Hoboken, N. J	Apr. 26, 1870	102, 367
Cut off-valve gear, Regulating	S. W. Robinson	Champaign, Ill	May 16, 1871	114, 861
Cut-off-valve gear, Steam-engine	H. Allen	New York, N. Y	Dec. 15, 1857	18, 837
Cut-off-valve gear, Steam-engine	W. L. Colborne	Elmira, N. Y	Mar. 1, 1864	41, 810
Cut-off-valve gear, Steam-engine	P. W. Gates, D. R. Fraser, and T. Chalmers.	Chicago, Ill	Oct. 5, 1858	21, 668
Cut-off-valve gear, Steam-engine	J. F. Hamilton	Pittsburgh, Pa	July 14, 1863	39, 228
Cut-off-valve gear, Steam-engine	J. Montfort	Newburgh, N. Y	Sept. 12, 1871	118, 873
Cut-off-valve gear, Steam-engine	C. W. Wailey	New Orleans, La	Sept. 15, 1868	82, 184
Cut-off valve gear, Steam-engine	W. Wright	New York, N. Y	Nov. 20, 1866	59, 886
Cut-off-valve gear, Steam-engine variable	M. W. Baldwin	Danvers Centre, Mass	Mar. 12, 1867	62, 797
Cut off-valve gear, Steam-engine variable	J. F. Hamilton	Philadelphia, Pa	Sept. 13, 1853	10, 007
Cut-off-valve gear, Steam-engine variable	H. O. Perry	Pittsburgh, Pa	July 14, 1863	39, 229
Cut-off-valve gear, Steam-engine variable	W. Wright	Buffalo, N. Y	Mar. 21, 1865	46, 932
Cut-off-valve gear, Variable	J. L. Dickinson	New York, N. Y	Nov. 20, 1866	59, 887
Cut-off-valve gear, Variable	G. E. Long	Dubuque, Iowa	Sept. 3, 1867	68, 492
Cut-off-valve gear, Variable	N. K. Lynch	Harrisburgh, Pa	June 7, 1870	103, 904
Cut-off valve, Steam-engine	H. Allen	New York, N. Y	Apr. 5, 1870	101, 634
Cut-off valve, Steam-engine	H. Allen	New York	Aug. 21, 1841	2, 227
Cut-off valve, Steam-engine	E. R. Arnold	New York	Apr. 30, 1842	2, 597
Cut-off valve, Steam-engine	J. Broughton	Providence, R. I	Nov. 1, 1859	25, 936
Cut-off valve, Steam-engine	B. Bunce	New York, N. Y	Dec. 3, 1861	33, 821
Cut-off valve, Steam-engine	A. S. Cameron	New York, N. Y	Oct. 19, 1858	21, 813
		New York, N. Y	Feb. 15, 1859	22, 935

Index of patents issued from the United States Patent Office from 1790 *to* 1873, *inclusive*—Continued.

Index of patents issued from the United States Patent Office from 1790 *to* 1873, *inclusive*—Continued.

Invention.	Inventor.	Residence.	Date.	No.
Cutter—Continued.				
See Glass-cutter.				
Grain-cutter.				
Hand-sole cutter.				
Harvester-cutter.				
Heel-cutter.				
Hemp-cutter.				
Hoop-lock cutter.				
Ice-cutter.				
Key-cutter.				
Key-hole cutter.				
Key-seat cutter.				
Lath-cutter.				
Leather-cutter.				
Lumber-cutter.				
Lumber tonguing and grooving cutter.				
Meat-cutter.				
Meat and vegetable cutter.				
Metal-cutter.				
Metallic-bar cutter.				
Molding-cutter.				
Molding-machine cutter.				
Nail-cutter.				
Operating cutter.				
Paper-cutter.				
Paper-box cutter.				
Pasteboard cutter.				
Pastry-cutter.				
Peach-cutter.				
Peg-cutter.				
Photographic-print cutter.				
Picket-cutter.				
Pipe-cutter.				
Pipe and rod cutter.				
Planing-cutter.				
Plug-cutter.				
Potato-cutter.				
Rag-cutter.				
Rattan-cutter.				
Relisher and wedge cutter.				
Revolving cutter.				
Ring-twine cutter.				
Rivet-cutter.				
Rock-channeling-machine cutter.				
Root-cutter.				
Rotary cutter.				
Sausage-cutter.				
Screw-cutter.				
Screw-thread cutter.				
Segment-cutter.				
Sewing-machine cutter.				
Sewing-machine-thread cutter.				
Shaft and pipe cutter.				
Sheet-iron cutter.				
Shingle-cutter.				
Shingle-band cutter.				
Shoe-lacing cutter.				
Slaw-cutter.				
Snuff-cutter.				
Soap-cutter.				
Sod-cutter.				
Sole-cutter.				
Stalk-cutter.				
Stave-cutter.				
Stencil-plate cutter.				
Stone-cutter.				
Stone-molding cutter.				
Straw-cutter.				
Stubble-cutter.				
Sugar-cutter.				
Thread-cutter.				
Tin-bottom cutter.				
Tobacco-cutter.				
Tobacco-plant cutter.				
Tonguing and grooving cutter.				
Tube-cutter.				
Tube-hole cutter.				
Twine-cutter.				
Twine or thread cutter.				
Vegetable-cutter.				
Veneer-cutter.				
Vine-cutter.				
Washer-cutter.				
Weed-cutter.				
Well-casing cutter.				
Welt and strap cutter.				
Wire cutter.				
Wrapper-cutter.				
Wrench and bolt cutter.				
Cutter-bars in grinding, Device for holding	W. H. Daniels	Bryan, Ohio	Apr. 22, 1873	138, 136
Cutter-bench and knee, Bent	E. Milner	Strathroy, Canada	Nov. 21, 1871	121, 058
Cutter-bits for rifling-machines, Apparatus for grinding.	D. Slate	Hartford, Conn	Feb. 16, 1864	41, 646
Cutter. grater, and sharpener	E. Culver	Shelburne Falls, Mass	May 28, 1867	65. 177
Cutter-head	D. C. Allen	Concord, N. H.	Aug. 12, 1873	141, 744
Cutter-head	E. Benjamin	Chicago, Ill	Sept. 9, 1873	142, 669
Cutter-head	M. W. Clark	Worcester, Mass	July 27, 1869	93, 056
Cutter-head	M. W. Clark	Worcester, Mass	Apr. 26, 1870	102, 222

Index of patents issued from the United States Patent Office from 1790 to 1873, inclusive—Continued.

Invention.	Inventor.	Residence.	Date.	No.
Cutter-head	J. F. W. Erdmann	Philadelphia, Pa	July 15, 1873	140, 905
Cutter-head	W. G. Farmer	Burlington, Vt	Oct. 6, 1868	82, 699
Cutter-head	S. Fawcett	Rochester, N. Y	Sept. 15, 1868	82, 211
Cutter-head	H. Fletcher	Louisville, Ky	Oct. 14, 1873	143, 565
Cutter-head	J. Gage	Henniker, N. H	Sept. 7, 1869	94, 590
Cutter-head	E. B. Hayes	Vergennes, Vt	Aug. 22, 1871	118, 365
Cutter-head	J. Kindleberger and W. A. Arnold.	San Francisco, Cal	Oct. 12, 1869	95, 809
Cutter-head	M. Lehman	Cincinnati, Ohio	Aug. 10, 1869	93, 628
Cutter-head	T. Long	Cerro Gordo, Ill	Nov. 26, 1872	133, 463
Cutter-head	C. E. McBeth	Hamilton, Ohio	Oct. 3, 1871	119, 632
Cutter-head	R. N. Meriam	Worcester, Mass	Dec. 28, 1869	98, 396
Cutter-head	D. W. Perry	Wilkesbarre, Pa	Nov. 15 1870	109, 244
Cutter-head	C. Richards and W. Curtiss	Cleveland, Ohio	June 15, 1869	91, 269
Cutter-head	G. Rowe and S. W. Nelson	Worcester, Mass	July 7, 1868	79, 690
Cutter-head	E. A. Rowley	Williamsport, Pa	Apr. 9, 1872	125, 618
Cutter-head	A. D. Sherar	Baltimore, Md	Apr. 2, 1872	125, 340
Cutter-head	A. H. Shipman	Arcadia, N. Y	Aug. 19, 1873	142, 048
Cutter-head	G. E. Somers	Waterbury, Conn	Aug. 9, 1870	106, 291
Cutter-head	A. L. Sweet	Norwich, Conn	Apr. 4, 1871	113, 464
Cutter-head	H. Thompson	Worcester. Mass	Dec. 21, 1869	98, 124
Cutter-head	A. Van Vleck	Jordan, N. Y	Feb. 20, 1872	123, 800
Cutter-head	J. V. Woolsey	Sandusky, Ohio	Aug. 8, 1871	117, 850
Cutter-head	E. S. Wright	New York, N. Y	Sept. 1, 1868	81, 859
Cutter-head and table-rest for cutting irregular forms.	J. P. Grosvenor	Lowell, Mass	May 25, 1858	20, 345
Cutter-head for dressing moldings	J. Temple	Terre Haute, Ind	Dec. 3, 1867	71, 817
Cutter-head for irregular form	D. Dunlap	Concord, N. H	Aug. 7, 1855	13, 386
Cutter-head for matching boards, Rotary	G. Greenwood	Lawrence, Mass	Feb. 18, 1873	135, 907
Cutter-head for moldings	J. Whitworth	Cleveland, Ohio	Dec. 3, 1867	71, 670
Cutter-head, Rotary	I. P. Tice	Baltimore, Md	Dec. 6, 1859	26, 383
Cutter-heads, Method of securing knife to	W. D. Hooker	Dedham, Mass	May 13, 1856	14, 871
Cutter-heads, pulleys, &c., Fastening	J. Du Bois and E. F. Bengler	Williamsport, Pa	Mar. 29, 1870	101, 238
Cutter-stock	J. Peace	Camden, N. J	Jan. 3, 1865	45, 742
Cutters for turning, Arrangement of	M. Roberts	South Levant, Me	Aug. 23, 1853	9, 957
Cutters in their heads for irregular forms, Method of operating.	J. Fear	Chicago, Ill	June 17, 1856	15, 153
Cutting and bending sheet-metal, Machine for	J. R. Maitland	Little Rock, Ark	May 2, 1871	114, 311
Cutting and coring knife, Apple	C. D. House	Lake Village, N. H	Feb. 2, 1869	86, 401
Cutting and grasping shears	S. W. Valentine	Boston, Mass	Aug. 1, 1865	49, 178
Cutting and grinding apparatus	P. Miles	New Haven, Conn	Jan. 8, 1861	31, 098
Cutting and polishing compound	J. P. Hall	New York, N. Y	July 27, 1869	92, 962
Cutting-blocks, Machine for leveling	A. Davis	Oxford, Mass	Sept. 12, 1871	118, 912
Cutting-board	R. C. Hussey	Milford, Mass	Aug. 28, 1866	57, 513
Cutting-box	W. Green	Kinzua, Pa	Mar. 6, 1866	52, 992
Cutting-box	C. Sewerkrop	Louisville, Ky	Aug. 7, 1860	29, 524
Cutting-box	W. Willis	South Carolina	Aug. 24, 1822	....
Cutting cavities, spherical, ellipsoidal, &c., Machine for.	I. B. Hartwell	Woodstock, Vt	July 3, 1855	13, 169
Cutting, grinding, and polishing tools, Manufacture of.	A. K. Eaton	Piermont, N. Y	July 12, 1870	105, 320
Cutting irregular forms, Feeding-apparatus for machines for.	C. P. Bailey	Zanesville, Ohio	Dec. 19, 1854	12, 088
Cutting irregular forms in wood	D. George and H. Robertson	Granville, Ohio	Oct. 24, 1848	5, 873
Cutting irregular forms, Machine for	A. Aldrich	Princeton, Mass	Sept. 26, 1854	11, 738
Cutting irregular forms, Machine for	A. Babbett	Auburn, N. Y	June 19, 1855	13, 076
Cutting irregular forms, Machine for	I. S. Barber	Boston, Mass	May 15, 1855	12, 884
Cutting irregular forms, Machine for	J. W. Campbell	New York, N. Y	June 13, 1871	115, 820
Cutting irregular forms, Machine for	W. N. Oakes	Dana, Mass	June 8, 1858	20, 505
Cutting irregular forms, Machine for	O. L. Reynolds	Dover, N. H	Aug. 15, 1854	11, 533
Cutting irregular forms, Machine for	J. Russell	Philadelphia, Pa	Jan. 3, 1854	10, 369
Cutting irregular forms, Machine for	C. Spofford	Amesbury, Mass	Oct. 7, 1856	15, 859
Cutting irregular forms, Machine for	H. D. Stover	Boston, Mass	Aug. 31, 1858	21, 379
Cutting irregular forms, Machine for	H. D. Stover and J. W. Ricknell	Boston, Mass	Mar. 11, 1856	14, 421
Cutting irregular forms, Machine for	I. P. Tice	Baltimore, Md	May 24, 1859	24, 163
Cutting irregular forms, Machine for	W. Wadleigh	Hill, N. H	Jan. 2, 1855	12, 174
Cutting irregular forms, Machine for	P. H. Wait	Barkersville, N. Y	Aug. 28, 1855	13, 511
Cutting-machine	W. H. Johnson	Springfield, Mass	May 19, 1868	77, 985
Cutting-machine	M. H. Merriam and E. L. Norton	Charlestown, Mass	June 5, 1866	55, 336
Cutting-machine, Reversing	S. D. Tripp	Lynn, Mass	Sept. 1, 1868	81, 707
Cutting-machine, Rotary	J. J. Butler	Cincinnati, Ohio	Dec. 11, 1866	60, 341
Cutting objects with straight sides and semicircular ends, Machine for.	C. W. Packer	Philadelphia, Pa	Dec. 20, 1864	45, 565
Cutting or lap board	N. O'Donnell	Cincinnati, Ohio	Sept. 13, 1870	107, 285
Cutting polygonal surfaces in timber, Machine for	E. Unger	Dayton, Ohio	Dec. 20, 1853	10, 346
Cutting-press	N. J. Simonds	Woburn, Mass	May 23, 1871	115, 118
Cutting, punching, and bending machine	C. Wright	Newark, N. J	Apr. 12, 1864	42, 325
Cutting, punching, and upsetting, Compound tool for.	A. A. Kent	Lyons, Iowa	Nov. 19, 1867	71, 181
Cutting-shears	R. O. Wood	Niagara Falls, N. Y	Aug. 22, 1871	118, 419
Cutting thin material, Method of	A. Delkescamp	Brooklyn, N. Y	Mar. 19, 1872	124, 669
Cutting-tool	B. F. Allen	Boston, Mass	Feb. 13, 1872	123, 604
Cutting-tool	G. L. Jaeger	New York, N. Y	Dec. 17, 1867	72, 400
Cutting-tool	M. J. Mellyn	Roxbury, Mass	Oct. 8, 1867	69, 689
Cyanogen and iron, Compound of	H. Halvorson	Cambridge, Mass	June 7, 1864	43, 022
Cylinder and cone, Casting chilled	J. Harley	Pittsburgh, Pa	Mar. 3, 1835	....
Cylinder, Apparatus for straightening	T. F. Hammer	Branford, Conn	Apr. 23, 1872	125, 892
Cylinder boat	A. Kendall	Washington, D. C	Sept. 2, 1834	....
Cylinder-boring apparatus	L. B. Flanders	Philadelphia, Pa	June 25, 1867	66, 011
Cylinder, Co-operating frictionless	J. P. Sawin	Roxbury, Mass	Jan. 24, 1811	....
Cylinder engine, Revolving	J. Allonas and W. Bauman	Mansfield, Ohio	Aug. 29, 1871	118, 503
Cylinder engine, Revolving	J. S. Foster	Virginia, Nev	Nov. 28, 1865	51, 166
Cylinder engine, Revolving	C. F. Ruset	Communipaw, N. J	Oct. 16, 1866	58, 891
Cylinder engine, Revolving	C. Scott and W. H. Morton	Hamilton, Ohio	Aug. 20, 1867	68, 008
Cylinder engine, Rotating	R. L. Cohen	Philadelphia, Pa	Mar. 14, 1871	112, 688
Cylinder engine, Triple	P. Brotherhood	Notting Hill, England	Dec. 23, 1873	145, 719
Cylinder, Expanding	P. Maltby	Kent, Ohio	Dec. 4, 1866	60, 214
Cylinder for drying paper, warp, thread, &c., Drying	M. A. Furbush	Philadelphia, Pa	Oct. 4, 1864	44, 524

Index of patents issued from the United States Patent Office from 1790 *to* 1873, *inclusive*—Continued.

Invention.	Inventor.	Residence.	Date.	No.
Cylinder for machine-cards	C. E. Brownell	East Haddam, Conn	Mar. 25, 1862	24, 734
Cylinder for smoothing walks, &c	J. Giles and C. B. Tompkins	Dryden and Ulysses, N. Y.	Nov. 15, 1859	26, 100
Cylinder for spreaders, cotton-gins, &c., Brush	A. M. Laupher	Gloucester, N. J	Sept. 31, 1858	21, 568
Cylinder-heads, Machine for grinding	J. Shirrell	Paterson, N. J	Sept. 16, 1873	142, 874
Cylinder, Hydraulic	D. Fitzgerald	New York, N. Y	Dec. 16, 1862	37, 195
Cylinder, Magnetic	S. Browning	Franconia, N. H	Nov. 25, 1814	....
Cylinder mill for grinding corn and other grain	H. W. Pitts	Wilsonville, Ala	July 1, 1840	1, 670
Cylinder mill for granulating corn, powder, bark, &c.	J. Wilson	New London, Conn	July 23, 1841	2, 192
Cylinder, Mixing and drying	L. B. Pitcher	Salina, N. Y	Nov. 12, 1867	70, 742
Cylinder, Multiplying reacting hollow	T. Powell	New York, N. Y	Dec. 3, 1813	....
Cylinder-press fan-fly attachment	P. A. Cotter	Boston, Mass	Apr. 8, 1873	137, 597
Cylinder, Pulverizing and mixing	W. Coggeshall and J. W. Stanley.	Springfield, Ohio., and Chicago, Ill.	July 13, 1869	92, 585
Cylinder, Steam	W. Corliss	Providence, R. I	Dec. 31, 1872	134, 362
Cylinder, Steam-drying	J. Booth	Pawtucket, R. I	Mar. 31, 1857	16, 905
Cylinder, Steam-drying	H. W. Butterworth	Philadelphia, Pa	June 25, 1867	66, 075
Cylinders, Apparatus for boring	S. R. Wilmot	Bridgeport, Conn	Sept. 20, 1864	44, 296
Cylinders, Apparatus for sizing glass	L. B. Flanders	Philadelphia, Pa	Nov. 10, 1868	84, 040
Cylinders, Arrangement for ports in steam	B. Eaton	Roanoke, Ind	Apr. 28, 1857	17, 142
Cylinders, Boring and squaring off	J. C. Chapman	Charlestown, Mass	Jan. 13, 1863	37, 384
Cylinders, Boring-bar for facing the ends of	J. Flower	Detroit, Mich	Dec. 26, 1871	122, 243
Cylinders, Casting copper	F. Adams	Somerville, Mass	Aug. 2, 1859	24, 915
Cylinders, Casting hollow	J. W. Brittin	Black Rock, Conn	Jan. 25, 1870	99, 144
Cylinders, Casting toothed	H. W. Cornell	Owego, N. Y	Sept. 5, 1871	118, 693
Cylinders, Device for cutting section of annular	J. O. Joyce	Dayton, Ohio	Dec. 8, 1868	84, 697
Cylinders, Lubricator for steam-engine	J. Henwood	New York, N. Y	Mar. 10, 1857	16, 796
Cylinders, Machine for boring and facing	J. MacDonald	Boston, Mass	Oct. 21, 1873	143, 917
Cylinders, Machine for boring curved	W. Wright	New York, N. Y	May 3, 1864	42, 616
Cylinders, Machine for cutting teeth on	S. B. Dean, W. T. Nicholson, and G. Ames.	Boston, Mass., Providence, R. I, and Oswego, N. Y.	Aug. 13, 1872	130, 361
Cylinders, Machine for facing	T. M. Henderson and F. L. McDonald.	Omaha, Nebr	Dec. 2, 1873	145, 103
Cylincers, Machine for making wire	C. H. Latham	Lowell, Mass	Apr. 18, 1871	113, 779
Cylinders, Machine for polishing the inner surfaces of.	T. Goodrum	Providence, R. I	July 16, 1872	129, 402
Cylinders, Machine for turning	J. M. Poole	Wilmington, Del	July 7, 1868	79, 683
Cylinders, Machine for turning	G. Sibley	Troy, N. H	July 3, 1860	29, 014
Cylinders, Manner of mounting hollow revolving	L. B. Pitcher	Salina, N. Y	Sept. 20, 1870	107, 534
Cylinders, Method of forming	H. Willard	Grand Rapids, Mich	May 10, 1870	103, 001
Cylinders, Method of lowering	G. R. Bramhall	Chicago, Ill	June 16, 1868	78, 860
Cylinders, Mode of making toothed	J. L. Tuttle	Lawrence, Mass	Oct. 30, 1849	6, 840
Cylinders, pulleys, &c., Mode of balancing	W. Kitson	Lowell, Mass	Dec. 20, 1870	110, 247
Cylinders, Repairing cast-iron	S. Falkenbury	Susquehannah Depot, Pa	June 22, 1858	20, 635
Cylinders to be polished, Clamp for holding	G. Williamson	Newark, N. J	Nov. 13, 1860	30, 660
Cylindrical boiler	P. M. Hackley	Hudson, N. Y	Oct. 9, 1813	....
Cylindrical box	E. Waters	Troy, N. Y	Jan. 30, 1855	12, 335
Cylindrical boxes, Machine for manufacturing	H. S. Smith, E. Hanson, and M. S. Richardson.	Rutland, Vt	Apr. 7, 1857	17, 001
Cylindrical brake	S. Cram	New York, N. Y	June 2, 1836	....
Cylindrical cutter for leather, &c	J. H. Golding	Liverpool, England	Dec. 22, 1868	85, 085
Cylindrical cutting-knives. Method of sharpening	J. H. and A. T. Goodell	New York, N. Y	May 1, 1860	28, 074
Cylindrical mold for rubber-goods, Machine for filling.	J. W. Cobb	Melrose, Mass	June 18, 1867	65, 794
Cylindrical or tapering sticks, Tool for cutting	G. Davis	Duquesne, Pa	July 13, 1858	20, 866
Cylindrical structures, Construction of double	D. L. Bartlett and G. H. Johnson.	Baltimore, Md	Aug. 28, 1866	57, 462
Cyphering-machine	S. J. Kelso	Detroit, Mich	Sept. 25, 1866	58, 347
D.				
Daguerreotype-apparatus	J. Brown	New York, N. Y	Nov. 15, 1853	10, 225
Daguerreotype-apparatus	W., W. H., and H. J. Lewis	New York, N. Y	Nov. 11, 1851	8, 513
Daguerreotype-apparatus for gilding plates	W. and W. H. Lewis	New York, N. Y	May 8, 1849	6, 431
Daguerreotype-apparatus for panoramic views	I. Van Bunschoten, J. G. Woodbridge, and W. Mann.	New York, N. Y	Apr. 17, 1849	6, 357
Daguerreotype-case	R. Hill	New York, N. Y	June 18, 1861	32, 562
Daguerreotype-case	I. F. Mascher	Philadelphia, Pa	Mar. 8, 1853	9, 611
Daguerreotype-case hinges, Fastening for	S. Peck	New Haven, Conn	Feb. 5, 1856	14, 202
Daguerreotype-case, Monumental	J. Bergstresser	Berrysburgh, Pa	Feb. 8, 1859	22, 850
Daguerreotype-cases, &c., Apparatus for the manufacture of.	H. T. Anthony and F. Phoebus	New York, N. Y	Mar. 23, 1854	10, 953
Daguerreotype-cases, Manufacture of	H. Halvorson	Cambridge, Mass	Aug. 7, 1855	13, 410
Daguerreotype-cases, Manufacture of	S. Peck	New Haven, Conn	Oct. 3, 1854	11, 758
Daguerreotype-cases, Mold for making	J. L. Baldwin	Newark, N. J	Feb. 11, 1862	34, 344
Daguerreotype-cases, &c., Process for ornamenting	J. F. Mascher	Philadelphia, Pa	Feb. 10, 1857	16, 600
Daguerreotype face-plates or mats, Machine for beveling and polishing the inner edges of.	E. Brown	Waterbury, Conn	July 3, 1855	13, 196
Daguerreotype-mat	J. Dean	Worcester, Mass	Dec. 24, 1861	34, 035
Daguerreotype-picture	C. J. Anthony	Pittsburgh, Pa	Jan. 1, 1851	7, 865
Daguerreotype-picture	H. E. Insley	New York, N. Y	Jan. 6, 1852	8, 633
Daguerreotype-picture case	A. F. Stiles	Southbury, Conn	Jan. 22, 1850	7, 041
Daguerreotype-pictures, Coloring	D. Davis, jr	Boston, Mass	Oct. 22, 1842	2, 826
Daguerreotype-pictures, Coloring	W. Thompson	Philadelphia, Pa	May 12, 1843	3, 085
Daguerreotype-plate buffing apparatus	W. and W. H. Lewis	New York, N. Y	July 22, 1851	8, 235
Daguerreotype-plate holder	P. H. Benedict	Syracuse, N. Y	Jan. 31, 1854	10, 466
Daguerreotype-plate holder	D. N. B. Coffin, jr	Lynn, Mass	Feb. 6, 1855	12, 344
Daguerreotype-plate holder	M. Finley	Canandaigua, N. Y	Oct. 4, 1853	10, 093
Daguerreotype-plate holder	J. Hill	Skaneateles, N. Y	Aug. 22, 1854	11, 565
Daguerreotype-plate holder	R. Knecht	Easton, Pa	Feb. 7, 1854	10, 508
Daguerreotype-plate holder	G. Mallory	New York, N. Y	Sept. 17, 1850	7, 655
Daguerreotype-plate holder	S. Peck	New Haven, Conn	Apr. 30, 1850	7, 326
Daguerreotype-plate holder	D. Shive	Philadelphia, Pa	Oct. 9, 1855	13, 665
Daguerreotype-plate-polishing machine	D. Shive	Philadelphia, Pa	Mar. 20, 1855	12, 560
Daguerreotype-plate vise	S. S. Day	New York, N. Y	Oct. 23, 1855	13, 701
Daguerreotype-plates, Apparatus for cleaning and buffing.	T. Longking	Brooklyn, N. Y	Jan. 31, 1854	10, 475
Daguerreotype-plates, Apparatus for holding	W. and W. H. Lewis	New York, N. Y	Oct. 23, 1849	6, 819

Index of patents issued from the United States Patent Office from 1790 *to* 1873, *inclusive*—Continued.

Invention.	Inventor.	Residence.	Date.	No.
Daguerreotype-plates, Apparatus for polishing	B. F. Upton	Bath, Me	Sept. 19, 1854	11, 709
Daguerreotype-plates, Block for holding	A. Beckers	New York, N. Y	Oct. 23, 1849	6, 812
Daguerreotype-plates, Box for coating	W. and W. H. Lewis	New York, N. Y	Nov. 15, 1853	10, 233
Daguerreotype-plates, Box for coating	J. H. Tompkins	Buffalo, N. Y	Jan. 15, 1856	14, 122
Daguerreotype-plates, Coloring	J. B. Isenring	Canton of St. Gall, Switzerland.	Jan. 30, 1846	4, 369
Daguerreotype-plates, Coloring	F. Langenheim	Philadelphia, Pa	Jan. 30, 1846	4, 370
Daguerreotype-plates, Machine for cleaning	C. Ketcham	Penn Yan, N. Y	June 29, 1858	20, 718
Daguerreotype-plates, Machine for polishing	T. Duryea	Williamsburgh, N. Y	June 15, 1852	9, 018
Daguerreotypes, Camera for taking stereoscopic and other.	S. A. Holmes	Brooklyn, N. Y	May 30, 1854	10, 987
Daguerreotypes, Coloring	W. A. Pratt	Alexandria, D. C	Mar. 14, 1846	4, 423
Daguerreoptyes, Coloring	B. R. Stevens and L. Morse	Lowell, Mass	Mar. 28, 1842	2, 522
Daguerreotypes for stereoscopes, Taking	A. S. Southworth and J. J. Hawes.	Boston, Mass	July 11, 1854	11, 304
Daguerreotypes, Gilding	C. L'Homdien	Charleston, S. C	Oct. 26, 1852	9, 354
Daguerreotypes in monumental stones, Securing	S. Jenkins	West Cambridge, Mass	Mar. 11, 1851	7, 974
Daguerreotypes, Taking	J. A. Whipple	Boston, Mass	Jan. 23, 1849	6, 056
Daguerreotyping	W. Yarnall	Newark, Ohio	Dec. 28, 1852	9, 511
Daguerreotyping, Mercury bath for	B. F. Upton	Bath, Me	Apr. 12, 1853	9, 666
Dairy-apparatus	J. A. Whitney	New York, N. Y	Aug. 17, 1869	93, 936
Dairy-can	L. A. Sunderland	Chagrin Falls, Ohio	Aug. 6, 1867	67, 607
Dairy-cooler	D. Whitmer	Paradise, Pa	Dec. 7, 1826	....
Dairy-heater for heating water	G. W. McCammon and D. H. Burrell.	Manheim and Little Falls, N. Y.	May 16, 1781	114, 836
Dairy-house	L. D. Yerks	Downington, Pa	Mar. 27, 1866	53, 516
Dam	J. Du Bois	Williamsport, Pa	Sept. 2, 1862	36, 342
Dam, Adjustable	J. A. Wood	Pittsburgh, Pa	July 11, 1871	117, 028
Dam and level	B. Britten	Galena, Ill	Sept. 11, 1866	57, 852
Dam or water-weir, Adjustable	M. S. Wheaton	Riga, N. Y	Apr. 3, 1849	6, 266
Dam, Portable and adjustable still-water	S. Lewis	Brooklyn, N. Y	July 28, 1868	80, 492
Dam, River	P. Magrath	Philadelphia, Pa	Oct. 10, 1829	....
Dam, Self-acting	H. Werner	Louisville, Ky	Jan. 21, 1873	135, 023
Dam, Still-water adjustable	S. Lewis	Brooklyn, E. D., N. Y	Jan. 5, 1869	85, 598
Dams, Building	S. J. Burr	Chambersburgh, Pa	Feb. 28, 1833	....
Dams, Mode of constructing	W. P. Craig	Newport, Ky	July 8, 1856	15, 317
Dams, Mode of constructing	J. M. Syme	Richmond, Va	Feb. 22, 1830	....
Damask hair-seating, Figured	S. Ross	Camden, N. J	Sept. 18, 1841	2, 262
Damper	F. Bleier	Pittsburgh, Pa	Jan. 15, 1860	7, 015
Damper	E. and J. Bourne	Pittsburgh, Pa	Mar. 24, 1863	37, J47
Damper	S. R. Brann	Hillsborough, Ill	June 10, 1862	35, 500
Damper	M. C. Burleigh	Somersworth, N. H	July 12, 1870	105, 167
Damper	G. Chilsom	Boston, Mass	Apr. 29, 1862	35, 080
Damper	D. B. Cox	Troy, N. Y	Aug. 11, 1868	80, 920
Damper	A. E. Elmer	Springfield, Mass	Oct. 24, 1865	50, 569
Damper	I. L. Frankem	Indianapolis, Ind	Oct. 15, 1867	69, 791
Damper	J. W. Freer	Pittston, Pa	Jan. 31, 1871	111, 336
Damper	W. Hailes	Albany, N. Y	July 28, 1868	80, 476
Damper	E. C. Harrison	New York, N. Y	May 26, 1863	38, 678
Damper	W. B. Hayden	Columbus, Ohio	Sept. 26, 1871	119, 270
Damper	F. M. Hubbard	Protection, N. Y	Mar. 31, 1863	38, 045
Damper	J. G. Jennings	Cleveland, Ohio	Nov. 15, 1864	45, 045
Damper	W. Johnson	Milwaukee, Wis	Sept. 26, 1865	50, 133
Damper	M. W. Kidder	Lowell, Mass	Feb. 26, 1867	62, 341
Damper	J. Kinckerbocker	Hartford, Conn	July 11, 1865	48, 695
Damper	J. W. Martin	Milton, Mass	Mar. 22, 1864	42, 049
Damper	S. Munson	Joliet, Ill	Feb. 16, 1864	41, 636
Damper	W. W. Paxson	Point Pleasant, Pa	Jan. 8, 1867	61, 093
Damper	N. R. Ramsey	Orange, Mass	June 27, 1865	48, 491
Damper	H. Raymond	Racine, Wis	Aug. 9, 1864	43, 794
Damper	J. C. Rhodes	Stillwater, Minn	July 21, 1868	80, 223
Damper	W. Sanford	Brooklyn, N. Y	May 6, 1862	35, 179
Damper	J. F. Stafford	North Granville, N. Y	Nov. 19, 1867	71, 077
Damper	C. C. F. Stender	Chicago, Ill	Mar. 8, 1864	41, 869
Damper	L. C. Taber	Eaton, N. Y	Feb. 23, 1869	87, 221
Damper	G. Tainter	Watertown, Mass	Oct. 14, 1862	36, 679
Damper	L. S Taylor	Sigel, Mo	Mar. 21, 1871	112, 863
Damper	H. B. Thomas	Racine, Wis	Apr. 8, 1862	34, 925
Damper	A. Tracy	Paris Hill, Me	May 23, 1871	115, 135
Damper	J. P. Tuttle	Warren, Ohio	Mar. 8, 1870	100, 692
Damper	P. Verbeck and O. T. Walker	Neenah, Wis	June 2, 1863	38, 776
Damper	N. W. Wheeler	Ripon, Wis	Feb. 2, 1864	41, 457
Damper and handle for anthracite coal stove	E. Nott	Schenectady, N. Y	June 29, 1833	....
Damper and register, Ventilating	W. J. Towne	Newton, Mass	Aug. 30, 1864	44. 030
Damper and shelf, Stove-pipe	E. H. Clinton and W. H. Lavinia	Chicago, Ill	May 23, 1871	115, 025
Damper and stove-pipe collar combined	A. L. Shontz	Quincy, Ill	Jan. 30, 1872	123, 298
Damper and valve regulator, Electro-magnetic	G. M. Sternberg	Fort Riley, Kans	Mar. 1, 1870	100, 462
Damper and ventilator	J. H. Littlefield	Cambridge, Mass	Sept. 12, 1865	49, 899
Damper and ventilator	J. L. Skinner	Amboy, Ill	Aug. 5, 1873	141, 601
Damper and ventilator, Combined	A. Anderson	London, Canada	Dec. 17, 1867	72, 261
Damper and ventilator, Stove-pipe	A. R. Burdick	Racine, Wis	Mar. 1, 1870	100, 369
Damper and ventilator, Thermo regulator for	J. N. Adams	Olathe, Kan	July 30, 1872	130, 002
Damper, Automatic	G. A. Townsend	Hornellsville, N. Y	Aug. 13, 1867	67, 687
Damper, Chimney	P. Baker	Oakland, Md	Apr. 4, 1871	113, 245
Damper, Chimney	A. Campbell	Philadelphia, Pa	Apr. 21, 1857	17, 077
Damper, Chimney	T. F. Engelbrecht	New York, N. Y	July 29, 1856	15, 458
Damper, Cooking-stove	H. L. Sheperd	Dayton, Ohio	Aug. 27, 1850	7, 607
Damper, Cooking-stove	L. L. Thomas	Dighton, Mass	July 19, 1859	24, 849
Damper die, Stove-pipe	J. Van Hagan	Chicago, Ill	Oct. 6, 1868	82. 771
Damper, Fire-place	J. Bradford	Charlotte, N. C	Oct. 8, 1872	131, 994
Damper, Fire-place	J. Bridgham	New York, N. Y	May 21, 1872	126, 927
Damper for air-tight stoves, Self-acting	S. P. Lyon	Farmington, Mich	Nov. 15, 1853	10, 234
Damper for chimneys, cranks, and wheels	J. Reilly and J. Flanagan	Waynesburgh, Pa	Mar. 10, 1827	....
Damper for fire-places, Adjustable	S. Albertson	New York, N. Y	Nov. 20, 1866	59, 789
Damper for flat-iron heater	G. R. Moore	Lyons, Iowa	Mar. 5, 1867	62, 557
Damper for flues	J. Liming	Philadelphia, Pa	Dec. 20, 1864	45, 508
Damper for heating-drums, Adjustable	A. J. Lovejoy	Augusta, Me	Nov. 12, 1872	132, 913
Damper for hot-air furnaces	A. H. Mershon	Philadelphia, Pa	May 5, 1868	77, 512

Index of patents issued from the United States Patent Office from 1790 *to* 1873, *inclusive*—Continued.

Invention.	Inventor.	Residence.	Date.	No.
Damper for hot-air furnaces	A. C. Warren	Concord, N. H	Feb. 25, 1873	136, 291
Damper for hot-air furnaces, Self-adjusting	E. Barrows, jr	Brooklyn, N. Y	Mar. 23, 1858	19, 678
Damper for multitubular steam-boilers	J. R. Robinson	Boston, Mass	Mar. 19, 1861	31, 744
Damper for steam-boilers	J. R. Robinson	Boston, Mass	June 11, 1861	32, 533
Damper for sugar-furnaces	J. Jordon	Boston, Mass	Dec. 27, 1833	....
Damper for windows, Fire	M. J. Barwick	New York, N. Y	June 17, 1873	139, 999
Damper, Grate	J. O'Brien	New York, N. Y	Dec. 15, 1857	18, 885
Damper guard, Stove	R. R. Howell	Fostoria, Ohio	July 9, 1872	128, 880
Damper, Illuminating	J. H. Keyser	New York, N. Y	Sept. 15, 1868	82, 228
Damper in steam-boiler	H. McDonough	New York, N. Y	June 25, 1867	66, 161
Damper, Operating stove	O. Paddock	Watertown, N. Y	Feb. 18, 1862	34, 446
Damper or draft-regulator for stove	S. Oppenheimer	Peru, Ind	Dec. 11, 1866	60, 412
Damper, Oven	J. P. Sherwood	Fort Edwards, N. Y	Oct. 10, 1854	11, 794
Damper-regulator	B. Fitts	Worcester, Mass	Jan. 21, 1873	135, 103
Damper-regulator	J. How and C. W. Copeland	Brooklyn, N. Y	Feb. 2, 1858	19, 249
Damper-regulator	J. H. Murrill	Baltimore, Md	Oct. 4, 1870	107, 948
Damper-regulator	J. F. Neall and W. Myers	Philadelphia, Pa	Nov. 6, 1866	59, 439
Damper-regulator	W. Noyes	New York, N. Y	Dec. 13, 1864	45, 429
Damper regulator and indicator, Steam	C. H. Ford	Baltimore, Md	Jan. 16, 1866	52, 114
Damper regulator, Fire	P. Lamb	San Francisco, Cal	June 21, 1864	41, 213
Damper regulator, Fire	W. Webster	Morrisania, N. Y	Aug. 19, 1862	36, 252
Damper regulator, Steam-boiler	W. Webster	Morrisania, N. Y	Feb. 17, 1857	16, 664
Damper regulator, Steam-boiler	P. White	Brooklyn, N. Y	June 9, 1857	17, 533
Damper regulator, Steam-pressure	E. B. Beach	West Meriden, Conn	Oct. 8, 1872	131, 992
Damper regularor, Stove	L. Boore	Buffalo, N. Y	Mar. 26, 1872	124, 931
Damper, Self-regulating	C. H. Lavis	Philadelphia, Pa	June 20, 1865	48, 291
Damper, Stove	G. Asmus	Portage, Mich	Oct. 3, 1865	50, 216
Damper, Stove	G. Cessford	Utica, N. Y	Mar. 11, 1862	34, 660
Damper, Stove	E. F. Cook	Omaha, Nebr	Dec. 30, 1873	145, 989
Damper, Stove	A. Cowan	Scranton, Pa	July 9, 1872	128, 859
Damper, Stove	E. T. Duke	Plattsmouth, Nebr	Nov. 26, 1867	71, 368
Damper, Stove	H. H. Huntley	Quincy, Ill	Nov. 18, 1873	144, 676
Damper, Stove	L. W. Langdon	Northampton, Mass	Nov. 30, 1869	97, 414
Damper, Stove	E. Lohsand	La Porte, Ind	May 2, 1871	114, 453
Damper, Stove	H. Mallory	Milwaukee, Wis	June 15, 1869	91, 245
Damper, Stove	J. Smith and S. C. Brown	Richmond, Ind	Mar. 12, 1867	62, 896
Damper, Stove	G. W. Walker	Boston, Mass	June 19, 1866	55, 451
Damper, Stove	J. D. Willoughby	Washington, D. C	Feb. 4, 1873	135, 611
Damper, Stove	R. Worster	Central Falls, R. I	Oct. 28, 1873	144, 043
Damper, Stove self-regulating	B. Owen	Seneca Falls, N. Y	June 19, 1849	6, 546
Damper, Stove-pipe	L. O. Allen	Gardiner, Me	Dec. 15, 1868	84, 930
Damper, Stove-pipe	I. C. and A. B. Allerton	Aztalan, Wis	Sept. 12, 1865	51, 409
Damper, Stove-pipe	T. K. Anderson	Hornellsville, N. Y	Mar. 26, 1867	63, 196
Damper, Stove-pipe	T. K. Anderson	Hornellsville, N. Y	June 8, 1869	90, 912
Damper, Stove-pipe	J. Ash	Sterling, Ill	Oct. 17, 1871	119, 911
Damper, Stove-pipe	H. Baker	Lancaster, Pa	Jan. 25, 1870	99, 048
Damper, Stove-pipe	J. M. Baker	Aurora, Ill	Oct. 19, 1869	95, 967
Damper, Stove-pipe	R. R. Ball	West Meriden, Conn	Sept. 24, 1872	131, 591
Damper, Stove-pipe	J. Barton	Battle Creek, Mich	Feb. 13, 1866	52, 514
Damper, Stove-pipe	N. A. Boynton	New York, N. Y	July 4, 1871	116, 675
Damper, Stove-pipe	J. Bradshaw and S. C. Wilson	Albion, N. Y	June 6, 1865	48, 046
Damper, Stove-pipe	R. M. Breckenridge	West Meriden, Conn	Aug. 7, 1866	56, 891
Damper, Stove pipe	H. C. Brown	Buffalo, N. Y	May 30, 1865	47, 923
Damper, Stove-pipe	E. C. Chapman	Lacon, Ill	Nov. 4, 1873	144, 192
Damper, Stove-pipe	D. A. Cleaveland	Providence, R. I	Sept. 3, 1872	130, 979
Damper, Stove-pipe	A. Colton	Sycamore, Ill	Mar. 24, 1868	75, 863
Damper, Stove-pipe	B. F. Cowan	New York, N. Y	July 17, 1866	56, 379
Damper, Stove-pipe	E. Cox and A. W. Potter	Monroe, Wis	Aug. 6, 1867	67, 415
Damper, Stove-pipe	E. K. Dean	Bangor, Me	Feb. 18, 1873	136, 042
Damper, Stove-pipe	D. De Garmo	Rochester, N. Y	Jan. 29, 1867	61, 613
Damper, Stove-pipe	W. H. Deily	Sycamore, Ill	Oct. 20, 1868	83, 139
Damper, Stove-pipe	N. J. Eldred	Chicago, Ill	Sept. 25, 1866	58, 233
Damper, Stove-pipe	J. Emmert	Chicago, Ill	May 20, 1873	138, 997
Damper, Stove-pipe	L. S. Enos	Almond, N. Y	July 14, 1868	79, 817
Damper, Stove-pipe	C. R. Everson	Palmyra, N. Y	Dec. 4, 1866	60, 160
Damper, Stove-pipe	A. V. and A. F. Fletcher	Athol, Mass	June 6, 1865	48, 058
Damper, Stove-pipe	J. Fowler	Watertown, Wis	Aug. 29, 1865	49, 619
Damper, Stove-pipe	J. Fraser and O. S. Garretson	Buffalo, N. Y	Dec. 18, 1866	60, 498
Damper, Stove-pipe	B. French	Rochester, N. Y	Aug. 8, 1871	117, 762
Damper, Stove-pipe	E. M. Gardner	Nantucket, Mass	Apr. 18, 1865	47, 293
Damper, Stove-pipe	W. T. Gray	Galesburgh, Ill	May 7, 1867	64, 415
Damper, Stove-pipe	D. L. Grover	Groton, N. Y	Apr. 3, 1866	53, 605
Damper, Stove-pipe	W. Hailes and P. Finkle	Albany, N. Y	Sept. 10, 1867	68, 738
Damper, Stove-pipe	E. R. Hall and D. A. Keyes	Norfolk, Conn	May 16, 1871	114, 806
Damper, Stove-pipe	G. B. Halsted	New York, N. Y	Nov. 12, 1867	70, 836
Damper, Stove-pipe	R. Ham	Troy, N. Y	Sept. 28, 1869	95, 340
Damper, Stove-pipe	B. I. Harris	Harrisburgh, Pa	Sept. 25, 1866	58, 251
Damper, Stove-pipe	J. Healey, jr	Detroit, Mich	July 7, 1868	79, 758
Damper, Stove-pipe	R. M. Hermance	Half Moon, N. Y	May 3, 1870	102, 539
Damper, Stove-pipe	T. Hessenbruch	Philadelphia, Pa	Jan. 19, 1869	85, 931
Damper, Stove-pipe	M. Hicks	Richmond, Ill	May 29, 1866	55, 101
Damper, Stove-pipe	R. Hillson	Albany, N. Y	Jan. 30, 1872	123, 171
Damper, Stove-pipe	G. Hollinger and S. H. Fry	Rothsville, Pa	Apr. 26, 1870	102, 266
Damper, Stove-pipe	J. L. Howard	Hartford, Conn	Mar. 7, 1865	46, 670
Damper, Stove-pipe	Z. Hunt	Hudson, N. Y	July 16, 1861	32, 834
Damper, Stove-pipe	J. A. Jacobs	Pittsfield, N. H	July 24, 1866	56, 566
Damper, Stove-pipe	J. Johnson	Atkinson, Ill	July 14, 1868	79, 836
Damper, Stove-pipe	E. Kabisius	Davenport, Iowa	Apr. 2, 1867	63, 395
Damper, Stove-pipe	C. Kathan	Hardin, Iowa	July 25, 1865	48, 952
Damper, Stove-pipe	W. J. Keep	Troy, N. Y	Mar. 8, 1870	100, 641
Damper, Stove-pipe	W. J. Keep	Troy, N. Y	Apr. 26, 1870	102, 407
Damper, Stove-pipe	W. J. Keep	Troy, N. Y	Dec. 12, 1871	121, 877
Damper, Stove-pipe	J. C. Kennedy	Chicago, Ill	Oct. 6, 1868	82, 848
Damper, Stove-pipe	H. Kipe	Thornbury, Pa	Apr. 4, 1865	47, 111
Damper, Stove-pipe	W. H. Lutz	Lancaster, Pa	Feb. 28, 1871	112, 260
Damper, Stove-pipe	E. Mackwitz and W. Frankfurth.	Milwaukee, Wis	Jan. 24, 1865	46, 007
Damper, Stove-pipe	D. Manuel	Boston, Mass	Dec. 17, 1867	72, 212

Index of patents issued from the United States Patent Office from 1790 *to* 1873, *inclusive*—Continued.

Invention.	Inventor.	Residence.	Date.	No.
Damper, Stove-pipe	M. S. Marshall	Melrose, Mass	Mar. 5, 1867	62, 655
Damper, Stove-pipe	S. Marvin	South Royalston, Mass	Apr. 10, 1866	53, 925
Damper, Stove-pipe	J. C. Merritt	Wyoming, N. Y	Mar. 6, 1866	53, 024
Damper, Stove-pipe	B. F. Miller	New York, N. Y	Aug. 7, 1866	56, 975
Damper, Stove-pipe	R. S. Miller	Battle Creek, Mich	Aug. 28, 1866	57, 544
Damper, Stove-pipe	M. F. Moody	Andover, N. H	Oct. 8, 1872	131, 963
Damper, Stove pipe	J. A. Noble	Florence, Mass	Sept. 6, 1864	44, 109
Damper, Stove-pipe	W. H. Nutting	Orange, Mass	July 31, 1866	56, 789
Damper, Stove-pipe	S. R. Nye	Barre, Mass	Apr. 24, 1866	54, 197
Damper, Stove-pipe	H. Ogborn	Richmond, Ind	Nov. 26, 1867	71, 524
Damper, Stove-pipe	H. Ogborn and A. T. Chapin	Richmond, Ind	Aug. 21, 1866	57, 368
Damper, Stove-pipe	E. Parker	Marlow, N. H	May 22, 1866	55, 017
Damper, Stove-pipe	E. Parker	Marlow, N. H	Apr. 23, 1867	64, 136
Damper, Stove-pipe	F. D. Pastorious	Philadelphia, Pa	Aug. 11, 1868	80, 832
Damper, Stove-pipe	W. V. Perry	Burnett, Wis	Mar. 5, 1867	62, 559
Damper, Stove-pipe	J. M. Read	Boston, Mass	Nov. 18, 1873	144, 634
Damper, Stove-pipe	C. Reed	Beaver Dam, Wis	June 4, 1872	127, 516
Damper, Stove-pipe	W. F. Rossman	Hudson, N. Y	Aug. 16, 1870	106, 410
Damper, Stove-pipe	R. S. Sanborn and W. Bennett	Ripon, Wis	Apr. 10, 1866	53, 886
Damper, Stove-pipe	D. Sanders	Milwaukee, Wis	Feb. 16, 1869	86, 943
Damper, Stove-pipe	D. Sanders	Milwaukee, Wis	June 28, 1870	104, 889
Damper, Stove-pipe	H. J. Sayers	Salina, Pa	Apr. 6, 1864	88, 744
Damper, Stove-pipe	G. E. Smith	Racine, Wis	Nov. 8, 1869	44, 981
Damper, Stove-pipe	J. Spear	Philadelphia, Pa	Mar. 12, 1872	124, 456
Damper, Stove-pipe	W. X. Stevens and W. E. Puffer	Worcester and Lexington, Mass.	Apr. 16, 1867	63, 959
Damper, Stove-pipe	W. W. St. John	Saint Louis, Mo	Feb. 6, 1866	52, 460
Damper, Stove-pipe	D. and S. Sweeney	Constableville, N. Y	May 22, 1866	54, 976
Damper, Stove-pipe	G. Tamkin	Newburgh, N. Y	Sept. 29, 1868	82, 565
Damper, Stove-pipe	W. Taylor	Lowell, Mass	May 31, 1870	103, 686
Damper, Stove-pipe	I. Van Hagen	Chicago, Ill	Sept. 21, 1869	94, 986
Damper, Stove-pipe	W. Wasson and G. W. Dungan	Genoa, Nev	Aug. 13, 1872	130, 456
Damper, Stove-pipe	D. A. White	Chagrin Falls, Ohio	June 16, 1868	79, 041
Damper, Stove-pipe	G. W. Wiseman	Sycamore, Ill	Dec. 17, 1867	72, 437
Damper valve, Stove-pipe	G. B. Wiseman	Sycamore, Ill	May 21, 1867	65, 033
Damper, Ventilating	N. A. Boynton	New York, N. Y	May 5, 1863	38, 370
Damper, Ventilating	G. Tainter	Watertown, Mass	May 13, 1862	35, 269
Damper, Ventilating and check-draft	J. A. Lawson	Troy, N. Y	Feb. 7, 1865	46, 249
Dampers for cooking-stoves, Arrangement of	W. E. Hayes	Geneva, N. Y	Apr. 22, 1856	14, 720
Dampers in rotary stoves, Arrangement of	W. W. Hill	Greenport, N. Y	June 6, 1854	11, 010
Dampers in steam-heating apparatus, Method of operating.	J. J. Smith	Elizabeth, N. J	Aug. 20, 1861	33, 107
Dampers of furnaces, Operating	D. Treadwell	Cambridge, Mass	Aug. 22, 1854	11, 583
Dampers of steam-boilers by the pressure of the steam, Regulating the.	P. Clark	Rahway, N. J	Jan. 3, 1854	10, 387
Damping-apparatus	P. Hill	Philadelphia, Pa	Feb. 5, 1867	61, 833
Damping-trough	H. T. Davis	New Cross, Great Britain	Dec. 1, 1868	84, 538
Dancer, Automatic	J. E. Atwood and C. P. Stimets	Trenton, N. J., and New York, N. Y.	Sept. 27, 1864	44, 378
Dancer, Automatic	C. Chinnock	Brooklyn, N. Y	Oct. 17, 1865	50, 452
Dancer, Automatic	T. N. and J. N. Crow	Mott Haven, N. Y	Dec. 1, 1863	40, 740
Dancer, Automatic tight-rope	W. Humans	Boston, Mass	Nov. 12, 1867	70, 850
Dancer, Automaton	I. S. Clough and V. Fountain, jr	Brooklyn and North Shore, N. Y.	Aug. 9, 1864	43, 759
Dancer, Automaton	W. Nunns	New York, N. Y	May 31, 1864	42, 962
Dash-board	J. S. Campbell	Bridgeport, Conn	Aug. 12, 1873	141, 758
Dash-board and rein-holder	P. Verplanck, jr	Binghamton, N. Y	Mar. 24, 1868	76, 007
Dash-board bag	S. Hipkiss	Charlestown, Mass	July 1, 1873	140, 416
Dash-wheel for washing and bleaching, Use of the.	J. Wallace, jr	Glasgow, North Britain	Sept 30, 1856	15, 836
Dating-machine	G. J. Hill	Buffalo, N. Y	June 26, 1860	28, 861
Davit and winch, Ship's	D. P. Nickerson	Cleveland, Ohio	Mar. 26, 1867	63, 289
Davit-fall-block hook	W. R. Satterly	Port Jefferson, N. Y	Sept. 18, 1866	58, 142
Davit-ring, Self-releasing	A. Guild and W. H. Pierce	Middletown, Conn	Oct. 18, 1870	108, 353
Davit, Ship's	W. Crosthwait	Buffalo, N. Y	Oct. 8, 1872	131, 941
Davit, Ship's	L. F. Frazee	South Amboy, N. J	July 30, 1867	67, 282
Davit, Ship's	S. Gill and D. C. Woods	San Pablo and San Francisco, Cal.	Dec. 8, 1868	84, 742
Davit, Ship's	C. G Meinhardt	Altoona, Pa	Aug. 11, 1868	80, 869
Davit, Ship's	C. Perley	New York, N. Y	Jan 27, 1852	8, 692
Davit, Ship's	J. A. Whiting	Baltimore, Md	Aug. 20, 1872	130, 678
Davits, Device for detaching boats from	E. H. Sheffield and E. P. Palmer.	Stonington, Conn., and New York, N. Y.	Apr. 10, 1866	53, 928
Davits, Tripping-block for boat's	C. Perley	New York, N. Y	Nov. 2, 1858	21, 979
Dead center, Device for overcoming	J. Coy	Oswego, N. Y	May 7, 1872	126, 526
Dead-center in cranks, Means for avoiding the	T. Williams	Providence, R. I	Apr. 2, 1861	31, 926
Dead-center lift	J. J. Gorman	Cincinnati, Ohio	Jan. 31, 1865	46, 102
Dead-eye blocks, Machine for making	T. Blanchard	United States	Aug. 10, 1836	7
Dead-point in cranks, Engine for overcoming	J. Griffin	Louisville, Ky	June 11, 1861	32, 575
Dead-point in cranks, Machine for overcoming	J. P. Cooper	Finleyville, Pa	Apr. 3, 1860	27, 698
Dead-point in cranks, Mode of arranging the cylinder in double-cylinder engines for overcoming.	C. S. Ferris	Courtland, N. Y	July 1, 1840	1, 667
Dead-spindle	H. G. Davis	Northborough, Mass	Sept. 9, 1835	......
Decanter-stopper, Self-acting	L. Loeffler	East Cambridge, Mass	May 29, 1866	55, 124
Decarbonizing furnace	S. H. Whitaker	Covington, Ky	Oct. 30, 1866	59, 301
Deck and side light for ship	E. S. Hidden	New York, N. Y	Mar. 29, 1864	42, 087
Deck and side light for vessel	C. Perley	New York, N. Y	July 18, 1865	48, 836
Deck and side light of vessels, Fastening for	S. Winter	Boston, Mass	Mar. 22, 1864	42, 058
Deck and side lights, Means of closing ship's	E. B. Vannevar	Boston, Mass	May 30, 1865	48, 033
Deck-light	I. M. Bearse	Boston, Mass	Feb. 25, 1873	136, 299
Deck-light	E. S. Hidden	New York, N. Y	Nov. 3, 1863	40, 479
Deck-light	H. Lanergan	East Cambridge, Mass	Jan. 29, 1861	31, 247
Deck-light for vessels	J. Sutton and J. Gregory	New York, N. Y	Apr. 28, 1863	38, 342
Deck light, Ship's	G. Gray	Lynn, Mass	Mar. 26, 1825	......
Decoctions, Apparatus for making	W. Adamson	Philadelphia, Pa	Sept. 6, 1859	25, 311
Decoloring compound, Artificial	F. Gerau	New York, N. Y	May 20, 1856	14, 911
Decolorizer, Compound used as	P. M. Belton	Brooklyn, N. Y	Aug. 14, 1860	29, 560
Decorative slab from plastic materials	E. Robbins	Somers Town, England	Feb. 1, 1870	99, 479
Decoy-duck	J. Foster	Philadelphia, Pa	Aug. 3, 1869	93, 293

Index of patents issued from the United States Patent Office from 1790 to 1873, inclusive—Continued.

Invention.	Inventor.	Residence.	Date.	No.
Delaine, Process of manufacturing	J. Marland	West Bridgewater, Mass	Sept. 9, 1856	15, 703
Delineator for giving fac-simile	J. Wickes	New York, N. Y	Sept. 9, 1825	....
Demijohn	G. W. and S. W. Foster	Charlestown, Mass	Apr. 2, 1872	125, 187
Dental anæsthetic-instrument	E. Cutter	Woburn, Mass	Nov. 13, 1866	59, 562
Dental and surgical instruments, Handle for	H. T. Fogg	San Paulo, Brazil	Sept. 1, 1868	81, 616
Dental apparatus	W. H. Dibble	Bordentown, N. J	Oct. 17, 1865	50, 461
Dental apparatus	W. H. Dibble	Bordentown, N. J	Apr. 9, 1867	63, 709
Dental apparatus for casting plates for artificial teeth.	J. A. Loomis and C. F. Moll	Carthage, Ill., and San Francisco, Cal.	Apr. 19, 1870	102, 019
Dental apparatus for relief of pain while operating	N. Washburn	Bridgewater, Mass	Dec. 6, 1859	26, 388
Dental articulating-cup	C. Von Bonhorst	Lancaster, Ohio	June 1, 1869	90, 706
Dental articulator	L. Hoffstadt	Philadelphia, Pa	Aug. 22, 1871	118, 242
Dental articulator	G. F. Schaffer	New York, N. Y	Feb. 8, 1870	99, 598
Dental blow-pipe	J. Thompson	North Bridgewater, Mass	Sept. 7, 1852	5, 254
Dental crystalline-gold, Manufacture of	A. J. Watts	Utica, N. Y	Dec. 4, 1860	30, 847
Dental drill	J. W. Baxter	Vevay, Ind	May 20, 1873	139, 039
Dental drill	G. V. Black	Jacksonville, Ill	Aug. 8, 1871	117, 732
Dental drill	H. F. Bryant	Marathon, N. Y	Nov. 26, 1867	71, 361
Dental drill	W. S. Elliott	Goshen, N. Y	Feb. 13, 1872	123, 559
Dental drill	W. S. Elliott	Goshen, N. Y	Sept. 10, 1872	131, 156
Dental drill	W. H. Gates	Louisville, Ky	Oct. 9, 1866	58, 631
Dental drill	A. Hartman	Murfreesborough, Tenn	Aug. 22, 1871	118, 237
Dental drill	C. Poor	Dubuque, Iowa	Aug. 22, 1871	118, 268
Dental drill	W. M. Reynolds	New York, N. Y	May 28, 1872	127, 269
Dental drill	J. J. Ross	Memphis, Tenn	June 4, 1872	127, 518
Dental drill	P. Soper	London, Canada	Mar. 13 1866	53, 245
Dental drill	K. Spencer	Athens, Ga	Oct. 17, 1848	5, 866
Dental drill	L. D. Walter	Fort Plain, N. Y	May 15, 1847	5, 121
Dental drill, Automatic	M. I. Gallager	Savannah, Ga	Jan. 21, 1873	135, 109
Dental drilling-machine	C. E. Edwards	Philadelphia, Pa	Dec. 2, 1873	145, 098
Dental engine	C. M. Curtis	Philadelphia, Pa	Apr. 29, 1873	138, 318
Dental engine	J. B. Morrison	Saint Louis, Mo	Feb. 7, 1871	111, 667
Dental file holder and polisher	H. Lawrence	New Orleans, La	May 23, 1871	115, 070
Dental filling	C. E. Blake	San Francisco, Cal	Dec. 9, 1873	145, 275
Dental fillings, Porcelain blocks for	C. H. Mack	New York, N. Y	May 20, 1873	139, 012
Dental fillings, Securing	C. H. Mack	Portland, Oreg	May 2, 1871	114, 454
Dental forceps	H. J. Batchelder	West Fairlee, Vt	July 1, 1856	15, 215
Dental forceps	E. Bourne	New Bedford, Mass	Sept. 25, 1849	6, 741
Dental forceps	J. C. Burch	Evansville, Ind	Sept. 9, 1851	8, 351
Dental forceps	J. D. Chevalier	New York, N. Y	Oct. 22, 1872	132, 354
Dental forceps	J. G. Coates	Big Lick, Va	Sept. 16, 1856	15, 730
Dental forceps	N. A. Durham	Duquoin, Ill	Dec. 5, 1871	121, 599
Dental forceps	N. A. Durham	Duquoin, Ill	Dec. 12, 1871	121, 858
Dental forceps	C. G. French	Springfield, Ill	Dec. 2, 1873	145, 058
Dental forceps	L. G. Haskins	Newport, N. Y	Dec. 19, 1871	122, 012
Dental forceps	P. N. Jacobus	Montague, N. J	Dec. 26, 1871	122, 254
Dental forceps	J. A. McClelland	Louisville, Ky	Apr. 21, 1857	17, 107
Dental forceps	S. Woolverton	Trenton, N. J	Mar. 17, 1868	75, 716
Dental gold, Manufacture of	R. S. Williams	New York, N. Y	July 15, 1873	140, 984
Dental hammer. *See* Hammer.				
Dental hydraulic-cup	J. Harrison	Jamestown, N. Y	Aug. 26, 1851	8, 312
Dental impression-cup	O. B. Buttles	Milwaukee, Wis	June 12, 1866	55, 465
Dental impression cup	R. V. Jenks	Paterson, N. J	Dec. 21, 1869	98, 066
Dental impression-cup	G. McDonald	Athens, Ga	Sept. 21, 1869	95, 126
Dental impression-cup	G. T. Schaffer	New York, N. Y	Mar. 20, 1866	53, 347
Dental impression-cup	C. L. Wuestenberg	Pittsburgh, Pa	June 1, 1869	90, 802
Dental impression-cup and suction-mold	G. H. Hurd	Memphis, Tenn	Apr. 27, 1869	89, 316
Dental impression-mold, Construction of	W. C. Smith	Warrensburgh, Mo	July 5, 1870	105, 008
Dental instrument	C. H. Dubs	Natchez, Miss	Oct. 17, 1848	5, 865
Dental instrument	C. H. Mack	Portland, Oreg	Aug. 8, 1871	117, 904
Dental instrument	W. R. Pomeroy	Millersburg, Ohio	May 25, 1869	90, 388
Dental instrument	I. A. Salmon	Boston, Mass	Feb. 26, 1867	62, 504
Dental instrument	S. C. Taylor	Toledo, Ohio	Feb. 18, 1868	74, 629
Dental jaw-brace	G. H. Hurd	Memphis, Tenn	Feb. 16, 1869	86, 922
Dental lathe	E. C. Rishel	White Haven, Pa	Jan. 10, 1871	110, 869
Dental mallet	J. A. Bidwell	Chicago, Ill	May 31, 1870	103, 706
Dental mallet	J. A. Harris	Pontiac, Mich	Oct. 17, 1865	50, 468
Dental mallet	C. Poor	Dubuque, Iowa	Sept. 18, 1866	58, 133
Dental mallet	J. N. Scranton and H. H. Parsons	Bennington, Vt., and Hoosick Falls, N. Y.	Oct. 24, 1865	50, 633
Dental mallet, Automatic	H. Hodge and J. P. Noyes	Binghamton, N. Y	July 23, 1867	67, 114
Dental models, Method of obtaining	L. Stuck	Bryan, Ohio	Sept. 29, 1868	82, 563
Dental mold	G. H. Hurd	Saint Louis, Mo	Sept. 4, 1866	57, 726
Dental mouth-meter	C. Von Bonhorst	Lancaster, Ohio	June 1, 1869	90, 705
Dental operations, Application of electricity in	W. G. A. Bomvill	Dover, Del	Feb. 8, 1859	22, 851
Dental operations, Applying electricity in	W. G. Oliver	Buffalo, N. Y	Apr. 19, 1859	23, 697
Dental pin	P. Crans, jr	Philadelphia, Pa	June 28, 1870	104, 835
Dental pins, Machine for heating	W. A. Duff and J. J. Griffith	Philadelphia, Pa	Mar. 14, 1865	46, 786
Dental pins, Manufacture of	F. W. Smith	Philadelphia, Pa	Mar. 4, 1862	34, 616
Dental plate	J. A. McClelland	Louisville, Ky	Aug. 11, 1863	39, 538
Dental plate	J. A. McClelland	Louisville, Ky	June 1, 1869	90, 765
Dental plate	E. J. Merick	Rochester, N. Y	July 23, 1867	67, 129
Dental plate	S. Purvine and H. Smith	Salem, Oreg	Aug. 29, 1871	118, 484
Dental plate	L. R. Streeter	Chelsea, Mass	Apr. 20, 1869	89, 253
Dental plate	G. W. Tripp	Auburn, N. Y	Aug. 4, 1862	39, 434
Dental plate	H. Twitchell	Pulaski, N. Y	Apr. 6, 1869	88, 682
Dental plate and teeth	R. E. Burlan	Lewisburgh, Pa	Feb. 6, 1872	123, 451
Dental plate, Atmospheric	N. T. Folsom	Laconia, N. H	Jan. 1, 1867	60, 871
Dental plate atmospheric attachment	J. P. Gillespie	San Francisco, Cal	June 13, 1871	115, 951
Dental plate, Atmospheric pressure	M. Levett	New York, N. Y	Apr. 6, 1858	19, 858
Dental plate atmospheric pressure attachment	J. B. Beers	San Francisco, Cal	Jan. 24, 1871	111, 168
Dental plate, Composition	A. Hill	Norwalk, Conn	July 27, 1869	93, 086
Dental plates and other articles, Composition for molds and models in casting.	S. Lawrence	Lowell, Mass	Aug. 31, 1869	94, 3[illegible]4
Dental plates and other purposes, Vulcanizing rubber for.	C. M. Kelsey	Mount Vernon, Ohio	Aug. 10, 1869	93, 623
Dental plates, Apparatur for molding	J. W., jr., and I. S. Hyatt	Albany, N. Y	Dec. 5, 1871	121, 522

Index of patents issued from the United States Patent Office from 1790 *to* 1873, *inclusive*—Continued.

Invention.	Inventor.	Residence.	Date.	No.
Dental plates, Casting	J. U. L. Fremster	Greencastle, Ind	Oct. 11, 1870	108. 245
Dental plates, Composition for	G. F. J. Colburn	Newark, N. J	July 23, 1867	67, 025
Dental plates, Composition for	L. R. Streeter	Chelsea, Mass	Mar. 23, 1869	88, 260
Dental plates, Flask for casting	S. Lawrence	Lowell, Mass	Aug. 31, 1869	94, 323
Dental plates from pyroxyline	I. S. J. W. Hyatt, jr., and J. A. Perkins.	Albany, N. Y	Mar. 28, 1871	113, 055
Dental plates from pyroxyline, Manufacture of	J. Brockway	Albany, N. Y	April 18, 1871	113, 736
Dental plates, Lathe-attachment for finishing	E. H. Danforth	Jamestown, N. Y	Jan. 11, 1859	22, 548
Dental plates, Machine for polishing	E. A. Mayor	Owego, N. Y	Aug. 6, 1872	130, 304
Dental plates, Manufacture of	F. M. Shields	Sacramento, Cal	Sept. 26, 1871	119, 245
Dental plates, &c., Material for	J. A. McClelland	Louisville, Ky	Apr. 28, 1868	77, 304
Dental plates, Mode of connecting artificial teeth with.	H. Crane	New York, N. Y	Feb. 16, 1869	86, 972
Dental plates, Mode of forming air-chamber in	M. A. Boughton	Norwalk, Conn	Jan. 31, 1871	111, 429
Dental plates, Preparing an application of pyroxyline for.	R. H. Winsborough	Saint Louis, Mo	April 25, 1871	114, 242
Dental plates, Vulcanized rubber for	C. M. Kelsey	Mount Vernon, Ohio	Nov. 21, 1871	121, 174
Dental plugger	J. T. Codman	Boston, Mass	Dec. 2, 1873	145, 051
Dental plugger	W. F. Griswold	Leavenworth, Kans	Apr. 9, 1872	125, 452
Dental plugger	C. M. Hooker	Hartford, Conn	Sept. 25, 1866	58, 257
Dental plugger	W. G. Redman	Louisville, Ky	Jan. 22, 1867	61, 460
Dental plugger	W. D. Stillman	West Winfield, N. Y	Aug. 27, 1872	130, 954
Dental plugger and burring-tool combined	I. M. Seaman	Buffalo, N. Y	Sept. 3, 1872	131, 123
Dental plugger, Pneumatic	E. A. Hyde	Ann Arbor, Mich	June 29, 1869	91, 849
Dental plugging-instrument	J. N. Scranton and H. A. Parsons	Bennington, Vt., and Hoosick Falls, N. Y.	June 1, 1869	90, 879
Dental plugging-instrument	G. B. Snow and T. G. Lewis	Buffalo, N. Y	Oct. 30, 1866	59, 284
Dental plugging-instrument	G. B. Snow and T. G. Lewis	Buffalo, N. Y	Nov. 20, 1866	59, 784
Dental plugging-instrument, Automatic	G. B. Snow and T. G. Lewis	Buffalo, N. Y	June 23, 1868	79, 270
Dental self-feeding plugging-tool	M. Burchardt	Berlin, Prussia	Oct. 10, 1865	50, 428
Dental substitute	G. W. Cool	Portland, Oreg	Dec. 3, 1867	71, 707
Dental swage, Adjustable	E. H. Danforth	Jamestown, N. Y	Mar. 22, 1859	23, 293
Dental tongue-holder	F. M. Osborn	Port Chester, N. Y	June 28, 1870	104, 874
Dental tongue-holder	F. M. Osborn	Port Chester, N. Y	Aug. 22, 1871	118, 386
Dental tool	C. P. Grout	New York, N. Y	Apr. 22, 1873	138, 150
Dental tools for separating teeth	R. Arthur	Baltimore, Md	Dec. 3, 1872	133, 617
Dental turnkey	H. Todd	Columbus, Ohio	May 9, 1846	4, 506
Dental use, Apparatus for packing rubber for	F. C. Brown	Palmyra, N. Y	Sept. 19, 1865	49, 969
Dental use, Compound for	F. Smith	Schenectady, N. Y	June 4, 1872	127, 656
Dental use, Diamond-pointed drill for	J. P. Gillespie	Louisville, Ky	Apr. 1, 1873	137, 434
Dental use, Gold for	R. S. Williams	New York, N. Y	Apr. 30, 1872	126, 242
Dental use, Making crystal shred gold for	E. Lamm	New Orleans, La	June 4, 1867	65, 398
Dental use, Manufacture of gold-foil for	U. K. Mayo	Boston, Mass	Nov. 1, 1870	108, 806
Dental use, Manufacture of rubber for	I. Woolworth	New Haven, Conn	Dec. 19, 1865	51, 642
Dental use, Metallic foil for	C. E. Blake	San Francisco, Cal	Dec. 9, 1873	145, 274
Dental use, Method of preparing gold for	E. G. Kearsing	New York, N. Y	Dec. 12, 1865	51, 459
Dental use, Method of preparing gold for	E. Lamm	New Orleans, La	July 31, 1866	56, 765
Dental use, Preparing gold for	G. J. Pack	New York, N. Y	May 17, 1870	103, 227
Dental use, Preparing gold for	G. J. Pack	New York, N. Y	May 14, 1872	126, 830
Dental use, Preparing gold for	R. S. Williams	New York, N. Y	June 21, 1870	104, 524
Dental use, Preparing gold for	R. S. Williams	New York, N. Y	Apr. 8, 1873	137, 747
Dental use, Process for preparing gold for	E. G. and L. Kearsing	Spring Valley, N. Y	Dec. 19, 1871	122, 029
Dental use, Rubber for	E. L. Simpson	Bridgeport, Conn	Oct. 16, 1866	58, 901
Dental vulcanizer, Safety-valve for	F. R. Moorhead	Chandlersville, Ohio	July 5, 1870	105, 115
Dental vulcanizing-heater	D. J. Peer	Rochester, N. Y	Apr. 3, 1866	53, 666
Dentifrice	J. G. Cook	Lewiston, Me	Jan. 22, 1867	61, 320
Dentifrice	O. Danforth	Bibb County, Ga	Feb. 14, 1871	111, 821
Dentifrice-p ste	G. F. J. Colburn	Newark, N. J	Dec. 22, 1868	85, 166
Dentifrice, Vegetable	E. H. Reid	Clarkesville, Ga	Mar. 1, 1834	....
Dentist's and barber's chair	R. W. and G. W. Archer	Rochester, N. Y	July 9, 1872	128, 775
Dentist's and barber's chair	A. T. Boon and J. B. Finchure	Galesburgh, Ill	Aug. 4, 1868	80, 589
Dentist's and barber's chair	F. J. Coates	Cincinnati, Ohio	Oct. 15, 1872	132, 255
Dentist's and barber's chair	A. Gebhard and J. Blodan	Indianapolis, Ind	June 27, 1871	116, 427
Dentist's and barber's chair	O. C. White	Hopkinton, Mass	Nov. 14, 1871	120, 919
Dentist's and barber's chair, Head-rest for	R. W. Archer	Rochester, N. Y	Aug. 17, 1869	93, 662
Dentist's and surgeon's chair	F. Searle	Springfield, Mass	Mar. 26, 1850	7, 224
Dentist's chair	A. M. Asay	Philadelphia, Pa	Apr. 23, 1850	7, 299
Dentist's chair	A. M. Asay	Philadelphia, Pa	Feb. 7, 1854	10, 490
Dentist's chair	W. M. Butler	San Francisco, Cal	Dec. 19, 1865	51, 553
Dentist's chair	W. M. Butler	Louisville, Ky	Aug. 16, 1870	106, 463
Dentist's chair	A. M. Holmes	Morrisville, N. Y	Sept. 14, 1858	21, 562
Dentist's chair	A. W. Morrison	Saint Louis, Mo	Aug. 13, 1872	130, 525
Dentist's chair	D. W. Perkins	Rome, N. Y	Aug. 7, 1855	13, 396
Dentist's chair	I. A. Salmon	Boston, Mass	June 5, 1866	55, 368
Dentist's chair	R. A. Stratton	Philadelphia, Pa	May 29, 1855	12, 972
Dentist's chair	V. Van Vleck	New York, N. Y	Nov. 20, 1860	30, 706
Dentist's chair	J. O. Whitcomb	New York, N. Y	Oct. 3, 1865	50, 291
Dentist's chair	O. C. White	Hopkinton, Mass	Mar. 12, 1867	62, 794
Dentist's chair	O. C. White	Hopkinton, Mass	July 30, 1872	130, 093
Dentist's chair	O. C. White and A. T. Ashmead	Hopkinton, Mass., and Hartford, Conn.	Oct. 6, 1868	82, 776
Dentist's chair, Folding	C. M. Adams	Canton, Miss	June 21, 1870	104, 398
Dentist's chairs, Construction of	J. Ask	Lyons, N. Y	Sept. 4, 1860	29, 936
Dentist's chairs, Construction of	N. C. Lewis, jr	Boston, Mass	Oct. 4, 1859	25, 702
Dentist's chairs, Head-rest for	W. W. Evans	Washington, D. C	Dec. 31, 1872	134, 424
Dentist's chairs, Stool for	T. B. Downs and W. D. Morris	Jefferson City, Mo	May 13, 1873	138, 796
Dentist's clamp	C. E. Blake	San Francisco, Cal	Apr. 29, 1873	138, 370
Dentist's flask	C. Bailey	Kinston, N. C	Apr. 8, 1873	137, 589
Dentist's flask	C. P. Bellows	Gloversville, N. Y	Dec. 15, 1868	84, 988
Dentist's flask	G. E. Donham	East Abington, Mass	Mar. 3, 1868	75, 000
Dentist's flask	G. E. Hays	Buffalo, N. Y	Feb. 4, 1868	73, 974
Dentist's flask	L. R. Streeter	Chelsea, Mass	Mar. 23, 1869	88, 229
Dentist's furnace	E. A. L. Roberts	New York, N. Y	July 19, 1859	24, 823
Dentist's grinding and polishing wheel	J. K. Merrick	Odell, Ill	Oct. 12, 1869	95, 823
Dentist's instrument-stand	J. J. Rop	Hernando, Miss	Apr. 4, 1871	113, 571
Dentist's lamp for vulcanizing	G. E. Hayes	Buffalo, N. Y	Dec. 1, 1863	40, 750
Dentist's lamp-heater	J. C. House	Lowville, N. Y	Oct. 8, 1867	69, 670
Dentist's lamp-stove	S. P. Hildreth	Mount Vernon, Ohio	Oct. 16, 1866	58, 822

Index of patents issued from the United States Patent Office from 1790 *to* 1873, *inclusive*—Continued.

Invention.	Inventor.	Residence.	Date.	No.
Dentists, Mold for metal die used by	F. Y. Clark	Savannah, Ga	May 22, 1860	28, 348
Dentist's operating-chair	G. W. Tripp	Auburn, N. Y	Jan. 5, 1858	19, 052
Dentist's pattern-plate	W. M. Wright	Pittsburgh, Pa	June 29, 1858	20, 754
Dentist's rest	R. F. Hunt	Washington, D. C	Nov. 22, 1870	109, 411
Dentist's saliva-pump	C. Newton	New York, N. Y	Sept. 27, 1825	......
Dentist's tool-holder	G. F. Foote	New York, N. Y	May 22, 1866	54, 882
Dentist's tool-rack	G. E. Hayes	Buffalo, N. Y	Nov. 11, 1873	144, 539
Dentist's tool-rack	I. A. Salmon	Boston, Mass	Feb. 26, 1867	62, 368
Dentist's use, Lathe-bath and sponge-holder for	D. Murlless	Holyoke, Mass	Nov. 2, 1869	96, 467
Dentist's use, Vulcanizing-vessel for	G. E. Hayes	Buffalo, N. Y	Apr. 3, 1866	53, 612
Dentist's vulcanizer	C. H. James	Cincinnati, Ohio	Aug. 7, 1866	56, 945
Dentists, Vulcanizing-apparatus for	J. L. McDermut	New York, N. Y	Oct. 4, 1864	44, 541
Dentist's vulcanizing-flask	H. F. Clark	Poughkeepsie, N. Y	Apr. 2, 1867	63, 470
Dentist's vulcanizing-flask	L. W. Dowlin	Sherbrooke, Canada	Jan. 15, 1867	61, 174
Dentist's vulcanizing-flask	L. Hoffstadt	Philadelphia, Pa	Aug. 14, 1866	57, 135
Dentist's vulcanizing-flask	E. T. Starr	Philadelphia, Pa	Aug. 14, 1866	57, 210
Dentistry	W. Ballard	New York, N. Y	Jan. 16, 1866	52, 012
Dentistry	J. Johnson	Saco, Me	July 19, 1864	43, 588
Dentistry	E. C. Smith and D. F. Wilcox	Greenville, N. Y	May 12, 1868	77, 773
Denture, Artificial	F. C. Brown	Palmyra, N. Y	May 7, 1872	126, 517
Denture, Artificial	S. C. Taylor	Monroe, Mich	Sept. 26, 1865	50, 186
Deodorizer for privy-seats	N. Clifford and A. N. Bell	Brooklyn, N. Y	May 21, 1867	64, 841
Deodorizing air and gases in fat-rendering, bone-boiling, &c.	A. and E. Lister and C. J. Eames.	Newark, N. J	Nov. 29, 1870	109, 638
Deodorizing and fertilizing compound	J. M. Loewenstein	New Orleans, La	Mar. 26, 1872	124, 901
Deodorizing and fertilizing materials	J. E. Dotch	Washington, D. C	Feb. 27, 1872	124, 041
Deodorizing and improving alcoholic and vinous liquids.	L. Bradley	Jersey City, N. J	Oct. 25, 1870	108, 558
Deodorizing and purifying sewerage	C. Rawson, P. Ovenden, J. Wylde, W. McCree, and H. Hill.	London, Surrey County, Leamington, Leyton, and Hastings, England.	Aug. 29, 1871	118, 485
Deodorizing and treating offal, Apparatus and process for.	J. J. Storer	Boston, Mass	Sept. 3, 1872	131, 131
Deodorizing-apparatus for water-closets	A. Jordan	Washington, D. C	Nov. 14, 1871	120, 979
Deodorizing-apparatus for water-closets	P. S. Luther	Hartford, Conn	June 29, 1869	91, 855
Deodorizing-apparatus for water-closets	H. Moule and H. J. Girdlestone	Fordington and London, England.	June 15, 1869	91, 474
Deodorizing composition	A. Rankin	New York, N. Y	Nov. 28, 1865	51, 216
Deodorizing compound	R. Fish	Washington, D. C	Aug. 10, 1869	93, 607
Deodorizing, desiccating, and mixing manure, Apparatus for.	H. S. Firman	New York, N. Y	Dec. 8, 1868	84, 686
Deodorizing excrement	P. N. Goux	Paris, France	Oct. 10, 1871	119, 756
Deodorizing gases from lard-boiling, &c	W. H. McNeill	West Orange, N. J	Dec. 26, 1871	122, 273
Deodorizing hydrocarbon oils, Process for	R. Gaggin	Erie, Pa	Aug. 22, 1871	118, 359
Deodorizing India rubber, gutta-percha, &c	E. De la Granja	Boston, Mass	Dec. 17, 1867	72, 272
Deodorizing offal	A. and E. Lister	Newark, N. J	Feb. 27, 1866	52, 863
Deodorizing oil obtained from acid-tar of oil-refineries.	R. Gaggin	Erie, Pa	May 6, 1873	138, 629
Deodorizing oils and fats, Process for	O. Loew and J. E. Dotch	New York, N. Y	Nov. 26, 1872	133, 322
Deodorizing night-soil, &c	J. E. Dotch	Washington, D. C	Apr. 23, 1872	125, 886
Deodorizing water-closets	G. E. Waring, jr	Newport, R. I	June 1, 1869	90, 709
Deoxodizing-furnace for iron ore	A. W. Honsinger	Rome, N. Y	Jan. 4, 1870	98, 496
Depilating animal carcasses, Apparatus for	D. H. Sherman	Jersey City, N. J	Oct. 28, 1873	144, 150
Depressor and atomizer	O. A. Shultz	Chicago, Ill	June 16, 1868	79, 015
Depth-gage	I. A. Hurd	Boston, Mass	July 2, 1867	66, 239
Depth of water indicator for vessels	H. B. Sommers	Ithaca, N. Y	Dec. 10, 1850	7, 826
Depth of water register for vessel's hold	N. Edwards	Chittenden County, Vt	Mar. 5, 1850	7, 138
Depthing-tool	A. Rush	Fairfield, Iowa	Dec. 6, 1864	45, 348
Depurator	W. Curran	Saint Louis, Mo	Feb. 23, 1869	87, 247
Depurator	W. Curran	Saint Louis, Mo	Mar. 15, 1870	100, 867
Depurator	S. C. Frink	Indianapolis, Ind	July 26, 1870	105, 666
Depurator	S. C. Frink and L. D. Harlan	Indianapolis, Ind	Dec. 14, 1869	97, 768
Depurator	F. C. Leland and F. W. Poland	Lowell, Mass	Apr. 4, 1871	113, 536
Depurator	L. W. Werner	Saint Louis, Mo	Dec. 31, 1872	134, 406
Depurator for the teeth and gums	J. D. Wingate	Carbondale, Pa	Feb. 4, 1873	135, 612
Depurator, Liquid	W. Needham and J. Kite	Vauxhall, England	July 13, 1869	92, 465
Dermopathic-instrument	J. Firmenich	Buffalo, N. Y	Mar. 18, 1862	34, 677
Derrick	H. S. Blood	New Orleans, La	Mar. 26, 1872	124, 878
Derrick	A. Campbell	Downierville, Cal	Dec. 29, 1868	85, 279
Derrick	J. S. Clarke	New Orleans, La	Oct. 22, 1872	132, 442
Derrick	F. Donald and G. S. Newton	Waupum, Wis	Feb. 11, 1873	135, 785
Derrick	J. B. Drake and W. H. Hutson	Montoursville, Pa	Nov. 10, 1868	83, 938
Derrick	E. Duchamp	Saint Martinsville, La	Sept. 20, 1859	25, 494
Derrick	F. M. Everingham	Lafayette, N. Y	Oct. 8, 1867	69, 552
Derrick	S. S. Fertig	Titusville, Pa	Aug. 20, 1872	130, 706
Derrick	C. Fricke	Mobile, Ala	Apr. 28, 1868	77, 271
Derrick	S. Hill and C. M. Dupuy, jr	Jersey City, N. J., and Rondout, N. Y.	June 15, 1852	9, 025
Derrick	J. B. Holmes	Boston, Mass	Feb. 21, 1854	10, 544
Derrick	W. M. and G. L. Howland	Topsham, Me	Oct. 3, 1871	119, 612
Derrick	J. S. Lester and L. G. Jennings	La Fayette, Ind	Apr. 31, 1867	64, 336
Derrick	C. S. Lockwood	Newburgh, N. Y	Aug. 25, 1868	81, 385
Derrick	N. Matlick	Williamstown, Mo	Apr. 27, 1869	89, 416
Derrick	D. J. McDonald	Gold Hill, Nev	Dec. 10, 1867	72, 064
Derrick	E. Mingay	Boston, Mass	July 17, 1855	13, 269
Derrick	J. W. Piper and W. J. and J. S. Hanger.	Chicago and Taylor, Ill	Nov. 23, 1869	97, 114
Derrick	W. J. Perkins	Louisville, Ky	May 7, 1872	126, 484
Derrick	A. K. Richmond	Boston, Mass	Dec. 23, 1873	145, 899
Derrick	J. E. Serrell	New York, N. Y	Sept. 9, 1873	142, 740
Derrick	J. Sheldon	Chicago, Ill	May 5, 1868	77, 541
Derrick	A. B. Sprout	Picture Rocks, Pa	Sept. 17, 1867	68, 909
Derrick	S. Turner	Onarga, Ill	Dec. 13, 1864	45, 445
Derrick	J. E. Walsh	New York, N. Y	Apr. 22, 1873	138, 215
Derrick	D. D. Winant	Brooklyn, N. Y	Feb. 20, 1866	52, 777
Derrick and crane	O. C. Brown	Iberia, Ohio	Oct. 8, 1872	131, 934
Derrick and crane	J. W. Kennedy	Saint Louis, Mo	Jan. 7, 1873	134, 677
Derrick and horse-power	D. Woodbury	Rochester, N. Y	Jan. 10, 1865	45, 890

Index of patents issued from the United States Patent Office from 1790 to 1873, inclusive—Continued.

Invention.	Inventor.	Residence.	Date.	No.
Derrick-boat	T. Saulpaugh	Rock Island, Ill	Sept. 20, 1870	107, 630
Derrick, Boom	A. D. Bishop	Brooklyn, N. Y	Aug. 22, 1846	4, 709
Derrick, Boom	G. E. Warner	Springfield, Mass	June 5, 1849	6, 515
Derrick, Double	A. M. Tomb	Owego, N. Y	Jan. 2, 1872	122, 416
Derrick, Field	I. J. Hattabough	Santa Clara County, Cal	Dec. 3, 1867	71, 749
Derrick, Floating	G. W. Brush	Brooklyn, N. Y	Apr. 9, 1861	31, 950
Derrick, Floating	J. V. Collins	Georgetown, D. C	May 3, 1864	42, 561
Derrick, Floating	G. W. Talcott	Buffalo, N. Y	Aug. 16, 1864	43, 873
Derrick, Floating and revolving	E. Jones	Brooklyn, N. Y	Feb. 1, 1859	22, 811
Derrick for hoisting and laying stone, &c., Boom	J. S. Savage	Boston, Mass	Feb. 15, 1838	607
Derrick for raising sunken vessels	S. Petrie	Buffalo, N. Y	Aug. 14, 1866	57, 183
Derrick, Hay	A. Barnes	Bloomington, Ill	Apr. 19, 1870	101, 971
Derrick, Hay	W. Denton	Iowa City, Iowa	Oct. 5, 1869	95, 439
Derrick, Hay	G. E. Johnston	Bloomington, Ill	Jan. 12, 1869	85, 831
Derrick, Hay	W. Minick	Kansas, Ill	Sept. 26, 1865	50, 150
Derrick, Hay	W. P. Squire	Paris, Ill	Nov 20, 1866	59, 870
Derrick, Hydraulic	D. L. Gorman	Michigan Bluff, Cal	Aug. 13, 1872	130, 367
Derrick or hoisting-crane	H. S. Blood	New Orleans, La	June 7, 1870	103, 834
Derrick, Portable	G. C. Brown	Philadelphia, Pa	Mar. 31, 1868	76, 155
Derrick, Portable	J. Greives	Brooklyn, N. Y	July 18, 1865	48, 808
Derrick, Portable	J. R. Hammond	Sedalia, Mo	Dec. 14, 1869	97, 911
Derrick, Portable	C. B. Wright	Belmont, Ohio	Dec. 3, 1867	71, 833
Derrick, Portable iron	S. Rider	Oil City, Pa	Aug. 16, 1870	106, 408
Desiccating-kiln	W. J. Rand	Brooklyn, N. Y	June 27, 1865	48, 492
Desiccating substances, Apparatus for	B. R. Smith and J. C. Harris	Philadelphia, Pa	Jan. 11, 1870	98, 808
Design-die for ornamenting	B. F. Calley	Saugus, Mass	July 26, 1870	105, 637
Desings, Method of molding	L. V. Hue and C. Roziére	Paris, France	Apr. 21, 1868	77, 042
Desk	F. Baltz	New York, N. Y	May 30, 1865	47, 915
Desk	J. T. Hammitt	Philadelphia, Pa	Nov. 4, 1851	8, 486
Desk	I. H. Norris and D. Flanders	Parishville, N. Y	Nov. 18, 1851	8, 530
Desk	G. and J. A. Schaff	Chicago, Ill	May 6, 1873	138, 585
Desk	C. W. Titus	Philadelphia, Pa	Aug. 30, 1870	106, 970
Desk, Adjustable	E. Dummer	Newburyport, Mass	Nov. 19, 1872	133, 087
Desk, Adjustable	M. Stahn	Baltimore, Md	July 23, 1872	129, 870
Desk, Adjustable school	C. H. Loomis	New Philadelphia, Ohio	June 21, 1870	104, 472
Desk, Advertising	H. W. Crotzer	Philadelphia, Pa	Apr. 19, 1870	102, 196
Desk and bed combined	F. Hopf	Chicago, Ill	Aug. 1, 1871	117, 632
Desk and book-case	J. A. Payne	New Orleans, La	May 27, 1873	139, 419
Desk and book-rack	G. Munger	Madison, Conn	Jan. 4, 1870	98, 614
Desk and chair	A. S. Vaughan	Elkton, Md	Jan. 3, 1871	110, 805
Desk and chair combined	A. A. Porter	Griffin, Ga	Oct. 31, 1871	120, 455
Desk and chair combined, Office	M. P. Norton	Troy, N. Y	Mar. 22, 1864	42, 016
Desk and chair combined, School	G. Buchanan	Hickory, Pa	Jan. 11, 1859	22, 545
Desk and musical instrument, School	W. O. Trowbridge	Newton Corner, Mass	Oct. 13, 1868	83, 008
Desk and seat	R. Cruikshank	Lawrenceville, N. J	Mar. 10, 1868	75, 379
Desk and seat	G. A. Stewart	Des Moines, Iowa	Oct. 6, 1868	82, 765
Desk and seat	H. W. White	Olney, Ill	Jan. 14, 1868	73, 416
Desk and seat, Folding	A. Chandler and S. F. Estell	Davenport, Iowa, and Richmond, Ind.	Feb. 16, 1869	87, 025
Desk and seat, Folding	H. E. Gillet	Oswego, N. Y	Oct. 26, 1869	96, 100
Desk and seat frame, School	J. L. Ritter	Brownsville, Ind	Oct. 10, 1871	119, 882
Desk and seat, School	J. P. Allen	Richmond, Ind	Mar. 28, 1865	46, 980
Desk and seat, School	B. W. Arnold	Des Moines, Iowa	Nov. 30, 1869	97, 338
Desk and seat, School	A. M. Bodwell	Ann Arbor, Mich	Aug. 22, 1871	118, 187
Desk and seat, School	A. Chandler	Davenport, Iowa	Sept. 7, 1869	94, 473
Desk and seat, School	C. T. Chase	Albany, N. Y	Dec. 17, 1867	72, 367
Desk and seat, School	W. Chase	Buffalo, N. Y	Mar. 8, 1870	100, 600
Desk and seat, School	M. B. Cochran	Davenport, Iowa	Feb. 18, 1868	74, 504
Desk and seat, School	R. Cruikshank	Pottstown, Pa	May 24, 1864	42, 859
Desk and seat, School	J. D. Diffenderfer	Lewisburgh, Pa	Mar. 8, 1870	100, 508
Desk and seat, School	W. P. Erwin and T. A. Dugdale	Richmond, Ind	Mar. 1, 1870	100, 276
Desk and seat, School	G. H. Grant	Richmond, Ind	Apr. 6, 1869	88, 563
Desk and seat, School	G. H. Grant	Richmond, Ind	May 23, 1871	115, 192
Desk and seat, School	C. J. Higgins	Indianapolis, Ind	Feb. 21, 1871	112, 038
Desk and seat, School	G. W. Hildreth	Lockport, N. Y	June 5, 1866	55, 293
Desk and seat, School	M. L. Holt and L. C. Abbott	Eaton, Ohio	Feb. 15, 1870	99, 895
Desk and seat, School	W. P. Hood	Winona, Minn	Nov. 28, 1871	121, 287
Desk and seat, School	A. Hutchinson	Philadelphia, Pa	Dec. 15, 1868	84, 947
Desk and seat, School	J. Ingels	Milton, Ind	May 10, 1870	102, 941
Desk and seat, School	N. Johnson	New York, N. Y	Nov. 22, 1870	109, 518
Desk and seat, School	W. H. Kline	Eaton, Ohio	Mar. 1, 1870	100, 416
Desk and seat, School	J. Monteith	New York, N. Y	Sept. 29, 1863	40, 117
Desk and seat, School	G. Munger	New York, N. Y	Sept. 4, 1866	57, 824
Desk and seat, School	J. S. Rankin	Minneapolis, Minn	Dec. 5, 1871	121, 470
Desk and seat, School	A. E. Roberts	Des Moines, Iowa	Apr. 22, 1873	138, 045
Desk and seat, School	J. P. Scott	Lewisburgh, Pa	Nov. 12, 1867	70, 904
Desk and seat, School	J. P. Scott and S. H. La Rue	Lewisburgh, Pa	Feb. 11, 1868	74, 433
Desk and seat, School	H. G. Sellman	North Newburgh, Mich	May 16, 1871	114, 869
Desk and seat, School	C. M. Sherwood	Chicago, Ill	Nov. 6, 1866	59, 466
Desk and seat, School	H. M. Sherwood	Chicago, Ill	Sept. 10, 1867	68, 660
Desk and seat, School	J. B. Sherwood	Chicago, Ill	Apr. 15, 1873	137, 866
Desk and seat, School	W. A. Slaymaker	Atlanta, Ga	Nov. 29, 1870	109, 770
Desk and seat, School	W. H. Soper	Baltimore, Md	Aug. 17, 1869	93, 919
Desk and seat, School	G. A. Stewart	Des Moines, Iowa	Jan. 5, 1869	85, 544
Desk and seat, School	J. Upham and W. H. Kline	Eaton, Ohio	June 11, 1872	127, 940
Desk and seat, School	W. Watson	Visalia, Cal	Nov. 21, 1871	121, 070
Desk and seat, School	T. Weaver	Harrisburgh, Pa	Nov. 29, 1864	45, 286
Desk and seat, School	J. F. Whiting and E. Fisher	Eaton, Ohio	Sept. 5, 1871	118, 765
Desk and seat, School	D. C. Wilson	Beaufort, S. C	Oct. 1, 1867	69, 523
Desk and seat, School	W. S. Wooton	Richmond, Ind	Mar. 2, 1869	87, 386
Desk and table	A. A. McMore	Brooklyn, N. Y	July 23, 1867	67, 063
Desk and table	A. Wibbert	Milwaukee, Wis	Nov. 7, 1871	120, 604
Desk and work-table, Combined	J. Trevor	Lockport, N. Y	Mar. 28, 1865	47, 058
Desk book-rest	L. A. Tarring	Washington, D. C	May 6, 1873	138, 728
Desk bracket, School	J. L. Riter	Brownsville, Ind	Apr. 9, 1872	125, 411
Desk, bureau, and churn-power	F. J. Slater	Nicholsville, Ohio	Oct. 18, 1870	108, 399
Desk, Cast-metal	D. Fitzgerald	New York, N. Y	Dec. 5, 1871	121, 504
Desk-cover	J. Heymann	Boston, Mass	Oct. 28, 1873	144, 027

Index of patents issued from the United States Patent Office from 1790 *to* 1873, *inclusive*—Continued.

Invention.	Inventor.	Residence.	Date.	No.
Desk, Fire-proof	M. B. Bigelow and A. Hardy	Boston, Mass	June 28, 1859	24, 534
Desk, Folding	A. Chandler	Davenport, Iowa	Apr. 6, 1869	88, 609
Desk, Folding	E. W. Gilles and J. Wendell	Oswego, N. Y	Mar. 22, 1870	101, 118
Desk, Folding	H. E. Gillet	Oswego, N. Y	Apr. 26, 1870	102, 249
Desk, Folding	T. W. Heitz	Quincy, Ill	Oct. 29, 1867	70, 339
Desk, Folding	J. F. McNel	Philadelphia, Pa	Aug. 15, 1865	49, 428
Desk, Folding	J. Milwain	Nashville, Tenn	Jan. 24, 1871	111, 230
Desk, Folding school	E. Convers	Oswego, N. Y	June 14, 1870	104, 116
Desk, Folding school	S. F. Estell	Richmond, Ind	Oct. 26, 1869	96, 096
Desk for chair, &c., Adjustable	S. C. Ripley	Portland, Me	Oct. 22, 1872	132, 494
Desk for school-rooms, Arrangement of	H. G. Eastman	Poughkeepsie, N. Y	Sept. 6, 1864	44, 082
Desk for school-teacher, &c	W. C. Carter and J. P. Emery	Galva, Ill	Sept. 26, 1871	119, 316
Desk, Hand	M. C. Stebbins	Springfield, Mass	Oct. 8, 1872	132, 030
Desk joint, School	C. U. Crandall	Sterling, Ill	May 9, 1871	114, 649
Desk joint, School	T. A. Galt and G. S. Tracy	Sterling, Ill	Sept. 5, 1871	118, 773
Desk joint, School	T. Gregg	Danville, Pa	Oct. 3, 1871	119, 515
Desk-lid, Prop	H. R. Russell	Woodbury, N. J	June 4, 1872	127, 519
Desk, Night-reading	J. Rogowiski	New York, N. Y	Apr. 23, 1861	32, 148
Desk, Office	H. Mott	Troy, N. Y	Sept. 26, 1871	119, 390
Desk or seat joint	L. and W. F. Sylla	Elgin, Ill	May 30, 1871	115, 383
Desk or table, Recording	W. B. C. Stirling	Batavia, Ohio	July 13, 1869	92, 550
Desk, Pendent reading	G. F. Perkins	San Francisco, Cal	Dec. 12, 1871	121, 810
Desk, Portable writing	W. Bothe	Williamsburgh, N. Y	June 14, 1870	104, 255
Desk, Portable writing	W. H. Lochman	York, Pa	Mar. 29, 1859	23, 379
Desk, Reading	E. Kast	Waterbury, Conn	May 22, 1866	54, 919
Desk, Reading	J. Moessinger	New York, N. Y	Sept. 28, 1869	95, 369
Desk, Reading and writing	C. Pretsch	Trenton, N. J	Apr. 10, 1860	27, 831
Desk, Revolving	C. Hope	Springfield, Mass	June 12, 1866	55, 492
Desk, School	H. L. Andrews	Chicago, Ill	Sept. 15, 1868	82, 061
Desk, School	W. Blackburn	Manchester, Great Britain	June 11, 1872	127, 839
Desk, School	J. E. Blair	New Haven, Conn	Apr. 16, 1867	63, 839
Desk, School	W. M. Brooke	Eaton, Ohio	Dec. 26, 1871	122, 221
Desk, School	J. D. Brown	Madisonville, Ohio	Feb. 15, 1870	99, 832
Desk, School	A. Chase	North Weare, N. H	Sept. 11, 1860	30, 020
Desk, School	J. W. Childs	Kansas City, Mo	June 3, 1873	139, 543
Desk, School	S. Cox and W. W. Fanning	Richmond, Ind	Jan. 21, 1873	135, 089
Desk, School	W. E. Doyle	Eaton, Ohio	Nov. 15, 1870	109, 304
Desk, School	W. P. Erwin and T. A. Dugdale	Richmond, Ind	Mar. 1, 1870	100, 275
Desk, School	W. N. Foster	Indianapolis, Ind	July 9, 1872	128, 871
Desk, School	J. H. French	Burlington, Vt	June 13, 1871	115, 839
Desk, School	E. W. Gilles and J. Wendell	Oswego, N. Y	Jan. 19, 1869	85, 924
Desk, School	B. N. Hemenway	Rockland, Me	July 16, 1872	129, 559
Desk, School	C. J. Higgins	Indianapolis, Ind	Oct. 21, 1873	143, 760
Desk, School	G. W. Hildreth	Lockport, N. Y	Feb. 18, 1868	74, 688
Desk, School	A. O. Huffman	Springfield, Ohio	Aug. 1, 1871	117, 633
Desk, School	A. J. Hull	Sterling, Ill	Feb. 1, 1870	99, 440
Desk, School	A. Hutchinson	Philadelphia, Pa	Apr. 27, 1869	89, 484
Desk, School	W. Johnson	Topsham, Me	Dec. 26, 1865	51, 727
Desk, School	J. Long and E. Convers	Oswego, N. Y	July 18, 1871	117, 090
Desk, School	H. S. McRae	Muncie, Ind	Dec. 24, 1867	72, 662
Desk, School	J. Mealey	Fairville, New Brunswick	Oct. 27, 1868	83, 521
Desk, School	H. B. Osborne and N. W. Hammon.	Des Moines, Iowa	June 27, 1871	116, 477
Desk, School	J. Peard	New York, N. Y	Feb. 2, 1869	86, 440
Desk, School	J. Peard	New York, N. Y	May 23, 1871	115, 232
Desk, School	J. F. Piehl	Richmond, Ind	Aug. 16, 1870	106, 403
Desk, School	I. N. Peirce	Philadelphia, Pa	Dec. 5, 1871	121, 543
Desk, School	J. F. Prehl	Richmond, Ind	Apr. 26, 1870	102, 317
Desk, School	C. H Bresbrey	Sterling, Ill	Jan. 21, 1873	135, 154
Desk, School	J. S. Rankin	Madison, Ind	Feb. 18, 1862	34, 448
Desk, School	A. E. Roberts	Des Moines, Iowa	Aug. 17, 1869	93, 907
Desk, School	A. E. Roberts	Des Moines, Iowa	Jan. 11, 1870	98, 801
Desk, School	C. H. Roberts	Geneva, Ohio	May 3, 1870	102, 713
Desk, School	J. Russel	Plymouth, Ind	July 25, 1871	117, 467
Desk, School	A. Schlag	Brooklyn, N. Y	Oct. 10, 1871	119, 794
Desk, School	C. W. Sherwood	Chicago, Ill	Apr. 5, 1870	101, 670
Desk, School	W. A. Slaymaker	Atlanta, Ga	Feb. 20, 1872	123, 796
Desk, School	W. A. Slaymaker	Atlanta, Ga	Feb 20, 1872	123, 797
Desk, School	J. Smith	Richmond, Ind	Feb. 26, 1867	62, 375
Desk, School	J. Smith	Richmond, Ind	Mar. 1, 1870	100, 334
Desk, School	J. Smith	Richmond, Ind	Aug. 15, 1871	118, 161
Desk, School	W. H. Soper	Baltimore, Md	Mar. 31, 1863	38, 064
Desk, School	E. I. Stearns	Cambridge, Md	Apr. 14, 1868	76, 839
Desk, School	R. A. Thompson	Beaver Falls, Pa	Dec. 3, 1872	133, 551
Desk, School	D. G. Venable	Christiansburgh, Ky	Mar. 25, 1873	137, 263
Desk, School	A. S. Vorse	Des Moines, Iowa	Nov. 16, 1869	97, 005
Desk, School	F. A. Wilber	Wabash, Ind	July 15, 1873	140, 979
Desk, School	E. Wilson	New Brighton, Pa	Dec. 13, 1870	110, 101
Desk, School	W. S. Wooton	Richmond, Ind	Nov. 10, 1868	83, 896
Desk seat, School	C. Perley	New York, N. Y	Mar. 30, 1858	19, 791
Desk seat, School	A. S. Vorse	Des Moines, Iowa	June 28, 1870	104, 906
Desk, table, and shelf combined	L. H. Morrill and E. G. P. Smith	Falmouth and Portland, Me	Dec. 12, 1871	121, 886
Desk, Wall	T. Cogswell	Boston, Mass	Oct. 29, 1872	132, 632
Desk, Writing	C. H Bergmann	New York, N. Y	Feb. 5, 1856	14, 217
Desk, Writing	J. W. Fiester	New Lexington, Ohio	Aug. 24, 1858	21, 249
Desk, Writing	J. H. Grimsley and P. I. Ankney.	New Lexington, Ohio	June 8, 1858	20, 487
Desk, Writing	W. Hoper	New Haven, Conn	Feb. 8, 1870	99, 672
Desk, Writing	E. Hughes	McCartysville, Cal	May 1, 1860	28, 086
Desk, Writing	S. Luther	Boston, Mass	June 19, 1838	786
Desk, Writing	C. Monson	New Haven, Conn	Feb. 4, 1862	34, 312
Desk, Writing	A. J. Ritter	Rahway, N. J	July 1, 1862	35, 781
Desk, Writing	F. Robbin	Hudson City, N. J	Nov. 16, 1869	96, 996
Desk, &c., Writing	D. Schafer	Parkersburgh, W. Va	Nov. 18, 1873	144, 636
Desk, Writing	E. Scheel	New York, N. Y	Oct. 25, 1870	108, 635
Desk, Writing	J. S. Watson	Newton, Mass	Aug. 16, 1864	43, 876
Desk, Writing	W. G. Wolf	Philadelphia, Pa	July 31, 1855	13, 371
Desk, Writing and drawing	W. W. Levering	New York, N. Y	Aug. 4, 1868	80, 749

Index of patents issued from the United States Patent Office from 1790 *to* 1873, *inclusive*—Continued.

Invention.	Inventor.	Residence.	Date.	No.
Desks, Attachment for school	D. I. Stagg	New York, N. Y	Oct. 1, 1867	69, 501
Desks, Device for supporting and connecting school	J. L. Rankin	Minneapolis, Minn	Dec. 5, 1871	121, 471
Desks in school-room, Arrangement of	V. Woodcock	Swanzey, N. H	Mar. 6, 1855	12, 497
Desks, seats, &c., to floors, Device for fastening	J. D. McAuliff	Saint Louis, Mo	Nov. 8, 1870	109, 031
Desks, System of school	S. L. Wilkinson	Cross Plains, Tenn	Sept. 18, 1860	30, 102
Desulphurizing and reducing furnace, Ore	J. Maunton	New York, N. Y	Nov. 12, 1867	70, 873
Desulphurizing furnace, Ore	W. Brückner	San Francisco, Cal	Dec. 4, 1866	60, 134
Desulphurizing furnace, Ore	W. Brückner	San Francisco, Cal	June 11, 1867	65, 538
Desulphurizing furnace, Ore	L. G. Marshall	Philadelphia, Pa	Sept. 1, 1863	39, 786
Desulphurizing furnace, Ore	C. A. Stelefeldt	New York, N. Y	June 14, 1864	43, 140
Desulphurizing furnace, Ore	T. D. Worrall	Central City, Colo	July 31, 1866	56, 842
Desulphurizing furnace, Steel and other wire	A. Cary	New York, N. Y	Dec. 1, 1868	84, 534
Desulphurizing metals and ores	L. G. Marshall	New York, N. Y	Apr. 19, 1864	42, 441
Detaching-block	T. J. Harte	New York, N. Y	May 24, 1870	103, 455
Detaching-hook	J. Bozorth	Camden, N. J	June 21, 1870	104, 547
Detaching-hook	J. O. Lord, jr	Middletown, Conn	Nov. 15, 1870	109, 328
Detaching-hook	J. W. Tuttle and J. Peterson	Rochester, N. Y	Apr. 19, 1870	102, 067
Detaching-hook for nautical use	J. M. Brooke	United States Navy	May 28, 1861	32, 408
Detective-register for watchman	P. H. Duffy	Somerset, Ohio	Nov. 29, 1859	26, 260
Detector: *See* Bank-note detector. Coin-detector. Egg-detector. Fire-detector. High and low water detector. Low-water detector. Pick-pocket detector. Steam-boiler detector.				
Detents, Tripping mechanical	S. Chester and C. T. Chester	New York, N. Y., and Englewood, N. J.	Mar. 23, 1869	88, 010
Detergent	S. Coburn	Stamford, Conn	July 5, 1864	43, 393
Detergent compound	W. Berrey	Boston, Mass	Mar. 15, 1870	100, 844
Detergent purposes, Composition for	A. Lovis	Boston, Mass	Nov. 8, 1859	26, 072
Detergent, Vegetable pulmonic	G. Rogers	Northampton, Mass	Dec. 31, 1808	....
Detergents, Manufacture of	S. Coburn	Stamford, Conn	Oct. 10, 1865	50, 336
Developing-stick	T. C. Roche	Williamsburgh, N. Y	Feb. 20, 1866	52, 748
Devulcanizing India rubber	S. Beer	New York, N. Y	May 29, 1855	12, 983
Dial, Sun	P. Fléchet	Paris, France	May 13, 1862	35, 225
Dial, Sun	J. P. Gardner	Columbia, Tenn	Dec. 5, 1843	3, 370
Dial, Sun	J. Scott	Portland, Me	June 5, 1849	6, 506
Dial, Steam-gage	E. H. Ashcroft	Lynn, Mass	July 23, 1867	67, 096
Dials to ascertain time, Combination of the needle and sun.	C. R. M. Pohlé	Richmond, Va	July 27, 1858	21, 020
Dials, Transmitting time-movement to distant	H. J. Wenzel	San Francisco, Cal	July 8, 1873	140, 661
Diamond-gage	W. Shumard	Richmond, Ind	Jan. 28, 1873	135, 249
Diamond-holder	G. D. Dunham	Chicago, Ill	Mar. 22, 1870	101, 111
Diamond-holder	J. E. Karelson	New York, N. Y	July 10, 1866	56, 231
Diamond-key	B. F. Southgate	Bridgewater, Vt	Sept. 3, 1867	68, 538
Diamond-setting	F. J. Herpers	Newark, N. J	Sept. 3, 1872	131, 058
Diamonds, Method of setting black	T. W. Baxter	Chicago, Ill	Jan. 21, 1873	134, 968
Diamonds, Setting	J. Dickinson	Phila elphia, Pa	Nov. 30, 183°	....
Diamonds, &c., Setting	I. Lindsley	Providence, R. I	Sept. 29, 1857	18, 288
Diamonds, Setting and adjusting glaziers'	M. Kleeman	Columbus, Ohio	May 9, 1865	47, 645
Diaper	J. C. Hempel	Baltimore, Md	June 15, 1869	91, 334
Diaper	A. M. Hughes	Hudson City, N. J	Nov. 22, 1870	109, 410
Diaper, Child's	M. A. Moore	Lisbon, Ill	Dec. 8, 1868	84, 703
Diaper or shawl pin	J. G. Klinger	Jersey City, N. J	Feb. 2, 1858	19, 280
Diaper-pin	J. Heilmann	New York, N. Y	July 21, 1857	17, 857
Diaper-pin	W. H. Hockensmith	Bridgeport, Conn	Aug. 29, 1871	118, 533
Diaper-pin	H. S. Lesher	Brooklyn, N. Y	Oct. 2, 1860	30, 232
Diaper-pin	J. Rabbeth	Mansfield, Conn	Jan. 21, 1847	4, 938
Diaper-pin	J. E. Rabbeth	East Hartford, Conn	July 11, 1854	11, 272
Diaper-pin	I. W. Stewart	New York, N. Y	Aug. 16, 1870	106, 422
Diaper-pin	A. Wanner	Hoboken, N. J	Aug. 30, 1864	44, 031
Diapers, Substitute for infants'	J. H. Hall	Kittaning, Pa	Feb. 23, 1858	19, 418
Dice apparatus, Mechanical	A. Klingenberg	Baltimore, Md	Nov. 21, 1871	121, 110
Dice-box	E. W. Quincy	Lacon, Ill	Dec. 11, 1866	60, 421
Dice-box	J. Twamly	Troy, N. Y	Oct. 14, 1873	143, 599
Dice-box	J. E. Zender	New York, N. Y	Nov. 12, 1867	70, 732
Dice, Instrument for throwing	B. F. Bee	Harwick, Mass	Sept. 5, 1865	49, 700
Dice, &c., Tool for marking	C. B. Rogers	Deep River, Conn	May 2, 1865	47, 571
Die: *See* Auger-die. Auger-bit die. Auger-blank die. Axle-clip trimming and straightening die. Axle-forming die. Axle-nut die. Bead-forming die. Bit and auger die. Blank-forming die. Blind-shaping die. Bolt-die. Bolt-heading die. Bolt-threading die. Box-stamping die. Buckle-closing die. Bullet-cutting die. Bullet-making die. Bunter-shape swaging and forming die. Button-molding die. Can-making die. Carriage-clip die. Carriage-clip-swaging die. Carriage-head block-plate die. Carriage-shackle die. Carriage-shackle-eye die. Carriage-shackle-eye-forging die. Carriage-shaft-shackle die. Carriage-spring-head-bar die. Cartridge-shell die. Chord-bar die. Chord-bar-head die. Clinch-ring die. Clinching-die. Clock-collet die. Collar-die. Cotton-tie-forming die. Damper die. Design-die. Draw-bar-followerdie. Draw-head-face-plate die. Electrotype-die. Elliptic-spring die. Eye-bolt-forging die. Eyelet-forming die. Felly-plate die. Felly-plate-making die. Fifth-wheel die. Fifth-wheel-head-forging die.				

Index of patents issued from the United States Patent Office from 1790 *to* 1873, *inclusive*—Continued.

Invention.	Inventor.	Residence.	Date.	No.
Die—Continued.				
See Finishing-die.				
Flange-turning die.				
Furniture-nail die.				
Hammer-die.				
Hammer-claw die.				
Hammer-forging die.				
Hammer-swaging die.				
Harness-blind die.				
Harness-blind-shaping die.				
Harness-tooth die.				
Hat-box-cutting die.				
Hat-forming die.				
Hat-pressing die.				
Hatter-iron-making die.				
Heel-die.				
Hoe and pick die.				
Horseshoe-die.				
Horseshoe-calk die.				
Kettle-die.				
Key-blank die.				
Key-blank-making die.				
King-bolt die.				
King-bolt-forging die.				
King-bolt-formingdie.				
King-bolt-head die.				
King-bolt-trimming die.				
Knife-die.				
Knob-die.				
Lead-die.				
Lead-pipe die.				
Link-welding die.				
Locket-die.				
Mattock, hoe, &c., swaging die.				
Metal-die.				
Metal cupping and raising die.				
Metal-raising die.				
Mold-board-bending die.				
Nail-machine-grinding die.				
Nozzle-bending die.				
Nut-machine die.				
Ox-shoe die.				
Ox-shoe-forging die.				
Paper-collar-cutting die.				
Perch-iron die.				
Pistol-frame die.				
Plier-joint die.				
Plier-making die.				
Plow-brace die.				
Plow mold-board die.				
Punching-die.				
Roller-die.				
Screw-capping die.				
Screw-cutting die.				
Screw-threading die.				
Shackle-die.				
Shears-die.				
Shears-bolt die.				
Shears-bow die.				
Shears-bow-forming die.				
Sheet-metal-box-cover die.				
Sheet-metal-ware die.				
Shovel-die.				
Shovel-making die.				
Shuttle-tip die.				
Skelping-die.				
Sleeve-die.				
Soap-bar die.				
Sole-die.				
Spike-head die.				
Spike-machine die.				
Spoon-blank-cutting die.				
Spring-calipers die.				
Square-forging die.				
Stay-end die.				
Straw-goods die.				
Tea-kettle-bail die.				
Thill-iron die.				
Thrashing-machine-teeth die.				
Tin-ware-stamping die.				
Tinners' die.				
Tube-punching die.				
Tube skelp die.				
Tube-thread die.				
Type-wheel die.				
Upsetting-die.				
Vise-box die.				
Washer-die.				
Washer-making die.				
Watch-die.				
Watch-case die.				
Watch-case-swaging die.				
Watch-hand-swaging die.				
Whiffletree T-boltdie.				
Wire-die.				
Wire-drawing die.				
Wrench-bar die.				
Wrench-head die.				
Wrench-head-forging die.				
Wrench-swaging die.				
Die	R. Brayton	Buffalo, N. Y	Jan. 13, 1857	16, 372
Die	L. Wetherell	Boston, Mass	Feb. 19, 1867	62, 174
Die and plunger	J. H. Cole	Millbury, Mass	Feb. 15, 1870	99, 848
Die and plunger	H. G. Williams	Providence, R. I	July 14, 1868	80, 042
Die and punch	T. G. Arnold	New York, N. Y	Nov. 3, 1868	83, 586
Die and punch, Method of holding and adjusting	P. E. Austin	New Haven, Conn	Dec. 1, 1868	84, 468
Die and tap	A. Lemont	Pittsburgh, Pa	Jan. 29, 1830	....
Die cutter stock	H. C. Meyer	Flushing, N. Y	May 28, 1872	127, 257
Die-cutting machine	W. Newsham	Philadelphia, Pa	July 5, 1870	104, 985
Die for stamping polished blanks or shells for buttons, &c., Bright.	W. B. Dunbar	Waterbury, Conn	Oct. 31, 1839	1, 392
Die-forming press	M. H. Stein	New York, N. Y	June 2, 1863	38, 767
Die-holder for screen-press	J. McWilliams	Providence, R. I	Aug. 3, 1869	93, 215
Die-mold for composition articles	I. Smith	New York, N. Y	Aug. 8, 1871	117, 826
Die-operating mechanism	H. Waters	Northbridge, Mass	Jan. 1, 1857	60, 973
Die-plate	F. D. Bliss	New Bedford, Mass	Jan. 28, 1873	135, 314
Die-plate	J. D. Driggs	New Bedford, Mass	Oct. 8, 1867	69, 644
Die-press	W. Burke	Brooklyn, N. Y	Jan. 25, 1870	99, 149
Die-press	J. Mays and E. W. Bliss	Brooklyn, N. Y	Jan. 28, 1868	73, 823
Die-rolling machine	H. Waters	Northbridge, Mass	Apr. 15, 1873	137, 984
Die spike-machinery, Reciprocating	M. Belknap	Canton, Mass	Nov. 9, 1852	9, 397
Die-stock	W. and J. Holroyd	Waterford, N. Y	Nov. 21, 1865	51, 048
Die-stock	J. Koberle	Bavaria	Aug. 29, 1869	49, 687
Die-stock	J. F. Sharer	New York, N. Y	June 2, 1857	17, 453
Die-stock	S. P. M. Tasker	Philadelphia, Pa	July 26, 1870	105, 861
Die-stock	J. Teachout	Waterford, N. Y	June 30, 1857	17, 707
Die-stock	J. Teachout	Waterford, N. Y	Mar. 27, 1860	27, 657
Die-work as a substitute for etching, &c	A. Brewster	Frankfort, Pa	July 15, 1813	....
Die-work on steel, Mode of impressing	J. Perkins and G. Murray	Newburyport, Mass., and Philadelphia, Pa.	June 25, 1813	....
Dies, Construction of	H. W. Hayden	Waterbury, Conn	June 17, 1851	8, 159
Dies, Formation of	T. Hollister	Cornwall, Conn	May 9, 1848	5, 554
Dies, Hardening	E. W. Sperry	Wolcottville, Conn	Feb. 2, 1869	86, 467
Dies, Manufacture of	F. W. Arvine	Fair Haven, Conn	Mar. 12, 1872	124, 414
Dies, Press for operating, bending, and shaping	W. D. Mendenhall	Farmington, Ill	July 27, 1869	93, 107
Diet, Compound to be used as an article of	C. G. Baylor	Quincy, Mass	Mar. 8, 1870	100, 587
Diet, Compound to be used as an article of	W. D. St. Clair	Chicago, Ill	Feb. 22, 1870	100, 081
Digester for treating bones, Steam	W. Perry	North Bridgewater, Mass	Feb. 26, 1867	62, 439

Index of patents issued from the United States Patent Office from 1790 to 1873, inclusive—Continued.

Invention.	Inventor.	Residence.	Date.	No.
Digger: *See* Garden-digger. Peanut-digger. Peat-digger. Post-hole digger. Potato-digger. Rotary-digger. Stone and root digger. Tree-digger.				
Digging-machine	N. Badger	Shelbyville, Ky	Feb. 25, 1862	34, 473
Digging-machine	A. A. Garver	Mechanicsburgh, Pa	May 22, 1860	28, 359
Digging-machine	J. Mitchell	Oceola, Iowa	Oct. 13, 1857	18, 404
Digging-machine	C. H. Stratton	Munroetown, Pa	Jan. 10, 1865	45, 877
Digging-machine	I. P. Tice	New York, N. Y	Jan. 31, 1871	111, 404
Digitorium	M. Marks	London, England	Aug. 8, 1871	117, 791
Dike-core	J. P. Culver	Jersey City, N. J	Nov. 14, 1871	120, 943
Dikes and embankments against the ravages of animals, &c., Protecting.	S. Gardner	New York, N. Y	Sept. 20, 1870	107, 608
Dikes and levees, Construction of	L. S. Robbins	New York, N. Y	Mar. 12, 1867	62, 889
Dikes, levees, and embankments, Construction of	J. C. Schooley	New York, N. Y	Dec. 31, 1867	72, 913
Diking-sheet frame	J. S. Pierson	New York, N. Y	Nov. 14, 1841	120, 896
Dining and ironing table and quilting-frame	T. Elkins	Albany, N. Y	Feb. 22, 1870	100, 020
Dining and other tables	A. Kinkead	Washington County, Ohio	Oct. 26, 1858	21, 885
Dining-chair, Child's	A. Smith	Perrysburgh, Ohio	Oct. 18, 1870	108, 526
Dining-room fixed at the end of vessels	B. Maillefert and L. Hayden	New York, N. Y	Jan. 5, 1864	41, 078
Dining-table	E.H.Bleebaum and C.H.Nagel	Saint Charles, Mo	Sept. 22, 1868	82, 375
Dining-table	D. Boardman	Columbus, Ind	Nov. 8, 1870	109, 107
Dining-table	M. P. Boyd	Unionville, S C	Aug. 16, 1870	106, 460
Dining-table	R. C. Case	Guilderland, N. Y	Nov. 1, 1870	108, 878
Dining-table	J. H. and L. Drane	Eminence, Ky	June 27, 1871	116, 419
Dining-table	S. R. Garner	Independence, Iowa	Mar. 7, 1871	112, 441
Dining-table	J. L. Knight	Tracy Creek, N. Y	May 31, 1870	103, 625
Dining table	A. Land	Hillsdale, Mich	Apr. 15, 1873	137, 779
Dining-table	G. L. Motter	Bloomington, Ill	Mar. 4, 1873	136, 381
Dining-table	J. C. Nichols	Woburn, Mass	Aug. 21, 1849	6, 665
Dining-table	M. Pirz	East New York, N. Y	Dec. 31, 1867	72, 896
Dining-table	B. Welteck	New York, N. Y	June 11, 1872	127, 720
Dining-table, Caloric	H. L. B. Lewis	New York, N. Y	Sept. 27, 1845	4, 207
Dining-table, Extension	D. T. Robinson	Boston, Mass	Aug. 10, 1869	93, 557
Dining-table, Revolving	R. Wilson	Rees Corners, Md	July 6, 1869	92, 413
Dining-table, Self-waiting	L. Pusey	Patterson, Pa	June 14, 1853	9, 783
Dinner-can	A. A. Arnold	Clyde, Ohio	June 17, 1873	139, 851
Dinner-can	J. H. Murphy	Boston, Mass	May 24, 1864	42, 865
Dinner-pot	A. F. Wolf	Beaver Falls, Pa	Mar. 19, 1872	124, 775
Dipper-handles, Means of attaching	J. B. Wood	Cranston, R. I	Jan. 7, 1868	73, 066
Direct-acting compound engine	W. M. Henderson	Philadelphia, Pa	Aug. 2, 1870	105, 941
Direct-acting engine	G. F. Blake	Boston, Mass	Jan. 1, 1867	60, 852
Direct-acting engine	A. S. Cameron	New York, N. Y	May 3, 1870	102, 486
Direct-acting engine	A. Campbell and C. B. Hardick	Brooklyn, N. Y	Apr. 26, 1864	42, 527
Direct-acting engine	C. P. Deane	Springfield, Mass	Apr. 26, 1870	102, 234
Direct-acting engine	T. Hanson	New York, N. Y	Dec. 1, 1863	40, 749
Direct-acting engine	T. Hanson	New York, N. Y	June 20, 1871	116, 051
Direct-acting engine	W. D. Hooker	San Francisco, Cal	Aug. 18, 1868	81, 168
Direct-acting engine	B. M. Johnson	Brooklyn, N. Y	Sept. 19, 1871	119, 088
Direct-acting engine	S.F.Shelbourne and C.E.Emery	New York and Brooklyn, N. Y.	Dec. 6, 1870	109, 951
Direct-acting engine	W. J. Stevens	New York, N. Y	Dec. 6, 1870	109, 963
Direct-acting engine	J. Storer	New York, N. Y	Sept. 7, 1869	94, 665
Direct-acting engine	L. Winterbauer	New York, N. Y	Apr. 14, 1863	38, 193
Direct-acting steam-engine	J. Bailey	Quincy, Ill	Mar. 18, 1873	136, 806
Direct-acting steam-engine	H. A. Benson and W. Avery	Warren, Mass	June 4, 1872	127, 451
Direct-acting steam-engine	C. P. Deane	Springfield, Mass	Dec. 5, 1871	121, 451
Direct-acting steam-engine	G. H. and C. P. Deane and J. B. Gardiner.	Springfield, Mass	Mar. 9, 1869	87, 550
Direct-acting steam-engine	W. B. Hayden	Columbus, Ohio	Sept. 28, 1869	95, 225
Direct-acting steam-engine	W. D. Hooker	San Francisco, Cal	July 16, 1872	129, 345
Direct-acting steam-engine	J. B. Smith	Dunmore, Pa	Mar. 19, 1872	124, 862
Direct-acting steam-engine	T. Thatcher	Danville, Pa	Dec. 1, 1863	40, 778
Dirk-knife	A. Henninger	New Haven, Conn	July 11, 1865	48, 685
Dirk-knife, Spring	C. Hibbard	Chicago, Ill	Sept. 11, 1866	57, 902
Discharge pipe and nozzle	A. Lovie	Albany, N. Y	Aug. 27, 1872	130, 927
Disengaging-hook	A. W. Roberts	Hartford, Conn	Oct. 25, 1859	25, 915
Dish and clothes washer	G. Williamson	Goldsborough, Pa	Mar. 23, 1869	88, 251
Dish and vegetable washer	J. N. Paddack	Oswego, N. Y	Sept. 24, 1867	69, 240
Dish, Baking	B. Harrington	China, Me	July 9, 1872	128, 725
Dish, Baking	H. C. Wilcox	West Meriden, Conn	Feb. 9, 1869	86, 791
Dish, Butter	N. Lawrence	Taunton, Mass	Apr. 14, 1868	76, 639
Dish, Butter	D. T. Lyon	West Meriden, Conn	Oct. 15, 1867	69, 824
Dish, Butter	W. M. Whittaker	Wallingford, Conn	Dec. 4, 1866	60, 308
Dish, Carving	E. P. Bernard	New York, N. Y	Apr. 30, 1872	126, 253
Dish, Carving	C. W. Sykes	Suffield, Conn	Aug. 17, 1869	93, 844
Dish-cleaner	R. T. Osgood	Orland, Me	Sept. 24, 1872	131, 555
Dish-cloth holder	W. J. Johnson	Newton Corner, Mass	Feb. 21, 1860	27, 225
Dish-cloth holder	J. T. Walker	Albany, N. Y	July 21, 1868	80, 100
Dish-cover	H. S. and F. M. Goff	Middletown, Conn	July 14, 1868	79, 824
Dish-cover	A. J. Ohmer	Hamilton, Ohio	Mar. 18, 1873	136, 856
Dish-cover	W. H. White	Kent Island, Md	June 11, 1867	65, 626
Dish-cover, Refrigerating	O. Marland	Boston, Mass	Jan. 12, 1864	41, 228
Dish-covers, Forming wire-cloth	L. H. Allen	Amherst, Mass	June 4, 1861	32, 457
Dish-covers, Machine for making wire	W. Lincoln	Oakham, Mass	Oct. 9, 1855	13, 651
Dish, Covered	W. A. Fenn	West Meriden, Conn	May 26, 1868	78, 366
Dish-drainer	C. Britain	Saint Joseph, Mich	Jan. 14, 1864	43, 088
Dish-drainer	W. H. Gould	Manchester, N. H	Aug. 12, 1873	141, 643
Dish-drainer	H. F. Pond	Franklin, Mass	Nov. 9, 1869	96, 725
Dish-drainer	H. R. Richmond	New Plymouth, New Zealand.	Feb. 27, 1872	124, 085
Dish drainer and drier	R.W. Stone and H A. Kendrick	Rowe, Mass	Mar. 6, 1866	53, 055
Dish, Earthenware	A. and T. Vail	Berlin, Wis	July 13, 1858	20, 906
Dish-fastener	C. R. Freet	Upper Strasburgh, Pa	July 26, 1870	105, 794
Dish, Glass	J. S. and T. B. Atterbury	Pittsburgh, Pa	Oct. 29, 1872	132, 558
Dish-heater	W. Brand	Burlington, Iowa	Aug. 11, 1863	39, 459
Dish-heater	S. S. Fitch	New York, N. Y	July 15, 1873	140, 769

Index of patents issued from the United States Patent Office from 1790 *to* 1873, *inclusive*—Continued.

Invention.	Inventor.	Residence.	Date.	No.
Dish-holder	T. M. Hubbard	Ripon, Wis	July 16, 1867	66, 842
Dish-mop	A. S. Hadley	Boston, Mass	Dec. 11, 1866	60, 364
Dish-pan	S. Lafferty	Chicago, Ill	May 27, 1873	139, 252
Dish-pan	J. Vandercar	Brooklyn, N. Y	July 5, 1864	43, 442
Dish pan and drainer	F. Bucknam	Portland, Me	Apr. 2, 1867	63, 361
Dish-rack	W. H. Duffett	Rochester, N. Y	Apr. 18, 1871	113, 862
Dish-rack	J. M. Rudiger	Brooklyn, N. Y	Jan. 16, 1872	122, 855
Dish-rack	E. A. Sawyer	Portland, Me	Oct. 27, 1868	83, 413
Dish, Soap	J. Cundy	Philadelphia, Pa	Feb. 16, 1864	41, 606
Dish-stand	W. F. Collier	Worcester, Mass	Aug. 23, 1870	106, 552
Dish-stand	I. W. Lamb	Rochester, N. Y	Feb. 13, 1866	52, 576
Dish-stand	D. Sherwood	Lowell, Mass	Apr. 27, 1869	89, 440
Dish-stand	H. C. Wilcox	West Meriden, Conn	Apr. 12, 1870	101, 797
Dish-stand, Table	F. Grosjean	New York, N. Y	July 8, 1873	140, 623
Dish-stand, Wire	G. D. Dudley	Lowell, Mass	Mar. 19, 1872	124, 798
Dish-stand, Wire	E. P. Woods and D. Sherwood	Lowell, Mass	Dec. 31, 1867	72, 770
Dish-stands, &c., Machine for making wire	E. P. Woods and D. Sherwood	Lowell, Mass	Nov. 12, 1867	70, 770
Dish-washer	C. Bartlett	Providence, R. I	Dec. 2, 1873	145, 042
Dish-washer	F. E. Clarke	Flint, Mich	Dec. 7, 1869	97, 604
Dish-washer	W. H. Emerson	Dixon, Ill	June 22, 1869	91, 529
Dish-washer	W. H. Emory	Ashburnham, Mass	Sept. 19, 1871	119, 085
Dish-washer	R. and R. D. S. Filson and J. D. Cope.	Xenia, Ill	Dec. 27, 1870	110, 559
Dish-washer	M. Hobson	Williamsburgh, Mo	Mar. 22, 1870	101, 129
Dish-washer	J. A. Hotchkiss and R. Eaves	Derby, Conn	July 18, 1865	48, 814
Dish-washer	M. E. Lewis	Mansfield, Ohio	Nov. 14, 1871	120, 886
Dish-washer, &c	W. M. Long	Moline, Ill	Sept. 17, 1872	131, 451
Dish-washer	C. Messinger	Cleveland, Ohio	Nov. 24, 1868	84, 431
Dish-washer	M. K. Morris	Louisville, Ky	May 7, 1872	126, 566
Dish-washer	S. D. Moxley	Keeseville, N. Y	May 14, 1872	126, 735
Dish-washer	E. and J. S. B. Norton	Boston, Mass	Nov. 12, 1867	70, 883
Dish-washer	M. S. Orton and P. B. Stiles	Galesburgh, Ill	Mar. 8, 1870	100, 657
Dish-washer	G. Richards and L. Alexander	Cummington and Shelburne Falls, Mass.	Oct. 13, 1863	40, 280
Dish-washer	J. J. Sawyer	Woodstock, Conn	Oct. 16, 1866	58, 895
Dish-washer	C. H. Sterling	Gambier, Ohio	Aug. 20, 1872	130, 761
Dish-washer	J. A. Strong	North Wolcott, Vt	Oct. 5, 1869	95, 618
Dish-washer	W. L. Thompson	Stanstead, Canada	Oct. 17, 1871	119, 953
Dish-washer	A. W. Thornton	Mendocino, Cal	Dec. 10, 1872	133, 811
Dish-washer	H. B. Todd	Plymouth, Conn	May 2, 1871	114, 367
Dish-washer	W. and S. B. Way and S. C. Pomeroy.	South Butler, N. Y	Apr. 9, 1867	63, 677
Dish washer	G. A. Wells	Luzerne, N. Y	June 14, 1870	104, 235
Dish washing and drying machine	A. W. Ward	Fishkill, N. Y	July 23, 1867	67, 088
Dish-washing machine	L. A. Alexander	Colerain, Mass	Nov. 21, 1865	51, 000
Dish-washing machine	A. M. and J. I. D. Bristol	Detroit, Mich	Nov. 3, 1863	40, 457
Dish-washing machine	D. Guptail	Elgin, Ill	Apr. 20, 1869	89, 216
Dish-washing machine	G. Richards	Cummington, Mass	Jan. 15, 1867	61, 256
Dish-washing machine	H. C. Robertson	East Saginaw, Mich	Aug. 1, 1871	117, 568
Dish-washing machine	S. S. Tupper	Churchville, N. Y	July 26, 1870	105, 742
Dish-washing machine	J. Usher	West Albany, N. Y	Jan. 21, 1873	135, 176
Dish-washing machine	J. Wheeler	Athol, Mass	Mar. 27, 1866	53, 511
Dish-washing machine	C. H. Williams	Syracuse, N. Y	Oct. 7, 1873	143, 484
Dish-washing machine	L. R. Witherell	Galesburgh, Ill	Sept. 28, 1869	95, 400
Dish-washing machine	L. R. and E. A. Witherell	Galesburgh, Ill	Dec. 22, 1868	85, 155
Dish-washing machine	C. Woodruff	Antioch, Cal	Feb. 6, 1872	123, 434
Dishes, Apparatus for drying	C. W. Schroeder	New York, N. Y	Oct. 23, 1866	59, 083
Dishes, knives and forks, &c., Machine for washing	C. M. Leland	Central City, Colo	July 6, 1869	92, 323
Disinfectant	L. Broad	Saint Louis, Mo	Sept. 18, 1866	58, 163
Disinfectant-compound	J. Hilton	New York, N. Y	July 29, 1873	141, 274
Disinfectant-package	W. P. Tenny	Boston, Mass	Sept. 16, 1873	142, 963
Disinfectant or ozone generator	W. H. Ford and S. Logan	New Orleans, La	Dec. 29, 1868	85, 297
Disinfectants, Apparatus for vaporizing and atomizing.	E. S. De Luce and H. M. Wells	United States Navy	June 25, 1872	128, 368
Disinfecting and antiseptic compound	J. Strigel	Louisville, Ky	July 23, 1867	67, 082
Disinfecting and deodorizing compound	H. G. Dayton	Dayton, Ohio	Nov. 30, 1869	97, 469
Disinfecting and embalming fluid	H. C. Coffman	Washington Court-House, Ohio.	May 28, 1867	65, 174
Disinfecting and preserving, Composition for	J. Hickson and L. L. Wilkinson	Auburn, N. Y	May 1, 1866	54, 345
Disinfecting and purifying hospitals, camps, &c., Composition for.	J. L. Kidwell	Georgetown, D. C	June 2, 1863	38, 748
Disinfecting and ventilating burial-vault	B. F. Lyford	San Francisco, Cal	Jan. 31, 1871	111, 358
Disinfecting-apparatus	J. M. Scudder	Cincinnati, Ohio	Apr. 24, 1866	54, 221
Disinfecting bone-boiling establishments, Mode of	D. E. Contaret	New York, N. Y	Aug. 14, 1866	57, 093
Disinfecting buildings, ships, &c., Process of	H. M. Wells	New York, N. Y	Oct. 24, 1871	120, 355
Disinfecting-composition	F. Hoffman and R. Wendler	Brooklyn, N. Y	May 22, 1866	54, 907
Disinfecting-compound	J. Gamage	London, England	Nov. 26, 1872	133, 430
Disinfecting-compound	J. Le Doyen	Paris, France	Dec. 24, 1844	5, 403
Disinfecting-compound	J. W. Lee and J. W. Davis	Washington, D. C	Dec. 9, 1873	145, 433
Disinfecting-compound	H. Napier	Elizabeth, N. J	Aug. 14, 1866	57, 174
Disinfecting-compound	O. Oldberg	Washington, D. C	Aug. 1, 1871	117, 671
Disinfecting compound	L. S. Robbins	New York, N. Y	Mar. 1, 1870	100, 327
Disinfecting-compound	C. A. Seely and C. J. Eames	New York, N. Y	Feb. 18, 1868	74, 608
Disinfecting-compound	E. Süvern	Halle, Prussia	Sept. 17, 1867	69, 043
Disinfecting-compound	H. A. Tilden	New Lebanon, N. Y	Sept. 26, 1871	119, 250
Disinfecting-compound	H. A. Tilden	New Lebanon, N. Y	Dec. 12, 1871	121, 913
Disinfecting-compound	G. Vique, sr	Bordeaux, France	Apr. 18, 1871	113, 820
Disinfecting fecal matter	D. E. Coutaret	Boston, Mass	Jan. 8, 1856	14, 053
Disinfecting foul air in vessels, Apparatus for	A. Peteler	New Brighton, N. Y	Apr. 9, 1861	31, 996
Disinfecting noxious vapors from rendering-houses, hospitals, &c., Mode of.	R. B. Lockwood	New York, N. Y	Dec. 11, 1866	60, 394
Disinfecting noxious vapors, Process for	W. Adamson	Philadelphia, Pa	Feb. 14, 1865	46, 317
Disinfecting-powder	H. G. Dayton	Maysville, Ky	Aug. 5, 1873	141, 546
Disinfecting rooms, ships, &c., Process of	E. Dumplemann	Washington, D. C	Nov. 26, 1867	71, 467
Disinfecting soap-boiling establishments	S. Strunz	Pittsburgh, Pa	May 3, 1873	138, 954
Disinfecting vessels, Apparatus for	E. H. Covel	New York, N. Y	Apr. 26, 1864	42, 464
Disinfecting water-closets, urinals, gutters, &c	A. G. Brown	Southwark, England	Sept. 17, 1872	131, 422
Disinfecter	J. D. C. Outwater	New York, N. Y	Aug. 28, 1866	57, 559

Index of patents issued from the United States Patent Office from 1790 *to* 1873, *inclusive*—Continued.

Invention.	Inventor.	Residence.	Date.	No.
Disintegrating-mill	G. B. Davis	Baltimore, Md	Dec. 14, 1869	97,890
Disintegrating substances, Apparatus for	C. G. C. Simpson	Montreal, Canada	Aug. 8, 1871	117,824
Disintegrator, Rotary	H. Dusch	Baltimore, Md	Oct. 25, 1870	108,692
Disk, bolt, and rivet machine, Rotating	J. G. Day	Brooklyn, N. Y	July 3, 1849	6,578
Disks, Machine for cutting out and flanging metallic	G. W. Bentley	Brooklyn, N. Y	Feb. 13, 1872	123,667
Disks, Machinery for cutting and bending metallic	E. Savage	Berlin, Conn	Aug. 30, 1853	9,982
Disks, Mode of securing cutters to rotary	J. Newton	New York, N. Y	June 19, 1855	13,096
Disks, rivets, &c., Machine for molding metal	P. L. Higley	Cincinnati, Ohio	May 11, 1869	89,866
Dislocations, Apparatus for reducing	G. O. Jarvis	Portland, Conn	Sept. 14, 1843	3,263
Distance between itself and the target without chaining, Instrument for ascertaining the.	B. De Villeroi	Philadelphia, Pa	Mar. 22, 1859	23,294
Distance-indicator for railway	S. O. Schoonmaker	Wright, N. Y	Sept. 22, 1863	40,066
Distillates, Apparatus for cooling	H. G. Dayton	Maysville, Ky	Oct. 18, 1864	44,712
Distillation	J. W. Alder	Pennsborough, Pa	Apr. 19, 1814	......
Distillation	J. F. Collins	New York, N. Y	Oct. 30, 1866	59,187
Distillation	J. Dimond	Philadelphia, Pa	Dec. 28, 1808	......
Distillation	W. Richardson	Kingston, N. Y	June 15, 1816	......
Distillation	H. B. Sinks	Cincinnati, Ohio	Mar. 29, 1870	101,388
Distillation and other purposes, Mode of fermenting liquids for.	R. D'Heureuse	San Francisco, Cal	Aug. 6, 1867	67,512
Distillation, Apparatus and process of	G. S. Williamson	Gallatin, Tenn	Sept. 6, 1870	107,147
Distillation, Apparatus for continuous	C. H. Budd	Philadelphia, Pa	Mar. 9, 1869	87,537
Distillation, Apparatus for continuous	W. G. W. Jaeger	Baltimore, Md	May 1, 1866	54,358
Distillation, Apparatus for cooling mash for	T. Whitemell	Halifax County, N. C	Sept. 2, 1815	......
Distillation, Apparatus for destructive	L. Atwood	Brooklyn, N. Y	Dec. 28, 1858	22,407
Distillation, Apparatus for destructive	L. Atwood	Brooklyn, N. Y	Feb. 22, 1859	23,006
Distillation, Apparatus for destructive	L. Atwood	Brooklyn, N. Y	Mar. 29, 1859	23,337
Distillation, Apparatus for treating grain for	A. Woolner	Louisville, Ky	June 10, 1873	139,749
Distillation, &c., Application of steam in	H. Kelly	Brooklyn, N. Y	Feb. 16, 1816	......
Distillation by steam in wooden or other stills	S. Brown, and E. and T. West	......	July 8, 1803	......
Distillation, Condensation of vapors in	N. Blanche	New York, N. Y	Mar. 2, 1811	......
Distillation for hydrocarbon oils	J. J. Johnston	Allegheny, Pa	Apr. 9, 1861	31,982
Distillation, Method of	V. B. Ryerson	New York, N. Y	May 7, 1861	32,254
Distillation, &c., Mode of cooling the mash used in the process of.	A. D. Ward	Minerva, Ky	Nov. 7, 1840	1,850
Distillation of bromine and iodine	D. Alter	Freeport, Pa	Mar. 19, 1867	62,988
Distillation of coal, Apparatus for	H. P. Gengembre	Allegheny, Pa	Mar. 20, 1860	27,542
Distillation of coal-oils	B. Garvey	New York, N. Y	July 17, 1860	29,218
Distillation of coal-oils	D. S. Stombs and J. Brace	Newport, Ky., and Point Creek, Va.	Apr. 10, 1860	27,842
Distillation of coal-oils, Apparatus for	L. Atwood	New York, N. Y	Apr. 10, 1860	27,768
Distillation of fresh water from salt water	A. Normandy	France	Oct. 5, 1858	21,693
Distillation of hydrocarbon oils	L. Atwood	New York, N. Y	Apr. 10, 1860	27,767
Distillation of hydrocarbon oils	H. Grogan and G. T. Lape	New York, N. Y	May 11, 1869	89,988
Distillation of hydrocarbon oils	F. McCarty	Smith's Ferry, Pa	June 29, 1869	91,953
Distillation of hydrocarbon oils	J. Merrill	Boston, Mass	July 2, 1861	32,704
Distillation of oils from resin	S. Frazer	Galena, Ill	June 12, 1860	28,663
Distillation of petroleum	A. Millochau	New York, N. Y	Mar. 13, 1866	53,167
Distillation of petroleum, &c., Apparatus for the continuous.	J. I. Vaughn	London, England	Aug. 29, 1869	49,689
Distillation of salt water, Ship's galley for	E. Hutchinson	Baltimore, Md	May 20, 1839	1,156
Distillation of spirits of turpentine, Apparatus for	J. Jennings	New York, N. Y	Aug. 27, 1835	......
Distillation of tar and other substances, Apparatus for.	L. S. Fales	Boston, Mass	Jan. 23, 1866	52,151
Distillation of wood, &c., Apparatus for destructive	L. Atwood	Brooklyn, N. Y	Dec. 28, 1858	22,408
Distillery	H. Becker	Warwick Township, Pa	June 29, 1833	......
Distillery	H. H. Kirk	Springfield, Tenn	Feb. 4, 1868	74,098
Distillery and brewery boiler	B. Eggleston, jr	Princetown, N. Y	May 27, 1805	......
Distillery and kitchen range	W. Gamble	Washington, D. C	Apr. 18, 1811	......
Distillery-boiler	T. Pierce	Hartwick, N. Y	Feb. 27, 1819	......
Distillery-furnace, Fire-proof	J. Miller	Northampton, Pa	Mar. 13, 1823	......
Distilleries, Mashing-machinery for	W. Whitney	Rochester, N. Y	Jan. 19, 1826	......
Distilleries, Refrigerator for	L. C. and P. Rodman	Baltimore, Md	Jan. 6, 1826	......
Distilling	J. M. Aiken	Philadelphia, Pa	Aug. 30, 1827	......
Distilling	E. Barnum and B. Brooks	New Marlborough, Mass	July 13, 1808	......
Distilling	J. Bates	Hartford, Conn	Jan. 26, 1813	......
Distilling	W. H. Bayless	New York, N. Y	May 29, 1847	5,136
Distilling	J. Cairon	New York, N. Y	Jan. 10, 1831	......
Distilling	J. Cairon	Boston, Mass	Aug. 22, 1831	......
Distilling	I. Carpenter and J. Roe	New York, N. Y	Jan. 29, 1824	......
Distilling	O. Carrier	Enfield, Conn	Jan. 18, 1813	......
Distilling	J. Carze	New York, N. Y	Mar. 18, 1819	......
Distilling	W. J. Cocke	Surry County, Va	Oct. 30, 1827	......
Distilling	W. Cook	Southport, N. Y	Dec. 7, 1829	......
Distilling	F. S. Cozzens	New York, N. Y	July 27, 1822	......
Distilling	A. Dunlap	Boston, Mass	May 3, 1811	......
Distilling	C. F. Fisher	York, Pa	Apr. 24, 1830	......
Distilling	T. Gallaher	Liverpool, Pa	Dec. 14, 1830	......
Distilling	M. Garber, sr	Staunton, Va	July 13, 1810	......
Distilling	J. J. Giraud	Baltimore, Md	June 14, 1811	......
Distilling	P. Griener	Brandywine, Pa	Apr. 17, 1822	......
Distilling	N. Hart	Easton, Pa	Oct. 29, 1814	......
Distilling	S. Harwood, 3d	Braintree, N. Y	Apr. 15, 1826	......
Distilling	J. L. Jencks	Sackett's Harbor, N. Y	May 2, 1831	......
Distilling	L. Johnston	Easton, Pa	Nov. 9, 1832	......
Distilling	A. R. Kerr and H. Hoover	Waynesborough, Pa	July 1, 1836	......
Distilling	T. Lawes and P. Grosjean	Louisville, Ky	July 30, 1828	......
Distilling	J. Lusk	Butler County, Ohio	Dec. 22, 1827	......
Distilling	M. McDonald	Fayette County, Pa	Feb. 19, 1823	......
Distilling	J. Miller	Lancaster County, Pa	Mar. 2, 1811	......
Distilling	J. Miller	Lancaster, Pa	Feb. 16, 1832	......
Distilling	S. Oliver	Northampton, Pa	Aug. 4, 1841	2,207
Distilling	A. Putnam	......	Jan. 29, 1791	......
Distilling	N. Read	......	Aug. 26, 1791	......
Distilling	G. Riley	New York, N. Y	May 29, 1847	5,133
Distilling	P. Swartz, jr	Muncy, Pa	June 25, 1836	......
Distilling	E. Talmage	Lancaster, Ky	July 8, 1816	......

Index of patents issued from the United States Patent Office from 1790 *to* 1873, *inclusive*—Continued.

Invention.	Inventor.	Residence.	Date.	No.
Distilling	W. M. B. Wollias and H. P. Barlow.	Oneida County, N. Y	May 6, 1814	....
Distilling	P. Woods	Frederick County, Md	Sept. 29, 1818	....
Distilling	L. W. Wright	London, England	Sept. 28, 1831	....
Distilling acid and naphtha from resin	L. S. Robbins	New York, N. Y	Nov. 4, 1851	8, 490
Distilling alcohol	J. T. Heard	Boston, Mass	Sept. 9, 1843	3, 256
Distilling alcohol, &c., Apparatus for	P. L. Howlett	Springfield, Ill	Dec. 16, 1862	37, 167
Distilling alcoholic liquors	S. F. Van Choat	Boston, Mass	Nov. 9, 1869	96, 640
Distilling alcoholic spirits	H. Fake	Chicago, Ill	Dec. 3, 1872	133, 523
Distilling alcoholic spirits and liquors	J. Dennis, jr	Washington, D. C	Aug. 10, 1869	93, 523
Distilling and bleaching oils	C. J. T. Burcey	Black Rock, Conn	Jan. 16, 1872	122, 810
Distilling and brewing, Treating grain for	A. Woolner	Louisville, Ky	Feb. 27, 1872	124, 188
Distilling and concentrating liquids, Apparatus for	C. C. Parsons	New York, N. Y	Jan. 17, 1871	110, 997
Distilling and condensing apparatus	J. K. Stafford	Brooklyn, N. Y	Apr. 25, 1854	10, 813
Distilling and employing the residuum advantageously.	I. Jennings	New York, N. Y	June 13, 1831	....
Distilling and evaporating extracts, saline solutions, &c., Apparatus for.	T. Steers, jr	Richmond, Va	July 12, 1870	105, 381
Distilling and evaporating liquids	P. T. Badoux	New York, N. Y	Jan. 29, 1867	61, 651
Distilling and evaporation	A. Matthews	Island Creek Township, Ohio.	June 13, 1831	....
Distilling and producing fresh potable water, Apparatus for.	W. A. Gray	New York, N. Y	May 25, 1869	90, 444
Distilling and purifying liquors, Apparatus for	F. M. Young	Nashville, Tenn	May 24, 1870	103, 535
Distilling and purifying spirits, Apparatus for	D. Woodard	Springfield, Tenn	May 11, 1869	89, 961
Distilling and purifying turpentine from wood	J. D. Stanley	Baltimore, Md	Aug. 20, 1872	130, 598
Distilling and rectifying apparatus	C. Gresinchna	New York, N. Y	Jan. 7, 1868	73, 092
Distilling and rectifying petroleum, Apparatus for.	C. Stott	San Francisco, Cal	Aug. 27, 1867	68, 257
Distilling and rectifying spirits	C. Falkman	Stockholm, Sweden	Nov. 20, 1849	6, 881
Distilling and rectifying spirits, Apparatus for	A. Foubert	New York, N. Y	Aug. 20, 1867	67, 863
Distilling and rectifying spirits, Apparatus for	G. Kaiser	New York, N. Y	Sept. 7, 1869	94, 611
Distilling and rectifying spirits, Apparatus for re	A. Noteman	Toledo, Ohio	Apr. 19, 1870	102, 033
Distilling and rectifying spirituous liquors, Apparatus for.	M. Newmann	New York, N. Y	Mar. 5, 1867	62, 670
Distilling and rectifying whisky and other spirits, Apparatus for.	E. F. Prentiss and R. A. Robertson.	Philadelphia, Pa., and Liverpool, England.	Mar. 1, 1864	41, 819
Distilling and rectifying whisky, Apparatus for	E. F. Prentiss and R. A. Robertson.	Philadelphia, Pa	June 27, 1865	48, 436
Distilling and refining apparatus	J. Sherer and A. Killian	Lancaster, Pa	June 17, 1811	....
Distilling and refining oils, Apparatus for	A. Foubert	Buffalo, N. Y	Aug. 29, 1871	118, 602
Distilling and refining oils, wines, and other liquids	A. Foubert	New York, N. Y	Dec. 4, 1866	60, 166
Distilling and refining petroleum, &c., Apparatus for.	J. Ellis and E. C. Kattell	New York and Binghamton, N. Y.	Apr. 16, 1867	63, 789
Distilling and refining petroleum, Apparatus for	C. G. Howell	Corning, N. Y	July 16, 1867	66, 841
Distilling and separating oils, fats, and glue, Apparatus for.	C. A. Seely	New York, N. Y	Feb. 23, 1869	87, 207
Distilling and treating alcoholic spirits	G. Clark	Buffalo, N. Y	Oct. 26, 1869	96, 201
Distilling and wooden still	J. J. Giraud	Baltimore, Md	Aug. 3, 1813	....
Distilling-apparatus	A. Blood	Boston, Mass	July 16, 1816	....
Distilling apparatus	A. Boucherie	Philadelphia, Pa	Jan. 30, 1809	....
Distilling-apparatus	W. Chamberlin	Vermont	Mar. 17, 1810	....
Distilling-apparatus	J. A. Coffey	London, England	Dec. 18, 1866	60, 477
Distilling-apparatus	W. Corfield	Philadelphia, Pa	Dec. 29, 1868	85, 286
Distilling-apparatus	W. Corfield	Philadelphia, Pa	Dec. 29, 1868	85, 287
Distilling-apparatus	H. G. Dayton	Maysville, Ky	Aug. 25, 1863	39, 635
Distilling-apparatus	C. Delescluze	New York, N. Y	Oct. 12, 1852	9, 317
Distilling-apparatus	S. De Witt	Albany, N. Y	Jan. 30, 1833	....
Distilling-apparatus	S. R. Divine and C. A. Seely	New York, N. Y	May 29, 1866	55, 071
Distilling-apparatus	J. C. Douglass	Troy, N. Y	Nov. 26, 1825	....
Distilling-apparatus	H. Drüding	Brieg, Prussia	June 25, 1872	128, 291
Distilling-apparatus	J. T. Dwyer	Nashville, Tenn	Nov. 19, 1833	....
Distilling-apparatus	J. T. Dwyer	Nashville, Tenn	Oct. 14, 1834	....
Distilling-apparatus	H. F. Fisher	Philadelphia, Pa	Sept. 12, 1821	....
Distilling-apparatus	C. L. Fleischmann	Washington, D. C	May 4, 1869	89, 749
Distilling-apparatus	A. Foubert	New York, N. Y	Nov. 19, 1867	71, 156
Distilling-apparatus	J. Frageser and I. Illofsky	New York, N. Y	Sept. 3, 1867	68, 470
Distilling apparatus	C. F. Frederici	New York, N. Y	July 11, 1865	48, 672
Distilling-apparatus	N. Goodsell	Paris, N. Y	June 13, 1816	....
Distilling-apparatus	T. Grundmann	Cleveland, Ohio	Dec. 3, 1867	71, 744
Distilling-apparatus	C. H. Hall and J. Ellis	New York, N. Y	Oct. 16, 1866	58, 813
Distilling-apparatus	H. Halvorson	Boston, Mass	May 2, 1854	10, 850
Distilling-apparatus	A. Heistand	York Town, Pa	Nov. 27, 1805	....
Distilling-apparatus	I. Hoffman	Shawangunk, N. Y	Dec. 31, 1813	....
Distilling-apparatus	G. Hunziker	Summit, Miss	Mar. 18, 1873	136, 921
Distilling-apparatus	F. Kalisky	Philadelphia, Pa	Dec. 14, 1819	....
Distilling-apparatus	P. Kessler	Belleville, Ill	Mar. 1, 1859	23, 089
Distilling-apparatus	J. Kunze	Chatsworth, Ill	Apr. 9, 1872	125, 463
Distilling-apparatus	L. Lecesne	Belleville, N. J	Dec. 4, 1811	....
Distilling-apparatus	A. A. Meyendorff	New York, N. Y	June 9, 1868	78, 678
Distilling-apparatus	C. Otis	Tunkhannock, Pa	Feb. 16, 1831	....
Distilling-apparatus	L. Pray	Charlestown, Mass	Jan. 8, 1867	61, 098
Distilling-apparatus	G. Riley	New York, N. Y	Apr. 17, 1849	6, 317
Distilling-apparatus	J. D. Riley	Cincinnati, Ohio	Apr. 28, 1868	77, 216
Distilling-apparatus	D. Savalle	Paris, France	June 23, 1868	79, 260
Distilling-apparatus	H. A. Schesch	Brooklyn, N. Y	Apr. 24, 1866	54, 218
Distilling-apparatus	J. Sloan	Pittsburgh, Pa	Jan. 17, 1860	26, 877
Distilling-apparatus	U. Swetland	Kingston, Pa	Nov. 18, 1822	....
Distilling-apparatus	A. H. Tait and J. W. Avis	New York, N. Y	Mar. 20, 1866	53, 359
Distilling-apparatus	A. Tobey	Alford, Mass	June 21, 1810	....
Distilling-apparatus	A. Webster	Seneca Falls, N. Y	June 2, 1868	78, 498
Distilling-apparatus	C. E. Werner	New Castle, Ill	Jan. 24, 1854	10, 463
Distilling-apparatus	E. Werner	Canton, Ill	Sept. 26, 1871	119, 436
Distilling-apparatus	J. M. Weyand	Saint Louis, Mo	May 3, 1870	102, 632
Distilling-apparatus	F. W. Willard	New York, N. Y	Mar. 13, 1860	27, 503
Distilling-apparatus and boiler	J. Lovatt	Philadelphia, Pa	Mar. 17, 1829	....
Distilling-apparatus, Combination	L. Wolff	Chicago, Ill	Apr. 5, 1870	101, 555
Distilling-apparatus, Construction of	J. Slean	Pittsburgh, Pa	Oct. 4, 1859	25, 705
Distilling, Apparatus for and mode of	A. Weaver	Adams County, Pa	Apr. 23, 1807	....

Index of patents issued from the United States Patent Office from 1790 *to* 1873, *inclusive*—Continued.

Invention.	Inventor.	Residence.	Date.	No.
Distilling-apparatus for spirits	J. C. Cookson	Lancaster, Pa	Sept. 1, 1868	81,605
Distilling-apparatus for spirits	D. Green	New York, N. Y	Sept. 1, 1868	81,624
Distilling-apparatus for spirits	F. Sonier	Springfield, Ill	Oct. 12, 1869	95,740
Distilling-apparatus, Wood	G. Hunziker	Summit, Miss	June 9, 1868	78,743
Distilling ardent spirits, Mode of	J. Shaw	Palmer, Mass	Nov. 13, 1811	
Distilling by steam	J. M. Bennett	Middletown, Conn	May 1, 1810	
Distilling by steam	W. Berkley	Lebanon, Ky	May 20, 1830	
Distilling by steam	D. Dewey	Sullivan, N. Y	July 5, 1814	
Distilling by steam	E. Geilhausen	Halifax Township, Pa	Jan. 14, 1823	
Distilling coal-oils	S. Hudson	Plainfield, N. J	Feb. 20, 1872	123,907
Distilling coal-oils and other substances, Apparatus for.	H. W. C. Tweddle	Great Britain	Feb. 4, 1862	34,324
Distilling coal-oils, Apparatus for	H. W. Adams	Brooklyn, N. Y	May 29, 1860	28,446
Distilling coal-oils, Apparatus for	J. L. Alberger	Buffalo, N. Y	Mar. 3, 1863	37,798
Distilling coal-oils, Apparatus for	J. Bullard	Stockbridge, Vt	Jan. 21, 1862	34,195
Distilling coal-oils, Apparatus for	H. P. Gengembre	Tarentum, Pa	Nov. 12, 1861	33,699
Distilling coal-oils, Apparatus for	G. W. Kirchhoffer	Cincinnati, Ohio	May 21, 1861	32,373
Distilling coal-oils, Apparatus for	F. W. Willard	New York, N. Y	Feb. 28, 1860	27,327
Distilling coal with hydrogen-gas	S. Meredith	Meadville, Pa	July 31, 1855	13,358
Distilling, Cold	A. V. H. Webb	Utica, N. Y	May 14, 1836	
Distilling, Combination of astringent woods and vegetables in.	F. Hall		Apr. 17, 1797	
Distilling corn-meal	A. Doolittle	Ann Harbor, Mich	Nov. 10, 1829	
Distilling crude oils from mineral coal, Apparatus for.	C. Cherry	Pittsburgh, Pa	Sept. 2, 1856	15,643
Distilling, Cylinder-heater for mash-water, &c., for	C. Jenks and A. Pease	East Windsor, Conn	July 12, 1813	
Distilling, Double boiler for	A. Gunn	New Haven, Conn	Dec. 19, 1808	
Distilling essential oils and spirits	D. Jewett	Pittsfield, Mass	Jan. 24, 1811	
Distilling, evaporating, and refining oils, &c., Apparatus for.	J. Ellis and E. C. Kattell	New York and Binghamton, N. Y.	Sept. 17, 1867	68,860
Distilling fresh from salt water	A. M. Ring	Saint John's, New Brunswick.	June 11, 1872	127,797
Distilling fresh from salt water, Caboose of a vessel for.	S. Rouse		Jan. 1, 1803	
Distilling from grain, &c	S. Shurtliff	Oneida County, N. Y	July 5, 1814	
Distilling gases and vapors, Apparatus for	J. Howarth	Salem, Mass	May 17, 1864	42,772
Distilling, Generating steam for	E. Brown	Oneida County, N. Y	Jan. 23, 1815	
Distilling, Heating liquor, mash, &c., for	T. Belden	Hartford, Conn	Aug. 18, 1813	
Distilling hydrocarbon-oils	W. Archer	New York, N. Y	Sept. 6, 1864	44,137
Distilling hydrocarbon-oils	C. H. Hall	Smith's Ferry, Pa	Feb. 2, 1869	86,535
Distilling hydrocarbon-oils	S. A. Hill and C. F. Thumm	Oil City, Pa	May 2, 1871	114,293
Distilling hydrocarbon-oils	H. W. C. Tweddle	Pittsburgh, Pa	Dec. 10, 1867	72,126
Distilling hydrocarbon-oils, Apparatus for	S. A. Hill and C. F. Thumm	Oil City, Pa	May 10, 1870	102,819
Distilling hydrocarbon-oils, Apparatus for	J. J. Johnston	Allegheny City, Pa	June 15, 1869	91,448
Distilling hydrocarbons, Apparatus for	H. W. Faucett and T. McGowan	Titusville and Meredith, Pa	Nov. 26, 1872	133,425
Distilling hydrocarbons, Apparatus for	J. J. Johnston	Allegheny City, Pa	June 15, 1869	91,447
Distilling hydrocarbons, Apparatus for	J. J. Johnston	Allegheny, Pa	July 25, 1871	117,425
Distilling hydrocarbons, Apparatus for	H. A. Stearns	Smithfield, R. I	May 24, 1870	103,385
Distilling in vacuo, Method of	F. Gutzkow	San Francisco, Cal	May 31, 1870	103,604
Distilling, Increasing the surface of evaporation for.	B. Henfrey		Mar. 2, 1801	
Distilling Indian-corn	J. R. Nance	Floyd County, Ind	Dec. 27, 1828	
Distilling liquids from coal-tar, Mode of	M. L. Keen	Royer's Ford, Pa	Sept. 20, 1859	25,552
Distilling liquors	N. and A. Wolcott	Bloomfield, N. Y	Dec. 8, 1826	
Distilling, Machinery used in	D. Jackson	Peru, N. Y	Dec. 28, 1824	
Distilling, &c., Mode of conducting off steam from cylinder of steam-engine in.	G. W. and G. Davis	Canal Fulton, Ohio	May 4, 1842	2,602
Distilling naphtha and other hydrocarbon liquids	H. H. Rogers	Brooklyn, N. Y	Oct. 31, 1871	120,539
Distilling oil	S. Andrews	Cleveland, Ohio	Sept. 25, 1866	58,197
Distilling oils and paraffine from heat and other substances.	S. L. Wiegand	Philadelphia, Pa	Aug. 18, 1863	39,607
Distilling oils, Apparatus and process for	W. G. Warden	Philadelphia, Pa	Jan. 3, 1871	110,806
Distilling oils, Apparatus for	J. B. Grant	New York, N. Y	Aug. 21, 1866	57,311
Distilling oils, Apparatus for	J. Howarth	Salem, Mass	June 15, 1858	20,562
Distilling oils, Apparatus for	J. J. Johnston	Allegheny City, Pa	June 20, 1865	48,285
Distilling oils, Apparatus for	E. G. Kelly and A. H. Tait	New York, N. Y., and Jersey City, N. J.	June 18, 1861	32,568
Distilling oils, Apparatus for	A. Quinn	New York, N. Y	Apr. 9, 1861	31,998
Distilling oils, Apparatus for	H. W. C. Tweddle	Pittsburgh, Pa	Dec. 10, 1867	72,125
Distilling oils from coal, Apparatus for	C. J. Van Wyck	New York, N. Y	Mar. 20, 1860	27,603
Distilling oils from coal, Method of	J. Howarth	Salem, Mass	Sept. 27, 1859	25,568
Distilling oils from coal, Process for	G. M. Mowbray	Green Point, N. Y	Sept. 27, 1859	25,575
Distilling oleaginous matter	A. M. Poisat and D. C. Knab	Paris, France	Feb. 26, 1850	7,124
Distilling or boiling water	E. Barnum and B. Brooks	New Marlborough, Mass	Mar. 21, 1808	
Distilling or making alcohol	L. J. Kilborn		June 4, 1803	
Distilling petroleum	H. Fleury	New York, N. Y	Oct. 24, 1865	50,571
Distilling petroleum	A. D. Higham	Jersey City, N. J	Apr. 24, 1866	54,157
Distilling petroleum	A. Kirk	Allegheny City, Pa	June 16, 1868	78,878
Distilling petroleum	W. H. Sangster and T. C. Spencer.	Buffalo, N. Y	July 10, 1866	56,276
Distilling petroleum	J. and J. Stuber and J. W. Mager.	Astoria, N. Y	Feb. 13, 1872	123,741
Distilling petroleum	A. Thirault	New York, N. Y	Jan. 8, 1867	61,120
Distilling petroleum	A. W. Wilkinson	New York, N. Y	Dec. 16, 1873	145,707
Distilling petroleum and other liquids	C. H. Hall	New York, N. Y	June 26, 1866	55,855
Distilling petroleum and other liquids	P. H. Vander Weyde	Philadelphia, Pa	Sept. 11, 1866	58,0[illegible]5
Distilling petroleum and other liquids, Apparatus for.	W. G. Warden	Philadelphia, Pa	Mar. 14, 1871	112,751
Distilling petroleum and other liquids, Apparatus for.	A. Quinn	New York, N. Y	Sept. 16, 1862	36,481
Distilling petroleum and other substances	O. Lugo	New York, N. Y	Jan. 2, 1866	51,843
Distilling petroleum, Apparatus for	E. Braggins	Titusville, Pa	Mar. 7, 1865	46,633
Distilling petroleum, &c., Apparatus for	J. F. Collins	New York, N. Y	Oct. 30, 1866	59,334
Distilling petroleum, Apparatus for	A. Dubreuil	Baltimore, Md	June 20, 1865	48,265
Distilling petroleum, Apparatus for	H. B. Everest	Rochester, N. Y	Sept. 3, 1867	68,426
Distilling petroleum, Apparatus for	M. P. Ewing	Rochester, N. Y	July 31, 1866	56,852
Distilling petroleum, Apparatus for	M. P. Ewing and H. B. Everest.	Rochester, N. Y	Sept. 11, 1866	58,021
Distilling petroleum, Apparatus for	L. V. Fichet	New York, N. Y	Apr. 17, 1866	53,964

Index of patents issued from the United States Patent Office from 1790 *to* 1873, *inclusive*—Continued.

Invention.	Inventor.	Residence.	Date.	No.
Distilling petroleum, Apparatus for	A. Kreusler	New Lebanon, N. Y	Oct. 10, 1865	50, 368
Distilling petroleum, &c., Apparatus for	O. Lugo	New York, N. Y	Sept 18, 1866	58, 113
Distilling petroleum, Apparatus for	J. Miller	Allegheny City, Pa	Apr. 21, 1868	77, 070
Distilling petroleum, &c., Apparatus for	A. Millochan	New York, N. Y	Mar. 21, 1865	46, 923
Distilling petroleum, Apparatus for	E. F. Prentiss and R. A. Robertson.	Philadelphia, Pa	June 27, 1865	48, 435
Distilling petroleum, Apparatus for	W. B. Snow	Brooklyn, N. Y	Aug. 20, 1872	130, 668
Distilling petroleum, &c., Apparatus for	A. H. Tait and J. W. Avis	New York, N. Y	Mar. 19, 1867	63, 115
Distilling petroleum, &c., Apparatus for	C. M. Warren	Boston, Mass	Apr. 11, 1865	47, 235
Distilling petroleum, oils, and other substances, Process of.	O. Lugo and T. O. L. Shrader	New York, N. Y	Dec. 11, 1866	60, 396
Distilling petroleum, Process for	R. A. Chesebrough	New York, N. Y	June 27, 1865	48, 367
Distilling petroleum, Process for	A. Dubreuil	Baltimore, Md	Nov. 28, 1865	51, 156
Distilling petroleum, Process of	L. S. Fales	Boston, Mass	Sept. 5, 1865	49, 739
Distilling petroleum, Retort for	G. H. S. Duffus	New Orleans, La	Jan. 31, 1865	46, 088
Distilling petroleum, Retort for	G. H. S. Duffus	New Orleans, La	Jan. 31, 1865	46, 089
Distilling petroleum, Retort for	G. H. S. Duffus	New Orleans, La	Jan. 31, 1865	46, 090
Distilling pine wood	J. A. Matlocks	Swansborough, N. C	Apr. 3, 1866	53, 641
Distilling pine wood	T. W. Wheeler	New Berne, N. C	July 5, 1870	105, 019
Distilling pine wood, &c., Apparatus for	S. L. Cole	Burlington, Vt	May 19, 1863	38, 560
Distilling pine wood, Apparatus for	D. F. Mellen	Manchester, N. H	May 5, 1868	77, 636
Distilling pine wood, Apparatus for	W. Messau	Atlanta, Ga	July 23, 1872	129, 849
Distilling pure alcoholic spirits	C. Andersen	Brooklyn, N. Y	Oct. 14, 1873	143, 654
Distilling, rectifying, &c	J. Whips	Jefferson County, Ky	Apr. 22, 1816	....
Distilling resin-oils, Process for	H. Halvorson	Boston, Mass	May 2, 1854	10, 849
Distilling resin-oils, Process of	M. Page	Williamsburgh, N. Y	May 24, 1853	9, 752
Distilling resin-oils, Process for	J. Riley and W. Allen	Southfield, N. Y	Mar. 22, 1853	9, 627
Distilling rock-oils	A. Thirault	New York, N. Y	Mar. 8, 1864	41, 871
Distilling rock-oils and other hydrocarbons, Apparatus for.	E. F. Prentiss and R. A. Robertson.	Philadelphia, Pa., and Liverpool, England.	Mar. 8, 1864	41, 858
Distilling salt-water, Caboose adapted for	M. Rocher	Nantes, France	Aug. 28, 1841	2, 232
Distilling sea-water, Apparatus for	J. Ericsson and R. B. Forbes	New York, N. Y., and Boston, Mass.	Oct. 23, 1849	6, 815
Distilling sea-water, Apparatus for	G. S. G. Spencer	Boston, Mass	July 3, 1860	29, 015
Distilling spirits	R. Blaikie	New York, N. Y	Apr. 27, 1821	....
Distilling spirits	D. Ilsley	....	May 14, 1803	....
Distilling spirits and other liquids, Apparatus for	F. Haeck	Brussels, Belgium	Dec. 5, 1865	51, 403
Distilling spirits and other liquids, Apparatus for condensing in.	H. G. Dayton and J. Christie	Maysville, Ky., and Atlanta, Ill.	Dec. 1, 1868	84, 479
Distilling spirits, Apparatus for	E. D. Bannister	Saint Louis, Mo	Mar. 30, 1869	88, 434
Distilling spirits, Apparatus for	H. G. Dayton	Maysville, Ky	Apr. 3, 1866	53, 581
Distilling spirits, Apparatus for	A. Foubert	New York, N. Y	Aug. 20, 1867	67, 864
Distilling spirits, Apparatus for	F. Haeck	Brussels, Belgium	Sept. 29, 1868	82, 519
Distilling spirits, Apparatus for	C. S. Hutchinson	Burlington, N. J	Aug. 4, 1868	80, 740
Distilling spirits, Apparatus for	G. Kleiner	Georgetown, Mo	Dec. 19, 1871	122, 032
Distilling spirits, Apparatus for	A. Kreusler and W. T. Pelton	New Lebanon, N. Y	Nov. 7, 1865	50, 876
Distilling spirits, Apparatus for	F. M. Young	Nashville, Tenn	May 3, 1870	102, 745
Distilling spirits by steam	W. Avery	Pompey, N. Y	May 13, 1825	....
Distilling spirits from corn-stalks	B. Allison	Philadelphia, Pa	July 6, 1809	....
Distilling spirits from grain	W. H. Ford, J. D. Burns, and L. C. Clarke.	New Orleans, La	Aug. 11, 1868	80, 860
Distilling spirits from grain and fruit	J. Hugus	Westmoreland County, Pa	May 17, 1828	....
Distilling spirits, Method of	A. Whiston	Staunton, Va	Sept. 28, 1825	....
Distilling spirits of turpentine	C. J. Meinicke	New York, N. Y	July 30, 1850	7, 528
Distilling spirits of turpentine, Apparatus for	D. Cashwell	Fayetteville, N. C	Nov. 19, 1867	70, 956
Distilling spirits of turpentine, Apparatus for	D. Reid	Washington, N. C	May 25, 1858	20, 371
Distilling spirits, Process of	J. Ellingwood	Owensborough, Ky	Oct. 5, 1869	95, 574
Distilling spirits, Process of	C. B. Jarvis	New York, N. Y	Apr. 27, 1869	89, 318
Distilling spirituous liquors	W. J. Cocke	Cabin Point, Va	Feb. 4, 1828	....
Distilling spirituous liquors	J. Simpson	....	Mar. 4, 1794	....
Distilling spirituous liquids, Apparatus for	O. Lugo	New York, N. Y	Jan. 21, 1868	73, 539
Distilling spirituous liquors, Apparatus for	W. Schilling	Baltimore, Md	Mar. 10, 1868	75, 470
Distilling, Steam	B. Barr	Strasburgh Township, Lancaster County, Pa.	Apr. 5, 1828	....
Distilling together spirits of turpentine and alcoholic liquids.	W. Corfield	Chicago, Ill	Feb. 25, 1873	136, 137
Distilling-tub for essential oils	B. P. Van Marter	Lyons, N. Y	Jan. 19, 1869	86, 115
Distilling turpentine	D. Cashwell	Wilmington, N. C	May 10, 1870	102, 774
Distilling turpentine	R. F. Smith	Charleston, S. C	Oct. 2, 1847	5, 320
Distilling turpentine, Apparatus for	L. Bellingrath, jr	Fayetteville, N. C	June 1, 1858	20, 465
Distilling turpentine, &c., Apparatus for	C. T. Burcey	Black Rock, Conn	May 30, 1871	115, 429
Distilling turpentine, Apparatus for	A. K. Lee	Galveston, Tex	Aug. 1, 1871	117, 549
Distilling turpentine, Apparatus for	A. K. Lee	Galveston, Tex	Sept. 24, 1872	131, 550
Distilling turpentine, Apparatus for	A. H. Van Bokkelen	Wilmington, N. C	Aug. 22, 1871	118, 305
Distilling turpentine, Apparatus for	J. E. Winants and J. F. Griffen	New York, N. Y	May 28, 1867	65, 147
Distilling turpentine from wood	G. R. H. Leffler	Baltimore, Md	Sept. 27, 1864	44, 435
Distilling turpentine, tar, &c	W. A. Rapp	New York, N. Y	Dec. 4, 1815	....
Distilling volatile hydrocarbons and other substances, Apparatus for.	C. M. James	New York, N. Y	June 26, 1869	86, 232
Distilling whisky and other spirits	J. S. Oliver and E. Harris	New York, N. Y	Oct. 19, 1869	96, 029
Distilling whisky and other spirits, Apparatus for	H. C. Dayton	Maysville, Ky	Feb. 16, 1869	87, 029
Distilling whisky from corn	D. Samson	Perrysburgh, N. Y	Oct. 1, 1830	....
Distilling with cast-iron stills, Mode of	J. Coppinger	Beaufort, S. C	Nov. 21, 1809	....
Distilling wood, &c., Portable apparatus for	G. E. Mills	New York, N. Y	Mar. 29, 1864	42, 101
Distributing-pipe, Hot-air	H. C. Crehore	Boston, Mass	Jan. 31, 1871	111, 483
Distributing table	G. Whitaker	Lewistown, Ill	Nov. 20, 1866	59, 883
Ditches, Construction of	T. K. Grube	Napoleon, Ohio	Aug. 23, 1870	106, 690
Ditches, Covering for blind	T. M. C. Lutes	New Mount Pleasant, Ind	Mar. 9, 1869	87, 694
Ditches, Machine for cutting open	J. N. Smith and W. O. Buckley	Washington, Ill	Aug. 11, 1868	80, 880
Ditcher	W. Robbins	Hancock County, Ill	June 22, 1869	91, 771
Ditcher	C. F. Woodruff	Newbern, Tenn	Feb. 22, 1870	100, 231
Ditcher and grader	E. L. Foreman	Rantoul, Ill	Nov. 23, 1869	97, 183
Ditcher and grader	J. H. Martin and C. D. Bradshaw.	Danville, Ill	Sept. 13, 1870	107, 393
Ditcher and grader	J. G. Sisson and L. Delana	Arcola, Ill	Feb. 22, 1870	100, 199
Ditcher and pipe-layer combined	J. I. Mettler	Mendota, Ill	July 25, 1871	117, 441
Ditching and boring machine	I. B. Jones	Xenia, Ohio	Oct. 6, 1868	82, 846
Ditching and embanking earth, Machine for	H. Cleveland	Fort Wayne, Ind	Dec. 31, 1842	2, 897

Index of patents issued from the United States Patent Office from 1790 *to* 1873, *inclusive*—Continued.

Invention.	Inventor.	Residence.	Date.	No.
Ditching and embanking machine	W. C. Bussey	Rock Grove, Ill	Sept. 3, 1846	4,740
Ditching and embanking machine	L. Thorn	Washington, D. C	Oct. 10, 1838	975
Ditching and grading machine	F. H. De Tray	Odin, Ill	Jan. 21, 1868	73,441
Ditching and grading machine	I. and J. Hoskins	Wilmington, Ohio	Oct. 25, 1859	25,899
Ditching and grading machine	F. M. Howard and D. W. Avery	Saint Paul, Ind	May 5, 1868	77,487
Ditching and grading machine	L. D. Noble	Cerro Gordo, Ill	Apr. 1, 1873	137,318
Ditching and grading machine	W. J. Wanchope	Brookfield, Ill	Feb. 19, 1867	62,171
Ditching and grading machine	H. A. Winter	Windsor, Ill	Nov. 8, 1870	109,091
Ditching and grading machine	W. S. Worley	Tuscola, Ill	Feb. 5, 1867	61,908
Ditching and hedging machine	W. Stacy	Hardin County, Iowa	July 26, 1870	105,858
Ditching and tile-laying machine	B. P. Foster and W. H. Chaffee	Flint, Mich	Oct. 15, 1861	33,506
Ditching and tile-laying machine	M. M. Hooton	Centralia, Ill	Feb. 6, 1872	123,479
Ditching and tile-laying machine, Combined	O. W. Voorhis, S. M. Mapes, and W. M. Voorhis.	Lawrence, Ill	May 31, 1870	103,801
Ditching, grading, &c., Machine for	W. Provines	Columbus, Mo	Jan. 4, 1859	22,512
Ditching-machine	I. V. Adair	Varick, N. Y	Nov. 12, 1867	70,772
Ditching-machine	P. W. Adaire	Hays' Creek, Miss	June 5, 1860	28,544
Ditching-machine	J. S. Anderson and J. B. Cooley	Clark's Hill, Ind	Sept. 7, 1869	94,539
Ditching-machine	A. S. Ballard	Mount Pleasant, Iowa	Aug. 14, 1860	29,647
Ditching-machine	J. Ballard and T. J. Magee	New Antioch and Cincinnati, Ohio.	Sept. 4, 1866	57,814
Ditching-machine	J. W. Barcroft	Friendship, Va	Oct. 25, 1859	25,875
Ditching-machine	J. W. Barcroft	Friendship, Va	Mar. 13, 1860	27,419
Ditching-machine	E. Barnes	Chelsea, Mich	May 19, 1868	77,949
Ditching-machine	C. Bartholomew	Ætna, N. Y	Sept. 14, 1869	94,696
Ditching-machine	I. A. Benedict and G. W. Cummings.	West Springfield, Pa., and Conneaut, Ohio.	May 15, 1860	28,247
Ditching-machine	I. A. Benedict and G. W. Cummings.	Conneaut, Ohio	Mar. 12, 1861	31,653
Ditching-machine	H. Benett	Linden, Cal	Nov. 9, 1869	96,540
Ditching-machine	S. Bentley	Green Oak, Mich	Aug. 10, 1869	93,401
Ditching-machine	U. Blickensderfer	Springfield, Pa	Aug. 23, 1870	106,653
Ditching-machine	I. Boas	New Orleans, La	May 2, 1871	114,395
Ditching-machine	J. C. Boyd	Rushville, Ind	Mar. 20, 1866	53,264
Ditching-machine	C. M. Brown	Baltimore, Md	June 10, 1873	139,765
Ditching-machine	M. E. Burtless	Seneca Falls, N. Y	May 25, 1869	90,495
Ditching-machine	E. T. Bussell	Indianapolis, Ind	Dec. 12, 1871	121,846
Ditching-machine	C. Bymer and J. Imlay	Greensburgh, Ind	Nov. 26, 1867	71,275
Ditching-machine	J. Calliham	Baton Rouge, La	Mar. 8, 1870	100,597
Ditching-machine	H. Carter	Aylmer, Canada	Nov. 23, 1869	97,164
Ditching-machine	H. Carter	Aylmer, Canada	Nov. 7, 1871	120,711
Ditching-machine	H. Carter	Aylmer, Canada	Apr. 2, 1872	125,167
Ditching-machine	W. Cline, jr	Clayton, Ind	Dec. 7, 1869	97,606
Ditching-machine	R. Conarroe	Camden, Ohio	Nov. 19, 1867	70,960
Ditching-machine	R. Conarroe	Camden, Ohio	July 20, 1869	92,707
Ditching-machine	T. R. Cook	West Newton, Ind	Apr. 17, 1866	53,950
Ditching-machine	A. W. Cox	Dublin, Ind	May 3, 1864	42,563
Ditching-machine	A. W. Cox	Dublin, Ind	Aug. 8, 1865	49,235
Ditching-machine	I. R. Crane	Warsaw, Mo	Dec. 27, 1859	26,566
Ditching-machine	W. Crawford	Streator, Ill	Aug. 10, 1869	93,419
Ditching-machine	R. Cummings	Lima, Ind	July 14, 1845	4,113
Ditching-machine	W. H. Dalbey	Clarksburgh, Ind	Aug. 8, 1865	49,239
Ditching-machine	A. C. Decker	Bushnell, Ill	Mar. 31, 1868	76,172
Ditching-machine	O. Doolittle	Dansville, N. Y	Sept. 23, 1862	36,511
Ditching-machine	O. Doolittle and H. Eldridge	Dansville, N. Y	Feb. 12, 1861	31,376
Ditching-machine	T. B. Fagan	Mendon, Ohio	May 26, 1868	78,198
Ditching-machine	J. W. Fauver	Augusta County, Va	Aug. 26, 1873	142,222
Ditching-machine	H. Felthoff and L. D. Tingley	Prince William, Ind	May 25, 1869	90,517
Ditching-machine	L. W. Fisher	Rockville, Ind	Dec. 17, 1872	133,974
Ditching-machine	E. L. Foreman	Rantoul, Ill	Apr. 21, 1868	77,024
Ditching-machine	R. G. Forsyth	Clayton, Ind	May 31, 1870	103,595
Ditching-machine	O. and O. S. Foster	Durhamville, N. Y	Oct. 8, 1872	131,947
Ditching-machine	C. Freese	Aurelius, N. Y	Apr. 3, 1866	53,737
Ditching-machine	A. A. Fuselier	Algiers, La	Feb. 4, 1868	74,074
Ditching-machine	S. F. Gard	New Orleans, La	Oct. 11, 1870	108,248
Ditching-machine	W. Gause	Greensborough, Ind	Feb. 11, 1868	74,338
Ditching-machine	A. Gifford and I. Seright	Milroy, Ind	Sept. 17, 1867	68,975
Ditching-machine	H. Gonellaz	Vermillionville, La	July 15, 1873	140,820
Ditching-machine	O. F. Hale	Irvington, Iowa	Oct. 10, 1871	119,759
Ditching-machine	H. L. Hall	Buffalo, N. Y	Dec. 7, 1869	97,504
Ditching-machine	E. Heath	Fowlerville, N. Y	Oct. 2, 1866	58,413
Ditching-machine	M. C. and G. B. Higgins	Flemington, N. J	June 14, 1870	104,153
Ditching-machine	E. and B. Holmes	Buffalo, N. Y	Oct. 13, 1857	18,398
Ditching-machine	W. T. Hoskins	Livia, Ky	Feb. 11, 1873	135,807
Ditching-machine	J. W. Humphreys	La Fayette, Ind	Jan. 21, 1873	134,991
Ditching-machine	J. W. Humphreys	Oxford, Ind	Dec. 23, 1873	145,736
Ditching-machine	H. C. Ingraham	Tecumseh, Mich	Oct. 23, 1866	59,028
Ditching-machine	H. C. Ingraham	Tecumseh, Mich	Dec. 17, 1867	72,399
Ditching-machine	G. E. Inman	Buffalo, N. Y	Nov. 22, 1859	26,189
Ditching-machine	G. Ives	Detroit, Mich	July 9, 1861	32,772
Ditching-machine	G. Jensen	Brooklyn, N. Y	Oct. 16, 1866	58,834
Ditching-machine	S. F. Jones	Saint Paul, Ind	Aug. 8, 1865	49,276
Ditching-machine	N. Kidder	Moscow, Iowa	June 24, 1862	35,736
Ditching-machine	J. King	Indianapolis, Ind	Feb. 18, 1868	74,540
Ditching-machine	A. La Tourrette	Waterloo, N. Y	Oct. 8, 1867	69,682
Ditching-machine	N. B. Lewis	Hopewell, N. Y	Sept. 14, 1869	94,901
Ditching-machine	P. Lugenbell	Greensburgh, Ind	Aug. 14, 1866	57,100
Ditching-machine	P. Lugenbeil	Greensburgh, Ind	Apr. 20, 1869	89,227
Ditching-machine	P. Lugenbell and J. S. Armstrong.	Greensburgh, Ind	Nov. 21, 1865	51,067
Ditching-machine	J. Marsh	Seneca, Ill	Oct. 20, 1868	83,296
Ditching-machine	J. H. Marshall and O. E. Mann	Lockport and Somerset, N. Y.	Nov. 27, 1866	60,025
Ditching-machine	E. C. Martin	Muscatine, Iowa	Feb. 26, 1861	31,551
Ditching-machine	J. Masters	Waukegan, Ill	May 22, 1860	28,387
Ditching-machine	J. Masters	Waukegan, Ill	Apr. 14, 1868	76,788
Ditching-machine	C. Metheny	Greensburgh, Ind	June 25, 1867	66,159
Ditching-machine	R. C. Mauck	Harrisonburgh, Va	Mar. 13, 1855	12,514

Index of patents issued from the United States Patent Office from 1790 *to* 1873, *inclusive*—Continued.

Invention.	Inventor.	Residence.	Date.	No.
Ditching-machine	J. W. McAlister and J. C. Roffenberger.	Jacksonville, Ill	Feb. 27, 1872	124, 074
Ditching-machine	J. W. McGehee	Fayetteville, Tex	Nov. 16, 1869	96, 824
Ditching-machine	S. P. McKelvy	Farmer City, Ill	Apr. 5, 1870	101, 640
Ditching-machine	W. D. McKinney	Marion, Ind	May 4, 1869	89, 676
Ditching-machine	J. W. Metz	Stouts, Ohio	Oct. 3, 1871	119, 633
Ditching-machine	H. E. Moon and J. Doan	Wilmington, Ohio	Oct. 16, 1866	58, 871
Ditching-machine	J. W. Morrill	Hampton Falls, N. H	May 10, 1853	9, 709
Ditching-machine	G. W. Nevill	Richmond, Va	Jan. 25, 1870	99, 102
Ditching-machine	G. W. Nevill	Richmond, Va	Jan. 30, 1872	123, 286
Ditching-machine	G. W. Nevill	Richmond, Va	Mar. 11, 1873	136, 752
Ditching-machine	W. A. Nichols	West Liberty, Iowa	Mar. 31, 1868	76, 237
Ditching-machine	L. D. Noble	Cerro Gordo, Ill	June 21, 1870	104, 631
Ditching-machine	P. O'Connor and M. Collins	Decatur, Ill	July 14, 1868	79, 851
Ditching-machine	J. C. Osgood	Troy, N. Y	Sept. 27, 1870	107, 710
Ditching-machine	R. R. Osgood	Troy, N. Y	Sept. 20, 1870	107, 529
Ditching-machine	A. B. Overbaugh	Newark, N. J	Mar. 2, 1869	87, 357
Ditching-machine	E. Owen	La Porte, Ind	Sept. 24, 1844	3, 757
Ditching-machine	J. A. Parsons	Cleveland, Ohio	July 12, 1870	105, 364
Ditching-machine	W. M. Perkins	La Fontaine, Ind	Jan. 2, 1866	51, 859
Ditching-machine	W. Pierce	Truxton, N. Y	Mar. 8, 1870	100, 664
Ditching-machine	R. C. Pratt	Canandaigua, N. Y	July 19, 1853	9, 860
Ditching-machine	J. P. Rees and R. A. Graham	Greensburgh, Ind	May 15, 1866	54, 823
Ditching-machine	J. R. Remington	Lowndes County, Ala	May 28, 1846	4, 539
Ditching-machine	B. Rhett	Abbeville, S. C	July 18, 1871	117, 235
Ditching-machine	L. Rickard	Danville, Ill	Nov. 17, 1868	84, 136
Ditching-machine	J. W. Roberts	Hartford City, Ind	Nov. 15, 1870	109, 251
Ditching-machine	D. Sawyer	Webster, Ohio	Oct. 23, 1866	59, 140
Ditching-machine	F. B. Scott	Buffalo, N. Y	Jan. 24, 1860	26, 934
Ditching-machine	W. R. Selfridge	Greensburgh, Ind	July 21, 1868	80, 092
Ditching-machine	W. Y. Singleton	Springfield, Ill	May 28, 1846	4, 535
Ditching-machine	J. H. Snyder	Killbuck, Ill	Apr. 11, 1865	47, 229
Ditching-machine	A. Spencer	Grampian Hills, Pa	Nov. 12, 1872	132, 929
Ditching-machine	R. Starbuck	Wilmington, Ohio	Feb. 4, 1873	135, 605
Ditching-machine	N. Starbuck	Wilmington, Ohio	Aug. 1, 1865	49, 166
Ditching-machine	N. Starbuck	Wilmington, Ohio	June 18, 1867	65, 841
Ditching-machine	A. J. Stephens	Milford, Ill	Aug. 2, 1870	105, 995
Ditching-machine	J. Stevens	Chicago, Ill	Aug. 7, 1866	57, 050
Ditching-machine	G. H. Stevenson	Washington, Ohio	May 12, 1868	77, 930
Ditching-machine	B. T. Stowell	Waddam's Grove, Ill	Aug. 22, 1848	5, 719
Ditching-machine	T. J. Stratton	Waterloo, N. Y	Jan. 15, 1856	14, 117
Ditching-machine	G. Sullivan	West Liberty, Ohio	Jan. 8, 1867	61, 114
Ditching-machine	F. Taylor	Indianapolis, Ind	Nov. 14, 1871	120, 836
Ditching-machine	E. W. Thomas	Chicago, Ill	Mar. 27, 1847	5, 042
Ditching-machine	S. Thrailkill	La Fayette, Ind	Sept. 10, 1846	4, 751
Ditching-machine	B. Tobias	Washington, Ill	May 27, 1862	35, 403
Ditching-machine	S. E. Todd	Brooklyn, N. Y	Nov. 12, 1872	133, 069
Ditching-machine	J. Valentine	Buffalo, N. Y	Dec. 12, 1871	121, 737
Ditching-machine	H. Vannatta	Jefferson, Ill	Oct. 11, 1870	108, 217
Ditching-machine	J. Vaughn and E. Chamness	Miami County and Grant County, Ind.	Apr. 7, 1868	76, 359
Ditching-machine	I. S. Weaver	Dayton, Ohio	July 10, 1866	56, 299
Ditching-machine	C. O., D., and E. West, J. R. Smith, J. Carey, G. Janney, R. Hunt, A. Hackett, and J. Garner.	Martinsville, Ohio	May 15, 1860	28, 320
Ditching-machine	J. W. and M. H. Weston	Windsor, Ill	Nov. 30, 1869	97, 464
Ditching-machine	D. Whisler	Union Township, Ohio	Dec. 31, 1867	72, 953
Ditching-machine	A. H. and T. S. Whitacre	Morrow, Ohio	Dec. 10, 1867	72, 140
Ditching-machine	D. Whitesell	Mattoon, Ill	Sept. 19, 1871	119, 209
Ditching-machine	G. W. Wiggin	Exeter, N. H	June 7, 1864	43, 060
Ditching-machine	I. M. Williams	Blanchester, Ohio	Feb. 18, 1868	74, 653
Ditching-machine	S. S. Wood	Brooklyn, N. Y	Nov. 22, 1870	109, 480
Ditching-machine	S. S. Wood	Brooklyn, N. Y	Aug. 8, 1871	117, 848
Ditching-machine	A. Woolfolk	Iberville Parish, La	Mar. 6, 1860	27, 404
Ditching-machine	E. Worth and C. A. Davis	Oswego, N. Y	Jan. 21, 1868	73, 560
Ditching machine, Mole	I. Hodgson	New Michigan, Ill	July 24, 1860	29, 269
Ditching machine, Mole	A. Peterson	Stockwell, Ind	Aug. 23, 1870	106, 721
Ditching machine, Railway	W. Chadwick and S. I. B. Anderson.	Terre Haute, Ind	Nov. 2, 1858	21, 942
Ditching-machine, Steam	A. J. Dye	New Orleans, La	Mar. 22, 1870	101, 112
Divan and bed	H. Buckler	New York, N. Y	June 11, 1867	65, 640
Dividers	A. A. Cook	Milford, Mass	Dec. 12, 1871	121, 850
Dividers	E. S. Fisher	Boston, Mass	Nov. 26, 1867	71, 370
Dividers	H. Gerecke	Carlstadt, N. J	June 11, 1872	127, 686
Dividers	C. M. Nichols	West Greenwich, R. I	Oct. 10, 1871	119, 875
Dividers	S. A. Shurtleff	Taunton, Mass	Sept. 26, 1865	50, 173
Dividers	O. E. Weatherhead	Winchendon, Mass	Dec. 17, 1867	72, 433
Dividers	H. Whiting	New York, N. Y	Apr. 30, 1867	64, 391
Dividers and calipers	G. W. Lane	Plantsville, Conn	Nov. 1, 1870	108, 799
Dividers and calipers	G. L. McKnight	Worcester, Mass	Apr. 30, 1867	64, 240
Dividers and calipers	S. Sawyer	Fitchburgh, Mass	Apr. 9, 1867	63, 656
Dividers and calipers	E. S. Scripture	Brooklyn, N. Y	Jan. 7, 1868	73, 126
Dividers, Extension	G. C. Miller	Detroit, Mich	Nov. 11, 1873	144, 468
Dividers, Mathematical	J. E. Earle	Leicester, Mass	Jan. 5, 1858	19, 058
Dividers, Mathematical	A. Schaefer	New York, N. Y	Mar. 9, 1858	19, 589
Dividers or compasses	D. H. Chamberlain	Boston, Mass	Apr. 9, 1850	7, 257
Dividers or measuring-compasses	C. Reiffel and N. Thorn	New York, N. Y	Nov. 21, 1848	5, 929
Dividers, Pencil-attachment for	W. Schollhorn	New Haven, Conn	June 17, 1873	140, 081
Dividers, Proportional	H. M Parkhurst	Perth Amboy, N. J	Dec. 25, 1855	13, 993
Dividers, Spring	J. W. Strange	Bangor, Me	Sept. 11, 1860	30, 009
Dividing-engine, Circular and straight-lined	R. Tyler	Philadelphia, Pa	Sept. 10, 1828	....
Dividing-engine, Variable dial for	W. H. Brown	Worcester, Mass	Jan. 15, 1856	14, 082
Diving-apparatus	T. C. McKeen	Dunkirk, N. Y	Sept. 29, 1863	40, 114
Diving-apparatus	T. C. McKeen	Irvington, N. J	June 11, 1867	65, 760
Diving-apparatus	W. H. Taylor	New York	Jan. 20, 1838	578
Diving-apparatus	G. Williamson	Brooklyn, N. Y	Sept. 22, 1857	18, 260
Diving-apparatus	C. Wilson	Bridgeport, Conn	Sept. 19, 1871	119, 210

Index of patents issued from the United States Patent Office from 1790 *to* 1873, *inclusive*—Continued.

Invention.	Inventor.	Residence.	Date.	No.
Diving-bell	H. C. Billings	Brooklyn, N. Y	Dec. 14, 1869	97, 862
Diving-bell	E. W. Foreman	New Rochelle, N. Y	Aug. 23, 1853	9, 965
Diving-bell	B. Maillefert	Astoria, N. Y	Mar. 30, 1858	19, 785
Diving-bell	J. R. Wooster	Baltimore, Md	Apr. 24, 1849	6, 397
Diving-bell, Deep-sea	J. A. Richards and I. W. Wolcott.	Boston, Mass	Apr. 3, 1849	6, 250
Diving-dress	C. Hall	Connecticut	Dec. 24, 1810	....
Diving-machine	R. Tripe	Dover, N. H	Apr. 1, 1806	....
Diving-suit	W. C. Davison and D. Clark	Pittsford, N. Y	July 26, 1824	....
Diving-suit	F. Smith and L. S. Steel	Norfolk, Va	Apr. 23, 1834	....
Diving suit	N. Wolcott	Sidney, N. Y	Apr. 17, 1834	....
Diving-suit, Elastic	L. Norcross	Dixfield, Me	June 14, 1834	....
Divitials and preventing forgery, Making	C. S. Rafinesque	Lexington, Ky	Aug. 23, 1825	....
Dock and wharf fender	J. Moomey	Clinton, Iowa	June 19, 1860	28, 765
Dock, Buoyant	M. Isnard	New York, N. Y	June 22, 1826	....
Dock, Dry	G. A. Albertson	New York, N. Y	Mar. 14, 1871	112, 525
Dock, Dry	D. Brown	New York, N. Y	Feb. 11, 1832	....
Dock, Dry	R. Bulkley	New York, N. Y	Aug. 28, 1821	....
Dock, Dry	G. H. Ferris	Brooklyn, N. Y	June 28, 1870	104, 843
Dock, Dry	J. L. Gilbert	New York, N. Y	May 12, 1840	1, 606
Dock, Dry	J. Houston, A. Hinman, and J. Ingraham.	Buffalo, N. Y	June 22, 1836	....
Dock, Dry	C. F. Johnson	Tioga, N. Y	Mar. 6, 1840	1, 506
Dock, Dry	W. Jones	New York, N. Y	May 3, 1870	102, 554
Dock, Dry	C. Krogh	Kroghsville, Wis	Sept. 8, 1863	39, 818
Dock, Dry	I. J. Merritt	New York, N. Y	Dec. 10, 1867	72, 066
Dock, Dry	J. W. Munger and W. O. Jones	Portland, Me	Aug. 28, 1866	57, 550
Dock, Dry	A. Place	New York, N. Y	Apr. 1, 1826	....
Dock, Dry	J. Richardson	Germantown, Va	Feb. 23, 1869	87, 291
Dock, Dry	J. Ryan	Saint Louis, Mo	Oct. 10, 1865	50, 390
Dock, Dry	E. Selden	Haddam, Conn	July 30, 1829	....
Dock, Dry	J. E. Simpson	East Boston, Mass	Dec. 5, 1854	12, 032
Dock, Dry	H. K. Wagner	Saint Louis, Mo	Nov. 10, 1846	4, 842
Dock, Floating	J. Campbell	London, England	Nov. 9, 1869	96, 672
Dock, Floating	S. Janicki	St. Petersburg, Russia	Feb. 6, 1872	123, 402
Dock, Floating	S. Janicki	St. Petersburg, Russia	Apr. 30, 1872	126, 146
Dock, Floating	A. R. McNair	Saint Louis, Mo	Sept. 17, 1872	131, 365
Dock, Floating	E. Turner	Baltimore, Md	Apr. 25, 1865	47, 501
Dock, Floating	H. Van Keuren	Jersey City, N. J	July 18, 1871	117, 223
Dock, Floating	O. T. Williams	Smithland, Ky	Feb. 24, 1852	8, 765
Dock, Floating dry	J. Adamson	Boston, Mass	Dec. 13, 1816	....
Dock, Floating dry	J. Barron	Philadelphia, Pa	July 17, 1837	272
Dock, Floating dry	A. Brux	Augusta, Ga	Nov. 6, 1818	....
Dock, Floating dry	J. R. Campbell and J. S. Withington.	Boston, Mass	Nov. 28, 1835	....
Dock, Floating dry	E. Covenhoven	Greenburgh, N. Y	Oct. 30, 1821	....
Dock, Floating dry	D. Dodge and P. Burgess	New York, N. Y	Oct. 9, 1841	2, 281
Dock, Floating dry	J. Floyd	Kittery, Me	Aug. 17, 1826	....
Dock, Floating dry	J. G. Gilbert	New York, N. Y	Mar. 25, 1840	1, 524
Dock, Floating dry	J. G. Gilbert	New York, N. Y	July 28, 1842	2, 742
Dock, Floating dry	J. Hawes	Cayuga County, N. Y	Apr. 22, 1835	....
Dock, Floating dry	S. Hill	Jersey City, N. J	Feb. 1, 1842	2, 442
Dock, Floating dry	S. Loveland	Oswego, N. Y	Feb. 5, 1847	4, 954
Dock, Floating dry	J. T. Martin	New York, N. Y	Mar. 19, 1840	1, 518
Dock, Floating dry	R. Moody and S. D. Dakin	New York, N. Y	Nov. 24, 1843	3, 351
Dock, Floating dry	R. Porter	Bellerica, Mass	Nov. 14, 1835	....
Dock, Floating dry	S. Rose	New York, N. Y	June 15, 1812	....
Dock, Floating dry	J. Thomas	Saint Louis, Mo	Mar. 26, 1834	....
Dock, Floating dry	J. Thomas	New York, N. Y	Dec. 20, 1837	522
Dock, Floating dry	J. Thomas	New York, N. Y	June 26, 1841	2, 149
Dock, Floating dry	W. Thomas	Saint Louis, Mo	May 19, 1840	1, 609
Dock, Floating graving	T. Cunningham	Pittsburgh, Pa	Oct. 1, 1830	....
Dock, Floating screw	S. F. Stinchfield	New Orleans, La	Apr. 27, 1832	....
Dock, Floating sectional dry	J. Seely	Buffalo, N. Y	Feb. 24, 1857	16, 696
Dock, Floating tide	W. Pickard	Jersey City, N. J	July 26, 1870	105, 725
Dock for cleansing the bottoms of vessels, Floating dry.	W. Rhodes	New York, N. Y	May 16, 1808	....
Dock for raising and removing vessels, Floating dry.	J. S. Gilbert	New York, N. Y	Sept. 19, 1840	1, 792
Dock for repairing vessels, Dry	C. Bergh	New York, N. Y	Apr. 29, 1825	....
Dock for repairing vessels, &c., Floating dry	W. R. Loweree	New York, N. Y	June 23, 1825	....
Dock, Fresh-water dry	L. H. Clarke and L. M. Wiss	New York, N. Y	Dec. 22, 1825	....
Dock, Graving	W. A. Kenrick and G. H. Whitcher.	Boston, Mass	Dec. 20, 1859	26, 501
Dock, Hydraulic	Z. Ring	New York, N. Y	Feb. 13, 1836	....
Dock, Hydrostatic suspension	T. Evans and J. Parsons	New York, N. Y	May 13, 1831	....
Dock indicator, Dry	P. F. Campbell	Jersey City, N. J	July 31, 1866	56, 702
Dock, Lifting	Z. P. Leach	Danbury, Conn	Jan. 3, 1865	45, 724
Dock, Marine dry	H. I. Crandall	New Bedford, Mass	June 12, 1860	28, 658
Dock, Marine elevating	E. Buck, H. Crehore, R. Gerry, and A. Bayard.	Boston, Mass	July 11, 1825	....
Dock, Rotary dry	J. Webb	New York, N. Y	May 3, 1817	....
Dock, Screw	J. C. Ely	New York, N. Y	Jan. 28, 1829	....
Dock, Screw	E. Turner	Rochester, N. Y	Oct. 10, 1829	....
Dock, Sectional dry	S. Loveland	Astoria, N. Y	Mar. 7, 1854	10, 600
Dock, Ship-railway	J. Thomas	New York, N. Y	Feb. 24, 1826	....
Docks and harbors, Machine for clearing	S. Colver		Dec. 14, 1798	....
Docks and marine-railway, Adjustable cradle for dry	W. Van Deusen	Philadelphia, Pa	Dec. 14, 1858	22, 317
Docks, Basin to be used in connection with floating dry.	R. Moody and S. D. Dakin	New York, N. Y	Sep. 17, 1844	3, 745
Docks, Bilge-supporter for holding vessels in	H. I. Crandall	Boston, Mass	Dec. 19, 1854	12, 093
Docks, Building	A. W. Thompson and J. Pattison	New York, N. Y	Mar. 4, 1873	136, 560
Docks, Carrying and lifting trunk	J. Barron	Norfolk, Va	Aug. 2, 1826	....
Docks, &c., Construction of	S. J. Seely	New York, N. Y	Jan. 31, 1865	46, 146
Docks, Erecting dry	J. Gardiner		Dec. 3, 1802	....
Docks, Machine for cleaning out	J. Greenleaf		July 13, 1802	....
Docks, Method of holding vessel by the keel in dry and other	J. Smith	Neponset Village, Mass	Dec. 19, 1854	12, 104

Index of patents issued from the United States Patent Office from 1790 *to* 1873, *inclusive*—Continued.

Invention.	Inventor.	Residence.	Date.	No.
Docks, Mode of adjusting vessel upon the keel-block of dry sectional or railway.	H. I. Crandall	East Boston, Mass	Nov. 14, 1854	11, 932
Docks, Mode of constructing projecting	M. Rogers	New York	Jan. 7, 1817	....
Docks or harbors, Machine for cleaning	P. Zacharie	....	Nov. 24, 1791	....
Document case	C. S. Trevitt	Washington, N. C	May 30, 1871	115, 542
Document holder and filer	M. L. M. Hussey	New York, N. Y	Aug. 7, 1866	56, 941
Dog kennel or house	S. S. Bent	Port Chester, N. Y	Jan. 5, 1869	85, 557
Dog-power	F. Cole	Conesus, N. Y	July 14, 1863	39, 263
Dog-power	A. W. Hager and J. H. S. Grove	Waverly, Iowa	Nov. 24, 1868	84, 354
Dog-power	A. Tieman	New York	Apr. 27, 1818	....
Dog-power, Operating machinery by	D. C. Slater	Laurens, N. Y	June 7, 1859	24, 338
Doll	M. Kintzback	Philadelphia, Pa	Oct. 5, 1869	95, 489
Doll	J. Lacman	Philadelphia, Pa	Apr. 4, 1871	113, 532
Doll	H. C. Work	Brooklyn, N. Y	July 8, 1873	140, 605
Doll, Creeping	G. P. Clarke	New York, N. Y	Aug. 29, 1871	118, 435
Doll, Creeping	R. J. Clay	New York, N. Y	Mar. 14, 1871	112, 550
Doll, Fancy	D. Checkeni	Marion, Conn	Feb. 20, 1866	52, 782
Doll-head	G. Benda	Coburg, Germany	July 16, 1872	129, 086
Doll-heads and other toys, Mode of constructing	L. E. Sallee	Decatur, Ill	Feb. 7, 1865	46, 270
Doll-heads, Constructing	L. Greiner	Philadelphia, Pa	Mar. 30, 1858	19, 770
Doll-heads, Construction of	G. H. Hawkins	New York, N. Y	Sept. 8, 1868	81, 999
Doll-heads, Manufacture of	G. H. Hawkins	New York, N. Y	Jan. 5, 1869	85, 589
Doll-joint	J. A. H. Ellis	Springfield, Vt	May 20, 1873	139, 130
Doll, Walking	A. W. Nicholson	Brooklyn, N. Y	Mar. 23, 1869	88, 197
Dolls, Arms for	C. F. Blakeslee	New York, N. Y	Jan. 3, 1865	45, 691
Dolls, Manufacture of	F. E. Darrow	Bristol, Conn	May 1, 1866	54, 301
Dolls, Process for manufacturing	W. Rumbold	Saint Louis, Mo	June 17, 1862	35, 630
Dome	I. F. Walker	Central Falls, R. I	Nov. 4, 1873	144, 373
Dome, bridge, &c., Mode of uniting timber for	W. Armesley	Albany, N. Y	Nov. 6, 1830	....
Domestic boiler	F. Till	Reading, Pa	Dec. 19, 1871	121, 973
Domino	J. W. Hyatt, jr	Albany, N. Y	June 15, 1869	91, 234
Domino	B. Louinean	Sables d'Olonnes, France	Oct. 14, 1873	143, 703
Domino	W. X. Stevens	East Brookfield, Mass	June 17, 1873	139, 928
Dominos, Manufacture of	J. W. Hyatt, jr	Albany, N. Y	June 15, 1869	91, 235
Dominos, &c., Treatment of wood for the manufacture of.	G. H. Chinnock	Brooklyn, N. Y	Dec. 10, 1872	133, 697
Donkey, Mechanical	A. and W. Shedlock	New York, N. Y	Feb. 25, 1868	74, 772
Door	E. R. Ball	Oswego, N. Y	Apr. 5, 1832	....
Door	B. F. Barker	Belfast, Me	June 20, 1871	116, 010
Door	O. C. Hill	Malone, N. Y	June 19, 1855	13, 090
Door	G. M. McMahan	Mount Sterling, Ky	Apr. 14, 1868	76, 790
Door-alarm	H. Behn	New York, N. Y	Oct. 2, 1860	30, 193
Door-alarm	H. Curtis and A. Tufts	Charlestown, Mass	Apr. 9, 1861	31, 957
Door and blind fastening	D. Bull	Amboy, Ill	Sept. 17, 1867	68, 947
Door and chest lock, Construction of	R. Wilson	Burdett, N. Y	May 10, 1838	734
Door and gate bolt	J. Smith	Southold, N. Y	Dec. 2, 1873	145, 131
Door and gate closer	H. N. Conklin	Indianapolis, Ind	Oct. 6, 1868	82, 804
Door and gate fastening, Barn	L. B. Hayes and W. Norris	Greene, N. Y	Nov. 5, 1867	70, 562
Door and gate spring	S. L. Bean and L. Mills	Galesburgh, Ill	Mar. 4, 1873	136, 482
Door and gate spring	J. T. Foster	Jersey City, N. J	Apr. 15, 1873	137, 710
Door and gate spring	J. C. Gould	Boonton, N. J	Oct. 2, 1866	58, 405
Door and gate spring	P. Kern	Dayton, Ohio	June 20, 1871	116, 196
Door and gate spring	C. N. Owen	Salem, Ohio	Apr. 8, 1873	137, 619
Door and gate spring	C. W. Saladee	Saint Catharine's, Canada	Feb. 20, 1872	123, 843
Door and gate spring	J. Stephenson	Canandaigua, N. Y	Jan. 1, 1867	60, 954
Door and gate spring	E. Stinson	Montpelier, Vt	Nov. 19, 1867	71, 081
Door and gate spring	M. F. Taber	Salem, Ohio	July 4, 1871	116, 644
Door and gate spring	S. D. Tuttle	Eaton, Ohio	Oct. 1, 1872	131, 914
Door and other locks	A. Prutzmann	Philadelphia, Pa	Sept. 5, 1840	1, 773
Door and other locks	C. F. Voorkies	Newark, N. J	June 9, 1843	3, 125
Door and other locks	R. Wilcox	New York	Apr. 20, 1824	....
Door and other locks, Manner of closing and opening the key-hole of.	D. Evans	Philadelphia, Pa	July 10, 1841	2, 166
Door and package holder	J. K. McDonald	Newark, N. J	Mar. 31, 1868	76, 224
Door and safe lock	J. Gillingham	Turner, Me	Sept. 8, 1868	81, 988
Door and sash clamp	E. F. Dunaway	Cincinnati, Ohio	Mar. 24, 1868	75, 880
Door and sash guide, Sliding	G. L. Waitt	Philadelphia, Pa	May 9, 1871	114, 731
Door and shutter	S. G. Cabell	Quincy, Ill	Nov. 13, 1866	59, 553
Door and shutter bolt	J. Feldman	New York, N. Y	Mar. 21, 1865	46, 891
Door and shutter bolt	G. B. Green	Philadelphia, Pa	June 21, 1870	104, 449
Door and shutter fastener	J. Anser	Mount Vernon, N. Y	Dec. 15, 1868	84, 984
Door and shutter fastening and buffer, Combined	W. C. Wendell	Philadelphia, Pa	June 27, 1871	116, 379
Door and shutter, Metallic	C. K. Marshall	New Orleans, La	May 26, 1868	78, 218
Door and shutter spring	L. Bommer	New York, N. Y	Mar. 31, 1863	38, 023
Door and trunk lock	A. Beals	Exeter, N. H	Jan. 26, 1830	....
Door and window bolt	G. Smith	Philadelphia, Pa	Dec. 18, 1839	1, 435
Door and window fastener	W. I. Corthell and P. Richards	Boston, Mass	Nov. 12, 1867	70, 699
Door and window fastener	J. Z. Davis	San Francisco, Cal	May 23, 1871	115, 176
Door and window fastener	H. Wilkinson	Newburgh, N. Y	Sept. 13, 1864	44, 245
Door and window frames, Machine for making	W. W. Maughlin	Baltimore, Md	May 14, 1867	64, 779
Door and window guard	J. W. White	Weymouth, Mass	Dec. 23, 1873	145, 772
Door and window locking device	A. H. Boyd	Rockville, Mass	Oct. 15, 1867	69, 753
Door and window tightener and fastener	D. Conlon	Portland, Me	Jan. 12, 1869	85, 729
Door apparatus, Sliding	J. Capron	New York, N. Y	May 17, 1870	103, 139
Door-bell	J. Harrison	East Hampton, Conn	Apr. 22, 1873	138, 151
Door-bell	W. M. Preston	Roxbury, N. Y	Sept. 2, 1873	142, 412
Door-bolt	G. T. Allamby	Bangor, Me	Aug. 28, 1866	57, 616
Door-bolt	S. D. Arnold	New Britain, Conn	Apr. 29, 1873	138, 360
Door, &c., bolt	W. F. Arnold and O. L. Steele	New Britain, Conn	Apr. 20, 1869	89, 191
Door-bolt	N. W. Bonney	Victoria, Tex	June 26, 1860	28, 827
Door-bolt	C. Chevallier	New York, N. Y	Aug. 1, 1865	49, 085
Door-bolt	J. H. Cohen, W. A. Pennoyer, and W. H. Crosby.	Catskill, N. Y	Aug. 13, 1872	130, 483
Door-bolt	J. M. Crosby	Norwalk, Ohio	May 5, 1857	17, 204
Door-bolt	F. Engelbrecht	Memphis, Tenn	May 16, 1871	114, 782
Door-bolt	L. J. Evans	Binghamton, N. Y	Nov. 7, 1871	120, 729
Door-bolt	A. P. Floyd	Niagara Falls, N. Y	Nov. 21, 1865	51, 034
Door-bolt	B. E. Fowler	Hartford, Conn	Jan. 15, 1867	61, 185

Index of patents issued from the United States Patent Office from 1790 *to* 1873, *inclusive*—Continued.

Invention.	Inventor.	Residence.	Date.	No.
Door-bolt	W. H. Hart	New Britain, Conn	July 4, 1865	48, 555
Door-bolt	W. H. Hart	New Britain, Conn	Sept. 26, 1865	50, 120
Door-bolt	W. H. Hart	New Britain, Conn	May 31, 1870	103, 606
Door-bolt	L. G. Hoffman	Waterford, N. Y	Sept. 6, 1859	25, 335
Door-bolt	A. G. Hofstatter	New York, N. Y	Apr. 15, 1873	137, 841
Door-bolt	C. C. Jones	Portland, Me	Dec. 17, 1867	72, 302
Door-bolt	J. Jones	New York, N. Y	Mar. 25, 1873	137, 211
Door-bolt	J. Levy	New York, N. Y	Feb. 1, 1870	99, 330
Door-bolt	D. McArthur	New Haven, Conn	Nov. 1, 1870	108, 920
Door-bolt	M. McGonnigle	Allegheny City, Pa	June 20, 1865	48, 297
Door-bolt	C. G. Page	Washington, D. C	Dec. 20, 1859	26, 518
Door-bolt	J. E. Parker	West Meriden, Conn	Aug. 8, 1865	49, 296
Door-bolt	A. Pelham	Plymouth, Mich	Mar. 18, 1873	136, 861
Door-bolt	G. A. Pridham	Newark, N. J	Nov. 12, 1867	70, 895
Door-bolt	R. B. Prindle	Norwich, N. Y	July 19, 1870	105, 489
Door-bolt	B. Russell	Brooklyn, N. Y	Apr. 9, 1861	32, 002
Door-bolt	J. B. Sargent	New Haven, Conn	June 6, 1871	115, 645
Door-bolt	J. B. Sargent	New Haven, Conn	June 18, 1872	128, 074
Door-bolt	A. H. Sherwood	Southport, Conn	Sept. 17, 1867	69, 033
Door-bolt	H. S. Smith	Brooklyn, N. Y	Feb. 10, 1863	37, 646
Door-bolt	S. B. Swartz and J. M. Opdycke	Lumberville, Pa	June 8, 1869	91, 180
Door-bolt	A. F. Warren	Brooklyn, N. Y	June 29, 1869	92, 130
Door-bolt	A. Westcott	Syracuse, N. Y	June 2, 1857	17, 464
Door-bolt	A. Westcott	Syracuse, N. Y	Mar. 7, 1865	46, 740
Door-bolt	W. Wheeler	Troy, N. Y	Oct. 23, 1847	5, 338
Door-bolt	S. R. Wilmot	Wartertown, Conn	June 2, 1857	17, 474
Door-bolt	G. M. Wood	Decatur, Ill	Aug. 14, 1866	57, 241
Door-bolt	J. Woolman	Philadelphia, Pa	Apr. 6, 1858	19, 891
Door-bolt	J. S. R. Wright	Nichols, N. Y	Apr. 15, 1873	137, 994
Door-bolt and lock	C. Sulzman	Waterford, N. Y	Apr. 14, 1868	76, 671
Door-bolt and weather-strip	A. Newcomb	Shipman, Ill	Jan. 26, 1869	86, 242
Door-bolt, Automatic	A. McKay	Montreal, Canada	Dec. 27, 1870	110, 486
Door bolt, Barn	W. Campbell	Belleville, Mich	May 21, 1872	127, 024
Door bolt, Barn	M. R. Green	Warwick, N. Y	Oct. 4, 1870	108, 017
Door-bolt, Cylindrical	C. G. Page	Washington, D. C	July 21, 1857	17, 843
Door-bolt for safes, &c	O. Gallagher	Boston, Mass	Sept. 14, 1869	94, 735
Door-bolt, Signal	C. Page	West Meriden, Conn	May 17, 1859	24, 051
Door-bolt, Spring	G. F. Atkinson	Seymour, Conn	Oct. 29, 1867	70, 149
Door-bolts, Adjustible keeper for	G. W. Davis	Brooklyn, N. Y	Jan. 26, 1869	86, 217
Door-bolts, Locking cylindrical	C. G. Page	Washington, D. C	July 14, 1857	17, 808
Door buffer and catch	A. A. Veer	Delaware, Ohio	Nov. 29, 1870	109, 692
Door-button	G. F. Beardsley	Binghamton, N. Y	Nov. 28, 1871	121, 271
Door-button	O. F. Fogelstrand	Kensington, Conn	Jan. 28, 1873	135, 215
Door-button	T. Lincoln and G. Hubbard	New Haven, Conn	Mar. 23, 1869	88, 186
Door-button	W. E. Sparks	New Haven, Conn	Mar. 24, 1868	75, 996
Door-button	H. Underwood	Tolland, Conn	Sept. 27, 1864	44, 474
Door-call	W. E. Sparks	New Haven, Conn	July 2, 1872	128, 565
Door-catch	A. Blood, jr	Independence, Iowa	Nov. 19, 1867	70, 945
Door-catch	J. M. Crosby	Norwalk, Ohio	June 9, 1857	17, 490
Door-catch	W. H. Fowler	Newburgh, N. Y	Nov. 6, 1866	59, 380
Door-catch	R. V. Phillips	Council Bluffs, Iowa	Nov. 15, 1870	109, 342
Door-catch and cushion	S. Wagner	Galion, Ohio	May 4, 1869	89, 819
Door-check	C. R. Anderson	Saint Louis, Mo	Mar. 14, 1871	112, 673
Door-check	J. Bader	Olathe, Kans	Nov. 4, 1873	144, 246
Door-check	C. Bird	Lower Merion Townsihp, Pa	Feb. 28, 1871	112, 210
Door-check	W. O. Clough	Lexington, Ky	May 7, 1872	126, 445
Door-check	L. Goodyear	Trumansburgh, N. Y	Aug. 5, 1873	141, 435
Door-check	A. Hanna	Dover, Ky	Jan. 21, 1873	135, 115
Door-check	L. S. Hicks	Peoria, Ill	Mar. 7, 1871	112, 455
Door-check	G. C. Idly	Burlingame, Kans	Mar. 11, 1873	136, 731
Door-check	J. P. Israel and D. W. McLane	Oil City, Pa	June 24, 1873	140, 139
Door-check	W. H. Johnson, sr	Clayton, Ind	Nov. 16, 1869	96, 924
Door-check	J. and A. Kolb	Oxford, Ind	Jan. 28, 1873	135, 346
Door-check	H. N. H. Lugrin	Chelsea, Mass	Jan. 17, 1871	110, 985
Door-check	J. A. McNabb	Olathe, Kans	Aug. 5, 1873	141, 577
Door-check	G. W. Pagett	Oxford, Ind	June 6, 1871	115, 768
Door-check	M. R. Perkins	Portsmouth, N. H	Nov. 4, 1873	144, 222
Door-check	S. Peters and C. D. Eisaman	Penn Station, Pa	Apr. 30, 1872	126, 328
Door-check	J. Pool	Elizabeth City, N. C	Oct. 31, 1871	120, 533
Door-check	B. Poulson	Forte Wayne, Ind	Mar. 25, 1873	137, 237
Door-check	G. Ramsay	Clyde, Ohio	Oct. 8, 1872	131, 970
Door-check	G. E. Rittenhouse	Akron, Ohio	May 20, 1873	139, 082
Door-check	G. Rohrbaker	Penn Station, Pa	Jan. 7, 1873	134, 612
Door-check	D. C. Smart	Cambridgeport, Mass	Aug. 15, 1871	118, 063
Door-check	H. D. Weever	Mount Vernon, Ind	July 23, 1872	129, 875
Door-check	C. P. Young	Attleborough, Miss	Nov. 8, 1870	109, 096
Door check and holder	S. L. Hart	Menasha, Wis	May 17, 1870	103, 044
Door check and retainer	G. H. Worster	Athol, Mass	May 6, 1873	138, 551
Door-clamp	H. O. Hooper	Diamond Springs, Cal	July 13, 1869	92, 531
Door-clamp	H. O. Hooper	Diamond Springs, Cal	Mar. 7, 1871	112, 342
Door-closing apparatus	M. T. Cooper	Ballston Spa, N. Y	Apr. 6, 1852	8, 855
Door-closing apparatus	H. F. Shaw	West Roxbury, Mass	Aug. 19, 1873	141, 896
Door-closing device	D. I. Stagg	New York, N. Y	Jan. 7, 1862	34, 075
Door-closing fixture	W. Wilson	Northampton, Mass	May 16, 1848	5, 579
Door-closing mechanism	H. Smith	Boston, Mass	Nov. 13, 1866	59, 667
Door-directory	L. Burger	Springfield, Ill	May 5, 1868	77, 450
Door, Double-acting	W. Rippon	Providence, R. I	July 6, 1852	9, 102
Door-fastener	L. C. Bignall	Lockport, N. Y	May 31, 1864	42, 924
Door-fastener	H. Burt	Newark, N. J	Aug. 16, 1859	25, 090
Door-fastener	J. M. Clark	Lancaster, Pa	Dec. 22, 1868	85, 165
Door-fastener	E. H. Crane	Jonesville, Mich	Feb. 21, 1865	46, 451
Door-fastener	L. Crofoot	Syracuse, N. Y	Nov. 11, 1856	16, 048
Door-fastener	W. F. Davis and C. E. Broad	Boston and Milton, Mass	Apr. 27, 1869	89, 295
Door-fastener	G. V. Hazard	Torrey, N. Y	Dec. 20, 1859	26, 494
Door-fastener	W. A. Howard	Dugway, N. Y	Sept. 12, 1871	118, 942
Door-fastener	H. M. Jones	West Meriden, Conn	Sept. 28, 1869	95, 231
Door-fastener	S. P. Kittle	Buffalo, N. Y	June 7, 1853	9, 765
Door-fastener	J. Letort	Wytheville, Va	Dec. 23, 1856	16, 282

Index of patents issued from the United States Patent Office from 1790 *to* 1873, *inclusive*—Continued.

Invention.	Inventor.	Residence.	Date.	No.
Door-fastener	F. C. Levalley	Warnerville, N. Y	Dec. 10, 1867	72, 053
Door-fastener	J. Lightfoot	Cold Spring, Ky	May 8, 1860	28, 186
Door-fastener	G. H. Lindner	Hoboken, N. J	Apr. 8, 1856	14, 594
Door-fastener	F. McDonough	Chicago, Ill	Nov. 6, 1866	59, 431
Door-fastener	D. E. McDougall	Troy, N. Y	May 31, 1853	9, 753
Door-fastener	J. H. McElroy and J. H. Holly	Warwick, N. Y	Dec. 29, 1868	85, 321
Door-fastener	G. W. McGill	Buffalo, N. Y	Apr. 13, 1858	19, 937
Door-fastener	J. A. Morris	Greenbush, N. Y	July 4, 1871	116, 621
Door-fastener	E. P. Moulton	Baltimore, Md	Apr. 29, 1856	14, 773
Door-fastener	H. Orcutt	Amherst, Wis	Aug. 12, 1873	141, 724
Door-fastener	G. F. Perkins and S. F. Gibbs	Holyoke, Mass	May 5, 1868	77, 651
Door-fastener	W. Quayle	Warsaw, Ill	May 11, 1869	89, 942
Door-fastener	J. O. Reilley	Baltimore, Md	Dec. 22, 1868	85, 243
Door-fastener	W. J. Ross	Worcester, Mass	Dec. 8, 1868	84, 767
Door-fastener	G. W. Sayre	Pisgah, Ohio	Aug. 29, 1865	49, 656
Door-fastener	W. H. Smith	New York, N. Y	May 6, 1856	14, 834
Door-fastener	H. N. Taft	Sag Harbor, N. Y	May 14, 1867	64, 808
Door-fastener	G. E. Thompson	New Haven, Conn	Feb. 6, 1866	52, 469
Door-fastener	G. Yates	West Dresden, N. Y	Dec. 28, 1858	22, 469
Door-fastener, Adjustable	L. P. Barnes	Fitchburgh, Mass	Jan. 10, 1871	110, 821
Door-fastener and key-ring combined	B. H. Melendy	Manchester, N. H	May 17, 1870	103, 068
Door-fastener and pocket-knife combined	B. F. Porter	Manchester, N. H	Nov. 5, 1867	70, 609
Door-fastener, Mortise-latch	L. Foster	Boston, Mass	Aug. 28, 1841	2, 231
Door-fastener, Portable	B. R. Eames	South Newry, Me	Apr. 10, 1855	12, 675
Door-fastener, Portable	G. W. Griswold	Carbondale, Pa	Sept. 5, 1854	11, 646
Door-fastener, Portable	E. H. Janney and E. J. Hamilton.	Fairfax, Va	June 5, 1866	55, 305
Door-fastener, Portable	L. C. Johnson	Buffalo, N. Y	June 28, 1859	24, 603
Door-fastener, Portable	J. Pepper	Lake Village, N. H	Aug. 6, 1867	67, 578
Door-fastening	M. Bennett	Kilbourne City, Wis	Jan. 1, 1867	60, 673
Door-fastening	A. Call	Springfield, Mass	Mar. 7, 1846	4, 407
Door-fastening	J. Decker	Sparta, N. J	Mar. 5, 1867	62, 532
Door-fastening	C. Frost	Waterbury, Conn	July 26, 1859	24, 859
Door-fastening	W. Hartsfield	Thomaston, Ga	July 5, 1859	24, 632
Door-fastening	L. G. Hoffman	Waterford, N. Y	Nov. 22, 1859	26, 185
Door-fastening	N. Hornady	West Elkton, Ohio	Nov. 12, 1867	70, 846
Door-fastening	F. H. and A. J. Illingworth	Weston, Mass	Jan. 16, 1866	52, 050
Door-fastening	J. C. Palmer	East Haddam, Conn	Oct. 11, 1845	4, 229
Door-fastening	R. Peck	Cortlandville, N. Y	Jan. 15, 1856	14, 112
Door-fastening	S. M. Pye	Acquackanocnk, N. J	Mar. 28, 1848	5, 492
Door-fastening	F. Reed	Brattleborough, Vt	Nov. 13, 1866	59, 653
Door-fastening	C. Washburn	Bridgewater, Mass	Apr. 18, 1832	....
Door-fastening and alarm	A. F. Kitchen	Shelton Depot, S. C	Nov. 24, 1868	84, 362
Door-fastening and alarm, Combined	O. Fisher	Smyrna, Del	June 6, 1871	115, 724
Door-fastening and knife	A. W. Hall	New York, N. Y	July 23, 1867	67, 044
Door fastening, Barn	A. D. Wright and A. Roys	Belleville, Mich	Jan. 21, 1873	135, 027
Door fastening, Chamber	J. P. Frazer	New York, N. Y	Sept. 30, 1862	36, 563
Door fastening, Double	G. H. Lindner	Hoboken, N. J	June 15, 1858	20, 570
Door, Fastening for folding	E. L. Roberts	Brooklyn, N. Y	Mar. 15, 1859	23, 264
Door fastening, Gate and barn	W. W. Peck	Cassopolis, Mich	Dec. 24, 1867	72, 536
Door-fastening, Mortise	L. Foster	Boston, Mass	Feb. 25, 1831	....
Door-fastening, Portable	D. Conlon	Portland, Me	June 19, 1866	55, 624
Door-fastening, Portable	L. S. Enos	Almond, N. Y	June 12, 1866	55, 477
Door-fastening, Spring-catch	J. Buck	Bucksport, Me	Sept. 28, 1831	....
Door-fastenings, Mode of operating the bolt of	A. Bingham	Boston, Mass	June 5, 1844	3, 611
Door-fender	W. Cole and I. Johnson	Randolph, Mass	Mar. 27, 1829	....
Door-fender	S. Durfee	Providence, R. I	Oct. 1, 1830	....
Door for buildings, Safety	C. Hesse	Champaign, Ill	Aug. 24, 1869	94, 109
Door for churches, Safety	A. H. Wagner	Pimlico, England	May 10, 1864	42, 719
Door for guarding against cold	D. Williams, jr	Colchester, Conn	Sept. 11, 1828	....
Door for steam-boilers, Water	I. W. Ayres	New York, N. Y	Oct. 31, 1848	5, 888
Door flush-bolt	S. D. Arnold	New Britain, Conn	June 10, 1873	139, 647
Door gong	C. W. Penfield	New Britain, Conn	Oct. 28, 1873	143, 996
Door guard	C. W. Gochwind	Port Republic, N. J	May 17, 1870	103, 042
Door guard	N. C. Perry and G. S. Gladding	Chester, Conn	Feb. 6, 1866	52, 497
Door guard	J. Tinkey	New Haven, Conn	Oct. 16, 1866	58, 916
Door guard, Elastic	W. N. Clark	Chester, Conn	Nov. 17, 1857	18, 627
Door guard, Elastic	D. E. Peck	Burlington, Conn	Apr. 26, 1870	102, 427
Door hanger	A. C. Arnold	Norwalk, Conn	July 23, 1872	129, 777
Door hanger	S. L. Bignall	Chicago, Ill	Oct. 14, 1873	143, 557
Door hanger	T. Kent and J. S. Barker	Waukegan, Ill	Aug. 8, 1871	117, 894
Door-hanger	A. K. Rider	Walden, N. Y	July 29, 1873	141, 382
Door hanger	S. P. Robinson	Canterbury, Conn	June 7, 1870	104, 064
Door hanger and rail	T. F. Hamilton	Geneseo, Ill	Jan. 31, 1871	111, 342
Door hanger, Barn	W. R. Axe	Rockton, Ill	Mar. 1, 1870	100, 351
Door hanger, Barn	J. A. Cauldwell	Watkins, N. Y	July 2, 1872	128, 590
Door hanger, Barn	W. W. Soden	Unadilla, N. Y	July 16, 1872	129, 065
Door hanger, Sliding	C. B. Clark	Buffalo, N. Y	June 11, 1872	127, 846
Door hanger, Sliding	G. Rumsey	Watkins, N. Y	Apr. 4, 1871	113, 697
Door, Hatchway	M. S. Driggs	Brooklyn, N. Y	May 22, 1866	54, 874
Door-holder	G. C. Bunsen	Belleville, Ill	Sept. 1, 1868	81, 746
Door-holder	J. K. Clark	Mount Pleasant, Iowa	July 28, 1868	80, 390
Door-holder	C. J. Fisher	Waukon, Iowa	Nov. 12, 1867	70, 823
Door-holder	J. Hale	Scranton, Pa	Oct. 29, 1867	70, 333
Door-holder	J. J. Harris and I. H. Mosher	Greene, N. Y	June 11, 1867	65, 566
Door-holder	E. Huddart	Prairie du Lac, Wis	July 9, 1867	66, 591
Door-holder	E. Manley	Marion, N. Y	Mar. 16, 1869	87, 785
Door-holder	W. A. Messler	Eureka, Ill	Sept. 24, 1867	69, 235
Door-holder	E. Morris	Burlington, N. J	June 19, 1849	6, 549
Door-holder	W. Quayle	Warsaw, Ill	May 14, 1867	64, 704
Door-holder	O. Sliker	Lincoln, Ill	June 29, 1869	91, 978
Door-holder	G. Wells	Bethel, Conn	Dec. 3, 1867	71, 828
Door-holder	J. S. White	Prescott, Wis	Apr. 20, 1869	89, 101
Door holder or check	O. Sliker	Lincoln, Ill	Nov. 29, 1870	109, 771
Door-holder, Spring	J. B. Okey	Indianapolis, Ind	Jan. 28, 1868	73, 829
Door, Illuminating	T. Hyatt	New York, N. Y	Dec. 2, 1873	145, 183
Door indicator	A. Hadley	Washington, D. C	May 26, 1868	78, 278
Door indicator	F. E. Mills	San Francisco, Cal	May 21, 1867	64, 894

Index of patents issued from the United States Patent Office from 1790 *to* 1873, *inclusive*—Continued.

Invention.	Inventor.	Residence.	Date.	No.
Door, Iron	L. Hover	Chicago, Ill	Oct. 13, 1868	83,062
Door-key	W. H. Andrews	New Haven, Conn	Mar. 23, 1869	88,113
Door-key	J. Brady	Branford, Conn	June 7, 1870	103,837
Door-key	J. Brady	Branford, Conn	June 28, 1870	104,823
Door-key	C. C. Carpenter	Huntsville, Ala	Feb. 28, 1871	112,123
Door-key	J. Collins	Hohokus Township, N. J	Dec. 9, 1873	145,401
Door-key	H. H. Elwell	South Norwalk, Conn	Oct. 18, 1870	108,466
Door-key	F. Greene	Troy, Pa	June 30, 1868	79,466
Door-key	E. Parker	New Britain, Conn	May 4, 1869	89,596
Door-key attachment	D. E. Campbell	Boston, Mass	Apr. 30, 1867	64,281
Door-key fastener	S. Macferran	Philadelphia, Pa	July 3, 1866	56,071
Door-key, Folding	J. Seibert	Pittsburgh, Pa	Feb. 4, 1873	135,609
Door-key guard	H. A. Adams	New York, N. Y	Feb. 9, 1869	86,797
Door-key guard, Safety	W. K. Marvin	New York, N. Y	May 5, 1868	77,633
Door-keys from being turned, Device for preventing	A. Westcott	Syracuse, N. Y	Feb. 10, 1863	37,665
Door-knob fastening	W. H. Mackrell	New York, N. Y	Oct. 30, 1847	5,345
Door-knobs, Manufacture of	J. Ottner	Bridgeport, Conn	May 13, 1873	138,758
Door-knobs, Shank for mineral	J. Laird	Cincinnati, Ohio	May 22, 1849	6,473
Door-knocker with brass plate, Cast-iron	I. Wilson	New London, Conn	Mar. 11, 1831	......
Door-latch	W. T. Munger	New Britain, Conn	Dec. 6, 1870	109,834
Door-latch	J. H. Vickers	Norwich, Conn	May 31, 1870	103,819
Door-lock	C. Adams	Pittsburgh, Pa	July 3, 1860	28,949
Door-lock	H. Ahrend	New York, N. Y	Jan. 12, 1869	85,776
Door-lock	E. Allen and J. Brady	Norwich, Conn	Sept. 3, 1867	68,334
Door-lock	W. H. Andrews	New Haven, Conn	Apr. 5, 1870	101,564
Door-lock	S. Andrews	Perth Amboy, N. J	Jan. 11, 1836	......
Door-lock	T. B. Atterbury	Pittsburgh, Pa	June 2, 1857	17,412
Door-lock	J. Atwater	New Haven, Conn	July 24, 1846	4,659
Door-lock	J. R. Baker	Jersey City, N. J	Apr. 30, 1872	126,248
Door-lock	D. Ball	Kingsbury, N. Y	July 12, 1838	833
Door-lock	F. H. Bartholomew	New York, N. Y	Mar. 24, 1867	75,838
Door-lock	T. Bartholomew	New York, N. Y	Apr. 3, 1829	......
Door-lock	C. H. Beatty	Wheeling, Va	Oct. 14, 1851	8,426
Door-lock	J. Behel	Rockford, Ill	Feb. 11, 1868	74,283
Door-lock	J. Brady	Branford, Conn	Nov. 7, 1871	120,704
Door-lock	J. Brady	Brooklyn, N. Y	Sept. 24, 1872	131,654
Door-lock	E. W. Brettell	Newark, N. J	Mar. 26, 1867	63,204
Door-lock	J. Brown and G. W. Robinson	Providence, R. I	Feb. 20, 1827	......
Door-lock	J. H. Butterworth	Dover, N. J	Apr. 11, 1846	4,452
Door-lock	W. Cayce	Franklin, Tenn	Aug. 1, 1854	11,412
Door-lock	S. L. Chase	New York, N. Y	Mar. 26, 1867	63,214
Door-lock	H. Clarke	Baltimore, Md	July 20, 1869	92,796
Door-lock	A. Conant	Lowell, Mass	Mar. 12, 1836	......
Door-lock	J. H. Davis	Lumberville, Pa	Mar. 7, 1846	4,403
Door-lock	W. P. Dodson	Philadelphia, Pa	Mar. 12, 1872	124,424
Door-lock	H. H. Elwell	South Norwalk, Conn	Feb. 18, 1873	136,049
Door-lock	J. B. Erb	Strasburgh, Pa	Apr. 22, 1856	14,714
Door-lock	P. S. Felter	Cincinnatus, N. Y	Mar. 31, 1868	76,066
Door-lock	W. D. Field	Providence, R. I	Apr. 28, 1868	77,267
Door-lock	E. Finney	Cleveland, Ohio	June 18, 1839	1,173
Door-lock	C. Fleischel and W. C. Bussey	San Francisco, Cal	Dec. 8, 1868	84,737
Door-lock	M. B. Foote	Northampton, Mass	Dec. 1, 1868	84,484
Door-lock	M. B. Foote	Northampton, Mass	May 4, 1869	89,753
Door-lock	E. P. Fowler and C. J. Clements	Brooklyn and Mott Haven, N. Y.	June 22, 1869	91,733
Door-lock	V. Frazee	San Francisco, Cal	Sept. 3, 1872	131,093
Door-lock	S. Frye	Monongahela City, Pa	Apr. 30, 1861	32,218
Door-lock	J. E. A. Gibbs	Steele's Tavern, Va	June 2, 1868	78,449
Door-lock	H. R. Gillingham	Baltimore, Md	Sept. 10, 1867	67,622
Door-lock	C. G. Gumpel	Leicester Square, England	Nov. 16, 1869	96,802
Door-lock	F. Gyss	New York, N. Y	Oct. 10, 1871	119,758
Door-lock	J. J. Hamilton	Newcastle, Ind	Apr. 27, 1858	20,063
Door-lock	R. C. Harrington	Newark, N. J	June 16, 1868	78,962
Door-lock	W. I. and J. W. Harris	Newport, N. Y	May 21, 1867	64,975
Door-lock	J. S. Hoard and V. O. Spencer	Mansfield, Pa	Feb. 8, 1859	22,872
Door-lock	E. B. Horn	Boston, Mass	Sept. 25, 1849	6,733
Door-lock	H. Jackson	New York, N. Y	Dec. 24, 1867	72,499
Door-lock	J. W. Jones	Paducah, Ky	May 11, 1869	89,931
Door-lock	D. C. Jordon, sr	Brooklyn, N. Y	Mar. 12, 1867	62,753
Door-lock	F. Just	Buffalo, N. Y	Feb. 12, 1867	61,036
Door-lock	J. Kinger	Pittsburgh, Pa	Nov. 13, 1860	30,654
Door-lock	J. H. Kinsman	Salem, Mass	June 27, 1871	116,452
Door-lock	J. Kinzer	Pittsburgh, Pa	Sept. 14, 1858	21,504
Door-lock	J. C. Kline	Pittsburgh, Pa	June 12, 1855	13,037
Door-lock	M. and J. Knapp	Hudson City, N. J	Nov. 5, 1867	70,576
Door-lock	C. Knauer	Birmingham, Pa	June 17, 1856	15,136
Door-lock	T. Kromer	Freiburg, Baden	July 1, 1873	140,377
Door-lock	J. Kyle	New York, N. Y	May 18, 1832	......
Door-lock	J. Kyle	Ramapo, N. Y	Sept. 16, 1833	......
Door-lock	J. Kyle	Ramapo, N. Y	Sept. 16, 1833	......
Door-lock	F. F. Landis	Lancaster, Pa	Mar. 5, 1867	62,545
Door-lock	A. Leich	Brooklyn, N. Y	Mar. 5, 1867	62,645
Door-lock	J. Linder	Seneca Falls, N. Y	June 28, 1870	104,749
Door-lock	H. Lockwood	New York, N. Y	May 1, 1860	28,092
Door-lock	W. Maguire	Cincinnati, Ohio	Jan. 15, 1850	7,026
Door-lock	J. R. Marston	New York, N. Y	June 15, 1858	20,571
Door-lock	W. Maurer	New York, N. Y	May 29, 1855	12,957
Door-lock	J. McClory	New York, N. Y	Mar. 18, 1836	......
Door-lock	J. McClory	New York, N. Y	July 2, 1836	......
Door-lock	R. McDowell	Lambertville, N. J	Apr. 6, 1869	88,577
Door-lock	A. McKinnon	New York, N. Y	Aug. 16, 1845	4,151
Door-lock	D. V. Miller	Weedsport, N. Y	Apr. 20, 1869	89,161
Door-lock	J. F. Milligan	Saint Louis, Mo	June 1, 1869	90,864
Door-lock	W. Moore	Williamsburgh, N. Y	Sept. 14, 1852	9,265
Door-lock	W. T. Munger	New Britain, Conn	Apr. 18, 1871	113,910
Door-lock	H. Norton	Delton, Wis	Mar. 17, 1868	75,569
Door-lock	G. Oates	Charleston, S. C	Sept. 2, 1845	4,175
Door-lock	T. A. Olson	Beloit, Wis	June 20, 1871	116,214

Index of patents issued from the United States Patent Office from 1790 to 1873, inclusive—Continued.

Invention.	Inventor.	Residence.	Date.	No.
Door-lock	E. Parker	New Britain, Conn	July 16, 1872	129, 587
Door-lock	J. E. Parker	West Meriden, Conn	Nov. 6, 1860	30, 586
Door-lock	W. Parr and J. F. Fowler	New York, N. Y	Nov. 19, 1833	....
Door-lock	A. Patterson	Pittsburgh, Pa	Apr. 8, 1856	14, 618
Door-lock	N. Petré	New York, N. Y	Sept. 7, 1869	94, 639
Door-lock	E. P. Porter and G. W. Hallett	Waterford, N. Y	Mar. 26, 1867	63, 295
Door-lock	E. P. Porter and G. W. Hallett	Waterford, N. Y	Nov. 26, 1867	71, 411
Door-lock	S. M. Pye	Acquackanonck, N. J	Mar. 13, 1849	6, 184
Door-lock	W. Pye	New York, N. Y	Apr. 28, 1818	....
Door-lock	J. M. Quimby and A. S. Dyer	Westbrook and Cape Elizabeth, Me.	Feb. 21, 1871	112, 076
Door-lock	W. Reynolds	Greenbriar County, Va	May 9, 1848	5, 557
Door-lock	B. F. Roberts	Lacona, Iowa	Jan. 17, 1871	111, 005
Door-lock	P. Rodgers	Philadelphia, Pa	Sept. 25, 1840	1, 796
Door-lock	A. Roff	New York, N. Y	July 1, 1836	....
Door-lock	R. Schade	New York, N. Y	Nov. 26, 1867	71, 534
Door-lock	I. P. Sherwood	Sandy Hill, N. Y	Dec. 17, 1842	2, 886
Door-lock	J. C. Silvy	New Orleans, La	June 12, 1855	13, 050
Door-lock	B. Smith	Canton, Conn	Apr. 13, 1836	....
Door-lock	D. M. Smith	Springfield, Vt	July 14, 1846	4, 635
Door-lock	J. G. Spathelf	Sandusky, Ohio	July 28, 1868	80, 369
Door-lock	M. R. Stephenson	Boston, Mass	July 13, 1852	9, 126
Door-lock	W. Stephenson	Cincinnati, Ohio	Sept. 12, 1848	5, 758
Door-lock	W. Stephenson	Cincinnati, Ohio	Sept. 12, 1848	5, 766
Door-lock	C. I. Stewart	Baltimore, Md	Sept. 12, 1871	118, 828
Door-lock	B. Van Dyke	Chicago, Ill	Aug. 20, 1872	130, 609
Door-lock	R. Vollschwitz	New York, N. Y	Jan. 1, 1867	60, 970
Door-lock	N. Warren	Wilmington, Del	Nov. 21, 1871	121, 142
Door-lock	W. Warwick	Birmingham, Pa	Mar. 13, 1855	12, 526
Door-lock	J. Weimar	New York, N. Y	Dec. 12, 1854	12, 081
Door-lock	J. Welsh	Canton, Ohio	Jan. 31, 1871	111, 499
Door-lock	W. P. Wentworth	Detroit, Mich	June 25, 1867	66, 059
Door-lock	J. L. Whetstone	Cincinnati, Ohio	July 20, 1846	4, 640
Door-lock	T. Whitehouse	Boston, Mass	Sept. 8, 1837	377
Door-lock	T. Whitehouse	Boston, Mass	June 14, 1838	783
Door-lock	A. F. Whiting	Bath, Me	Feb. 6, 1872	123, 531
Door-lock	L. Whitney	Toledo, Ohio	June 8, 1858	20, 524
Door-lock	J. T. Williams	Chicago, Ill	July 11, 1871	116, 907
Door-lock	A. Wolfe	Pittsburgh, Pa	Apr. 18, 1831	....
Door-lock	W. E. Woodbridge	New York, N. Y	July 14, 1846	4, 630
Door-lock	L. L. Woolley	Medford, Mass	May 19, 1868	78, 038
Door-lock	T. B. Worrell	Frankford, Pa	July 13, 1869	92, 502
Door-lock	L. Yale	Springfield, Mass	June 13, 1844	3, 630
Door-lock	L. Yale	Newport, N. Y	Oct. 18, 1853	10, 144
Door-lock-alarm attachment	J. Schneider	Rochester, N. Y	May 1, 1855	12, 784
Door lock and bolt, Sliding	T. Whaley	New York, N. Y	Nov. 2, 1821	....
Door lock and key	G. M. Zahm	Lancaster, Pa	May 26, 1842	2, 638
Door lock and latch	E. Halley	Branford, Conn	Sept. 1, 1863	39, 770
Door lock and latch	N. and C. Hunt	Cleveland, Ohio	July 29, 1839	1, 268
Door lock and latch	W. T. Mungeo	Branford, Conn	July 7, 1863	39, 203
Door lock and latch	G. W. Wilson	Nashua, N. H	June 11, 1841	2, 123
Door lock and latch, Sliding	E. J. Kehoe	New York, N. Y	Nov. 15, 1870	109, 220
Door-lock and other fastening	J. R. and H. C. Cambell	Charlestown, Mass	Dec. 28, 1835	....
Door-lock bolt	E. Parker	West Meriden, Conn	Feb. 7, 1860	27, 064
Door-lock, Burglar-alarm	E. Tracy	Kansas City, Mo	Dec. 3, 1867	71, 822
Door-lock by a combined key and gage, also a thief detector.	F. C. Goffin	Philadelphia, Pa	Mar. 10, 1849	6, 165
Door-lock by which one key-hole serves for two distinct keys.	A. Call	Springfield, Mass	Feb. 13, 1849	6, 109
Door-lock, Combination	M. R. Stephenson and O. Edwards.	Boston, Mass	Apr. 17, 1844	3, 543
Door-lock, Combination safety	S. Yale and C. Wilson	Springfield, Mass	Oct. 20, 1843	3, 312
Door-lock, Combination tumbler	R. M. Tuttle	Newark, N. J	Dec. 5, 1842	2, 869
Door-lock, Compound	G. Hubert	Lancaster, Pa	July 23, 1867	67, 116
Door-lock, Divided-bolt	J. Kyle	New York, N. Y	Aug. 15, 1848	5, 708
Door-look, Escutcheon latch and bolt	P. E. W. and J. A. Blake	New Haven, Conn	Dec. 31, 1833	....
Door-lock, Extension-bit guard-key for	W. Damerel	Brooklyn, N. Y	May 9, 1854	10, 883
Door-lock guard	W. Ballauff and F. Wurth	Cincinnati, Ohio	Mar. 6, 1855	12, 476
Door lock, Jail	E. Jacobs	Cincinnati, Ohio	Aug. 23, 1864	43, 915
Door-lock keeper	G. W. DaCunha	New York, N. Y	May 21, 1867	64, 847
Door-lock key	T. K. Webster	Lawrence, Mass	Nov. 17, 1857	18, 654
Door-lock key, Swivel-nibbed	A. C. Harig	Louisville, Ky	Mar. 1, 1853	9, 599
Door-lock, night-latch, &c	J. McClory	New York, N. Y	June 14, 1837	215
Door lock or latch	W. Hall	Chelsea, Mass	Oct. 18, 1843	3, 308
Door lock or latch	T. L. Littlefield	Philadelphia, Pa	Nov. 21, 1843	3, 343
Door lock or latch, Compound lever graduating	S. Noyes	Beverly, Mass	Feb. 20, 1843	2, 963
Door-lock or spring-catch	R. J. Byram	Boston, Mass	May 31, 1832	....
Door-lock, Portable	E. G. F. Arndt	Rondout, N. Y	Sept. 18, 1860	30, 035
Door-lock, Portable	J. W. Webb	Washington, D. C	Feb. 7, 1854	10, 511
Door lock, Prison	L. M. Ham	Boston, Mass	May 4, 1869	89, 830
Door-lock, Protector-slide for	G. F. I. Colburn	Newark, N. J	Feb. 27, 1849	6, 142
Door-lock, Safety	W. Stillman	Westerly, R. I	Sept. 14, 1839	1, 323
Door-lock, Sliding	A. W. Cram	Saint Louis, Mo	May 20, 1873	139, 118
Door-lock, Sliding	J. Davidson	Albany, N. Y	Nov. 1, 1870	108, 767
Door-lock strike or keeper	J. Scott	Philadelphia, Pa	Apr. 22, 1873	138, 048
Door-locks, Bolt-attachment to	W. H. Andrews	New Haven, Conn	Dec. 10, 1867	71, 944
Door-locks for banks, vaults, &c., Mode of constructing.	J. McClory	New York, N. Y	June 19, 1838	787
Door-locks, Keeper for	S. B. Williams	Leavenworth City, Kans	Dec. 19, 1865	51, 639
Door-locks, Keeper for right and left hand	C. Adams	Oak Hill, N. Y	Feb. 24, 1857	16, 676
Door locks or latches, Check-bolt of	A. Bingham	Boston, Mass	Oct. 17, 1842	2, 719
Door-locks, Safety key-holder for	F. J. May	Morrisania, N. Y	Mar. 12, 1867	62, 865
Door-mat	N. Berkeley	Aldie, Va	Apr. 4, 1871	113, 389
Door-mat	J. J. Diehl	Idaville, Pa	Feb. 2, 1869	86, 512
Door-mat	P. W. Neefus	New York, N. Y	May 30, 1871	115, 345
Door-mat	A. P. Noyes	Middleton, Mass	June 7, 1870	103, 916
Door-mat	F. V. Noyes	Mattoon, Ill	Sept. 5, 1865	49, 780
Door-mat	J. V. Oldaker	Atlanta, Ill	July 9, 1872	128, 897

Index of patents issued from the United States Patent Office from 1790 *to* 1873, *inclusive*—Continued.

Invention.	Inventor.	Residence.	Date.	No.
Door-mat	W. Young, jr	Franklin, Mass	July 28, 1868	80, 379
Door-mat, India-rubber	E. M. Chaffee	Providence, R. I	Feb. 16, 1858	19, 347
Door-mat, Metallic and elastic	P. W. Neefus	New York, N. Y	Nov. 29, 1870	109, 650
Door-numbers, Casting	J. T. Fuller	Louisville, Ky	Apr. 10, 1860	27, 861
Door or partition for separating rooms, &c., Drop	G. Kilburn	Walpole, N. H	Dec. 31, 1838	1, 052
Door or shutter fastener	W. Fields	Wilmington, Del	Feb. 20, 1872	123, 816
Door or shutter, Metallic	A. B. Mullett and B. Oertly	Washington, D. C	July 6, 1869	92, 205
Door-panel	L. W. Kimball	Pittsford, Vt	Aug. 11, 1868	80, 972
Door-panels, Composition for	R. B. Andrews	Poland, Me	Nov. 9, 1869	96, 655
Door-plate	J. W. Bliss	Hartford, Conn	July 13, 1858	20, 859
Door-plate	W. A. Caron	Springfield, Mass	Dec. 12, 1871	121, 703
Door-plate	J. H. Grout and F. M. Ray	New York, N. Y	Mar. 20, 1844	3, 496
Door-plate	J. S. Richardson	Boston, Mass	Oct. 15, 1836	56
Door-plate	J. S. Richardson	Boston, Mass	Oct. 15, 1836	....
Door-plate	E. I. Sause	New York	Aug. 11, 1843	3, 216
Door-plate and alarm	C. H. Miller	Buffalo, N. Y	Mar. 16, 1869	87, 958
Door-plate and bell-alarm	W. G. Brady	Lowell, Mass	June 29, 1869	91, 820
Door-plate and card-receiver	S. A. Baldwin	Waterbury, Conn	Sept. 9, 1862	36, 388
Door-plate and indicator	T. M. and W. F. Tobin	Middlebury, Vt	Apr. 1, 1873	137, 508
Door-plate and letter-box	E. A. Hopkins	Minneapolis, Minn	Dec. 24, 1867	72, 493
Door-plate and letter-box, Combined	J. T. Green	Marquette, Wis	Oct. 15, 1867	69, 800
Door-plate and letter-chute, Combined	B. Morningstar	Pittsburgh, Pa	Apr. 15, 1873	137, 946
Door-plate, Glass	D. Montague and J. Townsend	New Bedford, Mass., and Gardiner, Me.	May 22, 1860	28, 396
Door-plate, Index	N. Ames	Saugus Centre, Mass	July 31, 1860	29, 430
Door-plate, letter-box, and bell-pull combined	T. Thompson	Cleveland, Ohio	Oct. 22, 1872	132, 378
Door-plates and signs of separate types or cutters, Method of making.	E. Morris	Philadelphia, Pa	Apr. 25, 1844	3, 565
Door-plates, Casting	J. A. Pease	Philadelphia, Pa	Aug. 14, 1847	5, 243
Door-plates, Making and preparing	W. C. Austin	Greensville, Va	Jan. 31, 1837	113
Door-plates, &c., Number for	J. T. Smith	Washington, D. C	Jan. 18, 1870	99, 019
Door-retainer	C. T. Gravatt	Philadelphia, Pa	Mar. 8, 1870	100, 616
Door-retainer	G. W. Perry and J. D. Billins	Wilmington, Del	Dec. 8, [illegible]868	84, 756
Door-register	J. G. Miller	Swanton, Md	Mar. 16, 1858	19, 646
Door, safe, &c., Burglar-proof	G. Little	Rutherford Park, N. J	Apr. 1, 1873	137, 456
Door-securer	D. Arndt	Cleveland, Ohio	Dec. 19, 1871	121, 980
Door-securer	B. F. Baer	Lancaster, Pa	May 23, 1871	115, 012
Door-securer	W. H. Caldwell	Wheeling, W. Va	Mar. 7, 1871	112, 320
Door-securer	B. H. Melendy	Amherst, N. H	Mar. 12, 1872	124, 503
Door-securer	G. B. Pharo	Philadelphia, Pa	Nov. 12, 1872	132, 922
Door-securer	W. H. Phipps	Poquonock, Conn	Nov. 4, 1873	144, 223
Door-securer	S. A. Pool	Memphis, Tenn	May 21, 1872	126, 903
Door-securer and key-ring, Combined	J. P. Tuck	Houlton, Me	Nov. 18, 1873	144, 642
Door, Self-closing	J. C. Harkness	Washington, D. C	Oct. 12, 1858	21, 754
Door-sheave	R. G. Hatfield	New York, N. Y	June 21, 1864	43, 205
Door sheave, Sliding	M. L. Deering	New York, N. Y	Mar. 2, 1869	87, 329
Door sheave, Sliding	M. L. Deering	New York, N. Y	May 25, 1869	90, 345
Door sheave, Sliding	R. G. Hatfield	New York, N. Y	May 5, 1863	38, 387
Door-sill	D. Hitchcock and D. S. Trout	Arcola, Ill	Oct. 12, 1869	95, 799
Door-sill	J. Johnston and E. O. Marlow	Brodhead, Wis	June 14, 1870	104, 321
Door-sill, Adjustable	M. Armstrong	Girard, Ill	July 26, 1870	105, 623
Door-sill, Adjustable	J. H. Morris	Philadelphia, Pa	Mar. 30, 1869	88, 502
Door-sill gutter	J. A. Hemberger	Reading, Ohio	Oct. 26, 1869	96, 233
Door-sill, Self-adjusting	G. C. Bigelow	Worcester, Mass	Mar. 23, 1858	19, 673
Door-sills and door-strips, Arrangement of	H. Tryon	Steuben, Pa	May 5, 1857	17, 242
Door, Sliding	J. Capron	New York, N. Y	Mar. 1, 1870	100, 373
Door, Sliding	L. A. Cook	Concord, Mich	Nov. 1, 1870	108, 763
Door, Sliding	W. T. Forsyth	Philadelphia, Pa	Feb. 12, 1844	3, 428
Door, Sliding	J. More	New York, N. Y	Jan. 31, 1871	111, 468
Door, Sliding	E. Prudhomme and P. M. Leprohon.	New York, N. Y	May 18, 1869	90, 193
Door, Sliding	W. A. Waterhouse	Liverpool, N. Y	Mar. 22, 1870	101, 194
Door-spring	W. Allen	Oswego, N. Y	July 5, 1870	105, 025
Door-spring	W. Badger, jr	North Chelsea, Mass	Mar. 22, 1864	41, 971
Door-spring	G. L. Bailey	Portland, Me	Apr. 15, 1856	14, 686
Door-spring	G. L. Bailey	Portland, Me	Apr. 21, 1857	17, 070
Door-spring	A. T. Ballantine	Titusville, Pa	Apr. 19, 1870	102, 197
Door-spring	B. F. Barker	San Francisco, Cal	Mar. 27, 1866	53, 391
Door-spring	J. Barkley	Weston, Mo	Jan. 11, 1859	22, 540
Door-spring	W. S. Barlow	Paterson, N. J	Nov. 17, 186[illegible]	84, 1[illegible]9
Door-spring	O. D. Barrett	Cleveland, Ohio	Sept. 27, 1859	25, 602
Door-spring	A. Barton	Syracuse, N. Y	Apr. 1, 1856	14, 583
Door-spring	A. S. Blake	Waterbury, Conn	May 3, 1859	23, 819
Door-spring	J. O. Blythe	Germantown, Pa	Apr. 2, 1861	31, 862
Door-spring	J. Boyd and J. Belford	Philadelphia, Pa	Oct. 16, 1860	30, 383
Door-spring	J. Broughton	Chicago, Ill	Aug. 19, 1856	15, 555
Door-spring	J. Broughton	Chicago, Ill	Jan. 6, 1857	16, 324
Door-spring	I. Buckman, jr	Williamsburgh, N. Y	Nov. 5, 1872	132, 710
Door-spring	C. Burnham	Philadelphia, Pa	Dec. 24, 1867	72, 449
Door-spring	L. Burwell	Berlin, Conn	Nov. 25, 1829	....
Door-spring	A. F. Chatman	New York, N. Y	Jan. 27, 1857	16, 512
Door-spring	H. Cody	New York, N. Y	Nov. 4, 1873	144, 256
Door-spring	J. M. Connel	Newark, Ohio	Dec. 18, 1866	60, 481
Door-spring	J. M. Connel	Newark, Ohio	Feb. 9, 1869	86, 815
Door-spring	J. J. Corvell	Newark, N. J	May 28, 1872	127, 311
Door-spring	S. Cutler	Cincinnati, Ohio	Mar. 18, 1873	136, 974
Door-spring	D. Dick	New York, N. Y	June 3, 1873	139, 458
Door-spring	F. Dodge	Syracuse, N. Y	Sept. 21, 1869	94, 949
Door-spring	R. B. Donaldson	Washington, D. C	Oct. 30, 1860	30, 531
Door-spring	W. Duncan	Lebanon, N. H	Feb. 23, 1869	87, 095
Door-spring	W. Duryea	Glen Cove, N. Y	Aug. 11, 1868	80, 931
Door-spring	I. Eaton	Boston, Mass	Feb. 24, 1830	....
Door-spring	S. Elliott	Richmond, Ind	June 22, 1869	91, 528
Door-spring	C. Ellis	Boston, Mass	Dec. 15, 1843	3, 376
Door-spring	B. G. Fitzhugh	Frederick, Md	Mar. 1, 1870	100, 388
Door-spring	O. V. Flora	Madison, Ind	June 15, 1869	91, 222
Door-spring	J. T. Folwell	Camden, N. J	June 7, 1870	103, 859
Door-spring	H. S. Frost	Watertown, Conn	Aug. 6, 1867	67, 525

Index of patents issued from the United States Patent Office from 1790 *to* 1873, *inclusive*—Continued.

Invention.	Inventor.	Residence.	Date.	No.
Door-spring	F. Gardner	Boston, Mass	Jan. 2, 1872	122, 316
Door-spring	G. Geer	Plainville, Conn	Mar. 4, 1873	136, 371
Door-spring	W. Gilfillan	Syracuse, N. Y	July 30, 1867	67, 290
Door-spring	W. Gilfillan	Syracuse, N. Y	Apr. 6, 1869	88, 625
Door-spring	J. S. Gray	New York, N. Y	Jan. 11, 1859	22, 554
Door-spring	W. M. Gray	Brooklyn, N. Y	Dec. 5, 1871	121, 507
Door-spring	W. M. Gray	Brooklyn, N. Y	Dec. 5, 1871	121, 508
Door-spring	G. W. Griswold	Carbondale, Pa	Apr. 1[illegible], 1856	14, 691
Door-spring	L. J. Higgins	Mount Desert, Me	Apr. 1, 1873	137, 367
Door-spring	L. Hillebrand	Philadelphia, Pa	Dec. 3, 1867	71, 616
Door-spring	L. Hillebrand	Philadelphia, Pa	Jan. 26, 1869	86, 228
Door-spring	H. Hochstrasser and A. Masson	Philadelphia, Pa	Feb. 3, 1852	8, 704
Door-spring	A. D. Hoffman	Chicago, Ill	Feb. 18, 1873	135, 914
Door-spring	G. L. Hudson	Conneaut, Ill	Feb. 8, 1859	22, 874
Door-spring	H. Hughes	Utica, N. Y	July 16, 1867	66, 843
Door-spring	S. W. Huntington	Augusta, Me	June 23, 1868	79, 125
Door-spring	M. Jincks	Wallace, N. Y	Sept. 20, 1870	107, 616
Door-spring	W. M. Kellie	Chicago, Ill	Mar. 18, 1873	136, 923
Door-spring	W. F. Kells	San Francisco, Cal	Oct. 11, 1870	108, 268
Door spring	F. Kellsey	Middletown, Conn	June 27, 1848	5, 654
Door-spring	D. C. King	Boston, Mass	July 9, 1872	128, 891
Door-spring	J. Losee	Peekskill, N. Y	July 16, 1872	129, 574
Door-spring	J. J. Mackey	South Brooklyn, N. Y	Dec. 3, 1867	71, 772
Door-spring	J. F. Mason and J. Johnson	Newark, N. J., and Brooklyn, N. Y.	Apr. 16, 1867	63, 914
Door-spring	J. Maxon	De Ruyter, N. Y	Sept. 19, 1848	5, 784
Door-spring	T. J. Mayall	Roxbury, Mass	Dec. 28, 1858	22, 441
Door-spring	A. Mishler	New York, N. Y	Apr. 19, 1870	102, 029
Door-spring	E. L. Morse	Saint Louis, Mo	Mar. 16, 1869	87, 867
Door-spring	E. D. Morton	Cuba, N. Y	Nov. 2, 1869	96, 470
Door-spring	J. M. Newton	Norwich, Conn	July 16, 1867	66, 873
Door-spring	C. W. Oldham	Leipsic, Ind	July 8, 1873	140, 638
Door-spring	C. A. Peck	New York, N. Y	Dec. 29, 1857	18, 987
Door-spring	T. Peck	Syracuse, N. Y	May 22, 1847	5, 124
Door-spring	T. Peck	Syracuse, N. Y	Oct. 17, 1848	5, 864
Door-spring	J. Post	Newark, N. J	June 12, 1860	28, 689
Door-spring	I. Powars	Waukegan, Ill	Jan. 26, 1869	86, 318
Door-spring	L. C. Prindle	Chicago, Ill	Feb. 4, 1868	73, 999
Door-spring	F. H. Richards	New Britain, Conn	Nov. 25, 1873	144, 926
Door-spring	W. T. Root	Geneva, N. Y	July 21, 1863	39, 308
Door-spring	W. Ross	Baltimore, Md	Apr. 13, 1869	88, 985
Door-spring	W. T. Rossiter	Lincoln, Nebr	Dec. 2, 1873	145, 244
Door-spring	C. W. Saladee	Newark, Ohio	Apr. 30, 1867	64, 372
Door-spring	C. W. Saladee	Newark, Ohio	Nov. 12, 1867	70, 903
Door-spring	S. Sawyer	Boston, Mass.	Jan. 21, 1841	1, 938
Door-spring	R. Scheidler	Newark, Ohio	Oct. 23, 1866	59, 082
Door-spring	R. Schrader	Indianapolis, Ind	Dec. 10, 1867	72, 089
Door-spring	J. Shaw, jr	Hinckley, Ohio	July 20, 1846	4, 641
Door-spring	J. Simpson	Newark, Ohio	Nov. 2, 1869	96, 493
Door-spring	D. G. Smith	Carbondale, Pa	Feb. 26, 1856	14, 326
Door-spring	P. Smith	Newport, Ky	Aug. 3, 1869	93, 359
Door-spring	W. W. Smith and B. Mulliken, jr.	New York, N. Y	Nov. 13, 1840	1, 854
Door-spring	W. H. Stafford	New York, N. Y	Dec. 6, 1870	109, 962
Door-spring	F. Stemmler	East New York, N. Y	Apr. 23, 1872	125, 995
Door-spring	E. Stimson	Montpelier, Vt	July 13, 1869	92, 486
Door-spring	L. Thomas	Allegheny City, Pa	Mar. 3, 1857	16, 759
Door-spring	T. Thorpe	West Cambridge, Mass	Dec. 7, 1837	506
Door-spring	A. Tittman	Evansville, Ind	July 1, 1873	140, 446
Door-spring	E. P. Torrey and W. B. Tilton	New York, N. Y	Sept. 8, 1857	18, 164
Door-spring	J. L. Tucker	Laconia, N. H	June 8, 1869	91, 184
Door-spring	T. Van Kannel	Cincinnati, Ohio	July 9, 1867	66, 538
Door-spring	M. L. Webster	Albany, N. Y	May 20, 1873	139, 037
Door-spring	A. Westcott	Syracuse, N. Y	Oct. 1, 1850	7, 692
Door-spring	A. Westcott	Syracuse, N. Y	Nov. 6, 1855	13, 770
Door-spring	F. S. Willcox	Bridgeport, Conn	Feb. 2, 1869	86, 482
Door-spring	W. Wilson	Greenfield, Mass	July 17, 1838	845
Door-spring	A. Wiswall	New York, N. Y	Aug. 14, 1866	57, 240
Door-spring	L. R. Witherell	Galesburgh, Ill	May 2, 1871	114, 379
Door-spring	W. H. Worcester and E. F. Jones	Farmington, N. H	Aug. 30, 1864	44, 034
Door spring and hinge	A. Wiswall	New York, N. Y	July 16, 1867	66, 927
Door spring and retainer combined	C. Bird	Lower Merion Tp., Pa	Oct. 8, 1872	132, 046
Door-spring bracket	A. J. Walker	New York, N. Y	Oct. 7, 1856	15, 864
Door-spring for closing doors, gates, &c	G. Barton, jr	Shaftsbury, Vt	July 1, 1840	1, 668
Door-spring, India-rubber	B. G. Fitzhugh	Sykesville, Md	Jan. 7, 1868	72, 992
Door-spring, Rubber	J. B. Hayden	Schaghticoke, N. Y	June 7, 1870	104, 027
Door, Spring wire	J. Codman	Boston, Mass	Aug. 17, 1835	....
Door springs and levers, Arrangement of	W. B. Barnard	Bristol, Conn	Apr. 9, 1850	7, 255
Door-springs, Vertical wire for	J. Codman	Boston, Mass	Dec. 28, 1832	....
Door-stay	A. H. Platt	Yellow Springs, Ohio	Aug. 5, 1856	15, 493
Door-stop	M. V. Doyle and J. Behel	Rockford, Ill	May 23, 1871	115, 038
Door-stop	W. H. Fahrney	Rockford, Ill	May 31, 1870	103, 729
Door-stop	W. H. Fahrney	Rockford, Ill	Dec. 13, 1870	110, 026
Door-stop	T. Hazard and J. M. Richardson	Wilmington, Ohio	Feb. 5, 1867	61, 830
Door-stop	G. W. Hunt	Winchendon, Mass	May 28, 1867	65, 083
Door-stop	W. May	Binghamton, N. Y	Nov. 3, 1868	83, 650
Door-stop	A. G. Stevens	Manchester, N. H	Apr. 21, 1868	77, 122
Door-stop	P. L. Weimer	Lebanon, Pa	Oct. 29, 1867	70, 302
Door-stop	I. J. Wells	Willmar, Minn	July 4, 1871	116, 654
Door-stop	L. C. Wemple	Rockford, Ill	Nov. 15, 1870	109, 281
Door stop and latch	C. L. Willis	Washington, D. C	June 25, 1867	66, 198
Door-straightener	O. C. Ross	Penfield, N. Y	July 14, 1868	79, 862
Door-strip	G. W. Carpender	Jarvis, Ind	May 7, 1867	64, 486
Door-strip	J. W. Kramer	Bloomsburgh, Pa	Aug. 30, 1870	106, 8[illegible]5
Door-strip, Metallic	A. D. Smith	Grafton, Ohio	May 25, 1869	90, 400
Door, Temporary folding	W. M. O'Brien	San Francisco, Cal	Sept. 3, 1872	131, 114
Door-threshold	C. Loring	South Braintree, Mass	Nov. 21, 1865	51, 066
Door-threshold	O. G. Thomas	Taunton, Mass	Apr. 29, 1873	138, 300

Index of patents issued from the United States Patent Office from 1790 *to* 1873, *inclusive*—Continued.

Invention.	Inventor.	Residence.	Date.	No.
Door, Trap	G. Runge	Philadelphia, Pa	July 11, 1854	11, 254
Door, Ventilating	T. R. Timby	Saratoga Springs, N. Y	June 4, 1867	65, 451
Doors, Adjustable hasp and hook for	W. Bisbee and F. G. Hearn	Yreka, Cal	Dec. 31, 1867	72, 784
Doors and gates, Closing	S. Stewart	Philadelphia, Pa	Sept. 11, 1860	30, 005
Doors and gates, Device for closing	S. P. Griffin and B. M. Harrod	New Orleans, La	Nov. 19, 1872	133, 221
Doors and gates, Spring for shutting	I. S. Richardson	Boston, Mass	Oct. 20, 1837	433
Doors and gates, Weight and lever attachment for	A. S. Blake	Waterbury, Conn	July 29, 1862	35, 988
Doors and sashes, Machine for laying out	S. C. Ellis	Jersey City, N. J	Feb. 15, 1870	99, 866
Doors and windows, Mode of hanging	A. B. Carpenter	New York, N. Y	Sept. 19, 1845	4, 200
Doors and windows, Tightening	S. Hayward	Plainfield, Mass	Oct. 17, 1829	....
Doors and windows water-tight, Method of rendering.	W. C. Fuller	London, Great Britain	Apr. 18, 1865	47, 368
Doors, Apparatus for closing	D. Ball	Albany, N. Y	Feb. 1, 1847	4, 947
Doors, Arrangement of spring and spring-catch for closing.	J. Clark	North Hadley, Mass	July 25, 1854	11, 358
Doors, &c., Attachment for opening and closing	A. W. Webster	Waterbury, Conn	May 25, 1858	20, 381
Doors, Automatic catch for sliding	A. W. Gibbs	Norwich, Conn	Dec. 16, 1873	145, 644
Doors, &c., being too tight, Preventing	D. Fraser	New York, N. Y	Dec. 12, 1825	....
Doors, Bolt for double	D. B. Conklin	Hardwick, N. J	Mar. 12, 1872	124, 422
Doors, Bolt-mechanism for prison	T. R. and J. Pullis	Saint Louis, Mo	Apr. 22, 1873	138, 099
Doors, Bolt-work for	J. Sargent	Rochester, N. Y	July 9, 1872	128, 816
Doors, Closing	C. Perley	New York, N. Y	Feb. 24, 1863	37, 770
Doors, Combined latch and lock for sliding	J. Capron	New York, N. Y	Nov. 15, 1870	109, 298
Doors, Construction of	W. C. James	Fishersville, N. H	July 18, 1871	117, 080
Doors, Construction of	W. W. J. Toussaint	Boston, Mass	July 15, 1873	140, 799
Doors, Device for closing	T. Peck	Syracuse, N. Y	Apr. 10, 1847	5, 058
Doors, Device for closing	R. F. Stevens	Syracuse, N. Y	Apr. 10, 1847	5, 061
Doors, Device for holding open	J. B. Sweetland	Pontiac, Mich	Oct. 18, 1870	108, 534
Doors, Device for opening and closing double	G. C. Gooch and T. B. Jeffery	Chicago, Ill	Feb. 16, 1869	86, 981
Doors, Elastic stud for	A. Eliaers	Boston, Mass	Jan. 24, 1865	45, 984
Doors, Fastening for jail	E. Jacobs	Cincinnati, Ohio	Nov. 15, 1859	26, 108
Doors, Fastening for sliding	E. W. Staples, jr	Norwich, Conn	July 22, 1873	141, 019
Doors, gates, &c., Apparatus for closing	G. Barton	Waterford, N. Y	Sept. 23, 1843	3, 280
Doors, gates, &c., Closing	O. Davidson	Ballston Spa, N. Y	Mar. 30, 1835	....
Doors, gates, &c., Device for closing	L. S. Deming	Newington, Conn	Jan. 1, 1867	60, 703
Doors, gates, &c., Spring for	R. Espey	Saint Louis, Mo	Jan. 21, 1873	135, 192
Doors, Guard-fastener for	E. C. Cochrane	Buffalo, N. Y	July 7, 1868	79, 550
Doors, Guide for fitting locks to	F. P. Pfleghar	New Haven, Conn	May 18, 1869	90, 299
Doors Hanging	G. Heegler	San Francisco, Cal	Jan. 7, 1873	134, 542
Doors, Hanging	A. Hitchcock	New York, N. Y	Dec. 22, 1863	41, 006
Doors, Hanging	G. W. Holly	Low Moor, Iowa	Oct. 16, 1866	58, 827
Doors, Hanging	A. W. Morse	Eaton, N. Y	June 9, 1857	17, 513
Doors, Hanging	G. W. Wilson	Nashua, N. H	July 10, 1840	1, 678
Doors, Hanging and fastening	C. W. Earl	Elk River, Minn	Apr. 13, 1869	88, 857
Doors, Hanging double	C. E. Brown	New York, N. Y	Jan. 8, 1856	14, 047
Doors, Hanging sliding	A. J. Culver	White Hall, Ill	June 29, 1869	91, 916
Doors, Hanging sliding	G. M. Darley	Nebraska City, Nebr	Nov. 26, 1872	133, 359
Doors, Hanging sliding	G. F. Prindle	Stockbridge, Mass	July 9, 1872	128, 903
Doors, Hook-catch for	C. B. Richards	Brooklyn, N. Y	May 8, 1860	28, 241
Doors, &c., Machine for relishing	F. D. Green	Williamsport, Pa	Mar. 4, 1873	136, 505
Doors, &c., Machinery and art of locking	J. Dillingham	Turner, Me	June 2, 1825	....
Doors, Method of mounting porcelain roses for	J. Bell	New York, N. Y	Apr. 3, 1849	6, 280
Doors on the inside, Fastening	B. H. Green	Princeton, N. J	June 11, 1841	2, 121
Doors open, Device for holding	W. W. Green, jr	Janesville, Wis	Nov. 24, 1868	84, 352
Doors open, Holding	A. D. Witzleben	Washington, D. C	Aug. 28, 1860	29, 770
Doors, Operating bolt and lock for controlling series of.	D. I. Stagg	Hoboken, N. J	June 13, 1854	11, 093
Doors or shutters, Attachment for opening and closing.	W. Post	Flushing, N. Y	Feb. 18, 1851	7, 940
Doors or windows, Guard for nursery	A. Howard	Milford, Mass	May 28, 1872	127, 166
Doors, Rail for sliding	J. Collins	Bergen County, N. J	Dec. 16, 1873	145, 630
Doors, &c., Retaining-device for	L. F. Howell	Camden, N. J	Nov. 10, 1868	83, 967
Doors, safe-walls, vaults, &c., Burglar-proof plates for.	L. Yale, jr	Newport, N. Y	Oct. 19, 1852	9, 350
Doors, Sheave and rail for sliding	T. M. Lyons	New York, N. Y	May 30, 1871	115, 333
Doors, Spring catch and stop for	C. W. Barnes	Janesville, Wis	May 26, 1868	78, 355
Doors, Spring-catch for	W. Glue	Muskegon, Mich	Jan. 24, 1871	111, 196
Doors, Spring-catch for	J. C. Plumer	Boston, Mass	Mar. 7, 1865	46, 697
Doors, Spring for closing	L. Evans	Fayetteville, N. Y	Apr. 18, 1848	5, 515
Doors, Spring for closing	R. T. Knight	Philadelphia, Pa	Nov. 6, 1866	59, 414
Doors, stair-rails, lasts, &c., Fastening for	W. Sellers	New York, N. Y	May 31, 1864	42, 994
Doors, stiles, &c., Mortising	A. Whitmarsh	East Bridgewater, Mass	July 30, 1833	....
Doors to prevent rain from being forced under them by wind, &c., Method of hanging.	E. C. Tilson	Thomaston, Me	Jan. 9, 1838	571
Doors, &c., Turn-button for	P. Bradford	New Haven, Conn	Oct. 3, 1871	119, 499
Doors, Water-excluding attachment for	A. Kirkpatrick	Newark, N. J	Dec. 14, 1840	1, 901
Doors, Water-table for	N. Briggs	Windham, Vt	June 17, 1809	....
Doors, Water-table for	D. Dulkely	Hampton, Conn	June 14, 1834	....
Doors, Water-table for	J. Burt, jr	Tiverton, R. I	Jan. 24, 1834	....
Doors, windows, &c., Rolling screen for	W. G. Perkins	Walden, Vt	June 11, 1867	65, 692
Double and single tree fastening for carriages	J. Ingels	Milton, Ind	May 10, 1870	102, 940
Double and whiffle tree coupling, Open link for	N. C. Sample	Penningtonville, Pa	Jan. 17, 1871	111, 007
Double-centered joint, butt, or hinge	E. Hedge	Hartford, Conn	Nov. 23, 1837	473
Double-cooler	J. Van Duzer	Otisville, N. Y	Mar. 23, 1869	88, 098
Double-cylinder boiler for hot-water apparatus	W. Beebe	New York, N. Y	Nov. 14, 1854	11, 926
Double gate	J. S. Brown	Washington, D. C	Aug. 10, 1852	9, 176
Double-headed wrench	J. J. Love	New York, N. Y	Oct. 30, 1866	59, 242
Double sliding or turning gate	R. Fleming	Victory, N. Y	Apr. 28, 1868	77, 367
Double-tree	H. L. Bowen	Belvidere, Ill	Oct. 8, 1872	132, 048
Double-tree	C. C. Bradley	Brodhead, Wis	Dec. 3, 1861	33, 820
Double-tree	W. Dowell	Hicksville, Ohio	June 23, 1863	38, 952
Double-tree	M. Durnell and W. Milner	Leesburgh, Ohio	Apr. 25, 1871	114, 116
Double-tree	A. W. Green	South Dansville, N. Y	Oct. 15, 1872	132, 281
Double-tree	J. B. Hough	Lebanon, Ohio	Mar. 19, 1867	62, 957
Double-tree	D. W. Kauffman	Sterling, Ill	Apr. 18, 1871	113, 895
Double-tree	A. Lomax	La Porte, Ind	Oct. 10, 1871	119, 777
Double-tree	W. Martin	Clarence, Iowa	Sept. 25, 1873	143, 173
Double-tree	H. Palmer and A. N. Case	Kingsville, Ohio	Aug. 11, 1868	80, 872

Index of patents issued from the United States Patent Office from 1790 *to* 1873, *inclusive*—Continued.

Invention.	Inventor.	Residence.	Date.	No.
Double-tree	H. W. Palmer	Kingsville, Ohio	Apr. 13, 1869	88, 896
Double-tree	M. V. B. Williamson	Jamesport, N. Y	Feb. 11, 1868	74, 471
Double-tree, Carriage	J. Hoover	Manchester, Md	Mar. 21, 1865	46, 904
Double-tree-clevis hook	W. W. Bently	Lee Township, Mich	Mar. 21, 1871	112, 888
Double-tree, Compound	J. Wykoff	Grant City, Mo	Oct. 6, 1868	82, 781
Double-tree draft-equalizer	M. V. B. Williamson	Jamesport, N. Y	Feb. 11, 1868	74, 470
Double-tree, Equalizing	E. Griswold, J. B. Cramer, and W. Blay.	Helena, Mont	Feb. 11, 1868	74, 347
Double-tree fastener	D. Werst	South Bend, Ind	Dec. 13, 1870	110, 095
Double-walled coolers, Construction of	J. Oldfield	Thompsonville, Conn	Feb. 23, 1869	87, 283
Doubling and spooling machine, Stopping-mechanism for.	J. Arnold	Pawtucket, R. I	May 24, 1870	103, 277
Doubling and twisting machine, Stopping-mechanism for.	W. Cockcroft and R. Ackroyd	Chester, Pa	Sept. 2, 1873	142, 374
Doubling, twisting, and reeling	G. Levan	Gap, Pa	July 20, 1846	4, 649
Doubling, twisting, and reeling machine	H. E. and J. Haight	New York, N. Y	Aug. 30, 1813	....
Douche-bath	M. Mattson	Boston, Mass	June 3, 1862	35, 458
Douche, Electro-nasal	S. E. Adams	Springfield, Ohio	Nov. 22, 1870	109, 367
Douche, Nasal	W. B. Snyder	Bridgeport, Conn	June 25, 1872	128, 257
Dough and cake mixer	T. Holmes	Williamsburgh, N. Y	Sept. 6, 1870	107, 050
Dough and cutting crackers, &c., Machine for plaiting.	I. M. Heagle	New Haven, Conn	Apr. 13, 1838	689
Dough and paste, Breaking	D. Pettibone		Nov. 16, 1797	....
Dough and paste, Liquid for raising	A. and G. Sauer and L. Cachal	Paris, France	Aug. 9, 1870	106, 213
Dough and sponge raiser	W. Pearce	McLean County, Ill	July 7, 1868	79, 680
Dough, Apparatus for aërating	E. Fitzgerald	New York, N. Y	Jan. 3, 1865	45, 706
Dough, Apparatus for raising	T. Stockton	North Chenango, Pa	Mar. 28, 1865	47, 050
Dough, Apparatus for rolling and spreading	G. F. Fessenden	West Cambridge, Mass	Feb. 26, 1867	62, 323
Dough apportioning, expanding, and shaping apparatus for manufacture of bread.	J. Perry	Brooklyn, N. Y	May 26, 1868	78, 232
Dough-board	H. F. Morse	Manchester, N. H	Apr. 2, 1872	125, 208
Dough-board	N. B. Petterson	McGregor, Iowa	Oct. 31, 1871	120, 394
Dough-box for cake-machine	D. M. Holmes	Williamsburgh, N. Y	Feb. 9, 1869	86, 752
Dough, Breaking and kneading	I. Cole	Cincinnati, Ohio	Dec. 12, 1848	5, 959
Dough-breaking machine	C. Barnard	Nantucket County, Mass	Apr. 17, 1810	....
Dough-breaking machine	L. Folger		Nov. 1, 1804	....
Dough-breaking machine	L. Shaw	Bath, Me	May 26, 1843	3, 102
Dough, crushing sugar, &c., Machine for rolling	G. Perkins	Burlington, Vt	Apr. 23, 1867	64, 029
Dough-cutter and rolling-pin, Combined	F. E. Chadwick	Gardiner, Me	July 2, 1872	128, 462
Dough-dresser, Rotary	G. H. Cross	Montpelier, Vt	June 11, 1867	65, 649
Dough for biscuit, &c., Preparing	J. R. Treadwell	Brooklyn, N. Y	Sept. 15, 1863	39, 972
Dough for bread, Apparatus for raising	J. Johnson	New York, N. Y	Oct. 5, 1858	21, 683
Dough for molding crackers, Machine for preparing	F. C. Treadwell, jr	New York, N. Y	May 13, 1856	14, 889
Dough into loaves or crackers, Machine for molding and shaping.	W. Hotine	Brooklyn, N. Y	Sept. 11, 1866	57, 908
Dough-kneader	J. H. Barr and J. H. Smith	Roanoke, Ind	July 12, 1870	105, 293
Dough-kneader	L. Bell	Washington, D. C	Dec. 10, 1872	133, 823
Dough-kneader	H. H. Hawes	Mount Meridian, Va	Mar. 14, 1871	112, 707
Dough-kneader	C. C. Johnson	Springfield, Vt	July 2, 1867	66, 351
Dough-kneader	H. P. Jones	Davenport, Iowa	Jan. 15, 1867	61, 206
Dough-kneader	F. C. King	Portland, Me	July 25, 1865	48, 979
Dough-kneader	S. H. Lombard	Winona, Minn	May 19, 1868	78, 106
Dough-kneader	J. C. Loveland	Springfield, Vt	Jan. 16, 1866	52, 062
Dough-kneader	E. Mathers	Harrisville, W. Va	Jan. 23, 1872	122, 902
Dough-kneader	W. B. Morrison	Muskegon, Mich	Oct. 1, 1867	69, 467
Dough-kneader	A. J. Palmer	Brooklyn, N. Y	Aug. 9, 1864	43, 791
Dough-kneader	P. W. Robinson	New Bedford, Mass	July 30, 1867	67, 353
Dough-kneader	A. P. Routt	Orange Court-House, Va	Aug. 7, 1866	56, 995
Dough-kneader	G. W. Sanders	Springfield, Mass	June 19, 1866	55, 718
Dough-kneader	C. B. Sawyer	Fitchburgh, Mass	July 13, 1869	92, 479
Dough-kneader	F. Thrall	Oshkosh, Wis	June 8, 1869	91, 054
Dough-kneader	W. Tyler	Georgetown, D. C	May 16, 1871	114, 993
Dough-kneader and meat-pounder, &c	C. I. Wilmans and W. J. Wolfe	Olney, Ill	June 9, 1868	78, 778
Dough kneader and roller	P. Perry	Charlestown, Mass	July 3, 1866	56, 091
Dough-kneading machine	W. A. Snow	Killingworth, Conn	June 25, 1872	128, 333
Dough-kneading machine	G. R. Baker	Brooklyn, N. Y	Oct. 10, 1865	50, 322
Dough-kneading machine	H. Berdan	New York, N. Y	Sept. 15, 1857	18, 180
Dough-kneading machine	S. Emmore	Stouchsburgh, Pa	Dec. 3, 1867	71, 723
Dough-kneading machine	J. S. Fisher and S. W. Talbot	Dedham, Mass	Apr. 23, 1822	....
Dough-kneading machine	S. Nixon	Prince George's County, Va	Feb. 28, 1818	....
Dough-kneading machine	J. L. Rolland	Paris, France	Jan. 9, 1855	12, 217
Dough-kneading machine	N. C. Willett	Jonesborough, Tenn	Dec. 28, 1832	....
Dough-machine	O. B. Fuller	Newark, N. J	July 8, 1873	140, 695
Dough-machine	D. and T. Shackford	West Brook, N. Y	Mar. 2, 1836	
Dough, Machine for breaking or working	J. W. Post	Baltimore, Md	Mar. 14, 1840	1, 516
Dough, Machine for feeding the flour, mixing the materials, and kneading.	J. Hecker and W. Hotine	New York, N. Y	Nov. 24, 1857	18, 723
Dough, Machine for making	G. Richards	Stonington, Conn	Feb. 14, 1806	....
Dough, Machine for sheeting	D. B. Fuller	Philadelphia, Pa	Mar. 12, 1872	124, 431
Dough, Machine for sheeting	O. B. Fuller	Newark, N. J	Nov. 18, 1873	144, 666
Dough, Machine for sheeting	J. H. Shrote	Baltimore, Md	July 25, 1871	117, 474
Dough making and kneading machine	S. M. Ridgaway	Saint Michael's, Md	July 1, 1856	15, 254
Dough, Method of kneading	H. N. Rider	Adams, Mass	Feb. 19, 1850	7, 105
Dough-mixer	C. Chavanne	New Orleans, La	Oct. 9, 1866	58, 596
Dough-mixer	W. E. Damant	West Hoboken, N. J	June 25, 1872	128, 288
Dough-mixer	F. Grenier	Beroserac, France	Oct. 6, 1868	82, 710
Dough-mixer	R. Kent	Portland, Me	Feb. 6, 1872	123, 403
Dough-mixer	A. E. Muth	Cincinnati, Ohio	Mar. 26, 1872	125, 072
Dough-mixer	J. M. Stanyan	Milford, N. H	Oct. 9, 1866	58, 691
Dough-mixer	S. J. Talbott	Milford, N. H	Dec. 18, 1866	60, 632
Dough mixer and roller	J. Bailie and J. Gervers	Cincinnati, Ohio	Jan. 22, 1867	61, 307
Dough-mixer, Revolving	T. Holmes	Williamsburgh, N. Y	June 15, 1869	91, 335
Dough-mixer, Revolving	T. Holmes	Williamsburgh, N. Y	Sept. 21, 1869	95, 021
Dough-mixing	E. Fitzgerald	New York, N. Y	Oct. 8, 1861	32, 432
Dough mixing and kneading machine	E. Edwards	London, England	Feb. 25, 1873	136, 144
Dough mixing and kneading machine	W. Wells	Middleton, Mass	July 5, 1864	43, 445
Dough-mixing machine	G. Clark, jr., and L. P. Jenks	Dorchester and Boston, Mass.	Feb. 20, 1866	52, 678

Index of patents issued from the United States Patent Office from 1790 *to* 1873, *inclusive*—Continued.

Invention.	Inventor.	Residence.	Date.	No.
Dough-mixing machine	J. Davidson	Xenia, Ohio	May 25, 1869	90, 509
Dough-mixing machine	W. Hotine	Brooklyn, N. Y	Jan. 8, 1861	31, 082
Dough, &c., Mixing or breaking	H. Ball	Cincinnati, Ohio	Feb. 27, 1847	4, 981
Dough, paste, &c., Machine for making	E. Stevens	London, England	Oct. 20, 1863	40, 370
Dough-raiser, Automatic	J. Stark	Thomasville, Ga	May 25, 1869	90, 471
Dough, Raising	J. Perry and E. Fitzgerald	New York, N. Y	Mar. 9, 1858	19, 610
Dough raising and kneading machine	G. Scott	Steubenville, Ohio	Nov. 27, 1860	30, 764
Dough-raising apparatus	S. V. Dodge	De Soto, Iowa	Apr. 9, 1872	125, 445
Dough-raising apparatus	W. Leas	Kokomo, Ind	Jan. 9, 1866	51, 949
Dough-raising trough	H. L. McKean	Allegheny, Pa	Mar. 7, 1865	46, 688
Dough-roller	D. B. Baker	Rollersville, Ohio	Dec. 19, 1865	51, 539
Dough-roller	W. Long	Watertown, N. Y	May 26, 1868	78, 300
Dough roller and cutter	W. F. Rippon	Providence, R. I	May 14, 1872	126, 833
Dough, Roller for pressing	F. J. Rice and G. W. Hayward	Providence, R. I	July 3, 1860	29, 006
Dough rolling and cutting machine	I. S. Schuyler	New York, N. Y	Apr. 13, 1858	19, 968
Dough-rolling machine	H. Goodwin and C. H. Bennett, 2d.	South Berwick Junction, Me.	May 4, 1869	89, 653
Dough-rolling machine	J. Hecker and W. Hotine	New York, N. Y	Mar. 22, 1859	23, 301
Dough-rolling machine	J. McCollum	New York, N. Y	Dec. 1, 1857	18, 758
Dough under pressure, Apparatus for measuring out and discharging.	J. Perry	Brooklyn, N. Y	July 8, 1862	35, 837
Dough-worker	W. R. Pool	Havana, Ala	May 24, 1870	103, 366
Doughnut-cutter	J. F. Blondel	Thomaston, Me	July 9, 1872	128, 783
Doughnut-mold	G. Machlet	Newark, N. J	Mar. 25, 1873	137, 142
Dovetail and tenon cutting machine	T. J. Wells	New York, N. Y	July 8, 1841	2, 158
Dovetail-cutter	J. C. Hursell	Boston, Mass	Aug. 13, 1867	67, 654
Dovetail-joint for wood, &c	F. S. Barnard	New York, N. Y	Apr. 12, 1859	23, 542
Dovetail-joint, Sliding	G. B. Raymond		Mar. 26, 1816	
Dovetail key-cutter	A. P. Hughes	Philadelphia, Pa	Sept. 18, 1855	13, 574
Dovetail-marker	J. Evans	Philadelphia, Pa	Apr. 21, 1868	77, 018
Dovetail-marking device	G. Ashby	Decatur, Ill	Dec. 16, 1873	145, 477
Dovetail-molding device	A. L. Finch	Sing Sing, N. Y	Apr. 2, 1872	125, 186
Dovetail-molding tool	B. McEnally and E. Farrell	Detroit, Mich	Nov. 30, 1869	97, 308
Dovetail-moldings, Machine for cutting	S. Withington	Saint Louis, Mo	Nov. 30, 1858	22, 222
Dovetail-mortises, Machine for making	T. D. White	Cincinnati, Ohio	Apr. 21, 1868	77, 149
Dovetails, Adjustable gage for	J. S. L. Babbs and A. H. Ray	Boston, Mass., and Providence, R. I.	Dec. 15, 1857	18, 838
Dovetails and their grooves, Machine for cutting	E. G. Matthews	Clear Water, Minn	Feb. 10, 1857	16, 627
Dovetailing-board	A. Davis	Lancaster, Mass	Sept. 6, 1833	
Dovetailing-machine	J. S. L. Babbs	New Albany, Ind	June 5, 1866	55, 223
Dovetailing-machine	J. S. L. Babbs	New Albany, Ind	Jan. 1, 1867	60, 843
Dovetailing-machine	J. Bain and S. C. Brown	Richmond, Ind	Oct. 16, 1860	30, 382
Dovetailing-machine	H. H. Bashore	Philadelphia, Pa	Mar. 8, 1870	100, 704
Dovetailing-machine	J. Bell	Harlem, N. Y	Oct. 23, 1855	13, 696
Dovetailing-machine	M. T. Boult	Battle Creek, Mich	Oct. 8, 1872	131, 932
Dovetailing-machine	W. G. Branch, J. A. Haseltine, and W. and C. F. Branch.	Pomeroy, Ohio	Sept. 19, 1871	119, 113
Dovetailing-machine	T. H. Burley	Cincinnati, Ohio	Jan. 2, 1855	12, 122
Dovetailing-machine	T. H. Burley	New York, N. Y	Jan. 3, 1860	26, 647
Dovetailing-machine	E. Coffin	Indianapolis, Ind	Feb. 5, 1861	31, 301
Dovetailing-machine	T. Cullen	Blackstone, Mass	Apr. 15, 1873	137, 830
Dovetailing-machine	A. Davis	Lowell, Mass	Nov. 28, 1871	121, 278
Dovetailing-machine	A. Davis	Lowell, Mass	May 27, 1873	139, 299
Dovetailing-machine	A. and A. Davis	Lowell, Mass	Feb. 26, 1856	14, 307
Dovetailing-machine	J. Dill	Grand Rapids, Mich	Aug. 30, 1870	106, 922
Dovetailing-machine	J. Dill	Boston, Mass	Nov. 19, 1872	133, 145
Dovetailing-machine	H. H. Evarts	Chicago, Ill	Jan. 3, 1871	110, 642
Dovetailing-machine	H. H. Evarts	Chicago, Ill	June 7, 1870	104, 001
Dovetailing machine	J. P. Flanders	Vergennes, Vt	Nov. 15, 1870	109, 194
Dovetailing-machine	A. S. Gear and B. F. Dunklee	Boston, Mass., and Concord, N. H.	Apr. 2, 1872	125, 282
Dovetailing-machine	R. J. Gould	Newark, N. J	Aug. 11, 1868	80, 942
Dovetailing-machine	F. D. Green	Williamsport, Pa	Feb. 25, 1873	136, 236
Dovetailing-machine	C. S. Griffin and J. W. Wilkins	Stockton, Me	Nov. 8, 1870	109, 123
Dovetailing-machine	C. S. Griffin and J. W. Wilkins	Chelsea, Me	June 20, 1871	116, 178
Dovetailing-machine	H. J. and H. J. Hale, jr	Indianapolis, Ind	Mar. 3, 1863	37, 812
Dovetailing-machine	J. J. Haley	Philadelphia, Pa	Sept. 4, 1855	13, 522
Dovetailing-machine	W. T. Hamilton	Dublin, Ireland	Nov. 11, 1873	144, 536
Dovetailing-machine	S. E. Hartwell	New Albany, Ind	Mar. 22, 1864	41, 992
Dovetailing-machine	J. E. Haskell	Baltimore, Md	Mar. 4, 1873	136, 374
Dovetailing-machine	E. Heath	San Francisco, Cal	July 21, 1863	39, 290
Dovetailing-machine	E. Heath	San Francisco, Cal	May 16, 1871	114, 814
Dovetailing-machine	E. Heath	San Francisco, Cal	June 14, 1870	104, 147
Dovetailing-machine	H. W. Jelliff	Appleton, Ohio	June 12, 1860	28, 672
Dovetailing-machine	E. L. Jones and M. E. Carter	Saint Louis, Mo	Aug. 20, 1861	33, 087
Dovetailing-machine	D. Jordan	Charlestown, Mass	Feb. 1, 1870	99, 446
Dovetailing-machine	D. Jordan and J. Dill	Charlestown, Mass., and Grand Rapids, Mich.	Apr. 19, 1870	102, 011
Dovetailing-machine	D. A. King and T. Norris	Lexington, Ky	Jan. 22, 1861	31, 175
Dovetailing-machine	T. E., A., and E. King	Cherry Valley, Ohio	Sept. 14, 1858	21, 503
Dovetailing-machine	C. B. Knapp and N. S. Clement	Waterloo, Wis., and Northampton, Mass.	Jan. 2, 1872	122, 390
Dovetailing-machine	C. F. Kuhnle	Washington, D. C	Dec. 15, 1868	84, 884
Dovetailing-machine	W. A. McDonald	Mott Haven, N. Y	Feb. 15, 1859	22, 967
Dovetailing-machine	W. A. McDonald	Mott Haven, N. Y	Sept. 13, 1859	25, 431
Dovetailing-machine	W. F. Moody	Chicago, Ill	Apr. 1, 1873	137, 315
Dovetailing-machine	C. Ohlemacher and O. Kromer	Sandusky, Ohio	Mar. 9, 1869	87, 586
Dovetailing-machine	L. A. Orcutt	Albany, N. Y	July 8, 1856	15, 391
Dovetailing-machine	O. J. Pennell and G. Zimmer	Williamsport, Pa	Nov. 7, 1871	120, 594
Dovetailing-machine	J. Phillips, jr	Chicago, Ill	Aug. 3, 1869	93, 225
Dovetailing-machine	D. Pomroy	San Francisco, Cal	Apr. 4, 1871	113, 560
Dovetailing-machine	J. B. Ritchey	Pomeroy, Ohio	May 16, 1871	114, 859
Dovetailing-machine	J. B. Ritchey	Pomeroy, Ohio	July 25, 1871	117, 332
Dovetailing-machine	J. B. Schmid	Salem, Va	Nov. 29, 1870	109, 674
Dovetailing-machine	J. B. Schmid	Salem, Va	Nov. 29, 1870	109, 675
Dovetailing-machine	J. M. Seymour	Newark, N. J	Mar. 7, 1871	112, 502
Dovetailing-machine	W. E. Sibley	Weston, Mass	Dec. 6, 1864	45, 351
Dovetailing-machine	W. E. Sibley	Weston, Mass	May 31, 1870	103, 670

Index of patents issued from the United States Patent Office from 1790 *to* 1873, *inclusive*—Continued.

Invention.	Inventor.	Residence.	Date.	No.
Dovetailing-machine	P. A. B. Triacca	Stamford, Conn	Aug. 13, 1872	130, 548
Dovetailing-machine	A. C. Van Alstine	New York, N. Y	July 22, 1873	141, 096
Dovetailing-machine	J. R. Van Epps	Albany, N. Y	Apr. 19, 1870	102, 069
Dovetailing-machine	T. D. White	Cincinnati, Ohio	Apr. 21, 1868	77, 148
Dovetailing-machine	D. Whitlock	Newark, N. J	Jan. 30, 1872	123, 137
Dovetailing-machine	E. Wight	Philadelphia, Pa	Mar. 11, 1856	14, 427
Dovetailing-machine	R. Wolf	Burlington, Iowa	Mar. 10, 1868	75, 506
Dovetailing-machine	C. Young	Auburn, N. Y	July 16, 1867	66, 763
Dovetailing-machine, Cutter for	J. C. Hursell	Boston, Mass	June 19, 1866	55, 661
Dovetailing-machine, Cutter for	D. Jordan	Charlestown, Mass	May 31, 1870	103, 750
Dovetailing-machine cutters, Method of operating	J Bell	Harlem, N. Y	July 10, 1860	29, 051
Dovetailing machine, Miter	F. A. Gleason	Rome, N. Y	Mar. 17, 1863	37, 910
Dovetailing machine, Miter	F. A. Gleason	New York, N. Y	Aug. 9, 1864	43, 767
Dovetailing machine, Square-joint	W. Perrin	Lowell, Mass	Mar. 24, 1841	2, 014
Dovetailing-machine table	J. Dill	Grand Rapids, Mich	July 5, 1870	104, 943
Dovetailing rotary cutters in their heads, Method of.	G. H. Mallory	New York, N. Y	Jan. 5, 1858	19, 035
Dovetailing-tool	G. W. Billings	Cleveland, Ohio	Feb. 23, 1858	19, 406
Dovetailing-tool	A. P. Hughes	Philadelphia, Pa	Feb. 21, 1854	10, 548
Dowel-brad, Double headed and pointed	J. Sowle	Boston, Mass	June 27, 1871	116, 367
Dowel-gage	J. Er	Cleveland, Ohio	Feb. 6, 1872	123, 337
Dowel-hoop for cooperage	A. W. Ballow	Saint Charles, Mich	Mar. 19, 1872	124, 656
Dowel-pins, Machine for making	M. Everts	Girard Township, Pa	Feb. 4, 1873	135, 416
Dowel-pins, Machine for making wooden	A. H. Boyd	Saco, Me	Apr. 20, 1858	20, 016
Doweling	P. Jones	Newark, N. J	Apr. 1, 1862	34, 835
Doweling-bit used in ship-building	J. Thomas	Washington, D. C	Apr. 2, 1825	....
Doweling heading for cooperage, Machine for	R. Hunt	Ontario County, N. Y	Dec. 11, 1820	....
Doweling-instrument	W. Bowhay	Philadelphia, Pa	Oct. 14, 1816	....
Doweling-machine, Cooper's	J. German	Oriskany Falls, N. Y	Feb. 10, 1863	37, 619
Draft and hold-back attachment	W. H. Chamberlin	Medina, N. Y	Jan. 21, 1870	104, 557
Draft and shaft tug	H. Bowers	Albany, N. Y	May 5, 1868	77, 575
Draft-attachment	M. Adsit	Forest, N. Y	May 5, 1868	77, 433
Draft-attachment	C. W. Conolly	Rochester, N. Y	July 3, 1866	56, 012
Draft-attachment for horse	C. C. Cole	Phelps, N Y	Jan. 4, 1870	98, 563
Draft-attachment for horse	E. Sanford	Meriden, Conn	Nov. 26, 1867	71, 415
Draft-bar	D. and M. Foreman	Dalton, Ohio	Apr. 14, 1868	76, 737
Draft-bar, Railway	R. Fiegel	Montgomery County, Pa	Apr. 4, 1865	47, 098
Draft-chain, whiffletree, &c., Hook for	V. B. Paul	Morrill, Me	Sept. 10, 1872	131, 224
Draft-device for three horses	O. W. Jones	Columbus, Wis	Feb. 21, 1871	111, 944
Draft-equalizer	J. Averill and E. S. Fitch	Champlain and Movers, N. Y.	Jan. 28, 1868	73, 864
Draft-equalizer	J. P. Beckenbaugh	Springfield, Ill	Oct. 28, 1873	144, 052
Draft-equalizer	E. H. Blake	Coatsburgh, Ill	Dec. 30, 1873	145, 983
Draft-equalizer	O. N. Collins and P. B. Stiles	Cedar Rapids, Iowa	Dec. 9, 1873	145, 474
Draft-equalizer	W. W. Hinman	Elkhart, Ill	Oct. 14, 1873	143, 696
Draft-equalizer	W. McClelland, sr	Fowler, Ill	Dec. 30, 1873	146, 011
Draft-equalizer	G. A. Mosher	Champlain, N. Y	Sept. 1, 1868	81, 669
Draft-equalizer	J. Parker	Pontiac, Mich	Aug. 12, 1873	141, 813
Draft-equalizer	W. M. Perkins and T. F. Vandergift.	La Fontaine, Ind	Oct. 25, 1870	108, 727
Draft-equalizer	S. Shadduck	Elk River Township, Iowa.	Sept. 22, 1868	82, 443
Draft-equalizer	I. Starr	Prairieville, Mich	Feb. 9, 1869	86, 708
Draft-equalizer	W. Stevenson	Neponset, Ill	May 16, 1871	114, 988
Draft-equalizer	S. H. Smith	North Adams, Mich	Feb. 23, 1869	87, 305
Draft-equalizer	R. Walker and G. Trumbull	Batavia, N. Y	Dec. 15, 1868	85, 039
Draft-equalizer	S. H. Wheeler	Dowagiac, Mich	May 4, 1869	89, 614
Draft-equalizer for horse-power	A. Stewart	Troy, Wis	May 28, 1867	65, 297
Draft-equalizer for wagons	L. Kendle, G. Gates, and J. M. Orput.	Rochelle, Ill	Nov. 5, 1872	132, 725
Draft equalizer, Three-horse	H. F. Jerauld	Vandalia, Ill	Mar. 14, 1871	112, 715
Draft-equalizing attachment	J. Wilkinson	Prophetstown, Ill	Aug. 11, 1863	39, 517
Draft-evener	R. Bunker	Hudson, Wis	Apr. 1, 1873	137, 286
Draft for furnaces, Hot-air	E. Boughton	South Norwalk, Conn	Nov. 4, 1873	144, 249
Draft-hook, Detachable	J. Nicholson	Monticello, Ind	Dec. 5, 1871	121, 464
Draft in flues, furnaces, &c., Machine for creating	T. Fairbanks	Saint Johnsbury, Vt	May 26, 1843	3, 100
Draft in smoke-pipes, Using exhaust steam for increasing.	R. Winans	Baltimore, Md	Apr. 10, 1847	5, 056
Draft-link, Spring	B. Richards	North Industry, Ohio	Feb. 9, 1869	86, 863
Draft-plate	G. Smith	Providence, R. I	Apr. 16, 1867	63, 954
Draft-promoter	B. A. Haycock	Richland, Iowa	May 2, 1871	114, 291
Draft-regulator	J. W. Branham	Franklin, Ind	Mar. 31, 1868	76, 045
Draft regulator	C. A. Harper	Burlington, N. J	July 16, 1872	129, 551
Draft-regulator	E. Harrington	Boston, Mass	Sept. 6, 1864	44, 091
Draft-regulator	A. L. Ide	Springfield, Mass	July 11, 1871	116, 962
Draft-regulator	J. J. Smith and S. Wood	Cleveland, Ohio	Nov. 2, 1869	96, 494
Draft-regulator	H. E. Turner	Boston, Mass	Feb. 24, 1863	37, 781
Draft-regulator	J. Woodruff	Rahway, N. J	July 9, 1872	128, 836
Draft regulator and ventilator	G. Chilson	Boston, Mass	Aug. 23, 1864	43, 897
Draft-regulator for heating-apparatus	R. E. Deane	Brooklyn, N. Y	June 8, 1869	90, 934
Draft-regulator for warming-apparatus	J. Brown and C. R. Ellis	New York and Brooklyn, N. Y.	Apr. 3, 1860	27, 687
Draft-regulator, Hot-air-furnace	F. E. Chatard, jr	Baltimore, Md	May 7, 1872	126, 377
Draft-regulator, Hot-air-furnace	S. J. Gold	Cornwall, Conn	June 18, 1872	128, 035
Draft-regulator, Locomotive	P. H. Corlett	West Manchester, Pa	Nov. 10, 1863	40, 555
Draft-regulator, Steam-boiler	P. E. Chase	Philadelphia, Pa	Oct. 14, 1856	15, 870
Draft-regulator, Steam-boiler	H. N. Winans	New York, N. Y	Nov. 8, 1864	44, 992
Draft-regulator, Steam-boiler-furnace	E. B. Beach	West Meriden, Conn	Feb. 13, 1872	123, 608
Draft-regulator, Stove	E. Foote, jr	Seneca Falls, N. Y	May 26, 1842	2, 636
Draft-regulator, Stove and furnace	J. Briggs	Roxbury, Mass	Feb. 16, 1864	41, 598
Drafting-attachment	T. Evans	Newark, N. J	Dec. 3, 1867	71, 595
Drafting-instrument	J. C. Moore	Belmont County, Ohio	Dec. 30, 1833	....
Draftsman, Adjustable curve for use of	S. K. Kirby	East Saginaw, Mich	June 7, 1870	103, 896
Drag	J. W. Newton	Geneva, Wis	Nov. 16, 1869	96, 929
Drag and cultivator combined	H. H. Christie	Perch River, N. Y	Nov. 11, 1862	36, 892
Drag and hook, Dung	H. W. Weiss	Quakertown, Pa	Nov. 24, 1868	84, 452
Drag for removing stones, &c., from the bottoms of rivers, lakes, &c.	P. L. Lyme	New York, N. Y	Oct. 31, 1839	1, 386
Drag, Ground	B. J. McAfee and J. H. Wirt	Delphia, Ind	Sept. 11, 1866	57, 939
Drag-hook	J. Parish	Chicago, Ill	Dec. 24, 1867	72, 671

Index of patents issued from the United States Patent Office from 1790 *to* 1873, *inclusive*—Continued.

Invention.	Inventor.	Residence.	Date.	No.
Drag, Marine	J. Faunce	Washington, D. C	Feb. 22, 1870	100, 021
Drag-rake	G. H. Hackett	North Tunbridge, Vt	Jan. 18, 1870	98, 954
Drag-rake	H. F. Morton	West Sumner, Me	Mar. 30, 1869	88, 325
Drag-rake, Hand	E. Brown	Kennebunk, Me	Nov. 15, 1870	109, 294
Drag, Sea	S. Curtis	Lynn, Mass	Aug. 31, 1869	94, 402
Drag, Sea	J. Edson	Boston, Mass	Aug. 31, 1869	94, 191
Drag, Steam	G. Bradley	Paterson, N. J	Nov. 11, 1856	16, 044
Drain	G. R. Moore	Lyons, Iowa	Feb. 11, 1873	135, 723
Drain	G. W. Smith	Springfield, N. J	Feb. 9, 1864	41, 542
Drain	G. W. Smith	Springfield Township, N. J.	Nov. 15, 1864	45, 089
Drain and sewer pipe molding machine	D. Copeland, jr	Rochester, N. Y	May 23, 1871	115, 030
Drain and water pipe	W. P. Kirkland	San Francisco, Cal	Apr. 14, 1868	76, 778
Drain and water pipe	G. H. Titcomb and J. P. Culver	Jersey City, N. J	May 3, 1870	102, 623
Drain, Arched	G. R. Moore	Lyons, Iowa	July 16, 1872	129, 578
Drain-pipe	S. Carlton	Lynn, Mass	May 4, 1869	89, 630
Drain-pipe	T. J. Mayall	Roxbury, Mass	June 21, 1859	24, 476
Drain-pipe	W. G. Monk	Brooklyn, E. D., N. Y	Feb. 11, 1873	135, 721
Drain-pipe	G. Richardson	Milwaukee, Wis	Sept. 6, 1870	107, 103
Drain-pipe machine	H. Bissell	Hartford, Conn	July 11, 1871	116, 920
Drain-pipe machine	I. C. Bryant	Washington, D. C	Nov. 23, 1869	97, 035
Drain-pipe machine	A. J. Davis	Newark, N. J	Nov. 21, 1871	121, 160
Drain-pipe machine	C. W. Kennedy	Brooklyn, E. D., N. Y	July 11, 1871	116, 839
Drain-pipe machine	P. McIntyre	Norwick, Conn	Dec. 7, 1869	97, 668
Drain-pipe machine	B. S. Pierce and M. R. Pierce	New Bedford and Mansfield, Mass.	Apr. 19, 1859	23, 703
Drain-pipe machine	M. R. Pierce	New York, N. Y	Nov. 14, 1871	120, 999
Drain-pipe machine	G. Richardson	Milwaukee, Wis	Sept. 12, 1871	118, 880
Drain-pipe machine	J. W. Stockwell	Portland, Me	June 27, 1871	116, 371
Drain-pipe machine	J. W. Stockwell	Portland, Me	July 4, 1871	116, 643
Drain-pipe mold	J. S. Roake	Williamsburgh, N. Y	Sept. 17, 1872	131, 375
Drain-pipes, Apparatus for carrying	D. W. Brown	Woodbridge, N. J	Mar. 14, 1871	112, 538
Drain-pipes, &c., Apparatus for mixing and preparing material for manufacture of.	J. A. Middleton	Williamsburgh, N. Y	Sept. 17, 1872	131, 366
Drain-pipes, Device for the manufacture of	J. W. Stockwell	Portland, Me	Mar. 19, 1872	124, 702
Drain-pipes, Flask for	E. Dayton	Meriden, Conn	Nov. 29, 1864	45, 229
Drain-pipes, Machine and mold for	P. McIntyre	Hartford, Conn	Oct. 15, 1872	132, 169
Drain-pipes, Machine for making	A. Müller	Jersey City, N. J	Sept. 16, 1873	142, 930
Drain-pipes, Machine for making	A. Skinner	San Francisco, Cal	May 12, 1868	77, 925
Drain-pipes, Machine for making composition	R. Skinner and R. Gaines	San Francisco, Cal	Aug. 9, 1870	106, 290
Drain-pipes, Manufacture of	F. Dickerson, jr	Hartford, Conn	June 25, 1872	128, 369
Drain pipes, tiles, &c., Composition for the manufacture of.	E. J. Eames	New York, N. Y	Apr. 2, 1872	125, 125
Drain roller and molder combined	F. B. Fournier	Berea, Ohio	Feb. 4, 1862	34, 327
Drain, Sediment	G. R. Moore	Lyons, Iowa	Mar. 11, 1873	136, 608
Drain-tile	A. Newkumet	Philadelphia, Pa	Apr. 16, 1861	32, 079
Drain-tile	E. W. Rowe	Brewer, Me	July 3, 1860	29, 040
Drain-tile-laying apparatus	G. W. Nevill	Richmond, Va	Mar. 11, 1873	136, 751
Drain-tile machine	J. Daines	Birmingham, Mich	June 14, 1859	24, 379
Drain-tile machine	G. Graessle	Hamilton, Ohio	Aug. 14, 1860	29, 586
Drain-tile machine	J. Hotchkiss	Yellow Springs, Ohio	Mar. 27, 1860	27, 669
Drain-tile machine	H. Knight	Brooklyn, N. Y	Aug. 18, 1863	39, 579
Drain-tile machine	T. Maycock	Buffalo, N. Y	Nov. 4, 1856	16, 085
Drain-tile machine	S. M. Smith and C. Winegar	Union Springs, N. Y	Jan. 10, 1860	26, 793
Drain-tile mold	J. J. Alvord	Tecumseh, Mich	Aug. 25, 1863	39, 621
Drain-tiles, Apparatus for laying	B. B. Briggs	Sharon, Ohio	Oct. 4, 1859	25, 624
Drain-tiles, Bottom-plate for	J. Parsons	Cleveland, Ohio	Aug. 7, 1860	29, 516
Drain-tiles, Casting	J. Coy	Oswego, N. Y	May 10, 1864	42, 643
Drain-tiles, Mode of laying	H. F. Baker	Centreville, Ind	May 1, 1860	28, 050
Drain-trap	J. C. I. Sturm	Buffalo, N. Y	Jan. 16, 1872	122, 866
Drain-tube	J. McMillan	Perinton, N. Y	Aug. 23, 1870	106, 709
Drain, Underground	S. H. Warner	Darbyville, Ohio	Nov. 22, 1870	109, 474
Drain, Waste-water	C. H. Parker and G. N. Copeland.	Cortland, N. Y	Juiy 2, 1867	66, 380
Drain-water pipes, Machine for making	C. Collier	Charlestown, Mass	Apr. 16, 1867	63, 784
Drains, Apparatus for suppressing effluvia from	W. Bryent	Boston, Mass	Aug. 21, 1866	57, 284
Drains, Construction of cement	B. Livermore	Hartland, Vt	May 1, 1860	28, 091
Drains, Lining underground	J. C. Miller, S. A. Clemens, and G. H. Clemens.	Irwin, Ohio, Rockford, Ill., and Urbana, Ohio.	Oct. 4, 1859	25, 703
Drains, Pen-stock for underground	D. A. Parks	Fairview, Ill	Mar. 10, 1868	75, 295
Drains with cement, Lining underground	A. Watson	London, Ohio	Feb. 28, 1860	27, 324
Drainer, Pipe	A. Brigham	Lawrence, Mass	May 5, 1863	38, 371
Draining and pipe-laying machine	I. C. Pratt	Morton, Ill	Jan. 3, 1860	26, 708
Draining-apparatus	J. Roy	New Orleans, La	May 11, 1869	89, 946
Draining cellars	E. S. Dickinson and J. L. Peake	New York, N. Y	July 18, 1871	117, 156
Draining cellars, Sink-apparatus for	A. Knacker	Meadville, Pa	Apr. 12, 1870	101, 744
Draining-machine	M. Bucklin	Grafton, N. H	Aug. 2, 1859	24, 928
Draining-machine	J. Cole and A. L. O. Wall	De Witt, Ill	Nov. 11, 1856	16, 046
Draining-machine	A. P. Routt	Somerset, Va	July 14, 1857	17, 809
Draining-machine	A. P. Routt	Liberty Mills, Va	Jan. 15, 1867	61, 263
Draining-machine, Centrifugal	W. Richardson	New Orleans, La	Sept. 27, 1853	10, 060
Draining machine, Underground	A. Miller	Mount Pleasant, Iowa	Aug. 9, 1859	25, 027
Drapery-hook	E. Carrington	West Meriden, Conn	Oct. 15, 1867	69, 766
Draw-bar	W. Kotzum	Aurora, Ind	June 29, 1869	92, 062
Draw-bar and bumper, Elastic	T. J. Mayall	Roxbury, Mass	Jan. 11, 1859	22, 667
Draw-bar, Railway	D. S. Beals	Adrian, Mich	Aug. 10, 1869	93, 584
Draw-bars of railway-couplings, Die for making followers for.	J. T. Wilson	Pittsburgh, Pa	Aug. 8, 1871	117, 953
Draw-bars or buffers, Method of making	J. T. Wilson	Pittsburgh, Pa	Aug. 22, 1871	118, 322
Draw-bolts, Blank for	D. B. Oliver	McClure Township, Pa	Mar. 11, 1873	136, 757
Draw-bridge	N. Macneale	Cincinnati, Ohio	Mar. 15, 1870	100, 910
Draw-bridge signal, Railway	S. L. Spofford	Philadelphia, Pa	July 10, 1855	13, 235
Draw-bridge signals, Connection for	T. S. Hall	Stamford, Conn	June 1, 1869	90, 743
Draw-head face-plates for railway-couplings, Die for making.	J. T. Wilson	Pittsburgh, Pa	Aug. 8, 1871	117, 954
Draw-head plates, Bending and punching	P. L. Weimer	Lebanon, Pa	Oct. 31, 1865	50, 756
Draw-heads, Regulating the speed of	H. MacLean	Paterson, N. J	Oct. 23, 1847	5, 341
Draw-heads, Stop-motion for	B. O. Paige	Lowell, Mass	Mar. 3, 1863	37, 823
Draw-heads, Stop-motion for railway	B. O. Paige	Lowell, Mass	Feb. 12, 1861	31, 398

Index of patents issued from the United States Patent Office from 1790 *to* 1873, *inclusive*—Continued.

Invention.	Inventor.	Residence.	Date.	No.
Draw-heads while in motion, Mode of changing the gearing of.	A. Jenks	Bridesburgh, Pa	Dec. 11, 1849	6, 938
Draw-hook and buffer, Self-coupling	P. G. Gardiner	New York, N. Y	May 23, 1871	115, 050
Draw-plate	J. O. Kane	New York, N. Y	July 2, 1867	66, 248
Drawer	B. J. Greely	Boston, Mass	Apr. 22, 1873	138, 017
Drawer	I. A. Howe	Boston, Mass	Aug. 8, 1871	117, 885
Drawer-alarm	W. B. Card	Sag Harbor, N. Y	Jan. 10, 1860	26, 813
Drawer-alarm	F. H. Purington	Willimantic, Conn	Jan. 22, 1861	31, 188
Drawer, Bureau	M. Fitzgibbons	New York, N. Y	Aug. 24, 1869	94, 096
Drawer catch, Money	S. Hubbell, jr	West Salem, Ohio	July 14, 1868	79, 981
Drawer, Closet, bureau, &c	H. R. Taylor	Roxbury, Mass	Apr. 27, 1858	20, 100
Drawer-fastening	J. Schweinfurt	Philadelphia, Pa	Oct. 8, 1872	132, 028
Drawer-fastening	J. Serrill	Philadelphia, Pa	Dec. 19, 1865	51, 625
Drawer-guide	G. S. and H. Curtis	Chicago, Ill	July 20, 1869	92, 798
Drawer-handle	J. Smith	Brockport, N. Y	Apr. 13, 1869	88, 817
Drawer-handle	C. A. Stock	New York, N. Y	Dec. 6, 1870	109, 964
Drawer-lock	A. G. Burton	Rochester, N. Y	Mar. 24, 1868	75, 854
Drawer-lock	E. L. Gaylord	Terryville, Conn	Jan. 1, 1867	60, 874
Drawer-lock	E. G. Gory	Cincinnati, Ohio	Apr. 22, 1873	138, 148
Drawer-lock	I. Robbins	Hughesville, Pa	Mar. 2, 1869	87, 514
Drawer-lock	E. Skinner	Sandwich, N. H	Apr. 13, 1830	....
Drawer-locking device	L. Renker	Hoboken, N. J	Aug. 5, 1873	141, 461
Drawer or till alarm	F. H. Purington	Willimantic, Conn	Dec. 13, 1864	45, 435
Drawer-pull	C. S. and E. A. Barnard	West Meriden, Conn	Feb. 11, 1873	135, 754
Drawer-pull	H. Herlt	New York, N. Y	Feb. 6, 1872	123, 398
Drawer-pull	J. Kintz	West Meriden, Conn	Mar. 21, 1871	112, 816
Drawer-pull	C. H. Pierpent	West Meriden, Conn	Jan. 24, 1871	111, 243
Drawer-pull	W. E. Sparks	New Haven, Conn	Jan. 30, 1872	123, 301
Drawer pull and label, Combined	J. Quinlan	New York, N. Y	June 22, 1869	91, 810
Drawer pull and label, Combined	C. H. Robinson	Greenwich, N. Y	June 29, 1869	92, 102
Drawer, Slide	J. L. Chapman	West Roxbury, Mass	May 3, 1870	102, 489
Drawer, Slide	J. S. Gibbons	Philadelphia, Pa	Apr. 18, 1871	113, 871
Drawer-support	J. Baggs	Easton, Md	Nov. 12, 1872	133, 005
Drawers	J. Bellamy	New York, N. Y	Nov. 19, 1872	133, 138
Drawers	H. Heath	New York, N. Y	Feb. 4, 1868	73, 975
Drawers	O. Jenkins	Boston, Mass	June 22, 1842	2, 690
Drawers	F. Rose	Boston, Mass	Mar. 18, 1873	136, 869
Drawers	J. Ware	North Hoboken, N. J	Oct. 3, 1865	50, 308
Drawers, Cloth	J. Dyson	New Britain, Conn	Oct. 20, 1868	83, 141
Drawers, Cloth	H. Heath	New York, N. Y	Apr. 28, 1868	77, 189
Drawers, Cloth	H. P. Wetmore and J. G. Hitchcock.	Elizabeth, N. J., and New York, N. Y.	Nov. 17, 1868	84, 073
Drawers, Device for cutting out bands and fronts of.	L. Bennet	Amsterdam, N. Y	Mar. 3, 1868	74, 977
Drawers, Device for locking	C. E. Gould	Northampton, Mass	July 16, 1872	129, 404
Drawers, Device for locking a series of	W. F. Daly	Peru, Ind	July 1, 1873	140, 481
Drawers, Device for locking a series of	C. A. Stock	New York, N. Y	Apr. 1, 1873	137, 391
Drawers, Device for supporting	C. F. Langford	Brooklyn, N. Y	Dec. 26, 1871	122, 178
Drawers for wearing-apparel	H. Brewer	New York, N. Y	Feb. 28, 1822	....
Drawers, Locking	E. Brown	Lynchburgh, Va	Oct. 28, 1835	....
Drawers, Men's	E. L. Closse	Cleveland, Ohio	Aug. 19, 1873	142, 000
Drawers, Men's	K. V. R. Lansingh, jr	Albany, N. Y	Nov. 28, 1871	121, 250
Drawers-stretching frame	J. Dyson	New Britain, Conn	Aug. 18, 1868	81, 263
Drawers-supporter	C. F. Langford	Brooklyn, N. Y	Jan. 23, 1872	123, 027
Drawers, Supporter for men's	E. K. Cleaveland	Boston, Mass	Apr. 15, 1873	137, 761
Drawers, Supporting furniture	H. R. Taylor	Roxbury, Mass	Feb. 21, 1860	27, 247
Drawing and camera table	J. B. Stackpole	Boston, Mass	Nov. 17, 1863	40, 647
Drawing apparatus, Perspective	A. Richter	New York, N. Y	Nov. 16, 1852	9, 409
Drawing-board	H. W. Chamberlin	Pittsfield, Mass	Dec. 25, 1849	6, 967
Drawing-board	C. G. Collins	New York, N. Y	June 4, 1872	127, 573
Drawing-board	J. B. Franklin	New York, N. Y	Aug. 8, 1871	117, 761
Drawing-board	I. P. Hansell	Springfield, Ill	Aug. 31, 1858	21, 336
Drawing-board	A. Hitchcock	New York, N. Y	June 3, 1873	139, 507
Drawing-board	T. H. Kane	New York, N. Y	Jan. 7, 1873	134, 550
Drawing-board	W. Keuffel and H. Esser	Hoboken, N. J	May 13, 1873	138, 896
Drawing-board	H. H. Meyer	Denver, Colo	May 27, 1873	139, 411
Drawing-board	L. F. Schwenkel	New York, N. Y	Mar. 26, 1872	125, 089
Drawing-board, Symmetrical	C. D. Hillman	Louisville, Ky	Oct. 24, 1865	50, 583
Drawing-board trestle	J. H. Harden	Philadelphia, Pa	Dec. 2, 1873	145, 174
Drawing-boards, Ruler-attachment for	T. Bergner	Philadelphia, Pa	Apr. 18, 1871	113, 726
Drawing-clamp	J. Bouniol	Philadelphia, Pa	Feb. 4, 1868	73, 947
Drawing, Device for teaching	A. C. Covell	New York, N. Y	May 27, 1873	139, 237
Drawing-frame	S. Brooks and J. Standish	West Gorton, England	July 16, 1872	128, 999
Drawing-frame	G. Draper	Hopedale, Mass	Apr. 18, 1871	113, 752
Drawing-frame	G. Draper	Hopedale, Mass	Dec. 27, 1870	110, 556
Drawing-frame	J. T. Harris	Richmond, R. I	July 2, 1872	128, 618
Drawing-frame	O. P. Hussey	Nashua, N. H	Sept. 14, 1869	94, 752
Drawing-frame	C. P. Leavitt	New York, N. Y	Nov. 22, 1870	109, 430
Drawing-frame	J. McDonald	Concord, N. C	Oct. 1, 1872	131, 766
Drawing-frame	B. Moon	Coventry, R. I	July 18, 1871	117, 098
Drawing-frame	J. Pray and C. Stafford	East Killingly, Conn	Nov. 12, 1846	4, 847
Drawing-frame	G. E. Taft	Northbridge, Mass	Apr. 25, 1871	114, 064
Drawing-frame	T. S. Winn	Lawrence, Mass	Sept. 6, 1870	107, 146
Drawing-frame for drawing fibrous material	S. C. Durgin	Holyoke, Mass	Jan. 25, 1859	22, 768
Drawing-frame for hemp, &c	G. Davis and J. R. Hoover	Elizabethport, N. J	Aug. 9, 1870	106, 130
Drawing-frame for flax, hemp, &c	C. Wall and J. Stewart	New York and Brooklyn, N. Y.	Feb. 1, 1870	99, 503
Drawing-frame regulator	S. J. Whitton	Coleraine, Mass	Feb. 9, 1869	86, 719
Drawing-frame roll	J. M. Stone	North Andover, Mass	May 9, 1865	47, 667
Drawing-frames, Mode of operating rolls in	J. Chase and J. M. Stone	Worcester and North Andover, Mass.	Mar. 29, 1864	42, 076
Drawing-frames, Stop-motion for	J. S. Casey	Voluntown, Conn	Oct. 12, 1869	95, 767
Drawing-frames, Stop-motion for	L. Cutting	Lowell, Mass	May 15, 1834	....
Drawing-frames, Stop-motion for	C. Danforth	Paterson, N. J	Jan. 9, 1849	6, 014
Drawing-frames, &c., Stop-motion for	A. I. Earle	Valley Falls, R. I	May 13, 1873	138, 868
Drawing-frames, Stop-motion for	D. W. Hayden	Wauregan, Conn	Feb. 20, 1872	123, 826
Drawing-frames, Stop-motion for	H. G. Williams	Warren, R. I	May 21, 1861	32, 393
Drawing-frames, Stop-motion for	P. J. Wright and J. Thompson	Holyoke, Mass	May 27, 1873	139, 447
Drawing-frames, Stopping-mechanism for	J. C. Taft	Pawtucket, R. I	Oct. 28, 1873	144, 162

Index of patents issued from the United States Patent Office from 1790 *to* 1873, *inclusive*—Continued.

Invention.	Inventor.	Residence.	Date.	No.
Drawing-frames, &c., Top-roll for	H. T. Potter	Norwich, Conn	Sept. 2, 1873	142, 509
Drawing-instrument	E. W. Ellsworth	East Windsor, Conn	June 7, 1845	4, 070
Drawing-instrument	J. P. Jamison	New York, N. Y	June 18, 1861	32, 566
Drawing-instrument	W. W. Wythes	Philadelphia, Pa	July 27, 1858	21, 041
Drawing-knife	J. H. Atwater	Oshtemo, Mich	June 18, 1870	98, 903
Drawing-knife	C. C. Barton	Rochester, N. Y	June 21, 1870	104, 539
Drawing-knife	W. Brady	Mount Joy, Pa	Mar. 5, 1872	124, 246
Drawing-knife	M. Brooks	Chicago, Ill	Dec. 16, 1873	145, 485
Drawing-knife	T. M. Clarke	Winsted, Conn	Nov. 21, 1871	121, 084
Drawing-knife	E. Parker	Chambersburgh, Pa	Sept. 9, 1873	142, 720
Drawing-knife	E. Richards	Hingham, Mass	June 28, 1836	....
Drawing-knife	J. C. Smead	Huntington, Ohio	Nov. 19, 1872	133, 126
Drawing-knife	H. C. Smith	Westville, Conn	July 9, 1872	128, 821
Drawing-knife	A. M. Steele	Danbury, Conn	Jan. 30, 1872	123, 206
Drawing-knife	R. N. Watrous	Charlestown, Ohio	Dec. 15, 1857	18, 877
Drawing-machine roller	F. and G. G. Crague	Lewiston, Me	Apr. 28, 1868	77, 173
Drawing-machinery, Applying pressure to top-rolls of.	N. E. Hale	Nashua, N. H	Nov. 8, 1859	26, 031
Drawing metals, Apparatus for	L. Christoph, W. Hanksworth, and G. P. Harding.	Paris, France, Gartness, North Britain, and Chiswick, England.	July 21, 1868	80, 057
Drawing perspective mechanically	S. De Witt	Albany, N. Y	Apr. 30, 1813	....
Drawing rods and shafting, Machine for cold	G. H. Billings	Cincinnati, Ohio	Aug. 13, 1872	130, 465
Drawing-roller	D. Hussey	Nashua, N. H	May 15, 1866	54, 732
Drawing-roller	S. P. Spencer	Lancaster, Pa	June 29, 1858	20, 766
Drawing-roller cleaner	D. Crowley	Philadelphia, Pa	Feb. 11, 1868	74, 202
Drawing-roller regulator	W. Hayden	Windham, Conn	Mar. 12, 1850	7, 165
Drawing-rollers, Machine for rolling cots on	W. C. Joslin	Putnam, Conn	Mar. 31, 1868	76, 199
Drawing-rollers, Mode of lubricating	F. Hardenburg	Providence, R. I	June 25, 1872	128, 224
Drawing-rollers, Wiper for	J. M. Dunham	Holyoke, Mass	Nov. 5, 1872	132, 714
Drawing scrolls, Instrument for	E. E. Stebbins	Meadville, Pa	Apr. 22, 1873	138, 053
Drawing-stand	W. Bell	Buffalo, N. Y	Oct. 31, 1871	120, 364
Drawing-table	J. L. Ross and F. C. Hanson	Boston and Charlestown, Mass.	Dec. 26, 1871	122, 134
Drawing-table and desk, Combined	J. A. Wilkens	Indianapolis, Ind	Dec. 9, 1873	145, 378
Drawings, Instrument for hatch-lining	E. K. Haynes	Hanover, N. H	Dec. 10, 1867	72, 036
Drawings, Machine for making perspective	R. Lyons	New Rumley, Ohio	June 14, 1834	....
Dray	C. S. Martin	Milwaukee, Wis	Apr. 30, 1867	64, 346
Dray	E. Warren and T. Brangwin	Ceresco, Mich	Apr. 14, 1868	76, 856
Dray	F. Van Doren	Adrian, Mich	July 2, 1867	66, 426
Dray-skid	E. H. Boswell	Philadelphia, Pa	Jan. 5, 1864	41, 054
Dredge	E. B. Bishop	San Francisco, Cal	Apr. 22, 1873	137, 998
Dredge, Bail	N. A. Williams	Warwick, R. I	Jan. 2, 1872	122, 423
Dredge-box	H. M. Clark	Toledo, Ohio	Oct. 1, 1872	131, 810
Dredge-box	P. F. Cooley	Pittsfield, Mass	Oct. 1, 1872	131, 744
Dredge-box	J. Dalzell	Birmingham, Pa	Nov. 26, 1872	133, 307
Dredge-box	W. S. Potwin	Chicago, Ill	Feb. 20, 1872	123, 792
Dredge-box	G. W. Putnam	Peterborough, N. Y	June 4, 1867	65, 430
Dredge-box	T. Smith	Cleveland, Ohio	Jan. 21, 1873	135, 048
Dredge-box	C. F. Spencer	Cleveland, Ohio	Oct. 24, 1871	120, 339
Dredge-box	A. F. Tripp	Buffalo, N. Y	Aug. 16, 1870	106, 430
Dredge-box	T. Williams	Boston, Mass	July 9, 1867	66, 661
Dredge-box, cake-cutter, and grater	S. Cooke	Bayonne City, N. J	Apr. 26, 1870	102, 225
Dredge-box with grater and cake-cutter attached	G. D. Bayley	Lebanon, N. H	Feb. 17, 1863	37, 716
Dredge, Salt and pepper	D. C. Ripley	Pittsburgh, Pa	Sept. 2, 1873	142, 351
Dredger	A. C. Booth	Portland, Me	Apr. 29, 1873	138, 227
Dredger	T. J. Clepper	Columbia, Pa	Nov. 26, 1872	133, 303
Dredger	J. F. Morse	Oshkosh, Wis	Sept. 9, 1873	142, 577
Dredger	R. G. Packard	Brooklyn, N. Y	Jan. 14, 1873	134, 817
Dredger	T. Symonds	Portland, Me	Feb. 25, 1873	136, 187
Dredger	T. Symonds	Portland, Me	Apr. 29, 1873	138, 295
Dredging and excavating machine	W. Atkinson	Brooklyn, N. Y	July 7, 1863	39, 194
Dredging and excavating machine	W. H. Pearson	Philadelphia, Pa	Mar. 22, 1870	101, 157
Dredging-apparatus	E. Bazin	Paris, France	Sept. 30, 1873	143, 269
Dredging-apparatus	B. and D. Hughes	Rochester, N. Y	Sept. 21, 1869	95, 114
Dredging-apparatus	C. A. Scanlan	Charleston, S. C	Apr. 19, 1870	102, 049
Dredging-boat for excavating rivers	E. L. Brady	New Orleans, La	Dec. 17, 1867	72, 360
Dredging boat, Harbor	W. H. Nobles	Saint Paul, Minn	June 5, 1866	55, 346
Dredging-boat, Steam	W. H. Nobles	Saint Paul, Minn	June 5, 1866	55, 347
Dredging-can	D. W. Pepper	Philadelphia, Pa	Apr. 26, 1870	102, 428
Dredging-machine	E. B. Bishop	Shreveport, La	Apr. 13, 1858	19, 908
Dredging-machine	A. Blood, sr	Norfolk, Va	Aug. 11, 1857	17, 959
Dredging-machine	A. Boschke	Boston, Mass	Sept. 22, 1868	82, 376
Dredging-machine	A. Boschke	San Francisco, Cal	Aug. 12, 1873	141, 752
Dredging-machine	T. Byrne	New York, N. Y	Sept. 11, 1866	57, 859
Dredging-machine	D. Carmichael and J. C. Osgood	Brooklyn and Chittenango, N. Y.	May 30, 1846	4, 547
Dredging-machine	O. Chanute and G. L. Morison	Kansas City, Mo	Jan. 18, 1870	98, 848
Dredging-machine	W. A. Collins	Bloomfield, N. J	Dec. 23, 1873	145, 723
Dredging-machine	D. C. Cregier	Chicago, Ill	June 30, 1868	79, 448
Dredging-machine	J. Du Bois	Williamsport, Pa	Aug. 18, 1868	81, 076
Dredging-machine	W. Easby	Washington, D. C	Aug. 25, 1840	1, 727
Dredging-machine	J. Ebbert	Chicago, Ill	June 27, 1871	116, 282
Dredging-machine	C. H. Fondé	Mobile, Ala	Apr. 17, 1855	12, 720
Dredging-machine	C. H. Fondé and T. B. Lyons	Mobile, Ala	Mar. 21, 1854	10, 668
Dredging-machine	H. Gonellaz	Houma, La	Aug. 6, 1872	130, 213
Dredging-machine	H. Gonellaz	Vermillionville, La	July 15, 1873	140, 821
Dredging-machine	L. J. Gonyo	North Buffalo, N. Y	June 4, 1872	127, 595
Dredging-machine	A. J. Gove	San Francisco, Cal	Nov. 10, 1868	83, 845
Dredging-machine	A. J. Gove	San Francisco, Cal	Sept. 24, 1872	131, 608
Dredging-machine	J. Hamilton	New York, N. Y	Mar. 30, 1852	8, 840
Dredging-machine	D. S. Howard	Lyonsdale, N. Y	Jan. 9, 1855	12, 201
Dredging-machine	D. S. Howard	Lyons Falls, N. Y	Jan. 21, 1870	104, 594
Dredging-machine	J. Johnson	Saco, Me	Aug. 7, 1866	56, 948
Dredging-machine	S. H. Long	Alton, Ill	Mar. 26, 1861	31, 811
Dredging-machine	W. H. Lotz	Chicago, Ill	Jan. 9, 1872	122, 529
Dredging-machine	J. McClean	New Orleans, La	Apr. 12, 1870	101, 898
Dredging-machine	J. R. McClintock and J. K. Scott	New Orleans, La	Jan. 15, 1867	61, 227

Index of patents issued from the United States Patent Office from 1790 *to* 1873, *inclusive*—Continued.

Invention.	Inventor.	Residence.	Date.	No.
Dredging-machine	J. H. McLean	Saint Louis, Mo	July 9, 1867	66, 608
Dredging-machine	A. Menge	Point à la Hache, La	July 12, 1859	24, 750
Dredging-machine	A. Minard	New York, N. Y	Sept. 28, 1858	21, 613
Dredging-machine	G. Mitchell	Philadelphia, Pa	July 5, 1870	104, 980
Dredging-machine	J. H. Newcomb	Port Norris, N. J	May 20, 1862	35, 324
Dredging-machine	W. H. Nobles	Saint Paul, Minn	Aug. 7, 1866	56, 984
Dredging-machine	R. R. Osgood	Troy, N. Y	June 28, 1870	104, 757
Dredging-machine	R. G. Packard	Brooklyn, N. Y	Oct. 1, 1872	131, 825
Dredging-machine	E. Platt	Charleston, S. C	July 18, 1871	117, 109
Dredging-machine	W. D. Robertson	San Francisco, Cal	May 24, 1870	103, 373
Dredging-machine	J. Stewart	New London, Conn	Dec. 28, 1858	22, 458
Dredging-machine	W. T. Thelin	Baltimore, Md	Apr. 2, 1872	125, 351
Dredging-machine	I. D. Vandecar	Chicago, Ill	Apr. 23, 1867	64, 169
Dredging-machine	I. D. Vandecar	Chicago, Ill	Oct. 10, 1871	119, 901
Dredging-machine	J. L. Vergniais	Paris, France	July 30, 1867	67, 384
Dredging-machine	N. A. Williams	Warwick, R. I	May 28, 1872	127, 397
Dredging machine bucket	T. and A. Walsh	New York, N. Y	Jan. 21, 1868	73, 678
Dredging-machine-dipper handles, Mode of fastening racks of.	R. R. Osgood	Troy, N. Y	Sept. 30, 1873	143, 247
Dredging machine, Harbor and river	A. J. Gove	San Francisco, Cal	Feb. 21, 1865	46, 463
Dredging-machine, Hydraulic	R. S. Elliott	Saint Louis, Mo	Sept. 28, 1869	95, 213
Dredging-machine, Pneumatic	A. F. Du Faur and C. H. Campbell.	Newark, N. J., and New York, N. Y.	Nov. 12, 1872	133, 022
Dredging-machine, Rotary	J. Molyneux	Bordentown, N. J	Oct. 4, 1859	25, 704
Dredging machine, Sand	J. T. Clarkson	Chicago, Ill	Mar. 26, 1872	125, 023
Dredging-machines, Method of directing the scoops in.	J. Callaghan	New Bedford, Mass	Jan. 16, 1849	6, 029
Dredging or excavating machine	O. Allen	Norwich, Conn	May 1, 1845	4, 020
Dredging-scoop	A. C. Ellithorpe	Chicago, Ill	June 20, 1871	116, 171
Dredging, Spice and pepper box	G. W. Putnam	Peterborough, N. Y	Nov. 5, 1867	70, 613
Dress and satchel holder combined	G. McFadden	Thomaston, Conn	July 21, 1868	80, 196
Dress and skirt protector	A. H. Graton	Lawrence, Kans	Nov. 16, 1869	96, 797
Dress-body lining, Lady's	S. Moschcowitz	New York, N. Y	Jan. 24, 1871	111, 236
Dress, Boy's	C. Herwick	New York, N. Y	May 16, 1826	....
Dress-cutting, System of	H. M. Carpenter	Grand Rapids, Mich	Sept. 4, 1866	57, 837
Dress-cutting, System of	C. Dittenhafer	Canton, Ohio	July 17, 1866	56, 383
Dress, Diving	J. R. Campbell	Boston, Mass	Nov. 30, 1835	....
Dress-elevator	M. Dewey	New Albany, Ind	July 16, 1872	129, 399
Dress-elevator	H. Essex and J. Johnson	West Haverstraw and Brooklyn, N. Y.	June 26, 1866	55, 842
Dress-elevator	M. Fox	New York, N. Y	Feb. 11, 1873	135, 703
Dress-elevator	G. Kammerl and D. L. Bollermann.	New York, N. Y	Aug. 28, 1866	57, 519
Dress-elevator	G. W. Keller	Philadelphia, Pa	Nov. 5, 1867	70, 575
Dress-elevator	S. A. Moody	New York, N. Y	Sept. 11, 1866	57, 950
Dress-elevator	I. Nathan	New Haven, Conn	Mar. 20, 1866	53, 324
Dress-elevator	T. V. Phelps	Worcester, Mass	July 10, 1866	56, 341
Dress-elevator	W. E. Stein	New York, N. Y	Mar. 20, 1860	27, 578
Dress-facing	C. Tage	Philadelphia, Pa	Nov. 11, 1873	144, 576
Dress-facing	W. H. Gallup	Troy, N. Y	Apr. 29, 1873	138, 326
Dress-frame, Lady's	J. A. Mackee	Boston, Mass	Jan. 10, 1865	45, 840
Dress-goods holder	J. R. Palmenberg	New York, N. Y	Jan. 10, 1860	26, 781
Dress-hook	A. S. Grant	Rochester, N. Y	Mar. 26, 1872	124, 948
Dress-hook, Self-fastening	M. A. Keilig	Boston, Mass	Dec. 31, 1867	72, 862
Dress, Jubilee stock	J. W. Strange	Bangor, Me	May 8, 1860	28, 213
	W. J. Cantolo and R. M. Kerrison.	New York, N. Y	Aug. 30, 1826	....
Dress, Lady's	H. and E. S. Judson	Brooklyn, N. Y	June 10, 1873	139, 677
Dress, Lady's	A. Kelley	New York, N. Y	Nov. 14, 1871	120, 981
Dress-looping clasp, Lady's	H. H. Cole	New York, N. Y	Feb. 27, 1866	52, 822
Dress measuring and fitting apparatus	P. A. La Ment	New York, N. Y	Aug. 14, 1866	57, 254
Dress, Mode of attaching trimming to articles of	J. Sims	Liverpool Road, England	Oct. 5, 1869	95, 525
Dress-pattern	E. P. Smith	Chicago, Ill	Apr. 19, 1873	137, 967
Dress-protector	A. Herrmann	New York, N. Y	July 1, 1873	140, 415
Dress protector	E. Higgins	Boston, Mass	Aug. 5, 1862	36, 125
Dress-protector	T. Himes	New Albany, Ind	June 30, 1868	79, 350
Dress-protector	F. Wittram	San Francisco, Cal	Mar. 22, 1870	101, 203
Dress-protector, Lady's	T. D. Day	New York, N. Y	Feb. 7, 1865	46, 218
Dress-protector, &c., Fabric for	J. McMullen	Brooklyn, N. Y	Apr. 1, 1873	137, 311
Dress-shield	B. M. Hotchkiss	Naugatuck, Conn	Nov. 1, 1870	108, 908
Dress-shield	I. M. Post	Philadelphia, Pa	July 1, 1873	140, 538
Dresses, Apparatus for obtaining the measure for ladies'.	C. E. McDonald	Indianapolis, Ind	Nov. 27, 1866	60, 028
Dresses, Cutting ladies'	H. Segar	Macon, Ga	Jan. 28, 1846	4, 367
Dresses, Fastening for ornaments on	D. B. Howell	New York, N. Y	May 5, 1863	38, 390
Dresses, Fitting ladies'	S. S. Richardson	Baldwin, Me	Apr. 4, 1844	3, 522
Dresses, Frame for finishing ladies'	J. R. Williams	Baltimore, Md	Mar. 3, 1824	....
Dresses, Instrument for drafting ladies'	H. A. Fowler	Afton, N. Y	May 13, 1862	35, 226
Dresses, Method of finding waist and chest measurement of ladies'.	M. M. Turner	North Fairfield, Ohio	Feb. 14, 1865	46, 409
Dresses, &c., Taking measure and drafting for the cutting out of ladies'.	A. A. Tenther	Philadelphia, Pa	Jan. 23, 1841	1, 944
Dressing-case	H. W. Eastman	Baltimore, Md	June 29, 1869	91, 830
Dressing-case and bath-tub	C. A. Staples	Boston, Mass	Sept. 10, 1867	68, 802
Dressing-case. Portable	J. Russell	Boston, Mass	Apr. 7, 1863	38, 130
Dressing-combs of wood, Manufacturing	N. Bushnell	Middletown, Conn	Apr. 14, 1829	....
Dressing conical tapering surfaces, Machine for	P. Peckham	Petersham, Mass	July 3, 1855	13, 180
Dressing-frames, Composition roll for	R. Collier	Laurel, Md	Aug. 15, 1871	117, 985
Dressing irregular forms, Machine for	A. Cary	Worcester, Mass	Nov. 5, 1850	7, 750
Dressing-roller	B. R. Cotton	Lewiston, Me	Dec. 10, 1867	71, 991
Dressing sticks to polygonal forms, Machine for	J. W. Killam	East Wilton, N. H	Jan. 22, 1856	14, 136
Drier: *See* Dryer.				
Drift	J. Eglin	Manchester, England	Feb. 19, 1867	62, 123
Drill:				

See Boring-drill. Bow-drill. Brace-drill. Chuck-drill. Dental drill. Dentist's drill.

Index of patents issued from the United States Patent Office from 1790 *to* 1873, *inclusive*—Continued.

Invention.	Inventor.	Residence.	Date.	No.
Drill—Continued. *See* Expanding drill. Grain-drill. Guano and seed drill. Hand-drill. Jeweler's drill. Machine-drill. Metal-drill. Mining-drill. Oil-well drill. Planter-drill. Pneumatic drill. Power and hand drill. Ratchet-drill. Reservoir-drill. Rock-drill. Seed-drill. Self-feeding drill. Steam-drill. Stone-drill. Stone-channeling drill. Suction-drill. Tapering drill. Twist-drill. Upright drill. Well-drill. Well-boring drill. Wheat-drill.				
Drill	A. E. Allen	Perch River, N. Y	Sept. 3, 1872	131, 074
Drill	C. H. Amidon	Miller's Falls, Mass	Dec. 24, 1872	134, 237
Drill	L. Andrews	Biddeford, Me	Jan. 15, 1867	61, 134
Drill	H. and D. L. Bates	Dayton, Ohio	July 13, 1869	92, 419
Drill	T. V. Boyden	Bridgeport, Conn	Feb. 28, 1871	112, 115
Drill	G. F. Case	New York, N. Y	Feb. 12, 1867	62, 003
Drill	R. Dudgeon	New York, N. Y	Aug. 2, 1864	43, 673
Drill	F. Glasser	Mystic Bridge, Conn	Sept. 1, 1868	81, 769
Drill	H. Hufendeck	Saint Louis, Mo	Jan. 1, 1867	60, 895
Drill	C. W. Le Count	Norwalk, Conn	Dec. 22, 1868	85, 107
Drill	H. Loftic and E. Hinman	Syracuse, N. Y	Nov. 21, 1865	51, 065
Drill	L. G. Marshall	Mokena, Ill	July 18, 1865	48, 819
Drill	L. G. Marshall	Mokena, Ill	July 18, 1865	48, 872
Drill	J. M. May	Janesville, Wis	Aug. 1, 1865	49, 129
Drill	C. L. Noé	Bergen Point, N. J	Mar. 14, 1865	46, 815
Drill	M. Nottingham and W. Duncan.	Vinton, Iowa	Nov. 13, 1866	59, 631
Drill	A. Shiland	West Troy, N. Y	Mar. 21, 1865	46, 949
Drill	W. Stivers	New York, N. Y	Feb. 9, 1864	41, 546
Drill	G. C. Taft	Worcester, Mass	Feb. 28, 1860	27, 320
Drill	G. C. Taft	Worcester, Mass	June 28, 1864	43, 372
Drill	A. Thompson	Burlington, Vt	Apr. 19, 1870	102, 066
Drill	F. Vestor	Newark, N. J	Mar. 24, 1868	75, 816
Drill	D. Warner	Port Clinton, Ohio	Aug. 13, 1867	67, 825
Drill	R. G. Wells	Plumer, Pa	July 4, 1865	48, 609
Drill	E. S. Young and A. Whipple	Whitingsville, Mass	Sept. 10, 1867	68, 827
Drill and bit stock	C. H. Amidon	Miller's Falls, Mass	July 1, 1873	140, 451
Drill and blacksmith's tongs, Combined	F. Nevergold and G. Stackhouse.	Pittsburgh, Pa	June 19, 1866	55, 696
Drill and counter sink	P. A. Whitney	Woodstock, Vt	Feb. 25, 1868	74, 870
Drill and crane attachment	I. S. Lauback	New York, N. Y	Apr. 17, 1866	53, 992
Drill and cultivator teeth, Securing	J. H. Thomas and P. P. Mast	Springfield, Ohio	July 12, 1870	105, 387
Drill and fertilizer, Combined	G. E. Cooper	Baltimore, Md	June 5, 1866	55, 243
Drill and holder	I. McLaughlin	East Arlington, Vt	Oct. 18, 1870	108, 374
Drill and sand-pump	E. R. Morrison	New York, N. Y	Nov. 15, 1864	45, 068
Drill and saw-gummer, Combined	W. C. Marr	Peru, Wis	July 27, 1869	92, 983
Drill and seeder, Combined	J. E. Buxton	Owatonna, Minn	July 6, 1869	92, 262
Drill and seeder, Combined	J. E. Buxton	Owatonna, Minn	Mar. 14, 1871	112, 540
Drill and seeding-machine, Combined	J. E. Buxton	Owatonna, Minn	Aug. 16, 1870	106, 320
Drill-bit	W. W. Grier and R. H. Boyd	Hulton, Pa	May 23, 1865	47, 812
Drill-bit	S. A. Morse	East Bridgewater, Mass	Apr. 7, 1863	38, 119
Drill-bit and reamer	P. M. Ward	Cow Run, Ohio	Nov. 21, 1871	121, 223
Drill-coupling	E. E. and C. T. Packer	Philadelphia, Pa	June 26, 1866	55, 895
Drill, Cylinder	C. S. Pattison	North Adams, Mass	Apr. 25, 1871	114, 193
Drill-extractor	D. Kunkel	Oregon, Mo	Aug. 26, 1873	142, 244
Drill-frame	G. C. Taft	Worcester, Mass	Nov. 27, 1860	30, 774
Drill-gage	A. C. Trafton	Worcester, Mass	June 3, 1873	139, 487
Drill-gage	W. C. Wells	Newark, N. J	Nov. 28, 1865	51, 248
Drill-gage	W. C. Wells	Newark, N. J	Mar. 31, 1868	76, 123
Drill gage, Machine	E. G. Parkhurst	Hartford, Conn	Oct. 1, 1872	131, 775
Drill-guide	J. F. Case	Windsor, Vt	Mar. 22, 1870	101, 095
Drill-head	W. W. St. John	Syracuse, N. Y	Jan. 9, 1866	51, 980
Drill-holder	T. K. Bacon	Norwich, Conn	May 26, 1868	78, 251
Drill-holder	C. Burleigh	Fitchburgh, Mass	July 28, 1868	80, 386
Drill-holder	W. Hall, jr	North Adams, Mass	July 28, 1868	80, 406
Drill-holes, Device for charging	F. X. Lavalle	Auburn, Cal	Nov. 12, 1872	133, 040
Drill-jar	J. C. Eastman	Titusville, Pa	July 16, 1867	66, 816
Drill-jar	E. Guillod	Titusville, Pa	Nov. 24, 1868	84, 353
Drill-jar	R. McMullen	Titusville, Pa	Apr. 1, 1873	137, 378
Drill-jar	C. A. Read	Lockport, N. Y	Nov. 28, 1865	51, 217
Drill-jars, Construction of	E. Guillod	Titusville, Pa	June 16, 1868	78, 958
Drill, Machine	J. Conner and T. Newby	Richmond, Ind	Aug. 7, 1855	13, 383
Drill or pump-rod coupling	J. R. Cross	Chicago, Ill	Aug. 15, 1865	49, 386
Drill or well tube	J. Hutchins	Elmira, N. Y	Nov. 13, 1866	59, 608
Drill-press	J. M. Cullen and A. J. Baird	Pittsburgh, Pa	Oct. 27, 1868	83, 366
Drill-rest	C. Teachout	Waterford, N. Y	June 19, 1860	28, 791
Drill-rod attachment	R. S. Torrey	Bangor, Me	Nov. 7, 1865	50, 857
Drill-rod coupling	R. A. Clark	Petrolia City, Pa	Aug. 12, 1873	141, 694
Drill-rod coupling	W. H. Downing	Shamburgh, Pa	Nov. 5, 1872	132, 755
Drill, roller, and cultivator, Combined	T. S. Mills	Kendallsville, Ind	Jan. 14, 1868	73, 365
Drill-shaft, Feeding	G. C. Taft	Worcester, Mass	Mar. 24, 1857	16. 900
Drill-sharpener	E. W. Walton	Drytown, Cal	Mar. 3, 1868	75, 084
Drill-spring for quarrying stone, &c	E. G. Lamson	Shelburne Falls, Mass	Mar. 26, 1867	63, 262
Drill-stock	M. S. Brooks	Chester, Conn	May 17, 1859	24, 005
Drill-stock	C. M. Daboll	New London, Conn	Aug. 11, 1868	80, 812
Drill-stock	M. Hainque	San Francisco, Cal	June 23, 1868	79, 115
Drill-stock	D. F. Hartford	Boston, Mass	Apr. 17, 1866	54, 069
Drill-stock	W. H. Rand	Brooklyn, N. Y	Apr. 4, 1871	113, 344
Drill-stock, Geared	G. Page	Keene, N. H	May 8, 1838	730
Drill-teeth	S. Black	West Alexandria, Ohio	Mar. 4, 1873	136, 408
Drill-teeth setting	P. I and P. Schmitt	Waterloo, Ill	Feb. 16, 1869	87, 074
Drill, Well	W. Tingley	New York, N. Y	Mar. 26, 1867	63, 338
Drills, Feed-motion for	G. F. Case	New York, N. Y	Nov. 27, 1866	59, 963

Index of patents issued from the United States Patent Office from 1790 *to* 1873, *inclusive*—Continued.

Invention.	Inventor.	Residence.	Date.	No.
Drills, Gearing for operating	A. Blatchly	San Francisco, Cal	May 16, 1871	114, 754
Drills, Machine for operating	J. W. Stockwell	Portland, Me	Apr. 1, 1873	137, 504
Drills, Mode of mounting	H. Haupt	Cambridge, Mass	May 2, 1865	47, 541
Drilling and babbiting jig	J. Underwood	Muscatine, Iowa	Oct. 29, 1867	70, 294
Drilling and bolt-tapping machine	C. W. Coe	Fentonville, Mich	Jan. 19, 1869	86, 001
Drilling and boring machine	H. Haupt	Cambridge, Mass	May 23, 1865	47, 819
Drilling and boring machine	S. W. Shryock	Hopkinsville, Ky	Nov. 6, 1855	13, 766
Drilling and driving machine	I. M. Rose	New York, N. Y	Mar. 13, 1866	53, 189
Drilling and hoisting machine	W. C. McGill and A. J. Gibson	Cincinnati, Ohio	Mar. 6, 1866	53, 022
Drilling and milling machine	W. D. Sloan	New York, N. Y	Sept. 29, 1857	18, 305
Drilling and pumping apparatus	M. T. McCormick	Meadville, Pa	July 2, 1867	66, 372
Drilling and pumping machinery, Standard jack-post for.	W. W. Eastman and W. H. H. Morris.	Meadville, Pa	Sept. 9, 1873	142, 619
Drilling-and screw-cutting machine	J. P. Heacock	Marlborough, Ohio	Nov. 27, 1855	13, 845
Drilling-and screw-cutting machinery	C. W. Coe	Corunna, Mich	Jan. 20, 1863	37, 433
Drilling-apparatus	J. Fanning	New York, N. Y	July 2, 1867	66, 229
Drilling-apparatus	A. J. Fullam	Springfield, Vt	July 30, 1867	67, 284
Drilling-apparatus	S. Lauchli	Saint Louis, Mo	June 1, 1869	90, 670
Drilling-apparatus	J. Y. Smith	Pittsburgh, Pa	Mar. 26, 1867	63, 319
Drilling-apparatus, Submarine	S. Lewis	Williamsburgh, N. Y	June 7, 1870	103, 899
Drilling, Coupling-tool for	J. R. Moore	Brooklyn, N. Y	Mar. 28, 1865	47, 031
Drilling-instrument	N. Barnum and G. C. Schreiber	Saint Louis, Mo	June 25, 1867	65, 991
Drilling-machine	M. M. Ammidown	Boston, Mass	Dec. 4, 1866	60, 114
Drilling-machine	T. P. Bashford	Decatur, Mich	Feb. 27, 1866	52, 815
Drilling-machine	G. W. Bishop	Stamford, Conn	Oct. 8, 1867	69, 533
Drilling-machine	P. Blaisdell	Worcester, Mass	Jan. 28, 1873	135, 313
Drilling-machine	J. Button	Springfield, Mass	Nov. 14, 1865	50, 985
Drilling-machine	C. W. Coe	Ashtabula, Ohio	Dec. 7, 1852	9, 441
Drilling-machine	J. Cumming	Boston, Mass	Oct. 25, 1859	25, 886
Drilling-machine	W. H. Elliot	New York, N. Y	June 18, 1867	65, 802
Drilling-machine	W. and J. Gardam	Williamsburgh, N. Y	Febr 20, 1872	123, 886
Drilling-machine	A. Gordon	New York, N. Y	Apr. 19, 1864	42, 365
Drilling-machine	W. D. Grimshaw	Newark, N. J	Sept. 3, 1867	68, 362
Drilling-machine	J. Hale	Scranton, Pa	Oct. 26, 1869	96, 227
Drilling-machine	E. Hall	Byron, N. Y	Aug. 29, 1839	1, 305
Drilling-machine	A. Hatch	New Haven, Conn	Nov. 21, 1865	51, 113
Drilling-machine	J. G. Hirzel	Wilmington, Del	Mar. 13, 1866	53, 145
Drilling-machine	T. M. Howard	Charlestown, Pa	Apr. 17, 1866	53, 981
Drilling-machine	A. P. Jackson and L. Thompson.	Memphis, Ind	Nov. 28, 1865	51, 189
Drilling-machine	G. Koch	Cass, Pa	Sept. 12, 1871	118, 950
Drilling-machine	I. S. Lauback	New York, N. Y	Feb. 21, 1865	46, 478
Drilling-machine	J. Lemman	Cincinnati, Ohio	Aug. 8, 1871	117, 896
Drilling-machine	W. Lyon	New York, N. Y	Jan. 31, 1865	46, 123
Drilling-machine	H. Martin	Duncan, Pa	May 6, 1873	138, 570
Drilling-machine	A. Morgan	Wooster, Ohio	May 30, 1844	3, 609
Drilling-machine	A. Morgan	Massillon, Ohio	May 15, 1847	5, 116
Drilling-machine	N. A. Morlan	Salem, Ohio	Dec. 31, 1872	134, 392
Drilling-machine	W. Morton	Woodhull, N. Y	Mar. 22, 1864	42, 011
Drilling-machine	J. Moulton	Boston, Mass	Mar. 14, 1865	46, 813
Drilling-machine	R. Nutty	New York, N. Y	Sept. 18, 1866	58, 175
Drilling-machine	W. F. Parker	Andover, Mass	Apr. 30, 1867	64, 357
Drilling-machine	C. P. Philippi	Crown Point, Ind	June 23, 1863	38, 981
Drilling-machine	W. Sellers	Philadelphia, Pa	July 2, 1872	128, 665
Drilling-machine	F. Shaller	Hudson, N. Y	June 26, 1866	55, 918
Drilling-machine	J. Smith	New Haven, Conn	May 24, 1870	103, 515
Drilling-machine	R. H. Springsteed	Wooster, Ohio	Feb. 20, 1843	2, 961
Drilling-machine	G. C. Taft	Worcester, Mass	Apr. 18, 1871	113, 815
Drilling-machine	R. A. Thomas	Damascus, Cal	Aug. 7, 1866	57, 011
Drilling-machine	W. H. Thorne	Philadelphia, Pa	June 3, 1873	139, 629
Drilling-machine	W. H. Thorne and I. M. De Haven.	Philadelphia, Pa	Sept. 16, 1873	142, 966
Drilling-machine	A. and C. Van Haagen	Philadelphia, Pa	Apr. 23, 1872	126, 111
Drilling-machine	D. Warner, W. F. Peiffer, and A. F. Lepper.	Port Clinton, Ohio	Aug. 13, 1867	67, 826
Drilling-machine	W. C. Whipple	New Haven, Conn	July 13, 1869	92, 682
Drilling-machine	E. J. Worcester	Worcester, Mass	Feb. 15, 1870	99, 803
Drilling-machine	E. J. Worcester	Worcester, Mass	Feb. 4, 1873	135, 613
Drilling-machine	E. J. Worcester and A. F. Prentice.	Worcester, Mass	May 21, 1872	127, 132
Drilling-machine carriage	C. Burleigh	Fitchburgh, Mass	July 28, 1868	80, 387
Drilling-machine-cutter head	E. Weicke	Philadelphia, Pa	Aug. 26, 1873	142, 133
Drilling machine, Hand	R. Daniels	Woodstock, Vt	Sept. 21, 1852	9, 271
Drilling machine, Hand	T. J. Sloan	New York, N. Y	June 25, 1872	128, 430
Drilling machine, Metal	W. H. Judson	Titusville, Pa	July 23, 1872	129, 831
Drilling machine, Metal	J. C. Keller	Philadelphia, Pa	July 15, 1873	140, 985
Drilling-machine, Portable	W. H. Thorne	Philadelphia, Pa	Apr. 25, 1871	114, 229
Drilling machine, Radial	G. A. Gray, jr	Cincinnati, Ohio	Dec. 14, 1869	97, 908
Drilling-machine, Steam	J. W. Fowle	Boston, Mass	Mar. 11, 1851	7, 972
Drilling-machine, Subaqueous	S. Lewis	Brooklyn, N. Y	Jan. 5, 1869	85, 597
Drilling-machine table	T. Reaney	Chester, Pa	Jan. 17, 1871	111, 084
Drilling-machine, Adjustable feed-gear for	A. R. Morrill and H. Baldwin	Nashville, N. H	Aug. 27, 1850	7, 604
Drilling metal, Machine for	G. C. Taft	Worcester, Mass	Feb. 1, 1870	99, 495
Drilling or tapping instruments, Stock for	G. Bunch	Bonnot's Mill, Mo	May 20, 1873	139, 042
Drilling, riveting, and watch-jeweling apparatus	C. Hopkins	Philadelphia, Pa	Jan. 4, 1870	98, 497
Drilling-tool, Portable	C. Van Haagen	Philadelphia, Pa	Dec. 20, 1870	110, 405
Drinking-cup	B. Adler	New York, N. Y	Dec. 7, 1869	97, 583
Drinking-cup	L. Grosholz	Philadelphia, Pa	Nov. 2, 1858	21, 955
Drinking-cup	A. Millar	Roxbury, Mass	Jan. 8, 1867	61, 084
Drinking-cup	J. S. Ostrander	Albany, N. Y	July 1, 1862	35, 777
Drinking-cup, Invalid	M. P. Codman	Boston, Mass	Nov. 14, 1865	50, 906
Drinking-cup, Pocket	J. S. Towndrow	Moline, Ill	Oct. 20, 1868	83, 341
Drinking-cup, Telescopic	P. H. Niles	Boston, Mass	June 5, 1860	28, 597
Drinking-cups, Mustache-shield for	J. J. Greenough	Syracuse, N. Y	Nov. 18, 1873	144, 614
Drinking-fountain, Poultry	G. H. Lomax	Somerville, Mass	Oct. 1, 1867	69, 351
Drinking-glass holder	J. V. Snider	Philadelphia, Pa	Mar. 21, 1871	112, 974
Drinking-glasses, Manufacture of	J. L. and T. B. Atterbury	Pittsburgh, Pa	Oct. 17, 1865	50, 437
Drinking-tube, Invalid	E. Chapin	Saint Louis, Mo	May 24, 1870	103, 299

Index of patents issued from the United States Patent Office from 1790 *to* 1873, *inclusive*—Continued.

Invention.	Inventor.	Residence.	Date.	No.
Drinking-tube, Invalid	E. Chapin	Saint Louis, Mo	May 24, 1870	103, 300
Drinking vessel cover	J. Henermann	Davenport, Iowa	May 25, 1869	90, 539
Drinks, Apparatus for mixing	W. Harnett	Brooklyn, N. Y	Dec. 24, 1872	134, 274
Drinks from fruit, Manufacture of effervescent	A. Meucci	Clifton, N. Y	Jan. 3, 1872	122, 478
Drip-collecting device	W. A. Harris	Providence, R. I	Mar. 11, 1873	136, 597
Drip-conductor	J. G. Widmann	Yonkers, N. Y	June 3, 1873	139, 641
Dripping-pan	A. Reid	Buffalo, N. Y	Aug. 16, 1870	106, 506
Drive-well tube	J. M. Mott	Chicago, Ill	Feb. 22, 1870	100, 179
Driver, Friction	J. T. Jones	New York, N. Y	Mar. 3, 1868	75, 023
Driver, Friction	J. T. Jones	New York, N. Y	Mar. 3, 1868	75, 024
Driving-bit	E. S. Dawson	Syracuse, N. Y	Sept. 24, 1867	69, 079
Driving-bit	D. Hale	Roston, Mass	Sept. 10, 1867	67, 623
Driving-bit	C. Huie	New York, N. Y	May 14, 1872	126, 704
Driving-hoop	T. Lucey	Salem, Mass	Sept. 22, 1868	82, 425
Driving-lines	L. D. Woodmansee	Dayton, Ohio	Mar. 17, 1868	75, 614
Driving-power, Machinery	S. P. W. Douglass	Williamson, N. Y	May 22, 1841	2, 104
Driving-wheel of locomotive-engines	J. F. Elliott	New Haven, Conn	Apr. 20, 1858	19, 986
Driving-wheel of locomotives for ascending inclined planes.	J. Morss	Washington, D. C	Sept. 5, 1854	11, 650
Driving-wheel of portable steam-engines and agricultural implements, &c.	G. W. Barnett	Springfield, Ohio	May 25, 1858	28, 318
Driving-wheel of steam drags or propellers	G. W. N. Yost	Pittsburgh, Pa	June 3, 1856	15, 049
Driving-wheels, Method of connecting the wheels of locomotive steam-engines to render them.	H. Hinckley	Boston, Mass	Apr. 1, 1845	3, 981
Driving-wheels of locomotive-engines, Applying	A. M. Eastwich	Philadelphia, Pa	Nov. 20, 1837	471
Driving-wheels of locomotives, Method of arranging galvano-electro helices for magnetizing.	O. D. Vosmus	Boston, Mass	Apr. 12, 1859	23, 648
Driving-wheels of steam vehicles, plows, &c., Giving adhesion to.	J. T. Price	Rockville, Ind	Apr. 13, 1858	19, 947
Drop-light and hanger	G. F. Blaisse and W. F. Crites	Philadelphia, Pa	Nov. 4, 1873	144, 309
Drop-light and hanger	W. Staehlen	Brooklyn, N. Y	Dec. 30, 1873	146, 104
Drop-light, Center	C. Deavs	New York, N. Y	Dec. 16, 1873	145, 632
Drop-light, Center-slide	C. Deavs	New York, N. Y	May 20, 1873	139, 123
Drop-light, Chandelier	R. Cornelius	Philadelphia, Pa	July 1, 1873	140, 347
Drop-light, Chandelier	R. Cornelius	Philadelphia, Pa	July 1, 1873	140, 348
Drop light, Chandelier	N. W. Williames	Philadelphia, Pa	July 1, 1873	140, 565
Drop-light coupling	R. S. Roeschlaub	Quincy, Ill	May 21, 1872	127, 105
Drop-light fixture	S. B. H. Vance and E. M. Smith.	New York, N. Y	May 29, 1866	55, 212
Drop-light, Gasolier	C. Deavs	New York, N. Y	July 19, 1870	105, 435
Drop-light joint	T. G. Arnold	New York, N. Y	June 5, 1860	28, 548
Drop-light of gasaliers, Sustaining-device for center	C. Deaves	New York, N. Y	Aug. 22, 1871	118, 349
Drop-light standard	J. Cunningham	West Meriden, Conn	May 30, 1871	115, 442
Drop-light standard	J. Horton	New York, N. Y	June 14, 1870	104, 155
Drop-lights, Sliding gas-joints for extension	C. Deavs	New York, N. Y	May 30, 1871	115, 447
Drop-pipe, Rotating	G. C. Morgan	Chicago, Ill	Sept. 26, 1871	119, 388
Drop-press	J. Duff	Peoria, Ill	June 16, 1868	78, 867
Drop-press	T. Gaillard	Brooklyn, N. Y	Apr. 9, 1872	125, 386
Drop-press	H. C. Gladding	Providence, R. I	Jan. 3, 1865	45, 792
Drop-press	W. F. Goulding and F. Cheney	Providence, R. I., and Hartford, Conn.	Sept. 16, 1862	36, 459
Drop-press	W. C. Hicks	New York, N. Y	Jan. 20, 1863	37, 445
Drop-press	B. Hotchkiss	New Haven, Conn	Feb. 17, 1863	37, 692
Drop-press	N. P. Maker	Providence, R. I	May 25, 1869	90, 456
Drop-press	J. P. Noyes	Binghamton, N. Y	Oct. 16, 1866	58, 880
Drop-press	M. Peck	New Haven, Conn	Nov. 25, 1851	8, 548
Drop-press	M. Peck	New Haven, Conn	May 26, 1857	17, 411
Drop-press	M. and C. Peck	New Haven, Conn	Sept. 4, 1860	29, 910
Drop-press	S. Remington	Ilion, N. Y	Dec. 1, 1863	40, 769
Drop-press	J. C. Rhodes	South Abington, Mass	July 16, 1867	66, 884
Drop-press	N. C. Stiles and J. S. Miller	Meriden, Conn., and Springfield, Mass.	Nov. 19, 1867	71, 080
Drop press	W. Wilson, jr	Brandywine, Del	Aug. 11, 1857	18, 004
Drop press and hammer lifting device	C. Peck	New Haven, Conn	June 8, 1869	91, 159
Drop press and trip-hammer	L. L. Whitlock	New York, N. Y	Aug. 22, 1871	118, 317
Drop-press strap	C. McBurney	Roxbury, Mass	Oct. 20, 1863	40, 349
Drop-weight-lifting machine	L. L. Whitlock	New York, N. Y	Aug. 22, 1871	118, 316
Dropper: *See* Corn-dropper. Corn and fertilizer dropper. Harvester-dropper. Hoe seed-dropper. Medicine-dropper. Seed-dropper. Seed and manure dropper.				
Drub, Universal	S. Felton	Holmes County, Ohio	July 22, 1828	....
Drug-crusher, Portable	H. C. Becker	New York, N. Y	Aug. 14, 1866	57, 071
Drugs, &c., Machine for pulverizing	I. Eddy	Boston, Mass	Dec. 31, 1823	....
Drum	J. Mason	Louisville, Ky	June 30, 1863	39, 058
Drum and grate for stoves and furnaces, Hot-air	B. F. Campbell	Boston, Mass	July 9, 1872	128, 709
Drum and oven, Combined	G. F. Reinhardt	Lincoln, Ill	July 13, 1869	92, 543
Drum and oven, Stove-pipe	J. Beebe	Chicago, Ill	Mar. 6, 1866	53, 077
Drum attachment for furnace, Heating	W. Duryea and W. Ennis	Glen Cove and Hudson, N. Y.	Apr. 30, 1867	64, 293
Drum, Clip	F. Pelzer	Kohlscheid, Prussia	Oct. 8, 1872	132, 099
Drum evaporator	D. Wolf	Easton, Kans	Aug. 18, 1868	81, 236
Drum for coal-stoves, Hot-air	G. J. Bentley	Michigan City, Mich	Apr. 26, 1870	102, 210
Drum for heating apartments	W. Frazier	Brooklyn, N. Y	Apr. 30, 1840	1, 571
Drum for hot-air furnace	J. H. Keyser	New York, N. Y	Oct. 6, 1868	82, 723
Drum, Heating	J. E. Bartlett	Boston, Mass	Aug. 12, 1873	141, 621
Drum, Heating	S. M. Bayard	Ionia, Mich	June 15, 1869	91, 301
Drum, Heating	J. Hall	Pleasant Hill, Ill	Oct. 1, 1872	131, 875
Drum, Heating	T. Jefferis	Council Bluffs, Iowa	June 29, 1869	91, 940
Drum, Heating	M. W. Lester	Chicago, Ill	July 16, 1872	129, 147
Drum, Heating	F. Proudfoot	Toronto, Canada	Dec. 2, 1873	145, 237
Drum, Heating	J. Reynolds	Philadelphia, Pa	May 10, 1870	102, 866
Drum, Heating	C. W. Servoss	Chicago, Ill	Apr. 19, 1870	102, 053
Drum, Heating	C. W. Servoss	Chicago, Ill	Aug. 8, 1871	117, 820
Drum, Heating	O. D. Spalding	Mankato, Minn	Nov. 11, 1873	144, 573
Drum, Heating	I. P. Tice	New York, N. Y	Oct. 21, 1873	143, 795

Index of patents issued from the United States Patent Office from 1790 *to* 1873, *inclusive*—Continued.

Invention.	Inventor.	Residence.	Date.	No.
Drum, Heating and ventilating	R. Jennings	Shelbyville, Ind	Apr. 25, 1871	114, 013
Drum, Hot-air	W. Allchin	Newburgh, N. Y	Jan. 5, 1869	85, 554
Drum, Military	J. Dermond	Louisville, Ky	Feb. 3, 1863	37, 570
Drum or heat-radiator, Stove-pipe	F. A. Balch	Hingham, Wis	July 24, 1866	56, 514
Drum, pulley, and wheel, Double-griping flanged	W. Kelsey	London, England	July 12, 1870	105, 341
Drum, Stove	J. Adams	Findlay, Ohio	Dec. 22, 1868	85, 196
Drum, Stove	W. Allchin	Newburgh, N. Y	Oct. 5, 1869	95, 551
Drum, Stove	G. C. Benton	Port Huron, Mich	Apr. 25, 1871	113, 969
Drum, Stove	H. E. Blemker	Evansville, Ind	June 8, 1869	91, 073
Drum, Stove	J. L. Collins and H. C. Bergie	Chicago, Ill	Oct. 29, 1867	70, 168
Drum, Stove	D. W. Cutting	Cambridge, Vt	June 23, 1868	79, 212
Drum, Stove	N. Downes	Syracuse, N. Y	Apr. 28, 1868	77, 179
Drum, Stove	C. Fisher	Niles, Mich	Apr. 4, 1871	113, 413
Drum, Stove	T. F. Hamilton	Geneseo, Ill	Dec. 17, 1872	134, 054
Drum, Stove	W. P. Hepburn and W. Reiner	Clarinda, Iowa	June 15, 1869	91, 231
Drum, Stove	R. A. Hunter	Ionia, Mich	May 18, 1869	90, 267
Drum, Stove	M. S. Morgan	Clintonville, Ill	Feb. 4, 1868	73, 990
Drum, Stove	T. Roberts	Shelby, Ohio	Feb. 21, 1865	46, 499
Drum, Stove	T. Scantlin	Evansville, Ind	Mar. 29, 1870	101, 316
Drum, Stove	E. Selbach	Columbus, Ohio	Mar. 10, 1868	75, 472
Drum, Stove	A. F. Smith	Ellsworth, Me	Jan. 28, 1868	73, 762
Drum, Stove	C. D. F. Smith	Geneva, Ill	Oct. 6, 1868	82, 761
Drum, Stove	G. T. Spaulding	Brodhead, Wis	Oct. 24, 1871	120, 217
Drum, Stove	F. Städter	Plattsmouth, Nebr	May 4, 1869	89, 698
Drum, Stove	M. Turly and J. D. Bayliss	Council Bluffs, Iowa	July 6, 1869	92, 404
Drum, Stove	W. Twitchell	Syracuse, N. Y	Feb. 7, 1871	111, 589
Drum, Stove	J. Voegtle	Indianapolis, Ind	Apr. 7, 1868	76, 564
Drum, Stove-pipe	J. F. Beckwith	Albion, N. Y	July 30, 1867	67, 251
Drum, Stove-pipe	E. S. Blake	Pittsburgh, Pa	Oct. 30, 1866	59, 173
Drum, Stove-pipe	E. E. Blinn	Brewerton, N. Y	Oct. 17, 1871	120, 024
Drum, Stove-pipe	A. Cant	Galt, Canada	Feb. 2, 1869	86, 362
Drum, Stove-pipe	J. P. Chaplin	Valparaiso, Ind	Mar. 6, 1866	52, 966
Drum, Stove-pipe	J. T. Clymer	Galion, Ohio	Aug. 31, 1869	94, 398
Drum, Stove-pipe	P. Cocker	Philadelphia, Pa	Dec. 20, 1864	45, 556
Drum, Stove-pipe	L. R. Comstock	Macon, Mo	Nov. 6, 1866	59, 363
Drum, Stove-pipe	J. Conver	Philadelphia, Pa	Nov. 19, 1867	70, 961
Drum, Stove-pipe	H. Cook	Bluffton, Ind	Feb. 20, 1866	52, 685
Drum, Stove-pipe	T. R. Cook	Saratoga Springs, N. Y	Apr. 10, 1866	33, 789
Drum, Stove-pipe	J. Crimmuins	Detroit, Mich	Apr. 17, 1866	53, 954
Drum, Stove-pipe	L. Dowe and A. C. Colton	Sycamore, Ill	Feb. 18, 1868	74, 520
Drum, Stove-pipe	A. Ernst	Milwaukee, Wis	June 11, 1867	65, 554
Drum, Stove-pipe	A. W. Foote	Buffalo, N. Y	July 22, 1873	140, 999
Drum, Stove-pipe	M. Gage	Medina, N. Y	Apr. 10, 1866	53, 806
Drum, Stove-pipe	G. D. Greenleaf	Three-Mile Bay, N. Y	May 30, 1865	47, 945
Drum, Stove-pipe	S. A. Halladay	Marilla, N. Y	Apr. 9, 1867	63, 720
Drum, Stove-pipe	G. H. Hammond	Oneonta, N. Y	Mar. 5, 1867	62, 627
Drum, Stove-pipe	W. Hearle	Beamsville, Canada	Dec. 14, 1869	97, 915
Drum, Stove-pipe	J. Heger	Jefferson, Wis	Apr. 19, 1870	102, 119
Drum, Stove-pipe	B. J. Hobson	Covington, Ky	Oct. 3, 1871	119, 609
Drum, Stove-pipe	F. P. Hunt	Medina, N. Y	May 22, 1866	54, 911
Drum, Stove-pipe	J. B. Hyzer	Janesville, Wis	May 23, 1865	47, 833
Drum, Stove-pipe	D. Knowlton	Liberty, Me	Apr. 14, 1868	76, 779
Drum, Stove-pipe	W. Kroeger	Milwaukee, Wis	June 13, 1865	48, 184
Drum, Stove-pipe	W. Kroeger	Milwaukee, Wis	June 5, 1866	55, 419
Drum, Stove-pipe	J. A. Lakin	Thompsonville, Conn	July 21, 1868	80, 076
Drum, Stove-pipe	G. W. Lore	Dexter, Mich	Jan. 18, 1870	98, 873
Drum, Stove-pipe	J. A. Marvin	Red Wing, Minn	Aug. 13, 1867	67, 779
Drum, Stove-pipe	H. Meyer	Richmond, Ind	Sept. 8, 1868	82, 019
Drum, Stove-pipe	B. F. Miller	New York, N. Y	July 31, 1866	56, 774
Drum, Stove-pipe	E. P. Morse	Batavia, N. Y	Mar. 6, 1866	53, 029
Drum, Stove-pipe	H. Northup and J. H. Patterson	Sandy Hill and Glen's Falls, N. Y.	Jan. 30, 1866	52, 366
Drum, Stove-pipe	J. C. Paine	Dubuque, Iowa	July 11, 1865	48, 716
Drum, Stove-pipe	W. L. Phillips	Normal, Ill	Apr. 4, 1871	113, 694
Drum, Stove-pipe	L. Racine	Hillsdale, Mich	Nov. 14, 1871	121, 001
Drum, Stove-pipe	T. R. Renwick	Grand Rapids, Mich	Oct. 3, 1871	119, 646
Drum, Stove-pipe	E. D. Roberts	Hartford, Conn	Mar. 8, 1870	100, 557
Drum, Stove-pipe	C. E. Russell	Jacksonville, Ill	June 11, 1867	65, 770
Drum, Stove-pipe	F. E. Ruth and J. De Long	Upper Sandusky, Ohio	Aug. 28, 1866	57, 572
Drum, Stove-pipe	H. H. Senniksen	Richmond, Ind	Sept. 4, 1866	57, 778
Drum, Stove-pipe	C. D. F. Smith	Geneva, Ill	June 29, 1869	92, 114
Drum, Stove-pipe	J. G. Sorgen	Kenton, Ohio	Mar. 5, 1867	62, 699
Drum, Stove-pipe	G. S. Walker	Erie, Pa	Sept. 1, 1868	81, 714
Drum, Stove-pipe	J. C. Ward	Bergen, N. Y	Oct. 22, 1867	70, 138
Drum, Stove-pipe	G. Van Waters	Durand, Wis	Nov. 26, 1872	133, 506
Drum, Stove-pipe	C. C. Webber	Calmar, Iowa	July 17, 1866	56, 473
Drum, Stove-pipe	B. Webster	Kingfield, Me	July 11, 1871	116, 900
Drum, Stove-pipe	E. Webster	Holland, Mich	Nov. 5, 1867	70, 659
Drum, Stove-pipe	H. G. Whitaker	Brattleborough, Vt	June 26, 1866	59, 939
Drum, Stove-pipe	T. Whitson	Woodstock, Ill	May 23, 1865	47, 887
Drum-wheels, Fork for holding rope belts upon	C. Foster	Pompey, N. Y	Apr. 24, 1849	6, 377
Drums, Construction of military	C. H. Zimmermann	Philadelphia, Pa	Mar. 9, 1858	19, 602
Drums, kegs, casks, &c., Manufacture of	D. Cope	Liverpool, England	Apr. 21, 1863	28, 217
Drums to shafts, Mode of attaching	B. G. Brooks	Manchester, N. H	Oct. 3, 1871	119, 504
Dry-goods boxes, Device for packing	T. Webber	Chicago, Ill	Nov. 29, 1864	45, 304
Dry goods, Device for displaying	J. D. Chambers	West Lebanon, Ind	Apr. 27, 1869	89, 383
Dry-goods rack	C. N. Cadwallader	Philadelphia, Pa	Apr. 1, 1872	137, 416
Dry-goods rack	D. Keiser	Reading, Pa	Sept. 10, 1872	131, 221
Dry-goods rack	D. Keiser	Reading, Pa	Dec. 10, 1872	133, 781
Dry-goods stand	J. J. Bisel	Williamsport, Pa	Apr. 29, 1873	138, 226
Dry-house	J. K. Boswell	Richmond, Ind	Sept. 4, 1866	57, 669
Dry-house	J. Schultz	Ellenville, N. Y	May 7, 1867	64, 586
Dry-house flue	W. B. Hix	Rome, Ga	Mar. 8, 1870	100, 530
Dryer: *See* Bone-black dryer. Brick-dryer. Cardboard-dryer. Cigar-dryer. Cloth-dryer. Clothes-dryer. Collar and cuff dryer. Fish-dryer.				

Index of patents issued from the United States Patent Office from 1790 *to* 1873, *inclusive*—Continued.

Invention.	Inventor.	Residence.	Date.	No.
Dryer—Continued. *See*, Fruit-dryer. Glue-dryer. Grain-dryer. Hop-dryer. Lead-dryer. Lumber-dryer. Malt-dryer. Offal-dryer. Paper-dryer. Photograph-dryer. Poudrette-dryer. Salt-dryer. Sand and gravel dryer. Steam dryer. Sugar-dryer. Sugar and salt dryer. Tan-dryer. Tobacco-dryer. Towel dryer. Wool-Dryer. Yeast-cake dryer.				
Dryer	H. W. Adams and S. S. Bacon	Boston, Mass	July 4, 1871	116, 531
Dryer	H. W. Adams and S. S. Bacon	Boston, Mass	July 4, 1871	116, 532
Dryer	R. N. Allen	Pittsford, Vt	Dec. 24, 1867	72, 583
Dryer	C. E. Ashcroft	Boston, Mass	Feb. 9, 1869	86, 625
Dryer	E. H. Ashcroft	Boston, Mass	July 7, 1868	79, 623
Dryer	D. Barker	Northfleet, England	Mar. 3, 1868	75, 113
Dryer	J. Brakeley	Bordentown, N. J	May 12, 1868	77, 712
Dryer	J. Butterworth	Philadelphia, Pa	May 6, 1873	138, 475
Dryer	S. L. Cheyney	Wooster, Ohio	Apr. 28, 1868	77, 253
Dryer	S. L. Cheyney	Springfield, Ohio	Oct. 24, 1871	120, 242
Dryer	A. W. Cox	Indianapolis, Ind	Apr. 26, 1870	102, 230
Dryer	A. W. Cox and W. Gause	Indianapolis, Ind	Jan. 4, 1870	98, 470
Dryer	C. Crane	Boston, Mass	Sept. 1, 1868	81, 606
Dryer	L. J. Crow and G. Sanderson	Fredericksburgh, Ohio	Apr. 28, 1868	77, 362
Dryer	E. Drevet	New York, N. Y	June 27, 1871	116, 279
Dryer	E. Foote and M. P. Smith	East Bloomfield, N. Y., and Baltimore, Md.	Mar. 26, 1872	124, 944
Dryer	H. Garrett	Linneus, Mo	Jan. 25, 1870	99, 182
Dryer	W. A. Gibbs and A. Borwick	Sewardstone and London, England.	Jan. 7, 1873	134, 659
Dryer	T. S. Harrison	Philadelphia, Pa	Apr. 5, 1870	101, 458
Dryer	H. Henry	Halbert's Bluff, Ind	May 26, 1868	78, 285
Dryer	C. H. Hersey	Boston, Mass	Apr. 25, 1871	114, 137
Dryer	J. B. Johnson	Indianapolis, Ind	Mar. 10, 1868	75, 277
Dryer	C. Kaibel	Sacramento City, Cal	Feb. 9, 1869	86, 840
Dryer	R. Kidd	Montezuma, Ind	July 23, 1872	129, 736
Dryer	I. B. Kinkead	Watertown, Ohio	May 31, 1870	103, 622
Dryer	A. W. J. Mason	New Orleans, La	Oct. 31, 1871	120, 447
Dryer	O. F. Mayhew	Indianapolis, Ind	Sept. 21, 1869	94, 967
Dryer	P. Mickel	Milford, N. Y	June 14, 1870	104, 333
Dryer	J. P. Miller	Somerville, N. J	Nov. 15, 1870	109, 234
Dryer	C. A. Moffat	Indianapolis, Ind	Oct. 25, 1870	108, 616
Dryer	J. E. Rice	Moline, Ill	Aug. 9, 1870	106, 205
Dryer	E. Y. Robbins	Cincinnati, Ohio	July 19, 1864	43, 603
Dryer	M. P. Smith	Baltimore, Md	Feb. 1, 1870	99, 488
Dryer	M. P. Smith	Baltimore, Md	Sept. 13, 1870	107, 417
Dryer	M. P. Smith	Baltimore, Md	Apr. 2, 1872	125, 344
Dryer	E. Trowbridge and J. M. Jones	Detroit, Mich	Oct. 4, 1870	107, 982
Dryer	J. Turner	Chicago, Ill	Dec. 12, 1871	121, 915
Dryer	C. H. Wakelee	San Francisco, Cal	Oct. 10, 1871	119, 903
Dryer	J. M. Ward	Oxford, Ohio	Nov. 16, 1869	97, 006
Dryer	G. W. White and L. E. Wentworth.	Malden and Melrose, Mass	May 3, 1870	102, 742
Dryer	W. E. Wright	Rome, N. Y	July 23, 1872	129, 775
Dryer and evaporator	C. Alden	Newburgh, N. Y	Mar. 15, 1870	100, 835
Drying and burning kiln	J. Buhrer	Munich, Bavaria	Sept. 29, 1868	82, 488
Drying and desiccating apparatus	H. G. Bulkley	New York, N. Y	Oct. 22, 1867	69, 964
Drying and oxidizing colored goods	J. C. Kempton	Manayunk, Pa	Sept. 2, 1851	8, 334
Drying and preserving animal and vegetable substances, Process and apparatus for.	C. Alden	Newburgh, N. Y	Dec. 5, 1871	121, 569
Drying and ventilating apparatus	J. Royal	Rochelle, Ill	Feb. 4, 1868	74, 149
Drying-apparatus	D. K. Boswell	Corinth, Miss	Nov. 14, 1865	59, 895
Drying-apparatus	J. O. Mellen	Saint Louis, Mo	Apr. 2, 1867	63, 412
Drying-apparatus	F. I. Norton	Tremont, Ohio	Apr. 28, 1868	77, 205
Drying-apparatus	A. Stevens	West Milton, Ohio	July 31, 1866	56, 823
Drying-chamber	J. E. Tourné	New Orleans, La	Oct. 16, 1860	30, 436
Drying cloth	R. L. Hawes	Worcester, Mass	Oct. 3, 1854	11, 748
Drying-cylinder for fibrous manufactures	H. W. Peaslee	Malden Bridge, N. Y	Dec. 2, 1856	16, 149
Drying-cylinder, Steam	W. C. Collier	Salford, England	Jan. 14, 1873	134, 786
Drying-cylinders by steam, Mode of heating	A. P. Pitkin	Hartford, Conn	July 19, 1859	24, 819
Drying fibrous substances, Machine for	J. Essex	North Bennington, Vt	May 31, 1859	24, 205
Drying-frame, Lace-curtain, &c	S. Short	Cincinnati, Ohio	Dec. 23, 1873	145, 912
Drying furnace and oven	J. K. Caldwell	Allegheny City, Pa	Mar. 30, 1869	88, 444
Drying-house	J. J. Allen	Philadelphia, Pa	Mar. 26, 1872	124, 998
Drying-house	H. B. Gallup and C. Wood	Watertown, Wis	Aug. 14, 1866	57, 117
Drying house and oven	D. Stouder	Dayton, Ohio	Mar. 26, 1867	63, 327
Drying-house for drying fruit	S. Strong	Wilks Township, Ohio	Sept. 2, 1811	......
Drying-house for fruit, &c	L. A. Oellig	Martinsburgh, Pa	Jan. 21, 1868	73, 637
Drying-kiln	H. G. Bulkley	Kalamazoo, Mich	Feb. 13, 1866	52, 524
Drying-kiln	B. R. Hawley	Normal, Ill	Nov. 17, 1868	84, 117
Drying-machine	N. E. Chaffee	Ellington, Conn	Feb. 20, 1849	6, 119
Drying-room	R. Dalrymple	Galt, Canada	Oct. 10, 1871	119, 747
Duck-call	E. Fisher	Detroit, Mich	May 10, 1870	102, 799
Duck-shooting boat	R. Bogle	Rock Hall, Md	May 5, 1857	17, 192
Ducks, Mode of catching	W. Coffield	Norfolk, Va	Jan. 18, 1830	......
Ducts or natural passages for animal fluid, Instrument for obliterating strictures in.	G. N. Palmer	Greene, N. Y	Sept. 10, 1867	68, 647
Dulcimer	E. Durand	Norwich, Conn	Dec. 31, 1867	72, 824
Dulcimer	J. Low	Clinton, Mass	June 19, 1860	28, 811
Dumb-bell	E. Ballou	Zanesville, Ohio	Apr. 25, 1871	113, 966
Dumb-bell	D. P. Butler	Boston, Mass	July 4, 1865	48, 514
Dumb-bell	D. F. Savage	Boston, Mass	May 29, 1860	28, 505
Dumb-bell, Graduated	G. B. Windship	Boston, Mass	Feb. 14, 1865	46, 413
Dumb-jockey, the cross and saddle-tree being made of gutta-percha.	S. Blackwell	Middlesex County, England.	Mar. 18, 1856	14, 438
Dumb-waiter	T. B. Bryson	Newcastle, Pa	Oct. 15, 1867	69, 759

Index of patents issued from the United States Patent Office from 1790 *to* 1873, *inclusive*—Continued.

Invention.	Inventor.	Residence.	Date.	No.
Dumb-waiter	A. Cannon, jr	Poughkeepsie, N. Y	May 25, 1869	90, 422
Dumb-waiter	L. Carrier	New York, N. Y	Dec. 3, 1872	133, 561
Dumb-waiter	W. H. Elliott	Boston, Mass	May 24, 1870	103, 313
Dumb-waiter	R. W. Lawrence	New York, N. Y	Sept. 11, 1866	57, 928
Dumb-waiter	A. Murtaugh	New York, N. Y	Dec. 4, 1860	30, 831
Dumb-waiters, Pulley arrangement for	A. Murtaugh	New York, N. Y	Apr. 3, 1855	12, 640
Dummy-engine	J. Comly	Philadelphia, Pa	Apr. 1, 1873	137, 421
Dummy-engine	M. H. Kollock	Philadelphia, Pa	Mar. 25, 1873	137, 216
Dummy-engine	T. C Robinson	Mystic River, Conn	Jan. 9, 1872	122, 534
Dummy-engine	C. G. Spengler	Hoboken, Ga	Aug. 22, 1871	118, 291
Dummy or traction engine	A. A. Wilder	Detroit, Mich	Mar. 18, 1873	136, 886
Dumping and loading machine, Portable	W. Goff	Big Flats, N. Y	Oct. 29, 1867	70, 196
Dumping-apparatus	F. Ficht	Isle Royal Mine, Mich	Sept. 27, 1864	44, 412
Dumping-apparatus	R. M. McGrath	La Fayette, Ind	June 6, 1871	115, 759
Dumping-apparatus	G. W. Reed	Middlesex, Pa	June 3, 1873	139, 613
Dumping-apparatus	L. B. Stilson and J. G. Payson	Minneapolis, Minn	Nov. 29, 1870	109, 776
Dumping-box for agricultural purposes	J. Van Doren	Farm Ridge, Ill	Mar. 16, 1858	19, 663
Dumping-machine	G. W. and W. S. Hough	Galva, Ill	June 28, 1870	104, 736
Dumping-machine	B. Walton	Farisburgh, Ill	Nov. 23, 1869	97, 252
Dumping-platform	J. Atkinson	Cameron, Tex	Feb. 11, 1873	135, 683
Dumping platform	R. M. Fish	Glenwood, Iowa	Apr. 5, 1870	101, 451
Dumping-platform	W. W. Hinman	Elkhart, Ill	Apr. 29, 1873	138, 400
Dumping-platform	S. C. Kenago	Kankakee, Ill	Oct. 20, 1868	83, 288
Dumping-platform	F. Peteler	Bloomington, Minn	Sept. 9, 1873	142, 724
Dumping-platform	N. Swickard	Galva, Ill	July 16, 1872	129, 618
Dumping-platform	D. A. Wells	Sandwich, Ill	Aug. 13, 1872	130, 551
Dumping-tub	W. H. Brown	Erie, Pa	Mar. 3, 1863	37, 808
Dumping-tub	J. S. Evans	Berkley, Mass	Sept. 7, 1869	94, 483
Dumping-tub hook	J. R. Thorne	Waldoborough, Me	Nov. 15, 1870	109, 274
Dumping-wagon	R. W. Green	Bradford, Pa	Aug. 11, 1863	39, 478
Dung-forks, Device for convertible	E. H. Dawes	Litchfield, Me	Oct. 25, 1853	10, 148
Dung-forks, Making four-tined	M. Stevens	New Haven, Conn	Apr. 16, 1824	....
Dung-hook	J. G. Good	Raps, Pa	Apr. 16, 1867	63, 878
Duplex wrench	A. B. Davis	Philadelphia, Pa	Aug. 24, 1869	94, 085
Duplex wrench	A. B. Davis	Philadelphia, Pa	Aug. 9, 1870	106, 129
Durogel, Manufacture of	C. C. Peck and F. E. Engelhardt	New York, N. Y	June 4, 1867	65, 425
Dust-pan	A. Brigham	Cold Brook, Mass	July 23, 1867	67, 019
Dust-pan	D. C. Colby	Washington, D. C	Apr. 28, 1868	77, 258
Dust-pan	S. E. Condon	Brooklyn, N. Y	Feb. 25, 1868	74, 801
Dust-pan	W. M. Conger	Newark, N. J	Aug. 22, 1871	118, 343
Dust-pan	W. M. Conger	Newark, N. J	Feb. 6, 1872	123, 331
Dust-pan	W. M. Conger	Newark, N. J	July 30, 1872	129, 935
Dust-pan	F. L. Daniels	Boston, Mass	Nov. 30, 1869	97, 276
Dust-pan	F. L. Daniels	Boston, Mass	July 12, 1870	105, 315
Dust-pan	F. L. Daniels and J. Russell	Boston, Mass	Apr. 25, 1871	113, 985
Dust-pan	O. C. Forsyth, jr	Newburgh, N. Y	Dec. 23, 1873	145, 792
Dust-pan	T. G. Harold	Brooklyn, N. Y	Mar. 1, 1864	41, 771
Dust-pan	A. B. Hoffman	Roxbury, Mass	Oct. 16, 1866	58, 825
Dust-pan	G. H. Horstman	Philadelphia, Pa	May 21, 1861	32, 368
Dust-pan	R. S. Jennings	Philadelphia, Pa	Mar. 12, 1872	124, 584
Dust-pan	T. E. McNeil	Philadelphia, Pa	July 6, 1858	20, 811
Dust-pan	J. B. Morgan	Davenport, Iowa	Sept. 19, 1871	119, 042
Dust-pan	W. Musgrove	New York, N. Y	Oct. 22, 1872	132, 367
Dust-pan	W. S. Potwin	Chicago, Ill	Oct. 15, 1872	132, 176
Dust-pan	J. H. Rohrman	Philadelphia, Pa	June 7, 1859	24, 331
Dust-pan	T. F. Rooney	Chicago, Ill.	May 2, 1871	114, 341
Dust-pan	P. A. Schanck	Matawan, N. J	July 20, 1869	92, 754
Dust-pan	M. Vanderhoven	Utica, N. Y	July 30, 1867	67, 232
Dust-pan	W. Westlake	Chicago, Ill	Aug. 15, 1871	118, 084
Dust-pan	L. F. Wheaton	Madison, Conn	May 26, 1868	78, 406
Dust-pan	H. Whittemore	Orangetown, N. Y	Feb. 21, 1871	112, 102
Dust pan and brush	C. H. Parker and G. Burnham	Waltham, Mass	July 4, 1865	48, 582
Dust pan and brush	C. H. Parker and G. Burnham.	Waltham, Mass	Oct. 3, 1865	50, 270
Dust-receptacle	C. Butterfield	West Waterville, Me	Feb. 12, 1867	61, 999
Duster	G. M. Smith	Louisville, Ky	Jan. 28, 1873	135, 376
Duster, Feather	C. L. W. Baker	Camden, N. J	Nov. 18, 1873	144, 728
Duster, Feather	M. A. Goodenough	New York, N. Y	Apr. 28, 1868	77, 278
Duster, Feather	M. A. Goodenough	New York, N. Y	Jan. 11, 1870	98, 759
Dye, Aniline	B. Block	Soultz, France	July 14, 1868	79, 942
Dye, Aniline	J. Lambert, jr	New York, N. Y	Sept. 15, 1868	82, 129
Dye for coloring wool	G. W. Talbot	Providence, R. I	Feb. 1, 1870	99, 496
Dye-stuff	C. E. and M. E. Fox	Gilroy, Cal	Sept. 8, 1868	81, 992
Dye-stuffs, bark, &c., Cutting	A. Broad	Litchfield, Conn	June 9, 1814	....
Dye-stuffs from spent matter, Preparation of	F. Pfanner	Providence, R. I	Sept. 13, 1845	4, 192
Dye-stuffs, Machine for cutting and shaving	B. Swift	Washington, N. Y	Aug. 10, 1836	....
Dye-stuffs, Preparing vegetable	F. E. Schmidt	New York, N. Y	July 15, 1856	15, 361
Dye-tub	T. Sampson	Providence, R. I	Nov. 8, 1870	109, 059
Dye-vat	T. E. Rogers	Dexter, Me	Mar. 8, 1870	100, 670
Dye-wood and dye-stuff cutting and shaving machine.	B. Swift	United States	Aug. 10, 1836	10
Dye-wood cutter, Self-feeding	A. Foster	Rochester, N. Y	Dec. 9, 1828	....
Dye-wood, Cutting and separating	G. W. Pearson and D. Coburn	Lowell, Mass	Feb. 18, 1868	74, 586
Dye-wood-cutting machine	M. Hurd	Augusta, N. Y	July 22, 1831	....
Dye-wood-cutting machine	L. H. Moseley	Poughkeepsie, N. Y	June 22, 1837	242
Dye-woods and bark, Machine for reducing and cutting.	A. McMillen	Bedford, N. H	Dec. 26, 1837	532
Dye-woods, Machine for cutting	E. Converse	Dayton, Ohio	Jan. 10, 1821	....
Dye-woods, Machine for cutting	O. E. Pray	Portsmouth, N. H	Jan. 21, 1868	73, 647
Dye-woods, Machine for cutting	J. Richards, 3d	Norfolk, Conn	Aug. 13, 1810	....
Dye-woods, Machine for cutting and chipping	E. Morris	Bloomfield, N. J	Sept. 27, 1825	....
Dye-woods, Machine for grinding, sawing, or rasping.	J. Mathews	Philadelphia, Pa	Aug. 13, 1805	....
Dye-woods, Machine for grinding, sawing, and rasping.	J. Mathews	Philadelphia, Pa	Feb. 6, 1807	....
Dye-woods, Machine for rasping	J. Reed	Boston, Mass	June 3, 1808	....
Dye-woods, Machine for rasping and cutting	B. Swift	Washington, N. Y	Feb. 1, 1817	....
Dye-woods, Machinery for rasping	C. W. Roberts and J. Hambley.	Philadelphia, Pa	Aug. 21, 1847	5, 251
Dye-woods, Mode of extracting color from	L. Kent	Dorset, Vt	Sept. 27, 1838	951

Index of patents issued from the United States Patent Office from 1790 to 1873, inclusive—Continued.

Invention.	Inventor.	Residence.	Date.	No.
Dye-woods, &c., Rasping and grinding	J. Broad	Springfield, Mass	Mar. 1, 1814	....
Dye-woods, Rasping or sawing	S. Stone	Enosburgh, Vt	May 3, 1811	....
Dyes and colors, Manufacture of	G. H. Reed	Boston, Mass	May 22, 1866	54, 957
Dyes and colors, Preparation of	M. Howe and H. R. Stevens	Boston, Mass	Oct. 13, 1863	40, 263
Dyes and colors, Preparation and manufacture of	G. H. Reed	Boston, Mass	Oct. 17, 1865	50, 495
Dyes by steam, Mode of manufacturing	R. Wood	Erin, N. Y	Aug. 25, 1829	....
Dyes from aniline, Manufacture of	R. Pinkney	London, England	May 21, 1872	127, 102
Dyes, ink, &c., from aniline, Production of colors for.	R. Pinkney	London, England	Aug. 23, 1870	106, 616
Dyes made from aniline, &c., from portions of fabric, Removing.	J. Lambert, jr	New York, N. Y	Nov. 8, 1870	109, 025
Dyes, Manufacture of madder	T. Bristow	Cranston, R. I	Mar. 1, 1870	100, 365
Dyes, Preparation of	J. Reynolds	San Francisco, Cal	Jan. 28, 1868	73, 756
Dyes, Putting up bluing and other	E. L. Molineaux	New York, N. Y	Dec. 24, 1867	72, 524
Dyer's and bleacher's implement	H. W. Holly	Norwich, Conn	Jan. 14, 1868	73, 331
Dyer's vat	H. Champenois	New York, N. Y	Oct. 5, 1869	95, 425
Dyer's vat	W. Vine and W. H. Jubb	Norwalk, Conn	Mar. 3, 1868	75, 081
Dyers' wood, Machine for cutting	E. Hubbel	Middlebury, Vt	June 27, 1809	....
Dyeing	J. P. Derby	Amesbury, Mass	Oct. 21, 1856	15, 959
Dyeing	H. Hibbard	Attica, N. Y	May 2, 1843	3, 068
Dyeing	P. Magennis	Paterson, N. J	Apr. 21, 1838	703
Dyeing	S. Mallard	Staten Island, N. Y	Mar. 27, 1849	6, 242
Dyeing and bleaching machine	J. Young	New York, N. Y	Aug. 18, 1863	39, 609
Dyeing and bleaching yarn and thread, Apparatus for.	W. H. Elliot and I. Osgood	New York and Utica, N. Y	Mar. 31, 1868	76, 178
Dyeing and embossing table and piano covers, Mode of.	A. Jack	Milton, N. H	Sept. 24, 1867	69, 098
Dyeing and finishing reel for silk and velvet woven fabrics.	E. Mafat	New York, N. Y	Aug. 30, 1870	106, 844
Dyeing and finishing, Silk	A. Stearns and W. Barrett	Boston, Mass	Sept. 2, 1818	....
Dyeing and finishing silk goods, Machine for	W. Barrett	....	June 27, 1809	....
Dyeing and printing a black color on fabrics with aniline compounds.	J. Lightfoot	Accrington, England	May 19, 1863	38, 589
Dyeing and printing, Compound to be used as a mordant in.	F. S. Dumont	New York, N. Y	Sept. 7, 1869	94, 581
Dyeing and printing, Material called ole-izerine for.	A. Paraf	New York, N. Y	Jan. 17, 1871	110, 994
Dyeing and printing, Mode of producing black in	A. Paraf	New York, N. Y	Sept. 24, 1867	69, 121
Dyeing and printing, Preparing coloring-matters for.	A. W. Hoffmann	Middlesex, England	June 7, 1864	43, 066
Dyeing and printing textile fabrics	J. Lightfoot	Burnley, England	Feb. 7, 1871	111, 654
Dyeing and printing textile fabrics and in compounds therefor.	A. Paraf	Mulhouse, France	Mar. 19, 1867	63, 084
Dyeing and printing with aniline colors	R. H. Gratrix	Salford, England	Jan. 5, 1864	41, 066
Dyeing and printing with aniline colors	N. Loyd and J. G. Dale	Accrington, England	Apr. 1, 1862	34, 840
Dyeing and tanning, Preparing quercitron or black-oak bark for.	N. Harper	Frankfort, Pa	Oct. 30, 1810	....
Dyeing and washing fibrous materials, Apparatus for.	C. G. Sargent	Graniteville, Mass	Jan. 14, 1873	134, 935
Dyeing-apparatus	C. T. Appleton	Roxbury, Mass	Mar. 21, 1854	10, 677
Dyeing-apparatus	E. Brierley	Lowell, Mass	Dec. 11, 1849	6, 932
Dyeing-apparatus	G. M. Rohrbacher	Philadelphia, Pa	May 7, 1872	126, 578
Dyeing-apparatus	M. Waterhouse	Passaic, N. J	June 2, 1868	78, 559
Dyeing black, Process of	J. Gee	West New Brighton, N. Y	Dec. 7, 1869	97, 497
Dyeing black with aniline colors	L. Jarossen and J. J. Muller-Pack.	Lille, France, and Basel, Switzerland.	Dec. 17, 1872	134, 066
Dyeing blue	J. Percy	....	Mar. 3, 1800	....
Dyeing blue, Process for	E. Swiney	Andover, Mass	Nov. 4, 1851	8, 494
Dyeing-composition	F. Graupner	Evansville, Ind	Aug. 16, 1870	106, 479
Dyeing, Composition of the pastel-vat to be used in	H. N. Barrow	East Windsor, Conn	Dec. 5, 1848	5, 947
Dyeing-compound	F. Graupner	Evansville, Ind	Aug. 1, 1871	117, 620
Dyeing-compound	F. G. Vettercke	New York, N. Y	July 26, 1853	9, 890
Dyeing cotton, &c	H. Haigh and R. Heaton	Huddersfield, England	Jan. 5, 1864	41, 153
Dyeing cotton in the staple	J. R. Stewart	New York, N. Y	Nov. 11, 1831	....
Dyeing cotton, &c., Mode of	A. J. J. d'Andiran	Mulhouse, France	Jan. 19, 1869	85, 912
Dyeing door-mats	R. Shaler	Madison, Conn	July 22, 1851	8, 249
Dyeing, &c., Extract of bark for	S. Downing	....	June 12, 1801	....
Dyeing, Extracting the effective matter in bark, &c., for.	T. Bedwell and W. Mitchell	....	June 7, 1803	....
Dyeing, Extracting the essence of bark for	S. Downing	....	Apr. 19, 1802	....
Dyeing fabric	F. G. Graupner	Evansville, Ind	June 3, 1873	139, 573
Dyeing fabrics, Apparatus for	T. M. Drown	Philadelphia, Pa	Jan. 10, 1871	110, 907
Dyeing fabrics with naphtylamine colors	F. Lamy, jr	Deville-les-Rouen, France	Dec. 17, 1872	134, 076
Dyeing felt caps, &c., Device for	J. McFarlane	Matteawan, N. Y	May 24, 1864	42, 915
Dyeing felt hats and caps	J. T. Waring	Yonkers, N. Y	Aug. 19, 1873	142, 064
Dyeing fibrous and textile materials, Apparatus for.	C. Corron	St. Etienne, France	Aug. 6, 1872	130, 280
Dyeing hair on seal-skins, &c	J. Herring	New York, N. Y	Nov. 16, 1826	....
Dyeing hair, wool, &c	D. Williams	Albany, N. Y	Mar. 14, 1846	4, 416
Dyeing hats	I. Taylor	Trenton, N. J	Aug. 27, 1824	....
Dyeing hats, Apparatus used in	M. Burt	Warwick, N. Y	Nov. 8, 1825	....
Dyeing in various colors for manufacture of Scotch plaids, &c., Frame for.	C. Barton	Paterson, N. J	Feb. 15, 1870	99, 750
Dyeing, Iron liquor as a mordant for	J. D. Prince	Lowell, Mass	Apr. 24, 1841	2, 060
Dyeing machine, Silk	N. Mary, sr	Philadelphia, Pa	Aug. 11, 1857	17, 974
Dyeing madder-colors	F. A. Gatty	Accrington, England	May 20, 1873	139, 056
Dyeing, Manufacturing of prussiate of potash and soda for.	F. Fossard	Philadelphia, Pa	Nov. 7, 1835	....
Dyeing &c., Mode of treating fish-water for use in	J. B. Herreshoff	Bristol, R. I	Dec. 15, 1863	40, 933
Dyeing, Mordant for	R. O. Burgess and S. LaRhett	Providence, R. I	Nov. 25, 1873	144, 952
Dyeing, Mordant for	G. A. Hagemann	Copenhagen, Denmark	Oct. 7, 1873	143, 449
Dyeing nankeen, Mode of	A. G. Trott	New York, N. Y	Feb. 11, 1825	....
Dyeing piece-goods, Apparatus for	H. Burrows	Lowell, Mass	Mar. 3, 1868	74, 985
Dyeing, &c., Preparing quercitron or black-oak bark for.	T. Benger	....	Jan. 25, 1804	....
Dyeing-press	J. Holt	Lowell, Mass	July 19, 1870	105, 455
Dyeing-process	C. T. Appleton	Roxbury, Mass	May 30, 1854	10, 978
Dyeing-process	C. T. Appleton	Roxbury, Mass	Mar. 21, 1854	10, 664

Index of patents issued from the United States Patent Office from 1790 to 1873, inclusive—Continued.

Invention.	Inventor.	Residence.	Date.	No.
Dyeing silk	T. Harrison and R. Pierpont	New York, N. Y	Oct. 18, 1814	....
Dyeing textile fabrics with aniline colors	I. O. Iverson	Madison, Wis	Oct. 27, 1868	83, 502
Dyeing with alkaline prussiate	F. Fossard	Pittsburgh, Pa	Dec. 14 1832	....
Dyeing with alkaline prussiates and manufacturing potash.	F. Fossard	Philadelphia, Pa	Apr. 23 1834	....
Dyeing with aniline black	J. Higgen	Manchester, England	Apr. 22, 1873	138, 155
Dyeing with indigo	L. Scala	Genoa, Italy	July 23, 1872	129, 753
Dyeing with madder-colors	A. C. and A. Duncan	Manchester, England	June 18, 1872	128, 027
Dyeing wool, Process of	F. Fossard	Philadelphia, Pa	Apr. 21, 1838	702
Dyeing wool, silk, &c., with aniline colors	R. H. Gratix	New York, N. Y	Apr. 28, 1863	38, 306
Dyeing woolen goods	W. W. Sanborn	Lewiston, Me	Oct. 18, 1870	108, 523
Dyeing yarn, &c	A. Paraf	Mulhouse, France	Feb. 27, 1866	52, 942
Dyeing yarn from the beam, Machine for	W. Spencer	Lowell, Mass	Sept. 25, 1838	947
Dyeing yarn, Machinery for	W. A. Burke	Manchester, N. H	May 30, 1844	3, 608
Dyeing yarn parti-colored	D. B. Hinman	Philadelphia, Pa	July 26, 1853	9, 887
Dyeing yarn parti-colored	D. B. Kerr	New York, N. Y	Mar. 23, 1858	19, 701
Dynamic machine	L. C. Errani and R. Anders	Liege, Belgium	June 17, 1873	140, 021
Dynamometer	G. W. Bigelow	New Haven, Conn	Nov. 25, 1873	144, 945
Dynamometer	J. Emerson	Lowell, Mass	July 7, 1868	79, 561
Dynamometer	J. Emerson	Lowell, Mass	July 7, 1868	79, 562
Dynamometer	J. Emerson	Lowell, Mass	Dec. 21, 1869	98, 043
Dynamometer	J. Emerson	Lowell, Mass	June 27, 1871	116, 285
Dynamometer	G. and J. W. Gibbs	Canton, Ohio	Aug. 26, 1856	15, 608
Dynamometer	G. Juengst	New York, N. Y	Dec. 22, 1857	18, 908
Dynamometer	C. Neer	Albany, N. Y	July 12, 1859	24, 753
Dynamometer	C. Neer	Brooklyn, N. Y	Jan. 15, 1867	61, 237
Dynamometer	J. W. Sutton	Portland, Oreg	Dec. 21, 1869	98, 205
Dynamometer	W. Tucker	Blackstone, Mass	Dec. 21, 1858	22, 388
Dynamometer	A. Warren and E. Damon, jr	Boston, Mass	June 26, 1860	28, 925
Dynamometer, Recording	L. Miller	Akron, Ohio	May 21, 1872	127, 088
Dynamometer, Registering	W. B. Leonard	New York, N. Y	Dec. 19, 1854	12, 099
Dynamometer-wrench	H. R. Leonard	Portland, Oreg	June 11, 1872	127, 902
E.				
Eagle-chair	W. I. Weaver	New York, N. Y	June 14, 1816	....
Ear-cover	M. Stone	Vermillion, Ohio	June 10, 1873	139, 831
Ear-drop lobe-attachment	L. L. Northrup	Johnston, R. I	Oct. 3, 1871	119, 536
Ear-drop suspender	L. Weed	Norwalk, Conn	Apr. 27, 1869	89, 452
Ear-muff	S. Goge	New York, N. Y	Dec. 2, 1873	145, 168
Ear-muff	C. Sedgwick	New York, N. Y	Apr. 2, 1872	125, 339
Ear-protector	M. Isidor	New York, N. Y	May 13, 1873	138, 894
Ear-ring	G. Haberland	Bloomington, Ill	Sept. 27, 1870	107, 679
Ear-ring	L. L. Northrup	Olneyville, R. I	Jan. 2, 1872	122, 328
Ear-ring supporter	J. P. Tryner	Bloomington, Ill	Aug. 20, 1872	130, 771
Ear-rings, Device for piercing and lining ears for the reception of.	W. C. Edge	Newark, N. J	Mar. 28, 1871	113, 031
Ear-rings, Fastening for	G. E. Higgins	Syracuse, N. Y	May 16, 1854	10, 921
Ear-trumpet	T. H. Stilwell	New York, N. Y	June 2, 1868	78, 493
Ear-trumpets, Construction of	E. G. Hyde	Camptown, N. J	May 29, 1855	12, 951
Ears, Device for applying remedies to the	J. Fanyon	Bridgeport, Conn	June 13, 1871	115, 944
Ears, Method of attaching ornaments to the	W. B. Carpenter	Brooklyn, N. Y	June 8, 1858	20, 480
Ears, &c., Surgical instrument for examining the	H. Le Riemondie	New Orleans, La	Feb. 8, 1853	9, 581
Earth and sand, Raising and removing	C. Barnard	Newlin, Pa	July 14, 1826	....
Earth and stone, Machine for removing	S. Davis, jr	North Dartmouth, Mass	Feb. 22, 1831	....
Earth-borer	M. C. Chamberlin	Warsaw, N. Y	Jan. 30, 1866	52, 267
Earth borer and elevator	P. Dow	Philadelphia, Pa	Dec. 25, 1849	6, 970
Earth-borer for post-holes, &c	A. S. Ballard	Mount Pleasant, Iowa	Sept. 25, 1860	30, 175
Earth-boring implement	D. Gordon	Evansville, Ind	Oct. 25, 1859	25, 896
Earth-boring implement	C. N. White	Concord, N. Y	Apr. 10, 1855	12, 706
Earth-boring instrument	W. T. Huntington	Washington, D. C	Jan. 12, 1869	85, 741
Earth-boring machine	W. H. Bouser	Paris, Ill	June 10, 1873	139, 761
Earth-boring machine	J. Burns	Anamosa, Iowa	Aug. 19, 1873	141, 916
Earth-boring machine	J. R. Failing	Canajoharie, N. Y	June 13, 1827	....
Earth-boring machine	J. R. Failing	Canajoharie, N. Y	Aug. 25, 1828	....
Earth-boring machine	I. G. Manley and F. Wedge	Zanesville, Ohio	Feb. 19, 1861	31, 469
Earth-boring machine	S. R. Owen	Stewartville, Mo	Nov. 25, 1873	145, 007
Earth-boring machine	J. E. Race	Chicago, Ill	July 21, 1868	80, 217
Earth-boring machine	W. H. Salyer	Hamburgh, Iowa	July 8, 1873	140, 595
Earth-boring machine	C. Vernier	Stryker, Ohio	May 21, 1872	127, 124
Earth-cabinet	G. Wilson and G. H. Baisley	Hamilton, Mo	July 22, 1873	141, 097
Earth chamber-vessel	W. H. Bliss	Newport, R. I	Sept. 5, 1871	118, 681
Earth chamber-vessel	G. G. Baldwin	Milford, Conn	May 10, 1870	102, 750
Earth-closet	W. H. Bliss	Newport, R. I	May 3, 1870	102, 480
Earth-closet	J. A. August	Hot Springs, Va	Apr. 16, 1872	125, 781
Earth-closet	A. F. Baird	Pimlico, England	Nov. 16, 1869	96, 767
Earth-closet	W. J. Bradshaw	Cleveland, Ohio	June 20, 1871	116, 015
Earth-closet	W. J. Bradshaw	Cleveland, Ohio	Oct. 31, 1871	120, 367
Earth-closet	H. C. Bull	New Orleans, La	Dec. 26, 1871	122, 103
Earth-closet	R. A. Cannell	New Orleans, La	July 5, 1870	105, 139
Earth-closet	R. A. Cannell	New Orleans, La	May 23, 1871	115, 024
Earth-closet	H. Clark	Baltimore, Md	Nov. 5, 1872	132, 802
Earth-closet	W. R. C. Clark	New Orleans, La	Mar. 22, 1870	101, 098
Earth-closet	W. R. C. Clark	New Orleans, La	May 3, 1870	102, 494
Earth closet	W. R. C. Clark	New Orleans, La	May 3, 1870	102, 495
Earth-closet	W. R. C. Clark	New Orleans, La	July 5, 1870	105, 045
Earth-closet	L. G. Clock	Manchester, N. H	June 14, 1870	104, 270
Earth-closet	L. G. Clock	Manchester, N. H	Sept. 20, 1870	107, 451
Earth-closet	D. B. Collins	Richmond, Va	Dec. 5, 1871	121, 489
Earth-closet	R. A. Cowell	Cleveland, Ohio	Nov. 1, 1870	108, 885
Earth-closet	R. A. Cowell	Cleveland, Ohio	June 6, 1871	115, 580
Earth-closet	A. W. Davis	Wilmington, Del	Mar. 19, 1872	124, 725
Earth-closet	M. E. Doolittle	Hartford, Conn	Sept. 12, 1871	118, 795
Earth-closet	J. A. Drake	New Orleans, La	May 17, 1870	103, 029
Earth-closet	J. A. Drake	New Orleans, La	July 5, 1870	105, 053
Earth-closet	J. A. Drake	New Orleans, La	Aug. 23, 1870	106, 564
Earth-closet	J. A. Drake	New Orleans, La	Aug. 23, 1870	106, 565
Earth-closet	J. A. Drake	New Orleans, La	Aug. 23, 1870	106, 566

Index of patents issued from the United States Patent Office from 1790 *to* 1873, *inclusive*—Continued.

Invention.	Inventor.	Residence.	Date.	No.
Earth-closet	J. A. Drake and W. R. C. Clark	New Orleans, La	Apr. 12, 1870	101, 717
Earth-closet	B. Ferris	Wilmington, Del	Apr. 4, 1871	113, 411
Earth-closet	J. A. French	Milwaukee, Wis	June 14, 1870	104, 296
Earth-closet	B. A. G. Fuller	West Roxbury, Mass	Dec. 7, 1869	97, 495
Earth-closet	H. J. and J. W. Girdlestone	London, England	Sept. 13, 1870	107, 358
Earth-closet	E. Griffith	Wilmington, Del	Feb. 21, 1871	111, 926
Earth-closet	D. C. Hartman	Baltimore, Md	Feb. 14, 1871	111, 842
Earth-closet	C. C. Haskins	New Albany, Ind	Sept. 30, 1873	143, 239
Earth-closet	C. D. Holmes	Boston, Mass	Oct. 3, 1871	119, 610
Earth-closet	A. A. Jaqua	New York, N. Y	Feb. 20, 1872	123, 780
Earth-closet	A. A. Jaqua	New York, N. Y	June 4, 1872	127, 606
Earth-closet	G. B. Jewett	Salem, Mass	Apr. 19, 1870	102, 128
Earth-closet	G. B. Jewett	Salem, Mass	June 14, 1870	104, 319
Earth-closet	G. B. Jewett	Salem, Mass	July 26, 1870	105, 803
Earth-closet	W. A. Jordan	New Orleans, La	June 27, 1871	116, 320
Earth-closet	W. A. Jordan	New Orleans, La	July 16, 1872	129, 480
Earth-closet	B. L. Kent	West Chester, Pa	Jan. 16, 1872	122, 727
Earth-closet	B. L. Kent	Coatesville, Pa	Apr. 9, 1872	125, 575
Earth-closet	B. L. Kent	Coatesville, Pa	Feb. 11, 1873	135, 817
Earth-closet	C. Kieffer and J. R. D. Seeds	Wilmington, Del	Aug. 23, 1870	106, 593
Earth-closet	J. M. Loewenstein	New Orleans, La	Sept. 26, 1871	119, 376
Earth-closet	J. M. Loewenstein	New Orleans, La	Dec. 26, 1871	122, 123
Earth-closet	B. L. Mack	Essex, Conn	Mar. 18, 1873	137, 013
Earth-closet	A. Mallory	Mystic Bridge, Conn	Sept. 24, 1872	131, 694
Earth-closet	P. Malone	New Orleans, La	Jan. 17, 1871	110, 989
Earth-closet	J. Megratten	Wilmington, Del	Mar. 28, 1871	113, 186
Earth-closet	H. Moule and J. Brannehr	Fordington and Exeter, England.	May 16, 1871	114, 960
Earth-closet	S. D. Newbro	Lansing, Mich	Mar. 14, 1871	112, 620
Earth-closet	W. H. Newton	Newport, R. I	Sept. 27, 1870	107, 799
Earth-closet	N. P. Rider	Mandeville, La	May 14, 1872	126, 745
Earth-closet	G. W. Roberts	Wilmington, Del	Nov. 29, 1870	109, 667
Earth-closet	G. W. Roberts and J. H. Graham	Wilmington, Del	Apr. 30, 1872	126, 155
Earth-closet	I. S. and H. R. Russell	New Market, Md., and Woodbury, N. J.	Mar. 25, 1873	137, 099
Earth-closet	I. S. and H. R. Russell	New Market, Md., and Woodbury, N. J.	Dec. 23, 1873	145, 756
Earth-closet	H. Sherman	Waverly, Pa	Oct. 18, 1870	108, 396
Earth-closet	H. Sherman	Waverly, Pa	Jan. 16, 1872	122, 784
Earth-closet	J. G. Smith	Saint Louis, Mo	Dec. 10, 1872	133, 899
Earth-closet	R. R. Strain	San Francisco, Cal	Nov. 21, 1871	121, 068
Earth-closet	E. W. C. Vanderveer	Linden, N. J	Nov. 15, 1870	109, 276
Earth-closet	G. H. Vroom	Zanesville, Ohio	Nov. 12, 1872	133, 071
Earth closet	C. A. Wakefield	Pittsfield, Mass	May 17, 1870	103, 260
Earth-closet	C. A. Wakefield	Pittsfield, Mass	May 23, 1871	115, 256
Earth-closet	C. A. Wakefield	Pittsfield, Mass	Dec. 3, 1872	133, 685
Earth-closet	C. A. Wakefield	Pittsfield, Mass	June 17, 1873	140, 099
Earth-closet	R. S. Williams	Norristown, Pa	Apr. 2, 1872	125, 156
Earth closet	R. S. Williams	Norristown, Pa	Apr. 2, 1872	125, 370
Earth-closet	R. S. Williams	Norristown, Pa	Jan. 21, 1873	135, 026
Earth-closet	J. Wood	Martinsville, Ohio	Aug. 9, 1870	106, 244
Earth-closet	J. L. Young	New York, N. Y	June 4, 1872	127, 540
Earth-closet	J. L. Young	New York, N. Y	Feb. 4, 1873	135, 504
Earth-closet	J. L. Young	New York, N. Y	Dec. 30, 1873	146, 119
Earth closet, Dry	W. H. Grove	Philadelphia, Pa	May 2, 1871	114, 434
Earth-closet, Portable	A. Panyard	Massillon, Ohio	Apr. 1, 1873	137, 319
Earth-closets, Portable dumping-hod for	E. W. C. Vanderveer	Elizabeth, N. J	Apr. 4, 1871	113, 364
Earth-conveyor	N. Russell	Harrison, Ohio	Dec. 24, 1867	72, 542
Earth, Endless-conveyor for removing	S. Falwell	Memphis, Tenn	Jan. 11, 1859	22, 559
Earth, Excavating and removing	N. Currier	Methuen, Mass	May 29, 1835	....
Earth, Machine for pulverizing	W. M. Bush	Greensburgh, Ind	Jan. 5, 1869	85, 639
Earth, Machine for raising and removing	J. Platt	Albany, N. Y	May 4, 1805	....
Earth, Machine for raising and removing	J. Platt	Albany, N. Y	May 4, 1805	....
Earth, &c., Machine for removing	W. H. Doll	Harrisburgh, Pa	May 29, 1826	....
Earth, Machine for removing	P. Reading	Trenton, N. J	June 1, 1826	....
Earth, &c., Machine for removing	J. Ring	Ogden, N. Y	July 1, 1840	1, 669
Earth, Machine for removing	J. Welsh	Boston, Mass	May 6, 1805	....
Earth-moving machine	J. Cowdon	New Orleans, La	Nov. 10, 1857	18, 573
Earth, mud, &c., Machine for raising	S. Collins	New York, N. Y	Sept. 26, 1826	....
Earth-pulverizer	J. W. Pence	Clayton, Ohio	Sept. 13, 1870	107, 287
Earth-removing apparatus	I. L. Skinner and S. F. Tooker	Hartford, Conn	Mar. 23, 1820	....
Earth scraping and loading machine	A. Ward	New Michigan, Ill	Mar. 9, 1869	87, 736
Earth, stones, &c., Concentrating the volatile parts of calcareous.	J. Fowler		Nov. 16, 1796	....
Earth to be excavated or removed, Machine for loosening.	W. H. Butler	Chicago, Ill	Sept. 3, 1867	68, 411
Earth-works, Machine for building	H. Morey	Cameron, Tex	Dec. 2, 1873	145, 119
Earthen covers, Molding	W. S. Thompson	Rochester, N. Y	June 25, 1872	128, 437
Earthen or cement pipe mold	C. M. Peirce, jr	New Bedford, Mass	Aug. 27, 1861	33, 161
Earthen or cement pipes for conveying water, Construction of.	C. Stearns	Springfield, Mass	May 8, 1839	1, 147
Earthen-pipe mold	P. McIntyre	Hartford, Conn	Apr. 22, 1873	138, 035
Earthen-pipe mold	P. McIntyre	Hartford, Conn	Aug. 26, 1873	142, 168
Earthen pipes, Machine for making curved	R. Connable	Jackson, Mich	Oct. 14, 1873	143, 670
Earthenware kiln	J. Hassell	Elizabethtown, N. J	Oct. 21, 1815	....
Earthenware, Machine for making	C. W. Saladee	Putnam, Ohio	Mar. 28, 1865	47, 043
Earthenware, Manufacture of ornamental	C. L. Fleishmann	Washington, D. C	July 2, 1872	128, 479
Earthenware, Ornamenting baked	R. B. Beech	Kensington, Pa	June 3, 1851	8, 140
Earthenware pipes, Machine for making	J. Bingham	Philadelphia, Pa	Oct. 21, 1873	143, 870
Earthenware pipes, Process of making curved	R. Connable	Jackson, Mich	Oct. 14, 1873	143, 671
Earthenware-shaping apparatus	J. H. Baddeley	Greensborough, Pa	Nov. 9, 1869	96, 537
Earthenware, Stilts for burning	P. Painton	Trenton, N. J	July 28, 1863	39, 356
Easel	E. G. Chormann	Philadelphia, Pa	Dec. 10, 1872	133, 759
Easel	N. Johnson	New York, N. Y	Jan. 16, 1872	122, 725
Easel	G. Munger and J. W. Schermerhorn.	New York, N. Y	Mar. 29, 1870	101, 377
Easel, Artist's	F. W. Bacon	New York, N. Y	Mar. 23, 1869	88, 256
Easel, Artist's	G. W. King	Perth Amboy, N. J	Oct. 6, 1863	40, 171

Index of patents issued from the United States Patent Office from 1790 to 1873, inclusive—Continued.

Invention.	Inventor.	Residence.	Date.	No.
Easel, Carriage-painter's	B. Irons	Columbus, Wis	May 18, 1869	90, 105
Easel, Coach-painter's	D. R. Harris	South New Berlin, N. Y	Mar. 28, 1871	113, 049
Easel, Folding	R. Wright	Brooklyn, N. Y	June 20, 1871	116, 133
Easel, Painter's	P. Dechause	New York, N. Y	Dec. 10, 1867	71, 858
Easel, Painter's	G. Gillett	Little York, N. Y	Sept. 15, 1857	18, 198
Easel, Studio	J. C. Forbes	Toronto, Canada	Nov. 28, 1871	121, 240
Easy-chair	A. Clarke	New York, N. Y	May 1, 1849	6, 416
Easy-chair	J. H. Devereux	Alexandria, Va	May 17, 1864	42, 753
Easy-chair	J. D. Finney	Keene, N. H	May 8, 1823	....
Easy-chair	E. Foster	Hartford, Conn	Aug. 23, 1859	25, 188
Easy-chair	B. F. Hays	Pittsfield, Mass	Dec. 17, 1834	....
Easy-chair	D. Mareau	Hubbardstown, Mass	Aug. 8, 1868	81, 282
Easy-chair	W. C. Poppendieche	New York, N. Y	Apr. 12, 1870	10[illegible], 764
Easy-chair	D. S. Rice	Portland, Me	May 21, 1872	126, 984
Easy-chair	A. Wood	Charlestown, Va	June 28, 1836	....
Easy-chair, Folding	C. C. Schmitt and R. Wodrich	New York, N. Y	Oct. 6, 1868	82, 754
Easy-chair for invalids, &c	P. O'Neil	Brooklyn, N. Y	Sept. 23, 1851	8, 377
Easy-motion and pavement-saving wheel	F. Varela	New York, N. Y	July 20, 1831	....
Eave-gutter for houses	G. W. Wheatly	Harrodsburgh, Ky	Mar. 13, 1855	12, 527
Eave-protector	J. J. Lovell	New York, N. Y	May 25, 1869	90, 454
Eave-trough	J. P. Abbott	Cleveland, Ohio	June 19, 1866	55, 595
Eave-trough	W. Adel	Rockton, Ill	Dec. 28, 1869	98, 328
Eave-trough	J. R. Baker	Kendallville, Ind	Nov. 19, 1872	133, 186
Eave-trough	J. P. Dauth	Reading, Pa	Dec. 31, 1867	72, 813
Eave-trough	N. J. Eldred	Elk Horn, Wis	Aug. 20, 1861	33, 076
Eave-trough	F. J. Emery	Springfield, Ill	Jan. 31, 1865	46, 093
Eave-trough	J. Jacoby	Doylestown, Ohio	Apr. 17, 1866	53, 984
Eave-trough	P. F. Kiblinger	Millersburgh, Ind	Nov. 18, 1873	144, 770
Eave-trough	T. C. Moore	Marion, Ind	Feb. 6, 1866	52, 438
Eave-trough	J. Reinig	Fond du Lac, Wis	Sept. 1, 1868	81, 818
Eave-trough	G. M. Selden	Troy, N. Y	Dec. 18, 1860	30, 934
Eave-trough	W. Stine	Elmore, Ohio	June 2, 1868	78, 017
Eave-trough	W. Yapp	Cleveland, Ohio	Sept. 4, 1866	57, 813
Eave-trough and conductor	K. B. Miller	Utica, N. Y	May 30, 1871	115, 340
Eave trough and gutter machine	J. Lee	Wellsville, Ohio	Oct. 8, 1850	7, 704
Eave-trough and lightning-rod, Combined	J. W. Hankenson and W. Baker	Minneapolis, Minn	Oct. 18, 1870	108, 354
Eave-trough attachment	F. M. Buckles	Altona, Ill	June 29, 1869	91, 823
Eave-trough brace	W. H. Henderson	Franklin, Ind	May 4, 1858	20, 155
Eave-trough bracket	J. W. Gillespie	Alliance, Ohio	Apr. 6, 1869	88, 626
Eave-trough bracket	J. Marshall	Hartland, Mich	Sept. 3, 1867	68, 765
Eave-trough bracket	W. Yapp	Cleveland, Ohio	Aug. 15, 1865	49, 466
Eave-trough bracket and cornice	J. N. Ball	Buffalo, N. Y	July 30, 1867	67, 155
Eave-trough fastener	L. Cook	Shreveport, La	Jan. 3, 1871	110, 633
Eave-trough fastener	M. Henney and W. B. Rager	North Manchester, Ind	Nov. 1, 1870	108, 906
Eave-trough fastening	J. P. Abbott	Cleveland, Ohio	June 18, 1872	128, 089
Eave-trough fastening	P. Ahn	Brandon, Vt	Dec. 10, 1867	71, 941
Eave-trough fastening	W. H. Hammond	Syracuse, Mo	Aug. 19, 1873	142, 016
Eave-trough former	L. Granger and L. Phillips	Memphis, Mich	July 20, 1869	92, 720
Eave-trough hanger	J. P. Abbott	Cleveland, Ohio	July 19, 1870	105, 404
Eave-trough hanger	J. P. Abbott	Cleveland, Ohio	Jan. 21, 1873	135, 056
Eave-trough hanger	D. Arter	Ashland, Ohio	Dec. 1, 1868	84, 467
Eave-trough hanger	T. W. Bailey	Savannah, Ohio	Sept. 16, 1873	142, 758
Eave-trough hanger	H. S. Bishop	Cleveland, Ohio	July 2, 1872	128, 457
Eave-trough hanger	D. Dimmick	Orwell, Pa	June 18, 1872	128, 128
Eave-trough hanger	J. J. Kaufman	Ashland, Ohio	Aug. 2, 1870	105, 947
Eave-trough hanger	R. A. Lucas	Wooster, Ohio	Apr. 26, 1870	102, 412
Eave-trough hanger	T. F. Palm	Toledo, Ohio	Sept. 30, 1873	143, 375
Eave-trough hanger	H. B. Todd	Plymouth, Conn	Feb. 18, 1873	136, 109
Eave-trough hanger	T. G. Williams	Akron, Ohio	Nov. 4, 1873	144, 299
Eave-trough hanging	H. M. Gilbert and F. Elberson	Ada, Ohio	July 28, 1868	80, 473
Eave-trough-making machine	L. Mann	Ionia, Mich	Apr. 17, 1860	27, 916
Eave-trough-sawing machine	J. Wing	Hancock, Vt	Mar. 17, 1838	635
Eave-trough support	T. F. Morrison	Findley, Ohio	Dec. 30, 1873	146, 014
Eave-trough support	J. D. Pierce	Milwaukee, Wis	Dec. 24, 1872	134, 163
Eave-trough supporter	A. G. Perry	Clyde, Ohio	May 18, 1869	90, 122
Eave-trough supporter	B. Prugh and H. Austry	Grant City, Mo	Apr. 9, 1872	125, 483
Eave-trough suspender	T. F. Palm	Toledo, Ohio	Apr. 20, 1869	89, 068
Eave troughs and pipings, Process for manufacturing wooden.	S. T. Field	Worcester, Mass	Feb. 19, 1861	31, 449
Eave-troughs, Apparatus for suspending	J. A. Watrous	Green Spring, Ohio	Sept. 1, 1857	18, 113
Eave-troughs, Connecting sheet-metal for	W. M. Phelps	Marshall, Mich	Oct. 28, 1862	36, 855
Eave-troughs, Finishing	C. A. Codding	Battle Creek, Mich	Oct. 11, 1864	44, 605
Eave-troughs, Machine for forming	J. Brett	Memphis, Mich	Sept. 15, 1868	82, 077
Eave-troughs, Machine for forming	A. Calkins and W. Tower	Almont, Mich	Mar. 19, 1867	63, 012
Eave-troughs, Machine for making	O. E. Mallory	Castile, N. Y	Jan. 1, 1856	14, 025
Eave-troughs, Machine for making	O. W. Noble	Darlington, Wis	May 15, 1866	54, 758
Eave-troughs, Machine for making sheet-metal	S. A. Scofield and E. Churchill	Morenci, Mich	Dec. 15, 1863	40, 951
Eave-troughs, Machine for making wooden	A. T. Stearns	Dorchester, Mass	Apr. 2, 1867	63, 572
Eave-troughs, Machine for manufacturing	S. Yates	Clarence, Iowa	Aug. 28, 1866	57, 615
Eave-troughs, Making	L. Pierson	Jeffersonville, Ind	June 26, 1868	28, 901
Eave-troughs, Suspending	C. D. Woodruff	Toledo, Ohio	Mar. 7, 1854	10, 606
Eave-troughs to houses, Method of attaching	W. R. Wallis	Alliance, Ohio	June 11, 1867	65, 779
Eave-troughs, while soldering them, Device for holding.	E. Wilkinson, jr	Mansfield, Ohio	Aug. 14, 1866	57, 237
Eccentric	J. C. Butterfield	Chicago, Ill	Sept. 16, 1873	142, 898
Eccentric	J. C. Wells	Warren, Pa	Sept. 21, 1869	95, 174
Eccentric, Adjustable	J. B. Strickland	Scranton, Pa	May 21, 1867	64, 922
Eccentric, Adjustable	D. F. Walker	Bowling Green, Ky	Mar. 7, 1865	46, 737
Eccentric adjustment	J. M. Stone	North Andover, Mass	Feb. 7, 1865	46, 278
Eccentric, Variable	T. Keeler and G. S. Avery	Danbury, Conn	Apr. 28, 1868	77, 197
Eccentrics, Mechanism for actuating adjustable	M. Stubbs	Cincinnati, Ohio	Apr. 6, 1852	8, 863
Edge-burnishing machine for boot and shoe soles	A. C. Carey	Maiden, Mass	July 16, 1872	129, 457
Edge-burnishing machine for boot and shoe soles	F. Curtis	Boston, Mass	Aug. 6, 1872	130, 109
Edge-burnishing machine for boot and shoe soles	S. H. Hodges	Lynn, Mass	July 23, 1872	129, 825
Edge-burnishing machine, Sole	A. C. Carey	Malden, Mass	Sept. 17, 1872	131, 425
Edge-burnishing tool, Welt	M. J. Ferren	Stoneham, Mass	Apr. 15, 1873	137, 903
Edge-finishing machine for boot and shoe soles	L. Cote	Saint Hyacinth, Quebec, Canada.	June 21, 1870	104, 560

Index of patents issued from the United States Patent Office from 1790 *to* 1873, *inclusive*—Continued.

Invention.	Inventor.	Residence.	Date.	No.
Edge-finishing tool	J. B. Blanchard	Marlborough, Mass	Apr. 7, 1868	76, 391
Edge-finishing tool, Sole	F. M. Schmitt	Jamaica Plains, Mass	Dec. 18, 1866	60, 563
Edge-gage, Feather	G. G. Townsend	Rochester, N. Y	Sept. 23, 1856	15, 789
Edge-iron for shoemakers	E. D. Beales	Gallipolis, Ohio	June 21, 1870	104, 540
Edge-key	E. Campbell	Bath, Me	Apr. 30, 1861	30, 230
Edge-key for boots	G. C. Todd	Lynn, Mass	Dec. 7, 1858	22, 248
Edge-key for making and polishing the edge of boot and shoe soles.	G. C. Todd	Lynn, Mass	Nov. 25, 1856	16, 128
Edge-key handle	R. A. Woodberry	Beverly, Mass	May 31, 1870	103, 812
Edge-parer, Boot and shoe	W. Frederick	Ashland, Pa	Aug. 28, 1866	57, 493
Edge-plane	C. P. Bigelow	Clinton, Mass	Mar. 30, 1869	88, 264
Edge-plane	J. H. Conklin	Yorktown, N. Y	Dec. 27, 1864	45, 669
Edge-plane	C. D. McAuley	Carthage, Ohio	Feb. 18, 1868	74, 561
Edge-plane	S. Miller	Urbana, Ohio	Sept. 18, 1866	58, 123
Edge-plane	M. Newman	Unadilla, N. Y	May 30, 1865	47, 973
Edge-plane	J. E. Plummer	Binghamton, N. Y	Nov. 12, 1867	70, 894
Edge-plane	H. Sauerbier	Newark, N. J	July 19, 1859	24, 825
Edge-plane for boot-soles	C. Coti	Marlborough, Mass	Apr. 6, 1869	88, 551
Edge-plane for boots, &c	F. Buxton and G. Crosby	Lake Village, N. H	June 23, 1868	79, 203
Edge-plane for boots and shoes	W. Bayhouse	Portland, Oreg	Jan. 22, 1867	61, 384
Edge-plane for boots and shoes	A. B. Clark	Auburn, Mass	Feb. 5, 1867	61, 804
Edge-plane for boots and shoes	P. S. Foster	Richmond, Me	May 4, 1869	89, 754
Edge-plane for boots and shoes	J. C. Grün	Peoria, Ill	Feb. 27, 1866	52, 845
Edge-plane for boots and shoes	E. S. Snell	North Bridgewater, Mass	Apr. 10, 1860	27, 840
Edge-plane for boots and shoes	B. Tolman	Pembroke, Mass	June 21, 1859	24, 526
Edge-plane for trimming boot and shoe soles	I. A. Dunham	North Bridgewater, Mass	Sept. 22, 1857	18, 237
Edge-plane for trimming boot and shoe soles	C. Warren	Putnam, Conn	July 28, 1857	17, 905
Edge-plane for trimming soles and welts of boots and shoes.	B. J. Tayman	Philadelphia, Pa	May 20, 1873	139, 034
Edge-plane for trimming soles of boots, &c	J. Brooks and J. H. Sanford	North Bridgewater, Mass	Mar. 31, 1863	38, 026
Edge-plane, Shoemaker's	F. F. Baumann	Cambridge, Mass	Jan. 28, 1873	135, 308
Edge-plane, Shoemaker's	N. Bucher	Weedsport, N. Y	Nov. 2, 1852	9, 364
Edge-plane, Shoemaker's	I. A. Dunham	North Bridgewater, Mass	June 24, 1856	15, 176
Edge-plane, Shoemaker's	A. P. Hazard	North Bridgewater, Mass	July 19, 1870	105, 452
Edge-plane, Shoemaker's	A. P. Hazard	North Bridgewater, Mass	Sept. 20, 1870	107, 491
Edge-plane, Shoemaker's	C. E. Hersey	East Stoughton, Mass	Sept. 6, 1864	44, 093
Edge-plane, Shoemaker's	D. W. Horton	Petersburgh, Ind	Oct. 17, 1854	11, 808
Edge-plane, Shoemaker's	A. J. Parker	Lynn, Mass	Dec. 5, 1871	121, 465
Edge-plane, Shoemaker's	J. H. Sandford	North Bridgewater, Mass	Apr. 19, 1870	102, 048
Edge plane, Sole, (*See also* Sole)	M. Packard	North Easton, Mass	June 9, 1868	78, 758
Edge-setting tool, Sole	J. H. Morrison	Marlborough, Mass	Feb. 18, 1873	136, 087
Edge, Straight	S. Darling	Bangor, Me	Jan. 7, 1868	73, 082
Edge-tool sharpener	J. W. Wood	Watertown, N. Y	July 24, 1860	29, 336
Edge-tools, Composition for sharpening	G. L. Witsil	Philadelphia, Pa	Mar. 24, 1868	76, 025
Edge-tools, Grinding	G. C. Eaton	Lockport, N. Y	Feb. 10, 1863	37, 616
Edge-tools, Grinding	S. D. McLelland	McGregor, Iowa	July 12, 1864	43, 518
Edge-tools, Machine for holding and grinding	J. Richardson	Turner, Me	Apr. 3, 1866	53, 747
Edge-tools, Making	W. White	Newark, N. J	Dec. 28, 1858	22, 466
Edge-tools, Manufacture of	S. W. Collins	Canton, Conn	Nov. 20, 1860	30, 668
Edge-tools, Manufacture of	R. C. Grover	Newton, Mass	Oct. 6, 1868	82, 711
Edge-tools, Process of hardening	L. W. Stuart	Hawley, Pa	July 1, 1873	140, 390
Edge trimming and burnishing machine, Jack-center for.	T. M. Orr and A. R. Sears	North Bridgewater, Mass	Nov. 11, 1873	144, 560
Edge trimming and setting machine, Boot and shoe.	S. H. Hodges	Lynn, Mass	Mar. 19, 1872	124, 744
Edge-trimming machine for boots and shoes	R. C. Lambert	Raynham, Mass	Aug. 11, 1868	80, 828
Edge-trimming tool, Sole	I. R. Rogers	Lynn, Mass	Feb. 18, 1873	136, 096
Edges of boot and shoe soles, Burnishing-machine for setting the.	S. H. Hodges	Lynn, Mass	July 23, 1872	129, 663
Edging and ripping machine	E. C. Dicey	Whitehall, Mich	July 16, 1872	129, 400
Edging boards, Machine for	S. H. Richardson	Bangor, Me	Jan. 16, 1872	122, 780
Edging-gage for shoemakers, Feather	J. Jenkins	Andover, Mass	July 22, 1851	8, 247
Edging leather straps, Machine for	J. Barnes	Franklin, N. Y	Sept. 6, 1853	9, 986
Edging-machine	H. E. Bradt	Manistee, Mich	Apr. 4, 1871	113, 487
Edging-machine	L. Fay	Cincinnati, Ohio	Sept. 3, 1861	33, 190
Edging-machine	O. H. Gromberg and W. M. Ferry.	Ferrysburgh and Grand Haven, Mich.	May 20, 1873	139, 142
Edging-machine	E. H. Stearns	Erie, Pa	Jan. 17, 1871	111, 093
Edging-machine	E. H. Stearns	Erie, Pa	Sept. 30, 1873	143, 264
Edging machine, Feather	L. Goddu	Braintree, Mass	Sept. 20, 1864	44, 300
Edging-machine for turning the edges of tin, copper, or sheet-iron.	J. Woolley	New York, N. Y	Mar. 16, 1831	....
Edging metals, Machine for	W. Crossley	Chicago, Ill	Oct. 27, 1868	83, 365
Edging soles, Machine for feather	A. S. Libby	Lawrence, Mass	June 4, 1872	127, 618
Edging tin plates, sheet-iron, sheathing-copper, &c	C. F. Fisher	York, Pa	Feb. 28, 1833	....
Edging-tool for boot and shoe soles	T. Smiley	Albia, Iowa	Apr. 13, 1869	88, 920
Edging-tool, Harness-maker's	O. W. Morley	Tarrytown, N. Y	Sept. 7, 1869	94, 629
Edging-tool, Harness-maker's	F. M. Patterson and H. P. Miller.	Seymour, Ind	July 12, 1870	105, 242
Edging-tool, Tinman's	W. H. Henderson	Franklin, Ind	May 1, 1866	54, 342
Edible composition	D. Fobes	Boston, Mass	May 21, 1867	64, 856
Education-table	E. Allen	Windham, Conn	May 1, 1849	6, 407
Eel-pot	G. D. Allen	New York, N. Y	Oct. 13, 1868	82, 913
Egg and cake beater	J. W. Condon	Logansport, Ind	July 15, 1873	140, 891
Egg and cake beater	H. S. Maltby	Cincinnati, Ohio	Oct. 18, 1870	108, 498
Egg and cream beater	F. Oakley	London, England	Jan. 29, 1867	61, 679
Egg and fruit carrier	I. R. Amsden	Buffalo, N. Y	Jan. 26, 1869	86, 267
Egg and fruit carrier	W. J. Clark	Lena, Ill	Mar. 21, 1871	112, 784
Egg and fruit carrier	J. T. Cornforth	Kansas City, Mo	Aug. 15, 1871	117, 988
Egg-beater	L. B. Alden	Cincinnati, Ohio	July 31, 1866	56, 845
Egg-beater	W. N. Angus	Morristown, N. J	Apr. 21, 1868	76, 970
Egg-beater	V. G. Arnold	Providence, R. I	Oct 16, 1866	58, 750
Egg-beater	F. Ashley	New York, N. Y	May 1, 1860	28, 047
Egg-beater	F. Ashley	New York, N. Y	June 26, 1866	55, 802
Egg-beater	U. Baker	Brooklyn, N. Y	Oct. 23, 1860	30, 453
Egg-beater	L. T. Blake	New Haven, Conn	July 28, 1868	80, 440
Egg-beater	A. R. Blasse	Westminster, Md	Feb. 13, 1866	52, 518
Egg-beater	G. E. Bridger	Milwaukee, Wis	Aug. 27, 1867	68, 037

Index of patents issued from the United States Patent Office from 1790 *to* 1873, *inclusive*—Continued.

Invention.	Inventor.	Residence.	Date.	No.
Egg-beater	C. H. Butterfield	Sturbridge, Mass	Oct. 16, 1866	58, 770
Egg-beater	G. G. Carver	Roxbury, Mass	Dec. 3, 1867	71, 696
Egg-beater	D. B. Clayton	Columbia, S. C	Aug. 15, 1871	117, 982
Egg-beater	M. G. Crane	Boston, Mass	July 4, 1865	48, 525
Egg-beater	M. G. Crane	Boston, Mass	June 19, 1866	55, 625
Egg-beater	W. O. Crocker	Laconia, N. H	Sept. 24, 1872	131, 600
Egg-beater	J. Dane, jr	Newark, N. J	Jan. 23, 1872	122, 997
Egg-beater	J. Davis	Schenectady, N. Y	July 17, 1866	56, 382
Egg-beater	G. K. Dearborn	Pawtucket, R. I	July 26, 1870	105, 655
Egg-beater	C. M. Drennan	Boston, Mass	Sept. 18, 1866	58, 165
Egg-beater	H. F. Drott	Cumberland, Md	Sept. 18, 1860	30, 053
Egg-beater	T. Earle	Smithfield, R. I	July 7, 1863	39, 134
Egg-beater	T. Earle	Valley Falls and Smithfield, R. I.	Apr. 24, 1866	54, .34
Egg-beater	T. Earle	Valley Falls, R. I	Jan. 25, 1870	99, 173
Egg-beater	T. Earle	Smithfield, R. I	July 5, 1870	105, 057
Egg-beater	T. Earle and G. K. Dearborn	Valley Falls, Smithfield, and Pawtucket, R. I.	Sept. 27, 1870	107, 673
Egg-beater	H. D. and J. T. Felthouse	Philadelphia, Pa	Mar. 27, 1866	53, 429
Egg-beater	H G. and A. C. Fougen	Cape Girardeau, Mo	Nov. 30, 1869	97, 379
Egg-beater	E. Hadley	Chicopee Falls, Mass	May 6, 1873	138, 647
Egg-beater, &c	W. H. Haines	Newark, N. J	Feb. 15, 1870	99, 883
Egg-beater	W. Hart	Philadelphia, Pa	Mar. 12, 1861	31, 663
Egg-beater	J. B. Heich	Cincinnati, Ohio	Dec. 15, 1857	18, 849
Egg-beater	H. P. Hood	Indianapolis, Ind	Aug. 6, 1872	130, 297
Egg-beater	J. M. Jay	Canton, Ohio	Feb. 7, 1860	27, 054
Egg-beater	J. M. Jay and J. Danner	Canton, Ohio	Apr. 17 1860	27, 908
Egg-beater	H. F. Jenks and T. Marsh	Pawtucket, R. I	June 13, 1871	115, 862
Egg-beater	S. F. Jones	Saint Paul, Ind	May 3, 1859	23, 843
Egg-beater	F. L. King	Worcester, Mass	Jan. 2, 1866	51, 839
Egg-beater	P. Klepper	Centralia, Ill	July 16, 1867	66, 855
Egg-beater	F. Krandelt	San Francisco, Cal	Feb. 22, 1870	100, 155
Egg-beater	E. B. Kunkle	Fort Wayne, Ind	Sept. 5, 1871	118, 727
Egg-beater	L. Laurie	Washington, D. C	Dec. 21, 1869	98, 168
Egg-beater	H. W. Louden	Ephratah, Pa	Feb. 25 1866	74, 837
Egg-beater	D. D. Mackay	Whitestone, N. Y	Dec. 29, 1868	85, 460
Egg-beater	T. Marsh	Pawtucket, R. I	Sept. 20, 1870	107, 515
Egg-beater	T. Marsh and J. Berney	Pawtucket, R. I	June 14, 1870	104, 174
Egg-beater	T. Marsh and J. Berney	Pawtucket, R. I	Aug. 9, 1870	106, 182
Egg-beater	T. McBean	Fowlerville, N. Y	May 24, 1859	24, 134
Egg-beater	J. P. McLean and P. A. Morley	Brooklyn, N. Y	June 19, 1860	28, 760
Egg-beater	P. Mihan	Boston, Mass	Mar. 23, 1858	19, 738
Egg-beater	H. Miller	Cincinnati, Ohio	Dec. 1, 1857	18, 759
Egg-beater	H. Miller	Cincinnati, Ohio	Feb. 6, 1872	123, 353
Egg-beater	N. C. Miller	Stroudsburgh, Pa	Feb. 1, 1870	99, 337
Egg-beater	N. C. Miller	Stroudsburgh, Pa	Apr. 22, 1873	138, 094
Egg-beater	J. F. Monroe and E. P. Monroe	Fitchburgh, Mass., and New York, N. Y.	Apr. 19, 1859	23, 694
Egg-beater	D. Munson	Indianapolis, Ind	Mar. 5, 1872	124, 375
Egg-beater	D. Munson	Indianapolis, Ind	June 17, 1873	140, 067
Egg-beater	W. T. Nicholson	Providence, R. I	July 23, 1861	32, 886
Egg-beater	W. T. Nicholson and T. Earle	Providence and Smithfield, R. I.	Sept. 25, 1860	30, 152
Egg-beater	J. L. Nicolai	Chicago, Ill	Apr. 26, 1859	23, 814
Egg-beater	S. B. Pangborn and G. H. Griffin.	Boston, Mass	May 15, 1866	54, 761
Egg-beater	C. R. Peirce	Philadelphia, Pa	Aug. 20, 1872	130, 591
Egg-beater	W. H. Peirce	Bangor, Me	Mar. 24, 1868	75, 787
Egg-beater	C. Pinder	Lowell, Mass	Oct. 16, 1866	58, 882
Egg-beater	A. M. Robinson	Boston, Mass	May 15, 1866	54, 776
Egg-beater	J. F. Rote	Reading, Pa	Nov. 11, 1873	144, 567
Egg-beater	W. Saladee and E. M. Luckett	Philadelphia, Pa	May 3, 1864	42, 603
Egg-beater	P. Schildecker	Pittsburgh, Pa	Oct. 6, 1863	40, 188
Egg-beater	C. F. A. Seitz	Philadelphia, Pa	Apr. 6, 1869	88, 672
Egg-beater	D. Shive	Philadelphia, Pa	July 29, 1862	36, 030
Egg-beater	W. B. Smith	New York, N. Y	Dec. 26, 1865	51, 758
Egg-beater	D. M. Swartz	Lewisburgh, Pa	Oct. 11, 1870	108, 203
Egg-beater	H. Tilden	Boston, Mass	Aug. 1, 1865	49, 176
Egg-beater	S. Walker	Boston, Mass	Jan. 24, 1860	26, 958
Egg-beater	D. Webster	Washington, D. C	Dec. 10, 1867	72, 136
Egg-beater	W. Weckersham	Boston, Mass	Apr. 9, 1867	63, 774
Egg-beater	L. Williams	Terrysville, Conn	Jan. 4, 1870	98, 533
Egg-beater	M. T. Williams	Milwaukee, Wis	May 21, 1867	64, 928
Egg-beater	T. Williams	Providence, R. I	May 31, 1870	103, 811
Egg-beater	S. C. Wilson	Central Falls, R. I	Mar. 10, 1868	75, 340
Egg-beater and liquor-mixer	E. C. Strange	Taunton, Mass	Aug. 28, 1866	57, 596
Egg beater and stirrer	C. Lehman	Hartford, Conn	Mar. 29, 1870	101, 281
Egg-beater, cake-cutter, and nutmeg-grater	S. C. Moore	Boston, Mass	May 1, 1866	54, 388
Egg beater or agitator	J. W. Bliss	Hartford, Conn	Dec. 6, 1864	45, 309
Egg beater or churn	J. J. Parker	Marietta, Ohio	Aug. 9, 1859	25, 038
Egg-beater or ice cream freezer	J. Pyne and W. Burr	Harrisburgh, Pa	Nov. 15, 1859	26, 123
Egg-beater, Rotary	R. Collier	Baltimore, Md	Dec. 23, 1856	16, 267
Egg-boiler	F. Ashley	New York, N. Y	Aug. 1, 1865	49, 062
Egg-boiler	P. Malapert and E. A. Des Courtis.	Poitiers, France, and New York, N. Y.	Oct. 24, 1865	50, 669
Egg-boiler	A. Sharlow	Fort Lee, N. J	Apr. 11, 1865	47, 228
Egg-boiling apparatus	I. Dimock	Florence, Mass	Oct. 20, 1868	83, 264
Egg-box	J. D. Michael	Baltimore, Md	Jan. 4, 1870	98, 609
Egg-carrier	J. R. Asher	Oskaloosa, Iowa	Sept. 6, 1870	106, 983
Egg-carrier	J. A. Beam	California, Mo	July 15, 1873	140, 807
Egg-carrier	W. Bramwell	Toledo, Ohio	Aug. 13, 1872	130, 469
Egg-carrier	A. H. Bryant	Chicago, Ill	Mar. 22, 1870	101, 092
Egg-carrier	A. H. Bryant	Chicago, Ill	Mar. 22, 1870	101, 093
Egg-carrier	A. H. Bryant	Chicago, Ill	June 14, 1870	104, 261
Egg-carrier	A. H. Bryant	New York, N. Y	July 16, 1872	129, 313
Egg-carrier	A. H. Bryant	New York, N. Y	Dec. 17, 1872	134, 031
Egg-carrier	A. H. Bryant	Chicago, Ill	June 3, 1873	139, 540
Egg-carrier	W. J. Clark	Lena, Ill	Mar. 15, 1870	100, 857

Index of patents issued from the United States Patent Office from 1790 to 1873, inclusive—Continued.

Index of patents issued from the United States Patent Office from 1790 *to* 1873, *inclusive*—Continued.

Invention.	Inventor.	Residence.	Date.	No.
Ejector for oil-wells, Oil	W. Reed	Pittsburgh, Pa	Dec. 31, 1872	134, 397
Ejector for refuse matter from steam-vessels	J. Palmer	Sandisfield, Mass	Sept. 26, 1865	50, 153
Ejector for steam-boiler furnaces	J. N. Snowdon and H. Wilkins	Brownsville, Pa	June 13, 1865	48, 218
Ejector for steam-vessels, Ash	G. P. Hunt	United States Navy	Dec. 3, 1872	133, 535
Ejector for steam-vessels, Refuse	J. Brown, jr., and S. R. Brooks	New York, N. Y., and Saint Louis, Mo.	Jan. 12, 1864	41, 192
Ejector, Insect-powder	S. Rose and N. Goldsmith	Cincinnati, Ohio	Nov. 29, 1870	109, 762
Ejector, Liquid	A. Brear	Saugatuck, Conn	Aug. 20, 1865	49, 602
Ejector, Oil	R. Boeklen	Brooklyn, N. Y	Nov. 21, 1865	51, 008
Ejector, Oil	A. Brear	Saugatuck, Conn	May 23, 1865	47, 793
Ejector, Oil	E. Crooker	Buffalo, N. Y	Apr. 25, 1865	47, 399
Ejector, Oil	W. R. Greenleaf	Buffalo, N. Y	Mar. 28, 1865	47, 011
Ejector, Oil	W. R. Greenleaf	Buffalo, N. Y	June 13, 1865	48, 170
Ejector, Oil	W. W. Hubbell	Philadelphia, Pa	Oct. 31, 1865	50, 710
Ejector, Oil	G. E. Mills	New York, N. Y	Apr. 25, 1865	47, 444
Ejector, Oil	G. M. Mowbray	Titusville, Pa	Mar. 21, 1865	46, 925
Ejector, Oil	F. S. Pease	Buffalo, N. Y	Mar. 21, 1865	46, 931
Ejector, Oil	F. S. Pease	Buffalo, N. Y	Mar. 28, 1865	47, 034
Ejector, Oil	S. F. Schoonmaker	New York, N. Y	Apr. 11, 1865	47, 226
Ejector, Oil	S. F. Schoonmaker	New York, N. Y	Apr. 11, 1865	47, 227
Ejector, Oil	H. Searl	Rochester, N. Y	Mar. 14, 1865	46, 824
Ejector, Oil	H. Searle	Rochester, N. Y	Jan. 29, 1867	61, 571
Ejector, Oil	J. Y. Smith	Alexandria, Va	May 23, 1865	47, 871
Ejector, Oil	J. Y. Smith	Alexandria, Va	June 6, 1865	48, 106
Ejector, Oil	L. W. Turrell, S. Stanton, and L. C. Ward.	Newburgh, N. Y	Apr. 4, 1865	47, 174
Ejector, Oil	G. L. Witsil and E. Burke	Philadelphia, Pa	Apr. 4, 1865	47, 148
Ejector, Perfume	C. L. Fehrenson	New York, N. Y	Apr. 12, 1870	101, 848
Ejector, Sand	E. W. Poston	Fort Wayne, Ind	Aug. 6, 1867	67, 580
Ejector, Sand	J. Thayer	Topeka, Kans	Apr. 1, 1873	137, 507
Ejector, Sand and seed	M. Christiansen	Winneconne, Wis	May 23, 1871	115, 164
Ejector, Steam	N. L. Chappell	New York, N. Y	Aug. 6, 1867	67, 413
Ejector, Steam air	J. Y. Smith	Pittsburgh, Pa	Mar. 11, 1873	136, 780
Ejector, Steam and air	S. E. Hewes	Albany, N. Y	Oct. 30, 1866	59, 221
Ejector, Steam-operated water	H. Coll	Millville Borough, Pa	June 8, 1869	90, 930
Ejector, Steam water	R. D. and W. F. Cox	Philadelphia, Pa	May 7, 1872	126, 525
Ejector, Steam water	J. Habermehl and H. Kleiman	Allegheny, Pa	Mar. 26, 1872	124, 895
Ejector, Steam water	H. S. Ross	New York, N. Y	May 26, 1868	78, 395
Ejector, Water	N. L. Chappell	New York, N. Y	Aug. 29, 1865	49, 603
Ejector, Water	H. Coll	Millvale, Pa	Oct. 21, 1873	143, 884
Ejector, Water	T. J. Jones	Madison, N. J	May 7, 1867	64, 539
Ejector, Water	H. S. Lansdell	New York, N. Y	Mar. 26, 1867	63, 264
Elastic bands, Clasp for	R. P. Staats	New York, N. Y	Dec. 2, 1873	145, 073
Elastic chair	S. Gragg	Boston, Mass	Aug. 31, 1808	....
Elastic clamp for shoemakers, &c	T. R. Hoyt	Goffstown, N. H	Dec. 6, 1817	....
Elastic coupling for mill-shafting, &c	W. S. Reeder	Saint Louis, Mo	Oct. 13, 1857	18, 412
Elastic cups, dippers, &c., Manufacture of	T. Smith, jr	Boston, Mass	Dec. 16, 1862	37, 181
Elastic cushion for horses' feet	J. Haseltine and C. L. Wheeler	Melrose and Cambridge, Mass.	Sept. 21, 1869	95, 107
Elastic cushion for piercing implements	B. D. Sanders	Wellsburgh, W. Va	Apr. 5, 1864	42, 228
Elastic fabric	H. Hynes	New York, N. Y	Mar. 7, 1865	46, 674
Elastic fabric	J. W. Newell	New Brunswick, N. J	Aug. 24, 1858	21, 270
Elastic fabrics, Apparatus for the manufacture of	C. A. Ensign	Naugatuck, Conn	Aug. 30, 1870	106, 924
Elastic gate, Self-adjusting	P. S. Reist	Oregon, Pa	Oct. 8, 1867	69, 704
Elastic goods	E. Brown	Rindge, N. H	Sept. 26, 1854	11, 716
Elastic hollow molded articles, Manufacture of	D. D. Parmelee	Salem, Mass	Jan. 17, 1860	26, 864
Elastic mold	T. Taylor	Washington, D. C	Jan. 29, 1867	61, 641
Elastic or yielding roller, Driving gear of	L. M. Woodcock	Auburn, N. Y	Aug. 6, 1872	130, 349
Elastic roller	L. R. Streeter	Chelsea, Mass	Apr. 19, 1864	42, 415
Elastic-rubber straps, Manufacture of	C. Goodyear, jr	New York, N. Y	May 26, 1863	38, 674
Elastic seat, Portable adjustable	H. Perkins	Mansfield, Mass	Sept. 29, 1868	82, 636
Elastic shirred goods, Machinery for making	R. Solis	New Brunswick, N. J	Mar. 26, 1861	31, 832
Elastic-spring cushion, &c	S. Blair	New Haven, Conn	Aug. 1, 1823	....
Elastic springs, Manufacture of	S. G. Randall	Providence, R. I	Aug. 21, 1866	57, 377
Elastic strap for garments	J. W. Boughton	Chicago, Ill	Dec. 4, 1866	60, 131
Elastic supporter, Ladies'	G. C. Stillson	Derby, Conn	July 11, 1871	116, 882
Elastic-tube clasp	E. A. Day	Oberlin, Ohio	Dec. 17, 1872	134, 038
Elastic tubes, Regulating-clasp for	W. Bunce	Oberlin, Ohio	Mar. 28, 1871	113, 016
Elastic washer	C. Dittman	Leacock, Pa	July 2, 1872	128, 602
Elastic washer	J. P. Verre	Philadelphia, Pa	Dec. 26, 1871	122, 141
Elbow-joint band	S. M. Parse	Newark, N. J	Nov. 11, 1862	36, 911
Elbows, Manufacture of curved	F. G. and W. F. Niedringhaus	Saint Louis, Mo	Jan. 12, 1869	85, 843
Electric alarm	C. T. Mason	Sumter, S. C	May 11, 1869	89, 936
Electric and vapor chair	M. A. Hayward	Brooklyn, N. Y	Sept. 26, 1871	119, 230
Electric bath	M. W. House	Cleveland, Ohio	Feb. 18, 1862	34, 425
Electric bath	M. W. House	Cleveland, Ohio	May 5, 1863	38, 389
Electric battery	L. Bastet	Tarrytown, N. Y	Sept. 26, 1871	119, 298
Electric battery	T. Chutaux	Paris, France	July 5, 1870	105, 042
Electric battery	T. Chutaux	Paris, France	July 5, 1870	105, 043
Electric battery	V. Himmer	New York, N. Y	Oct. 10, 1871	119, 763
Electric battery for telegraphing and other purposes.	D. M. Cook	Mansfield, Ohio	Apr. 4, 1871	113, 399
Electric battery, Compound switch for	J. Kidder	New York, N. Y	July 26, 1870	105, 697
Electric currents, Apparatus for regulating and measuring the intensity of.	J. Lacassagne and R. Thiers	Lyons, France	Mar. 16, 1858	19, 642
Electric currents, Method of integrating inconstant.	C. Kirchhof	New York, N. Y	Feb. 26, 1861	31, 545
Electric fire and burglar alarm	E. Fontaine	Fort Wayne, Ind	Aug. 17, 1869	93, 816
Electric light	A. A. Meynial	New York, N. Y	Feb. 20, 1872	123, 923
Electric light	J. F. May	London, England	Oct. 8, 1861	33, 457
Electric light	J. F. May	London, England	Oct. 8, 1861	33, 458
Electric light, Producing the	E. L. C. d'Ivernois	Paris, France	May 9, 1871	114, 652
Electric machine	A. L. Fleury	New York, N. Y	July 14, 1868	79, 819
Electric machine, Frictional	R. Cornelius	Philadelphia, Pa	May 21, 1861	32, 354
Electric switch	S. Gardiner, jr	New York, N. Y	Nov. 19, 1867	71, 158
Electrical apparatus, Connecting-post for	T. Wishart	Clayton, Del	Apr. 15, 1873	137, 810
Electrical apparatus, Construction of	T. Brown	Albany, N. Y	Dec. 20, 1816	....

Index of patents issued from the United States Patent Office from 1790 *to* 1873, *inclusive*—Continued.

Invention.	Inventor.	Residence.	Date.	No.
Electrical apparatus, Keys for use with	J. Olmsted	Providence, R. I	Dec. 10, 1872	133, 797
Electrical apparatus, Pole-changer for	J. E. Smith	New York, N. Y	June 11, 1872	127, 809
Electrical apparatus, Portable	F. H. Varley	London, England	Apr. 29, 1873	138, 455
Electrical apparatus, Screw-post for	S. E. Field	San Francisco, Cal	July 8, 1873	140, 618
Electrical bath	C. Winterburn and W. Kent	Cincinnati, Ohio	July 14, 1868	79, 929
Electrical-current breaker, Automatic	C. Robinson and C. T. Chester.	New York, N. Y	Jan. 1, 1856	14, 032
Electrical heating apparatus	G. B. Simpson	Washington, D. C	Sept. 20, 1859	25, 532
Electrical indicator	S. S. Laws	New York, N. Y	Dec. 31, 1867	72, 742
Electrical indicator	G. Little	Rutherford Park, N. J	Apr. 9, 1872	125, 586
Electrical indicator for elevators	J. H. Corey	New York, N. Y	Apr. 1, 1872	137, 422
Electrical machine	E. S. Blake	Allegheny County, Pa	Mar. 28, 1848	5, 485
Electrical machine	H. J. Smith	Boston, Mass	Aug. 10, 1869	93, 563
Electrical machine, Holtz	E. S. Ritchie	Brookline, Mass	Feb. 4, 1868	74, 139
Electrical machines from moisture, Protecting frictional.	C. A. Seely	New York, N. Y	Dec. 13, 1859	26, 445
Electrical-pole changer	J. E. Smith	New York, N. Y	July 23, 1872	129, 867
Electrical regulator for transmitting-instruments	T. A. Edison	Newark, N. J	Sept. 9, 1873	142, 688
Electrical vacuum and vapor treatment, Apparatus for.	M. H. Utley	Montreal, Canada	May 23, 1871	115, 137
Electricity, Apparatus for obtaining light by frictional.	M. Wesolowski	New York, N. Y	Jan. 29, 1861	31, 273
Electricity as a remedial agent, Apparatus for using.	G. Munroe	Philadelphia, Pa	Feb. 28, 1871	112, 270
Electricity, Method of utilizing atmospheric	H. C. Vion	Paris, France	June 19, 1860	28, 793
Electricity, Remedial application of	M. W. House	Cleveland, Ohio	Sept. 1, 1863	39, 733
Electrifying-machine	E. Cole	Richmond, Ohio	July 22, 1843	3, 191
Electro and permanent magnets to render telegraphing-magnets easy of adjustment, Combination of.	E. F. Barnes	Brooklyn, N. Y	July 20, 1858	20, 930
Electro-bathing apparatus	W. W. Karshner	Cincinnati, Ohio	Jan. 25, 1859	22, 733
Electro-deposition of metals	I. Adams, jr	Boston, Mass	Mar. 22, 1870	100, 961
Electro-gilding iron	H. Tucker	Newton, Mass	June 1, 1869	90, 893
Electro-heating apparatus	W. L. Burton	Richmond, Va	Mar. 23, 1869	88, 006
Electro-magnet	S. G. Cabell	Quincy, Ill	June 25, 1867	66, 001
Electro-magnet	C. T. Chester	New York, N. Y	Sept. 4, 1860	29, 862
Electro-magnet	W. E. Davis	Jersey City, N. J	Nov. 9, 1869	96, 554
Electro-magnet	M. S. Frost	New York, N. Y	May 24, 1870	103, 440
Electro-magnet	A. G. Holcomb	New York, N. Y	June 4 1861	32, 478
Electro-magnet	J. S. Jenness	Bangor, Me	July 23, 1861	32, 874
Electro-magnet	H. M. Paine	Newark, N. J	May 10, 1870	102, 856
Electro-magnet	H. M. Paine	Newark, N. J	May 17, 1870	103, 230
Electro-magnet	E. Provost	New York, N. Y	Sept. 19, 1871	119, 176
Electro-magnet	W. W. Smith	Cincinnati, Ohio	Aug. 16, 1870	106, 418
Electro-magnet	I. P. Tice	New York, N. Y	Apr. 2, 1872	125, 151
Electro-magnet circuit	J. E. Smith	New York, N. Y	Apr. 8, 1873	137, 730
Electro-magnet sector	H. M. Paine	Newark, N. J	May 31, 1870	103, 768
Electro-magnets, Construction of	H. M. Paine	Newark, N. J	May 17, 1870	103, 231
Electro-magnets, Induction-coil for	J. B. Lyon and A. Doll	Cleveland, Ohio	Oct. 2, 1866	58, 439
Electro-magnets, Mode of changing the poles of	N. Walkly	Tuscaloosa, Ala	June 27, 1838	809
Electro magnetic adjuster	T. A. Edison	Newark, N. J	Jan. 14, 1873	134, 868
Electro-magnetic alarm	J. G. Butler	New York, N. Y	July 6, 1869	92, 260
Electro-magnetic alarm	M. G. Crane	Newton, Mass	Aug. 11, 1868	80, 922
Electro-magnetic alarm	W. B. Guernsey	Jersey City, N. J	Apr. 30, 1872	126, 287
Electro-magnetic alarm	T. S. Hall	Stamford, Conn	July 19, 1870	105, 447
Electro magnetic alarm	A. R. Pope	Somerville, Mass	June 21, 1853	9, 802
Electro-magnetic alarm for railway-car	W. Gillett	Ypsilanti, Mich	Aug. 31, 1869	94, 198
Electro-magnetic apparatus	H. N. Baker	Binghamton, N. Y	Sept. 4, 1860	29, 850
Electro-magnetic apparatus	S. Gardiner, jr	New York, N. Y	July 11, 1871	116, 944
Electro-magnetic apparatus	J. Kidder	New York, N. Y	Sept. 18, 1860	30, 068
Electro-magnetic apparatus	G. Little	Rutherford Park, N. J	Dec. 26, 1871	122, 266
Electro-magnetic apparatus	J. W. Powell	New York, N. Y	Mar. 26, 1872	125, 0.8
Electro-magnetic apparatus	R. Sayer	New York, N. Y	July 23, 1872	129, 752
Electro-magnetic apparatus, Circuit-closer for	M. G. Farmer	Salem, Mass	May 14, 1872	126, 627
Electro-magnetic apparatus with tooth-forceps, Mode of connecting.	J. J. Clark	Philadelphia, Pa	Nov. 16, 1858	22, 063
Electro-magnetic-bathing apparatus	J. Young	New York, N. Y	May 14, 1861	32, 332
Electro-magnetic battery	P. Bruso	Erie, Pa	Apr. 16, 1867	63, 848
Electro-magnetic batteries to car-brakes, Application of.	S. D. Carpenter	Madison, Wis	Jan. 19, 1858	19, 132
Electro-magnetic burglar-alarms, Automatic circuit-closer for.	J. P. Snyder	Brooklyn, N. Y	Sept. 13, 1870	107, 301
Electro-magnetic currents constant or intermittent, Device for making.	M. Marshall	Lowell, Mass	Aug. 9, 1859	25, 023
Electro-magnetic engine	T. C. Avery	New York, N. Y	Feb. 25, 1851	7, 950
Electro-magnetic engine	A. P. Berlioz	Paris, France	Oct. 16, 1866	58, 960
Electro magnetic engine	H. S. Daggett	La Fayette, Ind	Jan. 9, 1872	122, 572
Electro-magnetic engine	A. J. B. De Morat	Philadelphia, Pa	July 28, 1868	80, 463
Electro-magnetic engine	A. E. Dupas	New Orleans, La	Aug. 17, 1869	93, 689
Electro-magnetic engine	M. G. Farmer	Salem, Mass	May 14, 1872	126, 628
Electro-magnetic engine	C. J. B. Gaume	Davenport, Iowa	Apr. 2, 1867	63, 380
Electro-magnetic engine	C. J. B. Gaume	New York, N. Y	Mar. 16, 1869	87, 835
Electro-magnetic engine	C. V. Gaume	Williamsburgh, N. Y	Jan. 23, 1872	122, 944
Electro-magnetic engine	J. S. Gustin	Trenton, N. J	Sept. 28, 1852	9, 291
Electro-magnetic engine	J. S. Gustin	Trenton, N. J	Oct. 12, 1852	9, 320
Electro-magnetic engine	S. Jones	New Orleans, La	Nov. 21, 1871	121, 173
Electro-magnetic engine	J. H. Lillie	Joliet, Ill	Apr. 16, 1850	7, 287
Electro-magnetic engine	J. Neff	Philadelphia, Pa	Jan. 7, 1851	7, 889
Electro-magnetic engine	W. H. Odell	Yonkers, N. Y	July 23, 1872	129, 857
Electro-magnetic engine	C. G. Page	Washington, D. C	Jan. 31, 1854	10, 480
Electro-magnetic engine	H. M. Paine	Newark, N. J	May 17, 1870	103, 228
Electro-magnetic engine	L. C. Stuart	New York, N. Y	June 2, 1868	78, 619
Electro-magnetic engine	W. G. Thornton	Victoria, Tex	Apr. 9, 1872	125, 504
Electro-magnetic engine	J. P. Tirrel	Charlestown, Mass	Mar. 31, 1870	103, 798
Electro-magnetic engine	J. P. Tirrel	Charlestown, Mass	Aug. 29, 1871	118, 561
Electro-magnetic engine	M. Vergnes	New York, N. Y	Apr. 15, 1856	14, 682
Electro-magnetic engine	M. Vergnes	New York, N. Y	Oct. 2, 1860	30, 272
Electro-magnetic engine	W. Wickersham	Boston, Mass	Oct. 15, 1867	69, 880

Index of patents issued from the United States Patent Office from 1790 *to* 1873, *inclusive*—Continued.

Invention.	Inventor.	Residence.	Date.	No.
Electro-magnetic engine	W. Wickersham	Boston, Mass	June 2, 1868	78, 629
Electro-magnetic engine	W. Wickersham	Boston, Mass	Mar. 19, 1872	124, 868
Electro-magnetic engine	F. Yeiser	Lexington, Ky	Aug. 3, 1858	21, 105
Electro-magnetic gate and signal-apparatus for railways.	W. Robinson	Brooklyn, N. Y	July 19, 1870	105, 493
Electro-magnetic gate-operating apparatus	W. Robinson	Brooklyn, N. Y	July 19, 1870	105, 494
Electro-magnetic helix	M. Vergnes	New York, N. Y	Oct. 2, 1860	30, 271
Electro-magnetic indicator	W. D. Smith	Washington, D. C	Jan. 31, 1871	111, 394
Electro-magnetic indicator for burglar-alarms, &c	J. P. Snyder	Brooklyn, N. Y	May 24, 1870	103, 383
Electro-magnetic instrument	L. L. Duerden	Brooklyn, N. Y	July 25, 1871	117, 268
Electro-magnetic lock	H. Arden	Saint Louis, Mo	Nov. 8, 1870	108, 952
Electro-magnetic lock	J. C. Smith	Brooklyn, N. Y	Dec. 21, 1869	98, 114
Electro-magnetic machine	W. H. Burnap and J. A. Bradshaw.	Lowell, Mass	Mar. 8, 1859	23, 214
Electro-magnetic machine, &c	T. Cook	New York, N. Y	Aug. 25, 1840	1, 735
Electro-magnetic machine	A. E. Dupas	New Orleans, La	Feb. 1, 1870	99, 300
Electro-magnetic machine	T. Hall	Boston, Mass	June 21 1859	24, 459
Electro-magnetic machine	J. Kidder	New York, N. Y	Aug. 10 1869	93, 625
Electro-magnetic machine	J. W. Powell	New York, N. Y	May 30, 1871	115, 518
Electro-magnetic machine for shocks	S. B. Smith	New York, N. Y	June 4, 1850	7, 420
Electro-magnetic machines, Commutator for	V. Barjon	New York, N. Y	July 16, 1872	129, 085
Electro-magnetic meter	S. Gardiner, jr	New York, N. Y	Oct. 29, 1872	132, 569
Electro-magnetic motor	J. S. Camacho	Havana, Cuba	July 16, 1872	129, 000
Electro-magnetic motor	S. C. Carter	Richmond, Ind	May 13, 1873	138, 855
Electro-magnetic motor	L. Finger	Cambridge, Mass	July 26, 1870	105, 663
Electro-magnetic motor	L. T. Lindsey	Jackson, Tenn	June 29, 1869	92, 066
Electro-magnetic motor	L. T. Lindsey	Jackson, Tenn	May 3, 1870	102, 562
Electro-magnetic motor	L. T. Lindsey	Jackson, Tenn	Aug. 16, 1870	106, 493
Electro-magnetic motor	G. Little	Rutherford Park, N. J	Nov. 2, 1869	96, 332
Electro-magnetic motor	H. M. Paine	Newark, N. J	Mar. 21, 1871	112, 841
Electro-magnetic motor	W. H. Richardson	Albany, N. Y	May 28, 1872	127, 369
Electro-magnetic motor	A. Schreiber	Brooklyn, N. Y	Sept. 17, 1872	131, 377
Electro-magnetic motor	J. Taggart	Boston, Mass	June 11, 1872	127, 810
Electro-magnetic motor	M. H. Utley and A. Ross	Montreal, Canada	Oct. 10, 1871	119, 899
Electro-magnetic regulator	F. F. A. Achard	Paris, France	Sept. 5, 1865	49, 842
Electro-magnetic separator	A. H. Balch and W. D. E. Nelson.	Montreal, Canada	June 6, 1871	115, 561
Electro magnetic separator	A. Ross	Montreal, Canada	Jan. 2, 1872	122, 493
Electro-magnetic signal	S. F. B. Morse	New York, N. Y	June 20, 1840	1, 647
Electro-magnetic-signal apparatus	A. Hahl	Washington, D. C	Feb. 28, 1871	112, 242
Electro-magnetic-signal apparatus	G. B. Hicks	Cleveland, Ohio	Aug. 24, 1869	93, 993
Electro-magnetic-signal apparatus for railway-crossings, &c.	W. Robinson	Brooklyn, N. Y	Nov. 22, 1870	109, 549
Electro-magnetic watchman's register	H. D. Sheppard	New York, N. Y	Apr. 9, 1872	125, 624
Electro-medical battery	H. Fritz	Cleveland, Ohio	Feb. 25, 1868	74, 905
Electromotor escapement	T. A. Edison	New York, N. Y	May 17, 1870	103, 035
Electromotors, Armature for	S. Jones	New Orleans, La	Aug. 29, 1871	118, 538
Electrodes, Fibrous covering for spiral	E. Maynard	Brooklyn, N. Y	Feb. 8, 1859	22, 922
Electrophorus	R. Cornelius	Philadelphia, Pa	May 21, 1861	32, 353
Electrophorus	D. F. Launy	Philadelphia, Pa	Aug. 23, 1811	....
Electro-photo damaskening and enameling	A. G. Morvan	New York, N. Y	Oct. 4, 1870	108, 041
Electroplated rollers, Constructing	J. W. Wilcox	West Roxbury, Mass	Apr. 12, 1859	23, 633
Electroplating	J. P. Woodworth	Brooklyn, N. Y	Nov. 17, 1868	84, 243
Electroplating and gilding cast iron	H. Tucker	Newton, Mass	June 1, 1869	90, 894
Electroplating and platedware	H. W. Wright	Taunton, Mass	June 30, 1868	79, 427
Electroplating-apparatus	W. Gates	New Haven, Conn	July 28, 1868	80, 402
Electroplating-apparatus	H. W. Wright	Taunton, Mass	Mar. 22, 1870	101, 075
Electroplating-frame	W. H. Watrous	Hartford, Conn	June 2, 1868	78, 497
Electroplating iron and other metals with copper, Process of.	J. E. Walcott	Boston, Mass	Feb. 18, 1862	34, 470
Electroplating iron, steel, &c., Process of	C. Beslay	Paris, France	Oct. 21, 1862	36, 750
Electroplating iron with copper, &c., Apparatus for.	A. L. Freeman	South Boston, Mass	July 16, 1872	129, 124
Electroplating letters and numbers on metallic plates for signs, &c.	J. J. Pratt	New York, N. Y	Mar. 28, 1871	113, 090
Electroplating steel wires for piano-strings, &c., Mode of.	M. Miller, jr	Vienna, Austria	Mar. 11, 1862	34, 640
Electroplating the interior of pipes and tubes with silver, nickel, &c.	D. D. Parmelee	New York, N. Y	Apr. 25, 1871	114, 191
Electroplating with alloys of gold	A. Berthoud	New York, N. Y	Nov. 20, 1860	30, 663
Electroplating with antimony	J. S. Howard	Mansfield, Mass	Feb. 22, 1870	100, 038
Electroplating with brass and other alloys	S. Rust, jr	Cincinnati, Ohio	Dec. 21, 1869	98, 110
Electroplating with copper, &c, the surfaces of mirrors, &c.	J. Von Liebig	Munich, Bavaria	Nov. 12, 1861	33, 721
Electroplating with nickel	I. Adams, jr	Boston, Mass	Apr. 4, 1871	113, 612
Electroplating with nickel and other metals	G. W. Beardslee	Brooklyn, N. Y	July 30, 1872	129, 881
Electroplating with nickel, Mode of	I. Adams, jr	Boston, Mass	May 25, 1869	90, 332
Electroplating with nickel, Process of	W. H. Remington	Boston, Mass	Oct. 6, 1868	82, 877
Electroplating with tin	J. E. Bingham	Sheffield, Great Britain	June 13, 1871	115, 926
Electro-pneumatic dispatch-transmitter	C. F. Varley	London, England	June 9, 1868	78, 845
Electrotype and stereotype plates, Manufacture of curved.	H. and H. W. Lovejoy and J. H. Ferguson.	New York, N. Y	Jan. 19, 1869	86, 021
Electrotype-backing pan	W. A. Leggo	Quebec, Canada	Aug. 4, 1863	39, 410
Electrotype die or mold for molding plastic material.	D. Scrymgeour	Foxborough, Mass	Feb. 7, 1871	111, 688
Electrotype-mold	S. P. Knight	Brooklyn, N. Y	May 2, 1871	114, 447
Electrotype-molds, Coating	S. P. Knight	Brooklyn, N. Y	Sept. 10, 1872	131, 169
Electrotype-molds, Machine for coating	H. Loveloy and R. Wheeler	Brooklyn, N. Y	Sept. 14, 1858	21, 509
Electrotype-molds with plumbago, Machine for coating surface of.	S. D. Tucker	New York, N. Y	Dec. 29, 1868	85, 411
Electrotype-plates, Method of backing	A. H. Jocelyn	New York, N. Y	July 7, 1857	17, 741
Electrotype-plates, Mode of backing	W. Filmer and E. Bookhout	New York, N. Y	Dec. 2, 1856	16, 141
Electrotype-plates, Mode of forming curved	W. H. Elliot	Plattsburgh, N. Y	May 3, 1859	23, 836
Electrotype-plates, Production of	S. P. Knight	New York, N. Y	May 25, 1858	20, 353
Electrotype-printing block	T. Cressley	Rockville, Conn	Nov. 1, 1859	25, 953
Electrotypes, Forming curved	A. Judson	Brooklyn, N. Y	Aug. 6, 1872	130, 140
Electrotyping	J. A. Adams	Brooklyn, N. Y	Apr. 19, 1870	102, 077
Electrotyping	W. A. Leggo	Montreal, Canada	Oct. 3, 1871	119, 622

Index of patents issued from the United States Patent Office from 1790 *to* 1873, *inclusive*—Continued.

Invention.	Inventor.	Residence.	Date.	No.
Electrotyping	G. Mathiot	Washington, D. C	Dec. 10, 1850	7, 821
Electrotyping-battery	P. S. Hoe	New York, N. Y	Jan. 25, 1870	99, 192
Electrotyping iron and steel with silver	A. Law	Kingston, Canada	Nov. 29, 1870	109, 633
Electrotyping-machine	J. A. Adams	Brooklyn, N. Y	Sept. 4, 1855	13, 516
Electrotyping, Method of etching plates for	A. and H. T. Dawson	London, England	July 22, 1873	140, 995
Electrotyping purposes, Process of obtaining mold for.	W. A. Leggo	Montreal, Canada	May 30, 1871	115, 490
Electrotyping, Surface conductor for	S. Hallock	New York, N. Y	Apr. 2, 1867	63, 512
Elevated way	H. W. Farley	Oswego, Ill	Apr. 8, 1873	137, 667
Elevated way	H. W. Farley	Oswego, Ill	Dec. 16, 1873	145, 638
Elevating and conveying device	W. Louden	Fairfield, Iowa	Sept. 24, 1867	69, 107
Elevating and transporting device	E. B. Coffin	Olneyville, R. I	June 28, 1864	43, 291
Elevating and weighing apparatus	J. H. Brookmire	Saint Louis, Mo	Mar. 23, 1869	88, 124
Elevating-apparatus	P. R. Berry	Youngstown, Ohio	June 14, 1870	104, 102
Elevating-apparatus	M. A. Blakley and J. Cahill	Chattanooga, Tenn	Feb. 4, 1873	135, 511
Elevating-block	W. H. Hawley	Utica, N. Y	Sept. 17, 1867	68, 874
Elevating-jack	S. Lauchli	Saint Louis, Mo	June 28, 1864	43, 369
Elevating-machine	G. A. Dupuis	Detroit, Mich	July 1, 1873	140, 485
Elevating-machine	D. L. Miller	Madison, N. J	Feb. 18, 1862	34, 440
Elevating power-engine and compressing-machine for raising vessels, &c.	B. Bruff	Rochester, N. Y	Sept. 13, 1831	....
Elevator:				
See Balance-elevator. Barrel and box elevator. Beer-elevator. Bucket-elevator. Building-elevator. Car-elevator. Car-body elevator. Chain-elevator. Coal-elevator. Corn-elevator. Cultivator-bar elevator. Dress-elevator. Fireman's elevator. Fluid-elevator. Fluid or grain elevator. Freight-elevator. Gear-elevator. Grain-elevator. Hay-elevator. Hod-elevator. Hog-elevator. Hose-elevator. Hydraulic elevator. Hydraulic-power elevator. Ice-elevator. Inclined-plane elevator. Last-block elevator. Liquid-elevator. Merchandise-elevator. Millstone-elevator. Mortar and brick elevator. Passenger and freight elevator. Patient-elevator. Peat-elevator. Platform-elevator. Pneumatic elevator. Portable elevator. Rotary fluid-elevator. Sash-elevator. Scaffold-elevator. Self-elevator. Sirup-pan elevator. Skirt-elevator. Sorghum-elevator. Steam-elevator. Store-house elevator. Straw-elevator. Sucker-rod elevator. Tailing-elevator. Telescopic-platform elevator. Tobacco-elevator. Wall-elevator. Warehouse-elevator. Water-elevator. Wind-elevator. Wool-elevator.				
Elevator	E. H. Ashcroft	Lynn, Mass	Nov. 5, 1867	70, 387
Elevator	E. H. Ashcroft and J. R. Brown	Boston, Mass	Aug. 25, 1868	81, 325
Elevator	L. Atwood	New York, N. Y	May 28, 1872	127, 138
Elevator	C. W. Baldwin	Boston, Mass	Nov. 19, 1867	70, 941
Elevator	C. W. Baldwin	Boston, Mass	Nov. 15, 1870	109, 169
Elevator	C. W. Baldwin	Boston, Mass	Feb. 20, 1872	123, 761
Elevator	J. S. Baldwin	Newark, N. J	Nov. 12, 1867	70, 777
Elevator	J. S. Baldwin	Newark, N. J	Apr. 14, 1868	76, 692
Elevator	J. S. Baldwin	Newark, N. J	Apr. 14, 1868	76, 693
Elevator	J. S. Baldwin	Newark, N. J	May 20, 1873	138, 985
Elevator	J. M. Bean	Watertown, Mass	Mar. 11, 1873	136, 580
Elevator	A. Betteley	Boston, Mass	May 3, 1859	23, 818
Elevator	J. Bishop	Austin, Tex	June 13, 1871	115, 813
Elevator	V. C. Blair	Wheatland, Pa	June 6, 1871	115, 683
Elevator	A. Blass and D. Brown	Brooklyn, N. Y	Dec. 31, 1872	134, 416
Elevator	J. H. Blohm	New York, N. Y	Feb. 4, 1873	135, 398
Elevator	C. D. Bois	Fishkill Landing, N. Y	Mar. 18, 1873	136, 978
Elevator	E. Boyden	Lowell, Mass	Sept. 9, 1873	142, 555
Elevator	P. Brand	Springfield, Ill	July 28, 1868	80, 445
Elevator	W. D. Brooks	Bethany, Pa	Dec. 8, 1868	84, 794
Elevator	P. Byrne	Nashville, Tenn	Oct. 3, 1871	119, 572
Elevator	P. Byrne	Nashville, Tenn	Oct. 1, 1872	131, 809
Elevator	S. B. Castle	Cortland, N. Y	Mar. 10, 1868	75, 366
Elevator	D. H. Chamberlain	West Roxbury, Mass	June 20, 1871	116, 156
Elevator	T. F. Christman	Wilson, N. C	Oct. 9, 1866	58, 715
Elevator	G. R. Clarke	New York, N. Y	June 9, 1868	78, 646
Elevator	C. W. Collyer	Marblehead, Mass	July 1, 1873	140, 473
Elevator	A. W. Cram	Litchfield, Ill	June 17, 1873	140, 017
Elevator	A. B. Darling and J. Bones	New York, N. Y	May 21, 1872	126, 935
Elevator	J. G. Drew	Boston, Mass	Aug. 6, 1872	130, 112
Elevator	W. Edson	Boston, Mass	Dec. 7, 1869	97, 489
Elevator	W. Elmes	Georgetown, D. C	Oct. 22, 1872	132, 459
Elevator	Z. C. Favor	Chicago, Ill	Mar. 14, 1871	112, 573
Elevator	C. R. Foreman	West Branch, N. Y	Sept. 6, 1870	107, 028
Elevator	P. Griffhorn	Akron, Ohio	Feb. 6, 1872	123, 391
Elevator	W. Hamilton	Allegheny City, Pa	June 28, 1870	104, 731
Elevator	M. Hanford	Boston, Mass	Feb. 1, 1870	99, 433
Elevator	M. Hanford	Boston, Mass	Mar. 26, 1872	124, 952
Elevator	M. Hanford	Boston, Mass	Dec. 24, 1872	134, 273
Elevator	D. D. Hanson	Weare, N. H	July 19, 1870	105, 449
Elevator	O. P. Hatfield	New York, N. Y	Mar. 4, 1862	34, 572
Elevator	A. G. Hawkes	Baltimore, Md	July 16, 1872	129, 339
Elevator	A. C. Herron	Boston, Mass	Apr. 22, 1873	138, 154
Elevator	E. Hicks	North Hempstead, N. Y	Mar. 26, 1867	63, 248

Index of patents issued from the United States Patent Office from 1790 *to* 1873, *inclusive*—Continued.

Invention.	Inventor.	Residence.	Date.	No.
Elevator	N. D. Hinman	Pleasant Vale, Conn	Nov. 29, 1864	45, 245
Elevator	J. E. Hollister	Calais, Vt	Dec. 8, 1868	84, 744
Elevator	E. T. Hope	Philadelphia, Pa	Aug. 18, 1868	81, 276
Elevator	G. W. Hubbard	Philadelphia, Pa	Sept. 23, 1873	143, 013
Elevator	A. B. Hunt	Matteson, Mich	Oct. 27, 1868	83, 385
Elevator	G. Joslin	Boston, Mass	June 23, 1868	79, 129
Elevator	J. T. Ketchledge	Burns, Mich	Sept. 14, 1869	94, 896
Elevator	D. Knowles	Philadelphia, Pa	Sept. 10, 1872	131, 170
Elevator	A. Lawton	Philadelphia, Pa	Aug. 30, 1870	106, 841
Elevator	J. S. Lester	Knoxville, Tenn	July 20, 1869	92, 847
Elevator	W. Livingstone	Brooklyn, N. Y	Aug. 13, 1872	130, 509
Elevator	W. Livingstone and W. F. Holske.	Brooklyn, N. Y	Jan. 16, 1872	122, 839
Elevator	W. Lowden	Cedar Township, Iowa	Nov. 17, 1868	84, 127
Elevator	G. B. Lowe	Jamesville, N. Y	Aug. 8, 1871	117, 789
Elevator	J. Macomb	Chicago, Ill	May 30, 1871	115, 335
Elevator	J. Macomb	Chicago, Ill	Aug 29, 1871	118, 468
Elevator	M. Mageean and G. W. Marks.	Chicago, Ill	Apr. 29, 1873	138, 337
Elevator	J. F. Marsh	Littleton, N. H	Apr. 15, 1873	137, 940
Elevator	W. K. Marvin	New York, N. Y	Nov. 15, 1870	109, 331
Elevator	G. McKinzie	Zanesville, Ohio	Nov. 15, 1870	109, 232
Elevator	P. W. Mellen	Saint Louis, Mo	Apr. 1, 1873	137, 464
Elevator	N. L. Milburn	Saint Louis, Mo	Sept. 29, 1868	82, 540
Elevator	C. E. Moore	Boston, Mass	May 2, 1871	114, 460
Elevator	C. E. Moore	Boston, Mass	Oct. 3, 1871	119, 635
Elevator	W. A. Morrison	Cambridge, Mass	Apr. 15, 1873	137, 789
Elevator	A. B. Nimbs	Buffalo, N. Y	Jan. 10, 1865	45, 851
Elevator	C. R. Otis	Yonkers, N. Y	Oct. 1, 1872	131, 896
Elevator	C. R. Otis	Yonkers, N. Y	Oct. 1, 1872	131, 897
Elevator	F. B. Perkins	Boston, Mass	Sept. 26, 1871	119, 401
Elevator	H. G. Porter	Grand Rapids, Mich	June 29, 1869	92, 092
Elevator	J. J. Rea	Cadiz, Ohio	Oct. 20, 1868	83, 319
Elevator	J. Reedy	Chicago, Ill	July 30, 1872	130, 073
Elevator	W. R. Rightor	Helena, Ark	June 17, 1873	140, 076
Elevator	T. H. Rudiger	Lawrence, Kans	Sept. 27, 1870	107, 726
Elevator	W. Sawyer	Lowell, Mass	Oct. 31, 1865	50, 778
Elevator	G. Scott	Louisville, Ky	Aug. 18, 1868	81, 299
Elevator	G. Scott	Louisville, Ky	June 22, 1869	91, 775
Elevator	G. Scott	New Orleans, La	Nov. 29, 1870	109, 676
Elevator	G. Scott	New Orleans, La	Aug. 27, 1872	130, 824
Elevator	A. Shoemaker and J. R. Gearhart.	Marion, Pa	Feb. 9, 1869	86, 703
Elevator	T. B. Simonton	New York, N. Y	Nov. 24, 1868	84, 385
Elevator	D. F. Skinner and J. Arnold	Albany, N. Y	Nov. 7, 1871	120, 676
Elevator	H. Spiro	Knoxville, Tenn	July 7, 1868	79, 697
Elevator	F. Stein and H. Haering	New York, N. Y	Dec. 21, 1869	98, 203
Elevator	H. S. Stewart	Yreka, Cal	Dec. 12, 1871	121, 822
Elevator	H. C. Stouffer	Salem, Ohio	Mar. 29, 1870	101, 407
Elevator	J. B. Sweetland	Pontiac, Mich	Oct. 28, 1873	144, 161
Elevator	G. C. Timpe	New Orleans, La	Nov. 15, 1870	109, 275
Elevator	G. C. Timpe	New Orleans, La	Aug. 22, 1871	118, 300
Elevator	G. C. Timpe	New Orleans, La	Aug. 22, 1871	118, 301
Elevator	G. C. Timpe	New Orleans, La	Feb. 4, 1873	135, 454
Elevator	O. Tufts	Boston, Mass	May 28, 1861	32, 441
Elevator	O. Tufts	Boston, Mass	Dec. 11, 1866	60, 441
Elevator	O. Tufts	Boston, Mass	Dec. 11, 1866	60, 442
Elevator	O. Tufts	Boston, Mass	Oct. 21, 1873	143, 944
Elevator	O. Tufts, jr	Boston, Mass	Feb. 16, 1869	87, 083
Elevator	R. M. Van Sickler	New York, N. Y	Feb. 12, 1867	62, 098
Elevator	J. Walker	Kansas City, Mo	Oct. 15, 1867	69, 871
Elevator	J. C. and J. W. Wandell	New York, N. Y	Mar. 6, 1866	53, 064
Elevator	J. D. Warner	Brooklyn, N. Y	Jan. 16, 1872	122, 791
Elevator	J. J. Weber	Saint Clair, Pa	Jan. 24, 1871	111, 285
Elevator	J. Westover	Taylor, Ill	July 21, 1868	80, 103
Elevator	C. Whittier	Boston, Mass	Dec. 24, 1872	134, 179
Elevator	F. Wicks	Decatur, Ill	Nov. 29, 1870	109, 787
Elevator	G. Williams	Sterling, Cal	July 17, 1866	56, 481
Elevator	G. G. Winans	Scranton, Pa	Feb. 15, 1870	99, 802
Elevator	M. L. Wyman	Melrose, Mass	Mar. 15, 1870	100, 832
Elevator	J. Yost	White Deer Township, Pa.	June 8, 1869	91, 193
Elevator and carrier	S. W. Lackore	Worth, Ill	Mar. 7, 1871	112, 357
Elevator and carrier	H. E. Plumb	Monroe, Conn	Feb. 17, 1863	37, 704
Elevator and conveyer	L. Y. Myers	Roanoke, Ind	Nov. 24, 1868	84, 436
Elevator and conveyer	T. J. Powell	Naples, N. Y	May 24, 1870	103, 367
Elevator and conveyer	T. J. Powell	Naples, N. Y	Feb. 14, 1871	111, 869
Elevator and conveyer	W. F. Shanks	Louisville, Ky	July 16, 1872	129, 298
Elevator and conveyer, Combined	T. J. Powell	Naples, N. Y	Oct. 11, 1870	108, 183
Elevator and distributer	T. Shanks	Baltimore, Md	Dec. 6, 1870	109, 950
Elevator and dumper, Portable	W. H. Herriott	Brimfield, Ill	Dec. 30, 1873	146, 002
Elevator-belt tightener	W. Merson	Danbury, Conn	July 23, 1872	129, 745
Elevating-block	W. H. Hawley	Utica, N. Y	Aug. 27, 1867	68, 193
Elevator-boxes, Machine for making	W. L. Young	Marthasville, Mo	Sept. 24, 1872	131, 648
Elevator-brake	F. P. Canfield	Boston, Mass	Jan. 7, 1873	134, 640
Elevator-brake	M. G. and C. L. Gill	Baltimore, Md	July 16, 1872	129, 546
Elevator-brake	T. Thorn	Saint Clair, Pa	Oct. 10, 1871	119, 800
Elevator-brake	T. Thorn	Saint Clair, Pa	July 16, 1872	129, 625
Elevator-bucket	J. S. Brooks	Rochester, N. Y	Dec. 23, 1862	37, 215
Elevator-bucket	J. S. Brooks	Rochester, N. Y	Nov. 24, 1863	40, 671
Elevator-bucket	C. Camp	Buffalo, N. Y	Oct. 9, 1866	58, 595
Elevator-bucket	H. Dorer and J. Storms	Buffalo, N. Y	July 17, 1866	56, 386
Elevator-bucket	J. Magee	Chicago, Ill	Dec. 19, 1865	51, 603
Elevator-bucket	J. Pfitzinger	Buffalo, N. Y	May 18, 1869	90, 189
Elevator-bucket	N. C. Staples	Lynchburgh, Va	June 29, 1833	....
Elevator-buckets, Construction of	N. Hawkins	Chicago, Ill	Jan. 19, 1869	85, 930
Elevator-buckets, Manufacture of	A. B. Nimbs	Buffalo, N. Y	May 10, 1864	42, 734
Elevator-cup	J. H. Gibbens	Saint Paul, Minn	Oct. 15, 1872	132, 276
Elevator-cup	A. A. Vitt	Union, Mo	Aug. 15, 1871	118, 076
Elevator-cup for flouring mills	L. B. Prindle	Litchfield, Conn	Apr. 29, 1862	35, 111

Index of patents issued from the United States Patent Office from 1790 *to* 1873, *inclusive*—Continued.

Invention.	Inventor.	Residence.	Date.	No.
Elevator for the sick	J. Elliot	Newbury, Mass	Feb. 15, 1811	....
Elevator-guide	C. R. Otis	Yonkers, N. Y	Jan. 7, 1873	134, 698
Elevator-guide	O. Tufts	Boston, Mass	Dec. 11, 1866	60, 443
Elevator-indicator, Electric	E. Holmes	Brooklyn, N. Y	Dec. 5, 1871	121, 620
Elevator-leg	G. H. Johnson	Buffalo, N. Y	Feb. 16, 1869	86, 926
Elevator-leg	G. H. Johnson	Buffalo, N. Y	Feb. 16, 1869	86, 927
Elevator stop-motion	H. E. Bathrick	Boston, Mass	Apr. 2, 1872	125, 256
Elevators, Apparatus for indicating the position of	W. S. Auchincloss	New York, N. Y	June 21, 1870	104, 535
Elevators, Automatic safety-lock for	G. C. Timpe	New Orleans, La	Feb. 4, 1873	135, 453
Elevators, Safety-device for	P. P. Lane	Cincinnati, Ohio	June 18, 1872	128, 152
Elevators, Safety-device for	T. Stebins	San Francisco, Cal	Oct. 8, 1872	132, 112
Elevators, Safety-platform for	R. Dunbar	Buffalo, N. Y	June 24, 1873	140, 190
Elixir of life	J. Bucco	Baltimore, Md	Mar. 29, 1817	....
Elixir or cough-drop, Vegetable	S. Cooley		Nov. 12, 1803	....
Elixir, Restorative	H. Frink	Northampton, Mass	Jan. 6, 1812	....
Ellipses, Instrument for describing	J. B. Atwater	Chicago, Ill	June 2, 1863	38, 720
Ellipses, Instrument for drawing	F. Bowly	Winchester, Va	Jan. 14, 1868	73, 290
Ellipses, Instrument for drawing	A. Warr	Lockport, N. Y	Mar. 19, 1867	63, 123
Ellipses, Machine for cutting	W. G. Merrell	Auburn, N. Y	Jan. 10, 1854	10, 412
Ellipses, Trammel for	C. Crow	Onargo, Ill	Mar. 27, 1866	53, 524
Ellipsograph	H. M. Albee	Webster, Mass	Dec. 18, 1866	60, 609
Ellipsograph	A. W. Browne	Bloomfield, N. J	May 14, 1872	126, 782
Ellipsograph	J. W. Campbell	Brooklyn, N. Y	Dec. 9, 1856	16, 175
Ellipsograph	E. G. Chormann	Philadelphia, Pa	Dec. 21, 1858	22, 347
Ellipsograph	R. E. Harte	Marietta, Ohio	Dec. 4, 1860	30, 816
Ellipsograph	H. Smith	Dayton, Oreg	June 22, 1869	91, 681
Ellipsograph	T. Williams and W. C. Joslyn	Fisherville, Conn	Feb. 8, 1859	22, 910
Elliptic spring	R. Gray	Litchfield, Ill	July 16, 1867	66, 702
Elliptic spring	F. Keeler	Bridgeport, Conn	Feb. 15, 1870	99, 776
Elliptic spring	E. C. Lewis	Auburn, N. Y	Aug. 27, 1867	68, 217
Elliptic spring	W. A. Sweet	Syracuse, N. Y	Nov. 30, 1869	97, 458
Elliptic spring	R. Vose	New York, N. Y	May 5, 1863	38, 432
Elliptic-spring clip	W. Evans	Pittsburgh, Pa	Jan. 25, 1870	99, 074
Elliptic spring for carriages	E. M. Chaffee	Providence, R. I	Feb. 5, 1867	61, 712
Elliptic spring for vehicles	R. Halloran	New York, N. Y	Feb. 6, 1872	123, 396
Elliptic-spring support	E. Eves	Millville, Pa	Nov. 30, 1869	97, 370
Elliptic-spring support	J. J. Unsworth	Washington, D. C	Aug. 24, 1869	94, 150
Elliptic springs, Confining the ends of	R. Vose	New York, N. Y	Dec. 15, 1863	40, 968
Elliptic springs, Die for making heads for	J. Palmer	Concord, N. H	Sept. 3, 1867	68, 454
Elliptic springs, Ear-blank for	W. Evans	Pittsburgh, Pa	Apr. 13, 1869	88, 861
Elliptic springs for carriages, &c., Construction of	M. Tuells	Penn Yan, N. Y	July 9, 1838	828
Elliptic springs, Machine for bending	J. Gatchell	Rahway, N. J	Nov. 3, 1868	83, 625
Elliptic springs, Machine for forming	J. B. Cleveland and H. C. Guildersleive.	Newark, N. J	June 14, 1870	104, 114
Elliptic springs, Machine for forming and tempering.	G. S. Long	Bridgeport, Conn	Dec. 10, 1867	72, 055
Elliptic springs, Machine for punching the leaves of	G. Hopson	Bridgeport, Conn	Mar. 8, 1870	100, 630
Elliptic springs, Machine for setting and fitting	G. I. Neviel	Philadelphia, Pa	Sept. 20, 1839	1, 333
Elliptic springs, Machine for straightening	J. W. Gray and C. H. Curtis	Bridgeport, Conn	Feb. 14, 1865	46, 421
Elliptic springs, Machine for welding ears on carriage.	A. Richards and J. Jones	Concord, N. H	Apr. 9, 1872	125, 485
Elliptic springs, Making the eyes of	W. T. Richards	Bridgeport, Conn	Aug. 27, 1867	68, 113
Elliptic springs to vehicles, Connecting	J. W. Lawrence	New York, N. Y	Dec. 13, 1859	26, 461
Elliptic springs, Welding the ears of	J. Palmer	Concord, N. H	May 26, 1868	78, 230
Elliptical spindles, Grinder for	J. Wheaton	Providence, R. I	May 28, 1834	....
Elliptical spring	D. A. Edwards	Boston, Mass	Nov. 16, 1841	2, 360
Elliptical-spring joints, Machinery for forming	W. T. Richards	New Haven, Conn	Aug. 26, 1851	8, 325
Elliptical-spring sockets, Machine for forming	W. T. Richards	Poultney, Vt	Nov. 16, 1841	2, 367
Elliptical springs for carriages, &c., Mode of constructing.	G. B. Robinson	Pawlet, Vt	Oct. 3, 1838	961
Elliptical steel-spring	C. Hinkle	Lehigh County, Pa	Nov. 12, 1831	....
Embalming	T. Holmes	Williamsburgh, N. Y	Dec. 10, 1861	33, 885
Embalming	B. F. Lyford	San Francisco, Cal	Jan. 24, 1871	111, 222
Embalming and preserving animal substances, Mode of.	L. Brunetti	Rovigne, Italy	Oct. 1, 1867	69, 312
Embalming and preserving dead bodies	E. De la Granja	Boston, Mass	July 30, 1867	67, 170
Embalming birds	S. A. Morey	Grand Rapids, Mich	Apr. 16, 1872	125, 685
Embalming bodies	G. W. Scollay	Saint Louis, Mo	Jan. 22, 1867	61, 472
Embalming composition	J. C. Sickle	Philadelphia, Pa	Mar. 24, 1868	75, 992
Embalming dead bodies	W. Iddings	Warren, Ohio	Nov. 6, 1860	30, 576
Embalming dead bodies, Compound for	E. H. Crane	Burr Oak, Mich	Sept. 1, 1868	81, 755
Embalming-fluid	F. A. Hutton	Washington, D. C	June 2, 1863	38, 747
Embalming human bodies, Fluid compound for	W. E. Chenoweth	Baltimore County, Md	Nov. 18, 1873	144, 602
Embankment, levee, &c., Method of forming	D. Green	Dalton, Ga	June 11, 1850	7, 425
Embossing	R. J. Chute	Philadelphia, Pa	Jan. 9, 1872	122, 569
Embossing and compressing machine, Hydraulic	J. Robertson	Brooklyn, N. Y	Dec. 3, 1872	133, 543
Embossing and finishing woven fabrics	W. Ralston	Manchester, England	Nov. 1, 1859	25, 980
Embossing and perforating paper, &c., Machine for.	C. Lang	New York, N. Y	Mar. 26, 1867	63, 263
Embossing and printing machine	S. W. Lowe	Philadelphia, Pa	Apr. 29, 1856	14, 789
Embossing and printing press	S. J. Smith and C. Loekle	New York, N. Y	July 7, 1857	17, 703
Embossing and seal press	B. B. Hill	Chicopee, Mass	May 23, 1865	47, 821
Embossing and seal press	G. B. Sargent	Boston, Mass	Sept. 27, 1864	44, 459
Embossing articles of wearing-apparel, Machine for	G. W. Ray	Springfield, Mass	Mar. 26, 1867	63, 177
Embossing bonnets and hats, Apparatus for	H. E. West	Attleborough, Mass	July 25, 1865	49, 049
Embossing brass tubes, Machinery for	D. F. Sanford	New York, N. Y	Nov. 27, 1866	60, 067
Embossing capsules, Machine for	W. Betts	No. 1 Wharf Road, City Road, England.	Sept. 13, 1870	107, 329
Embossing cloth	J. E. Pollard	Franklin, Mass	Nov. 19, 1867	71, 213
Embossing consecutive numbers, Machine for	L. Henser	Boston, Mass	Jan. 1, 1867	60, 726
Embossing designs on metal for jewelry, Process of	W. Riker	Newark, N. J	Mar. 13, 1860	27, 474
Embossing fabrics, Implement for	J. W. McKee	Brooklyn, E. D., N. Y	Dec. 19, 1871	122, 044
Embossing-machine	J. C. Arms	Northampton, Mass	Dec. 4, 1866	60, 115
Embossing napkin-rings, Machine for	E. H. Eldridge and T. Leach	Taunton, Mass	May 8, 1866	54, 515
Embossing-press	G. Clisbee	Marlborough, Mass	Apr. 22, 1873	138, 130
Embossing-press	A. Komp	New York, N. Y	Jan. 23, 1872	123, 026
Embossing-press	J. C. Merriam and W. N. Weeden.	Boston, Mass	July 10, 1866	56, 244
Embossing-press	J. M. Willbur	Cleveland, Ohio	Aug. 31, 1869	94, 368

Index of patents issued from the United States Patent Office from 1790 *to* 1873, *inclusive*—Continued.

Invention.	Inventor.	Residence.	Date.	No.
Embossing-press for linen and other fabrics	L. H. Olmsted	New York, N. Y	Apr. 9, 1872	125, 610
Embossing-press gage	W. Richards, jr	Cincinnati, Ohio	May 19, 1868	78, 130
Embossing-press, Rotary	J. M. Willbur	Cleveland, Ohio	Sept. 22, 1868	82, 461
Embossing wood, Mode of	H. May	Bridgeport, Conn	Oct. 24, 1865	50, 608
Embroidering-machine	H. Berger	Marthalen, Switzerland	Sept. 28, 1869	95, 186
Embroidering-machine	J. Einhorn	New York, N. Y	July 14, 1868	79, 901
Embroidering-machine	A. Heaven and R. Smith	Manchester, England	Apr. 28, 1863	38, 358
Embroidery and sewing stand	W. H. Trowbridge	Saginaw, Mich	Sept. 28, 1858	21, 633
Embroidery-holder	H. Gundaker	Lancaster, Pa	May 3, 1870	102, 537
Embroidery, Imitation metal	J. S. Smith, jr	New York, N. Y	Apr. 8, 1862	34, 907
Embroidery, Imitation of	B. Müller	Fulda, Germany	Dec. 19, 1865	51, 668
Embroidery-patterns, Apparatus for transferring	C. Bordas	New York, N. Y	Oct. 21, 1873	143, 872
Emery-cleaning machine	J. H. Hawes and G. H. Bliss	Boston and West Stockbridge, Mass.	Apr. 28, 1868	77, 281
Emery-cloth, polishing-wheels, &c., Composition for the manufacture of.	L. Francis	New York, N. Y	Jan. 7, 1868	72, 994
Emery-cloth, Substitute for	T. J. Mayall	Roxbury, Mass	Apr. 10, 1860	27, 817
Emery for grinding and polishing tools, Composition of.	T. J. Mayall	Roxbury, Mass	Oct. 18, 1859	25, 841
Emery from foreign substances, Process of separating crude.	N. C. Sawyer	Boston, Mass	Oct. 2, 1866	58, 487
Emery-grinder, Table for	J. L. Jackson	New York, N. Y	Aug. 15, 1871	118, 017
Emery-grinder, Table for	J. L. Jackson	New York, N. Y	Dec. 26, 1871	122, 253
Emery-grinding machine	H. S. Lucas	Chester, Mass	Feb. 4, 1873	135, 482
Emery or sand-paper holder	E. R. Barnes	Brookfield, Conn	Oct. 13, 1863	40, 231
Emery polishing-stick	G. C. Taft	Worcester, Mass	May 30, 1871	115, 384
Emery polishing-wheel, &c	W. P. Van Kleeck	Charlestown, Mass	June 14, 1870	104, 231
Emery sticks and wheels, Composition for	T. J. Mayall	Roxbury, Mass	May 17, 1859	24, 039
Emery-vulcanite or hard-rubber compounds, Preparing.	T. J. Mayall	Boston, Mass	Apr. 9, 1872	125, 600
Emery-wheel	T. Harding	La Fayette, Ind	Jan. 17, 1871	111, 057
Emery-wheel	W. S. Jarboe	New York, N. Y	May 14, 1872	126, 706
Emery-wheel	W. A. Johnston	New York, N. Y	July 8, 1873	140, 630
Emery-wheel	T. Nelson	West Troy, N. Y	Nov. 1, 1864	44, 918
Emery-wheel	J. L. Otis	Leeds, Mass	Mar. 8, 1870	100, 659
Emery-wheel	A. M. Sawyer	Athol, Mass	Jan. 5, 1869	85, 537
Emery-wheel	J. Tyzick	Saint John's, Canada	Sept. 17, 1872	131, 481
Emery-wheel	W. T. Vose	Newtonville, Mass	May 7, 1872	126, 597
Emery-wheel clamp	E. A. Süwerkrop	Washington, D. C	Mar. 22, 1870	101, 158
Emery-wheel-cleaning machine	S. A. and H. Whipple	Shaftsbury, Vt., and Port Richmond, N. Y.	Oct. 14, 1856	15, 912
Emery-wheel cowl	C. Heaton	New York, N. Y	Feb. 27, 1872	124, 134
Emery-wheel for grinding and polishing saws, &c	G. L. Benton	Rochester, N. Y	Nov. 5, 1867	70, 393
Emery-wheels and artificial stone, Manufacture of.	E. C. Merrill	Charleston, Vt	Apr. 4, 1871	113, 439
Emery-wheels and grinding-surfaces, Forming	G. E. Vanderburgh	Mamaroneck, N. Y	Jan. 7, 1862	34, 123
Emery-wheels and oil-stones, Compound for	N. B. Hadley and R. J. Costain	Northampton, Mass	Sept. 21, 1869	94, 955
Emery-wheels and other abrading implements, Composition for.	E. C. Merrill and A. W. Willard.	Charleston, Vt	Oct. 12, 1869	95, 824
Emery wheels and sticks, Manufacture of	T. J. Mayall	Roxbury, Mass	Jan. 11, 1859	22, 606
Emery-wheels, Clamp for girding	H. S. Vrooman	Springfield, Mass	Nov. 5, 1850	7, 761
Emery-wheels, Machine for cleaning	J. Chandler	Collinsville, Conn	Jan. 7, 1862	34, 104
Emery-wheels, &c., Machine for molding	J. S. Elliott	Chelsea, Mass	July 23, 1872	129, 801
Emery-wheels, Making	J. D. Alvord	Bridgeport, Conn	Oct. 2, 1860	30, 189
Emery-wheels, Manufacture of	J. F. Wood	Boston, Mass	Dec. 13, 1870	110, 102
Emery-wheels, &c., Mode of finishing and turning	C. Marsh	Bridgeport, Conn	Apr. 9, 1869	88, 725
Emery-wheels, Mode of forming	C. G. Marshall	Florence, Mass	June 18, 1867	65, 924
Emery-wheels, Process of making	T. Nelson	West Troy, N. Y	Mar. 6, 1866	53, 032
Emery with caoutchouc, Method of combining	T. J. Mayall	Roxbury, Mass	Oct. 11, 1859	25, 747
Enamel	T. L. Oest	Berlin, Prussia	June 27, 1865	48, 500
Enamel, Composition of	J. G. Dunn and A. F. Howes	Lawrenceburgh, Ind	Sept. 7, 1852	9, 250
Enamel for clay gas-retorts, burners, tiles, &c	D. W. Clark	Chicago, Ill	Mar. 15, 1870	100, 725
Enamel for covering hoop-skirt springs	W. S. Ryerson	Philadelphia, Pa	Feb. 12, 1867	62, 074
Enamel for iron	C. Stumer	New York, N. Y	July 25, 1848	5, 681
Enamel for leather	C. W. Held	Brooklyn, N. Y	Feb. 11, 1862	34, 366
Enamel for leather	S. A. Hickel	Spencer, Pa	Nov. 26, 1867	71, 485
Enamel or glaze for pottery, bricks, tiles, &c	W. S. Thomas	Carbon Cliff, Ill	Nov. 9, 1869	96, 634
Enamel, paint, &c., Composition for	M. W. Brown	Morrisania, N. Y	Sept. 5, 1865	49, 708
Enamel to be applied to metals, &c	W. J. Cheyney and E. F. Dieterichs.	Wallingford and Philadelphia, Pa.	Oct. 1, 1867	69, 318
Enameled-metal comb	C. Foster	Wappinger's Falls, N. Y	July 7, 1868	79, 747
Enameled metal, Manufacture of articles of	G. A. Burrough	Providence, R. I	May 30, 1871	115, 278
Enameling and coating articles with rubber, gutta-percha, &c.	P. Finley	New York, N. Y	Apr. 30, 1872	126, 195
Enameling bricks, Compound for	D. W. Clark	Chicago, Ill	Jan. 31, 1871	111, 318
Enameling cast-iron	G. W. Holley	Niagara, N. Y	Mar. 10, 1857	16, 798
Enameling checkers, Apparatus for	I. S., J. W., and C. M. Hyatt	Albany, N. Y	July 9, 1872	128, 729
Enameling cloth, leather, &c., Machine for	L. L. Allen	Hallowell, Me	Sept. 3, 1872	130, 968
Enameling elliptical frames, Machine for	G. W. and H. H. Ferguson	New York, N. Y	June 21, 1864	43, 189
Enameling hard rubber, gutta-percha, &c., Process of.	A. W. and I. C. Monroe	Rahway, N. J., and Brooklyn, N. Y.	Apr. 2, 1867	63, 553
Enameling hollow ware, Composition for	C. E. and C. H. Paris	Paris, France	Mar. 5, 1850	7, 145
Enameling iron and steel	B. Baugh	Chadwick, England	Dec. 28, 1869	98, 336
Enameling iron and steel	J. G. Dreyfus	New York, N. Y	Mar. 2, 1869	87, 475
Enameling iron pipes and hollow ware	E. Pierce	Philadelphia, Pa	June 23, 1857	17, 635
Enameling iron, Process for	C. F. Thomin and C. Stumer	Cincinnati, Ohio	Oct. 17, 1854	11, 815
Enameling, japanning, and inlaying, Preparing metallic substances for.	D. Rait	New York, N. Y	Aug. 30, 1864	44, 046
Enameling jewelry and other articles	G. Krebs	New York, N. Y	Apr. 30, 1872	126, 308
Enameling-machine	H. R. Hall	Philadelphia, Pa	Apr. 28, 1868	77, 187
Enameling moldings, Apparatus for	B. McEachren	San Francisco, Cal	Sept. 26, 1865	58, 146
Enameling moldings, Machine for	G. Boos	New York, N. Y	Dec. 18, 1860	30, 954
Enameling moldings, Machine for	J. Johnson	Boston, Mass	June 9, 1868	78, 747
Enameling moldings, Machine for	J. Johnson	Boston, Mass	Dec. 5, 1871	121, 628
Enameling moldings, Machine for	R. Marcher	New York, N. Y	July 26, 1859	24, 879
Enameling moldings, Machine for	T. Moore	Boston, Mass	Nov. 7, 1871	120, 659
Enameling moldings, Machine for	R. Ten Eyck, jr	Brooklyn, N. Y	Apr. 24, 1860	28, 022
Enameling moldings, &c., Machinery for	R. Marcher	Cornwall, N. Y	Oct. 21, 1851	8, 448

Index of patents issued from the United States Patent Office from 1790 *to* 1873, *inclusive*—Continued.

Invention.	Inventor.	Residence.	Date.	No.
Enameling paper, cloth, card-board, &c., Composition for.	M. W. Brown	New York, N. Y	Feb. 9, 1869	86, 639
Enameling paper, Machine for	M. H. Gardner	New York, N. Y	May 19, 1868	78, 079
Enameling photographic pictures, Apparatus for	N. Weston	San Francisco, Cal	Dec. 10, 1867	72, 133
Enameling picture-frames, &c.	R. Marcher	New York, N. Y	Aug. 14, 1860	29, 609
Enameling picture-frames, &c., Tool for	J. Sperry	New York, N. Y	July 10, 1860	29, 122
Enameling sheet-iron, Apparatus for	W. Harris and D. Hall	Chicago, Ill	May 31, 1870	103, 741
Enameling stone and earthen ware	J. H. Giles	New York, N. Y	May 4, 1869	89, 574
Enameling wood and other solid materials with hard rubber.	P. Finley	Memphis, Tenn	Aug. 10, 1869	93, 606
Enema-chair	W. Laighton	Portsmouth, N. H	Aug. 8, 1846	4, 684
Enema-giving apparatus	B. T. Babbitt	New York, N. Y	Mar. 17, 1857	16, 822
Engine:				

See Air-engine.

Air-engine.	Hot-water engine.
Air and gas engine.	Hydraulic engine.
Air-compressing engine.	Hydrostatic engine.
Ammonia-engine.	Knapsack-engine.
Ammonia-gas engine.	Link-motion engine.
Ammoniacal engine.	Locomotive-engine.
Atmospheric engine.	Marine-engine.
Atmospheric steam-engine.	Motive-power engine.
Beam-engine.	Motor-engine.
Beating-engine.	Non-condensing engine.
Blowing-engine.	Oscillating engine.
Boring-engine.	Oscillating steam-engine.
Caloric engine.	Paper-pulp engine.
Carbonic-acid engine.	Paper-rag engine.
Carding-engine.	Piston-engine.
Chloride of calcium engine.	Pneumatic engine.
Compound-engine.	Pneumo-hydraulic engine.
Compound steam-engine.	Propeller-engine.
Compressed air engine.	Propulsion-engine.
Condensing-engine.	Pumping-engine.
Crank-engine.	Rag or paper engine.
Cylinder-engine.	Reacting engine.
Dental engine.	Reacting steam-engine.
Direct-acting engine.	Reciprocating-engine.
Direct-acting compound steam-engine.	Reciprocating steam-engine.
Direct-acting steam-engine.	Rose-engine.
Drum-engine.	Rosette-engine.
Dummy-engine.	Rotary engine.
Electro-magnetic engine.	Rotary-cylinder engine.
Exhaust-steam engine.	Rotary steam-engine.
Expansion engine.	Rotating cylinder-engine.
Explosive engine.	Stationary engine.
Fire-engine.	Steam-engine.
Fire-extinguishing engine.	Steam and air engine.
Garden-engine.	Steam and caloric engine.
Gas-engine.	Steam-hammer engine.
Gas-explosive engine.	Traction-engine.
Gear-cutting engine.	Traveling engine.
Gunpowder-engine.	Triple-cylinder engine.
Hand-engine.	Trunk-engine.
Heated-liquid engine.	Turning-engine.
High and low pressure engine.	Two-beam engine.
Hoisting-engine.	Vacuum-engine.
Horizontal engine.	Vapor-engine.
Horse-power engine.	Vertical engine.
	Water-engine.

Invention.	Inventor.	Residence.	Date.	No.
Engine	J. Idler	Philadelphia, Pa	July 31, 1819	....
Engine	T. C. Robinson and C. W. Clift.	Mystic Bridge, Conn	Mar. 11, 1873	136, 764
Engine and pump	O. P. Lewis	Cincinnati, Ohio	Nov. 28, 1871	121, 388
Engine-boiler and superheater, Combined	F. B. Blanchard	Spuyten Duyvil, N. Y	Mar. 1, 1870	100, 359
Engine by bi-sulphuret of carbon, Actuating	B. Hughes	Rochester, N. Y	Aug. 22, 1854	11, 553
Engine for raising weights	A. Holmes	Boston, Mass	Oct. 21, 1824	....
Engine-furnace Air-heating apparatus for	J. H. Shotwell	Rahway, N. J	Nov. 26, 1861	33, 801
Engine indicator	C. W. F. Krausch	Chicago, Ill	Sept. 9, 1862	36, 411
Engine operated by heated liquid	J. S. Baldwin	Newark, N. J	Dec. 5, 1871	121, 479
Engine operated by heated liquid	J. S. Baldwin	Newark, N. J	Dec. 5, 1871	121, 480
Engine operated by heated liquid	J. S. Baldwin	Newark, N. J	Dec. 5, 1871	121, 481
Engine-power, Augmenting	A. Ochler	Eastville, Va	Mar. 30, 1835	....
Engine smoke-stack	T. B. Phoebus	Memphis, Tenn	Nov. 21, 1871	121, 062
Engine-turning machine	A. Schwitler	New York, N. Y	Sept. 1, 1863	39, 755
Engines, Applying power to the cranks of	T. Stewart	Philadelphia, Pa	Oct. 26, 1858	21, 911
Engines, Combined quarter-connection and parallel motion for.	A. J. L. Loretz	Brooklyn, N. Y	Nov. 25, 1873	144, 852
Engines, Crank-motion for	J. Smith and G. Joithe	Newark, N. J	June 14, 1870	104, 367
Engines, Device for controlling	J. J. L. Farcot	Saint Ouen, (Seine,) France	Dec. 29, 1868	85, 439
Engines for using steam expansively, Arrangement of.	J. Ericsson	New York, N. Y	Nov. 6, 1849	6, 844
Engines, Holding-keys for strap-connections for	T. Cook	Washington, D. C	Apr. 19, 1859	23, 743
Engines, Mechanism for reversing	G. W. Bishop	Brooklyn, N. Y	Nov. 18, 1873	144, 731
Engines, Method of lubricating	J. Marks	Boston, Mass	Oct. 16, 1860	30, 416
Engines off their centers, Starting	A. Sims	New York, N. Y	July 9, 1867	66, 527
Engines, Utilizing the exhaust steam of	E. Körting	Berlin, Prussia	Mar. 26, 1872	125, 056
Engraver, Pantographic	J. Hope	Providence, R. I	Nov. 19, 1867	71, 010
Engraver's plate	T. Bardon	Brooklyn, N. Y	Aug. 25, 1868	81, 461
Engraver's ring-clamp	T. R. Hopkins	Petersburgh, Va	June 29, 1858	20, 711
Engraving	J. Hutton and G. Fairman		Jan. 8, 1802	....
Engraving	H. S. Ingersoll	Lansing, Mich	Dec. 10, 1872	133, 860

Index of patents issued from the United States Patent Office from 1790 *to* 1873, *inclusive*—Continued.

Invention.	Inventor.	Residence.	Date.	No.
Engraving and carving machine	T. W. Minter	New York, N. Y	Oct. 29, 1872	132, 594
Engraving and carving machine	T. W. Minter	New York, N. Y	Feb. 18, 1873	135, 998
Engraving and chasing articles of metal	T. Lippiatt	Orange, N. J	May 10, 1870	102, 950
Engraving and printing, Making and preparing metallic plates for.	D. H. Mason	Philadelphia, Pa	May 17, 1822	....
Engraving and printing to prevent counterfeits	G. Murray	Philadelphia, Pa	Feb. 13, 1810	....
Engraving bank-notes	B. Chambers	Washington, D. C	June 12, 1847	5, 155
Engraving bank-notes	W. L. Ormsby	Jersey City, N. J	May 16, 1865	47, 744
Engraving bank-notes, &c., Style of	J. Macdonough	New York, N. Y	Oct. 23, 1860	30, 488
Engraving calico-printers' rollers, Machine for	J. and T. Hope	Providence, R. I	Aug. 21, 1855	13, 462
Engraving copper, &c	V. Q. Wedikind	Philadelphia, Pa	Oct. 2, 1866	58, 516
Engraving copper cylinders, Machine for	B. L. Phillips	Providence, R. I	July 3, 1860	29, 002
Engraving copper, &c., Process for	A. F. Burson	Mount Blanchard, Ohio	Nov. 29, 1864	45, 224
Engraving cylinders, Machine for	R. Muckett and W. Rigby	Salford, Great Britain	May 12, 1857	17, 290
Engraving-machine	E. Allen	Grafton, Mass	Jan. 16, 1843	2, 912
Engraving-machine	J. B. Blair	Alton, Ill	May 24, 1853	9, 743
Engraving-machine	C. J. Coulter	Seville, Ohio	May 31, 1870	103, 576
Engraving-machine	J. C. Guerrant and B. J. Field	Leaksville, N. C	Nov. 5, 1867	70, 553
Engraving-machine	J. C. Guerrant and B. J. Field	Leaksville, N. C	Nov. 3, 1868	83, 708
Engraving-machine	J. S. Ives	Morrisania, N. Y	Dec. 23, 1862	37, 235
Engraving-machine	J. D. Lathrop	New York, N. Y	June 28, 1870	104, 744
Engraving-machine	J. C. Spencer	Phelps, N. Y	Feb. 15, 1870	99, 794
Engraving-machine, Apparatus for supporting and adjusting graver.	J. Hope	Providence, R. I	June 8, 1858	20, 528
Engraving-machine, Device for pentagraphic	J. Hope	Providence, R. I	Mar. 9, 1858	19, 607
Engraving-machinery	J. C. Guerrant and B. J. Field	Leaksville, N. C	Dec. 18, 1866	60, 506
Engraving metallic plates, Apparatus for	W. L. Ormsby	Jersey City, N. J	May 16, 1865	47, 745
Engraving or printing upon glass, Process of	M. D. and L. W. Whipple	Somerville, Mass	June 27, 1854	11, 189
Engraving rollers, &c., Machine for	W. Shields	Manchester, England	May 15, 1860	28, 332
Engraving-surface	I. Taylor	Stanford Rivers, England	June 1, 1852	8, 991
Engraving, Transfer	R. Meale	Brooklyn, N. Y	Aug. 31, 1869	94, 432
Engravings, &c., Binding	W. T. Anderson	Brooklyn, N. Y	Oct. 2, 1860	30, 277
Engravings, &c., Process for reducing copies of	J. A. F. Lair	Paris, France	Oct. 15, 1861	33, 485
Entrail-cleaning machine	J. A. Huss	Louisville, Ky	Nov. 24, 1868	84, 358
Entry-light	C. W. Felt	Salem, Mass	Jan. 9, 1855	12, 200
Envelope	E. L. Barrett	Springfield, Ohio	Oct. 24, 1865	50, 549
Envelope	S. D. Bates	Boston, Mass	Dec. 26, 1871	122, 212
Envelope	H. C. Bristol	Saint Clair, Mich	Mar. 8, 1864	41, 828
Envelope	R. S. Clark	Great Bend, Pa	Apr. 7, 1868	76, 303
Envelope	W. B. Coates	Philadelphia, Pa	Aug. 5, 1856	15, 475
Envelope	C. Crowell	Syracuse, N. Y	Apr. 16, 1872	125, 664
Envelope	B. C. Davis	Binghamton, N. Y	Sept. 6, 1870	107, 166
Envelope	F. W. Eberman	West Salem, Ill	Nov. 9, 1869	96, 566
Envelope	E. S. Ellis	Trenton, N. J	Mar. 14, 1871	112, 568
Envelope	H. Everdell	New York, N. Y	May 14, 1867	64, 753
Envelope	E. Fitzki	Washington, D. C	May 14, 1867	64, 653
Envelope	J. C. Gaston	Cincinnati, Ohio	Dec. 31, 1867	72, 836
Envelope	J. C. Gaston	Cincinnati, Ohio	Feb. 23, 1869	87, 253
Envelope	L. Giebrich	Ottumwa, Iowa	July 1, 1873	140, 496
Envelope	L. Giebrich	Ottumwa, Iowa	Sept. 23, 1873	143, 136
Envelope	J. Gray	Raymond, Miss	Mar. 26, 1861	31, 801
Envelope	H. K. Gregg	Baltimore, Md	May 9, 1871	114, 672
Envelope	J. W. Groomes	Portsmouth, Ohio	June 20, 1871	116, 180
Envelope	G. P. Hachenberg	Hudson, N. Y	June 15, 1869	91, 331
Envelope	E. Harmon	Washington, D. C	Nov. 20, 1855	13, 838
Envelope	W. E. Haskins	New York, N. Y	July 23, 1867	67, 111
Envelope	B. J. Hoffacker	Melrose, N. Y	June 1, 1869	90, 663
Envelope	J. W. Huffington and A. Buckman.	Brooklyn, N. Y	Dec. 24, 1872	134, 142
Envelope	R. S. Jennings	New York, N. Y	May 21, 1867	64, 879
Envelope	A. A. C. Klaucke and J. Fraser.	Washington, D. C., and New York, N. Y.	Apr. 6, 1869	88, 722
Envelope	R. T. Knight	Philadelphia, Pa	June 24, 1856	15, 185
Envelope	S. Kuh	Jefferson, Iowa	June 24, 1873	140, 204
Envelope	R. K. Kuhn and T. O. Atkinson	Doylestown, Pa	Feb. 11, 1873	135, 819
Envelope	P. Lockwood	Auburn, Ind	Apr. 25, 1871	114, 161
Envelope	R. Magee	Philadelphia, Pa	Nov. 20, 1866	59, 849
Envelope	G. H. Mathews	New York, N. Y	Dec. 1, 1868	84, 501
Envelope	W. F. McCrary	Baker City, Oreg	Apr. 30, 1872	126, 316
Envelope	C. E. McMahan	Elizabeth, N. J	Aug. 8, 1871	117, 795
Envelope	B. Morison	Philadelphia, Pa	June 19, 1860	28, 767
Envelope	J. B. Murray	New York, N. Y	Dec. 18, 1860	30, 980
Envelope	S. E. Pettee	Philadelphia, Pa	Mar. 22, 1859	23, 309
Envelope	C. Pink and M. Silverberg	Cairo, Ill	June 3, 1873	139, 475
Envelope	C. R. M. Pohlé	Richmond, Va	Apr. 28, 1868	77, 316
Envelope	C. R. M. Pohlé	Richmond, Va	Dec. 8, 1868	84, 840
Envelope	J. Reckendorfer and J. C. Richard.	New York and Brooklyn, N. Y.	July 9, 1861	32, 813
Envelope	C. W. Saladee	Circleville, Ohio	Apr. 27, 1869	89, 507
Envelope	F. M. Shields	Macon, Miss	Jan. 7, 1868	73, 203
Envelope	E. J. Smith	Washington, D. C	May 16, 1871	114, 871
Envelope	G. W. Starr	Washington, D. C	Dec. 22, 1868	85, 253
Envelope	R. S. Stubbs	Lisbon, N. H	Jan. 3, 1871	110, 798
Envelope	M. Saylor	Buffalo, N. Y	July 11, 1871	117, 015
Envelope	J. P. Thompson	Philadelphia, Pa	Apr. 25, 1871	114, 228
Envelope	S. Ullman	New York, N. Y	Feb. 25, 1868	74, 867
Envelope	S. Ullman	New York, N. Y	Sept. 8, 1868	81, 962
Envelope	S. Ullman	New York, N. Y	Sept. 8, 1868	81, 963
Envelope	J. B. Wheeler	Neville, Ohio	July 21, 1868	80, 257
Envelope	S. L. Wiegand	Philadelphia, Pa	Mar. 22, 1864	42, 035
Envelope	J. W. Wilcox	New York, N. Y	Apr. 8, 1862	34, 927
Envelope	C. C. Woolworth	New York, N. Y	Mar. 1, 1864	41, 804
Envelope and letter-sheet combined	J. B. Anderson	Wakefield, Mass	Jan. 16, 1872	122, 694
Envelope and letter sheet combined	C. Schweizer	Boston, Mass	Sept. 10, 1872	131, 184
Envelope and letter-sheet combined	C. Rowland	Clinton, Ill	Dec. 19, 1865	51, 623
Envelope-blanks, Gumming and feeding	J. B. Duff	Patchogue, N. Y	Dec. 13, 1864	45, 397
Envelope, Card	C. Rowland	Washington, D. C	Aug. 8, 1871	117, 818

Index of patents issued from the United States Patent Office from 1790 to 1873, inclusive—Continued.

Invention.	Inventor.	Residence.	Date.	No.
Envelope, Document	J. W. Wilcox	New York, N. Y	July 23, 1867	67,092
Envelope-drying apparatus	W. E. Lockwood	Philadelphia, Pa	Jan. 14, 1868	73,350
Envelope, Expansive	R. Van Buskirk	Newark, N. J	Aug. 16, 1864	43,875
Envelope, Express money	E. Hulbert	Atlanta, Ga	May 14, 1867	64,674
Envelope-fastener	M. E. Berolzheimer	New York, N. Y	Apr. 26, 1870	102,211
Envelope-feeding machine	E. Allen	Norwich, Conn	Nov. 19, 1872	133,184
Envelope-folding device	W. H. Hook	Walworth, England	June 21, 1864	43,272
Envelope-folding machine	E. Coleman	Philadelphia, Pa	Apr. 26, 1853	9,683
Envelope-folding machine	R. L. Hawes	Worcester, Mass	June 21, 1853	9,812
Envelope-folding machine	R. Parks	Philadelphia, Pa	Dec. 17, 1867	72,225
Envelope for rolled prints, &c	L. Prang	Boston, Mass	Sept. 3, 1872	131,117
Envelope for sewing-needles	G. P. Former	Philadelphia, Pa	June 17, 1862	35,596
Envelope, Glove	A. D. Foster	Sayville, N. Y	Aug. 6, 1872	130,205
Envelope, Glove	F. Hegle	New York, N. Y	Sept. 3, 1872	130,997
Envelope-gummer	E. Allen	Norwich, Conn	July 13, 1869	92,558
Envelope, Kid-glove	F. Hegle	New York, N. Y	July 30, 1872	129,890
Envelope, Letter	J. G. Arnold	Worcester, Mass	Dec. 28, 1858	22,405
Envelope, Letter	J. G. Arnold	Worcester, Mass	May 2, 1865	47,508
Envelope, Letter	H. C. Bristol	Chicago, Ill	Jan. 27, 1863	37,490
Envelope, Letter	C. Conkling	Ashland, Ohio	Feb. 12, 1867	62,011
Envelope, Letter	E. Harmon	Washington, D. C	Mar. 22, 1859	23,300
Envelope, Letter	T. S. Lambert	Peekskill, N. Y	Mar. 2, 1865	46,911
Envelope, Letter	J. D. McAnulty	Brooklyn, N. Y	Sept. 2, 1873	142,405
Envelope, Letter	A. M. Murphy	New York, N. Y	June 14, 1864	43,126
Envelope, Letter	W. Murphy	New York, N. Y	June 2, 1863	38,755
Envelope, Letter	C. Phelps	Salem, Mass	Apr. 27, 1858	20,087
Envelope, Letter	A. C. Richard	Newtown, Conn	Nov. 15, 1859	26,127
Envelope, Letter-sheet	C. Foster	Cortland Village, N. Y	Feb. 6, 1872	123,389
Envelope-machine	E. Allen	Norwich, Conn	Aug. 28, 1866	57,617
Envelope-machine	J. Ball	New York, N. Y	May 26, 1868	78,353
Envelope-machine	J. Ball	New York, N. Y	Jan. 31, 1871	111,420
Envelope-machine	T. Bergner	Philadelphia, Pa	Feb. 10, 1857	16,576
Envelope-machine	H. W. Chamberlin	New York, N. Y	Feb. 1, 1870	99,290
Envelope-machine	J. H. Cooper	Philadelphia, Pa	Mar. 19, 1867	62,939
Envelope-machine	W. W. Cotton	New York, N. Y	Apr. 8, 1856	14,625
Envelope-machine	J. B. Duff and T. W. Keating	New York, N. Y	Feb. 1, 1859	22,788
Envelope-machine	J. C. Gaston	Cincinnati, Ohio	Aug. 29, 1871	118,525
Envelope-machine	E. W. Goodale	Clinton, Mass	Oct. 9, 1855	13,647
Envelope-machine	J. M. D. Keating	New York, N. Y	June 30, 1863	39,053
Envelope-machine	J. M. D. Keating	New York, N. Y	Feb. 19, 1867	62,274
Envelope-machine	W. H. Low	Albany, N. Y	Apr. 8, 1856	14,643
Envelope-machine	J. K. Park and C. S. Watson	New York, N. Y	Jan. 23, 1849	6,055
Envelope-machine	R. Parks	Philadelphia, Pa	July 10, 1866	56,325
Envelope-machine	M. G. Puffer	Rockville, Conn	Nov. 23, 1858	22,149
Envelope-machine	G. H. Reay	New York, N. Y	Aug. 25, 1863	39,702
Envelope-machine	G. H. Reay	New York, N. Y	Jan. 29, 1867	61,687
Envelope-machine	A. A. Rheutan	Worcester, Mass	Feb. 4, 1868	74,140
Envelope-machine	A. A. Rheutan	Worcester, Mass	Dec. 10, 1872	133,800
Envelope-machine	D. M. Smyth	Orange, N. J	June 22, 1869	91,575
Envelope-machine	H. D. and D. W. Swift	Worcester, Mass	May 30, 1871	115,381
Envelope-machine	H. D. and D. W. Swift	Worcester, Mass	May 30, 1871	115,382
Envelope-machine	T. V. Waymouth	New York, N. Y	Sept. 25, 1866	58,327
Envelope-machine	T. V. Waymouth	New York, N. Y	Nov. 19, 1867	71,252
Envelope-machines, Apparatus for feeding paper to	G. H. Reay	New York, N. Y	Jan. 26, 1864	41,395
Envelope-machines, Counting-attachment for	G. H. Reay	Hudson, N. J	Dec. 16, 1862	37,199
Envelope-machines, Cutting and gumming apparatus for.	E. B. Olmsted	Washington, D. C	July 16, 1867	66,877
Envelope-machines, Feed-box for	E. Allen	Norwich, Conn	Sept. 9, 1873	142,606
Envelope-machines, Folding and printing bed for	E. B. Olmsted	Washington, D. C	July 16, 1867	66,878
Envelope-machines, Packing-apparatus for	E. B. Olmsted	Washington, D. C	July 16, 1867	66,876
Envelope, Metallic-seal	R. S. Jennings	New York, N. Y	June 26, 1866	55,955
Envelope, Oil-cake	C. Eskrett and H. Searle	Hull, England	Nov. 1, 1870	108,774
Envelope-opener	J. P. Gruger	Philadelphia, Pa	June 7, 1870	104,118
Envelope-opener	G. S. Hastings, jr	Plainville, Conn	May 27, 1873	139,312
Envelope-opener	A. Kline	Manheim, Pa	Dec. 10, 1872	133,783
Envelope-opener	W. H. Mantz	Centralia, Ill	May 4, 1869	89,675
Envelope-opener	L. A. Miller	Sewicklyville, Pa	July 12, 1870	105,356
Envelope-opener	C. B. Stephens	Riverton, Conn	Nov. 16, 1869	96,845
Envelope, Paper	R. Shepard	Brooklyn, N. Y	Nov. 29, 1864	45,301
Envelope, Paper-collar-packing	G. K. Snow	Watertown, Mass	July 18, 1865	48,848
Envelope, Picture	C. A. Kihlgren	Boston, Mass	Nov. 5, 1867	70,442
Envelope, Picture	R. Price	New York, N. Y	Feb. 22, 1870	100,067
Envelope, Picture	E. L. Wilson	Philadelphia, Pa	July 8, 1862	35,861
Envelope, Post-office way-bill	M. P. Norton	Troy, N. Y	June 10, 1862	35,539
Envelope, Postal-currency	F. Mills	Washington, D. C	June 15, 1869	91,473
Envelope, Prepaid	J. B. Murray	New York, N. Y	Sept. 25, 1860	30,151
Envelope, Registered-package	W. T. McCrary	Baker County, Oreg	Dec. 10, 1872	133,712
Envelope, Rouleau	A. J. Gove	San Francisco, Cal	Sept. 24, 1872	131,609
Envelope, Safety	W. S. Stetson	Baltimore, Md	Sept. 27, 1859	25,590
Envelopes, Apparatus for drying pasted	J. Armstrong	Philadelphia, Pa	Feb. 4, 1862	34,326
Envelopes, Apparatus for making	E. L. Barrett	Springfield, Ohio	Jan. 15, 1867	61,140
Envelopes, Construction of	R. T. Knight	Philadelphia, Pa	Apr. 8, 1856	14,614
Envelopes, Device for moistening and closing	R. W. Walker	Washington, D. C	Aug. 2, 1870	106,098
Envelopes for bottles and jars, Manufacture of	G. Clark	London, England	Apr. 17, 1866	54,080
Envelopes from sheets of paper, Mode of cutting	J. P. Herron	Washington, D. C	Aug. 30, 1864	43,989
Envelopes, Machine for gumming and printing	T. V. Waymouth	New York, N. Y	June 12, 1866	55,562
Envelopes, Manufacturing	J. Ball	New York, N. Y	Aug. 16, 1870	106,451
Envelopes, Opening	A. L. Seabury	Norfolk, Va	Mar. 29, 1864	42,113
Envelopes, pocket-books, &c., Slide for fastening	J. W. Wilcox	New York, N. Y	May 7, 1867	64,608
Envelopes, &c., Pressure adhesive gum for	P. S. P. Connor	Philadelphia, Pa	Aug. 31, 1869	94,399
Epaulet	W. H. Gray	Philadelphia, Pa	Nov. 12, 1861	33,702
Epaulet	J. S. Smith	New York, N. Y	Apr. 12, 1859	23,620
Epaulet-fastening	E. J. Helwig	Philadelphia, Pa	Sept. 20, 1870	107,614
Epaulet-fastening	W. J. Horstmann	Philadelphia, Pa	Jan. 31, 1871	111,456
Epergne for fruit and flowers	S. S. Barrie	Green Point, N. Y	Aug. 15, 1871	117,965
Epergne for fruit and flowers	S. S. Barrie	New York, N. Y	Mar. 26, 1872	125,003
Epilobium, Manufacture from the fibers of	R. B. Miller	Utica, N. Y	Mar. 21, 1865	46,922
Equalizer	A. Lafever and R. K. Laranay	Battle Creek, Mich	Apr. 28, 1868	77,200

Index of patents issued from the United States Patent Office from 1790 *to* 1873, *inclusive*—Continued.

Invention.	Inventor.	Residence.	Date.	No.
Equalizer	G. W. N. Yost	Corry, Pa	Mar. 9, 1869	87, 612
Equalizer for double-trees, Draft	R. F. Judson	Kalamazoo, Mich	Apr. 7, 1868	76, 466
Equalizer or power-regulator	A. Gregory	Brooklyn, N. Y	Sept. 9, 1851	8, 345
Equalizer, Spring	H. Davis	Berlin, Wis	Oct. 28, 1873	143, 966
Equalizer, Three-horse	E. F. Avery and H. B. Leckenby	Dowagiac, Mich	Feb. 2, 1869	86, 491
Equalizer, Three-horse	F. E. Barr	Albion, N. Y	June 8, 1869	90, 915
Equalizer, Three-horse	J. Blackwood	Lithopolis, Ohio	Mar. 12, 1872	124, 537
Equalizer, Three-horse	W. P. Brooks	Bloomington, Ill	Aug. 10, 1869	93, 593
Equalizer, Three-horse	J. E. Coveney	Buchanan, Mich	Mar. 2, 1869	87, 471
Equalizer, Three-horse	G. Cramton	Marshall, Mich	Feb. 4, 1868	74, 053
Equalizer, Three-horse	G. Cramton	Marshall, Mich	Sept. 14, 1869	94, 718
Equalizer, Three-horse	G. Cramton and P. A. Spicer	Marshall, Mich	July 21, 1868	80, 060
Equalizer, Three-horse	W. Haistings, sr	Fowler, Ill	June 24, 1873	140, 268
Equalizer, Three-horse	T. Hoadley	Toulon, Ill	Aug. 19, 1873	141, 875
Equalizer, Three-horse	A. March	Taylor, Ill	Mar. 19. 1872	124, 692
Equalizer, Three-horse	E. K. Parish	Shelbyville, Ind	July 16, 1872	129, 050
Equalizer, Three-horse	J. C. Pfeil	Arenzville, Ill	Oct. 11, 1870	108, 179
Equalizer, Three-horse	Z. B. Sims	Bonham, Tex	Aug. 31, 1869	94, 355
Equalizer, Three-horse	A. L., G. J., and T. N. Thomas	Lee's Summit, Mo	May 21, 1872	126, 998
Equalizer, Three-horse	J. T. Thornton	Kewanee, Ill	Mar. 1, 1870	100, 469
Equalizer, Three-horse	L. Warren	Dwight, Ill	July 11, 1871	117, 021
Equalizer, Three-horse	E. M. Wright	Castile, N. Y	July 13, 1869	92, 503
Equalizer, Three-horse-draft	E. J. Toof	Fort Madison, Iowa	Dec. 19, 1865	51, 632
Equalizer, Two-horse	Z. B. Sims	Bonham, Tex	Aug. 2, 1870	106, 084
Equalizing-bars, Die for bending	A. E. Barnard	Akron, Ohio	Feb. 9, 1869	86, 629
Eraser	H. T. Cushmann	North Bennington, Vt	May 17, 1870	103, 149
Eraser	J. M. Hicks	Boston, Mass	May 14, 1861	32, 288
Eraser	A. H. Hook and H. B. Adams	New York, N. Y	July 17, 1866	56, 415
Eraser	W. A. Morse and J. G. Powell	Philadelphia, Pa	June 16, 1868	78, 887
Eraser	A. G. Shaver	New Haven, Conn	Jan. 24, 1865	46, 032
Eraser	R. Smith	Bangor, Me	Aug. 7, 1866	57, 004
Eraser and burnisher	S. R. Andres	Troy, N. Y	Aug. 21, 1866	57, 273
Eraser and burnisher	A. G. Shaver	New Haven, Conn	Aug. 22, 1865	49, 558
Eraser and burnisher	A. G. Shaver	New Haven, Conn	Oct. 9, 1866	58, 684
Eraser and burnisher, Combined	A. G. Shaver	New Haven, Conn	Aug. 22, 1865	49, 559
Eraser and eraser-sharpener, Combination of	J. Gault	Boston, Mass	Sept. 17, 1861	33, 304
Eraser and letter-opener	G. C. Barney	Philadelphia, Pa	Oct. 1, 1867	69, 388
Eraser and pen-handle combined	W. A. Morse and J. G. Powell	Philadelphia, Pa	Feb. 26, 1867	62, 496
Eraser and pencil-sharpener	A. G. Shaver	Hartford, Conn	Mar. 8, 1859	23, 196
Eraser and pencil-sharpener combined	C. D. Smith	Washington, D. C	Oct. 2, 1866	58, 558
Eraser-attachment for pencil, &c	G. H. Richards	New York, N. Y	Feb. 22, 1870	100, 189
Eraser, Blackboard	F. G. Johnson	Brooklyn, N. Y	June 18, 1872	128, 149
Eraser, Blackboard	W. S. Read	Oakland, Cal	Feb. 4, 1873	135, 487
Eraser, Chalk	G. Munger	New Haven, Conn	Nov. 5, 1861	33, 684
Eraser for pencils, Detachable	W. K. Evans	New York, N. Y	May 14, 1872	126, 792
Eraser-holder	W. P. Patton	Harrisburgh, Pa	Aug. 21, 1866	57, 370
Eraser, Ink	J. W. Tallmage	New York, N. Y	Jan. 16, 1872	122, 867
Eraser, Rubber	W. N. Bartholomew	Newton Centre, Mass	Dec. 15, 1868	84, 985
Eraser, Rubber	W. N. Bartholomew	Newton Centre, Mass	Dec. 19, 1871	121, 982
Eraser, Rubber	M. D. Converse and F. A. Bates	London, Ohio, and Boston, Mass.	June 28, 1870	104, 831
Eraser, Rubber	E. W. Frost	Brooklyn, N. Y	Apr. 1, 1873	137, 299
Eraser, Rubber	S. D. Hovey	Brooklyn, N. Y	Aug. 17, 1869	93, 823
Eraser, Rubber	T. H. Müller	Yonkers, N. Y	Mar. 5, 1872	124, 374
Eraser, Rubber	G. Stackpole	Elizabeth, N. J	Nov. 4, 1873	144, 364
Eraser, Vulcanized rubber pencil-mark	T. H. Müller	Yonkers, N. Y	Jan. 23, 1872	122, 304
Escape-pipe, Condensing	B. F. Smith	New Orleans, La	Sept. 12, 1871	118, 885
Escapement	W. C. Kellum	San Francisco, Cal	Sept. 1, 1868	81, 789
Escapement	W. C. Kellum	San Francisco, Cal	Sept. 8, 1868	81, 910
Escapement	W. N. Manning	Rockport, Mass	May 19, 1863	38, 594
Escapement	J. V. D. Patch	Brownville, Nebr	July 21, 1868	80, 209
Escapement, Duplex	C. E. Jacot	New York, N. Y	July 27, 1852	9, 150
Escapements, Arrangement of dead-beat	D. J. Mozart	Yellow Springs, Ohio	Aug. 9, 1859	25, 034
Escritoir, Portable	W. G. Mitchell	Holliston, Mass	Jan. 24, 1871	111, 231
Esophagus-tube	N. Q. Munger	Brookfield Centre, Wis	July 10, 1860	29, 092
Essence-extracting apparatus	G. Bantz	Frederick, Md	Feb. 9, 1869	86, 803
Essences, Mode of extracting	J. C. Walker	Waco Village, Tex	Aug. 13, 1867	67, 824
Essences, Vessel for extracting	O. Ernst	New York, N. Y	July 1, 1862	35, 747
Etching-grounds, Mode of producing	R. Gottgetreu	Munich, Bavaria	Aug. 28, 1866	57, 651
Etching stones, Composition for	A. Hoen	Baltimore, Md	Apr. 24, 1860	27, 981
Ether, Apparatus for administering	L. Roper	Philadelphia, Pa	Oct. 10, 1848	5, 850
Ether, Process of manufacturing sulphuric	F. Renz	Poughkeepsie, N. Y	Aug. 11, 1868	80, 835
Evaporating alkaline solutions, &c., Furnace for	M. L. Keen	Jersey City, N. J	Nov. 18, 1873	144, 682
Evaporating and clarifying pan	C. Jacobs	Columbus, Ohio	Aug. 6, 1861	32, 998
Evaporating and distilling apparatus	W. R. King	Yellow Springs, Ohio	Jan. 7, 1862	34, 062
Evaporating and distilling apparatus	W. P. Wheeler	Louisville, Ky	Aug. 8, 1865	49, 328
Evaporating and distilling by solar heat	N. W. Wheeler and W. W. Evans.	Brooklyn and New Rochelle, N. Y.	May 3, 1870	102, 633
Evaporating and distilling liquids	J. Adair and H. W. C. Tweddle.	Pittsburgh, Pa	July 17, 1866	56, 343
Evaporating-furnace	J. Richardson		Sept. 13, 1802	
Evaporating-furnace	J. Richardson		Apr. 4, 1803	
Evaporating furnace, Salt	J. Colguhoun	Charleston, Va	Feb. 6, 1834	
Evaporating furnace, Sugar-juice	L. Lefebvre	New Orleans, La	Nov. 23, 1858	22, 126
Evaporating furnace, Sugar-juice	F. Roy	Saint Bernard Parish, La	Dec. 14, 1858	22, 307
Evaporating-kettle furnaces, Arch for	E. Laass	Syracuse, N. Y	June 2, 1868	78, 527
Evaporating-pan	H. O. Ames	New Orleans, La	Oct. 16, 1860	30, 443
Evaporating-pan	H. F. Bartlett	La Grange, Mo	Sept. 5, 1865	49, 694
Evaporating-pan	G. Bez	Mokena, Ill	May 2, 1865	47, 511
Evaporating-pan	S. P. Dyer	Prairie Depot, Ohio	June 11, 1867	65, 656
Evaporating-pan	J. J. Mapes and W. A. Cox	New York, N. Y	Jan. 7, 1846	4, 352
Evaporating-pan	S. Page	McAllisterville, Pa	June 25, 1867	66, 037
Evaporating-pan	G. W. Storer	Philadelphia, Pa	June 3, 1873	139, 625
Evaporating-pan	G. W. Storer	Philadelphia, Pa	Oct. 21, 1873	143, 939
Evaporating-pan	G. W. Storer	Philadelphia, Pa	Oct. 28, 1873	144, 158
Evaporating-pan	E. E. Stubbs and T. C. Davis	West Elkton, Ohio	July 16, 1867	66, 750
Evaporating-pan	D. Watson	Mexico, Mexico	Oct. 14, 1873	143, 649

Index of patents issued from the United States Patent Office from 1790 *to* 1873, *inclusive*—Continued.

Invention.	Inventor.	Residence.	Date.	No.
Evaporating-pan	J. A. Webb and C. Cory	Madison, N. J., and Lima, Ind.	Apr. 9, 1867	63, 678
Evaporating-pan	N. Willey	South Windsor, Conn	Sept. 10, 1867	68, 821
Evaporating-pan	J. B. Williams	Glastenbury, Conn	June 29, 1869	91, 890
Evaporating-pan	J. B. Williams	Glastenbury, Conn	June 27, 1871	116, 524
Evaporating pan and furnace	A. Kolb	Steubenville, Ohio	Jan. 14, 1825	....
Evaporating pan and furnace	L. Scott	Sinking Spring, Ohio	Sept. 21, 1869	95, 049
Evaporating pan and plant for salt-manufacture	G. W. Storer	Philadelphia, Pa	Oct. 28, 1873	144, 159
Evaporating pan, Cane-juice	D. M. Cook	Mansfield, Ohio	June 22, 1858	20, 631
Evaporating pan, Cane-juice	C. A. Desobry	Plaquemine, La	Nov. 8, 1859	26, 024
Evaporating-pan for saccharine juices	E. D. Foss	Maineville, Ohio	Feb. 25, 1862	34, 484
Evaporating-pan for saccharine juices	J. V. Harter	Plymouth, Ill	Dec. 17, 1861	33, 942
Evaporating-pan for saccharine juices	J. M. Moss	Waverly, Iowa	May 20, 1862	35, 350
Evaporating-pan for saccharine juices	D. B. Neal	Mount Gilead, Ohio	Mar. 11, 1862	34, 644
Evaporating-pan for saccharine juices	W. E. Prall	Maineville, Ohio	Feb. 18, 1862	34, 469
Evaporating-pan for saccharine liquids	O. N. Brainerd	Marion, Iowa	Jan. 28, 1862	34, 236
Evaporating-pan for saccharine liquids	F. G. Butler	Bellows Falls, Vt	Mar. 7, 1871	112, 319
Evaporating-pan for saccharine liquids	P. W. Gates	Chicago, Ill	Oct. 28, 1862	36, 881
Evaporating-pan for saccharine liquids	G. Stump	New York, N. Y	July 8, 1862	35, 844
Evaporating-pan for sorghum-juice	N. Z. Potter	Uniontown, Ill	June 9, 1863	38, 839
Evaporating-pan for sugar	H. F. Bartlett	La Grange, Mo	Oct. 18, 1864	44, 698
Evaporating-pan for sugar and sirup	L. Megowan	Upper Alton, Ill	June 21, 1864	43, 220
Evaporating-pan for sugar-juices	D. M. Cook	Mansfield, Ohio	Apr. 14, 1863	38, 152
Evaporating-pan for sugar, salt, &c	L. R. Cornell	Brooklyn, N. Y	Feb. 11, 1873	135, 630
Evaporating pan, Steam	T. C. Bartle and C. F. Putney	Independence, Iowa	Dec. 27, 1864	45, 667
Evaporating pan, Sugar	C. B. Darrow	Orland, Ind	Nov. 3, 1863	40, 460
Evaporating pan, Sugar	S. H. Gilman	New Orleans, La	Sept. 9, 1856	15, 694
Evaporating pan, Sugar	H. F. Schroder	Cincinnati, Ohio	Dec. 18, 1866	60, 564
Evaporating pan, Sugar-juice	W. Hedges	Chicago, Ill	Nov. 29, 1859	26, 325
Evaporating-pan with cellular boiler	D. M. Cook	Mansfield, Ohio	Feb. 24, 1863	37, 736
Evaporating-pan with tubular boiler	D. M. Cook	Mansfield, Ohio	Feb. 24, 1863	37, 737
Evaporating-pans, Apparatus for heating	H. O. Ames	New Orleans, La	Aug. 9, 1859	24, 978
Evaporating-pans, Apparatus for heating	E. Skelly	Plaquemine, La	Aug. 9, 1859	25, 052
Evaporating-pans, Cellular or tubular boiler for	D. M. Cook	Mansfield, Ohio	Feb. 24, 1863	37, 735
Evaporating-pans, Device to regulate the flow of sap to.	J. Thomas	Hardwick, Vt	Sept. 6, 1870	107, 124
Evaporating-pans for saccharine juices, &c., Construction of.	T. J. Price	Industry, Ill	Jan. 28, 1862	34, 264
Evaporating-pans furnaces, Arch bar and support for constructing.	G. S. Deane	Grand Rapids, Mich	Oct. 15, 1867	69, 781
Evaporating-pans, &c., Method of inserting tubes in.	G. H. Thomas	Kingston, Mass	Nov. 20, 1855	13, 832
Evaporating pans or boilers, Feeder for	S. A. Mitchell	Alstead Centre, N. H	Apr. 16, 1867	63, 925
Evaporating-vessel	J. P. Hale	Kanawha Court-House, Va	Nov. 22, 1859	26, 182
Evaporating-vessels, Arrangement of steam-coils in.	H. O. Ames	New Orleans, La	June 29, 1858	20, 687
Evaporation	T. Arnold	....	Mar. 5, 1806	....
Evaporation	T. Bedwell and B. Henfrey	....	Feb. 12, 1801	....
Evaporation and vaporization	M. S. Bringier	Ascension Parish, La	Nov. 12, 1867	70, 690
Evaporator:				
See Brine-evaporator. Cane-juice evaporator. Compound evaporator. Fluid-evaporator. Hot-air-pipe evaporator. Liquid-evaporator. Medicine-evaporator. Portable evaporator. Rotary evaporator. Saccharine-juice evaporator. Saccharine-liquid evaporator. Salt-evaporator. Salt-water evaporator. Sap-evaporator. Sorghum-evaporator. Steam-heated evaporator. Sugar-evaporator. Sugar-juice evaporator. Sugar-pan evaporator. Vacuum-evaporator.				
Evaporator	C. Alden	New York, N. Y	Oct. 4, 1859	25, 614
Evaporator	P. J. Badoux	New York, N. Y	June 9, 1868	78, 639
Evaporator	A. Balding	Madison, Ind	Aug. 28, 1866	57, 459
Evaporator	J. Beechley	Dayton, Ohio	Mar. 6, 1866	52, 952
Evaporator	J. C. Bell	Pawnee City, Nebr	Dec. 26, 1865	51, 683
Evaporator	J. Bogue	Auburn, Wis	Sept. 12, 1865	49, 851
Evaporator	S. Bowerman	Battle Creek, Mich	Dec. 20, 1864	45, 470
Evaporator	W. Canning	New York, N. Y	May 23, 1865	47, 795
Evaporator	B. F. Cauffman	Millerstown, Pa	Sept. 1, 1868	81, 752
Evaporator	E. Chitister	Chatham, Iowa	Jan. 12, 1869	85, 791
Evaporator	J. Cook	Wellsville, N. Y	Nov. 14, 1871	120, 859
Evaporator	J. Cooper	Mount Vernon, Ohio	Oct. 9, 1866	58, 607
Evaporator	S. P. Dyer	Ankney Town, Ohio	May 2, 1871	114, 278
Evaporator	A. S. Eddy	Smithville, N. Y	Aug. 21, 1866	57, 301
Evaporator	J. H. Elward	Polo, Ill	Feb. 6, 1866	52, 402
Evaporator	N. Evinger	Terre Haute, Ind	July 28, 1868	810, 342
Evaporator	F. Farquahar and R. E. Doan	Wilmington, Ohio	Sept. 25, 1866	58, 236
Evaporator	H. C. Gilbert	Cambridge, Vt	Feb. 12, 1867	61, 025
Evaporator	S. D. Gilson	Syracuse, N. Y	Oct. 25, 1870	108, 701
Evaporator	L. R. Gleason	Dundee, N. Y	Mar. 27, 1866	53, 438
Evaporator	F. M. Harris	Winnamac, Ind	Jan. 2, 1866	51, 828
Evaporator	J. Harris	Janesville, Wis	Nov. 24, 1868	84, 277
Evaporator	S. T. Harrison	Baltimore, Md	Dec. 31, 1838	1, 056
Evaporator	B. R. Hawley	Normal, Ill	May 1, 1866	54, 340
Evaporator	R. Hawley, jr	Detroit, Mich	Jan. 24, 1871	111, 121
Evaporator	G. W. L. Hazen	Franklin, Ind	Apr. 17, 1866	53, 975
Evaporator	W. H. Isaacs	Terre Haute, Ind	Aug. 21, 1866	57, 327
Evaporator	D. Kinzer	Sardinia, Ohio	Sept. 11, 1866	57, 923
Evaporator	H. Lighty	Attica, Ind	June 4, 1867	65, 404
Evaporator	J. and S. W. Little	Patoka, Ind	Aug. 28, 1866	57, 523
Evaporator	F. A. Lord	Aurora, N. Y	May 30, 1865	47, 966
Evaporator	F. M. Love	Waldron, Ind	July 4, 1865	48, 628
Evaporator	N. R. Martin	Canandaigua, N. Y	Aug. 9, 1870	106, 270
Evaporator	S. B. Maulsby	Muncie, Ind	Mar. 20, 1866	53, 318

Index of patents issued from the United States Patent Office from 1790 *to* 1873, *inclusive*—Continued.

Invention.	Inventor.	Residence.	Date.	No.
Evaporator	S. B. Maulsby	Muncie, Ind	July 24, 1866	56, 585
Evaporator	J. McCracken	Bloomfield, N. J	Mar. 13, 1855	12, 516
Evaporator	J. A. McKinney	Griggsville, Ill	Jan. 8, 1867	61, 081
Evaporator	C. and D. Mercer	Strickersville, Pa	Mar. 13, 1866	53, 229
Evaporator	F. Michael	Gratis, Ohio	Feb. 19, 1867	62, 147
Evaporator	J. E. Moeller	Terre Haute, Ind	Nov. 28, 1865	51, 206
Evaporator	R. C. Nourse	Corydon, Ind	Sept. 11, 1866	57, 954
Evaporator	S. Page	McAllisterville, Pa	Oct. 9, 1866	58, 671
Evaporator	M. Pake	Dorchester, Ill	Sept. 15, 1863	39, 948
Evaporator	I. H. Palmer	Lodi, Wis	July 25, 1865	48, 990
Evaporator	W. H. Parmelee	Hopkins, Mich	Nov. 15, 1864	45, 072
Evaporator	D. Reeves	South Pass, Ill	Apr. 10, 1866	53, 873
Evaporator	D. Reynolds	Albany, N. Y	June 15, 1869	91, 485
Evaporator	J. F. Riggs	Saint Joseph, Mo	July 3, 1866	56, 097
Evaporator	T. and J. M. Scantlin	Evansville, Ind	July 25, 1865	48, 982
Evaporator	I. Sherman	Cleveland, Ohio	May 22, 1860	28, 415
Evaporator	A. W. Shidler	South Bend, Ind	Aug. 31, 1869	94, 249
Evaporator	H. Staude and A. Traver	Troy, N. Y	Oct. 8, 1872	131, 979
Evaporator	H. Stoller	Watertown, Ohio	May 11, 1869	89, 951
Evaporator	J. J. Stout	Greensburgh, Ind	Jan. 8, 1867	61, 112
Evaporator	J. Sutton	New York, N. Y	Jan. 3, 1860	26, 717
Evaporator	E. W. Taylor	Franklin, Ind	July 17, 1866	56, 464
Evaporator	T. Thomasson	Keoga, Ill	Apr. 24, 1866	54, 231
Evaporator	S. M. Williams	Pine Village, Ind	June 11, 1867	65, 714
Evaporator	W. S. Worthington	Newtown, N. Y	Jan. 11, 1859	22, 601
Evaporator	J. E. Youngman	Rockford, Ill	Sept. 25, 1866	58, 331
Evaporator	A. Youngs	Berlin, Mich	May 3, 1870	102, 643
Evaporator and condenser	E. Lynch	Brooklyn, N. Y	July 18, 1848	5, 668
Evaporator and cooler	H. Koehly, jr	Potosi, Mo	Mar. 13, 1866	53, 226
Evaporator and drier	W. Kendrick	New York, N. Y	Apr. 24, 1866	54, 173
Evaporator and steam-engine	D. Heath	Philadelphia, Pa	July 26, 1816	....
Evaporator and still	M. H. Powers	Pike County, Ind	June 14, 1870	104, 351
Evaporator, Compound	J. A. Wadsworth	Portsmouth, R. I	Nov. 23, 1835	....
Evaporator-furnace	W. Barrett	Malden, Mass	Nov. 19, 1833	....
Evaporator-furnace	M. P. Hays	Seaforth, Canada	Dec. 24, 1872	134, 277
Evaporator-furnace	T. Spencer	Syracuse, N. Y	July 25, 1848	5, 676
Evaporator furnace and fire-place	P. Tomlinson	Derby, Conn	Mar. 7, 1815	....
Evaporator furnace or fire-arch	S. Berry	Grand Rapids, Mich	Nov. 17, 1863	40, 599
Evaporators, Adjuster for	J. A. De Tar	Lanesfield, Kans	Mar. 27, 1866	53, 422
Evaporators, Automatic feeder for	J. Smead	East Wallingford, Vt	May 5, 1868	77, 666
Evaporators, Construction of	N. Bourne	Peosto, Iowa	Oct. 4, 1859	25, 623
Evaporators, Construction of	J. B. Dagne	Ashley, Ohio	Oct. 18, 1859	25, 813
Evaporators, Construction of	T. J. Price	Industry, Ill	May 15, 1860	28, 304
Evener	M. Terrill	Star Prairie, Wis	July 18, 1871	117, 123
Evener, Three-horse	B. W. Sutherlen	Fillmore, Minn	Feb. 11, 1873	135, 861
Evener, Two-horse	J. Sturgis	Minneapolis, Minn	Jan. 14, 1873	134, 946
Earth, Machine for excavating and conveying	J. A. Sprague	Dayton, Ohio	Oct. 15, 1850	7, 723
Excavating and ditching machine	M. M. Stewart and J. D. M. Russell.	Wayfield, Ky	Jan. 30, 1872	123, 303
Excavating and ditching machine	B. F. Stowell	Quincy, Ill	Mar. 10, 1863	37, 878
Excavating and dredging	W. S. Smith	Oak Park, Ill	Apr. 9, 1867	63, 661
Excavating and dredging machine	O. Allen	Norwich, Conn	May 8, 1840	1, 591
Excavating and dredging machine	J. R. Anderson	Chicago, Ill	Feb. 3, 1857	16, 519
Excavating and dredging machine	C. Willey, jr	Chicago, Ill	Feb. 10, 1852	8, 731
Excavating and embanking ditches, Machine for	G. Page	Keene, N. H	Oct. 26, 1838	993
Excavating and grading machine	P. T. Mayne	Keosaugua, Iowa	July 17, 1860	29, 183
Excavating and grading machine	W. P. Miller	Marysville, Cal	Aug. 2, 1859	24, 946
Excavating and grading machine	Z. N. Morrel	Cameron, Tex	July 19, 1859	24, 816
Excavating and moving earth, Machine for	A. W. Cady	Sullivan, N. Y	Apr. 29, 1856	14, 762
Excavating-apparatus	P. W. Clark	Oblong, N. Y	May 3, 1870	102, 492
Excavating canals, ditches, &c	W. Graham	New Orleans, La	Sept. 5, 1831	....
Excavating canals, Machine for	L. Forrester	Bridgefield, Conn	May 10, 1826	....
Excavating-cart	J. King	Oswego, N. Y	Nov. 21, 1871	121, 108
Excavating earth, Inclined elevating-box wheel for	J. Rowe	Triana, Ala	Feb. 15, 1838	602
Excavating from rivers	S. Russell	Buffalo, N. Y	Apr. 28, 1836	....
Excavating-machine	S. Achenbach	Orangeville, Pa	June 11, 1872	127, 671
Excavating-machine	J. A. Bailey	Detroit, Mich	July 13, 1869	92, 508
Excavating-machine	J. A. Bailey	Detroit, Mich	Aug. 2, 1870	105, 883
Excavating-machine	C. Bishop	Norwalk, Ohio	Mar. 30, 1852	8, 836
Excavating-machine	D. Brown	Southampton, Ill	Apr. 30, 1872	126, 178
Excavating-machine	T. Claton	Shelbyville, Ind	Dec. 1, 1837	500
Excavating-machine	R. W. Davis	Sonora, N. Y	Aug. 15, 1871	117, 993
Excavating-machine	W. H. Elliot	New York, N. Y	Nov. 23, 1869	97, 068
Excavating-machine	P. W. Hamel	San Francisco, Cal	Feb. 23, 1869	87, 168
Excavating-machine	W. Hamilton	South Paris, Me	May 12, 1863	38, 497
Excavating-machine	W. M. Hansen	Stockton, Cal	May 31, 1870	103, 740
Excavating-machine	J. M. Hughes and A. J. Mapes	Independence, Mo	Feb. 21, 1871	111, 940
Excavating-machine	W. S. Lane	Cairo, Ill	July 30, 1872	130, 059
Excavating-machine	W. R. Maffet	Wilkesbarre, Pa	Aug. 17, 1858	21, 206
Excavating-machine	G. S. Manning	Springfield, Ill	July 17, 1860	29, 182
Excavating-machine	T. R. Markillie	Winchester, Ill	Dec. 27, 1859	26, 601
Excavating-machine	M. Newman, 2d	Lanesborough, Pa	Nov. 19, 1850	7, 787
Excavating-machine	B. W. Remy	Brookville, Ind	Oct. 21, 1851	8, 453
Excavating-machine	J. Robertson	Glasgow, Scotland	July 21, 1868	80, 225
Excavating-machine	N. Saunders and F. T. Sherman.	Chicago, Ill	June 8, 1858	20, 511
Excavating-machine	J. J. Savage	New York, N. Y	Jan. 8, 1856	14, 068
Excavating-machine	C. Schott and J. C. Baldwin	Nashville, Tenn	Apr. 19, 1859	23, 713
Excavating-machine	G. D. Stillson	Rochester, N. Y	Nov. 14, 1854	11, 949
Excavating-machine	G. D. Stillson	Rochester, N. Y	Apr. 19, 1859	23, 723
Excavating-machine	A. Taggart	Warrenton, Mo	June 16, 1857	17, 595
Excavating-machine	M. B. True	Newburyport, Mass	Sept. 18, 1860	30, 098
Excavating-machine	W. F. Wickersham	Springfield, Ill	June 25, 1861	32, 660
Excavating-machine	C. Williams	Jackson, Tenn	Apr. 3, 1855	12, 651
Excavating-machine	E. Williams	Covington, Ky	July 17, 1855	13, 282
Excavating-machine	E. H. Williams	Clermont, Iowa	June 13, 1871	115, 998
Excavating-machine	E. H. and D. R. W. Williams	Grand Meadow, Iowa, and Werner, Wis.	Oct. 1, 1867	69, 378

Index of patents issued from the United States Patent Office from 1790 *to* 1873, *inclusive*—Continued.

Invention.	Inventor.	Residence.	Date.	No.
Excavating-machine	C. Winegar	Union Springs, N. Y	Aug. 28, 1866	57, 614
Excavating machine, Coal	J. Borton	Antrim, Ohio	Aug. 14, 1868	76, 704
Excavating machine, Coal	H. Wilverth	Caseyville, Ky	Jan. 3, 1860	26, 726
Excavating machine, Ditch	H. Aiken	Franklin, N. H	Feb. 1, 1842	2, 440
Excavating-machine for ditches, canals, &c	J. Hanchett	Coldwater, Mich	Feb. 28, 1840	1, 502
Evcavating-machine for ditching, &c	G. W. Cherry	Washington, D. C	Jan. 16, 1839	1, 069
Excavating-machine, Mode of operating	J. A. H. Ellis and A. Gordon	Rochester, N. Y	July 18, 1854	11, 323
Excavating-machine, Submarine	J. C. Osgood	Troy, N. Y	May 22, 1855	12, 919
Excavating mud, Borer for	M. H. Ford	New York, N. Y	Oct. 4, 1859	25, 636
Excavating, plowing, and grading machines	E. H. Williams and D. R. W. Williams.	Clermont, Iowa, and Werner, Wis.	Apr. 15, 1862	35, 005
Excavating rock, Machinery for	J. C. Osgood	Troy, N. Y	Feb. 17, 1857	16, 652
Excavating-scoop	O. P. Kniffin	Clinton, Conn	Nov. 8, 1870	109, 132
Excavating-shovel	T. Pierce	Hartwick, N. Y	Mar. 3, 1825	....
Excavating, Softening frozen ground for	A. Derrom	Paterson, N. J	Feb. 6, 1872	123, 384
Excavating tunnels and cuttings, Sinking shafts and.	J. Shelley and M. C. Bullock	Mahanoy and Pottsville, Pa.	May 23, 1871	115, 115
Excavating under water, Method of	A. Duvall	Baltimore, Md	Mar. 3, 1868	75, 003
Excavating under water, Method of	A. Duvall	Baltimore, Md	Mar. 3, 1868	75, 004
Excavating-vehicle	J. P. Smith	Cherry Hill, Pa	Apr. 28, 1868	77, 329
Excavation, Curbing for	H. Whitestone	Louisville, Ky	Jan. 10, 1871	110, 882
Excavator	S. B. Alger	Oswego, N. Y	Dec. 24, 1872	134, 187
Excavator	A. J. Bartlett	Romulus, N. Y	Nov. 13, 1860	30, 610
Excavator	J. Bartoo	East Aurora, N. Y	June 4, 1861	42, 494
Excavator	H. H. Beard	Friar's Point, Miss	Aug. 17, 1869	93, 796
Excavator	J. Bourbin	San Francisco, Cal	Oct. 14, 1856	15, 875
Excavator	J. Bradley	Cedar Falls, Iowa	Aug. 7, 1866	57, 038
Excavator	H. Brown	Lowell, Mass	Aug. 25, 1868	81, 334
Excavator	Z. Butt	Lincolnton, N. C	Sept. 15, 1857	18, 185
Excavator	W. J. Carroll	Natchez, Miss	June 6, 1871	115, 700
Excavator	O. S. Chapman	Canton, Mass	Apr. 16, 1867	63, 857
Excavator	D. Close	Harmonsburgh, Pa	July 5, 1864	43, 392
Excavator	D. Comstock	Lockport, N. Y	Mar. 15, 1825	....
Excavator	J. Cowden and D. Brown	La Prairie Centre and Akron, Ill.	Oct. 12, 1869	95, 658
Excavator	J. M. Curless	Cedar Rapids, Iowa	May 6, 1873	138, 485
Excavator	S. S. Curtis	Croton Corners, N. Y	Dec. 14, 1858	22, 279
Excavator	J. P. T. Davis	New Trenton, Ind	Sept. 12, 1871	118, 793
Excavator	J. M. Dunn	Erin, Miss	Sept. 26, 1871	119, 334
Excavator	H. W. Farley	Hannibal, Mo	July 11, 1865	48, 669
Excavator	D. Gilmore and W. W. Forrest.	Peotone, Ill	Jan. 3, 1871	110, 759
Excavator	J. T. Ham	Liberty, Ind	Feb. 4, 1873	135, 475
Excavator	W. Hamilton	Saint Catherine, Mo	Sept. 13, 1859	25, 408
Excavator	B. Hancock	Troy, N. Y	Oct. 16, 1855	13, 680
Excavator	J. P. Hayes	Hennepin, Ill	Dec. 4, 1860	30, 819
Excavator	J. Hodges	Bagshot, England	May 23, 1865	47, 908
Excavator	M. M. Hodgman	Weymouth, Mass	Apr. 18, 1871	113, 883
Excavator	M. M. Hodgman	Weymouth, Mass	Oct. 8, 1872	131, 954
Excavator	T. Hutchinson	Green Point, N. Y	Aug. 6, 1861	32, 997
Excavator	D. Judd	Hinsdale, N. Y	July 10, 1866	56, 229
Excavator	D. Judd	Hinsdale, N. Y	Nov. 18, 1873	144, 679
Excavator	A. Kirby	Paris, Ill	Apr. 30, 1861	32, 194
Excavator	J. Lee	Galesburgh, Ill	Feb. 9, 1869	86, 678
Excavator	J. R. Lewis	Piper City, Ill	Dec. 14, 1869	97, 783
Excavator	A. P. Lomax	Warsaw, Ill	Mar. 18, 1873	137, 011
Excavator	C. A. Mann, jr	Pike, N. Y	May 27, 1856	14, 966
Excavator	D. McNabb	Moscow, Mich	Dec. 8, 1863	40, 850
Excavator	E. B. Meatyard	Chicago, Ill	May 14, 1872	126, 729
Excavator	S. Miller and E. Claringdon	Waterloo, N. Y	May 23, 1871	115, 084
Excavator	J. Milroy	Edinburgh, Great Britain	Apr. 23, 1867	64, 023
Excavator	J. Molyneux	Bordentown, N. J	Jan. 1, 1867	60, 774
Excavator	S. G. L. Morron	Linn, Mo	Sept. 30, 1856	15, 813
Excavator	B. A. Oliver	Bunker Hill, Ill	Mar. 18, 1873	136, 936
Excavator	J. C. Osgood	Troy, N. Y	July 4, 1865	48, 581
Excavator	J. C. Osgood	Troy, N. Y	Apr. 4, 1871	113, 691
Excavator	E. Phelps	Hendersonville, Ill	June 6, 1854	11, 028
Excavator	A. E. Pierce	Gilroy, Cal	Nov. 9, 1869	96, 724
Excavator	W. Provines	Columbia, Mo	Jan. 13, 1857	16, 397
Excavator	W. Randall	Uxbridge, Canada	Apr. 2, 1861	31, 905
Excavator	C. H. Sage and S. B. Alger	Norwich, N. Y	Nov. 29, 1870	109, 764
Excavator	F. W. Schultz	Mount Pleasant, Iowa	Jan. 21, 1873	135, 162
Excavator	B. Slusser	Sidney, Ohio	Dec. 4, 1866	60, 268
Excavator	B. Slusser	Sidney, Ohio	Dec. 10, 1867	72, 098
Excavator	B. Slusser	Sidney, Ohio	June 28, 1870	104, 782
Excavator	A. B. Smith	Denison, Tex	May 6, 1873	138, 539
Excavator	J. D. Smith	Panton, Vt	Jan. 12, 1858	19, 104
Excavator	S. W. Soule	Oswego, N. Y	July 10, 1855	13, 234
Excavator	S. W. Soule	Saint Louis, Mo	Apr. 28, 1857	17, 174
Excavator	B. T. Stowell	Quincy, Ill	Dec. 17, 1867	72, 336
Excavator	B. T. Stowell	Quincy, Ill	Nov. 17, 1868	84, 230
Excavator	B. T. Stowell	Quincy, Ill	Mar. 28, 1871	113, 221
Excavator	J. W. Swickard	Galva, Ill	July 6, 1869	92, 224
Excavator	T. Symonds	Portland, Me	Feb. 14, 1871	111, 787
Excavator	T. Symonds	Portland, Me	May 16, 1871	114, 878
Excavator	I. D. Vandecar	Chicago, Ill	Apr. 23, 1867	64, 168
Excavator	B. R. Wehner	Mankato, Minn	Oct. 19, 1869	95, 961
Excavator	J. F. Willey	Fredonia, N. Y	Nov. 11, 1856	16, 031
Excavator	G. W. Williams	Corning, N. Y	Jan. 2, 1872	122, 344
Excavator	C. F. Woodruff	Newbern, Tenn	Aug. 4, 1868	80, 696
Excavator and ditching-plow	C. A. Robbins	Iowa City, Iowa	Nov. 14, 1854	11, 948
Excavator and potato-digger	B. O. Warren	Elkhart, Ind	Feb. 12, 1867	61, 963
Excavator and self-loading cart	M. Alden and W. T. Boyd	Northumberland, Pa	July 20, 1831	....
Excavator, Canal	I. L. Skinner	Hartford, Conn	May 3, 1820	....
Excavator-carrier	J. Williams	Hemmingford, Canada	May 27, 1873	139, 446
Excavator, Channel	R. Gamble, jr	Tallahassee, Fla	Nov. 13, 1860	30, 626
Excavator, Ditch	G. W. Cherry	Washington, D. C	Mar. 26, 1841	2, 017
Excavator, Earth	J. P. Barker	Wayne, Ohio	Feb. 8, 1859	22, 847
Excavator, Earth	C. Colby	Wilson, N. Y	June 30, 1857	17, 669

Index of patents issued from the United States Patent Office from 1790 *to* 1873, *inclusive*—Continued.

Index of patents issued from the United States Patent Office from 1790 *to* 1873, *inclusive*—Continued.

Invention.	Inventor.	Residence.	Date.	No.
Exhaust of caloric engines, Utilizing the	P. Shaw	Boston, Mass	Oct. 15, 1861	33, 497
Exhaust of locomotive-engines, Variable	W. S. G. Baker	Chicago, Ill	May 17, 1859	23, 999
Exhaust of locomotive-engines, Variable	S. L. Hay	Reading, Mass	Oct. 9, 1855	13, 648
Exhaust of locomotives, Means for directing the	R. Hale	Roxbury, Mass	May 5, 1857	17, 215
Exhaust of steam-engines, Regulating	J. Thierry	Detroit, Mich	June 26, 1860	28, 918
Exhaust-pipe for steam-engines	R. Hale	Roxbury, Mass	Sept. 13, 1859	25, 407
Exhaust-pipe, Variable	W. P. Parrott and S. H. Head	Boston, Mass	Feb. 1, 1859	22, 819
Exhaust-pipe, Variable	J. Patrick	Chicago, Ill	Feb. 1, 1859	22, 820
Exhaust-regulator for locomotives, Rotary	E. R. Addison	Baltimore, Md	Oct. 13, 1857	18, 373
Exhaust-regulator, Steam	J. Shackleton	Rahway, N. J	Dec. 1, 1868	84, 586
Exhaust-valve for steam and air cylinders	G. Westinghouse, jr	Pittsburgh, Pa	Jan. 9, 1872	122, 544
Exhauster, Barometric vacuum	J. Gordon and J. Archbald	San Francisco, Cal	Sept. 17, 1867	68, 979
Exhausting ventilator and chimney-cowl	J. W. C. Anderson	New York, N. Y	Jan. 14, 1868	73, 280
Exhibiting case, Spool	J. D. Cutter	Brooklyn, N. Y	Feb. 11, 1873	135, 782
Exhibiting dry goods, Device for	J. J. Bisel	Williamsport, Pa	July 16, 1872	128, 943
Exhibiting hymns, &c., Apparatus for	H. V. Edmond	Norwich, Conn	July 30, 1867	67, 177
Exhibiting hymns, Apparatus for	H. V. Edmond	Norwich, Conn	Apr. 13, 1869	88, 947
Expanding and folding wheel	H. Steward	Clark County, Ind	Jan. 24, 1811	......
Expanding bit	C. L. Barnes	New York, N. Y	Nov. 16, 1852	9, 398
Expanding bit	H. Stone	Blackstone, Mass	Dec. 28, 1858	22, 474
Expanding die	B. T. Loomis	New York, N. Y	Aug. 20, 1867	67, 993
Expanding drill	F. Gleason	Philadelphia, Pa	July 4, 1865	48, 548
Expanding drill	T. Prosser	New York, N. Y	Jan. 25, 1853	9, 562
Expanding drill	J. L. Sayles	Gloucester, R. I	Dec. 29, 1868	85, 334
Expanding tool	J. Greenhalgh, jr	Burville, R. I	Feb. 23, 1858	19, 416
Expansible bit	C. L. Adancourt	Troy, N. Y	Aug. 27, 1850	7, 589
Expansion-bolt	D. L. Bartlett	Baltimore, Md	Apr. 1, 1873	137, 338
Expansion-drill	G. Mackinnon	Millbury, Mass	Oct. 15, 1867	69, 825
Expansion-engine	C. G. Wilson	Brooklyn, N. Y	June 21, 1870	104, 526
Expansive bit	W. A. Clark	Bethany, Conn	Sept. 28, 1858	21, 597
Expansive bit	W. A. Clark	New Haven, Conn	July 29, 1873	141, 324
Expansive bit	E. Ford	Westville, Conn	Oct. 8, 1872	131, 946
Expansive bit	A. Hall	New York, N. Y	Apr. 14, 1857	17, 038
Expansive bit	A. Weeks	South Boston, Mass	May 2, 1854	10, 872
Expansive bit for wood-boring	W. A. Ives	New Haven, Conn	Oct. 13, 1868	82, 956
Expansive bits, Method of seating the movable cutter in	W. A. Clark	Bethany, Conn	May 11, 1858	20, 192
Exploding by electricity, Apparatus for	J. B. Dowse	Lockport, Ill	Aug. 27, 1867	68, 055
Explosive agent, xyloglodine	C. Dittmar	Charlottenburg, Prussia	Jan. 25, 1870	99, 069
Explosive agents, Manufacture of xyloglodine and other	C. Dittmar	Charlottenburg, Prussia	Jan. 25, 1870	99, 070
Explosive composition	E. Harrison	New York, N. Y	Feb. 9, 1864	41, 578
Explosive compound	O. H. Bandisch	Berlin, Prussia	May 11, 1869	89, 910
Explosive compound	T. S. Beach	New York, N. Y	May 13, 1873	138, 841
Explosive compound	C. Dittmar	Charlottenburg, Prussia	Jan. 18, 1870	98, 854
Explosive compound	C. Dittmar	Boston, Mass	Dec. 9, 1873	145, 403
Explosive compound	E. Gomez	New York, N. Y	Feb. 16, 1869	86, 980
Explosive compound	E. Gomez	New York, N. Y	Dec. 26, 1871	122, 245
Explosive compound	J. Hafenegger	San Francisco, Cal	Sept. 8, 1868	81, 894
Explosive compound	J. Hafenegger	San Francisco, Cal	Dec. 20, 1870	110, 355
Explosive compound	J. Hafenegger	San Francisco, Cal	Feb. 7, 1871	111, 642
Explosive compound	H. Halvorson	Cambridge, Mass	Jan. 7, 1864	43, 621
Explosive compound	W. Mills	New York, N. Y	Oct. 26, 1869	96, 248
Explosive compound	W. Mills	New York, N. Y	Feb. 28, 1871	112, 163
Explosive compound	A. Nobel	New York, N. Y	Aug. 14, 1866	57, 175
Explosive compound	A. Nobel	Hamburg, Germany	May 26, 1868	78, 317
Explosive compound	A. Nobel	Hamburg, Germany	Aug. 5, 1873	141, 455
Explosive compound	J. H. Norrbin and J. Ohlsson	Stockholm, Sweden	Aug. 5, 1873	141, 585
Explosive compound	E. A. L. Roberts	Titusville, Pa	Nov. 7, 1871	120, 776
Explosive compound	T. P. Shaffner	Louisville, Ky	Aug. 17, 1869	93, 753
Explosive compound	T. P. Shaffner	Louisville, Ky	Aug. 17, 1869	93, 754
Explosive compound	T. P. Shaffner	Louisville, Ky	Dec. 28, 1869	98, 427
Explosive compound	T. P. Shaffner	Louisville, Ky	June 10, 1873	139, 738
Explosive compound	T. Varney	San Francisco, Cal	June 10, 1873	139, 746
Explosive compound	C. W. Volney	Boston, Mass	Mar. 5, 1872	124, 397
Explosive compound for use in fire-arm, blasting, &c.	T. P. Shaffner	Louisville, Ky	Aug. 17, 1869	93, 752
Explosive compound from gun-cotton	R. Punshon	Newcastle-upon Tyne, England.	Mar. 12, 1872	124, 510
Explosive compound or giant-powder	E. Judson	San Francisco, Cal	June 3, 1873	139, 468
Explosive engine	J. M. Welbourn	Caledonia, Ohio	Sept. 16, 1873	142, 880
Explosive powder	G. Designoble and J. Casthelaz	France	Mar. 31, 1868	76, 173
Explosive powder	F. M. Ruschhaupt	New York, N. Y	June 16, 1868	79, 010
Explosive substances and process of manufacturing	J. B. Muschamp	Kensington, Great Britain	June 25, 1872	128, 450
Express-bag, &c., Lock for	E. A. Locke and W. B. Mason	Boston, Mass	Feb. 2, 1869	86, 560
Express-signal	C. H. Seawell	Saint Louis, Mo	Oct. 6, 1868	82, 756
Extender	A. Bisbee	North Yarmouth, Me	Jan. 27, 1824	......
Extension-bit	H. P. Chapman	Essex, Conn	Nov. 23, 1869	97, 166
Extension-bit	J. P. Rollins	Boston, Mass	Dec. 25, 1855	13, 998
Extension-bracket	W. S. Elliott	Goshen, N. Y	June 18, 1872	127, 967
Extension-chair	S. J. Anderson and N. Richardson.	Erieville, N. Y	Feb. 17, 1857	16, 635
Extension-chair	N. C. O. and A. Collignon	Closter, N. J	June 27, 1871	116, 271
Extension-chair	F. W. Lange	Chicago, Ill	Sept. 7, 1869	94, 615
Extension-chair	L. F. Schwenkel	New York, N. Y	Nov. 7, 1871	120, 780
Extension-chair	J. H. Travis	Charlestown, Mass	July 23, 1872	129, 873
Extension-chair	B. H. Zinn	New York, N. Y	Apr. 7, 1868	76, 577
Extension-elevator	P. Porter	Hooksett, N. H	Oct. 6, 1857	18, 353
Extension-gate	C. S. Snead	Louisville, Ky	July 23, 1867	67, 143
Extension-platform	A. Morse	Portland, Me	May 7, 1861	32, 265
Extension-slide	J. Dourson	Columbus, Ohio	Mar. 26, 1872	125, 034
Extension-table	J. J. Arnaud	Boston, Mass	May 28, 1872	127, 291
Extension-table	H. Bachman	Philadelphia, Pa	Mar. 30, 1869	88, 433
Extension-table	A. Bader	New York, N. Y	Dec. 7, 1858	22, 224
Extension-table	M. Bancroft	Montague, Mass	Sept. 13, 1870	107, 215
Extension-table	W. Beale	Keyport, N. J	Apr. 10, 1866	53, 774
Extension-table	J. M. Blaisdell	Sanbornton, N. H	Apr. 20, 1869	89, 266
Extension-table	J. M. Blaisdell	Sanbornton, N. H	Jan. 16, 1872	122, 800

Index of patents issued from the United States Patent Office from 1790 *to* 1873, *inclusive*—Continued

Invention.	Inventor.	Residence.	Date.	No.
Extension-table	W. J. Boda	Dayton, Ohio	Jan. 4, 1870	98, 465
Extension-table	C. Briggs	Roxbury, Mass	Sept. 9, 1845	4, 185
Extension-table	C. Briggs	Roxbury, Mass	Sept. 1, 1843	3, 249
Extension-table	N. Carl	Cincinnati, Ohio	Jan. 8, 1867	60, 996
Extension-table	G. Chelini	Washington, D. C	Mar. 23, 1869	88, 132
Extension-table	A. F. Chubbuck	Casnovia, Mich	Dec. 2, 1873	145, 049
Extension-table	B. Clark	Oriskany Falls, N. Y	Jan. 6, 1857	16, 350
Extension-table	C. B. Clark	Mount Pleasant, Iowa	Dec. 1, 1857	18, 733
Extension-table	M. Coleman	Ionia, Mich	Mar. 11, 1873	136, 705
Extension-table	E. A. Curley	Westport, Conn	Jan. 15, 1856	14, 093
Extension-table	E. A. Curley	Westport, Conn	Dec. 22, 1857	18, 891
Extension-table	J. P. Curry	Vincennes, Ind	Sept. 21, 1869	95, 003
Extension-table	J. B. Curtis	Port Henry, N. Y	Nov. 6, 1866	59, 367
Extension-table	A. Dietsch	Frankfort Station, Ill	Apr. 26, 1870	102, 235
Extension-table	J. Dourson	Columbus, Ohio	June 8, 1869	91, 099
Extension-table	J. Dourson	Columbus, Ohio	Oct. 26, 1869	96, 209
Extension-table	T. E. Ellison	Albany, N. Y	Dec. 20, 1870	110, 220
Extension-table	W. Farson	Philadelphia, Pa	Jan. 18, 1870	98, 942
Extension-table	G. F. Folsom	Roxbury, Mass	May 21, 1867	64, 857
Extension-table	T. Franck	New York, N. Y	Apr. 10, 1849	6, 302
Extension-table	L. Freeman	New York, N. Y	Mar. 23, 1869	88, 028
Extension-table	T. Gray	Philadelphia, Pa	Mar. 8, 1859	23, 219
Extension-table	H. Gross	Tiffin, Ohio	Nov. 17, 1857	18, 636
Extension-table	J. Haines	West Middlebury, Ohio	Mar. 20, 1855	12, 5[illegible]8
Extension-table	T. Q. Hall	Fairfield, Iowa	Feb. 5, 1861	31, 312
Extension-table	F. Hamblin	Madrid Springs, N. Y	June 15, 1869	91, 332
Extension-table	E. Hambujer	Detroit, Mich	Apr. 15, 1873	137, 917
Extension-table	A. Hanson	San Francisco, Cal	Feb. 6, 1872	123, 397
Extension-table	W. Heerdt	New York, N. Y	June 8, 1858	20, 489
Extension-table	G. H. Henkel	Germantown, Ohio	Jan. 23, 1872	122, 946
Extension-table	A. Herzog	New York, N. Y	Feb. 25, 1873	136, 243
Extension-table	M. Hofmann	Darlington, Wis	Jan. 28, 1873	135, 274
Extension-table	F. Hoguet	Philadelphia, Pa	Feb. 25, 1851	7, 954
Extension-table	F. Huber	Franklin, Ind	May 28, 1872	127, 168
Extension-table	A. Iske	Lancaster, Pa	Feb. 7, 1865	46, 240
Extension-table	A. Iske	Lancaster, Pa	Apr. 5, 1859	23, 469
Extension-table	G. Lauter and J. Kautz	Vincennes, Ind	Sept. 11, 1866	57, 927
Extension-table	C. P. Lenz	Poughkeepsie, N. Y	Sept. 7, 1869	94, 617
Extension-table	C. P. Lenz	Poughkeepsie, N. Y	Mar. 15, 1870	100, 776
Extension-table	G. S. Manning	Danville, Ill	Jan. 12, 1869	85, 836
Extension-table	S. H. Martin	Mount Vernon, N. Y	May 6, 1873	138, 571
Extension-table	J. J. McKnight	Tarrytown, N. Y	Aug. 1, 1871	117, 659
Extension-table	F. Menzer	Flint, Mich	May 2, 1871	114, 319
Extension-table	E. Metz	Rochester, N. Y	Sept. 16, 1862	36, 496
Extension-table	L. Meyer	Columbus, Ga	Nov. 15, 1859	26, 116
Extension-table	C. Monson	New Haven, Conn	Jan. 14, 1862	34, 160
Extension-table	L. Munzer	Goodrich, Mich	Nov. 7, 1871	120, 700
Extension-table	G. Munzinger	Brooklyn, N. Y	Dec. 4, 1860	30, 823
Extension-table	F. R. Osgood	Roxbury, Mass	Jan. 2, 1866	51, 855
Extension-table	J. Pleukharp	Columbus, Ohio	May 14, 1872	126, 831
Extension-table	E. Pond, jr	Woonsocket, R. I	Mar. 12, 1872	124, 447
Extension-table	G. Pratt	Boston, Mass	Nov. 7, 1848	5, 905
Extension-table	G. Pratt	Boston, Mass	June 8, 1858	20, 530
Extension-table	A. E. Preston	Battle Creek, Mich	Oct. 15, 1867	69, 937
Extension-table	A. Reed	Chicago, Ill	Mar. 25, 1873	137, 148
Extension-table	W. Reichenbach and F. Roschdiantzky.	Chicago, Ill	Mar. 1, 1870	100, 446
Extension-table	E. Richards	New York, N. Y	Apr. 15, 1843	3, 050
Extension-table	C. Rieger	Morrisania, N. Y	Aug. 19, 1873	141, 954
Extension-table	S. M. Rounds	Somerset, Mass	July 23, 1861	32, 894
Extension-table	J. M. Sackman	Ridgeville, Ind	Sept. 9, 1873	142, 586
Extension-table	T. P. Sherborne	Philadelphia, Pa	June 26, 1849	6, 557
Extension-table	J. Sherman	Burlington, N. J	Aug. 31, 1869	94, 447
Extension-table	E. F. Shoenberger	Pittsburgh, Pa	Nov. 19, 1850	7, 789
Extension-table	L. Showalter	Beetown, Wis	June 25, 1872	128, 330
Extension-table	J. Simensen	Hermon, N. Y	June 29, 1869	91, 974
Extension-table	F. Souweine	New York, N. Y	May 9, 1871	114, 722
Extension-table	B. D. Sutlief	Wallingford, Conn	Jan. 19, 1869	86, 113
Extension-table	L. Thorn	Philadelphia, Pa	Mar. 25, 1851	7, 997
Extension-table	S. Tilton	Alton, N. H	Apr. 5, 1870	101, 545
Extension-table	W. Valentin	New York, N. Y	July 1, 1873	140, 392
Extension-table	S. E. Wales	Lebanon, N. H	Feb. 6, 1872	123, 372
Extension-table	S. E. Wales	Oxford, N. H	Apr. 1, 1873	137, 516
Extension-table	F. R. Wolfinger	Chicago, Ill	Nov. 5, 1867	70, 668
Extension-table	F. R. Wolfinger and J. Barrett	Chicago, Ill	Oct. 20, 1868	83, 350
Extension-table and settee	M. Debzell	Wabash, Ind	Sept. 20, 1870	107, 463
Extension-table, Convertible	M. Quigley	Watertown, Wis	Apr. 27, 1858	20, 088
Extension-table, Fall-leaf	J. Dourson	Columbus, Ohio	Mar. 16, 1869	87, 829
Extension-table rail	J. Lotz	Brooklyn, N. Y	May 21, 1872	126, 972
Extension-table, Revolving	F. Menzer	Flint, Mich	Apr. 16, 1872	125, 827
Extension-table roller slide	J. Dourson	Columbus, Ohio	Dec. 28, 1869	98, 360
Extension-table slide	E. P. Allyn	North Canaan, Conn	Apr. 25, 1865	47, 374
Extension-table slide	J. F. Birchard	Milwaukee, Wis	June 19, 1866	55, 606
Extension-table slide	M. E. Carter and E. Mets	Rochester, N. Y	Sept. 6, 1864	44, 073
Extension-table slide	W. Donoghue	Philadelphia, Pa	Aug. 16, 1870	106, 340
Extension-table slide	G. Finkbeiner	Syracuse, N. Y	Feb. 4, 1873	135, 418
Extension-table slide	M. Fleck	Milwaukee, Wis	Jan. 25, 1870	99, 075
Extension-table slide	S. R. Garner	Independence, Iowa	Mar. 26, 1872	124, 947
Extension-table slide	S. J. Genung	Waterloo, N. Y	Sept. 7, 1869	94, 486
Extension-table slide	N. Jenkins	New York, N. Y	Mar. 3, 1868	75, 022
Extension-table slide	J. King	Oswego, N. Y	Nov. 4, 1873	144, 276
Extension-table slide	P. Marx	Buffalo, N. Y	Aug. 20, 1872	130, 731
Extension-table slide	E. Metz	Rochester, N. Y	Nov. 12, 1867	70, 879
Extension-table slide	S. B. Nash	Buffalo, N. Y	May 23, 1871	115, 087
Extension-table slide	H. Olds	Syracuse, N. Y	Sept. 10, 1867	68, 782
Extension-table slide	H. Olds	Syracuse, N. Y	June 14, 1870	104, 340
Extension-table slide	J. Pleukharp	Columbus, Ohio	Dec. 26, 1871	122, 278
Extension-table slide	O. E. Sanford	La Porte, Ind	Aug. 31, 1869	94, 243

Index of patents issued from the United States Patent Office from 1790 *to* 1873, *inclusive*—Continued.

Invention.	Inventor.	Residence.	Date.	No.
Extension-table slide	A. H. Shipman	Arcadia, N. Y	Nov. 2, 1869	96, 533
Extension-table slide	A. P Shute and J. F. Jackson	Charlestown, Mass	Oct. 22, 1867	70, 127
Extension-table slide	S. Stilwell	Waterloo, N. Y	Sept. 20, 1870	107, 636
Extension-table slide	J. W. Tefft	Buffalo, N. Y	June 22, 1869	91, 579
Extension-table slide	G. W. Townsend	Hubbardston, Mich	Jan. 7, 1873	134, 711
Extension-table slide	A. Zeh, G. P. Livingston, and P. V. W. Cullings.	Gallupsville, N. Y	Oct. 7 1873	143, 554
Extension-table slides, Tool for making	S. J. Moore and G. A. Buckman.	Ogdensburgh, N. Y	May 10, 1870	102, 850
Extension-table tops, Machine for making	R. G. Utley	Boston, Mass	Feb. 25, 1873	136, 193
Extension-tables, Making slides of	C. F. Hobe	New York, N. Y	June 22, 1841	2, 135
Extract of barley-malt	M. Wesselhoeft	Baltimore, Md	Oct. 6, 1868	82, 909
Extract of fruit	R. Carpenter	Medford, Mass	Sept. 13, 1859	25, 384
Extract of sea-clams	B. G. Noble	New York, N. Y	July 9, 1867	66, 616
Extracts and decoctions from coffee, tea, &c., Apparatus for making.	L. Brauer	Washington, D. C	Mar. 15, 1870	100, 849
Extracts and essences, Apparatus for making	E. E. Burrough	Baltimore, Md	Nov. 10, 1868	83, 922
Extracts, Apparatus for concentrating	T. W. Johnson	New York, N. Y	Sept. 1, 1868	81, 643
Extracts, &c., Apparatus for evaporating and concentrating bark.	T. W. Johnson and A. W. Goodell.	New York, N. Y	Nov. 1, 1870	108, 793
Extracts, Apparatus for leaching and concentrating bark.	P. M. Church	Sault de Ste. Marie, Mich.	July 1, 1873	140, 469
Extracts, Apparatus for making	W. Class and E. Reubenow	Cincinnati, Ohio	Mar 22, 1864	41, 974
Extracts, Apparatus for making	J. Miller	Upton, Canada East	Jan. 23, 1866	52, 253
Extracts, Apparatus for making	L. Smith	Erie, Pa	June 6, 1865	48, 107
Extracts, Apparatus for making	A. Steers	Medina, N. Y	Mar. 11, 1856	14, 418
Extracts, Apparatus for making	A. Steers	New York, N. Y	Mar. 12, 1867	62, 901
Extracts from animal and vegetable substances, Apparatus for making.	J. Chilcott	Brooklyn, N. Y	Dec. 27, 1864	45, 584
Extracts from bark and other materials, Apparatus for making.	J. W. Jones	Cumberland, Md	Nov. 5, 1867	70, 439
Extracts from bark, &c., Apparatus for making	F. W. Perry and J. H. Pierce	Woburn, Mass	May 22, 1866	54, 945
Extracts from bark, Obtaining	W. G. Fessenden	Walpole, N. H	Oct. 20, 1812	....
Extracts from dye-woods, Apparatus for obtaining	I. B. Kirby	Providence, R. I	Jan. 2, 1872	122, 467
Extracts from sumac, Manufacture of	T. Steers, jr., and J. R. Sedgwick.	Richmond, Va	Feb. 13, 1872	123, 590
Extracts from tan-bark, Apparatus for making	S. W. Pingree	Lawrence, Mass	Oct. 24, 1865	50, 626
Extracts from vegetable and animal substances, Apparatus for making.	L. Brauer	Philadelphia, Pa	Apr. 18, 1871	113, 844
Extracts from vegetables, &c., Apparatus for obtaining.	J. Chilcott	Brooklyn, N. Y	Apr. 25, 1865	47, 393
Extracts from vegetables, Liquid	G. De Villepoix and J. F. Bonneterre.	Abbeville, Somme, and Paris, France.	May 12, 1868	77, 805
Extracts, Making	T. Close and J. C. Sandford	Rye, N. Y	Mar. 18, 1836	....
Extracts, Making vegetable	B. Brandeth	New York, N. Y	Jan. 20, 1843	2, 920
Extracts, Making vegetable	T. J. Covell	Bergen, N. J	Feb. 13, 1872	123, 679
Extracts, Manufacture of fluid	L. D. Puy	Grand Rapids, Mich	Feb. 6, 1872	123, 336
Extracts, Mode of obtaining	B. G. Martin	Philadelphia, Pa	Mar. 14, 1865	46, 809
Extracts of bark, &c., Apparatus for concentrating.	T. W. Johnson	New York, N. Y	July 13, 1869	92, 455
Extracts of bark, &c., Apparatus for evaporating	G. W. Klein	Boston, Mass	Dec. 15, 1838	1, 035
Extracts of bark, Making	J. Biddis and T. Bedwell		Aug. 10, 1791	....
Extracts of bark, &c., Process and apparatus for manufacture of.	J. Pickles	Wigan, England	Apr. 13, 1869	88, 807
Extracts, Preparation of desiccated vegetable	W. J. Rand	Brooklyn, N. Y	July 11, 1865	48, 719
Extracts, Process of making	J. Robert	Seelowitz, Austria	Oct. 30, 1866	59, 330
Extracts, Process of making	N. S. Thomas	Painted Post, N. Y	Aug. 14, 1866	57, 217
Extracts, Process of making concentrated fluid	N. S. Thomas	Painted Post, N. Y	Jan. 31, 1865	46, 156
Extracts under pressure, Apparatus for making	A. A. Burlingame	New York, N. Y	Apr. 9, 1861	31, 951
Extracting process and apparatus	G. W. Sylvester	Belleville, N. J	Apr. 16, 1872	125, 856
Extracting tanning matter from bark, Apparatus for.	N. S. Thomas	Painted Post, N. Y	Aug. 14, 1866	57, 218
Extractor: *See* Bolt-extractor. Bung-extractor. Cartridge-extractor. Cartridge-shell extractor. Cheese-skipper extractor. Cork-extractor. Drill-extractor. Hop-extractor. Kettle-molding extractor. Madder-extractor. Nail-extractor. Obstetrical extractor. Oil-tool extractor. Peach-stone extractor. Root-extractor. Spike-extractor. Stone extractor. Stump-extractor. Tack-extractor. Tooth-extractor.				
Eye and lung protector	G. A. Crofutt	New York, N. Y	Dec. 9, 1873	145, 337
Eye, Artificial	A. Boissonneau	Paris, France	June 19, 1866	55, 793
Eye-bolt die	C. Norton	New Haven, Conn	July 24, 1866	56, 599
Eye-bolt-forging die	F. Leonard	Cleveland, Ohio	June 16, 1868	78, 977
Eye-bolts or links without welding, Method of making.	G. H. Sellers	Phœnixville, Pa	Sept. 29, 1868	82, 559
Eye-cup	B. K. Maltby	Cincinnati, Ohio	July 14, 1868	79, 847
Eye-cup	B. F. Stephens	Brooklyn, N. Y	Dec. 10, 1867	72, 108
Eye-cup	J. M. Winslow	Rochester, N. Y	Aug. 13, 1867	67, 832
Eye-glass	A. D. Ansell	Hartford, Conn	May 27, 1873	139, 353
Eye-glass	J. Cadman	Chatham Village, N. Y	June 14, 1872	127, 459
Eye-glass	I. Clements	Fort Ann, N. Y	June 6, 1871	115, 574
Eye-glass	E. M. Cole	Southbridge, Mass	Jan. 7, 1868	72, 978
Eye-glass	J. J. Crispin	Providence, R. I	Dec. 3, 1867	71, 584
Eye-glass	G. M. Cummings	Providence, R. I	July 30, 1867	67, 167
Eye-glass	A. H. Daniels	Hartford Conn	May 26, 1868	78, 190
Eye-glass	S. Greacen	Rye, N. Y	July 15, 1873	140, 912
Eye-glass	E. K. Josselyn	Cambridge, Mass	Jan. 8, 1867	61, 073
Eye-glass	E. Maynard	Tarrytown, N. Y	Aug. 7, 1866	56, 967
Eye-glass	J. K. McDonald	Newark, N. J	Feb. 4, 1868	74, 109
Eye-glass	J. Prentice	New York, N. Y	Jan. 8, 1867	61, 099
Eye-glass	M. Risley	Springfield, Mass	Aug. 26, 1873	142, 277
Eye-glass	R. Stranbel	Williamsburgh, N. Y	July 27, 1869	93, 020

Index of patents issued from the United States Patent Office from 1790 *to* 1873, *inclusive*—Continued.

Index of patents issued from the United States Patent Office from 1790 *to* 1873, *inclusive*—Continued.

Invention.	Inventor.	Residence.	Date.	No.
Eyeleting-machine	A. Delkescamp	New York, N. Y	Dec. 5, 1871	121, 495
Eyeleting-machine	L. Goddu	Boston, Mass	July 23, 1872	129, 812
Eyeleting-machine	L. Hall	Boston, Mass	May 14, 1867	64, 761
Eyeleting-machine	P. Harlow	Hudson, Mass	Nov. 27, 1866	60, 000
Eyeleting-machine	H. Juge	New York, N. Y	Apr. 16, 1867	63, 904
Eyeleting-machine	J. Keith	New Bedford Mass	Jan. 19, 1864	41, 303
Eyeleting machine	A. Komp	New York, N. Y	June 30, 1868	79, 482
Eyeleting-machine	A. Komp	New York, N. Y	Jan. 3, 1871	110, 767
Eyeleting-machine	H. L. Lipman	Philadelphia, Pa	Jan. 3, 1865	45, 727
Eyeleting-machine	A. A. Reed	North Bridgewater, Mass	Apr. 10, 1866	53, 926
Eyeleting-machine	T. K. Reed	North Bridgewater, Mass	Aug. 9, 1864	43, 816
Eyeleting-machine	J. F. Sargent	Boston, Mass	Jan. 6, 1863	37, 306
Eyeleting-machine	J. F. Sargent	Boston, Mass	Aug. 25, 1863	39, 705
Eyeleting-machine	J. F. Sargent	Boston, Mass	Jan. 5, 1864	41, 125
Eyeleting-machine	J. F. Sargent	Melrose, Mass	Aug. 3, 1869	93, 353
Eyeleting-machine	E. Shaw	Milwaukee, Wis	Apr. 19, 1870	102, 054
Eyeleting-machine	G. Shipman	West Bridgewater, Mass	Aug. 21, 1866	57, 446
Eyeleting-machine	J. E. Wiggin	Stoneham, Mass	May 21, 1872	126, 916
Eyeleting-machine	G. Wunderlick	Philadelphia, Pa	Nov. 21, 1871	121, 224
Eyeleting-machine for attaching buttons to garments.	E. J. Warner	Newark, N. J	Sept. 24, 1867	69, 284
F.				
Fabric: *See* Bleached fabric. Carpet-fabric. Cellular fabric. Corrugated fabric. Elastic fabric. Felted fabric. Felted tufted fabric. Fiber and gum fabric. Figured fabric. Fire proof fabric. Flocked fabric. Fringed fabric. Fur-coated fabric. Gathered fabric. Hat and bonnet fabric. Hydrofuze fabric. India-rubber fabric. Knit fabric. Knitted medicated fabric. Metallic fabric. Netted or laced fabric. Palm-leaf fabric. Pile-fabric. Plain and figured fabric. Rubber fabric. Spring-fabric. Terry fabric. Textile fabric. Tubing-fabric. Tubular fabric. Tufted fabric. Warp-net fabric. Water-proof fabric. Water-proof furze fabric. Water-proofing fabric. Wire fabric. Woolen fabric. Woven fabric. Woven tuck fabric.				
Fabric	A. Forot	Paris, France	Aug. 16, 1859	25, 107
Fabric	C. Roder	Philadelphia, Pa	Oct. 1, 1872	131, 906
Fabric for bags, roofs, &c	H. W. Johns	New York, N. Y	Apr. 9, 1872	125, 573
Fabric for head-covering	E. Arnheim	New York, N. Y	Aug. 2, 1870	106, 024
Fabric for head-covering	H. Loewenberg	New York, N. Y	Mar. 12, 1872	124, 599
Fabric for wrapping and packing	J. A. Turner	Manchester, Great Britain	Aug. 6, 1872	130, 340
Fabric from fibrous sheets and hard rubber	R. O. Lowrey	Salem, N. Y	June 15, 1869	91, 458
Fabric opening, smoothing, and guiding machine	W. Birch	Manchester, England	Mar. 25, 1873	137, 169
Fabric or paper cloth, Imitation	J. H. Newton	Holyoke, Mass	July 20, 1869	92, 743
Fabric, Plaiting	W. Fuzzard	Chelsea, Mass	Jan. 19, 1869	85, 922
Fabric-pressing machine	G. H. Mussey and W. B. Leachman.	Leeds, England	Jan. 24, 1871	111, 141
Fabric-shirring machine	A. Hecht	New York, N. Y	June 7, 1870	104, 029
Fabric-stretching machine	I. E. Palmer	Hackensack, N. J	Mar. 30, 1869	88, 505
Fabric to be used as a substitute for japanned leather.	J. Fletcher	Newark, N. J	Aug. 21, 1866	57, 307
Fabrics and introducing flat skirt-wires therein, Instrument for puncturing.	M. Fishel	New York, N. Y	June 5, 1866	55, 267
Fabrics, &c., Apparatus for discharging extraneous coloring-matter from.	B. G. Brooks	Manchester, N. H	Oct. 3, 1871	119, 503
Fabrics, Coating and metallizing	R. Nöggerath	Paris, France	Dec. 24, 1867	72, 601
Pabrics, Composition for making designs upon	R. L. Jones	Sacramento, Cal	Sept. 15, 1868	82, 122
Fabrics, Fluid for extracting grease from	J. B. Wilson	Philadelphia, Pa	Dec. 19, 1871	122, 094
Fabrics from vegetable and animal fibers, Manufacture of useful and ornamental.	E. Pavy	Paris, France	June 21, 1870	104, 638
Fabrics impervious to gas, Composition to render	G. L. Burnham	Providence, R. I	Oct. 31, 1871	120, 368
Fabrics, Machine for cutting	W. J. Cussen	Richmond, Va	Dec. 10, 1872	133, 700
Fabrics, Machine for disintegrating	M. Marshall	Lowell, Mass	Sept. 30, 1873	143, 290
Fabrics, Machine for disintegrating	M. Marshall	Lowell, Mass	Nov. 18, 1873	144, 685
Fabrics, Machine for drying and stretching	O. C. Sweet	Albany, N. Y	Sept. 29, 1868	82, 564
Fabrics, Machine for scouring and washing	H. and F. Bates	Sowerby Bridge, Halifax, England.	Jan. 28, 1870	104, 818
Fabrics, Method of making dies for figures in press-dyed.	J. Holt	Lowell, Mass	Mar. 26, 1867	63, 159
Fabrics, Mode of ornamenting	W. Swan	New York, N. Y	May 19, 1868	78, 156
Fabrics, &c., Preventing mildew and decay in	W. A. Torrey	Mount Clair, N. J	Apr. 26, 1870	102, 450
Fabrics, Renovating faded	R. H. Klauder	Philadelphia, Pa	July 24, 1866	56, 574
Fabrics, Singeing-machine for	P. McEwen and W. McKenzie	Jersey City and Hudson City, N. J.	Oct. 27, 1868	83, 395
Fabrics to render them water-proof, Treating	C. Tappan	Wakefield, Mass	May 31, 1870	103, 799
Fabrics uninflammable, Rendering	D. Nicoll	London, England	Aug. 26, 1873	142, 267
Fabrics with button-hole therein, Mode of weaving	J. Connor	Boston, Mass	Apr. 4, 1865	47, 151
Fabrics with fluid or semi-fluid substances, Apparatus for coating.	G. Adamson	Philadelphia, Pa	Jan. 1, 1867	60, 660
Fabrics with parkesine, Coating	J. Lewthwaite	London, England	May 17, 1870	103, 209
Fagot for beams	W. W. Miller	Safe Harbor, Pa	Sept. 24, 1867	69, 237
Fagot for beams	G. Walters and T. Shaffer	Phœnixville, Pa	Mar. 17, 1868	75, 712
Fagot for wrought-metal cannon, hydraulic pumps, &c.	S. J. Reeves	Philadelphia, Pa	Dec. 9, 1862	37, 168
Fagot, Kindling	C. R. Jacobi	Green Point, N. Y	July 12, 1870	105, 213
Fagots, Construction of	F. W. Webb	Crewe, England	Mar. 28, 1865	47, 077

Index of patents issued from the United States Patent Office from 1790 *to* 1873, *inclusive*—Continued.

Invention.	Inventor.	Residence.	Date.	No.
Fall-leaf table	W. Caldwell	Bryan, Ohio	May 10, 1870	102, 771
Fall-leaf table	J. S. Rankin	Detroit, Mich	May 5, 1868	77, 655
Family-mill	D. Flagg, jr	Gardiner, Me	Feb. 23, 1826	....
Fan	S. Ainsworth	Saratoga Springs, N. Y	Apr. 21, 1868	76, 875
Fan	G. Anton	Philadelphia, Pa	Jan. 16, 1866	52, 109
Fan	O. Brück	New York, N. Y	Nov. 19, 1867	70, 952
Fan	O. Brück	New York, N. Y	Aug. 31, 1869	94, 390
Fan	O. Brück	New York, N. Y	Oct. 12, 1869	95, 764
Fan	O. Brueck	New York, N. Y	Oct. 25, 1870	108, 681
Fan	G. Brueck	New York, N. Y	Apr. 1, 1873	137, 413
Fan	A. B. Hancock	Suspension Bridge, N. Y	Nov. 24, 1868	84, 276
Fan	N. Harper	Newark, N. J	Sept. 25, 1866	58, 250
Fan	T. W. Hawkins	New Haven, Conn	Dec. 31, 1867	72, 372
Fan	T. W. Hawkins	New Haven, Conn	June 1, 1869	90, 748
Fan	L. Hill	North Brookfield, Mass	Apr. 23, 1867	64, 013
Fan	J. L. Jackson	New York, N. Y	Apr. 2, 1867	63, 389
Fan	F. Koppenfels and G. Brueck	New York, N. Y	Apr. 2, 1867	63, 396
Fan	W. Lucas	New Haven, Conn	Sept. 17, 1867	68, 890
Fan	C. C. Lusby	Philadelphia, Pa	Dec. 16, 1873	145, 666
Fan	G. Mallory	Watertown, Conn	Apr. 24, 1866	54, 184
Fan	J. McLoughlin	Morrisania, N. Y	July 23, 1872	129, 848
Fan	O. R. Nitsch	New York, N. Y	June 8, 1869	91, 156
Fan	O. R. Nitsch	New York, N. Y	Sept. 7, 1869	94, 633
Fan	A. K. Rich and J. B. Thurston	New York, N. Y	Sept. 3, 1872	131, 026
Fan	B. M. Smith	New York, N. Y	Nov. 28, 1865	51, 228
Fan	H. B. Smith	Essex, Conn	May 19, 1868	78, 145
Fan and air-circulator	J. D. Wright	Bergen, N. J	Jan. 26, 1869	86, 336
Fan and caster-stand, Table	N. H. Tilman and D. G. Good	Arcanum, Ohio	Apr. 28, 1868	77, 415
Fan and fly-brush combined	H. Mee	Crown Point, Ind	Sept. 12, 1871	118, 870
Fan and fly-brush, Rotating	O. Matcalf	Orleans, Ind	Sept. 13, 1870	107, 396
Fan and fly-driver	F. M. Hunt	Clinton, Ga	Mar. 21, 1871	112, 921
Fan and parasol	H. H. White	West River, Md	May 15, 1866	54, 799
Fan and parasol combined	J. Brooks	Boston, Mass	Mar. 15, 1870	100, 719
Fan and parasol combined	G. Mallory	Bridgeport, Conn	Oct. 29, 1867	70, 344
Fan and parasol, Combined	C. St. John	New York, N. Y	Nov. 11, 1873	144, 575
Fan and screen in cleaning grain for grinding	J. Peyton	Orange City, Va	Nov. 7, 1825	....
Fan attachment, Spring	J. Carey	Victoria, Mo	July 13, 1869	92, 513
Fan attachment, Table	S. S. Binkard and R. H. Boal	Urbana, Ohio	July 11, 1871	116, 919
Fan attachment, Table	J. E. Mansker	Clinton, La	Mar. 2, 1869	87, 417
Fan, Automatic	D. Aaron	San Francisco, Cal	June 15, 1869	91, 297
Fan, Automatic	J. Beck	New York, N. Y	May 12, 1868	77, 709
Fan, Automatic	J. M. Belcour	Paris, France	June 25, 1872	128, 279
Fan, Automatic	G. C. Bovey	Cincinnati, Ohio	Apr. 4, 1871	113, 620
Fan, Automatic	G. H. Briggs	Montgomery, Ala	July 26, 1870	105, 772
Fan, Automatic	C. F. Burleigh	Tuftonborough, N. H	Dec. 28, 1869	98, 347
Fan, Automatic	J. D. Bush	Elyton, Ala	Aug. 26, 1873	142, 081
Fan, Automatic	W. B. Campbell	Selma, Ala	Nov. 11, 1873	144, 505
Fan, Automatic	G. F. Case	Claremont, N. H	July 2, 1872	128, 589
Fan, Automatic	C. H. Clark	Hartford, Conn	Feb. 4, 1873	135, 524
Fan, Automatic	F. O. Degener	New York, N. Y	Oct. 25, 1859	25, 892
Fan, Automatic	C. Dieringer and M. Lindamann.	Cincinnati, Ohio	July 13, 1869	92, 589
Fan, Automatic	E. Duffner	Vicksburgh, Miss	May 13, 1873	138, 741
Fan, Automatic	J. R. Dunn and G. B. Burroughs.	Queens County and Brooklyn, N. Y.	May 4, 1869	89, 566
Fan, Automatic	W. Fay	New Orleans, La	Mar. 25, 1873	137, 189
Fan, Automatic	W. D. Hall	Memphis, Tenn	Sept. 26, 1871	119, 351
Fan, Automatic	W. D. Hall	Memphis, Tenn	Feb. 6, 1872	123, 476
Fan, Automatic	W. D. Hall	Memphis, Tenn	Apr. 30, 1872	126, 292
Fan, Automatic	W. D. Hall	Memphis, Tenn	Feb. 18, 1873	136, 044
Fan, Automatic	A. P. Heidt	New York, N. Y	Jan. 7, 1873	134, 543
Fan, Automatic	H. Hoffman	New York, N. Y	July 5, 1859	24, 638
Fan, Automatic	F. W. Jones	Orangeburgh, S. C	Apr. 8, 1873	137, 687
Fan, Automatic	W. Lawrence and J. Sanders	New Albany, Ind	Jan. 7, 1873	134, 605
Fan, Automatic	J. Lehner	Galena, Ill	May 6, 1873	138, 567
Fan, Automatic	M. Lochner	Newark, N. J	Oct. 24, 1871	120, 293
Fan, Automatic	W. O. Loeffler	New York, N. Y	July 9, 1867	66, 507
Fan, Automatic	J. A. W. Lundborg	San Francisco, Cal	Feb. 26, 1867	62, 431
Fan, Automatic	J. Maltry	Morrisania, N. Y	May 11, 1869	90, 009
Fan, Automatic	H. T. McCormick	Selma, Ala	Apr. 15, 1873	137, 785
Fan, Automatic	J. McLain	Saint Mary's, Ohio	Jan. 7, 1862	34, 066
Fan, Automatic	J. Naugle	Mooresville, Ind	Mar. 23, 1869	88, 062
Fan, Automatic	J. B. Powell	Philadelphia, Pa	Sept. 6, 1859	25, 368
Fan, Automatic	J. B. Powell	Philadelphia, Pa	Oct. 4, 1864	44, 553
Fan, Automatic	D. Ramler	Union Deposit, Pa	June 28, 1870	104, 768
Fan, Automatic	L. Rebstock and N. Reimel	Philadelphia, Pa	June 9, 1857	17, 517
Fan, Automatic	P. H. Reichhardt and M. Schnapp.	Brunswick, Mo	Apr. 29, 1873	138, 345
Fan, Automatic	J. Schuell and P. Schmidt	Waterloo, Ill	Oct. 25, 1870	108, 637
Fan, Automatic	C. B Smith	Griffin, Ga	Feb. 28, 1871	112, 293
Fan, Automatic	G. C. Steinhauer	Indianapolis, Ind	Mar. 30, 1869	88, 343
Fan, Automatic	E. D. Swartwout	Chicago, Ill	July 25, 1871	117, 346
Fan, Automatic	B. D. Thompson	New York, N. Y	May 2, 1871	114, 491
Fan, Automatic	S. J. Tucker and J. H. Rose	Richmond, Va	Mar. 22, 1870	101, 188
Fan, Automatic	O. Vanorman	Fond du Lac, Wis	Sept. 23, 1873	143, 206
Fan, Automatic	L. A. Walton	Byhalia, Miss	Sept. 12, 1871	118, 832
Fan, Automatic	T. Welch	Philadelphia, Pa	May 6, 1873	138, 720
Fan, Automatic	J. B. Williamson	Louisville, Ky	Feb. 27, 1872	124, 103
Fan, Automatic	S. E. Winslow	Kensington, Pa	Nov. 9, 1852	9, 395
Fan, Automatic	G. W. Zeigler	Tiffin, Ohio	Apr. 19, 1859	23, 739
Fan, Automatic or musquito	C. A. Gale	Boston, Mass	Apr. 10, 1860	27, 862
Fan, Bed and table	W. H. Downs	Jeffersonville, Ind	Sept. 19, 1871	119, 128
Fan blast-regulator	D. Garver	Ringgold, Md	May 9, 1871	114, 547
Fan, Blowing	T. Wallace and H. Backmaster	Philadelphia, Pa	June 6, 1854	11, 047
Fan, brush, and rack	B. Poulson	Fort Wayne, Ind	Dec. 18, 1866	60, 554
Fan, Buckwheat	A. Platt	Waterbury, Conn	Jan. 13, 1852	8, 659
Fan-chair	D. Linzie	Petersham, Mass	Apr. 10, 1849	6, 307
Fan, Electro-magnetic	G. Stevens and J. W. Moyle	Cincinnati, Ohio	July 4, 1871	116, 770

Index of patents issued from the United States Patent Office from 1790 *to* 1873, *inclusive*—Continued.

Invention.	Inventor.	Residence.	Date.	No.
Fan, Exhaust	R. Cook	Saratoga Springs, N. Y	Nov. 12, 1861	33, 693
Fan, Exhaust	J. V. Merrick	Philadelphia, Pa	June 13, 1854	11, 071
Fan, Fly	G. A. Goodrich and P. Miller	Hope, Ohio	Feb. 27, 1872	124, 054
Fan, Fly	J. G. Schwemmer and T. Mueller.	Philadelphia, Pa	Oct. 9, 1866	58, 683
Fan, Fly-expelling	W. R. Fowler	Anne Arundel County, Md	Dec. 1, 1863	40, 746
Fan, Folding	T. S. Thorn	South Amboy, N. J	Mar. 10, 1868	75, 316
Fan for apartments, Revolving	L. Stein	New York, N. Y	Dec. 19, 1854	12, 106
Fan for bed-chambers, dining-rooms, &c., moved by mechanism.	J. Barron	Norfolk, Va	Nov. 27, 1830	....
Fan for furnaces, gas-works, &c , Rotary blowing	H. Aland	London, England	Mar. 2, 1869	87, 456
Fan for ventilating mines, Cross-paddle propeller	D. Rissinger	Mahanoy City, Pa	May 27, 1873	139, 330
Fan, Grain	P. Bailey	Falls Township, Pa	Mar. 13, 1860	27, 417
Fan, Grain	G. Goewey	Philadelphia, Pa	Jan. 10, 1860	26, 760
Fan, Grain	O. Lindsay and R. F. Strean	Washington, Pa	Nov. 29, 1859	26, 273
Fan, Grain	J. McPhail	Charles City, Iowa	Mar. 1, 1870	100, 454
Fan, Grain	I. Pennington	Tiffin, Ohio	Feb. 1, 1870	99, 469
Fan, House	J. and A. Bloom	New Brunswick, N. J., and New York, N. Y.	Sept. 3, 1867	68, 483
Fan, House	E. Marquam		Mar. 28, 1803	....
Fan, Lady's	S. A. Grant	Burlington, N. J	Aug. 14, 1866	57, 253
Fan, Lady's	F. B. Scott	Lancaster, N. Y	Apr. 16, 1867	63, 817
Fan, Lady's	B. Todd	Reading, Pa	Aug. 28, 1866	57, 598
Fan, Lady's	J. and I. N. Todd	Madison, Ind	Sept. 13, 1864	44, 237
Fan, Lady's	T. Welham	Washington, D. C	Apr. 3, 1866	53, 714
Fan, Mechanical	J. Barron	Norfolk, Va	June 13, 1831	....
Fan-mill	I. T. Grant	Schaghticoke, N. Y	July 10, 1845	4, 105
Fan-mill	W. C. Hawley	Brookfield, Conn	Oct. 1, 1830	....
Fan-mill	F. H. Plymate	Blue Earth County, Minn	Sept. 19, 1871	119, 174
Fan, Musquito	J. M. Beverly	Chicago, Ill	Aug. 15, 1871	117, 969
Fan or brush, Automatic table	R. E. Tolar and W. D. Orr	Newnan, Ga	Aug. 9, 1870	106, 236
Fan-power	W. E. Wilcox	Saint Louis, Mo	Dec. 13, 1864	45, 447
Fan, Reversible screw	W. H. Goyne	Shamokin, Pa	Dec. 10, 1872	133, 772
Fan, Rocking-chair	O. Brueck	Union Hill, N. J	May 20, 1873	139, 112
Fan, Rocking-chair	O. Matzke	Durant, Miss	Nov. 19, 1872	133, 164
Fan, Rocking-chair	L. T. Simon	New York, N. Y	Aug. 17, 1869	93, 916
Fan, Rotary	I. S. Eastman	Madison, Wis	Jan. 14, 1873	134, 740
Fan, Rotary	G. Leach	Elmira, N. Y	May 23, 1865	47, 838
Fan, Rotating	I. S. Eastman	Madison, Wis	Mar. 31, 1868	76, 174
Fan, Rotating	J. H. Reynerson	Pleasant Plain, Iowa	Aug. 25, 1868	81, 539
Fan, Self-acting fly	W. D. Harrell and H. M. Hall	Osgood, Ind	Oct. 15, 1867	69, 802
Fan, Spring	O. Brueck	New York, N. Y	Oct. 11, 1870	108, 236
Fan-sticks, Machine for making	E. S. Hunt	Weymouth, Mass	Oct. 4, 1870	107, 913
Fan, Suction	C. R. Patterson	Pittston, Pa	Aug. 6, 1872	130, 310
Fan, Suction and blast	C. R. Patterson	Pittston, Pa	Oct. 24, 1871	120, 317
Fan, Table	J. C. McEwen	Micanopy, Fla	Dec. 20, 1870	110, 381
Fan, Table	W. A. McReynolds	Elkton, Ky	Aug. 6, 1867	67, 562
Fan, Table	A. R. Traber	Saint Martinsville, La	Jan. 29, 1861	31, 269
Fan, table-caster, and lamp-stand, Automatic	T. W. Carmichael	Indianapolis, Ind	Feb. 11, 1868	74, 303
Fan, Wheat	W. Adams	Virginia	Mar. 1, 1816	....
Fan, Wheat	J. Bamborough	Lancaster, Pa	Mar. 20, 1847	5, 031
Fan, Wheat	J. Deakyne	Petersburgh, Va	Sept. 13, 1833	....
Fan, Wheat	G. Hoffman	Fredericktown, Md	Aug. 21, 1829	....
Fan, Wheat	J. Hollingsworth	Zanesville, Ohio	Apr. 1, 1851	8, 008
Fan, Wheat	J. Montgomery and J. Montgomery.	Lancaster, Pa., and Baltimore, Md.	June 12, 1855	13, 062
Fan, Wheat	W. Stanley	Jamestown, N. C	Nov. 18, 1844	3, 829
Fan, Wheat	D. Watkins	Port Republic, Va	Feb. 2, 1844	3, 423
Fan, Wheat	P. H. Watson	Rockford, Ill	June 6, 1848	5, 616
Fan, Wheat	S. Weaver	Frederick County, Md	Mar. 11, 1816	....
Fan, Wheat	J. White	Barnesville, Ohio	Apr. 1, 1851	8, 013
Fan-wheel, Rotary	T. H. Walton	Ashland, Pa	Sept. 10, 1872	131, 238
Fans, Apparatus for binding circular paper or other	R. T. Smith	Nashua, N. H	Nov. 20, 1866	59, 783
Fans, Apparatus for operating	W. M. Bruton	Baltimore, Md	July 4, 1871	116, 788
Fans, Device for opening and closing	G. Bordes	New York, N. Y	Oct. 11, 1870	108, 096
Fans, Device for operating hand	W. A. Ireland	New York, N. Y	Sept. 20, 1870	104, 501
Fans, Escapement movement for automatic	D. J. Mozart	Xenia, Ohio	Oct. 14, 1856	15, 889
Fans, Machine for folding paper	R. T. Smith	Nashua, N. H	Nov. 20, 1866	59, 867
Fans, Machine for manufacturing webbing for ladies'.	J. W. White	Weymouth, Mass	June 1, 1869	90, 711
Fans, Manufacture of	E. S. Hunt	Weymouth, Mass	Dec. 15, 1868	84, 882
Fans, Manufacture of portable	J. C. Hall	Fayette, Miss	Dec. 21, 1858	22, 366
Fanning and rocking chair	T. Kerr	York, Pa	Aug. 11, 1868	80, 970
Fanning and smut mill	H. Littlefield	Lewis, Iowa	Mar. 15, 1870	100, 908
Fanning and winnowing mill	B. Atwell	Burlington, N. Y	Sept. 19, 1808	....
Fanning-apparatus	L. C. England	Philadelphia, Pa	Feb. 28, 1871	112, 232
Fanning-chair	A. R. Hobbs and N. F. Wright	Elizabethport, N. J	Apr. 7, 1868	76, 452
Fanning machine	E. P. Doremus	Washington, La	Feb. 21, 1871	111, 917
Fanning-machine	G. F. Evans	Chelsea, Mass	Aug. 25, 1868	81, 353
Fanning-machine	D. Flanders and C. Rathburn	Fort Covington, N. Y	Feb. 13, 1836	....
Fanning-mill	C. Altringer	Racine, Wis	May 7, 1872	126, 371
Fanning-mill	J. Ashton	Red Wing, Minn	July 20, 1869	92, 774
Fanning-mill	H. R. Averill	New Oregon, Iowa	Nov. 8, 1870	108, 953
Fanning-mill	B. Barney	Time, Ill	Mar. 7, 1871	112, 407
Fanning-mill	T. S. Barnum	Sharon, Conn	Jan. 15, 1812	....
Fanning-mill	E. Bless	Minerva, Ky	Dec. 17, 1850	7, 849
Fanning-mill	L. M. Boardman	Ottawa, Minn	Aug. 26, 1873	142, 197
Fanning-mill	F. E. Bowen	Knightstown, Ind	July 25, 1871	117, 374
Fanning-mill	J. J. Bradner	Pine Creek, N. Y	July 30, 1867	67, 260
Fanning-mill	L. Bronson	Buffalo, N. Y	Dec. 20, 1870	110, 196
Fanning-mill	H. Bruggeman	Petersburgh, Ind	June 22, 1869	91, 600
Fanning-mill	J. K. Buck	Winona, Minn	Apr. 11, 1865	47, 186
Fanning-mill	G. E. Clarke	Racine, Wis	Sept. 27, 1870	107, 759
Fanning-mill	G. E. Clarke	Racine, Wis	Mar. 14, 1871	112, 687
Fanning-mill	G. E. Clarke	Kenosha, Wis	Jan. 7, 1873	134, 643
Fanning-mill	D. Clinton	New Haven, Conn	Nov. 29, 1832	....
Fanning-mill	D. Clow	Port Byron, N. Y	June 16, 1846	4, 582
Fanning-mill	D. Collins	Zanesfield, Ohio	Nov. 21, 1871	121, 085

Index of patents issued from the United States Patent Office from 1790 *to* 1873, *inclusive*—Continued.

Invention.	Inventor.	Residence.	Date.	No.
Fanning-mill	B. Cortrite	Norwalk, Ohio	Apr. 12, 1870	101, 713
Fanning-mill	L. M. Crosby	Ashtabula, Ohio	May 2, 1871	114, 416
Fanning-mill	A. B. Culver	Westfield, N. Y	July 16, 1867	66, 803
Fanning-mill	H. W. Curtis	Lockport, Ill	Aug. 28, 1866	57, 483
Fanning-mill	L. H. Decker	Sauk Centre, Minn	Mar. 25, 1873	137, 185
Fanning-mill	L. H. Decker	Grove Lake, Minn	Sept. 23, 1873	143, 127
Fanning-mill	E. P. Dickey	Racine, Wis	June 8, 1869	90, 936
Fanning-mill	J. Drummond	Trenton, Mo	Feb. 20, 1872	123, 814
Fanning-mill	C. K. Ehle	Greenbush, Wis	Aug. 7, 1866	56, 912
Fanning-mill	L. Ensign	Millburn, N. Y	June 17, 1862	35, 594
Fanning-mill	A. Erwin	Jefferson, Md	Sept. 30, 1839	1, 354
Fanning-mill	F. Eves	Fountain City, Wis	Sept. 26, 1871	119, 340
Fanning-mill	A. Fanckboner	Schoolcraft, Mich	May 2, 1865	47, 533
Fanning-mill	L. Farnham and J. Mosher	Delta, Mich	May 4, 1869	89, 643
Fanning-mill	S. Fitch	Homer, N. Y	Oct. 29, 1814	....
Fanning-mill	S. Fitch	Edmeston, N. Y	Jan. 6, 1832	....
Fanning-mill	S. Foster, jr	Des Moines, Iowa	July 20, 1869	92, 814
Fanning-mill	J. Frary	Southampton, Mass	Feb. 17, 1829	....
Fanning-mill	J. W. Free	Richmond, Ind	Feb. 27, 1866	52, 840
Fanning-mill	D. A. Fulk and L. B. McLain	Hookstown and Lisbon, Ohio.	Oct. 6, 1843	3, 294
Fanning-mill	E. M. Gilbert	Kasson, Minn	Oct. 7, 1873	143, 506
Fanning-mill	R. Gipson	Shelby, Ohio	Mar. 14, 1871	112, 584
Fanning-mill	J. Goetz	West Bend, Wis	Aug. 1, 1871	117, 7[illegible]5
Fanning-mill	A. Hall and J. Faulkner	Dansville, N. Y	Feb. 25, 1862	34, 488
Fanning-mill	J. W. Hunter, M. L. Hunt, and S. Holman.	Oakland County, Mich	Dec. 18, 1832	....
Fanning-mill	J. Hutchins	Syracuse, Iowa	Nov. 7, 1871	120, 745
Fanning-mill	J. W. Johnson	Port Treverton, Pa	Apr. 29, 1873	138, 407
Fanning-mill	D. Kane	Tivoli, Iowa	Aug. 30, 1870	106, 8[illegible]
Fanning-mill	H. Kelly and W. Franklin	Decorah, Iowa	Dec. 15, 1863	40, 937
Fanning-mill	J. M. Kendall and J. Peel	Madelia, Minn	July 30, 1872	129, 965
Fanning-mill	T. B. Kirkwood	Dublin, Ind	Aug 17, 1869	93, 721
Fanning-mill	G. Leach	Elmira, N. Y	May 23, 1865	47, 839
Fanning-mill	E. Lindsley	Neenah, Wis	Sept. 8, 1868	82, 011
Fanning-mill	S. McMillan	Fletcher, Ohio	Feb. 18, 1868	74, 567
Fanning-mill	E. Michael	La Porte, Ind	May 25, 1869	90, 570
Fanning-mill	F. Miles	Rochester, N. Y	June 10, 1862	35, 536
Fanning-mill	J. Miller	Canton, Ohio	June 3, 1862	35, 461
Fanning-mill	J. Miller	Canton, Ohio	June 3, 1862	35, 462
Fanning-mill	J. Miller	Canton, Ohio	June 3, 1862	35, 463
Fanning-mill	S. Miller and I. J. Chase	Barrington, Ill	July 30, 1867	67, 206
Fanning-mill	J. Mumma	Middletown, Ohio	June 12, 1866	55, 526
Fanning-mill	D. Neefus	Flatlands, N. Y	Sept. 22, 1823	....
Fanning-mill	A. Niebel	Tiffin, Ohio	June 11, 1872	127, 912
Fanning-mill	H. Ogborn	Richmond, Ind	Aug. 16, 1870	106, 395
Fanning-mill	H. Ogborn	Richmond, Ind	Jan. 31, 1871	111, 375
Fanning-mill	L. Peck	Brookfield, Conn	Aug. 13, 1824	....
Fanning-mill	C. Peterson	Red Wing, Minn	Oct. 4, 1870	108, 049
Fanning-mill	O. Pinney and J. Olmested	Caledonia, N. Y	Nov. 7, 1826	....
Fanning-mill	J. F. Pool	Monroe, Wis	Jan. 19, 1869	86, 099
Fanning-mill	J. P. Preston	Monroe, Wis	Oct. 20, 1868	83, 312
Fanning-mill	P. Prine and L. Huestis	Auburn, N. Y	Feb. 23, 1832	....
Fanning mill	B. F. Randell	Des Moines, Iowa	Mar. 9, 1869	87, 706
Fanning-mill	W. C. Ray and G. Leigh	Pleasant Run and Clinton Station, N. J.	May 19, 1868	78, 008
Fanning-mill	J. E. Rice	Oneida, Ill	Feb. 1, 1859	22, 825
Fanning-mill	G. Richards and D. Stickland	Richland Centre, Wis	Oct. 13, 1868	83, 096
Fanning-mill	J. Roberts	Penn's Square Post-Office, Pa.	May 28, 1850	7, 400
Fanning-mill	B. F. Roe	Nebraska City, Nebr	Sept. 11, 1860	29, 998
Fanning-mill	J. L. Runk and B. H. Sharp	Nashville, Ill	Apr. 5, 1870	101, 514
Fanning-mill	F. Sauer and J. Coerver	Waterloo, Ill	Dec. 20, 1870	110, 291
Fanning-mill	D. Sears	Franklin, N. Y	Mar. 24, 1830	....
Fanning-mill	H. H. Seeley	Hudson, Mich	Sept. 29, 1868	82, 558
Fanning-mill	H. H. Seeley	Hudson, Mich	May 16, 1871	114, 977
Fanning-mill	H. F. Seibert	Brady's Bend, Pa	Apr. 13, 1869	88, 913
Fanning-mill	L. Shultz	Upper Sandusky, Ohio	June 19, 1866	55, 725
Fanning-mill	A. W. Smith	Dudleyville, Ala	Jan. 21, 1868	73, 469
Fanning-mill	S. Snedeker	Johnstown, N. Y	Apr. 27, 1810	....
Fanning-mill	H. A. Snyder	Shullsburgh, Wis	Sept. 8, 1868	81, 955
Fanning-mill	J. Soper	Windsor, Conn	Dec. 3, 1814	....
Fanning-mill	W. Stoddard	Winona, Minn	Sept. 8, 1868	81, 957
Fanning-mill	H. K. Stoner	West Lampeter, Pa	Mar. 26, 1872	124, 918
Fanning-mill	E. Tush	Manchester, Iowa	May 17, 1864	42, 808
Fanning-mill	H. W. Veregge	Richmond, Ind	Feb. 7, 1865	46, 312
Fanning-mill	W. W. Wait	Richmond, Ind	Mar. 23, 1869	88, 241
Fanning-mill	E. Walker	Springfield Four-Corners, Pa.	Sept. 20, 1827	....
Fanning-mill	J. Weidman	Littlestown, Pa	May 28, 1850	7, 405
Fanning-mill	N. Wells	Milford, Pa	Sept. 11, 1833	....
Fanning-mill	B. C. White	Richmond, Ind	Jan. 24, 1865	46, 039
Fanning-mill	L. L. Willson	Denmark, Mich	Nov. 5, 1872	132, 784
Fanning-mill	B. Wright and J. C. Hogaboam	Hudson, Mich	May 7, 1872	126, 616
Fanning-mill	P. Young	Austin, Minn	May 16, 1871	114, 898
Fanning-mill and grain-separator	H. Ogborn	Richmond, Ind	Jan. 3, 1865	45, 796
Fanning-mill and grain-separator	H. Ogborn	Richmond, Ind	Mar. 8, 1870	100, 552
Fanning-mill and grain-separator	D. Wilcox	Chariton, Iowa	Aug. 24, 1869	94, 052
Fanning-mill and grain-separator	D. C. Hill	Red Wing, Minn	Oct. 8, 1872	131, 953
Fanning-mill and wheelbarrow combined	J. Wormeldorff	Warsaw, Ind	Feb. 2, 1869	86, 618
Fanning-mill attachment	P. Austin	Medina, Wis	Aug. 3, 1869	93, 165
Fanning-mill-elevator attachment	N. Hinman	Sparta Centre, Mich	Apr. 27, 1869	89, 406
Fanning-mill feeding-device	N. Kibler	Milton, Ill	Oct. 11, 1870	108, 154
Fanning-mill for cleaning grain	C. O. Guernsey	Russia, N. Y	Oct. 12, 1844	3, 794
Fanning-mill for cleaning grain	T. W. Jessup	Fredericktown, Md	June 18, 1811	....
Fanning-mill for cleaning grain	E. Walker	Springville, Pa	Mar. 26, 1825	....
Fanning-mill for cleaning wheat, &c	B. D. Beecher	New Haven, Conn	May 30, 1816	....
Fanning-mill for grain, &c	S. Dodge	Cleveland, Ohio	Feb. 9, 1808	....

Index of patents issued from the United States Patent Office from 1790 *to* 1873, *inclusive*—Continued.

Invention.	Inventor.	Residence.	Date.	No.
Fanning-mill for grain, &c	D. Donoghoe	New York	Apr. 16, 1821	....
Fanning-mill for grain or clover-seed, &c	O. Barrett, jr	Sandy Hill, N. Y	Mar. 23, 1808	....
Fanning-mill, grain and seed separator	H. Ogborn	Richmond, Ind	Nov. 12, 1867	70, 885
Fanning-mill or machine for winnowing grain	D. Philips and A. Jackson	Georgetown, Pa., and Franklin Mills, Va.	May 4, 1841	2, 080
Fanning-mill shakers, Attachment to	J. Van Houten	Mount Morris, N. Y	Mar. 29, 1864	42, 132
Fanning-mill shoes, Device for balancing	W. Crain	Millgrove, Ind	Sept. 20, 1870	107, 459
Fanning-mill, Adjusting-screw for the legs of	H. Wolf	Avon, Pa	May 14, 1867	64, 728
Fanning-mills, Bag-filler for	J. C. Gephart	Dowagiac, Mich	June 14, 1870	104, 297
Fare-box	H. Baranger	Saint Louis, Mo	Jan. 9, 1872	122, 549
Fare-box	H. Baranger	Saint Louis, Mo	May 13, 1873	138, 839
Fare-box	H. Baranger	Saint Louis, Mo	July 8, 1873	140, 669
Fare-box	W. B. Bartram	Norwalk, Conn	Jan. 22, 1861	31, 204
Fare-box	A. H. Bugher	Cincinnati, Ohio	Aug. 1, 1865	49, 078
Fare-box	A. H. Bugher	Cincinnati, Ohio	Aug. 1, 1865	49, 079
Fare-box	A. H. Bugher	Cincinnati, Ohio	Aug. 1, 1865	49, 080
Fare-box	T. L. Johnson	Louisville, Ky	Oct. 14, 1873	143, 698
Fare-box	S. H. Little	Rosemond, Ill	Sept. 30, 1873	143, 360
Fare-box	W. H. McLellan	New Orleans, La	Apr. 16, 1867	63, 864
Fare-box	C. Newman	San Francisco, Cal	Apr. 16, 1872	125, 836
Fare-box	Z. I. Pratt	Chicago, Ill	Mar. 18, 1873	136, 865
Fare-box	J. B. Slawson	New Orleans, La	July 28, 1857	17, 899
Fare-box	J. B. Slawson	New Orleans, La	Jan. 29, 1861	31, 262
Fare-box	J. B. Slawson	New Orleans, La	Aug. 7, 1866	57, 000
Fare-box	J. B. Slawson	New York, N. Y	June 11, 1872	127, 808
Fare-box	J. B. Slawson	New York, N. Y	Oct. 29, 1872	132, 698
Fare-box	J. B. Slawson	New York, N. Y	Dec. 24, 1872	134, 321
Fare-box	R. D. O. Smith	Washington, D. C	July 15, 1873	140, 959
Fare-box	J. F. Winchell	Springfield, Ohio	June 13, 1871	115, 999
Fare-box	J. F. Winchell	Springfield, Ohio	Dec. 12, 1871	121, 920
Fare-box	J. F. Winchell	Springfield, Ohio	Feb. 18, 1873	136, 118
Fare-box	J. F. Winchell	Springfield, Ohio	July 15, 1873	140, 985
Fare-box, Car	B. F. Grimes	Memphis, Tenn	Sept. 12, 1871	118, 927
Fare-box, Car and omnibus	J. B. Slawson	New Orleans, La	Sept. 24, 1867	69, 263
Fare-box, City-car and omnibus	J. Blackadder	New Orleans, La	Sept. 24, 1867	69, 166
Fare-box, Omnibus	I. S. Reeves	New Orleans, La	Feb. 23, 1858	19, 471
Fare-box, Omnibus, &c	J. B. Slawson	New Orleans, La	Aug. 31, 1858	21, 372
Fare-box, Omnibus, &c	W. H. Young	Chicago, Ill	June 21, 1870	104, 530
Fare-box, Passenger-car	H. Baranger	Saint Louis, Mo	June 28, 1870	104, 816
Fare-box, Passenger-car	J. E. Woodruff	Buffalo, N. Y	Mar. 31, 1868	76, 290
Fare-box, Portable	A. Bradley	New Orleans, La	Dec. 27, 1870	110, 428
Fare-box, Portable	J. S. Hagerty	Baltimore, Md	June 11, 1872	127, 869
Fare-box, Portable	W. J. Hooper	Baltimore, Md	Aug. 13, 1872	130, 499
Fare-box, Portable	W. L. May	Philadelphia, Pa	June 27, 1871	116, 466
Fare-box, Portable	J. W. Prendergast	New York, N. Y	Mar. 5, 1872	124, 287
Fare-box, Portable	J. C. Schooley	New York, N. Y	Mar. 26, 1872	125, 088
Fare-box, Portable	J. C. Schooley	New York, N. Y	Sept. 24, 1872	131, 710
Fare-box, Railway	J. H. Phares	Zionsville, Ind	Apr. 27, 1869	89, 337
Fare-box, Railway-car	G. C. Hathorn	New York, N. Y	Mar. 15, 1870	100, 758
Fare-box, Railway-car	R. H. Long	Philadelphia, Pa	Apr. 6, 1869	88, 648
Fare-box, Railway-car	W. G. Raoul	Independence, La	Feb. 1, 1870	99, 351
Fare-box, Railway-car, &c	H. Tupper	Buffalo, N. Y	Mar. 19, 1861	31, 755
Fare-box, Registering	T. L. Johnson	Louisville, Ky	Oct. 29, 1872	132, 535
Fare-box, Street-car	W. W. Wormood	Dubuque, Iowa	Feb. 8, 1870	99, 616
Fare-box, Street-car	W. H. Young	Chicago, Ill	Feb. 15, 1870	99, 804
Fare-box, Vehicle	J. B. Slawson	New York, N. Y	July 5, 1870	105, 005
Fare on street-railway cars, Mode of collecting	J. H. Dennis	Louisville, Ky	Dec. 17, 1861	33, 928
Fare-receiver, Registering	W. G. Smoot	Washington, D. C	Aug. 18, 1868	81, 223
Fare-recorder for cars, &c	G. R. Metten and O. S. Pease	Cleveland and Xenia, Ohio	Jan. 3, 1871	110, 770
Fare-recorder for cars	O. S. Pease	Xenia, Ohio	June 7, 1870	104, 059
Fare-register	F. Blackburn and G. W. Woodside	Philadelphia, Pa	Mar. 15, 1870	100, 715
Fare-register	C. A. Calvert	Manchester, Great Britain	Dec. 20, 1870	110, 339
Fare-register, Railway	S. F. Covington	Cincinnati, Ohio	May 5, 1868	77, 593
Fare-registering machine	M. W. Helton	Bloomington, Ind	July 24, 1860	29, 274
Fare taken on public conveyances, Apparatus for ascertaining the	D. F. Haasz	Philadelphia, Pa	Feb. 19, 1861	31, 513
Fare-ticket	T. A. Jebb	Buffalo, N. Y	July 2, 1872	128, 629
Farinaceous substances, Apparatus for desiccating and torrefying	F. Huckins and E. C. R. Walker	Roxbury, Mass	Mar. 19, 1861	31, 763
Farm-gate	G. W. Baker	Neponset, Ill	Mar. 29, 1859	23, 339
Farm-gate	M. Barthel	San Francisco, Cal	July 18, 1871	117, 143
Farm-gate	J. Baughman	Suffield Township, Ohio	June 29, 1869	91, 816
Farm-gate	G. F. Bissell	Oneonta, N. Y	Nov. 16, 1869	96, 771
Farm-gate	G. W. Blackwell	Lebanon, Ind	July 28, 1868	80, 439
Farm-gate	C. S. Bonney	Penn Yan, N. Y	Sept. 15, 1868	82, 075
Farm-gate	G. C. Bovey	Chillicothe, Ohio	Nov. 27, 1860	30, 784
Farm-gate	J. W. Brewster	West Laurens, N. Y	July 9, 1867	66, 454
Farm-gate	W. Brown	Addison, Mich	Mar. 17, 1868	75, 517
Farm-gate	W. S. Brown	Norwood, Ill	Oct. 7, 1873	143, 493
Farm-gate	E. and A. Buckman	East Greenbush, N. Y	Jan. 29, 1867	61, 605
Farm-gate	M. Burtless	Seneca Falls, N. Y	July 8, 1873	140, 612
Farm-gate	A. L. Butler	Ripon, Wis	Aug. 25, 1868	81, 337
Farm-gate	H. F. Butler	Troy, Mich	Feb. 25, 1868	74, 748
Farm-gate	J. W. Byers	Mechanicsburgh, Pa	May 25, 1869	90, 496
Farm-gate	J. Campbell and A. D. Krewson	Harrison, Ohio	Apr. 2, 1867	63, 363
Farm-gate	E. P. H. Capron	Springfield, Ohio	Apr. 28, 1868	77, 251
Farm-gate	W. D. Carlton	Morrisville, N. C	July 25, 1871	117, 381
Farm-gate	J. Case	Springfield, Ohio	May 29, 1866	55, 058
Farm-gate	L. Charles	Clear Spring, Md	Sept. 8, 1868	81, 982
Farm-gate	J. A. Cheatham	Nashville, Tenn	Oct. 23, 1866	58, 983
Farm-gate	W. D. Clark	Ottawa, Ill	Jan. 7, 1868	72, 976
Farm-gate	C. N. Cole	Pleasant Valley, N. Y	May 13, 1856	14, 851
Farm-gate	T. Collier	Springfield, Ohio	Aug. 13, 1867	67, 723
Farm-gate	S. B. Cooper	Beloit, Wis	Apr. 16, 1867	63, 861
Farm-gate	A. W. Cox	Indianapolis Ind	July 2, 1867	66, 300
Farm-gate	W. M. Crawford	Ashland, Ohio	Apr. 16, 1867	63, 862
Farm-gate	J. Curtis	Truro, Ill	Apr. 28, 1868	77, 174

Index of patents issued from the United States Patent Office from 1790 *to* 1873, *inclusive*—Continued.

Invention.	Inventor.	Residence.	Date.	No.
Farm-gate	J. David	East Enterprise, Ind	June 23, 1868	79, 109
Farm-gate	E. B. Decker	Carrollton, Ill	Aug. 5, 1873	141, 497
Farm-gate	L. S. Deming	Newington, Conn	Jan. 2, 1866	51, 811
Farm-gate	J. Dickason	Vevay, Ind	Feb. 5, 1867	61, 725
Farm-gate	J. Dickason	Vevay, Ind	June 21, 1870	104, 435
Farm-gate	J. Dickason and G. W. D. Culp	Vevay and Moore's Hill, Ind.	Jan. 4, 1870	98, 573
Farm-gate	A. Dietz	Raritan, N. J	Mar. 2, 1858	19, 4[illegible]9
Farm-gate	E. Easton	Prairieville, Mich	Aug. 18, 1868	81, 0[illegible]7
Farm-gate	W. Elliott	Stockport, N. Y	June 12, 1866	55, 4[illegible]6
Farm-gate	J. W. Epperson	Woodhull, Ill	June 25, 1867	66, 008
Farm-gate	G. J. Fiedler	Danby, Ill	July 26, 1870	105, 6[illegible]2
Farm-gate	G. W. Fox	Saint Joseph County, Mich	July 16, 1867	66, 8[illegible]3
Farm-gate	D. C. Frazeur and W. D. Cocklin.	Sidonsburgh, Pa	June 22, 1869	91, 735
Farm-gate	F. Gay	Bedford, Ohio	July 24, 1866	56, 550
Farm-gate	F. Gay	Bedford, Ohio	Sept. 4, 1866	57, 6[illegible]8
Farm-gate	G. Gibbs	Fairview, Ind	Sept. 22, 1868	82, 3[illegible]8
Farm-gate	L. Gibbs and H. M. Shaw	Fremont, Ohio	Jan. 2, 1866	51, 8[illegible]4
Farm-gate	C. E. Gillespie	Edwardsville, Ill	Dec. 17, 1872	133, 980
Farm-gate	N. J. Glover	Waveland, Ind	Oct. 1, 1867	6[illegible], 357
Farm-gate	S. Goewey	Dormansville, N. Y	July 16, 1867	6[illegible], 8[illegible]9
Farm-gate	C. E. Goodwin	Portland, Mich	Feb. 18, 1868	74, 5[illegible]9
Farm-gate	O. Graham	Lima, N. Y	Nov. 5, 1867	70, 548
Farm-gate	H. Gray and W. A. Bury	Goose Isle, Mich	Mar. 26, 1861	31, 800
Farm-gate	M. Gunshenan	New York, N. Y	Apr. 19, 1870	101, 998
Farm-gate	T. F. Hall	Circleville, Ohio	June 1, 1869	90, 839
Farm-gate	W. D. Harrah	Davenport, Iowa	Sept. 20, 1864	44, 3[illegible]6
Farm-gate	C. L. Harsen and M. R. Brailey	Norwalk, Ohio	May 1, 1855	12, 790
Farm-gate	C. Hart	Farmington, Ill	June 29, 1869	92, 046
Farm-gate	C. Hart	Farmington, Ill	Mar. 8, 1870	100, 620
Farm-gate	H. P. Haskins	Roscoe, Ill	Apr. 4, 1871	113, 657
Farm-gate	A. Hathaway	Prairie du Sac, Wis	Aug. 9, 1870	106, 156
Farm-gate	W. C. Hender	Miami, Ohio	June 30, 1863	39, 044
Farm-gate	H. A. Henderson	Avoca, N. Y	Mar. 24, 1868	75, 7[illegible]6
Farm-gate	W. G. Hermance	Albany, N. Y	Feb. 20, 1866	52, 713
Farm-gate	E. Hill, C. Ostrander, and H. A. Spink.	Bainbridge, Mich	Apr. 6, 1869	88, 569
Farm-gate	V. W. Horton	Palmyra, N. Y	May 19, 1868	78, 095
Farm-gate	J. G. Hunt	Cincinnati, Ohio	Apr. 18, 1865	47, 307
Farm-gate	T. W. Johnson	Granger, Ohio	June 9, 1868	78, 670
Farm-gate	J. Kelsey and J. McLain	Saint Mary's, Ohio	Feb. 26, 1867	62, 426
Farm-gate	H. H. Kelty	Northfield, Ohio	July 9, 1867	66, 501
Farm-gate	D. L. Koons	Trimble, Ohio	Jan. 5, 1869	85, 674
Farm-gate	I. L. Landis	Lancaster, Pa	Jan. 7, 1868	73, 016
Farm-gate	A. Larowe	Cohocton, N. Y	Dec. 18, 1866	60, 527
Farm-gate	J. Lee	Galesburgh, Ill	May 23, 1865	47, 840
Farm-gate	J. Lee	Galesburgh, Ill	June 26, 1866	55, 876
Farm-gate	W. Lucus	Olive Green, Ohio	Nov. 25, 1873	144, 994
Farm-gate	H. B. Lum	Sandusky, Ohio	May 22, 1855	12, 914
Farm-gate	D. F. Luse	Spring Mills, Pa	May 29, 1866	55, 129
Farm-gate	J. Martin	New Oxford, Pa	Mar. 21, 1865	46, 920
Farm-gate	J. Mayben	Milroy, Pa	May 3, 1870	102, 569
Farm-gate	D. McCurdy	Ottawa, Ohio	Aug. 28, 1866	57, 537
Farm-gate	R. W. McFarland	Monticello, Wis	Nov. 13, 1866	59, 621
Farm-gate	W. McGuire	Chess Springs, Pa	Aug. 25, 1868	81, 521
Farm-gate	H. D. Mead	Wayne, N. Y	Sept. 11, 1866	57, 946
Farm-gate	A. T. Morris	Seal, Ohio	Mar. 30, 1869	88, 323
Farm-gate	D. Mougey	Marshallsville, Ohio	Oct. 20, 1868	83, 197
Farm-gate	J. T. Maxley	Owasso, Mich	Apr. 27, 1869	89, 424
Farm-gate	H. W. Mullennex	Alpine, N. Y	May 13, 1873	138, 918
Farm-gate	F. Müller	Mokena, Ill	Mar. 6, 1866	53, 084
Farm-gate	J. C. Murphy	Mendota, Ill	Apr. 10, 1866	53, 856
Farm-gate	J. H. and P. J. Murphy	Abingdon, Ill	Mar. 15, 1870	100, 787
Farm-gate	D. S. Neal	Lynn, Mass	Mar. 26, 1867	63, 288
Farm-gate	D. A. Neidig	Paris, Ohio	Dec. 30, 1873	146, 015
Farm-gate	W. Newlove	Penn Yan, N. Y	Nov. 23, 1858	22, 131
Farm-gate	E. Nicholson	East Rockport, Ohio	Aug. 11, 1863	39, 491
Farm-gate	J. H. Nonamaker	Middletown, Pa	Apr. 23, 1867	64, 133
Farm-gate	C. Ostrander	Lodi, Wis	Dec. 21, 1869	98, 185
Farm-gate	H. Palmerlee	Grand Rapids, Mich	Jan. 26, 1869	86, 315
Farm-gate	E. H. Peck	Brownhelm, Ohio	Dec. 11, 1866	60, 415
Farm-gate	D. Phillips	Shaftsbury, Vt	Nov. 7, 1854	11, 905
Farm-gate	H. Piper	Haskins, Ohio	May 10, 1870	102, 859
Farm-gate	J. Pool	Elizabeth City, N. C	July 12, 1870	105, 366
Farm-gate	A. J. Potter	Omaha, Nebr	Dec. 29, 1868	85, 474
Farm-gate	J. H. Reinhart	McKay, Ohio	Apr. 23, 1867	64, 034
Farm-gate	C. Rich	Marshallsville, Ohio	May 5, 1868	77, 658
Farm-gate	J. C. Rohrer	Gomer, Ohio	Nov. 4, 1873	144, 291
Farm-gate	J. Root	Cass County, Mich	Sept. 1, 1868	81, 820
Farm-gate	C. W. Saladee	Saint Catharine's, Canada	Aug. 16, 1870	106, 411
Farm-gate	S. Scott	Yane, Ohio	Dec. 21, 1869	98, 195
Farm-gate	E. C. Sears	Crystal Lake, Ill	Apr. 9, 1867	63, 757
Farm-gate	O. E. Seymour	Madison, Ind	Dec. 1, 1868	84, 512
Farm-gate	H. S. Shisler	Manheim Township, Pa	Aug. 11, 1868	81, 017
Farm-gate	D. Shockey	Waynesborough, Pa	Dec. 15, 1868	85, 033
Farm-gate	D. Shockey	Waynesborough, Pa	Oct. 5, 1869	95, 524
Farm-gate	S. A. Skinner	Derby, Vt	Oct. 9, 1855	13, 658
Farm-gate	H. R. Sloat	Sloatsburgh, N. Y	July 9, 1861	32, 793
Farm-gate	W. F. Smelley	Vevay, Ind	May 30, 1865	47, 994
Farm-gate	A. D. Smith	Grafton, Ohio	Mar. 19, 1867	62, 975
Farm-gate	A. J. Smith and G. S. Hudson	Ellisburgh, N. Y	Feb. 29, 1866	52, 759
Farm-gate	G. Smith	Providence, R. I	Jan. 26, 1869	86, 184
Farm-gate	R. M. Smith	La Fargeville, N. Y	Mar. 30, 1869	88, 339
Farm-gate	B. Snyder	Clinton, Wis	Dec. 1, 1868	84, 653
Farm-gate	H. Snyder	Wooster, Ohio	Mar. 6, 1866	53, 053
Farm-gate	G. W. States and A. W. Lutts	Norwalk, Ohio	Sept. 24, 1867	69, 138
Farm-gate	T. Steers	Milton, Wis	July 15, 1873	140, 960

Index of patents issued from the United States Patent Office from 1790 *to* 1873, *inclusive*—Continued.

Invention.	Inventor.	Residence.	Date.	No.
Farm-gate	J. G. Talbot	Sloansville, N. Y	June 11, 1867	65, 707
Farm-gate	J. G. Talbot	Sloansville, N. Y	Mar. 9, 1869	87, 600
Farm-gate	J. Tanner	Hebron, Ky	Feb. 2, 1869	86, 603
Farm-gate	L F. Tanner	Milan, Ind	Aug. 25, 1868	81, 558
Farm-gate	A. Taylor	New Hartford, Conn	June 16, 1868	79, 025
Farm-gate	G. Taylor	Richmond, Ind	June 24, 1856	15, 213
Farm-gate	A. C. Teel	Girard, Ill	Dec. 1, 1863	40, 777
Farm-gate	F. F. Terry	Port Gibson, N. Y	Nov. 20, 1866	59, 873
Farm-gate	C. W. Todd	Spring Arbor, Mich	June 8, 1869	90, 971
Farm-gate	J. A. Treat	Tallmadge, Ohio	Oct. 12, 1858	21, 785
Farm-gate	S. Vreeland	Cuba, N. Y	July 20, 1869	92, 914
Farm-gate	J. B. Webb	Muscatine, Iowa	June 4, 1867	65, 519
Farm-gate	B. Werrich and J. Smith	Middleburgh, Ind	Apr. 21, 1868	77, 142
Farm-gate	J. W. Westbrooks and A. J. Clemmons.	Aberdeen, Miss	May 20, 1873	139, 219
Farm-gate	T. B. Wickham	Granville, Ohio	Feb. 5, 1867	61, 785
Farm-gate	T. B. Wickham	Granville, Ohio	Sept. 29, 1868	82, 668
Farm-gate	W. H. Wright	Rochester, N. Y	Mar. 1, 1870	100, 484
Farm-gate	G. Yeomans	Romulus, Mich	Jan. 12, 1869	85, 886
Farm-gate	I. N. Young	Swann, Ind	July 16, 1867	66, 764
Farm-gate, Approach-opening	C. W. Smart	Watertown, N. Y	Dec. 22, 1857	18, 932
Farm-gate, Iron	W. F. Whitney	Milwaukee, Wis	Jan. 12, 1869	85, 882
Farm-gate, Self-acting	E. Dunbar	Philadelphia, Pa	Mar. 4, 1856	14, 351
Farm-gate, Sliding	T. Ellison	Abingdon, Ill	July 20, 1869	92, 805
Farm-gates by approaching vehicles, Method of operating.	A. J. Hamilton	Kewanee, Ill	Mar. 8, 1859	23, 169
Farm-gates by hand, Mode of opening and closing	G. Yates	West Dresden, N. Y	Apr. 19, 1859	23, 738
Farm-gates, Closing	T. B. Hand	Madison, Ind	Oct. 20, 1857	18, 449
Farm-gates, Device for raising or lowering	D. E. Fenn	Tallmadge, Ohio	Mar. 10, 1857	16, 791
Farm-gates, Hanging	J. Filson	Milroy, Pa	Feb. 8, 1853	9, 576
Farm-gates, Hinging	I. S. Rowland	West Earl, Pa	Jan. 13, 1857	16, 400
Farm-gates, Hanging and elevating	J. F. Downing	Erie, Pa	Apr. 15, 1856	14, 689
Farm-gates, Hanging and latching	J. Kinman	Freeport, Ill	Apr. 25, 1865	47, 429
Farm-gates, Opening and closing	J. A. Ayres	Hartford, Conn	Jan. 22, 1856	14, 131
Farm-gates, Opening and closing	W. F. C. Beattie	Cornwall, N. Y	May 18, 1858	20, 247
Farm-gates, Opening and closing	W. T. Boggs	Cincinnati, Ohio	Oct. 19, 1858	21, 811
Farm-gates, Opening and closing	B. M. Dorr	Kewanee, Ill	July 5, 1859	24, 619
Farm-gates, Opening and closing	D. E. Fenn	Tallmadge, Ohio	Oct. 14, 1856	15, 881
Farm-gates, Opening and closing	W. G. Hermance	Geneva, N. Y	Nov. 9, 1858	22, 023
Farm-gates, Opening and closing	W. Sherwood	Beloit, Wis	June 30, 1857	17, 699
Farm-gates, Opening and closing	D. Warren	Gettysburgh, Pa	June 21, 1859	24, 513
Farm-gates, Operating	J. H. Butler and P. G. Van Houten.	Cohocton, N. Y	July 12, 1859	24, 717
Farm-gates, Operating	J. K. Weber	Seneca Falls, N. Y	Oct. 9, 1855	13, 673
Farm-gates, Operating	C. Winegar	Union Springs, N. Y	Aug. 26, 1856	15, 631
Farm-gates, Raising, lowering, and operating	C. Hunter and N. Isham	Norwalk, Ohio	Aug. 12, 1856	15, 518
Farmer's boiler	G. H. Buckley	Quincy, Ill	July 25, 1871	117, 254
Farmer's boiler	P. Colvin	Pecatonica, Ill	Nov. 25, 1873	144, 890
Farmer's boiler	M. Ellis	South Carver, Mass	Apr. 28, 1863	38, 299
Farmer's boiler	J. H. Hunter	Pennington Point, Ill	Dec. 13, 1870	110, 139
Farmer's boiler	A. N. Merrill	Batavia, Ill	Jan. 18, 1870	98, 988
Farmer's boiler	A. N. Merrill	Batavia, Ill	May 31, 1870	103, 639
Farmer's boiler	A. B. Nott	Fairhaven, Mass	Jan. 21, 1873	135, 005
Farmer's boiler	W. M. Pryor and R. Ludwick	Kellogg Iowa	July 4, 1871	116, 753
Farmer's boiler	G. Stevenson	Zionsville, Ind	Aug. 15, 1871	118, 068
Farmer's boiler	R. W. Thickens	Batavia, Ill	July 26, 1870	105, 740
Farmer's boiler	R. W. Thickens and D. R. Sperry.	Batavia, Ill	Mar. 28, 1871	113, 222
Farrier's clinching-tool	S. Ogden	Sicily, Ohio	Apr. 23, 1872	126, 078
Farrier's tool	E. Warren and W. Johnston	Marshall, Mich	May 24, 1864	42, 918
Fat, lard, &c., Machine for cutting	C. Forschner	New York, N. Y	Aug. 14, 1866	57, 111
Fat, Machine for cutting	W. E. Boulger	Janesville, Wis	July 3, 1860	28, 960
Fat, Machine for cutting	J. M. Hunter	New York, N. Y	May 8, 1860	28, 179
Fat, Machine for cutting	A. Pfaff	Baltimore, Md	June 28, 1870	104, 878
Fat, Machine for cutting up	F. Miller and J. N. McIntire	New York, N. Y	Mar. 15, 1864	41, 963
Fats and oils from grains, &c., Process and apparatus for separating.	E. S. Hutchinson	Baltimore, Md	Apr. 2, 1872	125, 300
Fats and oils, Hardening	C. W. Schindler	New York, N. Y	Nov. 5, 1850	7, 760
Fats into fatty acids and glycerine, Method of decomposing.	R. A. Tilghman	Philadelphia, Pa	May 15, 1860	28, 315
Fats, Process for decomposing	R. A. Wright and L. J. Fouché	Paris, France	Jan. 25, 1859	22, 765
Fats, Purifying and separating	A. Paraf	New York, N. Y	Apr. 8, 1873	137, 564
Fats, Saponifying	G. F. Wilson and G. Payne	Belmont, Vauxhall, England.	June 17, 1856	15, 158
Fatty matter from residuum, Obtaining	D. Thain and W. Jackson	Philadelphia, Pa	Jan. 10, 1860	26, 799
Fatty matter from vegetable substances, Apparatus for separating.	T. Sim	Baltimore, Md	Dec. 12, 1871	121, 903
Fatty matter, Process for extracting	E. Deiss	Paris, France	Apr. 27, 1858	20, 048
Fatty matter, Treating	L. L. A. E. P. De la Peyrouse	Paris, France	Nov. 21, 1871	121, 162
Fatty material, Purifying	J. B. Moinier and P. H. Bontigny.	Paris, France	Feb. 8, 1853	9, 572
Fatty substances, Hardening	B. C. Tilghman	Philadelphia, Pa	Jan. 11, 1859	22, 593
Fatty substances, Process and apparatus for extracting.	C. F. A. Simonin	Philadelphia, Pa	Sept. 19, 1871	119, 188
Faucet	J. H. Alexander	Geneva, N. Y	Dec. 19, 1871	121, 978
Faucet	C. E. Allan	Boston, Mass	Nov. 27, 1866	59, 939
Faucet	L. L. Alrich	Carthage, Miss	Sept. 25, 1860	30, 110
Faucet	J. Ashcroft	New York, N. Y	Apr. 5, 1870	101, 410
Faucet	F. M. Bachman and S. Ricker	Fredericksburgh, Pa	Mar. 8, 1870	100, 489
Faucet	A. Bain	New York, N. Y	Nov. 15, 1864	45, 012
Faucet	W. Ball	Cabotville, Mass	Feb. 1, 1848	5, 431
Faucet	S. Barker	Hartford, Conn	May 3, 1859	23, 817
Faucet	T. W. Bartholomew	New York, N. Y	Feb. 7, 1871	111, 569
Faucet	D. L. and R. Bollerman	New York, N. Y	Feb. 2, 1869	86, 498
Faucet	R. M. Bouton	West Troy, N. Y	Apr. 18, 1854	10, 800
Faucet	C. K. Bradford	Linn, Mass	Apr. 19, 1859	23, 658
Faucet	A. Brinckmann	New York, N. Y	May 28, 1867	65, 160
Faucet	J. Broughton	Chicago, Ill	Sept. 8, 1863	39, 795

Index of patents issued from the United States Patent Office from 1790 *to* 1873, *inclusive*—Continued.

Invention.	Inventor.	Residence.	Date.	No.
Faucet	J. Broughton	New York, N. Y	Nov. 8, 1864	44, 933
Faucet	J. Broughton	New York, N. Y	Sept. 21, 1869	95, 077
Faucet	H. L. and J. A. Buckwalter	Kimberton, Pa	Oct. 9, 1866	58, 592
Faucet	J. Bulger	Port Sherman, Mich	Mar. 15, 1870	100, 721
Faucet	G. F. Burkhardt	Roxbury, Mass	Oct. 11, 1864	44, 596
Faucet	M. Burnett	Boston, Mass	June 18, 1867	65, 791
Faucet	M. Burnett	Boston, Mass	Dec. 5, 1871	121, 487
Faucet	E. Candler	London, England	Sept. 25, 1866	58, 356
Faucet	I. Carey	Morristown, N. J	May 10, 1870	102, 773
Faucet	W. P. Clark	Medford, Mass	Nov. 2, 1869	96, 394
Faucet	W. F. Class and L. W. Sapp	Cleveland, Ohio	Oct. 25, 1870	108, 568
Faucet	W. Cleveland	Orange, N. J	Sept. 25, 1860	30, 122
Faucet	W. Cleveland	Orange, N. J	Aug. 8, 1871	117, 866
Faucet	W. Cleveland	Orange, N. J	Mar. 12, 1872	124, 546
Faucet	J. S. Clute and H. W. and W. Frissler.	Cleveland, Ohio	Apr. 4, 1871	113, 630
Faucet	D. N. B. Coffin, jr	Newton, Centre, Mass	June 16, 1857	17, 604
Faucet	T. M. Coffin	Plymouth, Mass	Oct. 9, 1866	58, 660
Faucet	W. S. Cooper	Philadelphia, Pa	Oct. 23, 1866	59, 122
Faucet	W. S. Cooper	Philadelphia, Pa	May 17, 1870	103, 145
Faucet	G. T. Dalton	New York, N. Y	May 25, 1869	90, 343
Faucet	J. S. Davison	Cranberry, N. J	Apr. 16, 1861	32, 053
Faucet	J. M. A. Dew	Chicago, Ill	Feb. 22, 1870	100, 126
Faucet	A. Doll and C. Elling	Cleveland, Ohio	Nov. 11, 1873	144, 516
Faucet	J. H. Dorst	New Albany, Ind	Jan. 11, 1870	98, 749
Faucet	C. A. Douglas	Franklin, N. Y	Nov. 15, 1870	109, 191
Faucet	D. Drawbaugh	Eberly's Mill, Pa	Nov. 20, 1866	59, 792
Faucet	D. Drawbaugh	Eberly's Mill, Pa	Nov. 20, 1866	59, 793
Faucet	E. Duchamp	Saint Martinsville, La	Sept. 13, 1859	25, 397
Faucet	A. Eggleston	Fall River, Mass	June 14, 1870	104, 128
Faucet	J. Fahrney	Boonsborough, Md	Sept. 4, 1866	57, 818
Faucet	J. Farnan	Cleveland, Ohio	Sept. 18, 1860	30, 055
Faucet	J. Flattery	Brooklyn, N. Y	July 24, 1860	29, 263
Faucet	W. Fowler	New York, N. Y	May 8, 1855	12, 817
Faucet	A. Fuller	Boston, Mass	Oct. 16, 1855	13, 677
Faucet	A. Fuller	Cincinnati, Ohio	Aug. 30, 1859	25, 253
Faucet	A. Fuller	Cincinnati, Ohio	Feb. 18, 1862	34, 419
Faucet	H. Getty	Brooklyn, N. Y	July 6, 1858	20, 788
Faucet	J. Goodridge	Boston, Mass	July 29, 1856	15, 459
Faucet	J. Goodridge	Boston, Mass	Sept. 9, 1856	15, 719
Faucet	A. J. Gove	San Francisco, Cal	Apr. 29, 1862	35, 094
Faucet	R. Graham	Brooklyn, N. Y	Mar. 26, 1861	31, 799
Faucet	D. W. Green	Port Chester, N. Y	Mar. 23, 1869	88, 032
Faucet	J. Grundy	Stoneham, Mass	Apr. 12, 1864	42, 286
Faucet	O. Hanks	Cincinnati, Ohio	Nov. 29, 1870	109, 730
Faucet	A. S. Hart	Buffalo, N. Y	Jan. 17, 1860	26, 844
Faucet	I. W. Harvey	Norwich, Conn	June 7, 1870	104, 025
Faucet	J. Hayden, jr	Haydenville, Mass	Jan. 26, 1869	86, 225
Faucet	W. H. Hedges	Newark, N. J	Feb. 21, 1871	112, 835
Faucet	J. Heine and J. Vonficht	Toledo, Ohio	Apr. 19, 1870	102, 120
Faucet	T. Hersee and P. J. Bourgnon	Buffalo, N. Y	Oct. 4, 1859	25, 643
Faucet	W. B. Hersman	Richmond, Va	July 23, 1872	129, 823
Faucet	P. Hille	Union, N. J	Oct. 28, 1873	144, 097
Faucet	J. Hills	Haydensville, Mass	Dec. 6, 1870	109, 902
Faucet	N. Hotz	Green Point, Brooklyn, N. Y	June 11, 1872	127, 883
Faucet	W. H. Humphrey	Lansingburgh, N. Y	July 23, 1867	67, 056
Faucet	G. R. Huntley	Taunton, Mass	Sept. 11, 1866	57, 910
Faucet	G. R. Huntley	Taunton, Mass	Mar. 24, 1868	75, 918
Faucet	J. Jahrans and J. G. Bickel	Buffalo, N. Y	June 3, 1862	35, 450
Faucet	N. Jenkins	Boston, Mass	Aug. 22, 1865	49, 527
Faucet	T. J. Jones and T. L. Webster	Summit, N. J., and Brooklyn, N. Y.	May 7, 1867	64, 427
Faucet	A. D. King and J. W. King	Granville Corners, Mass., and Bridgeport, Conn.	Nov. 4, 1873	144, 208
Faucet	J. Knoche	Cincinnati, Ohio	June 21, 1870	104, 602
Faucet	J. Knoche	Cincinnati, Ohio	June 27, 1871	116, 326
Faucet	L. J. Knowles	Warren, Mass	June 2, 1857	17, 433
Faucet	B. F. Kraft	Reading, Pa	Sept. 15, 1868	82, 232
Faucet	A. W. P. Ladd	San Francisco, Cal	July 2, 1861	32, 699
Faucet	J. Laing	Hoboken, N. J	May 28, 1867	65, 238
Faucet	J. Leitch	Buffalo, N. Y	Dec. 15, 1863	40, 939
Faucet	J. H. Lord	San Francisco, Cal	July 26, 1870	105, 818
Faucet	J. R. Mackay	Meriden, Conn	Feb. 7, 1871	111, 659
Faucet	J. Marchbank	Lansingburgh, N. Y	Jan. 28, 1868	73, 822
Faucet	J. Marchbank and W. H. Humphrey.	Lansingburgh, N. Y	Nov. 6, 1866	59, 426
Faucet	J. Matthews, jr	New York, N. Y	June 27, 1865	48, 421
Faucet	W. McKay	Newburyport, Mass	June 7, 1870	104, 050
Faucet	J. McKenna	Pittsburgh, Pa	Sept. 16, 1862	36, 473
Faucet	J., A., and T. McKenna	Pittsburgh, Pa	Nov. 4, 1873	144, 213
Faucet	F. Meyrose	Saint Louis, Mo	June 14, 1864	43, 125
Faucet	W. Morgenstern	New York, N. Y	July 12, 1870	105, 360
Faucet	O. F. Morrill	Chelsea, Mass	Dec. 4, 1866	60, 223
Faucet	A. J. Morse	Melrose, Mass	Oct. 25, 1864	44, 813
Faucet	R. Murray	Boston, Mass	Jan. 17, 1865	45, 949
Faucet	J. Neumann	New York, N. Y	Apr. 30, 1861	32, 201
Faucet	C. B. O'Sullivan	New Orleans, La	Feb. 28, 1871	112, 173
Faucet	S. W. and J. F. Palmer	Auburn, N. Y	Aug. 3, 1869	93, 222
Faucet	E. Parks	Winchendon, Mass	July 23, 1861	32, 888
Faucet	G. G. Percival	Brooklyn, N. Y	July 25, 1865	48, 995
Faucet	S. Pfleger	Reading, Pa	Mar. 7, 1871	112, 487
Faucet	J. Phelps	Owego, N. Y	Sept. 24, 1867	69, 124
Faucet	W. Pinkerman	Bridgeport, N. Y	Aug. 18, 1863	39, 585
Faucet	J. Powell	Cincinnati, Ohio	Mar. 22, 1859	23, 310
Faucet	J. Powell	Cincinnati, Ohio	Sept. 6, 1859	25, 349
Faucet	A. D. Puffer	Boston, Mass	June 28, 1870	104, 880
Faucet	A. Putnam, jr	Saratoga Springs, N. Y	Nov. 1, 1870	108, 824
Faucet	G. W. Randall	Boston, Mass	Apr. 5, 1859	23, 533

Index of patents issued from the United States Patent Office from 1790 to 1873, inclusive—Continued.

Invention.	Inventor.	Residence.	Date.	No.
Faucet	J. Regester	Baltimore, Md	Apr. 9, 1867	63, 748
Faucet	E. Ripley	Troy, N. Y	Mar. 14, 1854	10, 641
Faucet	F. Roach	Boston, Mass	Sept. 8, 1868	81, 948
Faucet	M. Robbins and J. Powell	Cincinnati, Ohio	Dec. 21, 1858	22, 402
Faucet	A. Rooker	London, England	June 16, 1868	78, 897
Faucet	J. Schaefer and G. Schock	New York, N. Y	July 12, 1870	105, 258
Faucet	H. Schnautz and H. Bremenkamp.	Cincinnati, Ohio	Nov. 15, 1864	45, 085
Faucet	C. Schullian	New York, N. Y	July 3, 1866	56, 105
Faucet	E. Scrannage, jr	Melrose, Mass	May 6, 1873	138, 700
Faucet	M. Scrannage and W. H. Bate	Medford and East Somerville, Mass.	Dec. 20, 1870	110, 292
Faucet	A. D. Smith	Grafton, Ohio	Oct. 26, 1869	96, 158
Faucet	A. D. Smith	Grafton, Ohio	Nov. 28, 1871	121, 430
Faucet	D. P. Smith	Salem, N. J	Aug. 18, 1868	81, 305
Faucet	J. N. Smith	Jersey City, N. J	Mar. 14, 1865	46, 862
Faucet	J. N. Smith	Jersey City, N. J	Nov. 13, 1866	59, 672
Faucet	J. N. Smith	Jersey City, N. J	July 9, 1872	128, 762
Faucet	T. Somerville	Washington, D. C	Apr. 5, 1870	101, 533
Faucet	C. W. Stearns	Springfield, Mass	July 22, 1851	8, 237
Faucet	E. Stebbins	Chicopee, Mass	July 9, 1861	32, 797
Faucet	E. A. Sterry	Norwich Town, Conn	June 12, 1855	13, 047
Faucet	F. P. Stiker	Buffalo, N. Y	Aug. 20, 1867	67, 925
Faucet	J. T. Stilwell	Dowagiac, Mich	Aug. 6, 1867	67, 605
Faucet	S. C. Stokes	Manchester, N. H	Oct. 15, 1867	69, 859
Faucet	H. Strater, jr	Boston, Mass	Mar. 7, 1865	46, 726
Faucet	H. Strater, jr	Boston, Mass	Mar. 7, 1865	46, 727
Faucet	H. Strater, jr	Boston, Mass	Mar. 7, 1865	46, 728
Faucet	J. R. Street and S. M. Davis	Washington, D. C	Nov. 27, 1866	60, 082
Faucet	I. C. Tate	New London, Conn	June 12, 1860	28, 699
Faucet	G. Taylor	New York, N. Y	May 24, 1870	103, 523
Faucet	S. R. Thompson	Portsmouth, N. H	Nov. 11, 1873	144, 422
Faucet	T. J. Thorn and J. Dennis, jr	Skaneateles, N. Y., and Washington, D. C.	Aug. 14, 1866	57, 220
Faucet	D. H. Thorp	Chelsea, Mass	June 19, 1860	28, 819
Faucet	H. B. Tiffany	Medina, Ohio	Aug. 2, 1870	106, 000
Faucet	R. Tilly	Brooklyn, N. Y	June 21, 1870	104, 663
Faucet	J. W. Trafton	Springfield, Mass	July 23, 1872	129, 872
Faucet	W. A. Traver	Rhinebeck, N. Y	Oct. 8, 1872	132, 036
Faucet	W. H. Trissler	Cleveland, Ohio	May 3, 1870	102, 624
Faucet	J. E. Tucker	Boston, Mass	Nov. 12, 1867	70, 919
Faucet	J. E. Tucker	Boston, Mass	Dec. 22, 1868	85, 257
Faucet	C. Utter and G. E. Bruster	Newark, N. J	Aug. 1, 1871	117, 699
Faucet	H. Varwig	Cincinnati, Ohio	Apr. 15, 1873	137, 981
Faucet	C. T. Vose	Stapleton, N. Y	Aug. 6, 1872	130, 341
Faucet	C. Weaver	Easton, Pa	Oct. 8, 1867	69, 730
Faucet	W. Weaver	Nashua, N. H	Nov. 2, 1869	96, 519
Faucet	D. Wellington	Boston, Mass	Apr. 26, 1870	102, 341
Faucet	H. G. Whitaker	West Brattleborough, Vt	May 21, 1872	127, 128
Faucet	J. White	Philadelphia, Pa	Oct. 28, 1873	144, 175
Faucet	N. P. Whittelsey	Meriden, Conn	July 6, 1858	20, 853
Faucet	F. Wieshofer	Fremont, Ohio	Dec. 13, 1870	110, 099
Faucet	J. Wilcox	Thompsonville, Conn	Mar. 19, 1867	63, 127
Faucet	W. C. Wise and J. Ashman	Chelsea, Mass	Feb. 21, 1871	112, 000
Faucet	M. Woodbury	Boston, Mass	Mar. 11, 1856	14, 429
Faucet	C. T. Woodman	Boston, Mass	Feb. 16, 1864	41, 659
Faucet	W. E. Worthen	New York, N. Y	Apr. 22, 1862	35, 054
Faucet	E. Young	Cleveland, Ohio	June 28, 1870	104, 811
Faucet	M. Zimmerman	Earl Township, Pa	Jan. 5, 1869	85, 717
Faucet	O. Zwietusch	Milwaukee, Wis	July 16, 1872	129, 263
Faucet, Alarm	T. M. Biddle	Fort Wayne, Ind	Dec. 7, 1869	97, 594
Faucet and valve, Grinding	F. Shaw	Philadelphia, Pa	June 13, 1865	48, 216
Faucet and vent	J. Jahraus	Buffalo, N. Y	Sept. 22, 1863	40, 044
Faucet and vent, Combined beer	J. G. Bickel	Buffalo, N. Y	May 24, 1864	42, 828
Faucet and vent, Combined beer	J. Miller	Buffalo, N. Y	May 10, 1864	42, 676
Faucet, Anti-frost	F. H. Bartholomew	New York, N. Y	Nov. 11, 1856	16, 043
Faucet-attachment	J. Church	Saint Louis, Mo	Sept. 23, 1873	143, 062
Faucet-attachment	E. L. Dunbar	Bay City, Mich	July 22, 1873	141, 040
Faucet-attachment	G. Johnson and W. H. Milliken	San Francisco, Cal	Dec. 3, 1867	71, 622
Faucet-attachment, Self-closing	R. McConnell, W. Truesdell, and F. Mertsheimer.	Omaha, Nebr	Sept. 30, 1873	143, 291
Faucet-attachment to cans	J. H. Garrigan and F. L. Hall	Sacramento, Cal., and Reno, Nev.	Nov. 30, 1869	97, 290
Faucet, Balance-gate	E. Ripley	Troy, N. Y	May 8, 1855	12, 830
Faucet, Basin	J. Benson	Yonkers, N. Y	June 23, 1868	79, 190
Faucet, Basin	J. Chambers	Boston, Mass	Dec. 24, 1867	72, 453
Faucet, Basin	W. P. Clark	Charlestown, Mass	Aug. 2, 1864	43, 736
Faucet, Basin	W. Gordon	Philadelphia, Pa	May 28, 1872	127, 341
Faucet, Basin	J. Hills	Haydenville, Mass	Aug. 14, 1866	57, 134
Faucet, Basin	W. C. Marshall and H. W. Smith	Hartford, Conn	June 9, 1857	17, 511
Faucet, Basin	F. Roach	Boston, Mass	Dec. 3, 1867	71, 790
Faucet, Basin	R. P. Ross	Bethlehem, Pa	Mar. 10, 1868	75, 464
Faucet, Basin	E. Stebbins	Chicopee, Mass	May 19, 1857	17, 342
Faucet, Beer	T. Ahrens	Louisville, Ky	Aug. 2, 1870	106, 020
Faucet, Beer	J. C. Baer	Cincinnati, Ohio	Apr. 24, 1866	54, 091
Faucet, Beer	C. Bourgeois	Buffalo, N. Y	July 16, 1872	129, 090
Faucet, Beer	C. Brown and C. McGhie	Chicago, Ill	July 17, 1866	56, 357
Faucet, Beer	P. F. Donnelly	San Francisco, Cal	July 30, 1872	129, 938
Faucet, Beer	O. T. Earle	Norwalk, Conn	Aug. 26, 1873	142, 090
Faucet, Beer	J. Firmenich	Buffalo, N. Y	Jan. 2, 1866	51, 820
Faucet, Beer	J. Firmenich	Buffalo, N. Y	Mar. 20, 1866	53, 283
Faucet, Beer	J. Firmenich	Buffalo, N. Y	May 15, 1866	54, 707
Faucet, Beer	A. Hallowell	Lowell, Mass	June 20, 1865	48, 273
Faucet, Beer	J. Jahraus	Buffalo, N. Y	May 1, 1866	54, 359
Faucet, Beer	C. Jakob	New Orleans, La	Oct. 4, 1870	108, 025
Faucet, Beer	F. Manz	Allegheny City, Pa	Apr. 12, 1870	101, 966
Faucet, Beer	T. Marsh	Smithfield, R. I	Mar. 12, 1867	62, 864
Faucet, Beer	J. Miller	Buffalo, N. Y	Apr. 25, 1865	47, 442

Index of patents issued from the United States Patent Office from 1790 *to* 1873, *inclusive*—Continued.

Invention.	Inventor.	Residence.	Date.	No.
Faucet, Beer	L. Mullenhoff	Buffalo, N. Y	May 14, 1872	126, 736
Faucet, Beer	L. Poh	Buffalo, N. Y	July 4, 1865	48, 588
Faucet, Beer	A. D. Puffer	Somerville, Mass	Feb. 9, 1869	86, 861
Faucet, Beer	J. M. Stark	Buffalo, N. Y	Feb. 18, 1868	74, 625
Faucet, Beer	E. Stirret	Buffalo, N. Y	Apr. 3, 1866	53, 692
Faucet, Beer	F. J. Walz and C. Steck	Hudson, N. J	Mar. 3, 1868	75, 085
Faucet, Beer and ale	J. Deasey	Fall River, Mass	Nov. 11, 1873	144, 447
Faucet, Boring	J. R. Lawrence and J. G. Johnson.	Cutler, Me	Oct. 6, 1868	82, 851
Faucet, Boring	S. McGee	Madison, N. J	Mar. 3, 1868	75, 177
Faucet, Boring	A. Weed	Boston, Mass	Feb. 25, 1868	74, 961
Faucet, Boring	A. Weed	Boston, Mass	Feb. 16, 1869	86, 956
Faucet, Bottle	H. Fischer	Lanesville, Ind	Nov. 1, 1870	108, 898
Faucet, Bottle	W. I. Luther	Rochester, N. Y	Mar. 17, 1868	75, 557
Faucet, Bottle	J. Sargent and L. F. Munger	Rochester, N. Y	Nov. 21, 1871	121, 201
Faucet, Bung	G. D. Lee	Brookhaven, N. Y	Dec. 9, 1873	145, 432
Faucet, Casting	O. T. Wood	Pittsburgh, Pa	May 25, 1858	20, 395
Faucet, Compound	W. S. Bate	Philadelphia, Pa	July 22, 1873	141, 102
Faucet, Compression basin	W. Gordon	Philadelphia, Pa	Feb. 4, 1873	135, 541
Faucet-connection	T. H. Brady	New Britain, Conn	Nov. 9, 1869	96, 545
Faucet-coupling	O. Salgee	Brooklyn, N. Y	Sept. 26, 1871	119, 243
Faucet, Diaphragm	W. Blake	Boston, Mass	Nov. 6, 1866	59, 346
Faucet, Double	E. Sauter	Hartford, Conn	Aug. 10, 1869	93, 561
Faucet, Filter	R. B. Coar	Jersey City, N. J	Aug. 27, 1867	68, 170
Faucet, Filtering	L. Finger	Boston, Mass	Nov. 20, 1855	13, 836
Faucet, Filtering	G. H. Fox and H. J. Siller	Boston and East Cambridge, Mass.	Dec. 16, 1856	16, 232
Faucet, Filtering	G. Hillegass	Philadelphia, Pa	May 1, 1866	54, 347
Faucet, Flexible-lined	J. Broughton	New York, N. Y	Oct. 4, 1864	44, 510
Faucet, Fluid	J. Hollely	Brooklyn, N. Y	May 29, 1855	12, 950
Faucet for beer and other barrels	S. Thompson	Schaghticoke, N. Y	Jan. 17, 1871	111, 018
Faucet for bottles	J. Hiney	Hartford, Conn	Dec. 11, 1860	30, 920
Faucet for discharging liquids, Ventilating	J. W. McKee	Brooklyn, N. Y	Mar. 30, 1869	88, 402
Faucet for drawing liquor from casks	I. Ives	Bristol, Conn	Apr. 14, 1808	....
Faucet for filling barrels, Self-closing	S. C. Catlin	Cleveland, Ohio	Feb. 16, 1869	86, 970
Faucet for measuring liquids	J. Cross	New London, Ohio	Mar. 14, 1854	10, 633
Faucet for measuring liquids	J. B. Larwill and J. Cross	Bucyrus, Ohio	Mar. 14, 1854	10, 638
Faucet for oil, &c	J. D. Frary	New Britain, Conn	Feb. 28, 1865	46, 555
Faucet for stove-reservoirs	J. B. Crowley	Cincinnati, Ohio	Nov. 12, 1867	70, 815
Faucet-holes, Bushing for	S. R. Thompson	Portsmouth, N. H	Aug. 5, 1873	141, 473
Faucet, Liquid measuring and registering	E. W. Scott	Wauregan, Conn	Feb. 9, 1869	86, 700
Faucet-lock	V. L. Maxwell	Wilkesbarre, Pa	Apr. 26, 1870	102, 290
Faucet, Measure	G. Hubbard	Sandisfield, Mass	Aug. 25, 1863	39, 654
Faucet, Measure	M. W. Nalton	Utica, N. Y	Mar. 13, 1860	27, 465
Faucet Measure	A. Vrooman	Kokomo, Ind	Feb. 13, 1866	52, 628
Faucet measure and funnel combined	T. McMahon	Williamsburgh, N. Y	Feb. 1, 1870	99, 454
Faucet measure and indicator	V. Squarza	San Francisco, Cal	Jan. 19, 1864	41, 350
Faucet measure, Graduated	G. H. Henkel	Middletown, Ohio	July 4, 1865	48, 556
Faucet, Measuring	G. K. Babcock	Utica, N. Y	Apr. 3, 1860	27, 679
Faucet, Measuring	J. G. Baker	Philadelphia, Pa	July 10, 1866	56, 342
Faucet, Measuring	E. Bigelow	Springfield, Mass	Feb. 22, 1859	23, 008
Faucet, Measuring	E. Bigelow	Springfield, Mass	May 1, 1866	54, 282
Faucet, Measuring	T. B. Bishop	New York, N. Y	Sept. 28, 1869	95, 308
Faucet, Measuring	O. B. Blake and O. E. Colony	Keene, N. H	Aug. 11, 1868	80, 852
Faucet, Measuring	E. T. Bussell	Covington, Ky	Oct. 4, 1859	25, 627
Faucet, Measuring	J. R. Byler and G. W. Seusenich	Beartown, Pa	May 11, 1852	8, 937
Faucet, Measuring	J. F. De Navarro	New York, N. Y	July 6, 1869	92, 287
Faucet, Measuring	E. Fitzgerald	New York, N. Y	Dec. 10, 1867	71, 865
Faucet, Measuring	F. C. Heiser	Brooklyn, N. Y	Aug. 30, 1870	106, 816
Faucet, Measuring	W. W. Hollman	Eddyville, Ky	Mar. 1, 1859	23, 143
Faucet, Measuring	A. Houpt and J. K. Griffith	Reading, Pa	Aug. 6, 1861	32, 996
Faucet, Measuring	G. Hubbard	Montville, Mass	July 6, 1858	20, 799
Faucet, Measuring	G. Hubbard	Montville, Mass	Oct. 9, 1860	30, 321
Faucet, Measuring	I. Kinman	Freeport, Ill	May 3, 1859	23, 845
Faucet, Measuring	T. McGirr	Richmond, Ind	June 20, 1865	48, 296
Faucet, Measuring	E. A. Palmer	Clayville, N. Y	Apr. 22, 1856	14, 734
Faucet, Measuring	T. S. Reeve	Chicago, Ill	July 28, 1868	80, 307
Faucet, Measuring	F. Saunders	Aberdeen, Miss	Feb. 18, 1868	74, 605
Faucet Measuring	J. Schalk, jr	Guttenberg, N. J	Oct. 21, 1873	143, 830
Faucet Measuring	J. Smith and G. B. Griffin	Cincinnati, Ohio, and Madison, Wis.	June 21, 1859	24, 498
Faucet, Measuring	J. D. Smith	New York, N. Y	May 12, 1868	77, 928
Faucet, Measuring	J. N. Smith	Jersey City, N. J	June 20, 1865	48, 318
Faucet, Measuring	D. S. Spafford and G. Elsey	Morrison, Ill	July 2, 1867	66, 406
Faucet, Measuring	O. L. Wheeler	Lewiston, Me	Aug. 11, 1868	80, 847
Faucet, Measuring	S. H. Wheeler	Dowagiac, Mich	July 11, 1865	48, 778
Faucet, Measuring	S. S. Wiles	Santa Clara, Cal	Nov. 27, 1866	60, 104
Faucet, Molasses	H. D. Blake	New Britain, Conn	Jan. 23, 1866	52, 241
Faucet, Molasses	J Dudley	Fall River, Mass	Feb. 1, 1848	5, 430
Faucet, Molasses	J. and S. Fahrney	Boonsborough, Md	Sept. 22, 1863	40, 029
Faucet, Molasses	D. D. Hanson	Weare, N. H	Feb. 1, 1848	5, 427
Faucet, Molasses	W. H. Hartman	Fostoria, Ohio	Feb. 20, 1866	52, 712
Faucet, Molasses	C. W. Peckham	New Haven, Conn	Feb. 10, 1836	....
Faucet, Molasses	E. Stebbins	Cabotville, Mass	Feb. 1, 1848	5, 428
Faucet or gate, Molasses	C. Woodyear	Philadelphia, Pa	Feb. 5, 1834	....
Faucet or supply-cock	E. F. Shoenberger	Philadelphia, Pa	May 16, 1871	114, 981
Faucet or trap	A. L. Webster	New York, N. Y	Aug. 2, 1870	106, 007
Faucet, Permutation-lock	W. F. Jones	Baltimore, Md	Oct. 11, 1870	108, 153
Faucet-plug	G. Howland and E. T. Ford	Brunswick and Stillwater, N. Y.	July 20, 1869	92, 725
Faucet, Plugged	W. Jones	New York, N. Y	May 13, 1812	....
Faucet, Protected	P. A. Schwarz	Boston, Mass	Dec. 17, 1872	134, 102
Faucet, Rotary measuring	J. G. Baker and W. Harbster	Philadelphia and Reading, Pa.	Sept. 3, 1867	68, 546
Faucet, Rotary measuring	W. H. Laubach	Philadelphia, Pa	Aug. 8, 1871	117, 895
Faucet, Self-balancing and self closing	E. Osgood	Boston, Mass	Apr. 21, 1863	38, 269
Faucet, Self-closing	W. H. Bate	Charlestown, Mass	Aug. 26, 1873	142, 192
Faucet, Self-closing	A. Brinckmann	New York, N. Y	Sept. 21, 1869	94, 998

Index of patents issued from the United States Patent Office from 1790 to 1873, inclusive—Continued.

Invention.	Inventor.	Residence.	Date.	No.
Faucet, Self-closing	A. Brinckmann	New York, N. Y	June 24, 1873	140, 183
Faucet, Self-closing	A. and E. Buckman	Brooklyn, N. Y	Apr. 22, 1873	138, 127
Faucet, Self-closing	M. S. Clark	New York, N. Y	Feb. 4, 1873	135, 407
Faucet, Self closing	J. Hayden, jr	Haydensville, Mass	Aug. 15, 1871	118, 012
Faucet, Self-closing	H. F. King	New York, N. Y	Aug. 31, 1869	94, 419
Faucet, Self-closing	C. Schultz and T. Warker	New York, N. Y	Mar. 8, 1870	100, 561
Faucet, Self-closing	J. W. Trafton	Chicopee Falls, Mass	Sept. 23, 1873	143, 042
Faucet, Self-closing compression	E. Noble	North Haven, Conn	Aug. 16, 1870	106, 394
Faucet, Self-inserting	S. N. Haight	Bedford Station, N. Y	June 5, 1866	55, 286
Faucet, Siphon	T. W. Pium	London, England	Mar. 3, 1868	75, 191
Faucet, Spring-lever	C. Goodyear	Philadelphia, Pa	Mar. 16, 1832	....
Faucet, Stop	W. Cleveland	Orange, N. J	Mar. 20, 1866	53, 271
Faucet, Swing-nose basin	M. and W. Scrannage and W. H. Bate.	Boston, Mass	Aug. 27, 1867	68, 123
Faucet, Vented	H. and J. Schild	New York, N. Y	May 21, 1867	65, 016
Faucet, Ventilating	C. Cleveland	Ashfield, Mass	Aug. 14, 1855	13, 424
Faucet, Wash-basin	D. Wellington	Boston, Mass	Jan. 20, 1863	37, 478
Faucet, Wash-stand	W. S. Bate	Philadelphia, Pa	Dec. 16, 1873	145, 609
Faucet, Water	P. A. Mayor	New York, N. Y	Dec. 27, 1870	110, 484
Faucet, Weighing	D. Lesh, jr	Liverpool, Pa	Apr. 5, 1870	101, 633
Faucet, Weighing	C. H. McAleer and J. Shively	Canton, Ohio	June 10, 1862	35, 530
Faucet-weighing attachment	J. Brown	Lawn Ridge, Ill	Nov. 13, 1855	13, 776
Faucet-weighing attachment	L. C. Fisher and A. D. Holliday.	El Paso, Ill	Oct. 1, 1867	69, 425
Faucets, Bush for	O. Netzow and J. F. Heck	Baltimore, Md	Feb. 20, 1872	123, 789
Faucets, Device for locking	H. Getty	Brooklyn, N. Y	Sept. 1, 1857	18, 091
Faucets into fluids under pressure, Method of inserting.	P. Mihan	Boston, Mass	July 29, 1856	15, 430
Faucets, Machine for reaming	W. Read	New York, N. Y	May 23, 1846	4, 534
Faucets, Method of manufacturing	W. Westlake	Brooklyn, N. Y	July 9, 1867	66, 656
Faucets, Method of manufacturing	W. Westlake	Brooklyn, N. Y	July 9, 1867	66, 657
Faucets of water-coolers, Arrangement of means for making tight joints around.	J. S. Clark	Philadelphia, Pa	Oct. 19, 1858	21, 819
Faucets to pipes, Coupling	N. F. Weston	Boston, Mass	July 9, 1867	66, 659
Faucets, Tool for inserting and withdrawing	K. A. Wattendorf	New York, N. Y	May 27, 1873	139, 443
Faucets, Valvular arrangement for	E. Hamilton	Chicago, Ill	June 23, 1857	17, 624
Faucets, &c., Valvular arrangement in	J. C. Macdonald	Cincinnati, Ohio	Aug. 11, 1857	17, 973
Faucets, Waste-way in	J. E. Boyle	Richmond, Va	Apr. 21, 1857	17, 074
Feather-cleaner	W. Mills	New Athens, Ohio	Oct. 29, 1861	33, 594
Feather cleaning and dressing machine	S. Swett, jr	Readfield, Me	July 31, 1837	330
Feather cleaning and drying machine	R. B. Lewis	Hallowell, Me	June 27, 1840	1, 655
Feather cleaning and purifying machine	S. G. Ladd	Hallowell, Me	Sept. 22, 1838	939
Feather-cleaning machine	H. J. Beckwith	Chicopee, Mass	May 5, 1868	77, 571
Feather-cleaning machine	J. H. Sardam	Wellington, Ohio	June 4, 1867	65, 458
Feather-cleaning machine	T. Taylor	Washington, D. C	Sept. 1, 1868	81, 841
Feather-cleaning machine	C. Toupet	New York, N. Y	Oct. 15, 1829	....
Feather cleansing and renovating machine	H. B. Steele	West Winstead, Conn	Apr. 14, 1868	76, 840
Feather-dresser	S. Keplinger	Baltimore, Md	Feb. 12, 1836	....
Feather-dresser	F. P. Knowlton	Clermont, N. H	July 2, 1836	....
Feather-dresser	S. Orr	Springfield, Ohio	Apr. 9, 1861	31, 993
Feather-dressing	B. P. Coston	Philadelphia, Pa	June 28, 1836	....
Feather-dressing	G. Reynolds	East Hartford, Conn	May 23, 1836	....
Feather-dressing	B. and A. Todd	Marietta, Ohio	May 6, 1836	....
Feather-dressing	E. Wilbur	Geneva, N. Y	Mar. 18, 1836	....
Feather dressing and cleaning machine	O. Badger	Otsego, N. Y	Apr. 22, 1835	....
Feather dressing and purifying machine	B. Smith	Shodiack, N. Y	Nov. 23, 1835	....
Feather dressing and renovating machine	G. W. Peabody and O. L. Cowles.	East Hampton and Westfield, Mass.	Aug. 6, 1867	67, 449
Feather-dressing machine	A. Bailey	East Poultney, Vt	Oct. 27, 1857	18, 493
Feather-dressing machine	S. W. Bevans	Plymouth, Conn	Apr. 21, 1868	76, 982
Feather-dressing machine	A. B. Morey and W. Scarlett	Aurora, Ill	July 1, 1862	35, 773
Feather-dressing machine	G. Reynolds	East Hartford, Conn	Oct. 17, 1835	....
Feather-dressing machine	A. Washburn and J. N. Van Sickle.	Medina, Ohio	June 11, 1867	65, 782
Feather drying, whipping, and cleansing machinery	N. L. Manning	Boston, Mass	Apr. 16, 1841	2, 053
Feather dusters, Manufacture of	C. F. Shourds	New York, N. Y	Apr. 20, 1869	89, 176
Feather purifying and renovating machine	C. Turner and J. A. Jackson	Triangle, N. Y	Dec. 8, 1863	40, 871
Feather renovating and dressing	A. C. Sanford	Plymouth, Conn	July 14, 1868	80, 017
Feather-renovating machine	J. S. Peasley	Providence, R. I	Oct. 22, 1867	70, 110
Feather-renovator	T. J. Adams	Ansonia, Conn	Nov. 26, 1872	133, 348
Feather-renovator	C. E. Barber and W. Dean	Central Village, Conn	July 4, 1871	116, 535
Feather-renovator	W. T. Beauchamp	Grenada, Miss	Oct. 15, 1872	132, 236
Feather-renovator	J. A. Bell	Tyrone, Pa	Feb. 7, 1871	111, 606
Feather-renovator	E. Bickell and M. F. Noraconk	Milton, Pa	Oct. 4, 1870	107, 999
Feather-renovator	L. Blair	Painesville, Ohio	Mar. 9, 1869	87, 534
Feather-renovator	A. Bond	Chicopee, Mass	Sept. 15, 1868	82, 200
Feather-renovator	J. B. Booker and M. C. Green	Chicago and Paris, Ill	July 16, 1872	129, 390
Feather-renovator	S. A. Boyett	Grenada, Miss	Feb. 4, 1873	135, 516
Feather-renovator	E. Colvin	Paulet, Vt	Feb. 23, 1869	87, 145
Feather-renovator	A. D. Cook and J. Graves	New Madrid, Mo	Feb. 22, 1870	100, 120
Feather-renovator	R. B. Cooper	Monticello, N. Y	Oct. 7, 1873	143, 457
Feather-renovator	B. F. Cramer	Tyrone, Pa	May 30, 1871	115, 440
Feather-renovator	J. Cyester	Osborn, Ohio	May 4, 1869	89, 740
Feather-renovator	W. F. Daugherty	Wellington, Ohio	Aug. 4, 1868	80, 610
Feather-renovator	L. Dimick	Ellsworth, N. Y	July 16, 1872	129, 325
Feather-renovator	W. H. Elliot	New York, N. Y	Dec. 8, 1868	84, 809
Feather-renovator	L. S. Enos	Almond, N. Y	Sept. 23, 1873	143, 130
Feather-renovator	C. H. Farnham	Canterbury, Conn	June 22, 1869	91, 731
Feather-renovator	J. Garrett and J. E. Rauch	Selin's Grove, Pa	Mar. 28, 1871	113, 157
Feather-renovator	R. H. Graves and H. L. Crandell.	McGranville, N. Y	Jan. 5, 1869	85, 654
Feather-renovator	W. H. Groff	Mahanoy City, Pa	Apr. 29, 1873	138, 327
Feather-renovator	C. E. Hendrick	Chicopee, Mass	July 14, 1868	79, 975
Feather-renovator	C. E. Hendrick	Chicopee, Mass	Aug 25, 1868	81, 369
Feather-renovator	C. Kendall	Beloit, Wis	Nov. 5, 1872	132, 838
Feather-renovator	W. G. Lambard	Georgetown, Ill	Oct. 11, 1870	108, 161
Feather-renovator	C. E. Macon and G. F. Bell	Wellington, Ohio	July 7, 1868	79, 668
Feather-renovator	W. McArthur	Philadelphia, Pa	July 11, 1865	48, 702

Index of patents issued from the United States Patent Office from 1790 *to* 1873, *inclusive*—Continued.

Invention.	Inventor.	Residence.	Date.	No.
Feather-renovator	E. N. McKimm and J. R. Gearhart.	Lathrop, Mo	Jan. 14, 1873	134, 914
Feather-renovator	A. McKissick and C. M. French	Jordan and Weedsport, N. Y.	Apr. 15, 1862	34, 975
Feather-renovator	J. C. Moorehead and W. W. Elliott.	New Madrid, Mo	Apr. 21, 1868	77, 075
Feather-renovator	M. K. Morris	Council Bluffs, Iowa	Oct. 4, 1870	107, 944
Feather-renovator	O. J. Pennell	Williamsport, Pa	May 29, 1860	28, 500
Feather-renovator	J. W. Perry	Perrysburgh, N. Y	Aug. 17, 1869	93, 901
Feather-renovator	J. W. Post and R. Collier	Baltimore, Md	Mar. 2, 1836	......
Feather-renovator	J. B. Riley	Muncy, Pa	Mar. 25, 1873	137, 240
Feather-renovator	H. H. Robbins	Lynn, Mass	Aug. 18, 1868	81, 294
Feather-renovator	J. C. Rose and J. F. Silversmith.	Albany, N. Y	June 20, 1871	116, 101
Feather-renovator	J. E. Schooler	Chillicothe, Mo	Dec. 23, 1873	145, 759
Feather-renovator	J. T. Sheldon	Chicopee, Mass	Feb. 7, 1871	111, 576
Feather-renovator	S. B. Shoemaker	Willoughby, Ohio	Nov. 28, 1871	121, [illegible]57
Feather-renovator	S. G. Thanhauser	Baltimore, Md	May 20, 1873	139, 211
Feather-renovator	U. B. Waddle	Cleveland, Ohio	Nov. 26, 1867	71, 555
Feather-renovator	J. Wellfare and F. Champagne	Aurora, Ill	Oct. 24, 1871	120, 174
Feather-renovator and cleaner	O. C. Monroe	Poultney, Vt	Sept. 24, 1867	69, 114
Feather washer and drier	J. Mallory	Penn Yan, N. Y	Mar. 15, 1870	100, 911
Feathers, Apparatus for cleaning and renovating	J. R. Morrison	East Springfield, Ohio	Oct. 15, 1861	33, 489
Feathers, Cleaning	D. K. Hall	New York, N. Y	Feb. 12, 1836	......
Feathers, Machine for dressing	R. Glore	Nashville, Tenn	July 20, 1869	92, 818
Feathers, Machine for dressing	G. Reynolds	East Hartford, Conn	Feb. 7, 1834	......
Feathers, Machine for washing and dressing	J. W. Howlet	Greensborough, N. C	Oct. 5, 1838	965
Feathers, Made	A. E. Woolf	New York, N. Y	Mar. 18, 1873	136, 891
Feathers, Manufacture of ornamental	T. E. Schmidt	Hoboken, N. J	June 18, 1867	65, 955
Feathers, Mode of disinfecting	J. W. Howlett	Greensborough, N. C	Oct. 4, 1859	25, 647
Feed and delivery apron	P. R. Mansfield	Lancaster, Mass	Mar. 25, 1873	137, 224
Feed and other rollers, Wiping-apparatus for	L. Crawford	Holyoke, Mass	Sept. 27, 1870	107, 667
Feed and straw cutter	A. P. Chapin	Chicopee Falls, Mass	Sept. 3, 1867	68, 412
Feed and water trough, Combined	J. Sherman	Coopersville, Mich	June 7, 1870	103, 935
Feed-attachment for machinery	S. Brown	Philadelphia, Pa	Dec. 22, 1868	85, 064
Feed-bag	J. Becher and W. Tustin	Philadelphia, Pa	Feb. 23, 1864	41, 670
Feed-bag	C. Chinnock	Brooklyn, N. Y	Sept. 15, 1868	82, 088
Feed-bag	T. Fisler	Camden, N. J	Mar. 18, 1862	34, 715
Feed-bag	H. Pennie and C. Chinnock	New York, N. Y	Apr. 25, 1865	47, 498
Feed-bag	F. J. Simeon	Brooklyn, N. Y	June 19, 1866	55, 726
Feed-bag	J. Vandercar	Brooklyn, N. Y	June 25, 1867	66, 190
Feed-bag	W. B. Wait	Greenwood, Mass	Jan. 19, 1864	41, 346
Feed-bag for horses	T. Adams	Jersey City, N. J	Aug. 12, 1873	141, 686
Feed-bag for horses	A. T. Ballentine	New York, N. Y	Apr. 18, 1865	47, 270
Feed-bag for horses	W. A. Hough	South Butler, N. Y	Mar. 7, 1871	112, 458
Feed-bag for horses and other animals	R. B. Fitts	Philadelphia, Pa	Sept. 17, 1861	33, 318
Feed-bag for horses or other animals	W. B. Wait	Greenwood, Mass	Dec. 23, 1862	37, 262
Feed-box	J. Hawse	Wolcott, Vt	June 8, 1869	91, 013
Feed-box	W. S. Shaw and M. B. Gould	Buffalo, N. Y	Nov. 14, 1871	120, 906
Feed-box and stanchion, Combined	W. I. and J. W. Harris	Newport, N. Y	May 21, 1872	126, 886
Feed-box, Folding	J. L. Edwards	Cochranville, Pa	Jan. 14, 1873	134, 869
Feed-box for animals, Automatic	A. Goodyear, 2d	Hamden, Conn	Nov. 16, 1858	22, 076
Feed-box, Ration	J. Hayden	Exeter, Wis	July 24, 1866	56, 557
Feed-changing device	R. L. Nelson	Mexico, N. Y	Feb. 18, 1868	74, 578
Feed-cutter	T. P. Allen	Gowanda, N. Y	Mar. 17, 1868	75, 598
Feed-cutter	W. F. Altfather	Johnstown, Pa	Oct. 23, 1866	58, 967
Feed-cutter	W. S. Bartle and J. Garlock	Newark, N. Y	Mar. 25, 1873	137, 168
Feed-cutter	J. H. Bradley	Hillsborough, Ohio	Jan. 17, 1871	111, 035
Feed-cutter	H. A. Buck	Fredonia, N. Y	Feb. 28, 1871	112, 117
Feed cutter	C. H. Cain	Dayton, Ohio	May 3, 1870	102, 485
Feed-cutter	J. Criley and W. Gilbreath	Shiloh Hill and Rockwood, Ill.	Feb. 14, 1871	111, 819
Feed-cutter	W. Dahlem	Madison, Ind	Jan. 10, 1871	110, 834
Feed-cutter	C. R. Donner	Sonora, Cal	Oct. 11, 1870	108, 118
Feed-cutter	C. R. Donner	Sonora, Cal	Aug. 1, 1871	117, 521
Feed-cutter	M. Fletcher	Louisville, Ky	Aug. 27, 1861	33, 137
Feed-cutter	W. Gale	Peekskill, N. Y	Nov. 30, 1869	97, 384
Feed-cutter	F. E. Garner	Cornwall, Conn	Oct. 11, 1870	108, 249
Feed-cutter	G. W. Hathaway	Tioga, Pa	Dec. 4, 1860	30, 817
Feed-cutter	H. Helm	Pittsburgh, Pa	June 11, 1867	65, 571
Feed-cutter	S. Holl	Reading, Pa	July 16, 1872	129, 562
Feed-cutter	W. Hutchins	Paw Paw, Mich	Dec. 6, 1870	109, 822
Feed-cutter	C. Kemper	Hermann, Mo	Jan. 17, 1871	111, 065
Feed-cutter	J. Massey	New York, N. Y	Oct. 23, 1866	59, 043
Feed-cutter	R. McCain	Rootstown, Ohio	Jan. 14, 1862	34, 154
Feed-cutter	J. B. and W. McClinton	Galion, Ohio	Feb. 28, 1871	112, 265
Feed-cutter	N. McLeod	Clio, S. C	Nov. 10, 1868	83, 984
Feed-cutter	J. J. Parker	Marietta, Ohio	Oct. 11, 1864	44, 650
Feed-cutter	G. Parnell	Ontario, N. Y	Jan. 18, 1870	98, 997
Feed-cutter	H. Peterson	Red Wing, Minn	Dec. 7, 1869	97, 687
Feed-cutter	S. Pettibone	Corunna, Mich	Mar. 13, 1866	53, 179
Feed-cutter	J. P. Randolph	Marietta, Ohio	Dec. 28, 1869	98, 414
Feed-cutter	W. H. Rosser and J. Stiver	Mill Hall, Pa	Feb. 1, 1870	99, 352
Feed-cutter	J. Seaman	Chicago, Ill	Apr. 4, 1871	113, 579
Feed-cutter	F. Sims	Ridgeville, Ind	July 13, 1869	92, 548
Feed-cutter	G. Small	Clayton, Mich	Sept. 24, 1867	69, 133
Feed-cutter	J. R. Whittemore	Chicopee Falls, Mass	Feb. 7, 1871	111, 594
Feed-cutter and box	A. B. King	Camden, Ohio	Oct. 9, 1866	58, 649
Feed-cutter and thrasher	N. McLeod	Clio, S. C	Feb. 15, 1870	99, 931
Feed-cutters, Adjusting knife of	J. R. Whittemore	Chicopee Falls, Mass	July 30, 1861	32, 936
Feed for horses and cattle, Cutting	E. H. Porter	Lincolnton, N. C	Mar. 14, 18[illegible]4	......
Feed for horses, &c., Concentrated	G. Jaques, D. F. White, and J. Stowell.	Somerville and Charlestown, Mass.	Jan. 19, 1864	41, 301
Feed-gates for mills, Method of regulating	C. Dare	Cincinnati, Ohio	Feb. 12, 1856	14, 227
Feed-grinder	M. Gore	Ottawa, Ill	Feb. 7, 1871	111, 530
Feed-guide, Platen	E. L. Megill	Brooklyn, N. Y	Mar. 21, 1871	112, 827
Feed-hopper	J. Mattison	Oswego, N. Y	Feb. 15, 1870	99, 926
Feed-manger	C. E. Steller	McGregor, Iowa	Nov. 22, 1864	45, 190

Index of patents issued from the United States Patent Office from 1790 *to* 1873, *inclusive*—Continued.

Invention.	Inventor.	Residence.	Date.	No.
Feed-mill	Z. S. Cracraft	Ottawa, Ill	Oct. 31, 1871	120, 562
Feed-mill	G. W. Gibson	Dryden, N. Y	May 21, 1872	126, 883
Feed-mill	J. W. Myers	Lyons City, Iowa	Dec. 26, 1871	122, 129
Feed-rack	J. W. Blanchard	Rutland, Wis	Mar. 5, 1867	62, 522
Feed-rack	E. Gratten	Williamstown, Mich	May 14, 1867	64, 759
Feed-rack	J. L. Rhodeback	Newway, Ohio	Aug. 27, 1872	130, 867
Feed-rack	J. Robinson	Carmel, Me	Jan. 23, 1872	122, 967
Feed-rack	W. and H. Sias	Henderson, N. Y	July 19, 1864	43, 609
Feed-rack	F. G. L. Struve	Jefferson, Wis	Mar. 25, 1862	34, 793
Feed-rack	J. M. Van Nest	Clayton, Iowa	Apr. 25, 1865	47, 472
Feed-rack	M. Workman	Washington Township, Ohio.	July 14, 1868	79, 886
Feed rack and trough combined	D. Close	Harmonsburgh, Pa	May 31, 1864	42, 927
Feed-rack for stock	J. C. Ramsey	Le Roy, Ohio	Aug. 27, 1867	68, 111
Feed-rack, Portable	E. Farquhar	Ankenytown, Ohio	Feb. 28, 1871	112, 133
Feed-rack, Portable	C. W. Lamb	Columbia City, Ind	Apr. 3, 1866	53, 634
Feed-regulator	W. F. H. Daniels	Belleville, Ohio	Dec. 13, 1870	110, 015
Feed-regulator for liquids	H. Seyter	Vaihingen, Würtemberg	Nov. 14, 1871	121, 010
Feed-regulator, Steam-boiler	R. Berryman	Hartford, Conn	Nov. 19, 1872	133, 189
Feed-regulator, Steam-boiler	L. Thorn	New York, N. Y	May 25, 1858	20, 380
Feed-roller	H. R. Crowe	Carondelet, Mo	Mar. 12, 1867	62, 822
Feed-rollers, Method of operating yielding	J. H. Brinton	West Chester, Pa	Feb. 15, 1859	22, 931
Feed-trough	G. M. Goss	Bloomfield, Iowa	Jan. 23, 1872	123, 012
Feed-trough	A. Simkins	Algansee, Mich	July 23, 1872	129, 687
Feed trough and rack, Combined	J. Depue	California, Mich	Mar. 31, 1868	76, 061
Feed trough and rack for sheep, cattle, and horses	J. D. McBride	Mansfield, Ohio	Jan. 14, 1868	73, 426
Feed-trough, Folding	W. N. Shellabarger	Union, Ohio	May 18, 1869	90, 129
Feed-trough guard	E. Hovenden	Bushnell, Ill	Jan. 24, 1871	111, 122
Feed-water and blow-off apparatus for steam-boilers	J. Frick	Philadelphia, Pa	Dec. 14, 1858	22, 284
Feed-water apparatus	J. Kindel	Wilmington, Ohio	Feb. 7, 1871	111, 650
Feed-water apparatus	J. H. Pease	Reading, Pa	Apr. 10, 1866	53, 864
Feed-water apparatus	G. A. Reidel	Philadelphia, Pa	Sept. 19, 1865	50, 033
Feed-water apparatus	G. I. Washburn	Worcester, Mass	Sept. 12, 1865	49, 941
Feed-water apparatus	J. Wozniakowski	St. Petersburg, Russia	Sept. 17, 1872	131, 490
Feed-water apparatus for locomotive-engines	R. A. Wilder	Schuylkill Haven, Pa	Mar. 27, 1860	27, 662
Feed-water apparatus for steam-boilers, Heating	T. Sloan	Saint Louis, Mo	May 6, 1856	14, 835
Feed-water apparatus of steam-boilers, Means for controlling.	B. F. Bee	Wareham, Mass	July 15, 1856	15, 324
Feed-water apparatus, Steam-boiler	T. Armitage	Philadelphia, Pa	Oct. 4, 1859	25, 617
Feed-water apparatus, Steam-boiler	W. Barnes	Troy, N. Y	Sept. 6, 1859	25, 360
Feed-water apparatus, Steam-boiler	B. F. Bee	Harwich, Mass	Aug. 1, 1854	11, 433
Feed-water apparatus, Steam-boiler	R. Berryman	Hartford, Conn	Feb. 11, 1873	135, 757
Feed-water apparatus, Steam-boiler	W. P. Curry	Vincennes, Ind	Oct. 18, 1859	25, 811
Feed-water apparatus, Steam-boiler	J. Densmore	Blooming Valley, Pa	Sept. 18, 1855	13, 567
Feed-water apparatus, Steam-boiler	J. W. Doughty	New York, N. Y	July 16, 1861	32, 824
Feed-water apparatus, Steam-boiler	T. Firth	Cincinnati, Ohio	Feb. 5, 1856	14, 191
Feed-water apparatus, Steam-boiler	C. G. Fisher	Washington, D. C	Oct. 10, 1871	119, 835
Feed-water apparatus, Steam-boiler	B. Fitts	Worcester, Mass	Mar. 28, 1854	10, 695
Feed-water apparatus, Steam-boiler	L. X. Gargan	Paris, France	Apr. 16, 1861	32, 061
Feed-water apparatus, Steam-boiler	H. Giffard	Paris, France	Apr. 24, 1860	27, 979
Feed-water apparatus, Steam-boiler	J. Hibbard	Hermitage, N. Y	Jan. 17, 1860	26, 847
Feed-water apparatus, Steam-boiler	W. A. Lighthall	New York, N. Y	Apr. 9, 1861	31, 988
Feed-water apparatus, Steam-boiler	E. Pierce	Cincinnati, Ohio	June 12, 1860	28, 685
Feed-water apparatus, Steam-boiler	G. W. Rains	Newburgh, N. Y	Apr. 24, 1860	28, 011
Feed-water apparatus, Steam-boiler	H. C. Sargeant	Cincinnati, Ohio	June 13, 1854	11, 095
Feed-water apparatus, Steam-boiler	W. C. Selden	Brooklyn, N. Y	Sept. 19, 1871	119, 185
Feed-water apparatus, Steam-boiler	J. B. Thompson	Warrenton, Ga	Aug. 30, 1859	25, 290
Feed-water apparatus, Steam-boiler	A. J. Vandegrift	Delaware, Ohio	Feb. 17, 1857	16, 662
Feed-water apparatus, Steam-boiler	A. Winkler	St. Petersburgh, Russia	Sept. 17, 1872	131, 489
Feed-water apparatus, Steam-boiler	C. and G. M. Woodward	New York, N. Y	Apr. 10, 1860	27, 853
Feed-water apparatus, Steam-boiler	J. W. Youman	Mobile, Ala	Nov. 5, 1872	132, 888
Feed-water arrangement for steam-boilers	T. Snowden	Pittsburgh, Pa	Apr. 3, 1860	27, 743
Feed-water attachment to steam-engines	L. Martin	New York, N. Y	Dec. 22, 1857	18, 919
Feed-water connecting-pipe, Locomotive	C. A. Peavey	Joliet, Ill	May 24, 1870	103, 494
Feed-water device for boilers	W. F. Duerr	Newark, N. J	Apr. 20, 1869	89, 208
Feed-water device for boilers	J. D. Otis	Peoria, Ill	Feb. 2, 1869	86, 580
Feed-water device for boilers, Automatic	J. A. Davis	Watertown, N. Y	Feb. 2, 1869	86, 372
Feed-water filtering-heater for boilers	D. Wineland	McComb, Ohio	Apr. 27, 1869	89, 366
Feed-water for steam-boilers, Heating	W. A. Lighthall	New York, N. Y	July 4, 1871	116, 727
Feed-water for steam-engines, Heating	F. B. Stevens	New York, N. Y	July 1, 1862	35, 789
Feed-water for steam-engines, Method of heating	F. B. Stevens	New York, N. Y	June 24, 1862	35, 714
Feed-water for steam-engines, Valve for heating	F. B. Stevens	New York, N. Y	July 1, 1862	35, 788
Feed-water for steam-generators, Apparatus for heating and purifying.	J. S. Hooton	New Carlisle, Ind	Mar. 3, 1868	75, 163
Feed-water heater	H. Anderson	Peoria, Ill	Oct. 4, 1870	107, 850
Feed-water heater	J. V. Beekman	Brooklyn, N. Y	Sept. 23, 1873	142, 983
Feed-water heater	R. Berryman	Hartford, Conn	Oct. 29, 1872	132, 518
Feed-water heater	H. F. Boody and E. P. Merrill	Portland, Me	Jan. 7, 1873	134, 634
Feed-water heater	W. Crighton, W. Wills, and L. Bastetter.	Fort Wayne, Ind	June 11, 1867	65, 547
Feed-water heater	W. B. Cross	Sacramento City, Cal	Oct. 17, 1865	50, 459
Feed-water heater	J. Fairclough	Saint Joseph, Mo	Nov. 30, 1869	97, 371
Feed-water heater	C. E. Frazier	Baltimore, Md	June 10, 1873	139, 665
Feed-water heater	J. Gates	Portland, Oreg	June 18, 1872	128, 033
Feed-water heater	J. G. Goesel	Saint Louis, Mo	Sept. 3, 1867	68, 433
Feed-water heater	C. S. S. Griffing	Salem, Ohio	Apr. 25, 1871	114, 132
Feed-water heater	J. H. Hale and A. C. Whitney	Boston, Mass	Dec. 2, 1873	145, 059
Feed-water heater	R. W. Hale	Boston, Mass	Oct. 7, 1862	36, 614
Feed-water heater	G. Hasecoster and I. Stephens	Richmond, Ind	Sept. 18, 1866	58, 099
Feed-water heater	N. Jones	Buffalo, N. Y	Oct. 29, 1872	132, 585
Feed-water heater	P. M. Kafer and J. M. De Lacy	Trenton, N. J	May 7, 1867	64, 540
Feed-water heater	J. F. Morse	Oshkosh, Wis	June 14, 1870	104, 187
Feed-water heater	G. M. Mullen	Baltimore, Md	Mar. 5, 1872	124, 373
Feed-water heater	J. Nicholson	Allegheny City, Pa	May 22, 1866	54, 940
Feed-water heater	J. Nicholson	Allegheny City, Pa	Oct. 8, 1867	69, 576
Feed-water heater	R. Poole	Baltimore, Md	Apr. 25, 1865	47, 599
Feed-water heater	L. B. Powelson	Pittsburgh, Pa	Oct. 27, 1868	83, 543
Feed-water heater	G. W. Richardson	Troy, N. Y	Dec. 16, 1873	145, 591

Index of patents issued from the United States Patent Office from 1790 *to* 1873, *inclusive*—Continued.

Invention.	Inventor.	Residence.	Date.	No.
Feed-water heater	W. Sambrook	Liverpool, England	Dec. 30, 1873	146, 024
Feed-water heater	J. R. Sees	New York, N. Y	Apr. 26, 1864	42, 538
Feed-water heater	T. Shipton	Newark, N. J	May 19, 1868	78, 140
Feed-water heater	T. D. Simpson	Mount Vernon, Ohio	Apr. 4, 1871	113, 355
Feed-water heater	E. R. Stilwell	Dayton, Ohio	July 23, 1867	66, 998
Feed-water heater	S. Stucky	New Albany, Ind	July 28, 1868	80, 516
Feed-water heater	G. Wales	Brooklyn, N. Y	Jan. 21, 1873	135, 181
Feed-water heater	W. E. Walsh	Jersey City, N. J	Dec. 26, 1871	122, 204
Feed-water heater	G. I. Washburn	Worcester, Mass	Feb. 25, 1868	74, 959
Feed-water heater	G. Waters	Cincinnati, Ohio	Aug. 30, 1870	106, 972
Feed-water heater	P. C. Wortman	Meadville, Pa	Aug. 17, 1869	93, 790
Feed-water heater and circulator combined	E. L. Jones	Memphis, Tenn	Jan. 24, 1871	111, 215
Feed-water heater and condenser	R. Berryman	Hartford, Conn	Feb. 11, 1873	135, 756
Feed-water heater and condenser	T. E. McNeill and R. N. Pratt	New York, N. Y	June 17, 1873	139, 908
Feed-water heater and condenser	L. C. Wyatt	Baltimore, Md	Aug. 6, 1872	130, 351
Feed-water heater and filter	J. Armstrong	Bucyrus, Ohio	Sept. 8, 1868	81, 971
Feed-water heater and filter	J. Armstrong	Toledo, Ohio	Dec. 31, 1872	134, 450
Feed-water heater and filter	D. E. Shaw	Chatsworth, Ill	May 17, 1870	103, 243
Feed-water heater and filter	E. R. Stilwell	Dayton, Ohio	Oct. 4, 1864	44, 561
Feed-water heater and filter	E. R. Stilwell	Dayton, Ohio	June 28, 1870	104, 789
Feed-water heater and purifier	R. K. McMurry	West New Brighton, N. Y	Oct. 15, 1872	132, 170
Feed-water heater and purifier	S. J. Sadler and H. Volmar	Cleveland, Ohio	Aug 5, 1873	141, 517
Feed-water heater and regulator	R. Berryman	Hartford, Conn	July 30, 1872	130, 609
Feed-water heater and steam-condenser	B. N. and D. W. Payne	Corning, N. Y	Aug. 20, 1872	130, 656
Feed-water heater for boilers	H. O. Perry	Buffalo, N. Y	June 30, 1868	79, 385
Feed-water heater for boilers	E. R. Stilwell	Dayton, Ohio	Aug. 18, 1868	81, 117
Feed-water heater for boilers	C. Webster	New Haven, Conn	Apr. 21, 1868	76, 961
Feed-water heater for boilers	H. Wigley	New Albany, Ind	Dec. 7, 1869	97, 737
Feed-water heater for locomotives	S. F. Allen	Chicago, Ill	Feb. 16, 1864	41, 591
Feed-water heater for locomotives	B. Crawford	Pittsburgh, Pa	June 2, 1863	38, 730
Feed-water heater for locomotives	P. S. Ebbert	Chicago, Ill	Feb. 5, 1856	14, 187
Feed-water heater for locomotives	B. Garvin and R. J. Pettibone	Oshkosh, Wis	Dec. 3, 1867	71, 604
Feed-water heater for locomotives	J. S. Hooton	New Carlisle, Ind	July 12, 1870	105, 207
Feed-water heater for locomotives	I. P. Magoon	Saint Johnsbury, Vt	Mar. 29, 1870	101, 287
Feed-water heater for locomotives	E. Moyel	Wyandotte, Kans	Feb. 4, 1873	135, 438
Feed-water heater for locomotives	D. Pollock	Lancaster, Pa	Jan. 19, 1864	41, 319
Feed water heater for locomotives	J. R. Sees	New York, N. Y	Apr. 26, 1864	42, 539
Feed-water heater for locomotives	W. H. Standeford	Stewartsville, Mo	Aug. 20, 1872	130, 759
Feed-water heater for locomotive-engines	D. Matthew	Philadelphia, Pa	July 17, 1855	13, 270
Feed-water heater for locomotive-engines	J. R. Sees	New York, N. Y	Nov. 11, 1856	16, 071
Feed-water heater for steam-boilers	M. W. Baldwin and D. Clark	Philadelphia and Schuylkill Haven, Pa.	Oct. 12, 1852	9, 312
Feed-water heater for steam-boilers	R. Berryman	Hartford, Conn	June 6, 1871	115, 682
Feed-water heater for steam-boilers	R. Berryman	Hartford, Conn	Apr. 9, 1872	125, 526
Feed-water heater for steam-boilers	R. C. Bristol	Chicago, Ill	Feb. 24, 1863	37, 731
Feed-water heater for steam-boilers	J. T. Brooks	New Albany, Ind	May 24, 1859	24, 094
Feed-water heater for steam-boilers	J. Codling, jr., and J. McCunniff	Fairbanks, Iowa	Feb. 28, 1860	27, 271
Feed-water heater for steam-boilers	W. C. Drum	Belle Vernon, Pa	June 26, 1860	28, 840
Feed-water heater for steam-boilers	S. W. France	Brooklyn, N. Y	Aug. 6, 1872	130, 122
Feed-water heater for steam-boilers	B. Garvin	Oshkosh, Wis	Nov. 14, 1871	120, 959
Feed-water heater for steam-boilers	A. M. Granger	Saint Louis, Mo	Jan. 26, 1864	41, 374
Feed-water heater for steam-boilers	B. A. Hopkins	Sodus, N. Y	Oct. 20, 1868	83, 161
Feed-water heater for steam-boilers	W. A. Lighthall	New York, N. Y	Feb. 7, 1865	46, 252
Feed-water heater for steam-boilers	D. Lordon	Memphis, Tenn	Sept. 6, 1870	107, 185
Feed-water heater for steam-boilers	L. E. C. Martin	London, England	Aug. 2, 1864	43, 750
Feed-water heater for steam-boilers	W. W. Martin	Allegheny, Pa	June 9, 1863	38, 870
Feed-water heater for steam-boilers	J. Rodgers	Clarington, Ohio	May 21, 1872	126, 988
Feed-water heater for steam-boilers	J. R. Sees	New York, N. Y	Aug. 5, 1856	15, 494
Feed-water heater for steam-boilers	S. T. Taylor	Charleston, S. C	Sept. 26, 1871	119, 428
Feed-water heater for steam-boilers	J. F. Taylor	Baltimore, Md	May 13, 1873	138, 957
Feed-water heater for steam-boilers	J. M. White	New York, N. Y	Aug. 30, 1859	25, 296
Feed-water heater for steam-boilers	D. Whitlock	Newark, N. J	July 12, 1870	105, 397
Feed-water heater for steam-boilers	R. Wilson	Columbus City, Iowa	July 3, 1855	13, 193
Feed-water heater for steam-engines	J. Armstrong	Bucyrus, Ohio	June 22, 1869	91, 700
Feed-water heater for steam fire-engines	W. A. Brickill	New York, N. Y	Aug. 18, 1868	81, 132
Feed-water heater for steam-generators	G. Candee	Berlin Heights, Ohio	May 5, 1868	77, 454
Feed-water heater for steam-generators	R. Eaton	Kentish Town, England	Feb. 2, 1869	86, 515
Feed-water heater for steam-generators	R. R. Fenner	Urbana, Ill	Oct. 27, 1868	83 370
Feed-water heater for steam-generators	E. B. Johnson	Chicago, Ill	Feb. 8, 1870	99, 573
Feed-water heater for steam-generators	G. W. Mack	Hamtramck, Mich	Mar. 31, 1868	76, 217
Feed-water heater for steam-generators	D. F. McKim	Austin, Nev	Dec. 22, 1868	85, 116
Feed-water heater for steam-generators	H. S. Saroni	Marietta, Ohio	Aug. 28, 1866	57, 576
Feed-water heater for steam-generators	H. N. Waters	Hartford, Conn	Feb. 18, 1868	74, 735
Feed-water heater for steam-generators	H. N. Waters	Meriden, Conn	Apr. 23, 1872	125, 918
Feed-water heater, Hollow grate	G. W. Jones	Nashville, Tenn	July 18, 1871	117, 176
Feed-water heaters and generators, Series of	J. Goulding	Worcester, Mass	June 25, 1872	128, 300
Feed-water heaters for steam-generators, Filtering	B. F. Wilson	Geddes, N. Y	Apr. 20, 1869	89, 106
Feed-water of locomotives, &c., Apparatus for heating.	I. P. Magoon	Saint Johnsbury, Vt	Sept. 7, 1852	9, 251
Feed-water of steam-boilers, Apparatus for heating and purifying.	J. Guhmann	Rochester, N. Y	Mar. 15, 1859	23, 244
Feed-water of steam-boilers, Device for heating	S. Lamon and W. S. Gaskill	Van Wert, Ohio	Apr. 12, 1859	23. 588
Feed-water pipe	J. Doyle	Baltimore, Md	Jan. 3, 1871	110, 753
Feed-water pipes in the beds of steam-engines, Arrangement of.	H. W. Bill	Cuyahoga Falls, Ohio	Nov. 3, 1857	18, 532
Feed-water purifier for steam-generators	H. K. Hazlett	Saint Louis, Mo	Feb. 22, 1870	100, 144
Feed-water regulator	R. Berryman	Hartford, Conn	Apr. 9, 1872	125, 527
Feed-water regulator	G. E. Chenoweth	Baltimore, Md	Feb. 26, 1867	62, 474
Feed-water regulator	E. C. Fernald	Flatbush, N. Y	Apr. 8, 1873	137, 541
Feed-water regulator	N. Nolan	New York, N. Y	Nov. 21, 1871	121, 122
Feed-water regulator	J. B. Root	New York, N. Y	Aug. 5, 1873	141, 592
Feed-water regulator and low-water alarm	A. W. Morrell	Niles, Mich	Apr. 12, 1870	101, 756
Feed-water regulator and low-water alarm for steam-boilers.	W. Painter	Baltimore, Md	June 11, 1872	127, 917
Feed-water regulator and low-water alarm for steam-boilers.	J. B. Smith	Milwaukee, Wis	Apr. 23, 1872	125, 993
Feed-water regulator, Boiler	C. H. Gould	Cincinnati, Ohio	June 25, 1867	66, 018
Feed-water regulator, Boiler	J. K. P. Nourse	West Medway, Mass	Nov. 10, 1868	83, 993

Index of patents issued from the United States Patent Office from 1790 *to* 1873, *inclusive*—Continued.

Invention.	Inventor.	Residence.	Date.	No.
Feed-water regulator for boilers, Alarm	N. L. Smith	Derby, Conn	May 11, 1869	89, 949
Feed-water regulator for steam-boilers	R. Berryman	Hartford, Conn	Sept. 5, 1871	118, 679
Feed-water regulator for steam-boilers	R. E. Bond	Detroit, Mich	Oct. 18, 1864	44, 702
Feed-water regulator for steam-boilers	J. Stowell	Charlestown, Mass	Mar. 20, 1860	27, 601
Feed-water to steam-boilers, Admission of	T. Roberts	Baltimore, Md	Apr. 25, 1871	114, 202
Feed-water to steam-boilers, Apparatus for regulating the supply of.	A. Jacobs	Ithaca, N. Y	May 23, 1854	10, 963
Feed-wheel as substitute for ratchets and pawls	O. C. Phelps	New York, N. Y	Apr. 4, 1865	47, 126
Feed wheel-operating mechanism	H. A. M. Harris	Philadelphia, Pa	July 16, 1872	129, 131
Feeder, Calf	W. C. Dodge	Washington, D. C	Aug. 23, 1864	43, 902
Feeder, Fowl	J. Richardson	Ballston Spa, N. Y	Apr. 12, 1870	101, 766
Feeder, Portable stock	J. M. Spencer	La Plata, Mo	Apr. 29, 1873	138, 351
Feeder, Stock	C. M. Hall	Yates City, Ill	May 13, 1873	138, 801
Feeder, Stock	T. W. Peirce	Richfield, Minn	Feb. 21, 1865	46, 493
Feeder, Stock and poultry	G. Bowerman	Napoleon, Ohio	Nov. 5, 1867	70, 401
Feeder, Swine	N. Stockwell	Windsor, N. Y	June 29, 1869	91, 881
Feeder, Swing sheep	A. Putnam	Vernon, Big Bend Post-Office, Wis.	Apr. 14, 1868	76, 810
Feeding-apparatus, Automatic	R. B. Ruggles	Hartford, Conn	Jan. 23, 1872	122, 914
Feeding chickens, Apparatus for	S. W. Abbee	Walpole, N. H	Sept. 7, 1852	9, 244
Feeding-door for animal-pens	B. C. Scott	Paxton, Ill	Aug. 3, 1869	93, 234
Feeding rack and trough combined, Stock	A. Ralston	West Middletown, Pa	May 13, 1862	35, 259
Feeding rack, Cattle	G. W. Balding	Angola, Ind	Oct. 8, 1867	69, 608
Feeding rack, Sheep	J. C. Colflesh	Delaware, Ohio	Aug. 4, 1868	80, 603
Feeding rack, Sheep	M. S. Every	Bridgewater, Mich	July 3, 1866	56, 024
Feeding stock, Compound for	C. G. Otis	Troy, N. Y	Oct. 30, 1866	59, 255
Feeding swine, Apparatus for	N. Stockwell	Ouaquaga, N. Y	Jan. 11, 1870	98, 718
Feeding-trough	R. Chesnut	Richmond, Ind	Aug. 28, 1866	57, 477
Feeding trough, Animal	J. N. Gray	Lynn, Mass	May 24, 1864	42, 909
Feeding-trough for stock	B. Arnold	Arba, Ind	July 10, 1866	56, 158
Feeding trough, Hog	J. H. McElrath and L. M. Houghton.	Princeton, Ill	Aug. 20, 1872	130, 735
Feeding trough, Horse	A. D. Barrett	Cambridgeport, Mass	Apr. 13, 1869	88, 768
Feeding trough, Sheep	C. Aulls	Bridgewater, Mich	Apr. 9, 1867	63, 688
Feet, Taking impressions of	J. E. A. Rillot	San José, Cal	Nov. 22, 1864	45, 178
Feldspar to obtain useful products, Treating	G. E. Vanderburgh	Mamaroneck, N. Y	July 12, 1864	43, 534
Fellies and spokes in carriage-wheels, Mode of securing ends of.	J. Switzer	Lynn, Mass	Mar. 10, 1868	75, 314
Fellies and tightening spokes, Expanding	A. N. Towne	Chicago, Ill	Jan. 21, 1868	73, 476
Fellies, Bending	A. W. Johnson	Saint George's, Del	Nov. 11, 1851	8, 511
Fellies, Bending	T. Kimball	Salem, N. H	Jan. 16, 1832	....
Fellies, Bending	E. Reynolds	Haddonfield, N. J	July 17, 1835	....
Fellies by steaming and bending, Method of making.	M. Blakeslee	Canaan, Conn	July 29, 1829	....
Fellies for wheels, Machine for rounding	H. A. Gore	Goshen, Ind	Aug. 30, 1870	106, 931
Fellies, Guide for doweling wheel	W. C. Dean	Jacksonville, N. Y	Oct. 4, 1853	10, 092
Fellies in wheels, Mode of tightening	A. Stoner	Mount Joy, Pa	Jan. 13, 1857	16, 408
Fellies, Mode of securing	M. J. Mellyn	Roxbury, Mass	Dec. 24, 1867	72, 663
Fellies of carriage-wheels, Machinery for shaping the inner sides of.	R. A. Henry	Barre, Mass	Nov. 21, 1842	2, 860
Fellies of wagon-wheels, Machine for dressing the.	A. P. Odholm	Bridgeport, Conn	Nov. 27, 1866	60, 040
Fellies of wheels, Mode of connecting	J. Eveland	Elizabeth City, N. J	Oct. 28, 1862	36, 774
Fellies, Sawing and boring	W. W. Forwood	Abingdon, Md	Nov. 19, 1833	....
Fellies, Sawing and boring	J. Hamilton	New York, N. Y	Apr. 29, 1833	....
Fellies, &c., Sawing wood for	B. Ruggles	Poultney, Vt	June 29, 1826	....
Fellies, Securing the ends of	H. and H. Inman	Amsterdam, N. Y	Nov. 5, 1867	70, 438
Fellies, Tightening	H. Thompson	Palmyra, Wis	Aug. 2, 1864	43, 722
Fellies to wheels, Machine for fitting	F. H. Brinkkotter	Callahan's Ranch, Cal	Mar. 30, 1869	88, 267
Fellies with oil, Apparatus for saturating	J. R. Faus	Espy, Pa	Aug. 30, 1870	106, 927
Fellies, spokes, &c., Apparatus for oiling	P. E. Bromby	Espy, Pa	Feb. 7, 1871	111, 511
Felling apparatus, Tree	M. R. Fory	New York, N. Y	June 1, 1869	90, 834
Felling trees, &c	J. Hamilton	New York, N. Y	June 26, 1835	....
Felling trees	P. Johnson	Whitney's Point, N. Y	May 15, 1870	28, 280
Felling trees by hand, Machine for	E. T. Miller	Chelsea, Mass	June 9, 1857	17, 512
Felling trees by saws, Method of	G. C. Ehrsam	New York, N. Y	June 24, 1856	15, 178
Felling trees, Machine for	H. J. Beard	New Sharon, Me	Oct. 25, 1870	108, 675
Felling trees, Machine for	T. Durden	Montgomery, Ala	Sept. 18, 1855	13, 568
Felling trees, Machine for	A. Edwards	New Haven, Conn	Mar. 24, 1868	75, 882
Felling trees, Machine for	M. R. Fory	New York, N. Y	Sept. 22, 1868	82, 303
Felling trees, Machine for	J. S. Foster	Vallicita, Cal	July 2, 1861	32, 688
Felling trees, Machine for	E. Mathers	Morgantown, Va	Mar. 18, 1856	14, 462
Felling trees, Machine for	E. R. Morrison	New York, N. Y	June 13, 1871	115, 975
Felling trees, Method of	S. Ingersoll	Green Point, N. Y	Oct. 14, 1856	15, 913
Felling trees, Saw for	F. Bauschtliker	Washington, D. C	Oct. 6, 1868	82, 789
Felly	W. L. Perry and A. J. Mathews	Hartwell, Ga	Apr. 15, 1873	137, 955
Felly-bending machine	J. Compton	Elmira, N. Y	Sept. 15, 1863	39, 887
Felly-bending machine	T. Cox	Lancaster, Pa	July 4, 1854	11, 209
Felly-bending machine	A. Hemenway	Cleveland, Ohio	May 1, 1860	28, 080
Felly-bending machine	J. L. Mann	Ravenna, Ohio	Aug. 31, 1858	21, 351
Felly bending machine	S. Moyer	Shinersville, Pa	Mar. 20, 1860	27, 560
Felly-bending machine	D. A. Sprinkle	Leoti, Ind	Mar. 8, 1870	100, 685
Felly-bending machine, Wood	G. A. Brown	Newfane, N. Y	Mar. 15, 1859	23, 227
Felly-boring and spoke-tenoning bit	H. McGrath	Meig's Creek, Ohio	Jan. 15, 1856	14, 105
Felly, Carriage-wheel	M. Turley	Council Bluffs, Iowa	Feb. 13, 1872	123, 750
Felly-clamp and spoke-support for carriage-wheels	J. C. Plumer	Boston, Mass	Dec. 19, 1865	51, 620
Felly-clip for vehicle-wheels	D. Grim	Pittsburgh, Pa	June 20, 1871	116, 050
Felly coupling, Carriage	H. Austin	East Liberty, Ohio	May 8, 1866	54, 485
Felly coupling, Wheel	C. Cheney	Milford, Mass	Aug. 2, 1870	105, 904
Felly cutting and dressing machine	S. Church	Amherst, Mass	Apr. 25, 1810	....
Felly-cutting machine	J. and L. Adams	Hadley, Mass	July 9, 1850	7, 485
Felly-cutting machine	J. and L. Adams and L. H. Moore.	Hadley and Leverett, Mass	June 12, 1849	6, 521
Felly-cutting machine	Bradley and Worthley	Phillips, Me	Oct. 14, 1835	....
Felly-cutting machine	Brown and Barker	Phillips, Me	Oct. 14, 1835	....
Felly-cutting machine	J. Goulding	Leicester, Mass	Feb. 28, 1811	....
Felly-cutting machine	B. L. Greenough	Lebanon, N. H	Dec. 22, 1824	....
Felly-cutting machine	W. A. Lewis and G. W. Butler	Joliet, Ill	Nov. 16, 1869	96, 818
Felly-cutting machine	I. S. Moon	Bennington, Vt	Sept. 10, 1872	131, 292
Felly dowel-pin	E. D. Tyler	Gibson, Pa	Apr. 28, 1868	77, 337

Index of patents issued from the United States Patent Office from 1790 *to* 1873, *inclusive*—Continued.

Invention.	Inventor.	Residence.	Date.	No.
Felly-doweling machine	J. P. O'Brien	Kewanee, Ill	July 30, 1872	129, 979
Felly-dressing machine	W. M. Bullock	Marcy, Ind	Sept. 16, 1856	15, 727
Felly-dressing machine	H. Kurtz	Washington, D. C	Jan. 11, 1870	98, 696
Felly-dressing machine	D. J. Marston	Amesbury, Mass	Apr. 25, 1871	114, 169
Felly-dressing machine	W. H. Rodeheaver	Miamisburgh, Ohio	Oct. 11, 1870	108, 189
Felly-dressing machine	W. H. Rodeheaver	Miamisburgh, Ohio	July 14, 1868	80, 013
Felly-dressing machine	C. W. Wyatt	New York, N. Y	Dec. 12, 1854	12, 082
Felly-finishing tool	C. H. Denison	Guilford, Vt	Dec. 6, 1859	26, 392
Felly fitting and boring machine	W. L. Perry	Jonesville, S. C	Oct. 10, 1871	119, 880
Felly, Hollow metallic rim or	S. R. Bryant	Waterford, Pa	June 28, 1870	101, 824
Felly joint, Carriage-wheel	F. M. Gibson	Chelsea, Mass	Feb. 5, 1861	31, 309
Felly joint, clip, and ferrule	P. Jones	Newark, N. J	Jan. 28, 1868	73, 903
Felly-joint coupling	S. A. Garrison and D. C. Morey	Chelsea, Mass	Feb. 26, 1856	14, 311
Felly-joints, Mode of securing	J. W. Lawrence	New York, N. Y	Sept. 17, 1867	68, 999
Felly-machine	H. H. Dean	Adrian, Mich	May 9, 1854	10, 884
Felly-machine	C. H. Denison	Brattleborough, Vt	Feb. 19, 1861	31, 443
Felly-machine	F. H. Mallett	New Haven, Conn	July 17, 1860	29, 233
Felly-machine	R. Massey	Philadelphia, Pa	Oct. 6, 1863	40, 177
Felly-manufacturing machine	A. B. Richmond	Meadville, Pa	May 27, 1856	14, 987
Felly-mill	S. Church	Amherst, Mass	May 4, 1805	....
Felly of wheel for trucks, &c	G. Andrews	Tolland, Conn	Oct. 24, 1828	....
Felly-planing machine	A. W. Fox	Athens, Pa	Apr. 8, 1856	14, 604
Felly-planing machine	W. W. Johnson	Clifford, Pa	Dec. 18, 1855	13, 947
Felly-plate	C. D. Everett	Cleveland, Ohio	May 6, 1873	138, 624
Felly-plate	C. D. Everett and A. Baldwin	Cleveland, Ohio	Oct. 8, 1872	132, 065
Felly-plate	F. B. Morse	Plantsville, Conn	May 30, 1871	115, 504
Felly-plate	F. B. Morse	Plantsville, Conn	July 2, 1872	128, 558
Felly-plate	H. Sylvester	Saint Louis, Mo	Apr. 8, 1873	137, 571
Felly-plate for carriage-wheels	F. B. Morse	Plantsville, Conn	May 17, 1870	103, 273
Felly-plate-making die	A. N. Clark	Plainville, Conn	Jan. 24, 1871	111, 177
Felly-plates, Die for forming	F. B. Morse	Plantsville, Conn	Apr. 16, 1872	125, 751
Felly-plates, Die for the manufacture of	L. C. Clark	Plantsville, Conn	July 9, 1872	128, 790
Felly-sawing machine	D. Bowen	Wadesville, Va	July 1, 1856	15, 216
Felly-sawing machine	S. and W. H. Brook	Rushville, Ohio	May 6, 1856	14, 802
Felly-sawing machine	W. Carr	Logansport, Ind	Jan. 26, 1869	86, 279
Felly-sawing machine	A. Colby	Sharon, Vt	Oct. 15, 1830	....
Felly-sawing machine	P. Collier	Catskill, N. Y	May 20, 1826	....
Felly-sawing machine	D. D. Hansom	Weare, N. H	Oct. 20, 1830	....
Felly-sawing machine	E. Rhodes	Erie, Pa	Sept. 21, 1869	95, 140
Felly-sawing machine	W. E. Sheffield	New York, N. Y	Dec. 20, 1809	....
Felly sawing machine	J. B. Zimmerman	Fort Seneca, Ohio	May 14, 1872	126, 768
Felly-sawing machine, Wagon	N. W. Graves	Winnebago, Ill	Feb 26, 1867	62, 408
Felly-sawing machine, Wheel	S. Fahrney	Washington County, Md	Dec. 9, 1828	....
Felly-sawing mill	I. Sheetz	Taneytown, Md	May 20, 1842	2, 632
Felt fabrics, Machine for disintegrating waste	J. T. Greene	Brooklyn, N. Y	Apr. 12, 1859	23, 643
Felt for bottoms of vessels, &c	T. R. Williams	Newport, R. I	May 22, 1830	....
Felt for cloth, padding, carpets, &c., Manufacturing	J. Barker and L. Kinsley	Greene County, N. Y	Aug. 29, 1829	....
Felt for weather-strips, packings, &c., Process of preparing.	F. Siering	New York, N. Y	Dec. 23, 1873	145, 913
Felt goods, Manufacturing seamless	J. H. Bloodgood	New York, N. Y	Feb. 6, 1855	12, 343
Felt, Machinery for manufacturing sheathing	J. B. Hyde	Newark, N. J	May 29, 1860	28, 489
Felt, Machinery for the manufacture of	M. D. Whipple	Cambridge, Mass	Oct. 12, 1869	95, 862
Felt, Machinery for the manufacture of	M. D. Whipple	Cambridge, Mass	Oct. 12, 1869	95, 863
Felt, Manufacture of	S. M. Allen	Boston, Mass	Feb. 3, 1863	37, 559
Felt, Manufacture of roofing and sheathing	W. P. Arnold	New York, N. Y	Aug. 20, 1872	130, 615
Felt of wool without spinning or weaving, Making	L. Van Hoesen	Norwalk, Conn	June 27, 1829	....
Felt skirts, &c., Ornamenting	J. W. Blackham	Brooklyn, N. Y	Sept. 9, 1873	142, 554
Felt washer	W. Cole	Lee, Mass	Jan. 23, 1833	....
Felted and other goods and fabrics, Process of finishing.	T. Crossley	Bridgeport, Conn	Jan. 29, 1867	61, 520
Felted articles, Manufacture of hollow	C. W. and H. E. Palmer and C. Houghton.	Lynn and Boston, Mass	July 26, 1870	105, 836
Felted fabric	W. E. Bloodgood	Rahway, N. J	July 9, 1872	128, 844
Felted fabric	T. Crossley	Bridgeport, Conn	Feb. 2, 1869	86, 510
Felted fabric	T. Demuth	Danbury, Conn	Apr. 4, 1871	113, 274
Felted fabric	H. Hayward	New York, N. Y	July 21, 1868	80, 263
Felted fabric	M. A. Johnson	Lowell, Mass	Jan. 5, 1864	41, 073
Felted fabric	M. A. Johnson	Lowell, Mass	Jan. 28, 1868	73, 727
Felted fabric	J. E. Pollard	Franklin City, Mass	Feb. 21, 1871	111, 968
Felted fabric	S. P. Siver	Danbury, Conn	Mar. 8, 1870	100. 563
Felted fabric	E. Waite	Franklin City, Mass	Aug. 8, 1865	49, 353
Felted fabric	E. Waite	Franklin, Mass	June 11, 1872	127, 813
Felted fabric	J. T. Waring	Yonkers, N. Y	Oct. 6, 1868	82, 904
Felted fabric for bonnets, Woven	R. S. Gillespie and F. Ziesing	New York, N. Y	Feb. 27, 1872	124, 191
Felted fabrics, Manufacture of	E. D. McCracken	New York, N. Y	Aug. 1, 1865	49, 131
Felted fabrics, Manufacture of	E. Waite	South Natick, Mass	Mar. 14, 1865	46, 837
Felted fabrics, Removing vegetable fiber from	J. T. Waring	Yonkers, N. Y	Aug. 19, 1873	142, 063
Felted tufted fabric	J. T. Waring	Yonkers. N. Y	Nov. 24, 1868	84, 325
Felting, Bat for	L. W. Boynton	South Coventry, Conn	Apr. 29, 1851	8, 068
Felting bats, Machine for forming	J. H. Bloodgood	New York, N. Y	Jan. 20, 1857	16, 431
Felting, Composition	L. S. Stimson	Lowell, Mass	Sept. 19, 1871	119, 059
Felting-compound	L. E. Hopkins	Brooklyn, N. Y	Aug. 19, 1856	15, 563
Felting for coats, hats, &c	M. Osborne	New York, N. Y	May 28, 1842	2, 646
Felting-machine	W. J. Benedict and J. Wylie	South Norwalk, Conn	Aug. 18, 1868	81, 244
Felting-machine	J. W. Blackham	Brooklyn, N. Y	Sept. 21, 1869	94, 939
Felting-machine	J. H. Bloodgood and M. A. Johnson.	New York, N. Y., and Lowell, Mass.	Nov. 15, 1864	45, 017
Felting-machine	H. Brisco and J. W. Blackham	Brooklyn, N. Y	June 1, 1869	90, 633
Felting-machine	G. N. Bronson	New Milford, Conn	Apr. 16, 1861	32, 042
Felting-machine	A. Cattaneo	Newark, N. J	Aug. 18, 1868	81, 252
Felting-machine	R. Eickemeyer	Yonkers, N. Y	Mar. 16, 1869	87, 763
Felting-machine	R. Eickemeyer	Yonkers, N. Y	June 22, 1869	91, 729
Felting-machine	R. Eickemeyer	Yonkers, N. Y	Oct. 11, 1870	108, 121
Felting-machine	R. Eickemeyer	New York, N. Y	Nov. 21, 1871	121, 045
Felting-machine	F. S. Jennings	Danbury, Conn	Jan. 18, 1870	98, 871
Felting-machine	C. P. Ladd	New York, N. Y	Dec. 27, 1870	110, 574
Felting-machine	C. P. Ladd	Bloomfield, N. J	Mar. 19, 1872	124, 828
Felting-machine	W. B. Lodge and H. Platner	Danbury, Conn	Nov. 26, 1867	71, 516

Index of patents issued from the United States Patent Office from 1790 *to* 1873, *inclusive*—Continued.

Invention.	Inventor.	Residence.	Date.	No.
Felting-machine	W. A. Lyon	Danbury, Conn	July 9, 1867	66, 603
Felting-machine	S. S. Middlebrook	Sandy Hook, Conn	June 4, 1867	65, 413
Felting-machine	C. Mossant	Bourg du Péage, France	June 16, 1868	78, 991
Felting-machine	L. Robinson, L. Conine, and N. F. and D. W. Hyatt.	Matteawan, N. Y	June 22, 1869	91, 670
Felting-machine	E. Waite	Lawrence, N. Y	Sept. 22, 1863	40, 087
Felting-machine	E. Waite	South Natick, Mass	May 3, 1864	42, 611
Felting-machine	E. Waite	South Natick, Mass	June 14, 1864	43, 164
Felting-machine	E. Waite	South Natick, Mass	June 4, 1867	65, 455
Felting, Machine for crossing bats for	A. C. Arnold	Norwalk, Conn	Oct. 22, 1867	70, 065
Felting, Machine for crossing fibers in forming bats for.	L. Robinson	Matteawan, N. Y	July 20, 1869	92, 750
Felting-machinery	J. H. Bloodgood and M. A. Johnson.	New York, N. Y., and Lowell, Mass.	Mar. 6, 1860	27, 345
Felting, Machinery for forming bats for	T. B. Butler	Norwalk, Conn	Aug. 10, 1858	21, 164
Felting, wadding, &c., Machine for forming bats for	E. Waite	Franklin, Mass	June 4, 1867	65, 456
Female-supporter	A. D. Reeves	Portland, Me	Apr. 1, 1862	34, 845
Fence	G. W. Adams	Rochester, N. Y	Oct. 1, 1867	69, 299
Fence	G. Albro	Columbus, Pa	Aug. 5, 1873	141, 532
Fence	E. Allen	Honeove, N. Y	July 16, 1872	129, 305
Fence	B. F. Allison	West Dayton, Iowa	Feb. 24, 1863	37, 729
Fence	A. M. Alpin	Chetopah, Kans	Feb. 20, 1872	123, 854
Fence	W. Altick and F. P. Grimes	Dayton, Ohio	Mar. 14, 1871	112, 6[illegible]2
Fence	D. B. Ayres	Brooklyn, Mich	Aug. 10, 1869	93, 397
Fence	C. M. Ballard and M. Morehouse.	Johnsbury, N. Y	Jan. 10, 1871	110, 888
Fence	H. M. Barber	Franklin Station, Ohio	Feb. 7, 1871	111, 508
Fence	A. T. Barnes	Seneca, N. Y	Jan. 5, 1869	85, 506
Fence	T. Barnes	Wayne, Mich	July 13, 1869	92, 509
Fence	H. Bartholomew	Dover Centre, Ohio	Oct. 22, 1867	69, 961
Fence	W. Bartlett	Little Hocking, Ohio	Aug. 16, 1870	106, 455
Fence	A. C. Betts	Troy, N. Y	Mar. 12, 1872	124, 477
Fence	B. Billings	Lyons, Iowa	Dec. 4, 1863	60, 127
Fence	I. Blackaby	Civer, Ill	Dec. 21, 1869	98, 016
Fence	I. Boone	Troy, Ohio	July 23, 1867	67, 017
Fence	L. H. Bowlus	Knoxville, Tenn	Aug. 20, 1867	67, 944
Fence	W. J. Brainard	Hamilton, N. Y	Dec. 28, 1847	5, 406
Fence	R. W. Brockway and H. Frederick.	Akron, Ohio	July 21, 1868	80, 054
Fence	C. E. Brown	Pamelia, N. Y	Dec. 5, 1871	121, 448
Fence	F. G. Brown	Chapel Hill, Tex	June 4, 1872	127, 558
Fence	M. Brown and O. J. Shannon	Fond du Lac and Fairwater, Wis.	Feb. 28, 1865	46, 541
Fence	W. H. Brown	Stockwell, Ind	Aug. 7, 1866	56, 895
Fence	J. H. Buckman	Cartersville, Ga	Apr. 16, 1872	125, 845
Fence	J. Bundy	West Liberty, Iowa	Oct. 30, 1866	59, 176
Fence	J. S. Burch	Buffalo, N. Y	Oct 19, 1869	95, 982
Fence	T. L. Burk	Greensburgh, Ind	July 21, 1868	80, 131
Fence	H. Burrows	Georgetown, D. C	Sept. 11, 1860	30, 021
Fence	T. R. Byrnes	Washington, D. C	Dec. 26, 1865	51 691
Fence	J. Card	Gainesville, N. Y	Feb. 17, 1852	8, 735
Fence	J. Carr	Emerald Post-Office, Ohio	July 13, 1869	92, 582
Fence	I. D. Carter	Wilson, N. Y	Jan. 26, 1869	86, 280
Fence	J. M. Chaplin	Middleport, Conn	Oct. 13, 1868	83, 038
Fence	J. W. Cherry	Carthage, Ill	July 11, 1870	105. 303
Fence	J. W. Cherry	Carthage, Ill	Dec. 12, 1871	121, 752
Fence	E. S. Clapp and E. Blanchard	Montague and Amherst, Mass.	Apr. 5, 1864	42, 166
Fence	G. R. Clark	Livonia, N. Y	Mar. 12, 1867	62, 732
Fence	J. K. and C. B. Clark	Mount Pleasant, Ohio	Mar. 21, 1865	46, 881
Fence	L. N. Clark	Brighton, Mich	May 23, 1871	115, 165
Fence	A. J. Clemmons and J. W. Westbrooks.	Aberdeen, Miss	Dec. 10, 1872	133, 831
Fence	T. Coffield	Natrona, Pa	Mar. 29, 1870	101, 351
Fence	T. C. Collins	Little Hockhocking, Ohio	Jan. 10, 1871	110, 899
Fence	J. Comstock	Greenfield, Ind	Oct. 4, 1870	107, 879
Fence	S. P. Coon	Milwaukee, Wis	July 11, 1870	105, 307
Fence	F. K. Cosgrove and R. Westerman.	Fort Wayne, Ind	Jan. 20, 1863	37, 434
Fence	E. Cozad	East Cleveland, Ohio	Mar. 27, 1866	53, 418
Fence	E. Crandal	Marshalltown, Iowa	Nov. 26, 1872	133, 358
Fence	S. Crocker	Fredonia Post-Office, Iowa.	Sept. 11, 1866	57, 872
Fence	J. M. Crull	Noblesville, Ind	Aug. 17, 1869	93, 860
Fence	H. J. Culp	Goshen, Ind	Aug 4, 1868	80, 609
Fence	P. Davis	Newport News, Va	June 29, 1869	91, 918
Fence	H. H. Dennis	Steam Mill, Pa	Nov. 27, 1855	13, 844
Fence	S. Denton	Penn Yan, N. Y	Feb. 21, 1860	27, 209
Fence	B. G. Devoe	Fredericktown, Ohio	Dec. 12, 1871	121, 8[illegible]5
Fence	B. G. Devoe, T. Rogers, and J. C. Beals.	Fredericktown, Ohio, and Searsport, Me.	Mar. 5, 1872	124, 342
Fence	H. Deyoe	Machias, N. Y	June 13, 1871	115, 8[illegible]0
Fence	M. Dickle	Wellersburgh, Pa	Feb. 27, 1866	52, 828
Fence	B. K. Dorwart and W. J. Hines.	Lancaster, Pa., and Frederick, Md.	Sept. 24, 1867	69, 191
Fence	B. K. Dorwart and G. F. Rote, jr	Lancaster, Pa	Nov. 19, 1867	71, 146
Fence	B. K. Dorwart and G. F. Rote, jr	Lancaster, Pa	Nov. 19, 1867	71, 147
Fence	J. G. Downer	Auburn, N. Y	Nov. 9, 1869	96, 561
Fence	J. T. Drummond	Mount Pleasant, Iowa	Apr. 18, 1871	113, 753
Fence	C. E Easton	Cedarville, N Y	Oct. 7, 1862	36, 605
Fence	J. Edmunds	South Adams, Mass	Mar. 6, 1866	52, 981
Fence	O. L. Edwards and N. Gabel	Gratis, Preble County, Ohio	Jan. 28, 1862	34, 246
Fence	A. Ellis and O. Albertson	Salem, Ohio	Dec. 24, 1867	72, 469
Fence	F. Ellis	La Fayette, Ohio	May 7, 1867	64, 509
Fence	E. Ely	Lockport, N. Y	May 9, 1871	114, 543
Fence	D S. Evans	Richmond, Ind	Oct. 30, 1866	59, 199
Fence	O. J. Everson	Lake City, Minn	May 19, 1868	77, 967
Fence	E. Fales	Glenwood, Mo	Sept 6, 1870	107, 020
Fence	E. Fales	Glenwood, Mo	Apr. 16, 1872	1[illegible]5, 880

Index of patents issued from the United States Patent Office from 1790 *to* 1873, *inclusive*—Continued.

Invention.	Inventor.	Residence.	Date.	No.
Fence	A. Fanckboner	Schoolcraft, Mich	Oct. 13, 1863	40, 253
Fence	A. Faulkner	Mount Pleasant, Iowa	Jan. 10, 1871	110, 910
Fence	O. F. A. Faulkner	Mount Pleasant, Iowa	Feb. 14, 1871	111, 831
Fence	R. J. Flanner	Plainfield Township, Mich	Nov. 29, 1870	109, 606
Fence	B. Force	Mount Pleasant, Iowa	Aug. 27, 1867	68, 177
Fence	J. Fox	Bucyrus, Ohio	Jan. 11, 1870	98, 682
Fence	E. L. Fraker	Oshkosh, Wis	Mar. 2, 1869	87, 336
Fence	A. A. Garver	Albion, Iowa	May 21, 1872	126, 950
Fence	M. J. Gaskill	Pleasant Plain, Ohio	Feb. 25, 1868	74, 908
Fence	M. J. Gaskill	Pleasant Plain, Ohio	Mar. 9, 1869	87. 558
Fence	G. L. Gavett	Sandstone, Mich	July 16, 1867	66, 826
Fence	W. Gibson	Fort Wayne, Ind	May 13, 1862	35, 238
Fence	T. Glover	Woodbury, N. J	Dec. 11, 1866	60, 350
Fence	J. Godfrey	Allegheny City, Pa	Aug. 31, 1869	94, 199
Fence	S. Good	Greenville, Ohio	Dec. 3, 1867	71, 736
Fence	E. C. Gordon	Sevastopol, Ind	Dec. 4, 1866	60, 174
Fence	E. C. Gordon	Sevastopol, Ind	May 28, 1867	65, 209
Fence	J. Gray	Raymond, Miss	Nov. 11, 1873	144, 534
Fence	I. M. Greene, sr	Clinton, Ill	Nov. 5, 1867	70, 549
Fence	C. S. S. Griffing	Ashtabula, Ohio	May 24, 1864	42, 848
Fence	C. S. S. Griffing	Geneva, Ohio	Mar. 12, 1867	62, 840
Fence	H. A. Grover	North Cohocton, N. Y	July 13, 1869	92, 442
Fence	C. Guthrie	Peoria, N. Y	Sept. 3, 1872	130, 991
Fence	S. W. Hall	Elmira, N. Y	Dec. 10, 1872	133, 853
Fence	T. Harrison	La Fayette, Wis	Aug. 3, 1869	93, 198
Fence	D. Harvey	Jackson Township, Iowa	Apr. 11, 1865	47, 203
Fence	J. Hatfield	Ogden, Ind	Jan. 6, 1863	37, 372
Fence	C. A. Haviland	Davenport, Iowa	Dec. 3, 1867	71, 750
Fence	W. D. Hillis	Elgin, Ill	Apr. 3, 1866	53, 616
Fence	W. D. Hillis	Elgin, Ill	June 23, 1868	79, 121
Fence	W. D. Hillis	Elgin, Ill	Dec. 20, 1870	110, 234
Fence	C. N. Hitchcock, and H. P. T. Wilson.	Canton, Miss	Nov. 8, 1870	109, 127
Fence	E. Hollinger	New Haven, Conn	Mar. 14, 1865	46, 799
Fence	H. G. Hood	Harlan, Ind	July 14, 1863	39, 231
Fence	T. J. Hubbard	Hamilton, N. Y	Dec. 28, 1847	5, 407
Fence	W. D. Hunt	Scott, N. Y	July 23, 1867	67, 117
Fence	G. W. Hunter	Versailles, Ind	Nov. 2, 1869	96, 435
Fence	C. B. Hunting	Clinton, Ill	Feb. 20, 1866	52, 717
Fence	F. W. Huxford	Boonesborough, Iowa	July 30, 1867	67, 308
Fence	P. J. Hynes	Stoughton, Wis	Aug. 29, 1871	118, 535
Fence	A. Jackson	Clifton Springs, N. Y	Jan. 16, 1866	52, 051
Fence	J. B. Johnson	Laurel, Ind	Dec. 17, 1867	72, 402
Fence	N. Johnson	Jasper, N. Y	Oct. 1, 1867	69, 444
Fence	N. Johnson	Jasper, N. Y	Oct. 1, 1867	69, 445
Fence	D. Johnston	Cranberry, Ohio	Dec. 14, 1869	97, 926
Fence	L. P. Judson	Rose, N. Y	Aug. 15, 1871	118, 135
Fence	D. Kaufman	Boiling Springs, Pa	Sept. 17, 1867	68, 991
Fence	D. Kaufman	Boiling Springs, Pa	Jan. 21, 1868	73, 611
Fence	M. Kelly	New York, N. Y	Feb. 11, 1868	74, 379
Fence	J. E. Kendeigh	Amherst, Ohio	Mar. 13, 1866	53, 153
Fence	J. Kerr	Corsicana, Tex	Feb. 15, 1870	99, 911
Fence	G. Kicherer	Brooklyn, N. Y	Oct. 2, 1866	58, 432
Fence	T. E. King	Ashtabula, Ohio	May 5, 1863	38, 396
Fence	T. E. King	Painesville, Ohio	June 26, 1866	55, 874
Fence	T. E. King	Painesville, Ohio	Feb. 19, 1867	62, 209
Fence	W. F. King	Gonzales County, Tex	Sept. 23, 1873	143, 161
Fence	E. Kirk	Sharon, Pa	May 10, 1870	102, 834
Fence	J. J. Knight	Long Point, Ill	June 6, 1871	115, 742
Fence	A. Kull, jr	Bloomfield. Wis	Nov. 16, 1869	96, 814
Fence	E. Kyes	Nunda, N. Y	May 21, 1872	127, 071
Fence	J. A. Kysor	Leon, N. Y	June 27, 1871	116, 329
Fence	I. Lackey	Lebanon, Ohio	Dec. 10, 1867	71, 888
Fence	H. H. Landis	Lancaster, Pa	Aug. 29, 1871	118, 461
Fence	I. L. Landis	Lancaster, Pa	June 18, 1867	65, 819
Fence	I. L. Landis	Lancaster, Pa	Feb. 6, 1872	123, 487
Fence	I. L. Landis	Lancaster, Pa	Feb. 6, 1872	123, 488
Fence	I. L. Landis	Lancaster, Pa	July 23, 1872	129, 837
Fence	A. Lapham	Farmington, Mich	Aug. 25, 1868	81, 513
Fence	H. Latshaw	McKnightstown, Pa	Sept. 26, 1871	119, 372
Fence	J. E. Layton	New Wilmington, Pa	Oct. 22, 1867	70, 097
Fence	C. G. Lazear	Norwalk, Ohio	Nov. 27, 1866	60, 020
Fence	L. Leavenworth	Trumansburgh, N. Y	Oct. 30, 1849	6, 831
Fence	C. Lee	Winchester, Ohio	Sept. 4, 1866	57, 735
Fence	C. Lee	Sandy Post-Office, Ohio	Apr. 7, 1868	76, 475
Fence	J. C. Leonard	Union City, Mich	Feb. 12, 1867	61, 942
Fence	I. N. Lerick	San Antonio, Tex	July 23, 1872	129, 738
Fence	L. E. Lockling	Perrysburgh, N. Y	Oct. 16, 1866	58, 848
Fence	L. E. Lockling and N. N. Whitaker.	Perrysburgh and Sheridan, N. Y.	June 7, 1870	103, 903
Fence	O. Love	Saxenburgh, Pa	Oct. 20, 1868	83, 184
Fence	C. Lowe	Randolph, N. H	Nov. 16, 1869	96, 934
Fence	W. Mallary	Bucyrus, Ohio	Oct. 5, 1869	95, 596
Fence	C. Maltby	Glenwood, N. Y	June 18, 1872	128, 157
Fence	H. H. Margeson	San Francisco, Cal	Feb. 22, 1870	100, 169
Fence	J. Markley	Bucyrus, Ohio	Oct. 12, 1869	95, 820
Fence	C. C. Mather	Burlington, N. Y	Nov. 6, 1866	59, 427
Fence	J. B. Mattern	Matternsville, Pa	June 25, 1872	128, 409
Fence	N. Maxson	Wilmington, Ohio	Oct. 6, 1868	82, 733
Fence	J. McConnell	Tyro, Ohio	Mar. 1, 1870	100, 431
Fence	J. W. McCormick	Youngstown, N. Y	Mar. 30, 1869	88, 401
Fence	D. McCurdy	Ottawa, Ohio	Aug. 9, 1870	106, 183
Fence	R. McFarlane	Dane, Wis	May 18, 1869	90, 280
Fence	J. M. McKee	Mount Vernon, Ohio	July 6, 1869	92, 334
Fence	J. M. McKee	Hunt's Station, Ohio	May 24, 1870	103, 483
Fence	J. McKnight	Romulus, N. Y	Mar. 12, 1872	124, 502
Fence	H. McMullin	Batesville, Ark	Nov. 19, 1872	133, 165
Fence	B. F. Mears	Washington, Ind	Nov. 14, 1871	120, 989

Index of patents issued from the United States Patent Office from 1790 to 1873, inclusive—Continued.

Invention.	Inventor.	Residence.	Date.	No.
Fence	R. B. Meeker	Sanford's Corners, N. Y	Dec. 12, 1871	121, 797
Fence	M. D. Messler	New Lebanon, Ohio	Oct. 1, 1867	69, 462
Fence	W. A. Middleton	Harrisburgh, Pa	July 30, 1867	67, 332
Fence	W. A. Middleton	Harrisburgh, Pa	Nov. 1, 1870	108, 925
Fence	U. D Mihills	Fond du Lac, Wis	Nov. 8, 1870	109, 036
Fence	G. S. Mills	Johnson, Vt	Mar. 1, 1870	100, 310
Fence	C. Milner	Des Moines, Iowa	Nov. 9, 1869	96, 719
Fence	D. Mitchel	New Berlin, Pa	Mar. 15, 1870	100, 919
Fence	S. H. Mitchell	El Paso, Ill	June 26, 1866	55, 883
Fence	J. Moore	Pittsburgh, Pa	June 30, 1857	17, 692
Fence	L. Moore	Baraboo, Wis	Apr. 25, 1871	114, 178
Fence	L. Moore	Baraboo, Wis	Mar. 26, 1872	125, 071
Fence	T. Morris	McGregor, Iowa	June 4, 1867	65, 415
Fence	I. J. Morrow	Everton, Ind	Aug. 3, 1869	93, 332
Fence	J. Morton	Thornville Mills, Mich	Feb. 21, 1871	112, 064
Fence	H. B. Myers	Schoolcraft, Mich	Nov. 29, 1894	45, 298
Fence	W. Neely	Sandy, Ohio	Aug. 21, 1866	57, 366
Fence	W. Nevins	Lyons, N. Y	June 6, 1865	48, 086
Fence	A. M. Nichol	Granville, Ohio	Sept. 24, 1867	69, 116
Fence	E Odell	Winterset, Iowa	Aug. 7, 1866	56, 985
Fence	A. W. Olds	Green Oak, Mich	June 5, 1866	55, 352
Fence	A. W. Olds	Green Oak, Mich	July 17, 1866	56, 441
Fence	E. F. Olds and W. Clark	Brighton and Green Oak, Mich.	Jan. 14, 1868	73, 257
Fence	F. W. Owens	Granville, Ohio	Sept. 6, 1870	107, 095
Fence	C. H. Paine	Providence, R. I	June 11, 1867	65, 592
Fence	P. C. Pearson	Louisville, Ky	July 11, 1871	116, 989
Fence	D. L. Pettegrew	Claremont, N. H	July 11, 1865	48, 774
Fence	W. Pettengell	Painesville, Ohio	Aug. 20, 1867	67, 902
Fence	L. Philips	Eau Claire, Wis	Sept. 20, 1870	107, 533
Fence	T. E. Phillips	Coatesville, Ind	Mar. 2, 1869	87, 364
Fence	S. B. Pierce	Homer, N. Y	Aug. 4, 1868	80, 661
Fence	J. M. Pitts	Sumter, S. C	Oct. 30, 1860	30, 544
Fence	S. S. Porter	Broad Ford, Pa	July 25, 1871	117, 325
Fence	A. W. Pratt	Pultneyville, N. Y	Mar. 26, 1867	63, 296
Fence	M. D. Pratt	Copley, Ohio	Jan. 14, 1868	73, 260
Fence	S. M. Prentice	Aurora, Ill	Aug. 31, 1869	93, 238
Fence	R. Prouty	Springfield, Iowa	Feb. 20, 1866	52, 795
Fence	H. B. Ramsey	Rockville, Ind	Aug. 13, 1872	130, 441
Fence	T. D. Read	Aberdeen, Ind	Sept. 10, 1867	68, 790
Fence	W. Reed	Wayne, N. Y	Apr. 30, 1867	64, 363
Fence	J. J. Reicherts	Delaware, Ohio	Jan. 12, 1869	85, 852
Fence	J. B. Reyman	Dubuque, Iowa	May 22, 1855	12, 921
Fence	D. Rhodes	Fredericktown, Mo	Sept. 19, 1871	119, 047
Fence	J. Richard	Columbiaville, Mich	Mar. 30, 1869	88, 412
Fence	R. A. Riggs	Salem, Oreg	May 7, 1872	126, 413
Fence	S. Riley	Kenton, Ohio	Oct. 5, 1869	95, 518
Fence	J. Riordan	Six Mile, Ind	July 20, 1869	92, 886
Fence	E. C. Roberts	Salem, Mich	June 11, 1867	65, 694
Fence	T. D. Roberts	Nekimi, Wis	Aug. 27, 1872	130, 823
Fence	J. Rohrer	Springfield, Ohio	Aug. 15, 1871	118, 156
Fence	W. P. Rollo	Holland Patent, N. Y	Apr. 16, 1872	125, 761
Fence	S. H. Rose	Wheeler, N. Y	Sept. 25, 1866	58, 300
Fence	H. S. Ross	Cincinnati, Ohio	Sept. 13, 1853	10, 015
Fence	B. Rowell	Elmira, N. Y	Sept. 18, 1866	58, 193
Fence	B. Rowell	Byersville, N. Y	June 7, 1870	104, 065
Fence	J. R. Rowley	Fabius, N. Y	Mar. 19, 1867	63, 160
Fence	J. W. Sanders	Ripon, Wis	May 28, 1867	65, 284
Fence	S. Saunders	Fort Plain, N. Y	Feb. 13, 1866	52, 611
Fence	C Seabaugh	San Antonio, Tex	Aug. 28, 1866	57, 581
Fence	I. R. Shank	Buffalo, Va	Mar. 31, 1863	38, 063
Fence	J. W. Shankland	Summerfield, Ohio	Sept. 25, 1866	58, 308
Fence	N. L. Shaw	Wilson, N. Y	Apr. 21, 1868	77, 109
Fence	A. Sheldon	Greenwich Station, Ohio	Jan. 28, 1868	73, 761
Fence	W. B. Shelton	Congruity, Pa	Dec. 18, 1866	60, 576
Fence	P. L. Sherman	Geneseo Township, Iowa	Sept. 13, 1870	107, 297
Fence	W. W. Sherman	Saint Charles County, Mo	Jan. 17, 1871	111, 090
Fence	A. D. Smith and E. Stow	Townsend Station, Ohio	Mar. 24, 1868	75, 994
Fence	A. H. Smith	Elkhorn, Wis	Aug. 22, 1871	118, 398
Fence	D. N. Smith and E. F. Olds	Salem and Lyon, Mich	Dec. 24, 1867	72, 693
Fence	R A. Smith	Newburyport, Mass	Mar. 11, 1862	34, 650
Fence	F. F. Sommer	Detroit, Mich	June 29, 1869	92, 118
Fence	J. Southwick	Brant, N. Y	Sept. 1, 1868	81, 836
Fence	A. B Sprout	Picture Rocks, Pa	Sept. 3, 1872	131, 038
Fence	S. Stanbro	Northville, Mich	Aug. 8, 1865	49, 313
Fence	T. Stanford	Noblesville, Ind	Apr. 13, 1869	88, 819
Fence	J. Stevens	Northumberland, N. Y	July 20, 1846	4, 650
Fence	H. M. Stoker	Watson, Ill	Apr. 30, 1867	64, 379
Fence	J. Stone and S. Blocker, sr	Plattsburgh, Mo	July 31, 1866	56, 826
Fence	J. H. Stone	Chapel Hill, Tex	Mar. 12, 1872	124, 516
Fence	N. M. Stratton	New York, N. Y	Oct. 23, 1860	30, 507
Fence	C. H. Strowger	Webster, N. Y	Oct. 31, 1871	120, 546
Fence	C. H Strowger	Webster, N. Y	May 28, 1872	127, 282
Fence	I. Subers	Philadelphia County, Pa	Feb. 19, 1850	7, 108
Fence	W. W. Sullivan	Liberty, Ind	Dec. 13, 1870	110, 170
Fence	J. B. Tedrow	Chillicothe, Ohio	Mar. 9, 1869	87, 729
Fence	J. W. Teller and W. Townsend.	Lapeer, Mich	Feb. 28, 1871	112, 194
Fence	D. Terry	Wakeman, Ohio	Aug. 21, 1866	57, 408
Fence	J. Thomas	Mount Union, Ohio	May 21, 1867	65, 025
Fence	J. D. Tifft	Cuyahoga Falls, Ohio	Oct. 3, 1871	119, 670
Fence	A. Todd, jr	Pultneyville, N. Y	Nov. 26, 1867	71, 340
Fence	A. Todd, jr	Pultneyville, N. Y	Feb. 4, 1868	74, 017
Fence	A. Todd, jr	Pultneyville, N. Y	Feb. 4, 1868	74, 018
Fence	W. H. Trimble	Hillsborough, Ohio	Apr. 27, 1869	89, 521
Fence	M. I. Turck	Schodack, N. Y	Jan. 5, 1869	85, 710
Fence	S. P. Tuttle	Decatur, Mich	Dec. 4, 1866	60, 286
Fence	W. Tuttle, jr	Dowagiac, Mich	Sept. 11, 1866	58, 003

Index of patents issued from the United States Patent Office from 1790 *to* 1873, *inclusive*—Continued.

Invention.	Inventor.	Residence.	Date.	No.
Fence	D. M. Tyler	Union Township, Ind	July 5, 1870	105, 143
Fence	I. Vanamberg	Watertown, N. Y	Dec. 20, 1843	3, 388
Fence	J. H. Van Dorn	Akron, Ohio	Apr. 30, 1872	126, 167
Fence	M. Van Wormer	Troy, Ohio	Feb. 15, 1870	99, 796
Fence	M. Van Wormer	Troy, Ohio	June 6, 1871	115, 663
Fence	J. Waddle	Bakerstown, Pa	Feb. 28, 1871	112, 198
Fence	J. F. Warner	Oskaloosa, Iowa	Feb. 13, 1866	52, 629
Fence	D. H. Weaver	North Liberty, Ind	June 25, 1872	128, 343
Fence	R. M. Weider and J. Meals	Carthage, Ill	Sept. 12, 1871	118, 888
Fence	J. L. Wellington	Dansville, N. Y	July 20, 1869	92, 917
Fence	I. M. West	Wilmington, Ohio	Aug. 23, 1870	106, 640
Fence	T. Westerman	Clinton Township, Pa	Aug. 31, 1869	94, 262
Fence	R. H. White	Billingsville, Ind	Apr. 30, 1872	126, 362
Fence	J. Whiteside	Coesse, Ind	Sept. 11, 1866	58, 013
Fence	H. Wicker	Olean, N. Y	Oct. 19, 1869	95, 963
Fence	B. Williams	Williamsburgh, Ohio	June 4, 1872	127, 536
Fence	S. P. Williams	Sheridan, N. Y	May 22, 1866	54, 989
Fence	S. P. Williams	Sheridan, N. Y	Apr. 2, 1867	63, 594
Fence	W. T. Willie	Independence, Tex	Aug. 20, 1872	130, 682
Fence	S. A. Wood	Cardington, Ohio	Nov. 16, 1869	96, 859
Fence	T. C. Wood	Augusta, Mich	July 4, 1871	116, 657
Fence	W. D. Woodruff	Phelps, N. Y	Sept. 5, 1865	49, 820
Fence	O. H. Woodworth	Upper Marlborough, Md	Nov. 27, 1860	30, 780
Fence	O. H. Woodworth	Columbia City, Ind	Mar. 21, 1865	46, 964
Fence	N. Woolsey	Ottawa, Ill	Oct. 11, 1870	108, 226
Fence	W. P. Wright	Winamac, Ind	May 17, 1870	103, 270
Fence	E. York	Windsor, Ill	May 1, 1866	54, 459
Fence	P. C. Yost	Carthage, Ill	Oct. 24, 1871	120, 178
Fence	J. Q. A. Youkey and F. M. Rawson	Frankfort, Ind	Mar. 23, 1869	88, 252
Fence adaptable to uneven ground	G. R. McIlroy	Covington, Ky	Feb. 10, 1857	16, 598
Fence, Adjustable picket	J. S. Anderson	Flintville, Wis	Jan. 7, 1873	134, 624
Fence and gate	E. Muirheid	Greenfield, Ohio	May 3, 1870	102, 576
Fence and gate, Combined	C. S. S. Griffing	Geneva, Ohio	Apr. 23, 1867	64, 095
Fence and gate combined, Field	J. C. Lee	Seville, Ohio	Oct. 24, 1865	50, 605
Fence and pen, Worm	J. Will	Bryan, Ohio	Jan. 14, 1868	73, 420
Fence and trellis hook	D. S. Humphrey	East Townsend, Ohio	Dec. 31, 1867	72, 856
Fence and trellis post	S. Lamont	Pittsburgh, Pa	Mar. 17, 1868	75, 636
Fence-bars, Bracket for	P. Cope	Perryopolis, Pa	Dec. 24, 1872	134, 256
Fence-board-gage holder	D. Bordner	Canton, Ohio	Apr. 27, 1869	89, 281
Fence-builder	J. M. Kirkpatrick	Utica, Ohio	Jan. 19, 1869	86, 081
Fence, Combined wire and stone	H. C. Wire	Wilmington, Ohio	May 10, 1870	102, 902
Fence, Farm	C. Abbott	Iowa City, Iowa	June 22, 1869	91, 585
Fence, Farm	T. G. Beecher	Beaver Dam, N. Y	Nov. 1, 1859	25, 944
Fence, Farm	J. S. Bettis	Hanover, N. Y	Aug. 14, 1866	57, 073
Fence, Farm	V. Calkins	Varysburgh, N. Y	Jan. 26, 1869	86, 133
Fence, Farm	D. S. Humphrey	East Townsend, Ohio	Oct. 15, 1867	69, 809
Fence, Farm	W. Jasper	Columbia, Ohio	Feb. 16, 1869	86, 924
Fence, Farm	H. A. Kephart	Fletcher, Ohio	Dec. 24, 1867	72, 504
Fence, Farm	P. Kidney	Olmstead Falls, Ohio	July 12, 1870	105, 221
Fence, Farm	W. S. McKenzie	Rockwell, Tex	July 29, 1873	141, 284
Fence, Farm	E. G. McMillan	Norwalk, Ohio	Nov. 10, 1868	83, 871
Fence, Farm	T. W. Owens	Granville, Ohio	July 12, 1870	105, 239
Fence, Farm	C. W. Reeder	Trenton, Mo	July 6, 1869	92, 359
Fence, Farm	J. K. Staman	Mansfield, Ohio	Mar. 9, 1869	87, 597
Fence, Farm	E. Stiles	Cleveland, Ohio	Oct. 15, 1867	69, 858
Fence, Farm	S. P. Williams	Sheridan, N. Y	Dec. 10, 1867	72, 141
Fence, Farm	E. M. Woodruff	Washington, D. C	July 12, 1864	43, 542
Fence, Field	T. H. Bellard	Colbrook, Ohio	Oct. 29, 1867	70, 153
Fence, Field	S. H. Bennett	Belleville, Pa	Jan. 13, 1857	16, 369
Fence, Field	P. S. Carhartt	Collomer, N. Y	Mar. 2, 1858	19, 491
Fence, Field	T. J. Carleton and S. Post	York, Ohio	Jan. 29, 1856	14, 152
Fence, Field	S. Cheney	Kiantone, N. Y	Jan. 25, 1859	22, 702
Fence, Field	D. S. Curtiss	Madison, Wis	Jan. 25, 1859	22, 709
Fence, Field	F. Dickinson	Geneva, Ohio	May 14, 1867	65, 641
Fence, Field	J. Drown	Huron, N. Y	Feb. 16, 1858	19, 353
Fence, Field	A. Frankboner	Schoolcraft, Mich	Nov. 20, 1866	59, 900
Fence, Field	B. Gabriel	Elmira, N. Y	Apr. 20, 1858	19, 990
Fence, Field	I. A. Gormly	Bucyrus, Ohio	Jan. 7, 1868	73, 090
Fence, Field	I. A. Gormly	Bucyrus, Ohio	Apr. 14, 1868	76, 745
Fence, Field	J. Haines	West Middleburgh, Ohio	Oct. 18, 1859	25, 821
Fence, Field	J. Heacock	Marlborough, Ohio	June 11, 1867	65, 670
Fence, Field	D. M. Heikes	York County, Pa	Aug. 3, 1858	21, 073
Fence, Field	T. Hoge	Waynesburgh, Pa	June 15, 1858	20, 560
Fence, Field	C. Horton	Phelps, N. Y	Aug. 3, 1858	21, 074
Fence, Field	D. S. Humphrey	East Townsend, Ohio	Jan. 7, 1868	73, 011
Fence, Field	M. Ingersoll	Elyria, Ohio	June 25, 1867	66, 027
Fence, Field	I. G. Inskeep	West Middleburgh, Ohio	Mar. 12, 1861	31, 667
Fence, Field	J. H. Jones and N. W. Smith	Lebanon, Ohio	Mar. 9, 1858	19, 566
Fence, Field	D. W. Keefer	Leechburgh, Pa	June 15, 1869	91, 344
Fence, Field	E. King	Dunkirk, N. Y	Sept. 21, 1869	94, 963
Fence, Field	P. W. Kniskern	Monee, Ill	Jan. 24, 1865	46, 052
Fence, Field	J. W. Larmore	Harrison, Ohio	July 31, 1866	56, 766
Fence, Field	W. H. Leech	Dunlapville, Ind	Oct. 21, 1862	36, 719
Fence, Field	E. E. Lewis	Geneva, N. Y	Apr. 27, 1858	20, 071
Fence, Field	J. B. Mitchell	Wayne, N. Y	Oct. 19, 1858	21, 843
Fence, Field	A. W. Olds	Green Oak, Mich	Mar. 26, 1867	63, 290
Fence, Field	D. R. Prindle	East Bethany, N. Y	Apr. 25, 1854	10, 829
Fence, Field	S. Rains	Lancaster County, Va	Apr. 7, 1857	16, 996
Fence, Field	J. B. Reyman	Salem, Ind	Mar. 25, 1856	14, 518
Fence, Field	B. Rowells	Ossian, Ohio	Apr. 6, 1858	19, 873
Fence, Field	D. Sattler	Mifflin, Ohio	Jan. 25, 1870	99, 112
Fence, Field	F. L. Sexton	Wellington, Ohio	Nov. 22, 1864	45, 184
Fence, Field	S. G. Tufts	Maineville, Ohio	Apr. 1, 1856	14, 581
Fence, Field	J. L. Wentworth	Spread Eagle, Pa	Oct. 25, 1859	25, 925
Fence, Field	W. I. Young	Le Roy, N. Y	July 12, 1864	43, 544
Fence, Flexible	M. P. Coons	Lansingburgh, N. Y	July 29, 1851	8, 253

Index of patents issued from the United States Patent Office from 1790 to 1873, inclusive—Continued.

Invention.	Inventor.	Residence.	Date.	No.
Fence, Floating	J. Pitcher	Mount Vernon, Ind	Aug. 27, 1867	68, 232
Fence, Flood	W. F. Auxier	Mason City, Ill	Aug. 3, 1869	93, 161
Fence, Flood	W. C. Barber	Van Wert, Ga	Aug. 10, 1869	93, 398
Fence, Flood	D. P. Bird	Richwood, Ohio	Dec. 29, 1868	85, 271
Fence, Flood	Z. Dowden and C. T. Anderson	Clarksburgh, Md	Mar. 23, 1869	88, 147
Fence, Flood	G. W. Duncan	Columbia, Mo	Dec. 16, 1873	145, 493
Fence, Flood	R. S. Gilcrest	De Graff, Ohio	Feb. 20, 1872	123, 889
Fence, Flood	C. R. Hunter	Douglas, Ill	Feb. 28, 1871	112, 250
Fence, Flood	W. R. McFarland	Paris, Tenn	May 21, 1872	127, 083
Fence, Flood	D. J. Miller	Indianapolis, Ind	July 12, 1870	105, 230
Fence, Flood	H. W. Nichols	Worthfield, Ind	Dec. 10, 1872	133, 795
Fence, Flood	A. Orvis	Ellington, Iowa	May 27, 1873	139, 418
Fence, Flood	H. Reichert	Shippensburgh, Pa	Feb. 27, 1849	6, 138
Fence, Flood	J. L. Seat	Nashville, Tenn	Dec. 19, 1871	122, 067
Fence, Flood	J. Sourbeer	Mount Joy Township, Pa	Jan. 30, 1849	6, 075
Fence, Flood	J. A. Swope	Germany, Pa	Oct. 17, 1848	5, 857
Fence, Flood	D. C. Wilkinson	Sidney, Ohio	Nov. 6, 1860	30, 599
Fence, Flood	J. L. Wines	Hebardsville, Ohio	Sept. 19, 1871	119, 211
Fence, Flood	J. B. Wolfe	Marion, Ohio	Jan. 23, 1872	123, 068
Fence, Flood	V. Wood	Richmond, Ind	June 2, 1868	78, 633
Fence for collecting rain-water for stock	L. Howe	Alamo, Mich	Feb. 18, 1868	74, 690
Fence for crossing streams	W. McFarlin	Jackson, Ill	June 23, 1868	79, 141
Fence for poultry-yard	H. W. Rutt	Reedsburgh, Ohio	Apr. 21, 1868	77, 101
Fence for poultry-yard	W. P. Thomas	Whitewater, Ind	Aug. 25, 1857	18, 064
Fence for preventing injury by floods	W. Miller	Orrstown, Pa	Oct. 2, 1847	5, 313
Fence for rolling ground, Farm	E. D. Foss	Mainville, Ohio	Aug. 19, 1856	15, 561
Fence for sheep-fold	A. P. Hopkins	Bentleysville, Pa	Oct. 14, 1862	36, 651
Fence for stock-pens, Portable prairie	T. Hoge	Waynesburgh, Pa	Dec. 9, 1856	16, 181
Fence foundation	T. W. Welch and G. B. Starbird	Mechanicsburgh, Pa	July 28, 1868	80, 318
Fence-gage	W. V. Van Syckel	Joshua, Ill	May 18, 1869	90, 138
Fence-gate	J. Bailey	Leatherwood, Ohio	Jan. 15, 1850	7, 013
Fence-gate	A. G. Barnard	Seville, Ohio	May 8, 1866	54, 487
Fence-gate	J. S. Benedict	Bedford, Ohio	Mar. 27, 1866	53, 394
Fence-gate	J. S. Benedict	Bedford, Ohio	Oct. 2, 1866	58, 369
Fence-gate	W. W. Bratt	Ottawa, Ill	Dec. 11, 1866	60, 331
Fence-gate	N. Burch	North Fairfield, Ohio	Sept. 3, 1867	68, 553
Fence-gate	J. S. Jewett	Ottawa, Ill	May 5, 1868	77, 492
Fence-gate	J. S. Jewett	Ottawa, Ill	Sept. 1, 1868	81, 640
Fence-gate	F. Raymond and A. Miller	Cleveland, Ohio	Jan. 10, 1865	45, 858
Fence-gate	F. Raymond and A. Miller	Cleveland, Ohio	June 20, 1865	48, 309
Fence-gate	F. Raymond and A. Miller	Cleveland, Ohio	June 26, 1866	55, 906
Fence-gate	M. L. Salyards	Troy Grove, Ill	Mar. 19, 1867	63, 103
Fence-gate	W. F. Veber	Perrysburgh, Ohio	Apr. 23, 1867	64, 171
Fence-gate	B. J. Wheelock	Bedford, Ohio	Nov. 6, 1866	59, 492
Fence gate and gateway	A. Hood	New York, N. Y	Dec. 16, 1845	4, 307
Fence gate, Iron	C. T. Bush	Oneonta, N. Y	Apr. 22, 1873	138, 071
Fence gate post	E. Humphreys	Samsville, Ill	July 13, 1869	92, 451
Fence, Hedge	A. Belt	Newton, Iowa	Mar. 15, 1870	100, 843
Fence, Hedge	S. N. Caldwell	Pilot Grove, Ind	Mar. 30, 1869	88, 445
Fence hook, Wire	A. J. Gill	Denver, Colo	Mar. 5, 1872	124, 349
Fence, Hurdle	C. C. Cole	Rushville, N. Y	Dec. 2, 1851	8, 555
Fence, Iron	C. T. Bush	Middleburgh, N. Y	July 13, 1869	92, 579
Fence, Iron	C. T. Bush	Oneonta, N. Y	Mar. 18, 1873	136, 966
Fence, Iron	M. P. Coons	Brooklyn, N. Y	Apr. 18, 1854	10, 781
Fence, Iron	S. Crowell	Philadelphia, Pa	Oct. 11, 1870	108, 112
Fence, Iron	D. I. De Groat	Newburgh, N. Y	June 15, 1869	91, 310
Fence, Iron	B. G. Devoe	Fredericktown, Ohio	July 15, 1873	140, 897
Fence, Iron	A. Ferry	Waterbury, Conn	Nov. 24, 1868	84, 318
Fence, Iron	W. S. Fuller	Millbury, Mass	Apr. 7, 1857	17, 010
Fence, Iron	J. Gorr	Erie, Pa	Dec. 3, 1872	133, 528
Fence, Iron	P. Hedl	New York	Dec. 18, 1828	......
Fence, Iron	B. F. Miller	New York, N. Y	Nov. 29, 1853	10, 273
Fence, Iron	T. Rogers	Fredericktown, Ohio	Nov. 26, 1872	133, 489
Fence, Iron	T. Rogers	Kenton, Ohio	Dec. 23, 1873	145, 902
Fence, Iron	T. Rogers and B. G. Devoe	Kenton, Ohio	Dec. 23, 1873	145, 903
Fence, Iron	S. Sellers and W. Beschke	Philadelphia, Pa	May 7, 1872	126, 417
Fence, Iron	J. B. Wickersham	New York, N. Y	July 1, 1851	8, 189
Fence, Iron	J. B. Wickersham	New York, N. Y	Mar. 9, 1852	8, 793
Fence, Iron	J. B. Wickersham	Philadelphia, Pa	Jan. 21, 1873	135, 188
Fence, Iron	J. B. Wickersham and H. Jenkins.	Brooklyn, N. Y	Aug. 16, 1859	25, 166
Fence, Iron	O. Wilson	Middleburgh, N. Y	Feb. 27, 1872	124, 105
Fence, Iron hurdle	M. P. Coons	Lansingburgh, N. Y	Oct. 17, 1848	5, 863
Fence, Iron picket	M. Walker, sr	Philadelphia, Pa	July 11, 1854	11, 315
Fence-jack	C. Lewis	Nelson, Ohio	Nov. 6, 1866	59, 421
Fence key, Wire	S. B. Hewett, jr	Eagle Grove, Iowa	Mar. 24, 1868	75, 759
Fence, Lattice iron	A. Betteley	Boston, Mass	June 1, 1858	20, 400
Fence-machine	J. H. Evans	Dallas, Ky	Dec. 30, 1873	145, 996
Fence, Metal	H. Noblit	Philadelphia, Pa	Sept. 3, 1861	33, 211
Fence, Metallic	W. Bush	Harrisburgh, Pa	Aug. 3, 1858	21, 064
Fence, Metallic	M. Kelly	New York, N. Y	Nov. 17, 1868	84, 062
Fence, Multigrade iron	M. P. Coons	Brooklyn, N. Y	Nov. 14, 1854	11, 931
Fence of wire, cords, or twine, and seines or nets of wire.	C. Hall	Meriden, Conn	Nov. 27, 1829	......
Fence, Panel	C. W. Sproull	Rome, Ga	July 7, 1868	79, 698
Fence-panel adjuster	F. M. Ranous	Yreka, Cal	July 2, 1872	128, 563
Fence, Picket	G. W. T. Grant	Winona County, Minn	Mar. 5, 1861	31, 594
Fence, Picket	H. N. Hill	Pontiac, Mich	Sept. 14, 1869	94, 824
Fence, Picket	R. H. McGinty	Moulton, Tex	Oct. 14, 1873	143, 632
Fence, Picket	W. Most	Greenville, N. J	May 6, 1873	138, 573
Fence, Picket	J. Willhite	Pilot Point, Tex	Dec. 24, 1872	134, 235
Fence-picket, Corrugated	B. F. Miller	New York, N. Y	July 26, 1864	43, 644
Fence-pickets, Cutting	J. Tichenor, S. Goodrich, and G. A. Hart.	Ithaca, N. Y	June 22, 1836	......
Fence-pickets, Machine for cutting the heads of	A. Burnham	Revere, Mass	Nov. 4, 1873	144, 187
Fence-pickets, Machine for dressing	W. and J. F. Nuelle	Saint Louis, Mo	Dec. 27, 1870	110, 586
Fence-pickets, Machine for sharpening	H. F. Bond	Waltham, Mass	July 30, 1861	32, 960
Fence-pickets, Machine for sharpening	J. A. Montgomery	Columbus, Ohio	Feb. 26, 1867	62, 355

Index of patents issued from the United States Patent Office from 1790 to 1873, inclusive—Continued.

Invention.	Inventor.	Residence.	Date.	No.
Fence-pickets, &c., Machine for turning	E. Briggs	Perry, N. Y	July 20, 1842	2, 734
Fence, Portable	A. Armitage and J. H. Olmsted	Phelps Township and Arcadia, N. Y.	Nov. 9, 1869	96, 536
Fence, Portable	J. Augspurger	Trenton, Ohio	Apr. 16, 1867	63, 825
Fence, Portable	J. Augspurger	Trenton, Ohio	Apr. 16, 1867	63, 826
Fence, Portable	J. Augspurger	Trenton, Ohio	Apr. 16, 1867	63, 827
Fence, Portable	J. Augspurger	Trenton, Ohio	Apr. 16, 1867	63, 828
Fence, Portable	J. Augspurger	Trenton, Ohio	Apr. 16, 1867	63, 829
Fence, Portable	J. Augspurger	Trenton, Ohio	July 7, 1868	79, 624
Fence, Portable	G. F. Becher	Brooklyn, N. Y	Aug. 12, 1873	141, 750
Fence, Portable	J. Bickhart	Harlan, Ind	June 17, 1862	35, 581
Fence, Portable	J. W. Blodgett	Three Rivers, Mich	Apr. 21, 1863	28, 214
Fence, Portable	L. W. Bosart	Saint Marie, Ill	June 9, 1868	78, 783
Fence, Portable	L. W. Bosart	Saint Marie, Ill	Nov. 10, 1868	83, 911
Fence, Portable	B. F. Brattain	Noblesville, Ind	July 21, 1868	80, 052
Fence, Portable	J. Breneman	Mount Joy, Pa	July 31, 1866	56, 695
Fence, Portable	A. Brooker	Atalissa, Iowa	May 29, 1866	55, 051
Fence, Portable	P. M. Brown	Carrollton, Ill	Dec. 13, 1859	26, 413
Fence, Portable	R. J. Brown	Perry, Pa	Aug. 31, 1858	21, 315
Fence, Portable	S. Bryan	Jefferson, Wis	Oct. 17, 1865	50, 446
Fence, Portable	S. G. Burke	Chambers Court-House, Ala	July 29, 1873	141, 202
Fence, Portable	B. and V. Calkins	Varysburgh, N. Y	Apr. 2, 1867	63, 362
Fence, Portable	G. W. Campbell	Pendleton, Ind	Oct. 29, 1867	70, 162
Fence, Portable	J. T. Campbell	Rockville, Ind	Apr. 16, 1867	63, 853
Fence, Portable	G. S. Carlisle	Columbus City, Iowa	Sept. 4 1866	57, 672
Fence, Portable	Z. Castaline	Baconsburgh, Ohio	Dec. 3, 1867	71, 697
Fence, Portable	P. Chandler	Olney, Ill	June 25, 1867	66, 129
Fence, Portable	P. Chandler	Olney, Ill	Aug. 27, 1867	68, 166
Fence, Portable	J. M. Clark	Lancaster, Pa	Nov. 26, 1867	71, 278
Fence, Portable	F. C. Class	Roanoke, Ind	Oct. 30, 1866	59, 185
Fence, Portable	J. Closs	Decatur, Ind	Oct. 26, 1869	96, 084
Fence, Portable	W. F. Converse	Harrison, Ohio	May 5, 1868	77, 589
Fence, Portable	C. S. Coolidge and J. A. Rollins	Jersey Mills, Pa	Dec. 29, 1868	85, 285
Fence, Portable	W. Corson	Camden, Ohio	June 9, 1868	78, 789
Fence, Portable	W. Crumb	Coloma, Mich	Oct. 4, 1870	107, 883
Fence, Portable	H. M. Dake	Nunda, N. Y	Feb. 25, 1873	136, 217
Fence, Portable	T. Doneboo	Richmond, Mo	Oct. 31, 1871	120, 506
Fence, Portable	J. C. Duncan	Olney, Ill	May 14, 1867	64, 648
Fence, Portable	P. S. Dusenbury	Boscobel, Wis	Jan. 8, 1867	61, 057
Fence, Portable	J. S. Edgar	Janesville, Wis	Jan. 29, 1867	61, 524
Fence, Portable	E. Gale	Kendall, Ill	Nov. 25, 1862	36, 996
Fence, Portable	J. E. Garlington	Chambers Court-House, Ala.	Mar. 18, 1873	136, 989
Fence, Portable	L. Gates	Mount Vernon, Ohio	May 13, 1873	138, 747
Fence, Portable	M. F. Gibbs	Livonia, N. Y	June 4, 1867	65, 372
Fence, Portable	A. C. Gilmore	Centralia, Ill	Oct. 22, 1872	132, 461
Fence, Portable	N. J. Glover	Waveland, Ind	Oct. 8, 1867	69, 654
Fence, Portable	N. J. Glover	Waveland, Ind	Sept. 21, 1869	94, 953
Fence, Portable	W. P. Goff	Yorkville, Wis	Mar. 30, 1869	88, 469
Fence, Portable	C. S. S. Griffing	Unionville, Ohio	Apr. 28, 1868	77, 371
Fence, Portable	F. W. Groff	Indianapolis, Ind	Apr. 13, 1869	88, 782
Fence, Portable	R. Haynes	Oberlin, Ohio	Apr. 7, 1863	38, 108
Fence, Portable	L. and S. D. Hazlett	Winfield Township, Pa	Nov. 16, 1869	96, 805
Fence, Portable	L. F. Henderson	Freeport, Ill	Jan. 7, 1868	73, 098
Fence, Portable	C. R. Hight	Geneva, Ill	June 1, 1869	90, 751
Fence, Portable	J. Hoffman	Hixton, Wis	Sept. 3, 1872	131, 000
Fence, Portable	N. Hornaday	West Elkton, Ohio	Aug. 19, 1862	36, 215
Fence, Portable	O. Huffman	Lucieville, Nebr	June 24, 1873	140, 273
Fence, Portable	J. C. Hughes	Robinson, Ill	Sept. 24, 1867	69, 214
Fence, Portable	G. H. Hume	Paola, Kans	Oct. 29, 1872	132, 534
Fence, Portable	E. Hunt	Hartford, Ind	Aug. 30, 1870	106, 826
Fence, Portable	A. Jewett	Sanford's Corners, N. Y	Oct. 3, 1871	119, 614
Fence, Portable	C. E. Johnson	San Francisco, Cal	Sept. 14, 1869	94, 828
Fence, Portable	R. Johnson	Griffin, Ga	Dec. 23, 1873	145, 875
Fence, Portable	S. Keller	Elizabethtown, Pa	Mar. 17, 1868	75, 551
Fence, Portable	J. Kennel	Morton, Ill	July 5, 1864	43. 415
Fence, Portable	J. Killian	Marshall, Iowa	Oct. 20, 1863	40, 342
Fence, Portable	A. Labair	Pewaukee, Wis	Dec. 31, 1867	72, 863
Fence, Portable	P. Lambkin	Saint Albans, Vt	Jan. 5, 1869	85, 595
Fence, Portable	I. L. Landis	Lancaster, Pa	Apr. 30, 1867	64, 335
Fence, Portable	J. Lefeber	Wayne County, Ind	Apr. 16, 1867	63, 801
Fence, Portable	J. Leonard	Basil, Ohio	June 16, 1868	78, 979
Fence, Portable	S. Leonard, jr	Fairfield County, Conn	Dec. 3, 1867	71, 767
Fence, Portable	A. Love	Saxonburgh, Pa	Dec. 7, 1869	97, 538
Fence, Portable	J. M. May and E. B. Godfrey	Janesville and Oshkosh, Wis.	July 11, 1865	48, 701
Fence, Portable	P. McCollum	Fayette, Mo	June 22, 1869	91, 651
Fence, Portable	G. R. McIlroy	Oakdale, Ind	Sept. 30, 1856	15, 812
Fence, Portable	J. J. McMaken	Middletown, Iowa	Nov. 12, 1872	132, 916
Fence, Portable	G. McQuisten	Cardington, Ohio	Dec. 3, 1872	133, 658
Fence, Portable	S. Miles	Fabius, N. Y	Aug. 6, 1867	67, 564
Fence, Portable	E. D. Montrose	Nashua, Iowa	May 15, 1866	54, 754
Fence, Portable	W. Morrison	Carlisle, Pa	June 9, 1857	17, 514
Fence, Portable	A. Mosher	Canton, Mo	Aug. 5, 1873	141, 584
Fence, Portable	T. Nevison	Morgan, Ohio	May 18, 1869	90, 117
Fence, Portable	H. W. and R. P. Nichols	Northfield, Ind	Sept. 30, 1873	143, 372
Fence, Portable	J. W. Norman	Eugene, Ind	Aug. 4, 1868	80, 657
Fence, Portable	A. W. Olds	Green Oak Station, Mich	Apr. 15, 1873	137, 949
Fence, Portable	A. H. Palmer	Skaneateles, N. Y	Nov. 8, 1864	44, 967
Fence, Portable	L. P. Pease	McCordsville, Ind	Apr. 21, 1868	77, 087
Fence, Portable	D. K. Peterson	Williamsburgh, Ohio	Nov. 19, 1872	133, 116
Fence, Portable	D. Philips	Cordova, Ill	May 18, 1869	90, 300
Fence, Portable	W. W. Potts	Swedeland, Pa	Sept. 16, 1873	142, 865
Fence, Portable	P. M. Purdy	Haysville, Ohio	Feb. 19, 1850	7, 103
Fence, Portable	J. Reber	Shoemakerville, Pa	Apr. 21, 1863	38, 241
Fence, Portable	J. Reedy	Toledo, Iowa	May 15, 1866	54, 772
Fence, Portable	F. Richter	Orange, Ohio	Aug. 27, 1861	33, 163
Fence, Portable	J. R. Robbins	Buttsville, Mo	Dec. 3, 1872	133, 542

Index of patents issued from the United States Patent Office from 1790 *to* 1873, *inclusive*—**Continued.**

Invention.	Inventor.	Residence.	Date.	No.
Fence, Portable	W. E. Roberts	North Coventry, Pa	Mar. 31, 1868	76, 102
Fence, Portable	G. Robinson	Chillicothe, Mo	Nov. 4, 1873	144, 228
Fence, Portable	H. P. Ross	Hastings, N. Y	Nov. 10, 1863	40, 574
Fence, Portable	C. Rowland	Clinton, Ill	Jan. 16, 1866	52, 080
Fence, Portable	R. Samuel	Walden, N. Y	May 26, 1868	78, 329
Fence, Portable	A. H. Scott	Concord, N. C	Mar. 16, 1869	87, 976
Fence, Portable	H. Sinclair	New York, N. Y	Feb. 7, 1871	111, 578
Fence, Portable	H. D. Smalley	New Baltimore, Ohio	May 5, 1868	77, 544
Fence, Portable	J. O. Smith	Reiley, Ohio	Aug. 2, 1870	106, 088
Fence, Portable	S. C. H. Smith	Belpre, Ohio	Mar. 30, 1869	88, 523
Fence, Portable	F. Sproul	Doniphan, Kans	Mar. 28, 1871	113, 105
Fence, Portable	J. K. Staman	Mifflin, Ohio	June 26, 1866	55, 926
Fence, Portable	H. A. Stewart	Minneapolis, Minn	Feb. 4, 1868	74, 166
Fence, Portable	J. W. Suidter	Sharon, Wis	Apr. 9, 1867	63, 764
Fence, Portable	G. D. Sweigert	Martic Township, Pa	Dec. 10, 1867	72, 113
Fence, Portable	B. L. Taylor	Marlin, Tex	Nov. 18, 1873	144, 641
Fence, Portable	C. A. Thomas	Bellbuckle, Tenn	Oct. 28, 1873	144, 165
Fence, Portable	J. Thompson	Williamsburgh, N. Y	May 15, 1866	54, 794
Fence, Portable	J. B. Tillinghast	Racine, Ohio	Apr. 7, 1863	38, 131
Fence, Portable	D. Unthank	Spiceland, Ind	Jan. 8, 1867	61, 123
Fence, Portable	C. Van De Mark	Oak's Corners, N. Y	Mar. 24, 1863	37, 989
Fence, Portable	W. M. Wallace	Cameron, Ill	July 3, 1860	29, 024
Fence, Portable	J. M. Wallis	Milton, Iowa	July 1, 1862	35, 793
Fence, Portable	D. S. Watts	Canton, Miss	Mar. 11, 1873	136, 630
Fence, Portable	J. W. Westbrooks and A. J. Clemmons.	Aberdeen, Miss	Nov. 26, 1872	133, 510
Fence, Portable	T. B. Wickham	Granville, Ohio	May 12, 1868	77, 942
Fence, Portable	H. Willard	Ripon, Wis	July 14, 1868	79, 884
Fence, Portable	H. C. Wilson	West Elkton, Ohio	Mar. 30, 1869	88, 427
Fence, Portable	T. L. Wiswell	Olathe, Kans	July 8, 1873	140, 752
Fence, Portable	G. C. Wright	Westfield, Ohio	Oct. 23, 1866	59, 114
Fence, Portable	E. York	Windsor, Ill	Mar. 5, 1867	62, 718
Fence, Portable extension	A. W. Overholser	Duchoyquet Township, Ohio.	Feb. 11, 1873	135, 837
Fence, Portable farm	H. C. Foote	Jersey City, N. J	Dec. 17, 1861	33, 937
Fence, Portable field	W. B. Burnett	Lyons, N. Y	Dec. 23, 1856	16, 265
Fence, Portable field	P. S. Carhart	Collamer, N. Y	Sept. 21, 1858	21, 549
Fence, Portable field	E. Cole	Fairfield, Mich	May 5, 1857	17, 201
Fence, Portable field	D. Denham	Virden, Ill	Dec. 17, 1861	33, 927
Fence, Portable field	J. J. Friend	Altona, Ill	Jan. 1, 1867	60, 873
Fence, Portable field	T. B. Garside	Danville, Iowa	Aug. 2, 1859	24, 930
Fence, Portable field	E. C. Goddard	Unionville, Ohio	Sept. 17, 1867	68, 976
Fence, Portable field	C. S. S. Griffing	Ashtabula, Ohio	May 24, 1864	42, 849
Fence, Portable field	C. S. S. Griffing	Ashtabula County, Ohio	July 16, 1867	66, 705
Fence, Portable field	M. Hall	Richfield, Ohio	May 29, 1866	55, 091
Fence, Portable field	J. G. Hunt	Reading, Ohio	Dec. 16, 1856	16, 236
Fence, Portable field	J. B. Johnson	Linden, Ind	Aug. 24, 1858	21, 260
Fence, Portable field	B. F. Lyon	Pleasantville, Pa	June 24, 1856	15, 188
Fence, Portable field	G. B. Mallette	Millport, N. Y	Mar. 5, 1861	31, 611
Fence, Portable field	O. H. P. Orendorff	Bloomington, Ill	Nov. 26, 1861	33, 792
Fence, Portable field	T. B. Page	Laurel, Ohio	Dec. 15, 1857	18, 858
Fence, Portable field	L. S. Robison	Gypsum, N. Y	Apr. 20, 1858	20, 005
Fence, Portable field	J. Rowe	Tampa Bay, Fla	Mar. 25, 1856	14, 519
Fence, Portable field	H. F. Stanard	Wayne, Mich	Dec. 22, 1857	18, 934
Fence, Portable field	S. G. Tufts	Mainville, Ohio	May 12, 1857	17, 302
Fence, Portable field	A. B. and M. Vandemark	Phelps, N. Y	Sept. 14, 1858	21, 529
Fence, Portable field	W. Wilson, jr	Raymond, Ill	July 13, 1869	92, 553
Fence, Portable flood	J. B. Stoner	Lacon, Ill	Oct. 15, 1867	69, 861
Fence, Portable folding	A. M. Olds	New York, N. Y	Oct. 6, 1868	82, 865
Fence, Portable gate	J. W. Curtis	Bath, N. Y	Aug. 23, 1870	106, 670
Fence, Portable picket	G. Gross	Buffalo, N. Y	Dec. 20, 1864	45, 490
Fence, Portable picket	G. W. C. Jarvis and C. Graves	Lapeer, Mich	July 20, 1869	92, 727
Fence, Portable picket	A. L. Thorp	Vandalia, Mich	July 17, 1866	56, 465
Fence, Portable wire	L. W. Bosart	Saint Marie, Ill	Mar. 31, 1868	76, 151
Fence-post	C. Ayres	Farmington Centre, Wis	Aug. 15, 1871	117, 963
Fence-post	J. M. Beebe	Cassadaga, N. Y	Mar. 4, 1873	136, 406
Fence-post	R. E. Bowen	Colden, N. Y	Dec. 29, 1868	85, 360
Fence-post	M. K. Butterfield	Eddyville, N. Y	Sept. 21, 1869	95, 178
Fence-post	M. K. Butterfield	Eddyville, N. Y	Feb. 1, 1870	99, 288
Fence-post	S. A. Darrach	Newburgh, N. Y	Aug. 2, 1870	106, 035
Fence-post	H. T. Dewey	Sandusky, Ohio	May 10, 1859	23, 905
Fence-post	H. N. Dunbar	Mentor, Ohio	June 11, 1872	127, 749
Fence-post	H. O. Elmer	Sand Bank, N. Y	Aug. 5, 1873	141, 499
Fence-post	R. M. Filson	Willsborough, N. Y	Dec. 20, 1870	110, 221
Fence-post	D. Fisher	Fair Haven, Ohio	Aug. 31, 1869	94, 195
Fence-post	H. K. Flinchbaugh	Conestoga Centre, Pa	Mar. 3, 1868	75, 144
Fence-post	A. M. Freeman, C. P. Idell, and B. Vanderhoven.	Metuchen, N. J	Apr. 2, 1872	125, 189
Fence-post	J. Gibbs	Fremont, Ohio	Sept. 16, 1873	142, 910
Fence-post	A. J. Gill	Denver, Colo	Apr. 18, 1871	113, 872
Fence-post	E. S. Goodrich	Oakland, Wis	Apr. 13, 1869	88, 866
Fence-post	A. W. Gore	Manhattan, Kans	July 2, 1867	66, 321
Fence-post	B. S. Haviland	Fort Dodge, Iowa	Jan. 1, 1867	60, 889
Fence-post	W. D. Hopgood	Henderson, Ky	Jan. 23, 1872	122, 947
Fence-post	W. Hunter	Detroit, Mich	Oct. 23, 1866	59, 023
Fence-post	G. Ipe	Kent, Ohio	Apr. 9, 1867	63, 725
Fence-post	J. F. Keeler	Pittsburgh, Pa	Mar. 19, 1872	124, 686
Fence-post	J. E. and A. H. Kendeigh	Amherst, Ohio	Oct. 8, 1867	69, 680
Fence-post	R. Ketcham	South Dansville, N. Y	Feb. 5, 1867	61, 838
Fence-post	E. King	Dunkirk, N. Y	Sept. 16, 1873	142, 795
Fence-post	O. L. Larkin	Otto, N. Y	Aug. 16, 1870	106, 375
Fence-post	C. Losee	Perrysburgh, N. Y	Sept. 21, 1869	95, 123
Fence-post	W. S. Mayo	New York, N. Y	Sept. 17, 1861	33, 310
Fence-post	P. McDuff	Atchison, Kans	Apr. 21, 1868	77, 066
Fence-post	R. Merrill	Elmira, N. Y	Apr. 6, 1858	19, 863
Fence-post	W. A. Middleton	Harrisburgh, Pa	Aug. 24, 1869	94, 124
Fence-post	W. A. Middleton	Harrisburgh, Pa	Oct. 24, 1871	120, 308
Fence-post	F. Miller	Warsaw, N. Y	Sept. 24, 1872	131, 624

Index of patents issued from the United States Patent Office from 1790 *to* 1873, *inclusive*—Continued.

Invention.	Inventor.	Residence.	Date.	No.
Fence-post	E. G. Nichols	Beaufort, S. C	June 1, 1869	90, 773
Fence-post	D. Oliver	Oxford, Ohio	Aug. 6, 1867	67, 446
Fence-post	J. A. Otis	Watertown, N. Y	Apr. 25, 1871	114, 189
Fence-post	H. S. Palmer	Norvell, Mich	May 21, 1872	127, 099
Fence-post	J. Palmer	Painesville, Ohio	Aug. 9, 1864	43, 790
Fence-post	G. R. Patton	Juda, Wis	Apr. 7, 1868	76, 515
Fence-post	C. Putman and L. S. Totman	Cassadaga, N. Y	June 18, 1872	127, 987
Fence-post	R. Ramsey	New Wilmington, Pa	Dec. 10, 1867	72, 086
Fence-post	I. D. Richards and H. D. Snyder	Carbondale, Pa	May 12, 1868	77, 842
Fence-post	H. S. Ross	Millville, Ohio	Dec. 16, 1873	145, 528
Fence-post	J. Scott	Antwerp, N. Y	Oct. 28, 1873	144, 635
Fence-post	W. H. Shay	Sylvania, Ohio	Dec. 10, 1867	72, 094
Fence-post	C. J. Shuttleworth	Springfield, N. Y	Aug. 2, 1870	105, 988
Fence-post	C. R. Smith	Haverhill, N. H	Feb. 9, 1864	41, 541
Fence-post	H. Vail	Cortland, N. Y	Feb. 1, 1833	....
Fence-post	G. Vanauken	Phelps, N. Y	Sept. 15, 1863	39, 973
Fence-post	G. Webb	Lewiston, Me	Feb. 2, 1869	86, 476
Fence-post base	A. B. Drake	Painesville, Ohio	Oct. 15, 1872	132, 259
Fence-post, Cast-iron	P. Stewart	New Lebanon, N. Y	Mar. 1, 1859	23, 140
Fence-post driver	W. S. Graves	Oberlin, Ohio	Sept. 1, 1868	81, 771
Fence-post driver	J. D. Israel	Utica, Iowa	Aug. 11, 1868	80, 962
Fence-post driver	I. J. Parker	Buffalo Grove, Iowa	July 28, 1868	80, 364
Fence-post driver, Portable	W. Carns	New Cumberland, Ohio	May 31, 1870	103, 716
Fence-post excavator	R. P. Adams	Clinton, Ill	May 23, 1854	10, 952
Fence-post, Iron	J. W. Jenkins	Greenport, N. Y	Aug. 9, 1853	9, 921
Fence post, Iron	M. M. Manly	Philadelphia, Pa	May 7, 1872	126, 406
Fence-post, Iron	W. Merrell	Kent, Ohio	Nov. 10, 1868	83, 985
Fence-post, Metallic	W. A. Middleton	Harrisburgh, Pa	Feb. 16, 1869	86, 935
Fence-post pedestal	G. W. Hatch	Garrettsville, Ohio	May 7, 1867	64, 417
Fence-post pedestal	J. Robbins	Amherst, Ohio	Sept. 17, 1867	68, 902
Fence post, Picket	A. M. Freeman, C. P. Idell, and B. Vanderhoven.	Metuchen, N. J	Nov. 22, 1870	109, 402
Fence-post, Portable	D. M. Graham	Evansville, Ind	Dec. 10, 1867	72, 022
Fence-post socket	J. S. Blood and J. W. Miller	Newport, N. H	July 3, 1860	28, 958
Fence-post socket	G. W. and J. B. Durant	Bryan, Tex	Sept. 9, 1873	142, 687
Fence-post socket	G. Unger	Danville, Pa	May 25, 1869	90, 611
Fence-post stub	G. Lakins	East Norwalk, Ohio	July 29, 1873	141, 362
Fence posts and rails, Machine for mortising and sharpening	J. A. Snyder	Georgetown, Pa	Jan. 28, 1868	73, 844
Fence posts and ties, Construction of iron	J. B. Wickersham	New York, N. Y	Sept. 16, 1856	15, 750
Fence-posts, Boring-machine for	R. Weems		Mar. 16, 1801	....
Fence-posts, Construction of	M. Krumm, jr	Columbus, Ohio	Apr. 14, 1868	76, 637
Fence-posts, Device for bracing and ventilating	C. R. Smith	Haverhill, N. H	Oct. 11, 1859	25, 768
Fence-posts, Device for driving	J. Anderson	Waukesha, Wis	Feb. 27, 1866	52, 811
Fence-posts, Device for securing ends of wire in	W. G. Lavers	New York, N. Y	July 25, 1854	11, 378
Fence-posts, Driving-machine for	D. M. Wotson	Albia, Iowa	July 16, 1872	129, 442
Fence-posts, Machine for boring	J. Agnew	Bath, Pa	Dec. 26, 1865	51, 676
Fence-posts, Machine for boring	J. Temple	Birmingham, Pa	May 1, 1855	12, 808
Fence-posts, Method of attaching the cappings to	R. S. Cadwell	Andover, Ohio	June 14, 1859	24, 373
Fence-posts, Method of constructing	F. G. Johnson	Brooklyn, N. Y	Jan. 27, 1857	16, 486
Fence-posts, Method of driving	G. E. West	Indianapolis, Ind	Aug. 14, 1866	57, 231
Fence-posts, Method of fastening wires to	Z. Knapp	Pittston, Pa	Oct. 24, 1848	5, 884
Fence-posts, Mode of constructing iron	H. S. Brooks and J. S. Lehman	Martickville, Pa	June 2, 1868	78, 572
Fence-posts, Setting	H. W. Clarke	Newport, R. I	Oct. 16, 1866	58, 776
Fence-posts, Setting	W. Fulkerson	Three Rivers, Mich	Sept. 27, 1870	107, 770
Fence, Rail	A. Frayer	Ripley, Ohio	Dec. 24, 1867	72, 473
Fence, Rail	J. W. Mathews	Killbourne, Ohio	Apr. 8, 1873	137, 701
Fence, Rail	E. G. Warner	Union Township, Ohio	Sept. 15, 1868	82, 186
Fence-rails, Machine for pointing	C. Yost	Intercourse, Pa	July 3, 1860	29, 030
Fence-rails, Splicing	D. W. Knowles	Hotchkissville, Conn	Oct. 28, 1873	144, 113
Fence, Self-locking rail	A. Bagley	Ypsilanti, Mich	Jan. 21, 1873	134, 967
Fence, Self-supporting	T. Flinn	Birmingham, Mich	Nov. 2, 1869	96, 411
Fence-stake	J. N. Pease	Panama, N. Y	Oct. 19, 1869	95, 932
Fence, Stream	J. Fryling	Fletcher, Ohio	Aug. 6, 1867	67, 526
Fence, Stream or river	H. A. Kephart	Fletcher, Ohio	Jan. 14, 1868	73, 345
Fence, Suspended portable	A. Skinner	Warren, Mich	Mar. 23, 1869	88, 090
Fence, Wire	P. S. Clinger	Conestoga Centre, Pa	Sept. 13, 1859	25, 387
Fence, Wire	S. F. Dexter	Auburn, N. Y	Aug. 19, 1833	....
Fence, Wire	G. W. Ensminger	Richland, Iowa	Dec. 8, 1868	84, 810
Fence, Wire	E. Gale	Pavilion, Ill	Aug. 18, 1863	39, 563
Fence, Wire	T. Hill, jr	Sturgis, Mich	June 18, 1872	128, 145
Fence, Wire	H. Jenkins	Pottsville, Pa	Feb. 13, 1849	6, 106
Fence, Wire	I. Knapp	Medina, N. Y	June 7, 1864	43, 032
Fence, Wire	G. and L. N. Lakins	East Norwalk and Toledo, Ohio.	Aug. 26, 1873	142, 246
Fence, Wire	J. A. Little	Cartersburgh, Ind	June 24, 1873	140, 147
Fence, Wire	A. H. Mendell	Adams, N. Y	Aug. 3, 1869	93, 326
Fence, Wire	W. H. Meriwether	Comal County, Tex	Nov. 8, 1853	10, 211
Fence, Wire	Z. Nichelson	Haddonfield, N. J	Apr. 25, 1871	114, 029
Fence, Wire	J. W. Norcross	Middletown, Conn	Jan. 10, 1865	45, 872
Fence, Wire	J. W. Rappleye	Farmer Village, N. Y	July 4, 1871	116, 755
Fence, Wire	J. W. Rappleye	Farmer Village, N. Y	Feb. 18, 1873	136, 094
Fence, Wire	J. B. Reyman	Bloomington, Ill	Sept. 29, 1857	18, 301
Fence, Wire	H. M. Rose	Waterman Station, Ill	May 13, 1873	138, 763
Fence, Wire	A. C. Sisson	Factoryville, Pa	Apr. 25, 1871	114, 057
Fence, Wire	L. B. Smith	Kent, Ohio	June 25, 1867	66, 182
Fence, Wire	T. H. Speakman	Philadelphia, Pa	Jan. 16, 1872	122, 862
Fence, Wire	C. A. Wakefield	Pittsfield, Mass	Mar. 14, 1871	112, 658
Fence, Wire	M., sr., M., jr., and D. S. Walker.	Philadelphia, Pa	Mar. 29, 1853	9, 642
Fence, Wire	B. Wilson and F. P. Grimes	Dayton, Ohio	Dec. 6, 1870	109, 858
Fence, Wire and picket	E. C. Patterson	Rochester, N. Y	July 14, 1868	79, 854
Fence, Wood	W. Riley	Madison County, Miss	Apr. 8, 1873	137, 723
Fence, Wood	D. G. Temple	Farmersville, La	Mar. 11, 1873	136, 679
Fence, Wood	R. F. Ward	Senatobia, Miss	Oct. 7, 1873	143, 478
Fences, Apparatus for building sod	C. W. Smith	Morrisville, N. Y	Nov. 23, 1869	97, 130
Fences, Apparatus for stretching wire	J. T. Maugham	Gatesville, Tex	Mar. 4, 1873	136, 528
Fences, Board-holder in making board	J. Wallmer	Goshen, Ind	Nov. 15, 1864	45, 098

Index of patents issued from the United States Patent Office from 1790 *to* 1873, *inclusive*—Continued.

Invention.	Inventor.	Residence.	Date.	No.
Fences, Bore or support for posts of field	O. Spencer	Jacksonburgh, Ohio	Jan. 13, 1857	16, 406
Fences, Brace-post for field	C. Quackenbush	Huron, N. Y	Nov. 30, 1858	22, 202
Fences, Device for connecting panels of field	R. Merrill	Elmira, N. Y	Feb. 23, 1858	19, 434
Fences, Device for connecting panels of portable field.	J. Haines	West Middleburgh, Ohio	Oct. 4, 1859	25, 641
Fences, Device for holding together panels of portable.	O. P. Moran	Haynesville, Mo	Aug. 9, 1859	25, 030
Fences, Device for straining wire	F. H. Crandall	Ontario, N. Y	Jan. 14, 1868	73, 302
Fences, Device for stretching wires of	W. H. Robinson and J. Behel	Earlville, Ill	Sept. 13, 1864	44, 221
Fences, Device for tightening wire	P. S. Crawford	Union, Ill	Apr. 24, 1866	54, 119
Fences, Device for tightening wire	D. W. Eaton	North Ridge, N. Y	Aug. 24, 1869	93, 976
Fences, Device to allow for contraction and expansion in wire.	T. D. Burk	Chicago, Ill	Apr. 22, 1856	14, 751
Fences, Fastening for wire	W. E. Lockwood	Philadelphia, Pa	Jan. 14, 1868	73, 349
Fences, Flood-gate for	S. D. Hopkins	Brookville, Va	Nov. 20, 1849	6, 892
Fences in the posts, Method of fastening the rails of iron.	J. B. Wickersham	New York, N. Y	Jan. 13, 1857	16, 419
Fences, Iron post for wire	H. K. Flinchbaugh	Conestoga Centre, Pa	Mar. 3, 1868	75, 145
Fences, Iron railing for	S. Crowell	Philadelphia, Pa	May 23, 1865	47, 801
Fences, Machine for boring the posts for post-and-rail.	W. H. Shay	New York, N. Y	Nov. 25, 1839	1, 420
Fences, Machine for excavating and removing earth in making prairie.	J. Sawyer	New York, N. Y	Mar. 23, 1838	652
Fences, Machine for making picket	J. Moore and A. Kelly	Pittsburgh, Pa	Apr. 17, 1860	27, 923
Fences, Machine for making sod	H. L. F. Gavett	Jackson, Mich	Sept. 9, 1851	8, 341
Fences, Machine for making sod	J. Osgood	Blue Hill, Me	Aug. 24, 1869	94, 128
Fences, Machine for mortising posts and sharpening rails for.	W. Mace	New York, N. Y	Mar. 11, 1837	140
Fences, Method of allowing for expansion and contraction of wire.	O. Williams	Saint Louis, Mo	Sept. 7, 1858	21, 459
Fences, Method of clamping and unclamping panels of portable field.	W. B. Burnett	Lyons, N. Y	Apr. 21, 1857	17, 075
Fences, Method of compensating for expansion and contraction of metallic.	L. Eikenberry	Easton, Pa	May 10, 1859	23, 908
Fences, Method of connecting panels of field	S. F. Jones	Milford, Ind	July 7, 1857	17, 742
Fences, Method of connecting panels of field	W. D. Sheldon	Huron, N. Y	Jan. 19, 1858	19, 159
Fences, Method of connecting panels of portable field.	J. H. Bruen	Elmira, N. Y	Dec. 29, 1857	18, 952
Fences, Method of locking and supporting the panels of field.	E. West	Ogden, N. Y	Feb. 15, 1859	22, 994
Fences, Method of uniting and sustaining the panels of portable field.	I. D. Garlick	Lyons, N. Y	May 5, 1857	17, 210
Fences, Method of uniting panels of portable	C. Van De Mark	Oak's Corners, N. Y	June 2, 1857	17, 459
Fences, Mode of building and coating earth	C. A. Strong	Brooklyn, N. Y	May 13, 1862	35, 268
Fences, Mode of constructing prairie	I. Van Kersen	Kalamazoo, Mich	June 2, 1868	78, 622
Fences, Mode of fastening the paling to the rail in iron.	G. Hess	Easton, Pa	Nov. 23, 1852	9, 420
Fences, Mode of making combined wood and wire	G. Fletcher, sr	Greensburgh, Ind	Nov. 17, 1868	84, 180
Fences, Mode of securing to each other the panels of field.	C. P. Garlick and M. G. Blackstone.	Amador, Minn., and Mainville, Ohio.	Feb. 17, 1857	16, 675
Fences, &c., Molding and pressing clay for the construction of.	M. Wright	Tallytown, Pa	May 15, 1841	2, 093
Fences, Ornamental connection of the parts of iron	H. Jenkins	Cincinnati, Ohio	Jan. 13, 1852	8, 654
Fences, Post for field	J. Drown	Huron, N. Y	Jan. 25, 1859	22, 712
Fences, Post for field	H. G. Seekins	Elyria, Ohio	Mar. 23, 1858	19, 724
Fences, Rail for ornamental	E. M. Stigale	Philadelphia, Pa	Dec. 14, 1869	97, 984
Fences, Raising	T. J. Smith	Jackson, Mich	Nov. 17, 1868	84, 141
Fences, Repairing	E. Pitkin	East Hartford, Conn	Dec. 13, 1828	
Fences, Rivet-clamp for wire	M. P. Coons	Brooklyn, N. Y	June 6, 1854	11, 012
Fences, telegraph-poles, &c., Composition post for	J. L. Boone	San Francisco, Cal	June 4, 1872	127, 552
Fences, Tool for holding and driving staples for wire.	A. C. Betts	Troy, N. Y	Aug. 7, 1866	56, 884
Fences, Triangular brace for locking the panels of field.	C. Van De Mark	Oak's Corners, N. Y	July 27, 1858	21, 037
Fences, Wire-stretcher for	D. J. Denmark and P. P. Hill	Vergil and Alto, Ill	May 21, 1872	126, 939
Fences, Wire-stretcher for	W. B. Hayden	Columbus, Ohio	Aug. 25, 1868	81, 367
Fences, Wire-stretcher for picket	B. F. Alkire	Williamsport, Ohio	Oct. 8, 1872	132, 040
Fences, Wire-tightener for wire	F. Fanning	Atchison, Kans	Apr. 5, 1870	101, 601
Fences, Wire-work for	A. Algoever	New York, N. Y	Jan. 6, 1863	37, 266
Fences, Yielding joint for portable	R. J. Brown	Perry, Pa	Dec. 9, 1856	16, 172
Fencing grounds, Mode of	O. Church	Friendship, N. Y	July 3, 1823	
Fencing, Machine for making wire	B. Greening	Hamilton, Canada	July 25, 1871	117, 275
Fencing, Machine for manufacturing picket	W. W. Johnson	Clarksburgh, Va	May 17, 1859	24, 029
Fencing, Mode of making clay posts for	W. Cooley	Bolton, Conn	May 27, 1805	......
Fender: *See* Boot and shoe fender. Clod-fender. Cultivator-plow fender. Fire-fender. Gas-bracket fender. Grate-fender. Ice and snow fender. Iron fender. Locomotive-fender. Plow-fender. Ship-fender.				
Fender	G. F. Filley	Saint Louis, Mo	Nov. 4, 1873	144, 269
Fender and ash-sifter combined	W. C. Dobbins	Zanesville, Ohio	Aug. 26, 1873	142, 214
Fender and fire-screen, Combined	J. Miller and W. H. B. Flender	Washington, Pa	July 12, 1870	105, 231
Fender and gage-wheel, Combined	A. B. Thornton	New Berlin, Ill	Feb. 14, 1871	111, 790
Fermentation and distillation	J. E. Siebel	Chicago, Ill	Dec. 3, 1872	133, 677
Fermentation and purification of organic substances.	R. D'Heureuse	New York, N. Y	Feb. 28, 1871	112, 226
Fermentation, Vinous	C. C. Edday	Benton, Miss	Apr. 1, 1842	2, 535
Fermentation, Vinous	C. O. Wolpers	Cincinnati, Ohio	July 16, 1841	2, 169
Fermenting and distilling	J. Stowell	Manchester, N. H	Feb. 5, 1836	......
Fermenting and distilling spirits	I. Belnap	Millersburgh, Pa	July 20, 1827	......
Fermenting apple cider	W. Elder	New Brunswick, N. J	Aug. 21, 1822	......
Fermenting in close vessels, Vinous	A. Harvie and C. Guild	Cincinnati, Ohio	Apr. 25, 1854	10, 814
Fermenting mash	C. H. Frings	Centreton, Mo	June 20, 1871	116, 043
Fermenting-tanks, Preventing waste of alcohol in	C. H. Frings	Centreton, Mo	June 20, 1871	116, 044
Fernery	C. L. Osborn	New York, N. Y	Sept. 28, 1869	95, 255
Ferrotype card-mount	J. H. Daniels	Boston, Mass	Aug. 9, 1864	43, 761

Index of patents issued from the United States Patent Office from 1790 *to* 1873, *inclusive*—Continued.

Invention.	Inventor.	Residence.	Date.	No.
Ferrotype-plate	H. M. Hedden	Worcester, Mass	Mar. 1, 1870	100, 291
Ferrule	S. N. Chapin	New Britain, Conn	Aug. 31, 1869	94, 280
Ferrule	T. W. Detray	Montpelier, Vt	May 29, 1860	28, 458
Ferrule	C. E. Green and J. H. Bodwell	Newark, N. J	Sept. 2, 1873	142, 461
Ferrule	S. E. Jeralds	Cheshire, Conn	Dec. 30, 1873	145, 950
Ferrule	D. Moore	Brooklyn, N. Y	Nov. 7, 1871	120, 763
Ferrule	W. H. Rodden	Toronto, Canada	Feb. 7, 1871	111, 573
Ferrule	D. G. Smith	Columbus, Ohio	Apr. 18, 1871	113, 805
Ferrule, Brush	W. F. Lorey	Mendon, N. Y	Mar. 11, 1873	136, 605
Ferrule for boiler-tubes	E. Lawton and T. J. Jones	New York, N.Y., and Summit, N. J.	Apr. 2, 1867	63, 399
Ferrule-making machine, Wire	H. O. Lothrop	Milford, Mass	Oct. 19, 1869	95, 918
Ferrule, Metal socket	T. H. Windle	Westchester, Pa	May 7, 1867	64, 613
Ferrule-tapering apparatus	J. L. Parker	Worcester, Mass	June 27, 1871	116, 479
Ferrules, Machine for forming	W. L. Newsham	Philadelphia, Pa	Dec. 17, 1872	134, 092
Ferrules, Machine for making	R. Briggs	Philadelphia, Pa	Dec. 21, 1869	98, 020
Ferrules, Machine for making	J. Stever and J. A. Way	Bristol, Conn	May 28, 1867	65, 132
Ferrules, Machinery employed in the manufacture of coiled wire.	W. T. Richards	New Haven, Conn	Sept. 14, 1852	9, 262
Ferrules, Manufacture of	A. Shaw	Philadelphia, Pa	Dec. 10, 1867	72, 093
Ferrules, Manufacture of wire	W. T. Richards	New Haven, Conn	Nov. 2, 1852	9, 369
Ferrules, Manufacturing	S. E. Jeralds, H. A. Nettleton, and E. R. Lawton.	West Cheshire, Conn	Sept. 12, 1871	118, 858
Ferrules to handles, Mode of attaching	A. L. Carrier	Washington, D. C	Nov. 12, 1867	70, 694
Ferruling machine, Ring	J. Siddons	Rochester, N. Y	Nov. 24, 1868	84, 311
Ferry-boat	M. D. Brown	Buffalo, Va	Oct. 1, 1830	......
Ferry-boat	H. Clenny	Gallatin, Tenn	June 20, 1871	116, 024
Ferry-boat	T. Cohoon	Pittsfield, Mass	June 25, 1813	......
Ferry-boat for trains of cars	F. Cass	New Orleans, La	Oct. 21, 1873	143, 808
Ferry-boat guard	D. Fitzgerald, T. Rogers, and W. C. Walker.	New York, N. Y	Dec. 19, 1854	12, 094
Ferry-boat locking-apparatus	J. L. Canham	Newark, N. J	Apr. 16, 1867	63, 855
Ferry-boats, Means of stopping and starting	J. Rowland	Brooklyn, N. Y	May 14, 1867	64, 710
Ferry-boats, Operating	J. C. Day	Jersey City, N. J	Nov. 24, 1857	18, 683
Ferry-boats or flying bridges, Means of guiding line.	E. W. Quincy and W. Fisher	Lacon, Ill	Apr. 21, 1868	76, 943
Ferry-boats, Tram-way for	W. A. Jordan	Thibodeaux, La	Aug. 5, 1856	15, 487
Ferrying and draying	N. W. Wheeler	Brooklyn, N. Y	Apr. 25, 1865	47, 479
Fertilizer	L. C. Abernathy	Burlington, Ky	Aug. 18, 1829	......
Fertilizer	J. Althouse	Cross Roads, Pa	May 12, 1868	77, 860
Fertilizer	A. F. Andrews	New Haven, Conn	Aug. 19, 1873	141, 848
Fertilizer	J. W. Bitner	Downsville, Md	Dec. 3, 1867	71, 689
Fertilizer	J. R. Black	Ninety Six, S. C	Apr. 23, 1872	125, 927
Fertilizer	S. A. Burkholder and G. W. Wilson.	Bendersville, Pa	Mar. 30, 1869	88, 443
Fertilizer	W. G. Busey	Georgetown, D. C	Apr. 21, 1868	76, 991
Fertilizer	J. Commins	Charleston, S. C	Feb. 1, 1870	99, 294
Fertilizer	J. Commins	Charleston, S. C	Mar. 15, 1870	100, 729
Fertilizer	B. R. Croasdale	Philadelphia, Pa	Aug. 30, 1870	109, 918
Fertilizer	J. P. Crutchfield	Fayette Corner, Tenn	May 21, 1872	126, 933
Fertilizer	S. O. Doris	Philadelphia, Pa	Dec. 24, 1861	34, 039
Fertilizer	L. S. Fales	Tarrytown, N. Y	Dec. 3, 1867	71, 725
Fertilizer	L. S. Fales	New York, N. Y	Mar. 30, 1869	88, 466
Fertilizer	R. Fish	Washington, D. C	July 20, 1869	92, 810
Fertilizer	J. and A. Fox	Avoca, N. Y	Mar. 12, 1872	124, 487
Fertilizer	W. H. H. Glover	New York, N. Y	July 26, 1864	43, 639
Fertilizer	J. Gould	Lexington, Mass	Feb. 27, 1866	52, 844
Fertilizer	W. C. Grimes	Ladiesburgh, Md	Dec. 10, 1867	72, 026
Fertilizer	W. D. Hall	Hamden, Conn	Dec. 20, 1859	26, 548
Fertilizer	L. Harper	Riceville, N. J	Nov. 22, 1859	26, 184
Fertilizer	L. Harper	Riceville, N. J	Jan. 31, 1860	26, 985
Fertilizer	L. Harper	Brooklyn, N. Y	May 27, 1862	35, 417
Fertilizer	H. A. Hogel	Brooklyn, N. Y	Mar. 22, 1870	101, 131
Fertilizer	J. and A. Hursh	Philadelphia, Pa	June 26, 1866	55, 871
Fertilizer	W. Lalor	Utica, N. Y	May 25, 1869	90, 367
Fertilizer	O. Lugo	New York, N. Y	Nov. 14, 1865	50, 940
Fertilizer	J. J. Mapes	Newark, N. J	Nov. 22, 1859	26, 196
Fertilizer	J. J. Mapes	Newark, N. J	Dec. 20, 1859	26, 507
Fertilizer	J. McDougall	London, England	Feb. 18, 1873	135, 995
Fertilizer	J. K. Moore	Millville, N. J	July 30, 1867	67, 335
Fertilizer	H. E. Pond	Franklin, Mass	Nov. 5, 1867	70, 608
Fertilizer	J. S. Ramsburgh	New Market, Md	May 12, 1868	77, 840
Fertilizer	F. C. Renner	Ladiesburgh, Md	Feb. 5, 1867	61, 870
Fertilizer	F. C. Renner	Ladiesburgh, Md	June 22, 1869	91, 667
Fertilizer	A. Rolland	Toulouse, France	Feb. 7, 1860	27, 072
Fertilizer	A. Smith	Cincinnati, Ohio	Jan. 1, 1867	60, 948
Fertilizer	A. Smith	Cincinnati, Ohio	May 5, 1868	77, 667
Fertilizer	A. Smith	Baltimore, Md	Mar. 23, 1869	88, 223
Fertilizer	I. W. Speyer	Hamburg, Germany	Jan. 25, 1870	99, 255
Fertilizer	L. Stephens	Philadelphia, Pa	May 29, 1860	28, 516
Fertilizer	J. W. Stubbs	Cheraw, S. C	Feb. 25, 1873	136, 277
Fertilizer	T. Taylor	Washington, D. C	Mar. 14, 1871	112, 653
Fertilizer	J. D. Whelpley	Boston, Mass	Sept. 12, 1865	49, 943
Fertilizer	J. Whitehill	Frederick, Md	Apr. 29, 1873	138, 458
Fertilizer	J. W. York	Nolensville, Tenn	Apr. 8, 1873	137, 586
Fertilizer and corn-planter combined	S. H. Wallize	Washingtonville, Pa	Dec. 10, 1867	71, 930
Fertilizer and distributer	C. Salvo	Columbus, Ga	Feb. 7, 1871	111, 685
Fertilizer and planter, Corn	J. M. Stoner	Greenville Lodge, Pa	June 22, 1869	91, 786
Fertilizer and seed-drill	T. B. McConaughey	Newark, Del	Jan. 26, 1869	86, 313
Fertilizer and seeding-machine	A. Putnam	Owego, N. Y	Mar. 7, 1871	112, 490
Fertilizer, Artificial	H. E. Pond	Franklin, Mass	Aug. 6, 1867	67, 450
Fertilizer-attachment	C. C. Foster	Odessa, Del	Apr. 28, 1868	77, 269
Fertilizer-attachment	C. C. Foster	Odessa, Del	Sept. 28, 1869	95, 336
Fertilizer-desiccating machine	A. Smith	Baltimore, Md	May 4, 1869	89, 800
Fertilizer-distributer	M. W. Faubion	Parrottsville, Tenn	Sept. 2, 1873	142, 454
Fertilizer-distributer	B. Fickes	North Codorus Township, Pa.	Aug. 20, 1872	130, 707
Fertilizer-distributer	D. F. Halliburton	Rutherford Station, Tenn	Apr. 29, 1873	138, 398

Index of patents issued from the United States Patent Office from 1790 *to* 1873, *inclusive*—Continued.

Invention.	Inventor.	Residence.	Date.	No.
Fertilizer-distributer	J. Lytch	Laurinburgh, N. C	Nov. 11, 1873	144, 465
Fertilizer-distributer	J. A. Morton	New Orleans, La	May 30, 1871	115 505
Fertilizer-distributer	J. J. Singleton	Forsyth, Ga	Oct. 11, 1870	108, 196
Fertilizer-distributer	W. F. White	Belchertown, Mass	Apr. 23, 1872	126, 118
Fertilizer-distributer	J. M. Wiltsie	Pittsford, N. Y	Feb. 21, 1865	46, 515
Fertilizer-distributer, Broadcast	J. P. Machen	Centreville, Va	June 6, 1871	115, 749
Fertilizer-distributer, corn and cotton-seed planter	R. M. Brooks	Woodbury, Ga	May 10, 1870	102, 910
Fertilizer, Distributing	H. M. Keith	Pontiac, Mich	Oct. 11, 1864	44, 635
Fertilizer-distributing machine	J. F. Fisher	Greencastle, Pa	Dec. 15, 1868	84, 870
Fertilizer-distributing machine	S. Hoke	Mount Pleasant Township, Md.	Aug. 27, 1867	68, 077
Fertilizer-distributing machine	J. F. Keller	Hagerstown, Md	June 28, 1870	104, 854
Fertilizer-distributing machine	I. M. Reames	Oxford, N. C	Feb. 11, 1873	135, 846
Fertilizer-distributing machine	W. F. Weirick, J. C. Weller, and D. E. Rohr.	Charlestown, W. Va	Feb. 2, 1869	86, 477
Fertilizer from excrements	F. Wicke, J. Brönner, T. Petersen, and J. G. Zehfuss.	Frankfort, Prussia	Mar. 1, 1870	100, 347
Fertilizer from glue residuum	A. Van Haagen and W. Adamson.	Philadelphia, Pa	Feb. 15, 1870	99, 978
Fertilizer from offal	J. J. Storer	Boston, Mass	Sept. 30, 1873	143, 310
Fertilizer from sea-weed	J. G. Nickerson	Boston, Mass	July 20, 1869	92, 744
Fertilizer from sea-weed	U. S. Treat	Eastport, Me	Sept. 12, 1871	118, 987
Fertilizer of land, Process of preparing greens and marl as a.	C. Stearns	New York, N. Y	May 5, 1857	17, 237
Fertilizer or fish guano	O. Lugo	Baltimore, Md	Feb. 1, 1870	99, 452
Fertilizer or guano	O. Lugo	Baltimore, Md	Dec. 14, 1869	97, 939
Fertilizer or manure	G. F. Wilson	East Providence, R. I	Aug. 11, 1863	39, 519
Fertilizer-spreader	J. W. Thornburg	Woodstock, Va	Apr. 8, 1873	137, 736
Fertilizers and ammoniacal salts, Treating blood for the manufacture of.	J. J. Craven	Jersey City, N. J	Feb. 21, 1871	111, 910
Fertilizers and extracting oils and fats, Manufacture of.	O. Lugo	Baltimore, Md	Feb. 15, 1870	99, 924
Fertilizers and oil from fish, Manufacture of	O. Lugo	Baltimore, Md	May 3, 1870	102, 689
Fertilizers and other articles of manufacture, Machine for disintegrating.	T. Carr	Bristol, Great Britain	June 8, 1869	91, 085
Fertilizers and other materials, Machine for disintegrating, dispersing, and mixing.	T. Carr	Bristol, Great Britain	Aug. 10, 1869	93, 595
Fertilizers and other purposes, Treating blood for the preparation of.	H. A. Hogel	New York, N. Y	Mar. 8, 1870	100, 629
Fertilizers and to destroy offensive gases and vapors, Treating offal to produce.	J. J. Storer	Boston, Mass	Oct. 22, 1872	132, 498
Fertilizers, Apparatus for pulverizing animal matter for.	A. Smith	Cincinnati, Ohio	Apr. 2, 1872	125, 343
Fertilizers, Apparatus for treating animal matter for.	W. L. Bradley	Boston, Mass	Apr. 2, 1872	125, 260
Fertilizers, Apparatus for treating animal matter for.	H. A. Hogel	New York, N. Y	May 2, 1871	114, 508
Fertilizers, Composition for deodorizing and preparing.	T. Sewell	Washington, D. C	May 7, 1872	126, 418
Fertilizers, Composition of matter for disinfecting and preparing.	J. A. Thompson	Auburn, N. Y	July 9, 1867	66, 650
Fertilizers, Compound for	G. Bourgrade	New York, N. Y	Aug. 16, 1870	106, 313
Fertilizers, Cylinder for mixing	L. B. Pitcher	Salina, N. Y	July 26, 1870	105, 839
Fertilizers, &c., Deodorizing animal matter for	E. C. C. Stanford	Glasgow, Scotland	Apr. 15, 1873	137, 969
Fertilizers from animal matter, Process and apparatus for the manufacture of.	C. G. Bruce and M. J. Stein	New York, N. Y	Oct. 15, 1872	132, 244
Fertilizers from animal substances, Manufacture of.	O. Lugo	Baltimore, Md	Feb. 22, 1870	100, 163
Fertilizers from earth, &c., Preparing	S. Brown	New Oxford, Pa	Mar. 26, 1872	125, 017
Fertilizers from fish, &c., Manufacture of	O. Lugo	Baltimore, Md	June 14, 1870	104, 327
Fertilizers from night-soil, Manufacture of	D. J. Loewenstein	New Orleans, La	July 23, 1872	129, 739
Fertilizers, Machine for distributing	J. S. Edwards	Medford, N. J	Mar. 10, 1868	75, 252
Fertilizers, Machine for distributing	Z. N. Morrell	Cameron, Tex	Sept. 27, 1859	25, 574
Fertilizers, Machine for distributing	J. H. Thomas and P. P. Mast	Springfield, Ohio	Apr. 4, 1865	47, 138
Fertilizers, Machine for distributing	J. H. Thomas and P. P. Mast	Springfield, Ohio	Aug. 3, 1869	93, 368
Fertilizers, Machine for sowing	J. F. Keller	Greencastle, Pa	Jan. 8, 1861	31, 084
Fertilizers, Manufacture of	J. Commins	Charleston, S. C	Oct. 4, 1870	107, 878
Fertilizers, Manufacture of	J. Y. Diaz	Havana, Cuba	Mar. 15, 1870	100, 871
Fertilizers, Manufacture of	L. S. Fales	New York, N. Y	June 9, 1868	78, 730
Fertilizers, Manufacture of	D. A. T. Hoeven	Philadelphia, Pa	June 23, 1868	79, 160
Fertilizers, Manufacture of	C. P. Houghton	Georgetown, D. C	Nov. 1, 1870	108, 909
Fertilizers, Manufacture of	F. Klett	Philadelphia, Pa	Sept. 12, 1865	49, 891
Fertilizers, Manufacture of	W. I. Sapp	Baltimore, Md	Apr. 26, 1870	102, 438
Fertilizers, Manufacture of	T. Sim	Baltimore, Md	Aug. 23, 1870	106, 626
Fertilizers, Manufacture of	H. Stevens	Brazoria, Tex	Nov. 25, 1873	144, 877
Fertilizers, Manufacture of	E. Whitley	Murfreesborough, N. C	July 12, 1870	105, 283
Fertilizers, Manufacture of phosphatic	G. F. Wilson	East Providence, R. I	Mar. 10, 1868	75, 325
Fertilizers, Mode of treating mineral phosphates for the manufacture of.	J. Commins	Charleston, S. C	Feb. 25, 1868	74, 799
Fertilizers, Mode of treating mineral phosphates for the manufacture of.	J. Commins	Charleston, S. C	May 19, 1868	78, 061
Fertilizers, Preparation of	W. S. De Zeng	Geneva, N. Y	Nov. 5, 1867	70, 671
Fertilizers, Preparing	G. A. Leinau	Philadelphia, Pa	Mar. 12, 1867	62, 760
Fertilizers, Preparing ammoniated sulphuric acid for the manufacture of.	C. U. Shepard, jr	Charleston, S. C	Mar. 1, 1870	100, 457
Fertilizers, Preparing blood for	W. D. Craven	Jersey City, N. J	Feb. 18, 1873	136, 036
Fertilizers, Process for manufacturing	J. A. Manning	London, England	Jan. 16, 1872	122, 773
Fertilizers, Process for preparing	L. Reid	Barren Island, N. Y	Mar. 24, 1857	16, 882
Fertilizers, &c., Process for treating sewage and ammoniacal water for the production of.	T. Christy, jr., and A. Bobrownicki.	London, England	Mar. 25, 1873	137, 059
Fertilizers, Process of treating offal-gelatine, &c., for the manufacture of.	G. F. Wilson	East Providence, R. I	May 18, 1869	90, 328
Fertilizers to growing plants, Implement for distributing.	D. C. Colby	Claremont, N. H	June 13, 1865	48, 155
Fertilizers, Treating animal matter and manufacturing.	W. C. and R. G. Sillar and C. Rawson.	Blackheath, Bolton and London, England.	Nov. 19, 1872	133, 125
Fertilizers, Treating animal matter for	H. A. Hogel and A. Edwards	New York, N. Y., and New Haven, Conn.	Nov. 5, 1872	132, 723
Fertilizers, Treating blood for manufacture of	L. S. Fales	New York, N. Y	Feb. 14, 1871	111, 734

Index of patents issued from the United States Patent Office from 1790 to 1873, inclusive—Continued.

Invention.	Inventor.	Residence.	Date.	No.
Fertilizers, Treating bones, horns, hoofs, &c., for manufacture of.	W. B. Johns	Philadelphia, Pa	Feb. 14, 1871	111, 851
Fertilizers, Treating offal and manufacturing	J. J. Storer	Boston, Mass	Jan. 28, 1873	135, 383
Fertilizers, Treating phosphate of lime for manufacture of.	N. A. Pratt	Charleston, S. C	Apr. 9, 1872	125, 613
Fertilizers, Treating refuse animal matter for manufacture of.	W. L. Bradley	Boston, Mass	Feb. 27, 1872	124, 112
Fertilizers, &c., Treating sewage for	H. H. Parish	Rome, Italy	Mar. 26, 1872	126, 074
Fertilizers, &c., Treatment of phosphates for the manufacture of.	N. A. Pratt and G. F. Lewis	Charleston, S. C., and Philadelphia, Pa.	May 21, 1872	126, 904
Fertilizing-composition	J. M. Gallacher	Roxbury, Mass	Apr. 1, 1862	34, 825
Fertilizing-compound	J. M. Deering	Boston, Mass	Apr. 23, 1872	125, 939
Fertilizing-compound	L. S. Fales	Tarrytown, N. Y	Dec 3, 1867	71, 724
Fertilizing-compound	W. B. Hamilton	New Orleans, La	Apr. 25, 1871	114, 133
Fertilizing-compound	J. M. Loewenstein	New Orleans, La	Oct. 18, 1870	108, 369
Fertilizing-compound	J. M. Loewenstein	New Orleans, La	Jan. 31, 1871	111, 357
Fertilizing-compound	E. N. McKimm and H. W. Bender.	Benderville, Pa	June 29, 1869	92, 077
Fertilizing-compound	L. S. Robbins	Brooklyn, N. Y	May 26, 1857	17, 392
Fertilizing-compound, Phosphate	D. Stewart	Port Penn, Del	May 11, 1869	90, 057
Fertilizing distributer	H. N. Sweeney	Worcester, Mass	Apr. 8, 1873	137, 734
Fertilizing purposes, Method of preparing bones for.	D. Stewart	Annapolis, Md	Oct. 11, 1859	25, 772
Fertilizing soil	J. B. Wilson	Townsend's Inlet, N. J	Sept. 23, 1873	143, 213
Fetter for cow's tail	C. F. Tolles	Nashua, N. H	Oct. 21, 1873	143, 943
Fiber and gum fabric	A. F. Bishop, J. M. Pendleton, and J. H. Aiken.	Norwalk, Conn., New York, N. Y., and Norwalk, Conn.	Sept. 1, 1868	81, 740
Fiber and other products from the maize plant, Mode of obtaining.	J. T. Harris	Tyngsborough, Mass	Aug. 31, 1869	94, 411
Fiber, Apparatus and process for treating animal and vegetable.	W. Adamson	Philadelphia, Pa	Sept. 5, 1871	118, 668
Fiber-cleaning machine	J. Brown	Bay Ridge, N. Y	July 4, 1871	116, 546
Fiber-cleaning machine	C. A. Dean	Boston, Mass	Aug. 29, 1871	118, 591
Fiber-cleaning machine	S. T. Lamb	Boston, Mass	June 17, 1873	140, 050
Fiber, Composition for preserving and water-proofing vegetable.	G. A. Cowles, J. P. Case, and V. Vierow.	New York, N. Y	Sept. 20, 1864	44, 285
Fiber, Disintegrating	C. Heaton	New York, N. Y	Mar. 13, 1866	53, 247
Fiber, Disintegrating	C. Heaton	New York, N. Y	Mar. 13, 1866	53, 248
Fiber, Disintegrating and separating vegetable	G. E. Sellers	Seller's Landing, Ill	Feb. 9, 1864	41, 538
Fiber, Disintegrating vegetable	C. Heaton	New York, N. Y	Aug. 1, 1865	49, 106
Fiber, Disintegrating vegetable	R. W. Russell	New York, N. Y	Sept. 17, 1872	131, 465
Fiber, &c., Disintegration of vegetable substances for the separation of.	G. E. Sellers	Hardin County, Ill	Nov. 10, 1863	40, 576
Fiber-disintegrator	A. Berthet and P. Laberie	New Orleans, La	Dec. 16, 1873	145, 613
Fiber drier, Disintegrated	W. Adamson	Philadelphia, Pa	Nov. 29, 1870	109, 766
Fiber during manufacture, Mode of preventing mildew or injury to.	E. T. Rice	New York, N. Y	Oct. 27, 1868	83, 409
Fiber, Engine for reducing rags, &c., to	J. McCracken	Bloomfield, N. J	Aug. 21, 1866	57, 355
Fiber for stuffing mattress and cushion, Preparing vegetable.	W. Staufen	Prussia	Dec. 23, 1856	16, 293
Fiber for textile and other fabrics, Preparing vegetable.	H. Messmer	Newark, N. J	Aug. 20, 1867	67, 997
Fiber from bamboo, Preparing	P. Lichtenstadt	New York, N. Y	Feb. 16, 1864	41, 627
Fiber from bamboo, &c., Preparing	L. S. Robbins and J. A. Southmayd.	New York, N. Y., and Elizabeth, N. J.	Feb. 23, 1869	87, 295
Fiber from mulberry-tree, Method of separating	W. Holdmann	New York, N. Y	Aug. 4, 1868	80, 737
Fiber from pine-leaves for hygienic and other purposes.	A. Roque	Brire, France	Jan. 1, 1867	60, 940
Fiber from plants, Machinery for separating	J. A. De Brame	New York, N. Y	Dec. 2, 1862	37, 081
Fiber from plants, Machinery for separating	G. Sanford	New York, N. Y	June 24, 1862	35, 708
Fiber from seed, Machine for separating	A. B. Ely	Newton, Mass	Apr. 2, 1867	63, 373
Fiber from sisal grass and like substances, Machine for preparing.	G. E. Hopkins and W. B. Shedd	Boston, Mass	Dec. 20, 1870	110, 237
Fiber from waste felted fabrics, Obtaining	J. F. Greene	Brooklyn, N. Y	Apr. 12, 1859	23, 642
Fiber from wood and other substances, Separating	M. R. Fletcher	Cambridgeport, Mass	Feb. 11, 1868	74, 332
Fiber from woody portion of tropical plants, Machinery for separating.	J. Lilley	Birkenhead, England	Dec. 5, 1854	12, 025
Fiber, Machine for separating from the stalk and twisting woody.	H. W. Bowen	Providence, R. I	Oct. 20, 1863	40, 322
Fiber, Machine for treating vegetable	G. Sanford	New York, N. Y	Feb. 19, 1861	31, 479
Fiber, Machinery for cleaning vegetable	C. Brody	New York, N. Y	Sept. 24, 1861	33, 333
Fiber, Manufacturing Asclepias syriaca	M. Gerrish	Salem, Mass	Mar. 27, 1834	....
Fiber, Method of preserving vegetable	T. P. Shaffner	Louisville, Ky	June 5, 1866	55, 371
Fiber, Mode of treating and separating vegetable	J. R. Haskell	New York, N. Y	Mar. 19, 1867	63, 044
Fiber of banana, plantain, aloe, &c., Machine for preparing the.	F. Burke	Montserrat, West Indies	June 2, 1857	17, 420
Fiber of plants, Machine for preparing	E. J. y Patrullo	New York, N. Y	Jan. 15, 1867	61, 244
Fiber of straw, &c., Separating	A. S. Lyman	New York, N. Y	Nov. 24, 1863	40, 696
Fiber of tropical plants, Machinery for separating the.	E. J. y Patrullo	New York, N. Y	Apr. 28, 1863	38, 330
Fiber of various materials are united with adhesive mixtures, Machinery for the manufacture of stuffs in which.	T. R. Williams	Newport, R. I	Apr. 24, 1840	1, 559
Fiber, Plastic compound made from vegetable	W. H. Pierson	New Orleans, La	May 28, 1867	65, 267
Fiber, Preparation of vegetable	J. J. Stover	Portsmouth, N. H	Apr. 12, 1864	42, 319
Fiber, Preparation of vegetable	D. A. Wells	Cambridge, Mass	Apr. 4, 1854	10, 727
Fiber, Preparing and treating vegetable	C. Heaton	New York, N. Y	Sept. 3, 1867	68, 363
Fiber, Preparing vegetable	J. Blanc	New Orleans, La	Oct. 9, 1855	13, 636
Fiber, Process for preparing vegetable	C. J. Pownall	Addison Road, England	Apr. 5, 1853	9, 650
Fiber, Process for removing mineral, gummy, and resinous substances from vegetable.	A. Meucci	Clifton, N. Y	Mar. 28, 1865	47, 068
Fiber, Process of obtaining useful	D. Bickford	Boston, Mass	Oct. 15, 1867	69, 752
Fiber, Process of obtaining vegetable	W. Adamson	Philadelphia, Pa	Apr. 23, 1872	126, 006
Fiber, Process of obtaining vegetable	W. Adamson and C. F. A. Simonin.	Philadelphia, Pa	Aug. 8, 1871	117, 852
Fiber, Process of treating vegetable	P. Claussen	London, England	June 3, 1851	8, 134
Fiber, Process of treating vegetable	S. Knowles	Trenton, N. J	Feb. 14, 1854	10, 518
Fiber, Process of treating vegetable	E. T. Rice	New York, N. Y	June 29, 1869	92, 098

Index of patents issued from the United States Patent Office from 1790 *to* 1873, *inclusive*—Continued.

Invention.	Inventor.	Residence.	Date.	No.
Fiber, Process of treating vegetable substances to obtain.	B. C. Tilghman	Philadelphia, Pa	July 6, 1869	92, 229
Fiber, Reducing long staple	S. M. Allen	Woburn, Mass	Mar. 10, 1863	37, 846
Fiber, Separating animal from vegetable	J. Stuart	London, England	Sept. 27, 1870	107, 833
Fiber, Separating gummy and silicious matter from vegetable.	C. Heaton	New York, N. Y	Apr. 18, 1865	47, 301
Fiber, Separating vegetable	W. E. Woodbridge	Little Falls, N. Y	Sept. 15, 1863	39, 981
Fiber, Separating vegetable	W. E. Woodbridge	Little Falls, N. Y	June 24, 1873	140, 333
Fiber to pulp, Process of reducing wood	T. B. Armitage	New York, N. Y	June 10, 1873	139, 646
Fiber winding and twisting machine	H. Parmelee	Philadelphia, Pa	July 10, 1866	56, 258
Fibrous and porous materials, Impregnating	S. Gwynn	New York, N. Y	Feb. 21, 1865	46, 466
Fibrous and textile substances, Apparatus for washing and bleaching.	J. A. Jillson and H. Whinfield	Poughkeepsie and New York, N. Y.	Oct. 9, 1855	13, 650
Fibrous and textile substances in a vacuum for cleansing purposes, Method of treating.	J. A. Jillson and H. Whinfield	Poughkeepsie and New York, N. Y.	Sept. 15, 1857	18, 204
Fibrous-composition tube	R. W. Russell	New York, N. Y	Mar. 30, 1869	88, 518
Fibrous material	P. Higginbotham	Hernando, Fla	Nov. 19, 1867	71, 006
Fibrous materials, Combing	J. Heilmann	France	Nov. 29, 1853	10, 289
Fibrous materials from corn-stalks	J. A. Roth	Philadelphia, Pa	Feb. 16, 1864	41, 642
Fibrous materials, Lubricating-compound for use in the manufacture of.	C. Clark	Dayton, Ky	Jan. 5, 1869	85, 642
Fibrous materials, Machine for breaking, scutching, and separating.	E. Brasier	New Cross, England	May 17, 1870	103, 135
Fibrous materials, Machine for cleaning	C. G. Sargent	Graniteville, Mass	Dec. 3, 1861	33, 852
Fibrous materials, Machine for cleaning and blending.	A. J. Loiseau	Philadelphia, Pa	Aug. 6, 1867	67, 559
Fibrous materials, Machine for combing	C. Whipple	Providence, R. I	Apr. 19, 1859	23, 732
Fibrous materials, Machine for combing	M. D. Whipple	Charlestown, Mass	Mar. 17, 1857	16, 865
Fibrous materials, Machine for disintegrating and pulping.	J. C. Beach and J. Abbey	Bloomfield and Orange, N. J.	Apr. 16, 1867	63, 832
Fibrous materials, Machine for picking	O. Woodsworth, jr., and J. D. Page.	East Hartford, Conn	Mar. 9, 1858	19, 600
Fibrous materials, Machine for picking and opening.	S. Baxendale	Boston, Mass	Oct. 9, 1866	58, 711
Fibrous materials, Machine for reducing	R. Daniels	Woodstock, Vt	Jan. 22, 1861	31, 154
Fibrous materials, Machine for sizing	W. Fuzzard	Chelsea, Mass	Dec. 4, 1866	60, 167
Fibrous materials, Machine for surface-sizing	W. Fuzzard	Chelsea, Mass	Jan. 12, 1864	41, 214
Fibrous materials, Machine for surface-sizing	W. Fuzzard	Malden, Mass	Aug. 9, 1864	43, 765
Fibrous materials, Machinery for picking	R. Kitson	Lowell, Mass	Sept. 18, 1855	13, 578
Fibrous materials, Machinery for washing and drying.	C. G. Sargent	Graniteville, Mass	May 24, 1870	103, 506
Fibrous materials, Manufacturing sheets of	W. Fuzzard	Charlestown, Mass	Sept. 18, 1860	30, 058
Fibrous materials, Method of cleansing	J. Howarth	Salem, Mass	Mar. 17, 1857	16, 838
Fibrous materials, Method of reducing long staple	S. M. Allen	Niagara Falls, N. Y	Apr. 17, 1860	27, 878
Fibrous materials, Picker for	L. Ferguson	Lowell, Mass	Aug. 14, 1866	57, 109
Fibrous materials, Picking-cylinder of machine for disintegrating.	S. Boydon	Newark, N. J	Feb. 7, 1865	46, 294
Fibrous materials, Preparation of	J. Woodruff and F. Boyd	Quincy, Ill	May 17, 1870	103, 115
Fibrous materials, Surfacing	W. Fuzzard	Charlestown, Mass	June 19, 1860	28, 745
Fibrous materials to textile stock, Apparatus for reducing.	L. Dean	Fort Edward, N. Y	Aug. 9, 1870	106, 133
Fibrous materials to textile stock, Boiler for reducing.	L. Dean	Fort Edward, N. Y	Aug. 9, 1870	106, 135
Fibrous plants from pulpy matter, Machine for separating.	J. R. Beckwith	New Orleans, La	Sept. 12, 1865	51, 413
Fibrous plants, Machine for disintegrating	J. Evans	Newark, N. J	Sept. 26, 1865	50, 108
Fibrous plants, Method of utilizing waste extracts of.	G. E. Sellers	Seller's Landing, Ill	Jan. 22, 1867	61, 364
Fibrous plants, Treatment of	S. M. Allen	Niagara Falls, N. Y	Mar. 20, 1860	27, 507
Fibrous stock, Machine for working waste	A. W. Johnson	Worcester, Mass	Feb. 2, 1869	86, 409
Fibrous substances, Apparatus for washing	W. Adamson	Philadelphia, Pa	Mar. 5, 1867	62, 517
Fibrous substances, Drawing	J. B. Fuller	Norwich, Conn	Aug. 25, 1868	81, 489
Fibrous substances, Machine for cutting and working.	A. W. Hale	New York, N. Y	Dec. 17, 1867	72, 393
Fibrous substances, Machine for separating the pulp from.	G. Sanford	Bergen Point, N. J	Nov. 17, 1868	84, 070
Fibrous substances, Machinery for twisting	G. W. Pittman	Bushwick, N. Y	Jan. 25, 1859	22, 745
Fibrous substances, Mode of reducing vegetable	J. R. Haskell	New York, N. Y	Mar. 19, 1867	63, 043
Fid	H. H. Pember	New York, N. Y	Mar. 12, 1867	62, 880
Fid	S. H. Sagett	Eden, Me	Mar. 10, 1863	37, 879
Field-battery, Magazine	J. O. Whitcomb	New York, N. Y	Apr. 28, 1863	38, 350
Field-gates, Operating	A. J. Curtis	Frankfort, Me	Oct. 18, 1859	25, 812
Field-press	E. J. Marsters	Shaw's Flat, Cal	Nov. 16, 1869	96, 822
Field-roller	W. W. Andrew	Grand Rapids, Mich	Nov. 10, 1868	83, 818
Field-roller	A. W. Brinkerhoff and A. J. Failor.	Upper Sandusky, Ohio	Sept. 2, 1862	36, 336
Field-roller	A. L. Chubb	Grand Rapids, Mich	June 30, 1868	79, 456
Field-roller	O. B. Colcord	Greenville, Ill	May 14, 1867	64, 748
Field-roller	G. H. Dow	Freeport, Ill	Nov. 14, 1871	120, 950
Field-roller	C. Dunham	Bedford, Ohio	Oct. 20, 1863	40, 380
Field-roller	R. Glover	Tonawanda, N. Y	Dec. 14, 1869	97, 905
Field-roller	A. S. Keagy	Harristown, Ill	Apr. 5, 1870	101, 626
Field-roller	J. M. Kelly	Ellison, Ill	Aug. 29, 1871	118, 616
Field-roller	G. Lindley	Chicago, Ill	May 31, 1859	24, 219
Field-roller	R. A. McConaughy	Ripley, Ohio	Sept. 6, 1870	107, 074
Field-roller	E. F. Olds	South Lyon, Mich	May 7, 1867	64, 559
Field-roller	J. C. Pease	Sycamore, Ohio	Jan. 8, 1861	31, 087
Field-roller	D. R. Prindle	East Bethany, N. Y	May 5, 1868	77, 523
Field-roller	A. Rogers	Freeport, Ill	Nov. 28, 1871	121, 424
Field-roller	W. M. Sunderland	Highgate, Vt	Dec. 15, 1863	40, 961
Field-roller	W. Westfall	Chelsea, Mich	Nov. 17, 1868	84, 150
Field-roller and furrowing-machine	M. M. Robbins	Centreville, Ind	Oct. 4, 1870	107, 056
Field-roller for cutting stalks and weeds	J. H. Gest	Batavia, Ohio	Dec. 14, 1852	9, 463
Fife	J. Pfaff	Philadelphia, Pa	Nov. 29, 1864	45, 270
Fife	J. W. Tanner	New York, N. Y	July 5, 1864	43, 438
Fife and flute	A. H. Stratton	New York, N. Y	May 2, 1865	47, 582
Fifth-wheel	A. Finley	Bainbridge, Ind	Dec. 5, 1871	121, 603
Fifth-wheel	J. Irving	New York, N. Y	Dec. 18, 1866	60, 634

Index of patents issued from the United States Patent Office from 1790 *to* 1873, *inclusive*—Continued.

Invention.	Inventor.	Residence.	Date.	No.
Fifth-wheel	J. Le Roy	Marathon, N. Y	Nov. 29, 1870	109, 636
Fifth-wheel	W. Munson	Abington, Pa	Mar. 23, 1869	88, 196
Fifth-wheel	J. Skeen	Mound City, Ill	May 23, 1871	115, 245
Fifth-wheel	R. D. Wilson	Pittston, Pa	Jan. 2, 1872	122, 425
Fifth-wheel and attachment	J. Skeen	Mound City, Ill	May 23, 1871	115, 246
Fifth-wheel and coupling for vehicles	E. W. Silsby	Ottumwa, Iowa	Sept. 26, 1871	119, 417
Fifth-wheel bender	G. W. Heckart	Columbiana, Ohio	Sept. 8, 1868	81, 902
Fifth-wheel-bending machine	G. W. Heckart	New Lisbon, Ohio	Oct. 5, 1869	95, 585
Fifth-wheel-bending machine	W. J. Jordan	New Lisbon, Ohio	Mar. 15, 1870	100, 903
Fifth-wheel die, Carriage	D. Wilcox	Birmingham, Conn	Oct. 29, 1872	132, 705
Fifth-wheel for carriages	E. H. Adams	Detroit, Mich	June 30, 1868	79, 429
Fifth-wheel for carriages	W. A. Collins	Bloomfield, N. J	July 5, 1870	104, 932
Fifth-wheel for carriages	D. D. Decker	Saugerties, N. Y	Nov. 16, 1869	96, 898
Fifth-wheel for carriages	J. Deeble	Plantsville, Conn	July 23, 1867	66, 956
Fifth-wheel for carriages	P. S. Eastman	Washington Mills, N. Y	May 9, 1871	114, 539
Fifth-wheel for carriages	R. Hoadly	Toulon, Ill	Dec. 14, 1869	97, 919
Fifth-wheel for carriages	D. A. Johnson	Boston, Mass	Apr. 9, 1872	125, 397
Fifth-wheel for carriages	J. Lawrence	Palmyra, N. Y	Dec. 10, 1867	72, 051
Fifth-wheel for carriages	H. A. Luttgens	Paterson, N. J	Jan. 14, 1873	134, 907
Fifth-wheel for carriages	J. A. Peck	Taunton, Mass	Nov. 23, 1869	97, 111
Fifth-wheel for carriages	H. Poth	Pittsburgh, Pa	Mar. 2, 1869	87, 429
Fifth-wheel for carriages	H. W. Ransom	Lawrenceburgh, Ind	Feb. 11, 1868	74, 422
Fifth-wheel for carriages	H. Sayler	Saint Paris, Ohio	Aug. 23, 1870	106, 731
Fifth-wheel for carriages	R. M. Stivers and G. W. V. Smith	New York, N. Y	Aug. 26, 1862	36, 310
Fifth-wheel for carriages	C. St. James	Pittsfield, Mass	Nov. 16, 1869	96, 846
Fifth-wheel for carriages	D. Weaver	Dayton, Ohio	Aug. 3, 1869	93, 376
Fifth-wheel for carriages	D. Wilcox	Birmingham, Conn	Aug. 3, 1869	93, 261
Fifth-wheel for carriages	E. Yeiser	Sheridan, Pa	Aug. 25, 1868	81, 452
Fifth-wheel for vehicles	G. A. Brice	Townville, Pa	July 8, 1873	140, 676
Fifth-wheel for vehicles	E. G. Cameron	Tiffin, Ohio	Mar. 23, 1869	88, 007
Fifth-wheel for vehicles	P. H. Cummins	Adams, N. Y	Aug. 22, 1871	118, 347
Fifth-wheel for vehicles	J. Du Bois	Kingston, N. Y	May 14, 1872	126, 790
Fifth-wheel for vehicles	H. T. Goodale	Clinton, Mass	Mar. 9, 1858	19, 558
Fifth-wheel for vehicles	J. R. McGuire	Ansonia, Conn	Dec. 30, 1873	145, 959
Fifth-wheel for vehicles	S. Moreland	Covington, Ky	May 21, 1872	127, 093
Fifth-wheel for vehicles	U. Reynolds	New York, N. Y	May 22, 1866	54, 958
Fifth-wheel for vehicles	J. P. Smith	Hillsborough, Ohio	May 23, 1871	115, 122
Fifth-wheel for wagons	J. B. Stuart	Bunker Hill, Ill	Apr. 30, 1867	64, 380
Fifth-wheel for wagons, Machine for bending	C. Kieser	Baltimore, Md	July 30, 1861	32, 947
Fifth-wheel head for carriages	R. R. Miller	Plantsville, Conn	Feb. 22, 1870	100, 177
Fifth-wheel-head-forging die	F. Van Patten	Auburn, N. Y	Nov. 29, 1870	109, 781
Fifth-wheel heads, Die for manufacture of	F. Van Patten	Auburn, N. Y	Aug. 12, 1873	141, 682
Fifth-wheel-milling machine, Vehicle	F. Van Patten	Auburn, N. Y	Aug. 12, 1873	141, 683
Fifth-wheel of fire-engines and other vehicles	R. Poole	Baltimore, Md	Aug. 16, 1859	25, 164
Fifth-wheel of whiffletree attachment for carriages	F. Van Patten	Auburn, N. Y	Feb. 5, 1867	61, 776
Fifth-wheels, Bending	W. Morgey	Wilmington, Del	July 7, 1868	79, 677
Fifth-wheels for carriages, Bearing for	S. B. Smith	Salem, N. J	Apr. 14, 1868	76, 667
Fifth-wheels for carriages, &c., Manufacture of	D. Wilcox	Derby, Conn	Mar. 25, 1873	137, 274
Fifth-wheels, Keep and brace for	F. B. Morse	Plantsville, Conn	Mar. 21, 1871	112, 950
Fifth-wheels, Machine for forming	G. Feightner	Wooster, Ohio	Apr. 21, 1868	76, 899
Figured fabrics, Manufacturing	J. Smith	Shaefferstown, Pa	Aug. 4, 1834	....
File	P. Cinquini	West Meriden, Conn	Apr. 3, 1860	27, 691
File	J. H. Clark	Washington, D. C	July 5, 1870	105, 044
File	G. B. Cubberley	Milwaukee, Wis	June 2, 1868	78, 435
File	H. Disston	Philadelphia, Pa	Sept. 2, 1873	142, 445
File	H. A. Harvey	New York, N. Y	Oct. 18, 1864	44, 724
File	J. U. Houston	West Meriden, Conn	Jan. 19, 1858	19, 143
File	A. Hyland	Hingham, Mass	May 16, 1871	114, 824
File	J. B. Johnson	San Francisco, Cal	Apr. 29, 1873	138, 256
File	C. M. Nes	York, Pa	Dec. 10, 1872	133, 793
File	W. T. Nicholson	Providence, R. I	Sept. 11, 1866	58, 025
File	H. B. Nickerson	Boston, Mass	Mar. 17, 1868	75, 568
File	H. B. Nickerson	Boston, Mass	May 16, 1871	114, 962
File	I. N. Patten	Memphis, Tenn	Oct. 17, 1871	119, 947
File	G. M. Ramsay	New York, N. Y	Aug. 12, 1856	15, 525
File	A. Thompson	Norway, Me	Nov. 16, 1869	97, 002
File	A. Weed	Boston, Mass	Sept. 3, 1867	68, 584
File	M. D. Whipple	Cambridge, Mass	Apr. 1, 1862	34, 866
File	H. Young	Carey, Ohio	Sept. 17, 1867	68, 929
File and binder, Letter	E. Buell and J. W. Lilley	Columbus, Ohio	July 8, 1873	140, 610
File and binder, Paper	L. P. Keech	Brooklyn, N. Y	Sept. 2, 1873	142, 541
File and rasp	F. C. Curie	Lancaster, Pa	July 28, 1868	80, 460
File and rasp	H. Powers	Florence, Italy	Oct. 4, 1853	10, 088
File, Bill	C. W. Bond	Biddeford, Me	Nov. 10, 1868	83, 908
File, Bill and letter	A. J. Marshall	North Buffalo, N. Y	Jan. 7, 1873	134, 554
File, Bill and paper	W. H. Foye	Portland, Me	Dec. 16, 1873	145, 497
File, Bill and paper	T. Orton	Chicago, Ill	Dec. 2, 1873	145, 230
File, Bill and paper	G. T. Wolcott	Detroit, Mich	June 24, 1873	140, 177
File-blanks, Apparatus for grinding	S. O. Morse	Medford, Mass	May 5, 1863	38, 403
File-blanks, flyers, &c., Machine for rolling, shaping, and forging.	J. Dodge	Waterford, N. Y	May 1, 1866	54, 310
File-blanks, Grinding	A. B. Southwick	Ballard Vale, Mass	Feb. 3, 1863	37, 603
File-blanks, Machine for cutting teeth of	H. B. Nickerson	Boston, Mass	Nov. 5, 1867	70, 458
File-blanks, Machine for forging	W. T. Nicholson	Providence, R. I	Jan. 10, 1865	45, 850
File-blanks, Machine for grinding	J. S. Brown	Pawtucket, R. I	June 7, 1864	43, 005
File-blanks, Machine for grinding	E. Norton	Boston, Mass	May 5, 1868	77, 646
File-blanks, Machine for grinding	R. G. Pine	Newark, N. J	Feb. 3, 1857	16, 552
File-blanks, Machine for rolling	P. D. Cummings	Portland, Me	Nov. 13, 1866	59, 561
File-blanks, Machine for rolling	J. B. Mignault	Chelsea, Mass	Mar. 6, 1866	53, 025
File-blanks, Machine for rolling	C. Spofford and A. B. Southwick.	Ballard Vale, Mass	May 5, 1863	38, 454
File-blanks, Machine for rolling	M. D. Whipple	Cambridge, Mass	May 31, 1864	42, 981
File-blanks, Machine for stripping	A. Marshall and A. B. Southwick.	Lawrence and Ballard Vale, Mass.	Sept. 1, 1863	39, 775
File-blanks, Machinery for grinding the edge of	H. E. Grandy and S. O. Morse	Ballard Vale and Medford, Mass.	Sept. 15, 1863	39, 992
File-blanks, Rolling	J. N. Aspinwall	Newark, N. J	May 13, 1856	14, 897
File-box, Paper	J. H. Bentley	Cambridge, Mass	June 10, 1873	139, 653

Index of patents issued from the United States Patent Office from 1790 *to* 1873, *inclusive*—Continued.

Invention.	Inventor.	Residence.	Date.	No.
File-clasp for document, &c	I. S. Richardson	Boston, Mass	Jan. 30, 1866	52, 323
File-cutter	G. F. Card and C. A. Studley	Bridgeport, Conn	Nov. 26, 1867	71, 362
File-cutter	P. J. Carlsson	Andover, Mass	Jan. 17, 1871	110, 956
File cutting and making machine	C. Jackson	Butternuts, N. Y	May 28, 1813	....
File-cutting apparatus	S. A. Sutton	Pawtucket, R. I	July 30, 1867	67, 374
File-cutting engine	N. Starr	Middletown, Conn	Oct. 17, 1848	5, 869
File-cutting machine	L. Anderson	Kensington, Pa	Nov. 16, 1841	2, 365
File-cutting machine	J. K. Barker	Lawrence, Mass	Aug. 8, 1865	49, 214
File-cutting machine	M. B. Belknap	Brimfield, Mass	Jan. 16, 1812	....
File-cutting machine	M. B. Belknap	Worcester, Mass	June 13, 1812	....
File-cutting machine	E. Bernot	Paris, France	July 24, 1860	29, 236
File-cutting machine	E. Bucklin, jr	Pawtucket, R. I	Feb. 27, 1866	52, 926
File-cutting machine	E. Bucklin, jr	North Providence, R. I	Sept. 17, 1867	68, 946
File-cutting machine	T. Burr	Hastings, Mich	Mar. 20, 1860	27, 591
File-cutting machine	T. Burr	Battle Creek, Mich	Nov. 3, 1863	40, 534
File-cutting machine	A. Chambers	North Providence, R. I	Mar 13, 1866	53, 113
File-cutting machine	I. H. Coller	Poughkeepsie, N. Y	Feb. 24, 1857	16, 681
File-cutting machine	H. B. Corner	Pittsburgh, Pa	Sept. 4, 1866	57, 680
File-cutting machine	J. C. Cooke	Middletown, Conn	July 10, 1860	29, 130
File-cutting machine	J. C. Cooke	Middletown, Conn	Feb. 20, 1866	52, 804
File-cutting machine	M. G. Crane	Boston, Mass	Mar. 14, 1865	46, 780
File-cutting machine	J. E. Crisp	Charlestown, Mass	June 11, 1872	127, 854
File-cutting machine	J. D. Crocker	Norwich, Conn	Feb. 28, 1865	46, 545
File-cutting machine	G. Crosby	Baltimore, Md	Dec. 4, 1849	6, 922
File-cutting machine	J. Dodge	Manchester, England	Mar. 7, 1871	112, 430
File-cutting machine	A. B. Ely	Newton, Mass	Jan. 22, 1867	61, 410
File cutting machine	M. H. Fisher	Bridgeport, Conn	Jan. 13, 1863	37, 419
File-cutting machine	G. W. Fogg	South Dedham, Mass	Dec. 14, 1858	22, 329
File-cutting machine	I. Goodspeed	Norwich, Conn	Jan. 1, 1867	60, 878
File-cutting machine	W. Halliwell and L. Osborn	Poughkeepsie, N. Y	Aug. 4, 1857	17, 928
File-cutting machine	C. Hesser and A. Paxson	Philadelphia, Pa	Apr. 11, 1812	....
File-cutting machine	S. Hoke	Union City, Ind	Jan. 13, 1863	37, 398
File-cutting machine	H. Hotchkiss	Waterbury, Conn	July 24, 1855	13, 310
File-cutting machine	J. N. Jacobs	Worcester, Mass	Apr. 6, 1858	19, 854
File-cutting machine	W. F. James	Greenwich, N. Y	Nov. 19, 1812	....
File-cutting machine	J. Jervis	Baltimore, Md	Oct. 11, 1864	44, 633
File-cutting machine	J. Jervis	Baltimore, Md	Oct. 24, 1865	50, 595
File-cutting machine	J. Jervis	Baltimore, Md	Oct. 12, 1869	95, 804
File-cutting machine	A. F. Johnson and M. P. Griffin	Medford, Mass	Jan. 22, 1867	61, 341
File-cutting machine	W. McGregor	Manchester, England	Oct. 1, 1872	131, 767
File-cutting machine	J. B. Mignault	Boston, Mass	Jan. 28, 1868	73, 824
File-cutting machine	C. Miller and T. W. Decker	New York, N. Y	Apr. 12, 1859	23, 645
File-cutting machine	G. Miller	New York, N. Y	Apr. 3, 1866	53, 650
File-cutting machine	W. T. Nicholson	Providence, R. I	Apr. 5, 1864	42, 216
File-cutting machine	W. T. Nicholson	Providence, R. I	Apr. 5, 1864	42, 217
File-cutting machine	W. T. Nicholson	Providence, R. I	June 9, 1868	78, 681
File-cutting machine	J. L. Norton	Alum Bank, Pa	May 13, 1856	14, 881
File-cutting machine	E. O. Potter	Pawtucket, R. I	Nov. 8, 1864	44, 998
File-cutting machine	E. O. Potter	Pawtucket, R. I	Apr. 30, 1867	64, 396
File-cutting machine	A. N. Redman	Charlestown, Mass	Apr. 23, 1867	64, 033
File-cutting machine	J. Rieppel	Renovo, Pa	July 30, 1872	130, 076
File-cutting machine	J. Rotherham and J. Holden	Middletown, N. Y	May 8, 1866	54, 606
File-cutting machine	T. Schultz and C. Renne	Hoboken, N. J., and New York, N. Y.	June 1, 1869	90, 788
File-cutting machine	A. B. Southwick	Ballard Vale, Mass	Jan. 27, 1863	37, 552
File-cutting machine	S. A. Sutton	Dixon, Ill	Aug. 4, 1868	80, 679
File-cutting machine	J. H. Thompson	Paterson, N. J	Jan. 27, 1852	8, 697
File-cutting machine	T. E. Thurston	Newark, N. J	July 14, 1868	79, 878
File-cutting machine	W. Van Anden	Poughkeepsie, N. Y	July 7, 1857	17, 760
File-cutting machine	W. Van Anden	Poughkeepsie, N. Y	Aug. 14, 1860	29, 641
File-cutting machine	S. Vanstone	Providence, R. I	Apr. 29, 1862	35, 130
File-cutting machine	C. Vogel	New York, N. Y	Jan. 22, 1867	61, 486
File-cutting machine	C. Vogel	New York, N. Y	Jan. 21, 1873	135, 179
File-cutting machine	S. Walton	Ballard Vale, Mass	Sept. 10, 1867	68, 815
File-cutting machine	A. Weed	Boston, Mass	Mar. 14, 1865	46, 865
File-cutting machine	A. Weed	Boston, Mass	Mar. 13, 1866	53, 204
File-cutting machine	A. Weed	Boston, Mass	Feb. 20, 1872	123, 849
File-cutting machine	A. Weed	Boston, Mass	July 29, 1873	141, 300
File-cutting machine	M. D. Whipple	Charlestown, Mass	Feb. 1, 1859	22, 842
File-cutting machine	M. D. Whipple	Cambridge, Mass	Apr. 1, 1862	34, 865
File-cutting machine	M. D. Whipple	Cambridgeport, Mass	Jan. 27, 1863	37, 554
File-cutting machine	M. D. Whipple	Boston, Mass	Oct. 24, 1865	50, 647
File-cutting machine	S. Whipple	Albany, N. Y	Feb. 12, 1845	3, 914
File-cutting machines, Feeding-mechanism for	C. E. Moore	Boston, Mass	May 16, 1871	114, 957
File-cutting machinery	J. C. Blair	Pittsburgh, Pa	Apr. 27, 1852	8, 901
File-cutting machinery	J. W. Conklin, H. L. Sidman, and E. Whritner.	Ramapo, N. Y	Aug. 17, 1852	9, 194
File-cutting machinery	J. Crum	Ramapo, N. Y	July 1, 1851	8, 199
File-cutting machinery	M. H. Fisher	Bridgewater, Mass	Aug. 28, 1846	4, 728
File-cutting machinery	J. Hatch	Dedham, Mass	May 1, 1815	....
File-cutting machinery	J. Hatch	Roxbury, Mass	Oct. 11, 1828	....
File-cutting machinery	R. Walker	Portsmouth, N. H	June 12, 1847	5, 149
File-fastener	E. P. McCeney	Washington, D. C	Aug. 11, 1868	80, 986
File for dressing metal	W. T. Nicholson	Providence, R. I	Sept. 27, 1864	44, 443
File for grooving rolls	H. L. Butts	Norwich, Conn	Apr. 14, 1868	76, 599
File-guiding machine	I. H. Spencer	Pawtucket, R. I	July 25, 1865	49, 042
File-griping handle	A. Weed	Boston, Mass	Feb. 6, 1872	123, 433
File-handle	H. K. Austin	Charlestown, Mass	Feb. 11, 1873	135, 684
File-handle	B. Boardman	Norwich, Conn	Dec. 1, 1868	84, 529
File-handle	W. W. Draper	Greenfield, Mass	Jan. 18, 1859	22, 635
File-handle	C. F. Hunter	Adrian, Mich	Aug. 30, 1864	43, 994
File-handle	D. Pfouts	Empire Mills, near Winesburgh, Ohio.	Aug. 1, 1871	117, 677
File-handle	H. C. Storrs	New York, N. Y	Apr. 28, 1868	77, 332
File-handle	A. Weed	Boston, Mass	Sept. 3, 1867	68, 585
File handle, Surface	J. Weare	Fitchburgh, Mass	July 7, 1868	79, 709
File-holder	A. P. Brown	Worcester, Mass	Apr. 20, 1869	89, 019
File-holder	W. C. McGill	Cincinnati, Ohio	Dec. 12, 1865	51, 470

Index of patents issued from the United States Patent Office from 1790 *to* 1873, *inclusive*—**Continued.**

Invention.	Inventor.	Residence.	Date.	No.
File holder, Newspaper	W. Bright	Saint Louis, Mo	Jan. 21, 1873	134, 974
File-hook	G. W. Schramm	Brooklyn, N. Y	July 5, 1864	43, 432
File, Horse-tooth	J. P. Howell	Washingtonville, N. Y	May 5, 1863	38, 391
File, Letter	F. Ashley	New York, N. Y	Aug. 21, 1866	57, 274
File, Letter	F. Ashley	New York, N. Y	Oct. 1, 1867	69, 385
File, Letter	J. C. Hadden	New York, N. Y	Apr. 10, 1866	53, 817
File, Letter	J. W. Hauxhurst	New York, N. Y	Oct. 24, 1865	50, 580
File, Letter	J. H. Shipman	Yorkville, N. Y	Aug. 30, 1859	25, 282
File, Letter	A. E. Taylor	New Britain, Conn	Aug. 25, 1868	81, 559
File, Letter and music	R. B. Irwin	New York, N. Y	Dec. 6, 1864	45, 327
File, Letter and paper	W. C. Choate	Washington, D C	Mar. 13, 1866	53, 114
File, Letter-box	T.K.Sterrett and W. R. Farrell	Philadelphia, Pa	Oct. 30, 1866	59, 287
File, Letter or invoice	R. Boeklen	Brooklyn, N. Y	Oct. 24, 1865	50, 652
File-machine	E. B. Rollins	Poland, N. Y	May 17, 1870	103, 240
File-machine	G. S. Tiffany and H. Ingraham	Palmyra and Tecumseh, Mich.	Feb. 26, 1861	31, 581
File, Newspaper	J. H. Atwater	Providence, R. I	Nov. 10, 1863	40, 545
File, Newspaper	W. Burnet	Providence, R. I	Jan. 31, 1865	46, 072
File, Newspaper	G. Coope	New York, N. Y	Aug. 6, 1872	130, 197
File, Newspaper	J. W. Foard	San Francisco, Cal	May 1, 1866	54, 320
File, Newspaper	J. Frick	Philadelphia, Pa	Aug. 29, 1865	49, 620
File, Newspaper	L. P. McCarty	San Francisco, Cal	Feb. 18, 1868	74, 562
File, Newspaper	W. R. McNorton	Livingston, Ala	Aug. 20, 1872	130, 649
File, Newspaper	L. C. Prindle	Chicago, Ill	Aug. 10, 1869	93, 475
File, Newspaper	J. Robson	Chicago, Ill	Sept. 28, 1869	95, 380
File, Newspaper	R. K. Ville	New York, N. Y	July 2, 1872	128, 681
File, Newspaper	H. S. White	Newport, R. I	Jan. 22, 1861	31, 200
File, Newspaper	H. S. White	Newport, R. I	Apr. 2, 1861	31, 919
File or bill-holder	T. W. Brown	Boston, Mass	Apr. 25, 1854	10, 820
File or binder, Paper	B. J. Beck	Brooklyn, N. Y	July 6, 1869	92, 251
File or holder, Document	H. E. Woodbury	Washington, D. C	Aug. 8, 1854	11, 504
File or ready binder for filing pamphlets, letters, papers, &c.	I. Detterer	Philadelphia, Pa	May 22, 1841	2, 105
File, Pamphlet	R. M. Abercrombie	Rahway, N. J	Dec. 27, 1864	45, 570
File, Paper	J. Adair	Pittsburgh, Pa	Dec. 31, 1867	72, 711
File, Paper	S. Angell	Providence, R. I	Apr. 22, 1834	....
File, Paper	H. J. Asthalter	Pittsburgh, Pa	Oct. 24, 1871	120, 227
File, &c., Paper	A. Baker	Appleton, Wis	Apr. 18, 1871	113, 721
File, Paper	G. W. Billow	Shelby, Ohio	Jan. 14, 1873	134, 724
File, Paper	R. Boeklen	Brooklyn, N. Y	May 14, 1867	64, 624
File, Paper	J. W. Boughton	New York, N. Y	June 30, 1868	79, 441
File, Paper	W. Boyrer	New York, N. Y	June 21, 1870	104, 415
File, Paper	S. Brock	Brooklyn, N. Y	Jan. 21, 1868	73, 572
File, Paper	F. Bühle	Newark, N. J	Oct. 8, 1867	69, 539
File, Paper	H. Burgess	San Francisco, Cal	Oct. 12, 1869	95, 766
File, Paper	W. Burnet	New York, N. Y	Jan. 10, 1865	45, 813
File, Paper	J. and J. Cash, jr	Coventry, England	Aug. 30, 1870	102, 778
File, Paper	C. Chapman	Chicago, Ill	Oct. 17, 1871	119, 917
File, Paper	W. Z. W. Chapman	New York, N. Y	Mar. 30, 1858	19, 748
File, Paper	E. Clark	New York, N. Y	May 8, 1860	28, 234
File, Paper	W. R. Clough	Cambridge, Mass	Aug. 18, 1868	81, 253
File, Paper	L. Cohn	Montreal, Canada	Dec. 19, 1871	121, 993
File, Paper	W. A. Collord	Cincinnati, Ohio	Sept. 17, 1850	7, 648
File, Paper	L. B. Covert	Chicago, Ill	Apr. 23, 1867	64, 679
File, Paper	M. Craft	Bellaire, Ohio	Nov. 11, 1873	144, 512
File, Paper	E. H. Craige	Brooklyn, N. Y	Nov. 17, 1868	84, 172
File, Paper	G. Crandell	Washington, D. C	Apr. 16, 1867	63, 785
File, Paper	E. J. Crane	La Porte, Ind	May 10, 1870	102, 918
File, Paper	A. Day	Detroit, Mich	June 14, 1870	104, 122
File, Paper	P. W. Derham	Brooklyn, N. Y	July 20, 1869	92, 925
File, Paper	P. W. Derham	New York, N. Y	Aug. 26, 1873	142, 213
File, Paper	E. D. Dodd	Cincinnati, Ohio	Jan. 15, 1850	7, 018
File, Paper	G. W. Emerson	Chicago, Ill	Apr. 26, 1870	102, 237
File, Paper	G. W. Emerson	Chicago, Ill	Apr. 25, 1871	113, 991
File, Paper	T. C. Fahnestock	Cincinnati, Ohio	June 7, 1870	104, 002
File, Paper	W. Fallon	Washington, D. C	Nov. 12, 1867	70, 708
File, Paper	J. D. Field	Wataga, Ill	Dec. 22, 1868	85, 078
File, Paper	J. Fleischl	New York, N. Y	Jan. 1, 1867	60, 710
File, Paper	J. G. Floyd, jr	New York, N. Y	Jan. 31, 1871	111, 335
File, Paper	J. W. Foard	San Francisco, Cal	Jan. 2, 1872	122, 314
File, Paper	J. O. Foster	Pecatonica, Ill.	Feb. 23, 1869	87, 161
File, Paper	J. M. D. France	Washington, D. C	Dec. 8, 1868	84, 738
File, Paper	E. K. Godfrey	New York, N. Y	Dec. 21, 1858	22, 363
File, Paper	S. E. Harrison	Jersey City, N. J	Apr. 23, 1872	126, 053
File, Paper	R. Henning	Ottawa, Ill	Apr. 22, 1873	138, 086
File, Paper	B. F. Herr	Livingston, Ala	Nov. 30, 1869	97, 397
File, Paper	A. Hockett	Wilmington, Ohio	Mar. 22, 1870	101, 130
File, Paper	L. P. Keech	Washington, D. C	Mar. 26, 1872	124, 961
File, Paper	L. P. Keech	New York, N. Y	Nov. 5, 1872	132, 836
File, Paper	J. W. Kinsley	Boston, Mass	Oct. 8, 1872	132, 085
File, Paper	G. Lautenschlager	New York, N. Y	July 4, 1865	48, 568
File, Paper	A. Liebenroth	New York, N. Y	May 22, 1860	28, 438
File, Paper	C. D. Lindsey	Cincinnati, Ohio	Apr. 15, 1873	137, 851
File, Paper	C. Mason	New York, N. Y	June 3, 1873	139, 592
File, Paper	J. Matthias	New York, N. Y	Oct. 22, 1872	132, 478
File, Paper	T. E. Moore	Columbus, Ohio	July 5, 1870	105, 114
File, Paper	E. Motz	Woodward, Pa	Jan. 21, 1873	135, 144
File, Paper	B. Ney and H. Hofheimer	Alexandria, Va.	Oct. 27, 1868	83, 531
File, Paper	C. M. O'Hara	New York, N. Y	Jan. 30, 1872	123, 288
File, Paper	L. H. Olmsted	Brooklyn, N. Y	Dec. 1, 1868	84, 505
File, Paper	J. J. Parker	Marietta, Ohio	Nov. 29, 1859	26, 285
File-Paper	S. E. Pettee and C. A. Wolle	Bethlehem, Pa	Oct. 25, 1870	108, 623
File, Baper	J. P. Quarles	Richmond, Va.	Apr. 19, 1870	102, 153
File, Paper	R. Rathbone	New York, N. Y	July 22, 1873	141, 170
File, Paper	A. S. Richards	Montgomery County, Md	June 15, 1869	91, 268
File, Paper	G. Rowe	London, England	Aug. 1, 1871	117, 688
File, Paper	W. W. Russell	Malden, Mass	Nov. 15, 1870	109, 253
File, Paper	F. C. Senseman	Philadelphia, Pa	Aug. 2, 1870	105, 987

Index of patents issued from the United States Patent Office from 1790 *to* 1873, *inclusive*—Continued.

Invention.	Inventor.	Residence.	Date.	No.
File, Paper	J. A. Shannon	Perrysburgh, Ohio	Dec. 5, 1871	121, 671
File, Paper	E. J. Smith and P. H. Cheever	Washington, D. C	Apr. 14, 1868	76, 834
File, Paper	H. L. Smith	Cleveland, Ohio	June 7, 1853	9, 776
File, Paper	A. P. Stephens	Brooklyn, N. Y	Mar. 9, 1869	87, 598
File, Paper	D. A. Stiles	West Meriden, Conn	June 9, 1857	17, 526
File, Paper	M. Sullivan and J. Reedy	New York, N. Y	Mar. 2, 1869	87, 443
File, Paper	C. Tabor	Craftsbury, Vt	Sept. 27, 1870	107, 836
File, Paper	J. F. Tapley	Springfield, Mass	Feb. 14, 1871	111, 886
File, Paper	J. P. Terrell and S. G. Brett	Somerville, Mass	Feb. 9, 1869	86, 712
File, Paper	E. H. Thompson	Washington, D. C	Sept. 5, 1871	118, 757
File, Paper	S. Thompson	San Pedro, Cal	Nov. 3, 1863	40, 518
File, Paper	J. P. Tirrell and H. Whitney	Charlestown and Watertown, Mass.	Apr. 27, 1869	89, 358
File, Paper	C. W. West	Shiloh, N. J	Dec. 14, 1869	97, 999
File, Paper	F. W. Whitney	Brooklyn, N. Y	Nov. 8, 1870	109, 085
File, Paper	D. Winslow and P. D. Cummings.	Westbrook and Portland, Me.	Aug. 30, 1853	9, 979
File, Paper	J. F. Winter	Brooklyn, N. Y	July 15, 1873	140, 863
File, Paper	J. Wolfe	Washington, D. C	Dec. 10, 1867	72, 146
File, Paper	E. W. Woodruff and G. C. Green	Washington, D. C	Mar. 31, 1868	76, 287
File, Paper	E. W. Woodruff and G. C. Green	Washington, D. C	Apr. 14, 1868	76, 872
File, Paper	W. L. Woods	Washington, D. C	Aug. 12, 1862	36, 200
File, Paper	W. L. Woods	Washington, D. C	June 9, 1863	38, 868
File, Paper	W. L. Woods	Washington, D. C	Feb. 14, 1865	46, 415
File, Paper	W. L. Woods	Washington, D. C	Oct. 25, 1870	108, 746
File, Paper	T. P. Yates	Concord, N. H	Oct. 22, 1861	33, 554
File, Paper	S. W. Young	Providence, R. I	May 2, 1871	114, 504
File, Paper and letter	J. B. McEnally	Clearfield, Pa	Feb. 28, 1860	27, 302
File, Paper and letter	H. Riggs	Oxford, Conn	Dec. 12, 1865	51, 520
File, Portable	C. A. Cook	Chicago, Ill	Dec. 17, 1872	133, 923
File, Portfolio	P. W. Toy	New York, N. Y	June 22, 1858	20, 670
File, Portfolio paper	J. N. Jacobs	Worcester, Mass	June 19, 1860	28, 755
File, Prescription	A. D. Foster	Council Bluffs, Iowa	Apr. 29, 1873	138, 390
File, Rat-tail	R. Flint	Meriden, Conn	July 24, 1846	4, 660
File, Saw	W. Roberts	Blue Hill, Me	June 1, 1869	90, 784
File-stripping apparatus	T. Coldwell	Matteawan, N. Y	Sept. 11, 1866	57, 869
File-supporter	J. B. Woodruff and B. M. Townsend.	Washington, D. C., and Quincy, Ill.	Nov. 6, 1849	6, 858
Files and rasps, Composition for recutting	A. Van Camp	Washington, D. C	Mar. 9, 1869	87, 733
Files and rasps, Manufacture of	T. Sheehan	Dunkirk, N. Y	Feb. 5, 1861	31, 331
Files and rasps, Mode of hardening and repairing	C. A. Clavel	Paris, France	July 26, 1864	43, 658
Files, Compound for cleaning, biting, or sharpening damaged.	A. P. and E. Brittingham	New York, N. Y	July 27, 1829	....
Files, Construction of round and half-round	J. N. Jacobs	Worcester, Mass	Aug. 9, 1864	43, 775
Files, Cutting	M. B. Fisher	Sing Sing, N. Y	Feb. 5, 1856	14, 189
Files, Cutting	E. Gehbe	Northampton County, Pa	Oct. 10, 1811	....
Files, Cutting	S. Hawes	Bennington, Vt	May 11, 1814	....
Files, Cutting	C. Miller	New York, N. Y	Nov. 11, 1856	16, 064
Files, Cutting	C. and D. Platt	Sharon, Conn	Mar. 7, 1814	....
Files, Cutting round	M. D. Whipple	Charlestown, Mass	Oct. 7, 1856	15, 867
Files for papers, accounts, &c., Construction of screw-nuts for.	W. Mann	Philadelphia, Pa	July 3, 1840	1, 673
Files, Machine for grinding half-round	T. Smith and J. A. Stafford	Charlestown and Boston, Mass.	Sept. 1, 1863	39, 782
Files, Machine for manufacturing	L. and D. Anderson	Philadelphia, Pa	Apr. 20, 1812	....
Files, Manufacture of	M. D. Whipple	Charlestown, Mass	June 26, 1860	28, 947
Files, Method of making	T. S. Smith	Cincinnati, Ohio	June 21, 1870	104, 507
Files, Method of restoring	T. James	Newport, England	Feb. 13, 1872	123, 705
Files, &c., Method of sharpening	W. B. Gillett	Auburn, N. Y	Feb. 7, 1860	27, 046
Files, Mode of repairing	A. A. Dunk	Manchester, N. H	July 31, 1866	56, 734
Files, &c., Process and apparatus for hardening	S. Darling	Providence, R. I	July 12, 1870	105, 180
Files, Process of hardening	J. Russell	Sing Sing, N. Y	Feb. 6, 1866	52, 450
Files, Process of recutting	A. I. Ferguson	Sharon, Pa	Jan. 21, 1868	73, 590
Files, Process of recutting	X. Robert	Worcester, Mass	June 15, 1869	91, 271
Files, rasps, and sickles, Machine for cutting	S. G. White	Haverhill, Mass	July 11, 1812	....
Files, saw-blades, &c., Machine for tempering	J. Small	Saint Louis, Mo	Nov. 12, 1867	70, 755
Files, saws, &c., Machine for tempering	J. Small	Saint Louis, Mo	Aug. 15, 1865	49, 451
Files, Tip for saw	H. Disston	Philadelphia, Pa	Sept. 16, 1873	142, 900
Filing-case, Numerical	G. W. Bettesworth	Cedar Rapids, Iowa	Nov. 18, 1873	144, 593
Filing-machine	A. G. Burton and H. W. Covert	Rochester, N. Y	Nov 20, 1866	59, 741
Filing-machine	M. Fiedler	Rochester, N. Y	Mar. 10, 1868	75, 402
Filing-machine	T. H. Lindley	Taunton, Mass	June 18, 1867	65, 920
Filing papers, Safety-hook for	L. H. Beckwith	Port Jervis, N. Y	May 13, 1873	138, 782
Filings, Apparatus for separating metallic	J. Jonson	Baltimore, Md	Jan. 24, 1865	46, 005
Filling-frame, Cap for	J. Ripka	Manayunk, Pa	July 13, 1833	....
Filter	C. Andersen	Brooklyn, N. Y	Sept. 23, 1873	143, 113
Filter	E. Andries	Schaerbeek, Belgium	Mar. 28, 1865	47, 074
Filter	B. Arnold	East Greenwich, R. I	July 16, 1861	32, 815
Filter	W. W. Ayres	Worcester, Mass	Aug. 25, 1857	18, 031
Filter	T. Barrows	Brooklyn, N. Y	May 24, 1870	103, 280
Filter	C. F. Baxter	Boston, Mass	Jan. 31, 1860	26, 965
Filter	C. F. Baxter	Boston, Mass	Mar. 11, 1862	34, 620
Filter	J. Bean	Hudson, Mich	Nov. 15, 1870	109, 288
Filter	B. Best	Dayton, Ohio	Jan. 26, 1864	41, 357
Filter	B. Best	Dayton, Ohio	Mar. 27, 1866	53, 3 5
Filter	C. D. Birdseye	New York, N. Y	Mar. 26, 1850	7, 208
Filter	T. Bishop	Dobb's Ferry, N. Y	Dec. 31, 1842	2, 892
Filter	J. Brady	New York, N. Y	May 21, 1872	126, 926
Filter	J. Brown	San Francisco, Cal	Apr. 4, 1871	113, 491
Filter	J. R. Brown and W. A. Foskett	New Haven, Conn	Apr. 8, 1873	137, 653
Filter	A. Chabot	San Francisco, Cal	Dec. 8, 1863	40, 818
Filter	L. S. Chichester	New York, N. Y	May 29, 1860	28, 533
Filter	T. C. Clarke	Camden, N. J	June 12, 1855	13, 026
Filter	T. C. Clarke	Camden, N. J	Oct. 9, 1860	30, 300
Filter	W. T. Class	Cincinnati, Ohio	Sept. 13, 1864	44, 253
Filter	C. Cleminshaw	Troy, N. Y	Jan. 31, 1865	46, 079
Filter	D. N. B. Coffin, jr	Newton, Mass	Sept. 2, 1856	15, 646
Filter	M. Cooke and J. W. Watt	Sacramento, Cal	Oct. 22, 1872	132, 385

Index of patents issued from the United States Patent Office from 1790 *to* 1873, *inclusive*—Continued.

Invention.	Inventor.	Residence.	Date.	No.
Filter	G. Curtis	Springfield, Mass	Nov. 29, 1870	109, 592
Filter	G. Curtis	Springfield, Mass	Nov. 29, 1870	109, 593
Filter	F. J. Delker	Philadelphia, Pa	June 11, 1872	127, 681
Filter	J. Doering	Philadelphia, Pa	Jan. 22, 1867	61, 407
Filter	N. Downes	Syracuse, N. Y	July 10, 1866	56, 193
Filter	E. Duchamp	Saint Martinsville, La	Sept. 13, 1859	25, 398
Filter	J. B. Ellis	Washington, D. C	Aug. 26, 1873	142, 217
Filter	M. A. Espirat and E. Sause	Marseilles, France	Apr. 11, 1865	47, 251
Filter	J. M. Evarts	New Haven, Conn	July 15, 1873	140, 907
Filter	E. S. Farson	Philadelphia, Pa	May 31, 1870	103, 592
Filter	J. Fitch	Seneca Falls, N. Y	Mar. 22, 1859	23, 297
Filter	J. Fitch	Seneca Falls, N. Y	July 10, 1860	29, 069
Filter	H. Flad	Saint Louis, Mo	Mar. 20, 1866	53, 287
Filter	H. Flad	Saint Louis, Mo	Mar. 5, 1867	62, 538
Filter	J. Ford	Newburgh, N. Y	Feb. 23, 1869	87, 250
Filter	A. Fox	Poughkeepsie, N. Y	July 30, 1867	67, 281
Filter	L. W. W. Goodwyn	New Orleans, La	Dec. 24, 1867	72, 481
Filter	L. A. Gouch	Yonkers, N. Y	Oct. 25, 1864	44, 797
Filter	F. P. Griffiths	Philadelphia, Pa	May 22, 1866	54, 894
Filter	J. M. Hackney	Danville, Ky	May 17, 1870	103, 173
Filter	C. M. Halsted	Troy, N. Y	Nov. 17, 1863	40, 620
Filter	T. Hayes	Cambridge, Mass	Dec. 18, 1866	60, 512
Filter	F. Henshaw	Washington, D. C	Oct. 1, 1867	69, 433
Filter	B. Holly	Lockport, N. Y	Apr. 4, 1871	113, 516
Filter	L. Holms	Paterson, N. J	June 2, 1868	78, 524
Filter	P. Huerne	San Francisco, Cal	July 16, 1872	129, 136
Filter	P. Huerne	San Francisco, Cal	Jan. 28, 1873	135, 222
Filter	C. H. Jackson	Saint Louis, Mo	Dec. 29, 1868	85, 449
Filter	A. Jaminet	Saint Louis, Mo	Mar. 22, 1859	23, 302
Filter	L. P. Jenks	Boston, Mass	Feb. 5, 1861	31, 352
Filter	W. H. Jennison	New York, N. Y	Feb. 20, 1846	4, 386
Filter	W. H. Jennison	New York, N. Y	Mar. 7, 1846	4, 399
Filter	J. Kedzie	Rochester, N. Y	Jan. 10, 1854	10, 420
Filter	J. G. Lefler	Philadelphia, Pa	Jan. 5, 1864	41, 075
Filter	J. N. Lighthall	Joliet, Ill	Jan. 7, 1868	73, 105
Filter	W. Linton	Baltimore, Md	Nov. 29, 1859	26, 274
Filter	W. Linton	Baltimore, Md	June 27, 1871	116, 463
Filter	W. C. Long and H. A. Lowsbery	Lockport, N. Y	Aug. 25, 1868	81, 386
Filter	C. F. Mietzsch	Philadelphia, Pa	Oct. 16, 1866	58, 868
Filter	G. S. Neff	Rochester, N. Y	Oct. 21, 1873	143, 924
Filter	J. Nixon	Pittsburgh, Pa	July 5, 1870	104, 987
Filter	G. Norris	New York, N. Y	Apr. 30, 1861	32, 202
Filter	G. T. Palmer	Brooklyn, N. Y	Jan. 28, 1868	73, 751
Filter	G. O. Parkman and J. M. Trussell	Lincolnville and Belfast, Me.	May 25, 1869	90, 383
Filter	J. D. Parrot	Morristown, N. J	Aug. 3, 1869	93, 339
Filter	W. C. Pettijohn	Ozark, Mo	Dec. 10, 1872	133, 720
Filter	J. S. Phillips	Philadelphia, Pa	Aug. 27, 1828	....
Filter	J. Porée	New Orleans, La	Apr. 18, 1848	5, 512
Filter	L. Raecke	New York, N. Y	Jan. 17, 1871	111, 001
Filter	W. Rice	Philadelphia, Pa	June 21, 1859	24, 489
Filter	A. J. Robinson	Troy, N. Y	Aug. 6, 1872	130, 316
Filter	C. C. Savery	Philadelphia, Pa	Dec. 17, 1872	133, 953
Filter	C. C. Savery	Philadelphia, Pa	June 17, 1873	139, 922
Filter	F. Shickle and E. Randals	Saint Louis, Mo	Mar. 5, 1867	62, 696
Filter	T. Simmons	Chicago, Ill	Apr. 11, 1865	47, 261
Filter	T. Simmons	Chicago, Ill	Nov. 6, 1866	59, 467
Filter	T. Simmons	Brooklyn, N. Y	Sept. 15, 1868	82, 255
Filter	T. C. Simonton	Paterson, N. J	June 26, 1860	28, 942
Filter	T. Stewart	Philadelphia, Pa	May 25, 1869	90, 405
Filter	C. Süssner	United States	Oct. 20, 1863	40, 369
Filter	H. N. Taft	Sag Harbor, N. Y	Dec. 26, 1871	122, 291
Filter	J. A. Thompson	Cayuga, N. Y	Aug. 11, 1857	17, 987
Filter	J. A. Thompson	Geneva, N. Y	Apr. 21, 1863	38, 270
Filter	T. P. Thompson	Charlestown, Mass	May 10, 1870	102, 881
Filter	L. Tilliers	West Morrisania, N. Y	Mar. 22, 1859	23, 325
Filter	J. P. A. Vollman	Bingen, Germany	Aug. 22, 1871	118, 306
Filter	M. W. Warne	Saint Louis, Mo	Oct. 9, 1860	30, 366
Filter	C. Warner	Green Point, N. Y	June 3, 1856	15, 027
Filter	J. Watts	New Orleans, La	Mar. 5, 1867	62, 714
Filter	W. H. Wiley	Fredonia, N. Y	Sept. 10, 1867	68, 675
Filter and cooler	J. S. Brooks	Rochester, N. Y	July 21, 1863	39, 271
Filter and cooler	G. B. Davis	Chicago, Ill	Oct. 20, 1863	40, 329
Filter and cooler	W. P. Dickinson	Reading, Pa	Dec. 26, 1865	51, 703
Filter and cooler	N. Downes	Syracuse, N. Y	Dec. 1, 1868	84, 482
Filter and cooler	H. M. Seavey	Chicago, Ill	Oct. 4, 1864	44, 558
Filter and cooler	J. C. Whitehill	Saint Louis, Mo	Mar. 5, 1867	62, 582
Filter and cooler, Combined	J. E. Cheney	Rochester, N. Y	May 19, 1863	38, 559
Filter and cooler, Combined	L. Scharff	Spring Mill, Pa	Dec. 12, 1865	51, 485
Filter and cooler combined	J. E. White	New York, N. Y	Apr. 26, 1870	102, 455
Filter and cooler combined	H. T. Woodman	Dubuque, Iowa	Oct. 2, 1866	58, 530
Filter and cooler, Water	L. Scharff	Spring Mill, Pa	June 11, 1872	127, 926
Filter and refrigerator	J. T. Craddock	Baltimore, Md	Dec. 31, 1845	4, 344
Filter-attachment for faucet	J. H. Wright	New York, N. Y	July 1, 1856	15, 263
Filter, Barrel	G. Armstrong	Liverpool, England	May 31, 1870	103, 540
Filter, Bone, coal, and other	H. Forstrick	New York, N. Y	Dec. 22, 1868	85, 256
Filter, Cistern	N. Gonner and H. Bader	Cape Girardeau, Mo	Feb. 11, 1868	74, 215
Filter, Cistern	G. W. Lampson	Waterloo, N. Y	Mar. 12, 1867	62, 757
Filter, Cistern	P. Laughlin	Danville, Ky	Mar. 1, 1870	100, 418
Filter, Cistern	P. Laughlin	Danville, Ky	May 17, 1870	103, 206
Filter, Cistern	B. B Redfield	Lapeer, Mich	Mar. 1, 1870	100, 323
Filter, Cistern	R. W. Thompson	Mansfield, Ohio	Feb. 7, 1871	111, 701
Filter-cooler	S. M. Pike	Cincinnati, Ohio	Dec. 18, 1866	60, 553
Filter, Driving-pump	A. J. Sternberg	Butler, Ind	Aug. 13, 1872	130, 447
Filter, Faucet	H. G. Fuhrmann	Brooklyn, N. Y	July 28, 1868	80, 469
Filter, Faucet	M. P. Simons	Philadelphia, Pa	Feb. 22, 1870	100, 079
Filter, Feed-water	J. S. Hooton	New Carlisle, Ind	July 12, 1870	105, 209
Filter, Feed-water	G. Waters	Cincinnati, Ohio	Mar. 1, 1870	100, 344

Index of patents issued from the United States Patent Office from 1790 *to* 1873, *inclusive*—Continued.

Invention.	Inventor.	Residence.	Date.	No.
Filter for artesian wells	J. Clay and E. B. Torrey	Ithaca, N. Y	Oct. 31, 1865	50, 686
Filter for cane-juice	H. Bessemer	Baxter House, England	Mar. 8, 1853	9, 608
Filter for cane-juice, &c	E. Tuttle	Jeanerett, La	Aug. 6, 1872	130, 255
Filter for cane juice, oil, and other dense liquids	A. K. Lee	Galveston, Tex	Apr. 15, 1873	137, 849
Filter for drawing sirups, Siphon	C. N. Brock	Philadelphia, Pa	June 30, 1863	39, 031
Filter for hydrants, Self-cleaning	J. Raible, M. Reis, and J. Ritter	Chicago, Ill	Jan. 18, 1870	99, 002
Filter for oils	W. Cady	Marietta, Ohio	Jan. 1, 1867	60, 685
Filter for oils, &c	P. Halle	Philadelphia, Pa	June 13, 1865	48, 172
Filter for oils, acids, &c	J. Jowett	Brooklyn, N. Y	Oct. 21, 1873	143, 768
Filter for petroleum	G. W. Sylvester	Newark, N. J	Dec. 18, 1866	60. 593
Filter for pharmaceutists and others	A. B. Spencer	Rochester, N. Y	June 4, 1867	65, 515
Filter for refining sugar	R. W. Bender	New York, N. Y	Aug. 27, 1867	68, 157
Filter for saccharine and other liquids	H. Merrill	New York, N. Y	Oct. 6, 1868	82, 735
Filter for sirup	P. and H. A. Perdew	Seal, Ohio	Oct. 2, 1866	58, 472
Filter for sirup, &c., Bag	G. A. Gasper	Charlestown, Mass	Apr. 12, 1864	42, 291
Filter for steam-boilers, Arrangement of	E. Blunt	Brooklyn, N. Y	Aug. 14, 1849	6, 650
Filter for sugar, &c	J. Watson	Elizabethtown, N. J	Jan. 31, 1844	3, 419
Filter for wine, sirup, &c	J. Strauss	New York, N. Y	Nov. 22, 1870	109, 467
Filter, Hydrant	T. C. Clarke	Camden, N. J	June 12, 1855	13, 027
Filter, Liquid	D. N. Denman	Millburn, N. J	May 5, 1863	38, 378
Filter, liquor-cooler, and refrigerator, Water	W. F. Nickels	Philadelphia, Pa	Feb. 22, 1870	100, 182
Filter, Nozzle	H. Houston	Pittsburgh, Pa	Jan. 26, 1869	86, 158
Filter, Petroleum	J. H. Smith	Pittsburgh, Pa	July 9, 1867	66, 645
Filter, Portable	H. A. Hall	Boston, Mass	May 21, 1861	32, 361
Filter, Portable	J. P. A. Havard and J. B. Bourgoise.	Paris, France	Feb. 25, 1862	34, 492
Filter, Portable floating	N. Waterman	Boston, Mass	June 12, 1855	13, 054
Filter-press	L. P. R. De Massy	Paris, France	Aug. 14, 1866	57, 265
Filter, Pressure	C. N. Brock	Philadelphia, Pa	July 2, 1867	66, 293
Filter, Pump	J. Christman	Syracuse, N. Y	Nov. 28, 1865	51, 145
Filter-pump	S. D. Richardson and T. S. Hughes.	Syracuse, N. Y	Jan. 23, 1866	52, 204
Filter-rack	E. C. Andrews	Seneca Falls, N. Y	Nov. 16, 1869	96, 765
Filter-regulator	R. Berryman	Hartford, Conn	July 30, 1872	130, 008
Filter, Reversible	J. D. Parrot	Morristown, N. J	Apr. 30, 1872	126, 325
Filter, Sirup	J. L. Smith	Tuscola, Ill	Apr. 3, 1866	53, 693
Filter, Sugar	C. E. Bertrand	New York, N. Y	Oct. 30, 1855	13, 740
Filter, Sugar	W. H. Merrick	Philadelphia, Pa	Apr. 7, 1868	76, 495
Filter-supporter	F. C. Hughes	Frankfort, Ky	July 5, 1870	104, 959
Filter, Thermo-udoric	G. Weissenborn	New York, N. Y	Oct. 2, 1855	13, 628
Filter, Turpentine	F. G. Richardson	Brooklyn, N. Y	Aug. 26, 1873	142, 276
Filter, Vacuum	T. Simmons	Chicago, Ill	Feb. 26, 1867	62, 505
Filter, Water	W. M. Conger	Newark, N. J	Mar. 4, 1873	136, 36[illegible]
Filter, Water	G. B. Davis	Chicago, Ill	Feb. 18, 1862	34, 411
Filter, Water	N. Downes	Syracuse, N. Y	Mar. 7, 1865	46, 646
Filter, Water	A. T. Dunshee	McKeesport, Pa	Sept. 24, 1861	33, 340
Filter, Water	A. T. Dunshee	Pittsburgh, Pa	June 29, 1869	91, 924
Filter, Water	F. Henshaw	Washington, D. C	Mar. 19, 1872	124, 741
Filter, Water	A. Jaminet	Florisant, Mo	Apr. 13, 1858	19, 929
Filter, Water	J. C. Jewett	Buffalo, N. Y	Nov. 16, 1869	96, 923
Filter, Water	J. Pearson	Sacramento, Cal	Sept. 9, 1873	142, 722
Filter, Water	A. J. Robinson	Troy, N. Y	Nov. 26, 1872	133, 488
Filter, Water	G. B. White	New York, N. Y	July 7, 1829	....
Filter, Water-cock	N. C. Wilder	Hartford, Conn	May 3, 1870	102, 636
Filter, Water-cooler	C. Schneider	Newark, N. J	May 20, 1873	139, 085
Filter, Well	J. P. Spaulding	Williamsport, Pa	July 5, 1870	105, 009
Filter, Well and cistern	W. H. Wiley	Fredonia, N. Y	Dec. 4, 1866	60, 311
Filter, Well-point	D. A. Danforth and A. N. Chamberlain.	Elkhart, Ind	June 18, 1872	128, 126
Filter, Well-tube	W. A. Sharpe	Syracuse, N. Y	May 1, 1866	54, 417
Filter, Wine	C. W. Farciot	Kelley's Island, Ohio	Dec. 9, 1873	145, 288
Filter, Wine	A. F. Schmidt	Davenport, Iowa	Nov. 21, 1871	121, 065
Filters, Cleaning	J. Watson	Kingston, Jamaica	Dec. 28, 1847	5, 404
Filters, Composition for	W. H. Jennison	New York, N. Y	May 31, 1853	9, 760
Filters to supply-pipes, Method of attaching	J. Fernald	Boston, Mass	Jan. 6, 1857	16, 330
Filterer and purifier	R. A. Maingay	Pottsville, Pa	May 17, 1859	24, 036
Filtering and drinking tube, Pocket	A. Fessenden	Boston, Mass	Apr. 2, 1850	7, 238
Filtering and purifying spirits, Apparatus for	W. F. Bearns	Mount Pleasant, N. Y	Oct. 1, 1867	69, 390
Filtering and rectifying spirits, Apparatus for	C. W. Ackermann	Pekin, Ill	Jan. 21, 1873	135, 057
Filtering and ventilating apparatus for wells and cisterns.	B. B. Redfield	Pontiac, Mich	July 27, 1869	93, 121
Filtering-apparatus	L. Ayres	Baltimore, Md	July 8, 1834	....
Filtering-apparatus	C. Ballard	Worcester, Mass	Feb. 16, 1858	19, 335
Filtering-apparatus	J. P. Gruber	New York, N. Y	Apr. 3, 1866	53, 606
Filtering-apparatus	N. F. Rice	New Orleans, La	Feb. 1, 1859	22, 826
Filtering-apparatus	J. H. G. D. Wagner	Paris, France	May 29, 1860	28, 524
Filtering-apparatus, Pipe	T. S. Hudson	East Cambridge, Mass	Jan. 4, 1870	98, 596
Filtering burning-fluid	J. D. Kirkpatrick	Urbana, Ohio	Feb. 18, 1868	74, 698
Filtering collodion, Vial for	D. H. Cross	Bennington, Vt	June 21, 1870	104, 432
Filtering-cup	L. Jennings	Brooklyn, N. Y	Sept. 10, 1861	33, 281
Filtering-diaphragm, Self-regulating	W. H. Jennison	New York, N. Y	May 1, 1849	6, 408
Filtering liquids, Apparatus for	B. N. De Buffon	Paris, France	Apr. 14, 1857	17, 028
Filtering liquids, Apparatus for	T. R. Sinclaire	New York, N. Y	Nov. 14, 1871	120, 908
Filtering liquids, &c., Apparatus for	R. Stewart	Brooklyn, N. Y	Mar. 7, 1865	46, 724
Filtering liquids under pressure, Apparatus for	T. R. Sinclaire	New York, N. Y	July 6, 1869	92, 215
Filtering machine, Water	J. Wiseman	New York, N. Y	Aug. 20, 1834	....
Filtering-materials	L. Brandeis	Brooklyn, N. Y	Apr. 25, 1871	113, 973
Filtering-medium	W. Wickersham	Boston, Mass	July 15, 1856	15, 363
Filtering, Metallic medium for	O. C. Phelps	Boston, Mass	June 12, 1855	13, 070
Filtering petroleum, Apparatus for	R. A. Chesebrough	New York, N. Y	Aug. 8, 1865	49, 230
Filtering petroleum, Apparatus for	R. B. Douglass	Cleveland, Ohio	Sept. 24, 1867	69, 192
Filtering petroleum, sirup, and other liquids, Composition for.	W. Van Wyck	Belleville, N. J	May 28, 1867	65, 313
Filtering-pot	M. Freytag	Philadelphia, Pa	July 1, 1807	....
Filtering, Preparing charcoal for	G. W. Ferris	Quincy, Ill	July 10, 1866	56, 198
Filtering-press	L. P. R. De Massy	Paris, France	Nov. 21, 1865	51, 124
Filtering-press	L. P. R. De Massy	Paris, France	Nov. 28, 1865	51, 277
Filtering-press, Cylindrical	P. Du Rieux and E. Roettger	Lille, France	July 21, 1868	80, 157

Index of patents issued from the United States Patent Office from 1790 *to* 1873, *inclusive*—Continued.

Invention.	Inventor.	Residence.	Date.	No.
Filtering-sand for cider	I. Holmes	Leicester, N. Y	Aug. 12, 1856	15, 517
Filtering sirup and other liquid, Apparatus for	H. A. Tilden	New Lebanon, N. Y	July 25, 1865	49, 013
Filtering sirup, &c., Mode of treating bone-black for	J. O. Donner	Jersey City, N. J	Aug. 17, 1869	93, 865
Filtering-stopper	G. H. Mellen	Hartford, Conn	Oct. 8, 1861	33, 443
Filtering sugar, Treating bone-black for	C. B. Orvis	Saint Louis, Mo	Sept. 24, 1867	69, 118
Filtering-tube	D. and E. Moore	Brooklyn, N. Y	July 27, 1869	92, 988
Filtering volatile liquids, Apparatus for	W. Boyce	Tuscola, Ill	Dec. 28, 1869	98, 343
Filtering water, &c	J. Barron	United States Navy	Aug. 30, 1831	....
Filtering water	C. Hall	Norfolk, Va	Feb. 22, 1828	....
Filtering water, Apparatus and process for	J. Du Commun	New York, N. Y	Feb. 10, 1816	....
Filtering water, &c., Apparatus for	J. Mulhern	Saint Louis, Mo	July 31, 1849	6, 621
Filtering water, wine, &c	S. McCombs, J. Smith, and B. D. Galpin.	Baltimore, Md	June 24, 1806	....
Filtration-press for expressing liquids from substances.	W. Needham and J. Kite	Vauxhall, England	Aug. 11, 1857	17, 978
Fine-tooth comb, Making ivory	F. Bush and J. H. Pratt	Meriden, Conn	June 5, 1849	6, 492
Finger-bars, Machine for straightening	J. Corns	Akron, Ohio	Jan. 16, 1872	122, 815
Finger-exerciser	F. Wenzel	Saint Louis, Mo	Aug. 13, 1872	130, 457
Finger-exercising apparatus	A. C. Armengol	New York, N. Y	Feb. 9, 1869	86, 722
Finger-guard	W. J. Budington	Lancaster, Pa	July 16, 1872	129, 392
Finger-guard	L. Holmes	Green Point, N. Y	Apr. 4, 1871	113, 519
Finger-guard	G. H. Spencer	Cleveland, Ohio	Sept. 13, 1870	107, 420
Finger-guard for holding hot corn	H. Hebbard	New York, N. Y	Aug. 10, 1869	93, 436
Finger-nails, Instrument for cutting	C. C. E. Schuartz	Philadelphia, Pa	May 22, 1866	54, 966
Finger-ring	S. Bogert	New York, N. Y	Aug. 23, 1859	25, 172
Finger-ring	C. S. Ford	New York, N. Y	Apr. 27, 1869	89, 475
Finger-ring	J. S. Palmer	Providence, R. I	Dec. 27, 1870	110, 587
Finger-ring	W. H. Peckham	New York, N. Y	Sept. 29, 1868	82, 546
Finger-ring	G. W. Tinsley and E. W. Storer	Minneapolis, Minn	Aug. 15, 1871	118, 169
Finger-ring, Extension	S. Friend and G. Seiler	New York, N. Y	Dec. 21, 1858	22, 358
Finger-rings, Die for enlarging and sizing plain	G. Krementz	Newark, N. J	July 1, 1873	140, 422
Finger-rings, Die for manufacturing plain	G. Krementz	Newark, N. J	July 1, 1873	140, 423
Finger-rings, Instrument for expanding	L. Mason	Rochester, N. Y	Dec. 17, 1867	72, 410
Finger-rings, Machine for rolling stock for	J. S. Palmer	Providence, R. I	Mar. 26, 1872	124, 971
Finger-rings, Manufacture of sheet-metal	I. M. Porter	Providence, R. I	Oct. 2, 1860	30, 246
Finger, scarf, and napkin rings	L. Sauter	Jersey City, N. J	July 24, 1866	56, 618
Finishing-die in machine for making rings from sheet-metal.	C. W. Dickinson	Newark, N. J	July 25, 1854	11, 363
Fire-alarm	W. W. Andrews, J. Cummer, J. F. Gauweiler, and J. Stengel.	Croton, Mich	June 11, 1867	65, 526
Fire-alarm	W. A. Barnes	Bridgeport, Conn	May 6, 1873	138, 467
Fire-alarm	H. L. Brower	New York, N. Y	Oct. 17, 1871	120, 029
Fire-alarm	A. F. Cobb	Chapel Hill, Mo	Nov. 13, 1860	30, 613
Fire-alarm	J. Coulson	Oskaloosa, Iowa	Dec. 15, 1868	84, 997
Fire-alarm	J. Dinsmore	Millburn, Me	May 14, 1833	....
Fire alarm	C. Dion	Montreal, Canada	Aug. 29, 1869	49, 686
Fire-alarm	C. Dion	Montreal, Canada	Apr. 3, 1866	53, 757
Fire-alarm	C. Dion	New York, N. Y	Apr. 6, 1869	88, 698
Fire-alarm	E. Ehlin	Boston, Mass	Nov. 13, 1860	30, 624
Fire-alarm	A. Fontaine	Milwaukee, Wis	Nov. 6, 1866	59, 379
Fire-alarm	E. Fontaine and O. A. Simons	Fort Wayne, Ind	Nov. 13, 1866	59, 579
Fire-alarm	J. F. Gauweiler and J. Stengel	Croton, Mich	Feb. 11, 1868	74, 339
Fire-alarm	T. Goodwin	Exeter, N. H	Jan. 30, 1841	1, 956
Fire-alarm	W. D. Grimshaw	Newark, N. J	Feb. 28, 1860	27, 282
Fire-alarm	F. F. Herman	Grafton, Mich	Mar. 19, 1872	124, 680
Fire-alarm	C. D. Kubach	Philadelphia, Pa	June 26, 1866	55, 957
Fire-alarm	I. T. Pease	Thompsonville, Conn	Dec. 24, 1867	72, 672
Fire-alarm	I. T. Pease	Thompsonville, Conn	Feb. 23, 1869	87, 194
Fire-alarm	J. N. Pitts and J. E. Russell	Niagara Falls, N. Y	Oct. 10, 1871	119, 713
Fire-alarm	R. Porter	Billerica, Mass	Dec. 28, 1840	1, 915
Fire-alarm	R. Powers	Prescott, Mass	Apr. 24, 1866	54, 202
Fire-alarm	H. N. Shultz	Mechanicstown, Md	Apr. 4, 1871	113, 460
Fire-alarm	D. Tomlinson and H. S. Hopkins	Brookfield, Conn	June 12, 1847	5, 152
Fire-alarm	J. R. Tunnicliff	Van Hornesville, N. Y	Apr. 22, 1862	35, 050
Fire-alarm and circuit therefor, Electrical and thermostatic.	J. H. Guest	Brooklyn, N. Y	Oct. 14, 1873	143, 691
Fire-alarm and heat-detector	A. Ross	Brooklyn, N. Y	Mar. 10, 1863	37, 875
Fire-alarm and police-signal box	W. J. Philips	Philadelphia, Pa	Oct. 31, 1871	120, 395
Fire-alarm, Annunciating	A. and E. Fontaine	Fort Wayne, Ind	Oct. 8, 1867	69, 648
Fire-alarm apparatus	A. Allen	Rochester, N. Y	June 1, 1869	90, 806
Fire-alarm apparatus	S. D. Cooper	Hartford, Conn	Aug. 6, 1861	32, 980
Fire-alarm apparatus	B. Seymour and J. Whipple	Utica, N. Y	Apr. 2, 1835	....
Fire-alarm apparatus, Electro-magnetic	M. G. Farmer	Salem, Mass	Nov. 16, 1858	22, 071
Fire-alarm apparatus, Electro-magnetic	M. G. Farmer	Salem, Mass	Jan. 11, 1859	22, 553
Fire-alarm apparatus, Electro-magnetic	M. G. Farmer	Salem, Mass	Feb. 22, 1859	23, 060
Fire-alarm apparatus, Electro-magnetic	M. G. Farmer and W. F. Channing.	Salem and Boston, Mass	Mar. 8, 1859	23, 217
Fire-alarm, Automatic	H. L. Brown	Middletown, Conn	May 27, 1873	139, 292
Fire-alarm, Automatic	C. H. Lehnis	Philadelphia, Pa	Nov. 25, 1873	144, 991
Fire-alarm box, &c., Lock for	J. M. Fairchild	New Haven, Conn	Oct. 28, 1873	144, 079
Fire-alarm, Electric	E. A. Hill	Chicago, Ill	Dec. 12, 1871	121, 717
Fire-alarm, Electro-magnetic	H. Van Ausdall	Preble County, Ohio	Aug. 31, 1852	9, 240
Fire-alarm, Electro-magnetic	W. Whiting	Roxbury, Mass	Mar. 31, 1863	38, 074
Fire-alarm, Magnetic	A. Smith	Perrysburgh, Ohio	July 24, 1846	4, 661
Fire-alarm, Self-acting	I. S. Richardson	New Market, N. H	Aug. 15, 1831	....
Fire-alarm-signal box	C. E. Carpenter	Providence, R. I	Feb. 28, 1865	46, 542
Fire-alarm-signal box	G. Floyd	Cincinnati, Ohio	May 21, 1872	127, 041
Fire-alarm-signal box	F. Latta	Cincinnati, Ohio	Jan. 7, 1873	134, 552
Fire-alarm-signal box	G. W. Shawk	Cleveland, Ohio	Feb. 1, 1870	99, 358
Fire-alarm-signal box	G. Wright, M. Y. Holley, and H. R. Miles.	Washington, D. C	June 17, 1873	140, 106
Fire-alarm, Telegraph	J. B. Franz	Cleveland, Ohio	July 13, 1869	92, 598
Fire-alarm, Telegraphic	G. W. Shaw, S. C. Morley, and S. D. Cushman.	Cleveland and New Lisbon, Ohio.	Sept. 11, 1866	57, 981
Fire-alarm, Thermostat	W. B. Watkins	Jersey City, N. J	May 10, 1870	102, 982
Fire and burglar alarm	H. L. Bower	New York, N. Y	Aug. 22, 1871	118, 191
Fire and burglar alarm	J. L. Cheston	South Easton, Pa	Aug. 31, 1869	94, 282
Fire and burglar alarm	W. Henckler	Kirkwood, Mo	Apr. 30, 1872	126, 294

Index of patents issued from the United States Patent Office from 1790 *to* 1873, *inclusive*—Continued.

Invention.	Inventor.	Residence.	Date.	No.
Fire and burglar alarm	E. Middleton	Cleveland, Ohio	Oct. 9, 1866	58, 662
Fire and burglar alarm	O. E. Pickett and R. S. Luce	North Auburn and Lawsville Centre, Pa.	Feb. 11, 1868	74, 418
Fire and burglar alarm	D. Ward and R. S. Luce	Lawsville Centre, Pa	Nov. 7, 1865	50, 863
Fire and light, Producing	I. Jennings	New York, N. Y	June 11, 1829	......
Fire and ventilating apparatus for ships	J. W. Richards	New York, N. Y	Feb. 1, 1859	22, 897
Fire and water proof compositions for ceilings, walls, and floors.	L. A. Tartierè	New York, N. Y	May 13, 1873	138, 956
Fire and water proofing paper, cloth, &c., Process for.	R. O. Lowrey	Salem, N. Y	Dec. 10, 1867	71, 892
Fire and weather proof composition or artificial slate.	W. Blake	Sharon, Ohio	Mar. 28, 1848	5, 490
Fire-annihilator	H. Baragwanath and M. Van Wisker.	New York, N. Y	Dec. 5, 1865	51, 248
Fire-annihilator	C. T. Jerome	Minneapolis, Minn	July 9, 1867	66, 498
Fire, Apparatus for barns, stables, &c., for rescuing horses and other stock from.	J. E. Hall	Cleveland, Ohio	Oct. 13, 1857	18, 395
Fire, Apparatus to protect buildings from	S. M. Andrus	Bellevue, Mich	July 10, 1860	29, 044
Fire, Apparatus to protect buildings from	T. Odion	Portsmouth, N. H	Sept. 29, 1857	18, 295
Fire-arm	J. Adams	Cleveland, Ohio	Sept. 30, 1856	15, 797
Fire-arm	S. Adams	Springfield, Mass	Oct. 3, 1838	960
Fire-arm	E. Allen	Worcester, Mass	July 3, 1855	13, 154
Fire-arm	E. Allen	Worcester, Mass	Jan. 13, 1857	16, 367
Fire-arm	E. Allen	Worcester, Mass	Mar. 13, 1860	27, 415
Fire-arm	L. Bailey, J. B. Ripley, and W. B. Smith.	Portland, Me	Feb. 20, 1839	1, 084
Fire-arm	E. Baldwinn	Templeton, Mass	July 11, 1854	11, 283
Fire-arm	T. H. Barlow	Lexington, Ky	Jan. 16, 1855	12, 230
Fire-arm	F. Beals	New Haven, Conn	Sept. 26, 1854	11, 715
Fire-arm	F. Beals	New Haven, Conn	June 24, 1856	15, 167
Fire-arm	F. B. E. Beaumont	Barnsley, England	June 3, 1856	15, 032
Fire-arm	F. Beerstecher	Philadelphia, Pa	Sept. 25, 1855	13, 592
Fire-arm	J. E. Blake	Norwich, Conn	June 25, 1867	66, 072
Fire-arm	G. A. Blittkowskie	New York, N. Y	Apr. 28, 1857	17, 136
Fire-arm	C. C. Brand	Norwich, Conn	July 29, 1862	35, 989
Fire-arm	F. S. Brettell and J. B. Frisbie	Allegheny City, Pa	Feb. 10, 1857	16, 575
Fire-arm	C. F. Brown	Warren, R. I	Sept. 18, 1860	30, 045
Fire-arm	I. W. Brown	West Springfield, Mass	Aug. 8, 1854	11, 470
Fire-arm	C. Buss	Marlborough, N. H	Apr. 25, 1854	10, 821
Fire-arm	E. B. Butterfield	Brattleborough, Vt	Mar. 16, 1839	1, 106
Fire-arm	S. Colt	Hartford, Conn	Feb. 25, 1836	......
Fire-arm	S. Colt	Paterson, N. J	Aug. 29, 1839	1, 304
Fire-arm	S. Colt	Hartford, Conn	May 20, 1856	14, 905
Fire-arm	T. Cook	New York, N. Y	Feb. 14, 1854	10, 520
Fire-arm	A. and C. Daniels	Chester, Conn	Feb. 15, 1838	610
Fire-arm	J. C. Day	Hackettstown, N. J	Aug. 8, 1854	11, 477
Fire-arm	J. C. Day	Hackettstown, N. J	Dec. 18, 1855	13, 941
Fire-arm	J. C. Day	Hackettstown, N. J	Jan. 15, 1856	14, 095
Fire-arm	S. Day	New York, N. Y	Oct. 8, 1840	1, 810
Fire-arm	S. Day and S. Hall	New York, N. Y	Dec. 31, 1839	1, 461
Fire-arm	J. A. De Brame	New York, N. Y	July 2, 1861	32, 685
Fire-arm	R. Eastman	Brunswick, Me	Dec. 7, 1829	......
Fire-arm	D. Edwards	Zanesville, Ohio	Apr. 25, 1839	1, 134
Fire-arm	J. Ells	Pittsburgh, Pa	Apr. 28, 1857	17, 143
Fire-arm	E. Fisher and D. H. Chamberlain.	Springfield and Boston, Mass.	Apr. 17, 1837	168
Fire-arm	L. H. Gibbs	Oberlin, Ohio	Oct. 2, 1847	5, 316
Fire-arm	J. W. Goodale	Amherst, Mass	June 1, 1869	90, 741
Fire-arm	E. H. Graham	Biddeford, Me	Jan. 16, 1855	12, 235
Fire-arm	E. H. Graham	Manchester, N. H	Sept. 16, 1856	15, 734
Fire-arm	R. Graham	Brooklyn, N. Y	Aug. 16, 1864	43, 881
Fire-arm	H. Gross	Tiffin, Ohio	May 22, 1855	12, 906
Fire-arm	J. E. Halsey	New York, N. Y	July 8, 1856	15, 292
Fire-arm	H. Harrington	Southbridge, Mass	July 29, 1837	297
Fire-arm	A. V. Hill	Carrollton, N. Y	May 28, 1861	32, 421
Fire-arm	F. W. Hoffman	New York, N. Y	Aug. 12, 1856	15, 516
Fire-arm	A. Holcomb	Litchfield, N. Y	Oct. 31, 1822	......
Fire-arm	J. Hollingsworth and R. S. Mershon.	Zanesville, Ohio	Feb. 27, 1855	12, 470
Fire-arm	J. C. Howe	Milwaukee, Wis	Oct. 31, 1854	11, 862
Fire-arm	W. W. Hubbell	Philadelphia, Pa	July 1, 1844	3, 649
Fire-arm	W. Jenks	Columbia, S. C	May 25, 1838	747
Fire-arm	B. F. Joslyn	Worcester, Mass	Oct. 8, 1861	33, 435
Fire-arm	G. Kesling	Lebanon, Ohio	June 3, 1856	15, 041
Fire-arm	F. Klein	Newark, N. J	Apr. 10, 1855	12, 681
Fire-arm	D. Knight	Salem, Ind	Aug. 8, 1854	11, 483
Fire-arm	P. Lancaster	Burr Oak, Mich	Apr. 15, 1856	14, 667
Fire-arm	E. Lefauchaux	Paris, France	Mar. 26, 1861	31, 809
Fire-arm	A. Le Mat	New Orleans, La	Oct. 21, 1856	15, 925
Fire-arm	A. Le Mat	New Orleans, La	Nov. 25, 1856	16, 124
Fire-arm	G. Leonard	Shrewsbury, Mass	Aug. 9, 1853	9, 922
Fire-arm	E. Lindner	New York, N. Y	May 26, 1857	17, 382
Fire-arm	J. Lord	Minersville, Pa	June 12, 1860	28, 677
Fire-arm	S. K. Lovewell	Gardner, Mass	Mar. 17, 1857	16, 846
Fire-arm	W. W. Marston	New York, N. Y	Sept. 18, 1855	13, 581
Fire-arm	W. McCord	Sing Sing, N. Y	Apr. 2, 1861	31, 933
Fire-arm	L. A. Merriam	New York, N. Y	Jan. 19, 1869	86, 091
Fire-arm	J. H. Merrill	Baltimore, Md	Jan. 8, 1856	14, 077
Fire-arm	J. H. Merrill	Baltimore, Md	May 28, 1861	32, 450
Fire-arm	J. H. Merrill	Baltimore, Md	May 28, 1861	32, 451
Fire-arm	J. Merwin and E. P. Bray	New York, N. Y	Jan. 5, 1864	41, 166
Fire-arm	D. Moore	Brooklyn, N. Y	Feb. 19, 1861	31, 473
Fire-arm	W. H. Morrison	Marion County, Ind	Sept. 19, 1854	11, 698
Fire-arm	F. Newbury	Albany, N. Y	Mar. 20, 1855	12, 555
Fire-arm	F. Newbury	Albany, N. Y	Apr. 29, 1856	14, 774
Fire-arm	F. D. Newbury	Albany, N. Y	Aug. 12, 1856	15, 521
Fire-arm	F. D. Newbury	Albany, N. Y	Feb. 9, 1858	19, 327
Fire-arm	A. N. Newton	Richmond, Ind	Aug. 12, 1856	15, 522

Index of patents issued from the United States Patent Office from 1790 *to* 1873, *inclusive*—Continued.

Invention.	Inventor.	Residence.	Date.	No.
Fire-arm	C. V. Nickerson	Baltimore, Md	Jan. 27, 1852	8, 690
Fire-arm	H. S. North	Middletown, Conn	June 5, 1847	5, 141
Fire-arm	H. S. North	Middletown, Conn	June 17, 1856	15, 144
Fire-arm	G. F. and A. H. Palmié	Berlin, Prussia	Oct. 24, 1854	11, 835
Fire-arm	J. Peck	New Haven, Conn	May 16, 1854	10, 930
Fire-arm	A. D. Perry	Newark, N. J	Jan. 16, 1855	12, 244
Fire-arm	J. W. Plummer and J. Clark	Wayne Township, Ohio	May 17, 1832	......
Fire-arm	J. B. Porter	Girard, Pa	Aug. 4, 1832	......
Fire-arm	J. B. Read	Tuscaloosa, Ala	May 5, 1857	17, 233
Fire-arm	J. A. Reynolds	Elmira, N. Y	July 17, 1855	13, 292
Fire-arm	J. A. Reynolds	Elmira, N. Y	July 17, 1855	13, 293
Fire-arm	F. Ruggles	Hardwick, Mass	Nov. 24, 1826	......
Fire-arm	E. Savage and S. North	Middletown, Conn	July 30, 1844	3, 686
Fire-arm	J. Shaw, jr	Hinckley Township, Ohio	June 30, 1857	17, 698
Fire-arm	C. Slotterbek	San Francisco, Cal	Nov. 17, 1868	84, 224
Fire-arm	B. F. Smith	South Hadley, Mass	Dec. 5, 1839	1, 422
Fire-arm	H. Smith and D. B. Wesson	Norwich, Conn	Feb. 14, 1854	10, 535
Fire-arm	D. E. Somes	Washington, D. C	Aug. 18, 1863	39, 592
Fire-arm	G. H. Soule	Jersey City, N. J	July 15, 1856	15, 347
Fire-arm	N. Starr	Middletown, Conn	May 3, 1839	1, 141
Fire-arm	J. Stowell	Charlestown, Mass	May 8, 1855	12, 836
Fire-arm	J. L. Swan	Lowville, N. Y	Apr. 8, 1862	34, 911
Fire-arm	W. A. Sweet	Pompey, N. Y	Aug. 15, 1854	11, 536
Fire-arm	W. Thornton and J. Hatkill	Washington, D. C	May 21, 1811	......
Fire-arm	J. Warner	Springfield, Mass	June 24, 1856	15, 202
Fire-arm	E. Whitney	Whitneyville, Conn	Aug. 1, 1854	11, 447
Fire-arm	J. Widmer	Newark, N. J	Nov. 9, 1869	96, 751
Fire-arm	S. W. Wood	Cornwall, N. Y	Aug. 18, 1863	39, 619
Fire-arm	W. Wright	New York, N. Y	Nov. 7, 1854	11, 917
Fire-arm, Accelerating	A. S. Lyman	New York, N. Y	Feb. 3, 1857	16, 568
Fire-arm and ordnance	H. C. Fay	Lancaster, Mass	May 22, 1837	203
Fire-arm applicable alike to small-arms and cannon.	A. S. Thistle	New Orleans, La	Aug. 1, 1838	865
Fire-arm barrels, Method of making	J. Pannabecker	Elizabeth Township, Pa	Aug. 6, 1850	7, 547
Fire-arm, Breech-loading	F. J. Abbey and J. H. Foster	Chicago, Ill	Apr. 25, 1871	114, 081
Fire-arm, Breech-loading	G. T. Abbey	Chicago, Ill	Mar. 16, 1869	87, 814
Fire-arm, Breech-loading	H. W. Adams	New York, N. Y	Sept. 19, 1854	11, 685
Fire-arm, Breech-loading	J. S. Adams	Taunton, Mass	Aug. 11, 1863	39, 455
Fire-arm, Breech-loading	J. S. Adams	Taunton, Mass	Sept. 27, 1864	44, 377
Fire-arm, Breech-loading	S. Albright	Ottawa, Ohio	May 5, 1863	38, 366
Fire-arm, Breech-loading	W. Aldrich	Cleveland, Ohio	May 12, 1863	38, 455
Fire-arm, Breech-loading	E. Allen	Worcester, Mass	Sept. 18, 1860	30, 033
Fire-arm, Breech-loading	E. Allen	Worcester, Mass	Aug. 22, 1865	49, 491
Fire-arm, Breech-loading	E. Allen	Worcester, Mass	Dec. 15, 1868	84, 929
Fire-arm Breech-loading	E. S. Allin	Springfield, Mass	Sept. 19, 1865	49, 959
Fire-arm, Breech-loading	C. H. Alsop	Middletown, Conn	Apr. 7, 1868	76, 374
Fire-arm, Breech-loading	P. Altmair	Lewistown, Pa	July 12, 1859	24, 774
Fire-arm, Breech-loading	J. W. Armstrong and J. Taylor	Augusta, Ky	Nov. 25, 1862	37, 025
Fire-arm, Breech-loading	W. H. Arnold	Washington, D. C	Nov. 15, 1859	26, 076
Fire-arm, Breech-loading	J. N. Aronson	New York, N. Y	Nov. 13, 1866	59, 540
Fire-arm, Breech-loading	E. H. Ashcroft	Boston, Mass	May 26, 1863	38, 645
Fire-arm, Breech-loading	J. Aston	Hythe, England	May 13, 1873	138, 837
Fire-arm, Breech-loading	N. L. Babcock	New Haven, Conn	Mar. 20, 1860	27, 509
Fire-arm, Breech-loading	A. M. Bacon	Washington, D. C	July 31, 1866	56, 846
Fire-arm, Breech-loading	G. R. Bacon	Providence, R. I	July 21, 1863	39, 270
Fire-arm, Breech-loading	C. W. Baldwin	Boston, Mass	Jan. 19, 1869	85, 897
Fire-arm, Breech-loading	C. H. Ballard	Worcester, Mass	Nov. 5, 1861	33, 631
Fire-arm, Breech-loading	J. Barber and P. C. Reinfried	Bridesburgh, Pa	Mar. 15, 1859	23, 224
Fire-arm, Breech-loading	K. V. Barnekov	Cornwall, N. Y	June 14, 1870	104, 100
Fire-arm, Breech-loading	O. W. Bayley	Somerville, Mass	Apr. 22, 1862	35, 008
Fire-arm, Breech-loading	F. Beals	New Haven, Conn	June 28, 1864	43, 284
Fire-arm, Breech-loading	F. Beals	New Haven, Conn	Jan. 30, 1866	52, 258
Fire-arm, Breech-loading	S. Belden and J. F. Crabtree	Visalia, Cal	Dec. 29, 1868	85, 268
Fire-arm, Breech-loading	H. Berdan	New York, N. Y	Jan. 10, 1865	45, 899
Fire-arm, Breech-loading	H. Berdan	New York, N. Y	Jan. 9, 1866	51, 991
Fire-arm, Breech-loading	H. Berdan	New York, N. Y	Feb. 27, 1866	52, 925
Fire-arm, Breech-loading	H. Berdan	New York, N. Y	Dec. 22, 1868	85, 162
Fire-arm, Breech-loading	H. Berdan	New York, N. Y	Mar. 30, 1869	88, 436
Fire-arm, Breech-loading	H. Berdan	New York, N. Y	Apr. 5, 1870	101, 418
Fire-arm, Breech-loading	H. Berdan	New York, N. Y	Nov. 1, 1870	108, 869
Fire-arm, Breech-loading	H. Berg	Davenport, Iowa	Mar. 25, 1862	34, 729
Fire-arm, Breech-loading	A. J. Bergen	Brooklyn, N. Y	Feb. 26, 1867	62, 465
Fire-arm, Breech-loading	A. J. Bergen and D. Williamson	Brooklyn, N. Y	Nov. 22, 1864	45, 202
Fire arm, Breech-loading	C. E. Billings	Windsor, Vt	Apr. 24, 1866	54, 100
Fire-arm, Breech-loading	J. D. Blaker	Newtown, Pa	Aug. 10, 1869	93, 403
Fire-arm, Breech-loading	S. Bostwick and C. G. Sargent	Graniteville, Mass	Nov. 11, 1862	36, 891
Fire-arm, Breech-loading	P. Bourdereaux	New York, N. Y	Nov. 13, 1866	59, 706
Fire-arm, Breech-loading	P. Bourdereaux	New York, N. Y	July 29, 1873	141, 198
Fire-arm, Breech-loading	F. E. Boyd and P. S. Tyler	Boston, Mass	Jan. 21, 1868	73, 494
Fire-arm, Breech-loading	F. E. Boyd and P. S. Tyler	Boston, Mass	Apr. 6, 1869	88, 540
Fire-arm, Breech-loading	J. Boynton	East Hartford, Conn	Nov. 27, 1860	30, 714
Fire-arm, Breech-loading	I. Bradley	Hartford, Conn	Aug. 7, 1866	56, 899
Fire-arm, Breech-loading	C. C. Brand	Norwich, Conn	Apr. 28, 1863	38, 280
Fire-arm, Breech-loading	C. C. Brand	Norwich, Conn	June 23, 1863	38, 943
Fire-arm, Breech-loading	W. Briggs	Norristown, Pa	Apr. 6, 1869	88, 605
Fire-arm, Breech-loading	L. W. Broadwell	New Orleans, La	Aug. 22, 1865	49, 583
Fire-arm, Breech-loading	E. Brooks and G. Walker	Philadelphia, Pa	July 6, 1858	20, 776
Fire-arm, Breech-loading	J. Broughton	New York, N. Y	Apr. 14, 1868	76, 595
Fire-arm, Breech-loading	W. H. Brown	Worcester, Mass	Mar. 4, 1862	34, 561
Fire-arm, Breech-loading	J. Brownes	Prince Edward Island	Aug. 2, 1864	43, 733
Fire-arm, Breech-loading	H. Brügmann	Waterloo, Mich	July 16, 1872	129, 312
Fire-arm, Breech-loading	H. Büchner	New York, N. Y	May 2, 1871	114, 259
Fire-arm, Breech-loading	A. Burgess	New York, N. Y	Sept. 26, 1871	119, 218
Fire-arm, Breech-loading	A. Burgess	Owego, N. Y	June 11, 1872	127, 737
Fire-arm, Breech-loading	J. Burke	Courtland, Ill	May 15, 1866	54, 680
Fire-arm, Breech-loading	J. Burke	Syracuse, Ill	June 19, 1866	55, 613
Fire-arm, Breech-loading	A. E. Burnside	Bristol, R. I	Mar. 25, 1856	14, 491
Fire-arm, Breech-loading	B. Burton	Brooklyn, N. Y	Dec. 20, 1859	26, 475
Fire-arm, Breech-loading	B. Burton	Brooklyn, N. Y	Aug. 11, 1868	81, 059

Index of patents issued from the United States Patent Office from 1790 *to* 1873, *inclusive*—Continued.

Invention.	Inventor.	Residence.	Date.	No.
Fire-arm, Breech-loading	C. Callaghon	Great Britain	Feb. 25, 1868	74, 888
Fire-arm, Breech-loading	J. F. C. Carle	Hamburg, Germany	Feb. 27, 1866	52, 938
Fire-arm, Breech-loading	H. Carter and G. H. Edwards	Stepney, England	Jan. 19, 1869	85, 999
Fire-arm, Breech-loading	H. A. Castle	Ilion, N. Y	Nov. 4, 1873	144, 190
Fire-arm, Breech-loading	C. Chabot	Philadelphia, Pa	Apr. 4, 1865	47, 163
Fire-arm, Breech-loading	C. Chabot	Philadelphia, Pa	Sept. 5, 1865	49, 718
Fire-arm, Breech-loading	M. J. Chamberlin	Springfield, Mass	Feb. 14, 1871	111, 814
Fire-arm, Breech-loading	M. J. Chamberlin	Springfield, Mass	July 16, 1872	129, 393
Fire-arm, Breech-loading	M. J. Chamberlin	Springfield, Mass	Feb. 4, 1873	135, 405
Fire-arm, Breech-loading	M. J. and H. M. Chamberlin	Springfield, Mass	Jan. 8, 1867	60, 998
Fire-arm, Breech-loading	L. N. Chapin	New Lisbon, N. Y	May 17, 1864	42, 748
Fire-arm, Breech-loading	A. A. Chassepot	Paris, France	Nov. 23, 1869	97, 167
Fire-arm, Breech-loading	F. Clark	North Oxford, Mass	July 19, 1864	43, 571
Fire-arm, Breech-loading	F. Clark	Auburn, Mass	Jan. 3, 1865	45, 701
Fire-arm, Breech-loading	F. Clark	Auburn, Mass	Mar. 27, 1866	53, 522
Fire-arm, Breech-loading	N. S. Clement	Worcester, Mass	May 27, 1856	14, 949
Fire-arm, Breech-loading	N. S. Clement	New York, N. Y	Oct. 10, 1865	50, 334
Fire-arm, Breech-loading	J. J. Cloes	Liege, Belgium	Apr. 12, 1870	101, 826
Fire-arm, Breech-loading	J. W. Cochran	New York, N. Y	July 7, 1863	39, 120
Fire-arm, Breech-loading	J. W. Cochran	New York, N. Y	Dec. 22, 1863	40, 992
Fire-arm, Breech-loading	J. W. Cochran	New York, N. Y	Apr. 4, 1865	47, 088
Fire-arm, Breech-loading	J. W. Cochran	New York, N. Y	Apr. 25, 1865	47, 396
Fire-arm, Breech-loading	J. W. Cochran	New York, N. Y	Mar. 26, 1867	63, 217
Fire-arm, Breech-loading	J. W. Cochran	New York, N. Y	Mar. 17, 1868	75, 627
Fire-arm, Breech-loading	J. W. Cochran	New York, N. Y	Jan. 5, 1869	85, 645
Fire-arm, Breech-loading	J. W. Cochran	New York, N. Y	May 7, 1872	126, 446
Fire-arm, Breech-loading	C. C. Coleman	Worcester, Mass	May 13, 1862	35, 217
Fire-arm, Breech-loading	C. C. Coleman	Worcester, Mass	Nov. 6, 1866	59, 500
Fire-arm, Breech-loading	H. Conant	Hartford, Conn	Apr. 1, 1856	14, 554
Fire-arm, Breech-loading	I. B. Conklin	Baltimore, Md	Feb. 16, 1869	86, 971
Fire-arm, Breech-loading	J. A. Conover	New York, N. Y	July 24, 1869	56, 669
Fire-arm, Breech-loading	L. Conroy	New York, N. Y	Dec. 31, 1867	72, 803
Fire-arm, Breech-loading	L. Conroy	New York, N. Y	June 15, 1869	91, 421
Fire-arm, Breech-loading	L. Conroy	New York, N. Y	Dec. 2, 1873	145, 154
Fire-arm, Breech-loading	R. F. Cook	Potsdam, N. Y	July 24, 1860	29, 340
Fire-arm, Breech-loading	R. F. Cook	Watertown, N. Y	Mar. 10, 1863	37, 854
Fire-arm, Breech-loading	J. C. Cooke	Middletown, Conn	June 3, 1862	35, 488
Fire-arm, Breech-loading	J. R. Cooper	Birmingham, England	Dec. 15, 1868	84, 938
Fire-arm, Breech-loading	C. Cox	Coxville, N. C	Apr. 27, 1858	20, 041
Fire-arm, Breech-loading	C. Cox	Coxville, N. C	Apr. 10, 1860	27, 778
Fire-arm, Breech-loading	S. Crispin	New York, N. Y	Jan. 1, 1867	60, 698
Fire-arm, Breech-loading	S. Crispin	New York, N. Y	Feb. 5, 1867	61, 722
Fire-arm, Breech-loading	F. Curtis	Saugus Centre, Mass	Feb. 15, 1859	22, 940
Fire-arm, Breech-loading	F. Curtis	Newton, (Lower Falls,) Mass.	Sept. 17, 1861	33, 317
Fire-arm, Breech-loading	F. Curtis	Newton, (Lower Falls,) Mass.	Jan. 19, 1864	41, 281
Fire-arm, Breech-loading	F. S. Dangerfield	Auburn, N. Y	Sept. 3, 1872	130, 984
Fire-arm, Breech-loading	J. Davis	Buffalo, N. Y	Jan. 27, 1863	37, 544
Fire-arm, Breech-loading	J. Davis	Buffalo, N. Y	July 7, 1863	39, 198
Fire-arm, Breech-loading	J. Davis	Buffalo, N. Y	Apr. 26, 1864	42, 529
Fire-arm, Breech-loading	J. Davis	Buffalo, N. Y	Nov. 28, 1865	51, 258
Fire-arm, Breech-loading	J. Davis	Buffalo, N. Y	May 17, 1870	103, 154
Fire-arm, Breech-loading	J. M. Deprez	Liege, Belgium	Feb. 9, 1869	86, 739
Fire-arm, Breech-loading	M. L. M. Descontures	Paris, France	July 25, 1865	49, 057
Fire-arm, Breech-loading	F. V. Diaz	New York, N. Y	Sept. 7, 1869	94, 577
Fire-arm, Breech-loading	W. C. Dodge	Washington, D. C	Sept. 20, 1864	44, 290
Fire-arm, Breech-loading	W. C. Dodge	Washington, D. C	Feb. 13, 1866	52, 547
Fire-arm, Breech-loading	W. C. Dodge	Washington, D. C	Mar. 14, 1871	112, 763
Fire-arm, Breech-loading	W. C. Dodge	Washington, D. C	Apr. 4, 1871	113, 408
Fire-arm, Breech-loading	W. C. Dodge	Washington, D. C	May 9, 1871	114, 653
Fire-arm, Breech-loading	W. C. and P. T. Dodge	Washington, D. C	Mar. 14, 1871	112, 694
Fire-arm, Breech-loading	W. C. and P. T. Dodge	Washington, D. C	Aug. 22, 1871	118, 350
Fire-arm, Breech-loading	W. C. and P. T. Dodge	Washington, D. C	June 11, 1872	127, 683
Fire-arm, Breech-loading	J. D. Dougall	Westminster, Great Britain.	Sept. 5, 1865	49, 844
Fire-arm, Breech-loading	J. Duval	Laprairie, Canada	Mar. 14, 1871	112, 565
Fire-arm, Breech-loading	J. Duval	Laprairie, Canada	Jan. 30, 1872	123, 159
Fire-arm, Breech-loading	J. Duval	Montreal, Canada	Dec. 16, 1873	145, 494
Fire-arm, Breech-loading	G. H. Earnest	Springfield, Ohio	July 16, 1872	129, 115
Fire-arm, Breech-loading	W. H. Elliot	Plattsburgh, N. Y	May 13, 1862	35, 284
Fire-arm, Breech-loading	W. H. Elliot	Plattsburgh, N. Y	July 7, 1863	39, 136
Fire-arm, Breech-loading	W. H. Elliot	Plattsburgh, N. Y	Apr. 18, 1865	47, 372
Fire-arm, Breech-loading	W. H. Elliot	Plattsburgh, N. Y	May 23, 1865	47, 809
Fire-arm, Breech-loading	W. H. Elliot	New York, N. Y	Dec. 13, 1870	110, 024
Fire-arm, Breech-loading	W. H. Elliot	New York, N. Y	May 9, 1871	114, 540
Fire-arm, Breech-loading	W. H. Elliot	New York, N. Y	Dec. 5, 1871	121, 499
Fire-arm, Breech-loading	W. H. Elliot	New York, N. Y	Apr. 2, 1872	125, 127
Fire-arm, Breech-loading	W. C. Ellis	Springfield, Mass	Apr. 26, 1859	23, 762
Fire-arm, Breech-loading	J. Elson	Boston, Mass	May 14, 1867	64, 650
Fire-arm, Breech-loading	J. Elson	Boston, Mass	July 23, 1867	67, 033
Fire-arm, Breech-loading	J. Elson	Boston, Mass	Nov. 19, 1867	71, 149
Fire-arm, Breech-loading	J. Elson and W. R. Schaefer	Boston, Mass	Feb. 2, 1869	86, 378
Fire-arm, Breech-loading	A. B. Ely and E. C. Clay	Newton and Malden, Mass.	July 5, 1870	105, 058
Fire-arm, Breech-loading	F. H. Escherich	Baltimore, Md	June 2, 1868	78, 519
Fire-arm, Breech-loading	L. T. Fairbanks	Worcester, Mass	June 22, 1869	91, 616
Fire-arm, Breech-loading	G. H. Ferriss	Utica, N. Y	Oct. 10, 1871	119, 834
Fire-arm, Breech-loading	V. Fogerty	Boston, Mass	Feb. 2, 1869	86, 520
Fire-arm, Breech-loading	V. Fogerty	Boston, Mass	July 25, 1871	117, 398
Fire-arm, Breech-loading	G. P. Foster	Providence, R. I	Apr. 10, 1860	27, 874
Fire-arm, Breech-loading	G. P. and G. F. Foster	Mohawk, N. Y	July 17, 1866	56, 399
Fire-arm, Breech-loading	G. H. Fox	Boston, Mass	Jan. 4, 1870	98, 579
Fire-arm, Breech-loading	A. T. Freeman	Herkimer, N. Y	Jan. 16, 1872	122, 717
Fire-arm, Breech-loading	A. T. Freeman	Herkimer, N. Y	Dec. 10, 1872	133, 770
Fire-arm, Breech-loading	M. J. Gallager	Savannah, Ga	July 17, 1860	29, 157
Fire-arm, Breech-loading	M. J. Gallager and W. H. Gladding.	Savannah, Ga	July 12, 1859	24, 730
Fire-arm, Breech-loading	L. Geiger	Hudson, N. Y	Jan. 27, 1863	37, 501

Index of patents issued from the United States Patent Office from 1790 *to* 1873, *inclusive*—Continued.

Invention.	Inventor.	Residence.	Date.	No.
Fire-arm, Breech-loading	S. Gerngross	Saint Louis, Mo	Dec. 20, 1870	110, 353
Fire-arm, Breech-loading	L. H. Gibbs	New York, N. Y	Jan. 8, 1856	14, 057
Fire-arm, Breech-loading	W. Golcher	Saint Paul, Minn	Mar. 30, 1869	88, 470
Fire-arm, Breech-loading	W. Golcher	Saint Paul, Minn	Oct. 19, 1869	95, 998
Fire-arm, Breech-loading	R. Gesham	New York, N. Y	Feb. 27, 1872	124, 056
Fire-arm, Breech-loading	T. P. Gould	Niagara Falls, N. Y	Jan. 3, 1860	26, 734
Fire-arm, Breech-loading	J. Goulding	Worcester, Mass	May 3, 1864	42, 573
Fire-arm, Breech-loading	G. B. Gray and J. H. Romans	Mount Vernon, Ohio	Mar. 21, 1871	112, 803
Fire-arm, Breech-loading	J. Gray	Medford, Mass	Nov. 8, 1864	44, 995
Fire-arm, Breech-loading	C. Green	Rochester, N Y	Dec. 6, 1870	109, 890
Fire-arm, Breech-loading	J. D. Greene	Cambridge, Mass	Jan. 3, 1854	10, 391
Fire-arm, Breech-loading	J. D. Greene	Cambridge, Mass	June 27, 1854	11, 157
Fire-arm, Breech-loading	J. D. Greene	Cambridge, Mass	Nov. 17, 1857	18, 634
Fire-arm, Breech-loading	J. D. Greene	United States Army	Feb. 18, 1862	34, 422
Fire-arm, Breech-loading	J. D. Greene	Cambridge, Mass	Mar. 23, 1869	88, 161
Fire-arm, Breech-loading	A. Grillet	Philadelphia, Pa	Nov. 22, 1864	45, 152
Fire-arm, Breech-loading	H. Gross	Seneca County, Ohio	June 10, 1856	15, 072
Fire-arm, Breech-loading	H. Gross	Tiffin, Ohio	Aug. 30, 1859	25, 259
Fire-arm, Breech-loading	H. Gross	Tiffin, Ohio	Aug. 11, 1863	39, 479
Fire-arm, Breech-loading	H. Gross	Tiffin, Ohio	Aug. 25, 1863	39, 646
Fire-arm, Breech-loading	H. Gross	Tiffin, Ohio	May 31, 1864	42, 941
Fire-arm, Breech-loading	F. Gueury	Paris, France	Dec. 24, 1872	134, 200
Fire-arm, Breech-loading	G. Gundersen	Chicago, Ill	Dec. 30, 1873	145, 998
Fire-arm, Breech-loading	E. F. Gunn	Charleston, S. C	Sept. 10, 1867	68, 736
Fire-arm, Breech-loading	E. F. Gunn	Charleston, S. C	Dec. 29, 1868	85, 442
Fire-arm, Breech-loading	E. Gwyn and A. C. Campbell	Hamilton, Ohio	Oct. 21, 1862	36, 709
Fire-arm, Breech-loading	A. Hamilton	Broad Brook, Conn	Nov. 19, 1861	33, 769
Fire-arm, Breech-loading	H. Hammond	Providence, R. I	Oct. 25, 1864	44, 798
Fire-arm, Breech-loading	H. Hammond	Hartford, Conn	Mar. 14, 1871	112, 589
Fire-arm, Breech-loading	G. Hancock	Providence, R. I	Apr. 26, 1864	42, 471
Fire-arm, Breech-loading	J. Hanson	Raschliffe, near Huddersfield, England.	Nov. 29, 1870	109, 731
Fire-arm, Breech-loading	I. Hartshorn	Providence, R. I	Mar. 31, 1863	38, 042
Fire-arm, Breech-loading	H. W. Hayden	Waterbury, Conn	Dec. 20, 1864	45, 495
Fire-arm, Breech-loading	J. S. Heath	London, England	May 13, 1873	138, 887
Fire-arm, Breech-loading	J. A. Heckenbach.	Mayville, Wis	June 22, 1869	91, 624
Fire-arm, Breech-loading	A. Henry	Edinburgh, North Britain	Feb. 13, 1866	52, 654
Fire-arm, Breech-loading	A. Henry	Edinburgh, North Britain	Sept. 19, 1871	119, 145
Fire-arm, Breech-loading	A. Henry	Edinburgh, Scotland	Dec. 30, 1873	145, 944
Fire-arm, Breech-loading	W. C. Hicks	New York, N. Y	Mar. 1, 1864	41, 814
Fire-arm, Breech-loading	A. V. Hill	Hinsdale, N. Y	Aug. 2, 1859	24, 936
Fire-arm, Breech-loading	W. D. Hillis	Joliet, Ill	Sept. 20, 1864	44, 312
Fire-arm, Breech-loading	A. J. H. Hilton	Boston, Mass	July 16, 1867	66, 709
Fire-arm, Breech-loading	M. J. Hinden	Detroit, Mich	June 29, 1869	92, 048
Fire-arm, Breech-loading	C. B. Holden	Worcester, Mass	Apr. 1, 1862	34, 859
Fire-arm, Breech-loading	C. B. Holden	Worcester, Mass	Mar. 29, 1864	42, 139
Fire-arm, Breech-loading	G. L. Holt and J. C. Marshall	Springfield, Mass	Apr. 22, 1873	138, 157
Fire-arm, Breech-loading	H. H. Hopkins	Norwich, Conn	Sept. 23, 1873	143, 012
Fire-arm, Breech-loading	H. Hoppenau	Kansas City, Mo	Mar. 18, 1873	136, 998
Fire-arm, Breech-loading	B. B. Hotchkiss	New York, N. Y	Aug. 17, 1869	93, 822
Fire-arm, Breech-loading	B. B. Hotchkiss	New York, N. Y	Feb. 15, 1870	99, 898
Fire-arm, Breech-loading	B. B. Hotchkiss	New York, N. Y	Jan. 2, 1872	122, 465
Fire-arm, Breech-loading	C. Howard	New York, N. Y	Sept. 26, 1865	50, 125
Fire-arm, Breech-loading	C. Howard	New York, N. Y	Oct. 10, 1865	50, 358
Fire-arm, Breech-loading	C. W. Howard	Hammonton, N. J	July 14, 1863	39, 232
Fire-arm, Breech-loading	S. Howard	Elyria, Ohio	Oct. 28, 1862	36, 779
Fire-arm, Breech-loading	F. W. Howe	Providence, R. I	Sept. 16, 1862	36, 466
Fire-arm, Breech-loading	F. W. Howe	Providence, R. I	Mar. 7, 1865	46, 671
Fire-arm, Breech-loading	W. W. Hubbell	Philadelphia, Pa	June 18, 1867	65, 812
Fire-arm, Breech-loading	D. Hug	New York, N. Y	Sept. 2, 1873	142, 396
Fire-arm, Breech-loading	C. Jackson and T. Goodrem	Providence, R. I	Mar. 7, 1863	37, 937
Fire-arm, Breech-loading	A. E. and P. J. Jarre	Paris, France	Apr. 15, 1873	137, 927
Fire-arm, Breech-loading	B. H. Jenks	Bridesburgh, Pa	Feb. 25, 1868	74, 760
Fire-arm, Breech-loading	L. Jennings	New York, N. Y	Dec. 25, 1849	6, 973
Fire-arm, Breech-loading	W. Johnston	Cincinnati, Ohio	May 13, 1862	35, 241
Fire-arm, Breech-loading	W. Johnston	Cincinnati, Ohio	Nov. 1, 1864	44, 868
Fire-arm, Breech-loading	F. Jonas	McConnel's Grove, Ill	Oct. 2, 1860	30, 228
Fire-arm, Breech-loading	B. F. Joslyn	Worcester, Mass	Aug. 28, 1855	13, 507
Fire-arm, Breech-loading	B. F. Joslyn	Worcester, Mass	July 1, 1856	15, 240
Fire-arm, Breech-loading	B. F. Joslyn	Stonington, Conn	June 24, 1862	35, 688
Fire-arm, Breech-loading	B. F. Joslyn	Stonington, Conn	Aug. 4, 1863	39, 407
Fire-arm, Breech-loading	B. F. Joslyn	Stonington, Conn	Mar. 22, 1864	42, 000
Fire-arm, Breech-loading	B. F. Joslyn	Stonington, Conn	June 6, 1865	48, 073
Fire-arm, Breech-loading	B. F. Joslyn	Stonington, Conn	June 20, 1865	48, 288
Fire-arm, Breech-loading	B. F. Joslyn	Stonington, Conn	Jan. 2, 1866	51, 837
Fire-arm, Breech-loading	B. F. Joslyn	New York, N. Y	Nov. 15, 1870	109, 218
Fire-arm, Breech-loading	A. B. Kay	Newark, N. J	Nov. 22, 1870	109, 419
Fire-arm, Breech-loading	H. Kellogg	New Haven, Conn	May 20, 1862	35, 356
Fire-arm, Breech-loading	A. Krupp	Essen, Prussia	Aug. 9, 1864	43, 820
Fire-arm, Breech-loading	T. T. S. Laidley and C. A. Emery	United States Army and Springfield, Mass.	May 15, 1866	54, 743
Fire-arm, Breech-loading	W. R. Landfear	Hartford, Conn	Sept. 6, 1864	44, 099
Fire-arm, Breech-loading	R. S. Lawrence	Windsor, Vt	Jan. 6, 1852	8, 637
Fire-arm, Breech-loading	R. S. Lawrence	Hartford, Conn	Dec. 20, 1859	26, 504
Fire-arm, Breech-loading	R. S. Lawrence	Hartford, Conn	Apr. 6, 1869	88, 645
Fire-arm, Breech-loading	D. Leavitt	Chicopee, Mass	June 14, 1859	24, 394
Fire-arm, Breech-loading	J. Lee	Stevens Point, Wis	July 22, 1862	35, 941
Fire-arm, Breech-loading	J. Lee	Milwaukee, Wis	May 15, 1866	54, 744
Fire-arm, Breech-loading	J. Lee	Milwaukee, Wis	May 16, 1871	114, 951
Fire-arm, Breech-loading	J. Lee	Milwaukee, Wis	June 20, 1871	116, 068
Fire-arm, Breech-loading	J. Lee	Milwaukee, Wis	Jan. 2, 1872	122, 470
Fire-arm, Breech-loading	J. Lee	Milwaukee, Wis	Jan. 16, 1872	122, 772
Fire-arm, Breech-loading	T. Lee	New York, N. Y	Apr. 27, 1858	20, 073
Fire-arm, Breech-loading	T. Lee	Newark, N. J	Nov. 19, 1861	33, 745
Fire-arm, Breech-loading	J. Letort and H. S. Mathews	Wytheville, Va	Apr. 3, 1860	27, 723
Fire-arm, Breech-loading	E. Lindner	New York, N. Y	Mar. 29, 1859	23, 378
Fire-arm, Breech-loading	H. Lord	Hartford, Conn	Feb. 11, 1868	74, 387
Fire-arm, Breech-loading	O. D. Lull	Watkins, N. Y	June 16, 1863	38, 903

Index of patents issued from the United States Patent Office from 1790 *to* 1873, *inclusive*—Continued.

Invention.	Inventor.	Residence.	Date.	No.
Fire-arm, Breech-loading	J. Manton	Montreal, Canada	Aug. 1, 1871	117, 552
Fire-arm, Breech-loading	A. Marelli	Milan, Italy	May 27, 1873	139, 323
Fire-arm, Breech-loading	S. W. Marsh	Washington, D. C	Dec. 6, 1859	26, 362
Fire-arm, Breech-loading	S. W. Marsh	Washington, D. C	Nov. 5, 1861	33, 655
Fire-arm, Breech-loading	J. P. Marshall	Millbury, Mass	Oct. 4, 1859	25, 661
Fire-arm, Breech-loading	J. P. Marshall	Millbury, Mass	Apr. 29, 1862	35, 107
Fire-arm, Breech-loading	J. M. Mason	Washington, D. C	Mar. 7, 1871	112, 523
Fire-arm, Breech-loading	J. M. Mason	Washington, D. C	Aug. 8, 1871	117, 906
Fire-arm, Breech-loading	F. Maton	New York, N. Y	Nov. 14, 1854	11, 938
Fire-arm, Breech-loading	E. Maynard	Washington, D. C	May 27, 1851	8, 126
Fire-arm, Breech-loading	E. Maynard	Washington, D. C	Dec. 6, 1859	26, 364
Fire-arm, Breech-loading	E. Maynard	Washington, D. C	Oct. 30, 1860	30, 537
Fire-arm, Breech-loading	E. Maynard	Washington, D. C	June 27, 1865	48, 423
Fire-arm, Breech-loading	E. Maynard	Washington, D. C	Aug. 1, 1865	49, 130
Fire-arm, Breech-loading	E. Maynard	Washington, D. C	Feb. 2, 1869	86, 566
Fire-arm, Breech-loading	E. Maynard	New York, N. Y	Feb. 18, 1873	135, 928
Fire-arm, Breech-loading	J. E. McBeth	New Orleans, La	Jan. 14, 1868	73, 357
Fire-arm, Breech-loading	J. E. McBeth	New Orleans, La	Aug. 11, 1868	80, 985
Fire-arm, Breech-loading	R. McChesney	Ilion, N. Y	Oct. 2, 1866	58, 444
Fire-arm, Breech-loading	R. McChesney	Utica, N. Y	May 28, 1867	65, 103
Fire-arm, Breech-loading	J. McGoveren	New York, N. Y	Apr. 13, 1869	88, 890
Fire-arm, Breech-loading	J. V. Meigs	Washington, D. C	Oct. 21, 1862	36, 721
Fire-arm, Breech-loading	J. V. Meigs	Washington, D. C	May 22, 1866	54, 934
Fire-arm, Breech-loading	J. V. Meigs	Washington, D. C	Aug. 18, 1868	81, 100
Fire-arm, Breech-loading	D. F. Mellen	Manchester, N. H	Oct. 4, 1864	44, 545
Fire-arm, Breech-loading	J. Merlett	Boundbrook, N. J	Aug. 18, 1868	81, 283
Fire-arm, Breech-loading	L. A. Merriam	New York, N. Y	Feb. 16, 1869	87, 058
Fire-arm, Breech-loading	G. Merrill	East Orange, N. J	Oct. 17, 1871	119, 939
Fire-arm, Breech-loading	G. Merrill	East Orange, N. J	Oct. 17, 1871	119, 940
Fire-arm, Breech-loading	J. H. Merrill	Baltimore, Md	July 20, 1858	20, 954
Fire-arm, Breech-loading	J. H. Merrill	Baltimore, Md	Apr. 9, 1861	32, 032
Fire-arm, Breech-loading	J. H. Merrill	Baltimore, Md	Apr. 9, 1861	32, 033
Fire-arm, Breech-loading	J. H. Merrill	Baltimore, Md	Oct. 22, 1861	33, 536
Fire-arm, Breech-loading	J. H. Merrill	Baltimore, Md	Dec. 8, 1863	40, 884
Fire-arm, Breech-loading	I. M. Milbank	Greenfield Hill, Conn	Dec. 2, 1862	37, 048
Fire-arm, Breech-loading	I. M. Milbank	Greenfield Hill, Conn	Jan. 31, 1865	46, 125
Fire-arm, Breech-loading	I. M. Milbank	Greenfield Hill, Conn	Feb. 20, 1866	52, 734
Fire-arm, Breech-loading	I. M. Milbank	Greenfield Hill, Conn	June 12, 1866	55, 520
Fire-arm, Breech-loading	I. M. Milbank	Greenfield Hill, Conn	Jan. 8, 1867	61, 082
Fire-arm, Breech-loading	I. M. Milbank	Greenfield Hill, Conn	Feb. 5, 1867	61, 751
Fire-arm, Breech-loading	I. M. Milbank	Greenfield Hill, Conn	June 11, 1867	65, 585
Fire-arm, Breech-loading	I. M. Milbank	Greenfield Hill, Conn	Dec. 1, 1868	84, 566
Fire-arm, Breech-loading	I. M. Milbank	Greenfield Hill, Conn	Apr. 16, 1872	125, 829
Fire-arm, Breech-loading	I. M. Milbank	Greenfield Hill, Conn	Mar. 18, 1873	136, 850
Fire-arm, Breech-loading	W. H. Miller	West Meriden, Conn	Nov. 13, 1866	59, 723
Fire-arm, Breech-loading	W. H. and G. W. Miller	Meriden, Conn	May 23, 1865	47, 902
Fire-arm, Breech-loading	W. H. and G. W. Miller	West Meriden, Conn	Dec. 26, 1865	51, 739
Fire-arm, Breech-loading	W. H. and G. W. Miller	West Meriden, Conn	May 14, 1867	64, 786
Fire-arm, Breech-loading	J. K. Millner	New York, N. Y	Feb. 17, 1863	37, 723
Fire-arm, Breech-loading	E. M. Mix and H. B. Horton	Ithaca, N. Y	Jan. 19, 1864	41, 343
Fire-arm, Breech-loading	D. Moore	Brooklyn, N. Y	Dec. 3, 1861	33, 847
Fire-arm, Breech-loading	W. Morgenstern	Philadelphia, Pa	Nov. 29, 1864	45, 262
Fire-arm, Breech-loading	W. Morgenstern	Philadelphia, Pa	June 6, 1865	48, 133
Fire-arm, Breech-loading	W. Morgenstern	Hartford, Conn	Dec. 24, 1867	72, 526
Fire-arm, Breech-loading	W. Morgenstern	Philadelphia, Pa	Feb. 18, 1868	74, 712
Fire-arm, Breech-loading	W. Morgenstern	Hartford, Conn	June 23, 1868	79, 291
Fire-arm, Breech-loading	W. Morgenstern	New York, N. Y	Feb. 2, 1869	86, 434
Fire-arm, Breech-loading	W. Morgenstern	New York, N. Y	Feb. 23, 1869	87, 190
Fire-arm, Breech-loading	W. Morgenstern	New York, N. Y	Aug. 3, 1869	93, 330
Fire-arm, Breech-loading	W. Morgenstern and E. Morwitz.	Philadelphia, Pa	Nov. 10, 1863	40, 572
Fire-arm, Breech-loading	S. Morris and W. and P. Mauser	Springfield, Mass., and Oberndorf, Würtemberg.	June 2, 1868	78, 603
Fire-arm, Breech-loading	G. W. Morse	Baton Rouge, La	Oct. 28, 1856	15, 995
Fire-arm, Breech-loading	G. W. Morse	Baton Rouge, La	June 8, 1858	20, 503
Fire-arm, Breech-loading	M. Moses	Malone, N. Y	Sept. 30, 1862	36, 571
Fire-arm, Breech-loading	A. Muller	Paris, France	May 24, 1870	103, 488
Fire-arm, Breech-loading	F. Muller	Hartford, Conn	Feb. 4, 1868	74, 119
Fire-arm, Breech-loading	A. S. Munger	Chicopee Falls, Mass	Mar. 12, 1867	62, 873
Fire-arm, Breech-loading	S. and G. H. Needham	London, England	May 21, 1867	64, 999
Fire-arm, Breech-loading	R. Nenninger	Newark, N. Y	Mar. 28, 1871	113, 194
Fire-arm, Breech-loading	F. D. Newbury	New York, N. Y	Jan. 9, 1867	51, 959
Fire-arm, Breech-loading	J. D. S. Newell	Tensas Parish, La	Apr. 6, 1869	88, 730
Fire-arm, Breech-loading	J. D. S. Newell	Tensas Parish, La	May 25, 1869	90, 381
Fire-arm, Breech-loading	A. N. Newton	Richmond, Ind	June 27, 1854	11, 198
Fire-arm, Breech-loading	A. N. Newton	Richmond, Ind	Sept. 19, 1854	11, 700
Fire-arm, Breech-loading	F. E. M. Nillus	Havre, France	May 7, 1872	126, 568
Fire-arm, Breech-loading	P. J. J. Noël	Paris, France	Dec. 22, 1868	85, 120
Fire-arm, Breech-loading	J. C. Nye	Cincinnati, Ohio	Oct. 28, 1862	36, 852
Fire-arm, Breech-loading	J. C. Nye	Cincinnati, Ohio	Jan. 6, 1863	37, 356
Fire-arm, Breech-loading	J. Oliphant	Uniontown, Pa	Jan. 13, 1863	37, 407
Fire-arm, Breech-loading	W. Palmer	New York, N. Y	July 23, 1861	32, 887
Fire-arm, Breech-loading	W. Palmer	New York, N. Y	Dec. 22, 1863	41, 017
Fire-arm, Breech-loading	W. R. Pape	Newcastle-upon-Tyne, Great Britain.	Nov. 5, 1867	70, 463
Fire-arm, Breech-loading	W. R. Pape	Newcastle-upon-Tyne, England.	Nov. 24, 1868	84, 373
Fire-arm, Breech-loading	C. F. Payne	Gardner, Mass	May 10, 1864	42, 685
Fire-arm, Breech-loading	H. O. Peabody	Boston, Mass	July 22, 1862	35, 947
Fire-arm, Breech-loading	H. O. Peabody	Providence, R. I	Dec. 10, 1867	72, 076
Fire-arm, Breech-loading	H. O. Peabody	Providence, R. I	Apr. 14, 1868	76, 805
Fire-arm, Breech-loading	J. Percy	Albany, N. Y	Aug. 11, 1863	39, 494
Fire-arm, Breech-loading	C. Perley	New York, N. Y	Feb. 24, 1863	37, 764
Fire-arm, Breech-loading	A. D. Perry	Newark, N. J	Nov. 28, 1854	12, 001
Fire-arm, Breech-loading	M. Pidault and G. Elize dit Lagieze.	Paris, France	Sept. 10, 1867	68, 786
Fire-arm, Breech-loading	G. R. Pierce	Grand Rapids, Mich	Oct. 3, 1871	119, 474

Index of patents issued from the United States Patent Office from 1790 to 1873, inclusive—Continued.

Invention.	Inventor.	Residence.	Date.	No.
Fire-arm, Breech-loading	T. Poultney and S. Crispin	Baltimore, Md., and New York, N. Y.	May 14, 1867	64, 701
Fire-arm, Breech-loading	J. W. Preston	Newton, Mass	Feb. 5, 1867	61, 865
Fire-arm, Breech-loading	F. B. Prindle	New Haven, Conn	Nov. 28, 1865	51, 213
Fire-arm, Breech-loading	S. S. Rembert	Memphis, Tenn	Oct. 29, 1867	70, 264
Fire-arm, Breech-loading	S. S. Rembert	Memphis, Tenn	Feb. 18, 1868	74, 594
Fire-arm, Breech-loading	S. Remington	Ilion, N. Y	Feb. 9, 1869	86, 690
Fire-arm, Breech-loading	E. S. Renwick	New York, N. Y	Apr. 26, 1870	102, 434
Fire-arm, Breech-loading	T. Restell	London, England	Mar. 26, 1867	63, 303
Fire-arm, Breech-loading	T. Restell	Birmingham, England	May 20, 1873	139, 190
Fire-arm, Breech-loading	H. Reynolds	New Haven, Conn	May 8, 1866	54, 600
Fire-arm, Breech-loading	C. B. Richards	Hartford, Conn	Aug. 18, 1868	81, 290
Fire arm, Breech-loading	W. Richards	Birmingham, England	July 14, 1863	39, 246
Fire-arm, Breech-loading	W. Richards	Birmingham, England	Oct. 10, 1865	50, 432
Fire-arm, Breech-loading	W. Richards	Birmingham, England	May 11, 1869	89, 889
Fire-arm, Breech-loading	W. Richards	Birmingham, England	June 22, 1869	91, 668
Fire-arm, Breech-loading	W. Richards	Birmingham, England	Dec. 3, 1872	133, 665
Fire-arm, Breech-loading	W. Richards	Birmingham, England	May 27, 1873	139, 422
Fire-arm, Breech-loading	G. J. Richardson	Philadelphia, Pa	Aug. 23, 1864	43, 929
Fire-arm, Breech-loading	J. Rider	Newark, Ohio	Sept. 13, 1859	25, 470
Fire-arm, Breech-loading	J. Rider	Newark, Ohio	Nov. 15, 1864	45, 123
Fire-arm, Breech-loading	J. Rider	Newark, Ohio	Jan. 3, 1865	45, 797
Fire-arm, Breech-loading	J. Rider	Newark, Ohio	Feb. 21, 1865	46, 532
Fire-arm, Breech-loading	J. Rider	Newark, Ohio	Mar. 27, 1866	53, 543
Fire-arm, Breech-loading	J. Rider	Newark, Ohio	Feb. 11, 1868	74, 428
Fire-arm, Breech-loading	J. Rider	Newark, Ohio	July 29, 1873	141, 383
Fire-arm, Breech-loading	J. Rider	Newark, Ohio	July 29, 1873	141, 384
Fire-arm, Breech-loading	B. S. Roberts	United States Army	Sept. 23, 1862	36, 531
Fire-arm, Breech-loading	B. S. Roberts	United States Army	Feb. 27, 1866	52, 887
Fire-arm, Breech-loading	B. S. Roberts	United States Army	June 11, 1867	65, 607
Fire-arm, Breech-loading	B. S. Roberts	United States Army	May 11, 1869	90, 024
Fire-arm, Breech-loading	W. H. Robertson and G. W. Simpson.	Hartford, Conn	Feb. 12, 1856	14, 253
Fire-arm, Breech-loading	W. H. Robertson and G. W. Simpson.	Hartford, Conn	Mar. 13, 1866	53, 187
Fire-arm, Breech-loading	O. M. Robinson	Upper Jay, N. Y	May 24, 1870	103, 504
Fire-arm, Breech-loading	O. M. Robinson	Plattsburgh, N. Y	Apr. 23, 1872	125, 988
Fire-arm, Breech-loading	E. K. Root	Hartford, Conn	June 4, 1867	65, 509
Fire-arm, Breech-loading	A. H. Rowe	Hartford, Conn	Apr. 5, 1864	42, 227
Fire-arm, Breech-loading	J. Ruler	Newark, Ohio	Dec. 8, 1863	40, 887
Fire-arm, Breech-loading	C. F. Russell	New Britain, Conn	May 14, 1872	126, 748
Fire-arm, Breech-loading	S. Rydbeck	Red Wing, Minn	June 28, 1870	104, 775
Fire-arm, Breech-loading	C. G. Saez	Madrid, Spain	Jan. 3, 1865	45, 801
Fire-arm, Breech-loading	E. L. Sargent	Watertown, N. Y	Mar. 1, 1870	100, 455
Fire-arm, Breech-loading	E. L. Sargent	Watertown, N. Y	June 21, 1870	104, 502
Fire-arm, Breech-loading	E. L. Sargent	Watertown, N. Y	Nov. 15, 1870	109, 255
Fire-arm, Breech-loading	A. Sayer	Naubuc, Conn	June 19, 1866	55, 719
Fire-arm, Breech-loading	J. Scharffe	New York, N. Y	Nov. 11, 1856	16, 070
Fire-arm, Breech-loading	J. P. Schenkl	Boston, Mass	Aug. 16, 1853	9, 943
Fire-arm, Breech-loading	J. P. Schenkl	Boston, Mass	June 23, 1857	17, 642
Fire-arm, Breech-loading	J. P. Schenkl	Worcester, Mass	Oct. 12, 1858	21, 802
Fire-arm, Breech-loading	H. A. Schesch	Ilion, N. Y	Apr. 9, 1872	125, 620
Fire-arm, Breech-loading	G. W. Schofield	United States Army	June 14, 1870	104, 211
Fire-arm, Breech-loading	G. W. Schofield	United States Army	June 20, 1871	116, 225
Fire-arm, Breech-loading	F. Schopp	New York, N. Y	Nov. 28, 1865	51, 225
Fire-arm, Breech-loading	H. Schroder	New York, N. Y	June 25, 1861	32, 653
Fire-arm, Breech-loading	H. Schroeder, L. Salewski, and W. Schmidt.	Bloomington, Ill	Dec. 23, 1856	16, 288
Fire-arm, Breech-loading	C. D. Schubarth	Providence, R. I	July 23, 1861	32, 895
Fire-arm, Breech-loading	P. Schuler	Morris, Ind	Jan. 5, 1869	85, 616
Fire-arm, Breech-loading	P. Schuler	Morris, Ind	Nov. 1, 1870	108, 836
Fire-arm, Breech-loading	G. Schulz	Fort Madison, Iowa	May 11, 1869	89, 947
Fire-arm, Breech-loading	W. Scott and W. J. Matthews	Birmingham and Aston, England.	Nov. 25, 1873	144, 870
Fire-arm, Breech-loading	W. M. Scott	Birmingham, England	Nov. 1, 1870	108, 942
Fire-arm, Breech-loading	J. H. Sears	Brantford, Canada	Dec. 20, 1859	26, 526
Fire-arm, Breech-loading	J. H. Selwyn	Grasmere, England	Aug. 14, 1866	57, 269
Fire-arm, Breech-loading	C. Sharps	Philadelphia, Pa	Jan. 25, 1859	22, 752
Fire-arm, Breech-loading	C. Sharps	Philadelphia, Pa	July 9, 1861	32, 790
Fire-arm, Breech-loading	C. Sharps	Philadelphia, Pa	Oct. 29, 1861	33, 607
Fire-arm, Breech-loading	C. Sharps	Philadelphia, Pa	Feb. 12, 1867	62, 077
Fire-arm, Breech-loading	C. Sharps	Philadelphia, Pa	Apr. 8, 1873	137, 625
Fire-arm, Breech-loading	H. N. Sherman	Beloit, Wis	Nov. 25, 1873	144, 872
Fire-arm, Breech-loading	T. E. Shull	Millersburgh, Pa	Apr. 5, 1859	23, 505
Fire-arm, Breech-loading	B. F. Skinner and A. Plummer, jr.	Mystic Bridge, Conn	Feb. 18, 1862	34, 449
Fire-arm, Breech-loading	C. D. Skinner and D. Tryon	Haddam and Middletown, Conn.	Oct. 20, 1857	18, 472
Fire-arm, Breech loading	J. Smiles	Birmingham, England	Dec. 27, 1870	110, 505
Fire-arm, Breech-loading	D. Smith	Springfield, Mass	July 16, 1872	129, 433
Fire-arm, Breech-loading	D. Smith	Springfield, Mass	Apr. 22, 1873	138, 207
Fire-arm, Breech-loading	D. Smith and M. J. Chamberlin	Springfield, Mass	Mar. 7, 1871	112, 505
Fire-arm, Breech-loading	D. Smith and J. C. Marshall	Springfield, Mass	Aug. 5, 1873	141, 603
Fire-arm, Breech-loading	G. Smith	Buttermilk Falls, N. Y	Dec. 25, 1855	14, 001
Fire-arm, Breech-loading	G. Smith	Buttermilk Falls, N. Y	Aug. 5, 1856	15, 496
Fire-arm, Breech-loading	G. Smith	Buttermilk Falls, N. Y	June 23, 1857	17, 644
Fire-arm, Breech-loading	H. Smith	Norwich, Conn	Aug. 26, 1851	8, 317
Fire-arm, Breech-loading	I. Smith	New York, N. Y	Apr. 26, 1864	42, 542
Fire-arm, Breech-loading	W. H. Smith	Birmingham, Conn	Dec. 10, 1861	33, 907
Fire-arm, Breech-loading	W. H. Smith	Birmingham, Conn	Aug. 23, 1864	43, 957
Fire-arm, Breech-loading	W. S. Smoot	Washington, D. C	June 1, 1869	90, 792
Fire-arm, Breech loading	W. S. Smoot	Ilion, N. Y	June 20, 1871	116, 106
Fire-arm, Breech-loading	W. S. Smoot	Ilion, N. Y	Nov. 7, 1871	120, 788
Fire-arm, Breech-loading	W. S. Smoot	Ilion, N. Y	Nov. 12, 1872	133, 063
Fire-arm, Breech loading	W. T. Snedden	Johnstown, Pa	June 27, 1871	116, 363
Fire-arm, Breech-loading	W. T. Snedden	Johnstown, Pa	June 27, 1871	116, 364
Fire-arm, Breech-loading	C. E. Sneider	Baltimore, Md	Mar. 20, 1860	27, 600
Fire-arm, Breech-loading	C. E. Sneider	Baltimore, Md	Aug. 25, 1863	39, 707

Index of patents issued from the United States Patent Office from 1790 to 1873, inclusive—Continued.

Invention.	Inventor.	Residence.	Date.	No.
Fire-arm, Breech-loading	C. E. Sneider	Baltimore, Md	May 16, 1865	47, 755
Fire-arm, Breech-loading	C. E. Sneider	Baltimore, Md	Dec. 22, 1868	85, 252
Fire-arm, Breech-loading	J. Snider, jr	Philadelphia, Pa	Oct. 15, 1867	69, 941
Fire-arm, Breech-loading	W. Soper	Reading, England	June 14, 1870	104, 223
Fire-arm, Breech-loading	G. H. Soule	Jersey City, N. J	Apr. 3, 1855	12, 655
Fire-arm, Breech-loading	G. H. Soule	Jersey City, N. J	July 6, 1858	20, 825
Fire-arm, Breech loading	A. Spellerberg	Philadelphia, Pa	July 30, 1861	32, 929
Fire-arm, Breech-loading	C. M. Spencer	South Manchester, Conn	Feb. 4, 1862	34, 319
Fire-arm, Breech-loading	C. M. Spencer	Hartford, Conn	Feb. 11, 1873	135, 671
Fire-arm, Breech-loading	S. F. Stanton	Manchester, N. H	Apr. 29, 1856	14, 780
Fire-arm, Breech-loading	E. T. Starr	New York, N. Y	Sept. 14, 1858	21, 523
Fire-arm, Breech-loading	R. E. Stephens	Owen Sound, Canada	June 11, 1867	65, 704
Fire-arm, Breech-loading	G. R. Stetson	New Haven, Conn	July 4, 1871	116, 642
Fire-arm, Breech-loading	A. C. Stevens	Hudson, N. Y	May 4, 1869	89, 699
Fire-arm, Breech-loading	J. Stevens	Chicopee Falls, Mass	Sept. 6, 1864	44, 123
Fire-arm, Breech-loading	W. X. Stevens	Worcester, Mass	Jan. 12, 1864	41, 242
Fire-arm, Breech-loading	J. Stillman	Springfield, Mass	Oct. 17, 1865	50, 507
Fire-arm, Breech-loading	J. T. Stoakes	Champlain, N. Y	July 6, 1869	92, 393
Fire-arm, Breech-loading	W. M. Storm	New York, N. Y	July 8, 1856	15, 307
Fire-arm, Breech-loading	W. M. Storm	New York, N. Y	June 14, 1859	24, 414
Fire-arm, Breech-loading	W. M. Storm	New York, N. Y	Nov. 5, 1872	132, 740
Fire-arm, Breech-loading	S. Strong	Washington, D. C	Dec. 16, 1862	37, 208
Fire-arm, Breech-loading	S. Strong	Washington, D. C	May 19, 1863	38, 643
Fire-arm, Breech-loading	S. Strong	Washington, D. C	May 19, 1863	38, 644
Fire-arm, Breech-loading	T. L. Sturtevant	Boston, Mass	Sept. 19, 1865	50, 048
Fire-arm, Breech-loading	T. L. Sturtevant	Boston, Mass	Dec. 18, 1866	60, 592
Fire-arm, Breech-loading	I. Sutvan	Bridesburgh, Pa	Mar. 14, 1865	46, 866
Fire-arm, Breech-loading	J. F. Swinburn	Birmingham, England	Dec. 17, 1872	134, 014
Fire-arm, Breech-loading	J. C. Symmes	Watertown, Mass	Nov. 16, 1858	22, 094
Fire-arm, Breech-loading	T. D. Sympson, G. B. Gray, and J. H. Romans.	Mount Vernon, Ohio	Aug. 2, 1870	106, 083
Fire-arm, Breech-loading	J. P. Taylor	Elizabethton, Tenn	May 6, 1873	138, 711
Fire-arm, Breech loading	W. Terry	Birmingham, England	Oct. 14, 1862	36, 681
Fire-arm, Breech-loading	J. F. Thomas	Ilion, N. Y	Apr. 2, 1872	125, 229
Fire-arm, Breech-loading	J. F. Thomas	Ilion, N. Y	May 28, 1872	127, 386
Fire-arm, Breech-loading	D. C. Thrasher and B. F. Aiken	Freetown, Mass	July 16, 1867	66, 913
Fire-arm, Breech-loading	F. A. Thuer	East Hartford, Conn	July 12, 1870	105, 388
Fire-arm, Breech-loading	W. Tibbals	South Coventry, Conn	Nov. 28, 1865	51, 243
Fire-arm, Breech-loading	G. H. Tibbets	Augusta, Me	Feb. 13, 1872	123, 595
Fire-arm, Breech-loading	L. B. Tiebel	Hudson City, N. J	May 11, 1869	89, 955
Fire-arm, Breech-loading	F. Tiesing and C. Gerner	New Haven, Conn	Apr. 4, 1871	113, 470
Fire-arm, Breech-loading	F. Tiesing and C. Gerner	New Haven, Conn	Apr. 25, 1871	114, 230
Fire-arm, Breech-loading	G. H. Todd	Montgomery, Ala	July 27, 1869	93, 023
Fire-arm, Breech-loading	I. Townsend	Albany, N. Y	Jan. 29, 1861	31, 268
Fire-arm, Breech-loading	F. Townsend and N. S. Clement.	Albany, N. Y., and Worcester, Mass.	Sept. 6, 1864	44, 127
Fire-arm, Breech-loading	H. Underwood	Tolland, Conn	June 2, 1863	38, 772
Fire-arm, Breech-loading	H. Updegraff	Fort Laramie, Wyo	Sept. 19, 1871	119, 098
Fire-arm, Breech-loading	H. Updegraff	Smithfield, Ohio	Aug. 6, 1872	130, 165
Fire-arm, Breech-loading	S. F. Van Choate	Boston, Mass	May 11, 1869	89, 902
Fire-arm, Breech-loading	S. F. Van Choate	Boston, Mass	Aug. 24, 1869	94, 047
Fire-arm, Breech-loading	S. F. Van Choate	Boston, Mass	June 13, 1871	115, 911
Fire-arm, Breech-loading	S. F. Van Choate	Boston, Mass	Oct. 22, 1872	132, 505
Fire-arm, Breech-loading	A. L. Varney	Watertown, Mass	Mar. 30, 1869	88, 530
Fire-arm, Breech-loading	A. L. Varney	Watertown, Mass	Mar. 30, 1869	88, 531
Fire-arm, Breech-loading	A. L. Varney	Watertown, Mass	Sept. 28, 1869	95, 395
Fire-arm, Breech-loading	F. Vetterlin	Newhausen, Switzerland	Nov. 15, 1870	109, 277
Fire-arm, Breech-loading	F. J. Vittum and E. M. Stevens	Boston and Medford, Mass	Oct. 22, 1861	33, 569
Fire-arm, Breech-loading	J. Von der Poppenburg	Birmingham, England	Oct. 24, 1865	50, 670
Fire-arm, Breech loading	E. Von Jeinsen	New York, N. Y	Dec. 15, 1868	84, 922
Fire-arm, Breech-loading	E. Von Martini	Frauenfeld, Switzerland	May 25, 1869	90, 614
Fire-arm, Breech-loading	F. Von Martini	Frauenfeld, Switzerland	May 30, 1871	115, 546
Fire-arm, Breech-loading	F. Von Martini	Frauenfeld, Switzerland	Nov. 7, 1871	120, 800
Fire-arm, Breech-loading	F. Von Martini	Frauenfeld, Switzerland	Oct. 15, 1872	132, 222
Fire-arm, Breech-loading	H. Walker	Handsworth, England	Sept. 17, 1872	131, 484
Fire-arm, Breech-loading	J. M. Wampler	Loudoun County, Va	Mar. 6, 1860	27, 399
Fire-arm, Breech-loading	W. G. Ward	New York, N. Y	Aug. 31, 1869	94, 458
Fire-arm, Breech-loading	W. G. Ward	New York, N. Y	Dec. 7, 1869	97, 734
Fire-arm, Breech-loading	W. G. Ward	Edgewater, N. Y	Feb. 1, 1870	99, 504
Fire-arm, Breech-loading	W. G. Ward	Edgewater, N. Y	Feb. 21, 1871	111, 994
Fire-arm, Breech-loading	J. Warner	Springfield, Mass	Dec. 27, 1864	45, 660
Fire-arm, Breech-loading	J. Warner	Springfield, Mass	Feb. 23, 1864	47, 732
Fire-arm, Breech-loading	T. A. Washington	United States Army	Oct. 28, 1856	15, 990
Fire-arm, Breech-loading	A. T. Watson	Castleton, N. Y	Mar. 20, 1855	12, 567
Fire-arm, Breech-loading	J. B. Wayne	Birmingham, England	Aug. 15, 1871	118, 171
Fire-arm, Breech-loading	H. B. Weaver	South Windham, Conn	Oct. 16, 1855	13, 691
Fire-arm, Breech-loading	T. W. Webley	Birmingham, England	June 11, 1867	65, 783
Fire-arm, Breech-loading	J. Wernal	Steyer, Austria	Feb. 18, 1868	74, 737
Fire-arm, Breech-loading	D. B. Wesson	Springfield, Mass	Dec. 17, 1867	72, 434
Fire-arm, Breech-loading	D. B. Wesson	Springfield, Mass	June 9, 1868	78, 847
Fire-arm, Breech-loading	F. Wesson	Worcester, Mass	Nov. 11, 1862	36, 925
Fire-arm, Breech-loading	F. Wesson	Worcester, Mass	May 31, 1870	103, 694
Fire-arm, Breech-loading	F. Wesson and N. S. Harrington	Worcester, Mass	Oct. 25, 1859	25, 926
Fire-arm, Breech-loading	H. F. Wheeler	Boston, Mass	Oct. 31, 1865	50, 760
Fire-arm, Breech-loading	H. F. Wheeler	Boston, Mass	June 19, 1866	55, 752
Fire-arm, Breech-loading	L. Wheelock	New Haven, Conn	Oct. 22, 1867	70, 141
Fire-arm, Breech-loading	A. M. White	Port Chester, N. Y	Apr. 18, 1865	47, 350
Fire-arm, Breech-loading	G. W. White	New York, N. Y	Feb. 4, 1862	34, 325
Fire-arm, Breech-loading	G. W. White	New York, N. Y	Jan. 6, 1863	37, 369
Fire-arm, Breech-loading	L. S. White	Waterbury, Conn	Jan. 6, 1863	37, 376
Fire-arm, Breech-loading	R. White	Hartford, Conn	Mar. 13, 1855	12, 528
Fire-arm, Breech-loading	R. White	Hartford, Conn	Mar. 13, 1855	12, 529
Fire-arm, Breech-loading	R. White	Hartford, Conn	Apr. 3, 1855	12, 638
Fire-arm, Breech-loading	A. E. Whitmore	Ilion, N. Y	Aug. 8, 1871	117, 843
Fire-arm, Breech-loading	A. E. Whitmore	Ilion, N. Y	Apr. 16, 1872	125, 775
Fire-arm, Breech-loading	E. Whitney	New Haven, Conn	Nov. 8, 1864	44, 991
Fire-arm, Breech-loading	E. Whitney	New Haven, Conn	Nov. 26, 1867	71, 349
Fire-arm, Breech-loading	E. Whitney	New Haven, Conn	Mar. 21, 1871	112, 997

Index of patents issued from the United States Patent Office from 1790 *to* 1873, *inclusive*—Continued.

Invention.	Inventor.	Residence.	Date.	No.
Fire-arm, Breech-loading	E. Whitney	New Haven, Conn	June 13, 1871	115, 997
Fire-arm, Breech-loading	E. Whitney	New Haven, Conn	Mar. 26, 1872	124, 994
Fire-arm, Breech-loading	E. Whitney, C. Gerner, and F. Tiesing.	New Haven, Conn	July 27, 1869	93, 149
Fire-arm, Breech-loading	E. Whitney and F. Tiesing	New Haven, Conn	July 16, 1872	129, 637
Fire-arm, Breech-loading	J. M. Whittemore	Augusta, Me	June 14, 1870	104, 387
Fire-arm, Breech-loading	J. M. Whittemore	Augusta, Me	Sept. 17, 1872	131, 487
Fire-arm, Breech-loading	J. M. Whittemore	Augusta, Me	Oct. 1, 1872	131, 921
Fire arm, Breech-loading	J. A. Wickmann	Oldenburg, Germany	Sept. 29, 1863	40, 151
Fire-arm, Breech-loading	J. D. Wilkinson	Plattsburgh, N. Y	Aug. 29, 1871	118, 569
Fire-arm, Breech-loading	D. Williamson	Brooklyn, N. Y	Mar. 21, 1865	46, 977
Fire-arm, Breech-loading	D. Williamson	New York, N. Y	Oct. 2, 1866	58, 525
Fire-arm, Breech-loading	D. Williamson	New York, N. Y	Mar. 16, 1869	87, 997
Fire-arm, Breech-loading	T. Wilson	Birmingham, England	July 14, 1868	80, 043
Fire-arm, Breech-loading	W. F. Wilson and H. Flather	Bridesburgh, Pa	Aug. 15, 1865	49, 463
Fire-arm, Breech-loading	F. Wohlgemuth	New York, N. Y	May 18, 1869	90, 214
Fire-arm, Breech-loading	H. H. Wolcott	Yonkers, N. Y	Nov. 27, 1866	60, 106
Fire-arm, Breech-loading	C. O. Wood	Worcester, Mass	Oct. 9, 1860	30, 372
Fire-arm, Breech-loading	C. O. Wood	Worcester, Mass	Jan. 1, 1861	31, 050
Fire-arm, Breech-loading	S. W. Wood	Cornwall, N. Y	Apr. 1, 1862	34, 854
Fire-arm, Breech-loading	F. G. Woodward	Worcester, Mass	Jan. 7, 1862	34, 084
Fire-arm, Breech-loading	E. S. Wright	New York, N. Y	Nov. 15, 1864	45, 126
Fire-arm, Breech-loading	A. Wyley	Birmingham, England	Nov. 24, 1868	84, 459
Fire-arm, Breech-loading	T. Yates	Milwaukee, Wis	Dec. 18, 1866	60, 607
Fire-arm, Breech-loading	J. Yglesias	New York, N. Y	May 9, 1871	114, 742
Fire-arm, Breech-loading	L. V. Young	Saint Louis, Mo	June 21, 1870	104, 682
Fire-arm, Breech-loading	G. Zeller	Brussels, Belgium	Mar. 18, 1873	136, 894
Fire-arm, Breech-loading and magazine	J. S. Reeder	Canton, Ohio	Nov. 27, 1860	30, 760
Fire-arm, Breech-loading and other	J. W. Cochran	New York, N. Y	Nov. 29, 1859	26, 256
Fire-arm, Breech-loading magazine	J. Swyney	Charlestown, Mass	Aug. 21, 1855	13, 474
Fire-arm, Breech-loading repeating	C. Sharps	Philadelphia, Pa	Jan. 25, 1859	22, 753
Fire-arm, Breech-loading revolving	E. Claude	New York, N. Y	Dec. 21, 1858	22, 348
Fire-arm, Breech-loading revolving	F. A. Le Mat	New Orleans, La	Dec. 14, 1869	97, 780
Fire-arm, Double-barreled	E. Maynard	Tarrytown, N. Y	Oct. 20, 1868	83, 194
Fire-arm, Double-barreled	E. Whitney	New Haven, Conn	Oct. 23, 1866	59, 110
Fire-arm, Double-barreled breech-loading	W. H. Elliot	Plattsburgh, N. Y	Aug. 16, 1864	43, 840
Fire-arm, Double-barreled revolving	H. D. Ward	Pittsfield, Mass	Sept. 8, 1863	39, 850
Fire-arm lock	E. Allen	Grafton, Mass	Nov. 11, 1837	461
Fire-arm lock	W. H. Baker	Marathon, N Y	Dec. 8, 1863	40, 809
Fire-arm lock	O. Blunt	New York, N. Y	Dec. 25, 1849	6, 966
Fire-arm lock	J. S. Butterfield	Philadelphia, Pa	Jan. 2, 1855	12, 124
Fire-arm lock	E. S. Chapin	Stafford Springs, Conn	July 17, 1837	274
Fire-arm lock	N. B. Cook	Chicago, Ill	Aug. 27, 1850	7, 596
Fire-arm lock	W. Frank	Mount Sterling, Ill	Sept. 10, 1861	33, 245
Fire-arm lock	P. Hiller	Mattapoisett, Mass	Apr. 30, 1861	32, 188
Fire-arm lock	J. P. Lindsay	New York, N. Y	Oct. 9, 1860	30, 332
Fire-arm lock	D. Moore	Brooklyn, N. Y	Dec. 2, 1873	145, 118
Fire-arm lock	J. Post	Newark, N. J	May 15, 1849	6, 453
Fire-arm lock	E. T. Starr	New York, N. Y	May 10, 1864	42, 697
Fire-arm lock	A. P. Stephens	Brooklyn, N. Y	Apr. 21, 1863	38, 249
Fire-arm lock	A. Tonks	Boston, Mass	Jan. 13, 1857	16, 411
Fire-arm lock	M. Tromly	Mount Vernon, Ill	Oct. 13, 1857	18, 418
Fire-arm lock	M. Tromly	Mount Vernon, Ill	July 12, 1859	24, 768
Fire-arm lock	A. Young	Philadelphia, Pa	June 14, 1870	104, 394
Fire-arm lock, Double-barreled	D. B. Wesson	Springfield, Mass	May 2, 1871	114, 374
Fire-arm lock, Repeating	J. R. Mock	Elizabethtown, Ky	May 31, 1859	24, 228
Fire-arm, Magazine	A. Assmus	Chicago, Ill	Dec. 16, 1873	145, 478
Fire-arm, Magazine	A. Ball	Worcester, Mass	Dec. 6, 1864	45, 307
Fire-arm, Magazine	S. G. Bayes	Wauseon, Ohio	Feb. 9, 1869	86, 723
Fire-arm, Magazine	J. H. Bean	Cincinnati, Ohio	Aug. 12, 1873	141, 624
Fire-arm, Magazine	P. Boynton	Canton, N. Y	Mar. 15, 1859	23, 226
Fire-arm, Magazine	P. Boynton	Canton, N. Y	Jan. 3, 1860	26, 646
Fire-arm, Magazine	G. W. Briggs	New Haven, Conn	Oct. 16, 1866	58, 937
Fire-arm, Magazine	A. Burgess	New York, N. Y	Sept. 19, 1871	119, 115
Fire-arm, Magazine	A. Burgess	Hartford, Conn	Jan. 7, 1873	134, 589
Fire-arm, Magazine	T. Cullen	San Francisco, Cal	Apr. 13, 1869	88, 853
Fire-arm, Magazine	J. Davis	Limestoneville, Pa	Feb. 28, 1871	112, 127
Fire-arm, Magazine	W. C. Dodge	Washington, D. C	Oct. 16, 1866	58, 790
Fire-arm, Magazine	J. B. Doolittle	Seymour, Conn	July 29, 1862	35, 996
Fire-arm, Magazine	W. H. Elliot	New York, N. Y	May 28, 1872	127, 323
Fire-arm, Magazine	D. Ellis	Whitestone, N. Y	Apr. 12, 1870	101, 845
Fire-arm, Magazine	V. Fogerty	Boston, Mass	Feb. 21, 1865	46, 459
Fire-arm, Magazine	V. Fogerty	Boston, Mass	Oct. 23, 1866	59, 126
Fire-arm, Magazine	V. Fogerty	Roxbury, Mass	Oct. 6, 1868	82, 819
Fire-arm, Magazine	H. K. Forbis	Danville, Ky	Mar. 21, 1871	112, 795
Fire-arm, Magazine	W. Gardner	Toledo, Ohio	Feb. 16, 1869	87, 038
Fire-arm, Magazine	J. Gray	Medford, Mass	July 4, 1865	48, 622
Fire-arm, Magazine	J. Gray	Medford, Mass	Apr. 17, 1866	54, 068
Fire-arm, Magazine	H. W. Hayden	Waterbury, Conn	Aug. 7, 1866	56, 939
Fire-arm, Magazine	B. T. Henry	New Haven, Conn	Oct. 16, 1860	30, 446
Fire-arm, Magazine	G. W. Hughes and J. G. Pusey	Providence, R. I	Aug. 15, 1865	49, 409
Fire-arm, Magazine	N. King	Bridgeport, Conn	May 22, 1866	55, 012
Fire-arm, Magazine	N. King	Bridgeport, Conn	Aug. 28, 1866	57, 636
Fire-arm, Magazine	E. C. Kirk and E. Sneider	Baltimore, Md	July 9, 1867	66, 596
Fire-arm, Magazine	J. L. Kirk	Mattoon, Ill	June 20, 1871	116, 066
Fire-arm, Magazine	J. Kraffert	Berlin, Prussia	July 5, 1870	105, 093
Fire-arm, Magazine	T. W. Lane	Boston, Mass	Jan. 1, 1867	60, 910
Fire-arm, Magazine	G. D. Luce	New Orleans, La	Mar. 11, 1873	136, 660
Fire-arm, Magazine	J. Nichols	Limestone, N. Y	Sept. 2, 1862	36, 3[illegible]8
Fire-arm, Magazine	B. F. Parkinson	Washington, Pa	June 13, 1865	48, 201
Fire-arm, Magazine	J. Rider	Newark, Ohio	Aug. 15, 1871	118, 152
Fire-arm, Magazine	J. Rider	Newark, Ohio	Aug. 15, 1873	141, 590
Fire-arm, Magazine	L. C. Rodier	Springfield, Mass	May 25, 1862	34, 776
Fire-arm, Magazine	L. C. Rodier and F. G. Bates	Springfield, Mass	Apr. 29, 1873	138, 439
Fire-arm, Magazine	J. Q. A. Scott	Pittsburgh, Pa	Aug. 12, 1862	36, 174
Fire-arm, Magazine	P. Sheckler	Orangeville, Ill	Apr. 2, 1867	63, 564
Fire-arm, Magazine	L. Sibert	Mount Solon, Va	May 14, 1861	32, 316
Fire-arm, Magazine	J. D. Smith	Bridgeport, Conn	Feb. 27, 1866	52, 933

Index of patents issued from the United States Patent Office from 1790 *to* 1873, *inclusive*—Continued.

Invention.	Inventor.	Residence.	Date.	No.
Fire-arm, Magazine	J. D. Smith	Bridgeport, Conn	Feb. 27, 1866	52,934
Fire-arm, Magazine	J. N. Smith	Cincinnati, Ohio	Aug. 18, 1863	39,591
Fire-arm, Magazine	C. M. Spencer	Boston, Mass	May 26, 1863	38,702
Fire-arm, Magazine	C. M. Spencer	Boston, Mass	Oct. 9, 1866	58,737
Fire-arm, Magazine	C. M. Spencer	Boston, Mass	Oct. 9, 1866	58,738
Fire-arm, Magazine	E. Stabler	Sandy Springs, Md	Mar. 14, 1865	46,828
Fire-arm, Magazine	E. L. Sturtevant	Boston, Mass	July 16, 1867	66,751
Fire-arm, Magazine	A. Swingle and F. A. Huntington	San Francisco, Cal	Feb. 18, 1873	135,947
Fire-arm, Magazine	L. Z. Terrill	Chicopee, Mass	May 4, 1869	89,705
Fire-arm, Magazine	F. Vetterlin	Newhausen, Switzerland	Dec. 29, 1868	85,494
Fire-arm, Magazine	H. F. Wheeler	Boston, Mass	June 25, 1867	66,110
Fire-arm, Magazine	L. Wheelock	New Haven, Conn	Dec. 1, 1868	84,598
Fire-arm, Magazine	J. A. Whitney	Maryland, N. Y	July 30, 1867	67,242
Fire-arm, Magazine	O. F. Winchester	New Haven, Conn	Sept. 4, 1866	57,808
Fire-arm, Magazine breech-loading	H. F. Wheeler	Boston, Mass	Feb. 7, 1865	46,286
Fire-arm, Magazine for breech-loading	W. H. Elliot	New York, N. Y	Feb. 14, 1871	111,827
Fire-arm, Magazine for self-loading	J. F. Appleby	Mazo Manie, Wis	Dec. 20, 1864	45,466
Fire-arm, Magazine or self-loading	W. Fitzgerald	Boston, Mass	Jan. 17, 1865	45,919
Fire-arm, Magazine or self-loading	J. Gray	Medford, Mass	Dec. 20, 1864	45,560
Fire-arm, Magazine or self-loading	G. W. Hughes	Bloomington, Ill	Nov. 15, 1864	45,043
Fire-arm, Magazine or self-loading	R. Roberts	Utica, N. Y	Dec. 27, 1864	45,638
Fire-arm, Magazine or self-loading	L. Triplett	Columbia, Ky	Dec. 6, 1864	45,361
Fire-arm, Magazine revolving	E. J. Frost	New York, N. Y	June 11, 1867	65,742
Fire-arm, Many-barreled	W. J. Christy	Philadelphia, Pa	Sept. 18, 1866	58,064
Fire-arm, Many-barreled	W. H. Elliot	Plattsburgh, N. Y	May 10, 1864	42,648
Fire-arm, Many-barreled	W. H. Elliot	Plattsburgh, N. Y	May 10, 1864	42,649
Fire-arm, Many-barreled	W. H. Elliot	Ilion, N. Y	Oct. 3, 1865	50,232
Fire-arm, Many-barreled	W. H. Elliot	Ilion, N. Y	Dec. 12, 1865	51,440
Fire-arm, Many-chambered	C. A. Bennett and P. F. Haviland	Waterville, Me	Feb. 15, 1838	603
Fire-arm, Many-chambered	C. Parkhurst	Lawrenceville, Pa	Sept. 25, 1837	409
Fire-arm, Many-chambered cylinder	J. W. Cochran	New York, N. Y	Apr. 29, 1837	183
Fire-arm, Many-chambered cylinder	J. W. Cochran	New York, N. Y	Mar. 8, 1837	188
Fire-arm, Many-chambered cylinder	E. Jaquith	Brattleborough, Vt	July 12, 1838	832
Fire-arm, Many-chambered cylinder	D. Leavitt	Cabotsville, Mass	Apr. 29, 1837	182
Fire-arm, Many-chambered cylinder	R. Nichols and E. Childs	Conway, Mass	Apr. 24, 1858	707
Fire-arm, Many-chambered cylinder	M. Nutting	Portland, Me	Apr. 25, 1838	713
Fire-arm, Many-chambered cylinder	T. F. Strong	Northampton, Mass	Apr. 21, 1838	698
Fire-arm, Many-chambered cylinder	O. W. Whittier	Enfield, N. H	May 30, 1837	216
Fire-arm, Many-chambered rotating-breech	S. Colt	Hartford, Conn	Feb. 24, 1857	16,683
Fire-arm, Many-chambered rotating-breech	S. Colt	Hartford, Conn	Nov. 24, 1857	18,678
Fire-arm percussion-lock	J. H. B. Latrobe	Howard County, Md	Feb. 26, 1856	14,319
Fire-arm percussion-lock	O. Moses	Malone, N. Y	May 3, 1828	....
Fire-arm, Portable	A. O. H. P. Schorn	Murfreesborough, Tenn	Jan. 30, 1855	12,328
Fire-arm, Portable	C. L. Stanislaus, Baron Heurteloup.	France	July 29, 1841	2,203
Fire-arm, Portable	J. Tilton and W. Floyd	Rock House, Ohio	Mar. 3, 1857	16,761
Fire-arm, Repeating	S. Colt	Hartford, Conn	Sept. 3, 1850	7,613
Fire-arm, Repeating	C. Draeger	Indianapolis, Ind	Apr. 8, 1862	34,922
Fire-arm, Repeating	H. J. Drew	Dixon, Ill	Mar. 14, 1871	112,564
Fire-arm, Repeating	W. H. Elliot	Plattsburgh, N. Y	May 29, 1860	28,460
Fire-arm, Repeating	A. C. Faivre	Meadville, Pa	Mar. 9, 1858	19,553
Fire-arm, Repeating	H. Genhart	Liege, Belgium	Jan. 27, 1857	16,477
Fire-arm, Repeating	J. Gray	Boston, Mass	Jan. 26, 1864	41,375
Fire-arm, Repeating	B. Groomes	Cumberland Township, Pa	Jan. 1, 1856	14,017
Fire-arm, Repeating	A. Hall	New York, N. Y	June 10, 1856	15,110
Fire-arm, Repeating	P. J. Jarre	Paris, France	June 24, 1862	35,685
Fire-arm, Repeating	E. M. Judd	New Britain, Conn	Feb. 25, 1862	34,504
Fire-arm, Repeating	G. Leonard	Shrewsbury, Mass	May 6, 1856	14,820
Fire-arm, Repeating	E. Maher	New York, N. Y	May 6, 1862	35,167
Fire-arm, Repeating	W. W. Marston	New York, N. Y	May 26, 1857	17,386
Fire-arm, Repeating	H. H. McKenny and F. Goth	Biddeford, Me	Feb. 15, 1859	22,969
Fire-arm, Repeating	R. S. Mershon and J. Hollingsworth.	Zanesville, Ohio	Feb. 27, 1855	12,471
Fire-arm, Repeating	W. H. Morris and C. L. Brown	New York, N. Y	Jan. 24, 1860	26,919
Fire-arm, Repeating	C. S. Pettengill	New Haven, Conn	July 22, 1856	15,388
Fire-arm, Repeating	F. B. Prindle	New Haven, Conn	Aug. 10, 1858	21,149
Fire-arm, Repeating	J. N. Smith	Jersey City, N. J	June 10, 1862	35,548
Fire-arm, Repeating	W. S. Smoot	Washington, D. C	Dec. 14, 1869	97,821
Fire-arm, Repeating	E. T. Starr	New York, N. Y	May 10, 1864	42,698
Fire-arm, Repeating	J. Stevens	Chicopee, Mass	Aug. 9, 1853	9,929
Fire-arm, Repeating	J. Stevens	Chicopee Falls, Mass	Jan. 2, 1855	12,189
Fire-arm, Repeating	C. C. Terrel	Shullsburgh, Wis	Feb. 16, 1858	19,387
Fire-arm, Repeating	C. N. Tyler	Worcester, Mass	May 3, 1853	9,701
Fire-arm, Repeating	L. Wheelock	New Haven, Conn	Jan. 31, 1871	111,500
Fire-arm, Repeating	R. White	Hartford, Conn	Apr. 3, 1855	12,648
Fire-arm, Repeating	R. White	Hartford, Conn	Apr. 3, 1855	12,649
Fire-arm, Repeating magazine	J. C. Smith	Camden, N. J	Jan. 1, 1856	14,034
Fire-arm, Repeating or many-chambered	H. and C. Daniels	Chester, Conn	Apr. 5, 1838	677
Fire-arm, Repeating single-barreled	D. B. Neal	Mount Gilead, Ohio	Feb. 27, 1855	12,440
Fire-arm, Revolver	J Adams	Dalston, England	Nov. 6, 1860	30,692
Fire-arm, Revolving	J. Adams	Strand, England	Dec. 29, 1868	85,350
Fire-arm, Revolving	R. Adams	London, England	May 3, 1853	9,694
Fire-arm, Revolving	E. Allen	Worcester, Mass	Dec. 15, 1857	18,836
Fire-arm, Revolving	E. Allen	Worcester, Mass	Sept. 7, 1858	21,400
Fire-arm, Revolving	E. Allen	Worcester, Mass	Nov. 9, 1858	22,005
Fire-arm, Revolving	E. Allen	Worcester, Mass	July 3, 1860	28,951
Fire-arm, Revolving	E. Allen	Worcester, Mass	Sept. 24, 1861	33,328
Fire-arm, Revolving	E. Allen	Worcester, Mass	Oct. 22, 1861	33,509
Fire-arm, Revolving	E. Allen	Worcester, Mass	Apr. 29, 1862	35,067
Fire-arm, Revolving	C. H. Alsop	Middletown, Conn	Nov. 26, 1861	33,770
Fire-arm, Revolving	C. R. Alsop	Middletown, Conn	July 17, 1860	29,213
Fire-arm, Revolving	C. R. Alsop	Middletown, Conn	May 14, 1861	32,333
Fire-arm, Revolving	C. R. Alsop	Middletown, Conn	Jan. 21, 1862	34,226
Fire-arm, Revolving	C. R. Alsop	Middletown, Conn	Mar. 25, 1862	34,803
Fire-arm, Revolving	T. K. Austin	New York, N. Y	Oct. 12, 1858	21,730
Fire-arm, Revolving	T. Bailey	New Orleans, La	June 7, 1859	24,274
Fire-arm, Revolving	F. Beals	New Haven, Conn	May 26, 1857	17,359
Fire-arm, Revolving	F. Beals	New Haven, Conn	Sept. 14, 1858	21,478

Index of patents issued from the United States Patent Office from 1790 *to* 1873, *inclusive*—Continued.

Invention.	Inventor.	Residence.	Date.	No.
Fire-arm, Revolving	W. H Bell	Washington, D. C	Mar. 20, 1860	27, 518
Fire-arm, Revolving	G. A. Blittkowski and F. W. Hoffman.	New York, N. Y	Apr. 22, 1856	14, 710
Fire-arm, Revolving	C. C. Brand	Norwich, Conn	Sept. 23, 1862	36, 505
Fire-arm, Revolving	C. C. Brand	Norwich, Conn	Apr. 28, 1863	38, 279
Fire-arm, Revolving	H. A. Briggs and S. S. Hopkins.	Norwich, Conn	Jan. 5, 1864	41, 117
Fire-arm, Revolving	G. C. Bunsen	Belleville, Ill	Dec. 26, 1865	51, 690
Fire-arm, Revolving	G. W. H. Calver	Burlington, N. J	May 17, 1870	103, 013
Fire-arm, Revolving	A. Christ	California, Ohio	Sept. 11, 1866	57, 865
Fire-arm, Revolving	F. G. Cochran	Saint Louis, Mo	July 4, 1871	116, 559
Fire-arm, Revolving	J. W. Cochran	New York, N. Y	Dec. 28, 1858	22, 412
Fire-arm, Revolving	J. W. Cochran	New York, N. Y	Nov. 10, 1863	40, 553
Fire-arm, Revolving	S. Colt	Hartford, Conn	May 4, 1858	20, 144
Fire-arm, Revolving	C. A. Converse and S. S. Hopkins.	Norwich, Conn	Aug. 28, 1866	57, 622
Fire-arm, Revolving	J. M. Cooper	Pittsburgh, Pa	Mar. 20, 1860	27, 526
Fire-arm, Revolving	J. M. Cooper	Pittsburgh, Pa	Sept. 4, 1860	29, 864
Fire-arm, Revolving	J. M. Cooper	Pittsburgh, Pa	Sept. 22, 1863	40, 021
Fire-arm, Revolving	S. Crispin	New York, N. Y	Oct. 3, 1865	50, 224
Fire-arm, Revolving	G. G. Crowell	Lime Rock, Conn	Apr. 17, 1866	53, 955
Fire-arm, Revolving	C. F. and J. E. De Dartein	Strasbourg, France	May 10, 1870	102, 782
Fire-arm, Revolving	W. C. Dodge	Washington, D. C	Jan. 24, 1865	45, 983
Fire-arm, Revolving	J. B. Doolittle	New Haven, Conn	Apr. 17, 1866	54, 065
Fire-arm, Revolving	B. K. Dorwart	Rockland, R. I	Sept. 2, 1873	142, 376
Fire-arm, Revolving	R. W. Drew	Lowell, Mass	Apr. 2, 1867	63, 450
Fire-arm, Revolving	W. H. Elliot	Plattsburgh, N. Y	Aug. 17, 1858	21, 188
Fire-arm, Revolving	W. H. Elliot	Plattsburgh, N. Y	May 29, 1860	28, 461
Fire-arm, Revolving	W. H. Elliot	Plattsburgh, N. Y	Oct. 1, 1861	33, 382
Fire-arm, Revolving	W. H. Elliot	Plattsburgh, N. Y	Feb. 7, 1865	46, 225
Fire-arm, Revolving	W. C. Ellis and J. N. White	Springfield, Mass	July 12, 1859	24, 726
Fire-arm, Revolving	W. C. Ellis and J. N. White	Springfield, Mass	July 21, 1863	39, 318
Fire-arm, Revolving	J. Ells	Pittsburgh, Pa	Apr. 25, 1854	10, 812
Fire-arm, Revolving	J. Ells	Pittsburgh, Pa	Apr. 14, 1857	17, 032
Fire-arm, Revolving	C. Foehl	Philadelphia, Pa	June 3, 1873	139, 461
Fire-arm, Revolving	S. Forehand and H. C. Wadsworth.	Worcester, Mass	June 27, 1871	116, 422
Fire-arm, Revolving	S. Forehand and H. C. Wadsworth.	Worcester, Mass	Oct. 14, 1873	143, 566
Fire-arm, Revolving	A. T. Freeman	Binghamton, N. Y	Dec. 9, 1862	37, 091
Fire-arm, Revolving	G. H. Gardner	New York, N. Y	May 16, 1865	47, 712
Fire-arm, Revolving	G. W. B. Gedney	New York, N. Y	July 29, 1862	35, 999
Fire-arm, Revolving	M. F. Geraghty	Jersey City, N. J	Aug. 25, 1863	39, 642
Fire-arm, Revolving	A. J. Gibson	Worcester, Mass	May 22, 1860	28, 437
Fire-arm, Revolving	A. J. Gibson	Worcester, Mass	July 10, 1860	29, 126
Fire-arm, Revolving	A. J. Gibson	Worcester, Mass	Oct. 9, 1860	30, 309
Fire-arm, Revolving	J. Gordon	San Francisco, Cal	July 16, 1872	129, 334
Fire-arm, Revolving	E. H. Graham	Yonkers, N. Y	Nov. 24, 1863	40, 687
Fire-arm, Revolving	H. Gross	Tiffin, Ohio	Dec. 3, 1861	33, 836
Fire-arm, Revolving	H. Gross	Tiffin, Ohio	Aug. 25, 1863	39, 645
Fire-arm, Revolving	J. Gruler and A. Rebetey	Norwich, Conn	Dec. 27, 1859	26, 641
Fire-arm, Revolving	A. Guerriero	Genoa, Italy	Apr. 11, 1865	47, 252
Fire-arm, Revolving	A. Hall	Danville, Iowa	Mar. 24, 1863	37, 961
Fire-arm, Revolving	F. H. Harrington	Springfield, Mass	June 15, 1858	20, 607
Fire-arm, Revolving	C. W. Harris	Pittsburgh, Pa	Sept. 1, 1863	39, 771
Fire-arm, Revolving	P. Haughian	New York, N. Y	Feb. 28, 1865	46, 562
Fire-arm, Revolving	W. C. Haynes	Melrose, Tex	Mar. 1, 1859	23, 087
Fire-arm, Revolving	B. R. Hill	Cranston, R. I	Feb. 15, 1870	99, 893
Fire-arm, Revolving	G. Holman	Waterville, N. Y	Mar. 3, 1868	75, 016
Fire-arm, Revolving	F. W. Hood	Worcester, Mass	Nov. 8, 1864	44, 953
Fire-arm, Revolving	F. W. Hood	Boston, Mass	July 4, 1871	116, 593
Fire-arm, Revolving	C. W. Hopkins	Norwich, Conn	May 27, 1862	35, 419
Fire-arm, Revolving	S. S. Hopkins	Norwich, Conn	Mar. 28, 1871	113, 053
Fire-arm, Revolving	J. C. Howe	Worcester, Mass	Feb. 17, 1863	37, 693
Fire-arm, Revolving	J. Jenkinson	Brooklyn, N. Y	Dec. 2, 1862	37, 075
Fire-arm, Revolving	B. F. Joslyn	Worcester, Mass	May 4, 1858	20, 160
Fire-arm, Revolving	B. F. Joslyn	Stonington, Conn	Aug. 4, 1863	39, 405
Fire-arm, Revolving	B. F Joslyn	Stonington, Conn	Aug. 4, 1863	39, 406
Fire-arm, Revolving	B. F. Joslyn	Stonington, Conn	Apr. 19, 1864	42, 379
Fire-arm, Revolving	B. F. Joslyn	Stonington, Conn	Feb. 7, 1865	46, 243
Fire-arm, Revolving	B. F. Joslyn	Stonington, Conn	June 20, 1865	48, 287
Fire-arm, Revolving	B. F. Joslyn	Stonington, Conn	Jan. 2, 1866	51, 836
Fire-arm, Revolving	B. F. Joslyn	New York, N. Y	Nov. 22, 1870	109, 417
Fire-arm, Revolving	B. F. Joslyn	New York, N. Y	May 30, 1871	115, 483
Fire-arm, Revolving	H. S. Josselyn	Roxbury, Mass	Jan. 23, 1866	52, 248
Fire-arm, Revolving	J. Kerr	Southwark, England	Aug. 4, 1863	39, 439
Fire-arm, Revolving	C. A. King	Springfield, Mass	Aug. 24, 1869	94, 003
Fire-arm, Revolving	M. Kinsey	Newark, N. J	June 8, 1858	20, 496
Fire-arm, Revolving	B. Kittridge	Cincinnati, Ohio	Mar. 8, 1864	41, 848
Fire-arm, Revolving	E. S. Leaycroft	Brooklyn, N. Y	Mar. 7, 1871	112, 471
Fire-arm, Revolving	E. S. Leaycroft	Brooklyn, N. Y	Mar. 7, 1871	112, 472
Fire-arm, Revolving	S. C. Lewis and F. P. Pfleghar	Whitneyville, Conn	Aug. 2, 1859	24, 942
Fire-arm, Revolving	B. T. Loomis	New York, N. Y	Feb. 13, 1866	52, 582
Fire-arm, Revolving	C. J. Lunberg and W. J. Phillips.	Saint Louis, Mo	Dec. 6, 1870	109, 914
Fire-arm, Revolving	J. M. Marlin	New Haven, Conn	July 1, 1873	140, 516
Fire-arm, Revolving	W. Mason	Ilion, N. Y	Nov. 21, 1865	51, 117
Fire-arm, Revolving	W. Mason	Ilion, N. Y	Mar. 27, 1866	53, 539
Fire-arm, Revolving	W. Mason	Hartford, Conn	July 2, 1872	128, 644
Fire-arm, Revolving	T. J. Mayall	Roxbury, Mass	Nov. 25, 1862	37, 004
Fire-arm, Revolving	J. C. Miller	Danville, Ky	Feb. 8, 1870	99, 693
Fire-arm, Revolving	D. Moore	Brooklyn, N. Y	Sept. 18, 1860	30, 079
Fire-arm, Revolving	D. Moore	Brooklyn, N. Y	Jan. 7, 1862	34, 067
Fire-arm, Revolving	D. Moore	Brooklyn, N. Y	Apr. 28, 1863	38, 321
Fire-arm, Revolving	J. L. Morse and E. W. Johnson	Columbus, Miss	June 20, 1871	116, 078
Fire-arm, Revolving	A. L. Munson	New Haven, Conn	Nov. 13, 1866	59, 629
Fire-arm, Revolving	F. Newbury	Albany, N. Y	June 12, 1855	13, 039
Fire-arm, Revolving	F. Newbury	Albany, N. Y	Sept. 18, 1855	13, 582

Index of patents issued from the United States Patent Office from 1790 *to* 1873, *inclusive*—Continued.

Invention.	Inventor.	Residence.	Date.	No.
Fire-arm, Revolving	F. Newbury	Albany, N. Y	Mar. 11, 1856	14, 406
Fire-arm, Revolving	F. D. Newbury	Albany, N. Y	Mar. 23, 1858	19, 739
Fire-arm, Revolving	F. D. Newbury	Albany, N. Y	June 29, 1858	20, 765
Fire-arm, Revolving	F. D. Newbury	Albany, N. Y	Apr. 10, 1860	27, 868
Fire-arm, Revolving	F. D. Newbury	Albany, N. Y	Oct. 23, 1860	30, 494
Fire-arm, Revolving	H. S. North and E. Savage	Middletown and Cromwell, Conn.	Jan. 18, 1859	22, 666
Fire-arm, Revolving	W. I. Page	Boston, Mass	June 21, 1870	104, 636
Fire-arm, Revolving	W. Palmer	New York, N. Y	Sept. 28, 1858	21, 623
Fire-arm, Revolving	W. Palmer	New York, N. Y	Mar. 8, 1864	41, 857
Fire-arm, Revolving	C. S. Pettengill	New Haven, Conn	Jan. 4, 1859	22, 511
Fire-arm, Revolving	W. H. Philip	Brooklyn, N. Y	Aug. 26, 1873	142, 175
Fire-arm, Revolving	W. J. Pitt	Middletown, Conn	Jan. 7, 1862	34, 093
Fire-arm, Revolving	R. H. Plass	New York, N. Y	Jan. 24, 1865	46, 023
Fire-arm, Revolving	P. Polain	Brussels, Belgium	Mar. 27, 1866	53, 548
Fire-arm, Revolving	L. W. Pond	Worcester, Mass	June 17, 1862	35, 623
Fire-arm, Revolving	L. W. Pond	Worcester, Mass	June 16, 1863	38, 934
Fire-arm, Revolving	E. A. Prescott	Worcester, Mass	Oct. 2, 1860	30, 245
Fire-arm, Revolving	E. A. Raymond and C. Robitaille.	Brooklyn, N. Y	July 27, 1858	21, 054
Fire-arm, Revolving	J. Reid	New York, N. Y	Apr. 28, 1863	38, 336
Fire-arm, Revolving	J. Reid	Catskill, N. Y	Dec. 26, 1865	51, 752
Fire-arm, Revolving	H. Reynolds	Springfield, Mass	May 10, 1864	42, 688
Fire-arm, Revolving	C. B. Richards	Hartford, Conn	July 25, 1871	117, 461
Fire-arm, Revolving	C. B. Richards	Hartford, Conn	Sept. 19, 1871	119, 048
Fire-arm, Revolving	J. Rider	Newark, Ohio	Aug. 17, 1858	21, 215
Fire-arm, Revolving	J. Rider	Newark, Ohio	May 3, 1859	23, 861
Fire-arm, Revolving	J. Rider	Newark, Ohio	Nov. 28, 1865	51, 269
Fire-arm, Revolving	C. Robitaille and F. Dahis	Brooklyn, N. Y	Aug. 2, 1864	43, 709
Fire-arm, Revolving	L. C. Rodier	Springfield, Mass	July 11, 1865	48, 775
Fire-arm, Revolving	H. S. Rogers	Willow Vale, N. Y	Oct. 28, 1862	36, 861
Fire-arm, Revolving	M. L. Rood	Marshall, Mich	Nov. 22, 1853	10, 259
Fire-arm, Revolving	E. K. Root	Hartford, Conn	Dec. 25, 1855	13, 999
Fire-arm, Revolving	E. K. Root	Hartford, Conn	June 4, 1867	65, 510
Fire-arm, Revolving	S. H. Roper	Roxbury, Mass	Apr. 10, 1866	53, 881
Fire-arm, Revolving	J. Rupertus	Philadelphia, Pa	Apr. 19, 1859	23, 711
Fire-arm, Revolving	J. Rupertus	Philadelphia, Pa	Dec. 2, 1862	37, 059
Fire-arm, Revolving	J. Rupertus	Philadelphia, Pa	July 19, 1864	43, 606
Fire-arm, Revolving	J. Rupertus	Philadelphia, Pa	Nov. 21, 1871	121, 199
Fire-arm, Revolving	E. Savage and H. S. North	Cromwell and Middletown, Conn.	May 15, 1860	28, 331
Fire-arm, Revolving	O. Schneeloch	Brooklyn, N. Y	Dec. 31, 1872	134, 442
Fire-arm, Revolving	G. W. Schofield	United States Army	Apr. 22, 1873	138, 047
Fire-arm, Revolving	C. Sharps	Philadelphia, Pa	Sept. 5, 1871	118, 752
Fire-arm, Revolving	T. Shaw	Philadelphia, Pa	Dec. 24, 1861	34, 032
Fire-arm, Revolving	F. P. Slocum	Brooklyn, N. Y	Jan. 27, 1863	37, 551
Fire-arm, Revolving	F. P. Slocum	Brooklyn, N. Y	Apr. 14, 1863	38, 204
Fire-arm, Revolving	A. Smith	Hartford, Conn.	Dec. 24, 1861	34, 016
Fire-arm, Revolving	H. Smith and D. B. Wesson	Springfield, Mass	July 5, 1859	24, 666
Fire-arm, Revolving	H. Smith and D. B. Wesson	Springfield, Mass	June 16, 1863	38, 921
Fire-arm, Revolving	H. Smith and D. B. Wesson	Springfield, Mass	Nov. 21, 1865	51, 092
Fire-arm, Revolving	O. A. Smith	Middlefield, Conn	Jan. 28, 1873	135, 377
Fire-arm, Revolving	O. A. Smith	Middlefield, Conn	Jan. 28, 1873	135, 378
Fire-arm, Revolving	O. A. Smith	Middlefield, Conn	Apr. 15, 1873	137, 968
Fire-arm, Revolving	W. S. Smoot	Ilion, N. Y	Oct. 21, 1873	143, 855
Fire-arm, Revolving	C. E. Sneider	Baltimore, Md	Mar. 18, 1862	34, 703
Fire-arm, Revolving	C. E. Sneider	Baltimore, Md	Feb. 28, 1865	46, 612
Fire-arm, Revolving	A. Spellier	Philadelphia, Pa	Oct. 2, 1860	30, 260
Fire-arm, Revolving	E. T. Starr	New York, N. Y	Jan. 15, 1856	14, 118
Fire-arm, Revolving	E. T. Starr	New York, N. Y	Dec. 19, 1865	51, 628
Fire-arm, Revolving	W. M. Storm	New York, N. Y	Mar. 11, 1856	14, 420
Fire-arm, Revolving	F. A. Thuer	East Hartford, Conn	Sept. 15, 1868	82, 258
Fire-arm, Revolving	W. Tibbals	South Coventry, Conn	June 19, 1866	55, 743
Fire-arm, Revolving	W. Tibbals	South Coventry, Conn	July 17, 1866	56, 466
Fire-arm, Revolving	A. C. Vaughan	Bedford, Pa	May 27, 1862	35, 404
Fire-arm, Revolving	J. H. Vickers	Worcester, Mass	June 17, 1862	35, 657
Fire-arm, Revolving	J. H. Vickers	Worcester, Mass	May 16, 1865	47, 775
Fire-arm, Revolving	J. H. Vickers	Norwich, Conn	Aug. 21, 1866	57, 448
Fire-arm, Revolving	J. Walch	New York, N. Y	Feb. 8, 1859	22, 905
Fire-arm, Revolving	J. Warner	Springfield, Mass	July 28, 1857	17, 904
Fire-arm, Revolving	D. B. Wesson	Springfield, Mass	Feb. 25, 1873	136, 348
Fire-arm, Revolving	D. B. Wessen and C. A. King	Springfield, Mass	July 16, 1872	128, 991
Fire-arm, Revolving	F. Wesson	Worcester, Mass	Dec. 15, 1868	84, 976
Fire-arm, Revolving	F. Wesson	Worcester, Mass	June 13, 1871	115, 916
Fire-arm, Revolving	J. A. Whalen	Brooklyn, N. Y	Apr. 22, 1862	35, 052
Fire-arm, Revolving	R. White	Hartford, Conn	Apr. 13, 1858	19, 961
Fire-arm, Revolving	R. White	Lowell, Mass	Nov. 29, 1864	45, 290
Fire-arm, Revolving	R. White	Lowell, Mass	July 9, 1867	66, 542
Fire-arm, Revolving	R. White	Lowell, Mass	Aug. 10, 1869	93, 572
Fire-arm, Revolving	R. White	Lowell, Mass	Aug. 10, 1869	93, 653
Fire-arm, Revolving	R. White	Lowell, Mass	Feb. 1, 1870	99, 505
Fire-arm, Revolving	R. White	Lowell, Mass	Sept. 30, 1873	143, 394
Fire-arm, Revolving	E. Whitney	New Haven, Conn	Jan. 9, 1866	51, 985
Fire-arm, Revolving	E. Whitney	New Haven, Conn	May 23, 1871	115, 258
Fire-arm, Revolving	D. Williamson	Brooklyn, N. Y	Jan. 5, 1864	41, 184
Fire-arm, Revolving	D. Williamson	Brooklyn, N. Y	May 17, 1864	42, 823
Fire-arm, Revolving	D. Williamson	Greenville, N. J	Mar. 18, 1873	137, 043
Fire-arm, Revolving	D. Williamson	Greenville, N. J	Nov. 18, 1873	144, 814
Fire-arm, Revolving	D. Williamson	Greenville, N. J	Nov. 18, 1873	144, 815
Fire-arm, Revolving	S. W. Wood	Cornwall, N. Y	Nov. 18, 1862	36, 984
Fire-arm, Revolving	S. W. Wood	Cornwall, N. Y	Mar. 1, 1864	41, 803
Fire-arm, Revolving	S. W. Wood	Cornwall, N. Y	Sept. 20, 1864	44, 363
Fire-arm, Revolving-breech	J. Ells	Pittsburgh, Pa	Aug. 1, 1854	11, 419
Fire-arm, Revolving-breech	H. Iversen	New York, N. Y	Mar. 26, 1850	7, 218
Fire-arm, Revolving-breech	H. S. North and C. D. Skinner	Middletown and Haddam, Conn.	June 1, 1852	8, 982
Fire-arm, Revolving-breech	P. W. Porter	Memphis, Tenn	July 8, 1851	8, 210
Fire-arm, Revolving-breech	J. Warner	Springfield, Mass	July 15, 1851	8, 229

Index of patents issued from the United States Patent Office from 1790 *to* 1873, *inclusive*—Continued.

Invention.	Inventor.	Residence.	Date.	No.
Fire-arm, Revolving chambered	S. Colt	Hartford, Conn	Sept. 10, 1850	7, 629
Fire-arm, Revolving-hammer	G. Leonard, jr	Shrewsbury, Mass	July 9, 1850	7, 493
Fire-arm, Rifled	J. B. Atwater	Ripon, Wis	Mar. 6, 1860	27, 342
Fire-arm, Rifled	A. Henry	Edinburgh, North Britain	Oct. 10, 1871	119, 846
Fire-arm, Safety sliding-breech	C. Hartung	Beichlingen, Prussia	Nov. 13, 1849	6, 871
Fire-arm, Self-cocking	R. S. Mershon and J. Hollingsworth.	Philadelphia, Pa., and Zanesville, Ohio.	Sept. 8, 1863	39, 825
Fire-arm, Self-feeding breech-loading	A. Ball	Worcester, Mass	Aug. 16, 1864	43, 827
Fire-arm, Self-loading	A. Ball	Worcester, Mass	June 23, 1863	38, 935
Fire-arm, Self-loading	N. W. Brewer	Williamsport, Pa	June 12, 1860	28, 646
Fire-arm, Self-loading	J. D. Moore	Zanesville, Ohio	Mar. 6, 1860	27, 374
Fire-arm, Self-loading	W. H. Rice	Windsor, Conn	May 19, 1863	38, 604
Fire-arm, Self-loading	C. M. Spencer	South Manchester, Conn	Mar. 6, 1860	27, 393
Fire-arm, Self-loading	C. M. Spencer	Boston, Mass	Jan. 17, 1865	45, 952
Fire-arm, Self-loading	R. Wilson	Macomb, Ill	Nov. 15, 1864	45, 105
Fire-arm, Self loading and priming	S. Day	New York, N. Y	Aug. 31, 1837	364
Fire-arm, Self-loading and self-capping repeating	M. M. Cass	Utica, N. Y	Sept. 26, 1848	5, 814
Fire-arm, Self-priming	G. W. B. Gedney	New York, N. Y	Mar. 15, 1859	23, 241
Fire-arm with several stationary barrels and a revolving hammer.	G. Leonard, jr	Shrewsbury, Mass	Sept. 18, 1849	6, 723
Fire-arms, Adjustable back-sight for	C. Sharps	Philadelphia, Pa	July 23, 1861	32, 899
Fire-arms, Adjustable hammer for many-barreled	R. I. Howland and W. H. Elliot	Ilion, N. Y	Feb. 9, 1864	41, 510
Fire-arms, Adjustable hammer for revolving	A. Le Mat	New Orleans, La	June 7, 1859	24, 312
Fire-arms, Adjustable sight for	R. S. Lawrence	Hartford, Conn	Feb. 15, 1859	22, 958
Fire-arms, Adjustable stock of	C. W. Jones	Cheltenham, England	July 17, 1866	56, 506
Fire-arms and the lock and appurtenances of the same, Movable breech for.	B. Chambers	Washington, D. C	July 31, 1849	6, 612
Fire-arms, Apparatus for cooling repeating	J. A. Reynolds	Elmira, N. Y	July 17, 1855	13, 294
Fire-arms, Attached muzzle for	P. S. Newton	Hartford, Conn	June 1, 1843	3, 115
Fire-arms, Attaching bayonets to	H. Berdan	New York, N. Y	Jan. 10, 1865	45, 901
Fire-arms, Automatic primer for	J. Rupertus	Philadelphia, Pa	May 10, 1859	23, 952
Fire-arms, Automatic primer for	J. H. Wells	Brooklyn, N. Y	Dec. 24, 1861	34, 020
Fire-arms, Back-sight for	W. Conner	Rensselaerville, N. Y	Mar. 19, 1867	63, 022
Fire-arms, Back-sight for	H. Hammond	Hartford, Conn	Jan. 8, 1867	61, 007
Fire-arms, Back-sight for	J. B. Learock	Boston, Mass	Jan. 27, 1863	37, 512
Fire-arms, Back-sight for	E. Maynard	Washington, D. C	Oct. 4, 1859	25, 663
Fire-arms, Ball-screw for	J. J. Alises	Washington, D. C	Oct. 11, 1864	44, 586
Fire-arms, Ball-screw for	A. De Witzleben	Washington, D. C	Feb. 7, 1865	46, 220
Fire-arms, blasting, &c., Explosive compound for use in.	T. Taylor	Washington, D. C	Dec. 7, 1869	97, 566
Fire-arms, Cap-priming attachment to	E. D. Seely	Brookline, Mass	July 1, 1862	35, 783
Fire-arms, Capsule for preventing the soiling of	F. L. M. Dorvault	Paris, France	Apr. 25, 1865	47, 503
Fire-arms, Charger attached to	O. B. Percival and A. Smith	East Haddam, Conn, and New York, N. Y.	July 9, 1850	7, 496
Fire-arms, &c., Charger for	J. Johnson	Washington, D. C	Sept. 11, 1855	13, 547
Fire-arms, Charger for	T. H. Peavey	South Montaille, Me	June 27, 1854	11, 174
Fire-arms, Charger for	W. M. Storm	New York, N. Y	May 2, 1854	10, 846
Fire-arms, Combined back-sight and cartridge-retractor for.	G. W. Bowlby	Pontiac, Mich	May 21, 1867	64, 941
Fire-arms, Combined implement for detaching and replacing the parts of.	A. Grillet	Philadelphia, Pa	May 16, 1865	47, 715
Fire-arms, Concealed trigger for	J. Pecare and J. M. Smith	New York, N. Y	Dec. 4, 1849	6, 925
Fire-arms, Construction and mode of loading	T. McCarty	Elmira, N. Y	Mar. 11, 1837	147
Fire-arms, Continuous priming for	D. G. Rollin	New York, N. Y	Apr. 27, 1858	20, 129
Fire-arms, Cylinder-pin for revolving	W. H. Elliot	Plattsburgh, N. Y	May 16, 1865	47, 707
Fire-arms, Discharging breech-loading	H. Stanton	United States Army	Aug. 16, 1853	9, 950
Fire-arms, Expanding tampion for	T. K. Schermerhorn and J. Anderson.	Brooklyn, N. Y	Dec. 27, 1864	45, 641
Fire-arms, Expansible tampion for	G. R. Willmott	Meriden, Conn	Nov. 24, 1863	40, 720
Fire-arms, Feeder for repeating	L. W. Broadwell	Carlsruhe, Grand Duchy of Baden.	Dec. 20, 1870	110, 338
Fire-arms, Fitting lock-plate to stock of	G. Henry	Nazareth, Pa	Dec. 4, 1866	60, 188
Fire-arms, Fly tumbler-lock for	S. W. Marston	New York, N. Y	Jan. 7, 1851	7, 887
Fire-arms, Frame and stock for revolving	C. F. Galand	Paris, France	June 17, 1873	140, 028
Fire-arms, Gas-check for breech-loading	F. Curtis	Newton Lower Falls, Mass	Feb. 9, 1864	41, 489
Fire-arms, Gas-check for breech-loading	J. M. Seymour	Boston, Mass	May 20, 1862	35, 354
Fire-arms, Gas-check for breech-loading	J. C. Symmes	United States Ordnance Corps.	Sept. 8, 1863	39, 844
Fire-arms, Gas-check for revolving	J. Davis	Limestoneville, Pa	Oct. 22, 1872	132, 357
Fire-arms, Hammer for breech-loading	W. H. Elliot	New York, N. Y	Aug. 27, 1867	68, 292
Fire-arms, Hammer-guard to	C. Sharps	Philadelphia, Pa	Oct. 22, 1861	33, 546
Fire-arms, Hammer-guard for	B. Singleton	Portsmouth, Va	May 1, 1860	28, 109
Fire-arms, Handle-attachment to	E. Charlesworth	London, England	Oct. 3, 1865	50, 312
Fire-arms, Hook-attachment to bands of	I. Merrill	Springfield, Mass	Apr. 15, 1873	137, 786
Fire-arms, Loading and discharging	P. A. Morineau	Philadelphia, Pa	Oct. 10, 1832	....
Fire-arms, Locking-apparatus of repeating	J. Stevens	Chicopee, Mass	Nov. 26, 1850	7, 802
Fire-arms, Locking-cylinder of revolving	T. Gibson	Yonkers, N. Y	Apr. 19, 1864	42, 435
Fire-arms, Lubricating	S. Colt	Hartford, Conn	Mar. 3, 1857	16, 716
Fire-arms, Lubricating-pellet for	S. W. Wood	Cornwall, N. Y	May 7, 1872	126, 614
Fire-arms, Lubricating-wad for	S. W. Wood	Cornwall, N. Y	May 7, 1872	126, 615
Fire-arms, Machine for boring chambers in the cylinders of.	E. K. Root	Hartford, Conn	Nov. 28, 1854	12, 002
Fire-arms, Magazine-hammer for	J. N. Ward	United States Army	July 1, 1856	15, 262
Fire-arms, Manner of discharging	J. Shaw	Philadelphia, Pa	Jan. 30, 1841	1, 958
Fire arms, Manufacture of the metallic parts of	I. Adams, jr	Boston, Mass	Dec. 21, 1869	98, 006
Fire-arms, Means for actuating movable parts of	T. Bailey	New Orleans, La	June 14, 1859	24, 437
Fire-arms, Means for revolving the breeches of repeating.	J. Warner	Springfield, Mass	Jan. 7, 1851	7, 894
Fire-arms, Means of attaching cartridge magazine-blocks to.	H. Metcalfe	United States Army	Oct. 7, 1873	143, 415
Fire-arms, Method of attaching the cylinder in revolving.	D. H. Chamberlain	Boston, Mass	Apr. 23, 1850	7, 300
Fire-arms, Method of connecting the hammer with the cylinder of revolving.	E. Wesson	Hartford, Conn	Aug. 28, 1849	6, 669
Fire-arms, Method of converting muzzle into breech loading.	C. E. Sneider	Baltimore, Md	Jan. 24, 1865	46, 054
Fire-arms, Method of forming receiver for breech-block of.	E. G. W. Bartlett	Providence, R. I	Dec. 23, 1873	145, 717

Index of patents issued from the United States Patent Office from 1790 *to* 1873, *inclusive*—Continued.

Invention.	Inventor.	Residence.	Date.	No.
Fire-arms, Method of revolving the hammer of repeating.	C. Sharps	Washington, D. C	Dec. 18, 1849	6, 960
Fire-arms, Mode of altering flint-lock to percussion	J. N. Ward	United States Army	Jan. 27, 1857	16, 503
Fire-arms, Mode of loading	C. A. McEvoy	Richmond, Va	Mar. 26, 1861	31, 815
Fire-arms, Mode of overcoming the windage in	A. E. Burnside	Bristol, R. I	May 12, 1857	17, 261
Fire-arms, Nipple-guard of	J. M. Hill and R. D. Hay	Crooked Creek, N. C	Oct. 26, 1869	96, 112
Fire-arms, Nipple-guard of	B. Lilly	Birmingham, England	May 3, 1864	42, 621
Fire-arms, Nipple-guard of	D. W. Smith	Boston, Mass	Jan. 26, 1858	19, 213
Fire-arms, Nipple of	E. Maynard	Washington, D. C	Oct. 4, 1859	25, 664
Fire-arms, Nipple-primer for	L. H. Bradford	Boston, Mass	May 17, 1864	42, 741
Fire-arms, Patched ball for	M. Peck	New Haven, Conn	July 19, 1864	43, 601
Fire-arms, Percussion-cap holder for	R. S. Pickett	New Haven, Conn	Apr. 4, 1865	47, 127
Fire-arms, Percussion-cap holder for priming	L. R. Budd	Oskaloosa, Iowa	May 19, 1863	38, 557
Fire-arms, Percussion-pellet for	J. Rupertus	Philadelphia, Pa	Aug. 16, 1859	25, 142
Fire-arms, Powder and shot charger for	A. Hall	New York, N. Y	May 13, 1873	138, 751
Fire-arms, Primer for	A. N. Newton	Richmond, Ind	May 23, 1854	10, 950
Fire-arms, Priming	C. Sharps	Hartford, Conn	Oct. 5, 1852	9, 308
Fire-arms, Priming-repeating	G. R. Crooker	New York, N. Y	Oct. 20, 1857	18, 486
Fire-arms, Rammer connection for revolving	F. Beals	New Haven, Conn	Jan. 6, 1863	37, 329
Fire-arms, Rammer for many-chambered-breech	J. Kerr	London, England	Apr. 14, 1857	17, 044
Fire-arms, Rammer for revolving	C. R. Alsop	Middletown, Conn	Aug. 7, 1860	29, 538
Fire-arms, Rammer for revolving	F. D Newbury	Hudson City, N. J	Jan. 31, 1865	46, 131
Fire-arms, Rear-sight base for	F. W. Howe	Providence, R. I.	Jan. 24, 1865	46, 000
Fire-arms, Removable rammer for revolving	H. S. North	Middletown, Conn	Apr. 6, 1858	19, 868
Fire-arms, Retractor for revolving	T. Lee	Westport, Conn	Dec. 26, 1871	122, 182
Fire-arms, Revolving block of revolving	C. Sharps	Philadelphia, Pa	Nov. 27, 1860	30, 765
Fire-arms, Rifling	E. G. Allen	Boston, Mass	June 30, 1863	39, 024
Fire-arms, Rifling	A. Frauth	Chemnitz, Saxony	Oct. 10, 1865	50, 433
Fire-arms, Rifling breech-loading	H. Berdan	New York, N. Y	Jan. 10, 1865	45, 898
Fire-arms, Safety-device for lock of	E. T. Starr	New York, N. Y	Dec. 20, 1864	45, 532
Fire-arms, Safety-guard for	B. P. Cutler	Boston, Mass	Mar. 31, 1868	76, 058
Fire-arms, Safety-guard for the hammers of	H. E. Gibbon	Brooklyn, N. Y	Jan. 31, 1865	46, 100
Fire-arms, Safety-nipple for	W. N. Rowe	Washington, D. C	Jan. 1, 1867	60, 791
Fire arms, Safety nipple-guard	J. Oliphant	Uniontown, Pa	Jan. 13, 1863	37, 406
Fire-arms, Safety-stop for revolving	W. Tileston	Georgetown, D. C	Sept. 6, 1864	44, 126
Fire-arms, Screw driver for	G. W. Schofield	United States Army	Oct. 3, 1871	119, 656
Fire-arms, Securing the base-pin of revolving	S. Remington	Ilion, N. Y	Mar. 17, 1863	37, 921
Fire-arms, Self-primer for	W. H. Bell	Washington, D. C	Jan. 18, 1859	22, 618
Fire-arms, Self-priming hammer for	A. F. Tait	Morrisania, N. Y	June 2, 1863	38, 770
Fire-arms, Sight for	H. W. Colvin	Pendleton County, Ky	Sept. 13, 1859	25, 389
Fire-arms, Sight for	W. McKibbin	Buck Valley, Pa	Sept. 15, 1863	39, 941
Fire-arms, Sight for	C. E. Sneider	Baltimore, Md	Dec. 7, 1869	97, 717
Fire-arms, Sight for	J. Warner	Springfield, Mass	Feb. 24, 1863	37, 782
Fire-arms, Sighting	A. F. Garretson	Mount Pleasant, Iowa	Oct. 13, 1863	40, 256
Fire-arms, Snap-ring for	J. Rider	Newark, Ohio	Feb. 11, 1868	74, 427
Fire-arms so as to prevent oxidation and corrosion, Mode of finishing.	J. Allen and S. P. Townsend	New York, N. Y., and New Providence, N. J.	Jan. 1, 1867	60, 915
Fire-arms, Spring-power repeating	J. Gordon	New London, Conn	Dec. 31, 1867	72, 844
Fire-arms, Stand for testing	J. Lehnert	Louisville, Ky	Apr. 16, 1872	125, 743
Fire-arms, Swivel-loop for	E. Whitney, jr	New Haven, Conn	Apr. 15, 1873	137, 989
Fire-arms, Swivel-loop for	O. F. Winchester	New Haven, Conn	Dec. 12, 1871	121, 835
Fire-arms, Tape primer for	T. T. S. Laidley	United States Army	Feb. 15, 1859	22, 957
Fire-arms, Telescopic sight for	J. M. Trowbridge	United States Army	Mar. 8, 1864	41, 874
Fire-arms, Tige for	D. A. Hopkins	Brooklyn, N. Y	Sept. 16, 1862	36, 464
Fire-arms, Tompion for	R. Kingsley	Springfield, Mass	Feb. 18, 1862	34, 430
Fire-arms, Tompion for	T. T. S. Laidley	United States Army	May 19, 1868	77, 988
Fire-arms, Tompion for	R. G. Shurtliff	Springfield, Mass	Feb. 10, 1863	37, 645
Fire-arms, Tompion for	A. Tracy	United States Army	Mar. 18, 1862	34, 705
Fire-arms, Tool for manufacture of	A. Rebetey	Norwich, Conn	May 10, 1859	23, 944
Fire-arms, Toothed segment-lock for	D. H. Chamberlain	Boston, Mass	May 14, 1850	7, 360
Fire-arms, Trigger-cover for	J. Birkenhead	Ilion, N. Y	Feb. 9, 1864	41, 472
Fire-arms, Trigger-protector for	B. H. Westerhood	Philadelphia, Pa	July 22, 1856	15, 397
Fire-arms with tin, Overlaying	S. Ladd	Waltham, Mass	Oct. 14, 1835	....
Fire-arrester	G. W. Cook	Geneseo, Ill	Apr. 16, 1872	125, 663
Fire-back and grate setting	R. T. Collis	Webster, Ohio	Aug. 6, 1872	130, 195
Fire-back for warming rooms, Iron	C. Harris	Boston, Mass	May 15, 1815	....
Fire-balk	H. H. Baker	New Market, N. J	Jan. 17, 1865	45, 956
Fire-bar	W. Batchelor	Winchester, England	Jan. 17, 1871	111, 031
Fire-bar and connecting-bearing, Tubular	R. J. Ellis	Liverpool, England	Dec. 30, 1873	145, 995
Fire beds and tuyeres, Construction jointly of	P. H and F. M. Roots	Connersville, Ind	Oct. 18, 1870	108, 519
Fire-board	G. W. and W. H. Metcalf	Baltimore, Md	Jan. 26, 1869	86, 240
Fire-box	R. Gay	Richmond, Va	Nov. 30, 1869	97, 387
Fire box and grate	J. J. Folts	Buffalo, N. Y	Feb. 16, 1858	19, 358
Fire-box, Smoke-consuming	G. H. Smith	Galesburgh, Ill	Dec. 14, 1869	97, 819
Fire-boxes, Smoke-consuming apparatus for	J. Durand	Columbus, Ohio	Oct. 10, 1871	119, 832
Fire, Canvas conductor to be used when houses are on.	S. Green	....	Mar. 28, 1792	....
Fire-cavern	A. Quincy	Boston, Mass	Mar. 20, 1811	....
Fire-chamber, Apparatus for clearing	G. R. Moore	Lyons, Iowa	Mar. 29, 1864	42, 102
Fire-chamber clearer	G. R. Moore	Lyons, Iowa	Jan. 17, 1865	45, 933
Fire, Closed blomary	J. Renton and J. H. Crane	Newark, N. J	Aug. 10, 1848	5, 702
Fire, Composition for coating wooden structures to protect them against.	A. Pirz	Long Island City, N. Y	Jan. 24, 1871	111, 148
Fire, Conveying warm air to	W. R. Miller	Hinsdale, N. Y	Dec. 31, 1833	....
Fire-cracker holder	C. Most	Bergen City, N. J	Aug. 23, 1870	106, 714
Fire-cracker holder	A. E. Peck	Brooklyn, N. Y	Oct. 26, 1869	96, 143
Fire-cracker pistol or holder	R. Hutcheson	Newark, N. J	Nov. 8, 1870	109, 014
Fire-cracker pistol, Revolving	J. H. Hawes and O. W. Brock	Monroeton, Pa	Oct. 4, 1870	108, 022
Fire-detector	W. C. Grimes	Philadelphia, Pa	Apr. 21, 1868	76, 911
Fire-detector with the alarm, Connecting	H. L. Brower	New York, N. Y	Nov. 15, 1870	109, 292
Fire, Device in the walls of buildings to prevent damage to goods by water in case of.	T. Estlack	Philadelphia, Pa	June 3, 1856	15, 002
Fire, Distributer for extinguishing	G. J. Orr	New York, N. Y	Nov. 18, 1873	144, 696
Fire-draft or regulator by atmospheric pressure	R. Mayo	Washington, D. C	Sept. 9, 1835	....
Fire-engine	J. Achman	Washington, D. C	Dec. 30, 1812	....
Fire-engine	J. R. Adams	Port Jervis, N. Y	May 1, 1855	12, 773
Fire-engine	O. G. Adkins	Oswego, N. Y	July 20, 1852	9, 127
Fire-engine	A. F. Allen	Providence, R. I	Oct. 22, 1872	132, 426

Index of patents issued from the United States Patent Office from 1790 *to* 1873, *inclusive*—Continued.

Invention.	Inventor.	Residence.	Date.	No.
Fire-engine	J. B. Babcock	Marietta, Ohio	July 1, 1841	2, 150
Fire-engine	A. Barrett	Baltimore, Md	Feb. 18, 1841	1, 982
Fire-engine	G. Barton, jr	Waterford, N. Y	Aug. 16, 1844	3, 707
Fire-engine	G. Barton, jr., and L. Button	Waterford, N. Y	Sept. 19, 1846	4, 767
Fire-engine	J. Briggs, jr	St. Louis, Mo	Sept. 3, 1842	2, 770
Fire-engine	E. Bryant	New London, Conn	Mar. 28, 1820	....
Fire-engine	L. Button and R. Blake	Waterford, N. Y	Nov. 30, 1858	22, 162
Fire-engine	C. and W. C. Cleveland	New York and Ithaca, N. Y	Apr. 23, 1872	125, 883
Fire-engine	E. Daboll	Canaan, Conn	Apr. 12, 1828	....
Fire-engine	J. N. Dennisson	Newark, N. J	Feb. 7, 1765	46, 219
Fire-engine	J. J. Giraud	Baltimore, Md	Nov. 11, 1830	....
Fire-engine	C. W. Grannis	Collins, N. Y	Oct. 7, 1846	4, 796
Fire-engine	I. Hathaway	Walton, N. Y	Aug. 6, 1812	....
Fire-engine	E. Higgins	Saratoga, N. Y	Apr. 27, 1822	....
Fire-engine	J. Kersey		Apr. 13, 1797	....
Fire-engine	S. Lockwood and W. Loveland	Little Falls, N. Y	Apr. 16, 1822	....
Fire-engine	E. Marx	New York, N. Y	Dec. 20, 1845	4, 316
Fire-engine	A. Nudd	Exeter, N. H	Feb. 27, 1855	12, 441
Fire-engine	T. Odiorne	Portsmouth, N. H	Aug. 27, 1835	....
Fire-engine	J. Perkins	Newburyport, Mass	Mar. 23, 1813	....
Fire-engine	N. Pierce	Whitehall, N. Y	Feb. 23, 1831	....
Fire-engine	A. Pollock and J. Perkins	Boston, Mass	Aug. 6, 1812	....
Fire-engine	F. Ransom and D. L. Farnam	New York, N. Y	Apr. 17, 1847	5, 077
Fire-engine	J. Richardson, jr	Shrewsbury, Mass	July 24, 1818	....
Fire-engine	A. W. Roberts	Hartford, Conn	Jan. 23, 1855	12, 284
Fire-engine	G. Shalk and W. Tintoff	Lebanon, Pa	Sept. 23, 1826	....
Fire-engine	J. Smith, jr	New York, N. Y	May 20, 1807	....
Fire-engine	J. A. Sinclair	Bridgeport, Ohio	Oct. 28, 1873	144, 001
Fire-engine	S. Steward		Dec. 6, 1803	....
Fire-engine	J. B. Tarr	Albany, N. Y	Nov. 6, 1849	6, 859
Fire-engine	A. Warth	New York, N. Y	Aug. 1, 1854	11, 450
Fire-engine	J. Williams, jr	Salem, N. Y	May 11, 1839	1, 150
Fire-engine and lawn-sprinkler, Pneumatic	H. C. Neer	Park Ridge, N. J	Oct. 22, 1872	132, 407
Fire-engine and pump	J. N. Dennisson	Newark, N. J	Apr. 24, 1866	54, 247
Fire engine and water-pump	E. Williams and J. Collins	Falley Cross-Roads, Mass	June 20, 1823	....
Fire-engine and wrecking-pump, Floating	G. W. Talcott	Buffalo, N. Y	Oct. 6, 1868	82, 767
Fire-engine, Capstan	S. E. Hamlin		Aug. 30, 1799	....
Fire-engine, Chemical	I. H. Clark	Boston, Mass	June 1, 1869	90, 637
Fire-engine, Chemical	E. Gordon	Boston, Mass	Aug. 4, 1868	80, 542
Fire-engine, Chemical	R. Lapham and G. Clark, jr	Boston, Mass	May 28, 1867	65, 240
Fire-engine, Chemical	J. B. Stillson and J. A. Kley	Chicago, Ill	Sept. 17, 1872	131, 414
Fire-engine, Floating	J. C. Lang	Washington, D. C	Feb. 18, 1873	135, 922
Fire-engine heater, Steam	W. F. Shaw	Boston, Mass	Sept. 9, 1873	142, 589
Fire-engine, Horizontal	L. Holland	Belchertown, Mass	Mar. 11, 1824	....
Fire-engine, Horse-power	J. C. McCarthy	New York, N. Y	Nov. 2, 1869	96, 457
Fire-engine, Miniature steam	A. L. Dewey	Westfield, Mass	Aug. 22, 1871	118, 218
Fire-engine on locomotive	D. Williams	Syracuse, N. Y	Apr. 22, 1862	35, 053
Fire-engine pipe	J. Riley	Boston, Mass	Feb. 6, 1834	....
Fire-engine, Steam	N. S. Bean	Manchester, N. H	Jan. 15, 1861	31, 138
Fire-engine, Steam	L. Button and R. Blake	Waterford, N. Y	May 3, 1864	42, 557
Fire-engine, Steam	L. and T. E. Button	Waterford, N. Y	June 20, 1871	116, 151
Fire-engine, Steam	M. R. Clapp	Seneca Falls, N. Y	Jan. 7, 1862	34, 087
Fire-engine, Steam	E. R. and H. S. Cole	Pawtucket, R. I	July 7, 1868	79, 730
Fire-engine, Steam	E. R. and H. S. Cole	Pawtucket, R. I	Oct. 18, 1870	108, 452
Fire-engine, Steam	W. C. Davol, jr	Fall River, Mass	Oct. 21, 1873	143, 750
Fire-engine water-heater	P. M. Kafer and J. M. De Lacy	Trenton, N. J	Jan. 7, 1868	73, 015
Fire-engine water-heater	A. H. Perry	Boston, Mass	Oct. 22, 1872	132, 488
Fire-engines, Application of carbonic acid in	I. H. Clark	Boston, Mass	Dec. 1, 1868	84, 613
Fire-engines, Carriage for steam	A. B. Latta	Cincinnati, Ohio	May 22, 1855	12, 912
Fire-engines, Circulating water-heater for steam	H. V. Coleman	Chicago, Ill	Aug. 30, 1870	102, 782
Fire-engines, Coupling for uniting streams from	D. J. Tapley	Danvers, Mass	Dec. 17, 1872	134, 110
Fire-engines, Device for applying power to	G. Backstein	Philadelphia, Pa	Nov. 28, 1854	11, 987
Fire-engines, Electric signaling-apparatus for	W. H. Mumler	Boston, Mass	Feb. 6, 1872	123, 355
Fire-engines, Flue-blast for steam	J. Grabner	Warsaw, Ind	Jan. 14, 1873	134, 796
Fire-engines, Method of applying horse-power to	D. Russell	Lockport, N. Y	June 10, 1856	15, 069
Fire-engines, Method of working the lever of	S. Huse	Newburyport, Mass	Apr. 9, 1833	....
Fire-engines, Mode of applying water to	F. Ransom and U. Winman	New York, N. Y	Feb. 13, 1841	1, 980
Fire-engines, Operating	J. P., P., and G. Cowing	Seneca Falls, N. Y	Jan. 15, 1856	14, 089
Fire-engines, Operating	F. G. Smith	Columbia, Tenn	Sept. 12, 1854	11, 684
Fire-engines, Supplementary heater for steam	R. Gilbert and J. P. Topham	Rochester, N. Y	May 13, 1873	138, 880
Fire-engines, Supporting the carriage-bodies of	L. Button and R. Blake	Waterford, N. Y	Oct. 4, 1859	25, 628
Fire-escape	L. E. Ainsworth	Bolton, Canada	Sept. 23, 1873	143, 112
Fire-escape	E. Ale	Clearfield, Pa	July 4, 1871	116, 659
Fire-escape	W. B. Avery	Cambridge, Mass	June 5, 1860	28, 551
Fire-escape	H. O. Baker and J. McGill	New York, N. Y	May 22, 1860	28, 340
Fire-escape	A. T. Ballentine	New York, N. Y	June 28, 1864	43, 283
Fire-escape	P. W. Barnes	Albany, N. Y	Aug. 26, 1873	142, 141
Fire-escape	J. Bearns and A. Olsson	Brooklyn, N. Y	Mar. 19, 1872	124, 715
Fire-escape	G. W. Bishop and H. H. Smith	Baltimore, Md	Sept. 19, 1871	119, 008
Fire-escape	W. Breitenstein	New York, N. Y	Oct. 9, 1860	30, 291
Fire-escape	W. and G. H. Burditt	Boston, Mass	Feb. 1, 1870	99, 399
Fire-escape	O. F. Burton	New York, N. Y	Aug. 7, 1860	29, 542
Fire-escape	J. A. Chambers	Ogdensburgh, N. Y	July 3, 1866	56, 006
Fire-escape	R. Chandler	New York, N. Y	Nov. 8, 1864	44, 934
Fire-escape	L. M. Chipley	Saint Louis, Mo	Dec. 23, 1873	145, 844
Fire-escape	J. T. Commoss	New York, N. Y	Mar. 31, 1863	38, 078
Fire-escape	J. Cox	Brush Valley, Pa	June 27, 1846	4, 597
Fire-escape	C. W. Crosley	New York, N. Y	June 19, 1860	28, 740
Fire-escape	J. W. Davis and J. Vermillion	Washington, D. C	Oct. 31, 1871	120, 420
Fire-escape	J. Deckelmann and F. Spiess	New York, N. Y	Apr. 24, 1860	27, 972
Fire-escape	W. De Pew	Paris, Canada	May 2, 1871	114, 421
Fire-escape	T. S. Diblin	New York, N. Y	Sept. 3, 1867	68, 491
Fire-escape	C. Dieterich	Boston, Mass	Oct. 14, 1873	143, 677
Fire-escape	W. Gardner	Boston, Mass	Apr. 4, 1871	113, 509
Fire-escape	J. A. Griswold	Chicago, Ill	Aug. 30, 1870	106, 812
Fire-escape	J. C. Hancock and E. P. Richardson	Charlestown and Somerville, Mass	Aug. 8, 1871	117, 773
Fire-escape	J. M. Hancock	Lansing, Iowa	Apr. 5, 1859	23, 462

Index of patents issued from the United States Patent Office from 1790 *to* 1873, *inclusive*—Continued.

Invention.	Inventor.	Residence.	Date.	No.
Fire-escape	J. J. Hartmann	Saint Louis, Mo	July 16, 1872	129, 336
Fire-escape	E. Hawthorne	Mountain View, Cal	Jan. 7, 1868	73, 007
Fire-escape	W. Henley	Chicago, Ill	June 27, 1871	116, 441
Fire-escape	C. Herold	Pittsburgh, Pa	Oct. 8, 1872	132, 073
Fire-escape	J. Heuermann	Davenport, Iowa	July 23, 1867	66, 961
Fire-escape	J. Heuermann	Davenport, Iowa	Nov. 9, 1869	96, 698
Fire-escape	G. Heydrich	Philadelphia, Pa	Feb. 14, 1860	27, 127
Fire-escape	J. Hobbs	Columbus, Ind	Oct. 9, 1860	30, 318
Fire-escape	J. F. Hobson	Newburgh, Ohio	July 16, 1872	129, 410
Fire-escape	J. Hoeflinger	Carrollton, Mo	Apr. 4, 1871	113, 667
Fire-escape	J. J. Holwell	New York, N. Y	June 26, 1860	28, 862
Fire-escape	W. L. Horne	Batavia, Ill	July 17, 1866	56, 417
Fire-escape	E. J. Hudson	Golconda, Ill	Jan. 10, 1871	110, 851
Fire-escape	G. A. W. Hüttman and G. K. Kornelio.	Philadelphia, Pa	Mar. 10, 1849	6, 155
Fire-escape	J. Ivory	New York, N. Y	Mar. 14, 1871	112, 714
Fire-escape	L. Jewell	Stratham, N. Y	Jan. 22, 1861	31, 173
Fire-escape	C. J. and E. W. Jones	New Brighton, N. Y	Oct. 24, 1865	50, 596
Fire-escape	J. L. Jürgens	New Orleans, La	Sept. 22, 1868	82, 325
Fire-escape	C. P. Keenyon	Selma, N. C	Mar. 21, 1871	112, 928
Fire-escape	W. Kegg	Lasellsville, N. Y	Apr. 29, 1873	138, 410
Fire-escape	G. W. Keller	Philadelphia, Pa	Apr. 18, 1854	10, 807
Fire-escape	L. King	Bridgeport, Conn	May 22, 1860	28, 378
Fire-escape	L. Knocke	Davenport, Iowa	May 22, 1860	28, 380
Fire-escape	E. B. Larchar	New York, N. Y	Nov. 13, 1860	30, 633
Fire-escape	J. A. Law	Meredith, N. Y	Jan. 29, 1861	31, 248
Fire-escape	G. Laynor and H. Helmling	Baltimore, Md	Apr. 4, 1871	113, 311
Fire-escape	M. Lewis and J. C. Swenson	Williamsburgh, N. Y	Mar. 15, 1870	100, 777
Fire-escape	A. Lippmann	New York, N. Y	Apr. 24, 1860	27, 992
Fire-escape	I. Lohr	New York, N. Y	Feb. 16, 1864	41, 628
Fire-escape	R. Mackenzie and J. Cooper	New York, N. Y	June 4, 1867	65, 408
Fire-escape	H. Marshall	Atlanta, Ga	Dec. 10, 1872	133, 870
Fire-escape	J. Marx	Rochester, N. Y	Nov. 19, 1867	71, 030
Fire-escape	W. McCord	Sing Sing, N. Y	June 26, 1860	28, 880
Fire-escape	J. W. McKenzie	Brooklyn, N. Y	Nov. 12, 1861	33, 710
Fire-escape	I. Merritt	North Bridgewater, Mass	May 27, 1873	139, 410
Fire-escape	W. Miller	Boston, Mass	July 9, 1861	32, 778
Fire-escape	W. Miller	Boston, Mass	Sept. 6, 1870	107, 077
Fire-escape	C. C. Milton	San Francisco, Cal	May 6, 1873	138, 513
Fire-escape	H. Morohan	Brooklyn, N. Y	Nov. 13, 1860	30, 639
Fire-escape	E. R. Morrison	Washington, D. C	Jan. 28, 1873	135, 354
Fire-escape	S. Murset and E. Zuberbuhler	Philadelphia, Pa	Oct. 28, 1873	144, 122
Fire-escape	G. H. Nichols	Richmond, Va	Apr. 4, 1871	113, 329
Fire-escape	J. O'Brien	New York, N. Y	May 27, 1873	139, 416
Fire-escape	J. Paar	New York, N. Y	Mar. 31, 1868	76, 097
Fire-escape	J. W. Patten and G. P. Terry	Albany, N. Y	June 26, 1860	28, 898
Fire-escape	A. Pelham	Plymouth, Mich	July 29, 1873	141, 378
Fire-escape	S. Penfield	Hartford, Conn	Mar. 3, 1840	1, 505
Fire-escape	H. Poweison	New Brunswick, N. J	May 29, 1860	28, 501
Fire-escape	T. C. Rice	Worcester, Mass	Oct. 17, 1871	119, 997
Fire-escape	H. A. Richards	Chicago, Ill	Apr. 1, 1873	137, 483
Fire-escape	E. P. Richardson	Manchester, N. H	May 4, 1869	89, 686
Fire-escape	A. Rigny	New York, N. Y	May 7, 1867	64, 569
Fire-escape	A. Robinson	New York, N. Y	Apr. 7, 1868	76, 527
Fire-escape	W. Sawyer	Lowell, Mass	May 27, 1862	35, 413
Fire-escape	D. A. Scott and A. W. Hiltz	Calais, Me	Nov. 4, 1873	144, 361
Fire-escape	H. G. Sedgwick	Warsaw, Mo	May 7, 1872	126, 581
Fire-escape	H. T. Seely and W. M. Harrison.	Kenosha, Wis	July 16, 1872	129, 179
Fire-escape	F. Seymour	Cincinnati, Ohio	May 1, 1860	28, 095
Fire-escape	W. A. Sharp and G. Hollenbeck.	Tama City, Iowa	Nov. 14, 1871	121, 011
Fire-escape	G. H. Shaw	Phelps, N. Y	Apr. 22, 1873	138, 051
Fire-escape	A. Shute	Flushing, N. Y	Nov. 19, 1861	33, 757
Fire-escape	A. Shute	Flushing, N. Y	July 29, 1862	36, 032
Fire-escape	W. H. Simmons	Skaneateles, N. Y	May 18, 1869	90, 316
Fire-escape	G. C. Smith and F. M. Burrows	Baltimore, Md	Apr. 4, 1871	113, 357
Fire-escape	G. C. Smith and F. M. Burrows	Baltimore, Md	July 11, 1871	117, 007
Fire-escape	T. L. Sommeril	Juda, Wis	Mar. 21, 1871	112, 982
Fire-escape	O. Spear	Baltimore, Md	Feb. 2, 1869	86, 600
Fire-escape	J. W. Sprague	Rochester, N. Y	Jan. 22, 1861	31, 216
Fire-escape	J. Steger	New York, N. Y	Mar. 29, 1870	101, 393
Fire-escape	E. Sturel	New York, N. Y	Dec. 18, 1860	30, 939
Fire-escape	S. A. Swalm and C. C. Schmitt	New York, N. Y	Sept. 15, 1868	82, 176
Fire-escape	O. Sweeney	Brooklyn, N. Y	Aug. 24, 1858	21, 282
Fire-escape	J. A. Talpey	Somerville, Mass	June 3, 1873	139, 484
Fire-escape	J. A. Talpey	Somerville, Mass	July 22, 1873	141, 093
Fire-escape	T. Thompson	Baltimore, Md	Aug. 11, 1863	39, 505
Fire-escape	T. Thompson, jr	New York, N. Y	Aug. 25, 1868	81, 436
Fire-escape	J. J. Treanor	New York, N. Y	Dec. 12, 1871	121, 826
Fire-escape	W. W. Van Loan	Catskill, N. Y	Feb. 20, 1847	4, 971
Fire-escape	F. A. L. Von Ehren	New Orleans, La	Mar. 18, 1873	136, 883
Fire-escape	J. Wagner and J. Schmid	Philadelphia, Pa	June 25, 1867	66, 058
Fire-escape	L. Wahle	Davenport, Iowa	July 12, 1870	105, 280
Fire-escape	A. Warth	New York, N. Y	May 29, 1860	28, 525
Fire-escape	W. M. Watson	Tonica, Ill	Jan. 23, 1872	123, 066
Fire-escape	C. Wiedling	New York, N. Y	July 30, 1867	67, 239
Fire-escape	E. R. Wethered	London, England	Aug. 15, 1871	118, 172
Fire-escape	W. P. Withey	Hartford, Conn	May 12, 1840	1, 599
Fire-escape	R. Wyatt	Brooklyn, N. Y	June 20, 1865	48, 333
Fire escape and alarm	J. Fischer	New York, N. Y	May 28, 1867	65, 194
Fire escape and extinguisher	S. Welsh and T. Linacree	Albany, N. Y	Jan. 23, 1841	1, 943
Fire-escape and hose-elevator	W. H. Nobles	Saint Paul, Minn	Oct. 24, 1871	120, 211
Fire-escape attachment for horse-stalls	D. S. Neal	Lynn, Mass	July 24, 1860	29, 301
Fire-escape, Pocket	G. A. England	Ripon, Wis	Apr. 25, 1871	114, 119
Fire-escape, Portable	H. C. Carrigan	New York, N. Y	Sept. 28, 1869	95, 321
Fire-escape, Portable	I. Grafius	Alexandria, Pa	Oct. 2, 1860	30, 219
Fire-escape, Portable	G. D. McCullen	New Orleans, La	Dec. 12, 1871	121, 796

Index of patents issued from the United States Patent Office from 1790 *to* 1873, *inclusive*—Continued.

Invention.	Inventor.	Residence.	Date.	No.
Fire-extinguisher	B. T. Babbitt	New York, N. Y	Dec. 17, 1872	134, 024
Fire-extinguisher	J. F. Babcock	Boston, Mass	Aug. 4, 1868	80, 701
Fire-extinguisher	E. Barrett	New York, N. Y	Mar. 12, 1872	124, 534
Fire-extinguisher	W. H. Bate and G. F. Pinkham	Medford and Cambridge, Mass.	Mar. 30, 1869	88, 359
Fire-extinguisher	W. F. Beasley	Louisville, Ky	July 16, 1872	129, 519
Fire-extinguisher	C. Blake	Boston, Mass	May 2, 1871	114, 256
Fire-extinguisher	G. Booth	Toronto, Canada	Nov. 18, 1873	144, 595
Fire-extinguisher	J. F. Boynton	Syracuse, N. Y	Sept. 17, 1867	68, 940
Fire-extinguisher	J. F. Boynton	Syracuse, N. Y	June 15, 1869	91, 304
Fire-extinguisher	J. F. Boynton	Syracuse, N. Y	June 15, 1869	91, 411
Fire-extinguisher	J. F. Boynton	Syracuse, N. Y	Nov. 16, 1869	96, 876
Fire-extinguisher	Z. Breed	Weare, N. H	June 22, 1869	91, 709
Fire-extinguisher	P. F. Carlier and A. A. C. Vignon.	Paris, France	Apr. 13, 1869	88, 844
Fire-extinguisher	G. Clark, jr	Boston, Mass	Sept. 22, 1868	82, 471
Fire-extinguisher	G. Clark, jr	Boston, Mass	Mar. 9, 1869	87, 544
Fire-extinguisher	G. Cowing	Seneca Falls, N. Y	Apr. 26, 1870	102, 229
Fire-extinguisher	C. A. Cox	Jeffersonville, Ind	Apr. 15, 1873	137, 895
Fire-extinguisher	A. Crane	Fortress Monroe, Va	Feb. 10, 1863	37, 610
Fire-extinguisher	R. O. Doremus	New York, N. Y	Feb. 19, 1867	62, 118
Fire-extinguisher	W. L. Drake	Evanston, Ill	Dec. 9, 1873	145, 405
Fire-extinguisher	W. L. Drake	Evanston, Ill	Dec. 9, 1873	145, 406
Fire-extinguisher	J. Duffy	Quincy, Ill	Apr. 2, 1872	125, 275
Fire-extinguisher	W. L. Ellsworth	Brooklyn, N. Y	Aug. 20, 1872	130, 626
Fire-extinguisher	D. M. Ford and J. A. Kley	Chicago, Ill	June 20, 1871	116, 041
Fire-extinguisher	J. Gardner	Bedford, Ind	Feb. 14, 1871	111, 836
Fire-extinguisher	J. Gardner	Louisville, Ky	Dec. 26, 1871	122, 244
Fire-extinguisher	J. Gardner	Louisville, Ky	Mar. 12, 1872	124, 565
Fire-extinguisher	H. Henly	Bloomington, Ind	July 16, 1872	129, 134
Fire-extinguisher	L. Herman	Detroit, Mich	July 12, 1870	105, 203
Fire-extinguisher	C. T. Holloway	Baltimore, Md	June 11, 1872	127, 770
Fire-extinguisher	S B. Johnson	Philadelphia, Pa	Aug. 8, 1871	117, 891
Fire-extinguisher	E. Jones	Pierrepont, N. Y	Oct. 17, 1871	119, 931
Fire-extinguisher	E. Jones	Colton, N. Y	Jan. 30, 1872	123, 106
Fire extinguisher	W. Kitson	Lowell, Mass	Oct. 20, 1863	40, 343
Fire-extinguisher	F. Latta	Cincinnati, Ohio	Sept. 9, 1873	142, 637
Fire-extinguisher	W. H. Laubach	Philadelphia, Pa	Sept. 22, 1868	82, 421
Fire-extinguisher	C. L. Levey	New York, N. Y	Feb. 20, 1872	123, 783
Fire-extinguisher	S. M. Lillie	Elizabeth, N. J	Aug. 8, 1871	117, 900
Fire-extinguisher	C. Lippincott and J. Patterson	Philadelphia, Pa	July 1, 1873	140, 512
Fire-extinguisher	S. C. Main	Boston, Mass	June 15, 1869	91, 459
Fire-extinguisher	J. H. Manning	Chicago, Ill	Oct. 17, 1871	119, 936
Fire-extinguisher	T. J. Martin	Dowagiac, Mich	Mar. 26, 1872	125, 063
Fire-extinguisher	H. S. Maxim	Brooklyn, N. Y	July 22, 1873	141, 062
Fire-extinguisher	H. L. McAvoy	Baltimore, Md	June 17, 1873	139, 906
Fire-extinguisher	J. C. Meehan	Springfield, Mass	Apr. 1, 1873	137, 379
Fire-extinguisher	W. Mullally	Boston, Mass	June 2, 1868	78, 636
Fire-extinguisher	G. W. Nichols	Chicago, Ill	Oct. 3, 1871	119, 638
Fire-extinguisher	W. M. Parker	Boston, Mass	Apr. 6, 1869	88, 583
Fire-extinguisher	W. M. Parker	Boston, Mass	May 3, 1870	102, 582
Fire-extinguisher	G. F. Pinkham	Cambridge, Mass	Apr. 26, 1870	102, 431
Fire-extinguisher	W. K. Platt	Camden, N. J	Apr. 1, 1873	137, 481
Fire-extinguisher	P. W. Pratt	Abington, Mass	Sept. 17, 1872	131, 370
Fire-extinguisher	E. Sander	Saint Louis, Mo	Oct. 12, 1869	95, 840
Fire-extinguisher	E. Sander	Saint Louis, Mo	Feb. 14, 1871	111, 877
Fire-extinguisher	A. J. Sparrow	Potsdam, N. Y	Dec. 17, 1872	134, 107
Fire-extinguisher	J. H. Steiner	Cincinnati, Ohio	Aug. 12, 1873	141, 825
Fire-extinguisher	J. B. Stillson	Chicago, Ill	May 27, 1873	139, 275
Fire-extinguisher	D. J. Tapley	Danvers, Mass	Aug. 5, 1873	141, 607
Fire-extinguisher	J. S. Tibbets	Jeffersonville, Ind	July 8, 1873	140, 744
Fire-extinguisher	J. S. Tibbets	Jeffersonville, Ind	Sept. 23, 1873	143, 203
Fire-extinguisher	T. Tripp	Chicago, Ill	Aug. 16, 1870	106, 517
Fire-extinguisher	W. P. Van Deursen	Cincinnati, Ohio	Dec. 6, 1870	109, 975
Fire-extinguisher	J. B. Van Dyne	Covington, Ky	July 12, 1870	105, 279
Fire-extinguisher	J. B. Van Dyne	Covington, Ky	Dec. 6, 1870	109, 976
Fire-extinguisher	J. B. Van Dyne	Covington, Ky	May 9, 1871	114, 627
Fire-extinguisher	J. B. Van Dyne	Louisville, Ky	Apr. 9, 1872	125, 634
Fire-extinguisher	C. G. Wheeler	Chicago, Ill	Jan. 25, 1870	99, 270
Fire-extinguisher and chimney-regulator	D. Shoemaker	Philadelphia, Pa	Mar. 27, 1812	....
Fire-extinguisher and chimney-sweeper	E. Paine	Philadelphia, Pa	Dec. 11, 1817	....
Fire-extinguisher and lightning-rod, Combined	N. Carl	Cincinnati, Ohio	July 2, 1872	128, 534
Fire-extinguisher, Anti-freezing	W. C. Bruson	Chicago, Ill	Oct. 21, 1873	143, 746
Fire-extinguisher, Automatic	E. H. Ashcroft	Boston, Mass	Nov. 11, 1873	144, 493
Fire-extinguisher, Automatic	R. Lapham	New York, N. Y	June 11, 1867	65, 682
Fire-extinguisher, Automatic	I. P. Tice	New York, N. Y	Oct. 21, 1873	143, 794
Fire-extinguisher, Boiler	J. D. Bresnihan	Montgomery, Ala	Apr. 22, 1873	138, 068
Fire-extinguisher, Boiler	T. R. Sinclaire	New York, N. Y	Sept. 19, 1871	119, 190
Fire-extinguisher, Chemical	E. Bigelow	Springfield, Mass	Nov. 29, 1870	109, 576
Fire-extinguisher, Chemical	J. W. Douglas	Middletown, Conn	Dec. 14, 1869	97, 894
Fire-extinguisher, Chemical	A. E. Hughes	Louisville, Ky	Sept. 2, 1873	142, 340
Fire-extinguisher, Chimney	D. Berrien	New York, N. Y	Aug. 21, 1823	....
Fire-extinguisher for buildings, Steam	J. A. Coleman	Providence, R. I	July 22, 1873	141, 034
Fire-extinguisher for houses	S. P. Lord, jr	....	Mar. 10, 1804	....
Fire-extinguisher for railway-cars, &c., Automatic	W. P. Van Deursen and W. C. Davis.	Cincinnati, Ohio	Aug. 2, 1870	106, 002
Fire-extinguisher for steam-boilers	A. G. Davids	Baltimore, Md	Feb. 20, 1872	123, 874
Fire-extinguisher for steam-boilers	W. E. Prall	Washington, D. C	Jan. 2, 1872	122, 489
Fire-extinguisher for use on railway-cars, Automatic.	H. C. Stewart and R. T. Bradley.	Cincinnati, Ohio	Mar. 29, 1870	101, 404
Fire-extinguisher, Portable	I. C. Andrews	New York, N. Y	Oct. 14, 1873	143, 605
Fire-extinguisher, Steam	J. Souther	Boston, Mass	June 16, 1868	79, 021
Fire-extinguisher, Steamship	D. Spooner	Lowell, Ohio	Sept. 24, 1867	69, 267
Fire-extinguishers, Acid-bottle for	J. C. Davison	Chicago, Ill	May 16, 1871	114, 775
Fire-extinguishers, Composition for use in	J. M. Muterse and H. G. De Valory.	Guérande, France	Sept. 21, 1869	95, 036
Fire, Extinguishing	C. Alden	Newburgh, N. Y	Oct. 30, 1866	59, 151
Fire, Extinguishing	R. B. Armitage	Philadelphia, Pa	July 8, 1856	15, 276

Index of patents issued from the United States Patent Office from 1790 *to* 1873, *inclusive*—Continued.

Invention.	Inventor.	Residence.	Date.	No.
Fire, Extinguishing	J. Autenrieth	Newark, N. J	Sept. 19, 1871	119, 104
Fire, Extinguishing	I. Clowes	Norfolk, Va	July 2, 1836	....
Fire, Extinguishing	J. R. Laurent	Milford, Pa	Nov. 20, 1866	59, 847
Fire, Extinguishing	W. C. Marshall	New York, N. Y	May 15, 1866	54, 748
Fire, Extinguishing	L. Pusey	Philadelphia, Pa	Feb. 12, 1856	14, 252
Fire-extinguishing apparatus	O. F. Burton	New York, N. Y	Feb. 11, 1873	135, 772
Fire-extinguishing apparatus	I. H. Clark	Boston, Mass	Feb. 1, 1870	99, 291
Fire-extinguishing apparatus	M. Dunn	Chelsea, Mass	Dec. 24, 1872	134, 132
Fire-extinguishing apparatus	C. T. Holloway	Baltimore, Md	Aug. 20, 1872	130, 716
Fire-extinguishing apparatus	R. Lapham	Boston, Mass	Sept. 1, 1868	81, 653
Fire-extinguishing apparatus	W. H. Phillips	North Brixton, Surrey, England.	Apr. 9, 1850	7, 269
Fire-extinguishing apparatus for vessels	V. E. Campbell	Sterling Centre, N. Y	Feb. 2, 1869	86, 360
Fire-extinguishing apparatus for vessels	J. L. Stuart	New York, N. Y	May 21, 1861	32, 389
Fire-extinguishing composition	C. G. Mueller	New York, N. Y	Mar. 6, 1860	27, 375
Fire-extinguishing composition	E. F. Overdeer	Chattanooga, Tenn	Mar. 13, 1855	12, 519
Fire-extinguishing composition	W. P. Van Deursen	Cincinnati, Ohio	May 31, 1870	103, 800
Fire-extinguishing compound	R. Bulkley	New York, N. Y	Mar. 21, 1854	10, 658
Fire-extinguishing compound	T. Drew	Newton, Mass	Dec. 29, 1868	85, 434
Fire-extinguishing compound	E. A. Galbraith	Boston, Mass	Aug. 4, 1868	80, 720
Fire-extinguishing compound	E. Thayer	New York, N. Y	June 29, 1869	91, 882
Fire-extinguishing compound	J. Upham	Salem, Mass	Nov. 4, 1851	8, 495
Fire-extinguishing compound	W. P. Van Duersen	Cincinnati, Ohio	Aug. 16, 1870	106, 518
Fire-extinguishing device	J. Lamm	Port Deposit, Md	Aug. 6, 1872	130, 226
Fire-extinguishing engine	J. B. Van Dyne	Louisville, Ky	May 28, 1872	127, 390
Fire-extinguishing water-pipe attachment for buildings.	T. Miller	New York, N. Y	Nov. 18, 1873	144, 690
Fire-fender	M. Benson	Cincinnati, Ohio	May 19, 1823	....
Fire-fender	M. Morgan, jr	New York, N. Y	July 22, 1844	3, 675
Fire-fender, grate, and andiron, Combined adjustable	W. N. Hall	Mexia, Tex	May 13, 1873	138, 885
Fire fender or guard	C. H. S. Schultz	Cincinnati, Ohio	Nov. 19, 1867	71, 224
Fire-fender to prevent spread of fire	E. Ruggles	Rochester, Mass	Feb. 27, 1821	....
Fire-fenders, Shaping	T. Thomas and R. Fuller	New York, N. Y	Nov. 19, 1832	....
Fire-hearth, &c., for vessels	J. Lamb	Amboy, N. J	Mar. 18, 1820	....
Fire-hearth range or galley	G. Youle	New York, N. Y	Mar. 19, 1819	....
Fire-heater	F. Strattner	Wilmington, Del	May 13, 1873	138, 952
Fire-hook	J. G. Ernst	Harrisburgh, Pa	Jan. 6, 1857	16, 329
Fire-hydrant, Extension	J. Fricker, jr	Cincinnati, Ohio	June 4, 1872	127, 473
Fire in buildings, Apparatus for extinguishing	J. D. Sutter	Chicago, Ill	Oct. 28, 1873	144, 039
Fire in buildings, Extinguishing	O. Lugo	Baltimore, Md	Dec. 21, 1869	98, 175
Fire in high buildings, Extinguishing	A. Watson	Washington, D. C	Dec. 23, 1873	145, 920
Fire in holds of ships, vessels, &c., Extinguishing	C. Godfrey	Brooklyn, N. Y	Sept. 23, 1862	36, 515
Fire in inaccessible places, Method of extinguishing	A. Stone	Philadelphia County, Pa	Dec. 5, 1854	12, 038
Fire in inflammable liquids, Apparatus for extinguishing.	H. Baker	Chicago, Ill	Mar. 26, 1872	124, 875
Fire in oil-tanks, &c., Method of extinguishing	J. W. Stanton	Brooklyn, N. Y	Dec. 2, 1873	145, 134
Fire in steam-vessels, Arrangement for extinguishing.	W. Arthur	Brooklyn, N. Y	Apr. 5, 1859	23, 435
Fire indicator and alarm	J. Fawcett	East Boston, Mass	Nov. 25, 1873	144, 895
Fire-iron	R. Willingham	New York, N. Y	June 8, 1821	....
Fire-kindler	R. Adams	East Enterprise, Ind	Feb. 4, 1883	135, 395
Fire-kindler	W. Altick	Dayton, Ohio	Apr. 30, 1872	126, 245
Fire-kindler	C. Batcheller	Des Moines, Iowa	Apr. 12, 1870	101, 704
Fire-kindler	A. Bushnell	Iona Island, N. Y	May 13, 1873	138, 852
Fire-kindler	G. S. Coleman	Alexandria, Va	Sept. 17, 1872	131, 428
Fire-kindler	J. C. Crumpton	Allegheny City, Pa	Sept. 23, 1873	143, 066
Fire-kindler	J. C. Crumpton	Allegheny City, Pa	Sept. 30, 1873	143, 277
Fire-kindler	E. Everett	Walpole, Mass	Dec. 17, 1872	133, 973
Fire-kindler	M. E. Ezell	Hatchechubbee, Ala	Mar. 9, 1869	87, 654
Fire-kindler	J. E. Finley, G. H. Hurd, and B. F. Tatem.	Memphis, Tenn	Aug. 8, 1871	117, 757
Fire-kindler	J. A. Fuller	Rockford, Ill	Nov. 2, 1869	96, 532
Fire-kindler	M. Gernsey	Middleburgh, N. Y	Feb. 11, 1873	135, 706
Fire-kindler	J. R. Hice	Salem, Ohio	May 23, 1871	115, 205
Fire-kindler	H. K. Horton	Winfield, Mich	Feb. 9, 1869	86, 838
Fire-kindler	C. E. Imlay	Brooklyn, N. Y	Oct. 1, 1872	131, 759
Fire-kindler	J. Kenney, sr	Baltimore, Md	Nov. 15, 1870	109, 324
Fire-kindler	A. B. King	Camden, Ohio	May 23, 1871	115, 216
Fire-kindler	A. Lodeman and M. Desenberg	Kalamazoo, Mich	Sept. 7, 1869	94, 621
Fire-kindler	J. W. Lowe	Ottumwa, Iowa	Feb. 9, 1869	86, 767
Fire-kindler	J. Y. Marks	Rochester, Pa	Jan. 28, 1873	135, 280
Fire-kindler	W. H. McCrary	Kingston, Ga	June 6, 1871	115, 757
Fire-kindler	J. McFarland	Clinton, Ill	June 23, 1868	79, 140
Fire-kindler	I. Morris	Clinton, Ill	Nov. 10, 1868	83, 988
Fire-kindler	A. S. Morse and E. A. Jefferies	Fort Wayne, Ind	June 14, 1870	104, 336
Fire-kindler	C. A. Nisbett	Pontiac, Mich	Mar. 4, 1873	136, 451
Fire-kindler	B. Pickering	Dayton, Ohio	May 28, 1872	127, 365
Fire-kindler	N. Rogers	Thomasville, Ga	Nov. 14, 1871	120, 898
Fire-kindler	R. P. Smith	Dubuque, Iowa	Oct. 10, 1871	119, 893
Fire-kindler	V. G. Tansey	Quincy, Ill	June 4, 1867	65, 448
Fire-kindler	V. G. Tansey	Indianapolis, Ind	July 28, 1868	80, 429
Fire-kindler	W. H. Thomas	Goshen, Ind	Nov. 5, 1872	132, 780
Fire-kindler	D. W. Thompson	Saint Joseph, Mo	Apr. 30, 1872	126, 347
Fire-kindler	H. Van Ausdall	Keokuk, Iowa	July 30, 1867	67, 382
Fire-kindler	S. F. Watson	Richmond, Ind	Mar. 19, 1872	124, 705
Fire-kindler	E. P. Wheeler	Corinth, Miss	Sept. 23, 1873	143, 048
Fire-kindler case	D. W. Thompson	Saint Joseph, Mo	July 2, 1872	128, 676
Fire-kindling	I. Bicknell	Cincinnati, Ohio	Feb. 9, 1869	86, 635
Fire-kindling	E. Buss	Springfield, Ohio	Sept. 3, 1872	130, 975
Fire-kindling	L. Dodge	Waterford, N. Y	Sept. 5, 1871	118, 700
Fire-kindling	G. W. Eldridge	South Chatham, Mass	Apr. 5, 1870	101, 597
Fire-kindling	G. W. Eldridge	South Chatham, Mass	May 17, 1870	103, 163
Fire-kindling	C. Gaudin, and Z. and J Granier	San Francisco, Cal	July 14, 1868	79, 903
Fire-kindling	J. L. Hannum and S. H. Stebbins.	Berea, Ohio	June 14, 1870	104, 145
Fire-kindling	J. W. Kennedy	Plainfield, Conn	May 21, 1872	127, 068
Fire-kindling	J. W. Still	San Francisco, Cal	Apr. 25, 1871	114, 219
Fire-kindling	W. B. Tucker	Baltimore, Md	Feb. 4, 1873	135, 607
Fire-kindling	W. F. Wenisch	New York, N. Y	July 31, 1866	56, 867

Index of patents issued from the Unit d States Patent Office from 1790 *to* 1873, *inclusive*—Continued.

Invention.	Inventor.	Residence.	Date.	No.
Fire-kindling and fuel	I. Bicknell	Cincinnati, Ohio	Jan. 7, 1868	72, 964
Fire-kindling composition	D. B. Andrews	Fort Wayne, Ind	Jan. 22, 1867	61, 378
Fire-kindling composition	E. Bellinger	Mohawk, N. Y	Dec. 13, 1859	26, 408
Fire-kindling composition	M. Carstens	New York, N. Y	Mar. 20, 1866	53, 270
Fire-kindling composition	J. C. Crumpton	Allegheny City, Pa	Apr. 29, 1873	138, 382
Fire kindling composition	B. O'Reilly	New York, N. Y	Nov. 6, 1855	13, 758
Fire-kindling composition	W. C. Philbrick	Lynn, Mass	Jan. 28, 1868	73, 922
Fire-kindling composition	A. Reed	Louisville, Ky	Mar. 23, 1869	88, 075
Fire-kindling composition	J. Smith	Providence, R. I	Oct. 17, 1865	50, 503
Fire-kindling composition	A. Van Camp	Washington, D. C	Mar. 9, 1869	87, 734
Fire-kindling composition	W. P. Winkley	Des Moines, Iowa	Sept. 29, 1868	82, 506
Fire-kindling compound	J. S. Carroll	Covington, Ga	Oct. 8, 1872	131, 997
Fire-kindling compound	J. L. Hannum and S. H. Stebbins.	Berea, Ohio	Mar. 7, 1871	112, 337
Fire-kindling compound	S. L. Loomis and C. T. Shepherd.	Washington, D. C	Sept. 3, 1872	131, 011
Fire-kindling fagot	W. J. Wiggins and C. Stout	Saint Louis, Mo	June 7, 1870	103, 951
Fire-kindling material	L. T. Cheever	East Greenwich, R. I	Feb. 20, 1849	6, 125
Fire-kindling material	C. A. Rose	Columbus, Ga	Sept. 4, 1866	57, 774
Fire, Lehigh forge	J. Evans	New Haven, Conn	Jan. 5, 1864	41, 143
Fire-lighter	J. McCallum and J. Hartzell	Alliance, Ohio	Oct. 24, 1871	120, 206
Fire-lighter	J. R. Murphy	Chicago, Ill	Mar. 12, 1872	124, 505
Fire-lighter	L. H. Whitney	Vallejo, Cal	Feb. 25, 1868	74, 782
Fire, Means of detecting and signaling	W. B. Watkins	Jersey City, N. J	Mar. 28, 1871	113, 120
Fire, Method of lighting, heating, and extinguishing.	J. Kidd	New York, N. Y	Jan. 23, 1872	122, 895
Fire, Method of removing and protecting stores in case of.	J. H. Page	Whitewater, Wis	Jan. 19, 1869	85, 954
Fire, Method of removing stores, &c., in case of	A. Blood, sr., and R. W. Brown	Norfolk, Va., and Washington, D. C.	Sept. 1, 1857	18, 118
Fire, Mode of kindling	W. H. Towers	Boston, Mass	Mar. 5, 1867	62, 704
Fire, Mode of preventing the progress of	B. Taylor		Mar. 23, 1794	
Fire, Mode of protecting plastered walls and ceilings from.	P. Naylor	New York, N. Y	Feb. 22, 1839	1, 087
Fire, Mode of supplying water to buildings for the purpose of extinguishing.	I. Lowell	Pendleton, N. Y	Oct. 16, 1840	1, 489
Fire-place	T. C. Aldridge	Saint Louis, Mo	June 8, 1869	91, 059
Fire-place	A. C. Bacon and J. G. Jennings	Cleveland, Ohio	Dec. 22, 1863	40, 989
Fire-place	T. W. Baird	Bowling Green, Ky	Sept. 6, 1870	106, 985
Fire-place	T. F. Baker	Cincinnati, Ohio	June 15, 1869	91, 299
Fire-place	Bean and Skinner	Sandwich, N. H	June 12, 1835	
Fire-place	J. S. Blair	Boston, Mass	May 17, 1864	42, 739
Fire-place	P. Brecher	Louisville, Ky	June 20, 1871	116, 146
Fire-place	W. Bryant	Boston, Mass	Apr. 19, 1864	42, 342
Fire-place	R. Buck	Acton, Me	July 1, 1836	
Fire-place	W. Burgess	Middleborough, Mass	Mar. 12, 1836	
Fire-place	T. F. Card	Cincinnati, Ohio	Apr. 17, 1860	27, 886
Fire-place	A. Carson	New York, N. Y	May 22, 1860	28, 347
Fire-place	A. E. Chamberlain and J. B. Crowley.	Cincinnati, Ohio	Dec. 3, 1867	71, 698
Fire-place	G. A. Clark	Farmington, Conn	Oct. 4, 1859	25, 629
Fire-place	W. B. Coates	Philadelphia, Pa	July 28, 1868	80, 455
Fire-place	B. F. Conley	Funnelton, W. Va	Nov. 23, 1869	97, 050
Fire-place	C. D. Cooper	Albany, N. Y	June 27, 1810	
Fire-place	B. F. Cowan	Memphis, Tenn	Mar. 26, 1861	31, 785
Fire-place	J. M. Crockett	Newbern, Va	Oct. 12, 1869	95, 660
Fire-place	M. A. Cushing	Aurora, Ill	July 18, 1871	117, 053
Fire-place	A. D. Dailey	Terre Haute, Ind	Apr. 20, 1869	89, 203
Fire-place	T. C. Damborg	Philadelphia, Pa	July 16, 1872	128, 953
Fire-place	S. D. Dearman	Rock Hill, S. C	Apr. 23, 1872	125, 938
Fire-place	C. Dodge	Pittsburgh, Pa	Mar. 18, 1856	14, 447
Fire-place	J. Ervin, sr	Princeton, Ind	Nov. 3, 1868	93, 701
Fire-place	J. U. Fiester	Winchester, Ohio	June 4, 1867	65, 364
Fire-place	J. S. Ganson and C. T. Coit	Buffalo, N. Y	May 14, 1861	32, 283
Fire-place	J. L. Garlington	Snapping Shoals, Ga	July 8, 1873	140, 621
Fire-place	E. H. Gibbs	New York, N. Y	Dec. 14, 1869	97, 902
Fire-place	E. S. Greely	Dover, Me	Oct. 6, 1835	
Fire-place	C. B. Gregory	Beverly, N. J	Mar. 3, 1868	75, 155
Fire-place	C. B. Gregory	Beverly, N. J	Apr. 14, 1868	76, 750
Fire-place	G. Gridley	Newburyport, Mass	Mar. 14, 1808	
Fire-place	E. Griffiths	Philadelphia, Pa	Mar. 6, 1820	
Fire-place	W. D. Guseman	Morgantown, W. Va	Apr. 14, 1868	76, 751
Fire-place	J. Hackett	Louisville, Ky	Feb. 21, 1871	111, 929
Fire-place	J. Hackett	Louisville, Ky	July 25, 1871	117, 276
Fire-place	S. Hammond	Baltimore, Md	June 25, 1839	1, 197
Fire-place	E. Hampton	New York, N. Y	Feb. 23, 1869	87, 099
Fire-place	C. Harris and P. W. Zoiner	Cincinnati, Ohio	Apr. 16, 1867	63, 885
Fire-place	D. Hattan	Zanesville, Ohio	Aug. 4, 1868	80, 731
Fire-place	M. Haughey	Saint Louis, Mo	Aug. 13, 1872	130, 427
Fire-place	D. Hemingway	Leesburgh, Ky	Nov. 13, 1844	3, 821
Fire-place	J. R. Hendrickson	McKeesport, Pa	Sept. 24, 1861	33, 349
Fire-place	W. Hoyland	Newcastle, Pa	Nov. 18, 1873	144, 675
Fire-place	W. Hoyt	Brookville, Ind	June 3, 1825	
Fire-place	J. M. Irwin	Pittsburgh, Pa	Sept. 16, 1873	142, 791
Fire-place	W. H. James	Cincinnati, Ohio	Apr. 25, 1865	47, 423
Fire-place	E. Jenks	Colebrook, Conn	Sept. 2, 1818	
Fire-place	A. B. Johnson	Washington, Ind	July 26, 1870	105, 806
Fire-place	I. Kepler	Corry, Pa	Nov. 12, 1867	70, 858
Fire-place	W. M. Kepler	Cincinnati, Ohio	Aug. 26, 1873	142, 241
Fire-place	D. F. Launy	Philadelphia, Pa	Mar. 18, 1813	
Fire-place	J. E. Layton	Pittsburgh, Pa	July 2, 1861	32, 700
Fire-place	E. Lester	Killingsworth, Conn	June 21, 1830	
Fire-place	C. A. Littlefield	Covington, Ky	Sept. 11, 1860	29, 980
Fire-place	W. Lossie	Owensborough, Ky	Aug. 19, 1873	141, 937
Fire-place	J. Maggini	Baltimore, Md	Feb. 4, 1822	
Fire-place	S. Martin	Detroit, Mich	Nov. 30, 1869	97, 307
Fire-place	T. McCleary	Blairsville, Pa	Nov. 13, 1866	59, 620
Fire-place	R. D. McDonald	Jersey City, N. J	Sept. 21, 1869	95, 030

Index of patents issued from the United States Patent Office from 1790 *to* 1873, *inclusive*—Continued.

Invention.	Inventor.	Residence.	Date.	No.
Fire-place	G. H. McElevey	Newcastle, Pa	Aug. 25, 1868	81, 389
Fire-place	W. T. McMillen	Cincinnati, Ohio	Dec. 4, 1866	60, 220
Fire-place	J. McMurtry	Fayette County, Ky	Jan. 22, 1861	31, 212
Fire-place	W. P. Miller	San Francisco, Cal	Feb. 12, 1867	62, 053
Fire-place	E. Mix	New Haven, Conn	Apr. 25, 1812	......
Fire-place	G. Moody	Piqua, Ohio	Jan. 30, 1866	52, 375
Fire-place	A. Moon	Bethlehem, Ind	Feb. 20, 1866	52, 735
Fire-place	J. Moore	Erginetb, Ireland	Apr. 5, 1870	101, 645
Fire-place	M. Moore	Bartlett, Tenn	Jan. 16, 1872	122, 775
Fire-place	E. Nott	Schenectady, N. Y	Feb. 3, 1819	......
Fire-place	C. W. Peale		Nov. 16, 1797	......
Fire-place	T. Phillips	Cadiz, Ohio	Oct. 11, 1870	108, 180
Fire-place	W. R. Prescott	Hallowell, Me	Mar. 19, 1836	......
Fire-place	F. Proudfoot	Toronto, Canada	June 10, 1873	139, 813
Fire-place	C. S. Rankin	Cincinnati, Ohio	Mar. 30, 1869	88, 509
Fire-place	A. G. Redway	Cincinnati, Ohio	May 7, 1867	64, 446
Fire-place	I. I. Richardson	Delaware, Ohio	Mar. 19, 1867	63, 098
Fire-place	W. D. Robb	New Philadelphia, Ohio	July 1, 1873	140, 543
Fire-place	E. Y. Robbins	Cincinnati, Ohio	July 19, 1864	43, 604
Fire-place	E. Y. Robbins	Cincinnati, Ohio	Sept. 10, 1867	68, 653
Fire-place	P. M. Roche	Cleveland, Ohio	Apr. 25, 1871	114, 045
Fire-place	F. M. Rogers	Cincinnati, Ohio	Apr. 10, 1866	53, 880
Fire-place	D. A. Ross	Cincinnati, Ohio	Sept. 8, 1863	39, 836
Fire-place	J. K. Ross	Lebanon, Ohio	Apr. 3, 1860	27, 741
Fire-place	W. Salisbury	Derby, Vt	Aug. 16, 1817	......
Fire-place	H. Saxton	Eden, N. Y	July 8, 1834	......
Fire-place	G. R. Scriven	Hanging Rock, Ohio	Feb. 6, 1872	123, 518
Fire-place	L. Shepherd	Northampton, Mass	Jan. 25, 1817	......
Fire-place	E. A. Skeele	Saint Louis, Mo	June 23, 1863	38, 991
Fire-place	E. Skinner	Sandwich, N. H	Mar. 12, 1836	......
Fire-place	H. E. Smith	Cincinnati, Ohio	Sept. 13, 1864	44, 228
Fire-place	J. Smith	Brantford, Canada	Feb. 28, 1871	112, 188
Fire-place	J. Smith	Pittsburgh, Pa	Jan. 2, 1872	122, 410
Fire-place	J. W. Smith and J. S. Gallaher, jr.	Washington, D. C	Mar. 6, 1855	12, 491
Fire-place	A. Snyder	Richmond, Va	Oct. 5, 1869	95, 529
Fire-place	S. Spalding	Colchester, Conn	Apr. 11, 1822	......
Fire-place	A. J. Sprague	Springfield, Mo	Mar. 16, 1869	87, 884
Fire-place	P. Sternberg	Montgomery County, N. Y	Mar. 20, 1811	......
Fire-place	P. Sternberg	Montgomery County, N. Y	Oct. 28, 1811	......
Fire-place	P. Sternberg	Montgomery County, N. Y	Nov. 19, 1811	......
Fire-place	D. and J. Stoner	West Overton, Pa	May 17, 1864	42, 804
Fire-place	J. C. Strong and L. C. McNeal	Buffalo, N. Y	Sept. 21, 1869	94, 984
Fire-place	R. P. Thomas	Sciotoville, Ohio	Dec. 12, 1871	121, 911
Fire-place	I. H. Upton	New York, N. Y	Nov. 6, 1866	59, 484
Fire-place	W. R. Warden	Boston, Mass	Aug. 3, 1858	21, 094
Fire-place	M. D. Wellman	Pittsburgh, Pa	May 7, 1867	64, 465
Fire-place	M. D. Wellman and J. Old	Pittsburgh, Pa	July 11, 1865	48, 752
Fire-place	J. W. Wetmore	Erie, Pa	Feb. 28, 1871	112, 199
Fire-place	E. Whiting	Great Barrington, Mass	June 18, 1810	......
Fire-place	A. Wilkin	Lucas County, Ohio	June 8, 1869	91, 190
Fire-place	J. E. Wood	Webster, Ohio	Apr. 25, 1871	114, 075
Fire-place and chimney	R. Bacon and E. Harris	Boston, Mass	Apr. 8, 1835	......
Fire-place and chimney	C. Blood	Washington County, Md	Apr. 7, 1825	......
Fire-place and chimney	D. Hemingway	Covington, Ky	July 17, 1860	29, 168
Fire-place and chimney for saving fuel	S. Morey	Philadelphia, Pa	Jan. 18, 1813	......
Fire-place and chimney-stack in buildings	H. R. Sawyer	New York, N. Y	Mar. 26, 1841	2, 016
Fire-place and furnace	E. B. Wilson	London, England	Nov. 15, 1864	45, 111
Fire-place and grate	C. Lane	Hingham, Mass	Dec. 15, 1835	......
Fire-place and grate, Open	R. M. Sherman	Fairfield, Conn	June 30, 1837	249
Fire-place and oven	J. Knight	Stoddard, N. H	Dec. 15, 1831	......
Fire-place and stove	C. Guernsey	Poultney, Vt	July 7, 1813	......
Fire-place and stove	E. Haskell	New London, Conn	June 25, 1813	......
Fire-place and stove	C. Neer	Troy, N. Y	May 31, 1853	9, 756
Fire-place and stove	E. W. Newton	Middletown, Conn	Dec. 3, 1834	......
Fire-place and stove	A. Pollock	Boston, Mass	Apr. 23, 1807	......
Fire-place and stove	F. Stevens	Springfield, Mass	Mar. 2, 1836	......
Fire-place and throat of chimney	C. W. Russell	Washington, D. C	Mar. 5, 1850	7, 149
Fire-place and warming buildings	J. Cox	Philadelphia, Pa	May 2, 1834	......
Fire-place back	J. C. Bard	New Orleans, La	Feb. 15, 1870	99, 748
Fire-place back	G. W. Cummings	Franklin, Pa	Apr. 22, 1873	138, 135
Fire-place called "The defiance," Construction of	E. Bartholomew and W. Church	Boston, Mass	Aug. 18, 1815	......
Fire-place, Cast-iron	A. Hayward	Easton, Mass	Sept. 25, 1834	......
Fire-place, Chimney funneled	A. Gerrish	Shopleigh, Me	Jan. 23, 1835	......
Fire-place, Close	H. G. Spafford	Hudson, N. Y	May 3, 1805	......
Fire-place, Close and open	J. C. Brush	Washington, D. C	Aug. 19, 1819	......
Fire-place, Columbia	R. Heterick		June 30, 1803	......
Fire-place, Cooking and baking	J. C. Howard	Howard's Valley, Conn	Oct. 27, 1835	......
Fire-place, Doric	J. Pierpont	Boston, Mass	Jan. 8, 1830	......
Fire-place, Double-draft	T. Rose		Jan. 23, 1804	......
Fire-place fender	C. C. Algeo	Pittsburgh, Pa	May 30, 1871	115, 265
Fire-place fender	C. C. Algeo	Pittsburgh, Pa	Nov. 21, 1871	121, 073
Fire-place fender	W. N. Hall	Springfield, Tex	Feb. 27, 1872	124, 132
Fire-place fender	E. Skinner	Sandwich, N. H	Apr. 19, 1822	......
Fire-place fender	J. W. Truslow	Lewisburgh, Va	July 15, 1856	15, 362
Fire-place frame-work	J. L. Henderson	Covington, Ky	Oct. 22, 1867	70, 090
Fire-place front	E. A. Jackson	New York, N. Y	Nov. 11, 1873	144, 459
Fire-place, furnace, &c	J. W. Boot and W. Lyman	Boston, Mass	Dec. 29, 1826	......
Fire-place, Grate	J. Snyder	Philadelphia, Pa	June 12, 1835	......
Fire-place, Grated	A. Graham	Fredericktown, Md	Nov. 24, 1810	......
Fire-place heater	B. C. Bibb	Baltimore, Md	July 20, 1869	92, 781
Fire-place heater	S. S. Benk	New York, N. Y	Oct. 1, 1861	33, 412
Fire-place heater	F. L. Hedenberg	New York, N. Y	May 21, 1861	32, 364
Fire-place heater	W. C. Lesster	New York, N. Y	Jan. 29, 1867	61, 623
Fire-place heater	D. S. Quimby	Brooklyn, N. Y	Oct. 9, 1860	30, 349
Fire-place heater	D. S. Quimby, jr	Brooklyn, N. Y	June 8, 1869	91, 957
Fire-place heater	W. Sanford	Brooklyn, N. Y	Jan. 5, 1869	85, 614
Fire-place heater	W. Sanford	Brooklyn, N. Y	May 4, 1869	89, 693

Index of patents issued from the United States Patent Office from 1790 *to* 1873, *inclusive*—Continued.

Invention.	Inventor.	Residence.	Date.	No.
Fire-place heater	J. M. Thatcher	Bergen, N. J	Dec. 22, 1868	85, 255
Fire-place heater	J. M. Thatcher	Bergen, N. J	June 15, 1869	91, 287
Fire-place heater	J. M. Thatcher	Bergen, N. J	June 14, 1870	104, 376
Fire-place heater	H. H. Welch	Athens, Ohio	Aug. 8, 1865	49, 325
Fire-place heater, Base-burning	B. C. Bibb	Baltimore, Md	Sept. 14, 1869	94, 700
Fire-place heater, Base-burning	B. C. Bibb and P. Klotz	Baltimore, Md	Feb. 8, 1870	99, 525
Fire-place heater, Base-burning	J. H. Burtis	Brooklyn, N. Y	July 4, 1871	116, 680
Fire-place heater, Base-burning	E. S. Heath	Baltimore, Md	Jan. 3, 1871	110, 654
Fire-place heater, Base-burning	J. Jaeger	Tompkinsville, N. Y	Dec. 6, 1870	109, 969
Fire-place heater, Base-burning	P. Klotz	Baltimore, Md	Aug. 31, 1769	94, 320
Fire-place heater, Base-burning	P. Klotz	Baltimore, Md	Feb. 14, 1871	111, 753
Fire-place heater, Base-burning	W. Magill	Port Deposit, Md	Jan. 28, 1873	135, 278
Fire-place heater, Base-burning	J. Martino	Philadelphia, Pa	Mar. 8, 1870	100, 539
Fire-place heater, Base-burning	W. L. McDowell	Philadelphia, Pa	Nov. 21, 1871	121, 185
Fire-place heater, Base-burning	A. Murdock	Brooklyn, E. D., N. Y	Aug. 9, 1870	106, 191
Fire-place heater, Base-burning	J. S. Perry	Albany, N. Y	Jan. 14, 1873	134, 767
Fire-place heater, Base-burning	S. B. Sexton	Baltimore, Md	Dec. 14, 1869	97, 970
Fire-place heater, Base-burning	S. B. Sexton	Baltimore, Md	Mar. 19, 1872	124, 860
Fire-place heater, Base burning	G. B. Snider	New York, N. Y	June 27, 1871	116, 365
Fire-place heater, Base-burning	J. Spear	Philadelphia, Pa	Jan. 16, 1872	122, 863
Fire-place heater, Base-burning	W. E. Wood	Baltimore, Md	Aug. 1, 1871	117, 585
Fire-place-heater fender	D. Stuart and L. Bridge	Philadelphia, Pa	Feb. 8, 1870	99, 726
Fire-place heater or furnace	O. Collins	New York, N. Y	Apr. 30, 1867	64, 284
Fire-place lining	E. A. Jackson	New York, N. Y	Nov. 11, 1873	144, 460
Fire-place lining	A. J. Redwig	Cincinnati, Ohio	Feb. 4, 1873	135, 587
Fire-place lining	C. Truesdale	Cincinnati, Ohio	June 22, 1869	91, 689
Fire-place mantel and front	D. K. Innes and W. W. Magill	Cincinnati, Ohio	Oct. 31, 1871	4, 613
Fire-place mantel and front	D. K. Innes and W. W. Magill	Cincinnati, Ohio	July 25, 1871	117, 294
Fire-place, Metallic	J. F. Snider	Culpeper County, Va	July 18, 1854	11, 340
Fire-place, Open	M. D. Wellman	Pittsburgh, Pa	Dec. 18, 1866	60, 601
Fire-place or Franklin stove	J. Peckover	Cincinnati, Ohio	May 28, 1872	127, 265
Fire place or grate, Portable	J. Williamson	Washington, D. C	Nov. 8, 1836	74
Fire-place or open stove	J. F. Gould	Newburyport, Mass	Feb. 1, 1808	....
Fire-place or stove	J. E. Cayford	Milburn, Me	Dec. 31, 1833	....
Fire-place, Parlor	E. Backus	Brooklyn, N. Y	Sept. 25, 1847	5, 307
Fire-place, Parlor and kitchen	J. Hagerty	Monroe, Mich	Mar. 24, 1838	655
Fire-place, Portable	W. B. Johns	Georgetown, D. C	Dec. 24, 1861	33, 995
Fire-place, Reflecting	R. Jobson	Dudley, England	May 20, 1851	8, 101
Fire-place register	E. A. Tuttle	Brooklyn, N. Y	July 2, 1861	32, 728
Fire-place, Sheet-iron	J. Ingalls	Sanbornton, N. H	May 23, 1833	....
Fire-place, Sheet-iron	L. Mansfield	New Hartford, N. Y	May 29, 1833	....
Fire-place, Sheet-iron	G. Richards	Ashfield, Mass	Nov. 26, 1835	....
Fire-place, Steam	J. W. Cochran	Lowell, Mass	July 18, 1834	....
Fire-place to prevent smoking	H. Roberts	Delhi, N. Y	Apr. 30, 1840	1, 578
Fire-place warming-apparatus	C. Kalbfuss	New Richmond, Ohio	Aug. 13, 1867	67, 769
Fire-place with a flat funnel	A. Fisher		Feb. 19, 1807	
Fire-place with gridiron attachment, Movable	J. W. Wetmore	Erie, Pa	Apr. 4, 1865	47, 176
Fire places and grates, Back-plate for	F. E. Pitts	Nashville, Tenn	Mar. 3, 1857	16, 768
Fire-places, Back-plate and chimney-throat for	J. E. Layton	Pittsburgh, Pa	Oct. 29, 1861	33, 591
Fire-places, Constructing flue of open	T. Whitson	New York, N. Y	June 14, 1838	784
Fire-places, Fire-back for	S. M. Echols	La Fayette, Ind	July 15, 1856	15, 331
Fire-places for burning coal, Building and altering	A. Terrell	Boston, Mass	Oct. 25, 1832	....
Fire-places, stoves, &c., Composition slab for lining.	J. Putnam	Salem, Mass	Aug. 20, 1835	....
Fire-places, Ornamental back for	W. H. Jackson	New York, N. Y	July 27, 1869	92, 974
Fire-plug	J. W. Baker	Parkersburgh, W. Va	Sept. 24, 1867	69, 156
Fire-plug	J. P. Cummings	Newport, Ky	July 14, 1868	79, 960
Fire-plug	J. Curran	Baltimore, Md	Feb. 8, 1870	99, 646
Fire-plug	J. Fricker	Cincinnati, Ohio	Oct. 3, 1865	50, 235
Fire-plug	J. P. Gallagher	Saint Louis, Mo	July 22, 1873	141, 131
Fire-plug	R. A. Hill	Washington, D. C	May 25, 1869	90, 540
Fire-plug	F. Latta	Cincinnati, Ohio	Aug. 20, 1872	130, 726
Fire-plug	J. L. Lowry	Pittsburgh, Pa	Feb. 22, 1859	23, 034
Fire-plug	J. McClelland	Washington, D. C	Sept. 27, 1864	44, 439
Fire-plug	J. McClelland	Washington, D. C	May 30, 1871	115, 495
Fire-plug	L. Moss	Philadelphia, Pa	Nov. 10, 1857	18, 595
Fire-plug	J. H. Rhodes	Brooklyn, N. Y	May 26, 1868	78, 393
Fire-plug and cistern combined	F. Latta and W. H. Hughes	Cincinnati, Ohio	Mar. 18, 1873	137, 006
Fire-plug and hydrant	A. F. Allen	Providence, R. I	Dec. 26, 1871	122, 097
Fire-plug and hydrant	J. M. Jorden	Baltimore, Md	Sept. 8, 1838	909
Fire-plug and lamp-post, Combined	D. Y. and J. H. Johns	Cincinnati, Ohio	Nov. 19, 1872	133, 230
Fire-plugs, Casting	W. M. and J. B. Ellis	Washington, D. C	Apr. 17, 1860	27, 896
Fire-poker	G. R. Moore	Pittsburgh, Pa	July 24, 1860	29, 298
Fire-pot for coal-stoves	D. G. Littlefield	Albany, N. Y	June 25, 1861	32, 635
Fire-pot for stoves	N. O. Bond	Needham, Mass	May 24, 1864	42, 831
Fire-pot for stoves, furnaces, &c	P. P. Stewart	Troy, N. Y	Mar. 28, 1865	47, 049
Fire-pot, Tinners'	F. M. Campbell and L. W. Brown.	Cleveland, Ohio	Nov. 22, 1870	109, 383
Fire-pot, Tinners'	C. W. Johnson	Neponset, Ill	Sept. 1, 1868	81, 644
Fire-pot, Tinsmiths'	W. Yapp	Cleveland, Ohio	Aug. 11, 1863	39, 521
Fire-pots, Iron lining for	E. A. Tuttle	Williamsburgh, N. Y	Nov. 8, 1870	109, 080
Fire-proof chest	J. Matthews	New York, N. Y	June 16, 1836	....
Fire-proof chest	J. Scott	Philadelphia, Pa	Nov. 12, 1830	....
Fire-proof chest or safe	D. Fitzgerald	New York, N. Y	June 1, 1843	3, 117
Fire-proof chests, Use of asbestos in manufacture of	J. Scott	Philadelphia, Pa	July 21, 1835	....
Fire-proof column, Double-cylinder	J. B. Cornell	New York, N. Y	Mar. 20, 1860	27, 528
Fire-proof composition	N. E. Blake	Almond, N. Y	Apr. 18, 1865	47, 275
Fire-proof composition for crucibles, &c	L. Held	New York, N. Y	July 12, 1864	43, 548
Fire-proof compound	R. Spencer	New York, N. Y	Nov. 17, 1868	84, 143
Fire-proof compound for coating roofs, &c	R. W. Piper	Girard, Mich	Jan. 21, 1873	135, 152
Fire-proof, Compound for rendering substances	O. A. Tooker	Green Bay, Wis	July 2, 1872	128, 678
Fire-proof houses, Construction of	W. A. Berkey	Grand Rapids, Mich	Nov. 10, 1868	84, 044
Fire-proof iron chest	J. Delano	New York, N. Y	Aug. 13, 1834	....
Fire-proof iron chest	C. A. Gayler	New York, N. Y	Apr. 12, 1833	....
Fire-proof iron chest	C. A. Gayler	New York, N. Y	Apr. 3, 1835	....
Fire-proof packing for smoke and hot-air flues	J. B. Harris	Germantown, Ky	Nov. 26, 1867	71, 300
Fire-proof paper for roofs, &c	R. W. Piper	Girard, Mich	Jan. 21, 1873	135, 153
Fire-proof store, counting-room, &c	R. D. Curtis	Erie, Pa	Feb. 15, 1848	5, 445

Index of patents issued from the United States Patent Office from 1790 *to* 1873, *inclusive*—Continued.

Invention.	Inventor.	Residence.	Date.	No.
Fire-proof structure	H. H. Bryant	Boston, Mass	July 14, 1868	79, 809
Fire-proof wrought-iron chest	J. Delano	New York, N. Y	Mar. 7, 1826	....
Fire, Protecting buildings against	W. McAllister	Lawrence, Mass	Nov. 18, 1873	144, 625
Fire-regulator	J. Rozell	Brooklyn, N. Y	Sept. 29, 1863	40, 147
Fire-regulator, Automatic	G. M. Eldridge	Philadelphia, Pa	Jan. 30, 1872	123, 252
Fire-regulator for steam-boiler	J. M. Stiven	New York, N. Y	Aug. 5, 1862	36, 128
Fire safety-line	C. A. Gregory	Montreal, Canada	July 29, 1873	141, 269
Fire-screen	A. Chestnutwood	Davidsville, Pa	Aug. 8, 1871	117, 863
Fire-screen	H. P. Gengembre	Pittsburgh, Pa	Aug. 29, 1865	49, 622
Fire-screen	H. P. Gengembre	Pittsburgh, Pa	May 15, 1866	54, 713
Fire-screen	C. Pierce	Salem, Mass	Apr. 29, 1829	....
Fire-screen and stand	H. O. Fritsch	New York, N. Y	Jan. 10, 1871	110, 844
Fire shield	J. C. Clark	La Grange, Mich	Dec. 15, 1868	84, 859
Fire-shield	J. Lee	Galesburgh, Ill	Dec. 11, 1866	60, 390
Fire-shield	H. Rieger	Beaufort, N. C	Nov. 28, 1871	121, 421
Fire shield, Camp	M. Saviers	Kansas City, Mo	May 24, 1864	42, 879
Fire-shield, Portable	A. Dean	Ann Arbor, Mich	Aug. 8, 1871	117, 749
Fire-shovel	I. W. Denning	Allegheny City, Pa	Mar. 21, 1871	112, 789
Fire-shovel	F. M. Ketcham	Indianapolis, Ind	Sept. 23, 1873	143, 018
Fire-shovel	S. Kyle	Hyde Park, Ill	Nov. 26, 1872	133, 320
Fire-shovel	F. J. Niemöller	Rich Fountain, Mo	May 24, 1864	42, 867
Fire-stop	E. Putnam		Aug. 24, 1802	....
Fire, Water-gun for extinguishing	H. Strait	Covington, Ky	Feb. 24, 1852	8, 762
Fire-works, Machine for charging	T. Sharfenberg	Brooklyn, N. Y	Oct. 11, 1864	44, 662
Fireman's and builder's elevator	A. M. Patrick	Long Lane, Mo	Feb. 13, 1872	123, 642
Fireman's elevator	V. O'Bryan	Saint John's, Mich	May 19, 1868	78, 122
Fireman's protector	C. D. Woodruff	Toledo, Ohio	Mar. 8, 1859	23, 212
Fireman's signal	H. D. Treadwell	Elmira, N. Y	May 3, 1859	23, 873
Fireman's trumpet	W. Staehlen	Williamsburgh, N. Y	July 20, 1858	20. 961
Firemen, Machine for protecting	R. Bulkley	New York, N. Y	Mar. 29, 1821	....
Firing, Mechanical	S. Danks	Cincinnati, Ohio	June 27, 1871	116, 275
Firkin	C. Alvord	Westford, Wis	Aug. 19, 1873	141, 847
Firkins, barrels, &c., Device to attach to	A. T. Dunbar and A. McNaught	Alba, Pa	Aug. 13, 1867	67, 737
Fish and animal matter to obtain oil, fats, &c., Treating.	W. J. and T. Hooper	Baltimore, Md	Feb. 15, 1870	99, 896
Fish and animal trap	J. T. Harcourt and G. W. Cottingham.	Columbus, Tex	Sept. 17, 1872	131, 439
Fish, Apparatus for curing and drying	B. Robinson	East Gloucester, Mass	July 11, 1865	48, 723
Fish, Article of prepared cod	E. Crowell	New York, N. Y	Dec. 8, 1868	84, 801
Fish-bait	G. T. Thorp	Philadelphia, Pa	Oct. 26, 1869	96, 288
Fish-bait, Compound for	S. A. Goodman, jr	Jamestown, Tex	Jan. 21, 1873	135, 113
Fish-bait cutter	V. Doane, jr	Harwich Port, Mass	Oct. 13, 1868	83, 048
Fish-bait mill	W. McKay	Newburyport, Mass	July 16, 1872	129, 576
Fish-bars, Machine for punching, shearing, and straightening.	W. Morehouse	Buffalo, N. Y	Apr. 20, 1869	89, 062
Fish, Catching	J. Ellicott		Mar. 25, 1795	....
Fish-catching apparatus	J. Garl	Suffield, Ohio	May 25, 1858	20, 343
Fish-catching apparatus	A. Tiffany	Gibson, Pa	July 26, 1838	854
Fish, Conveying live	A. J. Smidth	Copenhagen, Denmark	Aug. 4, 1868	80, 775
Fish, Curing and preserving	R. A. Adams	Cambridge, Mass	Nov. 23, 1869	97, 145
Fish-cutter or bait-mill	N. Richardson	Gloucester, Mass	June 3, 1862	35, 472
Fish-decoy	I. B. Quimby	East Boston, Mass	Nov. 21, 1865	51, 120
Fish, Device for hatching the spawn of	S. Green	Rochester, N. Y	Sept. 17, 1867	68, 871
Fish-dressing machine	F. Kelsey and F. H. Hosmer	Sandusky, Ohio	Sept. 16, 1873	142, 918
Fish, &c., drier	W. J. and S. Hooper and O. Lugo.	Baltimore, Md	Feb. 8, 1870	99, 673
Fish-flake	J. Foster	Beverly, Mass	Aug. 20, 1867	67, 970
Fish for food, Mode of preparing	G. H. Heron	Washington, D. C	Nov. 5, 1867	70, 435
Fish for food, Process of preparing	J. Nickerson	Booth Bay, Me	Mar. 23, 1869	88, 064
Fish for use, Putting up cod	B. F. Stephens	Brooklyn, N. Y	Mar. 16, 1869	87, 986
Fish, Globe for	A. Ivers	New York, N. Y	Mar. 14, 1865	46, 801
Fish-grappling spear	J. W. Knapp	Cross River, N. Y	Oct. 28, 1873	144, 110
Fish-hatching apparatus	L. Stone	Charlestown, N. H	June 20, 1871	116, 112
Fish-hook	B. F. Allen	Boston, Mass	Aug. 19, 1873	141, 910
Fish-hook	F. T. Angers	Canastota, N. Y	Oct. 12, 1869	95, 755
Fish-hook	F. Angilard	Royan, France	Aug. 27, 1867	68, 027
Fish-hook	L. Arnold	Belchertown, Mass	Feb. 21, 1871	111, 898
Fish-hook	J. B. Christian	Mount Carroll, Ill	June 30, 1868	79, 446
Fish-hook	R. F. Cook	Troy, Ala	June 19, 1855	13, 081
Fish-hook	C. O. Crosby	New Haven, Conn	Nov. 20, 1866	59, 893
Fish-hook	W. Davis and J. Johnson	Brooklyn, N. Y	Dec. 19, 1865	51, 651
Fish-hook	T. F. Engelbrecht and G. F. Skiff.	Philadelphia, Pa	July 28, 1846	4, 670
Fish-hook	R. A. Fish	Worcester, Mass	Apr. 28, 1868	77, 365
Fish-hook	N. A. Gardiner, jr	Willett, N. Y	Sept. 20, 1864	44, 368
Fish-hook	W. C. Goodwin	Hamden, Conn	Oct. 2, 1866	58, 404
Fish-hook	M. Hiltz	Gloucester, Mass	Jan. 26, 1869	86, 154
Fish-hook	J. Johnson	Brooklyn, N. Y	Aug. 21, 1847	5, 256
Fish-hook	J. Johnson	Brooklyn, N. Y	Oct. 9, 1855	13, 649
Fish-hook	J. Johnson and H. Howarth	Brooklyn, N. Y	Apr. 24, 1866	54, 251
Fish-hook	F. Kemlo	Boston, Mass	Sept. 14, 1869	94, 893
Fish-hook	F. Kemlo	Boston, Mass	Sept. 14, 1869	94, 894
Fish-hook	F. Kemlo	Boston, Mass	Sept. 14, 1869	94, 895
Fish-hook	B. Lee, jr	Williamsburgh, N. Y	Feb. 12, 1867	61, 042
Fish-hook	A. I. Lenhart	New Brunswick, N. J	Nov. 12, 1867	70, 868
Fish-hook	H. B. Livermore	Ashland, Pa	Jan. 9, 1866	51, 951
Fish-hook	W. S. Morris	New York, N. Y.	Feb. 12, 1861	31, 396
Fish hook	S. Pendleton	New Haven, Conn	Aug. 21, 1847	5, 255
Fish-hook	E. Pitcher	Brooklyn, N. Y	July 16, 1872	129, 053
Fish-hook	H. Sigler	Houston, Tex	Apr. 11, 1854	10, 761
Fish-hook	E. Sterling	Cleveland, Ohio	Nov. 12, 1867	70, 913
Fish-hook, Double-lever	G. Crandell	Washington, D. C	Nov. 7, 1865	50, 799
Fish-hook holder	L. Arnold	Belchertown, Mass	Aug. 8, 1871	117, 719
Fish-hook, Spring	W. P. Blake	New York, N. Y	Aug. 15, 1848	5, 710
Fish-hook, Spring	D. Ellis and C. T. Grilley	Naugatuck, Conn	Aug. 15, 1848	5, 709
Fish-hook, Spring	D. Kidder	Franklin, N. H	Sept. 24, 1867	69, 221
Fish-hook, Spring	J. King, jr	Fort Wayne, Ind	Nov. 20, 1866	59, 844

Index of patents issued from the United States Patent Office from 1790 *to* 1873, *inclusive*—Continued.

Invention.	Inventor.	Residence.	Date.	No.
Fish-hook, Spring	E. Rhodes, jr., and J. W. Rhoades.	Clyde, Ohio	Jan. 1, 1867	60, 786
Fish-hooks and artificial bait, Attachment for	J. T. Buel	Whitehall, N. Y	Apr. 11, 1854	10, 771
Fish-hooks, Machine for making	C. O. Crosby	New Haven, Conn	Mar. 7, 1865	46, 644
Fish in warm weather, Mode of carrying	N. Robbins		Mar 11, 1802	......
Fish-nurseries	R. E. Sabin	West Springfield, Mass	July 11, 1871	116, 995
Fish-plates, Fastening for	R. Anthony	Kingston, Pa	Jan. 9, 1872	122, 547
Fish, Preparing and pres rving	W. Sharp	Portland, Me	Sept. 30, 1873	143, 386
Fish, Preparing, desiccating, and preserving	W. D. Cutler	PhiladelphiaPa.	Sept. 8, 1868	81, 987
Fish, Process of curing and put'ing up	J. Atwood, jr	Provincetown, Mass	May 25, 1869	90, 334
Fish, Process of extracting oil from	E. H. Woodward	New York, N. Y	Aug. 21, 1866	57, 427
Fish, Putting up salt mackerel and similar	E. A. Pharo	Philadelphia, Pa	Oct. 15, 1872	132, 316
Fish-scaling machine	N. B. White	Cecil County, Md	Nov. 26, 1867	71, 434
Fish, Serpentine spinner to catch	C. De Saxe	New York, N. Y	June 12, 1855	13, 068
Fish-spawn hatcher	M. G. Helton	Rochester, N. Y	Mar. 18, 1873	136, 834
Fish, Spinning bait for catching	J. T. Buel	Whitehall, N. Y	Apr. 6, 1852	8, 853
Fish to obtain oil and fertilizers, Process of treating.	A. Smith	Baltimore, Md	Jan. 25, 1870	99, 251
Fish to guano, &c., Apparatus for reducing	T. L. Robinson	Boston, Mass	Apr. 12, 1864	42, 310
Fish-trap	E. B. Beach	West Meriden, Conn	Dec. 22, 1868	85, 199
Fish-trap	D. Bowman	Tampico, Tenn	Mar. 8, 1859	23, 154
Fish-trap	R. Gray	Anson, Me	Jan. 18, 1859	22, 644
Fish-trap	F. Goodwin	Astoria, N. Y	Apr. 8, 1862	34, 887
Fish-trap	J. E Hammond	New Bedford, Mass	Apr. 4, 1871	113, 292
Fish-trap	C. E. Ketcham	Washington, D. C	Apr. 15, 1873	137, 930
Fish-trap	T. B McCaughan	Moscow, Tenn	Apr. 7, 1868	76, 489
Fish trap	S. Pavonarius and C. C. Michtle	Philadelphia, Pa	Aug. 5, 1873	141, 588
Fish-trap	L. Van Hoesen	Westville, Conn	Dec. 9, 1856	16, 217
Fish, Trolling-bait for catching	R. Haskell	Painesville, Ohio	Sept. 20, 1859	25, 507
Fish-way	E. A. Brackett	Winchester, Mass	Oct. 22, 1872	132, 349
Fish-way	A. Livermore	Ashland, Pa	Aug. 14, 1866	57, 159
Fish-way for dams	D. Steck	Hughesville, Pa	June 26, 1866	55, 929
Fish-weir	H. Emerson	Hancock	Aug. 14, 1811	......
Fishing-apparatus	P. H. Ferl and W. Larkins	Detroit, Mich	Aug. 7, 1866	56, 917
Fishing-apparatus	O. M. Fuller	Catasauqua, Pa	Jan. 30, 1872	123, 164
Fishing-apparatus	J. Koehler	New York, N. Y.	May 12, 1868	77, 893
Fishing-apparatus	J. Stetson	West Harwich, Mass	Nov. 3, 1868	83, 740
Fishing-bait, Mode of preserving	E E. Burnham	Gloucester, Mass	Dec. 15, 1868	84, 855
Fishing-bait, Preserving	T. D. Kellogg	New York, N. Y	Feb. 11, 1868	74, 3?8
Fishing-boat	J. Donn	Washington, D. C	Mar. 6, 1835	......
Fishing-implement	E. Horton	Bristol, Conn	Nov. 4, 1856	16, 014
Fishing-jack	M. D. Kirk and W. H. Belnap	Sturgis, Mich	July 20, 1869	92, 730
Fishing-jig	S. Albee	Wiscasset, Me	Nov. 3, 1868	83, 681
Fishing-lead	W. Smith	South Thomaston, Me	Apr. 1, 1856	14, 587
Fishing-line	F. E. Foster	New York, N. Y	Feb. 18, 1873	135, 900
Fishing-line float	E. Jewell	Louisville, Ky	July 9, 1872	128, 885
Fishing-line, Metallic	H. Camp	Covington, Ga	Sept. 5, 1871	118, 772
Fishing-line, Pivot	F. X. Monnier	Detroit, Mich	Feb. 23, 1869	87, 188
Fishing-line reel	A. Dougherty	Brooklyn, N. Y	Feb. 9, 1864	41, 494
Fishing-line reel	D. Ellis	Waterbury, Conn	July 12, 1864	43, 485
Fishing-line reel	A. B. Hartill	New York, N. Y	Aug. 7, 1866	56, 937
Fishing-line reel	W. M. Stuart	Newark, N. J.	Aug. 29, 1865	49, 663
Fishing-line reel	W. H. Van Gieson	New York, N. Y	July 5, 1864	43, 460
Fishing-line sinker	L. A. Burnham	Gloucester, Mass	Sept. 25, 1866	58, 211
Fishing-line sinker	E. F. Decker	Southport, Me	Feb. 21, 1865	46, 453
Fishing-line sinker	J. R. Martin	Booth Bay, Me	July 31, 1866	56, 857
Fishing-line sinker	J. R. Martin	Booth Bay, Me	Jan. 29, 1867	61, 625
Fishing-line sinker	R. T. Osgood	Orland, Me	Aug. 3, 1869	93, 220
Fishing-line sinker	W. H. Smith	New York, N. Y	May 12, 1868	77, 774
Fishing-line swivel	M. Hiltz	Gloucester, Mass	Dec. 10, 1867	71, 879
Fishing-lines and other small cords, Machine for making.	P. Brooks	New Haven, Conn	Oct. 23, 1866	59, 120
Fishing-lines, Attaching float and sinker to	J. Ingrams	Troy, N. Y.	Feb. 8, 1870	99, 572
Fishing-reel	P. A. Altmaeir	Harrisburgh, Pa	Nov. 9, 1869	96, 652
Fishing-reel	W. Billinghurst	Rochester, N. Y	Aug. 9, 1859	24, 987
Fishing-reel	A. H. Fowler	Batavia, N. Y	June 18, 1872	128, 137
Fishing-reel	A. Hatch	New Haven, Conn	June 19, 1866	55, 653
Fishing-reel	G. Mooney	Providence, R. I	Jan. 14, 1873	134, 917
Fishing-reel	M. S. Palmer	New Bedford, Mass	Feb. 28, 1860	27, 205
Fishing-reel	J. J. Ross	Buffalo, N. Y	Oct. 12, 1869	95, 839
Fishing-reel	S. B. Terry	Waterbury, Conn	Nov. 14, 1871	121, 020
Fishing-reel	J. Von Hofe	Brooklyn, E. D., N. Y	Nov. 26, 1867	71, 344
Fishing-rod	R. N. Isaacs	New York, N. Y	Oct. 16, 1866	58, 833
Fishing-rod	J. B. McHarg	Rome, N. Y	Mar. 18, 1873	137, 015
Fishing-rod	W. N. Smith	Reading, Pa	July 8, 1873	140, 656
Fishing-rod	T. Tout	Cambridge, Mass	Sept. 26, 1871	119, 251
Fishing-rod and float	C. De Saxe	New York, N. Y	Apr. 18, 1854	10, 795
Fishing-rod, Hinged	J. H. Montrose	New York, N. Y	Dec. 24, 1867	72, 667
Fishing-rod holder	F. Senieur	Mount Sterling, Ky	Aug. 26, 1873	142, 126
Fishing-rod joint	W. J. Hubbard	Ansonia, Conn	Mar. 15, 1870	100, 895
Fishing-rod reel	J. A. Bailey	Jersey City, N. J	Aug. 5, 1856	15, 466
Fishing-rod reel	E. Deacon	Brooklyn, N. Y	Feb. 10, 1857	16, 626
Fishing-rod tip	J. V. Hofe	New York, N. Y	May 20, 1862	35, 339
Fishing-rod tip	J. C. Underwood and T. J. Bargis.	Richmond, Ind	May 18, 1858	20, 309
Fishing-rods, Attaching reels to	T. W. Cummings	New York, N. Y	July 12, 1864	43, 546
Fishing-rods, Guide-ring for	H. Pritchard	Brooklyn, N. Y	Oct. 4, 1859	25, 693
Fishing-stake	J. O. Campbell	Alpena, Mich	Nov. 25, 1873	144, 888
Fishing-tackle	J. T. Buel	Whitehall, N. Y	Apr. 22, 1856	14, 706
Fishing-tackle	W. D. Chapman	Theresa, N. Y	May 15, 1866	54, 684
Fishing-tackle	A. A. Dennett	New Brunswick, N. J	July 21, 1868	80, 151
Fishing-tackle	J. D. Leach and S. Hutchings	Penobscot, Me	Dec. 15, 1868	84, 885
Fishing-tackle	A. J. Lenhart	Trenton, N. J	Aug. 2, 1864	43, 694
Fishing-tackle	L. D. Lothrop	Dover, N. H	May 5, 1868	77, 628
Fishing-tackle	D. C Talbot	Holden, Mass	Mar. 3, 1868	75, 675
Fishing-tackle	F. Tellgmann	Stamford, Conn	Feb. 9, 1869	86, 786
Fishing-tackle for deep-sea fishing	W. Woodbury	Gloucester, Mass	July 7, 1863	39, 192
Fishing-torch	S. Bartholomew	Sturgis, Mich	Apr. 15, 1873	137, 877

Index of patents issued from the United States Patent Office from 1790 *to* 1873, *inclusive*—Continued.

Invention.	Inventor.	Residence.	Date.	No.
Fishing vessel and net	J. Wills, jr	Galloway, N. J	Dec. 17, 1814	....
Fistula, Instrument for treating	E. F. Garvin	New York, N. Y	Dec. 8, 1868	84, 815
Fixing, puddling, and boiler furnace	M. Z. Evans	Ormsby, Pa	Oct. 5, 1869	95, 450
Flag	A. Watson	Washington, D. C	Feb. 16, 1864	41, 656
Flag-halyard	W. Albert	Brooklyn, N. Y	Sept. 26, 1871	119, 292
Flag-hoisting apparatus	J. W. Mackenzie	San Francisco, Cal	May 7, 1872	126, 469
Flag, Signal, &c	J. Holt	Lowell, Mass	Apr. 26, 1870	102, 267
Flag, Signal	H. J. Rogers	Baltimore, Md	Jan. 2, 1855	12, 140
Flags, Apparatus for press-dyeing stars for	D. C. Farrington	Lowell, Mass	Mar. 12, 1872	124, 428
Flags, Apparatus for press-dyeing stripes for	D. C. Farrington	Lowell, Mass	Mar. 12, 1872	124, 427
Flail-cap	T. J. Hubbard	Hamilton, N. Y	Jan. 11, 1859	22, 557
Flange	D. Paterson	Jersey City, N. J	July 10, 1866	56, 326
Flange-turning die	A. W. Sangster	Buffalo, N. Y	Mar. 10, 1863	37, 876
Flange-union	A. A. Dame	Titusville, Pa	June 17, 1873	139, 877
Flanges, Tool for upsetting	J. Connolly	Newburgh, N. Y	Apr. 2, 1867	63, 474
Flanged and beaded hoops, Machine for making	J. Blood	Watertown, N. Y	Sept. 13, 1870	107, 216
Flanged pipes, Mold for casting	G. T. Sheldon	Chelmsford, Mass	Apr. 3, 1866	53, 690
Flanging-apparatus	E. Regan	Indianapolis, Ind	Feb. 8, 1870	99, 707
Flanging-machine	G. A. Bowers	Chicago, Ill	Mar. 26, 1872	124, 879
Flanging-machine	S. Lowen	Temperanceville, Pa	Aug. 3, 1869	93, 320
Flannel and other cloth, Machine for drying	D. M. Cummings and A. Hayes	Enfield, N. H	Sept. 27, 1864	44, 402
Flannel and other cloth, Wetting	J. W. Hale	Haverhill, Mass	Apr. 2, 1841	2, 028
Flannel, Process of finishing	S. Archer	Globe Village, Mass	Aug. 18, 1863	39, 539
Flash-light signal	W. Mitchell and J. J. Mayo	London and Sutton, England.	July 22, 1873	141, 010
Flash-light-signal recorder	M. Gustin	Troy, Pa	May 6, 1873	138, 498
Flask: *See* Car-wheel flask. Column-flask. Dentists' flask. Founders' flask. Kettle-molding flask. Liquor-flask. Molders' flask. Molding-flask. Pipe flask. Pocket-flask. Powder-flask. Vulcanizing-flask.				
Flask	W. T. Fry	Philadelphia, Pa	Dec. 5, 1865	51, 303
Flask and bottle	W. T. Fry	Philadelphia, Pa	Oct. 20, 1863	40, 338
Flask clamp, Foundery	C. C. Stewart	Oneonta, N. Y	May 10, 1870	102, 987
Flask, Guide	T. S. Brown	Poughkeepsie, N. Y	Apr. [illegible], 1870	101, 426
Flask or bottle	W. T. Fry	New York, N. Y	Aug. 11, 1868	80, 939
Flask or mold fastening	O. H. Burdick	Auburn, N. Y	Apr. 26, 1864	42, 543
Flask or retort	S. L. Geer	Norwich, Conn	June 21, 1864	43, 193
Flask-pin	B. B. Whaley	Brooklyn, N. Y	Mar. 28, 1865	47, 059
Flasks or molds, Slide and guide for	S. A. Traugh	Cincinnati, Ohio	Apr. 19, 1864	42, 419
Flat-boats, Launching	J. and E. Davis	Matildaville, Pa	Nov. 22, 1859	26, 168
Flat-bottom boat with movable keel	S. Buel	Burlington, Vt	July 7, 1810	....
Flat-bottom boats, Construction of	H. Haupt and J. Y. Smith	Cambridge, Mass., and Alexandria, Va.	July 25, 1865	48, 935
Flat-iron	J. Alexander	Brooklyn, N. Y	Feb. 13, 1866	52, 512
Flat-iron	A. Bachelder	Pelham, N. H	Aug. 1, 1871	117, 592
Flat-iron	T. Butlers	Concord, N. H	Feb. 13, 1866	52, 529
Flat-iron	J. Grüss	Cleveland, Ohio	Mar. 3, 1868	75, 156
Flat-iron	D. Lithgow	Philadelphia, Pa	Oct. 26, 1858	21, 891
Flat-iron	E. B. Robinson	Portland, Me	Sept. 18, 1866	58, 138
Flat-iron	P. T. Wilbur	Boston, Mass	July 9, 1872	128, 936
Flat-iron, Gas	A. Hallowell and A. T. Atherton.	Lowell, Mass	Oct. 26, 1869	96, 228
Flat-iron guard	J. C. Briggs	Concord, N. H	Feb. 5, 1861	31, 295
Flat-iron heater	M. D. Birge	Grand Rapids, Mich	July 16, 1872	129, 453
Flat-iron heater	J. F. Hall	Albia, Iowa	Nov. 4, 1873	144, 335
Flat-iron heater	A. Hallowell	Lowell, Mass	Mar. 17, 1868	75, 631
Flat-iron heater	G. R. Moore	Lyons, Iowa	Aug. 9, 1864	43, 786
Flat-iron heater	S. W. Smith	Addison, Vt	Mar. 24, 1868	75, 995
Flat-iron heater	W. C. Wren	Brooklyn, N. Y	Sept. 10, 1872	131, 202
Flat-iron heater	J. C. Wright and J. W. Sursa	Moline, Ill	July 15, 1873	140, 987
Flat-iron heater	J. H. Yates	Batavia, N. Y	Aug. 20, 1867	68, 024
Flat-iron holder	F. W. Brocksieper	New Britain, Conn	Dec. 27, 1864	45, 581
Flat-iron holder	F. Bruns	Cleveland, Ohio	Nov. 28, 1871	121, 273
Flat-iron holder and polisher	I. F. Brown	New London, Conn	Oct. 15, 1872	132, 137
Flat-iron polisher and holder	W. B. Mason	Boston, Mass	Mar. 8, 1870	100, 649
Flat-iron, Steam	C. C. Walworth	Boston, Mass	Dec. 21, 1852	9, 494
Flat or sad iron	L. Lincoln	Norton, Mass	June 28, 1870	104, 748
Flat or sad iron	W. Wilson	Greenfield, Mass	Mar. 27, 1835	....
Flavoring compound	W. Zeigler and J. H. Seal	New York, N. Y	Nov. 15, 1870	109, 364
Flax and cotton, Treating	C. W. and O. G. Newton	Edinburgh, Mo	Feb. 27, 1866	52, 877
Flax and hemp, Bleaching unrotted	W. Cumberland	New York, N. Y	Aug. 27, 1822	....
Flax and hemp brake	J. L. Boorum and C. D. Palmiter.	Homer, N. Y	July 5, 1864	43, 452
Flax and hemp brake	J. A. Dana	Cazenovia, N. Y	July 23, 1811	....
Flax and hemp brake	A. W. Hall	New York, N. Y	Nov. 22, 1864	45, 154
Flax and hemp brake	E. Rumsey	Christian County, Ky	Apr. 26, 1811	....
Flax and hemp, Breaking	T. Pullen	Lexington, Ky	June 20, 1816	....
Flax and hemp, Breaking and cleaning	A. Zellnor	Pulaski, Tenn	July 26, 1833	....
Flax and hemp breaking and cleaning machine	J. Godden	Pittsburgh, Iowa	Jan. 20, 1843	2, 922
Flax and hemp breaking and cleaning machine	W. McMillen	Ripley, Ohio	Sept 30, 1842	2, 796
Flax and hemp breaking and cleaning machine	O. Roberts		Apr. 12, 1804	....
Flax and hemp breaking and cleaning machinery	G. Sanford and J. E. Mallory	New York, N. Y	June 24, 1862	35, 710
Flax and hemp breaking and cleaning machinery	G. Sanford and J. E. Mallory	New York, N. Y	Sept. 16, 1862	36, 485
Flax and hemp breaking and cleaning machinery	G. Sanford and J. E. Mallory	New York, N. Y	Oct. 14, 1862	36, 674
Flax and hemp breaking and corn-shelling machine	R. Gillman	New York, N. Y	Nov. 5, 1824	....
Flax and hemp, Breaking and dressing	E. C. Chase	Jay, Me	Jan. 6, 1832	....
Flax and hemp, Breaking and dressing	F. Cox	Somerville, N. J	Dec. 4, 1823	....
Flax and hemp, Breaking and dressing	A. Zellnor	Giles County, Tenn	Oct. 25, 1832	....
Flax and hemp breaking and dressing machine	E. Christian	Philadelphia, N. Y	Feb. 8, 1828	....
Flax and hemp breaking and dressing machine	G. H. Rickets and J. Kinney, jr	Mount Pleasant, N. J	July 12, 1814	....
Flax and hemp breaking and dressing machine	L. Tibbitts	New Glasgow, Va	Oct. 22, 1831	....
Flax and hemp breaking and dressing machinery	B. M. Smith	Massillon, Ohio	Oct. 7, 1845	4, 219
Flax and hemp breaking and swingling	W. C. Davison	Pittsford, N. Y	June 22, 1824	....
Flax and hemp breaking and thrashing	N. G. Hayden	Harrodsburgh, Ky	Apr. 12, 1833	....
Flax and hemp breaking machine	F. M. Barnes	Lexington, Mo	Jan. 10, 1843	2, 905

Index of patents issued from the United States Patent Office from 1790 *to* 1873, *inclusive*—Continued.

Invention.	Inventor.	Residence.	Date.	No.
Flax and hemp breaking machine	G. W. Billings	New York, N. Y	Mar. 15, 1864	41, 903
Flax and hemp breaking machine	J. P. Fry	Pulaski, Tenn	Dec. 21, 1842	2, 889
Flax and hemp breaking machine	A. Forsyth	Columbia, Tenn	Jan. 9, 1838	550
Flax and hemp breaking machine	J. Hinde	Schenectady, N. Y	Sept. 5, 1854	11, 647
Flax and hemp breaking machine	D Melville	Newport, R. I	Nov. 18, 1818	....
Flax and hemp breaking machine	N. P. Robinson	Fleming County, Ky	Aug. 31, 1810	....
Flax and hemp breaking machine	G. Sanford, J. E. Mallory, and C. P. Hayes.	New York, N. Y	Oct. 14, 1862	36, 675
Flax and hemp breaking machine	W. Van Duser	Warwick, N. Y	June 30, 1813	....
Flax and hemp breaking machinery	G. Sanford and J. E. Mallory	New York, N. Y	Mar. 25, 1862	34, 779
Flax and hemp breaking, scutching, or cleaning machine.	H. Johnson	Maysville, Ky	Mar. 4, 1843	2, 980
Flax and hemp by animal power, Machine for gathering or pulling.	W. Brittain and J. Silvers	New Hope, Pa	Nov. 25, 1838	1, 020
Flax and hemp by steam, Rotting	A. Chinn	Harrison County, Ky	Dec. 28, 1825	....
Flax and hemp, Cleaning and breaking	R. Miller	Glasgow, Ky	May 23, 1834	....
Flax and hemp cleaning and dressing machine	I. C., G. W., and C. E. Geisindorff.	Cincinnati, Ohio	Nov. 28, 1842	2, 864
Flax and hemp cleaning and hackling machine	C. Learned	Saint Louis, Mo	Oct. 22, 1842	2, 831
Flax and hemp, Cleaning and separating the fibers of.	G. W. Billings	New York, N. Y	Mar. 8, 1864	41, 826
Flax and hemp cleaning machine	A. Gärtner	New York, N. Y	Apr. 2, 1867	63, 499
Flax and hemp, Cleaning, softening, separating, and shortening the fibers of.	H. Hawley	Louisville, Ky	June 29, 1833	....
Flax and hemp dresser	E. H. Nichols and T. Fairbanks	Saint Johnsbury, Vt	Oct. 1, 1830	....
Flax and hemp, Dressing	L. S. Chichester	Brooklyn, N. Y	Apr. 11, 1854	10, 772
Flax and hemp dressing machine	T. S. and A. Barnum	New York, N. Y	Nov. 27, 1824	....
Flax and hemp dressing machine	R. S. Chappel	Pawlet, N. Y	Jan. 18, 1809	....
Flax and hemp dressing machine	W. Cumberland	New York, N. Y	July 9, 1822	....
Flax and hemp dressing machine	N. Goodsall	Paris, N. Y	Sept. 19, 1822	....
Flax and hemp dressing machine	H. Hawley	Louisville, Ky	June 4, 1833	....
Flax and hemp dressing machine	J Hines and W. Bain	Columbus, Ohio	Feb. 3, 1823	....
Flax and hemp dressing machine	C. G. Howard	Topeka, Kans	Sept. 6, 1864	44, 138
Flax and hemp dressing machine	E Kellogg	Palmyra, N. Y	May 11, 1824	....
Flax and hemp dressing machine	B. Lewis	Ligonia, Me	Jan. 28, 1824	....
Flax and hemp dressing machine	J. Macdonald	New York, N. Y	Aug. 31, 1822	....
Flax and hemp dressing machine	A. Smith and J. Olney	Westmoreland, N. Y	July 10, 1830	....
Flax and hemp dressing machine	J. C. Wenzle	Louisville, Ky	Jan. 17, 1828	....
Flax and hemp dressing machinery	F. Kellsey	Middletown, Conn	Nov. 26, 1824	....
Flax and hemp dressing machinery	G. Sanford	New York, N. Y	June 24, 1862	35, 709
Flax and hemp drying	G. W. Billings	New York, N. Y	Oct. 13, 1863	40, 236
Flax and hemp drying frame	G. W. Billings	New York, N. Y	Oct. 6, 1863	40, 155
Flax and hemp, Extracting vegetable matter from	S. Olcott	Harsimus, N. J	Dec. 3, 1828	....
Flax and hemp for spinning, Preparing	P. Antrim	Philadelphia, Pa	Nov. 25, 1814	....
Flax and hemp for spinning, Roving	G. Brown	Schatecoke, N. Y	June 16, 1809	....
Flax and hemp hackling machine	J. P. Arnold	Louisville, Ky	Jan. 4, 1853	9, 512
Flax and hemp hackling machine	F. Demasters	Shelbyville, Ky	Nov. 20, 1838	1, 012
Flax and hemp machine	D. Ball	Sandy Hill, N. Y	July 7, 1830	....
Flax and hemp machine	G. W. Billings	New York, N. Y	Mar. 15, 1864	41, 902
Flax and hemp machine	J. Dewey, jr	Troy, N. Y	Nov. 25, 1830	....
Flax and hemp machine	E. F. Hill	Fabias, N. Y	Feb. 11, 1809	....
Flax and hemp machine	R. McCormick	Rockbridge County, Va	Oct. 1, 1830	....
Flax and hemp machine	J. Rich, jr	Troy, N. Y	May 20, 1830	....
Flax and hemp machine	W. K. Scott	Sandy Hill, N. Y	Feb. 11, 1830	....
Flax and hemp, Machine for reducing the fibers of	S. Olcott	Philadelphia, Pa	Apr. 11, 1840	1, 550
Flax and hemp, Manufacture of	R. Patterson	New Hartford, N. Y	Dec. 18, 1849	6, 958
Flax and hemp mill	B. Tyler	....	Feb. 26, 1799	....
Flax and hemp preparatory to the various processes of cleaning and separating the fibers, &c., Mode of preparing.	S. Olcott	New Hope, Pa	Mar. 31, 1840	1, 532
Flax and hemp, Retting	G. W. Billings	New York, N. Y	Mar. 8, 1864	41, 825
Flax and hemp scutching machine	G. Sanford and J. E. Mallory	New York, N. Y	Mar. 25, 1862	34, 781
Flax and hemp to make them resemble cotton, Treating.	J. P. Comly	Dayton, Ohio	Mar. 4, 1862	34, 619
Flax and hemp without rotting by water, Preparing.	W. Hickingbottom	New York, N. Y	July 12, 1816	....
Flax-bleaching machine	I. A. Roth and J. Lea	Philadelphia County, Pa	Apr. 18, 1854	10, 780
Flax-boll brake	B. S. Burgan	Congress, Ohio	Jan. 12, 1869	85, 725
Flax-brake	A. G. Bill	Cuyahoga Falls, Ohio	Aug. 31, 1869	94, 177
Flax-brake	J. Boyce	Wooster, Ohio	Sept. 14, 1869	94, 706
Flax-brake	J. C. Kurtz	Wooster, Ohio	June 27, 1871	116, 328
Flax-brake	M. B. Southwick	Mont Saint Hillaire, Canada.	July 25, 1871	117, 342
Flax-brake and swingler	S. Cowan	Bloomfield, Iowa	Sept. 29, 1863	40, 091
Flax breaking and cleaning machine	S. A. Clemens	Chicago, Ill	Nov. 6, 1866	59, 503
Flax breaking and cleaning machine	G. Sanford and J. E. Mallory	England and New York, N. Y.	Aug. 25, 1863	39, 680
Flax breaking and cleaning machine	G. B. Turner	Cuyahoga Falls, Ohio	Sept. 6, 1864	44, 128
Flax breaking and dressing machine	S. M. Allen	Woburn, Mass	June 30, 1863	39, 025
Flax breaking and dressing machine	J. Briggs	Schoharie, N. Y	Oct. 1, 1825	....
Flax breaking and dressing machine	S. A. Clemens	Springfield, Mass	Mar. 8, 1853	9, 609
Flax breaking and dressing machine	L. Rundell	Pike, N. Y	Sept. 27, 1864	44, 458
Flax-breaking machine	J. Cannon	....	Jan. 17, 1801	....
Flax breaking machine	S. Davison	Romulus Township, N. Y	May 17, 1822	....
Flax-breaking machine	J. Haues and L. Kohler	Millersburgh, Ind	Nov. 17, 1863	40, 621
Flax-breaking machine	A. S. Miller	Bluffton, Ind	June 9, 1868	78, 811
Flax-breaking machine	P. Reading	Trenton, N. J	Jan. 3, 1825	....
Flax-breaking machine	H. Schoonhoven	Pultney, N. Y	Dec. 11, 1827	....
Flax cleaning and dressing machine	J. E. Crowell	Chelsea, Mass	Aug. 5, 1862	36, 075
Flax-cleaning machine	C. Beach	Penn Yan, N. Y	Sept. 27, 1864	44, 387
Flax-cleaning machine	W. W. Whiddit	Richmond, Ind	May 28, 1867	65, 319
Flax, &c., cleansing, dressing, and cutting machine	E. Phillips	Blackstone, Mass	May 2, 1865	47, 604
Flax, Dressing	E. Mack	....	Mar. 21, 1806	....
Flax-dressing machine	D. S. Abbott	Ischua, N. Y	June 12, 1866	55, 446
Flax-dressing machine	J. Barlow	Granville, Mass	Aug. 23, 1809	....
Flax-dressing machine	F. Burdick	Kortright, N. Y	Feb. 28, 1825	....
Flax-dressing machine	J. Haws	Hudson, N. Y	June 19, 1812	....
Flax-dressing machine	J. Hines and W. Bain	New York, N. Y	Aug. 12, 1826	....

Index of patents issued from the United States Patent Office from 1790 to 1873, inclusive—Continued.

Invention.	Inventor.	Residence.	Date.	No.
Flax-dressing machine	W. C. McBride	Somerville, N. J	Mar. 17, 1868	75, 559
Flax-dressing machine	W. Wheeler	Warren, N. Y	Mar. 26, 1825	......
Flax-dressing machinery	E. L. Norfolk	Salem, Mass	May 9, 1854	10, 876
Flax-dressing machinery	D. Warner, jr	South Hadley, Mass	Sept. 5, 1854	11, 654
Flax, &c., gathering and loading machine	G. W. Hatch	Parkman, Ohio	June 6, 1865	48, 064
Flax, &c., gathering and loading machine	G. W. Hatch	Parkman, Ohio	Dec. 4, 1866	60, 185
Flax, hemp, and Manila grass, Combing and separating fibers of.	S. Couillard, jr	Boston, Mass	Oct. 19, 1836	......
Flax, hemp, and other fibrous materials, Mode of treating.	J. W. Burton and G. Pye	Eye and Ipswich, England	Dec. 8, 1857	18, 801
Flax, hemp, and other fibrous plants, Disintegrating	R. M. Russell	New York, N. Y	Aug. 6, 1867	67, 455
Flax, hemp, &c., Apparatus for separating fibers of	J. B. Fuller	Claremont, N. H	Mar. 1, 1864	41, 812
Flax, hemp, &c., breaking and cleaning machine	G. Sanford and J. E. Mallory	New York, N. Y	June 16, 1863	38, 916
Flax, hemp, &c., breaking and cleaning machine	G. Sanford and J. E. Mallory	New York, N. Y	Sept. 15, 1863	39, 962
Flax, hemp, &c., breaking machine	J. Millard	Kent, Conn	May 6, 1813	......
Flax, hemp, &c., breaking machine	J. L. F. Roumage	New York, N. Y	Aug. 6, 1829	......
Flax, hemp, &c., cutting and gathering machine	R. M. Cooch	Lambertsville, N. J	July 16, 1841	2, 179
Flax, hemp, &c., Disintegrating and separating fibers of.	J. B. Fuller	Claremont, N. H	Mar. 1, 1864	41, 813
Flax, hemp, &c., Disintegrating or cottonizing	H. Burgess	Royer's Ford, Pa	Jan. 19, 1864	41, 276
Flax, hemp, &c., Fiber from	H. Burgess	Royer's Ford, Pa	Jan. 19, 1864	41, 275
Flax, hemp, &c., for spinning, Mode of preparing	J. Owings	Baltimore, Md	Mar. 5, 1812	......
Flax, hemp, &c., Mode of separating the fibers of	R. T. Shaw	New York N. Y	Jan. 19, 1864	41, 352
Flax, hemp, &c., Plant used as a substitute for	C. Whitlow	New York, N. Y	Jan. 11, 1812	......
Flax, hemp, &c., Retting and disintegrating	R. T. Shaw	New York, N. Y	Jan. 19, 1864	41, 350
Flax, hemp, silk, tow, and wool, Combing and hackling.	J. and J. Westerman	New York, N. Y	Aug. 14, 1833	......
Flax-machine	J. Halladay, T. Eldridge, C. Gibbs, and J. D. Smith.	Hartford, N. Y	Mar. 4, 1830	......
Flax-machine	O. Wilde	Canaan, N. Y	Apr. 16, 1810	......
Flax, &c., Machine for drawing	J. Good	Brooklyn, E. D., N. Y	Oct. 5, 1869	95, 462
Flax, Machine for scutching and thrashing	M. Jerome	Dixon, Ill	June 8, 1869	90, 946
Flax, Machine for scutching tangled	W. C. McBride	Raritan, N. J	Nov. 29, 1864	45, 259
Flax, Machine for straightening and thrashing tangled.	S. A. Clemens	Chicago, Ill	Nov. 6, 1866	59, 502
Flax, Machine for thrashing	S. A. Clemens	Chicago, Ill	Nov. 6, 1866	59, 360
Flax, Machinery for preparing	J. B. Fuller	Claremont, N. H	Dec. 19, 1865	51, 652
Flax, Mode of treating	A. F. and J. H. Andrews	Avon, Conn	Apr. 23, 1867	64, 058
Flax or hemp breaking and dressing machinery	G. Sanford	New York, N. Y	Mar. 18, 1862	34, 697
Flax or hemp, Cylindrical frame for roving or spinning.	W. Giffen	New York, N. Y	Jan. 11, 1812	......
Flax or hemp dressing wheel	C. and W. A. Benton	Amenia, N. Y	May 22, 1816	......
Flax or hemp, Machine for cleaning unrotted or rotted.	W. Cumberland	New York, N. Y	Nov. 29, 1822	......
Flax, Picking and cleaning	A. H. Caryl	Sandusky, Ohio	Feb. 14, 1854	10, 513
Flax, Process for bleaching	I. A. Roth and J. Lea	Philadelphia County, Pa	Apr. 18, 1854	10, 808
Flax-puller	J. H. Bennett	East Bennington, Vt	Jan. 23, 1845	3, 890
Flax-puller	L. S. Chichester	Brooklyn, N. Y	Nov. 16, 1852	9, 400
Flax-puller	B. M. Reeves	Franklin, Pa	Aug. 8, 1871	117, 926
Flax-puller	J. Smith	Troy, N. Y	May 10, 1870	102, 875
Flax-puller	S. W. Tyler	Troy, N. Y	July 18, 1871	117, 125
Flax-pulling machine	A. Burchard	Livingstonville, N. Y	Mar. 14, 1865	46, 857
Flax-pulling machine	J. Harrington	Menomonee, Wis	Jan. 22, 1867	61, 422
Flax-pulling machine	B. M. Reeves	Franklin, Pa	Mar. 28, 1871	113, 094
Flax-pulling machine	J. Silvers	Lambertsville, N. J	July 11, 1865	48, 731
Flax-pulling machine	S. W. Tyler	Troy, N. Y	Feb. 12, 1867	62, 090
Flax-pulling machine	S. W. Tyler	Troy, N. Y	Aug. 27, 1867	68, 132
Flax-scutching machine	H. Black	Lewisburgh, Ohio	June 5, 1866	55, 232
Flax-scutching machine	A. S. Miller	Bluffton, Ind	June 9, 1868	78, 810
Flax-scutching machinery	W. C. McBride	Raritan, N. J	Jan. 25, 1859	22, 738
Flax-seed-cleaning machine	P. Goltry		Apr. 24, 1804	......
Flax-seed, paint, &c., Grinding	A. Cross and E. Brown	Cazenovia, N. Y	Feb. 4, 1830	......
Flax-seeds, Machine for extracting	A. Hook	Searsmont, Me	Sept. 9, 1823	......
Flax-seeds, &c., Mill for grinding	G. L. Stearns	Boston, Mass	Aug. 17, 1838	886
Flax-spinning machine	L. Skeels	Worthington, Ohio	Feb. 6, 1866	52, 506
Flax-straw, Machine for preparing tow from tangled	G. F. Clemons	Springfield, Mass	July 7, 1863	39, 119
Flax-straw, Machine for straightening, breaking, and cleaning.	S. A. Clemens	Chicago, Ill	Nov. 6, 1866	59, 501
Flax-swingling machine	S. Achey	Heidelberg Township, Lebanon County, Pa.	Feb. 18, 1828	......
Flax, &c., to produce short fibers for spinning, Treating.	S. M. Allen	Woburn, Mass	Jan. 12, 1864	41, 185
Flea-powder	C. E. Jaycox	San Francisco, Cal	Mar. 30, 1869	88, 300
Fleece-bundling apparatus	J. Walton	Roseburgh, Oreg	July 20, 1869	92, 916
Fleece-folder	C. W. Rudgers	Brecksville, Ohio	Sept. 19, 1865	50, 040
Fleece-tying apparatus	S. Cooley	Caro, Mich	Nov. 16, 1869	96, 779
Flesh-fork	P. Fisher	Brooklyn, N. Y	June 27, 1871	116, 291
Flesh-fork	J. C. Klein	Birmingham, Pa	Feb. 19, 1867	62, 276
Flesh-fork and skimmer, Combination of	C. B. Bristol	Naugatuck, Conn	July 12, 1859	24, 716
Flesh, fruit, &c., Machinery for cutting	J. Buffington	Chester, Pa	July 5, 1833	......
Flesh hooks and forks, Manufacturing	M. V. Trask	Meriden, Conn	Aug. 11, 1863	39, 534
Flesh-offal, &c., Treating	W. Adamson and C. F. A. Simonin.	Philadelphia, Pa	Sept. 19, 1871	119, 000
Flexible boat, Division between the tubes of	E. T. Harr	New York, N. Y	Apr. 17, 1849	6, 363
Flexible coupling	N. W. Wheeler	Brooklyn, N. Y	June 25, 1867	66, 193
Flexible gate	J. M. Richardson	Elmira, N. Y	Nov. 12, 1872	133, 057
Flexible pipe for mining	A. Clark	Todd's Valley, Cal	Feb. 7, 1865	46, 295
Flexible platform for giving motion to wheels and machinery.	J. Johnson, jr	Charleston, S. C	Nov. 22, 1809	......
Flexible tube, hose, &c., Manufacture of	I. B. Harris	Edinburgh, Scotland	May 24, 1864	42, 900
Flexible tube or hose	D. K. Hoxsie and T. L. Reed	Providence, R. I	Sept. 27, 1864	44, 425
Flexible tubes, Device for making curvatures in	A. Beckers	New York, N. Y	Nov. 29, 1864	45, 218
Flexible tubing	E. P. Gleason	New York, N. Y	Feb. 28, 1865	46, 558
Flexible tubing for illuminating-gas	W. B. S. Taylor	New York, N. Y	Feb. 21, 1865	46, 507
Flexible tubing or hose	T. L. Reed	Providence, R. I	Apr. 30, 1867	64, 362
Flexible tubings, Elastic connection for	W. B. S. Taylor	New York, N. Y	Oct. 28, 1873	144, 163
Flexible tubings, Manufacture of	D. K. Hoxsie and T. L. Reed	Providence, R. I	Nov. 21, 1865	51, 052
Flexible tubings, Manufacture of	D. K. Hoxsie and T. L. Reed	Providence, R. I	May 22, 1866	54, 909

Index of patents issued from the United States Patent Office from 1790 *to* 1873, *inclusive*—Continued.

Invention.	Inventor.	Residence.	Date.	No.
Flexible tubings, Manufacture of	W. B. S. Taylor	New York, N. Y	June 19, 1866	55, 740
Flexomanus	H. A. Nathans	Philadelphia, Pa	Dec. 15, 1857	18, 857
Float-gage, feed-regulator, &c., for steam-boilers, &c.	T. J. Sloan	New York, N. Y	Apr. 27, 1852	8, 912
Float or raft	A. Carson	Memphis, Tenn	June 18, 1867	65, 792
Floating-bag	G. Mitchell	London, England	Feb. 16, 1864	41, 632
Floating battery	A. B. Cooley	Philadelphia, Pa	Apr. 1, 1862	34, 867
Floating battery	P. Martin	Philadelphia, Pa	Aug. 11, 1841	2, 217
Floating battery	J. P. Taylor	Little Compton, R. I	Oct. 17, 1848	5, 854
Floating battery	E. A. Willis	Cold Spring, N. Y	Nov. 22, 1859	26, 219
Floating battery and light-house	J. Moody	York, England	Mar. 12, 1867	62, 870
Floating battery for ships, &c	J. Hyde	New York, N. Y	Dec. 23, 1862	37, 232
Floating battery or steamship	G. Stiles	Baltimore, Md	Apr. 4, 1815	....
Floating house or bathing-machine	H. Chevens	New York, N. Y	July 13, 1813	....
Floating-mill	J. A. Dorman	New York, N. Y	May 17, 1864	42, 756
Floating-mill and silent fisher	E. Ford	Charleston, S. C	Dec. 21, 1809	....
Floating-wheel	J. Owings	Baltimore, Md	Sept. 9, 1816	....
Floating-wheel for vessels	J. Spilman	Tonawanda, N. Y	July 9, 1867	66, 532
Flock and paper-stock cutting machine	J. N. Pitts	Blackstone, Mass	Jan. 29, 1856	14, 165
Flock-cleaning machine	R. H. Dawson	Rockville, Conn	July 8, 1873	140, 684
Flock-cloth dyed or printed	T. Crossly	Bridgeport, Conn	Jan. 31, 1865	46, 199
Flock-cutter	M. and C. H. Morse	Franklin, Mass	May 9, 1871	114, 588
Flock-cutting machine	J. Chase	Lowell, Mass	Oct. 14, 1862	36, 642
Flock-cutting machine, Reversible	J. Pitts	Millerville, Mass	Nov. 28, 1871	121, 415
Flock-duster	M. Waterhouse	Passaic, N. J	Jan. 7, 1868	73, 144
Flock for felting, Machine for preparing	L. W. Boynton	South Coventry, Conn	July 18, 1854	11, 318
Flock-grinder	E. T. Marble	Worcester, Mass	Mar. 26, 1872	124, 903
Flock-grinder	J. Waterhouse	Passaic, N. J	Mar. 12, 1872	124, 463
Flock-grinding machine	J. C. Fonda	Albany, N. Y	July 29, 1851	8, 261
Flock-machine	W. McAllister	Lawrence, Mass	Dec. 6, 1870	109, 920
Flock, Machine for cleaning and opening	W. C. Geer	Rockville, Conn	Oct. 25, 1859	25, 895
Flock, Machine for opening and cleaning	E. C. Brett	Great Barrington, Mass	Oct. 7, 1851	8, 404
Flock, Machine for preparing	L. W. Boynton	Worcester, Mass	Aug. 7, 1855	13, 380
Flock, Machine for preparing	J. R. Peters, jr	New York, N. Y	June 22, 1852	9, 059
Flock, Machine for sifting	J. F. Greene	Brooklyn, N. Y	Feb. 26, 1861	31, 574
Flock, Machinery for cutting	A. Parber	Stephentown, N. Y	Nov. 18, 1846	4, 857
Flock, Machinery for cutting	J. Tilton and E. Ritson	Northfield and Sanbornton, N. H.	May 15, 1860	28, 316
Flock to cloth, Apparatus for applying	D. and R. Pratt	Elmira, N. Y	Oct. 7, 1851	8, 405
Flock with flannel, Mode of incorporating	J. M. Pratt	Dudley, Mass	Sept. 5, 1848	5, 757
Flocked cloth, Dyeing, printing, and manufacture of water-proof.	T. Crossly	Bridgeport, Conn	Jan. 31, 1865	46, 200
Flocked fabric for hats, bonnets, &c	C. L. Mitchell	Westborough, Mass	Oct. 17, 1871	120, 084
Flocking-machines, Feeding-apparatus for	H. Turner	New York, N. Y	Jan. 21, 1868	73, 676
Flocculent substances, Press for bundling	D. Kellogg	Pittsfield, Mich	Oct. 12, 1852	9, 324
Flood-gate	A. H. Adams	Piqua, Ohio	Mar. 28, 1871	113, 001
Flood-gate	T. H. Breed	Dundee, Mich	Sept. 6, 1870	107, 159
Flood-gate	J. Campbell and A. Watson	London, Ohio	Sept. 27, 1870	107, 659
Flood-gate	J. W. Edgarton	Thornton, Ind	June 20, 1871	116, 170
Flood-gate	G. W. Flanders	Lynn, Mass	Apr. 8, 1856	14, 602
Flood-gate	E. H. Hancock	Augusta, Ga	Sept. 6, 1859	25, 330
Flood-gate	N. Hinckley	Marston's Mills, Mass	Nov. 29, 1870	109, 734
Flood-gate	B. S. Hort	Georgetown, S. C	Oct. 9, 1813	....
Flood-gate	R. Keese, D. T. Ward, and J. G. Wilkinson.	Cardington, Ohio	Sept. 3, 1861	33, 201
Flood-gate	H. A. Kephart	Fletcher, Ohio	Apr. 28, 1868	77, 384
Flood-gate	J. J. Kimball	Napierville, Ill	Feb. 16, 1869	86, 985
Flood-gate	J. J. Kimball	Napierville, Ill	Jan. 18, 1870	98, 979
Flood-gate	A. L. King	Farmersville, Ohio	Mar. 16, 1869	87, 944
Flood-gate	A. L. King	Farmersville, Ohio	Dec. 21, 1869	98, 071
Flood-gate	J. Leatherman	Napoleon, Ohio	Aug. 18, 1868	81, 180
Flood-gate	J. Maeby	Portage, Wis	Aug. 7, 1866	56, 962
Flood-gate	A. Main	Delaware, Ohio	Mar. 30, 1869	88, 598
Flood-gate	C. B. McKinney	Houston, Ohio	Sept. 3, 1867	68, 767
Flood-gate	A. E. Noble	La Motte, Iowa	Apr. 5, 1870	101, 651
Flood-gate	A. Ralston	Carlisle, Pa	July 25, 1865	48, 977
Flood-gate	S. Rowland and T. C. Tipton	Williamsport, Ohio	Nov. 15, 1870	109, 252
Flood-gate	S. B. Shoup	Dayton, Ohio	Apr. 6, 1869	88, 673
Flood-gate	J. B. Stoner	Lacon, Ill	Oct. 15, 1867	69, 860
Flood-gate	H. O. Way	Thorntown, Ind	June 20, 1871	116, 247
Flood-gate, Adjustable	N. Tupper	Grand Blanc, Mich	Nov. 24, 1868	84, 321
Flood-gate for mill-dams	M. Cotton	Sardinia, N. Y	Mar. 21, 1865	46, 883
Flood-gate, Revolving	J. Du Bois	Williamsport, Pa	Jan. 17, 1865	45, 913
Flood-gate, Self-acting	G. B. Markham	Mead's Mills, Mich	Sept. 4, 1860	29, 896
Flood-gate, Self-operating	W. H. Allard and R. W. Thomas	Portage, Wis	Sept. 11, 1866	57, 839
Flood or waste gate	W. L. Clark	Cambria, Wis	July 16, 1867	66, 795
Flood-roller	W. J. Connell	West Unity, Ohio	June 22, 1869	91, 607
Floor	L. S. Wood	Marion, Iowa	July 15, 1873	140, 864
Floor and ceiling, Fire-proof	W. T. Butler	Chicago, Ill	July 29, 1873	141, 260
Floor and ceiling, Fire-proof	I. Hodgson and W. H. Brown	Indianapolis, Ind	July 30, 1872	129, 955
Floor and ceiling, Fire-proof	F. A. Petersen	New York, N. Y	Apr. 3, 1855	12, 642
Floor and chimney connection, Fire-proof	W. Neracher	Cleveland, Ohio	Apr. 22, 1873	138, 096
Floor and roof, Fire-proof	T. Hyatt	New York, N. Y	Dec. 2, 1873	145, 181
Floor and wall, Fire-proof	J. John	Chicago, Ill	Dec. 2, 1873	145, 211
Floor, Basement	E. S. Vaughan	Flatbush, N. Y	Apr. 8, 1873	137, 740
Floor-clamp	J. H. Baker	Saratoga Springs, N. Y	Apr. 28, 1868	77, 243
Floor-clamp	G. E. Banner	Newark, N. J	Apr. 14, 1868	76, 583
Floor-clamp	H. D. Barnes	Fair Haven, Conn	Nov. 26, 1867	71, 440
Floor-clamp	J. L. Bryant	Logansport, Ind	Sept. 12, 1865	49, 852
Floor-clamp	J. L. Clough	Suffield, Conn	Apr. 23, 1861	32, 120
Floor-clamp	W. Conner and C. W. Mitchell	Wilmington, Del	Dec. 10, 1867	71, 987
Floor-clamp	J. H. Ferriera	Newark, N. J	Oct. 6, 1868	82, 700
Floor-clamp	J. J. Foster	Belmont, Tex	Mar. 4, 1873	136, 428
Floor-clamp	J. A. Haase	Philadelphia, Pa	May 4, 1869	89, 577
Floor-clamp	E. A. Hall	Sugar Branch, Ind	May 5, 1868	77, 610
Floor-clamp	J. F. Hammond	Providence, R. I	July 30, 1867	67, 297
Floor-clamp	A. Lloyd	Mattoon, Ill	June 5, 1866	55, 424
Floor-clamp	F. S. Mack	Galesburgh, Ill	July 21, 1868	80, 079
Floor-clamp	D. Nettleton	Humbird, Wis	Sept. 20, 1870	107, 527

Index of patents issued from the United States Patent Office from 1790 *to* 1873, *inclusive*—Continued.

Invention.	Inventor.	Residence.	Date.	No.
Floor-clamp	D. Nevin	Georgetown, Colo	Feb. 7, 1871	111, 560
Floor-clamp	H. W. Oliver	Whitneysville, Conn	Apr. 15, 1856	14, 676
Floor-clamp	D. K. Peoples	Philadelphia, Pa	Nov. 24, 1863	40, 727
Floor-clamp	G. B. Perkins	Utica, N. Y	Aug. 25, 1868	81, 531
Floor-clamp	W. Partlock and J. H. Smith	Pleasant Grove and Toolesborough, Iowa.	Feb. 4, 1868	74, 128
Floor-clamp	H. J. O. Reed	Salem, Mass	Oct. 28, 1873	144, 140
Floor-clamp	J. G. Rogers	Miles, Mich	May 27, 1873	139, 265
Floor-clamp	J. B. Spencer	Norwich, Conn	Dec. 21, 1869	98, 116
Floor-clamp	O. Taff	Whitestone, N. Y	June 28, 1870	104, 790
Floor-clamp	T. S. Urie	Hubbardston, Mich	Sept. 19, 1871	119, 203
Floor-clamp	H. C. Wight	Worcester, Mass	July 13, 1858	20, 913
Floor-clamp	J. D. Winslow	Wilmington, Del	Oct. 30, 1866	59, 305
Floor-clamp	S. C. and S. Winslow	Worcester, Mass	May 15, 1866	54, 805
Floor-clamp	G. Wood	Philadelphia, Pa	Mar. 8, 1870	100, 700
Floor-cleaner	A. Armstrong	Newburgh, Ohio	Mar. 10, 1868	75, 236
Floor-cloth: *See* oil-cloth.				
Floor-cloth	S. M. Allen	Woburn, Mass	Nov. 17, 1863	40, 592
Floor-cloth	T. Griffin	Roxbury, Mass	June 18, 1867	65, 904
Floor-cloth	S. Hawksworth	Doncaster, England	Apr. 26, 1864	42, 479
Floor-cloth and carpeting	J. B. Meldrum	Paterson, N. J	July 30, 1867	67, 330
Floor cloth and carpeting	J. B. Meldrum	Paterson, N. J	Nov. 19, 1867	71, 034
Floor-cloth, Composition for	J. W. Harmon	Brooklyn, N. Y	Mar. 31, 1857	16, 918
Floor-cloth, &c., Composition for sizing and stiffening.	N. B. Powers	Lansingburgh, N. Y	Oct. 28, 1862	36, 794
Floor-cloth, &c., Composition for sizing for use in the manufacture of.	N. B. Powers	Lansingburgh, N. Y	Oct. 28, 1862	36, 793
Floor-cloth from straw-board, Manufacture of	F. N. Davis	Beloit, Wis	Nov. 19, 1872	133, 210
Floor-cloth, leather-cloth, &c., Mode of producing	R. O. Lowrey	Salem, N. Y	May 19, 1868	77, 993
Floor-cloth, &c., Leather-paper for	E. Richmond	Brookline, Mass	June 28, 1864	43, 340
Floor-cloth, Machine for printing	S. Savage	Lowell, Mass	Mar. 2, 1852	8, 778
Floor-cloth, Manufacture of	J. B. Hodgskin	New York, N. Y	Jan. 2, 1866	51, 830
Floor-cloth, Manufacture of	G. F. Hopper	New York, N. Y	Jan. 2, 1866	51, 832
Floor-cloth, Manufacture of	C. L. Laurence	New York, N. Y	Aug. 22, 1865	49, 537
Floor-cloth, Manufacture of	J. H. Williams	Essex, Conn	July 7, 1868	79, 712
Floor-cloth, Manufacture of rubber	H. W. Joslin	Jersey City, N. J	Feb. 7, 1871	111, 549
Floor-cloth, Method of printing	J. Albro	Elizabeth, N. J	Nov. 15, 1859	26, 075
Floor-cloth, &c., Printing	J. Marchbank	Lansingburgh, N. Y	Jan. 6, 1863	37, 298
Floor-cloth, Roller for	R. Hoskin	Brooklyn, N. Y	May 28, 1867	65, 223
Floor-covering	W. Howell, J. C. Finn, and C. A. Duy.	Philadelphia, Pa	Sept. 15, 1868	82, 119
Floor-covering	C. Keen	Philadelphia, Pa	Sept. 18, 1866	58, 171
Floor-covering	A. H. Platt	Yellow Springs, Ohio	Jan. 17, 1865	45, 937
Floor-covering	A. H. Platt	Ann Arbor, Mich	Sept. 4, 1866	57, 763
Floor-covering, roof, &c	D. L. Wolff	Chicago, Ill	June 20, 1871	116, 128
Floor-covering, wainscoting, &c., Fabric for	M. Flurscheim	New York, N. Y	Nov. 10, 1868	83, 948
Floor-coverings, Manufacture of	J. H. Spencer	Philadelphia, Pa	Oct. 2, 1866	58, 497
Floor, Fire-proof	N. Cheney	New York, N. Y	Apr. 1, 1873	137, 345
Floor, Fire-proof	J. B. Cornell	New York, N. Y	Feb. 15, 1859	22, 939
Floor, Fire-proof	J. Dunseith	New York, N. Y	Aug. 22, 1871	118, 220
Floor, Fire-proof	I. Hodgson and W. H. Brown	Indianapolis, Ind	July 23, 1872	129, 827
Floor, Fire-proof	G. H. Johnson and W. Freeborn	Chicago, Ill	Oct. 15, 1872	132, 292
Floor, Fire-proof tile	W. L. Drake	Evanston, Ill	July 1, 1873	140, 352
Floor for buildings	H. M. and W. C. Smith	New York, N. Y	Aug. 10, 1869	93, 646
Floor-gage	J. Fales	Cambridge, Mass	Sept. 14, 1869	94, 814
Floor, Hollow tile	G. H. Johnson and B. Kreischer	New York, N. Y	Mar. 21, 1871	112, 926
Floor-mat, India-rubber	T. J. Mayall	Roxbury, Mass	Dec. 24, 1861	34, 001
Floor, Mosaic	G. G. Garibaldi	Buffalo, N. Y	Apr. 14, 1868	76, 742
Floor, Mosaic	J. G. Kappes	New York, N. Y	Mar. 16, 1869	87, 853
Floor, Mosaic	J. Stegmiller	New York, N. Y	Nov. 15, 1870	109, 265
Floor-set	C. M. Shaw	Portland, Me	Nov. 2, 1869	96, 355
Floor, Skating	W. L. Nelson	Saint Louis, Mo	Jan. 19, 1869	86, 031
Floor, Tessellated	L. B. Hamilton	Boston, Mass	Nov. 1, 1870	108, 781
Floor, Tessellated	J. F. Worth	Brooklyn, N. Y	Oct. 19, 1869	96, 069
Floor-timber support	T. Voelckers	Boston, Mass	Mar. 18, 1873	136, 947
Floor-warmer	C. Britain	Saint Joseph, Mich	Mar. 10, 1863	37, 851
Floor, Water-proof	T. New	New York, N. Y	Apr. 18, 1871	113, 787
Floors, ceilings, roofs, &c., Mode and material for constructing.	G. C. Bouziat	Vincennes, France	May 31, 1870	103, 553
Floors, Clamp for laying	S. E. Parrish	New York, N. Y	Sept. 27, 1853	10, 061
Floors, Construction of marquetry	B. H. Shedaker	Philadelphia, Pa	Jan. 19, 1858	19, 174
Floors, Laying malting	C. Hollmann	Union Hill, N. J	May 24, 1870	103, 333
Floors, Mosaic covering for	S. H. Pearce	Boston, Mass	June 15, 1869	91, 364
Floors of houses, Machine for scrubbing or scouring.	S. B. Willard	Bloomfield, N. Y	Nov. 17, 1808	....
Floors, pavements, &c., Coating concrete and cement.	C. C. Hallock	Brooklyn, N. Y	Nov. 28, 1871	121, 284
Floors, pavements, &c., Composition for	A. Solari	Louisville, Ky	Jan. 31, 1871	111, 395
Floors, sidewalks, &c., Composition for	T. Landmann	Cincinnati, Ohio	May 11, 1869	89, 933
Floors, walls, &c., Ornamental covering for	H. Whittemore	New York, N. Y	Nov. 2, 1869	96, 523
Floors, Waxing	R. B. Walker	Claremont, N. H	May 12, 1868	77, 784
Flooring	W. Baum	Hoboken, N. J	Feb. 8, 1870	99, 522
Flooring	C. J. McAlister	Chicago, Ill	Apr. 5, 1870	101, 638
Flooring-clamp	A. S. Blinn and D. W. Hewitt	Dubuque, Iowa	Aug. 14, 1866	57, 078
Flooring-clamp	F. G. Lafayette	Middletown, Ohio	Jan. 19, 1869	85, 939
Flooring-clamp	D. D. Mackay	Whitestone, N. Y	Sept. 1, 1868	81, 803
Flooring-clamp	D. Nevin	Boulder City, Colo	Aug. 3, 1869	93, 217
Flooring-clamp	S. B. Wood and W. S., R. Y. H., and A. W. Terry.	Hamburgh, Ark	Aug. 5, 1873	141, 528
Flooring houses, Composition for	R. Robotham		Apr. 12, 1794	....
Flooring-sett	T. M. Richardson	Stockton, Me	Feb. 1, 1870	99, 477
Flooring, wainscoting, and other purposes, Composition for.	S. Whitmarsh	Northampton, Mass	Jan. 4, 1870	98, 652
Flooring, Wood-block for	F. Donbrava and J. Staib	Hartford, Conn	June 10, 1873	139, 662
Floral bracket	E. M. Stigale	Philadelphia, Pa	Mar. 14, 1871	112, 748
Floral ornament	H. J. Rogers	Saint Denis, Md	July 16, 1872	129, 173
Flour	T. Carr	Bristol, Great Britain	Aug. 15, 1871	118, 105

Index of patents issued from the United States Patent Office from 1790 *to* 1873, *inclusive*—Continued.

Invention.	Inventor.	Residence.	Date.	No.
Flour and grain elevator	H. Stanley	Saint Johnsbury, Vt	Oct. 13, 1863	40, 313
Flour and grist mill	A. J. Vandegrift	Covington, Ky	Sept. 13, 1870	107, 311
Flour and meal, Bolting	W. H. Akins	Berkshire, N. Y	Dec. 3, 1829	....
Flour and meal chest	E. M. Collis	Princeton, Ind	Aug. 27, 1872	130, 898
Flour and meal chest	T. J. Corr	Carlinville, Ill	Feb. 11, 1868	74, 200
Flour and meal chest	J. M. Dashiell	Decatur, Ill	Apr. 15, 1870	101, 438
Flour and meal for transportation, Method of preparing.	E. B. Larcher	New York, N. Y	July 11, 1865	48, 697
Flour and meal, Machine for manufacturing	O. Evans	Philadelphia, Pa	Jan. 22, 1808	....
Flour and meal, Manufacturing	O. Evans	....	Dec. 18, 1790	....
Flour and meal, Process for treating corn, &c., in the manufacture of.	W. Freudenau	Saint Louis, Mo	May 18, 1869	90, 253
Flour and middlings, Bolt for	B. F. Foreman and J. B. Speck	Upton, Pa	Apr. 1, 1873	137, 430
Flour and middlings purifier	E. N. Lacroix	Minneapolis, Minn	Sept. 9, 1873	142, 705
Flour and semolina, Manufacture of	G. A. Buchholz	London, England	Oct. 11, 1870	108, 102
Flour and semolina, Manufacture of	G. A. Buchholz	London, England	Oct. 11, 1870	108, 103
Flour, Apparatus for bolting	J. E. Madigan	Beloit, Wis	Mar. 21, 1865	46, 919
Flour, Apparatus for dressing	H. Cabanes	Bordeaux, France	Dec. 7, 1869	97, 600
Flour, &c., Apparatus for drying	J. Ballantine and A. Clark	Zanesville, Ohio	Mar. 4, 1839	1, 093
Flour-bolt	W. H. Akins	Ithaca, N. Y	Sept. 5, 1854	11, 637
Flour-bolt	R. H. Alexander	Plato, Ohio	Aug. 20, 1867	67, 836
Flour-bolt	W. H. Allen and W. Stoddard	Winona, Minn	May 11, 1869	89, 907
Flour-bolt	E. D. Auchey	Manheim, Pa	May 30, 1871	115, 412
Flour-bolt	W. Bashor	Johnson City, Tenn	Aug. 29, 1871	118, 508
Flour-bolt	E. Bateman	Frederick City, Md	June 18, 1867	65, 789
Flour-bolt	N. Baumann	Elmore, Ill	Sept. 15, 1857	18, 179
Flour-bolt	J. Beall and S. K. Shaffer	Decatur, Ill	Aug. 15, 1865	49, 365
Flour-bolt	S. H. Blossom and J. E. Huston	Buffalo, N. Y., and Hillsdale, Mich.	July 16, 1867	66, 783
Flour-bolt	J. C. Blythe	Perry, N. Y	Aug. 7, 1866	56, 887
Flour-bolt	A. T. Boon	Galesburgh, Ill	Dec. 6, 1864	45, 311
Flour-bolt	B. Boorman	Waukesha, Wis	May 14, 1867	64, 625
Flour-bolt	G. W. Brown	Jackson, Mich	Nov. 27, 1849	6, 897
Flour-bolt	J. Brown	Utica, N. Y	Dec. 19, 1865	51, 548
Flour-bolt	J. Brown	North Buffalo, N. Y	Dec. 31, 1872	134, 354
Flour-bolt	R. Buchanan	Sullivan County, Tenn	July 24, 1844	3, 680
Flour-bolt	D. T. Choat and H. Rich	Cedar Falls, Iowa	Aug. 26, 1873	142, 083
Flour-bolt	J. M. Clark	Philadelphia, Pa	July 26, 1859	24, 857
Flour-bolt	W. F. Cochrane	Springfield, Ohio	Jan. 1, 1867	60, 860
Flour-bolt	S. Cook	Adam's Basin, N. Y	July 22, 1851	8, 241
Flour-bolt	S. Cook	Adam's Basin, N. Y	Mar. 9, 1852	8, 786
Flour-bolt	J. C. Cookson	Lancaster, Pa	June 2, 1863	38, 783
Flour-bolt	M. Cosgro	Peoria, Ill	Dec. 3, 1867	71, 709
Flour-bolt	W. Craig	Uniontown, Pa	Jan. 7, 1868	73, 167
Flour-bolt	R. W. Cunningham	Chesterville, Ohio	Jan. 3, 1871	110, 6 5
Flour-bolt	E. Davies and J. Gerrard	Liverpool, Great Britain	May 21, 1872	126, 936
Flour bolt	W. Derwent, jr	Rockford, Ill	Mar. 19, 1867	62, 943
Flour-bolt	A. J. Dibble	Franklin, N. Y	Feb. 15, 1870	99, 861
Flour-bolt	E. V. Easley	Johnson City, Tenn	July 25, 1871	117, 395
Flour-bolt	J. Frickinger	Kingsville, Ohio	Oct. 11, 1870	108, 246
Flour-bolt	W. Gosshorn	Waterloo, Pa	Oct. 21, 1873	143, 821
Flour-bolt	E. Graham and I. N. Patten	Elizabethtown, Ky	Aug. 30, 1859	25, 256
Flour-bolt	H. Gross and J. B. Rumsey	Tiffin, Ohio	Apr. 2, 1867	63, 506
Flour-bolt	C. T. Hanna	Keokuk, Iowa	Dec. 27, 1870	110, 456
Flour-bolt	S. Heflebower	Alexandria, Va	July 11, 1865	48, 684
Flour-bolt	C. B. Horton	Sand Bank, N. Y	Aug. 18, 1868	81, 086
Flour-bolt	J. G. Humes	Gravios Mills, Mo	June 30, 1868	79, 354
Flour-bolt	J. E. Huston	Hillsdale, Mich	Aug. 29, 1865	49, 629
Flour-bolt	J. W. Johnson	Evansport, Ohio	Nov. 12, 1872	132, 965
Flour-bolt	J. H. Jones	Yellow Springs, Ohio	Aug. 23, 1870	106, 697
Flour-bolt	J. G. Kaufman	Nevada, Mo	Oct. 21, 1873	143, 910
Flour-bolt	E. H. Kellogg	Mukwonago, Wis	Feb. 4, 1868	73, 981
Flour-bolt	I. B. Ketchum	Rochester, Minn	Mar. 16, 1869	88, 792
Flour-bolt	I. B. Lewis	Belvidere, Ill	Apr. 16, 1872	125, 819
Flour-bolt	S. Lewis	Tiffin, Ohio	Oct. 10, 1865	50, 371
Flour-bolt	S. Lewis	Tiffin, Ohio	Nov. 26, 1867	71, 395
Flour-bolt	S. Lewis	Tiffin, Ohio	June 4, 1872	127, 617
Flour-bolt	J. Mallin	Chicago, Ill	Oct. 4, 1870	107, 935
Flour-bolt	D. Marsh	Fairfield, Conn	May 25, 1852	8, 970
Flour-bolt	S. G. McMurtry	West Urbana, Ill	Feb. 9, 1858	19, 303
Flour-bolt	S. C. Mendenhall	Richmond, Ind	Feb. 5, 1856	14, 199
Flour-bolt	S. C. Mendenhall and I. Conner	Richmond, Ind	July 29, 1856	15, 455
Flour-bolt	R. S. Mitchell and G. Z. Kessinger.	Elizabeth, Ind	May 11, 1869	90, 011
Flour-bolt	R. Mohler and J. Becker	Lancaster County, Pa	Sept. 8, 1863	39, 826
Flour-bolt	T. G. Morgan	Murfreesborough, Tenn	Oct. 10, 1871	119, 783
Flour-bolt	E. and A. H. Nordyke	Richmond, Ind	July 5, 1859	24, 656
Flour-bolt	D. N. M. Peregoy	Elizabethtown, Tenn	May 16, 1871	114, 966
Flour-bolt	H. N. Peterson	Peoria, Ill	Dec. 19, 1871	122, 054
Flour-bolt	J. M. Reed and W. B. Willis	Charlestown, Va	Jan. 29, 1850	7, 063
Flour-bolt	H. Reichert	Shippensburgh, Pa	July 22, 1862	35, 953
Flour-bolt	F. Bender	Etchells, England	Feb. 4, 1873	135, 588
Flour-bolt	A. N. Shultz	Sabillasville, Md	Sept. 1, 1868	81, 831
Flour-bolt	J. Skinner	Hadley, Mich	Aug. 9, 1870	106, 223
Flour-bolt	C. B. Slater	Blanchester, Ohio	Dec. 10, 1872	133, 805
Flour-bolt	A. R. Smith	Delaware, Ohio	Nov. 27, 1866	60, 073
Flour-bolt	G. T. and A. Smith	Minneapolis, Minn	Feb. 4, 1873	135, 496
Flour-bolt	J. C. Smith	Zanesfield, Ohio	Aug. 8, 1871	117, 938
Flour-bolt	S. A. Smith	Philadelphia, Pa	Mar. 10, 1868	75, 307
Flour-bolt	R. H. St. John	Bellefontaine, Ohio	Dec. 5, 1871	121, 681
Flour-bolt	F. S. Thayer	Troy, N. Y	Apr. 11, 1865	47, 233
Flour-bolt	T. M. Walker	Bellefontaine, Ohio	Mar. 28, 1871	113, 226
Flour-bolt	J. W. Walters	Tiffin, Ohio	May 7, 1867	64, 603
Flour-bolt	J. Wernz	Epolzheim, Bavaria	Oct. 1, 1872	131, 919
Flour-bolt	J. Wister	Greencastle, Pa	Aug. 26, 1862	36, 317
Flour-bolt and bran-duster	A. Burbank	Rochester, N. Y	Mar. 20, 1866	53, 268
Flour bolt and reel	J. R. Bradfield	Ada, Mich	Apr. 25, 1871	113, 972
Flour-bolt as applied to grinding-mills	S. E. Fitch and T. Sharp	Greenbush, N. Y	Jan. 13, 1857	16, 383

Index of patents issued from the United States Patent Office from 1790 *to* 1873, *inclusive*—Continued.

Invention.	Inventor.	Residence.	Date.	No.
Flour-bolt attachment	R. Turner	Rochester, N. Y	July 19, 1870	105, 618
Flour-bolt feeder	J. Boehm	Peru, Ohio	June 11, 1872	127, 673
Flour-bolt feeder	J. Cornwell	Kalamazoo, Mich	Nov. 12, 1867	70, 806
Flour-bolt feeder	J. Mallin	Chicago, Ill	June 29, 1869	92, 070
Flour-bolt feeder	E. J. Weaver	Sterling, Ill	Oct. 18, 1870	108, 417
Flour-bolt knocker	J. W. Bradley	Rocheport, Mo	July 5, 1870	105, 034
Flour-bolt knocker	E. B. Lowe	Bellefontaine, Ohio	Apr. 30, 1872	126, 312
Flour-bolt screen	D. Landis	Lancaster, Pa	Oct. 23, 1860	30, 483
Flour-bolt vibrator	C. E. Canan	Coldwater, Mich	Dec. 5, 1871	121, 585
Flour-bolt, Wire-cloth	F. B. Hunt and E. Nordyke	Richmond, Ind	Feb 6, 1855	12, 349
Flour-bolt, Feeding	S. Taggart	Indianapolis, Ind	Feb. 20, 1855	12, 422
Flour, Bolting	E. Bradfield	Rochester, N. Y	Sept. 15, 1846	4, 758
Flour, Bolting	J. H. Burk and T. W. Trussell	New Market, Va	Jan. 30, 1872	123, 082
Flour-bolting	W. F. Cochrane	Springfield, Ohio	Jan. 6, 1863	37, 317
Flour-bolting	W. F. Cochrane	Springfield, Ohio	Jan. 6, 1863	37, 318
Flour-bolting	W. F. Cochrane	Springfield, Ohio	Jan. 6, 1863	37, 319
Flour-bolting	W. F. Cochrane	Springfield, Ohio	Jan. 6, 1863	37, 320
Flour-bolting	W. F. Cochrane	Springfield, Ohio	Jan. 6, 1863	37, 321
Flour-bolting	D. Geib	Mifflintown, Pa	Jan. 5, 1858	19, 024
Flour, Bolting	E. S. Snyder	Charlestown, Va	May 2, 1846	4, 490
Flour, &c., Bolting	J. Wright	Saint Louis, Mo	Feb. 23, 1864	47, 737
Flour bolting and dressing machine	J. Hean	Annville, Pa	Feb. 28, 1842	2, 471
Flour-bolting apparatus	L. Fagin and H. C. Hayman	Cincinnati, Ohio	Apr. 8, 1851	8, 025
Flour-bolting apparatus	L. C. A. Schmidt	Tiffin, Ohio	May 31, 1870	103, 784
Flour-bolting apparatus	D. Speck	Weaverton, Md	July 16, 1872	129, 611
Flour-bolting device	W. Halderman	Freeport, Ill	Aug. 7, 1860	29, 483
Flour-bolting machine	M. H. Collins	Chelsea, Mass	Oct. 25, 1859	25, 884
Flour-bolting machine	J. C. Cookson	Lancaster, Pa	Dec. 12, 1871	121, 708
Flour-bolting machinery	R. Mauck	Honeyville, Va	Dec. 26, 1845	4, 329
Flour-bolting machinery	B. D. Sanders	Halliday's Cove, Va	Aug. 24, 1858	21, 277
Flour-bolting process	L. G. Binkley	Baughman, Ohio	June 27, 1871	116, 260
Flour-bolting reel	F. B. Lewis	Tiffin, Ohio	May 30, 1871	115, 331
Flour-bolting reel	D. Shamp	Perry, N. Y	May 16, 1871	114, 980
Flour-bolting reel	A. L. Williams	Orth, Ind	May 10, 1870	103, 000
Flour-bolting reel	A. L. Williams	Orth, Ind	Jan. 24, 1871	111, 292
Flour-box	F. Monroe	Charlestown, Mass	July 16, 1867	66, 867
Flour by means of heated air, Kiln-drying	N. Tyson	Baltimore, Md	Aug. 3, 1831	....
Flour-chest	M. W. Hill	New York, N. Y	Aug. 15, 1865	49, 404
Flour-chest	I. R. Shank	Buffalo, Va	July 24, 1860	29, 319
Flour-chest and cupboard combined	T. J. Corr	Bloomington, Ill	Oct. 31, 1871	120, 496
Flour-cooler	H. B. Allis	Little Rock, Ark	Nov. 9, 1858	22, 006
Flour-cooler	W. W. Goff	Avoca, N. Y	Aug. 3, 1869	93, 297
Flour-cooler	Hebard, Catlin & Abell	Pomfret, N. Y	May 22, 1835	....
Flour-cooler	P. Johnson	Wauconda, Ill	Apr. 27, 1869	89, 485
Flour-cooler	J. Pope	Windham, Me	Feb. 5, 1836	....
Flour-cooler	J. S. Reynolds	Wauconda, Ill	Dec. 8, 1868	84, 842
Flour-cooler	A. and P. Staffer	Salt Creek, Ind	June 15, 1869	91, 381
Flour, Cooling and drying	G. B. Willison	Elizabeth, Pa	July 3, 1855	13, 192
Flour cooling and drying machine	S. N. Park and J. A. Staats	Somerville, N. J	May 28, 1861	32, 433
Flour dipper, measurer, and sifter	G. C. Carver	Roxbury, Mass	May 22, 1866	54, 857
Flour-dispenser	D. L. Johnson	Yorkville, Mich	Oct. 13, 1868	82, 958
Flour-dredge	E. A. Goodes	Philadelphia, Pa	Sept. 1, 1868	81, 770
Flour, Drying	H. Ely	Rochester, N. Y	June 6, 1854	11, 046
Flour-drying reel	W. H. Dole and D. R. Frazer	Chicago, Ill	June 9, 1863	38, 816
Flour, Elevating, cooling, and conveying	J. White and J. Bundy	Barnesville, Ohio	Dec. 10, 1850	7, 828
Flour, feed, and grain elevator	J. Raney	Newcastle, Pa	May 5, 1868	77, 654
Flour, Filling barrels with	B. Bowman and A. Kauffman	Franklin County, Pa	Nov. 7, 1848	5, 907
Flour for bread-making, Preparation of	H. Jones	Bristol, Great Britain	May 1, 1849	6, 418
Flour for confectionery, &c., Preparing	W. G. Dean	Brooklyn, N. Y	Apr. 18, 1871	113, 859
Flour for exportation, Preserving	E. Hord and B. Browning	Fleming County, Ky	Mar. 12, 1825	....
Flour from becoming sour, Preserving	W. Wood, jr	Saint Clairsville, Ohio	Nov. 4, 1824	....
Flour from bran, Machine for separating	W. W. Huntley and A. Babcock	Silver Creek, N. Y	Apr. 12, 1870	101, 736
Flour from bran, &c., Machinery for separating	E. and J. M. Clark	Lancaster, Pa	Apr. 17, 1849	6, 335
Flour from bran, Machinery for separating	I. Frost and J. Monroe	Albion, Mich	Feb. 27, 1849	6, 148
Flour from bran, Machinery for separating	J. Johnston	Wilmington, Del	Apr. 17, 1849	6, 366
Flour, &c., Hopper-filler for bolting	A. Hershe	Lancaster, Pa	Jan. 15, 1814	....
Flour in barrels, Machine for packing	J. Bartholomew	Union, N. Y	Nov. 1, 1859	25, 940
Flour in casks, &c., Machine for packing	J. Hinman	Hartley Township, Pa	Mar. 2, 1835	....
Flour, Machine for manufacture of	T. Carr	Bristol, Great Britain	Nov. 25, 1873	144, 830
Flour, Machine for preparing	L. S. Chichester	Brooklyn, N. Y	Jan. 7, 1873	134, 513
Flour, Machinery for dressing	C. Learned and S. Hughes	Indianapolis, Ind	Nov. 27, 1849	6, 902
Flour-making	G. Motley	Rochester, N. Y	Mar. 28, 1871	113, 079
Flour, Manner of mixing the middlings with the chops in the process of making.	A. D. Worman	Fredericktown, Md	July 23, 1841	2, 189
Flour, Manufacture of	L. G. Binkly	Baughman, Ohio	Feb. 6, 1872	123, 446
Flour, Manufacture of	C. Colgate	Lancaster, Ohio	Mar. 31, 1863	38, 033
Flour, Manufacture of	W. Freudenau	Saint Louis, Mo	Feb. 1, 1870	99, 427
Flour, Manufacture of	E. S. Hutchinson	Baltimore, Md	Feb. 21, 1871	112, 046
Flour, Manufacture of corn	C. Jones and W. Standing	De Soto, Ill	Jan. 5, 1869	85, 592
Flour, Manufacturing	D. R. Clem	Edinburgh, Va	Oct. 19, 1869	95, 987
Flour, Manufacturing	J. Rich	Madison County, New York	Jan. 27, 1812	....
Flour, meal, and other farinaceous substances, Process and apparatus for treating.	J. J. Ridge	St. John's, Southwark, England.	Aug. 31, 1869	94, 341
Flour, meal, &c., Treating cereals for the manufacture of.	E. Weise	Basle, Switzerland	Dec. 22, 1868	85, 261
Flour, Method of dressing	B. Barter	Faribault, Minn	Apr. 9, 1872	125, 518
Flour-mill	W. B. Allen	Winona, Minn	Dec. 10, 1872	133, 819
Flour-mill	J. M. Clark	Philadelphia, Pa	Jan. 10, 1860	26, 751
Flour-mill	H. Cutler	Ashland, Mass	June 22, 1869	91, 612
Flour-mill	J. Dean	New Baltimore, N. Y	June 3, 1873	139, 553
Flour-mill	J. B. Gendebien and A. Honyet	Brussels, Belgium	Jan. 5, 1847	4, 920
Flour-mill	R. M. Thomas	Brady, Pa	Apr. 23, 1872	126, 101
Flour-mill bran-duster	J. E. Weaver	Nether Providence Township, Pa.	Sept. 28, 1869	95, 293
Flour-mill, Cutting	J. Burdge	Cincinnati, Ohio	June 10, 1856	15, 060
Flour-mill, Friction-bolt	L. S. Reynolds	Indianapolis, Ind	Mar. 29, 1859	23, 394
Flour-mill ventilator	C. Moegling	Milwaukee, Wis	Dec. 17, 1867	72, 317
Flour-mills, Hopper-boy for	W. F. Cochrane	Springfield, Ohio	Sept. 30, 1862	36, 583

Index of patents issued from the United States Patent Office from 1790 *to* 1873, *inclusive*—Continued.

Invention.	Inventor.	Residence.	Date.	No.
Flour, &c., mixing-machine	J. B. Peterson	Brooklyn, E. D., N. Y	June 30, 1868	79, 386
Flour-packer	J. Beall	Berlin, Ill	May 19, 1863	38, 546
Flour-packer	I. Cook	Saint Louis, Mo	Apr. 26, 1864	42, 463
Flour-packer	I. Cook	Saint Louis, Mo	Sept. 12, 1865	49, 859
Flour-packer	I. Cook	Saint Louis, Mo	May 29, 1866	55, 063
Flour-packer	C. Custer	Philadelphia, Pa	June 23, 1868	79, 211
Flour-packer	S. Hewit	Seneca Falls Village, N. Y	Apr. 28, 1868	77, 374
Flour-packer	N. Kinman	Buffalo, N. Y	Oct. 30, 1849	6, 830
Flour-packer	N. Kinman	Lewiston, N. Y	Mar. 23, 1852	8, 826
Flour-packer	J. Musser	Frizellburgh, Md	June 26, 1866	55, 885
Flour-packer	A. H. Nordyke	Richmond, Ind	Dec. 13, 1870	110, 155
Flour-packer	O. Place	Brooklyn, N. Y	July 24, 1866	56, 605
Flour-packer	S. Taggart	Indianapolis, Ind	Dec. 30, 1873	146, 107
Flour-packer	L. W. Teeter	Hagerstown, Ind	Nov. 20, 1866	59, 872
Flour-packer	C. E. Timmerman	Frederick, Md	Aug. 13, 1872	130, 461
Flour-packer	G. A. Warner	Portland, Oreg	Sept. 25, 1866	58, 325
Flour packing and pressing machine	O. Jewell	Rochester, N. Y	Mar. 30, 1837	156
Flour-packing machine	J. Banta, jr	Utica, N. Y	Aug. 2, 1839	1, 271
Flour-packing machine	H. A. Barnard	Moline, Ill	July 30, 1867	67, 249
Flour-packing machine	W. Bashoe	Johnson City, Tenn	June 28, 1870	104, 817
Flour-packing machine	B. F. Bashor	Carter's Depot, Tenn	Aug. 30, 1870	106, 912
Flour-packing machine	A. Brown	Mifflinville, Pa	Apr. 29, 1862	35, 075
Flour-packing machine	S. A. Clapp	Hamilton, Ill	June 5, 1860	28, 561
Flour-packing machine	J. Mattison	Oswego, N. Y	Mar. 9, 1858	19, 572
Flour, Preparation of	T. Byrne	Baton Rouge, La	Sept. 4, 1860	29, 859
Flour-pressing machine	S. Hathaway	Massillon, Ohio	Feb. 13, 1832	....
Flour, Process of manufacturing middlings	G. T. Smith	Minneapolis, Minn	Apr. 1, 1873	137, 495
Flour-sack	J. M. Hurd	Auburn, N. Y	Dec. 26, 1865	51, 722
Flour-sack packer, Hand	J. Detweiler	West Liberty, Ohio	Jan. 23, 1872	123, 000
Flour-sacks, Apparatus for filling	J. C. Philbrook	East Sanbornton, N. H	July 29, 1862	36, 023
Flour-sacks, Machine for making	J. M. Hurd	Auburn, N. Y	Aug. 9, 1864	43, 773
Flour-sacker	C. B. Horton	Sand Bank, N. Y	Mar. 16, 1869	87, 930
Flour, Self-raising	W. C. Hughes	Scio, Mich	Feb. 25, 1868	74, 824
Flour-separator	S. Hughes	Hamilton, Ohio	Apr. 17, 1860	27, 907
Flour-separating machine, Power vibrating	B. Calver	Glastonbury, Conn	Oct. 6, 1830	....
Flour, starch, and spirit from wheat, &c., Making	E. Perkins	Shrewsbury, N. J	Feb. 26, 1811	....
Flour, Stone facing to admit cool air to	R. C. Stephens	Hackettstown, N. J	Feb. 18, 1834	....
Flour, Sweet-potato	C. K. Marshall	New Orleans, La	June 22, 1869	91, 554
Flour, Worm for cooling	J. Smelzer	Adams County, Pa	May 6, 1813	....
Flouring and bolting	J. Stouffer, P. Brough, and J. W. Barr.	Chambersburgh, Pa	July 11, 1854	11, 312
Flouring and grist mill	E. Clark	Lancaster, Pa	Jan. 14, 1862	34, 129
Flouring and grist mills, Alarm-indicator for	E. Clark	Lancaster, Pa	Jan. 14, 1862	34, 130
Flouring and grist mills, Mode of preventing the destruction of bolting-cloth in.	W. McKain	Conoy Township, Pa	Dec. 24, 1861	34, 002
Flouring-apparatus	J. M. Clark	Lancaster, Pa	May 13, 1851	8, 089
Flouring-mill	W. Bowman	Craig's Mills, Va	May 10, 1870	103, 003
Flouring-mill	R. J. Brown	Perry, Pa	Aug. 3, 1858	21, 062
Flouring-mill	E. Clark	Lancaster, Pa	Jan. 5, 1858	19, 016
Flouring-mill	E. Clark	Lancaster, Pa	May 25, 1858	20, 329
Flouring-mill	E. and J. M. Clark	Lancaster, Pa	June 6, 1854	11, 015
Flouring-mill	T. Crane	Fort Atkinson, Wis	Jan. 22, 1856	14, 132
Flouring-mill	J. Durling	Sparta, N. J	Apr. 21, 1842	2, 579
Flouring-mill	A. F. Menefee	Rappahannock County, Va	Mar. 12, 1850	7, 174
Flouring-mill	C. Rands	Peoria, Ill	May 25, 1858	20, 370
Flouring-mill	J. Thompson	Ashtabula, Ohio	July 7, 1830	....
Flouring-mill	D. S. Wagener	Penn Yan, N. Y	Sept. 25, 1855	13, 610
Flouring-mill	J. Weis	Bordentown, N. J	Jan. 29, 1856	14, 179
Flouring-mill	S. Wolff	Vicksburgh, Miss	July 20, 1858	20, 972
Flouring-mill bolt	J. L. Yule	New Orleans, La	Jan. 9, 1855	12, 223
Flouring-mill bolt	S. Godfrey	Peoria, Ill	Oct. 17, 1865	50, 466
Flouring-mill cleaning-bolt	S. Godfrey	Peoria, Ill	Jan. 9, 1866	51, 939
Flouring-mills, Distributing-apparatus in	W. Cann	Black Rock, N. Y	June 6, 1854	11, 039
Flouring-mills, Distributing-apparatus in	A. T. Clark	Lancaster, Pa	June 30, 1857	17, 667
Flouring, Process of	J. M. Clarke	Lancaster, Pa	Nov. 10, 1857	18, 574
Flower, Artificial	D. P. Bonnell	Tecumseh, Mich	Aug. 14, 1849	6, 648
Flower-bracket	C. A. Schaller	Philadelphia, Pa	Sept. 2, 1873	142, 353
Flower-casket	W. Hichborn	Charlestown, Mass	July 4, 1871	116, 708
Flower-casket	J. M. Hess	Philadelphia, Pa	Oct. 25, 1870	108, 592
Flower-frame	J. M. Hess	Philadelphia, Pa	Mar. 5, 1872	124, 207
Flower-pot	C. Hochbrunn	New York, N. Y	June 23, 1868	79, 122
Flower-pot	E. Baldwin	Washington, D. C	July 16, 1872	129, 451
Flower-pot	O. Eberhardt	Brooklyn, N. Y	Feb. 5, 1861	31, 306
Flower-pot	M. B. Gould	Buffalo, N. Y	Jan. 28, 1873	135, 271
Flower-pot	T. Griffin	Roxbury, Mass	July 30, 1861	32, 942
Flower-pot	J. Hively	Dayton, Ohio	Jan. 17, 1860	26, 849
Flower pot	A. D. Judd	New Haven, Conn	Sept. 26, 1871	119, 272
Flower-pot	M. Ludlum	Williston, Vt	Mar. 1, 1870	100, 300
Flower-pot	M. Ludlum	Middlebury, Vt	July 30, 1872	129, 972
Flower-pot	T. J. Mayall	Roxbury, Mass	Sept. 3, 1861	33, 205
Flower-pot	C. R. Penfield	Lockport, N. Y	Apr. 26, 1870	102, 311
Flower-pot	W. H. Pugh	Chicago, Ill	Oct. 15, 1872	132, 178
Flower-pot	B. W. Putnam	Boston, Mass	Oct. 26, 1869	96, 144
Flower-pot	C. L. Steele	Boston, Mass	Sept. 20, 1870	107, 558
Flower-pot	F. C. A. Von Levetzow	Kiel, Prussia	Apr. 23, 1872	126, 112
Flower-pot	E. Whitman	Fitchburgh, Mass	Sept. 8, 1868	82, 052
Flower pot and tub	J. Booher	Dayton, Ohio	Mar. 12, 1867	62, 728
Flower-pot bracket	M. D. Jones	Boston, Mass	Jan. 14, 1873	134, 751
Flower-pot support	T. Prince	Roxbury, Mass	Jan. 23, 1866	52, 200
Flower pot, vase, and basket	T. McClunie	Hartford, Conn	Sept. 6, 1870	107, 073
Flower-pots and other pottery, Machine for molding.	F. Herrmann	Milwaukee, Wis	June 18, 1872	128, 039
Flower-pots, Green-glaze for	J. E. Brooks	Yarmouth, Me	Mar. 21, 1871	112, 776
Flower-pots, Hook for hanging	I. A. Lovejoy	Lynn, Mass	May 27, 1873	139, 254
Flower-pots, Machine for cleaning	S. W. Curtis	Stoughton, Mass	July 25, 1865	48, 912
Flower-pots or vases for plants, &c., Construction of.	J. Adams	Boston, Mass	Sept. 3, 1840	1, 761
Flower-shears	J. W. Barbour	Winooski Falls, Vt	Oct. 11, 1870	108, 091

Index of patents issued from the United States Patent Office from 1790 to 1873, inclusive—Continued.

Invention.	Inventor.	Residence.	Date.	No.
Flower-stand	E. T. Cobb	Conway, Mass	Sept. 12, 1871	118, 844
Flower-stand	E. D. Castelow	Meriden, Conn	Feb. 7, 1871	111, 608
Flower-stand	H. J. Coster	Chicago, Ill	July 17, 1860	29, 145
Flower-stand	J. Crawford	Roxbury, Mass	Dec. 2, 1862	37, 070
Flower-stand	A. C. Dakin	Clinton, Mass	Feb. 11, 1873	135, 696
Flower-stand	G. Erkson	Brooklyn, N. Y	Mar. 28, 1871	113, 035
Flower-stand	C. J. Hauck	Williamsburgh, N. Y	Jan. 7, 1868	73, 005
Flower-stand	H. H. Kendrick	Fulton, N. Y	Nov. 8, 1870	109, 130
Flower-stand	C. T. Lee	Taunton, Mass	June 27, 1871	116, 459
Flower-stand	P. B. Sheldon	Prattsburgh, N. Y	May 9, 1865	47, 664
Flower stand and holder, Metallic	H. Miller	Cranston, R. I	July 6, 1869	92, 203
Flower-stand f r windows	F. W. Test	Chicago, Ill	Dec. 19, 1871	121, 972
Flower-stand, Revolving	T. Leslie	Brooklyn, N. Y	Oct. 17, 1871	120, 078
Flowers and fruit, Composition for wax	A. A. Hinkley	Boston, Mass	Feb. 16, 1869	86, 921
Flowers, Artificial stem for cut	J. B. Craig	Perrysville, Pa	Nov. 11, 1873	144, 446
Flowers, Case for preserving	E. M. Stigale	Philadelphia, Pa	Nov. 24, 1868	84, 445
Flowers, Compound for the manufacture of wax	M. J. McColl	Chicago, Ill	Mar. 8, 1870	100, 540
Flowers, Cutting wax for artificial	M. J. McColl	Chicago, Ill	Feb. 15, 1870	99 928
Flowers, Machine for branching artificial	A. Giraudat	New York, N. Y	Nov. 9, 1869	96, 582
Flowers, Manufacture of artificial	O. E. Fillion	Paris, France	May 10, 1870	102, 798
Flowers, Manufacture of artificial	C. E. Howard	San Gabriel, Cal	Mar. 1, 1870	100, 293
Flowers, Manufacture of artificial	C. C. Nichols	Providence, R. I	Oct. 17, 1848	5, 867
Flowers, Manufacture of artificial and preservation of natural.	E. S. Harris	Philadelphia, Pa	Nov. 9, 1869	96, 587
Flowers, Preserving natural	P. T. Vining	Springfield, Mass	June 27, 1871	116, 375
Flowers, Process for preserving and restoring natural.	A. J. Knox	Boston, Mass	Feb. 7, 1865	46, 247
Flowers, Process of bleaching, &c., dried natural	J. W. and M. L. Ware	Philadelphia, Pa	Jan. 7, 1873	134, 714
Flowers, &c., Transplanting	E. G. Nichols	Beaufort, S. C	Nov. 5, 1867	70, 457
Flue	N. Fouché	New Orleans, La	Oct. 4, 1870	108, 014
Flue and setting of open boilers	J. Chilcott	Brooklyn, N. Y	Jan. 9, 1866	51, 922
Flue-block	J. Binns	Oskaloosa, Iowa	June 9, 1868	78, 717
Flue-bottom of steam-boilers, Adjustable	A. M. Sprague	Mobile, Ala	May 9, 1854	18, 897
Flue-brazier, Portable upright	W. Grunert and G. E. Bingham	Milwaukee, Wis	Mar. 24, 1868	75, 899
Flue-cleaner	N. W. Wheeler	Brooklyn, N. Y	Nov. 28, 1865	51, 249
Flue-cleaner, Revolving	G. E. Bingham	Milwaukee, Wis	Dec. 4, 1866	60, 128
Flue-cleaner, Steam	W. Doty	Circleville, Ohio	Aug. 29, 1871	118, 596
Flue-contractor and chimney-valve for fire-places and grates.	H. Batchelder	Beverly, Mass	Jan. 8, 1842	2, 418
Flue-expander	I. S. Hamilton	Hamilton, Ohio	Dec. 5, 1871	121, 514
Flue for buildings	E. T. Gould	Mont Clair, N. J	July 29, 1873	141, 219
Flue-stopper	H. Smith	Southington, Conn	May 25, 1869	90, 401
Flues and dampers of cooking-apparatus, Arrangement of.	J. A. Elder and W. J. Thorn	Westbrook, Me., and Hollister, Mass.	Dec. 4, 1855	13, 868
Flues, Construction of	R. F. O'Brien	Boonville, Mo	May 15, 1860	28, 295
Fluid and mode of generating the same, Imponderable.	M. Ziegler	Mulhouse, France	Jan. 1, 1867	60, 986
Fluid, Burning	E. Howe	Brooklyn, N. Y	Sept. 24, 1850	7, 667
Fluid-can	J. Clements	Ann Arbor, Mich	Aug. 12, 1862	36, 190
Fluid, Compound illuminating	N. A. Dyar and J. F. Augustus	Medford and Boston, Mass	Sept. 6, 1859	25, 362
Fluid-distributing apparatus	J. W. Middleton	Philadelphia, Pa	Sept. 12, 1854	11, 675
Fluid-elevator, Rotary	S. P. Ruggles	Boston, Mass	July 7, 1868	79, 781
Fluid-evaporator	C. S. Wheeler	Flowerfield, Mich	Mar. 8, 1859	23, 209
Fluid-forcing apparatus	T. W. Malone	Mason City, W. Va	Sept. 5, 1871	118, 776
Fluid-gate or faucet	J. W. Smith	Hartford, Conn	Mar. 10, 1857	16, 810
Fluid-indicator	J. Cordnan	Brooklyn, N. Y	Mar. 31, 1868	76, 166
Fluid-meter	L. Allen	New York, N. Y	Mar. 10, 1868	75, 342
Flui -meter	N. Aubin	Albany, N. Y	July 8, 1862	35, 806
Fluid-meter	E. J. Baker	Philadelphia, Pa	Apr. 19, 1864	42, 337
Fluid-meter	C. W. Baldwin	Boston, Mass	May 25, 1869	90, 479
Fluid-meter	P. Ball and B. Fitts	Worcester, Mass	July 20, 1869	92, 691
Fluid-meter	S. J. Burr	Brooklyn, N. Y	Apr. 21, 1857	17, 127
Fluid-meter	L. F. Buschmann	New York, N. Y	Jan. 17, 1871	111, 040
Fluid-meter	E. A. Chameroy	Paris, France	Dec. 1, 1868	84, 476
Fluid-meter	J. Cochrane	New York, N. Y	Mar. 31, 1857	16, 945
Fluid-meter	R. Crewzbaur	Williamsburgh, N. Y	Jan. 24, 1871	111, 179
Fluid-meter	H. A. Desper	Worcester, Mass	Nov. 18, 1873	144, 747
Fluid-meter	J. Dooling	Boston, Mass	May 14, 1872	126, 626
Fluid-meter	J. H. A. Gericke	Hoboken, N. J	Apr. 16, 1872	125, 805
Fluid-meter	W. Hamilton	Toronto, Canada	Oct. 26, 1869	96, 105
Fluid-meter	J. Harris	Boston, Mass	Jan. 5, 1869	85, 656
Fluid-meter	A. Heaton	Bridgeton, Conn	Apr. 6, 1869	88, 709
Fluid meter	H. B. Herbert	New York, N. Y	Aug. 1, 1871	117, 628
Fluid-meter	J. O. Johnson	New York, N. Y	May 6, 1873	138, 657
Fluid-meter	H. B. Leach	Boston, Mass	Feb. 23, 1869	87, 104
Fluid-meter	H. B. Leach	Boston, Mass	Nov. 23, 1869	97, 097
Fluid-meter	W. B. Leonard	New York, N. Y	Nov. 1, 1853	10, 198
Fluid-meter	W. H. Lindsay	New York, N. Y	Feb. 20, 1849	6, 130
Fluid-meter, &c	W. H. Lindsay	New York, N. Y	June 22, 1852	9, 060
Fluid-meter	J. R. Maxwell	Cincinnati, Ohio	June 2, 1857	17, 443
Fluid-meter	C. Moore	New York, N. Y	June 29, 1869	92, 082
Fluid-meter	C. Moore	New York, N. Y	Jan. 24, 1871	111, 233
Fluid-meter	C. E. Moore	Elizabethport, N. J	Apr. 28, 1868	77, 307
Fluid meter	W. Park	Norwich, Conn	Aug. 11, 1868	80, 996
Fluid-meter	W. C. Perrine	New York, N. Y	May 4, 1858	20, 169
Fluid-meter	T. Poore	Scranton, Pa	May 18, 1869	90, 123
Fluid-meter	J. Powers and J. B. Van Deusen.	New York, N. Y	Mar. 17, 1868	75, 644
Fluid-meter	H. C. Sergeant	Newark, N. J	May 24, 1870	103, 509
Fluid-meter	J. Sheffield	Williamson, N. Y	Nov. 3, 1863	40, 504
Fluid-meter	J. Sheffield	Buffalo, N. Y	Feb. 23, 1869	87, 118
Fluid-meter	G. Sickles	Boston, Mass	Feb. 16, 1869	86, 949
Fluid-meter.	C. W. Siemens	London, England	Dec. 14, 1858	22, 315
Fluid-meter	D. B. Spooner	Syracuse, N. Y	Sept. 5, 1871	118, 755
Fluid-meter	C. Stein	Philadelphia, Pa	Mar. 10, 1868	75, 313
Fluid-meter	J. Taggart	Roxbury, Mass	May 15, 1855	12, 887
Fluid-meter	J. B. Van Deusen	New York, N. Y	June 29, 1869	91, 989
Fluid-meter	F. Wagner	New York, N. Y	June 29, 1869	91, 991

Index of patents issued from the United States Patent Office from 1790 *to* 1873, *inclusive*—Continued.

Invention.	Inventor.	Residence.	Date.	No.
Fluid-meter	F. Wagner	New York, N. Y	Nov. 15, 1870	109, 359
Fluid-meter	E. D. Weatherbee	Worcester, Mass	July 20, 1858	20, 979
Fluid meter	J. W. Weller	Cleveland, Ohio	Feb. 25, 1868	74, 778
Fluid-meter, Diaphragm	J. H. Darlington and W. Piper	New York, N. Y	Nov. 11, 1856	16, 049
Fluid-meter, Diaphragm	R. L. Hawes	Worcester, Mass	Apr. 15, 1856	14, 692
Fluid-meters by hand, Device for operating	W. Mason	Warren, Mass	Dec. 23, 1856	16, 284
Fluid-meters, Propeller-wheel for	W. Van Anden	Poughkeepsie, N. Y	Apr. 9, 1872	125, 506
Fluid-motor, Rotary	H. Q. Hawley	Albany, N. Y	July 1, 1873	140, 367
Fluid-motor, Rotary	H. Q. Hawley and E. Anthony	Albany, N. Y	July 30, 1872	129, 943
Fluid or grain elevator	R. Christy	Jefferson County, Va	Nov. 21, 1812	......
Fluid-pressure regulator	W. J. Fay and T. A. Cairns	Denver, Colo	July 16, 1872	129, 015
Fluid-regulator	W. A. Simonds	Boston, Mass	May 1, 1866	54, 424
Fluids, Apparatus for compressing elastic	J. Jameson	Gateshead, England	Feb. 15, 1859	22, 956
Fluids, Apparatus for expanding and compressing elastic.	J. Jameson	Gateshead, England	Dec. 11, 1860	30, 924
Fluids. Apparatus for raising	J. Parker	Camberwell, Great Britain	Nov. 19, 1867	71, 050
Fluids, Apparatus for raising and forcing	M. Alden	Philadelphia, Pa	Feb. 1, 1859	22, 773
Fluids, Cooling	W. Thornton	Washington, D. C	July 31, 1827	......
Fluids, Gage for measuring the pressure of	V. Beaumont	New York, N. Y	June 14, 1859	24, 365
Fluids, Machine for raising or projecting	D. Tomlinson	Brookfield, Conn	May 27, 1805	......
Fluids, Pipe for transmission of	E. P. Vaux	Washington, D. C	Oct. 15, 1867	69, 869
Fluids, Retention and discharge of	B. Seymour	Westmoreland, N. Y	Nov. 10, 1827	......
Fluids under pressure, Device for controlling	J. Atkins	Washington, D. C	Mar. 30, 1869	88, 431
Flush-bolt	W. B. Carness	Roxbury, Mass	May 22, 1860	28, 346
Flush-bolt	J. M. Hopkins	New York, N. Y	May 1, 1866	54, 350
Flush-bolt	C. E. Lattin	Birmingham, Conn	Oct. 29, 1867	70, 340
Flute	T. Berteling	New York, N. Y	Apr. 7, 1868	76, 389
Flute	C. G. Christman	New York, N. Y	Dec. 25, 1849	6, 968
Flute, &c	S. C. Goodsell	New Haven, Conn	Sept. 11, 1866	58, 022
Flute	J. Pfaff	Philadelphia, Pa	Apr. 14, 1857	17, 054
Flute	O. J. C. Wardrum	Chicago, Ill	Sept. 14, 1869	94, 795
Flute, German	W. Schaffer	New York, N. Y	June 6, 1814	......
Fluted cutters, Machine for grinding cylindrical	A. G. Coes	Worcester, Mass	Oct. 19, 1869	95, 988
Fluted rollers, Machinery for making	W. Weild	Manchester, Great Britain	Nov. 14, 1865	50, 997
Fluting and plaiting machine	M. E. Wilson and A. J. Perry	Galesburgh, Ill	June 24, 1873	140, 331
Fluting and puffing iron	A. M. Thorne	Syracuse, N. Y	Feb. 9, 1869	86, 711
Fluting and quilling, Apparatus for	G. R. Houghton	Flint, Mich	Aug. 31, 1869	94, 211
Fluting and sad iron	C. Hyatt	Buffalo, N. Y	Aug. 30, 1870	106, 828
Fluting and sad iron	M. H. Knapp	Bay City, Mich	Aug. 2, 1870	105, 953
Fluting and sad iron, Combined	A. S. Mann	Pittsburgh, Pa	Oct. 29, 1872	132, 590
Fluting-apparatus	G. B. Arnold, A. H. Price, and A. S. Urner.	New York, N. Y	July 24, 1860	29, 337
Fluting-apparatus	F. B. Perkins	Elgin, Ill	June 18, 1872	128, 168
Fluting-iron	S. D. Hubbard	Pittsburgh, Pa	Nov. 19, 1872	133, 104
Fluting-machine	H. B. Adams	New York, N. Y	Apr. 6, 1869	88, 598
Fluting-machine	H. B. Adams	New York, N. Y	Dec. 14, 1869	97, 751
Fluting-machine	S. W. Babbitt	West Meriden, Conn	Apr. 4, 1871	113, 242
Fluting-machine	A. Burbank	Rochester, N. Y	Jan. 10, 1871	110, 895
Fluting-machine	S. G. Cabell	Quincy, Ill	Nov. 10, 1868	83, 924
Fluting-machine	S. G. Cabell	Washington, D. C	Mar. 1, 1870	100, 371
Fluting-machine	S. G. Cabell	Washington, D. C	Sept. 27, 1870	107, 657
Fluting-machine	M. P. Carpenter	San Francisco, Cal	June 28, 1870	104, 825
Fluting-machine	H. H. Cole	New York, N. Y	June 12, 1866	55, 469
Fluting-machine	C. F. Corbett	Boston, Mass	Apr. 20, 1869	89, 028
Fluting-machine	W. D. Corrister	New York, N. Y	Dec. 8, 1868	84, 799
Fluting-machine	E. M. Deey	New York, N. Y	Mar. 22, 1870	100, 989
Fluting-machine	E. M. Deey	New York, N. Y	Apr. 4, 1871	113, 271
Fluting-machine	E. M. Deey	New York, N. Y	Nov. 28, 1871	121, 235
Fluting-machine	E. M. Deey	New York, N. Y	Aug. 13, 1872	130, 413
Fluting-machine	E. M. Deey	New York, N. Y	Aug. 19, 1873	141, 923
Fluting-machine	C. Dion	New York, N. Y	Oct. 18, 1870	108, 336
Fluting-machine	J. F. Hayen	Buffalo, N. Y	Jan. 9, 1872	122, 607
Fluting-machine	F. Hewitt	Bloomfield, N. J	Dec. 14, 1869	97, 774
Fluting-machine	E. P. Holley	Lockport, N. Y	May 6, 1873	138, 650
Fluting-machine	C. E. L. Holmes	New York, N. Y	Mar. 1, 1870	100, 405
Fluting-machine	A. Y. Hubbell	Elmira, N. Y	Nov. 29, 1870	109, 738
Fluting-machine	G. E. King	New York, N. Y	Feb. 26, 1867	62, 492
Fluting-machine	S. Knox and W. D. Corriston	New York, N. Y	Apr. 3, 1866	53, 633
Fluting-machine	S. R. Knox	New York, N. Y	Nov. 20, 1866	59, 913
Fluting-machine	S. Leavitt and E. L. Howard	Malden, Mass	Mar. 8, 1870	100, 645
Fluting-machine	H. Luchs	Washington, D. C	Nov. 16, 1869	96, 820
Fluting-machine	J. W. Madden	Buffalo, N. Y	Nov. 7, 1871	120, 591
Fluting-machine	J. W. Madden and J. Dodsworth	Buffalo, N. Y	June 18, 1872	128, 156
Fluting-machine	E. J. Manville	Waterbury, Conn	Feb. 23, 1869	87, 182
Fluting-machine	E. J. Manville	Waterbury, Conn	May 17, 1870	103, 063
Fluting-machine	W. P. McKee	Cincinnati, Ohio	Nov. 22, 1870	109, 535
Fluting-machine	H. Sauerbier	Newark, N. J	Nov. 22, 1870	109, 456
Fluting-machine	D. R. Saunders	Houston, Tex	Jan. 14, 1873	134, 936
Fluting-machine	C. Schortau	New York, N. Y	July 6, 1869	92, 374
Fluting-machine	H. C. Sergeant	Newark, N. J	Nov. 23, 1869	97, 125
Fluting-machine	H. Sommer and C. Bauer	Newark, N. J	May 23, 1871	115, 251
Fluting-machine	C. A. Sterling	New York, N. Y	Aug. 21, 1866	57, 403
Fluting-machine	T. Stockmarr	New York, N. Y	Mar. 25, 1873	137, 258
Fluting-machine	C. W. Thompson	Chicago, Ill	Aug. 1, 1871	117, 578
Fluting-machine	T. M. Tucker	Newark, N. J	Feb. 7, 1871	111, 588
Fluting-machine	T. M. Tucker	Newark, N. J	July 25, 1871	117, 487
Fluting machine	W. Weitling	New York, N. Y	Aug. 27, 1872	130, 962
Fluting, Machine for straight and spiral	J. January	Greenville, Ky	Mar. 3, 1829	......
Fluting-machine roller	H. G. Pearson	New York, N. Y	Dec. 21, 1869	98, 103
Fluting-rollers, Machine for	W. Weild	Manchester, Great Britain	Oct. 4, 1864	44, 585
Fluting sad-iron	E. A. Franklin	Brenham, Tex	Nov. 14, 1871	120, 869
Fluting-tongs	H. Gerecke	Carlstadt, N. J	May 17, 1870	103, 039
Fluting-tongs	J. B. Mersereau	Binghamton, N. Y	May 17, 1870	103, 224
Fluting-tongs	J. C. Robie	Binghamton, N. Y	Jan. 31, 1871	111, 475
Fluting-tongs	E. R. Shepard	Binghamton, N. Y	Nov. 28, 1871	121, 428
Fluting trimmings, Apparatus for	S. E. Totten	Brooklyn, N. Y	Jan. 31, 1865	46, 157
Flutter-wheels, Application of water to	J. Stewart	Muloy's Post Office, Tenn	Oct. 24, 1828	......
Flutter-wheels, Letting water on	J. Smith	Queen's County, N. Y	July 1, 1817	......

Index of patents issued from the United States Patent Office from 1790 to 1873, inclusive—Continued.

Invention.	Inventor.	Residence.	Date.	No.
Flutter-wheels. Method of applying water to compound buckets of.	D. Rankin	Augusta County, Va	July 11, 1854	11,255
Flux for working metals and minerals	S. W. Kirk	Philadelphia, Pa	Nov. 8, 1870	109,021
Fly and musquito bar	J. H. Jennings	New Bedford, Mass	Mar. 13, 1866	53,149
Fly and musquito bar for windows	F. D. Wright	Jordan, N. Y	July 2, 1867	66,272
Fly-catcher	L. Bartlett	Yarmouth, Me	Feb. 18, 1873	135,961
Fly-catcher	H. A. Farnam	South Bend, Ind	Jan. 31, 1871	111,332
Fly-catcher	J. Olson	Chicago, Ill	Nov. 5, 1872	132,733
Fly-catcher	C. E. Penny	Fort Wayne, Ind	Nov. 12, 1872	132,977
Fly-catcher	W. H. Rice	Oberlin, Ohio	Nov. 12, 1872	133,056
Fly-driver	R. H. Fauntleroy	New Harmony, Ind	Aug. 13, 1834	......
Fly-expelling compound	J. Baquoi	Baltimore, Md	Feb. 25, 1873	136,203
Fly-exterminator	A. K. Walker	Hampden, Me	Jan. 31, 1871	111,497
Fly-flap	T. Royer	Lancaster, Pa	Aug. 21, 1866	57,384
Fly-frame	J. G. Luscomb	Taunton, Mass	June 8, 1869	91,032
Fly-frame flyer	J. S. Streeter	Providence, R. I	Aug. 4, 1868	80,783
Fly-killing machine	A. Glendening	Loudoun County, Va	Sept. 9, 1824	......
Fly or balance wheel	R. Rice	Mineral, Ill	Sept. 3, 1867	68,530
Fly-protector	I. Daugherty	Rocky Station, Va	June 3, 1873	139,552
Fly-screen for animals	B. F. Adams	Bangor, Me	Feb. 2, 1869	86,342
Fly-string cutter and rounder	S. W. Sheller	Mount Carroll, Ill	July 6, 1869	92,377
Fly-switch	L. S. Welch	Canaan, N. H	May 6, 1873	138,719
Fly-trap	S. Arnold	Wilson County, Tenn	Aug. 5, 1856	15,464
Fly-trap	S. Arnold	Green Hill, Tenn	Sept. 23, 1856	15,752
Fly-trap	B. Atwater	Berlin, Conn	Oct. 5, 1858	21,646
Fly-trap	S. B. Bacheller	McDonough, N. Y	Jan. 14, 1873	134,833
Fly-trap	N. P. Bassett	Fulton, N. Y	Apr. 28, 1863	38,277
Fly-trap	V. D. Beach	Battle Creek, Mich	May 28, 1872	127,297
Fly-trap	E. D. Blakeman	New Lebanon, N. Y	Mar. 29, 1859	23,422
Fly-trap	A. Bristol	Constantine, Mich	Dec. 24, 1867	72,446
Fly-trap	M. V. Bulla	South Bend, Ind	May 14, 1872	126,671
Fly-trap	P. A. Burgess	Butler, Mo	Apr. 30, 1872	126,180
Fly-trap	A. M. Chapel and J. G. Hubbard	Pittsfield, Mass	Nov. 26, 1872	133,354
Fly-trap	H. L. Chapman	Marcellus, Mich	Dec. 2, 1873	145,093
Fly-trap	J. B. Childs	Lee Centre, Ill	July 16, 1872	129,317
Fly-trap	W. B. Clark	South Bend, Ind	Oct. 1, 1872	131,848
Fly-trap	I. S. Clough and S. R. Burrell	Brooklyn and New York, N. Y.	June 14, 1859	24,375
Fly-trap	J. J. Craig	Knoxville, Tenn	Sept. 7, 1869	94,569
Fly-trap	E. N. Cummings	Colebrook, N. H	Jan. 16, 1866	52,034
Fly-trap	A. J. Davis	Hartford, Mich	Aug. 5, 1873	141,495
Fly-trap	S. S. Day	New York, N. Y	Nov. 13, 1860	30,619
Fly-trap	W. De Puy	Polk Station, Pa	Oct. 1, 1872	131,813
Fly-trap	A. Eames	Kalamazoo, Mich	Nov. 1, 1859	25,961
Fly-trap	G. W. Eicholtz	New Berlin, Ill	Dec. 23, 1873	145,791
Fly-trap	W. Elwell	Gardiner, Me	Nov. 1, 1859	25,997
Fly-trap	J. Emery and J. E. Gott	Bucksport, Me	May 22, 1866	55,007
Fly-trap	G. W. Ernst and I. Lucas	Fremont, Ohio	May 13, 1873	138,797
Fly-trap	D. and S. K. Flanders	Parishville, N. Y	Apr. 18, 1854	10,801
Fly-trap	J. Fletcher	Cleveland, Ohio	Dec. 17, 1872	133,931
Fly-trap	S. Friend	Decatur, Ill	Dec. 3, 1872	133,579
Fly-trap	J. B. Fuller and G. W. Pierce	Worcester, Mass	Apr. 16, 1850	7,288
Fly-trap	G. Gilbert	Westville, Conn	Oct. 7, 1856	15,843
Fly-trap	B. Glasscock	Hillsborough, Ohio	Apr. 27, 1869	89,305
Fly-trap	L. M. Gould	Chicago, Ill	Mar. 26, 1872	124,951
Fly-trap	J. O. Greene	New Albany, Ind	July 9, 1872	128,874
Fly-trap	L. Grim	Fort Branch, Ind	Sept. 12, 1871	118,852
Fly-trap	C. R. Hardy	Lexington, Ind	Apr. 4, 1871	113,515
Fly-trap	J. M. Harper	El Paso, Ill	Sept. 3, 1872	131,098
Fly-trap	D. Henderson	North Bridgewater, Mass	Feb. 20, 1872	123,897
Fly-trap	J. Hoover	Gratis, Ohio	Dec. 8, 1868	84,825
Fly-trap	P. D. Horton and E. T. Bryan	Marengo Township, Mich	Feb. 21, 1871	111,936
Fly-trap	J. Hyter	Kent, Ind	July 22, 1856	15,378
Fly-trap	J. R. Johnson	Benton Harbor, Mich	Aug. 19, 1873	141,931
Fly-trap	J. H. and J. Kiplinger	North Manchester, Ind	Mar. 23, 1869	88,044
Fly-trap	D. Lake	Smith's Landing, N. J	June 21, 1864	41,212
Fly-trap	D. Lake	Smith's Landing, N. J	Jan. 10, 1865	45,839
Fly-trap	D. Lake	Smith's Landing, N. J	Mar. 20, 1866	53,310
Fly-trap	T. Lane	Whitesburgh, Tenn	Dec. 16, 1873	145,507
Fly-trap	G. E. Le Roy	South Bend, Ind	July 16, 1872	129,350
Fly-trap	B. J. Leslie	Irvine, Ky	Aug. 22, 1871	118,373
Fly-trap	M. Little	Ashley, Ill	Jan. 18, 1870	98,984
Fly-trap	S. F. McGown	Rockville, Ind	Oct. 31, 1871	120,449
Fly-trap	M. Merk	Rochester, N. Y	Aug. 2, 1870	105,963
Fly-trap	R. P. and W. I. Miller	Lineville Station, Pa	May 27, 1873	139,412
Fly-trap	A. C. Mills	Oaktown, Ind	Aug. 3, 1869	93,329
Fly-trap	C. Packard and S. Standish	Eureka, Nev., and Pacheco, Cal.	Nov. 25, 1873	144,860
Fly-trap	J. Parker	Dubuque, Iowa	Nov. 22, 1870	109,444
Fly-trap	M. Patzaner	New York, N. Y	May 3, 1870	102,585
Fly-trap	N. Pike	Brooklyn, N. Y	Feb. 20, 1866	52,794
Fly-trap	H. H. Potter	Carthage, N. Y	Jan. 8, 1867	61,096
Fly-trap	M. M. Preble	Kokomo, Ind	Jan. 22, 1867	61,358
Fly-trap	W. Riley	Madison County, Miss	Apr. 27, 1858	20,091
Fly-trap	H. H. Robertson	Kingston, Mo	July 3, 1860	29,008
Fly-trap	C. S. Rouse	Dowagiac, Mich	Oct. 7, 1873	143,423
Fly-trap	G. T. Savary	Reading, Mass	Sept. 27, 1864	44,486
Fly-trap	T. M. Scott	La Grange, Ga	Feb. 16, 1858	19,382
Fly-trap	R. Shaler	Madison, Conn	Feb. 1, 1859	22,840
Fly-trap	W. T. Shannon	Greensborough, Ga	July 14, 1857	17,811
Fly-trap	W. Shreve	Elkton, Ky	Aug. 22, 1848	5,726
Fly-trap	J. C. Skeen	Barton Township, Ind	July 16, 1872	128,982
Fly-trap	S. W. Smith and H. Bigelow	Brooklyn and New York, N. Y.	Feb. 15, 1859	22,986
Fly-trap	W. Smith	Alexandria, Ind	May 25, 1869	90,402
Fly-trap	H. Snow and L. T. Smart	Lowell, Mass., and Campton, N. H.	July 29, 1851	8,252
Fly-trap	A. Snyder	Jackson, Mich	Mar. 10, 1868	75,480

Index of patents issued from the United States Patent Office from 1790 *to* 1873, *inclusive*—Continued.

Invention.	Inventor.	Residence.	Date.	No.
Fly-trap	A. Standinger	Saint Louis, Mo	Oct. 19, 1869	96, 050
Fly-trap	F. Stengel	New York, N. Y	July 23, 1872	129, 763
Fly-trap	J. E. Stone	Erving, Mass	Mar. 16, 1869	87, 801
Fly-trap	G. J. Swingle	Knoxville Ill	Apr. 14, 1868	76, 673
Fly-trap	A. Veling	Pana, Ill	Nov. 26, 1872	133, 507
Fly-trap	E. Victor	Fort Branch, Ind	July 25, 1871	117, 489
Fly-trap	T. H. Whiting	Grand Haven, Mich	June 24, 1873	140, 176
Fly-trap	T. Whitmore and J. Beebe	Waterloo, Iowa	June 25, 1872	128, 444
Fly-trap	S. R. Wilmot	Watertown, Conn	Sept. 22, 1857	18, 261
Fly-trap, Automatic	F. L. Rosentrater	Des Moines, Iowa	Sept. 5, 1871	118, 747
Fly-wheel	D. Eldridge	Philadelphia, Pa	Mar. 21, 1865	46, 890
Fly-wheel	H. Gerner	New York, N. Y	Sept. 23, 1873	143, 009
Fly-wheel	J. W. Keely	Philadelphia, Pa	Aug. 15, 1871	118, 022
Fly-wheel	C. Root	Cleveland, Ohio	July 18, 1871	117, 115
Fly-wheel and crank-shaft, Combined	E. B. Jucket	Roxbury, Mass	Oct. 23, 1866	59, 032
Fly-wheel, Compensating	A. H. Smith	Charlton, N. Y	June 30, 1868	79, 403
Fly-wheel for rolling-mill machinery	J. Geyser	Allegheny, Pa	Mar. 22, 1859	23, 299
Fly-wheel press	D. Evans	Philadelphia, Pa	Aug. 26, 1831	....
Fly-wheel, Sectional	H. L. Farr	Indianapolis, Ind	Feb. 8, 1870	99, 659
Fly-wheels, Construction of	J. Bryant	Brooklyn, N. Y	Sept. 22, 1863	40, 011
Fly-wheels, Mechanical movement for regulating the action of.	A. C. Frederick	Clarendon, N. Y	Dec. 1, 1857	18, 734
Fly-wheels, Method of arranging	H. B. Peck	Wolcott, N. Y	Apr. 2, 1861	31, 935
Fly-wheels to multiply their motion, Manner of actuating.	C. Johnson	Amity, Ill	Oct. 11, 1841	2, 295
Flyer	E. C. Johnson	Lowell, Mass	Dec. 12, 1854	12, 063
Flyer	D. F. Smith	Manchester, N. H	Dec. 3, 1861	33, 866
Flyer for roving, &c.	T. Marquis	New York, N. Y	Oct. 31, 1848	5, 892
Flyer-guides, Machine for manufacturing	D. L. Hill	Lowell, Mass	Jan. 19, 1864	41, 297
Flyers and spindles, Arrangement of	J. Dermond	Paterson, N. J	Feb. 12, 1850	7, 081
Flyers, Compresser for	W. H. Thompson and R. H. Plummer.	Biddeford, Me	July 19, 1853	9, 865
Flying-apparatus	W. F. Quinby	Wilmington, Del	Sept. 10, 1867	68, 789
Flying-apparatus	W. F. Quinby	Wilmington, Del	Oct. 8, 1872	132, 022
Flying-horse	E. S. Scripture	Green Point, N. Y	June 4, 1850	7, 419
Flying-machine	W. F. Quinby	Wilmington, Del	Oct. 5, 1869	95, 513
Flying-machine	J. Wootton	Boonton, N. J	May 22, 1866	54, 992
Fodder-cutter	J. Eiberweiser	Cincinnati, Ohio	July 20, 1869	92, 946
Fodder-cutter	T. Hazard	Wilmington, Ohio	May 22, 1866	54, 899
Fodder-cutter	A. R. Reese	Phillipsburgh, N. J	Nov. 20, 1860	30, 697
Fodder-cutter	E. F. Varner	Harveysburgh, Ohio	Oct. 27, 1868	83, 571
Fodder-cutter and corn-husking machine, Combined	M. C. Jeffers	New York, N. Y	Feb. 11, 1868	74, 370
Fodder, Cutting and crushing corn	R. Miller	York, Pa	Oct. 3, 1844	3, 775
Fodde , Cutting and crushing corn	H. A. Pitts	Winthrop, Me	May 1, 1845	4, 024
Fodder cutting and grinding machine	I. Fulton	Madison, Pa	Aug. 15, 1865	49, 396
Fodder cutting and grinding machine	J. Jessop, J. Wanbaugh, G. W. Ilgenfritz, and J. C. Baker.	York, Pa	Feb. 20, 1846	4, 391
Fodder cutting and grinding machine	J. Royer	Uniontown, Md	Mar. 12, 1845	3, 938
Fodder-cutting machine	P. S. Clinger	Conestoga Centre, Pa	Aug. 14, 1860	29, 572
Fodder-cutting machine, Corn	J. Elgar	Baltimore, Md	June 19, 1847	5, 159
Fodder-grating machine	J. Lusk	Fredonia, Mich	Mar. 9, 1869	87, 573
Fodder, hay, straw, and potato cutter	W. A. Hamilton	Albany, N. Y	Nov. 19, 1833	....
Fodder, hay, straw, and potato cutter	A. Russell and N. Davis	Syracuse, N. Y	Nov. 19, 1833	....
Fodder-stand	J. Antram and E. B. Mullin	Franklin, Ohio	Jan. 31, 1871	111, 301
Fog-alarm	J. R. Anderson	Brooklyn, N. Y	July 7, 1868	79, 538
Fog-alarm	S. G. Cabell	Quincy, Ill	Aug. 8, 1871	117, 861
Fog-alarm	C. L. Daboll	New London, Conn	June 26, 1860	28, 837
Fog-alarm	G. Hull	Wallingford, Conn	Sept. 19, 1865	50, 004
Fog-alarm	J. C. Lyons	New York, N. Y	Sept. 3, 1867	68, 763
Fog-alarm	C. and G. M. Stevens	Boston, Mass	July 29, 1873	141, 396
Fog-horn	J. R. Anderson	Brooklyn, N. Y	Aug. 8, 1871	117, 717
Fog-signal	F. Brown	New York, N. Y	July 23, 1867	66, 942
Fog-signal	G. P. Goulding, D. Clark, and T. Dickinson.	Buffalo, N. Y	Nov. 13, 1866	59, 589
Fog-signal	G. C. Pattison	Baltimore, Md	Nov. 25, 1873	145, 008
Fog-signal	W. A. Stewart	Bethel, Me	Dec. 23, 1873	145, 915
Fog signal	I. Van Trump	Wilmington, Del	Jan. 29, 1867	61, 643
Fog-signal machine	J. D. Custer	Morristown, Pa	Oct. 5, 1858	21, 656
Folding and extension table	G. Mayer	Sullivan, Ill	Dec. 21, 1869	98, 079
Folding and reclining chair	G Hunzinger	Brooklyn, N. Y	Sept. 15, 1863	39, 995
Folding arm-chair	H. S. Golightly and C. S. Twitchell.	New Haven, Conn	Oct. 6, 1863	40, 210
Folding-bench	T. S. Lewis	Kendall's Mills, Me	Dec. 21, 1858	22, 371
Folding-boat	J. Hegeman	Vischer's Ferry, N. Y	July 16, 1872	129, 026
Folding-box	M. and J. Erpelding	Chicago, Ill	May 23, 1871	115, 183
Folding-box	C. C. Moore	New York, N. Y	June 6, 1871	115, 764
Folding-case	J. C. Barker	Chicago, Ill	Feb. 18, 1873	135, 878
Folding-chair	E. Bartels	New York, N. Y	Oct. 18, 1870	108, 314
Folding-chair	B. Beach	Meriden, Conn	Dec. 21, 1869	98, 141
Folding-chair	P. Born	New York, N. Y	Oct. 6, 1868	82, 791
Folding-chair	A. C. Boyd	Grafton, Mass	Oct. 13, 1868	83, 034
Folding-chair	A. C. Boyd	Worcester, Mass	Aug. 2, 1870	105, 898
Folding-chair	A. C. Boyd	Worcester, Mass	Apr. 22, 1873	138, 124
Folding-chair	C. Brada	New York, N. Y	July 4, 1871	116, 676
Folding-chair	H. E. Braunfeld	Philadelphia, Pa	Aug. 1, 1871	117, 509
Folding-chair	W. E. Cameron	Taunton, Mass	Aug. 2, 1870	105, 903
Folding-chair	A. Collignon	Closter, N. J	Sept. 29, 1868	82, 493
Folding-chair	A. Collignon	Closter, N. J	Nov. 24, 1868	84, 344
Folding-chair	A., C. O., and N. Collignon	Closter, N. J	June 10, 1873	139, 776
Folding-chair	C. O. and N. Collignon	Closter, N. J	Sept. 29, 1868	82, 494
Folding-chair	N. and C. O. Collignon	Closter, N. J	Nov. 16, 1869	96, 778
Folding-chair	N., C. O., and A. Collignon	Closter, N. J	July 11, 1871	116, 811
Folding-chair	F. Colton	Brooklyn, N. Y	Dec. 27, 1870	110, 438
Folding-chair	T. B. Comins, jr.	Lowell, Mass	Jan. 11, 1870	98, 743
Folding-chair	T. R. Comins, jr.	Lowell, Mass	Nov. 8, 1870	108, 973
Folding-chair	J. C. Compton	Clarksville, N. J	May 28, 1872	127, 221
Folding-chair	J. Cram	Boston, Mass	Aug. 21, 1855	13, 479
Folding-chair	N. Cross	New York, N. Y	July 29, 1862	35, 993

Index of patents issued from the United States Patent Office from 1790 to 1873, inclusive—Continued.

Invention.	Inventor.	Residence.	Date.	No.
Folding-chair	J. C. Crummy and A. A. Parsons	Pittsburgh, Pa	Mar. 15, 1870	100, 733
Folding-chair	I. N. Dann	New Haven, Conn	Feb. 6, 1866	52, 488
Folding-chair	I. N. Dann	New Haven, Conn	Oct. 29, 1867	70, 323
Folding-chair	I. N. Dann	New Haven, Conn	Dec. 6, 1870	109, 876
Folding-chair	J. A., W. F., and I. A. Dann	New Haven, Conn	Jan. 6, 1863	37, 277
Folding-chair	C. Dietrich	West Roxbury, Mass	Apr. 13, 1869	88, 776
Folding-chair	A. Eliaers	Boston, Mass	Nov. 22, 1864	45, 146
Folding-chair	F. B. Fabri	New York, N. Y	Dec. 3, 1872	133, 576
Folding-chair	W. Gardner	Glen Gardner, N. J	Oct. 10, 1871	119, 754
Folding-chair	K. Geisler	New York, N. Y	Aug. 12, 1873	141, 784
Folding-chair	H. S. Golightly and C. S. Twitchell.	New Haven, Conn	Mar. 10, 1863	37, 864
Folding-chair	H. S. Golightly and C. S. Twitchell.	New Haven, Conn	Feb. 25, 1868	74, 910
Folding-chair	E. Hambujer	Detroit, Mich	Sept. 4, 1866	57, 709
Folding-chair	P. J. Hardy	New York, N. Y	Dec. 22, 1863	41, 001
Folding-chair	P. J. Hardy	New York, N. Y	Aug. 13, 1867	67, 759
Folding-chair	B. J. Harrison	New York, N. Y	July 29, 1873	141, 271
Folding-chair	B. J. Harrison and J. Condie	New York, N. Y	July 17, 1866	56, 410
Folding-chair	F. M. Holmes	Boston, Mass	June 28, 1870	104, 851
Folding-chair	F. M. Holmes	Boston, Mass	July 5, 1870	105, 079
Folding-chair	F. M. Holmes	Boston, Mass	Jan. 24, 1871	111, 209
Folding-chair	F. M. Holmes	Boston, Mass	Jan. 31, 1871	111, 455
Folding-chair	F. M. Holmes	Boston, Mass	Feb. 21, 1871	112, 040
Folding-chair	F. M. Holmes	Boston, Mass	Feb. 21, 1871	112, 041
Folding-chair	D. Howarth	Portland, Me	Apr. 16, 1867	63, 897
Folding-chair	G. Hunzinger	New York, N. Y	May 13, 1873	138, 892
Folding-chair	J. Hyde	Troy, N. Y	May 7, 1867	64, 535
Folding-chair	J. Hyde	Troy, N. Y	July 23, 1867	67, 119
Folding-chair	C. A. Jackson	Boston, Mass	Oct. 10, 1871	119, 849
Folding-chair	C. A. Jackson	Boston, Mass	Apr. 23, 1872	126, 059
Folding-chair	H. James	North Adams, Mass	Apr. 2, 1872	125, 195
Folding-chair	T. Johnson	Portland, Me	Jan. 16, 1872	122, 834
Folding-chair	F. Kilian	New York, N. Y	Apr. 5, 1870	101, 628
Folding-chair	B. Koechling	New York, N. Y	Feb. 26, 1867	62, 428
Folding-chair	M. Lechler	New York, N. Y	July 5, 1870	104, 965
Folding-chair	R. M. Lytle, W. I. Ashton, and L. W. True.	Williamson County, Tenn	Dec. 14, 1858	22, 297
Folding-chair	C. Marcher	New York, N. Y	Feb. 20, 1872	123, 832
Folding-chair	S. E. Mason and E. Downe	Bangor, Me	Aug. 13, 1867	67, 780
Folding-chair	G. McAleer	Worcester, Mass	Nov. 3, 1868	83, 720
Folding-chair	G. McAleer	Worcester, Mass	Apr. 19, 1870	102, 022
Folding-chair	G. McAleer	Worcester, Mass	July 19, 1870	105, 472
Folding-chair	J. D. Merriam	Ashburnham, Mass	Jan. 6, 1863	37, 324
Folding-chair	G. Miller and P. Hannah	Rochester, N. Y	Mar. 28, 1871	113, 077
Folding-chair	G. F. Mitchell and A. A. Sheafe	South Boston, Mass	Feb. 6, 1872	123, 412
Folding-chair	E. P. More and S. J. Anderson	Cazenovia, N. Y	Sept. 6, 1870	107, 094
Folding-chair	W. Morstatt	New York, N. Y	May 24, 1870	103, 358
Folding-chair	W. Morstatt and F. Kipps	New York, N. Y	Sept. 20, 1870	107, 525
Folding-chair	J. Nicolai	Boston, Mass	Nov. 19, 1867	71, 045
Folding-chair	J. Nicolai	Boston, Mass	Aug. 4, 1868	80, 656
Folding-chair	J. Nicolai	Boston, Mass	Sept. 27, 1870	107, 708
Folding-chair	J. Nicolai and J. P. Rinn	Boston, Mass	Oct. 26, 1869	96, 139
Folding-chair	P. W. Nolan	New York, N. Y	Aug. 5, 1873	141, 456
Folding-chair	P. W. Nolan	New York, N. Y	Aug 5, 1873	141, 457
Folding-chair	H. T. Pratt	Fitchburgh, Mass	Dec. 4, 1860	30, 858
Folding-chair	E. C. Ranks	Boston, Mass	Dec. 28, 1869	98, 415
Folding-chair	F. W. Richardson	New York, N. Y	Apr. 25, 1871	114, 199
Folding-chair	A. M. Rodgers	Brooklyn, N. Y	Mar. 30, 1869	88, 513
Folding-chair	A. M. Rodgers	Brooklyn, N. Y	July 12, 1870	105, 253
Folding-chair	C. C. Schmitt and R. Wodrich	New York, N. Y	Oct. 6, 1868	82, 755
Folding-chair	D. N. Selleg	Newburgh, N. Y	Oct. 14, 1873	143, 721
Folding-chair	A. W. Stewart	Glasgow, Scotland	Feb. 22, 1870	100, 209
Folding-chair	A. W. Stewart	Boston, Mass	Apr. 19, 1870	102, 178
Folding-chair	A. W. Stewart	Boston, Mass	Apr. 19, 1870	102, 179
Folding-chair	A. W. Stewart	Boston, Mass	Apr. 19, 1870	102, 180
Folding-chair	A. W. Stewart	Boston, Mass	Aug. 23, 1870	106, 633
Folding-chair	A. W. Stewart	Boston, Mass	Aug. 6, 1872	130, 327
Folding-chair	J. H. Swan	New York, N. Y	Sept. 13, 1859	25, 452
Folding-chair	J. H. Swan	New York, N. Y	Aug. 21, 1860	29, 750
Folding-chair	J. H. Travis	Charlestown, Mass	Mar. 22, 1870	101, 186
Folding-chair	J. H. Travis and E. H. Mahoney	Charlestown and Boston, Mass.	Nov. 26, 1872	133, 503
Folding-chair	G. Trinks	New York, N. Y	July 30, 1867	67, 378
Folding-chair	G. Trinks	New York, N. Y	June 29, 1869	92, 125
Folding-chair	C. S. Twitchell	New Haven, Conn	Sept. 24, 1867	69, 276
Folding-chair	E. W. Vaill	Worcester, Mass	Apr. 7, 1863	38, 132
Folding chair	E. W. Vaill	Worcester, Mass	Sept. 17, 1867	69, 050
Folding-chair	E. W. Vaill	Worcester, Mass	Sept. 24, 1867	69, 145
Folding-chair	E. W. Vaill	Worcester, Mass	Nov. 3, 1868	83, 744
Folding chair	E. W. Vaill	Worcester, Mass	Oct. 12, 1869	95, 856
Folding-chair	P. B. Viele	Rochester, N. Y	May 7, 1872	126, 595
Folding-chair	P. B. Viele	Rochester, N. Y	July 9, 1872	128, 767
Folding-chair	N. Waterman	Toledo, Ohio	Apr. 23, 1867	64, 173
Folding-chair	E. Watkins and A. McConnell	Philadelphia, Pa	Aug. 29, 1871	118, 500
Folding-chair	A. D. Whitmore	Housatonic, Mass	Nov. 3, 1863	40, 526
Folding-chair	G. E. Whitmore	Housatonic, Mass	June 4, 1867	65, 520
Folding chair	C. A. Wiedemann	New York, N. Y	June 29, 1869	92, 133
Folding-chair and life-preserver	J. M. Rudiger	New York, N. Y	July 18, 1871	117, 209
Folding-chair, bed, and ottoman combined	E. Fauh	New York, N. Y	Jan. 14, 1873	134, 875
Folding-chair for pews	M. S. Beach	Brooklyn, N. Y	Oct. 13, 1857	18, 377
Folding-chair, Iron	J. Lauer	Chicago, Ill	June 11, 1872	127, 899
Folding-chair, Iron	G. Wilson	Chicago, Ill	June 27, 1871	116, 384
Folding-chair, Iron	G. Wilson	Chicago, Ill	July 4, 1871	116, 784
Folding-chair, Iron	G. Wilson	Chicago, Ill	Nov. 14, 1871	121, 034
Folding-chair or lounge	J. Sutter	New York, N. Y	Nov. 3, 1863	40, 512
Folding-chair or table	F. Lüdke	New York, N. Y	Jan. 31, 1865	46, 121
Folding convertible chair	W. B. Kimball	Peterborough, N. H	Apr. 5, 1870	101, 472

Index of patents issued from the United States Patent Office from 1790 *to* 1873, *inclusive*—Continued.

Invention.	Inventor.	Residence.	Date.	No.
Folding-frame for boats, Sectional	W. M. Ducker	Brooklyn, N. Y	Aug. 19, 1873	141, 860
Folding-gate	C. A. Ackerson and W. D. Harrah.	Bath, N.Y., and Davenport, Iowa.	July 16, 1867	66, 668
Folding-gate	I. Meritt	North West Bridgewater, Mass.	Dec. 18, 1849	6, 957
Folding gates, Method of opening and closing vertico-lateral.	F. Thrasher and H. B. Horton	Akron, Ohio	Sept. 29, 1857	18, 308
Folding-guide	A. Douglas	New York, N. Y	Oct. 5, 1858	21, 659
Folding ironing-table	J. H. Mallory	La Porte, Ind	Mar. 1, 1870	100, 430
Folding knife and fork	H. Schumacher	Philadelphia, Pa	Mar. 4, 1873	136, 552
Folding or ironing table	M. G. Briggs	Boston, Mass	June 9, 1868	78, 642
Folding or knock-down chair	J. K. Coolidge and N. H. Hill	Cincinnati, Ohio	Sept. 20, 1870	107, 455
Folding or lunch box	G. B. Mershon	Philadelphia, Pa	Nov. 26, 1867	71, 511
Folding or spiral spring for carriages	J. Armour, jr	Baltimore, Md	June 27, 1808	....
Folding-screen	M. G. Lazarus	Pittston, Pa	May 6, 1873	138, 507
Folding-seat	C. J. Everickx	Paris, France	Apr. 9, 1872	125, 556
Folding-seat	C. A. French	Lowell, Mass	Aug. 30, 1870	106, 805
Folding-seat	J. H. Hall	Chicago, Ill	Oct. 28, 1873	143, 979
Folding-seat	G. W. King	Oswego, N. Y	July 12, 1870	105, 222
Folding-seat	D. H. Nation	Saint Louis, Mo	May 18, 1869	90, 183
Folding-seat	A. M. Olds	New York, N. Y	May 14, 1867	64, 694
Folding-seat	W. W. Parker	Minneapolis, Minn	May 20, 1873	139, 021
Folding-seat	T. Reeve and M. B. Swezey	Brooklyn, N. Y	Apr. 19, 1859	23, 708
Folding-seat	C. W. Sherwood	Chicago, Ill	Mar. 26, 1867	63, 315
Folding-seat	G. Sherwood	Chicago, Ill	Oct. 15, 1867	69, 830
Folding-seat	J. M. Swain	Peekskill, N. Y	Sept. 19, 1871	119, 060
Folding-seat and arm	F. J. Dibble	Chicago, Ill	July 23, 1867	67, 107
Folding-seat for carriage-bodies	R. F. Briggs	Amesbury, Mass	Oct. 1, 1867	69, 397
Folding-table	A. C. Ballard	Winooski, Vt	Feb. 20, 1872	123, 806
Folding-table	J. F. Birchard	Milwaukee, Wis	Feb. 18, 1873	135, 881
Folding-table	J. P. Brown	Boston, Mass	Aug. 6, 1867	67, 406
Folding-table	J. H. Bush	Bone Creek, W. Va	Feb. 26, 1867	62, 391
Folding-table	S. V. Cornell	New Haven, Conn	Apr. 8, 1873	137, 658
Folding-table	D. Doty	Detroit, Mich	July 16, 1867	66, 813
Folding-table	F. W. Heublein	Hartford, Conn	June 10, 1873	139, 673
Folding-table	J. Johnson	New York, N. Y	Feb. 5, 1861	31, 353
Folding-table	C. Lammrich	New York, N. Y	Dec. 14, 1858	22, 294
Folding-table, &c	H. Lorth	Philadelphia, Pa	May 13, 1873	138, 904
Folding-table, &c	T. P. I. Magoun	Worcester, Mass	Feb. 25, 1873	136, 167
Folding-table	F. Mohr	Fond du Lac, Wis	May 1, 1866	54, 386
Folding-table	P. Munde	East Hampton, Mass	Aug. 5, 1873	141, 453
Folding-table	G. W. Nellis	Richmondville, N. Y	Oct. 2, 1866	58, 462
Folding-table	J. Juevedo	Brooklyn, N. Y	Apr. 9, 1872	125, 484
Folding-table	G. G. Small	New York, N. Y	Oct. 24, 1865	50, 664
Folding-table	J. W. Smith	Charlestown, Mass	Apr. 13, 1869	88, 988
Folding-table	W. Smith	Cincinnati, Ohio	Sept. 22, 1868	82, 360
Folding-table	J. Sutter	New York, N. Y	Nov. 3, 1863	40, 513
Folding-table	J. Sutter	New York, N. Y	Dec. 24, 867	72, 562
Folding-table	J. W. Wayne and J. R. Miller	Cincinnati, Ohio	July 12, 1864	43, 538
Folding-table	C. H. Wheeler	Saint Louis, Mo	Aug. 26, 1873	142, 306
Folding-table	A. Windeck	Peoria, Ill	Aug. 30, 1870	106, 906
Folding-table and toilet-glass	H. W. Eastman	Baltimore, Md	Oct. 18, 1864	44, 714
Folding table or bench	R. Carter, jr	Lawrence, Mass	Aug. 22, 1865	49, 500
Folding-table, Portable	C. D. Barnitz	Baltimore, Md	July 29, 1856	15, 407
Folding-table support	J. Daly	Troy, N. Y	June 20, 1871	116, 164
Folding-table top	C. W. Mills	Chicago, Ill	Apr. 4, 1871	113, 548
Folding-tub	C. H. Hudson	New York, N. Y	Jan. 26, 1869	86, 159
Fomentations, Preparing and applying	T. L. Jennison	Cambridge, Mass	Aug. 30, 1832	....
Food and diet from cerealines, Article of	J. E. Brown	Philadelphia, Pa	June 14, 1864	43, 091
Food, Article of	G. G. Campbell	Oswego, N. Y	Jan. 28, 1873	135, 204
Food, Article of	W. J. and A. Coleman	Bury St. Edmunds and London, England.	Dec. 22, 1868	85, 070
Food, Article of	L. S. and M. G. Johnson	Cortland, N. Y	Feb. 25, 1873	136, 165
Food, Article of	J. R. Weed	New York, N. Y	July 30, 1872	130, 091
Food, Artificial	H. C. Morris	Richmond, Va	Mar. 4, 1873	136, 447
Food, Cattle	J. Christie and H. G. Dayton	Atlanta, Ill., and Maysville, Ky.	Mar. 2, 1869	87, 397
Food, Cattle	J. T. Harris	Tyngsborough, Mass	Nov. 2, 1869	96, 317
Food, Composition or paste for articles of	R. M. Livingston	Mobile, Ala	June 4, 1867	65, 494
Food-compound	L. S. Chichester	Brooklyn, N. Y	Feb. 25, 1873	136, 302
Food-compound	L. S. Chichester	Brooklyn, N. Y	Feb. 25, 1873	136, 303
Food-compound	L. S. Chichester	Brooklyn, N. Y	Feb. 25, 1873	136, 304
Food, Compound for cattle	W. Marsden	Newburgh, N. Y	May 31, 1870	103, 761
Food, Concentrated	J. H. Schenck	Saint Louis, Mo	Nov. 22, 1864	45, 180
Food, Cooked vegetable	E. C. Frost	Highland Nurseries, N. Y.	Mar. 31, 1863	38, 039
Food for animals	J. S. Kirk	Pittsburgh, Pa	Dec. 29, 1868	85, 313
Food for animals, Machine for cutting	R. Fanslan	New Haven, Conn	Jan. 28, 1873	135, 326
Food for cattle, Composition for	E. Payne	London, England	Dec. 1, 1868	84, 574
Food for domestic animals	M. S. Bringier	Ascension Parish, La	Mar. 16, 1869	87, 821
Food for infants from cereals	L. S. Chichester	Brooklyn, N. Y	Nov. 11, 1873	144, 508
Food for stock	D. Embree	Dayton, Ohio	Dec. 31, 1867	72, 827
Food for the sick, Article of	A. Meyer-Berck	Frankfort-on-the-Maine, Prussia.	Aug. 18, 1868	81, 190
Food for the table, Device for forming articles of	J. Blaettler	Philadelphia, Pa	Mar. 19, 1872	124, 782
Food from algae or sea mosses, Article of	W. J. Rand, jr	Brooklyn, E. D., N. Y	Apr. 4, 1871	113, 562
Food from cereals	L. S. Chichester	Brooklyn, N. Y	Feb. 25, 1873	136, 305
Food from cider, Article of	S. W. Mahan	Middlebury, Vt	Oct. 21, 1873	143, 918
Food from oyster-juice, Article of	B. G. Noble	New York, N. Y	July 16, 1867	66, 732
Food from potatoes, Article of	C. K. Marshall	New Orleans, La	May 19, 1868	77, 995
Food from pumpkins and squashes, Article of	E. W. and M. C. Ayer	South Waterford, Me	Aug. 31, 1869	94, 375
Food from wheat and process of preparing the same	E. H. Murray	Saint Paul, Minn	June 3, 1873	139, 600
Food, Horse and cattle	E. H. Clowser	Boston, Mass	Jan. 18, 1870	98, 849
Food made from beans, Article of	S. R. Andres	Troy, N. Y	July 23, 1861	32, 853
Food made of sweet potatoes, Article of	S. R. and S. Andres, and M. Bucklin.	Troy and New York, N. Y	Feb. 11, 1862	34, 389
Food, Method of preparing and preserving	J. M. Call and B. G. Sloper	London and Walthamstow, England.	Dec. 20, 1864	45, 569
Food or beef-tea, Concentrated	E. S. Muringer	Philadelphia, Pa	Mar. 11, 1862	34, 642

Index of patents issued from the United States Patent Office from 1790 *to* 1873, *inclusive*—**Continued.**

Invention.	Inventor.	Residence.	Date.	No.
Food or pearl wheat, Article of	J. E. Weaver	Havelock, Pa	Jan. 9, 1872	122, 543
Food or sauce	F. A. Friscia	Brooklyn, N. Y	Dec. 16, 1873	145, 562
Food, Preparation of wheat for	W. S. Brewster	Chicago, Ill	Jan. 28, 1870	104, 700
Food prepared from fish and potatoes, Article of	W. D. Cutler	Philadelphia, Pa	Nov. 10, 1868	83, 836
Food, Preparing	J. Bouis	Baltimore, Md	Jan. 19, 1826	....
Food, Preparing farinaceous	E. N. Horsford	Cambridge, Mass	Mar. 16, 1869	87, 850
Food, Preparing tomatoes and other fruits and vegetables to be used as.	C. Alden	Newburgh, N. Y	Apr. 5, 1870	101, 562
Food, Preparing wheat and other cereals for	G. W. Waitt	Philadelphia, Pa	June 25, 1872	128, 342
Food, Preparing wheat for	M. H. Kollock	Philadelphia, Pa	July 23, 1872	129, 835
Food, Process for preparing fish for	I. L. Stanley	New York, N. Y	May 21, 1872	127, 115
Food, Process of preparing Iceland and Irish moss for use in.	W. J. Rand	Brooklyn, N. Y	Aug. 10, 1869	93, 477
Food, Sauce for	A. P. Agresta and A. Meuecci	New York and Clifton, N. Y	Aug. 26, 1873	142, 071
Food, Treating vegetables for	W. Adamson and C. F. A. Simonin.	Philadelphia, Pa	Apr. 2, 1872	125, 247
Food-warmer	E. Bevan and A. Fleming	Birkenhead and Shipston, England.	Apr. 7, 1868	76, 390
Foot-ball covering	H. A. Alden	Matteawan, N. Y	Dec. 31, 1867	72, 772
Foot-bath	I. A. Isaacs	Cleveland, Ohio	Dec. 18, 1866	60, 521
Foot-boards to splints, Attachment of adjustable	J. Gruol	New York, N. Y	Jan. 5, 1858	19, 025
Foot-cleaner	A. McKeachnie	New York, N. Y	May 25, 1858	20, 359
Foot-cleaner	R. Shaler and C. B. Rogers	Madison and Deep River, Conn.	July 31, 1860	29, 443
Foot-comforter	G. W. Rothrock	Mifflin, Pa	June 22, 1869	91, 567
Foot-light for theaters	C. Defries	London, England	June 2, 1868	78, 580
Foot-lights, Mode of constructing and arranging theater.	G. W. Lloyd	Detroit, Mich	June 24, 1862	35, 696
Foot-mill, Vertical cylindrical	S. Fowks	Catskill, N. Y	Sept. 13, 1824	....
Foot or bed warmer	L. M. Roby	Leesville, Ohio	Oct. 12, 1869	95, 728
Foot-power	E. Hutson	Brockport, N. Y	Dec. 17, 1867	72, 300
Foot-power	G. B. Kirkham	New York, N. Y	Dec. 5, 1871	121, 525
Foot-power apparatus	E. and H. Harding	Delevan, Minn	Aug. 19, 1873	141, 928
Foot-power machine	F. S. Stoddard	Litchfield, Conn	May 17, 1859	24, 064
Foot-press	J. N. Smith	Jersey City, N. J	Nov. 13, 1866	59, 669
Foot-press	N. C. Stiles	West Meriden, Conn	Mar. 27, 1866	53, 499
Foot-press	W. M. Warren	Watertown, Conn	Sept. 16, 1862	36, 491
Foot-rest	C. Adams	Pittsburgh, Pa	Oct. 1, 1867	69, 298
Foot-rest	C. S. Adams	Hillsdale, Mich	July 11, 1865	48, 638
Foot-rest	I. Van Hagen	Chicago, Ill	Feb. 19, 1867	62, 167
Foot-rest	F. M. Watson and H. H. Clough	Warner, N. H	May 26, 1863	38, 718
Foot-rest	E. Wilkins	Warner, N. H	May 19, 1863	38, 619
Foot-rest and boot-jack, Combined	D. H. James	Cincinnati, Ohio	Dec. 20, 1870	110, 240
Foot-rest and kneeling-board	H. Morrison	Steubenville, Ohio	Feb. 26, 1867	62, 434
Foot-rest for chairs	T. A. Johnson	Portland, Me	Mar. 11, 1873	136, 734
Foot-rest for chairs, &c., Extension	J. H. Travis	Charlestown, Mass	Oct. 22, 1872	132, 503
Foot-rest, grate, and fire-brick base, Combined	J. H. Keyser	New York, N. Y	Aug. 11, 1868	80, 971
Foot-shield	D. C. Winans	New York, N. Y	Feb. 23, 1864	47, 735
Footstool for churches, &c	W. G. Brown	Monmouth, Me	Mar. 22, 1870	100, 971
Footstool, Nursery	L. Burnell	Milwaukee, Wis	Oct. 4, 1870	107, 869
Foot-straightening reel	G. Goodrich	Portland Harbor, N. Y	July 22, 1833	....
Foot-warmer	J. J. Andrews	Clyde, Ill	Dec. 19, 1865	51, 534
Foot-warmer	D. Austin	Sheldon, Vt	Aug. 24, 1869	94, 060
Foot-warmer	N. H. Bruce	Forge Valley, Westford, Mass.	Sept. 17, 1867	68, 944
Foot-warmer	J. B. Craig	Perrysville, Pa	Nov. 11, 1873	144, 445
Foot-warmer	A. Eckert	Trenton, Ohio	Mar. 7, 1865	46, 649
Foot-warmer	C. R. Everson	Palmyra, N. Y	Nov. 6, 1866	59, 374
Foot-warmer	H. Forncrook	Elbridge, N. Y	Dec. 4, 1855	13, 871
Foot-warmer	W. Heissenbuttel	Brooklyn, N. Y	Oct. 28, 1873	144, 095
Foot-warmer	H. Hock and J. Zilz	Philadelphia, Pa	Sept. 26, 1865	50, 123
Foot-warmer	S. Hunt	Danville, Ind	July 22, 1862	35, 936
Foot-warmer	J. W. Merrill and J. H. Rowe	Boston, Mass	June 10, 1862	35, 532
Foot-warmer	A. Palmer	Hudson, Mich	Aug. 15, 1865	49, 433
Foot-warmer	C. L. Palmer	Brookline, Mass	Apr. 18, 1865	47, 327
Foot-warmer	J. H. Parsons	Quincy, Mich	Feb. 13, 1866	52, 593
Foot-warmer	H. G. Seekins, sr., and H. G. Seekins, jr.	Elyria, Ohio	Dec. 15, 1857	18, 868
Foot-warmer	G. W. Smith	Aurora, Ind	Aug. 31, 1858	21, 376
Foot-warmer	N. Waterman	Boston, Mass	Nov. 27, 1855	13, 859
Foot-warmer and reflecting-lamp, Combined	C. S. Merwin and C. A. Metcalf	Dubuque, Iowa	Oct. 2, 1866	58, 453
Foot-warmer, Electric	C. V. Littlepage	Austin, Tex	May 17, 1870	103, 061
Forage-ration	M. Fletcher	Louisville, Ky	Dec. 15, 1863	40, 922
Forceps, Surgical	J. G. Loomis	Philadelphia, Pa	Nov. 21, 1854	11, 982
Forceps, Tubular	G. W. Wooley	New York, N. Y	Nov. 25, 1862	37, 023
Forces, Machine for applying and measuring	H. T. Stanard	Wayne, Mich	July 23, 1867	67, 144
Fore-arm, Artificial	B. F. Palmer	Philadelphia, Pa	Jan. 11, 1859	22, 575
Fore-iron for shoemakers	S. A. Shurtleff	North Carver, Mass	Nov. 15, 1859	26, 129
Fore-part irons	J. Blease	Richmond, Ind	Apr. 28, 1868	77, 246
Forebays, Gate for	J. J. Meeker and L. Cortright	Columbia County, Pa	June 26, 1866	55, 882
Forebays, Regulating	H. Mallow	Pendleton County, Va	Mar. 13, 1849	6, 172
Forge	G. Campbell	North Buffalo, N. Y	Mar. 22, 1870	100, 975
Forge	J. Howe	Alna, Me	Apr. 16, 1834	....
Forge	W. T. Kosinski	Philadelphia, Pa	July 25, 1871	117, 303
Forge	C. V. Queen	Peekskill, N. Y	Nov. 18, 1845	4, 273
Forge	C. N. Taylor and E. J. Holmes	Upton and Dedham, Mass	Apr. 25, 1865	47, 467
Forge and applying bellows thereto	R. Green	Brunswick, N. Y	May 31, 1834	....
Forge and bellows for blacksmiths, jewelers, &c	J. C. Conklin	Peekskill, N. Y	June 19, 1835	....
Forge and furnace blowing apparatus	J. A. Stewart	Springfield, Tenn	Aug. 18, 1840	1, 724
Forge and furnace water-back	W. McEwen	Norristown, Pa	Nov. 10, 1841	2, 349
Forge and furnace water-back and tuyere	A. Blood	Cohocton, N. Y	Sept. 29, 1825	....
Forge and other furnaces, Application of air to	S. W. Watson and C. Robinson	Ashtabula, Ohio	Sept. 26, 1835	....
Forge-back	F. Avery	Hamilton, N. Y	Feb. 12, 1833	....
Forge-back	T. Bush, jr	Springhort, N. Y	Apr. 12, 1833	....
Forge-back	A. Graham	Hamilton, N. Y	Feb. 26, 1833	....
Forge back, Blacksmith's	I. Sawyer	Hallowell, Me	Feb. 13, 1835	....
Forge-bellows	C. McMurtry	New Marlborough, Mass	July 8, 1811	....
Forge, Blacksmith's	T. S. Clark	Lena, Ill	Dec. 29, 1868	85, 283

Index of patents issued from the United States Patent Office from 1790 *to* 1873, *inclusive*—Continued

Invention.	Inventor.	Residence.	Date.	No.
Forge, Blacksmith's	J. H. Gould	Cincinnati, Ohio	Mar. 14, 1865	46, 792
Forge, Blacksmith's	J. R. Hobbs	Huntsville, Ala	Sept. 15, 1846	4, 760
Forge, Blacksmith's	W. T. Kellogg	Troy, N. Y	Nov. 25, 1873	144, 986
Forge, Blacksmith's	J. Knickerbacer	La Porte, Ind	Sept. 26, 1835	....
Forge, Blacksmith's	C. Richardson	Greenfield, N. H	Mar. 12, 1836	....
Forge, Blacksmith's	M. Scott	Fairfield, Iowa	Jan. 30, 1872	123, 202
Forge, Blacksmith's	F. A. Stuart	Catharine, N. Y	July 1, 1844	3, 647
Forge, Blacksmith's portable	J. F. Bridge	Deadwood, Cal	Mar. 29, 1864	42, 068
Forge, Blacksmith's portable	M. De La Montanya	San Francisco, Cal	Oct. 15, 1861	33, 479
Forge-bonnet	W. Dunkerly	Woonsocket, R. I	May 17, 1870	103, 160
Forge-bonnet	W. Dunkerly	Woonsocket, R. I	Apr. 25, 1871	114, 115
Forge-chimneys, Flue-casing for	H. S. Wilcox	Nevada, Ohio	Aug. 26, 1873	142, 311
Forge-fire	J. Evans	New Haven, Conn	Mar. 24, 1863	37, 958
Forge-fire	W. Rodgers and A. Bannon	Bellefonte, Pa	Apr. 1, 1856	14, 575
Forge fire, Hot-blast blomary	P. A. Sabbaton	Reading, Pa	Feb. 24, 1845	3, 922
Forge for manufacturing wrought iron, Blomary	S. Guilford	Lebanon, Pa	Dec. 27, 1843	3, 390
Forge-furnace	B. Hotchkiss and H. Shattuck	New Haven and Hamden, Conn.	Sept. 25, 1866	58, 346
Forge-hearth	R. W. Clark	Pittsburgh, Pa	Jan. 21, 1868	73, 502
Forge-hearth	J. Caubach	Middletown, Pa	Nov. 10, 1841	2, 348
Forge-hearth, Hot-air	L. V. Badger and R. Walker	Portsmouth, N. H	Nov. 30, 1835	....
Forge, Hollow-fire	S. Collins	Springfield, Mass	Sept. 11, 1822	....
Forge, Mode of producing the blast for blacksmith's	W. and H. W. Sharp	Catharine, N. Y	Aug. 2, 1839	1, 270
Forge, Portable	J. B. Bolinger	Detroit, Mich	Nov. 17, 1868	84, 048
Forge, Portable	J. M. Cayce	Franklin, Tenn	Aug. 20, 1867	67, 845
Forge, Portable	J. H. Dickerson	Cincinnati, Ohio	Jan. 17, 1865	45, 911
Forge, Portable	W. G. Hyndman	Cincinnati, Ohio	Nov. 17, 1857	18, 643
Forge, Portable	S. Rohrer	Palmyra, N. Y	Dec. 27, 1864	45, 678
Forge, Portable	P. H. and F. M. Roots	Connersville, Ind	Nov. 1, 1870	108, 941
Forge, Queen's portable	C. V. Queen	Peekskill, N. Y	Sept. 19, 1848	5, 782
Forge, Smith's	A. Bissey	Point Pleasant, Pa	June 23, 1838	803
Forge, Smith's	J. W. Crannel	Olivet, Mich	Feb. 3, 1857	16, 531
Forge, Smith's	W. E. Risher	Austin, Tex	Feb. 12, 1867	62, 072
Forges and furnaces, Back for	L. Wilder	Leominster, Mass	Mar. 17, 1838	645
Forges, Bellows-box for blacksmith's	C. Foster	Rochester, N. Y	May 11, 1841	2, 088
Forges, Blowing wind for	D. Strobel, jr	Washington, D. C	May 9, 1831	....
Forges, Driving-gear for portable	L. R. Fitch	Chicago, Ill	Nov. 14, 1871	120, 957
Forges, Fire-box for	H. Sayler and J. Bair	Saint Paris, Ohio	Sept. 10, 1867	68, 657
Forges, Gas-chamber and valve for	J. V. Karr	Goshen, Ind	June 4, 1867	65, 490
Forges, &c., where a blast is required, Back for	A. Graham	Hamilton, N. Y	Dec. 15, 1832	....
Forgery, Means to prevent	R. Willcox	New York, N. Y	Sept. 19, 1823	....
Forging and punching machines, Drop and die	S. Andrews	Perth Amboy, N. J	Apr. 4, 1824	10, 220
Forging and stamping machine	W. Ball	Chicopee, Mass	Jan. 18, 1870	98, 839
Forging-apparatus	J. G. N. Alleyne	Alfreton, Great Britain	Nov. 27, 1866	59, 940
Forging-apparatus	D. Davies	Crumlin, England	Sept. 15, 1868	82, 093
Forging-apparatus	T. B. De Forest	Birmingham, Conn	Sept. 24, 1867	69, 080
Forging-apparatus	A. J. Grainger	Wilmington, Ill	Feb. 26, 1867	62, 407
Forging-apparatus	R. S. Lawrence	Hartford, Conn	Feb. 18, 1862	34, 434
Forging-apparatus	E. F. McFarland	Worcester, Mass	Jan. 17, 1865	45, 928
Forging-apparatus	N. Peterson and G. W. Jones	Antioch, Cal	June 9, 1868	78, 825
Forging-apparatus	J. Price	New York, N. Y	June 18, 1867	65, 941
Forging-apparatus	E. A. Raymond	Brooklyn, N. Y	Nov. 28, 1865	51, 267
Forging-apparatus	A. Rix	San Francisco, Cal	Feb. 6, 1866	52, 447
Forging-apparatus, Blacksmith's	L. and I. Andrews	Biddeford, Me	Dec. 18, 1866	60, 458
Forging bolts and round iron	C. Rentgin		Nov. 17, 1796	....
Forging, drawing, swaging, or forming spindles, rollers, bolts, &c., in metal, Machine for.	W. Rider	Bolton, England	Aug. 26, 1843	3, 235
Forging, Furnace for heating wrought-iron wheels for.	W. R. Thomson	Cleveland, Ohio	May 1, 1855	12, 800
Forging, Heating wrought-iron wheels for	W. R. Thomson	Cleveland, Ohio	June 12, 1855	13, 053
Forging iron, &c., Blacksmith's striker for	E. Brooks	Ridgeville, Va	Feb. 16, 1843	2, 956
Forging-machine	J. Copley, jr	Allegheny City, Pa	June 16, 1868	78, 862
Forging-machine	L. L. Crane	Cleveland, Ohio	May 7, 1867	64, 408
Forging-machine	W. H. Defrees	Andover, Mass	May 26, 1868	78, 266
Forging-machine	J. C. Jewell	Boston, Mass	July 4, 1865	48, 560
Forging-machine	W. T. Leach	East Wareham, Mass	Mar. 20, 1860	27, 550
Forging-machine	D. Noyes	Abington, Mass	Aug. 14, 1855	13, 434
Forging-machine	J. Pipes	Ripley, W. Va	Apr. 18, 1871	113, 922
Forging-machine	S. S. Putnam	Boston, Mass	May 29, 1855	12, 961
Forging-machine	G. H. Richards	West Roxbury, Mass	Sept. 14, 1852	9, 266
Forging-machine	E. Schlanker	Buffalo, N. Y	Mar. 29, 1859	23, 400
Forging-machine	J. Stone	Chicago, Ill	Feb. 5, 1867	61, 888
Forging machine, Iron	S. S. Putnam	Boston, Mass	Dec. 9, 1856	16, 186
Forging metal ic sockets, Machine for	G. Fitch and P. Shaffer	Etna, Pa	Dec. 17, 1872	134, 045
Forging metals, Apparatus for	T. Shaw	Philadelphia, Pa	Feb. 15, 1870	99, 958
Forging metals, Drop for	E. K. Root	Hartford, Conn	Nov. 9, 1858	22, 034
Forging metals, &c., Machinery for	W. Field	Providence, R. I	Dec. 14, 1852	9, 471
Forging, shearing, and punching device	J. E. Emerson	Trenton, N. J	June 5, 1866	55, 253
Forging wrench-bars, Die for	L. Coes	Worcester, Mass	Nov. 25, 1873	144, 832

Fork:

See Agricultural fork.
Band-cutting fork.
Barley-fork.
Barley or gavel fork.
Convertible fork.
Cook-fork.
Culinary fork.
Dung-fork.
Flesh-fork.
Four-tined fork.
Fruit-fork.
Gavel-fork.
Gaveling-fork.
Gold-digger's fork.
Grain-fork.
Grappling-fork.
Grate-fork.
Green-corn fork.
Hay-fork.
Hay-elevating fork.
Hoisting-fork
Horse-bag fork.
Loom-filling fork.
Manure-fork.
Parer-fork.
Pickle-fork.
Pitch-fork.
Plate-handling fork.
Potato-fork.
Steel-spring fork.
Table-fork.
Toasting-fork.
Tuning-fork.
Turning-fork.
Wire fork.

Index of patents issued from the United States Patent Office from 1790 *to* 1873, *inclusive*—Continued.

Invention.	Inventor.	Residence.	Date.	No.
Fork	S. C. Blodgett	Philadelphia, Pa	Feb. 19, 1856	14, 275
Fork and band-cutter combined	J. H. Beffemmyer	Enterprise, Pa	Feb. 13, 1872	123, 671
Fork and band-cutter, Combined	M. Wenger	Upper Leacock Post-Office, Pa.	Aug. 23, 1870	106, 758
Fork and cake-lifter	P. L. Suine	Shirleysburgh, Pa	Dec. 18, 1866	60, 650
Fork and pick, Combined	D. Mater, jr	Bellmore, Ind	Aug. 13, 1872	130, 517
Fork and sharpener, Combined	F. C. Beach	Stratford, Conn	Dec. 5, 1865	51, 285
Fork-bending machine	I. G. Batcheller	Wallingford, Vt	Aug. 30, 1870	106, 769
Fork-blank	J. C. Richardson	Ilion, N. Y	Nov. 24, 1868	84, 377
Fork-blank	J. C. Richardson	Ilion, N. Y	Dec. 29, 1868	85, 400
Fork-blank	J. C. Richardson	Ilion, N. Y	June 15, 1869	91, 367
Fork-guard	H. D. P. Cunningham	Alverstroke, England	July 23, 1872	129, 796
Fork, hoe, &c	S. Sheble	Philadelphia, Pa	Mar. 28, 1871	113, 212
Fork-making machinery	H. B. Kinney	Leonardsville, N. Y	Aug. 7, 1866	56, 954
Fork, shovel, and hoe combined	J. A. Heald	Columbus, Miss	Aug. 18, 1868	81, 166
Fork-tine	W. H. Rodden	Toronto, Canada	Oct. 31, 1871	120, 458
Fork-tines, Die for punching	L. S. White	Hartford, Conn	June 2, 1857	17, 475
Fork-wrench	D. Witt	Hubbardston, Mass	July 16, 1872	129, 302
Forks and rakes to their handles, Mode of securing the tines of.	J. P. W. Riley	Montrose, Pa	May 28, 1867	65, 277
Forks and spoons to knives, Means of attaching	N. Ames	Saugus Centre, Mass	Nov. 19, 1861	33, 730
Forks, Machine for forming shoulders on	J. C. Batcheller	Wallingford, Vt	Mar. 20, 1866	53, 258
Forks, Making	H. Whipple and E. Denio	Baldwinsville, N. Y	Sept. 8, 1868	81, 965
Forks, Making four-tined	G. Martin	Waterbury, Conn	June 13, 1823	....
Forks, Manufacture of	S. H. Gilman	Boston, Mass	Oct. 16, 1845	4, 235
Forming and planing machines, Carriage for	E. Allen	Newark, N. J	Nov. 11, 1862	36, 884
Forming and punching articles of irregular form	L. Dodge	Waterford, N. Y	Mar. 12, 1861	31, 660
Forms, Machine for cutting tapering	L. E. Phillips	Bristol, N. Y	Apr. 26, 1864	42, 504
Fort	E. Clark	Philadelphia, Pa	Apr. 2, 1812	....
Forts, &c., by means of inflammable liquid, Destroying.	A. Berney	Jersey City, N. J	Apr. 26, 1864	42, 458
Forts, Construction of iron	E. G. Woodman	Lowell, Mass	Jan. 27, 1843	2, 928
Fortification	J. W. Reid	New York, N. Y	Jan. 27, 1863	37, 519
Fortification	A. Roff	Bridgeport, Conn	Apr. 7, 1863	38, 127
Fortifications, Constructing field	P. Jamain	Bordeaux, France	Oct. 6, 1863	40, 213
Foundations, Constructing subaqueous	J. B. Eads	Saint Louis, Mo	Feb. 13, 1872	123, 685
Foundations, Construction of subaqueous	J. B. Eads	Saint Louis, Mo	Jan. 23, 1872	123, 002
Foundations, Forming sub	C. Pontez and C. L. McAlpine	New York, N. Y	Nov. 11, 1862	36, 913
Founder's cleansing-mill	S. D. Horton	Peekskill, N. Y	Sept. 11, 1866	57, 907
Founder's flask	A. L. Robinson, M. M. Donnelly, and D. H. Kruna.	Cincinnati, Ohio	Apr. 21, 1863	38, 243
Founding	A. H. Emery	New York, N. Y	Jan. 27, 1863	37, 497
Founding-apparatus	C. Warner	Louisville, Ky	Dec. 2, 1851	8, 570
Foundery-press, Double	W. Blake	Boston, Mass	Apr. 29, 1833	....
Foundery purposes, Venting-core for	H. Tucker	Newton, Mass	Nov. 13, 1866	59, 684

Fountain:

See Crystal fountain.
Drinking-fountain.
Ink-fountain.
Lamp-fountain.
Parlor-fountain.
Pen-fountain.
Portable fountain.
Portable refresh-ment-fountain.
Rotating fountain.
Self-feeding fountain.
Soda-fountain.
Soda-water fountain.
Wash-boiler fountain.

Invention.	Inventor.	Residence.	Date.	No.
Fountain	H. Broezel	Mauston, Wis	Dec. 5, 1871	121, 580
Fountain	G. Finley	Pittsburgh, Pa	Aug. 27, 1872	130, 798
Fountain	G. Polyblank and J. Parkin	Cleveland, Ohio	Sept. 19, 1871	119, 046
Fountain	J. Ross	Greenville, Mich	Mar. 19, 1867	63, 099
Fountain	H. H. Sawtell	Randolph, N. Y	Nov. 14, 1871	120, 902
Fountain	L. Schoeney	New York, N. Y	May 9, 1871	114, 718
Fountain	J. Storer	Hammersmith, England	Feb. 7, 1871	111, 697
Fountain	A. P. Yates	Syracuse, N. Y	May 31, 1870	103, 814
Fountain and aquarium	J. Moore	Brooklyn, N. Y	Oct. 7, 1873	143, 456
Fountain and cooler combined	C. Lauby	Brodhead, Wis	Jan. 9, 1872	122, 617
Fountain and evaporator, Combined	G. H. Thatcher	Albany, N. Y	July 22, 1851	8, 232
Fountain and pump combined	J. J. Sink	Philadelphia, Pa	Aug. 5, 1873	141, 600
Fountain-comb	W. Kerr, jr., and J. A. Robbins	Boston, Mass	Dec. 29, 1868	85, 311
Fowl fetter or clog	S. J. Baker	Madison Centre, Me	Jan. 3, 1871	110, 621
Fowl-perch	L. T. Stetson	Randolph, Mass	Nov. 21, 1871	121, 067
Fowl, Trapping wild	A. Brear	Saugatuck, Conn	Nov. 27, 1866	59, 953
Fractured thigh, Apparatus for	S. Wolston	Vincentown, N. J	July 2, 1836	....
Fractures and dislocations, Apparatus for treating	E. H. Barnes	Marathon, N. Y	Apr. 7, 1868	76, 382
Fractures and displacements, Apparatus for treating.	J. S. McClelland	Crawfordsville, Ind	Dec. 17, 1867	72, 215
Fractures, Apparatus for	T. Burr	Battle Creek, Mich	Apr. 30, 1861	30, 229
Fractures, Apparatus for	W. Mills and M. Hoar	New Athens, Ohio	Nov. 8, 1845	4, 255
Fractures, Apparatus for	L. Post	Lodi, N. Y	Dec. 16, 1844	3, 859
Fractures, Apparatus for	L. Roe	White Plains, N. Y	Nov. 6, 1844	3, 810
Fractures, Apparatus for	J. Whitten	Boston, Mass	Feb. 19, 1861	31, 501
Fractures, Apparatus for the cure of	O. M. Allaben	Middletown, N. Y	Dec. 12, 1839	1, 428
Fractures, Apparatus for treatment of	Z. Hussey	Chillicothe, Ohio	Dec. 14, 1852	9, 467
Fractures, Instrument for supporting	S. B. Tucker	Saint Louis, Mo	Mar. 19, 1867	62, 982
Fractures of the lower jaw, Apparatus for the treatment of.	J. Stowe	Lawrence, Mass	Aug. 2, 1870	106, 091

Frame:

See Advertising-frame.
Alphabet-frame.
Arithmetical frame.
Artist's stretching-frame.
Awning-frame.
Bag-frame.
Baking-frame.
Barrow-frame.
Bed-frame.
Bed-bottom frame.
Bridge-truss frame.
Bonnet-frame.
Canopy-frame.
Car-frame.
Car-body frame.
Car-seat frame.
Car-seat end frame.
Carpet-bag frame.
Carriage-top frame.
Caster-frame.
Chair-frame.
Chair-seat frame.
Cheese-frame.
Clothes-frame.
Comb-frame.
Diking-sheet frame.
Door-frame.

Index of patents issued from the United States Patent Office from 1790 *to* 1873, *inclusive*—Continued.

Invention.	Inventor.	Residence.	Date.	No.
Frame—Continued. *See* Drawing-frame. Dress-frame. Dressing-frame. Drying-frame. Electroplating-frame. Fender-frame. Filling-frame. Flax and hemp drying frame. Flower-frame. Fly-frame. Folding-frame. Fruit-drying frame. Furnace-door frame. Gilt-frame. Glass-frame. Grate-frame. Grate-bar frame. Grindstone-frame. Harrow-frame. Harvester-frame. Hat-frame. Heddle-frame. Hinged frame. Hoop-skirt frame. Hop-frame. Hop-vine frame. Ivory frame. Kite-frame. Lace-drying frame. Looking-glass frame. Mattress-frame. Mica-frame. Milk-stool frame. Mirror-frame. Mirror and picture frame. Necktie-frame. Net-frame. Oval frame. Pants-stretching frame. Photographic frame. Photographic-paper frame. Photographic-printing frame. Piano-frame. Picture-frame. Pier-glass frame. Pillow-frame. Pillow-spread frame. Porcelaine-printing frame. Printing-frame. Quilting-frame. Revolving-frame. Rock-drilling-machine frame. Roller-frame. Roofing-frame. Roving-frame. Sample-frame. Sash-frame. Saw-frame. Settee-frame. Ship-frame. Shutter-blind frame. Skirt-frame. Slate-frame. Spectacles-frame. Spinning-frame. Spool-frame. Staging-frame. Stool or ottoman frame. Stove-door frame. Stripper-frame. Tatting-frame. Tender-frame. Tent-frame. Throstle-frame. Traveling-bag frame. Trunk-caster frame. Tub-washer frame. Umbrella-frame. Vehicle-seat frame. Vertical-engine frame. Wardrobe-frame. Warp-setting frame. Warping-frame. Watch-frame. Wheelbarrow-frame. Winding-frame. Window-frame. Window and door frame. Window-screen frame.				
Frame	B. A. Stevens	Toledo, Ohio	Oct. 8, 1872	132, 114
Frame, bottle, &c., Composition for	J. T. Peet	Cincinnati, Ohio	Aug. 1, 1865	49, 147
Frame-joints, Method of securing	C. F. Linscott	New York, N. Y	Jan. 24, 1871	111, 128
Frame or caissons of breakwater, &c	E. H. Tracy	New York, N. Y	Apr. 27, 1859	20, 105
Frame, Standing-press	B. F. Brown	New York, N. Y	Mar. 2, 1822	....
Framing-joint	J. Newton	New York, N. Y	July 4, 1871	116, 623
Frecing-tool, screw, and jewel-setter combined	C. E. Evard	Leesburgh, Va	May 4, 1869	89, 568
Freezer, Artificial	N. M. Johnson	Philadelphia, Pa	Sept. 9, 1843	3, 254
Freezer, Ice-cream	E. P. Torrey	New York, N. Y	Sept. 19, 1861	31, 497
Freezing-apparatus	J. B. Toselli	Paris, France	May 19, 1868	78, 159
Freezing-apparatus, Process of forming vacuum in	E. S. Boynton	Meriden, Conn	Aug. 9, 1870	106, 251
Freezing-box for fish, &c	W. Davis	Detroit, Mich	Jan. 19, 1869	85, 913
Freezing fish, meat, &c., Device for	D. Y. Howell	Toledo, Ohio	Dec. 6, 1870	109, 820
Freezing liquids, Apparatus for	J. Baptiste, J. Mignon, and S. H. Rouart.	Paris, France	Sept. 26, 1865	50, 212
Freight, Device for unloading or storing	H. A Whitney	Brooklyn, N. Y	June 20, 1865	48, 330
Freight-elevator	W. F. Morrow	Shippensburgh, Pa	Jan. 30, 1872	123, 189
Freight or merchandise, Unloading	R. Bragg	San Francisco, Cal	June 2, 1863	38, 725
Freight, Receiving and discharging	N. A. Patterson	Kingston, Tenn	Sept. 4, 1866	57, 825
Freight-transferring mechanism	J. H. James, jr	Urbana, Ohio	Jan. 12, 1869	85, 829
Friction-brake	R. Kitson	Lowell, Mass	June 2, 1863	38, 749
Friction-brake for machinery	J. C. Clapp	Boston, Mass	Aug. 15, 1871	118, 107
Friction brake or clutch	A. G. Waldo	Milwaukee, Wis	June 6, 1871	115, 792
Friction-coupling	A. and F. Brown	New York, N. Y	Nov. 4, 1862	36, 832
Friction-coupling	J. Mattison	Oswego, N. Y	Apr. 10, 1866	53, 846
Friction-coupling, Double	J. Hendy	San Francisco, Cal	Oct. 11, 1859	25, 733
Friction-furnace for generating heat without the consumption of fuel.	J. W. Cochran	Lowell, Mass	Nov. 19, 1833	....
Friction-heater for mill, &c	D. H. Ball	Enterprise, Pa	Jan. 26, 1869	86, 270
Friction in machinery, Reducing	J. J. Reekers	Baltimore, Md	June 13, 1831	....
Friction in mill-gudgeons, Reducing	F. Eichelberger	Creagerstown, Md	June 29, 1833	....
Friction-joint, Hydrodynamic	T. E. Sandgren	Wilmington, Del	July 24, 1855	13, 324
Friction, Mode of diminishing	C. Badin	Paris, France	Aug. 29, 1869	49, 685
Friction of axles, mandrils, &c., Reducing	I. Cooper	Baltimore, Md	Sept. 28, 1831	....
Friction-roller	D. Baldwin	Queensbury, N. Y	Oct. 1, 1830	....
Friction-roller	W. Brown	Portsmouth, England	Dec. 6, 1870	109, 866
Friction-roller	J. Patterson	Franklinville, N. Y	Aug. 9, 1853	9, 926
Friction-roller	C. W. Pierce	Albany, N. Y	Jan. 7, 1868	73, 117
Friction-roller	J. M. Totten	Peoria, Ill	Sept. 10, 1850	7, 639
Friction-roller, Adjustable	J. and T. Sweeney	Birmingham, Pa	Apr. 10, 1855	12, 701
Friction-roller bearing	C. M. Daboll	New London, Conn	May 10, 1870	102, 779
Friction-roller, Elastic	A. A. Moss	Philadelphia, Pa	July 5, 1859	24, 653
Friction-roller for machinery	I. Clowes	Norfolk, Va	Sept. 9, 1834	....
Friction-roller for machinery	B. Stancliffe	Philadelphia, Pa	Dec. 31, 1833	....
Friction-roller support or bearing	S. W. Eaton	Farmington, Me	May 21, 1867	64, 958
Friction-rollers, Mode of applying	R. Dickson and S. G Merriman	Southington, Conn	Mar. 26, 1838	657
Friction-rollers to gudgeons, Applying	M. C. Forrist	Foxborough, Mass	Nov. 16, 1841	2, 364

Index of patents issued from the United States Patent Office from 1790 *to* 1873, *inclusive*—Continued.

Invention.	Inventor.	Residence.	Date.	No.
Friction rolling	G. Danforth	Centreville, Ill	May 17, 1833	....
Friction-spindles, Reducing	F. Fredley	Logan, Pa	Mar. 10, 1834	....
Friction-wheel	T. Hotchkiss	Stratford, Conn	Apr. 2, 1867	63, 520
Friction-wheel and oil-chamber	A. I. Ambler	Chicago, Ill	Mar. 21, 1865	46, 867
Friction-wheel for driving machinery	J. Hinkley	Norwalk, Ohio	Apr. 14, 1863	38, 165
Friction-wheel for gearing	W. D. Andrews	New York, N. Y	Mar. 15, 1870	100, 839
Friction-wheel or roller for machinery	J. D. Cobb	Lebanon, Ohio	June 22, 1832	....
Frictional-brake	P. W. Yarrell	Garysburgh, N. C	Apr. 2, 1872	125, 158
Frilling and crimping machine	C. O. Crosby and H. Kellogg	Hartford, Conn	Dec. 2, 1862	37, 033
Frilling, Double	C. O. Crosby and H. Kellogg	New Haven, Conn	Apr. 28, 1863	38, 289
Fringe	J. E. Gillespie	Norwich, Conn	Feb. 27, 1872	124, 051
Fringe-cutting machine	W. J. Horstmann	Philadelphia, Pa	Apr. 7, 1857	16, 977
Fringe, Machinery for twisting shawl	M. D. Whipple	Lowell, Mass	Nov. 27, 1849	6, 911
Fringe-making machine	E. Doran	Philadelphia, Pa	Aug. 18, 1868	81, 073
Fringe-making machine	L. D. Valetton	New York, N. Y	Apr. 2, 1861	31, 916
Fringe of shawls, &c., Machine for twisting	J. Nesmith and W. Sawyer	Lowell and Dracut, Mass	Oct. 13, 1851	8, 424
Fringe on textile fabrics, Machine for forming	T. Henderson	Glasgow, North Britain	Jan. 23, 1872	123, 017
Fringe-twister	H. Howard	Leeds, N. Y	May 9, 1871	114, 562
Fringe-twisting machine, Bullion	E. Barton	Paterson, N. J	Mar. 9, 1869	87, 620
Fringed fabric, Woven	P. Cocker	Philadelphia, Pa	Dec. 13, 1870	110, 012
Frog. Self-acting guard	C. A. Postley	Spring Garden, Pa	June 24, 1851	8, 171
Frowers, Die for forging	C. Konold	Pittsburgh, Pa	Jan. 21, 1873	134, 993
Fruit and berry basket	G. W. Swan and W. Finch	San Francisco, Cal	Aug. 26, 1873	142, 299
Fruit and berry box or basket	J. H. Marvil	Laurel, Del	May 2, 1871	114, 456
Fruit and clothes drier	T. B. Carroll	Indianapolis, Ind	July 20, 1869	92, 700
Fruit and ice house combined	J. S. Ross	Hiram, Ohio	Jan. 8, 1867	61, 107
Fruit and lard press	W. H. Davis	Lexington, Ind	Sept. 2, 1873	142, 329
Fruit and meat drying apparatus	H. Endemann	New York, N. Y	Oct. 24, 1871	120, 253
Fruit and poultry box	D. B. Spinning	Brooklyn, N. Y	June 25, 1867	66, 183
Fruit and preserve jar	J. S. and T. B. Atterbury	Pittsburgh, Pa	June 30, 1863	39, 027
Fruit and vegetable crate	E. B. Georgia	Clifton Station, Va	July 16, 1872	129, 332
Fruit and vegetable cutter	W. H. Trissler	Lima, Ind	May 22, 1860	28, 419
Fruit and vegetable desiccating apparatus	J. A. Miller	New York, N. Y	Sept. 13, 1864	44, 260
Fruit and vegetable preserver	P. Kephart	Uniontown, Md	Sept. 24, 1844	3, 758
Fruit and vegetable press	C. Parham	Philadelphia, Pa	Oct. 18, 1864	44, 742
Fruit annually, Method for growing	F. Clymer	Galion, Ohio	Feb. 9, 1869	86, 733
Fruit, Apparatus for cooking and preserving	W. Janney	Martinsville, Ohio	Mar. 19, 1867	62, 960
Fruit, Apparatus for packing	R. Law	Lockport, N. Y	July 12, 1859	24, 779
Fruit-bag	W. I. Ludlow	Cleveland, Ohio	Apr. 28, 1868	77, 201
Fruit-basket	R. S. Bartlett	Northampton, Mass	Jan. 23, 1872	122, 984
Fruit-basket	S. D. Bedell	New York, N. Y	May 6, 1873	138, 605
Fruit-basket	L. W. Beecher	New Haven, Conn	Apr. 29, 1862	35, 073
Fruit-basket	L. W. Beecher	New Haven, Conn	May 31, 1864	42, 923
Fruit-basket	L. W. Beecher	New Haven, Conn	Dec. 17, 1872	134, 027
Fruit-basket	W. H. Burridge	Cleveland, Ohio	Jan. 31, 1865	46, 177
Fruit-basket	H. Carpenter	New York, N. Y	Sept. 20, 1864	44, 280
Fruit-basket	H. Carpenter	Williamsburgh, N. Y	Nov. 7, 1871	120, 620
Fruit-basket	N. S. Clement	New Britain, Conn	Sept. 29, 1868	82, 601
Fruit-basket	W. Clement	Charlestown, Ill	May 29, 1866	55, 061
Fruit-basket	E. Colby	Brockport, N. Y	May 31, 1870	103, 570
Fruit-basket	S. B. Conover	New York, N. Y	Oct. 1, 1872	131, 851
Fruit-basket	C. C. Converse	Dubuque, Iowa	Aug. 30, 4864	43, 974
Fruit-basket	C. C. Converse	Dubuque, Iowa	Sept. 27, 1864	44, 400
Fruit-basket	D. Cook	New Haven, Conn	July 12, 1859	24, 723
Fruit-basket	A. Dean	Otto, Ind	July 30, 1872	130, 025
Fruit-basket	J. H. Doolittle	Derby, Conn	Aug. 15, 1865	49, 390
Fruit-basket	I. C. Gleason	Middletown, Conn	June 26, 1866	55, 851
Fruit-basket	J. S. Hoard and C. M. Milese	Vineland, N. J	May 23, 1865	47, 822
Fruit-basket	E. A. Jeffery	New Haven, Conn	Aug. 22, 1865	49, 590
Fruit-basket	H. C. Jones	Dowagiac, Mich	Aug. 6, 1872	130, 299
Fruit-basket	O. Macdaniel	New York, N. Y	May 19, 1868	78, 110
Fruit-basket	R. Mitchell	Smyrna, Del	Aug. 23, 1870	106, 605
Fruit-basket	C. Moore	Stratford, Conn	Nov. 17, 1868	84, 207
Fruit-basket	G. Munger	New Haven, Conn	Oct. 17, 1865	50, 486
Fruit-basket	O. A. North	New Britain, Conn	Aug. 27, 1872	130, 937
Fruit-basket	J. K. Park	Marlborough, N. Y	June 12, 1866	55, 530
Fruit-basket	S. D. Payne	Kasota, Minn	Aug. 23, 1870	106, 719
Fruit-basket	E. F. Percival and N. S. True	Hammonton, N. J	Aug. 18, 1868	81, 205
Fruit-basket	D. H. Priest	Watertown, Mass	May 8, 1866	54, 594
Fruit-basket	P. B. Sheldon	Prattsburgh, N. Y	May 1, 1866	54, 422
Fruit-basket	D. Sherwood and G. D. Dudley	Lowell, Mass	Mar. 16, 1869	87, 797
Fruit-basket	M. L. and O. A. Stray	Willoughby, Ohio	Mar. 25, 1862	34, 792
Fruit-basket	M. L. and O. A. Stray	Willoughby, Ohio	Oct. 30, 1866	59, 290
Fruit-basket	O. Stoddard	Busti, N. Y	Oct. 27, 1863	40, 436
Fruit-basket	W. D. Trissler	Dunkirk, N. Y	Feb. 11, 1873	135, 737
Fruit-basket	W. H. Trissler	Dunkirk, N. Y	July 9, 1872	128, 927
Fruit-basket	P. B. Viele	Rochester, N. Y	Feb. 11, 1868	74, 456
Fruit-basket	C. Wheat and C. Bunge	Geneva, N. Y	Aug. 7, 1866	57, 024
Fruit-basket	G. H. White	Huntington, N. Y	Dec. 6, 1870	109, 983
Fruit-basket	W. R. Wilcox	Saint Joseph, Mich	Nov. 4, 1873	144, 376
Fruit basket and crate	A. F. Newell	Warren, Ohio	Oct. 21, 1862	36, 723
Fruit-basket cover	A. S. Dyckman	South Haven, Mich	Nov. 28, 1871	121, 238
Fruit-basket, Paper	A. Adams and J. F. Jewett	Chagrin Falls and Cincinnati, Ohio.	July 26, 1870	105, 619
Fruit-basket platform or handle	J. Knapp	Coloma, Mich	Nov. 21, 1871	121, 055
Fruit-baskets, Cutter for cutting the bodies of	B. D. Whitney	Winchendon, Mass	May 24, 1870	103, 533
Fruit-baskets, Machine for bottoming	B. D. Whitney	Winchendon, Mass	May 24, 1870	103, 532
Fruit-baskets, Machine for cutting out the bodies of	H. Mellish	Walpole, N. H	Apr. 2, 1867	63, 414
Fruit-baskets, Machine for cutting the bottoms of	H. Mellish	Walpole, N. H	Mar. 26, 1867	63, 280
Fruit-baskets, Machine for making	H. Mellish	Walpole, N. H	Mar. 26, 1867	63, 279
Fruit-baskets, Tool for fitting bottoms to	A. S. Parks	Winchendon, Mass	May 24, 1870	103, 493
Fruit-box	C. A. Blair	New Britain, Conn	Dec. 19, 1871	121, 926
Fruit-box	C. A. Blair	New Britain, Conn	May 7, 1872	126, 440
Fruit-box	E. P. Bothwell	McArthur, Ohio	Jan. 21, 1873	134, 973
Fruit-box	I. F. Brown	New London, Conn	Sept. 17, 1867	68, 943
Fruit-box	C. Colby and H. C. Ward	Benton Harbor, Mich	Aug. 23, 1870	106, 603
Fruit-box	I. Copeland	North Bridgewater, Mass	Jan. 14, 1868	73, 300
Fruit-box	T. B. Doolittle	Ansonia, Conn	June 19, 1866	55, 631

Index of patents issued from the United States Patent Office from 1790 *to* 1873, *inclusive*—Continued.

Invention.	Inventor.	Residence.	Date.	No.
Fruit-box	W. H. Earle	Vineland, N. J	Oct. 16, 1866	58, 939
Fruit-box	J. H. Fisher	Chicago, Ill	Dec. 7, 1869	97, 493
Fruit-box	L. L. Gilliland	Dayton, Ohio	June 7, 1870	104, 116
Fruit-box	H. H. and M. L. Gridley	Auburn, N. Y., and Burlington, N. J.	Apr. 9, 1867	63, 629
Fruit-box	N. Hallock	Flushing, N. Y	Sept. 7, 1858	21, 415
Fruit-box	J. W. Hayes	Newark, N. J	Aug 12, 1858	15, 514
Fruit-box	W. Huey	Galena, Md	July 2, 1867	66, 346
Fruit-box	T. B. Jones	Radnor, Ohio	Oct. 27, 1868	83, 504
Fruit-box	J. H. Marvil	Laurel, Del	Oct. 10, 1871	119, 872
Fruit-box	E. Morris	Burlington, N. J	Jan. 10, 1865	45, 848
Fruit-box	E. Morris	Burlington, N. J	Sept. 12, 1865	49, 908
Fruit-box	A. F. Newell	Warren, Ohio	Jan. 5, 1864	41, 168
Fruit-box	W. Nicklin	Marlborough, N. Y	July 16, 1872	129, 047
Fruit-box	J. M. Perkins	Plainfield, N. J	Sept. 22, 1868	82, 438
Fruit-box	S. W. Phelps	Sandusky, Ohio	Nov. 28, 1871	121, 412
Fruit-box	C. Reese	Baltimore, Md	June 19, 1866	55, 709
Fruit-box	A. T. Robinson and J. Shepard	Bristol, Conn	Mar. 16, 1869	87, 875
Fruit-box	J. T. Severns	Burlington, N. J	Oct. 16, 1866	58, 899
Fruit-box	J. Shepard	Bristol, Conn	Feb. 20, 1873	123, 941
Fruit-box	J. Shepard and B. B. Lewis	Bristol, Conn	May 8, 1866	54, 614
Fruit-box	J. S. Shields	Medora, Ind	Nov. 17, 1868	84, 139
Fruit-box	J. Vincent	Newfane, N. Y	Feb. 9, 1869	86, 886
Fruit-box	C. W. Weston	San Francisco, Cal	Oct. 10, 1871	119, 806
Fruit-box	C. W. Weston	San Francisco, Cal	Aug. 27, 1872	130, 963
Fruit-box	C. W. Weston	San Francisco, Cal	Nov. 12, 1872	132, 942
Fruit-box	C. W. Weston	San Francisco, Cal	Aug. 12, 1873	141, 842
Fruit-box	J. White	Cleveland, Ohio	Aug. 20, 1867	68, 019
Fruit-box	J. F. and O. B. Whitney	Milton, N. Y	Sept. 3, 1867	68, 401
Fruit-box	W. R. Wilcox	Saint Joseph, Mich	July 16, 1867	66, 925
Fruit-box	H. B. Willcox	Troy Mills, Pa	Nov. 26, 1867	71, 435
Fruit-box crate	J. White	Cleveland, Ohio	Nov. 16, 1869	96, 856
Fruit-box fastening	H. T. Barker	Napa, Cal	June 12, 1866	55, 455
Fruit-box, Folding	J. H. Hollingsworth	Philadelphia, Pa	Jan. 26, 1869	86, 157
Fruit-box handle, Adjustable	C. Peebles	Santa Clara, Cal	Mar. 19, 1867	63, 086
Fruit-box lock	C. Colby	Madison, Wis	Mar. 10, 1868	75, 372
Fruit box or basket	J. H. Doolittle	Ansonia, Conn	Sept. 12, 1865	49, 866
Fruit-boxes, Case for conveying	R. H. Baker	Jamestown, N. Y	Jan. 5, 1864	41, 049
Fruit-can	M. Bray	Newton, Mass	Nov. 11, 1873	144, 380
Fruit-can, &c	C. Becker, J. A. Ross, and J. Steuernagel.	Allegheny City, Pa	July 14, 1868	79, 889
Fruit-can	H. Callahan	Dayton, Ohio	Aug. 29, 1871	118, 586
Fruit-can	B. F. Ells	Dayton, Ohio	July 17, 1866	56, 390
Fruit-can	R. G. Farnham	Elbridge, N. Y	Jan. 3, 1871	110, 757
Fruit-can	O. F. Fitch	Morristown, Ind	Mar. 11, 1862	34, 627
Fruit-can	J. Grimes	Portsmouth, Ohio	Jan. 2, 1872	122, 455
Fruit-can	J. Haines	West Middleburgh, Ohio	Aug. 13, 1867	67, 754
Fruit-can	G. E. Heinig	Louisville, Ky	July 19, 1870	105, 572
Fruit-can	L. E. Holden	Cleveland, Ohio	Apr. 21, 1868	76, 917
Fruit-can	I. Kling	Seymour, Ind	Mar. 22, 1870	101, 024
Fruit-can	S. Lutz	Philadelphia, Pa	May 10, 1870	102, 952
Fruit-can	W. W. Lyman	West Meriden, Conn	Dec. 28, 1858	22, 436
Fruit-can	W. W. Lyman	West Meriden, Conn	June 10, 1862	35, 529
Fruit-can	A. J. McMillen	Ravenswood, W. Va	Dec. 7, 1869	97, 669
Fruit-can	J. F. Merrill	Cincinnati, Ohio	July 5, 1870	105, 109
Fruit-can	H. Mitchell	Osborne, Ohio	Apr. 27, 1869	89, 327
Fruit-can	J. Neuberger and P. J. Illig	Buffalo, N. Y	Apr. 11, 1865	47, 216
Fruit-can	P. Nemsen	Baltimore, Md	Mar. 3, 1868	75, 189
Fruit-can	M. O'Connor	New Brunswick, N. J	Sept. 2, 1873	142, 503
Fruit-can	W. and G. Pollyblank	Cleveland, Ohio	Feb. 27, 1872	124, 156
Fruit-can	J. S. Shaneman and J. S. Hoon	Beaver Falls, Pa	Feb. 21, 1871	112, 082
Fruit-can	C. F. Spencer	Cleveland, Ohio	Feb. 27, 1872	124, 171
Fruit-can	L. J. Wicks	Bridgeton, N. J	Oct. 23, 1866	59, 111
Fruit-can	A. A. Wilcox	Philadelphia, Pa	Mar. 19, 1872	124, 710
Fruit-can	J. R. Williamson	Bethlehem, N. J	May 26, 1868	78, 346
Fruit-can	J. B. Wilson	Fislerville, N. J	Jan. 26, 1864	41, 417
Fruit can and jar	J. K. Chase	Brooklyn, N. Y	June 18, 1872	128, 112
Fruit-can bodies, Manufacturing	J. Pfau	Cincinnati, Ohio	June 2, 1868	78, 605
Fruit-can cover	T. Earle	Smithfield, R. I	Dec. 22, 1863	40, 996
Fruit-can fastening	G. H. Hammer	Newville, Pa	June 26, 1866	55, 953
Fruit-can manufacture	C. Becker, J. A. Ross, and J. Steuernagel.	Allegheny City, Pa	July 14, 1868	79, 890
Fruit-can opener	W. W. Lyman	West Meriden, Conn	May 22, 1866	54, 929
Fruit-can opener	A. McClanahan	Bath, Ill	Nov. 19, 1872	133, 108
Fruit-can-opening tool	H. Holt	New York, N. Y	Dec. 10, 1867	72, 042
Fruit-can-sealing apparatus	T. M. Ferguson	Rainsborough, Ohio	July 28, 1868	80, 401
Fruit-can, Self-sealing	T. Parker	Germantown, Pa	Dec. 18, 1866	60, 547
Fruit-can, Sheet-metal	C. F. Sturgis	Buena Vista, Ala	Mar. 28, 1871	113, 111
Fruit-can tongs	A. T. Atherton	Lowell, Mass	Oct. 17, 1871	119, 912
Fruit-can top	A. Stewart	Cincinnati, Ohio	Oct. 2, 1866	58, 502
Fruit-cans and cleaning off the tops, Machine for filling.	L. J. Wicks	Bridgeton, N. J	Dec. 26, 1871	122, 298
Fruit-cans and other cylindrical packages, Machine for varnishing and labeling.	A. C. Platt	Sandusky, Ohio	Aug. 23, 1870	106, 724
Fruit-cans, Apparatus for exhausting the air from	L. C. Cooley	Albany, N. Y	June 24, 1873	140, 247
Fruit-cans by steam, Mode of exhausting air from	D. Beardsley	Ithaca, N. Y	Jan. 29, 1867	61, 506
Fruit-cans, Closing	N. S. Gilbert	Lockport, N. Y	Aug. 25, 1863	39, 643
Fruit-cans, Closing	W. Webster	Middletown, Ohio	Sept. 8, 1863	39, 853
Fruit-cans, Closing	W. Webster	Middletown, Ohio	Feb. 16, 1864	41, 657
Fruit-cans, Machine for closing	G. C. Grodhaus	Jamestown, Ohio	Sept. 3, 1872	130, 989
Fruit-cans, Machine for making	I. Kling	Seymour, Ind	Mar. 15, 1870	100, 775
Fruit-cans, Machine for making tin	A. Kollenberg	Owensborough, Kentucky	May 22, 1866	54, 923
Fruit-cans, Making tin	J. M. Francis	Waldo, Ohio	June 26, 1866	55, 845
Fruit-cans, Sealing	E. K. and J. M. Bruce	Wilkins, Pa	Nov. 20, 1866	59, 816
Fruit-cans, Sealing	J. M. Cooper and W. L. Haller	Chambersburgh and Carlisle, Pa.	Aug. 7, 1860	29, 544
Fruit-cans, Sealing	H. S. Fisher	Newburgh, Pa	Mar. 22, 1864	41, 985
Fruit-cans, &c., Sealing	W. K. Lewis and J. W. Bailey	Boston and West Brookfield, Mass.	Apr. 24, 1866	54, 182

Index of patents issued from the United States Patent Office from 1790 *to* 1873, *inclusive*—Continued.

Invention.	Inventor.	Residence.	Date.	No.
Fruit-cans, Sealing	C. F. Spencer	Rochester, N. Y	Nov. 19, 1867	71, 239
Fruit-cans, Tongs for lifting	A. H. Johnson	Bridgeport, Conn	Jan. 7, 1873	134, 604
Fruit, Canning	J. R. Matthews	Cincinnati, Ohio	Feb. 18, 1873	135, 927
Fruit, Canning	C. J. Paine	Young America, Ill	Sept. 3, 1867	68, 384
Fruit-canning ladle	R. D. Baker	Syracuse, N. Y	Aug. 8, 1871	117, 723
Fruit-canning machine	J. H. Ellis	Peoria, Ill	June 21, 1870	104, 438
Fruit-carrier	J. W. Weston	New York, N. Y	Nov. 3, 1868	83, 678
Fruit-carrying box	G. A. Lloyd	San Francisco, Cal	Sept. 6, 1870	107, 070
Fruit-case	C. Doolittle and A. Carson	Oswego and New York, N. Y.	Oct. 9, 1860	30, 303
Fruit collecting and drying apparatus	S. N. Thomas	Auburn, N. Y	June 9, 1863	38, 848
Fruit-coring knife	A. L. Harris	Kent, Ohio	July 5, 1870	104, 954
Fruit-cover	O. W. Alexander	Osage, Tex	Dec. 17, 1872	133, 914
Fruit-crate	G. Clapp	Geneva, N. Y	Nov. 15, 1870	109, 175
Fruit-crate	W. H. Clarkson	Bridgeville, Del	Mar. 18, 1873	136, 968
Fruit-crate	D. Crane	Saginaw, Mich	Aug. 26, 1873	142, 148
Fruit-crate	G. M. Fenley	Medora, Ind	Aug. 17, 1869	93, 693
Fruit-crate	W. G. Goodale	Centralia, Ill	Aug. 18, 1868	81, 153
Fruit-crate	H. and E. W. Humphreys	Salisbury, Md	Dec. 24, 1872	134, 205
Fruit-crate	L. A. Lindsey and J. F. O'Sullivan.	Jackson, Miss	Apr. 6, 1869	88, 647
Fruit-crate	T. Mabbett, sr	Vineland, N. J	Apr. 21, 1868	77, 060
Fruit-crate	J. H. Marvil	Laurel, Del	Mar. 5, 1872	124, 366
Fruit-crate	L. Selling	Detroit, Mich	July 29, 1873	141, 389
Fruit-crate	F. R. Van Dake	Jackson, Miss	Sept. 21, 1869	95, 169
Fruit, Crate for carrying	W. Gilbert	Catskill, N. Y	Mar. 27, 1866	53, 437
Fruit-crates, Hasp for fastening	F. R. Baird	Norfolk, Va	Feb. 8, 1870	99, 620
Fruit, Device for boiling and stirring	M. G. Collins	Baltimore, Md	May 12, 1868	77, 870
Fruit, Device for quartering, coring, and stringing	S. T. Sanford	Fall River, Mass	Sept. 15, 1863	39, 963
Fruit-dryer	J. Allen	Everett, Mo	July 18, 1871	117, 029
Fruit-dryer	J. Allen	Everett, Mo	July 2, 1872	128, 521
Fruit-dryer	R. N. Allen	Cleveland, Ohio	Nov. 20, 1866	59, 806
Fruit-dryer	S. C. Barth	Indianapolis, Ind	Mar. 19, 1872	124, 658
Fruit-dryer	H. Beamer	Smithburg, Pa	Oct. 9, 1860	30, 285
Fruit-dryer	J. S. Biddle and T. S. Lingenfelter.	Pattonville, Pa	Mar. 14, 1871	112, 532
Fruit-dryer	J. I. Boone	West Milton, Ohio	Jan. 2, 1866	51, 792
Fruit-dryer	D. K. Boswell	Columbus, Ohio	Jan. 28, 1868	73, 691
Fruit-dryer	J. K. Boswell	Richmond, Ind	Apr. 23, 1867	64, 065
Fruit-dryer	C. A. Boynton	Vineland, N. J	Apr. 23, 1872	125, 929
Fruit-dryer	J. W. Brooks and H. Rudoff	Ashley, Ill	July 27, 1869	93, 048
Fruit-dryer	W. R. Clark	Indianola, Ill	Nov. 23, 1869	97, 168
Fruit-dryer	N. C. Cooley	Wyoming, Del	June 21, 1870	104, 559
Fruit dryer	I. B. Cottrell	Bridgeville, Del	Aug. 26, 1873	142, 147
Fruit-dryer	E. Dilday	South Pass, Ill	Nov. 23, 1869	97, 174
Fruit-dryer	E. Duncan	West Hilton, Ohio	June 18, 1861	32, 597
Fruit-dryer	D. Durfee	Fort Seneca, Ohio	Dec. 17, 1867	72, 377
Fruit-dryer	A. Edwards	New Haven, Conn	Jan. 7, 1873	134, 528
Fruit-dryer	B. F. Ells	Dayton, Ohio	Feb. 21, 1871	112, 022
Fruit-dryer	O. P. Fence	Des Moines, Iowa	Oct. 4, 1870	108, 048
Fruit-dryer	W. D. Fisher and W. Holly	Freeport, Ill	Sept. 11, 1866	57, 885
Fruit-dryer	M. W. Florer	Bracken County, Ky	Dec. 24, 1867	72, 470
Fruit-dryer	J. Harvey	Martinsville, Ind	Oct. 19, 1869	95, 900
Fruit-dryer	W. Heaton	Greene County, Pa	Mar. 16, 1858	19, 635
Fruit-dryer	H. Henley	Shoals Station, Ind	June 22, 1869	91, 626
Fruit-dryer	H. Henley	Shoals, Ind	Aug. 16, 1870	106, 484
Fruit-dryer	J. Hildebrand	Taneytown, Md	May 31, 1870	103, 744
Fruit-dryer	I. Hogeland	Indianapolis, Ind	Aug. 17, 1869	93, 885
Fruit-dryer	C. Leavitt	Cleveland, Ohio	May 14, 1867	64, 678
Fruit-dryer	A. C. Lewis	Burlington, Mich	July 31, 1860	29, 390
Fruit-dryer	D. Lippy	Mansfield, Ohio	July 4, 1865	48, 570
Fruit-dryer	M. E. Lloyd	New York, N. Y	May 25, 1869	90, 369
Fruit-dryer	R. A. Lucas and L. S. Lehman	Wooster, Ohio	Dec. 29, 1868	85, 459
Fruit-dryer	C. H. Martin	Chapinville, N. Y	Dec. 12, 1871	121, 795
Fruit-dryer	J. B. May	Magnolia, Ohio	Mar. 15, 1870	100, 914
Fruit-dryer	J. V. R. Miller	Richmond, Ind	Jan. 31, 1871	111, 467
Fruit-dryer	J. Mongene	Vincennes, Ind	Apr. 26, 1870	102, 298
Fruit-dryer	R. S. Morse	East Dixfield, Me	Jan. 2, 1855	12, 167
Fruit-dryer	G. R. Nebinger	Lewisberry, Pa	Aug. 7, 1866	56, 982
Fruit-dryer	G. R. Nebinger	Philadelphia, Pa	May 6, 1873	138, 516
Fruit-dryer	J. B. Okey	Indianapolis, Ind	Oct. 11, 1870	108, 289
Fruit-dryer	A. Paige and G. Wilkinson	Chicago, Ill	Aug. 27, 1872	130, 861
Fruit-dryer	J. L. Post	Ashley, Ill	Mar. 11, 1873	136, 670
Fruit-dryer	J. W. Pyne	Danville, Ill	Oct. 20, 1868	83, 313
Fruit-dryer	I. Randall, 2d	Claremont, N. H	June 19, 1860	28, 775
Fruit-dryer	S. D. Rogers and F. C. Selby	Allegan, Mich	June 22, 1869	91, 671
Fruit-dryer	B. L. Ryder	London, Pa	Nov. 12, 1872	133, 060
Fruit-dryer	M. A. Shepard	Bridgeport, Ill	Apr. 7, 1868	76, 532
Fruit-dryer	R. H. Sipes and D. Defibaugh	Bloody Run, Pa	Oct. 19, 1869	95, 942
Fruit-dryer	A. J. Smith	Decorah, Iowa	June 29, 1869	92, 113
Fruit-dryer	M. P. Smith	Baltimore, Md	Sept. 3, 1872	131, 125
Fruit-dryer	A. Snyder	Clyde, Ohio	July 11, 1865	48, 733
Fruit-dryer	J. Spear	Carbondale, Ill	Sept. 29, 1868	82, 654
Fruit-dryer	H. Speer and J. L. Harlow	Chelsea, Mich	Nov. 17, 1863	40, 646
Fruit-dryer	G. M. Sternberg	Fort Riley, Kans	June 28, 1870	104, 788
Fruit-dryer	G. W. Stevens and J. Gray	San Francisco, Cal	Apr. 16, 1872	125, 703
Fruit-dryer	J. Stevenson	Sparta, Ill	Apr. 8, 1873	137, 634
Fruit-dryer	J. H. L. Tuck	Saint Charles, Ill	Aug. 11, 1863	39, 5 9
Fruit-dryer	W. Vogel	Chelsea, Mich	June 28, 1864	43, 354
Fruit-dryer	T. L. West	West Salem, Wis	June 7, 1864	43, 058
Fruit-dryer	J. Williams	South Haven, Mich	Oct. 21, 1873	143, 949
Fruit-dryer and food-warmer	S. Richmond	Annapolis, Md	Mar. 21, 1871	112, 963
Fruit, Drying	I. Lynde	Marathon, N. Y	Apr. 19, 1864	42, 386
Fruit, Drying and preserving	J. Lowe	Guilford County, N. C	July 4, 1871	116, 729
Fruit-drying frame	C. Gardner	Freedom, Ohio	Sept. 26, 1865	50, 113
Fruit-drying house	J. Billings	Dayton, Ohio	Oct. 10, 1865	50, 414
Fruit-drying house	J. F. Winchell	Springfield, Ohio	Jan. 8, 1867	61, 130
Fruit-drying machine	D. Rice	Shelburne Falls, Mass	Sept. 3, 1867	68, 4[illegible]9

Index of patents issued from the United States Patent Office from 1790 *to* 1873, *inclusive*—Continued.

Invention.	Inventor.	Residence.	Date.	No.
Fruit-fork	I. C. Draper	New York, N. Y	Feb. 18, 1873	135, 975
Fruit-frame	S. T. Emerson	Seville, Ohio	July 28, 1868	80, 468
Fruit-frame	C. Stone	Ravenna, Ohio	Feb. 11, 1868	74, 252
Fruit-gatherer	G. Aldridge	Henderson City, Ky	July 11, 1871	116, 789
Fruit-gatherer	J. S. and A. M. Barry	Sheridan Township, Mich	Aug. 15, 1871	117, 966
Fruit-gatherer	J. Bowles	Augusta, Ga	Aug. 17, 1869	93, 667
Fruit-gatherer	A. W. Brinkerhoff and A. T. Barnes.	Upper Sandusky, Ohio	Sept. 2, 1862	36, 334
Fruit-gatherer	N. H. Bruce	Shirley, Mass	Aug. 2, 1861	32, 973
Fruit-gatherer	M. L. Byrn and G. Clark	New York, N. Y	Feb. 26, 1861	31, 527
Fruit-gatherer	I. T. Carpenter	Thompsontown, Pa	Aug. 28, 1866	57, 473
Fruit-gatherer	D. P. Chamberlin	Hudson, Mich	Jan. 3, 1860	26, 651
Fruit-gatherer	P. Collyer	Hunter, N. Y	June 12, 1847	5, 156
Fruit-gatherer	J. H. Colthar	Hamersville, Ohio	Oct. 29, 1872	132, 634
Fruit-gatherer	P. Conver	Farmington, Ill	Mar. 5, 1872	724, 334
Fruit-gatherer	J. C. Counts	Cross Roads, Ohio	May 14, 1867	64, 749
Fruit-gatherer	M. Darling	Blodgett's Mills, N. Y	July 16, 1867	66, 807
Fruit-gatherer	J. A. Daum	Canton, Ohio	May 15, 1866	54, 696
Fruit-gatherer	O. R. Dinsmore	Auburn, N. H	Sept. 4, 1866	57, 685
Fruit-gatherer	W. Doty	South Hartford, N. Y	Oct. 27, 1857	18, 503
Fruit-gatherer	J. Evans	Newark, N. J	Nov. 4, 1862	36, 834
Fruit-gatherer	T. Flagler	Grass Lake, Mich	Oct. 22, 1867	69, 980
Fruit-gatherer	L. Fleckenstine	Manor Township, Pa	July 30, 1867	67, 279
Fruit-gatherer	J. Franz	Selbysport, Md	Mar. 5, 1867	62, 623
Fruit-gatherer	C. W. Gage	Homer, N. Y	Apr. 7, 1868	76, 320
Fruit-gatherer	E. Gilliam	Allegheny City, Pa	May 8, 1866	54, 527
Fruit-gatherer	F. Goodwin	Astoria, N. Y	Nov. 10, 1857	18, 582
Fruit-gatherer	J. B. Haines	Millersville, Pa	Apr. 23, 1867	64, 096
Fruit-gatherer	R. S. Hall	Hamburgh, Mich	June 9, 1868	78, 662
Fruit-gatherer	O. C. Hamilton and H. McKinney.	Turtle Creek, Pa	July 14, 1868	79, 827
Fruit-gatherer	F. L. Hicks	New Haven, Conn	Feb. 13, 1872	123, 630
Fruit-gatherer	N. E. Hinds	Cooperstown, N. Y	Mar. 19, 1872	124, 743
Fruit-gatherer	J. C. Kearns	Lewistown, Pa	July 4, 1871	116, 718
Fruit-gatherer	H. Kelsey	Ottawa, Kans	May 6, 1873	138, 565
Fruit-gatherer	Z. S. Kelsey	Huntington, Ohio	Sept. 4, 1866	57, 731
Fruit-gatherer	J. Kisor	Nevada, Ohio	Feb. 18, 1873	136, 072
Fruit-gatherer	J. Lane	Chicago, Ill	Feb. 27, 1872	124, 000
Fruit-gatherer	C. F. Lang	Venedy, Ill	Oct. 20, 1868	83, 178
Fruit-gatherer	J. A. Little	Danville, Ind	Mar. 21, 1865	46, 916
Fruit-gatherer	M. Lounsbury	Seymour, Conn	Dec. 3, 1861	33, 843
Fruit-gatherer	V. H. Lyon	Plainfield, Ind	Mar. 3, 1868	75, 035
Fruit-gatherer	F. A. Maxfield	East Spring Hill, Pa	Apr. 19, 1864	42, 387
Fruit-gatherer	A. McWilliams	Washington, D. C	Mar. 13, 1844	3, 478
Fruit-gatherer	S. Mellinger, jr	Mount Pleasant, Pa	Aug. 7, 1866	56, 972
Fruit-gatherer	M. A. and D. F. Morton	Angola, N. Y	June 4, 1867	65, 416
Fruit-gatherer	I. L. Myers and I. Corl	Agricultural College, Pa	July 2, 1872	128, 648
Fruit-gatherer	H. W. Neal	Sidney, Ohio	Apr. 28, 1868	77, 400
Fruit-gatherer	E. H. Newcomb	New York, N. Y	May 29, 1866	55, 150
Fruit-gatherer	C. Ostrander	Poughkeepsie, N. Y	Mar. 14, 1871	112, 622
Fruit-gatherer	L. M. Parker	Shirley Village, Mass	Feb. 19, 1861	31, 475
Fruit-gatherer	O. B. Parker	Hopkinston, Mass	Nov. 17, 1863	40, 638
Fruit-gatherer	H. Perry	Manlius, N. Y	Mar. 23, 1869	88, 071
Fruit-gatherer	C. H. Phillips and D. F. Briggs	Providence, R. I	Mar. 5, 1872	124, 384
Fruit-gatherer	F. I. Rausschert	Buffalo, N. Y	Sept. 4, 1866	57, 766
Fruit-gatherer	C. R. Roberts and J. S. Hartzell	Addison, Pa	Dec. 4, 1866	60, 249
Fruit-gatherer	H. L. Scott	Plessis, N. Y	Dec. 18, 1866	60, 565
Fruit-gatherer	W. Sedgwick and J. Brooks	Poughkeepsie, N. Y	Oct. 3, 1846	4, 792
Fruit-gatherer	A. Selover	Brooklyn, Ohio	Mar. 28, 1865	47, 044
Fruit-gatherer	Y. W. Smith	Bristol, N. Y	Sept. 4, 1866	57, 785
Fruit-gatherer	W. H. Stone	Saint John's, Mich	Aug. 7, 1866	57, 008
Fruit-gatherer	G. Tanner	Freetown, N. Y	Sept. 15, 1868	82, 177
Fruit-gatherer	D. W. Thompson	Saint Joseph, Mo	Aug. 20, 1872	130, 769
Fruit-gatherer	B. Tukey	Fairfield, Me	June 24, 1862	35, 717
Fruit-gatherer	J. Vail	Newfield, N. Y	Nov. 15, 1864	45, 095
Fruit-gatherer	W. H. Wakeman	North Fairfield, Ohio	Oct. 15, 1872	132, 190
Fruit-gatherer	J. Waters	West Sutton, Mass	Mar. 12, 1872	124, 464
Fruit-gathering bag	N. B. Dixon and M. W. Sprague	Rochester, N. Y	Nov. 23, 1869	97, 061
Fruit, grain, &c., Mill for grinding	M. N. Whitely, J. Tassler, and O. S. Kelly.	Springfield, Ohio	Dec. 15, 1863	40, 987
Fruit-house	N. Cope	New Waterford, Ohio	Oct. 12, 1869	95, 657
Fruit-house	E. C. Roberts	Salem, Mich	Oct. 13, 1863	40, 282
Fruit-house ventilator	J. S. Houghton and C. B. Rees	Philadelphia, Pa	Oct. 20, 1868	83, 163
Fruit in barrels, &c., Machine for pressing	E. S. Holmes	Wilson, N. Y	Mar. 12, 1861	31, 689
Fruit, Instrument to loosen dried	J. F. Schmeltzer and J. M. Roberts.	Winona, Minn	Oct. 28, 1873	144, 147
Fruit-jar	H. U. Allen	North Bennington, Vt	Dec. 26, 1871	122, 143
Fruit-jar	J. C. Baker	Mechanicsburgh, Ohio	Aug. 14, 1860	29, 557
Fruit-jar	T. J. Bargis and J. C. Underwood.	Richmond, Ind	Dec. 7, 1869	97, 588
Fruit-jar	T. E. Batterson	Rochester, N. Y	May 5, 1868	77, 570
Fruit-jar	E. Bennett	Philadelphia, Pa	Feb. 6, 1866	52, 379
Fruit-jar	E. Bennett	Baltimore, Md	Nov. 16, 1869	96, 869
Fruit-jar	L. F. Betts	Philadelphia, Pa	July 23, 1872	129, 780
Fruit-jar	M. R. Bissell	Kalamazoo, Mich	Feb. 7, 1871	111, 607
Fruit-jar	J. Borden	Bridgeton, N. J	Dec. 18, 1866	60, 468
Fruit-jar	I. Buckman, jr	Williamsburgh, N. Y	Mar. 8, 1870	100, 496
Fruit-jar	G. W. Buffington	Mechanicsburgh, Ohio	Mar. 5, 1867	62, 603
Fruit-jar	N. C. Burnap	Argusville, N. Y	Dec. 3, 1872	133, 518
Fruit-jar	T. B. Carroll	Indianapolis, Ind	May 20, 1873	139, 115
Fruit-jar	T. A. Clark and H. C. Mascroft	Worcester, Mass	Sept. 20, 1870	107, 449
Fruit-jar	W. F. Corpe	Windsor Locks, Conn	Apr. 24, 1866	54, 117
Fruit-jar	E. Croft	Philadelphia, Pa	Sept. 20, 1870	107, 598
Fruit-jar	R. M. Dalbey	Springfield, Ohio	June 5, 1866	55, 248
Fruit-jar	W. H. Daniels	Bryan, Ohio	Feb. 15, 1870	99, 763
Fruit-jar	E. M. Davis	Pittsburgh, Pa	Oct. 27, 1868	83, 367
Fruit-jar	H. S. Draper	Rochester, N. Y	June 22, 1869	91, 726
Fruit-jar	T. Earle	Valley Falls, R. I	June 28, 1870	104, 839

Index of patents issued from the United States Patent Office from 1790 *to* 1873, *inclusive*—Continued

Invention.	Inventor.	Residence.	Date.	No.
Fruit-jar	A. E. Fife	Rochester, N. Y	Feb. 8, 1870	99, 5[illegible]0
Fruit-jar	J. Focer	Glassborough, N. J	June 12, 1866	55, 581
Fruit-jar	W. Galloway	Philadelphia, Pa	Feb. 8, 1870	99, 662
Fruit-jar	H. C. Gaskill	Mount Holly, N. J	Feb. 25, 1873	*136, 148
Fruit-jar	T. P. Gibbons	Philadelphia, Pa	June 21, 1870	104, 574
Fruit-jar	T. Gibson	Rochester, N. Y	July 28, 1868	80, 472
Fruit-jar	W. T. Gillinder and E. Bennett	Philadelphia, Pa	Aug. 8, 1865	49, 256
Fruit-jar	C. A. Gregory	Stratford, Conn	Aug. 17, 1869	93, 820
Fruit-jar	J. F. Griffen	New York, N. Y	Oct. 7, 1862	36, 612
Fruit-jar	J. Haines	West Middleburgh, Ohio	Mar. 1, 1870	100, 396
Fruit-jar	E. G. Haller	Carlisle, Pa	Feb. 25, 1873	136, 240
Fruit-jar	E. G. Haller	Carlisle, Pa	July 15, 1873	140, 916
Fruit-jar	W. L. Haller	Carlisle, Pa	Feb. 5, 1867	61, 827
Fruit-jar	W. L. Haller	Philadelphia, Pa	Jan. 4, 1870	98, 586
Fruit-jar	L. B. Harberger	Philadelphia, Pa	Nov. 30, 1869	97, 293
Fruit-jar	B. I. Harris	Harrisburgh, Pa	Aug. 7, 1866	56, 934
Fruit-jar	E. Harris	Boston, Mass	Feb. 9, 1864	41, 575
Fruit-jar	R. Hemingray	Cincinnati, Ohio	June 27, 1865	48, 399
Fruit-jar	G. W. Hempell	Millville, N. J	Mar. 18, 1873	136, 996
Fruit-jar	H. Hering	Philadelphia, Pa	Mar. 4, 1873	136, 513
Fruit-jar	W. Hicks	Trenton, N. J	Mar. 24, 1863	39, 962
Fruit-jar	P. M. Hinman	Rochester, N. Y	Jan. 19, 1869	85, 932
Fruit-jar	D. I. Holcomb	Henry County, Iowa	Dec. 14, 1869	97, 920
Fruit-jar	E. D. Holman	Buffalo, N. Y	July 2, 1867	66, 237
Fruit-jar	T. and H. H. Houghton	Philadelphia, Pa	Dec. 14, 1869	97, 922
Fruit-jar	H. Howson	Philadelphia, Pa	Sept. 3, 1872	131, 003
Fruit-jar	H. Howson	Philadelphia, Pa	Feb. 4, 1873	135, 430
Fruit-jar	D. Hughes, H. E. Shaffer, and W. S. Thompson.	Rochester, N. Y	Mar. 30, 1869	88, 296
Fruit-jar	W. Hunt	New York, N. Y	July 31, 1866	56, 755
Fruit-jar	C. G. Imlay	Philadelphia, Pa	May 23, 1865	47, 834
Fruit-jar	C. G. and W. L. Imlay	Philadelphia, Pa	Nov. 29, 1870	109, 625
Fruit-jar	C. G. and W. L. Imlay	Camden, N. J	July 16, 1872	129, 235
Fruit-jar	J. Johnson	New York, N. Y	May 16, 1865	47, 731
Fruit-jar	J. Johnson	New York, N. Y	Mar. 13, 1866	53, 151
Fruit-jar	J. Johnson	New York, N. Y	July 10, 1866	56, 226
Fruit-jar	A. T. Jones	Clinton, Wis	June 3, 1873	139, 580
Fruit-jar	A. T. Jones	Clinton, Wis	July 1, 1873	140, 508
Fruit-jar	J. M. W. Kitchen	New York, N. Y	Jan. 7, 1868	73, 187
Fruit-jar	J. M. W. Kitchen	Brooklyn, N. Y	Sept. 21, 1869	95, 025
Fruit-jar	J. Letchworth	Philadelphia, Pa	June 12, 1866	55, 512
Fruit-jar	W. A. Loder	Rochester, N. Y	July 14, 1868	79, 843
Fruit-jar	J. R. and N. E. Lupton	Stafford, Ohio	May 29, 1866	55, 128
Fruit-jar	W. W. Lyman	West Meriden, Conn	July 28, 1868	80, 296
Fruit-jar	W. W. Lyman	Meriden, Conn	Oct. 12, 1869	95, 819
Fruit-jar	J. B. Lyon	East Cleveland, Ohio	Aug. 4, 1868	80, 554
Fruit-jar	J. L. Mason	New York, N. Y	Jan. 19, 1869	86, 089
Fruit-jar	J. L. Mason	New York, N. Y	Jan. 19, 1869	86, 090
Fruit-jar	J. L. Mason	New York, N. Y	Feb. 23, 1869	87, 274
Fruit-jar	J. L. Mason	New York, N. Y	Mar. 1, 1870	100, 306
Fruit-jar	J. L. Mason	New York, N. Y	May 10, 1870	102, 913
Fruit-jar	J. L. Mason	New York, N. Y	Sept. 24, 1872	131, 695
Fruit-jar	J. L. Mason	New Brunswick, N. J	Apr. 1, 1873	137, 461
Fruit-jar	R. McCully	Philadelphia, Pa	Apr. 19, 1870	102, 024
Fruit-jar	A. D. McMaster	Rochester, N. Y	Mar. 17, 1868	75, 639
Fruit-jar	T. McSpedon and J. Steger	New York, N. Y	May 7, 1872	126, 561
Fruit-jar	H. B. Morton	Rochester, N. Y	June 2, 1868	78, 474
Fruit-jar	G. H. Myers	Philadelphia, Pa	Dec. 22, 1868	85, 236
Fruit-jar	T. G. Otterson	Millville, N. J	Oct. 28, 1862	36, 853
Fruit-jar	P. Pallissard	Saint Anne, Ill	Sept. 19, 1865	50, 027
Fruit-jar	S. J. Parker	Ithaca, N. Y	Feb. 9, 1864	41, 532
Fruit-jar	T. C. Purdy	Janesville, Wis	Apr. 5, 1870	101, 505
Fruit-jar	N. Raymer	New Sterling, N. C	Nov. 17, 1868	84, 214
Fruit-jar	G. A. Reynolds	Rochester, N. Y	Oct. 9, 1866	58, 674
Fruit-jar	F. Rohrbacher and F. Hormann.	Philadelphia, Pa	Aug. 18, 1868	81, 296
Fruit-jar	S. B. Rowley	Philadelphia, Pa	Dec. 14, 1869	97, 964
Fruit-jar	G. H. Russell	Dickinson, Pa	Apr. 3, 1866	53, 685
Fruit-jar	H. E. Shaffer	Rochester, N. Y	Aug. 31, 1869	94, 248
Fruit-jar	H. E. Shaffer	Rochester, N. Y	Nov. 2, 1869	96, 490
Fruit-jar	A. Sherwood	Auburn, N. Y	Sept. 26, 1865	50, 172
Fruit-jar	C. S. Siddons	Rochester, N. Y	Mar. 24, 1868	75, 993
Fruit-jar	C. F. Spencer	Rochester, N. Y	Aug. 27, 1867	68, 319
Fruit-jar	C. F. Spencer	Rochester, N. Y	Jan. 28, 1868	73, 846
Fruit-jar	C. F. Spencer	Rochester, N. Y	Aug. 18, 1868	81, 116
Fruit-jar	J. J. Squire	Windsor Locks, Conn	Mar. 7, 1865	46, 720
Fruit-jar	J. J. Squire	New London, Conn	Sept. 26, 1865	50, 181
Fruit-jar	D. Stone	Rochester, N. Y	Apr. 14, 1868	76, 846
Fruit-jar	H. Stone	Cleveland, Ohio	July 21, 1868	80, 234
Fruit-jar	L. C. Straub	Pittsburgh, Pa	July 18, 1871	117, 121
Fruit-jar	W. S. Thompson	Rochester, N. Y	Aug. 31, 1869	94, 452
Fruit-jar	W. S. Thompson	Rochester, N. Y	Apr. 12, 1870	101, 944
Fruit-jar	J. Weidig	Philadelphia, Pa	May 5, 1868	77, 555
Fruit-jar	E. B. Whitmore	Rochester, N. Y	Jan. 14, 1868	73, 271
Fruit-jar	B. B. Wilcox	New Haven, Conn	Mar. 26, 1867	63, 193
Fruit-jar	B. B. Wilcox	New Haven, Conn	June 21, 1870	104, 676
Fruit-jar	E. R. Williams	Rochester, N. Y	Sept. 1, 1868	81, 856
Fruit-jar	G. Williams	West Middlebury, Ohio	Nov. 13, 1866	59, 699
Fruit-jar	J. F. Winchell	Springfield, Ohio	Sept. 18, 1866	58, 185
Fruit-jar	L. Worcester and J. S. Brown	Lowell, Mass	May 5, 1868	77, 559
Fruit-jar and clamp	A. J. H. Hilton	Boston, Mass	Apr. 21, 1868	76, 915
Fruit-jar and preserving-vessel	A. Warner	Williamsburgh, N. Y	Mar. 24, 1868	76, 008
Fruit-jar cap	R. Gray and R. Hemingray	Covington, Ky	June 9, 1863	38, 820
Fruit-jar cap	N. P. Whittelsey	West Meriden, Conn	May 19, 1863	38, 617
Fruit-jar cap	H. Wright	Pittsburgh, Pa	Mar. 22, 1870	101, 074
Fruit-jar clamp	I. Buckman, jr	Williamsburgh, N. Y	Nov. 28, 1871	121, 232
Fruit-jar cover	T. B. and J. S. Atterbury	Pittsburgh, Pa	Nov. 20, 1860	30, 662
Fruit-jar cover	F. A. Bunnell	Syracuse, N. Y	Feb. 13, 1866	52, 525

Index of patents issued from the United States Patent Office from 1790 *to* 1873, *inclusive*—Continued.

Invention.	Inventor.	Residence.	Date.	No.
Fruit-jar cover	T. A. Clark and H. C. Mascroft.	Worcester, Mass	Sept. 6, 1870	107, 003
Fruit-jar cover	J. L. Mason	New York, N. Y	Apr. 1, 1873	137, 462
Fruit-jar cover	H. C. Nicholson	Mount Washington, Ohio	Apr. 15, 1862	34, 976
Fruit-jar cover	T. G. Otterson	Port Elizabeth, N. J	Dec. 19, 1865	51, 613
Fruit-jar cover	S. B. Rowley	Philadelphia, Pa	Dec. 31, 1872	134, 400
Fruit-jar cover	J. Siddons	Rochester, N. H	Oct. 6, 1868	82, 884
Fruit-jar covers, Screw-fastening for	S. B. Rowley	Philadelphia, Pa	Mar. 2, 1869	87, 515
Fruit-jar fastening	E. Boorse	Philadelphia, Pa	Nov. 16, 1869	96, 873
Fruit-jar fastening	W. Galloway	Philadelphia, Pa	Apr. 5, 1870	101, 606
Fruit-jar fastening	T. F. Woodward	Winslow, N. J	Apr. 12, 1870	101, 958
Fruit-jar fastening	T. F. Woodward	Winslow, N. J	Nov. 29, 1870	109, 703
Fruit-jar holder	C. G. Imlay	Philadelphia, Pa	Oct. 3, 1865	50, 301
Fruit-jar holder	W. P. Walter	Philadelphia, Pa	Apr. 26, 1870	102, 453
Fruit-jar lifter	S. R. Pinckney	New York, N. Y	Mar. 21, 1871	112, 955
Fruit-jar safety-case	J. Carter	Towanda, Pa	Dec. 20, 1870	110, 340
Fruit-jar screw-cap and ring	J. L. Mason	New York, N. Y	June 6, 1871	115, 754
Fruit-jar stopper	W. Chrysler	Lockport, N. Y	Nov. 21, 1865	51, 020
Fruit-jar stopper	C. R. Doane	Spotswood, N. J	Nov. 7, 1865	50, 806
Fruit-jar stopper	W. W. Lyman	West Meriden, Conn	Jan. 2, 1866	51, 844
Fruit-jar stopper	T. T. Prosser	Fond du Lac, Wis	May 19, 1863	38, 600
Fruit-jar stopper	J. M. Whitall	Philadelphia, Pa	Apr. 11, 1865	47, 238
Fruit-jars and other vessels, Process of manufacturing enameled.	H. Reed	Jersey City, N. J	Feb. 3, 1863	37, 591
Fruit-jars, &c., Closing	R. M. Dalbey	Springfield, Ohio	Jan. 12, 1864	41, 202
Fruit-jars, &c., Closing	J. Harbster	Reading, Pa	Aug. 4, 1863	39, 441
Fruit-jars, Closing	W. L. Imlay	Philadelphia, Pa	Jan. 28, 1868	73, 724
Fruit-jars, Closing	W. L. Imlay	Philadelphia, Pa	Mar. 10, 1868	75, 275
Fruit-jars, Closing	C. Newman	Pittsburgh, Pa	May 12, 1863	38, 536
Fruit-jars, Constructing	G. M. Ramsy	New York, N. Y	Apr. 17, 1866	54, 015
Fruit-jars for receiving written labels, Compound to be applied to.	G. W. Doty	Wooster, Ohio	Aug. 31, 1869	94, 405
Fruit-jars, Instrument for extracting air from	J. H. Parrish	Talladega, Ala	Mar. 19, 1872	124, 758
Fruit-jars, Instrument for tightening and loosening screw-caps of.	A. French	Philadelphia, Pa	Sept. 21, 1869	95, 102
Fruit-jars, Machine for grinding glass	M. Neckermann	Pittsburgh, Pa	Dec. 15, 1868	85, 022
Fruit-jars, Manufacture of locking-rings for closing	L. R. Boyd	New York, N. Y	Jan. 21, 1868	73, 571
Fruit-jars, Sealing	R. S. Manning	Trenton, N. J	May 20, 1873	139, 170
Fruit-jars, Sealing	E. Purdy	Ithaca, N. Y	July 30, 1867	67, 215
Fruit-jars, Securing the tops of	W. Galloway	Philadelphia, Pa	Nov. 22, 1870	109, 507
Fruit-jelly	C. Alden	Newburgh, N. Y	July 11, 1871	116, 912
Fruit-knife	H. Soggs	Columbus, Pa	Dec. 17, 1872	134, 106
Fruit-knife	G. W. Stevens	San Francisco, Cal	Apr. 1, 1873	137, 501
Fruit-knife and nut-picker	G. Mayland	Brooklyn, N. Y	June 13, 1865	48, 194
Fruit, Knife for seeding stone	S. S. Marques	Mason, Ill	Aug. 17, 1869	93, 896
Fruit-knife gage	C. R. Howe	Minneapolis, Min	Nov. 12, 1872	133, 031
Fruit lifter, Pressed	E. A. Raymond	Waterloo, Iowa	Mar. 21, 1871	112, 959
Fruit loosener, Dried	H. W. Holman	Waterloo, Iowa	Nov. 11, 1873	144, 542
Fruit loosener, Dried	C. Ragan	Waterloo, Iowa	Apr. 8, 1873	137, 622
Fruit-masher and lemon-squeezer	W. W. Armington	Lowell, Mass	Sept. 26, 1865	50, 199
Fruit-mill	E. Greenlee	Rundells, Pa	Feb. 2, 1869	86, 391
Fruit-mill	G. S. Hull	Washington, Iowa	Feb. 11, 1868	74, 365
Fruit-mill and press	H. A. Holderman	North Manchester, Ind	Feb. 11, 1868	74, 361
Fruit or grain dryer	C. W. Davis	Newark, N. J	Mar. 25, 1856	14, 494
Fruit-package	D. Snedeker	New York, N. Y	Aug. 5, 1873	141, 605
Fruit-packer	M. E. Lewis	Mansfield, Ohio	July 1, 1873	140, 511
Fruit-packing box	E. D. Lewelling	San Lorenzo, Cal	Oct. 10, 1871	119, 776
Fruit-packing implement	C. C. Roberts	Perrysburgh, Ohio	Apr. 2, 1872	125, 146
Fruit-packing press	T. J. Jones	Rochester, Mich	Feb. 18, 1868	74, 543
Fruit-paring machine	W. H. Williams	Canton, Ohio	Mar. 1, 1870	100, 548
Fruit-peeling machine	C. Lehman	Hartford, Conn	June 15, 1869	91, 238
Fruit-picker	J. H. Adams	Martinsville, Ind	June 22, 1869	91, 698
Fruit-picker	J. Bally	Deposit, N. Y	May 7, 1867	64, 472
Fruit-picker	A. T. Barnes	Tiffin, Ohio	Aug. 27, 1867	68, 030
Fruit-picker	W. Brown	Worcester, Mass	Feb. 5, 1867	61, 709
Fruit-picker	L. D. Cogswell	Lowell, Mass	Nov. 30, 1869	97, 275
Fruit-picker	P. and S. Conver	Yates City, Ill	Dec. 31, 1867	72, 804
Fruit-picker	E. Evans	Sparta, Ohio	Aug. 14, 1866	57, 106
Fruit-picker	R. Evans	Brant, N. Y	June 30, 1868	79, 455
Fruit-picker	E. W. Gurnee	Haverstraw, N. Y	July 30, 1867	67, 296
Fruit-picker	L. Hotchkiss	Torrington, Conn	July 14, 1868	79, 906
Fruit-picker	N. G. Hughes	Waynesburgh, Pa	Aug. 11, 1868	80, 864
Fruit-picker	J. A. Knight	Durham, Me	July 28, 1868	80, 288
Fruit-picker	J. H. and D. A. Lament	Troy, Pa	June 23, 1868	79, 292
Fruit-picker	C. M. and W. F. Lunt	Biddeford, Me	Sept. 4, 1866	57, 741
Fruit-picker	J. Melendy	Southbridge, Mass	June 27, 1854	11, 170
Fruit-picker	H. Mewes	Binghamton, N. Y	May 20, 1873	139, 015
Fruit-picker	O. B. Moore	Walled Lake, Mich	Feb. 4, 1868	73, 989
Fruit-picker	J. Neff, jr	Pultney, N. Y	Aug. 28, 1866	57, 553
Fruit-picker	T. Nutting	Georgiaville, R. I	June 15, 1869	91, 256
Fruit-picker	S. Page	McAllisterville, Pa	July 9, 1867	66, 517
Fruit-picker	G. J. Parham	Harrodsburgh, Ind	Aug. 3, 1869	93, 338
Fruit-picker	B. C. Phelps	Weathersfield, Conn	Apr. 18, 1865	47, 364
Fruit-picker	L. Richards and D. Lincoln	Orangeville, N. Y	Nov. 13, 1866	59, 655
Fruit-picker	G. S. Richardson	Stow, Ohio	Jan. 15, 1867	61, 257
Fruit-picker	O. P. Rogers	Roxbury, Mass	May 28, 1867	65, 279
Fruit-picker	J. F. Saiger and A. Davis	Shelby, Ohio	Dec. 18, 1866	60, 561
Fruit-picker	J. Schroy	Fortville, Ind	Oct. 13, 1868	83, 098
Fruit-picker	W. L. Shaw	Etna, Pa	Jan. 17, 1871	111, 059
Fruit-picker	D. F. Slane	Chillicothe, Ohio	Jan. 4, 1870	98, 639
Fruit-picker	A. S. Stevens	Attica, N. Y	June 2, 1868	78, 549
Fruit-picker	I. Tower	Milford, Mich	Sept. 24, 1867	69, 144
Fruit-picker	S. W. Valentine	Bristol, Conn	July 6, 1869	92, 235
Fruit-picker	E. Ward	Urbana, Ohio	Aug. 31, 1869	94, 457
Fruit-picker	J. T. Warren	Lee Roy, N. Y	Jan. 26, 1869	86, 333
Fruit-preserving can	W. H. Hammond	Syracuse, Mo	Jan. 14, 1873	134, 799
Fruit-press	G. Jenkins	Queensburgh, N. Y	May 26, 1863	38, 682
Fruit-press	J. Maurow	Sacramento, Cal	Dec. 1, 1863	40, 762
Fruit-press	O. Metcalf	Mitchell, Ind	Apr. 15, 1873	137, 787

Index of patents issued from the United States Patent Office from 1790 *to* 1873, *inclusive*—Continued.

Invention.	Inventor.	Residence.	Date.	No.
Fruit-press	D. T. Robinson	Boston, Mass	Nov. 27, 1866	60, 060
Fruit-press	G. R. Rust	Chester, Ill	July 15, 1862	35, 896
Fruit-press	G. J. Wells	South Vineland, N. J	Feb. 28, 1871	112, 304
Fruit-pulp, Apparatus for separating seed from	R. H. Mayo	Paris, Tex	Mar 29, 1870	101, 289
Fruit-renovator	J. E. Weaver	Lancaster, Pa	Aug. 6, 1872	130, 167
Fruit-seeder	S. A. Russell	Huntington, Conn	Nov. 19, 1867	71, 069
Fruit-stand	A. Greenman	East Kendall, N. Y	June 15, 1869	91, 330
Fruit stem cutter	A. Strong and J. A. Dadmun	South Boston, Mass	Aug. 20, 1867	68, 011
Fruit-stoning machine	R. McCormick	Greenville, Va	June 26, 1869	28, 881
Fruit-straining apparatus	M. E. Grigsby	Putnamville, Ind	July 12, 1870	105, 196
Fruit-transportation box	W. Shaw	Newfane, N. Y	Mar. 8, 870	100, 676
Fryer and boiler, Combined	G. Smith	Williamsburgh, N. Y	Oct. 28, 1873	144, 151
Frying-pan	G. H. Knight	Cincinnati, Ohio	July 9, 1867	66, 503
Frying-pan	S. B. Sexton	Baltimore, Md	Sept. 20, 1864	44, 346
Frying-pan	F. P. Warren	East Court Cosham, Great Britain.	Feb. 13, 1872	123, 753
Frying-pan	R. C. Whitehouse	Booth Bay, Me	Mar. 3, 1868	75, 095
Frying-pan and kettle	C. Avery	Ashtabula, Ohio	Aug. 30, 1864	43, 965
Frying-rack, Reversible	F. Bucknam	Portland, Me	Jan. 28, 1868	73, 692
Fuel	I. Bicknell	Cincinnati, Ohio	June 29, 1869	92, 003
Fuel, also improved stove, Mode of preparing	T. Joice	Camberwell, England	Nov. 12, 1838	1, 005
Fuel and for compressing conglomerate substances into masses, Machine for the manufacture of artificial.	A. Dietz	New York, N. Y	Nov. 30, 1869	97, 279
Fuel, and method of burning same, Artificial	A. Berney	Jersey City, N. J	May 27, 1873	139, 288
Fuel and other purposes, Composition of matter for	J. S. Lipps	Brooklyn, N. Y	Jan. 1, 1867	60, 755
Fuel, Apparatus for economizing	W. Levin	Saint Louis, Mo	Apr. 8, 1862	34, 891
Fuel, Apparatus for the combustion of	J. Bührer	Munich, Bavaria	Aug. 6, 1867	67, 492
Fuel, Apparatus for using mineral oils as	D. Dick	Meadville, Pa	Oct. 28, 1862	36, 769
Fuel, Artificial	J. E. Atwood	Trenton, N. J	Nov. 19, 1867	71, 119
Fuel, Artificial	R. B. Bayard	Philadelphia, Pa	July 4, 1865	48, 506
Fuel, Artificial	W. D. Beaumont	Mobile, Ala	June 12, 1855	13, 056
Fuel, Artificial	A. W. Buckland and A. M. Daniels.	Hartford, Conn	Apr. 21, 1868	76, 887
Fuel, Artificial	W. Budd and J. L. Husband	Philadelphia, Pa	July 12, 1864	43, 475
Fuel, Artificial	P. Binson	New York, N. Y	Apr. 5, 1864	42, 163
Fuel, Artificial	J. F. Bulkley	Brooklyn, N. Y	Apr. 10, 1866	53, 779
Fuel, Artificial	J. Cornell	Brooklyn, N. Y	Nov. 8, 1864	44, 994
Fuel, Artificial	R. Courtney	Albany, N. Y	Sept. 9, 1856	15, 688
Fuel, Artificial	H. Cutler	Springfield, Mass	Mar. 12, 1872	124, 553
Fuel, Artificial	A. M. Daniels	Hartford, Conn	Dec. 24, 1867	72, 460
Fuel, Artificial	A. De Lentilhac	Tamaqua, Pa	Feb. 6, 1866	52, 397
Fuel, Artificial	F. C. Epting	Schuylkill Haven, Pa	Jan. 2, 1872	122, 447
Fuel, Artificial	M. A. Febrey and T. Smith	Washington, D. C	Jan. 9, 1872	122, 590
Fuel, Artificial	T. M. Fell	Brooklyn, N. Y	Dec. 15, 1863	40, 920
Fuel, Artificial	R. Fish	Washington, D. C	Mar. 27, 1866	53, 431
Fuel, Artificial	G. R. Gladding	Providence, R. I	Apr. 18, 1865	47, 296
Fuel, Artificial	G. Gray	Temperanceville, Pa	Jan. 8, 1867	61, 006
Fuel, Artificial	W. Halsted	Trenton, N. J	Jan. 17, 1865	45, 922
Fuel, Artificial	W. and O. S. Halsted, jr	Washington, D. C., and Newark, N. J.	June 14, 1864	43, 112
Fuel, Artificial	T. Hooker and W. D. Beaumont	New Orleans, La	May 8, 1855	12, 847
Fuel, Artificial	S. D. Hovey	Chicago, Ill	Oct. 24, 1865	50, 588
Fuel, Artificial	S. D. Hovey	Chicago, Ill	Jan. 2, 1866	51, 833
Fuel, Artificial	J. H. Hubbard	Hartford, Conn	Dec. 1, 1863	40, 753
Fuel, Artificial	J. Kircher	New York, N. Y	Mar. 5, 1872	124, 363
Fuel, Artificial	C. Korff	New York, N. Y	July 4, 1865	48, 564
Fuel, Artificial	E. Loiseau and C. F. Reguin	Nashville, Tenn	June 16, 1868	78, 982
Fuel, Artificial	E. F. Loiseau and C. F. Reguin	Nashville, Tenn	June 21, 1870	104, 471
Fuel, Artificial	H. S. Lucas	Chester, Mass	Aug. 2, 1864	43, 695
Fuel, Artificial	M. Mann	Syracuse, N. Y	May 27, 1862	35, 427
Fuel, Artificial	P. M. McGill	Washington, D. C	Jan. 11, 1870	98, 786
Fuel, Artificial	P. M. McGill	Washington, D. C	Nov. 15, 1870	109, 231
Fuel, Artificial	E. Miannay	New York, N. Y	Dec. 22, 1857	18, 920
Fuel, Artificial	F. C. Payne	New York, N. Y	Jan. 17, 1865	45, 935
Fuel, Artificial	B. F. Penny	Rochester, N. Y	Feb. 7, 1871	111, 565
Fuel, Artificial	H. Realich	Chicago, Ill	June 27, 1865	48, 439
Fuel, Artificial	J. A. Roth	Philadelphia, Pa	Apr. 18, 1865	47, 337
Fuel, Artificial	H. Seymour	Dartmoor, Great Britain	Aug. 26, 1862	36, 305
Fuel, Artificial	E. Sharpe	Paris, France	May 15, 1866	54, 832
Fuel, Artificial	J. and R. C. Walrath	Chittenango, N. Y	Feb. 5, 1867	61, 778
Fuel, Artificial	C. D. Williams	Saint Paul, Minn	May 17, 1870	103, 266
Fuel, Artificial	C. D. Williams	Saint Paul, Minn	Feb. 20, 1872	123, 962
Fuel, Artificial	H. Wilverth	Caseyville, Ky	Dec. 20, 1859	26, 541
Fuel, Artificial	H. Wurtz	New York, N. Y	Feb. 8, 1870	99, 738
Fuel-box and washing-apparatus with settees, Combination of.	M. Bishop	Washington, D. C	May 13, 1862	35, 210
Fuel-burning furnace, Metal-heating	T. J. Leigh	London, England	May 12, 1868	77, 822
Fuel, Burning oil for	R. A. Chesebrough	New York, N. Y	Oct. 2, 1866	58, 379
Fuel, Burning wet	J. F. Manahan	Lowell, Mass	Jan. 8, 1856	14, 063
Fuel by the aid of calcarious cement, Preparing a composition for.	J. Allen	New York, N. Y	Mar. 17, 1838	643
Fuel, Combustion of	R. Coad	Kensington, England	May 8, 1849	6, 446
Fuel, Composition	E. Bellenger	Mohawk, N. Y	Dec. 1, 1857	18, 729
Fuel, Composition	F. N. Hopkins	Baltimore, Md	Nov. 23, 1869	97, 087
Fuel, Composition	J. Newton	Providence, R. I	Mar. 28, 1871	113, 083
Fuel, Composition	E. H. Richter	Taunton, Mass	June 27, 1871	116, 489
Fuel, Composition	H. Slatter	Covington, Ky	Jan. 15, 1867	61, 271
Fuel, Composition for	S. O. Doris	Philadelphia, Pa	Jan. 2, 1855	12, 159
Fuel, Composition for	W. Stickney and N. B. Chase	Lockport, N. Y	Jan. 21, 1868	73, 553
Fuel, Composition for	F. Stoker	Lyons, France	May 1, 1866	54, 477
Fuel, Consolidating coal-dust for	W. Footner	Chicago, Ill	May 19, 1868	77, 970
Fuel-dumper	J. Haines	West Middleburgh, Ohio	Sept. 25, 1866	58, 248
Fuel-economizing and smoke-consuming furnace	F. P. Dimpfel	New York, N. Y	May 9, 1839	1, 148
Fuel for glass-furnaces, Drying	J. Lee	Philadelphia, Pa	Jan. 14, 1811	......
Fuel for metallurgic purposes	C. E. Lester	New York, N. Y	Oct. 26, 1869	96, 243
Fuel for producing heat and light, Process and apparatus for using liquid.	T. S. Dickerson and R. M. Whipple.	Chicago, Ill	May 3, 1870	102, 662

Index of patents issued from the United States Patent Office from 1790 to 1873, inclusive—Continued.

Invention.	Inventor.	Residence.	Date.	No.
Fuel from anthracite coal-dust, Manufacture of	E. G. Markley and G. H. Bardwell	Sunbury and Philadelphia, Pa.	Apr. 30, 1867	64, 243
Fuel from brush-wood and twigs, Manufacture of granular.	R. Daniels	Woodstock, Vt	June 15, 1852	9, 015
Fuel from coal-dust and fresh-water peat, Preparing.	H. S. Lucas	Chester, Mass	Apr. 30, 1867	64, 233
Fuel from coal-dust, &c., Manufacture of	A. Kloezewski and D. Mindeleff.	Washington, D. C	July 2, 1872	128, 636
Fuel from coal dust or slack, Manufacture of	H. M. Smith	Chicago, Ill	Oct. 1, 1872	131, 794
Fuel from coal-dust, Solidified	E. D. Williams	Philadelphia, Pa	Mar. 6, 1860	27, 401
Fuel from coal-waste, Manufacture of	J. R. Hayes	Washington, D. C	Mar. 4, 1873	136, 375
Fuel from spent tan-bark	B. Irving	New York, N. Y	Oct. 13, 1868	83, 066
Fuel, Generating and burning vapor	A. J. Works and A. J. Daniels	New York, N. Y., and Washington, D. C.	Jan. 10, 1871	110, 946
Fuel into blocks or bricks, Machine for pressing	J. B. Collen	Philadelphia, Pa	Jan. 22, 1867	61, 318
Fuel, Machine for making peat	G. Weissenborn	New York, N. Y	Dec. 24, 1867	72, 573
Fuel, Machine for preparing peat for	J. H. Ames	New York, N. Y	July 2, 1867	66, 277
Fuel, Machine for preparing peat for	S. Marden	Newton, Mass	Dec. 5, 1865	51, 390
Fuel, Machine for preparing peat for	A. M. Sawyer	Athol, Mass	Oct. 1, 1867	69, 370
Fuel, Machinery for manufacturing artificial	M. L. Keen	Royer's Ford, Pa	Aug. 16, 1859	25, 124
Fuel, Manufacture of	D. E. Breinig	Brooklyn, N. Y	Apr. 16, 1872	125, 656
Fuel, Manufacture of	M. Rae	Upball, North Britain	Feb. 25, 1873	136, 263
Fuel, Manufacture of artificial	D. Barker	Northfleet, England	July 1, 1873	140, 452
Fuel, Manufacture of artificial	W. A. Bradley and J. Bigelow	Washington, D. C	Dec. 28, 1868	22, 410
Fuel, Manufacture of artificial	G. H Bronson	New York, N. Y	Sept. 15, 1868	82, 079
Fuel, Manufacture of artificial	J. F. Cranston, J. H. Banks, and J. M. Ingersoll.	Springfield, Mass	Apr. 12, 1870	101, 828
Fuel, Manufacture of artificial	H. G. Dayton	Maysville, Ky	Nov. 20, 1866	59, 791
Fuel, Manufacture of artificial	E. J. De Smedt	New York, N. Y	May 31, 1870	103, 580
Fuel, Manufacture of artificial	C. Du Lin	Mans, France	Mar. 9, 1869	87, 650
Fuel, Manufacture of artificial	S. D. Gilson	Oswego Falls, N. Y	Sept. 11, 1866	57, 890
Fuel, Manufacture of artificial	W. Holsted	Trenton, N. J	Sept. 4, 1866	57, 708
Fuel, Manufacture of artificial	C. Saffray	New York, N. Y	June 5, 1866	55, 367
Fuel, Manufacture of peat	J. Webster	Malden, Mass	Jan. 29, 1867	61, 586
Fuel, Mears of utilizing fine	E. Thayer	Worcester, Mass	Sept. 13, 1864	44, 262
Fuel, Non-fusible	W. Budd	Philadelphia, Pa	Dec. 1, 1863	40, 791
Fuel, Pine resin as	R. L. Wood	Philadelphia, Pa	Sept. 10, 1828	....
Fuel, Preparation of artificial	H. A. Archereau	Paris, France	Feb. 22, 1859	23, 005
Fuel, Preparation of peat for	A. Betteley	Boston, Mass	Nov. 21, 1865	51, 004
Fuel, Preparing artificial	D. E. Contardt	New York, N. Y	Mar. 8, 1864	41, 832
Fuel, Preparing artificial	W. H. and E. Cory	Notting Hill, England	Nov. 5, 1872	132, 751
Fuel, Preparing peat	E. H. Ashcroft and A. Betteley	Lynn and Boston, Mass	Sept. 6, 1864	44, 062
Fuel, Preparing peat and other substances for	L. S. Robbins	New York, N. Y	Dec. 4, 1866	60, 248
Fuel, Preparing peat for	A. Betteley	Boston, Mass	Aug. 8, 1865	49, 218
Fuel, Preparing peat for	L. W. Boynton	Hartford, Conn	Feb. 26, 1867	62, 469
Fuel, Preparing peat for	N. F. Potter	Providence, R. I	May 2, 1865	47, 564
Fuel, Preparing peat for	N. C. Sawyer	Boston, Mass	Sept. 20, 1864	44, 342
Fuel, Preparing peat for	H. Winter and F. W. Newton	Williamsburgh, N. Y., and South Orange, N. J.	Oct. 2, 1866	58, 528
Fuel-press	T. M. Mitchell	Philadelphia, Pa	June 7, 1870	103, 913
Fuel, &c., Process of forming blocks and slabs from green grasses for.	C. L. Fleishmann	Washington, D. C	July 2, 1872	128, 478
Fuel, Saving	Z. Phinney	Cairo, N. Y	Oct. 31, 1812	....
Fuel size-indicator	J. Gibson, jr	Albany, N. Y	Aug. 25, 1868	81, 490
Fullers' nap and card machine	J. Groff	Rapho, Pa	Feb. 10, 1830	....
Fulling-mill	W. and F. Bates	Sowerby Bridge, Halifax, England.	June 28, 1870	104, 819
Fulling-mill	G. M. Coppo	Paris, France	May 12, 1863	38, 469
Fulling-mill	C. P. Ladd	Bloomfield, N. J	June 4, 1872	127, 489
Fulling-mill	R. Hunt and J. H. Waite	Orange, Mass	July 29, 1862	36, 005
Fulling-mill	S. E. Coleman	West Haven, Vt	June 11, 1841	2, 126
Fulling-mill	E. Dams	Newark, N. J	Sept. 21, 1869	95, 094
Fulling-mill	A. Danielson	Otsego, N. Y	May 16, 1815	....
Fulling-mill	R. Eickemeyer	Yonkers, N. Y	Nov. 25, 1873	144, 841
Fulling-mill	E. Gessner	Aue, Saxony	June 9, 1868	78, 660
Fulling-mill	W. B. Lodge	Danbury, Conn	July 27, 1869	93, 100
Fulling-mill	T. J. Mayall	Roxbury, Mass	Nov. 24, 1863	40, 700
Fulling-mill	J. McKenzie and J. C. Millar	Troy, N. Y	July 29, 1862	36, 013
Fulling-mill	J. C. Millar	Springfield, Mass	Aug. 20, 1850	7, 579
Fulling-mill	S. M. Pike	Providence, R. I	May 28, 1872	127, 267
Fulling-mill	W. E. Underwood	Middlefield, Mass	Dec. 21, 1852	9, 492
Fulling-mill	J. H. Waite	Orange, Mass	Mar. 16, 1869	87, 806
Fulling-mill and power-loom	A. Atwood	Salem, N. Y	Oct. 1, 1830	....
Fulling-mill and washing-machine	J. Kennion	Dutchess County, N. Y	Mar. 26, 1806	....
Fulling-mill called the double-crank, Construction of the.	L. Osborne	....	July 12, 1804	....
Fulling-mill stock, Propelling	E. S. Norris	Monmouth, Me	July 7, 1835	....
Fulling-stock	W. B. Lodge	Danbury, Conn	June 25, 1867	66, 095
Fulling vegetable and other textures, Chemical process for.	J. Mercer	Oakenshaw, England	Aug. 19, 1851	8, 303
Fulminating-composition	G. Boldt	Chicago, Ill	July 10, 1866	56, 167
Fulminating-compound	J. Goldmark	Brooklyn, N. Y	Sept. 24, 1867	69, 206
Fulminating-compound	H. B. Stockwell	Brooklyn, N. Y	June 27, 1865	48, 469
Fulminating-powder for needle-guns	H. Büchner and F. Ebertz	New York, N. Y	Aug. 13, 1867	67, 714
Fulminating-powder, Manufacture of	J. S. Lipps	Brooklyn, N. Y	Sept. 6, 1864	44, 143
Fumiductor	W. A. Baron	New York, N. Y	Sept. 10, 1812	....
Fumigating-apparatus	K. P. Kidder	Burlington, Vt	Apr. 19, 1864	42, 380
Fumigating plants, Machine for	D. S. Brown	Surrey County, England	Sept. 24, 1850	7, 663
Fumigator	S. Andrews	Perth Amboy, N. J	Sept. 3, 1872	131, 075
Fumigator	A. R. Delinger	Gordonville, Pa	Nov. 25, 1873	144, 962
Fumigator	I. Hutchins, jr	Wellington, Me	Oct. 23, 1866	59, 025
Fumigator	R. Kerr	Boston, Mass	Apr. 21, 1868	76, 923
Fumigator	A. W. Todd	Chicago, Ill	Aug. 2, 1864	43, 723
Fumigator	S. Vanstone	Providence, R. I	Oct. 31, 1865	50, 752
Fumigator for destroying vermin	J. R. Hamilton	Portland, Oreg	Aug. 27, 1867	68, 188
Fumigator for smoking bees, &c	G. W. Hughes	Bloomington, Ill	Sept. 27, 1864	44, 427
Fumigator, Hospital, &c	T. J. Mayall	Boston, Mass	Oct. 14, 1873	143, 583
Funnel	J. I. Beaumont	Saint Paul, Minn	June 29, 1869	91, 899

Index of patents issued from the United States Patent Office from 1790 *to* 1873, *inclusive*—Continued.

Invention.	Inventor.	Residence.	Date.	No.
Funnel	F. Catlin	Watertown, Conn	June 25, 1867	66, 127
Funnel	W. B. Crowther	Philadelphia, Pa	July 1, 1873	140, 350
Funnel	L. P. Edwards	Hamlinton, Pa	Feb. 7, 1871	111, 525
Funnel	B. French	Rochester, N. Y	Nov. 7, 1871	120, 637
Funnel	J. Goudoin	New York, N. Y	Nov. 26, 1867	71, 378
Funnel	G. F. Herrick	Sharon, Minn	June 7, 1864	43, 075
Funnel	J. Q. Hill	Worcester, Mass	Aug. 22, 1865	49, 522
Funnel	C. Jones	Brooklyn, N. Y	Dec. 12, 1865	51, 515
Funnel	H. F. Lawrence	Vallejo, Cal	July 6, 1869	92, 321
Funnel	H. F. Lawrence	New York, N. Y	Aug. 6, 1872	130, 302
Funnel	W. E. Ledmun	Bridgeville, Del	Mar. 1, 1870	100, 420
Funnel	C. L. Lochman	Carlisle, Pa	Feb. 7, 1865	46, 255
Funnel	C. L. Lochman	Carlisle, Pa	Aug. 21, 1866	57, 347
Funnel	P. H. Niles	Boston, Mass	May 10, 1870	102, 855
Funnel	S. W. O'Laughlen	Baltimore, Md	June 4, 1872	127, 506
Funnel	N. L. Price	Lynchburgh, Va	Aug. 15, 1871	118, 051
Funnel	C. Ruf	New York, N. Y	May 10, 1870	102, 974
Funnel	C. Schneider	Washington, D. C	May 4, 1852	8, 928
Funnel	T. W. Slade	Manchester, Mass	Mar. 3, 1868	75, 065
Funnel	J. D. Smedley	Chicago, Ill	Oct. 6, 1863	40, 193
Funnel	A. Warth	Stapleton, N. Y	Nov. 15, 1870	109, 360
Funnel	C. G. Way	Boston, Mass	Jan. 28, 1873	135, 391
Funnel, Air-escape	J. I. Beaumont	Saint Paul, Minn	July 7, 1868	79, 799
Funnel, Air-escape	H. F. Hildebrand	Baltimore, Md	Sept. 14, 1869	94, 742
Funnel, Alarm	J. Sholl and J. Collins	Burlington, N. J	June 12, 1866	55, 544
Funnel and cock, Combined	A. G. Wilson	Chicago, Ill	Mar. 6, 1866	53, 074
Funnel and grater	E. A. Goodes	Philadelphia, Pa	Mar. 17, 1868	75, 541
Funnel, Beer	W. Golden	Flint, Mich	June 21, 1870	104, 575
Funnel, Bung	J. Buck	Baltimore, Md	Aug. 22, 1871	118, 337
Funnel can-filler	N. L. Brundage and B. Downing	Pittston, Pa	Mar. 15, 1870	100, 851
Funnel can-filler	T. Scantling	Evansville, Ind	Apr. 19, 1870	102, 050
Funnel, Filtering	J. H. Goodfellow	Troy, N. Y	May 25, 1869	90, 523
Funnel for canning fruit	S. B. Hertwig	Cincinnati, Ohio	Nov. 26, 1872	133, 445
Funnel for filling vessels with fluid	P. Apple	Philadelphia, Pa	July 13, 1808	....
Funnel, Indicating	J. D. Smedley	Chicago, Ill	Mar. 22, 1864	42, 024
Funnel-measure	C. W. Fuller	Earlville, Ill	Aug. 28, 1866	57, 494
Funnel-measure	G. A. Keene	Newburyport, Mass	Aug. 5, 1862	36, 095
Funnel, Measuring	C. Chinnock	Brooklyn, N. Y	Sept. 15, 1868	82, 087
Funnel, Measuring	J. W. Clarke	Kingston, Wis	July 31, 1866	56, 710
Funnel, Measuring	W. B. Cleves	Binghamton, N. Y	Jan. 22, 1867	61, 398
Funnel, Measuring	J. K. Cohic and D. H. Shartzer	West Hempfield Township, Pa.	June 3, 1873	139, 496
Funnel, Measuring	T. E. Cropper	Suffolk, Va	Mar. 25, 1873	137, 182
Funnel, Measuring	J. H. Elward	Mendota, Ill	May 14, 1867	64, 825
Funnel, Measuring	J. M. Estabrook	Worcester, Mass	Sept. 5, 1865	49, 739
Funnel, Measuring	R. F. Fisher and G. F. Waldron	Boston, Mass	July 4, 1871	116, 580
Funnel, Measuring	W. E. Keene	Lynn, Mass	Oct. 23, 1866	59, 034
Funnel, Measuring	J. Lloyd	Springfield, Ohio	Apr. 10, 1866	53, 841
Funnel, Measuring	G. B. Massey	New York, N. Y	May 25, 1869	90, 457
Funnel, Measuring	G. W. McCann	Springfield, Ohio	Mar. 12, 1867	62, 866
Funnel, Measuring	N. Otis	Cook County, Ill	May 10, 1864	42, 684
Funnel, Measuring	W. H. Rodgers	Brooklyn, N. Y	Sept. 24, 1867	69, 129
Funnel, Measuring	A. H. Whitney	Portland, Me	June 26, 1866	55, 941
Funnel, Measuring	H. J. Wolters	Salem, Mass	Mar. 9, 1869	87, 745
Funnel, Pendent measuring	G. A. Keene	Newburyport, Mass	June 17, 1862	35, 613
Funnel, Safety measuring	F. H. Smith	Burke, Vt	July 26, 1870	105, 857
Funnel, Sorghum	A. Bare	Mexico, Ohio	May 1, 1866	54, 279
Funnel with a vent	D. Clark	West Stockbridge, Mass	Apr. 9, 1810	....
Funnels, Manufacture of	L. R. Comstock	Baltimore, Md	June 17, 1873	139, 873
Funnels, Mode of connecting the sections of metallic	J. W. Cochran	New York, N. Y	Mar. 23, 1858	19, 685
Fur-band, Elastic	J. W. Gay	Brooklyn, N. Y	Jan. 5, 1864	41, 145
Fur-blower	W. B. Rotch	New Bedford, Mass	Mar. 9, 1824	....
Fur-blower	R. Wildman	Danbury, Conn	Mar. 6, 1866	53, 071
Fur-blowing machine	S. Whitman	New Albany, Ind	Mar. 22, 1834	....
Fur-box	H. Braunhold	New York, N. Y	June 20, 1871	116, 145
Fur-box	M. Fitzgibbons	New York, N. Y	Mar. 2, 1869	87, 334
Fur-box	S. C. Nichols	Buffalo, N. Y	Sept. 19, 1871	119, 168
Fur-cleaning machine	W. Woodworth	New York, N. Y	Apr. 19, 1831	....
Fur-coated fabrics, Manufacture of	H. Kellogg	Milford, Conn	Sept. 23, 1873	143, 163
Fur cuff	E. H. Hart	New York, N. Y	May 5, 1868	77, 483
Fur-cutting and cloth-shearing machine	B. Martial and D. Johnson	Concord, N. H	Dec. 7, 1816	....
Fur-cutting machine	D. Beard	Guildford, N. C	Apr. 30, 1816	....
Fur-cutting machine	M. Petre	Womelsdorf, Pa	Dec. 20, 1827	....
Fur-cutting machine, Reciprocating	C. M. Sampson	New York, N. Y	Feb. 5, 1835	....
Fur, Dyeing	P. Norden and H. Mischo	New York, N. Y	Sept. 27, 1870	107, 800
Fur, Dyeing and coloring	A. Müller	San Francisco, Cal	June 20, 1871	116, 211
Fur for the use of hatters, Machine for cutting	N. Young		May 14, 1802	....
Fur from hair, Machine for separating	C. and R. Lockwood and J. Arnold.	Fairfield County, Conn	Dec. 4, 1832	....
Fur from pelt, Machine for cutting	D. Williams	New York, N. Y	Oct. 25, 1832	....
Fur from peltry, Cutting	J. Witherle	Boston, Mass	May 28, 1811	....
Fur from peltry, Machine for cutting	E. Cutter	Walpole, N. H	Feb. 4, 1813	....
Fur from peltry, Machine for cutting	W. Jackson	Burlington, N. J	Feb. 2, 1815	....
Fur from peltry, Machine for cutting	E. Sprague	Danbury, Conn	May 9, 1810	....
Fur from peltry, Machine for pulling	W. Jackson	Burlington, N. J	Feb. 2, 1815	....
Fur from skins and pelts, Machine for cutting	M. Clark	New York, N. Y	Apr. 21, 1810	....
Fur from skins, Machine for cutting	C. C. K. Beach	Portland, Me	Nov. 10, 1827	....
Fur from skins, Machine for cutting	J. Hubbard	Jamestown, N. C	Nov. 4, 1826	....
Fur, Machine and process for blowing or cleaning	I. W. Cochran	New York, N. Y	Nov. 4, 1842	2, 842
Fur, &c., Manufacture of vegetable	P. Baumgras	Syracuse, N. Y	Apr. 12, 1864	42, 326
Fur muffler	G. W. and W. F. Parker	Kalamazoo, Mich	Apr. 21, 1868	77, 085
Fur, Process of treating	A. C. Brush	Darien, Conn	Jan. 23, 1866	52, 133
Fur-set box	J. Crane	Bloomfield, N. J	Feb. 6, 1872	123, 333
Fur-set box	J. Crane	Bloomfield, N. J	Feb. 6, 1872	123, 334
Fur-set box	C. W. Frazer	Brooklyn, N. Y	Dec. 27, 1870	110, 615
Fur-set box	B. F. Porter	Nashau, N. H	Nov. 14, 1871	120, 897
Fur-set box	E. H. Smith	Bergen Heights, N. J	Sept. 19, 1871	119, 095
Fur-skins, Machine for extracting hair from	S. Graham	Farmington, Me	Mar. 27, 1835	....

Index of patents issued from the United States Patent Office from 1790 *to* 1873, *inclusive*—Continued.

Invention.	Inventor.	Residence.	Date.	No.
Fur-substitute for dressed fur-skin	A. Belden	Hudson, N. Y	Apr. 22, 1835	....
Fur tassel	J. Schmid	Boston, Mass	Mar. 12, 1872	124, 633
Fur, wool, &c., Mode of cleaning	D. Vail	Philadelphia, Pa	Feb. 21, 1821	....
Furs, clothes, &c., from injury by moths, &c., Preserving.	J. Crane	Bloomfield, N. J	July 6, 1869	92, 274
Furnace:				

See Air-furnace.
Air-heating furnace.
Annealing-furnace.
Bagasse-furnace.
Barrel-pitching furnace.
Blast-furnace.
Blast-heating furnace.
Boiler-furnace.
Boiler and stove furnace.
Boiling-furnace.
Bone-black furnace.
Bone-black reburning and ore-reducing furnace.
Broiling-furnace.
Charcoal-furnace.
Chimney-furnace.
Cider-clarifying furnace.
Coal-furnace.
Coking-furnace.
Cooking-furnace.
Crucible and metal-heating furnace.
Cupola furnace.
Decarbonizing-furnace.
Dentist's furnace.
Deodorizing-furnace.
Deoxidizing-furnace.
Desulphurizing-furnace.
Drying-furnace.
Evaporating-pan furnace.
Evaporator-furnace.
Fire-proof furnace.
Fixing, puddling, and boiler furnace.
Forge-furnace.
Friction-furnace.
Fuel-burning furnace.
Fuel-economizing and smoke-consuming furnace.
Gas-furnace.
Gas-burning furnace.
Gas-generating furnace.
Gas-producing furnace.
Gas-retort furnace.
Gypsum-calcining furnace.
Glass-furnace.
Glass-melting furnace.
Hatter's furnace.
Hatter's iron-furnace.
Heating-furnace.
Heating and puddling furnace.
Hot-air furnace.
House-warming furnace.
Iron furnace.
Iron and steel furnace.
Iron boiling and puddling furnace.
Iron-finishing furnace.
Iron-galvanizing furnace.
Iron-heating furnace.
Iron heating and puddling furnace.
Iron-manufacturing furnace.
Iron-melting furnace.
Iron melting and refining furnace.
Iron-plating furnace.
Iron-remelting furnace.
Iron-smelting furnace.
Iron-working furnace.
Kiln-furnace.
Lead-smelting furnace.
Lime-burning furnace.
Lime-burning and ore-smelting furnace.
Locomotive-furnace.
Locomotive-boiler furnace.
Melting-furnace.
Melting and smelting furnace.
Metal-heating furnace.
Metal-melting furnace.
Metal-plate-heating furnace.
Metal-treating furnace.
Metallurgic furnace.
Metallurgic gas-furnace.
Ore-furnace.
Ore-calcining furnace.
Ore-oxidizing furnace.
Ore-reducing furnace.
Ore-roasting furnace.
Ore roasting and treating furnace.
Ore-smelting furnace.
Ore-treating furnace.
Oven-heating furnace.
Pipe-welding furnace.
Portable furnace.
Puddling-furnace.
Pyrite-burning furnace.
Quicksilver-furnace.
Radiating-furnace.
Refuse-burning furnace.
Regenerator-furnace.
Retort-furnace.
Reverberatory furnace.
Reverberatory and cupola furnace.
Roasting-furnace.
Salt-furnace.
Salt-boiling furnace.
Sand-drying furnace.
Saw-dust-burning furnace.
Saw-heating furnace.
Scythe-heating furnace.
Shaft-furnace.
Small-coal furnace.
Small coal or culm furnace.
Smelting-furnace.
Smoke-burning furnace.
Smoke-condensing furnace.
Smoke-consuming furnace.
Soldering-furnace.
Spider-furnace.
Steam-furnace.
Steamboat-furnace.
Steam-engine furnace.
Steam-generating furnace.
Steam-generator furnace.
Steel heating and tempering furnace.
Steel, iron, &c., melting furnace.
Steel-melting furnace.
Steel-tempering furnace.
Stove-furnace.
Sugar-juice-evaporating-furnace.
Summer-furnace.
Tailor's furnace.
Tank-furnace.
Tinman's furnace.
Tinner's furnace.
Tinsmith's furnace.
Tire-heating furnace.
Tire sprinkling and removing furnace.
Tube-heating furnace.
Zinc-furnace.

Index of patents issued from the United States Patent Office from 1790 to 1873, inclusive—Continued.

Invention.	Inventor.	Residence.	Date.	No.
Furnace	J. Albee	Boston, Mass	Oct. 4, 1864	44, 497
Furnace	T. and J. Aldridge	Hudson City, N. J	Oct. 27, 1857	18, 491
Furnace	M. Battel	Albany, N. Y	Sept. 18, 1860	30, 040
Furnace	C. F. Baxter	Boston, Mass	June 19, 1860	28, 730
Furnace	B. F. Blood	Port Jackson, N. Y	Dec. 29, 1857	18, 951
Furnace	J. Case and I. Soules	Amsterdam, N. Y	Oct. 20, 1857	18, 432
Furnace	E. B. Cherevy	New York, N. Y	June 28, 1859	24, 599
Furnace	I. Clute and J. Seabury	Cohoes, N. Y	Dec. 5, 1842	2, 875
Furnace	C. F. Cory	Lebanon, Ill	Nov. 27, 1860	30, 724
Furnace	N. F. B. De Chodzko	Paris, France	June 16, 1863	38, 885
Furnace	J. H. Duhme	Cincinnati, Ohio	June 28, 1859	24, 547
Furnace	B. W. Dunklee	Boston, Mass	Oct. 11, 1859	25, 729
Furnace	W. Duryea and W. Ennis	Glen Cove, N. Y., and Hudson, N. J.	Mar. 24, 1868	75, 881
Furnace	J. Ekin	Xenia, Ohio	Mar. 18, 1862	34, 714
Furnace	W. Ennis	New York, N. Y	May 29, 1866	55, 079
Furnace	W. Ennis	Hudson, N. J	Jan. 1, 1867	60, 867
Furnace	B. D. Evans	Mount Vernon, Ohio	Mar. 20, 1860	27, 541
Furnace	S. E. Foster	Fitchburgh, Mass	Oct. 25, 1864	44, 835
Furnace	D. Hargar	Des Moines, Iowa	Nov. 5, 1867	70, 557
Furnace	M. L. Horton	Worcester, Mass	Feb. 18, 1868	74, 536
Furnace	J. B. Hoyt	Stamford, Conn	Feb. 8, 1870	99, 570
Furnace	J. Leeds	Philadelphia, Pa	May 30, 1865	47, 963
Furnace	D. G. Littlefield	Albany, N. Y	Oct. 9, 1860	30, 333
Furnace	P. Low	Cincinnati, Ohio	Apr. 3, 1860	27, 724
Furnace	M. F. Magliocco	Philadelphia, Pa	Oct. 18, 1864	44, 772
Furnace	J. McCracken	Bloomfield, N. J	Apr. 13, 1858	19, 942
Furnace	J. T. McDougall	San Francisco, Cal	Dec. 4, 1866	60, 219
Furnace	J. H. Mearns	Philadelphia, Pa	Oct. 16, 1866	58, 863
Furnace	J. R. Morris	Houston, Tex	Apr. 12, 1870	101, 965
Furnace	B. Palazot	Bordeaux, France	July 7, 1863	39, 168
Furnace	S. Pierce	Troy, N. Y	Jan. 18, 1859	22, 670
Furnace	R. B. Pullan	Cincinnati, Ohio	Jan. 24, 1860	26, 920
Furnace	G. E. Reynolds	Philadelphia, Pa	May 1, 1866	54, 409
Furnace	C. Schacher and L. Feret	Paris, France	Dec. 16, 1873	145, 689
Furnace	G. Schreyer	Columbus, Ohio	Mar. 10, 1868	75, 300
Furnace	T. P. Scripter	Des Moines, Iowa	May 23, 1871	115, 111
Furnace	J. Sholl	Burlington, N. J	Aug. 6, 1867	67, 457
Furnace	J. W. Smith	Chicago, Ill	Jan. 30, 1872	123, 300
Furnace	L. Solomon	New York, N. Y	July 19, 1859	24, 828
Furnace	T. S. Speakman	Camden, N. J	Feb. 22, 1870	100, 205
Furnace	J. L. Stevens	Kennington, England	Nov. 21, 1854	11, 975
Furnace	W. Stevens	Bloomington, Ill	Oct. 1, 1867	69, 505
Furnace	R. R. Taylor	Reading, Pa	May 8, 1860	28, 214
Furnace	W. D. Thomas	Morristown, Vt	Sept. 4, 1860	29, 925
Furnace	C. W. Trotter	Rochester, N. Y	June 18, 1867	65, 967
Furnace	J. I. Vinton and E. John	Ironton, Ohio	Oct. 4, 1859	25, 694
Furnace	B. H. Washington	Hannibal, Md	July 6, 1858	20, 836
Furnace	R. Wells	Baltimore, Md	Sept. 30, 1856	15, 832
Furnace	D. T. Woodrow	Cincinnati, Ohio	Aug. 28, 1860	29, 840
Furnace air and boiler-water heater	Z. C. Robbins	Saint Louis, Mo	Oct. 16, 1844	3, 796
Furnace and bake-oven	C. E. Russell	Philadelphia, Pa	Feb. 13, 1835	....
Furnace and boiler	H. McClure	Terre Haute, Ind	Oct. 2, 1866	58, 552
Furnace and boiler combined	A. Harrison	New Haven, Conn	Mar. 2, 1836	....
Furnace and cooking-range	J. Brown and L. Bridges	Chicago, Ill	Oct. 2, 1860	30, 203
Furnace and cupola	Z. Ellis	Philadelphia, Pa	Nov. 15, 1864	45, 030
Furnace and furnace-bar	E. R. Austin	Norwalk, Conn	Oct. 31, 1871	120, 361
Furnace and furnace-door	B. R. Hawley	Normal, Ill	Mar. 4, 1873	136, 509
Furnace and range, Combined	E. Edwards	South Boston, Mass	Sept. 13, 1870	107, 347
Furnace and steam-boiler, Combined	L. Disbrow	New York, N. Y	May 6, 1833	....
Furnace and steam-generator combined	O. W. Ketchum	Toronto, Canada	June 10, 1873	139, 795
Furnace and steam-generator, Combined	F. Morth	Vienna, Austria	Aug. 10, 1869	93, 468
Furnace and steam-generator for heating, &c	B. R. Hawley	Normal, Ill	Mar. 4, 1873	136, 510
Furnace and stove	C. B. Sawyer	Fitchburgh, Mass	June 7, 1859	24, 332
Furnace and stove draft-regulator	F. E. Chatard, jr	Baltimore, Md	Dec. 26, 1871	122, 105
Furnace and tea-kettle combined	C. L. Taylor and G. C. Setchell	Greenville, Conn	Aug. 30, 1870	106, 889
Furnace and ventilator	C. B. Sawyer	Fitchburgh, Mass	Aug. 30, 1859	25, 279
Furnace baker	G. Chilson	Boston, Mass	June 10, 1840	1, 627
Furnace-blower tube	B. W. Reynolds	Evansville, Ind	July 18, 1871	117, 206
Furnace-boiler	J. Seabury	New York, N. Y	July 31, 1847	5, 213
Furnace, Combined	L. Disbrow	New York, N. Y	May 6, 1833	....
Furnace dead-plate	A. C. Rand	Aurora, Ill	Mar. 26, 1872	125, 080
Furnace-door	D. Auld, jr	Glasgow, North Britain	Dec. 16, 1873	145, 608
Furnace-door	J. M. Clark	Louisville, Ky	Dec. 2, 1873	145, 150
Furnace-door	W. W. Crane	Philadelphia, Pa	Sept. 22, 1868	82, 294
Furnace-door	L. F. Johnson	Buffalo, N. Y	Oct. 6, 1868	82, 721
Furnace-door	J. Philips and D. Keeley	Phœnixville, Pa	Sept. 28, 1869	95, 263
Furnace-door	N. L. Sibley and B. Shiverick	Waltham and Weston, Mass.	July 25, 1865	49, 041
Furnace-door	J. Wickersham	Baltimore, Md	Dec. 19, 1865	51, 638
Furnace-door fastening	P. E. Shear	Saugerties, N. Y	Apr. 21, 1868	77, 112
Furnace-door for boilers	J. Penketh	Chicago, Ill	Feb. 14, 1865	46, 423
Furnace-door frame	P. A. Sabbaton	Albany, N. Y	June 21, 1859	24, 493
Furnace-door frame	D. H. Young	Manchester, N. H	Oct. 5, 1869	95, 550
Furnace-door latch	J. L. Reilly	Chester, Pa	Mar. 3, 1868	75, 053
Furnace-door-opening device	H. Fessler and H. Maxell	Canton, Ohio	Oct. 16, 1866	58, 799
Furnace door-regulator	O. A. Dodge	Burlington, Vt	Feb. 20, 1866	52, 695
Furnace-doors, Constructing	J. Watson	Buffalo, N. Y	Nov. 26, 1867	71, 433
Furnace draft-regulator	A. H. Mershon	Philadelphia, Pa	May 27, 1866	55, 138
Furnace draft-regulator	J. F. Neall and W. Myers	Philadelphia, Pa	Apr. 21, 1868	77, 081
Furnace draft-regulator, Boiler	W. E. Pearson	East Boston, Mass	May 16, 1871	114, 965
Furnace draft-regulator, Steam-boiler	T. H. Moore	Alexandria, Va	Nov. 28, 1871	121, 398
Furnace for burning as fuel tar, oil, &c	J. C. Love	Pittsburgh, Pa	June 30, 1863	39, 057
Furnace for burning coal-dust, small fuel, &c	J. Tissot	Paris, France	Jan. 28, 1873	135, 298
Furnace for burning edgings, saw-dust, &c	I. O. Smith	Muskegon, Mich	Feb. 15, 1870	99, 793
Furnace for burning pulverized fuel	T. R. Crampton	London, Great Britain	Feb. 7, 1871	111, 616
Furnace for burning pulverized fuel under steam-boilers, evaporators, &c	J. D. Whelpley and J. J. Storer	Boston, Mass	Feb. 7, 1871	111, 705

Index of patents issued from the United States Patent Office from 1790 to 1873, inclusive—Continued.

Invention.	Inventor.	Residence.	Date.	No.
Furnace for burning shavings	L. Crandell	New York, N. Y	Dec. 13, 1870	110, 118
Furnace for burning wet fuel	M. Thompson	Henrico County, Va	Apr. 10, 1855	12, 678
Furnace for heating metallic bars, tubes, &c	S. P. M. Tasker	Philadelphia, Pa	Feb. 11, 1873	135, 674
Furnace for liberating and using the gaseous products of coal.	T. T. Prosser	Chicago, Ill	Apr. 20, 1869	89, 239
Furnace for producing hydrocarbon and treating ores.	J. R. Morris	Houston, Tex	Mar. 28, 1871	113, 191
Furnace-grate bar	J. Cone and W. K. Kelly	Bristol, Pa	Feb. 8, 1870	99, 642
Furnace-grate bar	C. Hoffman	New York, N. Y	Jan. 23, 1872	123, 018
Furnace-heater	G. W. Wilson	Chelsea, Mass	Sept. 13, 1864	44, 246
Furnace, Heating	M. F. Magliocco	Philadelphia, Pa	Oct. 8, 1867	69, 568
Furnace, Heating	W. Von Faber du Faur	Würtemberg	Apr. 16, 1842	2, 558
Furnace-lining	B. A. Haycock	Richland, Iowa	Apr. 7, 1868	76, 442
Furnace-lining, retorts, &c., Plastic compound for	S. J. Payne	Charlton, England	May 6, 1873	138, 68
Furnaceman's clamp	A. P. Briggs	Taunton, Mass	Mar. 6, 1866	52, 95
Furnace-mold clamp	E. Card	Providence, R. I	Apr. 2, 1867	63, 46
Furnace or oven and tempering-die combined	H. Disston	Philadelphia, Pa	Sept. 20, 1870	107, 599
Furnace or stove, Economical	J. Poudrell	New York, N. Y	Apr. 6, 1808	....
Furnace-pot	J. Ballou	Boston, Mass	June 6, 1871	115, 677
Furnace-raking-attachment	J. Rice	Owego, N. Y	July 12, 1870	105, 249
Furnace-register	E. Barrows	New York, N. Y	May 28, 1846	4, 538
Furnace-regulator for hot-water apparatus	T. T. Tasker	Philadelphia, Pa	Dec. 5, 1854	12, 039
Furnace-shield	E. S. Collins	United States Navy	Jan. 22, 1867	61, 401
Furnace smoke-stack	B. F. Smith	New Orleans, La	Sept. 12, 1871	118, 883
Furnace, stove, &c	S. Rolph	Wales, N. Y	Nov. 8, 1826	....
Furnace with reflectors	J. Hurd, jr	Boston, Mass	Nov. 10, 1829	....
Furnaces, Apparatus for burning liquid fuel to heat	W. C. Wren	Brooklyn, N. Y	Oct. 1, 1872	131, 804
Furnaces, Apparatus for distributing and feeding powdered fuel to.	T. R. Crampton	London, Great Britain	Feb. 7, 1871	111, 615
Furnaces, Apparatus for increasing the draft of	I. B. Martin	Wilmington, N. C	Feb. 1, 1859	22, 814
Furnaces, Apparatus for supplying gases to	S. C. Salisbury	New York, N. Y	Aug. 27, 1867	68, 119
Furnaces, Apparatus for supplying saw-dust to	H. Doty	Cardington, Ohio	Nov. 22, 1859	26, 169
Furnaces, Apparatus for utilizing heat from	S. Bennett	Newcastle, Pa	Oct. 24, 1865	50, 55
Furnaces, &c., Blasts by steam-jets for	P. W. Mackenzie	Blauveltville, N. Y	Apr. 25, 1871	114, 16
Furnaces, Bridge-wall for	S. C. Sturtevant	Cleveland, Ohio	Nov. 11, 1873	144, 42
Furnaces, Checking draft in	J. C. Bagnall	Saint Louis, Mo	June 11, 1867	65, 6
Furnaces, crucibles, &c., Composition for lining	A. E. Bates	Monroe, Mich	Dec. 9, 1873	145, 3
Furnaces, Door for reverberatory and other	C. H. Burgess, 2d	Sandwich, Mass	Feb. 18, 1862	34, 4
Furnaces, Feeding	S. M. Fales	Baltimore, Md	May 16, 1846	4, 5
Furnaces, Feeding fuel to	G. F. Deacon	Liverpool, England	June 20, 1871	116, 1
Furnaces, Feeding fuel to	H. Delano	Syracuse, N. Y	Mar. 20, 1855	12, 5
Furnaces, Feeding fuel to	J. Henington	Richmond, Ind	Apr. 14, 1857	17, 0
Furnaces, Feeding fuel to	C. P. Leavitt	Newark, N. J	Aug. 12, 1873	141, 6
Furnaces, Feeding fuel to	J. G. McCormick	Louisville, Ky	July 14, 1868	79, 91
Furnaces, Feeding fuel to	J. Y. Smith	Pittsburgh, Pa	Oct. 17, 1871	120, 00
Furnaces, Feeding fuel to	J. Y. Smith	Pittsburgh, Pa	Nov. 7, 1871	120, 68
Furnaces, Feeding fuel to	A. M. Sprague	Mobile, Ala	Apr. 15, 1856	14, 6
Furnaces, Feeding fuel to	J. D. Whelpley and J. J. Storer	Boston, Mass	Nov. 13, 1866	59, 69
Furnaces, Feeding pulverized fuel to	T. R. Crampton	London, Great Britain	Feb. 7, 1871	111, 61
Furnaces, Feeding pulverized fuel to metallurgical and other.	J. Y. Smith	Pittsburgh, Pa	Oct. 17, 1871	120, 00
Furnaces, Feeding saw-dust to	A. J. Emlaw	Grand Haven, Mich	Oct. 17, 1870	120, 0
Furnaces, Feeding saw-dust to	J. A. McClelland	Vernon, Ind	Aug. 4, 1868	80, 6
Furnaces, Feeding saw-dust to	J. P. Wigal	Henderson, Ky	May 8, 1860	28, 2
Furnaces, Fire-chamber for	S. Smith	Worcester, Mass	Apr. 16, 1867	63, 9
Furnaces, Machine for feeding fuel to	A. Rogers and E. Tarrant	Muskegon, Mich	Mar. 14, 1871	112, 6
Furnaces, Means for directing blast in	B. H. Washington	Hannibal, Mo	Nov. 7, 1854	11, 9
Furnaces, Method of promoting combustion in	C. Fisher	Trenton, N. J	Sept. 25, 1866	58, 25
Furnaces, Mode of making	G. W. Robinson	Attleborough, Mass	Aug. 19, 1822	....
Furnaces, Mode of producing hot-blast in	T. McDowell	Light Street, Pa	Mar. 10, 1868	75, 44
Furnaces of steam-engines, &c., Supplying air to	P. Robinson	Waterloo, N. Y	Apr. 20, 1844	3, 555
Furnaces, Protecting hearths of	H. W. Ellicott	Baltimore, Md	Jan. 3, 1871	110, 639
Furnaces, Protecting heated parts of	E. G. Scovil	St. John's, New Brunswick	Oct. 22, 1867	70, 031
Furnaces, Removing slag from	J. Thomas	Hokendauqua, Pa	Feb. 14, 1871	111 789
Furnaces, Restoring waste heat from	W. Gorman and J. Paton	Glasgow, Great Britain	July 16, 1872	129, 403
Furnaces, safes, &c., Lining for	A. C. Hamlin	Bangor, Me	Apr. 23, 1872	125, 951
Furnaces, Saw-dust feeder for	M. Garland	West Eau Claire, Wis	May 28, 1872	127, 335
Furnaces, Saw-dust feeder for	S. Sykes	Chippewa Falls, Wis	June 15, 1869	91, 384
Furnaces, Self-regulating hot-blast for	C. Schinz	Camden, N. J	Dec. 4, 1855	13, 887
Furnaces, Straw-feeding attachment for	D. Morey	Watsonville, Cal	Feb. 11, 1873	135, 659
Furnaces, Straw-feeding attachment to	D. Morey	Watsonville, Cal	May 20, 1873	139, 075
Furnaces, Treating cinder for fixing	A. McDonald	Allegheny, Pa	Mar. 31, 1868	76, 223
Furnaces, Utilizing slag of	W. H. Smith	Philadelphia, Pa	Dec. 7, 1852	9, 459
Furnaces, Ventilating	H. Ruttan	Coburg, Canada	May 20, 1851	8, 109
Furnaces with hot-air, Apparatus for supplying	C. Fletcher	Cincinnati, Ohio	Aug. 16, 1859	25, 159
Furnaces with hot-air, Supplying	C. Fletcher	Cincinnati, Ohio	June 12, 1855	13, 031
Furnaces with hydrocarbon liquids, Heating	J. K. Caldwell	Philadelphia, Pa	Mar. 19, 1872	124, 791
Furnaces with pulverized metal, Apparatus for supplying.	E. Schmitz	New York, N. Y	Jan. 23, 1855	12, 286
Furniture and other purposes, Composition for	J. J. Wiggin	Cincinnati, Ohio	Dec. 4, 1866	60, 309
Furniture-attachments	J. B. Haupt	Philadelphia, Pa	July 16, 1872	129, 556
Furniture, Barber's	O. Stoelker	Montgomery, Ala	Aug. 16, 1870	106, 425
Furniture-button	J. H. Howcroft	Cincinnati, Ohio	Mar. 12, 1872	124, 577
Furniture, Cabinet	J. D. Browne	Cincinnati, Ohio	Aug. 16, 1859	25, 089
Furniture, Construction of	A. D. Brown	Glasgow, North Britain	Jan. 19, 1858	19, 127
Furniture-drawer	C. Brada	Newton, Mass	Sept. 24, 1867	69, 068
Furniture-drawer	E. B. Johnson	Milwaukee, Wis	May 25, 1869	90, 45
Furniture-drawer	G. F. Joyce	Boston, Mass	May 10, 1870	102, 82
Furniture-drawer	J. W. Warth	New York, N. Y	July 16, 1872	129, 3
Furniture-drawer and fastening	J. Koch	Brookline, Mass	Mar. 17, 1868	75, 5
Furniture, Drawer for cabinet	L. Burnell	Milwaukee, Wis	Nov. 8, 1870	109, 1
Furniture-edge	F. Hütwohl	New York, N. Y	Mar. 24, 1868	75, 9
Furniture, &c., Elastic tip for	W. S. Paddock	Albany, N. Y	July 13, 1869	92, 63
Furniture, Extensible	J. M. Dennis	Galesburgh, Ill	Dec. 5, 1871	121, 59
Furniture-fastening	L. A. Johnson	Candor, N. Y	Aug. 23, 1870	106, 59
Furniture fender	G. R. Willmot	Meriden, Conn	Feb. 20, 1872	123, 96
Furniture fender or cushion	G. R. Willmot	Meriden, Conn	Apr. 11, 1871	113, 60
Furniture Floor-attachment for	H. Cole	Cincinnati, Ohio	Mar. 27, 1866	53, 41

Index of patents issued from the United States Patent Office from 1790 *to* 1873, *inclusive*—Continued.

Invention.	Inventor.	Residence.	Date.	No.
Furniture, Folding	A. Tracy	United States Army	June 26, 1860	28, 921
Furniture for vessels	S. F. Pratt	Roxbury, Mass	May 21, 1867	65, 007
Furniture, House, ship, and camp	E. Clibborn	Cincinnati, Ohio	Sept. 12, 1831	....
Furniture, Joint for folding	N. Thompson	Brooklyn, N. Y	Dec. 12, 1871	121, 735
Furniture-leg	J. C. Orr and J. M. Baird	Wheeling, W. Va	July 22, 1873	141, 071
Furniture-legs, Elastic tip for	E. S. Torrey	New York, N. Y	Jan. 15, 1867	61, 282
Furniture-lock, Automatic	J. T. Brown and I. Hird	Cincinnati, Ohio	Nov. 1, 1870	108, 874
Furniture, Marine	J. Foster	Sandwich, Mass	May 21, 1867	64, 962
Furniture, Marine	E. Gallier	Saint Louis, Mo	Apr. 6, 1869	88, 622
Furniture, Marine	L. D. Newell	New York, N. Y	Apr. 5, 1870	101, 495
Furniture, Means for attaching elastic tips to legs of	E. S. Torrey	New York, N. Y	Jan. 1, 1867	60, 967
Furniture, Means for attaching hooks to	A. Moutleart and W. Tent	Mildmay Park and London, England.	Feb. 9, 1864	41, 528
Furniture, Method of manufacturing	J. H. Belter	New York, N. Y	Feb. 23, 1858	19, 405
Furniture-nail die	W. H. Van Gieson	Waterbury, Conn	Sept. 5, 1865	49, 808
Furniture, Ornamenting	J. E. Rogers	Chelsea, Mass	Mar. 12, 1872	124, 630
Furniture-pad	D. H. Clark and H. Winson	New Haven, Conn	Mar. 14, 1871	112, 686
Furniture-pad	J. E. Earle	New Haven, Conn	Nov. 6, 1866	59, 371
Furniture-pad	W. B. Moses	Washington, D. C	Apr. 5, 1870	101, 648
Furniture-polish	J. L. Brabyn	New York, N. Y	Aug. 19, 1856	15, 550
Furniture-polish	C. A. Libby	La Cygne, Kans	Apr. 29, 1873	138, 262
Furniture-polish	A. Sabatier	New York, N. Y	June 19, 1860	28, 779
Furniture polish and restorer	G. Bricker, sr	Newville, Pa	Jan. 1, 1867	60, 679
Furniture, Printers'	T. J. House	Pittsburgh, Pa	Mar. 2, 1869	87, 339
Furniture-protector	G. F. Boyden	Providence, R. I	Dec. 29, 1868	85, 361
Furniture-rocker	A. Strawbridge	Covington, Ky	Nov. 29, 1864	45, 302
Furniture rubber or brush	B. Y. Conklin	Brooklyn, N. Y	Oct. 29, 1872	132, 522
Furniture, School	J. M. Johnson	Chillicothe, Ohio	June 14, 1870	104, 161
Furniture, School	J. L. Riter	Brownsville, Ind	Mar. 4, 1873	136, 548
Furniture-spring	W. T. Doremus	New York, N. Y	Dec. 17, 1872	133, 971
Furniture-spring	W. T. Doremus	New York, N. Y	Apr. 8, 1873	137, 600
Furniture-spring	W. T. Doremus	New York, N. Y	June 24, 1873	140, 188
Furniture-spring	W. T. Doremus	New York, N. Y	June 24, 1873	140, 189
Furniture-spring	D. Hawkins	Derby, Conn	Oct. 17, 1865	50, 471
Furniture-spring	J. M. Kirkpatrick	Utica, Ohio	Jan. 19, 1869	85, 938
Furniture-spring	C. F. and J. W. Tillman	La Crosse, Wis	Dec. 15, 1863	40, 963
Furniture-spring	F. Tylee	Cleveland, Ohio	Mar. 22, 1864	42, 053
Furniture-spring	W. C. Wyckoff	Brooklyn, N. Y	Dec. 19, 1865	51, 644
Furniture, Spring-guard for legs of	D. F. Hartman	Charles City, Iowa	May 16, 1871	114, 812
Furniture-spring, Japanned	J. J. Eagleton	New York, N. Y	Dec. 19, 1871	122, 001
Furniture-springs, Method of manufacturing	J. L. Haigh	New York, N. Y	Feb. 13, 1872	123, 627
Furniture, Stationary	A. D. Hibbs	Trenton, N. J	Apr. 26, 1870	102, 398
Furniture-tip	J. F. Akerstein	Chicago, Ill	June 27, 1871	116, 392
Furniture-tip	O. B. Collins	Carthage Landing, N. Y	Dec. 22, 1868	85, 071
Furniture, Triangular stand for	T. W. Currier	Lawrence, Mass	Feb. 22, 1859	23, 013
Furniture, Upholstering	F. Mathesius	New York, N. Y	May 17, 1853	9, 730
Furniture, Wall-guard for	J. L. Brander	Boston, Mass	Feb. 6, 1872	123, 377
Furrow in land, Machine for dressing	J. Whitehead	Manchester, Va	June 2, 1857	17, 467
Furrow-staff	G. H. Comer	Indiana, Canada	Nov. 12, 1872	132, 952
Furrowing and dropping corn, Machine for	M. and I. Nichols	Otsego County, N. Y	Oct. 12, 1809	....
Furrowing-device	W. S. Riggs	Hightstown, N. J	Oct. 11, 1864	44, 659
Furrowing-machine	F. M. Birdsall	Martinsville, Ohio	May 22, 1866	54, 842
Furrowing-machine	P. Jotter	Monroe, Ohio	Dec. 3, 1872	133, 537
Furrowing-machine	W. H. Rutledge	Hamilton, Ohio	Aug. 26, 1873	142, 283
Furrowing-machine	J. I. Wilson	Abingdon, Ill	Feb. 6, 1866	52, 482
Fuse, Blasting, &c	R. Bacon	Simsbury, Conn	Feb. 5, 1847	4, 957
Fuse, Blasting, &c	J. S. Bickford	Tuckingmill, Great Britain	May 9, 1865	47, 677
Fuse, Blasting	S. H. Daddow	Saint Clair, Pa	Oct. 8, 1872	132, 061
Fuse, Blasting	S. H. Daddow	Saint Clair, Pa	Dec. 24, 1872	134, 128
Fuse, Blasting	T. P. Shaffner	Louisville, Ky	Aug. 17, 1869	93, 755
Fuse, Blasting	T. P. Shaffner	Louisville, Ky	Dec. 28, 1869	98, 428
Fuse, Blasting	T. H. Walton	Ashland, Pa	Jan. 3, 1865	45, 774
Fuse, Blasting	R. Wren	Santa Cruz, Cal	Apr. 25, 1871	114, 233
Fuse, Combined time and concussion	C. Arick	Saint Clairsville, Ohio	Sept. 6, 1864	44, 061
Fuse, Combined time and concussion	W. F. Goodwin	Powhatan, Ohio	Nov. 1, 1864	44, 861
Fuse, Combined time, percussion, and concussion	A. Merriman	New York, N. Y	Oct. 4, 1864	44, 546
Fuse-composition	W. H. Rogers	Brooklyn, N. Y	June 7, 1870	103, 931
Fuse composition, Safety	E. Gomez and W. Mills	New York, N. Y	Sept. 15, 1857	18, 199
Fuse, Concussion	E. A. Dana	Brookline, Mass	Apr. 20, 1869	89, 204
Fuse, Electric	C. A. and I. S. Browne	Adams, Mass	July 16, 1872	128, 945
Fuse, Electric	G. M. Mowbray	North Adams, Mass	June 10, 1873	139, 686
Fuse, Electric	T. P. Shaffner	Louisville, Ky	Dec. 18, 1866	60, 569
Fuse, Electric	H. J. Smith	Boston, Mass	June 23, 1868	79, 268
Fuse, Electric	H. J. Smith	Boston, Mass	Nov. 23, 1869	97, 241
Fuse, Electric	H. J. Smith	Boston, Mass	Mar. 21, 1871	112, 859
Fuse, Electric blasting	C. A. and I. S. Browne	North Adams, Mass	Oct. 18, 1870	108, 324
Fuse for conveying fire in blasting rocks, &c., Machine for making safety.	W. Lewis	New York, N. Y	Oct. 26, 1838	991
Fuse for explosive projectiles, Concussion	J. L. Henry	United States Army	June 2, 1863	38, 797
Fuse for explosive projectiles, Safety concussion	J. P. Schenckl	Boston, Mass	Oct. 15, 1861	33, 495
Fuse for explosive shell	F. Alger	Boston, Mass	Sept. 2, 1862	36, 329
Fuse for explosive shell	C. Arick	Saint Clairsville, Ohio	Oct. 11, 1864	44, 588
Fuse for explosive shell, &c	J. F. Clew	New York, N. Y	Nov. 15, 1864	45, 024
Fuse for explosive shell	J. McIntyre	New York, N. Y	Oct. 20, 1863	40, 350
Fuse for explosive shell	J. McIntyre	New York, N. Y	Oct. 4, 1864	44, 581
Fuse for explosive shell	B. F. Sturtevant	Boston, Mass	July 29, 1862	36, 037
Fuse for explosive shell	B. F. Sturtevant	Boston, Mass	July 29, 1862	36, 038
Fuse for explosive shell	B. F. Sturtevant	Boston, Mass	July 29, 1862	36, 039
Fuse for explosive shell, Combined time and percussion.	S. and A. M. Sawyer	Fitchburgh, Mass	Aug. 12, 1862	36, 172
Fuse for explosive shell, Concussion	W. S. Beebe	Philadelphia, Pa	Apr. 16, 1867	63, 834
Fuse for explosive shell, Concussion	G. P. Ganster	New York, N. Y	June 13, 1865	48, 167
Fuse for explosive shell, Concussion	B. B. Hotchkiss	Sharon, Conn	June 17, 1862	35, 611
Fuse for explosive shell, Concussion	J. P. Schenckl	Boston, Mass	Aug. 19, 1862	36, 236
Fuse for explosive shell, Concussion	A. J. Simpson and J. J. Janezeck.	Philadelphia, Pa., and Washington, D. C.	July 9, 1867	66, 644
Fuse for explosive shell, Percussion	J. W. Cochran	New York, N. Y	Jan. 6, 1863	37, 275
Fuse for explosive shell, Percussion	J. W. Cochran	New York, N. Y	Feb. 17, 1863	37, 675

Index of patents issued from the United States Patent Office from 1790 *to* 1873, *inclusive*—Continued.

Invention.	Inventor.	Residence.	Date.	No.
Fuse for explosive shell, Percussion	J. A. Curran	United States Army	May 23, 1865	47, 803
Fuse for explosive shell, Percussion	B. B. Hotchkiss	Sharon, Conn	Sept. 16, 1862	36, 465
Fuse for explosive shell, Percussion	R. P. Parrott	Cold Spring, N. Y	Mar. 15, 1864	41, 937
Fuse for explosive shell, Percussion	J. P. Schenkl	Boston, Mass	Sept. 30, 1862	36, 576
Fuse for explosive shell, Percussion	J. F. Shearman	Brooklyn, E. D., N. Y	Apr. 17, 1866	54, 027
Fuse for explosive shell, Time	M. McDevitt	Beloit, Wis	Aug. 23, 1864	43, 922
Fuse for explosive shell, Time	G. Wright	Washington, D. C	Dec. 6, 1864	45, 381
Fuse for explosive shell, Time	G. Wright	Washington, D. C	Mar. 21, 1865	46, 965
Fuse for explosive shell, Time and percussion	B. B. Hotchkiss	New York, N. Y	Dec. 24, 1867	72, 494
Fuse for shell	E. Estabrook	Jersey City, N. J	Jan. 24, 1865	45, 986
Fuse for shell, Combined time and concussion	C. Arick	Saint Clairsville, Ohio	Nov. 22, 1864	45, 128
Fuse for shell, Combined time and percussion	F. Alger	Boston, Mass	Sept. 30, 1862	36, 553
Fuse for shell, Combined time and percussion	J. P. Schenkl	Boston, Mass	Aug. 25, 1863	39, 682
Fuse for shell, Combined time and percussion	W. S. Smoot	Washington, D. C	Oct. 28, 1862	36, 806
Fuse for shell, Combined time and percussion	C. W. Stafford	New York, N. Y	Aug. 30, 1864	44, 023
Fuse for shell, Concussion	W. F. Goodwin	New York, N. Y	No 15, 1864	45, 035
Fuse for shell, Concussion	W. H. Hubbell	Philadelphia, Pa	Sept. 30, 1862	36, 566
Fuse for shell, Concussion	S. R. Russell	Middletown, Ohio	Dec. 16, 1862	37, 200
Fuse for shell, Concussion	I. P. Tice	New York, N. Y	June 23, 1863	38, 994
Fuse for shell, Igniting	C. W. Smith	Evans, N. Y	Aug. 9, 1864	43, 801
Fuse for shell, Igniting time	A. M. Sawyer	Fitchburgh Mass	May 26, 1863	38, 699
Fuse for shell, Percussion	A. H. Emery	New York, N. Y	Dec. 8, 1863	40, 828
Fuse for shell, Percussion	G. P. Ganster	New York, N. Y	Apr. 5, 1864	42, 185
Fuse for shell, Percussion	B. B. Hotchkiss	Sharon, Conn	Feb. 24, 1863	37, 756
Fuse for shell, Percussion	W. F. Patterson	Somerset, Ky	Dec. 8, 1863	40, 885
Fuse for shell, Time	B. B. Hotchkiss	Sharon, Conn	May 10, 1864	42, 660
Fuse for shell, Time and concussion	J. P. Schenkl	Boston, Mass	July 15, 1862	35, 897
Fuse head, Electric	H. J. Smith	Boston, Mass	June 14, 1870	104, 217
Fuse heads, Electric and other	E. A. L. Roberts	Titusville, Pa	Mar. 21, 1871	112, 850
Fuse-hood for explosive shell	T. Taylor	Washington, D. C	Apr. 11, 1865	47, 231
Fuse-hood for shell	S. Sawyer	Fitchburgh, Mass	Dec. 24, 1861	34, 040
Fuse-lock, Miners'	G. Hagenmeyer	Big River, Cal	Oct. 9, 1866	58, 639
Fuse-making machine	A. F. Andrews	Avon, Conn	July 28, 1857	17, 863
Fuse-manufacturing machine	T. Richards	Medford, Mass	Aug. 18, 1868	81, 291
Fuse of shells and other projectiles	N. Scholfield	Norwich, Conn	Dec. 15, 1857	18, 866
Fuse or slow-match for igniting under water	T. K. Anderson	Hornellsville, N. Y	Jan. 28, 1862	34, 233
Fuse, Percussion	W. Gardner	San Francisco, Cal	May 31, 1870	103, 599
Fuse, Safety	A. F. and J. H. Andrews	Avon, Conn	Dec. 26, 1865	51, 679
Fuse, Safety	J. H. Andrews	Avon, Conn	Dec. 27, 1864	45, 572
Fuse, Safety	J. E. Chase and J. Foy	Simsbury, Conn	Nov. 22, 1864	45, 140
Fuse, Shell	J. D. Bacon	New York, N. Y	June 22, 1869	91, 701
Fuse, Shell	J. G. Butler	United States Army	July 30, 1872	129, 929
Fuse, Shell	J. G. Butler	United States Army	July 30, 1872	129, 930
Fuse, Shell	E. Drake	Stoughton, Mass	Feb. 14, 1871	111, 823
Fuse, Shell	J. Eggo	Jersey City, N. J	Dec. 20, 1870	110, 219
Fuse-stock for bomb-shells	A. Powell, jr	Mare Island, Cal	June 26, 1855	13, 138
Fuse, Tape	P. Bacon	Simsbury, Conn	May 8, 1855	12, 810
Fuse, Tape	J. E. Chase and J. Foy	Simsbury, Conn	June 30, 1863	39, 033
Fuse to shell, Applying	R. P. Parrott	Cold Spring, N. Y	Nov. 5, 1861	36, 662
Fuses, Apparatus for casting	G. Wright	Washington, D. C	Apr. 28, 1863	38, 352
Fuses by electricity, Firing	F. E. Beardslee	College Point, N. Y	Aug. 18, 1863	39, 542
Fuses, Compound for priming electric	G. M. Mowbray	Titusville, Pa	Nov. 2, 1869	96, 465
Fuses, Compound for priming electric	G. M. Mowbray	North Adams, Mass	June 25, 1872	128, 241
Fuses, Concussion-bulb for	G. P. Gauston and I. S. Schuyler	New Nork, N. Y	Mar. 29, 1864	42, 082
Fuses, Construction of trains or	E. Gomez	New York, N. Y	Jan. 7, 1862	34, 057
Fuses for explosive shell, Percussion-igniter of time	B. B. Hotchkiss	Sharon, Conn	Aug. 30, 1864	43, 993
Fuses, Machine for making	A. F. Andrews	Avon, Conn	May 25, 1852	8, 963
Fuses, Machinery for manufacturing safety	R. Uren, T. Dunstone, and J. Blight	Eagle Harbor, Mich	Dec. 2, 1862	37, 079
Fuses, Manufacture of	G. F. James	Manchester, England	May 7, 1872	126, 462
G.				
Gaff-chock for vessels	J. C. Cottingham	Philadelphia, Pa	July 6, 1869	92, 271
Gaff-fastening	J. H. David	Damariscotta, Me	Sept. 24, 1872	131, 508
Gaff for ships' spars	G. C. Pattison	Baltimore, Md	Apr. 2, 1867	63, 555
Gaff for vessels' sails	W. H Conway	Harrison, Md	Sept. 17, 1861	33, 295
Gag-runner	W. Robotham	Newark, N. J	Feb. 14, 1860	27, 158
Gag runner and buckle	S. Siebold	Syracuse, N. Y	Mar. 29, 1870	101, 386
Gag-swivel	W. Greacen	Newark, N. J	Oct. 4, 1864	44, 579
Gag-swivel	C. B. Payne	Clinton, Ill	Nov. 3, 1868	83, 787

Gage:

See Auger-gage.
Axle-gage.
Belt-lacing gage.
Bilge-water gage.
Bill-folding gage.
Boiler-gage.
Boiler alarm-gage.
Boiler safety-gage.
Boot and shoe gage.
Boot and shoe heel gage.
Button hole-cutter gage.
Button-shank gage.
Carpenter's gage.
Carriage-axle gage.
Cart-gage.
Changeable gage.
Clapboard-gage.
Clock-work-turning gage.
Cloth-gage.
Cock-gage.
Cock and last gage.
Collar-gage.
Compound gage.
Countersink-gage.
Depth-gage.
Diamond-gage.
Dowel-gage.
Drill-gage.
Edge gage.
Embossing-press gage.
Feather-edging gage.
Feed-gage.
Floor-gage.
Fruit-knife gage.
Gas-fitter's gage.
Grain-gage.
Gun-hammer gage.
Harness-maker's gage.
Index-gage.
Joiner's gage.
Key-way gage
Lamp-wick gage.
Leather-gage.
Leather-splitting gage.
Liquid-gage.
Lumber-slitting gage.

Index of patents issued from the United States Patent Office from 1790 *to* 1873, *inclusive*—Continued.

Invention.	Inventor.	Residence.	Date.	No.
Gage—Continued.				
See Machine-drill gage. Machinist's gage. Marking and cutting gage. Metal-planing gage. Meter-gage. Micrometer-gage. Millstone-dressing gage. Mortising-machine gage. Paper-cutter gage. Platen-gage. Plow-gage. Pressure-gage. Printing-press feed-gage. Railway-gage. Railway-track gage. Reading-gage. Ruling-gage. Safety-gage. Sash-mortising gage. Saw-gage. Saw-table gage. Saw-tooth gage. Screw-cutting-tool gage. Sewing-gage. Sewing-machine gage. Shingling-gage. Siding gage. Sliding-gage. Speed-gage. Splitting-gage. Spooling-gage. Steam-gage. Steam-pressure gage. Stencil-cutting gage. Surface-gage. Surface and depth gage. Taper-gage. Tide-gage. Track-gage. Watch-maker's gage. Water-gage. Water-level gage. Water-plow gage. Weather-board gage. Wire-gage.				
Gage	J. W. Butler	New Britain, Conn	Jan. 14, 1873	134,729
Gage	J. A. Marden	Veazie, Me	Apr. 16, 1872	125,823
Gage	E. W. Mathewson	Norwich, Conn	May 19, 1868	77,997
Gage	L. H. Mayott	Springfield, Mass	Oct. 13, 1868	82,969
Gage	B. F. Merrill	West Lebanon, N. H	Aug. 4, 1868	80,649
Gage, Adjustable	T. E. Warren	Shelburne Falls, Mass	Aug. 2, 1870	106,099
Gage and dividers	G. Kenny	Nashua, N. H	Jan. 4, 1870	98,504
Gage and try-square, Combined	F. Castle	Montana, Iowa	Mar. 21, 1871	112,780
Gage and try-square combined	F. Castle	Montana, Iowa	July 11, 1871	116,805
Gage, calipers, and rule, Connecting	N. H. Bundy	Boston, Mass	Aug. 8, 1865	49,337
Gage-knife for cutting leather	C. T. Durham, J. Wood, and T. Hair.	Norristown, Pa	June 21, 1870	104,436
Gage-rod	E. S. Prime	Baltimore, Md	Mar. 29, 1870	101,309
Gaging-rod	W. J. Tait	Bergen, N. J	Apr. 24, 1866	54,229
Gaging-rod for liquor-cask	D. C. Hyde	New York, N. Y	May 26, 1863	38,681
Graining-machine	J. Richards	Philadelphia, Pa	Oct. 7, 1873	143,531
Gaiter	P. E. Clark	Hartford, Conn	Apr. 4, 1871	113,259
Gaiter	H. B. Goodyear	New Haven, Conn	Jan. 14, 1862	34,140
Gaiter	W. Love	McConnellsville, Ohio	May 27, 1873	139,321
Gaiter	G. W. Ludlow	Elizabeth, N. J	Apr. 21, 1863	38,235
Gaiter	W. E. Putnam	Boston, Mass	Apr. 2, 1872	125,219
Gaiter	W. E. Putnam	Boston, Mass	Sept. 23, 1873	143,186
Gaiter	M. M. Wheeler	Galena, Ill	Oct. 28, 1873	144,173
Gaiter, Ankle-supporting	T. G. Rich	Milton, Mass	Jan. 29, 1861	31,259
Gaiter, Congress	H. S. Holmes	Lynn, Mass	May 14, 1861	32,291
Gaiter, Man's	E. R. Ware	Worcester, Mass	Mar. 28, 1871	113,119
Gaiter-pattern	S. Hodgins	Saint Louis, Mo	Nov. 28, 1871	121,367
Gaiter, Spring	F. M. Blodgett	Boston, Mass	July 25, 1865	48,897
Gaiters, Hand-jack for Congress	H. F. Dougherty	Monmouth, Ill	Aug. 21, 1866	57,297
Gaiters, Instrument for stretching elastic	A. Werden	Ypsilanti, Mich	Oct. 8, 1861	33,456
Gaiters, Machine for fastening lacing-hook to	L. S. Wiswell	Utica, N. Y	June 2, 1868	78,503
Gaiters, Machine for goring	J. Walden	Newark, N. J	May 7, 1872	126,598
Galley	F. M. Wildman	Bethel, Conn	Sept. 6, 1870	107,144
Galley-rest	H. H. Gale	Eugene City, Oreg	Sept. 13, 1870	107,246
Galley-rest	M. T. Lincoln	Washington, D. C	Nov. 16, 1869	96,929
Galley-rest	E. Morgan	Washington, D. C	July 27, 1869	93,109
Galley-rest, Printers'	J. M. Murphy	Olympia, Wash	July 11, 1871	116,857
Galloon, Method of weaving	J. Kemper	Philadelphia, Pa	Aug. 10, 1869	93,539
Galvanic apparatus	E. Smith	Normal, Ill	Oct. 19, 1869	96,044
Galvanic bath for treating disease	J. B. Hatting	New York, N. Y	Apr. 9, 1872	125,567
Galvanic batteries, Carbon-plate for	C. T. Chester	New York, N. Y	Apr. 14, 1863	38,146
Galvanic batteries, Combining, generating, and secondary or accumulating.	G. L. Leclanché	Paris, France	Apr. 23, 1867	64,113
Galvanic batteries, Connecting-clamp for the plates of.	C. T. Chester	New York, N. Y	May 15, 1855	12,855
Galvanic batteries, Device for preventing corrosion of the binding-screws in.	G. Doyle	Ottawa, Ill	Mar. 30, 1858	19,759
Galvanic batteries, Liquid for	V. Barjon	New York, N. Y	July 11, 1871	116,917
Galvanic batteries, Liquid for	D. H. Fitch, jr	Litchfield, Ill	June 16, 1863	38,894
Galvanic batteries, Liquid for	E. Prevost	New York, N. Y	Sept. 27, 1870	107,811
Galvanic batteries, Liquids for exciting	M. Vergnes	New York, N. Y	Mar. 11, 1862	34,656
Galvanic batteries, Method of attaching the electrodes to the poles of.	J. Elmendorf	Penn Yan, N. Y	Feb. 2, 1858	19,245
Galvanic batteries, Porous cell for	M. G. Farmer	Salem, Mass	Jan. 11, 1853	9,530
Galvanic batteries, Zinc for	E. A. Hill	Chicago, Ill	Apr. 25, 1871	114,005
Galvanic battery	T. C. Avery	New York, N. Y	Oct. 11, 1859	25,710
Galvanic battery	L. Bastet	New York, N. Y	May 6, 1873	138,602
Galvanic battery	F. E. Beardslee	Brooklyn, N. Y	Aug. 13, 1872	130,353
Galvanic battery	J. Blackie	Washington, D. C	Sept. 5, 1865	49,824
Galvanic battery	C. Boulay	Paris, France	Nov. 12, 1867	70,791
Galvanic battery	D. Brooks	Philadelphia, Pa	May 16, 1871	114,911
Galvanic battery	D. J. Browne and C. W. Baldwin.	Boston, Mass	Apr. 25, 1865	47,389
Galvanic battery	J. A. Calland	Nantes, France	Oct. 26, 1869	96,199
Galvanic battery	J. A. Calland	Nantes, France	Oct. 17, 1871	120,034
Galvanic battery	C. T. Chester	Englewood, N. J	Jan. 26, 1869	86,134
Galvanic battery	D. M. Cook	Mansfield, Ohio	Dec. 20, 1870	110,206
Galvanic battery	A. G. Davis	Baltimore, Md	Dec. 31, 1872	134,364
Galvanic battery	J. Dixon	Jersey City, N. J	Sept. 4, 1866	57,687
Galvanic battery	L. Drescher	New York, N. Y	Dec. 7, 1852	9,447
Galvanic battery	L. Drescher	New York, N. Y	July 18, 1871	117,057

Index of patents issued from the United States Patent Office from 1790 *to* 1873, *inclusive*—Continued.

Invention.	Inventor.	Residence.	Date.	No.
Galvanic battery	T. A. Edison	Newark, N. J	Sept. 23, 1873	142, 999
Galvanic battery	E. J. Fraser	Chicago, Ill	Nov. 3, 1863	40, 537
Galvanic battery	G. W. Freed	Lancaster, Pa	June 7, 1864	43, 014
Galvanic battery	A. C. Garratt	Boston, Mass	July 7, 1868	79, 567
Galvanic battery	E. Grenet, jr	Paris, France	Sept. 20, 1859	25, 503
Galvanic battery	H. Highton	Putney, England	Oct. 1, 1872	131, 878
Galvanic battery	E. A. Hill	Galesburgh, Ill	Aug. 18, 1863	39, 571
Galvanic battery	E. A. Hill	Chicago, Ill	Apr. 25, 1871	114, 006
Galvanic battery	J. Hill	New York, N. Y	May 24, 1870	103, 331
Galvanic battery	J. Kidder	New York, N. Y	Oct. 25, 1870	108, 602
Galvanic battery	J. Kidder	New York, N. Y	June 27, 1871	116, 451
Galvanic battery	J. Kidder	New York, N. Y	Oct. 10, 1871	119, 855
Galvanic battery	G. L. Leclanché	Paris, France	June 5, 1866	55, 441
Galvanic battery	E. J. Leland	Worcester, Mass	Nov. 7, 1871	120, 651
Galvanic battery	C. A. Linke	Pittsburgh, Pa	July 16, 1872	129, 148
Galvanic battery	R. M. Lockwood	New York, N. Y	Apr. 8, 1873	137, 556
Galvanic battery	E. D. McCracken	New York, N. Y	Aug. 16, 1870	106, 383
Galvanic battery	J. R. McPherson	Beloit, Wis	Dec 14, 1869	97, 949
Galvanic battery	A. L. Nolf	New York, N. Y	Sept. 2, 1873	142, 502
Galvanic battery	A. Olmstead	Easton, Pa	Apr. 17, 1849	6, 362
Galvanic battery	G. M. Phelps	Brooklyn, N. Y	Aug. 20, 1872	103, 593
Galvanic battery	J. Plumbe, jr	Boston, Mass	Mar. 4, 1843	2, 984
Galvanic battery	J. W. Powell	New York, N. Y	May 30, 1871	115, 519
Galvanic battery	E. Provost	New York, N. Y	Sept. 19, 1871	119, 175
Galvanic battery	J. A. Robbins	Medford, Mass	Oct. 24, 1871	120, 327
Galvanic battery	J. A. Robbins	Medford, Mass	July 2, 1872	128, 660
Galvanic battery	E. Seaver	Boston, Mass	Jan. 26, 1858	19, 209
Galvanic battery	W. J. Wilder	Cleveland, Ohio	Nov. 12, 1872	132, 997
Galvanic battery adapted to medical purposes	P. Coad	Philadelphia, Pa	Mar. 28, 1842	2, 521
Galvanic-battery connection	F. Splitdorf	New York, N. Y	Feb. 13, 1866	52, 617
Galvanic battery for remedial uses	H. Fitz	Cleveland, Ohio	Mar. 19, 1867	63, 032
Galvanic-battery switch	G. Little	Rutherford, Park, N. J	June 13, 1871	115, 966
Galvanic-battery switch	E. M. Pierson	Newark, N. J	Oct. 18, 1870	108, 513
Galvanic chains, bands, &c., Electro	I. L. Pulvermacher	London, England	Nov. 7, 1871	120, 772
Galvanic chains, Fluid for exciting	F. T. Bakker	Chicago, Ill	June 30, 1868	79, 432
Galvanic electricity applied to cure disease	D. Harrington	Philadelphia, Pa	Mar. 31, 1835	....
Galvanic electricity applied to surface of the body	D. Harrington	Philadelphia, Pa	Apr. 8, 1835	....
Galvanic fluid for curing certain diseases	D. Harrington	Philadelphia, Pa	July 22, 1833	....
Galvanic fluid to cure diseases	D. Harrington	Philadelphia, Pa	Jan. 27, 1834	....
Galvanic instrument	D. Harrington	Philadelphia, Pa	Sept. 2, 1845	4, 176
Galvanic plate for medical use	J. Hill	Brooklyn, N. Y	Oct. 9, 1860	30, 317
Galvanic plate for remedial purposes	C. Maray	New York, N. Y	Jan. 19, 1869	86, 088
Galvanic register for steam-boilers	A. Dunn	Dalston, England	May 28, 1850	7, 394
Galvanic ring, belt, &c	D. C. Moorhead	New York, N. Y	Aug. 26, 1845	4, 167
Galvanism, Apparatus for treating diseases by means of.	J. B. Hatting	New York, N. Y	June 21, 1870	104, 587
Galvanism for salivating, &c	W. Phœbus	New York, N. Y	June 9, 1807	....
Galvanized nails, Cleansing and separating	W. Blake	Boston, Mass	Aug. 21, 1860	29, 664
Galvanizing metals, Process for	C. B. Miller	Wilmington, Del	May 30, 1854	10, 976
Galvano-electric machine	J. R. Palmenberg	New York, N. Y	Nov. 9, 1858	22, 029
Galvano-plastic process for precipitating iron on molds, &c.	M. H. Jacobi and E. Klein	St. Petersburg, Russia	Sept. 29, 1868	82, 525
Galvanometer, Reflecting	W. Thomson	Glasgow College, Scotland	July 6, 1869	92, 228
Game	C. B. Barlow	Portsmouth, N. H	Oct. 19, 1869	95, 969
Game	J. Brandl	New York, N. Y	Dec. 24, 1872	134, 125
Game	E. J. Brooks	New York, N. Y	Mar. 19, 1872	124, 786
Game	J. M. Fletcher	Clifton, N. Y	Apr. 16, 1872	125, 671
Game	D. F. Hale	Chicopee, Mass	Oct. 25, 1870	108, 587
Game	H. Jackson	New York, N. Y	Feb 11, 1868	74, 368
Game	G. Krotzinger	New York, N. Y	Dec. 3, 1867	71, 626
Game	B. E. Mead	Peekskill, N. Y	Jan. 3, 1865	45, 731
Game	J. B. Murray	Newark, N. J	Aug. 6, 1872	130, 234
Game	R. Patterson	New Santa Fé, Mo	July 30, 1872	129, 983
Game	J. M. Rix	Warren, N. H	Mar. 26, 1872	124, 915
Game	A. G. Slagle	Memphis, Tenn	Mar. 30, 1869	88, 421
Game	N. J. Vander Weyde	Franklin, N. H	June 14, 1870	104, 379
Game	W. H. Wilson	Providence, R. I	Sept. 22, 1868	82, 370
Game and rat trap	B. F. Tatem	Memphis, Tenn	Nov. 7, 1871	120, 683
Game-apparatus	R. E. Bean	Franklin, N. H	Dec. 9, 1873	145, 385
Game-apparatus	J. N. Sawkins	Langport, England	Dec. 23, 1873	145, 757
Game-apparatus	A. M. Smith	Chicago, Ill	Feb. 4, 1873	135, 494
Game-apparatus	F. W. Smith	Bridgeport, Conn	Dec. 23, 1873	145, 914
Game-apparatus	P. West and G. S. Lee	Worcester, Mass	Sept. 23, 1873	143, 047
Game-apparatus	P. West and G. S. Lee	Worcester, Mass	Oct. 21, 1873	143, 799
Game-apparatus	J. C. Wilhelm	Shaver's Creek, Pa	Dec. 16, 1873	145, 706
Game, Arithmetical	S. A. Emery	Boston, Mass	Nov. 10, 1868	83, 946
Game ball	E. A. Barrett	New York, N. Y	Feb. 6, 1872	123, 442
Game-board	J. C. Arms	Northampton, Mass	Feb. 25, 1873	136, 200
Game-board	A. F. R. Arndt	Detroit, Mich	Dec. 9, 1873	145, 271
Game-board	H. C. Drexel	Baltimore, Md	Jan. 23, 1872	123, 001
Game-board	J. T. Edson	Stowe, Mass	Apr. 25, 1865	47, 491
Game-board	G. R. Elliott	Boston, Mass	Feb. 18, 1873	136, 048
Game-board	F. P. Holmes	North Bridgewater, Mass	July 22, 1873	141, 053
Game-board	G. W. Kintz	Rochester, N. Y	Aug. 22, 1871	118, 249
Game-board	C. Krath and G. H. Moll	Saint Louis, Mo	Sept. 6, 1870	107, 065
Game-board	J. Smith and E. M. Nutter	Feltonville, Mass	May 23, 1865	47, 867
Game-board	G. Wentz	Cincinnati, Ohio	Aug. 19, 1873	142, 066
Game board, Top	D. Wight	New London, Conn	Apr. 7, 1868	76, 572
Game-box	S. F. Brooks	Weston, Mass	June 19, 1860	28, 733
Game-box for ten-pins	G. B. Fowler	Brooklyn, N. Y	Sept. 6, 1870	107, 030
Game called vinco	J. Carlin	New York, N. Y	Dec. 29, 1868	85, 281
Game-card	C. W. Saladee	Paducah, Ky	Feb. 9, 1864	41, 587
Game-counter	A. A. Griffing	Lexington, Mass	Feb. 16, 1869	86, 919
Game-counter	F. H. Richardson	Brooklyn, N. Y	July 9, 1872	128, 910
Game-counter	E. Schellhorn	Urbana, Ohio	Feb. 9, 1869	86, 781
Game entitled "Talisman"	J. W. Wilson	New York, N. Y	Apr. 27, 1869	89, 535
Game for pastime	C. Richardson	Richmond, Va	Apr. 23, 1867	64, 149

Index of patents issued from the United States Patent Office from 1790 *to* 1873, *inclusive*—Continued.

Index of patents issued from the United States Patent Office from 1790 *to* 1873, *inclusive*—Continued.

Invention.	Inventor.	Residence.	Date.	No.
Garment-fastening	F. I. Palmer	Springfield, Mass	Mar. 27, 1866	53, 472
Garment-fastening	R. S. Pickett	New Haven, Conn	Mar. 20, 1866	53, 331
Garment for ladies, Interlined under	C. L. Morehouse	Cleveland, Ohio	Feb. 28, 1865	46, 575
Garment for travelers, Sleeping	V. P. Corbett	Washington, D. C	Apr. 5, 1864	42, 168
Garment-hanger and size-ticket holder combined	I. Desky and S. A. Jennings	Seneca Falls, N. Y	Jan. 4, 1870	98, 476
Garment having body and sleeves	H. Osler	Philadelphia, Pa	June 2, 1863	38, 757
Garment-holder	B. V. Taylor	Newark, N. J	Nov. 8, 1870	109, 156
Garment, Hooded	A. Keating	Boston, Mass	Dec. 20, 1870	110, 242
Garment-hook	M. Fowler	North Branford, Conn	Jan. 12, 1864	41, 213
Garment, Lady's accouching	H. R. Esterling	Bennettsville, S. C	Jan. 30, 1872	123, 249
Garment-pattern, Adjustable	S. Shawcross	Freeport, Ill	Mar. 22, 1870	101, 052
Garment-pressing apparatus	J. W. Thorp	South Weare, N. H	Dec. 16, 1851	8, 598
Garment-pressing machine	W. B. Walker	Boston, Mass	Feb. 27, 1872	124, 177
Garment, Skin under	J. H. Andrus	New York, N. Y	Mar. 28, 1871	113, 128
Garment-strap, Adjustable and detachable	S. L. Clemens	Hartford, Conn	Dec. 19, 1871	121, 992
Garment-strap, Adjustable and detachable	H. C. Lockwood	Baltimore, Md	Dec. 19, 1871	122, 038
Garment-support	J. L. Kendall	Foxborough, Mass	Apr. 20, 1869	89, 152
Garment-supporter	H. M. Clemence	Worcester, Mass	Nov. 19, 1867	71, 136
Garment, Under	O. P. Flynt	Boston, Mass	Nov. 18, 1873	144, 609
Garments, Apparatus for cutting	T. H. Moore	Boston, Mass	Mar. 25, 1862	34, 766
Garments, Apparatus for fitting and laying out	W. H. Mayer	Newark, N. J	Dec. 5, 1871	121, 642
Garments, Apparatus for laying off the scye in cutting.	P. Spilman	Richmond, Va	July 18, 1854	11, 339
Garments, Applying measurement to and laying out.	J. J. Miller	McAlevy's Fort, Pa	July 12, 1870	105, 355
Garments, Attachment for fastening overlapping parts of.	B. J. Greely	Boston, Mass	Dec. 7, 1869	97, 629
Garments, Attachment for protecting cuff and sleeve of.	J. C. Reed	Boston, Mass	Oct. 24, 1871	120, 324
Garments, Brace for supporting	D. Minthorn	New York, N. Y	June 5, 1855	13, 011
Garments, Cutting	O. Madison	Troy, N. Y	Aug. 18, 1829	....
Garments, Cutting	D. N. Sipperly	Troy, N. Y	July 8, 1843	3, 160
Garments, Cutting	S. B. Stilwell	Brooklyn, N. Y	June 20, 1845	4, 083
Garments, Drafting	W. L. Leete	Oswego, N. Y	Sept. 6, 1870	107, 068
Garments, Drafting and cutting	J. H. Chappell	Chillicothe, Ohio	Jan. 18, 1834	....
Garments, Drafting and measuring	C. Lucas	Charlottsville, Va	June 13, 1848	5, 635
Garments, Hook for fastening	G. A. Watkins	Springfield, Vt	Feb. 24, 1863	37, 783
Garments, Machine for cutting out	W. A. McLaughlin	New York, N. Y	Nov. 10, 1868	83, 870
Garments, Machine for drafting	J. M. Weston	Chesterfield, N. Y	Feb. 2, 1858	19, 271
Garments, Manufacturing seamless felt	D. W. Gitchell and L. W. Badger	Matteawan, N. Y	Oct. 20, 1857	18, 487
Garments, Marking off and cutting out	G. Beard, jr	West Whiteland Township, Pa.	Aug. 5, 1833	....
Garments, Marking off and cutting out	B. J. Lewis	Mount Vernon, Ohio	Nov. 19, 1833	....
Garments, Mode of constructing	T. S. Lambert	Peekskill, N. Y	Feb. 18, 1862	34, 464
Garments, Pattern for cutting	S. A. Millwee	Greenwood, S. C	May 14, 1872	126, 825
Garments, Pattern for laying out	W. M. Michael	Indiana, Pa	Dec. 7, 1869	97, 672
Garments, Sleeve of knitted	W. H. Abel	Greenville, R. I	Aug. 4, 1868	80, 697
Garments, Spring-hook fastening for	D. M. Smith	Springfield, Vt	June 16, 1863	38, 920
Garments, Spring hook for fastening	A. Putnam, jr	Chester, Vt	Nov. 11, 1862	36, 933
Garments, Square for cutting	J. G. Wilson	New York, N. Y	Feb. 28, 1827	....
Garments, System of cutting	A. Wiswell	Exeter, N. H	July 11, 1837	256
Garments, System of laying out	J. A. Shreckengaust	Chillicothe, Ohio	July 16, 1872	129, 603
Garments, Tailor's instrument and mode of measuring	L. Flenner	Philadelphia, Pa	Nov. 10, 1841	2, 341
Garments, Tailor's instrument for drafting	E. J. Axford	Philadelphia, Pa	Mar. 30, 1839	1, 113
Garments to hooks, Securing	G. Woodward and F. S. Hathaway.	Keene, N. H	Feb. 15, 1859	22, 996
Garments, Uniting bats for making seamless felt	D. W. Gitchell	Rahway, N. J	Sept. 26, 1854	11, 739
Garments without spinning or weaving, Manufacturing woolen.	J. Foster and J. Stoudenburgh	Greene County, N. Y	Apr. 20, 1830	....
Garter	E. F. Burrows	Mystic River, Conn	July 24, 1866	56, 524
Garter	C. Coester, jr., and J. L. Moore	Bridgeport, Conn	Aug. 16, 1870	106, 328
Garter	J. P. Fuller	New York, N. Y	June 26, 1860	38, 841
Garter	H. A. House	Bridgeport, Conn	Nov. 29, 1870	109, 617
Garter	H. A. House	Bridgeport, Conn	Nov. 29, 1870	109, 736
Garter	H. A. House	Bridgeport, Conn	Nov. 29, 1870	109, 737
Garter	H. A. House	Bridgeport, Conn	Mar. 11, 1873	136, 599
Garter and belt, Coil-spring	J. M. Ellis	Bridgeport, Conn	Dec. 2, 1873	145, 160
Garter, Metallic	H. A. House	Bridgeport, Conn	Mar. 7, 1871	112, 344
Garter, Metallic	W. H. McCoy and A. Wheeler	Charlestown, Mass	Oct. 29, 1867	70, 238
Gas-alarm	P. H. Vander Weyde	New York, N. Y	Mar. 22, 1864	42, 032
Gas and heating buildings, Apparatus for generating.	S. C. Salisbury	New York, N. Y	Mar. 23, 1869	88, 079
Gas and heating dwellings, Apparatus for generating.	T. S. C. Lowe	Morristown, Pa	Aug. 13, 1872	130, 381
Gas and illuminating street and other cars, Production of.	A. Barbarin	New Orleans, La	Sept. 22, 1868	82, 273
Gas and in application of the same, Process of generating.	J. T. Rich	Philadelphia, Pa	Feb. 4, 1868	74, 001
Gas and liquid meter	H. H. Stuart	Jamaica, N. Y	Dec. 11, 1866	60, 437
Gas and liquid regulator	T. H. Dodge	Nashua, N. H	June 27, 1854	11, 154
Gas and oil, Apparatus and process for obtaining light from.	J. Kidd	New York, N. Y	Aug. 31, 1869	94, 219
Gas and oil from coal, Apparatus for manufacturing.	J. Shoemaker	Putneyville, Pa	Nov. 19, 1867	71, 233
Gas and other heater	J. Q. Birkey	Philadelphia, Pa	Mar. 17, 1868	75, 516
Gas and other lights, Apparatus for igniting	W. Klinkerfues	Göttingen, Prussia	May 16, 1871	114, 950
Gas and other liquids, Device for igniting	W. H. Smith	New York, N. Y	May 30, 1871	115, 373
Gas and other products, Manufacture of illuminating.	J. Absterdam	New York, N. Y	Dec. 15, 1868	84, 981
Gas and smoke consuming apparatus	G. Marlow	Chicago, Ill	Oct. 24, 1871	120, 299
Gas and steam fittings	A. Hallowell	Lowell, Mass	Nov. 17, 1868	84, 112
Gas and vapor from petroleum, Apparatus for generating.	A. I. Ambler	Washington, D. C	May 14, 1872	126, 770
Gas and water fittings, Machine for tapping	A. D. Laws	Bridgeport, Conn	Sept. 23, 1873	143, 168
Gas and water mains, Coupling for	W. Kilburn	Napa, Cal	July 15, 1873	140, 834
Gas and water meter	G. R. Moore	Philadelphia, Pa	June 1, 1869	90, 678
Gas and water meter register	J. J. Squire	New Haven, Conn	Mar. 3, 1863	37, 832

Index of patents issued from the United States Patent Office from 1790 *to* 1873, *inclusive*—Continued.

Invention.	Inventor.	Residence.	Date.	No.
Gas and water pipe	E. H. Austin	New York, N. Y	Mar. 18, 1873	136, 952
Gas and water pipe	W. B. Guy	Boston, Mass	Dec. 20, 1864	45, 491
Gas and water pipe	A. Wyckoff	Elmira, N. Y	Nov. 22, 1864	45, 201
Gas and water pipe coating	N. Clute	Schenectady, N. Y	Mar. 21, 1871	112, 786
Gas and water pipe joint	C. W. Isbell	New Haven, Conn	July 24, 1860	29, 281
Gas and water pipe joint	J. E. Quinn	Chicago, Ill	Apr. 26, 1859	23, 811
Gas and water pipe plug	E. P. Schutt	Cortland, N. Y	Mar. 1, 1870	100, 456
Gas and water pipes, Casting packing-rings in	R. C. Robbins	New York, N. Y	Jan. 26, 1864	41, 413
Gas-apparatus	A. L. Bogart	New York, N. Y	Jan. 3, 1871	110, 729
Gas apparatus	J. W. Brown and H. B. West	Butlerville, Ind., and Put in Bay, Ohio.	Apr. 29, 1873	138, 377
Gas-apparatus	C. H. Childs	Cleveland, Ohio	May 26, 1868	78, 185
Gas-apparatus	R. T. Coverdale	Circleville, Ohio	July 23, 1867	66, 950
Gas-apparatus	C. Deavs	New York, N. Y	June 25, 1867	66, 004
Gas-apparatus	W. Foster, jr., and G. P. Ganster.	New York, N. Y	Mar. 9, 1869	87, 556
Gas-apparatus	J. S. Gallaher, jr, and J. W. Smith.	Washington, D. C	Dec. 11, 1855	13, 904
Gas-apparatus	D. M. Graham	Evansville, Ind	Sept. 3, 1867	68, 435
Gas-apparatus	E. J. Manville and S. G. Blackman.	Waterbury, Conn	June 1, 1858	20, 438
Gas-apparatus	H. S. Maxim	Brooklyn, N. Y	Jan. 9, 1872	122, 625
Gas-apparatus	J. McCleish	New York, N. Y	May 26, 1868	78, 307
Gas-apparatus	R. Morton	London, England	Sept. 12, 1871	118, 967
Gas-apparatus	A. Myers	Philadelphia, Pa	June 7, 1864	43, 079
Gas-apparatus	G. Olney	Brooklyn, N. Y	Jan. 16, 1872	122, 733
Gas-apparatus	A. Pierce	Philadelphia, Pa	Feb. 27, 1849	6, 145
Gas-apparatus	E. A. Pond and M. S. Richardson.	Rutland, Vt	Mar. 27, 1866	53, 482
Gas-apparatus	J. Ponton	Buffalo, N. Y	May 19, 1868	78, 006
Gas-apparatus	R. Porter and T. Lane	London, England	Aug. 13, 1872	130, 388
Gas-apparatus	J. E. Richard	Columbia, S. C	Nov. 1, 1870	108, 937
Gas-apparatus	C. Seeger	Quincy, Ill	Mar. 19, 1872	124, 766
Gas-apparatus	J. F. Spence	Williamsburgh, N. Y	Feb. 5, 1867	61, 887
Gas-apparatus	J. H. Steiner	Cincinnati, Ohio	Apr. 9, 1872	125, 496
Gas-apparatus	O. Tirrill	Burlington, Mass	June 5, 1866	55, 395
Gas-apparatus	H. H. Wainwright	Philadelphia, Pa	Feb. 4, 1873	135, 455
Gas-apparatus	A. Walker, jr	Burke, Vt	Aug. 7, 1849	6, 626
Gas-apparatus	W. Warner and E. S. Redstreake.	Philadelphia, Pa	Aug. 7, 1866	57, 020
Gas apparatus and carbureter, Portable	J. MacDougall	New York, N. Y	Nov. 26, 1867	71, 514
Gas apparatus and carbureter, Portable	M. A. Root and J. D. Custer	Philadelphia and Norristown, Pa.	Nov. 23, 1869	97, 122
Gas-apparatus, Automatic air-holder for	H. L. McAvoy	Baltimore, Md	Nov. 22, 1864	45, 205
Gas apparatus, Coal	E. Jones	Boston, Mass	Dec. 19, 1871	122, 024
Gas, Apparatus for broiling or roasting by	W. F. Shaw	Boston, Mass	June 19, 1860	28, 781
Gas, Apparatus for carbonizing	W. H. Laubach	Philadelphia, Pa	Apr. 21, 1868	77, 156
Gas, Apparatus for carbureting and regulating the pressure of.	T. G. Springer	New York, N. Y	Sept. 2, 1873	142, 525
Gas, Apparatus for collecting marsh and other	C. S. Hunt and J. B. Knight	Terre Bonne Parish and New Orleans, La.	May 19, 1868	77, 982
Gas, Apparatus for compressing	W. H. Gwynne	New York, N. Y	Aug. 7, 1860	29, 481
Gas, Apparatus for condensing and purifying	A. Hendrickx	New York, N. Y	Aug. 3, 1858	21, 072
Gas, Apparatus for domestic manufacture of	J. W. Brown	Wooster, Ohio	Sept. 15, 1868	82, 080
Gas-apparatus for domestic use	W. Mills and O. H. Burdett	New Athens, Ohio	July 7, 1863	39, 159
Gas, &c., Apparatus for exhausting	R. Laidlaw and J. Thomson	Glasgow, Great Britain	Apr. 25, 1871	114, 156
Gas, Apparatus for exhausting and purifying	P. Munzinger	Philadelphia, Pa	Aug. 29, 1871	118, 472
Gas, Apparatus for exhausting and washing	C. H. Brown and J. J. Thomas	Philadelphia, Pa	Mar. 25, 1873	137, 174
Gas, Apparatus for generating and burning	W. Stewart	Steubenville, Ohio	July 5, 1870	105, 012
Gas, Apparatus for generating and carbureting	C. F. Dunderdale	New York, N. Y	June 1, 1869	90, 644
Gas, Apparatus for generating and carbureting hydrogen.	J. B. Terry	Brooklyn, N. Y	Sept. 12, 1871	118, 983
Gas, Apparatus for generating carbonic-acid	S. T. Bacon	Boston, Mass	Jan. 29, 1867	61, 596
Gas, Apparatus for generating carbonic-acid	J. Matthews	New York, N. Y	June 25, 1872	128, 234
Gas, Apparatus for generating carbonic-acid	J. D. O'Donnell	Washington, D. C	Aug. 8, 1871	117, 805
Gas, Apparatus for generating carbonic-acid	T. Warker	New York, N. Y	Apr. 27, 1859	20, 110
Gas, Apparatus for generating carbonic-acid	O. Zwietusch	Milwaukee, Wis	July 30, 1872	130, 001
Gas, Apparatus for generating carbonic-acid and other.	F. M. Ruschhaupt	Philadelphia, Pa	Apr. 19, 1864	42, 447
Gas, Apparatus for generating, carbureting, and burning hydrogen.	B. Sloper	Saint Louis, Mo	May 30, 1871	115, 369
Gas, Apparatus for generating illuminating	M. P. Coons	Brooklyn, N. Y	Nov. 22, 1859	26, 163
Gas, Apparatus for generating illuminating	D. Davison	New York, N. Y	Feb. 15, 1870	99, 860
Gas, Apparatus for generating illuminating	C. N. Tyler	Washington, D. C	Dec. 28, 1858	22, 463
Gas, Apparatus for generating oxygen	B. R. Smithson	New York, N. Y	Dec. 3, 1867	71, 657
Gas, Apparatus for making illuminating	O. Collins	New York, N. Y	Aug. 22, 1865	49, 505
Gas, Apparatus for making water	W. H. Gwynne	Brooklyn, N. Y	Mar. 11, 1862	34, 632
Gas, Apparatus for manufacturing	J. Absterdam	Boston, Mass	June 15, 1858	20, 534
Gas, Apparatus for manufacturing	W. Beaumont	Paterson, N. J	June 15, 1858	20, 541
Gas, Apparatus for manufacturing	R. Carkhuff	Lewisburgh, Pa	Aug. 2, 1864	43, 668
Gas, Apparatus for manufacturing	B. F. Caston	Washington, D. C	July 26, 1847	5, 210
Gas, Apparatus for manufacturing	J. C. Clapp	Homer, N. Y	July 16, 1867	66, 676
Gas, Apparatus for manufacturing	E. Jones	Boston, Mass	Dec. 19, 1871	122, 025
Gas, Apparatus for manufacturing	W. Maynard	Salem, Mass	Feb. 18, 1873	136, 081
Gas, Apparatus for manufacturing	S. C. Salisbury	New York, N. Y	May 30, 1865	47, 986
Gas, Apparatus for manufacturing	W. H. Spencer	New York, N. Y	July 22, 1873	141, 090
Gas, Apparatus for manufacturing	G. Symes	London, England	Aug. 6, 1872	130, 164
Gas, Apparatus for manufacturing coal	S. B. Darwin	Shrewsbury, England	Feb. 13, 1872	123, 682
Gas, Apparatus for manufacturing coal	W. Gibson	Cambridge, Mass	Sept. 19, 1871	119, 135
Gas, Apparatus for manufacturing coal	A. M. Giles	Boston, Mass	Mar. 26, 1872	125, 043
Gas, Apparatus for manufacturing coal	D. H. Irland	Fayette, N. Y	May 26, 1868	78, 209
Gas, Apparatus for manufacturing coal	E. Jones	Boston, Mass	Feb. 11, 1873	135, 647
Gas, Apparatus for manufacturing heating and illuminating.	J. A. Bassett	Salem, Mass	Sept. 8, 1868	81, 974
Gas, Apparatus for manufacturing hydrocarbon	E. J. L. Caillot	Barcelona, Spain	Dec. 10, 1872	133, 829
Gas, Apparatus for manufacturing hydrocarbon	J. Kidd	New York, N. Y	Aug. 29, 1871	118, 459
Gas, Apparatus for manufacturing hydrogen	W. L. Imlay	Philadelphia, Pa	Nov. 11, 1873	144, 543
Gas, Apparatus for manufacturing illuminating	R. Alsop	Philadelphia, Pa	June 22, 1869	91, 588

Index of patents issued from the United States Patent Office from 1790 *to* 1873, *inclusive*—Continued.

Index of patents issued from the United States Patent Office from 1790 *to* 1873, *inclusive*—Continued.

Invention.	Inventor.	Residence.	Date.	No.
Gas, Converting waste into combustible	B. Todd	Newcastle-upon-Tyne, England.	Dec. 24, 1872	134, 332
Gas cooking-apparatus	M. Germann	Cincinnati, Ohio	Apr. 7, 1868	76, 321
Gas cooler and washer	R. Salter	Cincinnati, Ohio	Mar. 30, 1869	88, 520
Gas-engine	G. B. Brayton	Boston, Mass	Apr. 2, 1872	125, 166
Gas-engine	P. Hugon	Paris, France	Jan. 19, 1864	41, 299
Gas-engine	P. Hugon	Paris, France	Aug. 8, 1865	49, 346
Gas-engine	M. Isnard	New York, N. Y	Dec. 11, 1824	....
Gas-engine	O. H. Kratze	Leipsic, Saxony	Aug. 4, 1863	39, 448
Gas-engine	W. H. Laubach	Philadelphia, Pa	Jan. 28, 1868	73, 816
Gas-engine	C. P. Leavitt	New York, N. Y	Aug. 15, 1871	118, 028
Gas-engine	F. Milion	Paris, France	Apr. 2, 1867	63, 416
Gas-engine	S. Perry	New York, N. Y	Oct. 7, 1846	4, 800
Gas-engine	S. L. Wiegand	Philadelphia, Pa	Oct. 4, 1864	44, 572
Gas-engine, Explosive	A. Drake	Philadelphia, Pa	Apr. 17, 1855	12, 715
Gas, Engines for the utilization of ammoniacal	J. Frot	Orleans, France	Dec. 18, 1866	60, 500
Gas-engine without steam	S. Brown	London, England	Mar. 2, 1824	....
Gas-engines, Arrangement of	J. C. F. Saloman	Baltimore, Md	May 4, 1858	20, 172
Gas-engines, Method of actuating	O. Hammel	Jersey City, N. J	June 18, 1867	65, 808
Gas-engines, Electrical apparatus for lighting	W. M. Storm	New York, N. Y	Sept. 25, 1855	13, 598
Gas-exhausters, &c., Governor for	R. Koch	New York, N. Y	Mar. 25, 1873	137, 215
Gas explosive engine for condensing air	B. T. Babbitt	New York, N. Y	July 14, 1868	79, 938
Gas explosive engine for condensing air	B. T. Babbitt	New York, N. Y	July 14, 1868	79, 939
Gas-extinguisher	D. M. Reynolds	Port Deposit, Md	Dec. 2, 1873	145, 243
Gas-extinguisher and cut-off, Automatic	G. R. Pierce	Grand Rapids, Mich	Aug. 27, 1872	130, 940
Gas-fitter's clamp	J. Peace	Camden, N. J	June 27, 1865	48, 431
Gas-fitter's gage-attachment	H. Getty	Hoboken, N. J	Aug. 29, 1871	118, 603
Gas-fitter's hook-blank	E. P. Gleason	New York, N. Y	Aug. 8, 1865	49, 257
Gas-fitter's pressure-gage	R. B. Donaldson	Washington, D. C	July 2, 1867	66, 227
Gas-fitter's tool	J. Himmer	Hartford, Conn	Sept. 29, 1868	82, 620
Gas-fitter's use, Rod of connected hook-blanks for	J. Fellows and J. W. Lyon	Brooklyn, N. Y	Mar. 16, 1869	87, 919
Gas-fitter's wrench	G. B. Phillips	Albany, N. Y	May 3, 1859	23, 857
Gas-fittings	G. Rosenthal	Pittsburgh, Pa	Oct. 24, 1871	120, 329
Gas-fittings, Device for boring and turning	M. Walty	Buffalo, N. Y	Apr. 29, 1873	138, 456
Gas-fittings, Machine for drilling and tapping	R. T. Crane	Chicago, Ill	Mar. 15, 1870	100, 731
Gas-fittings, Machine for finishing	J. W. Lyon	Brooklyn, N. Y	Oct. 30, 1860	30, 536
Gas-fittings, Machine for reaming and tapping	H. A. Chapin	Springfield, Mass	July 1, 1856	15, 219
Gas-fixture	N. L. Bradley and J. A. Evarts	West Meriden, Conn	Oct. 20, 1868	83, 246
Gas-fixture	E. McClintock	New Brunswick, N. J	Mar. 31, 1868	76, 222
Gas-fixture	D. Milne	Norwich, N. Y	Sept. 6, 1870	107, 191
Gas-fixture	C. A. Shaw	Biddeford, Me	Feb. 16, 1864	41, 644
Gas-fixture	G. G. Sheldon	Chicago, Ill	May 13, 1873	138, 766
Gas-fixture attachment	B. Jacobs	New York, N. Y	May 22, 1866	54, 915
Gas-fixture, Extension	H. Coester	New York, N. Y	Aug. 2, 1870	105, 911
Gas-fixture, Extension	H. Krüger	New York, N. Y	Mar. 29, 1870	101, 369
Gas-fixture, Oxyhydrogen	A. W. Wilkinson	New York, N. Y	Feb. 6, 1872	123, 535
Gas-fixture socket-coupling	T. L. Reed	Providence, R. I	July 17, 1866	56, 448
Gas-fixtures, Device for raising and lowering	J. F. Pond	Cleveland, Ohio	June 14, 1870	104, 197
Gas-fixtures, Extension-slide for	J. Horton	New York, N. Y	June 15, 1869	91, 443
Gas-fixtures, Lock-coupling for	T. L. Reed	Providence, R. I	Jan. 7, 1868	73, 059
Gas-fixtures, Trap-attachment for	J. B. Hyde	New York, N. Y	Sept. 13, 1870	107, 264
Gas-flame expander	H. Whitney	Watertown, Mass	Oct. 17, 1871	120, 139
Gas for fuel, illumination, &c., Manufacture of	W. Elmer	New York, N. Y	Feb. 23, 1869	87, 156
Gas for fuel, &c., Manufacture of inflammable	W. Elmer	New York, N. Y	Aug. 23, 1870	106, 569
Gas for head-lights, Apparatus for generating	H. S. Maxim	New York, N. Y	Feb. 15, 1870	99, 927
Gas for heating and illuminating, Apparatus for the manufacture of.	J. H. Connelly and J. McLure	Wheeling, W. Va	May 23, 1871	115, 028
Gas for heating and illuminating buildings and for other purposes, Manufacture of.	T. Arnold	New York, N. Y	Aug. 15, 1865	49, 358
Gas for heating and illuminating purposes, Method of applying.	C. Pepper	Albany, N. Y	Dec. 14, 1858	22, 331
Gas for heating and illumination, Generating	J. Arbos	Barcelona, Spain	Dec. 8, 1863	40, 804
Gas for heating and lighting, Producing	T. G. Springer	Clinton, Iowa	May 4, 1869	89, 802
Gas for illuminating and heating, Manufacture of	G. Eveleigh	Peckham, Great Britain	Jan. 30, 1872	123, 255
Gas for illuminating, heating, &c., Manufacture of	R. M. Whipple and J. G. Blunt	Chicago, Ill., and Leavenworth, Kans.	Apr. 1, 1873	137, 521
Gas for illuminating, &c., Manufacture of pneumatic.	J. W. Stow	San Francisco, Cal	Apr. 4, 1871	113, 702
Gas for illumination, Apparatus for purifying	D. Garnet	Richfield, N. Y	June 27, 1840	1, 656
Gas for illumination, &c., Manufacture of	J. Kidd	New York, N. Y	Aug. 23, 1870	106, 699
Gas for illumination, Manufacture of coal	J. W. Smith	Washington, D. C	Dec. 13, 1864	45, 460
Gas for illumination, Treating	W. A. Simonds and S. Warner	Boston and East Hampton, Mass.	Mar. 24, 1863	38, 017
Gas for illumination, &c., Treating	L. Stevens	Fitchburgh, Mass	July 3, 1866	56, 116
Gas for light-houses, Generating and burning	B. F. Coston	Washington, D. C	Jan. 31, 1845	3, 894
Gas for lighting and heating purposes, Manufacture and purification of.	G. Eveleigh	London, Great Britain	Jan. 30, 1872	123, 093
Gas for lighting railway-cars, Apparatus for generating and carbureting.	J. H. Steiner	Saint Louis, Mo	July 6, 1869	92, 391
Gas for metallurgic and other purposes, Apparatus for burning.	C. M. Tessié du Motay	Paris, France	Mar. 2, 1869	87, 476
Gas for motive-power	J. Arbos	Barcelona, Spain	Dec. 8, 1863	40, 805
Gas for motive-power for extinguishing fire and for other purposes, Generating.	C. G. Wheeler	Chicago, Ill	Aug. 24, 1869	94, 157
Gas for motive-power, Generating	D. E. Somes	Washington, D. C	Sept. 4, 1866	57, 787
Gas for the production of heat, light, &c., Process of burning.	S. Stevens	New York, N. Y	July 24, 1866	56, 629
Gas for various purposes, Manufacture and application of.	P. Salmon	London, England	Dec. 15, 1868	84, 967
Gas from anthracite coal, Generating inflammable	M. Ward and R. W. Hall	Baltimore, Md	Jan. 19, 1828	....
Gas from bitumen, Manufacture of illuminating	A. Gesner	Halifax, Nova Scotia	Jan. 29, 1850	7, 052
Gas from bituminous coal, Purifying and condensing illuminating.	W. H. St. John and P. Cartwright.	New York, N. Y	Aug. 13, 1872	130, 545
Gas from blast and puddling furnaces, Apparatus for collecting and forcing.	D. H. Geiger	Saint Clair, Pa	May 25, 1869	90, 440
Gas from coal and other materials, Manufacture of illuminating.	G. McKenzie	Glasgow, Scotland	Mar. 1, 1870	100, 432
Gas from coal, Apparatus for producing	T. O'Meara	Brooklyn, N. Y	Aug. 30, 1870	106, 863

Index of patents issued from the United States Patent Office from 1790 *to* 1873, *inclusive*—Continued.

Invention.	Inventor.	Residence.	Date.	No.
Gas from coal-tar, Manufacture of illuminating	J. Kidd	New York, N. Y	Mar. 12, 1872	124, 441
Gas from fuel, Process of producing	J. H. Burgin	Philadelphia, Pa	Sept. 17, 1867	68, 840
Gas from gasoline, Apparatus for making illuminating.	H. S. Maxim and J. Radley	New York, N. Y	May 4, 1869	89, 588
Gas from hydrocarbon liquids, Generating	M. S. Richardson and E. A. Pond.	Rutland, Vt	Oct. 1, 1867	69, 483
Gas from hydrocarbon vapors, Method of generating fixed.	J. Butler	Brooklyn, N. Y	Nov. 24, 1868	84, 259
Gas from hydrocarbons, Apparatus for making	J. T. Tyler and J. J. Johnston	Pittsburgh and Allegheny City, Pa.	June 15, 1869	91, 499
Gas from hydrocarbons, Apparatus for the manufacture of.	J. Rigby	Portsmouth, Ohio	Feb. 11, 1873	135, 666
Gas from hydrocarbons, Apparatus for the manufacture of.	J. H. Smith	Newark, Ohio	Dec. 6, 1870	109, 848
Gas from hydrocarbons, Generating	J. R. Smedberg	San Francisco, Cal	Nov. 22, 1870	109, 460
Gas from hydrocarbons, Manufacture of illuminating.	H. H. Eames and C. J. Eames	Philadelphia, Pa., and New York, N. Y.	Oct. 15, 1872	132, 265
Gas from hydrocarbons, Producing	X. Moussard	Paris, France	Mar. 28, 1871	113, 192
Gasfrom hydrocarbons, Retort for the manufacture of.	B. Sloper and R. M. Potter	New York, N. Y., and Jersey City, N. J.	Apr. 23, 1872	126, 098
Gas from naphtha, &c., Apparatus for generating and burning.	D. H. Lowe	Boston, Mass	Feb. 23, 1869	87, 272
Gas from oil, Apparatus for the manufacture of	M. J. Barry	Washington, D. C	May 7, 1872	126, 510
Gas from oil, Apparatus for the manufacture of	C. and F. A. G. Gearing	Pittsburgh, Pa., and Houston, Tex.	June 18, 1872	128, 199
Gas, &c., from oil, Generating	S. Andrews	Perth Amboy, N. J	Apr. 15, 1831	....
Gas from oil-vapor, hydrogen, and air, Manufacture of illuminating.	T. G. Springer	New York, N. Y	Dec. 17, 1872	134, 109
Gas from oils, Apparatus for making	S. H. Goldthorp	Pittsburgh, Pa	July 15, 1873	140, 911
Gas from peat, Manufacture of illuminating	J. B. Hyde	Newark, N. J	July 19, 1864	43, 585
Gas from peat, Method of making	J. B. Hyde	Newark, N. J	Oct. 18, 1859	25, 866
Gas from petroleum and other hydrocarbons, Apparatus for generating.	G. W. Thompson and J. Foster	Eordentown, N. J	May 5, 1863	38, 429
Gas from petroleum and other wells, Device for collecting.	H. M. Hamilton	Franklin, Pa	June 21, 1864	43, 201
Gas from petroleum, Apparatus for generating	A. I. Ambler	Washington, D. C	Aug. 29, 1871	118, 575
Gas from petroleum, Apparatus for generating	A. I. Ambler	Washington, D. C	Sept. 10, 1872	131, 240
Gas from petroleum, Apparatus for generating	D. M. Graham	Evansville, Ind	May 9, 1865	47, 634
Gas from petroleum, Apparatus for manufacturing	H. H. Edgerton	Fort Wayne, Ind	Apr. 23, 1872	125, 941
Gas from petroleum, Apparatus for manufacturing	W. C. Wren and W. Barker	Brooklyn, N. Y	July 31, 1866	56, 843
Gas from petroleum, Apparatus for producing	H. Hirzel	Leipsic, Saxony	May 14, 1867	64, 672
Gas from petroleum, Apparatus for separating	J. Smith and A. Greig	Tarville, Pa	Mar. 29, 1864	42, 121
Gas from petroleum, Apparatus for separating and collecting.	I. H. Hobbs	Philadelphia, Pa	Sept. 27, 1864	44, 421
Gas from petroleum-distilleries, Mode of utilizing waste.	H. W. C. Tweddle	Pittsburgh, Pa	Dec. 6, 1864	45, 363
Gas from petroleum, Generating	W. M. Sloane	Buffalo, N. Y	Feb. 23, 1869	87, 210
Gas from petroleum, Manufacture and application of.	T. S. Dickerson	Chicago, Ill	Oct. 12, 1869	95, 665
Gas from petroleum, Manufacture of	L. Stevens	Washington, D. C	Feb. 22, 1870	100, 208
Gas from petroleum, Manufacture of	G. W. Wren	Brooklyn, N. Y	Feb. 9, 1869	86, 793
Gas from petroleum, Manufacture of illuminating	R. H. Smith	Pittsburgh, Pa	Aug. 12, 1873	141, 732
Gas from petroleum, Manufacture of illuminating	W. H. Spencer	Brooklyn, N. Y	Feb. 20, 1872	123, 950
Gas from petroleum, Manufacture of illuminating	W. C. and G. W. Wren	Brooklyn, N. Y	Oct. 31, 1871	120, 409
Gas from petroleum, Method of generating	C. Carpenter	Buffalo, N. Y	Aug. 18, 1868	81, 136
Gas from petroleum, naphtha, &c., Apparatus for generating and burning.	M. E. Hanson	Newport, Me	Aug. 28, 1866	57, 502
Gas from resin, Apparatus for making	A. Schmidt	New York, N. Y	Sept. 27, 1859	25, 610
Gas from retorts, &c., Apparatus for exhausting	J. Kidd	New York, N. Y	Jan. 23, 1872	122, 894
Gas from retorts, Apparatus for exhausting	S. Trumbore	Easton, Pa	June 6, 1871	115, 788
Gas from solid carbons, Producing	L. Stevens	Washington, D. C	Feb. 21, 1871	112, 088
Gas from the ignition of explosive compounds applied to machinery.	H. Rogers	New York, N. Y	June 29, 1833	....
Gas from volatile liquids, Manufacture of	C. A. Seely	New York, N. Y	Nov. 17, 1868	84, 219
Gas from volatile oils, &c., Machine for making	J. E. Schwippel	Saint Joseph, Mo	Feb. 23, 1869	87, 299
Gas from wood, Apparatus for making	A. Schmidt	New York, N. Y	Sept. 27, 1859	25, 609
Gas from wood, Making	L. R. Breisach	New York, N. Y	Sept. 6, 1859	25, 316
Gas from wood, Making	A. Schmidt	New York, N. Y	Sept. 4, 1860	29, 941
Gas from wood, Making illuminating	W. P. McConnell	Washington, D. C	Sept. 26, 1854	11, 740
Gas from wood, Retort for the manufacture of	M. Levy	New York, N. Y	Apr. 9, 1861	31, 986
Gas-furnace	J. Jordan	Liverpool, England	Sept. 12, 1871	118, 947
Gas-furnace	L. Stevens	Washington, D. C	July 22, 1873	141, 179
Gas-furnace	L. Stevens	Washington, D. C	July 22, 1873	141, 180
Gas furnace and forge	J. R. Morris	Houston, Tex	Nov. 7, 1871	120, 764
Gas-furnace for heating metals, &c	Z. S. Durfee	New York, N. Y	Sept. 17, 1872	131, 333
Gas, &c., Furnace for the manufacture of	F. Carroll	New Orleans, La	Nov. 11, 1873	144, 506
Gas-furnace, Metallurgic	H. Frank	Pittsburgh, Pa	Feb. 11, 1873	135, 639
Gas-furnace, Metallurgic	J. M. Hartman	Philadelphia, Pa	July 22, 1873	141, 002
Gas-furnace, Metallurgic	C. W. Siemens	Westminster, England	Nov. 8, 1870	109, 064
Gas-furnace, Metallurgic	J. Thomas	Middlesborough, England	June 10, 1873	139, 834
Gas-furnace, Metallurgic	W. C. Wren	Brooklyn, N. Y	July 8, 1873	140, 606
Gas, &c., Gage for measuring the pressure of explosive.	T. Shaw	Philadelphia, Pa	May 3, 1864	42, 627
Gas, Gas-works for making coal	A. Babbett and W. W. Binney	Auburn, N. Y	June 8, 1869	91, 066
Gas-generating and blast-heating apparatus for metallurgic and other purposes.	J. D. Whelpley and J. J. Storer	Boston, Mass	Jan. 24, 1871	111, 288
Gas generating and carbureting machine, Hydrogen	H. I. Hoyt	Norwalk, Conn	May 24, 1870	103, 465
Gas, Generating and charging liquid with	J. H. Laning	Davidson County, Tenn	Dec. 23, 1824	
Gas, Generating and supplying illuminating	E. A. Pond and M. S. Richardson.	Rutland, Vt	June 5, 1866	55, 359
Gas-generating apparatus	Z. S. Durfee	New York, N. Y	Jan. 2, 1872	122, 311
Gas-generating furnace, Portable	W. M. Sloane	Buffalo, N. Y	Nov. 5, 1867	70, 477
Gas-generating furnace, Steam	J. M. Sanders	New York, N. Y	July 6, 1869	92, 212
Gas, Generating heat by hydrogen	L. A. Hall	Newark, N. J	June 3, 1843	3, 121
Gas, Generating hydrogen and hydrocarbon	J. S. Wood	Philadelphia, Pa	Dec. 7, 1869	97, 580
Gas, Generating hydrogen and hydrocarbon	J. S. Wood	Philadelphia, Pa	Apr. 5, 1870	101, 558
Gas, Generating illuminating	A. Hamar	Philadelphia, Pa	Dec. 7, 1869	97, 632
Gas, Generating illuminating	F. King	Richmond, Va	Jan. 14, 1868	73, 252

Index of patents issued from the United States Patent Office from 1790 to 1873, inclusive—Continued.

Invention.	Inventor.	Residence.	Date.	No.
Gas, Generating steam	E. A. Lester	Boston, Mass	Mar. 10, 1828	... ...
Gas-generator	N. Aubin	Albany, N. Y	Sept. 26, 1854	11, 714
Gas-generator	N. Aubin	Albany, N. Y	June 23, 1857	17, 614
Gas-generator	J. Bagot	New York, N. Y	Oct. 24, 1865	50, 548
Gas-generator	J. A. Bassett	Salem, Mass	Oct. 13, 1868	83, 026
Gas-generator	C. F. Brown	Baltimore, Md	Feb. 26, 1850	7, 115
Gas-generator	J. A. Bruce	Baltimore, Md	May 12, 1857	17, 309
Gas-generator	J. Butler	Brooklyn, N. Y	Sept. 15, 1857	18, 184
Gas-generator	J. Butler	New York, N. Y	Jan. 24, 1871	111, 174
Gas-generator	S. Chamberlaine	Philadelphia, Pa	Apr. 10, 1860	27, 775
Gas-generator	M. P. Coons	Brooklyn, N. Y	May 3, 1859	23, 828
Gas-generator	M. P. Coons	Brooklyn, N. Y	Sept. 4, 1866	57, 682
Gas-generator	E. Doty	Janesville, Wis	May 10, 1870	102, 784
Gas-generator	J. J. Ensley	New York, N. Y	Aug. 28, 1866	57, 491
Gas-generator	M. Faloon	Bloomington, Ill	Oct. 29, 1867	70, 184
Gas-generator	T. B. Fogarty	New York, N. Y	May 17, 1870	103, 036
Gas-generator	J. W. Fox and D. H. Irland	Chicago, Ill	Apr. 23, 1872	125, 888
Gas-generator	A. M. Giles	Boston, Mass	Mar. 17, 1857	16, 830
Gas-generator	J. Hansor	Wardsworth Road, England.	Feb. 3, 1857	16, 544
Gas-generator	A. A. Hayes	Boston, Mass	June 16, 1857	17, 574
Gas-generator	J. G. Hock	Newark, N. J	Mar. 30, 1858	19, 777
Gas-generator	C. A. Howard	Pontiac, Mich	Oct. 28, 1856	15, 973
Gas-generator	D. H. Irland	Hudson, Mich	Apr. 22, 1873	138, 160
Gas-generator	P. Kelly	Dayton, Ohio	July 6, 1869	92, 317
Gas-generator	F. King	Richmond, Va	Feb. 11, 1868	74, 230
Gas-generator	H. Lyles	Washington, D. C	Aug. 10, 1858	21, 142
Gas-generator	R. J. Malcolm	Cincinnati, Ohio	June 2, 1868	78, 600
Gas-generator	A. L. McKay	Bolton, Miss	Apr. 26, 1870	102, 293
Gas generator	J. McWilliams	Pittsburgh, Pa	June 5, 1866	55, 328
Gas-generator	M. Pettenkofer and C. Ruland	Munich, Bavaria	May 20, 1856	14, 926
Gas-generator	A. Pierce	Philadelphia, Pa	June 27, 1846	4, 604
Gas-generator	A. Pollock	Washington, D. C	Aug. 11, 1857	17, 981
Gas-generator	D. W. Ranke	Limestoneville, Pa	June 9, 1868	78, 828
Gas-generator	S. C. Salisbury	New York, N. Y	Aug. 6, 1872	130, 318
Gas-generator	D. F. Scheaf	Dayton, Ohio	Jan. 7, 1868	73, 124
Gas-generator	A. Schwaninger	Milwaukee, Wis	July 24, 1860	29, 318
Gas-generator	G. W. R. Seal	Winchester, Va	July 13, 1858	20, 897
Gas-generator	S. Skinner	Yonkers, N. Y	Oct. 13, 1857	18, 414
Gas-generator	W. M. Sloan	Buffalo, N. Y	Apr. 7, 1868	76, 535
Gas-generator	J. W. Smith	Washington, D. C	June 30, 1857	17, 704
Gas-generator	J. H. Steiner	Kansas City, Mo	Jan. 19, 1869	85, 972
Gas-generator	C. A. Stevens	New York, N. Y	Jan. 2, 1872	122, 499
Gas-generator	W. N. Taylor	Philadelphia, Pa	May 4, 1858	20, 177
Gas-generator	A. K. Tupper	Milford, Mich	May 22, 1860	28, 421
Gas-generator	P. H. Vander Weyde	Philadelphia, Pa	Feb. 12, 1867	62, 095
Gas-generator	J. Watson and E. Cart	Hull, England	Sept. 18, 1849	6, 729
Gas-generator	E. W. Whitehead and J. L. Conklin.	Newark, N. J	June 16, 1857	17, 599
Gas-generator	A. B. Wilson	Waterbury, Conn	Oct. 26, 1858	21, 914
Gas-generator	J. S. Wood	Brooklyn, N. Y	May 13, 1873	138, 778
Gas-generator and burner	C. B. Loveless	Syracuse, N. Y	Dec. 21, 1869	98, 174
Gas-generator and burner	C. N. Tyler	Washington, D. C	July 24, 1860	29, 328
Gas-generator and carbureter	C. F. Dunderdale	New York, N. Y	Nov. 9, 1869	96, 565
Gas-generator and carbureter	J. B. Olney	New York, N. Y	Dec. 22, 1868	85, 239
Gas-generator and carbureter	J. H. Steiner	Cincinnati, Ohio	May 14, 1872	126, 652
Gas-generator and carbureter	A. Stevens	Fitchburgh, Mass	Dec. 21, 1869	98, 118
Gas-generator and carbureter	W. Thompson	Cleveland, Ohio	Apr. 30, 1867	64, 382
Gas generator and carbureter, Hydrogen	J. S. Wood	Philadelphia, Pa	Apr. 5, 1870	101, 557
Gas generator and carbureter, Illuminating	A. Stevens	Fitchburgh, Mass	Nov. 30, 1869	97, 457
Gas generator, Carbonic-acid	P. and F. Hinkel	New York, N. Y	Apr. 11, 1865	47, 205
Gas generator, Carbonic-acid	J. W. Stanton	Brooklyn, N. Y	Apr. 1, 1873	137, 330
Gas generator, Carbonic-acid	F. W. Wiesebrock	New York, N. Y	Nov. 18, 1873	144, 719
Gas-generator, Compressed	J. Gros	Paris, France	Apr. 3, 1866	53, 758
Gas-generator for heating purposes	L. Stevens	Washington, D. C	Oct. 11, 1870	108, 302
Gas generator, Illuminating	S. N. Chamberlin	Abington, Mass	Feb. 27, 1866	52, 946
Gas generator, Illuminating	G. P. Ganster	New York, N. Y	Mar. 31, 1868	76, 182
Gas generator, Illuminating	H. B. Myer	Philadelphia, Pa	Aug. 28, 1866	57, 551
Gas-generator, Portable	C. B. Loveless	Syracuse, N. Y	Feb. 23, 1869	87, 271
Gas-generator, Portable	W. A. Simonds	Boston, Mass	Sept. 1, 1857	18, 109
Gas-generator, Portable	W. Snodgrass	Macomb, Ill	Feb. 22, 1870	100, 080
Gas generator, Steam	H. S. Maxim	New York, N. Y	Nov. 26, 1867	71, 400
Gas generator, Steam petroleum	R. M. Whipple and A. I. Ambler.	Chicago, Ill	July 13, 1869	92, 687
Gas generator, Wood	W. D. Porter	New York, N Y	Aug. 22, 1854	11, 560
Gas generator, Wood	C. F. Werner	New York, N. Y	June 2, 1857	17, 465
Gas-generators, Feed-apparatus to	S. Meredith	Erie, Pa	Sept. 6, 1853	9, 994
Gas-generators, Feeding	C. B. Loveless	Syracuse, N. Y	June 2, 1857	17, 435
Gas-generators, Method of cleansing	S. Coates	New York, N. Y	Mar. 23, 1858	19, 686
Gas-governor	E. Beggs	San Francisco, Cal	May 5, 1868	77, 572
Gas-governor and by-pass	P. Munzinger	Philadelphia, Pa	June 3, 1873	139, 514
Gas-heater	A. Adams	Sturgis, Mich	June 21, 1870	104, 531
Gas-heater	B. Allen	Boston, Mass	Sept. 7, 1869	94, 538
Gas-heater	J. Bannihr	Hempstead, N. Y	Nov. 9, 1869	96, 539
Gas-heater	E. Barnes	London, England	Nov. 5, 1872	132, 793
Gas-heater	J. Q. Birkey	Philadelphia, Pa	Oct. 31, 1865	50, 678
Gas-heater	A. L. Bogart	New York, N. Y	July 27, 1869	92, 931
Gas-heater	C. Burnham	Philadelphia, Pa	May 19, 1868	78, 053
Gas-heater	C. C. Burt	Jackson, Mich	May 24, 1870	103, 297
Gas-heater	L. Bush	Boston, Mass	Apr. 26, 1870	102, 258
Gas-heater	J. Comly	Philadelphia, Pa	May 17, 1870	103, 017
Gas-heater	J. E. Cone	Chicago, Ill	July 12, 1870	105, 305
Gas heater	S. G. Dare	New York, N. Y	Nov. 1, 1870	108, 888
Gas-heater	S. Darling	Bangor, Me	Feb. 4, 1868	73, 957
Gas-heater	L. A. and W. G. Duval	Charleston, S. C	Jan. 12, 1869	85, 804
Gas-heater	O. C. Fox	Georgetown, D. C	June 16, 1868	78, 951
Gas-heater	C. Geisse	Taycheedah, Wis	July 10, 1866	56, 204
Gas-heater	R. George	Kilburn, England	Oct. 25, 1870	108, 583

Index of patents issued from the United States Patent Office from 1790 *to* 1873, *inclusive*—Continued.

Invention.	Inventor.	Residence.	Date.	No.
Gas-heater	J. T. Greenwood	Beloit, Wis	Sept. 8, 1868	81, 892
Gas-heater	D. G. Haskins	Cambridge, Mass	Nov. 19, 1867	71, 002
Gas-heater	D. G. Haskins	Cambridge, Mass	Jan. 11, 1870	98, 767
Gas-heater	D. G. Haskins	Cambridge, Mass	Nov. 21, 1871	121, 170
Gas-heater	J. P. Hayes	Philadelphia, Pa	Oct. 10, 1871	119, 761
Gas-heater	J. S. Hull	Cincinnati, Ohio	Mar. 31, 1868	76, 195
Gas-heater	J. H. Jones	New York, N. Y	Apr. 25, 1865	47, 426
Gas-heater	W. Jones	Chelsea, Mass	June 23, 1868	79, 230
Gas-heater	D. Kellogg	Jackson, Mich	Mar. 10, 1868	75, 429
Gas-heater	A. Komp	New York, N. Y	Sept. 10, 1872	131, 280
Gas-heater	H. Y. Lazear	New York, N. Y	July 14, 1868	79, 989
Gas-heater	H. Y. Lazear and J. L. Sharp	New York, N. Y	July 6, 1869	92, 322
Gas-heater	D. H. Lowe	Boston, Mass	Dec. 15, 1868	84, 889
Gas-heater	J. Lundgren	New York, N. Y	Aug. 2, 1870	106, 070
Gas-heater	S. T. McDougall	Brooklyn, N. Y	May 5, 1868	77, 635
Gas-heater	G. F. Meiggs	Boston, Mass	May 17, 1870	103, 222
Gas-heater	A. H. Mershon	Philadelphia, Pa	May 7, 1872	126, 473
Gas-heater	H. B. Musgrave	Cincinnati, Ohio	Oct. 26, 1869	96, 260
Gas-heater	C. H. Prentiss	Cleveland, Ohio	Mar. 4, 1873	136, 383
Gas-heater	O. M. Reynolds and D. T. Kitchell.	Oil City, Pa	Apr. 4, 1871	113, 567
Gas-heater	E. O. Schartan	Philadelphia, Pa	Nov. 9, 1869	96, 623
Gas-heater	P. Schreyer	New York, N. Y	Apr. 14, 1868	76, 828
Gas-heater	W. F. Shaw	Boston, Mass	Jan. 23, 1855	12, 267
Gas-heater	J. Sheedy	New York, N. Y	Apr. 12, 1870	101, 928
Gas-heater	G. Smith	Ayer, Mass	Apr. 16, 1872	125, 850
Gas-heater	W. C. Trowbridge	New York, N. Y	Feb. 16, 1869	87, 014
Gas-heater	H. F. W. Wesche	New York, N. Y	June 6, 1871	115, 666
Gas-heater and petroleum-stove	A. T. Boon	Galesburgh, Ill	Mar. 19, 1867	63, 004
Gas-heater boiler	C. Geisse	Taycheedah, Wis	Oct. 16, 1866	58, 807
Gas-heater, Drum	W. H. Towers	New York, N. Y	Dec. 27, 1864	45, 655
Gas-heater for cooking, &c	J. S. Hull	Cincinnati, Ohio	July 3, 1866	56, 052
Gas heating and cooking apparatus	R. S. Andrews	Baltimore, Md	May 12, 1857	17, 251
Gas heating and cooking apparatus	F. A. Jacquet	Paris, France	June 1, 1869	90, 846
Gas heating and lighting apparatus for locomotive and railway cars.	J. H. Connelly	Wheeling, W. Va	May 23, 1871	115, 027
Gas-heating apparatus	P. S. Devlan	Camden, N. J	Jan. 26, 1858	19, 185
Gas-heating apparatus	C. M. Guild	Brooklyn, N. Y	July 11, 1854	11, 244
Gas-heating apparatus	D. G. Haskins	Cambridge, Mass	Apr. 10, 1866	53, 820
Gas-heating apparatus	W. F. Shaw	Boston, Mass	Feb. 26, 1856	14, 325
Gas-heating apparatus	S. L. Wiegand	Philadelphia, Pa	Nov. 10, 1863	40, 591
Gas heating or cooking apparatus	W. F. Shaw	Boston, Mass	Mar. 11, 1856	14, 414
Gas heating or cooking apparatus	W. F. Shaw	Boston, Mass	Nov. 4, 1856	16, 031
Gas, Heating, warming, and cooking by	W. Boggett and G. B. Pettit	Westminster, England	Apr. 18, 1854	10, 793
Gas-holder	S. Hill and W. I. Wood	Rochester, N. Y	Nov. 6, 1855	13, 754
Gas-holder	C. A. Stebbins	Springfield, Mass	Apr. 20, 1869	89, 090
Gas-holder	T. R. White	Philadelphia, Pa	Aug. 6, 1872	130, 346
Gas-holder, Dry	N. Aubin	Albany, N. Y	July 2, 1861	32, 676
Gas-holder, Portable	O. L. Lawson	New York, N. Y	June 26, 1860	28, 873
Gas-holder, Portable	J. McFarlan	Brooklyn, N. Y	Sept. 27, 1859	25, 607
Gas, Illuminating	W. Elmer	New York, N. Y	Jan. 17, 1865	45, 915
Gas, Illuminating	C. S. Hunt and J. B. Knight	Terre Bonne Parish and New Orleans, La.	May 19, 1868	77, 983
Gas-illuminator	M. Andrew	New York, N. Y	Sept. 5, 1871	118, 669
Gas-illuminator	J. B. Van Patten	Newark, N. J	Mar. 4, 1873	136, 565
Gas in apparatus for diving, Regulating the flow of.	B. Bouquayrol	Paris, France	Nov. 6, 1866	59, 529
Gas in fire-extinguishers, &c., Composition for generating.	J. F. Babcock	Boston, Mass	Sept. 29, 1868	82, 582
Gas in fu naces, Method of burning	J. C. Tiffany	Portsmouth, N. H	Dec. 2, 1873	145, 257
Gas, In the preparation of materials to be used in the purification of.	A. A. Croll	London, England	Apr. 4, 1865	47, 160
Gas into furnaces, retorts, converters, &c., Introducing oxygen.	O. M. Phillips	New York, N. Y	Jan. 26, 1869	86, 247
Gas, iron, and steel from the ore, Process and furnace for making.	T. J. Chubb	Williamsburgh, N. Y	Nov. 19, 1872	133, 202
Gas-jets, Device for igniting	G. W. Parke and J. G. Stowe	Bloomington, Ill	July 16, 1872	129, 363
Gas-lamps, &c., Apparatus for lighting street	J. W. Beard	St. John's, New Brunswick	Dec. 10, 1867	71, 842
Gas-light and pressure-indicator	W. W. Goodwin	Camden, N. J	Apr. 22, 1873	138, 016
Gas-light apparatus	J. Crutchett	Cincinnati, Ohio	May 6, 1844	3, 573
Gas-light globe	C. Collier	Selma, Ala	Nov. 29, 1870	109, 589
Gas light globe	T. Trudeau	Ottawa, Canada	July 25, 1871	117, 486
Gas-light-governor case	N. Tufts	Boston, Mass	May 25, 1869	90, 609
Gas-light multiplier	J. F. Boynton	Syracuse, N. Y	Sept. 5, 1865	49, 705
Gas-light multiplier	J. F. Boynton	Syracuse, N. Y	Oct. 8, 1867	69, 621
Gas-light shade	W. Fullagar	Brooklyn, N. Y	June 7, 1870	103, 865
Gas-lights, Apparatus for lighting and extinguishing.	N. S. Manross	Forestville, Conn	Feb. 5, 1861	31, 322
Gas-lights from cotton-seed, Manufacture of	D. Olmstead	New Haven, Conn	July 21, 1827	....
Gas-lights, Means of extinguishing	C. H. Harwood	Salem, Mass	Aug. 22, 1871	118, 239
Gas-lights, Means of extinguishing	O. S. Judd	New Britain, Conn	June 10, 1862	35, 525
Gas-lights, Mode of extinguishing	H. K. Symmes	Newton, Mass	Feb. 14, 1860	27, 170
Gas-lights, Packing for sliding	G. Clay	New York, N. Y	Nov. 8, 1859	26, 015
Gas-lighter	W. B. Johns	United States Army	Aug. 16, 1859	25, 122
Gas-lighter	E. D. McCracken	New York, N. Y	Feb. 18, 1873	136, 084
Gas-lighter	W. Wiler and L. Moss	Philadelphia, Pa	June 26, 1855	13, 134
Gas-lighter, Automatic	A. N. Allen and R. H. Dewey	Pittsfield, Mass	Oct. 24, 1871	120, 226
Gas-lighter, Electric	A. N. Allen and R. H. Dewey	Pittsfield, Mass	Nov. 28, 1871	121, 308
Gas-lighter, Electric	E. J. Frost and G. A. Lawrence	Springfield, Mass	Aug. 8, 1865	49, 254
Gas-lighter, Fulminate	H. B. Stockwell	Brooklyn, N. Y	June 27, 1865	48, 459
Gas, Lighting	A. Barbarin	New Orleans, La	Nov. 20, 1866	59, 754
Gas, Lighting	A. Barbarin	New Orleans, La	Apr. 30, 1867	64, 188
Gas, Lighting	W. W. Batchelder	New York, N. Y	May 14, 1867	64, 738
Gas, Lighting	W. W. Batchelder	New York, N. Y	Nov. 3, 1868	83, 593
Gas, Lighting	E. D. McCracken	New York, N. Y	Oct. 17, 1871	120, 082
Gas, Lighting and extinguishing	M. G. Farmer	Salem, Mass	Dec. 24, 1867	72, 616
Gas lighting and extinguishing apparatus	A. N. Allen and R. H. Dewey	Pittsfield, Mass	Apr. 12, 1870	101, 806
Gas lighting and extinguishing apparatus, Automatic.	E. P. Russell	Manlius, N. Y	June 25, 1867	66, 044
Gas lighting and extinguishing apparatus, Electric	J. Vansant	San Francisco, Cal	Apr. 4, 1871	113, 370

Index of patents issued from the United States Patent Office from 1790 *to* 1873, *inclusive*—Continued.

Invention.	Inventor.	Residence.	Date.	No.
Gas lighting and extinguishing, Automatic machine for.	H. R. Richardson	Manlius, N. Y	June 26, 1866	55, 909
Gas lighting and extinguishing by electricity, Apparatus for.	E. E. Bean	Boston, Mass	June 1, 1869	90, 629
Gas lighting and extinguishing by electricity, Apparatus for.	F. Bean	Somerville, Mass	Oct. 18, 1870	108, 434
Gas lighting and extinguishing by electricity, Apparatus for.	J. P. Tirrell	Charlestown, Mass	Nov. 28, 1871	121, 301
Gas lighting and extinguishing by electricity, Apparatus for.	J. P. Tirrell	Charlestown, Mass	Nov. 28, 1871	121, 302
Gas lighting and extinguishing by electricity, Apparatus for.	J. Vansant	San Francisco, Cal	Mar. 19, 1872	124, 773
Gas-lighting apparatus	A. Barbarin	New Orleans, La	Sept. 1, 1868	81, 735
Gas-lighting apparatus	H. G. Fiske	Springfield, Mass	May 25, 1869	90, 519
Gas-lighting apparatus	E. D. McCracken	New York, N. Y	Oct. 17, 1871	120, 083
Gas-lighting apparatus	A. Potter	Philadelphia, Pa	July 8, 1873	140, 591
Gas-lighting apparatus, Automatic	F. Korwan	Mannheim, Germany	Mar. 11, 1873	136, 739
Gas-lighting apparatus, Electric	C. N. Ealer	Opelousas, La	Sept. 20, 1870	107, 465
Gas-lighting apparatus, Electrical	J. Vansant	San Francisco, Cal	June 10, 1873	139, 692
Gas-lighting apparatus, Electro-magnetic	A. N. Allen and R. H. Dewey	Pittsfield, Mass	Oct. 3, 1871	119, 492
Gas-lighting apparatus, Electro-magnetic	F. Heyl and P. Diehl	East New York, N. Y	June 20, 1871	116, 054
Gas-lighting apparatus, Electro-magnetic	A. W. Scharit, L. A. Hudson, and D. Lyman.	Saint Louis, Mo., Syracuse, N. Y., and Parkman, Ohio.	Nov. 2, 1869	96, 488
Gas-lighting apparatus, Magneto-electric	A. N. Allen and R. H. Dewey	Pittsfield, Mass	July 4, 1871	116, 660
Gas-lighting by electrical sparks, Apparatus for	J. Vansant	San Francisco, Cal	May 21, 1872	127, 000
Gas-lighting by electricity, &c	A. N. Allen	Pittsfield, Mass	Feb. 6, 1872	123, 439
Gas-lighting by electricity	E. E. Bean and W. H. Mumler	Boston, Mass	June 4, 1867	65, 466
Gas-lighting by electricity	A. L. Bogart	New York, N. Y	Oct. 3, 1871	119, 561
Gas-lighting by electricity	R. Cornelius	Philadelphia, Pa	June 4, 1861	32, 471
Gas-lighting by electricity	R. Cornelius	Philadelphia, Pa	May 19, 1863	38, 562
Gas-lighting by electricity	R. Cornelius	Philadelphia, Pa	May 19, 1863	38, 563
Gas-lighting by electricity	S. Gardiner, jr	New York, N. Y	Dec. 22, 1857	18, 945
Gas-lighting by electricity	S. Gardiner, jr	New York, N. Y	Mar. 30, 1858	19, 766
Gas-lighting by electricity	S. Gardiner, jr	New York, N. Y	Nov. 29, 1864	45, 239
Gas-lighting by electricity	S. Gardiner, jr	New York, N. Y	Nov. 29, 1864	45, 240
Gas-lighting by electricity	S. Gardiner, jr	New York, N. Y	Nov. 29, 1864	45, 241
Gas-lighting by electricity	S. Gardiner, jr	New York, N. Y	June 19, 1866	55, 641
Gas-lighting by electricity	J. M. Higgins	Saint Louis, Mo	Nov. 19, 1867	71, 005
Gas-lighting by electricity	G. G. Percival	Brooklyn, N. Y	Jan. 15, 1867	61, 247
Gas-lighting by electricity	R. G. Pike	New York, N. Y	Sept. 4, 1866	57, 761
Gas-lighting by electricity	W. A. Pitt	New York, N. Y	June 10, 1873	139, 811
Gas-lighting by electricity	A. Wilson	New York, N. Y	Aug. 16, 1859	25, 167
Gas-lighting by electricity, Apparatus for	A. Barbarin	New Orleans, La	June 1, 1869	90, 626
Gas-lighting by electricity, Apparatus for	W. W. Batchelder	Boston, Mass	Sept. 7, 1869	94, 545
Gas-lighting by electricity, Apparatus for	W. W. Batchelder	Boston, Mass	May 17, 1870	103, 127
Gas-lighting by electricity, Apparatus for	W. W. Batchelder	Boston, Mass	June 7, 1870	103, 831
Gas-lighting by electricity, Apparatus for	S. Gardiner, jr	New York, N. Y	Feb. 13, 1867	62, 125
Gas-lighting by electricity, Apparatus for	W. H. Kelly	New York, N. Y	Mar. 15, 1870	100, 773
Gas-lighting by electricity, Apparatus for	W. Klinkerfues	Göttingen, Germany	Jan. 2, 1872	122, 389
Gas-lighting by electricity, Apparatus for	W. J. Morris and W. J. Reid	New York, N. Y	Apr. 5, 1870	101, 491
Gas-lighting by electricity, Apparatus for	H. T. Robbins	Boston, Mass	Sept. 6, 1870	107, 105
Gas-lighting by electricity, Apparatus for	T. P. Tirrell	Charlestown, Mass	Aug. 20, 1872	130, 770
Gas-lighting by electricity, Apparatus for	J. Vansant	San Francisco, Cal	Oct. 31, 1871	120, 469
Gas-lighting by electricity, Instrument for	R. Cornelius	Philadelphia, Pa	Oct. 18, 1864	44, 708
Gas-lighting by electro-galvanic battery, Method of	A. Wilson	Boston, Mass	Feb. 23, 1858	19, 460
Gas-lighting by frictional electricity, Method of	W. W. Batchelder	New York, N. Y	Jan. 31, 1860	26, 964
Gas-lighting by galvanic electricity, Method of	W. W. Hopkins	Amelia, Ohio	Feb. 15, 1859	22, 952
Gas-lighting, Construction of instrument for	A. Damarin and G. C. Brower	New Orleans, La	Apr. 24, 1860	27, 971
Gas-lighting device	E. P. Gleason	New York, N. Y	Dec. 8, 1868	84, 689
Gas-lighting device	J. G. Harper	New York, N. Y	July 25, 1865	48, 932
Gas-lighting, &c., Distributing electricity for	S. Gardiner, jr	New York, N. Y	Apr. 9, 1872	125, 387
Gas-lighting, Electric apparatus for	F. Bean	Boston, Mass	Jan. 28, 1868	73, 868
Gas-lighting, Electric torch for	W. W. Batchelder	New York, N. Y	Apr. 30, 1872	126, 251
Gas-lighting, Electrical	C. G. Mueller and H. Meier	Hanover, Germany	June 3, 1873	139, 599
Gas-lighting, Electrical apparatus for	R. Cornelius	Philadelphia, Pa	May 24, 1864	42, 840
Gas-lighting, Electrical apparatus for	S. B. H. Vance	New York, N. Y	Feb. 5, 1861	31, 359
Gas-lighting, Implement for	T. W. Houchin	Morrisania, N. Y	Dec. 10, 1867	71, 880
Gas-lighting instrument	T. W. Houchin	Morrisania, N. Y	Apr. 11, 1865	47, 207
Gas-lighting instrument	A. Shipley and W. T. Mersereau	Newark, N. J	May 12, 1868	77, 771
Gas, Lighting street	E. P. Russell	Manlius, N. Y	Oct. 29, 1867	70, 272
Gas-lighting torch	E. P. Gleason	Brooklyn, N. Y	June 24, 1873	140, 191
Gas-lighting torch	C. H. Roberts	Troy, N. Y	Feb. 18, 1873	136, 006
Gas-lighting torch, Electric	W. W. Batchelder	New York, N. Y	Mar. 5, 1872	124, 317
Gas-lime, &c., Obtaining bisulphide or carbon from	J. Kircher	New York, N. Y	May 6, 1873	138, 504
Gas-lime, &c., Obtaining sulphur, sulphuric acid, and sulphurets of sodium and potassium from.	J. Kircher	New York, N. Y	Apr. 8, 1873	137, 692
Gas-machine	S. R. Ball	Hyde Park, Ill	Jan. 4, 1870	98, 462
Gas-machine	N. W. Bancroft	Worcester, Mass	Oct. 6, 1868	82, 786
Gas-machine	W. W. Binny	Auburn, N. Y	June 14, 1870	104, 253
Gas-machine	E. H. and W. H. Covel	New York, N. Y	Nov. 24, 1868	84, 460
Gas-machine	T. B. Fogarty	Brooklyn, N. Y	June 6, 1871	115, 591
Gas-machine	T. B. Fogarty	Kings County, N. Y	June 6, 1871	115, 592
Gas-machine	T. B. Fogarty	Brooklyn, N. Y	June 6, 1871	115, 593
Gas-machine	T. B. Fogarty	Brooklyn, N. Y	June 6, 1871	115, 594
Gas-machine	T. B. Fogarty	Brooklyn, N. Y	June 6, 1871	115, 595
Gas-machine	T. B. Fogarty	Brooklyn, N. Y	June 6, 1871	115, 596
Gas-machine	T. B. Fogarty	Brooklyn, N. Y	June 6, 1871	115, 597
Gas-machine	T. B. Fogarty	Brooklyn, N. Y	Sept. 26, 1871	119, 227
Gas-machine	T. B. Fogarty	New York, N. Y	Feb. 18, 1873	135, 980
Gas-machine	J. P. Gallagher	Saint Louis, Mo	July 12, 1870	105, 190
Gas-machine	D. M. Graham	Evansville, Ind	July 28, 1868	80, 404
Gas-machine	J. W. Groat	Fremont, Ohio	May 25, 1869	90, 445
Gas-machine	C. C. Hare	Kansas City, Mo	May 18, 1869	90, 259
Gas-machine	J. Kaufmann	Jackson, Miss	May 28, 1872	127, 245
Gas-machine	P. H. Lawler and W. H. Gibson	Rochester, N. Y	Dec. 22, 1868	85, 104
Gas-machine	H. S. Maxim	New York, N. Y	Sept. 8, 1868	81, 922
Gas-machine	H. S. Maxim	Brooklyn, N. Y	Oct. 24, 1871	120, 302

Index of patents issued from the United States Patent Office from 1790 to 1873, inclusive—Continued.

Invention.	Inventor.	Residence.	Date.	No.
Gas-machine	H. S. Maxim and J. F. Lockwood.	New York, N. Y	June 2, 1868	78, 465
Gas-machine	W. T. McMillen	Richmond, Ind.	Nov. 14, 1871	120, 824
Gas machine	I. Prichard	Terre Haute, Ind.	Feb. 4, 1868	74, 132
Gas-machine	W. A. Simonds	Boston, Mass	May 2, 1871	114, 358
Gas-machine	A. R. Spang and D. F. Scheaf	Dayton, Ohio	Sept. 21, 1869	94, 982
Gas-machine	J. D. Spang	Dayton, Ohio	Dec. 22, 1868	85, 185
Gas-machine	T. G. Springer	Clinton, Iowa	Dec. 7, 1869	97, 748
Gas-machine	T. G. Springer	Saint Louis, Mo	Mar. 8, 1870	100, 684
Gas-machine	T. G. Springer	Saint Louis, Mo	Mar. 21, 1871	112, 975
Gas-machine	T. G. Springer	Fayette City, Pa	Oct. 3, 1871	119, 663
Gas-machine	W. Thompson	Cleveland, Ohio	Nov. 10, 1868	84, 021
Gas-machine	H. Tilden	Boston, Mass	Nov. 23, 1869	97, 247
Gas-machine	H. Tilden	Philadelphia, Pa	May 6, 1873	138, 715
Gas-machine	J. H. Van Houten	Newark, N. J	Nov. 22, 1870	110, 568
Gas-machine	H. Wain	Ravenna, Ohio	Oct. 20, 1868	83, 344
Gas machine, Air	P. Werni	Newark, N. J	Aug. 19, 1873	141, 973
Gas machine, Electro-hydrocarbon	H. J. Smith	Boston, Mass	Mar. 22, 1870	101, 171
Gas, Machine for generating and carbureting	C. F. Dunderdale	New York, N. Y	July 20, 1869	92, 943
Gas machine, Pneumatic	D. Boyle	San Francisco, Cal	June 7, 1870	103, 836
Gas-machine, Portable	B. Sloper	New York, N. Y	Oct. 7, 1873	143, 426
Gas-machines, Adjustable pulley for	E. F. Van Houten	Newark, N. J	Aug. 19, 1873	141, 967
Gas machines and carbureters, Safety-apparatus for	J. F. Boynton	Syracuse, N. Y	Sept. 18, 1866	58, 055
Gas-main	I. N. Stanley	Brooklyn, N. Y	Aug. 7, 1866	57, 006
Gas-main pipes, Condensing liquids in	J. Walton	Louisville, Ky	June 30, 1857	17, 711
Gas-mains, Leveling hydraulic	P. Munzinger	Philadelphia, Pa	May 25, 1869	90, 379
Gas-mains, Straddle-pipe for hydraulic	P. Munzinger	Philadelphia, Pa	May 4, 1869	89, 592
Gas, Making	J. Hansor	Wardsworth Road, England.	Feb. 10, 1857	16, 591
Gas, Making	J. McGeary	Salem, Mass	Apr. 7, 1868	76, 338
Gas, Making and application of combustible	W. Garnet	New York, N. Y	Dec. 27, 1822	....
Gas, Making and using nitrous-oxide	P. H. Vander Weyde	Philadelphia, Pa	Feb. 12, 1867	62, 094
Gas, Making carbureted hydrogen	B. Kugler	Philadelphia, Pa	Apr. 23, 1816	....
Gas, Making illuminating	H. W. Adams	New York, N. Y	Aug. 10, 1852	9, 175
Gas, Making illuminating	N. Aubin	Albany, N. Y	Jan. 8, 1856	14, 045
Gas, Making illuminating	G. Danré, P. Nicolas, and F. Loperz.	Marseilles, France	Dec. 28, 1852	9, 501
Gas, Making illuminating	R. Grant	Brooklyn, N. Y	Jan. 27, 1857	16, 480
Gas, Making illuminating	J. L. Graves	Springfield, Mass	June 4, 1872	127, 411
Gas, Making illuminating	L. L. Hill	Hudson, N. Y	June 17, 1862	35, 610
Gas, Making illuminating	J. Howarth	Salem, Mass	May 17, 1864	42, 771
Gas, Making illuminating	A. C. Rand	Union Mills, Pa	June 25, 1867	66, 041
Gas, Manufacture of	J. H. Connelly	Wheeling, W. Va	Sept. 18, 1866	58, 221
Gas, Manufacture of	J. H. Connelly	Wheeling, W. Va	May 21, 1867	64, 845
Gas, Manufacture of	J. Crutchett	Stroud, England	Mar. 12, 1867	62, 823
Gas, Manufacture of	L. D. Gale	Washington, D. C	Nov. 8, 1859	26, 028
Gas, Manufacture of	L. D. Gale	Washington, D. C	Nov. 8, 1859	26, 030
Gas, Manufacture of	D. C. Knab	Paris, France	Mar. 9, 1858	19, 575
Gas, Manufacture of	T. S. C. Lowe	Norristown, Pa	Aug. 13, 1872	130, 382
Gas, Manufacture of	S. T. McDougall	New York, N. Y	Mar. 13, 1860	27, 460
Gas, Manufacture of	G. Michiels	Paris, France	Oct. 3, 1846	4, 789
Gas, Manufacture of	C. Noble	New York, N. Y	July 11, 1865	48, 714
Gas, Manufacture of	R. H. Patterson	Hammersmith, England	Apr. 8, 1873	137, 713
Gas, Manufacture of	J. T. Rich	Philadelphia, Pa	July 30, 1867	67, 217
Gas, Manufacture of	F. A. Sabbaton	Troy, N. Y	May 20, 1873	139, 083
Gas, Manufacture of	H. Skoines	London, Great Britain	Nov. 25, 1873	145, 021
Gas, Manufacture of	J. L. Stewart	East Boston, Mass	June 1, 1858	20, 453
Gas, Manufacture of	C. N. Tyler	Washington, D. C	June 21, 1859	24, 506
Gas, Manufacture of carbon-black for natural carbureted hydrogen.	J. Howarth	Salem, Mass	Sept. 17, 1872	131, 446
Gas, Manufacture of coal	J. A. Bassett	Salem, Mass	Aug. 29, 1871	118, 579
Gas, Manufacture of coal	D. Davison	New York, N. Y	Oct. 24, 1871	120, 151
Gas, Manufacture of coal	W. Gibson	Cambridge, Mass	Oct. 5, 1869	95, 459
Gas, Manufacture of coal	W. H. Gwynne	New York, N. Y	Dec. 15, 1863	40, 925
Gas, Manufacture of hydrocarbon	J. A. Bassett	Salem, Mass	Jan. 3, 1860	26, 644
Gas, Manufacture of hydrocarbon	J. Calkins	Hudson, N. Y	Jan. 3, 1860	26, 649
Gas, Manufacture of hydrocarbon	F. A. Sabbaton	Troy, N. Y	July 11, 1871	116, 994
Gas, Manufacture of illuminating	J. A. Bassett	Salem, Mass	July 8, 1862	35, 807
Gas, Manufacture of illuminating	J. A. Bassett	Salem, Mass	Aug. 19, 1862	36, 254
Gas, Manufacture of illuminating	J. A. Bassett	Salem, Mass	June 25, 1867	66, 070
Gas, Manufacture of illuminating	J. A. Bassett	Salem, Mass	June 25, 1867	66, 071
Gas, Manufacture of illuminating	J. A. Bassett	Salem, Mass	Oct. 20, 1868	83, 239
Gas, Manufacture of illuminating	J. Boston	New York, N. Y	Feb. 11, 1831	....
Gas, Manufacture of illuminating	A. S. Cameron and W. E. Everett.	New York and Rye, N. Y	May 9, 1871	114, 642
Gas, Manufacture of illuminating	C. Carpenter	Buffalo, N. Y	Mar. 2, 1869	87, 466
Gas, Manufacture of illuminating	D. Davison	New York, N. Y	Sept. 26, 1871	119, 329
Gas, Manufacture of illuminating	C. F. Dieterich and A. Schüssler.	New York, N. Y	July 23, 1872	129, 720
Gas, Manufacture of illuminating	C. F. Dunderdale	New York, N. Y	Aug. 31, 1869	94, 297
Gas, Manufacture of illuminating	W. Elmer	New York, N. Y	Sept. 15, 1863	39, 905
Gas, Manufacture of illuminating	W. Elmer	New York, N. Y	Feb. 2, 1864	41, 431
Gas, Manufacture of illuminating	W. Elmer	New York, N. Y	June 11, 1867	65, 733
Gas, Manufacture of illuminating	W. Elmer	New York, N. Y	Aug. 5, 1873	141, 500
Gas, Manufacture of illuminating	J. J. Ensley	New York, N. Y	Feb. 4, 1868	74, 063
Gas, Manufacture of illuminating	R. M. and A. R. Fryer	New York, N. Y	Jan. 21, 1868	73, 521
Gas, Manufacture of illuminating	G. P. Ganster	New York, N. Y	June 16, 1868	78, 870
Gas, Manufacture of illuminating	C. Gearing	Pittsburgh, Pa	June 24, 1873	140, 264
Gas, Manufacture of illuminating	H. D. Green	Portland, Oreg	Aug. 14, 1866	57, 123
Gas, Manufacture of illuminating	H. D. Green	Portland, Oreg	Oct. 23, 1866	59, 004
Gas, Manufacture of illuminating	W. H. Gwynne	White Plains, N. Y	July 29, 1862	36, 002
Gas, Manufacture of illuminating	W. H. Gwynne	White Plains, N. Y	Jan. 6, 1863	37, 289
Gas, Manufacture of illuminating	W. H. Gwynne	White Plains, N. Y	July 14, 1863	39, 227
Gas, Manufacture of illuminating	A. W. Hall	New York, N. Y	Nov. 22, 1870	109, 510
Gas, Manufacture of illuminating	G. W. Harris and H. P. Allen	Elizabeth, N. J., and New York, N. Y.	July 30, 1872	129, 951
Gas, Manufacture of illuminating	G. W. Harris and H. Holdrege	Elizabeth, N. J., and New York, N. Y.	Sept. 7, 1869	94, 596

Index of patents issued from the United States Patent Office from 1790 *to* 1873, *inclusive*—Continued.

Invention.	Inventor.	Residence.	Date.	No.
Gas, Manufacture of illuminating	S. Hevner	San Francisco, Cal	May 9, 1871	114, 559
Gas, Manufacture of illuminating	E. R. Hopkins	Newark, N. J	Dec. 16, 1873	145, 655
Gas, Manufacture of illuminating	J. Kidd	New York, N. Y	Sept. 19, 1871	119, 034
Gas, Manufacture of illuminating	F. King	Washington, D. C	Aug. 1, 1871	117, 645
Gas, Manufacture of illuminating	M. Levy	New York, N. Y	Dec. 9, 1862	37, 101
Gas, Manufacture of illuminating	J. S. Lipps	New York, N. Y	Dec. 29, 1868	85, 458
Gas, Manufacture of illuminating	C. B. Loveless	Syracuse, N. Y	Dec. 1, 1868	84, 636
Gas, Manufacture of illuminating	W. L. Lowrey	Saratoga Springs, N. Y	July 23, 1867	67, 127
Gas, Manufacture of illuminating	W. L. Lowrey	Saratoga Springs, N. Y	Aug. 4, 1868	80, 642
Gas, Manufacture of illuminating	E. P. McCarthy	San Francisco, Cal	May 17, 1870	103, 218
Gas, Manufacture of illuminating	W. P. McConnell	Washington, D. C	July 28, 1863	39, 350
Gas, Manufacture of illuminating	E. D. McCracken	New York, N. Y	July 12, 1870	105, 351
Gas, Manufacture of illuminating	E. D. McCracken	New York, N. Y	Aug. 13, 1872	130, 437
Gas, Manufacture of illuminating	G. A. McIlhenny	Washington, D. C	June 18, 1867	65, 927
Gas, Manufacture of illuminating	G. McKenzie	Glasgow, Great Britain	July 9, 1867	66, 511
Gas, Manufacture of illuminating	T. N. Miller	Pittsburgh, Pa	June 17, 1873	139, 910
Gas, Manufacture of illuminating	A. Millochau	New York, N. Y	Oct. 2, 1866	58, 553
Gas, Manufacture of illuminating	E. A. Pond	Rutland, Vt	Oct. 10, 1865	50, 385
Gas, Manufacture of illuminating	E. A. and G. H. Pond and M. S. Richardson.	Rutland, Vt	Sept. 27, 1870	107, 809
Gas, Manufacture of illuminating	A. C. Rand	Union Mills, Pa	Feb. 26, 1867	62, 363
Gas, Manufacture of illuminating	A. C. Rand	New York, N. Y	Apr. 13, 1869	88, 982
Gas, Manufacture of illuminating	A. C. Rand	New York, N. Y	Mar. 8, 1870	100, 668
Gas, Manufacture of illuminating	W. D. Seal	Washington, D. C	Feb. 27, 1866	52, 893
Gas, Manufacture of illuminating	B. Silliman	New Haven, Conn	Oct. 19, 1869	95, 941
Gas, Manufacture of illuminating	B. Sloper and R. M. Potter	New York, N. Y., and Jersey City, N. J.	Jan. 23, 1872	123, 052
Gas, Manufacture of illuminating	J. W. Smith and T. H. Phillips	Washington, D. C	Oct. 8, 1867	69, 594
Gas, Manufacture of illuminating	J. Somerville and R. Elsdon	Maidstone, Great Britain	Apr. 14, 1868	76, 837
Gas, Manufacture of illuminating	A. P. Southwick	Buffalo, N. Y	Aug. 26, 1873	142, 289
Gas, Manufacture of illuminating	W. H. Spencer	Brooklyn, N. Y	July 9, 1872	123, 918
Gas, Manufacture of illuminating	W. H. Spencer	Brooklyn, N. Y	Sept. 3, 1872	131, 635
Gas, Manufacture of illuminating	T. G. Springer	Fayette City, Pa	Dec. 5, 1871	121, 679
Gas, Manufacture of illuminating	T. G. Springer	Fayette City, Pa	Dec. 17, 1872	134, 108
Gas, Manufacture of illuminating	J. W. Stanley	Brooklyn, N. Y	Jan. 31, 1871	111, 486
Gas, Manufacture of illuminating	L. Stevens	Washington, D. C	Jan. 7, 1868	73, 057
Gas, Manufacture of illuminating	L. Stevens	Washington, D. C	Jan. 26, 1869	86, 187
Gas, Manufacture of illuminating	L. Stevens	Washington, D. C	Feb. 23, 1869	87, 123
Gas, Manufacture of illuminating	J. B. Terry	Hartford, Conn	Dec. 10, 1867	72, 118
Gas, Manufacture of illuminating	J. E. Thomson	Buffalo, N. Y	May 20, 1862	35, 336
Gas, Manufacture of illuminating	P. H. Vander Weyde	New York, N. Y	Apr. 9, 1872	125, 633
Gas, Manufacture of illuminating	A. Walton and J. L. Kite	Philadelphia, Pa	Jan. 14, 1862	34, 185
Gas, Manufacture of illuminating	S. L. Wiegand	Philadelphia, Pa	May 5, 1863	38, 438
Gas, Manufacture of illuminating	S. L. Wiegand	Philadelphia, Pa	Aug. 18, 1863	39, 605
Gas, Manufacture of illuminating	S. L. Wiegand	Philadelphia, Pa	Aug. 18, 1863	39, 606
Gas, Manufacture of illuminating	A. W. Wilkinson	New York, N. Y	Feb. 6, 1872	123, 538
Gas, Manufacture of illuminating	A. W. Wilkinson	New York, N. Y	June 17, 1873	140, 104
Gas, Manufacture of illuminating	W. C. Wren	Brooklyn, N. Y	Apr. 8, 1873	137, 750
Gas, Manufacture of illuminating	W. Young and P. Brash	Magdalen Bridge and Leith, Scotland.	Sept. 27, 1870	107, 848
Gas, Manufacture of oxygen	O. M. Phillips	New York, N. Y	Jan. 26, 1869	86, 248
Gas, Manufacture of oxygen	J. Webster	Birmingham, England	June 24, 1862	35, 742
Gas, Manufacture of pneumatic	H. Bloomfield	San Francisco, Cal	June 6, 1871	115, 684
Gas, Manufacture of vapor	C. L. Cohen	Philadelphia, Pa	July 22, 1873	141, 119
Gas, Manufacture of vapor	C. A. Seely	New York, N. Y	June 29, 1869	92, 105
Gas, Manufacture of water	W. H. Gwynne	White Plains, N. Y	Aug. 11, 1863	39, 480
Gas, Manufacture of water	E. J. Jerzmanowski	New York, N. Y	Dec. 9, 1873	145, 350
Gas-meter	T. Brattan	Birkenhead, England	Feb. 1, 1870	99, 285
Gas-meter	C. F. Brown	Baltimore, Md	June 22, 1842	2, 687
Gas-meter	S. Clegg	Great Britain	Sept. 22, 1838	942
Gas-meter	S. Clegg	Putney, County Surrey, England.	Apr. 16, 1861	32, 049
Gas-meter	A. A. Croll	London, England	Feb. 22, 1853	9, 591
Gas-meter	D. De Castro and R. Burton	Mortlake, Surrey, and London, England.	July 29, 1873	141, 263
Gas-meter	J. S. Elliott	Philadelphia, Pa	Sept. 15, 1863	39, 904
Gas-meter	J. E. Fisk	Salem, Mass	Oct. 5, 1858	21, 663
Gas-meter	T. B. Fogarty	Charleston, S. C	Dec. 13, 1859	26, 423
Gas-meter	D. Forrest	Eastport, Me	Nov. 26, 1867	71, 375
Gas-meter	E. R. Hallam and T. B. Barnard	New Haven, Conn., and Brooklyn, N. Y.	Feb. 8, 1853	9, 580
Gas-meter	J. Hemming	Great Britain	Nov. 4, 1842	2, 837
Gas-meter	H. J. Hyams	Pittsburgh, Pa	Sept. 13, 1870	107, 380
Gas-meter	H. H. and J. F. G. Kromschroeder.	Regent's Park, Hanover	Oct. 13, 1863	40, 266
Gas-meter	G. Laidlaw	New York, N. Y	Nov. 2, 1852	9, 367
Gas-meter	T. W. Lane	Woburn, Mass	Oct. 8, 1850	7, 703
Gas-meter	J. Long	Chicago, Ill	Mar. 5, 1850	7, 154
Gas-meter	J. W. Mahlon	Brooklyn, N. Y	Apr. 28, 1868	77, 298
Gas-meter	H. Robinson	Boston, Mass	Mar. 10, 1831	....
Gas-meter	J. Schatt	Philadelphia, Pa	July 24, 1860	29, 323
Gas-meter	J. T. Scholte	Paris, France	Aug. 14, 1866	57, 268
Gas-meter	T. Shaw	Philadelphia, Pa	Apr. 27, 1858	20, 130
Gas-meter	A. W. Smith	Birmingham, Pa	Mar. 17, 1868	75, 653
Gas-meter	W. A. Telling and S. Johnson	London, England	Sept. 30, 1873	143, 390
Gas-meter	N. Tufts, jr	Boston, Mass	Aug. 14, 1860	29, 639
Gas-meter	G. J. Willson and D. H. Fox	Reading, Pa	Aug. 21, 1860	29, 740
Gas meter and regulator	C. C. Lloyd	West Philadelphia, Pa	June 20, 1854	11, 128
Gas-meter, Dry	D. Alcorn	New York, N. Y	July 21, 1863	39, 266
Gas-meter, Dry	A. W. Almqvist and F. W. Ofeldt.	New York, N. Y	June 6, 1871	115, 557
Gas-meter, Dry	L. Boore	Buffalo, N. Y	July 9, 1861	32, 747
Gas-meter, Dry	F. Darracott	Boston, Mass	Dec. 5, 1854	12, 045
Gas-meter, Dry	S. Down	New York, N. Y	May 24, 1859	24, 108
Gas-meter, Dry	J. E. Fisk	Salem, Mass	Mar. 18, 1862	34, 678
Gas-meter, Dry	J. E. Fisk	Salem, Mass	Sept. 15, 1863	39, 907
Gas-meter, Dry	W. W. Goodwin	Camden, N. J	Apr. 21, 1868	76, 908
Gas-meter, Dry	R. H. Gratz and C. C. Lloyd	Philadelphia, Pa	Oct. 23, 1860	30, 524

Index of patents issued from the United States Patent Office from 1790 *to* 1873, *inclusive*—Continued.

Invention.	Inventor.	Residence.	Date.	No.
Gas-meter, Dry	T. Grodjinski	New York, N. Y	July 12, 1859	24,736
Gas-meter, Dry	A. Harris	Philadelphia, Pa	Dec. 16, 1873	145,650
Gas-meter, Dry	P. Hodge	Seneca Falls, N. Y	Oct. 11, 1870	108,263
Gas-meter, Dry	T. C. Hopper	Philadelphia, Pa	Jan. 14, 1868	73,332
Gas-meter, Dry	H. J. Hyams	Hampton Road, England	Mar. 31, 1857	16,922
Gas-meter, Dry	C. Kirk	New Haven, Conn	Feb. 18, 1862	34,465
Gas-meter, Dry	C. C. Lloyd	Philadelphia, Pa	May 5, 1863	38,448
Gas-meter, Dry	C. C. Lloyd	Philadelphia, Pa	Nov. 26, 1867	71,518
Gas-meter, Dry	H. Logue	Philadelphia, Pa	May 31, 1859	24,263
Gas-meter, Dry	G. Lowen and S. H. Goldthorp	Pittsburgh, Pa	Nov. 15, 1870	109,230
Gas-meter, Dry	W. Lyon and C. W. Dickinson	Newark, N. J	Apr. 29, 1856	14,770
Gas-meter, Dry	E. Meredith	Philadelphia, Pa	Dec. 17, 1867	72,217
Gas-meter, Dry	J. A. Sabbaton	New York, N. Y	July 9, 1861	32,788
Gas-meter, Dry	J. Schatt	Philadelphia, Pa	Apr. 22, 1862	35,065
Gas-meter, Dry	J. Schatt	Philadelphia, Pa	July 2, 1867	66,256
Gas-meter, Dry	C. L. Vasquez	Philadelphia, Pa	Apr. 10, 1860	27,847
Gas-meter indicator	T. I. Pitt	New York, N. Y	Nov. 17, 1857	18,648
Gas-meter, Paper dry	A. Schüssler and C. F. Dieterich	New York, N. Y	Nov. 25, 1873	144,868
Gas-meter register	H. H. Edgerton	Fort Wayne, Ind	Aug. 22, 1871	118,221
Gas-meter register	J. S. Elliott	Philadelphia, Pa	Sept. 17, 1861	33,300
Gas-meter register	W. McDonald	Albany, N. Y	Oct. 9, 1866	58,731
Gas-meter, Wet	J. Cochrane	Harburn, North Britain	Mar. 18, 1862	34,670
Gas-meter, Wet	W. Richards	Barcelona, Spain	Mar. 20, 1860	27,567
Gas meter, Diaphragm-bellows for dry	A. Tufts	Malden, Mass	Oct. 13, 1868	83,113
Gas-meters, Gearing for	J. S. Elliott	Philadelphia, Pa	Mar. 26, 1861	31,791
Gas-meters in the wall of building, Setting	A. Potts	Philadelphia, Pa	July 5, 1859	24,660
Gas-meters, Liquid for	H. P. Gengembre	Rock Island, Ill	Apr. 27, 1858	20,058
Gas-meters, Liquid for fluid	J. Taylor	South Dartmouth, Mass	June 5, 1860	28,617
Gas-meters, Movement of	J. Thatcher	Philadelphia, Pa	Oct. 3, 1854	11,767
Gas-meters, Packing the shafts of dry	N. Tufts, jr	Boston, Mass	Sept. 12, 1865	49,935
Gas, Method and apparatus for separating hydrocarbon vapors from illuminating.	H. H. Edgerton	Fort Wayne, Ind	Aug. 15, 1871	117,998
Gas, Method of burning	J. C. McNulty	San Francisco, Cal	June 19, 1866	55,777
Gas, Method of compressing and liquefying nitrous-oxide and other.	W. F. and W. A. Johnston	Brooklyn, N. Y	Nov. 14, 1871	120,978
Gas, Method of obtaining benzole and its homologues from coal.	F. Engelhorn, A. Clemm, H. Caro, and C. Clemm.	Mannheim, Germany	July 6, 1869	92,176
Gas, Method of preserving and storing carbonic-acid.	J. S. Baldwin	New York, N. Y	May 29, 1866	55,039
Gas, Method of producing illuminating	J. F. Boynton	Syracuse, N. Y	Dec. 27, 1870	110,427
Gas, Method of producing water	W. H. Gwynne	White Plains, N. Y	Aug. 4, 1863	39,393
Gas, Method of relieving liquids of	N. W. Wheeler	Brooklyn, N. Y	Feb. 6, 1866	52,477
Gas, Mode of obtaining antiseptic	P. De la Bigarre and J. B. M. Picornell.	....	Mar. 24, 1803	....
Gas, Mode of producing hydrogen	C. M. E. Du Motay and C. R. Maréchal.	Metz, France	Nov. 12, 1867	70,704
Gas, Mode of producing oxygen	E. M. E. Du Motay and C. R. Maréchal.	Metz, France	Nov. 12, 1867	70,705
Gas, Mode of treating the spent oxide of iron used for purifying.	S. R. Divine	New York, N. Y	June 15, 1869	91,313
Gas of mineral springs, Method of exhibiting the	J. P. Haskins	Saratoga Springs, N. Y	Nov. 27, 1866	60,001
Gas of stone-coal, Light by a flame produced from the vapor on.	L. Enters and W. Zigler	Georgetown, D. C	Sept. 28, 1815	....
Gas on steamboats and other vessels, Apparatus for supplying.	N. Treadwell	New York, N. Y	Sept. 25, 1866	58,321
Gas or air with hydrocarbon vapors, Apparatus for charging.	W. H. Burridge	Cleveland, Ohio	Jan. 1, 1867	60,857
Gas or other retort	J. Chilcote	Brooklyn, N. Y	Jan. 17, 1865	45,908
Gas or vapor engine	S. Morey	Orford, N. H	Apr. 1, 1826	....
Gas or vapor engine	S. Perry	Newport, N. Y	May 25, 1844	3,597
Gas-pendant, Extensible	G. Warner	Des Moines, Iowa	Aug. 30, 1870	106,895
Gas-pipe	W. Stephenson	Galion, Ohio	July 10, 1860	29,111
Gas pipe and fixture	J. and T. D. Richardson	New York, N. Y	June 24, 1873	140,165
Gas-pipe and stair-rod	M. Gould	Newark, N. J	July 12, 1864	43,491
Gas-pipe coupling	D. Freeborn	New York, N. Y	Sept. 26, 1865	50,111
Gas-pipe cutter	J. E. Stanwood	Malden, Mass	Apr. 26, 1859	23,792
Gas-pipe drill	W. Daggett	Troy, N. Y	June 1, 1858	20,469
Gas-pipe fittings, Machine for finishing	C. C. Walworth	Boston, Mass	Oct. 7, 1856	15,860
Gas-pipe joint	C. Bruso, jr	Worcester, Mass	July 30, 1867	67,161
Gas-pipe joint	R. Wilson	Cold Spring, N. Y	Nov. 13, 1866	59,702
Gas-pipe nipples, Device for holding	L. W. Stockwell	Ravenna, Ohio	July 29, 1873	141,244
Gas-pipe safety-attachment	L. W. Turrell, S. Stanton, and L. C. Ward.	Newburgh, N. Y	July 31, 1866	56,836
Gas-pipe wrench	G. A. Jenks	Worcester, Mass	June 24, 1856	15,184
Gas-pipe, Apparatus for proving	A. C. Beardslee	Brooklyn, N. Y	Oct. 15, 1867	69,894
Gas-pipes, Apparatus for removing obstructions from.	G. Goldsmith and J. Dilkes	Leicester, England	Nov. 25, 1873	144,843
Gas-pipes, Apparatus for temporarily increasing the pressure in.	G. S. Dunbar	Pittsfield, Mass	May 7, 1872	126,451
Gas-pipes, Conduit-joint for	C. Monson	New Haven, Conn	Apr. 13, 1858	19,944
Gas-pipes, Connection for flexible	W. B. S. Taylor	New York, N. Y	June 25, 1872	128,336
Gas-pipes, Die for forming	J. Fieldhouse	Dighton, Mass	May 8, 1866	54,643
Gas-pipes, Dirt-chamber for	M. N. Dial	Painesville, Ohio	Apr. 18, 1871	113,749
Gas-pipes, Drip or water-trap for	J. H. Vansteenburgh	Ballston Spa, N. Y	Nov. 12, 1872	132,939
Gas-pipes, Hydrostatic stop for	J. Battin	Philadelphia, Pa	Sept. 18, 1847	5,295
Gas-pipes, Machine for	C. C. Walworth	Boston, Mass	Feb. 17, 1857	16,663
Gas-pipes, Mode of preventing the freezing up of	J. H. Mulhall	Albany, N. Y	Aug. 14, 1866	57,172
Gas-pipes, Street-box for	E. McMillin	Ironton, Ohio	Dec. 28, 1869	98,287
Gas-pipes, &c., Tool for cutting	W. S. Howarth	New Haven, Conn	Feb. 20, 1866	52,715
Gas-pipes, Tool for cutting	J. F. Howland	New York, N. Y	Oct. 4, 1859	25,646
Gas-pliers	H. Richter	New York, N. Y	Mar. 16, 1869	87,972
Gas, Portable apparatus for generating illuminating	J. O. Halsey	Essex County, N. J	Apr. 21, 1857	17,090
Gas, Portable apparatus for manufacturing illuminating.	G. H. Kitchen	New York, N. Y	Aug. 5, 1862	36,096
Gas, Preparing illuminating	S. White	Manchester, England	Jan. 22, 1850	7,043
Gas, Preparing petroleum for the manufacture of illuminating.	T. Clough	New York, N. Y	Sept. 16, 1862	36,453
Gas-pressure regulator	C. C. Ramsay	Chicago, Ill	Sept. 8, 1868	81,944

Index of patents issued from the United States Patent Office from 1790 *to* 1873, *inclusive*—Continued.

Invention.	Inventor.	Residence.	Date.	No.
Gas, Process and apparatus for generating combustible.	C. M. Tessié du Motay	Paris, France	Mar. 2, 1869	87, 478
Gas, Process and apparatus for making	T. S. C. Lowe	Norristown, Pa	Aug. 13, 1872	130, 383
Gas. Processes and compositions for the manufacture of illuminating.	W. H. Sterling	San Francisco, Cal	Sept. 16, 1873	142, 959
Gas producer or furnace	C. W. Siemens and F. Siemens.	Westminster, England, and Berlin, Prussia.	May 10, 1864	42, 717
Gas-producing apparatus for use in metallurgy and glass-making and for other purposes.	J. Green	Norristown, Pa	Feb. 2, 1869	86, 529
Gas-producing furnace	W. Gerhardt	New York, N. Y	Feb. 8, 1870	99, 554
Gas, Producing illuminating	W. Elmer	New York, N. Y	Aug. 4, 1863	39, 388
Gas, Production of illuminating	J. M. Sanders	Cincinnati, Ohio	July 27, 1858	21, 027
Gas, Purification of coal	A. A. Croll	London, England	June 18, 1867	65, 880
Gas, Purification of coal	E. McMillen	Ironton, Ohio	Mar. 1, 1870	100, 309
Gas, Purification of illuminating	W. H. St. John	New York, N. Y	Sept. 2, 1873	142, 358
Gas-purifier	R. C. Bocking	Indianapolis, Ind	July 3, 1866	55, 994
Gas-purifier	P. Fontain	Philadelphia, Pa	June 28, 1859	24, 551
Gas-purifier	J. T. Goodfellow and F. A. Sabbaton.	Troy, N. Y	June 17, 1873	140, 034
Gas-purifier	P. Munzinger	Philadelphia, Pa	Mar. 30, 1869	88, 326
Gas purifier and regulator	J. A. Enos	Peabody, Mass	May 20, 1873	139, 131
Gas-purifier, Dry	E. Duffee	Haverhill, Mass	May 2, 1871	114, 275
Gas purifier, Dry lime	C. F. Werner and C. Deutschmann.	New York and Buffalo, N. Y.	July 1, 1856	15, 267
Gas-purifier screen	T. G. Arnold	New York, N. Y	Oct. 16, 1866	58, 749
Gas-purifier screen	E. Duffee	Haverhill, Mass	July 17, 1866	56, 490
Gas-purifier screen	E. Duffee	Haverhill, Mass	Aug. 6, 1872	130, 190
Gas purifier screen, Coal	G. W. Day	Haverhill, Mass	June 11, 1872	127, 856
Gas purifier screen, Coal	E. Duffee	Haverhill, Mass	Dec. 12, 1871	121, 764
Gas purifier screen, Coal	J. Hale	Georgetown, Mass	June 18, 1872	127, 973
Gas-purifier screen, Dry	E. Duffee	Haverhill, Mass	Oct. 2, 1866	58, 393
Gas-purifiers, Mode of forming connection of	R. Briggs	Philadelphia, Pa	June 8, 1869	90, 917
Gas-purifiers, Tray for	P. Muszinger	Philadelphia, Pa	Oct. 11, 1870	108, 167
Gas-purifiers, Tray or grating for	J. L. Cheesman	New York, N. Y	June 10, 1873	139, 702
Gas, Purifying	W. T. Kosinski	Philadelphia, Pa	May 14, 1872	126, 717
Gas-purifying apparatus	W. F. Danowsky	Allentown, Pa	Aug. 10, 1858	21, 121
Gas-purifying apparatus	A. F. Havens	Brooklyn, N. Y	Aug. 29, 1871	118, 609
Gas-purifying apparatus	A. Longbottom	New York, N. Y	Feb. 3, 1852	8, 705
Gas-purifying apparatus	A. Walker	Claremont, N. H	Aug. 3, 1858	21, 095
Gas-purifying apparatus	W. T. Walker	London, England	July 30, 1872	129, 997
Gas-purifying apparatus	J. Waterhouse	Little Falls, N. Y	Aug. 3, 1858	21, 096
Gas, Purifying carbonic-acid	A. P. Meylert	New Britain, Conn	Jan. 21, 1873	135, 001
Gas, Purifying carbonic-acid	A. P. Meylert	New Britain, Conn	Apr. 8, 1873	137, 615
Gas, Purifying coal	J. A. Sabbaton	Albany, N. Y	July 30, 1850	7, 534
Gas, Purifying illuminating	F. J. De Cavaillon	Paris, France	May 6, 1851	8, 081
Gas, Purifying illuminating	R. J. Everett	Bridgeport, Conn	Apr. 23, 1872	125, 943
Gas, Purifying illuminating	A. Ruthel	Savona, Italy	Oct. 29, 1872	132, 642
Gas, Purifying illuminating	P. Spence	Manchester, England	Oct. 27, 1868	83, 417
Gas, Purifying illuminating	W. H. St. John and P. Cartwright.	New York, N. Y	Nov. 15, 1870	109, 268
Gas-purifying machine	W. C. Turnbull	New York, N. Y	Sept. 11, 1866	58, 002
Gas-reflector	W. J. McLea	Buffalo, N. Y	Apr. 14, 1868	76, 789
Gas-regulator	L. Abbott	Boston, Mass	Dec. 10, 1861	33, 873
Gas-regulator	S. D. Baldwin	Milwaukee, Wis	Apr. 12, 1859	23, 540
Gas-regulator	J. Battin	Philadelphia, Pa	Sept. 18, 1847	5, 300
Gas-regulator	S. Bidwell	Chicago, Ill	Sept. 21, 1858	21, 544
Gas-regulator	J. E. Boyle	Brooklyn, N. Y	Apr. 19, 1864	42, 430
Gas-regulator	F. W. Brocksieper	Bridgeport, Conn	Nov. 17, 1863	40, 602
Gas-regulator	F. W. Brocksieper	Bridgeport, Conn	Feb. 16, 1864	41, 599
Gas-regulator	F. H. Brown	Chicago, Ill	Nov. 18, 1862	36, 940
Gas-regulator	S. W. Brown	Lowell, Mass	May 29, 1855	12, 943
Gas-regulator	S. W. Brown	Lowell, Mass	July 10, 1855	13, 210
Gas-regulator	S. W. Brown	Lowell, Mass	Aug. 7, 1855	13, 377
Gas-regulator	T. Champion and C. Champion.	Washington, D. C., and El Dorado County, Cal.	July 24, 1860	29, 246
Gas-regulator	J. S. Conant	Lowell, Mass	Dec. 9, 1851	8, 578
Gas-regulator	J. B. Coolidge	Boston, Mass	Oct. 31, 1871	120, 416
Gas-regulator	J. H. Cooper	Philadelphia, Pa	Feb. 17, 1857	16, 639
Gas-regulator	J. H. Cooper	Philadelphia, Pa	June 30, 1857	17, 671
Gas-regulator	J. H. Cooper	Philadelphia, Pa	Aug. 18, 1857	18, 008
Gas-regulator	J. H. Cooper	Philadelphia, Pa	June 22, 1858	20, 625
Gas-regulator	R. Cornelius	Philadelphia, Pa	Apr. 21, 1857	17, 079
Gas-regulator	R. Cornelius	Philadelphia, Pa	May 19, 1857	17, 317
Gas-regulator	R. Cornelius	Philadelphia, Pa	May 19, 1863	38, 561
Gas-regulator	E. H. Covel	New York, N. Y	May 31, 1859	24, 199
Gas-regulator	C. M. Cresson	Philadelphia, Pa	Apr. 11, 1865	47, 189
Gas-regulator	O. Dean	Richmond, Va	May 2, 1871	114, 272
Gas-regulator	J. C. Dickey	Saratoga Springs, N. Y	Oct. 23, 1855	13, 700
Gas-regulator	G. B. Dixwell and J. A. Dorr	Boston, Mass., and New York, N. Y.	Apr. 18, 1854	10, 786
Gas-regulator	J. Edson	Boston, Mass	May 6, 1862	35, 146
Gas-regulator	C. H. Edwards	Chicago, Ill	Aug. 26, 1873	142, 155
Gas-regulator	G. M. Eldridge	Philadelphia, Pa	Apr. 23, 1872	125, 942
Gas-regulator	J. Foster	Richmond, Va	Oct. 16, 1860	30, 399
Gas-regulator	H. Gerner	New York, N. Y	June 20, 1871	116, 176
Gas-regulator	H. Giroud	Paris, France	Nov. 3, 1868	83, 706
Gas-regulator	W. Grover	Holyoke, Mass	Sept. 2, 1862	36, 343
Gas-regulator	C. J. Halsted and J. Coeyman	New York, N. Y	Mar. 31, 1857	16, 951
Gas-regulator	C. L. Herring	Saint Louis, Mo	Jan. 22, 1861	31, 168
Gas-regulator	J. W. Hoard	Providence, R. I	Mar. 13, 1855	12, 509
Gas-regulator	C. F. Holzer	Philadelphia, Pa	July 27, 1858	21, 048
Gas-regulator	H. G. Hubert	New York, N. Y	Dec. 24, 1867	72, 643
Gas-regulator	C. T. Judkins	Boston, Mass	Feb. 25, 1862	34, 507
Gas-regulator	J. Keeling	New York, N. Y	Oct. 24, 1871	120, 280
Gas-regulator	P. Keller	New York, N. Y	Jan. 11, 1870	98, 776
Gas-regulator	P. Keller	New York, N. Y	Oct. 4, 1870	108, 029
Gas-regulator	P. Keller	New York, N. Y	June 27, 1871	116, 321
Gas-regulator	P. Keller	New York, N. Y	Apr. 1, 1873	137, 454

Index of patents issued from the United States Patent Office from 1790 *to* 1873, *inclusive*—Continued.

Invention.	Inventor.	Residence.	Date.	No.
Gas-regulator	W. Kidder	Lowell, Mass	Oct. 12, 1852	9,325
Gas-regulator	W. Kidder	Lowell, Mass	Oct. 12, 1852	9,326
Gas-regulator	W. Kidder	Lowell, Mass	Oct. 12, 1852	9,327
Gas-regulator	G. H. Kitchen	New York, N. Y	Oct. 18, 1859	25,834
Gas-regulator	J. G. Leffingwell	Newark, N. J	Oct. 16, 1860	30,415
Gas-regulator	W. B. Leonard	New York, N. Y	Feb. 10, 1852	8,725
Gas-regulator	C. C. Loyd and R. M. Potter	Philadelphia, Pa., and New York, N. Y.	Mar. 15, 1864	41,930
Gas-regulator	W. Mallerd	Bridgeport, Conn	Oct. 12, 1858	21,765
Gas-regulator	S. F. Mathews	Mechanicsburgh, Pa	May 26, 1868	78,305
Gas-regulator	J. S. Merriken	Baltimore, Md	May 9, 1871	114,699
Gas-regulator	S. P. Mervine, jr	Philadelphia, Pa	Jan. 19, 1869	86,027
Gas-regulator	J. H. Norton	Boston, Mass	June 10, 1862	35,538
Gas-regulator	E. T. Orne	Boston, Mass	Mar. 27, 1860	27,659
Gas-regulator	S. P. Parham	Trenton, N. J	Apr. 10, 1855	12,692
Gas-regulator	A. H. Phillippi	Reading, Pa	Sept. 24, 1872	131,701
Gas-regulator	C. C. Place	Boston, Mass	June 27, 1871	116,349
Gas-regulator	R. M. Potter	Jersey City, N. J	Oct. 8, 1872	132,103
Gas-regulator	J. H. Powers	Newark, N. J	Sept. 1, 1857	18,103
Gas-regulator	J. H. Powers	Newark, N. J	July 27, 1858	21,022
Gas-regulator	T. Powers	Philadelphia, Pa	May 14, 1861	32,312
Gas-regulator	S. P. Ruggles	Boston, Mass	Aug. 14, 1855	13,437
Gas-regulator	H. Schutte	Kansas City, Mo	Apr. 25, 1871	114,206
Gas-regulator	H. C. Sergeant	Cincinnati, Ohio	Apr. 15, 1862	34,985
Gas-regulator	W. A. Simonds	Boston, Mass	Aug. 15, 1865	49,449
Gas regulator	W. A. Simonds	Boston, Mass	Dec. 19, 1865	51,664
Gas-regulator	W. A. Simonds	Boston, Mass	Mar. 3, 1868	75,210
Gas-regulator	A. Smith	Brooklyn, N. Y	Aug. 26, 1862	36,306
Gas-regulator	G. H. Smith	Rochester, N. Y	June 25, 1861	32,670
Gas-regulator	W. G. Sterling	Bridgeport, Conn	Nov. 11, 1856	16,073
Gas-regulator	W. G. Sterling	Bridgeport, Conn	Aug. 24, 1858	21,281
Gas-regulator	S. O. Trudell	Detroit, Mich	Sept. 6, 1870	107,127
Gas-regulator	H. Waterman	Hudson, N. Y	May 13, 1856	14,893
Gas-regulator	D. Wheeler and I. Little	Fairfield and Bridgeport, Conn.	May 31, 1859	24,253
Gas-regulator	M. Wheeler	Honesdale, Pa	June 3, 1856	15,028
Gas-regulator	H. Wheelock	Boston, Mass	June 5, 1855	13,020
Gas-regulator	S. H. Whitaker	Cincinnati, Ohio	June 12, 1860	28,722
Gas-regulator	A. H. Wood	Boston, Mass	Jan. 29, 1867	61,593
Gas-regulator	J. S. Wood	Philadelphia, Pa	Aug. 1, 1865	49,188
Gas-regulator	J. S. Wood	Philadelphia, Pa	Aug. 21, 1866	57,425
Gas-regulator	J. S. Wood	Philadelphia, Pa	Aug. 13, 1867	67,694
Gas-regulator	G. B. Woodruff and J. N. Palmer	New Haven, Conn	May 8, 1855	12,844
Gas regulator and purifier	P. Fontain	New York, N. Y	Jan. 27, 1863	37,500
Gas-regulator, Dry	J. B. Hoffman	Philadelphia, Pa	Dec. 3, 1867	71,757
Gas-regulator for blow-pipes, Automatic	J. H. Snow	Providence, R. I	June 15, 1869	91,376
Gas-regulator for nitrous-oxide apparatus, Mercurial.	J. B. Coolidge	Boston, Mass	Nov. 23, 1869	97,051
Gas-regulator for railway-cars	A. H. Phillippi	Reading, Pa	July 31, 1860	29,401
Gas-regulator, Stop-cock	O. L. Lawson and A. A. Starr	New York, N. Y	Aug. 4, 1857	17,936
Gas-retort	W. P. Battey and W. N. Taylor	Utica, N. Y., and Philadelphia, Pa.	Aug. 5, 1862	36,065
Gas-retort	W. Beaumont	Paterson, N. J	June 28, 1859	24,531
Gas-retort	H. P. M. Birkinbine	Philadelphia, Pa	Apr. 17, 1855	12,711
Gas-retort	J. Butler	New York, N. Y	Mar. 5, 1872	124,250
Gas-retort	J. Butler	New York, N. Y	Aug. 27, 1872	130,842
Gas-retort	M. L. Callender	New York, N. Y	Dec. 8, 1868	84,679
Gas-retort	S. Coates	New York, N. Y	Sept. 8, 1857	18,134
Gas-retort	C. M. Cresson	Philadelphia, Pa	Oct. 3, 1854	11,742
Gas-retort	J. Davis and S. Chaddock	Boston, Mass	Feb. 14, 1860	27,113
Gas-retort	D. Davison	New York, N. Y	Oct. 4, 1870	108,009
Gas-retort	D. Davison	New York, N. Y	Mar. 14, 1871	112,557
Gas-retort	D. Davison	New York, N. Y	Aug. 1, 1871	117,610
Gas-retort	D. Davison	New York, N. Y	Oct. 14, 1873	143,675
Gas-retort	E. Duffee	Haverhill, Mass	Feb. 27, 1866	52,929
Gas-retort	H. H. Edgerton	Fort Wayne, Ind	Oct. 3, 1871	119,586
Gas-retort	J. M. Gallacher	Roxbury, Mass	Apr. 30, 1861	32,219
Gas-retort	J. Green	Saint Louis, Mo	July 16, 1872	129,276
Gas-retort	R. C. Harrington	Newark, N. J	Mar. 20, 1860	27,543
Gas-retort	G. W. Harris	Elizabeth, N. J	Dec. 4, 1866	60,183
Gas-retort	G. W. Harris	Elizabeth, N. J	Mar. 14, 1871	112,593
Gas-retort	J. D. Higgins	Rome, N. Y	Sept. 12, 1871	118,936
Gas-retort	J. J. Holden and S. J. Best	London, England	Dec. 3, 1867	71,617
Gas-retort	A. S. King	Commerce, Mich	Feb. 4, 1862	34,333
Gas-retort	W. H. Lauback	Philadelphia, Pa	Dec. 28, 1858	22,434
Gas-retort	A. Marsh	Detroit, Mich	Aug. 10, 1858	21,169
Gas-retort	A. Marsh	Detroit, Mich	May 17, 1859	24,038
Gas-retort	E. D. McCracken	New York, N. Y	July 11, 1870	105,350
Gas-retort	G. A. McIlhenny	Washington, D. C	Dec. 16, 1873	145,672
Gas-retort	M. I. Miller	Saint Louis, Mo	Feb. 17, 1857	16,651
Gas-retort	J. D. Patton	Trevorton, Pa	June 3, 1873	139,605
Gas-retort	J. Rigby and P. A. Palmer	Marietta, Ohio	June 7, 1870	103,929
Gas-retort	C. A. Robbe	Augusta, Ga	Aug. 9, 1859	25,046
Gas-retort	W. A. Simonds	Chelsea, Mass	June 1, 1858	20,448
Gas-retort	I. T. Sloan, V. Smith, M. Hoover, and R. M. Briggs.	Jackson, Cal	May 25, 1858	20,375
Gas-retort	J. H. Smith	Newark, Ohio	May 16, 1871	114,872
Gas-retort	J. W. Smith	Washington, D. C	June 21, 1859	24,524
Gas-retort	T. G. Springer	Fayette City, Pa	May 28, 1872	127,381
Gas-retort	W. Stratton	Philadelphia, Pa	July 5, 1859	24,670
Gas-retort	H. K. Symmes	Newton, Mass	Aug. 16, 1859	25,225
Gas-retort	J. C. Tiffany	Boston, Mass	Aug. 13, 1872	130,453
Gas-retort	J. C. Tiffany	Portsmouth, N. H	Dec. 2, 1873	145,256
Gas-retort	A. K. Tupper	Pontiac, Mich	July 9, 1861	32,802
Gas-retort	A. K. Tupper	Pontiac, Mich	Dec. 10, 1861	33,921
Gas-retort	C. N. Tyler	Washington, D. C	June 22, 1858	20,671
Gas-retort	E. Walcott	Providence, R. I	June 21, 1859	24,510

Index of patents issued from the United States Patent Office from 1790 *to* 1873, *inclusive*—Continued.

Invention.	Inventor.	Residence.	Date.	No.
Gas-retort	J. H. Walker and J. Walker	Milwaukee, Wis., and South Bend, Ind.	Dec. 10, 1872	133, 731
Gas-retort	S. H. and M. C. Walker	Boston, Mass	Apr. 9, 1861	32, 019
Gas-retort	D. L. Weatherhead	Philadelphia, Pa	Mar. 29, 1859	23, 433
Gas-retort	C. Wooster	New York, N. Y	May 22, 1860	28, 431
Gas-retort	J. Wotherspoon and W. Foulis	Glasgow, North Britain	Dec. 2, 1873	145, 267
Gas-retort and apparatus for charging	L. F. Blair	Painesville, Ohio	May 28, 1872	127, 144
Gas-retort and heating-furnace	G. H. Pond	Rutland, Vt	Dec. 27, 1870	110, 496
Gas-retort benches, Arrangement of	J. G. Hock	Newark, N. J	June 3, 1856	15, 010
Gas-retort charger	T. F. Rowland	Green Point, N. Y	Apr. 1, 1873	137, 487
Gas-retort charges, Apparatus for filling	T. F. Rowland	Brooklyn, N. Y	Dec. 31, 1872	134, 399
Gas-retort, Clay	J. P. Kennedy	Trenton, N. J	Mar. 13, 1860	27, 451
Gas-retort cleaner	S. H. and M. C. Walker	Lancaster, Pa	May 20, 1856	14, 934
Gas-retort closings	G. A. McIlhenney	Washington, D. C	Apr. 8, 1873	137, 704
Gas-retort, Composite	J. Cochrane	New York, N. Y	Sept. 14, 1869	94, 714
Gas-retort cover	A. F. Havens	Brooklyn, N. Y	Dec. 26, 1871	122, 250
Gas-retort fastening	J. G. Hock	Newark, N. J	May 20, 1856	14, 913
Gas-retort fastening, Copper-ring	W. H. St. John	New York, N. Y	Nov. 11, 1856	16, 075
Gas, Retort for generating	J. H. Irwin	Chicago, Ill	Sept. 11, 1866	57, 915
Gas, Retort for generating	W. H. Laubach	Philadelphia, Pa	June 15, 1858	20, 567
Gas, Retort for the manufacture of coal	W. J. Cochran	Baltimore, Md	June 1, 1869	90, 819
Gas, Retort for the manufacture of illuminating	A. Millochau	New York, N. Y	Dec. 18, 1866	60, 537
Gas-retort furnace	W. C. Wren	Brooklyn, N. Y	Apr. 15, 1873	137, 811
Gas-retort mouth-piece	P. Munzinger	Philadelphia, Pa	Nov. 25, 1873	144, 857
Gas-retort mouth-piece	T. F. Rowland	Brooklyn, N. Y	Oct. 28, 1873	143, 998
Gas-retort, Portable	J. W. Smith	Washington, D. C	Mar. 16, 1858	19, 655
Gas-retort, Portable	D. L. Weatherhead and J. T. Henry	Philadelphia, Pa	Apr. 6, 1858	19, 900
Gas-retorts, Air-tight joint for	C. F. Dieterich and A. Schüssler	New York, N. Y	June 6, 1871	115, 586
Gas-retorts, Air-tight joint for	C. F. Dieterich and A. Schüssler	New York, N. Y	June 18, 1872	127, 965
Gas-retorts and other like purposes, Apparatus for charging and drawing	S. J. Best and J. J. Holden	London, England	Jan. 15, 1867	61, 144
Gas-retorts, Apparatus for charging	A. F. Havens	Brooklyn, N. Y	Dec. 17, 1872	134, 055
Gas-retorts, Apparatus for charging	A. F. Havens	Brooklyn, N. Y	Dec. 17, 1872	134, 056
Gas-retorts, Apparatus for charging	A. F. Havens	Brooklyn, N. Y	July 8, 1873	140, 624
Gas-retorts, Apparatus for charging	T. F. Rowland	Green Point, N. Y	Apr. 1, 1873	137, 485
Gas-retorts, Apparatus for charging	J. F. Snediker and W. F. Bailey	Bristol, Pa	Mar. 10, 1868	75, 308
Gas-retorts, Apparatus for charging	J. Somerville and J. Robinson	Dublin, Ireland	Sept. 23, 1873	143, 039
Gas-retorts, Apparatus for charging and discharging	J. J. Holden	Bermondsey, England	May 9, 1871	114, 682
Gas-retorts, Apparatus for charging and drawing	W. Foulis	Glasgow, North Britain	Nov. 11, 1873	144, 526
Gas-retorts, Apparatus for raking	T. F. Rowland	Green Point, N. Y	Apr. 1, 1873	137, 486
Gas-retorts, Binder for closing	A. Fulton	Albany, N. Y	May 18, 1869	90, 162
Gas-retorts, Casting	A. Pevey	Lowell, Mass	Dec. 22, 1857	18, 926
Gas-retorts, Casting	A. Pevey	Lowell, Mass	Apr. 15, 1862	34, 980
Gas-retorts, Cleaning	C. A. Bush	New London, Conn	Oct. 26, 1869	96, 197
Gas-retorts, Cleaning	S. W. Carpenter	Yonkers, N. Y	Dec. 1, 1857	18, 791
Gas retorts, Closing	N. Aubin	Albany, N. Y	Apr. 21, 1857	17, 068
Gas-retorts, Closing	G. A. McIlhenney	Washington, D. C	Aug. 26, 1873	142, 253
Gas-retorts, Coating	A. R. Terry	Detroit, Mich	June 20, 1854	11, 142
Gas-retorts, Composition for luting	J. Chilcott	Brooklyn, N. Y	Aug. 30, 1864	43, 970
Gas-retorts, Compound for	G. F. Kreischer and L. A. Tartière	New York, N. Y	Apr. 15, 1873	137, 934
Gas-retorts, Compound for cleaning	W. L. Pitkin and C. J. Gleason	Montpelier, Vt	July 8, 1873	140, 590
Gas-retorts, Cover for	B. H. Bartol	Philadelphia, Pa	July 9, 1867	66, 444
Gas-retorts, &c., Enameling	D. W. Clark	Chicago, Ill	May 16, 1871	114, 763
Gas-retorts, Feeding-apparatus for	N. Aubin	Albany, N. Y	June 3, 1856	14, 996
Gas-retorts, Feeding hydrocarbon liquids to hot	T. G. Springer	Saint Louis, Mo	Nov. 15, 1870	109, 264
Gas retorts, furnaces, &c., Cover or door for	W. Matthews and J. Moore	Philadelphia, Pa	Dec. 29, 1868	85, 463
Gas-retorts, Machine for charging	T. F. Rowland	Green Point, N. Y	Sept. 24, 1872	131, 564
Gas-retorts, Machine for charging	N. O. J. Tisdale	New Orleans, La	Aug. 24, 1869	94, 045
Gas-retorts, Machine for making fire-clay	J. S. Heartt and S. English	Troy, N. Y	Oct. 29, 1861	33, 623
Gas-retorts, Method of preventing deposition of carbon in	A. Marsh	Detroit, Mich	Oct. 11, 1859	25, 790
Gas-retorts, Method of removing carbon from	J. A. Bassett and O. C. Smith	Salem, Mass	July 28, 1868	80, 438
Gas-retorts, Method of removing carbon from	B. E. Chollar	Leavenworth, Kans	Oct. 29, 1867	70, 166
Gas-retorts, Method of removing incrustation from	A. J. White	New York, N. Y	Aug. 8, 1865	49, 329
Gas-retorts, Mode of constructing head, neck, and connection of	J. Chilcott	Brooklyn, N. Y	Aug. 8, 1865	49, 231
Gas-retorts, Mode of securing lid on	J. R. Thomas	Williamsburgh, N. Y	May 29, 1860	28, 519
Gas-retorts, Securing lids to	S. Holman	London, England	Aug. 1, 1871	117, 631
Gas-retorts, Setting	T. Curley	Wilmington, Del	May 31, 1870	103, 579
Gas-retorts, Setting	A. Weber	New York, N. Y	Aug. 5, 1862	36, 119
Gas-retorts, Stand-pipe of	M. Combs, jr	Youngstown, Ohio	Nov. 28, 1871	121, 276
Gas-retorts with liquids, Device for supplying	E. L. Norfolk	Salem, Mass	Sept. 22, 1863	40, 058
Gas-seal, &c	R. M. Caffall	Alton, England	Nov. 11, 1873	144, 504
Gas service-pipes for automatically cutting off the gas in case of fire, Attachment to	W. Humphreys	Waterford, N. Y	Apr. 25, 1871	114, 144
Gas-socket	G. Mooney	Providence, R. I	Dec. 15, 1868	84, 961
Gas-socket	T. L. Reed	Providence, R. I	Jan. 7, 1868	73, 040
Gas-stand, Portable	B. Allen and J. Riddell	Boston, Mass	June 12, 1866	55, 447
Gas, steam, and fluid regulator	E. C. Maldant	Paris, France	Mar. 9, 1869	87, 575
Gas, steam, &c., Method of drawing from manufacturing inclosures waste	R. F. Brower	Bloomfield, N. J	Sept. 9, 1856	15, 716
Gas stopper or cork	N. D. Whitin	New York	Apr. 29, 1842	2, 593
Gas-tar and ammoniacal liquors, Apparatus for separating	E. D. McCracken	New York, N. Y	Oct. 18, 1870	108, 499
Gas-tar, Apparatus for burning	A. Smith	Niagara Falls, N. Y	Sept. 2, 1873	142, 357
Gas through pipes, Mode of facilitating the flow of illuminating	J. F. Russell	Washington, D. C	Oct. 29, 1867	70, 365
Gas to be used for motive-power, Process of generating	D. E. Somes	Washington, D. C	Apr. 3, 1866	53, 695
Gas-torch	W. A. Lawton	New York, N. Y	Feb. 4, 1868	74, 039
Gas-tube	E. L. Perry	New York, N. Y	Apr. 28, 1868	77, 314
Gas-tube, Extension	C. Monson	New Haven, Conn	Sept. 8, 1857	18, 154
Gas-tube, Flexible	J. Butler	Brooklyn, N. Y	June 12, 1860	28, 647
Gas-tube for chandelier, Extensible	H. L. McAvoy	Baltimore, Md	Feb. 22, 1870	100, 171
Gas-tube joint	C. Monson	New Haven, Conn	Jan. 19, 1858	19, 150
Gas-tube support, Flexible	A. Housinger	Chicago, Ill	Mar. 26, 1872	124, 897

Index of patents issued from the United States Patent Office from 1790 *to* 1873, *inclusive*—Continued.

Invention.	Inventor.	Residence.	Date.	No.
Gas-tubes, Flexible joint for	A. Stratton	Brooklyn, N. Y	July 10, 1860	29, 113
Gas-tubes, Manufacture of flexible	T. J. Mayall	Roxbury, Mass	Sept. 4, 1860	29, 898
Gas-tubing, Flexible	A. W. Hale	New York, N. Y	July 16, 1872	129, 405
Gas-tubing, Manufacture of flexible	G. L. Burnham	Providence, R. I	Dec. 3, 1872	133, 560
Gas, water, &c., Machine for making laminated pipes for.	J. S. Patric	Rochester, N. Y	May 13, 1873	138, 814
Gas with vapor of hydrocarbon liquids, Charging	C. M. Williams	New York, N. Y	Nov. 3, 1868	83, 748
Gas-works, Apparatus for regulating the exhauster in.	J. A. Harris	Philadelphia, Pa	Mar. 26, 1867	63, 155
Gas-works, By-pass for	P. Munzinger	Philadelphia, Pa	Nov. 11, 1873	144, 408
Gas-works, Center-seal for	P. Munzinger	Philadelphia, Pa	Nov. 22, 1870	109, 540
Gas-works, Center-seal for	P. Munzinger	Philadelphia, Pa	June 6, 1871	115, 632
Gas-works, Condensing-apparatus for	H. H. Edgerton	Fort Wayne, Ind	July 23, 1872	129, 723
Gas-works, Dip-pipe in hydraulic main for	J. Hannan	Gallipolis, Ohio	Apr. 16, 1872	125, 733
Gas-works, Exhaust-apparatus for	P. W. McKenzie	Blauveltville, N. Y	May 30, 1871	115, 334
Gas-works, Exhausting-apparatus for	S. R. Brick	Philadelphia, Pa	Mar. 29, 1870	101, 348
Gas-works, Hydraulic main for	P. T. Burtis	Chicago, Ill	Apr. 8, 1873	137, 655
Gas-works, Hydraulic main for	C. Collier	Selma, Ala	Oct. 17, 1871	119, 919
Gas-works, Hydraulic main for	W. Farmer	New York, N. Y	Jan. 14, 1873	134, 874
Gas-works, Hydraulic main for	J. R. Floyd and J. A. Sabbaton.	New York, N. Y	Apr. 21, 1868	76, 904
Gas-works, Mode of deodorizing the spent lime of	A. Millochau	New York, N. Y	Apr. 6, 1869	88, 658
Gas-works, Purifier, condenser, &c., for	A. Fulton	Albany, N. Y	Feb. 14, 1871	111, 737
Gas-works, Regulator for exhauster in	A. S. Cameron	New York, N. Y	Oct. 3, 1871	119, 505
Gas-works, Seal for dip-pipes in	W. Cartwright	Oswego, N. Y	Dec. 10, 1872	133, 756
Gas-works, Seal for dip-pipes in	G. Douty	Columbus, Ohio	Jan. 31, 1871	111, 439
Gas-works, Seal for dip-pipes in	G. Douty	Columbus, Ohio	Jan. 31, 1871	111, 440
Gas-works, Seal for dip-pipes in	A. F. Havens	Brooklyn, N. Y	Aug. 29, 1871	118, 608
Gas-works, Seal for hydraulic mains for	A. Odierne	Springfield, Ill	June 6, 1871	115, 634
Gas-works, Seal for hydraulic mains of	S. Trumbore	Easton, Pa	Apr. 18, 1871	113, 946
Gas-works, Sealing and unsealing dip-pipes in hydraulic mains of.	W. E. Grenelle	New York, N. Y	Aug. 8, 1871	117, 880
Gas-works, Sealing dip-pipes of	H. H. Edgerton	Fort Wayne, Ind	Apr. 1, 1872	137, 425
Gas-works, Sealing dip-pipes of	E. Jones	Boston, Mass	Nov. 18, 1873	144, 619
Gas-works, &c., Superheater for	M. L. Callender	Brooklyn, N. Y	Oct. 18, 1870	108, 328
Gas-works, &c., Treating ammoniacal liquor of	H. H. Eames and C. J. Eames	Philadelphia, Pa., and New York, N. Y.	Oct. 15, 1872	132, 264
Gasalier	J. W. Kerr	Pittsburgh, Pa	May 22, 1860	28, 377
Gasalier	J. H. Seaman	Brooklyn, N. Y	June 27, 1871	116, 358
Gasalier	J. H. Seaman	Brooklyn, N. Y	June 27, 1871	116, 359
Gasalier	J. H. Seaman	Brooklyn, N. Y	June 27, 1871	116, 360
Gasalier and cigar-lighter combined	W. C. Lester	New York, N. Y	Nov. 13, 1866	59, 617
Gasalier drop-light	J. Horton	New York, N. Y	Jan. 16, 1872	122, 721
Gasalier extension	L. Hull	Charlestown, Mass	Oct. 21, 1873	143, 765
Gasalier-extension apparatus	L. Hull	Charlestown, Mass	Jan. 23, 1872	123, 022
Gasalier, Sliding center-light extension	S. B. H. Vance	New York, N. Y	Apr. 30, 1872	126, 241
Gasalier smoke-bell	C. Deans	New York, N. Y	Aug. 12, 1873	141, 634
Gasaliers and drop-lights, Means of suspending	S. B. H. Vance	New York, N. Y	Nov. 5, 1867	70, 653
Gaseous bodies, Machinery for compressing	W. A. Royce	Newburgh, N. Y	May 26, 1857	17, 394
Gases, Apparatus for liquefying nitrous-oxide and other.	W. F. and W. A. Johnston	Brooklyn, N. Y	Nov. 14, 1871	120, 977
Gases, Apparatus for mixing	W. D. Parrish	Philadelphia, Pa	May 12, 1863	38, 499
Gases, Apparatus for naphthalizing	E. H. Ashcroft	Boston, Mass	June 5, 1860	28, 549
Gases for the purpose of elevating water, Method of applying heat to dilate.	J. W. Middleton	Philadelphia, Pa	Sept. 12, 1854	11, 671
Gases from furnaces, Collecting, separating, and purifying the.	J. H. Connelly and J. McLure	Wheeling, W. Va	May 31, 1870	103, 572
Gases from oils, Retort for generating	S. C. Salisbury	New York, N. Y	Apr. 24, 1866	54, 214
Gases from petroleum and water, Apparatus for producing and burning.	A. J. Griffin	Lowell, Mass	July 3, 1866	56, 143
Gases, &c., in furnaces, Apparatus for separating	W. D. Jones	Hagaman's Mills, N. Y	May 25, 1858	20, 351
Gases, Method of collecting and separating carbonic acid from mixtures of.	J. S. Baldwin	New York, N. Y	May 29, 1866	55, 038
Gases of smelting-furnaces, Apparatus for collecting the waste.	B. Branon and G. H. Baldwin	Sharon, Pa	June 29, 1869	92, 006
Gasoline for heating and illuminating, Vaporizing and burning.	H. Gilbert	Philadelphia, Pa	Feb. 12, 1867	61, 024
Gasometer	J. Bogardus	New York, N. Y	Oct. 7, 1834	......
Gasometer	J. Butler	New York, N. Y	Feb. 27, 1872	123, 977
Gasometer	J. E. Hobbs	North Berwick, Me	Jan. 26, 1869	86, 155
Gasometer	G. W. Kraft	Philadelphia, Pa	Oct. 26, 1858	21, 887
Gasometer	T. F. Rowland	Green Point, N. Y	Mar. 8, 1870	100, 559
Gasometer	H. B. Williams	Baltimore, Md	June 11, 1830	......
Gasometer or gas-holder	J. C. Tiffany	Portsmouth, N. H	Jan. 21, 1873	135, 172
Gasometer, Portable	J. H. Hayward	New York, N. Y	Mar. 29, 1864	42, 085
Gasometer, Portable	J. H. Hayward	New York, N. Y	Aug. 10, 1869	93, 435
Gasometers, Method of counterpoising	P. T. Burtis	Chicago, Ill	July 27, 1858	20, 988

Gate:

See Approach-opening gate.
Automatic gate.
Balance-gate.
Bridge-gate.
Bridge turning gate.
Canal-lock gate.
Cattle-gate.
Change-gate.
Chute-gate.
Double gate.
Double sliding gate.
Elastic gate.
Farm-gate.
Fence-gate.
Field-gate.
Flexible gate.
Flood-gate.
Fluid-gate.
Folding gate.
Grain-gate.
Hanging gate.
Hatchway safety-gate.
Molasses-gate.
Molder's gate.
Nursery-gate.
Operating-gate.
Railway-gate.
Railway cattle-gate.
Railway-crossing gate.
Rising gate.
Road-gate.
Safety-gate.
Saw-mill gate.
Self-acting gate.
Self-operating gate.
Sliding gate.
Sliding and swinging gate.
Sluice-gate.

Index of patents issued from the United States Patent Office from 1790 *to* 1873, *inclusive*—Continued.

Invention.	Inventor.	Residence.	Date.	No.
Gate—Continued.				
See Snow-gate. Stock guard-gate. Suspension-gate. Swing-bridge gate. Swinging gate. Tilting gate. Turbine-gate. Turbine-motor gate. Turnpike-gate. Wagon end-gate. Waste-gate. Water-gate. Water-wheel gate.				
Gate	P. M. Ackerman	Webster, N. Y	Oct. 9, 1866	58, 566
Gate	P. M. Ackerman	Webster, N. Y	June 4, 1867	65, 326
Gate	H. Adams	Seattle, Wash	Oct. 23, 1866	58, 965
Gate	J. Adams	Pontiac, Mich	May 5, 1868	77, 564
Gate	R. Adams	Ottawa, Ill	Jan. 14, 1868	73, 277
Gate	J. B. Alexander	Washington, D. C	June 4, 1867	65, 327
Gate	W. J. Alexander	Rolling Prairie, Ind	Dec. 3, 1867	71, 675
Gate	S. S. Allen	Belvidere, N. Y	Jan. 5, 1869	85, 555
Gate	S. Allington	West Dresden, N. Y	Oct. 5, 1858	21, 645
Gate	T. Andrews	Warren, Wis	Aug. 30, 1870	106, 763
Gate	S. E. Anthony	Stillwater, N. Y	Apr. 2, 1867	63, 455
Gate	W. D. and W. I. Armstrong	Harlem, Ill	Apr. 9, 1867	63, 687
Gate	W. I. Armstrong	Rockford, Ill	Aug. 21, 1866	57, 432
Gate	J. Atkins	Mokena, Ill	Feb. 4, 1868	74, 034
Gate	V. Babcock	Marshall, Mich	July 31, 1866	56, 693
Gate	D. B. Baker	Rollerville, Ohio	May 8, 1866	54, 641
Gate	D. D. Baker	West Alexandria, Ohio	June 25, 1867	66, 113
Gate	G. E. Baker	Waukegan, Ill	July 24, 1860	29, 234
Gate	F. Ball	Cleona, Iowa	Mar. 28, 1865	46, 985
Gate	H. F. Balschmiter	Davenport, Iowa	Sept. 10, 1867	68, 684
Gate	H. Barber	Milpitas, Cal	Feb. 19, 1861	31, 435
Gate	J. F. Barrow	Baltimore, Md	Sept. 17, 1872	131, 417
Gate	U. N. Beardsley	Goshen, Ind	July 16, 1867	66, 778
Gate	A. Beeraft	Jacksonville, Ill	Apr. 5, 1870	101, 570
Gate	J. Behel	Rockford, Ill	Apr. 13, 1869	88, 770
Gate	J. S. Benedict	Bedford, Ohio	Sept. 19, 1871	119, 110
Gate	I. C. Bennett	Clinton, Ill	Jan. 18, 1870	98, 841
Gate	J. H. H. Bennett	Hunt's Hollow, N. Y	Oct. 23, 1860	30, 460
Gate	E. L. Bergstresser	Sunbury, Pa	Jan. 1, 1867	60, 828
Gate	A. H. Betts	Cleveland, Ohio	Nov. 14, 1871	120, 932
Gate	J. Bickhart	Harlan, Ind	Dec. 9, 1862	37, 083
Gate	G. W. Bishop	Stamford, Conn	Mar. 3, 1868	75, 116
Gate	J. Blinn	Rockford, Ill	Feb. 4, 1873	135, 463
Gate	A. Blood, sr	Janesville, Wis	Nov. 22, 1864	45, 132
Gate	F. F. Blood	Janesville, Wis	Dec. 15, 1863	40, 902
Gate	C. S. Borney	Syracuse, N. Y	Feb. 15, 1870	99, 822
Gate	A. T. Boon	Galesburgh, Ill	Mar. 20, 1866	53, 263
Gate	D. Bordner	Canton, Ohio	Apr. 30, 1867	64, 190
Gate	R. A. Boulware	Doniphan, Kans	Feb. 6, 1872	123, 447
Gate	G. C. Bovey	Cincinnati, Ohio	Oct. 31, 1871	120, 484
Gate	J. Bowers	Clinton, Wis	Mar. 10, 1868	75, 356
Gate	R. T. Bowne	Fallston, Md	Apr. 25, 1871	114, 101
Gate	C. D. Brewer	Williamsport, Pa	May 18, 1869	90, 074
Gate	J. R. Brisel	Middletown, N. Y	Jan. 1, 1867	60, 678
Gate	L. D. Brooks	Syene, Wis	Apr. 13, 1869	88, 773
Gate	H. Brouse	Wellington, Ohio	May 21, 1872	126, 928
Gate	F. C. Brown and C. Allen	Palmyra, N. Y	May 25, 1869	90, 421
Gate	L. F. Brown	Keokuk, Iowa	Oct. 8, 1867	69, 535
Gate	R. T. Browne	Fallston, Md	Nov. 29, 1870	109, 713
Gate	E. and A. Buckman	Greenbush, N. Y	Dec. 8, 1868	84, 729
Gate	J. Budd	Pittsford, N. Y	Apr. 30, 1867	64, 193
Gate	J. A. Burchard	Beloit, Wis	May 21, 1867	64, 945
Gate	J. A. Burchard and R. Tatters-hall.	Beloit, Wis	June 23, 1868	79, 201
Gate	T. I. Burhyte	Fond du Lac, Wis	Mar. 19, 1867	63, 008
Gate	J. P. Butz and A. McFarland	Enterprise, Ind	May 19, 1868	77, 956
Gate	J. P. Cadman	Freeport, Ill	July 10, 1866	56, 174
Gate	S. W. Chamberlain	Three Oaks, Mich	Dec. 20, 1859	26, 477
Gate	J. D. Chambers	Williamsport, Ind	June 21, 1870	104, 555
Gate	J. D. Chambers	Williamsport, Ind	June 21, 1870	104, 556
Gate	I. A. Clark	Marion, N. Y	Sept. 21, 1869	95, 083
Gate	B. Clough	Amherst, Ohio	Sept. 18, 1866	58, 218
Gate	D. M. Cochran and L. A. Haw-kins.	Richmond, Ind	May 23, 1871	115, 167
Gate	R. Conway	Velga, Ind	Feb. 26, 1867	62, 396
Gate	R. R. Cool	Milien's Bay, N. Y	Mar. 1, 1864	41, 700
Gate	W. Cooley	Tafton, Wis	Nov. 19, 1867	70, 963
Gate	J. S. Corbin	Ann Arbor, Mich	Nov. 17, 1863	40, 609
Gate	J. B. Cottom	Dayton, Ohio	Mar. 28, 1871	113, 020
Gate	J. S. Covell	Salt River, Mich	Nov. 1, 1870	108, 764
Gate	W. H. Cowley	Cleveland, Ohio	Dec. 22, 1868	85, 072
Gate	H. B. Crandall	Brocton, N. Y	Jan. 31, 1871	111, 322
Gate	P. S. Crawford	Union, Ill	Mar. 21, 1871	112, 904
Gate	P. S. Crawford	Union, Ill	July 25, 1871	117, 263
Gate	D. Creighton	Vacaville, Cal	Nov. 7, 1871	120, 573
Gate	J. Curry	South Bend, Ind	Nov. 5, 1872	132, 809
Gate	G. G. Curtis	Rochester, N. Y	Mar. 12, 1867	62, 735
Gate	A. Dambacher	New York, N. Y	July 16, 1872	129, 538
Gate	S. A. Darrach	Newburgh, N. Y	Oct. 3, 1871	119, 509
Gate	J. R. Davis	Covington, Ga	Nov. 16, 1869	96, 783
Gate	S. S. Davis	Edgerton, Wis	Oct. 13, 1868	83, 046
Gate	L. S. Deming	Newington, Conn	June 22, 1869	91, 723
Gate	D. A. Denison	Troy, Mich	Apr. 2, 1867	63, 370
Gate	S. M. Denniston	Hudson, Wis	Oct. 8, 1867	69, 640
Gate	J. Dickason	Vevay, Ind	June 4, 1867	65, 354
Gate	B. F. Dickey	Marshall Township, Mich	May 10, 1870	102, 783
Gate	B. F. Dickey	Marshall, Mich	May 31, 1870	103, 583
Gate	A. J. Dimick	Berlinville, Ohio	Jan. 3, 1871	110, 750
Gate	A. O. Divine	Cambria Mills, Mich	Apr. 30, 1867	64, 289
Gate	C. Dixon and S. H. Close	Port Byron, N. Y	July 3, 1866	56, 019

Index of patents issued from the United States Patent Office from 1790 *to* 1873, *inclusive*—Continued.

Invention.	Inventor.	Residence.	Date.	No.
Gate	E. R. Dobbs	Poughkeepsie, N. Y	Apr. 9, 1867	63, 710
Gate	O. Dowd	Dansville, N. Y	Nov. 12, 1867	70, 819
Gate	W. R. Dugdale	Penn Township, Ind	July 9, 1867	66, 572
Gate	E. Duncan	West Milton, Ohio	Oct. 30, 1866	59, 196
Gate	R. R. Earnest	Springfield, Ohio	June 5, 1866	55, 259
Gate	C. Edgar	Dayton, Ohio	Sept. 10, 1872	131, 257
Gate	C. H. Eggleston	Marshall, Mich	Feb. 22, 1870	100, 129
Gate	S. Elliott	Richmond, Ind	Aug. 24, 1869	93, 977
Gate	T. Ellison	Abingdon, Ill	Mar. 15, 1870	100, 739
Gate	C. H. Embree	West Dresden, N. Y	Jan. 28, 1868	73, 704
Gate	S. F. Emerson	Seville, Ohio	July 17, 1866	56, 392
Gate	J. B. Erwin	Pittsburgh, Pa	Apr. 4, 1871	113, 282
Gate	L. Essig	Canton, Ohio	Jan. 14, 1868	73, 312
Gate	F. Ewer	Honeoye Falls, N. Y	Jan. 29, 1867	61, 526
Gate	B. H. Fairchild and E. Sadler	Farmington, Mich	June 4, 1867	65, 359
Gate	S. G. Farnham	East Hartford, Conn	Nov. 10, 1863	40, 557
Gate	J. Fausett	Leonardtown, Md	Dec. 19, 1871	122, 003
Gate	D. E. Fenn	Tallmadge, Ohio	Aug. 28, 1860	29, 774
Gate	S. L. Fisher	Brimfield, Ill	Aug. 29, 1865	49, 615
Gate	B. F. Fisk	Fredonia Township, Mich	Apr. 27, 1869	89, 301
Gate	J. E. Fitts	Candia Village, N. H	Nov. 6, 1866	59, 378
Gate	G. C. Flagg	Tanktown, Ohio	Apr. 9, 1861	31, 965
Gate	D. Flint	Sacramento, Cal	Oct. 15, 1867	69, 788
Gate	W. G. Franklin	Shelbina, Mo	Jan. 24, 1871	111, 190
Gate	P. Freeman	Benton County, Iowa	June 12, 1866	55, 481
Gate	S. Freet	Upper Strasburgh, Pa	Nov. 30, 1869	97, 288
Gate	F. Friedrich and C. H. Allen	Platteville, Wis	Apr. 23, 1872	125, 947
Gate	D. Fuller	Oakwood, Mich	Sept. 10, 1867	68, 726
Gate	A. Gaskill	Neoga, Ill	Jan. 16, 1872	122, 764
Gate	W. Gause and J. C. Curryer	Indianapolis and Thorntown, Ind.	Nov. 15, 1870	109, 198
Gate	E. Gemberling	Elkhart, Ind	Nov. 29, 1870	109, 609
Gate	R. Gidley	La Grange, N. Y	Dec. 24, 1867	72, 477
Gate	R. Gidley	La Grangeville, N. Y	Jan. 3, 1871	110, 646
Gate	R. Gidley	Moore's Mill, N. Y	July 25, 1871	117, 403
Gate	R. J. Gilbert	Hanover, Wis	Dec. 4, 1866	60, 171
Gate	R. J. Gilbert	Hanover, Wis	Oct. 1, 1867	69, 336
Gate	S. M. Gillett	Homer, N. Y	Sept. 12, 1865	49, 873
Gate	J. and W. B. Goff	Hornellsville, N. Y	Apr. 30, 1867	64, 303
Gate	W. H. Goodale	Colton, N. Y	July 13, 1869	92, 524
Gate	S. A. Gould	Aberdeen, Ind	Feb. 16, 1869	87, 040
Gate	G. S. Granger and W. Northop	Wayland, N. Y	May 28, 1867	65, 210
Gate	J. H. Graves	Rochester, N. Y	July 3, 1866	56, 039
Gate	I. J. Gray	Seville, Ohio	Nov. 20, 1866	59, 743
Gate	R. D. Green	Columbia, Mo	May 21, 1867	64, 860
Gate	W. W. Green, jr	Janesville, Wis	Sept. 15, 1868	82, 112
Gate	B. Greenside	Fort-Dodge, Iowa	July 16, 1867	66, 832
Gate	A. L. Grinnell	Des Moines, Iowa	Jan. 9, 1866	51, 940
Gate	S. Grinnell, G. Bez, and H. C. Stoll.	Mokena, Ill	Sept. 5, 1865	49, 753
Gate	W. H. Griscom	Salem, N. J	Mar. 16, 1869	87, 841
Gate	I. H. Gustin	Middletown, Ind	July 5, 1870	105, 067
Gate	D. Hague	Balville Township, Ohio	Feb. 26, 1867	62, 412
Gate	U. W. Hardy	Abingdon, Ill	Dec. 29, 1868	85, 306
Gate	U. W. Hardy	Abingdon, Ill	Apr. 4, 1871	113, 294
Gate	J. H. Hardy	Penn Township, Pa	Sept. 13, 1870	107, 254
Gate	J. Harpster	Clyde, Ohio	Oct. 19, 1869	96, 002
Gate	N. Harrier	Muscatine, Iowa	June 29, 1869	91, 932
Gate	J. K. Harris	Springfield, Ohio	Jan. 24, 1871	111, 201
Gate	H. P. Haskin	Roscoe, Ill	Dec. 22, 1868	85, 091
Gate	H. P. Haskin	Roscoe, Ill	May 18, 1869	90, 261
Gate	H. R. Haskins	Harlan, Ind	Dec. 20, 1870	110, 363
Gate	R. Hatch, jr	Traverse City, Mich	Oct. 26, 1869	96, 106
Gate	W. Hathaway	Northbridge, Mass	Sept. 19, 1871	119, 141
Gate	C. P. Hawley and E. B. Murdock	East Galway, N. Y	July 17, 1866	56, 412
Gate	J. Hays	Fostoria, Ohio	Mar. 15, 1870	100, 759
Gate	B. S. Healy	Cohocton, N. Y	Sept. 4, 1866	57, 713
Gate	N. B. Helm	Aldin, Ill	May 21, 1872	126, 958
Gate	H. A. Henderson	Avoca, N. Y	Aug. 25, 1868	81, 501
Gate	S. Henry	Chenoa, Ill	Apr. 19, 1870	102, 003
Gate	L. Hermance	Hudson, N.Y	Oct. 15, 1867	69, 914
Gate	J. Hibbard	Prospect Lake, Mich	Feb. 5, 1867	61, 832
Gate	E. Hickman	Red Bluff, Cal	Dec. 2, 1873	145, 104
Gate	F. M. Hickman	Rolling Prairie, Ind	Mar. 31, 1868	76, 190
Gate	E. Higgins	Sacramento, Cal	July 25, 1871	117, 418
Gate	W. Holdredge	Oxford, N. Y	Apr. 30, 1867	64, 317
Gate	W. C. Hooker	Abingdon, Ill	June 2, 1868	78, 452
Gate	W. C. Hooker	Abingdon, Ill	Nov. 1, 1870	108, 786
Gate	J. J. and J. T. Hoss	Tipton County, Ind	Feb. 20, 1872	123, 902
Gate	H. A. House	Brooklyn, N. Y	Aug. 21, 1860	29, 696
Gate	R. H. Hudgin	Fairfield, Canada	Jan. 24, 1871	111, 124
Gate	R. H. Hudgin	Fairfield, Canada	Feb. 13, 1872	123, 700
Gate	R. H. Hudgin	Fairfield, Canada	July 2, 1872	128, 490
Gate	G. Hungerford	Union, Ill	July 23, 1872	129, 733
Gate	H. Hunt	Delavan, Wis	Dec. 24, 1867	72, 495
Gate	H. Hunt	Delavan, Wis	Apr. 21, 1868	76, 980
Gate	H. Hunt	Delavan, Wis	May 12, 1868	77, 735
Gate	J. H. Hunter	Versailles, Ind	Nov. 2, 1869	96, 436
Gate	A. B. Hurd	Watkins, N. Y	Apr. 23, 1867	64, 108
Gate	C. B. Huixthal and J. Lee	Bolivar, Ohio	Oct. 2, 1860	30, 223
Gate	J. L. Janeway	Flemington, N. J	Mar. 10, 1868	75, 276
Gate	J. L. Janeway	Flemington, N. J	Mar. 31, 1868	76, 081
Gate	H. J. Johnson	Saint Peter, Minn	Oct. 1, 1867	69, 443
Gate	J. Johnson	Geneseo, N. Y	Mar. 6, 1860	27, 367
Gate	S. S. Kappel	Woodhull, Ill	Nov. 6, 1866	59, 405
Gate	C. Kark	Huntington, Ohio	Aug. 27, 1867	68, 084
Gate	R. Kelly	Tuscola, Ill	Jan. 31, 1865	46, 112
Gate	E. Kemper	Thorn Township, Ohio	June 24, 1862	35, 689

Index of patents issued from the United States Patent Office from 1790 *to* 1873, *inclusive*—**Continued.**

Invention.	Inventor.	Residence.	Date.	No.
Gate	E. Kemper	Thornville, Ohio	July 16, 1867	66, 852
Gate	M. F. Kent	West Union, Iowa	Aug. 11, 1868	80, 969
Gate	S. Kepner	Pottstown, Pa	Oct. 4, 1870	108, 031
Gate	J. H. King	Smithfield, Ind	Sept. 15, 1868	82, 229
Gate	W. H. Kosht	Ashland, Ohio	July 25, 1871	117, 431
Gate	S. A. Kroner	Doylestown, Pa	Jan. 15, 1867	61, 215
Gate	A. Larrowe	Cohocton, N. Y	Nov. 6, 1866	59, 416
Gate	A. Larrowe	Cohocton, N. Y	May 21, 1867	64, 987
Gate	J. Lee	Massillon, Ohio	June 2, 1868	78, 598
Gate	W. Leonard	Orleans, Ind	Sept. 14, 1869	94, 900
Gate	F. Livings	East Enterprise, Ind	Oct. 18, 1870	108, 368
Gate	F. Livingston	Marathon, N. Y	Apr. 21, 1868	76, 930
Gate	M. E. Livingston	Manchester, Ill	Aug. 8, 1871	117, 901
Gate	J. C. Long	Eaton, Ind	Mar. 12, 1872	124, 600
Gate	N. Long	Muncie, Ind	June 20, 1871	116, 202
Gate	M. Loomis	Liberty Centre, Ohio	Dec. 9, 1873	145, 434
Gate	W. I. Ludlow	Cleveland, Ohio	Nov. 21, 1871	121, 181
Gate	C. Mack	Leipsic, Ohio	Oct. 1, 1867	69, 455
Gate	C. Mack	Leipsic, Ohio	July 6, 1869	92, 329
Gate	G. W. and G. S. Mackey	Lathrop, Pa	Nov. 18, 1873	144, 780
Gate	H. Mansfield	Warsaw, Ind	Sept. 24, 1867	69, 108
Gate	S. L. Marsden	New Haven, Conn	Apr. 6, 1869	88, 575
Gate	J. Martin	New Oxford, Pa	May 1, 1866	54, 381
Gate	J. W. Martin	Washington, D. C	Mar. 19, 1867	63, 067
Gate	M. Martin	Rockford, Ill	Dec. 26, 1871	122, 125
Gate	H. Z. Mast	Fork Meeting-House, Md	Dec. 10, 1872	133, 788
Gate	E. A. Matthiessen	Cornwall, N. Y	July 2, 1867	66, 368
Gate	C. F. Mawbey	Woodbridge, N. Y	Dec. 24, 1867	72, 657
Gate	H. Maxell	Canton, Ohio	June 5, 1866	55, 322
Gate	J. M. May	Janesville, Wis	Aug. 8, 1865	49, 286
Gate	J. M. May	Janesville, Wis	Dec. 17, 1867	72, 213
Gate	W. McAfee	Summerville, Mich	Aug. 21, 1860	29, 706
Gate	O. C. McCarty	Haysville, Ohio	Oct. 23, 1866	59, 045
Gate	P. McCollum	Fayette, Md	Nov. 10, 1868	83, 867
Gate	J McCreary	Middletown, Pa	May 7, 1867	64, 548
Gate	F. H. McGeorge	Corning, N. J	Dec. 14, 1869	97, 947
Gate	E. R. McKinney	Lacon, Ill	Apr. 30, 1867	64, 239
Gate	G. McKnight	Hebron, N. Y	Jan. 29, 1867	61, 676
Gate	J. H. McKnight	Oakwood, Mich	Sept. 1, 1868	81, 662
Gate	I. H. McOmber	El Paso, Ill	Sept. 3, 1867	68, 376
Gate	A. W. Meek	Waterloo City, Ind	Sept. 22, 1868	82, 336
Gate	J. L. Meredith	Bloomingsburgh, Ind	Apr. 25, 1871	114, 177
Gate	R. C. Mighell	Plano, Ill	Sept. 8, 1868	82, 021
Gate	L. Miller	Johnsville, Ohio	May 26, 1868	78, 389
Gate	P. L. Miller	Mechanicsburgh, Pa	May 7, 1867	64, 551
Gate	P. L. Miller	Mechanicsburgh, Pa	Mar. 17, 1868	75, 642
Gate	J. H. Morgan	Warsaw, Ind	Aug. 6, 1872	130, 306
Gate	J. D. Morrison	Richfield, Ill	Nov. 21, 1871	121, 188
Gate	G. Mott	Big Run, Ohio	June 13, 1871	115, 883
Gate	G. Mott, J. Morris, jr., and D. Lupton.	Big Run, Ohio	Sept. 14, 1869	94, 764
Gate	T. Munger	Cedar Falls, Iowa	July 16, 1867	66, 870
Gate	T. Munger	Jaynesville, Iowa	Nov. 26, 1867	71, 503
Gate	E. M. Naramore	North Underhill, Vt	Apr. 14, 1868	76, 800
Gate	M. Neudgent	Hartland, Mich	Jan. 12, 1869	85, 755
Gate	V. C. Newland	Sparta, Wis	Dec. 5, 1871	121, 654
Gate	W. E. Nichols	Baldwin, Mo	Aug. 18, 1868	81, 196
Gate	P. O'Neil	Murfreesborough, Tenn	Jan. 24, 1871	111, 238
Gate	H. S. Otis	Prescott, Wis	Jan. 5, 1869	85, 607
Gate	C. N. Owen	Salem, Ohio	Feb. 28, 1871	112, 174
Gate	M. Packard	Clarendon, N. Y	Oct. 23, 1866	59, 060
Gate	C. H. Paine	Providence, R. I	Apr. 16, 1867	63, 932
Gate	C. H. Paine	Providence, R. I	June 11, 1867	65, 591
Gate	B. F. Palmer	Baraboo, Wis	Apr. 4, 1871	113, 334
Gate	N. Parker	Trimble County, Ky	Mar. 21, 1871	112, 843
Gate	J. M. Peirce	Mokena, Ill	Apr. 28, 1863	38, 331
Gate	J. J. Pellett	Oconomowoc, Wis	Apr. 21, 1868	77, 088
Gate	W. A. Penny	Morrisville, N. C	Nov. 7, 1871	120, 770
Gate	O. and C. Perry	Ortonville, Mich	Feb. 18, 1868	74, 716
Gate	E. Petteys	Chestertown, N. Y	July 23, 1867	67, 135
Gate	E. Petteys and T. C. Leggett	Chestertown, N. Y	Dec. 18, 1866	60, 551
Gate	A. H. Phillips	Pontiac, Mich	Aug. 27, 1872	130, 938
Gate	W. H. Phillips	Staunton, Md	Mar. 12, 1872	124, 508
Gate	C. H. Platt	North Fairfield, Ohio	Dec. 31, 1867	72, 897
Gate	N. M. Platt	North Fairfield, Ohio	Dec. 22, 1868	85, 127
Gate	J. Pool	Elizabeth City, N. C	Feb. 7, 1871	111, 676
Gate	J. Potter	Pierceton, Ind	July 14, 1868	79, 857
Gate	J. B. Powell and S. H. Everett	Macedon, N. Y	June 25, 1867	66, 171
Gate	S. Puffer	Oxford, N. Y	Sept. 11, 1866	57, 967
Gate	C. Purdy	Bedford, Ohio	May 8, 1866	54, 596
Gate	H. R. Raub	Pymatuning, Pa	July 21, 1868	80, 220
Gate	F. Raymond	Cleveland, Ohio	Apr. 12, 1870	101, 765
Gate	F. Raymond and A. Miller	Cleveland, Ohio	Nov. 10, 1868	83, 881
Gate	C. D. Reed	Polo, Ill	Oct. 25, 1870	108, 628
Gate	S. Regan and E. Mensy	La Motte, Iowa	Oct. 11, 1870	108, 186
Gate	E. Reynolds	Metomen, Wis	Sept. 17, 1867	69, 024
Gate	W. W. Riley	Columbus, Ohio	Mar. 24, 1868	75, 792
Gate	J. F. Rodgers	South Bend, Ind	Oct. 27, 1868	83, 550
Gate	W. H. Rogers	Harlem, Ill	Jan. 7, 1868	73, 049
Gate	H. Root	Union City, Ind	Aug. 22, 1871	118, 393
Gate	E. Roth	New Oxford, Pa	Mar. 31, 1868	76, 251
Gate	E Roth and G. Shane	New Oxford, Pa	Apr. 23, 1867	64, 150
Gate	N. Rue	Harrodsburgh, Ky	Dec. 29, 1868	85, 479
Gate	C. W. Saladee	St. Catharine's, Canada	Feb. 14, 1871	111, 780
Gate	C. Saxton	Fredonia, Ohio	Dec. 21, 1869	98, 194
Gate	E. B. Scattergood	Saint John's, Mich	Feb. 23, 1869	87, 206
Gate	E. B. Scattergood	Saint John's, Mich	June 14, 1870	104, 210
Gate	M. Schneider	Rochester, Minn	Aug. 27, 1872	130, 946

Index of patents issued from the United States Patent Office from 1790 *to* 1873, *inclusive*—Continued.

Invention.	Inventor.	Residence.	Date.	No.
Gate	P. Schwebel	Quincy, Ill	Mar. 19, 1872	124, 858
Gate	L. M. Scothorn	Findley, Ohio	Nov. 5, 1867	70, 626
Gate	C. Seefeld	Lomira, Wis	Oct. 23, 1866	59, 084
Gate	W. Serviss	Sidney, Ohio	Oct. 1, 1867	69, 494
Gate	J. Shartle	Lima, Ind	Dec. 24, 1867	72, 549
Gate	C. Shepard	Binghamton, N. Y	Sept. 10, 1867	68, 658
Gate	F. R. Sherman	Dowagiac, Mich	Jan. 7, 1868	73, 129
Gate	L. W. Sibley	Ames, Iowa	June 22, 1869	91, 677
Gate	L. W. Sibley	Ames, Iowa	Jan. 24, 1871	111, 266
Gate	G. W. Sigerfoos, J. J. Sands, and G. Fry.	Potsdam, Ohio	Sept. 4, 1866	57, 782
Gate	S. H. Sill	Geneva, N. Y	May 8, 1860	28, 205
Gate	J. W. Singleton	Quincy, Ill	Jan. 1, 1867	60, 798
Gate	G. W. Sizer	Springvale, Wis	May 14, 1867	64, 802
Gate	G. A. Slater	Benton Harbor, Mich	Jan. 3, 1871	110, 687
Gate	L. and D. Slinger	Carmi, Ill	June 18, 1872	128, 183
Gate	J. A. Smith	Lacon, Ill	Sept. 29, 1868	82, 648
Gate	L. Smith	Easton, Mass	May 29, 1849	6, 488
Gate	S. Smith and A. Persels	Rockford, Ill., and Beloit, Wis.	Sept. 10, 1867	68, 801
Gate	W. B. Smith	Copper Creek, Ill	Dec. 10, 1872	133, 807
Gate	S. Smyth	East Bridgewater, Pa	Oct. 3, 1871	119, 661
Gate	J. Snell	Evans' Mills, N. Y	June 22, 1869	91, 783
Gate	B. Snyder	Clinton, Wis	Sept. 26, 1871	119, 421
Gate	W. W. Sowles	Manlius, N. Y	Apr. 13, 1869	88, 818
Gate	S. Spoor	Phelps, N. Y	Nov. 20, 1866	59, 930
Gate	S. Spoor	Phelps, N. Y	Dec. 11, 1866	60, 435
Gate	G. S. Spragg and G. Mott	Tabor, Iowa	Jan. 23, 1872	122, 973
Gate	G. P. Stebbins	Sparta Centre, Mich	Jan. 7, 1868	73, 206
Gate	E. Stevens	New Carlisle, Ind	June 25, 1872	128, 259
Gate	G. Stovel	Chicago, Ill	Mar. 21, 1865	46, 954
Gate	W. W. Sutliff	Town Line, Pa	Jan. 8, 1867	61, 115
Gate	J. M. Swift	Shelbyville, Ill	Sept. 10, 1867	68, 804
Gate	W. Tallman	Manteno, Ill	June 20, 1865	48, 322
Gate	A. Tandy	Columbia, Mo	Oct. 1, 1867	69, 509
Gate	R. Tattershall and J. A. Burchard.	Beloit, Wis	June 25, 1867	66, 108
Gate	T. D. Taylor	Hartleton, Pa	Dec. 3, 1872	133, 682
Gate	G. L. Templeton	Pierceton, Ind	Nov. 12, 1867	70, 762
Gate	J. W. Thompson	Greenfield, Mass	Dec. 10, 1867	72, 120
Gate	W. Tobey	Naples, N. Y	Sept. 14, 1858	21, 526
Gate	J. J. Tofflemire and J. D. Linnell.	Rockford, Ill	Sept. 20, 1870	107, 638
Gate	C. Trexler	La Grange, Ill	July 2, 1867	66, 421
Gate	G. W. Tucker	Elba Township, Ill	May 5, 1868	77, 682
Gate	I. C. Tunison and A. Reeve	Roodhouse and Pike County, Ill.	Sept. 3, 1872	131, 039
Gate	H. Turner	Ripon, Wis	July 2, 1867	66, 423
Gate	J. Vail	Beloit, Wis	May 21, 1867	64, 924
Gate	C. H. Van Eps	Farmington, Iowa	Apr. 2, 1867	63, 444
Gate	H. B. Van Voorhis	Pittsburgh, Pa	June 11, 1867	65, 619
Gate	W. B. Waldo	Johnsville, N. Y	Apr. 19, 1864	42, 422
Gate	W. S. Wandell	Battle Creek, Mich	Mar. 17, 1868	75, 607
Gate	H. M. Ward	Stone Church, N. Y	May 15, 1866	54, 796
Gate	D. R. Warfield	Muscatine, Iowa	Jan. 27, 1863	37, 536
Gate	D. R. Warfield	Muscatine, Iowa	Mar. 15, 1864	41, 951
Gate	N. Waterbury	Fond du Lac, Wis	Nov. 8, 1859	26, 063
Gate	S. H. Wheeler	Dowagiac, Mich	June 18, 1867	65, 974
Gate	E. B. Whitaker	Bel Air, Md	Dec. 17, 1872	133, 960
Gate	S. Whitman	Wayland, N. Y	Sept. 6, 1864	44, 132
Gate	A. D. Wilcox	Manlius, N. Y	Apr. 13, 1869	88, 828
Gate	T. P. Wilcox	Hebron, Ind	Apr. 9, 1872	125, 511
Gate	M. S. G. Wilde	Somerville, Mass	July 27, 1869	93, 027
Gate	S. P. Williams	Sheridan, N. Y	Mar. 23, 1869	88, 249
Gate	J. F. Winchell	Springfield, Ohio	Sept. 18, 1866	58, 161
Gate	J. Wineganer	Gratiot, Ohio	June 25, 1872	128, 269
Gate	G. Winteringer	Fredericktown, Ohio	Apr. 21, 1868	76, 963
Gate	D. D. Wisell	Zanesville, Ind	July 23, 1872	129, 702
Gate	H. S. Wolf	Rolling Prairie, Ind	Apr. 16, 1867	63, 974
Gate	E. R. Wolfe	Plymouth, Pa	July 23, 1867	67, 094
Gate	E. J. Wolfgang and J. M. Kenreigh.	Salem, Ohio	Mar. 23, 1869	88, 108
Gate	J. A. Wood	Crosswicks, N. J	June 27, 1871	116, 526
Gate	J. A. Wood and E. V. Marbaker.	Crosswicks, N. J	Dec. 7, 1869	97, 745
Gate	R. J. Wood	Hancock, Mich	Oct. 24, 1871	120, 358
Gate	S. G. Wood	Rochester, N. Y	Sept. 19, 1871	119, 101
Gate	J. Woodward	Wilmot, N. H	Oct. 28, 1862	36, 817
Gate	W. J. Wooster	Harvard, Ill	Sept. 3, 1872	131, 043
Gate	W. Worth	Jackson, Mich	Aug. 11, 1863	39, 520
Gate	C. Wright	Mason, Ill	Mar. 14, 1871	112, 762
Gate	J. Wright	Versailles, Ill	Nov. 29, 1870	109, 704
Gate	E. Young	Camden Centre, Mich	July 9, 1867	66, 663
Gate	I. N. Young	Swan, Ind	Dec. 31, 1867	72, 957
Gate, Adjustably-hinged	E. M. George	Three Rivers, Mich	Feb. 25, 1873	136, 149
Gate and barn-door fastening	E. Pea	Mechanicsburgh, Ill	Dec. 3, 1867	71, 636
Gate and door closing device	G. Turner	Lansing, Mich	Mar. 15, 1864	41, 948
Gate and door spring	J. Bliss	Cleveland, Ohio	Oct. 15, 1872	132, 133
Gate and door spring	A. T. Boon and L. Mills	Galesburgh, Ill	July 11, 1871	116, 800
Gate and door spring	J. C. Gould	Oxford, N. J	Nov. 26, 1867	71, 478
Gate and door spring	M. Harrington	Waupun, Wis	Nov. 22, 1870	109, 512
Gate and door spring	J. Palmer	Cincinnati, Ohio	July 16, 1872	129, 584
Gate and door spring	B. F. Whitaker	Whitestown, Ind	Mar. 10, 1868	75, 324
Gate and door swing	F. W. Kroeber	Forbestown, Cal	Oct. 9, 1860	30, 327
Gate and fence combined	J. W. Sanford	Bath, N. Y	Aug. 20, 1867	67, 915
Gate and gate-post	G. W. Balding	Pleasant, Ind	Nov. 20, 1866	59, 753
Gate, Approach-opening	A. Iske and J. B. Erbe	Lancaster and Conestoga Township, Pa.	Aug. 9, 1859	25, 018
Gate, Approach-opening	J. S. Lloyd	Philadelphia, Pa	Sept. 20, 1859	25, 515

Index of patents issued from the United States Patent Office from 1790 *to* 1873, *inclusive*—Continued.

Invention.	Inventor.	Residence.	Date.	No.
Gate, Approach-opening	E. C. Rowland	Phelps, N. Y	Oct. 19, 1858	21, 851
Gate, Approach-opening	C. Winegar	Union Springs, N. Y	Dec. 7, 1858	22, 261
Gate-attachment	H. M. Long	Williamsville, N. Y	Dec. 29, 1868	85, 390
Gate-catch	W. W. Robinson	Ripon, Wis	Oct. 6, 1863	40, 187
Gate-closing device	W. Gilfillan	Syracuse, N. Y	Oct. 14, 1862	36, 647
Gate-closing device	L. Kolloff	Brooklyn, N. Y	Nov. 25, 1862	37, 001
Gate-fastener	J. H. Nichols	La Fayette, Ill	Dec. 9, 1873	145, 362
Gate-fastener	P. Philippi	Beardstown, Ill	Sept. 18, 1866	58, 130
Gate-fastener	H. S. Shisler	Manheim Township, Pa	May 11, 1869	89, 893
Gate-fastening	J. D. Bourne	De Witt, Iowa	Feb. 5, 1867	61, 800
Gate-fastening	W. Broomhall	Circleville, Ohio	Sept. 12, 1865	51, 417
Gate-fastening	J. Bull	Galesburgh, Ill	Dec. 3, 1867	71, 694
Gate-fastening	S. E. James	Smithfield Station Post-Office, Ohio.	Feb. 19, 1867	62, 136
Gate-fastening	J. C. Kellogg	Thorntown, Ill	Jan. 2, 1866	51, 838
Gate-fastening	W. Kimball	Salem, Ohio	Sept. 10, 1867	68, 752
Gate-fastening	H. Last	West Lebanon, Ind	Nov. 22, 1864	45, 162
Gate-fastening	R. A. Leeds	Stamford, Conn	Sept. 25, 1866	58, 268
Gate-fastening	J. Lintner	Indianapolis, Ind	May 31, 1870	103, 629
Gate-fastening	M. B. Markham	Grass Lake, Mich	Sept. 29, 1868	82, 537
Gate-fastening	G. McCoy	Antioch, Cal	Nov. 19, 1867	71, 195
Gate-fastening	C. Seymour	La Porte, Ind	Apr. 23, 1867	64, 153
Gate-fastening	J. P. Woodcock	Bedford, N. Y	Sept. 5, 1865	49, 819
Gate-fastening	S. Young	Milton, N. Y	Oct. 14, 1856	15, 911
Gate-flume	H. Frink	Chautauqua County, N. Y	Feb. 25, 1836	......
Gate opening and closing apparatus	S. G. Dugdale	Richmond, Ind	Oct. 11, 1853	10, 105
Gate opening and closing apparatus	E. Woolman	Damascoville, Ohio	Dec. 2, 1851	8, 572
Gate opening and closing device	J. H. Nevin	Ogdensburgh, N. Y	Feb. 28, 1860	27, 304
Gate opening and closing device	C. Winegar	Union Springs, N. Y	Aug. 14, 1855	13, 445
Gate-operating device	D. G. Goodall	Beloit, Wis	Mar. 16, 1869	87, 839
Gate or door spring	J. W. Briggs	Cleveland, Ohio	Sept. 24, 1861	33, 332
Gate-post	A. Fishburn	Carlisle, Pa	Feb. 2, 1869	86, 383
Gate-post	G. O. Hutson	Iowa City, Iowa	Sept. 19, 1855	50, 005
Gate-post	W. F. Veber	Perrysburgh, Ohio	Mar. 3, 1868	75, 224
Gate-post attachable to any panel of its corresponding field-fence.	J. G. Hunt	Cincinnati, Ohio	Apr. 7, 1857	16, 978
Gate-post, Non-swagging	I. Johnson	Alliance, Ohio	Oct. 29, 1861	33, 590
Gate-post, Portable	W. A. Dillon	Middletown, Mo	Dec. 3, 1872	133, 571
Gate purchase, Labor-saving rotary	J. Musser and L. B. Gitchel	Canton, Ohio	Feb. 11, 1832	....
Gate-spring	W. W. Sutliff	Town Line, Pa	Nov. 19, 1867	71, 083
Gates and presses, Locking device for	W. Welch	Bridgeport, Conn	Sept. 3, 1867	68, 540
Gates, Arrangement of weight and pulley for closing.	W. Twitchell	Syracuse, N. Y	Aug. 7, 1849	6, 632
Gates, &c., Closing and opening	C. Winegar	Union Springs, N. Y	May 22, 1855	12, 930
Gates, Construction and hanging of	J. Healy	South Dansville, N. Y	Mar. 7, 1865	46, 667
Gates, doors, &c., Hanging	S. Oberholzer	Terre Hill, Pa	May 6, 1856	14, 826
Gates, &c., Hanging	N. W. Cilley	Nottingham Township, N. Y.	May 16, 1854	10, 928
Gates, Hanging	A. Hotchkin	Schenevus, N. Y	Mar. 7, 1854	10, 619
Gates, Hanging	T. S. Minniss	Meadville, Pa	Feb. 14, 1865	46, 377
Gates, Hanging and operating	T. Parkinson	Naples, N. Y	Aug. 6, 1850	7, 548
Gates, Hanging for	P. Rasar and D. J. Mayes	Illiopolis, Ill	July 28, 1868	80, 305
Gates, Opening	J. A. Ayres	Hartford, Conn	Feb. 8, 1859	22, 845
Gates, Opening	F. B. Betts	Brownhelm, Ohio	Mar. 29, 1859	23, 345
Gates, Opening and closing	B. R. Cole	Geneva, N. Y	Oct. 18, 1859	25, 808
Gates, Opening and closing	W. T. Merritt	Hart's Village, N. Y	Nov. 1, 1853	10, 199
Gates, Opening and closing	W. G. Philips	Newport, Del	Mar. 7, 1854	10, 593
Gates, Operating	J. L. Betts	Archbold, Ohio	Apr. 9, 1872	125, 375
Gathered fabrics	H. S. Brown	New York, N. Y	Aug. 23, 1864	43, 943
Gavel-fork	T. R. George	West Dryden, N. Y	Nov. 16, 1869	96, 794
Gavel-fork	B. Wright and W. C. Park	Cardiff, N. Y	Apr. 25, 1871	114, 080
Gear, Changing	W. M. Davis	Gardiner, Me	July 17, 1847	5, 192
Gear, Chill for casting toothed	R. T. Davis	Canton, Ohio	May 10, 1870	102, 781
Gear-cutter	C. Evotte	Paris, France	Mar. 16, 1869	87, 766
Gear-cutter	H. N. Keables	Worcester, Mass	July 2, 1867	66, 354
Gear-cutter	E. Parker	Middletown, Conn	July 3, 1866	56, 149
Gear-cutter	W. Sellers	Philadelphia, Pa	Feb. 1, 1870	99, 356
Gear-cutters, Machine for shaping and cutting	F. A. Pratt	Hartford, Conn	Mar. 7, 1871	112, 379
Gear-cutters, Making	L. F. Grant	Thomaston, Conn	July 17, 1866	56, 495
Gear-cutting	A. T. Boon	Galesburgh, Ill	Nov. 10, 1863	40, 550
Gear-cutting engine	H. Pfarrer	New York, N. Y	Oct. 11, 1859	25, 759
Gear-cutting machine	R. Cairns	Waterbury, Conn	Aug. 6, 1872	130, 188
Gear-cutting machine	J. W. Foster	Pawtucket, R. I	Mar. 4, 1873	136, 429
Gear-cutting machine	J. J. Greenough	Syracuse, N. Y	Jan. 28, 1873	135, 218
Gear-cutting machine	C. Van Haagen	Philadelphia, Pa	Dec. 20, 1870	110, 406
Gear-cutting machine, Adjustable scroll-index for	W. M. Galusha	Arlington, Vt	Apr. 14, 1868	76, 741
Gear-cutting tool	J. Harrington	New London, Conn	Sept. 22, 1868	82, 402
Gear-cutting tool	H. R. Taylor	Westport, Conn	Mar. 3, 1868	75, 219
Gear-cutting wheel	T. B. Russell	Salem, Mass	July 30, 1867	67, 219
Gear elevator, Differential	D. Hussey	Nashua, N. H	Nov. 17, 1868	84, 123
Gear for vehicles, Running	S. J. Edwards	New Berlin, N. Y	Nov. 19, 1867	70, 980
Gear for wagons, Front	A. Finley	Bainbridge, Ind	July 20, 1869	92, 948
Gear, Friction	F. Simmons	New Orleans, La	Apr. 29, 1873	138, 444
Gear, Friction driving	W. W. Tunis	Easton, Md	Jan. 14, 1873	134, 825
Gear, Machine for cutting teeth of beveled	G. H. Corliss	Providence, R. I	Mar. 10, 1849	6, 161
Gear, Machine for cutting wood	T. F. Freeman	Brooklyn, N. Y	Oct. 29, 1867	70, 192
Gear, Machinery for cutting bevel	C. E. Roper	Canton, Ohio	Aug. 6, 1867	67, 587
Gear or pulley mechanism	A. M. Damon	Lowell, Mass	Feb. 13, 1872	123, 618
Gear or universal joint, Spherical	T. Weaver	Harrisburgh, Pa	Dec. 2, 1873	145, 075
Gear power machine, Differential	H. F. Shaw	West Roxbury, Mass	Mar. 10, 1868	75, 304
Gear press, Power	W. and R. Skene	Louisville, Ky	June 28, 1859	24, 585
Gear, Proportional scales for the construction of toothed.	J. Walker	Philadelphia, Pa	Feb. 21, 1871	112, 097
Gear, Reversing	G. Juengst	New York, N. Y	Sept. 9, 1856	15, 697
Gear, Shifting	C. D. Rogers	Utica, N. Y	Jan. 10, 1865	45, 864
Gear, Spherical	T. Weaver	Harrisburgh, Pa	Aug. 1, 1871	117, 706
Gear, Universal bevel	J. Lewis	Burlington, Vt	Nov. 29, 1838	1, 025
Gear-wheel	W. H. Ward	Auburn, N. Y	May 24, 1870	103, 398

Index of patents issued from the United States Patent Office from 1790 *to* 1873, *inclusive*—Continued.

Invention.	Inventor.	Residence.	Date.	No.
Gear-wheel, Adjustable	W. H. Ward	Auburn, N. Y	Oct. 18, 1870	108, 539
Gear-wheel and pulley	G. I. Washburn	Worcester, Mass	Mar. 8, 1864	41, 896
Gear-wheel, Stop-motion	J. H. Whitney	Rochester, Minn	Nov. 8, 1870	109, 164
Gear-wheel teeth	J. Letzkus	Allegheny City, Pa	July 28, 1868	80, 291
Gear-wheels, Cutter for cutting	J. R. Brown	Providence, R. I	Nov. 29, 1864	45, 294
Gear-wheels, Cutting teeth of	G. W. Bigelow	New Haven, Conn	Nov. 20, 1855	13, 813
Gear-wheels, Manufacture of	J. Comly	Williamsburgh, N. Y	Nov. 12, 1872	132, 899
Gear-wheels, Molding	W. Rowell	New York, N. Y	Nov. 24, 1868	84, 305
Gear-wheels, Pattern for curved-tooth	R. Oliver	Buffalo, N. Y	Nov. 26, 1872	133, 384
Gear-worm	D. Harrigan and J. Whitney	Winchester, Mass	Jan. 7, 1868	73, 095
Gearing	B. Arnold	East Greenwich, R. I	Oct. 30, 1849	6, 821
Gearing	A. M. Beard	Hillsborough, N. H	Feb. 18, 1868	74, 488
Gearing	E. G. Chormann	Philadelphia, Pa	Jan. 7, 1873	134, 642
Gearing	P. D. Cummings	Portland, Me	June 5, 1860	28, 631
Gearing	F. Emmerich	New York, N. Y	July 8, 1873	140, 690
Gearing	G. P. Ganster	Reading, Pa	Nov. 23, 1858	22, 116
Gearing	E. A. Goodes	Philadelphia, Pa	Nov. 23, 1858	22, 118
Gearing	J. J. Greenough	Syracuse, N. Y	Dec. 16, 1873	145, 499
Gearing	S. W. and J. F. Palmer	Auburn, N. Y	May 8, 1866	54, 589
Gearing	A. Sibley	Pawtucket, R. I	July 7, 1868	79, 605
Gearing	H. M. Street	Denmark, Tenn	Aug. 14, 1860	29, 634
Gearing and self-adjusting shaft	G. B. Hamlin	Willimantic, Conn	Aug. 16, 1870	106, 360
Gearing and ungearing pinions and wheels	J. Skinner	Hadley, Mich	Feb. 28, 1871	112, 187
Gearing, Anti-friction	T. Chase	Washington, D. C	July 19, 1870	105, 426
Gearing, Belted friction	J. D. Whelpley	Boston, Mass	Apr. 3, 1866	53, 717
Gearing, Changeable	J. Evans and J. H. Thompson	Paterson, N. J	Apr. 24, 1847	5, 085
Gearing, Cog	J. A. Bazin	Canton, Mass	Aug. 22, 1854	11, 585
Gearing, Cog	L. R. Faught	Philadelphia, Pa	Oct. 3, 1871	119, 683
Gearing, Device for adjusting	G. S. Barton	Worcester, Mass	Apr. 6, 1869	88, 691
Gearing, Device for increasing the friction of helical.	G. Lindsay	Belfast, Ireland	July 22, 1873	141, 149
Gearing, Differential	W. D. Andrews	Brookhaven, N. Y	June 25, 1872	128, 352
Gearing, Differential wheel	C. F. Cooke and J. Standfield.	York and Lambeth, Great Britain.	Nov. 5, 1867	70, 416
Gearing, Driving	G. B. Hamlin	Willimantic, Conn	Apr. 12, 1870	101, 872
Gearing for connecting feed or pressure rollers	W. E. Cornell and C. W. Brown	Boston, Mass	July 17, 1847	5, 191
Gearing for driving agricultural machinery, Arrangement of.	T. Salisbury	Albion, Pa	Sept. 14, 1869	94, 844
Gearing for driving machinery	M. Young, jr	Frederick, Md	July 2, 1861	32, 733
Gearing for driving machinery, Arrangement of	J. Urmy	Wilmington, Del	Feb. 6, 1838	593
Gearing for grinding-rollers	W. E. Sibley	Weston, Mass	Mar. 13, 1866	53, 236
Gearing for twisting-apparatus, &c., Differential	J. and J. H. Webster	Boston, Mass	Mar. 25, 1873	137, 267
Gearing for machinery	S. P. Gary	Oshkosh, Wis	Dec. 16, 1862	37, 165
Gearing for machinery	T. M. Millet, sr	Savannah, Ga	Dec. 13, 1870	110, 151
Gearing for machinery	W. Webster	Jefferson County, Wash	June 22, 1858	20, 672
Gearing for machinery, Extension	H. P. Minot	Chicago, Ill	Dec. 24, 1872	134, 304
Gearing for machinery, Friction	F. P. Dimpfel	Philadelphia, Pa	Mar. 12, 1861	31, 659
Gearing, Frictional	D. H. Chamberlain	West Roxbury, Mass	Mar. 22, 1870	100, 977
Gearing, Gin	H. R. Easterling	Bennettsville, S. C	Jan. 30, 1872	123, 248
Gearing, Intermittent and expansive	L. B. Potter	Putnam, Conn	Oct. 23, 1866	59, 064
Gearing, Machine	R. J. Gatling	Indianapolis, Ind	Sept. 18, 1860	30, 059
Gearing, Machine	A. Hitchcock	New York, N. Y	Feb. 12, 1867	61, 938
Gearing, Mill	J. C. Cunningham	Oglethorpe, Ga	Nov. 19, 1867	70, 970
Gearing, Mill	J. H. Glover	Glasgow, Ky	Jan. 8, 1861	31, 078
Gearing, Mill	J. Shickel	Harrisonburgh, Va	Sept. 10, 1867	68, 799
Gearing, Mill	H. Shoemaker and J. A. McClintock.	Perry, Ill	Nov. 8, 1870	109, 063
Gearing, Mill	H. M. Smith	Peoria, Ill	June 28, 1870	104, 793
Gearing mills	C. Coleman	Barry's Bridge, Va	Aug. 27, 1835	....
Gearing, Mode of increasing, reducing, and communicating motion in machinery to be used as a substitute for cog.	F. S. Barnard	New York, N. Y	Sept. 10, 1839	1, 318
Gearing, Multiplying	F. Dibben and L. Bollman	New York, N. Y	Aug. 9, 1853	9, 914
Gearing, Pivot	J. J. Duchesne	Lacon, Ill	Dec. 10, 1867	72, 003
Gearing, Spiral	U. Haskins, jr	Pittsburgh, Pa	Mar. 17, 1868	75, 544
Gearing, Spur or bevel	C. Neer	Waterford, N. Y	Mar. 9, 1827	....
Gelatine-factories, Mode of recovering useful products from the waste liquor of.	F. Bihn and W. Schrader	Frankford, Pa	Dec. 15, 1868	84, 934
Gelatine from algæ, Manufacture of vegetable	W. J. Rand, jr	Brooklyn, N. Y	June 25, 1872	128, 248
Gelatine from bones, horn, pith, &c., Obtaining	N. B. Rice	East Saginaw, Mich	May 9, 1871	114, 602
Gelatine from fish-heads, Process of obtaining	B. Robinson	East Gloucester, Mass	May 19, 1868	78, 016
Gelatine from impure solution, Manufacture of	B. F. Shaw	Cambridge, Mass	Dec. 10, 1872	133, 896
Gelatine or glue, Treating liquors containing	O. Lugo	Baltimore, Md	Mar. 8, 1870	100, 647
Gelatine, Preparation of portable	P. Cooper	New York, N. Y	June 20, 1845	4, 084
Gelatine, tannin, and cellulose, Compound of	A. K. Eaton	Brooklyn, N. Y	July 16, 1872	129, 217
Gelatine, Treating marine plants to obtain	G. Bourgade	New York, N. Y	Mar. 14, 7871	112, 535
Genealogy, chronology, &c., Representing by lines	A. Bostwick	New York, N. Y	Aug. 2, 1826	....
Generating coal-oil, Apparatus for	W. T. Barnes	Buffalo, N. Y	Aug. 2, 1859	24, 921
Generator: *See* Acid-generator. Gas-generator. Hydrogen-generator Ozone-generator. Steam-generator. Steam and vapor generator. Vapor-generator.				
Geometrical interest-table	G. Cobb	Boston, Mass	Sept. 7, 1808	....
Geometrical interest-table	G. Cobb	Boston, Mass	Apr. 24, 1809	....
Geometrical lines, centers, &c., Machinist's instrument for determining.	R. B. Light	Dunkirk, N. Y	Apr. 9, 1861	31, 987
Geometrical lines, Instrument for drawing	W. Ritchie	Wilmington, Ill	Feb. 22, 1870	100, 072
Geometrical protractor and tablet	J. Pool, jr	Easton, Mass	June 16, 1830	....
Gib	A. H. Rhodes	Fall River, Mass	June 9, 1868	78, 761
Gib and self-oiler	C. B. White	Port Richmond, N. Y	July 28, 1868	80, 376
Gib for cross-heads	W. R. Bishop	Sherwood, Wis	July 3, 1866	55, 992
Gib, Self-lubricating	W. A. Devon	Port Richmond, N. Y	Nov. 19, 1867	71, 145
Gig for cloth	O. M. Stillman	Westerly, R. I	Feb. 11, 1868	74, 443
Gig-mill	E. Gessner	Aue, Saxony	May 21, 1867	64, 859
Gig-mill	R. Kershaw	Philadelphia, Pa	Dec. 24, 1861	34, 038
Gig-mill	J. C. Miller	Troy, N. Y	Sept. 23, 1862	36, 550
Gig-mill	J. Shaw	Manayunk, Pa	Aug. 28, 1860	29, 828

Index of patents issued from the United States Patent Office from 1790 *to* 1873, *inclusive*—Continued.

Index of patents issued from the United States Patent Office from 1790 *to* 1873, *inclusive*—Continued.

Invention.	Inventor.	Residence.	Date.	No.
Glass-cutting, Graved rule and self-adjusting diamond for.	J. Dickinson	Brooklyn, N. Y	Oct. 1, 1861	33, 380
Glass-cutting machine	W. Vom Hofe	New York, N. Y	Dec. 31, 1872	134, 498
Glass-cutting machinery	J. P. Colné	New York, N. Y	Aug. 26, 1851	8, 323
Glass-cutting, polishing, dressing, and ornamenting machine.	T. J. W. Robertson	New York, N. Y	June 7, 1864	43, 042
Glass-cutting tool	S. G. Monce	Bristol, Conn	June 8, 1869	91, 150
Glass-decoration	W. Nielson	New York, N. Y	July 23, 1872	129, 856
Glass, Device for cutting	A. S. McClure	New Buffalo, Pa	Sept. 20, 1864	44, 331
Glass, Embellishment for	E. Ingraham	Bristol, Conn	Nov. 29, 1870	109, 626
Glass, Fire-polishing	J. S. Gilliland	Brooklyn, N. Y	Jan. 11, 1853	9, 533
Glass, Flattening	J. J. Adams	Washington, D. C	Apr. 21, 1832	....
Glass, Flattening and annealing	S Rickards	Philadelphia, Pa	Apr. 6, 1843	3, 035
Glass, Flattening and tempering window	J. J. Adams	Winslow N. J	Oct. 17, 1842	2, 820
Glass-flattening furnace and leer	J. Clabby	Lenox, Mass	Mar. 7, 1871	112, 322
Glass, Flexible fastening for	W. H. Taylor	Stamford, Conn	Feb. 28, 1871	112, 193
Glass for windows, Flattening-cylinder for	W. Coffan, jr	Hammonton, N. J	Oct. 1, 1830	....
Glass frame for looking-glass, &c	J. Scott	Philadelphia, Pa	Apr. 6, 1831	....
Glass from cylinders, Manufacturing plate	W. P. Walter	Philadelphia, Pa	July 10, 1855	13, 245
Glass, Frosting plates of	I. Taylor	New York, N. Y	Nov. 11, 1851	8, 519
Glass-furnace	E. Bayard and B. A. Mason	New York, N. Y	Mar. 18, 1873	136, 957
Glass-furnace	J. Carroll	Longacoming, N. J	July 25, 1865	48, 903
Glass-furnace	H. R. De Monthureux	Birmingham, Pa	Nov. 19, 1872	133, 083
Glass-furnace	J. W. Ells and J. J. Deabold	Pittsburgh, Pa	July 9, 1872	128, 716
Glass-furnace	G. W. and C. W. Foster	Charlestown, Mass	July 23, 1872	129, 657
Glass-furnace	N. Granger	Saratoga, N. Y	Aug. 4, 1868	80, 623
Glass-furnace	J. Green	Philadelphia, Pa	Nov. 7, 1854	11, 890
Glass-furnace	J. Green	Norristown, Pa	Mar. 26, 1867	63, 241
Glass-furnace	A. K. Hay	Winslow, N. J	Nov. 7, 1854	11, 893
Glass-furnace	A. K. Hay	Winslow, N. J	May 20, 1862	35, 314
Glass-furnace	J. B. Hay	Winslow, N. J	July 24, 1860	29, 342
Glass-furnace	J. Henderson	Wheeling, W. Va	Jan. 18, 1870	98, 963
Glass-furnace	D. Jarves	Boston, Mass	Oct. 3, 1846	4, 783
Glass-furnace	T. Leighton	Cambridge, Mass	Sept. 19, 1846	4, 770
Glass-furnace	W. Leighton, jr	Wheeling, W. Va	Aug. 19, 1873	141, 935
Glass-furnace	S. Oakman	Boston, Mass	Mar. 23, 1869	88, 066
Glass-furnace	A. Pocheron	Paris, France	Aug. 12, 1873	141, 666
Glass-furnace	S. Richards	Philadelphia, Pa	June 3, 1856	15, 018
Glass-furnace	S. Richards	Philadelphia, Pa	July 22, 1856	15, 389
Glass-furnace	S. Richards	Philadelphia, Pa	Aug. 25, 1857	18, 059
Glass-furnace	S. Richards	Philadelphia, Pa	Mar. 12, 1867	62, 887
Glass-furnace	F. Rohrbacher and F. Hormann	Philadelphia, Pa	Aug. 6, 1872	130, 154
Glass-furnace	F. Schaum	Baltimore, Md	Apr. 25, 1854	10, 830
Glass-furnace	C. W. and F. Siemens	Westminster, England, and Dresden, Saxony.	June 11, 1872	127, 806
Glass-furnace	J. Stanger and D. H. Miller	Philadelphia, Pa	Aug. 16, 1815	....
Glass-furnace, kiln, &c	J. Green	Norristown, Pa	Feb. 23, 1869	87, 255
Glass furnaces, and pots. Manufacture of	E. Wells	Covington, Pa	Dec. 21, 1858	22, 393
Glass-furnaces, Mode of feeding resin in the fires of	B. Shiverick	North Sandwich, Mass	Feb. 1, 1853	9, 569
Glass furniture, Method of making	J. P. Bakewell	Pittsburgh, Pa	Sept. 9, 1825	....
Glass, Gilding on	V. Schwarzenbach	Berne, Switzerland	Aug. 12, 1873	141, 670
Glass-globe-cutting apparatus	J. T. H. Richardson	Tutbury, England	July 18, 1871	117, 208
Glass-goblet mold	A. J. Sweeney	Wheeling, Va	July 3, 1860	29, 017
Glass goblets, glasses, &c., Mold for making	J. Magoun	East Cambridge, Mass	Sept. 10, 1867	68, 633
Glass grinding and polishing machine	A Hathaway	Lenox, Mass	May 28, 1872	127, 344
Glass grinding, cutting, and engraving machine	T. J. W. Robertson	New York, N. Y	June 7, 1864	43, 041
Glass-heating table for silvering	H. B. Walker	New York, N. Y	Oct. 18, 1870	108, 413
Glass-holder	B. H. Badger	New York, N. Y	Dec. 31, 1867	72, 776
Glass-house pot	T. Scanlon	Birmingham, Pa	Aug. 17, 1869	93, 839
Glass in molds, Pressing	J. Magoun	East Cambridge, Mass	Dec. 6, 1845	4, 297
Glass into molds, Pressing melted	D. Jarves	Boston, Mass	Dec. 1, 1828	....
Glass-jar	W. M. Kirchner	Pittsburgh, Pa	May 30, 1871	115, 326
Glass-jar mold	R. Hemingway	Covington, Ky	Sept. 18, 1860	30, 063
Glass jars, Machine for grinding	A. W. Kelly and J. B. Samuel.	Philadelphia, Pa	Dec. 28, 1869	98, 270
Glass jars, Machine for removing the "blow-over" on	J. Chambers	Birmingham, Pa	Jan. 25, 1870	99, 160
Glass jars, Means for attaching covers to	D. C. Ripley	Pittsburgh, Pa	June 29, 1869	91, 871
Glass jars, Mode of grinding the mouth-edges of	O. O. G. Elliott	Philadelphia, Pa	Mar. 13, 1866	53, 218
Glass jars, Process of forming smooth tops on	F. Rohrbacher and F. Hormann	Philadelphia, Pa	Aug. 23, 1870	106, 624
Glass jars, Tool for forming mouths of	G. M. Keeffer	East Birmingham, Pa	Dec. 20, 1870	110, 243
Glass jars, Tool for forming screw-threads on	W. M. Kirchner	Pittsburgh, Pa	Dec. 6, 1870	109, 825
Glass, Kiln for annealing	T. B. Atterbury	Pittsburgh, Pa	June 28, 1864	43, 278
Glass, Kiln for annealing	T. Lowry	Pittsburgh, Pa	Mar. 1, 1870	100, [illegible]24
Glass, Ladling of molten	W. P. Walter and J. Green	Philadelphia, Pa	May 6, 1856	14, 838
Glass-lamp mold	D. C. Ripley	Birmingham, Pa	June 14, 1870	104, 205
Glass letters, numbers, &c., Making	O. Stietz	New York, N. Y	May 14, 1867	64, 807
Glass light	W. A. Demuth	New York, N. Y	Sept. 22, 1868	82, 389
Glass light for ship and building	J. Oakes	Boston, Mass	May 11, 1822	....
Glass, &c., Machine for cutting and polishing	J. P. and C. Colné	Washington, D. C	Nov. 19, 1872	133, 204
Glass, Machine for grinding	A. H. Hook	New York, N. Y	Nov. 15, 1859	26, 103
Glass, &c., Machine for marking designs on	H. A. Wassell	Stourbridge, England	Feb. 4, 1873	135, 610
Glass, Machine for making sheet	S. S. Ferries	New York, N. Y	Nov. 10, 1868	83, 841
Glass, Machine for mixing "batch" for	W. T. Gillinder	Philadelphia, Pa	Dec. 13, 1870	110, 029
Glass, &c., Machinery for grinding and polishing	A. Broughton	Malone, N. Y	May 29, 1860	28, 532
Glass, Machinery for polishing	P. Burgess	New York, N. Y	Nov. 18, 1856	16, 085
Glass, &c., Machinery for polishing	A. Lindsay	Malone, N. Y	Mar. 9, 1858	19, 569
Glassmaker's mold	D. Jarves	Boston, Mass	May 28, 1830	....
Glassmaker's pot	W. H. Capewell	Westville, N. J	Dec. 3, 1867	71, 578
Glass, Making	W. Price	Pennsylvania	Mar. 4, 1816	....
Glass-making pot	E. Dithridge	Pittsburgh, Pa	Jan. 1, 1861	31, 011
Glass-making, Tank-furnace for	G. Leuffgen	Charlottenburg, Prussia	May 17, 1870	103, 208
Glass, Manufacture of	W. H. Balmain	St. Helen's, Great Britain	June 20, 1871	116, 137
Glass, Manufacture of	L. Bémelmans and L. De Give.	Atlanta, Ga	July 14, 1868	79, 892
Glass, Manufacture of	L. Bémelmans and L. De Give.	Atlanta, Ga	July 14, 1868	79, 893
Glass, Manufacture of	J. Best	Pittsburgh, Pa	Jan. 10, 1865	45, 809
Glass, Manufacture of	J. L. Gilliland	Brooklyn, N. Y	Jan. 5, 1864	41, 065
Glass, Manufacture of	C. H. Jenkins	Boston, Mass	Apr. 12, 1870	101, 737
Glass, Manufacture of	H. Napier and J. J. Hollins	Elizabeth, N. J	Dec. 5, 1865	51, 343
Glass, Manufacture of	D. C. Ripley	Pittsburgh, Pa	Mar. 17, 1868	75, 577

Index of patents issued from the United States Patent Office from 1790 to 1873, inclusive—Continued.

Invention.	Inventor.	Residence.	Date.	No.
Glass, Manufacture of	H. Trumbull	Jersey City, N. J	July 3, 1860	29, 020
Glass, Manufacture of	R. Washburn	Monsey, N. Y	Feb. 27, 1866	52, 917
Glass, Manufacture of	S. Wetherill	Bethlehem, Pa	Oct. 16, 1860	30, 439
Glass, Manufacture of	T. A. Zellers	East Birmingham, Pa	July 2, 1867	66, 274
Glass, Manufacture of black bottle	J. F. McCully	Gonzales, Tex	Sept. 2, 1856	15, 665
Glass, Manufacture of crystal	O. Wuth	Pittsburgh, Pa	Oct. 12, 1869	95, 752
Glass, Manufacture of plate	J. J. Greenough	Boston, Mass	June 14, 1853	9, 791
Glass, Manufacture of plate	P. Stenger	Philadelphia, Pa	Aug. 7, 1855	13, 411
Glass, Manufacture of plate and window	T. Clark	Pittsburgh, Pa	June 8, 1852	8, 994
Glass, Manufacture of window	C. F. Carstens and T. H. Schwenck.	Chicago, Ill	Mar. 5, 1872	124, 195
Glass, Manufacturing	J. M. Brookfield and E. V. White	Honesdale, Pa	June 14, 1853	9, 789
Glass-manufacturing process	S. Richards	Philadelphia, Pa	May 29, 1866	55, 159
Glass, Material for the manufacture of	P. E. Minor	Schenectady, N. Y	Dec. 31, 1867	72, 878
Glass, Melting and fusing	T. W. Dyott	Philadelphia, Pa	Sept. 10, 1828	....
Glass-melting furnace	L. B. Goodhue	Saint Louis, Mo	May 5, 1868	77, 479
Glass-melting pot	J. Dougherty	Somerville, Mass	Jan. 14, 1873	134, 789
Glass-melting pot	C. Newman	San Francisco, Cal	May 7, 1867	64, 558
Glass, Metallic molds for molten	W. Pountney	Brooklyn, N. Y	June 21, 1864	43, 224
Glass, Method of lettering and ornamenting	J. B. Shaw	Pittsburgh, Pa	Aug. 12, 1856	15, 532
Glass, Method of preventing the corrosion or staining of the surface of.	W. B. Richards	New York, N. Y	Mar. 28, 1865	47, 040
Glass, Mode of applying crystal frosting to	H. B. Kimball	Charlotte, Mich	Dec. 8, 1868	84, 699
Glass, Mode of frosting	J. Levy and C. Jones	New York, N. Y	Dec. 7, 1852	9, 453
Glass-mold	H. Brooke	New York, N. Y	Jan. 5, 1869	85, 637
Glass-mold	H. Brooke	New York, N. Y	Aug. 10, 1869	93, 591
Glass-mold	W. Brooke	Jersey City, N. J	Aug. 15, 1854	11, 515
Glass-mold	E. W. Cooper	Williamstown, N. J	Nov. 24, 1868	84, 412
Glass-mold	T. G. Otterson	Philadelphia, Pa	Mar. 2, 1869	87, 506
Glass-mold	C. H. Over, J. Robinson, and H. Faupel.	Bellaire, Ohio	July 23, 1872	129, 679
Glass-mold	J. H. Reighard	Wheeling, W. Va	Nov. 19, 1867	71, 216
Glass-mold	J. C. Schaffer	Rochester, N. Y	Mar. 16, 1869	87, 795
Glass-mold	G. W. Scollay	Saint Louis, Mo	Sept. 18, 1860	30, 093
Glass-mold	C. Stadelmann	Pittsburgh, Pa	May 30, 1871	115, 377
Glass-mold	C. Stadelmann	Pittsburgh, Pa	Apr. 23, 1872	126, 099
Glass-mold	M. Sweeney, J. E. Mathews, and T. Hartley.	Wheeling, W. Va	Jan. 30, 1866	52, 338
Glass-mold	D. Turpie	Sandwich, Mass	Mar. 17, 1868	75, 604
Glass-mold	C. H. Warner	Brooklyn, N. Y	Apr. 10, 1860	27, 873
Glass-mold	F. A. Weise	Baltimore, Md	Apr. 2, 1872	125, 234
Glass-mold	H. Wickham	Chicago, Ill	Feb. 4, 1873	135, 502
Glass-mold and press	W. M. Kirchner	Birmingham, Pa	July 16, 1872	129, 036
Glass-mold carriage	D. Bennett, M. Krebs, and J. Haley.	Baldwin Township and Pittsburgh, Pa.	Apr. 29, 1873	138, 368
Glass-mold for making argand-lamp fountain	E. B. Horn	Boston, Mass	Aug. 1, 1848	5, 684
Glass, Mold for pressing	H. Dillaway	Boston, Mass	Aug. 21, 1841	2, 226
Glass, Mold for pressing	J. Magoun	East Cambridge, Mass	Sept. 25, 1847	5, 303
Glass, Mold for pressing	T. Shaw	Philadelphia, Pa	June 7, 1859	24, 360
Glass, Mold for working	A. N. N. Aubin and J. P. Colné	Portland, Conn., and Washington, D. C.	Dec. 3, 1872	133, 515
Glass-mold from gas-carbon or graphite	J. B. Lhote	Paris, France	June 29, 1869	91, 946
Glass-mold, Machine for operating	W. C. King	Pittsburgh, Pa	May 9, 1871	114, 569
Glass-mold plunger	J. McCord	East Birmingham, Pa	Apr. 5, 1870	101, 485
Glass-molds, Method of treating	S. Lydiatt	Montreal, Canada	Sept. 9, 1873	142, 573
Glass-molder's press	F. Klinck	Philadelphia, Pa	Aug. 8, 1871	117, 785
Glass, Molding	J. Magoun	Cambridge, Mass	Oct. 24, 1848	5, 875
Glass, Molding and pressing	J. Magoun	East Cambridge, Mass	Sept. 25, 1847	5, 302
Glass-molding press	W. O. Davis	Pittsburgh, Pa	Jan. 31, 1854	10, 470
Glass-molding tool	D. Challinor	Birmingham, Pa	June 6, 1871	115, 569
Glass, Ornamental window	C. Frederici	New York, N. Y	Aug. 12, 1873	141, 782
Glass-ornamentation	W. G. Webb	Wordsley, Great Britain	Dec. 5, 1871	121, 696
Glass, Ornamenting	S. M. Adams	New York, N. Y	July 29, 1873	141, 305
Glass, Ornamenting	J. S. Miles	Ann Arbor, Mich	Oct. 26, 1858	21, 896
Glass, Ornamenting	J. C. Millward	New York, N. Y	Oct. 29, 1867	70, 240
Glass, Ornamenting	C. Schüssler	New York, N. Y	Aug. 26, 1873	142, 125
Glass-ornamenting composition	J. J. H. Brainchon	Paris, France	Jan. 18, 1859	22, 620
Glass-ornamenting machine	T. J. W. Robertson	New York, N. Y	June 7, 1864	43, 043
Glass pipes, Mold for making	G. Scott	Albany, N. Y	Sept. 4, 1849	6, 694
Glass plates, &c., Machine for grinding	J. Kendall and A. Hathaway	South Framingham and Lenox, Mass.	June 23, 1868	79, 131
Glass-polish	J. M. Warren	Boston, Mass	Oct. 10, 1865	50, 406
Glass-polishing machine	A. H. Hook	New York, N. Y	July 26, 1859	24, 908
Glass-polishing machine	J. Kendall and A. Hathaway	South Framingham and Lenox, Mass.	June 23, 1868	79, 132
Glass, porcelain, &c., Attaching metal caps to	C. B. Jenkins	New York, N. Y	June 17, 1873	139, 961
Glass-pot	D. McAfee	Pittsburgh, Pa	Nov. 21, 1865	51, 068
Glass-press	A. P. Brooke	New York, N. Y	July 11, 1871	116, 926
Glass-press	A. P. Brooke	New York, N. Y	June 25, 1872	128, 280
Glass-press	A. P. Brooke	New York, N. Y	July 16, 1872	129, 095
Glass-press	W. O. Davis	Pittsburgh, Pa	Aug. 25, 1863	39, 698
Glass-press	H. Dillaway	Sandwich, Mass	Aug. 21, 1866	57, 296
Glass-press	J. Haley	Allentown, Pa	July 30, 1872	130, 039
Glass-press	J. Haley	Pittsburgh, Pa	May 13, 1873	138, 750
Glass-press	W. C. King	Pittsburgh, Pa	Dec. 17, 1872	134, 070
Glass-press	W. C. King	Pittsburgh, Pa	Dec. 17, 1872	134, 071
Glass-press	H. J. Leasure	Wheeling, W. Va	Mar. 5, 1872	124, 364
Glass-press	A. M. Smith	Brooklyn, N. Y	Jan. 28, 1873	135, 292
Glass-press, Cooling	H. J. Leasure and J. S. Gill	Wheeling, W. Va	Dec. 4, 1866	60, 203
Glass-pressing machine	G. J. Capewell	West Chesire, Conn	July 7, 1868	79, 635
Glass-pressing machine	F. McKee and C. Ballinger	Pittsburgh, Pa	Mar. 29, 1864	42, 143
Glass-pressing machine	M. Sweeney	Wheeling, W. Va	July 1, 1868	79, 786
Glass-pressing mold	P. C. Dummer	Jersey City, N. J	Oct. 16, 1827	....
Glass, process for the manufacture of	H. M. Baker	Rochester, N. Y	Dec. 4, 1866	60, 119
Glass, Process of forming letters, characters, and ornaments on.	C. M. Strauss	Memphis, Tenn	Oct. 5, 1869	95, 617
Glass, Removing stains from	J. G. Pohle and J. N. Crow	Morrisania and Mott Haven, N. Y.	Mar. 24, 1863	37, 975
Glass shades and globes, Mode of ornamenting	R. Guthrie and J. Shearer	New York, N. Y	Nov. 5, 1867	70, 554

Index of patents issued from the United States Patent Office from 1790 *to* 1873, *inclusive*—Continued.

Invention.	Inventor.	Residence.	Date.	No.
Glass shades, Machine for grinding	C. C. P. Waterman	Sandwich, Mass	Aug. 27, 1861	33, 175
Glass, silver-ware, &c., Composition for cleaning	W. H. Rosevelt	Rochester, N. Y	Aug. 12, 1873	141, 819
Glass, Silvering	H. B. Walker	New York, N. Y	Dec. 20, 1870	110, 408
Glass-smelting pot	R. Combs and H. J. Leasure	Wheeling, W. Va	May 3, 1870	102, 502
Glass, soluble silicates, hydrochloric acid, and bleaching-powder, Process to be used in the manufacture of.	W. R. Stace and H. M. Baker	Rochester, N. Y	Aug. 27, 1867	68, 254
Glass, Staining	E. A. Goodes	Philadelphia, Pa	Mar. 14, 1871	112, 703
Glass-staining composition	H. V. Edmond	Norwich, Conn	July 6, 1869	92, 292
Glass stoppers for bottles, &c., Apparatus for making.	T. R. Hartell	Philadelphia, Pa	Oct. 19, 1858	21, 831
Glass, Tool for manufacture of	G. Matthewman	Williamsburgh, N. Y	Oct. 10, 1865	50, 373
Glass, Transparent and flexible materials, designed as a partial substitute for.	C. Sussegger	New York, N. Y	July 14, 1863	39, 265
Glass tubes for philosophical apparatus, Joint around.	A. B. Latta	Cincinnati, Ohio	Mar. 16, 1852	8, 801
Glass, Use of the alkalies obtained from the spent lye of soap-makers as a flux in the manufacture of.	G. H. Burgin	Philadelphia, Pa	Apr. 3, 1829	....
Glass-vessel covers, Screw-collar for	J. Cook	New York, N. Y	Feb. 23, 1869	87, 146
Glass vessels, Attachment of covers to	R. D. Bryce	East Birmingham, Pa	Aug. 21, 1860	29, 066
Glass vessels, Attachment of hinged covers of	J. H. Reighard and C. L. Knecht.	Birmingham, Pa	Aug. 21, 1860	29, 718
Glass vessels, Fastening for metallic covers to	J. Bird	Birmingham, Pa	Aug. 21, 1860	29, 743
Glass vessels to metal base, Mode of securing	A. French	Philadelphia, Pa	July 5, 1870	105, 061
Glass-ware and metal stem connections	C. L. Knecht and T. Adams	Saint Clair and Stow Township, Pa.	Dec. 13, 1870	110, 049
Glass-ware, Annealing	E. Dithridge	Pittsburgh, Pa	June 16, 1863	38, 930
Glass-ware, Cooling-stand for	J. Osterling	Wheeling, W. Va	Mar. 5, 1872	124, 378
Glass-ware holder	H. Dillaway	Sandwich, Mass	Aug. 11, 1857	17, 9[illegible]0
Glass-ware in bas-relief, Manufacture of hollow	J. S. and T. B. Atterbury and J. Reddick.	Pittsburgh, Pa	June 3, 1862	35, 429
Glass-ware makers, Snap for	O. B. Brigham	Cambridge, Mass	Nov. 26, 1867	71, 447
Glass-ware, Manufacture of	J. S. and T. B. Atterbury	Pittsburgh, Pa	Mar. 3, 1868	75, 110
Glass-ware, Manufacture of	J. S. and T. B. Atterbury	Pittsburgh, Pa	Apr. 20, 1869	89, 005
Glass-ware, Manufacture of	A. H. Baggs	Bridgeport, Ohio	Dec. 2, 1873	145, 144
Glass-ware, Manufacture of	J. P. Pears	Birmingham, Pa	Oct. 13, 1868	82, 983
Glass-ware, Manufacture of	D. C. Ripley	Birmingham, Pa	Oct. 20, 1868	83, 210
Glass-ware, Manufacture of graduated	J. H. Hobbs	Wheeling, W. Va	Oct. 15, 1872	132, 208
Glass-ware, Manufacture of hollow	J. S. and T. B. Atterbury and J. Reddick.	Pittsburgh, Pa	Feb. 11, 1862	34, 345
Glass-ware, Manufacture of hollow	J. S. and T. B. Atterbury and J. Reddick.	Pittsburgh, Pa	June 3, 1862	35, 430
Glass-ware, Manufacture of silvered	J. W. Haines	Somerville, Mass	May 30, 1865	47, 948
Glass-ware, Manufacture of stemmed	J. Oesterling	Wheeling, W. Va	Oct. 15, 1872	132, 216
Glass-ware, Manufacture of stemmed	J. Oesterling	Wheeling, W. Va	Oct. 21, 1873	143, 778
Glass-ware, Method and mold for manufacturing	J. S. and T. B. Atterbury	Pittsburgh, Pa	June 17, 1873	139, 993
Glass-ware mold	D. Ashworth	East Cambridge, Mass	July 16, 1872	129, 306
Glass-ware mold	J. S. Atterbury, J. Reddick, and T. B. Atterbury.	Pittsburgh, Pa	Mar. 4, 1862	34, 555
Glass-ware mold	H. Dillaway	Sandwich, Mass	July 7, 1868	79, 737
Glass-ware mold	R. E. Haines	Cambridge, Mass	Apr. 30, 1867	64, 312
Glass-ware mold	G. H. Lomax	Somerville, Mass	Oct. 14, 1873	143, 629
Glass-ware mold	J. B. Lyon	Pittsburgh, Pa	Dec. 13, 1870	110, 056
Glass-ware mold	J. H. Reighard	Wheeling, W. Va	Mar. 26, 1872	125, 083
Glass-ware mold	A. E. Young	Dorchester, Mass	May 11, 1869	90, 040
Glass-ware, Mold for forming blown	H. Dillaway	Sandwich, Mass	Feb. 22, 1870	100, 127
Glass-ware mold, Sectional	R. D. Haines	Cambridge, Mass	June 8, 1869	91, 118
Glass-ware mold, Solid	R. D. Haines	Cambridge, Mass	June 8, 1869	91, 119
Glass-ware molds, Cooling	H. Dillaway	Sandwich, Mass	July 7, 1868	79, 738
Glass-ware-molding press	J. Bird	Philadelphia, Pa	Sept. 14, 1869	94, 863
Glass-ware press	H. Dillaway	Sandwich, Mass	Oct. 1, 1867	69, 325
Glass-ware press	J. Haley	Cambridge, Mass	Oct. 1, 1867	69, 426
Glass-ware press	C. H. Hersey and W. E. Hawes	Boston, Mass	June 8, 1869	91, 132
Glass-ware press	J. Myers	Dorchester, Mass	Jan. 15, 1867	61, 233
Glass-ware, Process of manufacturing silvered	E. Dithridge	Pittsburgh, Pa	Oct. 29, 1867	70, 325
Glass-ware with handles, Manufacturing	J. S. and T. B. Atterbury	Pittsburgh, Pa	June 30, 1868	79, 298
Glass with platinum, Method of coating	L. P. Angenard	New York, N. Y	Mar. 14, 1865	46, 767
Glasses for orreries, &c., Preparing luminous	M. J. Gardiner	York, Pa	Apr. 30, 1832	....
Glasses for plants, hot-beds, &c., Folding	J. St. John	Stamford, Conn	Dec. 3, 1867	71, 812
Glazier's diamonds, Mounting	J. Dickinson	Brooklyn, N. Y	Sept. 3, 1861	33, 188
Glazier's diamonds, Mounting	P. Sinsz	Baltimore, Md	Apr. 23, 1867	64, 157
Glazier's pins, Machine for cutting	J. G. Baker	New Brunswick, N. J	June 15, 1858	20, 539
Glazier's points, Machine for cutting	W. Eaton	Carbondale, Pa	May 16, 1854	10, 925
Glazier's points, Machine for cutting	J. M. Jay	Canton, Ohio	Feb. 18, 1868	74, 693
Glazier's points, Tool for driving	A. Woodworth and E. W. Warren.	Cambridge, N. Y	June 28, 1870	104, 915
Glazier's tool	S. G. Monce	Bristol, Conn	July 1, 1873	140, 426
Glazing, Composition for	T. and E. Parker	Orangeville, Pa	Feb. 20, 1844	3, 443
Glazing, Heating calender-rollers for	J. Brimhall and T. Keyes, jr	West Boylston, Mass	Apr. 1, 1826	....
Glazing pottery-ware	C. W. Fenton	Bennington, Vt	Nov. 27, 1849	6, 907
Globe	G. D. Abbott	New York, N. Y	Aug. 11, 1868	80, 891
Globe	E. Bascom	New York, N. Y	Oct. 16, 1866	58, 757
Globe	J. Monteith	New York, N. Y	July 24, 1860	29, 291
Globe	G. Vale	New York	Oct. 28, 1843	3, 318
Globe, Automatic terrestrial time	F. S. Barnard	New York, N. Y	Aug. 28, 1860	29, 755
Globe, Clock-work	L. J. N. Mouret	Paris, France	June 17, 1873	140, 065
Globe, Concentric celestial and terrestrial	H. Williamson	New York, N. Y	Dec. 3, 1867	71, 830
Globe for teaching geography	E. Oram	New York	Jan. 12, 1831	....
Globe, Geographical	E. Perce	Brooklyn, N. Y	Mar. 15, 1864	41, 938
Globe-holder	T. Hay	Newark, N. J	Jan. 26, 1869	86, 224
Globe-holders, Manufacture of	C. H. Barney	Providence, R. I	Oct. 14, 1873	143, 656
Globe-joint	J. F. Hollister	Plano, Ill	June 23, 1868	79, 124
Globe, School	J. R. Agnew	Philadelphia, Pa	July 24, 1860	29, 229
Globe, School	J. R. Agnew	Mercersburgh, Pa	June 10, 1862	35, 498
Globe, School	J. R. Agnew	Mercersburgh, Pa	Sept. 9, 1862	36, 387
Globe, School	J. R. Agnew	Mercersburgh, Pa	June 16, 1863	38, 875
Globe, School	E. Weissenborn	Hudson City, N. J	Nov. 24, 1868	84, 398

Index of patents issued from the United States Patent Office from 1790 *to* 1873, *inclusive*—**Continued.**

Invention.	Inventor.	Residence.	Date.	No.
Globe, Terrestrial	G. P. Clarke	New York, N. Y	Oct. 1, 1867	69, 408
Globes, Apparatus for roughening the outer surface of.	T. W. Mellor	Philadelphia, Pa	Aug. 13, 1872	130, 520
Globes, Making artificial	W. B. Annin	Boston, Mass	May 4, 1826	......
Globes, maps, &c., for schools, Construction of	J. D. Brinkerhoff and J. Duthie.	Morrisania, N. Y	Nov. 29, 1870	109, 581
Globes, Method of relieving geographical outlines on molded elastic.	H. B. Goodyear	New Haven, Conn	Feb. 5, 1861	31, 311
Globes, Method of suspending school	J. Monteith	New York, N. Y	July 24, 1860	29, 292
Globes, Mold for making school	J. R. Agnew	Lancaster, Pa	July 5, 1864	43, 379
Globes, Mounting	S. Cornell	Rochester, N. Y	July 5, 1845	4, 098
Glossing and fluting iron	C. W. Monroe	Chicago, Ill	Dec. 12, 1871	121, 885
Glove	E. W. and A. A. Avery	Plymouth, N. H	Apr. 25, 1871	114, 091
Glove	O. Bartlit and G. D. Edson	Rockford, Ill	Aug. 31, 1869	94, 380
Glove	C. J. Brown	Plymouth, N. H	May 14, 1872	126, 781
Glove	C. J. Brown	Plymouth, N. H	July 2, 1872	128, 536
Glove	R. D. Burr	Kingsborough, N. Y	Aug. 4, 1868	80, 707
Glove	R. D. Burr	Kingsborough, N. Y	May 30, 1871	115, 277
Glove	R. D. Burr	Kingsborough, N. Y	July 29, 1873	141, 203
Glove	G. Chant	Port Jervis, N. Y	Oct. 8, 1872	131, 998
Glove	S. J. Clute and D. M. Durfee	Rockwood, N. Y	July 30, 1872	129, 934
Glove	T. G. Foster	Gloversville, N. Y	Nov. 18, 1873	144, 665
Glove	S. Goge	Brooklyn, N. Y	June 8, 1869	91, 113
Glove	D. S. Hulett	Gloversville, N. Y	July 26, 1870	105, 687
Glove	H. Z. and A. J. Kasson	Gloversville, N. Y	Dec. 5, 1871	121, 630
Glove	T. Kehoe	Poughkeepsie, N. Y	Nov. 26, 1872	133, 319
Glove	J. F. Mason	Johnstown, N. Y	Nov. 26, 1872	133, 376
Glove	J. I. McMartin	Johnstown, N. Y	Dec. 17, 1872	133, 998
Glove	J. H. Putnam	Gloversville, N. Y	Apr. 5, 1870	101, 506
Glove	J. H. Putnam	Gloversville, N. Y	Apr. 4, 1871	113, 695
Glove	A. J. Stevens	Rumney, N. H	Jan. 2, 1872	122, 413
Glove	W. S. Tooker	Kingsborough, N. Y	Nov. 1, 1870	108, 849
Glove	W. S. Tooker	Kingsborough, N. Y	Nov. 1, 1870	108, 850
Glove	E. V. Whitaker	Gloversville, N. Y	Dec. 19, 1871	122, 091
Glove	E. V. Whitaker	Gloversville, N. Y	Feb. 18, 1873	136, 115
Glove	W. W. Whitaker	Gloversville, N. Y	Mar. 28, 1871	113, 230
Glove	W. W. Whitaker	Gloversville, N. Y	Dec. 12, 1871	121, 832
Glove	W. W. Whitaker	Gloversville, N. Y	Dec. 12, 1871	121, 833
Glove	J. L. Whitten	Essex, Vt	May 31, 1870	103, 696
Glove-clasp	J. L. Weir	Indianapolis, Ind	May 14, 1872	126, 657
Glove-cleaning composition	M. J. Pulte	Cincinnati, Ohio	June 24, 1862	35, 706
Glove, Corn-husking	P. N. Harts	French Grove, Ill	Apr. 15, 1873	137, 918
Glove, Corn-husking	A. W. Preston	Mazon, Ill	June 18, 1867	65, 827
Glove-fastener	I. Cole	Brooklyn, N. Y	June 9, 1868	78, 649
Glove-fastener	P. Courvoisier	Paris, France	Nov. 10, 1868	83, 935
Glove-fastener	L. Ferris	San Francisco, Cal	Nov. 4, 1873	144, 197
Glove-fastener	M. B. Foote	Northampton, Mass	Aug. 1, 1871	117, 530
Glove-fastener	L. Meyers	New York, N. Y	Aug. 5, 1873	141, 580
Glove-fastener	H. R. Minns	Bristol, England	July 30, 1872	129, 900
Glove-fastening	H. P. Carver	Binghamton, N. Y	July 8, 1873	140, 613
Glove-fastening	D. A. Cooper	Worcester, England	June 10, 1873	139, 703
Glove-fastening	T. Deschamps	Paris, France	May 25, 1869	90, 348
Glove-fastening	M. B. Foote	Northampton, Mass	July 7, 1868	79, 746
Glove-fastening	C. H. Hall and R. Knott	Trenton, N. J., and Brooklyn, N. Y.	Aug. 5, 1873	141, 555
Glove, Gauntlet	F. Farrant	Gloversville, N. Y	Dec. 2, 1873	145, 099
Glove, Gauntlet	V. Price	New York, N. Y	Oct. 3, 1871	119, 642
Glove, Gauntlet	I. B. Whipple	Gloversville, N. Y	Apr. 15, 1873	137, 988
Glove, Husking	J. H. Titus	Independence, Iowa	Jan. 5, 1869	85, 707
Glove, Husking and shelling	E. Cohen	Washington. D. C	Jan. 5, 1858	19, 018
Glove, Swimming	D. Campbell	New York, N. Y	Aug. 21, 1855	13, 455
Glove-turning machine	F. Vanderpool	Mayfield, N. Y	Dec. 30, 1873	146, 111
Gloves, Dyeing kid	J. T. Reed	Charlestown, Mass	Mar. 29, 1864	42, 111
Gloves, Dyeing kid	J. T. Reed	Charlestown, Mass	Apr. 11, 1865	47, 221
Gloves, Machine for cutting out	H. J. Dickerson	Gloversville, N. Y	Nov. 22, 1864	45, 145
Gloves, Machine for cutting out	J. H. Harlan and T. Pomeroy	Denver City, Colo	Dec. 17, 1867	72, 197
Gloves, Manufacturing knit	J. Peatfield	Ipswich, Mass	July 13, 1858	20, 893
Gloves, &c., Mode of coloring kid	S. C. Chase	Charlestown, Mass	Jan. 19, 1864	41, 279
Glucose and white lead, Manufacture of	R. Rowland	New York, N. Y	June 6, 1865	48, 099
Glucose, Manufacture of	W. H. Keyt	Madison, Ind	June 4, 1872	127, 418
Glucose, Manufacture of	G. R. Percy	New York, N. Y	Feb. 28, 1865	46, 585
Glucose-manufacture, Process in	G. Riley	New York, N. Y	Mar. 5, 1850	7, 148
Glue	W. Adamson	Philadelphia, Pa	June 18, 1867	65, 785
Glue	E. Goddard	New York, N. Y	Oct. 4, 1864	44, 528
Glue and paint pot	J. J. Wilson	New York, N. Y	Nov. 18, 1870	109, 090
Glue and water heater	J. Edgecomb	Worcester, Mass	Sept. 20, 1864	44, 294
Glue-can	W. F. Muchmore	Astoria, N. Y	Mar. 31, 1868	76, 233
Glue-ement	S. Krewson	Springfield, Ohio	Feb. 9, 1869	86, 843
Glue, Clarifying	W. Adamson	Philadelphia, Pa	Jan. 30, 1855	12, 304
Glue-cutting machine	T. Brown, jr	South Danvers, Mass	Aug. 7, 1860	29, 541
Glue-dryer	J. J. Manning	Rockport, Mass	Aug. 8, 1871	117, 905
Glue-drying apparatus	M. Newbauer and P. Adelmann.	New York, N. Y	June 7, 1859	24, 325
Glue-drying apparatus	C. Wahl	Chicago, Ill	July 21, 1868	80, 248
Glue-drying, Holder for	W. Adamson	Philadelphia, Pa	Jan. 14, 1868	73, 220
Glue-drying machine	C. Wahl	Chicago, Ill	July 21, 1868	80, 249
Glue-drying machine	C. Wahl	Chicago, Ill	July 21, 1868	80, 250
Glue from the pith of horns, Manufacture of	J. Winward	East Cambridge, Mass	July 13, 1869	92, 683
Glue, gelatine, and size, Bleaching and clarifying	N. J. Wells	Holyoke, Mass	June 18, 1872	128, 000
Glue, Heating	R. Jackman	New York, N. Y	Aug. 9, 1864	43, 774
Glue into grooves, Machine for putting	J. A. Danforth	Potsdam, N. Y	Oct. 15, 1872	132, 257
Glue, Liquid	C. F. Binder	Philadelphia, Pa	July 24, 1866	56, 665
Glue, Liquid	W. Horwitz	New York, N. Y	Nov. 8, 1870	109, 010
Glue, Manufacture of	W. Adamson	Philadelphia, Pa	June 18, 1867	65, 787
Glue, Manufacture of	W. Adamson	Philadelphia, Pa	Aug. 16, 1870	106, 448
Glue, Manufacture of	C. W. Cooper	Brooklyn, N. Y	Aug. 6, 1861	32, 979
Glue, Manufacture of	A. Dietz	New York, N. Y	June 7, 1870	103, 852
Glue, Manufacture of	H. Fleck	Dresden, Saxony	May 18, 1869	90, 160
Glue, Manufacture of	O. S. Follett	Mont Clair, N. J	Apr. 15, 1873	137, 835

Index of patents issued from the United States Patent Office from 1790 to 1873, inclusive—Continued.

Invention.	Inventor.	Residence.	Date.	No.
Glue, Manufacture of	G. Guenther	Chicago, Ill	Dec. 14, 1869	97,771
Glue, Manufacture of	J. A. and R. Lighthall	Brooklyn, N. Y	Apr. 28, 1868	77,388
Glue, Manufacture of	J. H. Mark	Philadelphia, Pa	Oct. 3, 1822	....
Glue, Manufacture of	H. and C. McDougall	Chicago, Ill	Aug. 6, 1872	130,230
Glue, Manufacture of	W. Plumer	Lexington, Mass	July 29, 1873	141,379
Glue, Manufacture of	L. Reid and J. Rogers	New York, N. Y	Feb. 28, 1860	27,310
Glue, Manufacture of	B. F. Shaw	Cambridge, Mass	Dec. 9, 1873	145,454
Glue, Manufacture of	G. Upton	South Danvers, Mass	Jan. 1, 1867	70,968
Glue, Manufacture of aërated	W. Adamson	Philadelphia, Pa	June 18, 1867	65,786
Glue, Manufacture of liquid	W. C. Watson	Paterson, N. J	Oct. 30, 1866	59,326
Glue, Manufacturing	P. Cooper	New York, N. Y	Apr. 29, 1830	....
Glue, Manufacturing	J. Morgan	Portland, Me	Sept. 18, 1835	....
Glue, Manufacturing	H. G. C. Paulsen	New York, N. Y	Feb. 28, 1842	2,474
Glue, Mode of drying	G. Guenther	New York, N. Y	June 4, 1867	65,377
Glue or cement	N. S. Whipple	Detroit, Mich	June 21, 1870	104,520
Glue or gelatine and other materials called durogel, Composition of.	H. Wurtz	New York, N. Y	Jan. 1, 1867	60,984
Glue-pot	J. I. Baringer	Germantown, N. Y	Sept. 19, 1871	119,002
Glue-pot	J. Bragdon	Boston, Mass	July 14, 1868	79,945
Glue-pot	T. C. Howes	Troy, N. Y	Apr. 16, 1872	125,680
Glue-pot	H. C. Stewart	Cincinnati, Ohio	Jan. 1, 1867	60,958
Glue-pot	J. Tinney	Westfield, N. Y	Feb. 9, 1869	86,798
Glue-pot	J. Turner	Cambridgeport, Mass	Jan. 24, 1860	26,957
Glue, Preparation of	J. M. Hunter	New York, N. Y	Aug. 14, 1860	29,596
Glue, Prepared	J. F. Peting	Rochester, Ind	Nov. 14, 1871	120,996
Glue, Process of making	D. A. James	Cincinnati, Ohio	July 26, 1853	9,877
Glue, Process of treating	A. Dietz	New York, N. Y	July 7, 1868	79,736
Glue-stock and other products from animal substances, Preparation of.	D. K. Tuttle and O. Lugo	Baltimore, Md	May 4, 1869	89,709
Glue-stock, Preparation of	W. Adamson	Philadelphia, Pa	June 23, 1868	79,178
Glue-stock, Preparing	O. Rich	Cambridge, Mass	Nov. 24, 1857	18,724
Glue-stock, Treating	C. W. Cooper	Brooklyn, N. Y	Aug. 15, 1865	49,383
Gluing and cementing machine	M. H. Merriam and E. L. Norton.	Charlestown, Mass	June 5, 1866	55,335
Gluing-hopper	J. W. Campbell	New York, N. Y	Nov. 22, 1870	109,494
Gluing-hopper	J. W. Campbell and W. J. Miller.	New York, N. Y	June 15, 1869	91,414
Gluing-press	M. H. Merriam and E. L. Norton.	Charlestown, Mass	June 5, 1866	55,334
Gluing-table	S. P. Groocock and W. G. Brasington.	Clifton, N. J., and Brooklyn, N. Y.	Dec. 5, 1871	121,510
Glycerine from soap-makers' spentlyes, Process of obtaining.	B. T. Babbitt	New York, N. Y	Sept. 13, 1870	107,324
Glycerine, Manufacture of	O. Laist	Cincinnati, Ohio	Mar. 24, 1868	75,929
Glycerine, Manufacture of nitro	T. P. Shaffner	Louisville, Ky	Dec. 28, 1869	98,425
Glycerine, Preparation of	J. F. Wisnewski	Cincinnati, Ohio	Aug. 9, 1859	25,072
Goblet	T. Leach	Taunton, Mass	Jan. 12, 1869	85,934
Goggles, Strabismus	A. Lake	Flatbush, N. Y	Nov. 4, 1842	2,838
Gold, amalgam, and quicksilver, Apparatus for saving.	G. R. Evans	Virginia City, Nev	Feb. 2, 1869	86,379
Gold, amalgam, and quicksilver saving device	O. H. Young and D. J. Vaughn.	Wisconsin Hill, Cal	Feb. 7, 1871	111,713
Gold and amalgam saving apparatus	G. R. Evans	Virginia City, Nev	July 18, 1871	117,162
Gold and other metals from ores, Reducing and separating.	C. M. Nes	Baltimore, Md	Apr. 16, 1872	125,835
Gold and other precious metals from foreign substances, Machine for separating.	E. N. Kent	New York, N. Y	Feb. 26, 1856	14,316
Gold and other precious metals from their ores, Extracting.	H. Wurtz	New York, N. Y	June 27, 1865	48,499
Gold and silver, Amalgamated plate for collecting	M. Attwood and J. Roach	San Francisco, Cal	Feb. 23, 1869	87,132
Gold and silver amalgamator	W. M. Fuller	Chicago, Ill	Sept. 1, 1868	81,767
Gold and silver from base metals, Process for separating.	E. Lundquist	Grass Valley, Cal	Apr. 1, 1873	137,375
Gold and silver from lead, Separating	C. Roswag and A. N. De Pauville.	Paris, France	May 13, 1873	138,938
Gold and silver from mineral and other substances, Process for separating.	G. N. Jennings	Virginia City, Nev	Mar. 21, 1865	46,909
Gold and silver from ores by means of the vapor of mercury, Mode of extracting.	R. Spencer	New York, N. Y	Nov. 22, 1864	45,188
Gold and silver from ores, earth, &c., Mill for grinding, washing, and separating.	W. H. Folger	Charlotte, N. C	Feb. 13, 1828	....
Gold and silver from ores, Separating	E. Brown	Chicago, Ill	Nov. 27, 1866	59,955
Gold and silver from quartz, Separating	E. A. Hyde	Ann Arbor, Mich	Feb. 8, 1870	99,571
Gold and silver from solutions, Apparatus for precipitating.	W. S. Laighton	Norwich, Conn	Oct. 11, 1870	108,158
Gold and silver from sweepings, washings, &c., Method of collecting.	J. H. Rae	Syracuse, N. Y	Mar. 12, 1867	62,776
Gold and silver from their ores, Extracting	J. W. Kidwell	Washington, D. C	Oct. 27, 1868	83,582
Gold and silver from their ores, Method of extracting.	J. Tunbridge	Newark, N. J	Dec. 22, 1868	85,258
Gold and silver, Machine for collecting and amalgamating.	E. B. Prater	Washoe County, Nev	Dec. 6, 1864	45,341
Gold and silver, Machine for grinding and amalgamating.	W. H. Hepburn and G. K. Peterson.	San Francisco, Cal	Apr. 19, 1864	42,371
Gold and silver ores, Apparatus for roasting and treating.	A. B. Crosby and R. L. Thompson.	Gilpin County, Colo	Oct. 18, 1864	44,767
Gold and silver ores, Machinery for washing	D. Asbury	Colburn's Post-Office, N.C	May 2, 1846	4,496
Gold and silver ores, Method and means for treating.	L. E. Rivot	Paris, France	Jan. 28, 1868	73,838
Gold and silver ores, Process of treating	L. E. Rivot	Paris, France	June 9, 1868	78,831
Gold and silver ores, Treating	J. Kallmes	San Francisco, Cal	May 6, 1873	138,500
Gold and silver, Process of collecting	A. F. W. Partz	Oakland, Cal	June 8, 1869	90,955
Gold and silver, Refining	N. A. F. Brewer	Camden, S. C	July 11, 1848	5,663
Gold and silver, Sluice and blanket for collecting	E. Coleman	San Francisco, Cal	July 27, 1869	93,060
Gold and silver washing apparatus	J. Hendy	San Francisco, Cal	Feb. 27, 1866	52,850
Gold, &c., Apparatus for disintegrating gravel containing.	I. B. Cox	San Francisco, Cal	Aug. 4, 1868	80,606
Gold, Apparatus for saving	C. Schofield	Kernville, Cal	May 10, 1870	102,870
Gold, Apparatus for saving floating	D. Gay, jr	Vallejo, Cal	Feb. 2, 1869	86,388

Index of patents issued from the United States Patent Office from 1790 *to* 1873, *inclusive*—Continued.

Invention.	Inventor.	Residence.	Date.	No
Gold, Apparatus for washing and amalgamating	J. C. Dickey	Saratoga Springs, N. Y	July 19, 1864	43,574
Gold-beater, Mechanical	R. B. Ruggles and L. W. Serrell	New York, N. Y	Jan. 6, 1852	8,642
Gold-beating apparatus	T. C. Robbins	Philadelphia, Pa	Aug. 21, 1866	57,380
Gold beating machine	M. Hastings	Philadelphia, Pa	June 27, 1865	48,394
Gold-beating machine	W. Vine and J. H. Ashmead	Hartford, Conn	Aug. 6, 1850	7,552
Gold-beating machinery	W. Vine	Hartford, Conn	May 11, 1852	8,945
Gold bullion to toughen and refine it, Process of treating.	F. B. Miller	Sidney, Colony of New South Wales.	Mar. 10, 1868	75,289
Gold, &c., by amalgamation, Method of obtaining	M. A. Bertolet, L. Kirk, and A. M. De Hart.	Reading, Pa	Dec. 28, 1852	9,499
Gold by a vibrating circular cylindrical trough, Separating.	R. King	McIntosh County, Ga	Oct. 1, 1830	......
Gold-cleaner	J. Bogardus	New York, N. Y	Apr. 7, 1834	......
Gold-collector	J. Perry	New York, N. Y	Dec. 12, 1854	12,069
Gold-digger's fork	L. Teese and Son	San Francisco, Cal	Feb. 20, 1855	12,453
Gold, Digging and procuring	H. Jordan	Lexington, Ky	Oct. 1, 1830	......
Gold-digging apparatus	D. Tisdale	Des Moines, Iowa	Sept. 7, 1869	94,669
Gold, Double-acting rocker for washing	A. Buffum and P. Thorp	New York, N. Y	Oct. 1, 1850	7,678
Gold-dust, &c., Machine for washing, cleaning, and separating.	R. Lee	Erwinsville, N. C	July 8, 1829	......
Gold from earth, &c., Machine for separating	S. H. Mead	Saint Andrew's, N. Y	Sept. 10, 1861	33,257
Gold from earth, &c., Machine for separating	G. B. Palmer	Pendleton, N. C	Apr. 23, 1831	......
Gold from foreign matter, Apparatus for extracting	A. W. Hall	New York, N. Y	Dec. 27, 1864	45,670
Gold from impurities, Machine for separating	J. Sullivan	Gold Region, N. C	Feb. 27, 1847	4,988
Gold from its ores, Machine for extracting	N. Bosworth	Philadelphia, Pa	June 6, 1835	......
Gold from its ores, Process of extracting	R. D'Heureuse	San Francisco, Cal	Apr. 7, 1868	76,413
Gold from ores, Collecting	A. F. W. Partz	Oakland, Cal	Jan. 4, 1870	98,518
Gold from pyrites, Apparatus for extracting	G. F. Deetken	Nevada, Cal	Jan. 6, 1863	37,278
Gold from river-bottoms, Collecting	J. Johnson	Saco, Me	Apr. 2, 1867	63,394
Gold from sand and gravel, Machine for washing	C. Bechtler	Rutherfordton, N. C	Oct. 5, 1831	......
Gold from sand and pounded ores, Machine for washing.	C. Bechtler	Rutherfordton, N. C	Oct. 5, 1831	......
Gold from sand, Apparatus for separating	D. F. Hawkes	Timbuctoo, Cal	Oct. 5, 1869	95,472
Gold-leaf condenser	J. F. Adams	Worcester, Mass	Apr. 20, 1869	89,002
Gold-leaf condenser	J. F. Adams	Worcester, Mass	May 18, 1869	90,062
Gold, Machine for collecting and amalgamating fine particles of.	A. B. Crosby and J. Ladd	Greene, Me., and Boston, Mass.	Sept. 30, 1862	36,557
Gold, Machine for excavating and washing	S. Johnson	New York, N. Y	Apr. 13, 1858	19,930
Gold, Machine for separating and collecting	W. H. Folger	Spartanburgh District, S. C	Feb. 13, 1828	......
Gold, Machinery for separating	I. Babbitt	Boston, Mass	May 2, 1848	5,545
Gold, Magnetic machine for washing and separating.	S. Gardiner, jr	New York, N. Y	Mar. 8, 1853	9,610
Gold mineral, Reducing	W. Longmaid	London, England	Aug. 10, 1852	9,187
Gold-ores and alluvial auriferous earth, Machine for washing.	J. Powell	Salisbury, N. C	Apr. 1, 1831	......
Gold-ores and alluvial soil, Washing-machine for	V. De Rivafinoli, C. Harslebin, and W. Davis.	London, England, and Great Britain.	Nov. 1, 1830	......
Gold, Process for dissolving	C. F. Spieker	New York, N. Y	Feb. 10, 1852	8,729
Gold, Process for preparing	A. J. Watts	Utica, N. Y	Apr. 26, 1853	9,691
Gold, Process for refining	J. C. Booth	Philadelphia, Pa	Sept. 24, 1850	7,661
Gold, &c., roasting and smelting furnace	J. W. Shaeffer	Red Wing, Minn	July 14, 1868	80,020
Gold, &c., Saving float or fine	E. J. Fraser	San Francisco, Cal	Jan. 30, 1872	123,096
Gold, Separating and grinding	S. Whisler and J. Smith	Rockingham County, Va	Jan. 15, 1830	......
Gold, Separating and grinding flint-rock	A. Carson	Kingston, Tenn	Jan. 29, 1833	......
Gold-separator	G. Aughinbaugh	Portland, Oreg	Apr. 23, 1867	63,981
Gold-separator	J. Dobson	Burke County, N. C	Aug. 21, 1834	......
Gold-separator	R. H. Dunning	North San Juan, Cal	Feb. 7, 1860	27,041
Gold-separator	W. T. Duvall	Georgetown, D. C	Oct. 15, 1867	69,784
Gold-separator	M. C. Gritzner	Washington, D. C	Nov. 29, 1853	10,272
Gold-separator	D. Jones	Granville, N. C	Nov. 3, 1830	......
Gold-separator	W. H. Long	Mountain City, Colo	May 2, 1865	47,557
Gold-separator	D. Pierce	Woodstock, Vt	Jan. 10, 1854	10,414
Gold-separator	R. L. Reaney	Philadelphia, Pa	Jan. 10, 1860	26,784
Gold-separator	F. D. Sanno	Philadelphia, Pa	Dec. 24, 1830	......
Gold-separator	T. Seay	Columbia County, Ga	May 4, 1841	2,078
Gold-separator	E. L. Seymour	New York, N. Y	Sept. 19, 1854	11,712
Gold-separator	E. L. Seymour	New York, N. Y	Apr. 7, 1857	16,999
Gold-separator	W. C. Stiles	Nevada City, Cal	Nov. 12, 1867	70,914
Gold-separator	C. F. Testman	Portland, Oreg	July 24, 1866	56,636
Gold-separator	J. A. Veatch	San Francisco, Cal	Oct. 2, 1860	30,273
Gold separator, amalgamator, and washer	J. L. Montandevert	Charlotte, N. C	Apr. 30, 1831	......
Gold, silver, and copper, Process for parting	G. A. Scherpf	Hoboken, N. J	Jan. 30, 1866	52,325
Gold, silver, and other ores, Smelting	W. O. Davis	Pittsburgh, Pa	Dec. 17, 1872	133,927
Gold, silver, and plated ware, Mode of manufacturing.	H. G. Reed	Taunton, Mass	Feb. 21, 1871	112,077
Gold-size, Composition for	C. Bartholomae	New York, N. Y	May 8, 1866	54,488
Gold-swinging riddle	O. Willis	Morgantown, N. C	Sept. 5, 1832	......
Gold-washer	W. Ball	Chicopee, Mass	June 19, 1849	6,535
Gold-washer	H. Barnard	Morristown, N. Y	Feb. 16, 1858	19,338
Gold-washer	M. Bradley	Empire Ranch, Cal	Oct. 1, 1861	33,376
Gold-washer	J. Canfield	Sabula, Iowa	Mar. 25, 1862	34,738
Gold-washer	N. D. Clark	Bentonsport, Iowa	Mar. 7, 1865	46,641
Gold-washer	W. Davis	Fauquier County, Va	Nov. 22, 1832	......
Gold-washer	M. English	Lagro, Ind	Aug. 28, 1849	6,676
Gold-washer	L. Jennings	New York, N. Y	May 1, 1849	6,410
Gold-washer	W. H. Jennison	New York, N. Y	Apr. 3, 1849	6,267
Gold-washer	L. Lacharme	St. Leger de Feugeret, France.	Oct. 2, 1849	6,771
Gold-washer	M. Nelson	New York, N. Y	Oct. 4, 1859	25,667
Gold-washer	I. F. Quinby	Rochester, N. Y	July 16, 1867	66,736
Gold-washer	C. Ringel	San Francisco, Cal	Dec. 20, 1859	26,524
Gold-washer	H. M. Ritterband	New York, N. Y	Nov. 1, 1853	10,190
Gold-washer	T. Rives, sr	Hall County, Ga	Feb. 3, 1832	......
Gold-washer	H. Roberts	Mormon Island, Cal	July 26, 1859	24,889
Gold-washer	P. Summey	Lincoln, N. C	Jan. 30, 1830	......
Gold-washer	J. H. Ward	Sonora, Cal	Oct. 4, 1853	10,090
Gold-washer	O. Willis	Burke County, N. C	Dec. 21, 1831	......

Index of patents issued from the United States Patent Office from 1790 *to* 1873, *inclusive*—Continued.

Index of patents issued from the United States Patent Office from 1790 to 1873, inclusive—Continued.

Invention.	Inventor.	Residence.	Date.	No.
Governor-cut-off gear	H. H. Meyer	Denver, Colo	Sept. 2, 1873	142, 491
Governor, Electro-magnetic speed	G. M. Phelps	Troy, N. Y	Jan. 5, 1858	19, 042
Governor, Engine	E. Buss	Buckan Magdeburg, Prussia.	July 1, 1873	140, 467
Governor, Engine	W. J. Hallefas	Brooklyn, N. Y	May 1, 1866	54, 336
Governor, Engine	J. F. Haskins	Fitchburgh, Mass	June 4, 1872	127, 602
Governor, Engine	J. Hendy	San Francisco, Cal	July 1, 1873	140, 414
Governor, Engine	W. S. Henson	New York, N. Y	Oct. 13, 1868	83, 060
Governor, Engine	F. W. Howe	Newark, N. J	July 28, 1857	17, 879
Governor, Engine	R. K. Huntoon	Boston, Mass	May 2, 1871	114, 296
Governor, Engine	B. S. Lawson	Brooklyn, N. Y	Nov. 14, 1871	120, 982
Governor, Engine	J. D. Lynde	Philadelphia, Pa	Feb. 21, 1871	112, 058
Governor, Engine	R. Stewart	Elmira, N. Y	June 4, 1867	65, 444
Governor for direct-acting engines	A. S. Cameron	New York, N. Y	Mar. 7, 1871	112, 415
Governor for direct-acting engines	A. S. Cameron	New York, N. Y	Mar. 7, 1871	112, 416
Governor for electro-motors	T. A. Eddison	Newark, N. J	Jan. 24, 1871	111, 112
Governor for engines, water-wheels, &c	J. P. Sibley and A. Walsh	Bennington, Vt	Oct. 6, 1868	82, 758
Governor for machinery	G. M. Phelps	Troy, N. Y	Dec. 22, 1857	18, 927
Governor for marine and other engines, Differential	C. N. Clow	Port Byron, N. Y	June 17, 1856	15, 122
Governor for marine or other enginery	J. Atkins	Washington, D. C	Jan. 12, 1869	85, 779
Governor for mill-wheels, steam-engines, &c	H. Burt	Boston, Mass	Aug. 31, 1844	3, 722
Governor for regulating the speed of steam-engines	T. Silver	Philadelphia, Pa	Apr. 26, 1859	23, 790
Governor for regulating the work of windmills, &c	E. Allen	Worcester, Mass	July 28, 1857	17, 862
Governor for side-wheel ocean-steamers, Engine	W. B. Godfrey	Auburn, Iowa	May 27, 1856	14, 954
Governor for steam and other enginery	J. Bell	Cincinnati, Ohio	Feb. 15, 1870	99, 751
Governor for steam and other enginery	T. Gill	Waltham, Mass	July 27, 1869	92, 956
Governor for steam and other enginery	W. J. and C. A. Kesselmeyer and E. H. Nacke.	Manchester, England, and Als-Schoenfeld, Saxony.	Oct. 5, 1869	95, 481
Governor for steam and other enginery	J. T. Lassen	Wurzburg, Bavaria	Oct. 27, 1868	83, 516
Governor for steam and other enginery	B. Mackerley	Paint, Ohio	Dec. 22, 1868	85, 113
Governor for steam and other enginery	D. F. Mosman	Chelsea, Mass	Mar. 30, 1869	88, 406
Governor for steam and other enginery	M. Murphy	Charlotte, N. C	Nov. 9, 1869	96, 608
Governor for steam and other engines	C. P. Bowen	Silver City, Idaho	Oct. 31, 1871	120, 366
Governor for steam and other engines	W. Clark	Plumstead, England	June 4, 1872	127, 572
Governor for steam and other engines	J. A. Marden	Boston, Mass	May 4, 1869	89, 780
Governor for steam and other engines	H. D. Snow	Rochester, N. Y	Oct. 11, 1859	25, 769
Governor for steam-engines, Centrifugal	C. T. Porter	New York, N. Y	June 18, 1861	32, 583
Governor for steam-engines, Centrifugal	J. Wheelock	Worcester, Mass	July 3, 1860	29, 025
Governor for steam-engines, Fan	I. Y. Chubbuck	Roxbury, Mass	Dec. 13, 1859	26, 414
Governor for steam-engines, Instantaneous	W. W. H. Mead	Chestertown, N. Y	May 27, 1856	14, 967
Governor for steam-engines, Marine	J. L. Cathcart	Georgetown, D. C	Apr. 26, 1859	23, 755
Governor for steam-engines, Throttle	H. Camp and G. W. McIntosh	Rouseville, Pa	Sept. 26, 1871	119, 315
Governor for steam-engines, &c., Vane	F. Gustine	Medford, Mass	June 23, 1857	17, 623
Governor for steam-engines, Vane	C. Whittier	Roxbury, Mass	Nov. 3, 1857	18, 563
Governor for steam-engines, Water	G. Aab	Brooklyn, N. Y	Oct. 3, 1871	119, 490
Governor for steam-heating apparatus	S. J. Olsson	Chicago, Ill	June 24, 1873	140, 156
Governor for water, steam, and other powers	N. Scholfield	Norwich, Conn	July 21, 1857	17, 847
Governor for water-wheels, &c	W. M. Bailey	Matteawan, N. Y	July 23, 1872	129, 707
Governor, Hydraulic	J. E. Gillespie	Trenton, N. J	Jan. 7, 1862	34, 055
Governor, Hydraulic	J. E. Gillespie	Boston, Mass	Aug. 28, 1866	57, 630
Governor, Hydraulic	S. M. Hunter	Terrysville, Conn	Dec. 18, 1866	60, 632
Governor, Hydro-pneumatic	A. Harris	Philadelphia, Pa	Dec. 21, 1869	98, 162
Governor, Marine	H. J. Behrens	New York, N. Y	Mar. 24, 1863	37, 944
Governor, Marine-engine	J. Sullivan	South Boston, Mass	Dec. 17, 1867	72, 239
Governor, Marine steam	J. B. Cullen	Philadelphia, Pa	Mar. 24, 1868	75, 868
Governor, Marine steam-engine	O. Marland	Boston, Mass	Apr. 29, 1873	138, 420
Governor, Marine steam-engine	T. Silver	Philadelphia, Pa	July 3, 1855	13, 202
Governor, Marine steam-engine	E. M. Troth	New York, N. Y	Dec. 31, 1867	72, 939
Governor or hydraulic regulator	S. T. Thomas	Gilford, N. H	Dec. 24, 1872	134, 328
Governor, Power	J. Judson	New York, N. Y	Mar. 4, 1851	7, 960
Governor, Steam	G. W. Clark	Council Bluffs, Iowa	Nov. 14, 1871	120, 856
Governor, Steam	W. L. Collamore	Warren, R. I	Jan. 21, 1868	73, 578
Governor, Steam	R. D. Jacobus	Newark, N. J	Apr. 20, 1858	19, 995
Governor, Steam	A. Matson	Quincy, Ill	July 16, 1872	129, 287
Governor, Steam	G. E. Noyes	Washington, D. C	Nov. 6, 1866	59, 442
Governor, Steam	E. H. Parker	Bucksport, Me	Nov. 3, 1868	83, 655
Governor, Steam	A. F. Reeder	Normal, Ill	July 26, 1870	105, 847
Governor, Steam	R. Sanderson	Cleveland, Ohio	Jan. 15, 1867	61, 266
Governor, Steam	R. Sanderson	Cleveland, Ohio	May 14, 1867	64, 712
Governor, Steam	H. D. Snow	Bennington, Vt	Oct. 1, 1867	69, 499
Governor, Steam	J. Tremper	Wilmington, Del	Feb. 11, 1868	74, 454
Governor, Steam-engine	C. D. Allen	New York, N. Y	Dec. 28, 1869	98, 216
Governor, Steam-engine	T. Alsop	Elkhart City, Ill	Aug. 25, 1868	81, 455
Governor, Steam-engine	A. Anderson	Lancaster, Ohio	Aug. 3, 1858	21, 056
Governor, Steam-engine, &c	J. and E. Arthur	New Brunswick, N. J	Aug. 14, 1855	13, 415
Governor, Steam-engine	W. Ashby	Timber Township, Ill	Jan. 5, 1869	85, 632
Governor, Steam-engine	C. H. Bacon	Boston, Mass	May 19, 1868	77, 948
Governor, Steam-engine	H. Bilgram	Philadelphia, Pa	Nov. 19, 1872	133, 192
Governor, Steam-engine	H. Boardman	Port Richmond, N. Y	Mar. 15, 1870	100, 847
Governor, Steam-engine, &c	J. M. Bottum	New York, N. Y	Apr. 8, 1873	137, 652
Governor, Steam-engine	J. Broughton	New York, N. Y	Apr. 19, 1859	23, 660
Governor, Steam-engine	A. Brown	New York, N. Y	Nov. 7, 1865	50, 793
Governor, Steam-engine	C. P. Buckingham	Mount Vernon, Ohio	Aug. 7, 1860	29, 459
Governor, Steam-engine	J. S. Camac	Shickshinny, Pa	May 28, 1872	127, 305
Governor, Steam-engine	D. L. F. Chase	Boston, Mass	Mar. 17, 1868	75, 523
Governor, Steam-engine	J. C. Clime	Philadelphia, Pa	Apr. 17, 1866	54, 057
Governor, Steam-engine	C. A. Condé	Indianapoks, Ind	June 29, 1869	91, 915
Governor, Steam-engine	C. A. Condé	Indianapolis, Ind	Mar. 7, 1871	112, 420
Governor, Steam-engine	C. G. Cross	Chicago, Ill	Aug. 4, 1868	80, 533
Governor, Steam-engine	G. W. Davis and G. A. Rollins	Nashua, N. H	May 7, 1867	64, 500
Governor, Steam-engine	T. S. Davis	Jersey City, N. J	June 18, 1867	65, 885
Governor, Steam-engine	R. Defrees	Newark, N. J	Nov. 23, 1869	97, 056
Governor, Steam-engine	J. L. Dickinson	Dubuque, Iowa	Aug. 18, 1868	81, 072
Governor, Steam-engine	F. Douglass	Norwich, Conn	Sept. 8, 1863	39, 799
Governor, Steam-engine	C. Duclos	New Harmony, Ind	Aug. 24, 1869	94, 091
Governor, Steam-engine	J. B. Duff	Patchogue, N. Y	Apr. 23, 1872	126, 039
Governor, Steam-engine	J. Eddy	Barnesville, Ohio	Sept. 3, 1867	68, 495
Governor, Steam-engine	E. C. Edmonds	Buffalo, N. Y	Oct. 23, 1866	58, 996

Index of patents issued from the United States Patent Office from 1790 *to* 1873, *inclusive*—Continued.

Invention.	Inventor.	Residence.	Date.	No.
Governor, Steam-engine	R. Eickemeyer	Yonkers, N. Y	Aug. 28, 1866	57, 490
Governor, Steam-engine	L. Eikenberry	Philadelphia, Pa	Sept. 15, 1863	39, 903
Governor, Steam-engine	J. Farcot	St. Ouen, (Seine,) France	Feb. 16, 1869	87, 034
Governor, Steam-engine	W. C. Freeman	Louisiana, Mo	Aug. 13, 1872	130, 422
Governor, Steam-engine	L. F. Fuller	Providence, R. I	Mar. 31, 1868	76, 071
Governor, Steam-engine	R. W. Gardner	Quincy, Ill	Aug. 14, 1860	29, 579
Governor, Steam-engine	R. W. Gardner	Quincy, Ill	Dec. 27, 1864	45, 599
Governor, Steam-engine	R. W. Gardner	Quincy, Ill	Sept. 16, 1873	142, 846
Governor, Steam-engine	W. W. Gilbert	New York, N. Y	Jan. 5, 1869	85, 580
Governor, Steam-engine	J. W. Gray	Clermont County, Ohio	Apr. 7, 1868	76, 322
Governor, Steam-engine	J. L. Hastings	Towanda, Pa	Apr. 26, 1864	42, 478
Governor, Steam-engine	A. A. Henderson	Norfolk, Va	Dec. 18, 1866	60, 629
Governor, Steam-engine	J. C. Hoadley	Lawrence, Mass	Oct. 28, 1873	144, 098
Governor, Steam-engine	W. H. Howland	San Francisco, Cal	June 29, 1869	92, 051
Governor, Steam-engine	J. E. Hugon	Richmond, Ind	Nov. 4, 1873	144, 204
Governor, Steam-engine	J. D. Humphreys	London, England	Dec. 1, 1863	40, 754
Governor, Steam-engine	R. K. Huntoon	Boston, Mass	Dec. 4, 1866	60, 192
Governor, Steam-engine	R. K. Huntoon	Boston, Mass	Nov. 19, 1867	71, 015
Governor, Steam-engine	R. K. Huntoon	Boston, Mass	Dec. 3, 1867	71, 761
Governor, Steam-engine	R. K. Huntoon	Boston, Mass	May 4, 1869	89, 581
Governor, Steam-engine	R. K. Huntoon	Wakefield, Mass	Nov. 8, 1870	109, 013
Governor, Steam-engine	R. K. Huntoon	Boston, Mass	July 4, 1871	116, 596
Governor, Steam-engine	J. Judson	Rochester, N. Y	Nov. 19, 1861	33, 743
Governor, Steam-engine	O. A. Kelly	Slatersville, R. I	Dec. 17, 1867	72, 204
Governor, Steam-engine	O. A. Kelly and E. Lamb	Slatersville, R. I	Jan. 31, 1865	46, 111
Governor, Steam-engine	A. K. Kline	Readington, N. J	Nov. 21, 1871	121, 109
Governor, Steam-engine	T. S. La France	Elmira, N. Y	Aug. 7, 1866	56, 956
Governor, Steam-engine	H. H. Lamont	San Francisco, Cal	Sept. 10, 1872	131, 282
Governor, Steam-engine	J. P. T. Lang	Washington, D. C	May 14, 1867	64, 775
Governor, Steam-engine	C. M. Langley	Lowell, Mass	Aug. 1, 1871	117, 548
Governor, Steam-engine	A. P. and B. F. Lanterman	Prairie City, Ill	Jan. 30, 1866	52, 298
Governor, Steam-engine	A. Lawrence	Lowell, Mass	Feb. 20, 1866	52, 789
Governor, Steam-engine	B. S. Lawson	Brooklyn, N. Y	July 11, 1871	116, 968
Governor, Steam-engine	J. F. Letellier	Grand Rapids, Mich	July 20, 1869	92, 848
Governor, Steam-engine	T. J. Lovegrove	Philadelphia, Pa	June 20, 1865	48, 344
Governor, Steam-engine	J. A. Lynch	Boston, Mass	Feb. 28, 1871	112, 159
Governor, Steam-engine	J. A. Lynch and R. K. Huntoon	Boston, Mass	Sept. 22, 1868	82, 332
Governor, Steam-engine	J. D. Lynde	Philadelphia, Pa	Oct. 11, 1870	108, 162
Governor, Steam-engine	J. D. Lynde	Philadelphia, Pa	Aug. 8, 1871	117, 903
Governor, Steam-engine	J. A. Marden	Boston, Mass	Oct. 6, 1868	82, 729
Governor, Steam-engine	J. A. Marden	Boston, Mass	Mar. 30, 1869	88, 495
Governor, Steam-engine	J. A. Marden	Boston, Mass	Mar. 30, 1869	88, 496
Governor, Steam-engine	J. A. Marden	Boston, Mass	Mar. 30, 1869	88, 497
Governor, Steam-engine	J. A. Marden and C. E. Abbott	Boston, Mass	Mar. 30, 1869	88, 498
Governor, Steam-engine	A. Matson	Quincy, Ill	Aug. 26, 1863	142, 167
Governor, Steam-engine	H. D. McMaster and A. Dale	Guilford, Ireland	Apr. 5, 1870	101, 487
Governor, Steam-engine	S. H. Miller	Hanoverton, Ohio	Sept. 11, 1860	29, 986
Governor, Steam-engine	T. Moore	Brooklyn, N. Y	Mar. 15, 1870	100, 785
Governor, Steam-engine	D. F. Mosman	Cambridge, Mass	Dec. 31, 1867	72, 882
Governor, Steam-engine	W. Nichols	Elmira, N. Y	Sept. 11, 1866	57, 953
Governor, Steam-engine	W. Ord	Brooklyn, Ohio	May 6, 1873	138, 521
Governor, Steam-engine	J. Parlane	Brooklyn, N. Y	May 18, 1869	90, 294
Governor, Steam-engine	G. T. Parry and H. W. Evans	Philadelphia, Pa	Mar. 1, 1859	23, 110
Governor, Steam-engine	A. J. Peavey	Boston, Mass	Aug. 16, 1870	106, 400
Governor, Steam-engine	T. R. Pickering	New York, N. Y	Oct. 24, 1865	50, 624
Governor, Steam-engine	W. H. Place	New York, N. Y	July 26, 1870	105, 841
Governor, Steam-engine	J. H. Pomeroy	Jordan, N. Y	Dec. 10, 1861	33 903
Governor, Steam-engine	C. T. Porter	New York, N. Y	July 13, 1858	20, 894
Governor, Steam-engine	G. F. Pottle	Boston, Mass	Feb. 7, 1871	111, 677
Governor, Steam-engine	G. T. Pracy	San Francisco, Cal	Jan. 12, 1869	85, 763
Governor, Steam-engine	G. W. Rains	Newburgh, N. Y	May 29, 1860	28, 502
Governor, Steam-engine	J. H. and C. E. Randall	Boston, Mass	Oct. 6, 1868	82, 874
Governor, Steam-engine	J. M. Rees	Scott, Ohio	Oct. 17, 1865	50, 496
Governor, Steam-engine	E. Reynolds and N. G. Herreshoff.	Providence, R. I	Mar. 26, 1872	125, 084
Governor, Steam-engine	S. P. Ruggles	Boston, Mass	Sept. 16, 1873	142, 869
Governor, Steam-engine	C. R. Rungvist	Stockholm, Sweden	Oct. 14, 1873	143, 642
Governor, Steam-engine	I. S. Schuyler	Brooklyn, E. D., N. Y	Feb. 20, 1872	123, 939
Governor, Steam-engine	H. C. Sergeant	Columbus, Ohio	Dec. 21, 1858	22, 380
Governor, Steam-engine	J. W. Shirley and W. H. Fasig	Terre Haute, Ind	June 11, 1867	65, 697
Governor, Steam-engine	D. Shive	Philadelphia, Pa	May 1, 1866	54, 423
Governor, Steam-engine	T. Silver	New York, N. Y	Oct. 2, 1866	58, 491
Governor, Steam-engine	B. B. Smith	Nashville, Tenn	Apr. 8, 1873	137, 573
Governor, Steam-engine	W. Smith	Philadelphia, Pa	June 1, 1869	90, 696
Governor, Steam-engine	W. Smith	Philadelphia, Pa	Feb. 22, 1870	100, 203
Governor, Steam-engine	H. D. Snow	Rochester, N. Y	Sept. 27, 1864	44, 466
Governor, Steam-engine	J. H. Springer	Philadelphia, Pa	July 21, 1868	80, 096
Governor, Steam-engine	G. S. Stearns and W. Hodgson	Cincinnati, Ohio	Aug. 31, 1852	9, 236
Governor, Steam-engine	F. Taggart	Brooklyn, N. Y	Nov. 27, 1866	60, 084
Governor, Steam-engine	H. N. Throop	Pultneyville, N. Y	Dec. 29, 1857	18, 997
Governor, Steam-engine	S. Trumbore	Easton, Pa	Nov. 24, 1868	84, 396
Governor, Steam-engine	J. H. Wait	Portsmouth, Ohio	Dec. 26, 1865	51, 764
Governor, Steam-engine	A. F. Ward	Louisville, Ky	July 14, 1857	17, 817
Governor, Steam-engine	C. Waters	Boston, Mass	Jan. 3, 1871	110, 703
Governor, Steam-engine	H. B. Weaver	Hartford, Conn	June 20, 1871	116, 248
Governor, Steam-engine	J. V. Weitz	Cleveland, Ohio	Feb. 26, 1867	62, 457
Governor, Steam-engine	M. Wheeler	Honesdale, Pa	June 10, 1856	15, 095
Governor, Steam-engine	W. Wickersham	Boston, Mass	Oct. 5, 1869	95, 543
Governor, Steam-engine	J. Wood	Brooklyn, N. Y	Apr. 16, 1867	63, 976
Governor, Steam-engine	W. W. W. Wood	Philadelphia, Pa	Sept. 7, 1858	21, 475
Governor, Steam-engine	J. H. Wooster	Strykersville, N. Y	Aug. 20, 1867	67, 936
Governor stop-valve	J. Crawley	Perrysville, Ind	July 10, 1866	56, 186
Governor stop-valve for steam-engines	C. A. Condé	Indianapolis, Ind	May 25, 1869	90, 427
Governor-valve	R. Andrews and E. Armstrong	Allegheny, Pa	June 2, 1868	78, 410
Governor-valve	W. W. Gilbert	New York, N. Y	Apr. 27, 1869	89, 399
Governor-valve for steam-engines	W. A. Cogswell	Rochester, N. Y	Aug. 13, 1872	130, 480
Governor-valve for steam-engines	W. A. Cogswell	Rochester, N. Y	Aug. 13, 1872	130, 481
Governor-valve for steam-engines	W. H. Cowles	Erie, Pa	Apr. 23, 1872	125, 937

Index of patents issued from the United States Patent Office from 1790 *to* 1873, *inclusive*—Continued.

Index of patents issued from the United States Patent Office from 1790 *to* 1873, *inclusive*—Continued.

Invention.	Inventor.	Residence.	Date.	No.
Grain and straw separator	S. Huson	Jacksonville, N. Y	Sept. 16, 1862	36, 467
Grain and straw separator	A. Major	Lebanon, Pa	Jan. 27, 1863	37, 549
Grain and straw shaker	W. H. and P. Bott	West Manchester Township, Pa.	July 18, 1871	117, 040
Grain, &c., Apparatus for collecting and transporting.	J. and G. Richards	Philadelphia, Pa	Sept. 30, 1873	143, 254
Grain, Apparatus for dampening	J. Shellabarger	Decatur, Ill	Aug. 11, 1868	81, 016
Grain, Apparatus for decorticating	S. Dodson	Jersey City, N. J	Nov. 26, 1872	133, 309
Grain, &c., Apparatus for decorticating, separating, and drying.	A. J. Glas	London, England	Sept. 20, 1870	107, 480
Grain, Apparatus for preserving	L. M. F. Doyére	Paris, France	Dec. 20, 1859	26, 481
Grain, &c., Apparatus for preventing overheating of	T. A. Hoffmann	Beardstown, Ill	Mar. 8, 1870	100, 628
Grain, Apparatus for washing, boiling, and fermenting.	E. F. Prentiss and R. A. Robertson.	Philadelphia, Pa	Aug. 1, 1865	49, 151
Grain-bag	J. W. H. Campbell	San Francisco, Cal	July 19, 1864	43, 567
Grain-bag	W. B. Carlock	Worshiam, Va	Oct. 21, 1873	143, 876
Grain-bag holder	W. P. Leland	Mendon, Mich	May 16, 1871	114, 952
Grain-band and bag-tie	E. Foote	East Bloomfield, N. Y	Feb. 18, 1873	135, 899
Grain-bands, Device for securing	J. Nelson	Rockford, Ill	May 2, 1865	47, 603
Grain-basket	G. P. Coan	Wyandotte, Mich	Oct. 28, 1873	144, 018
Grain-bin	D. D. Badger and W. S. Sampson.	New York, N. Y	June 14, 1869	24, 424
Grain-bin	L. S. Chichester	Brooklyn, N. Y	Apr. 2, 1872	125, 169
Grain-bin	G. H. Johnson	New York, N. Y	Dec. 9, 1862	37, 134
Grain-bin	S. Marsh	Roxbury, Mass	Oct. 23, 1860	30, 486
Grain-bin	C. W. Mills	Brooklyn, N. Y	May 18, 1869	90, 116
Grain-bin	O. I. Porter	Hudson, Ohio	Jan. 8, 1867	61, 095
Grain-bin	F. Raymond	Cleveland, Ohio	July 12, 1870	105, 367
Grain bin	F. Raymond	Cleveland, Ohio	Aug. 30, 1870	106, 961
Grain-bin	F. Raymond and A. Miller	Cleveland, Ohio	Dec. 21, 1869	98, 190
Grain-bin	W. S. Sampson	New York, N. Y	Jan. 14, 1862	34, 164
Grain-bin	W. S. Sampson	New York, N. Y	Dec. 9, 1862	37, 139
Grain-bin	W. S. Sampson	New York, N. Y	Aug. 3, 1869	93, 352
Grain-bin	C. D. Woodruff	Toledo, Ohio	Aug. 31, 1869	94, 265
Grain-bin, Fire-proof	G. H. Johnson	Buffalo, N. Y	Mar. 9, 1869	87, 568
Grain-bin for storing and drying	J. Royal	White Rock, Ill	Feb. 6, 1872	123, 424
Grain-bins, Operating discharge-valve for	G. H. Johnson	Buffalo, N. Y	Feb. 16, 1869	86, 928
Grain-binder	J. F. Appleby	Mazo Manie, Wis	June 1, 1869	90, 807
Grain-binder	J. Baker	Fairbury, Ill	July 27, 1869	93, 035
Grain-binder	W. P. Barker	Grand Rapids, Mich	Dec. 9, 1862	37, 123
Grain-binder	W. P. Barker	Grand Rapids, Mich	Mar. 21, 1865	46, 869
Grain-binder	C. L. Beamer	Cambria, N. Y	June 15, 1869	91, 202
Grain-binder	J. Behel	Earlville, Ill	Feb. 16, 1864	41, 661
Grain-binder	J. Behel	Rockford, Ill	Aug. 3, 1869	93, 165
Grain-binder	C. W. Bowron	Chicago, Ill	Jan. 16, 1872	122, 803
Grain-binder	H S. L. Bryan	Kearney, Mo	Oct. 22, 1872	132, 436
Grain-binder	J. K. Bull	Buckingham, Iowa	Sept. 8, 1868	81, 875
Grain-binder	W. W. Burson	Yates, Ill	June 26, 1860	28, 830
Grain-binder	W. W. Burson	Atkinson, Ill	Aug. 11, 1863	39, 463
Grain-binder	W. W. Burson	Rockford, Ill	Oct. 4, 1864	44, 513
Grain-binder	W. W. Burson	Rockford, Ill	July 25, 1865	48, 900
Grain-binder	S. D. Carpenter	Madison, Wis	Dec. 22, 1868	85, 209
Grain-binder	S. D. Carpenter	Madison, Wis	Dec. 22, 1868	85, 210
Grain-binder	E. Chapman	Rochester, Minn	May 7, 1872	126, 520
Grain-binder	E. H. Clinton	Iowa City, Iowa	July 13, 1869	92, 517
Grain-binder	E. H. Clinton	Iowa City, Iowa	Apr. 30, 1872	126, 129
Grain-binder	C. G. Dickenson	Poughkeepsie, N. Y	May 30, 1871	115, 284
Grain-binder	C. H. Durkee	Hartford, Wis	Nov. 22, 1859	26, 171
Grain-binder	N. B. Fassett	Saint Louis, Mo	May 21, 1872	127, 036
Grain-binder	N. F. Gilman	Rochester, Minn	Aug. 24, 1869	93, 985
Grain-binder	W. F. Goodwin	Washington, D. C	Mar. 13, 1866	53, 138
Grain-binder	J. F. Gordon	Kalamazoo, Mich	May 12, 1868	77, 878
Grain-binder	J. F. Gordon	Rochester, N. Y	July 15, 1873	140, 822
Grain-binder	J. B. Greenhut	Chicago, Ill	Sept. 8, 1868	81, 891
Grain-binder	W. D. Harrah	Davenport, Iowa	Dec. 22, 1863	41, 002
Grain-binder	W. D. Harrah, I. M. Gifford, and E. T. Johnston.	Davenport, Iowa	May 10, 1870	102, 815
Grain-binder	H. Harrier	Indianapolis, Ind	Sept. 4, 1866	57, 710
Grain-binder	V. Hayes, C. G. Waldo, and H. A. Main.	Tekonsha, Mich	Dec. 8, 1868	84, 693
Grain-binder	J. F. Hemperly and C. Barns	West Liberty, Iowa	Sept. 5, 1865	49, 756
Grain-binder	J. S. Hickey	Pike, Ill	July 17, 1860	29, 170
Grain-binder	S. T. Holly	Rockford, Ill	Aug. 14, 1866	57, 138
Grain-binder	S. T. Holly	Rockford, Ill	Sept. 11, 1866	57, 904
Grain-binder	S. T. Holly	Rockford, Ill	June 11, 1867	65, 746
Grain-binder	A. S. Hoyt	Winona, Minn	Dec. 13, 1870	110, 138
Grain-binder	A. S. Hoyt	Winona, Minn	Aug. 5, 1873	141, 507
Grain-binder	S. Judevine and Z. Shaw	Roxbury, Wis	July 14, 1863	39, 234
Grain-binder	H. Kaller	Perry, Ill	June 5, 1860	28, 584
Grain-binder	I. Lancaster	Baltimore, Md	Jan. 23, 1866	52, 175
Grain-binder	I. Lancaster	Baltimore, Md	Apr. 24, 1866	54, 177
Grain-binder	I. Lancaster	Baltimore, Md	Nov. 19, 1867	71, 023
Grain-binder	S. D. Locke	Janesville, Wis	Dec. 19, 1865	51, 599
Grain-binder	S. D. Locke	Janesville, Wis	Dec. 19, 1865	51, 600
Grain-binder	S. D. Locke	Janesville, Wis	July 24, 1866	56, 580
Grain-binder	S. D. Locke	Janesville, Wis	Sept. 4, 1866	57, 739
Grain-binder	S. D. Locke	Janesville, Wis	Dec. 4, 1866	60, 208
Grain-binder	S. D. Locke	Janesville, Wis	Dec. 7, 1869	97, 531
Grain-binder	S. D. Locke	Janesville, Wis	Dec. 7, 1869	97, 532
Grain-binder	S. D. Locke	Janesville, Wis	Dec. 7, 1869	97, 533
Grain-binder	S. D. Locke	Janesville, Wis	Dec. 7, 1869	97, 534
Grain-binder	S. D. Locke	Janesville, Wis	Dec. 7, 1869	97, 535
Grain-binder	S. D. Locke	Janesville, Wis	Dec. 7, 1869	97, 536
Grain-binder	S. D. Locke	Janesville, Wis	Jan. 17, 1871	111, 069
Grain-binder	S. D. Locke	Janesville, Wis	Nov. 28, 1871	121, 290
Grain-binder	S. D. Locke	Hoosick Falls, N. Y	Feb. 11, 1873	135, 826
Grain-binder	S. D. Locke	Hoosick Falls, N. Y	May 20, 1873	139, 008
Grain-binder	W. Lottridge	Charles City, Iowa	Aug. 10, 1869	93, 458

Index of patents issued from the United States Patent Office from 1790 *to* 1873, *inclusive*—Continued.

Invention.	Inventor.	Residence.	Date.	No.
Grain-binder	W. Lottridge	Charles City, Iowa	Oct. 4, 1870	107, 933
Grain-binder	G. W. Loveless and C. H. Shaffer.	Charles Hill, Ind	Nov. 16, 1869	96, 933
Grain-binder	W. D. May	Baltimore, Md	Feb 16, 1869	87, 056
Grain-binder	J. M. McMaster	Rochester, N. Y	June 29, 1869	91. 860
Grain-binder	D. McPherson	Caledonia, N. Y	Sept. 27, 1870	107, 797
Grain-binder	J. H. Morse	Peoria, Ill	Dec. 17, 1872	134, 089
Grain-binder	J. H. Mudgett	Camanche, Iowa	Oct. 12, 1869	95, 714
Grain-binder	W. B. Oglesby	Ridge Prairie, Ill	June 21, 1870	104, 487
Grain-binder	W. H. Paine	Janesville, Wis	Aug. 2, 1870	105, 974
Grain-binder	L. Parker	Davenport, Iowa	Dec. 27, 1870	110, 492
Grain-binder	L. F. Parker	Davenport, Iowa	Aug. 31, 1869	94, 436
Grain-binder	J. Pearson	West Milton, Ohio	Apr. 9, 1872	125. 481
Grain-binder	P. A. Perry	Perth Amboy, N. J	Apr. 18, 1871	113, 792
Grain-binder	A. Philippi	Saint Louis, Mo	Sept. 26, 1871	119, 241
Grain-binder	A. Phillippi	Saint Louis, Mo.	Aug. 19, 1873	142, 040
Grain-binder	F. W. Randall	Tekonsha, Mich	Dec. 27, 1870	110, 498
Grain-binder	M. T. Ridout	Sun Prairie, Wis	Nov. 14, 1871	120, 828
Grain-binder	O. B. Ross	Bowen's Prairie, Iowa	Sept. 12, 1871	118, 975
Grain-binder	Q. A. Scott	Pittsburgh, Pa	June 17, 1873	140, 083
Grain-binder	G. A. Scribner	Rochester, N: Y	Sept. 27, 1870	107, 823
Grain-binder	G. B. Shafer	Delta, Ohio	Mar. 23, 1869	88, 087
Grain-binder	W. Siverd, jr	Geneseo, N. Y	Feb. 27, 1872	124, 020
Grain-binder	A. J. Smith	New York, N. Y	Jan. 15, 1867	61, 272
Grain-binder	W. W. Snell	Bushford, Minn	Feb. 23, 1869	87, 215
Grain-binder	G. H. Spaulding	Rockford, Ill	May 31, 1870	103, 673
Grain-binder	A. S. Stone	Plain View, Minn	Mar. 3, 1868	75, 215
Grain-bidder	O. O. Storle	Norway, Wis	Dec. 3, 1867	71, 660
Grain-binder	O. O. Storle	Norway, Wis	June 15, 1869	91, 282
Grain-binder	M. Summers and S. B. Lane	Zionsville, Ind	Oct. 3, 1871	119, 668
Grain-binder	J. Waddington	New Philadelphia, Ohio	Aug. 26, 1873	142, 304
Grain-binder	S. J. Wallace	Carthage, Ill	Apr. 12, 1864	42, 322
Grain-binder	S. J. Wallace	Carthage, Ill	Jan: 10, 1865	45, 885
Grain-binder	G. Warner	West Liberty, Iowa	July 30, 1867	67, 389
Grain-binder	J. H. Whitney	Rochester, Minn	Apr. 2, 1872	125, 366
Grain-binder	C. B. Withington	Janesville, Wis	Feb. 20, 1872	123, 967
Grain-binder	J. Youll	Manchester, Iowa	Aug. 15, 1871	118, 091
Grain-binder, Automatic	R. D. Brown	Covington, Ind	Apr. 7, 1863	38, 093
Grain-binders, Band-tuck for	J. Beall	Defiance, Ohio	Dec. 24, 1872	134, 192
Grain-binders' reel	S. D. Locke	Janesville, Wis	Feb. 20, 1866	52, 722
Grain-binders' reel	S. D. Locke	Janesville, Wis	Nov. 17, 1868	84, 065
Grain-binding apparatus	J. F. Barrett	North Granville, N. Y	Apr. 28, 1857	17, 135
Grain-binding apparatus	G. W. Chandler	Mason, N. H	Feb. 1, 1870	99, 404
Grain-binding machine	D. W. Ayres	Middleport, Ill	May 22, 1860	28, 338
Grain-binding machine	J. F. Black	Lancaster, Ill	Sept. 22, 1857	18, 231
Grain-binding machine	W. W. Burson	Yates City, Ill	Feb. 26, 1861	31, 526
Grain-binding machine	T. Courser	Princeton, Ill	July 10, 1860	29, 063
Grain-binding machine	T. Fowler	Tottensville, N. Y	June 7, 1870	103, 861
Grain-binding machine	T. Fowler	Tottensville, N. Y	June 7, 1870	103, 862
Grain-binding machine	A. Goodyear, 2d	New Haven, Conn	Nov. 7, 1865	50, 814
Grain-binding machine	S. T. Holly	Rockford, Ill	Jan. 26, 1864	41, 377
Grain-binding machine	S. T. Holly	Rockford, Ill	Jan. 26, 1864	41, 378
Grain-binding machine	I. Keyes	Putney, Vt	July 17, 1839	1, 251
Grain-binding machine	W. F. Pagett	Stone Bridge, Va	July 29, 1856	15, 436
Grain-binding machine	L. D. Phillips	Chicago, Ill	Dec. 29, 1857	18, 988
Grain-binding machine	S. Reynolds	Richmond, R. I	Feb. 12, 1861	31, 402
Grain-binding mechanism	A. Sherwood	Auburn, N. Y	Aug. 30, 1859	25, 308
Grain, Bushing-screen for cleaning	E. Tisdall	Windham, Conn	July 17, 1815	....
Grain by blowing, Machine for raising	J. Bailey	Chester County, Pa	Mar. 5, 1810	....
Grain by hand or horse power, Grinding or cracking	B. Hinkley	New York, N. Y	Mar. 10, 1834	....
Grain by horse-power, Machine for cutting	L. Durham and J. S. Pleasants	Halifax County, Va	July 28, 1827	....
Grain by sulphurous acid, Bleaching	J. M. Clark	Philadelphia, Pa	Apr. 17, 1860	27, 887
Grain-carrier	O. M. Gould	Montreal, Canada	Dec. 12, 1871	121, 867
Grain-carriers, Construction of	A. Linhart and S. McClain	Fulton, Ohio	Sept. 25, 1849	6, 749
Grain-cleaner	W. Ager	Washington, D. C	July 13, 1869	92, 556
Grain-cleaner	J. E. Anderson	Boiling Springs, Pa	Oct. 27, 1868	83, 438
Grain-cleaner	G. B. Bailey	Greenfield, Ind	Dec. 2, 1862	37, 031
Grain-cleaner	C. F. Baylor	Clinton, N. J	Aug. 7, 1866	56, 880
Grain-cleaner	E. Bradfield	Rochester, N. Y	Sept. 19, 1840	1, 794
Grain-cleaner	G. Clark	Sandusky, Ohio	Apr. 21, 1863	38, 258
Grain-cleaner	J. H. De Force	Healdsburgh, Cal.	June 11, 1872	127, 680
Grain-cleaner	W. S. Deisher	Hamburgh, Pa	Feb. 16, 1864	41, 609
Grain-cleaner	P. C. Fritz	Barrytown, N. Y	May 1, 1860	28, 071
Grain-cleaner	A. Gaar	Richmond, Ind	Mar. 13, 1860	27, 497
Grain-cleaner	J. A. Hall	Greenfield, Ind	May 28, 1867	65, 074
Grain-cleaner	T. Hancock and J. H. Leaman	Richmond, Va	Jan. 28, 1868	73, 803
Grain-cleaner	W. Houghton	Great Grimsby, England	Nov. 11, 1873	144, 455
Grain-cleaner	A. Hunter	Rockford, Ill	Oct. 29, 1872	132, 664
Grain-cleaner	J. Hutchison	Three Rivers, Mich	Dec. 8, 1863	40, 840
Grain-cleaner	C. Jones	De Soto, Ill	Jan. 1[illegible], 1870	98, 975
Grain-cleaner	J. A. Krake	Alden, N. Y	May 7, 1872	126, 400
Grain-cleaner	G. Leach	Owego, N. Y	Mar. 13, 1855	12, 512
Grain-cleaner	J. P. Leonard	Five Mile, Ohio	Nov. 12, 1872	133, 041
Grain-cleaner	W. A. Lewis	Joliet, Ill	June 22, 1869	91, 549
Grain-cleaner	E. Mantz	Frederick, Md	Feb. 9, 1864	41, 519
Grain-cleaner	J. McTaggart	Rochester, N. Y	Apr. 2, 1867	63, 410
Grain-cleaner	L. and J. Miller	Canton, Ohio	Dec. 17, 1861	33, 956
Grain-cleaner	J. Montgomery	Baltimore, Md	Dec. 29, 1868	85, 325
Grain-cleaner	S. Moore	Rochelle, Ill	May 27, 1873	139, 257
Grain-cleaner	M. T. Nesbitt	Colora, Md	Apr. 4, 1871	113, 686
Grain-cleaner	G. W. Osborn	Centreville, Mich	Mar. 13, 1860	27, 467
Grain-cleaner	C. Perry	Rochester, N. Y	June 7, 1870	103, 922
Grain-cleaner	C. Perry and J. E. Wheat	Rochester, N. Y	Mar. 1, 1870	100, 441
Grain-cleaner	P. Provost	Rochester, Minn	Oct. 15, 1872	132, 319
Grain-cleaner	N. Reece	Hemlock Township, Pa	Aug. 1, 1871	117, 681
Grain-cleaner	H. D. Reynolds	Pendleton, Ind	June 19, 1855	13, 103
Grain-cleaner	F. P. Root	Sweden, N. Y	Nov. 15, 1843	3, 341
Grain-cleaner	J. Simpson and W. Hayden	Tecumseh, Mich	Oct. 14, 1862	36, 676

Index of patents issued from the United States Patent Office from 1790 *to* 1873, *inclusive*—Continued.

Invention.	Inventor.	Residence.	Date.	No.
Grain-cleaner	W. B. Smith	Clayton, Ill	July 6, 1869	92, 385
Grain-cleaner	J. Sternberg	Webster City, Iowa	Feb. 21, 1871	112, 087
Grain-cleaner	W. H. Stevens	Salem, Ill	July 7, 1846	4, 624
Grain-cleaner	G. Stevenson	Zionsville, Ind	Apr. 9, 1867	63, 763
Grain-cleaner	G. Stevenson and J. J. Crider	Zionsville and Greenfield, Ind.	May 29, 1866	55, 174
Grain-cleaner	J. Stroop	Joliet, Ill	Nov. 9, 1869	96, 630
Grain-cleaner	B. T. Trimmer	Parma, N. Y	Apr. 3, 1855	12, 657
Grain-cleaner	I. Wait	Watertown, N. Y	Oct. 11, 1859	25, 777
Grain-cleaner	A. T. Waldo	Dryden, N. Y	Mar. 13, 1860	27, 489
Grain-cleaner	J. H. Weaver	Gap, Pa	June 20, 1871	116, 122
Grain-cleaner	W. B. Webster	Foxville, Va	May 1, 1860	28, 119
Grain-cleaner	J. E. Wheat	Rochester, N. Y	Jan. 26, 1869	86, 194
Grain-cleaner and fertilizer-sifter	J. A. Green	Mill Dale, Va	Mar. 7, 1871	112, 334
Grain cleaner and scourer	E. H. Kellogg	Milwaukee, Wis	May 9, 1871	114, 688
Grain cleaner and separator	G. S. Carter	Lewisburgh, Pa	Jan. 2, 1872	122, 436
Grain cleaner and separator	W. A. Cockrill	Zanesville, Ohio	June 6, 1871	115, 707
Grain cleaner and separator	W. Gardner	Catalpa, Ky	Jan. 28, 1870	104, 728
Grain cleaner and separator	J. W. Patterson	Monticello, Minn	Dec. 15, 1863	40, 942
Grain cleaner and separator	J. H. Redfield	Salem, Ind	Feb. 6, 1872	123, 361
Grain cleaner and separator	B. T. Trimmer	Rochester, N. Y	May 29, 1866	55, 179
Grain cleaner and separator	R. Ward	Edinburgh, Ind	Aug. 26, 1856	15, 630
Grain cleaner and separator	F. Wegmann	Naples, Italy	Oct. 18, 1870	108, 541
Grain cleaner and smut-machine	J. Ferguson	Fall River, Mass	July 20, 1869	92, 809
Grain, Cleaning	R. Redfield	Salem, Ind	Feb. 27, 1866	52, 882
Grain, Cleaning	M. Urffer	Upper Milford, Pa	Oct. 1, 1830	....
Grain, Cleaning and dressing	G. Arnold	Angelica, N. Y	June 17, 1830	....
Grain cleaning and drying machine	H. N. Black	Philadelphia, Pa	Aug. 1, 1854	11, 458
Grain cleaning and hulling machine	J. M. Mayer	New York, N. Y	Feb. 14, 1865	46, 374
Grain cleaning and winnowing machine	Z. Rice	Lyons, N. Y	Apr. 24, 1841	2, 067
Grain-cleaning device	P. G. B. Westmacott	Elswick, Newcastle-upon-Tyne, England.	July 16, 1867	66, 759
Grain, Cleaning, hulling, and drying	E. Cole	Dryden, N. Y	June 23, 1863	38, 948
Grain-cleaning machine	B. Barney	Tune, Ill	Sept. 18, 1866	58, 048
Grain-cleaning machine	J. D. Beers	Philadelphia, Pa	Dec. 10, 1841	2, 384
Grain-cleaning machine	S. Bentz	Boonsborough, Md	July 23, 1841	2, 193
Grain-cleaning machine	J. N. Bird and E. D. Weld	Trenton, N. J	Mar. 28, 1842	2, 513
Grain-cleaning machine	J. L. Booth	Cuyahoga Falls, Ohio	Dec. 18, 1855	13, 937
Grain-cleaning machine	G. W. Bowers	Leitersburgh, Md	Sept. 24, 1850	7, 662
Grain cleaning machine	D. H. Cole	Portland, Me	July 17, 1838	846
Grain-cleaning machine	P. Cook	Westfield, N. Y	Jan. 17, 1842	2, 423
Grain-cleaning machine	W. Crotzer	Spruce Creek, Pa	Nov. 20, 1860	30, 704
Grain-cleaning machine	J. De Rush	Saint Mary's, Ohio	May 11, 1858	20, 196
Grain-cleaning machine	M. Eckley	Olney, Ill	May 5, 1868	77, 468
Grain-cleaning machine	W. T. Fisher	Cleaveland, Tenn	Oct. 5, 1858	21, 662
Grain-cleaning machine	H. Fitts	Somerset, Mich	Apr. 12, 1859	23, 641
Grain-cleaning machine	D. Flickenger and S. Krim	Hanover, Pa	May 28, 1842	2, 649
Grain-cleaning machine	J. R. Gates	Eckmansville, Ohio	May 19, 1857	17, 325
Grain-cleaning machine	T. C. Gleason	Rochester, N. Y	May 3, 1859	23, 838
Grain-cleaning machine	I. G. Goshen and W. Bowers	Mercersburgh and Chambersburgh, Pa.	June 1, 1858	20, 422
Grain-cleaning machine	M. L. Hall	Bridgeport, Conn	June 1, 1858	20, 425
Grain-cleaning machine	C. B. Horton	Elmira, N. Y	Nov. 18, 1856	16, 088
Grain-cleaning machine	G. H. Johnson	Johnsonburgh, N. Y	Dec. 15, 1838	1, 039
Grain-cleaning machine	G. Mann, jr	Lockport, N. Y	June 29, 1839	1, 211
Grain-cleaning machine	R. C. Mauck	Harrisonburgh, Va	Dec. 18, 1860	30, 978
Grain-cleaning machine	T. McCrea	Anne Arundel County, Md	Aug. 9, 1839	1, 280
Grain-cleaning machine	J. Nott	West Hempfield, Pa	Aug. 11, 1841	2, 220
Grain-cleaning machine	W. H. Orr	Martin's Ferry, Ohio	July 13, 1858	20, 923
Grain-cleaning machine	J. Outram	Elmira, N. Y	Oct. 30, 1860	30, 540
Grain-cleaning machine	W. Partridge and G. W. Shaw	Ellicott's Mills, Md	June 15, 1858	20, 581
Grain-cleaning machine	T. Reese	Baltimore, Md	July 20, 1831	....
Grain-cleaning machine	J. Rood	Beaver Dam, Wis	Sept. 28, 1869	95, 269
Grain-cleaning machine	N. H. Sherburne	Campton, Ill	July 13, 1858	20, 899
Grain-cleaning machine	S. Spangler	Stoney Creek, Pa	May 19, 1840	1, 608
Grain-cleaning machine	E. Sweetland	South Bend, Ind	Mar. 27, 1866	53, 502
Grain-cleaning machine	G. E. Throop	Syracuse, N. Y	Feb. 11, 1873	135, 675
Grain-cleaning machine	B. T. Trimmer	Rochester, N. Y	July 27, 1858	21, 036
Grain-cleaning machine	J. P. Tunison	Ovid, N. Y	Sept. 18, 1860	30, 099
Grain-cleaning machine	J. Tyler	Claremont, N. H	May 18, 1827	....
Grain-cleaning machine	W. H. Waldby	Cooperstown, N. Y	Aug. 19, 1862	36, 250
Grain-cleaning machine	H. Wallace and W. Mellon	North Sewickley, Pa	Mar. 8, 1859	23, 205
Grain cleaning, scouring, and decorticating machine	J. A. Maloney	Georgetown, D. C	Nov. 14, 1871	120, 987
Grain-cleaning screen, Revolving	E. P. Fitzpatrick	Mount Morris, N. Y	Nov. 14, 1835	....
Grain, &c., Cleansing	I. Honeywells	Hillsdale, N. Y	Apr. 27, 1816	....
Grain, coffee, and rice cleaner	A. and I. D. Crawford	Wilkesbarre, Pa., and Bloomington, Ill.	Jan. 31, 1871	111, 323
Grain-conveyer	D. L. Bartlett	Rockford, Ill	Jan. 12, 1869	85, 782
Grain-conveyer	D. W. Bryant	Chicago, Ill	Mar. 21, 1865	46, 876
Grain-conveyer	O. C. Dodge	New York, N. Y	Feb. 10, 1863	37, 615
Grain-conveyer	J. Gardiner	Philadelphia, Pa	Nov. 23, 1869	97, 070
Grain-conveyer	O. Johnson and C. O. Wall	Galva, Ill	Feb. 22, 1870	100, 154
Grain-conveyer	C. Lazarevitch	Brooklyn, N. Y	Dec. 2, 1873	145, 112
Grain-conveyer	J. M. Rush	Marengo, Iowa	Apr. 7, 1868	76, 529
Grain-conveyer	S. W. Wood	Cornwall, N. Y	Dec. 4, 1866	60, 313
Grain-conveyer for elevators	D. C. Chester	Ogdensburgh, N. Y	Apr. 13, 1869	88, 849
Grain-conveying apparatus	H. G. Yates	Kalamazoo, Mich	Nov. 11, 1873	144, 589
Grain-conveying machine	S. W. Wood	Cornwall, N. Y	May 28, 1867	65, 149
Grain-cooler	S. A. Stebbins	Toledo, Ohio	Apr. 19, 1870	102, 175
Grain-cooler	G. D. Woodworth	Chicago, Ill	July 3, 1866	56, 136
Grain cooler and sifter	Armstrong and King	New York, N. Y	May 9, 1835	....
Grain cooler and ventilator	C. D. Clark	Chicago, Ill	Aug. 24, 1858	21, 245
Grain-cradle	J. Babcock	Rensselaerville, N. Y	Mar. 24, 1823	....
Grain-cradle	D. D Devoe	Ilion, N. Y	Apr. 5, 1864	42, 173
Grain-cradle	D. Duesler	Coldwater, Mich	Dec. 30, 1873	145, 933
Grain-cradle	M. R. Flanders	Parishville, N. Y	June 28, 1859	24, 550
Grain-cradle	C. Goss	Madrid, N. Y	Feb. 16, 1843	2, 955
Grain-cradle	I. T. Grant and D. H. Viall	Schaghticoke, N. Y	Oct. 15, 1850	7, 720

Index of patents issued from the United States Patent Office from 1790 *to* 1873, *inclusive*—Continued.

Invention.	Inventor.	Residence.	Date.	No.
Grain-cradle	A. P. Grover	Eureka, Wis	Jan. 12, 1864	41, 215
Grain-cradle	C. P. Kelsey	Livingstonville, N. Y	Nov. 1, 1853	10, 186
Grain-cradle	C. P. Kelsey	Livingstonville, N. Y	Apr. 2, 1867	63, 530
Grain-cradle	J. Leidy	Lamar, Pa	Mar. 16, 1858	19, 643
Grain-cradle	W. H. Locke	Canton, Pa	Apr. 6, 1869	88, 724
Grain-cradle	D. Miffleton	King George Court-House, Va.	Oct. 20, 1857	18, 464
Grain-cradle	D. H. Vaill	Schaghticoke, N. Y	Apr. 30, 1861	32, 215
Grain-cradle	S. D. Warren	Lebanon, Ala	Mar. 10, 1857	16, 815
Grain-cradle	E D. Wilcox	Bellingham, Mass	July 24, 1860	29, 349
Grain-cradle	W. A. Wood and J. C. Loveland	Hoosick Falls, N. Y	Nov. 13, 1844	3, 814
Grain-cradle finger	J. Houghton	Ogden, N. Y	Dec. 17, 1850	7, 840
Grain-cradles, Method of securing braces in the snaths of.	W. W. Bryan	Schaghticoke, N. Y	Dec. 23, 1856	16, 264
Grain, Curing and preserving	F. Dorsett	Chicago, Ill	Aug. 27, 1867	68, 291
Grain-cutter	J. Reeder	Alton, Ill	Jan. 20, 1843	2, 919
Grain-cutter	E. H. Woodsum	Harrison, Me	Apr. 25, 1871	114, 079
Grain cutter and binder	S. D. Locke	Janesville, Wis	Aug. 30, 1870	106, 945
Grain cutting and binding machine	H. Kellogg	McHenry, Ill	May 5, 1857	17, 223
Grain cutting and binding machine	F. Meyer	Naperville, Ill	Aug. 2, 1859	24, 945
Grain, Cutting and cleaning	C. Brown and F. S. Craus	Barton, N. Y	Apr. 6, 1842	2, 542
Grain-dampening apparatus	I. Shellabarger	Decatur, Ill	June 9, 1868	78, 838
Grain-decorticating machine	W. P. Robinson	Buffalo, N. Y	July 9, 1872	128, 912
Grain, Destroying weevil in	W. Watson	Chicago, Ill	May 8, 1849	6, 434
Grain-distilling apparatus	J. W. Beebe and T. F. Lloyd	Albany, N. Y	Dec. 11, 1866	60, 327
Grain-distributer	L. P. Haradon	Independence, Iowa	Jan. 6, 1863	37, 291
Grain-distributer	G. H. Johnson	Buffalo, N. Y	Feb. 2, 1869	86, 410
Grain, Distributer for sowing and drilling	G. A. Titus	Mantorville, Minn	July 20, 1869	92, 904
Grain-distributer, Rotary	A. D. Foote	Berlin, Wis	Nov. 18, 1873	144, 753
Grain-drill	J. B. Amos	Lower Chanceford, Pa	June 13, 1865	48, 145
Grain-drill	T. D. Aylsworth	Pine Bend, Minn	Apr. 28, 1863	38, 275
Grain-drill	G. S. Ball	Dayton, Ohio	Nov. 26, 1872	133, 292
Grain-drill	W. Ballard	Homer, N. Y	Sept. 30, 1862	36, 554
Grain-drill	A. P. Barry	Martinsville, Miss	Oct. 3, 1871	119, 495
Grain-drill	A. G. Bartlett	Oxford, Ohio	Mar. 10, 1849	5, 992
Grain-drill	P. D. Beckwith	Dowagiac, Mich	Nov. 17, 1863	40, 597
Grain-drill	P. D. Beckwith	Dowagiac, Mich	Apr. 30, 1867	64, 274
Grain-drill	H. Beitzell	Centreville, Ind	Oct. 6, 1857	18, 317
Grain-drill	L. Bickford	Macedon, N. Y	Mar. 19, 1867	63, 001
Grain-drill	S. Binkley	Dublin, Ind	Oct. 12, 1858	21, 736
Grain-drill	A. R. Blood and B. Uhrich	Independence, Iowa	Jan. 23, 1872	122, 986
Grain-drill	G. G. Blunt	Bath, Ill	June 17, 1873	139, 860
Grain-drill	A. Brooks	Factoryville, N. Y	Mar. 29, 1864	42, 069
Grain-drill	A. L. and C. P. Brown	Shortsville, N. Y	June 10, 1862	35, 501
Grain-drill	J. Bucknell	Decorah, Iowa	Feb. 9, 1864	41, 474
Grain-drill	W. Bullock	Philadelphia, Pa	Jan. 8, 1850	6, 996
Grain-drill	T. Burns and H. S. Jamison	Greensburgh, Ind	Jan. 31, 1871	111, 421
Grain-drill	C. E. Burton	Gillespie, Ill	Aug. 6, 1872	130, 187
Grain-drill	N. N. Conger	Sugar Valley, Ohio	Apr. 19, 1864	42, 350
Grain-drill	T. R. Cormick	Independence, Iowa	Jan. 12, 1864	41, 199
Grain-drill	J. W. Craven and A. Clark	Centre Valley, Ind	Mar. 28, 1871	113, 022
Grain-drill	C. F. Davis	Auburn, N. Y	Feb. 18, 1868	74, 515
Grain-drill	J. Davis	Allegheny City, Pa	Nov. 14, 1865	50, 914
Grain-drill	H. B. Dean and S. A. Baker	Ludlowville, N. Y	Apr. 27, 1869	89, 386
Grain-drill	J. S. De Haven	North Springfield, Ohio	June 10, 1862	35, 510
Grain-drill	D. L. Dickson	Durham, Ill	June 9, 1868	78, 792
Grain-drill	M. A. Dilley	Mendon, Mich	June 2, 1863	38, 733
Grain-drill	J. W. H. Doubler	Warren, Ill	Mar. 1, 1864	41, 811
Grain-drill	J. W. H. Doubler	Warren, Ill	Feb. 21, 1865	46, 523
Grain-drill	J. Easterday and J. B. Crowell	Frederick County, Md., and Greencastle, Pa.	July 20, 1869	92, 945
Grain-drill	J. B. Edgell, G. P. Martin, H. C. Kellogg, and E. A. Alexander.	Quasqueton, Iowa	Oct. 13, 1863	40, 314
Grain-drill	G. W. Farley	Manchester, N. H	Dec. 26, 1865	51, 707
Grain drill	J. R. Finch	Dayton, Ohio	Mar. 3, 1868	75, 142
Grain-drill	J. P. Floom	Canton, Ohio	Sept. 19, 1871	119, 022
Grain-drill	J. P. Floom	Canton, Ohio	Sept. 19, 1871	119, 023
Grain-drill	J. Ford	Wabash, Ind	Apr. 12, 1859	23, 561
Grain-drill	A. Fravel	La Porte, Ind	June 24, 1856	15, 210
Grain-drill	J. P. Fulghum	Dublin, Ind	Mar. 22, 1870	100, 998
Grain-drill	J. P. Fulghum	Dublin, Ind	Dec. 23, 1873	145, 795
Grain-drill	S. N. Gallup	Macedon, N. Y	Mar. 14, 1871	112, 700
Grain-drill	C. O. Gardiner	Springfield, Ohio	Oct. 6, 1868	82, 822
Grain-drill	J. F. Gibson	Chesnut Level, Pa	Dec. 14, 1869	97, 903
Grain-drill	B. W. Gould	Winterset, Iowa	Feb. 6, 1872	123, 394
Grain-drill	S. Grenell	Mokena, Ill	Oct. 13, 1863	40, 257
Grain-drill	W. N. Hamilton	Odessa, Del	Oct. 27, 1868	83, 491
Grain-drill	W. N. Hamilton	Odessa, Del	Nov. 30, 1869	97, 394
Grain-drill	S. Hart	Fulton, N. Y	Nov. 11, 1873	144, 452
Grain-drill	M. Hayden	Dowagiac, Mich	Feb. 22, 1870	100, 033
Grain-drill	P. F. Hodges	Saint Paul, Minn	Jan. 12, 1869	85, 824
Grain-drill	P. F. Hodges	Saint Paul, Minn	Apr. 6, 1869	88, 710
Grain-drill	A. Hoffman	Half Day, Ill	Apr. 5, 1864	42, 192
Grain-drill	G. Huntington	Norwichville, Canada	Apr. 12, 1864	42, 290
Grain-drill	A. Ingalls	Independence, Iowa	May 12, 1863	38, 485
Grain-drill	J. Ingels	Fayette County, Ind	Nov. 10, 1857	18, 590
Grain-drill	J. Ingels	Milton, Ind	Jan. 6, 1863	37, 345
Grain-drill	J. Ingels	Milton, Ind	June 6, 1865	48, 069
Grain-drill	J. Ingels	Milton, Ind	Mar. 6, 1866	53, 005
Grain-drill	J. Ingels	Milton, Ind	Oct. 4, 1870	107, 915
Grain-drill	S. Jolly	Ripley, Ohio	Nov. 10, 1863	40, 565
Grain-drill	D. J. Jones	Sadsbury, Pa	Aug. 1, 1865	49, 117
Grain-drill	J. D. Jones	Pittsburgh, Pa	Sept. 12, 1865	49, 882
Grain-drill	J. D. Jones	Pittsburgh, Pa	Sept. 26, 1865	50, 134
Grain-drill	J. F. Keller	Hagerstown, Md	Apr. 15, 1873	137, 929
Grain-drill	J. King	Indianapolis, Ind	Apr. 22, 1873	138, 166
Grain-drill	S. L. King and W. Ogden	Owego, N. Y	Aug. 23, 1870	106, 594
Grain-drill	B. Kuhns	Dayton, Ohio	Feb. 16, 1869	87, 052

Index of patents issued from the United States Patent Office from 1790 to 1873, inclusive—Continued.

Invention.	Inventor.	Residence.	Date.	No.
Grain-drill	B. Kuhns	Dayton, Ohio	Nov. 16, 1869	96, 926
Grain-drill	B. Kuhns	Dayton, Ohio	Feb. 14, 1871	111, 853
Grain-drill	B. Kuhns	Dayton, Ohio	July 4, 1871	116, 719
Grain-drill	E. Lake	Davisburgh, Mich	Sept. 17, 1867	68, 996
Grain-drill	G. Leigh	Clinton Station, N. J	Oct. 13, 1863	40, 270
Grain-drill	M. T. Lowth and T. J. Howe	Owatonna, Minn	Oct. 6, 1868	82, 853
Grain-drill	J. T. Lynam	Jeffersonville, Ind	Oct. 27, 1868	83, 517
Grain-drill	S. Markham	Flat Rock, Kans	Aug. 29, 1871	118, 621
Grain-drill	R. Marks and A. C. Behne	Connersville, Ind	July 4, 1871	116, 612
Grain-drill	A. Maschka	Chicago, Ill	Feb. 23, 1864	41, 713
Grain-drill	P. P. Mast and C. O. Gardiner	Springfield, Ohio	Apr. 2, 1872	125, 314
Grain-drill	J. McDonald	Hardin, Ill	Sept. 25, 1866	58, 274
Grain-drill	D. E. McSherry	Dayton, Ohio	Jan. 24, 1871	111, 132
Grain-drill	D. E. McSherry	Dayton, Ohio	July 18, 1871	117, 191
Grain-drill	D. E. McSherry	Dayton, Ohio	Jan. 16, 1872	122, 729
Grain-drill	D. E. McSherry and J. H. Landis.	Dayton, Ohio	Dec. 5, 1871	121, 648
Grain-drill	T. S. Mills	Iberia, Ohio	Apr. 19, 1864	42, 390
Grain-drill	J. Milton	Hillsborough, Va	May 14, 1872	126, 826
Grain-drill	W. H. Moore, jr	Blooming Grove, Ind	Nov. 16, 1869	96, 825
Grain-drill	H. A. and L. B. Myers	Elmore, Ohio	June 25, 1861	32, 641
Grain-drill	H. B. and G. A. Myers	Schoolcraft, Mich	Nov. 3, 1863	40, 495
Grain-drill	M. L. Nickels	Dunlapsville, Ind	June 9, 1868	78, 682
Grain-drill	S. R. Nye	Winchendon, Mass	Feb. 18, 1873	135, 932
Grain-drill	H. Paddack and C. Hollar	Abington, Ind	Apr. 27, 1869	89, 333
Grain-drill	A. Palmer	Brockport, N. Y	June 19, 1849	6, 536
Grain-drill	C. E. Patric	Macedon, N. Y	Dec. 29, 1868	85, 472
Grain-drill	C. E. Patric	Springfield, Ohio	Nov. 5, 1872	132, 889
Grain-drill	C. E. Patric	Springfield, Ohio	Feb. 25, 1873	136, 260
Grain-drill	C. E. Patric and L. Bickford	Macedon, N. Y	Nov. 26, 1867	71, 528
Grain-drill	W. P. Penn	Belleville, Ill	May 24, 1864	42, 871
Grain-drill	W. P. Penn, J. Geiss, and J. Brosius.	Belleville, Ill	June 27, 1865	48, 432
Grain-drill	O. M. Pond	Independence, Iowa	Oct. 20, 1863	40, 359
Grain-drill	D. J. Powers	Madison, Wis	Nov. 18, 1862	36, 966
Grain drill	A. Pritz	Dayton, Ohio	Aug. 17, 1858	21, 212
Grain-drill	H. Pulse	Saint Paul, Ind	Feb. 12, 1867	62, 066
Grain-drill	H. Pulse	Saint Paul, Ind	Nov. 12, 1867	70, 896
Grain-drill	H. Pulse	Waldron, Ind	Mar. 8, 1870	100, 555
Grain-drill	G. A. Pursly	Pittsfield, Ill	June 3, 1873	139, 610
Grain-drill	A. Putnam	Owego, N. Y	June 19, 1866	55, 707
Grain-drill	B. Regan	Miamisburgh, Ohio	Mar. 14, 1865	46, 821
Grain-drill	D. Rentchler	Belleville, Ill	Apr. 4, 1871	113, 564
Grain-drill	M. Rich	Horicon, Wis	Dec. 1, 1863	40, 770
Grain-drill	J. L. Riter	Brownsville, Ind	May 25, 1869	90, 582
Grain-drill	J. L. Riter	Brownsville, Ind	Oct. 11, 1870	108, 294
Grain-drill	J. L. Riter	Brownsville, Ind	May 30, 1871	115, 523
Grain-drill	W. Ross	Oshkosh, Wis	Mar. 29, 1864	42, 112
Grain-drill	J. R. Rude	Liberty, Ind	Feb. 5, 1867	61, 765
Grain-drill	J. R., S. B., and G. W. Rude	Liberty, Ind	Oct. 5, 1869	95, 607
Grain-drill	P. J. Schmitt	Carlinville, Ill	Apr. 20, 1869	89, 247
Grain-drill	P. J. Schmitt	Carlinville, Ill	July 13, 1869	92, 658
Grain-drill	P. and P. J. Schmitt	Waterloo, Ill	Apr. 24, 1866	54, 219
Grain-drill	P. and P. J. Schmitt	Waterloo, Ill	Feb. 5, 1867	61, 879
Grain-drill	S. C. Schofield	Freeport, Ill	Oct. 20, 1863	40, 365
Grain-drill	J. Scovil	Hamburgh, N. Y	Dec. 22, 1863	41, 023
Grain-drill	U. H. Shockley	Litchfield, Ill	Mar. 29, 1864	42, 116
Grain-drill	J. H. Shreiner	Camp Hill, Pa	May 12, 1868	77, 924
Grain-drill	J. H. Shreiner	Camp Hill, Pa	Aug. 25, 1868	81, 547
Grain-drill	B. F. Smith	Unity, Iowa	Oct. 27, 1863	40, 435
Grain-drill	J. Smith	Tiffin, Ohio	June 17, 1862	35, 634
Grain-drill	J. Smith	Tiffin, Ohio	Nov. 18, 1862	36, 969
Grain-drill	E. Steacy	Strasburgh, Pa	June 5, 1849	6, 497
Grain-drill	C. Street	Barre Centre, N. Y	May 29, 1855	12, 973
Grain-drill	J. H. Thomas	Springfield, Ohio	July 26, 1870	105, 866
Grain-drill	J. H. Thomas	Springfield, Ohio	Feb. 28, 1871	112, 299
Grain-drill	J. H. Thomas and P. P. Mast	Springfield, Ohio	Oct. 17, 1865	50, 545
Grain-drill	J. H. Thomas and P. P. Mast	Springfield, Ohio	June 19, 1866	55, 742
Grain-drill	J. H. Thomas and P. P. Mast	Springfield, Ohio	Dec. 7, 1869	97, 729
Grain-drill	J. H. Thomas and P. P. Mast	Springfield, Ohio	Feb. 21, 1871	112, 093
Grain-drill	J. H. Thomas, P. P. Mast, and C. O. Gardiner.	Springfield, Ohio	Aug. 3, 1869	93, 369
Grain-drill	J. H. Thomas, P. P. Mast, and C. O. Gardiner.	Springfield, Ohio	Aug. 3, 1869	93, 370
Grain-drill	W. H. Trimmer	Round Hill, Pa	Oct. 5, 1869	95, 619
Grain-drill	W. W. Tuttle	Gratiot, Wis	Dec. 1, 1863	40, 780
Grain-drill	S. E. Tyler	Horicon, Wis	May 31, 1864	42, 975
Grain-drill	J. G. Vale	Cumberland County, Pa	Dec. 10, 1867	72, 127
Grain-drill	E. Wagoner	Westminster, Md	Nov. 27, 1866	60, 096
Grain-drill	L. N. Warren	Milwaukee, Wis	May 29, 1866	55, 184
Grain-drill	T. W. Watts	Rushville, Ill	Apr. 19, 1864	42, 424
Grain-drill	W. Weusthoff	Dayton, Ohio	July 13, 1869	92, 678
Grain-drill	W. Weusthoff	Dayton, Ohio	July 13, 1869	92, 679
Grain-drill	W. Weusthoff	Dayton, Ohio	Oct. 19, 1869	96, 064
Grain-drill	W. Weusthoff and T. G. Troup	Dayton, Ohio	Mar. 19, 1872	124, 707
Grain-drill	S. H. Wheeler and W. Tuttle, jr	Dowagiac, Mich	Feb. 5, 1867	61, 903
Grain-drill	L. Wight and O. G. Ewings	Whitewater, Wis	Dec. 26, 1871	122, 299
Grain-drill and broadcast sower	J. Ingels	Milton, Ind	Jan. 25, 1870	99, 198
Grain-drill and corn-cultivator, Combined	D. B. Platt	Jeffersonville, Ind	Feb. 7, 1871	111, 675
Grain-drill and corn-dropper	J. D. Sater and T. Barns	Greensburgh, Ind	Mar. 9, 1869	87, 716
Grain-drill and corn-planter, Combined	T. Short	Fairmount, Ill	Dec. 6, 1864	45, 350
Grain-drill and cultivator combined	E. Badlam	Ogdensburgh, N. Y	Oct. 17, 1865	50, 438
Grain-drill and land-roller, Combined	H. D. Palmer	Volena Township, Mich	Sept. 28, 1869	95, 375
Grain-drill attachment	S. Bowman	Camp Hill, Pa	Feb. 2, 1869	86, 354
Grain-drill feed	J. L. Riter	Liberty, Ind	Jan. 25, 1870	99, 238
Grain-drill feeder	M. L. Nickels	Dunlapsville, Ind	Aug. 30, 1870	106, 950
Grain-drill grass-seed attachment	J. P. Fulghum	Dublin, Ind	July 8, 1873	140, 694
Grain-drill-lifting apparatus	C. E. Patric	Macedon, N. Y	Dec. 17, 1867	72, 323

Index of patents issued from the United States Patent Office from 1790 *to* 1873, *inclusive*—Continued.

Invention.	Inventor.	Residence.	Date.	No.
Grain-drill roller	O. F. Momany	Dowagiac, Mich	June 1, 1869	90, 770
Grain-drill, seed-sower, and corn-planter, Combined.	D. Evans	Newton, Iowa	June 29, 1869	91, 925
Grain-drill shoe	P. and P. J Schmitt	Waterloo, Ill	Aug. 4, 1868	80, 671
Grain-drill teeth	L. Brickford	Macedon, N. Y	Apr. 19, 1870	102, 081
Grain-drill teeth	I. H. Palmer	Lodi, Wis	Nov. 3, 1863	40, 497
Grain-drill teeth	C. E. Patric	Macedon, N. Y	Feb. 1, 1870	99, 467
Grain-drill teeth	F. Villard	Mount Eaton, Ohio	Feb. 15, 1870	99, 981
Grain-drill teeth	J. D. Welsh	Farmland, Ind	Jan. 21, 1873	135, 186
Grain-drill tube	S. K. Lighter and T. Harding	Hamilton, Ohio	Nov. 6, 1866	59, 422
Grain-drill tube	S. K. Lighter, T. Harding, and J. Curtis.	Hamilton, Ohio	Aug. 13, 1867	67, 660
Grain-drills, Adjustable drag-bar for	J. D. Jones	Pittsburgh, Pa	Sept. 12, 1865	49, 883
Grain-drills, &c., Cup for	J. P. Zeller	South Bend, Ind	May 2, 1871	114, 384
Grain-drills, Drag-bar and teeth for	J. D. Jones	Pittsburgh, Pa	Sept. 26, 1865	50, 135
Grain-drills, Drag-bar for	J. H. Thomas and P. P. Mast	Springfield, Ohio	June 6, 1865	48, 112
Grain-drills, Fastening tubes in	F. Gardner	Carlisle, Pa	May 3, 1870	102, 531
Grain-drills, Feed-roller for	J. H. Thomas	Springfield, Ohio	Apr. 25, 1871	114, 227
Grain-drills, Seed-box for	C. H. Godfrey	Stewartsville, N. J	June 4, 1867	65, 374
Grain-drills, Seed-coverer for	J. S. Gage	Dowagiac, Mich	Aug. 26, 1862	36, 291
Grain-drills, &c., Seed-feeding device for	A. Schopp	Belleville, Ill	Sept. 21, 1869	95, 149
Grain-drills, Seed-wheel for	W. Weusthoff	Dayton, Ohio	Oct. 19, 1869	96, 063
Grain-drills, Seed-wheel for	W. Weusthoff and C. Schmidt	Dayton, Ohio	July 13, 1869	92, 680
Grain-drills, Vulcanized-rubber tube for	J. R. Bird	Brooklyn, N. Y	May 2, 1871	114, 394
Grain-drilling machine	J. Ingels	Milton, Ind	July 26, 1870	105, 960
Grain-dryer	E. A. Abbott	Baltimore, Md	Mar. 9, 1869	87, 613
Grain-dryer	L. Abbott and J. A. Sherburne	Lewiston, Me., and Boston, Mass.	Jan. 24, 1871	111, 163
Grain-dryer	J. W. Adams	Elyria, Ohio	Oct. 8, 1867	69, 530
Grain-dryer	S. V. Appleby	Spotswood, N. J	Oct. 3, 1871	119, 493
Grain-dryer	J. Babillion	Detroit, Mich	Sept. 13, 1864	44, 150
Grain-dryer	A. H. C. Barber	Clinton, Ill	Jan. 1, 1867	60, 666
Grain-dryer	H. H. Beach	Philadelphia, Pa	June 2, 1863	38, 723
Grain-dryer	H. H. Beach	Rome, N. Y	Nov. 7, 1865	50, 789
Grain-dryer	H. H. Beach	Rome, N. Y	July 17, 1866	56, 348
Grain-dryer	A. Bigelow	Adrian, Mich	May 16, 1848	5, 574
Grain-dryer	H. H. Bingham and J. C. Hunt	Terre Haute, Ind	Mar. 24, 1868	75, 844
Grain-dryer	W. Blakey	Baltimore, Md	Oct. 5, 1869	95, 556
Grain-dryer	W. Blakey	Brooklyn, N. Y	Apr. 18, 1871	113, 840
Grain-dryer	H. Boden	Olney, Ill	July 30, 1867	67, 256
Grain-dryer	D. Bonnell	Oswego, N. Y	Sept. 29, 1868	82, 589
Grain-dryer	A. T. Boon and C. L. Stevens	Galesburgh, Ill	Nov. 7, 1865	50, 792
Grain-dryer	C. H. Booth	Dubuque, Iowa	Jan. 19, 1864	41, 273
Grain-dryer	J. Buckingham	Wethersfield, Conn	Feb. 28, 1871	112, 118
Grain-dryer	J. S. Buell and S. A. W. Marsh	Buffalo, N. Y	Oct. 4, 1864	44, 512
Grain-dryer	J. S. Buell and S. A. W. Marsh	Buffalo, N. Y	Dec. 20, 1864	45, 471
Grain-dryer	H. G. Bulkley	Kalamazoo, Mich	Mar. 2, 1852	8, 769
Grain-dryer	H. G. Bulkley	New York, N. Y	Aug. 20, 1872	130, 564
Grain-dryer	J. Burns	New York, N. Y	May 30, 1871	115, 431
Grain-dryer	J. Burt	Westport, Mass	July 9, 1867	66, 559
Grain-dryer	W. L. Card	Kankakee, Ill	July 9, 1872	128, 707
Grain-dryer	C. F. Chichester	Brooklyn, N. Y	Dec. 5, 1871	121, 588
Grain-dryer	L. S. Chichester	New York, N. Y	Aug. 25, 1863	39, 630
Grain-dryer	L. S. Chichester	Brooklyn, N. Y	Mar. 15, 1864	41, 905
Grain-dryer	L S. Chichester	Brooklyn, N. Y	May 2, 1865	47, 596
Grain-dryer	L. S. Chichester	Brooklyn, N. Y	Aug. 15, 1865	49, 470
Grain-dryer	L. S. Chichester	Brooklyn, N. Y	June 18, 1867	65, 793
Grain-dryer	L. S. Chichester	Brooklyn, N. Y	Feb. 25, 1868	74, 893
Grain-dryer	L. S. Chichester	Brooklyn, N. Y	Oct. 20, 1868	83, 255
Grain-dryer	L. S. Chichester	Brooklyn, N. Y	Jan. 10, 1871	110, 896
Grain-dryer	G. Clark	Buffalo, N. Y	Apr. 19, 1864	42, 348
Grain-dryer	G. Clark	Buffalo, N. Y	Apr. 3, 1866	53, 575
Grain-dryer	G. Clark	Buffalo, N. Y	July 31, 1866	56, 708
Grain-dryer	C. R. Coe	Bloomington, Ill	Nov. 26, 1872	133, 410
Grain-dryer	M. C. Cogswell and A. G. Williams.	Buffalo, N. Y	Feb. 18, 1862	34, 405
Grain-dryer	M. C. Cogswell and A. G. Williams.	Buffalo, N. Y	Aug. 11, 1863	39, 466
Grain-dryer	J. De Bary	Offenbach, Germany	Mar. 10, 1868	75, 386
Grain-dryer	W. H. Dole and D. R. Fraser	Chicago, Ill	Sept. 1, 1863	39, 722
Grain-dryer	M. R. Dudley	New Orleans, La	Apr. 30, 1850	7, 321
Grain-dryer	T. W. Eaton	Kankakee, Ill	June 13, 1871	115, 833
Grain-dryer	J. R. Evertson	Mount Vernon, Ind	Nov. 27, 1866	59, 989
Grain-dryer	J. R. Evertson	Mount Vernon, Ind	Oct. 15, 1867	69, 785
Grain-dryer	E. S. Forgy	Dayton, Ohio	Apr. 8, 1873	137, 606
Grain-dryer	S. Godfrey	Peoria, Ill	Mar. 20, 1866	53, 293
Grain-dryer	R. Grotz	Chicago, Ill	Aug. 31, 1869	94, 308
Grain-dryer	J. Guardiola	Chocola, Central America	May 7, 1872	126, 455
Grain-dryer	H. B. Hebert	New York, N. Y	Dec. 28, 1869	98, 374
Grain-dryer	R. Heneage	Buffalo, N. Y	Aug. 29, 1865	49, 626
Grain-dryer	R. Heneage	Buffalo, N. Y	Sept. 25, 1866	58, 254
Grain-dryer	H. Henley and J. A. Reinhart	Shoals and Loogootee, Ind	Dec. 15, 1868	85, 007
Grain-dryer	J. Hollingsworth	New York, N. Y	Sept. 24, 1872	131, 682
Grain-dryer	J. R. Hoopes	West Philadelphia, Pa	Aug. 27, 1850	7, 601
Grain-dryer	W. Hull and C. W. Hammond	Baltimore, Md	Sept. 28, 1869	95, 351
Grain-dryer	G. H. Johnson and G. Milsom	Buffalo, N. Y	Jan. 1, 1869	90, 847
Grain-dryer	S. C. Kenaga	Kankakee, Ill	Dec. 7, 1869	97, 650
Grain-dryer	E. Knaur and S. Beaver, jr	Valley Forge and Great Valley, Pa.	Apr. 18, 1848	5, 517
Grain-dryer	C. W. T. Krausch	Chicago, Ill	Mar. 10, 1863	37, 869
Grain-dryer	S. Marsh	Chicago, Ill	May 31, 1864	42, 957
Grain-dryer	S. Marsh	Chicago, Ill	July 4, 1865	48, 573
Grain-dryer	A. W. J. Mason	New Orleans, La	Sept. 12, 1871	118, 954
Grain-dryer	H. Merrill	Brooklyn, N. Y	Nov. 22, 1870	109, 536
Grain-dryer	F. H. C. Mey	Buffalo, N. Y	Apr. 12, 1864	42, 333
Grain-dryer	F. H. C. Mey	Buffalo, N. Y	May 28, 1872	127, 256
Grain-dryer	F. H. C. Mey and H. W. Dopp	Buffalo, N. Y	Dec. 3, 1872	133, 590
Grain-dryer	R. Milbourne and T. Browning	Whitechapel, Great Britain	July 26, 1870	105, 709
Grain-dryer	C. W. Mills and L. S. Chichester	Brooklyn, N. Y	Sept. 22, 1868	82, 432
Grain-dryer	W. F. Morgan	Buffalo, N. Y	July 22, 1873	141, 160

Index of patents issued from the United States Patent Office from 1790 *to* 1873, *inclusive*—Continued.

Invention.	Inventor.	Residence.	Date.	No.
Grain-dryer	J. J. Munger	Syracuse, N. Y	Feb. 2, 1869	86, 436
Grain-dryer	I. Y. Munn	Chicago, Ill	Apr. 12, 1870	101, 759
Grain-dryer	J. Norton, 3d	Chicago, Ill	July 10, 1866	56, 255
Grain-dryer	J. H. Pattee and E. S. Cleveland	Galva, Ill	Dec. 5, 1865	51, 348
Grain-dryer	R. S. Reynolds	New Haven, Conn	June 23, 1863	38, 985
Grain-dryer	A. Robert	Paris, France	May 23, 1871	115, 240
Grain dryer	A. W. Roper	Glasgow, Mo	Sept. 17, 1872	131, 464
Grain-dryer	P.C. Schuyler and S.W. Warren	New York, N. Y	June 2, 1863	38, 790
Grain-dryer	S. Schuyler	Brooklyn, N. Y	Jan. 22, 1861	31, 191
Grain-dryer	H. H. Scoville	Chicago, Ill	Sept. 26, 1848	5, 816
Grain-dryer	C. S. Snead	Louisville, Ky	Oct. 1, 1850	7, 690
Grain-dryer	A. Soper	New York, N. Y	Nov. 18, 1873	144, 710
Grain-dryer	J. Souter	Chicago, Ill	June 28, 1859	24, 589
Grain-dryer	H. Spendelow and R. Heneage	Buffalo, N. Y	Sept. 15, 1868	82, 170
Grain-dryer	J. R. Stafford	Cleveland, Ohio	Apr. 18, 1848	5, 518
Grain-dryer	J. R. Stafford	Cleveland, Ohio	Apr. 18, 1848	5, 519
Grain-dryer	J. R. Stafford	Cleveland, Ohio	Apr. 18, 1848	5, 524
Grain-dryer	W. Standing	Cairo, Ill	July 7, 1868	79, 699
Grain-dryer	W. Stark and J. G. Fisher	Toledo, Ohio	Mar. 28, 1871	113, 219
Grain-dryer	S. M. Stevens	Elwood, Ill	Oct. 5, 1869	95, 616
Grain-dryer	J. E. Strode	Litchfield, Ill	Feb. 19, 1867	62, 233
Grain-dryer	B. F. Sturtevant	Jamaica Plain, Mass	Feb. 22, 1870	100, 212
Grain-dryer	B. F. Sturtevant	Jamaica Plain, Mass	Feb. 22, 1870	100, 235
Grain-dryer	R. T. Sutton	Rochester, N. Y	Dec. 22, 1863	41, 028
Grain-dryer	R. T. Sutton	Rochester, N. Y	Mar. 15, 1864	41, 947
Grain-dryer	R. T. Sutton	Rochester, N. Y	Jan. 2, 1866	51, 880
Grain-dryer	W. H. Sutton and J. J. Gibson	Bantford, Canada	Sept. 29, 1863	40, 130
Grain-dryer	E. Trenholm	Washington, D. C	May 19, 1863	38, 614
Grain-dryer	H. Walker	Detroit, Mich	Jan. 26, 1869	86, 192
Grain-dryer	T. Wallace	Chicago, Ill	Oct. 6, 1863	40, 219
Grain-dryer	T. E. Weed	Williamsburgh, N. Y	Feb. 24, 1852	8, 764
Grain-dryer	J. B. Wheeler	Bolton, Mass	Apr. 14, 1863	38, 191
Grain-dryer	R. J. Williams	Ottumwa, Iowa	Aug. 26, 1873	142, 184
Grain-dryer	H. Wood, G. A. Fourdinier, and R. Haselden.	Montreal, Lyn, and Montreal, Canada.	Feb. 23, 1864	47, 748
Grain-dryer	M. H. Wright	Chicago, Ill	Feb. 5, 1867	61, 909
Grain dryer, cooler, and scourer	L. S. and C. F. Chichester	Brooklyn, N. Y	Dec. 12, 1871	121, 847
Grain-dryer furnace	M. C. Coggswell and A. G. Williams.	Buffalo, N. Y	Sept. 29, 1863	40, 090
Grain-dryer, Vacuum	G. Clark	Buffalo, N. Y	Dec. 10, 1867	71, 981
Grain-dryers, Endless bands for	J. Massey	New York, N. Y	Apr. 17, 1849	6, 322
Grain, Drying	J. Babillion	Detroit, Mich	Mar. 3, 1863	37, 800
Grain, Drying	J. H. Patten	New York, N. Y	June 19, 1849	6, 550
Grain, Drying	H. Quinn	New Alexandria, N. J	Mar. 10, 1849	6, 163
Grain drying and cleaning machine	A. Tonnar	Prussia, Germany	July 31, 1866	56, 868
Grain drying and cooling machine	J. B. Wheeler	Chicago, Ill	Oct. 23, 1860	30, 512
Grain drying and renovating process	G. H. Johnson and G. Milsom	Buffalo, N. Y	Feb. 9, 1869	86, 758
Grain-drying apparatus	C. W. T. Krausch	Chicago, Ill	Mar. 31, 1863	38, 048
Grain, &c., drying apparatus	S. Marsh	Chicago, Ill	Feb. 10, 1863	37, 632
Grain-drying apparatus	T. F. Rowland J. Stephens, and W. H. Mason.	Brooklyn, N. Y	May 22, 1855	12, 922
Grain-drying machine	E. I. Bodrio	Saint Louis, Mo	Oct. 2, 1860	30, 196
Grain, &c., drying machine	C. C. Custer	Philadelphia, Pa	Sept. 8, 1859	18, 137
Grain-drying machine	R. Else	Middlesex County, England	Aug. 14, 1839	1, 286
Grain-drying machine	T. H. McCulloch	Peoria, Ill	Apr. 17, 1860	27, 917
Grain-drying machine	S. B. Robinson	Oswego, N. Y	June 20, 1854	11, 140
Grain-drying machine, Wet	S. V. Appleby	New York, N. Y	Apr. 8, 1856	14, 588
Grain-dump	W. M. Hall, jr	Bloomington, Ill	Sept. 6, 1870	107, 040
Grain, Dumping	J. Sypes	Fairbury, Ill	Oct. 10, 1871	119, 799
Grain elevating and bagging apparatus	J. S. Hasbrouck	Tyre, N. Y	Mar. 15, 1870	100, 887
Grain elevating, cleaning, and bagging machine	I. A. Stafford	Essex, N. Y	Oct. 23, 1860	30, 506
Grain, Elevating, conveying, storing, and shipping	H. I., P. F., and E. D. Chase	Peoria, Ill	May 20, 1873	138, 994
Grain-elevating machine	J. Bruckshaw, H. Bruckshaw, and W. S. Underhill.	Oakley, Market Drayton, and Newport, England.	Nov. 5, 1861	33, 634
Grain elevating, measuring, registering, and bagging machine.	P. Barker	North Adams, Mich	June 1, 1858	20, 399
Grain-elevator	T. H. Green	Fond du Lac, Wis	Sept. 4, 1860	29, 881
Grain-elevator	C. W. T. Krausch	Chicago, Ill	Mar. 24, 1863	38, 004
Grain-elevator	H. Merrill	Brooklyn, N. Y	Apr. 22, 1873	138, 175
Grain-elevator	J. F. Moulton	Chicago, Ill	June 21, 1864	43, 222
Grain-elevator	J. Nichols	Fond du Lac, Wis	Dec. 20, 1864	45, 515
Grain-elevator	A. B. Nimbs	Buffalo, N. Y	Dec. 6, 1864	45, 336
Grain-elevator	A. B. Nimbs	Buffalo, N. Y	Dec. 6, 1864	45, 337
Grain-elevator	J. T. Parlour	Buffalo, N. Y	May 15, 1866	54, 822
Grain-elevator	F. Taggart, L. S. Chichester, and C. W. Mills.	Brooklyn, N. Y	June 27, 1865	48, 495
Grain elevator and dryer, Floating	J. W. Sykes	Chicago, Ill	Apr. 15, 1862	34, 992
Grain elevator and dumping apparatus	M. Eldridge and F. A. Reed	Alexandria, Va	Apr. 23, 1867	64, 002
Grain elevator and feeder	B. F. Sherman	San Francisco, Cal	Oct. 22, 1867	70, 126
Grain-elevator, Floating	A. B. Nimbs	Buffalo, N. Y	Aug. 12, 1862	36, 193
Grain elevator, Pneumatic	S. W. Wood	Cornwall, N. Y	June 16, 1868	79, 044
Grain elevator, Portable wet	D. W. Kellogg and J. W. McKee.	Buffalo, N. Y	May 10, 1864	42, 733
Grain-elevator, Adjustable gate for	S. Brown	Utica, Ohio	Apr. 27, 1869	89, 284
Grain-elevators, Clearing guard of	G. Mann, jr	Ottawa, Ill	May 12, 1857	17, 289
Grain-elevators, Yoke for	T. E. Jewell	Brooklyn, N. Y	Nov. 12, 1867	70, 855
Grain-fan and corn-sheller	H. E. Smith	Philadelphia, Pa	Jan. 18, 1859	22, 689
Grain-fan blast	J. and W. H. Butterworth	Trenton, N. J	Apr. 13, 1869	88, 843
Grain fanning and assorting machine	R. Nutting	Randolph, Vt	Aug. 10, 1858	21, 144
Grain fanning and separating machine	J. I. Smith and W. H. Nicodemus.	Frederick, Md	June 28, 1870	104, 784
Grain-feeder, Revolving	M. Decamp	South Bend, Ind	Apr. 24, 1866	54, 126
Grain feeding and scouring apparatus	J. W. Ardinger	Mount Pulaski, Ill	Jan. 26, 1869	86, 199
Grain, Fermenting alcoholic liquors from	J. Ellinwood	Owensborough, Ill	Nov. 1, 1870	108, 773
Grain, Filling bags with	L. Eggleston	Battle Creek, Mich	Oct. 28, 1862	36, 772
Grain, Finger for lifting lodged	S. Manning	San José, Cal	Sept. 17, 1867	69, 004
Grain, flour, &c., Apparatus for drying	J. R. Stafford	Cleveland, Ohio	Aug. 14, 1847	5, 238
Grain for distillation, Mashing	M. Thompson	Saint Louis, Mo	July 20, 1869	92, 901

Index of patents issued from the United States Patent Office from 1790 to 1873, inclusive—Continued.

Invention.	Inventor.	Residence.	Date.	No.
Grain for distillation, Preparing	T. Sim and E. S. Hutchinson	Baltimore, Md	Mar. 16, 1869	87, 980
Grain for distillation, Preparing	H. Tauszky	New York, N. Y	Nov. 16, 1869	96, 998
Grain for distillation, Preparing and treating	M. Thompson	Saint Louis, Mo	June 7, 1864	43, 047
Grain for distillation, Process for preparing	J. Chilcott	Brooklyn, N. Y	Apr. 25, 1865	47, 394
Grain for distillation, Process of preparing	J. Fleischman	New York, N. Y	Jan. 3, 1865	45, 793
Grain for flouring, Preparing	I. W. Howlet and F. M. Walker	Greensborough, N. C	Oct. 14, 1846	4, 813
Grain for food, Preparing	R. B. Fitts	Philadelphia, Pa	June 13, 1871	115, 947
Grain for food. Preparing wheat, corn and other	R. B. Fitts	Philadelphia, Pa	July 18, 1871	117, 165
Grain for grinding, Apparatus and process for preparing.	A. S. Sackett	Rochester, Minn	July 30, 1872	129, 906
Grain for grinding, Preparing	L. S. Chichester	New York, N. Y	Apr. 2, 1872	125, 170
Grain for grinding, Process of preparing	A. E. Wells	Jamestown, N. Y	Sept. 17, 1872	131, 379
Grain, &c, for malting, Preparing	S. Weidenfield	New York, N. Y	Apr. 28, 1863	38, 347
Grain for manufacture into flour, Process for preparing.	O. F. Cook	Grand Island, Cal	Dec. 23, 1873	145, 846
Grain for mashing, Process of preparing	F. W. De Spessbourg	Normandy, France	Sept. 7, 1869	94, 479
Grain-fork	W. W. Bryan	Schaghticoke, N. Y	Aug. 23, 1870	106, 656
Grain-fork	E. G. Bullis	Manchester, Iowa	Dec. 10, 1867	71, 971
Grain-fork	H. M. and W. W. Bruson	Atkinson, Ill	Aug. 11, 1863	39, 464
Grain-fork	A. Clow	Port Byron, N. Y	Sept. 3, 1867	68, 352
Grain-fork	M. C. Remington	Auburn, N. Y	July 10, 1866	56, 328
Grain fork and rake	M. M. Wells	Hartwick, N. Y	Feb. 25, 1873	136, 347
Grain fork, Loose	H. Gary	Croton, N. J	Nov. 23, 1869	97, 187
Grain from garlic, Machine for separating	H. Staub	Shepherdstown, Va	Nov. 11, 1837	460
Grain from smut, Machine for cleansing	L. Lee and C. Maston	Penn Yan, N. Y	Apr. 25, 1826	....
Grain from straw, &c., Machine for clearing	W. Tunstall	....	June 30, 1804	....
Grain from straw, Machine for extracting	C. Hoxie	....	Aug. 20, 1801	....
Grain from straw, Machine for separating	M. and C. B. Packard	Clarendon, N. Y	Nov. 18, 1844	3, 832
Grain, &c., from vessels, Apparatus for discharging	J. Pagin	Buffalo, N. Y	May 8, 1847	5, 101
Grain from wagons, Apparatus for unloading	C. S. Dole	Chicago, Ill	Jan. 5, 1869	85, 573
Grain-gage	H. Haak	Myerstown, Pa	Oct. 30, 1866	59, 213
Grain-gate	G. Seitzinger	Ottawa, Ill	Aug. 21, 1866	57, 445
Grain-gatherer	W. Herries	Fayette, N. Y	Mar. 13, 1849	6, 183
Grain gathering and binding machine	M. L. Baker	Mannsville, N. Y	Apr. 15, 1862	34, 934
Grain, grass, &c., Machine for cutting	C. Taylor	McKeesport, Pa	June 26, 1855	13, 143
Grain, grass-seed, rice, &c., from straw, Separating	A. Look and W. Coleman, jr	Fredericktown, Md	Mar. 21, 1832	....
Grain-grinder, Conical	D. and J. Fitzgerald	New York, N. Y	Oct. 30, 1834	....
Grain grinder, scourer, and huller	D. Parmelee and J. Mooers	New Troy, Pa	June 12, 1832	....
Grain, Grinding	J. Bicknell	Buckfield, Me	Dec. 11, 1822	....
Grain, Grinding	M. W. Chapman	Girard, Pa	June 1, 1833	....
Grain, Grinding	B. and J. C. Langdon, and J. Trash	Troy, N. Y	Sept. 26, 1823	....
Grain, Grinding	D. Parmelee and J. C. Shoemaker	Kingston, Pa	Apr. 16, 1833	....
Grain-grinding machine	O. Wyman	East Cambridge, Mass	Dec. 20, 1837	531
Grain, hemp, &c., Band for binding	A. Ralston	West Middletown, Pa	Apr. 5, 1859	23, 496
Grain-huller	W. C. Grimes	Philadelphia, Pa	Feb. 20, 1866	52, 706
Grain-huller	M. Hoffmann	Munich, Bavaria	Apr. 18, 1871	113, 770
Grain-huller	B. Maltby and M. Fowler	New Haven, Conn	Feb. 28, 1818	....
Grain-huller	J. H. Thompson	Hoboken, N. J	June 20, 1865	48, 326
Grain-huller	R. W. Van Peyma	Lancaster, N. Y	Feb. 20, 1866	52, 774
Grain huller and cleaner	O. Lull	Otsego, N. Y	Aug. 25, 1832	....
Grain hulling and cleaning machine	J. Cross	Centre Liste, N. Y	Oct. 8, 1838	968
Grain hulling and cleaning machine	G. H. Rice	Kingston, Pa	June 25, 1872	128, 424
Grain hulling and scouring machine	T. F. Wagoner	Trenton, N. J	June 28, 1859	24, 595
Grain hulling, cleaning, and grinding machine	C. F. Campbell	Chenango, N. Y	Sept. 24, 1861	33, 336
Grain hulling cotton-seed, rubbing down whetstones, &c., Grinding.	B. N. Fyler	Bradford, Vt	Apr. 15, 1830	....
Grain-hulling machine	C. O. Bullot	Santa Rosa de los Andes, Chili.	Aug. 1, 1865	49, 205
Grain-hulling machine	S. Gardner and A. B. Howe	New York, N. Y	Dec. 12, 1865	51, 445
Grain-hulling machine	W. J. Hart	Talmadge, Ohio	May 21, 1824	....
Grain-hulling machine	F. Henckel and W. Seck	Munich, Bavaria	Nov. 20, 1866	59, 838
Grain-hulling machine	A. Hubbell	Sharon, Conn	Mar. 24, 1868	75, 761
Grain-hulling machine	J. A. Welsh	Xenia, Ohio	Aug. 28, 1866	57, 605
Grain in a vessel's hold, Apparatus for leveling	H. Milsom, H. Spendelow, and G. V. Watson.	Buffalo, N. Y	Mar. 28, 1865	47, 030
Grain in bulk, Treating and preserving	C. F. Parrott	New York, N. Y	July 26, 1870	105, 720
Grain in bundles, Machine for binding	J. D. Osborn	Constantine, Mich	June 14, 1859	24, 400
Grain in bundles or sheaves, Mode of securing	J. P. Manny	Rockford, Ill	July 6, 1858	20, 809
Grain in elevator-bins, Device for distributing	W. O. Strong	Detroit, Mich	Feb. 11, 1862	34, 379
Grain in mills, &c., Distributing	C. S. Hamilton	Fond du Lac, Wis	June 21, 1864	43, 200
Grain in process of grinding, Steaming	J. F. Lawton	Venedy, Ill	Sept. 20, 1864	44, 322
Grain in the mass, Apparatus for drying	J. C. Pedrick	Washington, D. C	Dec. 16, 1856	16, 259
Grain into flour and meal, Machinery of mills for the manufacture of.	W. Parkinson	Wheeling, Va	July 31, 1829	....
Grain into gavels, Device for gathering	W. M. Waggoner	Middletown, Ind	Jan. 18, 1859	22, 682
Grain-kiln	R. Andrews	Fleming County, Ky	Nov. 7, 1822	....
Grain-kiln	I. S. Stover	Erwinna, Pa	Nov. 11, 1851	8, 518
Grain, Kiln for drying	J. Abraham	Middlesex, N. J.	Dec. 20, 1805	....
Grain, Kiln for drying	W. W. Alcott	Boston, Mass	Sept. 26, 1846	4, 774
Grain, Kiln for drying	T. Cook	New Hope, Pa	Nov. 30, 1835	....
Grain, Kiln for drying	J. Deneale, jr	....	July 10, 1800	....
Grain, Kiln for drying	A. Greenleaf, jr., and T. C. Vice	Brooklyn, N. Y., and New Haven, Conn.	May 5, 1863	38, 382
Grain, Kiln for drying	H. Y. and A. Haupt, jr	Bucks County, Pa	Dec. 5, 1842	2, 867
Grain, Kiln for drying	S. Howard	Adrian, Mich	Sept. 26, 1846	4, 777
Grain, Kiln for drying	W. Poole	Wilmington, Del	Dec. 20, 1805	....
Grain, Kiln for drying	N. F. Potter	Providence, R. I	Dec. 11, 1847	5, 391
Grain, Kiln for drying	J. H. Tower	Kirkland, N. Y	June 20, 1848	5, 638
Grain, Kiln for drying	N. Wallaster	Detroit, Mich	Oct. 23, 1866	59, 102
Grain, Machine for cleaning and separating garlic, &c., from.	J. Heygel	Salisbury, Pa	Sept. 25, 1841	2, 270
Grain, Machine for cutting the bands of	A. J. Luckey	Bradford, Wis	Apr. 15, 1862	34, 973
Grain, Machine for decorticating and drying	E. Skelly	Plaquemine, La	Aug. 23, 1870	106, 627
Grain by wind power, Machine for grinding	D. McColler	Hudson, Ohio	May 2, 1829	....
Grain, Machine for registering measured	A. Rakestraw and W. Colwell	Chillicothe, Ill	Dec. 13, 1864	45, 436
Grain, Machine for separating	L. Butler	Cobleskill, N. Y	June 3, 1837	225
Grain, Machine for separating and scouring	D. S. Mackey	Batavia, N. Y	Sept. 18, 1860	30, 072
Grain, Machine for separating garlic from	P. C. Fritz	Barrytown, N. Y	Dec. 21, 1858	22, 359

Index of patents issued from the United States Patent Office from 1790 *to* 1873, *inclusive*—Continued.

Invention.	Inventor.	Residence.	Date.	No.
Grain, Machine for separating garlic from	W. C. Grimes	York, Pa	Sept. 13, 1840	1, 763
Grain, Machine for separating straw from	E. S. Snyder	Charlestown, Va	June 3, 1848	5, 630
Grain, Machine for sowing and drilling	G. A. Titus	Mantorville, Minn	July 20, 1869	92, 903
Grain, malt, &c., Apparatus for drying	S. R. Andres, S. Andres, and M. Bucklin.	Troy and New York, N. Y	Jan. 18, 1859	22, 614
Grain, malt, &c., Drying	S. Marsh	Chicago, Ill	Jan. 13, 1863	37, 403
Grain, Malting	R. D'Heureuse	New York, N. Y	Feb. 8, 1870	99, 541
Grain mash, wort, and beer after fermentation	N. Eisendrath	Chicago, Ill	June 1, 1869	90, 827
Grain, meal, &c., Elevating	J. Baily	Kennett's Square, Pa	June 19, 1812	......
Grain-meter	T. Brockett and J. J. Brown	Davenport, Iowa	Nov. 16, 1869	96, 878
Grain-meter	F. G. Chesman	Lemont, Ill	July 19, 1870	105, 427
Grain-meter	J. T. Keeling	Hibernia, Mo	May 24, 1870	103, 470
Grain-meter	A. McBride	Fayette, Pa	July 16, 1872	129, 045
Grain-meter	H. Pooley, T. Roberts, and E. O'Brien.	Liverpool, England	Apr. 9, 1872	125, 612
Grain-meter	R. Rutherford	Belleville, Ill	May 14, 1872	126, 749
Grain-meter	W. S. Sampson	New York, N. Y	Apr. 17, 1866	54, 072
Grain-meter	W. Schnebley	Hackensack, N. J	Aug. 28, 1866	57, 560
Grain meter	J. C. Walker	Waco Village, Tex	Sept. 3, 1867	68, 472
Grain, Method of unloading	I. H. Palmer	Lodi, Wis	June 18, 1867	65, 826
Grain-mill	E. H. Austin	Scott's Hill, Tenn	Sept. 5, 1871	118, 672
Grain-mill	G. W. Grader and B. F. Cowan	Memphis, Tenn	Dec. 19, 1854	12, 096
Grain-mill	J. J. Johnston	Allegheny, Pa	Feb. 2, 1858	19, 251
Grain-mill	H. Mellish	Walpole, N. H	Aug. 22, 1854	11, 558
Grain-mill	J. Norman	New Orleans, La	Sept. 7, 1869	94, 635
Grain-mill	P. Perry	Troy, N. Y	Feb. 23, 1858	19, 441
Grain-mill	J. C. Smith	Lacon, Ill	Sept. 25, 1837	397
Grain-mill	E. G. Ward	New York	Feb. 20, 1841	1, 988
Grain-mill	W. Westrup	Wapping, England	June 6, 1854	11, 011
Grain-mill	J. W. Wheeler	Cleveland, Ohio	Feb. 21, 1860	27, 250
Grain-mill	O. Wyman	Watertown, Mass	July 1, 1836	......
Grain, Mill for cutting	W. Gerrish	Portsmouth, N. H	Jan. 11, 1836	......
Grain, Mill for hulling, grinding, and sifting	T. Ellicot	Pennsylvania	Feb. 17, 1816	......
Grain-mill, Portable	C. Leavitt	Quincy, Ill	July 6, 1852	9, 096
Grain-mill, Portable	C. Leavitt	Quincy, Ill	Feb. 27, 1855	12, 461
Grain-mills, Dress of grinding-surfaces for	O. W. Stanford	Cincinnati, Ohio	Aug. 11, 1857	17, 985
Grain-mills, Feeder for	M. Decamp	South Bend, Ind	Dec. 10, 1867	71, 857
Grain-mills, Feeding and cooling device for	J. Nairn	Milton, Ind	Oct. 6, 1868	82, 741
Grain-mills, Shoe for	H. Mellish	Walpole, N. H	May 29, 1855	12, 959
Grain-moistener	L. J. Adams and J. H. Esale	Avon, Ill	Sept. 8, 1868	81, 861
Grain-mover	W. Moses	Buffalo, N. Y	Jan. 30, 1866	52, 309
Grain, nuts, &c., Machine for cleaning	J. Johnson	New York, N. Y	May 27, 1873	139, 248
Grain of wood, Transferring the natural	J. R. Cross	Morrisania, N. Y	July 2, 1872	128, 468
Grain, paints, &c., Machine for grinding	H. Averill	Richland, N. Y	Apr. 30, 1829	......
Grain, plaster, &c., for sowing, Apparatus for carrying.	O. W. Smith	Flint, Mich	Aug. 9, 1870	106, 216
Grain, Portable mill for grinding	E. Arnold	Otego, N. Y	Mar. 9, 1844	3, 468
Grain preparatory to grinding, Steaming	B. F. Broomell	London Grove, Pa	May 14, 1850	7, 358
Grain, Preparing and washing	C. H. Frings	Centreton, Mo	Apr. 25, 1871	113, 997
Grain, Process for mashing	E. J. Hainault	Belgium	Dec. 9, 1856	16, 179
Grain-rake	W. Baldwin	Plymouth, Conn	July 7, 1868	79, 720
Grain-rake	W. Gates	Barre Centre, N. Y	Dec. 20, 1843	3, 386
Grain-rake	D. Harkness	Raisin, Mich	Mar. 14, 1846	4, 422
Grain-rake	E. Palmer	Solon, N. Y	Aug. 27, 1867	68, 227
Grain-rake	B. F. Partridge	Onondaga, N. Y	Mar. 13, 1844	3, 482
Grain-rake	E. G. Warner	Union Township, Ohio	July 4, 1865	48, 606
Grain-rake	R. Warnock and C. Abbey, 2d	Ridgeville, Ohio	July 28, 1868	80, 523
Grain-rake	M. M. Wells	Hartwick, N. Y	May 5, 1868	77, 556
Grain-rake	M. M. Wells	Hartwick, N. Y	Dec. 27, 1870	110, 613
Grain rake and loader	G. S. Dudley	Dixon, Cal	May 27, 1873	139, 303
Grain raking and binding apparatus	T. K. Griffith	Redstone, Pa	Aug. 16, 1870	106, 357
Grain raking and binding machine	A. S. Harding	Mount Hope, N. Y	Apr. 15, 1862	34, 999
Grain raking and binding machine	J. E. Heath	Warren, Ohio	July 22, 1850	7, 520
Grain, Reaping	C. H. McCormick	Rockbridge County, Va	June 21, 1834	......
Grain, Reaping and cutting	O. Hussey	Cincinnati, Ohio	Dec. 31, 1833	......
Grain reaping and thrashing machine	L. B. Lathrop	San José, Cal	May 25, 1869	90, 561
Grain-reaping machine	S. Adams		Dec. 28, 1805	
Grain-register	G. D. Denison	Troy, Ohio	Jan. 7, 1862	34, 052
Grain-register	L. O. Hayworth	New Cumberland, Ind	Mar. 12, 1867	62, 845
Grain-register	L. W. Hines, G. Saterlee, and S. W. Harden.	Quasqueton, Iowa	Sept. 30, 1862	36, 565
Grain-register	W. C. Howard	Belle Plaine, Iowa	July 27, 1869	92, 971
Grain-register	S. Hudson	Milford, Mich	May 26, 1868	78, 208
Grain-register	C. Lamb	Binghamton, N. Y	Nov. 22, 1870	109, 522
Grain-register	G. W. Nesmith	Metamora, Ill	Oct. 7, 1873	143, 527
Grain-register	M. W. and G. W. Nesmith	Metamora, Ill	Mar. 3, 1868	75, 187
Grain-register	J. S. Price and P. Miller	Champaign County and Clarke County, Ohio.	Jan. 14, 1873	134, 928
Grain-register	S. Ramer	Winona, Minn	Feb. 11, 1873	135, 845
Grain-register	B. Taylor	Forestville, Minn	July 28, 1868	80, 372
Grain-register	W. Z. Taylor	Burlington, Iowa	Dec. 27, 1870	110, 605
Grain-register	T. N. Wheeler	Rio, Wis	July 29, 1862	36, 045
Grain-register	J. T. Wiley	Clayton, Iowa	Sept. 5, 1865	49, 817
Grain-registering device	W. H. Stinson	Newbern, Iowa	May 9, 1871	114, 723
Grain-regulator	E. W. Hitchings	Potsdam, N. Y	Feb. 25, 1868	74, 824
Grain, rice, coffee, &c., Machine for hulling and scouring.	L. H. Whitney	Washington, D. C	Sept. 20, 1870	107, 644
Grain-riddle	A. Rowe	Atalissa, Iowa	Dec. 4, 1860	30, 840
Grain, Riddle for separating	M. M. Cooper and J. W. Donaldson.	Fairfield, Cal	Nov. 29, 1870	109, 716
Grain-sacking device	P. Van Lackum	Saint Charles, Minn	Apr. 16, 1867	63, 967
Grain-sampler	J. J. Bois	Rantoul, Ill	Oct. 28, 1873	143, 953
Grain-scale, Automatic	A. Stevely	Fond du Lac, Wis	Apr. 14, 1863	38, 185
Grain-scourer	I. N. Harshbarger	Bloomington, Wis	Sept. 20, 1870	107, 486
Grain-scourer	C. B. Horton	Waterloo, N. Y	Nov. 25, 1873	144, 980
Grain-scourer	G. S. Newman	Liberty Mill, Va	May 7, 1872	126, 567
Grain-scourer	D. M. Richardson	Detroit, Mich	Aug. 6, 1872	130, 315
Grain-scourer	T. Shively	North Lansing, Mich	Feb. 4, 1873	135, 602

Index of patents issued from the United States Patent Office from 1790 *to* 1873, *inclusive*—Continued.

Invention.	Inventor.	Residence.	Date.	No.
Grain-scourer	A. Smith	Valmont, Colo	Dec. 20, 1870	110, 301
Grain-scourer	G. E. Throop	Chicago, Ill	Dec. 22, 1863	41, 031
Grain-scourer	V. Weismantel	Belleville, Ill	Apr. 12, 1870	101, 950
Grain scourer and cleaner	A. B. Paige	Washington, D. C	Jan. 2, 1872	122, 330
Grain scourer and huller	L. S. Chichester	Brooklyn, N. Y	Apr. 2, 1872	125, 171
Grain scourer and separator	S. Canby	Ellicott's Mills, Md	May 26, 1857	17, 363
Grain scourer and separator	S. H. Hinsdell, H. W. Drake, and C. B. Way.	Camillus, N. Y	Apr. 4, 1871	113, 296
Grain scourer and separator	C. B. Horton	Waterloo, N. Y	July 16, 1872	129, 346
Grain scourer, smutter, and separator	J. C. Hunt and W. W. Ingraham	Chicago, Ill	Aug 27, 1872	130, 804
Grain, Scouring	P. Pettis	Middlebury, Ohio	July 2, 1830	......
Grain, Scouring and cleaning	J. Sitzenberger	Philadelphia, Pa	Nov. 24, 1823	......
Grain, Scouring and cleaning	B. T. Trimmer	Rochester, N. Y	Sept. 23, 1862	36, 541
Grain scouring and hulling machine	J. N. Treadwell	Reading, Conn	Mar. 1, 1859	23, 127
Grain scouring and separating machine	M. Bartholomew	Enterprise, Pa	July 24, 1860	29, 237
Grain scouring and thrashing machines, Cylinder for.	J. A. Welsh	Xenia, Ohio	Mar. 4, 1862	34, 617
Grain-scouring apparatus	P. H. Massey	South Bend, Ind	Apr. 7, 1868	76, 486
Grain scouring, cleaning, and polishing machine	G. P. Plant and J. Raith	Saint Louis, Mo	Sept. 18, 1860	30, 090
Grain-scouring machine	J. Bergstresser	Berrysburgh, Pa	Jan. 20, 1863	37, 430
Grain-scouring machine	W. McLaughlin	Jersey City, N. J	Aug. 9, 1870	106, 184
Grain-scouring machine	W. P. Robinson	Buffalo, N. Y	Jan. 16, 1872	122, 782
Grain-scouring machine	B. T. Trimmer	Rochester, N. Y	Apr. 12, 1870	101, 946
Grain-screen	J. E. Anderson	Boiling Springs, Pa	Sept. 28, 1869	95, 303
Grain-screen	G. B. Bailey	Greenfield, Ind	May 10, 1864	42, 635
Grain-screen	C. F. Baylor	Trenton, N. J	Sept. 18, 1866	58, 050
Grain-screen	S. Blair	New Wilmington, Pa	Feb. 23, 1869	87, 239
Grain-screen	D. A. and S. Church	Friendship, N. Y	June 27, 1846	4, 600
Grain-screen	J. J. Crider	Greenfield, Ind	Nov. 24, 1868	84, 345
Grain-screen	J. Hatfield and J. Wall	Ogden, Ind	Oct. 11, 1864	44, 625
Grain-screen	R. Hawkins	Palestine, Ind	July 2, 1867	66, 333
Grain-screen	W. G. Hoag	Hoosick, N. Y	Mar. 24, 1863	39, 963
Grain-screen	J. Long	Morristown, Ind	Oct. 2, 1866	58, 551
Grain-screen	C. F. and J. B. Messinger	Logansport, Ind	Aug. 15, 1865	49, 429
Grain-screen	H. Ogborn	Richmond, Ind	Jan. 3, 1865	45, 740
Grain-screen	H. Ogborn	Richmond, Ind	May 29, 1866	55, 152
Grain-screen	W. Rowan and J. M. H. Gill	Freeport, Pa	Oct. 21, 1862	36, 733
Grain-screen	D. D. Schamp	Pleasant Run, N. J	May 7, 1872	126, 493
Grain-screen	W. Tullock	Lantern Mills, Orange Court-House, Va.	Oct. 28, 1806	......
Grain-screen	J. H. H. Wiseheart	Shawneetown, Ill	Aug. 25, 1868	81, 567
Grain screen and sieve	H. B. Thomas	Chicago, Ill	Nov. 15, 1864	45, 125
Grain-screen, Revolving	D. Loeffel	Mount Vernon, Ind	July 6, 1869	92, 327
Grain-screen, Revolving	F. Mills	Mount Vernon, Ind	Aug. 30, 1870	106, 949
Grain-screen, Rotary	D. Pease, jr	Floyd, N. Y	Oct. 29, 1850	7, 744
Grain, seed, &c., Machine for hulling and scouring	O. P. Stevens	Cleveland, Ohio	July 1, 1856	15, 269
Grain-separator	S. Adams	Boston, Mass	Feb. 16, 1864	41, 660
Grain-separator	A. J. Alexander	Chicago, Ill	June 11, 1867	65, 716
Grain-separator	J. R. Allen	Edinburgh, Ind	Dec. 7, 1869	97, 584
Grain-separator	W. B. Allen	Winona, Minn	Feb. 11, 1873	135, 746
Grain-separator	J. Allonas	Mansfield, Ohio	Apr. 9, 1872	125, 513
Grain-separator	P. I. Aukney and D. McGreery	New Lexington, Ohio	Sept. 20, 1859	25, 478
Grain-separator	S. K. Ayers	Delton, Wis	June 13, 1865	48, 233
Grain-separator	S. K. Ayres	Delton, Wis	May 2, 1871	114, 251
Grain-separator	C. F. Babcock	Chicago, Ill	June 8, 1869	91, 067
Grain-separator	F. A. Balch	Hingham, Wis	Aug. 15, 1871	118, 094
Grain-separator	S. Ballard, sr	Sullivan, Ind	June 8, 1869	90, 981
Grain-separator	J. B. Barcalo	Mount Morris, N. Y	Dec. 9, 1862	37, 082
Grain-separator	M. J Barcalo	Mount Morris, N. Y	June 23, 1863	38, 937
Grain-separator	M. J. Barcalo	Mount Morris, N. Y	Jan. 12, 1864	41, 188
Grain-separator	P. Barker	Battle Creek, Mich	Apr. 6, 1869	88, 690
Grain-separator	A., C., and H. K. Barkholder	Clear Spring, Pa	June 15, 1869	91, 412
Grain-separator	H. A. Barnard	Moline, Ill	July 25, 1865	48, 890
Grain-separator	C. R. Barnes	Muncy, Pa	Aug. 4, 1863	39, 374
Grain-separator	C. Bates	Hardin, Iowa	Nov. 24, 1863	40, 670
Grain-separator	H. H. Beach	Philadelphia, Pa	Sept. 17, 1861	33, 231
Grain-separator	F. A. Begole	Jackson, Mich	Oct. 4, 1870	107, 856
Grain-separator	J. Benner	Allegheny, Pa	Apr. 12, 1859	23, 546
Grain-separator	C. Bergen	Farmer, N. Y	July 31, 1860	29, 352
Grain-separator	C. Bergen	Covert, N. Y	May 6, 1862	35, 134
Grain-separator	J. Bergey	Wadsworth, Ohio	Oct. 5, 1852	9, 293
Grain-separator	D. Best	Yuba, Cal	Apr. 25, 1871	114, 097
Grain-separator	C. A. Bikle	Hagerstown, Md	Aug. 10, 1869	93, 586
Grain-separator	J. Blue	Covert, N. Y	Nov. 8, 1853	10, 202
Grain-separator	J. S. Bodge	Bath, N. Y	Nov. 10, 1863	40, 548
Grain-separator	J. S. Bodge	Bath, N. Y	Dec. 26, 1865	51, 687
Grain-separator	J. S. Bodge	La Porte, Ind	Sept. 18, 1866	58, 052
Grain-separator	J. L. Booth	New York, N. Y	Mar. 8, 1859	23, 153
Grain-separator	J. L. Booth	Cuyahoga Falls, Ohio	July 12, 1859	24, 714
Grain-separator	J. L. Booth	Cuyahoga Falls, Ohio	Sept. 20, 1859	25, 484
Grain-separator	J. L. Booth	Rochester, N. Y	Nov. 4, 1862	36, 830
Grain-separator	J. C. Bowden	Farmington, Cal	Mar. 5, 1872	124, 244
Grain-separator	J. W. Breese	Canandaigua, Mich	Sept. 9, 1873	142, 672
Grain-separator	J. Brightbill	Lebanon, Pa	June 22, 1869	91, 710
Grain-separator	G. E. Bringman	Gettysburgh, Pa	Jan. 19, 1869	85, 992
Grain-separator	J. D. Brunner	Doylestown, Pa	July 4, 1871	116, 549
Grain-separator	J. D. Brunner and E. R. J. Ueberroth.	Doylestown, Pa	Sept. 6, 1870	106, 994
Grain-separator	M. Bugher	New Philadelphia, Ohio	Sept. 18, 1860	30, 047
Grain-separator	H. Burdick	Monroe, Wis	Feb. 15, 1870	99, 835
Grain-separator	H. K. Burkholder	Clear Spring, Pa	Dec. 28, 1869	98, 346
Grain-separator	M. Burr	Plymouth, Mich	Dec. 16, 1862	37, 148
Grain-separator	W. T. Chaffee	Richmond, Va	Mar. 17, 1868	75, 520
Grain-separator	W. C. Chamberlain	Dubuque, Iowa	May 17, 1864	42, 747
Grain-separator	J. H. Chase and J. M. Tiffany	Montgomery, Ill	July 10, 1866	56, 177
Grain-separator	D. Claude, jr	Annapolis, Md	Feb. 24, 1843	2, 974
Grain-separator	J. N. Clees	Darbyville, Ohio	July 5, 1864	43, 390
Grain-separator	J. Clum and G. A. Fisher	Shelby and Alabama, N. Y	July 12, 1864	43, 478

Index of patents issued from the United States Patent Office from 1790 *to* 1873, *inclusive*—Continued.

Invention.	Inventor.	Residence.	Date.	No.
Grain-separator	J. A. Cohoon	Toledo, Ohio	Mar. 19, 1872	124, 793
Grain-separator	H. Cook	Dillsburgh, Pa	Sept. 13, 1870	107, 338
Grain-separator	L. B. Corbin	Dryden, N. Y	Feb. 28, 1860	27, 274
Grain-separator	J. W. Cormack and F. C. Walker	Quincy, Ill	Oct. 14, 1856	15, 879
Grain separator	W. R. Cox	Delhi, Iowa	Nov. 2, 1858	21, 945
Grain-separator	J. B. Crist	Evansville, Ind	May 10, 1859	23, 903
Grain-separator	J. F. and H. D. Cummings	Fremont, N. Y	Dec. 22, 1863	40, 995
Grain-separator	A. Curtis	Lena, Ill	July 7, 1857	17, 728
Grain-separator	A. Curtis	Warren, Ill	Oct. 8, 1867	69, 545
Grain-separator	J. L. Custer	Bonaparte, Iowa	May 14, 1872	126, 683
Grain-separator	G. A. Dabney	San José, Cal	Mar. 26, 1872	125, 027
Grain-separator	E. Davis	Almond, N. Y	Mar. 1, 1870	100, 379
Grain-separator	E. Davis and A. Palmer	Hudson, Mich	Mar. 5, 1861	31, 591
Grain-separator	J. Davis	Allegheny City, Pa	Nov. 14, 1865	50, 913
Grain-separator	L. H. Davis	Newark, Del	July 14, 1868	79, 813
Grain-separator	S. Daw	Corvallis, Oreg	Nov. 25, 1873	144, 836
Grain-separator	M. Decamp	South Bend, Ind	Jan. 27, 1857	16, 471
Grain-separator	J. A. Denton, D. W. Shannon, and E. Lucas.	Winslow, Ind	Jan. 12, 1869	85, 801
Grain-separator	W. W. Dingee and A. B. Farquhar.	York, Pa	June 17, 1862	35, 590
Grain-separator	A. H. Dixson	San Francisco, Cal	Nov. 5, 1861	33, 636
Grain-separator	J. W. Donaldson	Fairfield, Cal	Dec. 27, 1870	110, 445
Grain-separator	E. L. Dorsey	Winslow, Ind	Mar. 3, 1868	75, 133
Grain-separator	E. Doud	Oshkosh, Wis	June 9, 1857	17, 498
Grain-separator	J. F. Dunham	Fayette, Iowa	July 10, 1866	56, 194
Grain-separator	A. Durrs	San Francisco, Cal	Oct. 11, 1870	108, 120
Grain-separator	T. Earheart	Donelson, Tenn	July 24, 1860	29, 256
Grain-separator	J. Esse	Redwood City, Cal	Feb. 6, 1872	123, 387
Grain-separator	J. G. Evans	Orville, Ohio	Oct. 25, 1870	108, 577
Grain-separator	O. J. Everson	Lake City, Minn	July 22, 1873	141, 130
Grain-separator	J. H. Fairchild	East Highgate, Vt	Apr. 3, 1866	53, 735
Grain-separator	A. Fancboner	Schoolcraft, Mich	Jan. 15, 1861	31, 114
Grain-separator	J. Fargusson	Dubuque, Iowa	Nov. 5, 1861	33, 639
Grain-separator	J. Fargusson and C. S. Burt	Dubuque, Iowa, and Dunleith, Ill.	June 30, 1863	39, 093
Grain-separator	J. Faulkner	Dansville, N. Y	Feb. 20, 1863	37, 440
Grain-separator	J. Felsing	Granville, Wis	Apr. 1, 1862	34, 824
Grain-separator	P. Flickinger	Hanover, Pa	May 16, 1871	114, 789
Grain-separator	A. Foster	Quincy, Ill	Oct. 26, 1858	21, 877
Grain-separator	F. R. Foster	Brandon, Wis	Oct. 12, 1869	95, 674
Grain-separator	S. W. Foster	Scio, Mich	Dec. 4, 1849	6, 920
Grain-separator	J. W. Free	Richmond, Ind	Feb. 27, 1866	52, 841
Grain-separator	A. Gaar	Richmond, Ind	June 21, 1870	104, 571
Grain-separator	D. Garver	Ringgold, Md	May 9, 1871	114, 546
Grain-separator	W. E. Gaunt and B. B. Hinman	Keokuk, Iowa	Mar. 19, 1861	31, 711
Grain-separator	D. Geiser	Waynesborough, Pa	Aug. 9, 1864	43, 766
Grain-separator	P. Geiser	Smithsburgh, Md	Oct. 19, 1852	9, 341
Grain-separator	P. Geiser	Smithsburgh, Md	Oct. 9, 1855	13, 644
Grain-separator	H. N. Goodrich	Aurora, Ill	Jan. 24, 1865	45, 993
Grain-separator	A. Gordon	Rochester, N. Y	Oct. 7, 1862	36, 611
Grain-separator	J. Gray	Milwaukee, Wis	Dec. 22, 1863	41, 000
Grain-separator	J. Green	Norwalk, Ohio	June 26, 1866	55, 952
Grain-separator	J. C. Gregg	Hillsborough, Ohio	July 17, 1860	29, 161
Grain-separator	P. Griswold	Hudson, Mich	May 22, 1866	54, 895
Grain-separator	P. Griswold	Hudson, Mich	Mar. 7, 1871	112, 447
Grain-separator	P. Griswold and H. H. Seeley	Hudson, Mich	Nov. 22, 1859	26, 181
Grain-separator	A. Hall	Dansville, N. Y	Jan. 19, 1858	19, 140
Grain-separator	C. S. Hall	Rochester, N. Y	Nov. 7, 1871	120, 582
Grain-separator	J. H. Hamaker	Frease's Store, Ohio	Apr. 25, 1865	47, 417
Grain-separator	G. F. Harlan	Elkton, Md	May 18, 1869	90, 260
Grain-separator	S. Harris	Springfield, Mass	Mar. 10, 1868	75, 417
Grain-separator	T. Harrison and W. C. Buchanan.	Belleville, Ill	Sept. 12, 1865	49, 876
Grain-separator	G. Harrowsmith	Lockport, N. Y	July 10, 1860	29, 045
Grain-separator	D. Hathaway	Wyoming, Wis	Nov. 3, 1868	83, 630
Grain-separator	G. Heberling	Quincy, Ill	Jan. 20, 1857	16, 439
Grain-separator	S. Heflebower and J. M. Reed	Alexandria and Loudoun County, Va.	Dec. 18, 1866	60, 513
Grain-separator	A. Higley	Sand Creek, Minn	Dec. 3, 1861	33, 838
Grain-separator	A. Higley	Chicago, Ill	Jan. 7, 1864	43, 026
Grain-separator	O. Holmes	New Lenox, Ill	Jan. 21, 1868	73, 605
Grain-separator	H. A. Hummer	Frenchtown, N. J	Feb. 15, 1870	99, 903
Grain-separator	A. Hunter	Solano County, Cal	Apr. 9, 1861	31, 981
Grain-separator	A. Hunter	Solano County, Cal	Oct. 20, 1863	40, 340
Grain-separator	A. Hunter	San Francisco, Cal	June 1, 1869	90, 844
Grain-separator	S. S. Hurlbut	Cordova, Ill	May 5, 1868	77, 490
Grain-separator	C. B. Hutchings	Rochester, N. Y	Nov. 20, 1860	30, 679
Grain-separator	C. B. Hutchings	Rochester, N. Y	Nov. 15, 1864	45, 120
Grain-separator	B. S. Hyers	Pekin, Ill	Dec. 22, 1863	41, 041
Grain-separator	T. W. Irvin	Marion County, Ind	Nov. 18, 1862	36, 980
Grain-separator	L. P. Josse	Paris, France	Dec. 1, 1863	40, 756
Grain-separator	C. Kathan	Hardin, Iowa	Oct. 14, 1862	36, 653
Grain-separator	J. Kefer	Hamilton, Ohio	May 17, 1864	42, 820
Grain-separator	E. L. Kelly	Reading, Mich	Aug. 1, 1871	117, 545
Grain-separator	J. B. King and O. N. Cronkite	Sacramento, Cal	Dec. 3, 1872	133, 648
Grain-separator	J. Koons	New Auburn, Minn	Sept. 23, 1873	143, 081
Grain-separator	D. Ladd	Chicago, Ill	Jan. 25, 1870	99, 093
Grain-separator	G. Landers and H. Lampman	Atton, N. Y	Oct. 16, 1860	30, 413
Grain-separator	J. L. La Rose	Leavenworth, Kans	Oct. 29, 1872	132, 674
Grain-separator	W. H. Lawrence	Baltimore, Md	Oct. 24, 1871	120, 285
Grain-separator	G. Leach	Elmira, N. Y	May 27, 1862	35, 421
Grain-separator	S. F. Lefler	Racine, Wis	Apr. 5, 1864	42, 205
Grain-separator	S. Lessig, sr	Reading, Pa	Nov. 14, 1871	120, 983
Grain-separator	S. Lessig, sr	Reading, Pa	Nov. 4, 1873	144, 211
Grain-separator	A. J. and H. Linebarger	Jackson, Ill	July 10, 1866	56, 237
Grain-separator	A. W. Lockhart	Sacramento, Cal	July 16, 1867	66, 859
Grain-separator	L. Low	San Francisco, Cal	June 28, 1864	43, 318

Index of patents issued from the United States Patent Office from 1790 *to* 1873, *inclusive*—Continued.

Invention.	Inventor.	Residence.	Date.	No.
Grain-separator	L. Low	Peoria, Ill	Mar. 30, 1869	88,396
Grain-separator	E. Lucas	Winslow, Ind	July 14, 1868	79,993
Grain-separator	G. Lull	Hardin, Iowa	Apr. 16, 1861	32,071
Grain-separator	C. B. Martin	Fond du Lac, Wis	May 28, 1861	32,429
Grain-separator	J. Matthews	Middletown, Pa	Feb. 26, 1861	31,577
Grain-separator	F. I. May	Beverly, N. J	July 19, 1859	24,812
Grain-separator	T. H. McCulloch	Monmouth, Ill	Jan. 28, 1868	73,736
Grain-separator	W. T. McGahey and H. C. Foote.	McGaheysville, Va., and Fredericktown, Ohio.	June 12, 1860	28,679
Grain-separator	I. H. McGehee	Athens, Ala	Apr. 17, 1860	27,919
Grain-separator	J. W. McKinstry	Columbus, Wis	June 4, 1872	127,626
Grain-separator	A. P. Merrill	Ypsilanti, Mich	July 23, 1861	32,883
Grain-separator	L. Merrill	Janesville, Wis	Apr. 9, 1861	31,989
Grain-separator	F. H. C. Mey	Buffalo, N. Y	Aug. 30, 1864	44,011
Grain-separator	J. Meyer	Chicago, Ill	Oct. 22, 1861	33,537
Grain-separator	E. Michael	La Porte, Ind	Jan. 8, 1861	31,096
Grain-separator	J. Miller	Detroit, Mich	Apr. 25, 1871	114,023
Grain-separator	J. A. and J. W. Miller	La Grange, Ind	Dec. 1, 1863	40,767
Grain-separator	W. K. Miller	Canton, Ohio	July 13, 1869	92,629
Grain-separator	D. Y. Milligan	Shelbyville, Ill	May 28, 1872	127,258
Grain-separator	C. W. Mills and L. S. Chichester	Brooklyn, N. Y	Sept. 22, 1868	82,431
Grain-separator	J. R. Moffitt	Saint Louis, Mo	Dec. 1, 1857	18,761
Grain-separator	J. R. Moffitt	Piqua, Ohio	Nov. 29, 1859	26,283
Grain-separator	H. Montgomery and S. Howes	Silver Creek, N. Y	Feb. 22, 1859	23,039
Grain-separator	J. and J. Montgomery and E. Davis.	Baltimore, Md	May 16, 1865	47,772
Grain-separator	J. Nash	Janesville, Wis	Sept. 27, 1859	25,608
Grain-separator	J. Nash	Janesville, Wis	Feb. 12, 1861	31,428
Grain-separator	M. T. Nesbitt	Colora, Md	June 11, 1872	127,788
Grain-separator	J. and E. C. Nichols and D. Shepard.	Battle Creek, Mich	Jan. 7, 1862	34,071
Grain-separator	G. S. and H. Nutter	Bunker Hill, Ill	June 11, 1872	127,915
Grain-separator	G. S., H., and C. H. Nutter	Hunker Hill, Ill	July 16, 1872	129,583
Grain-separator	H. Ogborn	Richmond, Ind	May 18, 1869	90,291
Grain-separator	H. Ogborn and J. W. Free	Green's Fork, Ind	Jan. 12, 1864	41,232
Grain-separator	S. E. Oviatt	Richfield, Ohio	Apr. 28, 1868	77,206
Grain-separator	S. E. Oviatt	Richfield, Ohio	Dec. 8, 1868	84,705
Grain-separator	J. J. Palmer and A. Plamondon	Chicago, Ill	June 24, 1862	35,701
Grain-separator	J. J. Palmer and A. Plamondon	Chicago, Ill	June 30, 1863	39,063
Grain-separator	J. C. Parmater and E. H. Bowen	Vinton, Iowa	Apr. 27, 1869	89,334
Grain-separator	L. Patric	Victor, N. Y	June 24, 1862	35,739
Grain-separator	E. C. Patterson	Chicago, Ill	Dec. 20, 1864	45,517
Grain-separator	E. C. Patterson	Chicago, Ill	Jan. 16, 1866	52,071
Grain-separator	E. C. Patterson	Rochester, N. Y	June 30, 1868	79,497
Grain-separator	N. A. Patterson	Kingston, Tenn	Aug. 28, 1860	29,812
Grain-separator	S. and O. Pettibone	Corunna, Mich	Nov. 20, 1860	30,690
Grain-separator	C. T. Phillips	Jordan, N. Y	June 22, 1869	91,561
Grain-separator	J. F. Plum	Near Greencastle, Pa	Sept. 21, 1869	95,136
Grain-separator	J. F. Pool	Monroe, Wis	Aug. 18, 1868	81,288
Grain-separator	A. Potter	Williamson, N. Y	June 28, 1859	24,605
Grain-separator	A. W. Powers	Barrington, Ill	Feb. 9, 1869	86,687
Grain-separator	T. J. Price	South Union, Ky	May 1, 1866	54,403
Grain-separator	H. Raymond	Tecumseh, Mich	Dec. 6, 1870	109,839
Grain-separator	P. N. and J. Recker	Dayton, Ohio	Oct. 13, 1868	82,989
Grain-separator	W. M. Redd	Hazel Green, Wis	Oct. 18, 1870	108,391
Grain-separator	B. Rice	Schuyler, N. Y	Mar. 12, 1861	31,675
Grain-separator	O. N. Rich	Geneva, Ill	Aug. 4, 1868	80,666
Grain-separator	H. Richmann	Cincinnati, Ohio	June 15, 1869	91,270
Grain-separator	C. Roberts	Belleville, Ill	July 20, 1852	9,140
Grain-separator	C. Robers and J. Cox	New Hope, Pa	Mar. 25, 1856	14,517
Grain-separator	F. W. Robinson	Richmond, Ind	Dec. 18, 1860	30,932
Grain-separator	F. W. Robinson	Richmond, Ind	Feb. 16, 1869	86,942
Grain-separator	A. Rowe	Atalissa, Iowa	Dec. 9, 1862	37,111
Grain-separator	L. Royer	Royerton, Ind	Apr. 5, 1870	101,664
Grain-separator	L. C. Royer	Royerton, Ind	May 16, 1871	114,974
Grain-separator	J. F. Russell	Franklin, Ohio	Aug. 10, 1869	93,559
Grain-separator	B. D. Sanders	Wellsburgh, Va	Apr. 19, 1864	42,409
Grain-separator	J. Schaeffer	Henderson, Ky	Apr. 10, 1860	27,834
Grain-separator	A. Schenck	Ovid, Mich	Dec. 3, 1867	71,796
Grain-separator	J. and D. Schildhauer	New Holstein, Wis	June 9, 1863	38,844
Grain-separator	F. Schunko	York, Pa	Apr. 6, 1858	19,877
Grain-separator	J. A. Scott	Rochester, N. Y	May 28, 1861	32,454
Grain-separator	J. Seebold	New Berlin, Pa	Oct. 4, 1859	25,677
Grain-separator	H. H. Seeley and P. Griswold	Hudson, Mich	June 29, 1858	20,735
Grain-separator	H. H. Seeley and P. Griswold	Hudson, Mich	July 2, 1861	32,723
Grain-separator	H. H. Seely	Hudson, Mich	Mar. 5, 1867	62,694
Grain-separator	D. Shannon and W. Spencer	Winslow, Ind	Sept. 8, 1868	81,951
Grain-separator	N. H. Sherburne	Elgin, Ill	Nov. 3, 1863	40,505
Grain-separator	M. Shreiner	Carlisle, Pa	June 20, 1871	116,229
Grain-separator	H. Siddall	San Francisco, Cal	Jan. 6, 1863	37,362
Grain-separator	H. Siddall	San Francisco, Cal	June 30, 1863	39,073
Grain-separator	A. Smith	Birmingham, Mich	Apr. 11, 1848	5,509
Grain-separator	A. B. Smith	Rochester, Pa	Dec. 22, 1868	85,141
Grain-separator	H. Smith	Hector, N. Y	May 15, 1849	6,452
Grain-separator	H. E. Smith	Philadelphia, Pa	Sept. 23, 1856	15,785
Grain-separator	M. H. Smith	Lawton, Mich	Mar. 29, 1870	101,324
Grain-separator	D. Spencer	Cortland, N. Y	Nov. 22, 1859	26,208
Grain-separator	W. Spencer	Winslow, Ind	Aug. 24, 1869	94,144
Grain-separator	O. W. Stanford	Lebanon, Ohio	Aug. 9, 1864	43,817
Grain-separator	O. W. Stanford and A. W. Crane.	Mason and Lebanon, Ohio	May 5, 1863	38,425
Grain-separator	C. G. and W. Stoll	East New York, N. Y	Apr. 24, 1866	54,263
Grain-separator	O. Stone	Ionia, Mich	May 31, 1870	103,680
Grain-separator	F. Swift	Hudson, Mich	Oct. 25, 1859	25,922
Grain-separator	F. Swift	Hudson, Mich	Dec. 24, 1867	72,696
Grain-separator	F. Swift and M. Standish	Hudson, Mich	Nov. 8, 1870	109,155
Grain-separator	A. Sylvester	Dubuque, Iowa	Apr. 21, 1863	38,251
Grain-separator	H. B. Thomas	Cascade, Iowa	July 14, 1863	39,262

Index of patents issued from the United States Patent Office from 1790 *to* 1873, *inclusive*—Continued.

Invention.	Inventor.	Residence.	Date.	No.
Grain-separator	H. B. Thomas	Chicago, Ill	May 3, 1864	42, 628
Grain-separator	J. Thompson	Chili, N. Y	Apr. 6, 1852	8, 864
Grain-separator	J. D. Tiff	Cuyahoga, Ohio	Aug. 31, 1858	21, 3-3
Grain-separator	W. Todd	Ottawa, Ill	Dec. 27, 1864	45, 654
Grain-separator	J. Tomlinson	Newburgh, Wis	July 4, 1865	48, 603
Grain-separator	J. P. Tostevin	Racine, Wis	Aug. 17, 1869	93, 926
Grain-separator	B. T. Trimmer	Rochester, N. Y	Feb. 14, 1860	27, 171
Grain-separator	B. T. Trimmer	Rochester, N. Y	Jan. 3, 1865	45, 772
Grain-separator	B. T. Trimmer	Rochester, N. Y	Apr. 18, 1865	47, 346
Grain-separator	G. B. Turner and J. A. Vaughn	Cuyahoga Falls, Ohio	Apr. 9, 1861	32, 017
Grain-separator	J. Turner	Sunapee, N. H	Apr. 6, 1858	19, 899
Grain-separator	W. E. Turner	Neosho Falls, Kans	Aug. 12, 1873	141, 676
Grain-separator	E. R. J. Weberroth	Doylestown, Pa	Nov. 12, 1872	133, 070
Grain-separator	J. S. Upton	Battle Creek, Mich	Apr. 28, 1868	77, 419
Grain-separator	A. J. Vandegrift	Lexington, Ky	June 8, 1858	20, 522
Grain-separator	A. J. Vandegrift	Saint Louis, Mo	July 24, 1860	29, 335
Grain-separator	A. J. Vandegrift	Cincinnati, Ohio	Dec. 12, 1865	51, 495
Grain-separator	J. Van Horne	Magnolia, Ill	Sept. 4, 1855	13, 532
Grain-separator	J. Van Houten	Mount Morris, N. Y	Dec. 22, 1863	41, 032
Grain-separator	A. J. Vantuyl	Hector, N. Y	July 2, 1861	32, 729
Grain-separator	J. Vaughn	Magnolia, Ill	May 10, 1859	23, 965
Grain-separator	J. A. Vaughn	Cuyahoga Falls, Ohio	Apr. 24, 1860	28, 026
Grain-separator	D. S. Wagener	Penn Yan, N. Y	Nov. 29, 1859	26, 311
Grain-separator	W. M. Watson	Tonica, Ill	Oct. 29, 1861	33, 617
Grain-separator	J. V. A. Wemple	Chicago, Ill	Nov. 1, 1853	10, 195
Grain-separator	J. V. A. Wemple and G. Westinghouse.	Mohawk and Schoharie, N. Y.	July 13, 1844	3, 663
Grain-separator	G. Westinghouse	Schenectady, N. Y	Apr. 17, 1860	27, 941
Grain-separator	H. L. Whitman	Saint Louis, Mo	May 2, 1871	114, 500
Grain-separator	W. Wicken	Muscoda, Wis	Jan. 28, 1862	34, 279
Grain-separator	L. Wilcox	Hudson, Mich	Aug. 17, 1858	21, 227
Grain-separator	J. N. Williams, jr	Saint Paul, Minn	Oct. 22, 1867	70, 060
Grain-separator	M. D. Williams	Lawton, Mich	Aug. 25, 1868	81, 448
Grain-separator	W. Wilmington	South Bend, Ind	May 23, 1848	5, 595
Grain-separator	W. Wilmington	Toledo, Ohio	Nov. 29, 1859	26, 316
Grain-separator	S. M. Wirts	Hudson, Mich	Sept. 11, 1860	30, 017
Grain-separator	S. M. Wirts and F. Swift	Medina, Mich	Oct. 9, 1866	58, 707
Grain-separator	S. M. Wirts and L. Swift	Hudson, Mich	June 22, 1869	91, 805
Grain-separator	S. M. Wirtz and F. Swift	Hudson, N. Y	Apr. 16, 1861	32, 099
Grain-separator	D. Woodbury	Perkinsville, Vt	Mar. 27, 1849	6, 235
Grain-separator	J. Woolerver	Peoria, Ill	July 10, 1866	56, 316
Grain-separator	B. Wright and J. Bean	Hudson, Mich	Oct. 16, 1855	13, 688
Grain-separator	D. S. Yeakel	Dillingersville, Pa	July 7, 1868	79, 715
Grain-separator	E. Youngs	Tuscarora, N. Y	June 30, 1863	39, 090
Grain-separator	E. Youngs	Tuscarora, N. Y	Jan. 24, 1865	46, 044
Grain-separator	W. Zimmerman	Quincy, Ill	July 21, 1857	17, 853
Grain-separator	A. Zwiebel	Burlington, Wis	June 18, 1872	128, 197
Grain separator and bagger	J. J. Bradner	Pine Creek, N. Y	Mar. 26, 1872	125, 014
Grain separator and cleaner	W. M. Arnall	Sperryville, Va	May 1, 1860	28, 046
Grain separator and cleaner	W. M. Arnall	Sperryville, Va	June 26, 1866	55, 800
Grain separator and cleaner	D. W. Harshbarger	Myersburgh, Pa	July 7, 1863	39, 146
Grain separator and cleaner	S. Howes and G. E. Throop	Silver Creek, N. Y., and Chicago, Ill.	Mar. 16, 1858	19, 637
Grain-separator and clover-cleaning machine	E. L. Kelly	Reading, Mich	Aug. 10, 1869	93, 450
Grain separator and conveyer	J. Lyndall	Santa Clara, Cal	Nov. 18, 1856	16, 103
Grain separator and fan	R. T. Merrill	Bloomfield, Mich	Apr. 8, 1851	8, 031
Grain separator and fanning mill	S. S. Hammond and J. S. Paden	North East, Pa	Feb. 28, 1871	112, 141
Grain separator and mixer	J. J. Crowley	San Francisco, Cal	June 8, 1869	90, 932
Grain separator and scourer	S. W. Andrews and L. Godfrey	Greeneville, Tenn	Nov. 4, 1873	144, 180
Grain separator and scourer	W. P. Clifford	West Jersey, Ill	Sept. 30, 1873	143, 331
Grain separator and scourer	W. C. Knox	Jacksonville, Ill	June 29, 1869	92, 061
Grain separator and scourer	W. C. Knox	Jacksonville, Ill	Aug. 9, 1870	106, 177
Grain-separator and straw-carrier	C. Van Derzee	Albany, N. Y	Feb. 24, 1857	16, 698
Grain-separator and straw-carrier, Combined	A. T. Dunbar and A. McNaught	Alba, Pa	Feb. 5, 1867	61, 816
Grain, Separator for cleaning	G. B. Turner	Cuyahoga Falls, Ohio	Aug. 5, 1873	141, 524
Grain-separator, Rotary	H. Moore	Green Lake, Wis	Oct. 1, 1872	131, 771
Grain-separators, Blast-governor for	I. Hoge	Washington, D. C	May 6, 1873	138, 563
Grain-separators, Blast-regulator for	A. W. Fox	McConnellsburgh, Pa	Oct. 1, 1872	131, 870
Grain-separators, Elevator and table for feeding	D. C. Matteson and T. P. Williamson.	Stockton, Cal	Feb. 2, 1869	86, 564
Grain-separators, Governor attachment to	L. D. Lane	Freeport, Ill	July 17, 1860	29, 178
Grain-separators, Hopper for	F. H. Schroeder	Bushnell, Ill	Mar. 21, 1865	46, 947
Grain-separators, Metallic sieve for	J. A. Maloney	Georgetown, D. C	Oct. 3, 1871	119, 626
Grain-separators, Platform-attachment for	J. Whiteside	Salina, Cal	Jan. 11, 1870	98, 829
Grain-separators, Riddle for	G. A. Wells	Oskaloosa, Iowa	Dec. 15, 1868	85, 043
Grain-separators, Screen for	A. Gaar	Richmond, Ind	Oct. 20, 1857	18, 444
Grain-separators, Shoe for	H. Aldridge	Michigan City, Ind	May 24, 1859	24, 084
Grain-separators, Sieve for	E. Dond	Oshkosh, Wis	Aug. 12, 1873	141, 637
Grain-shovel	B. E. Miles	Washington, D. C	July 12, 1864	43, 519
Grain-shovel	D. B. Rogers	Pittsburgh, Pa	May 10, 1859	23, 949
Grain-shovel	G. V. Watson, G. Milson, and Spendelow.	Buffalo, N. Y	Nov. 22, 1864	45, 197
Grain-shovel	E. P. Williams	Buffalo, N. Y	Aug. 8, 1865	49, 350
Grain-shovel handler	F. Clark	Marcellus, N. Y	Mar. 26, 1867	63, 139
Grain shoveling apparatus	T. D. Hawley	Detroit, Mich	June 15, 1869	91, 440
Grain Sifting and bagging	E. L. Seymour	New York, N. Y	Aug. 12, 1862	36, 176
Grain smutter, scourer, and separator	J. C. Hunt and W. W. Ingraham	Terre Haute, Ind., and Chicago, Ill.	Dec. 28, 1869	98, 266
Grain, snuff, paint, &c., Grinding	W. S. Johnson	New York	June 26, 1835	......
Grain-spout	A. D. Foote	Berlin, Wis	May 1, 1866	54, 322
Grain-spout	J. O. Frost	Candor, N. Y	Aug. 1, 1871	117, 531
Grain-spout	G. H. Johnson	Buffalo, N. Y	Feb. 16, 1869	86, 925
Grain-spreading device	H. H. Beach	Philadelphia, Pa	July 22, 1862	35, 916
Grain-steaming apparatus	J. H. Ebert and E. Pitts	Decatur, Ill	Aug. 19, 1873	142, 008
Grain stirring and delivering apparatus	S. Marsh	West Roxbury, Mass	Oct. 11, 1859	25, 745
Grain stirring and drying apparatus	T. C. Vice	New Haven, Conn	Jan. 20, 1863	37, 467
Grain, Stirring, conveying, and cooling	J. B. Wheeler	Bolton, Mass	May 13, 1862	35, 271
Grain-storehouse	G. Clark	Buffalo, N. Y	Mar. 9, 1869	87, 634

Index of patents issued from the United States Patent Office from 1790 *to* 1873, *inclusive*—Continued.

Invention.	Inventor.	Residence.	Date.	No.
Grain-storer	R. M. Mitchell	Fort Atkinson, Wis	Sept. 22, 1868	82, 339
Grain-stripper	J. O. King and H. A. Rice	Louisiana, Mo	Dec. 14, 1869	97, 932
Grain-tallying machine	A. Harter	Delphi, Ind	Oct. 30, 1866	59, 217
Grain, Tallying-machine for measured	S. Hudson	Milford, Mich	Jan. [illegible], 1865	45, 829
Grain, Teeth for lifting lodged	W. M. Jackson	Woodland, Cal	June 11, 1867	65, 750
Grain-tester, Pocket	B. Martin	Prairie du Chien, Wis	Jan. 14, 1868	73, 356
Grain to burr-millstones, Mode of cleaning and feeding.	S. G. Morrison	Williamsport, Pa	June 18, 1861	32, 578
Grain to flour or meal, Apparatus for reducing	J. J. Webster	Magog, Canada	Sept. 17, 1872	131, 485
Grain to millstones, Cleaning and feeding in	S. Shearman	Goshen, Ind	Aug. 1, 1854	11, 443
Grain to millstones, Feeding	M. and C. Painter	Owing's Mills, Md	June 2, 1857	17, 446
Grain to mills, Feeding	M. H. Ferguson	Sunfish, Ohio	Sept. 11, 1860	29, 961
Grain to obtain extractive matter for coloring and flavoring spirits, Treating.	S. H. Gilman	Galveston, Tex	June 14, 1870	104, 138
Grain to prevent smut, Composition to be applied to	A. N. Overton	Knoxville, Iowa	Oct. 1, 1861	33, 400
Grain-toller	S. R. and B. A. Peden	Franklin, Ky	Jan. 5, 1869	85, 610
Grain-tolling apparatus	J. Armstrong	Bucyrus, Ohio	Sept. 15, 1868	82, 193
Grain-tolling apparatus	W. S. Widger and W. M. Reed	Fairfield, Iowa	Oct. 27, 1868	83, 428
Grain-transferer, Railway	J. W. Sykes	Chicago, Ill	Oct. 12, 1869	95, 747
Grain, &c., Transporting and storing	S. F. Schoomaker	New York, N. Y	May 12, 1868	77, 768
Grain, &c., Treating and storing	C. Alden	Newburgh, N. Y	Nov. 3, 1868	83, 753
Grain, Ventilating	B. Dunwiddie	Monroe, Wis	May 20, 1873	139, 126
Grain, Ventilating	J. Shone	Blackwoodtown, N. J	May 22, 1866	54, 969
Grain-vessels, Ventilating	W. S. Sampson	New York, N. Y	Oct. 17, 1871	120, 108
Grain-washer	G. and G. W. Feaga	Frederick, Md	Jan. 4, 1853	9, 517
Grain-washing machine	G. Copeland	Denver, Colo	May 7, 1872	126, 523
Grainer's tool	R. A. Adams	Indianapolis, Ind	Feb. 12, 1861	31, 363
Graining and ornamental painting, Process for	W. J. Potter	Chicago, Ill	May 5, 1863	38, 412
Graining-apparatus	W. H. Kay	Lemont, Ill	May 11, 1869	90, 004
Graining-instrument	W. Russell	Beloit, Wis	Apr. 18, 1865	47, 339
Graining-machine	W. H. Berger	Pittsburgh, Pa	Oct. 12, 1869	95, 760
Graining-machine	G. Geiger	Cleveland, Ohio	Nov. 26, 1872	133, 311
Graining-machine	B. M. Hall	South Bend, Ind	Aug. 13, 1867	67, 755
Graining-machine	W. H. Smith	New York, N. Y	May 11, 1869	90, 056
Graining, Method of	J. J. Callow	Cleveland, Ohio	July 5, 1870	104, 929
Graining or imitating wood	J. Johnston	New York, N. Y	Jan. 21, 1873	135, 039
Graining, printing, &c., Flexible forms for	H. Tubesing	Pittsburgh, Pa	Mar. 7, 1865	46, 736
Graining process	J. R. Cross	Morrisania, N. Y	Oct. 29, 1872	132, 638
Graining-tool	W. J. Potter and W. H. Arnold	Chicago, Ill	Sept. 15, 1863	39, 953
Grains of wood, Process and composition for printing.	J. Bougardt	New York, N. Y	Dec. 8, 1868	84, 728
Grammatical key	J. Brown	Newark, N. J	Mar. 3, 1817	....
Granaries and other buildings, Cooling, drying, and ventilating.	D. E. Somes	Washington, D. C	Mar. 21, 1865	46, 950
Granary	A. C. L. Devaux	London, England	June 30, 1863	39, 030
Granary	E. Ford	Spring Cottage, Miss	Oct. 24, 1854	11, 830
Granary	T. Hermans	Mitchellsville, Tenn	Oct. 13, 1863	40, 261
Granary	C. T. Moorman, jr	Jamestown, Ohio	Nov. 19, 1872	133, 244
Granary	B. M. Nice	Cleveland, Ohio	Oct. 3, 1865	50, 267
Granary	J. Walsh	Valley Town, Ill	Jan. 1, 1867	60, 971
Granary and fruit-house	S. R. Beckwith	Cleveland, Ohio	Nov. 7, 1865	50, 790
Granary, Fire-proof	G. H. Johnson	Buffalo, N. Y	Mar. 9, 1869	87, 679
Granary, fruit-house, &c	S. R. Beckwith	Cleveland, Ohio	Mar. 13, 1866	53, 101
Granary to preserve wheat, &c., Construction of	J. Harmony	Chambersburgh, Pa	Aug. 20, 1835	....
Granite and stone cutting and dressing machine	W. C. Poland and E. Blossom	Portland, Me	Nov. 11, 1837	458
Granite-cutting machine	J. W. Maloy	Boston, Mass	Oct. 16, 1866	58, 853
Granite-dressing machine	J. D. Buzzell	Cape Elizabeth, Me	June 2, 1836	....
Granite, Hammering and dressing	W. Morse	Corinna, Me	Sept. 5, 1836	....
Granite, Machine for dressing or hammering	W. Morse	United States	Sept. 5, 1836	21
Grape and flower picker	C. W. H. Delano	Cedar Rapids, Iowa	Nov. 25, 1873	144, 837
Grape-box	O. Mallory	Rochester, N. Y	June 13, 1865	48, 192
Grape-box	S. P. Talman	Perrysburgh, Ohio	Jan. 23, 1872	123, 060
Grape-crusher	F. B. Schoenstein and A. Klein	San Francisco, Cal	Oct. 31, 1871	120, 400
Grape crusher and stem separator	T. C. Purington	Lincoln, Cal	Mar. 30, 1869	88, 329
Grape crusher and stem separator	T. C. Purington	Lincoln, Cal	Nov. 30, 1869	97, 313
Grape crusher and stemmer	O. Hyde	Oakland, Cal	Sept. 17, 1872	131, 352
Grape crusher and stoner	G. Johnston and W. F. Johnson	Sacramento and Folsom, Cal.	Apr. 2, 1872	125, 197
Grape-frame	S. O. Cross	Kingsbury, N. Y	June 27, 1854	11, 153
Grape-gatherer, Self-relieving	C. Wadsworth	Madison, Ohio	Sept. 3, 1867	68, 399
Grape-mill	A. Hemminger	Sandusky, Ohio	Nov. 22, 1864	45, 156
Grape-picker	G. A. Warner	San Francisco, Cal	Oct. 3, 1871	119, 485
Grape-pressing machine	H. Krause	New York, N. Y	July 20, 1858	20, 947
Grape-vine cutter	L. W. Mayer	Sonoma, Cal	Feb. 9, 1869	86, 681
Grape-vine protector	J. S. Davis	La Porte, Ind	May 1, 1866	54, 302
Grape-vine protector	J. Walter	Princeton, Ill	Nov. 27, 1866	60, 099
Grape-vine support	F. B. Green	Seneca Falls, N. Y	Apr. 25, 1865	47, 415
Grape-vines, Mode of cultivating	G. Perry	Georgetown, Conn	Feb. 11, 1868	74, 417
Grape-vines, Propagating	W. Griffeth	North East, Pa	May 14, 1867	64, 760
Grape-vines, Training	G. S. Salsbury	Clarendon, N. Y	Sept. 4, 1866	57, 777
Grapes, apples, &c., Grinding and pressing	C. B. Hutchinson	Auburn, N. Y	Feb. 3, 1863	37, 579
Grapes, Implement for harvesting	J. F. Single	Painesville, Ohio	May 26, 1868	78, 241
Grapes, Machine for stemming and crushing	C. Wadhams	Los Angeles, Cal	Sept. 10, 1872	131, 316
Grapes, Mill for compressing and grinding	Z. C. Robbins	Washington, D. C	Aug. 14, 1855	13, 448
Graphite in reducing metals, Using	J. Weisman	Philadelphia, Pa	Mar. 16, 1858	19, 668
Graphodometer, Automatic mechanism for operating surveyor's.	J. M. Wampler	Baltimore, Md	July 13, 1858	20, 908
Graphotype	J. McElheran	Brooklyn, N. Y	Mar. 23, 1858	19, 707
Grapnel	N. P. and N. E. Allen	Salem, Mass	Dec. 20, 1870	110, 329
Grapple	J. H. Brinton	Thornbury, Pa	Aug 10, 1869	93, 590
Grapple	J. Burkhart	Brookville, Ind	Apr. 8, 1873	137, 592
Grapple	E. G. Crandal	Belfast, N. Y	Mar. 14, 1871	112, 690
Grapple	S. B. Dexter	Mason City, Iowa	Mar. 25, 1873	137, 130
Grapple	C. La Dow	South Galway, N. Y	Sept. 19, 1871	119, 157
Grapple	S. Rogers	Pittsburgh, Pa	Mar. 1, 1870	100, 330
Grapple	J. F. Thomas	Ilion, N. Y	Aug. 3, 1869	93, 247
Grapple	H. Whitehall and J. Burson	Philadelphia, Pa., and Yates City, Ill.	Oct. 10, 1871	119, 729

Index of patents issued from the United States Patent Office from 1790 to 1873, inclusive—Continued.

Invention.	Inventor.	Residence.	Date.	No.
Grapple and excavator	A. T. Morris	Bloomfield, N. J	Jan. 5, 1869	85, 602
Grapple for raising stone	M. M. Cass and L. R. Bigelow	Watkins, N. Y	Feb. 19, 1856	14, 282
Grapple for raising sunken bodies	G. A. Wilbur	Skowhegan, Me	Aug. 19, 1856	15, 586
Grapple for raising sunken vessels	J. T. Martin	New York, N. Y	July 11, 1854	11, 291
Grapple, Spring	O. Warner and C. S. Gaylord	Gaylord's Bridge, Conn	Oct. 8, 1850	7, 709
Grapple, Store and household	J. Seltzer	Philadelphia, Pa	Apr. 12, 1870	101, 772
Grappling-apparatus, Naval	S. Riggs	Blanchester, Ohio	Sept. 5, 1865	49, 790
Grappling-fork	G. Hunziker	Summit, Miss	July 16, 1872	129, 032
Grappling-hook	E. Harter	Dowagiac, Mich	Dec. 28, 1869	98, 259
Grappling-hook	H. H. Hatheway	Clockville, N. Y	Dec. 7, 1869	97, 635
Grappling-hook	E. J. Riker	Lewiston, Me	Dec. 1, 1868	84, 510
Grappling-hook	E. Ziegler and C. Cable	Harmony, Pa	Mar. 28, 1871	113, 126
Grappling-iron	J. H. Chapman	Utica, N. Y	Aug. 28, 1866	57, 476
Grappling-iron	W. H. Hawley	Utica, N. Y	Oct. 22, 1867	69, 992
Grappling-iron	W. H. Hawley	Utica, N. Y	Oct. 6, 1868	82, 832
Grappling or dredging machine	A. Stoner	Mount Joy, Pa	Mar. 24, 1857	16, 886
Grappling-tool	N. W. Green	Cortland Village, N. Y	Dec. 11, 1866	60, 361
Grass and grain cutting machine	W. Manning	Plainfield, N. J	May 3, 1831	....
Grass-burner	J. A. Craig	Columbia, Ark	Feb. 3, 1852	8, 701
Grass-cutter	S. W. Sears	New York, N. Y	Sept. 20, 1870	107, 632
Grass-cutter	Sturdivant and Holmes	Portland, Me	June 19, 1835	....
Grass cutter and cradling machine	E. Badlam, jr	Chester, Vt	Sept. 18, 1835	....
Grass-cutting machine	W. Boone	New Hope, Mo	Nov. 21, 1848	5, 931
Grass-cutting machine	C. H. McCormick	Chicago, Ill	Sept. 21, 1858	21, 573
Grass-cutting machine, Flower-maker's	T. and J. Millot	New York, N. Y	Nov. 25, 1873	144, 916
Grass, &c., Extracting fiber from China	J. Steart	Bermondsey, England	Oct. 30, 1866	59, 331
Grass, Implement for destroying quack	C. W. Moseley	Onondaga, N. Y	Nov. 2, 1869	96, 464
Grass, Mode of destroying cocoa	H. B. Kenner	New Orleans, La	Oct. 6, 1843	3, 296
Grass. Mowing	J. P. Chandler	Wilton, Me	Aug. 17, 1835	....
Grass or grain cutter and rake	A. Rundel	Verona, N. Y	Apr. 22, 1835	....
Grass-renovator	J. Gould	Lexington, Mass	May 11, 1869	89, 983
Grass-seed-cleaning machine	H. Child	Canton, Me	Jan. 11, 1823	....
Grass-seed huller	H. P. Byram	Louisville, Ky	Oct. 11, 1853	10, 102
Grass-seed, &c., Machine for hulling	T. Blake	Paris, Me	July 12, 1822	....
Grass-seed separator for mangers	D. B. Dixon	Unionville, Mo	Aug. 23, 1870	106, 562
Grass under water, Apparatus for cutting	P. J. Stone	Athens, Pa	Aug. 2, 1870	105, 996
Grass under water, Machine for cutting	J. Hinds	Hindsburgh, N. Y	Sept. 5, 1840	1, 772
Grasshopper-exterminating composition	S. Green	Arapahoe County, Colo	Aug. 8, 1865	49, 258
Grate	J. L. Babbitt	Glen Cove, N. Y	Mar. 24, 1868	75, 830
Grate	J. Bennett	New York, N. Y	May 14, 1836	....
Grate	E. Bosdevex	Philadelphia, Pa	June 23, 1868	79, 195
Grate	J. S. Clark	Philadelphia, Pa	Feb. 12, 1867	62, 005
Grate	J. C. Cochrane	Rochester, N. Y	Feb. 26, 1867	62, 316
Grate	B. W. Dunklee	Boston, Mass	June 1, 1869	90, 645
Grate	F. Glick and U. Keck	Allentown, Pa	Oct. 4, 1870	107, 899
Grate	G. W. Griswold	Carbondale, Pa	July 25, 1854	11, 371
Grate	R. Ham	Troy, N. Y	Aug. 17, 1869	93, 712
Grate	C. Hires	Salem, N. J	May 15, 1866	54, 725
Grate	C. Hunt	New York, N. Y	Mar. 17, 1812	....
Grate	W. Keiser	Stroudsburgh, Pa	June 4, 1867	65, 392
Grate	S. J. Kelly	Pemberton, N. J	Oct. 20, 1863	40, 341
Grate	S. Kepner	Pottstown, Pa	Dec. 2, 1873	145, 212
Grate	D. Lister	Glasgow, Great Britain	Jan. 12, 1864	41, 225
Grate	C. B. Loveless	Syracuse, N. Y	Aug. 5, 1862	36, 098
Grate	L. G. Marshall	Mokena, Ill	Jan. 17, 1865	45, 948
Grate	W. L. McDowell	Philadelphia, Pa	Apr. 28, 1863	38, 317
Grate	J. H. Meissner	New York, N. Y	Oct. 22, 1867	70, 102
Grate	J. Miller	Saint Louis, Mo	June 27, 1865	48, 425
Grate	H. Mulford	Philadelphia, Pa	Feb. 8, 1870	99, 584
Grate	P. Murray	Milwaukee, Wis	Dec. 12, 1865	51, 472
Grate	P. Murray	Philadelphia, Pa	Dec. 29, 1868	85, 470
Grate	J. Old	Pittsburgh, Pa	Mar. 16, 1869	87, 790
Grate	P. A. Palmer	Troy, N. Y	Nov. 5, 1867	70, 462
Grate	J. F. Phelps	Huntsville, Ala	May 30, 1871	115, 515
Grate	B. Pike	New York, N. Y	Mar. 11, 1835	....
Grate	N. Pramer	Troy, N. Y	July 23, 1867	67, 136
Grate	E. S. Renwick	New York, N. Y	Feb. 23, 1869	87, 112
Grate	G. L. Smith	Brooklyn, N. Y	Mar. 29, 1864	42, 118
Grate	G. L. Smith	Brooklyn, N. Y	Aug. 21, 1866	57, 396
Grate	P. Smith	Fall River, Mass	Apr. 12, 1859	23, 621
Grate	S. Smyth	East Bridgewater, Pa	Oct. 21, 1873	143, 856
Grate	S. Stevens and J. P. Smith	Pittsburgh, Pa	Sept. 15, 1863	39, 969
Grate	J. Tiberi	Saint Louis, Mo	Sept. 6, 1859	25, 352
Grate	T. Tomkinson	Philadelphia, Pa	Aug. 17, 1869	93, 925
Grate	W. J. Towne	Newton, Mass	May 30, 1865	48, 000
Grate	J. Watson, jr	Louisville, Ky	Oct. 13, 1863	40, 315
Grate	M. D. Wellman	Allegheny City, Pa	Nov. 19, 1867	71, 253
Grate	J. S. Williams	Saint Louis, Mo	Sept. 13, 1859	25, 462
Grate	C. J. Woolson	Cincinnati, Ohio	Apr. 14, 1863	38, 194
Grate	C. J. Woolson	Cleveland, Ohio	Jan. 1, 1867	60, 981
Grate	W. Wright	South River, N. J	Mar. 17, 1863	37, 933
Grate	J. H. Yocum	Ashland, Pa	Jan. 28, 1868	73, 767
Grate, Adjustable stove	R. Moss	Philadelphia, Pa	Oct. 24, 1865	50, 613
Grate, Agitating coal	A. Keeney	Carlisle, Pa	Apr. 2, 1850	7, 245
Grate and andiron, Portable fire	J. H. Coate	West Milton, Ohio	Nov. 27, 1866	59, 968
Grate and ash-sifter in cooking-stoves	D. E. Paris	Troy, N. Y	June 9, 1868	78, 685
Grate and back for cooking-stoves	J. B. Slusser and W. H. Meech	Roanoke, Ind	June 7, 1870	104, 070
Grate and blower for coal	S. C. Mott and W. Holmes	New York	Nov. 26, 1826	....
Grate and cooking-stove	J. J. Giraud	Baltimore, Md	Feb. 10, 1836	....
Grate and door of open-grate or parlor stove	E. Brown	Philadelphia, Pa	Mar. 5, 1872	124, 325
Grate and fire-place	E. Conger	Newark, N. J	July 22, 1837	288
Grate and fire-place, Adjustable hood for coal	W. T. Foster	Jeffersonville, Ind	Oct. 22, 1867	69, 982
Grate and fire-place for anthracite coal	W. M. Russell	Boston, Mass	Feb. 8, 1833	....
Grate and fire-place register and air-box	A. Pollock	Boston, Mass	Mar. 3, 1837	137
Grate and fire-pot for heating-stoves	P. J. Schopp	Louisville, Ky	June 14, 1870	104, 212
Grate and grate-bar	J. C. Grant	Salem, Mass	Aug. 29, 1871	118, 604
Grate and grate-bar	W. B. Treadwell	Albany, N. Y	Mar. 6, 1866	53, 059
Grate and stove-bar	J. L. Mott	New York	Oct. 14, 1835	....

Index of patents issued from the United States Patent Office from 1790 to 1873, inclusive—Continued.

Invention.	Inventor.	Residence.	Date.	No.
Grate and stove, Coal	C. K. Marshall	New Orleans, La	July 14, 1868	79, 845
Grate and stove for warming apartments	E. H. Dixon	New York, N. Y	Aug. 8, 1837	347
Grate and stove, Parlor and kitchen	E. Ingalls	Providence, R. I	Apr. 22, 1835	
Grate, Anthracite-coal stove	E. Nott	Schenectady, N. Y	June 29, 1833	
Grate-bar	H. Ball	New York, N. Y	Nov. 29, 1870	109, 711
Grate-bar	H. K. Bates	North La Crosse, Wis	June 19, 1866	55, 603
Grate-bar	H. L. Budd	New York, N. Y	Apr. 30, 1867	64, 191
Grate-bar	J. Buzby	Moorestown, N. J	Oct 11, 1859	25, 720
Grate-bar	G. H. Clarke	Brooklyn, N. Y	Aug. 26, 1873	142, 146
Grate-bar	G. H. Clarke and C. Van Wagenen.	New York, N. Y	Oct. 30, 1866	59, 311
Grate-bar	H. Collinson	Boston, Mass	May 5, 1868	77, 458
Grate-bar	H. Collinson	Dorchester, Mass	Feb. 8, 1870	99, 640
Grate-bar	H. Collinson	Boston, Mass	July 11, 1871	116, 812
Grate-bar	A. F. Crowell	Hyannis, Mass	Sept. 30, 1873	143, 224
Grate-bar	T. S. Davis	Lancaster, Pa	Feb. 22, 1870	100, 016
Grate-bar	J. Easterly	Albany, N. Y	Apr. 5, 1859	23, 456
Grate-bar	W. Ferrel	Charleston, S. C	Apr. 25, 1843	3, 059
Grate-bar	A. Fickett and C. C. Benton	Rochester, N. Y	Jan. 17, 1871	111, 047
Grate-bar	A. C. Fletcher	New York, N. Y	Apr. 17, 1866	53, 966
Grate-bar	A. C. Fletcher	New York, N. Y	Jan. 29, 1867	61, 615
Grate-bar	A. C. Fletcher	New York, N. Y	Feb. 18, 1868	74, 526
Grate-bar	A. C. Fletcher	New York, N. Y	Apr. 14, 1868	76, 734
Grate-bar	A. C. Fletcher	New York, N. Y	July 7, 1868	79, 743
Grate-bar	A. C. Fletcher	New York, N. Y	Aug. 18, 1868	81, 155
Grate-bar	A. C. Fletcher	New York, N. Y	Nov. 7, 1871	120, 733
Grate-bar	A. W. Foster, jr	Pittsburgh, Pa	Oct. 24, 1871	120, 188
Grate-bar	C. A. Greenleaf	Indianapolis, Ind	Nov. 16, 1869	96, 798
Grate-bar	P. Griffith	Philadelphia, Pa	Aug. 5, 1862	36, 083
Grate-bar	J. W. Griswold and E. L. Thomson.	Philadelphia, Pa	Aug. 18, 1868	81, 161
Grate-bar	S. Harrison	Philadelphia, Pa	Sept. 25, 1866	58, 252
Grate-bar	A. Hartupee	Pittsburgh, Pa	Apr. 21, 1868	77, 038
Grate-bar	H. Hartwig	Covington, Ky	Jan. 2, 1872	122, 460
Grate-bar	M. Helbling	Allegheny, Pa	Mar. 9, 1869	87, 564
Grate-bar	W. E. Hill	Brooklyn, N. Y	Dec. 12, 1865	51, 455
Grate-bar	M. Hodgson	East Saginaw, Mich	Aug. 2, 1870	105, 942
Grate-bar	D. Housten	New York, N. Y	May 1, 1866	54, 349
Grate-bar	R. A. Hutchinson	Jersey City, N. J	July 30, 1872	130, 053
Grate-bar	R. A. Hutchinson	Bergen, N. J	Sept. 2, 1873	142, 471
Grate-bar	W. Kearney	Belleville, N. J	Aug. 2, 1870	105, 948
Grate-bar	H. C. Kerstine	Cleveland, Ohio	Sept. 23, 1873	143, 079
Grate-bar	H. King	Waterbury, Conn	May 26, 1868	78, 292
Grate bar	C. Kingsland	Allegheny, Pa	Oct. 16, 1849	6, 799
Grate-bar	J. S. Kirk and W. H. Elliot	Plattsburgh, N. Y	Mar. 27, 1855	12, 597
Grate-bar	L. F. Lakey and W. B. Hayte	Quincy, Ill	Oct. 12, 1869	95, 812
Grate-bar	D. Lasher	Brooklyn, N. Y	Sept. 11, 1860	29, 977
Grate-bar	W. S. Mackintosh	Allegheny City, Pa	Oct. 13, 1868	82, 965
Grate-bar	W. W Marsh	Alton, Ill	Sept. 18, 1860	30, 074
Grate-bar	J. T. Marshall	Wilmington, Del	Jan. 24, 1871	111, 223
Grate-bar	J. McDonald	Cincinnati, Ohio	Mar. 29, 1864	42, 099
Grate-bar	W. McMannies	Brooklyn, N. Y	Feb. 8, 1870	99, 582
Grate-bar	J. A. Miller	Providence, R. I	Sept. 12, 1871	118, 962
Grate-bar	R. Montgomery	New York, N. Y	Aug. 21, 1866	57, 358
Grate-bar	W. Müir and P. Butler	Archibald and Carbondale, Pa.	June 20, 1871	116, 080
Grate-bar	J. T. Osborn	New Orleans, La	Jan. 22, 1856	14, 142
Grate-bar	B. F. Penny and J. Jones	Rochester, N. Y	July 6, 1869	92, 351
Grate-bar	C. W. Pierce	Albany, N. Y	Oct. 2, 1866	58, 474
Grate-bar	A. D. Puffer	Somerville, Mass	Mar. 21, 1865	46, 939
Grate-bar	W. Randall	Salem, Mass	Feb. 9, 1869	86, 862
Grate-bar	M. L. Roucout	Paris, France	May 31, 1853	9, 758
Grate-bar	S. T. Savage	Albany, N. Y	Nov. 23, 1858	22, 134
Grate-bar	H. B. Scofield	New York, N. Y	May 1, 1866	54, 415
Grate-bar	J. Sherman	Burlington, N. J	May 28, 1867	65, 289
Grate-bar	J. A. Sinclair	Bridgeport, Ohio	Feb. 4, 1873	135, 447
Grate-bar	J. R. Smith	Salem, Mass	Nov. 5, 1867	70, 638
Grate-bar	S. Smith	Salem, Mass	Oct. 2, 1866	58, 494
Grate-bar	J. W. Stanton	Barnesville, Ohio	Feb. 13, 1872	123, 739
Grate-bar	O. H. Taylor	Brooklyn, N. Y	Aug. 4, 1868	80, 574
Grate-bar	L. B. Tupper	New York, N. Y	June 14, 1864	43, 143
Grate-bar	L. B. Tupper	New York, N. Y	Feb. 27, 1866	52, 913
Grate-bar	L. B. Tupper	New York, N. Y	Aug. 21, 1866	57, 411
Grate-bar	P. Umholtz	Tremont, Pa	Nov. 4, 1873	144, 372
Grate-bar	P. and A. Umholtz	Tremont, Pa	Sept. 24, 1872	131, 641
Grate-bar	S. Vansyckel	Little York, N. J	Aug. 23, 1853	9, 958
Grate-bar	S. Vansyckel	Little York, N. J	Oct. 31, 1854	11, 879
Grate-bar	J. Yocum, jr	Philadelphia, Pa	Mar. 30, 1869	88, 355
Grate-bar	J. Yocom, jr	Philadelphia, Pa	Nov. 15, 1870	109, 285
Grate-bar, Air-conducting	R. G. Orwig	Des Moines, Iowa	Apr. 9, 1872	125, 404
Grate-bar and bearer	W. Kearney	Belleville, N. J	Nov. 1, 1870	108, 796
Grate bar and grating, Furnace	J. Cuthbert	Pittsburgh, Pa	Apr. 26, 1870	102, 233
Grate-bar for boiler and other furnaces	G. A. Eyars	Meriden, Conn	May 18, 1869	90, 248
Grate-bar for boilers	D. Byard	Sharon, Pa	Sept. 7, 1869	94, 561
Grate-bar for boilers	E. Thayer	Worcester, Mass	Mar. 28, 1865	47, 052
Grate-bar for furnaces and heaters	J. Reynolds	Philadelphia, Pa	May 11, 1869	89, 888
Grate-bar for furnaces, Hollow	G. S. Nevins	Bushnell, Ill	Jan. 23, 1866	52, 189
Grate-bar for steam and other enginery	M. and C. H. Morse	Franklin, Mass	Oct. 26, 1869	96, 138
Grate-bar for steam-boilers, Hollow	B. L. Griffith	Reading, Pa	Aug. 30, 1859	25, 258
Grate-bar for steam-generators	E. G. Blakeslee and A. Manser	Sing Sing, N. Y	Oct. 10, 1865	50, 327
Grate-bar frame, Hollow	G. E. Turner	Chicago, Ill	Dec. 28, 1869	98, 321
Grate bar, Furnace	F. Armstrong	New Orleans, La	Feb. 17, 1852	8, 732
Grate-bar, Furnace	C. C. Bemis	San Francisco, Cal	Dec. 4, 1866	60, 125
Grate bar, Furnace	W. C. Cambridge	Bristol, Great Britain	Nov. 3, 1868	83, 689
Grate bar, Furnace	G. H. Clarke	Brooklyn, N. Y	Jan. 21, 1868	73, 436
Grate bar, Furnace	E. Dugdale	Burlington, N. J	Aug. 18, 1857	18, 010
Grate bar, Furnace	J. H. Fellows	Cincinnati, Ohio	Mar. 27, 1847	5, 035
Grate bar, Furnace	H. Gerner	New York, N. Y	Oct. 11, 1864	44, 622

Index of patents issued from the United States Patent Office from 1790 *to* 1873, *inclusive*—Continued.

Invention.	Inventor.	Residence.	Date.	No.
Grate bar, Furnace	R. C. Graves	Barnesville, Ohio	Feb. 11, 1873	135, 799
Grate bar, Furnace	J. Kymer	Caermarthen, South Wales	July 18, 1844	3, 671
Grate bar, Furnace	W. S. Low	Albany, N. Y	June 7, 1859	24, 316
Grate bar, Furnace	W. Mellor	Paterson, N. J	Mar. 26, 1872	124, 966
Grate bar, Furnace	A. Rawson	Des Moines, Iowa	June 4, 1872	127, 428
Grate bar, Furnace	A. Rawson	Des Moines, Iowa	July 23, 1872	129, 862
Grate bar, Furnace	A. Rawson	Des Moines, Iowa	Aug. 19, 1873	141, 890
Grate bar, Furnace	J. C. Schlough	Easton, Pa	May 29, 1855	12, 967
Grate bar, Furnace	W. H. Settle	Louisville, Ky	July 23, 1872	129, 756
Grate bar, Furnace	G. O. Tupper	New York, N. Y	Apr. 10, 1865	53, 903
Grate bar, Furnace	J. Vandercar	Brooklyn, N. Y	May 24, 1864	42, 888
Grate bar, Furnace	B. C. Vanduzen	Cincinnati, Ohio	May 30, 1854	10, 996
Grate bar, Furnace	R. Wicks	New York, N. Y	May 15, 1855	12, 883
Grate bar, Furnace	J. Williams, J. Forgie, and J. Edwards.	New York, N. Y	Mar. 24, 1868	76, 023
Grate-bar, Hollow	N. Shaw	West Eau Claire, Wis	Aug. 31, 1869	94, 345
Grate-bar, Revolving	G. C. Waggonner	Hamilton, Ill	Oct. 22, 1872	132, 423
Grate-bar, Rocking	J. Jones	Rochester, N. Y	May 30, 1871	115, 482
Grate-bar, Rotary	D. Byard	Sharon, Pa	Oct. 18, 1870	108, 327
Grate bar, Steam-generator	J. Mahony	Newport, R. I	June 29, 1869	91, 949
Grate bar, Steam-generator	M. D. Wellman	Allegheny County, Pa	Apr. 12, 1870	101, 795
Grate bar, Stove	G. W. Gardner	Albany, N. Y	Nov. 18, 1851	8, 523
Grate-bar supporter	W. F. Morgan and F. C. Bartlett.	New York, N. Y	Sept. 19, 1865	50, 023
Grate bar, Tubular	W. H. Farris	Cairo, Ill	May 2, 1871	114, 280
Grate bar, Water	W. H. Farris	Cairo, Ill	Mar. 28, 1871	113, 037
Grate-bars, Agitating	A. D. Spoor	Troy, N. Y	Apr. 15, 1851	8, 043
Grate-bars, Construction of	F. P. Dimpfel	Philadelphia, Pa	Mar. 16, 1852	8, 799
Grate-bars for furnaces, Casting	J. A. Miller	New York, N. Y	Jan. 31, 1865	46, 127
Grate-bars, Interlocking	S. Vansyckel	Jersey City, N. J	Oct. 9, 1855	13, 669
Grate, Brick-kiln	J. Maltpress	Edgerton, Wis	Dec. 1, 1868	84, 639
Grate called caloret, Fire	C. M. Graham	New York	Nov. 10, 1821	....
Grate, Coal	C. O. Greene	West Troy, N. Y	Jan. 1, 1850	6, 985
Grate, Coal	W. T. McMillen	Saint Louis, Mo	May 22, 1860	28, 390
Grate, Coal	I. Pickard	Lena, Ill	Dec. 13, 1870	110, 157
Grate, Coal-stove	S. G. Morrison	Williamsport, Pa	Aug. 5, 1873	141, 452
Grate, Coal-vault	G. W. Scott	Greencastle, Ind	Oct. 11, 1870	108, 298
Grate combination	J. Wallace	Louisville, Ky	Nov. 12, 1867	70, 924
Grate, Common fire	J. Atwater	New Haven, Conn	Oct. 25, 1832	....
Grate, Cooking	T. Vinton	Philadelphia, Pa	Oct. 31, 1829	....
Grate, Drip-receiving	W. N. Reed	Arlington, Va	Nov. 22, 1870	109, 452
Grate-fender	S. S. Bent	Port Chester, N. Y	July 12, 1870	105, 295
Grate-fender	G. Buchanan	Washington, Pa	Mar. 9, 1869	87, 629
Grate-fender	G. Buchanan	Washington, Pa	Nov. 7, 1871	120, 571
Grate, Fire	J. Atwater	New Haven, Conn	Jan. 9, 1833	....
Grate, Fire	W. S. Bronson	Hartford, Conn	May 5, 1868	77, 445
Grate, Fire	W. H. Farris	Cairo, Ill	Aug. 12, 1873	141, 780
Grate, Fire	W. D. Guseman	Morgantown, W. Va	June 30, 1863	39, 043
Grate, Fire	J. Habermehl	Wheeling, W. Va	May 30, 1865	47, 947
Grate, Fire	G. R. Moore	Philadelphia, Pa	Apr. 18, 1871	113, 908
Grate, Fire	M. Peckham	Utica, N. Y	Aug. 16, 1870	106, 502
Grate, Fire	W. H. Pulver	Troy, N. Y	Aug. 7, 1847	5, 232
Grate, Fire	H. Speeler	Trenton, N. J	Jan. 21, 1868	73, 662
Grate, Fire	J. Tiberi	Saint Louis, Mo	Apr. 2, 1861	31, 914
Grate, Fire	J. Vandercar	Brooklyn, N. Y	Aug. 11, 1868	80, 844
Grate, Fire	R. F. Weller	Albany, N. Y	Sept. 5, 1871	118, 762
Grate, Fire	G. Wellhouse	Akron, Ohio	Mar. 7, 1871	112, 399
Grate, Fire	G. Williamson	Milwaukee, Wis	May 4, 1869	89, 720
Grate, Fire-place	S. S. Bent	Port Chester, N. Y	July 12, 1870	105, 296
Grate, Fire-place	S. S. Bent	Port Chester, N. Y	July 12, 1870	105, 297
Grate, Fire-place	L. Bertsche, jr	Allegheny City, Pa	June 22, 1869	91, 597
Grate, Fire-place	F. S. Bissell	Pittsburgh, Pa	Aug. 30, 1870	106, 772
Grate, Fire-place	J. Caven	Indianapolis, Ind	Oct. 31, 1871	120, 490
Grate, Fire-place	G. Chilson	Boston, Mass	June 4, 1850	7, 411
Grate, Fire-place	J. A. Crawford	Newcastle, Pa	Jan. 28, 1870	104, 836
Grate, Fire-place	G. W. Everhart	Louisville, Ky	Dec. 21, 1869	98, 045
Grate, Fire-place	W. H. Garrett	Cannonsburgh, Pa	Apr. 18, 1871	113, 757
Grate, Fire-place	J. Goldenburgh	Cincinnati, Ohio	Oct. 7, 1846	4, 801
Grate, Fire-place	A. Greenaway and H. J. Needham.	New Albany, Ind	June 27, 1871	116, 438
Grate, Fire-place	J. Hackett	Louisville, Ky	Sept. 20, 1870	107, 483
Grate, Fire-place	G. W. Hinman	Paducah, Ky	June 29, 1869	92, 049
Grate, Fire-place	M. C. Hull	New York, N. Y	Sept. 15, 1863	39, 928
Grate, Fire-place	G. H. McElevey	Newcastle, Pa	Feb. 8, 1870	99, 581
Grate, Fire-place	E. D. Merrick	New Brighton, Pa	Mar. 29, 1870	101, 296
Grate, Fire-place	E. Norman	New Orleans, La	July 12, 1870	105, 234
Grate, Fire-place	M. Obmer	Dayton, Ohio	June 11, 1872	127, 916
Grate, Fire-place	J. Old	Pittsburgh, Pa	Nov. 2, 1869	96, 342
Grate, Fire-place	D. Pangle	Belmont, Ohio	Dec. 19, 1871	122, 050
Grate, Fire-place	W. Pulsfort	Louisville, Ky	Feb. 21, 1871	111, 971
Grate, Fire-place	C. S. Rankin	Cincinnati, Ohio	June 30, 1868	79, 391
Grate, Fire-place	J. S. Runyan	Columbus, Ohio	Nov. 22, 1870	109, 550
Grate, Fire-place	J. S. Runyan	Columbus, Ohio	Sept. 2, 1873	142, 514
Grate, Fire-place	R. P. Sause	Louisville, Ky	Mar. 19, 1872	124, 699
Grate, Fire-place	J. W. Thorniley	New Brighton, Pa	Mar. 21, 1871	112, 986
Grate, Fire-place	M. S. Watkins	Memphis, Tenn	Nov. 14, 1871	120, 917
Grate, Fire-place	W. E. Whitehurst	Norfolk, Va	Feb. 1, 1870	99, 384
Grate, Fire-place	W. S. Withers	Atlanta, Ga	Aug. 17, 1869	93, 939
Grate, Fire-place	J. E. Wood	Huntington, W. Va	Dec. 2, 1873	145, 141
Grate, Fire-place	J. M. Woodcock	Bridgeport, Ohio	Apr. 15, 1873	137, 993
Grate, fire-place, and furnace	W. Young	London, England	Nov. 19, 1867	71, 260
Grate, fire-place, and hot-air furnace	J. H. Sherwood	New York, N. Y	Mar. 11, 1873	136, 775
Grate, Fire-place and stove	C. S. Rankin	Cincinnati, Ohio	Feb. 13, 1872	123, 727
Grate for admitting air to sides and body of the fuel-stove.	B. Franklin	Indianapolis, Ind	Mar. 26, 1872	125, 040
Grate for base-burning stoves, Plate or	L. W. Harwood and C. D. Newton.	Troy, N. Y	Jan. 19, 1869	86, 069

Index of patents issued from the United States Patent Office from 1790 to 1873, inclusive—Continued.

Invention.	Inventor.	Residence.	Date.	No.
Grate for burning coal, Open	J. Atwater	New Haven, Conn	Nov. 9, 1838	1,001
Grate for burning coal-screenings	G. W. Hildreth	Lockport N. Y	Feb. 28, 1871	112,246
Grate for burning fuel, Suspension	T. Pearson	Monroe Works, N. Y	July 17, 1837	270
Grate for burning petroleum and other liquid fuel	J. D. Smedley	Chicago, Ill	Oct. 28, 1862	36,805
Grate for coal-stoves	J. Easterly	Albany, N. Y	Sept. 7, 1858	21,410
Grate for coal-stoves	C. Isbister	Allegheny City, Pa	Apr. 17, 1849	6,358
Grate for coal-stoves and furnaces	W. Hailes	Albany, N. Y	Nov. 18, 1862	36,950
Grate for cooking-stoves	J. G. Clarke	Cincinnati, Ohio	July 11, 1865	48,773
Grate for cooking-stoves	J. T. Davy	Troy, N. Y	Apr. 16, 1850	7,284
Grate for cooking-stoves	L. Emmons	Hamilton, Ohio	Jan. 28, 1868	73,705
Grate for cooking-stoves	D. P. Foster	Shelburne Falls, Mass	July 25, 1865	48,923
Grate for cooking-stoves	J. H. Roelker	Evansville, Ind	Mar. 23, 1869	88,213
Grate for cooking-stoves	W. H. Whitehead	Chicago, Ill	Oct. 15, 1867	69,879
Grate for cooking-stoves, Portable summer	S. Spoor	Troy, N. Y	July 30, 1872	129,907
Grate for fire-places, Revolving	A. Brase and L. Salladey	Sciotoville, Ohio	May 17, 1870	103,134
Grate for furnaces, Basket	J. J. Heindl	New York, N. Y	Apr. 7, 1868	76,444
Grate for furnaces, Circular	G. L. Smith	Brooklyn, N. Y	Sept. 24, 1867	69,134
Grate for furnaces, Drop	W. Shepherd	New York, N. Y	Aug. 15, 1871	118,062
Grate for furnaces, stoves, &c	P. W. Pratt	Abington, Mass	June 4, 1872	127,514
Grate for furnaces, Water	R. L. Walker	Globe Village, Mass	Nov. 19, 1867	71,251
Grate for heating stoves	G. Nimmo	Jersey City, N. J	Oct. 23, 1866	59,057
Grate for locomotive-engines	J. W. Pole	Philadelphia, Pa	July 27, 1858	21,021
Grate for steam-boiler furnaces	G. L. Smith	Brooklyn, N. Y	June 27, 1865	48,455
Grate for steam-boiler furnaces	J. Zeh	Vienna, Austria	June 13, 1865	48,247
Grate for steam-boilers	F. A. Hull	Belvidere, Ill	Sept. 11, 1860	29,974
Grate for steam boilers	J. Montgomery	Brooklyn, N. Y	July 27, 1858	21,013
Grate for steam-boilers, Fire	R. Eaton	Lee, England	Jan. 1, 1867	60,706
Grate for steam-boilers, Hollow	C. E. Hutson	Commerce, Mo	Mar. 8, 1870	100,631
Grate for steam-boilers, Tubular	S. C. Sturtevant	Cleveland, Ohio	June 11, 1861	32,541
Grate for steam-boilers, Water	T. Stone	Carbondale, Ill	Oct. 24, 1871	120,342
Grate for steam-engines	R. Winans	Baltimore, Md	Apr. 6, 1858	19,890
Grate for steam-generator and other furnaces	B. T. Babbitt	New York, N. Y	June 29, 1869	91,999
Grate for steam-generators, Circulating	J. Braden	Indianapolis, Ind	Nov. 2, 1869	96,354
Grate for stoves and fire-places	W. B. Sutor	Indiana, Pa	Apr. 8, 1873	137,733
Grate for stoves and furnaces	N. A. Boynton	New York, N. Y	May 14, 1872	126,777
Grate for stoves and furnaces	J. B. McIntosh	Erie, Pa	Sept. 28, 1869	95,250
Grate for stoves and furnaces	J. J. Richardson	Brooklyn, N. Y	Aug. 19, 1873	141,891
Grate for stoves and furnaces	G. Vander Heyden	Troy, N. Y	Dec. 1, 1863	40,781
Grate for stoves, Basket	F. H. Root	Buffalo, N. Y	Apr. 27, 1869	89,347
Grate for stoves, ranges, and heaters	R. Whiting and A. Hamilton	New York, N. Y	Aug. 4, 1868	80,689
Grate for stoves, Revolving	E. Draper	Oskaloosa, Iowa	Feb. 13, 1866	52,548
Grate-frame, Base for summer-pieces and	A. Brown and W. Patterson	Brooklyn, E. D., N. Y	Feb. 8, 1870	99,631
Grate-front and oven, Combined	J. W. Gillespie and W. Hughes	Alliance, Ohio	June 14, 1870	104,136
Grate, Furnace	A. Adams	Stamford, Conn	Apr. 2, 1867	63,354
Grate, Furnace	J. Alexander	Green Point, N. Y	Jan. 1, 1867	60,842
Grate, Furnace	A. J. Allen and W. S. Hudson	Paterson, N. J	May 25, 1858	20,316
Grate, Furnace	I. R. Barbour	Hannibal, Mo	July 13, 1869	92,562
Grate, Furnace	C. F. Cory	Lebanon, Ill	July 24, 1860	29,248
Grate, Furnace	R. W. and D. Davis	Flushing and Long Island City, N. Y.	Apr. 17, 1866	54,064
Grate, Furnace	R. C. Graves	Barnesville, Ohio	Apr. 30, 1872	126,285
Grate, Furnace	T. T. Holdsworth	Brooklyn, N. Y	Aug. 18, 1863	39,573
Grate, Furnace	R. A. Hutchinson	Bergen, N. J	Sept. 27, 1870	107,780
Grate, Furnace	C. Kugler	Barnesville, Ohio	Apr. 23, 1872	126,060
Grate, Furnace	E. Langen	Cologne, Prussia	Nov. 14, 1865	50,996
Grate, Furnace	D. Lasher	Brooklyn, N. Y	Jan. 7, 1862	34,064
Grate, Furnace	W. B. LeVan	Philadelphia, Pa	Aug. 19, 1873	141,936
Grate, Furnace	E. I. McCarthy	Saugerties, N. Y	July 19, 1859	24,815
Grate, Furnace	G. R. Moore	Philadelphia, Pa	Apr. 4, 1871	113,324
Grate, Furnace	C. M. Northrup	New York, N. Y	July 25, 1871	117,319
Grate, Furnace	A. L. Pennock	Upper Darby, Pa	Sept. 27, 1870	107,715
Grate, Furnace	T. E. Purchase	Danville, Pa	Feb. 14, 1860	27,155
Grate, Furnace	J. Reynolds	Philadelphia, Pa	July 9, 1867	66,628
Grate, Furnace	M. M. Rounds	New Haven, Conn	Mar. 12, 1861	31,676
Grate, Furnace	H. Ryder	Somerville, Mass	Mar. 11, 1873	136,673
Grate, Furnace	W. H. Short	Brooklyn, N. Y	Jan. 3, 1865	45,760
Grate, Furnace	B. F. Smith	New Orleans, La	Sept. 12, 1871	118,884
Grate, Furnace	G. L. Smith	Brooklyn, N. Y	Nov. 24, 1863	40,712
Grate, Furnace	E. F. Steele	Wallingford, Conn	July 11, 1871	117,009
Grate, Furnace	T. Von Bolzano	Schlan, Austria	Jan 28, 1873	135,390
Grate, Furnace	R. Van Velthoven	Philadelphia, Pa	Apr. 19, 1859	23,729
Grate, Furnace	C. Van Wagenen	New York, N. Y	May 28, 1872	127,203
Grate, Furnace	A. B. Weeks	Rockland, Me	Oct. 4, 1870	107,987
Grate, Furnace	C. Whittier	Roxbury, Mass	Jan. 23, 1866	52,235
Grate, Furnace	S. L. Wiegand and W. B. Le Van.	Philadelphia, Pa	Jan. 5, 1864	41,181
Grate, Furnace	W. A. Wilson and J. Smith	Liverpool, England	Aug. 27, 1867	68,139
Grate, Furnace	A. Winterburn	Albany, N. Y	Nov. 3, 1863	40,532
Grate, Furnace	J. Withington	Blossburgh, Pa	Aug. 19, 1873	142,068
Grate, Furnace and fire-place	G. Warriner	Little Ilford, England	Feb. 12, 1867	61,964
Grate furnace for locomotive-boilers, Multiple	F. Harbach	Cleveland, Ohio	Jan. 30, 1849	6,076
Grate, Hot-air	T. F. Randolph	Cincinnati, Ohio	Dec. 22, 1863	41,019
Grate, Hot-air-furnace	E. S. Reunick	New York, N. Y	Aug. 19, 1868	81,109
Grate in cooking-stoves, Apparatus for raising the	B. K. Maltby	Cleveland, Ohio	June 18, 1850	7,442
Grate in heating-stoves	G. Moody	Falmouth, Me	May 11, 1869	89,880
Grate, Kitchen	L. Disbrow	New York	Nov. 22, 1830	......
Grate, Locomotive and furnace	M. E. Brown	Buffalo, N. Y	Apr. 26, 1864	42,526
Grate, Locomotive-fire	R. Eaton	Lee, England	Oct. 30, 1866	59,329
Grate, Locomotive-furnace	G. R. Comstock	Manheim, N. Y	Jan. 29, 1856	14,153
Grate, Open	J. Easterly	Troy, N. Y	May 16, 1842	2,627
Grate, Open coal	C. Crawford	East Cleveland, Ohio	May 29, 1866	55,065
Grate or furnace, Anthracite coal-stove	J. Atwater	New Haven, Conn	Aug. 23, 1833	......
Grate or lining of fire-pot, Fire	D. H. Dean	Lowell, Mass	June 9, 1857	17,540
Grate or parlor-stove, Open	O. Jenks	Albany, N. Y	Nov. 16, 1841	2,357
Grate or portable fire-place	J. Williamson	Washington, D. C	Nov. 8, 1836	......
Grate, Parlor	W. Anderson	New York, N. Y	June 25, 1836	......
Grate, Parlor	D. Lamoureux	New York, N. Y	Sept. 6, 1859	25,338
Grate, Parlor	J. Wilson	New York, N. Y	Sept. 13, 1845	4,191

Index of patents issued from the United States Patent Office from 1790 *to* 1873, *inclusive*—Continued.

Invention.	Inventor.	Residence.	Date.	No.
Grate, Pendulum	N. Winslow	Portland, Me	July 2, 1836	....
Grate, Portable camp	L. D. Gavitt	Los Angeles, Cal	Jan. 17, 1871	110, 967
Grate, Quadrant hinged	G. H. Thatcher	Albany, N. Y	Aug. 5, 1851	8, 277
Grate, Reverberating	R. Mayo	Richmond, Va	Nov. 1, 1825	....
Grate, Revolving	P. J. Boris	Fairfax, Nova Scotia	Jan. 10, 1865	45, 895
Grate, Revolving	C. Evans	Charlestown, Mass	Dec. 25, 1855	14, 007
Grate, Revolving and sifting	E. B. M. Hughes	New Haven, Conn	July 29, 1837	312
Grate, Revolving coal	J. B. Chollar	West Troy, N. Y	Aug. 27, 1850	7, 593
Grate, Revolving horizontal coal	J. F. Weishampel	Baltimore, Md	June 19, 1849	6, 541
Grate, Rocking	J. Seddon	Susquehanna Depot, Pa	Sept. 23, 1873	143, 034
Grate, Rotary stove	A. Harrison	Philadelphia, Pa	Oct. 5, 1852	9, 297
Grate, Sliding	J. C. Howard	Hampton, Conn	Feb. 13, 1836	....
Grate, Sliding	G. W. Walker	Boston, Mass	Feb. 25, 1862	34, 536
Grate, Sliding-flue	D. Desmond	New York, N. Y	Dec. 26, 1837	541
Grate, Steam-boiler	A. M. Searles	Cincinnati, Ohio	Sept. 30, 1856	15, 825
Grate, Steam-furnace	E. H. Jones	West Albany, N. Y	June 20, 1865	48, 286
Grate, Stove	A. J. Arnold	Columbus, Nebr	Dec. 24, 1872	134, 238
Grate, Stove	D. S Baker	West Bloomfield, N. Y	May 12, 1868	77, 707
Grate, Stove	M. R. Barr	Erie, Pa	June 27, 1871	116, 396
Grate, Stove	P. Boyden	Amsterdam, N. Y	May 23, 1871	115, 021
Grate, Stove	L. Bridge	Philadelphia, Pa	June 14, 1870	104, 258
Grate, Stove	A. Brown	Troy, N. Y	Mar. 25, 1862	34, 731
Grate, Stove	A. Brown	Troy, N. Y	Dec. 11, 1866	60, 334
Grate, Stove	E. Card	Pawtucket, R. I	Dec. 7, 1869	97, 477
Grate, Stove	G. Chilson	Boston, Mass	Apr 7, 1868	76, 302
Grate, Stove	G. Chilson	Boston, Mass	June 30, 1868	79, 312
Grate, Stove	J. Churchman	Burlington, N. J	May 16, 1871	114, 762
Grate, Stove	W. Craven	Cincinnati, Ohio	Aug. 18, 1868	81, 139
Grate, Stove	S. P. Davis	Constantine, Mich	June 13, 1871	115, 827
Grate, Stove	P. Dempsey	Erie, Pa	May 24, 1870	103, 307
Grate, Stove	E. Dewitt	Belvidere, N. J	Jan. 21, 1873	135, 095
Grate, Stove	E. Dewitt	Belvidere, N. J	Jan. 21, 1873	135, 096
Grate, Stove	W. Doyle	Albany, N. Y	Nov. 28, 1871	121, 236
Grate, Stove	W. Doyle	Albany, N. Y	June 11, 1872	127, 684
Grate, Stove	S. Duncan	Syracuse, N. Y	Feb. 4, 1873	135, 471
Grate, Stove	G. W. Eltonhead	Saint Louis, Mo	Oct. 7, 1873	143, 443
Grate, Stove	M. P. Farnham	Janesville, Wis	Mar. 26, 1867	63, 150
Grate, Stove	D. A. Flood and D. W. Brown	Woodbridge, N. J	June 6, 1871	115, 590
Grate, Stove	C. O. Foley	Troy, N. Y	Sept. 26, 1871	119, 343
Grate, Stove	C. Fulton	Rochester, N. Y	Apr. 27, 1869	89, 304
Grate, Stove	J. Glass	Troy, N. Y	Aug. 8, 1865	49, 344
Grate, Stove	W. A. Green	Brooklyn, N. Y	Nov. 21, 1871	121, 097
Grate, Stove	W. Hagerty	Philadelphia, Pa	Dec. 2, 1873	145, 172
Grate, Stove	W. Hailes	Albany, N. Y	May 11, 1869	89, 863
Grate, Stove	L. F. Hake	Salem, Ohio	Feb. 19, 1867	62, 265
Grate, Stove	R. Ham	Troy, N. Y	Aug. 9, 1870	106, 154
Grate, Stove	E. Harrington	Boston, Mass	Nov. 18, 1862	36, 952
Grate, Stove	A. M. Harris	Columbus, Ga	Feb. 18, 1873	135, 909
Grate, Stove	T. Hartley	Bridgeport, Ohio	Nov. 28, 1871	121, 363
Grate, Stove	C. R. Harvey and J. H. Foote	New York, N. Y	May 2, 1871	114, 298
Grate, Stove	L. W. Harwood	Troy, N. Y	Nov. 13, 1860	30, 620
Grate, Stove	L. W. Harwood	Troy, N. Y	Sept. 8, 1863	39, 811
Grate, Stove	D. Hathaway	Troy, N. Y	Sept. 13, 1864	44, 188
Grate, Stove	D. Hathaway	Troy, N. Y	Nov. 9, 1869	96, 588
Grate, Stove	D. Hathaway	Troy, N. Y	Dec. 21, 1869	98, 059
Grate, Stove	B. F. Holbrook and E. B. Rumrill.	Boston, Mass	Sept. 8, 1868	81, 904
Grate, Stove	W. J. Keep	Troy, N. Y	June 3, 1873	139, 583
Grate, Stove	H. G. Leonard	Taunton, Mass	May 7, 1867	64, 430
Grate, Stove	E. C. Loud	Springfield, Mass	Mar. 8, 1870	100, 536
Grate, Stove	J. Magee	Boston, Mass	Dec. 10, 1861	33, 919
Grate, Stove	A. J. Magoon	Providence, R. I	Aug. 4, 1868	80, 645
Grate, Stove	H. C. March	Lawrenceville, Pa	July 29, 1862	36, 055
Grate, Stove	W. L. McDowell	Philadelphia, Pa	Oct. 14, 1862	36, 662
Grate, Stove	W. McIlvain	Philadelphia, Pa	Oct. 3, 1865	50, 258
Grate, Stove	H. Miner	Green Island, N. Y	Apr. 19, 1870	102, 143
Grate, Stove	E. Mingay	Boston, Mass	May 1, 1860	28, 102
Grate, Stove	E. Moncuse and L. Duparquet	New York, N. Y	July 7, 1868	79, 590
Grate, Stove	G. R. Moore	Philadelphia, Pa	July 18, 1871	117, 194
Grate, Stove	G. R. Moore	Philadelphia, Pa	Sept. 12, 1871	118, 874
Grate, Stove	G. R. Moore	Philadelphia, Pa	Dec. 9, 1873	145, 360
Grate, Stove	J. Moran	Washington, D. C	Sept. 7, 1869	94, 627
Grate, Stove	R. Moss	Philadelphia, Pa	Nov. 14, 1871	120, 992
Grate, Stove	D. H. Nation	Albany, N. Y	July 24, 1860	29, 300
Grate, Stove	D. E. Paris	Troy, N. Y	Dec. 31, 1867	72, 754
Grate, Stove	J. W. Parnell	Troy, N. Y	Sept. 25, 1860	30, 182
Grate, Stove	H. W. Pell	Rome, N. Y	Nov. 26, 1872	133, 477
Grate, Stove	G. A. Philip	New York, N. Y	Sept. 4, 1847	5, 277
Grate, Stove	G. H. Philips and W. H. Johnson.	Troy, N. Y	Sept. 29, 1863	40, 146
Grate, Stove	J. A. Price	Scranton, Pa	Oct. 10, 1871	119, 787
Grate, Stove	W. Quay and E. M. Hinsdale	Troy, N. Y	Sept. 6, 1870	107, 197
Grate, Stove	S. H. Ransom and B. Burton	Albany, N. Y	June 11, 1872	127, 795
Grate, Stove	H. J. Ruggles	West Poultney, Vt	Nov. 8, 1851	8, 535
Grate, Stove	W. Sanford	Brooklyn, N. Y	Dec. 12, 1871	121, 733
Grate, Stove	J. J. Savage	Troy, N. Y	July 16, 1867	66, 743
Grate, Stove	I. Smith	Albany, N. Y	Nov. 27, 1860	30, 769
Grate, Stove	I. Smith	Albany, N. Y	Dec. 16, 1862	37, 204
Grate, Stove	J. C. Smith	Troy, N. Y	July 21, 1868	80, 233
Grate, Stove	S. Smyth	East Bridgewater, Pa	June 24, 1873	140, 224
Grate, Stove	R. Solliday	Allentown, Pa	May 17, 1870	103, 098
Grate, Stove	S. B. Stewart	Brush Valley, Pa	Dec. 3, 1867	71, 659
Grate, Stove	J. R. Stone	Round Grove, Kans	Feb. 11, 1873	135, 735
Grate, Stove	W. H. Stryker	Syracuse, N. Y	Oct. 7, 1873	143, 541
Grate, Stove	D. Stuart and L. Bridge	Philadelphia, Pa	Dec. 2, 1873	145, 253
Grate, Stove	L. Stuck	Bryan, Ohio	May 10, 1870	102, 880
Grate, Stove	W. Teamer	Evansville, Ind	July 12, 1870	105, 384
Grate, Stove	J. M. Thatcher	Bergen, N. J	Dec. 6, 1870	109, 968

Index of patents issued from the United States Patent Office from 1790 *to* 1873, *inclusive*—**Continued.**

Invention.	Inventor.	Residence.	Date.	No.
Grate, Stove	W. S. Thomas	Kendallville, Ind	Jan. 31, 1871	111, 493
Grate, Stove	G. Troh	Philadelphia, Pa	Feb. 22, 1870	100, 024
Grate, Stove	C. Truesdale	Cincinnati, Ohio	July 20, 1869	92, 906
Grate, Stove	C. Truesdale	Cincinnati, Ohio	Nov. 9, 1869	96, 743
Grate, Stove	J. S. Van Buren	Troy, N. Y	June 17, 1873	139, 934
Grate, Stove	G. Vander Hayden	Troy, N. Y	Jan. 10, 1865	45, 881
Grate, Stove	G. W. Walker	Boston, Mass	Jan. 10, 1865	45, 883
Grate, Stove	R. Ward	Edinburgh, Ind	Aug. 16, 1870	106, 521
Grate, Stove	C. Waters and H. A. Brown	Poughkeepsie, N. Y	Nov. 26, 1867	71, 345
Grate, Stove	M. D. Welman	Allegheny City, Pa	Jan. 18, 1870	99, 041
Grate, Stove	R. Wilson	Half Moon, N. Y	June 27, 1848	5, 652
Grate, Stove	G. A. Wing	Albany, N. Y	Feb. 2, 1869	86, 616
Grate, Stove	A. Winterburn	Albany, N. Y	June 14, 1870	104, 242
Grate, Stove	G. D. Woodworth	Chicago, Ill	Oct. 20, 1868	83, 232
Grate, Stove	J. H. Yocum	Ashland, Pa	July 7, 1868	97, 716
Grate, Stove and furnace	J. V. B. Carter	Albany, N. Y	July 24, 1860	29, 245
Grate, Stove and furnace	W. T. Coggeshall	Fall River, Mass	Oct. 20, 1857	18, 434
Grate, Stove and furnace	J. W. Griswold	Philadelphia, Pa	Jan. 28, 1868	73, 800
Grate, Stove and furnace	B. Gommenginger and C. W. Trotter.	Rochester, N. Y	July 28, 1868	80, 280
Grate, Stove and furnace	H. W. Pell	Rome, N. Y	Oct. 29, 1872	132, 688
Grate, Stove-furnace	G. A. Wing	Albany, N. Y	Aug. 17, 1869	93, 781
Grate, Stove or furnace	G. A. Wing	Albany, N. Y	Aug. 16, 1870	106, 442
Grate, Stove-oven	W. G. James	Richland Centre, Wis	Feb. 14, 1871	111, 850
Grate, Stove-pipe	J. O. Malley	Montreal, Canada	Nov. 8, 1870	109, 043
Grate, Tubular	B. Garvin and R. J. Pettibone	Oshkosh, Wis	Feb. 12, 1867	61, 022
Grate, Tubular	B. Garvin and R. J. Pettibone	Oshkosh, Wis	Sept. 13, 1870	107, 356
Grate, Tubular	E. Thayer	Worcester, Mass	Mar. 28, 1865	47, 053
Grate, Union or double	J. Wilson	New York	Sept. 12, 1834	......
Grate, Water	W. A. L. Kirk	Hamilton, Ohio	Oct. 23, 1866	59, 130
Grate, Water	J. Ryan	Saint Louis, Mo	Jan. 31, 1871	111, 387
Grates and dampers for chimneys, Arrangement of	J. Cohen	New York, N. Y	Apr. 15, 1856	14, 650
Grates and stools, Fire-back for	J. Habermehl	Wheeling, W. Va	Nov. 26, 1867	71, 297
Grates, Apparatus for raising	J. W. Allyn	Philadelphia, Pa	Nov. 1, 1864	44, 844
Grates, Ash-sifting device for	J. Beesley	Philadelphia, Pa	Jan. 22, 1867	61, 386
Grates, Casting fire	H. Ansbutz	Allegheny, Pa	Jan. 14, 1873	134, 832
Grates, Coal-stirrer for furnace	W. R. Nichols and B. C. Boyes	Philadelphia, Pa	Apr. 23, 1850	7, 306
Grates, Constructing and arranging stove	G. Chilson	Boston, Mass	Sept. 11, 1841	2, 246
Grates, &c., Draft and ventilating device for open	J. M. Crockett	Newbern, Va	Dec. 3, 1867	71, 711
Grates, Construction of grates for lime-kilns	W. B. Hill	Bellevue, Mich	July 30, 1841	2, 206
Grates, Hoisting	G. Stowe	Braceville, Ohio	July 14, 1868	79, 873
Grates, Raising and lowering stove	J. Dutcher	New York, N. Y	June 30, 1838	822
Grates, Shade and draft-blower for	F. McCarthy	Chicago, Ill	Jan. 7, 1873	134, 556
Grates, Shaker-bar of stove	G. W. Gardner	Troy, N. Y	Feb. 3, 1857	16, 538
Grater	N. Amas	Saugus, Mass	Oct. 13, 1857	18, 422
Grater	G. Booth	Philadelphia, Pa	Apr. 22, 1873	138, 122
Grater	J. A. Hard	Lawrence, Kans	July 9, 1872	128, 796
Grater	W. Kinyon and J. Maxson	Scott, N. Y	Mar. 22, 1870	101, 023
Grater	J. A. Latham	New Haven, Conn	Mar. 24, 1868	75, 931
Grater	L. Schmidt	Lancaster, Pa	Mar. 15, 1870	100, 809
Grater	J. M. Smith	Seymour, Conn	June 7, 1870	104, 071
Grater	H. Stone	Williamsburgh, N. Y	Aug. 27, 1867	68, 256
Grater	E. B. Strong	Buffalo, N. Y	Aug. 27, 1835	......
Grater	H. C. White	Philadelphia, Pa	Apr. 25, 1871	114, 070
Grater and egg-beater	W. A. Bemis	Spencer, Mass	June 26, 1866	55, 807
Grater and slicer	C. W. Saladee and J. H. Hall	Newark, Ohio, and Pittsburgh, Pa.	June 2, 1868	78, 612
Grater, Apple	U. Emmons	New York	Feb. 13, 1828	......
Grater, Apple, &c., revolving	D. Flagg, jr	New York	Dec. 20, 1830	......
Grater, Cider-mill	W. Barr	Ypsilanti, Mich	May 20, 1873	139, 105
Grater, Cocoa-nut	W. H. McCall	Philadelphia, Pa	July 16, 1872	129, 288
Grater, Corn	G. C. Richards, jr	Philadelphia, Pa	Apr. 30, 1872	126, 333
Grater, Corn	B. Taylor	Philadelphia, Pa	Feb. 12, 1856	14, 259
Grater, Corn and apple	C. Moore	Guilford County, N. C	Dec. 30, 1826	......
Grater for carrots, &c	E. Stimson	Montpelier, Vt	Aug. 14, 1866	57, 213
Grater for spice and fruit	H. S. Shepardson	Shelburne Falls, Mass	Sept. 5, 1865	49, 836
Grater, Green-corn	G. L. Witsil	Philadelphia, Pa	Nov. 16, 1869	97, 019
Grater, Nutmeg	H. J. Amerling	Erie, Pa	Apr. 9, 1872	125, 428
Grater, Nutmeg	L. V. Badger	Chicago, Ill	Jan. 8, 1867	61, 037
Grater, Nutmeg	H. H. Barstow	Chicago, Ill	July 21, 1868	80, 116
Grater, Nutmeg	G. Blanchard	New York, N. Y	July 15, 1856	15, 325
Grater, Nutmeg	W. Bradley	Lynn, Mass	July 25, 1854	11, 351
Grater, Nutmeg	W. Bradley	Lynn, Mass	Jan. 29, 1867	61, 511
Grater, Nutmeg	H. Carsley	Lynn, Mass	Nov. 20, 1855	13, 834
Grater, Nutmeg	R. H. Chinn	Washington, D. C	July 16, 1867	66, 675
Grater, Nutmeg	J. L. and D. H. Coles	New York, N. Y	July 28, 1868	80, 456
Grater, Nutmeg	C. A. Durgin	New York, N. Y	July 16, 1867	66, 689
Grater, Nutmeg	C. L. Gilpatric	South Dedham, Mass	Oct. 9, 1866	58, 632
Grater, Nutmeg	E. A. Goodes	Philadelphia, Pa	July 28, 1868	80, 347
Grater, Nutmeg	H. H. Herrick	Boston, Mass	Nov. 1, 1864	44, 865
Grater, Nutmeg	J. A. Hooper	South Berwick, Me	Mar. 5, 1867	62, 542
Grater, Nutmeg	J. Lofoendahl	Boston, Mass	June 13, 1865	48, 237
Grater, Nutmeg	T. Marriatt	Detroit, Mich	Dec. 23, 1873	145, 884
Grater, Nutmeg	W. W. Owen and D. Kelly	Muskegon, Mich	Dec. 17, 1867	72, 223
Grater, Nutmeg	A. L. Platt	Bloomington, Ill	Apr. 22, 1873	138, 191
Grater, Nutmeg	G. C. Richards, jr	Philadelphia, Pa	July 16, 1872	129, 426
Grater, Nutmeg	J. Riddle and B. Allen	Boston, Mass	Oct. 30, 1866	59, 266
Grater, Nutmeg	J. G. Roth	New York, N. Y	Apr. 14, 1868	76, 820
Grater, Nutmeg	A. S. Skillin and G. W. Reed	Portland, Me	Sept. 3, 1867	68, 463
Grater, Nutmeg	L. Von Froben	Washington, D. C	Aug. 6, 1867	67, 616
Grater, Nutmeg	D. C. Warner	Chicago, Ill	Jan. 24, 1871	111, 284
Grater, Nutmeg	R. W. Whitney and J. P. Davis	South Berwick, Me	July 23, 1867	67, 010
Grater, Nutmeg	G. L. Witsil	Philadelphia, Pa	May 6, 1862	35, 192
Grater, Nutmeg	C. Worden	Binghamton, N. Y	Sept. 24, 1867	69, 150
Grater, Revolving	S. S. Wilcox and E. J. Colegrove.	Lincklaen, N. Y	Jan. 18, 1870	99, 042
Grater, Rotary vegetable	W. E. Knight	Shrewsbury, Vt	Oct. 19, 1869	95, 913
Grater, Spice	H. W. Oliver	New Haven, Conn	July 9, 1867	66, 619

Index of patents issued from the United States Patent Office from 1790 to 1873, inclusive—Continued.

Invention.	Inventor.	Residence.	Date.	No.
Grater, Vegetable	G. Booth	Philadelphia, Pa	Apr. 22, 1873	138, 123
Grater, Vegetable	T. W. Houchin	Morrisania, N. Y	Apr. 30, 1872	126, 144
Grater, Vegetable	J. Keagy	Newark, Ohio	Dec. 3, 1872	133, 585
Grater, Vegetable	S. A. McGill	Cincinnati, Ohio	Mar. 6, 1866	53, 021
Grater, Vegetable	A. S. McNeir and W. D. Stockton.	Philadelphia, Pa	May 20, 1873	139, 173
Grater, Vegetable	C. R. Peirce	Philadelphia, Pa	Aug. 8, 1871	117, 924
Grater, Vegetable	J. Wehrle and W. Wittlinger	Cincinnati, Ohio	Mar. 21, 1871	112, 872
Grater, Vegetable	S. M. Wilson	New York, N. Y	Jan. 11, 1870	98, 832
Grating for stove, furnace, &c., Supplementary	B. F. Foering	Philadelphia, Pa	Mar. 4, 1856	14, 356
Grating, Illuminating	J. K. Ingalls	Brooklyn, N. Y	Mar. 18, 1856	14, 456
Grating machine, Vegetable	H. Arthur	Martinsburgh, N. Y	Oct. 8, 1867	69, 607
Grating, sifting, and slicing machine	J. F. Shepard	Hampton Falls, N. H	May 24, 1870	103, 378
Grave-cover, Terra-cotta	C. C. W. Morgan	Holly Springs, Miss	Nov. 5, 1872	132, 851
Grave-covering	J. R. Abrams	Greenville, Ala	Nov. 4, 1873	144, 300
Grave-guard	A. Rank	Salem, Ohio	July 8, 1873	147, 728
Grave-mound	J. Meley	Trenton, Tenn	Oct. 13, 1868	83, 077
Grave-mound cover	B. Hunter	Philadelphia, Pa	Nov. 11, 1873	144, 392
Grave-stone, monument, &c., Composition	E. A. Locke	Boston, Mass	Sept. 3, 1872	131, 010
Gravel and sand heater	P. Le Goullon	Pittsburgh, Pa	Jan. 10, 1871	110, 861
Gravel-heater	W. A. Gay	Newark, N. J	Jan. 25, 1870	99, 077
Gravel-heating apparatus	W. D. Andrews	Brookhaven, N. Y	June 25, 1872	128, 351
Gravel-pan	H. Franke	Brooklyn, N. Y	Mar. 15, 1870	100, 745
Gravel-spreader, Railway	J. C. Casement and J. Elliott	Cleveland, Ohio, and Erie, Pa.	Mar. 9, 1869	87, 542
Graver	R. S. Mershon	Zanesville, Ohio	Dec. 24, 1867	72, 521
Graver	F. R. Stockton	New York, N. Y	Feb. 20, 1866	52, 769
Gravimotometer	J. W. Whetmore	Erie, Pa	Feb. 16, 1858	19, 392
Grease and oils from animal and vegetable substances, Extracting.	J. Besso	Philadelphia, Pa	June 25, 1867	66, 119
Grease-cup	G. Hagenmeyer	Big River, Cal	Aug. 8, 1865	49, 259
Grease-cup	G. Hagenmeyer	Big River, Cal	Apr. 17, 1866	53, 973
Grease from cloth, &c., Composition for removing	S. Guess	Boston, Mass	Aug. 26, 1845	4, 161
Grease from slush, Apparatus for separating	D. H. Kaufman	Kokomo, Ind	July 4, 1865	48, 559
Grease or sizing	G. and R. Birtwistle	Fall River, Mass	Feb. 11, 1868	74, 288
Grease-trap	E. Whiteley	Cambridge, Mass	Aug. 3, 1869	93, 258
Green-brier, Obtaining useful products from the berries of the.	P. Baumgras	Syracuse, N. Y	May 17, 1864	42, 814
Green-houses, Portable steam-apparatus for	E. Whiteley	Cambridge, Mass	Aug. 3, 1869	93, 380
Green, Manufacture of Paris	T. Schwartz	New York, N. Y	Apr. 17, 1849	6, 327
Grenade, Hand	G. P. Ganster	New York, N. Y	Apr. 19, 1864	42, 363
Grenade, Hand	G. P. Ganster and I. S. Schuyler	New York, N. Y	Feb. 16, 1864	41, 615
Grenade, Hand	W. F. Ketchum	Buffalo, N. Y	Aug. 20, 1861	33, 089
Grenade, Igniting hand	J. S. Adams	Taunton, Mass	Jan. 10, 1865	45, 806
Griddle	W. Bennett	New York, N. Y	May 26, 1857	17, 361
Griddle	T. D. Gail	Waukegan, Ill	Feb. 14, 1871	111, 835
Griddle	B. Gilbert	Pittsburgh, Pa	Oct. 11, 1853	10, 118
Griddle	E. A. Jeffery	Trappe, Md	Aug. 27, 1867	68, 202
Griddle	S. Kennedy	Allegheny City, Pa	Aug. 26, 1873	142, 164
Griddle	G. W. Pittock	Union Mills, N. Y	Sept. 11, 1860	30, 028
Griddle	C. Wilson	New York, N. Y	Feb. 20, 1855	12, 419
Griddle-greaser	W. H. Bixler	Easton, Pa	July 22, 1873	140, 992
Griddle-greaser	L. Ward	Poughkeepsie, N. Y	July 18, 1871	117, 130
Griddle-lifter	M. D. Murphy	Watkins, N. Y	July 4, 1871	116, 739
Griddle or cooking-utensil	E. J. Smith	Washington, D. C	Jan. 22, 1867	61, 478
Gridiron	W. Andrews	Alleghany County, Md	Mar. 3, 1868	75, 107
Gridiron	D. Ball	Ballston Spa, N. Y	Aug. 29, 1822	......
Gridiron	A. Bataille	New York, N. Y	Mar. 18, 1873	136, 955
Gridiron	W. Bennett	New York, N. Y	Sept. 30, 1856	15, 799
Gridiron	E. C. Brewster	Bristol, Conn	Feb. 9, 1864	41, 475
Gridiron	J. S. Brooks and L. B. Grover	Rochester, N. Y	July 24, 1860	29, 238
Gridiron	F. Bush and L. Pratt	Meriden, Conn	July 21, 1832	......
Gridiron	A. Clark	Berlin, Conn	Mar. 28, 1834	......
Gridiron	I. Damon	Northampton, Mass	Jan. 30, 1841	1, 960
Gridiron	C. Denn	Frankford, Pa	June 23, 1868	79, 056
Gridiron	W. A. Greene and J. G. Treadwell.	Albany, N. Y	Mar. 29, 1859	23, 367
Gridiron	J. Hawkins	West Windsor, N. J	Mar. 26, 1845	3, 973
Gridiron	G. B. Isham	Burlington, Vt	Feb. 16, 1869	86, 923
Gridiron	C. H. Mock	Quincy, Ill	Aug. 11, 1868	80, 831
Gridiron	C. Noble	Philadelphia, Pa	Oct. 31, 1871	120, 528
Gridiron	M. V. Nobles	Elmira, N. Y	Nov. 23, 1869	97, 217
Gridiron	J. T. Page	Rochester, N. Y	Feb. 26, 1867	62, 437
Gridiron	E. B. Phelps	New York, N. Y	Dec. 8, 1868	84, 758
Gridiron	J. Powers	Lansingburgh, N. Y	Aug. 10, 1829	......
Gridison	E. P. Russell	Manlius, N. Y	Apr. 9, 1867	63, 753
Gridiron	J. Shavor and A. C. Corse	Troy, N. Y	Feb. 19, 1861	31, 482
Gridiron	J. G. Treadwell	Albany, N. Y	Dec. 20, 1859	26, 539
Gridiron	N. Waterman	Boston, Mass	May 1, 1847	5, 092
Gridiron	E. Webster	Hartford, Conn	Feb. 14, 1860	27, 176
Gridiron and spider	A. and G. Sizer	Meriden, Conn	Nov. 14, 1836	......
Gridiron and spider, Combined	A. and G. Sizer	Meriden, Conn	Nov. 14, 1836	78
Gridiron-case	M. V. Nobles	Elmira, N. Y	Apr. 5, 1870	101, 652
Gridiron-cover	R. Shaler	Madison, Conn	June 11, 1867	65, 614
Gridiron, Folding	J. H. Thomas	Newark, N. J	Aug. 10, 1858	21, 157
Gridiron, Gas-burning	E. D. Willard	Washington, D. C	July 17, 1855	13, 283
Gridiron, Hollow or guttered	T. Mussey	New London, Conn	Mar. 6, 1820	......
Gridiron, Rotary	K. Strong	Meriden, Conn	Oct. 28, 1820	......
Gridiron, Sheet-metal	G. Booth	Philadelphia, Pa	Aug. 22, 1865	49, 581
Gridirons, Construction of	F. Bush and L. Pratt	Meriden, Conn	Sept. 17, 1834	......
Grinder, Card	J. S. Dronsfield	Olham, Great Britain	Apr. 26, 1870	102, 236
Grinder, Cutter	N. F. Stone	Chicago, Ill	Apr. 17, 1866	54, 036
Grinder, Flock	R. Aldrich	Forestdale, R. I	Jan. 28, 1870	104, 813
Grinder or hone for razors, penknives, &c., Metallic.	C. Veltenair		Aug. 10, 1804	......
Grinder, Percussion	A. P. Stevens	Brooklyn, N. Y	Feb. 28, 1865	46, 597
Grinders, Method of forming teeth upon cast-iron	E. Ripley	Troy, N. Y	Aug. 12, 1851	8, 293

Index of patents issued from the United States Patent Office from 1790 *to* 1873, *inclusive*—Continued.

Invention.	Inventor.	Residence.	Date.	No.
Grinding and abrading surfaces, Method of employing Franklinite pig-metal for making.	T. Selleck	Greenwich, Conn	May 1, 1860	28, 107
Grinding and amalgamating machine, Shoe and die for.	F. G. Belknap	Washoe, Nev	Jan. 16, 1866	52, 015
Grinding and bolting machines, Combining	J. M. Clark	Lancaster, Pa	Jan. 1, 1850	6, 982
Grinding and bolting mill	J. Hollingsworth	Zanesville, Ohio	Nov. 18, 1851	8, 527
Grinding and chopping mill, Grain	P. Pope	Washington Township, Pa	Aug. 17, 1825	....
Grinding and crushing mill	W. Frost	New York, N. Y	Nov. 19, 1850	7, 782
Grinding and crushing mill	P. Perry	Troy, N. Y	Apr. 19, 1859	23, 702
Grinding and crushing mill	G. Sanford	Poughkeepsie, N. Y	Jan. 4, 1859	22, 515
Grinding and crushing mills, Ore-feeder for	C. P. Stanford	San Francisco, Cal	June 3, 1873	139, 523
Grinding and cutting machine	F. B. Hunt	Richmond, Ind	June 8, 1858	20, 490
Grinding and hulling mill, Universal	J. Bogardus	New York	July 29, 1841	2, 194
Grinding and polishing composition	N. A. Buhle	New York, N. Y	Aug. 7, 1866	56, 896
Grinding and polishing cylindrical concave surfaces, Apparatus for.	W. C. Hicks	New York, N. Y	Dec. 24, 1867	72, 636
Grinding and polishing knives	J. Dodge	Waterford, N. Y	Oct. 12, 1858	21, 746
Grinding and polishing machine	D. Lovejoy and G. F. Butterfield.	Lowell, Mass	Oct. 27, 1857	18, 509
Grinding and polishing machine	T. Prosser	New York, N. Y	June 18, 1867	65, 942
Grinding and polishing machine, Metal	J. H. and J. S. Lane	Akron, Ohio	Nov. 1, 1870	108, 822
Grinding and polishing machine, Metal	J. A. Whelpley	Greenwich, New Brunswick.	Dec. 31, 1867	72, 957
Grinding and polishing machine, Tool	J. A. Hendrick	Providence, Pa	Dec. 8, 1863	40, 838
Grinding and polishing metal	M. A. and J. H. Diedrichs	Baltimore, Md	Sept. 17, 1867	68, 854
Grinding and polishing metal	J. Dodge	Waterford, N. Y	May 2, 1865	47, 527
Grinding and polishing metallic surfaces, Machinery for.	R. M. Hoe	New York	May 30, 1842	2, 656
Grinding and polishing tools, Machine for	J. Vaughan, jr	Union, Me	Dec. 11, 1849	6, 949
Grinding and pounding mill	A. D. Moore	New Haven County, N. C	Aug. 7, 1813	....
Grinding and shaping metal, Apparatus for	S. Darling	Bangor, Me	Aug. 30, 1853	9, 976
Grinding-apparatus	A. Assman	Linden, N. J	Oct. 15, 1872	132, 231
Grinding-apparatus	W. L. Washburn	New York, N. Y	July 12, 1870	105, 392
Grinding artificial granite, &c., Facing-beds for	B. Hardingo	New York, N. Y	May 8, 1855	12, 820
Grinding-cylinder	J. Bridges	Troy, N. Y	July 24, 1832	....
Grinding cylinder, Toothed	B. Mackerley	New Petersburgh, Ohio	May 18, 1858	20, 282
Grinding device, Tool	D. W. Ayres	Sheldon, Il	June 23, 1868	79, 049
Grinding edge-tools	L. Zimmerman	Wakeshma, Mich	July 27, 1869	93, 031
Grinding edged tools, Device for	H. K. Trask	Beaver Dam, Wis	Apr. 6, 1869	88, 679
Grinding edge-tools, Machine for	J. D. Smith	Greig, N. Y	Feb. 23, 1869	87, 212
Grinding ledger and fly blades, planer-knives, &c	C. Hardy	Biddeford, Me	June 21, 1870	104, 584
Grinding-machine	W. Battell and M. E. Worrell	Quincy, Ill	Apr. 9, 1872	125, 434
Grinding-machine	G. T. Chattaway and J. Dickinson.	Brooklyn and New York, N. Y.	Dec. 21, 1869	98, 028
Grinding-machine	P. D. Cummings	Portland, Me	Dec. 16, 1873	145, 488
Grinding-machine	J. Flint	Rochester, N. Y	Dec. 22, 1868	85, 081
Grinding-machine	G. W. Libbey	Philadelphia, Pa	Dec. 24, 1872	134, 296
Grinding-machine	J. L. Otis	Leeds, Mass	Oct. 27, 1868	83, 405
Grinding-machine	T. H. Worrall	East Blackstone, Mass	Apr. 5, 1870	101, 559
Grinding-machine for the knives of mowing-machines.	B. F. Davis	Auburn, N. Y	Feb. 23, 1869	87, 150
Grinding machine, Metal-plate	A. R. Reynolds	Auburn, N. Y	Mar. 5, 1867	62, 566
Grinding-machinery for knives which have warped surfaces.	W. Hovey	Worcester, Mass	Dec. 18, 1847	5, 398
Grinding metal articles, Machine for	J. P. Curtiss	New Britain, Conn	Sept. 1, 1868	81, 608
Grinding metal plates, Device for	E. A. White	Rock Island, Ill	Oct. 19, 1869	96, 066
Grinding, Method of removing and discharging dust produced by.	T. Fairbanks	Saint Johnsbury, Vt	May 26, 1843	3, 105
Grinding-mill	H. Albright	Cranesville, W. Va	June 8, 1869	90, 978
Grinding-mill	G. N. Annan	Buffalo, N. Y	Sept. 1, 1868	81, 725
Grinding-mill	A. Arnold	Troy, N. Y	Sept. 15, 1857	18, 178
Grinding-mill	J. Aubin	Paris, France	June 19, 1866	55, 792
Grinding-mill	J. G. Baker	Philadelphia, Pa	June 21, 1870	104, 537
Grinding-mill	J. G. Baker	Philadelphia, Pa	Oct. 21, 1873	143, 867
Grinding-mill	J. W. Baldwin	Sidney, N. J	May 8, 1866	54, 486
Grinding-mill	S. A. Bantz and W. Andrew	Frederick, Md	Dec. 4, 1849	6, 916
Grinding-mill	A. Barber	Stephentown, N. Y	June 5, 1847	5, 143
Grinding-mill	L. B. Bealay	Lebanon, N. H	Sept. 17, 1872	131, 385
Grinding-mill	J. H. Bear	York, Pa	Jan. 29, 1867	61, 599
Grinding-mill	B. A. Beardsley	Waterville, N. Y	June 29, 1858	20, 692
Grinding-mill	T. Bennett	New York, N. Y	Jan. 4, 1859	22, 479
Grinding-mill	C. P. Benoit	Detroit, Mich	Oct. 23, 1866	58, 977
Grinding-mill	C. Bollinger	Harrisburgh, Pa	Sept. 8, 1863	39, 793
Grinding-mill	C. Bollinger	Harrisburgh, Pa	Sept. 4, 1866	57, 667
Grinding-mill	R. W. Bowman	Orangeville, Pa	Oct. 5, 1869	95, 557
Grinding-mill	D. E. Breinig	Philadelphia, Pa	Apr. 6, 1858	19, 826
Grinding-mill	A. Briggs	Harrison, Ohio	Sept. 14, 1869	94, 808
Grinding-mill	E. Brisson	Orleans, France	Feb. 10, 1863	37, 605
Grinding-mill	J. Broughton	New York, N. Y	Dec. 13, 1859	26, 412
Grinding-mill	W. Broughton	London, England	Apr. 25, 1846	4, 473
Grinding-mill	C. W. Brown	Boston, Mass	Jan. 15, 1850	7, 016
Grinding-mill	C. W. Brown	Boston, Mass	Aug. 23, 1859	25, 173
Grinding-mill	J. Bryant	Brooklyn, N. Y	Feb. 14, 1860	27, 102
Grinding-mill	C. P. Buckingham	Mount Vernon, Ohio	Dec. 6, 1859	26, 332
Grinding-mill	J. Burns	New York, N. Y	Sept. 24, 1867	69, 174
Grinding-mill	J. Burns	New York, N. Y	Jan. 16, 1872	122, 811
Grinding-mill	N. Burr	Batavia, Ill	Aug 24, 1869	94, 072
Grinding-mill	T. B. Burtis	Chicago, Ill	Dec. 19, 1865	51, 551
Grinding-mill	D. A. Caldwell	Jacksonville, Ill	Dec. 23, 1873	145, 786
Grinding-mill	J. Carl	Grenada, Miss	Aug. 23, 1859	25, 176
Grinding-mill	T. A. Chandler	Rockford, Ill	July 10, 1849	6, 583
Grinding-mill	E. Clark	Lancaster, Pa	Nov. 11, 1862	36, 893
Grinding-mill	G. W. Clark and E. Butterfield	Lowell, Mass	Oct. 3, 1848	5, 824
Grinding-mill	J. M. Clark	Lancaster, Pa	Dec. 16, 1862	37, 151
Grinding-mill	D. R. Clem	Front Royal, Va	Feb. 5, 1867	61, 714
Grinding-mill	E. Coleman	Philadelphia, Pa	June 23, 1857	17, 619
Grinding-mill	W. P. Coleman	New Orleans, La	Oct. 1, 1850	7, 679
Grinding-mill	M. Cosgro	Peoria, Ill	Apr. 2, 1867	63, 479

Index of patents issued from the United States Patent Office from 1790 *to* 1873, *inclusive*—Continued.

Invention.	Inventor.	Residence.	Date.	No.
Grinding-mill	E. H. Cotton	Manchester, N. H	Nov. 21, 1865	51, 022
Grinding-mill	A. Crease	Cleveland, Ohio	Apr. 16, 1872	125, 791
Grinding-mill	H. P. Crouse	Hartland, Mich	Nov. 8, 1864	44, 942
Grinding-mill	J. Culbertson	Covington, Ky	Jan. 6, 1857	16, 325
Grinding-mill	W. H. Culver	West Troy, N. Y	Oct. 10, 1871	119, 746
Grinding-mill	C. F. Dean	Saint Johnsbury, Vt	Oct. 9, 1866	58, 719
Grinding-mill	R. Denison and J. B. Moon	Grand Rapids, Mich	Nov. 7, 1865	50, 805
Grinding-mill	J. Donaldson	Rockford, Ill	Oct. 15, 1867	69, 903
Grinding-mill	J. F. Drummond	New York, N. Y	Dec. 10, 1867	72, 002
Grinding-mill	H. V. Duryea	Fult n, N. Y	Feb. 9, 1858	19, 289
Grinding-mill	G. Eberius and F. A. Heinig	Washington, Mo	Aug. 18, 1863	39, 558
Grinding-mill	S. T. Eck	Taneytown, Md	July 5, 1870	104, 946
Grinding-mill	J. Ellston	Cleveland, Ohio	Mar. 3, 1868	75, 135
Grinding-mill	G. Erkson	New York, N. Y	July 20, 1858	20, 941
Grinding-mill	P. Evans	New York, N. Y	Oct. 21, 1873	143, 816
Grinding-mill	L. Fagin	Cincinnati, Ohio	Oct. 30, 1849	6, 827
Grinding-mill	A. Felton	Troy, N. Y	Jan. 2, 1855	12, 181
Grinding-mill	J. Fickinger	Kingsville, Ohio	Apr. 9, 1867	63, 712
Grinding-mill	J. A. Forsman	Jamestown, Ohio	Nov. 3, 1863	40, 467
Grinding-mill	J. A. Forsman	Chicago, Ill	Sept. 13, 1870	107, 353
Grinding-mill	M. Foster	Cleveland, Ohio	Sept. 21, 1869	94, 952
Grinding-mill	S. Godfrey	Peoria, Ill	Apr. 3, 1866	53, 603
Grinding-mill	R. D. Granger	Philadelphia, Pa	Mar. 9, 1858	19, 559
Grinding-mill	T. Grey	Clarence, N. Y	Dec. 3, 1867	71, 606
Grinding-mill	Z. Griffin	Montgomery, Ala	Oct. 3, 1848	5, 829
Grinding-mill	D. Halladay	Batavia, Ill	July 19, 1870	105, 448
Grinding-mill	C. A. Harper	Canterbury, N. H	Nov. 8, 1864	44, 950
Grinding-mill	E. Harrison	New York, N. Y	Nov. 20, 1847	5, 374
Grinding-mill	E. Harrison	New Haven, Conn	June 6, 1854	11, 040
Grinding-mill	E. Harrison	New Haven, Conn	May 8, 1866	54, 535
Grinding-mill	E. Harrison	New Haven, Conn	Sept. 1, 1868	81, 780
Grinding-mill	E. Harrison	New Haven, Conn	Dec. 29, 1868	85, 444
Grinding-mill	J. T. Harvey	Murrysville, Pa	Feb. 25, 1873	136, 321
Grinding-mill	P. Hauser	Cincinnati, Ohio	Nov. 7, 1835	....
Grinding-mill	M. W. Helton and J. H. Redfield.	Bloomington, Ind	Jan. 1, 1867	60, 724
Grinding-mill	F. M. Hemphill	Newport, Ky	Mar. 29, 1859	23, 372
Grinding-mill	F. M. Hemphill and R. H. Knox.	Washington, Ohio	Oct. 10, 1848	5, 836
Grinding-mill	H. Hensley	Elysianfield, Tex	May 9, 1871	114, 558
Grinding-mill	S. S. Howard	Milton, N. Y	June 18, 1861	32, 565
Grinding-mill	S. S. Howard	Milton, N. Y	Sept. 22, 1863	40, 078
Grinding-mill	G. W. Hubbard and S. A. Smith.	Philadelphia, Pa	Mar. 10, 1868	75, 273
Grinding-mill	T. E. Hunt	Louisville, Ky	Jan. 3, 1860	26, 738
Grinding-mill	J. Hutchison	Three Rivers, Mich	June 26, 1866	55, 872
Grinding-mill	E. J. Hyde	Philadelphia, Pa	Oct. 16, 1860	30, 407
Grinding-mill	S. Hyde	Malone, N. Y	Nov. 26, 1835	....
Grinding-mill	C. and A. Kaestner	Chicago, Ill	Apr. 17, 1866	53, 989
Grinding-mill	F. Kaiser	Buffalo, N. Y	Sept. 29, 1868	82, 532
Grinding-mill	J. Kemp	Brooklyn, N. Y	Dec. 19, 1865	51, 596
Grinding-mill	B. Kenoyer	Edina, Mo	Sept. 1, 1863	39, 735
Grinding-mill	F. Klinkerman	Farmer's Retreat, Ind	Aug. 30, 1864	44, 000
Grinding-mill	P. Kraus	Augusta, Ga	May 31, 1870	103, 626
Grinding-mill	F. W. Krause and G. W. Strong	Chicago, Ill	June 18, 1861	32, 570
Grinding-mill	J. G. and W. J. Lane	Mill-brook, N. Y	Apr. 9, 1872	125, 399
Grinding-mill	F. H. La Port	Clarinda, Iowa	Apr. 4, 1871	113, 534
Grinding-mill	J. A. Lechler	Philadelphia, Pa	Feb. 2, 1864	41, 439
Grinding-mill	B. W. Leonard	Bridgeport, Conn	Jan. 12, 1858	19, 093
Grinding-mill	E. D. and E. C. Little	Shabbonas Grove, Ill	Mar. 28, 1871	113, 179
Grinding-mill	G. W. Lay and F. C. Baker	Jefferson, Tex	July 23, 1872	129, 741
Grinding-mill	J. C. Lyon and H. F. Phillips	Auburn and Seneca Falls, N. Y.	May 3, 1859	23, 850
Grinding-mill	J. C. Lyon and H. F. Phillips	Auburn and Seneca Falls, N. Y.	June 21, 1859	24, 473
Grinding-mill	D. Marsh and E. B. Nichols	Fairfield, Conn	Oct. 30, 1849	6, 834
Grinding-mill	I. and W. D. Mayfield	Mayfield, Ky	Jan. 10, 1871	110, 927
Grinding-mill	P. G. McCulla	Philadelphia, Pa	Aug. 14, 1860	29, 612
Grinding-mill	J. W. Miles	Hubbardstown Village, Mich.	July 10, 1866	56, 246
Grinding-mill	J. M. Miller	Hamilton, Ohio	Dec. 11, 1866	60, 404
Grinding-mill	S. M. Miller	Beaver Mill, Pa	Nov. 6, 1866	59, 435
Grinding-mill	G. Mitchell	Philadelphia, Pa	Aug. 9, 1870	106, 271
Grinding-mill	C. Moegling	Milwaukee, Wis	Mar. 19, 1867	62, 966
Grinding-mill	J. A. Montgomery	Crawford, N. J	Aug. 25, 1868	81, 526
Grinding-mill	J. P. Moore	Morning View, Ky	June 9, 1868	78, 814
Grinding-mill	S. Moore	Wellsburgh, Va	June 12, 1860	28, 681
Grinding-mill	J. R. Morrison	East Springfield, Ohio	Dec. 22, 1857	18, 923
Grinding-mill	E. Munson	Utica, N. Y	Apr. 3, 1860	27, 735
Grinding-mill	W. Newlove	Utica, N. Y	Oct. 14, 1851	8, 425
Grinding-mill	O. Nichols	Lowell, Mass	Oct. 12, 1852	9, 330
Grinding-mill	A. L. Norcross	Hallowell, Me	Aug. 20, 1835	....
Grinding-mill	A. H. Nordyke and D. W. Marmon.	Richmond, Ind	Aug. 1, 1871	117, 667
Grinding-mill	A. H. Nordyke and D. W. Marmon.	Richmond, Ind	Aug. 1, 1871	117, 668
Grinding-mill	E. and A. H. Nordyke	Richmond, Ind	Sept. 4, 1866	57, 754
Grinding-mill	W. B. North	Jersey City, N. J	Oct. 3, 1848	5, 831
Grinding-mill	L. Norton	Madison, Conn	Oct. 3, 1848	5, 826
Grinding-mill	H. P. and P. Nuchols	Barren County, Ky	Jan. 21, 1835	....
Grinding-mill	G. and H. O'Connor	Mishawaka, Ind	July 19, 1870	105, 592
Grinding-mill	F. Olds	Providence, R. I	Dec. 29, 1857	18, 985
Grinding-mill	J. Ormsby and T. Cohoon	Washington, D. C	Jan. 4, 1816	....
Grinding-mill	A. Orvis	Niagara, N. Y	May 31, 1859	24, 266
Grinding-mill	J. F. Ostrander	New York, N. Y	Apr. 25, 1846	4, 478
Grinding-mill	D. Paddack	Pontiac, Mich	Nov. 28, 1848	5, 944
Grinding-mill	L. Paige	Cavendish, Vt	Jan. 29, 1856	14, 164

Index of patents issued from the United States Patent Office from 1790 *to* 1873, *inclusive*—Continued.

Invention.	Inventor.	Residence.	Date.	No.
Grinding-mill	R. and S. Patterson	Newcastle-upon-Tyne, England.	Oct. 14, 1873	143, 710
Grinding-mill	W. Peck	Rockford, Ill	May 3, 1870	102, 588
Grinding-mill	J. Peirce	Buffalo, N. Y	Sept. 18, 1847	5, 298
Grinding-mill	E. Peugeot and J. B. C. Laurent.	Paris, France	Nov. 6, 1866	59, 528
Grinding-mill	J. Platt	Bridgeport, Conn	Dec. 26, 1845	4, 334
Grinding-mill	S. W. Powell	Tuscarora Valley, Pa	Dec. 4, 1849	6, 926
Grinding-mill	A. Prosens	Philadelphia, Pa	Jan. 11, 1859	22, 580
Grinding-mill	D. Read	Hamilton, N. Y	Aug. 7, 1860	29, 521
Grinding-mill	J. C. Reed	Mount Vernon, Ohio	June 6, 1854	11, 045
Grinding-mill	E. Ripley	Troy, N. Y	Apr. 21, 1857	17, 116
Grinding mill	C. H. Roberts	Evansville, Ind	Apr. 12 1870	101, 917
Grinding-mill	J. C. Roberts	Adamstown, Md	Oct. 3, 1865	50, 274
Grinding-mill	J. C. Roberts	Adamstown, Md	Jan. 1, 1867	60, 789
Grinding-mill	J. Rogers, jr	Jackson, Mich	Dec. 17, 1850	7, 844
Grinding-mill	S. G. Rollins	Boston, Mass	Aug. 20, 1872	130, 750
Grinding-mill	J. Ruof, A. Heupel, and F. Leuthy.	Lancaster, Pa	July 3, 1860	29, 010
Grinding-mill	J. H. Rusk	Philadelphia, Pa	Dec. 20, 1870	110, 397
Grinding-mill	G. Sanford	Poughkeepsie, N. Y	Mar. 9, 1858	19, 587
Grinding-mill	W. Scarlett	Kenosha, Wis	June 29, 1858	20, 734
Grinding-mill	A. H. Searfoss	Newark, N. J	Feb. 24, 1863	37, 793
Grinding-mill	G. Selsor	Philadelphia, Pa	Mar. 29, 1859	23, 431
Grinding-mill	G. Selsor	Philadelphia, Pa	Feb. 21, 1871	112, 080
Grinding-mill	H. Shaw	Cincinnati, Ohio	Aug. 17, 1869	93, 913
Grinding-mill	C. W. Shedd	Addison, Ala	Oct. 30, 1860	30, 552
Grinding-mill	O. Sherwood, jr	Independence, Iowa	Oct. 28, 1862	36, 804
Grinding-mill	N. Shoemaker	Montrose, Pa	Nov. 28, 1865	51, 227
Grinding-mill	T. J. Sloan	New York, N. Y	Apr. 24, 1866	54, 260
Grinding-mill	T. J. Sloan	New York, N. Y	Apr. 24, 1866	54, 261
Grinding-mill	T. J. Sloan	New York, N. Y	Apr. 24, 1866	54, 262
Grinding-mill	F. Smith	Evans, N. Y	Mar. 18, 1835	....
Grinding-mill	J. B. Smith	Bowensburgh, Ill	Dec. 12, 1871	121, 819
Grinding-mill	T. F. Smith	Ohio County, W. Va	Nov. 13, 1866	59, 674
Grinding-mill	E. S. Snyder	Charlestown, Va	Feb. 10, 1846	4, 382
Grinding-mill	J. Snyder	Hart's Mills, Ind	Feb. 4, 1868	74, 161
Grinding-mill	H. Southwick	Little Cooley, Pa	Mar. 2, 1858	19, 521
Grinding-mill	J. R. Stafford	Cleveland, Ohio	June 18, 1850	7, 447
Grinding-mill	O W. Stanford	Cincinnati, Ohio	Feb. 14, 1860	27, 164
Grinding-mill	W. H. Starry	Franklin, Ohio	Aug. 9, 1864	43, 802
Grinding-mill	W. Stauffer	Middlebury, Ind	Sept. 8, 1857	18, 163
Grinding-mill	W Stewart	Philadelphia, Pa	Mar. 20, 1860	27, 579
Grinding-mill	G. M. Stone	Saint Louis, Mo	Nov. 12, 1867	70, 916
Grinding-mill	J. J. Storer and J. D. Whelpley	Boston, Mass	Sept. 30, 1862	36, 580
Grinding-mill	T. B. Stout	Keyport, N. J	Dec. 16, 1856	16, 249
Grinding-mill	H. P. Straub	Cincinnati, Ohio	Mar. 20, 1866	53, 358
Grinding-mill	I. Straub	Cincinnati, Ohio	June 27, 1854	11, 181
Grinding-mill	I. Straub	Kenton County, Ky	May 28, 1867	65, 133
Grinding-mill	B. Swift	Washington, N. Y	Aug. 16, 1845	4, 149
Grinding-mill	C. N. Taylor	Cookstown, N. J	Sept. 3, 1867	68, 539
Grinding-mill	A. Thwing and C. H. Fowler	Hopedale and West Roxbury, Mass.	May 4, 1869	89, 609
Grinding-mill	G. Todd	Saint Louis, Mo	Apr. 26, 1859	23, 799
Grinding-mill	C. Tripp	Ann Arbor, Mich	Nov. 10, 1857	18, 610
Grinding mill	C. T. Umfrid	Stuttgart, Würtemberg	Aug. 27, 1867	68, 263
Grinding-mill	A. J. and G. W. M. Vandegrift	Cincinnati, Ohio	May 17, 1870	103, 106
Grinding-mill	S. Vascou and A. Guirand	Cincinnati, Ohio	May 18, 1858	20, 310
Grinding-mill	A. Verbeck	Sterling, Ill	Sept. 20, 1870	107, 640
Grinding-mill	E. H. Vining	Govington, Ga	Mar. 15, 1870	100, 824
Grinding-mill	A. H. Wagner	Chicago, Ill	Feb. 24, 1863	37, 796
Grinding-mill	A. H. Wagner	Chicago, Ill	June 2, 1863	38, 795
Grinding-mill	A. H. Wagner	Staunton, Va	June 2, 1868	78, 624
Grinding-mill	J. N. Walker	Cincinnati, Ohio	Dec. 24, 1850	7, 862
Grinding-mill	G. T. Walters	Nicholasville, Ky	Feb. 12, 1844	3, 434
Grinding-mill	J. W. Webb	Ledyard, N. Y	Jan. 29, 1850	7, 062
Grinding-mill	D. Weimer	Gettysburgh, Ohio	Sept. 28, 1869	95, 396
Grinding-mill	J. M. Westmoreland	Danville, Tex	Nov. 15, 1870	109, 361
Grinding-mill	C. T. Weston	Scranton, Pa	Oct. 24, 1865	50, 646
Grinding-mill	J. D. Whelpley and J. J. Storer	Boston, Mass	Nov. 8, 1864	44, 990
Grinding-mill	G. L. Witsil	Philadelphia, Pa	Nov. 27, 1866	60, 105
Grinding-mill	T. B. Woodward	Kensington, Pa	July 11, 1854	11, 309
Grinding-mill	C. D. Young and J. McLean	Waterloo, N. Y	July 24, 1866	56, 659
Grinding mill, Barley	J. Mackay	Dundas, Canada	Mar. 4, 1873	136, 527
Grinding-mill bolt	R. Denison	Grand Rapids, Mich	Nov. 22, 1864	45, 142
Grinding mill, Bone	A. and E. Lister	Newark, N. J	Sept. 6, 1870	107, 184
Grinding-mill, Cast-iron	J. Russell	Troy, N. Y	May 17, 1859	24, 058
Grinding-mill, Conical	A. W. Sweet	Cincinnati, Ohio	Feb. 21, 1860	27, 246
Grinding-mill cooler	P. C. Fritz	Barrytown, N. Y	Nov. 6, 1866	59, 382
Grinding mill, Corn	J. J. Johnston	Allegheny, Pa	Jan. 25, 1870	99, 205
Grinding mill, Corn	R. Medley	Bloomfield, Ky	Sept. 5, 1828	....
Grinding mill, Cracker and	C. Parker	Meriden, Conn	Sept. 21, 1869	95, 131
Grinding-mill crushing and hulling attachment	G. C. Hohenstein and C. T. Glaeser.	Cincinnati, Ohio	Jan. 17, 1871	111, 061
Grinding-mill, Family	P. M. Wright	New York	Apr. 14, 1835	
Grinding mill, Feed	M. S. Harsha	Batavia, Ill	Mar. 19, 1872	124, 737
Grinding-mill feed	G. Parker	Poughkeepsie, N. Y	Oct. 12, 1869	95, 829
Grinding-mill feed-apparatus	J. D. Whelpley and J. J. Storer	Boston, Mass	May 16, 1870	102, 987
Grinding-mill feed-mechanism	H. L. Bennett	Geneva, Ill	Oct. 19, 1869	95, 973
Grinding-mill feed-regulator	E. H. Austin	Scott's Hill, Tenn	Sept. 30, 1873	143, 315
Grinding-mill feed-regulator	J. C. Dunlap	Moffat's Creek, Va	Apr. 29, 1873	138, 384
Grinding-mill feeding-shoe	J. C. Andrew	Seventy-Six, Ky	Mar. 9, 1869	87, 615
Grinding-mill for coffee, spices, &c	J. Garfield	Ayer's, Mass	Aug. 8, 1871	117, 879
Grinding-mill for grain	J. Brown	Utica, N. Y	Dec. 19, 1865	51, 549
Grinding-mill for grain	P. Davis	North Providence, R. I	May 17, 1838	738
Grinding-mill for grain	V. Fouchier	Paris, France	Sept. 17, 1861	33, 301
Grinding-mill for grain	J. Grant	Providence, R. I	Aug. 19, 1834	....
Grinding-mill for grain	E. Holton	Westminster, Vt	Apr. 3, 1835	....

Index of patents issued from the United States Patent Office from 1790 to 1873, inclusive—Continued.

Invention.	Inventor.	Residence.	Date.	No.
Grinding-mill for grain	W. H. Hope	Washington, D. C	Feb. 1, 1859	22, 807
Grinding-mill for grain, &c	J. Ives	New York	Dec. 2, 1825	
Grinding-mill for grain	E. B. Nichols and D. Marsh	Fairfield, Conn	Mar. 13, 1844	3, 479
Grinding-mill for grain	H. Pearce	Cincinnati, Ohio	July 29, 1839	1, 266
Grinding-mill for grain	F. Price	New York, N. Y	Feb. 20, 1843	2, 962
Grinding-mill for grain	E. Ripley	Troy, N. Y	Dec. 25, 1855	13, 996
Grinding-mill for grain	J. R. Sleeper	Philadelphia, Pa	Jan. 27, 1835	
Grinding-mill for grain	J. Ware	Farmington, Ohio	July 20, 1842	2, 733
Grinding-mill for grain, Cast-iron	H. Bailey	Hartford, Conn	May 14, 1825	
Grinding-mill for grain, Cylindrical	J. Groat	Troy, N. Y	July 11, 1844	3, 658
Grinding-mill for mustard, &c	C. Walker	Brooklyn, N. Y	Aug. 28, 1847	5, 262
Grinding-mill hopper	M. Replogle	Moulton, Iowa	June 17, 1873	139, 973
Grinding mill hopper	G. S. Thompson	Philadelphia, Pa	Apr. 22, 1873	138, 212
Grinding-mill hopper and casing	J. M. Finch	Cedar Falls, Iowa	Apr. 1, 1873	137, 429
Grinding-mill-hopper attachment	D. C. Walters	Warsaw, Ind	Jan. 7, 1868	73, 142
Grinding-mill, Metallic	G. Palmer	Littlestown, Pa	Apr. 22, 1862	35, 036
Grinding-mill, Metallic	H. B. Stevens	Buffalo, N. Y	Apr. 1, 1873	137, 503
Grinding-mill, Portable	L. Scott	Saint Louis, Mo	May 16, 1854	10, 931
Grinding-mill, Portable	A. W. Straub	Philadelphia, Pa	Feb. 25, 1868	74, 774
Grinding-mill regulator	A. B. Hamaker	Salunga, Pa	Aug. 25, 1863	39, 648
Grinding-mill spindles, Bush for	J. H. Teahl	Eberly's Mills, Pa	May 14, 1867	64, 809
Grinding mill, Sumac	R. G. Chase	Alexandria, Va	Jan. 12, 1869	85, 726
Grinding-mills, Apparatus for delivering grain, ore, &c., into.	J. D. Whelpley and J. J. Storer	Boston, Mass	Feb. 7, 1871	111, 593
Grinding-mills, Attachment for preparing corn in the ear for.	J. M. Seely and W. E. Tomlinson.	Lockport, Ohio	May 21, 1850	7, 386
Grinding-mills, Dress of metallic hemispherical	A. Atwood	Troy, N. Y	Sept. 9, 1856	15, 680
Grinding-mills, Feeding-device for	S. Middleton	Dunkirk, Ind	Oct. 28, 1873	143, 989
Grinding-mills, Flour-distributing bolt for	W. W. Hamer	Cincinnati, Ohio	Nov. 17, 1857	18, 637
Grinding-mills, Operating	J. Holben	Allentown, Pa	June 3, 1862	35, 444
Grinding-mills, Tram-staff for	W. Ring	Gosport, Ind	Oct. 29, 1867	70, 266
Grinding-mills, Tramming-bolt for	J. T. Noye	Buffalo, N. Y	Sept. 16, 1873	142, 809
Grinding objects cylindrical, Machine for	J. M. Poole	Wilmington, Del	Aug. 20, 1872	130, 741
Grinding objects cylindrically, Machine for	J. M. Poole	Wilmington, Del	Jan. 18, 1870	99, 000
Grinding or hulling mills, Grinding or hulling plates for.	H. Shaw	Cincinnati, Ohio	Nov. 9, 1869	96, 736
Grinding or polishing implement	N. A. Buhle	New York, N. Y	Feb. 19, 1867	62, 110
Grinding-pan and amalgamator	W. H. Thoss	West Point, Cal	Mar. 7, 1871	112, 395
Grinding-plate	E. S. Howland	Batavia, Ill	Apr. 4, 1871	113, 432
Grinding-plate	E. S. Howland	Batavia, Ill	Aug. 20, 1872	130, 719
Grinding-plate	P. M. Randall	San Francisco, Cal	Nov. 22, 1864	45, 175
Grinding-plate, Iron	D. Halladay and B. H. Ruggles	Batavia, Ill	May 25, 1869	90, 531
Grinding, scouring, and polishing machine	A. B. Walters	Philadelphia, Pa	Feb. 4, 1873	135, 456
Grinding-surface in mills	C. Ross	Rochester, N. Y	Oct. 17, 1854	11, 811
Grinding-surfaces, Roll for	O. I. Foster	Salem, N. H	Dec. 10, 1872	133, 843
Grinding-tool, hammer, sadiron, &c	D. Welch	Canaan, N. H	Jan. 10, 1873	
Grinding tools, Device for	S. Bennett	Wilkesbarre, Pa	Feb. 2, 1869	86, 352
Grinding tools, &c., Machine for	N. B. Reynolds	Auburn, N. Y	Aug. 28, 1866	57, 571
Grinding tools, Machinery for	W. Hovey	Worcester, Mass	Sept. 23, 1845	4, 204
Grinding tools, Rest for	J. H. Brown	Berea, Ohio	Mar. 26, 1867	63, 207
Grindstone	C. Burkholder	Sterling, Ill	July 30, 1872	130, 014
Grindstone	J. F. and S. H. Green	Haverstraw, N. Y	July 15, 1873	140, 823
Grindstone	T. Loring	Blackwoodtown, N. J	Nov. 23, 1869	97, 098
Grindstone	W. P. Miller	New York, N. Y	May 21, 1867	64, 893
Grindstone	J. F. Shillaber	Portsmouth, N. H	Sept. 12, 1865	49, 928
Grindstone, Artificial	G. G. Griswold	Chester, Conn	Nov. 29, 1864	45, 243
Grindstone-axle	J. W. Ligon	Marion, Ky	June 4, 1872	127, 492
Grindstone-cutting machine	J. P. Gillespie	Louisville, Ky	Apr. 1, 1873	137, 435
Grindstone-dresser	R. Barkley and L. Semple	Philadelphia, Pa	May 17, 1864	42, 738
Grindstone-frame	B. Bisbee	North Waterford, Me	Oct. 12, 1869	95, 639
Grindstone-frame	G. L. Cummings	New York, N. Y	Dec. 31, 1872	134, 363
Grindstone-frame	J. W. Douglas	Middletown, Conn	Sept. 1, 1868	81, 611
Grindstone-frame	I. L. Lord	Chester, Conn	Aug. 15, 1854	11, 526
Grindstone-hanger	S. L. Bignall	Chicago, Ill	Oct. 28, 1873	143, 952
Grindstone journal-box	T. W. Brown	Reading, Pa	June 22, 1869	91, 514
Grindstone journal-box	P. P. Child	Saint Louis, Mo	Sept. 5, 1871	118, 688
Grindstone journal-box	J. L. Haven	Cincinnati, Ohio	Jan. 12, 1869	85, 819
Grindstone journal-box	J. B. Sargent	New Haven, Conn	Aug. 16, 1870	106, 510
Grindstone or tool-sharpener	U. Farris and A. Miller	Red Rock, Iowa	Sept. 27, 1870	107, 768
Grindstone, Parlor	J. M. Simpson	Oshkosh, Wis	Aug. 16, 1870	106, 413
Grindstone-rest	F. M. Stearns	Berea, Ohio	Dec. 27, 1864	45, 650
Grindstone, Self-sharpening	J. Pannabecker	Elizabeth Township, Pa	Jan. 20, 1852	8, 676
Grindstone trough and support	C. P. Richardson	Groveton, N. H	Mar. 18, 1873	136, 939
Grindstone turning and grinding machine	P. Leonard	Sharon, Pa	June 29, 1869	91, 944
Grindstones, Apparatus for supplying water to	F. Blauss	New York, N. Y	Mar. 24, 1868	75, 725
Grindstones, Device for holding tools against	E. Fernald	Turner, Me	July 28, 1868	80, 344
Grindstones, Device for wetting	D. Cumming, jr	Jersey City, N. J	Nov. 25, 1873	144, 891
Grindstones, Dressing	G. C. Howard	Philadelphia, Pa	Feb. 5, 1867	61, 738
Grindstones, Dressing	J. F. Schuyler	Philadelphia, Pa	June 5, 1860	28, 639
Grindstones, Gearing for	F. Howlett and C. R. Sherman	West Rupert, Vt., and Salem, N. Y.	Oct. 13, 1868	83, 064
Grindstones, Hanging	H. M. Church	Brunswick, Ohio	Sept. 6, 1870	107, 002
Grindstones, Hanging	D. B. Herrington	Detroit, Mich	Sept. 17, 1867	68, 876
Grindstones, Hanging	D. Hinman	Berea, Ohio	June 24, 1856	15, 180
Grindstones, Machine for cutting out	J. E. Mitchell	Philadelphia, Pa	June 7, 1870	103, 912
Grindstones, Machine for facing	J. Bidwell	New York, N. Y	June 2, 1863	38, 782
Grindstones, Machine for facing	R. B. Matthews	Fitchburgh, Mass	Apr. 28, 1868	77, 303
Grindstones, Machine for making	J. Baldwin, jr	Berea, Ohio	Jan. 14, 1862	34, 124
Grindstones, Machine for making	C. Foss	Painsville, Ohio	Apr. 24, 1849	6, 399
Grindstones, Machine for splitting	J. McDermott	Cleveland, Ohio	May 3, 1870	102, 693
Grindstones, Machine for turning off	J. Thierry	Aurora, Ill	June 10, 1862	35, 553
Grindstones Method of holding edged tools on	P. V. Dunn	Calamus, Wis	Mar. 19, 1867	62, 945
Grindstones, minerals, &c., Machine for dressing	P. Leonard	Sharon, Pa	July 14, 1868	79, 990
Grindstones, Mode of forming the center for the shaft of.	D. Hinman	Berea, Ohio	July 31, 1860	29, 375
Grindstones, Mode of packing	F. M. Stearns	Berea, Ohio	June 20, 1865	48, 319
Grindstones, Operating	H. Pray	Sharon, Conn	Nov. 30, 1869	97, 441
Grindstones, Tools for dressing	C. Arthur	Keeseville, N. Y	Aug. 26, 1845	4, 160

Index of patents issued from the United States Patent Office from 1790 *to* 1873, *inclusive*—**Continued.**

Invention.	Inventor.	Residence.	Date.	No.
Grindstones, Tool-holder for	P. Leonard	Sharon, Pa	Dec. 7, 1869	97,656
Grindstones, Tool-rest for	W. H. Mosteller	Sharonville, Ohio	Apr. 26, 1870	102,420
Grindstones, Tool-rest for	W. H. Straham	Philadelphia, Pa	Oct. 2, 1866	58,504
Grindstones, Truing	C. E. Wilson	Boston, Mass	July 6, 1869	92,242
Griping-block for press	G. W. Swinebroad	Bolivar, Tenn	Sept. 24, 1872	131,640
Grist and coffee mill	J. Stroub	Charleston, S. C	Sept. 6, 1833	
Grist and flour mill	I. Straub	Northumberland County, Pa.	May 9, 1834	
Grist and saw mill combined	N. Porter	Boston, Mass	Nov. 17, 1831	
Grist-mill	R. W. Adams	Marlborough, Vt	Dec. 8, 1808	
Grist-mill	W. Adams	Guilford, N. C	July 18, 1827	
Grist-mill	J. Amell and J. Gallery	Brooklyn, N. Y	Aug. 21, 1844	3,711
Grist-mill	J. F. Anderson	Louisville, Ky	May 3, 1831	
Grist-mill	A. Bacon	Windsor, N. Y	June 4, 1832	
Grist-mill	C. Badger	Edgerton, Wis	May 22, 1860	28,434
Grist-mill	A. Barnes	Colesville, N. Y	Apr. 27, 1832	
Grist-mill	J. Demis	Worcester, Mass	Nov. 17, 1826	
Grist-mill	W. Benbow	Guilford County, N. C	Jan. 19, 1827	
Grist-mill	A. Bencine	Caswell County, N. C	Jan. 16, 1827	
Grist-mill	E. Bigelow	Georgetown, D. C	Feb. 11, 1831	
Grist-mill	E. Brees	Kingston, Pa	Sept. 28, 1831	
Grist-mill	M. L. Chase	Frankfort, Me	Dec. 23, 1829	
Grist-mill	A. and S. Coe	Guilford, N. C	July 21, 1827	
Grist-mill	W. Coleman	Euclid, Ohio	Feb. 25, 1830	
Grist-mill	J. Copes	Georgetown, Del	May 2, 1805	
Grist-mill	J. Crail	Warren, Ohio	June 19, 1828	
Grist-mill	J. W. Dart and W. and H. Webster.	Truxton, N. Y	July 20, 1831	
Grist-mill	A. Delap and A. Coe	Guilford County, N. C	May 31, 1827	
Grist-mill	W. W. Forwood	Harford County, Md	June 15, 1827	
Grist-mill	J. C. Gentry	Philadelphia, Pa	July 14, 1832	
Grist-mill	E. Goodell and O. P. Stevens	Port Lawrence, Ohio	Oct. 10, 1838	974
Grist-mill	E. Gray	Ulysses, N. Y	Mar. 25, 1840	1,523
Grist-mill	C. B. Gregory	Danbury, Conn	Feb. 16, 1843	2,952
Grist-mill	E. Griswold	Truxton, N. Y	Oct. 16, 1830	
Grist-mill	E. Hale	Haverhill, Mass	Feb. 22, 1807	
Grist-mill	S. M. Handy, J. Albee, and E. B. Cutts.	Hallowell, Me	June 6, 1834	
Grist-mill	F. Harris and J. Wilson	Albany, N. Y	June 9, 1826	
Grist-mill	E. Harrison	New Haven, Conn	Jan. 9, 1872	122,605
Grist-mill	E. Harrison	New Haven, Conn	Jan. 16, 1872	122,827
Grist-mill	J. and D. Hascall	Sikay, N. Y	Mar. 9, 1822	
Grist-mill	S. Holland	Hanover, Ohio	Mar. 1, 1828	
Grist-mill	J. W. Holly		Mar. 27, 1802	
Grist-mill	G. Hotchkiss	Windsor, N. Y	Jan. 27, 1832	
Grist-mill	C. Kaestner	Chicago, Ill	July 4, 1871	116,717
Grist-mill	R. Kendall	Wilkes County, Ga	Nov. 19, 1833	
Grist-mill	E. C. L. Kunnecke	Dayton, Ohio	Aug. 15, 1871	118,136
Grist-mill	C. Sangford	Claridon, Ohio	July 8, 1830	
Grist-mill	S. Lawing and J. Monteith	Statesville, N. C	June 11, 1827	
Grist-mill	O. N. May	Hancock, N. Y	May 5, 1832	
Grist-mill	W. and J. McCreight	Winnsborough, S. C	Feb. 5, 1836	
Grist-mill	M. Mendenhall	Greensborough, N. C	Oct. 20, 1826	
Grist-mill	T. Newman	Guilford County, N. C	Feb. 6, 1827	
Grist-mill	A. L. Norcross	Hallowell, Me	Dec. 21, 1833	
Grist-mill	J. Northorp	Woodbridge, Conn	July 31, 1826	
Grist-mill	L. Olds	Otsego, N. Y	May 18, 1833	
Grist-mill	B. Overman	Greensborough, N. C	Feb. 28, 1827	
Grist-mill	J. P. Phipps and J. Holliday	Wilmington, Del	May 12, 1831	
Grist-mill	I. Platt	Weston, Conn	Sept. 28, 1843	3,283
Grist-mill	A. Porter	New London, Ind	Mar. 10, 1830	
Grist-mill	J. Robinson	Buckskin Township, Ohio	Dec. 11, 1827	
Grist-mill	I. Ryon	Bridgewater, Mass	June 29, 1826	
Grist-mill	A. Sawyer	Chester, Vt	July 20, 1813	
Grist-mill	J. Smith and W. Sapp	Mount Vernon, Ohio	Sept. 9, 1828	
Grist-mill	J. C. Smith	Wheeling, Va	Jan. 9, 1830	
Grist-mill	M. Stockman	Hampton, N. H	Apr. 18, 1826	
Grist-mill	W. L. Taylor	McMinn County, Tenn	Jan. 28, 1828	
Grist-mill	R. S. Thomas	Rockingham, N. C	June 4, 1827	
Grist-mill	W. A. Turner	Plymouth, N. C	June 27, 1827	
Grist-mill	B. Tyler		Feb. 20, 1800	
Grist-mill	D. D. Wagener	Pittsburgh, Pa	Jan. 8, 1839	1,065
Grist-mill	A. H. Wagner	Chicago, Ill	June 6, 1871	115,791
Grist-mill	A. Warren	Saugerties, N. Y	Mar. 12, 1828	
Grist-mill	C. T. Weston, S. Broadbent, and W. B. Culver.	Scranton, Pa	June 14, 1870	104,384
Grist-mill	B. Whitney	New Brunswick, N. J	Oct. 27, 1868	83,574
Grist-mill	J. Wickersham and T. Crozer	East Fairfield, Ohio	Aug. 29, 1829	
Grist-mill	I. Wilson	New London, Conn	Feb. 23, 1830	
Grist-mill	O. Wyman	East Cambridge, Mass	Apr. 18, 1839	1,125
Grist-mill	J. Yeamans	Ashtabula, Ohio	Mar. 8, 1831	
Grist-mill alarm	J. D. Irwin and B. I. Seward	Corydon and Bloomington, Ind.	Jan. 28, 1868	73,896
Grist-mill and cotton-seed huller combined	J. W. Smith	Columbus, Ga	Apr. 25, 1871	114,215
Grist-mill and horizontal spiral water-wheel	A. Temple	Brookfield, Ohio	Dec. 11, 1829	
Grist-mill bush	G. M. Copeland	Genoa, Ohio	Oct. 11, 1841	2,313
Grist-mill, Conical	S. Sheldon	Cincinnati, Ohio	Sept. 11, 1841	2,251
Grist mill, crusher, and sheller	J. G. Morse	Randolph County, N. C	Mar. 20, 1827	
Grist-mill for grain, &c	J. Ambler, jr., and D. C. Ambler.	New Berlin, N. Y	June 13, 1831	
Grist-mill for grinding grain	J. Platt	Weston, Conn	Oct. 9, 1841	2,282
Grist-mill-gate pressure	S. Prentiss	New York	June 4, 1832	
Grist-mill grinding-plate	H. Shaw and W. D. Leavitt	New Orleans, La	July 14, 1868	79,865
Grist-mill-hopper shoe	W. E. Wyche and Y. P. Dickson.	Brookville, N. C	July 7, 1868	79,620
Grist-mill, Portable	M. Riech	Wheeling, W. Va	Nov. 5, 1872	132,777
Grist-mill, Portable	S. Sheldon	Cincinnati, Ohio	Sept. 12, 1854	11,681
Grist-mill, Portable	I. Straub	Cincinnati, Ohio	Feb. 18, 1840	1,491

Index of patents issued from the United States Patent Office from 1790 *to* 1873, *inclusive*—Continued.

Invention.	Inventor.	Residence.	Date.	No.
Grist-mill, Portable	H. Weed	Sandwich, N. H	Jan. 10, 1831	
Grist-mill, Portable iron-husk	H. W. Shipley and Z. Blair	Mount Vernon, Ohio	Aug. 16, 1859	25, 144
Grist-mill, Pressure or weighted	J. , jr., and D. C. Ambler	New Berlin, N. Y	Dec. 6, 1830	
Grist-mill, Screw	J. L. Miller	Brookfield, N. Y	Mar. 28, 1810	
Grist-mill, "Self-stopper"	R. McCormick, jr	Augusta County, Va	Apr. 20, 1831	
Grist-mill, spindle, and bush	J. Barber	Phelps, N. Y	Feb. 28, 1834	
Grist-mill spindle, Cast-iron	W. Rice	Athens, Pa	Oct. 1, 1823	
Grist-mill spindle, Apparatus for lubricating	C. Brown, sr	Richmond, Ind	May 6, 1856	14, 791
Grist-mill toll-collector	F. Klinkerman	Aurora, Ind	Feb. 23, 1869	87, 176
Grist-mill toll-collector	T. R. Van Gelder	Damascus, Pa	Mar. 20, 1860	27, 582
Grist-mill toll-gathering machine	J. Bartholomew	Dundee, N. Y	June 15, 1858	20, 540
Grist-mill toll-taker	W. W. McCauley	Fancy Farm, Ky	Oct. 15, 1872	132, 302
Grist-mill toll-taking machine	C. F. Keller	Nevada, Ohio	May 19, 1868	78, 099
Grist-mills, Automatic alarm for	M. W. Helton and J. H. Redfield	Bloomington, Ind	Jan. 14, 1868	73, 247
Grist-mills, Mode of constructing bales and divers for.	E. R. Benton	Ohio City, Ohio	Mar. 31, 1840	1, 533
Grist-mills, Motive-power cut-off in	J. Hough	West Chester, Pa	July 5, 1864	43, 409
Grist-mills, &c., "Wry-fly" which may be applied by wind or water to	B. Tyler		Mar. 19, 1804	
Grist, paint, and plaster mill	C. Manning	Littleton, Mass	Apr. 2, 1835	
Grocer's can	J. Adair	Kilbuck Township, Pa	July 9, 1872	128, 837
Grocer's canister	W. H. Smith	Portland, Conn	May 2, 1871	114, 360
Grommet	W. Brown	Middletown, Conn	July 2, 1867	66, 296
Grommet	J. W. Norcross	Boston, Mass	Dec. 15, 1868	84, 900
Grommet, Metallic	J. Allender	New London, Conn	June 20, 1854	11, 108
Grommet, Metallic	J. Mair	Philadelphia, Pa	July 6, 1869	92, 199
Grommet, Metallic	E. H. Penfield	Middletown, Conn	Sept. 19, 1848	5, 779
Grommet, Strap	E. F. Southward	Wellfleet, Mass	Oct. 8, 1850	7, 7[illegible]8
Grooved-roll-shaping die	A. Reese	Pittsburgh, Pa	Nov. 19, 1867	71, 062
Grooved rollers, Pattern for casting	J. Herald	Unadilla, N. Y	Feb. 21, 1871	111, 931
Grooved wheel or pulley	J. H. Barnes	Brooklyn, N. Y	Jan. 1, 1867	60, 668
Grooves and slots, Machine for cutting	R. F. Underhill	Indianapolis, Ind	May 26, 1857	17, 398
Grooves, Device for casting circle-plates, roses, &c., with dovetailed.	N. Matthews	Pittsburgh, Pa	Apr. 20, 1852	8, 898
Grooves in the necks of cans, Machine for forming.	J. D. Willoughby	Petersburgh, Va	Aug. 21, 1860	29, 739
Grooves, Machine for boring circular	W. Wright	Rochester, N. Y	June 10, 1845	4, 076
Grooves on metallic rods, Machine for forming spiral.	T. T. Prosser	Chicago, Ill	June 20, 1871	116, 218
Grooving boards, Device for	G. C. Fisk	Dansville, N. Y	Aug. 8, 1854	11, 478
Grooving-machine	W. H. Bond and G. G. Lee	Syracuse, N. Y	Oct. 20, 1868	83, 128
Grooving-machine	D. D. Hanson	Weare, N. H	June 29, 1833	
Grooving-machine	T. Holt	Brooklyn, N. Y	June 22, 1869	91, 635
Grooving-machine	W. G. Raoul	Independence, La	Feb. 1, 1870	99, 350
Grooving-machine	J. W. Smith	Hartford, Conn	Nov. 1, 1870	108, 943
Grooving-machine	J. L. Taylor	New York, N. Y	Oct. 15, 1861	33, 501
Grooving-machine and planing-bit	J. Shugert	Lawrenceville, Pa	Nov. 19, 1833	
Grooving machine, Sheet-metal	C. H. Raymond	Southington, Conn	Feb. 8, 1870	99, 592
Grooving-tool	J. Dill	Grand Rapids, Mich	Dec. 13, 1870	110, 123
Grooving-tool	C. Van Hagen	Philadelphia, Pa	Dec. 20, 1870	110, 407
Ground-roller	C. D. Roberts	Jacksonville, Ill	Sept. 4, 1866	57, 773
Ground-roller and stalk-cutter	P. H. Tompkins and E. Dougal	Van Buren, Iowa	June 22, 1869	91, 581
Grout-conductor	W. S. Folensbee	Janesville, Wis	Nov. 19, 1867	71, 155
Grub and stump machine	A. McKinney	Maumee, Ohio	Apr. 17, 1866	54, 000
Grub-hook	J. W. Goodall	Eldred, Pa	Dec. 21, 1869	98, 159
Grubbing-implement	J. Sattazahn, jr	Pine Grove Township, Pa	May 17, 1870	103, 086
Grubbing-machine	J. B. Ash	Elkton, Md	Aug 21, 1860	29, 659
Grubbing-machine	C. Ball	Augusta, Mich	Nov. 29, 1864	45, 216
Grubbing-machine	C. Ball and J. W. Houghtelin	Detroit, Mich	Aug. 14, 1866	57, 667
Grubbing-machine	C. E. Chase and B. T. Devendorff.	Wyoming Township, Mich	Dec. 22, 1868	85, 069
Grubbing-machine	O. A. Cheney	Orleans Township, Mich	Mar. 6, 1866	52, 967
Grubbing-machine	J. H. Flanagan and W. Lanning	Chicago, Ill., and Stoughton, Wis.	Jan. 19, 1864	41, 287
Grubbing-machine	J. Frey	Battle Creek, Mich	Feb. 3, 1863	37, 574
Grubbing-machine	D. I. Hall	Dowagiac, Mich	Dec. 3, 1867	71, 608
Grubbing-machine	A. McKenney	Maumee City, Ohio	Feb. 18, 1862	34, 429
Grubbing-machine	C. R. Moffett	Philadelphia, Pa	June 20, 1865	48, 298
Grubbing-machine	D. C. Payne	Elkhart, Ind	July 29, 1862	36, 022
Grubbing machine	Y. W. Short	Oglethorpe County, Ga	Jan. 28, 1840	1, 480
Grubbing-machine	W. W. St. John	Saint Louis, Mo	Aug. 28, 1866	57, 593
Grubbing-machine	T. C. Wood	Charleston, Mich	Nov. 9, 1858	22, 047
Grubbing-machine and for working on roads	E. Fraser	Hector, N. Y	Dec. 31, 1833	
Grubbing-machine capstan	B. B. Newell	Centreville, Mich	June 30, 1868	79, 380
Guano and other fertilizers, Machine for distributing	E. Wagner	Westminster, Md	Nov. 30, 1858	22, 212
Guano and seed distributer	J. H. Boyd	Plain, S. C	Nov. 4, 1873	144, 250
Guano and seed drill	L. M. Rhodes	Warrenton, Ga	Apr. 18, 1871	113, 797
Guano and the manufacture of fertilizer, Treatment of.	C. Morfit	New York, N. Y	Apr. 30, 1870	106, 851
Guano-distributer	J. D. Coxwell	Gibson, Ga	June 15, 1869	91, 434
Guano-distributer	N. Foster	Palmyra, N. Y	Apr. 14, 1868	76, 621
Guano-distributer	W. E. Martin	Oconee, Ga	Oct. 4, 1870	107, 937
Guano-distributer	E. R. Stedman	Sparta, Ga	Oct. 4, 1870	107, 975
Guano-distributer and seed-sower	B. F. Hinkley	Baltimore, Md	Oct. 11, 1870	108, 262
Guano-distributing machine	H. L. and C. P. Brown	Manchester, N. Y	May 25, 1869	90, 493
Guano-distributing machine	G. W. Sizer and W. M. Owen	New Orleans, La	Sept. 6, 1870	107, 111
Guano, &c., distributing machine	J. F. Thomas	Adamstown, Md	Feb. 11, 1868	74, 448
Guano, Drying	E. P. Baugh	Philadelphia, Pa	May 3, 1870	102, 648
Guano, phosphate, &c., bag	B. R. Croasdale	Philadelphia, Pa	Nov. 23, 1869	97, 169
Guano, Process of treating Navassa	R. B. Potts	Camden, N. J	Mar. 7, 1865	46, 700
Guano. Restoring de-ammoniated	A. A. Hayes	Boston, Mass	Feb. 16, 1864	41, 663
Guano, Restoring phosphate	L. Harper	Brooklyn, Mass	Feb. 2, 1864	41, 428
Guano-s reader	I J. Saunders	Sparta, Ga	Feb. 12, 1861	31, 409
Guano, Treating and preparing "Navassa"	G. A. Liebig	Baltimore, Md	Mar. 22, 1864	42, 006
Guano, Treating phosphatic	L. D. Gale	Washington, D. C	Mar. 31, 1863	38, 040
Guard: *See* Ankle or knee guard. Bed or cradle guard. Ax-handle guard. Boot-strap guard.				

Index of patents issued from the United States Patent Office from 1790 *to* 1873, *inclusive*—Continued.

Invention.	Inventor.	Residence.	Date.	No.
Guard—Continued.				
See Bridge-guard. Candle-guard. Car safety-guard. Carriage-guard. Carriage-spring guard. Carriage-wheel guard. Catamenial guard. Cattle-guard. Damper-guard. Door-guard. Door and window guard. Door-key guard. Door-knob guard. Dress-guard. Ferry-boat guard. Finger-guard. Fork-guard. Grave-guard. Hame-guard. Harvester-guard. Hat and coat guard. Hatchway-guard. Hatchway safety-guard. Heel-guard. Heel-shaving guard. Horse's elastic heel-guard. Insect-guard. Key-guard. Key-hole guard. Key-hole key-guard. Knife-guard. Knob-guard. Lantern-guard. Lock-guard. Loom-shuttle guard. Mining-shaft safety-guard. Mowing-machine guard. Musquito-guard. Mustache-guard. Paint-guard. Pistol safety-guard. Plow-guard. Pocket-book guard. Railway cattle-guard. Railway safety-guard. Rein-guard. Roof-guard. Roof snow-guard. Safety-guard. Saw-guard. Sawing-machine guard. Screen-guard. Shoe-knife guard. Spool-thread guard. Spring-guard. Steam-boiler safety-guard. Thill-guard. Thread-guard. Trunk-guard. Watch-guard. Weather-guard. Window-guard. Window safety-guard.				
Gudgeon-box, Self-oiling	J. Shugert	Elizabeth, Pa	Apr. 1, 1842	2, 527
Gudgeon, Wing	M. Wilder	Princeton, Mass	Apr. 9, 1850	7, 278
Gudgeons, journals, and wheel-boxes, Iron	S. Reynolds	Guilford, N. Y	June 13, 1829	......
Gudgeons, Oiling	P. S. Devlin, J. Hancock, and C. S. Wood.	Philadelphia, Pa	Aug. 8, 1846	4, 686
Guide and sign board	S. Carter	Raynham, Mass	Nov. 25, 1829	......
Guide-post	R. C. Maners	Boston, Mass	Aug. 13, 1833	......
Guitar	R. Knaffl	Nashville, Tenn	Jan. 7, 1873	134, 679
Guitar	G. D. Reed	Springfield, Vt	Dec. 2, 1873	145, 241
Guitar	E. N. Sherr	Philadelphia, Pa	Oct. 6, 1831	......
Guitar	W. B. Tilton	New York, N. Y	Jan. 3, 1854	10, 380
Guitar	W. B. Tilton	New York, N. Y	Mar. 1, 1856	14, 378
Guitar	W. H. Towers	Philadelphia, Pa	May 16, 1854	10, 934
Guitar-banjo	L. Brown	Baltimore, Md	Oct. 17, 1865	50, 444
Guitar-head and capio d'astra	J. Ashborn	Wolcottville, Conn	Apr. 16, 1850	7, 279
Guitars, Bracing the sounding-board of	J. E. Bini	Mount Vernon, N. Y	Dec. 24, 1867	72, 591
Guitars, Tuning-attachment for	H. Seehausen	Memphis, Tenn	Sept. 24, 1867	69, 259
Gum and palate, Artificial	J. A. Cummings	Boston, Mass	June 7, 1864	43, 009
Gum, Chewing	T. Adams	Hudson City, N. J	Feb. 14, 1871	111, 798
Gum, Chewing	W. W. Kilbourn	Sanford, N. Y	Sept. 27, 1870	107, 693
Gum, Chewing	W. F. Semple	Mount Vernon, Ohio	Dec. 28, 1869	98, 304
Gum, Chewing	N. Wood	Portland, Me	Dec. 17, 1872	134, 022
Gum compound, Chewing	A. H. Tyler	Toledo, Ohio	July 27, 1869	93, 141
Gum-elastic: (*See* India rubber.)				
Gum-elastic braces	R. Warner	New York, N. Y	Mar. 18, 1836	......
Gum-elastic-cloth making	H. G. Tyer and J. Helm	Ballard Vale, Mass., and New Brunswick, N. J.	May 6, 1856	14, 814
Gum for coating and water-proofing, Artificial	R. O. Lowrey	Salem, N. Y	Aug. 4, 1868	80, 641
Gum from machinery, Composition for removing	S. Maxwell	Baltimore, Md	Apr. 24, 1860	28, 001
Gum, Process of purifying spruce	H. B. Esty	Houlton, Me	Aug. 21, 1866	57, 334
Gum, Treating waste and inferior	J. Murphy	New York, N. Y	Jan. 3, 1860	26, 698
Gun	A. Boyden	Newark, N. J	Jan. 10, 1824	......
Gun, Air	G. W. B. Gedney	New York, N. Y	Sept. 24, 1861	33, 344
Gun, Air	P. Giffard	Paris, France	Feb. 9, 1864	41, 500
Gun, Air	E. Lindner	New York, N. Y	Dec. 16, 1862	37, 173
Gun, Air repeating	P. Giffard	Paris, France	Feb. 25, 1873	136, 315
Gun, Alarm	C. J. Beasley	Petersburgh, Va	Feb. 22, 1870	100, 106
Gun, Alarm	M. C. Heptinstall	Enfield, N. C	June 29, 1869	91, 935
Gun, Alarm	A. Johnson and S. E. Allen	Raleigh, N. C	Jan. 1, 1867	60, 897
Gun, Alarm	T. A. Marable and G. Utley	Petersburgh, Va	May 14, 1867	64, 682
Gun and bayonet battery	J. W. Andrews	Norristown, Pa	Nov. 19, 1861	33, 731
Gun and blasting powder	C. F. Fuchs and A. Clement	Boston, Mass	Aug. 6, 1872	130, 123
Gun and blasting powder	G. A. Neumeyer	Altenburg, Germany	July 2, 1867	66, 378
Gun and blasting powder	G. B. Weistling	Oxford Furnace, N. J	Sept. 22, 1863	40, 070
Gun and fire-arm	J. Shaw	Philadelphia, Pa	May 7, 1829	......
Gun and percussion-lock, Revolving four-barreled	H. Rogers	Middletown, Ohio	May 7, 1829	......
Gun and pistol, Air	E. H. Hawley	Kalamazoo, Mich	June 1, 1869	90, 749
Gun and pistol bayonet, Combined	R. K. Colvin	Lancaster, Pa	Oct. 25, 1864	44, 784
Gun and pistol lock	J. Marsh	East Dorset, Vt	July 1, 1836	......
Gun and pistol, Percussion	J. Finch	New York, N. Y	Apr. 12, 1823	......
Gun and rifle	F. Oswan	Harper's Ferry, Va	Feb. 25, 1815	......
Gun and rifle, Convertible shot	S. McCulloch	Yellow Springs, Ohio	Dec. 10, 1867	72, 063
Gun, Arrow	C. Robinson	Boston, Mass	Aug. 17, 1869	93, 908
Gun-barrel	E. Allen	Worcester, Mass	June 20, 1865	48, 249
Gun-barrel	R. R. Moore	Cortland, N. Y	Dec. 26, 1871	122, 187
Gun-barrel scraper	E. L. Pratt	Boston, Mass	Apr. 11, 1865	47, 260
Gun-barrels, Cleaning the bores of	P. F. Carr	Wyalusing, Pa	Feb. 9, 1864	41, 481
Gun-barrels, Coloring	J. S. Howard	Mansfield, Mass	Aug. 30, 1870	106, 823
Gun-barrels, Expanding scraper for	A. L. Bausman	Minneapolis, Minn	Sept. 11, 1866	57, 846
Gun-barrels, &c., from Bessemer steel, Process of making.	J. Thompson	Bilston, England	Nov. 28, 1865	51, 281

Index of patents issued from the United States Patent Office from 1790 *to* 1873, *inclusive*—Continued.

Invention.	Inventor.	Residence.	Date.	No.
Gun-barrels, Heating and soldering	E. Allen	Worcester, Mass	June 19, 1866	55, 596
Gun-barrels, Instrument for cleaning	W. E. Turner	Fort Snelling, Minn	June 3, 1873	139, 633
Gun barrels, locks, &c., Plating	J. Fowler	Lancaster, Pa	Jan. 4, 1810	....
Gun-barrels, Machine for boring	N. Fobes	....	Dec. 31, 1804	....
Gun-barrels, Machine for draw-grinding	T. Dakin	Harper's Ferry, Va	July 28, 1820	....
Gun-barrels, Machine for rifling	W. and C. Sellers	Philadelphia, Pa	Mar. 7, 1865	46, 714
Gun-barrels, Machine for rolling	J. Yates	Mott Haven, N. Y	Nov. 7, 1865	50, 869
Gun-barrels, Machine for turning	S Nash	Harper's Ferry, Va	Apr. 11, 1818	....
Gun-barrels, Machine for turning	S. Youngs	Hartford, Conn	May 1, 1810	....
Gun-barrels, Manufacture of	W. Baker	Ilion, N. Y	Feb. 23, 1864	41, 669
Gun-barrels, Manufacture of	J. H. Burton	Jefferson County, Va	Mar. 20, 1860	27, 539
Gun-barrels, Manufacture of	R. A. Douglas	Orange, N. J	Apr. 14, 1863	38, 201
Gun-barrels, Manufacture of	S. Hiler	Harlem, N. Y	Apr. 15, 1862	34, 961
Gun-barrels, Manufacture of	J. Thompson	Bilston, England	May 10, 1864	42, 718
Gun-barrels, Manufacturing	A. Waters	Millbury, Mass	Oct. 25, 1817	....
Gun-barrels, Manufacturing plates for	H. Mills	Springfield, Mass	July 12, 1834	....
Gun-barrels, Means of attaching the fore-end stock to.	J. Deeley and J. S. Edge, jr	Yardley, England	July 1, 1873	140, 482
Gun-barrels, Process for making twisted	T. Warner	Chicopee, Mass	Sept. 6, 1853	9, 999
Gun-barrels, Scarfing and rolling the scalp for	J. Bruce	Springfield, Mass	Oct. 15, 1824	....
Gun-barrels, Scraper for cleaning	M. G. Crane	Charlestown, Mass	July 19, 1864	43, 573
Gun-barrels, System of rifling	O. F. Winchester	New Haven, Conn	June 25, 1872	128, 446
Gun-barrels to stocks, Attaching	N. R. Davis	Freetown, Mass	Aug. 25, 1868	81, 348
Gun-barrels, Turning	A. Waters	Millbury, Mass	Dec. 19, 1818	....
Gun, Battery	W. Douglas	Corry, Pa	Aug. 23, 1864	43, 903
Gun, Battery	W. Fields	Wilmington, Del	Apr. 25, 1871	113, 996
Gun, Battery	R. J. Gatling	Indianapolis, Ind	May 9, 1865	47, 631
Gun, Battery	A. E. Miltimore	United States Army	Dec. 2, 1873	145, 224
Gun battery, Many-barreled	J. Brett	Matteawan, N. Y	May 3, 1864	42, 552
Gun battery, Repeating	E. Ripley	Troy, N. Y	Oct. 22, 1861	33, 544
Gun, Blast	C. Kirchhof	Newark, N. J	Apr. 21, 1868	76, 925
Gun-boat	S. D. Carpenter	Madison, Wis	Jan. 31, 1865	46, 074
Gun-boat	L. M. Van Sickle	Woodbridge, N. J	Apr. 28, 1862	38, 344
Gun-boats, Drilling bolt-holes in turrets of	T. F. Rowland	Green Point, N. Y	May 19, 1863	38, 605
Gun, Breech-loading	E. Lindner	New York, N. Y	May 6, 1856	14, 819
Gun, Breech-loading	C. Sharps	Philadelphia, Pa	Nov. 11, 1856	16, 072
Gun, Breech-loading and magazine	B. and W. G. Burton	Brooklyn, N. Y	Oct. 14, 1873	143, 614
Gun, Burglar-alarm	P. Swisher	Versailles, Ohio	Jan. 1, 1867	60, 960
Gun, Burglar-alarm	J. Wilson	Anderson Court-House, S. C.	July 9, 1867	66, 662
Gun, Cane	A. Davis	Shelbyville, Ind	Aug. 21, 1860	29, 676
Gun, Cane	A. Karutez	Brooklyn, N. Y	Apr. 9, 1872	125, 460
Gun, Cane	J. F. Thomas	Ilion, N. Y	Feb. 9, 1858	19, 328
Gun, Cane	W. H. Werner	Nazareth, Pa	June 7, 1870	104, 087
Gun-cap	J. C. Bandle and E. J. Christner	Cincinnati, Ohio	July 25, 1871	117, 367
Gun-cap	J. Talbott	Albany, Ga	July 6, 1869	92, 398
Gun-capper	H. Buffington	South Coventry, Conn	June 1, 1869	90, 634
Gun-capping implement	E. D. Seely	Brookline, Mass	Oct. 29, 1861	33, 626
Gun-carriage	G. Birkbeck, jr	New York, N. Y	Oct. 4, 1864	44, 506
Gun-carriage	L. W. Broadwell	New Orleans, La	June 1, 1869	90, 814
Gun-carriage	T. Coughlan	Newton, Mass	Nov. 15, 1870	109, 181
Gun-carriage	J. B. Eads	Saint Louis, Mo	Aug. 17, 1869	93, 691
Gun-carriage	J. B. Eads	Saint Louis, Mo	May 23, 1871	115, 181
Gun-carriage	J. B. Eads	Saint Louis, Mo	June 18, 1872	128, 130
Gun-carriage	J. Ericsson	New York, N. Y	June 28, 1864	43, 298
Gun-carriage	J. Ericsson	New York, N. Y	Mar. 8, 1870	100, 514
Gun-carriage	J. Ericsson	New York, N. Y	July 23, 1872	129, 804
Gun-carriage	W. C. Fuller	London, England	July 9, 1861	32, 762
Gun-carriage	S. Hawkins	New York, N. Y	Feb. 19, 1806	....
Gun-carriage	J. C. Honel and F. L. F. Caillet	Paris, France	Jan. 4, 1870	98, 595
Gun-carriage	J. R. Kelso	Freedom, Mo	June 15, 1869	91, 345
Gun-carriage	J. Laurens	Charleston, S. C	July 1, 1856	15, 244
Gun-carriage	J. B. Lyons	Litchfield, Conn	June 9, 1863	38, 831
Gun-carriage	A. Moncrieff	Woolwich, Conn	Nov. 10, 1868	83, 873
Gun-carriage	A. F. Potter	Oakland, Cal	Sept. 6, 1870	107, 099
Gun-carriage	T. R. Timby	Tarrytown, N. Y	Oct. 31, 1871	120, 553
Gun-carriage	C. S. Tyson	Old Point Comfort, Va	May 19, 1868	78, 030
Gun-carriage	G. J. Van Brunt	Dedham, Mass	June 15, 1858	20, 597
Gun-carriage	J. J. Walsh	New York, N. Y	Sept. 25, 1860	30, 186
Gun-carriage	M. Wappick	Sacramento, Cal	Mar. 10, 1863	37, 882
Gun-carriage	G. R. Wilson	Washington, D. C	Feb. 9, 1869	86, 720
Gun-carriage	J. W. Wilson	New York, N. Y	Mar. 1, 1870	100, 482
Gun-carriage, Counterpoise	J. G. Foster	Nashua, N. H	July 13, 1869	92, 597
Gun-carriage, Counterpoise	J. G. Foster	Boston, Mass	May 2, 1871	114, 430
Gun-carriage, Counterpoise	J. G. Foster	Boston, Mass	Oct. 17, 1871	119, 924
Gun-carriage for naval and other purposes	W. Smith	Washington, Ky	July 19, 1838	847
Gun-carriage for ships of war	J. Bubier	Marblehead, Mass	Oct. 20, 1837	432
Gun carriage, Pivot	L. M. Laighton	Portsmouth, N. H	Nov. 29, 1814	....
Gun-carriages, Operating	S. J. Ashley	San Francisco, Cal	Dec. 15, 1863	40, 893
Gun-carriages, Operating	A. L. Caswell	Lansingburgh, N. Y	Oct. 18, 1859	25, 806
Gun-carriages, Operating	J. Ericsson	New York, N. Y	Dec. 15, 1863	40, 919
Gun-carriages, Operating	J. Ericsson	New York, N. Y	Jan. 23, 1866	52, 150
Gun-carriages, Operating	R. H. Long	Philadelphia, Pa	Aug. 4, 1863	39, 449
Gun-carriages, Operating	I. Rindge	Cincinnati, Ohio	June 21, 1864	43, 228
Gun-carriages, Pile or fagot for shoe-rails for	J. L. Lewis	Pittsburgh, Pa	Aug. 19, 1862	36, 223
Gun-carriages, Quoins for	D. D. Porter	United States Navy	Dec. 21, 1858	22, 377
Gun-carriages, &c., Wheel for	C. F. Brown	Warren, R. I	July 10, 1860	29, 055
Gun, Centrifugal	C. S. Dickinson	Cleveland, Ohio	Aug. 9, 1859	24, 997
Gun, Centrifugal	G. C. Eaton and S. W. Turner	Cleveland, Ohio	Dec. 16, 1862	37, 159
Gun, Centrifugal	W. Joslin	Cleveland, Ohio	May 17, 1859	24, 031
Gun, Centrifugal	C. B. Thayer	Boston, Mass	Aug. 3, 1858	21, 109
Gun, Centrifugal spring	R. Shaler	Madison, Conn	Nov. 18, 1862	36, 968
Gun-charger	H. Kahn	Troy, N. Y	Oct. 18, 1870	108, 359
Gun-cleaner	M. G. Crane	Charlestown, Mass	Nov. 15, 1864	45, 027
Gun-cleaner	C. G. Gould	Windham, Vt	Sept. 23, 1873	143, 139
Gun, Combined piston-breech and firing-cock repeating.	W. Hunt	New York, N. Y	Aug. 21, 1849	6, 663
Gun-cotton	B. W. Lenk	Vienna, Austria	June 14, 1864	43, 166

Index of patents issued from the United States Patent Office from 1790 *to* 1873, *inclusive*—Continued.

Invention.	Inventor.	Residence.	Date.	No.
Gun-cotton and lint, Manufacture of	J. P. McLean	Brooklyn, N. Y	Apr. 18, 1865	47, 316
Gun-cotton, Making solutions of	M. Newton	Boston, Mass	June 25, 1872	128, 416
Gun-cotton, Manufacture of	F. A. Abel	Woolwich, Great Britain	Nov. 20, 1866	59, 888
Gun-cotton, Manufacture of	S. J. Mackie	Westminster, England	Aug. 12, 1873	141, 654
Gun-cotton, Manufacture of	J. J. Révy	Vienna, Austria	Sept. 19, 1865	50, 082
Gun-cotton, Manufacture of	J. J. Révy	London, England	Sept. 19, 1865	50, 083
Gun-cotton, Manufacture of	T. P. Shaffner	Louisville, Ky	Dec. 18, 1866	60, 571
Gun, Eight-barreled percussion	S. L. Faries	Middletown, Ohio	Oct. 10, 1829	....
Gun, Faucet-breech	A. D. Perry	New York, N. Y	Dec. 11, 1849	6, 945
Gun for driving stock from railroad-track, Steam	F. G. Smith	Columbia, Tenn	Nov. 28, 1865	51, 229
Gun-hammer gage	W. T. Round	Middletown, Conn	July 2, 1867	66, 399
Gun-lock	J. Albright	Pleasant View, Mo	June 27, 1871	116, 252
Gun-lock	S. Belden and J. F. Crabtree	Visalia, Cal	Apr. 14, 1868	76, 587
Gun-lock	W. N. Bennett	Illyria, Iowa	July 25, 1871	117, 245
Gun-lock	W. Briggs	Norristown, Pa	Aug. 30, 1859	25, 244
Gun-lock	E. F. Burrows	Mystic River, Conn	Oct. 1, 1872	131, 845
Gun-lock	J. J. Byers	Delta, N. Y	June 18, 1872	128, 015
Gun-lock	J. J. Byers	Delta, N. Y	Sept. 24, 1872	131, 598
Gun-lock	P. F. Charpie	Mount Vernon, Ohio	Aug. 16, 1853	9, 934
Gun-lock	J. F. and W. N. Crabtree	Visalia, Cal	Mar. 10, 1868	75, 248
Gun-lock	S. Cromwell	Edgecomb, Me	Feb. 3, 1827	....
Gun-lock	J. C. Dane	La Crosse, Wis	Mar. 26, 1872	124, 939
Gun-lock	W. Dashner	Point Pleasant, W. Va	Aug. 10, 1869	93, 420
Gun-lock	J. Deutz	San Antonio, Tex	Aug. 16, 1870	106, 337
Gun-lock	W. R. Evans	Thomaston, Me	Sept. 19, 1871	119, 020
Gun-lock	R. H. Fauntleroy	Posey County, Ind	Aug. 17, 1833	....
Gun-lock	E. Gilbert	Rochester, N. Y	Apr. 3, 1829	....
Gun-lock	C. Gordon	Goswell Road, England	Apr. 22, 1873	138, 145
Gun-lock	J. B. Gray	Fredericksburgh, Va	Oct. 7, 1834	....
Gun-lock	L. Hailer	Washington, D. C	July 26, 1870	105, 799
Gun-lock	R. D. Hay	Crooked Creek, N. C	Nov. 22, 1870	109, 514
Gun-lock	R. D. Hay and J. M. Hill	Crooked Creek, N. C	June 8, 1869	91, 014
Gun-lock	E. B. Hendee	San Francisco, Cal	Dec. 20, 1870	110, 360
Gun-lock	C. B. Holden	Worcester, Mass	Nov. 8, 1870	109, 128
Gun-lock	P. W. Hoyt	Danbury, Conn	Mar. 10, 1838	629
Gun-lock	J. Hults	Berlin Township, Ohio	May 16, 1854	10, 927
Gun-lock	A. Judson	New Lebanon, N. Y	June 13, 1831	....
Gun-lock	J. Kelsay	Richmond, Mo	Jan. 5, 1869	85, 672
Gun-lock	E. C. Kirk	Baltimore, Md	Mar. 20, 1866	53, 306
Gun-lock	N. C. Lock	Salem, Mass	Jan. 11, 1870	98, 781
Gun-lock	J. A. Sowe	New York, N. Y	Feb. 8, 1859	22, 881
Gun-lock	W. W. Marston	New York, N. Y	June 5, 1849	6, 514
Gun-lock	T. J. Massie	Arrington, Va	Oct. 31, 1871	120, 388
Gun-lock	A. Miller	Daleville, Ala	Sept. 26, 1871	119, 386
Gun-lock	J. C. Miller	Danville, Ky	Mar. 30, 1869	88, 319
Gun-lock	E. Monroe	Charlestown, Mass	Mar. 25, 1856	14, 513
Gun-lock	W. J. Morris	New York, N. Y	Apr. 1, 1873	137, 381
Gun-lock	S. Morrison	Milton, Pa	Feb. 10, 1836	....
Gun-lock	J. Phin	Rochester, N. Y	Nov. 20, 1855	13, 825
Gun-lock	W. Rudolph and A. Braun	San Francisco, Cal	Apr. 17, 1866	54, 021
Gun-lock	W. Rudolph and A. Braun	San Francisco, Cal	June 19, 1866	55, 716
Gun-lock	N. Saltonstall	New London, Conn	May 29, 1828	....
Gun-lock	F. Schenck	San Antonio, Tex	Sept. 11, 1866	57, 978
Gun-lock	C. W. Scott	Constantia, Ohio	Oct. 25, 1864	44, 827
Gun-lock	J. H. Shapley	Exeter, N. H	Aug. 11, 1863	39, 501
Gun-lock	J. H. Smith	Brewster's Station, N. Y	Sept. 24, 1861	33, 371
Gun-lock	E. T. Starr	New York, N. Y	Dec. 19, 1865	51, 629
Gun-lock	J. Stokes	Springfield, Mass	Nov. 24, 1868	84, 314
Gun-lock	M. Tromly	Mount Vernon, Ill	Aug. 14, 1855	13, 442
Gun-lock	M. Tromly	Washington, D. C	Nov. 17, 1868	84, 233
Gun-lock	G. B. Warren	Prospect, Pa	June 13, 1871	115, 996
Gun-lock	F. Wesson	Worcester, Mass	Apr. 9, 1872	125, 640
Gun-lock	E. West	....	July 6, 1802	....
Gun-lock	A. Young	Philadelphia, Pa	Feb. 23, 1869	87, 316
Gun-lock cover	J. B. Leverich	New York, N. Y	June 3, 1862	35, 456
Gun-lock, Double-acting	E. Brey	Pennsburgh, Pa	June 29, 1858	20, 757
Gun lock, Lever percussion	J. Amtler, jr	South New Berlin, N. Y	Oct. 16, 1827	....
Gun lock, Magazine percussion	J. B. Lowry	Mayville, N. Y	Sept. 11, 1827	....
Gun lock, Percussion	J. Caswell	Manlius, N. Y	May 8, 1828	....
Gun lock, Percussion	M. Davis	Mayville, N. Y	July 10, 1827	....
Gun lock, Percussion	W. A. Hart	Fredonia, N. Y	Feb. 20, 1827	....
Gun lock, Percussion	J. Lawrence	New Berlin, N. Y	May 24, 1828	....
Gun lock, Percussion	J. Shattuck	Jefferson County, Ohio	Nov. 10, 1827	....
Gun-lock, Rotary-tumbler	T. W. Harvey	New York, N. Y	June 19, 1849	6, 537
Gun-lock, Safety	C. Bowlen	Milwaukee, Wis	Dec. 10, 1867	71, 963
Gun-lock, Safety	W. F. Kussmaul	Baltimore, Md	Oct. 15, 1867	69, 919
Gun-lock, Safety	D. H. Mapother	Louisville, Ky	Jan. 28, 1873	135, 233
Gun-lock, Safety	J. E. McBeth	New Orleans, La	Oct. 2, 1866	58, 443
Gun-lock, Self-priming	F. H. Bell	Washington, D. C	Apr. 12, 1859	23, 545
Gun-lock, Self-priming	J. S. Butterfield and S. Marshall	Philadelphia, Pa	June 14, 1859	24, 372
Gun-lock, Self-priming	M. I. Gallagher	Savannah, Ga	July 7, 1857	17, 733
Gun-lock, Self-priming	R. S. Lawrence	Hartford, Conn	Apr. 12, 1859	23. 590
Gun-lock, Self-setting hair-triggered	J. Altman	Armstrong County, Pa	Feb. 17, 1857	16, 634
Gun-lock springs, Forging	G. P. Foster	Bristol, R. I	Feb. 3, 1857	16, 537
Gun lock, Water-proof self-priming percussion	S. L. Faries	Middletown, Ohio	Aug. 29, 1828	....
Gun-locks, Safety-stop for	W. G. Oliver	Buffalo, N. Y	June 12, 1866	55, 588
Gun, Machine	L. Christopher and J. Montigney.	Brussels, Belgium	Nov. 28, 1871	121, 277
Gun, Machine	W. B. Farwell	New York, N. Y	Apr. 1, 1873	137, 428
Gun, Machine	W. A. Miles	Salisbury, Conn	July 30, 1872	129, 976
Gun, Machine	C. Stensland	Negaunee, Mich	Oct. 14, 1873	143, 729
Gun, Machine	M. Wood	Lewisburgh, W. Va	July 30, 1872	130, 098
Gun, Magazine	N. H. Ambler	East Cleveland, Ohio	Aug. 9, 1870	106, 246
Gun, Magazine	F. Brady and J. C. Noble	Washington, Pa	Jan. 14, 1862	34, 126
Gun, Magazine	A. Burgess	Owego, N. Y	June 25, 1872	128, 208
Gun, Magazine	A. Burgess	Owego, N. Y	July 16, 1872	129, 523
Gun, Magazine	H. J. Drew	Dixon, Ill	Mar. 14, 1871	112, 563
Gun, Magazine	W. R. Evans	Thomaston, Me	Dec. 8, 1868	84, 685

Index of patents issued from the United States Patent Office from 1790 *to* 1873, *inclusive*—Continued.

Invention.	Inventor.	Residence.	Date.	No.
Gun, Magazine	E. H. Graham	Biddeford, Me	Oct. 4, 1853	10,084
Gun, Magazine	E. H. Graham	Biddeford, Me	May 16, 1854	10,944
Gun, Magazine	C. R. Stickney	Ilion, N. Y	July 2, 1872	128,671
Gun, Magazine	A. Swingle	San Francisco, Cal	Apr. 1, 1873	137,392
Gun, Magazine, repeating, and needle	E. Lindner	New York, N. Y	June 27, 1854	11,197
Gun, Many-barreled	E. A. Mejia	Mexico	Feb. 19, 1867	62,281
Gun, Method of manufacturing ribs and bolsters for double-barrel.	D. B. Wesson and J. H. Blaze	Springfield, Mass	Dec. 31, 1867	72,949
Gun, Needle	G. A. Blittkowski and F. W. Hoffman.	New York, N. Y	Mar. 25, 1856	14,488
Gun, Needle	W. Burghart	Lawrence, Mass	Jan. 12, 1858	19,068
Gun, Needle	A. A. Chassepot	Paris, France	Jan. 1, 1867	60,832
Gun, Needle	T. Twickeler	Boston, Mass	Mar. 18, 1862	34,706
Gun-nipple protector	J. Haskins	Roxbury, Mass	May 27, 1862	35,418
Gun-nipples, Safety-guard for	C. T. Moore	Gilmanton, N. H	Aug. 17, 1869	93,733
Gun or rifle	O. Stith	Brunswick County, Va	Mar. 16, 1819	....
Gun or rifle, Double-shooting	S. Mosher and N. White	Hamilton, N. Y	May 5, 1828	....
Gun, Percussion	J. Shaw	Philadelphia, Pa	June 19, 1822	....
Gun, Percussion	J. Tyrer	Petersburgh, Va	June 3, 1825	....
Gun, Repeating	C. H. Palmer	Lakeville, Conn	Dec. 2, 1862	37,052
Gun, Repeating	A. Wheeler	Concord, Mass	June 10, 1818	....
Gun-rest, Hinged	J. Hawkins	West Windsor, N. J	May 14, 1867	64,765
Gun, Revolving battery	R. J. Gatling	Indianapolis, Ind	Nov. 4, 1862	36,836
Gun, Revolving battery	R. J. Gatling	Indianapolis, Ind	Feb. 28, 1871	112,138
Gun, Revolving battery	R. J. Gatling	Hartford, Conn	Apr. 9, 1872	125,563
Gun, Rifle or plain barrel	A. Wheeler	Massachusetts	Feb. 19, 1819	....
Gun-scraper, Adjustable	E. L. Pratt	Boston, Mass	Jan. 31, 1865	46,140
Gun-scrubber holder	L. Fleckenstine	Manor Township, Pa	Jan. 7, 1868	72,993
Gun, Self-loading battery	T. J. Cranmer	Vallicita, Cal	Mar. 3, 1868	74,994
Gun, Shot	G. Buckel and E. Dorsch	Monroe, Mich	Apr. 8, 1856	14,597
Gun, Shot	D. Smith	Springfield, Mass	May 30, 1871	115,370
Gun-shot wounds and extracting bullets, Instrument for treating.	S. J. Howell	Orange, Mass	Jan. 19, 1869	86,016
Gun-sight	H. B. Barber	Scott, N. Y	Apr. 5, 1870	101,568
Gun-sight	F. A. Churchill	Pittsfield, Mass	Feb. 28, 1871	112,124
Gun-sight	M. W. Harrington	Homestead, Iowa	Feb. 25, 1873	136,159
Gun-sight	J. T. La Rue	Pleasant Post-Office, Ind	June 18, 1872	128,049
Gun-sight	F. N. Martin	Covington, Ky	Jan. 5, 1864	41,162
Gun-sight	H. A. Schottky and T. Simendinger.	Hartford, Conn	Jan. 14, 1873	134,772
Gun-sight, Telescopic	D. Wood	Rochester, N. Y	May 31, 1864	42,983
Gun, Sighting	E. A. Stevens	Hoboken, N. J	Jan. 6, 1863	37,365
Gun, Spring	J. E. Blythe	New York, N. Y	Jan. 3, 1865	45,789
Gun, Spring	A. Gemunder	Springfield, Mass	Jan. 12, 1858	19,086
Gun, Spring	W. W. Hannah	Hudson, N. Y	June 11, 1872	127,873
Gun-stock	C. R. Alsop	Middletown, Conn	May 22, 1860	28,433
Gun-stock	W. Burnett	Boston, Mass	Jan. 7, 1862	34,103
Gun stock and barrel attachment	D. C. Thresher	Freetown, Mass	May 8, 1866	54,624
Gun-stock and spade combined	W. E. Blake	New York, N. Y	Feb. 11, 1873	135,624
Gun-stock, Canteen	S. Colt	Hartford, Conn	Jan. 18, 1859	22,627
Gun-stock machine	E. S. Wright and E. Allen	New York, N.Y., and Newark, N. J.	Nov. 17, 1863	40,665
Gun-stock machine	E. S. Wright and E. Allen	New York, N.Y., and Newark, N. J.	Nov. 17, 1863	40,666
Gun-stock machine	E. S. Wright and E. Allen	New York, N.Y., and Newark, N. J.	Nov. 17, 1863	40,667
Gun-stocks, Machine for making	H. W. Oliver	New Haven, Conn	July 1, 1862	35,775
Gun-stocks, &c., Machine for turning	T. Blanchard	Middlebury, Mass	Sept. 6, 1819	....
Gun-stocks, &c., Machine for turning	A. Woolworth	Waterbury, Conn	June 15, 1820	....
Gun-stocks, Machinery for carving and drilling	J. G. Pusey	New York, N. Y	Feb. 17, 1863	37,705
Gun-stocks, Mode of crooking	J. Schirer	Charleston, S. C	Apr. 12, 1826	....
Gun-stocks, Sawing	A. Myers	Boonsborough, Md	Nov. 9, 1832	....
Gun-stocks to pistols, Mode of attaching	E. B. Savage	Cromwell, Conn	Apr. 9, 1861	32,003
Gun-stocks with pistols, Mode of coupling	S. Colt	Hartford, Conn	Jan. 18, 1859	22,626
Gun, Submarine steam	W. W. W. Wood and J. L. Lay	Philadelphia, Pa., and Buffalo, N. Y.	July 18, 1865	48,862
Gun-swab	P. M. Hendrick and J. J. Chattaway.	Springfield, Mass	July 24, 1866	56,559
Gun traversing-mechanism, Machine	R. J. Gatling	Hartford, Conn	Dec. 16, 1873	145,563
Gun-trucks, Machine for cutting	R. Rose	Washington, D. C	Aug. 6, 1821	....
Gun-wad	C. W. Lovett, jr.	Boston, Mass	June 25, 1872	128,231
Gun, Walking-stick	R. R. Beckwith	New York, N. Y	Mar. 23, 1858	19,674
Gun, Walking-stick	I. Buckman, jr	New York, N. Y	Aug. 4, 1857	17,915
Gun-wiper	H. Berdan	New York, N. Y	Sept. 12, 1865	49,848
Gun-wiper	C. F. Gillette	Sparta, Wis	Feb. 8, 1870	99,557
Gun-wiper	H. Greve	Sparta, Wis	Apr. 19, 1870	101,997
Gun-worm	H. C. Bascom	La Crosse, Wis	Mar. 19, 1867	62,994
Guns and gun-towers, Operating	J. B. Eads	Saint Louis, Mo	Mar. 31, 1863	38,038
Guns and gun-turrets, Operating	J. B. Eads	Saint Louis Mo	Feb. 7, 1865	46,223
Guns, Apparatus for unspiking	T. J. Dobbs	Weehawken, N. J	Apr. 6, 1869	88,615
Guns by adjustable ports, Directing	R. Trussell	Brooklyn, N. Y	Jan. 12, 1864	41,245
Guns, &c., Combining barrels of	J. Griswold	Chambers County, Ala	Feb. 1, 1842	2,441
Guns, Constructing large	R. F. Loper	Philadelphia, Pa	July 30, 1844	3,685
Guns, Detachable muzzle for shot	S. H. Roper	Roxbury, Mass	July 14, 1868	79,861
Guns exhausted of air, Muzzle-cap for	R. Gibbons	Oakland, Cal	Aug. 2, 1864	43,680
Guns, Flexible rammer for turret	J. B. Walker	Elizabeth, Pa	Sept. 10, 1867	68,813
Guns in turrets, Operating	J. B. Eads	Saint Louis, Mo	Feb. 7, 1865	46,222
Guns, Lock of double-barreled	H. Barnes	Wilson, N. C	Jan. 19, 1858	19,121
Guns, Machine for battery	B. B. Hotchkiss	New York, N. Y	Aug. 13, 1872	130,501
Guns, Machine for elevating heavy	L. and W. H. Bell	Fortress Monroe, Va	Dec. 8, 1829	....
Guns, Machine for finishing the cascabel of	E. Kaylor	Pittsburgh, Pa	Sept. 26, 1865	50,139
Guns, Machinery for rifling	W. H. Gwynne	Brooklyn, N. Y	Dec. 10, 1861	33,884
Guns, Making rifle	E. Strong and C. Killogg	New Hartford, Conn	Aug. 31, 1810	....
Guns, Maneuvering heavy	K. Stewart	San Francisco, Cal	Nov. 26, 1861	33,803
Guns, Manufacture of breech-plates for	W. Baker	Ilion, N. Y	Mar. 23, 1869	88,003
Guns, Manufacturing	D. G. Colburn	Canton Canal, N. Y	June 29, 1833	....
Guns, Means for operating heavy	A. Moncrieff	Culfargie, Scotland	Oct. 28, 1873	144,120

Index of patents issued from the United States Patent Office from 1790 *to* 1873, *inclusive*—Continued.

Invention.	Inventor.	Residence.	Date.	No.
Guns, &c., Method of attaching bits eccentrically to arbors for stocking.	S. D. Sizer	Springfield, Mass	May 12, 1843	3,090
Guns, Method of preventing accidental discharge in the Prussian.	J. Wurfflein	Philadelphia, Pa	Apr. 30, 1850	7,334
Guns, Mounting heavy	J. B. Newman	Milford, Pa	June 28, 1864	43,330
Guns, Muzzle for shot	J. Fry	Latrobe, Pa	Dec. 15, 1868	84,942
Guns, Muzzle for shot	J. A. McKenzie	Galesburgh, Ill	Jan. 19, 1869	85,949
Guns, Operating heavy	P. F. Jones	New York, N. Y	Dec. 10, 1861	33,891
Guns, Operating machine	G. O. Kinne	Hartford, Conn	Nov. 7, 1871	120,588
Guns, Operating ships'	J. H. Ward	United States Navy	May 3, 1864	42,630
Guns, Percussion-lock for	R. Beale	Washington, D. C	Feb. 20, 1835	......
Guns, Percussion-lock for	T. Daplyn	Dover, Ohio	Feb. 20, 1835	......
Guns, pistols, and other fire-arms	G. Stocker and J. Bentley	Birmingham, England	Apr. 8, 1840	1,544
Guns, Piston for muzzle-loading	J. T. Foster and J. J. Banta	Jersey City, N. J	Mar. 17, 1857	16,860
Guns, Planing the rim-bases of	E. Kaylor	Pittsburgh, Pa	July 12, 1864	43,510
Guns, Priming for needle	C. H. F. Thieme	North Vernon, Ind	Aug. 11, 1868	81,035
Guns, Removing spikes from	A. Lafever	Battle Creek, Mich	Sept. 22, 1863	40,051
Guns, Rifled muzzle for smooth-bored	C. R. Alsop	Middletown, Conn	Dec. 16, 1862	37,193
Guns, Rifling of	J. B. Atwater	Ripon, Wis	Sept. 30, 1862	36,592
Guns, Shot-charge for measuring shot in charging	G. W. Dobbin	Baltimore, Md	Mar. 23, 1838	654
Guns, Target for air	C. A. Demling	New York, N. Y	Jan. 1, 1867	60,862
Gunnery, Repeating	J. C. Chambers	West Middletown, Pa	Mar. 23, 1813	......
Gunpowder	E. B. Dodson	Reading, Pa	Feb. 10, 1857	16,580
Gunpowder	L. Du Pont	Wilmington, Del	May 19, 1857	17,321
Gunpowder	L. H. G. Ehrhardt	Baywater, England	June 19, 1866	55,795
Gunpowder	L. H. G. Ehrhardt	London, England	Jan. 28, 1868	73,786
Gunpowder	A. Molfino	New York, N. Y	Aug. 15, 1871	118,040
Gunpowder	J. F. E. Schultze	Potsdam, Prussia	June 2, 1863	38,789
Gunpowder	T. Taylor	Washington, D. C	Dec. 7, 1869	97,567
Gunpowder, &c	C. M. Wetherill	La Fayette, Ind	Mar. 22, 1864	42,056
Gunpowder and blasting-powder, Manufacture of	G. A. Neumeyer	Altenburg, Saxe-Altenburg.	Sept. 1, 1868	81,670
Gunpowder and lamp-black, Application of Grahamite in the manufacture of.	P. H. Vander Weyde	New York, N. Y	Mar. 2, 1869	87,382
Gunpowder and other explosive substances, Method of blasting with.	T. P. Shaffner	Louisville, Ky	Aug. 17, 1869	93,757
Gunpowder, Apparatus for graining	P. A. Oliver	Wilkesbarre, Pa	Dec. 31, 1872	134,438
Gunpowder, Apparatus for the manufacture of	C. M. Wetherill	La Fayette, Ind	Mar. 22, 1864	42,057
Gunpowder-canister	O. Scott	Bennington, Vt	Dec. 31, 1867	72,916
Gunpowder, &c, Composition for	E. Harrison	New York, N. Y	Feb. 9, 1864	41,576
Gunpowder, Composition for	H. Leibert	Norristown, Pa	Jan. 6, 1863	37,296
Gunpowder corning or graining machine	L. T. Swett	Canton, Conn	Nov. 16, 1841	2,362
Gunpowder, Cylinder for polishing	R. H. Cunningham	Schaghticoke, N. Y	Aug. 26, 1862	36,320
Gunpowder, Drying	E. E. Hendrick	Carbondale, Pa	Nov. 19, 1867	71,004
Gunpowder, Drying and glazing	J. Smith	Kingston, N. Y	Feb. 7, 1865	46,275
Gunpowder-engine	J. S. Foster	Salem, Mass	Jan. 30, 1872	123,095
Gunpowder, Granulating	R. Crosbie	Newark, N. J	Feb. 17, 1812	......
Gunpowder, Granulating	C. Dalmas	Wilmington, Del	Apr. 26, 1816	......
Gunpowder, Granulating	G. Fulton	Philadelphia, Pa	Aug. 21, 1819	......
Gunpowder, Granulating	D. Rogers	Newburgh, N. Y	Aug. 26, 1818	......
Gunpowder, Granulating	J. Worseley	Berne, N. Y	Feb. 21, 1815	......
Gunpowder, Machine for granulating	E. J. Du Pont de Nemours		Nov. 23, 1804	......
Gunpowder, Machine for granulating	W. H. Richardson	Baltimore, Md	Aug. 3, 1822	......
Gunpowder, Machinery for the manufacture of	P. A. Oliver	New York, N. Y	Apr. 7, 1868	76,510
Gunpowder-making	J. M. Merrow and R. McKee, jr	East Hartford, Conn	Apr. 19, 1822	......
Gunpowder, Making charcoal for manufacturing	F. Parker	Prince George's, Md	July 25, 1815	......
Gunpowder, Manufacture of	F. S. Allen	New York, N. Y	Oct. 9, 1866	58,567
Gunpowder, Manufacture of	F. Burney	Faversham, Great Britain	Dec. 2, 1873	145,149
Gunpowder, Manufacture of	C. W. Curtis	London, England	Nov. 14, 1871	120,862
Gunpowder, Manufacture of	H. E. Drayson	Edgeworth Lodge, England	Sept. 13, 1864	44,269
Gunpowder, Manufacture of	L. and E. Du Pont	New Castle County, Del	Dec. 3, 1872	133,522
Gunpowder, Manufacture of	L. H. G. Ehrhardt	London, England	Jan. 5, 1869	85,576
Gunpowder, Manufacture of	W. H. Jackson	Salem, Mass	June 23, 1868	79,229
Gunpowder, Manufacture of	V. L. Maxwell	Wilkesbarre, Pa	Dec. 27, 1859	26,602
Gunpowder, Manufacture of	F. G. Murray	Washington, D. C	June 20, 1865	48,303
Gunpowder, Manufacture of	R. I. L. Witty	Lowell, Mass	Apr. 2, 1838	669
Gunpowder, Manufacturing	T. Ewell	Georgetown, D. C	Dec. 7, 1813	......
Gunpowder, Manufacturing	H. Keyser		Aug. 10, 17 1	......
Gunpowder-mill	B. Potter, jr	Hubbardston, Mass	Jan. 24, 1860	26,932
Gunpowder, mining-powder, &c	H. Hochstalter	Hesse-Darmstadt, Germany.	Mar. 22, 1864	42,047
Gunpowder, Mode of keeping	J. Gale, jr	Devonshire Terrace, England.	Oct. 3, 1865	50,313
Gunpowder, Mode of preparing vessels for holding	A. T. Rand	New York, N. Y	Mar. 13, 1866	53,184
Gunpowder, Plate for pressing	L. Du Pont	Wilmington, Del	Sept. 26, 1865	50,104
Gunpowder, Preparation of cotton-wool, &c., as substitutes for.	C. F. Schoenbein	Basel, Switzerland	Dec. 5, 1846	4,877
Gunpowder, Press for pressing	L. Du Pont	Wilmington, Del	Oct. 24, 1865	50,568
Gunpowder, Substitute for	A. Noble	Hamburg	Oct. 24, 1865	50,617
Gunpowder to flat projectiles, giving them rotation, Application of.	E. O. C. Ord	United States Army	Feb. 14, 1860	27,147
Gunpowder to form cartridges, Treating	R. O. Doremus and B. L. Budd	New York, N. Y	Mar. 18, 1862	34,724
Gunpowder to serve as charge for fire-arms, Preparation of granulated.	J. H. Brown	Ramsey, England	Aug. 20, 1861	33,069
Gutta-percha and caoutchouc, Covering with	C. Goodyear	New Haven, Conn	Oct. 11, 1853	10,106
Gutta-percha and caoutchouc, Dissolving and softening.	C. F. Durant	Jersey City, N. J	Apr. 25, 1848	5,539
Gutta-percha and India rubber, Manufacturing	C. Goodyear and R. Haering	New Haven, Conn	Apr. 12, 1853	9,668
Gutta-percha, Apparatus for cleaning	J. Reynolds	New York, N. Y	May 27, 1856	14,972
Gutta-percha, Apparatus for covering wire with	J. Reynolds	New York, N. Y	July 29, 1856	15,439
Gutta-percha boats, Making	E. B. Larchar	Baltimore, Md	July 24, 1855	13,315
Gutta-percha by molding, stamping, or embossing, Making articles of.	R. A. Brooman	London, England	May 23, 1848	5,592
Gutta-percha cord, Making	J. Reynolds	New York, N. Y	Dec. 9, 1856	16,215
Gutta-percha, &c., Desulphurizing	W. E. Rider and J. Murphy	New York, N. Y	Nov. 7, 1854	11,906
Gutta-percha fabrics in imitation of patent-leather, Preparing.	H. H. Day	Jersey City, N. J	May 2, 1848	5,543

Index of patents issued from the United States Patent Office from 1790 *to* 1873, *inclusive*—Continued.

Invention.	Inventor.	Residence	Date.	No.
Gutta-percha, Feed-apparatus for working	J. Reynolds	New York, N. Y	June 10, 1856	15, 087
Gutta-percha, Process for purifying	R. Haering	New York, N. Y	May 5, 1857	17, 214
Gutta-percha, Process for treating	J. Murphy	New York, N. Y	May 30, 1854	10, 977
Gutta-percha, Process of manufacturing	J. Rider	New York, N. Y	June 1, 1852	8, 992
Gutta-percha, Process of working	S. T. Armstrong and C. J. Gilbert.	New York, N. Y	Sept. 17, 1850	7, 643
Gutta-percha tubing and covering wire. Machine for.	J. Reynolds	New York, N. Y	Apr. 22, 1851	8, 051
Gutter-fastening	W. L. Rogers	Rochefort, Mo	Aug. 29, 1871	118, 556
Gutter, Iron	E. Whitehead	Cincinnati, Ohio	Nov. 20, 1866	59, 804
Gutter, Metallic	T. McClunie	Hartford, Conn	May 27, 1873	139, 324
Gutter-strap	A. B. Schulz	Baltimore, Md	Aug. 15, 1871	118, 061
Gutter, Street	H. O. Ames	New Orleans, La	Sept. 24, 1872	131, 588
Gutter-sweeping machine	W. H. King	Philadelphia, Pa	Aug. 19, 1856	15, 566
Gutter-sweeping machine	R. I. Smith	Philadelphia, Pa	Oct. 23, 1855	13, 709
Gutter-trough, Tinners'	J. N. Woolwin	Mechanicsburgh, Ohio	July 20, 1869	92, 922
Gutters, Beading rain	W. H. Henderson	Franklin, Ind	May 28, 1861	32, 420
Gutters for buildings, Machine for making	A. K. P. Buffum	Gardiner, Me	July 25, 1871	117, 255
Gutters, Machine for forming tubular beads on sheet-metal.	O. W. Stow	Plantsville, Conn	Mar. 10, 1868	75, 487
Gutters to buildings, Mode of sustaining	W. Yapp	Cleveland, Ohio	Oct. 28, 1862	36, 876
Gymnasium, Equestrian	E. S. Scripture	Williamsburgh, N. Y	July 30, 1872	129, 988
Gymnasium, Portable	W. G. Hanlon	New York, N. Y	Oct. 29, 1861	33, 583
Gymnastic apparatus	G. W. Bacon	London, England	Oct. 22, 1867	69, 956
Gymnastic apparatus	D. P. Butler	Boston, Mass	Jan. 19, 1869	85, 996
Gymnastic apparatus	W. Hanlon	New York, N. Y	Jan. 4, 1870	98, 587
Gymnastic apparatus	W. Hanlon	New York, N. Y	Jan. 4, 1870	98, 588
Gymnastic apparatus	J. Smith	Saint Paul, Minn	Oct. 18, 1870	108, 401
Gymnastic apparatus	J. Smith	Philadelphia, Pa	May 6, 1873	138, 590
Gymnastic apparatus	F. Veerkamp and F. Leopold	Philadelphia, Pa	Jan. 1, 1861	31, 059
Gypsum-calcining furnace	B. Fowler	Lubec, Me	June 18, 1850	7, 439
Gypsum-calcining furnace	F. Godfrey	Grand Rapids, Mich	Sept. 11, 1866	57, 892

AMERICA IN TWO CENTURIES:
An Inventory

An Arno Press Collection

American Association of Museums. **A Statistical Survey of Museums in the United States and Canada.** 1965

Andrews, Israel D. **On the Trade and Commerce of the British North American Colonies, and Upon the Trade of the Great Lakes and Rivers.** 1853

Audit Bureau of Circulations. **Scientific Space Selection.** 1921

Austin, E. L. and Odell Hauser. **The Sesqui-Centennial International Exposition.** 1929

Barnett, James H. **The American Christmas.** 1954

Barton, L[eslie] M. **A Study of 81 Principal American Markets.** 1925

Bennitt, Mark, comp. **History of the Louisiana Purchase Exposition.** 1905

Bowen, Eli. **The United States Post-Office Guide.** 1851

Bureau of Applied Social Research, Columbia University. **The People Look at Radio.** 1946

Burlingame, Roger. **Engines of Democracy:** Inventions and Society in Mature America. 1940

Burlingame, Roger. **March of the Iron Men:** A Social History of Union Through Invention. 1938

Burnham, W. Dean. **Presidential Ballots, 1836-1892.** 1955

Cochrane, Rexmond C. **Measures for Progress:** A History of the National Bureau of Standards. 1966

Cohn, David L. **The Good Old Days.** 1940

Cozens, Frederick W. and Florence Scovil Stumpf. **Sports in American Life.** 1953

Day, Edmund E. and Woodlief Thomas. **The Growth of Manufactures, 1899 to 1923.** 1928

Edwards, Richard Henry. **Popular Amusements.** 1915

Evans, Charles H., comp. **Exports, Domestic and Foreign, From the American Colonies to Great Britain, From 1697 to 1789, Inclusive;** Exports, Domestic, From the U.S. to All Countries, From 1789 to 1883, Inclusive. 1884

Federal Reserve System, Board of Governors. **All-Bank Statistics, United States, 1896-1955.** 1959

Flexner, Abraham. **Funds and Foundations:** Their Policies, Past and Present. 1952

Flint, Henry M. **The Railroads of the United States.** 1868

Folger, John K. and Charles B. Nam. **Education of the American Population.** 1967

Handel, Leo A. **Hollywood Looks At Its Audience:** A Report of Film Audience Research. 1950

Harlow, Alvin F. **Old Waybills:** The Romance of the Express Companies. 1934

Harrison, Shelby M. **Social Conditions in an American City:** A Summary of the Findings of the Springfield Survey. 1920

Homans, J. Smith, comp. **An Historical and Statistical Account of the Foreign Commerce of the United States.** 1857

Ingram, J. S. **The Centennial Exposition.** 1876

Institute of American Meat Packers and the School of Commerce and Administration of the University of Chicago. **The Packing Industry:** A Series of Lectures. 1924

Leech, D[aniel] D. T[ompkins]. **The Post Office Department of the United States of America.** 1879

Leggett, M. D., comp. **Subject-Matter Index of Patents for Inventions Issued by the United States Patent Office From 1790 to 1873, Inclusive.** 1874. Three vols.

Magazine Marketing Service. **M.M.S. County Buying Power Index.** 1942

Martin, Robert F. **National Income in the United States, 1799-1938.** 1939

McCullough, Edo. **World's Fair Midways.** 1966

Melish, John. **Surveys for Travellers, Emigrants and Others.** 1976

National Advertising Company. **America's Advertisers.** 1893

Peters, Harry T. **America On Stone:** The Other Printmakers to the American People. 1931

Peters, Harry T. **California On Stone.** 1935

Peters, Harry T. **Currier & Ives:** Printmakers to the American People. 1929/1931. Two vols.

Pownall, T[homas]. **A Topographical Description of the Dominions of the United States of America.** Edited by Lois Mulkearn. 1949

Reed, Alfred Zantzinger. **Present-Day Law Schools in the United States and Canada.** 1928

Reed, Alfred Zantzinger. **Training for the Public Profession of the Law.** 1921

Rogers, Meyric R. **American Interior Design.** 1947

Romaine, Lawrence B. **A Guide to American Trade Catalogs, 1744-1900.** 1960

Scammon, Richard M., comp. **America at the Polls:** A Handbook of American Presidential Election Statistics, 1920-1964. 1965

Smillie, Wilson G. **Public Health:** Its Promise for the Future. 1955

Thompson, Warren S. **Population: The Growth of Metropolitan Districts in the United States, 1900-1940.** 1947

Thorndike, E[dward] L. **Your City.** 1939

Truman, Ben[jamin] C. **History of the World's Fair.** 1893

U.S. Bureau of the Census, Department of Commerce. **Housing Construction Statistics: 1889 to 1964.** 1966

U.S. Census Office (12th Census). **Street and Electric Railways.** 1903

Urban Statistical Surveys. 1976

Wayland, Sloan and Edmund de S. Brunner. **The Educational Characteristics of the American People.** 1958

Woytinsky, W. S. **Employment and Wages in the United States.** 1953

U.S. Census Office (1st Census, 1790). **Return of the Whole Number of Persons Within the Several Districts of the United States.** 1802

U.S. Census Office (2nd Census, 1800). **Return of the Whole Number of Persons Within the Several Districts of the United States.** 1802

U.S. Census Office (3rd Census, 1810). **Aggregate Amount of Each Description of Persons Within the United States of America.** 1811

U.S. Census Office (4th Census, 1820). **Census for 1820.** 1821

U.S. Census Office (5th Census, 1830). **Abstract of the Returns of the Fifth Census.** 1832

U.S. Census Office (6th Census, 1840). **Compendium of the Enumeration of the Inhabitants and Statistics of the United States.** 1841

U.S. Census Office (7th Census, 1850). **The Seventh Census of the United States.** 1853

U.S. Census Office (8th Census, 1860). **Statistics of the United States in 1860.** 1866

U.S. Census Office (9th Census, 1870). **A Compendium of the Ninth Census.** 1872

U.S. Census Office (10th Census, 1880). **Compendium of the Tenth Census.** Parts I and II. 1883. Two vols.

U.S. Census Office (11th Census, 1890). **Abstract of the Eleventh Census.** 1894

U.S. Bureau of the Census (12th Census, 1900). **Abstract of the Twelfth Census of the United States.** 1904

U.S. Bureau of the Census (13th Census, 1910). **Thirteenth Census of the United States: Abstract of the Census.** 1913

U.S. Bureau of the Census (14th Census, 1920). **Abstract of the Fourteenth Census of the United States.** 1923

U.S. Bureau of the Census (15th Census, 1930). **Fifteenth Census of the United States: Abstract of the Census.** 1933

U.S. Bureau of the Census (16th Census, 1940). **Sixteenth Census of the United States: United States Summary.** 1943

U.S. Bureau of the Census (17th Census, 1950). **A Report of the Seventeenth Decennial Census of the United States: United States Summary.** 1953

U.S. Bureau of the Census (18th Census, 1960). **The Eighteenth Decennial Census of the United States: United States Summary.** 1964

U.S. Bureau of the Census (19th Census, 1970). **1970 Census of Population: United States Summary.** 1973. Two vols.

AMERICA IN TWO CENTURIES:
An Inventory

This is a volume in the Arno Press collection

AMERICA IN TWO CENTURIES: An Inventory

SUBJECT-MATTER INDEX

OF

PATENTS FOR INVENTIONS

ISSUED BY THE

UNITED STATES PATENT OFFICE

FROM

1790 TO 1873, INCLUSIVE

VOLUME II

ARNO PRESS
A New York Times Company
1976

Editorial Supervision: ANDREA HICKS

Reprint Edition 1976 by Arno Press Inc.

Reprinted from a copy in
The New York Public Library

AMERICA IN TWO CENTURIES: An Inventory
ISBN for complete set: 0-405-07666-5
See last pages of this volume for titles.

Manufactured in the United States of America

Library of Congress Cataloging in Publication Data

United States. Patent Office.
Subject-matter index of patents for inventions issued by the United States Patent Office from 1790 to 1873, inclusive.

(America in two centuries, an inventory)
On spine: Patents for inventions: 1790-1873.
Reprint of the 1874 ed. published by Government Printing Office, Washington.
1. Patents--United States--Indexes. I. Leggett, Mortimer Dormer, 1821-1896. II. Title. III. Title: Patents for inventions: 1790-1873. IV. Series.
T223.D7A45 1976 608'.7'73 75-24110
ISBN 0-405-07737-8

SUBJECT-MATTER INDEX

OF

PATENTS FOR INVENTIONS

ISSUED BY THE

UNITED STATES PATENT OFFICE

FROM

1790 TO 1873, INCLUSIVE.

VOLUME II.

COMPILED AND PUBLISHED UNDER THE DIRECTION OF

M. D. LEGGETT,

COMMISSIONER OF PATENTS.

WASHINGTON:
GOVERNMENT PRINTING OFFICE.
1874.

INDEX OF PATENTS

ISSUED FROM

THE UNITED STATES PATENT-OFFICE

FROM 1790 TO 1873, INCLUSIVE.

VOLUME II.

Invention.	Inventor.	Residence.	Date.	No.
H.				
Hackling-machine	L. F. Lannay and J. Webb	Baltimore County, Md	Mar. 12, 1872	124, 497
Hackling-machine	W. Montgomery	West Roxbury, Mass	July 2, 1872	128, 647
Hæmagatactiphorus	D. Houghtaling and A. Meneely	Watervliet, N. Y	June 11, 1829	....
Hair and cutting fur from peltry, Machine for extracting.	E. Cutter	Brattleborough, Vt	Mar. 1, 1809	....
Hair and grass cloth, Machine for preparing woof for the manufacture of.	J. Downie	Paterson, N. J	July 18, 1865	48, 796
Hair and oakum, Picking	R. B. Lewis	Hallowell, Me	June 25, 1836	....
Hair, Apparatus for arranging	G. Lieberknecht	Wiemar, North German Confederation.	Aug. 1, 1871	117, 648
Hair, Application for restoring the	B. Harris	New Orleans, La	Mar. 1, 1859	23, 086
Hair, Barber's apparatus for brushing	C. P. Kroll	New York, N. Y	Sept. 12, 1865	49, 893
Hair-clasp	C. Rowland	Washington, D. C	Feb. 16, 1869	87, 072
Hair-clipping device	F. C. Richardson	New York, N. Y	May 2, 1871	114, 477
Hair-clipping device	J. C. Wilson	New York, N. Y	Nov. 22, 1870	109, 479
Hair-clipping machine	G. H. Pratt	Boston Highlands, Mass	Feb. 6, 1872	123, 508
Hair-cloth	I. Lindsley	Pawtucket, R. I	Apr. 7, 1868	76, 476
Hair-cloth	J. Noblit	Philadelphia, Pa	Jan. 28, 1868	73, 920
Hair-cloth	W. Rossnagel	Newark, N. J	Oct. 24, 1871	120, 330
Hair-cloth, Imitation	W. Rossnagel	Newark, N. J	Oct. 24, 1871	120, 331
Hair-cloth, Imitation	C. Sherriff	Newark, N. J	Dec. 5, 1871	121, 674
Hair-cloth, Imitation	I. N. Tichenor	Newark, N. J	Jan. 16, 1872	122, 868
Hair-cloth, Machine for trimming and shearing	O. Arnold and I. Lindsay	North Providence, R. I	Aug. 9, 1870	106, 106
Hair-cloth, Making	W. Shotwell and A. Kinder	New York	July 23, 1813	....
Hair-cloth, Manufacture of imitation	J. J. Comstock and J. Aborn	Providence, R. I	May 17, 1870	103, 018
Hair-cloth, Process of separating the hair from	D. Whiteley	Providence, R. I	Dec. 6, 1870	109, 857
Hair-comb	G. Hooker	Bristol, Conn	Apr. 3, 1835	....
Hair-comb	M. Moss	Farmington, Conn	Jan. 10, 1809	....
Hair-comb	T. Stanley	Southington, Conn	July 6, 1809	....
Hair-comb, Metal	L. North	Otsego, N. Y	May 28, 1818	....
Hair-comb, Metallic	N. Bushnell	Middletown, Conn	Oct. 31, 1835	....
Hair-comb, Socket	E. Parsons	Bristol, Conn	Aug. 16, 1810	....
Hair-combs, Machinery for pointing the teeth of	C. B. Rogers	Deep River, Conn	Jan. 4, 1859	22, 513
Hair-combs, Manufacturing	I. Ives	Bristol, Conn	Apr. 5, 1823	....
Hair, Composition for coloring	J. C. Ayer and E. Haeffely	Lowell, Mass	Jan. 28, 1868	73, 865
Hair, Composition for preserving and curing the	H. Rose	Milford, Mass	Apr. 12, 1864	42, 311
Hair, Composition for the	J. Mackay	New York	Apr. 6, 1842	2, 541
Hair-compound	B. F. Atwood	New York, N. Y	Oct. 27, 1868	83, 440
Hair, Compound for cleansing	A. Miles	Toledo, Ohio	Dec. 5, 1871	121, 536
Hair, Compound for coloring	G. Smith	Groton Junction, Mass	Mar. 23, 1869	88, 224
Hair, Compound for coloring and preserving the	G. Smith	Groton Junction, Mass	Feb. 1, 1870	99, 487
Hair, Compound for curling	E. L. Parsons	Grand Ledge, Mich	June 20, 1871	116, 215
Hair, Compound for treating human	W. R. Lount	Austin, Tex	Oct. 26, 1869	96, 245
Hair-crimper	H. H. Adams	New York, N. Y	Mar. 24, 1863	38, 010
Hair-crimper	H. Christian	New York, N. Y	Mar. 27, 1866	53, 413
Hair-crimper	S. F. Conant	Skowhegan, Me	Apr. 7, 1868	76, 401
Hair-crimper	W. F. George	New York, N. Y	Jan. 29, 1861	31, 234
Hair-crimper	E. Ivins	Waterbury, Conn	May 3, 1859	23, 842
Hair-crimper	B. Mannon	Newport, Ky	Aug. 21, 1866	57, 441
Hair-crimper	J. B. Siccardi	New York, N. Y	Feb. 19, 1867	62, 294
Hair-crimper	G. W. Wood	Philadelphia, Pa	Apr. 14, 1868	76, 683
Hair-crimping pin	E. A. Tyler	Buffalo, N. Y	Mar. 6, 1866	53, 060
Hair-curler	C. H. Barney	Providence, R. I	Oct. 17, 1871	119, 960
Hair-curler	S. A. Early	Philadelphia, Pa	July 9, 1867	66, 476
Hair-curler	J. W. Kenny and J. H. Adams	Albany, N. Y	Oct. 4, 1870	107, 920
Hair-curler	C. H. Lavis and J. McMillan	Philadelphia, Pa	July 9, 1867	66, 599
Hair-curler	I. S. Marcy	Nashua, N. H	July 4, 1871	116, 731
Hair-curler	C. Markley	New York, N. Y	July 25, 1871	117, 439
Hair curler and crimper, Combined	E. Matteson	Jersey City, N. J	July 26, 1870	105, 822
Hair, Curling	E. T. Crain	Kansas City, Mo	May 20, 1873	138, 995
Hair-curling apparatus	M. Adkins	Oswego, N. Y	July 27, 1869	92, 926
Hair-curling clamp	F. Arnold	Middle Haddam, Conn	Feb. 2, 1858	19, 228
Hair-curling device	S. L. Tibbals	Dutch Flat, Cal	July 19, 1870	105, 519

Index of patents issued from the United States Patent Office from 1790 to 1873, inclusive—Continued.

Index of patents issued from the United States Patent Office from 1790 *to* 1873, *inclusive*—Continued.

Invention.	Inventor.	Residence.	Date.	No.
Hair-roll	T. Schionbeck	Saint Louis, Mo	Nov. 26, 1872	133, 490
Hair-roll	J. H. Vogt and G. Dietzel	New York, N. Y	Sept. 6, 1870	107, 134
Hair, Roll for ladies'	J. Edwards	New York, N. Y	May 19, 1863	38, 574
Hair-rope, Machine for opening	P. Wisdom	New York, N. Y	Mar. 24, 1868	76, 025
Hair-rope picker	H. R. Hildreth	Lynn, Mass	Apr. 27, 1869	89, 400
Hair-seating, Weaving	C. R. Harvey	Poughkeepsie, N. Y	Nov. 25, 1837	490
Hair spinning and curling machine	P. Wisdom	Brooklyn, N. Y	Dec. 21, 1869	98, 135
Hair spinning, twisting, and kinking machine	H. Burnham	New York	June 22, 1842	2, 680
Hair-switch	B. F. Burgess, jr	Boston, Mass	Dec. 26, 1871	122, 153
Hair-switch	G. H. Cutter	Arlington, Mass	Apr. 9, 1872	125, 382
Hair-tonic	S. A. Reybert	Plainfield, N. J	Mar. 14, 1871	112, 632
Hair tonic and restorer	C. Stock and W. B. Ross	Pawling, N. Y	Feb. 6, 1872	123, 370
Hair-turning hatchel	L. Wilbur	Watertown, N. Y	May 14, 1872	126, 661
Hair twisting and curling machine	P. Wisdom and J. H. Wilcox	Brooklyn and New York, N. Y.	Apr. 12, 1870	101, 800
Hair, Vegetable fiber as a substitute for	W. Staufen	Paris, France	Sept. 1, 1868	81, 702
Hair, Vegetable fiber to imitate	A. Couturier	Trinidad, Cuba	May 11, 1869	89, 855
Hair-wash	R. Travis	Elkton, Ky	Sept. 2, 1873	142, 532
Hair weaving and mounting apparatus	N. Demongeot	Washington, D. C	Dec. 19, 1871	121, 997
Hair-whipping machinery	I. Davis	Erie, Pa	Dec. 25, 1855	13, 981
Hall, church, &c., Seats for	J. L. Kapple	Mechanicsburgh, Ohio	June 11, 1872	127, 891
Halter	J. Carpenter	Wilmington, Ohio	Dec. 12, 1871	121, 704
Halter	S. K. Dexter	Lowell, Mass	June 25, 1872	128, 215
Halter	W. Garrison and C. H. Stevens	Syracuse, N. Y	Jan. 28, 1868	73, 711
Halter	W. M. Harris	Dixon, Ill	Feb. 22, 1870	100, 031
Halter	R. Porter	Ottumwa, Iowa	Dec. 6, 1870	109, 937
Halter	C. H. Sawyer	Buxton, Me	Oct. 30, 1866	59, 277
Halter	C. A. Steinbrocke	Louisville, Ill	Aug. 25, 1868	81, 552
Halter	J. Thornton	Wellsville, N. Y	Apr. 4, 1871	113, 468
Halter	J. Thornton and E. G. Latta	Genesee, N. Y	Aug. 20, 1872	130, 004
Halter	L. Whitehead	Nunda, N. Y	May 1, 1860	28, 121
Halter	S. C. Hawkins	Patchogue, N. Y	Oct. 5, 1858	21, 674
Halter and bridle for horses	E. Howe	Marlborough, Mass	May 15, 1866	54, 729
Halter apparatus, Tethering	J. H. Plumstead	Lynn, Mass	June 20, 1865	48, 307
Halter-clasp	G. H. Stewart	Philadelphia, Pa	Oct. 26, 1869	96, 164
Halter-clasp	J. C. Ford	Montreal, Canada	Nov. 4, 1873	144, 198
Halter for hitching horses, Pastern	A. T. Boon and L. Mills	Galesburgh, Ill	June 1, 1869	90, 812
Halter-hitch	G. Race	Norwich, N. Y	Feb. 2, 1869	86, 587
Halter-hitch	G. F. Jerome	Hempstead, N. Y	July 26, 1864	43, 643
Halter-holder	E. H. Le Baron	Mattepoisett, Mass	Aug. 9, 1864	43, 781
Halter, Horse	A. Le Plongeon	San Francisco, Cal	Nov. 22, 1870	109, 431
Halter, Horse	J. E. Marshall	West Chester, Pa	July 16, 1861	32, 837
Halter, Horse	I. P. Osborn and W. A. Bayhan	Wilmington, Ohio	Sept. 12, 1871	118, 877
Halter, Horse	M. Wesson	Springfield, Mass	Feb. 13, 1872	123, 653
Halter, Horse	S. F. Cross	Canton, Ohio	Feb. 23, 1869	87, 246
Halter-loop	K. Gibbs	Berwick, Me	Dec. 19, 1865	51, 582
Halter, Neck-rope	L. C. Chase	Boston, Mass	Apr. 30, 1861	32, 180
Halter-ring	E. M. Kinne	Cuba, N. Y	Feb. 18, 1873	136, 071
Halter-rings, &c., Fastening for	H. B. Ware	Burlington, Iowa	Sept. 12, 1865	49, 939
Halters, Mode of manufacturing	J. W. Briggs	Cleveland, Ohio	May 22, 1849	6, 465
Hame	T. G. Brooks	Oneida, Ill	Aug. 25, 1868	81, 469
Hame	R. D. Brown	Covington, Ind	Feb. 16, 1864	41, 600
Hame	W. H. Bustin	Watertown, Mass	Apr. 27, 1869	89, 287
Hame	H. Cogswell	Greenwich, N. Y	Feb. 15, 1859	22, 937
Hame	B. Crawford	Allegeny City, Pa	Mar. 10, 1868	75, 250
Hame	C. D. and R. Hazard	Cleveland Ohio	Apr. 16, 1872	125, 675
Hame	R. Hazard	New London, Ohio	Sept. 10, 1861	33, 280
Hame	F. X. Kaffer	Champaign City, Ill	Apr. 10, 1866	53, 830
Hame	E. F. Lacy and D. K. Woodbury	Danville, Ill	May 5, 1868	77, 625
Hame	G. J. Letchworth	Auburn, N. Y	Jan. 11, 1870	98, 834
Hame	G. J. Letchworth	Auburn, N. Y	July 4, 1871	116, 725
Hame	D. M. Nixon	Danville, Ill	May 5, 1868	77, 645
Hame	I. B. Smith and H. C. Burr	Springfield, Vt	Sept. 15, 1868	82, 168
Hame	J. Thornton and E. G. Latta	Wellsville and Friendship, N. Y.	Nov. 4, 1873	144, 371
Hame	L. Whitehead	Nunda, N. Y	June 28, 1870	104, 806
Hame, Adjustable	D. M. Cummings	Enfield, N. H	Oct. 28, 1862	36, 767
Hame and collar	J. G. Smith	Oregon, Wis	Dec. 28, 1869	98, 311
Hame and strap fastener	J. B. Waterman	Summit, Mich	Sept. 22, 1868	82, 367
Hame-attachment	C. H. Easte	South Boston, Mass	Apr. 19, 1870	102, 100
Hame-attachment	J. Thornton and E. G. Latta	Wellsville, N. Y	Apr. 4, 1871	113, 469
Hame-bending apparatus	A. Gardner	Buffalo, N. Y	Dec. 11, 1849	6, 934
Hame cap, Harness	L. Bonine	Vandalia, Mich	Aug. 9, 1870	106, 111
Hame-clasp	J. H. Snyder	Rockford, Ill	Aug. 13, 1867	67, 681
Hame, Coach and chaise	H. Aiken	Brighton, Mass	Mar. 12, 1823	
Hame-coupling	G. W. Heckart	Columbiana, Ohio	June 23, 1868	79, 224
Hame, Extension	F. M. Schaeffer	Blooming Grove, Kans	Aug. 3, 1869	93, 233
Hame-fastener	H. W. Austin and E. C. Perry	Portage Township, Mich	Oct. 5, 1869	95, 410
Hame-fastener	E. Bradley	Farm Village, N. Y	Mar. 5, 1872	124, 323
Hame-fastener	T. L. Booker	Shady Grove, Va	Aug. 12, 1873	141, 690
Hame-fastener	J. Clendening	Rockford, Ill	Oct. 5, 1869	95, 430
Hame-fastener	E. A. Cooper	Lancaster, N. Y	Mar. 24, 1868	75, 864
Hame-fastener	E. A. Cooper	Buffalo, N. Y	Jan. 26, 1869	86, 137
Hame-fastener	R. A. Daniels	Wayne, Ohio	Feb. 11, 1862	34, 351
Hame-fastener	A. M. Dorman	Philadelphia, Pa	Aug. 24, 1869	94, 090
Hame-fastener	M. L. Drake	Rockford, Ill	Sept. 21, 1869	95, 007
Hame-fastener	J. G. Fleming	Cochranton, Pa	Dec. 2, 1873	145, 163
Hame-fastener	A. J. Foster	Lake Mills, Wis	Sept. 17, 1867	68, 866
Hame-fastener	A. Gale and H. R. Johnson	Shelby County, Ky	Dec. 20, 1870	110, 226
Hame-fastener	J. Harding	Schoolcraft, Mich	June 23, 1868	79, 223
Hame-fastener	J. V. Hutschler	Keyport, N. J	Feb. 9, 1869	86, 756
Hame-fastener	J. Koch and D. Seachrist	Columbiana, Ohio	Sept. 1, 1868	81, 650
Hame-fastener	J. T. McDivitt	Fayetteville, Ohio	Aug. 31, 1869	93, 230
Hame-fastener	R. R. McDonald	Syracuse, N. Y	Sept. 15, 1868	82, 142
Hame-fastener	J. H. McKinley	New York, N. Y	Dec. 29, 1868	85, 464
Hame-fastener	C. H. Miller	Buffalo, N. Y	Apr. 5, 1870	101, 642
Hame-fastener	J. D. Miller	Enon, Ohio	Aug. 4, 1868	80, 757
Hame-fastener	A. Palmer	Hudson, Mich	Nov. 19, 1867	71, 209

Index of patents issued from the United States Patent Office from 1790 *to* 1873, *inclusive*—Continued.

Invention.	Inventor.	Residence.	Date.	No.
Hame-fastener	W. W. Palmer	Hudson, Mich	Apr. 16, 1867	63, 933
Hame-fastener	W. H. Payne	Janesville, Wis	Oct. 13, 1868	83, 087
Hame-fastener	A. J. Preston	Dryden, N. Y	Nov. 7, 1865	50, 842
Hame-fastener	J. R. Richardson	Newcastle, Pa	Sept. 3, 1867	68, 575
Hame-fastener	W. A. Robinson	Grand Rapids, Mich	Jan. 21, 1868	73, 651
Hame-fastener	M. R. Shalters	Alliance, Ohio	July 24, 1866	56, 621
Hame-fastener	T. Skelton	Rockford, Ill	Dec. 22, 1868	85, 250
Hame-fastener	A. Sweetland	Syracuse, N. Y	Mar. 3, 1868	75, 074
Hame-fastener	A. Sweetland	Syracuse, N. Y	June 13, 1871	115, 908
Hame-fastener	W. S. Thompson and R. V. Love	Montgomery, Ala	Sept. 8, 1868	82, 045
Hame-fastener	W. H. Tillon	Lee Roy, N. Y	Dec. 28, 1869	98, 315
Hame-fastener	J. Tingley	Potter County, Pa	Nov. 16, 1858	22, 096
Hame-fastener	A. J. Tompkins and J. M. Wegand.	Clarksville, Iowa	Oct. 12, 1869	95, 855
Hame-fastener	A. B. and S. A. Woodard	Alfred Centre and Hornellsville, N. Y.	July 14, 1868	79, 930
Hame-fastener	I. B. Woolsey	Bloomfield, Iowa	May 16, 1865	47, 762
Hame-fastening	W. J. Alexander	Manchester, Iowa	Dec. 11, 1866	60, 319
Hame-fastening	E. Covert	Farmer Village, N. Y	July 1, 1873	140, 475
Hame-fastening	W. Fawcett	New York, N. Y	Mar. 2, 1869	87, 402
Hame-fastening	H. A. Fowler	East Guilford, N. Y	Jan. 27, 1857	16, 475
Hame-fastening	H. B. Grumling	Grant, Pa	Jan. 10, 1871	110, 848
Hame-fastening	W. W. Kittleman	Bloomfield, Iowa	May 16, 1865	47, 735
Hame-fastening	J. McLain	Saint Mary's, Ohio	Mar. 15, 1864	41, 962
Hame-fastening	C. Morgan	Waumandee, Wis	Nov. 10, 1868	83, 987
Hame-fastening	L. D. G. Niles	Helena, Ark	Sept. 20, 1870	107, 528
Hame-fastening	M. R. Shalters and T. Catern	Alliance, Ohio	Dec. 18, 1866	60, 574
Hame-fastening	T. Taylor	Loudoun County, Va	Mar. 11, 1835	....
Hame-fastening	J. B. Tinkelpaugh	Hastings, Minn	May 10, 1864	42, 651
Hame-fastening	E. Turner	Baltimore, Md	Aug. 8, 1854	11, 501
Hame fastening, Harness	H. Compton	Wells' Corner, Pa	Feb. 3, 1857	16, 530
Hame fastening, Harness	T. Taylor	Loudoun County, Va	Jan. 15, 1850	7, 030
Hame guard, Harness	C. H. Allen	Saint Louis, Mo	Mar. 4, 1873	136, 403
Hame, Harness	M. E. Abbey	Sardis, Miss	Oct. 8, 1872	131, 986
Hame, Harness	J. E. Brown	Lansingburgh, N. Y	Nov. 28, 1865	51, 256
Hame, Harness	W. H. Bustin	Watertown, Mass	Apr. 27, 1869	89, 288
Hame, Harness	D. Clemons	Scranton, Pa	Oct. 13, 1868	83, 041
Hame, Harness	A. Dietz	New York, N. Y	Apr. 2, 1850	7, 235
Hame, Harness	C. H. Drury	Oceola, Ill	Oct. 10, 1871	119, 697
Hame, Harness	W. Duncan	Spring Hill, Ind	Aug. 16, 1870	106, 341
Hame, Harness	D. Foreman	Milton, Ill	Apr. 2, 1872	125, 278
Hame, Harness	K. Frazer	Fayetteville, N. Y	Sept. 26, 1846	4, 773
Hame, Harness	J. Holt and S. G. Cheever	Chelsea and Boston, Mass	Apr. 30, 1867	64, 222
Hame, Harness	R. W. Jones	Syracuse, N. Y	Feb. 6, 1872	123, 401
Hame, Harness	J. Letchworth	Buffalo, N. Y	Apr. 19, 1870	102, 138
Hame, Harness	J. Low	New Britain, Conn	Apr. 9, 1850	7, 265
Hame, Harness	C. Pope	Syracuse, N. Y	Nov. 6, 1849	6, 850
Hame, Harness	C. Robinson	Eau Claire, Wis	Mar. 25, 1873	137, 241
Hame, Harness	J. Thornton and E. G. Latta	Wellsville, N. Y	Apr. 18, 1871	113, 945
Hame, Harness	S. G. Tufts	Mainville, Ohio	July 30, 1867	67, 379
Hame, Harness	P. B. Watson	Belvidere, N. J	Aug. 22, 1871	118, 310
Hame hook and clevis	A. Strever	Albany, N. Y	Jan. 26, 1869	86, 188
Hame, Horse	N. Post	Madrid, N. Y	June 15, 1844	3, 634
Hame, Horse	J. K. Slater and S. G. Pratt	Boston, Mass	Sept. 20, 1844	3, 754
Hame, Horse-collar	J. E. Brown	Lansingburgh, N. Y	Apr. 10, 1866	53, 916
Hame, Horse-collar	M. Drew	Saint Paul, Minn	Sept. 18, 1860	30, 052
Hame-lock	J. A. Stausbury	Marion, Iowa	Apr. 7, 1868	76, 356
Hame-making machine	H. Burt and J. T. Hedden	Newark City, N. J	Feb. 17, 1857	16, 638
Hame-ring	J. Letchworth	Buffalo, N. Y	Mar. 9, 1869	87, 686
Hame-staple	J. Letchworth	Buffalo, N. Y	Feb. 6, 1872	123, 350
Hame-staple socket	R. J. Alger	Kalamazoo, Mich	Mar. 18, 1873	136, 803
Hame-strap	M. T. Briggs	Schoolcraft, Mich	June 4, 1867	65, 334
Hame-strap	G. Paddington and W. F. Crew	Waubeck, Iowa	June 15, 1869	91, 258
Hame-tug	J. C. Covert	Townsendville, N. Y	Dec. 10, 1867	71, 993
Hame-tug	L. Hall	Henrietta, Mich	July 7, 1863	39, 143
Hame-tug	J. Hovey	Bedford, Mich	May 8, 1860	28, 178
Hame-tug	J. McInnes	Waverly, Ill	Sept. 11, 1866	57, 944
Hame-tug	R. B. Whipple	Cleveland, Ohio	Apr. 20, 1852	8, 896
Hame-tug and breast-collar	A. McMullen	Sterling, Ill	Nov. 27, 1866	60, 031
Hame tug and buckle	L. W. Heelan	Petersburgh, Ill	Sept. 11, 1866	57, 899
Hame tug and buckle	J. S. Topham	Washington, D. C	June 12, 1860	28, 700
Hame tug fastener	E. D. Lockwood	Churchville, N. Y	May 18, 1858	20, 278
Hame-tug fastening	J. E. Ball	Newark, N. J	May 18, 1858	20, 246
Hame-tug fastening	J. W. Church	Cold Water, Mich	Dec. 18, 1860	30, 911
Hame-tug fastening	W. J. Lockwood	Sturgis, Mich	Sept. 29, 1857	18, 290
Hame tug, Harness	N. Botsford	Somerset, N. Y	June 14, 1870	104, 106
Hame tug, Harness	P. R. Dawson	Brenham, Tex	Aug. 15, 1871	117, 994
Hame tug, Harness	S. A. Summers	Trappe, Md	Aug. 27, 1872	130, 955
Hame-tug, Metallic	J. M. Curran and J. C. Baxter	Washington, D. C	Oct. 1, 1867	69, 413
Hames and collars, Stretching and blocking	M. Eddy	Adams, N. Y	Apr. 22, 1835	....
Hames and means of hitching horses to vehicles, Appliance to.	J. L. Kreider	Chestnut Level, Pa	June 1, 1869	90, 669
Hames, Breast-loop for	W. B. Hayden	Columbus, Ohio	Sept. 7, 1869	94, 597
Hames, Mode of attaching hold-back rings to	W. B. Hayden	Columbus, Ohio	July 26, 1870	105, 679
Hames to supersede use of collars, Wooden	S. Norton	East Bloomfield, N. Y	June 12, 1835	....
Hammer	S. Anderson	Garrettsville, N. Y	Aug. 20, 1845	4, 155
Hammer	C. Ball	Augusta, Mich	Sept. 9, 1862	36, 389
Hammer	R. Boeklen	Brooklyn, N. Y	Aug. 28, 1860	29, 760
Hammer	R. Boeklen and G. W. Schramm	Brooklyn and New York, N. Y.	Nov. 4, 1862	36, 829
Hammer	H. Cheney	Little Falls, N. Y	July 2, 1867	66, 298
Hammer	J. P. Clark	Portland, Me	Oct. 19, 1858	21, 820
Hammer	T. S. Coffin	Harrington, Me	Oct. 27, 1868	83, 463
Hammer	D. T. Crockett	Newark, N. J	Dec. 18, 1866	60, 482
Hammer	R. Devereux	Buffalo, N. Y	Mar. 26, 1867	63, 228
Hammer	J. H. Godwin	Scotland Neck, N. C	June 23, 1868	79, 066
Hammer	C. Hammond	Philadelphia, Pa	Jan. 19, 1847	4, 934
Hammer	P. C. Havely and W. W. Coggshall.	Rensselaerville, N. Y	June 30, 1868	79, 346

Index of patents issued from the United States Patent Office from 1790 *to* 1873, *inclusive*—Continued.

Invention.	Inventor.	Residence.	Date.	No.
Hammer	G. W. Hubbard	Lowville, N. Y	Sept. 29, 1868	82, 621
Hammer	J. H. Littlefield	Cambridge, Mass	Sept. 12, 1865	49, 898
Hammer	G. H. Mills and J. M. Hanscom	Boston, Mass	July 15, 1862	35, 885
Hammer	C. Monson	New Haven, Conn	Oct. 3, 1865	50, 262
Hammer	J. P. Radley	Albany, N. Y	Feb. 8, 1870	99, 705
Hammer	O. Rock	Hudson, Mass	May 31, 1870	103, 660
Hammer	O. Shepard	Alton, Ill	Feb. 18, 1868	74, 615
Hammer	S. B. Smith	New Haven, Conn	Dec. 7, 1869	97, 715
Hammer	H. H. W. Wright	South Boston, Mass	Mar. 17, 1868	75, 615
Hammer	G. M. Young	El Paso, Ill	May 16, 1871	115, 008
Hammer	W. Zimmerman	Quincy, Ill	Nov. 10, 1868	83, 897
Hammer and anvil-stock, Trip	J. C. Higgins	Skowhegan, Me	Oct. 4, 1870	108, 023
Hammer and anvil, Trip	D. A. Morris	Pittsburgh, Pa	Oct. 5, 1858	21, 691
Hammer and carpet-stretcher, Tack	A. Hicks	Factoryville, N. Y	June 28, 1864	43, 309
Hammer and drop, Adjustable	T. P. Keeler	Worcester, Mass	May 26, 1868	78, 291
Hammer and mallet	W. S. McNeil	Springfield, Mass	Sept. 8, 1868	82, 017
Hammer and nail-holder combined	R. W. Green	Bradford, Pa	Nov. 24, 1868	84, 351
Hammer and turning nipple locks, Concealed	A. Wurfflein	Philadelphia, Pa	Dec. 18, 1849	6, 964
Hammer, Atmospheric	W. D. Grimshaw	Birmingham, England	Jan. 10, 1865	45, 896
Hammer, Atmospheric	M. Peck	New Haven, Conn	Apr. 17, 1860	27, 925
Hammer, Atmospheric	J. Robertson	New York, N. Y	May 1, 1866	54, 413
Hammer, Atmospheric	H. Shattuck	Hamden, Conn	Aug. 22, 1865	49, 593
Hammer, Atmospheric trip	B. Hotchkiss	New Haven, Conn	Sept. 15, 1863	39, 924
Hammer, Automatic steam	C. Sellers	Philadelphia, Pa	Oct. 22, 1872	132, 374
Hammer, Automatic steam	W. Sellers	Philadelphia, Pa	Oct. 22, 1872	132, 375
Hammer, Blacksmith's	G. B. Cubberley	Milwaukee, Wis	Feb. 25, 1873	136, 139
Hammer, Blacksmith's spring	J. Rainey	Orange County, N. C	Mar. 7, 1827	....
Hammer, Bush	J. W. Maloy	Boston, Mass	Dec. 7, 1869	97, 663
Hammer, Bush	W. C. Peckham	Troy, Ohio	Oct. 29, 1872	132, 539
Hammer, Bush	A. Wheeler	Gloucester, Mass	Nov. 6, 1866	59, 491
Hammer, Bush	E. J. Worcester	Worcester, Mass	May 13, 1873	138, 975
Hammer, Carpenter's	J. O. Montignani	Albany, N. Y	Feb. 28, 1865	46, 574
Hammer, Cast-iron	T. Jones	Glastenbury, Conn	July 31, 1828	....
Hammer, Claw	L. P. Edmeston	New York	Aug. 1, 1848	5, 697
Hammer claw-die	H. Hammond	Hartford, Conn	Nov. 19, 1867	70, 998
Hammer, Compressed-air forge	C. Vogel	New York, N. Y	Aug. 30, 1870	106, 892
Hammer, Steam-cylinder	R. Mitchell	Wolverhampton, Great Britain.	Nov. 5, 1867	70, 595
Hammer, Dental	J. C. Dean	Chicago, Ill	July 11, 1865	48, 708
Hammer-die	H. W. Bailey	Springfield, Mass	Mar. 17, 1868	75, 662
Hammer-die	J. Yerkes	Philadelphia, Pa	Aug. 19, 1873	142, 070
Hammer, Driect-action steam	J. H. Towne	Philadelphia, Pa	Sept. 3, 1850	7, 623
Hammer, Drop	J. Aughe	Dayton, Ohio	July 3, 1866	55, 976
Hammer, Drop	J. Blackadder	New Orleans, La	Aug. 6, 1872	130, 183
Hammer, Drop	J. E. Crisp	Charlestown, Mass	June 11, 1872	127, 853
Hammer, Drop	J. Evans	New Haven, Conn	Apr. 25, 1865	47, 405
Hammer, Drop	B. Hotchkiss	New Haven, Conn	Mar. 1, 1864	41, 775
Hammer, Drop	E. Kaylor	Pittsburgh, Pa	Oct. 12, 1869	95, 693
Hammer, Drop	J. McGeorge	Bellaire, Ohio	July 8, 1873	140, 717
Hammer, Drop	L. H. Olmsted	Stamford, Conn	July 16, 1867	66, 879
Hammer, Drop	E. K. Root	Hartford, Conn	Aug. 16, 1853	9, 941
Hammer, Drop	N. C. Stiles	Middletown, Conn	Aug. 27, 1872	130, 877
Hammer, Drop	J. Tobin	Newark, N. J	Dec. 16, 1873	145, 601
Hammer, Drop	W. H. Waters	Springfield, Mass	Oct. 1, 1867	69, 518
Hammer, Drop	J. Wool	Boston, Mass	Apr. 21, 1868	77, 155
Hammer, Fly or trip	J. Hines	Providence, R. I	June 16, 1810	....
Hammer, Foot-trip	E. Pierce and J. Hathaway	Pultney, N. Y	Oct. 19, 1827	....
Hammer, Foot-trip	C. V. Statler and G. W. Wilson	Walnut Grove, Ill	Jan. 11, 1859	22, 589
Hammer for bending couplings	J. T. Bruen	New York, N. Y	June 26, 1866	55, 813
Hammer for cutting and dressing stone	J. Richards	Braintree, Mass	Feb. 20, 1828	....
Hammer for forging bloom	G. D. Manley	Cogan Station, Pa	Nov. 6, 1866	59, 425
Hammer for pecking or working mill and other stones, Peck.	B. Gardner	Ashfield, Mass	Aug. 3, 1838	873
Hammer for planishing sheet-metal	W. D. Wood	McKeesport, Pa	Sept. 9, 1873	142, 754
Hammer for plating out scythes, Gage to be attached to the anvil of trip.	B. Richardson	Western, Mass	May 17, 1814	....
Hammer, Forge	J. Comstock	New London, Conn	Apr. 24, 1855	12, 770
Hammer, Forge	E. Paye	New York, N. Y	Oct. 9, 1860	30, 343
Hammer, Forge	B. Shiverick	Pittsburgh, Pa	Nov. 16, 1858	22, 092
Hammer, Forge	E. Spaulding	Worcester, Mass	Nov. 22, 1864	45, 187
Hammer, Forge	B. Walker	Green Point, N. Y	Dec. 3, 1867	71, 823
Hammer, Forge and trip	J. Goulding	Leicester, Mass	Mar. 1, 1811	....
Hammer, Forge arm	J. Sharp	Colerain, Pa	June 14, 1834	....
Hammer, Forging	S. B. Dodge	Roslyn, N. Y	Jan. 1, 1869	90, 642
Hammer, Forging	W. D. Grimshaw	Newark, N. J	May 20, 1862	35, 309
Hammer, Forging	L. Wetherell	Boston, Mass	Mar. 12, 1867	62, 793
Hammer forging-die	T. Daffin	Philadelphia, Pa	Oct. 7, 1873	143, 406
Hammer, Furrowing	L. Sauers	Mount Joy, Pa	Apr. 24, 1866	54, 258
Hammer, Hand	A. Gregory	Washington, D. C	Nov. 16, 1858	22, 073
Hammer-handle	T. Phillips	Ann Arbor, Mich	Apr. 9, 1861	31, 997
Hammer-handle	D. Weiser	Philadelphia, Pa	May 19, 1868	78, 164
Hammer, Hatchet	J. Howe	Princeton, Mass	July 8, 1834	....
Hammer-head	R. Black	Holyoke, Mass	July 16, 1867	66, 781
Hammer-head	R. Dawes	Washington, D. C	Oct. 19, 1858	21, 823
Hammer, Lasting	J. W. Warner	Dover, N. H	Oct. 19, 1869	96, 061
Hammer lifter, Drop	F. M. Hodge	Shelburne Falls, Mass	Jan. 3, 1871	110, 655
Hammer lifter, Drop	C. Peck	New Haven, Conn	Sept. 9, 1873	142, 723
Hammer, Machine	D. Noyes	Abington, Mass	Oct. 25, 1853	10, 170
Hammer, Meat	A. H. Brown	Springfield, Vt	Apr. 30, 1867	64, 279
Hammer, Nail	C. Carlisle	Woodstock, Vt	Oct. 16, 1860	30, 386
Hammer, Nail	H. Cheney	Little Falls, N. Y	July 4, 1871	116, 553
Hammer, Nail	C. G. Dodge, jr	Providence, R. I	July 10, 1866	56, 191
Hammer, Nail	E. S. Morton	Plymouth, Mass	June 7, 1870	103, 914
Hammer, Nail	G. Selsor	Philadelphia, Pa	Mar. 19, 1867	63, 106
Hammer, Nail	H. M. Stocum	Painted Post, N. Y	Dec. 5, 1871	121, 682
Hammer, Nail	W. G. Ward	Savona, N. Y	July 31, 1866	56, 838
Hammer, Portable trip	S. Kilburn	Sterling, Md	Aug. 18, 1829	....
Hammer, Power	I. Althouse	Columbus, Ohio	Oct. 17, 1871	120, 017

Index of patents issued from the United States Patent Office from 1790 *to* 1873, *inclusive*—Continued.

Invention.	Inventor.	Residence.	Date.	No.
Hammer, Power	J. C. Butterfield	Chicago, Ill	Oct. 28, 1873	144, 058
Hammer, Power	J. C. Butterfield	Chicago, Ill	Oct. 28, 1873	144, 059
Hammer, Power	A. Cunningham and A. Sharp	Salem, Ohio	July 2, 1867	66, 222
Hammer, Power	T. B. Harrison	Maquoketa, Iowa	Sept. 10, 1867	68, 740
Hammer, Power	M. Hunkley	Rochester, N. Y	Aug. 27, 1867	68, 301
Hammer, Power	J. M. Long	Hamilton, Ohio	Feb. 25, 1873	136, 253
Hammer, Power	J. Palmer	Concord, N. H	Jan. 9, 1872	122, 647
Hammer, Power	S. Pennock	Kennett's Square, Pa	Mar. 25, 1873	137, 092
Hammer, Power	T. F. Preston	Pawtucket, R. I	May 7, 1867	64, 567
Hammer, Power	T. T. Prosser	Chicago, Ill	Apr. 13, 1869	88, 980
Hammer, Power	T. Scott and J. Clarridge	Madison Mills and Pancoastburgh, Ohio.	Sept. 10, 1867	68, 797
Hammer, Power	T. Shaw	Philadelphia, Pa	Feb. 27, 1866	52, 894
Hammer, Power	T. Shaw	Philadelphia, Pa	Nov. 17, 1868	84, 221
Hammer, Power	A. Strube	Frederick, Md	Mar. 24, 1868	75, 808
Hammer, Revolving	G. Stacy	Nanuet, N. Y	Nov. 18, 1873	144, 638
Hammer, Riveting	H. A. Seymour and W. H. Nettleton.	Bristol, Conn	May 10, 1870	102, 871
Hammer saw-set, Trip	W. Duesler	Saint Joseph, Mo	Sept. 28, 1869	95, 331
Hammer, screw-driver, and wrench, Combined	W. D. Gold	Philadelphia, Pa	May 14, 1867	64, 662
Hammer shafts or stamps, Device for lubricating, cooling, and washing vertical.	Z. E. Coffin	Newton, Mass	Sept. 6, 1870	107, 005
Hammer, Shoemaker's	A. Clarke	Boston, Mass	Mar. 21, 1871	112, 902
Hammer, Shoemaker's	A. and A. Clarke	Philadelphia, Pa	July 2, 1872	128, 463
Hammer, Shoemaker's	J. Vigeant	Marlborough, Mass	Jan. 7, 1868	73, 141
Hammer. Spring-balance	J. Collins	Parma, Ohio	Apr. 27, 1869	89, 292
Hammer standards, Construction of steam	O. C. Ferris and F. B. Miles	Philadelphia, Pa	Apr. 4, 1871	113, 508
Hammer, Steam	T. Beach	Freeport, Pa	Sept. 4, 1860	29, 854
Hammer, Steam	P. Danvers	New York, N. Y	Sept. 14, 1858	21, 489
Hammer, Steam	O. C. Ferris and F. B. Miles	Philadelphia, Pa	July 5, 1870	104, 948
Hammer, Steam	D. Joy	Middlesboro', Great Britain.	Aug. 4, 1868	80, 550
Hammer, Steam	E. L. Kinsley	Cambridge, Mass	May 8, 1866	54, 560
Hammer, Steam	L. Kirk	Reading, Pa	Sept. 19, 1848	5, 774
Hammer, Steam	J. L. L. Morris	Reading, Pa	Jan. 31, 1854	10, 479
Hammer, Steam	R. Morrison	Newcastle-upon-Tyne, England.	Oct. 29, 1861	33, 595
Hammer, Steam	D. R. Quick and G. A. Gardner	New York, N. Y	Aug. 12, 1873	141, 667
Hammer, Steam	W. Sellers	Philadelphia, Pa	Oct. 22, 1872	132, 373
Hammer, Steam	T. Sumner	Paterson, N. J	June 27, 1854	11, 182
Hammer, Steam	R. R. Taylor	Reading, Pa	Jan. 1, 1861	31, 041
Hammer, Steam	P. L. Weimer	Reading, Pa	Jan. 31, 1854	10, 486
Hammer, Steam	C. W. and J. P. Willard	Chicago, Ill	Oct. 4, 1864	44, 573
Hammer, Steam drop	F. B. Miles	Philadelphia, Pa	Apr. 1, 1873	137, 313
Hammer, Steam trip	L. Kirk	Reading, Pa	Apr. 3, 1847	5, 044
Hammer, Stone	G. H. Pierce	Middlebury, Mass	Dec. 17, 1834	....
Hammer-strap	W. J. Lewis	Pittsburgh, Pa	Oct. 4, 1870	107, 929
Hammer-straps for wagons, Blank for	W. J. Lewis	Pittsburgh, Pa	Oct. 4, 1870	107, 931
Hammock-support	F. Park	Deerfield, Mass	Nov. 4, 1873	144, 219
Hammock-support	O. Tufts	Boston, Mass	Oct. 28, 1873	144, 168
Hammer, Swage spring	W. Field	North Providence, R. I	Sept. 14, 1833	....
Hammer-swaging die	H. Hammond	Hartford, Conn	Nov. 19, 1867	70, 997
Hammer, Tack	G. J. Capewell	Cheshire, Conn	Nov. 25, 1873	144, 953
Hammer, Tack	J. Crandell and N. P. Braman	Chicopee Falls, Mass., and Bridgeport, Conn.	Jan. 5, 1869	85, 568
Hammer, Tack	T. Evans	Newark, N. J	Nov. 26, 1867	71, 290
Hammer, Tack	A. Pond	Southington, Conn	June 19, 1866	55, 704
Hammer, Tack	T. B. Stout	Keyport, N. J	June 11, 1867	65, 706
Hammer, Tilt	H. Baines	Toronto, Canada	Aug. 22, 1871	118, 181
Hammer, Tilt	P. Breen	Auburn, N. Y	Oct. 31, 1871	120, 412
Hammer, Tilt	J. F. Gould	....	Mar. 8, 1806	....
Hammer, Trip	O. Ames	Plymouth, Mass	Apr. 25, 1810	....
Hammer, Trip	J. Beal	Ontario County, N. Y	Oct. 5, 1811	....
Hammer, Trip	J. Briggs	Louisville, Ky	Oct. 27, 1863	40, 390
Hammer, Trip	H. Bushnell	New Haven, Conn	Mar. 3, 1857	16, 714
Hammer, Trip	J. C. Butterfield and J. Hay	Chicago, Ill	Oct. 18, 1870	108, 326
Hammer, Trip	E. Downer	De Ruyter, N. Y	Apr. 11, 1810	....
Hammer, Trip	H. E. Fessel and F. Krautwadel	Chicago, Ill	Feb. 7, 1865	46, 229
Hammer, Trip	J. C. Forrest and G. Baker	Schenectady, N. Y	Dec. 14, 1852	9, 462
Hammer, Trip	B. Hershey	Erie, Pa	Feb. 21, 1871	112, 036
Hammer, Trip	B. Hotchkiss	New Haven, Conn	June 14, 1859	24, 428
Hammer, Trip	B. Hotchkiss	New Haven, Conn	May 3, 1864	42, 581
Hammer, Trip	D. Howell	Louisville, Ky	Apr. 10, 1860	27, 804
Hammer, Trip	B. Hughes	Rochester, N. Y	May 16, 1854	10, 923
Hammer, Trip	L. Kinsley	Cambridgeport, Mass	Feb. 10, 1863	37, 628
Hammer, Trip	J. Kreshbiel	Williamsville, N. Y	July 25, 1871	117, 432
Hammer, Trip	S. Morrill	Readfield, Mass	Sept. 28, 1812	....
Hammer, Trip	J. Neuberger	Buffalo, N. Y	Oct. 15, 1872	132, 173
Hammer, Trip	E. Osborn	Berkshire County, Mass	Jan. 8, 1810	....
Hammer, Trip	U. Palmer	Spencer, N. Y	July 31, 1815	....
Hammer, Trip	M. Peck	New Haven, Conn	July 3, 1855	13, 179
Hammer, Trip	J. W. Peer	Schenectady, N. Y	Nov. 29, 1853	10, 274
Hammer, Trip	P. P. Read	Mercer, Me	Dec. 7, 1826	....
Hammer, Trip	H. Redfield	Grafton, Mass	Mar. 5, 1835	....
Hammer, Trip	T. J. Root	Galena, Ill	Feb. 7, 1865	46, 267
Hammer, Trip	T. J. and R. L. Root	Andover, Ohio	Aug. 20, 1867	67, 913
Hammer, Trip	L. Rosenkrans	Big Flats, N. Y	Mar. 17, 1828	....
Hammer, Trip	J. Smith	Otsego County, N. Y	Oct. 14, 1809	....
Hammer, Trip	J. Tandler	Grand Rapids, Mich	Aug. 27, 1867	68, 295
Hammer, Trip	W. Van Anden	Poughkeepsie, N. Y	Aug. 16, 1853	9, 942
Hammer, Trip	C. Vogel	New York, N. Y	Feb. 25, 1868	74, 868
Hammer, Vertical trip	P. L. Weimer	Reading, Pa	Sept. 25, 1855	13, 608
Hammer, &c., with craw and peen, Nail	P. Stebbins and J. Holmes	Schenectady, N. Y	June 1, 1852	8, 989
Hammers and hatchets, Mode of manufacturing socket.	C. A. Strong	South Glastonbury, Conn	Mar. 3, 1829	....
	P. Eastman	Canaan, N. H	July 17, 1839	1, 247
Hammers and stamps, Operating	C. K. James	Jersey City, N. J	June 19, 1866	55, 772
Hammers, Arrangement of valves, ports, and passages for operating steam.	R. R. Taylor	Reading, Pa	Nov. 29, 1853	10, 276

Index of patents issued from the United States Patent Office from 1790 *to* 1873, *inclusive*—Continued.

Invention.	Inventor.	Residence.	Date.	No.
Hammers, Cylinder for steam	T. R. Morgan	Pittsburgh, Pa	Apr. 23, 1872	125, 976
Hammers, Machine for forging	W. Evans	Long Island City, N. Y	Nov. 25, 1873	144, 969
Hammers, Machine for making	R. B. Perkins	Meriden, Conn	Oct. 13, 1857	18, 408
Hammers, Making	J. Yerkes	Fox Chase, Pa	Feb. 12, 1867	62, 105
Hammers, Manner of working forge	G. E. Sellers	Cincinnati, Ohio	Jan. 10, 1845	3, 882
Hammers, Manufacture of tack	T. A. Conklin	New Britain, Conn	Dec. 10, 1867	71, 986
Hammers, Method of adjusting the stroke of trip	L. Briggs, jr	Braintree, Mass	Mar. 11, 1851	7, 965
Hammers, Mode of hanging forge	J. McNeel	Coleman Township, Pa	Jan. 21, 1829	....
Hammers, Moving trip	E. Crowell	Gardiner, Me	Aug. 4, 1826	....
Hammers, Operating blacksmith's	J. W. Kerr	Rochester, N. Y	Apr. 20, 1858	19, 997
Hammers, Operating steam trip	J. S. Bonney and C. W. Willard	Hanson and Bridgewater, Mass.	Aug. 17, 1858	21, 183
Hammers, spikes, gimlets, &c., Machine for shaping	E. L'Hommedieu	Saybrook, Conn	Apr. 28, 1830	....
Hammers to helves, Attachment of	D. Hicks	Duncansville, Pa	Apr. 2, 1850	7, 242
Hammers, Valve arrangement for steam	J. Watt	South Boston, Mass	Dec. 6, 1853	10, 297
Hammers, Valve-gear for steam	F. B. Miles	Philadelphia, Pa	July 21, 1868	80, 082
Hammers with perpendicular handles, Machine for working trip.	J. Fox, jr	Middlesex, Mass	Feb. 25, 1807	....
Hammock	W. H. Forbush	Buffalo, N. Y	Nov. 5, 1861	33, 678
Hammock	A. Woods	Liverpool, England	Sept. 17, 1867	68, 927
Hammock, Knapsack	G. Abbott	United States Army	Feb. 2, 1864	41, 418
Hammock, Knapsack	A. W. Süs	New York, N. Y	Mar. 8, 1864	41, 870
Hammock-supporter	J. M. K. Southwick	Newport, R. I	May 23, 1871	115, 128
Hammocks, Device for suspending	O. Tufts	Boston, Mass	Sept. 30, 1873	143, 392
Hams, Apparatus for extracting marrow from	W. W. Macqueen	Saint Louis, Mo	Mar. 26, 1872	125, 062
Hams, Composition for covering	H. A. Amelung	New York, N. Y	Dec. 8, 1863	40, 803
Hams, Composition for covering	H. Billings	Beardstown, Ill	Apr. 9, 1850	7, 256
Hams, Holder for slicing	J. Baumgartner and L. Angster	Newark, N. J	Nov. 12, 1867	70, 780
Hams, Putting up	S. E. Kelly	Philadelphia, Pa	Mar. 7, 1871	112, 467
Hand and arm, Artificial	J. S. Drake	Boston, Mass	July 22, 1856	15, 372
Hand, Apparatus for development of the muscles of the.	D. P. D. Kops	New York, N. Y	Nov. 28, 1871	121, 289
Hand, Artificial	O. Lindsay and I. Vance	Washington, D. C	Sept. 19, 1865	50, 014
Hand, Artificial	J. F. Maguire	East Boston, Mass	July 17, 1866	56, 427
Hand, Artificial	W. Selpho	New York, N. Y	Aug. 18, 1857	18, 021
Hand, Artificial	W. Selpho and J. Walber	New York, N. Y	Dec. 6, 1859	26, 378
Hand clamp	S. L. Thompson	Lowell, Mass	May 7, 1872	126, 425
Hand-drill	E. Beeman	Owego, N. Y	Aug. 12, 1862	36, 135
Hand-drill	W. Bushnell	New York, N. Y	Dec. 2, 1851	8, 554
Hand-drill	W. C. Burch	Gloucester, N. J	May 26, 1868	78, 183
Hand-drill	L. Hillebrand	Philadelphia, Pa	Mar. 19, 1872	124, 681
Hand-drill	J. Johnson	Somerville, Mass	Oct. 31, 1848	5, 894
Hand-drill	F. McNair	Fultonham, Ohio	Nov. 16, 1858	22, 085
Hand-drill	S. H. Ostrom	Schenectady, N. Y	May 18, 1869	90, 292
Hand-drill	H. H. Packer	Boston, Mass	June 29, 1858	20, 728
Hand-drill	J. H. Parker	Boston, Mass	May 15, 1860	28, 327
Hand-drill	J. H. Parker	Boston, Mass	May 15, 1860	28, 328
Hand-drilling machine	J. E. Hunter	North Adams, Mass	Apr. 5, 1870	101, 466
Hand-engine	C. T. Ulmann and M. Bockman	New York and Brooklyn, N. Y.	Mar. 17, 1868	75, 657
Hand-fork	F. W. Palmer	West Richmondville, N. Y.	May 12, 1868	77, 906
Hand griping-tool	E. A. Alpress	Bristol, Conn	Dec. 31, 1867	72, 712
Hand-hole plate, Crab for	J. S. Hooton	New Carlisle, Ind	July 12, 1870	105, 208
Hand-indicator for showing course of vessels	H. P. Tuttle	Brooklyn, N. Y	Mar. 15, 1870	100, 821
Hand-mill	E. Alsop	New York, N. Y	Sept. 1, 1868	81, 574
Hand-motor, Mechanical	H. Fowler	Washington, D. C	Sept. 20, 1870	107, 474
Hand-motor, Mechanical	H. Fowler	Washington, D. C	Sept. 20, 1870	107, 475
Hand-preserving compound	J. W. Osborn	Brooklyn, N. Y	Jan. 31, 1871	111, 376
Hand-press	B. F. Day	Philadelphia, Pa	Aug. 15, 1854	11, 520
Hand-press, Jeweler's	J. McWilliams	Providence, R. I	Sept. 23, 1873	143, 023
Hand-press, Self-inking	D. Zuern and L. L. Bevan	Shamokin, Pa	Nov. 2, 1858	21, 997
Hand-presses, Rack-motion for	C. Wells	Cincinnati, Ohio	Dec. 4, 1866	60, 301
Hand-protector	J. Turnbull	Simsbury, Conn	Dec. 19, 1871	121, 974
Hand-rake	C. Bissell	Aurora, Ohio	May 7, 1867	64, 475
Hand-rake	H. M. Clark	Brewer, Me	Sept. 15, 1868	82, 207
Hand-rake	M. Ellis	Greenfield, Mass	Sept. 6, 1870	107, 018
Hand-rake	D. Johnston	Sulphur Springs, Ohio	Aug. 24, 1869	94, 118
Hand-rake	G. W. Stearns	Lebanon, N. H	Feb. 6, 1872	123, 369
Hand-rake	A. Winters	Washington, Pa	Dec. 22, 1868	85, 154
Hand-rake for hay, &c	A. Foster	Auburn, N. Y	Dec. 19, 1827	....
Hand-rake, Revolving	S. C. Rundlett	Portland, Me	Oct. 25, 1864	44, 824
Hand-rest	I. Belford	Caldwell, Ohio	Mar. 4, 1873	136, 357
Hand-shears	J. N. Wallis	Fleming, N. Y	Dec. 27, 1864	45, 659
Hand-shears	C. Witte	Brooklyn, N. Y	Dec. 29, 1868	85, 500
Hand shears and nippers	T. Wallis and T. Witbeck	Scipio, N. Y	Mar. 21, 1865	46, 960
Hand-spring for machinery	P. Howe	Saffordville, Conn	Apr. 30, 1867	64, 318
Hand-support, Penman's	D. A. Sanborn	Brooklyn, N. Y	July 5, 1870	105, 001
Hand-wrench	G. B. Phillips	Albany, N. Y	July 21, 1857	17, 844
Hands, Substitute for artificial	J. Reichenbach	Pittsburgh, Pa	June 27, 1865	48, 440
Handcuff	W. H. Kimball	Augusta, Me	Aug. 7, 1860	29, 495
Handcuff, Adjustable	O. C. Phelps	New York, N. Y	July 17, 1866	56. 463
Handcuff-lock	H. H. Cheney	East Saginaw, Mich	Apr. 4, 1871	113, 499
Handcuff-lock	J. A. Wisner and M. Hoyt	East Saginaw, Mich	July 6, 1869	92, 414
Handkerchief and fan holder	G. D. Stevens	New York, N. Y	July 23, 1872	129, 871

Handle:
See Augur-handle.
Awl-handle.
Ax-handle.
Basket-handle.
Boat-hook handle.
Bone handle.
Bottle-handle.
Brad-awl handle.
Broom-handle.
Brush-handle.
Brush and broom handle.
Burial-casket handle.
Can-handle.
Cane-handle.
Carriage-handle.
Casket-handle.
Chisel-handle.
Coffee and spice mill handle.
Coffin-handle.
Comb-handle.
Cross-cut-saw handle.
Crutch-handle.
Cutlery-handle.
Dipper-handle.

Index of patents issued from the United States Patent Office from 1790 *to* 1873, *inclusive*—Continued.

Invention.	Inventor.	Residence.	Date.	No.
Handle—Continued. *See* Drawer-handle. File-handle. File-griping handle. Fork and spoon handle. Gimlet-handle. Hammer-handle. Implement-handle. Kettle-handle. Knife-handle. Knife and fork handle. Lamp-chimney handle. Latch-handle. Lifting-handle. Milk-can handle. Mop-handle. Mucilage-brush handle. Organ-handle. Pick-handle. Plow-handle. Plow-beam handle. Pocket-knife handle. Pump-handle. Rake-handle. Sadiron-handle. Satchel-handle. Sauce-pan handle. Saw-handle. Screw-driver handle. Sheet-metal-ware handle. Shovel-handle. Smoothing-iron handle. Spade-handle. Tea and coffee pot handle. Tool-handle. Traveling-bag handle. Trunk-handle. Umbrella-handle. Whip-handle.				
Handles, &c., Finishing and bleaching	D. H. Taylor	Westfield, N. Y	Nov. 1, 1870	108, 846
Handles, molds, &c., Composition for making	C. Branwhite	New York, N. Y	Jan. 23, 1846	4, 362
Hanger: *See* Adjustable hanger. Bracket-hanger. Broom-hanger. Car-hanger. Clothes-hanger. Door-hanger. Eave-trough hanger. Garment-hanger. Grindstone-hanger. Harness-check-hook hanger. Journal or shaft hanger. Lamp-hanger. Mirror-hanger. Picture-hanger. Pulley-hanger. Shaft-hanger. Shafting-hanger. Sword-hanger. Tobacco-hanger. Whip-hanger.				
Hanger	P. P. Lane and E. Myers	Cincinnati, Ohio	Oct. 12, 1869	95, 813
Hanger	J. M. Stone	North Andover, Mass	Sept. 17, 1867	69, 041
Hanger-bar socket	T. A. Summers	Rochester, N. Y	Dec. 22, 1863	41, 027
Hanging bracket	H. Polley	San Francisco, Cal	Sept. 10, 1872	131, 298
Hanging bracket or shelf	A. C. Brown	Chicago, Ill	Mar. 18, 1873	136, 900
Hanging gate	E. Secor	Niles, Mich	Sept. 2, 1873	142, 522
Harbor and river channels, Means of defending	R. H. Jewett	Mount Sterling, Ill	Sept. 8, 1863	39, 816
Harbor defense, Submarine	J. S. Gilbert	New York, N. Y	Apr. 7, 1863	38, 106
Harbor obstructions, Method of removing	J. D. Hall	Philadelphia, Pa	Dec. 20, 1864	45, 562
Harbors, &c., Mud-machine for deepening	J. Watchman and J. Bratt	Baltimore, Md	Feb. 18, 1825	....
Hard-rubber molds, Apparatus for filling	J. C. Robie	Binghamton, N. Y	Jan. 2, 1866	51, 865
Harmonicons, Glass or metal	J. Koppe	New York, N. Y	Oct. 2, 1860	30, 230
Harmonicons, Musical glass called the grand	F. H. Smith	Northampton, Va	Apr. 7, 1825	....
Harmonium	J. Gilmour	Glasgow, North Britain	July 18, 1865	48, 881
Harmonium	H. W. Smith	Boston, Mass	July 3, 1860	29, 041
Harmonium	C. F. Uhlig	Chemnitz, Saxony	Aug. 14, 1866	57, 270
Harness	M. E. Abbey	Sardiz, Miss	Oct. 8, 1872	131, 987
Harness	E. S. Basnett	Morgantown, W. Va	Apr. 5, 1870	101, 569
Harness	S. L. Bond	Greenwood, S. C	Oct. 23, 1860	30, 457
Harness	J. Calkins	Hudson, Mich	July 18, 1865	48, 790
Harness	S. G. Cheever	Boston, Mass	June 8, 1869	91, 087
Harness	E. Covert	Farmer Village, N. Y	Aug. 12, 1873	141, 698
Harness	C. Daniels	Fremont, Ohio	Feb. 19, 1867	62, 257
Harness	G. B. Durkee	Alden, N. Y	Oct. 8, 1867	69, 550
Harness	G. W. Dutton	Tomales, Cal	Oct. 17, 1871	120, 049
Harness	I. Ellis	Tyler, Tex	June 25, 1872	128, 377
Harness	J. Fassnacht	New Milltown, Pa	Sept. 13, 1859	25, 401
Harness	E. E. Hardy	New York, N. Y	May 30, 1865	47, 949
Harness	G. M. Harnisch	Chicago, Ill	July 6, 1869	92, 188
Harness	J. K. Harris	Springfield, Ohio	May 26, 1868	78, 282
Harness	J. K. Harris	Springfield, Ohio	June 8, 1869	91, 127
Harness	A. Harroun, jr	South Onondaga, N. Y	Mar. 17, 1868	75, 632
Harness	N. T. Healy	Medina, N. Y	Nov. 5, 1867	70, 434
Harness	R. Huyck	Sheboygan Falls, Wis	Aug. 14, 1866	57, 143
Harness	W. A. Jordan	New Orleans, La	Sept. 1, 1868	81, 788
Harness	A. D. Kendig	Safe Harbor, Pa	Oct. 30, 1866	59, 316
Harness	F. D. Ladenberger	Glenbeulah, Wis	Nov. 21, 1865	51, 062
Harness	A. W. Lawton	Rochester, N. Y	Oct. 31, 1871	120, 445
Harness	J. B. Lindeman	Manor Township, Pa	Mar. 19, 1867	63, 064
Harness	A. McCracken	South Byron, N. Y	May 2, 1871	114, 318
Harness	A. McMullen and J. H. Rock	Sterling, Ill	May 5, 1868	77, 508
Harness	B. Mirick	Rutland County, N. Y	Apr. 21, 1809	....
Harness	F. Monroe	Romeo, Mich	Aug. 24, 1858	21, 267
Harness	R. O'Brien	Fowlersville, N. Y	Apr. 7, 1868	76, 507
Harness	E. C. Patterson	Rochester, N. Y	Nov. 5, 1872	132, 855
Harness	W. Peabody	Dixmont Centre, Me	July 17, 1866	56, 462
Harness	G. Renton	Boston, Mass	May 6, 1873	138, 530
Harness	H. B. Richmond	New Britain, Conn	Aug. 20, 1872	130, 747
Harness	H. T. Robbins	Hyde Park, Mass	Oct. 22, 1872	132, 413
Harness	J. J. Smokey	Natchez, Miss	Feb. 18, 1868	74, 623
Harness	J. C. Spooner	Houlton, Me	Oct. 17, 1871	120, 118
Harness	J. Stanbrough	Newark, N. J	Mar. 22, 1853	9, 628
Harness	D. Waldhauer	New York, N. Y	Nov. 23, 1869	97, 251
Harness	L. Whitehead	Nunda, N. Y	May 17, 1870	103, 264
Harness	W. S. Wood	Haltborough, Pa	Apr. 7, 1868	76, 369
Harness	G. M. Zell	Waynesville, Ohio	Apr. 14, 1868	76, 874
Harness adapted to horse-rakes	W. Parker	Putney, Vt	June 26, 1849	6, 551
Harness and carriage to prevent accident from runaway horses, Making safety.	E. Eastlack	Greenwich, N. J	Jan. 10, 1840	1, 467
Harness and other articles of leather, Composition for dressing.	M. T. Boyd	New York, N. Y	June 1, 1869	90, 813

Index of patents issued from the United States Patent Office from 1790 *to* 1873, *inclusive*—Continued.

Invention.	Inventor.	Residence.	Date.	No.
Harness and saddle pad	G. Elsey	Sterling, Ill	Feb. 11, 1873	135, 790
Harness and thills of vehicles, Safety-apparatus to be applied to.	J. H. Wilson, jr	Nashville, Tenn	Mar. 18, 1856	14, 484
Harness, Attaching rosettes to	W. L. Denio	Rochester, N. Y	Oct. 13, 1868	82, 927
Harness-attachment	W. W. Beebee	Dubuque, Iowa	Jan. 14, 1868	73, 286
Harness-attachment	W. A. Blundell	Saint Louis, Mo	Mar. 21, 1871	112, 889
Harness-attachment	E. Croughwell, A. Tellman, and E. L. Vanderburg.	San Francisco, Cal	Aug. 26, 1873	142, 085
Harness-attachment for hitching and unhitching horses.	T. J. Dobbs	Weehawken, N. J	Oct. 14, 1873	143, 678
Harness-attachment for supporting driving-lines	T. D. Brown	Montville, Ohio	Mar. 1, 1859	23, 073
Harness back-band	A. J. Rutledge	Jefferson Parish, La	July 16, 1872	129, 498
Harness back-band hook	S. Reynolds	Pittsburgh, Pa	July 11, 1871	116, 993
Harness back-pad	J. Umbach	Kankakee, Ill	July 28, 1868	80, 433
Harness-bell	C. T. Liebold	New York, N. Y	Jan. 16, 1866	52, 058
Harness-bell	R. Schmidt	New York, N. Y	July 10, 1866	56, 277
Harness-bit	B. S. Roberts	New Haven, Conn	Nov. 8, 1870	109, 145
Harness, Bitting	B. F. Brewster	Norwich, Conn	July 27, 1869	93, 047
Harness, Blacking for	S. Sherwood	New York, N. Y	May 8, 1866	54, 616
Harness-blinds, Die for shaping	E. Ward	Newark, N. J	July 16, 1872	129, 630
Harness breast-collar	A. B. Morrison	Le Roy, N. Y	Apr. 22, 1873	138, 039
Harness breast-plate	O. Ramsdell	Westfield, Vt	Feb. 19, 1850	7, 106
Harness, Breast-plate for the breast-collar of double	C. Graham	New York, N. Y	Mar. 1, 1870	100, 283
Harness breast-strap attachment	J. Letchworth	Buffalo, N. Y	Oct. 19, 1869	96, 016
Harness, Breast-strap fastening for	J. H. Martin	Columbus, Ohio	June 21, 1870	104, 475
Harness, Breast-strap protector for	L. R. Ward	Ward's Corners, Iowa	Sept. 6, 1870	107, 134
Harness breast-straps, Shield for	M. W. Pond and H. E. Mussey	Elyria, Ohio	Sept. 6, 1864	44, 113
Harness breast-straps, Slide for	D. Pettengill	Delhi, N. Y	Nov. 25, 1862	37, 006
Harness breast-straps, Slide for	M. W. Pond and L. W. Darling	Elyria, Ohio	Jan. 19, 1864	41, 320
Harness breeching and breast plate, Brace for	E. Brickett	Minot, Me	Feb. 14, 1860	27, 098
Harness-breeching holder	F. L. Churchill and R. C. Beach	Minneapolis, Minn	May 16, 1871	114, 925
Harness, Breeching-stay for	W. R. Knowles	Columbiana, Ohio	Apr. 23, 1872	125, 963
Harness-breeching to wagon-thills, Mode of attaching.	A. Parker	Coventry, N. Y	May 3, 1859	23, 854
Harness, bridles, &c., Machine for the manufacture of.	S. Wilson	Dansville, N. Y	May 9, 1846	4, 509
Harness-buckle	J. McLean	Meadville, Pa	Apr. 8, 1873	137, 558
Harness-buckle	I. Roraback	South Bend, Ind	July 7, 1868	79, 779
Harness, Carriage	E. Wilder	South Hingham, Mass	Jan. 7, 1868	73, 149
Harness, Cart	P. K. Curll	Elkridge Landing, Md	Jan. 14, 1868	73, 236
Harness, Cart	E. E. Hardy	New York, N. Y	Sept. 13, 1864	44, 185
Harness-check hook	C. H. Bassett	Derby, Conn	Nov. 17, 1868	84, 080
Harness-check hook	A. B. Buell	Westmoreland, N. Y	Mar. 13, 1844	3, 484
Harness-check hook	H. A. Carlton	Chelsea, Mass	Apr. 22, 1873	138, 003
Harness-check hook and terret	W. V. Kay	Chicago, Ill	May 23, 1871	115, 061
Harness, Check-hook guard for	W. C. Shipherd	Cleveland, Ohio	Oct. 1, 1872	131, 793
Harness-check loop	W. Parsons	Palmyra, N. Y	Mar. 4, 1873	136, 454
Harness-clamp	T. B. Hodge	Francistown, N. H	Feb. 12, 1867	61, 939
Harness-clamp	A. W. Porter and S. A. Drake	Farmer Village, N. Y	Aug. 27, 1872	130, 942
Harness-clasp	M. A. Penn	Washington, D. C	Apr. 15, 1873	137, 954
Harness-clip	J. S. Hays	Williamsport, Pa	July 12, 1870	105, 202
Harness-clip	C. M. Tripp	Pittsburgh, Pa	Feb. 8, 1870	99, 731
Harness-clips, Mode of attaching	R. S. B. Milliken	South Bend, Ind	June 4, 1872	127, 631
Harness coach-pad	A. Gilliam	Pittsburgh, Pa	Aug. 19, 1873	141, 873
Harness, Cock-eye for	S. D. Bingham	Maumee City, Ohio	Jan. 12, 1869	85, 784
Harness, Cock-eye for	J. W. Briggs	Cleveland, Ohio	June 5, 1849	6, 490
Harness, Cock-eye for	T. J. Magruder	Marion, Ohio	Dec. 7, 1869	97, 662
Harness, Cock-eye for	L. Worster	Syracuse, N. Y	Aug. 17, 1869	93, 788
Harness, Crupper for	J. J. Flack	Joliet, Ill	July 20, 1852	9, 133
Harness, Draft	G. Chamberlin	Olean, N. Y	Sept. 13, 1870	107, 335
Harness, Elastic draft-attachment for single and double.	J. Barrow	Cincinnati, Ohio	Sept. 15, 1868	82, 072
Harness-fastening	A. and G. Bernd and A. White	Macon, Ga	Sept. 14, 1869	94, 861
Harness-fastening	E. Covert	Farmer Village, N. Y	July 1, 1873	140, 474
Harness-fastening	W. Hayden	New Milford, Pa	Jan. 21, 1837	111
Harness-fastening	T. Henderson	Harford County, Md	July 27, 1852	9, 149
Harness-fastening	S. S. Little and H. P. Westcott	Macon, Ga	Mar. 22, 1870	101, 141
Harness-fastening	J. Shepard	New Britain, Conn	Apr. 25, 1865	47, 500
Harness for breaking horses	T. Hammond	Petersburgh, Ind	Apr. 14, 1868	76, 626
Harness for carriages and saddlery, Making	B. Foster, D. Washburn, N. Rockwell, and J. O. Walker.	Castleton, Vt	Aug. 16, 1810	
Harness for horses	J. Palen	Lockport, N. Y	Dec. 14, 1869	97, 799
Harness for horses	J. Rouse	Port Gibson, N. Y	Sept. 27, 1859	25, 587
Harness for horses	H. C. Smith	Dublin, Ind	July 10, 1866	56, 283
Harness for horses	J. Smith	Delaware, Ohio	Feb. 17, 1857	16, 660
Harness for horses, Breast-plate	A. Post	Henrietta, N. Y	Apr. 8, 1840	1, 545
Harness for horses in shafts	R. Beale	Washington, D. C	June 15, 1837	237
Harness for preventing the fore legs of horses from interfering.	D. G. Kettell	Worcester, Mass	Jan. 22, 1861	31, 174
Harness for shoeing horses	J. P. Reynolds	Mirabile, Mo	Jan. 22, 1861	31, 213
Harness for shoeing horses	W. P. Thomas	Hillsborough, Ind	Mar. 18, 1856	14, 477
Harness for vicious horses	S. L. Gray	Chillicothe, Ohio	May 19, 1868	78, 082
Harness-frame, Guide or clamp for	A. S. Villee	Lancaster, Pa	Nov. 12, 1867	70, 923
Harness from horses, Apparatus for detaching	S. Hunt	Baltimore, Md	July 11, 1854	11, 262
Harness from horses, Detaching	G. Yellott	Bel Air, Md	June 15, 1852	9, 044
Harness-hook	S. P. Babcock	Jordan, N. Y	May 10, 1864	42, 634
Harness-hook	G. and A. Bernd	Macon, Ga	Feb. 22, 1870	100, 233
Harness-hook	J. U. Fiester	Winchester, Ohio	Feb. 6, 1872	123, 340
Harness-hook	M. Reilly	Covington, Ky	June 21, 1870	104, 646
Harness-hook	I. C. Richmond	West Meriden, Conn	July 2, 1867	66, 391
Harness-hook	J. Seaman	New York, N. Y	Feb. 15, 1870	99, 957
Harness-hook	W. H. Taylor	Baldwinsville, N. Y	Oct. 25, 1870	108, 739
Harness-hook	A. White	Macon, Ga	Feb. 22, 1870	100, 226
Harness-hook	J. J. Wilkins	Virden, Ill	June 26, 1866	55, 942
Harness hook and bit	A. A. Hotchkiss and E. Garnsey.	Sharon, Conn	Jan. 23, 1834	....
Harness hook and loop shield	W. H. Morgan	Alliance, Ohio	Sept. 6, 1870	107, 087
Harness-hook, Safety	J. L. Dickinson	Dubuque, Iowa	Feb. 4, 1868	73, 958

Index of patents issued from the United States Patent Office from 1790 *to* 1873, *inclusive*—Continued.

Invention.	Inventor.	Residence.	Date.	No.
Harness, Horse yoke	B. F. Baker	Milton, N. Y	Apr. 2, 1867	63, 357
Harness-leather-cutting machine	J. Wehr	Roanoke, Ind	Aug. 15, 1865	49, 459
Harness, Leather-fastening for	T. G. Moore	Albia, Iowa	Jan. 4, 1870	98, 514
Harness-loop	G. L. Du Laney and J. S. Huston.	Mechanicsburgh, Pa	May 24, 1870	103, 311
Harness-loop	F. Hickman	Reading, Pa	Dec. 9, 1873	145, 299
Harness-loop	A. M. Osborn	Girard, Pa	Aug. 15, 1871	118, 046
Harness-loop	G. W. Roland	Salem, Ohio	May 12, 1868	77, 918
Harness-loop	A. Worster	Hannibal, N. Y	June 22, 1869	91, 809
Harness-loops, Clamp for embossing	O. H. Morris	New Haven, Conn	Dec. 7, 1869	97, 675
Harness-maker's bench-iron	J. Thornton and E. G. Latta	Genesee and Friendship, N. Y.	Sept. 16, 1873	142, 967
Harness-maker's clamp	D. Eighme	Chicago, Ill	Sept. 2, 1873	142, 384
Harness-maker's clamp	J. F. Johnson	Monrovia, Ind	Sept. 1, 1868	81, 787
Harness-maker's gage	D. G. Garlock	White Cloud, Kans	Feb. 4, 1873	135, 472
Harness, Metallic layer loops for	R. J. Algeo	Kalamazoo, Mich	Mar. 26, 1872	124, 996
Harness, Mode of dressing side-strap for	J. F. Valentine	Union County, Ohio	June 18, 1867	65, 845
Harness-mounting	T. Fawcett	New York, N. Y	Oct. 21, 1873	143, 817
Harness-mounting	J. F. Knorr	New York, N. Y	Feb. 18, 1873	136, 073
Harness-mounting	S. S. Sargent	Newark, N. J	July 16, 1872	129, 254
Harness-mounting	C. M. Theberath	Newark, N. J	Dec. 3, 1867	71, 819
Harness-mounting	C. H. Thornton	Newark, N. J	Nov. 14, 1871	121, 022
Harness-mountings, Mode of covering	W. Fawcett	New York, N. Y	Dec. 26, 1871	122, 163
Harness-nails, Die for forming metal heads on	F. Reynold and F. L. Hilbright.	Newark, N. J	June 12, 1866	55, 536
Harness-nails, Mode of manufacturing	F. Reynold	Newark, N. J	July 24, 1866	56, 611
Harness-operating mechanism	R. Blakie	Hyde Park, Mass	May 31, 1870	103, 708
Harness or vehicles, Breaking the surge on	F. P. Hart	Strasburgh, Pa	Dec. 29, 1868	85, 307
Harness-ornament	W. Ulrich and C. Hackmeister.	Newark, N. J	May 25, 1869	90, 610
Harness, Ornamenting	W. W. Kitch	Crestline, Ohio	Feb. 13, 1872	123, 709
Harness, Ornamenting	F. Meinberg	New York, N. Y	Mar. 12, 1872	124, 446
Harness-pad	C. Drew	Newark, N. J	Feb. 2, 1869	86, 376
Harness-pad	C. Drew	Newark, N. J	Nov. 9, 1869	96, 563
Harness-pad	J. H. Garrett	Greencastle, Ind	Dec. 12, 1871	121, 863
Harness pad	A. Gilliam	Cincinnati, Ohio	Apr. 30, 1867	64, 215
Harness-pad	G. W. Graves	Chicago, Ill	Sept. 13, 1870	107, 361
Harness-pad	J. D. Groff and K. H. Wilson	Oxford, Pa	Feb. 25, 1873	136, 155
Harness-pad	J. Hosford	Monroeville, Ohio	June 25, 1867	66, 086
Harness-pad	J. Hughes	Newark, N. J	Feb. 18, 1868	74, 691
Harness-pad	J. Hughes	Newark, N. J	June 22, 1869	91, 747
Harness-pad	J. Ives	Mount Carmel, Conn	Apr. 24, 1860	27, 984
Harness-pad	J. F. Knorr	New York, N. Y	Oct. 22, 1872	132, 472
Harness-pad	J. P. Lovett and F. P. Lefevre	Oxford, Pa	Dec. 10, 1872	133, 786
Harness pad	J. Maclure	Newark, N. J	Dec. 24, 1867	72, 514
Harness-pad	S. A. Marker	Indianapolis, Ind	Nov. 25, 1873	144, 853
Harness-pad	M. W. Pond	Elyria, Ohio	Feb. 21, 1871	112, 073
Harness-pad	M. W. Pond	Elyria, Ohio	July 1, 1873	140, 384
Harness-pad	M. W. Pond	Elyria, Ohio	July 1, 1873	140, 385
Harness-pad	W. A. Reddick	Niles, Mich	Sept. 30, 1873	143, 253
Harness-pad	H. R. Ridgley	Mansfield, Ohio	Jan. 14, 1868	73, 389
Harness-pad	W. H. Rutan	Newark, N. J	Nov. 26, 1872	133, 338
Harness-pad	R. M. Selleck	New York, N. Y	June 15, 1858	20, 588
Harness-pad	H. C. Swift	Fond du Lac, Wis	Feb. 16, 1869	87, 078
Harness-pad	J. H. Van Riper	Chelsea, Mich	Mar. 26, 1872	125, 101
Harness-pad	P. H. Wiedersum	New York, N. Y	Feb. 27, 1872	124, 185
Harness-pad block	W. H. Rannels	Oakland Mills, Pa	June 4, 1867	65, 433
Harness-pad block	P. Shaw	Syracuse, N. Y	Dec. 3, 1867	71, 653
Harness-pad block	J. Zimmer, jr	Half Day, Ill	Nov. 16, 1869	96, 860
Harness-pad brace or stiffener	S. Samson	Sterling, Mass	Jan. 5, 1833	......
Harness-pad clamp	J. H. Garrett	Greencastle, Ind	Feb. 13, 1872	123, 689
Harness-pad crimp and last	R. W. Calvert and W. Michael.	Mansfield, Ind	July 30, 1872	129, 931
Harness-pad, Elastic water-proof	A. Deitz	Albany, N. Y	Oct. 14, 1836	......
Harness-pad former or mold	T. H. Girard	Batavia, N. Y	Nov. 3, 1863	40, 473
Harness-pad plate	C. Gahr	Newark, N. J	May 24, 1870	103, 321
Harness-pad plate	P. Shaw and E. S. Dawson	Syracuse, N. Y	Mar. 19, 1867	63, 108
Harness-pad press	G. W. Lawbaugh	Geneseo, Ill	Feb. 25, 1868	74, 921
Harness-pad press	G. W. Lawbaugh	Geneseo, Ill	Dec. 15, 1868	85, 013
Harness-pad tree	J. W. Hinman	Berlin, Wis	Feb. 25, 1868	74, 823
Harness-pad tree	W. V. Kay	Chicago, Ill	May 23, 1871	115, 062
Harness-pad tree	P. Shaw and E. S. Dawson	Syracuse, N. Y	Nov. 5, 1867	70, 633
Harness-pad tree	W. H. Taylor	Baldwinsville, N. Y	Aug. 12, 1873	141, 835
Harness-pads, Manufacturing	J. Maclure	Newark, N. J	Feb. 11, 1868	74, 390
Harness-pads, Method of making	C. Gahr	Newark, N. J	Nov. 28, 1871	131, 352
Harness-padding	J. Carmac and A. Stobbs	Hector, N. Y	Aug. 16, 1870	106, 321
Harness, Padding or stuffing for	H. Hauer	Philadelphia, Pa	Nov. 17, 1868	84, 116
Harness-receptacle	H. E. Pond	Franklin, Mass	Aug. 13, 1867	67, 672
Harness-rein guide	L. Richmond	Derby, Vt	Dec. 14, 1869	97, 814
Harness, &c., Renovating	D. McCleary	Hollidaysburgh, Pa	July 23, 1867	66, 982
Harness-ring	S. Aldud	Avoca, Wis	July 19, 1870	105, 406
Harness-ring	G. H. Buckius	Canton, Ohio	July 7, 1868	79, 725
Harness-ring	A. Dietz	New York, N. Y	Apr. 9, 1850	7, 259
Harness-ring	H. N. Eames	Newport, N. Y	Mar. 1, 1870	100, 384
Harness-ring	W. Yates	Canton, Ohio	July 7, 1868	79, 797
Harness, Riveting	W. Dukehart	Baltimore, Md	Mar. 3, 1836	
Harness round-knife	J. H. Quackenbush	Springfield, Mass	Aug. 4, 1868	80, 765
Harness-rosette	G. S. Caldwell	Syracuse, N. Y	Nov. 19, 1867	71, 132
Harness-rosette	W. D. Davis, C. W. Blakeslee, and J. C. Peck.	Watertown, Conn	Mar. 21, 1871	112, 906
Harness-rosette	J. W. Dayton	Waterbury, Conn	Apr. 23, 1872	125, 885
Harness-rosette	F. F. Reynold	Newark, N. J	Nov. 12, 1872	132, 983
Harness-rosette	C. F. Richers	New York, N. Y	Aug. 18, 1868	81, 110
Harness-rosette	L. C. Wilson	Albany, N. Y	Mar. 16, 1869	87, 896
Harness-saddle	M. E. Abbey	Sardis, Miss	Oct. 8, 1872	131, 988
Harness-saddle	V. Borst	New York, N. Y	June 25, 1867	65, 995
Harness-saddle	V. Borst	New York, N. Y	Apr. 20, 1869	89, 196
Harness-saddle	V. Borst	New York, N. Y	May 17, 1870	103, 132
Harness-saddle	V. Borst	New York, N. Y	Nov. 8, 1870	109, 108
Harness-saddle	V. Borst	New York, N. Y	Mar. 28, 1871	113, 136

Index of patents issued from the United States Patent Office from 1790 *to* 1873, *inclusive*—Continued.

Invention.	Inventor.	Residence.	Date.	No.
Harness-saddle	V. Borst	New York, N. Y	Sept. 19, 1871	119, 112
Harness-saddle	V. Borst	New York, N. Y	Mar. 26, 1872	125, 011
Harness-saddle	P. Böttyer	Newark, N. J	Apr. 18, 1865	47, 276
Harness-saddle	J. W. Briggs	Cleveland, Ohio	June 12, 1849	6, 518
Harness-saddle	A. D. Brown	New York, N. Y	Dec. 28, 1847	5, 408
Harness-saddle	A. D. Brown	New York, N. Y	Oct. 3, 1848	5, 822
Harness-saddle	G. H. Buckius	Canton, Ohio	Aug. 25, 1868	81, 336
Harness-saddle	J. T. Denniston	Lynde, N. Y	Nov. 20, 1846	4, 860
Harness-saddle	E. Dixon	Emporia, Kans	June 18, 1872	128, 025
Harness-saddle	G. W. Dutton	Tomales, Cal	Aug. 27, 1872	130, 906
Harness-saddle	J. Fye	Hamilton, Ohio	Dec. 4, 1866	60, 168
Harness-saddle	A. H. Gaylay	Saratoga Springs, N. Y	Mar. 14, 1848	5, 476
Harness-saddle	A. Gilliam	Cincinnati, Ohio	Apr. 30, 1867	64, 214
Harness-saddle	A. Gilliam	Pittsburgh, Pa	Oct. 19, 1869	95, 895
Harness-saddle	A. Gilliam	Pittsburgh, Pa	Jan. 17, 1871	110, 969
Harness-saddle	A. Gilliam	Pittsburgh, Pa	July 11, 1871	116, 946
Harness-saddle	A. Gilliam	Pittsburgh, Pa	Nov. 18, 1873	144, 613
Harness-saddle	E. E. Hardy	Joliet, Ill	June 7, 1870	104, 023
Harness-saddle	W. G. Hull	Sing Sing, N. Y	Apr. 13, 1869	88, 961
Harness-saddle	W. Leonard	Boston, Mass	Dec. 31, 1867	72, 743
Harness-saddle	W. and W. E. Leonard	Chelsea, Mass	Aug. 15, 1871	118, 137
Harness-saddle	C. K. Marshall	New Orleans, La	May 10, 1870	102, 953
Harness-saddle	G. Marshall	Indiana, Pa	Nov. 5, 1867	70, 589
Harness-saddle	J. H. Martin	Columbus, Ohio	Apr. 9, 1872	125, 589
Harness-saddle	J. McLain	Circleville, Ohio	Dec. 23, 1851	8, 610
Harness-saddle	O. B. North	New Haven, Conn	Apr. 11, 1865	47, 244
Harness-saddle	C. Pope and K. Frazer	Syracuse, N. Y	Feb. 5, 1847	4, 959
Harness-saddle	H. Sanders	Utica, N. Y	Jan. 5, 1858	19, 048
Harness-saddle	A. V. Sargeant	Newark, N. J	Mar. 18, 1873	136, 870
Harness-saddle	R. M. Selleck	New York, N. Y	Nov. 7, 1854	11, 911
Harness-saddle	P. Shaw	Syracuse, N. Y	Apr. 21, 1857	17, 117
Harness-saddle	R. Spencer	Brooklyn, N. Y	Aug. 6, 1850	7, 551
Harness-saddle	R. Spencer	New York, N. Y	Aug. 22, 1854	11, 576
Harness-saddle	R. Swift	New Haven, Conn	Dec. 29, 1857	18, 996
Harness-saddle	J. C. Tobias	Middleport, Ill	June 17, 1862	35, 656
Harness-saddle	S. E. Tompkins	Newark, N. J	Nov. 17, 1863	40, 653
Harness-saddle	S. E. Tompkins	Sing Sing, N. Y	Apr. 16, 1872	125, 770
Harness-saddle	S. E. Tompkins	Sing Sing, N. Y	July 15, 1873	140, 968
Harness-saddle	S. E. Tompkins	Sing Sing, N. Y	July 15, 1873	140, 969
Harness-saddle	S. E. Tompkins	Sing Sing, N. Y	July 15, 1873	140, 970
Harness-saddle	J. L. Van Wert	Tolland, Mass	Jan. 18, 1870	98, 899
Harness-saddle	J. Waite	Palmer, Mass	Mar. 9, 1869	87, 606
Harness saddle, Cart	H. A. Rains	Nashville, Tenn	Nov. 13, 1866	59, 651
Harness-saddle check-hook	A. V. Sargeant	Newark, N. J	Sept. 2, 1873	142, 415
Harness-saddle check-hook	P. H. Wiedersum	New York, N. Y	Mar. 5, 1872	124, 408
Harness-saddle hook	R. Hart	Dunkirk, N. Y	Jan. 25, 1870	99, 185
Harness-saddle hook and terret	S. E. Tompkins	Newark, N. J	Aug. 19, 1862	36, 248
Harness-saddle loop	J. M. Roe	Worcester, Mass	Mar. 16, 1869	87, 973
Harness-saddle mounting	J. J. Carrel	Petersburgh, Va	Oct. 17, 1848	5, 860
Harness-saddle pad	H. Lerew	Panora, Iowa	May 27, 1873	139, 403
Harness-saddle pad	R. C. Sturges	Boston, Mass	Jan. 19, 1869	86, 112
Harness-saddle pad or housing	C. B. Hogg	Boston, Mass	Nov. 30, 1869	97, 295
Harness-saddle plate	J. H. Garrett	Mount Pleasant, Iowa	Nov. 1, 1870	108, 900
Harness-saddle seat	P. Burns	Syracuse, N. Y	May 10, 1870	103, 004
Harness-saddle seat	O. B. North	New Haven, Conn	June 5, 1866	55, 348
Harness-saddle tree	J. C. Dickey	Saratoga Springs, N. Y	Mar. 6, 1855	12, 481
Harness-saddle tree	J. Fonda	Albany, N. Y	Apr. 7, 1863	38, 104
Harness-saddle tree	G. D. Gillett	Meridian, N. Y	July 20, 1869	92, 717
Harness-saddle tree	J. F. Knorr	Orange, N. J	Oct. 13, 1868	82, 961
Harness-saddle tree	A. Koehler	Holyoke, Mass	May 9, 1865	47, 647
Harness-saddle tree	J. A. Lawrence	New Haven, Conn	July 22, 1851	8, 238
Harness-saddle tree	T. Mardock and W. C. Kellar	Cincinnati, Ohio	Oct. 12, 1852	9, 328
Harness-saddle tree	J. H. Martin	Columbus, Ohio	Feb. 21, 1871	112, 061
Harness-saddle tree	J. H. Martin	Columbus, Ohio	Apr. 4, 1871	113, 541
Harness-saddle tree	J. H. Martin	Columbus, Ohio	July 25, 1871	117, 309
Harness-saddle tree	T. G. Moore	Albia, Iowa	Aug. 29, 1871	118, 547
Harness-saddle tree	O. B. North	New Haven, Conn	Feb. 21, 1865	46, 489
Harness-saddle tree	C. J. Paine	Young America, Ill	Apr. 23, 1872	126, 080
Harness-saddle tree	G. Pennoyer	New York, N. Y	May 24, 1870	103, 495
Harness-saddle tree	R. Spencer	Southport, Conn	Aug. 22, 1854	11, 573
Harness-saddle tree	S. E. Tompkins	Newark, N. J	Jan. 10, 1865	45, 880
Harness-saddle tree	S. E. Tompkins and J. Maclure	Newark, N. J	Feb. 1, 1859	22, 841
Harness-saddle tree	P. Washburn	Taunton, Mass	Mar. 8, 1832	....
Harness-saddle tree	P. H. Wiedersum	New York, N. Y	Sept. 27, 1870	107, 843
Harness-saddle tree	P. H. Wiedersum	New York, N. Y	Jan. 31, 1871	111, 502
Harness-saddle tree	P. H. Wiedersum	New York, N. Y	June 27, 1871	116, 518
Harness-saddle tree	P. H. Wiedersum	New York, N. Y	June 27, 1871	116, 519
Harness-saddle tree	P. H. Wiedersum	New York, N. Y	Apr. 30, 1872	126, 304
Harness-saddle tree, Metallic	S. E. Tompkins	New York, N. Y	Nov. 27, 1855	13, 858
Harness-saddle tree, Safety	J. S. Hall	Pittsburgh, Pa	Sept. 1, 1868	81, 629
Harness-saddle tree, Wooden	F. P. Ambler, jr	Trumbull, Conn	June 1, 1858	20, 463
Harness-saddles, Back-band for	S. F. Marshall	Wilmington, Del	Feb. 11, 1873	135, 831
Harness-saddles, Buckle-attachment to	J. B. Welpton	Tabor, Iowa	Apr. 1, 1873	137, 519
Harness-saddles, Fastening of terrets in	P. B. Cool	Columbus, Ohio	Mar. 12, 1850	7, 157
Harness-saddles, Revolving back-band hook for	G. Elbet	Pittsburgh, Pa	May 31, 1870	103, 588
Harness, Safety	J. L. Arnold	Lowndesville, S. C	June 4, 1872	127, 403
Harness, Safety	I. Ellis	Tyler, Tex	Oct. 1, 1872	131, 747
Harness, Safety	C. Rogers	Baltimore, Md	July 22, 1833	....
Harness safety-hook	H. Beagle, jr	Philadelphia, Pa	Apr. 9, 1861	31, 946
Harness, Safety-line for	A. H. Rockwell	Harpersville, N. Y	Dec. 4, 1866	60, 253
Harness safety-loop	C. H. Trumbull	Marion, N. Y	Mar. 5, 1872	124, 298
Harness, Safety-trap for	C. W. Nichols	Chicago, Ill	Nov. 25, 1873	145, 0[illegible]6
Harness, Securing buckles and rings to	R. B. Anderson	Oneida, Ill	June 23, 1868	79, 180
Harness shaft-loop	B. J. Aurand	Mount Gilead, Ohio	July 16, 1867	66, 771
Harness shaft-tug	J. L. Hoyt	Port Jervis, N. Y	June 19, 1847	5, 160
Harness, Shaft-tug lug for	T. J. Magruder	Marion, Ohio	Dec. 14, 1869	97, 786
Harness-shield	M. W. Pond	Elyria, Ohio	Feb. 23, 1869	87, 110
Harness, Single	O. V. Flora	Madison, Ind	July 7, 1868	79, 745
Harness, Single	J. S. Reid	Orange, Ind	Jan. 28, 1868	73, 835

Index of patents issued from the United States Patent Office from 1790 *to* 1873, *inclusive*—Continued.

Invention.	Inventor.	Residence.	Date.	No.
Harness, Single	C. R. Stuart	Winslow, Me	Oct. 24, 1871	120, 344
Harness-slide	J. S. Topham	Washington, D. C	Apr. 22, 1862	35, 049
Harness-slide snap	J. and D. C. Kerr	Ames, Iowa	June 9, 1868	78, 807
Harness-snap	P. Beckman	Naperville, Tenn	Oct. 6, 1863	40, 152
Harness-snap	E. A. Cooper	Lancaster, N. Y	Apr. 2, 1867	63, 478
Harness-snap	E. A. Cooper	Buffalo, N. Y	June 23, 1868	79, 107
Harness-snap	W. F. Davidson, O. A. Bates, S. M. Wilson, and A. P. Russell.	Janesville, Wis	Dec. 24, 1867	72, 462
Harness-snap	G. W. Devin	Ottumwa, Iowa	May 15, 1866	54, 700
Harness-snap	G. B. Durkee	Alden, N. Y	Oct. 8, 1867	69, 548
Harness-snap	G. B. Durkee	Alden, N. Y	Oct. 8, 1867	69, 549
Harness-snap	H. Harris	Newark, N. J	Feb. 21, 1865	46, 468
Harness-snap	H. Harris	Newark, N. J	Mar. 14, 1865	46, 796
Harness-snap	B. B. Hotchkiss	Sharon, Conn	Dec. 14, 1858	22, 290
Harness-snap	P. W. Jones	Syracuse, N. Y	Nov. 3, 1863	40, 486
Harness-snap	C. H. Palmer	Newark, N. J	Nov. 29, 1864	45, 299
Harness-snap	S. Reynolds	Allegheny, Pa	Mar. 12, 1872	124, 628
Harness-snap	J. B. Tibbits	Palmyra, N. Y	Aug. 11, 1863	39, 506
Harness-snap	A. B. Woodward and W. Bruen	Alfred Centre, N. Y., and Newark, N. J.	Feb. 7, 1871	111, 710
Harness steel spring-worm	G. Pritchard	Clarksburgh, Va	May 3, 1832	....
Harness-tree	T. Dempsey	Newark, N. J	Jan. 12, 1858	19, 078
Harness-tree	F. B. Kuehnhold and D. B. Sturges.	Newark, N. J	Feb. 16, 1858	19, 371
Harness-tree	C. M. Sturgess	Washington, Iowa	Feb. 5, 1867	61, 891
Harness-tree	J. H. Whissemore	Mansfield, Ohio	Aug. 11, 1868	81, 041
Harness-tree pad	W. A. Sharp and J. A. Shannon	Tama City, Iowa	Nov. 17, 1868	84, 071
Harness-trimming	J. Bauer	Newark, N. J	July 4, 1871	116, 536
Harness-trimming	J. Bauer	Newark, N. J	Oct. 17, 1871	119, 914
Harness-trimming	M. A. Fisk	Springfield, Mass	Mar. 18, 1868	75, 257
Harness-trimming	T. J. Magruder	Marion, Ohio	Feb. 11, 1868	74, 391
Harness-trimming	S. Wiener	Newark, N. J	July 11, 1871	116, 904
Harness-trimmings and carriage-ornaments, Manufacture of.	W. S. Robinson	Taunton, Mass	Sept. 22, 1829	....
Harness-trimmings, Manufacture of rubber-coated	A. Albright	Newark, N. J	Feb. 13, 1872	123, 603
Harness-trimmings, Manufacture of rubber-coated	A. Albright	Newark, N. J	Apr. 15, 1873	137, 873
Harness-trimmings, Mode of covering	C. M. Theberath	Newark, N. J	Jan. 18, 1870	99, 032
Harness-tug	I. H. Alexander	Newfield, N. Y	Oct. 10, 1871	119, 684
Harness-tug	N. Botsford	Lockport, N. Y	June 11, 1872	127, 840
Harness-tug	J. S. H. Dickinson	Jackson, Pa	June 15, 1869	91, 425
Harness-tug clasp	L. D. Cowles	Armada, Mich	Feb. 17, 1863	37, 678
Harness-tug clasp	L. D. Cowles	Armada, Mich	Dec. 15, 1863	40, 911
Harness-tug clasp	L. D. Cowles	Armada, Mich	Dec. 15, 1863	40, 912
Harness-tug clasp	J. S. Gilman	Tecumseh, Mich	July 29, 1862	35, 995
Harness-tug or trace-supporter	C. W. Terpening	Geneseo, Ill	June 29, 1869	91, 986
Harness, Water-hook bolt for	J. W. Bishop	New Haven, Conn	Mar. 16, 1869	87, 755
Harness, Wire	A. R. Reese	Phillipsburgh, N. J	Mar. 8, 1864	41, 860
Harnessing horses to vehicles, Mode of	R. McHardy	Edinburgh, Great Britain	Oct. 3, 1871	119, 526
Harp, Keyed	A. Kuhn	Baltimore, Md	Jan. 27, 1857	16, 489
Harpoon	M. Adams	Chilmark, Mass	Apr. 21, 1863	28, 207
Harpoon	W. Carsley	New Bedford, Mass	July 29, 1841	2, 195
Harpoon	D. H. Chamberlain	Boston, Mass	Aug. 17, 1835	....
Harpoon	G. Doyle	Provincetown, Mass	Nov. 2, 1858	21, 949
Harpoon	J. Holmes and A. West	Tisbury, Mass	Nov. 24, 1846	4, 865
Harpoon	J. Q. Kelley	Sag Harbor, N. Y	Oct. 20, 1857	18, 458
Harpoon	Z. Kelley	New Bedford, Mass	Dec. 3, 1867	71, 763
Harpoon	A. Moor	Hampden, Me	Mar. 16, 1844	3, 490
Harpoon	C. Randall	Palmyra, Ga	Dec. 3, 1846	4, 872
Harpoon	J. Sizer, 2d	New London, Conn	Oct. 16, 1829	....
Harpoon and lance	J. D. B. Stillman	New York, N. Y	Apr. 6, 1852	8, 862
Harpoon and lance, Gun	H. W. Harkness	Bristol, Conn	Feb. 16, 1858	19, 363
Harpoon, Bomb	R. Brown	New London, Conn	Aug. 20, 1850	7, 572
Harpoon, Exploding	C. Freeman	Brewster, Mass	May 7, 1872	126, 388
Harpoon, Explosive	C. Burt	Belfast, Me	May 6, 1851	8, 073
Harpoon, Gun	T. Briggs	Philadelphia, Pa	Dec. 11, 1860	30, 869
Harpoon, Gun	O. Allen	Norwich, Conn	Dec. 5, 1848	5, 949
Harpoon, Gun	R. Brown	New London, Conn	June 4, 1850	7, 410
Harpoon, Gun	J. P. Rechten	New York, N. Y	Dec. 7, 1869	97, 693
Harpoon, Gun	R. E. Smith	Provincetown, Mass	Apr. 23, 1867	64, 045
Harpoon, Hinged gun	T. W. Roys	Southampton, N. Y	Jan. 22, 1861	31, 190
Harpoon, Rocket	W. Albertson	New London, Conn	Nov. 19, 1850	7, 777
Harpoon, Rocket	T. W. Roys	Southampton, N. Y	June 3, 1862	35, 474
Harpoons, Method of attaching lines to	T. W. Roys and G. A. Lilliendahl.	New York, N. Y	Apr. 24, 1866	54, 211
Harrow	C. F. Brown	Warren, R. I	Sept. 3, 1850	7, 610
Harrow	J. Adams	Monroe, Mich	Sept. 24, 1861	33, 327
Harrow	J. Aiken	Warner, N. H	Dec. 17, 1867	72, 260
Harrow	J. Allen	Union, N. Y	Dec. 18, 1860	30, 905
Harrow	W. Anderson	Jacksonville, N. Y	Aug. 1, 1854	11, 452
Harrow	M. Atwood	New Sharon, Iowa	Nov. 24, 1868	84, 339
Harrow	J. Ayers	Canterbury, N. H	July 10, 1866	56, 159
Harrow	D. C. Ayres	Lumberland, N. Y	Sept. 7, 1858	21, 403
Harrow	A. W. Ball	Delaware Grove, Pa	June 8, 1869	90, 980
Harrow	A. H. Ballagh	Westport, Md	Mar. 16, 1869	87, 819
Harrow	W. H. Barry	Rabbit River, Mich	Oct. 20, 1868	83, 238
Harrow	L. Beckelshymen	Leavenworth, Kans	Aug. 20, 1867	67, 940
Harrow	R. N. Bennett	Union Mills, Ind	Oct. 26, 1869	96, 191
Harrow	J. Benson	Belle Plains, Iowa	Aug. 1, 1871	117, 595
Harrow	A. Berdan	Macon, Mich	Aug. 10, 1858	21, 113
Harrow	W. Berlin	Berryville, Va	Apr. 12, 1853	9, 657
Harrow	A. W. Bohaker	Napa, Cal	Dec. 3, 1872	133, 623
Harrow	M. Böshenz	Chili, Ill	Feb. 11, 1868	74, 194
Harrow	E. Bradley	Clyde, N. Y	Sept. 11, 1866	57, 851
Harrow	T. E. C. Brinly	Louisville, Ky	May 14, 1872	126, 779
Harrow	T. S. and T. A. Brown	Brooklyn, N. Y	Aug. 24, 1869	93, 959
Harrow	W. Brown	Butte des Morts, Wis	Feb. 18, 1873	136, 027
Harrow	R. W. Buckles	Grayville, Ill	May 24, 1859	24, 095

Index of patents issued from the United States Patent Office from 1790 *to* 1873, *inclusive*—Continued.

Invention.	Inventor.	Residence.	Date.	No.
Harrow	M. Bucklin	Grafton, N. H	June 4, 1861	32,462
Harrow	J. G. Burcham	Noblesville, Ind	Mar. 29, 1870	101,350
Harrow	W. J. Burdick	Alfred, N. Y	July 26, 1870	105,775
Harrow	S. P. Campbell	Rochester, Minn	Feb. 9, 1864	41,479
Harrow	A. C. Carnes	Smithville, Tenn	July 1, 1873	140,403
Harrow	H. Cartwright	Owatonna, Minn	Feb. 4, 1873	135,464
Harrow	J. C. Center	Mason City, Ill	July 25, 1871	117,258
Harrow	V. M. Chafee	Grayville, Ill	May 25, 1858	20,325
Harrow	J. Chase	Farmington, Pa	Nov. 24, 1868	84,342
Harrow	J. Chellew	Glasford, Ill	July 18, 1871	117,149
Harrow	C. Clareni	New York, N. Y	Apr. 3, 1855	12,659
Harrow	J. Click	Springfield, Ohio	Aug. 27, 1867	68,169
Harrow	O. Coe	Port Washington, Wis	Mar. 2, 1858	19,489
Harrow	D. C. Colby	Newport, N. H	June 26, 1860	28,938
Harrow	L. Coleman	New Orleans, La	Aug. 13, 1867	67,634
Harrow	T. B. Collins	Noank, Conn	June 29, 1869	91,914
Harrow	J. C. Concklin	Peekskill, N. Y	Apr. 21, 1836	....
Harrow	J. C. Concklin	Yorktown, N. Y	June 3, 1862	35,487
Harrow	J. C. Conkey	Washington, Ohio	May 11, 1858	20,195
Harrow	D. B. Conover	New Bedford, N. J	Dec. 20, 1870	110,344
Harrow	G. Cooke	Bristol Station, Ill	June 10, 1862	35,563
Harrow	W. J. Cordill	Blue Earth City, Minn	Apr. 30, 1872	126,186
Harrow	A. J. Craig	Ashmore Station, Ill	Oct. 13, 1868	82,923
Harrow	I. Crum	West Chester, Ohio	Mar. 24, 1868	75,867
Harrow	I. Crum	West Chester, Ohio	June 22, 1869	91,525
Harrow	J. S. Davis	Washington, Ohio	June 1, 1858	20,410
Harrow	L. P. Davis and O. P. Beagle	Altona, Ill	Nov. 26, 1872	133,308
Harrow	N. A. Davis	Sutton, N. H	Apr. 10, 1866	53,792
Harrow	J. Dawson	Greenwood, Ill	Oct. 25, 1870	108,690
Harrow	J. Dawson	Greenwood, Ill	July 2, 1872	128,601
Harrow	J. Delong	Covert, N. Y	Oct. 14, 1862	36,644
Harrow	A. B. Depeny	Hamburgh, Iowa	Nov. 19, 1872	133,212
Harrow	W. De Witt and O. D. Barrett	Cleveland, Ohio	Mar. 2, 1858	19,494
Harrow	D. L. Dickson	Durham, Ill	Feb. 18, 1868	74,673
Harrow	J. Dingman	Decatur, Ill	Sept. 13, 1870	107,344
Harrow	S. Donaldson	Fenton, Mich	Jan. 16, 1872	122,711
Harrow	M. Easterbrook, jr	Geneva, N. Y	Sept. 12, 1865	49,867
Harrow	O. W. Edmunds	Bluffdale, Ill	Dec. 8, 1868	84,807
Harrow	G. M. Ellis	Huntertown, Ind	Aug. 23, 1870	106,677
Harrow	J. H. Elward	Ottawa, Ill	Nov. 17, 1863	40,614
Harrow	H. L. Eshelman	Elizabethtown, Pa	July 30, 1867	67,178
Harrow	T. H. Enlass	Mason City, Ill	Jan. 26, 1869	86,218
Harrow	D. S. Fisher	Cedar Spring, Ind	May 28, 1867	65,196
Harrow	J. Foltz	Valley Mills, Ind	June 8, 1869	90,939
Harrow	J. N. Fordyce	Cambridge, Ohio	Aug. 13, 1867	67,835
Harrow	E. Forney	Fishersville, Pa	Jan. 28, 1868	73,792
Harrow	J. T. Foster	Jersey City, N. J	June 18, 1861	32,599
Harrow	W. Fox	Beaver Dam, Wis	Apr. 4, 1871	113,286
Harrow	J. T. Frankeberger	Hensly, Ill	Dec. 24, 1867	72,623
Harrow	J. H. French	Syracuse, N. Y	May 10, 1859	23,914
Harrow	A. Friedemann	Waverly, Ohio	Aug. 2, 1870	106,046
Harrow	A. Friedemann	Waverly, Iowa	Jan. 31, 1871	111,445
Harrow	D. L. Garver	Hart Township, Mich	Oct. 5, 1869	95,458
Harrow	J. F. Gazley	Canyonville, Oreg	May 27, 1873	139,309
Harrow	L. C. Gifford	Monticello, Ill	Mar. 11, 1873	136,717
Harrow	L. C. Gillaspie	Denmark, Tenn	July 10, 1860	29,073
Harrow	E. A. Goodes	Philadelphia, Pa	July 20, 1869	92,719
Harrow	W. Gourley	Clarke County, Va	July 17, 1855	13,257
Harrow	W. Grange	Augusta, Ky	Sept. 2, 1862	36,345
Harrow	F. Granger	Lockport, Ill	July 17, 1866	56,405
Harrow	P. S. Graves and P. B. Parcell	Ashmore, Ill	Oct. 5, 1869	95,467
Harrow	O. C. Green	Dublin, Ind	Nov. 22, 1859	26,179
Harrow	M. W. Gunn	La Salle, Ill	Feb. 11, 1868	74,348
Harrow	E. L. Hagan	Frankfort, N. Y	Dec. 12, 1854	12,059
Harrow	C. Hairgrove	Jacksonville, Ill	Jan. 16, 1872	122,765
Harrow	D. Haldeman	Morgantown, Va	Feb. 20, 1855	12,415
Harrow	S. W. Hamsher	Decatur, Ill	Oct. 18, 1859	25,824
Harrow	C. Hanson	Owatonna, Minn	Dec. 1, 1868	84,547
Harrow	J. M. Harper	El Paso, Ill	Dec. 13, 1870	110,133
Harrow	J. Harris	San Francisco, Cal	Aug. 27, 1872	130,916
Harrow	G. Heffner	Homer, Iowa	Mar. 16, 1869	87,845
Harrow	J. H. Hendee	Jackson, Mich	June 21, 1864	43,207
Harrow	J. Herald and C. B. Tomkins	Trumansburgh, N. Y	June 7, 1859	24,303
Harrow	E. W. Herendeen	Geneva, N. Y	Oct. 24, 1871	120,195
Harrow	E. S. Herrington	Emmett, Ohio	Nov. 14, 1871	120,873
Harrow	H. M. Hickman and B. G. Devoe.	Vandalia, Ill	Nov. 10, 1868	83,962
Harrow	A. Hockstein	Williamsville, N. Y	July 14, 1868	79,829
Harrow	S. S. Hogle	York, Ohio	Mar. 17, 1857	16,866
Harrow	W. R. Hollingsworth	Mount Pleasant, Iowa	Sept. 28, 1869	95,348
Harrow	T. C. Hooker	Kendall, N. Y	June 18, 1861	32,564
Harrow	T. C. Hooker	Brockport, N. Y	Nov. 12, 1872	132,964
Harrow	M. W. House	Cleveland, Ohio	Sept. 4, 1860	29,885
Harrow	A. Jacobson	Logan, Nebr	Apr. 9, 1872	125,395
Harrow	J. Jacobson	Austin, Minn	Nov. 1, 1870	108,910
Harrow	D. L. Jaques	Hudson, Mich	Dec. 13, 1870	110,142
Harrow	D. L. Jaques	Hudson, Mich	Apr. 15, 1873	137,926
Harrow	J. Jay and J. Coppock	Jonesborough, Ind	Aug. 3, 1869	93,306
Harrow	C. Jillson	Worcester, Mass	Oct. 17, 1865	50,478
Harrow	B. Johnston	Sterling, Ill	July 4, 1871	116,716
Harrow	A. Jones	Randolph, Wis	Mar. 12, 1872	124,586
Harrow	S. G. Jones	Niantic, Ill	June 8, 1869	91,024
Harrow	W. P. Jones	Arcade, N. Y	Sept. 6, 1870	107,057
Harrow	A. and N. Kane	Newport, N. Y	July 13, 1869	92,614
Harrow	J. Kelsey	Yardleyville, Pa	Dec. 23, 1862	37,234
Harrow	T. M. King	Murfreesborough, Tenn	May 20, 1873	139,065
Harrow	J. Kinhart	Athens, Ill	Jan. 23, 1872	123,025
Harrow	J. Knickerbocker	Dunning, Pa	July 6, 1869	92,320

Index of patents issued from the United States Patent Office from 1790 *to* 1873, *inclusive*—Continued.

Invention.	Inventor.	Residence.	Date.	No.
Harrow	C. Lane and J. M. Healy	Jamestown, N. Y	Mar. 23, 1869	88, 046
Harrow	O. J. Leabo	Forest Grove, Oreg	Mar. 19, 1872	124, 752
Harrow	A. Lewis	Hastings, Minn	Sept. 13, 1870	107, 269
Harrow	J. S. Lewis	Elkader, Iowa	Oct. 26, 1869	96, 244
Harrow	H. C. Lezott	Osage, Iowa	July 20, 1869	92, 849
Harrow	S. H. Lintan	Burrows, Ind	Nov. 30, 1869	97, 416
Harrow	P. H. Lang	Palmyra, Me	Dec. 31, 1872	134, 433
Harrow	I. Low	East Fairfield, Ohio	Aug. 31, 1869	94, 326
Harrow	L. Lupton	Winchester, Va	May 24, 1853	9, 747
Harrow	C. R. Macy	Bedminster, N. J	Dec. 28, 1869	98, 281
Harrow	M. R. Marcell	Dansville, N. Y	June 29, 1869	92, 071
Harrow	N. B. Marsh	Marengo, Ill	Mar. 9, 1869	87, 577
Harrow	B. T. Martin	Charlotte, Mich	Dec. 15, 1868	84, 956
Harrow	J. Mathison	Fremont, Nebr	Oct. 24, 1871	120, 204
Harrow	D. B. Maze	West Buffalo, Ohio	Oct. 7, 1873	143, 520
Harrow	J. G. McCauley	Stone Bridge, Va	Aug. 29, 1854	11, 614
Harrow	N. McCuen	South Potsdam, N. Y	Feb. 25, 1862	34, 516
Harrow	W. McFishback	Union, Ohio	Aug. 14, 1866	57, 166
Harrow	J. Mellinger	Greensburgh, Pa	Oct. 11, 1870	108, 279
Harrow	S. Mendenhall	Muncy Station, Pa	Mar. 23, 1869	88, 195
Harrow	J. Mercier	Detroit, Mich	Jan. 7, 1868	73, 024
Harrow	M. D. Meriweather	Denmark, Tenn	Aug. 28, 1860	29, 804
Harrow	A. Mero	New Haven, Conn	Apr. 10, 1866	53, 850
Harrow	J. H. Miller and F. A. Pickering	Niantic, Ill	Nov. 23, 1869	97, 104
Harrow	N. B. Miller	Orion, Mich	Jan. 21, 1868	73, 626
Harrow	S. Miller	Winchester, Ohio	Mar. 19, 1861	31, 731
Harrow	J. E. Morgan	Deerfield, N. Y	Mar. 31, 1857	16, 933
Harrow	J. E. Morgan	Deerfield, N. Y	July 16, 1867	66, 869
Harrow	S. A. and C. C. Morgan	Auburn, N. Y	June 26, 1860	28, 890
Harrow	D. C. Myers	South Bend, Ind	Mar. 12, 1867	62, 769
Harrow	J. Myers	Powhatan Point, Ohio	Aug. 15, 1854	11, 528
Harrow	D. B. Neal	Mount Gilead, Ohio	Aug. 24, 1858	21, 269
Harrow	S. Neill	Chillicothe, Ill	Sept. 4, 1860	29, 903
Harrow	S. J. Newell	South China, Me	Nov. 5, 1872	132, 853
Harrow	A. A. Nuquist	Oneida, Ill	July 27, 1869	93, 115
Harrow	W. S. O'Brien	Brimfield, Ill	June 5, 1866	55, 351
Harrow	G. Ogg	Lacon, Ill	Mar. 12, 1867	62, 875
Harrow	R. W. O'Neal	Punta Arenas, Cal	Jan. 30, 1872	123, 193
Harrow	S. J. Orange	Grayville, Ill	Feb. 2, 1858	19, 259
Harrow	S. J. Orange and G. Beidelman	Grayville, Ill	Sept. 7, 1858	21, 439
Harrow	W. F. Osborn	Mount Pleasant, Pa	Nov. 10, 1868	83, 878
Harrow	G. Paddington	Springville, Iowa	Dec. 21, 1869	98, 100
Harrow	O. D. Padrick	Shelbyville, Ind	July 3, 1866	56, 088
Harrow	W. F. Pagett	Stone Bridge, Va	May 2, 1854	10, 863
Harrow	D. A. Parkman	Union City, Tenn	Sept. 2[illegible], 1871	119, 397
Harrow	J. A. B. Patterson	Springfield, Ill	Apr. 20, 1869	89, 238
Harrow	N. A. Patterson	Kingston, Tenn	Feb. 14, 1860	27, 150
Harrow	J. M. Payne	Benton, Ill	Sept. 26, 1871	119, 398
Harrow	G. W. Pense and C. E. Lykke	Franklin Grove, Ill	Mar. 16, 1869	87, 872
Harrow	W. H. Platt	Dayton, Ohio	Apr. 8, 1873	137, 569
Harrow	E. J. Preston	Eureka, Mo	Nov. 16, 1869	96, 936
Harrow	J. Rankin	Taunton, Mass	Apr. 21, 1868	77, 091
Harrow	G. Reed	Greencastle, Ind	June 18, 1872	128, 067
Harrow	W. Rennyson	Norristown, Pa	Sept. 12, 1871	118, 973
Harrow	W. Rennyson	Norristown, Pa	Nov. 14, 1871	121, 003
Harrow	J. W. Richardson	Sligo, Ohio	Jan. 15, 1867	61, 258
Harrow	W. E. Robbins and G. Enderton	Sterling, Ill	May 30, 1871	115, 525
Harrow	J. Robins	Leominster, Mass	Feb. 2, 1858	19, 281
Harrow	H. W. Robinson	Woodstock, Ill	Jan. 23, 1872	122, 911
Harrow	L. C. Rumberger	Cass, Pa	Nov. 25, 1873	145, 016
Harrow	J. Russell	Grampian Hills, Pa	Jan. 17, 1860	26, 864
Harrow	J. Ruth	Defiance, Ohio	Mar. 3, 1868	75, 202
Harrow	J. Ruth and A. Vaughn	Grayville, Ill	Aug. 10, 1858	21, 153
Harrow	A. Safley	Mount Vernon, Iowa	Apr. 19, 1864	42, 408
Harrow	S. C. Schofield	Freeport, Ill	Mar. 13, 1860	27, 477
Harrow	D. W. Shares	Hamden, Conn	Jan. 27, 1857	16, 498
Harrow	M. Sheldon, jr	Calais, Vt	July 1, 1862	35, 784
Harrow	B. F. Shellabarger	Mifflintown, Pa	Aug. 1, 1848	5, 687
Harrow	H. and C. Shirk	Lebanon, Pa	Feb. 8, 1870	99, 719
Harrow	V. Sippel	Niagara City, N. Y	June 29, 1869	91, 977
Harrow	C. Smith	Irwin's Station, Pa	Feb. 4, 1873	135, 448
Harrow	J. Smith	Woodland, Mich	Mar. 25, 1873	137, 254
Harrow	J. Snowdin and J. Kent	Westford and Beaver Dam, Wis.	Mar. 23, 1869	88, 225
Harrow	N. Starr, jr	Homer, N. Y	Dec. 17, 1867	72, 331
Harrow	C. E. Steller	Genesee, Wis	Dec. 2, 1862	37, 061
Harrow	C. E. Steller	Chicago, Ill	Feb. 18, 1868	74, 626
Harrow	A. J. Stewart	Chillicothe, Mo	Nov. 4, 1873	144, 236
Harrow	W. D. Summers	Hopkinsville, Ky	Aug. 19, 1873	142, 055
Harrow	S. Svanson	Sweede Point, Iowa	Mar. 12, 1867	62, 904
Harrow	C. Svenzen	Blair, Nebr	Oct. 21, 1873	143, 941
Harrow	H. N. Swift	Matteawan, N. Y	Aug. 13, 1872	130, 451
Harrow	F. Y. Tavenner, J. W. Galbraith, and A. Smith.	Sedalia, Mo	Aug. 24, 1869	94, 044
Harrow	R. L. Taylor	Golden Prairie, Iowa	Sept. 9, 1873	142, 593
Harrow	J. Temple	Bellefonte, Pa	Nov. 27, 1860	30, 789
Harrow	C. and G. P. Tharp	Bryan, Ohio	Nov. 14, 1871	121, 021
Harrow	C. Thayer and L. L. Thomas	Otego, N. Y	Aug. 23, 1870	106, 742
Harrow	J. J. Thomas	Union Springs, N. Y	Sept. 22, 1868	82, 451
Harrow	G. W. Tolhurst	Cleveland, Ohio	Apr. 21, 1857	17, 121
Harrow	W. Tuttle	Fayette, Miss	Apr. 5, 1870	101, 547
Harrow	M. Vanbibber	Tipton, Ind	Jan. 4, 1870	98, 531
Harrow	J. E. Van Riper	Dearborn, Mich	Aug. 6, 1867	67, 466
Harrow	J. A. Walker	Nashville, Tenn	July 22, 1873	141, 021
Harrow	J. N. and T. Wallis	Fleming, N. Y	Jan. 7, 1868	73, 061
Harrow	J. Walsh	Stark County, Ill	July 30, 1867	67, 387
Harrow	J. N. Wamsley	Galena, Ohio	Jan. 21, 1873	135, 022
Harrow	E. S. Warner	Union Township, Ohio	Aug. 24, 1869	94, 153

Index of patents issued from the United States Patent Office fram 1790 *to* 1873, *inclusive*—Continued.

Invention.	Inventor.	Residence.	Date.	No.
Harrow	C. Watson	Cascade, Va	Oct. 2, 1860	30, 280
Harrow	G. Watt	Richmond, Va	July 1, 1873	140, 563
Harrow	H. Wendover	Norway, N. Y	Apr. 1, 1873	137, 520
Harrow	M. D. Wells	Morgantown, W. Va	Jan. 23, 1866	52, 233
Harrow	S. White	Penfield, Ohio	Sept. 14, 1858	21, 542
Harrow	J. Wigle	West Point, Ill	Oct. 17, 1871	120, 140
Harrow	J. P. Wile	Nashville, Mich	June 1, 1869	90, 713
Harrow	B. B. Williams	Laclede, Mo	Apr. 13, 1869	88, 931
Harrow	J. H. Williams	Tontzville, Kan	Sept. 14, 1869	94, 855
Harrow	J. L. Willoughby	Bowling Green, Ky	Aug. 29, 1871	118, 570
Harrow	F. R. Willson	Columbus, Ohio	June 16, 1868	79, 042
Harrow	L. L. Willson	Denmark, Mich	Nov. 5, 1872	132, 885
Harrow	S. and A. Woodard	Saratoga, Ind	Aug. 30, 1870	106, 908
Harrow	P. N. Woodworth	Stony Point, Cal	June 17, 1873	139, 987
Harrow	G. Workman	Rochester, N. Y	Dec. 21, 1869	98, 137
Harrow	M. K. Young	Glen Haven, Wis	Oct. 28, 1873	144, 044
Harrow, Adjustable	J. C. Center	Bath, Ill	Apr. 28, 1868	77, 254
Harrow, Adjustable	A. Havens	Trenton, N. J	Oct. 18, 1870	108, 478
Harrow, Adjustable	J. Kinhart	Athens, Ill	Sept. 21, 1869	94, 964
Harrow, Adjustable	H. Pulse	Saint Paul, Ind	Dec. 26, 1865	51, 750
Harrow and clod-crusher, Combined	A. C. Tower	Troy Grove, Ill	Aug. 20, 1872	130, 606
Harrow and cultivator	W. A. Estes	China, Me	Aug. 3, 1869	93, 187
Harrow and cultivator	J. F. Jaquess	Commerce, Miss	Jan. 24, 1871	111, 125
Harrow and cultivator	J. Lerch	Uhlersville, Pa	Aug. 16, 1870	106, 376
Harrow and cultivator	E. S. Rice	Paw-Paw, Mich	Apr. 6, 1869	88, 740
Harrow and cultivator	A. S. White	Malone, N. Y	Sept. 29, 1868	82, 573
Harrow and cultivator, Combined	G. W. Bressler	La Fayette, Iowa	Mar. 30, 1869	88, 363
Harrow and cultivator, Combined	H. Culver	Dansville, N. Y	Mar. 9, 1869	87, 642
Harrow and cultivator, Combined	W. R. Cummings	McCutchanville, Ind	June 29, 1869	91, 917
Harrow and cultivator, Combined	W. J. Funk	Portland, Oreg	July 20, 1869	92, 714
Harrow and cultivator, Combined	D. M. Harkrader	Chili, Ill	Oct. 13, 1868	82, 942
Harrow and cultivator, Combined	J. Lerch	Uhlersville, Pa	Nov. 23, 1869	97, 205
Harrow and cultivator, Combined	J. K. Minich	Mount Joy, Pa	May 8, 1866	54, 648
Harrow and cultivator, Combined	J. R. Ross and W. D. Mitchell	Centralia, Ill	Apr. 6, 1869	88, 669
Harrow and cultivator, Combined	A. H. Sheffer	West Donegal, Pa	May 1, 1866	54, 421
Harrow and cultivator, Combined	N. W. Wheeler	Ripon, Wis	Oct. 6, 1868	82, 775
Harrow and cultivator combined, Revolving	S. D. Carpenter	Madison, Wis	Dec. 29, 1868	85, 282
Harrow and cultivator, Revolving	A. Newell	New York, N. Y	July 2, 1867	66, 245
Harrow and cultivator, Sulky	W. and J. Whait	Independence, Iowa	July 23, 1867	67, 009
Harrow and earth-cutter	J. Schroeffel and W. Dell	Allegheny City, Pa	Apr. 5, 1870	101, 522
Harrow and field-roller	A. S. Keagy	Harristown, Ill	Feb. 28, 1871	112, 149
Harrow and marker, Combined	W. Addleton	Mottville, N. Y	July 6, 1869	92, 139
Harrow and marker, Combined	B. F. Barney	Pontiac, Ill	July 27, 1869	93, 038
Harrow and marker, Combined	S. Hutchinson	Griggsville, Ill	July 18, 1871	117, 079
Harrow and marker, Combined	A. Pierce	Olmsted, Ohio	Sept. 11, 1866	57, 961
Harrow and plow-hoe	I. Cheatham	Providence Inn, Va	July 31, 1827	....
Harrow and roller, Combined	T. Bailey	Wyoming, Ill	Aug. 30, 1870	106, 765
Harrow and roller, Combined	J. M. Blankenbeker	Power's Station, Ind	Aug. 23, 1870	106, 639
Harrow and roller, Combined	W. H. Converse	New Castle, Me	Apr. 4, 1865	47, 090
Harrow and roller, Combined	F. A. Dann and J. McKibben	Wellsville, Mo	Sept. 20, 1870	107, 461
Harrow and roller, Combined	J. Steiner and J. P. Miller	Harrisburgh, Ohio	May 15, 1866	54, 790
Harrow and roller, Revolving	E. K. Harvey	Quincy, Ohio	May 26, 1868	78, 283
Harrow and seed-drill	J. A. Underwood	Grand River, Iowa	Oct. 21, 1862	36, 739
Harrow and seed-sower, Combined	E. A. Barton	Boonville, Ind	Dec. 28, 1869	98, 335
Harrow and seeder	D. L. and J. M. Barlow	Coboctah, Mich	Jan. 17, 1865	45, 903
Harrow and seeder, Combined	L. G. Jackson	Marion, Wis	May 13, 1873	138, 806
Harrow and seeder, Combined	B. Randall	Adams, N. Y	Nov. 12, 1867	70, 898
Harrow and seeder, Combined	E. Stoner	Greencastle, Pa	Nov. 15, 1870	109, 269
Harrow and seeder combined, Revolving sulky	A. L. Taveau	Chaptico, Md	May 4, 1869	89, 812
Harrow and seeder combined, Revolving sulky	A. L. Taveau	Chaptico, Md	Apr. 5, 1870	101, 540
Harrow and seeder, Combined wheel	F. Bramer	Little Falls, N. Y	June 24, 1873	140, 182
Harrow and sower, Combined	J. Brewer	Albany, Ill	Oct. 1, 1861	33, 377
Harrow and stone-remover, Combined	E. Babcock	Scott, N. Y	Feb. 4, 1868	74, 035
Harrow, Carriage	T. J. Marinus, and S. and W. Whait.	Independence, Iowa	Oct. 16, 1866	58, 854
Harrow, Circular	G. W. Hills	Dodge Centre, Minn	Nov. 25, 1873	144, 978
Harrow, Corn, &c	P. Clark	Aurora, N. Y	Aug. 20, 1835	....
Harrow, Corn	G. Hill	Galva, Ill	May 23, 1871	115, 055
Harrow, Corn and grain cultivator	N. Stanton, jr	Florida, N. Y	July 9, 1832	....
Harrow, Cultivating	J. Slocum	Syracuse, N. Y	Nov. 27, 1860	30, 767
Harrow, cultivator, and planter combined	J. G. S Garwood	Vermillion, Ill	Jan. 21, 1868	73, 593
Harrow, cultivator, wheelbarrow, and sled combined	J. R. Wagner	Manada Hill, Pa	Aug. 2, 1870	106, 004
Harrow, Double-rotary	G. W. Hall	New Haven, Mich	Oct. 1, 1867	69, 340
Harrow, drill, grass-seeder, and roller, Combined	J. P. Long	Osage, Iowa	May 12, 1863	38, 495
Harrow, drill, planter, and roller combined	D. B. Platt	Madison, Ind	June 16, 1868	78, 998
Harrow, Expanding	I. J. Halsted	Springfield, Ill	Jan. 5, 1869	85, 582
Harrow, Flexible	C. D Blim	Port Huron, Mich	May 5, 1868	77, 574
Harrow, Flexible	W. B. and G. M. Ramsay	South Strabane, Pa	May 2, 1854	10, 865
Harrow for clearing ground, Cleave	L. Tam	Sussex County, Del	July 11, 1826	....
Harrow-frame	J. Brainerd, E. F. Olds, and A. W. Olds.	Cleveland, Ohio, and Lyon and Green Oak, Mich.	Apr. 2, 1861	31, 865
Harrow, Hoe	A. Newbern	Bristol Township, Pa	June 1, 1824	....
Harrow, Plow	S Vinton	Wellington, Conn	Nov. 18, 1807	....
Harrow, Portable field	J. D. Johnson	Tylersville, Pa	Mar. 30, 1869	88, 301
Harrow, Reciprocating	J. Densmore	Holley, N. Y	July 16, 1867	66, 684
Harrow, Revolving	G. S. Bartholomew and A. R. Chapman.	Reading, Mich	May 18, 1869	90, 222
Harrow, Revolving	C. Bates	Kingston, Mass	June 1, 1869	90, 628
Harrow, Revolving	M. G. Cass	Utica, N. Y	Sept. 10, 1839	1, 317
Harrow, Revolving	J. Hill and J. G. Stewart	Jonesville, Ind	Dec. 26, 1871	122, 251
Harrow, Revolving	W. A. Horrall and R. G. Sirwell	Grayville, Ill	Feb. 16, 1858	19, 365
Harrow, Revolving	M. W. House	Cleveland, Ohio	Jan. 18, 1859	22, 651
Harrow, Revolving	B. Lyon	Mount Pleasant, Ill	Sept. 29, 1868	82, 623
Harrow, Revolving	H. Mittendorf	Stony Hill, Mo	Feb. 25, 1873	136, 169
Harrow, Revolving	H. H. Monroe	Thomaston, Me	Mar. 8, 1870	100, 547
Harrow, Revolving	G. J. Olendorf	Middlefield, N. Y	June 17, 1856	15, 151
Harrow, Revolving	S. Rugg	Lancaster, Mass	Jan. 11, 1831	....
Harrow, Revolving	H. C. Stoll	Mokena, Ill	Feb. 26, 1867	62, 379

Index of patents issued from the United States Patent Office from 1790 *to* 1873, *inclusive*—Continued.

Invention.	Inventor.	Residence.	Date.	No.
Harrow, Revolving	W. R. Toby and M. J. Barcalo	Nunda and Mount Morris, N. Y.	Dec. 1, 1868	84, 656
Harrow, Revolving cylindrical	J. D. Woodside	Washington, D. C	July 28, 1832	....
Harrow, Revolving disk	J. F. Pond	Cleveland, Ohio	May 9, 1871	114, 707
Harrow, roller, and drill, Combined	S. Bradbury	Dresden, Mo	Aug. 24, 1869	93, 956
Harrow, roller, and seed-planter, Combined	R. J. Colvin	Lancaster, Pa	May 23, 1871	115, 026
Harrow, roller, and seed-sower combined	C. E. Pierce	New York, N. Y	Feb. 15, 1870	99, 946
Harrow, roller, and seeder, Combined	A. Powell	New Baltimore, N. Y	Jan. 4, 1870	98, 627
Harrow, Rotary	A. H. Acken	Griggstown, N. J	Apr. 23, 1867	64, 055
Harrow, Rotary	O. D. Barrett	Cleveland, Ohio	Oct. 4, 1859	25, 620
Harrow, Rotary	J. W. Barton and T. Holmes	Emporia, Kans	Mar. 14, 1871	112, 676
Harrow, Rotary	C. Bates	Kingston, Mass	Mar. 26, 1867	63, 199
Harrow, Rotary	M. L. Bawder	Cleveland, Ohio	Dec. 4, 1860	30, 795
Harrow, Rotary	E. Bennett	Nankin, Mich	Apr. 9, 1872	125, 523
Harrow, Rotary	L. Brainard and L. Newton	Attica, N. Y	Apr. 10, 1855	12, 667
Harrow, Rotary	J. Brainerd	Cleveland, Ohio	Oct. 2, 1860	30, 198
Harrow, Rotary	M. B. Champion	Sturgis, Mich	July 13, 1869	92, 422
Harrow, Rotary	J. F. Chase	Westbrook, Me	Mar. 28, 1871	113, 141
Harrow, Rotary	G. Collins	Fremont, Nebr	Sept. 13, 1870	107, 337
Harrow, Rotary	G. Cook	Paris, Ill	Sept. 13, 1859	25, 390
Harrow, Rotary	C. Daniel	Sigel, Mo	Dec. 1, 1863	40, 741
Harrow, Rotary	C. P. Fisher	Leesville, Ohio	June 1, 1869	90, 652
Harrow, Rotary	H. A. Gaston	Stockton, Cal	Oct. 12, 1869	95, 788
Harrow, Rotary	P. Gevin and E. Foreman	Summer Hill, Pa	Nov. 27, 1860	30, 728
Harrow, Rotary	C. and J. K. Gingrich	Annville, Pa	June 28, 1859	24, 554
Harrow, Rotary	W. P. Goolman	Dublin, Ind	Aug. 30, 1859	25, 301
Harrow, Rotary	W. T. Hildrup	Harrisburgh, Pa	Feb. 1, 1859	22, 805
Harrow, Rotary	W. T. Hildrup	Harrisburgh, Pa	Apr. 12, 1859	23, 578
Harrow, Rotary	S. S. Hogle	Cleveland, Ohio	Oct. 11, 1859	25, 736
Harrow, Rotary	S. S. Hogle	Cleveland, Ohio	Oct. 30, 1860	30, 554
Harrow, Rotary	C. Hood	Seneca Falls, N. Y	Dec. 5, 1871	121, 621
Harrow, Rotary	C. Howell	Cleveland, Ohio	Jan. 4, 1859	22, 502
Harrow, Rotary	A. C. Jenson and J. Mathison	Fremont, Nebr	Apr. 9, 1872	125, 572
Harrow, Rotary	M. C. Kilgore	Washington, Iowa	Oct. 4, 1859	25, 651
Harrow, Rotary	S. Lubolt and J. Trout	Lykens, Pa	Feb. 9, 1869	86, 768
Harrow, Rotary	W. H. Main	Liverpool, Ohio	Aug. 24, 1858	21, 265
Harrow, Rotary	W. H. Main	Liverpool, Ohio	Nov. 9, 1858	22, 026
Harrow, Rotary	W. H. Main	Liverpool, Ohio	Mar. 26, 1861	31, 813
Harrow, Rotary	J. W. McLean	Lebanon, Ind	June 7, 1859	24, 322
Harrow, Rotary	H. H. Monroe	Thomaston, Me	Aug. 2, 1864	43, 700
Harrow, Rotary	D. Morris	Bartlett, Ohio	Jan. 1, 1867	60, 776
Harrow, Rotary	A. W. Olds	Green Oak, Mich	Sept. 3, 1861	33, 212
Harrow, Rotary	A. W. Olds	Green Oak, Mich	Dec. 24, 1861	34, 007
Harrow, Rotary	A. W. Olds	Green Oak Station, Mich	Oct. 1, 1872	131, 893
Harrow, Rotary	J. F. Ostrander	New York, N. Y	Aug. 26, 1851	8, 313
Harrow, Rotary	J. D. Parrot	Morristown, N. J	Oct. 31, 1865	50, 729
Harrow, Rotary	R. Rakestraw	Wyoming, Ill	July 11, 1871	116, 991
Harrow, Rotary	R. Rakestraw	Wyoming, Ill	June 10, 1873	139, 814
Harrow, Rotary	F. Raymond	Sandusky, Ohio	Nov. 27, 1860	30, 759
Harrow, Rotary	J. Robins	Boston, Mass	Sept. 21, 1858	21, 577
Harrow, Rotary	H. W. Robinson	Woodstock, Ill	Aug. 20, 1872	130, 748
Harrow, Rotary	P. B. B. Stiles	Galesburgh, Ill	July 30, 1867	67, 372
Harrow, Rotary	H. Thompson	Rockland, Me	May 22, 1866	55, 024
Harrow, Rotary	S. S. Thompson	Heller's Corners, Ind	Sept. 21, 1858	21, 580
Harrow, Rotary	G. W. Toleman	Augusta, Ky	May 10, 1859	23, 962
Harrow, Rotary	L. S. Tyler	Lineville, Pa	Feb. 5, 1861	31, 341
Harrow, Rotary	S. M. Wade	Andover, Ohio	Apr. 12, 1859	23, 627
Harrow, Rotary	B. A., and M. H. Welds and H. W. Strong.	Reading, Mich	Feb. 2, 1869	86, 478
Harrow, Rotating	S. M. Garver	Monticello, Ill	Sept. 8, 1863	39, 806
Harrow, Rotating	J. B. Glascock	Fancy Creek, Ill	July 21, 1857	17, 831
Harrow, Rotating	S. Grenell	Mokena, Ill	Sept. 18, 1866	58, 191
Harrow, Seed-sowing	J. Davis	Allegheny City, Pa	Nov. 11, 1862	36, 898
Harrow, seeder, and roller combined	J. Vessot, sr., and S. Vessot, jr	Joliette, Canada	Mar. 21, 1871	112, 869
Harrow, Seeding	E. Badlam	Ogdensburgh, N. Y	June 18, 1861	32, 551
Harrow, Seeding	W. Finlay	Schoolcraft, Mich	Apr. 10, 1860	27, 789
Harrow, Seeding	H. Hewitt	San Francisco, Cal	Mar. 27, 1860	27, 668
Harrow, Seeding	A. E. Jerome	Monroeville, Ohio	Nov. 15, 1859	26, 110
Harrow, Seeding	M. S. Root	Medina, Ohio	May 22, 1860	28, 411
Harrow, Self-sharpening	H. Hopkins and J. Pratt	Pompey, N. Y	Nov. 9, 1824	....
Harrow, Sulky	W. C. Bye	Collette, Ind	Feb. 25, 1873	136, 131
Harrow, &c., Sulky	J. A. Casey	Maysville, Ky	June 21, 1870	104, 423
Harrow, Sulky	J. E. Cheasebro	Marilla, N. Y	Feb. 9, 1869	86, 731
Harrow, Sulky	W. H. H. Frye	North Fryeburgh, Me	Sept. 7, 1869	94, 589
Harrow, Sulky	E. W. Hewitt	Pecatonica, Ill	Dec. 15, 1868	84, 879
Harrow, Sulky	B. Randall	Adams, N. Y	Apr. 14, 1868	76, 657
Harrow, Sulky	F. P. L. Reimer and H. Asbahr	Davenport, Iowa	July 25, 1871	117, 330
Harrow, Sulky	P. Speelmon	Knox, Ind	Nov. 4, 1873	144, 231
Harrow, Sulky	G. W. Van Goeder	Warren, Ohio	Mar. 4, 1873	136, 564
Harrow, Sward-cutting	D. Rice	Rowe, Mass	May 17, 1844	3, 589
Harrow-teeth	S. W. Corbin	Bainbridge, N. Y	Nov. 29, 1870	109, 719
Harrow-teeth	D. M. Cummings	Enfield, N. H	Nov. 22, 1859	26, 167
Harrow-teeth	D. France	Lakeville, Minn	Dec. 24, 1872	134, 137
Harrow-teeth	E. N. Higley	Lake Village, N. H	July 12, 1864	43, 500
Harrow-teeth	P. V. Hixson	Tioga, Pa	Oct. 6, 1868	82, 835
Harrow-teeth	A. Jones	Westford, Wis	May 12, 1868	77, 737
Harrow-teeth	H. R. Kinney	Portsmouth, Ohio	Sept. 20, 1870	107, 506
Harrow-teeth	W. McConaughey	New Garden, Pa	Dec. 8, 1823	....
Harrow-teeth	A. Patterson	Birmingham, Pa	Nov. 3, 1868	83, 656
Harrow-teeth	W. H. Platt	Dayton, Ohio	Oct. 14, 1873	143, 713
Harrow-teeth	P. Prettyman	Georgetown, Del	Apr. 3, 1834	....
Harrow-teeth	F. R. Willson	Columbus, Ohio	Sept. 24, 1867	69, 290
Harrow-teeth, Apparatus for making	W. McConaughey	New Garden, Pa	Feb. 16, 1827	....
Harrow-teeth, Die for forging	J. Pedder and G. Abel	Pittsburgh, Pa	Mar. 12, 1872	124, 619
Harrow-teeth, Fluke	L. C. Tann	Milton, Del	Mar. 30, 1830	....
Harrow-teeth, Machine for making	J. Morgan	Pittsburgh, Pa	June 14, 1870	104, 186
Harrow-teeth, Machine for rolling	A. H. Holmes	Allegheny, Pa	Feb. 27, 1872	124, 062
Harrow-teeth, Machine for rolling	S. C. Murdoch	Pittsburgh, Pa	Dec. 5, 1871	121, 653

Index of patents issued from the United States Patent Office from 1790 *to* 1873, *inclusive*—Continued.

Invention.	Inventor.	Residence.	Date.	No.
Harrow-teeth, Mode of fastening	G. Fry	Newaygo, Mich	Feb. 28, 1871	112, 236
Harrow-teeth, Reversible	G. W. Hurst	Avon, Ohio	June 24, 1873	140, 198
Harrow, Wheel	T. M. Brintnall	Medina, Ohio	Aug. 27, 1872	130, 839
Harrow, Wheel	E. R. Powell	Jeffersonville, Vt	Nov. 29, 1870	109, 754
Harrow, Wheeled	A. C. Baker and N. O. Hoyt	La Fayette, N. Y	Oct. 29, 1867	70, 314
Harrow, Wheeled	A. L. P. Vairin	Ripley, Miss	Jan. 4, 1870	98, 647
Harrow with cultivator-attachment	A. M. Bakewell	Normal, Ill	Apr. 20, 1869	89, 192
Harrows, Combined	C. Wickoff, jr	Fairview, Ill	Apr. 25, 1871	114, 244
Harrows, Riding-attachment for	J. M. Freeman	Belleville, N. Y	Aug. 27, 1867	68, 179
Harrows, Riding-attachment to	A. D. Shaver	Belleville, N. Y	July 7, 1868	79, 604
Harrows to land-rollers, Attachment of	D. Hill	Bartonia, Ind	Oct. 11, 1853	10, 109
Harrows to plows, Attachment of	J. Stroop	Philadelphia, Pa	June 26, 1849	6, 562
Harrows to seed-planters, Attachment of	M. and S. S. Sage	Windsor, N. Y	Apr. 16, 1850	7, 293
Harvester	A. S. Acker	Albion, N. Y	Feb. 27, 1866	52, 809
Harvester	H. A. Adams	Sandwich, Ill	Nov. 4, 1873	144, 179
Harvester	J. H. Adamson	Auburn, South Australia	Aug. 29, 1871	118, 574
Harvester	H. Adkins	Plymouth, Ill	Sept. 2, 1856	15, 638
Harvester	R. L. Allen	New York, N. Y	Sept. 7, 1858	21, 401
Harvester	S. S. Allen	Bristol, Pa	Apr. 7, 1857	16, 957
Harvester	N. Allstatter	Hamilton, Ohio	Sept. 28, 1869	95, 301
Harvester	J. Anthony	Ledyard, N. Y	Oct. 8, 1867	69, 606
Harvester	S. Augspurger	Trenton, Ohio	Feb. 16, 1864	41, 593
Harvester	E. M. Awrey	Caistor, Canada	July 15, 1873	140, 804
Harvester	A. P. Ayers	Chicago, Ill	June 21, 1870	104, 407
Harvester	D. Babcock	Warsaw, Ill	June 30, 1868	79, 300
Harvester	G. Bailey	Wiscotta, Iowa	Sept. 6, 1864	44, 064
Harvester	N. A. Baker	Covington, Ky	May 21, 1872	126, 919
Harvester	M. and W. P. Bales	London, Ohio	Sept. 13, 1870	107, 214
Harvester	E. Ball	Canton, Ohio	Oct. 18, 1859	25, 797
Harvester	E. Ball and M. L. Ballard	Canton, Ohio	Mar. 20, 1860	27, 511
Harvester	V. S. Barber	Alliance, Ohio	Aug. 23, 1870	106, 535
Harvester	S. and J. H. Barley	Longwood, Mo	June 19, 1860	28, 727
Harvester	A. B. Barnard, S. R. Nye, and R. L. Hewett.	West Fitchburgh, Mass	Aug. 18, 1868	81, 129
Harvester	J. Barnes	Rockford, Ill	Dec. 13, 1864	45, 383
Harvester	J. Barnes	Rockford, Ill	Jan. 12, 1869	85, 722
Harvester	J. Barnes	Rockford, Ill	Sept. 28, 1869	95, 304
Harvester	J. Barnes	Rockford, Ill	May 16, 1871	114, 747
Harvester	L. Barnes	Islip Township, N. Y	Jan. 8, 1856	14, 046
Harvester	T. J. Barnes	Corry, Pa	Jan. 17, 1871	110, 949
Harvester	J. A. Barrington	Fredericktown, Ohio	Dec. 21, 1858	22, 341
Harvester	S. S. Bartlett	Providence, R. I	Feb. 25, 1862	34, 545
Harvester	S. S. Bartlett	Providence, R. I	Nov. 10, 1863	40, 585
Harvester	S. S. Bartlett	Providence, R. I	May 16, 1865	47, 691
Harvester	S. O. Bartow	Bethel, Conn	Feb. 4, 1868	74, 036
Harvester	S. D. Bates	Lewisburgh, Pa	Nov. 9, 1869	96, 660
Harvester	L. M. Batty	Canton, Ohio	Dec. 28, 1869	98, 220
Harvester	L. M. Batty	Canton, Ohio	Jan. 31, 1871	111, 424
Harvester	L. M. Batty	Canton, Ohio	Mar. 14, 1871	112, 530
Harvester	J. Beach	De Ruyter, N. Y	Mar. 19, 1872	124, 781
Harvester	J. H. Beam	Springfield, Ill	Mar. 28, 1871	113, 134
Harvester	W. G. Beels	Independence, Iowa	Jan. 18, 1870	98, 910
Harvester	S. Bell	Marseilles, Ill	Feb. 28, 1854	10, 581
Harvester	O. Billings	Elyria, Ohio	Sept. 12, 1871	118, 837
Harvester	J. Birch, A. Crosby, and T. Birch.	Westfield, N. Y., and Meadville, Pa.	June 28, 1870	104, 695
Harvester	V. W. Blanchard	Bridgeport, Vt	Dec. 8, 1863	40, 812
Harvester	V. W. Blanchard	Bridgeport, Vt	Oct. 17, 1865	50, 441
Harvester	V. W. Blanchard	Bridgeport, Vt	Jan. 1, 1867	60, 676
Harvester	J. Blue	Covert, N. Y	Dec. 18, 1860	30, 953
Harvester	J. Blue	Covert, N. Y	Mar. 26, 1861	31, 774
Harvester	J. Blue	Trumansburgh, N. Y	Nov. 24, 1868	84, 349
Harvester	O. Bonney, jr	San Francisco, Cal	July 21, 1868	80, 124
Harvester	O. Bonney, jr	San Francisco, Cal	Sept. 28, 1869	95, 188
Harvester	J. W. Bosse	Saint Louis, Mo	Jan. 10, 1865	45, 810
Harvester	J. W. Bosse	Saint Louis, Mo	Jan. 17, 1865	45, 905
Harvester	J. W. Bosse	Saint Louis, Mo	Jan. 17, 1865	45, 906
Harvester	J. Bardwell	Brockport, N. Y	Jan. 2, 1872	122, 432
Harvester	J. B. Bowen and J. E. Baker	Madison, Wis	Apr. 22, 1862	35, 009
Harvester	W. F. Brabrook	South Hardwick, Vt	Aug. 6, 1867	67, 490
Harvester	C. J. Brackebush and C. E. Merrifield.	Indianapolis, Ind	Sept. 14, 1869	94, 865
Harvester	H. Brackett	Valley Falls, N. Y	May 4, 1869	89, 730
Harvester	C. Bradfield	Philadelphia, Pa	Aug. 21, 1855	13, 480
Harvester	I. W. Bragg	Kidder, Mo	July 8, 1873	140, 675
Harvester	F. Bramer	Fabius, N. Y	Dec. 19, 1865	51, 546
Harvester	T. Brett	Geneva, Ohio	Nov. 17, 1863	40, 601
Harvester	H. H. Bridenthall, jr	New Derby, Pa	Oct. 17, 1871	120, 027
Harvester	A. D. Briggs	Springfield, Mass	Feb. 16, 1858	19, 344
Harvester	A. B. Briggs	South Deerfield, Mass	Feb. 9, 1864	41, 566
Harvester	C. R. Brinkerhoff	Batavia, N. Y	July 10, 1860	29, 053
Harvester	C. R. Brinkerhoff	Rochester, N. Y	Jan. 11, 1870	98, 737
Harvester	A. Brown	Worcester, Mass	Aug. 24, 1869	94, 167
Harvester	A. Brown, L. G. Kniffen, and T. H. Dodge.	Worcester, Mass., and Nashua, N. H.	Nov. 17, 1863	40, 658
Harvester	C. B. Brown	Alton, Ill	May 11, 1858	20, 191
Harvester	L. D. Brown	Saint Louis, Mo	Nov. 20, 1860	30, 664
Harvester	R. Brown	Frederick, Md	June 18, 1861	32, 556
Harvester	R. Brown	Frederick, Md	Oct. 27, 1863	40, 395
Harvester	R. D. Brown	Covington, Ky	Feb. 9, 1864	41, 477
Harvester	T. S. Brown	Poughkeepsie, N. Y	Aug. 25, 1868	81, 472
Harvester	T. S. Brown	Poughkeepsie, N. Y	July 2, 1872	128, 584
Harvester	A. C. Brownlich	Buffalo, N. Y	Jan. 4, 1859	22, 483
Harvester	C. Brownlich	Buffalo, N. Y	Mar. 1, 1859	23, 075
Harvester	S. T. Bruce	Marshall, Mo	May 1, 1860	28, 061
Harvester	R. Bryson	Schenectady, N. Y	June 5, 1860	28, 557
Harvester	R. Bryson	Schenectady, N. Y	Mar. 31, 1863	38, 030
Harvester	R. Bryson	Schenectady, N. Y	July 31, 1866	56, 698

Index of patents issued from the United States Patent Office from 1790 *to* 1873, *inclusive*—Continued.

Invention.	Inventor.	Residence.	Date.	No.
Harvester	A. Bullock	Busti, N. Y	Nov. 11, 1856	16, 079
Harvester	C. Bullock	Jamestown, N. Y	Dec. 21, 1858	22, 345
Harvester	O. H. Burdick	Auburn, N. Y	Feb. 27, 1866	52, 927
Harvester	J. Burke	Sycamore, Ill	Nov. 12, 1867	70, 798
Harvester	W. H. Burkhart	Bucyrus, Ohio	May 30, 1865	47, 926
Harvester	W. H. Burkhart	Bucyrus, Ohio	Feb. 13, 1866	52, 526
Harvester	H. K. Burnett	Poughkeepsie, N. Y	July 28, 1868	80, 272
Harvester	J. P. Burnham	Rockford, Ill	Nov. 29, 1859	26, 251
Harvester	G. E. Burt	Harvard, Mass	Dec. 29, 1868	85, 274
Harvester	S. and J. H. Buser	Warren, Ill	Feb. 14, 1860	27, 105
Harvester	J. Butter	Buffalo, N. Y	Nov. 15, 1859	26, 142
Harvester	J. Butter	Buffalo, N. Y	June 16, 1863	38, 882
Harvester	J. S. Butterfield	Philadelphia, Pa	Mar. 2, 1858	19, 483
Harvester	C. Cadwell	Waukegan, Ill	Jan. 16, 1866	52, 026
Harvester	C. Cadwell	Waukegan, Ill	Jan. 16, 1866	52, 027
Harvester	C. Cadwell	Waukegan, Ill	Aug. 14, 1866	57, 082
Harvester	C. Cadwell	Waukegan, Ill	May 12, 1868	77, 714
Harvester	J. M. Canfield, A. T. Still, and E. P. Wheeler.	Lawrence, Kans	Mar. 27, 1866	53, 406
Harvester	C. W. Cardot	Fredonia, N. Y	May 14, 1867	64, 749
Harvester	S. D. Carpenter	Madison, Wis	Nov. 4, 1873	144, 189
Harvester	W. D. Carpenter	South Berwick, Me	Jan. 21, 1868	73, 501
Harvester	H. Carter	Greene, N. Y	July 19, 1859	24, 793
Harvester	H. M. Carter	La Fayette, Ind	Feb. 4, 1873	135, 403
Harvester	J. A. Cauldwell	Watkins, N. Y	Aug. 8, 1871	117, 862
Harvester	W. Chapman	Salisbury, Conn	Feb. 23, 1869	87, 244
Harvester	C. H. Charlesworth and J. H. Short.	Avoca, N. Y	May 17, 1870	103, 136
Harvester	G. E. Chenoweth	Baltimore, Md	Mar. 30, 1858	19, 749
Harvester	G. E. Chenoweth	Baltimore, Md	Feb. 8, 1859	22, 855
Harvester	G. E. Chenoweth	Baltimore, Md	Mar. 1, 1859	23, 077
Harvester	G. E. Chenoweth	Baltimore, Md	Nov. 15, 1859	26, 091
Harvester	G. E. Chenoweth	Baltimore, Md	Mar. 13, 1860	27, 617
Harvester	W. L. Childs	Piermont, N. Y	Mar. 2, 1858	19, 486
Harvester	C. Clapp	Trumansburgh, N. Y	May 10, 1870	102, 776
Harvester	C. Clapp	Trumansburgh, N. Y	Aug. 1, 1871	117, 514
Harvester	H. Clark	Rochester, N. Y	Mar. 10, 1857	16, 788
Harvester	S. A. Clemens	Rockford, Ill	Mar. 3, 1857	16, 721
Harvester	E. H. Clinton	Iowa City, Iowa	Dec. 2, 1873	145, 050
Harvester	D. Clow	Janesville, Wis	Feb. 15, 1859	22, 936
Harvester	D. Clow	Janesville, Wis	July 22, 1862	35, 923
Harvester	N. Clute	Dunnsville, N. Y	June 16, 1857	17, 555
Harvester	N. Clute	Dunnsville, N. Y	Nov. 30, 1358	22, 163
Harvester	W. F. Cochrane	Springfield, Ohio	Jan. 31, 1865	46, 178
Harvester	W. F. Cochrane	Springfield, Ohio	Jan. 31, 1865	46, 179
Harvester	W. F. Cochrane	Springfield, Ohio	Jan. 31, 1865	46, 180
Harvester	W. F. Cochrane	Springfield, Ohio	Jan. 31, 1865	46, 181
Harvester	W. F. Cochrane	Springfield, Ohio	Jan. 31, 1865	46, 182
Harvester	W. F. Cochrane	Springfield, Ohio	Jan. 31, 1865	46, 183
Harvester	W. F. Cochrane	Springfield, Ohio	Jan. 1, 1867	60, 692
Harvester	W. F. Cochrane	Springfield, Ohio	Jan. 1, 1867	60, 693
Harvester	J. F. Coddington	Boundbrook, N. J	July 7, 1868	79, 551
Harvester	J. T. Coddington	Newark, N. J	Mar. 9, 1869	87, 637
Harvester	W. Cogswell	Ottawa, Ill	Aug. 29, 1865	49, 607
Harvester	W. Cogswell	Ottawa, Ill	Aug. 29, 1865	49, 608
Harvester	W. Cogswell	Ottawa, Ill	Nov. 2, 1869	96, 311
Harvester	W. and I. Cogswell, jr	Ottawa, Ill	Dec. 6, 1859	26, 338
Harvester	L. H. Colborn	Baltimore, Md	Apr. 12, 1859	23, 552
Harvester	I. H. Caller	Poughkeepsie, N. Y	Sept. 1, 1863	39, 719
Harvester	I. H. Caller	Poughkeepsie, N. Y	Nov. 24, 1863	40, 675
Harvester	J. Collins	Guelph, Canada	Feb. 19, 1867	62, 252
Harvester	R. Conarroe	Camden, Ohio	Dec. 14, 1869	97, 884
Harvester	I. H. Conklin	Rockford, Ill	May 25, 1858	20, 334
Harvester	J. M. Connel	Newark, Ohio	Sept. 14, 1869	94, 871
Harvester	R. F. Cooke	Newark, N. J	Dec. 31, 1867	72, 722
Harvester	G. T. Coolman and C. M. Young	Corry, Pa	Mar. 30, 1869	88, 453
Harvester	G. T. Coolman and C. M. Young	Corry, Pa	May 11, 1869	89, 973
Harvester	G. T. Coolman and C. M. Young	Corry, Pa	Feb. 15, 1870	99, 850
Harvester	G. E. Cooper	Baltimore, Md	Oct. 12, 1858	21, 741
Harvester	F. C. Coppage	Terre Haute, Ind	June 19, 1866	55, 763
Harvester	F. C. Coppage	Terre Haute, Ind	July 10, 1866	56, 320
Harvester	F. C. Coppage	Terre Haute, Ind	Dec. 10, 1867	71, 989
Harvester	W. R. Cory	Springfield, Ill	Jan. 3, 1871	110, 742
Harvester	J. F. Crawford	New York Mills, N. Y	Jan. 11, 1870	98, 671
Harvester	S. Crawford	London, Canada	May 21, 1872	126, 932
Harvester	T. Crumbling	Hellam, Pa	Sept. 27, 1859	25, 559
Harvester	G. S. Curtis	Chicago, Ill	Feb. 23, 1858	19, 411
Harvester	G. S. and H. Curtis	Chicago, Ill	May 31, 1864	42, 930
Harvester	J. Curtis	Hackettstown, N. J	Nov. 29, 1864	45, 228
Harvester	T. Curtis	New Hudson, Mich	May 3, 1864	42, 564
Harvester	J. D. Custer	Norristown, Pa	Sept. 27, 1859	25, 561
Harvester	R. Daniels	Woodstock, Vt	Oct. 13, 1857	18, 390
Harvester	A. L. Darby	White Creek, N. Y	Dec. 23, 1862	37, 253
Harvester	J. S. Davis	Tiffin, Ohio	Oct. 31, 1865	50, 693
Harvester	J. S. Davis	Tiffin, Ohio	Mar. 10, 1868	75, 383
Harvester	J. S. Davis	Toledo, Ohio	May 20, 1873	139, 122
Harvester	R. M. Davis	Eaton, N. Y	July 22, 1862	35, 970
Harvester	T. D. Davis and J. M. Waldron.	Syracuse and South Otselic, N. Y.	Jan. 7, 1862	34, 097
Harvester	C. Denton	Pekin, Ill	Oct. 9, 1866	58, 720
Harvester	C. Denton	Decatur, Ill	June 30, 1868	79, 452
Harvester	W. De Witt	Cleveland, Ohio	Jan. 7, 1862	34, 090
Harvester	J. Dick, jr	Canton, Ohio	Oct. 9, 1866	58, 617
Harvester	C. G. Dickinson	Poughkeepsie, N. Y	Feb. 1, 1859	22, 786
Harvester	C. G. Dickinson	Poughkeepsie, N. Y	Jan. 29, 1861	31, 229
Harvester	S. P. Doane	San Francisco, Cal	Feb. 2, 1869	86, 375
Harvester	J. A. Dodge	Auburn, N. Y	May 23, 1865	47, 807
Harvester	J. A. Dodge	Auburn, N. Y	June 26, 1866	55, 837

Index of patents issued from the United States Patent Office from 1790 *to* 1873, *inclusive*—Continued.

Invention.	Inventor.	Residence.	Date.	No.
Harvester	J. A. Dodge	Auburn, N. Y	Aug. 20, 1867	67, 852
Harvester	J. A. Dodge	Auburn, N. Y	Dec. 22, 1868	85, 168
Harvester	J. A. Dodge	Auburn, N. Y	Aug. 6, 1872	130, 285
Harvester	T. H. Dodge	Worcester, Mass	Jan. 2, 1872	122, 445
Harvester	A. G. Donnelly	Breesport, N. Y	Nov. 17, 1868	84, 051
Harvester	J. W. H. Doubler	West Alden, N. Y	Sept. 16, 1873	142, 903
Harvester	W. Dripps	Coatesville, Pa	Oct. 7, 1856	15, 843
Harvester	J. J. Duchesne	Lacon, Ill	July 7, 1868	79, 740
Harvester	I. A. Dufield	McHenry, Ill	Nov. 1, 1859	25, 960
Harvester	R. Dutton	Dayton, Ohio	Apr. 27, 1858	20, 050
Harvester	R. Dutton	New York, N. Y	Feb. 11, 1868	74, 208
Harvester	R. Dutton	New York, N. Y	Feb. 11, 1868	74, 209
Harvester	R. Dutton	New York, N. Y	Feb. 11, 1868	74, 210
Harvester	P. Dyer, jr	Unadilla, Mich	May 18, 1869	90, 247
Harvester	J. F. Earl	San Francisco, Cal	Nov. 10, 1868	83, 839
Harvester	M. Easterbrook, jr	Geneva, N. Y	May 22, 1866	54, 876
Harvester	M. Easterbrook, jr	Geneva, N. Y	June 11, 1867	65, 732
Harvester	J. Ebner and F. Leuthy	Lancaster, Pa	Oct. 18, 1859	25, 815
Harvester	R. Eickemeyer	Yonkers, N. Y	Apr 12, 1870	101, 719
Harvester	G. S. Ellard	Westerly, R. I	July 6, 1869	92, 294
Harvester	M. E. Ellsworth	Hudson, Ohio	Aug. 10, 1858	21, 125
Harvester	J. H. Elward	Ottawa, Ill	Apr. 26, 1864	42, 532
Harvester	J. H. Elward	Ottawa, Ill	Nov. 1, 1864	44, 857
Harvester	J. H. Elward	Polo, Ill	June 14, 1870	104, 290
Harvester	J. H. Elward	Polo, Ill	Apr. 25, 1871	113, 9[illegible]0
Harvester	J. H. Elward	Polo, Ill	May 28, 1872	127, 324
Harvester	J. H. Elward	Saint Paul, Minn	Aug. 19, 1873	141, 863
Harvester	J. H. Elward	Polo, Ill	Aug. 19, 1873	141, 864
Harvester	J. H. Elward	Saint Paul, Minn	Aug. 19, 1873	141, 865
Harvester	J. H. Elward	Saint Paul, Minn	Aug. 19, 1873	141, 866
Harvester	J. H. Elward	Saint Paul, Minn	Aug. 19, 1873	141, 867
Harvester	J. H. Elward	Saint Paul, Minn	Aug. 19, 1873	141, 868
Harvester	P. Embree	West Chester, Pa	May 19, 1863	38, 626
Harvester	D. L. Emerson	Rockford, Ill	Dec. 10, 1861	33, 918
Harvester	D. L. Emerson	Rockford, Ill	Aug. 11, 1863	39, 526
Harvester	D. L. Emerson	Rockford, Ill	Aug. 18, 1863	39, 559
Harvester	D. L. Emerson	Rockford, Ill	Sept. 6, 1864	44, 084
Harvester	D. L. Emerson	Rockford, Ill	Sept. 10, 1867	68, 615
Harvester	R. Emerson, jr., and F. Graham	Rockford, Ill	Jan. 14, 1862	34, 181
Harvester	H. L. Emery	Albany, N. Y	Oct. 25, 1859	25, 888
Harvester	J. Emery	Cedar Falls, Iowa	June 25, 1867	66, 135
Harvester	E. Emmert	Franklin Grove, Ill	Jan. 19, 1858	19, 137
Harvester	E. Emmert	Franklin Grove, Ill	Jan. 28, 1868	73, 788
Harvester	E. Emmert	Franklin Grove, Ill	July 13, 1869	92, 435
Harvester	G. Engle	Bunker Hill, Wis	Mar. 3, 1863	37, 813
Harvester	D. W. Entrikin and L. H. Davis	West Chester, Pa	Apr. 13, 1858	19, 919
Harvester	G. Esterly	White Water, Wis	Jan. 22, 1861	31, 158
Harvester	G. Esterly	White Water, Wis	Jan. 22, 1861	31, 159
Harvester	G. Esterly	White Water, Wis	Jan. 13, 1863	37, 391
Harvester	G. Esterly	White Water, Wis	Feb. 18, 1868	74, 676
Harvester	F. Ewer	Mendon Centre, N. Y	Sept. 8, 1863	39, 804
Harvester	E. W. Fairman	Orfordville, Wis	Mar. 10, 1868	75, 401
Harvester	J. A. Falk, A. Johnson, and G. A. Erickson.	Altona, Ill	Aug. 30, 1859	25, 251
Harvester	G. C. Fanckboner	Schoolcraft, Mich	Nov. 28, 1865	51, 163
Harvester	G. Farmer	Oceola, Ill	Aug. 21, 1860	29, 685
Harvester	J. Farrington	Corry, Pa	June 20, 1871	116, 037
Harvester	J. Farrington	Corry, Pa	Jan. 7, 1873	134, 531
Harvester	D. S. Fisher	Cedar Spring, Ind	Jan. 14, 1868	73, 314
Harvester	H. Fisher	Alliance, Ohio	July 26, 1859	24, 860
Harvester	H. Fisher	Alliance, Ohio	Jan. 26, 1864	41, 411
Harvester	R. H. Fisher	Claremont, N. H	May 4, 1858	20, 152
Harvester	R. H. Fisher	Claremont, N. H	Oct. 19, 1858	21, 827
Harvester	B. G. Fitzhugh	Baltimore, Md	May 1, 1866	54, 466
Harvester	P. Flickinger	Hanover, Pa	Feb. 14, 1860	27, 120
Harvester	A. B. J. Flowers	Greenfield, Ind	Sept. 22, 1857	18, 240
Harvester	A. B. J. Flowers	Greenfield, Ind	June 1, 1858	20, 416
Harvester	A. Foot	Earlville, Ill	Apr. 13, 1869	88, 951
Harvester	E. T. Ford	Stillwater, N. Y	Jan. 6, 1863	37, 371
Harvester	E. T. Ford	Stillwater, N. Y	July 17, 1866	56, 397
Harvester	T. N. Foster	Watertown, N. Y	Oct. 2, 1860	30, 215
Harvester	J. L. and H. K. Fountain	Rockford, Ill	May 15, 1849	6, 450
Harvester	H. E. Fowler	North Branford, Conn	Aug. 18, 1868	81, 158
Harvester	J. Fox	Baltimore, Md	Oct. 15, 1867	69, 908
Harvester	J. Fox and S. C. Ridgway	Baltimore, Md	Apr. 17, 1866	53, 969
Harvester	H. H. Faye	Ottawa, Ill	Sept. 25, 1860	30, 135
Harvester	J. Q. A. Frazier	Piqua, Ohio	July 16, 1861	32, 828
Harvester	W. Gage	Buffalo, N. Y	Sept. 16, 1856	15, 735
Harvester	J. L. Garber	Hamilton, Ohio	Dec. 19, 1865	51, 579
Harvester	C. B. and G. B. Garlinghouse	Allensville, Ind	July 19, 1859	24, 803
Harvester	G. B. and C. B. Garlinghouse	Allensville, Ind	Oct. 28, 1862	36, 775
Harvester	J. Geiss, J. Brosius, and W. P. Penn.	Belleville, Ill	July 8, 1862	35, 815
Harvester	D. D. Gitt	Arendtsville, Pa	Apr. 12, 1864	42, 282
Harvester	D. D. Gitt	Arendtsville, Pa	Jan. 10, 1865	45, 823
Harvester	D. D. Gitt	Arendtsville, Pa	Feb. 27, 1866	52, 843
Harvester	C. W. Glover	Roxbury, Conn	Oct. 14, 1856	15, 882
Harvester	R. Glover	Grayville, Ill	May 12, 1863	38, 533
Harvester	W. H. Goble and A. Stuart	Urbana, Ohio	Nov. 21, 1848	5, 933
Harvester	W. F. Goodwin	Washington, D. C	Mar. 13, 1866	53, 137
Harvester	W. F. Goodwin	East New York, N. Y	Jan. 14, 1868	73, 424
Harvester	W. F. Goodwin	East New York, N. Y	June 9, 1868	78, 800
Harvester	W. F. Goodwin	East New York, N. Y	Mar. 30, 1869	88, 381
Harvester	W. F. Goodwin	Metuchen, N. J	Mar. 22, 1870	101, 120
Harvester	W. F. Goodwin	Metuchen, N. J	Feb. 7, 1871	111, 633
Harvester	W. F. Goodwin	Metuchen, N. J	Aug. 29, 1871	118, 526
Harvester	W. F. Goodwin and A. M. Smith	East New York, N. Y., and Washington, D. C.	Mar. 17, 1868	75, 677

Index of patents issued from the United States Patent Office from 1790 *to* 1873, *inclusive*—Continued.

Invention.	Inventor.	Residence.	Date.	No.
Harvester	H. S. Goudon	Steel's Mills, Ill	Oct. 8, 1872	132, 003
Harvester	J. F. Gordon	Rochester, N. Y	Aug. 27, 1872	130, 852
Harvester	J. Gore	Brattleborough, Vt	Dec. 27, 1859	26, 582
Harvester	A. B. Graham	Waukegan, Ill	July 23, 1867	67, 041
Harvester	A. B. Graham	Waukegan, Ill	Feb. 11, 1868	74, 342
Harvester	A. B. Graham	Waukegan, Ill	Apr. 7, 1868	76, 436
Harvester	A. B. Graham	Waukegan, Ill	Mar. 9, 1869	87, 561
Harvester	G. S. Grier	Milford, Del	Oct. 31, 1871	120, 431
Harvester	J. M. Groh	Shaefferstown, Pa	Oct. 28, 1862	36, 780
Harvester	C. P. Gronberg	Aurora, Ill	Mar. 16, 1869	87, 770
Harvester	S. Gumaer	Chicago, Ill	Aug. 4, 1857	17, 927
Harvester	S. B. Haines	Lewistown, Pa	Mar. 8, 1864	41, 842
Harvester	S. B. Haines	Lewistown, Pa	Jan. 7, 1868	72, 999
Harvester	J. C. Hall	Monroe, Wis	Mar. 2, 1869	87, 488
Harvester	L. Hall	Metamora, Ill	Sept. 6, 1870	107, 039
Harvester	M. Hallenbeck	Albany, N. Y	May 18, 1858	20, 271
Harvester	M. Hallenbeck	Albany, N. Y	May 18, 1858	20, 272
Harvester	M. Hallenbeck	Albany, N. Y	Nov. 3, 1868	83, 627
Harvester	M. Hallenbeck	Phillipsburgh, N. Y	July 23, 1872	129, 817
Harvester	M. Hallenbeck	Albany, N. Y	July 30, 1872	130, 041
Harvester	G. Hamel	Jenkintown, Pa	Sept. 3, 1872	130, 993
Harvester	H. D. Hammond	Batavia, N. Y	June 30, 1857	17, 676
Harvester	R. Hance	Piscatonica, Ill	Sept. 16, 1862	36, 460
Harvester	W. H. Harman	Westminster, Md	July 4, 1871	116, 588
Harvester	D. S. Harner	Xenia, Ohio	Nov. 19, 1867	71, 164
Harvester	J. Harris	Janesville, Wis	July 28, 1868	80, 479
Harvester	J. Harris	Janesville, Wis	July 9, 1872	128, 726
Harvester	J. K. Harris	Allensville, Ind	Oct. 12, 1858	21, 804
Harvester	L. W. Harris	Waterville, N. Y	Mar. 3, 1857	16, 730
Harvester	W. Harrison	Linneus, Mo	June 10, 1873	139, 787
Harvester	W. O. Harrison	Chittenden, Vt	July 21, 1868	80, 170
Harvester	A. J. Haswell and J. W. Irwin	Circleville, Ohio	Sept. 20, 1870	107, 487
Harvester	B. G. H. Hathaway	Rock Stream, N. Y	May 28, 1867	65, 218
Harvester	B. G. H. Hathaway	Rock Stream, N. Y	Feb. 4, 1868	74, 085
Harvester	B. G. H. Hathaway	Rock Stream, N. Y	Feb. 18, 1868	74, 685
Harvester	B. G. H. and G. M. Hathaway	Rock Stream, N. Y	May 28, 1867	65, 219
Harvester	J. S. and R. Hawkins	Greenfield, Ind	Aug. 23, 1859	25, 194
Harvester	G. F. Hawley	Grand Rapids, Mich	Dec. 27, 1870	110, 568
Harvester	A. A. Heath	West Greenville, Pa	May 9, 1865	47, 639
Harvester	B. Hess	Manor Township, Pa	Feb. 27, 1866	52, 851
Harvester	J. C. Henermann and J. Reeves	Camden, N. J	July 1, 1856	15, 236
Harvester	A. Hilts, jr	Springdale, Ohio	Oct. 1, 1867	69, 342
Harvester	I. I. Hite	White Post, Va	July 10, 1855	13, 226
Harvester	P. F. Hodges	Massillon, Ohio	Aug. 26, 1873	142, 231
Harvester	R. Hoffheins	Dover, Pa	May 20, 1862	35, 315
Harvester	R. Hoffheins	Dover, Pa	Nov. 3, 1863	40, 481
Harvester	L. B. Hoit and M. Laflin	Cedar Falls, Iowa, and Chicago, Ill.	July 20, 1869	92, 723
Harvester	C. Holloway	Petersburgh, Va	Mar. 17, 1857	16, 836
Harvester	S. T. Holly	Rockford, Ill	Mar. 25, 1862	34, 804
Harvester	S. T. Holly	Rockford, Ill	Apr. 29, 1862	35, 096
Harvester	S. T. Holly	Rockford, Ill	June 23, 1863	39, 014
Harvester	C. C. Holman	Clayville, N. Y	Jan. 23, 1866	52, 168
Harvester	G. W. Holmes	Council Bluffs, Iowa	Feb. 13, 1872	123, 631
Harvester	H. L. Hopkins	Lebanon, N. Y	Dec. 17, 1861	33, 943
Harvester	H. L. Hopkins	Eaton, N. Y	Dec. 12, 1871	121, 784
Harvester	J. Howard and E. T. Bonsfield	Bedford, England	Nov. 19, 1867	71, 175
Harvester	H. Howe	Oneonta, N. Y	Oct. 20, 1868	83, 164
Harvester	C. Howell	Cleveland, Ohio	Feb. 23, 1858	19, 422
Harvester	M. G. Hubbard	Penn Yan, N. Y	Jan. 20, 1857	16, 441
Harvester	M. G. Hubbard	Penn Yan, N. Y	Jan. 27, 1857	16, 484
Harvester	M. G. Hubbard	Penn Yan, N. Y	Mar. 17, 1857	16, 840
Harvester	M. G. Hubbard	Penn Yan, N. Y	May 12, 1857	17, 277
Harvester	M. G. Hubbard	Penn Yan, N. Y	May 12, 1857	17, 280
Harvester	M. G. Hubbard	Penn Yan, N. Y	Feb. 9, 1858	19, 298
Harvester	M. G. Hubbard	Penn Yan, N. Y	Feb. 15, 1859	22, 953
Harvester	M. G. Hubbard	Penn Yan, N. Y	Feb. 22, 1859	23, 029
Harvester	M. G. Hubbard	New York, N. Y	Nov. 15, 1859	26, 105
Harvester	M. G. Hubbard	Syracuse, N. Y	Sept. 13, 1864	44, 192
Harvester	M. G. Hubbard	Syracuse, N. Y	Sept. 22, 1868	82, 412
Harvester	M. G. Hubbard	Syracuse, N. Y	Sept. 22, 1868	82, 413
Harvester	M. G. Hubbard	Syracuse, N. Y	Sept. 22, 1868	82, 414
Harvester	M. G. Hubbard	Syracuse, N. Y	Sept. 22, 1868	82, 415
Harvester	S. Hull	Poughkeepsie, N. Y	Nov. 16, 1858	22, 077
Harvester	S. Hull	Poughkeepsie, N. Y	Dec. 2, 1862	37, 073
Harvester	S. Hull	Poughkeepsie, N. Y	Dec. 22, 1863	41, 040
Harvester	S. Hull	Poughkeepsie, N. Y	June 14, 1864	43, 116
Harvester	J. P. Humphreys	Nashville, Tenn	Apr. 16, 1872	125, 816
Harvester	S. R. Hunter	Cortland, N. Y	July 22, 1856	15, 377
Harvester	T. S. Hunter	Cross Plains, Iowa	Jan. 5, 1864	41, 071
Harvester	O. Hussey	Baltimore, Md	Aug. 23, 1859	25, 201
Harvester	J. H. Irwin	Beardstown, Ill	Aug. 21, 1860	29, 698
Harvester	G. M. Jackson	North Hector, N. Y	Dec. 10, 1867	72, 046
Harvester	G. M. Jackson	North Hector, N. Y	Mar. 30, 1869	88, 483
Harvester	J. Jann	New Windsor, Md	June 5, 1866	55, 304
Harvester	B. A. Jenkins	White Water, Wis	July 10, 1860	29, 084
Harvester	G. T. and M. Jerome	Mineola, N. Y	Oct. 5, 1858	21, 681
Harvester	I. Jewell	Wheaton, Ill	May 21, 1861	32, 400
Harvester	W. B. Johns and W. J. Read	Georgetown, D. C., and Cumberland, Md.	June 28, 1870	104, 852
Harvester	I. A. Johnson	Rockford, Ill	Feb. 28, 1871	112, 147
Harvester	L. H. Johnson	Rochester, Minn	Feb. 16, 1869	87, 048
Harvester	S. and L. Johnson, jr	Avon, N. Y	Dec. 8, 1857	18, 813
Harvester	W. N. Johnson	Oxford, Wis	Aug. 1, 1871	117, 639
Harvester	S. Johnston	Buffalo, N. Y	Oct. 28, 1862	36, 843
Harvester	S. Johnston	Buffalo, N. Y	Jan. 31, 1865	46, 190
Harvester	S. Johnston	Syracuse, N. Y	July 7, 1868	79, 575
Harvester	S. Johnston and C. H. Jenner	Brockport, N. Y	June 20, 1871	116, 063

Index of patents issued from the United States Patent Office from 1790 *to* 1873, *inclusive*—Continued.

Invention.	Inventor.	Residence.	Date.	No.
Harvester	E. Jones	Chester Cross Roads, Ohio	Mar. 27, 1860	27, 636
Harvester	J. H. Jones	Rockford, Ill	Feb. 23, 1869	87, 263
Harvester	J. H. Jones	Rockford, Ill	Feb. 23, 1869	87, 264
Harvester	J. H. Jones	Rockford, Ill	June 1, 1869	90, 666
Harvester	J. H. Jones	Rockford, Ill	Mar. 12, 1872	124, 587
Harvester	L. A. and G. J. Jones	Barrington, N. Y	May 21, 1867	64, 984
Harvester	W. Jones	Saint Louis, Mo	Dec. 8, 1863	40, 883
Harvester	A. and H. Kane	Newport, N. Y	Apr. 24, 1860	27, 989
Harvester	H. R. Keese	Bridport, Vt	Mar. 29, 1859	23, 376
Harvester	J. M. and M. L. Kellar	Buckeye, Iowa	Oct. 8, 1867	69, 679
Harvester	M. A. Keller	Littletown, Pa	Oct. 22, 1867	70, 001
Harvester	D. A. Kellogg	Valparaiso, Ind	July 4, 1871	116, 601
Harvester	P. H. Kells	Hudson, N. Y	Mar. 21, 1854	10, 671
Harvester	W. A. Kerby	Auburn, N. Y	Apr. 19, 1870	102, 131
Harvester	J. Kershaw	Kent, Ohio	June 15, 1869	91, 347
Harvester	W. F. Ketchum	Buffalo, N. Y	June 29, 1858	20, 719
Harvester	W. F. Ketchum	Buffalo, N. Y	Jan. 18, 1859	22, 655
Harvester	M. C. Kilgore	Washington, Iowa	Nov. 8, 1864	44, 956
Harvester	O. H. King	Salem, Iowa	Aug. 2, 1859	24, 941
Harvester	J. F. Kingwill	Chicago, Ill	Feb. 25, 1873	136, 325
Harvester	D. P. Kinyon	Raritan, N. J	Mar. 1, 1859	23, 090
Harvester	W. A. Kirby	Buffalo, N. Y	Nov. 15, 1859	26, 114
Harvester	W. A. Kirby	Buffalo, N. Y	May 15, 1860	28, 284
Harvester	W. A. Kirby	Auburn, N. Y	Apr. 5, 1864	42, 255
Harvester	W. A. Kirby	Buffalo, N. Y	Apr. 19, 1864	42, 440
Harvester	W. A. Kirby	Auburn, N. Y	May 18, 1869	90, 273
Harvester	W. A. Kirby	Auburn, N. Y	June 25, 1872	128, 402
Harvester	J. Kline	Mechanicsburgh, Pa	May 10, 1870	102, 835
Harvester	G. S. Knapp	Dubuque, Iowa	Dec. 24, 1861	33, 997
Harvester	I. Knauer	Valley Forge, Pa	Apr. 28, 1857	17, 157
Harvester	J. M. Knepley	Jersey Shore, Pa	Nov. 10, 1868	83, 975
Harvester	L. G. Kniffen	North Salem, N. Y	Dec. 13, 1859	26, 437
Harvester	L. G. Kniffen	Worcester, Mass	Apr. 12, 1864	42, 296
Harvester	L. G. Kniffen	Worcester, Mass	Apr. 12, 1864	42, 331
Harvester	L. G. Kniffen	Worcester, Mass	Aug. 20, 1867	67, 882
Harvester	L. G. Kniffen	Worcester, Mass	Aug. 20, 1867	67, 883
Harvester	L. G. Kniffen	Worcester, Mass	Aug. 20, 1867	67, 884
Harvester	L. G. Kniffen	Worcester, Mass	Aug. 20, 1867	67, 885
Harvester	A. E. Kroger	Norwalk, Conn	Sept. 4, 1855	13, 524
Harvester	E. M. Krum	Nassau, N. Y	May 1, 1866	54, 372
Harvester	M. Laflin	Chicago, Ill	July 2, 1872	128, 638
Harvester	S. I. Lamb and W. N. Whiteley	Springfield, Ohio	July 12, 1870	105, 343
Harvester	I. Lancaster	Baltimore, Md	Feb. 18, 1868	74, 552
Harvester	I. Lancaster	Baltimore, Md	Mar. 24, 1868	75, 773
Harvester	I. Lancaster	Baltimore, Md	Mar. 24, 1868	75, 774
Harvester	I. Lancaster	Baltimore, Md	Aug. 10, 1869	93, 627
Harvester	F. Landon	Brockport, N. Y	Nov. 13, 1860	30, 655
Harvester	F. Landon	Brockport, N. Y	May 28, 1861	32, 446
Harvester	C. Lee	Sandy, Ohio	June 25, 1867	66, 154
Harvester	G. W. Leisher	New Wilmington, Pa	June 17, 1873	139, 900
Harvester	E. E. Lewis	Geneva, N. Y	Feb. 25, 1868	74, 835
Harvester	E. I. Leyburn	Lexington, Va	May 12, 1868	77, 897
Harvester	C. Lidren	La Fayette, Ind	Sept. 29, 1868	82, 535
Harvester	C. Lidren	La Fayette, Ind	May 28, 1872	127, 252
Harvester	C. Lidren	La Fayette, Ind	Feb. 4, 1873	135, 480
Harvester	S. K. Lighter	Hamilton, Ohio	Nov. 3, 1863	40, 488
Harvester	S. K. Lighter and J. Curtis	Hamilton, Ohio	Dec. 15, 1868	84, 887
Harvester	S. A. Lindsay	Unionville, Md	Aug. 2, 1859	24, 944
Harvester	S. D. Locke	Janesville, Wis	Aug. 6, 1872	130, 227
Harvester	J. Long	Massillon, Ohio	Dec. 29, 1857	18, 981
Harvester	J. M. Long	Hamilton, Ohio	Sept. 4, 1860	29, 938
Harvester	J. M. Long	Hamilton, Ohio	Apr. 21, 1863	38, 232
Harvester	J. M. Long, P. Black, and R. Allstatter.	Hamilton, Ohio	Mar. 23, 1858	19, 703
Harvester	G. Lord	Watertown, N. J	Apr. 12, 1859	23, 592
Harvester	I. S. Love	Beloit, Wis	Oct. 7, 1856	15, 855
Harvester	W. R. Low	Sandwich, Ill	Nov. 12, 1872	132, 970
Harvester	W. R. Low and A. Adams	Sandwich, Ill	Feb. 1, 1870	99, 451
Harvester	A. Lowmiller	Jewett, Ohio	Feb. 20, 1866	52, 724
Harvester	S. Luce	Syracuse, N. Y	Aug. 19, 1873	141, 938
Harvester	D. B. Luckey	Bloomingburgh, N. Y	Jan. 6, 1863	37, 351
Harvester	D. B. Luckey	Bloomingburgh, N. Y	Sept. 13, 1864	44, 259
Harvester	C. M. Lufkin	Acworth, N. H	Sept. 8, 1857	18, 173
Harvester	G. G. Lyman	Independence, Iowa	Apr. 21, 1868	77, 058
Harvester	G. G. Lyman	Independence, Iowa	June 22, 1869	91, 757
Harvester	A. J. Manny	Freeport, Ill	Nov. 28, 1865	51, 203
Harvester	F. H. Manny	Rockford, Ill	Aug. 28, 1860	29, 800
Harvester	F. H. Manny	Rockford, Ill	Oct. 2, 1860	30, 235
Harvester	F. H. Manny	Rockford, Ill	June 23, 1863	38, 969
Harvester	F. H. Manny	Rockford, Ill	June 23, 1863	38, 970
Harvester	J. H. Manny	Waddam's Grove, Ill	Nov. 23, 1852	9, 423
Harvester	J. H. Manny	Rockford, Ill	Jan. 1, 1856	14, 026
Harvester	J. P. Manny	Rockford, Ill	July 14, 1857	17, 779
Harvester	J. P. Manny	Rockford, Ill	July 6, 1858	20, 806
Harvester	J. P. Manny	Rockford, Ill	Mar. 25, 1862	34, 760
Harvester	J. P. Manny	Rockford, Ill	Mar. 25, 1862	34, 761
Harvester	J. P. Manny	Rockford, Ill	Mar. 25, 1862	34, 762
Harvester	J. P. Manny	Rockford, Ill	Mar. 31, 1863	38, 053
Harvester	J. P. Manny	Rockford, Ill	Mar. 31, 1863	38, 054
Harvester	J. P. Manny	Rockford, Ill	June 15, 1869	91, 246
Harvester	J. P. Manny	Rockford, Ill	Mar. 7, 1871	112, 361
Harvester	J. P. Manny	Rockford, Ill	Mar. 7, 1871	112, 362
Harvester	J. P. Manny	Rockford, Ill	Mar. 21, 1871	112, 942
Harvester	J. P. Manny	Rockford, Ill	Apr. 16, 1872	125, 746
Harvester	J. P. Manny	Rockford, Ill	Jan. 21, 1873	135, 041
Harvester	P. Manny	Stephenson County, Ill	June 26, 1849	6, 560
Harvester	P. Manny	Waddam's Grove, Ill	Oct. 21, 1856	15, 927
Harvester	P. Manny	Waddam's Grove, Ill	Jan. 20, 1857	16, 445

Index of patents issued from the United States Patent Office from 1790 to 1873, inclusive—Continued.

Invention.	Inventor.	Residence.	Date.	No.
Harvester	P. Manny	Waddam's Grove, Ill	Apr. 7, 1857	16, 984
Harvester	P. Manny	Waddam's Grove, Ill	Apr. 7, 1857	16, 985
Harvester	P. Manny	Waddam's Grove, Ill	Mar. 27, 1860	27, 642
Harvester	H. Marcellus	Amsterdam, N. Y	Apr. 20, 1858	19, 999
Harvester	C. W. and W. W. Marsh	Clinton, Ill	Feb. 14, 1865	46, 373
Harvester	C. W. and W. W. Marsh	Shabbona, Ill	Nov. 12, 1867	70, 730
Harvester	C. W. and W W. Marsh	Sycamore, Ill	June 18, 1872	127, 981
Harvester	J. S. Marsh	Lewisburgh, Pa	Nov. 16, 1858	22, 084
Harvester	J. S. Marsh	Lewisburgh, Pa	May 21, 1861	32, 375
Harvester	J. S. Marsh	Lewisburgh, Pa	Feb. 10, 1863	37, 630
Harvester	J. S. Marsh	Lewisburgh, Pa	Feb. 12, 1867	61, 944
Harvester	J. S. Marsh	Lewisburgh, Pa	Feb. 28, 1871	112, 263
Harvester	J. S. Marsh, E. and C. C. Shorkley, and P. Beaver.	Lewisburgh, Pa	Feb. 10, 1863	37, 631
Harvester	C. E. Mason	Elgin, Ill	Sept. 21, 1869	94, 966
Harvester	N. Maxson	Wilmington, Ohio	Nov. 13, 1860	30, 636
Harvester	J. McAleer	Chambersburgh, Pa	Oct. 11, 1859	25, 748
Harvester	J. McCafferty	Waterloo, Iowa	Feb. 22, 1870	100, 051
Harvester	C. N. McCandlish and J. C. Naginey.	West Rushville, Ohio	Apr. 8, 1873	137, 614
Harvester	R. W. McClelland	Springfield, Ill	Mar. 22, 1864	42, 009
Harvester	S. S. McColloch	Wheeling, W. Va	Apr. 22, 1873	138, 172
Harvester	J. A. McCormick	Versailles, Ky	Apr. 27, 1858	20, 080
Harvester	J. B. McCormick	Saint Louis, Mo	May 19, 1863	38, 596
Harvester	J. B. McCormick	Dayton, Ohio	Feb. 15, 1870	99, 929
Harvester	J. B. McCormick	Dayton, Ohio	Mar. 22, 1870	101, 031
Harvester	L. J. McCormick and W. R. Baker.	Chicago, Ill	June 21, 1870	104, 618
Harvester	L. J. McCormick and W. R. Baker.	Chicago, Ill	Nov. 1, 1870	108, 922
Harvester	L. J. McCormick and W. R. Baker.	Chicago, Ill	Oct. 10, 1871	119, 712
Harvester	L. J. McCormick and W. R. Baker.	Chicago, Ill	Oct. 17, 1871	119, 937
Harvester	L. J. McCormick, W. R. Baker, and L. Erpelding.	Chicago, Ill	Oct. 27, 1868	83, 520
Harvester	L. J. McCormick, W. B. Baker, and L. Erpelding.	Chicago, Ill	Apr. 27, 1869	89, 324
Harvester	L. J. McCormick, W. R. Baker, and L. Erpelding.	Chicago, Ill	June 14, 1870	104, 331
Harvester	L. J. McCormick and L. Erpelding.	Chicago, Ill	Jan. 15, 1867	61, 228
Harvester	L. J. McCormick, L. Erpelding, and W. R. Baker.	Chicago, Ill	Dec. 28, 1869	98, 394
Harvester	J. C. McDougal	Black Rock, N. Y	Jan. 26, 1864	41, 384
Harvester	W. McIntosh	Wilmington, Ill	Oct. 22, 1861	33, 533
Harvester	W. McLagan	Cuylerville, N. Y	June 15, 1852	9, 031
Harvester	D. S. McNamara	North Hoosick, N. Y	June 30, 1857	17, 691
Harvester	D. S. McNamara	North Hoosick, N. Y	Sept. 28, 1858	21, 612
Harvester	W. Meigs	Waynesville, Ohio	July 22, 1862	35, 945
Harvester	E. W. Merrill	West Buxton, Me	Mar. 17, 1868	75, 640
Harvester	Q. F. Messinger	Easton, Pa	Jan. 9, 1872	122, 630
Harvester	H. Merves	Binghamton, N. Y	July 5, 1870	104, 978
Harvester	W. Michael	Murrysville, Pa	Feb. 9, 1869	86, 774
Harvester	J. C. Miles	Bloomington, Wis	June 21, 1870	104, 482
Harvester	C. G. Miller	Springfield, Ohio	Sept. 10, 1867	68, 775
Harvester	C. G. Miller	Springfield, Ohio	Mar. 10, 1868	75, 444
Harvester	C. G. Miller and B. Kersting	Springfield, Ohio	Jan. 11, 1870	98, 700
Harvester	L. Miller	Canton, Ohio	May 4, 1858	20, 180
Harvester	L. Miller	Canton, Ohio	May 4, 1858	20, 181
Harvester	L. Miller	Canton, Ohio	May 4, 1858	20, 182
Harvester	L. Miller	Akron, Ohio	Nov. 24, 1868	84, 432
Harvester	L. and J. Miller	Canton, Ohio	July 5, 1859	24, 700
Harvester	L. and J. Miller	Canton, Ohio	July 8, 1862	35, 830
Harvester	W. K. Miller	Canton, Ohio	Apr. 6, 1858	19, 864
Harvester	W. K. Miller	Canton, Ohio	Feb. 8, 1859	22, 885
Harvester	W. K. Miller	Canton, Ohio	July 2, 1861	32, 707
Harvester	W. K. Miller	Canton, Ohio	Sept. 10, 1867	68, 642
Harvester	W. K. Miller	Canton, Ohio	June 3, 1873	139, 596
Harvester	W. K. Miller	Canton, Ohio	June 3, 1873	139, 597
Harvester	W. T. Miller	New Geneva, Pa	Oct. 25, 1870	108, 716
Harvester	J. Mitchell	Gosport, N. Y	July 6, 1858	20, 813
Harvester	J. A. Moore and A. H. Patch	Louisville, Ky	Mar. 29, 1869	23, 383
Harvester	S. M. Moore	Beloit, Wis	Nov. 15, 1864	45, 067
Harvester	J. Moran and C. D. Wallace	Auburn, N. Y., and Corry, Pa.	Feb. 7, 1871	111, 557
Harvester	D. S. Morgan	Brockport, N. Y	Jan. 22, 1861	31, 183
Harvester	R. Morris	Salem, Ind	Nov. 9, 1869	96, 606
Harvester	W. Morrison	Carlisle, Pa	Nov. 22, 1859	26, 198
Harvester	C. Moul	Hanover, Pa	Jan. 28, 1868	73, 914
Harvester	D. Mulock	Mount Hope, N. Y	Feb. 28, 1871	112, 166
Harvester	W. Needham	Rockford, Ill	Aug. 28, 1866	57, 552
Harvester	W. Needham and J. Nelson	Rockford, Ill	Feb. 1, 1865	46, 486
Harvester	W. Neff	Centre Hall, Pa	Feb. 18, 1868	74, 577
Harvester	H. H. Nestestu	Deerfield, Wis	Sept. 6, 1870	107, 193
Harvester	C. Newhall	Hinsdale, N. H	July 29, 1862	36, 017
Harvester	P. Nicola	Massillon, Ohio	Nov. 21, 1871	121, 061
Harvester	F. Nishwitz	Brooklyn, N. Y	Feb. 16, 1858	19, 377
Harvester	F. Nishwitz	Brooklyn, N. Y	Feb. 21, 1865	46, 488
Harvester	F. Nishwitz	Brooklyn, N. Y	Jan. 28, 1868	73, 826
Harvester	F. Nishwitz	Brooklyn, N. Y	Jan. 28, 1868	73, 827
Harvester	F. Nishwitz	Brooklyn, N. Y	Sept. 28, 1869	95, 372
Harvester	F. Nishwitz	Brooklyn, N. Y	Sept. 28, 1869	95, 373
Harvester	D. G. Norris	Saint Johnsbury, Vt	June 19, 1866	55, 780
Harvester	H. Opp	Belleville, Ill	Dec. 7, 1858	22, 237
Harvester	D. M. Osborne	Auburn, N. Y	June 19, 1866	55, 781
Harvester	C. N. Owen	Salem, Ohio	May 24, 1870	103, 490
Harvester	E. F. Page	Brooklyn, N. Y	July 25, 1865	48, 988

Index of patents issued from the United States Patent Office from 1790 *to* 1873, *inclusive*—Continued.

Invention.	Inventor.	Residence.	Date.	No.
Harvester	D. R. Paiste	Williamstown, Pa	June 5, 1866	55, 427
Harvester	A. Palmer	Brockport, N. Y	May 1, 1866	54, 396
Harvester	I. H. Palmer	Lodi, Wis	Dec. 22, 1863	41, 016
Harvester	I. H. Palmer	Lodi, Wis	Oct. 20, 1868	83, 304
Harvester	I. H. Palmer	Lodi, Wis	Aug. 2, 1870	105, 973
Harvester	J. and E. Panton	London, Canada	Oct. 22, 1872	132, 408
Harvester	L. F. Parker	Davenport, Iowa	Dec. 1, 1868	84, 507
Harvester	H. A. Parkhurst	Fairfield, N. Y	Feb. 23, 1858	19, 442
Harvester	E. G. Passmore, jr	Philadelphia, Pa	Apr. 10, 1866	53, 862
Harvester	E. G. Passmore, jr	Philadelphia, Pa	Sept. 15, 1868	82, 156
Harvester	J. W. Patterson and L. H. Colborn.	Baltimore, Md	Mar. 8, 1859	23, 190
Harvester	N. A. Patterson	Kingston, Tenn	Oct. 13, 1857	18, 405
Harvester	J. L. Paxson	Norristown, Pa	Apr. 10, 1860	27, 826
Harvester	H. Pease	Brockport, N. Y	July 30, 1867	67, 211
Harvester	E. Peck	San José, Cal	Feb. 14, 1860	27, 151
Harvester	E. A. Peck	Sycamore, Ill	July 16, 1872	128, 974
Harvester	H. W. Pell	Rome, N. Y	Dec. 10, 1867	71, 906
Harvester	H. W. Pell	Rome, N. Y	Jan. 21, 1868	73, 641
Harvester	H. W. Pell	Rome, N. Y	Jan. 21, 1868	73, 642
Harvester	W. P. Penn	Belleville, Ill	Sept. 10, 1861	33, 263
Harvester	W. P. Penn	Belleville, Ill	Sept. 27, 1864	44, 452
Harvester	W. P Penn, J. Geiss, and J. Brosius.	Belleville, Ill	July 8, 1862	35, 836
Harvester	W. P. Penn, J. Geiss, and J. Brosius.	Belleville, Ill	Oct. 27, 1863	40, 427
Harvester	S. Pennock	Kennett's Square, Pa	Sept. 22, 1857	18, 267
Harvester	J. G. Perry	South Kingston, R. I	May 22, 1866	54, 946
Harvester	J. G. Perry	Kingston, R. I	Feb. 2, 1869	86, 584
Harvester	J. G. Perry	Kingston, R. I	May 1, 1869	90, 02[illegible]
Harvester	J. G. Perry	Kingston, R. I	June 21, 1870	104, 489
Harvester	J. G. Perry	Kingston, R. I	June 21, 1870	104, 490
Harvester	J. G. Perry	Kingston, R. I	Aug. 13, 1872	130, 386
Harvester	J. G. Perry	Kingston, R. I	Sept. 16, 1873	142, 811
Harvester	S. Persons and A. M. Cone	Panama, N. Y	Feb. 1, 1859	22, 824
Harvester	G. M. Peters, jr	Granville, Ohio	Sept. 17, 1867	69, 018
Harvester	C. J. C. Petersen	Port Chester, N. Y	Sept. 19, 1871	119, 172
Harvester	J. J. Piggott and C. L. Crowell.	Belleville and Peoria, Ill	Apr. 7, 1863	38, 123
Harvester	J. Pine	Troy, N. Y	May 7, 1861	32, 267
Harvester	J. Pine	Troy, N. Y	Oct. 22, 1861	33, 557
Harvester	J. Pine	Troy, N. Y	Oct. 27, 1863	40, 448
Harvester	J. Pine	Troy, N. Y	Apr. 26, 1864	42, 506
Harvester	J. Pine	Troy, N. Y	Mar. 5, 1867	62, 681
Harvester	N. Platt	Ottawa, Ill	June 12, 1849	6, 517
Harvester	T. E Platt	Newtown, Conn	Dec. 31, 1867	72, 756
Harvester	J. C. and L. C. Pluche	Cape Vincent, N. Y	June 17, 1856	15, 146
Harvester	T. Plumleigh	Dundee, Ill	Jan. 30, 1872	123, 291
Harvester	J. Powers and E. M. Smith	New York, N. Y	Jan. 14, 1862	34, 186
Harvester	J. W. Prentiss	Pultney, N. Y	Nov. 28, 1865	51, 211
Harvester	J. W. Prentiss and E. M. Birdsall.	Penn Yan, N. Y	June 21, 1864	43, 225
Harvester	J. B. Pressey, M. A. Wheaton, and D. Sheets.	Dubuque County, Iowa, Suisun City, Cal., and Pike County, Mo.	July 12, 1864	43, 525
Harvester	K. H. C. Preston	Manlius, N. Y	Dec. 17, 1861	33, 975
Harvester	K. H. C. Preston	Manlius, N. Y	Oct. 15, 1867	69, 838
Harvester	K. H. C. Preston	Manlius, N. Y	Sept. 1, 1868	81, 686
Harvester	A. Pritz	Dayton, Ohio	Apr. 16, 1861	32, 080
Harvester	S. N. Purse	Ashley, Mo	Dec. 27, 1859	26, 616
Harvester	G. Pye	Boston, Mass	June 15, 1869	91, 482
Harvester	O. Qwick, W. S. Opie, and A. J. Farrand.	Raritan, N. J	Dec. 24, 1867	72, 678
Harvester	W. K. Rairigh	Plumville, Pa	June 3, 1873	139, 611
Harvester	W. K. Rairigh	Plumville, Pa	June 3, 1873	139, 612
Harvester	A. Ralston	West Middletown, Pa	Mar. 8, 1859	23, 194
Harvester	H. B. Ramsey	Indianapolis, Ind	Oct. 11, 1859	25, 761
Harvester	J. Ramsey	Milford, Tex	May 5, 1868	77, 524
Harvester	A. Rank	Salem, Ohio	Apr. 4, 1865	47, 128
Harvester	A. Rank	Salem, Ohio	Feb. 12, 1867	61, 952
Harvester	A. Rank	Salem, Ohio	Feb. 12, 1867	61, 953
Harvester	A. Rank	Salem, Ohio	Nov. 5, 1867	70, 614
Harvester	A. Rank	Salem, Ohio	May 5, 1868	77, 525
Harvester	A. Rank	Salem, Ohio	May 5, 1868	77, 526
Harvester	A. Rank	Salem, Ohio	June 30, 1868	79, 500
Harvester	A. Rank	Salem, Ohio	Mar. 23, 1869	88, 209
Harvester	A. Rank	Salem, Ohio	Nov. 2, 1869	96, 350
Harvester	A. Rank	Salem, Ohio	Nov. 2, 1869	96, 351
Harvester	A. Rank	Salem, Ohio	Nov. 2, 1869	96, 352
Harvester	A. Rank	Salem, Ohio	July 19, 1870	105, 597
Harvester	A. Rank	Salem, Ohio	Nov. 7, 1871	120, 669
Harvester	A. Rank and J. H. Cox	Salem, Ohio	Oct. 13, 1868	82, 987
Harvester	A. Rank and J. H. Cox	Salem, Ohio	Feb. 2, 1869	86, 449
Harvester	D. Rauck	Intercourse, Pa	Mar. 1, 1859	23, 115
Harvester	S. and I. Rawson	Almond, N. Y	Nov. 29, 1870	109, 755
Harvester	S. and I. Rawson	Almond, N. Y	July 25, 1871	117, 460
Harvester	B. F. Ray	Baltimore, Md	Feb. 5, 1856	14, 205
Harvester	B. F. Ray	Baltimore, Md	Feb. 15, 1859	22, 977
Harvester	B. F. Ray	Baltimore, Md	Jan. 21, 1862	34, 215
Harvester	S. Ray and E. Grant	Alliance, Ohio	Nov. 14, 1865	50, 956
Harvester	H. W. Read	West Windsor Vt	Nov. 9, 1858	22, 032
Harvester	W. T. B. Read	Alton, Ill	June 2, 1857	17, 451
Harvester	G. Rector	Sodus, Mich	June 2, 1868	78, 482
Harvester	J. H. Redfield and W. J. Cox	Salem, Ind	Feb. 25, 1868	74, 847
Harvester	O. Redmond	Rochester, N. Y	Apr. 10, 1866	53, 872
Harvester	J. H. and A. E. Redstone	Indianapolis, Ind	Aug. 19, 1862	36, 258
Harvester	C. A. Reed and J. M. Campbell	Madison and Beaver Dam, Wis.	Aug. 10, 1869	93, 641
Harvester	A. R. Reese	Phillipsburgh, N. J	May 14, 1867	64, 797

Index of patents issued from the United States Patent Office from 1790 *to* 1873, *inclusive*—Continued.

Invention.	Inventor.	Residence.	Date.	No.
Harvester	L. C. Reese	Phillipsburgh, N. J	Apr. 10, 1860	27, 832
Harvester	J. Reily	Hart Prairie, Wis	Jan. 8, 1856	14, 079
Harvester	B. W. S. and C. H. Ressler	Tipton, Ind	June 10, 1873	139, 817
Harvester	S. Rex	Orefield, Pa	Apr. 7, 1868	76, 522
Harvester	G. S. Reynolds	Tunbridge, Vt	July 23, 1861	32, 915
Harvester	I. Reynolds	Republic, Ohio	June 13, 1854	11, 086
Harvester	E. B. Rice	Madison, Wis	Apr. 5, 1870	101, 659
Harvester	E. B. Rice	Madison, Wis	Nov. 14, 1871	121, 005
Harvester	G. W. Richardson	Grayville, Ill	June 25, 1861	32, 669
Harvester	G. W. Richardson	Grayville, Ill	Mar. 4, 1862	34, 614
Harvester	G. W. Richardson	Grayville, Ill	Dec. 6, 1864	45, 345
Harvester	G. W. Richardson	Grayville, Ill	Nov. 7, 1865	50, 844
Harvester	G. W. Richardson and R. Glover.	Grayville, Ill	Jan. 25, 1859	22, 772
Harvester	A. Rickart	Schoharie, N. Y	Feb. 11, 1873	135, 729
Harvester	G Riemer	Fayette, N. Y	Nov. 12, 1867	70, 744
Harvester	M. H. Ripley	Minneapolis, Minn	Jan. 4, 1870	98, 635
Harvester	M. H. Ripley and W. N. Temple.	Minneapolis, Minn	Mar. 30, 1869	88, 511
Harvester	C. Roberts	Livonia, N. Y	Feb. 23, 1858	19, 447
Harvester	R. A. Roberts	Salisbury, Mo	Mar. 18, 1873	136, 940
Harvester	S. Rockafellow	Muscatine, Iowa	May 26, 1868	78, 236
Harvester	J. L. Rohrer and C. E. Roper	Upper Leacock, Pa., and Canton, Ohio.	Dec. 19, 1871	122, 062
Harvester	W. Rose	Lake, Ill	Feb. 18, 1868	74, 600
Harvester	A. M. Ross	Ilion, N. Y	Jan. 2, 1872	122, 492
Harvester	N. Rowe	Emmittsburgh, Md	Dec. 11, 1866	60, 427
Harvester	J. S. Royce	Cuylerville, N. Y	Apr. 1, 1862	34, 849
Harvester	G. H. Rugg	South Ottawa, Ill	June 8, 1852	9, 005
Harvester	C. Russell and W. K. Miller	Massillon, Ohio	Mar. 24, 1868	75, 797
Harvester	E. P. Russell	Manlius, N. Y	Nov. 12, 1861	33, 712
Harvester	E. P. Russell	Manlius, N. Y	Dec. 24, 1861	34, 014
Harvester	E. P. Russell	Manlius, N. Y	Mar. 4, 1862	34, 594
Harvester	E. P. Russell	Manlius, N. Y	Nov. 14, 1865	50, 959
Harvester	F. Russell	Boston, Mass	Aug. 14, 1855	13, 438
Harvester	H. R. Russell	New Market, Md	Sept. 1, 1863	39, 780
Harvester	I. S. and H. R. Russell	New Market, Md	Mar. 29, 1859	23, 399
Harvester	I. S. and H. R. Russell	New Market, Md	Aug. 20, 1867	67, 914
Harvester	I. S. and H. R. Russell	New Market, Md., and Woodbury, N. J.	Aug. 18, 1868	81, 215
Harvester	E. H. Rurve	Walcott, Iowa	Feb. 13, 1872	123, 645
Harvester	J. Rynearson	Farmington, Ill	Oct. 28, 1862	36, 800
Harvester	A. Saltsman, C. H. Charlesworth, and R. F. Osgood.	Avoca and Rochester, N. Y	Apr. 20, 1869	89, 082
Harvester	D. Sandford	Taylor, Ill	Dec. 6, 1859	26, 376
Harvester	W. and T. Schnebly	New York, N. Y	June 15, 1852	9, 038
Harvester	W. and T. Schnebly	Hackensack, N. J	Nov. 30, 1858	22, 203
Harvester	J. W. Schuckers	Philadelphia, Pa	Jan. 31, 1871	111, 482
Harvester	J. W. Schuckers	Philadelphia, Pa	Feb. 28, 1871	112, 288
Harvester	H. H. Scoville	Syracuse, N. Y	Apr. 12, 1859	23, 613
Harvester	J. Scoville	Buffalo, N. Y	Jan. 31, 1860	27, 010
Harvester	J. Scuder and E. Miller	Litiz, Pa	May 30, 1871	115, 376
Harvester	T. C. Sebring	Milford, Mich	Sept. 15, 1868	82, 253
Harvester	J. Seibel	Manlius, Ill	Dec. 27, 1864	45, 642
Harvester	J. Seibel	Manlius, Ill	Feb. 21, 1865	46, 502
Harvester	J. Seibel	Manlius, Ill	May 15, 1866	54, 784
Harvester	J. Seibel	Manlius, Ill	Nov. 10, 1868	84, 006
Harvester	J. Seibel	Manlius, Ill	June 7, 1870	104, 067
Harvester	J. F. Seiberling	Doylestown, Ohio	Oct. 15, 1861	33, 496
Harvester	J. F. Seiberling	Doylestown, Ohio	Nov. 1, 1864	44, 893
Harvester	J. F. Seiberling	Doylestown, Ohio	Nov. 1, 1864	44, 894
Harvester	J. F. Seiberling	Doylestown, Ohio	Dec. 5, 1865	51, 359
Harvester	J. F. Seiberling	Doylestown, Ohio	Jan. 30, 1866	52, 327
Harvester	J. F. Seiberling	Doylestown, Ohio	May 8, 1866	54, 611
Harvester	J. F. Seiberling	Akron, Ohio	Aug. 28, 1866	57, 583
Harvester	J. F. Seiberling	Akron, Ohio	June 8, 1869	91, 172
Harvester	J. F. Seiberling	Akron, Ohio	Aug. 1, 1871	117, 692
Harvester	J. F. Seiberling	Akron, Ohio	Jan. 21, 1873	135, 013
Harvester	J. F. Seiberling	Akron, Ohio	Apr. 1, 1873	137, 490
Harvester	J. F. Seiberling	Akron, Ohio	May 27, 1873	139, 268
Harvester	W. H. Seymour	Brockport, N. Y	Mar. 28, 1854	10, 707
Harvester	W. H. Seymour and D. S. Morgan.	Brockport, N. Y	Feb. 11, 1873	135, 731
Harvester	W. H. Seymour and H. Pease.	Brockport, N. Y	Sept. 9, 1856	15, 721
Harvester	W. H. Seymour and H. Pease.	Brockport, N. Y	Sept. 9, 1856	15, 722
Harvester	W. H. Seymour and H. Pease.	Brockport, N. Y	May 25, 1858	20, 394
Harvester	M. R. Shalters	Alliance, Ohio	Jan. 9, 1866	51, 975
Harvester	W. A. Sharpe	Syracuse, N. Y	July 27, 1869	93, 013
Harvester	J. W. Shaw	Concord, N. H	Apr. 27, 1869	89, 509
Harvester	M. and S. Shawver	Bellefontaine, Ohio	May 27, 1862	35, 397
Harvester	A. Shebanck	Cleveland, Ohio	Mar. 31, 1868	76, 255
Harvester	A. Sheline and E. Burke	Edon, Ohio	Dec. 7, 1869	97, 556
Harvester	J. A. Shephard	Portsmouth, Ohio	Sept. 10, 1872	131, 307
Harvester	N. C. Sherman and S. Lightcap	Hazle Green, Wis	Aug. 4, 1857	17, 942
Harvester	A. Sherwood	Auburn, N. Y	Oct. 5, 1869	95, 522
Harvester	A. and C. Sherwood	Auburn, N. Y	Nov. 8, 1870	109, 062
Harvester	J. H. Shireman	New York, N. Y	Aug. 5, 1873	141, 597
Harvester	H. J. Silvernale	Golden Lake, Wis	May 20, 1873	139, 199
Harvester	W. H. H. Sisum	Cherry Valley, N. Y	Aug. 24, 1869	94, 141
Harvester	E. W. Skinner	Madison, Wis	Aug. 18, 1868	81, 221
Harvester	W. D. Slack	Lewisburgh, Pa	Mar. 30, 1869	88, 334
Harvester	W. D. Slack	Lewisburgh, Pa	Oct. 5, 1869	95, 526
Harvester	J. Smalley	Boundbrook, N. J	Apr. 12, 1859	23, 618
Harvester	A Smith	Vienna Cross Roads, Ohio	Aug. 25, 1868	81, 423
Harvester	A. Smith	Springfield, Ohio	Oct. 12, 1869	95, 738
Harvester	A. B. Smith	Rochester, Pa	Sept. 1, 1868	81, 832
Harvester	A. L. Smith	Bristol Centre, N. Y	Dec. 21, 1869	98, 113
Harvester	B. H. Smith and G. W. Archer.	Ipswich, Mass	June 24, 1862	35, 712

Index of patents issued from the United States Patent Office from 1790 *to* 1873, *inclusive*—Continued.

Invention.	Inventor.	Residence.	Date.	No.
Harvester	E. Smith	Cold Spring Harbor, N. Y	Jan. 21, 1862	34, 219
Harvester	E. Smith	Cold Spring Harbor, N. Y	Apr. 22, 1862	35, 044
Harvester	E. H. Smith and H. G. Smith	Saint Albans, Vt., and Goshen, Conn.	May 19, 1868	78, 144
Harvester	E. M. Smith	New York, N. Y	Feb. 23, 1864	47, 746
Harvester	H. C. Smith	Cleveland, Ohio	Mar. 9, 1858	19, 590
Harvester	H. C. Smith	Cleveland, Ohio	May 11, 1858	20, 225
Harvester	J. B. Smith	Winfield, N. Y	Feb. 19, 1861	31, 486
Harvester	J. D. Smith	Lancaster, Ohio	Oct. 19, 1858	21, 854
Harvester	S. H. Smith	Magnolia, Ill	June 15, 1858	20, 593
Harvester	D. Snell	Little Falls, N. Y	July 29, 1862	36, 034
Harvester	L. D. Snook	Barrington, N. Y	June 2, 1868	78, 490
Harvester	J. H. Snyder	Killbuck, Ill	Mar. 29, 1864	42, 148
Harvester	W. W. Sowles	Manlius, N. Y	Aug. 7, 1866	57, 005
Harvester	G. H. Spaulding	Rockford, Ill	Dec. 17, 1867	72, 238
Harvester	G. H. Spaulding	Rockford, Ill	Feb. 21, 1871	111, 987
Harvester	G. H. Spaulding	Rockford, Ill	July 8, 1873	140, 738
Harvester	W. Sprague	Farnham, N. Y	July 14, 1868	79, 871
Harvester	P. H. Standish	Pacheco, Cal	Feb. 19, 1861	31, 490
Harvester	C. T. Stetson	Amherst, Mass	June 30, 1857	17, 705
Harvester	C. T. Stetson	Amherst, Mass	Nov. 2, 1858	21, 993
Harvester	W. S. Stetson	Baltimore, Md	Apr. 5, 1859	23, 508
Harvester	W. S. Stetson	Baltimore, Md	Aug. 14, 1860	29, 632
Harvester	W. H. Stevenson	Auburn, N. Y	Mar. 3, 1868	75, 070
Harvester	G. W. Stewart	Adairsville, Ga	Dec. 22, 1868	85, 254
Harvester	L. B. Stilson	Woodland, Minn	Oct. 6, 1868	82, 890
Harvester	O. Stoddard	Busti, N. Y	May 11, 1858	20, 221
Harvester	O. Stoddard	Busti, N. Y	Dec. 14, 1858	22, 312
Harvester	A. Stoler and S. A. Sisson	Bristol, Pa	May 15, 1860	28, 313
Harvester	C. S. Stone	Indianapolis, Ind	Nov. 4, 1873	144, 367
Harvester	G. Stone and J. P. Bullock	Beloit, Wis	Mar. 14, 1865	46, 830
Harvester	J. W. Stout	Raritan, N. J	June 3, 1862	35, 479
Harvester	J. W. Street	Salem, Ohio	July 15, 1862	35, 901
Harvester	H. Stuckey	Bucyrus, Ohio	Jan. 24, 1871	111, 272
Harvester	J. M. Swain	Howard, Ind	Feb. 26, 1867	62, 452
Harvester	D. M. Swartz and J. Kreamer	Millheim, Pa	May 12, 1863	38, 514
Harvester	A. J. Sweeney	Wheeling, W. Va	Sept. 3, 1872	131, 134
Harvester	W. A. Sweet	Syracuse, N. Y	Dec. 8, 1863	40, 864
Harvester	L. Swenson	North Cape, Wis	Aug. 17, 1869	93, 765
Harvester	Z. Swope	Lancaster, Pa	Feb. 25, 1873	136, 279
Harvester	C. R. and J. O. Taber	Salem, Ohio	June 30, 1868	79, 518
Harvester	E. Taylor	Memphis, Tenn	Apr. 1, 1873	137, 506
Harvester	T. H. Taylor	Jeffersonville, Ill	Oct. 5, 1869	95, 537
Harvester	D. H. Thayer	Ludlowville, N. Y	Mar. 24, 1868	76, 001
Harvester	D. H. Thayer	Ludlowville, N. Y	Feb. 13, 1872	123, 650
Harvester	J. B. Thomison	Lynchburgh, Tenn	Dec. 26, 1871	122, 201
Harvester	J. W. Thompson	Greenfield, Mass	Feb. 4, 1868	74, 016
Harvester	T. J. Tindall	New York, N. Y	Oct. 11, 1864	44, 679
Harvester	C. Tinker and J. A. Sprague	Mantua, Ohio	Aug. 4, 1857	17, 945
Harvester	J. B. Tinker	Plymouth, N. Y	Feb. 5, 1861	31, 339
Harvester	W. Tinker	Kelloggsville, Ohio	Aug. 19, 1856	15, 582
Harvester	W. Tinker	Kelloggsville, Ohio	Dec. 9, 1856	16, 194
Harvester	B. Titcomb	Baltimore, Md	Apr. 19, 1864	42, 416
Harvester	J. Toay	Mineral Point, Wis	Oct. 27, 1868	83, 568
Harvester	P. A. Tobey	Caton, N. Y	Mar. 15, 1870	100, 817
Harvester	C. Tompkins	Troy, N. Y	Mar. 6, 1866	53, 057
Harvester	J. S. Troxel	Mount Pleasant, Pa	May 11, 1858	20, 227
Harvester	I. V. Trump	Somerville, N. Y	Aug. 3, 1858	21, 093
Harvester	J. S. Truxell	Greensburgh, Pa	Aug. 30, 1870	106, 891
Harvester	J. S. Truxell	Greensburgh, Pa	Nov. 14, 1871	120, 911
Harvester	J. S. Truxell	Greensburgh, Pa	Mar. 11, 1873	136, 628
Harvester	W. O. Tubbs	Spring, Pa	Apr. 19, 1864	42, 420
Harvester	A. W. Tucker	Waxahachie, Tex	Nov. 12, 1867	70, 918
Harvester	B. H. Turner	Fremont, Nebr	May 2, 1871	114, 493
Harvester	S. W. Tyler	Greenwich, N. Y	Jan. 26, 1858	19, 218
Harvester	S. W. Tyler	Greenwich, N. Y	Mar. 29, 1859	23, 413
Harvester	S. W. Tyler	Greenwich, N. Y	Nov. 13, 1860	30, 651
Harvester	J. Urmy	Wilmington, Del	Mar. 10, 1863	37, 881
Harvester	W. Van Anden	Poughkeepsie, N. Y	Mar. 30, 1858	19, 803
Harvester	W. Van Anden	Poughkeepsie, N. Y	Dec. 24, 1861	34, 019
Harvester	W. Van Anden	Poughkeepsie, N. Y	July 1, 1862	35, 792
Harvester	W. Van Anden	Poughkeepsie, N. Y	Sept. 9, 1862	36, 430
Harvester	W. Van Anden	Poughkeepsie, N. Y	Aug. 11, 1863	39, 511
Harvester	J. Van De Water	Whitewater, Wis	Mar. 28, 1871	113, 225
Harvester	J. Van Doren	Somerville, N. J	Mar. 2, 1858	19, 522
Harvester	J. Van Doren	Somerville, N. J	Apr. 6, 1858	19, 884
Harvester	D. Van Kleeck	Cohocton, N. Y	Mar. 20, 1860	27, 583
Harvester	J. P. Van Sickle	Geneva, N. Y	Oct. 26, 1869	96, 170
Harvester	N. T. Veatch	Camden, Ill	July 16, 1872	129, 501
Harvester	H. C. Velie	Poughkeepsie, N. Y	July 19, 1870	105, 612
Harvester	H. C. Velie	Poughkeepsie, N. Y	Jan. 2, 1872	122, 501
Harvester	W. A. Vertrees	Winchester, Mo	Feb. 28, 1860	27, 323
Harvester	A. H. Wagner	Chicago, Ill	Oct. 4, 1870	108, 069
Harvester	A. H. Wagner	Chicago, Ill	Mar. 14, 1871	112, 750
Harvester	C. B. Wagner	Philadelphia, Pa	June 24, 1856	15, 204
Harvester	D. B. Waite	Spring Water, N. Y	June 1, 1858	20, 456
Harvester	L. Wallace	Leavenworth, Kans	Mar. 17, 1868	75, 606
Harvester	W. Wallace	Syracuse, N. Y	Aug. 19, 1873	141, 970
Harvester	W. J. Wallis and W. E. Huttmann.	Chicago, Ill	June 15, 1869	91, 501
Harvester	T. and I. W. Ward	Lane Depot, Ill	July 7, 1863	39, 186
Harvester	R. Warner	Brattleborough, Vt	Mar. 1, 1859	23, 132
Harvester	D. Warren	Gettysburgh, Pa	Dec. 16, 1862	37, 187
Harvester	D. Watson	Petersburgh, Va	July 17, 1855	13, 285
Harvester	D. Watson	New York, N. Y	Mar. 3, 1857	16, 763
Harvester	W., jr., and J. Webber	Rockton, Ill	Dec. 22, 1857	18, 938
Harvester	G. Weiland	Buffalo, N. Y	May 26, 1868	78, 245

Index of patents issued from the United States Patent Office from 1790 *to* 1873, *inclusive*—Continued.

Invention.	Inventor.	Residence.	Date.	No.
Harvester	T. Welch	Churchville, N. Y	Aug. 1, 1865	49, 183
Harvester	T. Welch	Churchville, N. Y	June 16, 1868	78, 904
Harvester	H. Wells	Erieville, N. Y	Apr. 16, 1872	125, 711
Harvester	J. V. A. Wemple	Chicago, Ill	Apr. 19, 1859	23, 730
Harvester	J. Werner, jr	Prairie du Lac, Wis	June 17, 1873	139, 986
Harvester	P. Werni	Chicago, Ill	Nov. 19, 1867	71, 097
Harvester	C. Wheeler, jr	Poplar Ridge, N. Y	May 26, 1863	38, 708
Harvester	C. Wheeler, jr	Auburn, N. Y	Oct. 8, 1867	69, 732
Harvester	C. Wheeler, jr	Auburn, N. Y	Apr. 7, 1868	76, 571
Harvester	C. Wheeler, jr	Auburn, N. Y	Apr. 21, 1868	77, 145
Harvester	C. Wheeler, jr	Auburn, N. Y	Nov. 17, 1868	84, 151
Harvester	C. Wheeler, jr., and C. Young	Auburn, N. Y	Apr. 4, 1871	113, 475
Harvester	J. Whitehead	Manchester, Va	Jan. 5, 1858	19, 055
Harvester	A. Whiteley	Springfield, Ohio	Apr. 24, 1855	12, 769
Harvester	A. Whiteley	Springfield, Ohio	Aug. 30, 1859	25, 297
Harvester	A. Whiteley	Springfield, Ohio	Mar. 3, 1868	75, 229
Harvester	W. N. Whiteley	Springfield, Ohio	Jan. 30, 1866	52, 351
Harvester	W. N. Whiteley	Springfield, Ohio	Aug. 13, 1867	67, 828
Harvester	W. N. Whiteley	Springfield, Ohio	Nov. 19, 1867	71, 255
Harvester	W. N. Whiteley	Springfield, Ohio	May 5, 1868	77, 691
Harvester	W. N. Whiteley	Springfield, Ohio	July 21, 1868	80, 258
Harvester	W. N. Whiteley	Springfield, Ohio	Aug. 11, 1868	81, 046
Harvester	W. N. Whiteley	Springfield, Ohio	Nov. 24, 1868	84, 399
Harvester	W. N. Whiteley	Springfield, Ohio	Dec. 15, 1868	85, 045
Harvester	W. N. Whiteley	Springfield, Ohio	June 14, 1870	104, 386
Harvester	W. N. Whiteley	Springfield, Ohio	July 12, 1870	105, 394
Harvester	W. N. Whiteley	Springfield, Ohio	July 12, 1870	105, 395
Harvester	W. N. Whiteley	Springfield, Ohio	Sept. 12, 1871	118, 996
Harvester	W. N. Whiteley, jr	Springfield, Ohio	Feb. 23, 1864	47, 734
Harvester	W. N. Whiteley, jr	Springfield, Ohio	Dec. 5, 1865	51, 374
Harvester	W. N. Whiteley, jr	Springfield, Ohio	Jan. 30, 1866	52, 349
Harvester	W. N. Whiteley, jr	Springfield, Ohio	Jan. 30, 1866	52, 350
Harvester	W. N. Whiteley, jr	Springfield, Ohio	Aug. 21, 1866	57, 419
Harvester	W. N. Whiteley, jr	Springfield, Ohio	Oct. 30, 1866	59, 302
Harvester	W. N. Whiteley, jr	Springfield, Ohio	Oct. 30, 1866	59, 303
Harvester	W. N. Whiteley, jr	Springfield, Ohio	Mar. 26, 1867	63, 192
Harvester	W. N. Whiteley, jr	Springfield, Ohio	May 14, 1867	64, 818
Harvester	W. N. Whiteley and S. T. Lamb.	Springfield, Ohio	Aug. 22, 1871	118, 315
Harvester	W. N. and A. Whiteley	Springfield, Ohio	Jan. 18, 1859	22, 684
Harvester	W. N., jr., and A. Whiteley	Springfield, Ohio	Dec. 17, 1867	72, 349
Harvester	J. H. Whitenack	Somerville, N. J	Mar. 6, 1866	53, 069
Harvester	D. D. Whitker	New York, N. Y	Aug. 30, 1870	106, 903
Harvester	J. A. Whitney	Maryland, N. Y	July 8, 1862	35, 852
Harvester	J. H. Whitney	Rochester, N. Y	Jan. 16, 1872	122, 750
Harvester	B. Wieland	Orangeville, Ill	Apr. 18, 7865	47, 351
Harvester	J. D. Wilber	Poughkeepsie, N. Y	Oct. 22, 1867	70, 059
Harvester	J. D. Wilber	Poughkeepsie, N. Y	Oct. 31, 1871	120, 407
Harvester	J. D. Wilber	Poughkeepsie, N. Y	May 13, 1873	138, 830
Harvester	J. D. Wilber	Poughkeepsie, N. Y	Sept. 16, 1873	142, 828
Harvester	D. H. Wilgus and S. M. Denny	Harveysburgh, Ohio	July 13, 1869	92, 500
Harvester	H. Willard and R. Ross	Vergennes, Vt	Nov. 3, 1857	18, 562
Harvester	J. S. Williams	Chicago, Ill	Dec. 20, 1864	45, 550
Harvester	R. M. Williams	Rockville, Md	Oct. 12, 1869	95, 864
Harvester	S. Williams	Stockton, Cal	June 15, 1858	20, 600
Harvester	J. C. and T. G. Wilson	Cedar Hill, Tex	Apr. 21, 1857	17, 123
Harvester	T. Windell	New Albany, Ind	June 8, 1858	20, 525
Harvester	M. J. Wine	Long Glade, Va	Oct. 27, 1868	83, 578
Harvester	C. P. Wing	Fayetteville, N. Y	May 28, 1861	32, 443
Harvester	C. P. Wing	Fayetteville, N. Y	Apr. 5, 1864	42, 248
Harvester	C. P. Wing	Fayetteville, N. Y	June 28, 1864	43, 361
Harvester	J. Winters and C. C. Gapen	Lacon, Ill	Dec. 10, 1867	71, 937
Harvester	C. B. Withington	Rock, Wis	Feb. 21, 1860	27, 252
Harvester	C. W. and B. F. Witt	Indianapolis, Ind	Aug. 4, 1868	80, 694
Harvester	C. W. and B. F. Witt	Indianapolis, Ind	July 12, 1870	105, 399
Harvester	D. Wolf	Lebanon, Pa	May 7, 1867	64, 615
Harvester	C. B. Wolgemuth	Mount Joy, Pa	Apr. 9, 1872	125, 512
Harvester	A. Wood	Henrietta, N. Y	Jan. 31, 1865	46, 170
Harvester	W. A. Wood	Hoosick Falls, N. Y	Dec. 29, 1857	19, 001
Harvester	W. A. Wood	Hoosick Falls, N. Y	Dec. 29, 1857	19, 002
Harvester	W. A. Wood	Hoosick Falls, N. Y	Feb. 22, 1859	23, 056
Harvester	W. A. Wood	Hoosick Falls, N. Y	Dec. 3, 1861	33, 862
Harvester	W. A. Wood	Hoosick Falls, N. Y	Apr. 15, 1862	34, 995
Harvester	W. A. Wood	Hoosick Falls, N. Y	June 10, 1862	35, 558
Harvester	W. A. Wood	Hoosick Falls, N. Y	Dec. 3, 1867	71, 832
Harvester	W. A. Wood	Hoosick Falls, N. Y	Feb. 25, 1873	136, 295
Harvester	W. A. Wood and J. M. Rosebrooks.	Hoosick Falls, N. Y	July 19, 1859	24, 836
Harvester	W. A. and W. A. Wood and J. M. Rosebrooks.	Hoosick Falls, N. Y	Jan. 3, 1871	110, 713
Harvester	W. A. and W. A. Wood and J. M. Rosebrooks.	Hoosick Falls, N. Y	Jan. 3, 1871	110, 714
Harvester	W. A. and W. A. Wood and J. M. Rosebrooks.	Hoosick Falls, N. Y	Jan. 3, 1871	110, 715
Harvester	J. Woodruff	Rahway, N. J	June 5, 1860	28, 625
Harvester	J. Woody	Mount Vernon, Ind	Oct. 12, 1858	21, 792
Harvester	J. Woody	Mount Vernon, Ind	Aug. 20, 1861	33, 120
Harvester	T. Y. Woolford	Romney, W. Va	Feb. 11, 1873	135, 741
Harvester	H. M. Yale	Cold Water, Mich	Mar. 25, 1873	137, 162
Harvester	T. Yates	Milwaukee, Wis	Sept. 23, 1873	143, 110
Harvester	B. Yeakel	Allentown, Pa	Feb. 23, 1858	19, 463
Harvester	G. W. N. Yost	Corry, Pa	June 9, 1868	78, 708
Harvester	G. W. N. Yost	Corry, Pa	June 9, 1868	78, 709
Harvester	G. W. N. Yost	Corry, Pa	June 9, 1868	78, 710
Harvester	G. W. N. Yost	Corry, Pa	Aug. 18, 1868	81, 241
Harvester	G. W. N. Yost	Corry, Pa	Aug. 25, 1868	81, 569
Harvester	G. W. N. Yost	Corry, Pa	Oct. 6, 1868	82, 782
Harvester	G. W. N. Yost	Corry, Pa	Dec. 15, 1868	84, 927

Index of patents issued from the United States Patent Office from 1790 *to* 1873, *inclusive*—Continued.

Invention.	Inventor.	Residence.	Date.	No.
Harvester	G. W. N. Yost	Corry, Pa	Dec. 22, 1868	85, 159
Harvester	G. W. N. Yost	Corry, Pa	Dec. 29, 1868	85, 349
Harvester	G. W. N. Yost	Corry, Pa	Jan. 5, 1869	85, 628
Harvester	G. W. N. Yost	Corry, Pa	Jan. 12, 1869	85, 887
Harvester	G. W. N. Yost	Corry, Pa	Mar. 23, 1869	88, 254
Harvester	G. W. N. Yost	Corry, Pa	Mar. 30, 1869	88, 533
Harvester	G. W. N. Yost	Corry, Pa	May 25, 1869	90, 621
Harvester	G. W. N. Yost	Corry, Pa	June 1, 1869	90, 909
Harvester	G. W. N. Yost	Corry, Pa	Apr. 15, 1873	137, 812
Harvester	G. W. N. Yost	Corry, Pa	Apr. 15, 1873	137, 813
Harvester	G. W. N. Yost	Corry, Pa	Apr. 15, 1873	137, 814
Harvester	G. W. N. Yost	Corry, Pa	Apr. 15, 1873	137, 815
Harvester	C. M. Young	Meadville, Pa	Jan. 3, 1871	110, 718
Harvester	C. M. Young	Meadville, Pa	Jan. 28, 1873	135, 394
Harvester	E. Young	Fayetteville, Mo	Oct. 1, 1867	69, 529
Harvester	M. Young, jr	Frederick, Md	Sept. 21, 1858	21, 587
Harvester	M. Young, jr	Frederick, Md	July 9, 1861	32, 807
Harvester	W. Zimmerman	Oskaloosa, Iowa	June 19, 1866	55, 758
Harvester	D. Zug	Shaefferstown, Pa	Oct. 4, 1859	25, 697
Harvester	D. Zug	Shaefferstown, Pa	June 3, 1862	35, 485
Harvester and binder	J. E. Nesen	Buffalo, N. Y	Dec. 13, 1853	10, 314
Harvester and binder, Combined	J. M. McKesson	Lincoln, Nebr	Feb. 28, 1871	112, 267
Harvester and binder, Combined grain	J. H. Whitney	Rochester, Minn	Dec. 6, 1870	109, 985
Harvester and binder, Grain	L. B. Stilson	Minneapolis, Minn	Feb. 15, 1870	99, 968
Harvester and binder, Grain	P. H. Watson and S. Renwick	Washington, D. C	Aug. 16, 1853	9, 930
Harvester and binder, Grain	W. Watson, E. S. Renwick, and P. H. Watson.	Chicago, Ill., and Washington, D. C.	May 13, 1851	8, 083
Harvester and husker combined, Corn	L. Devore	Victor, Iowa	Aug. 22, 1871	118, 217
Harvester and husker, Combined corn	J. McLeish	Chicago, Ill	Apr. 2, 1872	125, 318
Harvester and husker, Combined corn	S. Patton	Chatsworth, Ill	May 24, 1870	103, 362
Harvester and husker, Combined corn	A. Smith	Pontiac, Ill	Oct. 11, 1870	108, 197
Harvester and husker, Combined corn	T. F. Vincent	Rock Island, Ill	June 18, 1872	127, 999
Harvester and shocker, Corn	E. Culp	Hilliard Station, Ohio	Oct. 24, 1871	120, 249
Harvester and shocker, Corn	W. H. Karicofe	Harrisonburgh, Va	Aug. 3, 1869	93, 311
Harvester and thrasher, Combined	J. H. Adamson	Auburn, Australia	July 1, 1873	140, 396
Harvester and thrasher, Combined	W. G. and L. T. Davis	McMinnville, Oreg	Nov. 16, 1869	96, 896
Harvester and thrasher, Combined	J. D. Field	Keokuk, Iowa	Apr. 12, 1870	101, 849
Harvester and thrasher combined	D. Howell	Saint Helena, Cal	Jan. 25, 1870	99, 089
Harvester and thrasher combined	W. B. Hubard	Arrington Depot, Va	Dec. 31, 1872	134, 476
Harvester and thrasher, Combined	J. H. Robbins	Bethel, Oreg	July 16, 1872	129, 171
Harvester-attachment for dumping and shocking sheaves.	W. A. Sharp and C. L. McClung.	Tama City, Iowa	May 6, 1873	138, 536
Harvester-attachment for raking and binding grain.	H. A. Reid	Beaver Dam, Wis	June 28, 1870	104, 883
Harvester, Bean	N. Chappell	East Avon, N. Y	Feb. 7, 1865	46, 216
Harvester, Bean	J. Hinds	Hindsburgh, N. Y	Apr. 1, 1873	137, 370
Harvester, Bean	A. Hutchinson and S. I. Lynd	Albion, N. Y	Oct. 15, 1872	132, 210
Harvester, Bean	J. Ketcham and J. Waterman	Orangeport, N. Y	Nov. 24, 1857	18, 697
Harvester, Bean	D. B. Munger	Mumford, N. Y	Apr. 18, 1865	47, 320
Harvester, Bean	D. C. Rosier	Clarkson, N. Y	Sept. 18, 1866	58, 140
Harvester, Bean	J. A. and T. Wood	Chemung, Ill	Mar. 28, 1871	113, 125
Harvester-binder	J. Barta	La Crosse, Wis	May 16, 1871	114, 749
Harvester binder-attachment	J. H. Garnhart	Madison, Wis	Sept. 2, 1873	142, 331
Harvester binder-attachment	J. H. Garnhart	Madison, Wis	Oct. 7, 1873	143, 505
Harvester binder-attachment	S. Mills	Madison, Wis	Nov. 26, 1872	133, 473
Harvester binder-attachment	H. Porter	Polo, Ill	Nov. 18, 1873	144, 789
Harvester binder-attachment	C. G. Price	Amana, Iowa	Aug. 27, 1872	130, 865
Harvester binder-attachment	W. L. Sanford	Ashton, Ill	Sept. 16, 1873	142, 870
Harvester binder-attachment	T. Urdahl	Cross Plains, Wis	Nov. 18, 1873	144, 715
Harvester binder-attachment	T. Webb	Freeport, Ill	May 13, 1873	138, 771
Harvester binder, Grain	C. A. McPhetridge	Saint Louis, Mo	Nov. 18, 1856	16, 097
Harvester-binder stand and table	W. R. Low	Sandwich, Ill	Feb. 4, 1873	135, 481
Harvester binding-apparatus	C. H. McAleer	Chambersburgh, Pa	Aug. 16, 1859	25, 131
Harvester binding-apparatus	D. W. Travis and C. M. Clinton	Enfield and Ithaca, N. Y	Sept. 6, 1870	107, 126
Harvester binding-attachment	C. Alvord	Westford, Wis	Mar. 5, 1861	31, 584
Harvester binding-attachment	J. W. Bates	Saint Paul, Minn	Apr. 20, 1869	89, 194
Harvester binding-attachment	J. Behel and W. Hedges	Earlville, Ill	Sept. 6, 1864	44, 139
Harvester binding-attachment	R. D. Brown	Covington, Ind	June 27, 1865	48, 363
Harvester binding-attachment	H. M. and W. W. Burson	Atkinson, Ill	Mar. 10, 1863	37, 852
Harvester binding-attachment	A. Freed and J. Snook	La Porte, Ind	Jan. 10, 1871	110, 841
Harvester binding-attachment	W. Grey	Nicholsville, Ohio	Nov. 16, 1858	22, 074
Harvester binding-attachment	L. P. Harris	Mansfield, Ohio	Feb. 26, 1861	31, 540
Harvester binding-attachment	F. B. Isett	Hollidaysburgh, Pa	Aug. 2, 1870	105, 946
Harvester binding-attachment	C. Powers and P. Lancaster	Bronson, Mich	Oct. 29, 1861	33, 601
Harvester binding-attachment	B. W. Squires and C. M. Clinton	Ithaca, N. Y	Jan. 25, 1870	99, 256
Harvester binding-attachment	A. Underwood	Kenosha, Wis	Aug. 11, 1863	39, 510
Harvester binding-device	G. Notman	Deerfield, Ohio	May 11, 1858	20, 215
Harvester, Broom-corn	J. O. Norton	Wilton, Ill	Oct. 14, 1862	36, 668
Harvester, Cane	W. B. Robertson	West Baton Rouge, La	Feb. 12, 1861	31, 406
Harvester, Cane	A. St. Dizier	Plaquemine, La	Oct. 9, 1860	30, 361
Harvester caster-wheel	J. S. and H. R. Russell	New Market, Md., and Woodbury, N. J.	Sept. 28, 1869	95, 383
Harvester, Clover	J. R. Bruton	Concord, N. C	May 13, 1873	138, 786
Harvester, Clover	F. Decker	Ostrander, Ohio	May 28, 1861	32, 416
Harvester, Clover	F. Decker	Ostrander, Ohio	Feb. 28, 1865	46, 547
Harvester, Clover	P. Dismukes	Gallatin, Tenn	Nov. 24, 1868	84, 416
Harvester, Clover	P. Dismukes	Gallatin, Tenn	Mar. 23, 1869	88, 144
Harvester, Clover	J. S. Gage	Dowagiac, Mich	Dec. 19, 1854	12, 095
Harvester, Clover	M. Garretson	Bermudian, Pa	Jan. 6, 1852	8, 628
Harvester, Clover	D. Hinkle	New Pittsburgh, Ohio	July 2, 1861	32, 692
Harvester, Clover	J. Hinton	Monroe County, Va	May 22, 1849	6, 475
Harvester, Clover	J. W. Hull and A. G. Stiffler	Alquina, Ind	Feb. 7, 1871	111, 544
Harvester, Clover	E. Kramer	Alvira, Pa	May 1, 1866	54, 370
Harvester, Clover	J. Krauser	Reading, Pa	June 22, 1852	9, 055
Harvester, Clover	S. Krauser	Reading, Pa	Dec. 18, 1849	6, 954
Harvester, Clover	J. Lamburn	Boundary City, Ind	Sept. 26, 1871	119, 369
Harvester, Clover	C. McCombie	Carrolltown, Pa	May 14, 1872	126, 726
Harvester, Clover	T. S. Steadman	Murray, N. Y	May 23, 1854	10, 967

Index of patents issued from the United States Patent Office from 1790 *to* 1873, *inclusive*—Continued.

Invention.	Inventor.	Residence.	Date.	No.
Harvester, Clover	S. L. Stockstill and W. H. H. Scarff.	Medway, Ohio	Feb. 25, 1868	74,861
Harvester, Clover	J. A. Wagener	Pultney, N. Y.	May 24, 1853	9,750
Harvester, Clover	J. Westhafer	Quincy, Ohio	Jan. 12, 1869	85,880
Harvester, Clover-seed	C. B. Wheeler and A. Bascom	Steuben, Ohio	July 29, 1856	15,449
Harvester, Corn	I. V. Adair	Varick, N. Y	Apr. 6, 1858	19,822
Harvester, Corn	M. and W. P. Bales	London, Ohio	May 12, 1868	77,708
Harvester, Corn	M. and W. P. Bales	London, Ohio	Jan. 26, 1869	86,203
Harvester, Corn	B. F. Barney	Pontiac, Ill	May 4, 1869	89,546
Harvester, Corn	W. Beach	Baltimore, Md	Oct. 4, 1859	25,699
Harvester, Corn	J. H. Beam	Woodside, Ill	May 11, 1869	89,968
Harvester, Corn	G. W. S. Bell	Tallula, Ill	Sept. 7, 1869	94,549
Harvester, Corn	J. H. Besant and M. B. Atkinson	Point of Rocks, Md., and Georgetown, D. C.	Mar. 19, 1872	124,661
Harvester, Corn	W. M. Bonwill	Camden, Del	Mar. 4, 1856	14,344
Harvester, Corn	J. W. Bope	Saint Louis, Mo	Jan. 10, 1865	45,811
Harvester, Corn	E. Boswell	Highland, Ohio	May 3, 1868	75,119
Harvester, Corn	J. Bowers	Iola, Kans	Nov. 19, 1872	133,141
Harvester, Corn	T. Boyd	Des Moines, Iowa	Feb. 27, 1866	52,819
Harvester, Corn	I. V. Brower	Millstone, N. J	Aug. 13, 1861	33,026
Harvester, Corn	A. R. Burdick and C. D. Read	Racine, Wis., and Elgin, Ill	July 22, 1862	35,921
Harvester, Corn	J. Burke	Sycamore, Ill	Jan. 9, 1872	122,561
Harvester, Corn	T. Butterworth	Shelbyville, Mo	Feb. 6, 1866	52,384
Harvester, Corn	J. F. Byland	Walton, Ky	Dec. 7, 1869	97,599
Harvester, Corn	J. and W. Carrothers	Newton, Iowa	Dec. 24, 1872	134,251
Harvester, Corn	J. and W. Carrothers	Newton, Iowa	May 27, 1873	139,367
Harvester, Corn	J. L. Chapman	Kinmundy, Ill	Mar. 1, 1859	23,076
Harvester, Corn	O. N. Chase	Boston, Mass	Oct. 13, 1863	40,243
Harvester, Corn	W. Cogswell and C. A. Mathewson.	Ottawa, Ill	Aug. 9, 1859	24,992
Harvester, Corn	R. B. Corbin and J. Morris	Saint Augustine, Ill	June 22, 1858	20,628
Harvester, Corn	C. U. Crandall	Galesburgh, Ill	Dec. 31, 1867	72,811
Harvester, Corn	J. Crossland	Spencer Grove, Iowa	July 25, 1871	117,264
Harvester, Corn	J. H. Crip	Quincy, Ohio	Nov. 12, 1867	70,816
Harvester, Corn	B. T. Currier	Bath, Me	Aug. 9, 1859	24,994
Harvester, Corn	T. E. Curtis	Titusville, Pa	Aug 22, 1871	118,207
Harvester, Corn	J. Dodenhoff	Bloomington, Ill	July 7, 1857	17,729
Harvester, Corn	A. Dyson and W. N. Macqueen.	Saint Louis, Mo	June 26, 1866	55,840
Harvester, Corn	E. J. Eno	Jacksonville, Ill	Dec. 15, 1863	40,979
Harvester, Corn	E. J. Eno	Springfield, Ill	Sept. 28, 1869	95,214
Harvester, Corn	J. H. Fisher and C. Holcomb	Mendota, Ill	May 11, 1869	89,979
Harvester, Corn	A. W. Fleming	Springfield, Ill	Apr. 27, 1869	89,396
Harvester, Corn	H. Flescher	Springfield, Ill	Apr. 12, 1870	101,721
Harvester, Corn	D. C. Flint	Bardolph, Ill	May 23, 1871	115,046
Harvester, Corn	A. Ford	Toledo, Ohio	June 8, 1869	91,005
Harvester, Corn	B. M. Fowler	Brooklyn, N. Y	July 12, 1864	43,487
Harvester, Corn	F. F. Fowler	Upper Sandusky, Ohio	Aug. 14, 1866	57,113
Harvester, Corn	W. N. Gates	Manchester Centre, N. Y	July 30, 1867	67,287
Harvester, Corn	G. Geer	Douglas, Ill	June 2, 1863	38,738
Harvester, Corn	S. Gesley	Beloit, Wis	Aug. 6, 1872	130,211
Harvester, Corn	M. A. Getzendaner	Polo, Ill	May 3, 1870	102,532
Harvester, Corn	H. Gortner and J. McCann	Nashport, Ohio	Mar. 6, 1860	27,360
Harvester, Corn	L. Hamilton	Panora, Iowa	July 22, 1873	141,138
Harvester, Corn	L. B. Hamilton	Boston, Mass	Dec. 31, 1867	72,848
Harvester, Corn	S. H. Hamilton	Macomb, Ill	Aug. 27, 1861	33,142
Harvester, Corn	G. B. Hamlin	Willimantic, Conn	June 28, 1870	104,846
Harvester, Corn	J. D. Hampshire	Paper Mills Post-Office, Md	Oct. 20, 1868	83,155
Harvester, Corn	H. Harris	Circleville, Ohio	Jan. 5, 1869	85,585
Harvester, Corn	E. K. Harvey	Quincy, Ohio	Nov. 11, 1868	83,851
Harvester, Corn	G. D. Haworth	Mechanicsburgh, Ill	July 21, 1857	17,832
Harvester, Corn	A. Henlings	Philadelphia, Pa	Mar. 31, 1857	16,921
Harvester, Corn	W. Hill and J. A. Harpham	Havana, Ill	Oct. 29, 1867	70,335
Harvester, Corn	N. Hospers and M. Sellers	Pella and Keokuk, Iowa	Jan. 12, 1869	85,828
Harvester, Corn	W. B. Hubard	Arrington Depot, Va	June 15, 1869	91,444
Harvester, Corn	A. Humberger	Somerset, Ohio	Dec. 1, 1857	18,747
Harvester, Corn	A. Humberger	Somerset, Ohio	Apr. 27, 1858	20,067
Harvester, Corn	A. Humberger	Somerset, Ohio	Apr. 3, 1860	27,720
Harvester, Corn	J. Johnson	Mott Haven, N. Y	Dec. 26, 1871	122,176
Harvester, Corn	S. Johnston	Buffalo, N. Y	Oct. 28, 1862	36,848
Harvester, Corn	W. H. Kenyon	North Providence, R. I	Jan. 7, 1873	134,678
Harvester, Corn	J. Kilborn	Chicago, Ill	Sept. 16, 1873	142,920
Harvester, Corn	J. H. Kite	Conrad's Store, Va	Mar. 8, 1859	23,174
Harvester, Corn	D. Landon	Wyandot, Ohio	June 22, 1858	20,645
Harvester, Corn	C. M. Lightner	Harrisburgh, Pa	July 10, 1866	56,236
Harvester, Corn	R. B. Linthicum	Lexington, Ill	Aug. 21, 1866	57,345
Harvester, Corn	C. B. Maclay	Delavan, Ill	Nov. 9, 1869	96,602
Harvester, Corn	J. Mains	Olean, Ill	Feb. 4, 1868	74,104
Harvester, Corn	G. H. Manlove and J. P. Green.	Chicago, Ill	Sept. 25, 1866	58,348
Harvester, Corn	G. H. Manlove and J. P. Green.	Chicago, Ill	Sept. 25, 1866	58,349
Harvester, Corn	W. M. Mason	Polo, Ill	Mar. 29, 1864	42,097
Harvester, Corn	R. C. Mauck	Conrad's Store, Va	Jan. 4, 1859	22,508
Harvester, Corn	R. C. Mauck and W. T. McGahey.	Conrad's Store and McGaheysville, Va.	Apr. 22, 1856	14,730
Harvester, Corn	J. McLeish	Chicago, Ill	Sept. 21, 1869	95,031
Harvester, Corn	G. Meader	Prairie Centre, Ill	Apr. 28, 1868	77,394
Harvester, Corn	T. Merrell	Dixon, Ill	Mar. 19, 1872	124,757
Harvester, Corn	T. Merrell	Dixon, Ill	Nov. 26, 1872	133,327
Harvester, Corn	Q. F. Messinger	Easton, Pa	Feb. 25, 1868	74,928
Harvester, Corn	J. I. Mettler	Mendota, Ill	July 4, 1871	116,735
Harvester, Corn	H. Molby	Davisburgh, Mich	Jan. 21, 1868	73,629
Harvester, Corn	A. Moravek	Rosnyo, Hungary	Feb. 5, 1867	61,852
Harvester, Corn	C. G. Moremen	Brandenburgh, Ky	Aug. 21, 1866	57,360
Harvester, Corn	B. Murray and J. Van Doren	Ottawa and Farm Ridge, Ill	Dec. 7, 1858	22,259
Harvester, Corn	W. Murray	Alexandria, Va	Feb. 1, 1870	99,463
Harvester, Corn	R. L. Nelson	Orange Court-House, Va	Apr. 25, 1871	114,182
Harvester, Corn	R. L. Nelson	Orange Court-House, Va	June 13, 1871	115,978
Harvester, Corn	R. L. Nelson	Orange Court-House, Va	July 30, 1872	130,065
Harvester, Corn	N. Newman	Springfield, Ill	Sept. 15, 1868	82,148

Index of patents issued from the United States Patent Office from 1790 *to* 1873, *inclusive*—Continued.

Invention.	Inventor.	Residence.	Date.	No.
Harvester, Corn	N. Newman	Springfield, Ill	Nov. 30, 1869	97,429
Harvester, Corn	N. Newman	Springfield, Ill	July 26, 1870	105,716
Harvester, Corn	J. Oadhoudt	Saint Anthony's Falls, Minn	May 14, 1867	64,789
Harvester, Corn	S. Patton	Chatsworth, Ill	Aug. 8, 1868	81,202
Harvester, Corn	S. Patton	Chatsworth, Ill	Apr. 22, 1873	138,181
Harvester, Corn	J. and I. M. Poffenberger	Urbana, Ohio	June 29, 1869	91,964
Harvester, Corn	W. E. Prall	Knoxville, Tenn	May 22, 1866	54,951
Harvester, Corn	E. W. Quincy	Lacon, Ill	July 7, 1868	79,775
Harvester, Corn	E. W. Quincy	Peoria, Ill	June 24, 1873	140,305
Harvester, Corn	I. Reamer	Conrad's Store, Va	Jan. 11, 1859	22,583
Harvester, Corn	I. Reamer and H. Miller	Conrad's Store, Va	Sept. 14, 1858	21,516
Harvester, Corn	I. Reamer and H. Miller	Conrad's Store, Va	Apr. 26, 1859	23,783
Harvester, Corn	J. H. Rible	Somerset, Ohio	Dec. 1, 1857	18,769
Harvester, Corn	J. E. Rice	Moline, Ill	May 31, 1870	103,779
Harvester, Corn	G. W. Richardson and J. W. White.	Grayville, Ill	June 14, 1859	24,434
Harvester, Corn	T. A. Risher	Circleville, Ohio	Mar. 23, 1858	19,716
Harvester, Corn	T. A. Risher	Oskaloosa, Iowa	Nov. 17, 1863	40,641
Harvester, Corn	A. F. Roberts	Lexington, Ky	Sept. 19, 1871	119,181
Harvester, Corn	J. A. C. Rose	Carrollton, Ill	Feb. 16, 1869	87,071
Harvester, Corn	D. Sarvar and R. Coons	Greensburgh, Pa	May 5, 1868	77,660
Harvester, Corn	A. Scott	Rochester, N. Y	Aug. 5, 1873	141,466
Harvester, Corn	S. Secrist	West Liberty, Ohio	Sept. 18, 1866	58,143
Harvester, Corn	J. W. Sharrard, S. Bryan, and H. Hunt.	Janesville, Wis	Aug. 31, 1869	94,446
Harvester, Corn	J. Shobe	Principio, Md	Mar. 19, 1867	63,111
Harvester, Corn	H. L. Smith	Watkins, N. Y	Dec. 17, 1867	72,236
Harvester, Corn	J. W. Smith	Iowa Point, Kans	Aug. 2, 1864	43,716
Harvester, Corn	J. W. Smith	Iowa Point, Kans	May 30, 1865	47,995
Harvester, Corn	M. A. Smith	Middlefield, N. Y	Oct. 4, 1870	107,972
Harvester, Corn	J. H. Spears and K. and R. Wells.	Piper City, Ill	June 3, 1873	139,623
Harvester, Corn	A. Sprague	Cold Water, Mich	Aug. 12, 1856	15,533
Harvester, Corn	L. Stadler	Bowensburgh, Ill	Sept. 7, 1869	94,662
Harvester, Corn	A. Stoddard	Tecumseh, Mich	July 27, 1858	21,031
Harvester, Corn	T. H. Storms and J. C. Poffenberger.	Jacksonville, Ill	Nov. 29, 1864	45,282
Harvester, Corn	C. S. Stull	Poolesville, Md	Feb. 28, 1871	112,191
Harvester, Corn	J. B. Sweetland	Pontiac, Mich	Dec. 3, 1867	71,815
Harvester, Corn	J. B. Sweetland	Pontiac, Mich	Feb. 11, 1868	74,447
Harvester, Corn	M. Thorp	Warterloo, Iowa	Sept. 12, 1871	118,984
Harvester, Corn	W. S. Tilton	Boston, Mass	June 17, 1856	15,152
Harvester, Corn	D. J. Tittle	Albany, N. Y	Sept. 25, 1866	58,320
Harvester, Corn	G. M. Tope	Leesville, Mo	July 15, 1873	140,971
Harvester, Corn	J. H. L. Tuck	Ottawa, Ill	Feb. 1, 1870	99,501
Harvester, Corn	E. Turk	Manchester, Ill	Jan. 18, 1870	99,034
Harvester, Corn	J. M. Wallis and E. Miller	Milton, Iowa	Apr. 3, 1866	53,710
Harvester, Corn	S. Ward	Lane, Ill	July 11, 1865	48,750
Harvester, Corn	J. H. Whitney	Rochester, Minn	Nov. 7, 1871	120,688
Harvester, Corn	L. C. Wilder	Lexington, N. C	Jan. 26, 1858	19,221
Harvester, Corn	C. W. Williams	Wyandotte, Mich	Oct. 8, 1867	69,600
Harvester, Corn	H. Williams	Chicago, Ill	July 29, 1862	36,046
Harvester, Corn	J. S. Williams	Chicago, Ill	Dec. 12, 1865	51,505
Harvester, Corn	J. F. Winchell	Springfield, Ohio	Oct. 9, 1866	58,706
Harvester, Corn	J. F. Winchell	Springfield, Ohio	May 5, 1868	77,696
Harvester, Corn	A. N. Woodard	Fentonville, Mich	Apr. 25, 1871	114,076
Harvester, Corn	J. Wright and J. J. Johnson	Cold Water, Mich	May 28, 1867	65,324
Harvester, Corn	T. Yates	Dubuque, Iowa	Dec. 4, 1866	60,318
Harvester, Corn	R. Yeilding	Ypsilanti, Mich	July 15, 1862	35,907
Harvester, Corn	G. W. N. Yost	Port Gibson, Miss	Jan. 8, 1856	14,076
Harvester, Corn	P. C. Yost	Hamilton, Ill	May 5, 1868	77,562
Harvester, Corn and cane	E. H. Clinton, W. Prather, and H. O. Hutchinson.	Iowa City, Iowa	Aug. 20, 1867	67,955
Harvester, Corn and cane	D. P. Flynn and R. S. Hayes	Le Roy, N. Y	Mar. 6, 1860	27,356
Harvester, Corn and cane	W. M. Mason	Polo, Ill	Sept. 3, 1861	33,204
Harvester, Corn and cane	H. D. McGeorge and D. S. Greer.	Morgantown, Va	June 21, 1859	24,477
Harvester, Corn and sugar-cane	G. W. N. Yost	Cincinnati, Ohio	July 10, 1870	29,131
Harvester, Cotton	W. Apperly and C. P. Johnson	Memphis, Tenn	Dec. 11, 1860	30,862
Harvester, Cotton	L. Bishop	Talladega, Ala	July 5, 1859	24,609
Harvester, Cotton	J. Davis, jr., and D. Scott, jr	Greensborough, N. C	Apr. 5, 1870	101,439
Harvester, Cotton	J. Griffin	Louisville, Ky	Mar. 8, 1859	23,168
Harvester, Cotton	J. Griffin	Louisville, Ky	Nov. 22, 1859	26,180
Harvester, Cotton	M. Hosford and J. C. Avery	Macon, Miss	Apr. 27, 1858	20,066
Harvester, Cotton	D. C. Hubbard	Point Coupee Parish, La	July 12, 1870	105,211
Harvester, Cotton	W. H. Irving	Philadelphia, Pa	Feb. 13, 1872	123,704
Harvester, Cotton	J. W. Leigh	Norfolk, Va	Mar. 18, 1873	136,924
Harvester, Cotton	O. P. Myers	Canton, Ohio	Dec. 24, 1872	134,157
Harvester, Cotton	W. H. Pedrick	Richmond, Ind	Nov. 18, 1873	144,629
Harvester, Cotton-stalk	S. Bowerman	Detroit, Mich	Oct. 1, 1850	7,677
Harvester, Cradling	M. Barlow	Lexington, Ky	Apr. 29, 1856	14,761
Harvester-crank	H. L. Wanzer	Lanesville, Conn	Nov. 9, 1869	96,643
Harvester crank and pin	J. Kline	Mechanicsburgh, Pa	Mar. 22, 1870	101,134
Harvester-crank motion	A. Palmer	Brockport, N. Y	May 1, 1866	54,397
Harvester-cutter	E. M. Allen	Darlington, Mich	Mar. 31, 1868	76,135
Harvester-cutter	J. Atkins	Mokena, Ill	Feb. 11, 1768	74,272
Harvester-cutter	H. C. Aydelot	Carthage, Ind	Aug. 16, 1870	106,305
Harvester-cutter	T. D. Aylesworth	Ilion, N. Y	Sept. 6, 1859	25,312
Harvester-cutter	E. M. Birdsall	Penn Yan, N. Y	Oct. 12, 1869	95,638
Harvester-cutter	J. A. Bonham and A. J. Harrington.	Lovely Dale, Ind	Jan. 16, 1872	122,801
Harvester-cutter	H. Bonholtzer and J. S. Shopp	Cumberland County, Pa	Sept. 14, 1869	94,705
Harvester-cutter	C. Cadwell	Waukegan, Ill	Jan. 22, 1867	61,394
Harvester-cutter	G. W. Chapman, jr	Iowa Falls, Iowa	Aug. 4, 1868	80,598
Harvester-cutter	T. J. Christy	Olney, Ill	June 2, 1868	78,515
Harvester-cutter	T. J. and G. M. Clark	Higganum, Conn	Oct. 25, 1870	108,686
Harvester-cutter	J. M. Connel	Newark, Ohio	July 19, 1870	105,550
Harvester-cutter	H. Cutler	Central Village, Conn	Mar. 14, 1871	112,692

Index of patents issued from the United States Patent Office from 1790 *to* 1873, *inclusive*—Continued.

Invention.	Inventor.	Residence.	Date.	No.
Harvester-cutter	J. W. H. Doubler	West Alden, N. Y	July 29, 1873	141, 210
Harvester-cutter	G. L. Du Laney	Mechanicsburgh, Pa	May 18, 1869	90, 086
Harvester-cutter	G. L. Du Laney	Mechanicsburgh, Pa	June 8, 1869	91, 102
Harvester-cutter	G. L. Du Laney	Mechanicsburgh, Pa	Nov. 23, 1869	97, 062
Harvester-cutter	G. Fetter	Philadelphia, Pa	Apr. 3, 1860	27, 757
Harvester-cutter	S. B. Fisk	Rising Sun, Ind	Oct. 1, 1872	131, 869
Harvester-cutter	J. L. Fountain	Rockford, Ill	Nov. 17, 1857	18, 659
Harvester-cutter	T. Garrick	Providence, R. I	Mar. 14, 1871	112, 582
Harvester-cutter	S. Gillam	New York, N. Y	May 29, 1866	55, 084
Harvester-cutter	J. Gore	Fredonia, N. Y	Feb. 16, 1858	19, 360
Harvester-cutter	P. Gregg	Brownsville, Mich	Apr. 25, 1871	114, 003
Harvester-cutter	M. Hallenbeck	Albany, N. Y	Nov. 3, 1868	83, 628
Harvester-cutter	M. Harrison	Laclede, Mo	Sept. 19, 1871	119, 029
Harvester-cutter	M. P. Hathaway	Mankato, Minn	Aug. 14, 1866	57, 130
Harvester-cutter	I. A. Hebbard	Rochester, N. Y	Apr. 5, 1870	101, 460
Harvester-cutter	H. L. Hervey	Quincy, Ill	Mar. 18, 1856	14, 453
Harvester-cutter	M. G. Hubbard	Penn Yan, N. Y	Jan. 20, 1857	16, 442
Harvester-cutter	M. G. Hubbard	Penn Yan, N. Y	June 16, 1857	17, 575
Harvester-cutter	S. Hull	Poughkeepsie, N. Y	Aug. 4, 1863	39, 401
Harvester-cutter	S. Hull	Poughkeepsie, N. Y	Sept. 14, 1869	94, 826
Harvester-cutter	J. Irwin	Frankfort, Ohio	July 7, 1857	17, 739
Harvester-cutter	W. G. Kenyon	Wakefield, R. I	Mar. 28, 1871	113, 175
Harvester-cutter	T. R. and S. Knowles	Jersey City, N. J	Feb. 18, 1868	74, 550
Harvester-cutter	A. G. Laughlin and K. G. Rice	Providence, Ky	Oct. 22, 1872	132, 513
Harvester-cutter	M. Lewis	Odell, Ill	June 2, 1868	78, 460
Harvester-cutter	H. A. Link	Columbus, Ohio	Aug. 3, 1869	93, 319
Harvester-cutter	I. S. Love	Beloit, Wis	Mar. 11, 1856	14, 402
Harvester-cutter	J. H. Manny	Waddam's Grove, Ill	June 21, 1853	9, 800
Harvester-cutter	J. H. Manny	Rockford, Ill	June 26, 1855	13, 149
Harvester-cutter	J. H. Manny	Rockford, Ill	Mar. 25, 1856	14, 544
Harvester-cutter	J. P. Manny	Rockford, Ill	July 7, 1857	17, 745
Harvester-cutter	H. W. Mason	Hagerstown, Md	Aug. 30, 1870	106, 948
Harvester-cutter	W. McKeever	Staunton, Va	Apr. 8, 1873	137, 705
Harvester-cutter	H. Mewes	Binghamton, N. Y	Dec. 17, 1872	134, 085
Harvester-cutter	R. J. Morrison	Richmond, Ind	Jan. 13, 1857	16, 393
Harvester-cutter	B. Murray	Farm Ridge, Ill	June 13, 1854	11, 076
Harvester-cutter	C. K. Myers	Pekin, Ill	Feb. 8, 1870	99, 695
Harvester-cutter	T. Neys	Menomonee, Wis	Dec. 21, 1869	98, 182
Harvester-cutter	J. T. Norris	Tiffin, Ohio	June 16, 1868	78, 889
Harvester-cutter	J. H. Owen	Houston Township, Ill	Apr. 20, 1869	89, 236
Harvester-cutter	W. J. Oxer	Williamsport, Ind	Sept. 19, 1871	119, 170
Harvester-cutter	J. T. Polson	Laclede, Mo	Oct. 3, 1871	119, 640
Harvester-cutter	C. Pomeroy	Mattoon, Ill	Apr. 22, 1873	138, 192
Harvester-cutter	G. F. Quick	Philadelphia, Pa	July 19, 1864	43, 602
Harvester-cutter	S. H. Reed and T. Thompson	Fredericktown, Ohio	Feb. 25, 1873	136, 266
Harvester-cutter	F. E. Rogers	Paw Paw, Ill	June 28, 1870	104, 773
Harvester-cutter	B. T. Roney	Philadelphia, Pa	Apr. 29, 1856	14, 777
Harvester-cutter	B. T. Roney	Philadelphia, Pa	Sept. 11, 1860	29, 999
Harvester-cutter	D. Russell	Drewersburgh, Ind	Jan. 30, 1855	12, 327
Harvester-cutter	L. Russell	Otsego, Mich	Feb. 13, 1872	123, 731
Harvester-cutter	J. Schneider	Canton, Ohio	Dec. 28, 1869	98, 303
Harvester-cutter	H. F. Shaw	West Roxbury, Mass	Sept. 14, 1869	94, 918
Harvester-cutter	W. E. Shoales	Sherburne Four Corners, N. Y.	Sept. 2, 1873	142, 417
Harvester-cutter	J. S. Smith and J. Coder	Swanton, Ohio	Aug. 31, 1869	94, 251
Harvester-cutter	W. G. Smith	Elizabethport, N. J	Nov. 6, 1860	30, 595
Harvester-cutter	L. E. Stillwell	Franklinville, N. Y	July 16, 1872	129, 068
Harvester-cutter	D. Stukey	Lancaster, Ohio	Sept. 6, 1870	107, 118
Harvester-cutter	F. R. and W. O. Sutton	Wellington, Ill	July 29, 1873	141, 298
Harvester-cutter	J. M. Taft	Arcadia, Wis	Nov. 2, 1869	96, 503
Harvester-cutter	P. Thayer	Lansingburgh, N. Y	Mar. 11, 1856	14, 423
Harvester-cutter	H. Unger	Germantown, Ohio	Sept. 24, 1872	131, 580
Harvester-cutter	C. B. Wagner	Philadelphia, Pa	June 24, 1856	15, 205
Harvester-cutter	W. Wasson, P. F. Powers, and G. W. Dungan.	Genoa, Nev	Feb. 4, 1873	135, 500
Harvester-cutter	J. M. Wehrly	Somerville, N. J	Aug. 7, 1866	57, 023
Harvester-cutter	S. H. Wellings and S. Soules	Bridgeville, Mich	June 10, 1873	139, 695
Harvester-cutter	W. N. Whiteley	Springfield, Ohio	Aug. 11, 1868	81, 044
Harvester-cutter	W. N. Whiteley, jr., J. Fassler, and O. S. Kelly.	Springfield, Ohio	Aug. 13, 1867	67, 829
Harvester-cutter	F. Wittram	San Francisco, Cal	Dec. 28, 1869	98, 456
Harvester-cutter	W. A. Wood	Hoosick Falls, N. Y	Dec. 2, 1862	37, 066
Harvester-cutter	E. L. Yancy	Batavia, N. Y	Aug. 6, 1872	130, 266
Harvester cutter and raker, Grain and grass	H. Knowles and H. C. Bevington.	Washington, D. C., and Holmes County, Ohio.	July 2, 1850	7, 475
Harvester-cutter bar	R. Allstatter	Hamilton, Ohio	May 31, 1870	103, 699
Harvester-cutter bar	B. Johnson and W. Johnson	Carrollton and Hanover, Ohio.	Aug. 10, 1869	93, 448
Harvester-cutter bar	J. H. Manny	Rockford, Ill	Jan. 22, 1856	14, 149
Harvester-cutter bar	W. J. Oxer	Williamsport, Ind	May 24, 1870	103, 361
Harvester-cutter bar	E. G. Short and C. Oberly	Carthage, N. Y	Apr. 12, 1870	101, 930
Harvester-cutter bar	T. Welch	Churchville, N. Y	Mar. 19, 1867	62, 985
Harvester-cutter-bar connection	G. D. Culp and W. J. Keeney	Allensville and Florence, Ind.	Aug. 25, 1863	39, 633
Harvester-cutter-bar heads, Manufacture of	A. Padgham	Syracuse, N. Y	June 20, 1871	116, 088
Harvester cutter-bars, Machine for grinding	J. H. and J. S. Griffin	Harvard, Ill	Feb. 20, 1872	123, 890
Harvester-cutter blades to sickle-bars, Attaching	W. H. Hovey	Springfield, Mass	Apr. 29, 1856	14, 768
Harvester cutter, Corn	J. J. De Frietas	Springfield, Ill	July 26, 1870	105, 784
Harvester-cutter finger	J. H. Manny	Waddam's Grove, Ill	Apr. 19, 1853	9, 675
Harvester cutter, Grain and grass	M. G. Hubbard	Penn Yan, N. Y	Nov. 11, 1856	16, 057
Harvester cutter, Grain and grass	W. Pierpoint	Salem, N. J	Nov. 22, 1853	10, 258
Harvester-cutter grinder	J. P. Barker	Vienna, Ohio	Sept. 19, 1871	119, 108
Harvester-cutter grinder	W. B. Denel	Ithaca, N. Y	Nov. 2, 1869	96, 402
Harvester-cutter grinder	E. F. Keeling	Milton, Ohio	Aug. 13, 1861	33, 043
Harvester-cutter grinder	E. R. McCall	Simcoe, Canada	Oct. 24, 1871	120, 205
Harvester-cutter grinder	S. O. and P. W. Vaughn	De Kalb, Ill	Oct. 4, 1870	107, 985
Harvester-cutter holder	S. H. Wilson	Auburn, N. Y	Jan. 5, 1869	85, 627
Harvester-cutter holder for grinders	T. E. King and G. C. Dolph	Cleveland and West Andover, Ohio.	Sept. 27, 1870	107, 784

Index of patents issued from the United States Patent Office from 1790 *to* 1873, *inclusive*—Continued.

Invention.	Inventor.	Residence.	Date.	No.
Harvester-cutter sharpener	E. L. Bushnell	Poughkeepsie, N. Y	July 14, 1868	79, 948
Harvester-cutter sharpener	I. H. Coller	Poughkeepsie, N. Y	Mar. 8, 1864	41, 831
Harvester-cutter sharpener	J. McKnight	Pomeroy, Ohio	July 24, 1866	56, 589
Harvester-cutter sharpener	J. K. Staman	Mifflin, Ohio	Sept. 8, 1863	39, 866
Harvester-cutters, Apparatus for grinding	J. H. Curran	Rochester, N. Y	Jan. 30, 1872	123, 088
Harvester-cutters, Apparatus for grinding	J. F. and R. G. Kirkwood	Ellicott City, Md	July 2, 1872	128, 634
Harvester-cutters, Hand-grinder for	H. C. Fisk	Wellsville, N. Y	Oct. 24, 1871	120, 258
Harvester-cutters, Machine for grinding	J. Weichhart	San Francisco, Cal	Oct. 12, 1869	95, 751
Harvester-cutters, Machine for grinding	H. Whitall	Woodbury, N. J	May 14, 1867	64, 817
Harvester-cutters, Machine for sharpening	W. S. Ingraham	Evanston, Ill	Aug. 9, 1870	106, 167
Harvester-cutters, Rest for grinding	T. Brett	Geneva, Ohio	Apr. 21, 1868	76, 883
Harvester-cutters, Sharpening-bench for	J. R. Cliffton	West Unity, Ohio	Nov. 26, 1872	133, 409
Harvester-cutters while being sharpened, Holder for.	W. H. Daniels	Bryan, Ohio	Feb. 27, 1872	124, 121
Harvester cutting-apparatus	J. L. Abell	Cummington, Mass	Nov. 29, 1870	109, 569
Harvester cutting-apparatus	S. W. Beach	Ypsilanti, Mich	July 2, 1872	128, 456
Harvester cutting-apparatus	S. Comfort, jr	Morrisville, Pa	Apr. 7, 1857	16, 968
Harvester cutting-apparatus	G. W. Cook and F. M. Duncan	Geneseo and Aledo, Ill	Mar. 14, 1871	112, 689
Harvester cutting-apparatus	J. M. Connel	Newark, Ohio	Aug. 20, 1872	130, 568
Harvester cutting-apparatus	R. Dutton	Brooklyn, N. Y	Feb. 11, 1868	74, 211
Harvester cutting-apparatus	R. Dutton	Brooklyn, N. Y	Feb. 11, 1868	74, 212
Harvester cutting-apparatus	R. Dutton	New York, N. Y	Feb. 11, 1868	74, 213
Harvester cutting-apparatus	T. C. Hargrave	Schenectady, N. Y	Dec. 17, 1861	33, 941
Harvester cutting-apparatus	C. Howell	Cleveland, Ohio	Sept. 14, 1858	21, 499
Harvester cutting-apparatus	W. G. Kenyon	Wakefield, R. I	July 18, 1871	117, 178
Harvester cutting-apparatus	L. G. Kniffen	Worcester, Mass	Dec. 24, 1861	33, 999
Harvester cutting-apparatus	P. Manny	Waddam's Grove, Ill	Aug. 25, 1857	18, 052
Harvester cutting-apparatus	H. Mewes	Binghamton, N. Y	July 5, 1870	105, 111
Harvester cutting-apparatus	H. Mewes	Binghamton, N. Y	July 16, 1872	129, 357
Harvester cutting-apparatus	R. J. Morrison	Richmond, Va	Aug. 14, 1855	13, 433
Harvester cutting-apparatus	C. D. Read	Ayer's, Mass	June 25, 1872	128, 326
Harvester cutting-apparatus	A. R. Reese	Phillipsburgh, N. J	Apr. 22, 1862	35, 040
Harvester cutting-apparatus	A. Stoler and S. A. Sisson	Bristol, Pa	Dec. 17, 1861	33, 966
Harvester cutting-apparatus	D. H. Thayer	Lansing, N. Y	Mar. 26, 1861	31, 854
Harvester cutting-apparatus	W. A. Wood	Hoosick Falls, N. Y	Dec. 28, 1858	22, 468
Harvester cutting-apparatus, Attachment for	A. Anderson and L. Johnston	London, Canada	May 30, 1871	115, 411
Harvester cutting-apparatus, Corn and cane	J. W. Batson	Triadelphia, Md	July 29, 1856	15, 409
Harvester cutting-apparatus, Grain and grass	J. W. Baltzly and W. Hobson	Pana, Ill	Sept. 22, 1857	18, 229
Harvester cutting-device	D. W. Entrikin and L. H. Davis	West Chester, Pa	Sept. 13, 1858	19, 920
Harvester cutting-device	C. W. Glover	Roxbury, Conn	July 15, 1856	15, 334
Harvester cutting-device	C. P. Gronberg	Montgomery, Ill	Sept. 7, 1858	21, 414
Harvester cutting-device	C. Wheeler, jr	Poplar Ridge, N. Y	Sept. 2, 1856	15, 677
Harvester discharging-apparatus	A. J. Cook	Enon, Ohio	Mar. 28, 1854	10, 691
Harvester discharging-apparatus	J. S. Fowler	Davenport, Iowa	July 14, 1868	79, 967
Harvester-divider	J. J. Lurvey	North Prairie, Wis	Oct. 20, 1868	83, 185
Harvester-divider	J. H. Shireman	East Berlin, Pa	Oct. 4, 1859	25, 679
Harvester divider-attachment	J. H. Keller and D. F. Luse	Boalsburgh and Centre Hall, Pa.	Apr. 2, 1872	125, 198
Harvester drive-wheel	W. A. Wood	Hoosick Falls, N. Y	Apr. 13, 1869	88, 938
Harvester driving-wheel	D. L. Emerson	Rockford, Ill	Oct. 25, 1864	44, 834
Harvester driving-wheel	E. P. Russell	Manlius, N. Y	Aug. 15, 1865	49, 439
Harvester driving-wheel	W. N. Whiteley and T. Harding	Springfield, Ohio	June 23, 1868	79, 171
Harvester dropper	W. Allstatter and F. Schurger	Hamilton, Ohio	Mar. 5, 1872	124, 311
Harvester-dropper	J. J. and D. N. Barnhill	Vincennes, Ind	Aug. 8, 1871	117, 726
Harvester-dropper	W. G. Beels	Independence, Iowa	Jan. 18, 1870	98, 909
Harvester-dropper	C. Bickel and J. F. Leibold	Delaware, Ohio	Aug. 22, 1871	118, 331
Harvester-dropper	H. Brackett	Valley Falls, N. Y	May 4, 1869	89, 732
Harvester-dropper	H. Brackett	Valley Falls, N. Y	May 4, 1869	89, 733
Harvester-dropper	J. Case	La Fayette, Ind	Feb. 9, 1869	86, 730
Harvester-dropper	G. R. Clements	Prescott, Wis	Oct. 13, 1868	83, 040
Harvester-dropper	O. Dorsey	Newark, Ohio	July 23, 1872	129, 798
Harvester-dropper	W. H. Forker	Meadville, Pa	Dec. 17, 1872	133, 933
Harvester-dropper	J. S. Fowler	Davenport, Iowa	Dec. 16, 1873	145, 640
Harvester-dropper	J. B. Gathright	Louisville, Ky	Mar. 4, 1873	136, 370
Harvester-dropper	A. Goodyear, 2d	Hamden, Conn	Nov. 12, 1872	133, 025
Harvester-dropper	O. C. Green	Dublin, Ind	Jan. 11, 1870	98, 689
Harvester-dropper	D. S. Harner	Xenia, Ohio	July 5, 1870	104, 953
Harvester-dropper	O. M. Harrison	Glasgow, Mo	Aug. 22, 1871	118, 236
Harvester-dropper	I. Hedges	Radnor, Ohio	May 23, 1871	115, 199
Harvester-dropper	A. J. Hodges	Peoria, Ill	Dec. 9, 1873	145, 419
Harvester-dropper	J. W. Irwin	Circleville, Ohio	Jan. 23, 1872	122, 892
Harvester-dropper	C. D. Jeffries	Wooster, Ohio	Feb. 1, 1870	99, 444
Harvester-dropper	N. Johnson	Newburgh, Minn	May 27, 1873	139, 395
Harvester-dropper	N. S. Ketchum	Marshalltown, Iowa	Mar. 29, 1870	101, 273
Harvester-dropper	T. W. S. Kidd	Springfield, Ill	Dec. 20, 1870	110, 246
Harvester-dropper	T. F. Lippincott	Conemaugh, Pa	Dec. 21, 1869	98, 075
Harvester-dropper	J. Miller	Canton, Ohio	Mar. 27, 1866	53, 551
Harvester-dropper	J. Miller	Canton, Ohio	Dec. 24, 1867	72, 523
Harvester-dropper	L. Miller	Akron, Ohio	Nov. 12, 1867	70, 735
Harvester-dropper	T. C. Moore	Dublin, Ind	Sept. 20, 1870	107, 522
Harvester-dropper	E. Myers	Creagerstown, Md	Dec. 21, 1869	98, 090
Harvester-dropper	E. Myers	Creagerstown, Md	Feb. 22, 1870	100, 181
Harvester-dropper	H. Nisen	Brighton, Wis	June 4, 1872	127, 636
Harvester-dropper	A. L. and G. M. Peters	Lancaster, Ohio	Nov. 17, 1868	84, 133
Harvester-dropper	G. M. Peters	Lancaster, Ohio	Sept. 8, 1868	81, 941
Harvester-dropper	W. J. Plecker	Bushnell, Ill	May 6, 1873	138, 526
Harvester-dropper	A. Rank	Salem, Ohio	May 4, 1869	89, 792
Harvester-dropper	R. A. Roberts	Salisbury, Mo	July 2, 1872	128, 564
Harvester-dropper	B. Seneff	Chillicothe, Ohio	Feb. 27, 1872	124, 166
Harvester-dropper	B. W. Steschult	Glandorf, Ohio	Dec. 29, 1868	85, 487
Harvester-dropper	G. Stevenson	Zionsville, Ind	July 6, 1869	92, 392
Harvester-dropper	J. O. Taber	Salem, Ohio	Feb. 2, 1869	86, 602
Harvester-dropper	J. O. Taber	Salem, Ohio	May 31, 1870	103, 796
Harvester-dropper	J. S. Truxell	Greensburgh, Pa	Sept. 30, 1873	143, 391
Harvester-dropper	J. N. and T. Wallis	Fleming, N. Y	June 20, 1871	116, 120
Harvester-dropper	A. Ward	Dublin, Ind	May 10, 1870	102, 890
Harvester-dropper	W. N. Whiteley	Springfield, Ohio	Sept. 28, 1869	95, 397
Harvester-dropper	W. N. Whiteley	Springfield, Ohio	July 12, 1870	105, 396

Index of patents issued from the United States Patent Office from 1790 *to* 1873, *inclusive*—Continued.

Invention.	Inventor.	Residence.	Date.	No.
Harvester dropping-platform	E. and M. Ball	Canton, Ohio	July 21, 1868	80,112
Harvester dropping-platform	E. P. Cady	Trenton, Wis	May 10, 1870	102,914
Harvester dropping-platform	J. Miller	Canton, Ohio	Nov. 24, 1868	84,368
Harvester dropping-platform	G. M. Peters	Lancaster, Ohio	Jan. 14, 1868	73,380
Harvester dropping-platform	B. T. Reynolds	Centreville, Ind	May 24, 1870	103,370
Harvester dropping-platform	J. B. Sawyer	Templeton, Mass	Apr. 26, 1870	102,439
Harvester dropping-platform	A. H. Shreffler	Joliet, Ill	Nov. 3, 1868	83,735
Harvester dropping-platform	C. R. and J. O. Taber	Salem, Ohio	Mar. 31, 1868	76,269
Harvester dropping-platform	P. Warner	Gilead, Mich	Nov. 4, 1873	144,239
Harvester dropping-platform	G. Wellhouse	Akron, Ohio	Nov. 10, 1868	84,033
Harvester dumping-platform	J. W. Bope	Saint Louis, Mo	Apr. 14, 1868	76,703
Harvester dumping-platform	D. M. Swartz	Lewisburgh, Pa	Aug. 20, 1867	67,929
Harvester dumping-reel	E. P. Russell	Manlius, N. Y	Sept. 3, 1867	68,461
Harvester-finger	J. P. Manny	Rockford, Ill	July 6, 1858	20,808
Harvester-finger	J. Reily	Heart Prairie, Wis	Apr. 29, 1856	14,790
Harvester-finger	E. P. Russell	Manlius, N. Y	Mar. 19, 1861	31,745
Harvester-finger	E. P. Russell	Manlius, N. Y	Oct. 13, 1863	40,283
Harvester-finger	H. C. Smith	Cleveland, Ohio	Mar. 2, 1858	19,518
Harvester finger-bar	J. J. Barnes	Monticello, Ind	Feb. 11, 1868	74,191
Harvester finger-bar	R. Dutton	New York, N. Y	May 29, 1866	55,074
Harvester finger-bar	R. Dutton	New York, N. Y	May 26, 1868	78,197
Harvester finger-bar	E. F. Ford	Stillwater, N. Y	Nov. 3, 1868	83,619
Harvester finger-bar	J. M. Long, P. Black, and R. Allstatter.	Hamilton, Ohio	Dec. 1, 1857	18,754
Harvester finger-bar	W. H. Seymour	Brockport, N. Y	Dec. 16, 1856	16,253
Harvester finger-bar	C. A. Smith	Napa, Cal	Feb. 4, 1873	135,495
Harvester finger-bars, Manufacture of	R. Dutton	New York, N. Y	May 26, 1868	78,196
Harvester finger-beam	A. Rank	Salem, Ohio	Mar. 14, 1871	112,629
Harvester finger-beam	A. R. Reese	Phillipsburgh, N. J	Apr. 22, 1862	35,041
Harvester finger-guard	C. T. Bush	Rensselaerville, N. Y	Sept. 5, 1865	49,713
Harvester finger-guard	R. Dutton	New York, N. Y	May 15, 1866	54,702
Harvester finger-guards, Machine for slotting	J. Fassler	Springfield, Ohio	May 28, 1867	65,193
Harvester finger-guards, Making	M. L. Ballard	Canton, Ohio	Jan. 29, 1861	31,218
Harvester finger-guards, Making	M. L. Ballard	Canton, Ohio	Jan. 29, 1861	31,219
Harvester finger-guards, Making	M. L. Ballard	Canton, Ohio	Jan. 29, 1861	31,220
Harvester finger-guards, Making	M. L. Ballard	Canton, Ohio	Jan. 29, 1861	31,284
Harvester finger-guards, Making	L. Miller	Canton, Ohio	Jan. 29, 1861	31,285
Harvester finger or guard	L. Miller	Canton, Ohio	May 11, 1858	20,243
Harvester, Flax	S. W. Tyler	Greenwich, N. Y	Sept. 1, 1863	39,764
Harvester, Flax	S. W. Tyler	Troy, N. Y	June 13, 1871	115,910
Harvester-frame	M. G. Hubbard	Penn Yan, N. Y	Apr. 28, 1857	17,151
Harvester-frame	H. F. Mann	Westville, Ind	June 3, 1856	15,013
Harvester frame, Grass	A. Palmer	Brockport, N. Y	Jan. 30, 1855	12,323
Harvester friction-gear	H. F. and G. F. Shaw	West Roxbury, Mass	Dec. 3, 1872	133,675
Harvester gaveling-attachment	J. W. Harvey	Marshalltown, Iowa	Dec. 27, 1864	45,604
Harvester gaveling-attachment	C. C. Shults	Waverly, Iowa	May 7, 1872	126,582
Harvester gaveling-fork	O. Webster	Murray, N. Y	Nov. 7, 1871	120,602
Harvester-gearing	J. Farrington	Corry, Pa	June 20, 1871	116,038
Harvester-gearing	M. G. Hubbard	Syracuse, N. Y	Sept. 22, 1868	82,410
Harvester-gearing	M. G. Hubbard	Syracuse, N. Y	Sept. 22, 1868	82,411
Harvester-gearing	J. P. Manny	Rockford, Ill	Dec. 12, 1871	121,881
Harvester-gearing	A. H. Wagner	Chicago, Ill	Feb. 22, 1870	100,221
Harvester-gearing	A. Warner	Ontario, N. Y	Nov. 14, 1865	50,969
Harvester-gearing for changing speed	M. G. Hubbard	Syracuse, N. Y	Feb. 1, 1870	99,439
Harvester, Grain	N. T. Allen	Ludlowville, N. Y	June 10, 1851	8,157
Harvester, Grain	T. D. Burrall	Geneva, N. Y	Apr. 5, 1853	9,644
Harvester, Grain	E. Danford, jr	Geneva, Ill	Sept. 17, 1850	7,649
Harvester, Grain	B. Densmore	Sweden, N. Y	Feb. 10, 1852	8,720
Harvester, Grain	A. Elliott	San Francisco, Cal	Apr. 1, 1856	14,556
Harvester, Grain	D. Fitzgerald	New York County, N. Y	Sept. 7, 1852	9,247
Harvester, Grain	B. G. Fitzhugh	Frederick, Md	Mar. 28, 1854	10,693
Harvester, Grain	J. Haines	Pekin, Ill	May 22, 1855	12,907
Harvester, Grain	S. S. Hurlbut	Racine, Wis	Feb. 4, 1851	7,928
Harvester, Grain	T. N. Lupton	Winchester, Va	May 8, 1855	12,824
Harvester, Grain	C. Marston	Viroqua, Wis	Aug. 14, 1860	29,610
Harvester, Grain	D. S. Middlekauff	Hagerstown, Md	Mar. 14, 1854	10,652
Harvester, Grain	J. E. Newcomb	Whitehall, N. Y	Jan. 9, 1855	12,215
Harvester, Grain	J. F. Nicholson	Davidsonville, Md	May 15, 1855	12,888
Harvester, Grain	F. Nishwitz	Williamsburgh, N. Y	Aug. 30, 1853	9,975
Harvester, Grain	A. Palmer and S. G. Williams	Brockport, N. Y	July 1, 1851	8,192
Harvester, Grain	A. Palmer and S. G. Williams	Brockport, N. Y., and Janesville, Wis.	Jan. 24, 1854	10,459
Harvester, Grain	J. Phillips	Harrisburgh, Pa	Feb. 12, 1856	14,250
Harvester, Grain	W. H. Start	Smyrna, Del	June 24, 1851	8,182
Harvester, Grain	M. Vanderpool	Polk County, Oreg	Oct. 6, 1868	82,896
Harvester, Grain	T. Van Fossen	Lancaster, Ohio	Jan. 20, 1852	8,667
Harvester, Grain and grass	O. Billings	La Grange, Ohio	Nov. 12, 1861	33,691
Harvester, Grain and grass	C. B. Brown	Griggsville, Ill	Dec. 7, 1852	9,446
Harvester, Grain and grass	J. E. Brown and S. S. Bartlett	Woonsocket, R. I	Dec. 20, 1853	10,320
Harvester, Grain and grass	J. E. Brown and S. S. Bartlett	Woonsocket, R. I	Jan. 2, 1855	12,121
Harvester, Grain and grass	R. Bryson	Schenectady, N. Y	Aug. 3, 1858	21,063
Harvester, Grain and grass	M. Burnett and C. Vander Woerd.	Boston, Mass	Jan. 2, 1855	12,123
Harvester, Grain and grass	T. D. Burrall	Geneva, N. Y	Mar. 18, 1856	14,441
Harvester, Grain and grass	J. Case	Springfield, Ohio	Apr. 17, 1855	12,714
Harvester, Grain and grass	S. Colburn	Ansonia, Conn	July 3, 1855	13,160
Harvester, Grain and grass	A. Dietz and J. G. Dunham	Raritan, N. J	Mar. 27, 1855	12,584
Harvester, Grain and grass	D. Fitzgerald and J. H. Smith	New York, N. Y	Aug. 10, 1852	9,182
Harvester, Grain and grass	E. B. Forbush	Buffalo, N. Y	July 20, 1852	9,134
Harvester, Grain and grass	E. B. Forbush	Buffalo, N. Y	Apr. 17, 1855	12,721
Harvester, Grain and grass	E. B. Forbush	Buffalo, N. Y	Mar. 18, 1856	14,448
Harvester, Grain and grass	U. H. Goble	Springfield, Ohio	Dec. 20, 1853	10,323
Harvester, Grain and grass	H. Green	Ottawa, Ill	Mar. 21, 1854	10,657
Harvester, Grain and grass	J. H. Heyser and E. M. Mobley	Hagerstown, Md	May 19, 1857	17,328
Harvester, Grain and grass	W. H. Hovey	Springfield, Mass	July 3, 1855	13,173
Harvester, Grain and grass	W. H. Hovey	Springfield, Mass	Apr. 15, 1856	14,661
Harvester, Grain and grass	M. G. Hubbard	New York, N. Y	June 5, 1855	13,004
Harvester, Grain and grass	M. G. Hubbard	Penn Yan, N. Y	Dec. 7, 1858	22,251

Index of patents issued from the United States Patent Office from 1790 *to* 1873, *inclusive*—**Continued.**

Invention.	Inventor.	Residence.	Date.	No.
Harvester, Grain and grass	W. G. Huyett	Williamsburgh, Pa	June 14, 1853	9,779
Harvester, Grain and grass	O. B. Judd	Little Falls, N. Y	Jan. 16, 1855	12,252
Harvester, Grain and grass	W. A. Kerby	Buffalo, N. Y	Apr. 15, 1856	14,694
Harvester, Grain and grass	W. F. Ketchum	Buffalo, N. Y	Dec. 19, 1854	12,113
Harvester, Grain and grass	W. F. Ketchum	Buffalo, N. Y	Jan. 15, 1856	14,102
Harvester, Grain and grass	J. H. Manny	Freeport, Ill	Oct. 17, 1854	11,810
Harvester, Grain and grass	J. H. Manny	Rockford, Ill	May 8, 1855	12,825
Harvester, Grain and grass	J. H. Manny	Rockford, Ill	Jan. 22, 1856	14,148
Harvester, Grain and grass	J. H. Manny and H. Marcellus	Rockford, Ill., and Amsterdam, N. Y.	Mar. 6, 1855	12,499
Harvester, Grain and grass	H. Marcellus	Amsterdam, N. Y	Apr. 13, 1858	19,938
Harvester, Grain and grass	W. P. Maxson	Albion, Wis	Sept. 9, 1856	15,701
Harvester, Grain and grass	L. Miller	Canton, Ohio	Dec. 10, 1861	33,895
Harvester, Grain and grass	R. T. Osgood	Orland, Me	Feb. 17, 1852	8,743
Harvester, Grain and grass	B. T. Roney	Philadelphia, Pa	Mar. 11, 1856	14,409
Harvester, Grain and grass	G. Sanford and T. and S. Hull	Poughkeepsie, N. Y	Jan. 15, 1856	14,127
Harvester, Grain and grass	W. and T. Schnebly	New York, N. Y	Dec. 20, 1853	10,326
Harvester, Grain and grass	W. H. Seymour	Brockport, N. Y	Dec. 14, 1852	9,476
Harvester, Grain and grass	O Stoddard	Busti, N. Y	Sept. 2, 1856	15,672
Harvester, Grain and grass	J. Swartz	Buffalo, N. Y	Nov. 14, 1854	11,951
Harvester, Grain and grass	P. Sylla	Elgin, Ill	Apr. 17, 1855	12,745
Harvester, Grain and grass	P. Sylla and A. Adams	Elgin, Ill	Sept. 20, 1873	10,038
Harvester, Grain and grass	J. Thompson	Clifton, N. Y	Aug. 28, 1855	13,508
Harvester, Grain and grass	J. Urmy	Wilmington, Del	July 24, 1855	13,330
Harvester, Grain and grass	A. Van Duzer	Goshen, N. Y	Feb. 9, 1858	19,319
Harvester, Grain and grass	J. J. Weeks	Oyster Bay, N. Y	Sept. 26, 1854	11,735
Harvester, Grain and grass	C. Wheeler, jr	Poplar Ridge, N. Y	Dec. 5, 1854	12,044
Harvester, Grain and grass	C. Wheeler, jr	Venice, N. Y	Feb. 6, 1855	12,367
Harvester, Grain and grass	A. Whiteley	Springfield, Ohio	Sept. 19, 1854	11,710
Harvester, Grain and grass	A. Whiteley	Clark County, Ohio	July 10, 1855	13,246
Harvester, Grain and grass	A. Whiteley	Springfield, Ohio	Feb. 5, 1856	14,212
Harvester, Grain and grass	A. Whiteley	Springfield, Ohio	Mar. 11, 1856	14,428
Harvester, Grain and grass	A. Whiteley	Springfield, Ohio	Mar. 25, 1856	14,541
Harvester, Grain and grass	A. P. Wilson	Waterbury, Conn	June 3, 1856	15,029
Harvester, Grain and grass	W. A. Wood	Hoosick Falls, N. Y	Mar. 20, 1855	12,570
Harvester, Grain and grass	G. W. N. Yost	Pittsburgh, Pa	Feb. 12, 1856	14,266
Harvester, Grain and grass	G. W. N. Yost	Pittsburgh, Pa	Apr. 1, 1856	14,582
Harvester, Grain and grass	J. T. Youart	Troy, Ohio	May 27, 1856	14,980
Harvester, Grain and maize	E. Quincy	Lacon, Ill	Oct. 8, 1850	7,705
Harvester, Grain-binding	H. H. Bridenthall, jr	Youngstown, Pa	Dec. 10, 1872	133,744
Harvester, Grain binding	C. F. Goddard	Saint Ansgar, Iowa	Feb. 18, 1873	135,985
Harvester grain-platform	A. E. Gleason	Newton, Mich	Mar. 17, 1868	75,674
Harvester, Grass	G. Esterly	Heart Prairie, Wis	June 27, 1854	11,155
Harvester, Grass	J. Haines	Pekin, Ill	Sept. 4, 1855	13,523
Harvester, Grass	W. K. Hall	Philippi, Va	Nov. 22, 1853	10,267
Harvester, Grass	M. Hallenbeck	Albany, N. Y	Apr. 18, 1854	10,802
Harvester, Grass	W. F. Ketchum	Buffalo, N. Y	Feb. 10, 1852	8,724
Harvester, Grass	J. S. and D. Lake	Smith's Landing, N. J	July 20, 1852	9,137
Harvester, Grass	W. Manning	South Trenton, N. J	July 20, 1852	9,138
Harvester, Grass	J. H. Maydole and A. W. Morse	Eaton, N. Y	Feb. 6, 1855	12,363
Harvester, Grass	R. J. Morrison	Richmond, Va	Feb. 13, 1855	12,393
Harvester, Grass	E. Neely	Savannah, Mo	Jan. 7, 1851	7,888
Harvester, Grass	F. Peabody	Salem, Mass	May 29, 1855	12,963
Harvester-guard	J. Birch	Corry, Pa	Sept. 28, 1869	95,187
Harvester-guard	J. H. Jones and M. S. Prentice	Rockford, Ill	Sept. 7, 1869	94,608
Harvester guard-finger	W. Allen and L. Russ	Worcester, Mass	Oct. 29, 1867	70,311
Harvester guard-finger	R. Beans	Johnsville, Pa	Mar. 20, 1860	27,515
Harvester guard-finger	J. W. Brokaw	Springfield, Ohio	Sept. 14, 1858	21,533
Harvester guard-finger	A. Brown, H. D. Worcester, A. M. Griswold.	Ganier, Ill	Oct. 22, 1867	70,071
Harvester guard-finger	S. Copeland	Worcester, Mass	May 16, 1865	47,702
Harvester guard-finger	A. Crosby	Westfield, N. Y	July 12, 1870	105,311
Harvester guard-finger	G. Fyfe and G. Hard	Ottawa, Ill	Mar. 7, 1871	112,440
Harvester guard-finger	A. A. Hotchkiss	Sharon, Conn	Mar. 15, 1864	41,922
Harvester guard-finger	A. A. Hotchkiss and J. P. Adriance.	Sharon, Conn., and New York, N. Y.	June 21, 1859	24,461
Harvester guard-finger	W. F. Ketchum	Buffalo, N. Y	Apr. 25, 1854	10,841
Harvester guard-finger	J. H. Manny	Rockford, Ill	June 26, 1855	13,150
Harvester guard-finger	K. H. C. Preston	Manlius, N. Y	Sept. 10, 1861	33,265
Harvester guard-finger	A. R. Reese	Phillipsburgh, N. J	July 28, 1857	17,894
Harvester guard-finger	M. B. Riggs	New York, N. Y	Jan. 28, 1862	34,267
Harvester guard-finger	M. B. Riggs	New York, N. Y	Oct. 28, 1862	36,796
Harvester guard-finger	W. A. Wood	Hoosick Falls, N. Y	July 1, 1856	15,264
Harvester guard-finger	W. A. Wood	Hoosick Falls, N. Y	Nov. 19, 1867	71,257
Harvester guard-fingers, Blank for	J. Fassler	Springfield, Ohio	Sept. 28, 1869	95,334
Harvester guard-fingers, Machine for drilling	J. Fassler	Springfield, Ohio	May 28, 1867	65,191
Harvester guard-fingers, Machine for making	J. Fassler	Springfield, Ohio	Mar. 15, 1870	100,878
Harvester guard-fingers, Machine for milling	J. Fassler	Springfield, Ohio	May 28, 1867	65,192
Harvester guard-fingers, Machine for sawing the slot in.	J. Fassler	Springfield, Ohio	July 12, 1870	105,323
Harvester guard-fingers, Making	L. Miller	Akron, Ohio	Nov. 24, 1868	84,433
Harvester guard fingers, Making	G. W. Slough	Canton, Ohio	May 8, 1860	28,206
Harvester guard-fingers, Manufacture of	W. N. Whiteley, jr., J. Fassler, and O. S. Kelly.	Springfield, Ohio	June 18, 1867	65,977
Harvester guard or finger	A. Wissler	Clay Township, Pa	Dec. 5, 1865	51,380
Harvester-guards, Method of renovating the cutting-edge of.	J. Rummel, jr	New Middletown, Ohio	Nov. 23, 1869	97,232
Harvester guiding-wheel	W. A. and W. A. Wood	Hoosick Falls, N. Y	Jan. 3, 1871	110,712
Harvester, Hemp	T. Berry	Louisburgh, Ky	June 22, 1858	20,618
Harvester, Hemp	W. B. Coates	Big Lick, Va	Oct. 15, 1850	7,719
Harvester, Hemp	O. Farrar	Jessamine County, Ky	May 14, 1872	126,793
Harvester, Hemp	P. S. Fitch	Hanly, Ky	Jan. 7, 1873	134,656
Harvester, Hemp	R. C. Wrenn	Waverly, Mo	Mar. 3, 1868	75,103
Harvester-knife	G. J. Wardwell	Rutland, Vt	Feb. 15, 1870	99,797
Harvester-knife grinder	J. Y. Hoagland	Auburn, N. Y	Mar. 17, 1868	75,633
Harvester-knife grinder	T. Loring	Blackwoodtown, N. J	May 31, 1870	103,633
Harvester-knife grinder	J. A. Thompson	Auburn, N. Y	Mar. 17, 1868	75,709
Harvester-knife grinder	P. W. Vaughan	De Kalb, Ill	Dec. 19, 1871	122,083

Index of patents issued from the United States Patent Office from 1790 *to* 1873, *inclusive*—Continued.

Invention.	Inventor.	Residence.	Date.	No.
Harvester-knife grinder	E. L. Yancey	Batavia, N. Y	Dec. 14, 1869	97,847
Harvester-knife head	H. A. Soliday	Doylestown, Ohio	Mar. 23, 1869	88,091
Harvester-knife sharpener	H. Fisher	Canton, Ohio	July 16, 1872	128,957
Harvester-knives, Apparatus for sharpening	I. C. Saunders	Trenton, Mich	Dec. 19, 1871	122,064
Harvester-knives, Machine for grinding	J. M. Connel	Newark, Ohio	July 15, 1873	140,810
Harvester-knives, Machine for grinding	J. M. Connel	Newark, Ohio	July 15, 1873	140,811
Harvester-knives, Machine for grinding	J. Murphy	New York, N. Y	Feb. 25, 1873	136,171
Harvester-knives, Machine for sharpening	J. S. Elliott	Chelsea, Mass	May 13, 1873	138,743
Harvester-knives, Sharpening	A. P. Taylor	Darien, N. Y	Mar. 22, 1870	101,061
Harvester-knives, Tempering	J. F. Shippey	Valley Falls, N. Y	Apr. 29, 1873	138,349
Harvester, Maize	G. A. Bruce	Mechanicsburgh, Ill	Aug. 29, 1854	11,595
Harvester, Maize	J. S. Burnham	West Jefferson, Ohio	Sept. 19, 1854	11,688
Harvester, Maize	S. Lapham	Salem, Ohio	Apr. 18, 1854	10,811
Harvester, Maize	C. B. Matthews	Oquawka, Ill	Oct. 19, 1858	21,840
Harvester, Maize	J. L. Ream	Mount Pulaski, Ill	Dec. 21, 1852	9,487
Harvester, Maize	W Watson	Chicago, Ill	Oct. 15, 1850	7,725
Harvester, &c., pitman	J. W. Doty	Lockport, N. Y	Oct. 30, 1866	59,192
Harvester-pitman	O. P. Drury	Niles, Mich	Aug. 4, 1868	80,614
Harvester-pitman	B. E. J. Eils	Washington, D. C	Sept. 24, 1872	131,671
Harvester-pitman	J. R. Finley	Delphi, Ind	Apr. 13, 1869	88,863
Harvester-pitman	G. B. and C. B. Garlinghouse	North Madison, Ind	June 23, 1868	79,112
Harvester-pitman	G. W. Harrison	Lansing, Mich	Sept. 26, 1871	119,356
Harvester-pitman	B. J. Hunter	Ledyard, N. Y	Aug. 14, 1866	57,140
Harvester-pitman	W. J. Keeney	Florence, Ind	Aug. 27, 1867	68,207
Harvester-pitman	J. L. Kintner	Harrison County, Ind	May 21, 1867	64,985
Harvester-pitman	J. M. Mourer	Millheim, Pa	Jan. 30, 1872	123,283
Harvester-pitman	C. E. Roper	Canton, Ohio	Apr. 27, 1869	89,348
Harvester-pitman	T. C. Sebring	Rochester, N. Y	June 25, 1867	66,047
Harvester-pitman	G. L. Squier	Buffalo, N. Y	Sept. 3, 1867	68,581
Harvester-pitman connection	J. Dixon and M. B. Sampson	Eddyville, Iowa	Aug. 16, 1870	106,339
Harvester-pitman connection	W. Ferris	Pleasant Plain, Ohio	Nov. 26, 1872	133,363
Harvester-pitman connection	S. J. Green	Syracuse, N. Y	July 25, 1871	117,409
Harvester-pitman connection	H. Howe	Houston, Minn	May 28, 1872	127,240
Harvester-pitman connection	W. Loucks	Lowville, N. Y	Apr. 30, 1872	126,220
Harvester-pitman connection	D. M. Osborne	Auburn, N. Y	Nov. 7, 1865	50,879
Harvester-pitman connection	D. M. Osborne	Auburn, N. Y	Nov. 7, 1865	50,880
Harvester-pitman connection	H. L. Wanzer	Clyde, Ohio	July 2, 1867	66,430
Harvester-pitman connection	B. F. Waters	Garrettsville, Ohio	Jan. 14, 1873	134,954
Harvester-pitman connection	C. Wheeler, jr	Auburn, N. Y	Feb. 11, 1868	74,463
Harvester-pitman connection	R. C. Wood	Le Roy, Kans	Oct. 5, 1869	95,548
Harvester-pitman coupling	H. Bishop	Batavia, Iowa	Mar. 20, 1866	53,261
Harvester-pitman head	S. H. Wilson	Auburn, N. Y	June 5, 1866	55,435
Harvester-pitman joint	S. T. Lamb	New Albany, Ind	Sept. 28, 1869	95,359
Harvester-pitman joint	S. T. Lamb	New Albany, Ind	Feb. 22, 1870	100,158
Harvester-pitman joint	S. L. Lamb	New Albany, Ind	Feb. 22, 1870	100,159
Harvester-platform	J. P. Adriance and T. S. Brown	Poughkeepsie, N. Y	Feb. 8, 1870	99,617
Harvester-platform	G. B. Deardorff	Canal Dover, Ohio	July 18, 1871	117,055
Harvester-platform	C. Lidren	La Fayette, Ind	May 10, 1870	102,839
Harvester platform, Grain	J. Atkins	Chicago, Ill	Apr. 24, 1855	12,756
Harvester-platforms, Adjustable grain-wheel for	G. W. Patten	Auburn, N. Y	Jan. 7, 1873	134,610
Harvester, Potato	J. E. Hardenbergh	Fultonville, N. Y	Nov. 15, 1859	26,102
Harvester, Potato	J. J. Hill	Xenia, Ohio	May 15, 1866	54,724
Harvester, Potato	J. D. Otstot	Springfield, Ohio	Nov. 15, 1859	26,118
Harvester-rake	A. Adams	Boston Station, Ky	June 15, 1869	91,401
Harvester-rake	N. Allstatter	Hamilton, Ohio	Apr. 26, 1864	42,454
Harvester-rake	E. Ames	Austin, Minn	Mar. 16, 1869	87,817
Harvester-rake	P. Ammerman	Cynthiana, Ky	Oct. 27, 1868	83,355
Harvester-rake	C. J. Arlington	Auburn, N. Y	July 7, 1868	79,539
Harvester-rake	J. K. Augspurger	Trenton, Ohio	Aug. 26, 1873	142,139
Harvester-rake	C. Aultman	Canton, Ohio	Dec. 31, 1867	72,775
Harvester-rake	J. Bacon	Medina, Wis	Oct. 17, 1865	50,529
Harvester-rake	J. Baldwin	Saint Paris, Ohio	June 28, 1864	43,282
Harvester-rake	T. P. Barbour	Louisville, Ky	Jan. 14, 1873	134,722
Harvester-rake	J. Barnes	Rockford, Ill	Apr. 17, 1866	53,942
Harvester-rake	J. Barnes	Rockford, Ill	Dec. 10, 1867	71,952
Harvester-rake	J. Barnes	Rockford, Ill	Jan. 12, 1869	85,723
Harvester-rake	J. Barnes	Rockford, Ill	Apr. 25, 1871	114,094
Harvester-rake	C. Barns	Oskaloosa, Iowa	Mar. 30, 1869	88,358
Harvester-rake	C. L. Barritt	Richland, Mich	Apr. 28, 1868	77,244
Harvester-rake	T. Baylis and D. Williams	Tecumseh, Mich	Jan. 11, 1853	9,528
Harvester-rake	E. R. Bergstresser	Hublersburgh, Pa	Feb. 5, 1867	61,704
Harvester-rake	G. Blake	Whitby, Canada	Sept. 8, 1868	81,978
Harvester-rake	J. B. Bowen, C. A. Reed, and C. A. Whelan.	Madison, Wis	Aug. 18, 1868	81,130
Harvester-rake	J. B. Bowen, C. A. Whelan, and C. A. Reed.	Madison, Wis	July 20, 1869	92,783
Harvester-rake	H. Brackett	Valley Falls, N. Y	May 4, 1869	89,731
Harvester-rake	C. B. Brown	Upper Alton, Ill	July 11, 1854	11,249
Harvester-rake	C. B. Brown	Springfield, Mo	July 5, 1870	104,927
Harvester-rake	J. O. Brown, A. Ingham, and F. T. Lomont.	Massillon, Ohio	July 31, 1866	56,696
Harvester-rake	R. D. Brown	Covington, Ind	July 24, 1866	56,521
Harvester-rake	R. D. Brown	Covington, Ind	Dec. 24, 1867	72,599
Harvester-rake	T. S. Brown	Poughkeepsie, N. Y	Aug. 25, 1868	81,473
Harvester-rake	T. S. Brown	Poughkeepsie, N. Y	Sept. 28, 1869	95,191
Harvester-rake	F. Brua	Gordonsville, Pa	Jan. 8, 1867	60,993
Harvester-rake	R. Bryson	Schenectady, N. Y	Apr. 8, 1862	34,875
Harvester-rake	R. Bryson	Schenectady, N. Y	Mar. 3, 1868	74,983
Harvester-rake	F. M. Buckles	Altona, Ill	July 27, 1869	93,049
Harvester-rake	C. Buckwalter	Davenport, Iowa	Feb. 1, 1870	99,397
Harvester-rake	C. Bullock	Jamestown, N. Y	Apr. 23, 1861	32,160
Harvester-rake	O. H. Burdick	Auburn, N. Y	June 7, 1864	43,007
Harvester-rake	O. H. Burdick	Auburn, N. Y	Dec. 8, 1868	84,730
Harvester-rake	C. T. Burgess	Brentwood, England	Nov. 3, 1868	83,599
Harvester-rake	I. P. Cadman	Mendota, Ill	Dec. 17, 1867	72,163
Harvester-rake	R. Carkhuff	Lewisburgh, Pa	Dec. 15, 1868	84,857
Harvester-rake	C. L. Carter	Union City, Ind	May 1, 1866	54,294
Harvester-rake	H. J. Case	Auburn, N. Y	Feb. 16, 1869	86,904

Index of patents issued from the United States Patent Office from 1790 *to* 1873, *inclusive*—Continued.

Invention.	Inventor.	Residence.	Date.	No.
Harvester-rake	W. J. and R. Case	Pittstown, N. J	Sept. 19, 1865	49, 976
Harvester-rake	W. J. and R. Case	Pittstown, N. J	May 29, 1866	55, 059
Harvester-rake	P. C. Chipron	Highland, Ill	May 29, 1866	55, 060
Harvester-rake	S. Clevenger	Vibbard, Mo	Aug. 12, 1873	141, 695
Harvester-rake	D. Clow	Janesville, Wis	Sept. 6, 1870	107, 004
Harvester-rake	H. Clymo	Galena, Ill	May 1, 1866	54, 298
Harvester-rake	W. F. Cochrane	Springfield, Ohio	Sept. 19, 1865	50, 066
Harvester-rake	O. B. Colcord	Greenville, Ill	Apr. 4, 1871	113, 632
Harvester-rake	J. Collins	Guelph, Canada	Feb. 11, 1873	135, 691
Harvester-rake	S. Comfort, jr	Morrisville, Pa	Oct. 20, 1857	18, 437
Harvester-rake	S. Comfort, jr	Morrisville, Pa	Jan. 5, 1858	19, 019
Harvester-rake	F. E. Cook	Seville, Ohio	May 28, 1867	65, 059
Harvester-rake	A. G. Crane and W. T. Johnston.	Ottumwa, Iowa	Nov. 15, 1870	109, 301
Harvester-rake	M. Crossman and P. A. Spicer.	Marengo and Marshall, Mich.	Oct. 1, 1867	69, 322
Harvester-rake	J. S. Crump	Williamsburgh, Mo	Dec. 10, 1867	71, 994
Harvester-rake	J. D. Custer	Norristown, Pa	Feb. 2, 1869	86, 511
Harvester-rake	C. F. Davis	Auburn, N. Y	May 22, 1866	54, 871
Harvester-rake	N. H. Dederer	Greene, N. Y	Feb. 6, 1866	52, 396
Harvester-rake	C. C. Dennis	Auburn, N. Y	Apr. 21, 1863	38, 219
Harvester-rake	H. F. W. Deterding	Alton, Ill	June 23, 1868	79, 057
Harvester-rake	J. Dick, jr	Canton, Ohio	Jan. 28, 1868	73, 878
Harvester-rake	J. Dick, jr	Oshawa, Ontario	Nov. 3, 1868	83, 610
Harvester-rake	W. P. Dillman	Joliet, Ill	Apr. 17, 1866	53, 958
Harvester-rake	W. P. Dillman	Joliet, Ill	Nov. 19, 1867	70, 976
Harvester-rake	J. A. Dodge	Auburn, N. Y	Aug. 20, 1867	67, 850
Harvester-rake	J. A. Dodge	Auburn, N. Y	Aug. 20, 1867	67, 851
Harvester-rake	J. A. Dodge and G. Perry	Auburn, N. Y	Dec. 3, 1867	71, 718
Harvester-rake	J. A. Dodge and W. H. and H. S. Stevenson.	Auburn, N. Y	Aug. 20, 1867	67, 853
Harvester-rake	O. Dorsey	Howard County, Md	Mar. 4, 1856	14, 350
Harvester-rake	J. C. Durborow	Ellicott City, Md	June 9, 1868	78, 654
Harvester-rake	J. C. Durborow	Ellicott City, Md	June 15, 1869	91, 315
Harvester-rake	W. T. Eastes	Summitville, Ind	Aug. 16, 1870	106, 342
Harvester-rake	W. T. Eastes	Summitville, Ind	July 22, 1873	141, 127
Harvester-rake	B. Ellingworth	Le Roy, Minn	May 2, 1871	114, 441
Harvester-rake	R. Emerson	Sycamore, Ill	Sept. 19, 1871	119, 130
Harvester-rake	J. Farrington	Corry, Pa	Mar. 29, 1870	101, 246
Harvester-rake	V. H. Felt	Rochester, N. Y	Jan. 7, 1868	72, 990
Harvester-rake	V. H. Felt	Kendall, N. Y	May 3, 1870	102, 525
Harvester-rake	J. R. Finley	Delphi, Ind	July 20, 1869	92, 713
Harvester-rake	E. T. Ford	Batavia, N. Y	Nov. 24, 1857	18, 686
Harvester-rake	E. T. Ford	Stillwater, N. Y	Feb. 20, 1866	52, 700
Harvester-rake	J. S. Fowler	Davenport, Iowa	Mar. 23, 1869	88, 156
Harvester-rake	J. French	Independence, Iowa	Jan. 18, 1870	98, 861
Harvester-rake	M. Gibbs	Homer, Mich	Dec. 8, 1868	84, 818
Harvester-rake	J. H. and A. J. Glass	McGregor, Iowa	Dec. 10, 1867	72, 016
Harvester-rake	W. F. Goodwin	Washington, D. C	Mar. 13, 1866	53, 136
Harvester-rake	W. F. Goodwin	Washington, D. C	June 12, 1866	55, 487
Harvester-rake	W. F. Goodwin	Washington, D. C	Mar. 12, 1867	62, 838
Harvester-rake	W. F. Goodwin	Washington, D. C	May 7, 1867	64, 520
Harvester-rake	W. F. Goodwin	Washington, D. C	May 7, 1867	64, 521
Harvester-rake	W. F. Goodwin	East New York, N. Y	Dec. 31, 1867	72, 839
Harvester-rake	W. F. Goodwin	East New York, N. Y	Dec. 31, 1867	72, 840
Harvester-rake	W. F. Goodwin	East New York, N. Y	Dec. 31, 1867	72, 841
Harvester-rake	W. F. Goodwin and A. W. Browne.	Washington, D. C., and Brooklyn, N. Y.	May 7, 1867	64, 522
Harvester-rake	W. F. Goodwin and A. W. Browne.	Washington, D. C., and Brooklyn, N. Y.	May 7, 1867	64, 523
Harvester-rake	W. F. Goulding	Providence, R. I	Mar. 8, 1870	100, 615
Harvester-rake	O. C. Green	Belleville, Ill	Aug. 28, 1855	13, 490
Harvester-rake	G. S. Grier	Milford, Del	Feb. 27, 1872	124, 130
Harvester-rake	G. S. Grier	Milford, Del	July 9, 1872	128, 875
Harvester-rake	C. P. Gronberg	Geneva, Ill	Dec. 1, 1857	18, 736
Harvester-rake	C. P. Gronberg	Elgin, Ill	Aug. 6, 1872	130, 295
Harvester-rake	G. Hall	Baltimore, Md	June 2, 1863	38, 796
Harvester-rake	J. C. Hall	Monroe, Wis	July 9, 1867	66, 584
Harvester-rake	J. C. Hall	Monroe, Wis	Sept. 1, 1868	81, 628
Harvester-rake	T. Harding	Springfield, Ohio	Mar. 10, 1868	75, 263
Harvester-rake	J. H. Harnly	Penn Township, Pa	Dec. 9, 1862	37, 133
Harvester-rake	H. A. M. Harris	Philadelphia, Pa	Dec. 17, 1867	72, 198
Harvester-rake	H. A. M. Harris	Philadelphia, Pa	Dec. 17, 1867	72, 199
Harvester-rake	J. Harris	Janesville, Wis	May 23, 1871	115, 052
Harvester-rake	G. Heuerman, W. Sternberg, and J. Stuhr.	Davenport, Iowa	Nov. 28, 1871	121, 247
Harvester-rake	G. W. Hines	Brookfield, Wis	Jan. 31, 1871	111, 344
Harvester-rake	J. F. Hirschy and W. M. McDonald.	Wooster, Ohio	Apr. 16, 1867	63, 890
Harvester-rake	S. T. Holly	Rockford, Ill	Mar. 7, 1871	112, 341
Harvester-rake	S. Hull	Poughkeepsie, N. Y	Aug. 18, 1868	81, 090
Harvester-rake	S. Hull	Poughkeepsie, N. Y	May 11, 1869	89, 867
Harvester-rake	W. R. G. Humphrey	Chillicothe, Mo	Feb. 13, 1872	123, 701
Harvester-rake	W. H. H. Hunter	Versailles, Ind	Apr. 26, 1864	42, 488
Harvester-rake	W. H. Hurlbut	Mirabile, Mo	May 14, 1872	126, 705
Harvester-rake	J. Irvine	Parkersburgh, Iowa	Sept. 9, 1873	142, 632
Harvester-rake	W. B. Johns	Cumberland, Md	Dec. 24, 1867	72, 644
Harvester-rake	I. A. Johnson	Middletown, Conn	Nov. 22, 1870	109, 415
Harvester-rake	S. Johnston	Buffalo, N. Y	Dec. 22, 1863	41, 009
Harvester-rake	S. Johnston	Syracuse, N. Y	Feb. 2, 1869	86, 554
Harvester-rake	J. H. Jones	Rockton, Ill	Aug. 22, 1865	49, 530
Harvester-rake	J. H. Jones	Rockford, Ill	Mar. 27, 1866	53 450
Harvester-rake	R. V. Jones and H. Fessler	Canton, Ohio	Feb. 12, 1867	61, 035
Harvester-rake	M. A. Keller	York County, Pa	Feb. 28, 1865	46, 565
Harvester-rake	D. A. Kellogg	Valparaiso, Ind	May 16, 1871	114, 827
Harvester-rake	W. King	Richmond, Va	July 16, 1872	129, 035
Harvester-rake	W. W. Kingsbury	Kalamazoo, Mich	June 29, 1869	92, 058
Harvester-rake	W. E. Kinnear	Bucyrus, Ohio	Oct. 29, 1872	132, 669

Index of patents issued from the United States Patent Office from 1790 *to* 1873, *inclusive*—Continued.

Invention.	Inventor.	Residence.	Date.	No.
Harvester-rake	D. R. Kinyon	Raritan, N. J	Mar. 14, 1871	112, 605
Harvester-rake	W. A. Kirby	Auburn, N. Y	Dec. 8, 1868	84, 833
Harvester-rake	W. A. Kirby	Auburn, N. Y	Apr. 2, 1872	125, 308
Harvester-rake	I. Lancaster	Baltimore, Md	Aug. 20, 1867	67, 888
Harvester-rake	I. Lancaster	Baltimore, Md	July 21, 1868	80, 188
Harvester-rake	I. Lancaster	Baltimore, Md	Apr. 27, 1869	89, 414
Harvester-rake	C. Ledren	La Fayette, Ind	Feb. 25, 1868	74, 836
Harvester-rake	E. J. Leyburn	Lexington, Va	June 26, 1866	55, 877
Harvester-rake	E. J. Leyburn	Lexington, Va	Oct. 22, 1867	70, 007
Harvester-rake	E. J. Leyburn	Lexington, Va	Dec. 10, 1867	71, 890
Harvester-rake	C. Lidren	La Fayette, Ind	Mar. 30, 1869	88, 394
Harvester-rake	S. K. Lighter	Hamilton, Ohio	Dec. 11, 1866	60, 392
Harvester-rake	S. A. Lindsay	Unionville, Md	Dec. 11, 1860	30, 882
Harvester-rake	E. Lippoldt	Brighton, Ill	Nov. 18, 1873	144, 683
Harvester-rake	R. Little and L. Gibbs	Canton, Ohio	Dec. 17, 1867	72, 208
Harvester-rake	S. D. Locke	Hoosick Falls, N. Y	Sept. 16, 1873	142, 708
Harvester-rake	J. M. Long	Hamilton, Ohio	May 21, 1867	64, 888
Harvester-rake	J. D. Loveland	Wheatfield, N. Y	Apr. 3, 1866	53, 740
Harvester-rake	F. H. Manny	Rockford, Ill	Dec. 28, 1869	98, 284
Harvester-rake	J. P. Manny	Rockford, Ill	Mar. 7, 1871	112, 363
Harvester-rake	J. P. Manny	Rockford, Ill	Mar. 21, 1871	112, 940
Harvester-rake	J. P. Manny	Rockford, Ill	Mar. 21, 1871	112, 941
Harvester-rake	J. P. Manny	Rockford, Ill	Apr. 25, 1871	114, 165
Harvester-rake	J. P. Manny	Rockford, Ill	Apr. 25, 1871	114, 166
Harvester-rake	J. P. Manny	Rockford, Ill	Apr. 16, 1872	125, 747
Harvester-rake	J. S. Marsh	Lewisburgh, Pa	Jan. 21, 1868	73, 619
Harvester-rake	A. J. Martin	Rockford, Ill	Apr. 30, 1867	64, 345
Harvester-rake	A. McArthur	Boonville, Mo	Apr. 4, 1871	113, 321
Harvester-rake	E. R. McCall	Simcoe, Canada	May 17, 1870	103, 217
Harvester-rake	J. B. McCormick	Saint Louis, Mo	Jan. 21, 1868	73, 623
Harvester-rake	L. J. McCormick, W. R. Baker, and L. Erpelding.	Chicago, Ill	Sept. 15, 1868	82, 141
Harvester-rake	J. B. McMillan	North Vernon, Ind	Sept. 2, 1873	142, 490
Harvester-rake	J. H. Mears	Oshkosh, Wis	Mar. 4, 1862	34, 584
Harvester-rake	J. H. Mears	Oshkosh, Wis	June 17, 1862	35, 653
Harvester-rake	D. Mendenhall	Fairfield, Iowa	Apr. 7, 1868	76, 339
Harvester-rake	W. G. Merrell	Auburn, N. Y	Jan. 21, 1868	73, 452
Harvester-rake	J. Miller	Canton, Ohio	July 9, 1867	66, 609
Harvester-rake	J. Miller	Canton, Ohio	July 23, 1867	66, 983
Harvester-rake	L. iller	Akron, Ohio	May 28, 1867	65, 106
Harvester-rake	L. Miller	Akron, Ohio	Feb. 25, 1868	74, 929
Harvester-rake	L. Miller	Akron, Ohio	May 28, 1872	127, 181
Harvester-rake	S. Miller	Urbana, Ohio	July 15, 1851	8, 225
Harvester-rake	W. W. Miller	Zionsville, Ind	June 21, 1870	104, 483
Harvester-rake	J. B. Morse and L. L. Carter	La Fayette, Ind	Mar. 9, 1869	87, 699
Harvester-rake	C. Moul	Hanover, Pa	Apr. 3, 1866	53, 653
Harvester-rake	J. Mumma	Middletown, Ohio	Mar. 13, 1866	53, 171
Harvester-rake	T. Murphy	Pine Bluffs, Wis	Aug. 26, 1873	142, 171
Harvester-rake	M. L. Nix	Chetopah, Kans	July 22, 1873	141, 164
Harvester-rake	G. Oerllein	Utica, Minn	Nov. 22, 1870	109, 441
Harvester-rake	E. Ogden	Lynchburgh, Va	Mar. 17, 1868	75, 570
Harvester-rake	J. L. Owens	Cambria, Wis	Sept. 9, 1873	142, 644
Harvester-rake	A. and C. W. Palmer	Brockport, N. Y	Mar. 16, 1869	87, 963
Harvester-rake	A. and C. W. Palmer	Brockport, N. Y	Feb. 1, 1870	99, 340
Harvester-rake	A. and C. W. Palmer	Brockport, N. Y	Mar. 15, 1870	100, 922
Harvester-rake	W. Partridge	Philadelphia, Pa	Mar. 22, 1870	101, 156
Harvester-rake	H. Pease	Brockport, N. Y	Dec. 20, 1870	110, 279
Harvester-rake	W. P. Penn	Belleville, Ill	Apr. 24, 1860	28, 010
Harvester-rake	G. M. Peters	Granville, Ohio	July 9, 1867	66, 625
Harvester-rake	G. M. Peters	Granville, Ohio	Oct. 1, 1867	69, 476
Harvester-rake	H. F. Phillips	Auburn, N. Y	Jan. 26, 1869	86, 178
Harvester-rake	W. Pimlott	Springfield, Ohio	Jan. 25, 1870	99, 108
Harvester-rake	W. Pimlott	Brockport, N. Y	Nov. 22, 1870	109, 449
Harvester-rake	J. Pine	Troy, N. Y	May 7, 1861	32, 266
Harvester-rake	J. Poulson, jr	Pittstown, N. J	Sept. 26, 1865	50, 159
Harvester-rake	D. J. Powers	Madison, Wis	Oct. 6, 1868	82, 872
Harvester-rake	K. H. C. Preston	Manlius, N. Y	Sept. 1, 1868	81, 685
Harvester-rake	A. Pritz	Dayton, Ohio	Dec. 4, 1860	30, 835
Harvester-rake	A. Quick, W. S. Opie, and A. J. Farrand.	Raritan, N. J	Feb. 25, 1868	74, 845
Harvester-rake	J. M. Randle	Brighton, Ill	Oct. 10, 1865	50, 388
Harvester-rake	A. Rank	Salem, Ohio	Aug. 18, 1868	81, 208
Harvester-rake	A. Rank	Salem, Ohio	Nov. 2, 1869	96, 353
Harvester-rake	A. Rank	Salem, Ohio	Sept. 27, 1870	107, 721
Harvester-rake	A. Rank	Salem, Ohio	Aug. 12, 1873	141, 668
Harvester-rake	C. D. Read	Hamilton, Ohio	Aug. 11, 1863	39, 497
Harvester-rake	O. Redmond	Rochester, N. Y	Mar. 10, 1868	75, 297
Harvester-rake	A. R. Reese	Phillipsburgh, N. J	Feb. 16, 1864	41, 641
Harvester-rake	A. R. Reese	Phillipsburgh, N. J	May 1, 1866	54, 406
Harvester-rake	A. R. Reese, W. Gould, and N. Lake.	Phillipsburgh, N. J	Nov. 5, 1861	33, 663
Harvester-rake	T. S. Reeve and C. D. Smith	Chicago, Ill	Oct. 13, 1868	82, 991
Harvester-rake	G. H. Reister	Washington, Iowa	Aug. 31, 1869	94, 440
Harvester-rake	L. M. Rice	Oregon, Wis	Apr. 2, 1867	63, 427
Harvester-rake	S. C. Ridgaway	Baltimore, Md	Nov. 26, 1867	71, 327
Harvester-rake	S. C. Ridgaway	Baltimore, Md	Jan. 14, 1868	73, 388
Harvester-rake	C. Roberts	Belleville, Ill	May 16, 1854	10, 904
Harvester-rake	A. B. Rodman	Lyons, Iowa	Oct. 18, 1864	44, 749
Harvester-rake	J. L. Rohrer	Upper Leacock Township, Pa.	Dec. 3, 1867	71, 649
Harvester-rake	J. M. Rosebrook	Hoosick Falls, N. Y	Aug. 1, 1871	117, 687
Harvester-rake	J. M. Rosebrooks	Hoosick Falls, N. Y	Nov. 7, 1871	120, 777
Harvester-rake	J. W. Schuckers	Philadelphia, Pa	Aug. 16, 1870	106, 513
Harvester-rake	F. Schürger and N. Allstatter	Hamilton, Ohio	Aug. 11, 1868	80, 879
Harvester-rake	F. B. Scott	Lancaster, N. Y	Feb. 28, 1871	112, 289
Harvester-rake	J. Seibel	Manlius, Ill	Apr. 16, 1872	125, 696
Harvester-rake	W. H. Seymour	Brockport, N. Y	July 8, 1851	8, 212
Harvester-rake	W. H. Seymour and A. Palmer	Brockport, N. Y	Apr. 15, 1862	34, 986

Index of patents issued from the United States Patent Office from 1790 *to* 1873, *inclusive*—Continued.

Invention.	Inventor.	Residence.	Date.	No.
Harvester-rake	W. H. Seymour and A. Palmer	Brockport, N. Y	Mar. 24, 1868	75, 801
Harvester-rake	S. S. and J. G. Sherman	McHenry, Ill	Mar. 6, 1866	53, 051
Harvester-rake	S. S. and J. G. Sherman	McHenry, Ill	Dec. 22, 1868	85, 139
Harvester-rake	A. Sherwood	Auburn, N. Y	Aug. 20, 1872	130, 597
Harvester-rake	D. B. Shirk	Brunersville, Pa	Feb. 22, 1870	100, 078
Harvester-rake	D. B. Shirk	Brunersville, Pa	Nov. 1, 1870	108, 839
Harvester-rake	E. W. Skinner	Madison, Wis	Feb. 22, 1870	100, 200
Harvester-rake	E. M. Smith	New York, N. Y	June 30, 1868	79, 507
Harvester-rake	E. H. Smith	Westminster, Md	Feb. 23, 1869	87, 211
Harvester-rake	E. M. Smith	New York, N. Y	May 14, 1872	126, 839
Harvester-rake	W. T. Smith	Bellefontaine, Ohio	Nov. 5, 1872	132, 870
Harvester-rake	A. D. Sprague and A. Dockum	Caledonia, Minn	May 15, 1866	54, 789
Harvester-rake	G. L. Squire	Buffalo, N. Y	Apr. 8, 1862	34, 909
Harvester-rake	E. Stewart	Fort Madison, Iowa	Dec. 10, 1867	72, 109
Harvester-rake	O. O. Storle	North Cape, Wis	June 12, 1866	55, 551
Harvester-rake	O. O. Storle	North Cape, Wis	May 5, 1868	77, 675
Harvester-rake	O. O. Storle	Norway, Wis	Jan. 12, 1869	85, 870
Harvester-rake	O. O. Storle	North Cape, Wis	July 5, 1870	105, 139
Harvester-rake	H. A. Stringer and A. F. Ward	Chatham, Canada	Nov. 22, 1870	109, 468
Harvester-rake	W. Sutliff	Burns, Wis	Nov. 22, 1870	109, 469
Harvester-rake	J. B. Sweetland	Pontiac, Mich	Dec. 3, 1867	71, 816
Harvester-rake	J. B. Sweetland	Pontiac, Mich	Mar. 24, 1868	76, 000
Harvester-rake	J. Taylor	Magnolia, Ill	May 13, 1862	35, 270
Harvester-rake	T. Taylor	Baltimore, Md	Apr. 10, 1866	53, 929
Harvester-rake	J. Tray	Mineral Point, Wis	July 19, 1870	105, 608
Harvester-rake	J. S. Truxell	Greenburgh, Pa	Apr. 1, 1873	137, 510
Harvester-rake	J. Underwood	Muscatine, Iowa	June 9, 1868	78, 700
Harvester-rake	I. Van Doren	Somerville, N. J	Sept. 22, 1857	18, 256
Harvester-rake	I. Van Doren	Somerville, N. J	Mar. 2, 1858	19, 523
Harvester-rake	J. Walmsley	London, Canada	Feb. 13, 1872	123, 752
Harvester-rake	W. H. Ward	Auburn, N. Y	June 22, 1869	91, 582
Harvester-rake	G. H. Weller	New Village, N. J	May 23, 1871	115, 141
Harvester-rake	G. Wellhouse	Akron, Ohio	Apr. 24, 1866	54, 235
Harvester-rake	H. Wells	Erieville, N. Y	Apr. 16, 1872	125, 710
Harvester-rake	P. Werni	Chicago, Ill	Nov. 19, 1867	71, 098
Harvester-rake	C. Wheeler, jr	Auburn, N. Y	Apr. 21, 1868	77, 146
Harvester-rake	C. Wheeler, jr	Auburn, N. Y	July 14, 1868	79, 881
Harvester-rake	C. Wheeler, jr	Auburn, N. Y	Dec. 15, 1868	85, 044
Harvester-rake	C. Wheeler, jr	Auburn, N. Y	Apr. 4, 1871	113, 474
Harvester-rake	W. N. Whiteley	Springfield, Ohio	Feb. 11, 1868	74, 464
Harvester-rake	W. N. Whiteley	Springfield, Ohio	June 23, 1868	79, 170
Harvester-rake	W. N. Whiteley	Springfield, Ohio	Aug. 11, 1868	81, 045
Harvester-rake	W. N. Whiteley	Springfield, Ohio	Jan. 26, 1869	86, 265
Harvester-rake	W. N. Whiteley	Springfield, Ohio	Feb. 1, 1870	99, 506
Harvester-rake	W. N. Whiteley, jr	Springfield, Ohio	May 14, 1867	64, 819
Harvester-rake	W. N. Whiteley, jr	Springfield, Ohio	May 21, 1867	65, 030
Harvester-rake	W. N. Whiteley and J. Fassler	Springfield, Ohio	Feb. 11, 1868	74, 465
Harvester-rake	T. S. Whitenack	Easton, Pa	Feb. 5, 1861	31, 345
Harvester-rake	J. R. Whittemore	Chicopee Falls, Mass	May 20, 1862	35, 343
Harvester-rake	J. Wilcke	Chicago, Ill	Aug. 3, 1869	93, 260
Harvester-rake	W. S. Wilmot	New York, N. Y	June 4, 1861	32, 493
Harvester-rake	W. A. Wood	Hoosick Falls, N. Y	Nov. 19, 1861	33, 761
Harvester-rake	W. A. Wood	Hoosick Falls, N. Y	Jan. 3, 1871	110, 716
Harvester rake and reel	A. B. Hitchcock	Juneau, Wis	Sept. 14, 1869	94, 887
Harvester rake and reel combined	P. F. Hodges	Moline, Ill	May 17, 1870	103, 185
Harvester rake and reel combined	P. F. Hodges	Moline, Ill	May 17, 1870	103, 186
Harvester rake and reel combined	S. Johnston	Buffalo, N. Y	Feb. 7, 1865	46, 300
Harvester rake and reel, Combined	W. A. Kirby	Auburn, N. Y	June 25, 1872	128, 403
Harvester rake and reel, Combined	J. B. Whiting	Ripon, Wis	May 16, 1871	115, 004
Harvester rake and reel, Combined	W. A. Wood	Hoosick Falls, N. Y	Dec. 8, 1863	40, 876
Harvester rake and reel, Combined	M. Young, jr	Frederick, Md	Sept. 18, 1860	30, 103
Harvester rake-attachment	C. Barns	West Liberty, Iowa	Aug. 9, 1870	106, 108
Harvester rake-attachment	L. M. Batty	Canton, Ohio	Sept. 19, 1865	49, 964
Harvester rake-attachment	R. D. Brown	Covington, Ind	Dec. 19, 1865	51, 550
Harvester rake-attachment	H. Fisher	Canton, Ohio	Nov. 21, 1865	51, 111
Harvester rake-attachment	L. Miller	Akron, Ohio	Nov. 21, 1865	51, 070
Harvester rake-attachment	L. Miller	Akron, Ohio	Nov. 21, 1865	51, 071
Harvester rake-attachment	W. B. Parsons	Granger, N. Y	July 18, 1865	48, 834
Harvester rake-attachment	E. Smith	Clinton, Pa	June 5, 1866	55, 386
Harvester-rake, Automatic	J. W. Brokaw	Springfield, Ohio	Aug. 25, 1857	18, 070
Harvester-rake, Automatic	S. Comfort, jr	Morrisville, Pa	Dec. 23, 1856	16, 307
Harvester-rake, Automatic	T. F. Kums	Rockford, Ill	Oct. 1, 1861	33, 418
Harvester-rake, Automatic	T. F. Kums	Rockford, Ill	Oct. 1, 1861	33, 419
Harvester-rake, Automatic	T. F. Kums	Rockford, Ill	Oct. 1, 1861	33, 420
Harvester-rake, Automatic	S. T. Lamb	New Washington, Ind	Apr. 29, 1856	14, 769
Harvester-rake, Automatic	J. S. Manning	Philadelphia, Pa	June 30, 1857	17, 687
Harvester-rake, Automatic	J. Ollis	Bloomington, Ill	June 26, 1860	28, 894
Harvester-rake, Automatic	I. C. Twining	Wrightstown, Pa	Aug. 14, 1860	29, 640
Harvester-rake, Automatic	W. N. Whiteley, jr	Springfield, Ohio	Aug. 29, 1865	49, 669
Harvester rake, Grain	J. Atkins	Chelsea, Ill	Dec. 21, 1852	9, 479
Harvester-rake, Self-acting	S. T. Lamb	New Washington, Ind	June 30, 1857	17, 685
Harvester-rake, Self-acting	J. Whitehead	Manchester, Va	Sept. 16, 1856	15, 751
Harvester-rake, Self-acting	J. Whitehead	Manchester, Va	Dec. 2, 1856	16, 156
Harvester-rakes, &c., Producing intermittent acceleration of motion in.	J. Richardson	Buckeystown, Md	June 19, 1855	13, 102
Harvester-raker	J. W. Haggard and G. Bull	Bloomington, Ill	Dec. 11, 1855	13, 933
Harvester-raker	W. A. Wood	Hoosick Falls, N. Y	Sept. 11, 1860	30, 018
Harvester raking and binding apparatus	A. Sherwood	Auburn, N. Y	Sept. 14, 1858	21, 540
Harvester raking and binding apparatus	A. B. Smith	Clinton, Pa	June 19, 1860	28, 783
Harvester raking and binding attachment	J. Barta	La Crosse, Wis	May 15, 1866	54, 672
Harvester raking and binding attachment	J. P. Manny	Rockford, Ill	July 6, 1858	20, 805
Harvester raking and binding attachment	T. Palmer	Catskill, N. Y	Jan. 13, 1863	37, 408
Harvester raking and binding device	A. Sherwood	Auburn, N. Y	Jan. 26, 1858	19, 212
Harvester raking and delivering attachment	W. A. Wood	Hoosick Falls, N. Y	Apr. 27, 1859	20, 119
Harvester raking-apparatus	G. A. Clarke	Philadelphia, Pa	Jan. 1, 1856	14, 043
Harvester raking-apparatus	I. Dodenhoff	Bloomington, Ill	Aug. 18, 1857	18, 009
Harvester raking-apparatus	S. R. Hunter	Cortland, N. Y	Dec. 2, 1856	16, 145
Harvester raking-apparatus	J. P. Manny	Rockford, Ill	July 14, 1857	17, 798

Index of patents issued from the United States Patent Office from 1790 *to* 1873, *inclusive*—Continued.

Invention.	Inventor.	Residence.	Date.	No.
Harvester raking-apparatus	D. C. Smith	Tecumseh, Mich	June 30, 1857	17,703
Harvester raking-apparatus	J. Urmy	Wilmington, Del	Mar. 31, 1857	10,941
Harverter raking-apparatus, Automatic	W. H. Wilson	Denton, Md	May 8, 1860	28,228
Harvester raking-apparatus, Corn and cane	J. W. Batson	Triadelphia, Md	July 29, 1856	15,408
Harvester raking-attachment	J. A. Barrington	Fredericktown, Ohio	June 8, 1858	20,475
Harvester raking-attachment	W. W. Burson	Yates City, Ill	Nov. 2, 1858	21,940
Harvester raking-attachment	A. H. Caryl	Sandusky, Ohio	Feb. 5, 1856	14,183
Harvester raking-attachment	G. E. Chenoweth	Baltimore, Md	Apr. 9, 1861	31,954
Harvester raking-attachment	I. H. Conklin	Rockford, Ill	Apr. 14, 1857	17,025
Harvester raking-attachment	I. C. Crane	Edgerton, Ohio	May 17, 1864	42,752
Harvester raking-attachment	P. S. Crawford	Marengo, Ill	Sept. 21, 1858	21,552
Harvester raking-attachment	H. Curtis	New York, N. Y	Oct. 18, 1864	44,709
Harvester raking-attachment	D. Davis, J. Heibeler, and S. A. Porter.	Prescott, Wis	Apr. 5, 1864	42,171
Harvester raking-attachment	D. O. De Wolf	New York, N. Y	June 1, 1858	20,411
Harvester raking-attachment	J. L. Fountain	Rockford, Ill	Jan. 12, 1858	19,085
Harvester raking-attachment	J. Fox and J. W. Van Hook	Uniontown, D. C	Dec. 27, 1864	45,597
Harvester raking-attachment	G. V. Griffith	Sandusky, Ohio	Apr. 27, 1858	20,061
Harvester raking-attachment	C. P. Gronberg	Montgomery, Ill	June 28, 1859	24,555
Harvester raking-attachment	C. P. Gronberg	Aurora, Ill	Sept. 11, 1860	29,966
Harvester raking-attachment	D. Guptail	Elgin, Ill	June 26, 1860	28,854
Harvester raking-attachment	W. H. Hovey	Springfield, Mass	Apr. 15, 1856	14,693
Harvester raking-attachment	R. V. Jones	Canton, Ohio	Dec. 27, 1864	45,614
Harvester raking-attachment	D. W. and H. A. Lafetra	Eatontown, N. J	Apr. 14, 1857	17,045
Harvester raking-attachment	A. E. McGaughey	Red Wing, Minn	Feb. 9, 1864	41,522
Harvester raking-attachment	J. McIntosh	Geneva, Ill	June 30, 1857	17,690
Harvester raking-attachment	W. G. Merrell	Auburn, N. Y	Apr. 26, 1864	42,498
Harvester raking-attachment	J. Nelson	Rockford, Ill	Sept. 7, 1858	21,437
Harvester raking-attachment	S. N. Page	Salona, Pa	Mar. 21, 1865	46,930
Harvester raking-attachment	J. W. Patterson	Philadelphia, Pa	Feb. 16, 1858	19,370
Harvester raking-attachment	A. R. Reese	Phillipsburgh, N. J	Oct. 19, 1858	21,847
Harvester raking-attachment	I. S. and H. R. Russell and J. I. D. Bristol.	New Market, Md., and Detroit, Mich.	July 21, 1863	39,329
Harvester raking-attachment	W. T. Shaw and J. Manz	Wilmington, Del	Nov. 22, 1864	45,185
Harvester raking-attachment	A. B. Smith	Clinton, Pa	May 14, 1861	32,317
Harvester raking-attachment	B. Smith	Batavia, Ill	July 14, 1863	39,249
Harvester raking-attachment	I. B. Snyder	Lancaster County, Pa	Jan. 20, 1863	37,463
Harvester raking-attachment	J. A. St. John	Janesville, Wis	May 25, 1858	20,378
Harvester raking-attachment	O. Stoddard	Busti, N. Y	Apr. 13, 1858	19,958
Harvester raking-attachment	A. Stoler and S. A. Sisson	Bristol, Pa	Mar. 1, 1864	41,793
Harvester raking-attachment	H. K. Taylor	Racine, Wis	July 12, 1864	43,531
Harvester raking-attachment	J. V. A. and A. Wemple	Chicago, Ill	Feb. 16, 1858	19,393
Harvester raking-attachment	C. Wheeler, jr	Poplar Ridge, N. Y	July 8, 1856	15,311
Harvester raking-attachment	C. Wheeler, jr	Poplar Ridge, N. Y	Apr. 26, 1864	42,521
Harvester raking-attachment	W. Whiteley, jr	Springfield, Ohio	Nov. 25, 1856	16,131
Harvester raking-attachment	J. Young	Marshallton, Pa	Dec. 14, 1858	22,326
Harvester raking-device	S. R. Hunter	Cortland, N. Y	Sept. 1, 1857	18,096
Harvester raking-device	C. W. and W. W. Marsh	Shabbona, Ill	Jan. 5, 1864	41,080
Harvester raking-device	J. Nelson	Rockford, Ill	Jan. 5, 1864	41,123
Harvester ratchet-attachment	G. E. Burt and S. B. Hildreth	Harvard, Mass	Dec. 1, 1868	84,475
Harvester ratchet-attachment	W. Cogswell	Ottawa, Ill	May 15, 1866	54,689
Harvester-reel	J. M. Bope	South Bend, Ind	June 7, 1870	103,969
Harvester-reel	J. M. Bope	South Bend, Ind	June 7, 1870	103,970
Harvester-reel	G. A. Bruce	Mechanicsburgh, Ill	July 3, 1855	13,159
Harvester-reel	G. S. Curtis	Chicago, Ill	Sept. 27, 1859	25,560
Harvester-reel	C. F. Goddard	Saint Ansgar, Iowa	Sept. 2, 1873	142,389
Harvester-reel	G. S. Grier	Milford, Del	Dec. 19, 1871	122,007
Harvester-reel	G. S. Grier	Milford, Del	July 2, 1872	128,616
Harvester-reel	H. B. Hurford and E. Miller	Canton, Ohio	Feb. 13, 1872	123,702
Harvester-reel	J. R. Jones	Clarksville, Iowa	Mar. 23, 1869	88,043
Harvester-reel	J. H. Keller	Boalsburgh, Pa	Nov. 1, 1870	108,912
Harvester-reel	J. Miller	Swan, Ind	Aug. 8, 1871	117,911
Harvester-reel	D. C. Nutting and J. D. Allen	Bowling Green, Ky	Aug. 15, 1871	118,045
Harvester-reel	C. N. Owen	Salem, Ohio	Sept. 27, 1870	107,804
Harvester-reel	J. L. Rohrer	Upper Leacock Twnsp., Pa	Nov. 17, 1868	84,138
Harvester-reel	W. F. Rundell	Genoa, N. Y	Dec. 31, 1867	72,910
Harvester-reel	W. F. Rundell	Genoa, N. Y	June 16, 1868	79,009
Harvester-reel	E. P. Russell	Manlius, N. Y	Nov. 12, 1861	33,713
Harvester-reel	E. P. Russell	Manlius, N. Y	Apr. 18, 1865	47,338
Harvester-reel	J. A. Shephard	Portsmouth, Ohio	July 30, 1872	129,990
Harvester-reel	J. Werner, jr	Prairie du Sac, Wis	Sept. 16, 1873	142,881
Harvester-reel	W. W. and C. C. Wright	Canton, Pa	Dec. 7, 1852	9,458
Harvester-reel	G. W. N. Yost	Corry, Pa	June 9, 1868	78,711
Harvester reel and rake	W. H. Seymour	Brockport, N. Y	Jan. 31, 1871	111,483
Harvester reel and rake	W. N. Whiteley	Springfield, Ohio	Sept. 8, 1868	82,050
Harvester reel and rake combined	S. W. Bingman	Laurelton, Pa	Feb. 15, 1870	99,820
Harvester reel-rake	M. K. Church	Stamford, Canada	Dec. 17, 1872	133,966
Harvester reel-rake	C. A. Werden	Waukegan, Ill	Mar. 18, 1873	136,885
Harvester reel-rake	C. Young	Auburn, N. Y	Nov. 19, 1872	133,285
Harvester-reels, Mode of supporting	T. I. Stealey	Middlebourne, Va	Dec. 15, 1857	18,871
Harvester, Rice	J. J. Herndon	Marlborough District, S. C.	Oct. 1, 1850	7,683
Harvester scroll-wheel	C. D. Rogers	Utica, N. Y	July 7, 1857	17,749
Harvester-seat	T. S. Brown	Poughkeepsie, N. Y	Apr. 24, 1866	54,245
Harvester-seat	W. J. Ludlow	Cleveland, Ohio	June 2, 1868	78,531
Harvester-seat	E. A. Smith	Saint Albans, Vt	Mar. 10, 1868	75,306
Harvester-seat	G. W. N. Yost	Corry, Pa	Apr. 15, 1873	137,816
Harvester, Seed	F. Decker	Delaware, Ohio	Aug. 2, 1870	105,920
Harvester sheaf-dropper	N. S. Ketchum	Marshalltown, Iowa	June 4, 1872	127,611
Harvester-sickle	I. C. Crane	Edgerton, Ohio	Apr. 5, 1864	42,169
Harvester-sickle	P. Manny	Waddam's Grove, Ill	Oct. 21, 1856	15,926
Harvester sickle-guard	A. Shogren	Chicago, Ill	Aug. 2, 1859	24,957
Harvester sickle-head	M. H. Hilburn	Wilmington, Ill	July 31, 1866	56,749
Harvester-sickles, Tool for shaping	J. Jann	New Windsor, Md	Nov. 20, 1866	59,770
Harvester, Steam	R. C. Parvin	Philadelphia, Pa	Apr. 16, 1872	125,843
Harvester, Sugar-cane	W. T. Johnson and J. Doyle	Philadelphia, Pa., and Wetumpka, Ala.	Oct. 2, 1860	30,226
Harvester, Sugar-cane	R. R. Taylor	Reading, Pa	May 24, 1859	24,164
Harvester, Thrashing	S. S. Rembert	Memphis, Tenn	Mar. 26, 1850	7,222

Index of patents issued from the United States Patent Office from 1790 *to* 1873, *inclusive*—Continued.

Invention.	Inventor.	Residence.	Date.	No.
Harvester-tongue	M. Rohrer	Polo, Ill	Apr. 25, 1871	114, 203
Harvester-tongue	J. Wadleigh	Chebanse, Ill	Dec. 22, 1868	85, 150
Harvester track-clearer	O. Du Bois	San José, Cal	Apr. 1, 1873	137, 351
Harvester track-clearer	W. F. Ketchum	Buffalo, N. Y	May 17, 1853	9, 737
Harvester track-clearer	J. P. Manny	Rockford, Ill	July 6, 1858	20, 807
Harvester track-clearer	C. N. Owen	Salem, Ohio	Nov. 21, 1871	121, 123
Harvester track-clearer	N. Stonecipher	Cambridge City, Ind	July 7, 1868	79, 609
Harvester track-clearer	J. O. Taber	Salem, Ohio	Mar. 24, 1863	38, 007
Harvester track-clearer	G. W. N. Yost	Corry, Pa	Dec. 8, 1868	84, 786
Harvester track-clearer, Grass	A. Whiteley	Springfield, Ohio	Aug. 22, 1854	11, 579
Harvester-wheel	P. Lincoln	Cold Water, Mich	Mar. 24, 1863	37, 969
Harvester-wheel	A. J. Sweeney	Wheeling, W. Va	Sept. 3, 1872	131, 133
Harvester wheel and axle	B. G. Turner	Fremont, Nebr	Aug. 19, 1873	141, 965
Harvesters, Adjustable reel and cut-off for	J. H. Keller and D. F. Luse	Boalsburgh and Centre Hill, Pa.	Nov. 21, 1871	121, 107
Harvesters, Adjustable reel-post for	J. W. Davis	Dublin, Ohio	July 26, 1870	105, 654
Harvesters, Apparatus for removing grain from	S. Comfort, jr	Morrisville, Pa	Apr. 1, 1856	14, 553
Harvesters, Arrangement of cutters in grain and grass.	J. Peirson	Wilmington, Del	July 2, 1850	7, 481
Harvesters, Attaching connecting-bars to cutters of	J. and W. Little	Princeton, Ind	July 3, 1855	13, 176
Harvesters, Attaching raker's seat to	S. Hull	Poughkeepsie, N. Y	June 26, 1855	13, 147
Harvesters, Attaching teeth to sickle-bars of	I. C. and L. C. Pluche	Cape Vincent, N. Y	June 10, 1856	15, 084
Harvesters, &c., Attaching wheels to	A. Whiteley	Springfield, Ohio	Apr. 24, 1855	12, 768
Harvesters, Binder's platform for	C. Miller	Lincoln, Pa	June 27, 1871	116, 471
Harvesters, Binder's table for	P. M. Donohoo	Saint Rose, Wis	Nov. 21, 1871	121, 090
Harvesters, Binder's table for	J. H. Smith	Orford, Iowa	Jan. 21, 1873	135, 168
Harvesters, Clutch-mechanism for	W. Michael	Murrysville, Pa	June 7, 1870	103, 910
Harvesters, Combined rake and reel attachment to	R. Hoffheins	Dover, Pa	July 4, 1865	48, 557
Harvesters, Cut-off for	W. R. G. Humphrey	Chillicothe, Mo	Jan. 23, 1872	123, 023
Harvesters, Cutting-gear of grain and grass	S. S. Allen	Salem, N. J	Nov. 8, 1853	10, 201
Harvesters, Cutting-off apparatus for	J. S. Schoonover and P. S. Bacon.	Corry, Pa	Apr. 30, 1873	126, 236
Harvesters, Delivering-apparatus of grain	E. A. Morrison	Lawrenceville, Pa	Jan. 30, 1855	12, 339
Harvesters, Draft-device for	J. M. Smith	Jerseyville, Ill	Nov. 15, 1870	109, 353
Harvesters, Finger-bar arrangement for	J. A. Moore and A. H. Patch	Louisville, Ky	Nov. 25, 1856	16, 134
Harvesters, Finger-fan for	W. A. Wood	Hoosick Falls, N. Y	Dec. 10, 1867	72, 149
Harvesters from reapers to mowers, &c., Mechanism for changing.	R. Beans	Johnsville, Pa	Aug. 28, 1855	13, 504
Harvesters, Gleaner-attachment for	J. V. Bryson	Greenup, Ky	Oct. 1, 1872	131, 844
Harvesters, Grain-binder for	G. W. N. Yost	Port Gibson, Miss	Jan. 1, 1856	14, 036
Harvesters, Grain-binding attachment for	T. S. Minnis	Meadville, Pa	Sept. 27, 1870	107, 703
Harvesters, Grain-discharger for	J. D. Congue	Lyndon, Ill	Nov. 24, 1863	40, 676
Harvesters, Grain-lifter for	W. M. Jackson	Woodland, Cal	June 21, 1870	104, 460
Harvesters, Lifting-finger for	A. Hughes	Brampton Ash, England	Nov. 11, 1873	144, 457
Harvesters, Means of adjusting the grain-wheel of	J. D. Wright	Columbia, Tenn	Nov. 22, 1870	109, 481
Harvesters, Method of gathering grain upon, and discharging it from, the platform of.	O. Hussey	Baltimore, Md	Dec. 21, 1858	22, 368
Harvesters, Method of raising and lowering the cutters of.	J. F. Barrett	North Granville, N. Y	July 10, 1855	13, 205
Harvesters, Running-gear for	M. G. Hubbard	Syracuse, N. Y	June 4, 1867	65, 386
Harvesters, &c., Seat for	J. G. Perry	Kingston, R. I	July 22, 1873	141, 075
Harvesters, Self-rake for	R. D. Brown	Covington, Ind	Apr. 7, 1863	38, 094
Harvesters, Self-raker for	S. G. Randall	Rockford, Ill	July 22, 1856	15, 387
Harvesters, Self-raker for	J. T. Whitaker	Saint Charles, Ill	Apr. 29, 1856	14, 784
Harvesters, Self-raking attachment to	H. Foresman	Enon, Ohio	May 13, 1856	14, 861
Harvesters, Spring-seat for	T. Brett	Geneva, Ohio	Feb. 27, 1872	124, 113
Harvesters, Swathing-apparatus for	S. C. Longshore	Lahaska, Pa	Oct. 20, 1857	18, 462
Harvesters, &c, Tempering knife sections for	A. R. Reynolds	Auburn, N. Y	Aug. 28, 1866	57, 655
Harvesting and seeding machine, Frame for	J. H. Snyder	Killbuck, Ill	Dec. 20, 1864	45, 531
Harvesting grain	A. Churchill	Geneva, Ill	Mar. 16, 1841	2, 007
Harvesting grain and grass, Machinery for	W. and T. Schnebly	Hackensack, N. Y	Feb. 17, 1857	16, 658
Harvesting, husking, and shelling machine, Corn	D. A. Dickinson	Baltimore, Md	Oct. 16, 1866	58, 785
Harvesting-machine	W. Allen	Worcester, Mass	Oct. 4, 1864	44, 500
Harvesting-machine	O. Barr	Aurora, Ill	Jan. 16, 1849	6, 034
Harvesting-machine	J. F. Barrett	North Granville, N. Y	Apr. 28, 1857	17, 134
Harvesting-machine	L. M. Batty	Canton, Ohio	Sept. 19, 1865	49, 963
Harvesting-machine	E. Bayliss, J. O. Brown, and F. T. Lomont.	Massillon, Ohio	Oct. 14, 1873	143, 609
Harvesting-machine	A. Belchambers	Ripley, Ohio	Mar. 7, 1865	46, 628
Harvesting-machine	W. B. Birdsall and E. H. Coggswell.	Hudson, Mich	Nov. 1, 1864	44, 847
Harvesting-machine	J. M. Bowman	Brockport, N. Y	July 10, 1860	29, 119
Harvesting-machine	J. Bradley	Cedar Falls, Iowa	Sept. 20, 1864	44, 277
Harvesting-machine	C. R. Brinckerhoff	Batavia, N. Y	May 24, 1859	24, 093
Harvesting machine	C. R. Brinckerhoff	Rochester, N. Y	Aug. 22, 1865	49, 497
Harvesting-machine	W. H. Burkhart	Bucyrus, Ohio	Jan. 30, 1866	52, 264
Harvesting-machine	T. B. Butler	Newark, Conn	July 5, 1859	24, 613
Harvesting-machine	J. Carpenter	Yorktown, N. Y	Dec. 23, 1856	16, 313
Harvesting-machine	G. and W. Chamberlin	Olean, N. Y	Apr. 26, 1859	23, 756
Harvesting-machine	H. Chandler	New Gloucester, Me	Aug. 22, 1865	49, 501
Harvesting-machine	D. A. Church, L. H. Obert, and W. W. and O. F. Willoughby.	Friendship, N. Y., and Chicago, Ill.	Aug. 7, 1847	5, 228
Harvesting-machine	F. Clark	Charlotte, N. Y	June 18, 1861	32, 602
Harvesting-machine	W. F. Cochrane	Springfield, Ohio	July 25, 1865	49, 024
Harvesting-machine	I. H. Coller	Poughkeepsie, N. Y	July 11, 1865	48, 658
Harvesting-machine	J. C. Cox and R. Newton	Greenville, N. C	Nov. 17, 1857	18, 628
Harvesting-machine	D. Cushing	Aurora, Ill	Nov. 21, 1848	5, 934
Harvesting machine	J. S. Davis	Tiffin, Ohio	Aug. 22, 1865	49, 506
Harvesting-machine	N. A. Dederer	Greene, N. Y	Feb. 6, 1866	52, 395
Harvesting-machine	J. Dunlap	Geneva, Wis	June 26, 1847	5, 174
Harvesting-machine	R. Dutton	Dayton, Ohio	Feb. 12, 1861	31, 378
Harvesting-machine	R. Dutton	New York, N. Y	July 17, 1866	56, 389
Harvesting-machine	G. Esterly	Heart Prairie, Wis	Oct. 22, 1844	3, 803
Harvesting-machine	G. Esterly	Heart Prairie, Wis	Mar. 24, 1857	16, 873
Harvesting-machine	G. Esterly	Whitewater, Wis.	Apr. 12, 1859	23, 666
Harvesting-machine	B. G. Fitzhaugh	Frederick, Md	June 28, 1859	24, 549
Harvesting-machine	G. W. Folhurst	Cleveland, Ohio	Sept. 16, 1856	15, 748
Harvesting-machine	E. T. Ford	Stillwater, N. Y	Feb. 20, 1866	52, 701

Index of patents issued from the United States Patent Office from 1790 *to* 1873, *inclusive*—Continued.

Invention.	Inventor.	Residence.	Date.	No.
Harvesting-machine	C. Foster	La Porte, Ind	Jan. 1, 1847	4, 916
Harvesting-machine	L. J. Fountain	New Milford, Ill	June 20, 1865	48, 335
Harvesting-machine	J. Haines	Union Grove, Ill	Mar. 27, 1849	6, 245
Harvesting-machine	M. Hallenbeck	Albany, N. Y	Apr. 17, 1860	27, 901
Harvesting-machine	T. C. Hargraves	Schenectady, N. Y	Apr. 30, 1861	30, 233
Harvesting-machine	J. K. Harris	Allensville, Ind	June 30, 1857	17, 678
Harvesting-machine	I. Hawley and D. S. Stafford	Pekin and Decatur, Ill	Nov. 8, 1864	44, 996
Harvesting-machine	J. E. Heath	Warren, Ohio	Jan. 15, 1850	7, 020
Harvesting-machine	J. E. Heath	Geneva, Ohio	Sept. 11, 1855	13, 545
Harvesting-machine	M. G. Hubbard	Penn Yan, N. Y	May 10, 1859	23, 921
Harvesting-machine	A. H. Inskeep	Middleburgh, Ohio	Sept. 13, 1859	25, 415
Harvesting-machine	J. Jann	New Windsor, Md	Aug. 4, 1863	39, 403
Harvesting-machine	W. Jones	Bradford, Vt	July 8, 1851	8, 207
Harvesting-machine	W. H. Jordan	Roseville, Ind	Apr. 5, 1864	42, 197
Harvesting-machine	H. G. Kaufmann	Saint Louis, Mo	Aug. 31, 1858	21, 343
Harvesting-machine	W. A. Kirby	Buffalo, N. Y	Sept. 2, 1856	15, 659
Harvesting-machine	W. A. Kirby	Buffalo, N. Y	July 2, 1861	32, 736
Harvesting-machine	S. T. Lamb	New Washington, Ind	Oct. 9, 1860	30, 330
Harvesting-machine	J. Little	Chambersburgh, Pa	May 24, 1859	24, 128
Harvesting-machine	H. H. Luther	Warren, R. I	June 7, 1859	24, 318
Harvesting-machine	F. H. Manny	Rockport, Ill	Apr. 30, 1861	32, 197
Harvesting-machine	H. Marcellus	Amsterdam, N. Y	May 3, 1859	23, 851
Harvesting-machine	D. J. Marvin	Stockton, Cal	Nov. 15, 1864	45, 061
Harvesting-machine	J. C. and C. N. Mayberry	White Rock, Ill	Nov. 19, 1861	33, 748
Harvesting-machine	J. H. Maydole	Eaton, N. Y	Aug. 7, 1860	29, 504
Harvesting-machine	R. W. McClelland	Springfield, Ill	Aug. 2, 1864	43, 697
Harvesting-machine	J. McPherson	Pennington, N. J	May 31, 1859	24, 226
Harvesting-machine	B. Mertz	Burlington, Iowa	Sept. 11, 1860	29, 985
Harvesting-machine	L. Miller	Canton, Ohio	Dec. 3, 1861	33, 845
Harvesting-machine	L. and J. Miller	Canton, Ohio	Dec. 2, 1862	37, 049
Harvesting-machine	H. Moore and J. Hascall	Kalamazoo, Mich	June 28, 1836	......
Harvesting-machine	L. L. Moore	Petersburgh, Va	Aug. 19, 1856	15, 569
Harvesting-machine	R. J. Morrison	Richmond, Va	Dec. 16, 1856	16, 244
Harvesting-machine	G. Murphy	Lewisburgh, Pa	Sept. 26, 1865	50, 206
Harvesting-machine	W. Neff	Centre Hall, Pa	Oct. 22, 1861	33, 539
Harvesting-machine	J. M. Orfut	Malta, Ill	Aug. 27, 1861	33, 158
Harvesting-machine	S. K Paden	Volant, Pa	June 28, 1864	43, 331
Harvesting-machine	S. N. Page	Salona, Pa	Mar. 21, 1865	46, 929
Harvesting machine	F. S. Pease	Buffalo, N. Y	Nov. 14, 1848	5, 925
Harvesting-machine	A. J. Purviance	Updegraff's, Ohio	May 22, 1849	6, 476
Harvesting-machine	S. Ray and M. R. Shalters	Alliance, Ohio	Apr. 19, 1859	23, 707
Harvesting-machine	J. Reily	Heart Prairie, Wis	Nov. 20, 1855	13, 828
Harvesting-machine	J. H. Rible	Somerset, Ohio	July 10, 1860	29, 129
Harvesting-machine	G. W. Richardson	Grayville, Ill	July 19, 1859	24, 848
Harvesting machine	G. W. Richardson and R. Glover	Grayville, Ill	Sept. 6, 1859	25, 370
Harvesting-machine	J. S. Royce	Cuylerville, N. Y	Dec. 18, 1860	30, 933
Harvesting-machine	E. P. Russell	Manlius, N. Y	Feb. 14, 1865	46, 394
Harvesting-machine	J. Y. Schelly and J. Stauffer	Hereford and Hosensack, Pa.	Sept. 2, 1856	15, 669
Harvesting-machine	W. and T. Schnebly	Hackensack, N. J	June 21, 1859	24, 495
Harvesting-machine	W. and T. Schnebly	Hackensack, N. J	June 21, 1859	24, 496
Harvesting-machine	J. Seibel	Manlius, Ill	Apr. 19, 1864	42, 410
Harvesting-machine	W. H. Seymour and D. S. Morgan.	Brockport, N. Y	June 8, 1858	20, 516
Harvesting-machine	J. B. Smith	Winfield, N. Y	Nov. 22, 1864	45, 186
Harvesting-machine	J. D. Smith	Lancaster, Ohio	May 3, 1859	23, 869
Harvesting-machine	W. S. Stetson	Baltimore, Md	May 17, 1859	24, 062
Harvesting-machine	W. S. Stetson and R. F. Maynard.	Baltimore, Md	May 17, 1859	24, 063
Harvesting-machine	A. G. Stipher	Richmond, Ind	June 28, 1859	24, 591
Harvesting-machine	S. Thomas	Burnett, Wis	May 10, 1859	23, 961
Harvesting-machine	B. Titcomb	Baltimore County, Md	June 5, 1860	28, 618
Harvesting-machine	J. A. Wagener	Pultney, N. Y	May 22, 1860	28, 427
Harvesting-machine	W., jr., and J. Webber	Rockton, Ill	Mar. 29, 1859	23, 417
Harvesting-machine	T. Welch	Churchville, N. Y	Apr. 3, 1866	53, 713
Harvesting-machine	J. Werner, jr	Prairie du Sac, Wis	Sept. 26, 1865	50, 191
Harvesting-machine	E. C. West	Bradford, Vt	June 25, 1845	4, 092
Harvesting-machine	M. A. Wheaton	Suisun City, Cal	Aug. 1, 1865	49, 186
Harvesting-machine	C. Wheeler, jr	Poplar Ridge, N. Y	Feb. 9, 1864	41, 556
Harvesting-machine	C. Wheeler, jr	Poplar Ridge, N. Y	Feb. 9, 1864	41, 557
Harvesting-machine	C. Wheeler, jr	Poplar Ridge, N. Y	Feb. 9, 1864	41, 558
Harvesting-machine	C. Wheeler, jr	Poplar Ridge, N. Y	Feb. 9, 1864	41, 559
Harvesting-machine	E. H. Wheeler	Keokuk, Iowa	Jan. 29, 1861	31, 275
Harvesting-machine	J. Whitehead	Manchester, Va	May 24, 1859	24, 170
Harvesting-machine	T. S. Whitenack	Easton, Pa	Apr. 30, 1861	32, 216
Harvesting-machine	H. Willard and R. Ross	Vergennes, Vt	May 10, 1859	23, 971
Harvesting-machine	W. Wilmington	Toledo, Ohio	July 3, 1860	29, 026
Harvesting-machine	W. H. Wilson	Denton, Md	May 10, 1859	23, 972
Harvesting-machine	B. F. Witt	Dublin, Ind	Apr. 10, 1860	27, 852
Harvesting-machine	W. A. Wood	Hoosick Falls, N. Y	Feb. 10, 1857	16, 618
Harvesting-machine	W. A. Wood	Hoosick Falls, N. Y	Feb. 10, 1857	16, 619
Harvesting-machine	W. A. Wood	Hoosick Falls, N. Y	Feb. 10, 1857	16, 620
Harvesting-machine	W. A. Wood	Hoosick Falls, N. Y	May 3, 1859	23, 878
Harvesting-machine	W. A. Wood	Hoosick Falls, N. Y	July 29, 1862	36, 047
Harvesting-machine	M. Young, jr	Frederick, Md	June 28, 1859	24 598
Harvesting-machine	M. Young, jr	Frederick, Md	Dec. 18, 1860	30, 999
Harvesting-machine, Bean	S. V. Newman	Covington, N. Y	June 21, 1859	24, 482
Harvesting-machine cutter	S. T. Lamb	New Washington Ind	Oct. 2, 1860	30, 231
Harvesting-machine for cutting, thrashing, and winnowing grain.	D. A. Church	Friendship, N. Y	May 4, 1841	2, 074
Harvesting machine, Grain	G. R. Crane	Caldwell, N. J	Mar. 10, 1857	16, 789
Harvesting machine, Grain	G. F. Foote	Buffalo, N. Y	Nov. 11, 1856	16, 052
Harvesting-machine guards, Manufacture of	J. E. Sweet	Syracuse, N. Y	Nov. 28, 1871	121, 261
Harvesting-machine header-attachment	G. Esterly	Whitewater, Wis	Jan. 13, 1863	37, 392
Harvesting machine, Hemp	J. B. McCormick	Versailles, Ky	June 2, 1857	17, 438
Harvesting-machine, Rake	L. P. Brady	Mount Joy, Pa	Dec. 18, 1860	30, 908
Harvesting-machine raking-attachment	G. Tatlock	Salem, Ind	May 10, 1859	23, 959
Harvesting-machines, Automatic rake for	J. P. Greene and I. Dodenhoff	Bloomington, Ill	Apr. 21, 1857	17, 088

Index of patents issued from the United States Patent Office from 1790 *to* 1873, *inclusive*—Continued.

Invention.	Inventor.	Residence.	Date.	No.
Harvesting-machines, Crank-wrist for	J. O. Taber	Salem, Ohio	June 7, 1870	104, 080
Harvesting-machines, Divider for	W. A. Woods	Hoosick Falls, N. Y	Mar. 24, 1863	37, 995
Harvesting-machines, Form of teeth in	E. B. Forbush	Buffalo, N. Y	Nov. 27, 1849	6, 903
Harvesting-machines, Grain-carrier for	J. J. and H. F. Mann	Clinton, Ind	June 19, 1849	6, 540
Harvesting-machines, Grain-discharging attachment to.	J. F. Black	Lancaster, Ill	Oct. 26, 1858	21, 869
Harvesting-machines, Machine for grinding the cutters of.	W. H. Stevenson	Auburn, N. Y	Feb. 18, 1868	74, 730
Harvesting-machines, Machine for grinding the cutters of.	W. H. Stevenson	Auburn, N. Y	Feb. 18, 1868	74, 731
Harvesting-machines, Method of manufacturing cutter-bars for.	G. G. Taylor	Worcester, Mass	Jan. 10, 1865	45, 879
Harvesting-machines, Method of operating the cutters of.	J. Haviland	Milton, N. Y	Sept. 1, 1857	18, 093
Harvesting-machines, Pitman-head and crank-wrist box for.	W. N. Whiteley, jr	Springfield, Ohio	July 10, 1866	56, 304
Harvesting-machines, Tongue and caster-plate for	R. Emerson, jr	Rockford, Ill	May 26, 1857	17, 367
Harvesting-rake	P. Sylla	Elgin, Ill	May 12, 1863	38, 515
Harvesting standing corn, Machine for	J. H. Frampton	Hopewell, Ohio	Mar. 3, 1857	16, 726
Hasp and staple	O. J. Smith	Wauwatosa, Wis	Dec. 24, 1872	134, 322
Hasp and staple fastening	E. P. Jones	Shell Mound, Miss	Mar 15, 1870	100, 899
Hasp-lock	J. H. Beauregard	Kingsbury, N. Y	Oct. 6, 1868	82, 681
Hasp-lock	W. N Chamberlain	Denton, Mich	Apr. 4, 1871	113, 496
Hasp-lock	E. R. Colver	New London, Conn	Dec. 21, 1869	98, 036
Hasp-lock	G. Crompton	Jersey City, N. J	Oct. 24, 1871	120, 247
Hasp-lock	J. S. Elliott	Philadelphia, Pa	May 12, 1868	77, 723
Hasp-lock	J. Fisler and G. Crompton	Jersey City, N. J	July 4, 1871	116, 581
Hasp-lock	E. L. Gaylord	Terrysville, Conn	July 27, 1869	93, 078
Hasp-lock	W. D. Heister	Newton Hamilton, Pa	Dec. 19, 1871	122, 013
Hasp-lock	M. G. Imbach	Hartford, Conn	Oct. 27, 1868	83, 500
Hasp-lock	J. Kinzer	Pittsburgh, Pa	July 15, 1873	140, 926
Hasp-lock	W. C. McGill	Cincinnati, Ohio	June 27, 1871	116, 469
Hasp-lock	C. W. A. Romer	Newark, N. J	Nov, 16, 1869	96, 968
Hasp-lock	T. Slaight	Newark, N. J	May 7, 1872	126, 583
Hasp-lock	A. B. Smith	Memphis, Mich	May 27, 1873	139, 430
Hasp-lock	E. P. Stimens and T. C. Martin	Perryville, Ohio	Mar. 18, 1873	136, 877
Hasp-lock	C. Walsh, J. F. Connelly, and A. Bratt.	Newark, N. J	Dec. 21, 1869	98, 212
Hasp or latch, Safety	E. C. Robinson	Boston, Mass	Mar. 14, 1871	112, 635
Hassock-machine	C. F. Anthony	Chicago, Ill	May 26, 1868	78, 350
Hassocks or stools, Device for forming	J. G. Flagg	Philadelphia, Pa	Jan. 15, 1867	61, 183
Hat	D. K. Albright and L. H. De Lange.	Philadelphia, Pa., and Burlington, N. J.	June 13, 1865	48, 222
Hat	D. K. Albright and L. H. De Lange.	Philadelphia, Pa., and Bordentown, N. J.	Nov. 5, 1867	70, 493
Hat	J. P. Beatty	Norwalk, Conn	June 27, 1865	48, 357
Hat	J. P. Beatty	Norwalk, Conn	Mar. 23, 1869	88, 116
Hat	A. Bogardus	Matteawan, N. Y	Sept. 19, 1871	119, 073
Hat	J. W. Corey	Newark, N. J	Jan. 24, 1871	111, 178
Hat	P. Curtis	Amesbury, Mass	June 7, 1864	43, 072
Hat	F. Degen	New York, N. Y	Dec. 7, 1852	9, 442
Hat	W. E. Doubleday	Brooklyn, N. Y	Aug. 26, 1862	36, 284
Hat	R. Dunlap	New York, N. Y	June 27, 1865	48, 384
Hat	J. H. Earle	Fall River, Mass	Nov. 7, 1865	50, 809
Hat	J. R. Ender	Trenton, La	Apr. 17, 1860	27, 897
Hat	F. P. Flanagan	Newark, N. J	June 16, 1863	38, 895
Hat	D. W. Gitchell	New York, N. Y	Nov. 28, 1865	51, 172
Hat	H. Hayward	New York, N. Y	Aug. 10, 1869	93, 616
Hat	J. M. Heard	Aberdeen, Miss	May 24, 1870	103, 457
Hat	C. A. Hopkins, C. H. Reid, G. N. Raymond, and J. S. Meeker.	Danbury, Conn	July 4, 1871	116, 710
Hat	F. Howard	Boston, Mass	Jan. 12, 1864	41, 260
Hat	H. C. Hulbert and A. Tollett	Brooklyn, N. Y	Aug. 18, 1868	81, 089
Hat	F. Huot and C. Bandouin	New York, N. Y	June 9, 1868	78, 668
Hat	R. B. Hurd and W. Halladay	Paterson, N. J., and Brooklyn, N. Y.	Jan. 12, 1864	41, 261
Hat	G. Johnson	Philadelphia, Pa	Dec. 27, 1870	110, 570
Hat	J. Keith	Charlton, Mass	Jan. 10, 1871	110, 921
Hat	H. Kellogg	New Haven, Conn	July 28, 1868	80, 490
Hat	W. B. Lodge	Danbury, Conn	Oct. 15, 1867	69, 921
Hat	G. Mallory	Bridgeport, Conn	Feb. 11, 1868	74, 392
Hat	W. H. Mallory	Watertown, Conn	Sept. 8, 1863	39, 822
Hat	J. W. McGill	Washington, D. C	Nov. 29, 1870	109, 641
Hat	E. Morris	Philadelphia, Pa	Jan. 6, 1857	16, 338
Hat	E. Morris	Philadelphia, Pa	July 18, 1865	48, 828
Hat	W. S. Nelson	Saint Louis, Mo	Aug. 28, 1866	57, 554
Hat	R. S. Nickerson and J. Wallace	Philadelphia, Pa	May 30, 1865	47, 974
Hat	J. O'Donnell	Saint Louis, Mo	Apr. 29, 1873	138, 430
Hat	W. H. Plumb	New York, N. Y	Sept. 17, 1867	68, 899
Hat	C. L. Rahmer	Brooklyn, N. Y	June 6, 1865	48, 092
Hat	C. L. Rahmer	Brooklyn, N. Y	July 17, 1866	56, 447
Hat	C. E. Richards	North Attleborough, Mass	Apr. 23, 1867	64, 145
Hat	B. Sherwood	New York County, N. Y	Aug. 10, 1852	9, 190
Hat	B. B. Taggart and C. W. Brown	Watertown, N. Y	June 23, 1868	79, 278
Hat	W. H. Towers	New York, N. Y	Aug. 22, 1865	49, 570
Hat	P. W. Vail	Newark, N. J	Aug. 3, 1869	93, 251
Hat	W. F. Warburton	Philadelphia, Pa	Jan. 2, 1855	12, 173
Hat	W. F. Warburton	Philadelphia, Pa	May 31, 1864	42, 977
Hat	W. F. Warburton	Philadelphia, Pa	Aug. 7, 1866	57, 019
Hat	M. S. Watkins	Mansfield, Tex	Sept. 1, 1868	81, 715
Hat	W. H. White	Kent Island, Md	Sept. 25, 1866	58, 328
Hat	W. H. White	Kent Island, Md	Sept. 25, 1866	58, 329
Hat	W. H. White	New York, N. Y	May 19, 1868	78, 036
Hat	W. H. White	Baltimore, Md	Mar. 14, 1871	112, 661
Hat	D. Wilcox	Boston, Mass	Mar. 23, 1869	88, 107
Hat	D. Wilcox	Boston, Mass	June 8, 1869	91, 189
Hat	S. Wilks and J. E. Dow	Poland, Russia, and Boston, Mass.	July 16, 1872	129, 194

Index of patents issued from the United States Patent Office from 1790 *to* 1873, *inclusive*—Continued.

Invention.	Inventor.	Residence.	Date.	No.
Hat, Adjustable	I. Y. Cassiano	San Antonio, Tex	Dec. 2, 1873	145, 092
Hat and block for forming, Ventilated	W. F. Warburton	Philadelphia, Pa	May 14, 1872	126, 854
Hat and bonnet	J. Dennis, jr	Portsmouth, R. I	May 2, 1843	3, 065
Hat and bonnet box	O. A. Dailey	Washington, D. C	Mar. 14, 1865	46, 782
Hat and bonnet pressing apparatus	W. E. Doubleday	Brooklyn, N. Y	Sept. 20, 1864	44, 293
Hat and bonnet pressing machine	H. E. West	Attleborough, Mass	July 25, 1865	49, 048
Hat and cap	W. H. White	West River, Md	Apr. 17, 1866	54, 049
Hat and cap	W. H. White	West River, Md	May 22, 1866	54, 986
Hat and cap bands, Manufacture of India-rubber	W. H. Halsey	Hoboken, N. J	Apr. 10, 1866	53, 920
Hat and cap bracket	T. Lawrence	Peoria, Ill	Sept. 13, 1870	107, 389
Hat and cap ear covering	B. R. Morse	Galesburgh, Ill	Apr. 22, 1873	138, 040
Hat and cap holder	C. Beeny	Albany, N. Y	Mar. 21, 1871	112, 767
Hat and cap holder and bracket	J. R. Hartman and M. Kais	Peoria, Ill	Aug. 2, 1870	105, 938
Hat and cap linings, Sweat-knife for cutting	E. R. Pye	New York, N. Y	Nov. 8, 1859	26, 052
Hat and cap ventilator	W. Dale	New York, N. Y	Nov. 7, 1871	120, 629
Hat and clothes hook	T. Bowerman	Newark, N. J	Apr. 30, 1872	126, 256
Hat and coat guard	S. M. Robbins	New York, N. Y	Feb. 18, 1873	136, 005
Hat and coat holder	E. J. Post	Vienna, N. J	Jan. 29, 1867	61, 633
Hat and coat hook	C. G. Cole	Bennington, Vt	Feb. 18, 1873	135, 968
Hat and coat hook	J. Danner	Canton, Ohio	Mar. 18, 1873	136, 905
Hat and coat hook	A. J. Goodrich	Wolcottville, Conn	Oct. 24, 1871	120, 156
Hat and coat rack	G. T. Benson	Jersey City N. J	Nov. 8, 1870	108, 961
Hat and coat rack	T. W. Brown	New York, N. Y	Oct. 1, 1867	69, 311
Hat and coat rack	G. F. J. Colburn	Newark, N. J	Feb. 7, 1865	46, 296
Hat and coat rack	J. E. Osborn	Chicago, Ill	June 22, 1869	91, 559
Hat and coat rack, Safety	F. W. Roth	Washington, D. C	Apr. 20, 1869	89, 171
Hat and dish rack	W. W. Horton	Katonah, N. Y	Apr. 28, 1868	77, 377
Hat and parcel holder	A. Clarke	Philadelphia, Pa	Mar. 11, 1873	136, 587
Hat and velvet polisher	J. A. Thompson	Auburn, N. Y	Mar. 8, 1864	41, 895
Hat bats, Machine for forming felt	W. G. Hagaman	Philadelphia, Pa	Feb. 3, 1857	16, 543
Hat-block	J. B. Brown	Middletown, N. Y	Sept. 24, 1867	69, 172
Hat-block	W. W. Cumberland	Newark, N. J	Feb. 19, 1861	31, 442
Hat-block	W. W. Jemeson	Wheeling, Ohio	Mar. 30, 1836	....
Hat-block	J. H. Masker	Newark, N. J	July 21, 1863	39, 299
Hat-block	H. A. Robison	Cleveland, Ohio	May 8, 1866	54, 601
Hat-block, Expansible	R. Eickemeyer	Yonkers, N. Y	July 29, 1873	141, 338
Hat-block to color hats	A. and S. Chichester	Wilton, Conn	May 29, 1835	....
Hat-blocks, Apparatus for holding	J. White	Cleveland, Ohio	Aug. 10, 1869	93, 571
Hat-blocks, Machine for turning	J. H. Masker	Newark, N. J	Jan. 3, 1860	26, 691
Hat blocking and shaping machine	R. Eickemeyer	Yonkers, N. Y	Dec. 31, 1867	72, 726
Hat blocking and shaping machine	G. Osterheld and R. Eickemeyer	Yonkers, N. Y	Apr. 3, 1866	53, 661
Hat blocking and stretching machine	J. De La Mar	Brooklyn, N. Y	Oct. 11, 1870	108, 116
Hat blocking and stretching machine	G. H. Hawkins	New York, N. Y	Aug. 4, 1868	80, 733
Hat blocking and stretching machine	G. A. Mandeville and W. E. Pine.	Newark, N. J	Jan. 21, 1868	73, 541
Hat-blocking machine	J. A. Ball and P. B. Lawson	Cold Spring, N. Y	Nov. 3, 1868	83, 590
Hat-blocking machine	C. H. Berry	Brooklyn, N. Y	Nov. 7, 1871	120, 699
Hat-blocking machine	S. Boyden	Newark, N. J	Nov. 13, 1866	59, 707
Hat-blocking machine	J. De La Mar	Brooklyn, N. Y	Feb. 26, 1867	62, 321
Hat-blocking machine	J. Eberhardt	Newark, N. J	June 23, 1868	79, 059
Hat-blocking machine	E. C. Fales	Foxborough, Mass	May 20, 1873	139, 133
Hat-blocking machine	W. C. Griswold	New York, N. Y	Jan. 7, 1868	72, 998
Hat-blocking machine	W. C. Griswold, A. Pelisse, and A. H. Hook.	Brooklyn, N. Y., Newark, N. J., and New York, N. Y.	Aug. 25, 1868	81, 364
Hat-blocking machine	S. S. Middlebrook	Sandy Hook, Conn	Apr. 2, 1867	63, 549
Hat-blocking machine	J. Sheldon	New York, N. Y	Mar. 19, 1867	63, 109
Hat-blocking machine	J. Sheldon	New York, N. Y	Apr. 2, 1867	63, 434
Hat-blocking machine	J. Sheldon	New York, N. Y	May 4, 1869	89, 601
Hat-blocking machine	J. Sheldon	New York, N. Y	May 31, 1870	103, 787
Hat-blocking machine	J. Surerus and W. H. Behrens	Newark, N. J	May 3, 1870	102, 613
Hat-blocking machines, Adjustable banding-ring for.	G. Hayden	Newark, N. J	Nov. 7, 1871	120, 740
Hat-blocking machines, Adjustable crown-blocks for.	W. C. Griswold	Newark, N. J	Nov. 7, 1871	120, 739
Hat-bodies, Alarm-attachment for machine for forming.	W. C. Waring	Yonkers, N. Y	Jan. 3, 1871	110, 700
Hat-bodies, Apparatus for forming	G. and C. Beatty	Norwalk, Conn	May 21, 1861	32, 346
Hat-bodies, Apparatus for stretching	T. G. Oakley and W. R. Finch	Brooklyn, N. Y	Oct. 20, 1863	40, 383
Hat-bodies, Apparatus for stretching	W. H. Stevens	Danbury, Conn	Sept. 27, 1864	44, 468
Hat-bodies, Batting or web for	T. Blanchard	New York, N. Y	June 14, 1837	230
Hat-bodies by machinery at one operation, Manufacturing.	G. L. Thatcher	Brooklyn, N. Y	May 31, 1828	....
Hat-bodies by machinery, Making	L. Van Hoesen	Norwalk, Conn	June 19, 1829	....
Hat bodies by machinery, Roll for felting or sizing	P. W. Somers	Danbury, Conn	Dec. 4, 1866	60, 275
Hat-bodies by sewing, Manner of cutting fabrics to be made into.	I. L. Chapman	New York, N. Y	May 1, 1845	4, 021
Hat-bodies, Clearing	W. F. Young	Buffalo, N. Y	Aug. 20, 1872	130, 781
Hat-bodies, Composition for stiffening	J. M. Bottum	New York, N. Y	June 21, 1864	43, 176
Hat-bodies, Composition for stiffening	T. Trowbridge	Danbury, Conn	June 20, 1865	48, 328
Hat-bodies, Conductor in machine for forming	L. E. Hopkins	New York, N. Y	Aug. 30, 1853	9, 970
Hat-bodies, Device for printing	A. Barnes	Newark, N. J	Apr. 14, 1868	76, 696
Hat-bodies, &c., Dyeing	A. C. Brush and G. C. White	Norwalk and Danbury, Conn.	Oct. 9, 1866	58, 591
Hat-bodies, Felted	R. Johnson	Danbury, Conn	Jan. 28, 1868	73, 812
Hat-bodies, Felting	E. R. Barnes and J. B. Blakslee	Brookfield and Newton, Conn.	Aug. 12, 1856	15, 508
Hat-bodies, Felting	W. W. Cumberland	Newark, N. J	Oct. 23, 1855	13, 698
Hat-bodies, Felting	I. Searles	Newark, N. J	Dec. 18, 1855	13, 963
Hat-bodies, Forming	S. Graham	Fayette, Me	May 15, 1832	....
Hat-bodies, Forming	M. Osborne	New York, N. Y	Nov. 18, 1845	4, 275
Hat-bodies, Forming	A. H. Stevens	Richland, N. Y	Jan. 27, 1830	....
Hat-bodies, Forming	H. F. West and A. H. Stevens	Richland, N. Y	Oct. 29, 1828	....
Hat-bodies, Forming	J. Wharton	Newark, N. J	Dec. 2, 1873	145, 262
Hat-bodies, Forming and hardening	A. Spencer and A. Loeschner	New York and Brooklyn, N. Y.	Apr. 25, 1854	10, 832
Hat-bodies, Implement for alluring	J. Northrop and J. F. Emmons	Bridgeport, Conn	May 2, 1871	114, 327
Hat-bodies, Machine for blocking	W. A. Fenn	New Milford, Conn	Apr. 21, 1857	17, 083
Hat-bodies, Machine for blocking and stretching	W. Humphreys, jr	Cold Spring, N. Y	Apr. 23, 1867	64, 107

Index of patents issued from the United States Patent Office from 1790 *to* 1873, *inclusive*—Continued.

Invention.	Inventor.	Residence.	Date.	No.
Hat-bodies, Machine for felting	J. W. Blackham	Brooklyn, N. Y	Nov. 20, 1866	59, 758
Hat-bodies, Machine for felting	J. Bowden and R. Shaw	Marple, Great Britain	Feb. 4, 1873	135, 399
Hat-bodies, Machine for felting	C. Faure	New York, N. Y	Feb. 6, 1866	52, 489
Hat-bodies, Machine for felting	W. Fuzzard	Cambridgeport, Mass	July 8, 1856	15, 290
Hat-bodies, Machine for felting	W. Fuzzard	Cambridgeport, Mass	Apr. 7, 1857	16, 973
Hat-bodies, Machine for felting	S. H. Gray	Bridgeport, Conn	June 3, 1856	15, 008
Hat-bodies, Machine for felting	L. E. Hopkins	Brooklyn, N. Y	July 22, 1856	15, 375
Hat-bodies, Machine for felting	J. Kirk, S. Shelmerdine, and C. Froggart.	Stockport, Great Britain	Mar. 29, 1870	101, 276
Hat-bodies, Machine for felting	R. Northrop	New Milford, Conn	Mar. 19, 1861	31, 732
Hat-bodies, Machine for felting	H. L. Randall	Roxbury, Conn	Apr. 21, 1857	17, 115
Hat-bodies, Machine for felting	R. Smith	Danbury, Conn	June 4, 1861	32, 500
Hat-bodies, Machine for felting	P. W. Somers	Danbury, Conn	Dec. 4, 1866	60, 274
Hat-bodies, Machine for felting	J. Thomas	Brooklyn, N. Y	Aug. 26, 1856	15, 627
Hat-bodies, Machine for felting	J. Vero	Dewsbury near Leeds, England.	Feb. 28, 1871	112, 197
Hat-bodies, Machine for felting	J. Walber	Washington, D. C	Apr. 23, 1867	64, 049
Hat-bodies, Machine for felting and hardening	J. T. Earle	Danbury, Conn	Oct. 4, 1870	107, 887
Hat-bodies, Machine for forming	F. Degen	Newark, N. J	Dec. 10, 1867	71, 996
Hat-bodies, Machine for forming	R. Fitzgerald	Newark, N. J	Dec. 6, 1859	26, 395
Hat-bodies, Machine for forming	I. Gill	Walpole, Mass	Jan. 13, 1857	16, 426
Hat-bodies, Machine for forming	I. Nutt	Newark, N. J	Oct. 11, 1870	108, 175
Hat-bodies, Machine for forming	O. Root	Montgomery County, N. Y	Dec. 19, 1829	....
Hat-bodies, Machine for forming and constructing	B. Lapham and E. Cady	Canaan, N. Y	Apr. 15, 1829	....
Hat-bodies, Machine for forming and hardening	A. C. Arnold	Norwalk, Conn	Oct. 6, 1857	18, 316
Hat-bodies, Machine for fulling	A. Cattaneo	Newark, N. J	Mar. 3, 1868	75, 122
Hat-bodies, Machine for fulling and felting	W. M. Storm and G. H. Ennis	New York, N. Y., and Hudson County, N. J.	Nov. 3, 1868	83, 743
Hat-bodies, Machine for hardening	J. Booth	Newark, N. J	Sept. 15, 1857	18, 181
Hat-bodies, Machine for hardening	S. Boyden	Newark, N. J	Aug. 30, 1859	25, 300
Hat-bodies, Machine for making	D. Barnum	Philadelphia County, Pa	July 1, 1851	8, 195
Hat-bodies, Machine for making	A. Clark and H. Chase	Falmouth, Mass	June 11, 1829	....
Hat-bodies, Machine for making	E. F. Condit and A. Taylor	Springfield, N. J	Mar. 19, 1850	7, 182
Hat-bodies, Machine for making	J. F. Greene	Brooklyn, N. Y	Jan. 1, 1861	31, 053
Hat-bodies, Machine for manufacturing	J. Booth	Newark, N. J	Aug. 25, 1857	18, 034
Hat-bodies, Machine for manufacturing	L. E. Hopkins	New York, N. Y	Dec. 7, 1852	9, 450
Hat-bodies, Machine for molding, felting, and fulling	M. A. Johnson and W. Murkland.	Lowell, Mass	Oct. 23, 1866	59, 031
Hat-bodies, Machine for planking	P. Emmons	New York, N. Y	Apr. 19, 1853	9, 672
Hat-bodies, Machine for pouncing and napping	R. Eickemeyer	Yonkers, N. Y	Feb. 28, 1865	46, 552
Hat-bodies, Machine for shrinking	P. V. W. Bishop	Morristown, N. J	Nov. 7, 1871	120, 617
Hat-bodies, Machine for shrinking	J. S. Taylor	Danbury, Conn	May 3, 1853	9, 700
Hat-bodies, Machine for sizing	P. Emmons	New York, N. Y	July 24, 1855	13, 304
Hat-bodies, Machine for sizing	S. H. Gray	Bridgeport, Conn	Dec. 23, 1856	16, 305
Hat-bodies, Machine for sizing	W. E. Hatfield	Newark, N. J	Dec. 3, 1861	33, 867
Hat-bodies, Machine for sizing	W. H. Hoyt	Bethel, Conn	July 14, 1868	79, 833
Hat-bodies, Machine for sizing	S. C. Ketchum	Brooklyn, N. Y	May 27, 1856	14, 960
Hat-bodies, Machine for sizing	W. B. Lodge and H. Platner	Danbury, Conn., and Hudson, N. Y.	June 25, 1867	66, 097
Hat-bodies, Machine for sizing	A. Spencer	New York, N. Y	Mar. 25, 1856	14, 521
Hat-bodies, Machine for sizing	J. Thomas	Brooklyn, N. Y	June 17, 1856	15, 154
Hat-bodies, Machine for sizing	H. Warner	Newark, N. J	Jan. 14, 1868	73, 411
Hat-bodies, Machine for sizing	S. W. Wood	Washington, D. C	June 15, 1858	20, 602
Hat-bodies, Machine for sizing and felting	G. A. Mandeville	Newark, N. J	Oct. 25, 1870	108, 715
Hat-bodies, Machine for sizing or planking	W. B. Lodge and H. Platner	Danbury, Conn	June 25, 1867	66, 096
Hat-bodies, Machine for stretching	R. Eickemeyer	Yonkers, N. Y	Feb. 28, 1865	46, 553
Hat-bodies, Machine for stretching	R. Eickemeyer	Yonkers, N. Y	Jan. 21, 1873	135, 033
Hat-bodies, Machine for stretching	W. C. Griswold	Brooklyn, N. Y	Dec. 8, 1868	84, 690
Hat-bodies, Machine for stretching	P. Keffer	Reading, Pa	Mar. 17, 1868	75, 684
Hat-bodies, Machinery for felting	J. B. Blakslee and E. R. Barnes	Newtown and Brookfield, Conn.	Mar. 17, 1857	16, 823
Hat-bodies, Machinery for felting	J. B. Blakslee and S. S. Middlebrook.	Newtown, Conn	Jan. 10, 1860	26, 744
Hat-bodies, Machinery for felting	W. Fuzzard	Newark, N. J	Feb. 13, 1855	12, 386
Hat-bodies, Machinery for felting	M. R. Lemman	Columbus, Miss	May 8, 1860	28, 185
Hat-bodies, Machinery for felting	S. S. Middlebrook and J. B. and C. F. Blakslee.	Newtown, Conn	Jan. 30, 1855	12, 321
Hat-bodies, Machinery for felting	J. S. Taylor	Danbury, Conn	May 6, 1856	14, 845
Hat-bodies, Machinery for felting	J. Thomas	Brooklyn, N. Y	July 1, 1856	15, 261
Hat-bodies, Machinery for forming	P. Arneson	Newark, N. J	Aug. 23, 1859	25, 168
Hat-bodies, Machinery for forming	D. Barnum	New York, N. Y	Oct. 17, 1854	11, 805
Hat-bodies, Machinery for forming	S. Boyden	Newark, N. J	Jan. 25, 1859	22, 698
Hat-bodies, Machinery for forming	S. Boyden	Newark, N. J	Jan 10, 1860	26, 811
Hat-bodies, Machinery for forming	W. Fuzzard	Charlestown, Mass	Aug. 7, 1860	29, 478
Hat-bodies, Machinery for forming	M. Hardy	New York, N. Y	Dec. 14, 1858	22, 288
Hat-bodies, Machinery for forming	L. Platt and R. Wildman	Danbury, Conn	July 31, 1860	29, 441
Hat-bodies, Machinery for forming	A. B. Taylor	Newark, N. J	July 29, 1856	15, 443
Hat-bodies, Machinery for forming	J. S. Taylor	Danbury, Conn	Oct. 14, 1856	15, 903
Hat-bodies, Machinery for forming	J. S. Taylor	Danbury, Conn	Sept. 13, 1859	25, 454
Hat-bodies, Machinery for forming	T. Walber	New York, N. Y	Aug. 17, 1852	9, 205
Hat-bodies, Machinery for forming	D. G. Wells	New York, N. Y	Sept. 9, 1856	15, 715
Hat-bodies, Machinery for forming	R. Wildman	Hartford, Conn	July 20, 1846	4, 643
Hat-bodies, Machinery for hardening	G. E. Cowperthwait	Danbury, Conn	May 17, 1859	24, 007
Hat-bodies, Machinery for making	L. W. Boynton	South Coventry, Conn	Oct. 10, 1854	11, 773
Hat-bodies, Machinery for making	W. Fosket	Ware, Mass	Jan. 23, 1846	4, 363
Hat-bodies, Machinery for making	I. Searles	Newark, N. J	Nov. 7, 1854	11, 910
Hat-bodies, Machinery for making	A. B. Taylor	Newark, N. J	Mar. 18, 1856	14, 476
Hat-bodies, Machinery for making	A. B. Taylor and H. A. Burr	New York, N. Y	Feb. 9, 1847	4, 962
Hat-bodies, Machinery for making	H. A. Wells	New York, N. Y	Apr. 25, 1846	4, 472
Hat-bodies, Machinery for manufacturing	P. Arnesen, J. Pederson, and H. Rees.	New York, N. Y	Oct. 2, 1855	13, 614
Hat-bodies, Machinery for manufacturing	L. E. Hopkins	New York, N. Y	Dec. 21, 1852	9, 484
Hat-bodies, Machinery for perforating	W. F. Warburton and W. B. Aitken.	Philadelphia, Pa	Sept. 20, 1859	25, 551
Hat-bodies, Making	A. Rankin	Newark, N. J	Oct. 3, 1854	11, 756
Hat-bodies, Making a batting or web for	H. A. Wells, J. James, and R. W. Peck.	Brooklyn, N. Y	Sept. 22, 1837	392
Hat-bodies, Manufacture of	S. Lyon	Roxbury, Mass	Feb. 5, 1847	4, 951

Index of patents issued from the United States Patent Office from 1790 *to* 1873, *inclusive*—Continued.

Invention.	Inventor.	Residence.	Date.	No.
Hat-bodies, Manufacture of	H. Moore	New York, N. Y	Apr. 20, 1837	173
Hat-bodies, Manufacture of	A. B. Taylor	Newark, N. J	Aug. 12, 1856	15, 534
Hat-bodies, Pouncing	W. H. Tupper	New York, N. Y	Nov. 15, 1859	26, 132
Hat-bodies, Preparing material for	P. Arnesen, J. Pederson, and H. Rees.	New York, N. Y	Oct. 2, 1855	13, 613
Hat-bodies, Rolling and felting	J. Crisp and V. V. Dodd	Bloomfield and Orange, N. J.	Oct. 17, 1871	119, 970
Hat-bodies, Sizing	J. McCracken	Brooklyn, N. Y	Oct. 21, 1856	15, 929
Hat-bodies, Steaming-on	S. Polley	Brooklyn, N. Y	Apr. 2, 1867	63, 556
Hat-bodies, Stiffening	H. Blynn	Newark, N. J	May 9, 1835	....
Hat-bodies, Stiffening	J. P. Kettell and J. Wright	Worcester County, Mass	June 11, 1836	....
Hat-bodies, Stiffening	J. McCracken	Brooklyn, N. Y	Sept. 2, 1856	15, 664
Hat-bodies, Stiffening	E. P. Spear	Lexington, Mass	July 2, 1836	....
Hat-bodies, Winding-machine for setting up	J. Grant	Providence, R. I	Apr. 28, 1825	....
Hat-bodies, Winding-machine for setting up	J. Grant	Providence, R. I	Apr. 10, 1827	....
Hat-bodies with fur, Process of facing wool	R. D. Hine	Matteawan, N. Y	May 19, 1868	78, 090
Hat-body	G. Johnson	Camden, N. J	Dec. 24, 1872	134, 206
Hat-body	W. Nunns	New York	Aug. 14, 1834	....
Hat-body	J. A. Pease	New York, N. Y	Jan. 22, 1867	61, 451
Hat-body bat	S. Hurlbut	Glastenbury, Conn	Feb. 14, 1831	....
Hat-body formers, Cover for cone of	F. Degen and E. R. Parsil	Newark, N. J	Aug. 22, 1871	118, 213
Hat-body machinery	A. B. Taylor	Newark, N. J	Aug. 31, 1858	21, 382
Hat-body machinery	D. Tenny	Plattsburgh, N. Y	Oct. 1, 1830	....
Hat-box	C. W. Packer	Philadelphia, Pa	July 16, 1867	66, 880
Hat-box and valise	Z. S. Durfee	Philadelphia, Pa	Oct. 16, 1866	58, 792
Hat-box-cutting die	F. Schoettle	Philadelphia, Pa	Jan. 30, 1872	123, 127
Hat-brim, Extension	I. Y. Cassiano	San Antonio, Tex	July 8, 1873	140, 614
Hat-brim, Extension	I. Y. Cassiano	San Antonio, Tex	July 8, 1873	140, 615
Hat brim, Soft	J. W. Corey	Newark, N. J	Nov. 1, 1870	108, 884
Hat-brim spring	G. Mallory	Bridgeport, Conn	Dec. 4, 1866	60, 213
Hat-brim spring	S. Peck	New Haven, Conn	Dec. 4, 1866	60, 236
Hat-brims, Apparatus for forming	F. Degan	New York, N. Y	Apr. 17, 1847	5, 065
Hat-brims, Block for cutting and shaping	F. R. Going	Boston, Mass	Mar. 24, 1868	75, 897
Hat-brims, Block for setting	S. Billings	Spring Garden, Pa	Aug. 7, 1849	6, 627
Hat-brims, Curling	F. Degen	New York, N. Y	Nov. 13, 1849	6, 867
Hat-brims, Curling	F. S. Sibley	Brooklyn, N. Y	Oct. 9, 1860	30, 379
Hat-brims, Curved spring for	E. S. Nichols	New Haven, Conn	Feb. 5, 1867	61, 856
Hat-brims, Machine for forming	C. T. Rowe	Norwalk, Conn	Jan. 30, 1872	123, 200
Hat-brims, Method of setting or ironing	F. Degen	New York	Aug. 31, 1842	2, 765
Hat-brims, Pattern for trimming	C. M. Hawes	New York, N. Y	Aug. 4, 1868	80, 732
Hat-brushing machine	C. Faure	New York, N. Y	June 13, 1865	48, 235
Hat-brushing machine	J. W. Hopkins	Brooklyn, E. D., N. Y	Feb. 23, 1869	87, 103
Hat-brushing machine	J. D. Parsons	Yonkers, N. Y	July 20, 1869	92, 877
Hat brushing and napping machine	J. W. Hopkins	Brooklyn, N. Y	Mar. 2, 1869	87, 495
Hats, Case for ladies' calash	H. Cook	Onondaga, N. Y	Feb. 20, 1833	....
Hat, Cassimere	O. Brooks and J. A. Sloan	Philadelphia, Pa	Nov. 9, 1842	2, 847
Hat-check	J. E. Smith	Waterbury, Conn	May 10, 1870	102, 982
Hat, Child's	S. B. Pratt	Boston, Mass	Sept. 17, 1872	131, 371
Hat-conformator	J. Dickson, jr	Brooklyn, N. Y	June 5, 1860	28, 564
Hat-conformator	B. Hogan	Albany, N. Y	Mar. 24, 1868	75, 760
Hat-conformator	J. F. Klein	Trenton, N. J	Feb. 24, 1863	37, 759
Hat-conformator	J. J. McKnight	Tarrytown, N. Y	Nov. 25, 1873	144, 855
Hat-conformator	E. Z. Webster	Norwich, Conn	Dec. 15, 1868	85, 041
Hat, Cork	A. C. Crondal	New York, N. Y	Nov. 8, 1864	44, 941
Hat-cushion	A. D. Purinton	Dover, N. H	June 26, 1860	28, 902
Hat, Elastic ventilating	D. Greenleaf	Vicksburgh, Miss	Nov. 26, 1836	90
Hat, Elastic water-proof	A. Buffum and J. Kelly	Smithfield, R. I	Feb. 17, 1820	
Hat, Enameled	E. S. Cheney and G. P. Perry	Providence, R. I	July 20, 1869	92, 704
Hat, Felt	J. T. Waring	Yonkers, N. Y	Aug. 6, 1872	130, 342
Hat, Felt	J. W. Whittal and W. W. Pendleton.	Greenwich, Conn	Aug. 1, 1854	11, 462
Hat, Felted	E. Sealy	Newark, N. J	May 18, 1869	90, 198
Hat-felting apparatus	W. H. Hoyt	Bethel, Conn	Apr. 15, 1862	34, 963
Hat-felting machine	J. F. Badoye	New York, N. Y	Oct. 22, 1867	70, 066
Hat-felting machine	J. W. Blackham	Brooklyn, N. Y	Dec. 10, 1867	71, 959
Hat-felting machine	A. Cataneo	Newark, N. J	June 21, 1870	104, 424
Hat-felting machine	A. C. Fuller	Danbury, Conn	May 13, 1856	14, 862
Hat-felting machinery	J. S. Taylor	Danbury, Conn	Jan. 15, 1856	14, 121
Hat-finishing apparatus	J. A. Roche and J. J. Stewart	Williamsburgh, N. Y	Sept. 26, 1865	50, 165
Hat-finishing machine	J. Cooper and T. Barnett	Philadelphia, Pa	May 31, 1826	....
Hat-finishing machine	J. H. La Ban	New York, N. Y	Jan. 22, 1861	31, 177
Hat finishing machine	J. C. Richardson	Newark, N. J	Jan. 19, 1869	85, 964
Hat-finishing machine	G. W. Stout	Newark, N. J	Nov. 8, 1870	109, 072
Hat-finishing press	W. W., jr., and M. J. Walsh	Brooklyn, N. Y	Oct. 2, 1866	58, 513
Hat, Fireman's	J. Brown	Newark, N. J	Nov. 1, 1845	4, 246
Hat-joiner	L. Simonet	Paris, France	Apr. 2, 1867	63, 435
Hat-forming die	R. T. Wilde and S. H. Lyon	Brooklyn, N. Y	Nov. 11, 1862	36, 927
Hat-forming die	R. T. Wilde and S. H. Lyon	Brooklyn, N. Y	Mar. 24, 1863	37, 993
Hat-forming die and block	L. Bommer	New York, N. Y	Nov. 3, 1868	83, 597
Hat-forming machine	D. Dennis	Barre, Mass	Sept. 12, 1871	118, 914
Hat-forming machine	H. N. Swift	Matteawan, N. Y	Mar. 13, 1866	53, 196
Hat-forming mold	J. E. Ward	Bredbury, Great Britain	Dec. 22, 1868	85, 191
Hat-frame	A. Komp	New York, N. Y	June 27, 1865	48, 412
Hat-frame, Metallic skeleton	A. Komp	New York, N. Y	Apr. 25, 1865	47, 431
Hat-frames, Machine for forming	B. Ertola and A. Caselli	New York, N. Y	Oct. 21, 1873	143, 814
Hat-frames, Machine for pressing	L. P Faught	Foxborough, Mass	July 30, 1872	130, 031
Hat-holder	Z. Waters	Bloomington, Ill	July 28, 1868	80, 374
Hat-holder for pews and seats	J. M. Cain	La Fayette, Wis	Mar. 16, 1869	87, 756
Hat-hook	J. Harvey	Scranton, Pa	Mar. 3, 1868	75, 159
Hat-hook	C. L. Spencer	Providence, R. I	Oct. 27, 1868	83, 562
Hat-hook for pews	R. W. Whitney and J. W. Shaw.	South Berwick, Me., and Concordia, N. H.	Aug. 6, 1867	67, 620
Hat-ironing machine	W. Best	Newark, N. J	Mar. 27, 1866	53, 401
Hat-ironing machine	W. Best and M. S. Drake	Newark, N. J	Mar. 27, 1866	53, 518
Hat-ironing machine	M. S. Drake	Newark, N. J	Aug. 21, 1866	57, 437
Hat-ironing machine	R. Eickemeyer	Yonkers, N. Y	Dec. 30, 1873	145, 934
Hat-ironing machine	G. W. Stout and J. C. Richardson.	Newark, N. J	Dec. 3, 1867	71, 662

Index of patents issued from the United States Patent Office from 1790 *to* 1873, *inclusive*—Continued.

Invention.	Inventor.	Residence.	Date.	No.
Hat-ironing machine	G. W. Stout and J. C. Richardson.	Newark, N. J	Dec. 15, 1868	84, 915
Hat-ironing machine	J. T. Waring	Yonkers, N. Y	Oct. 15, 1867	69, 950
Hat-ironing machinery	S. A. Kinsman and S. Field	Barre, Mass	Mar. 11, 1856	14, 401
Hat, Leather	A. Bogardus	Newburgh, N. Y	June 4, 1872	127, 550
Hat lining and tips, Ornamenting	T. W. Bracher	New York, N. Y	June 6, 1871	115, 689
Hat-linings, Embossing	T. W. Bracher	New York, N. Y	Jan. 10, 1871	110, 891
Hat-linings, Press for printing	W. Moultrie	New York, N. Y	Dec. 16, 1856	16, 245
Hat-machine	C. M. Osgood	Amherst, Mass	Jan. 17, 1871	111, 079
Hat machine, Paper	H. Kellogg	New Haven, Conn	June 4, 1867	65, 393
Hat, Man's	T. W. Adams	Baltimore, Md	Dec. 24, 1861	34, 043
Hat, Military	C. L. Pascal	Philadelphia, Pa	Dec. 10, 1861	33, 900
Hat, Military	W. F. Warburton	Philadelphia, Pa	Feb. 25, 1862	34, 538
Hat or bonnet	D. Scrymgeour	Foxborough, Mass	Dec. 5, 1871	121, 668
Hat or cap ventilator	W. Dale	New York, N. Y	Dec. 19, 1871	121, 995
Hat or hat-body felting machinery	J. B. Laville	Paris, France	Sept. 26, 1854	11, 731
Hat, Palm-leaf	F. Groening	Brooklyn, N. Y	June 25, 1836	....
Hat, Paper-pulp	J. A. Pease	New York, N. Y	Oct. 18, 1864	44, 744
Hat-planking machine	R. Bacon	Boston, Mass	Jan. 31, 1827	....
Hat-pouncing apparatus	E. Nougaret	Newark, N. J	Feb. 20, 1866	52, 739
Hat-pouncing machine	J. B. Brown	Danbury, Conn	Oct. 25, 1870	108, 562
Hat-pouncing machine	M. S. Drake	Newark, N. J	Feb. 5, 1867	61, 814
Hat-pouncing machine	R. Eickemeyer	Yonkers, N. Y	Nov. 23, 1869	97, 178
Hat-pouncing machine	J. L. Labiaux	Newark, N. J	Mar. 26, 1867	63, 261
Hat-pouncing machine	E. Mougaret	Newark, N. J	Sept. 18, 1866	58, 126
Hat-pouncing machine	A. Pelisse and G. W. Stoute	Newark, N. J	Apr. 4, 1871	113, 446
Hat-pouncing machine	J. C. Richardson	Newark, N. J	Jan. 7, 1868	73, 044
Hat-pouncing machine	J. C. Richardson	Newark, N. J	Aug. 17, 1869	93, 836
Hat-pouncing machine	J. Rosencranz	Boston, Mass	Sept. 28, 1869	95, 270
Hat-pouncing machine	P. W. Vail	Newark, N. J	Sept. 24, 1867	69, 277
Hat-pouncing machine	C. H. Van Houten and J. M. Crane.	Newark, N. J	Apr. 9, 1867	63, 674
Hat-pouncing machine	S. S. Wheeler	Danbury, Conn	Aug. 24, 1869	94, 158
Hat-pouncing machine	S. S. Wheeler and D. B. Manley	Danbury, Conn	Aug. 14, 1866	57, 232
Hat-press	G. Bartel	New York, N. Y	Aug. 29, 1871	118, 578
Hat-press	S. Beatty	Norwalk, Conn	Mar. 14, 1871	112, 677
Hat-press	R. Eickemeyer	Yonkers, N. Y	May 13, 1873	138, 871
Hat-press	R. Eickemeyer	Yonkers, N. Y	June 24, 1873	140, 335
Hat-press	G. C. Howard	Philadelphia, Pa	May 4, 1869	89, 834
Hat pressing and ironing apparatus	H. A. Robison	Cleveland, Ohio	July 22, 1862	35, 954
Hat-pressing apparatus	J. S. Giles, W. Halladay, and J. A. Rue.	New York and Brooklyn, N. Y.	Dec. 23, 1862	37, 255
Hat-pressing apparatus	M. Morse	Franklin, Mass	Mar. 6, 1866	53, 083
Hat-pressing apparatus	H. L. Sweet	Foxborough, Mass	Mar. 31, 1863	38, 067
Hat-pressing device	C. M. Osgood	Amherst, Mass	Feb. 4, 1873	135, 557
Hat-pressing die	J. Desper	Barre, Mass	June 10, 1873	139, 708
Hat-pressing die	J. Desper	Barre, Mass	Nov. 4, 1873	144, 262
Hat-pressing die	W. E. Doubleday and S. H. Lyon	Brooklyn, N. Y	Mar. 5, 1861	31, 592
Hat-pressing machine	S. G. Congdon and D. C. Mowry	Mansfield and Milford, Mass.	Dec. 27, 1864	45, 588
Hat-pressing machine	J. De La Mar	Brooklyn, N. Y	Jan. 12, 1867	62, 018
Hat-pressing machine	W. E. George	Wrentham, Mass	June 30, 1868	79, 338
Hat-pressing machine	I. T. Green	Milford, Conn	Oct. 2, 1866	58, 544
Hat-pressing machine	E. B. Hastings and J. Desper	Palmer and Barre, Mass	Feb. 4, 1873	135, 428
Hat-pressing machine	N. B. Hooper	Newark, N. J	Apr. 10, 1866	53, 826
Hat-pressing machine	S. Howard	Luton, Great Britain	July 21, 1868	80, 181
Hat-pressing machine	M. Morse	Franklin, Mass	Aug. 28, 1866	57, 638
Hat-pressing machine	M. and C. H. Morse	Franklin, Mass	Oct. 1, 1867	69, 358
Hat-pressing machine	J. Stearns	Templeton, Mass	July 8, 1851	8, 213
Hat-pressing machine	G. L. Thompson	New York, N. Y	Nov. 27, 1866	60, 088
Hat-pressing machine	F. Wolfram	New York, N. Y	Mar. 9, 1869	87, 744
Hat-pressing machinery	D. Brown	Wrentham, Mass	July 22, 1873	140, 993
Hat-pressing machinery	B. Potter, jr	Templeton, Mass	Oct. 29, 1850	7, 745
Hat-protector	G. W. Bidwell	New Haven, Conn	Aug. 14, 1866	57, 074
Hat-protector	H. Hodgson	New York, N. Y	Oct. 23, 1866	59, 019
Hat-rack	N. Clarke	Boston, Mass	Feb. 18, 1873	136, 032
Hat-rack	C. H. Keener	Baltimore, Md	July 24, 1866	56, 569
Hat-rack	T. S. Lambert	Peekskill, N. Y	Oct. 6, 1863	40, 173
Hat-rack and seat	E. S. Blake	Pittsburgh, Pa	Aug. 14, 1866	57, 076
Hat-rack and umbrella-stand combined	J. W. Currier	Springfield, Mass	Mar. 12, 1872	124, 552
Hat-rack, Extension	N. Hayden	Chicago, Ill	Oct. 25, 1870	108, 590
Hat-rack, Folding	N. Hayden	Chicago, Ill	Dec. 26, 1871	122, 168
Hat-rack, Folding	M. L. Roddy	Cincinnati, Ohio	Oct. 29, 1872	132, 692
Hat-rack for cigars, &c	C. Branwhite	Williamsburgh, N. Y	Apr. 24, 1860	28, 034
Hat-rack for seats	P. Sylla	Elgin, Ill	Feb. 25, 1868	74, 953
Hat-rounder and hat-gage	S. Jones	Bridgeport, Pa	June 2, 1819	....
Hat, Safety	J. J. Giltenan	Cincinnati, Ohio	Feb. 23, 1869	87, 164
Hat-shade	M. L. Battle	Bainbridge, Ga	July 9, 1872	128, 780
Hat-shaper	A. and F. Brown	New York, N. Y	July 11, 1854	11, 302
Hat-shaping apparatus	S. H. Apple, B. Lindheim, and H. Baer.	New York, N. Y	Dec. 4, 1860	30, 791
Hat-shaping machine	G. W. Gallagiere and E. W. Ruby.	New Milford, N. Y	Apr. 13, 1869	88, 865
Hat-shaping machine	J. P. Morlat	Paris, France	Feb. 1, 1870	99, 458
Hat-shaping machine	S. Wing	Munson, Mass	Aug. 17, 1869	93, 783
Hat-shaping machine	R. Cooke	New Hartford, N. Y	Dec. 2, 1873	145, 052
Hat-shell iron	J. P. Kettell	Worcester, Mass	Apr. 1, 1862	34, 838
Hat sizing and felting machine	E. R. Gardiner	Brooklyn, N. Y	May 21, 1867	64, 966
Hat-sizing machine	J. H. Hopkins	Newark, N. J	Feb. 21, 1871	111, 935
Hat-sizing machine	A. Pelisse and F. Degen	Newark, N. J	Feb. 8, 1870	99, 700
Hat-sizing machine	J. S. Taylor	Danbury, Conn	May 28, 1867	65, 300
Hat, Spring-brim	S. Collins	New Haven, Conn	Nov. 8, 1864	44, 939
Hat-stand	J. B. Wickersham	New York, N. Y	June 2, 1857	17, 461
Hat steaming and hardening machine	N. Wildman	Danbury, Conn	Apr. 2, 1824	....
Hat-steaming apparatus	C. W. Banta	Orange, N. J	July 1, 1873	140, 341
Hat stretching and blocking machine	S. Polley	Brooklyn, N. Y	Oct. 6, 1868	82, 748
Hat-stretching apparatus	J. W. Blackham	Brooklyn, N. Y	Dec. 22, 1863	41, 036
Hat-stretching machine	R. Eickemeyer	Yonkers, N. Y	June 22, 1869	91, 730

Index of patents issued from the United States Patent Office from 1790 *to* 1873, *inclusive*—Continued.

Invention.	Inventor.	Residence.	Date.	No.
Hat-stretching machine	R. Eickemeyer	Yonkers, N. Y	July 15, 1873	140, 903
Hat-stretching machine	R. Eickemeyer	Yonkers, N. Y	July 29, 1873	141, 337
Hat-stretching machine	S. Goodman	Natick, Mass	Jan. 21, 1873	135, 112
Hat-stretching machine	A. Pelisse	Newark, N. J	Jan. 4, 1870	98, 624
Hat supporter and ventilator, Combined	J. A. Borthwick	Philadelphia, Pa	Jan. 31, 1871	111, 309
Hat-sweat	B. W. Fay	Boston, Mass	Sept. 2, 1862	36, 344
Hat-sweat	G. W. Thompson	Brooklyn, N. Y	Sept. 22, 1868	82, 364
Hat sweat-band	W. Burgyes	Chelsea, Mass	Apr. 22, 1862	35, 011
Hat-sweat from water-proof paper	W. M. Waterbury	New York, N. Y	Dec. 12, 1871	121, 829
Hat-tip	S. H. Greenbaum	New York, N. Y	Oct 26, 1869	96, 222
Hat tips and linings, Ornamentation of	T. W. Bracher	New York, N. Y	Apr. 9, 1872	125, 534
Hat-tips, Cutting	T. W. Bracher	New York, N. Y	Jan. 8, 1869	91, 077
Hat-tips, Lining	W. H. M. Pye	Brooklyn, N. Y	Aug. 12, 1873	141, 727
Hat-tips, Machine for pulling out	J. Stevens	Chicopee Falls, Mass	Feb. 5, 1867	61, 773
Hat-tips, Machinery for pulling out	W. Selpho	New York, N. Y	Apr. 24, 1847	5, 082
Hat-tips, Printing	M. J. Duffy	New York, N. Y	Oct. 1, 1872	131, 858
Hat-trimming, Machine for covering reeds for	G. A. Shephard	Bethel, Conn	May 3, 1870	102, 722
Hat, Ventilated	J. N. Genin	New York, N. Y	July 22, 1845	4, 118
Hat, Ventilated	W. F. Warburton	Philadelphia, Pa	Dec. 11, 1860	30, 900
Hat-ventilation	J. McMannus	New York, N. Y	Jan. 3, 1860	26, 692
Hat-ventilator	D. K. Albright and L. H. De Laage.	Philadelphia, Pa., and Bordentown, N. J.	Nov. 21, 1865	50, 999
Hat-ventilator	S. Beatty	Norwalk, Conn	May 31, 1870	103, 703
Hat-ventilator	S. Beatty	Norwalk, Conn	Aug. 1, 1871	117, 506
Hat-ventilator	F. H. Bell	Washington, D. C	July 10, 1860	29, 050
Hat-ventilator	T. W. Bracher	New York, N. Y	July 20, 1869	92, 785
Hat-ventilator	T. W. Bracher	New York, N. Y	Sept. 21, 1869	95, 074
Hat-ventilator	B. J. Burnett	Mount Vernon, N. Y	Mar. 27, 1866	53, 406
Hat-ventilator	C. H. Coffin	San Francisco, Cal	Oct. 15, 1867	69, 900
Hat-ventilator	W. Dale	New York, N. Y	Apr. 16, 1872	125, 725
Hat-ventilator	G. Deas	New York, N. Y	Apr. 21, 1868	77, 007
Hat-ventilator	P. N. Horsley	Jersey City, N. J	June 3, 1862	35, 446
Hat-ventilator	H. A. House	Bridgeport, Conn	Mar. 28, 1871	113, 054
Hat-ventilator	W. M. Irwin and A. H. Moses	Montgomery, Ala	Sept. 1, 1868	81, 784
Hat-ventilator	J. Jenkinson	Brooklyn, N. Y	Apr. 24, 1860	27, 985
Hat-ventilator	A. O. Knapp	Danbury, Conn	Apr. 23, 1872	125, 962
Hat-ventilator	A. Komp	New York, N. Y	Sept. 5, 1865	49, 767
Hat-ventilator	G. Munro	Philadelphia, Pa	July 6, 1869	92, 340
Hat-ventilator	E. G. Nichols	Beaufort, S. C	Oct. 24, 1871	120, 163
Hat-ventilator	J. Pollock	Morrisania, N. Y	Apr. 3, 1860	27, 736
Hat-ventilator	G. B. Smith	Detroit, Mich	July 23, 1872	129, 866
Hat-ventilator	W. Smith	Philadelphia, Pa	Oct. 3, 1865	50, 285
Hat-ventilator	A. Solmans	New York, N. Y	Apr. 18, 1871	113, 941
Hat-ventilator	C. C. Stremme	Austin, Tex	Mar. 30, 1869	88, 527
Hat-ventilator	W. F. Warburton	Philadelphia, Pa	Nov. 24, 1868	84, 323
Hat-ventilator	W. F. Warburton	Philadelphia, Pa	Nov. 22, 1870	109, 473
Hat-ventilator	A. H. Young	Milford, Mass	Aug. 6, 1872	130, 173
Hat, Water-proof	R. Mills	Baltimore, Md	Feb. 27, 1832	....
Hat, Water-proof silk	G. B. Dexter	Boston, Mass	Jan. 6, 1836	....
Hat wiring and binding machine	H. A. Whiting	New York, N. Y	Apr. 29, 1873	138, 459
Hats and bonnets, Apparatus for forming	S. H. Lyon	New York, N. Y	Jan. 19, 1864	41, 342
Hats and bonnets, Fabric for	E. Copleston	Wrentham, Mass	Nov. 20, 1866	59, 822
Hats and bonnets, Fabric for	P. and A. Erhard	New York, N. Y	June 22, 1869	91, 614
Hats and bonnets, Fabric for	H. Loenberg	Boston, Mass	Aug. 5, 1862	36, 097
Hats and bonnets, Fabric for	H. Lowenberg	New York, N. Y	Feb. 28, 1865	46, 568
Hats and bonnets, Finishing leghorn	J. Snyder	Philadelphia, Pa	Aug. 21, 1834	....
Hats and bonnets, Manufacture of palm-leaf	D. C. Perrin	Roxbury, Mass	Jan. 13, 1863	37, 409
Hats and bonnets of horse-hair, Manufacture of	C. L. Noe	New York, N. Y	July 20, 1843	3, 188
Hats and bonnets, Pressing	S. E. Pettee	Foxborough, Mass	Feb. 6, 1855	12, 353
Hats and caps, Fabric for the manufacture of	T. Garceau and E. De La Granja	Boston, Mass	Mar. 10, 1868	75, 406
Hats and caps, Shaping and embossing	A. L. Bayley	Amesbury, Mass	Apr. 9, 1861	31, 944
Hats and caps, Ventilating	J. O. Blythe	Germantown, Pa	Aug. 20, 1861	33, 121
Hats and caps, Ventilating sweat-leather for	O. A. Taylor	San Francisco, Cal	Oct. 11, 1864	44, 697
Hats and caps, Water-proof sweat-band for	P. F. Lenhart	Brooklyn, N. Y	June 28, 1870	104, 861
Hats, and frame-block for, Dyeing and airing	N. L. Canfield	Wilton, Conn	Mar. 14, 1834	....
Hats and making cloth without yarn, Machine for planking.	J. Pitkin and T. Kimball	Hartford, Conn	Oct. 19, 1807	....
Hats, Apparatus for attaching mourning-badge to	T. H. Lowerre	New York, N. Y	June 27, 1865	48, 417
Hats, Apparatus for making brush	J. C. Raake	Brooklyn, N. Y	Apr. 26, 1864	42, 508
Hats, Attaching wire to brims of	C. F. Bosworth	Milford, Conn	Aug. 18, 1868	81, 246
Hats, Blocking	D. Beard	Guilford, N. C	May 28, 1816	....
Hats, Blocking	W. Wright	Philadelphia, Pa	Oct. 10, 1848	5, 849
Hats, bonnets, &c., Compound for coating textile fabrics for manufacture of.	J. L. Kendall and R. H. Trested	Foxborough, Mass., and Jamaica, N. Y.	Feb. 9, 1869	86, 841
Hats, bonnets, &c., Machine for ironing and pressing	R. Murdock	Baltimore, Md	June 17, 1840	1, 635
Hats, bonnets, &c., Machine to cut chips or strips of wood to make chip.	J. Roberts, A. D. Allen, and E. Kelsey.	....	Sept. 5, 1804	....
Hats, bonnets, &c., Machine to cut strips or chips of wood to make chip.	A. D. Allen	....	May 10, 1804	....
Hats, bundles, &c., Rack for	C. A. Young	Providence, R. I	Mar. 12, 1872	124, 469
Hats by application of hot water, Napping	H. Wortham and G. S. Petty	Lancaster, Ky	July 9, 1832	....
Hats by application of steam, Making	W. Harkins	Wheeling, Va	Dec. 29, 1826	....
Hats, &c., by machinery, Mode of planking wool	L. L. Macomber	Bath, Me	Dec. 31, 1825	....
Hats by steam, Manufacture of	D. Sutton	Lancaster, Ky	June 13, 1831	....
Hats, caps, &c., Application of bark for	S. G. Whipple	Hallowell, Mass	Apr. 17, 1807	....
Hats, caps, bonnets, &c., Making	E. Pratt	Cambridge, Mass	May 16, 1835	....
Hats, Carding fur and wool for	A. Buffum	....	Feb. 21, 1807	....
Hats, cloths, &c., from a fleece or sheet taken from a carding-engine without bowing, Manufacturing.	R. Pitkin	East Hartford, Conn	Apr. 23, 1808	....
Hats, &c., Coloring	J. Taylor	Danbury, Conn	Nov. 20, 1823	....
Hats, Coloring woolen	G. D. Foote	Danbury, Conn	Sept. 27, 1859	25, 564
Hats, Combined sweat-band, fastener, and size-mark for.	T. A. Lawrence and J. H. Murfey.	New York, N. Y	Mar. 17, 1868	75, 685
Hats, Composition for stiffening	C. Bent and F. Bush	Chelmsford, Mass	Nov. 3, 1831	....
Hats, Composition for stiffening wool and fur	L. L. Macomber	Gardiner, Me	Jan. 19, 1831	....

Index of patents issued from the United States Patent Office from 1790 *to* 1873, *inclusive*—Continued.

Invention.	Inventor.	Residence.	Date.	No.
Hats, Composition of materials for the manufacture of.	F. Messer	Guildhall, Vt	Jan. 23, 1812	....
Hats, Cooling and coloring	R. Pike	Wilton, Conn	Jan. 23, 1834	....
Hats, Dyeing	A. Gould	Washington, Conn	Jan. 11, 1836	....
Hats, Embossing	R. Eickemeyer	Yonkers, N. Y	Feb. 8, 1870	99, 544
Hats, Felting	M. D. Whipple	Cambridge, Mass	Dec. 1, 1868	84, 599
Hats, Felting and napping	T. J. Cornell	Randolph, Vt	Feb. 11, 1831	....
Hats, Finishing palm-leaf	D. Dennis	Barre, Mass	July 4, 1854	11, 214
Hats, Forming bell-crown	W. E. Doubleday	Brooklyn, N. Y	Feb. 17, 1863	37, 682
Hats from a mixture of clothier's flocks with sheep's wool, Manufacturing felt.	J. Carleton	Salem, N. H	Jan. 21, 1814	....
Hats, fur, &c., Process of coloring	H. Hibbard	Attica, N. Y	May 25, 1838	746
Hats, Gearing of cones for bowing	T. F. Mayhew	Boston	Aug. 22, 1827	....
Hats, Machine for basoning, hardening, and steaming.	J. Sizer	New London, Conn	Nov. 6, 1811	....
Hats, Machine for carding, winding, and making.	I. Sanford	Blockley, Pa	Oct. 11, 1828	....
Hats, Machine for coating	J. F. Mathias and D. M. Legat.	Paris, France	June 18, 1867	65, 926
Hats, Machine for coloring	G. M. Johnson	Port Deposit, Md	Dec. 31, 1839	1, 454
Hats, Machine for forming the brims of felt	W. A. Fenn	New Milford, Conn	Apr. 14, 1857	17, 033
Hats, Machine for lustering	R. Eickemeyer	Yonkers, N. Y	July 29, 1873	141, 339
Hats Machine for making bats or frames for wool.	R. Gookins	Hampshire, Mass	Jan. 24, 1806	....
Hats, Machine for making bodies of fur	H. A. Wells, J. James, and R. W. Peck.	Brooklyn, N. Y	Sept. 20, 1837	383
Hats, Machine for making felt in the manufacture of fur.	H. S. Miller	Philadelphia, Pa	Mar. 5, 1839	1, 094
Hats, Machine for making leather and India-rubber.	S. Gleason	New York, N. Y	Nov. 24, 1843	3, 357
Hats, Machine for napping	A. Rankin	Newark, N. J	Sept. 20, 1839	1, 328
Hats, Machine for polishing and equalizing the nap or pile on the surface of.	J. Loudon and T. Shaw	New York, N. Y	May 7, 1845	4, 033
Hats, Machine for pressing bell-crowned	L. P. Faught	Foxborough, Mass	June 25, 1872	128, 293
Hats, Machine for pressing palm-leaf	C. Gorham	Barre, Mass	Mar. 3, 1840	1, 503
Hats, Machine for pressing straw or chip	S. Prince	New York, N. Y	Aug. 31, 1808	....
Hats, Machine for printing	T. Byrne and T. Henry	New York, N. Y	May 2, 1865	47, 516
Hats, Machine for raising and drying the nap of	M. Esselen	Roxbury, Mass	Aug. 13, 1867	67, 740
Hats, Machine for stretching the brims of	J. Sheldon	Newark, N. J	May 23, 1871	115, 114
Hats, Machine for washing and cleaning	W. Carlock	Baltimore, Md	Mar. 12, 1830	....
Hats, Machinery for forming brims of felt	W. A. Fenn	Brookfield, Conn	Jan. 19, 1858	19, 138
Hats, Machinery for hardening	R. Wildman	Charlestown, Mass	Feb. 26, 1856	14, 330
Hats, Machinery for setting up	J. Grant	Providence, R. I	Aug. 11, 1821	....
Hats, Making	W. E. Doubleday	Brooklyn, N. Y	Oct. 2, 1866	58, 391
Hats, Making	W. Goodrich	Philadelphia, Pa	Aug. 20, 1816	....
Hats, Making paper water-proof	B. Grut	New York, N. Y	Oct. 1, 1830	....
Hats, Making printed felt	A. Bailey	Amesbury, Mass	Apr. 18, 1865	47, 269
Hats, Manufacturing	A. Bancker and C. F. Alvord.	New York, N. Y	Jan. 9, 1849	6, 010
Hats, Manufacturing	J. W. Beebe	Brooklyn, N. Y	Mar. 4, 1856	14, 343
Hats, Manufacturing	J. and J. Bryan	Saratoga, N. Y	Aug. 26, 1815	....
Hats, Manufacturing	E. Corning, jr		June 13, 1803	....
Hats, Manufacturing	L. E. Hopkins	Brooklyn, N. Y	June 5, 1855	13, 005
Hats, Manufacturing	S. Hurlbut	Glastenbury, Conn	June 13, 1831	....
Hats, Manufacturing	J. Johnson	New Orleans, La	Jan. 8, 1856	14, 062
Hats, Manufacturing	H. Kellogg	New Haven, Conn	Feb. 12, 1867	61, 037
Hats, Manufacturing	J. Maguire	Washington, D. C	Dec. 7, 1844	3, 851
Hats, Manufacturing	A. Rankin	Newark, N. J	June 5, 1855	13, 013
Hats, Manufacturing	T. Sealy	Newark, N. J	Dec. 24, 1867	72, 546
Hats, Manufacturing	J. Taylor and C. Brown	Danbury, Conn	Jan. 13, 1832	....
Hats, Manufacturing	S. Tibballs, jr	Tyringham, Mass	Dec. 26, 1810	....
Hats, Manufacturing	W. F. Warburton	Philadelphia, Pa	Aug. 2, 1859	24, 968
Hats, Manufacturing	W. F. Warburton and C. C. Lovett.	Philadelphia, Pa	Apr. 10, 1860	27, 848
Hats, Manufacturing cemented	N. Weston	Reading, Mass	May 24, 1816	....
Hats, Manufacturing felt	J. Monach	Rahway, N. J	May 10, 1859	23, 934
Hats, Manufacturing felt	G. H. Scribner	Natick, Mass	Jan. 21, 1870	104, 652
Hats, Manufacturing leather	J. S. and W. Wibirh	Eden, N. Y	Sept. 4, 1841	2, 241
Hats, Manufacturing light	H. Lainhart, jr	Baltimore, Md	July 6, 1809	....
Hats, Manufacturing military	S. Hoskin	Catskill, N. Y	Apr. 30, 1810	....
Hats, Manufacturing napped	L. Lyon, 2d	Needham, Mass	May 16, 1834	....
Hats, Manufacturing soft	B. McNamee	Boston, Mass	July 16, 1872	129, 356
Hats, Method of extracting alcohol used in stiffening	W. McCay	Northumberland, Pa	June 22, 1832	....
Hats, Method of surfacing felt	A. Hurd	Danbury, Conn	Feb. 10, 1857	16, 588
Hats, &c., Method of treating straw-braid for	G. Cornwall, 2d	Milford, Conn	May 19, 1857	17, 318
Hats, Mold for forming wool and rorams, &c., into hats for.	J. Werely	Albany, N. Y	Oct. 14, 1813	....
Hats, Napping	J. Dart, jr., W. Wells, and D. Olmsted.	Buffalo, N. Y	May 12, 1832	....
Hats, Napping	J. Long		Aug. 5, 1799	....
Hats, Napping	L. Lyon, 2d	Needham, Mass	Apr. 13, 1836	....
Hats of leather, skins, &c., Machinery for forming, &c.	R. Fish	New York, N. Y	Oct. 12, 1844	3, 790
Hats of paper, Making	J. Ford and J. White	Boston, Mass	Nov. 25, 1811	....
Hats, Planking	J. Gregg	Greene County, Pa	Mar. 14, 1810	....
Hats, Planking	G. and E. Page	Manchester, Conn	Nov. 9, 1832	....
Hats, Planking and felting	J. Swan	Haverhill, N. H	Mar. 7, 1811	....
Hats, Preparing seal's fur for	G. Cleveland		Sept. 9, 1803	....
Hats, Pressing	J. F. Mathias	Paris, France	Jan. 19, 1864	41, 311
Hats, Pressing leghorn, straw, and other	R. Tyler and B. P. Coston	Philadelphia, Pa	May 23, 1826	....
Hats, Process of treating muslin for sweat-linings, &c., for.	W. P. Wright	Philadelphia, Pa	Apr. 14, 1868	76, 685
Hats, Rounding-jack for trimming brims of	S. Polley	Brooklyn, N. Y	Mar. 26, 1867	63, 173
Hats, Scalding and napping	D. Baldwin	Ithaca, N. Y	Oct. 15, 1829	....
Hats, Scalding and napping	A. P. Gregory	Ithaca, N. Y	May 13, 1830	....
Hats, Size-mark for	L. C. Woehming	New York, N. Y	Apr. 20, 1869	89, 108
Hats, Sizing and napping	G. Henning	Ithaca, N. Y	Apr. 28, 1830	....
Hats, &c., Sizing for	H. E. Pond	Franklin, Mass	Dec. 8, 1863	40, 886
Hats, Stiffening for	A. and G. V. Raymond	Baltimore, Md	Feb. 7, 1818	....
Hats, Stiffening for	J. D. Wilson	New York	Aug. 6, 1829	....
Hats, Stiffening for water-proof	G. W. Downs	Circleville, Ohio	Oct. 25, 1832	....
Hats, Sweat-leather for	D. Shive	Philadelphia, Pa	May 14, 1867	64, 717
Hats, Sweat-leather ventilator for	H. Curtis and A. Tufts	Charlestown, Mass	June 24, 1862	35, 672

Index of patents issued from the United States Patent Office from 1790 *to* 1873, *inclusive*—Continued.

Invention.	Inventor.	Residence.	Date.	No.
Hats, Ventilating	J. W. Beebe	New York, N. Y	Mar. 16, 1858	19,616
Hats, Ventilating	G. W. Cherry	Alexandria, D. C	July 26, 1845	4,122
Hats, Ventilating	R. Halloran	New York, N. Y	Mar. 21, 1846	4,427
Hats, Ventilating	T. R. Johnson	Montreal, Canada	Aug. 11, 1868	80,963
Hats, Ventilating	A. Maginnis	Philadelphia, Pa	May 17, 1859	24,035
Hats, Ventilating	A. Maginnis	Philadelphia, Pa	Sept. 11, 1860	29,984
Hats, Ventilating	W. Sellers	New York, N. Y	July 31, 1855	13,361
Hats, Ventilating sweat-leather for	J. Pollock	Morrisania, N. Y	Nov. 13, 1860	30,643
Hats, Washing	S. Dinkhouse	Eastonborough, Pa	Jan. 30, 1832	....
Hats, Water-proof stiffening for	S. Hempstead, jr	Saint Charles County, Mo	May 25, 1827	....
Hats, Water-proof stiffening for	S. Hempstead, jr	Saint Charles County, Mo	Oct. 26, 1827	....
Hats, Water-proof stiffening for	J. H. Tilge	Washington, D. C	Sept. 23, 1819	....
Hats, Water-proof stiffening for fur and wool	W. Buckles	Baltimore, Md	Nov. 28, 1817	....
Hats, &c., Weaving grass for making	L. Burnap	Merrimack, N. H	Feb. 16, 1823	....
Hats with rabbit's fur, Napping	D. Evans	Alexandria, Va	Oct. 2, 1820	....
Hatch	D. Talcot	New York, N. Y	May 9, 1854	10,899
Hatch for vessels	R. C. Holmes	Cape May, N. J	Mar. 7, 1846	4,397
Hatch for warehouses, Safety	A. D. Medlicott	Windsor Locks, Conn	Nov. 19, 1867	71,199
Hatch of vessels, Securing	E. S. Keyser	New York, N. Y	Dec. 8, 1857	18,816
Hatch, Safety	G. N. Creamer	Trenton, N. J	May 26, 1868	78,265
Hatch, Safety	R. H. Martin	Staten Island, N. Y	July 12, 1868	77,744
Hatch, Self-closing	C. H. Reynolds	Williamsburgh, N. Y	Feb. 11, 1873	135,728
Hatch, Ship's	J. R. Rich	Tremont, Me	Mar. 30, 1869	88,330
Hatchet	S. Daugherty	Belle Vernon, Pa	Sept. 10, 1872	131,211
Hatchet	N. F. English	Hartland, Vt	Apr. 27, 1858	20,052
Hatchet	J. Jenkins	Philadelphia, Pa	July 13, 1869	92,532
Hatchet	M. E. Rudasill	Shelby, N. C	Nov. 20, 1860	30,692
Hatchet	R. H. Thompson	Troy, N. Y	Nov. 26, 1872	133,343
Hatchet	D. E. Weaver	Cheviot, Ohio	Nov. 12, 1872	132,996
Hatchet-blanks, Die for shaping	J. Yerkers	Philadelphia, Pa	Mar. 4, 1873	136,479
Hatchet, Carpenter's	J. T. Shank	Martinsburgh, W. Va	Nov. 26, 1867	71,416
Hatchet, hammer, and scraper, Combined	A. Barbarin	New Orleans, La	Oct. 13, 1868	83,021
Hatchet-heads, Machine for swaging	L. Dodge	Cohoes, N. Y	Sept. 1, 1857	18,087
Hatching chickens by artificial heat	N. E. Guerin	New York, N. Y	Mar. 30, 1843	3,019
Hatchway	G. Follett and A. Brummel	Brooklyn, N. Y	May 27, 1873	139,307
Hatchway	C. I. Russell	Chicago, Ill	Dec. 24, 1872	134,169
Hatchway	J. D. Sinclair	Brooklyn, N. Y	Nov. 24, 1868	84,387
Hatchway-door, Metallic	T. Hyatt	New York, N. Y	Dec. 2, 1873	145,184
Hatchway for elevators	J. Wayland	Jersey City, N. J	Dec. 24, 1872	134,334
Hatchway-guard	E. H. Ball	New York, N. Y	July 2, 1872	128,523
Hatchway-guard	H. H. Covert	New York, N. Y	Nov. 2, 1869	96,398
Hatchway-guard	M. J. Hinden	Detroit, Mich	Feb. 25, 1873	136,161
Hatchway-guard, Automatic	G. E. Berry and F. C. Pingree	Detroit, Mich	Nov. 11, 1873	144,432
Hatchway opening and closing apparatus	E. C. Ford	New York, N. Y	Sept. 25, 1860	30,134
Hatchway-protector, Automatic	J. B. Waring and J. W. Wilson	New York, N. Y	Nov. 26, 1872	133,345
Hatchway protector, Elevator	J. W. Meaker	Detroit, Mich	May 20, 1873	139,014
Hatchway safety-gate	J. W. Meaker	Detroit, Mich	Sept. 3, 1872	131,109
Hatchway safety guard	Z. S. B. Weeks and C. L. Kohler	New York, N. Y	Mar. 5, 1872	124,301
Hatchway, Self-closing	W. P. Cherrington	Boston, Mass	Dec. 16, 1873	145,626
Hatchway, Self-closing	A. B. Lee	Yonkers, N. Y	Sept. 9, 1873	142,738
Hatchway, Self-closing	A. Reid	New York, N. Y	July 22, 1873	141,171
Hatchways, Apparatus for opening and closing	H. Sizer and E. Stone	New York, N. Y., and Lowell, Mass.	Oct. 2, 1855	13,624
Hatchways, Closing	J. S. Baldwin	Newark, N. J	May 20, 1873	138,984
Hatchways, Closing	J. M. Bradner	Williamsport, Pa	July 23, 1872	129,646
Hatchways, Closing	E. M. Hackett	New York, N. Y	July 1, 1873	140,499
Hatchways, Operating shutters for	J. H. McKernan	Indianapolis, Ind	May 10, 1870	102,954
Hatchways, Safety-attachment for	J. Bridge	Augusta, Me	June 16, 1857	17,551
Hatchways, Safety-device for	A. Fries	Cincinnati, Ohio	Feb. 7, 1871	111,628
Hatters' cards or jacks	J. C. Seeley	Dutchess County, N. Y	Mar. 15, 1827	....
Hatters' finishing-irons, &c., Furnaces for heating	G. V. Raymond	Richmond, Va	Sept. 29, 1825	....
Hatters' iron-making die	B. Hull	Westport, Conn	May 1, 1866	54,356
Hatters' irons, Furnace for heating	D. L. Tuthill	New York, N. Y	Nov. 8, 1831	....
Hatters, Machine for bowing wool for	H. S. Holladay and E. G. Griffen	Lyme, Conn	June 6, 1812	....
Hatters, Machine for bowing wool or fur for	E. C. Griffing	Lyme, Conn	Aug. 10, 1815	....
Hatters, Making bow-strings for	D. Nichols	New Milford, Conn	Mar. 1, 1809	....
Hatters, Mode of preparing fur for	A. Buffum	Troy, Mass	Sept. 28, 1825	....
Hatters' plank-kettle	S. Fisher	Herkimer County, N. Y	July 15, 1816	....
Haversack	T. Keech	New York, N. Y	Sept. 22, 1863	40,047
Hawse-hole, Ship's	R. R. Osgood	Troy, N. Y	May 5, 1857	17,228
Hawse-pipe	P. Burnham	Gloucester, Mass	Oct. 8, 1867	69,625
Hawse-pipe	P. Moody	Gloucester, Mass	May 21, 1867	64,997
Hawse-pipe	C. Perley	New York, N. Y	Nov. 27, 1860	30,755
Hawse-pipe	T. J. Southard	Richmond, Me	Mar. 13, 1860	27,482
Hawse-pipe for ships	A. S. Phillips	Boston, Mass	July 5, 1859	24,702
Hawse-pipe stopper	O. B. Bearse	Hyannis, Mass	June 14, 1870	104,101
Hawse-pipe stopper	C. Perley	New York, N. Y	Aug. 9, 1864	43,793
Hawse-pipe stopper	J. Stewart	Bangor, Me	Mar. 12, 1867	62,902
Hawser-clamp	H. H. Ellis and C. F. Gladding	Norwich, Conn	June 9, 1868	78,794
Hay and cotton press	A. Adams	Jerseyville, Ill	Nov. 28, 1854	11,986
Hay and cotton press	H. A. Ashley	Springfield, Ohio	Jan. 6, 1863	37,370
Hay and cotton press	S. T. Baker	West Gorham, Me	Mar. 20, 1835	....
Hay and cotton press	J. C. Baldwin	Stanton, Va	Feb. 22, 1839	1,086
Hay and cotton press	R. Ball	Petersburgh, Va	June 28, 1870	104,689
Hay and cotton press	W. C. Banks	Como Depot, Miss	Oct. 12, 1869	95,631
Hay and cotton press	H. Barnes	Blairsville, Pa	Nov. 30, 1858	22,216
Hay and cotton press	C. J. Beasley	Petersburgh, Va	July 5, 1870	104,922
Hay and cotton press	J. Berkeley	Washington, Tex	Oct. 5, 1869	95,416
Hay and cotton press	B. R. Brown and J. Toone, jr	Jackson, Tenn	May 31, 1870	103,557
Hay and cotton press	S. W. Bullock	Catskill, N. Y	Mar. 23, 1842	2,507
Hay and cotton press	S. Q. Carey	Waxahatchie, Tex	Nov. 10, 1868	83,926
Hay and cotton press	N. Chapman	Hopedale, Mass	Dec. 13, 1870	110,011
Hay and cotton press	N. Chapman	Hopedale, Mass	Jan. 17, 1871	111,043
Hay and cotton press	R. H. and G. F. Cole	Greenport, N. Y	Mar. 15, 1870	100,861
Hay and cotton press	L. L. Cummings	Munnsville, N. Y	June 15, 1858	20,551
Hay and cotton press	M. G. Cunningham	Corsicana, Tex	Dec. 27, 1870	110,442
Hay and cotton press	J. K. Davis	Monticello, S. C	Dec. 27, 1870	110,443
Hay and cotton press	E. M. Day and J. T. Noel	Lower Lake, Cal	Feb. 20, 1866	52,786

Index of patents issued from the United States Patent Office from 1790 *to* 1873, *inclusive*—Continued.

Invention.	Inventor.	Residence.	Date.	No.
Hay and cotton press	L. Dederick	Albany, N. Y	May 16, 1854	10, 920
Hay and cotton press	L. Dederick	New York, N. Y	May 25, 1869	90, 431
Hay and cotton press	G. N. Doolittle	Louisville, Ky	June 23, 1863	38, 950
Hay and cotton press	B. F. Dunning	Galesburgh, Ill	Mar. 14, 1865	46, 785
Hay and cotton press	E. Evans	Montgomery, Ala	Jan. 12, 1869	85, 808
Hay and cotton press	C. D. Findlay and D. D. Craig.	Macon, Ga	Feb. 21, 1871	111, 923
Hay and cotton press	D. Flagg and C. Peck	Gardiner, Me	Dec. 30, 1826	....
Hay and cotton press	J. Garfield	Groton, Mass	Apr. 23, 1867	64, 093
Hay and cotton press	J. Grosvenor	New York, N. Y	Aug. 31, 1837	369
Hay and cotton press	J. K. Harris	Allensville, Ind	Aug. 20, 1861	33, 083
Hay and clothes press	G. W. Hart	Aurora, Ind	Dec. 15, 1863	40, 930
Hay and cotton press	C. W. Hawkes	Brunswick, Me	May 29, 1841	2, 111
Hay and cotton press, &c	G. A. Hotchkiss	Pleasant Township, Ind	July 15, 1862	35, 913
Hay and cotton press	P. C. Ingersoll	Green Point, N. Y	July 28, 1863	39, 365
Hay and cotton press	S. Ingersoll	Green Point, N. Y	Apr. 15, 1856	14, 663
Hay and cotton press	I. James	Mattoon, Ill	Feb. 16, 1864	41, 625
Hay and cotton press	J. H. Johnson	Griffin, Ga	Oct. 25, 1870	108, 599
Hay and cotton press	D. Knowles	Philadelphia, Pa	July 11, 1871	116, 965
Hay and cotton press	N. J. Lampman	Coxsackie, N. Y	Oct. 17, 1846	4, 818
Hay and cotton press	C. Lent	Washington, D. C	Feb. 18, 1868	74, 703
Hay and cotton press	E. G. Maxey and W. R. Mason.	Troy, Ind., and Lewisport, Ky.	Jan. 21, 1868	73, 542
Hay and cotton press	W. H. McBurney	Sacramento City, Cal	May 13, 1873	138, 909
Hay and cotton press	M. McElroy	Springfield, Mass	Sept. 3, 1867	68, 568
Hay and cotton press	J. A. McGillivrae and C. O. Wheeler.	Mattison, Ill	Oct. 12, 1869	95, 707
Hay and cotton press	J. W. McIntyre	Philadelphia, Pa	Apr. 9, 1872	125, 471
Hay and cotton press	D. L. Miller	Madison, N. J	Mar. 20, 1860	27, 556
Hay and cotton press	S. Miller	Mount Union, Pa	Mar. 8, 1870	100, 546
Hay and cotton press	M. D. Moore	Brooklyn, N. Y	Feb. 6, 1866	52, 495
Hay and cotton press	J. Müller	Nacogdoches, Tex	Oct. 14, 1873	143, 586
Hay and cotton press	P. L. Negley	Castleton, Ind	Oct. 3, 1871	119, 528
Hay and cotton press	C. H. Parshall	Detroit, Mich	Dec. 19, 1865	51, 615
Hay and cotton press	I. N. Patten and D. G. Marden.	Memphis, Tenn	Jan. 25, 1870	99, 228
Hay and cotton press	J. Peevy	Passadumkeag, Me	Jan. 1, 1856	14, 031
Hay and cotton press	J. B. Pugh	Champaign, Ill	Dec. 17, 1872	134, 097
Hay and cotton press	W. Read	Greenwood, Cal	Sept. 22, 1863	40, 064
Hay and cotton press	T. D. and F. R. Reed and C. Dorrel.	Rising Sun, Ind	May 22, 1866	54, 956
Hay and cotton press	W. Ridemour and M. K. Biser.	Springfield, Ohio	Mar. 8, 1864	41, 893
Hay and cotton press	J. Robertson	Gosport, Ind	Sept. 8, 1868	82, 036
Hay and cotton press	C. A. Robinson	Florence, Ind	Mar. 21, 1871	112, 966
Hay and cotton press	C. H. Robinson	Bath, Me	Aug. 18, 1863	39, 588
Hay and cotton press	E. Rock	Greenvale, N. Y	May 23, 1871	115, 106
Hay and cotton press	J. S. Schofield	Macon, Ga	Sept. 3, 1867	68, 462
Hay and cotton press	P. A. Shearer	Morristown, Tenn	Jan. 28, 1873	135, 291
Hay and cotton press	W. E. Sheffield	Saint Joseph, Mo	Aug. 17, 1869	93, 914
Hay and cotton press	J. Simpson	Chester, S. C	Aug. 31, 1869	94, 347
Hay and cotton press	J. J. Sivley	Clarksville, Tex	Nov. 21, 1871	121, 133
Hay and cotton press	J. N. Smith	Jersey City, N. J	Jan. 16, 1866	52, 084
Hay and cotton press	G. W. Stewart	Atlanta, Ga	Apr. 22, 1873	138, 210
Hay and cotton press	G. W. Swift	Memphis, Tenn	Mar. 8, 1870	100, 687
Hay and cotton press	G. W. Swinebroad	Bolivar, Tenn	Aug. 16, 1870	106, 427
Hay and cotton press	B. F. Taft	Groton Junction, Mass	Aug. 4, 1868	80, 681
Hay and cotton press	E. Thomas	Craigsville, Va	June 8, 1869	90, 970
Hay and cotton press	G. Utley	Charlotte, N. C	May 12, 1868	77, 852
Hay and cotton press	G. Utley	Charlotte, N. C	Aug. 23, 1870	106, 638
Hay and cotton press	M. Wallace	Little Rock, Ark	Aug. 30, 1870	106, 893
Hay and cotton press	H. R. Walton	Philadelphia, Pa	Apr. 26, 1870	102, 463
Hay and cotton press	C. O. Wheeler	Mattison, Ill	Feb. 13, 1872	123, 754
Hay and cotton press	W. H. Whetstone	Lowndesborough, Ala	June 4, 1872	127, 666
Hay and cotton press	J. L. White	Hernando, Miss	Feb. 7, 1871	111, 706
Hay and cotton presses, Mode of fastening door of.	P. C. Ingersoll	Green Point, N. Y	June 16, 1863	38, 928
Hay and dung forks, &c., Making	C. Lawrence	Chatham, N. Y	July 23, 1823	....
Hay and grain, Apparatus for stacking	W. F. Browne	Washington, D. C	June 4, 1867	65, 471
Hay and grain cock cover	E. R. Whitney	Plattsburgh, N. Y	Nov. 24, 1868	84, 453
Hay and grain covering	F. Wingate	Hallowell, Me	Mar. 19, 1833	....
Hay and grain elevator	J. Dennis	Oswego, N. Y	Dec. 21, 1869	98, 042
Hay and grain loader	W. J. Webber	Hollister, Cal	Aug. 12, 1873	141, 840
Hay and grain protector	A. J. Frisbie	Saint Mary's, Ohio	Oct. 2, 1866	58, 401
Hay and grain rack	R. Cobb	Hadley, Mich	Oct. 7, 1862	36, 604
Hay and grain rake, Revolving hand	S. Coats	Shoreham, Vt	Apr. 17, 1837	163
Hay and grain roof, Portable	J. J. Naylon	Brighton, Mich	Feb. 4, 1868	74, 120
Hay and grain, Thatching for stacks of	R. McLarn	Shirland, Pa	Mar. 2, 1869	87, 425
Hay and hemp press	G. W. Penniston	North Vernon, Ind	Sept. 23, 1862	36, 525
Hay and lime elevator	A. and G. D. Thomas	Saint Thomas, Pa	May 3, 1870	102, 729
Hay and manure fork	J. K. Babcock	Shortsville, N. Y	Apr. 30, 1867	64, 272
Hay and manure fork	G. L. Barton and A. E. Roberts.	Albany, N. Y	Nov. 24, 1857	18, 670
Hay and manure fork	W. Jones	Speedsville, N. Y	Aug. 25, 1857	18, 047
Hay and manure fork	A. J. Martin	Catskill, N. Y	July 7, 1868	79, 586
Hay and manure fork	E. Moore	Brooklyn, N. Y	July 27, 1869	92, 989
Hay and manure fork	R. A. Peet	Caledonia Township, Mich	July 5, 1870	104, 993
Hay and manure fork	L. D. Pitcher	Pitcherville, Ill	May 25, 1869	90, 463
Hay and manure fork	F. Villard	Mount Eaton, Ohio	Feb. 15, 1870	99, 982
Hay and manure fork, Combined	F. Villard	Mount Eaton, Ohio	Aug. 22, 1871	118, 410
Hay and manure forks, Fastening for	A. Clark	Southfield, N. Y	Mar. 5, 1850	7, 134
Hay and manure forks, Shank of	R. M. Hine	Mentz, N. Y	Jan. 3, 1854	10, 384
Hay and other forks, Manufacture of	G. B. Ely	Saint Johnsbury, Vt	Sept. 24, 1867	69, 196
Hay and pruning knife	J. Fasig	West Salem, Ohio	Jan. 30, 1866	52, 279
Hay and pruning knife combined	J. Fasig	West Salem, Ohio	May 3, 1870	102, 524
Hay and straw cutter	A. C. Both	Hesse-Cassel, Germany	May 31, 1870	103, 552
Hay and straw cutter	C. D. and W. S. Brewer	Lewisburgh, Pa	Dec. 13, 1864	45, 385
Hay and straw cutter	J. H. Dickinson	Chicopee Falls, Mass	June 4, 1867	65, 355
Hay and straw cutter	J. D. Felthousen	Michigan City, Ind	Sept. 18, 1860	30, 056
Hay and straw cutter	H. R. Hawkins	Akron, Ohio	Nov. 20, 1860	30, 677
Hay and straw cutter	T. Hazard	Wilmington, Ohio	Aug. 30, 1864	43, 988
Hay and straw cutter	W. Henshaw	Spencer, Mass	Apr. 3, 1866	53, 614
Hay and straw cutter	J. N. Neff	Strasburgh, Pa	Nov. 27, 1860	30, 750

Index of patents issued from the United States Patent Office from 1790 *to* 1873, *inclusive*—Continued.

Invention.	Inventor.	Residence.	Date.	No.
Hay and straw cutter	F. Pohlman	Coxsackie, N. Y	June 3, 1862	35, 468
Hay and straw cutter	D. H. Whittemore	Lynchburgh, Va	Sept. 4, 1860	29, 930
Hay and straw cutting machine	C. Brown	Buffalo, N. Y	May 26, 1868	78, 181
Hay and straw elevator	J. H. Gill	Mount Pleasant, Ohio	Jan. 12, 1858	19, 087
Hay and straw for feed for cattle and horses, Preparing.	C. Brown	Buffalo, N. Y	Sept. 3, 1867	68, 345
Hay and straw stacker	C. Rundell	Chicago, Ill	Apr. 24, 1866	54, 213
Hay and straw stacking apparatus	D. W. Baird	Lebanon, Tenn	Sept. 24, 1872	131, 590
Hay-binder	A. R. Clark	Onondaga, N. Y	Nov. 15, 1870	109, 176
Hay-binder	M. A. Dilley	Le Roy, Kans	Sept. 24, 1872	131, 509
Hay-carrier	C. S. Boothby	Saco, Me	May 3, 1870	102, 651
Hay-carrier	J. B. Drake	Indianapolis, Ind	Apr. 2, 1872	125, 183
Hay-carrier	T. J. Powell	Naples, N. Y	July 26, 1870	105, 723
Hay-carrier	H. C. Stouffer	Beaver Township, Ohio	Dec. 1, 1868	84, 591
Hay-carrier	E. B. Tanner	Attica, Ohio	May 13, 1873	138, 955
Hay, Carrier for unloading	G. Smith	Rochester, N. Y	June 20, 1871	116, 230
Hay, cider, and oil press	N. Whitney	Augusta, Me	Feb. 15, 1826	....
Hay-cock protector	O. R. Dinsmore	Auburn, N. H	Mar. 23, 1858	19, 689
Hay-crane	M. Simms and J. V. Chambers	Wheeling, W. Va	Jan. 14, 1868	73, 263
Hay-crane or carrier	B. P. Barackman	Lineville, Pa	Feb. 8, 1870	99, 520
Hay-cutter	F. Gerfen	West Hempfield Township, Pa.	May 20, 1868	78, 369
Hay-cutter	H. Kinsey, F. W. Kissell, and J. E. and J. M. Smith.	Ligonier, Pa	Nov. 17, 1868	84, 063
Hay-cutter	J. G. Perry	South Kingston, R. I	July 14, 1863	39, 242
Hay, &c., Cutting and pressing	R. Wakeman and J. L. Ballance	Port Deposit, Md	May 23, 1865	47, 883
Hay, &c., cutting machine	T. H. and D. T. Willson	Harrisburgh, Pa	Nov. 6, 1860	30, 597
Hay-derrick	I. Van Voorhis	Hillsborough, Pa	Oct. 1, 1867	69, 375
Hay, Derrick for stacking	T. G. Palmer	Greenville, N. Y	Aug. 23, 1864	43, 925
Hay, Derrick for stacking	S. Turner	Onargo, Ill	Aug. 23, 1864	43, 936
Hay, Device for mowing away	L. E. Palmer	Le Roy, Pa	Apr. 6, 1869	88, 664
Hay, Double box for pressing	E. Waterhouse	Gardiner, Me	Nov. 14, 1826	....
Hay-drag	D. and A. Eddelman	Madison, Ind	Feb. 15, 1870	99, 766
Hay elevating and unloading apparatus	H. Buck	Polo, Ill	Feb. 18, 1873	136, 029
Hay-elevating apparatus	J. Bolles	Jackson, Ohio	Oct. 31, 1871	120, 365
Hay-elevating apparatus	J. F. Nugent	Cannonsburgh, Mich	Apr. 7, 1868	76, 506
Hay-elevating device	F. Calvert	Wabash City, Ind	Sept. 12, 1871	118, 791
Hay-elevating fork	L. A. Beardsley	South Edmeston, N. Y	Sept. 11, 1860	29, 948
Hay-elevating fork	L. A. Beardsley	South Edmeston, N. Y	Sept. 17, 1861	33, 288
Hay-elevating fork	T. T. Calkins and D. E. Wing	Coxsackie, N. Y	Aug. 16, 1864	43, 832
Hay-elevating fork	C. E. and J. N. Gladding	Troy, Pa	May 11, 1858	20, 241
Hay-elevating fork	J. B. Hawley	Albany, N. Y	Aug. 23, 1864	43, 946
Hay-elevating fork	G. C. Howard and I. N. Wilfong	Philadelphia, Pa	Mar. 24, 1863	39, 965
Hay-elevating fork	D. F. Neikirk	Republic, Ohio	Feb. 23, 1864	41, 715
Hay-elevating fork	W. S. Newton	Norwich, Conn	Jan. 24, 1865	46, 018
Hay-elevating fork	E. Reynolds	Corunna, Mich	Dec. 27, 1864	45, 634
Hay-elevating fork	L. Rundell	Coxsackie, N. Y	Apr. 7, 1863	38, 129
Hay-elevating fork	L. Rundell	New Baltimore, N. Y	May 5, 1863	38, 420
Hay-elevating fork	E. W. Seymour and G. W. Gregory.	Centre Lisle and Binghamton, N. Y.	June 9, 1863	38, 872
Hay-elevating fork	R. J. Stanley	Mount Morris, N. Y	Dec. 8, 1863	40, 860
Hay-elevating fork	S. Wheeler	Albany, N. Y	Feb. 23, 1864	47, 733
Hay-elevating fork	J. A. Whitney	Maryland, N. Y	Aug. 23, 1864	43, 939
Hay-elevating machine	J. C. McGrew	Smithfield, Ohio	Dec. 21, 1858	22, 372
Hay-elevator	T. H. Arnold	Troy, Pa	Aug. 10, 1869	93, 579
Hay-elevator	T. V. Bagly	Jones' Station, Ind	Apr. 18, 1871	113, 837
Hay-elevator	A. S. Brown	Delphi, Ind	Mar. 19, 1872	124, 787
Hay-elevator	E. H. Carpenter	Dexter, Mich	Dec. 10, 1867	71, 850
Hay-elevator	A. Chapman	Delta, N. Y	July 2, 1872	128, 537
Hay-elevator	F. A. Crane	Zanesville, Ohio	Oct. 27, 1868	83, 479
Hay-elevator	W. Derr	Tiffin, Ohio	Aug. 10, 1869	93, 423
Hay-elevator	J. M Dick	Buffalo, N. Y	July 7, 1863	39, 128
Hay-elevator	F. R. Dufour	Vevay, Ind	Nov. 30, 1869	97, 282
Hay-elevator	W. E. Durkee	Fort Edward, N. Y	June 12, 1860	28, 713
Hay-elevator	P. C. Ellsworth	Venice, N. Y	Oct. 2, 1866	58, 541
Hay-elevator	F. F. Fowler	Crane Township, Ohio	Apr. 17, 1860	27, 899
Hay-elevator	R. Furnas	Friendswood, Ind	Dec. 20, 1870	110, 224
Hay-elevator	J. Gifford, jr	Watertown, N. Y	Nov. 13, 1866	59, 585
Hay-elevator	C. E. Gladding	Towanda, Pa	June 27, 1871	116, 300
Hay-elevator	E. C. Green	Plainfield, Ind	Mar. 19, 1867	63, 037
Hay-elevator	T. Harter	Ilion, N. Y	Oct. 24, 1871	120, 273
Hay-elevator	G. F. Hipp and J. B. Fast	Nova, Ohio	Sept. 3, 1867	68, 508
Hay-elevator	J. L. Hubbell and E. Sherman	Fairfield, Conn	Sept. 25, 1866	58, 259
Hay-elevator	T. T. Jarrett	Horsham, Pa	May 30, 1854	10, 989
Hay-elevator	J. H. Jenkins	Upper Sandusky, Ohio	Nov. 28, 1865	51, 190
Hay-elevator	J. L. Kintner	Harrison, Ind	Apr. 23, 1867	64, 111
Hay-elevator	C. H. Kirkpatrick	La Fayette, Ind	Sept. 9, 1873	142, 635
Hay-elevator	J. Linderman	Bullville, N. Y	Sept. 27, 1870	107, 695
Hay-elevator	W. Louden	Fairfield, Iowa	Mar. 17, 1868	75, 686
Hay-elevator	O. E. Mabie	Camden, N. Y	May 4, 1869	89, 673
Hay-elevator	H. and L. M. McCown	Enon Valley, Pa	Jan. 7, 1868	73, 190
Hay-elevator	H. and L. M. McCown	Enon Valley, Pa	Sept. 22, 1868	82, 427
Hay-elevator	M. Mitchell	Crown Point, Ind	Jan. 1, 1867	60, 773
Hay elevator	A. Mott	Scott, N. Y	Jan. 21, 1868	73, 544
Hay-elevator	M. D. Myers	Ilion, N. Y	July 15, 1862	35, 887
Hay-elevator	J. W. Odaniel	Cloverdale, Ind	Dec. 21, 1869	98, 097
Hay-elevator	W. L. Overhiser	Stockton, Cal	July 23, 1867	66, 990
Hay-elevator	N. Palmer	Greenville, N. Y	Sept. 30, 1862	36, 572
Hay-elevator	H. G. Porter	Grand Rapids, Mich	Mar. 3, 1868	75, 193
Hay-elevator	G. W. Prout	Ashland, N. Y	Feb. 16, 1864	41, 640
Hay-elevator	H. H. Quint	Brookfield, Mo	Aug. 22, 1871	118, 270
Hay-elevator	E. M. Rees	Norristown, Pa	Aug. 10, 1858	21, 150
Hay-elevator	S. Rogers	Pittsburgh, Pa	Jan. 24, 1865	46, 027
Hay-elevator	S. B. Secrist and I. Seyster	Ogle County, Ill	Aug. 3, 1869	93, 236
Hay-elevator	A. Smith	Shellsburgh, Pa	Oct. 5, 1869	95, 611
Hay-elevator	H. C. Stouffer	East Lewiston, Ohio	Nov. 16, 1869	96, 996
Hay-elevator	J. B. Summerill	Penn's Grove, N. J	Mar. 24, 1868	75, 810
Hay-elevator	J. M. Taber	Greenwich, N. Y	Apr. 22, 1862	35, 045

Index of patents issued from the United States Patent Office from 1790 *to* 1873, *inclusive*—Continued.

Invention.	Inventor.	Residence.	Date.	No.
Hay-elevator	N. C. Thomas and J. H. Coe	Brighton, Mich	Oct. 13, 1868	83, 007
Hay-elevator	E. J. Toof	Fort Madison, Iowa	May 9, 1865	47, 671
Hay-elevator	J. M. Van Demark and N. Barlow.	Phelps, N. Y	May 2, 1871	114, 496
Hay-elevator	E. L. Walker	Bendford's Store, Pa	Sept. 6, 1864	44, 129
Hay-elevator	G. Wilbur	Macedon, N. Y	July 17, 1837	271
Hay-elevator	E. L. Yancy	Batavia, N. Y	Nov. 9, 1869	96, 647
Hay elevator and carrier	W. S. Neil	Greensborough, Pa	July 4, 1871	116, 622
Hay elevator and carrier	J. H. White	Columbus City, Iowa	May 21, 1872	127, 003
Hay elevator and carrier, Track-rope	E. H. Carpenter	Dexter, Mich	Feb. 20, 1866	52, 676
Hay elevator and conveyer	T. E. Haymond	Morris, Ill	Dec. 13, 1870	110, 135
Hay elevator and conveyer	N. D. Hinman	Stepney Depot, Conn	Oct. 23, 1866	59, 018
Hay elevator and conveyer	C. A. Miller	Marengo Township, Mich	July 19, 1870	105, 477
Hay elevator and conveyer	T. I. Powell	Naples, N. Y	Jan. 26, 1869	86, 179
Hay elevator and loader	J. H. Bean	Macon, Ill	Sept. 16, 1873	142, 836
Hay-elevator and manure-drag combined	P. H. Stauffer	Lehighton, Pa	Jan. 19, 1869	86, 109
Hay elevator and stacker	A. W. Tooker	Harvard, Ill	July 11, 1865	48, 742
Hay-elevator clutch	N. D. Hinman	Stepney Depot, Conn	Nov. 12, 1867	70, 717
Hay-elevator fork	B. G. Fox	Pricetown, Pa	Aug. 15, 1871	118, 121
Hay-elevator or stack-builder	F. Wicks	Kansas, Ill	Oct. 18, 1864	44, 761
Hay-elevator sling	G. Smith	Rochester, N. Y	June 20, 1871	116, 231
Hay-elevators, Tackle-block for	F. Wicks and F. F. Fowler	Upper Sandusky, Ohio	Aug. 28, 1866	57, 610
Hay fed to stock, Device for saving the seed from	R. A. Campbell	Salem, Ind	Nov. 16, 1858	22, 062
Hay for baling, Device for preparing	G. H. Nye	Monmouth, Ill	Jan. 10, 1865	45, 853
Hay for market, Apparatus for preparing	S. Colahan	Cleveland, Ohio	May 22, 1866	54, 864
Hay for pressing, Machine for cutting	O. and C. Waste	Cameron, Ill	Jan. 26, 1864	41, 416
Hay-fork	C. S. Ambruster	Woodstown, N. J	Sept. 1, 1868	81, 726
Hay-fork	H. F. Bernendefer and D. H. Finch.	Attica, Ohio	Apr. 21, 1868	76, 980
Hay-fork	A. Clark	Southfield, N. Y	Oct. 8, 1850	7, 697
Hay-fork	J. Dampman	Lebanon, Pa	Dec. 28, 1869	98, 236
Hay-fork	E. G. Dorchester and U. Scott	Geneva, N. Y	Dec. 8, 1868	84, 712
Hay-fork	C. L. Driesslein	Chicago, Ill	July 11, 1865	48, 665
Hay-fork	W. H. Elliot	New York, N. Y	Nov. 13, 1866	59, 570
Hay-fork	T. Foster	Coxsackie, N. Y	Sept. 29, 1863	40, 098
Hay-fork	J. G. Hitchcock	New York, N. Y	Jan. 7, 1868	73, 099
Hay-fork	W. Jones	Speedsville, N. Y	Jan. 13, 1857	16, 388
Hay-fork	J. E. Lobdell and L. H. Smith	Centre Lisle, N. Y	Feb. 1, 1870	99, 449
Hay-fork	J. A. Montgomery	Williamsport, Pa	Jan. 23, 1866	52, 186
Hay-fork	F. L. Morrison	New Albany, Ind	June 29, 1869	91, 959
Hay-fork	G. W. Robinson	New Wilmington, Pa	Aug. 25, 1868	81, 541
Hay-fork	L. Rogers	Pittsburgh, Pa	Jan. 11, 1870	98, 711
Hay-fork	W. F. Rundell	Genoa, N. Y	Mar. 28, 1865	47, 042
Hay-fork	W. F. Rundell	Genoa, N. Y	Nov. 27, 1866	60, 065
Hay-fork	J. B. Stewart	South Paris, Me	May 1, 1866	54, 435
Hay-fork	E. W. Walton and A. J. Brown	Stockton, Cal	Nov. 2, 1869	96, 517
Hay fork and carrier	J. B. Drake	Indianapolis, Ind	Aug. 6, 1872	130, 286
Hay fork hook	T. W. Peirce	Minneapolis, Minn	Sept. 20, 1870	107, 531
Hay-fork, Elevator	J. F. H. Brown	Greenup, Ill	Apr. 25, 1871	113, 975
Hay-fork grapple	J. K. O'Neil and J. F. Thomas	Kingston and Herkimer County, N. Y.	Jan. 12, 1869	85, 844
Hay-fork, Horse	B. F. Alexander	Glen Hope, Pa	June 29, 1869	91, 998
Hay-fork, Horse	D. W. Amos	Broad Top City, Pa	Jan. 14, 1868	73, 222
Hay-fork, Horse	T. H. Arnold	Troy, Pa	May 29, 1866	55, 035
Hay-fork, Horse	T. H. Arnold	Troy, Pa	Nov. 13, 1866	59, 539
Hay-fork, Horse	T. H. Arnold	Troy, Pa	Sept. 10, 1867	68, 683
Hay-fork, Horse	G. Atherton	Sparta Township, Mich	Feb. 18, 1873	136, 016
Hay-fork, Horse	L. Atwater	Ithaca, N. Y	Jan. 14, 1868	73, 223
Hay-fork, Horse	C. N. Baldwin	Wilmington, Conn	Oct. 25, 1870	108, 551
Hay-fork, Horse	W. D. Ballard	Davisburgh, Mich	Aug. 31, 1869	94, 174
Hay-fork, Horse	C. Bean	Pawtucket, R. I	May 2, 1871	114, 391
Hay-fork, Horse	S. Bebout	Waterford, Ohio	Oct. 11, 1870	108, 094
Hay-fork, Horse	H. J. Beemer	Honesdale, Pa	Nov. 10, 1868	83, 821
Hay-fork, Horse	J. R. Benedict	Marion, N. Y	May 21, 1872	126, 866
Hay-fork, Horse	W. H. Berdan	York, Mich	Mar. 16, 1869	87, 902
Hay-fork, Horse	D. S. Bigler	Siddonstown, Pa	May 18, 1869	90, 149
Hay-fork, Horse	M. D. Birge	Grand Rapids, Mich	Sept. 10, 1867	68, 690
Hay-fork, Horse	C. C. Blodgett	Watertown, N. Y	Dec. 18, 1866	60, 467
Hay-fork, Horse	C. C. Blodgett	Watertown, N. Y	Aug. 13, 1867	67, 711
Hay-fork, Horse	D. S. Blue	Tremont, Ohio	May 7, 1867	64, 403
Hay-fork, Horse	L. L. Bond	Chicago, Ill	Dec. 3, 1867	71, 575
Hay-fork, Horse	S. B. Bowen	Stockton, Cal	Aug. 19, 1873	141, 851
Hay-fork, Horse	G. W. Bowlsby	Monroe, Mich	July 28, 1868	80, 442
Hay-fork, Horse	J. Bradley	Racine, Wis	June 8, 1869	91, 078
Hay-fork, Horse	A. C. Briggs	North Easton, N. Y	July 10, 1866	56, 170
Hay-fork, Horse	W. D. Brooks	Bethany, Pa	Sept. 29, 1868	82, 485
Hay-fork, Horse	A. S. Brown	Lebanon, Pa	May 18, 1869	90, 233
Hay-fork, Horse	B. F. Brown	Catlin, Ind	Mar. 21, 1871	112, 777
Hay-fork, Horse	J. S. Brown	Washington, D. C	July 17, 1866	56, 359
Hay-fork, Horse	J. S. Brown and W. F. Browne	Washington, D. C	Apr. 9, 1867	63, 608
Hay-fork, Horse	E. and A. Buckman	East Greenbush, N. Y	Jan. 23, 1866	52, 134
Hay-fork, Horse	J. H. Bunton	Thornbury, Pa	Nov. 17, 1868	84, 084
Hay-fork, Horse	B. S. Burgan	Congress, Ohio	Nov. 3, 1868	83, 598
Hay-fork, Horse	J. R. Cadwell	Dexter, Mich	Dec. 20, 1864	45, 472
Hay-fork, Horse	J. R. Cadwell	Dexter, Mich	Aug. 1, 1865	49, 083
Hay-fork, Horse	W. Carlton	Adrian, Mich	Nov. 26, 1867	71, 453
Hay-fork, Horse	J. B. Carothers	Pine Grove Mills, Pa	Sept. 19, 1871	119, 120
Hay-fork, Horse	W. Carroll	Hillsdale, Mich	Nov. 12, 1867	70, 695
Hay-fork, Horse	A. Y. Case	Dexter, Mich	Dec. 19, 1865	51, 555
Hay-fork, Horse	A. M. Cheney and H. B. Kimball.	Charlotte, Mich	Dec. 1, 1868	84, 477
Hay-fork, Horse	S. Clark	Howard, N. Y	Feb. 1, 1870	99, 292
Hay-fork, Horse	D. B. Clement	Brooklyn, N. Y	Feb. 9, 1864	41, 484
Hay-fork, Horse	D. B. Clement	Brooklyn, N. Y	Jan. 24, 1865	46, 047
Hay-fork, Horse	A. Coates	Watertown, N. Y	Nov. 19, 1867	71, 137
Hay-fork, Horse	M. Coffin	Milton, Ky	Nov. 5, 1867	70, 527
Hay-fork, Horse	W. S. Coffman	Coldwater, Mich	June 23, 1868	79, 106
Hay-fork, Horse	A. J. Cook	Guilford, Conn	Nov. 19, 1867	70, 962

Index of patents issued from the United States Patent Office from 1790 *to* 1873, *inclusive*—Continued.

Invention.	Inventor.	Residence.	Date.	No.
Hay-fork, Horse	S. Cook	Harrison, Ill	Sept. 17, 1872	131, 394
Hay-fork, Horse	A. J. Cooley	Chardon, Ohio	July 17, 1866	56, 375
Hay-fork, Horse	J. A. Cowles	Chicago, Ill	Feb. 21, 1865	46, 450
Hay-fork, Horse	F. Cramer	Chess Springs, Pa	Aug. 10, 1869	95, 520
Hay-fork, Horse	E. G. Crandell	Belfast, N. Y	Jan. 18, 1870	98, 930
Hay-fork, Horse	J. Crandell	Ilion, N. Y	Oct. 25, 1864	44, 788
Hay-fork, Horse	T. C. Craven	Albany, N. Y	June 26, 1866	55, 824
Hay-fork, Horse	C. N. Culver	Bowling Green, Ohio	June 5, 1866	55, 246
Hay-fork, Horse	F. Culver	Elkland, Pa	Dec. 17, 1867	72, 270
Hay-fork, Horse	J. S. Culver	Springport, N. Y	Nov. 12, 1867	70, 700
Hay-fork, Horse	J. Cummins	Perry, Mich	Mar. 9, 1869	87, 643
Hay-fork, Horse	J. and E. C. Cummins	Perry, Mich	Nov. 16, 1869	96, 894
Hay-fork, Horse	J. J. De Grummond	Knoxville, Ill	Nov. 23, 1869	97, 171
Hay-fork, Horse	M. Dennis	Barton, N. Y	Dec. 22, 1868	85, 167
Hay-fork, Horse	W. E. Derrick	Jordan, N. Y	Nov. 24, 1868	84, 269
Hay-fork, Horse	W. E. Derrick	Jordan, N. Y	Jan. 12, 1869	85, 733
Hay-fork, Horse	W. E. Derrick	Jordan, N. Y	July 6, 1869	92, 175
Hay-fork, Horse	W. E. Derrick	Jordan, N. Y	June 14, 1870	104, 124
Hay-fork, Horse	H. E. Dewey	Aurora, Ill	Jan. 21, 1868	73, 515
Hay-fork, Horse	H. L. Doane	Green Oak, Mich	June 23, 1868	79, 058
Hay-fork, Horse	L. M. Doudna	Elmira, N. Y	Apr. 23, 1867	64, 083
Hay-fork, Horse	G. H. Dow	Freeport, Ill	Nov. 24, 1868	84, 271
Hay-fork, Horse	J. B. Drake	Picture Rocks, Pa	July 31, 1866	56, 851
Hay-fork, Horse	J. Drinkwater	Adams, Ohio	June 30, 1868	79, 328
Hay-fork, Horse	A. T. Dunbar and J. H. Fellows.	Alba, Pa	Nov. 15, 1870	109, 305
Hay-fork, Horse	J. A. Eberly and H. Wechter	Reamstown Station, Pa	July 13, 1869	92, 434
Hay-fork, Horse	F. Ebert	Saxonburgh, Pa	May 11, 1869	89, 920
Hay-fork, Horse	J. T. Elliott	Grand Rapids, Mich	Dec. 29, 1868	85, 437
Hay-fork, Horse	H. A. Estes	Jersey City, N. J	Feb. 4, 1868	74, 064
Hay-fork, Horse	O. Evans	Alliance, Ohio	Sept. 24, 1867	69, 084
Hay-fork, Horse	E. J. Fenn	Medina, Ohio	July 7, 1868	79, 644
Hay-fork, Horse	E. J. Fenn	Medina, Ohio	Sept. 21, 1869	95, 012
Hay-fork, Horse	B. Field	Three Rivers, Mich	Dec. 24, 1872	134, 265
Hay-fork, Horse	D. Figge	Jenner's Cross Roads, Pa	Apr. 14, 1866	57, 110
Hay-fork, Horse	H. Fisher	Canton, Ohio	July 31, 1866	56, 739
Hay-fork, Horse	H. Fisher and M. Ball	Canton, Ohio	Mar. 20, 1866	53, 286
Hay-fork, Horse	E. Forney and J. Swab	Elizabethville, Pa	Jan. 18, 1870	98, 946
Hay-fork, Horse	F. Foster	Coxsackie, N. Y	Mar. 20, 1866	53, 288
Hay-fork, Horse	F. F. Fowler	Upper Sandusky, Ohio	Aug. 14, 1866	57, 114
Hay-fork, Horse	R. S. Frame	Washington, D. C	Dec. 1, 1868	84, 540
Hay-fork, Horse	R. S. Frame	Washington, Ohio	Mar. 8, 1870	100, 518
Hay-fork, Horse	W. B. Gabel	East Cocalico Township, Pa.	Dec. 17, 1867	72, 282
Hay-fork, Horse	J. S. Gage	Dowagiac, Mich	Feb. 9, 1864	41, 499
Hay-fork, Horse	D. M. Garrett	Shelby, Ohio	Aug. 29, 1865	49, 621
Hay-fork, Horse	S. L. Gates	Verona, N. Y	Aug. 25, 1863	39, 641
Hay-fork, Horse	S. L. Gates	Verona, N. Y	Sept. 6, 1864	44, 087
Hay-fork, Horse	C. H. Gifford	Philadelphia, Pa	Oct. 8, 1867	69, 651
Hay fork, Horse	W. M. Gillan	Mount Parnel, Pa	Nov. 3, 1868	83, 705
Hay-fork, Horse	E. Gilliam	Allegheny, Pa	June 5, 1866	55, 278
Hay-fork, Horse	J. Gilmore	Phœnixville, Pa	Dec. 31, 1867	72, 837
Hay-fork, Horse	J. Gilmore	Phœnixville, Pa	Oct. 27, 1868	83, 482
Hay-fork, Horse	B. F. Gladding	Providence, R. I	Feb. 9, 1869	86, 661
Hay-fork, Horse	C. E. Gladding	Troy, Pa	Jan. 7, 1868	73, 179
Hay-fork, Horse	J. A. Glenn	West Middlesex, Pa	Dec. 1, 1868	84, 544
Hay-fork, Horse	J. A. Glenn	West Middlesex, Pa	May 18, 1869	90, 256
Hay-fork, Horse	G. D. S. Gochnauer	Mulberry, Pa	Feb. 9, 1869	86, 831
Hay-fork, Horse	J. S. Gochnauer	York, Pa	Dec. 17, 1867	72, 286
Hay-fork, Horse	A. Gordon	Rochester, N. Y	Oct. 9, 1866	58, 635
Hay-fork, Horse	H. N. Green	Fort Wayne, Ind	Nov. 22, 1870	109, 509
Hay-fork, Horse	C. S. Grimes	Lancaster, Ky	Mar. 22, 1870	101, 122
Hay-fork, Horse	J. T. Hall	Trenton, N. Y	Apr. 9, 1867	63, 719
Hay-fork, Horse	A. M. Halsted	Rye, N. Y	Mar. 7, 1865	46, 664
Hay-fork, Horse	J. D. Halsted	Rye, N. Y	Jan. 5, 1864	41, 155
Hay-fork, Horse	W. Hannah	Middlefield Centre, N. Y	July 7, 1868	79, 655
Hay-fork, Horse	O. J. Hardgrove	Canton, Ohio	Mar. 19, 1867	63, 042
Hay-fork, Horse	S. and D. A. Harris	Shippensburgh, Pa	Aug. 7, 1866	56, 936
Hay-fork, Horse	S. and D. A. Harris	Shippensburgh, Pa	Apr. 23, 1867	64, 100
Hay-fork, Horse	A. J. Harrison	Cadiz, Ohio	Jan. 5, 1869	85, 658
Hay-fork, Horse	E. Harter	Dowagiac, Mich	Apr. 27, 1869	89, 478
Hay-fork, Horse	E. Harter	Dowagiac, Mich	Dec. 28, 1869	98, 258
Hay-fork, Horse	H. H. Hatheway	Clockville, N. Y	May 21, 1867	64, 863
Hay-fork, Horse	I. J. Hattabaugh	Santa Clara County, Cal	Nov. 19, 1867	71, 167
Hay-fork, Horse	L. Haverstick	Manor Township, Pa	Sept. 17, 1867	68, 873
Hay-fork, Horse	L. Haverstick	Manor Township, Pa	Jan. 12, 1869	85, 739
Hay-fork, Horse	L. Haverstick	Manor Township, Pa	Apr. 29, 1873	138, 249
Hay-fork, Horse	S. Z. Hawbecker and A. Thomas.	Upton and Saint Thomas, Pa.	May 18, 1869	90, 098
Hay-fork, Horse	G. W. Heath	Burlington, Pa	Mar. 26, 1867	63, 244
Hay-fork, Horse	G. W. Heath	Burlington, Pa	Apr. 9, 1867	63, 634
Hay-fork, Horse	G. W. Heath	Burlington, Pa	Aug. 11, 1868	80, 950
Hay-fork, Horse	G. W. Heath	Burlington, Pa	Mar. 30, 1869	88, 477
Hay-fork, Horse	J. S. Henry and A. H. Reist	Manheim, Pa	Aug. 27, 1867	68, 195
Hay-fork, Horse	P. Hill	Millport, N. Y	Dec. 6, 1864	45, 324
Hay-fork, Horse	N. Hinman	Sparta, Mich	Aug. 10, 1869	93, 439
Hay-fork, Horse	N. D. Hinman	Pleasant Dale, Conn	Aug. 15, 1865	49, 405
Hay-fork, Horse	B. F. Hisert	Norton Hill, N. Y	Sept. 1, 1863	39, 732
Hay-fork, Horse	B. F. Hisert	Norton Hill, N. Y	Feb. 20, 1866	52, 805
Hay-fork, Horse	S. M. Hoagland	Catawissa, Pa	Apr. 14, 1868	76, 758
Hay-fork, Horse	A. Houghton	Seville, Ohio	June 23, 1868	79, 228
Hay-fork, Horse	C. A. Howard	Pontiac, Mich	Oct. 13, 1868	83, 063
Hay-fork, Horse	C. A. Howard	Pontiac, Mich	Oct. 31, 1871	120, 440
Hay-fork, Horse	D. J. Howenstine	Marshallville, Ohio	Feb. 18, 1868	74, 538
Hay-fork, Horse	J. W. Hull	Connersville, Ind	May 17, 1870	103, 192
Hay-fork, Horse	A. B. Hunt	Matteson, Mich	Nov. 16, 1869	96, 810
Hay-fork, Horse	E. Huoncker	Bethel, Pa	Mar. 16, 1869	87, 979
Hay-fork, Horse	J. Huy	Bakerstown, Pa	Aug. 1, 1871	117, 541

Index of patents issued from the United States Patent Office from 1790 *to* 1873, *inclusive*—Continued

Invention.	Inventor.	Residence.	Date.	No.
Hay-fork, Horse	T. D. Ingersoll	Monroe, Mich	Mar. 15, 1870	100, 767
Hay-fork, Horse	T. H. and H. James	Stockport, N. Y	Jan. 5, 1864	41, 072
Hay-fork, Horse	J. Jennings	Jacksonville, Ill	Jan. 5, 1869	85, 669
Hay-fork, Horse	J. Jones	Burtonville, N. Y	July 16, 1872	129, 237
Hay-fork, Horse	R. V. Jones	Canton, Ohio	Dec. 31, 1867	72, 860
Hay-fork, Horse	H. Kauffman	York, Pa	Oct. 23, 1866	59, 033
Hay-fork, Horse	C. H. B. Kellogg	Tontogany, Ohio	Oct. 13, 1868	83, 068
Hay-fork, Horse	T. C. Kelly	West Liberty, Pa	July 27, 1869	92, 975
Hay-fork, Horse	J. H. Kendrick	Dexter, Mich	Oct. 22, 1867	70, 002
Hay-fork, Horse	J. G. Kimberlin	West Dryden, N. Y	Feb. 16, 1869	87, 051
Hay-fork, Horse	N. W. Kingsley	Swansea, Mass	Aug. 31, 1869	94, 319
Hay-fork, Horse	G. Kinney	Bristol, Ind	Mar. 10, 1868	75, 430
Hay-fork, Horse	A. Knapp	North Fairfield, Ohio	Dec. 10, 1867	72, 050
Hay-fork, Horse	L. G. Kniffen	Worcester, Mass	Oct. 25, 1864	44, 805
Hay-fork, Horse	H. Krafft	Mulberry, Pa	May 5, 1868	77, 496
Hay-fork, Horse	E. M. Krum	Nassau, N. Y	May 28, 1867	65, 237
Hay-fork, Horse	J. B. Kurtz	Davisburgh, Pa	Oct. 20, 1868	83, 175
Hay-fork, Horse	A. J. Laird	Middletown, Pa	Aug. 21, 1866	57, 337
Hay-fork, Horse	A. J. Laird	Middletown, Pa	Dec. 17, 1867	72, 304
Hay-fork, Horse	H. Laird	Mechanicsburgh, Pa	May 26, 1868	78, 294
Hay-fork, Horse	J. C. Lampman	Baltimore, Md	Oct. 3, 1871	119, 619
Hay-fork, Horse	S. F. Leavitt	North Fairfield, Ohio	Sept. 12, 1865	49, 897
Hay-fork, Horse	C. E. Lins	Ashland, Pa	Aug. 25, 1868	81, 517
Hay-fork, Horse	D. Lippy	Mansfield, Ohio	Jan. 9, 1866	51, 950
Hay-fork, Horse	D. Lippy and J. H. Palm	Mansfield, Ohio	Sept. 6, 1864	44, 102
Hay-fork, Horse	R. W. Liscomb	Smithfield, Pa	Jan. 16, 1866	52, 060
Hay-fork, Horse	T. Lloyd	Muncy, Pa	Sept. 18, 1866	58, 110
Hay-fork, Horse	J. E. Lobdell	Centre Lisle, N. Y	Mar. 10, 1868	75, 282
Hay-fork, Horse	J. W. Lowe	Ashland, Ohio	Sept. 23, 1873	143, 170
Hay-fork, Horse	A. W. Lozier	New York, N. Y	Dec. 1, 1868	84, 559
Hay-fork, Horse	A. W. Lozier	New York, N. Y	Mar. 30, 1869	88, 397
Hay-fork, Horse	J. R. Lyons	Montrose, Pa	Feb. 4, 1868	74, 103
Hay-fork, Horse	J. M. Mansfield	Watertown, N. Y	Oct. 1, 1867	69, 353
Hay-fork, Horse	H. C. Mapes	Rushville, N. Y	Apr. 28, 1868	77, 299
Hay-fork, Horse	A. M. Martin and J. C. Blocher	Bloomville, Ohio	Apr. 14, 1868	76, 787
Hay-fork, Horse	L. S. Mason	Middlefield Centre, N. Y	Feb. 4, 1868	74, 106
Hay-fork, Horse	N. F. Mathewson	Barrington, R. I	Feb. 16, 1869	86, 934
Hay-fork, Horse	D. C. Matteson and T. P. Williamson.	Stockton, Cal	Feb. 2, 1869	86, 565
Hay-fork, Horse	S. S. Mattis	Curtin, Pa	Jan. 12, 1869	85, 840
Hay-fork, Horse	D. C. McDonald	Rushford, N. Y	May 6, 1873	138, 572
Hay-fork, Horse	J. M. McDonald	McCoysville, Pa	Mar. 8, 1870	100, 650
Hay-fork, Horse	W. W. McFaddin	Ennisville, Pa	Nov. 8, 1870	109, 032
Hay-fork, Horse	E. McKenzie	Watertown, N. Y	Feb. 18, 1868	74, 565
Hay-fork, Horse	S. P. Mecay	Kilbourne, Ohio	May 24, 1870	103, 485
Hay-fork, Horse	J. Milholland	New Concord, Ohio	May 19, 1868	78, 114
Hay-fork, Horse	J. A. Miller	Shippensburgh, Pa	Mar. 9, 1869	87, 696
Hay-fork, Horse	J. A. Miller	Shippensburgh, Pa	Dec. 28, 1869	98, 397
Hay-fork, Horse	S. Miller	Mohawk, N. Y	Feb. 9, 1869	86, 775
Hay-fork, Horse	D. Morton	Mount Vernon, Ohio	Jan. 5, 1869	85, 686
Hay-fork, Horse	P. A. Mowers	Cleversburgh, Pa	Mar. 24, 1868	75, 956
Hay-fork, Horse	J. H. Mullin	Schellsburgh, Pa	Feb. 11, 1868	74, 405
Hay-fork, Horse	C. E. Murray	Sugar Valley, Pa	July 14, 1868	80, 002
Hay-fork, Horse	M. D. Myers	Dion, N. Y	Mar. 14, 1865	46, 814
Hay-fork, Horse	D. B. Neal	Mount Gilead, Ohio	May 11, 1869	89, 938
Hay-fork, Horse	D. F. Neikirk	Republic, Ohio	July 16, 1867	66, 730
Hay-fork, Horse	A. J. Nellis	Pittsburgh, Pa	Oct. 21, 1873	143, 775
Hay-fork, Horse	H. Newmeyer	Millerstown, Pa	Nov. 20, 1866	59, 859
Hay-fork, Horse	S. T. Nigh	Leitersburgh, Md	May 4, 1869	89, 683
Hay-fork, Horse	F. Nishwitz	Williamsburgh, N. Y	July 17, 1866	56, 439
Hay-fork, Horse	F. Nishwitz and B. F. Hyers	Williamsburgh, N. Y., and Pekin, Ill.	July 17, 1866	56, 440
Hay-fork, Horse	J. K. O'Neil	Kingston, N. Y	Apr. 3, 1866	53, 660
Hay-fork, Horse	J. K. O'Neil	Kingston, N. Y	Apr. 17, 1866	54, 007
Hay-fork, Horse	J. K. O'Neil	Kingston, N. Y	June 12, 1866	55, 528
Hay-fork, Horse	J. K. O'Neil	Kingston, N. Y	Nov. 13, 1866	59, 633
Hay-fork, Horse	J. K. O'Neil	Kingston, N. Y	Oct. 13, 1868	82, 979
Hay-fork, Horse	J. O'Rourke	Columbiana, Ohio	Apr. 7, 1868	76, 511
Hay-fork, Horse	O. Paddock	Watertown, N. Y	Mar. 24, 1868	75, 786
Hay-fork, Horse	O. Paddock	Watertown, N. Y	Oct. 6, 1868	82, 867
Hay-fork, Horse	O. Paddock	Watertown, N. Y	Mar. 23, 1869	88, 067
Hay-fork, Horse	S. K. Padden	Pulaski, Pa	Mar. 5, 1872	124, 381
Hay-fork, Horse	S. Page	McAllisterville, Pa	Dec. 17, 1867	72, 321
Hay-fork, Horse	S. Page	McAllisterville, Pa	Mar. 10, 1868	75, 452
Hay-fork, Horse	S. Page	McAllisterville, Pa	June 30, 1868	79, 384
Hay-fork, Horse	H. D. Palmer	Leonidas, Mich	June 9, 1868	78, 816
Hay-fork, Horse	N. Palmer	Albany, N. Y	Feb. 25, 1868	74, 934
Hay-fork, Horse	N. Palmer	Albany, N. Y	May 12, 1868	77, 907
Hay-fork, Horse	N. Palmer	Athens, N. Y	Dec. 31, 1872	134, 482
Hay-fork, Horse	J. A. Park	Lansing, Mich	Apr. 13, 1869	88, 804
Hay-fork, Horse	J. A. Park	Lansing, Mich	Sept. 14, 1869	94, 909
Hay-fork, Horse	J. A. Park	Lansing, Mich	Nov. 2, 1869	96, 474
Hay-fork, Horse	E. M. Parker	Zion, Md	June 9, 1868	78, 820
Hay-fork, Horse	J. H. Parker, J. T. Hall, and I. Pierce.	Trenton, N. Y	Mar. 19, 1867	62, 970
Hay-fork, Horse	S. W. Patterson and S. Dewey	Mainesburgh, Pa	July 9, 1867	66, 621
Hay-fork, Horse	G. C. Perry	Ortonville, Mich	June 8, 1869	91, 160
Hay-fork, Horse	J. F. Pierce	Holland Patent, N. Y	Mar. 6, 1866	53, 085
Hay-fork, Horse	M. H. Pope	Susquehanna Depot, Pa	May 26, 1868	78, 324
Hay-fork, Horse	H. G. Porter	Grand Rapids, Mich	July 26, 1870	105, 722
Hay-fork, Horse	I. T. Price	Leesville, Ohio	June 26, 1866	55, 902
Hay-fork, Horse	E. Raber	Lake, Ohio	May 12, 1868	77, 839
Hay-fork, Horse	E. Raber	Roanoke, Ind	Sept. 7, 1869	94, 644
Hay-fork, Horse	A. Ream and C. L. Bush	Reading, Pa	Apr. 2, 1872	125, 330
Hay-fork, Horse	C. W. Reed	Chagrin Falls, Ohio	Dec. 24, 1867	72, 540
Hay-fork, Horse	C. W. Reed	Chagrin Falls, Ohio	Dec. 8, 1868	84, 708
Hay-fork, Horse	A. Reynolds	Rock Spring, Md	Apr. 14, 1868	76, 661
Hay-fork, Horse	E. Reynolds	Corunna, Mich	Oct. 18, 1864	44, 748

Index of patents issued from the United States Patent Office from 1790 *to* 1873, *inclusive*—Continued.

Invention.	Inventor.	Residence.	Date.	No.
Hay-fork, Horse	R. Reynolds and C. Young	Stockport, N. Y	May 30, 1865	47,982
Hay-fork, Horse	R. Reynolds and C. Young	Stockport, N. Y	May 30, 1865	48,029
Hay-fork, Horse	E. Rhoades, sr	Clyde, Ohio	Jan. 21, 1868	73,650
Hay-fork, Horse	S. H. Rhoades and W. Carroll	Clyde, Ohio, and Hillsdale, Mich.	May 12, 1868	77,760
Hay-fork, Horse	T. Rhoades	Ottawa, Ill	Apr. 29, 1873	138,286
Hay-fork, Horse	E. Rhoades, jr	Clyde, Ohio	Apr. 16, 1867	63,812
Hay-fork, Horse	E. Rhoades, jr	Clyde, Ohio	Nov. 3, 1868	83,794
Hay-fork, Horse	J. L. Ripley	Fremont, Ohio	Sept. 6, 1864	44,116
Hay-fork, Horse	J. L. Ripley	Fremont, Ohio	Feb. 14, 1865	46,393
Hay-fork, Horse	G. M. Robinson	New Wilmington, Pa	May 12, 1868	77,843
Hay-fork, Horse	L. M. Roby	Leesville, Ohio	Apr. 10, 1866	53,878
Hay-fork, Horse	B. B. Rockwell	Waterford Mills, Ind	Oct. 28, 1873	144,143
Hay-fork, Horse	L. Rogers	Pittsburgh, Pa	Jan. 18, 1870	98,887
Hay-fork, Horse	S. Rogers	Pittsburgh, Pa	Mar. 20, 1866	53,345
Hay-fork, Horse	S. Rogers	Pittsburgh, Pa	Oct. 23, 1866	59,139
Hay-fork, Horse	J. W. Row	Lewisburgh, Pa	Aug. 17, 1869	93,750
Hay fork, Horse	E. D. Rundell	Hudson, N. Y	Dec. 13, 1864	45,439
Hay-fork, Horse	A. V. Ryder	Germano, Ohio	Aug. 14, 1866	57,193
Hay-fork, Horse	W. Scholl	Marion Township, Pa	Mar. 24, 1868	75,800
Hay-fork, Horse	P. Schwitzer	Robeson Township, Pa	Apr. 7, 1868	76,352
Hay-fork, Horse	E. U. Scoville	Manlius, N. Y	Apr. 5, 1870	101,524
Hay-fork, Horse	O. P. Secor	Chicago, Ill	Sept. 5, 1865	49,834
Hay-fork, Horse	G. W. Shade	Shippensburgh, Pa	Dec. 17, 1867	72,232
Hay-fork, Horse	G. W. Shade	Shippensburgh, Pa	June 28, 1870	104,891
Hay-fork, Horse	J. Shearer	Reading, Pa	Nov. 26, 1867	71,334
Hay-fork, Horse	R. S. Sheldon	Chicago, Ill	Jan. 5, 1869	85,702
Hay-fork, Horse	A. Shellenberger	Versailles, Ohio	June 14, 1870	104,365
Hay-fork, Horse	H. L. Sheperd	Osborn, Ohio	July 28, 1868	80,513
Hay-fork, Horse	E. Shorkley	Lewisburgh, Pa	Aug. 14, 1866	57,199
Hay-fork, Horse	S. G. Simpson	Mill Creek, Pa	Nov. 30, 1869	97,449
Hay-fork, Horse	S. Slack	Dowagiac, Mich	Aug. 24, 1869	94,142
Hay-fork, Horse	A. Smith	Schellsburgh, Pa	Aug. 11, 1868	80,839
Hay-fork, Horse	R. A. Smith	Washington Mills, N. Y	May 4, 1869	89,603
Hay-fork, Horse	F. Snyder	Hinkletown, Pa	July 9, 1867	66,530
Hay-fork, Horse	I. E. Snyder	Woodward, Pa	Sept. 23, 1873	143,192
Hay-fork, Horse	I. C. Spear	New Wilmington, Pa	Jan. 14, 1868	73,403
Hay-fork, Horse	W. S. Spratt	West Manchester, Pa	Sept. 4, 1866	57,789
Hay-fork, Horse	A. B. Sprout	Hughesville, Pa	May 1, 1866	54,431
Hay-fork, Horse	A. B. Sprout	Hughesville, Pa	Apr. 23, 1867	64,163
Hay-fork, Horse	A. B. Sprout	Hughesville, Pa	May 26, 1868	78,335
Hay-fork, Horse	N. Starr, jr	Homer, N. Y	Apr. 21, 1868	77,120
Hay-fork, Horse	G. M. Stearns	Syracuse, N. Y	Jan. 25, 1870	99,120
Hay-fork, Horse	H. B. Steele	West Winsted, Conn	Mar. 10, 1868	75,481
Hay-fork, Horse	W. W. Stevens and J. Patcher, jr	Tontogany, Ohio	June 30, 1868	79,406
Hay-fork, Horse	B. F. Stewart	Freeport, Ohio	Oct. 15, 1867	69,857
Hay-fork, Horse	D. P. Stewart	Spruce Creek, Pa	Oct. 12, 1869	95,743
Hay-fork, Horse	O. O. Storle	North Cape, Wis	Apr. 23, 1867	64,164
Hay-fork, Horse	O. O. Storle	Norway, Wis	Nov. 10, 1868	84,017
Hay-fork, Horse	H. C. Stouffer	Canfield, Ohio	Apr. 22, 1873	138,211
Hay-fork, Horse	H. C. Stouffer, J. Heaton, and A. A. Bushong.	Columbiana, Ohio	Dec. 3, 1867	71,661
Hay-fork, Horse	H. C. and A. Stouffer	Beaver Township, Ohio	Mar. 10, 1868	75,486
Hay-fork, Horse	G. F. Strong	Onondaga, N. Y	Dec. 12, 1865	51,490
Hay-fork, Horse	G. H. Strough	Watertown, N. Y	June 30, 1868	79,512
Hay-fork, Horse	L. W. Stuart	Narrowsburgh, N. Y	Aug. 9, 1870	106,230
Hay-fork, Horse	J. W. Summers	Sandy Hill, N. Y	Mar. 24, 1868	75,997
Hay-fork, Horse	J. B. Sweetland	Pontiac, Mich	Apr. 21, 1868	77,125
Hay-fork, Horse	J. B. Sweetland	Pontiac, Mich	Apr. 13, 1869	88,991
Hay-fork, Horse	J. B. Sweetland	Pontiac, Mich	May 4, 1869	89,810
Hay-fork, Horse	E. B. Tanner	Attica, Ohio	Sept. 27, 1864	44,470
Hay-fork, Horse	L. H. Tears	Troy, Pa	Oct. 22, 1867	70,044
Hay-fork, Horse	T. H. Tears	Le Roy, Pa	Apr. 7, 1868	76,550
Hay-fork, Horse	A. G. Thomas	Sandy Spring, Md	Apr. 6, 1869	88,754
Hay-fork, Horse	J. T. Thomas	Ilion, N. Y	July 20, 1869	92,762
Hay-fork, Horse	F. W. Thorla	Hoskinville, Ohio	Jan. 18, 1870	99,033
Hay-fork, Horse	L. N. Tinkham	Sylvania, Pa	Dec. 24, 1867	62,567
Hay-fork, Horse	W. H. T. Tomlin	Mullica Hill, N. J	Apr. 23, 1867	64,167
Hay-fork, Horse	H. Totten	Leesville, Ohio	Oct. 17, 1865	50,516
Hay-fork, Horse	J. F. Troxel	Bloomville, Ohio	Apr. 6, 1869	88,593
Hay-fork, Horse	M. I. Turck	Schodack, N. Y	May 7, 1867	64,462
Hay-fork, Horse	P. Vanderbelt, jr	Hughesville, Pa	Nov. 26, 1867	71,343
Hay-fork, Horse	F. Van Doren	Adrian, Mich	Jan. 23, 1872	123,063
Hay-fork, Horse	O. Vanorman	Ripon, Wis	Dec. 10, 1867	72,131
Hay-fork, Horse	G. H. Waldo	Prattsburgh, N. Y	Dec. 10, 1867	71,928
Hay-fork, Horse	E. L. Walker	Benford's Store, Pa	Apr. 17, 1866	54,044
Hay-fork, Horse	E. L. Walker	Benford's Store, Pa	Apr. 21, 1868	77,134
Hay-fork, Horse	J. H. Walker	Grand Rapids, Mich	Jan. 7, 1868	73,211
Hay-fork, Horse	J. H. Walker	Grand Rapids, Mich	Apr. 21, 1868	77,136
Hay-fork, Horse	J. M. Walker	Rossville, Pa	July 13, 1869	92,495
Hay-fork, Horse	C. E. Warner	Syracuse, N. Y	Mar. 16, 1869	87,807
Hay-fork, Horse	C. E. Warner	Syracuse, N. Y	Oct. 26, 1869	96,173
Hay-fork, Horse	J. L. Wells	Stockbridge, N. Y	Oct. 11, 1864	44,682
Hay-fork, Horse	G. F. Weymouth	Dresden, Me	Mar. 1, 1870	100,476
Hay-fork, Horse	S. H. Wheeler	Dowagiac, Mich	Sept. 26, 1865	50,210
Hay-fork, Horse	E. I. White	Locke, N. Y	July 7, 1868	79,711
Hay-fork, Horse	A. Whitmarsh	East Bridgewater, Mass	Mar. 6, 1866	53,070
Hay-fork, Horse	A. F. Wicke and O. Evans	Alliance, Ohio	Mar. 26, 1867	63,345
Hay-fork, Horse	F. Wicks	Kansas, Ill	Oct. 18, 1864	44,760
Hay-fork, Horse	F. Wicks	Upper Sandusky, Ohio	Aug. 28, 1866	57,609
Hay-fork, Horse	J. H. Wilder	Farmington, Ohio	Apr. 7, 1868	76,573
Hay-fork, Horse	H. Willard	Grand Rapids, Mich	Dec. 3, 1867	71,829
Hay-fork, Horse	H. Willard	Grand Rapids, Mich	June 23, 1868	79,173
Hay-fork, Horse	C. W. Willis and A. Stephenson	Cambridge City, Ind	Apr. 17, 1866	54,050
Hay-fork, Horse	F. M. Willson	Whitney's Point, N. Y	Apr. 14, 1868	76,868
Hay-fork, Horse	W. D. Wilson	Watertown, N. Y	May 26, 1868	78,347
Hay-fork, Horse	M. Winsler, W. Campbell, and L. Hardman.	Tuscarawas County, Ohio	May 21, 1867	65,031

Index of patents issued from the United States Patent Office from 1790 *to* 1873, *inclusive*—Continued.

Invention.	Inventor.	Residence.	Date.	No.
Hay-fork, Horse	P. A. Wise	Stockbridge, N. Y	June 5, 1866	55,408
Hay-fork, Horse	J. Wolfrom	York, Pa	June 2, 1868	78,504
Hay-fork, Horse	E. B. Woodruff	Morristown, N. J	Jan. 29, 1867	61,697
Hay-fork, Horse	L. Woodworth	Troy, Pa	May 19, 1868	78,169
Hay-fork, Horse	E. Yancey	Utica, N. Y	Sept. 3, 1867	68,405
Hay-fork, Horse	E. Yeiser and J. S. Sheetz	Sheridan, Pa	Jan. 15, 1867	61,302
Hay-fork, Horse	C. Yengst	Annville, Pa	June 13, 1871	116,000
Hay-fork, Horse	J. S. Yinger	Manchester Township, Pa	Dec. 29, 1868	85,416
Hay-fork, Horse	J. S. Yinger	Manchester, Pa	June 21, 1870	104,529
Hay-fork, Horse	W. Zimmerman	Colfax, Iowa	Mar. 24, 1868	76,029
Hay-fork, Spiral	A. Neumeyer	Millerstown, Pa	Apr. 2, 1867	63,554
Hay-forks, Apparatus for operating horse	G. M. Robinson	New Wilmington, Pa	Feb. 4, 1868	74,142
Hay-forks, Clutch for	G. D. Melotte	Watertown, N. Y	Oct. 29, 1867	70,239
Hay-forks, Clutch for operating horse	D. B. Baker	Rollersville, Ohio	Dec. 31, 1867	72,778
Hay-forks, Device for suspending	W. S. Salisbury	Adams Centre, N. Y	May 28, 1867	65,283
Hay-forks, &c., Device for suspending horse	G. B. and C. Lewis	Adams Centre, N. Y	Jan. 1, 1867	60,758
Hay-forks, Elevating	J. H. Chapman	Utica, N. Y	Mar. 26, 1867	63,212
Hay-forks, Grapple for suspending	W. E. Derrick	Jordan, N. Y	Jan. 12, 1869	85,734
Hay-forks, Machine for bending	N. Brand	Leonardsville, N. Y	June 17, 1856	15,118
Hay-forks, Operating horse	H. Maycock	Verona, N. Y	Feb. 20, 1866	52,808
Hay-forks, Operating horse	A. J. Purviance	Keosauqua, Iowa	Oct. 1, 1867	69,479
Hay-forks, Self-tightening band for	J. H. Melick	Albany, N. Y	Nov. 24, 1863	40,701
Hay-forks, Shank and socket for hand	F. F. Terry	Port Gibson, N. Y	Apr. 30, 1867	64,381
Hay-forks, Shank of	D. Anthony, jr	Spring Port, N. Y	Oct. 8, 1850	7,694
Hay-forks, Suspending claw for horse	E. S. Kershaw	Sherburne, N. Y	July 30, 1867	67,199
Hay-gatherer	W. E. Phelps	Elmwood, Ill	Feb. 8, 1870	99,589
Hay-gatherer	A. Stream	Harrodsburgh, Ind	June 6, 1871	115,783
Hay-gatherer	A. Stream	Harrodsburgh, Ind	May 21, 1872	127,116
Hay-gatherer	C. Waste	Galesburgh, Ill	June 14, 1870	104,233
Hay gatherer and carrier	G. Meaders	Jeffersonville, Ind	June 21, 1870	104,480
Hay gatherer and shocker	C. M. Terrell	Oskaloosa, Iowa	Mar. 22, 1870	101,183
Hay-gathering machine	F. F. Fowler	Crane Township, Ohio	Jan. 22, 1861	31,162
Hay, grain, &c., Curing and drying	T. Heermans	Mitchellsville, Tenn	July 5, 1864	43,406
Hay, grain, &c., Mode of ventilating	A. Post	Henrietta, N. Y	June 18, 1861	32,584
Hay-grinding machine	W. P. Miller	New York, N. Y	June 17, 1873	139,911
Hay-hook, Horse	A. Kratz	Plumsteadville, Pa	Nov. 17, 1863	40,628
Hay-knife	J. S. Ball	Mishawaka, Ind	Nov. 18, 1873	144,59
Hay-knife	C. Dittman	Upper Leacock Township, Pa.	July 5, 1870	105,050
Hay-knife	J. Fasig	Jackson, Ohio	Apr. 13, 1858	19,921
Hay-knife	J. Fasig	West Salem, Ohio	Feb. 11, 1862	34,356
Hay-knife	G. Fenton	Streetsborough, Ohio	Mar. 10, 1868	75,255
Hay-knife	F. J. Fischer	Hamilton, Ohio	June 19, 1866	55,765
Hay-knife	C. A. Fisher	Geneseo, Ill	Jan. 7, 1868	73,177
Hay-knife	C. A. Fisher	Geneseo, Ill	Sept. 29, 1868	82,505
Hay-knife	M. Merrill	Oneida County, N. Y	Jan. 18, 1870	98,876
Hay-knife	J. Offineer	Ashland, Ohio	Sept. 15, 1868	82,152
Hay-knife	G. W. Parsons and W. S. Finney.	Harrisburgh, Ohio	July 16, 1872	129,247
Hay-knife	G. Reiber	Dover, Pa	Aug. 22, 1871	118,390
Hay-knife	J. L. Ripley	Fremont, Ohio	Sept. 29, 1868	82,641
Hay-knife	H. M. Smith	Kalamazoo, Mich	Sept. 3, 1867	68,536
Hay-knife	P. O. Soper	San Francisco, Cal	Jan. 29, 1867	61,576
Hay-knife	A. N. Staley	Waynesborough, Pa	July 18, 1869	90,202
Hay-knife	G. F. Weymouth	Dresden, Me	Mar. 7, 1871	112,400
Hay-knife	S. Whalen	West Milton, N. Y	Mar. 2, 1854	10,874
Hay-knife	D. H. Wile	New Pittsburgh, Ohio	Oct. 4, 1870	108,076
Hay-knife	J. H. Wolfe	Lytle City, Iowa	Aug. 29, 1871	118,571
Hay-knife and pruner combined	D. Fasig	Rowsburgh, Ohio	Feb. 16, 1869	87,035
Hay-knife and pruning-hook, Combined	D. Fasig	Rowsburgh, Ohio	Mar. 8, 1870	100,611
Hay-loader	I. Anderson	Poland, Ohio	Oct. 5, 1869	95,408
Hay-loader	J. Bachelder	Norwich, Conn	Mar. 15, 1870	100,710
Hay-loader	J. R. Bailey	Woonsocket, R. I	Aug. 10, 1869	93,581
Hay-loader	H. Baker	Cortland, N. Y	July 18, 1871	117,033
Hay-loader	A. Barker	Camanche, Iowa	June 2, 1868	78,566
Hay-loader	J. M. Boorman	Scarborough, N. Y	Apr. 12, 1870	101,705
Hay-loader	A. Buck	Hebron, Ind	July 19, 1870	105,418
Hay loader	J. Bullis	Macedon, N. Y	May 7, 1867	64,482
Hay-loader	J. Capen	Charlton, Mass	Dec. 21, 1869	98,027
Hay-loader	D. Clagett	Hagerstown, Md	Aug. 30, 1870	106,916
Hay-loader	E. Cooper	Theresa, N. Y	Apr. 13, 1869	88,945
Hay-loader	E. Cooper	Theresa, N. Y	July 12, 1870	105,308
Hay-loader	E. N. Curtice	Spring Water, N. Y	Feb. 4, 1868	74,055
Hay-loader	T. P. Davidson and A. Parvis	Liberty, Ind	Apr. 8, 1873	137,751
Hay-loader	L. De Lacee	Springfield, Ill	May 21, 1867	64,848
Hay-loader	W. Denton	Iowa City, Iowa	Apr. 27, 1869	89,388
Hay-loader	M. A. Dilley	Mendon, Mich	Sept. 17, 1867	68,855
Hay-loader	N. B. Douglas	Cornwall, Vt	June 30, 1868	79,327
Hay-loader	W. H. Elliot	Plattsburgh, N. Y	Apr. 26, 1864	42,466
Hay-loader	W. H. Elliot	New York, N. Y	July 3, 1866	56,023
Hay-loader	W. H. Elliot	New York, N. Y	Apr. 23, 1867	64,003
Hay-loader	W. .H Elliot	New York, N. Y	June 4, 1867	65,478
Hay-loader	J. E. Fell and J. H. Mattocks	Maquokita, Iowa	Feb. 14, 1871	111,735
Hay-loader	A. Fuller	Plymouth, Ind	Aug. 7, 1866	56,924
Hay-loader	C. W. Gage	Homer, N. Y	July 16, 1867	66,699
Hay-loader	C. B. Garlinghouse	Carpentersville, Ind	Jan. 25, 1870	99,181
Hay-loader	C. B. Garlinghouse and J. Dickason.	Allensville, Ind	July 14, 1863	39,264
Hay-loader	A. Garver	Lime Spring Station, Iowa	Nov. 12, 1872	132,960
Hay-loader	C. Gibbs	Pittsfield, Vt	May 29, 1866	55,083
Hay-loader	T. Giffin	erryville, Ohio	Jan. 30, 1872	123,166
Hay-loader	C. H. Gifford	Philadelphia, Pa	Oct. 8, 1867	69,652
Hay-loader	H. L. Gockley	Jackson, Ill	Sept. 6, 1870	107,038
Hay-loader	W. H. Gray	Ashfield, Mass	Aug. 16, 1870	106,355
Hay-loader	C. P. Hale	Calhoun, Ky	Oct. 12, 1869	95,680
Hay-loader	W. M. Hall and J. Johnson	Barrington, N. Y	Apr. 2, 1867	63,382
Hay-loader	H. W. Hamilton	Brandon, Vt	Dec. 12, 1871	121,777
Hay-loader	N. L. Hatch	Cape Elizabeth, Me	Dec. 22, 1868	85,092

Index of patents issued from the United States Patent Office from 1790 to 1873, inclusive—Continued.

Invention.	Inventor.	Residence.	Date.	No.
Hay-loader	S. R. Higgins	Parma, Mich	May 15, 1866	54, 723
Hay-loader	H. C. Howland and T. L. Moore	Oxford, N. Y	Aug. 26, 1873	142, 234
Hay-loader	W. L. Hubbell	Brooklyn, N. Y	Aug. 25, 1863	39, 655
Hay-loader	H. Hull	West Exeter, N. Y	Feb. 11, 1868	74, 226
Hay-loader	D. C. Jewett and A. C. Bowen	Sand Spring and Bowen's Prairie, Iowa.	Jan. 9, 1866	51, 947
Hay-loader	M. K. Lewis and J. C. Durbin	Iowa City, Iowa	Jan. 3, 1865	45, 726
Hay-loader	A. W. Lozier	New York, N. Y	June 23, 1868	79, 135
Hay-loader	A. W. Lozier	New York, N. Y	Dec. 1, 1868	84, 560
Hay-loader	A. W. Lozier	New York, N. Y	June 15, 1869	91, 242
Hay-loader	O. T. Nanny	Mamakating, N. Y	Mar. 27, 1866	53, 470
Hay-loader	A. J. and S. D. McKee	Beaver Dam, Ohio	June 8, 1869	91, 147
Hay-loader	B. J. Moore	Dresbach, Minn	Nov. 2, 1869	96, 462
Hay-loader	J. Morgan	West Springfield, Mass	Apr. 16, 1867	63, 806
Hay-loader	G. G. Park	Xenia, Nebr	Aug. 23, 1870	106, 611
Hay-loader	I. J. Parker	Buffalo Grove, Iowa	July 26, 1870	105, 837
Hay-loader	W. Platt and A. G. Burnham	Greenfield, Mass	Jan. 31, 1865	46, 139
Hay-loader	A. C. Preston	Battle Creek, Mich	Feb. 15, 1870	99, 947
Hay-loader	G. Race	Norwich, N. Y	Apr. 17, 1866	54, 084
Hay-loader	F. M. Reynolds	Mile Strip, N. Y	July 21, 1868	80, 222
Hay-loader	J. P. Rideout	Bowdoinham, Me	Jan. 24, 1871	111, 252
Hay-loader	J. G. Sehorn	Iowa City, Iowa	Sept. 14, 1869	94, 846
Hay-loader	R. Shaw	Warsaw, Ind	Apr. 23, 1872	125, 992
Hay-loader	W. H. Straub	Danville, Pa	Dec. 7, 1869	97, 724
Hay-loader	E. Sweet	Whitney's Point, N. Y	Mar. 26, 1867	63, 332
Hay-loader	L. D. Taylor	Granville Centre, Pa	May 7, 1867	64, 459
Hay-loader	L. D. Taylor	Granville Centre, Pa	Apr. 4, 1871	113, 363
Hay-loader	F. Van Etten	Jackson, Mich	May 8, 1866	54, 627
Hay-loader	A. Vose	Pittsfield, Vt	Nov. 19, 1867	71, 250
Hay-loader	C. E. Warner	Burlington, Kans	Dec. 9, 1873	145, 321
Hay, Loading	R. Cobb	Hadley, Mich	Dec. 5, 1865	51, 294
Hay, Loading	W. Dixon	Chicago, Ill	June 26, 1860	28, 839
Hay, Loading	T. J. Jolly	Olean, Ind	June 26, 1860	28, 868
Hay loading and pitching machine	J. S. De Haven	Akron, Ohio	Aug. 19, 1862	36, 208
Hay loading and unloading apparatus	G. W. Long	Delaware Centre, Iowa	Feb. 11, 1873	135, 720
Hay-loading device	A. B. Clark	Grass Lake, Mich	June 7, 1870	103, 844
Hay-loading machine	J. B. Atwater	Chicago, Ill	Aug. 30, 1864	43, 964
Hay-loading machine	W. H. Bentley	Westford, N. Y	Nov. 17, 1863	40, 598
Hay-loading machine	G. Constable	Cannonsville, N. Y	July 14, 1863	39, 215
Hay-loading machine	L. De Lacie	Newark, N. Y	Dec. 13, 1864	45, 394
Hay-loading machine	S. V. Essick	Moultrie, Ohio	May 10, 1859	23, 911
Hay-loading machine	J. W. Foust	Hammondsburgh, Pa	Nov. 18, 1862	36, 947
Hay-loading machine	J. F. Hall and I. Pierce	Holland Patent, N. Y	June 28, 1864	43, 307
Hay-loading machine	J. F. Hall and I. Pierce	Holland Patent, N. Y	Dec. 13, 1864	45, 405
Hay-loading machine	S. R. Higgins	Parma, Mich	Jan. 10, 1865	45, 826
Hay-loading machine	H. Maycock	Verona, N. Y	Oct. 4, 1864	44, 540
Hay-loading machine	J. B. McIntosh	Girard, Pa	Feb. 19, 1861	31, 470
Hay-loading machine	L. Mishler	Mogadore, Ohio	Sept. 24, 1861	33, 359
Hay-loading machines, Tongue for	E. Sweet	Whitney's Point, N. Y	Jan. 21, 1868	73, 474
Hay, Machine for spreading and turning	J. Center	Hudson, N. Y	June 27, 1834	....
Hay-making machine	G. A. Brown	Middletown, R. I	Jan. 23, 1855	12, 269
Hay-making machine	T. J. Goff	Warren, R. I	June 14, 1859	24, 386
Hay-making machine	F. Peabody	Salem, Mass	Apr. 17, 1855	12, 728
Hay-making machine	J. C. Stoddard	Worcester, Mass	June 28, 1859	24, 588
Hay-making machine	J. C. Stoddard	Worcester, Mass	Dec. 6, 1859	26, 380
Hay-mow, Ventilating	G. Race	Norwich, N. Y	Dec. 10, 1867	72, 081
Hay or cotton press	F. Horton	Silver Creek, N. Y	Nov. 12, 1867	70, 720
Hay or cotton press	D. A. Nelson	Tyler, Tex	Sept. 9, 1873	142, 718
Hay or feed cutter	J. R. Whittemore	Chicopee Falls, Mass	Aug. 26, 1862	36, 316
Hay or straw cutter	S. S. Clark	Manchester, N. H	Jan. 17, 1860	26, 835
Hay, &c., Preserving	A. D. Ditman	Chester, Pa	Feb. 15, 1838	599
Hay-press	D. R. Allen	Cumberland, Me	Apr. 25, 1846	4, 484
Hay-press	J. S. Arnold	South Milan, Ind	July 3, 1860	28, 952
Hay-press	G. H. Aylworth	Brighton, Ill	May 21, 1867	64, 828
Hay-press	G. H. Aylworth	Brighton, Ill	Apr. 5, 1870	101, 566
Hay-press	H. F. Blank	Liberty, Ill	Nov. 18, 1873	144, 733
Hay-press	M. B. Bliss	Pittston, Me	Jan. 26, 1828	....
Hay-press	I. Bisbee and A. Bisbee	East Pharsalia, N. Y., and Polk Township, Mo.	Apr. 8, 1862	34, 872
Hay-press	O. Bossée	Millbrae, Cal	Aug. 29, 1871	118, 582
Hay-press	J. Briggs	Homewood, Ill	Mar. 19, 1872	124, 717
Hay-press	A. R. Chamberlin and A. Cleflin	Richmond, Me	Mar. 30, 1836	....
Hay-press	S. Colahan	Cleveland, Ohio	Mar. 11, 1862	34, 622
Hay-press	G. W. D. Culp	Allensville, Ind	July 12, 1864	43, 480
Hay-press	G. W. D. Culp	East Enterprise, Ind	Feb. 4, 1868	74, 054
Hay-press	G. W. D. Culp	East Enterprise, Ind	Oct. 22, 1872	132, 448
Hay-press	J. H. Davison and T. J. Sasher	Mound Station, Ill	Nov. 1, 1870	108, 829
Hay-press	L. Dederick	Kinderhook, N. Y	June 24, 1843	3, 148
Hay-press	L. Dederick	Albany, N. Y	June 6, 1854	11, 043
Hay-press	L. Dederick	Albany, N. Y	July 14, 1863	39, 258
Hay-press	W. Deering	Louisville, Ky	May 26, 1863	38, 659
Hay-press	E. A. Field	Sidney, Me	Dec. 26, 1865	51, 709
Hay-press	M. Fletcher	Louisville, Ky	Sept. 27, 1864	44, 413
Hay-press	F. Frey	Liberty, Ill	July 5, 1870	105, 062
Hay-press	G. Gibbs	Sugar Branch, Ind	Oct. 25, 1864	44, 795
Hay-press	J. H. Gove	San Francisco, Cal	Mar. 6, 1860	27, 361
Hay-press	F. F. Hamilton	Green Bay, Wis	Jan. 10, 1871	110, 849
Hay-press	F. F. Hamilton	Green Bay, Wis	Nov. 28, 1871	121, 244
Hay-press	C. W. Hawkes	Brunswick, Me	Apr. 18, 1839	1, 124
Hay-press	A. Hayford and A. Strout	Belfast, Me	May 30, 1865	47, 950
Hay-press	S. Hewitt	Allensville, Ind	Dec. 30, 1843	3, 394
Hay-press	H. C. Hunt	Indianapolis, Ind	Dec. 6, 1870	109, 905
Hay-press	I. H. Johnson	Long Reach, W. Va	Aug. 21, 1866	57, 331
Hay-press	G. W. Lockhart	Charlestown, Ind	Apr. 7, 1868	76, 478
Hay-press	S. Mahurin	Liberty, Ill	July 16, 1867	66, 860
Hay-press	P. Manny	Waddam's Grove, Ill	Apr. 17, 1855	12, 740
Hay-press	C. Martine	Greensburgh, N. Y	May 30, 1842	2, 657
Hay-press	W. McCord	Sing Sing, N. Y	Jan. 24, 1860	26, 917
Hay-press	M. R. Molleur	Chicago, Ill	Feb. 18, 1868	74, 571

Index of patents issued from the United States Patent Office from 1790 *to* 1873, *inclusive*—Continued.

Invention.	Inventor.	Residence.	Date.	No.
Hay-press	M. V. Northrup	Hornitos, Cal	Dec. 27, 1870	110, 585
Hay-press	G. Noyes	Pownal, Me	Dec. 17, 1867	72, 221
Hay-press	C. F. Paine	Winslow, Me	Apr. 25, 1844	3, 561
Hay-press	P. C. Pearson	Harrison County, Ind	July 19, 1864	43, 600
Hay-press	W. R. and H. M. Peavey	Swanville, Me	July 15, 1862	35, 889
Hay-press	S. G. Randall	New Britain, Conn	July 17, 1860	29, 194
Hay-press	W. H. Reynolds	New Orleans, La	July 11, 1871	116, 869
Hay-press	A. C. Richard	Point Lookout, Tenn	Dec. 24, 1867	72, 681
Hay-press	W. Ridonour	Springfield, Ohio	Mar. 10, 1863	37, 873
Hay-press	B. Roberts	Highland, N. Y	Apr. 24, 1866	54, 209
Hay-press	C. Rundlett and J. W. Drummond.	Alden and Winslow, Me	July 19, 1859	24, 824
Hay-press	J. Scott	Brooklyn, N. Y	Feb. 6, 1872	123, 517
Hay-press	A. Spanier	Saint John, Ind	Aug. 15, 1871	118, 164
Hay-press	C. S. Stevens and A. T. Boon	Galesburgh, Ill	Oct. 11, 1864	44, 673
Hay-press	S. Stevens	Sacramento, Cal	July 24, 1860	29, 320
Hay-press	R. V. W. Thorn	New York, N. Y	Apr. 29, 1813	
Hay-press	E. A. Ward	Gallipolis, Ohio	Aug. 14, 1866	57, 227
Hay-press pawl	L. D. Rundell	South Westerlo, N. Y	Feb. 10, 1863	37, 643
Hay-press, Portable	J. Mauck	Cheshire, Ohio	May 28, 1867	65, 249
Hay-press, Wagon	H. Brocksmith	Saint Louis, Mo	Dec. 30, 1873	145, 986
Hay-presses, Capstan for working beater	P. K. Dederick	Albany, N. Y	Aug. 29, 1865	49, 678
Hay-presses, Method of securing the door of	C. Martratt	New Baltimore, N. Y	Apr. 21, 1857	17, 108
Hay, Pressing and baling	G. H. Aylworth	Brighton, Ill	Mar. 21, 1871	112, 766
Hay-rack	G. Baldwin	Bluffton, Ind	Apr. 25, 1865	47, 379
Hay-rack	A. C. Bayley	Battle Creek, Mich	May 4, 1869	89, 619
Hay-rack	C. G. Boyer	Greenfield, Ind	July 2, 1867	66, 292
Hay-rack	S. Brownell	Irving, N. Y	Sept. 22, 1868	82, 288
Hay-rack	G. Denis and G. Grassel	Oceola, Iowa	Nov. 23, 1869	97, 172
Hay-rack	W. E. Grimes	Allen, Mo	Apr. 10, 1866	53, 814
Hay-rack	W. P. Jones	Arcade, N. Y	Jan. 2, 1872	122, 387
Hay-rack	J. Mandigo	Wayland, Mich	Oct. 3, 1871	119, 627
Hay-rack	D. Matteson	Pittsfield, Mich	Aug. 17, 1869	93, 897
Hay-rack	A. Naramor	Berlin Heights, Ohio	Jan. 24, 1865	46, 017
Hay-rack	W. Nash	Corning, N. Y	Apr. 30, 1867	64, 349
Hay-rack	W. Smith	Tod Township, Ohio	Oct. 28, 1868	83, 561
Hay-rack	A. T. Soule	Savannah, N. Y	July 21, 1868	80, 095
Hay-rack	D. N. Webster	Geneva, Ohio	Sept. 18, 1866	58, 158
Hay-rack and wagon-box, Combined	F. G. McClellan	Attica, Ohio	Apr. 19, 1870	102, 023
Hay-rack, Feeding	T. Moore	New York, N. Y	Apr. 17, 1866	54, 005
Hay-rack, Hinged	M Carpenter	Wayland, Mich	June 20, 1871	116, 153
Hay-rake	D. Armel	Somerset, Pa	Jan. 20, 1863	37, 429
Hay-rake	E. P. Barton and R. W. Towle	Batavia and Bath, N. Y	July 9, 1861	32, 745
Hay-rake	H. Brown	Cape Elizabeth, Me	July 3, 1866	56, 000
Hay-rake	L. A. C. Brown	Sparta, Ill	Dec. 1, 1857	18, 731
Hay-rake	R. I. Burbank	Boston, Mass	Mar. 23, 1869	88, 127
Hay-rake	C. P. Carpenter	Saint Johnsbury, Vt	Aug. 26, 1856	15, 601
Hay-rake	C. P. Carpenter	Saint Johnsbury, Vt	S pt. 15, 1863	39, 989
Hay-rake	W. Deckman	Canton, Ohio	Apr. 16, 1861	32, 054
Hay-rake	C. J. Fay	Hammonton, N. J	Sept. 25, 1860	30, 132
Hay-rake	P. Fitzgerald	Constantia, Ohio	Oct. 5, 1858	21, 664
Hay-rake	A. J. Greene	Sterling, Mass	Nov. 5, 1867	70, 550
Hay-rake	T. C. Hance	Palmyra, N. Y	June 25, 1821	
Hay-rake	H. Heaberlin	Scipio, Ind	Sept. 2, 1856	15, 653
Hay-rake	S. J. Homan	Walden, N. Y	Oct. 2, 1860	30, 222
Hay-rake	W. King	Springfield, Ill	Sept. 1, 1868	81, 791
Hay-rake	R. G. Lamson	Brownsville, Vt	Oct. 16, 1866	58, 841
Hay-rake	W. H. Misner and G. E. Marker.	Heyworth, Ill	Aug. 2, 1870	105, 967
Hay-rake	J. I. Monroe	Burlington, Mass	Mar. 19, 1867	63, 079
Hay-rake	F. L. Nagler	Irving Township, Mich	Jan. 24, 1871	111, 138
Hay-rake	G. Notman	Deerfield, Ohio	Aug. 8, 1871	117, 917
Hay-rake	F. B. Parker	Queensville, Ind	Aug. 23, 1853	9, 956
Hay-rake	M. Raezer	Reading, Pa	Oct. 5, 1858	21, 698
Hay-rake	T. R. Roach	West Needham, Mass	Nov. 4, 1856	16, 025
Hay-rake	I. I. Robbins	Penn's Manor, Pa	Sept. 23, 1856	15, 777
Hay-rake	D. Robinson	Reading Centre, N. Y	Apr. 1, 1862	34, 848
Hay-rake	Z. Sanders	West Windsor, Vt	June 8, 1852	9, 007
Hay-rake	D. Smith	Vincennes, Ind	June 7, 1838	773
Hay-rake	C. R. Soule	Fairfield, Vt	May 18, 1852	8, 965
Hay-rake	C. R. Soule	Fairfield, Vt	July 9, 1861	32, 791
Hay-rake	J. J. Squire	Saint Louis, Mo	Dec. 23, 1856	16, 318
Hay-rake	J. C. Stoddard	Lockport, N. Y	June 4, 1872	127, 659
Hay-rake	J. S. Sturges	Litchfield, Ohio	Mar. 9, 1852	8, 791
Hay-rake	J. Wallace and D. Carpenter	Goshen, N. Y	Dec. 1, 1863	40, 782
Hay-rake	G. Whitcomb	Port Chester, N. Y	Oct. 5, 1858	21, 712
Hay-rake	F. G. Willson	Ontario, Canada	July 30, 1861	32, 935
Hay-rake	S. W. Wood	Washington, D. C	July 7, 1857	17, 772
Hay rake and feeder, Combined	S. J. Taylor	Rome, N. Y	May 24, 1870	103, 524
Hay rake and fork	T. W. Peirce	Minneapolis, Minn	July 18, 1871	117, 201
Hay-rake and grain-harrow	G. Davis	Belmont, Ohio	July 22, 1839	1, 258
Hay rake and loader	J. Adams	Transfer, Pa	June 16, 1868	78, 912
Hay rake and loader	H. P. Burdick	Buffalo, N. Y	June 1, 1869	90, 818
Hay rake and loader	R. Chesnut	Richmond, Ind	Mar. 23, 1869	88, 009
Hay rake and loader	A. Clark	Cadiz, Ohio	May 25, 1869	90, 425
Hay rake and loader	W. A. Dean	New Lexington, Ohio	July 16, 1872	129, 009
Hay rake and loader	N. Gabel	Minneapolis, Kans	Jan. 4, 1870	98, 581
Hay rake and loader	F. W. Harlow	Hannibal, Mo	June 22, 1869	91, 623
Hay rake and loader	J. W. Henry	Pecatonica, Ill	Oct. 1, 1867	69, 432
Hay rake and loader	H. Kimmell	Waynesburgh, Ohio	Aug. 13, 1867	67, 771
Hay rake and loader	O. Luce	Virgil, N. Y	Mar. 31, 1868	76, 215
Hay rake and loader	F. Marion and N. E. Wilson	Groveland, Ill	Jan. 21, 1873	134, 997
Hay rake and loader	A. J. Preston	East Guilford, N. Y	July 18, 1871	117, 204
Hay rake and loader	A. Sheline	Edon, Ohio	Sept. 14, 1869	94, 782
Hay rake and loader	F. Terwilleger and J. R. Isdell	Wyanet, Ill	Sept. 14, 1869	94, 791
Hay rake and loader	J. J. Thompson	Richwood, Ohio	Oct. 13, 1868	83, 111
Hay rake and loader	M. Webb	Chenango Forks, N. Y	Sept. 3, 1867	68, 587
Hay rake and loader	G. A. Wing	Albany, N. Y	May 3, 1870	102, 744

Index of patents issued from the United States Patent Office from 1790 *to* 1873, *inclusive*—Continued.

Invention.	Inventor.	Residence.	Date.	No.
Hay rake and seeder combined, Horse	J. H. Bear	York, Pa	Jan. 1, 1867	60, 826
Hay rake and tedder	J. C. Mills	Palmyra, N. Y	Dec. 7, 1869	97, 543
Hay rake and tedder combined	C. La Dow	South Galway, N. Y	Mar. 28, 1871	113, 066
Hay rake and tedder combined	F. A. Thayer	Sheldonville, Mass	Apr. 25, 1871	114, 065
Hay rake and tedder, Combined horse	H. C. Varnum	Hartford, Vt	July 19, 1870	105, 611
Hay-rake fastening	O. W. Hogle	Somerset, N. Y	Sept. 3, 1850	7, 618
Hay-rake, Flexible horse	W. Buckminster	Framingham, Mass	July 12, 1838	838
Hay-rake, Horse	N. E. Allen	Trenton, Wis	Apr. 20, 1858	19, 975
Hay-rake, Horse	A. Amos	Potsdam Junction, N. Y	Nov. 4, 1873	144, 305
Hay-rake, Horse	A. Amos	Potsdam Junction, N. Y	Nov. 4, 1873	144, 306
Hay-rake, Horse	F. Andrews	Galesburgh, Ill	June 27, 1871	116, 254
Hay-rake, Horse	O. B. Austin	Potsdam Junction, N. Y	Dec. 24, 1872	134, 239
Hay-rake, Horse	L. B. Ball	Dayton, Ohio	Feb. 27, 1872	124, 110
Hay-rake, Horse	A. B. Barnard	Worcester, Mass	Feb. 22, 1870	100, 104
Hay-rake, Horse	N. M. Barnes	Tiffin, Ohio	July 19, 1870	105, 542
Hay-rake, Horse	N. M. Barnes	Tiffin, Ohio	May 28, 1872	127, 295
Hay-rake, Horse	J. H. Bean	Macon, Ill	July 8, 1873	140, 671
Hay-rake, Horse	C. D. Blinn	Port Huron, Mich	Aug. 27, 1867	68, 160
Hay-rake, Horse	O. Bonney, jr	San Francisco, Cal	Jan. 31, 1871	111, 308
Hay-rake, Horse	L. S. Bortree	Grand Rapids, Mich	June 21, 1870	104, 545
Hay-rake, Horse	C. L. and F. Brough	Delphi, Ind	Dec. 31, 1872	134, 353
Hay-rake, Horse	A. Brown	Worcester, Mass	July 12, 1870	105, 165
Hay-rake, Horse	F. Brown	Russell, Ohio	Jan. 30, 1872	123, 081
Hay-rake, Horse	J. H. Bullard	Chicopee, Mass	Nov. 12, 1872	133, 012
Hay-rake, Horse	G. E. Burt	Harvard, Mass	Dec. 17, 1872	133, 920
Hay-rake, Horse	M. Butler	Vernon, Ind	Mar. 19, 1867	63, 011
Hay-rake, Horse	G. E. Carleton	Old Town, Me	Sept. 27, 1870	107, 660
Hay-rake, Horse	A. L. Chubb	Grand Rapids, Mich	Sept. 26, 1871	119, 320
Hay-rake, Horse	J. M. Colson	Morrill, Me	Aug. 23, 1870	106, 554
Hay-rake, Horse	J. Comly	York, Pa	Dec. 26, 1871	122, 109
Hay-rake, Horse	I. N. Condra	Genoa, Iowa	Jan. 31, 1871	111, 320
Hay-rake, Horse	J. H. Cook	Hagerstown, Md	Feb. 8, 1870	99, 643
Hay-rake, Horse	A. Cowley	Harpersfield, N. Y	Mar. 30, 1858	19, 753
Hay-rake, Horse	J. Crellin	Marshalltown, Iowa	May 16, 1865	47, 703
Hay-rake, Horse	L. A. Crockett	Wythe County, Va	Nov. 21, 1871	121, 159
Hay-rake, Horse	N. W. Curtis	Johnsburgh, N. Y	Jan. 25, 1870	99, 066
Hay-rake, Horse	S. L. Denny	Christiana, Pa	Aug. 16, 1870	106, 336
Hay-rake, Horse	B. J. Downing	Mitchell, Iowa	Sept. 30, 1873	143, 278
Hay-rake, Horse	A. T. Dunbar	Alba, Pa	Dec. 15, 1868	84, 999
Hay-rake, Horse	H. Eastman	Indianapolis, Ind	Mar. 6, 1860	27, 407
Hay-rake, Horse	C. Edgar	Dayton, Ohio	Sept. 16, 1873	142, 844
Hay-rake, Horse	D. S. Emrick	Fayette, N. Y	Apr. 26, 1870	102, 238
Hay-rake, Horse	W. P. Ewing	Fancy Hill, Va	Sept. 21, 1869	95, 010
Hay-rake, Horse	J. F. Faust	Lebanon, Ohio	July 6, 1858	20, 844
Hay-rake, Horse	W. H. Fay	Camden, N. J	June 14, 1870	104, 131
Hay-rake, Horse	H. Gale	Albion, Mich	Nov. 9, 1869	96, 576
Hay-rake, Horse	O. L. Genung and W. H. Blackman.	Caroline, N. Y	Nov. 19, 1872	133, 091
Hay-rake, Horse	C. N. Goss	Claremont, N. H	Jan. 28, 1873	135, 330
Hay-rake, Horse	P. M. Gundlach	Belleville, Ill	Nov. 2, 1869	96, 425
Hay-rake, Horse	E. R. Hall	Utica, N. Y	June 9, 1868	78, 735
Hay-rake, Horse	G. N. Hall	Mamakating, N. Y	Dec. 6, 1859	26, 396
Hay-rake, Horse	J. Harris	Mount Vernon, Ind	June 11, 1872	127, 876
Hay-rake, Horse	W. H. Hartley	Philadelphia, Pa	Oct. 14, 1873	143, 572
Hay-rake, Horse	G. Hauck	Mechanicsburgh, Pa	Sept. 13, 1870	107, 256
Hay-rake, Horse	J. Hayward	Greene, N. Y	Oct. 2, 1866	58, 412
Hay-rake, Horse	W. A. Heath	Apalachin, N. Y	May 19, 1868	78, 086
Hay-rake, Horse	J. Heidy	Martinsburgh, Ohio	Nov. 12, 1872	133, 028
Hay-rake, Horse	T. N. Henderson	Jackson, Mich	Apr. 23, 1867	64, 104
Hay-rake, Horse	J. F. Henkley	Saint Louis, Mo	Mar. 28, 1871	113, 165
Hay-rake, Horse	M. S. Holman and C. S. Farrer	Armada, Mich	Aug. 15, 1871	118, 130
Hay-rake, Horse	W. Holmes	Clarksville, N. Y	Sept. 8, 1868	82, 002
Hay-rake, Horse	B. F. Horton	Ithaca, N. Y	Feb. 15, 1870	99, 897
Hay-rake, Horse	C. Howard	West Hurley, N. Y	Feb. 1, 1870	99, 319
Hay-rake, Horse	J. Howard and E. S. Bousfield	Bedford, England	Mar. 7, 1871	112, 461
Hay-rake, Horse	J. G. Huntington	Atkinson, Me	Oct. 19, 1869	96, 005
Hay-rake, Horse	A. K. Hurst	Chambersburgh, Pa	July 10, 1860	29, 082
Hay-rake, Horse	G. L. Ives	Rome, N. Y	Aug. 13, 1872	130, 431
Hay-rake, Horse	O. S. Jarvis	Xenia, Ill	Apr. 25, 1871	114, 150
Hay-rake, Horse	A. B. Johnson	Washington, Ind	Mar. 6, 1860	27, 366
Hay-rake, Horse	J. F. Keller	Hagerstown, Md	Mar. 25, 1873	137, 212
Hay-rake, Horse	J. G. Kimberlin	Dryden, N. Y	Apr. 30, 1867	64, 331
Hay-rake, Horse	W. S. Kincaid	Leavenworth County, Kans	Jan. 19, 1869	86, 080
Hay-rake, Horse	G. W. King	Greenville, N. Y	May 2, 1865	47, 553
Hay-rake, Horse	C. Kugler	Barnesville, Ohio	Sept. 5, 1871	118, 726
Hay-rake, Horse	G. O. Lackey	Akron, Ohio	Sept. 6, 1870	107, 183
Hay-rake, Horse	D. Z. Lantz	Salisbury Township, Pa	Dec. 17, 1872	133, 944
Hay-rake, Horse	S. Lessig	Reading, Pa	Dec. 27, 1859	26, 599
Hay-rake, Horse	L. Litchfield and J. S. Corbin	Gouverneur, N. Y	July 1, 1873	140, 513
Hay-rake, Horse	W. H. Locke	Canton, Pa	Apr. 18, 1871	113, 895
Hay-rake, Horse	J. G. Lockwood	West Davenport, N. Y	May 23, 1871	115, 071
Hay-rake, Horse	J. G. Lockwood	West Davenport, N. Y	Dec. 19, 1871	122, 039
Hay-rake, Horse	W. H. Long	Eddyville, Iowa	Feb. 1, 1870	99, 331
Hay-rake, Horse	C. M. Lufkin and C. G. Allen	Barre, Mass	Aug. 19, 1873	142, 031
Hay-rake, Horse	S. I. Lynd and E. M. Tousley	Albion, N. Y	July 23, 1872	129, 742
Hay-rake, Horse	S. I. Lynd and E. M. Tousley	Albion, N. Y	July 23, 1872	129, 743
Hay-rake, Horse	R. H. Martindale	Hobbieville, Ind	Apr. 30, 1872	126, 314
Hay-rake, Horse	W. C. Martindale	Philadelphia, Pa	July 8, 1873	140, 634
Hay-rake, Horse	W. Matthews	Vinton, Ohio	Aug. 31, 1869	94, 425
Hay-rake, Horse	R. W. McClelland	Springfield, Ill	Apr. 12, 1870	101, 752
Hay-rake, Horse	J. A. McGee	Sharpsville, Ind	May 17, 1870	103, 065
Hay-rake, Horse	G. M. L. McMillen	Drayton, Ohio	Aug. 16, 1870	106, 385
Hay-rake, Horse	J. H. Mears	Oshkosh, Wis	July 29, 1873	141, 370
Hay-rake, Horse	C. R. Merriam	Middlebury, Vt	Jan. 3, 1871	110, 771
Hay-rake, Horse	J. Mills	Milan, Ind	Aug. 23, 1870	106, 604
Hay-rake, Horse	P. J. Moose and J. Kuhn	Dansville, N. Y	Oct. 24, 1871	120, 160
Hay-rake, Horse	J. Morgan and W. Cline, jr	Clayton, Ind	Feb. 1, 1870	99, 457
Hay-rake, Horse	J. H. Morris	Maquoketa, Iowa	Apr. 5, 1870	101, 646

Index of patents issued from the United States Patent Office from 1790 *to* 1873, *inclusive*—Continued.

Invention.	Inventor.	Residence.	Date.	No.
Hay-rake, Horse	B. Morse	Ithaca, N. Y	Aug. 19, 1873	141, 884
Hay-rake, Horse	J. I. Munroe	Woburn, Mass	Apr. 12, 1870	101, 904
Hay-rake, Horse	L. Myers	Ludlowville, N. Y	July 5, 1870	105, 117
Hay-rake, Horse	W. A. Myers	York, Pa	Apr. 4, 1871	113, 552
Hay-rake, Horse	O. T. Nanny	Amity, N. Y	Nov. 22, 1870	109, 439
Hay-rake, Horse	D. B. Neal	Mount Gilead, Ohio	Aug. 17, 1869	93, 829
Hay-rake, Horse	G. Notman	Deerfield, Ohio	May 3, 1870	102, 578
Hay-rake, Horse	G. Notman	Deerfield, Ohio	May 9, 1871	114, 591
Hay-rake, Horse	S. R. Nye	Barre, Mass	July 26, 1870	105, 833
Hay-rake, Horse	J. K. O'Neil	Kingston, N. Y	Feb. 15, 1870	99, 939
Hay-rake, Horse	L. A. Paddock	Pecatonica, Ill	Feb. 6, 1872	123, 357
Hay-rake, Horse	L. A. Paddock	Pecatonica, Ill	Apr. 2, 1872	125, 216
Hay-rake, Horse	M. Pennock	East Marlborough, Pa	June 26, 1822	....
Hay-rake, Horse	P. Pfeifer	Durhamville, N. Y	Oct. 3, 1871	119, 532
Hay-rake, Horse	R. Pratt	Marple Township, Pa	Jan. 8, 1856	14, 067
Hay-rake, Horse	P. Prescott	Boonville, N. Y	Oct. 6, 1868	82, 749
Hay-rake, Horse	M. Raezer	Reading, Pa	Nov. 29, 1859	26, 294
Hay-rake, Horse	M. C. Remington	Weedsport, N. Y	Sept. 28, 1869	95, 268
Hay-rake, Horse	M. C. Remington	Weedsport, N. Y	Oct. 10, 1871	119, 721
Hay-rake, Horse	G. S. Reynolds	East Bethel, Vt	May 10, 1859	23, 943
Hay-rake, Horse	M. J. Robinson	Ashley, Ill	Sept. 13, 1870	107, 411
Hay-rake, Horse	S. Rockafellow	Moline, Ill	Jan. 24, 1871	111, 150
Hay-rake, Horse	S. Rockafellow	Moline, Ill	Dec. 26, 1871	122, 282
Hay-rake, Horse	S. Rockafellow	Moline, Ill	Mar. 19, 1872	124, 855
Hay-rake, Horse	R. S. Ryker	Canaan, Ind	Sept. 10, 1872	131, 183
Hay-rake, Horse	C. W. Sanborn	Morrill, Me	Jan. 4, 1870	98, 637
Hay-rake, Horse	W. Sharkey	Chico, Cal	Dec. 7, 1869	97, 708
Hay-rake, Horse	D. P. Sharp	Ithaca, N. Y	Mar. 5, 1872	124, 225
Hay rake, Horse	R. B. Sheldon	Canastota, N. Y	Apr. 1, 1873	137, 389
Hay-rake, Horse	J. H. Shoonmaker	Bethlehem, N. Y	Mar. 21, 1871	112, 854
Hay-rake, Horse	J. S. Shrawder	Fairview Village, Pa	May 17, 1870	103, 092
Hay-rake, Horse	A. J. Skunk	Des Moines, Iowa	July 26, 1870	105, 733
Hay-rake, Horse	S. P. Smith	Waterford, N. Y	Apr. 19, 1870	102, 172
Hay-rake, Horse	S. P. Smith	Waterford, N. Y	May 31, 1870	103, 789
Hay-rake, Horse	S. P. Smith	Waterford, N. Y	Mar. 7, 1871	112, 507
Hay-rake, Horse	S. P. Smith	Waterford, N. Y	Jan. 23, 1872	123, 053
Hay-rake, Horse	S. Stevenson	Dansville, N. Y	Mar. 11, 1873	136, 785
Hay-rake, Horse	J. C. Stoddard	Worcester, Mass	Aug. 6, 1861	33, 023
Hay-rake, Horse	J. C. Stoddard	Worcester, Mass	Sept. 27, 1870	107, 832
Hay-rake, Horse	J. C. Stoddard	Worcester, Mass	Jan. 10, 1871	110, 876
Hay-rake, Horse	O. O. Storle	North Cape, Wis	Apr. 18, 1871	113, 813
Hay-rake, Horse	G. Sweet	Dansville, N. Y	Oct. 3, 1871	119, 669
Hay-rake, Horse	G. Sweet	Dansville, N. Y	Nov. 25, 1873	145, 029
Hay-rake, Horse	B. C. Taylor	Dayton, Ohio	Sept. 1, 1868	81, 706
Hay-rake, Horse	R. M. Treat	Morris, Conn	May 2, 1871	114, 368
Hay-rake, Horse	M. W. Trescott	North Canaan, Conn	June 13, 1871	115, 909
Hay-rake, Horse	C. Tyler	Dryden, N. Y	Mar. 22, 1870	101, 189
Hay-rake, Horse	J. H. Walker	Grand Rapids, Mich	Aug. 12, 1873	141, 684
Hay-rake, Horse	C. W. Warner	New Haven, Vt	Aug. 18, 1868	81, 234
Hay-rake, Horse	D. H. Weaver	South Bend, Ind	May 31, 1870	103, 691
Hay-rake, Horse	G. Weiland	Dayton, Ohio	Mar. 22, 1870	101, 196
Hay-rake, Horse	J. R. Whittemore	Chicopee Falls, Mass	Nov. 19, 1872	133, 280
Hay-rake, Horse	J. E. Wisner	Friendship, N. Y	Oct. 25, 1870	108, 744
Hay-rake, Horse	J. E. Wisner	Friendship, N. Y	Nov. 29, 1870	109, 700
Hay-rake, Horse	J. E. Wisner	Friendship, N. Y	Nov. 14, 1871	121, 035
Hay-rake, Horse	E. Zimerman	Pamelia Four Corners, N. Y.	May 31, 1870	103, 815
Hay-rake, Revolving	E. Calderwood	Portland, Me	Jan. 21, 1865	46, 447
Hay-rake, Revolving	H. and E. Hoopes	Wilmington, Del	July 26, 1824	....
Hay-rake, Revolving	E. B. and M. D. Wells	Morgantown, Va	Sept. 20, 1839	1, 329
Hay-rake spring-teeth	Z. Breed	Weare, N. H	June 18, 1859	7, 434
Hay rake, thrasher, loader, and stacker	J. R. Hammond	Sedalia, Mo	Nov. 30, 1869	97, 395
Hay-rakes, Tooth-fastening for horse	B. Pickering	Dayton, Ohio	July 9, 1872	128, 749
Hay raker and loader	J. Armstrong and J. Jeffcoat	Onawa, Iowa	Feb. 2, 1869	86, 345
Hay raker and loader	H. Baker	Cortland, N. Y	July 3, 1866	55, 979
Hay raker and loader	J. M. Boorman	Scarborough, N. Y	Apr. 28, 1868	77, 247
Hay raker and loader	C. C. Brandt and J. C. Shillook	New Ulm, Minn	June 18, 1872	128, 012
Hay raker and loader	W. F. Brown and J. N. Smith	Washington, D. C., and Jersey City, N. J.	Dec. 3, 1867	71, 577
Hay raker and loader	A. Campbell	Peoria, Ill	Oct. 15, 1867	69, 764
Hay raker and loader	J. F. Chandler	Concord, N. H	Aug. 17, 1869	93, 806
Hay raker and loader	L. Clarke	Candor, N. Y	Feb. 11, 1868	74, 307
Hay raker and loader	J. S. Coffman and M. Graybill	Greenville, Ind	Mar. 17, 1868	75, 528
Hay raker and loader	L. D. Copeland	Chenango Forks, N. Y	Sept. 24, 1867	69, 187
Hay raker and loader	E. N. Curtice	Springwater, N. Y	Nov. 26, 1867	71, 366
Hay raker and loader	M. A. Dilly	Mendon, Mich	May 12, 1868	77, 721
Hay raker and loader	R. Elarton and E. A. Thomas	Hillsborough, Iowa	Jan. 5, 1869	85, 651
Hay raker and loader	N. Farlow and J. A. Ham	Sullivan, Ill	Oct. 5, 1869	95, 451
Hay raker and loader	J. W. Foust	Evansburgh, Pa	Oct. 16, 1866	58, 804
Hay raker and loader	M. B. Fuller	Sanborn, N. Y	Feb. 25, 1868	74, 814
Hay raker and loader	G. B. Garlinghouse and J. C. Moore.	Madison, Ind	Apr. 14, 1868	76, 743
Hay raker and loader	A. Godfrey	Dupage, Ill	Oct. 15, 1867	69, 794
Hay raker and loader	J. Harper	Hillsborough, Iowa	Mar. 10, 1868	75, 416
Hay raker and loader	S. R. Higgins	Parma, Mich	May 12, 1868	77, 817
Hay raker and loader	W. H. Hiteshew	Perrysburgh, Ind	Sept. 22, 1868	82, 316
Hay raker and loader	G. F. Holmes	Cortland, N. Y	Feb. 16, 1869	87, 045
Hay raker and loader	H. Kewley	Perry, Ohio	Dec. 3, 1867	71, 764
Hay raker and loader	N. S. Kinyon and B. F. Smith	Chenango Forks, N. Y	Sept. 17, 1867	68, 887
Hay raker and loader	J. C. Leonard, S. B. Holcomb, and W. B. Wright.	Clinton, Mo	Oct. 5, 1869	95, 493
Hay raker and loader	S. M. Livingston	Claverack, N. Y	Jan. 28, 1868	73, 731
Hay raker and loader	A. W. Lozier	New York, N. Y	June 15, 1869	91, 243
Hay raker and loader	G. C. Mead	Guilford, N. Y	Nov. 19, 1867	71, 198
Hay raker and loader	J. C. Moore and C. B. and G. B. Garlinghouse.	Madison and North Madison, Ind.	Feb. 23, 1869	87, 106
Hay raker and loader	S. K. Morse	Commerce, Mich	May 5, 1868	77, 641
Hay raker and loader	W. T. Nichols	Rutland, Vt	Jan. 28, 1868	73, 747
Hay raker and loader	W. T. Nichols	Rutland, Vt	Mar. 10, 1868	75, 450

Index of patents issued from the United States Patent Office from 1790 *to* 1873, *inclusive*—Continued.

Invention.	Inventor.	Residence.	Date.	No.
Hay raker and loader	C. Plaisted	Kenockee, Mich	Jan. 21, 1868	73, 460
Hay raker and loader	A. J. Preston	Guilford, N. Y	Mar. 24, 1868	75, 790
Hay raker and loader	G. H. Reister	Washington, Iowa	Mar. 15, 1870	100, 804
Hay raker and loader	F. M. Robinson and T. G. Springer.	Conneautville, Pa	July 16, 1867	66, 741
Hay raker and loader	A. Royse and M. K. Morris	Le Roy, Pa	Dec. 3, 1867	71, 792
Hay raker and loader	J. Ruhl and E. S. Herrington	Defiance, Ohio	May 19, 1868	78, 136
Hay raker and loader	B. Shirley	Moravia, N. Y	Oct. 22, 1867	70, 032
Hay raker and loader	G. W. Swartz	Newburgh, Pa	July 2, 1867	66, 412
Hay raker and loader	L. Underwood	Ottawa, Ill	Oct. 6, 1868	82, 770
Hay raker and loader	A. Vose	Pittsfield, Vt	Dec. 10, 1867	72, 133
Hay raker and loader	H. Willard	Vergennes, Vt	Oct. 15, 1867	69, 881
Hay raking and bunching machine	L. Wallace	Leavenworth City, Kans	Dec. 12, 1865	51, 500
Hay raking and cocking machine	F. Granger	Homer, Ill	Jan. 1, 1867	60, 880
Hay raking and cocking machine	Z. C. Steele	Pana, Ill	May 17, 1870	103, 099
Hay raking and cocking machine	S. Stewart	Lowell, Mass	May 17, 1870	103, 100
Hay raking and cocking machine	L. R. Stone	Owassa, Mich	Feb. 26, 1861	31, 568
Hay raking and cocking machine	J. Wadleigh	Chebanse, Ill	May 4, 1869	89, 611
Hay raking and loading machine	J. A. Althouse	Phillipstown, Ill	Nov. 22, 1859	26, 225
Hay raking and loading machine	W. W. Barnes	Lyme, N. H	Feb. 27, 1866	52, 814
Hay raking and loading machine	J. B. Benton, J. F. Behn, and G. Bastian.	Buffalo, N. Y	July 6, 1858	20, 772
Hay raking and loading machine	M. and D. Cole	Colebrook and Kent, Conn	Oct. 9, 1866	58, 601
Hay raking and loading machine	M. G. Couch	Odessa, N. Y	July 23, 1861	32, 860
Hay raking and loading machine	W. A. Duncan	Syracuse, N. Y	Mar. 7, 1865	46, 647
Hay raking and loading machine	W. A. Duncan	Syracuse, N. Y	Apr. 24, 1866	54, 133
Hay raking and loading machine	W. A. Duncan	Syracuse, N. Y	July 30, 1867	67, 176
Hay raking and loading machine	T. Frame	Barnesville, Ohio	Oct. 30, 1866	59, 205
Hay raking and loading machine	J. K. Harris	Allensville, Ind	Dec. 11, 1855	13, 908
Hay raking and loading machine	A. W. Heaney	Doylestown, Pa	May 29, 1866	55, 098
Hay raking and loading machine	G. Howard	Westfield Township, Ohio	July 19, 1859	24, 806
Hay raking and loading machine	M. K. Lewis, J. C. Durbin, and L. P. Lewis.	Iowa City, Iowa	Aug. 15, 1865	49, 423
Hay raking and loading machine	W. D. Mayfield	Ashley, Ill	Mar. 27, 1866	53, 464
Hay raking and loading machine	F. Nevergold and G. Stackhouse.	Pittsburgh, Pa	May 1, 1866	54, 392
Hay raking and loading machine	H. S. Palmer	Norvell, Mich	Sept. 25, 1866	58, 282
Hay raking and loading machine	J. S. Preston	Mechanicsville, Pa	Feb. 13, 1866	52, 601
Hay raking and loading machine	M. S. and C. B. Rawson	Winhall and Londonderry, Vt.	Feb. 19, 1867	62, 224
Hay raking and loading machine	J. Smith	Condit, Ohio	June 3, 1856	15, 046
Hay raking and loading machine	J. N. Smith	Jersey City, N. J	Mar. 14, 1865	46, 863
Hay raking and loading machine	J. P. Smith	Hudson, N. Y	July 10, 1866	56, 331
Hay raking and loading machine	D. H. Thompson	Fitchburgh, Mass	Mar. 25, 1856	14, 538
Hay raking and loading machine	B. M. Townsend	Quincy, Ill	Apr. 9, 1850	7, 277
Hay raking and loading machine	T. J. Wallace	Cameron, Ill	Sept. 13, 1859	25, 459
Hay raking and loading machine	M. Webb	Chenango Forks, N. Y	Oct. 23, 1866	59, 105
Hay raking and loading machine	H. Willard	Vergennes, Vt	Aug. 21, 1866	57, 421
Hay raking and loading machine	R. and J. Wright	Franklin Township, Pa	Oct. 23, 1866	59, 116
Hay raking and pitching machine	A. I. Preston	East Guilford, N. Y	Oct. 16, 1860	30, 423
Hay-raking machine	D. Ramler	Union Deposit, Pa	Apr. 26, 1859	23, 782
Hay-rigging	O. Reed	Paris, Mich	May 20, 1862	35, 352
Hay-shocker	C. M. Terrell	Oskaloosa, Iowa	Sept. 17, 1867	69, 045
Hay-shocking machine	N. W. Plymate	Freeland, Iowa	Mar. 8, 1864	41, 859
Hay, &c., Sling for loading and unloading	G. W. Long	Delaware Centre, Iowa	Sept. 9, 1873	142, 638
Hay-spreader	R. Adams and J. D. Sheetz	Heidelburgh Township, Pa	June 22, 1869	91, 699
Hay-spreader	A. B. Barnard	Worcester, Mass	Dec. 8, 1868	84, 725
Hay-spreader	H. Beers	Brookfield, Conn	May 30, 1865	47, 919
Hay-spreader	M. Bowker	Newark, N. J	Oct. 19, 1869	95, 875
Hay-spreader	N. Brand	Ilion, N. Y	Mar. 2, 1869	87, 392
Hay-spreader	H. M. Burdick	Ilion, N. Y	Mar. 30, 1869	88, 270
Hay-spreader	J. M. Burdick	Ilion, N. Y	Nov. 17, 1868	84, 087
Hay-spreader	G. E. Burt	Harvard, Mass	June 25, 1867	66, 074
Hay-spreader	G. E. Burt	Harvard, Mass	Oct. 15, 1867	69, 761
Hay-spreader	G. E. Burt and E. A. Hildreth	Harvard, Mass	Nov. 24, 1868	84, 257
Hay-spreader	W. H. Butterworth	Trenton, N. J	Nov. 24, 1868	84, 409
Hay-spreader	W. H. Butterworth	Trenton, N. J	May 18, 1869	90, 077
Hay-spreader	A. H. Caryl	Groton, Mass	Aug. 11, 1868	80, 806
Hay-spreader	N. Chapman	Milford, Mass	Sept. 22, 1868	82, 382
Hay-spreader	D. B. Clement	Brighton, Mass	July 6, 1869	92, 166
Hay-spreader	T. C. Craven	Albany, N. Y	Dec. 8, 1868	84, 800
Hay-spreader	N. Eaton	Woburn, Mass	Sept. 7, 1869	94, 480
Hay-spreader	W. H. Elliot	New York, N. Y	Feb. 2, 1869	86, 516
Hay-spreader	W. H. Elliot	New York, N. Y	June 8, 1869	91, 001
Hay-spreader	C. R. Frink	Norwich, N. Y	May 8, 1866	54, 524
Hay-spreader	C. R. Frink	Norwich, N. Y	Oct. 23, 1866	59, 001
Hay-spreader	C. R. Frink	Norwich, N. Y	Dec. 24, 1867	72, 624
Hay-spreader	C. R. Frink	Norwich, N. Y	Jan. 4, 1870	98, 486
Hay-spreader	W. H. H. Frye	North Fryeburgh, Me	May 24, 1870	103, 441
Hay-spreader	J. Garfield	Groton, Mass	May 7, 1867	64, 518
Hay-spreader	W. C. Gifford	Jamestown, N. Y	Jan. 24, 1865	45, 992
Hay-spreader	D. Kennedy	New York, N. Y	July 28, 1868	80, 412
Hay-spreader	W. H. Kenyon	Wakefield, R. I	Aug. 3, 1869	93, 313
Hay-spreader	L. L. Knight	Barre, Mass	Apr. 3, 1866	53, 739
Hay-spreader	G. G. Knowles	Wakefield, R. I	May 12, 1868	77, 739
Hay-spreader	D. Lyman	Middlefield, Mass	Apr. 25, 1865	47, 437
Hay-spreader	N. F. Matheuson	Barrington, R. I	Mar. 30, 1869	88, 314
Hay-spreader	M. D. Myers	Frankfort, N. Y	Feb. 9, 1869	86, 777
Hay-spreader	E. W. Nichols	Worcester, Mass	Dec. 28, 1869	98, 292
Hay-spreader	H. L. Perkins	Kingsman, Ohio	May 31, 1870	103, 771
Hay-spreader	E. D. and O. B. Reynolds	North Bridgewater, Mass	Feb. 16, 1869	87, 066
Hay-spreader	G. T. Savary	Byfield, Mass	Jan. 12, 1869	85, 858
Hay-spreader	J. C. Stoddard	Worcester, Mass	June 29, 1869	92, 122
Hay-spreader	B. F. Taft	Groton Junction, Mass	Apr. 13, 1869	88, 822
Hay-spreader	J. A. Talpey	Somerville, Mass	June 22, 1869	91, 790
Hay-spreader	J. F. Thomas	Ilion, N. Y	May 11, 1869	89, 899
Hay-spreader	J. F. Thomas and D. H. McLane	Ilion, N. Y	July 5, 1870	105, 141

Index of patents issued from the United States Patent Office from 1790 *to* 1873, *inclusive*—Continued.

Invention.	Inventor.	Residence.	Date.	No.
Hay-spreader	M. N. Ward	Bangor, Me	Mar. 2, 1869	87, 383
Hay-spreader	G. I. Wooster	Plymouth, Conn	Aug. 16, 1870	106, 444
Hay spreader and cocker combined	R. T. Gill	Poughkeepsie, N. Y	Mar. 31, 1868	76, 073
Hay spreader and elevator combined	T. C. Craven	Albany, N. Y	Dec. 19, 1865	51, 650
Hay-spreader fork	G. E. Burt and E. A. Hildreth	Harvard, Mass	June 8, 1869	90, 991
Hay-spreader teeth	G. A. Squier	Syracuse, N. Y	Apr. 21, 1868	76, 952
Hay spreading and turning machine	G. Moody	Falmouth, Me	Oct. 28, 1862	36, 791
Hay-spreading machine	C. Willard	Newtown, Pa	July 25, 1865	49, 021
Hay-stacker	J. T. Breneman	Springfield, Ohio	Jan. 8, 1867	61, 044
Hay-stacker	T. N. Bunnell	Reynolds, Ind	Oct. 19, 1869	95, 879
Hay-stacker	J. Forshee and J. C. McCland	Unionville Centre, Ohio	Oct. 29, 1867	70, 189
Hay-stacker	W. S. Nichols	Rutland, Vt	Jan. 28, 1868	73, 748
Hay-stacker	C. Rundell	Chicago, Ill	July 10, 1866	56, 273
Hay-stacker	S. J. Wallace	Keokuk, Iowa	Apr. 23, 1867	64, 050
Hay-stacking apparatus	C. H. Kirkpatrick	Sugar Grove, Ind	July 30, 1872	129, 967
Hay-stacking apparatus	C. H. Tryon	Greenwood, Ill	Jan. 29, 1867	61, 642
Hay-stacking device	W. Louden	Fairfield, Iowa	Aug. 28, 1866	57, 525
Hay-stacking machine	W. M. Mason	Polo, Ill	Feb. 25, 1862	34, 513
Hay-stacks, &c., Portable roof for	T. Munson	Canandaigua, N. Y	July 28, 1868	80, 301
Hay, straw, and vegetables, Machine for cutting	R. I. Burbank	Boston, Mass	Dec. 22, 1868	85, 067
Hay-tedder	O. A. Benton	Amenia, N. Y	Dec. 6, 1870	109, 863
Hay-tedder	A. Brown	Worcester, Mass	Jan. 10, 1871	110, 893
Hay-tedder	R. Bryson	Schenectady, N. Y	Feb. 15, 1870	99, 833
Hay-tedder	E. W. Bullard	Barre, Mass	Jan. 31, 1871	111, 314
Hay-tedder	H. M. Burdick	Ilion, N. Y	Oct. 12, 1869	95, 647
Hay-tedder	H. M. Burdick	Ilion, N. Y	Apr. 18, 1871	113, 848
Hay-tedder	H. M. Burdick	Ilion, N. Y	Apr. 18, 1871	113, 849
Hay-tedder	J. M. Burdick	Ilion, N. Y	Mar. 22, 1870	100, 972
Hay-tedder	W. H. Butterworth	Trenton, N. J	Jan. 17, 1871	111, 041
Hay-tedder	A. H. Caryl	Groton, Mass	Aug. 17, 1869	93, 805
Hay-tedder	T. J. and G. M. Clark	Higganum, Conn	Mar. 15, 1870	100, 856
Hay-tedder	T. J. and G. M. Clark	Higganum, Conn	Aug. 19, 1873	141, 999
Hay-tedder	J. K. Collins	Hartford, Vt	Oct. 12, 1869	95, 656
Hay-tedder	J. K. Collins	Hartford, Vt	Nov. 7, 1871	120, 626
Hay-tedder	T. C. Craven	Albany, N. Y	Nov. 16, 1869	96, 892
Hay-tedder	C. R. Frink	Norwich, N. Y	Sept. 24, 1872	131, 513
Hay-tedder	L. Hale	Hollis, N. H	Feb. 21, 1871	112, 030
Hay-tedder	M. B. Harvey	Stafford, Conn	Sept. 13, 1870	107, 365
Hay-tedder	J. B. Kelley	Brandon, Vt	Sept. 28, 1869	95, 234
Hay-tedder	W. G. Kenyon	Wakefield, R. I	Mar. 14, 1871	112, 602
Hay-tedder	J. Lee	Pottstown, Pa	Feb. 13, 1872	123, 700
Hay-tedder	W. H. Mickle	Utica, N. Y	Sept. 12, 1871	118, 810
Hay-tedder	M. D. Myers	Frankfort, N. Y	Dec. 7, 1869	97, 677
Hay-tedder	J. G. Perry	Kingston, R. I	Nov. 30, 1869	97, 437
Hay-tedder	J. G. Perry	Kingston, R. I	Nov. 30, 1869	97, 438
Hay-tedder	J. G. Perry	Kingston, R. I	Aug. 1, 1871	117, 561
Hay-tedder	J. G. Perry	Kingston, R. I	Aug. 1, 1871	117, 562
Hay-tedder	J. G. Perry	Kingston, R. I	Dec. 19, 1871	121, 958
Hay-tedder	H. F. and G. F. Shaw	West Roxbury, Mass	July 16, 1872	129, 431
Hay-tedder	F. R. Smith	Ilion, N. Y	Aug. 2, 1870	106, 087
Hay-tedder	W. Smith	Weston, Mass	Apr. 19, 1870	102, 173
Hay-tedder	J. A. Talpey	Somerville, Mass	Nov. 9, 1869	96, 741
Hay-tedder	H. Warren	Leicester, Mass	June 21, 1870	104, 669
Hay tedder and rake, Combined	S. Perry	Newport, N. Y	Jan. 2, 1872	122, 487
Hay-tedders, Device for securing the tines of	D. Kennedy	New York, N. Y	Jan. 10, 1871	110, 856
Hay, Teeth for scattering	J. C. Stoddard	Worcester, Mass	July 24, 1860	29, 327
Hay turning and spreading machine	E. W. Bullard	Barre, Mass	July 21, 1861	32, 350
Hay-unloader	H. H. Ensminger	Buffalo, N. Y	July 5, 1870	104, 947
Hay-unloader	J. Wood	Smyrna, N. Y	May 14, 1867	64, 729
Hay, Unloading	H. H. Angell	Clermont, Iowa	June 26, 1860	28, 822
Hay unloading and mowing apparatus	A. Smith	Hoosick Four Corners, N. Y	May 23, 1871	115, 120
Hay unloading and stacking device	W. R. Waldron	Webster, Mich	May 28, 1867	65, 316
Hay-unloading apparatus	E. Harrison	Mountain View, Cal	Aug. 29, 1871	118, 531
Hay-unloading apparatus	G. Smith	Providence. R. I	Oct. 26, 1869	96, 161
Hay-unloading device	J. Backus	Greenvale, Ill	Oct. 13, 1868	83, 018
Hay-unloading device	G. Smith	Providence, R. I	Nov. 3, 1868	83, 667
Hay-unloading device	G. Van Sickle	Auburn, N. Y	Sept. 29, 1868	82, 570
Head and shoulder rest	A. V. W. Van Vechten	New York, N. Y	June 9, 1868	78, 702
Head-band	D. McKinnon	Wappinger's Falls, N. Y	Aug. 26, 1873	142, 169
Head-block	A. A. Adams	Felchville, Vt	Sept. 22, 1868	82, 269
Head-block	N. F. Beckwith	Omro, Wis	Dec. 26, 1871	122, 215
Head-block	C. R. Bushnell	Saint Anthony's Falls, Minn	Nov. 10, 1868	83, 830
Head-block	T. L. Clark	Mount Vernon, Ohio	Mar. 10, 1868	75, 371
Head-block	J. M. Comins	Boscawen, N. H	Dec. 26, 1871	122, 228
Head-block	J. F. Cook	Baltimore, Md	Dec. 15, 1868	84, 937
Head-block	P. M. Cummings	Cincinnati, Ohio	Dec. 13, 1870	110, 119
Head-block	P. M. Cummings	Lyons, Iowa	Nov. 11, 1873	144, 388
Head-block	T. Douglass	Warren, Ohio	Dec. 19, 1871	121, 934
Head-block	T. Douglass	Warren, Ohio	Jan. 30, 1872	123, 091
Head-block	C. R. Ely	Northfield, Vt	Feb. 2, 1869	86, 517
Head-block	P. Estes	Leavenworth, Kans	Dec. 31, 1867	72, 828
Head-block	H. H. Gridley	Auburn, N. Y	Aug. 30, 1870	106, 932
Head-block	N. Hunt	Salem, Ohio	July 20, 1869	92, 828
Head-block	W. A. L. Kirk	Hamilton, Ohio	Sept. 22, 1868	82, 328
Head-block	M. C. Lewis	Glasgow, Mo	Dec. 24, 1867	72, 510
Head-block	A. C. Martin and W. Ritchie	Hamilton, Ohio	Nov. 10, 1868	83, 982
Head-block	H. R. Martin	Hillsborough, N. H	Aug. 30, 1871	115, 492
Head-block	D. Parkhurst	Saint Louis, Mo	Feb. 6, 1872	123, 507
Head-block	H. M. Popple	Warren, Pa	May 23, 1871	115, 237
Head-block	E. H. Stearns	Erie, Pa	Sept. 1, 1868	81, 837
Head-block	W. Tauney	Jeromesville, Ohio	July 16, 1872	129, 620
Head-block, Shoemaker's	J. H. Morton	Deering, Me	Mar. 26, 1872	124, 908
Head-comb, Lady's	D. Warring	Boston, Mass	July 16, 1872	129, 441
Head-covering	T. Bracher	New York, N. Y	July 4, 1865	48, 510
Head-covering	M. D. Cohen	Philadelphia, Pa	Dec. 24, 1861	33, 984
Head-covering	E. Copleston	Wrentham, Mass	Jan. 22, 1867	61, 403
Head-covering	T. S. Lambert	Peekskill, N. Y	Oct. 20, 1863	40, 382
Head-dress for ladies	J. Edwards	New York, N. Y	Jan. 5, 1864	41, 062

Index of patents issued from the United States Patent Office from 1790 *to* 1873, *inclusive*—Continued.

Invention.	Inventor.	Residence.	Date.	No.
Head-dress for ladies	H. S. C. Iwerson	New York, N. Y	Aug. 21, 1866	57, 328
Head-dress for ladies, Waterfall	P. Walter	New York, N. Y	Mar. 21, 1865	46, 961
Head-dress for smoke, Safety	W. N. Ball	La Porte, Ind	Apr. 9, 1872	125, 515
Head dress, Infant's	L. E. Love	New York, N. Y	Nov. 28, 1871	121, 391
Head-dresses, Band for	N. Grant and G. Downs	Providence, R. I	July 11, 1865	48, 677
Head-light	H. S. Maxim and J. Radley	New York, N. Y	Sept. 28, 1869	95, 248
Head-light	M. M. Rounds	New Haven, Conn	Apr. 17, 1866	54, 020
Head-light burner, Locomotive	A. C. Vaughan	Philadelphia, Pa	Aug. 30, 1870	106, 971
Head-light, Engine	P. Budenbach	New York, N. Y	May 22, 1866	54, 850
Head-light, Engine	S. M. Davies	Chicago, Ill	Mar. 28, 1865	46, 998
Head-light, Engine	J. Radley	New York, N. Y	June 26, 1866	55, 904
Head-light, Gasoline locomotive	J. B. Terry	Hartford, Conn	Dec. 24, 1867	72, 697
Head-light lens or glass	H. C. Felthousen	Buffalo, N. Y	Dec. 27, 1870	110, 449
Head-light, Locomotive	J., W. H., and A. E. Briggs	Newport, Ky	Aug. 26, 1873	142, 204
Head-light, Locomotive	S. E. Cleveland	Buffalo, N. Y	June 3, 1862	35, 486
Head-light, Locomotive	A. Collins and J. Hardy	Galesburgh, Ill	June 27, 1871	116, 272
Head-light, Locomotive	S. M. Davis	Chicago, Ill	Dec. 6, 1870	109, 877
Head-light, Locomotive	E. L. Hall	Utica, N. Y	July 30, 1867	67, 192
Head-light, Locomotive	R. W. Love	Windsor, Vt	June 7, 1870	103, 905
Head-light, Locomotive	H. S. Maxim and J. Radley	New York, N. Y	Oct. 5, 1869	95, 498
Head-light, Locomotive	L. Michaels	Cincinnati, Ohio	Nov. 23, 1869	97, 211
Head-light, Locomotive	L. Michaels	Covington, Ky	Apr. 29, 1873	138, 426
Head-light, Locomotive	T. J. Newland	Utica, N. Y	Feb. 14, 1865	46, 380
Head-light, Locomotive	J. Radley	New York, N. Y	Nov. 13, 1866	59, 650
Head-light, Locomotive	T. S. Ray and S. E. Cleveland	Buffalo, N. Y	Dec. 5, 1865	51, 351
Head-light, Locomotive	T. S. Ray and S. E. Cleveland	Buffalo, N. Y	Apr. 10, 1866	53, 871
Head-light, Locomotive	P. F. Stout	Philadelphia, Pa	July 22, 1873	141, 183
Head-light, Locomotive	A. C. Vaughan	Philadelphia, Pa	Nov. 26, 1867	71, 431
Head light, Locomotive	W. Westlake	Chicago, Ill	Oct. 7, 1873	143, 550
Head lights, Mechanism for adjustment of	C. D. Gibson	New York, N. Y	May 15, 1866	54, 714
Head-lights, Reflector for locomotive	C. S. Lee and W. M. Baldwin	Troy, N. Y	July 18, 1871	117, 089
Head-rest	J. S. Bartlett	Warsaw, N. Y	Nov. 8, 1870	108, 956
Head-rest	E. P. Cook	Cartersville, Ga	Feb. 22, 1870	100, 014
Head-rest	A. Dunlap	Clyde, Ohio	Dec. 24, 1867	72, 467
Head-rest	F. Fabrici	New York, N. Y	Oct. 1, 1872	131, 816
Head-rest	D. R. V. Goetchins	Little Falls, N. Y	Dec. 28, 1869	98, 250
Head-rest	R. Hale	Chicago, Ill	Nov. 5, 1867	70, 555
Head-rest	H. C. Hunt	Amboy, Ill	Apr. 21, 1868	77, 043
Head-rest	C. V. Littlepage	Austin, Tex	May 31, 1870	103, 631
Head-rest	J. W. Lockwood	New York, N. Y	Feb. 7, 1860	27, 058
Head-rest	C. B. Loveless	Syracuse, N. Y	Dec. 21, 1869	98, 077
Head-rest	B. Lyon and C. M. Curtis	Springfield, Mass	June 21, 1870	104, 473
Head-rest	D. M. Mitchell	Marathon, N. Y	Mar. 4, 1873	136, 531
Head-rest	T. B. Stout	Keyport, N. J	Mar. 16, 1869	87, 987
Head-rest	T. D. Thompson	Providence, R. I	Oct. 17, 1871	120, 128
Head-rest	M. Warne	Philadelphia, Pa	Oct. 20, 1868	83, 227
Head-rest	M. Warne	Philadelphia, Pa	Oct. 11, 1870	108, 308
Head-rest	E. Waters	Troy, N. Y	Dec. 29, 1857	18, 998
Head-rest for chairs	H. Snowden	Baltimore, Md	Sept. 24, 1867	69, 135
Head-rest for chairs, Portable	C. P. Bailey	Zanesville, Ohio	Mar. 28, 1854	10, 698
Head-rest for seats	J. R. Chiles	Richmond, Va	Jan. 28, 1868	73, 782
Head-rest for travelers in railway-cars, &c	G. Utley	Chapel Hill, N. C	Mar. 6, 1860	27, 395
Head-rest, Portable	H. E. Churchill	Portland, Me	Mar. 9, 1869	87, 543
Head-rest, Portable	J. M. Whitney	Boston, Mass	Jan. 9, 1866	51, 987
Head-rest, Portable	A. B. Wilson	Waterbury, Conn	Dec. 23, 1856	16, 300
Head-stall overcheck	J. B. Loomis	Ypsilanti, Mich	Sept. 10, 1872	131, 284
Head-supporting apparatus for travelers	H. C. Howells	New York, N. Y	Apr. 10, 1860	27, 806
Hearse	E. Allen	Norwich, Conn	Sept. 21, 1869	95, 066
Hearse	M. Jincks and F. Altmeyer	Danville, N. Y	Oct. 16, 1866	58, 835
Hearse	J. F. Kinback	Carbondale, Pa	Feb. 23, 1869	87, 175
Hearth, grate, and fender, Combined	G. Buchanan	Washington, Pa	June 28, 1870	104, 703
Hearth, Metallic	O. Keels	Steubenville, Ohio	Feb. 15, 1870	99, 777
Heat, Apparatus for accumulating and retaining	T. J. Chubb	Brooklyn, N. Y	May 21, 1867	64, 947
Heat, Apparatus for applying and conducting	J. C. Douglass	New York	Dec. 23, 1834	......
Heat, Apparatus for utilizing	B. F. Sturtevant	Jamaica Plain, Mass	Oct. 12, 1869	95, 849
Heat by friction, Apparatus for generating	P. Vera	Bogota, United States of Colombia.	Jan. 19, 1869	86, 046
Heat by the circulation of hot water, Apparatus for transmitting.	A. M. Perkins	London, England	Sept. 4, 1840	1, 768
Heat by the combustion of fuel of various kinds, Mode of raising.	J. F. Bennett	Pittsburgh, Pa	Mar. 13, 1866	53, 104
Heat-controller attachment	M. S. Hitchcock	Springfield, Mass	July 25, 1865	49, 029
Heat, Evolution and management of	E. Nott	Schenectady, N. Y	Feb. 23, 1826	......
Heat, Evolution and management of	E. Nott	Schenectady, N. Y	Dec. 29, 1826	......
Heat, Evolution and management of	E. Nott	Schenectady, N. Y	May 30, 1827	......
Heat, Evolution and management of	E. Nott	Schenectady, N. Y	Mar. 26, 1828	......
Heat, Evolution and management of	E. Nott	Schenectady, N. Y	Sept. 17, 1832	......
Heat, Evolution and management of	E. Nott	Schenectady, N. Y	Sept. 17, 1832	......
Heat, Evolution, &c., of	E. Nott	Schenectady, N. Y	June 21, 1826	......
Heat from lime, &c., Applying	P. Wenn	Philadelphia, Pa	Mar. 8, 1836	......
Heat from smoke by basket-grate, Abstracting	D. Steinhauer	Philadelphia, Pa	Dec. 13, 1832	......
Heat-generating apparatus for cooking-purposes	O. F. Morrill	Chelsea, Mass	June 5, 1866	55, 340
Heat generator and radiator	G. Chilson	Boston, Mass	Sept. 26, 1854	11, 718
Heat-governor or draft-regulator	F. E. Chatard, jr	Baltimore, Md	May 13, 1873	138, 736
Heat in chemical and other processes, Method of conveying and using.	C. A. Seely	New York, N. Y	Feb. 23, 1869	87, 208
Heat indicator, Distant	J. A. Read	Enfield, New South Wales	Sept. 9, 1873	142, 727
Heat, Machinery for regulating	L. W. Wright	London, England	Aug. 25, 1831	......
Heat, Management of	L. Lewis	Lewistown, N. Y	Mar. 2, 1836	......
Heat, Management of combustion in producing	S. A. Bille	New York	Nov. 8, 1828	......
Heat, Material for transmitting	W. C. Baker	New York, N. Y	Dec. 10, 1867	71, 948
Heat, Mode of generating	W. Hartell and J. Lancaster	Kensington and Spring Garden, Pa.	Nov. 23, 1852	9, 419
Heat, Mode of obtaining	J. Johnson	Saco, Me	Apr. 7, 1868	76, 463
Heat, Non-conductor of	J. Chalmers	London, England	Aug. 4, 1868	80, 709
Heat-regulator	H. Boyle	London, England	Sept. 2, 1873	142, 371
Heat-regulator	S. M. Kellogg	Battle Creek, Mich	June 28, 1864	43, 313
Heat-regulator for hot-air furnaces	A. H. Tingley	Providence, R. I	Feb. 15, 1870	99, 972

Index of patents issued from the United States Patent Office from 1790 *to* 1873, *inclusive*—Continued.

Invention.	Inventor.	Residence.	Date.	No.
Heat-regulator for hot-water apparatus	H. L. McAvoy	Baltimore, Md	July 27, 1869	93, 104
Heat retainer and imparter	J. K. Mitchell	Philadelphia, Pa	Nov. 19, 1818	....
Heat to fluids, Application of	S. Applegate, T. Howard, and P. H. Wharton.	Philadelphia, Pa	Nov. 22, 1809	....
Heater: *See* Air-heater. Barrel-heater. Boiler-heater. Building-heater. Car-heater. Cask-heater. Cheese-vat heater. Chimney-heater. Coal-oil heater. Cotton-seed heater. Cream-heater. Curling-tongs heater. Cylinder-heater. Dairy-heater. Dentist's lamp-heater. Dish-heater. Feed-water heater. Fire-engine heater. Fire-place heater. Flat-iron heater. Friction-heater. Furnace-heater. Gas-heater. Gravel-heater. Gravel and sand heater. Hot-water heater. Hot-water and hot-air heater. Hot-water or steam heater. Hydrocarbon-heater. Hydraulic heater. Kerosene-heater. Lamp-heater. Lamp-chimney heater. Laundry-heater. Laundry-iron heater. Liquid-heater. Nursery-lamp heater. Parlor-heater. Peanut-heater. Portable heater. Ruling-machine heater. Sad-iron heater. Service-heater. Sleigh-heater. Smoke-consuming heater. Spirituous-liquor heater. Stamp-heater. Steam-heater. Steam-generating heater. Steam-generator heater. Stove and water heater. Sugar-sirup heater. Superheater. Tea-pot heater. Tire-heater. Topical heater. Tubular heater. Water-heater.				
Heater	O. Abbott	New York, N. Y	Feb. 9, 1864	41, 467
Heater	L. F. Betts	Albion, Mich	Jan. 5, 1864	41, 051
Heater	C. T. Boardman	Bergen Point, N. J	Apr. 5 1864	42, 161
Heater	N. A. Boynton	New York, N. Y	Nov. 25, 1862	36, 989
Heater	S. A. Briggs	Providence, R. I	May 20, 1862	35, 294
Heater	H. L. Budd	New York, N. Y	Jan. 12, 1864	41, 256
Heater	J. Carton	Utica, N. Y	May 13, 1862	35, 216
Heater	J. Chilcott	Brooklyn, N. Y	Aug. 22, 1865	49, 504
Heater	T. S. Clogston	Boston, Mass	Dec. 13, 1864	45, 389
Heater	T. Dowling	Lynn, Mass	Aug. 4, 1863	39, 385
Heater	C. H. Frost	Peekskill, N. Y	June 3, 1862	35, 440
Heater	C. C. Hare	Louisville, Ky	Oct. 10, 1865	50, 354
Heater	J. C. Henderson	Albany, N. Y	Dec. 23, 1862	37, 228
Heater	G. Joslin	Boston, Mass	Feb. 25, 1862	34, 503
Heater	J. Magee	Chelsea, Mass	Nov. 25, 1862	37, 027
Heater	J. A. Marvin	Red Wing, Minn	Nov. 26, 1867	71, 399
Heater	O. F. Morrill	Chelsea, Mass	Oct. 4, 1864	44, 548
Heater	O. F. Morrill	Chelsea, Mass	Oct. 4, 1864	44, 549
Heater	G. W. Oakeley	Reading, Pa	Dec. 3, 1861	33, 849
Heater	M. Pond	Boston, Mass	June 10, 1862	35, 541
Heater	J. S. Reid	Muncie, Ind	Sept. 1, 1863	39, 750
Heater	J. S. Reid	Muncie, Ind	Apr. 5, 1864	42, 225
Heater	E. C. Robinson	Troy, N. Y	Aug. 29, 1865	49, 652
Heater	W. H. Seymour	West Hartford, Conn	Mar. 25, 1862	34, 785
Heater	S. B. Sexton	Baltimore, Md	June 17, 1862	35, 632
Heater	T. Shipton	Newark, N. J	Oct. 29, 1867	70, 276
Heater	J. H. Simonds	New York, N. Y	Apr. 22, 1862	35, 043
Heater	E. Sprague	Allegheny City, Pa	Oct. 20, 1863	40, 366
Heater	J. C. Underwood	Richmond, Ind	Dec. 27, 1864	45, 656
Heater	A. Weller	Albany, N. Y	Oct. 20, 1863	40, 375
Heater and blower	W. M. Jackson	Wooland, Cal	Sept. 2, 1873	142, 397
Heater and boiler	S. A. Willett	Philadelphia, Pa	July 2, 1861	32, 731
Heater and condenser	P. W. Mackenzie	Blauveltville, N. Y	Feb. 13, 1872	123, 570
Heater and condenser combined	B. Holly	Lockport, N. Y	Apr. 25, 1871	114, 142
Heater and evaporator	F. Farquahar and R. E. Doan	Wilmington, Ohio	Aug. 13, 1867	67, 742
Heater and filter	L. F. Meunier	Fires-Lille, France	July 8, 1873	140, 719
Heater and filter, Combined	W. T. Bate	Norristown, Pa	Apr. 15, 1873	137, 878
Heater and ventilator	G. W. Blake	New York, N. Y	Feb. 22, 1870	100, 110
Heater and ventilator	B. F. Sturtevant	Jamaica Plain, Mass	Feb. 22, 1870	100, 211
Heater-boiler	C. R. Ellis	Brooklyn, N. Y	Jan. 7, 1868	73, 084
Heater, condenser, and filter combined	W. A. Lighthall	New York, N. Y	July 29, 1862	36, 010
Heater, cooler, and ventilator	D. E. Somes	Washington, D. C	Nov. 28, 1865	51, 236
Heater or cooler	J. C. Hoadley	Lawrence, Mass	Jan. 26, 1858	19, 197
Heater-screen	W. C. Baker	New York, N. Y	Apr. 21, 1863	28, 210
Heating and cooking apparatus	E. Conway	Dayton, Ohio	Aug. 28, 1860	29, 769
Heating and fuel-saving device	D. C. Colby	Claremont, N. H	Apr. 25, 1865	47, 489
Heating and illuminating apparatus	L. Ruel	Saint Johnsbury, Vt	Nov. 11, 1873	144, 414
Heating and puddling furnace	M. S. Ridgeway and C. Lewis	Danville and Harrisburgh, Pa.	May 7, 1867	64, 617
Heating and puddling furnace, Iron, &c	J. Morrison	East Saint Louis, Ill	Jan. 30, 1872	123, 282
Heating and puddling furnaces, Boshes for	W. Batty	Lawrenceburgh, Pa	Sept. 24, 1867	69, 063
Heating and puddling furnaces, Slide-door for	S. M. Laughlin	Philadelphia, Pa	Oct. 18, 1870	108, 501
Heating and ventilating	G. F. Schulze	Janesville, Wis	Mar. 22, 1870	101, 168
Heating and warming furnace, Apartment	J. A. Page	Boston, Mass	Sept. 25, 1841	2, 269
Heating and welding furnace	G. Nimmo	Jersey City, N. J	May 28, 1867	65, 110
Heating and welding furnace	G. Nimmo and R. S. Stenton	Jersey City, N. J., and New York, N. Y.	June 14, 1864	43, 127
Heating apartments	T. H. Parker	York, Pa	Aug. 1, 1848	5, 692
Heating apartments, Apparatus for	S. Whitmarsh	Northampton, Mass	Feb. 20, 1849	6, 118
Heating-apparatus	J. Armstrong, jr	San Francisco, Cal	Oct. 12, 1869	95, 756
Heating-apparatus	W. C. Baker	New York, N. Y	Nov. 18, 1873	144, 650

Index of patents issued from the United States Patent Office from 1790 *to* 1873, *inclusive*—Continued.

Invention.	Inventor.	Residence.	Date.	No.
Heating-apparatus	L. Bridges	Chicago, Ill	June 19, 1860	28,805
Heating-apparatus	M. G. Fagan	Troy, N. Y	Feb. 28, 1871	112,233
Heating-apparatus	E. C. Gillette	San Francisco, Cal	June 30, 1863	39,041
Heating-apparatus	R. D. Granger	Philadelphia, Pa	Jan. 18, 1859	22,642
Heating-apparatus	M. Greenebaum	Chicago, Ill	Oct. 15, 1861	33,483
Heating-apparatus	F. L. Hedenberg	New York, N. Y	Mar. 30, 1858	19,775
Heating-apparatus	J. R. Jenness	Norwich, Conn	Sept. 22, 1868	82,322
Heating-apparatus	J. Johnson	Saco, Me	Apr. 7, 1868	76,329
Heating-apparatus	A. Kohler	Boston, Mass	May 28, 1861	32,427
Heating-apparatus	U. D. Mihills	Hartford, Wis	Jan. 25, 1859	22,740
Heating-apparatus	P. Mitchell	Greenfield, Mass	June 25, 1861	32,639
Heating-apparatus	G. R. Osbrey	Providence, R. I	Nov. 8, 1859	26,048
Heating-apparatus	C. Pepper	Albany, N. Y	Dec. 20, 1859	26,521
Heating-apparatus	S. Smith	Greenfield, Mass	July 31, 1866	56,816
Heating-apparatus	G. W. Williamson	Scranton, Pa	Jan. 25, 1859	22,760
Heating-apparatus regulator	E. L. Brown	Brooklyn, N. Y	Nov. 27, 1860	30,716
Heating-boiler	Z. Ellis	Philadelphia, Pa	Nov. 15, 1864	45,031
Heating-boiler	J. A. Maynard	Newtonville, Mass	Mar. 21, 1871	112,824
Heating-boiler	J. G. Wilson	New York, N. Y	Jan. 10, 1871	110,884
Heating-boiler for buildings	A. Gay	Rochester, N. Y	Sept. 12, 1871	118,799
Heating-boiler for buildings	A. E. Hitchings	New York, N. Y	Nov. 24, 1857	18,693
Heating-boiler for buildings	C. F. Hitchings	New York, N. Y	May 15, 1860	28,272
Heating-boiler for buildings	C. F. Hitchings	New York, N. Y	July 23, 1867	67,053
Heating-boiler for buildings	T. J. Myers	Philadelphia, Pa	July 25, 1871	117,448
Heating buildings and evaporating fluids	A. M. Perkins	United States	Aug. 20, 1838	888
Heating buildings by the combination of and burning gas, air, and steam, Apparatus for.	C. H. Johnson	Boston, Mass	Mar. 4, 1856	14,360
Heating buildings, &c., Warm air for	J. L. Dutton	Philadelphia, Pa	Mar. 13, 1834	....
Heating, cooking, and ventilating apparatus	C. B. Sawyer	Fitchburgh, Mass	June 28, 1859	24,581
Heating-furnace	W. D. Bartlett	Amesbury, Mass	Mar. 22, 1870	101,082
Heating-furnace	N. E. Cornnall	New York, N. Y	Feb. 4, 1868	74,051
Heating-furnace	E. W. Crane	Roselle, N. J	Apr. 8, 1873	137,598
Heating-furnace	C. Harkinson	Philadelphia, Pa	May 17, 1870	103,176
Heating-furnace	T. Hydes and J. Bennett	Sheffield, England	Mar. 22, 1870	101,017
Heating-furnace	J. Maunton	New York, N. Y	Nov. 12, 1867	70,874
Heating-furnace	J. G. Porter and E. Davis	New York, N. Y., and Jersey City, N. J.	June 10, 1873	139,812
Heating-furnace	C. R. Rand	Dubuque, Iowa	Aug. 14, 1866	57,187
Heating furnace	C. R. Rand	Saint Louis, Mo	Dec. 17, 1867	72,229
Heating-furnace	W. A. Sweet	Syracuse, N. Y	July 26, 1870	105,738
Heating-furnace	J. Watts	Birmingham, Great Britain	Mar. 24, 1868	76,010
Heating-furnace and cooking-stove combined	B. McConnell	Philadelphia, Pa	May 21, 1872	127,082
Heating-furnace for buildings	B. W. Dunklee	Boston, Mass	May 10, 1859	23,907
Heating-furnace for buildings	F. A. Frickhardt	Easton, Pa	Sept. 8, 1836	23
Heating-furnace for buildings	J. Leeds	Philadelphia, Pa	Nov. 7, 1854	11,896
Heating-furnace for buildings	J. Leeds	Philadelphia, Pa	Aug. 14, 1855	13,432
Heating-furnace for buildings	J. Leeds	Philadelphia, Pa	Apr. 2, 1861	31,892
Heating-furnace for buildings	J. Plant	Washington, D. C	Oct. 5, 1858	21,724
Heating-furnace for buildings	S. B. Sexton	Baltimore, Md	Aug. 14, 1855	13,439
Heating-furnace for buildings	G. Walker	New Haven, Conn	June 10, 1844	3,623
Heating furnaces, Consuming escape steam as an adjunct in.	T. Maskell	Franklin, La	Sept. 4, 1855	13,526
Heating-furnaces, Device for drawing iron from	W. Stevens	New Albany, Ind	Aug. 15, 1871	118,165
Heating-furnaces, Hot-air flue for	G. R. Barker	Philadelphia, Pa	July 22, 1873	141,101
Heating-pipe, Jacket for	M. Ber	New York, N. Y	Aug. 23, 1870	106,652
Heating plates for hot-pressing	M. A. Quigley	Washington County, Md	June 15, 1819	....
Heating-purposes, Metallic base and molding for	C. E. Fenkle	New York, N. Y	Apr. 21, 1868	77,022
Heating rooms	A. McAllester and J. Iggett	Salem, N. Y	Dec. 15, 1827	....
Heating rooms, &c	A. Pollock	Boston, Mass	Apr. 24, 1807	....
Heating rooms	B. Russell	Hartford, Conn	Mar. 4, 1811	....
Heating rooms	R. B. Varden	Baltimore, Md	Feb. 6, 1832	....
Heating rooms and ovens	J. A. Pitts	Winthrop, Me	July 2, 1836	....
Heating rooms, &c., by rarefied air	D. Pettibone	Philadelphia, Pa	May 12, 1812	....
Heating rooms, Device for	L. Newson	Gallipolis, Ohio	Aug. 21, 1860	29,710
Heddle-actuating mechanism	J. Crawshaw	Cobourg, Canada	July 23, 1872	129,718
Heddle-actuating mechanism	W. Gminder	Reutlingen, Germany	Oct. 15, 1872	132,279
Heddle-actuating mechanism	R. B. Goodyear	Wilmington, Del	July 16, 1872	129,126
Heddle-eye opener	J. H. Crowell	Providence, R. I	June 17, 1873	139,875
Heddle-frame	J. Dyson	Philadelphia, Pa	July 23, 1872	129,722
Heddle-machine	P. Philip	Stockport, N. Y	Aug. 9, 1870	106,277
Heddle, Metallic	J. Senneff	Philadelphia, Pa	Apr. 13, 1852	8,662
Hedge and tree trimmer	C. E. Healy	New London, Ohio	Feb. 22, 1870	100,145
Hedge and tree trimmer	S. McElhaney	Milledgeville, Ill	Sept. 3, 1872	131,108
Hedge-cutting machine	D. Oliver	Carthage, Ill	Mar. 6, 1866	53,033
Hedge-fastener	C. D. Brown	Tampico, Ill	July 2, 1867	66,209
Hedge-hook	E. S. Turner	Palmyra, Wis	May 20, 1873	139,215
Hedge-plant grab	H. Ferris	Galesburgh, Ill	Apr. 17, 1866	53,963
Hedge-plants and weeds, Implement for extracting	J. L. Knick	Lexington, Ill	Apr. 28, 1868	77,385
Hedge-planting device	C. D. Brown	Sterling, Ill	Jan. 15, 1867	61,150
Hedge-plashing machine	D. Gore	Carlinville, Ill	June 22, 1869	91,620
Hedge-setter	J. H. Holbert	Ottawa, Ill	July 27, 1869	92,961
Hedge-shears	J. O. Minor	Wapello, Iowa	July 30, 1867	67,207
Hedge-trimmer	C. W. Aikin	Decatur, Ill	Apr. 30, 1872	126,243
Hedge-trimmer	F. Binder	Baltimore, Md	Aug. 27, 1867	68,280
Hedge-trimmer	J. Block	Eureka, Ill	Jan. 30, 1872	123,224
Hedge-trimmer	G. and S. P. Clark	Dover, Ill	Feb. 28, 1871	112,221
Hedge-trimmer	J. S. Crum	Scottville, Ill	Apr. 23, 1872	126,032
Hedge-trimmer	W. E. Horne	Decatur, Ill	July 5, 1870	104,956
Hedge-trimmer	A. H. Hussey	Mount Pleasant, Ohio	Dec. 14, 1869	97,925
Hedge-trimmer	S. T. Hyde	Piasa, Ill	May 4, 1869	89,769
Hedge-trimmer	T. C. Mathews	Yates City, Ill	June 30, 1868	79,372
Hedge-trimmer	J. McAnulty	Bentley, Ill	Sept. 20, 1870	107,517
Hedge-trimmer	T. V. Nichols	Olena, Ill	Sept. 9, 1862	36,420
Hedge-trimmer	D. Oliver	Galesburgh, Ill	Feb. 15, 1870	99,938
Hedge-trimmer	R. T. Payne	Eureka, Ill	July 16, 1872	129,493
Hedge-trimmer	A. Selover	Brooklyn, Ohio	Aug. 8, 1865	49,311
Hedge-trimmer	W. and H. Perry	Tivoli, Ill	Aug. 28, 1866	57,590
Hedge-trimmer	S. B. Turner	Adams County, Ill	Apr. 22, 1873	138,106

Index of patents issued from the United States Patent Office from 1790 *to* 1873, *inclusive*—Continued.

Invention.	Inventor.	Residence.	Date.	No.
Hedge-trimmer	J. M. and O. W. Van Nosdall	Newark, Ill	Jan. 25, 1870	99, 121
Hedge-trimmer	J. M. and O. W. Van Nosdall	Newark, Ill	Feb. 28, 1871	112, 300
Hedge-trimmer	L. Woods	Quincy, Ill	July 25, 1854	11, 400
Hedge-trimmer and mower	G. Waddington	Le Roy, Ill	Mar. 15, 1870	100, 955
Hedge-trimmer and mower, Combined	S. H. Hepperly	Elmira, Ill	Feb. 26, 1867	62, 487
Hedge-trimmer and stalk-cutter	J. G. Sprague	Lexington, Ill	Feb. 1, 1870	99, 490
Hedge-trimmer and corn-cutting machine	J. W. Hull	Connersville, Ill	Dec. 17, 1867	72, 299
Hedge-trimming machine	E. M. Benfield	Maquin, Ill	July 29, 1862	36, 049
Hedge-trimming machine	S. Bradbury	Griggsville, Ill	Jan. 27, 1857	16, 462
Hedge-trimming machine	W. C. Hooker	Abingdon, Ill	July 25, 1865	48, 946
Hedge-trimming machine	A. Tweedy	Collinsville, Ohio	July 30, 1867	67, 380
Hedge-trimming machine	W. Wimmer	Billingsville, Ind	July 7, 1857	17, 764
Hedges, Device for bending down plants to form	C. D. Brown	Tampico, Ill	July 2, 1867	66, 210
Heel	S. A. Brackett	Boston, Mass	Nov. 5, 1872	132, 747
Heel	W. T. Downs	Saint Louis, Mo	Mar. 30, 1869	88, 461
Heel and sole edge trimming machine, Boot and shoe.	R. C. Lambert	Quincy, Mass	May 23, 1871	115, 219
Heel and spoke shave	J. A. Perley	Lynn, Mass	Feb. 8, 1859	22, 889
Heel and spur, Boot	C. F. Woodruff	Newbern, Tenn	Jan. 19, 1869	86, 049
Heel and toe irons for boots and shoes, Coating	H. O. Lothrop	Milford, Mass	Jan. 21, 1868	73, 451
Heel and toe plates for boots and shoes	I. E. Loughborough	Pittsford, N. Y	Apr. 4, 1865	47, 115
Heel-attachment for boots and shoes	G. C. Aiken	Worcester, Mass	Feb. 5, 1861	31, 286
Heel, Blank boot	H. H. Bigelow	Worcester, Mass	Sept. 7, 1869	94, 552
Heel blank manufacturing machine, Boot and shoe	C. W. Glidden	Lynn, Mass	May 27, 1873	139, 382
Heel blanks, Machine for compressing and punching.	R. C. Lambart	Quincy, Mass	Oct. 21, 1873	143, 914
Heel, Boot	F. Closs	New Haven, Conn	Sept. 18, 1866	58, 067
Heel, Boot	G. W. Davis	Milford Centre, Ohio	Feb. 12, 1867	62, 017
Heel, Boot	E. Dunbar	Buffalo, N. Y	June 20, 1865	48, 266
Heel, Boot	J. Fearn	Tompkinsville, N. Y	Feb. 18, 1868	74, 525
Heel, Boot	J. B. Gebhard and C. G. Schwarz.	Philadelphia, Pa	Oct. 14, 1873	143, 688
Heel, Boot	J. H. Greenleaf	New Haven, Conn	Nov. 20, 1866	59, 908
Heel, Boot	F. D. Hayward and P. Stone	Malden and Charlestown, Mass.	July 11, 1865	48, 682
Heel, Boot	G. Henning and H. P. Willie	Buffalo, N. Y	Jan. 14, 1868	73, 328
Heel, Boot	S. Hodgins	Saint Louis, Mo	July 25, 1865	48, 943
Heel, Boot	L. and A. A. Hoffman	Buffalo, N. Y	June 4, 1867	65, 487
Heel, Boot	J. Hubbell	Buffalo, N. Y	Sept. 3, 1867	68, 364
Heel, Boot	G. Lane	Hamilton, Ohio	Apr. 21, 1868	77, 055
Heel, Boot	G. A. Mitchell	Turner, Me	June 25, 1861	32, 638
Heel, Boot	E. Newhall	Lynn, Mass	Sept. 19, 1865	50, 026
Heel, Boot	A. T. Perrine	Louisville, Ky	Mar. 16, 1869	88, 000
Heel, Boot	F. Richardson	New Bedford, Mass	May 5, 1868	77, 529
Heel, Boot	F. Richardson and F. Hacker	Providence, R. I	May 7, 1872	126, 489
Heel, Boot	F. Richardson and F. Hacker	Providence, R. I	May 7, 1872	126, 490
Heel, Boot	C. Robinson and J. C. Marshall	Springfield, Mass	Aug. 7, 1866	56, 992
Heel, Boot	H. S. Snow	West Meriden, Conn	Mar. 12, 1867	62, 783
Heel, Boot	S. Thorp and W. Thorp	Portland and Turner, Me	Jan. 25, 1859	22, 755
Heel, Boot	R. Vandevort	Pittsburgh, Pa	Sept. 17, 1867	69, 051
Heel, Boot and shoe	I. Banister	Newark, N. J	May 21, 1872	127, 013
Heel, Boot and shoe	G. Beaty	Middlebury, Ohio	Jan. 7, 1868	72, 963
Heel, Boot and shoe	H. H. Bigelow	Worcester, Mass	Nov. 1, 1870	108, 970
Heel, Boot and shoe	H. H. Bigelow	Worcester, Mass	June 27, 1871	116, 400
Heel, Boot and shoe	L. R. Blake and A. S. Libby	Brooklyn, N. Y., and Lawrence, Mass.	Dec. 9, 1873	145, 389
Heel, Boot and shoe	F. Bleisewick	New York, N. Y	Mar. 22, 1870	100, 968
Heel, Boot and shoe	M. Bray	Newton, Mass	Sept. 12, 1871	118, 839
Heel, Boot and shoe	M. Bray	Newton, Mass	Aug. 26, 1873	142, 203
Heel, Boot and shoe	H. Cordtz	Chicago, Ill	Sept. 21, 1869	95, 087
Heel, Boot and shoe	A. O. Crane	Boston, Mass	Dec. 15, 1868	84, 861
Heel, Boot and shoe	A. O. Crane	Boston, Mass	June 6, 1871	115, 581
Heel, Boot and shoe	A. O. Crane	Boston, Mass	Aug. 22, 1871	118, 203
Heel, Boot and shoe	S. Dodge, jr., and B. Potter, jr.	Marblehead, Mass	Sept. 13, 1859	25, 395
Heel, Boot and shoe	C. Dyer, jr., and E. Drake	Stoughton, Mass	Aug. 6, 1867	67, 422
Heel, Boot and shoe	C. Dyer, jr., and E. Drake	Stoughton, Mass	Oct. 22, 1867	70, 078
Heel, Boot and shoe	C. H. Eggleston	Marshall, Mich	Dec. 26, 1871	122, 112
Heel, Boot and shoe	T. S. Engledow	Cedar Falls, Iowa	Mar. 3, 1868	75, 137
Heel, Boot and shoe	G. F. Fling	Portland, Me	Sept. 19, 1871	119, 021
Heel, Boot and shoe	S. Flint and R. S. Rogers	Lynn, Mass	Dec. 14, 1858	22, 328
Heel, Boot and shoe	H. F. Harris and G. P. Pinney	Chicago, Ill	Aug. 30, 1870	106, 814
Heel, Boot and shoe	C. H. Helms	Poughkeepsie, N. Y	June 7, 1864	43, 024
Heel, Boot, &c	R. Herr	Brooklyn, N. Y	June 9, 1868	78, 667
Heel, Boot and shoe	W. Hunt	New York, N. Y	June 21, 1859	24, 517
Heel, Boot and shoe	J. M. Hunter	New York, N. Y	Mar. 19, 1872	124, 682
Heel, Boot and shoe	J. M. Hunter	Morristown, N. J	Aug. 27, 1872	130, 805
Heel, Boot and shoe	G. W. Keene	Lynn, Mass	Mar. 6, 1860	27, 369
Heel, Boot and shoe	G. W. Keene	Lynn, Mass	Dec. 24, 1861	34, 037
Heel, Boot and shoe	F. Kettler	Milwaukee, Wis	May 27, 1862	35, 377
Heel, Boot and shoe	R. Lapham	Boston, Mass	Mar. 30, 1869	88, 310
Heel, Boot and shoe	G. W. Martin	Boston, Mass	Jan. 12, 1869	85, 748
Heel, Boot and shoe	J. Miller	Antwerp, N. Y	Mar. 15, 1864	41, 933
Heel, Boot and shoe	N. Mole	Bay City, Mich	Dec. 17, 1872	134, 087
Heel, Boot and shoe	G. E. Newcomb	Bucksport Me	May 10, 1870	102, 961
Heel, Boot and shoe	E. Newhall	Lynn, Mass	Dec. 10, 1867	72, 073
Heel, Boot and shoe	C. B. Norton	Crittenden, N. Y	Apr. 2, 1872	125, 214
Heel, Boot and shoe	S. Oliver, jr	Lynn, Mass	Apr. 14, 1857	17, 051
Heel, Boot and shoe	W. H. Peckham	Hoboken, N. J	Oct. 30, 1860	30, 542
Heel, Boot and shoe	A. T. Perrine	Boston, Mass	Oct. 3, 1871	119, 473
Heel, Boot and shoe	M. H. Prescott	Ottawa, Ill	May 18, 1869	90, 192
Heel, Boot and shoe	C. A. Read	Bridgeport, Conn	Sept. 26, 1871	119, 407
Heel, Boot and shoe	J. Read	Philadelphia, Pa	Mar. 22, 1859	23, 312
Heel, Boot and shoe	J. Read	Philadelphia, Pa	June 14, 1870	104, 200
Heel, Boot and shoe	J. R. Ryerson	St. Albans, Me	July 16, 1872	129, 369
Heel, Boot and shoe	P. Shaw	Boston, Mass	Jan. 3, 1860	26, 712
Heel, Boot and shoe	H. S. Smythe	Darlington, Wis	Feb. 21, 1871	111, 986
Heel, Boot and shoe	D. E. Somes	Biddeford, Me	July 24, 1860	29, 324
Heel, Boot and shoe	D. E. Somes	Biddeford, Me	Sept. 18, 1860	30, 096

Index of patents issued from the United States Patent Office from 1790 *to* 1873, *inclusive*—Continued.

Invention.	Inventor.	Residence.	Date.	No.
Heel, Boot and shoe	J. W. Tull and G. Stevenson	Zionsville, Ind	Sept. 29, 1868	82, 662
Heel, Boot and shoe	O. Underwood	Milford, Mass	Nov. 11, 1873	144, 580
Heel, Boot and shoe	A. B. Wilton	Dorchester, Mass	May 31, 1859	24, 269
Heel, Boot and shoe	J. Woodley	Quebec, Canada	Mar. 29, 1870	101, 399
Heel-breasting machine	C. H. Helms	Poughkeepsie, N. Y	Mar. 6, 1866	52, 997
Heel-breasting machine	A. B. Jaquith and J. Wolohan	Rochester, N. Y	Dec. 2, 1873	145, 061
Heel-breasting machine	V. K. Spear	Lynn, Mass	May 27, 1873	139, 273
Heel breasting machine, Boot and shoe	W. C. Butler	Baltimore, Md	Dec. 16, 1873	145, 625
Heel-breasting machinery	V. K. Spear	Lynn, Mass	Aug. 12, 1873	141, 735
Heel-burnisher	M. Burnham 2d	Wenham, Mass	Aug. 23, 1870	106, 657
Heel-burnishing machine	E. H. Downing and W. Joint	Lynn, Mass	Feb. 11, 1873	135, 786
Heel-burnishing machine	C. W. Glidden	Lynn, Mass	Oct. 7, 1873	143, 507
Heel-burnishing machine	C. W. Glidden	Lynn, Mass	Oct. 14, 1873	143, 690
Heel-burnishing machine	G. C. Hawkins	Boston, Mass	Dec. 30, 1873	145, 943
Heel-burnishing machine	R. C. Lambert	Quincy, Mass	Aug. 8, 1871	117, 955
Heel-burnishing machine	H. S. Vrooman	Boston, Mass	June 3, 1873	139, 636
Heel burnishing machine, Boot and shoe	C. W. Glidden	Lynn, Mass	Oct. 21, 1873	143, 899
Heel burnishing machine, Boot and shoe	G. C. Hawkins	Boston, Mass	June 4, 1872	127, 414
Heel burnishing machine, Boot and shoe	G. C. Hawkins, A. G. Mead, and V. K. Spear.	Boston and Lynn, Mass	Apr. 4, 1871	113, 658
Heel burnishing machine, Boot and shoe	R. C. Lambart	Quincy, Mass	Jan. 30, 1872	123, 112
Heel burnishing machine, Boot and shoe	R. C. Lambart	Quincy, Mass	Dec. 9, 1873	145, 430
Heel burnishing machine, Boot and shoe	A. C. McKnight	Philadelphia, Pa	Feb. 20, 1872	123, 785
Heel burnishing machine, Boot and shoe	J. H. Sawyer and C. Keniston	Boston and Somerville, Mass.	Mar. 22, 1870	101, 167
Heel burnishing machine, Boot and shoe	V. K. Spear	Lynn, Mass	June 6, 1871	115, 651
Heel burnishing machinery, Boot and shoe	J. Beasley	Lynn, Mass	Jan. 28, 1873	135, 259
Heel burnishing machinery, Boot and shoe	I. Van Nouhuys	Albany, N. Y	Aug. 12, 1873	141, 678
Heel-calk	G. F. Clemons	Springfield, Mass	May 19, 1868	77, 960
Heel-calk	J. F. Richardson and G. F. Morse.	Portland, Me	Jan. 16, 1866	52, 077
Heel-calk	T. Symonds	Portland, Me	Nov. 28, 1865	51, 241
Heel-calk	M. F. Walter and C. Standinger.	Hartford, Conn	Sept. 6, 1870	107, 132
Heel-casing	J. R. Moffitt	Chelsea, Mass	Apr. 14, 1868	76, 794
Heel, Composition boot and shoe	F. Marquard	Newburyport, Mass	Nov. 23, 1869	97, 206
Heel, Composition for the manufacture of boot and shoe.	C. Brocking	Boston, Mass	Nov. 29, 1870	109, 578
Heel-cutter	J. H. Bean	Marietta, Ohio	Mar. 16, 1869	87, 753
Heel-cutter	J. H. Bean	Marietta, Ohio	June 22, 1869	91, 508
Heel-cutter	S. Keen	East Bridgewater, Mass	Oct. 26, 1869	96, 120
Heel-cutter	A. D. Kelley	Rochester, N. H	June 27, 1854	11, 196
Heel-cutter	O. H. and J. A. Marston	Centre Sandwich, N. H	Feb. 6, 1866	52, 430
Heel cutter, Boot	B. F. Goddard	Charlton Depot, Mass	Nov. 19, 1867	70, 992
Heel-cutter for cutting out heels of boots and shoes.	J. Shaw	Natick, Mass	June 2, 1857	17, 455
Heel cutting and finishing machine, Shoe	W. F. Edson	Philadelphia, Pa	Sept. 5, 1859	25, 326
Heel-cutting machine	O. G. Critchett	Stoneham, Mass	June 21, 1864	43, 183
Heel-cutting machine	R. C. Lambart	Raynham, Mass	Mar. 31, 1868	76, 207
Heel-cutting machine	R. C. Lambart	Raynham, Mass	Nov. 16, 1869	96, 817
Heel cutting machine, Boot and shoe	M. Bray and E. Edmands	Newton, Mass	July 16, 1872	129, 093
Heel cutting machine, Boot and shoe	E. T. Green	Stoneham, Mass	Oct. 9, 1860	30, 376
Heel cutting machine, Boot and shoe	R. C. Lambart	South Abington, Mass	Oct. 25, 1870	108, 605
Heel-cutting machine, Knife-holder and guide for	S. Keen	East Bridgewater, Mass	Oct. 25, 1870	108, 601
Heel, Detachable boot and shoe	C. W. Bailey	Boston, Mass	Nov. 9, 1869	96, 538
Heel, Detachable boot and shoe	C. W. Bailey	Boston, Mass	Sept. 6, 1870	106, 984
Heel, Detachable boot and shoe	A. C. Wyman	Boston, Mass	July 16, 1872	129, 447
Heel-die	B. F. Fisk and M. B. Stone	Haverhill, Mass	Apr. 25, 1871	114, 121
Heel, Elastic boot and shoe	T. P. Monzani	Columbusville, N. Y	Dec. 17, 1872	134, 088
Heel enameling, Boot and shoe	C. H. Orcutt	Leominster, Mass	May 2, 1871	114, 470
Heel-faces for boots and shoes, Machine for dressing	T. K. Reed	East Bridgewater, Mass	Feb. 27, 1866	52, 883
Heel finishing machine, Boot and shoe	H. H. Bigelow	Worcester, Mass	Oct. 25, 1870	108, 678
Heel finishing machine, Boot and shoe	J. L. Joyce	New Haven, Conn	Feb. 9, 1864	41, 512
Heel-finishing machine, Shave and burnisher for	J. L. Joyce	New Haven, Conn	Aug. 13, 1872	130, 375
Heel finishing machinery, Boot and shoe	L. P. Hawkins	Lynn, Mass	Apr. 1, 1873	137, 304
Heel finishing machinery, Boot and shoe	L. P. Hawkins	Lynn, Mass	Aug. 12, 1873	141, 712
Heel forming, smoothing, and polishing machine	G. P. Merriam	Lynn, Mass	July 29, 1862	36, 015
Heel gage, Boot and shoe	S. Ward	Westfield, N. Y	Aug. 29, 1871	118, 499
Heel guard, Boot and shoe	W. A. Harris	Providence, R. I	June 5, 1860	28, 575
Heel, India-rubber boot and shoe	P. Boisset	Paris, France	Apr. 22, 1862	35, 056
Heel, Interchangeable boot and shoe	J. Norburn	Pittsburgh, Pa	Dec. 7, 1869	97, 681
Heel, Interchangeable boot and shoe	J. C. Woodhead	Pittsburgh, Pa	Dec. 14, 1869	98, 004
Heel-iron	W. M. Rice	Boston, Mass	Oct. 9, 1866	58, 675
Heel-iron and ice-calk	W. Weaver	Nashua, N. H	Aug. 11, 1863	39, 514
Heel, Metallic	E. P. Bray	Elizabeth, N. J	Aug. 22, 1871	118, 335
Heel, Metallic	A. S. Mann	Saint Louis, Mo	Jan. 3, 1871	110, 768
Heel, Metallic	A. S. Mann	Saint Louis, Mo	July 11, 1871	116, 973
Heel, Metallic boot	P. S. Devlan	Reading, Pa	July 24, 1849	6, 610
Heel, Metallic boot and shoe	E. P. Bray	Elizabeth, N. J	Oct. 10, 1871	119, 691
Heel, Metallic boot and shoe	J. V. Dinsmore	Auburn, Me	Apr. 3, 1860	27, 702
Heel, Metallic boot and shoe	M. C. Easterly	Antwerp, N. Y	Oct. 6, 1863	40, 162
Heel, Metallic boot and shoe	H. Green	Antwerp, N. Y	July 1, 1862	35, 751
Heel, Metallic boot and shoe	A. L. Holbrook	Fremont, Nebr	May 23, 1871	115, 206
Heel, Metallic boot and shoe	E. T. Miller	Albany, N. Y	Aug. 9, 1870	106, 186
Heel, Metallic boot and shoe	A. T. Perrine	Louisville, Ky	June 7, 1870	103, 921
Heel nailing and shaving machine	L. P. Hawkins	Lynn, Mass	Feb. 18, 1873	135, 911
Heel-nailing machine	C. W. Glidden	Lynn, Mass	July 23, 1872	129, 811
Heel of rubber boots and shoes	F. M. Shepard	New York, N. Y	Mar. 8, 1870	100, 677
Heel-pattern, Metallic	J. Brobst	Fort Wayne, Ind	Aug. 11, 1868	80, 904
Heel-plate	H. L. Drake	Lynn, Mass	June 3, 1873	139, 500
Heel plate, Boot and shoe	W. E. Hamlin, jr	Providence, R. I	Sept. 1, 1868	81, 777
Heel plate, Boot and shoe	W. H. and W. Lewis	New York, N. Y	Oct. 31, 1839	1, 391
Heel plate, Boot and shoe	G. B. Massey	New York, N. Y	Sept. 17, 1872	131, 452
Heel plate, Boot and shoe	G. Rohr	Williamsburgh, N. Y	Oct. 14, 1873	143, 719
Heel-plate, Skate	G. Havell	Newark, N. J	Apr. 8, 1873	137, 678
Heel polisher, Boot	V. K. Spear	Lynn, Mass	Dec. 10, 1867	72, 105
Heel polisher, Boot and shoe	C. H. Helms	Poughkeepsie, N. Y	Oct. 10, 1871	119, 707
Heel-polishing machine	B. Q. Budding	Milford, Mass	Aug. 18, 1863	39, 546

Index of patents issued from the United States Patent Office from 1790 *to* 1873, *inclusive*—Continued.

Invention.	Inventor.	Residence.	Date.	No.
Heel-polishing machine	J. M. Thompson and S. D. Tripp	Stoneham and Lynn, Mass	Mar. 7, 1865	46, 761
Heel-polishing machine	J. M. Thompson and S. D. Tripp	Stoneham and Lynn, Mass	Feb. 6, 1866	52, 498
Heel-polishing machine	J. M. Thompson and S. D. Tripp	Stoneham and Lynn, Mass	Feb. 6, 1866	52, 499
Heel-polishing machine	S. D. Tripp	Lynn, Mass	Feb. 6, 1866	52, 470
Heel-polishing machine, Boot	J. M. Thompson, G. P. French, and S. D. Tripp.	Stoneham and Lynn, Mass	Sept. 27, 1864	44, 489
Heel polishing machine, Boot and shoe	B. Q. Budding	Milford, Mass	May 3, 1864	42, 555
Heel polishing machine, Boot and shoe	B. Q. Budding	Worcester, Mass	Oct. 6, 1868	82, 798
Heel polishing machine, Boot and shoe	C. H. Helms	Poughkeepsie, N. Y	June 13, 1871	115, 853
Heel polishing machine, Boot and shoe	C. H. Helms	Poughkeepsie, N. Y	July 25, 1871	117, 283
Heel polishing machine, Boot and shoe	C. H. Helms	Poughkeepsie, N. Y	May 28, 1872	127, 234
Heel polishing machine, Boot and shoe	V. K. Spear	Lynn, Mass	Apr. 7, 1868	76, 543
Heel polishing machine, Shoe and boot	S. W. Chamberlin	Stoneham, Mass	July 23, 1861	32, 859
Heel-press	H. H. Bigelow	Worcester, Mass	July 5, 1870	105, 030
Heel press, Boot	C. H. Helms	Poughkeepsie, N. Y	Aug. 13, 1867	67, 650
Heel-protector	H. H. Southwick	Mahaska County, Iowa	Mar. 31, 1868	76, 262
Heel-randing machine, Boot and shoe	H. F. Packard	North Bridgewater, Mass	Aug. 22, 1871	118, 263
Heel-rands, Machine for forming	S. Holmes and J. F. Sargent	North Brookfield and Melrose, Mass.	Apr. 4, 1871	113, 432
Heel-rands, Machine for turning	J. B. Johnson and G. W. Moulton	Lynn, Mass	Apr. 29, 1873	138, 257
Heel, Reversible boot	F. Richardson and F. Hacker	Providence, R. I	Mar. 21, 1871	112, 962
Heel, Reversible boot	F. Richardson and F. Hacker	Providence, R. I	July 4, 1871	116, 758
Heel, Revolving boot	T. Walker	Birmingham, England	June 29, 1852	9, 088
Heel, Revolving boot and shoe	J. H. Roome	New York, N. Y	June 8, 1858	20, 510
Heel, Rotary boot and shoe	A. O. Crane	Boston, Mass	Mar. 19, 1872	124, 723
Heel-rounding machine	J. C. White	Auburn, N. Y	Aug. 30, 1870	106, 901
Heel, Rubber boot	J. F. Barrett	Concord, Mass	Sept. 24, 1867	69, 163
Heel screw, Boot	W. Ackerman	Flint, Mich	Aug. 19, 1873	141, 908
Heel seat, Boot and shoe	G. I. Mason	Lewistown, Ill	Mar. 24, 1868	75, 779
Heel-seat-cutting machine	A. B. Keith	North Bridgewater, Mass	May 26, 1868	78, 211
Heel-shave	J. G. Ross	Philadelphia, Pa	June 27, 1865	48, 448
Heel-shave	J. H. Sanford	North Bridgewater, Mass	July 9, 1872	128, 757
Heel-shave	D. E. Somes	Biddeford, Me	Nov. 13, 1860	30, 648
Heel shave, Boot	E. S. Snell	North Bridgewater, Mass	July 29, 1862	36, 035
Heel shave, Boot and shoe	F. S. Dawes	Hudson, Mass	Dec. 31, 1867	72, 724
Heel shave, Boot and shoe	V. Snell	North Bridgewater, Mass	July 20, 1858	20, 960
Heel-shaving guard	J. E. Wiggin and D. G. Crosby	Stoneham, Mass	Mar. 26, 1867	63, 346
Heel shells with wood, Machine for filling metallic boot.	F. Richardson, F. Hacker, and J. A. Blake.	Providence, R. I	May 7, 1872	126, 491
Heel, sole, &c., Composition boot and shoe	F. Marquard	Newburyport, Mass	May 23, 1871	115, 075
Heel-spur	M. Young, jr	Frederick, Md	Jan. 31, 1860	27, 019
Heel-stiffener	L. C. Crowell	Boston, Mass	Jan. 14, 1873	134, 736
Heel-stiffener	L. C. Crowell	Boston, Mass	May 27, 1873	139, 238
Heel-stiffener	A. B. Ely	Newton, Mass	Feb. 27, 1872	123, 985
Heel-stiffener	J. W. Hatch	Rochester, N. Y	July 16, 1872	129, 338
Heel-stiffener	J. R. Moffitt	Chelsea, Mass	Oct. 19, 1869	96, 023
Heel-stiffener	I. R. Rogers	Lynn, Mass	Jan. 14, 1873	134, 933
Heel-stiffener	E. M. Stevens	Chelsea, Mass	Apr. 2, 1867	63, 573
Heel-stiffener	J. M. Watson	Sharon, Mass	July 30, 1872	130, 089
Heel stiffener, Boot and shoe	S. Moore	Sudbury, Mass	Nov. 11, 1873	144, 556
Heel stiffener, Boot and shoe	G. V. Sheffield	Northbridge, Mass	May 3, 1870	102, 720
Heel stiffener or counter, Boot and shoe	D. E. Hayward	Melrose, Mass	Oct. 27, 1863	40, 409
Heel-stiffener, Rubber	E. M. Stevens	Chelsea, Mass	Apr. 16, 1867	63, 958
Heel-stiffeners for boots and shoes, Manufacture of.	J. R. Moffitt	Chelsea, Mass	May 21, 1872	127, 090
Heel-stiffeners, Machine for making	S. Moore and H. Rogers	Sudbury, Mass	Nov. 5, 1872	132, 849
Heel-stiffeners, Machine for making	J. Samuels	Vineland, N. J	Nov. 25, 1873	145, 017
Heel stiffeners, Machine for shaping boot and shoe	J. Kimball	Boston, Mass	Oct. 8, 1872	131, 957
Heel-stiffening	A. B. Ely	Newton, Mass	Mar. 16, 1869	87, 916
Heel-stiffening for boots and shoes	F. E. Darrow and O. E. Wait	Rockport, Mass	Apr. 22, 1873	138, 010
Heel-stiffening for boots and shoes	E. M. Stevens	Boston, Mass	Aug. 18, 1863	39, 594
Heel-tap for boots and shoes	A. Warner	Brooklyn, E. D., N. Y	Jan. 3, 1871	110, 701
Heel-tread-leveling machine, Boot	A. B. Keith	Braintree, Mass	May 16, 1871	114, 949
Heel trimmer	C. H. Helms	Poughkeepsie, N. Y	Aug. 11, 1868	80, 953
Heel trimmer, Boot and shoe	L. Coté	Quebec, Canada	Nov. 15, 1870	109, 300
Heel trimming and burnishing machine, Attachment for.	S. H. Hodges	Lynn, Mass	Mar. 18, 1873	136, 916
Heel trimming and burnishing machine, Boot and shoe.	T. K. Reed	East Bridgewater, Mass	July 11, 1871	116, 867
Heel trimming and burnishing machines, Holding device for.	J. R. Folsom	Boston, Mass	Nov. 12, 1872	132, 957
Heel trimming and finishing machine, Boot and shoe.	G. C. Hawkins	Boston, Mass	Jan. 2, 1872	122, 462
Heel-trimming machine	C. J. Addy	Boston, Mass	Aug. 20, 1872	130, 686
Heel-trimming machine	C. J. Addy	Boston, Mass	Nov. 12, 1872	132, 944
Heel-trimming machine	C. J. Addy	Boston, Mass	Feb. 11, 1873	135, 745
Heel-trimming machine	C. C. Ballou	Albany, N. Y	Sept. 23, 1873	142, 980
Heel-trimming machine	E. U. Jones	Woodhaven, N. Y	July 1, 1873	140, 420
Heel-trimming machine	G. W. Moore	Newark, N. J	Oct. 15, 1872	132, 309
Heel-trimming machine	J. F. Sargent	Boston, Mass	Apr. 18, 1865	47, 341
Heel trimming machine, Boot and shoe	C. H. Helms	Poughkeepsie, N. Y	Mar. 12, 1861	31, 666
Heel trimming machine, Boot and shoe	C. H. Helms	Poughkeepsie, N. Y	Dec. 7, 1869	97, 638
Heel trimming machine, Boot and shoe	E. P. Richardson	Lawrence, Mass	Apr. 19, 1870	102, 159
Heel trimming machine, Boot and shoe	C. S. Stearns	Marlborough, Mass	June 25, 1867	66, 184
Heel-turning machine	J. Q. Molton	Lynn, Mass	Apr. 4, 1871	113, 684
Heels and soles for boots and shoes, Machine for cutting.	A. Warren	Jefferson, Ohio	July 19, 1859	24, 835
Heels and soles for boots and shoes, Machine for rasping and dressing.	J. P. Molliere	Lyons, France	Nov. 27, 1855	13, 854
Heels, Attaching boot and shoe	B. Giroux	Chicago, Ill	Dec. 13, 1870	110, 030
Heels, Breasting	W. H. Pitkin	Providence, R. I	Sept. 27, 1864	44, 453
Heels, Cutting boot	P. Shaw	Abington, Mass	Feb. 6, 1849	6, 095
Heels for boots and shoes, Machine for polishing and grinding the edges of.	J. M. Thompson and G. P. French.	Stoneham, Mass	May 3, 1864	42, 629
Heels for boots and shoes, Machine for preparing	J. Jenkins	Lynn, Mass	Sept. 30, 1862	36, 586
Heels for boots and shoes, Machine for shaping	J. H. Belser	Marlborough, Mass	Nov. 28, 1865	51, 254
Heels for boots and shoes, Machine for shaping	L. Hall	Boston, Mass	Sept. 27, 1859	25, 605
Heels for boots and shoes, Machine for shaping	J. Samuels	Lynn, Mass	Apr. 12, 1864	42, 334
Heels for boots and shoes, Process of forming pierced	H. H. Bigelow	Worcester, Mass	Apr. 9, 1872	125, 528

Index of patents issued from the United States Patent Office from 1790 *to* 1873, *inclusive*—Continued.

Invention.	Inventor.	Residence.	Date.	No.
Heels for shoes, Machine for finishing wooden	R. Hubler	New York, N. Y	Jan. 21, 1873	135, 038
Heels, Hand-tool for manufacture of	S. L. Ricker	New York, N. Y	Oct. 14, 1873	143, 718
Heels, Machine for attaching and finishing boot	H. Saloshinsky	Boston, Mass	Jan. 31, 1860	27, 008
Heels, Machine for cutting down boot and shoe	J. L. Joyce	New Haven, Conn	Jan. 8, 1869	91, 136
Heels, Machine for cutting leather for boot and shoe.	M. H. Hall	Philadelphia, Pa	Dec. 17, 1873	145, 647
Heels, Machine for dressing boot and shoe	J. S. Stuart and A. L. Corson	Marblehead, Mass	Oct. 23, 1860	30, 508
Heels, Machine for forming boot	C. W. Glidden	Lynn, Mass	May 6, 1873	138, 634
Heels, Machine for making boot and shoe	H. H. Bigelow	Worcester, Mass	Apr. 9, 1872	125, 529
Heels, Machine for making boot and shoe	C. W. Glidden	Lynn, Mass	Mar. 14, 1871	112, 702
Heels, Machine for making boot and shoe	A. Knowlton	Boston, Mass	Jan. 14, 1873	134, 899
Heels, Machine for making boot and shoe	A. C. McKnight	Philadelphia, Pa	June 17, 1873	140, 060
Heels, Machine for pricking and nail-loading boot and shoe.	C. W. Glidden	Lynn, Mass	Mar. 4, 1873	136, 503
Heels, Machine for punching the lifts of boot	G. W. Ellis	Lynn, Mass	Dec. 22, 1863	41, 037
Heels, Machine for shaping boot	G. W. Warfield	Hudson, Mass	Oct. 22, 1867	70, 139
Heels, Machine for shaping the edges of boot and shoe.	C. W. Glidden	Lynn, Mass	Feb. 4, 1873	135, 538
Heels, Machine for turning wooden	R. Hubler	New York, N. Y	Jan. 21, 1873	135, 037
Heels, Manufacture of boot and shoe	J. Blukey	Leeds, England	June 17, 1873	139, 859
Heels, Manufacture of boot and shoe	C. W. Glidden	Lynn, Mass	Jan. 7, 1873	134, 538
Heels, Metallic stiffener for boot and shoe	J. Barsaloux, J. S. James, and N. Lyon.	Sandy Hill and Hope Falls, N. Y.	July 9, 1872	128, 843
Heels of boots and shoes, Apparatus for trimming.	L. Coté	St. Hyacinthe, Canada	Apr. 1, 1873	137, 293
Heels of boots and shoes, Constructing	S. Warner, J. Hodgkins, and W. E. Traver.	Lowell and Westfield, Vt., and Watervliet, N. Y.	July 1, 1840	1, 665
Heels of boots and shoes, Machine for dressing	W. H. Bush	New Haven, Conn	Mar. 12, 1872	124, 482
Heels of boots and shoes, Machine for dressing	H. Guild and L. Hall	Boston, Mass	Mar. 15, 1859	23, 245
Heels of boots and shoes, Machine for pricking and cutting.	E. S. Snell	North Bridgewater, Mass	Mar. 9, 1858	19, 611
Heels of boots and shoes, Machine for shaving	A. B. Keith and T. K. Reed	North Bridgewater and East Bridgewater, Mass.	June 21, 1870	104, 399
Heels of boots and shoes, Metallic brace for	G. W. Griswold	Carbondale, Pa	Sept. 23, 1856	15, 762
Heels of boots and shoes, Shaping	E. J. Beane	Providence, R. I	June 14, 1864	43, 082
Heels to boots and shoes and polishing the same, Machine for attaching.	G. W. Ellis and C. W. Glidden	Stoneham, Mass	Oct. 7, 1862	36, 607
Heels to boots and shoes, Machine for applying	J. Jenkins	Lynn, Mass	Mar. 12, 1861	31, 690
Heels to boots and shoes, Machine for nailing	G. W. Ellis	Lynn, Mass	Dec. 22, 1863	41, 038
Heels to boots and shoes, Mode of attaching composition.	C. H. Helms	Poughkeepsie, N. Y	Dec. 13, 1870	110, 037
Heels to boots, Attaching	C. D. Ulmer	Boston, Mass	Feb. 16, 1869	86, 955
Heels to India-rubber soles, Attaching metallic	S. T. Parmelee	New Brunswick, N. J	July 17, 1855	13, 272
Heels, Tool for trimming boot and shoe	S. C. Bedell	Red Bank, N. J	Apr. 3, 1866	53, 558
Heels, Vulcanized-rubber rings for boot and shoe	W. H. Towers	Boston, Mass	Jan. 16, 1872	122, 869
Heeling-machine	C. W. Glidden and H. P. Fairfield.	Lynn and Boston, Mass	May 20, 1873	139, 058
Heeling machine, Boot	C. W. Glidden	Lynn, Mass	May 31, 1870	103, 735
Heeling machine, Boot	T. K. Reed and A. B. Keith	East Bridgewater and North Bridgewater, Mass.	Nov. 29, 1870	109, 756
Heeling machine, Boot and gaiter	L. Graf	Newark, N. J	Dec. 16, 1873	145, 645
Heeling machine, Boot and shoe	H. H. Bigelow	Worcester, Mass	Oct. 25, 1870	108, 677
Heeling machine, Boot and shoe	J. Gilson	Stoneham, Mass	May 31, 1870	103, 734
Heeling machine, Boot and shoe	C. W. Glidden	Lynn, Mass	Feb. 25, 1873	136, 233
Heeling machine, Boot and shoe	W. F. Spinney	Lynn, Mass	May 31, 1870	103, 792
Heeling machine, Shoe	C. H. Trask and H. Eldridge	Lynn, Mass	Jan. 5, 1869	85, 625
Heeling machines, Bed-plate for boot and shoe	C. L. Frye	Marlborough, Mass	June 11, 1872	127, 861
Heeling-plate	N. A. Swett	Westbrook, Me	Sept. 3, 1867	68, 466
Heliogeastrum or planetarium	T. Newell	Poultney, Vt	Apr. 12, 1820	....
Heliographic instrument for taking the sun's altitude.	J. Oakes	New York, N. Y	June 8, 1858	20, 506
Helical spring	W. R. Nichols	Philadelphia, Pa	Apr. 26, 1870	102, 421
Helical spring	J. W. Peck, jr	Brooklyn, N. Y	Jan. 24, 1860	26, 954
Helical spring	R. Vose	New York, N. Y	Mar. 15, 1864	41, 950
Heliographic and photographic spectrum for producing line-engravings.	F. Von Egloffstein	New York, N. Y	Nov. 21, 1865	51, 103
Heliometer	C. F. L. Kisch	Huntingburgh, Ind	Apr. 28, 1868	77, 324
Heliometer	C. F. L. Kisch	Huntingburgh, Ind	May 19, 1868	78, 133
Heliotrope, Automatic	L. F. Morawetz and C. Volkmar	Baltimore, Md	June 12, 1866	55, 523
Helm-wheel, Spiral	R. P. Cunningham	Pomfret, Conn	Aug. 3, 1820	....
Hem-folder	L. Clark	Monticello, N. Y	Oct. 18, 1859	25, 807
Hem, Gathered	H. A. Ellis	Chicopee Falls, Mass	Apr. 30, 1872	126, 138
Hemmer	O. L. Brown	Hopkinton, Mass	Oct. 7, 1873	143, 433
Hemmer	E. P. Davis	North Attleborough, Mass	Apr. 14, 1868	76, 720
Hemmer	H. A. Ellis	Chicopee Falls, Mass	Apr. 30, 1872	126, 139
Hemmer	H. M. Hall	Philadelphia, Pa	July 9, 1872	128, 876
Hemmer	E. S. Yentzer	Ottawa, Ill	Apr. 22, 1873	138, 064
Hemmer and folder	J. and F. Stevens	New York, N. Y	Mar. 26, 1861	31, 833
Hemmer-guide	S. Perry	New York, N. Y	Oct. 8, 1872	132, 101
Hemming and cording guide	H. B. Odiorne	Philadelphia, Pa	May 8, 1855	12, 826
Hemming and tucking guide	G. W. Downes	New York, N. Y	May 26, 1863	38, 662
Hemming-guide	W. Gaskill	Cincinnati, Ohio	May 9, 1865	47, 629
Hemming-guide	W. Gaskill and G. H. Knight	Cincinnati, Ohio	May 9, 1865	47, 630
Hemming-guide	J. Howell	Sacramento, Cal	Mar. 5, 1861	31, 602
Hemming-guide	G. L. Jencks	Providence, R. I	June 11, 1861	32, 519
Hemorrhage from internal organs or cavities, Instruments for arresting.	A. B. Haile	Norwich, Conn	Oct. 16, 1849	6, 796
Hemorrhoidian	R. A. Cameron	Valparaiso, Ind	Aug. 27, 1867	68, 042
Hemostatic and anti-septic agents, Preparation to serve as.	P. A. F. Boboeuf	Paris, France	June 23, 1863	38, 940
Hemp and cotton stalks, Machine for pulling	A. W. Goddard	Clinton, Mass	Dec. 11, 1866	60, 359
Hemp and flax and reducing the length of the fibers, Machine for breaking.	J. S. Treat and S. Randall	Voluntown, Conn	Sept. 16, 1851	8, 360
Hemp and flax brake	A. F. Bruce	Marshall Post-Office, Mo	June 24, 1844	3, 641
Hemp and flax brake	J. Bryant	Akron, Ind	Aug. 9, 1864	43, 812
Hemp and flax, Breaking	J. C. Johnston	Fayette County, Ky	May 16, 1811	....
Hemp and flax, Breaking	R. Miller	Lexington, Ky	July 13, 1811	....
Hemp and flax breaking and cleaning machine	R. Medley	Bloomfield, Ky	June 11, 1829	....
Hemp and flax breaking and cleaning machine	W. N. Stewart	Mayslick, Ky	Feb. 4, 1843	2, 943

Index of patents issued from the United States Patent Office from 1790 *to* 1873, *inclusive*—Continued.

Invention.	Inventor.	Residence.	Date.	No.
Hemp and flax breaking and dressing	H. Lull	Ithaca, N. Y	June 14, 1837	235
Hemp and flax, Breaking and dressing	A. Salisbury and J. C. Langdon	Troy, N. Y	July 29, 1829	....
Hemp and flax breaking and dressing machine	C. Warner, A. T. Mixell, and E. I. Horn.	Belvidere, N. J	July 31, 1837	319
Hemp and flax, Breaking and swingling	A. Tanner	Trumbull County, Ohio	Oct. 24, 1811	....
Hemp and flax breaking and swingling machine	T. Cohoon	Cazenovia, N. Y	Apr. 13, 1808	....
Hemp and flax breaking machine	A. Kyes	Crittenden, Ky	Apr. 4, 1838	671
Hemp and flax breaking machine	O. A. Leavitt	Maysville, Ky	Aug. 30, 1853	9, 973
Hemp and flax breaking machine	A. Salisbury and J. C. Langdon	Troy, N. Y	Apr. 18, 1829	....
Hemp and flax breaking machine	F. Stith	Franklin, Tenn	Apr. 22, 1835	....
Hemp and flax breaking machine	J. Warren	Westbrook, Me	June 14, 1837	234
Hemp and flax breaking machine	J. Y. Watson, J. Blossom, and A. Burnett.	Salem, N. Y	Apr. 21, 1829	....
Hemp and flax, Breaking, swingling, and beating	W. Harper	Richmond, Va	Oct. 15, 1811	....
Hemp and flax cleaning and dressing machine	A. Salisbury	Troy, N. Y	Apr. 15, 1829	....
Hemp and flax, Drawing-frame for	O. S. Leavitt	Marcellus, N. Y	Sept. 20, 1853	10, 034
Hemp and flax dressing machine	H. S. Barnum and M. Stevenson	Cambridge, N. Y	July 8, 1829	....
Hemp and flax dressing machine	W. and R. Brittain	Amwell, N. J	Oct. 12, 1837	423
Hemp and flax dressing machine	H. Burden	Albany, N. Y	June 15, 1822	....
Hemp and flax dressing machine	J. Macdonald	New York, N. Y	Jan. 18, 1823	....
Hemp and flax dressing machine	G. H. Ricketts and J. Kinney, jr	Morristown, N. J	Oct. 24, 1814	....
Hemp and flax dressing machinery	W. W. Grant	Providence, R. I	Mar. 5, 1850	7, 139
Hemp and flax fiber for the manufacture of dusters, &c., Preparing.	E. G. Wayman	Louisville, Ky	Feb. 18, 1868	74, 736
Hemp and flax for carding, Machinery for preparing.	C. Beach	Penn Yan, N. Y	Sept. 23, 1862	36, 551
Hemp and flax for spinning and weaving, Preparation of.	L. Tibbitts	New Glasgow, Va	Aug. 1, 1829	....
Hemp and flax hackling machine	J. Rinek	Easton, Pa	Sept. 24, 1872	131, 708
Hemp and flax hatcheling and cleaning machine	G. M. Billings and J. Harrison	Glasgow, Mo	May 1, 1845	4, 022
Hemp and flax machine	R. Button	Cazenovia, N. Y	Feb. 11, 1807	....
Hemp and flax machine	T. Cohoon	Troy, N. Y	Apr. 25, 1829	....
Hemp and flax, Machine and mode of preparing	F. Hall	Tennessee	Oct. 7, 1813	....
Hemp and flax, Machinery for breaking	G. Sanford and J. E. Mallory	England and New York, N. Y.	Apr. 28, 1863	38, 340
Hemp and flax, Machinery for preparing	E. Davy	Crediton, England	Mar. 11, 1856	14, 394
Hemp and flax, Preparing and hatcheling	J. Goulding	Boston, Mass	Aug. 17, 1835	....
Hemp and flax, Process of bolting	G. W. Billings	New York, N. Y	Oct. 6, 1863	40, 154
Hemp and flax scutching and hackling machine	O. W. Grimes	Paducah, Ky	Sept. 23, 1851	8, 374
Hemp and flax softening machinery	R. Boyack	Poughkeepsie, N. Y	Oct. 24, 1854	11, 825
Hemp and making oakum of same, Machinery for tarring slivers or bands of.	W. Montgomery	Boston, Mass	Aug. 28, 1840	1, 747
Hemp and other fibrous material, Apparatus and process of rotting.	L. W. Wright	Plainfield, N. H	Dec. 25, 1849	6, 980
Hemp and parting its fibers, Machinery for drawing	O. S. Leavitt	Maysville, Ky	Sept. 24, 1850	7, 668
Hemp-brake	J. F. Adams	Lexington, Ky	Nov. 4, 1873	144, 301
Hemp-brake	W. W. Austin and F. Creasy	Carrollton, Mo	Mar. 5, 1861	31, 585
Hemp-brake	J. Barkley	Saint Louis, Mo	Apr. 7, 1857	17, 015
Hemp-brake	C. A. Biermann	Waterloo, Ill	Feb. 5, 1867	61, 797
Hemp-brake	R. W. Bewen	Marshall, Mo	June 24, 1856	15, 166
Hemp-brake	G. M. Campbell	Lewistown, Ill	Feb. 27, 1866	52, 821
Hemp-brake	L. S. Chichester	Williamsburgh, N. Y	Feb. 3, 1852	8, 700
Hemp-brake	E. Christianson	Saint Joseph, Mo	July 30, 1867	67, 268
Hemp-brake	E. M. Crandal	Alton, Ill	Oct. 12, 1869	95, 659
Hemp-brake	J. Crane and F. H. Hamilton	Schenectady, N. Y	Jan. 1, 1850	6, 984
Hemp-brake	R. Dodsworth	Saint Louis, Mo	Feb. 12, 1861	31, 375
Hemp-brake	Z. Feagan	Palmyra, Mo	Aug. 9, 1859	24, 999
Hemp-brake	P. S. Fitch	Hanly, Ky	June 18, 1872	128, 030
Hemp-brake	T. L. Fortune	Liberty, Mo	Sept. 3, 1846	4, 741
Hemp-brake	T. L. Fortune	Weston, Mo	July 26, 1864	43, 654
Hemp-brake	P. G. Gardiner	New York, N. Y	Feb. 28, 1845	3, 936
Hemp-brake	R. J. Gatling	Murfreesborough, N. C	Apr. 17, 1847	5, 073
Hemp-brake	R. J. Gatling	Indianapolis, Ind	Sept. 4, 1860	29, 875
Hemp-brake	J. T. Gillman	Walnut Fork, Iowa	May 24, 1864	42, 847
Hemp-brake	H. Guild	Cincinnati, Ohio	Sept. 26, 1848	5, 811
Hemp-brake	E. Guile	Waverly, Mo	Oct. 16, 1860	30, 402
Hemp-brake	W. W. Hampton	Winchester, Va	Apr. 21, 1857	17, 092
Hemp-brake	J. S. Hardeman	Arrow Rock, Mo	May 12, 1857	17, 274
Hemp-brake	R. Heneage	Buffalo, N. Y	Dec. 21, 1858	22, 399
Hemp-brake	R. Heneage	Buffalo, N. Y	Jan. 22, 1861	31, 167
Hemp-brake	J. Hindman	Haynesville, Mo	Jan. 25, 1859	22, 725
Hemp-brake	F. P. Holcomb	New Castle, Del	Mar. 13, 1847	5, 010
Hemp-brake	D. W. Hughes	New London, Mo	Feb. 27, 1855	12, 456
Hemp-brake	W. C. Hutchinson	Saint Joseph, Mo	Oct. 5, 1858	21, 680
Hemp-brake	W. Jones	Saint Louis, Mo	Jan. 29, 1861	31, 245
Hemp-brake	J. Kaye	Louisville, Ky	Sept. 15, 1868	82, 123
Hemp-brake	E. W. Lacy	Oak Park, Va	Dec. 23, 1856	16, 279
Hemp-brake	E. W. Lacy	Oak Park, Va	July 31, 1860	29, 388
Hemp-brake	W. S. Lawrence	Winchester, Ky	Apr. 9, 1872	125, 466
Hemp-brake	W. S. Lawrence	Winchester, Ky	Aug. 5, 1873	141, 569
Hemp-brake	O. S. Leavitt	Marcellus, N. Y	Sept. 20, 1853	10, 033
Hemp-brake	S. H. Little	Saint Louis, Mo	Jan. 6, 1857	16, 365
Hemp-brake	S. H. Little	Saint Louis, Mo	Aug. 24, 1858	21, 264
Hemp-brake	H. F. Mann	La Porte, Ind	May 31, 1859	24, 264
Hemp-brake	R. Mansley	Philadelphia, Pa	Jan. 18, 1859	22, 661
Hemp-brake	W. Mason	Washington County, Ohio	Dec. 28, 1809	....
Hemp-brake	J. C. Matherly	Irvine, Ky	June 24, 1873	140, 289
Hemp-brake	J. B. McCormick and W. R. Baker.	Saint Louis, Mo	Jan. 22, 1861	31, 181
Hemp-brake	J. R. McDonald	Fayette, Mo	Apr. 2, 1861	31, 899
Hemp-brake	H. D. McGeorge	Morgantown, Va	Sept. 14, 1858	21, 513
Hemp-brake	J. Mills, jr	Quincy, Ill	May 29, 1860	28, 497
Hemp-brake	S. P. Moore	Arrow Rock, Mo	Feb. 2, 1858	19, 255
Hemp-brake	T. H. Murphy	New Orleans, La	Oct. 14, 1862	36, 667
Hemp-brake	G. M. Newell	Lexington, Mo	July 13, 1858	20, 890
Hemp-brake	J. H. Phillips	Waverly, Mo	June 11, 1861	32, 530
Hemp-brake	J. W. Rinehart	Lexington, Mo	May 10, 1859	23, 946
Hemp-brake	G. Rynel	Paris, Ky	May 3, 1870	102, 596

Index of patents issued from the United States Patent Office from 1790 *to* 1873, *inclusive*—Continued.

Invention.	Inventor.	Residence.	Date.	No.
Hemp-brake	S. H. Sample	Fayette, Mo	Aug. 18, 1846	4, 698
Hemp-brake	I. Scoville	Chicago, Ill	Sept. 25, 1860	30, 161
Hemp-brake	W. Shelby	Waverly, Mo	Nov. 2, 1858	21, 983
Hemp-brake	S. Sherman	Weston, Mo	Mar. 26, 1867	63, 183
Hemp-brake	C. Simon	Louisville, Ky	Sept. 29, 1857	18, 303
Hemp-brake	A. Smith	Mobile, Ala	Nov. 20, 1849	6, 888
Hemp-brake	S. Stafford	Carroll County, Mo	July 28, 1857	17, 900
Hemp-brake	M. Thompson	Saint Joseph, Mo	Aug. 5, 1856	15, 498
Hemp-brake	W. A. Vertrees	Winchester, Mo	Aug. 30, 1859	25, 292
Hemp-brake	P. M. Walker	Marshall, Mo	May 27, 1851	8, 122
Hemp-brake	D. F. Wallace	Ripley, Ohio	May 5, 1868	77, 553
Hemp-brake	A. Wilson and G. C. Fletcher	Saint Thomas, Mo	Aug. 2, 1859	24, 966
Hemp-brake	C. Witt and A. Sina	Davenport, Iowa	Sept. 3, 1867	68, 477
Hemp-brake	G. F. S. Zimmerman and A. Beattie.	Saint Joseph, Mo	Nov. 17, 1857	18, 657
Hemp brake and cleaner	C. B. Butler	Petersburgh, Tenn	Jan. 6, 1844	3, 402
Hemp-breaker	S. A. Clemens	Rockford, Ill	Feb. 11, 1862	34, 349
Hemp, Breaking and cleaning	G. W. Billings and J. Harrison	Glasgow, Mo	June 7, 1845	4, 071
Hemp breaking and cleaning	H. Zellner	Columbia, Tenn	July 16, 1867	66, 767
Hemp breaking and cleaning machine	J. R. Booton	Luray, Va	Oct. 11, 1859	25, 717
Hemp, &c., breaking and cleaning machine	W. Y. Singleton	Springfield, Ill	May 7, 1845	4, 034
Hemp breaking and cleaning machine	G. T. Tate and W. English	Frankford, Mo	July 11, 1842	2, 717
Hemp breaking and dressing machinery	L. W. Colver	Glasgow, Mo	Sept. 26, 1848	5, 801
Hemp, Breaking and swingling	S. Mulliken		Mar. 11, 1791	
Hemp-breaking machine	L. W. Colver	Louisville, Ky	Mar. 29, 1853	9, 632
Hemp-breaking machine	W. Greathouse	Mason County, Ky	July 30, 1831	
Hemp-breaking machine	J. S Hoskins	Spring Hill, Mo	May 12, 1868	77, 815
Hemp-breaking machine	J. Pursell	Perryville, Ky	Mar. 18, 1836	
Hemp-breaking machine	S. Stafford	Carrollton, Mo	Dec. 13, 1859	26, 449
Hemp-breaking machine	W. Stone, jr	Williamson County, Tenn	Sept. 8, 1831	
Hemp-breaking machine	J. S. Van de Graaf	Scott County, Ky	May 12, 1828	
Hemp cleaner and broom-corn stripper, Sisal	G. D. Allen	Key West, Fla	Mar. 6, 1855	12, 472
Hemp-cradle	G. Reynolds, jr	Washington, Ky	May 30, 1844	3, 607
Hemp-cutter	J. L. Hardeman	Arrow Rock, Mo	Aug. 21, 1855	13, 460
Hemp-cutter	J. L. Hardeman	Arrow Rock, Mo	Nov. 17, 1857	18, 638
Hemp-cutting cradle	W. L. Larimore	Paris, Ky	Apr. 25, 1837	178
Hemp-cutting machine	J. L. Hardeman	Arrow Rock, Mo	Aug. 20, 1850	7, 578
Hemp-drawing machine	S. Lownds	Brooklyn, N. Y	July 14, 1857	17, 795
Hemp-drawing machine	G. W. Pittman	Dartmouth, Canada	Nov. 28, 1871	121, 414
Hemp-dresser	B. Langdon and A. Salisbury	Troy, N. Y	Apr. 25, 1846	4, 469
Hemp, flax, and other fibers, Bleaching and scouring.	L. Jarosson	Lille, France	Dec. 24, 1867	72, 500
Hemp, flax, and wool, Machine for hatcheling	D. Treadwell	Boston, Mass	Aug. 18, 1834	
Hemp, flax, &c., Belted roving-machine for	D. Treadwell	Boston, Mass	Feb. 3, 1834	
Hemp, flax, &c., breaking	S. Mulliken		Jan. 15, 1795	
Hemp, flax, &c., breaking and cleaning machine	G. Sanford and J. E. Mallory	England and New York, N. Y.	Sept. 22, 1863	40, 063
Hemp, flax, &c., breaking and cleaning machine	T. Tebow	Lexington, Ky	May 30, 1871	115, 387
Hemp, flax, &c., Curing	W. D. Monk	Williamsburgh, N. Y	Nov. 6, 1866	59, 436
Hemp, flax, &c., for spinning, Preparing	J. B. Fuller and J. P. Upham	Claremont, N. H	May 2, 1865	47, 538
Hemp, flax, jute-grass, &c., Method of treating	S. M. Allen	Woburn, Mass	July 18, 1865	48, 782
Hemp, flax, &c., New manufacture from	S. M. Allen	Woburn, Mass	Apr. 12, 1864	42, 265
Hemp, flax, &c., Preparing short-staple fibers from	J. B. Fuller and J. P. Upham	Claremont, N. H	June 7, 1864	43, 015
Hemp, flax, &c., Process for separating fiber of	J. B. Fuller and J. P. Upham	Claremont, N. H	May 2, 1865	47, 539
Hemp, flax, &c., Separating fibers of	J. B. Fuller	Claremont, N. H	June 7, 1864	43, 073
Hemp, flax, &c., Separating fibers of	J. B. Fuller and J. P. Upham	Claremont, N. H	Sept. 27, 1864	44, 415
Hemp, flax, &c., to a fibrous condition, Reducing	R. Sherwood	Fort Edward, N. Y	Nov. 10, 1863	40, 577
Hemp, flax, tow, &c., Preparing and hatcheling	J. Goulding	Boston, Mass	Oct. 10, 1835	
Hemp, flax, &c., Treating	T. Gray	Wadsworth, England	Dec. 4, 1866	60, 177
Hemp from okra, Making	J. Blanc	New Orleans, La	June 24, 1851	8, 184
Hemp, &c., hackling machine	P. G. and G. Rice	Danville, Ky	July 11, 1837	261
Hemp-hatcheling machine	G. W. Pittman	Brooklyn, N. Y	June 20, 1871	116, 092
Hemp into rope, drawing and hackling, Spinning	J. Lang	Greenock, Scotland	Jan. 16, 1834	
Hemp-leaves, Machine for separating the fiber from the pulp in.	S. S. Mills	Charleston, S. C	Oct. 12, 1858	21, 771
Hemp-machine	J. Anderson	Louisville, Ky	Nov. 13, 1849	6, 860
Hemp-machine	G. Ehrsam	New York, N. Y	July 14, 1863	39, 219
Hemp-machine	C. C. Estes	Maury County, Tenn	July 14, 1845	4, 114
Hemp-machine	J. E. Neill	Key West, Fla	Sept. 23, 1873	143, 179
Hemp, Machine for dressing Manila	C. E. Potter	Portsmouth, N. H	Dec. 23, 1834	
Hemp, Machine for dressing Sisal	S. A. Clemens	Springfield, Mass	July 15, 1851	8, 218
Hemp, &c., Machine for lowering and separating the boon from the fibers of.	R. Deering, sr	Louisville, Ky	June 10, 1845	4, 075
Hemp, &c., Machine for preparing and spinning	M. Day	Roxbury, Mass	Apr. 30, 1840	1, 580
Hemp, Machinery for breaking and dressing	A. Eldred	Oppenheim, N. Y	Apr. 24, 1849	6, 371
Hemp, &c., Machinery for hackling or preparing and spinning.	W. Montgomery	Boston, Mass	Feb. 28, 1844	3, 452
Hemp, &c., Machinery for spinning	W. C. Hibbard	Boston, Mass	Apr. 24, 1849	6, 388
Hemp or flax breaking and cleaning machinery	G. Sanford and J. E. Mallory	New York, N. Y	Mar. 18, 1862	34, 698
Hemp or flax dressing machinery	G. F. Schaffer	New York, N. Y	Oct. 29, 1861	33, 625
Hemp or flax waste within a sliver of longer staple, Machine for enveloping.	J. Leinweber	Louisville, Ky	May 1, 1866	54, 375
Hemp, Preparation of	R. Deering, sr	Louisville, Ky	June 25, 1845	4, 093
Hemp, Process of preparing	T. H. Barlow	Lexington, Ky	June 25, 1845	4, 094
Hemp, Process of rotting	W. Watt	Glasgow, Great Britain	Nov. 21, 1854	11, 981
Hemp, Process of treating	L. C. Suggett	Lexington, Ky	May 23, 1854	10, 968
Hemp, ramie, &c., cleaning machine	B. Roezl	Santo Comapam, Mexico	Sept. 17, 1867	68, 905
Hemp, Rotting	G. W. Billings and J. Harrison	Glasgow, Mo	May 10, 1845	4, 041
Hemp-rotting apparatus	C. H. Van Dorn	Saint Louis, Mo	Sept. 19, 1848	5, 776
Hemp-scutchers	F. H. Hamilton and T. Bullock	Schenectady, N. Y	Jan. 1, 1850	6, 987
Hemp, Sisal, and other grass cleaning machine	C. A. Dean	Boston, Mass	June 13, 1871	115, 828
Hemp, Treatment of fiber of Tampico	W. Staufen	London, England	July 6, 1858	20, 827
Hens from scratching, Device for preventing	E. Rice	West Northfield, Mass	Nov. 24, 1868	84, 303
Herbs for cookery, &c., Preparing	J. Culin	Philadelphia, Pa	July 16, 1872	129, 323
Hernia by means of injections, Cure of	C. T. Sage	New York, N. Y	Jan. 10, 1843	2, 906
Hernia-pad	W. Pomeroy	Brooklyn, N. Y	Aug 25, 1868	81, 533
Hernia-pad	W. Pomeroy	New York, N. Y	Aug. 29, 1871	118, 483
Hernia, Treating reducible	Z. Jayne	Carrollton, Ill	Apr. 2, 1841	2, 032

Index of patents issued from the United States Patent Office from 1790 *to* 1873, *inclusive*—Continued.

Invention.	Inventor.	Residence.	Date.	No.
Hide and fulling mill	S. Hussey	Gowanda, N. Y	Apr. 2, 1872	125, 135
Hide and leather dressing machine	J. R. Bumgarner and L. White	Davenport, Iowa	July 13, 1858	20, 861
Hide and leather softening machine	L. H. Stanley and G. B. Draper	Attleborough, Mass	May 23, 1871	115, 131
Hide and leather washing machine	H. M. Meeker	Smith's Mills, N. Y	May 4, 1869	89, 590
Hide and leather working machine	E. Fitzhenry	Somerville, Mass	Apr. 2, 1872	125, 276
Hide and skin cleaner and scourer	J. Turney	Nottingham, England	Sept. 17, 1872	131, 480
Hide-drying apparatus	M. H. Merriam and E. L. Norton	Charlestown, Mass	June 5, 1866	55, 333
Hide-handling cylinders, Beater in	J. R. Innis	Easton, Pa	Mar. 19, 1850	7, 192
Hide-mill	J. P. Friend and B. A. Annable	Peabody and Salem, Mass	Mar. 8, 1870	100, 519
Hide-mill or machine for softening dry hides	W. Edwards	Northampton, Mass	Dec. 30, 1812	
Hide-mills, Operating	J. G. Curtis	Emporium, Pa	Nov. 22, 1870	109, 393
Hide scouring or dressing machinery	C. and F. E. Holmes	Boston, Mass	May 24, 1870	103, 463
Hide-treating device	H. Royer	San Francisco, Cal	June 22, 1869	91, 772
Hide-washing machine	A. Ross	Maine, N. Y	May 7, 1867	64, 575
Hide-working, leather-scouring, &c., Machinery for	T. R. Williams	Salem, Mass	Aug. 16, 1870	106, 439
Hides, Adjustable frame for stretching	W. Dunn	Newark, N. J	Oct. 1, 1867	69, 327
Hides and furs, Dressing	C. I. Weston	Cummington, Mass	Mar. 24, 1868	76, 015
Hides and leather, Frame for stretching	T. P. Howell and C. P. Oliver	Newark, N. J	May 5, 1868	77, 615
Hides and manufacturing leather, Treating	W. H. Fuller	Brockport, N. Y	Dec. 27, 1870	110, 562
Hides and preserving wood, Depilating	J. E. Siebel	Chicago, Ill	July 4, 1871	116, 638
Hides and shaving leather, Machine for splitting	S. Boyden	Foxborough, Mass	Jan. 7, 1809	
Hides and skins, Apparatus for fleshing and stoning	J. S. Wheat	South Wheeling, W. Va	Nov. 13, 1866	59, 692
Hides and skins, Breaking, unhairing, fleshing, and scraping.	T. E. Barker and S. S. Fordham	Hudson, N. Y	Oct. 15, 1812	
Hides and skins, Compound for bating	C. F. Panknin	Charleston, S. C	Nov. 29, 1870	109, 656
Hides and skins, Compound for bating	L. F. Robertson	Morrisania, N. Y	Feb. 23, 1869	87, 202
Hides and skins, Compound for bating	L. F. Robertson	New York, N. Y	Dec. 13, 1870	110, 161
Hides and skins, Compound for treating	L. F. Robertson	West Farms, N. Y	Apr. 21, 1868	77, 099
Hides and skins, Curing	H. Napier	Elizabeth, N. J	Oct. 30, 1866	59, 251
Hides and skins, Depilating and tanning	J. Henry	New York, N. Y	Jan. 28, 1870	104, 734
Hides and skins, Machine for dressing	J. Schiffer	New York, N. Y	July 9, 1867	66, 640
Hides and skins, Machine for scouring	E. Fitzhenry	Boston, Mass	Oct. 3, 1871	119, 513
Hides and skins, Machine for splitting	W. Brown	Philadelphia, Pa	Nov. 20, 1809	
Hides and skins, Machine for splitting	T. O. Harrison		May 19, 1800	
Hides and skins, Process for treating	W. Maynard	Salem, Mass	Feb. 18, 1873	136, 082
Hides and skins, Process of bating	W. Zollikoffer	Middleburgh, Md	Feb. 3, 1838	592
Hides and skins, Reel for handling	W. Brown	Frankford and Philadelphia, Pa.	Apr. 17, 1828	
Hides and skins, Removing hair from	W. Anderson	Inverkeithing, Scotland	Oct. 8, 1872	131, 927
Hides and skins, Stuffing	W. Panton	Quincy, Mass	Jan. 29, 1867	61, 560
Hides and skins, Treatment of	J. Armstrong	Woburn, Mass	Feb. 15, 1859	22, 925
Hides and working leather, Machine for breaking	E. D. Taylor and W. Rude	Hornellsville, N. Y	Sept. 20, 1870	107, 562
Hides, Apparatus for handling	C. Weston	Salem, Mass	Feb. 22, 1859	23, 053
Hides, Apparatus for liming	S J. Miller, A. B. Barnett, and W. H. Study.	Economy, Ind	July 4, 1865	48, 738
Hides, Apparatus for removing hair and lime from	S. S. Weed	Stoneham, Mass	Sept. 5, 1865	49, 811
Hides, Apparatus for removing hair from	M. Bray	Boston, Mass	Aug. 22, 1865	49, 496
Hides, Apparatus for removing hair from	S. S. Weed	Stoneham, Mass	Sept. 5, 1865	49, 839
Hides, Apparatus for tanning	H. Liebermann	Louisville, Ky	June 21, 1864	43, 258
Hides, Apparatus for treating green	L. Mackey	Dunmore, Pa	May 8, 1866	54, 568
Hides, Apparatus for working	A. Adler	Paris, France	July 31, 1866	56, 687
Hides, Apparatus for working	H. Lampert	Rochester, N. Y	Mar. 15, 1870	100, 907
Hides, Bating and tanning	W. B. Milligan	Edinburgh, Va	Nov. 4, 1851	8, 500
Hides, Breaking and scraping	A. Smith	Cumberland Valley, Pa	Aug. 26, 1846	4, 712
Hides, Breaking, fleshing, and hairing	J. Dunaway	Woodville, Va	May 13, 1834	
Hides, Breaking, working, and scouring	J. Bonney	Cornwall, Conn	May 24, 1834	
Hides, Composition for unhairing	A. H. Ward, jr	Boston, Mass	Jan. 2, 1855	12, 151
Hides, Compound for bating and raising	G. W. Adler	Philadelphia, Pa	Feb. 1, 1870	99, 387
Hides, Compound for stuffing and tanning	E. England	Mossy Creek, Tenn	June 28, 1870	104, 719
Hides, Depilating composition for	P. H. Schlosser	Middletown, Md	Jan. 2, 1868	78, 543
Hides, Depilating compound for	A. K. Eaton	New York, N. Y	June 16, 1857	17, 562
Hides for belts, Preparing	J. J. Travis	Franklin, Conn	Oct. 22, 1828	
Hides for tanning, Process of scraping flesh and hair from.	T. Williams	Rochester, N. Y	Nov. 4, 1830	
Hides for tanning, Scraping the flesh and hair from	T. Williams	Rochester, N. Y	Oct. 31, 1831	
Hides for the manufacture of various articles, Process of preparing raw.	W. H. Towers	New York, N. Y	Aug. 21, 1866	57, 409
Hides from animals, Apparatus for removing	P. Lull	Norwich, N. Y	Apr. 14, 1868	76, 642
Hides from cattle, Machine for stripping the	C. Brühl	Green Point, N. Y	July 9, 1867	66, 558
Hides, Gage-frame for slitting raw	J. Hoffman	Belvidere, N. J	May 19, 1868	78, 092
Hides, Handling	A. H. Beschormann	New York, N. Y	Nov. 14, 1846	4, 851
Hides in tanning, Apparatus for handling	D. A. Haviland and A. S. Phillips	Fort Dodge, Iowa	Nov. 5, 1861	33, 645
Hides in the lime, Softening, breaking, and stoning	B. Aylsworth	Masonville, N. Y	Feb. 13, 1833	
Hides in the vats, Handling	S. Stem and D. Wierman	Mechanicstown, Md	Jan. 9, 1834	
Hides into hard transparent material for making combs, &c., Manufacturing raw or green.	S. Pike	Providence, R. I	June 11, 1829	
Hides, Knife for removing hair from	B. Harrington	Boston, Mass	July 3, 1866	56, 144
Hides, leather, &c., Apparatus for drying	A. W. Roberts	Hartford, Conn	June 11, 1867	65, 767
Hides, Machine for beaming	P. Lennox	Lynn, Mass	Aug. 11, 1868	80, 829
Hides, Machine for beaming	P. Lennox	Lynn, Mass	Nov. 8, 1870	109, 135
Hides, Machine for beaming	P. Lennox, H. H. Robbins, and E. Hayes.	Lynn, Mass	May 26, 1868	78, 380
Hides, Machine for breaking	C. Bauchman	North Whitehall, Pa	Apr. 16, 1850	7, 281
Hides, Machine for breaking	H. S. Clark	Randolph County, N. C	Dec. 22, 1826	
Hides, Machine for breaking	I. S. Hershey	Hagerstown, Md	Sept. 11, 1849	6, 710
Hides, Machine for breaking and scouring	O. W. Bean	Farmington, Tex	July 20, 1869	92, 776
Hides, Machine for breaking, hairing, and fleshing	N. Kirk and S. C. Clark	Saint Clairsville, Ohio	July 18, 1812	
Hides, Machine for cleaning hair from	A. Hasbrouck	Ithaca, N. Y	May 11, 1869	89, 864
Hides, Machine for cutting	J. C. Flint	Boston, Mass	Nov. 11, 1851	8, 510
Hides, Machines for dressing	H. L. Arnold	Elk Horn, Wis	Sept. 10, 1861	33, 229
Hides, Machines for dressing	E. Fitzhenry	Boston, Mass	Aug. 15, 1871	118, 002
Hides, Machine for handling	J. Hammond	Lattisburgh, Ohio	Aug. 11, 1868	80, 947
Hides, Machine for handling	B. B. Mereness	Georgetown, N. Y	Sept. 8, 1863	39, 824
Hides, Machine for handling	J. Snell, jr	Pottsville, Pa	Nov. 6, 1866	59, 469
Hides, Machine for handling	J. S. Wheat	South Wheeling, W. Va	Jan. 19, 1864	41, 336
Hides, Machine for preparing	T. W. Jones	Philomath, Ga	Feb. 4, 1851	7, 924
Hides, Machine for scraping	R. Shaler	Haddam, Conn	June 19, 1837	239
Hides, Machine for shaving	T. F. Weston	Salem, Mass	Mar. 26, 1867	63, 191

Index of patents issued from the United States Patent Office from 1790 *to* 1873, *inclusive*—Continued.

Index of patents issued from the United States Patent Office from 1790 *to* 1873, *inclusive*—Continued.

Invention.	Inventor.	Residence.	Date.	No.
Hinge	L. T. Howell	Burlington, N. J	Apr. 26, 1859	23, 808
Hinge	A. Huffer	Hagerstown, Md	Oct. 11, 1870	108, 143
Hinge	G. S. Hurford	Canton, Ohio	June 14, 1870	104, 316
Hinge	W. Johnson	Milwaukee, Wis	Aug. 9, 1870	106, 169
Hinge	W. Johnston	Cincinnati, Ohio	Sept. 7, 1869	94, 606
Hinge	F. B. Jones	Louisville, Ky	Aug. 9, 1870	106, 170
Hinge	J. Jones	Rochester, N. Y	Feb. 7, 1860	27, 056
Hinge	D. C. Jordan, sr	Brooklyn, N. Y	Apr. 24, 1866	54, 171
Hinge	A. C. Kasson	Milwaukee, Wis	Dec. 31, 1872	134, 383
Hinge	C. M. Lane	Cincinnati, Ohio	June 4, 1861	32, 482
Hinge	G. W. Lane	Plantsville, Conn	Aug. 20, 1867	67, 889
Hinge	M. R. Lemman	Hamilton, Ohio	Mar. 21, 1871	112, 934
Hinge	E. Lindsley	Neenah, Wis	Aug. 4, 1868	80, 639
Hinge	J. R. Lomas	West Haven, Conn	Apr. 3, 1866	53, 637
Hinge	D. W. Long	Parkersburgh, W. Va	May 31, 1870	103, 758
Hinge	D. W. Long	Parkersburgh, W. Va	Aug. 20, 1872	130, 729
Hinge	W. I. Ludlow	Washington, D. C	Apr. 1, 1873	137, 458
Hinge	P. P. Lynch	Newark, N. J	Apr. 29, 1873	138, 336
Hinge	H. Manneck	New York, N. Y	Oct. 14, 1873	143, 704
Hinge	J. C. Mason	New Hartford Centre, Conn	Feb. 16, 1858	19, 374
Hinge	T. D. McCall and S. Bushnell	Walton, N. Y	Sept. 1, 1868	81, 805
Hinge	W. C. McGill	Cincinnati, Ohio	Dec. 19, 1865	51, 604
Hinge	J. Miller	Watkins, N. Y	Nov. 7, 1871	120, 762
Hinge	A. L. Mora	New York, N. Y	Nov. 5, 1867	70, 597
Hinge	J. A. Morrell	Chicago, Ill	Sept. 10, 1867	68, 644
Hinge	F. Müsser	Cincinnati, Ohio	Jan. 7, 1873	134, 697
Hinge	G. R. Nebringer	Lewisberry, Pa	July 16, 1867	66, 729
Hinge	E. D. Norton	Cuba, N. Y	May 30, 1871	115, 509
Hinge	A. A. Oat	Philadelphia, Pa	Oct. 17, 1871	119, 944
Hinge	J. K. Otis	East Cambridge, Mass	May 7, 1872	126, 571
Hinge	J. K. Otis	East Cambridge, Mass	May 14, 1872	126, 647
Hinge	G. F. Outten	Norfolk, Va	Apr. 12, 1870	101, 908
Hinge	J. Parker	Meriden, Conn	Feb. 19, 1867	62, 219
Hinge	J. Plant	Washington, D. C	Aug. 9, 1870	106, 278
Hinge	J. Plegar	Birmingham, Pa	July 24, 1855	13, 321
Hinge	E. N. Porter	Morrisville, Vt	Dec. 12, 1865	51, 476
Hinge	J. W. Powers	Evanston, Ill	June 11, 1867	65, 601
Hinge	G. M. Ramsay	New York, N. Y	Jan. 29, 1856	14, 167
Hinge	A. Rankin	Philadelphia, Pa	Apr. 8, 1873	137, 623
Hinge	V. Rathknecht	Chicago, Ill	Mar. 12, 1872	124, 627
Hinge	D. Reed	San Francisco, Cal	July 16, 1872	129, 591
Hinge	R. Reiber	Lebanon, Pa	Jan. 7, 1868	73, 120
Hinge	S. M. Richardson	New York, N. Y	Dec. 6, 1859	26, 375
Hinge	J. M. Riley	Newark, N. J	Oct. 10, 1865	50, 423
Hinge	J. Schafer	New York, N. Y	Feb. 5, 1867	61, 766
Hinge	W. R. Searle	East Hampton, Mass	May 14, 1872	126, 650
Hinge	N. Sehner	Hagerstown, Md	July 4, 1865	48, 630
Hinge	S. Selden	Fairview Township, Pa	Feb. 13, 1866	52, 649
Hinge	S. Selden and W. J. F. Liddell	Erie, Pa	July 16, 1867	66, 393
Hinge	A. P. Seymour, jr	Hecla Works, N. Y	June 1, 1869	90, 880
Hinge	R. Seymour	Waterford, N. Y	Aug. 6, 1872	130, 248
Hinge	W. Shannon	Allegheny City, Pa	Dec. 8, 1868	84, 771
Hinge	W. Shannon	Allegheny City, Pa	Dec. 8, 1868	84, 772
Hinge	J. D. Shepard	Buffalo, N. Y	Sept. 3, 1872	131, 032
Hinge	J. D. Shepard and R. W. English	Buffalo, N. Y	July 13, 1869	92, 662
Hinge	T. Smith	Boston, Mass	Feb. 15, 1870	99, 999
Hinge	J. Sow[illegible]	Boston, Mass	Jan. 21, 1868	73, 661
Hinge	S. S. S[illegible]aire and T. Scharfenberg	Brooklyn, N. Y	Feb. 1, 1859	22, 830
Hinge	S. Stiger	New York, N. Y	May 14, 1872	126, 850
Hinge	E. Swan	New Bedford, Mass	May 14, 1872	126, 654
Hinge	A. W. Sweeny	Washington, D. C	June 26, 1860	28, 944
Hinge	L. D. Tice	New York, N. Y	Sept. 17, 1872	131, 478
Hinge	L. Upham	Pawtucket, R. I	Nov. 7, 1865	50, 861
Hinge	S. D. Van Pelt	Anderson, Ind	July 26, 1870	105, 743
Hinge	A. Velguth	Milwaukee, Wis	Sept. 28, 1869	95, 290
Hinge	W. Wakenshaw	Newark, N. J	Apr. 2, 1872	125, 354
Hinge	F. H. Walker	Boston, Mass	Mar. 30, 1869	88, 351
Hinge	B. D. Washburn	Roxbury, Mass	Oct. 29, 1867	70, 299
Hinge	W. Webb	Waterbury, Conn	Aug. 2, 1864	43, 746
Hinge	W. T. Wells	Decatur, Ill	Oct. 23, 1866	59, 107
Hinge	W. T. Wells	Decatur, Ill	Dec. 19, 1871	122, 088
Hinge	D. Werner	Saint Louis, Mo	Jan. 11, 1870	98, 827
Hinge	S. M. Wheeler	Dowagiac, Mich	Apr. 4, 1871	113, 378
Hinge	A. Whelan	Washington, D. C	Jan. 19, 1869	85, 977
Hinge	A. Wirth	New York, N. Y	Aug. 26, 1873	142, 315
Hinge	J. W. Wood	Conshohocken, Pa	July 1, 1873	140, 449
Hinge	H. Young and M. Stachelin	Port Chester, N. Y	Oct. 10, 1865	50, 411
Hinge	H. M. Zimmerman	Washington, D. C	June 26, 1860	28, 937
Hinge	J. H. Zinn	Idaville, Pa	Mar. 5, 1867	62, 72[illegible]
Hinge, Adjustable	J. G. Ralph	Aurora, Ill	Aug. 20, 1867	67, 904
Hinge and blind support	W. W. S. Orbeton	Haverhill, Mass	Aug. 20, 1867	67, 999
Hinge and fastener	N. B. Spooner	Plymouth, Mass	Oct. 29, 1867	70, 282
Hinge and fastener combined	L. Felker	Tewkesbury, Mass	Oct. 27, 1868	83, 369
Hinge and fastening, Combined shutter	W. S. Gerard	Newburgh, N. Y	Oct. 3, 1865	50, 239
Hinge and fastening, Window-blind	W. Baker	Utica, N. Y	Sept. 17, 1842	2, 776
Hinge and fastening, Window-blind	C. Reed	Cambridge, Mass	Aug. 1, 1848	5, 683
Hinge and hook	F. H. Cuypers	Newark, N. J	Mar. 4, 1862	34, 565
Hinge and k[illegible]ob, Shutter	J. W. Tripp	Gallipolis, Ohio	Jan. 12, 1869	85, 772
Hinge and ring for lockets	G. Hartje	New York, N. Y	July 12, 1870	105, 201
Hinge and spring combined	J. J. Cowell	Newark, N. J	May 25, 1869	90, 504
Hinge and spring, Combined door	W. Gilfillan	Syracuse, N. Y	Sept. 30, 1873	143, 234
Hinge and spring, Combined double	A. B. Taft	New York, N. Y	Jan. 23, 1849	6, 052
Hinge, Angular strap	C. F. Hawley	Kansas City, Mo	June 13, 1871	115, 851
Hinge, Bag	G. Havell	Newark, N. J	June 25, 1867	66, 150
Hinge, Blind	M. Adler	Buffalo, N. Y	Nov. 3, 1868	83, 585
Hinge, Blind	W. L. Barnes	Kingston, N. Y	June 16, 1863	38, 876
Hinge, Blind	D. Bull	Amboy, Ill	Nov. 5, 1872	132, 748

Index of patents issued from the United States Patent Office from 1790 *to* 1873, *inclusive*—Continued.

Invention.	Inventor.	Residence.	Date.	No.
Hinge, Blind	J. L. Cathcart	Georgetown, D. C	Feb. 23, 1869	87, 243
Hinge, Blind	C. B. Clark	Buffalo, N. Y	Mar. 24, 1868	75, 862
Hinge, Blind	C. B. Clark	Buffalo, N. Y	Oct. 13, 1868	82, 920
Hinge, Blind	C. B. Clark	Buffalo, N. Y	Nov. 3, 1868	83, 603
Hinge, Blind	C. B. Clark	Buffalo, N. Y	May 25, 1869	90, 500
Hinge, Blind	C. B. Clark	Buffalo, N. Y	Feb. 15, 1870	99, 844
Hinge, Blind	A. Eccard	Washington, D. C	July 19, 1870	105, 441
Hinge, Blind	W. T. Freligh	Jersey City, N. J	Sept. 21, 1869	95, 101
Hinge, Blind	O. S. Garretson	Buffalo, N. Y	Apr. 9, 1872	125, 450
Hinge, Blind	O. S. Garretson	Buffalo, N. Y	Dec. 23, 1873	145, 797
Hinge, Blind	L. Gathmann	Chicago, Ill	Nov. 25, 1873	144, 898
Hinge, Blind	A. Huffer	Rochester, N. Y	Feb. 18, 1873	136, 061
Hinge, Blind	T. C. Lord	Chicago, Ill	Sept. 28, 1869	95, 363
Hinge, Blind	W. McKee and C. H. Jordan	Washington, D. C	Nov. 10, 1868	83, 869
Hinge, &c., Blind	J. Plant	Washington, D. C	Apr. 24, 1847	5, 080
Hinge, Blind	R. B. Prindle	Norwich, N. Y	Sept. 15, 1868	82, 245
Hinge, Blind	D. C. Sage	Middletown, Conn	Mar. 1, 1870	100, 453
Hinge, Blind	D. C. Sage	Middletown, Conn	June 14, 1870	104, 359
Hinge, Blind	A. L. Seabury and D. Morris	Norfolk, Va	Oct. 26, 1869	96, 273
Hinge, Blind	C. G. Shepard	Buffalo, N. Y	Sept. 24, 1872	131, 571
Hinge, Blind	E. R. Shepard	Scranton, Pa	Oct. 2, 1860	30, 281
Hinge, Blind	A. Velguth	Milwaukee, Wis	Oct 26, 1869	96, 171
Hinge, Blind-fastening	J. H. Nevins	Brooklyn, E. D., N. Y	Aug. 9, 1870	106, 272
Hinge, Box	R. Frisbie	Cromwell, Conn	Sept. 21, 1869	95, 103
Hinge, Butt	M. C. Ames	Hartford, Conn	June 11, 1867	65, 717
Hinge, Butt	M. A. Avery	Groton, N. Y	Feb. 25, 1868	74, 788
Hinge, Butt	B. F. Barker	San Francisco, Cal	Aug. 27, 1867	68, 150
Hinge, Butt	E. Boileau and C. Mesnier	Saint Louis, Mo	Mar. 23, 1869	88, 122
Hinge, Butt	M. Bush	Bloomington, Ill	Apr. 27, 1869	89, 464
Hinge, Butt	J. H. Carkeet	Montgomery, Ala	Nov. 3, 1868	83, 760
Hinge, Butt	J. W. Carleton	New Britain, Conn	Mar. 7, 1871	112, 417
Hinge, Butt	G. F. J. Colburn	Newark, N. J	Aug. 2, 1864	43, 670
Hinge, Butt	C. Cole	Ithaca, N. Y	Jan. 12, 1869	85, 728
Hinge, Butt	G. H. Fayman	Washington, D. C	Apr. 10, 1860	27, 788
Hinge, Butt	C. H. Foster	San Francisco, Cal	Nov. 2, 1869	96, 414
Hinge, Butt	C. Goodrich	Plainsville, Conn	Aug. 13, 1867	67, 747
Hinge, Butt	R. Hoadley	Derby, Conn	Feb. 16, 1864	41, 664
Hinge, Butt	H. Hockemeyer	Toledo, Ohio	Oct. 6, 1868	82, 717
Hinge, Butt	J. C. Hyde	West Haven, Conn	June 26, 1866	55, 954
Hinge, Butt	G. A. Jenks	Chicago, Ill	Dec. 1, 1868	84, 630
Hinge, Butt	J. Johnson	Geneseo, N. Y	Dec. 18, 1860	30, 970
Hinge, Butt	J. W. Jordan	Lexington, Va	Dec. 29, 1868	85, 310
Hinge, Butt	B. C. Pole	Richmond, Va	Nov. 3, 1868	83, 659
Hinge, Butt	A. Rankin	Philadelphia, Pa	Jan. 22, 1867	61, 458
Hinge, Butt	J. Rouse	Troy, N. Y	May 17, 1836	....
Hinge, Butt	D. C. Sage	Middletown, Conn	Nov. 12, 1867	70, 748
Hinge, Butt	J. E. Shields	Washington, D. C	May 29, 1860	28, 510
Hinge, Butt	C. Sholl	Mount Joy, Pa	July 30, 1861	32, 927
Hinge, Butt	I. L. Thompson	Sardis, Ohio	Jan. 7, 1873	134, 617
Hinge, Butt	J. F. Townsend	Westfield, N. Y	Oct. 28, 1862	36, 883
Hinge, Butt	S. Van Gilder	Knoxville, Tenn	June 17, 1873	139, 982
Hinge, &c., Butt	W. Whittaker	Troy, N. Y	Mar. 12, 1836	....
Hinge butts, Finishing loose	E. Luther, P. Lyon, and W. Edwards.	West Troy, N. Y	Dec. 22, 1868	85, 112
Hinge, Carriage	G. W. Beers	Bridgeport, Conn	July 3, 1866	55, 987
Hinge, Carriage	C. E. Schwind	New York, N. Y	Aug. 27, 1867	68, 121
Hinge, Carriage-door	G. W. Beers	Bridgeport, Conn	Mar. 7, 1871	112, 313
Hinge, Carriage-door	E. Wells	New Haven, Conn	Apr. 2, 1872	125, 236
Hinge, Carriage-top	H. Killam	New Haven, Conn	Oct. 12, 1869	95, 808
Hinge, Cast	N. Gates	Cincinnati, Ohio	Aug. 29, 1854	11, 606
Hinge-clamp	E. L. Seger and S. L. Smith	Yonkers, N. Y	Sept. 3, 1867	68, 533
Hinge, Compound	H. Loth	Philadelphia, Pa	July 15, 1873	140, 930
Hinge, Compound safe-door	P. J. Stuhltrager	Philadelphia, Pa	Sept. 29, 1868	82, 658
Hinge, Concealed	G. R. Cady and W. H. Cooper	New Haven, Conn	Nov. 26, 1867	71, 452
Hinge, Concealed	F. Wood	Bridgeport, Conn	Mar. 15, 1864	41, 956
Hinge, Door	S. M. Bullard	Holliston, Mass	June 2, 1857	17, 419
Hinge, Door	C. Dupré	Louisville, Ky	Jan. 7, 1868	73, 174
Hinge, Door	J. Elgar	Baltimore, Md	Apr. 11, 1854	10, 774
Hinge, Door	O. Judd	Cherry Valley, N. Y	Apr. 17, 1847	5, 066
Hinge, Door	G. Lane	New York, N. Y	Feb. 11, 1868	74, 384
Hinge, Door	C. H. Robison	Syracuse, N. Y	May 30, 1848	5, 608
Hinge, Door	A. W. Walter	Canton, Ohio	Oct. 27, 1868	83, 572
Hinge, Double	J. S. Jenness	Bangor, Me	Jan. 3, 1871	110, 659
Hinge, Double-acting	J. H. Carkeet	Montgomery, Ala	July 14, 1868	79, 950
Hinge, Double-acting	G. Dumbolton	Baltimore, Md	Apr. 7, 1868	76, 422
Hinge, Double-acting spring	H. E. Canfield	New York, N. Y	May 15, 1855	12, 853
Hinge, Double-acting spring	T. F. Engelbrecht	New York, N. Y	Feb. 25, 1851	7, 953
Hinge, Double-acting spring	T. F. Engelbrecht	New York, N. Y	Jan. 3, 1854	10, 389
Hinge, Elliptical	M. and R. Reeve	Burlington, N. J	Oct. 31, 1812	....
Hinge-fastener and shutter-opener, Combined	A. S. Pelton	Clinton, Conn	Jan. 23, 1849	6, 059
Hinge-finishing machine, Butt	J. J. Crooke	New York, N. Y	Feb. 26, 1867	62, 479
Hinge for blinds, Lock	T. Clucas, jr	Saint Louis, Mo	Sept. 9, 1873	142, 679
Hinge for carriage-doors, Concealed	E. Wells	New Haven, Conn	Jan. 23, 1872	122, 977
Hinge for daguerreotype-cases	E. G. Kingsley and S. A. W. Parker, jr.	Stoughton, Mass	June 1, 1858	20, 436
Hinge for doors, Helical-spring-joint	D. A. Hoyt and P. W. Bulkeley	Danbury, Conn	Apr. 28, 1838	721
Hinge for doors or window-frames of stoves	T. J. Coulston	Springfield, Pa	Nov. 9, 1869	96, 551
Hinge for doors, shutters, &c., Crane	E. Ripley	Troy, N. Y	May 13, 1851	8, 090
Hinge for doors, Spring	R. Adams	Borough of Southwark, Great Britain.	June 20, 1871	116, 002
Hinge for doors, &c., Spring	A. Rankin	Philadelphia, Pa	Sept. 27, 1870	107, 813
Hinge for doors, &c., Spring	A. F. Schiffling	Evansville, Ind	Sept. 27, 1870	107, 819
Hinge for fastening blinds, shutters, and doors	R. B. Varden	Baltimore, Md	Feb. 12, 1845	3, 903
Hinge for folding bedsteads	J. Binder	Chelsea, Mass	Aug. 16, 1853	9, 933
Hinge for gates and doors	D. S. Esten	Monson, Mass	Apr. 14, 1868	76, 727
Hinge for inkstand-covers	J. Nock	Philadelphia, Pa	Dec. 13, 1853	10, 310
Hinge for ladder-brackets	J. H. Balsley	Dayton, Ohio	Feb. 8, 1870	99, 622
Hinge for lamp-burners	W. Webb	Waterbury, Conn	June 18, 1867	65, 972

Index of patents issued from the United States Patent Office from 1790 *to* 1873, *inclusive*—Continued.

Invention.	Inventor.	Residence.	Date.	No.
Hinge for landau carriages, Concealed	E. Wells	New Haven, Conn	June 7, 1870	103, 948
Hinge for molders' flasks	G. Grant	Troy, N. Y	Dec. 7, 1852	9, 448
Hinge for molders' flasks	E. C. Little	Saint Louis, Mo	Oct. 16, 1866	58, 948
Hinge for piano-fortes, &c	R. Perkins	Boston, Mass	June 3, 1843	3, 122
Hinge for picture-cases	A. P. Critchlow	Florence, Mass	Oct. 14, 1856	15, 915
Hinge for reflectors of stereoscopes, &c	A. Beckers	New York, N. Y	Mar. 29, 1859	23, 342
Hinge for safe and vault doors	H. Gross	Cincinnati, Ohio	Nov. 18, 1873	144, 669
Hinge for safe-doors, Sliding	W. H. Butler	Brooklyn, N. Y	May 28, 1872	127, 147
Hinge for school-desks	A. E. Roberts	Des Moines, Iowa	Sept. 27, 1870	107, 724
Hinge for school-desks and settee-seats	N. O. Tiffany	Buffalo, N. Y	Dec. 20, 1870	110, 402
Hinge for sewing-machine tables	S. C. Brinser	Middletown, Pa	Aug. 13, 1872	130, 472
Hinge for sewing-machine tables	J. C. Gove	Cleveland, Ohio	Oct. 8, 1872	132, 069
Hinge for sheet-metal boxes	C. J. Hauck	Williamsburgh, N. Y	Mar. 24, 1868	75, 903
Hinge for show-cases	J. McAdams	Brooklyn, N. Y	Feb. 1, 1870	99, 453
Hinge for sleeping-car berths	J. I. Fogg	Chicago, Ill	Apr. 14, 1868	76, 620
Hinge for step-ladders	C. G. Udell	Chicago, Ill	Oct. 1, 1872	131, 798
Hinge for stove-lids	T. J. Coulston	Springville, Pa	Aug. 10, 1869	93, 600
Hinge for stoves, Quadrant	E. Barrows	Boston, Mass	Aug. 31, 1837	366
Hinge for sustaining, operating, closing, and fastening window-blinds.	L. T. Talbot	Taunton, Mass	Apr. 18, 1848	5, 521
Hinge for table-leaves	D. A. Garver	Bryan, Ohio	Apr. 2, 1872	125, 281
Hinge for table-leaves, &c	P. Hires	Columbus, Ky	Apr. 25, 1871	114, 009
Hinge for tables, &c., Lock	E. E. Hendrick	Carbondale, Pa	Mar. 12, 1872	124, 438
Hinge for tea-kettle covers, &c	J. Easterly	Albany, N. Y	Apr. 16, 1867	63, 874
Hinge for window-blinds	J. London and H. Iversen	New York, N. Y	Aug. 31, 1858	21, 347
Hinge for window-blinds	T. E. Williams	Washington, D. C	Nov. 30, 1858	22, 214
Hinge for window-shutters	P. P. Child	Saint Louis, Mo	Dec. 24, 1867	72, 603
Hinge for window-shutters	W. H. Foulds	Henderson, Ky	Nov. 13, 1866	59, 581
Hinge-forming die	J. O'Kane	New York, N. Y	Apr. 9, 1867	63, 742
Hinge, Friction-roller door	J. Bailey, jr	Plymouth County, Mass	May 16, 1808	....
Hinge, Gate	C. E. Burnham	Binghamton, N. Y	Nov. 2, 1858	21, 939
Hinge, Gate	E. A. Bushnell	Burnett, Wis	July 16, 1872	129, 209
Hinge, &c., Gate	P. P. Child	Saint Louis, Mo	Oct. 31, 1871	120, 492
Hinge, Gate	C. B. Clark	Buffalo, N. Y	Feb. 7, 1871	111, 515
Hinge, Gate	P. Dennis	Schuylerville, N. Y	Feb. 25, 1868	74, 805
Hinge, Gate	J. W. Everham	Pitt's Grove, N. J	May 11, 1869	89, 978
Hinge, Gate	J. B. Farmer	Indianapolis, Ind	Feb. 9, 1869	86, 658
Hinge, Gate	W. Field and R. Carruthers	Bergen, N. J	Oct. 30, 1866	59, 202
Hinge, Gate	W. G. Franklin	Shelbina, Mo	Apr. 22, 1873	138, 081
Hinge, Gate	G. Garrett	Elkhart, Ill	Apr. 25, 1871	113, 998
Hinge, Gate	D. H. Gould	Troy, N. Y	Sept. 3, 1867	68, 434
Hinge, Gate	B. Greenside	Fort Dodge, Iowa	Oct. 30, 1866	59, 211
Hinge, Gate	A. T. Hendrick	Clyde, N. Y	Sept. 14, 1858	21, 496
Hinge, Gate	C. Hermance	Schuylerville, N. Y	Feb. 19, 1867	62, 196
Hinge, Gate	W. Hull	Fort Wayne, Ind	Oct. 21, 1873	143, 764
Hinge, Gate	H. Last	West Lebanon, Ind	June 12, 1866	55, 505
Hinge, Gate	A. D. McMaster	Rochester, N. Y	July 11, 1871	116, 851
Hinge, Gate	J. Miller	Penn Yan, N. Y	Dec. 10, 1872	133, 715
Hinge, Gate	E. D. Norton	Cuba, N. Y	Apr. 18, 1871	113, 788
Hinge, Gate	S. J. Olmsted	Binghamton, N. Y	Mar. 26, 1861	31, 852
Hinge, Gate	W. W. Robinson	Ripon, Wis	June 9, 1863	38, 843
Hinge, Gate	I. J. Ryerson	Princeton, Ind	June 8, 1869	91, 167
Hinge, Gate	J. H. Stone	Chapel Hill, Tex	Apr. 15, 1873	137, 972
Hinge, Gate	J. Sweet	Decatur, Mich	Sept. 28, 1869	95, 393
Hinge, Gate	D. Wadsworth, jr	Nashua, N. H	Nov. 6, 1860	30, 600
Hinge, Gate	J. T. Watson	Richmond, Ind	Oct. 13, 1863	40, 314
Hinge, Gate	J. T. Watson	Richmond, Ind	Jan. 1, 1867	60, 808
Hinge, Gate	L. E. Woodard	Cohocton, N. Y	Oct. 16, 1866	58, 930
Hinge hooks, Machine for making blind	G. Orr	Needham, Mass	Oct. 17, 1871	119, 945
Hinge-hooks, Manufacture of	C. Kanebeck	Boston, Mass	Jan. 21, 1868	73, 533
Hinge-hooks, Manufacture of	W. J. Lewis	Pittsburgh, Pa	Aug. 13, 1872	130, 508
Hinge hooks, Manufacture of blind	W. H. McIntosh	Needham Upper Falls, Mass.	Apr. 30, 1872	126, 150
Hinge-joints or coupling-shafts of wagons, Shifting	E. W. Seymour	Centre Lisle, N. Y	Feb. 25, 1862	34, 553
Hinge, Latch	J. Plant	Washington, D. C	May 3, 1859	23, 859
Hinge, Lock	C. B. Clark	Buffalo, N. Y	July 29, 1873	141, 323
Hinge, Lock	M. C. Lee	New York, N. Y	Apr. 22, 1873	138, 092
Hinge, Loose-joint butt	M. Perry	New York, N. Y	Oct. 1, 1867	69, 365
Hinge, Lounge	A. Spiegel	Indianapolis, Ind	Sept. 16, 1873	142, 957
Hinge-machine	J. H. Baird	Oakville, Conn	Mar. 29, 1870	101, 210
Hinge-machine	E. Brown	New York, N. Y	Nov. 24, 1868	84, 256
Hinge-making machine	E. Brown	Waterbury, Conn	May 16, 1854	10, 943
Hinge-making machine	J. B. Evrard and J. P. Boyer	Paris, France	June 19, 1866	55, 794
Hinge-making machine	D. Morse		June 10, 1803	....
Hinge-making machine	L. P. Summers	New Britain, Conn	Dec. 10, 1872	133, 903
Hinge-making machine, Butt	H. D. Blake	New Britain, Conn	Mar. 24, 1868	75, 724
Hinge-making machine, Butt	E. Brown and W. H. Van Gieson.	Waterbury, Conn., and Paterson, N. J.	Apr. 9, 1861	31, 949
Hinge-making machine, Butt	C. Miller	New York, N. Y	Feb. 20, 1855	12, 443
Hinge-making machine, Butt	A. Rais	Waterbury, Conn	June 25, 1867	66, 172
Hinge-making machine, Butt	A. Rais	Waterbury, Conn	July 9, 1867	66, 626
Hinge-making machine, Butt	A. Rais	Waterbury, Conn	Sept. 3, 1867	68, 529
Hinge-making machine, Butt	W. H. Van Gieson and J. J. Crooke.	Passaic, N. J., and New York, N. Y.	Mar. 10, 1868	75, 320
Hinge-making machine, Strap	W. J. Lewis	Pittsburgh, Pa	Apr. 25, 1871	114, 019
Hinge of rolling iron shutters	A. L. Johnson	Baltimore, Md	June 25, 1850	7, 457
Hinge, Pin	A. Lyons and I. B. Abrahams	New York, N. Y	Apr. 29, 1873	138, 418
Hinge-pintle	W. Wells	Ashtabula, Ohio	Mar. 9, 1869	87, 739
Hinge-pivots, Machine for forging solid felted	J. J. and L. Crooke	Southfield and New York, N. Y.	Sept. 16, 1873	142, 774
Hinge, Plated butt	J. J. and L. Crooke	Southfield and New York, N. Y.	Apr. 4, 1871	113, 501
Hinge, Quadrant	N. Jones	Lockport, Ind	Aug. 24, 1869	94, 000
Hinge, Reversible	P. P. Child	Saint Louis, Mo	Sept. 26, 1871	119, 319
Hinge, Reversible	F. W. Marston	Philadelphia, Pa	May 9, 1871	114, 579
Hinge, Reversible	A. P. Seymour	Hecla Works, N. Y	Mar. 1, 1870	100, 332
Hinge, Reversible	H. J. Wolters	Salem, Mass	May 18, 1869	90, 330
Hinge, Reversible butt	D. F. Hale	Chicopee, Mass	May 3, 1870	102, 538

Index of patents issued from the United States Patent Office from 1790 *to* 1873, *inclusive*—Continued.

Invention.	Inventor.	Residence.	Date.	No.
Hinge, Reversible butt	W. Hancock	Saco, Me	Jan. 22, 1867	61,421
Hinge, Reversible butt	W. S. Weed	Auburn, N. Y	Sept. 24, 1867	69,287
Hinge-riveting machine	H. D. Blake	New Britain, Conn	Mar. 17, 1868	75,663
Hinge-riveting machine	E. H. Lacy and P. Lyon	West Troy, N. Y	Jan. 16, 1872	122,837
Hinge-riveting machine	H. M. Ritter	Covington, Ky	Dec. 15, 1868	84,907
Hinge-riveting machine	I. W. Valance and H. Littlejohn.	Lansingburgh and New York, N. Y.	Aug. 27, 1861	33,174
Hinge, Safe-door	J. L. Hall	Cincinnati, Ohio	Feb. 25, 1873	136,239
Hinge, Safe-door	W. F. Stevens	Melrose, Mass	Dec. 30, 1873	145,973
Hinge, Seat	M. W. Chase	Buffalo, N. Y	July 22, 1873	141,115
Hinge, Seat	W. A. Slaymaker	Atlanta, Ga	Nov. 28, 1871	121,259
Hinge, Self-closing	C. Vetter	Cincinnati, Ohio	Nov. 12, 1867	70,921
Hinge, Self-fastening shutter	A. Nicholson	Poland, N. Y	Mar. 21, 1854	10,673
Hinge, Self-lock window	J. R. Murphy	Pittsburgh, Pa	Sept. 29, 1863	40,145
Hinge, Self-locking	G. E. Boisselier	Saint Louis, Mo	Sept. 20, 1870	107,437
Hinge, Self-locking blind	O. S. Garretson	Buffalo, N. Y	Jan. 9, 1872	122,594
Hinge, Self-locking shutter	E. H. Benjamin	Oak Hill, N. Y	June 23, 1868	79,051
Hinge, Sheet-metal	W. H. Hart	New Britain, Conn	June 3, 1873	139,464
Hinge, Shutter	D. G. Coppin	Cincinnati, Ohio	Sept. 26, 1865	50,099
Hinge, Shutter	R. Covington	Washington, D. C	Nov. 6, 1866	59,365
Hinge, Shutter	I. Davis	Groton, N. Y	Mar. 4, 1856	14,349
Hinge, Shutter	H. F. Drott	Cumberland, Md	Oct. 11, 1859	25,728
Hinge, Shutter	S. Drum	Allegheny City, Pa	June 20, 1865	48,264
Hinge, Shutter	T. F. Duley	Olney, Md	Oct. 22, 1872	132,512
Hinge, Shutter	C. H. Forbes and W. F. Rutter	Philadelphia, Pa	Apr. 26, 1870	102,242
Hinge, Shutter	W. Fosket	Meriden, Conn	Nov. 25, 1873	144,971
Hinge, Shutter	O. S. Garretson	Buffalo, N. Y	June 19, 1866	55,644
Hinge, Shutter	J. Hof and P. Brenners	Baltimore, Md	Apr. 19, 1870	102,125
Hinge, Shutter	C. F. Knauer	Pittsburgh, Pa	June 6, 1865	48,075
Hinge, Shutter	R. Lee	Cincinnati, Ohio	Feb. 7, 1865	46,251
Hinge, Shutter	H. Lull	South Coventry, Conn	Jan. 31, 1854	10,477
Hinge, Shutter	H. L. Norton	Middletown, Conn	June 4, 1872	127,425
Hinge, Shutter	J. Slusser	Cincinnati, Ohio	May 8, 1866	54,618
Hinge, Shutter	J. Smith, jr	Allegheny City, Pa	Oct. 28, 1862	36,865
Hinge, Shutter	E. L. Tevis	Philadelphia, Pa	Apr. 16, 1867	63,962
Hinge, Sliding	J. M. Eveleth and G. P. Moore	Oroville, Cal	Feb. 17, 1863	37,684
Hinge, Spiral or worm-joint	P. G. Bates	Waterbury, Conn	Jan. 17, 1854	10,425
Hinge, Spring	A. Acker	Ramapo, N. Y	July 17, 1860	29,212
Hinge, Spring	J. Bidwell	San Francisco, Cal	June 14, 1870	104,252
Hinge, Spring	N. Birdsall	Port Jervis, N. Y	June 28, 1859	24,535
Hinge, Spring	L. Bommer	New York, N. Y	Dec. 8, 1863	40,879
Hinge, Spring	A. Crawford	Boston, Mass	Dec. 10, 1872	133,699
Hinge, Spring	J. T. Garlick	New York, N. Y	Dec. 23, 1856	16,272
Hinge, Spring	D. B. Hawkes	Providence, R. I	May 16, 1871	114,813
Hinge, Spring	H. N. Hemingway	Rochester, N. Y	Sept. 3, 1872	190,998
Hinge, Spring	F. Herrmann	Newport, Ky	Mar. 12, 1867	62,747
Hinge, Spring	V. S. Herzog	Baltimore, Md	Apr. 6, 1869	88,567
Hinge, Spring	W. Hoar	Floyd, Iowa	Sept. 2, 1873	142,392
Hinge, Spring	S. Joyce	New York, N. Y	Sept. 16, 1873	142,855
Hinge, Spring	C. W. Lawrence	Milton, Ind	Feb. 19, 1867	62,140
Hinge, Spring	J. Maxson	De Ruyter, N. Y	Nov. 10, 1857	18,593
Hinge, Spring	H. B. Middaugh	Mansfield, Pa	June 1, 1869	90,769
Hinge, Spring	A. M. Olds	New York, N. Y	June 25, 1867	66,103
Hinge, Spring	J. Palmer	Cincinnati, Ohio	Apr. 15, 1878	137,951
Hinge, Spring	D. Renshaw	Brooklyn, N. Y	Jan. 11, 1870	98,709
Hinge, Spring	C. I. Rooney and D. Renshaw	New York, N. Y	Sept. 27, 1859	25,585
Hinge, Spring	H. W. Sabin and G. Drew	Canandaigua, N. Y	Feb. 25, 1851	7,957
Hinge, Spring	N. Sehner	Hagerstown, Md	May 9, 1871	114,719
Hinge, Spring	N. Sehner	Hagerstown, Md	Aug. 1, 1871	117,691
Hinge, Spring	J. S. Smith	New York, N. Y	May 19, 1857	17,354
Hinge, Spring	B. Turner	London, England	May 7, 1872	126,593
Hinge, Spring	W. Wells	Cleveland, Ohio	Dec. 6, 1870	109,855
Hinge, Spring	A. Wiswall	New York, N. Y	June 18, 1867	65,980
Hinge, Spring butt	A. Acker	Ramapo, N. Y	Nov. 18, 1862	36,976
Hinge, Spring butt	H. A. Clark	Boston, Mass	Mar. 25, 1873	137,178
Hinge-stock, Composition	G. R. Sillibridge	New York, N. Y	Oct. 10, 1829	....
Hinge, Stop	G. F. J. Colburn	Newark, N. J	Sept. 6, 1864	44,076
Hinge, Stop	H. Pennie	Buffalo, N. Y	Nov. 6, 1860	30,587
Hinge, Stop	G. C. Thomas	Brooklyn, N. Y	Dec. 9, 1873	145,315
Hinge-stop	M. Umstadter	Norfolk, Va	Jan. 12, 1869	85,876
Hinge, Stove-door, &c	C. J. Woolson	Cleveland, Ohio	Mar. 16, 1852	8,814
Hinge, Table	E. C. Brinser	Middletown, Pa	Aug. 26, 1873	142,077
Hinge, Table	M. C. Brinser	Lancaster, Pa	Apr. 16, 1872	125,657
Hinge, Table	W. F. Daly	Peru, Ind	Aug. 26, 1873	142,210
Hinge, Table	J. W. Palmer	Pana, Ill	July 15, 1873	140,947
Hinge, Telescopic quadrant	A. S. Blake	Waterbury, Conn	Aug. 10, 1869	93,402
Hinge, Trunk	A. M. Darrell	Washington, D. C	Feb. 18, 1873	136,040
Hinge, Trunk	O. Luce	Cortland, N. Y	Nov. 3, 1868	83,647
Hinge, Trunk	W. Wakenshaw and T. R. Dunham.	Newark, N. J	Apr. 10, 1866	53,908
Hinge, Trunk	L. Wheaton	Auburn, N. Y	Nov. 19, 1867	71,254
Hinge, Trunk and box	B. Hogan	Albany, N. Y	Nov. 26, 1872	133,313
Hinge, Wrought-iron blind	W. R. Goodrich	Utica, N. Y	Mar. 11, 1873	136,650
Hinges and tubes, Machine for making	W. Shaw	Buffalo, N. Y	Oct. 22, 1835	....
Hinges, Arrangement of friction-roller in inclined-plane.	E. Woolman	Damascoville, Ohio	May 9, 1854	10,900
Hinges, Casting	N. A. Fenner	Providence, R. I	Dec. 22, 1857	18,896
Hinges, Casting	C. M. Lane	Cincinnati, Ohio	July 20, 1858	20,948
Hinges, Casting	G. A. Rogers	Newburyport, Mass	July 3, 1810	....
Hinges, Casting	D. Stuart and C. C. Lloyd	Danville, Pa	Jan. 20, 1843	2,917
Hinges, Casting auger gate	B. Roux	Cincinnati, Ohio	Feb. 8, 1870	99,712
Hinges, Casting butt	R. W. Belson	Danville, Pa	July 26, 1843	3,195
Hinges, Casting butt, &c	T. Shepherd	Philadelphia, Pa	Oct. 9, 1841	2,288
Hinges, Casting iron	C. Alger	Boston, Mass	May 18, 1814	....
Hinges, Cutting out strap	S. M. Richardson	New York, N. Y	June 7, 1859	24,330
Hinges, Device for inserting the pins in butt	L. P. Summers	New Britain, Conn	July 19, 1870	105,513
Hinges, Fastening metal plates upon door	W. Whiting	Brooklyn, N. Y	Oct. 29, 1867	70,304
Hinges, Flask for molding	T. Loring	Gloucester, N. J	Feb. 7, 1844	3,427

Index of patents issued from the United States Patent Office from 1790 *to* 1873, *inclusive*—Continued.

Invention.	Inventor.	Residence.	Date.	No.
Hinges, Implement for expelling the joint-pin from	W. H. H. Barton	Uxbridge, Mass	Dec. 10, 1872	133, 821
Hinges, Machine for countersinking holes in butt	L. P. Summers	New Britain, Conn	Dec. 6, 1870	109, 851
Hinges, Machine for forming the eyes of	D. W. Lyon	West Troy, N. Y	Sept. 11, 1849	6, 704
Hinges, Machine for forming the eyes of butt	W. Whitaker	Albany, N. Y	July 8, 1843	3, 156
Hinges, Machine for grinding butt	C. Kenney and W. Gurley	Troy, N. Y	July 1, 1856	15, 241
Hinges, Machine for manufacturing	W. F. Lewis	Waterbury, Conn	Jan. 17, 1871	110, 983
Hinges, Machine for manufacturing butt and other kinds of.	C. R. Macy	Hyde Park, N. Y	Oct. 5, 1838	966
Hinges, Machine for milling the knuckles of butt	E. Parker	New Britain, Conn	July 19, 1870	105, 487
Hinges, Machine for wiring	E. Luther	West Troy, N. Y	Oct. 12, 1869	95, 700
Hinges, Machinery for manufacturing	G. H. Horton and L. Armstrong	Hartford, Conn	Sept. 11, 1847	5, 293
Hinges made of wrought-iron, &c., Machine for bending the knuckles of butt.	C. Kenney	Troy, N. Y	Aug. 23, 1844	3, 717
Hinges, Manufacture of	L. Crooke	New York, N. Y	Jan. 9, 1872	122, 516
Hinges, Manufacturing butt	H. P. Anderson and J. Sawyer	Waterford, N. Y	June 18, 1833	....
Hinges, Manufacturing butt	D. Ellis	Attleborough, Mass	Aug. 8, 1834	....
Hinges, Manufacturing table butt	H. Treadwell	Hyde Park, N. Y	Dec. 31, 1833	....
Hinges, Material for manufacture of butt	O. S. Judd	New Britain, Conn	June 14, 1864	43, 117
Hinges, Mold for butt	B. F. Harley and J. D. Morris	Philadelphia, Pa	Feb. 12, 1844	3, 435
Hinges, Mold for casting butt	T. Shepherd and T. Loring	Philadelphia, Pa	Mar. 16, 1841	2, 005
Hinges, Mold for casting metallic butt	J. Kendall, jr	Watertown, Mass	July 21, 1831	....
Hinges of wrought-iron, Machinery for making butt	C. Kenney	West Troy, N. Y	Aug. 7, 1844	3, 690
Hinges on the inside, Machine for planing or dressing the knuckles of butt.	G. Stickney	Blackwoodtown, N. J	Dec. 19, 1844	3, 863
Hinges on their axes, Casting	S. Wilkes	Darleston, Great Britain	Apr. 10, 1841	2, 043
Hinges, Pintle of butt	L. Maschauer and W. Frankfurth.	Milwaukee, Wis	Oct. 27, 1868	83, 393
Hinges, &c., Plating iron for the manufacture of	J. J. and L. Crooke	Southfield and New York, N. Y.	Dec. 28, 1869	98, 354
Hinges, Recess-cutter for setting	D. Snow	Cleveland, Ohio	Feb. 16, 1869	87, 008
Hinges, Spike for	H. Garrett	Richmond, Mo	Aug. 7, 1860	29, 480
Hinges, &c., Spring for	A. French	Waterbury, Conn	June 19, 1855	13, 085
Hinges with tin, Coating	H. M. Myers	Allegheny City, Pa	June 15, 1869	91, 477
Hinged-claw wrench	A. Hay	Newark, N. J	Jan. 30, 1849	6, 077
Hinged, sliding, and other bodies, Fastening	L. R. Streeter	Chelsea, Mass	Feb. 2, 1864	41, 451
Hitch-hook	G. W. Chandler	Mason, N. H	Aug. 23, 1870	106, 549
Hitch, Horse	W. A. Middleton and J. A. Haller.	Harrisburgh, Pa	Aug. 11, 1868	80, 870
Hitch, Safety horse	T. Weaver	Harrisburgh, Pa	Nov. 24, 1868	84, 450
Hitching and sign post	C. F. Barnard	Victoria, British Columbia	Mar. 12, 1872	124, 532
Hitching-clamp or hold-fast	J. P. Sinclair	Mill Port, N. Y	Oct. 23, 1866	59, 085
Hitching-device	C. H. Sawyer	Hollis, Me	Oct. 30, 1866	59, 276
Hitching device, Horse	W. H. Bell	Spring Lake, Mich	May 24, 1870	103, 283
Hitching device, Horse	F. Deming	Bridgeport, Conn	Apr. 28, 1868	77, 364
Hitching device, Horse	S. Galbraith	New Orleans, La	Dec. 10, 1867	72, 012
Hitching device, Horse	C. Rogers	Bergen, N. J	Nov. 13, 1866	59, 660
Hitching device, Horse	J. B. Thornton	Madison, Wis	Sept. 17, 1867	69, 048
Hitching device, Horse	W. G. Ward	Steuben County, N. Y	Sept. 17, 1867	68, 919
Hitching device, Horse	M. Warne and W. H. Pearce	Philadelphia, Pa	Aug. 13, 1867	67, 689
Hitching hook, Cam	C. Starbuck	Philadelphia, Pa	Apr. 9, 1867	63, 762
Hitching horses	J. L. Kreider	Drumore Township, Pa	June 29, 1869	91, 854
Hitching horses, clothes-lines, &c., Apparatus for	E. S. Boynton	East Hartford, Conn	July 8, 1856	15, 279
Hitching horses to seeding-machines	D. C. Baughman	Tiffin, Ohio	Aug. 14, 1866	57, 068
Hitching horses to vehicles and plows	J. Graham	Ceresco, Mich	May 3, 1864	42, 574
Hitching-post	J. Gibson, jr	Albany, N. Y	Dec. 28, 1869	98, 247
Hitching-post	J. Melchers	Detroit, Mich	May 21, 1872	127, 085
Hitching-post	J. Oothoudt	Minneapolis, Minn	June 28, 1870	104, 875
Hitching-post	W. S. Owings	Pan Handle Post-Office, W. Va.	Dec. 13, 1870	110, 066
Hitching-post	W. W. Powers	Belleville, N. Y	Apr. 5, 1870	101, 657
Hitching-post	J. S. Rohrer	Lancaster, Pa	Apr. 5, 1870	101, 662
Hitching-post	E. F. Shoenberger	Germantown, Pa	May 16, 1871	114, 982
Hitching-post	B. and P. Walther	New York, N. Y	May 5, 1868	77, 686
Hitching-post	V. Ward	San Francisco, Cal	June 5, 1866	55, 445
Hitching post, Animal	D. Newton	Southington, Conn	Aug. 27, 1867	68, 308
Hitching-post cap	R. N. Roberts	New Britain, Conn	Feb. 18, 1868	74, 596
Hitching post, Horse	C. Bush	Newburgh, N. Y	June 18, 1861	32, 601
Hitching post, Horse	T. Garrish	Providence, R. I	Dec. 24, 1872	134, 266
Hitching-ring	J. P. Howell	New York, N. Y	Sept. 24, 1867	69, 212
Hitching-strap	A. M. Brubaker	Millersville, Pa	Mar. 18, 1873	136, 011
Hitching-strap	T. P. Chambers	Newtown, Pa	Dec. 3, 1867	71, 699
Hitching-strap	P. Conover	Kingsessing, Pa	Jan. 26, 1864	41, 366
Hitching-strap	J. Dubree	Drumore Township, Pa	July 24, 1866	56, 537
Hitching-strap	B. Hunter	Philadelphia, Pa	Apr. 30, 1872	126, 145
Hitching-strap	J. D. Miller	Russellville, Pa	Sept. 3, 1867	68, 569
Hitching-strap	A. J. Ross	Rochester, N. Y	Nov. 5, 1867	70, 473
Hitching-straps, Clasp for	M. R. Margerum and T. P. Marshall.	Trenton, N. J	Dec. 20, 1859	26, 508
Hobby-horse	L. Anderson	Chicago, Ill	Sept. 14, 1869	94, 858
Hobby-horse	J. A. Crandall	Brooklyn, N. Y	Apr. 15, 1873	137, 896
Hobby-horse	W. E. and J. A. Crandall	New York and Brooklyn, N. Y.	Nov. 5, 1872	132, 806
Hobby-horse	C. Hitzelberger	South Orange, N. J	Sept. 6, 1870	107, 049
Hobby-horse	J. Liming	Philadelphia, Pa	June 14, 1870	104, 170
Hobby-horse	P. J. Marqua	Cincinnati, Ohio	Feb. 7, 1865	46, 258
Hobby-horse	J. Reinhart	New York, N. Y	Oct. 21, 1873	143, 930
Hobby-horse and carriage-motor combined	J. Liming	Philadelphia, Pa	July 26, 1870	105, 701
Hobby-horse, Child's	W. L. Williams	New York, N. Y	Jan. 3, 1871	110, 709
Hobby-horse-seat guard	W. L. Williams	New York, N. Y	Dec. 27, 1870	110, 527
Hod	J. Short	Roxbury, Mass	Oct. 31, 1865	50, 741
Hod, Brick and mortar	C. Roehl	Chicago, Ill	Mar. 12, 1872	124, 448
Hod-elevator	L. Atwood	New York, N. Y	July 18, 1871	117, 140
Hod-elevator	G. W. Brown	New York, N. Y	Sept. 26, 1871	119, 305
Hod-elevator	P. Dehlinger	Buffalo, N. Y	Nov. 24, 1868	84, 268
Hod-elevator	E. H. Garrigues	Saint Louis, Mo	Oct. 24, 1871	120, 189
Hod, &c., elevator	E. Harlow	New York, N. Y	June 24, 1873	140, 195
Hod-elevator	T. M. Pelham	New York, N. Y	Sept. 28, 1869	95, 262
Hod-elevator	J. Powers	Chicago, Ill	July 12, 1870	105, 246

Index of patents issued from the United States Patent Office from 1790 *to* 1873, *inclusive*—**Continued.**

Invention.	Inventor.	Residence.	Date.	No.
Hod-elevator	W. H. Totten	Academia, Pa	Sept. 5, 1865	49, 805
Hod-iron, Mason's	L. W. Lake	Providence, R. I	Apr. 23, 1872	126, 061
Hod, Mason's	B. C. Monroe	Newburgh, N. Y	Dec. 16, 1873	145, 675
Hoe	S. W. Adams	Wethersfield, Conn	Feb. 13, 1866	52, 510
Hoe	S. W. Adams	Wethersfield, Conn	Oct. 23, 1866	58, 966
Hoe	L. J. Aderhold	Wedowee, Ala	July 19, 1870	105, 405
Hoe	A. Allen	Ramapo, N. Y	Jan. 23, 1836	
Hoe	J. M. Baird	Wheeling, W. Va	July 29, 1873	141, 308
Hoe	H. R. Barnes	Rock Stream, N. Y	Feb. 4, 1873	135, 462
Hoe	H. R. Barnes	Rock Stream, N. Y	Apr. 15, 1873	137, 755
Hoe	W. G. Beckwith	Lowndesborough, Ala	Sept. 28, 1869	95, 306
Hoe	N. D. Beecroft	Bangor, Me	Feb. 18, 1873	136, 020
Hoe	L. Billings	Gallipolis, Ohio	Dec. 14, 1869	97, 863
Hoe	L. L. Bond	Chicago, Ill	Feb. 15, 1870	99, 821
Hoe	R. P. Buttles	Mansfield, Pa	Oct. 29, 1872	132, 520
Hoe	H. Camp	Covington, Ga	Jan. 21, 1873	135, 076
Hoe	J. S. Carroll	Covington, Ga	Apr. 9, 1872	125, 438
Hoe	I. Cook and J. T. Bever	Haynesville, Mo	May 2, 1871	114, 266
Hoe	M. Cookerly	Baxter Springs, Kans	Oct. 22, 1872	132, 444
Hoe	P. and H. E. Crapo	Bridgewater, Mass	Aug. 6, 1861	32, 982
Hoe	J. Dodge	Grass Valley, Cal	June 9, 1868	78, 724
Hoe	A. Ellis	Canton, Ind	Oct. 17, 1871	120, 050
Hoe	A. Ellis and O. Albertson	Salem, Ind	Oct. 18, 1870	108, 465
Hoe	J. Ells	Pittsburgh, Pa	Dec. 8, 1863	40, 826
Hoe	J. and A. Fairley	Birmingham, England	Dec. 13, 1870	110, 126
Hoe	W. C. Finney	Fayette County, Tenn	Nov. 30, 1852	9, 430
Hoe	S. H. Folsom	Winchester, Mass	Dec. 17, 1872	133, 932
Hoe	J. L. Fountain	New Milford, Ill	June 30, 1868	79, 462
Hoe	G. Freeborn	New York	July 7, 1829	
Hoe	M. Gates	Gallipolis, Ohio	Dec. 12, 1854	12, 057
Hoe	J. Glasson	Brooklyn, N. Y	Aug. 14, 1866	57, 119
Hoe	J. H. Gould	Rutland, Ohio	Aug. 26, 1873	142, 159
Hoe	G. Harper	Montgomery County, Pa	Nov. 19, 1833	
Hoe	B. Hinkley	Fayette, Me	June 23, 1832	
Hoe	J. Hinman	Hamden, Conn	Apr. 22, 1835	
Hoe	C. P. Howell	Covington, Ky	Dec. 29, 1868	85, 384
Hoe	G. S. Jobson	Macon, Ga	Jan. 4, 1870	98, 598
Hoe	M. Johnson	Three Rivers, Mich	Mar. 11, 1873	136, 733
Hoe	A. C. Kasson	Milwaukee, Wis	Dec. 4, 1866	60, 195
Hoe	T. B. Lockwood	Smyrna, Del	July 29, 1873	141, 365
Hoe	H. A. Lothrop	Sharon, Mass	Dec. 29, 1857	18, 979
Hoe	J. F. Lowe	Louisville, Ky	Mar. 23, 1869	88, 051
Hoe	C. Mahan	Grand Island, Cal	Sept. 3, 1867	68, 764
Hoe	E. W. McLendon	Griffin, Ga	June 24, 1873	140, 291
Hoe	P. H. Merrill	Hinsdale, N. H	Apr. 6, 1820	
Hoe	J. C. Miller	Danville, Ky	Feb. 6, 1872	123, 502
Hoe	E. C. and J. W. Newland	Lawrence County, Ind	Jan. 13, 1863	37, 405
Hoe	G. H. Owens	Maysville, Ky	Apr. 21, 1868	77, 083
Hoe	H. Parkman	Philadelphia, Pa	Oct. 28, 1873	144, 127
Hoe	J. C. Plumer	Boston, Mass	Mar. 20, 1866	53, 332
Hoe	H. Porter	Cummington, Mass	Mar. 6, 1860	27, 382
Hoe	A. Prentiss	Prentiss Vale, Pa	May 14, 1867	64, 792
Hoe	E. Riston and W. W. Brigg	Maltaville, N. Y	June 5, 1866	55, 365
Hoe	H. C. Rogers	Scranton, Pa	June 17, 1862	35, 627
Hoe	S. Rogers	Pittsburgh, Pa	Jan. 16, 1866	52, 079
Hoe	C. W. Saladee and J. S. Hall	Newark, Ohio, and Pittsburgh, Pa.	June 2, 1868	78, 611
Hoe	S. Springstead	Westchester, N. Y	Nov. 9, 1869	96, 739
Hoe	J. Stilwell	Griffin, Ga	May 23, 1871	115, 132
Hoe	C. H. Strong and E. A. Sterry	Norwich, Conn	Jan. 5, 1833	
Hoe	T. R. Timby	Saratoga Springs, N. Y	Mar. 5, 1867	62, 574
Hoe	E. Warren	Ceresco, Mich	May 10, 1870	102, 891
Hoe	P. Whitin	Northbridge, Mass	Feb. 22, 1825	
Hoe	J. F. Wilson	Athens, Ga	Aug. 24, 1869	94, 054
Hoe	I. N. Wood	Fall River, Mass	June 2, 1868	78, 561
Hoe, Adjustable	T. Drake	Hartford, Conn	Jan. 28, 1868	73, 702
Hoe, Adjustable weeding	A. C. Arnold	Norwalk, Conn	Oct. 17, 1865	50, 436
Hoe and corn-planter, Combined	J. A. Burchard	Beloit, Wis	June 19, 1866	55, 612
Hoe and cotton-cultivator, Combined revolving	M. E. Davis	Rome, Ga	Nov. 30, 1869	97, 277
Hoe and cultivator, Horse	G. D. Rowell	Menamonee Falls, Wis	Apr. 18, 1871	113, 934
Hoe and fork, Combined	B. L. Tibbetts	South China, Me	Apr. 6, 1869	88, 591
Hoe and garden-rake combined	R. H. Gordon, sr	Brooklyn, Ohio	Jan. 21, 1868	73, 525
Hoe and garden-trimmer, Scuffle	I. Pardee	Vineland, N. J	Mar. 31, 1868	76, 288
Hoe and potato-digger, Combined	E. F. Morris and R. J. Greed	Cicero, N. Y	Oct. 15, 1867	69, 928
Hoe and rake	A. Ellis and O. Albertson	Salem, Ind	June 15, 1869	91, 317
Hoe and rake, Combined	I. Cook	Haynesville, Mo	Sept. 17, 1867	68, 958
Hoe and rake, Combined	W. M. McLendon	Greenville, Ga	May 14, 1872	126, 821
Hoe and rake, Combined	F. L. Senour	Eaton, Ohio	June 1, 1869	90, 691
Hoe and rake, Combined garden	G. W. Lockwood	Fairport, N. Y	May 21, 1872	127, 077
Hoe and roller combined, Garden	E. Blanchard	Poolesville, Md	Mar. 25, 1873	137, 124
Hoe and seed-dropper	W. Snyder	Ypsilanti, Mich	Aug. 20, 1872	130, 757
Hoe and seed-planter	C. H. Wolcott	Randolph, N. Y	Aug. 30, 1864	44, 032
Hoe-blank	H. Waters	Boston, Mass	Feb. 26, 1867	62, 456
Hoe-blank	H. Waters	Boston, Mass	Feb. 25, 1868	74, 960
Hoe-blank-rolling machine	S. A. Millard	Clayville, N. Y	July 21, 1868	80, 083
Hoe-blank-rolling machine	H. Waters	Boston, Mass	May 2, 1871	114, 498
Hoe-blanks, Machine for manufacture of	N. Brand	Leonardsville, N. Y	Feb. 26, 1861	31, 525
Hoe-blanks, Machine for rolling	W. T. Clement and E. V. Foster	Northampton, Mass	Aug. 17, 1869	93, 679
Hoe, Cane-stubble	G. H. Wright and A. K. Johnson	New Orleans, La	Oct. 17, 1871	120, 143
Hoe, Cast-iron	B. F. Boyden	Boston, Mass	Mar. 31, 1836	
Hoe-cleansing attachment	R. R. and H. T. Spedden	Astoria, Oreg	Nov. 1, 1870	108, 843
Hoe, Compound adjustable garden	M. H. Fletcher	Richmond, Ind	Nov. 3, 1868	83, 771
Hoe, Cotton	J. S. Carroll	Covington, Ga	June 6, 1871	115, 568
Hoe, Cotton	Z. B. Sims	Bonham, Tex	Aug. 31, 1869	94, 353
Hoe, Cotton	Z. B. Sims	Bonham, Tex	Sept. 26, 1871	119, 418
Hoe, Cultivating	J. J. Ray and J. R. Young	New Orleans, La	Sept. 21, 1869	95, 844
Hoe, Double	A. Burchard	New Brenton, Ill	Mar. 3, 1868	74, 984
Hoe, Extension weeding	M. Renz	Naugatuck, Conn	Dec. 18, 1866	60, 641

Index of patents issued from the United States Patent Office from 1790 *to* 1873, *inclusive*—Continued.

Invention.	Inventor.	Residence.	Date.	No.
Hoe for weeding, Cast-iron horse	W. Carmichael	Sand Lake, N. Y	July 28, 1827	....
Hoe, Garden	J. R. Albertson	East Deer Township, Pa	Apr. 3, 1860	27, 675
Hoe, Garden	I. W. Averill	Plymouth, Mich	June 11, 1836	....
Hoe, Garden	J. P. Avery	Norwich, Conn	Mar. 14, 1871	112, 527
Hoe, Garden	C. Belden	Akron, Ohio	Dec. 17, 1872	134, 028
Hoe, Garden	J. H. Brower	Atlas, Mich	May 28, 1867	65, 053
Hoe, Garden	D. E. Eaton	Boston, Mass	Jan. 2, 1866	51, 893
Hoe, Garden	C. R. Gilbert	High Point, N. C	Aug. 22, 1871	118, 360
Hoe, Garden	C. S. Homer	Boston, Mass	May 8, 1840	1, 593
Hoe, Garden	J. Howard	West Bridgewater, Mass	Apr. 9, 1861	31, 977
Hoe, Garden	A. C. Judson	Grand Rapids, Ohio	Sept. 6, 1870	107, 060
Hoe, Garden	S. V. Kimball	Indianapolis, Ind	Oct. 17, 1871	119, 986
Hoe, Garden	A. M. Ross	Ilion, N. Y	Mar. 2, 1869	87, 368
Hoe, Garden	S. Shetter	Allegheny, Pa	July 21, 1857	17, 848
Hoe, Garden	L. Streeter	Chicopee, Mass	Sept. 1, 1868	81, 839
Hoe, Garden	F. Trigalet	Astoria, N. Y	Dec. 5, 1871	121, 691
Hoe-handle	S. Reynolds	Duquesneborough, Pa	Apr. 23, 1861	32, 146
Hoe-handles, &c., Fastening	R. Dare	Greenwich, N. J	Feb. 14, 1832	....
Hoe-handles, Fastening	G. Hight	Gorham, Me	May 25, 1838	751
Hoe, Hand	C. A. Rose	Columbus, Ga	Jan. 1, 1867	60, 941
Hoe, Horse	M. Chandler	East Corinth, Me	Jan. 14, 1862	34, 128
Hoe, Horse	M. Chandler and J. B. Nickels	Corinth and Kenduskeag, Me.	Feb. 19, 1867	62, 181
Hoe, Horse	G. W. Cooper	Ogeechee, Ga	June 11, 1867	65, 728
Hoe, Horse	I. Copeland	North Bridgewater, Mass	Sept. 20, 1870	107, 456
Hoe, Horse	T. P. Fish	Litchfield, N. Y	May 1, 1866	54, 315
Hoe, Horse	J. Gifford, jr	Watertown, N. Y	Nov. 13, 1866	59, 584
Hoe, Horse	D. Harris	Canaan, Me	July 24, 1866	56, 555
Hoe, Horse	H. W. Hasslock	Nashville, Tenn	Mar. 22, 1870	101, 010
Hoe, Horse	N. H. Lindley	Bridgeport, Conn	Feb. 9, 1869	86, 846
Hoe, Horse	N. H. Lindley	Bridgeport, Conn	Jan. 18, 1870	98, 983
Hoe, Horse	C. Lobdell	Fort Hill, Ill	Mar. 2, 1869	87, 347
Hoe, Horse	D. C. Matteson and T. P. Williamson.	Stockton, Cal	Nov. 3, 1868	83, 719
Hoe, Horse	W. Muir	Wauconda, Ill	Feb. 16, 1869	87, 060
Hoe, Horse	T. Rice and L. R. Hitchcock	Caneadea, N. Y	Nov. 5, 1867	70, 472
Hoe, Horse	G. W. Sawin	Nashua, N. H	Aug. 20, 1867	67, 917
Hoe, Horse	E. W. Walton	Stockton, Cal	June 9, 1868	78, 774
Hoe, Horse	C. Whitlock	Bridport, Vt	Oct. 25, 1870	108, 741
Hoe-machine for land	M. Spofford	Georgetown, Mass	Sept. 11, 1847	5, 290
Hoe-necks, Swage-die, &c., for manufacturing	N. Brand	Leonardsville, N. Y	Apr. 1, 1845	3, 980
Hoe, Planting	C. N. Jones	Galway, N. Y	Sept. 29, 1863	40, 107
Hoe, Planting	W. G. Sterling	Bridgeport, Conn	Feb. 14, 1854	10, 525
Hoe, Planting	A. Williams	Sebec, Me	June 25, 1867	66, 196
Hoe-plates, Die for forging	L. T. Richardson	Auburn, N. Y	Mar. 11, 1873	136, 672
Hoe, Plowing	T. J. Mason	Harmony, Me	Dec. 15, 1868	84, 957
Hoe, Prong	J. Wilson	Marlborough, N. H	Sept. 20, 1827	....
Hoe, rake, &c	J. S. Hall	West Manchester, Pa	Feb. 9, 1864	41, 503
Hoe-rolling machine	W. F. Stillman	Ilion, N. Y	July 16, 1867	66, 905
Hoe, Scuffle	T. R. Peck	Waterloo, N. Y	Sept. 24, 1872	131, 627
Hoe seed-dropper	T. Nevison, jr	Morgan, Ohio	Mar. 31, 1868	76, 235
Hoe, Seed-planting	S. Woodruff	Sparta, N. J	Apr. 20, 1858	20, 014
Hoe, Seeding	L. B. Brown	Simsbury, Conn	Oct. 30, 1860	30, 526
Hoe, Seeding	J. A. Pease	New York, N. Y	May 17, 1853	9, 731
Hoe seeding-attachment	E. L. Berdstresser	Hublersburgh, Pa	Mar. 24, 1868	75, 843
Hoe, Shuffle	J. D. Jenkins	Jackson, Ill	May 22, 1866	54, 916
Hoe, Weed-puller	N. Ames	Saugus Centre, Mass	Aug. 16, 1864	43, 869
Hoe, Weeding	A. B. Adams	Westport, Conn	Aug. 30, 1864	44, 035
Hoe, Weeding	J. M. Adams	Canton, Mass	Oct. 18, 1859	25, 862
Hoe, Weeding	G. P. Allen	Woodbury, Conn	Apr. 24, 1866	54, 087
Hoe, Weeding	H. H. Baker	New Market, N. J	Oct. 4, 1859	25, 619
Hoe, Weeding	A. Coleman	Red Bank, N. J	Apr. 28, 1868	77, 258
Hoe, Weeding	E. M. Conkling	Parma, N. Y	Aug. 27, 1867	68, 047
Hoe, Weeding	C. Crofut	Weston, Conn	Sept. 19, 1865	49, 985
Hoe, Weeding	H. S. Crossland	Dresden, Tex	Aug. 26, 1873	142, 149
Hoe, Weeding	J. G. Harriman	Gardiner, Me	Nov. 19, 1872	133, 151
Hoe, Weeding	L. King	Oriskany Falls, N. Y	Oct. 20, 1868	83, 290
Hoe, Weeding	A. E. Lyman	Northampton, Mass	Nov. 10, 1868	83, 866
Hoe, Weeding	J. S. Munger	Olean, N. Y	Apr. 6, 1869	88, 659
Hoe, Weeding	J. Nawgle	Mooresville, Ind	June 20, 1865	48, 304
Hoe, Weeding	E. Wilcox	East Cleveland, Ohio	Oct. 31, 1871	120, 408
Hoe, Wheel	E. D. and O. B. Reynolds	North Bridgewater, Mass	Mar. 10, 1868	75, 460
Hoes and picks, Die for making	D. Carr	Allegheny City, Pa	Feb. 13, 1872	123, 552
Hoes and picks, Manufacture of grubbing	J. Lee	Cleveland, Ohio	Jan. 23, 1872	122, 897
Hoes, Attaching eyes to the blades of	H. Havell	Newark, N. J	Feb. 10, 1857	16, 584
Hoes by rolling cast-steel, Making	C. Bulkley	Colchester, Conn	Jan. 10, 1872	....
Hoes, chisels, &c., Socket for	G. Banister	Hartford, Vt	July 4, 1865	48, 505
Hoes, Corn-dropping attachment for	E. L. Staples	Chillicothe, Ohio	Feb. 14, 1871	111, 883
Hoes, Corn-dropping attachment for	C. W. Cotton and E. L. Staples	Cincinnati, Ohio	Sept. 10, 1867	68, 607
Hoes, Garden and other	J. T. Sargent	Sutton, N. H	Oct. 25, 1853	10, 162
Hoes, Machine for making	H. Sauerbier	Newark, N. J	Apr. 5, 1859	23, 502
Hoes, Machinery for manufacturing	W. T. Clement and E. V. Foster	Northampton, Mass	Sept. 3, 1867	68, 414
Hoes, Making	J. Abbott	Boston, Mass	Jan. 25, 1848	5, 417
Hoes, Making	M. Depuy	Pittsburgh, Pa	Nov. 6, 1860	30, 569
Hoes, Making	J. Watson	New Canaan, Conn	July 26, 1823	....
Hoes, Manufacture of	W. Acheson and W. H. Ridley	Pittsburgh, Pa	May 28, 1872	127, 288
Hoes, Manufacture of	W. Baker	West Winsted, Conn	May 21, 1861	32, 344
Hoes, Manufacture of	J. A. Black	Columbia, S. C	Oct. 13, 1828	....
Hoes, Manufacture of	S. Boyd	New York, N. Y	Jan 13, 1857	16, 371
Hoes, Manufacture of	S. Boyd	Brooklyn, N. Y	Dec. 27, 1859	26, 500
Hoes, Manufacture of	W. T. Clement	Northampton, Mass	Jan. 21, 1868	73, 437
Hoes, Manufacture of	J. S. Craig	Guilford, Kans	Feb. 13, 1872	123, 680
Hoes, Manufacture of	R. Harper	Philadelphia, Pa	Nov. 20, 1866	59, 744
Hoes, Manufacture of	J. Knight	Newark, N. J	Mar. 30, 1858	19, 812
Hoes, Manufacture of	A. Patterson	Birmingham, Pa	July 5, 1859	24, 658
Hoes, Manufacture of	A. Patterson	Birmingham, Pa	Sept. 27, 1859	25, 580
Hoes, Manufacture of	A. Patterson	Birmingham, Pa	Aug. 29, 1865	49, 640
Hoes, Manufacture of	L. T. Richardson	Clayville, N. Y	May 26, 1868	78, 326

Index of patents issued from the United States Patent Office from 1790 *to* 1873, *inclusive*—Continued.

Invention.	Inventor.	Residence.	Date.	No.
Hoes, Manufacture of	E. C. Tuttle	Naugatuck, Conn	Oct. 18, 1859	25, 856
Hoes, Manufacturing	S. Boyd	Brooklyn, N. Y	Aug. 2, 1859	24, 926
Hoes, Process of rolling	S. A. Millard	Clayville, N. Y	Feb. 18, 1868	74, 570
Hoes, Roll for the manufacture of planters'	J. T. Tyler	Pittsburgh, Pa	Mar. 21, 1871	112, 867
Hoes, Securing handles to	W. T. Clement	Northampton, Mass	May 14, 1861	32, 275
Hoes, Seed-dropping attachment for	A. T. Large	Chicago, Ill	Mar. 5, 1867	62, 642
Hoes to handles, Attaching	O. Clark and W. D. Hillis	Hudson, Ohio	Sept. 9, 1843	3, 255
Hoes to handles, Attaching	W. H. Hartzman	Big Lick, Va	Mar. 12 1867	62, 900
Hoes to handles, Attaching	J. A. Marine	Mooresville, Ind	Apr. 13, 1869	88, 884
Hoeing-machine	H. C. Briggs	West Auburn, Me	Nov. 17, 1868	84, 165
Hoeing-machine	H. C. Briggs	West Auburn, Me	Mar. 9, 1869	87, 627
Hoeing-machine	H. W. Clapp	Northampton, Mass	Nov. 9, 1869	96, 549
Hog-cleaning machine	N. Silverthorn	Prescott, Wis	Aug. 30, 1864	44, 021
Hog-elevator	A. J. Chambers and T. Jackson	New Washington, Ohio	Aug. 24, 1869	94, 076
Hog-elevator	D. Kaufman	Boiling Spring, Pa	Mar. 6, 1860	27, 368
Hog-feeder	I. S. Pope	Napoleon, Ohio	May 21, 1867	65, 004
Hog-holder	W. and C. Leffingwell	Clarksburgh, Ohio	May 21, 1867	64, 885
Hog-holding clamp	R. T. Leaverton	Holden, Mo	Aug. 20, 1872	130, 727
Hog-lifting machine	J. Dagley	Gosport, Ind	Nov. 14, 1871	120, 946
Hog-ring	J. J. Hutson	South Solon, Ohio	Aug. 13, 1867	67, 765
Hog-ring	W. S. and E. S. Shoemaker	Towsontown, Md., and Lancaster, Ohio.	May 11, 1869	90, 028
Hog-ringing device	G. W. Clark	Frankfort, Ohio	Oct. 1, 1867	69, 406
Hog-ringing device	W. S. Houston	Mansfield, Ohio	Nov. 14, 1871	120, 876
Hog-ringing implement	H. W. Hill	Decatur, Ill	Aug. 27, 1872	130, 853
Hog-scalder, Portable	A. Clarke	Philadelphia, Pa	Dec. 26, 1865	51, 783
Hog-scalding machine	M. Stricker	Vincennes, Ind	Oct. 23, 1866	59, 093
Hog-scalding tank	G. King	Eminence, Ky	Apr. 28, 1868	77, 198
Hog-scraper, Rotary	R. Fyfe	New York, N. Y	July 1, 1873	140, 409
Hog-scraping apparatus	O. McNeil and P. W. Dalton	Jersey City, N. J	July 15, 1873	140, 838
Hog-singeing apparatus	P. Kenney	Chicago, Ill	July 16, 1872	129, 481
Hog-snout cutter	S. Hair	Kankakee City, Ill	Aug. 23, 1864	43, 909
Hog-snout slitter	J. J. Gisb	Milton, Ohio	Mar. 3, 1868	75, 148
Hog-snouter	J. C. Campbell and W. S. Bruce	Good Hope, Ill	Jan. 3, 1871	110, 629
Hog-stock	W. C. Hays and J. D. Scott	Sharonville, Ohio	June 20, 1871	116, 183
Hog-tamer	T. G. Telton	Lyons, Iowa	Apr. 18, 1865	47, 329
Hog-trap	M. Caywood	Farmington, Ill	June 14, 1872	127, 461
Hog-trap	J. J. Cox	Woodhull, Ill	Apr. 2, 1872	125, 120
Hog-trap	T. Darby	Duncan, Ill	Aug. 12, 1873	141, 633
Hog-trap	A. B. De Vore	Talkington Township, Ill	July 15, 1873	140, 816
Hog-trap	H. W. Hill	Forsythe, Ill	June 11, 1872	127, 769
Hog-trap	R. Keiler	Kent, Ill	July 23, 1872	129, 832
Hog-trap	C. B. Weeks	Galesburgh, Ill	June 29, 1869	91, 887
Hog-trough	E. Johnson	Manchester, Mass	July 21, 1857	17, 839
Hog-trough	W. H. Robbins	Richmond, Ind	Sept. 4, 1866	57, 772
Hog-trough	C. F. Rolfe	Laconia, N. H	July 7, 1868	79, 688
Hogs, Apparatus for cutting up	W. C. Marshall	New York, N. Y	Aug. 13, 1872	130, 515
Hogs by steam, Scalding	T. J. Godman	Baltimore, Md	Feb. 13, 1835	....
Hogs, Device for catching and holding	F. Voss, J. P. Hax, and H. Krug.	Saint Joseph, Mo	Sept. 13, 1870	107, 312
Hogs, Forceps for snouting	G. Stephenson	Zionsville, Ind	Oct. 14, 1873	143, 730
Hogs from rooting, Attachment for preventing	J. Butts	Evansville, Wis	Aug. 20, 1867	67, 952
Hogs from rooting, Device for preventing	H. Judson	Galesburgh, Ill	July 3, 1866	56, 061
Hogs from rooting, Device for preventing	D. K. Nixon	Sandyville, Ohio	Dec. 18, 1860	30, 928
Hogs from rooting, Device for preventing	G. and H. L Nixon	Sandyville, Ohio	Feb. 18, 1868	74, 582
Hogs from rooting, Device to prevent	D. P. Heflebower	Champaign County, Ohio	Aug. 8, 1871	117, 883
Hogs from rooting, Device to prevent	R. Little	Middle Branch, Ohio	Mar. 5, 1861	31, 610
Hogs from rooting, Device to prevent	L. L. Rinehart and P. Phillips	Evansburgh, Ohio	Feb. 4, 1868	74, 002
Hogs from rooting, Preventing	J. D. Kirkpatrick	Urbana, Ohio	May 12, 1868	77, 891
Hogs in veterinary operations, Device for holding	A. M. Scott and T. Wilson	Clearmont, Mo	Jan. 7, 1873	134, 707
Hogs, Instrument for ringing	J. and G. Heesen and H. Nyland.	Tecumseh, Mich	June 27, 1871	116, 440
Hogs, Instrument for taming and marking	S. Long	Ogle County, Ill	Aug. 3, 1869	93, 210
Hogs, Nose-ring for	O. P. Goodman	Chillicothe, Ohio	Feb. 8, 1870	99, 559
Hogsheads, &c., Apparatus for moving and rolling	F. O. J. Burr	Brewer, Me	Mar. 4, 1873	136, 486
Hoist, Air	L. Taws and J. M. Hartman	Philadelphia, Pa	Oct. 19, 1869	95, 952
Hoist, Cellar	P. Coble	Mechanicsburgh, Pa	July 23, 1872	129, 790
Hoist or elevator, Cellar	H. L. Bowers	Harrisburgh, Pa	Feb. 22, 1870	100, 006
Hoist, Pneumatic	J. C. Kent and H. C. Rich	Phillipsburgh, N. J	Apr. 26, 1870	102, 275
Hoist, Portable vertical	J. E. Walsh	New York, N. Y	Oct. 8, 1872	131, 983
Hoist, Self-acting hatchway	G. N. Creamer	Trenton, N. J	Mar. 1, 1870	100, 265
Hoist, Vertical	W. E. Worthen	New York, N. Y	Nov. 14, 1871	120, 844
Hoists, Safety-device for	P. J. Borger	Cincinnati, Ohio	May 2, 1871	114, 399
Hoists, Safety-device for	P. J. Borger	Cincinnati, Ohio	Feb. 13, 1872	123, 672
Hoisting and conveying apparatus	O. T. McIntosh	New York, N. Y	May 13, 1873	138, 910
Hoisting and conveying apparatus	G. Stancliff	New York, N. Y	May 13, 1873	138, 950
Hoisting and conveying apparatus	C. B. Stough	Monticello, Ill	July 29, 1873	141, 297
Hoisting and dumping apparatus	W. B. Culver	Scranton, Pa	Dec. 15, 1868	84, 939
Hoisting and dumping apparatus	G. Martz	Pottsville, Pa	Sept. 7, 1858	21, 431
Hoisting and dumping apparatus	G. Martz	Pottsville, Pa	Sept. 28, 1869	95, 366
Hoisting and lowering apparatus	R. L. Fitch	Sing Sing, N. Y	May 30, 1871	115, 292
Hoisting and lowering apparatus	I. J. Lancaster	Vancouver, Wash	June 27, 1865	48, 414
Hoisting and lowering apparatus	H. E. Towle	New York, N. Y	Apr. 5, 1870	101, 681
Hoisting and lowering apparatus, Self-braking	W. Hart	Philadelphia, Pa	Aug. 19, 1873	142, 017
Hoisting and lowering goods, &c., Machinery for	G. Thompson	Cincinnati, Ohio	June 1, 1858	20, 455
Hoisting and lowering weights	W. C. McGill	Cincinnati, Ohio	May 30, 1865	47, 969
Hoisting and storing apparatus, Ice	H. Van Steenburgh and J. Egnor.	Catskill, N. Y	Nov. 30, 1858	22, 210
Hoisting and weighing machine	J. L. Dutton, sr	Philadelphia, Pa	Mar. 13, 1860	27, 623
Hoisting-apparatus	G. Ambrose	New York, N. Y	May 23, 1865	47, 782
Hoisting-apparatus	W. D. Andrews	New York, N. Y	Oct. 1, 1867	69, 384
Hoisting-apparatus	W. D. Andrews	Brookhaven, N. Y	Dec. 20, 1870	110, 331
Hoisting-apparatus	C. W. Baldwin	Boston, Mass	Jan. 25, 1870	99, 049
Hoisting-apparatus	J. Beattie	Chicago, Ill	July 6, 1869	92, 250
Hoisting-apparatus	J. R. Benedict	Williamson, N. Y	June 13, 1871	115, 810
Hoisting-apparatus	E. H. Bermier	Paris, France	June 1, 1869	90, 810
Hoisting-apparatus	H. Black	Carrollton, Ill	Sept. 11, 1866	57, 850
Hoisting-apparatus	J. F. Bodley, J. Reynolds, and S. Van Emon.	Cincinnati, Ohio	July 5, 1864	43, 451

Index of patents issued from the United States Patent Office from 1790 to 1873, inclusive—Continued.

Invention.	Inventor.	Residence.	Date.	No.
Hoisting-apparatus	J. N. B. Bond	New York, N. Y	Feb. 4, 1873	135, 513
Hoisting-apparatus	G. W. Brown	New York, N. Y	May 24, 1870	103, 292
Hoisting-apparatus	W. H. Brown	New York, N. Y	May 10, 1870	102, 763
Hoisting-apparatus	L. Burnell	Milwaukee, Wis	Oct. 4, 1870	107, 868
Hoisting-apparatus	F. P. Canfield	Brighton, Mass	Oct. 6, 1868	82, 799
Hoisting-apparatus	M. K. Carpenter	Cincinnati, Ohio	Aug. 30, 1870	106, 914
Hoisting-apparatus	D. H. Chamberlain	West Roxbury, Mass	Feb. 23, 1869	87, 143
Hoisting-apparatus	H. Chatfield	Wolcottville, Conn	Nov. 5, 1867	70, 412
Hoisting-apparatus	J. Christiansen	New York, N. Y	June 16, 1868	78, 927
Hoisting-apparatus	C. Clark	Unadilla Forks, N. Y	Feb. 25, 1873	136, 212
Hoisting-apparatus	C. S. Crane	Chicago, Ill	July 2, 1872	128, 465
Hoisting-apparatus	W. W. Crapster	Mechanicsburgh, Pa	Aug. 18, 1868	81, 256
Hoisting-apparatus	L. Cutting	San Francisco, Cal	June 1, 1869	90, 823
Hoisting-apparatus	J. A. Dayton	New London, Conn	May 28, 1867	65, 180
Hoisting-apparatus	W. Dyatt	New York, N. Y	Dec. 7, 1869	97, 487
Hoisting-apparatus	F. Farrell	Ansonia, Conn	Aug. 5, 1873	141, 551
Hoisting-apparatus	R. Finnegan	New York, N. Y	Oct. 22, 1867	69, 978
Hoisting-apparatus	H. Flad and J. B. Eads	Saint Louis, Mo	Jan. 24, 1871	111, 188
Hoisting-apparatus	O. H. Flook	Middletown, Md	June 21, 1870	104, 567
Hoisting-apparatus	D. Frisbie	New Haven, Conn	Sept. 25, 1866	58, 340
Hoisting-apparatus	E. R. Gard	Chicago, Ill	Aug. 29, 1871	118, 447
Hoisting-apparatus	S. C. Goodsell and D. Frisbie	New Haven, Conn	May 1, 1866	54, 467
Hoisting-apparatus	E. G. Green	East Gloucester, Mass	Sept. 14, 1869	94, 817
Hoisting-apparatus	F. H. Hambleton	Baltimore, Md	May 21, 1872	127, 049
Hoisting-apparatus	D. Head	Medusa, N. Y	June 23, 1868	79, 071
Hoisting-apparatus	P. Higdon	Lewisport, Ky	Nov. 5, 1867	70, 565
Hoisting-apparatus	J. Hoffman	Tewkesbury, N. J	Aug. 8, 1871	117, 778
Hoisting-apparatus	J. Hoffman	Lebanon, N. J	Dec. 24, 1872	134, 141
Hoisting-apparatus	G. L. Howland	Topsham, Me	Sept. 18, 1866	58, 103
Hoisting-apparatus	G. L. and W. M. Howland	Topsham, Me	Mar. 26, 1867	63, 251
Hoisting-apparatus	W. M. and G. L. Howland	Topsham, Me	Oct. 12, 1869	95, 691
Hoisting-apparatus	H. H. Hunt	Saratoga Springs, N. Y	Dec. 20, 1870	110, 238
Hoisting-apparatus	J. L. Isbell	Naugatuck, Conn	May 6, 1873	138, 655
Hoisting-apparatus	B. H. Jenks	Bridesburgh, Pa	Aug. 7, 1866	56, 947
Hoisting-apparatus	G. H. Kamnacher	Columbus, Ohio	Sept. 21, 1869	95, 022
Hoisting-apparatus	G. B. Keeler	Portchester, N. Y	Sept. 17, 1867	68, 992
Hoisting-apparatus	D. Knowles	Philadelphia, Pa	July 1, 1873	140, 421
Hoisting-apparatus	T. Krausch	New York, N. Y	Nov. 28, 1871	121, 386
Hoisting-apparatus	W. H. Kuntz	Mount Rock, Pa	Nov. 26, 1867	71, 393
Hoisting-apparatus	E. Learned	Boston, Mass	Feb. 6, 1849	6, 085
Hoisting-apparatus	A. R. Lemen	Kalamazoo, Mich	Mar. 5, 1867	62, 647
Hoisting-apparatus	J. Lemman	Cincinnati, Ohio	July 17, 1860	29, 180
Hoisting-apparatus	A. B. Lipsey	New York, N. Y	Jan. 31, 1871	111, 356
Hoisting-apparatus	S. M. Longley	Hudson, N. Y	Oct. 24, 1865	50, 607
Hoisting-apparatus	J. E. Marshall and J. W. Schroeder.	Baltimore, Md	June 23, 1868	79, 138
Hoisting-apparatus	W. K. Marvin	New York, N. Y	Feb. 7, 1865	46, 257
Hoisting-apparatus	L. S. Mason	Middlefield Centre, N. Y	Apr. 30, 1872	126, 222
Hoisting-apparatus	G. B. Massey	New York, N. Y	Oct. 7, 1873	143, 518
Hoisting-apparatus	G. B. Massey and A. B. Darling	New York, N. Y	Oct. 7, 1873	143, 519
Hoisting-apparatus	N. S. McFarland	Brooklyn, N. Y	Mar. 26, 1872	125, 064
Hoisting-apparatus	N. S. McFarland	Williamsburgh, N. Y	Oct. 14, 1873	143, 706
Hoisting-apparatus	J. V. and W. H. Merrick	Philadelphia, Pa	June 9, 1868	78, 675
Hoisting-apparatus	J. V. and W. H. Merrick	Philadelphia, Pa	Aug. 31, 1869	94, 429
Hoisting-apparatus	J. H. Mills	Boston, Mass	Jan. 19, 1869	85, 952
Hoisting-apparatus	J. Nicholson	Monticello, Ind	Dec. 5, 1871	121, 463
Hoisting-apparatus	D. A. Noble	Boston, Mass	June 11, 1872	127, 789
Hoisting-apparatus	C. R. Otis	Yonkers, N. Y	Feb. 28, 1865	46, 580
Hoisting-apparatus	C. R. Otis	Yonkers, N. Y	Mar. 31, 1868	76, 240
Hoisting-apparatus	C. R. Otis	Yonkers, N. Y	Dec. 30, 1873	146, 090
Hoisting-apparatus	C. R. and N. P. Otis	Yonkers, N. Y	Nov. 3, 1868	83, 725
Hoisting-apparatus	C. R. and N. P. Otis	Yonkers, N. Y	Jan. 17, 1871	110, 993
Hoisting-apparatus	C. R. and N. P. Otis	Yonkers, N. Y	Apr. 4, 1871	113, 555
Hoisting-apparatus	C. R. and N. P. Otis	Yonkers, N. Y	Jan. 16, 1872	122, 846
Hoisting-apparatus	E. G. Otis	Yonkers, N. Y	Jan. 15, 1861	31, 128
Hoisting-apparatus	S. K. Paden	Pulaski, Pa	Mar. 5, 1872	124, 380
Hoisting apparatus	H. Pennepacker	Kimberton, Pa	Sept. 1, 1863	39, 745
Hoisting-apparatus	T. Pollard	Philadelphia, Pa	Aug. 9, 1870	106, 284
Hoisting-apparatus	J. L. Pott	Pottsville, Pa	Aug. 16, 1859	25, 215
Hoisting-apparatus	J. L. Pott	Pottsville, Pa	Jan. 24, 1860	26, 921
Hoisting-apparatus	W. R. Reaney	Chester, Pa	Oct. 12, 1869	95, 837
Hoisting-apparatus	J. F. Rochow	New York, N. Y	Dec. 9, 1862	37, 110
Hoisting-apparatus	J. F. Rochow	New York, N. Y	Oct. 22, 1867	70, 122
Hoisting-apparatus	W. Rung	New York, N. Y	Sept. 10, 1867	68, 794
Hoisting-apparatus	J. Sanderson	Fredericksburgh, Ohio	May 19, 1868	78, 138
Hoisting-apparatus	H. A. Schneekloth	New York, N. Y	May 3, 1870	102, 597
Hoisting apparatus	E. U. and W. L. Scoville	Manlius and West Bloomfield, N. Y.	Sept. 24, 1867	69, 257
Hoisting-apparatus	E. U. and W. L. Scoville	Manlius, N. Y	Aug. 4, 1868	80, 569
Hoisting-apparatus	H. F. Shaw	West Roxbury, Mass	May 1, 1866	54, 419
Hoisting-apparatus	M. Shepard	Pontiac, Mich	Feb. 13, 1872	123, 737
Hoisting-apparatus	T. Silver	New York, N. Y	Apr. 4, 1871	113, 585
Hoisting-apparatus	W. Smith	Philadelphia, Pa	Nov. 1, 1864	44, 895
Hoisting-apparatus	G. Sprague	South Addison, N. Y	July 4, 1871	116, 769
Hoisting-apparatus	H. D. Stover	New York, N. Y	June 15, 1869	91, 283
Hoisting-apparatus	L. W. Stuart	Narrowsburgh, N. Y	Apr. 4, 1871	113, 362
Hoisting-apparatus	A. L. Sweet	Norwich, Conn	Sept. 6, 1864	44, 124
Hoisting-apparatus	J. A. Talpey	Somerville, Mass	May 23, 1865	47, 878
Hoisting-apparatus	T. Terrell	Yonkers, N. Y	Dec. 12, 1871	121, 910
Hoisting-apparatus	T. R. Vestal	Santa Fé, Tenn	July 18, 1871	117, 224
Hoisting-apparatus	J. H. Violett	Goshen, Ind	Nov. 15, 1870	109, 278
Hoisting-apparatus	J. D. Warner	Brooklyn, N. Y	Apr. 9, 1872	125, 507
Hoisting-apparatus	W. S. Watson	New York, N. Y	Nov. 20, 1866	59, 880
Hoisting-apparatus	A. Weide	Chicago, Ill	Aug. 8, 1871	117, 840
Hoisting-apparatus	T. A. Weston	Ridgewood, N. J	Dec. 14, 1869	98, 000
Hoisting-apparatus	J. Wheelock	Worcester, Mass	Aug. 8, 1871	117, 948
Hoisting-apparatus	W. C. Williamson	Philadelphia, Pa	June 11, 1872	127, 726
Hoisting-apparatus and derrick	E. Osgood	Boston, Mass	Nov. 23, 1869	97, 223

Index of patents issued from the United States Patent Office from 1790 *to* 1873, *inclusive*—Continued.

Invention.	Inventor.	Residence.	Date.	No.
Hoisting-apparatus and safety-hatch	J. W Osgood	New York, N. Y	Mar. 12, 1872	124, 614
Hoisting-apparatus, Animal	J. Newton	Marengo Township, Mich.	Feb. 7, 1871	111, 671
Hoisting-apparatus, Automatic stop for steam	W. H. Lavinia	Chicago, Ill	July 2, 1872	128, 495
Hoisting-apparatus, Brake-clutch for	T. A. Weston	Ridgewood, N. J	Jan. 14, 1873	134, 957
Hoisting-apparatus, Brake of	C. R. Otis	Yonkers, N. Y	Oct. 18, 1864	44, 740
Hoisting apparatus, Coal	L. S. Chichester	Brooklyn, N. Y	Mar. 1, 1870	100, 259
Hoisting apparatus, Coal	G. Martz	Pottsville, Pa	Aug. 15, 1856	14, 671
Hoisting apparatus, Coal	J. C. White and R. Hay	Tuckerville, Pa	Nov. 25, 1856	16, 133
Hoisting-apparatus for bricks, &c	J. Cranshaw	Rochester, N. Y	Nov. 10, 1857	18, 575
Hoisting-apparatus for builders	L. Atwood	New York, N. Y	Dec. 29, 1868	85, 419
Hoisting-apparatus for miners, Swivel	T. M. Martin	San Francisco, Cal	Mar. 11, 1873	136, 661
Hoisting-apparatus, Hydraulic	P. Hinkle	San Francisco, Cal	July 1, 1873	140, 502
Hoisting-apparatus, Hydraulic	A. Lucius	New York, N. Y	Aug. 15, 1871	118, 031
Hoisting-apparatus, Hydraulic	J. R. Ritter	Reading, Pa	Nov. 19, 1872	133, 256
Hoisting-apparatus, Hydraulic	J. R. Ritter	Reading, Pa	Nov. 19, 1872	133, 257
Hoisting-apparatus, Portable and adjustable	G. A. Myers	Williamsburgh, N. Y	July 1, 1873	140, 427
Hoisting apparatus, Rafter-hook for	D. Hicks and S. Doty	Pontiac, Mich	Apr. 28, 1868	77, 375
Hoisting-apparatus, Safety	C. W. Copeland	New York, N. Y	July 12, 1870	105, 177
Hoisting-apparatus, Safety-mechanism for	W. H. Merrick	Philadelphia, Pa	Mar. 8, 1870	100, 651
Hoisting-apparatus, Safety-pawl for	R. T. Crane and W. H. Lavinia	Chicago, Ill	July 2, 1872	128, 467
Hoisting-apparatus, Ship's	D. J. Wilcoxson	Milan, Ohio	Aug. 2, 1859	24, 967
Hoisting-apparatus, Steam	J. Beggs	Chicago, Ill	July 29, 1873	141, 197
Hoisting-apparatus, Steam	C. S. Crane and W. H. Lavinia	Chicago, Ill	July 2, 1872	128, 466
Hoisting-apparatus, Steam	J. S. Neal	Madison, Ind	Feb. 2, 1864	41, 446
Hoisting-apparatus, Steam	C. R. Otis	Yonkers, N. Y	May 16, 1865	47, 773
Hoisting-apparatus, Steam	C. R. Otis	Yonkers, N. Y	Oct. 1, 1872	131, 894
Hoisting-apparatus, Steam	C. R. Otis	Yonkers, N. Y	Oct. 1, 1872	131, 895
Hoisting-apparatus, Steam	C. R. and N. P. Otis	Yonkers, N. Y	Nov. 21, 1865	51, 076
Hoisting-apparatus, Steam	C. R. and N. P. Otis	Yonkers, N. Y	Jan. 7, 1873	134, 699
Hoisting-apparatus, Steam	O. Tufts	Boston, Mass	Apr. 10, 1866	53, 904
Hoisting apparatus, Water	H. Waterman	Haverhill, Mass	Mar. 6, 1860	27, 400
Hoisting-attachment for the shaft of well-augers	H. H. Russell	Maysville, Mo	Aug. 13, 1872	130, 442
Hoisting-block	J. A. Burr	Brooklyn, N. Y	Oct. 18, 1870	108, 446
Hoisting-block	W. Duchermin	Charlottetown, Prince Edward Island.	Dec. 17, 1861	33, 929
Hoisting-block	W. H. Merrill	Taunton, Mass	Dec. 25, 1855	13, 991
Hoisting-blocks, Anti-friction bearing of	W. Duchermin	Charlottetown, Prince Edward Island.	Feb. 4, 1862	34, 295
Hoisting-bucket	G. Focht	Reading, Pa	Nov. 3, 1857	18, 540
Hoisting-bucket for coal, &c	G. Focht	Reading, Pa	Apr. 14, 1857	17, 034
Hoisting-cages, Safety-attachment for	W. C. Williamson	Philadelphia, Pa	July 4, 1871	116, 656
Hoisting-crane	C. Bassett	Massillon, Ohio	Oct. 7, 1862	36, 598
Hoisting-device	J. J. Doyle	New York, N. Y	Jan. 8, 1861	31, 077
Hoisting-device	S. C. Goodsell	New Haven, Conn	June 25, 1867	66, 079
Hoisting-drum	G. W. La Baw	Jersey City, N. J	Apr. 15, 1856	14, 666
Hoisting-drum	H. Strickler	Carlisle, Pa	Feb. 11, 1868	74, 445
Hoisting drum and grain-fork, Hay	H. Strickler	Carlisle, Pa	Apr. 21, 1868	76, 954
Hoisting-engine	F. Murgatroyd	Cleveland, Ohio	Nov. 4, 1873	144, 217
Hoisting-fork	E. Magruder	Cap au Gris, Mo	Jan. 31, 1871	111, 361
Hoisting-gear	G. Duerre	Williamsburgh, N. Y	Nov. 14, 1871	120, 951
Hoisting-gear	W. O. Jones	Portland, Me	Aug. 11, 1868	80, 968
Hoisting-gear	H. F. Shaw	West Roxbury, Mass	May 21, 1867	64, 915
Hoisting-gear	H. F. Shaw	West Roxbury, Mass	May 3, 1870	102, 719
Hoisting-gear, Attachment for	G. Duerre	Williamsburgh, N. Y	Jan. 16, 1872	122, 712
Hoisting-grapple	J. A. Burgess	Plymouth, Mass	Sept. 20, 1870	107, 444
Hoisting-grapple	T. McGrath	Albany, N. Y	Mar. 2, 1869	87, 349
Hoisting ice, Apparatus for	A. Hunt	Philadelphia, Pa	Feb. 2, 1858	19, 250
Hoisting in mines, Cage for	H. Berry, H. Hochholzer, and F. Denver.	Virginia City, Nev	May 8, 1866	54, 490
Hoisting-jack	B. Beuchel	Kalamazoo, Mich	Nov. 2, 1869	96, 299
Hoisting-jack	E. A. Castellaw	Savannah, Ga	May 31, 1870	103, 563
Hoisting-jack	J. H. Churchill	Cross River, N. Y	Aug. 19, 1873	141, 919
Hoisting-jack	F. Hollenberry	Frizellburgh, Md	Aug. 16, 1870	106, 485
Hoisting-jack	W. Kearney	Newark, N. J	Oct. 19, 1858	21, 837
Hoisting-jack	S. B. Rittenhouse	Plymouth, Ind	Dec. 17, 1867	72, 329
Hoisting-jack	J. W. Steed	Minneapolis, Minn	Nov. 27, 1866	60, 080
Hoisting-machine	C. Abel	New York, N. Y	Jan. 24, 1865	45, 962
Hoisting-machine	W. C. Allison	Philadelphia, Pa	Apr. 2, 1850	7, 230
Hoisting-machine	J. Ashcroft	Brooklyn, N. Y	Nov. 26, 1872	133, 289
Hoisting-machine	D. W. Barr	Lancaster, Pa	Nov. 9, 1858	22, 008
Hoisting-machine	J. Bates	Baltimore, Md	Apr. 18, 1871	113, 836
Hoisting-machine	A. Betteley	Boston, Mass	Dec. 20, 1859	26, 469
Hoisting-machine	J. Bird	New York, N. Y	June 6, 1865	48, 044
Hoisting-machine	D. A. Bolt	Harrisburgh, Pa	Aug. 15, 1871	117, 971
Hoisting-machine	M. K. Carpenter	Cincinnati, Ohio	May 18, 1869	90, 236
Hoisting-machine	G. R. Clarke	New York, N. Y	June 9, 1868	78, 645
Hoisting-machine	S. Croghan	Flatbush, N. Y	Nov. 25, 1862	37, 026
Hoisting-machine	J. Darling	Cincinnati, Ohio	Dec. 23, 1873	145, 727
Hoisting-machine	J. Drummond	New York, N. Y	Mar. 10, 1834	....
Hoisting-machine	J. Edson	Boston, Mass	Oct. 29, 1867	70, 180
Hoisting-machine	W. Eppelsheimer	San Francisco, Cal	Oct. 5, 1869	95, 448
Hoisting-machine	D. Evans	Philadelphia, Pa	Mar. 19, 1834	....
Hoisting-machine	H. T. Goodling	York, Pa	July 21, 1868	80, 067
Hoisting-machine	C. H. Hersey	Boston, Mass	July 22, 1873	141, 051
Hoisting-machine	P. Higdon	Cropper's Depot, Ky	Dec. 26, 1865	51, 719
Hoisting-machine	J. Jewsbury	Brook Fields, England	Mar. 15, 1870	100, 769
Hoisting-machine	G. Johnson	Cincinnati, Ohio	Dec. 28, 1869	98, 388
Hoisting-machine	G. Johnson	Cincinnati, Ohio	Mar. 29, 1870	101, 367
Hoisting-machine	J. Jones	Providence, R. I	Dec. 16, 1873	145, 657
Hoisting-machine	R. A. and W. Kendall	Mineral Point, Wis	Apr. 26, 1870	102, 274
Hoisting-machine	J. Kennedy	Chicago, Ill	Dec. 1, 1868	84, 496
Hoisting-machine	G. R. Long and S. King	Lanark, Ill	Oct. 17, 1871	119, 988
Hoisting-machine	S. M Longley	Hudson City, N. Y	Mar. 1, 1864	41, 779
Hoisting-machine	S. L. Lord	New York, N. Y	Nov. 21, 1871	121, 113
Hoisting-machine	W. M. Loyd	New York, N. Y	July 4, 1871	116, 607
Hoisting-machine	M. Lynch	New York, N. Y	Feb. 8, 1870	99, 578
Hoisting-machine	H. Malley	Chicago, Ill	Mar. 5, 1867	62, 549
Hoisting-machine	J. McCalvey	Philadelphia, Pa	Aug. 28, 1866	57, 535

Index of patents issued from the United States Patent Office from 1790 *to* 1873, *inclusive*—Continued.

Invention.	Inventor.	Residence.	Date.	No.
Hoisting-machine	D. McIntyre and G. C. Reeves	Central City, Colo	Nov. 15, 1864	45, 062
Hoisting-machine	P. W. Mellen	Saint Louis, Mo	July 15, 1873	140, 786
Hoisting-machine	W. Miller	Cincinnati, Ohio	May 12, 1863	38, 497
Hoisting-machine	W. Miller	Cincinnati, Ohio	July 4, 1865	48, 579
Hoisting-machine	W. Miller	Cincinnati, Ohio	May 14, 1867	64, 785
Hoisting-machine	W. Murray	Chicago, Ill	Nov. 19, 1867	71, 043
Hoisting-machine	W. Neal	Louisville, Ky	Apr. 23, 1872	125, 978
Hoisting-machine	R. Packard	Rockland, Me	May 4, 1858	20, 170
Hoisting-machine	J. S. Park	Montandon, Pa	June 29, 1869	92, 088
Hoisting-machine	N. Parrish	Kalamazoo, Mich	Sept. 27, 1870	107, 807
Hoisting-machine	F. B. Perkins	Roxbury, Mass	May 3, 1864	42, 595
Hoisting-machine	S. B. Phelps and C. A. Slack	Norwich, Vt	Mar. 28, 1865	47, 036
Hoisting-machine	G. H. Pittman	McConnellsburgh, Pa	Oct. 1, 1872	131, 902
Hoisting-machine	W. Purden	Baltimore, Md	Mar. 19, 1806	....
Hoisting-machine	H. J. Reedy	Cincinnati, Ohio	June 9, 1868	78, 829
Hoisting-machine	H. J. Reedy	Cincinnati, Ohio	June 20, 1871	116, 096
Hoisting-machine	H. J. and J. Reedy	Cincinnati, Ohio	Oct. 23, 1866	59, 137
Hoisting-machine	G. H. Reynolds	New York, N. Y	Aug. 18, 1868	81, 289
Hoisting-machine	S. Reynolds	Pierrepont Manor, N. Y	June 3, 1873	139, 614
Hoisting-machine	J. Scott	Pontiac, Mich	Mar. 21, 1871	112, 855
Hoisting-machine	W. Sellers	Philadelphia, Pa	Mar. 21, 1871	112, 856
Hoisting-machine	H. M. Smith	Richmond, Va	Apr. 16, 1867	63, 955
Hoisting-machine	I. Smith	Tonhannock, N. Y	Mar. 25, 1873	137, 152
Hoisting-machine	W. M. Smith	Augusta, Ga	Aug. 10, 1869	93, 490
Hoisting-machine	J. W. Tucker	New York, N. Y	May 28, 1867	65, 307
Hoisting-machine	S. Van Emon	Cincinnati, Ohio	Mar. 12, 1867	62, 788
Hoisting-machine	R. A. Wilder	Cressona, Pa	Mar. 18, 1862	34, 710
Hoisting-machine	N. J. Wilkinson	Kalamazoo, Mich	Apr. 26, 1870	102, 345
Hoisting-machine	M. Willard	Cincinnati, Ohio	May 16, 1865	47, 761
Hoisting-machine	W. C. Williamson	Philadelphia, Pa	Dec. 29, 1868	85, 415
Hoisting-machine	W. C. Williamson	Philadelphia, Pa	Aug. 23, 1870	106, 642
Hoisting-machine	I. W. Willson	Syracuse, N. Y	Oct. 3, 1848	5, 828
Hoisting-machine	N. Wonlarlarsky	St. Petersburg, Russia	Dec. 6, 1870	109, 988
Hoisting-machine	J. M. Yates	Princeville, Ill	Sept. 23, 1873	143, 214
Hoisting-machine capstan	C. S. Houck	Greenport, N. Y	Apr. 28, 1868	77, 191
Hoisting-machine for cellars	J. Ingram	New York, N. Y	Feb. 5, 1867	61, 835
Hoisting-machine for stacking hay	W. Louden	Fairfield, Iowa	Dec. 3, 1867	71, 771
Hoisting-machine for vessels	H. O. Perry	Buffalo, N. Y	July 2, 1867	66, 251
Hoisting-machine grapple	J. H. Hecker	Lebanon, Pa	June 25, 1872	128, 309
Hoisting-machine hatch	R. H. Martin and C. R. Otis	Brooklyn and Yonkers, N. Y.	Oct. 1, 1872	131, 888
Hoisting machine, Hay, &c	J. S. Lloyd	Salem, N. J	Apr. 24, 1860	27, 994
Hoisting-machine, Hydrostatic	S. Van Emon	Cincinnati, Ohio	Dec. 17, 1872	134, 019
Hoisting machine, Ice	W. G. Brower	Staatsburgh, N. Y	May 25, 1858	28, 322
Hoisting machine, Ice	J. Wagner	Philadelphia, Pa	Oct. 4, 1859	25, 691
Hoisting machine, Marl, &c	T. A. Granger	Wilson County, N. C	Nov. 8, 1859	26, 029
Hoisting-machine, Pneumatic	J. R. Anderson and P. C. Harlan	Freeport, Pa	July 11, 1871	116, 916
Hoisting-machines, Brake-clutch for	T. A. Weston	Ridgewood, N. J	Jan. 14, 1873	134, 958
Hoisting-machine, Safety-apparatus for	W. T. Sands	New York, N. Y	June 20, 1871	116, 103
Hoisting-machines, Safety-attachment for	W. D. Andrews	Brookhaven, N. Y	Mar. 25, 1873	137, 165
Hoisting-machinery	J. B. Holmes	Boston, Mass	June 7, 1841	2, 115
Hoisting-machinery	R. A. Wilder	Cressona, Pa	May 1, 1860	28, 122
Hoisting-machinery, Horse-power	A. Jackson	New York, N. Y	Aug. 1, 1854	11, 457
Hoisting persons, &c., from one story in a building to another, Machine for.	J. H. Hobbs	Philadelphia, Pa	May 15, 1860	28, 273
Hoisting-protector	J. W. Wilson	New York, N. Y	Oct. 1, 1872	131, 923
Hoisting-power, Tread-wheel	J. B. Cassel	Worcester Township, Pa	Nov. 20, 1866	59, 818
Hoisting-tackle	J. W. Norcross	Middletown, Conn	Dec. 26, 1865	51, 742
Hoisting-tackle	J. C. Price	New Philadelphia, Ohio	Jan. 1, 1867	60, 935
Hoisting-wheels, Operating	J. McMurtry	Fayette County, Ky	Apr. 3, 1860	27, 760
Hoisting-whim	J. C. Marshall	San Francisco, Cal	May 13, 1873	138, 811
Holdback	A. C. Beckwith and J. H. Graham.	Oriskany, N. Y	Jan. 19, 1869	85, 898
Holdback	R. W. Carrier	Sherburne, N. Y	Dec. 20, 1859	26, 476
Holdback	G. D. Cleaveland	Flint, Ind	Oct. 28, 1873	143, 964
Holdback	J. C. Covert	Townsendville, N. Y	Oct. 1, 1867	69, 412
Holdback	G. F. De Vine	Hamilton, Canada	Mar. 26, 1872	124, 888
Holdback	W. Garrison and C. H. Stevens	Syracuse, N. Y	May 4, 1869	89, 573
Holdback	D. A. Gorham	Norway, Me	Mar. 23, 1869	88, 159
Holdback	L. Kruse	Sabula, Iowa	Sept. 1, 1868	81, 651
Holdback	G. B. Lumpkin	Lexington, Ga	Feb. 27, 1872	124, 144
Holdback	R. Moody	Monmouth, Me	Oct. 27, 1868	83, 400
Holdback	M. D. Myers	Schuyler, N. Y	Feb. 11, 1873	135, 833
Holdback	J. G. Rogers	Cass County, Mich	July 16, 1872	129, 253
Holdback	B. E. Sperry	Aurora, Ill	Sept. 3, 1872	131, 036
Holdback-hook	N. W. Robinson	Moriah, N. Y	Feb. 4, 1868	74, 143
Holdback-hook	U. B. Winchell	Oak Hill, N. Y	Dec. 10, 1861	33, 915
Holdback-iron for carriages	J. P. Simmons	Fulton, N. Y	May 24, 1864	42, 884
Holdback-irons	E. F. Howe	Dowagiac, Mich	Aug. 20, 1872	130, 640
Holdback-straps, Hook for	W. A. Bagley	Ansonia, Conn	Nov. 19, 1867	70, 938

Holder:

See Amunition-holder.
Bag-holder.
Barber's check-holder.
Bed-clothes holder.
Bill-holder.
Billiard-table ball-holder.
Billiard-table chalk-holder.
Bit-holder.
Blacking and brush holder.
Blacking-box holder.
Blacking-brush holder.
Blotter-holder.
Blower-holder.
Bolt-holder.
Book-holder.
Book and manuscript holder.
Bottle-holder.
Bouquet-holder.
Brace-bit holder.
Breeching-holder.
Broom-holder.
Brush-holder.
Brush and mop holder.
Bucket-holder.
Button-holder.
Camera plate-holder.
Candle-holder.

Index of patents issued from the United States Patent Office from 1790 *to* 1873, *inclusive*—Continued.

Invention.	Inventor.	Residence.	Date.	No.

Holder—Continued.

See Canopy-holder.
Cap-holder.
Car-stake holder.
Card-holder.
Carriage-tongue holder.
Cartridge-holder.
Carving knife and fork holder.
Chain-holder.
Chalk-holder.
Chalk-line holder.
Cigar-holder.
Cigar-ash holder.
Cigar-cap holder.
Clothes-holder.
Clothes-line holder.
Clothes-pin holder.
Coil-holder.
Coin-holder.
Colter-holder.
Comb-holder.
Copy-holder.
Cork or stopper holder.
Corn-holder.
Cow's tail holder.
Crayon-holder.
Cuff holder.
Currency and postage-stamp holder.
Curtain-holder.
Curtain-roller holder.
Daguerreotype-plate holder.
Diamond-holder.
Document-holder.
Door-holder.
Dress-goods holder.
Drill-holder.
Drinking-glass holder
Egg-holder.
Elastic-bosom holder.
Embroidery-holder.
Eraser-holder.
Eye-glass holder.
Fan and handkerchief holder.
Faucet-holder.
Fence-board-gage holder.
File-holder.
Fire-cracker holder.
Fire-set holder.
Fish-hook holder.
Flat-iron holder.
Fruit-jar holder.
Gas-holder.
Gimlet-holder.
Glass-holder.
Glass-blower's tool-holder.
Globe-holder.
Grain-bag holder.
Hammer-holder.
Handkerchief-holder.
Hat-holder.
Hat and cap holder.
Hat and coat holder.
Hat and parcel holder.
Hog-holder.
Horse-holder.
Horse-tail holder.
Hose-holder.
Ice-cream holder.
Insulator-holder.
Iron-holder.
Jar or vase holder.
Knife-holder.
Knife and fork holder.
Knife-blade holder.
Knife, fork, and spoon holder.
Knitting-holder.
Kite-string holder.
Label-holder.
Lamp-holder.
Lamp-chimney holder.
Lamp-shade holder.
Last-holder.
Last-block holder.
Lead-pencil holder.
Line and strap holder.
Lint and basket holder.
Mail-pouch holder.
Map-holder.
Match-holder.
Measure-holder.
Meat-holder.
Mirror-holder.
Mop-holder.
Mucilage-holder.
Music-leaf holder.
Napkin-holder.
Neck-tie holder.
Needle and tool holder.
Oil-can holder.
Oil-stone holder.
Painter's holder.
Painter's brush-holder.
Painter's drill-holder.
Paper-holder.
Paper-ribbon holder.
Pen-holder.
Pencil-holder.
Percussion-cap holder.
Photographic-paper holder.
Photographic-plate holder.
Pipe-holder.
Pipe-coupling holder.
Pitcher-holder.
Planer's tool-holder.
Plate-holder.
Portfolio-holder.
Postage-holder.
Pounce-holder.
Press-holder.
Railway-rail-bolt holder.
Razor-strop holder.
Rein-holder.
Rosette-holder.
Rubber-block holder.
Ruche-holder.
Sad-iron holder.
Salt-holder.
Sand-paper holder.
Sash-holder.
Sash-spring holder.
Scythe-holder.
Scythe-stone holder.
Seat-holder.
Sewing-machine-bobbin holder.
Sewing-machine cloth-holder.
Sewing-machine work-holder.
Sewing-thread holder.
Sewing-work holder.
Shade-holder.
Shaft-holder.
Shaft and pole holder.
Sheep-holder.
Shoe-holder.
Shook-holder.
Shutter-holder.
Siding-holder.
Signal-holder.
Slat-holder.
Slate-pencil holder.
Sliding-holder.
Soap-holder.
Soda-water-tumbler holder.
Spice-holder.
Spinning-machine bobbin holder.
Spinning-ring holder.
Spittoon-holder.
Sponge-holder.
Spool-holder.
Spoon-holder.
Stalk-holder.
Stamp-holder.
Stereoscopic-picture holder.
Stereotype-block holder.
Stereotype-plate holder.
Stitching-work holder.
Stove-holder.
Stove-utensil holder.

Index of patents issued from the United States Patent Office from 1790 *to* 1873, *inclusive*—Continued.

Invention.	Inventor.	Residence.	Date.	No.
Holder—Continued.				
See Strap-holder. Umbrella-holder.				
Tack-holder. Vine-holder.				
Tag-holder. Violin-holder.				
Tailor's pressing-board holder. Wash-tub holder.				
Watch-wheel holder.				
Taper-holder. Wax-candle holder.				
Thill-holder. Wax candle or taper holder.				
Ticket-holder.				
Tongue-holder. Whip-holder.				
Tool-holder. Whip and line holder.				
Towel-holder. Wire jar-holder.				
Tray-holder. Work-holder.				
Twine-holder. Yoke-holder.				
Holding and soldering cans, Machine for	W. J., jr., and T. G. Redhefter, and L. A. Smith.	Kansas City, Mo., and Pawnee, Kans.	Mar. 18, 1873	137, 027
Hollow bolt	D. F. Fetter, M. D	New York, N. Y	Feb. 16, 1869	86, 976
Hollow drill, Adjustable	C. E. Butler	Hudson, N. Y	Apr. 19, 1870	102, 090
Hollow ring	K. Frazer	Syracuse, N. Y	Jan. 1, 1867	60, 712
Hollow-ware, &c., Apparatus for pressing	N. Thompson	Brooklyn, N. Y	Oct. 10, 1871	119, 725
Hollow-ware from tinned plates, sheet-iron, &c., Manufacturing.	E. M. Converse	Southington, Conn	July 20, 1831	....
Hollow-ware from zinc, &c., Manufacturing	S. Davis	New York	Apr. 18, 1831	....
Hollow-ware, Making gutta-percha	S. T. Armstrong	New York, N. Y	June 24, 1851	8, 180
Hollow-ware, Manufacture of sheet-iron	J. and J. D. Gray	Pittsburgh, Pa	May 5, 1863	38, 446
Hollow-ware, Mode of hanging covers to baled metallic.	E. Ripley	Troy, N. Y	Jan. 1, 1861	31, 035
Hollow-ware molding	J. I. Johnston	Cincinnati, Ohio	June 29, 1852	9, 074
Holster	W. Tileston	Georgetown, D. C	June 21, 1864	43, 275
Hominy and pearling mill	E. A. Duer	Decatur, Ill	Aug. 4, 1868	80, 713
Hominy and smut mill	W. Wright	Saint Louis, Mo	Dec. 8, 1868	84, 723
Hominy-machine	R. Campbell	Martinsburgh, Va	Apr. 9, 1827	....
Hominy-machine	W. Davis	Middleburgh, Md	May 24, 1859	24, 104
Hominy-machine	E. Fahrney	Mount Morris, Ill	May 15, 1855	12, 860
Hominy-machine	J. Gehr	Clear Spring, Md	July 17, 1860	29, 159
Hominy-machine	D. Haines	Union Bridge, Md	June 11, 1861	32, 516
Hominy-machine	O. F. Mayhew	Indianapolis, Ind	June 2, 1857	17, 469
Hominy-machine	J. Nesbitt, jr., and T. J. Crosley	Clear Spring, Md	Sept. 11, 1855	13, 549
Hominy-machine	S. Null	Carroll County, Md	May 25, 1852	8, 972
Hominy-machine	P. Siemers	Saint Louis, Mo	Oct. 13, 1857	18, 413
Hominy-machine	R. H. Taylor	Goose Creek, Va	Aug. 21, 1866	57, 406
Hominy, Making	W. Y. Singleton	Springfield, Mass	Apr. 25, 1846	4, 486
Hominy-mill	B. Bridendolph	Clear Spring, Md	Aug. 22, 1854	11, 547
Hominy-mill	J. M. Clark	Lancaster, Pa	May 25, 1858	20, 327
Hominy-mill	E. M. Coombes	Memphis, Ind	July 19, 1870	105, 430
Hominy-mill	W. C. Coombs and J. M. Gray	Memphis, Ind	Jan. 4, 1870	98, 565
Hominy-mill	F. E. Drake and J. W. Teal	Indianapolis, Ind	Mar. 23, 1858	19, 691
Hominy-mill	E. Fahrney	Deep River, Iowa	Jan. 5, 1858	19, 060
Hominy-mill	J. B. Gowdy and J. A. Welsh	Xenia, Ohio	Aug. 11, 1857	17, 966
Hominy-mill	J. Hayden	Philadelphia, Pa	Apr. 3, 1866	53, 611
Hominy-mill	J. D. Heatwole and R. C. Mauck	Harrisonburgh, Va	May 1, 1860	28, 079
Hominy-mill	P. Honarighaus	Royalton, Ohio	Feb. 9, 1858	19, 297
Hominy-mill	T. Rudnut	Terre Haute, Ind	Dec. 26, 1871	122, 172
Hominy-mill	J. Hughes	Cambridge City, Ind	Oct. 12, 1852	9, 323
Hominy-mill	B. Kemp	Knoxville, Md	Dec. 11, 1866	60, 385
Hominy-mill	J. R. Leedy	Maurertown, Va	Oct. 4, 1870	108, 036
Hominy-mill	N. Read	Belfast, Me	May 8, 1820	....
Hominy-mill	I. Speight	Woodville, Miss	Dec. 21, 1858	22, 384
Hominy-mill	G. Strause	Boonsborough, Md	Sept. 20, 1859	25, 536
Hominy-mill	J. L. Toner	Edinburgh, Ind	Nov. 11, 1873	144, 486
Hominy-mill	W. Wright	Springfield, Ohio	Sept. 6, 1864	44, 136
Hominy-mill	W. Wright	Springfield, Ohio	June 20, 1865	48, 332
Hominy-mill burr	A. P. Jackson	Memphis, Ind	June 15, 1869	91, 343
Hominy-mills, Separator for	J. Donaldson	Rockford, Ill	Jan. 24, 1860	26, 893
Homomotive	W. S. Hall	Quincy, Mass	June 8, 1869	91, 123
Hone	I. Babbitt	Roxbury, Mass	Mar. 23, 1854	10, 954
Hone, Boxed	J. Potter and O. Abell	Whitehall, N. Y	Jan. 14, 1868	73, 383
Hone, Glass	A. Gordon and J. P. Bakewell	Pittsburgh, Pa	Dec. 8, 1832	....
Hone, Metallic	J. Houston	Williamsburgh, Va	May 4, 1805	....
Hone, Metallic	W. H. Webb, jr	Chelsea, Mass	Apr. 17, 1855	12, 748
Hones, Manufacture of artificial	T. Deming	East Hartford, Conn	Oct. 13, 1857	18, 392
Honey, Artificial	Z. Corbin and G. Marlett	Syracuse, N. Y	May 12, 1857	17, 264
Honey, Artificial	G. H. Smitson	Ripley, Ohio	July 13, 1869	92, 666
Honey-comb, Artificial	S. Wagner	York, Pa	May 7, 1861	32, 258
Honey-comb knife	H. O. Peabody	Boston, Mass	May 28, 1872	127, 187
Honey-combs, Device for making artificial	J. Williams	Bean's Station, Tenn	Dec. 31, 1872	134, 411
Honey-cutting knife	A. W. Todd	Chicago, Ill	May 17, 1864	42, 807
Honey from the comb, Apparatus for extracting	L. L. Langstroth and S. Wagner	Butler County, Ohio, and Washington, D. C.	Jan. 15, 1867	61, 216
Honey from the comb, Centrifugal apparatus for extracting.	W. Probasco	Lawrenceburgh, Ind	Jan. 7, 1873	134, 561
Honey from the comb, Centrifugal machine for extracting.	H. O. Peabody	Boston, Mass	Oct. 26, 1869	96, 142
Honey from the comb, Centrifugal machine for extracting.	H. O. Peabody	Boston, Mass	Dec. 7, 1869	97, 684
Hood, Lady's	E. Hill	Philadelphia, Pa	July 18, 1865	48, 867
Hood, Lady's	M. Landenberger	Philadelphia, Pa	July 19, 1864	43, 591
Hoof-expander	C. B. and S. Galentine and A. J. Russell.	Nunda, N. Y	Dec. 23, 1856	16, 273
Hoof-expander	J. Tipton	Malaga, Ohio	Jan. 21, 1868	73, 675
Hoof-hooks, Joining and fitting	F. Stanley	Austin, Tex	Sept. 15, 1868	82, 172
Hoof-pad, Elastic	F. B. Carleton	Cambridge, Vt	Jan. 1, 1867	60, 687
Hoof-parer	I. Baker	Long Branch, Mo	Oct. 10, 1871	119, 734
Hoof parer, Horse	A. Baker	Shenandoah County, Va	Aug. 7, 1860	29, 543
Hoof parer, Horse	A. Crafts and E. Weeks	Auburn, Ohio	Jan. 1, 1850	6, 983
Hoof parer, Horse	S. D. Freet	McCutchenville, Ohio	Mar. 5, 1867	62, 624
Hoof parer, Horse	F. Lehmann	Riverside, Ill	Nov. 16, 1869	90, 928
Hoof parer, Horse	V. N. Mitchell	Concord, N. C	June 2, 1857	17, 441

Index of patents issued from the United States Patent Office from 1790 *to* 1873, *inclusive*—Continued.

Invention.	Inventor.	Residence.	Date.	No.
Hoof parer, Horse	W. Tansley	Salisbury Centre, N. Y	Apr. 22, 1862	35, 046
Hoof parer, Horse	J. Temple	Van Buren, Ohio	June 15, 1869	91, 286
Hoof-pairing tool, Horse	J. C. Johnson	Sulphur Springs, Ind	Oct. 8, 1872	132, 010
Hoof-shears	M. C. Malone	Palmyra, Ill	Apr. 30, 1872	126, 221
Hoof-spreader	T. Armstrong	Hamilton, Canada	Mar. 11, 1873	136, 689
Hoofs, Machine for paring horses'	G. W. Schafer	Saint Charles, Mo	Apr. 22, 1873	138, 101

Hook	W. M. Knight and J. H. Orne	Marblehead, Mass	Aug. 22, 1865	49, 535
Hook	R. V. Laney	Cumberland, Md	July 19, 1870	105, 582
Hook	J. B. Sweetland	Pontiac, Mich	June 23, 1868	79, 159
Hook	J. H. Tracy	Mayville, Mich	Sept. 14, 1869	94, 926
Hook	R. Weiss	Brooklyn, N. Y	Feb. 18, 1873	136, 114
Hook and button, Combined	W. H. Shurtleff	Providence, R. I	June 5, 1866	55, 431
Hook and eye	A. Bennett	Brooklyn, N. Y	Feb. 12, 1867	61, 989
Hook and eye	J. P. Culver	New York, N. Y	Jan. 24, 1865	45, 979
Hook and eye	M. R. L. Davis	Kansas City, Mo	Feb. 13, 1872	123, 619
Hook and eye	J. H. Fairchild	Jericho, Vt	May 30, 1854	10, 984
Hook and eye	J. C. Howells	New York, N. Y	Dec. 13, 1864	45, 411
Hook and eye	W. Law	Birmingham, England	Jan. 28, 1868	73, 817
Hook and eye	J. B. Martindale	Newcastle, Ind	Nov. 27, 1866	60, 026
Hook and eye	J. B. Martindale	Newcastle, Ind	Mar. 5, 1867	62, 554
Hook and eye	A. C. Mason	Springfield, Vt	Apr. 9, 1861	32, 031
Hook and eye	H. McEroy	Birmingham, England	Apr. 17, 1849	6, 329
Hook and eye	H. Nickolds	Providence, R. I	Oct. 25, 1864	44, 815
Hook and eye	E. P. Roche	Bath, Me	Oct. 19, 1869	95, 936
Hook and eye	E. C. Savage	Hartford, Conn	Mar. 26, 1844	3, 512
Hook and eye	D. M. Smith	Springfield, Vt	May 22, 1860	28, 443
Hook and eye	I. Weinberg	Philadelphia, Pa	Dec. 18, 1866	60, 600
Hook and eye card	M. Fowler	Northport, Conn	Oct. 29, 1867	70, 190
Hook and eye for connecting cords	A. Codding, jr	North Attleborough, Mass	July 7, 1863	39, 121
Hook and eye for fastening garments	C. Atwood	Derby, Conn	Feb. 24, 1843	2, 978
Hook and eye making machine	J. S. Ford	San Francisco, Cal	Mar. 7, 1871	112, 331
Hook and eye making machine	J. Stewart	Boston, Mass	July 20, 1831	....
Hook and eye, Wire	C. Atwood	Derby, Conn	July 1, 1851	8, 198

Index of patents issued from the United States Patent Office from 1790 *to* 1873, *inclusive*—Continued.

Invention.	Inventor.	Residence.	Date.	No.
Hook and hasp, Combined	B. H. Brooks	New Haven, Conn	Aug. 12, 1873	141, 691
Hook and ladder apparatus	J. S. Hunt	Richmond, Ind	July 9, 1872	128, 882
Hook and terret	W. R. Ferguson	Marseilles, Ill	Aug. 20, 1867	67, 861
Hook bending and punching	P. L. Weimer	Lebanon, Pa	Oct. 31, 1865	50, 757
Hook-bending machine	W. Malick	Erie, Pa	Sept. 10, 1872	131, 289
Hook-bending machine	R. B. Sears	Providence, R. I	Dec. 24, 1867	72, 547
Hook-eye for wearing-apparel and other purposes	J. C. Howells	Washington, D. C	May 12, 1863	38, 483
Hook-pin for fastening wearing-apparel	A. Lindsay and M. Moses	Malone, N. Y	Oct. 23, 1866	59, 040
Hooks and eyes, Machine for separating	M. Fowler	Northford, Conn	Nov. 29, 1864	45, 238
Hooks and eyes to cards, Attaching	C. Atwood	Birmingham, Conn	Sept. 25, 1849	6, 745
Hooks and eyes to cards, Attaching	C. Atwood	Birmingham, Conn	Dec. 20, 1853	10, 330
Hooks and eyes to cards, Attaching	T. B. De Forest	Birmingham, England	Sept. 15, 1863	39, 893
Hooks and eyes to cards, Machine for attaching	A. Capron and J. S. Dennis	Attleborough and Somerville, Mass.	Aug. 7, 1855	13, 409
Hooks and eyes to cards, Securing	J. D. Franklin	Attleborough, Mass	Feb. 4, 1868	74, 072
Hooks and eyes to paper cards, Attaching	P. Kirkham	Waterbury, Conn	July 30, 1850	7, 526
Hooks and eyes to paper cards, Fastening	C. J. Darrington	Waterbury, Conn	Sept. 9, 1851	8, 354
Hooks and eyes to tapes and dresses, Securing	C. Atwood	Derby, Conn	Aug. 7, 1849	6, 628
Hooks and eyes upon cards, Fastening	E. J. Warner	Waterbury, Conn	Nov. 5, 1850	7, 762
Hooks and staples, Machinery for bending	S. Brooks	Chester, Conn	Nov. 19, 1861	33, 733
Hooks, Elastic mousing for	E. E. Stone	United States Navy	Aug. 8, 1865	49, 316
Hooks, staples, &c., Machine for pointing	W. Malick	Tidioute, Pa	Dec. 1, 1868	84, 562
Hooker-up, Mechanical	D. J. Happersett	Downingtown, Pa	July 8, 1851	8, 204
Hoop:				
See Automatic hoop. Bale hoop. Barrel-hoop. Bucket-hoop. Cheese-hoop. Dowel-hoop. Driving-hoop. Flanged and beaded hoop. Grace-hoop. Guide-rolling hoop. Iron hoop. Jumping-hoop. Mast-hoop. Pail-hoop. Propelling-hoop. Skirt-hoop. Steam coiled hoop. Toy-hoop. Trundling-hoop. Truss-hoop.				
Hoop and barrel dressing machine	Kinbell, Perry, and Spalding	Peterborough, N. H	Sept. 18, 1835	....
Hoop and lath sawing machine	I. Price, jr	Lockport, N. Y	Aug. 19, 1828	....
Hoop and train skirt	L. Guelle	Hoboken, N. J	Mar. 4, 1873	136, 434
Hoop-bending machine	E. Coapman	Rochester, N. Y	Dec. 9, 1873	145, 398
Hoop-bending machine	G. E. Smith	Middleport, N. Y	Oct. 12, 1869	95, 845
Hoop-bolts, Machine for pointing and checking	A. A. Wilder	Detroit, Mich	Mar. 31, 1863	38, 085
Hoop-box, Wooden	S. S. Barrie	Green Point, N. Y	Nov. 22, 1870	109, 373
Hoop cutting and bending	J. Dobbins	Litchfield, Mich	July 4, 1865	48, 535
Hoop cutting and bending machine	J. Dobbins	Litchfield, Mich	Feb. 4, 1873	135, 413
Hoop cutting and dressing machine	F. Ellis	Sylvania, Ohio	Dec. 28, 1869	98, 240
Hoop-cutting machine	W. H. Davis	Lexington, Ind	Aug. 23, 1870	106, 671
Hoop-cutting machine	A. G. Parkhurst	Appleton, Wis	Apr. 1, 1873	137, 384
Hoop-cutting machine	P. Welch	Saint Louis, Mo	July 29, 1873	141, 246
Hoop-dressing machine	S. L. Heywood	Minneapolis, Minn	May 27, 1873	139, 387
Hoop-dressing machine	J. Penney	Rochester, N. Y	Nov. 1, 1870	108, 820
Hoop-dressing machine	A. Prenatt	Buffalo, N. Y	Jan. 11, 1859	22, 581
Hoop-driving and barrel-crozing machine	E. Holmes	Buffalo, N. Y	Feb. 17, 1863	37, 719
Hoop-irons, Machinery for rolling	B. Norton	Boonton, N. J	Sept. 26, 1846	4, 772
Hoop-lock	G. N. Beard	Saint Louis, Mo	Oct. 16, 1866	58, 760
Hoop-lock	C. Hughes	New Orleans, La	Feb. 5, 1861	31, 319
Hoop-lock	D. Hughes	Rochester, N. Y	Dec. 13, 1859	26, 435
Hoop-lock	E. A. Jeffery	Corning, N. Y	Dec. 21, 1858	22, 369
Hoop-lock	T. E. Lucas	Chesterfield, S. C	Oct. 28, 1873	144, 117
Hoop-lock	A. P. Merrill, jr	Natchez, Miss	Jan. 3, 1860	26, 694
Hoop-lock clasp	T. Hanvey	Elma, N. Y	July 12, 1864	43, 495
Hoop-lock cutter	J. G. Logginsand P. P. Wilkins	Williston, Vt	Dec. 30, 1873	145, 956
Hoop-lock cutter	W. and H. A. Tripp	Williamson, N. Y	Oct. 1, 1872	131, 835
Hoop-lock-cutting machine	T. Conklin	Fond du Lac, Wis	Apr. 5, 1870	101, 436
Hoop-lock-cutting machine	A. Cutter	Boston, Mass	Oct. 11, 1870	108, 114
Hoop-lock-cutting machine	J. McGren	Ravenswood, W. Va	June 3, 1873	139, 594
Hoop-lock for metal bale-hoops	C. A. Dubs	Natchez, Miss	May 8, 1860	28, 164
Hoop-machine	J. J. Alvord	Tecumseh, Wis	Apr. 30, 1867	64, 270
Hoop-machine	S. F. Atherton	Fitchburgh, Mass	Nov. 13, 1860	30, 609
Hoop-machine	J. B. Dougherty	Rochester, N. Y	Nov. 17, 1863	40, 613
Hoop-machine	G. W. Holmes	Buckfield, Me	Apr. 22, 1856	14, 753
Hoop machine	H. C. Peirson	Philadelphia, Pa	Mar. 29, 1859	23, 389
Hoop-machine	J. and S. Sawyer	Fitchburgh, Mass	May 6, 1856	14, 833
Hoop-machine	J. and S. Sawyer	Fitchburgh, Mass	Sept. 23, 1856	15, 780
Hoop-machine	J. Thompson	Rochester, N. Y	Nov. 17, 1863	40, 664
Hoop-machine	W. B. Wood	Fitchburgh, Mass	Oct. 7, 1856	15, 865
Hoop-making	M. Granger	Utica, N. Y	Dec. 31, 1833	....
Hoop-making machine	W. Lawyer	Macomb, N. Y	Nov. 2, 1869	96, 448
Hoop-making machine	A. A. Wilder	Detroit, Mich	Mar. 31, 1863	38, 084
Hoop manufacturing machine	J. B. Dougherty	Rochester, N. Y	Aug. 9, 1870	106, 260
Hoop-manufacturing machine	J. Peirson	Alexandria, Va	Feb. 20, 1855	12, 429
Hoop-notching machine	D. Lamson	East Weymouth, Mass	Sept. 23, 1856	15, 768
Hoop notching and trimming machine	S. Littlefield	West Troy, N. Y	Sept. 14, 1858	21, 507
Hoop-planing machine	S. Sawyer	Fitchburgh, Mass	Jan. 27, 1857	16, 513
Hoop-planing machine	T. S. Scoville	Elmira, N. Y	Apr. 28, 1857	17, 171
Hoop-poles, Machine for splitting	J. Penney	Grand Rapids, Mich	May 21, 1872	126, 983
Hoop-poles, Machine for splitting	J. Penney	Grand Rapids, Mich	July 2, 1872	128, 561
Hoop-poles, Machine for splitting	J. and S. Sawyer	Fitchburgh, Mass	Apr. 7, 1857	17, 014
Hoop-racking machine	R. M. Shaner	Genoa, Ohio	Dec. 20, 1870	110, 295
Hoop-riving machine	G. J. Bentley	Michigan City, Ind	Feb. 21, 1865	46, 439
Hoop-sawing machine	N. Bratt	Lockport, N. Y	July 31, 1828	....
Hoop-sawing machine	A. Lutz	Orangeville, Ill	Jan. 21, 1868	73, 615
Hoop, &c., sawing machine	P. Slayton	Lockport, N. Y	May 1, 1828	....
Hoop-sawing machine	E. Strange and T. B. Smith	Taunton, Mass	Sept. 4, 1855	13, 531
Hoop-sawing machine	E. C. Strange	Taunton, Mass	Jan. 13, 1857	16, 407
Hoop sawing machine, Cooper's	J. O. Woodward	Taunton, Mass	July 22, 1856	15, 399
Hoop-shaving machine	J. A. Douglass and J. Wagner	Chest Springs, Pa	Feb. 11, 1873	135, 637
Hoop-shaving machine	G. Heatwoal	Harrisburgh, Pa	July 27, 1831	....
Hoop-shaving machine	S. Parker	Altoona, Pa	Sept. 20, 1870	107, 622

Index of patents issued from the United States Patent Office from 1790 *to* 1873, *inclusive*—Continued.

Invention.	Inventor.	Residence.	Date.	No.
Hoop-skirt	E. Adams, jr	Attleborough, Mass	June 18, 1867	65, 854
Hoop-skirt	H. B. Ames	Brooklyn, N. Y	Jan. 21, 1862	34, 225
Hoop-skirt	H. B. Aurel	Brooklyn, N. Y	Mar. 1, 1864	41, 809
Hoop-skirt	J. E. Atwood	New York, N. Y	July 24, 1866	56, 512
Hoop-skirt	A. M. Bardwell	Amherst, Mass	May 22, 1866	54, 998
Hoop-skirt	M. Berliner	New York, N. Y	Oct. 1, 1872	131, 840
Hoop-skirt	G. Brering	New York, N. Y	Dec. 7, 1869	97, 750
Hoop-skirt	M. P. Bray	Birmingham, Conn	Mar. 14, 1871	112, 681
Hoop-skirt	F. A. Brewster	Springfield, Mass	July 30, 1867	67, 158
Hoop-skirt	O. R. Burnham	New York, N. Y	Mar. 25, 1862	34, 737
Hoop-skirt	J. P. Buzzell	Clinton, Mass	Feb. 13, 1866	52, 637
Hoop-skirt	C. E. Carpenter	New York, N. Y	June 27, 1871	116, 411
Hoop-skirt	A. Carter	New York, N. Y	Apr. 18, 1871	113, 854
Hoop-skirt	J. F. Chase	Augusta, Me	June 15, 1869	91, 418
Hoop-skirt	W. Coe	New Haven, Conn	Dec. 23, 1873	145, 845
Hoop-skirt	S. Collins	New Haven, Conn	Dec. 20, 1864	45, 557
Hoop-skirt	H. Cook	Manchester, England	Oct. 22, 1861	33, 517
Hoop-skirt	C. Daniels	Birmingham, Conn	Apr. 2, 1867	63, 366
Hoop-skirt	T. D. Day	New York, N. Y	Dec. 13, 1864	45, 392
Hoop-skirt	T. D. Day	New York, N. Y	July 3, 1866	56, 016
Hoop-skirt	T. D. Day	New York, N. Y	Feb. 16, 1869	86, 910
Hoop-skirt	T. B. De Forest	Birmingham, Conn	Jan. 15, 1867	61, 169
Hoop-skirt	T. B. De Forest	Birmingham, Conn	Jan. 15, 1867	61, 170
Hoop-skirt	T. B. De Forest	Birmingham, Conn	Mar. 26, 1867	63, 145
Hoop-skirt	D. H. Fanning	Worcester, Mass	Mar. 31, 1868	76, 064
Hoop-skirt	A. Fellheimer	New York, N. Y	Mar. 26, 1867	63, 233
Hoop-skirt	L. Fellheimer	New York, N. Y	Mar. 26, 1867	63, 234
Hoop-skirt	M. Fishel	New York, N. Y	June 5, 1866	55, 266
Hoop-skirt	E. Fleischer	Cincinnati, Ohio	July 20, 1869	92, 811
Hoop-skirt	L. H. Foy	Worcester, Mass	Aug. 21, 1866	57, 309
Hoop-skirt	H. Freudenburg	New York, N. Y	July 25, 1865	48, 924
Hoop-skirt	T. S. Gilbert	Derby, Conn	Jan. 5, 1864	41, 148
Hoop-skirt	T. S. Gilbert	Derby, Conn	Jan. 5, 1864	41, 149
Hoop-skirt	T. S. Gilbert	New Haven, Conn	June 19, 1866	55, 768
Hoop-skirt	D. Hawkins	Birmingham, Conn	Aug. 4, 1863	39, 396
Hoop-skirt	C. F. Hendee	Waterbury, Conn	Feb. 25, 1862	34, 496
Hoop-skirt	H. A. Horn	New York, N. Y	Jan. 26, 1869	86, 340
Hoop-skirt	W. E. Houston	Birmingham, Conn	May 21, 1867	64, 870
Hoop-skirt	T. Hull	Birmingham, Conn	Nov. 26, 1867	71, 487
Hoop-skirt	J. Levy	Philadelphia, Pa	May 16, 1865	47, 738
Hoop-skirt	L. Lobenstein	New York, N. Y	Feb. 6, 1866	52, 427
Hoop-skirt	H. Loewenberg	New York, N. Y	Nov. 25, 1862	37, 003
Hoop-skirt	G. Mallory	Watertown, Conn	Oct. 19, 1858	21, 839
Hoop-skirt	J. Mann and A. McDonald	Waltham and Cambridge, Mass.	June 9, 1863	38, 832
Hoop-skirt	J. Mayer	Brooklyn, N. Y	June 1, 1869	90, 673
Hoop-skirt	J. McKeever	New York, N. Y	Nov. 19, 1867	71, 033
Hoop-skirt	K. McRae	New York, N. Y	Oct. 5, 1869	95, 625
Hoop-skirt	I. T. Meyer and J. F. J. Gunning	New York, N. Y	Nov. 10, 1868	83, 986
Hoop-skirt	S. A. Moody	New York, N. Y	May 10, 1864	42, 677
Hoop-skirt	E. L. Morris	Boston, Mass	Aug. 14, 1866	57, 169
Hoop skirt	C. Neumann	New York, N. Y	Oct. 20, 1863	40, 355
Hoop-skirt	C. Neumann	New York, N. Y	July 31, 1866	56, 787
Hoop-skirt	C. Neumann	New York, N. Y	Jan. 8, 1867	61, 088
Hoop-skirt	F. S. Otis	Brooklyn, N. Y	Nov. 15, 1864	45, 070
Hoop-skirt	C. E. Pratt	Rahway, N. J	July 5, 1870	105, 124
Hoop-skirt	W. S. Ryerson	Philadelphia, Pa	Aug. 27, 1867	68, 238
Hoop-skirt	L. Sanders	New York, N. Y	Oct. 20, 1863	40, 363
Hoop-skirt	L. Sanders	New York, N. Y	May 10, 1864	42, 736
Hoop-skirt	L. Sanders	New York, N. Y	Aug. 2, 1864	43, 710
Hoop-skirt	J. Schleisinger	New York, N. Y	July 31, 1866	56, 803
Hoop-skirt	L. S. Scofield	Belmont, Mass	Oct. 13, 1863	40, 285
Hoop-skirt	P. E. Sheffield	Pontiac, Mich	May 15, 1866	54, 785
Hoop-skirt	S. S. Sherwood	Acquackamack, N. J	Dec. 15, 1863	40, 985
Hoop-skirt	W. D. Sloan	New York, N. Y	Oct. 14, 1862	36, 677
Hoop-skirt	T. S. Sperry	Chicago, Ill	June 27, 1871	116, 507
Hoop-skirt	A. R. Stanley	Boston, Mass	Mar. 31, 1868	76, 291
Hoop-skirt	W. H. Towers	Boston, Mass	Nov. 17, 1868	84, 149
Hoop-skirt	A. Trager	New York, N. Y	June 18, 1867	65, 986
Hoop-skirt	H. Vincent	New York, N. Y	Jan. 1, 1867	60, 969
Hoop-skirt	E. C. Walker	Newark, N. J	Feb. 26, 1867	62, 381
Hoop-skirt	J. Waterman	New York, N. Y	Oct. 25, 1864	44, 841
Hoop-skirt	J. Waterman	New York, N. Y	Mar. 22, 1870	101, 195
Hoop-skirt	J. O. West	New York, N. Y	Feb. 21, 1871	111, 998
Hoop-skirt	J. Whitehead and J. McKeever	New York, N. Y	Aug. 10, 1869	93, 573
Hoop-skirt	S. R. Wilmot	Brooklyn, N. Y	June 11, 18 1	32, 547
Hoop-skirt and bustle combined	A. R. Young	Boston, Mass	Aug. 18, 1868	81, 319
Hoop-skirt bands, Buckle or clasp for	H. P. Brooks	Waterbury, Conn	Aug. 10, 1869	93, 592
Hoop-skirt clasp	A. J. Cross	Greenport, N. Y	June 23, 1868	79, 210
Hoop-skirt clasp	T. B. De Forest	Birmingham, Conn	Oct. 2, 1866	58, 540
Hoop-skirt clasp	A. Delkescamp	Brooklyn, N. Y	Apr. 16, 1867	63, 867
Hoop-skirt clasp	G. P. Evans	Malden, Mass	Dec. 11, 1860	30, 872
Hoop-skirt clasp	L. Fellheimer	New York, N. Y	July 28, 1868	80, 343
Hoop-skirt clasp	J. F. J. Gunning	New York, N. Y	Feb. 11, 1868	74, 349
Hoop-skirt clasp	W. E. Houston	Birmingham, Conn	May 21, 1867	64, 871
Hoop-skirt clasp	J. Ingraham	New York, N. Y	Dec. 1, 1868	84, 492
Hoop-skirt clasp	M. Samuels	New York, N. Y	Dec. 29, 1868	85, 480
Hoop-skirt clasp	A. Smart	New York, N. Y	Aug. 31, 1858	21, 373
Hoop-skirt clasp	S. R. Wilmot	Bridgeport, Conn	Oct. 2, 1866	58, 527
Hoop-skirt clasp	S. R. Wilmot	Bridgeport, Conn	Sept. 10, 1867	68, 677
Hoop-skirt clasps, Forceps for fastening	G. D., S. A., and C. L. Russell	Birmingham, Conn	Dec. 14, 1858	22, 308
Hoop-skirt clasps, Machine for making	A. Delkescamp	Brooklyn, N. Y	June 11, 1867	65, 652
Hoop-skirt clasps, Machinery for forming	F. J. Terrell	Ansonia, Conn	May 3, 1864	42, 608
Hoop-skirt clasps, Making	J. H. Doolittle	Ansonia, Conn	Nov. 15, 1859	26, 144
Hoop-skirt clasps, Making	J. H. Doolittle	Ansonia, Conn	Sept. 26, 1865	50, 102
Hoop-skirt fastening	F. E. Day	New York, N. Y	Sept. 28, 1869	95, 203
Hoop-skirt fastening	A. W. Hale	New York, N. Y	Aug. 30, 1859	25, 260
Hoop-skirt former	A. Raffel	New York, N. Y	Feb. 13, 1866	52, 602

Index of patents issued from the United States Patent Office from 1790 to 1873, inclusive—Continued.

Invention.	Inventor.	Residence.	Date.	No.
Hoop-skirt former	A. R. Stanley	Boston, Mass	Mar. 31, 1868	76, 263
Hoop-skirt former	C. C. Wilson	Baltimore, Md	May 3, 1870	102, 638
Hoop-skirt joint	S. J. Sherman	Brooklyn, N. Y	Aug. 15, 1865	49, 447
Hoop-skirt, Lady's	S. Beberdy	Philadelphia, Pa	Sept. 14, 1858	21, 479
Hoop-skirt, Lady's	J. Holmes	Boston, Mass	May 3, 1859	23, 841
Hoop-skirt, Lady's	F. Hull	Derby, Conn	Feb. 8, 1859	22, 875
Hoop-skirt, Lady's	S. Peberdy	Philadelphia, Pa	Nov. 30, 1858	22, 197
Hoop-skirt-making machine	E. R. Hopkins	New York, N. Y	Nov. 27, 1866	60, 010
Hoop-skirt-making machine	C. Neumann	New York, N. Y	June 26, 1860	28, 893
Hoop-skirt-making machine	S. H. Perkins and T. S. Gilbert.	New Haven, Conn	Feb. 26, 1867	62, 359
Hoop-skirt-making machine	S. H. Perkins and T. S. Gilbert.	New Haven, Conn	Apr. 16, 1867	63, 936
Hoop-skirt, Skeleton	R. J. Mann	Brooklyn, N. Y	Nov. 9, 1858	22, 051
Hoop-skirt, Skeleton	C. Neumann	New York, N. Y	Nov. 1, 1859	25, 976
Hoop-skirt, Spiral	D. G. Rollin	New York, N. Y	Dec. 4, 1860	30, 838
Hoop-skirt spring	T. B. DeForest and T. S. Gilbert	Birmingham, Conn	Feb. 18, 1868	74, 672
Hoop-skirt supporter	E. Blakeslee	Plymouth, Conn	Aug. 14, 1866	57, 077
Hoop-skirt tape	C. C. Carpenter	New York, N. Y	July 22, 1873	141, 112
Hoop-skirt tape	W. E. Houston	Birmingham, Conn	May 21, 1867	64, 872
Hoop-skirt tape	W. E. Houston	Birmingham, Conn	May 21, 1867	64, 873
Hoop-skirt wire	T. B. De Forest	Derby, Conn	Nov. 13, 1866	59, 711
Hoop-skirt wire	J. W. Stiles	New York, N. Y	Dec. 15, 1863	40, 960
Hoop-skirt wires, Attaching	J. E. Earle	New Haven, Conn	June 5, 1866	55, 258
Hoop-skirt wires, Coating	W. H. Towers	Boston, Mass	July 6, 1869	92, 403
Hoop-skirts, Adjustable frame for forming	H. S. Loper	New Haven, Conn	Oct. 16, 1866	58, 949
Hoop-skirts, Apparatus for clasping	C. L. Olmstead	Brooklyn, N. Y	Oct. 31, 1865	50, 728
Hoop-skirts, Apparatus for clinching clasp on	G. F. Wright	Clinton, Mass	Sept. 12, 1865	49, 957
Hoop-skirts, Clasp for bottom hoop of	L. W. McFarland	Brooklyn, N. Y	July 3, 1866	56, 146
Hoop-skirts, Covered clasp for	F. E. Day	New York, N. Y	July 27, 1869	93, 068
Hoop-skirts, Frame for forming	J. F. J. Gunning	New York, N. Y	June 9, 1863	38, 821
Hoop-skirts, Punch for making slides for	H. Schenerle	New York, N. Y	Jan. 1, 1861	31, 038
Hoop-splitting machine	S. F. Atherton	Fitchburgh, Mass	May 1, 1860	28, 048
Hoop-splitting machine	J. W. Chittenden and W. C. Mead.	Vevay, Ind	Feb. 28, 1854	10, 563
Hoop-splitting machine	W. Rose	Philadelphia, Pa	Sept. 3, 1844	3, 727
Hoops from edges of boards, Machine for cutting	S. C. Blinn, J. J. Alvord, and H. Brewer.	Tecumseh, Mich	Mar. 10, 1868	75, 352
Hoops, Machine for cutting and distancing locks on	A. H. Crozier	Oswego, N. Y	Feb. 21, 1860	27, 206
Hoops, Machine for cutting the ends of	L. N. Hewes	Swanzey, N. H	Dec. 27, 1870	110, 462
Hoops on casks, Machine for driving	H. C. Sherman	Buffalo, N. Y	July 18, 1865	48, 843
Hoops out of split timber, Machine for shaving	D. Shriver	Petersburgh, Pa	Mar. 17, 1810	....
Hoops, Tool for riving	W. Baker	East Templeton, Mass	Aug. 2, 1859	24, 917
Hooped skirt	C. H. De Forest	Birmingham, Conn	Jan. 6, 1863	37, 336
Hooped skirt	C. S. Hutchinson	Burlington, N. J	Nov. 18, 1862	36, 979
Hooped skirt	S. S. Sherwood	Acquackanonck, N. J	Jan. 6, 1863	37, 374
Hooped skirt, Lady's	J. Holmes	Boston, Mass	Dec. 28, 1858	22, 426
Hooped skirts, Apparatus for attaching clasps to	T. B. De Forest	Birmingham, Conn	Oct. 28, 1862	36, 877
Hooped skirts, Machine for making	C. Neumann	New York, N. Y	Aug. 16, 1859	25, 163
Hop and hay press	J. W. Shipman	Springfield Centre, N. Y	Nov. 26, 1861	33, 798
Hop and hay press	J. Wilson	Otsego, N. Y	Feb. 17, 1826	....
Hop-box	W. R. Crandall	Deansville, N. Y	June 2, 1868	78, 433
Hop-curing apparatus	N. E. Hinds	Cooperstown, N. Y	Nov. 19, 1872	133, 152
Hop-curing apparatus	J. S. Saudt	Saint Joseph, Mo	June 2, 1868	78, 485
Hop-curing process	C. C. Mason	Brookfield, N. Y	Nov. 27, 1866	60, 027
Hop-dryer	W. Loofbourow	Fayette, Wis	Mar. 30, 1869	88, 395
Hop-dryer	W. F. Waterhouse	Weyauwega, Wis	Aug. 4, 1868	80, 578
Hop-dryer	J. Whitney	Fort Winnebago, Wis	Mar. 10, 1868	75, 503
Hop-frame	T. D. Aylsworth	Frankfort, N. Y	Feb. 13, 1855	12, 375
Hop-frame	T. D. Aylsworth	Ilion, N. Y	Mar. 29, 1859	23, 338
Hop-frame	L. A. Beardsley	South Edmeston, N. Y	Jan. 10, 1860	26, 743
Hop-frame	L. A. Beardsley	South Edmeston, N. Y	Sept. 3, 1861	33, 177
Hop-frame	L. A. Beardsley	South Edmeston, N. Y	May 20, 1862	35, 291
Hop-frame	J. C. and D. P. Leonard	Union City, Mich	May 14, 1867	64, 679
Hop-frame	L. S. Mason	Middlefield Centre, N. Y	Sept. 6, 1864	44, 104
Hop-frame	J. B. Van Dewerker	Cobleskill, N. Y	Mar. 19, 1867	63, 119
Hop-hook	E. and E. C. Denio	Baldwinsville and New Hartford, N. Y.	July 28, 1868	80, 277
Hop-liquor for distillers and brewers, Preparing	A. S. Rollins	Albany, N. Y	Mar. 15, 1859	23, 266
Hop-picker	H. Farncrook, F. J. Shepperd, and A. Garton.	Watertown, Wis	June 30, 1868	79, 336
Hop-picker	M. Mases	Malone, N. Y	Apr. 5, 1870	101, 494
Hop-picking tool	J. Dean	Baraboo, Wis	Mar. 10, 1868	75, 385
Hop-pole	L. B. Clark	Bainbridge, N. Y	July 7, 1868	79, 636
Hop-pole	W. Conner	Rensselaerville, N. Y	June 9, 1868	78, 787
Hop-pole	J. G. Wilber	Kilburn City, Wis	July 21, 1868	80, 104
Hop-pole	D. F. Wilcox	Greenville, N. Y	May 19, 1868	78, 037
Hop-pole cleaner	H. Forncrook	Watertown, Wis	July 28, 1868	80, 468
Hop-pole puller	O. B. Hale	Malone, N. Y	Mar. 16, 1869	87, 843
Hop-pole puller	A. L. and W. A. Hatch	Loyd, Wis	Jan. 5, 1869	85, 587
Hop-pole puller	I. W. Legg	Long Eddy, N. Y	May 19, 1868	78, 103
Hop-pole puller	J. R. Woodworth	Munda, N. Y	Jan. 21, 1878	73, 482
Hop-pole, Pulling-machine for	W. Smith	Munda, N. Y	July 30, 1867	67, 365
Hop-pole sharpener	S. V. Barns	Triangle, N. Y	Aug. 3, 1869	93, 269
Hop-pole-sharpening machine	T. S. Angel	Watertown, N. Y	July 21, 1868	80, 050
Hop-pole-sharpening machine	A. H. West	Hamilton, N. Y	July 19, 1864	43, 614
Hop-poles, Jack for pulling	O. S. Foster	Durhamville, N. Y	Dec. 17, 1867	72, 277
Hop-poles, Machine for dressing	C. D. Brown	Bainbridge, N. Y	Nov. 10, 1868	83, 918
Hop-press	J. Brockway	Richland Centre, Wis	Aug. 9, 1870	106, 252
Hop-press	N. Carpenter and J. Hutchinson.	White Creek, Wis	July 16, 1867	66, 793
Hop-press	M. Coryell	New York, N. Y	Jan. 26, 1869	86, 215
Hop-press	J. Hutchinson	Fond du Lac, Wis	Mar. 3, 1868	75, 166
Hop-press	F. Momburg	Burlington, Wis	June 18, 1872	128, 059
Hop-press	H. Taylor	Middletown, Wis	June 30, 1868	79, 413
Hop-press power	N. Carpenter	White Creek, Wis	Jan. 21, 1868	73, 500
Hop-stripper	S. Holt	Baraboo, Wis	June 23, 1868	79, 072
Hop-trellis	L. D. Snook	Barrington, N. Y	May 12, 1868	77, 775
Hop-vine frame	A. Shoemaker and W. Phelps	Conesville, N. Y	Mar. 10, 1868	75, 474
Hop-vine support	C. T. Bush	Rensselaerville, N. Y	Nov. 19, 1867	71, 128

Index of patents issued from the United States Patent Office from 1790 to 1873, inclusive—Continued.

Invention.	Inventor.	Residence.	Date.	No.
Hop-vine support	P. J. Fuller	Clarksville, N. Y	July 7, 1868	79, 648
Hop-vine support	N. C. Roberts and E. W. Badger	Burlington and Otsego, N. Y	Aug. 21, 1866	57, 381
Hop-vine support	N. C. Roberts and E. W. Badger	Fly Creek, N. Y	July 2, 1867	66, 393
Hop-vine trellis	Y. W. Smith	Bristol, N. Y	Aug. 25, 1868	81, 426
Hop-vines by charring the stems, Process of preserving roots of.	S. Cummings	Middlefield, N. Y	Aug. 8, 1865	49, 238
Hop-vines, Training	B. C. Rogers	Stockbridge, Vt	May 29, 1866	55, 163
Hops, Apparatus for extracting	A. Hammer	Philadelphia, Pa	Jan. 9, 1855	12, 204
Hops, Apparatus for producing extract of	J. Schneider	Williamsburgh, N. Y	June 26, 1866	55, 915
Hops, Cultivating	N. Baker	Algansee, Mich	June 23, 1868	79, 097
Hops, Cultivating	E. A. Wightman and W. C. Williams.	Livingstonville, N. Y	Jan. 1, 1867	60, 976
Hops, Extract of	H. Burgess	Royer's Ford, Pa	Sept. 28, 1869	95, 317
Hops, Extract of	C. A. Seely	New York, N. Y	Aug. 22, 1871	118, 396
Hops in brewing, Preserving and using	J. Seeger and J. Boyd	Baltimore, Md	Dec. 12, 1871	121, 902
Hops, Kiln for drying and curing	W. M. Haynie	Sacramento County, Cal	Apr. 30, 1867	64, 220
Hops, Manufacture of extracts of	C. A. Seely	New York, N. Y	May 17, 1870	103, 091
Hops, Poling	A. Ingalls	Hartwick, N. Y	Dec. 31, 1867	72, 858
Hops, Poling	G. J. Olendorf	Middlefield, N. Y	Nov. 24, 1868	84, 299
Hops, Poling	G. J. Olendorf and A. O. Parshall.	Middlefield, N. Y	Jan. 28, 1868	73, 750
Hops, Preserving	S. F. Schoonmaker	New York, N. Y	Mar. 3, 1868	75, 203
Hops, Preserving the aromatic principle of	H. Bartholomay	Rochester, N. Y	Mar. 1, 1870	100, 352
Hops, Preserving the aromatic principle of	E. D. Brainard	Albany, N. Y	July 27, 1869	92, 934
Hops, Process for obtaining a condensed extract of	S. R. Percy and W. S. Wells	New York, N. Y	Mar. 21, 1865	46, 973
Hops, Training	F. W. Collins	Morris, N. Y	Dec. 1, 1863	40, 792
Hops, Training	W. C. Dunn	Greene County, N. Y	Oct. 30, 1866	59, 197
Hops, Training	L. H. Whitney	Vallejo, Cal	Dec. 4, 1866	60, 306
Hopper-boy rake	A. C. Pry	Keedysville, Md	Dec. 28, 1869	98, 412
Hoppers, Feed-regulating mechanism for	J. S. Bodge	Bath, N. Y	July 11, 1865	48, 646
Hopple	G. W. Hyatt	Auburn, N. Y	July 16, 1867	66, 846
Hopple, Animal	G. L. Brent	Gordonsville, Va	June 6, 1871	115, 691
Hopple for horses	R. N. Eagle	Washington, D. C	Mar. 19, 1867	62, 946
Hopple for horses and other animals	R. N. Eagle	Washington, D. C	Apr. 7, 1863	38, 100
Hopple, Training	W. M. Greenwood	Cincinnati, Ohio	July 23, 1867	67, 042
Horizon for quadrants, Artificial	J. Rahill	Ramsgate, England	Aug. 20, 1861	33, 104
Horizontal engines, Expansion-gear for	S. H. Gilman	Cincinnati, Ohio	Dec. 17, 1850	7, 838
Horizontal engines, Frame for	W. Wright	Newburgh, N. Y	Nov. 18, 1873	144, 818
Horizontal mill	J. Kelsey	Hartford, Conn	Aug. 25, 1809	......
Horizontal mill	T. Reynolds	America, N. Y	Mar. 15, 1827	......
Horizontal revolving jointed screen	P. Broughton	Charleston, S. C	Aug. 5, 1831	......
Horn and shell, Machine for splitting	J. Robins	Boston, Mass	June 24, 1851	8, 174
Horn and tortoise-shell combs, Machine for twining	D. E. Noyes	Philadelphia, Pa	Sept. 6, 1817	......
Horn and tortoise-shell combs, Manufacturing	M. J. Littleboy	Philadelphia, Pa	Nov. 19, 1830	......
Horn comb	M. H. Fairchild	Newtown, Conn	May 25, 1869	90, 516
Horn combs with tortoise-shell, Plating	J. P. Spies	Baltimore, Md	Jan. 8, 1810	......
Horn, Composition resembling	W. M. Welling	New York, N. Y	Jan. 11, 1870	98, 727
Horn for lantern-lights, combs, &c., Splitting and shearing.	J. Pulsifer	Newbury, Mass	May 24, 1814	......
Horn handles to knives, Attaching	M. Bradley	Westport, Conn	Aug. 8, 1865	49, 222
Horn, hard rubber, &c., Composition substitute for	R. Haering	New York, N. Y	Sept. 9, 1862	36, 406
Horn, Machine for shaving	W. Noyes, jr	Newburyport, Mass	July 2, 1867	66, 247
Horn, Machine for splitting	E. Prescott	Leominster, Mass	Oct. 10, 1854	11, 793
Horn, Manufacture of articles from pulverized	W. F. Niles and S. G. Pitts	Leominster, Mass	June 20, 1871	116, 213
Horn, Molding and clarifying	E. F. Coffin	Newburyport, Mass	Apr. 5, 1870	101, 587
Horn, Process of whitening	A. Scheller	New York, N. Y	Apr. 16, 1867	63, 949
Horn-splitting and comb-stamp-twining machine	S. Hills and M. Emery, jr	Newburyport, Mass	Oct. 21, 1814	......
Horn to resemble whalebone, Treating	W. Birch	Walworth, Great Britain	Dec. 2, 1873	145, 045
Horns, hoof, &c., Treatment of	E. P. and D. Baugh	Philadelphia, Pa	July 16, 1872	129, 517
Horometer	A. Abbott	Manchester, N. H	Sept. 11, 1855	13, 560
Horoscope	M. Eble	Ellwangen, Würtemberg	Sept. 8, 1863	39, 860
Horse and cattle tie, Self-loosening	J. J. Eshleman	Lancaster, Pa	Mar. 30, 1858	19, 761
Horse and mule shoe	J. Wonderlin	Louisville, Ky	Nov. 29, 1870	109, 701
Horse and ox shoe	N. E. Hinds	Cooperstown, N. Y	Feb. 22, 1859	23, 027
Horse and ox shoe	P. Sternberg	Danube, N. Y	Apr. 10, 1824	......
Horse and vehicle, Auto-propelling	P. W. Mackenzie	Jersey City, N. J	Jan. 19, 1864	41, 310
Horse and wagon brake, Combined	G. Haberland	Pontiac, Ill	Dec. 24, 1867	72, 484
Horse and water wheel, Inclined	L. Wheeler and T. Powell	Coxsackie, N. Y	Dec. 6, 1822	......
Horse-blanket adjuster	A. Z. Neff	Amsterdam, N. Y	Mar. 12, 1872	124, 506
Horse-blinder	J. Dunlap	Madison Township, Pa	Dec. 22, 1868	85, 075
Horse-blinder	S. Ferris	New York, N. Y	Sept. 24, 1867	69, 199
Horse-blinder	R. Lehmicke	Stillwater, Minn	Apr. 30, 1872	126, 218
Horse-blinder	S. B. Rumery	Plainwell, Mich	Sept. 3, 1872	131, 029
Horse-block and hitching-post	G. W. Preston	Corning, N. Y	Nov. 19, 1867	71, 059
Horse-boat	M. Isaacs	Philadelphia, Pa	Mar. 17, 1818	......
Horse-boats, Machinery for	D. Dunham	New York, N. Y	June 28, 1814	......
Horse-boot	P. Murray and F. Koch	Morrisania, N. Y	May 7, 1872	126, 479
Horse-boot	R. Williams	Philadelphia, Pa	Oct. 10, 1871	119, 906
Horse-bracket	H. B. Davis	Lexington, Mass	June 14, 1859	24, 380
Horse bristle-boot	J. J. Davy	Newark, N. J	Jan. 29, 1867	61, 655
Horse-cage	W. G. Hughes	Memano, Ind	Sept. 6, 1864	44, 095
Horse cant-hook	E. W. Gale	Monroeton, Pa	Oct. 12, 1869	95, 675
Horse-checking apparatus	S. S. Ingalls	Chicago, Ill	Mar. 26, 1872	125, 052
Horse-checking apparatus	J. Jackson	Hartford, Conn	Sept. 30, 1873	143, 244
Horse clipping or shearing instrument	H. Knight	Ryde, Isle of Wight, England.	Oct. 18, 1870	108, 489
Horse-collar fastener	A. Steinbach	Evansville, Ind	July 11, 1865	48, 735
Horse-collar, &c., Lining for	I. Lindsley and J. Mackintosh	Pawtucket, R. I	Oct. 28, 1873	143, 987
Horse controller, Carriage	F. Marlow	Cleveland, Ohio	Feb. 5, 1867	61, 748
Horse-controlling device	N. P. Slade	Franklin Grove, Ill	Jan. 11, 1870	98, 715
Horse-cover	E. D. French	Camden, N. J	Sept. 17, 1872	131, 433
Horse-cover	A. H. Morris	Boston, Mass	Jan. 14, 1873	134, 921
Horse-cover	E. L. Perry	New York, N. Y	Oct. 11, 1864	44, 654
Horse-cover, Ventilated	C. P. Eager	Boston, Mass	Apr. 26, 1870	102, 381
Horse-cover, Ventilated	S. Sibley	Chelsea, Mass	July 18, 1871	117, 118
Horse-covers, Ventilating	C. P. Eager	Boston, Mass	Dec. 14, 1869	97, 896
Horse, Extension	R. Hammill	Mineral Point, Wis	June 11, 1867	65, 666
Horse-fastener	F. Vogeli	Newburgh, N. Y	June 13, 1865	48, 225

Index of patents issued from the United States Patent Office from 1790 *to* 1873, *inclusive*—Continued.

Invention.	Inventor.	Residence.	Date.	No.
Horse-fastening	J. Bolton	Richmond, Va	Dec. 23, 1856	16,312
Horse-fetter	A. P. Mason	Gowanda, N. Y	June 15, 1869	91,462
Horse-fetter	J. Watson	New Canaan, Conn	Jan. 15, 1818	....
Horse foot-cushion	L. Hale	Milford, Mass	May 29, 1860	28,473
Horse foot-pad	J. Johnson	Lowell, Mass	Feb. 11, 1873	135,814
Horse foot-pad	J. Johnson	Lowell, Mass	July 5, 1870	104,960
Horse-frame for facilitating the shoeing of horses, Swinging.	J. Kirk	Weston, Vt	Nov. 8, 1820	....
Horse-grooming apparatus	R. Beaumont and W. Clark, jr.	Albany, N. Y	Dec. 1, 1868	84,528
Horse-hair woven garments	P. L. Slayton	New York, N. Y	Nov. 22, 1864	45,209
Horse head-pad	D. K. Humphrey	Buffalo, N. Y	Mar. 16, 1869	87,938
Horse head-protector	W. H. Wilson	Cynthiana, Ky	June 11, 1872	127,820
Horse-holder	W. B. Chapman	La Salle, Ill	Nov. 13, 1866	59,710
Horse-holder	S. Hague	Utica, N. Y	Aug. 30, 1864	43,986
Horse-holder	J. P. Reynolds	Mirabile, Mo	Dec. 3, 1867	71,788
Horse-holder	T. S. W. Russell	Kansas, Ky	July 9, 1872	128,814
Horse-holder	J. M. Whiting	Providence, R. I	Mar. 26, 1867	63,344
Horse hoof-armor	H. Splitdorf	Brighton, Mass	Oct. 8, 1867	69,720
Horse hoof-pad	J. Haseltine	Methuen, Mass	May 19, 1868	77,979
Horse hoof-pad	J. Johnson	Lowell, Mass	June 22, 1869	91,541
Horse hoof-pad	W. H. Southworth	Mittineague, Mass	Aug. 11, 1868	81,025
Horse hoof-pad	J. L. Wetherell	Attleborough, Mass	July 28, 1868	80,319
Horse interfering-boot	A. D. Westbrook	Astoria, N. Y	July 2, 1872	128,687
Horse interfering pad	W. H. Hall	Boston, Mass	Feb. 14, 1871	111,740
Horse interfering-pad	J. Smith	Boston, Mass	Aug. 2, 1870	105,990
Horse interfering-pad	A. D. Westbrook	Buffalo, N. Y	Apr. 16, 1867	63,821
Horse leg-fender	S. Rossman	Hudson, N. Y	June 6, 1865	48,097
Horse machine, Flying	G. L. Newhall and J. F. Cummings.	Chelmsford, Mass	Aug. 31, 1869	93,234
Horse-mill	E. Brown	....	Apr. 10, 1810	....
Horse-mill	W. Cornwell	Accomack, Va	Dec. 14, 1816	....
Horse-mill	J. Grannis	Cheshire, Conn	May 12, 1812	....
Horse-mill	A. Warren	Saugerties, N. Y	Apr. 15, 1826	....
Horse-mills, Lever for working	J. Galbraith	Maury County, Tenn	Mar. 6, 1828	....
Horse-nail machine	E. W. Kelly	Hamilton, Scotland	Oct. 10, 1871	119,854
Horse neck-pad	C. J. Fisher	Waukon, Iowa	May 4, 1869	89,646
Horse neck-protector	J. P. Myer	Waukesha, Wis	Jan. 8, 1867	61,016
Horse-nicking apparatus	H. Dodge	Buffalo, N. Y	Mar. 17, 1863	37,904
Horse-power	J. Abbott	South Reading, Mass	June 30, 1836	....
Horse-power	A. Adams	Augusta, Me	May 6, 1836	....
Horse-power	A. Adams	Sandwich, Ill	July 2, 1861	32,673
Horse-power	J. M. Albertson	New London, Conn	May 4, 1869	89,825
Horse-power	J. M. Albertson	New London, Conn	Sept. 12, 1871	118,893
Horse-power	H. Aldridge and W. Bedford	Goshen, Ind., and Chicago, Ill.	May 19, 1868	77,945
Horse-power	S. S. Allen	Saratoga Springs, N. Y	June 29, 1833	....
Horse-power	S. S. Ammons	Winona, Miss	Oct. 17, 1871	120,018
Horse-power	D. Anthony	Sharon, N. Y	Aug. 7, 1847	5,215
Horse-power	W. E. Arnold	Chatham, Conn	Mar. 13, 1835	....
Horse-power	J. E. Atwood	Willimantic, Conn	Mar. 1, 1870	100,350
Horse-power	C. Aultman and J. Allonas	Mansfield, Ohio	July 13, 1869	92,560
Horse-power	C. Avery	Tunkhannock, Pa	June 3, 1851	8,136
Horse-power	J. F. Barrett	North Granville, N. Y	Oct. 30, 1847	5,348
Horse-power	S. Basket	Crittenden County, Ark	Oct. 24, 1871	120,232
Horse-power	B. D. Beecher	Cheshire, Conn	Dec. 23, 1833	....
Horse-power	W. N. Berkeley	Cedar Rapids, Iowa	Oct. 26, 1869	96,192
Horse-power	J. Bish	Dayton, Ohio	Sept. 10, 1867	68,691
Horse-power	J. M. Blake	Madison, Wis	Mar. 4, 1862	34,559
Horse-power	B. Bogue	Trenton, Iowa	Sept. 4, 1860	29,855
Horse-power	T. J. Bottoms	Thomasville, Ga	Feb. 5, 1861	31,294
Horse-power	G. M. Branch	Winona, Miss	Mar. 18, 1873	136,899
Horse-power	L. H. Bratt	South Bend, Ind	Jan. 7, 1873	134,510
Horse-power	G. Brodie	Jefferson County, Ark	Jan. 30, 1872	123,229
Horse-power	L. Bronson	Buffalo, N. Y	Jan. 7, 1873	134,511
Horse-power	N. B. Brown	Antwerp, N. Y	Jan. 23, 1872	122,989
Horse-power	H. L. and J. A. Buckwalter	Kimberton, Pa	July 17, 1866	56,362
Horse-power	G. E. Burt	Harvard, Mass	Aug. 13, 1861	33,028
Horse-power	G. E. Burt	Harvard, Mass	Mar. 17, 1863	37,892
Horse-power	G. E. Burt and A. and G. F. Wright.	Harvard, Mass	Sept. 22, 1857	18,232
Horse-power	D. Cary	Clarkson, N. Y	June 27, 1846	4,595
Horse-power	J. L. Cathcart	Washington, D. C	July 16, 1850	7,504
Horse-power	R. J. Cheney	Petaluma, Cal	Sept. 26, 1871	119,221
Horse-power	T. C. Churchman	Sacramento, Cal	Feb. 13, 1872	123,614
Horse-power	P. W. Clark	Oblong, N. Y	May 3, 1870	102,491
Horse-power	S. Coin	Cazenovia, N. Y	Sept. 17, 1867	68,957
Horse-power	J. F. Collins	Lodi, Miss	Aug. 1, 1871	117,516
Horse-power	M. H. Cornell	Feasterville, Pa	Feb. 10, 1852	8,717
Horse-power	C. Coster and D. Pennypacker	Upper Providence, Pa	May 23, 1836	....
Horse-power	J. C. Cox	Greenville, N. C	July 30, 1867	67,273
Horse-power	D. D. Craig	Macon, Ga	June 6, 1871	115,711
Horse-power	A. D. Crane	Newark, N. J	Apr. 8, 1851	8,028
Horse-power	A. D. Crane	Newark, N. J	June 22, 1852	9,049
Horse-power	E. Crockett	Macon, Ga	Apr. 1, 1873	137,294
Horse-power	P. C. Curtis and L. Yale	Otsego, N. Y	Sept. 17, 1833	....
Horse-power	C. Custer and D. Ponnepacker	Providence Township, Pa.	Mar. 30, 1835	....
Horse-power	J. Darling	Cincinnati, Ohio	May 18, 1858	20,257
Horse-power	S. W. Davis	Plattsburgh, N. Y	Mar. 6, 1866	52,975
Horse-power	S. W. Davis	Brasher Falls, N. Y	Jan. 19, 1869	86,061
Horse-power	P. K. Dederick	Albany, N. Y	June 25, 1872	128,214
Horse-power	W. Deering	Louisville, Ky	Dec. 19, 1871	121,932
Horse-power	A. J. Detrick	Dryden, N. Y	Feb. 9, 1864	41,493
Horse-power	J. Diffendal and S. Hughes	Westminster, Md	Oct. 5, 1869	95,440
Horse-power	J. G. Dillaha	Waco, Tex	May 18, 1869	90,245
Horse-power	W. W. Dingee	Racine, Wis	Apr. 18, 1871	113,750
Horse-power	N. S. Dodge	Indianapolis, Ind	Feb. 28, 1860	27,278
Horse-power	C. L. Drury	Rockingham, Vt	Sept. 21, 1869	95,008

Index of patents issued from the United States Patent Office from 1790 to 1873, inclusive—Continued.

Invention.	Inventor.	Residence.	Date.	No.
Horse-power	J. A. Eberly, J. Lutz, and H. Becker	East Cocalico Township, Pa	Feb. 9, 1869	86, 656
Horse-power	R. S. Eddy	La Crosse, Wis	Apr. 21, 1868	77, 015
Horse power	G. Eichenseer	Waterloo, Ill	Dec. 4, 1866	60, 159
Horse-power	W. P. Emerson	Pleasantville, Ky	May 13, 1873	138, 873
Horse-power	W. Emmons	New York, N. Y	Dec. 24, 1834	
Horse-power	C. M. Erwin	Winona, Miss	Oct. 3, 1871	119, 587
Horse-power	T. A. E. Evans	Albany, Ga	Apr. 4, 1871	113, 410
Horse-power	A. B. Farquhar	York, Pa	May 7, 1872	126, 386
Horse-power	J. B. Fasset	Irasburgh, Vt	Aug. 17, 1869	93, 813
Horse-power	L. R. Faught	Philadelphia, Pa	Nov. 21, 1871	121, 048
Horse-power	L. R. Faught	Philadelphia, Pa	Feb. 20, 1872	123, 774
Horse-power	L. R. Faught	Philadelphia, Pa	Oct. 28, 1873	143, 974
Horse-power	J. A. Fay	Baltimore, Md	Oct. 10, 1838	978
Horse-power	S. M. Feezler	Seneca Falls, N. Y	July 22, 1862	35, 973
Horse-power	M. Fiske	Sparta, Tenn	Nov. 24, 1868	84, 348
Horse-power	D. Fitzgerald	New York, N. Y	Oct. 19, 1836	
Horse-power	J. Fitzgerald	New York, N. Y	June 16, 1846	4, 579
Horse-power	J. Fraser and W. Thomas	New York, N. Y	Jan. 18, 1870	98, 947
Horse-power	J. Frezer	Newberry, Pa	Jan. 25, 1859	22, 721
Horse-power	H. H. Fultz	Lexington, Mass	July 3, 1855	13, 167
Horse-power	C. F. Gay	Albany, Oreg	Jan. 5, 1869	85, 578
Horse-power	P. Geiser	Waynesborough, Pa	Sept. 29, 1868	82, 617
Horse-power	C. H. Gifford	Potsdam Junction, N. Y	Feb. 8, 1870	99, 555
Horse-power	C. G. Gilbert	Leeds, Me	July 2, 1836	
Horse-power	W. Gilfillan	Speier, Minn	July 8, 1873	140, 622
Horse-power	M. Gillam	Troy, Pa	June 1, 1858	20, 421
Horse-power	H. Glass	Racine, Wis	May 9, 1871	114, 549
Horse-power	S. Gleason	Westville, N. Y	Sept. 18, 1834	
Horse-power	W. A. Glidden and A. Starkweather	Alvaretta, Wis	May 29, 1860	28, 468
Horse-power	W. F. Goodwin	Metuchen, N. J	Feb. 7, 1871	111, 634
Horse-power	C. S. Graves	Elyria, Ohio	Apr. 10, 1860	27, 795
Horse-power	J. A. Green	North Waterford, Me	June 16, 1868	78, 957
Horse-power	M. G. Groff	Vogansville, Pa	July 21, 1868	80, 164
Horse-power	J. A. Hafner	Commerce, Mo	Oct. 13, 1868	83, 059
Horse-power	A. G. Hagerstrom	Red Wing, Minn	Jan. 28, 1873	135, 331
Horse-power	S. B. Haines	Lancaster, Pa	Jan. 17, 1865	45, 921
Horse-power	W. G. Hulbert	Columbus, Miss	June 11, 1872	127, 763
Horse-power	J. and L. Hale	Hollis, N. H	Dec. 24, 1833	
Horse-power	A. H. Halton	Tyro, Miss	Apr. 9, 1872	125, 455
Horse-power	G. B. Hamlin	Willimantic, Conn	Aug. 16, 1870	106, 483
Horse-power	T. Harrison	Belleville, Ill	Feb. 12, 1867	61, 030
Horse-power	T. Harrison and W. C. Buchanan	Belleville, Ill	May 9, 1871	114, 555
Horse-power	T. Harrison and W. C. Buchanan	Belleville, Ill	July 16, 1872	129, 552
Horse-power	J. Haw	Hanover County, Va	Mar. 15, 1845	3, 953
Horse power	J. R. Hedges	Glenwood, N. Y	July 26, 1870	105, 680
Horse-power	D. S. and J. D. Heebner	Norristonville, Pa	Oct. 17, 1871	120, 066
Horse-power	G. Hely	Rochester, Wis	Sept. 14, 1858	21, 495
Horse-power	W. Heston	Bedford, Ohio	Apr. 22, 1862	35, 060
Horse-power	J. Heuermann	Davenport, Iowa	Aug. 10, 1869	93, 438
Horse-power	D. W. Hunt	San Francisco, Cal	May 23, 1865	47, 830
Horse-power	R. Hunt	Freeport, Ill	Mar. 25, 1856	14, 547
Horse-power	J. W. Huntoon	Montgomery, Ala	Sept. 13, 1870	107, 379
Horse-power	J. W. Huntoon	Saint Louis, Mo	Feb. 28, 1871	112, 251
Horse-power	H. M. Irwin	Davidson College, N. C	Apr. 8, 1873	137, 684
Horse-power	T. S. Johnson	Winona, Miss	Oct. 8, 1872	132, 011
Horse-power	I. Keller	Randolph, Ohio	June 30, 1868	79, 476
Horse-power	W. King	Richmond, Mo	June 9, 1868	78, 808
Horse-power	J. H. Kleppinger	Cherryville, Pa	June 23, 1868	79, 074
Horse-power	P. Kline	Johnsville, Ohio	June 22, 1869	91, 546
Horse-power	J. W. Knox	Winona, Miss	Oct. 10, 1871	119, 774
Horse-power	R. W. Krouse	Westminster, Md	Nov. 8, 1870	109, 114
Horse-power	G. Kuenne, J. N. Cole, and D. F. Rath	Fond du Lac, Wis	Oct. 13, 1863	40, 267
Horse-power	D. L. Lamon	Boston, Ga	Feb. 21, 1871	111, 949
Horse power	J. S. Lamphier	Dresden, N. Y	Jan. 23, 1872	122, 896
Horse-power	Z. P. Landrum	Columbus, Miss	Feb. 18, 1873	136, 075
Horse-power	Z. P. Landrum	Columbus, Miss	July 29, 1873	141, 363
Horse-power	C. Lane	Hamilton, Ohio	Apr. 26, 1859	23, 809
Horse-power	H. B. Larzelere	Doylestown, Pa	May 21, 1872	127, 072
Horse-power	W. Lanver	Peru Mills, Pa	June 15, 1869	91, 456
Horse-power	J. A. Leibey	Davenport, Iowa	Nov. 26, 1867	71, 522
Horse power	S. Leonard	Hampden, Me	Sept. 23, 1834	
Horse-power	G. Lewis	Panama, N. Y	Aug. 2, 1864	43, 738
Horse-power	W. J. F. Liddell	Madison, Wis	Mar. 5, 1872	124, 212
Horse-power	J. H. Lighter	Freedom Township, Ill	Jan. 6, 1863	37, 350
Horse-power	F. M. C. Liles	Roanoke, Ala	July 11, 1871	116, 843
Horse-power	B. F. Love and J. H. Frazee	Shelbyville, Ind	May 18, 1858	20, 279
Horse-power	J. E. Lutz	East Cocalico Township, Pa	June 1, 1869	90, 671
Horse-power	D. G. Marden	Memphis, Tenn	Apr. 25, 1871	114, 168
Horse-power	M. H. Marmaduke and B. F. Stewart	Santa Fé, Mo	Dec. 12, 1871	121, 882
Horse-power	H. J. Marnette	Chicago, Ill	June 10, 1873	139, 682
Horse-power	J. Marshall	New Orleans, La	Sept. 26, 1871	119, 279
Horse-power	D. Marvin	New York, N. Y	Feb. 5, 1836	
Horse-power	W. McCord	Sing Sing, N. Y	July 11, 1854	11, 258
Horse-power	E. G. McMillan	Berlin, Ohio	Aug. 26, 1873	142, 115
Horse-power	C. L. Merrill	Watertown, N. Y	May 4, 1869	89, 678
Horse-power	D. Michaels and J. H. Croskey	Hopedale, Ohio	May 18, 1869	90, 182
Horse-power	J. Milbourn	Millport, Ohio	June 10, 1873	139, 725
Horse-power	S. B. Minnech	Landsville, Pa	Nov. 24, 1868	84, 434
Horse-power	T. Mitchel	Newburgh, N. Y	July 7, 1835	
Horse-power	L. B. Morris	Hopefield, Ark	Oct. 15, 1872	132, 313
Horse-power	G. W. Moyers	Gordonsville, Va	Oct. 12, 1869	95, 826
Horse-power	J. Müller	Nacogdoches County, Tex	Feb. 4, 1873	135, 439
Horse-power	J. W. Murrell	El Dorado, Ark	Aug. 9, 1870	106, 194
Horse-power	J. Musten	Franklin County, Vt	Feb. 12, 1836	

Index of patents issued from the United States Patent Office from 1790 *to* 1873, *inclusive*—Continued.

Invention.	Inventor.	Residence.	Date.	No.
Horse-power	S. Newton	Dayton, Ohio	Jan. 20, 1836	....
Horse-power	J. Nichols and E. D. Brown	Battle Creek, Mich	Mar. 11, 1873	136,754
Horse-power	F. I. Norton	Lower Sandusky, Ohio	Apr. 25, 1848	5,530
Horse-power	G. Oerllein	Utica, Minn	July 14, 1868	79,852
Horse-power	G. Oerllein	Utica, Minn	Aug. 17, 1869	93,737
Horse-power	G. Oerllein	Utica, Minn	Aug. 17, 1869	93,738
Horse-power	G. Oerllein	Utica, Minn	Apr. 5, 1870	101,498
Horse-power	W. R. Palmer	Elizabeth City, N. C	July 18, 1854	11,328
Horse-power	G. Partridge and D. M. Johnson	Saint Louis, Mo., and Coshocton, Ohio.	Apr. 3, 1866	53,664
Horse-power	J. N. Pease	Harmony, N. Y	Sept. 16, 1862	36,477
Horse-power	J. N. Pease	Panama, N. Y	Nov. 3, 1863	40,499
Horse-power	W. L. Peet	Maple Rapids, Mich	Jan. 1, 1867	60,933
Horse-power	S. Pelton	New Windsor, Md	Dec. 18, 1855	13,955
Horse-power	S Pelton	Marysville, Cal	Feb. 6, 1872	123,416
Horse-power	W. P. Penn	Belleville, Ill	Feb. 21, 1860	27,235
Horse-power	T. D. Pennington	Forsyth, Ga	Mar. 28, 1871	113,088
Horse-power	S. Perry	Newport, N. Y	June 10, 1862	35,567
Horse-power	S. Perry	Newport, N. Y	July 21, 1863	39,324
Horse-power	S. Perry	Newport, N. Y	July 21, 1863	39,325
Horse-power	L. A. Peter	Neffs, Pa	Oct. 11, 1870	108,178
Horse-power	W. Pierpont	Salem, N. J	Dec. 16, 1862	37,178
Horse-power	J. A. Pitts	Buffalo, N. Y	July 4, 1854	11,232
Horse-power	R. Porter	Billerica, Mass	June 11, 1836	....
Horse-power	D. J. Powers and H. B. Stevens	Madison, Wis	May 14, 1867	64,791
Horse-power	T. B. Pyron	Springfield, Mo	Aug. 16, 1870	106,505
Horse-power	R. Quinn	Whitefield, Miss	July 26, 1870	105,724
Horse-power	J. M. Rand	Chicago, Ill	Feb. 8, 1870	99,591
Horse-power	J. M. Randle	Brighton, Ill	Mar. 20, 1866	53,378
Horse-power	J. W. Reid	New York, N. Y	Apr. 25, 1865	47,456
Horse-power	J. R. Remington	Lowndes County, Ala	June 20, 1846	4,590
Horse-power	E. B. Regna	Jersey City, N. J	Apr. 17, 1860	27,928
Horse-power	L. Price	Attleborough, Pa	July 17, 1837	276
Horse-power	L. Price and D. Congdon	West Chester, Pa	July 17, 1837	277
Horse-power	D. Richardson	Onowa, Iowa	Dec. 24, 1872	134,167
Horse-power	M. A. Richardson	Sherman, N. Y	Apr. 14, 1868	76,816
Horse-power	M. A. Richardson	Sherman, N. Y	July 28, 1868	80,308
Horse-power	W. Rider	Almont, Mich	July 20, 1858	20,978
Horse-power	W. Rider	Almont, Mich	July 1, 1862	35,780
Horse-power	C. Roberts and J. A. Throp	Three Rivers, Mich	July 13, 1869	92,473
Horse-power	C. Roberts and J. A. Throp	Three Rivers, Mich	July 13, 1869	92,474
Horse-power	C. Roberts and J. A. Throp	Three Rivers, Mich	July 13, 1869	92,475
Horse-power	C. Roberts and J. A. Throp	Three Rivers, Mich	Feb. 15, 1870	99,951
Horse-power	F. W. Robinson	Richmond, Ind	Jan. 10, 1860	26,786
Horse-power	F. W. Robinson	Richmond, Ind	Mar. 22, 1870	101,045
Horse-power	J. S. Rowell and W. F. Lowth	Beaver Dam, Wis	Sept. 30, 1862	36,575
Horse-power	C. Russell	Massillon, Ohio	May 1, 1855	12,782
Horse-power	D. Russell	Saint Louis, Mo	Aug. 24, 1852	9,221
Horse-power	E. P. Russell	Manlius, N. Y	May 9, 1865	47,662
Horse-power	W. J. Sage	Steubenville, Ohio	May 5, 1863	38,421
Horse-power	G. Sanford	New York, N. Y	Jan. 8, 1861	31,090
Horse-power	G. Sanford	New York, N. Y	June 13, 1865	48,212
Horse-power	G. W. Sanor and J. Stoffer	Hanoverston and New Chambersburgh, Ohio.	Feb. 20, 1866	52,753
Horse-power	J. Schley	Savannah, Ga	July 2, 1867	66,401
Horse-power	W. Schuyler	Orangeville, Pa	Sept. 17, 1872	131,413
Horse-power	W. Schuyler	Orangeville, Pa	Dec. 3, 1872	133,546
Horse-power	C. J. Shuttleworth	Springville, N. Y	July 13, 1869	92,664
Horse-power	J. Simpson	Atlanta, Ga	Apr. 17, 1855	12,731
Horse-power	R. Skinner	Williamson, N. Y	Mar. 31, 1836	....
Horse-power	A. Smith	Dayton, Oreg	July 21, 1868	80,228
Horse-power	H. Smith	Shelby Station, Tenn	Nov. 7, 1871	120,678
Horse-power	F. M. Sofge	Columbus, Ga	Jan. 18, 1859	22,680
Horse-power	O. W. Stanford	Cincinnati, Ohio	Apr. 24, 1860	28,018
Horse-power	I. Starr	Wooster, Ohio	Aug. 22, 1871	118,292
Horse-power	N. Starr, jr	Homer, N. Y	May 1, 1866	54,433
Horse-power	J. G. Stephenson	Bucyrus, Ohio	Sept. 14, 1869	94,921
Horse-power	H. B. Stevens	Buffalo, N. Y	May 16, 1871	114,874
Horse-power	W. S. Stone	Pitt's Point, Ky	Oct. 8, 1872	132,033
Horse-power	S. Stone	Canton, Ohio	July 12, 1870	105,276
Horse-power	O. O. Storle	Milwaukee, Wis	May 21, 1872	126,912
Horse-power	I. Straub	Lewiston, Pa	May 14, 1836	....
Horse-power	J. B. Streeter	Middlesex, N. Y	Aug. 26, 1834	....
Horse-power	J. B. Sweetland	Pontiac, Mich	July 4, 1865	48,600
Horse-power	J. B. Sweetland	Pontiac, Mich	Sept. 11, 1866	57,995
Horse-power	J. S. Tadlock	Belmont, Tex	Oct. 28, 1873	144,040
Horse-power	H. Tarpley	Wesley, Ky	Mar. 17, 1868	75,597
Horse-power	R. B. Tatum	Helena, Ark	Feb. 13, 1872	123,745
Horse-power	L. P. Teed	Mechanicsburgh, Pa	Dec. 6, 1870	109,967
Horse-power	D. J. Tittle	Albany, N. Y	June 23, 1868	79,282
Horse-power	P. A. Tobey	Caton, N. Y	Mar. 15, 1870	100,818
Horse-power	F. Tobias	Covington, Ohio	Oct. 15, 1867	69,865
Horse-power	A. W. Tooker	Chemung, Ill	Dec. 20, 1864	45,540
Horse-power	S. E. Tooly	Delphi, N. Y	July 16, 1872	129,627
Horse-power	A. Trahern, H. and J. Heberling, and W. E. Lukens.	Harrison County, Ohio	Oct. 28, 1835	....
Horse-power	J. S. Upton	Battle Creek, Mich	Mar. 15, 1859	23,280
Horse-power	J. S. Upton	Battle Creek, Mich	Feb. 5, 1861	31,342
Horse-power	J. S. Upton	Battle Creek, Mich	Sept. 21, 1869	95,168
Horse-power	J. Urmy	Wilmington, Del	Mar. 6, 1847	5,001
Horse-power	J. Valentine	Buffalo, N. Y	Oct. 17, 1871	119,954
Horse-power	D. Van Houten	Fuller's Corners, Ind	Feb. 17, 1863	37,713
Horse power	B. Wales	Hallowell, Me	Aug. 17, 1835	
Horse power	W. Ward	Zanesville, Ohio	Sept. 11, 1849	6,713
Horse-power	A. G. Waterhouse	San Francisco, Cal	Apr. 22, 1873	138,108
Horse-power	S. Wheeler	Albany, N. Y	Mar. 2, 1869	87,450
Horse-power	S. Wheeler	Albany, N. Y	June 14, 1870	104,240
Horse-power	S. Wheeler and E. Jerome	Albany, N. Y	Sept. 27, 1864	44,479

Index of patents issued from the United States Patent Office from 1790 *to* 1873, *inclusive*—Continued.

Invention.	Inventor.	Residence.	Date.	No.
Horse-power	W. Whitman	Haverhill, N. H	June 20, 1836	....
Horse-power	F. Wicks	Kansas, Ill	Nov. 29, 1864	45, 291
Horse-power	B. H. Wilcox	Petroleum Centre, Pa	Oct. 6, 1868	82, 777
Horse-power	J. W. Wilcox	Macon, Ga	Dec. 20, 1870	110, 320
Horse-power	W. H. Wiley	Fredonia, N. Y	Nov. 12, 1867	70, 768
Horse-power	T. H., J. E., J. F., and R. J. Wilson.	Athens, Ga	June 1, 1858	20, 461
Horse-power	T. Wiltse, jr	Panama, N. Y	Oct. 8, 1867	69, 736
Horse-power	A. Wissler and J. Gamber	Brunersville and Petersburgh, Pa.	Dec. 12, 1871	121, 740
Horse-power	D. Woodbury	Weathersfield, Vt	Aug. 26, 1846	4, 718
Horse-power	D. Woodbury	Rochester, N. Y	Apr. 27, 1869	89, 368
Horse-power	D. Woodbury	Rochester, N. Y	Oct. 5, 1869	95, 549
Horse-power	D. Woodbury	Rochester, N. Y	June 4, 1872	127, 537
Horse-power	D. Woodbury	Rochester, N. Y	Dec. 9, 1873	145, 473
Horse-power	W. R. Wright and D. R. Warnock.	Barnwell County and Beaufort County, S. C.	Oct. 10, 1871	119, 909
Horse-power, Adjustment of gearing for	J. S. Everitt	Oshkosh, Wis	June 2, 1868	78, 445
Horse-power and accelerating machine	R. S. Schevenell	Orangeburgh, S. C	Nov. 13, 1832	....
Horse-power and baling-press, Combined	C. A. Wright	Rodney, Miss	Oct. 4, 1870	107, 990
Horse-power and truck	C. Roberts and J. A. Throp	Three Rivers, Mich	Dec. 8, 1868	84, 766
Horse-power apparatus	A. Gaar	Richmond, Ind	Aug. 30, 1870	106, 979
Horse-power apparatus	E. O. and C. B. Thompson	Thomasville, Ga	Sept. 13, 1870	107, 309
Horse-power, Apparatus for confining	R. Knott	Suisun, Cal	Feb. 9, 1869	86, 677
Horse-power apparatus for elevating hay, &c	F. Wicks	Upper Sandusky, Ohio	Aug. 28, 1866	57, 608
Horse-power, Attaching the arms of	C. Roberts	Belleville, Ill	Feb. 10, 1857	16, 612
Horse-power brake	T. Harvey and N. J. Becky	Amsterdam, N. Y	July 24, 1860	29, 280
Horse-power brake	J. Hull and W. P. Anderson	Hackettstown, N. J	Oct. 2, 1866	58, 547
Horse-power brake	T. G. Palmer	Shultville, N. Y	Nov. 10, 1868	83, 995
Horse-power brake	A. Pursell	New Village, N. J	Sept. 27, 1864	44, 454
Horse-power brake	W. F. Rundell	Geneva, N. Y	Oct. 3, 1865	50, 277
Horse-power brake	A. D. Tingley	Adrian, Mich	Sept. 10, 1867	68, 806
Horse-power, Chain-band for	J. S. and H. S. Pitts	Winthrop and Livermore, Me.	Aug. 15, 1834	....
Horse-power, Chain-wheel for chain	S. Perry	Newport, N. Y	Sept. 24, 1867	69, 243
Horse-power, Circuit	S. Perry	Newport, N. Y	Apr. 22, 1862	35, 063
Horse-power, Circuit	S. Perry	Newport, N. Y	Aug. 26, 1862	36, 327
Horse-power, Connecting the parts of endless floors for.	S. Hale	Hollis, N. H	Jan. 16, 1845	3, 887
Horse-power connection	J. A. Hafner	Commerce, Mo	Feb. 2, 1869	86, 533
Horse-power connection	J. A. Hafner	Commerce, Mo	Oct. 26, 1869	96, 225
Horse-power, Coupling of endless-chain	W. E. Arnold	Rochester, N. Y	May 2, 1854	10, 852
Horse-power, Device for securing	F. W. Randall	Burlington, Mich	May 16, 1871	114, 853
Horse-power draft	J. H. Jones	Rockton, Ill	Nov. 16, 1858	22, 079
Horse-power elevator	W. H. Hawley	Utica, N. Y	Nov. 26, 1867	71, 303
Horse-power elevator and excavator	S. T. Bishop and A. Steveley	Fond du Lac, Wis	Jan. 24, 1865	45, 965
Horse-power elevator and excavator	S. T. Bishop and A. Steveley	Fond du Lac, Wis	Jan. 24, 1865	45, 966
Horse-power elevator and excavator	S. T. Bishop and A. Steveley	Fond du Lac, Wis	Jan. 24, 1865	45, 967
Horse-power elevator and excavator	S. T. Bishop and A. Steveley	Fond du Lac, Wis	Jan. 24, 1865	45, 968
Horse-power, Endless band for	J. G. Stanley and J. C. Howard	Winthrop, Me	Jan. 16, 1835	
Horse-power, Endless-chain	H. C. Dodge	North Danville, Vt	Mar. 5, 1872	124, 208
Horse-power, Endless-chain	H. L. Emery	Albany, N. Y	Feb. 24, 1852	8, 754
Horse-power, Endless-chain	W. Furbush	Hallowell, Me	Apr. 14, 1835	
Horse-power, Endless-chain	A. W. Gray	Middletown, Vt	Oct. 26, 1842	2, 833
Horse-power, Endless-chain	J. Kelly	Lewistown, Pa	Mar. 4, 1842	2, 480
Horse-power, Endless-chain	I. R. Lawrence	Green Island, N. Y	Sept. 2, 1862	36, 353
Horse-power, Endless-chain	A. Parker	Windham, Me	Aug. 30, 1838	898
Horse-power, Endless-chain	S. Richards	East Poultney, Vt	Oct. 28, 1843	3, 322
Horse-power, Endless-chain	T. Sharp	Albany, N. Y	Mar. 2, 1852	8, 779
Horse-power, Endless-chain	G. Westinghouse	Central Bridge, N. Y	June 13, 1854	11, 104
Horse-power, Endless chain and tread of	G. E. Burt	Harvard, Mass	Aug. 6, 1861	33, 024
Horse-power, Endless chain for	A. and G. F. Wright	Clinton, Mass	Jan. 26, 1869	86, 197
Horse-power, Endless-chain inclined-plane	O. Badger	Cooperstown, N. Y	Oct. 15, 1836	....
Horse-power engine, Portable endless-screw	I. Van Doren	Hopewell Township, N. J	June 13, 1831	....
Horse-power equalizer	G. Hely	La Porte, Ind	Feb. 15, 1859	22, 950
Horse-power, Equalizing draft in	J. Wilkinson	Prophetstown, Ill	Aug. 11, 1863	39, 516
Horse-power, Equalizing the action of gearing in	C. Caples	Savannah, Mo	July 31, 1849	6, 614
Horse-power, Equalizing the draft of	W. P. Dunlap	Maquoketa, Iowa	July 16, 1867	66, 688
Horse-power fastener	R. Knott	Suisun, Cal	Nov. 10, 1868	83, 862
Horse-power fastening	H. B. Hassler	New Berlin, Ohio	Dec. 12, 1871	121, 874
Horse-power fastening	C. Ross	Hartland, Mich	Aug. 4, 1863	39, 424
Horse-power for driving machinery	S. S. Allen	Miamisburgh, Ohio	Jan. 10, 1840	1, 466
Horse-power for driving machinery	W. E. Arnold	Rochester, N. Y	Apr. 26, 1839	1, 135
Horse-power for driving machinery	A. D. Childs	Rochester, N. Y	Apr. 6, 1844	3, 572
Horse-power for driving machinery	M. Davenport	Pittsburgh, Pa	Sept. 4, 1841	2, 239
Horse-power for driving machinery	S. B. Haines	Greensburgh, Pa	Dec. 31, 1844	3, 871
Horse-power for driving machinery	C. Hibbard	Gilford, N. H	Sept. 14, 1840	1, 788
Horse-power for driving machinery	B. Hinkley	Fayette, Me	Nov. 25, 1837	487
Horse-power for driving machinery	S. H. Little	Gettysburgh, Pa	June 11, 1841	2, 124
Horse-power for driving machinery	M. C. Mix	Danby, N. Y	June 23, 1838	802
Horse-power for driving machinery	R. Montgomery	Waterville, N. Y	Apr. 10, 1845	3, 988
Horse-power for driving machinery	E. Piper	Camden, Me	June 10, 1840	1, 629
Horse-power for driving machinery	L. S. Rand	Townshend, Vt	Mar. 4, 1843	2, 992
Horse-power for driving machinery	J. Secor	New York, N. Y	Apr. 28, 1838	714
Horse-power for driving machinery	O. Straight	Lycoming County, Pa	Aug. 3, 1838	874
Horse-power for driving machinery	G. Strenge and J. Rohrer	Lancaster, Pa	May 8, 1840	1, 586
Horse-power for driving machinery	E. Warren	New York	Jan. 5, 1841	1, 924
Horse-power for driving machinery	T. J. Wells	New York	July 1, 1841	2, 152
Horse-power for driving machinery, Endless-chain	M. Davenport	Phillips, Me	July 10, 1839	1, 229
Horse-power for driving machinery, Endless-chain	J. G. Hall	Zanesville, Ohio	Sept. 21, 1837	395
Horse-power for driving machinery, Endless-chain	H. G. Hall	Putnam, Ohio	Sept. 28, 1837	417
Horse-power for driving machinery, Endless-chain	I. R. Lawrence	Chatham, N. Y	Oct. 7, 1842	2, 799
Horse-power for driving machinery, Endless-chain	A. Palmer	Akron, N. Y	Sept. 22, 1837	398
Horse-power for driving machinery, Endless-chain	A. and W. C. Wheeler	Chatham, N. Y	July 8, 1841	2, 157
Horse-power for driving machinery, Endless-floor	J. M. Reed	Middlefield, N. Y	Jan. 30, 1841	1, 963
Horse-power for driving machinery, Endless-floor	O. Badger	Cooperstown, N. Y	Dec. 14, 1840	1, 898
Horse-power for machinery	H. Smith	Bethel, Ohio	Oct. 6, 1837	418
Horse-power for propelling machinery	M. Davenport	Phillips, Me	Oct. 10, 1835	....

Index of patents issued from the United States Patent Office from 1790 to 1873, inclusive—Continued.

Invention.	Inventor.	Residence.	Date.	No.
Horse-power for propelling machinery	C. Emmons	New York	Aug. 2, 1831	....
Horse-power for propelling machinery	B. Langdon	Troy, N. Y	July 19, 1837	286
Horse-power for propelling saw-mills, &c	I. Jones	Natchez, Miss	Jan. 13, 1835	....
Horse-power, Frame-work of portable	J. A. Nelson and J. P. Ross	Lewisburgh, Pa	June 30, 1837	252
Horse-power, Gearing for	C. Avery	Tunkhannock, Pa	Apr. 20, 1858	19, 976
Horse-power, &c., Gearing for	C. Avery	Tunkhannock, Pa	June 5, 1860	28, 550
Horse-power, Gearing for	S. Gardiner	Auburn, N. Y	May 14, 1833	....
Horse-power governor	T. B. McConaughey	Newark, Del	Sept. 11, 1866	57, 941
Horse-power governor	L. Pusey	Wilmington, Del	May 25, 1858	20, 368
Horse-power, Hoisting-attachment for portable	P. Cary	Coeymans, N. Y	May 25, 1869	90, 423
Horse-power, Inclined plane	W. G. Johnston	Bridgeton, N. J	Oct. 4, 1834	....
Horse-power jack	S. Perry	Newport, N. Y	Jan. 2, 1872	122, 488
Horse-power link	S. Wheeler	Albany, N. Y	Sept. 2, 1862	36, 378
Horse-power, Link for	A. W. Gray	Middletown, Conn	Sept. 9, 1856	15, 693
Horse-power, Link for endless-chain	G. E. Burt	Harvard, Mass	Oct. 25, 1870	108, 564
Horse-power, Link for endless-chain	J. Casho	Newark, Del	Aug. 18, 1868	81, 251
Horse-power, Link for endless-chain	A. W. Gray	Middletown, Vt	July 29, 1873	141, 268
Horse-power, Link for endless-chain	G. L. Sheldon	Hartsville, Mass	July 12, 1870	105, 263
Horse-power, Link for railway	S. Wheeler	Albany, N. Y	June 2, 1863	38, 778
Horse-power, Link-gearing for	T. D. Burk	Chicago, Ill	Apr. 22, 1856	14, 750
Horse-power machine	J. F. Axtell	Geneva, N. Y	Mar. 13, 1834	....
Horse-power machine	E. Briggs	Fort Covington, N. Y	July 12, 1834	....
Horse-power machine	A. B. Colton	Athens, Ga	May 24, 1859	24, 101
Horse-power machine	J. Eastman	Elbridge, N. Y	May 31, 1834	....
Horse-power machine	W. Field	Providence, R. I	June 7, 1859	24, 291
Horse-power machine	L. R. Fraught	Atlanta, Ga	Sept. 20, 1859	25, 497
Horse-power machine	J. Grant	Rochester, N. Y	Mar. 30, 1858	19, 769
Horse-power machine	E. J. Keep and W. H. Briggs	Stockton, Cal	Sept. 10, 1861	33, 252
Horse-power machine	H. B. Middaugh and A. Clarke	Mansfield, Pa	Sept. 24, 1861	33, 358
Horse-power machine	W. Phelps and W. H. Hanford	Sycamore, Ill	Nov. 22, 1859	26, 203
Horse-power machine	G. Sandford	Poughkeepsie, N. Y	Sept. 20, 1859	25, 528
Horse-power machine	J. A. Stone	Rochester, N. Y	July 27, 1858	21, 032
Horse-power machine	F. B. Williams	Freeport, Ill	Aug. 16, 1859	25, 155
Horse-power-machine brake	D. G. Terrell	Wakefield, Pa	June 23, 1868	79, 281
Horse-power machine, Endless chain for	I. R. Lawrence and G. E. Gould	Green Island, N. Y	Oct. 11, 1859	25, 743
Horse-power machine, Mode of applying and constructing.	W. Zeller	Lebanon County, Pa	Apr. 12, 1859	23, 638
Horse-power machine, Mode of connecting the draft-lever for.	H. Shaw	New Orleans, La	Aug. 25, 1868	81, 544
Horse-power, Master-wheel of	J. A. Taplin	Fishkill, N. Y	June 12, 1849	6, 525
Horse-power, Master-wheel of portable	J. A. Taplin	Hammond, N. Y	Dec. 30, 1841	2, 407
Horse-power, Mode of constructing	T. Schankwiler	Fayette, N. Y	Nov. 19, 1861	33, 755
Horse-power, Mode of fastening	H. C. Drew	Jamestown, Mich	Nov. 28, 1871	121, 281
Horse-power, Mounted	M. B. Erskine	Racine, Wis	Sept. 13, 1870	107, 237
Horse-power, Platform	F. J. Culver	Hartford, Vt	Jan. 31, 1871	111, 326
Horse-power, Portable	H. Aldridge	Goshen, Ind	Sept. 17, 1867	68, 829
Horse-power, Portable	S. S. Allen	Saratoga Springs, N. Y	Apr. 3, 1835	....
Horse-power, Portable	J. Brandon	Williamsport, Pa	Apr. 8, 1835	....
Horse-power, Portable	W. Burk	Whitemarsh Township, Pa	June 25, 1834	....
Horse-power, Portable	J. H. Corey	Bethel Township, Ohio	Feb. 27, 1832	....
Horse-power, Portable	D. Flagg, jr	New York	Dec. 28, 1832	....
Horse-power, Portable	S. Gardner	Rochester, N. Y	Nov. 25, 1834	....
Horse-power, Portable	E. H. Jaques	Springfield, Vt	May 16, 1845	4, 051
Horse-power, Portable	C. M. Keller	Washington, D. C	June 3, 1833	....
Horse-power, Portable	I. J. Richardson	New York	Aug. 26, 1843	3, 229
Horse-power, Portable	I. J. Richardson	New York, N. Y	Feb. 10, 1846	4, 374
Horse-power, Portable	G. W. Swift	Oxford, Miss	May 3, 1859	23, 871
Horse-power, Portable	D. Woodbury	Rochester, N. Y	Aug. 18, 1857	18, 028
Horse-power, Propelling	G. Griffen	Greece, N. Y	May 31, 1833	....
Horse-power, Propelling machinery by	T. Showerman	Covington, N. Y	June 27, 1832	....
Horse-power, Propelling machinery by	I. Stoddard	Great Bend, Pa	Oct. 9, 1860	30, 362
Horse-power, Reversible	P. H. Kells	Hudson, N. Y	July 8, 1856	15, 296
Horse-power, Safety-brake for	D. M. Reynolds	Rising Sun, Md	Dec. 22, 1863	41, 020
Horse-power, &c., Speed-regulator for	S. Perry	Newport, N. Y	Apr. 22, 1862	35, 064
Horse-power, Staking	W. Gregg	Dansville, Mich	Aug. 22, 1871	118, 232
Horse-power staking-apparatus	B. F. Osgood	Coloma, Mich	July 12, 1870	105, 238
Horse-power, Sun and planet	J. Bogardus	New York, N. Y	Aug. 29, 1848	5, 735
Horse-power, Suspending band-wheel in	I. Straub	Lewistown, Pa	July 11, 1837	267
Horse-power, Sweep	R. J. M. King	Ypsilanti, Mich	Apr. 19, 1870	102, 015
Horse-power to machinery, Applying	B. Mattby	New York	Apr. 11, 1825	....
Horse-power to mills, Applying	B. E. Orton	Lyndon, Ill	Mar. 13, 1860	27, 468
Horse-power to the ground, Fastening	W. Buchanan, jr	Maine Prairie, Cal	Nov. 10, 1868	83, 919
Horse-power to the ground, Securing	W. H. Buell	Union City, Mich	Oct. 6, 1868	82, 688
Horse-power to the ground, Securing	F. W. Randall	Burlington, Mich	Apr. 5, 1870	101, 507
Horse-power, Tread	S. Perry	Newport, N. Y	Apr. 22, 1862	35, 062
Horse-power, Water-supported	A. James	Vinton, Ohio	July 23, 1872	129, 734
Horse-power, &c., Wheel and axle attachment of	G. E. Burt and G. F. Wright	Harvard, Mass	Feb. 23, 1858	19, 408
Horse-propelling	J. H. Brown	New York, N. Y	Dec. 11, 1866	60, 335
Horse-pulleying apparatus	A. H. Trego	Lambertville, N. J	July 9, 1861	32, 801
Horse-radish for use, Preparing	J. D. Husbands, jr	Saint Louis, Mo	Apr. 30, 1872	126, 212
Horse-rake	J. W. Acker	Copenhagen, N. Y	Sept. 29, 1868	82, 474
Horse-rake	J. Q. Adams	High Spire, Pa	June 3, 1862	35, 428
Horse-rake	D. G. Adelsberger	Emmetsburgh, Md	Jan. 23, 1866	52, 122
Horse-rake	H. Albright	Lewisburgh, Pa	Feb. 21, 1865	46, 435
Horse-rake	M. Alden	Auburn, N. Y	Apr. 7, 1868	76, 480
Horse-rake	F. M. Allerton	Alliance, Ohio	July 6, 1869	92, 140
Horse-rake	S. E. Ament	Oswego, Ill	Feb. 19, 1867	62, 107
Horse-rake	D. W. Amos	Bedford, Pa	Jan. 12, 1864	41, 187
Horse-rake	H. A. Bailey and A. R. Burdick	Racine, Wis	June 26, 1866	55, 804
Horse-rake	M. W. Baldwin and A. S. Lyman.	Philadelphia, Pa	May 16, 1848	5, 587
Horse-rake	L. B. Ball	Dayton, Ohio	Aug. 13, 1867	67, 705
Horse-rake	L. B. Ball	Dayton, Ohio	Oct. 8, 1867	69, 610
Horse-rake	A. T. Barnes	Tiffin, Ohio	Mar. 24, 1868	75, 836
Horse-rake	W. H. Barton	Olney, Ill	July 28, 1868	80, 383
Horse-rake	J. N. Baxter	Greensburgh, Ind	July 24, 1866	56, 517
Horse-rake	H. L. Beach	Montrose, Pa	Dec. 10, 1867	71, 841
Horse-rake	H. L. Beach	Montrose, Pa	Mar. 17, 1868	75, 514

Index of patents issued from the United States Patent Office from 1790 *to* 1873, *inclusive*—Continued.

Invention.	Inventor.	Residence.	Date.	No.
Horse-rake	H. L. Beach	New York, N. Y	Mar. 31, 1868	76, 040
Horse-rake	L. Beach	Montrose, Pa	Apr. 10, 1860	27, 771
Horse-rake	L. Beach	Montrose, Pa	Jan. 19, 1864	41, 270
Horse-rake	E. L. Bergstresser	Berrysburgh, Pa	Aug. 26, 1862	36, 268
Horse-rake	S. Bingham	Troy, N. Y	Dec. 5, 1865	51, 287
Horse-rake	R. H. Blair and A. W. Beatty	Saltsburgh, Pa	Mar. 4, 1862	34, 558
Horse-rake	I. W. Boatman	Seven Mile, Ohio	Sept. 10, 1867	68, 693
Horse-rake	J. Bohner	Alden, N. Y	May 4, 1869	89, 555
Horse-rake	O. Bonney, jr	San Francisco, Cal	Nov. 23, 1869	97, 157
Horse-rake	L. S. Bortree	Grand Rapids, Mich	Nov. 24, 1868	84, 254
Horse-rake	W. L. Bostwick	Ithaca, N. Y	Mar. 5, 1867	62, 524
Horse-rake	M. Bradley	Dundee, Ill	Aug. 16, 1859	25, 088
Horse-rake	B. Bridendolph	Clear Spring, Md	Jan. 4, 1859	22, 526
Horse-rake	N. Briggs	New Hartford, N. Y	Feb. 11, 1835	......
Horse-rake	S. C. Brinser	Middletown, Pa	Feb. 17, 1863	37, 672
Horse-rake	S. C. Brinser	Middletown, Pa	Apr. 28, 1863	38, 281
Horse-rake	S. C. Brinser	Middletown, Pa	Oct. 27, 1868	83, 452
Horse-rake	H. L. Brown	Adrian, Mich	Nov. 17, 1868	84, 085
Horse-rake	N. H. Brown	Derby, N. H	Apr. 14, 1868	76, 597
Horse-rake	N. T. Brown	Ononwa, Iowa	June 10, 1862	35, 502
Horse-rake	W. H. Brown	Middletown, N. Y	Jan. 4, 1859	22, 482
Horse-rake	E. Buckman	East Greenbush, N. Y	Sept. 3, 1861	33, 183
Horse-rake	F. M. Buckmaster	Galesburgh, Ill	Sept. 28, 1869	95, 315
Horse-rake	D. Bull	Amboy, Ill	July 6, 1869	92, 156
Horse-rake	E. W. Bullard	Barre, Mass	Aug. 27, 1867	68, 283
Horse-rake	E. W. Bullard	Barre, Mass	Feb. 25, 1868	74, 794
Horse-rake	I. L. Bullock	Marcy, Ind	May 28, 1867	65, 164
Horse-rake	I. C. Burget	Davenport Centre, N. Y	Mar. 8, 1859	23, 155
Horse-rake	G. E. Burt	Harvard, Mass	Mar. 14, 1865	46, 776
Horse-rake	G. E. Burt	Harvard, Mass	Oct. 24, 1865	50, 557
Horse-rake	G. E. Burt	Harvard, Mass	Sept. 17, 1867	68, 950
Horse-rake	G. E. Burt	Harvard, Mass	Dec. 29, 1868	85, 275
Horse-rake	C. Carlisle	Norwich, Vt	May 15, 1847	5, 119
Horse-rake	I. Carman	Sandwich, Ill	May 11, 1869	89, 972
Horse-rake	S. Carpenter	Brookfield, Ill	Dec. 3, 1867	71, 579
Horse-rake	P. S. Carver	Honeoye Falls, N. Y	May 2, 1865	47, 519
Horse-rake	A. H. Chaplin	Adrian, Mich	Nov. 25, 1862	36, 992
Horse-rake	A. Chapman	Delta, N. Y	Feb. 4, 1868	73, 952
Horse-rake	J. Chappel	Greene, N. Y	Nov. 20, 1860	30, 666
Horse-rake	L. Clinton	North Haven, Conn	Nov. 17, 1868	84, 090
Horse-rake	L. Clinton	North Haven, Conn	Feb. 2, 1869	86, 366
Horse-rake	L. Clinton and E. S. Munson	North Haven, Conn	May 22, 1866	54, 862
Horse-rake	A. W. Coates	Alliance, Ohio	Aug. 27, 1867	68, 288
Horse-rake	C. B. Cogswell	Essex, Mass	June 17, 1862	35, 586
Horse-rake	W. H. Cook	Bridgehampton, N. Y	Sept. 22, 1868	82, 292
Horse-rake	E. Crandal	Northville, Mich	Apr. 20, 1869	89, 029
Horse-rake	J. Crites	Orrville, Ohio	Apr. 2, 1861	31, 875
Horse-rake	A. J. Curtis, D. J. Roberts, and W. Curtis.	Swanville and Monroe, Me	Aug. 28, 1866	57, 482
Horse-rake	G. H. Dailey and R. M. Treat	Morris, Conn	Nov. 11, 1862	36, 897
Horse-rake	M. Davenport	Minerva, Ohio	Sept. 6, 1864	44, 079
Horse-rake	B. W. Davis	Fort Madison, Iowa	Mar. 17, 1868	75, 532
Horse-rake	G. Deal	Wilmot, Ohio	Apr. 7, 1868	76, 307
Horse-rake	G. Deal	Wayne Township, Ohio	July 10, 1866	56, 189
Horse-rake	N. C. Decker	Saint Louis, Mo	Sept. 18, 1866	58, 073
Horse-rake	C. Delano	East Livermore, Me	Feb. 27, 1849	6, 151
Horse-rake	L. S. Deming	Newington, Conn	Jan. 18, 1859	22, 632
Horse-rake	S. L. Denney	Christiana, Pa	Aug. 4, 1863	39, 383
Horse-rake	S. L Denney and J. L. Chalfant	Christiana and Chester County, Pa.	Nov. 10, 1868	83, 937
Horse-rake	D. Dewey	Poultney, Vt	Nov. 23, 1837	472
Horse-rake	J. Dillier	Greensburgh, Ind	Sept. 18, 1866	58, 075
Horse-rake	T. H. Dodge	Worcester, Mass	Dec. 20, 1864	45, 481
Horse-rake	H. S. Doolittle	Kortright, N. Y	Aug. 22, 1846	4, 708
Horse-rake	E. Dorr	Rockford, Ill	July 13, 1869	92, 521
Horse-rake	J. B. Drake	Picture Rocks, Pa	Feb. 23, 1864	41, 740
Horse-rake	D. M. Dunham, and J. and A. Webb.	Bangor, Me	Aug. 28, 1866	57, 488
Horse-rake	O. D. Dunham	Plainville, Mich	Mar. 29, 1864	42, 079
Horse-rake	S. Eberly	Mechanicsburgh, Pa	Aug. 30, 1864	43, 980
Horse-rake	S. Eberly and G. Hauck	Mechanicsburgh, Pa	July 2, 1867	66, 314
Horse-rake	S. Eberly and S. Hauck	Mechanicsburgh, Pa	Oct. 29, 1867	70, 327
Horse-rake	C. Edgar	Dayton, Ohio	Dec. 23, 1873	145, 852
Horse-rake	J. Eliot	Vermillion, Ill	Mar. 3, 1868	75, 006
Horse-rake	W. Emmons	Sandwich, Ill	Apr. 13, 1869	88, 858
Horse-rake	R. M. Ewing	Clinton, Ill	Jan. 30, 1866	52, 360
Horse-rake	J. Farmwalt	German Township, Ohio	Dec. 1, 1863	40, 793
Horse-rake	H. V. Farris	Richmond, Ind	July 16, 1867	66, 696
Horse-rake	E. A. Field	Sidney, Me	Dec. 10, 1867	72, 006
Horse-rake	C. H. Finson	Bangor, Me	Jan. 23, 1866	52, 152
Horse-rake	L. W. Frederick	Ray, Ind	Feb. 16, 1864	41, 613
Horse-rake	L. W. Frederick	Gosport, Ind	Aug. 6, 1867	67, 523
Horse-rake	S. Freet	Upper Strasburgh, Pa	Oct. 1, 1867	69, 332
Horse-rake	C. Furst, D. Bradley, and J. Lacey.	Chicago, Ill	Apr. 15, 1862	34, 953
Horse-rake	C. Garver	Londonderry, Pa	Dec. 7, 1858	22, 232
Horse-rake	E. Geiger	Lancaster, Pa	May 24, 1859	24, 114
Horse-rake	J. Ginther	Mier, Ill	June 23, 1868	79, 065
Horse-rake	D. D. Gitt	Arendtsville, Pa	Jan. 17, 1865	45, 920
Horse-rake	C. N. Goss	Claremont, N. H	June 9, 1868	78, 801
Horse-rake	R. A. Graham	Greensburgh, Ind	Dec. 12, 1865	51, 450
Horse-rake	J. S. Grant	Sidney Centre, Me	Sept. 3, 1867	68, 500
Horse-rake	S. H. Grinnell	Charlestown, N. H	Feb. 20, 1849	6, 123
Horse-rake	J. W. Hadcock and P. Wilcox	Norway, N. Y	Dec. 7, 1858	22, 235
Horse-rake	E. R. Hall	Ilion, N. Y	Sept. 4, 1866	57, 706
Horse-rake	S. J. Halstead	Margarettville, N. Y	May 25, 1869	90, 446
Horse-rake	O. J. Hardgrove	Massillon, Ohio	Mar. 22, 1864	41, 990
Horse-rake	E. Harris	Princeton, Ill	July 5, 1859	24, 631
Horse-rake	A. L. Haskell	Amity, Pa	Oct. 15, 1867	69, 913

Index of patents issued from the United States Patent Office from 1790 *to* 1873, *inclusive*—Continued.

Invention.	Inventor.	Residence.	Date.	No.
Horse-rake	J. V. Hawkey	Greensburgh, Pa	Dec. 10, 1867	71, 875
Horse-rake	H. R. Hawkins	Akron, Ohio	Dec. 27, 1864	45, 605
Horse-rake	H. R. Hawkins	Akron, Ohio	Apr. 24, 1866	54, 151
Horse-rake	H. R. Hawkins	Akron, Ohio	Mar. 24, 1868	75, 905
Horse-rake	J. D. Heebner	Norrittonville, Pa	Aug. 27, 1867	68, 075
Horse-rake	G. L. Heidler	York, Pa	July 10, 1866	56, 222
Horse-rake	H. Hersh	Lancaster, Pa	June 14, 1859	24, 389
Horse-rake	J. N. Hicks	Barre Centre, N. Y	Mar. 17, 1868	75, 548
Horse-rake	T. Himmelberger	Heidelburgh Township, Pa	Feb. 11, 1868	74, 358
Horse-rake	J. B. Hoag	Oxford, Ill	July 23, 1867	67, 054
Horse-rake	H. W. Holcomb	Northville, Mass	Nov. 16, 1869	96, 917
Horse-rake	C. B. Holden	Worcester, Mass	Apr. 14, 1868	76, 760
Horse-rake	F. Holden	Clyde, Ill	Aug. 8, 1865	49, 269
Horse-rake	F. Holden	Litchfield, Ill	Oct. 8, 1867	69, 667
Horse-rake	J. Hollingsworth	Chicago, Ill	Feb. 2, 1864	41, 433
Horse-rake	J. Hollingsworth	Chicago, Ill	July 25, 1865	48, 944
Horse-rake	J. Hollingsworth	Chicago, Ill	June 11, 1867	65, 573
Horse-rake	J. Hollingsworth	Chicago, Ill	May 25, 1869	90, 362
Horse-rake	W. Horning	New Lebanon, Ohio	Feb. 23, 1858	19, 420
Horse-rake	B. F. Horton	Ithaca, N. Y	Jan. 14, 1868	73, 248
Horse-rake	A. Hovey	Brookfield, Vt	Feb. 12, 1850	7, 084
Horse-rake	C. Howard	Bearsville, N. Y	July 30, 1867	67, 305
Horse-rake	E. Huber	Kelso, Ind	Jan. 24, 1865	46, 001
Horse-rake	J. Hudson	Charleston, Ill	Sept. 14, 1869	94, 889
Horse-rake	J. Hunsberger	Worcester Township, Pa	June 16, 1868	78, 969
Horse-rake	J. P. Hunter	Williamsport, Ind	July 16, 1867	66, 679
Horse-rake	C. S. Huntington	Black River, N. Y	Oct. 16, 1866	58, 832
Horse-rake	E. Huson	Ithaca, N. Y	Feb. 12, 1867	61, 034
Horse-rake	D. G. Hussey	Nantucket, Mass	June 23, 1863	38, 965
Horse-rake	D. G. Hussey	Nantucket, Mass	Oct. 18, 1864	44, 728
Horse-rake	D. G. Hussey	Nantucket, Mass	June 13, 1865	48, 179
Horse-rake	D. G. Hussey	Nantucket, Mass	Nov. 28, 1865	51, 187
Horse-rake	V. Hyatt	Westfield, Mass	Dec. 15, 1857	18, 850
Horse-rake	J. M. Jay	Canton, Ohio	Apr. 18, 1865	47, 371
Horse-rake	C. Jennings	Easton, Conn	May 10, 1864	42, 664
Horse-rake	R. B. and A. C. Jennings	Livermore, Me	Oct. 3, 1848	5, 833
Horse-rake	A. B. Johnson	Washington, D. C	Dec. 1, 1868	84, 551
Horse-rake	S. Johnson	New Harmony, Ind	July 2, 1867	66, 352
Horse-rake	W. H. Johnson	Northborough, Mass	Oct. 10, 1865	50, 363
Horse-rake	E. O. Jones	Oakwood, Mich	July 10, 1866	56, 227
Horse-rake	J. D. Jones	Pittsburgh, Pa	May 9, 1865	47, 644
Horse-rake	S. A. and L. M. Kays	Independence, Iowa	Apr. 30, 1867	64, 329
Horse-rake	C. P. Kelly	Phelps, N. Y	Apr. 6, 1869	88, 719
Horse-rake	A. S. Kendall	Guilford, Me	Jan. 28, 1868	73, 904
Horse-rake	J. E. Kendall	Plymouth, Ind	Dec. 17, 1867	72, 404
Horse-rake	G. Kimball	Springfield, Vt	Dec. 20, 1864	45, 502
Horse-rake	G. W. King	Schoharie, N. Y	Oct. 22, 1867	70, 004
Horse-rake	J. King	Omaha, Nebr	June 19, 1866	55, 672
Horse-rake	W. King	Springfield, Ill	Dec. 5, 1865	51, 322
Horse-rake	W. King	Springfield, Ill	Apr. 9, 1867	63, 729
Horse-rake	W. King	Springfield, Ill	June 22, 1869	91, 640
Horse-rake	G. S. Kinsey	Reading, Pa	Aug. 28, 1860	29, 795
Horse-rake	F. C. Kneeland	Hartford, Wis	Mar. 1, 1859	23, 091
Horse-rake	S. D. Knight, J. W. Smith, and G. W. Bercaw.	Bryan, Ohio	Feb. 23, 1869	87, 267
Horse-rake	J. B. Koon	Aurelius, N. Y	July 20, 1869	92, 732
Horse-rake	A. Krause	West Liberty, Ohio	Jan. 29, 1867	61, 670
Horse-rake	C. Kugler	Cadiz, Ohio	Aug. 28, 1866	57, 520
Horse-rake	J. La Croix	Chicopee, Mass	Mar. 16, 1869	87, 782
Horse-rake	J. Lacy	Chicago, Ill	May 16, 1865	47, 736
Horse-rake	S. Ladd	Danville, Vt	Mar. 14, 1846	4, 421
Horse-rake	E. E. Sauer and H. W. Eisenhart.	York, Pa	Mar. 2, 1869	87, 345
Horse-rake	A. S. Lazier	North Parma, N. Y	Sept. 28, 1869	95, 361
Horse-rake	P. Lebzelter	Lancaster, Pa	July 5, 1859	24, 645
Horse-rake	W. A. Lewis	Joliet, Ill	June 22, 1869	91, 548
Horse-rake	J. M. Long	Hamilton, Ohio	Oct. 22, 1867	70, 008
Horse-rake	W. H. Long	Lancaster, Pa	Mar. 1, 1859	23, 098
Horse-rake	R. Lownsbury and F. G. Willson.	Fulton and Ontario, N. Y	Jan. 31, 1860	27, 000
Horse-rake	C. O. Luce	Brandon, Vt	Jan. 28, 1868	73, 820
Horse-rake	J. B. Luce	Earlville, Ill	Aug. 13, 1867	67, 778
Horse-rake	P. Lugenbell and T. Barns	Greensburgh, Ind	Aug. 7, 1866	56, 960
Horse-rake	E. Luther	West Troy, N. Y	Feb. 25, 1868	74, 922
Horse-rake	E. Luther	West Troy, N. Y	Oct. 6, 1868	82, 855
Horse-rake	S. P. Macay	Killbourne, Ohio	Apr. 21, 1868	76, 932
Horse-rake	J. Maltby	North Branford, Conn	Dec. 5, 1843	3, 369
Horse-rake	E. C. Martin	West Liberty, Iowa	Jan. 24, 1865	46, 009
Horse-rake	N. Martz	Briar Creek Township, Pa	Feb. 26, 1856	14, 321
Horse-rake	R. W. McClelland	Springfield, Ill	June 29, 1869	91, 858
Horse-rake	W. McCord	Sing Sing, N. Y	Aug. 7, 1866	56, 968
Horse-rake	W. McCord	Sing Sing, N. Y	Feb. 25, 1868	74 924
Horse-rake	G. M. L. McMillen	Dayton, Ohio	Oct. 13, 1868	82, 972
Horse-rake	G. M. L. McMillen	Dayton, Ohio	Mar. 2, 1869	87, 504
Horse-rake	W. H. McPherson	Danby, N. Y	May 7, 1867	64, 550
Horse-rake	W. H. McPherson	Danby, N. Y	July 16, 1867	66, 865
Horse-rake	B., S., jr., and J. Mellinger	Mount Pleasant, Pa	May 13, 1862	35, 245
Horse-rake	B., S., jr., and J. Mellinger	Mount Pleasant, Pa	Nov. 18, 1862	36, 961
Horse-rake	G. W. Middlecoff	Atlanta, Ill	Apr. 20, 1869	89, 160
Horse-rake	S. D. Milam	Leoti, Ind	May 18, 1869	90, 286
Horse-rake	M. Miles	Middlesex, N. Y	Dec. 17, 1867	72, 315
Horse-rake	A. Miller	Hagerstown, Md	Feb. 9, 1869	86, 855
Horse-rake	J. Miller	Paris, Ohio	Feb. 18, 1862	34, 441
Horse-rake	M. Morgan	Lancaster, Pa	Aug. 24, 1858	21, 268
Horse-rake	L. H. Morrill	West Cumberland, Me	Nov. 10, 1868	83, 875
Horse-rake	C. H. Mosey	Mansfield, Ohio	July 6, 1869	92, 339
Horse-rake	W. Mosher	Pittstown, N. Y	May 8, 1866	54, 582
Horse-rake	S. Mowry	Womelsdorf, Pa	May 14, 1861	32, 307

Index of patents issued from the United States Patent Office from 1790 *to* 1873, *inclusive*—Continued.

Invention.	Inventor.	Residence.	Date.	No.
Horse-rake	C. E. Murray	Sugar Valley, Pa	June 30, 1868	79, 378
Horse-rake	G. D. Neal	Mount Vernon, Ohio	Oct. 6, 1868	82, 864
Horse-rake	O. Nivison	Hector, N. Y	Jan. 21, 1868	73, 457
Horse-rake	H. B. Noble	South Windsor, Conn	Sept. 29, 1868	82, 631
Horse-rake	S. R. Nye	Barre, Mass	Mar. 13, 1866	53, 172
Horse-rake	S. R. Nye	Barre, Mass	Jan. 12, 1869	85, 757
Horse-rake	G. Palmer	Littlestown, Pa	Oct. 3, 1865	50, 269
Horse-rake	L. H. Parson and G. Houston	Middletown, N. Y	Aug. 31, 1858	21, 358
Horse-rake	J. Partridge	Pittsfield, Mass	Aug. 2, 1864	43, 702
Horse-rake	G. Peirce	Ercildoun, Pa	Nov. 29, 1859	26, 293
Horse-rake	L. D. Pennington and J. G. Woodfill.	Vernon, Ind	June 9, 1868	78, 824
Horse-rake	J. Pennypacker	Charlestown, Pa	Feb. 28, 1865	46, 583
Horse-rake	C. B. Perkins	Kenduskeag, Me	Apr. 7, 1868	76, 518
Horse-rake	J. Perry	Vernon, Ind	Mar. 2, 1869	87, 363
Horse-rake	W. Phillips	New Brunswick, N. Y	Aug. 15, 1831	....
Horse-rake	W. M. Piatt	West Liberty, Ohio	May 22, 1866	54, 947
Horse-rake	O. Pier	Ludlow, Vt	Sept. 13, 1859	25, 441
Horse-rake	O. Pier	Winhall, Vt	Feb. 5, 1867	61, 862
Horse-rake	C. H. Poage	Perry, Mo	Sept. 22, 1868	82, 345
Horse-rake	L. L. Pollard	Worcester, Mass	Dec. 27, 1864	45, 626
Horse-rake	D. Prest	Marlborough, N. J	May 30, 1865	47, 980
Horse-rake	D. Prest	Marlborough, N. J	Aug. 13, 1867	67, 673
Horse-rake	J. Pudney	Stanford, N. Y	Nov. 7, 1835	....
Horse-rake	J. S. Randall	Grand Rapids, Mich	May 28, 1867	65, 271
Horse-rake	O. E. Randall	Lewiston, Me	May 23, 1865	47, 857
Horse-rake	O. E. Randall	Lewiston, Me	Feb. 4, 1868	74, 136
Horse-rake	W. Read	Vernon, Ind	Aug. 11, 1868	81, 003
Horse-rake	P. Reading	Trenton, N. J	Aug. 1, 1818	....
Horse-rake	S. B. Reed	Stuyvesant, N. Y	Jan. 9, 1866	51, 969
Horse-rake	H. Reese	Petersburgh, Ind	Apr. 27, 1869	89, 343
Horse-rake	A. R. Reese	Phillipsburgh, N. J	Dec. 8, 1868	84, 760
Horse-rake	G. M. and C. C. Richardson	Dana, Mass	Apr. 20, 1869	89, 169
Horse-rake	A. H. Robbins	Copenhagen, N. Y	Aug. 25, 1868	81, 409
Horse-rake	I. Robbins and J. Old	Pittsburgh, Pa	Dec. 6, 1864	45, 346
Horse-rake	R. I. Robeson	Chicago, Ill	July 4, 1865	48, 629
Horse-rake	J. Robinson	Lawrence, Pa	Apr. 12, 1864	42, 309
Horse-rake	E. S. Root	Mount Morris, N. Y	May 17, 1836	....
Horse-rake	A. V. Ryder	Germano, Ohio	Oct. 10, 1865	50, 391
Horse-rake	A. V. Ryder	Germano, Ohio	Jan. 8, 1867	61, 023
Horse-rake	H. W. Sabin	Canandaigua, N. Y	Dec. 3, 1850	7, 813
Horse-rake	C. W. Sanborn	Morrill, Me	Sept. 21, 1869	95, 147
Horse-rake	C. Satterlee	Paris, Ill	Nov. 13, 1866	59, 663
Horse-rake	E. Saunders	Wethersfield, Vt	Oct. 24, 1848	5, 883
Horse-rake	W. and T. Schnebly	Hackensack, N. J	July 10, 1860	29, 105
Horse-rake	S. C. Schofield and A. J. Wise	Chicago, Ill	Oct. 26, 1869	96, 272
Horse-rake	J. Seely	North Java, N. Y	May 26, 1868	78, 239
Horse-rake	F. Seidle	Mechanicsburgh, Pa	Sept. 20, 1864	44, 344
Horse-rake	F. Seidle	Mechanicsburgh, Pa	Sept. 12, 1865	49, 926
Horse-rake	F. Seidle and S. Eberley	Mechanicsburgh, Pa	July 3, 1860	29, 012
Horse-rake	N. Selby	Flora, Ill	Oct. 20, 1868	83, 326
Horse-rake	G. C. Shaler and H. Barlow	Gilboa and Hobart, N. Y	Oct. 13, 1868	83, 000
Horse-rake	D. P. Sharp	Ithaca, N. Y	Jan. 19, 1864	41, 326
Horse-rake	D. P. Sharp	Ithaca, N. Y	Dec. 12, 1865	51, 486
Horse-rake	D. P. Sharp	Ithaca, N. Y	July 16, 1867	66, 894
Horse-rake	S. M. Sherman	Fort Dodge, Iowa	Aug. 1, 1865	49, 164
Horse-rake	J. H. Shireman	East Berlin, Pa	Sept. 9, 1862	36, 426
Horse-rake	J. H. Shireman	York, Pa	Oct. 8, 1867	69, 713
Horse-rake	J. H. Shireman	York, Pa	Mar. 16, 1869	87, 977
Horse-rake	T. J. Shreves	Greenbush, Ill	Aug. 4, 1868	80, 772
Horse-rake	A. J. Shunk	Shanesville, Ohio	Aug. 30, 1864	44, 020
Horse-rake	A. J. Shunk	Millersburgh, Ohio	May 25, 1869	90, 592
Horse-rake	E. Smith and S. Cowles	Northford, Conn	Sept. 10, 1861	33, 267
Horse-rake	F. Smith	Highgate, Vt	Sept. 8, 1868	82, 040
Horse-rake	F. M. Smith and E. Brumfield	Albion, N. Y	Dec. 4, 1866	60, 270
Horse-rake	J. B. Smith	Newton, Ill	Oct. 13, 1868	83, 002
Horse-rake	M. Smith	Worcester, Mass	May 23, 1865	47, 872
Horse-rake	M. Smith	Worcester, Mass	May 19, 1868	78, 147
Horse-rake	S. P. Smith	Waterford, N. Y	May 4, 1869	89, 697
Horse-rake	G. W. Snyder	Kalamazoo, Mich	Sept. 17, 1867	69, 036
Horse-rake	E. R. and W. P. Spear	Orland, Ind	Apr. 24, 1866	54, 228
Horse-rake	E. R. and W. P. Spear	Orland, Ind	May 11, 1869	89, 950
Horse-rake	J. A. Spear	Braintree, Vt	Oct. 9, 1866	58, 689
Horse-rake	A. B. Sprout	Hughesville, Pa	Nov. 25, 1862	37, 012
Horse-rake	A. B. Sprout	Hughesville, Pa	Jan. 17, 1865	45, 942
Horse-rake	A. B. Sprout	Hughesville, Pa	June 6, 1865	48, 109
Horse-rake	W. Squire	Edon, Ohio	Feb. 15, 1870	99, 966
Horse-rake	J. M. Stafford	Pike, N. Y	Sept. 11, 1847	5, 291
Horse-rake	C. Starrett	Chicago, Ill	July 21, 1868	80, 097
Horse-rake	T. J. Steffe	Lancaster, Pa	Sept. 20, 1859	25, 535
Horse-rake	T. Stewart	Pittsburgh, Pa	Feb. 6, 1866	52, 458
Horse-rake	H. Stimmel	Canton, Ohio	June 26, 1866	55, 930
Horse-rake	W. Stinson	Cool Spring Township, Pa	Jan. 12, 1869	85, 869
Horse-rake	J. C. Stoddard	Worcester, Mass	Sept. 11, 1860	30, 007
Horse-rake	J. C. Stoddard	Worcester, Mass	Jan. 3, 1865	45, 769
Horse-rake	J. C. Stoddard	Worcester, Mass	Jan. 5, 1869	85, 620
Horse-rake	J. C. Stoddard	Worcester, Mass	Mar. 15, 1870	100, 950
Horse-rake	A. C. Stone	Steeleville, Pa	Nov. 7, 1865	50, 853
Horse-rake	A. C. Stone	Steeleville, Pa	Nov. 21, 1865	51, 097
Horse-rake	H. K. Stoner	Lancaster, Pa	June 2, 1863	38, 769
Horse-rake	H. K. Stoner	Lancaster, Pa	Nov. 5, 1867	70, 643
Horse-rake	O. O. Storle and L. Swenson	Norway, Wis	Nov. 24, 1868	84, 315
Horse-rake	S. Stoughton	Windsor, Ohio	May 4, 1869	89, 605
Horse-rake	H. A. Streeter	Worcester, Mass	May 4, 1869	89, 808
Horse-rake	D. Strock	Chambersburgh, Pa	Sept. 11, 1860	30, 010
Horse-rake	G. E. Sutphen	Louisiana, Mo	Dec. 8, 1868	84, 777
Horse-rake	E. Sweet	Triangle, N. Y	Jan. 12, 1869	85, 872
Horse-rake	L. Swift	Clarkson, N. Y	Jan. 19, 1847	4, 933

Index of patents issued from the United States Patent Office from 1790 *to* 1873, *inclusive*—Continued.

Invention.	Inventor.	Residence.	Date.	No.
Horse-rake	J. F. Swinnerton	Marion, Ohio	June 25, 1867	66, 186
Horse-rake	B. C. Taylor	Dayton, Ohio	May 22, 1866	54, 977
Horse-rake	B. C. Taylor	Dayton, Ohio	Aug. 6, 1867	67, 609
Horse-rake	P. A. Thayer	Theresa, N. Y	Sept. 21, 1869	95, 164
Horse-rake	J. Thompson	Bridgeport, Ill	Oct. 12, 1869	95, 852
Horse-rake	C. M. Titus and L. Mood	Ithaca, N. Y	Nov. 5, 1872	132, 741
Horse-rake	E. J. Toof	Fort Madison, Iowa	Dec. 15, 1868	84, 917
Horse-rake	R. M. Treat	Morris, Conn	June 10, 1862	35, 572
Horse-rake	A. Tschop and J. Hartman	East Berlin, Pa	May 11, 1869	89, 956
Horse-rake	H. Tunison	White Hall Grove, Ill	May 2, 1865	47, 587
Horse-rake	T. J. Turner	Richmond County, Ill	Aug. 13, 1867	67, 688
Horse-rake	J. A. Varney	Alton, N. H	May 1, 1866	54, 448
Horse-rake	J. E. Voiles	Madison, Ind	Nov. 10, 1868	84, 028
Horse-rake	C. F. Walker	Benford's Store, Pa	July 29, 1862	36, 043
Horse-rake	S. W. Walker	Anson, Me	June 11, 1867	65, 778
Horse-rake	M. N. Ward	Bangor, Me	Feb. 9, 1869	86, 717
Horse-rake	M. N. Ward	Linneus, Me	Apr. 20, 1869	89, 257
Horse-rake	C. W. Warner	Williston, Vt	Nov. 15, 1864	45, 099
Horse-rake	C. W. Warner and H. N. Tracy	Williston and Essex, Vt	Oct. 17, 1865	50, 525
Horse-rake	B. Webb	Unadilla Forks, N. Y	Mar. 28, 1871	113, 121
Horse-rake	J. W. Webb	Mount Morris, N. Y	Feb. 5, 1836	....
Horse-rake	A. Wells	Morgantown, W. Va	Jan. 16, 1866	52, 098
Horse-rake	M. D. Wells	Morgantown, Va	Aug. 15, 1854	11, 538
Horse-rake	M. D. Wells	Morgantown, W. Va	Dec. 12, 1865	51, 503
Horse-rake	T. J. West	Alfred Centre, N. Y	Aug. 17, 1869	93, 777
Horse-rake	W. H. White	Garrattsville, N. Y	Mar. 1, 1859	23, 133
Horse-rake	H. C. Whitney	Coxsackie, N. Y	Sept. 19, 1865	50, 060
Horse-rake	J. R. Whittemore	Chicopee Falls, Mass	Feb. 4, 1868	74, 023
Horse-rake	F. Wicks	Upper Sandusky, Ohio	June 23, 1868	79, 284
Horse-rake	T. Witmer	Williamsville, N. Y	Aug. 29, 1865	49, 673
Horse-rake	H. Wood	East Henrietta, N. Y	Feb. 23, 1869	87, 315
Horse-rake	J. Wood	North Bloomfield, N. Y	Oct. 23, 1866	59, 113
Horse-rake	J. M. Woodcock	Bridgeport, Ohio	Nov. 11, 1862	36, 928
Horse-rake	M. Woodman and L. Atwood	Farmington, Me., and Norwich, Conn.	Jan. 29, 1867	61, 646
Horse-rake	D. B. Woodward	Ercildoun, Pa	Feb. 19, 1861	31, 507
Horse-rake	J. Zimmerman	Powhatan, Md	Aug. 27, 1867	68, 140
Horse-rake and hay-spreader	G. N. Palmer	Greene, N. Y	May 28, 1867	65, 113
Horse-rake and hay-spreader, Combined	J. M. Low	Portlandville, N. Y	Nov. 12, 1867	70, 870
Horse-rake and hay-spreader, Combined	F. E. Nearing	Brookfield, Conn	Jan. 7, 1868	73, 192
Horse-rake and hay-spreader, Combined	F. E. Nearing	Brookfield, Conn	Jan. 8, 1869	91, 037
Horse-rake and hay-spreader, Combined	G. N. Palmer	Greene, N. Y	Dec. 12, 1865	51, 473
Horse-rake and hay-spreader, Combined	C. Rogers	Barker, N. Y	Oct. 1, 1867	69, 487
Horse-rake and hay-spreader, Combined	E. E. Seymour and S. J. Taylor	Rome, N. Y	July 30, 1867	67, 221
Horse-rake and hay-spreader, Combined	J. M. Spangler	Canton, Ohio	Mar. 16, 1869	87, 982
Horse-rake and hay-spreader, Combined	H. C. Varnum	Hartford, Vt	Aug. 24, 1869	94, 048
Horse-rake and tedder combined	G. L. Ives	Rome, N. Y	Jan. 28, 1873	135, 276
Horse-rake clearer-bar	D. E. Bristol	Albany, N. Y	Dec. 10, 1872	133, 745
Horse-rake, harrow, and roller combined	H. Cousins	Whiting, Me	Dec. 24, 1872	134, 127
Horse-rake, Hay, &c	J. Baily	Philadelphia, Pa	Mar. 30, 1827	....
Horse-rake, Hay and grain	M. and S. Pennock	East Marlborough, Pa	Feb. 17, 1827	....
Horse-rake, Revolving	L. S. Edleblute	Tullahoma, Tenn	Oct. 22, 1872	132, 457
Horse-rake, Revolving	M. K. Flory	Viola, Ill	Oct. 14, 1873	143, 684
Horse-rake, Revolving	H. Hunt	Bridgewater, N. Y	Dec. 10, 1836	104
Horse-rake, Revolving	G. Penisten, sr	Buchanan, Ohio	Sept. 9, 1873	142, 581
Horse-rake, Revolving	W. Wells	Salem, Mass	Apr. 15, 1873	137, 987
Horse-rake, Revolving hay and grain	M. Pennock	Kennett's Square, Pa	Nov. 23, 1824	....
Horse-rake, Revolving hay and grain	M. Pennock	East Marlborough, Pa	Feb. 8, 1825	....
Horse-rake, Spring-tooth	L. M. Whitman	Pike, N. Y	Apr. 4, 1846	4, 444
Horse-rake teeth	I. L. and J. B. L. Bartlett	North Jay, Me	Sept. 21, 1869	95, 070
Horse-rake teeth	J. B. L. Bartlett	North Jay, Me	July 25, 1871	117, 243
Horse-rake teeth	C. Coleman	Allegheny City, Pa	Feb. 20, 1866	52, 680
Horse-rake teeth	A. B. Sprout	Hughesville, Pa	Sept. 5, 1865	49, 800
Horse-rake teeth	A. B. Sprout	Hughesville, Pa	May 1, 1866	54, 430
Horse-rake teeth, Machine for bending and hardening.	G. F. Simmonds	Fitchburgh, Mass	June 6, 1871	115, 775
Horse-rake teeth, &c., Tempering	J. A. Ferson	Fitchburgh, Mass	Dec. 24, 1872	134, 264
Horse-rake-tooth fastening	B. C. Taylor	Dayton, Ohio	Apr. 9, 1872	125, 630
Horse-rakes, Bending teeth for	H. Brandt	Columbia, Pa	June 5, 1860	28, 555
Horse-releasing apparatus	J. Harrison	New York, N. Y	Oct. 24, 1871	120, 272
Horseshoe	D. Anthony, jr	Adams, Mass	Apr. 8, 1831	....
Horseshoe	H. Armstrong	Sparta, Wis	June 5, 1866	55, 220
Horseshoe	J. Austin	Rockford, Ill	Jan. 2, 1866	51, 786
Horseshoe	J. Austin	Rockford, Ill	Aug. 21, 1866	57, 433
Horseshoe	J. Austin	Rockford, Ill	Feb. 12, 1867	61, 982
Horseshoe	H. H. Baker	New Market, N. J	July 10, 1866	56, 161
Horseshoe	H. H. Baker	New Market, N. J	Oct. 9, 1866	58, 573
Horseshoe	S. J. Baker	Madison Centre, Me	June 14, 1870	104, 097
Horseshoe	J. Barker	Champlain, N. Y	Dec. 22, 1868	85, 052
Horseshoe	J. Behel and J. M. Buell	Rockport, Ill	May 29, 1866	55, 045
Horseshoe	J. Behel, J. Perrine, and J. M. Buell.	Rockford and Ogle County, Ill.	Nov. 12, 1867	70, 783
Horseshoe	T. B. Bishop	Baltimore, Md	Feb. 12, 1867	61, 990
Horseshoe	T. B. Bishop	Baltimore, Md	Dec. 24, 1867	72, 594
Horseshoe	G. Bonnet	New York, N. Y	Aug. 7, 1866	56, 888
Horseshoe	J. Bracket	Lynn, Mass	Mar. 21, 1871	112, 773
Horseshoe	J. B. Brown and J. Farmer	Boston, Mass	Mar. 11, 1822	....
Horseshoe	J. H. Brown	Watertown, Minn	Jan. 9, 1866	51, 916
Horseshoe	J. Carlin	Cumminsville, Ohio	Jan. 17, 1860	26, 832
Horseshoe	I. Carman	Schoolcraft, Mich	Nov. 19, 1867	70, 954
Horseshoe	N. B. Carpenter	New York, N. Y	May 13, 1856	14, 852
Horseshoe	E. Cate	Franklin, N. H	Feb. 5, 1861	31, 296
Horseshoe	E. Cate	Watertown, Mass	Apr. 5, 1870	101, 581
Horseshoe	G. T. Chapman	New York, N. Y	Feb. 25, 1868	74, 892
Horseshoe	M. Chittenden	Danbury, Conn	Oct. 11, 1864	44, 603
Horseshoe	J. K. Christopher	Dayton, Ohio	Oct. 25, 1870	108, 684
Horseshoe	M. C. Clark	Manchester, N. H	Apr. 15, 1873	137, 891
Horseshoe	J. N. Clarke	Cincinnati, Ohio	July 10, 1866	56, 181

Index of patents issued from the United States Patent Office from 1790 *to* 1873, *inclusive*—Continued.

Invention.	Inventor.	Residence.	Date.	No.
Horseshoe	J. N. Clarke	Cincinnati, Ohio	Aug. 18, 1868	81, 142
Horseshoe	N. Clouse	Valley Grove, W. Va	July 30, 1872	130, 019
Horseshoe	W. Coes	Worcester, Mass	Apr. 25, 1865	47, 397
Horseshoe	T. M. Coleman	Philadelphia, Pa	June 12, 1860	28, 656
Horseshoe	R. F. Cooke	Brooklyn, N. Y	Nov. 25, 1873	144, 833
Horseshoe	W. Cooper	Brooklyn, N. Y	June 30, 1857	17, 672
Horseshoe	G. Copeland	Denver, Colo	May 3, 1870	102, 504
Horseshoe	D. Cumming	Sorrel Horse, Pa	May 12, 1857	17, 265
Horseshoe	G. Custer	Monroe, Mich	June 28, 1864	43, 293
Horseshoe	G. Custer	Monroe, Mich	July 4, 1865	48, 618
Horseshoe	J. M. Cuykendal	Metomen, Wis	June 2, 1868	78, 436
Horseshoe	J. Deitz	New York	Apr. 22, 1831	......
Horseshoe	W. Disbrow	San Francisco, Cal	Jan. 31, 1865	46, 087
Horseshoe	H. Dawnie and I. B. Harris	Corstophine and Edinburgh, Scotland.	Jan. 25, 1870	99, 072
Horseshoe	F. W. Edison	Port Huron, Mich	Dec. 8, 1868	84, 684
Horseshoe	W. L. Edwards	Ellison, Ill	Jan. 23, 1872	123, 007
Horseshoe	J. W. Faulkner	Rockford, Ill	Apr. 10, 1866	53, 801
Horseshoe	H. B. Ferren	Batavia, N. Y	Aug. 20, 1872	130, 629
Horseshoe	W. H. Freleigh	Troy, N. Y	Mar. 5, 1872	124, 203
Horseshoe	E. C. Gers	Galesburgh, Mich	July 17, 1866	56, 401
Horseshoe	C. Goodenough	New York, N. Y	Aug. 26, 1873	142, 097
Horseshoe	R. A. Goodenough	Brooklyn, N. Y	May 29, 1860	28, 469
Horseshoe	D. Grim	Pittsburgh, Pa	Nov. 14, 1871	120, 813
Horseshee	L. M. Guiteau	Batavia, N. Y	May 9, 1865	47, 635
Horseshoe	P. Hanly	New York, N. Y	Dec. 24, 1867	72, 485
Horseshoe	W. T. Harmer	New York, N. Y	Feb. 4, 1868	73, 971
Horseshoe	J. Haseltine	Warren, N. H	July 25, 1865	49, 028
Horseshoe	J. Henderson	Elmira, N. Y	May 20, 1856	14, 915
Horseshoe	J. Henderson	Albion, N. Y	Apr. 19, 1870	102, 121
Horseshoe	J. A. Heyl	Boston, Mass	July 21, 1868	80, 074
Horseshoe	J. A. Heyl	Boston, Mass	Mar. 9, 1869	87, 565
Horseshoe	N. E. Hinds	Cooperstown, N. Y	Nov. 8, 1859	26, 036
Horseshoe	W. Hinds	Worcester, Mass	Oct. 23, 1866	59, 016
Horseshoe	W. Hinds	Worcester, Mass	Apr. 7, 1868	76, 449
Horseshoe	H. S. Hitner	Marble Hall, Pa	Jan. 19, 1869	86, 014
Horseshoe	J. W. Hodges	Baltimore, Md	July 4, 1865	48, 623
Horseshoe	J. W. Hodges	Baltimore, Md	Oct. 22, 1867	70, 091
Horseshoe	A. S. Hopson	Plain View, Minn	Mar. 31, 1868	76, 078
Horseshoe	O. A. Howe	Fort Plain, N. Y	Sept. 15, 1863	39, 926
Horseshoe	N. W. Hubbard	New York, N. Y	Feb. 18, 1868	74, 539
Horseshoe	W. E. Hubbard	Randolph, N. Y	June 29, 1858	20, 713
Horseshoe	B. P. Hutchinson	Cady, Mich	Nov. 19, 1872	133, 105
Horseshoe	H. Ingraham	Armada, Mich	Mar. 15, 1870	100, 768
Horseshoe	R. G. Jameson and W. H. Chamberlain.	Bristol, N. H	Dec. 8, 1868	84, 745
Horseshoe	J. Johnson	Lowell, Mass	May 4, 1869	89, 774
Horseshoe	J. Johnson	Lowell, Mass	May 4, 1869	89, 775
Horseshoe	J. M. Johnson	Washington, D. C	May 24, 1864	42, 857
Horseshoe	P. C. Johnson	Central City, Colo	Aug. 10, 1869	93, 447
Horseshoe	P. C. Johnson and E. Froggott.	Central City, Colo	Sept. 29, 1868	82, 528
Horseshoe	J. B. Johnston	New York, N. Y	Sept. 30, 1873	143, 353
Horseshoe	R. K. Jordan	Oakland, Cal	Dec. 2, 1873	145, 108
Horseshoe	J. Jorey	Rocky Hill, Conn	July 5, 1859	24, 643
Horseshoe	J. Jorey	Westville, Conn	Feb. 25, 1868	74, 829
Horseshoe	J. Jorey	Westville, Conn	Mar. 30, 1869	88, 304
Horseshoe	L. H. Kellogg	Monroe, Ohio	Nov. 24, 1868	84, 360
Horseshoe	J. B. Kendall	Boston, Mass	Nov. 12, 1861	33, 709
Horseshoe	T. and R. Kinghorn	Morgan, Ohio	Feb. 6, 1872	123, 404
Horseshoe	A. E. Kroger	Norwalk, Conn	Aug. 25, 1868	81, 512
Horseshoe	B. Ladd	Ottumwa, Iowa	Sept. 1, 1868	81, 796
Horseshoe	P. A. La France	Elmira, N. Y	July 3, 1866	56, 065
Horseshoe	R. Láportà	New York, N. Y	Dec. 8, 1868	84, 834
Horseshoe	A. Leach	Lynn, Mass	Aug. 5, 1873	141, 509
Horseshoe	G. W. Lewis	Providence, R. I	Oct. 29, 1867	70, 231
Horseshoe	W. Litzenberg	Macomb, Ill	Aug. 21, 1866	57, 346
Horseshoe	H. D. Lyman	Kalamazoo, Mich	Sept. 8, 1868	82, 013
Horseshoe	O. P. Macgill	Brooklandville, Md	Apr. 11, 1865	47, 242
Horseshoe	J. Maddock	Bloomington, Ill	Sept. 21, 1858	21, 571
Horseshoe	J. F. Mallett	New York, N. Y	Aug. 30, 1864	44, 009
Horseshoe	S. Mason	Newark, N. J	Nov. 24, 1868	84, 294
Horseshoe	D. L. McDonell	Detroit, Mich	Mar. 12, 1867	62, 867
Horseshoe	J. McPherson	Rockford, Ill	Dec. 26, 1865	51, 738
Horseshoe	J. J. Mervesp	New York, N. Y	Dec. 15, 1868	84, 958
Horseshoe	G. B. Milligan	Baltimore, Md	July 21, 1868	80, 199
Horseshoe	S. A. Moore	Bloomfield, Iowa	July 18, 1865	48, 827
Horseshoe	I. D. Mott	Cannonsville, N. Y	Mar. 25, 1873	137, 128
Horseshoe	A. L. Murphy	Philadelphia, Pa	Feb. 22, 1870	100, 180
Horseshoe	P. Murray	East Morrisania, N. Y	Apr. 7, 1868	76, 501
Horseshoe	H. H. Palmer	Rockford, Ill	Nov. 21, 1865	51, 078
Horseshoe	C. Parish	Elkhart, Ind	July 16, 1872	129, 586
Horseshoe	G. A. Parker	Westford, Mass	Jan. 5, 1869	85, 689
Horseshoe	R. H. Parks	Columbus, Ohio	Apr. 20, 1869	89, 165
Horseshoe	I. Peacock	Shortsville, N. Y	May 5, 1863	38, 451
Horseshoe	C. Peillard	France	Sept. 21, 1869	95, 133
Horseshoe	J. I. Peyton	Washington, D. C	Nov. 13, 1866	59, 725
Horseshoe	E. J. Pleyel	Dallas County, Iowa	Oct. 6, 1863	40, 215
Horseshoe	I. R. Potter	Dartmouth, Mass	July 21, 1868	80, 213
Horseshoe	Z. V. Purdy	Washington, D. C	June 30, 1868	79, 389
Horseshoe	D. Roberge	Mooers, N. Y	May 25, 1869	90, 394
Horseshoe	D. Roberge	Moore's Forks, N. Y	Mar. 1, 1870	100, 328
Horseshoe	D. Roberge	Moore's Forks, N. Y	Mar. 1, 1870	100, 329
Horseshoe	J. S. Robertson	Cathcart, Scotland	Nov. 8, 1870	109, 055
Horseshoe	J. Rowe, jr., and F. B. Brown.	Wilmington and Boston, Mass.	Aug. 26, 1873	142, 282
Horseshoe	H. Schreiner	Philadelphia, Pa	May 22, 1866	55, 019
Horseshoe	R. Seiffert	Chicago, Ill	Sept. 26, 1871	119, 415
Horseshoe	G. Sewell	Poughkeepsie, N. Y	Feb. 12, 1867	62, 076

Index of patents issued from the United States Patent Office from 1790 *to* 1873, *inclusive*—Continued.

Invention.	Inventor.	Residence.	Date.	No.
Horseshoe	F. Shinn	Rockford, Ill	June 5, 1866	55, 374
Horseshoe	S. Short	New London, Conn	July 8, 1856	15, 306
Horseshoe	W. H. Shurtleff	Providence, R. I	Sept. 10, 1867	68, 800
Horseshoe	T. Skelton	Rockford, Ill	July 18, 1865	48, 845
Horseshoe	G. W. Skinner	Rockford, Ill	Feb. 18, 1868	74, 620
Horseshoe	S. Sloat	Morgan, Ohio	Oct. 22, 1867	70, 035
Horseshoe	S. Sloat	Morgan, Ohio	Jan. 23, 1872	122, 972
Horseshoe	A. W. Smith	Manchester, N. H	June 11, 1872	127, 932
Horseshoe	A. W. Smith	Manchester, N. H	Sept. 9, 1873	142, 590
Horseshoe	H. Smith and J. H. Evans	Sandyville and Bolivar, Ohio.	June 22, 1869	91, 571
Horseshoe	L. A. Smith	Pekin, Ill	Aug. 4, 1868	80, 778
Horseshoe	L. T. C. Smith	Scott County, Ky	Dec. 15, 1831	....
Horseshoe	E. Sneider	Baltimore, Md	Oct. 2, 1866	58, 495
Horseshoe	U. Snyder	Conshohocken, Pa	Nov. 11, 1873	144, 483
Horseshoe	A. Soles	Fonda, N. Y	June 4, 1872	127, 657
Horseshoe	C. O. Stevens	Auburn, Me	Aug. 18, 1868	81, 307
Horseshoe	J. Stickney	Manchester, N. H	May 7, 1872	126, 497
Horseshoe	J. Stickney	Manchester, N. H	Dec. 3, 1872	133, 549
Horseshoe	P. Thiry	Paris, France	July 22, 1862	35, 958
Horseshoe	J. S. Toan	King's Ferry, N. Y	July 14, 1868	80, 033
Horseshoe	G. H. Todd	Montgomery, Ala	Dec. 9, 1873	145, 463
Horseshoe	W. H. Towers	Philadelphia, Pa	Dec. 20, 1853	10, 345
Horseshoe	W. H. Towers	Philadelphia, Pa	July 25, 1854	11, 392
Horseshoe	J. H. Tyler	Martin, N. C	Apr. 20, 1869	89, 096
Horseshoe	A. Tyrrell	Batavia, N. Y	July 26, 1864	43, 650
Horseshoe	A. Tyrrell	Batavia, N. Y	Apr. 25, 1865	47, 502
Horseshoe	S. Ward and L. J. Munger	Cambridge and Charlestown, Mass.	May 24, 1864	42, 889
Horseshoe	T. Waterhouse and C. F. McKenney.	West Gorham and Saco, Me.	Dec. 28, 1869	98, 323
Horseshoe	B. R. Watson	New Bedford, Mass	Feb. 19, 1867	62, 170
Horseshoe	W. R. Watson	Stockton, Cal	Mar. 2, 1869	87, 384
Horseshoe	A. Weitman	West Union, Iowa	Aug. 8, 1865	49, 324
Horseshoe	A. Weitman	West Union, Iowa	Sept. 5, 1865	49, 812
Horseshoe	C. Weitman	Hazelton, Iowa	May 7, 1867	64, 604
Horseshoe	C. M. Werner	Rockford, Ill	Apr. 10, 1866	53, 930
Horseshoe	C. M. Werner	Rockford, Ill	Sept. 25, 1866	58, 355
Horseshoe	C. M. Werner	Rockford, Ill	Feb. 19, 1867	62, 242
Horseshoe	E. Wheeler	Marlborough, Mass	Mar. 2, 1858	19, 526
Horseshoe	J. Wheeler	Huntington, Ind	Nov. 5, 1867	70, 660
Horseshoe	W. C. Whitmore	Macon City, Mo	Nov. 24, 1868	84, 462
Horseshoe	A. S. Wilkinson	Pawtucket, R. I	July 3, 1866	56, 134
Horseshoe	A. S. Wilkinson	Pawtucket, R. I	July 10, 1866	56, 307
Horseshoe	A. S. Wilkinson	Pawtucket, R. I	July 10, 1866	56, 308
Horseshoe	A. S. Wilkinson	Pawtucket, R. I	July 10, 1866	56, 309
Horseshoe	A. S. Wilkinson	Pawtucket, R. I	July 10, 1866	56, 310
Horseshoe	A. S. Wilkinson	Pawtucket, R. I	July 17, 1866	56, 475
Horseshoe	A. S. Wilkinson	Pawtucket, R. I	July 17, 1866	56, 476
Horseshoe	A. S. Wilkinson	Pawtucket, R. I	July 17, 1866	56, 477
Horseshoe	A. S. Wilkinson	Pawtucket, R. I	Aug. 7, 1866	57, 029
Horseshoe	A. S. Wilkinson	Pawtucket, R. I	Aug. 7, 1866	57, 030
Horseshoe	A. S. Wilkinson	Pawtucket, R. I	Aug. 14, 1866	57, 235
Horseshoe	A. S. Wilkinson	Pawtucket, R. I	Aug. 21, 1866	57, 420
Horseshoe	A. S. Wilkinson	Pawtucket, R. I	Feb. 12, 1867	61, 971
Horseshoe	A. S. Wilkinson	Pawtucket, R. I	Feb. 12, 1867	61, 972
Horseshoe	A. S. Wilkinson	Pawtucket, R. I	Feb. 12, 1867	61, 973
Horseshoe	A. S. Wilkinson	Pawtucket, R. I	Feb. 12, 1867	61, 974
Horseshoe	A. S. Wilkinson	Pawtucket, R. I	Feb. 12, 1867	61, 975
Horseshoe	A. S. Wilkinson	Pawtucket, R. I	May 28, 1867	65, 144
Horseshoe	A. S. Wilkinson	Pawtucket, R. I	May 28, 1867	65, 145
Horseshoe	A. S. Wilkinson	Pawtucket, R. I	Sept. 24, 1867	69, 146
Horseshoe	A. A. York	De Lancey, N. Y	June 29, 1869	91, 996
Horseshoe, Adjustable	H. and P. Moran	New York, N. Y	Mar. 5, 1872	124, 215
Horse shoe and boot, Combined	H. G. Haedrich	Philadelphia, Pa	Oct. 26, 1869	96, 104
Horse shoe and boot, Combined	H. G. and E. M. Haedrich	Philadelphia, Pa	July 19, 1870	105, 566
Horseshoe and calk	C. H. Johnson	Boston, Mass	Apr. 25, 1865	47, 495
Horseshoe and other nails, Making	B. W. Pierce	New Bedford, Mass	Feb. 10, 1863	37, 640
Horseshoe-attachment	P. H. O'Neill	Saint Louis, Mo	Apr. 15, 1873	137, 794
Horseshoe-bar	E. Cate	Watertown, Mass	Nov. 29, 1870	109, 586
Harseshoe-bars, Machine for rolling	A. Reese	Pittsburgh, Pa	July 30, 1867	67, 348
Horseshoe-bars, Roll for rolling	E. Cate	East Woburn, Mass	Apr. 4, 1871	113, 255
Horseshoe bending and channeling machine	A. Oehme	Rock Island, Ill	Oct. 17, 1871	120, 087
Horseshoe-bending machine	H. J. Batchelder	Boston, Mass	May 29, 1866	55, 042
Horseshoe-bending machine	H. F. Goulding	Providence, R. I	Aug. 28, 1866	57, 631
Horseshoe-bending machine	E. Wheeler	Marlborough, Mass	Dec. 22, 1857	18, 940
Horseshoe-beveler	E. Quinby	Comstock, Mich	Dec. 14, 1869	97, 811
Horseshoe-blank	J. Day	Buffalo, N. Y	Apr. 19, 1870	102, 095
Horseshoe-blank	E. B. Turner	Providence, R. I	Nov. 30, 1869	97, 327
Horseshoe-blank bar	E. Cate	Watertown, Mass	May 10, 1870	102, 775
Horseshoe-blank bar	J. Montgomery	Sing Sing, N. Y	Dec. 28, 1869	98, 290
Horseshoe-blanks, Forming	C. A. Perkins	Providence, R. I	Mar. 10, 1868	75, 456
Horseshoe-blanks, Machine for making	H. and J. Reese	Pittsburgh, Pa	July 19, 1870	105, 490
Horseshoe-blanks, Machine for swaging	C. H. Perkins and R. W. Comstock.	Providence, R. I	May 21, 1867	64, 903
Horseshoe-blanks, Manufacture of	H. J. Batchelder	Boston, Mass	Nov. 20, 1866	59, 755
Horseshoe-blanks, Roll for	J. W. Kingsbury	New Bedford, Mass	Sept. 1, 1863	39, 773
Horseshoe-blanks, Roll for forming	W. Acheson	Pittsburgh, Pa	Mar. 14, 1871	112, 668
Horseshoe-blanks, Roll for forming	C. L. Fitzhugh	Pittsburgh, Pa	Dec. 20, 1870	110, 349
Horseshoe-blanks, Rolling	A. Reese	Pittsburgh, Pa	Oct. 20, 1868	83, 207
Horseshoe-calk	T. M. Clarke	Winsted, Conn	Jan. 21, 1873	135, 081
Horseshoe-calk	G. Custer	Monroe, Mich	May 6, 1873	138, 618
Horseshoe-calk	J. H. Jennings	New Bedford, Mass	May 21, 1861	32, 371
Horseshoe-calk	J. Jorey	North Manchester, Conn	June 15, 1869	91, 237
Horseshoe-calk	F. Judson	Jersey City, N. J	May 22, 1866	54, 918
Horseshoe-calk	E. Maynard	Williamsburgh, N. Y	Feb. 24, 1857	16, 691
Horseshoe-calk	J. J. Mervesp	New York, N. Y	Aug. 29, 1871	118, 470
Horseshoe-calk	J. J. Mervesp	New York, N. Y	Mar. 4, 1873	136, 379

Index of patents issued from the United States Patent Office from 1790 *to* 1873, *inclusive*—Continued.

Invention.	Inventor.	Residence.	Date.	No.
Horseshoe-calk	C. H Perkins	Providence, R. I	June 16, 1868	78, 892
Horseshoe-calk	J. L. Pike	Lynn, Mass	Feb. 7, 1865	46, 262
Horseshoe-calk	I. R. Potter	Dartmouth, Mass	Nov. 14, 1865	50, 952
Horseshoe-calk	A. S. Wilkinson	Pawtucket, R. I	Aug. 14, 1866	57, 236
Horseshoe-calk, Adjustable	W. J. Berne	Cincinnati, Ohio	Nov. 5, 1867	70, 507
Horseshoe calk and toe, Movable	S. Lloyd	Washington, D. C	Jan. 31, 1865	46, 192
Horseshoe-calk blank	R. B. Caswell	Palmer, Mass	June 11, 1867	65, 723
Horseshoe-calk, Detachable	W. J. Berne	Cincinnati, Ohio	May 18, 1869	90, 224
Horseshoe-calk, Detachable	J. E. Byers	Butler, Pa	June 11, 1872	127, 843
Horseshoe-calk, Detachable	C. H. Johnson	Boston, Mass	Apr. 18, 1865	47, 360
Horseshoe-calk, Detachable	P. Laflin	Warren, Mass	July 27, 1869	93, 098
Horseshoe-calk die	S. Stone	North Manchester, Conn	Oct. 12, 1869	95, 745
Horseshoe-calk-grinding machine	J. Little	Newburgh, N. Y	Apr. 15, 1873	137, 852
Horseshoe-calk, Removable	J. D. Barnum	Amenia Union, N. Y	June 7, 1870	103, 827
Horseshoe-calk sharpener	R. Crocker	Marshalltown, Iowa	June 30, 1868	79, 319
Horseshoe-calk sharpener	W. Duncan	Vinton, Iowa	Oct. 6, 1868	82, 813
Horseshoe-calk sharpening	A. W. Payne	Morris, N. Y	Apr. 26, 1859	23, 780
Horseshoe-calk-sharpening device	J. Johnson	Barrington, N. Y	Jan. 28, 1868	73, 901
Horseshoe-calk-sharpening device	W. M. Jones	Horicon, Wis	Jan. 7, 1868	73, 102
Horseshoe-calk-sharpening device	P. Winegard	Cold Water, Mich	July 14, 1868	80, 044
Horseshoe-calk-sharpening implement	H. M. Close	Chariton, Iowa	June 30, 1868	79, 316
Horseshoe-calk-sharpening machine	E. A. Bushnell	Dodge County, Wis	June 19, 1866	55, 617
Horseshoe-calk-sharpening machine	E. A. Bushnell	Horicon, Wis	Aug. 11, 1868	80, 805
Horseshoe-calk-sharpening machine	W. M. Butler	Waukegan, Ill	Nov. 5, 1867	70, 519
Horseshoe-calk-sharpening machine	H. Howell	Salem, Ohio	Apr. 4, 1871	113, 670
Horseshoe-calk-sharpening machine	J. B. McClanathan	Horicon, Wis	Feb. 12, 1867	62, 048
Horseshoe-calk-sharpening tool	O. E. Butler, S. P. Dunham, and G. K. Wann.	Marshalltown, Iowa	Jan. 19, 1869	85, 997
Horseshoe-calk-sharpening tool	N. Hays, W. Duncan, and E. H. Bowen.	Winton, Iowa	Dec. 24, 1867	72, 491
Horseshoe-calk-turning machine	A. T. Boon, D. M. Osborn, and G. Geer.	Galesburgh, Ill	Aug. 1, 1871	117, 508
Horseshoe-calks, Attachable and removable	G. S. Norris	Baltimore, Md	Dec. 14, 1869	97, 798
Horseshoe-calks, Attaching	W. J. Berne	Cincinnati, Ohio	July 16, 1867	66, 780
Horseshoe-calks, Clamp for forming	J. Rhoads	Niles, Mich	Feb. 1, 1870	99, 476
Horseshoe-calks, Die for swaging	J. Allen	Palmer, Mass	Apr. 16, 1867	63, 823
Horseshoe-calks, Die for swaging	P. A. Page	Palmer, Mass	May 7, 1867	64, 561
Horseshoe-calks, Swage for	E. D. Withers	Parkton, Md	May 30, 1871	115, 405
Horseshoe-cork, Removable	G. W. Griswold	New York, N. Y	Aug. 2, 1864	43, 747
Horseshoe-cushion	G. E. Rust	Boston, Mass	Jan. 7, 1868	73, 050
Horseshoe, Elastic	J. O. Jones	Newton, Mass	Aug. 3, 1852	9, 173
Horseshoe for cure of hoof-bound horses	D. Homer	Alton, Ill	Apr. 16, 1842	2, 564
Horseshoe interfering-attachment	H. S. Davis	Camden, N. J	Apr. 8, 1873	137, 660
Horseshoe-irons, Formation of	E. Cate	Franklin, N. H	Mar. 26, 1861	31, 781
Horseshoe-irons, Machine for rolling	C. Richardson	Auburn, N. Y	Oct. 8, 1861	33, 452
Horseshoe-jack	J. Shimer	Scranton Station, Iowa	Mar. 12, 1872	124, 452
Horseshoe-machine	D. N. Allard	McConnellsville, Ohio	Apr. 9, 1861	31, 941
Horseshoe-machine	F. D. Althans and J. F. Allen	Morrisania and Tremont, N. Y.	Nov. 16, 1869	96, 762
Horseshoe-machine	W. Anderson	Pittsburgh, Pa	Nov. 23, 1869	97, 021
Horseshoe-machine	R. Y. Babcock	New London, Conn	Apr. 29, 1851	8, 067
Horseshoe-machine	H. J. Bachelder and G. E. Woods.	Marlborough, Mass	Apr. 21, 1868	76, 975
Horseshoe-machine	H. J. Batchelder	Boston, Mass	June 19, 1866	55, 602
Horseshoe-machine	L. H. Bigelow	Philadelphia, Pa	Oct. 17, 1865	50, 440
Horseshoe-machine	U. Billings	New Bedford, Mass	Nov. 12, 1861	33, 692
Horseshoe-machine	U. Billings	New Bedford, Mass	Sept. 9, 1862	36, 390
Horseshoe-machine	U. Billings	New Bedford, Mass	June 9, 1863	38, 804
Horseshoe-machine	U. Billings	Cambridgeport, Mass	Mar. 21, 1871	112, 769
Horseshoe-machine	H. Burden	Troy, N. Y	Nov. 23, 1835	....
Horseshoe-machine	H. Burden	Troy, N. Y	Sept. 14, 1843	3, 261
Horseshoe machine	H. Burden	Troy, N. Y	June 30, 1857	17, 665
Horseshoe-machine	H. Burden	Troy, N. Y	July 1, 1862	35, 746
Horseshoe-machine	J. Christie	Pittsburgh, Pa	May 4, 1869	89, 737
Horseshoe-machine	J. B. Collen	Philadelphia, Pa	May 10, 1859	23, 976
Horseshoe-machine	E. Davis and S. Field	Barre and Oakum, Mass	Oct. 8, 1834	....
Horseshoe-machine	S. W. Davis	Wilmington, Del	Nov. 17, 1863	40, 611
Horseshoe-machine	S. W. Davis	Wilmington, Del	Aug. 30, 1864	43, 977
Horseshoe-machine	S. Decatur and W. Tatham		Oct. 11, 1809	....
Horseshoe-machine	J. Desnos	Troy, N. Y	July 31, 1860	29, 365
Horseshoe-machine	M. Hardaway	Saint Louis, Mo	Nov. 24, 1863	40, 689
Horseshoe-machine	R. E. Hobart	Pottstown, Pa	May 24, 1828	....
Horseshoe-machine	O. A. Howe	Jersey City, N. J	Sept. 22, 1868	82, 409
Horseshoe-machine	W. R. Justus	Pittsburgh, Pa	Mar. 28, 1871	113, 173
Horseshoe-machine	W. R. Justus	Pittsburgh, Pa	Jan. 23, 1872	122, 893
Horseshoe-machine	J. W. Kingsbury	New Bedford, Mass	May 21, 1867	64, 883
Horseshoe-machine	W. W. Lewis	Cincinnati, Ohio	June 22, 1858	20, 646
Horseshoe-machine	W. W. Lewis	Cincinnati, Ohio	Feb. 1, 1859	22, 812
Horseshoe-machine	W. W. Lewis	Cincinnati, Ohio	Aug. 31, 1869	94, 225
Horseshoe-machine	J. McCarty	Philadelphia, Pa	Apr. 27, 1858	20, 079
Horseshoe-machine	J. McCarty	Philadelphia, Pa	Oct. 16, 1860	30, 448
Horseshoe-machine	J. McCarty	Philadelphia, Pa	Nov. 13, 1860	30, 656
Horseshoe machine	B. Mee	Troy, N. Y	June 9, 1863	38, 873
Horseshoe-machine	J. Neff	Prattsburgh, N. Y	Sept. 16, 1862	36, 475
Horseshoe-machine	C. H. Perkins	Putnam, Conn	June 1, 1858	20, 441
Horseshoe-machine	C. H. Perkins	Providence, R. I	June 25, 1861	32, 645
Horseshoe-machine	C. H. Perkins	Providence, R. I	Oct. 25, 1864	44, 838
Horseshoe-machine	C. H. Perkins	Providence, R. I	Oct. 25, 1864	44, 839
Horseshoe-machine	C. H. Perkins	Providence, R. I	Oct. 25, 1864	44, 840
Horseshoe-machine	C. H. Perkins	Providence, R. I	Oct. 21, 1873	143, 781
Horseshoe-machine	C. H. Perkins	Providence, R. I	Oct. 21, 1873	143, 782
Horseshoe-machine	C. H. Perkins and R. W. Comstock.	Providence, R. I	May 28, 1867	65, 265
Horseshoe-machine	D. I. Pruner	Bellefonte, Pa	May 30, 1871	115, 520
Horseshoe-machine	P. P. Read	Durham, Me	Nov. 13, 1847	5, 364
Horseshoe-machine	C. B. Reed	West Bridgewater, Mass	Oct. 20, 1834	....
Horseshoe-machine	A. Reese	Pittsburgh, Pa	July 23, 1867	66, 991
Horseshoe-machine	J. and A. Reese	Pittsburgh, Pa	Nov. 23, 1869	97, 118

Index of patents issued from the United States Patent Office from 1790 to 1873, inclusive—Continued.

Invention.	Inventor.	Residence.	Date.	No.
Horseshoe-machine	W. D. Rinehart	Pittsburgh, Pa	June 8, 1869	91, 165
Horseshoe-machine	A. J. Roberts	Boston, Mass	Sept. 1, 1863	39, 778
Horseshoe-machine	A. J. Roberts	Boston, Mass	Mar. 28, 1865	47, 071
Horseshoe-machine	A. J. Roberts	Boston, Mass	Oct. 19, 1869	95, 935
Horseshoe-machine	L. D. Roberts	Cleveland, Ohio	Sept. 2, 1862	36, 370
Horseshoe-machine	L. D. Roberts	Cleveland, Ohio	Dec. 22, 1863	41, 022
Horseshoe-machine	L. D. Roberts	Cleveland, Ohio	Apr. 5, 1870	101, 513
Horseshoe-machine	W. Roberts	Cleveland, Ohio	Sept. 19, 1871	119, 051
Horseshoe-machine	D. T. Robinson	Boston, Mass	Apr. 30, 1867	64, 254
Horseshoe-machine	T. H. Russell and A. Morrill	Northfield and Strafford, Vt	Oct. 12, 1858	21, 779
Horseshoe-machine	E. Shaw and C. Carpenter, jr	Providence, R. I., and Pawtucket, Mass.	Apr. 20, 1858	20, 023
Horseshoe-machine	S. Shetter	Allegheny, Pa	Jan. 18, 1859	22, 677
Horseshoe-machine	J. Shinneller and J. Brislin	Temperanceville, Pa	July 10, 1866	56, 280
Horseshoe-machine	J. H. Snyder	Troy, N. Y	Oct. 18, 1870	108, 530
Horseshoe-machine	J. H. Snyder	Richmond, Va	Aug. 22, 1871	118, 399
Horseshoe-machine	G. Stiles, jr., and S. Kneass	Philadelphia, Pa	Apr. 13, 1858	19, 957
Horseshoe-machine	J. G. Stowe	Providence, R. I	May 31, 1870	103, 681
Horseshoe-machine	T. R. Taylor	Cleveland, Ohio	Apr. 3, 1860	27, 750
Horseshoe-machine	T. R. Taylor	Cleveland, Ohio	July 29, 1862	36, 041
Horseshoe-machine	W. H. Thompson	Cleveland, Ohio	Dec. 2, 1862	37, 062
Horseshoe-machine	E. Tolles	Winchester, Conn	Oct. 24, 1834	....
Horseshoe-machine	E. B. Turner	Providence, R. I	Oct. 20, 1868	83, 342
Horseshoe-machine	J. T. Walker	Albany, N. Y	Feb. 22, 1870	100, 222
Horseshoe-machine	W. Wallick	Philadelphia, Pa	May 1, 1866	54, 452
Horseshoe-machine	E. Wassell	Pittsburgh, Pa	Dec. 29, 1868	85, 414
Horseshoe-machine	E. Wassell	Wood's Run, Pa	Mar. 8, 1870	100, 570
Horseshoe-machine	H. L. Watts	Chester, Mass	June 28, 1859	24, 596
Horseshoe-machine	T. J. West	Alfred, N. Y	Dec. 27, 1864	45, 661
Horseshoe-machine	C. W. Wettengel	Pittsburgh, Pa	Mar. 8, 1870	100, 574
Horseshoe-machine	C. P. Williamson	Louisville, Ky	Aug. 17, 1869	93, 778
Horseshoe-machine	H. A. Wills	Keeseville, N. Y	Mar. 2, 1858	19, 528
Horseshoe-machine	H. A. Wills	Keeseville, N. Y	July 3, 1860	29, 042
Horseshoe-machine	B. Young and S. Titus	Killingly and Brooklyn, Conn.	July 29, 1837	301
Horseshoe-machine	J. Zepf	Troy, N. Y	Nov. 10, 1868	84, 043
Horseshoe-machinery	S. S. Greene	Lowell, Mass	Nov. 12, 1850	7, 772
Horseshoe-machinery	S. Shetter	Allegheny, Pa	Nov. 9, 1852	9, 390
Horseshoe-nail	T. Fowler	Seymour, Conn	June 5, 1866	55, 270
Horseshoe-nail	J. Jorey	Norwich, Conn	May 14, 1872	126, 712
Horseshoe-nail	A. Reese	McClure Township, Pa	Feb. 2, 1869	86, 450
Horseshoe-nail	A. S. Wilkinson	Pawtucket, R. I	Feb. 12, 1867	61, 970
Horseshoe-nail blank	M. D. Whipple	Cambridge, Mass	Mar. 8, 1864	41, 881
Horseshoe-nail blanks from plate-metal, Machine for punching.	G. L. Hall	Boston, Mass	Mar. 26, 1872	125, 045
Horseshoe-nail clincher	E. E. Fisher and W. H. Mack	Indianola, Ill	Mar. 30, 1869	88, 378
Horseshoe-nail clincher	W. H. Lyman	Springfield, Mo	July 2, 1872	128, 642
Horseshoe-nail clincher	D. Mater, jr	Bellmore, Ind	Aug. 8, 1871	117, 793
Horseshoe-nail clincher	A. Morley	Addison, Mich	Sept. 28, 1869	95, 254
Horseshoe-nail clincher	N. Repp	Waterloo, Iowa	June 8, 1869	91, 044
Horseshoe, &c., nail cutter	E. Rolph	North Providence, R. I	Sept. 16, 1833	....
Horseshoe-nail machine	A. D. Bingham	Vergennes, Vt	Jan. 14, 1873	134, 841
Horseshoe-nail machine	E. L. Brundage	Middletown, N. Y	Aug. 31 1869	94, 391
Horseshoe-nail machine	M. Burnett	Boston, Mass	Apr. 1 1851	8, 006
Horseshoe-nail machine	C. Carpenter, jr	Providence, R. I	June 9, 1857	17, 491
Horseshoe-nail machine	T. Carpenter	Providence, R. I	May 4, 1858	20, 141
Horseshoe-nail machine	A. H. Caryl	Groton, Mass	Sept. 2, 1873	142, 437
Horseshoe-nail machine	H. D. Cowles and G. Stacy	Montreal, Canada	May 2, 1871	114, 413
Horseshoe-nail machine	D. Dodge	Keeseville, N. Y	Jan. 5, 1864	41, 141
Horseshoe-nail machine	L. H. Dwelley	Dorchester, Mass	July 18, 1865	48, 798
Horseshoe-nail machine	L. H. Dwelley	Dorchester, Mass	Feb. 26, 1867	62, 322
Horseshoe-nail machine	D. J. Farmer	Wheeling, W. Va	Aug. 24, 1869	94, 095
Horseshoe-nail machine	T. Fowler	Seymour, Conn	May 21, 1867	64, 964
Horseshoe-nail machine	P. N. Gallaer and I. C. Tate	New London, Conn	July 12, 1870	105, 189
Horseshoe-nail machine	W. C. Grimes	Philadelphia, Pa	July 21, 1863	39, 287
Horseshoe-nail machine	E. W. Kelley	Hamilton, Scotland	May 23, 1871	115, 064
Horseshoe-nail machine	E. W. Kelley	Lowell, Mass	May 27, 1873	139, 250
Horseshoe-nail machine	J. M. Laughlin	Boston, Mass	Feb. 18, 1873	135, 993
Horseshoe-nail machine	B. A. Mason	Newport, R. I	Jan. 18, 1859	22, 663
Horseshoe-nail machine	J. Mills	Keeseville, N. Y	Jan. 21, 1873	135, 140
Horseshoe-nail machine	S. S. Putnam	Dorchester, Mass	June 11, 1861	32, 531
Horseshoe-nail machine	S. S. Putnam	Dorchester, Mass	Dec. 9, 1862	37, 107
Horseshoe-nail machine	S.S. Putnam and L. H. Dwelley	Dorchester, Mass	Mar. 28, 1865	47, 070
Horseshoe-nail machine	S.S. Putnam and L. H. Dwelley	Dorchester, Mass	Apr. 24, 1866	54, 256
Horseshoe-nail machine	S.S. Putnam and L. H. Dwelley	Dorchester, Mass	Mar. 5, 1867	62, 684
Horseshoe-nail machine	S.S. Putnam and L. H. Dwelley	Dorchester, Mass	Mar. 5, 1867	62, 685
Horseshoe-nail machine	G. W. Sargent and B. P. Rider	Chelsea, Mass	Jan. 16, 1866	52, 081
Horseshoe-nail machine	G. W. Sargent and B. P. Rider	Fairhaven and Chelsea, Mass.	June 26, 1866	55, 912
Horseshoe-nail machine	H. F. Sehnders	Buffalo, N. Y	Aug. 29, 1865	49, 657
Horseshoe-nail machine	W. Sharts	Hudson, N. Y	July 9, 1867	66, 642
Horseshoe-nail machine	A. Shaw	Westford, Mass	Dec. 24, 1867	72, 550
Horseshoe-nail machine	A. St. Louis	Keeseville, N. Y	Oct. 29, 1867	70, 374
Horseshoe-nail machine	J. Stone	Batavia, Ill	Mar. 19, 1872	124, 771
Horseshoe-nail machine	W. Tallman	Providence, R. I	Mar. 27, 1860	27, 656
Horseshoe-nail machine	E. Thomas	Middleborough, Mass	Mar. 19, 1867	62, 980
Horseshoe-nail machine	G. D. Walcott	Jackson, Mich	June 30, 1868	79, 417
Horseshoe-nail machine	M. D. Whipple	Cambridge, Mass	Mar. 15, 1864	41, 955
Horseshoe-nail machine	M. D. and L. W. Whipple	Brighton and Boston, Mass	Apr. 9, 1872	125, 423
Horseshoe-nail machine	J. White and J. Madden	Cleveland and Youngstown, Ohio.	May 10, 1864	42, 709
Horseshoe-nail machine	A. Whittemore	Cambridgeport, Mass	Aug. 14, 1860	29, 643
Horseshoe-nail machine	A. Whittemore	Cambridgeport, Mass	Jan. 12, 1864	41, 251
Horseshoe-nail machine	J. A. Whitney	Maryland, N. Y	Jan. 9, 1866	51, 986
Horseshoe-nail machine	W. Wickersham	Boston, Mass	Oct. 8, 1867	69, 734
Horseshoe-nail machine	W. Wickersham	Boston, Mass	Nov. 23, 1869	97, 141
Horseshoe-nail machine	D. Wilson, jr	North Chelmsford, Mass	Jan. 21, 1851	7, 913
Horseshoe-nail machine	C. W. Woodford	Montreal, Canada	Mar. 5, 1872	124, 411

Index of patents issued from the United States Patent Office from 1790 *to* 1873, *inclusive*—Continued.

Invention.	Inventor.	Residence.	Date.	No.
Horseshoe-nail machine	H. E. and C. W. Woodford	Keeseville, N. Y	Oct. 30, 1866	59, 327
Horseshoe-nail machine	S. Wootton	Boonton, N. J	Nov. 10, 1857	18, 617
Horseshoe-nail machine	C. D. Wrightington and B. P. Rider.	Fairhaven and Chelsea, Mass.	Sept. 10, 1867	68, 824
Horseshoe-nail manufacture	A. H. Caryl	Groton, Mass	July 18, 1871	117, 148
Horseshoe-nail pointer	W. T. Vann	Macomb, Ill	June 22, 1869	91, 797
Horseshoe-nail-pointing device	C. H. Perkins	Providence, R. I	Aug. 1, 1871	117, 560
Horseshoe-nail-pointing device	A. H. Wills	Vergennes, Vt	Aug. 1, 1871	117, 584
Horseshoe-nail, Steel-headed	J. H. Smith	Allegheny City, Pa	June 28, 1870	104, 785
Horseshoe-nails, Apparatus for pointing	C. H. Perkins	Providence, R. I	Sept. 26, 1871	119, 402
Horseshoe-nails, Clinching	J. Houck	Greencastle, Ind	Apr. 20, 1858	19, 993
Horseshoe-nails from plates, Die for punching	A. H. Caryl	Forge Village, Mass	Sept. 9, 1873	142, 678
Horseshoe-nails, Hammer-roll device for forming	T. Harris and B. Dunn	Coté St. Paul, Canada	Jan. 7, 1873	134, 664
Horseshoe-nails, Machine for cutting off	C. F. Johnson, jr	Oswego, N. Y	May 30, 1871	115, 480
Horseshoe-nails, Machine for feeding	S. S. Putnam	Boston, Mass	Nov. 4, 1873	144, 226
Horseshoe-nails, Machine for finishing	J. Huggett and J. A. Huggett	Eastbourne and Clapham, England.	Oct. 8, 1872	132, 082
Horseshoe-nails, Machine for finishing	C. H. Perkins and C. E. Sheridan.	Providence, R. I	Oct. 8, 1872	131, 967
Horseshoe-nails, Machine for finishing	S. S. Putnam	Boston, Mass	Nov. 25, 1873	144, 922
Horseshoe-nails, Machine for finishing	R. Ross	Vergennes, Vt	May 27, 1873	139, 331
Horseshoe-nails, Machine for finishing	R. Ross	Vergennes, Vt	Sept. 2, 1873	142, 352
Horseshoe-nails, Machine for finishing	H. A. Wills	Vergennes, Vt	Jan. 16, 1872	122, 876
Horseshoe-nails, Machine for forging	R. Cook	South Abington, Mass	June 9, 1857	17, 542
Horseshoe-nails, Machine for forging	J. and J. A. Huggett	Eastbourne, England	Apr. 20, 1869	89, 224
Horseshoe-nails, Machine for forging	C. Parkhurst and C. Weed	Boston, Mass	Aug. 19, 1856	15, 571
Herreshoe-nails, Machine for forging	S. S. Putnam	Dorchester, Mass	Jan. 11, 1870	98, 707
Horseshoe-nails, Machine for forging	F. H. Richards	New Britain, Conn	Aug. 19, 1873	142, 044
Horseshoe-nails, Machine for forging	J. Roy	Boston, Mass	Oct. 2, 1866	58, 485
Horseshoe-nails, Machine for manufacturing	G. D. Hall	Boston, Mass	Dec. 5, 1871	121, 511
Horseshoe-nails, Machine for pointing	D. Armstrong	Chicago, Ill	Sept. 29, 1868	82, 476
Horseshoe-nails, Machine for pointing	A. H. Caryl	Groton, Mass	May 16, 1871	114, 920
Horseshoe-nails, Machine for pointing extremities of.	H. A. Wills	Vergennes, Vt	July 16, 1872	129, 077
Horseshoe-nails, Machine for shaping heads of	N. Dexter	Bower Hill, Pa	Sept. 27, 1870	107, 670
Horseshoe-nails, Machine for stamping, pressing, and pointing.	F. Sandham	Montreal, Canada	Dec. 6, 1870	109, 844
Horseshoe-nails, Machine for swaging and trimming points of.	R. Ross	Vergennes, Vt	May 27, 1873	139, 332
Horseshoe-nails, Machinery for clinching	D. Kirk	Orleans, N. Y	Jan. 26, 1869	86, 166
Horseshoe-nails, Machinery for making	A. Whittemore	Cambridgeport, Mass	Oct. 2, 1866	58, 520
Horseshoe-nails, Manufacture	S. E. Chase	Boston, Mass	June 9, 1868	78, 644
Horseshoe-nails, Making	A. M. Polsey	Boston, Mass	July 6, 1869	92, 355
Horseshoe-nails, Making	A. H. Caryl	Groton, Mass	Sept. 24, 1872	131, 664
Horseshoe-nails, Manufacture of	H. D. Cowles	Bridgeport, Conn	Dec. 3, 1867	71, 710
Horseshoe-nails, Manufacture of	H. D. Cowles	Montreal, Canada	Apr. 15, 1873	137, 762
Horseshoe-nails, Manufacture of	H. D. Cowles	Montreal, Canada	Dec. 9, 1873	145, 336
Horseshoe-nails, Manufacture of	J. C. Paige	Stoneham, Mass	Sept. 2, 1873	142, 506
Horseshoe-nails, Manufacture of	D. Turbayne	Boston, Mass	Sept. 9, 1873	142, 597
Horseshoe-nails, Manufacture of	D. Turbayne and G. M. Wyman	Boston, Mass	Jan. 14, 1873	134, 948
Horseshoe-nails, Manufacture of	H. R. Underhill	Derry, N. H	Dec. 31, 1872	134, 448
Horseshoe-nails, Roll for rolling iron for	J. Huggett and J. A. Huggett	Eastbourne and Clapham, England.	May 6, 1873	138, 499
Horseshoe-nails, Tool for clinching	J. E. Draper	Northville, Mich	Aug. 6, 1861	32, 989
Horseshoe-nails, Tool for clipping and clinching	J. E. Giles	Marshall, Mich	Feb. 23, 1864	41, 695
Horseshoe-rubber pad	D. L. Corbin	Friendship, N. Y	Jan. 10, 1871	110, 828
Horseshoe, Sectional	T. P. Randall	Adrian, Mich	Nov. 26, 1872	133, 337
Horseshoe-sharpener	H. Thompson	Philadelphia, Pa	Feb. 6, 1872	123, 523
Horseshoe-stretcher	J. D. Abbott	Manchester, N. H	Sept. 9, 1873	142, 547
Horseshoe, Supplemental	T. P. Handy and C. B. Kleibacker.	Baltimore, Md	May 11, 1869	89, 991
Horseshoe, Temporary	W. H. Halsey	Philadelphia, Pa	May 30, 1871	115, 311
Horseshoe toe-calk	G. Custer	Monroe, Mich	Feb. 12, 1867	62, 016
Horseshoe toe-calk	G. Custer	Monroe, Mich	Apr. 12, 1870	101, 831
Horseshoe toe-calk die	P. F. Burke	Worcester, Mass	July 6, 1869	92, 259
Horseshoe toe-calk die	E. R. Cheney	South Boston, Mass	Aug. 11, 1868	80, 807
Horseshoe toe-calks, Die for welding and forming	J. Flynn	Portland, Conn	Jan. 30, 1872	123, 256
Horseshoe-tool	R. Stout	Matteawan, N. Y	Apr. 5, 1870	101, 538
Horseshoes, Apparatus for swaging and welding toe-calks to.	A. Cook	Hillsdale, Mich	Apr. 27, 1869	89, 384
Horseshoes, Attaching	H. Schreiner	Philadelphia, Pa	Oct. 29, 1867	70, 274
Horseshoes, Attaching	J. Wagner	Washington, D. C	Mar. 10, 1868	75, 494
Horseshoes, Attaching calks to	J. Forbes	Plainwell, Mich	Sept. 24, 1867	69, 201
Horseshoes, Attaching calks to	P. Laflin	Warren, Mass	Dec. 7, 1869	97, 525
Horseshoes, Attaching soles to	A. H. Knapp	Coxsackie, N. Y	May 29, 1866	55, 119
Horseshoes, Attaching elastic soles to	W. Somerville	Buffalo, N. Y	Sept. 29, 1857	18, 306
Horseshoes, chain-links, &c., Machine for forming	J. F. Winslow and T. Bigwood	Troy, N. Y	Aug. 11, 1843	3, 217
Horseshoes, Detachable calk for	K. Goddard	Richmond, N. Y	May 25, 1869	90, 442
Horseshoes, Device for attaching	H. B. Ferren	Batavia, N. Y	Sept. 15, 1868	82, 215
Horseshoes, Device for forming	E. Cate	Franklin, N. H	Mar. 26, 1861	31, 780
Horseshoes, Device for forming	U. Stewart	Berlin, Wis	Jan. 9, 1866	51, 979
Horseshoes, Expanding	B. P. Sargent	Sutton, N. H	Oct. 25, 1853	10, 161
Horseshoes, Machine for forging	R. Griffiths and G. Shield	Allegheny City, Pa., and Cincinnati, Ohio.	Dec. 19, 1854	12, 112
Horseshoes, Machine for forming	D. F. Hofstatter and W. H. Shaw.	New London, Ohio	Apr. 15, 1873	137, 923
Horseshoes, Machine for punching	C. H. Perkins	Providence, R. I	Nov. 20, 1860	30, 689
Horseshoes, Machine for punching	S. D. Turner	Providence, R. I	Sept. 5, 1865	49, 838
Horseshoes, Machine for punching holes in, and bending the calks on.	A. Oehme	Rock Island, Ill	Oct. 17, 1871	120, 088
Horseshoes, Machine for rolling bars for	A. Reese	McClure Township, Pa	Nov. 23, 1869	97, 117
Horseshoes, Machine for the manufacture of	A. J. Roberts	Boston, Mass	Aug. 20, 1867	67, 910
Horseshoes, Machinery for manufacturing	W. Gibson	Troy, N. Y	Dec. 27, 1843	3, 392
Horseshoes, Making	W. Morehouse	Buffalo, N. Y	Dec. 22, 1868	85, 119
Horseshoes, Making steel	I. Peacock	Shortsville, N. Y	Jan. 20, 1863	37, 458
Horseshoes, Manufacturing	S. King	Sullivan, N. Y	Apr. 1, 1830	......
Horseshoes, &c., Metal for	J. Kennelly	Hartford, Conn	Mar. 31, 1863	38, 046
Horseshoes, Mold for casting	J. Kennelly	Hartford, Conn	Mar. 31, 1863	38, 047

Index of patents issued from the United States Patent Office from 1790 *to* 1873, *inclusive*—Continued.

Invention.	Inventor.	Residence.	Date.	No.
Horseshoes, &c., Process of treating cast iron for the manufacture of.	J. R. Speer	Pittsburgh, Pa	July 6, 1869	92, 220
Horseshoes, Removable calk for	E. Whitehead	Cincinnati, Ohio	Aug. 10, 1869	93, 654
Horseshoes, Securing	L. Carpenter	Lake City, Minn	June 21, 1864	43, 179
Horseshoes, Securing	W. H. Forker	Meadville, Pa	Oct. 4, 1864	44, 522
Horseshoes, Securing	W. C. Stickney and H. B. Taylor.	Putnam, Ohio	Oct. 18, 1864	44, 754
Horseshoes, Securing	C. Weitman	Hazleton, Iowa	Oct. 4, 1864	44, 569
Horseshoes, Toe-calk for	C. H. Perkins	Providence, R. I	Apr. 9, 1861	31, 995
Horseshoes, Tool for making	T. G. Thompson	Oswego, N. Y	Nov. 6, 1866	59, 479
Horseshoeing	P. Charlier	Paris, France	Apr. 23, 1867	64, 073
Horseshoeing	G. R. Stevens	Clarksville, Mo	July 3, 1860	29, 016
Horseshoeing-apparatus	J. B. Brusoe	Lula, Ill	Jan. 30, 1872	123, 231
Horseshoeing-apparatus	N. Warlick	La Fayette, Ala	Aug. 29, 1854	11, 624
Horseshoeing rest	S. S. Blackburn	Fredericktown, Ohio	Apr. 27, 1869	89, 379
Horseshoeing-rest	G. Stansel	Saint Johnsville, N. Y	Oct. 17, 1871	120, 119
Horseshoeing, Tool for clinching nails in	D. W. Bush	Clarence, Mo	Aug. 27, 1867	68, 040
Horseshoer's rest	J. Legg	Rushville, N. Y	Aug. 12, 1873	141, 651
Horse, Spring	H. F. Metzler	New York, N. Y	Feb. 21, 1865	46, 529
Horse, Spring	H. F. Metzler	New York, N. Y	May 5, 1868	77, 513
Horse, Spring rocking	J. A. Crandell	New York, N. Y	Feb. 9, 1864	41, 570
Horse, Spring rocking	C. B. Northrup	New York, N. Y	May 31, 1864	42, 960
Horse-stocking	W. Lewis	Astoria, N. Y	Dec. 12, 1871	121, 880
Horse-tail holder	G. Gray	West Chester, Pa	July 31, 1846	4, 680
Horse-tail holder	F. A. Roberts	North Vassalborough, Me.	Mar. 22, 1870	101, 043
Horse-tail protector	C. A. Warren	Watertown, Conn	May 21, 1872	127, 002
Horse-taming apparatus	J. M. Lanier	Eufaula, Ala	Nov. 29, 1859	26, 272
Horse-taming apparatus	D. Springer	Pontiac, Mich	Oct. 8, 1872	131, 978
Horse-taming bit	A. L. Weymouth	Boston, Mass	July 1, 1862	35, 796
Horse-tie, Stable	E. D. Cramer	Hackettstown, N. J	Oct. 19, 1869	95, 883
Horse-troughs, &c., Float-valve for	J. Johnson	New York, N. Y	May 6, 1873	138, 659
Horse-wheel, Hollow inclined	J. Keller and S. B. Burt	Cincinnati, Ohio	Apr. 23, 1824	....
Horse-wheel, Inclined	J. Bradford	Lexington, Ky	May 14, 1824	....
Horse-wheel, Inclined	E. Holliday	Schoharie, N. Y	May 5, 1826	....
Horse-wheel, Inclined	J. B. Robinson	Cincinnati, Ohio	Oct. 14, 1816	....
Horse-wheel, Vertical	J. A. Morton	Baltimore, Md	July 21, 1815	....
Horses, &c., across rivers, Floating	S. P. Heintzelman	Newport Barracks, Ky	Nov. 24, 1857	18, 691
Horses, Administering balls to	C. P. Moyer	Womelsdorf, Pa	Nov. 5, 1867	70, 600
Horses and cattle, Breeding-instrument for	C. Addle	Winthrop, Me	May 9, 1835	....
Horses and cattle, Breaking	D. Cooley	Orange County, N. Y	June 23, 1810	....
Horses and cattle, Condition-powder for	J. H. Woolrich	Woburn, Mass	June 7, 1870	104, 094
Horses and other animals, Apparatus for clipping	J. Tidmarsh	Twickenham, England	Nov. 9, 1869	96, 742
Horses and shafts from carriages, Apparatus for discharging.	G. Hubbard	Sandisfield, Mass	May 26, 1857	17, 376
Horses, Anti-kicking attachment for	O. H. P. Fancher	New York, N. Y	Oct. 29, 1867	70, 185
Horses, Apparatus for breaking	S. Tomlinson	Washington Hollow, N. Y.	Sept. 3, 1850	7, 625
Horses, Apparatus for cleaning	O. W. Allison and P. Homeliers	Buffalo, N. Y	Mar. 25, 1873	137, 045
Horses, Apparatus for clipping hair from	S. H. Salom and J. Field	London and Westminster, Great Britain.	June 1, 1869	90, 877
Horses, Apparatus for detaching	J. L. Hamilton	Saint Joseph, Mo	May 9, 1871	114, 553
Horses, Apparatus for detaching	J. C. Hancock and E. P. Richardson.	Charlestown and Somerville, Mass.	July 12, 1870	105, 198
Horses, Apparatus for detaching	H. Latshaw	McKnightstown, Pa	May 20, 1873	139, 068
Horses, Apparatus for grooming	J. J. Greenough	Syracuse, N. Y	Aug. 29, 1871	118, 449
Horses, Apparatus for handling vicious	O. Turner	Clinton, Wis	Oct. 11, 1864	44, 680
Horses, Apparatus for shearing	W. Harrison	Chicago, Ill	Mar. 18, 1873	136, 993
Horses, Apparatus for stopping runaway	S. M. Cooper	Fairfax County, Va	Sept. 10, 1867	68, 606
Horses, Apparatus for throwing	S. Shaw	Russellville, Ohio	Oct. 11, 1864	44, 665
Horses, Apparatus for treating	D. Sullivan	Washington, D. C	May 20, 1873	139, 207
Horses attached to carriages, Device for holding	A. Moore and J. Aylwerd	Mission of San José, Cal	June 1, 1869	90, 865
Horses attached to vehicles, Apparatus for checking.	M. O'Connell	Boston, Mass	July 21, 1868	80, 085
Horses, &c., Attaching and detaching	J. K. Harris	Madison, Ind	May 14, 1867	64, 666
Horses, Attaching shoes to	T. H. Ince	Westminster, England	May 29, 1866	55, 215
Horses, Awning for	W. McCormick	Philadelphia, Pa	Nov. 22, 1870	109, 532
Horses, Boot for	J. Fennell	Cynthiana, Ky	Dec. 10, 1872	133, 768
Horses, Breeching-attachment for	W. F. Schatz	Columbus, Ohio	Aug. 23, 1870	106, 732
Horses, Bridle for stopping	P. Laporte	Louisa County, Va	Mar. 20, 1821	....
Horses, Clipping the hair of	G. F. Evans	Norway, Me	Sept. 13, 1864	44, 171
Horses, Combined boot and shoe for	H. G. and E. M. Haedrich	Philadelphia, Pa	July 5, 1870	105, 068
Horses, Crib for	H. Eddy	North Bridgewater, Mass.	Aug. 26, 1862	36, 287
Horses, Curing blindness in	J. Sater	Huntsville, N. C	Aug. 14, 1821	....
Horses, Detaching	P. T. Share	Baltimore, Md	Mar. 18, 1836	....
Horses, Device for attaching and detaching	G. L. Du Laney	Mechanicsburgh, Pa	May 25, 1869	90, 352
Horses, Device for checking	T. Itzstein	New York, N. Y	Mar. 2, 1869	87, 497
Horses, Device for detaching	E. P. Jones	Shell Mound, Miss	Dec. 9, 1873	145, 426
Horses, Device for detaching	R. R. Jones	Pillar Point, N. Y	Sept. 30, 1873	143, 286
Horses, Device for detaching	I. L. Landis	Lancaster, Pa	Dec. 16, 1863	145, 662
Horses, Device for detaching	A. Schmitt	Cincinnati, Ohio	Sept. 27, 1870	107, 727
Horses, Device for detaching	S. Sykes	Portland, Oreg	Oct. 7, 1873	143, 428
Horses, Device for detaching runaway	E. P. Connor	Jeffersonville, Ohio	Jan. 1, 1867	60, 836
Horses, Device for fettering	A. H. Lewis	North Greenbush, N. Y	Apr. 14, 1868	76, 640
Horses, Device for forming the curve in tails of	I. B. Phillips	Philadelphia, Pa	Aug. 27, 1872	130, 939
Horses, Device for moistening the legs of	G. J. Harris	New York, N. Y	June 11, 1872	127, 875
Horses, Device for preventing and curing cribbiting in.	G. H. White	Huntington, N. Y	May 5, 1868	77, 689
Horses, Device for singeing	J. Alexander	Nashua, N. H	Dec. 15, 1868	84, 983
Horses, Device for spurring or driving	J. Davis	Northampton, Ill	Nov. 7, 1865	50, 803
Horses, Device for training	O. S. Pratt	Batavia, N. Y	Oct. 29, 1872	132, 689
Horses, Dock-holder for	E. H. Penfield	Middletown, Conn	Dec. 12, 1854	12, 068
Horses, Dock-holder for	S. Tomlinson	Pleasant Valley, N. Y	July 18, 1854	11, 341
Horses' ears, Improving or setting	S. Janes	....	Dec. 31, 1804	....
Horses, Elastic heel-guard for	W. H. Hall	New Gloucester, Me	June 8, 1869	91, 122
Horses, Fastening for	P. Ricord and J. R. Scattergood	Newark, N. J	May 18, 1869	90, 196
Horses, Feeding	A. Carman	Hyde Park, N. Y	Nov. 27, 1828	....
Horses' feet, Device for measuring	M. S. Woodward	Marshalltown, Pa	Oct. 29, 1867	70, 307
Horses' feet, Elastic cushion and guard for	W. H. Hall	New Gloucester, Me	May 1, 1866	54, 335

Index of patents issued from the United States Patent Office from 1790 *to* 1873, *inclusive*—Continued.

Invention.	Inventor.	Residence.	Date.	No.
Horses' feet, Elastic protector for	W. H. Hall	Boston, Mass	Mar. 8, 1870	100, 618
Horses' feet, Stopping for	A. S. Wilkinson	Pawtucket, R. I	Aug. 28, 1866	57, 611
Horses, Foot-rest for	J. E. Tucker	Montfort, Wis	Feb. 5, 1867	61, 897
Horses from biting and crib-biting, Device for preventing.	B. D. Howe	Hanover, N. H	July 16, 1867	66, 840
Horses from carriages, and drag for arresting the motion thereof, Manner of combining an apparatus for disengaging.	J. Harlacher	Lancaster, Pa	Nov. 4, 1842	2, 839
Horses from carriages, Apparatus for detaching	T., jr., and J. Byrd	Williamsville, Mich	Mar. 19, 1872	124, 790
Horses from carriages, Apparatus for detaching	J. Davis	Elmira, N. Y	Nov. 6, 1860	30, 566
Horses from carriages, Apparatus for detaching	G. Gabriel	Pittsburgh, Pa	Sept. 15, 1868	82, 217
Horses from carriages, Apparatus for detaching	S. Kepner	Pottstown, Pa	June 15, 1869	91, 346
Horses from carriages, Apparatus for releasing	T. B. Pyron	Salina, Tenn	Oct. 22, 1850	7, 733
Horses from carriages, Danger-escape to disengage	W. Start	Greensborough, Md	May 3, 1816	....
Horses from carriages, Detaching	O. R. Broyles	Anderson Court-House, S.C	Mar. 17, 1834	
Horses from carriages, Detaching	J. W. Harrison	Logansport, Ind	Aug. 27, 1850	7, 600
Horses from carriages, Detaching	E. P. Jones	Shell Mound, Miss	Nov. 9, 1869	96, 703
Horses from carriages, Detaching	S. Kepner	North Coventry, Pa	Dec. 18, 1860	30, 972
Horses from carriages, Detaching	J. Madden	Warren, Ohio	July 9, 1844	3, 652
Horses from carriages, Detaching	E. Myers	New Holland, Pa	Apr. 10, 1847	5, 055
Horses from carriages, Detaching	J. Summers	Kokomo, Ind	Mar. 19, 1872	124, 703
Horses from carriages, Detaching	H. C. Wiatt	Weldon, N. C	Oct. 29, 1846	4, 828
Horses from carriages, Device for detaching	C. McElroy	New Baltimore, Mich	Apr. 20, 1869	89, 159
Horses from carriages, Device for detaching	C. McElroy	New Baltimore, Mich	July 6, 1869	92, 333
Horses from carriages, Disengaging	D. Little	Gettysburgh, Pa	Mar. 30, 1843	3, 020
Horses from carriages, Disengaging	J. S. Shnell	Shiremantown, Pa	Apr. 13, 1844	3, 536
Horses from carriages, Releasing	J. T. Guthrie	Leesburgh, Ohio	Jan. 9, 1866	51, 941
Horses from carriages, Releasing	G. Kintzi	Exeter, Pa	May 4, 1842	2, 601
Horses from carriages suddenly, Splinter-bar and swingle-tree to disengage.	H. Dunlap	Georgetown, D. C	Apr. 23, 1807	....
Horses from carts, Device for releasing	G. H. Meier	Cincinnati, Ohio	July 1, 1873	140, 521
Horses from cribbing, Apparatus to prevent	S. S. Bent	Port Chester, N. Y	June 30, 1868	79, 437
Horses from cribbing, Device for preventing	W. H. Bishop and A. H. Low	Warren, Mass	July 31, 1860	29, 354
Horses from cribbing, Device for preventing	G. G. Hickman	Coatesville, Pa	Oct. 1, 1867	69, 341
Horses from cribbing, Device for preventing	R. Jennings and J. A. Marshall	Bordentown, N. J	Jan. 29, 1867	61, 666
Horses from cribbing, Device for preventing	A. Stillwell	Dwaar's Kill, N. Y	Aug. 19, 1873	141, 962
Horses from crowding, Device for preventing	O. C. Ross	Penfield, N. Y	July 29, 1873	141, 386
Horses from dragging weights, Device for preventing.	A. Pope, jr	Dorchester, Mass	Apr. 28, 1868	77, 317
Horses from fire, Apparatus for rescuing	E. French	Braintree, Mass	July 31, 1860	29, 370
Horses from flies, Device for protecting the necks of	E. L. Kurtz	New York, N. Y	Aug. 4, 1857	17, 934
Horses from interfering, Device to prevent	W. Somerville	New York, N. Y	Apr. 24, 1860	28, 017
Horses from kicking, Apparatus to prevent	E. H. Gammon	Batavia, Ill	May 22, 1866	54, 887
Horses from kicking in the stable, Apparatus for preventing.	W. Itschner	Philadelphia, Pa	Nov. 16, 1869	96, 812
Horses from running away, Apparatus for preventing.	W. Hall	Indianapolis, Ind	Oct. 18, 1859	25, 822
Horses from running away, Apparatus for preventing.	C. C. Moore and R. Blair	Philadelphia, Pa	July 16, 1872	129, 358
Horses from running away, Preventing	E. E. Burnham and G. Brown	Gloucester, Mass	Sept. 10, 1867	68, 600
Horses from stables in case of fire, Releasing	E. N. Moore	Boston, Mass	May 16, 1848	5, 583
Horses from sun-stroke, Shield for protecting	J. Anderson	Brooklyn, N. Y	July 6, 1869	92, 142
Horses from vehicles, Apparatus for detaching	T. Y. Vancleve	Cornersville, Tenn	Jan. 21, 1873	135, 177
Horses from vehicles, Apparatus for unhitching	J. K. Harris	Madison, Ind	Feb. 19, 1867	62, 132
Horses from vehicles, Attaching and detaching	G. Switzer	Washington, D. C	Aug. 9, 1870	106, 293
Horses from vehicles, Detaching	F. M. English	Hopkinsville, Ky	June 24, 1856	15, 179
Horses from vehicles, Detaching	A. B. Johnson, M. H. Vaughan, and J. Stinnett.	Clarksville, Ark., and Shelby County, Tenn.	July 3, 1860	28, 988
Horses from vehicles, Detaching	W. D. Mayfield	Bloomington, Ill	Oct. 13, 1857	18, 426
Horses from vehicles, Detaching	J. M. Roberds	Plaisance, La	July 24, 1860	29, 313
Horses from vehicles, Device for detaching	G. B. Lumpkin	Lexington, Ga	July 25, 1871	117, 306
Horses from vehicles, Device for detaching	B. A. McConnaughey	New Market, Ohio	Dec. 18, 1866	60, 533
Horses from vehicles, Device for detaching	J. McNeill	Wells, Minn	June 10, 1873	139, 803
Horses from vehicles, Disconnecting	J. Rancevan	Carthage, Ohio	Nov. 12, 1867	70, 897
Horses from vehicles, Unhitching	O. N. Weaver	Dover, Ky	Feb. 5, 1867	61, 782
Horses from whiffletrees, Mechanism for disengaging.	J. B. Goldsmith	Rockport, Mass	June 11, 1872	127, 760
Horses, Harnessing	J. Murray, jr	Worcester, Mass	Apr. 16, 1872	125, 752
Horses, Head-gear for stopping runaway	J. Koehler	New York, N. Y	Apr. 17, 1860	27, 952
Horses, Heel-boot for	W. Mathis	New York, N. Y	Nov. 18, 1873	144, 782
Horses in carriages from falling, Apparatus to prevent.	R. D. Dwyer	Richmond, Va	Mar. 11, 1856	14, 395
Horses in navigating canals, Disengaging	G. Shultz	Philadelphia, Pa	Sept. 9, 1835	....
Horses, India-rubber fender for interfering	C. Brinckerhoff	Fishkill, N. Y	Dec. 22, 1868	85, 205
Horses, Insect-guard for	B. Treadwell	Reading, Conn	Jan. 28, 1868	73, 851
Horses, Knee-boot for	J. Finlay	New York, N. Y	Dec. 8, 1868	84, 813
Horses, &c., Knee-clasp for	J. A. Warden	Minnesota Station, Wis	Aug. 10, 1869	93, 570
Horses, Making pads, caps, blinds, &c., for	S. Boyden	Newark, N. J	Apr. 10, 1824	....
Horses, Means for stopping	N. Fountain	New York, N. Y	Aug. 18, 1868	81, 157
Horses, Nose-bag for	H. D. McGovern	Brooklyn, N. Y	June 17, 1873	140, 058
Horses of cribbing, Apparatus to cure	G. W. Metcalfe	Hummelstown, Pa	July 2, 1867	66, 375
Horses or mules to rack, Apparatus for training	C. Daniels	Barnwell Court-House, S.C	Apr. 30, 1861	32, 185
Horses, Preventing interfering in	E. M. Gardner	Nantucket, Mass	May 31, 1864	42, 937
Horses, Quarter-boot for	W. H. Jones	Boston, Mass	June 23, 1868	79, 128
Horses, Remedy for spavin in	S. E. Thayer	Manchester, Vt	Aug. 27, 1867	68, 260
Horses, Rubber boot for	J. H. Jennings	Cambridgeport, Mass	May 28, 1867	65, 231
Horses, Safety-check for	J. W. Domet	Boston, Mass	Jan. 28, 1873	135, 208
Horses, Securing and releasing	C. Wills	New York, N. Y	Nov. 19, 1867	71, 107
Horses, Shaving	F. J. Eldred	Webster, N. Y	Oct. 3, 1871	119, 512
Horses, Sun-bonnet for	J. Anderson	Brooklyn, N. Y	Feb. 22, 1870	100, 000
Horses, Sun-bonnet for	J. Anderson	Brooklyn, N. Y	June 3, 1873	139, 491
Horses, Sun-shade for	E. T. Balch	Philadelphia, Pa	Feb. 1, 1870	99, 282
Horses, Sun-shade for	J. L. McIntosh	Brooklyn, N. Y	Aug. 3, 1869	93, 325
Horses, Sun-shade for	S. Ruth	Philadelphia, Pa	Aug. 25, 1868	81, 412
Horses, Sun-shade for	S. Ruth	Philadelphia, Pa	Jan. 7, 1873	134, 564
Horses, Sun-shield for	H. D. McGovern	Brooklyn, N. Y	Oct. 24, 1871	120, 208
Horses, Sun-shield for	H. D. McGovern	Brooklyn, N. Y	July 2, 1872	128, 554
Horses, Suspender for fly-guards upon	G. Shelton	Normal, Ill	Apr. 22, 1873	138, 205

Index of patents issued from the United States Patent Office from 1790 *to* 1873, *inclusive*—Continued.

Invention.	Inventor.	Residence.	Date.	No.
Horses, Throat-check for	S. French	Boston, Mass	Dec. 26, 1871	122, 114
Horses, Throwing	O. S. Pratt	Batavia, N. Y	Feb. 13, 1872	123, 726
Horses to and from carriages, Apparatus for attaching and detaching.	P. W. Hardwick	Williamsburgh, Ind	Jan. 28, 1862	34, 253
Horses to and from vehicles, Attaching and detaching.	A. Colburn	Leominster, Mass	Mar. 18, 1862	34, 671
Horses to carriages, Attaching	L. W. Boynton	Hartford, Conn	July 5, 1864	43, 388
Horses to carriages, Attaching	W. H. Townsend	Camden, Ohio	Sept. 24, 1867	69, 273
Horses to machinery, Applying draft	W. Hart	Accomack, Va	Jan. 22, 1817	....
Horses to one-wheeled carriages, Attaching	J. R. Remington	Lowndes County, Ala	Apr. 4, 1846	4, 436
Horses to plows, Attaching	J. D. Filkins and W. H. De Puy	Lima, Ind	Oct. 25, 1853	10, 153
Horses to shafts of vehicles, Attaching	G. B. Kaign	Lumberton, N. J	June 10, 1856	15, 103
Horses to shoe them, Apparatus for raising and securing the legs of.	J. P. Champion	Phelps, N. Y	Nov. 5, 1867	70, 523
Horses to two-wheeled vehicles, Attaching	J. W. Barnes	Murfreesborough, N. C	Oct. 23, 1860	30, 455
Horses to vehicles, Apparatus for attaching	C. Leroy	Mexico, N. Y	Oct. 6, 1868	82, 852
Horses to vehicles, Attaching	J. K. Andrews	East Lampeter Township, Pa.	Feb. 1, 1870	99, 278
Horses to vehicles, Attaching	A. Avery	South Windham, Conn	Sept. 18, 1860	30, 037
Horses to vehicles, Attaching	G. H. Gray	Clinton, Miss	Sept. 9, 1856	15, 692
Horses to vehicles, Attaching	E. D. Lockwood	Penfield, N. Y	Apr. 5, 1859	23, 476
Horses to vehicles, Device for connecting	A. Shaler	New York, N. Y	Feb. 1, 1870	99, 357
Horses to vehicles, Mechanism for connecting	D. Belcher	Easton, Mass	Mar. 30, 1869	88, 263
Horses to wagons, Attaching	M. Knapp	Springwater, N. Y	Sept. 5, 1846	4, 743
Horses, Training and breaking	T. Cock	New York, N. Y	Mar. 1, 1811	....
Horses, Weight and hitch-strap for fastening	D. S. Bartlett	Roxbury, Mass	Jan. 15, 1861	31, 104
Horses while in harness, Refreshing	W. Compton	Newark, N. J	July 13, 1869	92, 425
Hose	A. Carney	New York, N. Y	Mar. 31, 1868	76, 049
Hose	H. H. Day	Jersey City, N. J	Aug. 16, 1845	4, 150
Hose	B. F. Lee	New York, N. Y	Jan. 4, 1870	98, 603
Hose	A. S. Libby	Lawrence, Mass	Aug. 6, 1812	130, 303
Hose	A. S. Libby	Lawrence, Mass	Nov. 12, 1872	133, 044
Hose	E. L. Perry	New York, N. Y	Dec. 29, 1868	85, 395
Hose	J. Sharp	Cincinnati, Ohio	Feb. 6, 1872	123, 365
Hose	G. C. Smith	Matteawan, N. Y	May 11, 1869	90, 031
Hose and belting, Machine for cutting India-rubber sheets into bands for.	T. J. Mayall	Boston, Mass	Apr. 9, 1872	125, 597
Hose and flexible tubes, Manufacture of	H. A. Alden	Fishkill, N. Y	Dec. 16, 1862	37, 192
Hose and machine for making	G. Coles, J. A. Jaques, and J. A. Fanshawe.	London and Tottenham, England.	Oct. 20, 1868	83, 132
Hose and pipe connection	G. E. Nutting	New York, N. Y	Dec. 24, 1872	134, 158
Hose and pipe coupling	T. E. Button	Waterford, N. Y	Apr. 15, 1873	137, 823
Hose and pipe coupling	M. S. Curtis and W. D. Tewksbury.	New York, N. Y	Mar. 31, 1868	76, 057
Hose and pipe coupling	F. Shaller	Hudson, N. Y	Apr. 12, 1870	101, 927
Hose and pipe coupling	J. C. Thompson	Charlestown, Mass	Oct. 26, 1869	96, 286
Hose and pipe coupling	C. H. Weston	Lowell, Mass	Mar. 16, 1869	87, 993
Hose and tubing	T. J. Mayall	Roxbury, Mass	July 14, 1863	39, 237
Hose and tubing for water, steam, &c., Manufacture of.	S. P. Cook	San Francisco, Cal	Apr. 2, 1872	125, 175
Hose and tubing, Machinery for making India-rubber.	T. J. Mayall	Boston, Mass	Apr. 9, 1872	125, 598
Hose and tubing, Manufacture of elastic	T. J. Mayall	Roxbury, Mass	July 14, 1863	39, 238
Hose and tubing, Vulcanized rubber and combination.	J. Murphy	New York, N. Y	July 25, 1871	117, 447
Hose and tubes with India rubber, &c., Lining flexible and other.	J. B. Forsyth	Boston, Mass	June 23, 1868	79, 220
Hose, Apparatus for oiling, testing, and washing fire-engine.	P. Noyes	Lowell, Mass	Aug. 19, 1873	141, 885
Hose, belting, &c., Machine for making India-rubber.	J. B. Forsyth	Roxbury, Mass	Sept. 13, 1864	44, 254
Hose-bridge	E. Batzel	Philadelphia, Pa	Oct. 7, 1873	143, 400
Hose-bridge	W. E. Cameron	Taunton, Mass	Oct. 4, 1870	107, 870
Hose-bridge	P. Daily	New York, N. Y	Mar. 7, 1871	112, 427
Hose-bridge	W. Donoghue and F. L. Charlton.	Philadelphia, Pa	Nov. 9, 1869	96, 559
Hose-bridge	J. S. Hagerty	Baltimore, Md	Oct. 29, 1872	132, 660
Hose-bridge	L. W. Hodges	Wilmington, N. C	May 14, 1872	126, 635
Hose-bridge	J. P. Maxwell	Baltimore, Md	Nov. 12, 1872	133, 045
Hose-bridge	J. E. McCaullay	Philadelphia, Pa	Mar. 11, 1873	136, 745
Hose-bridge	G. J. Orr and C. E. Gildersleeve	New York, N. Y	Jan. 28, 1873	135, 285
Hose-bridge, Portable	A. L. Wilkinson and E. Y. Beggs	Huntsville, Ala., and Nashville, Tenn.	Dec. 3, 1867	71, 671
Hose-bridge, Railway	J. Burson	Yates City, Ill	Mar. 30, 1869	88, 364
Hose-carriage	N. S. Bean	Manchester, N. H	June 11, 1867	65, 634
Hose-carriage	W. Boate	Philadelphia, Pa	June 22, 1869	91, 705
Hose-carriage	A. A. Justin	Buffalo, N. Y	Aug. 29, 1871	118, 615
Hose-carriage	K. Perrine and S. M. Stewart	Rochester, N. Y	Apr. 18, 1865	47, 330
Hose carriage	J. W. Wiler, S. B. Sturges, and G. McFall.	Mansfield, Ohio	Oct. 13, 1857	18, 421
Hose-cart	W. E. Shaw and C. A. Ashley	Stockton, Cal	May 28, 1872	127, 277
Hose, Clamp for closing ruptures in fire	P. H. Collins	Philadelphia, Pa	Apr. 30, 1867	64, 285
Hose-clasp	T. J. Pettit	Brooklyn, N. Y	Aug. 1, 1871	117, 563
Hose, Compound for lining textile	J. Dolman and F. W. Claessens	Boston, Mass	Dec. 7, 1869	97, 486
Hose connection	P. Schemmals	Chicago, Ill	Sept. 27, 1864	44, 487
Hose-coupling	A. F. Allen	Providence, R. I	Apr. 27, 1869	89, 455
Hose-coupling	A. F. Allen	Providence, R. I	Jan. 30, 1872	123, 070
Hose-coupling	A. E. Barnard	Paterson, N. J	Jan. 15, 1861	31, 103
Hose-coupling	W. G. Bedford	Philadelphia, Pa	Nov. 26, 1867	71, 266
Hose-coupling	W. H. Bliss	Newport, R. I	Feb. 25, 1862	34, 476
Hose-coupling	J. R. Buchanan	Chicago, Ill	Sept. 3, 1867	68, 348
Hose-coupling	H. C. Bull and S. T. Shelley	Louisville, Ky	Oct. 27, 1868	83, 455
Hose-coupling	L. Button and R. Blake	Waterford, N. Y	Jan. 10, 1860	26, 749
Hose-coupling	A. W. Cary	Brockport, N. Y	Feb. 10, 1852	8, 716
Hose-coupling	J. T. Condon and F. Jeffers	New Orleans, La., and Pawtucket, R. I.	Dec. 24, 1872	134, 195
Hose-coupling	J. C. Cooke	Middletown, Conn	Nov. 30, 1858	22, 166
Hose-coupling	M. S. Curtis	New York, N. Y	Oct. 15, 1867	69, 778

Index of patents issued from the United States Patent Office from 1790 to 1873, inclusive—Continued.

Invention.	Inventor.	Residence.	Date.	No.
Hose, &c., coupling	M. S. Curtis and W. D. Tewksbury.	New York, N. Y	July 16, 1867	66, 804
Hose-coupling	C. H. Cushman	Alexandria, Va	Mar. 29, 1870	101, 403
Hose-coupling	C. H. Cushman	Alexandria, Va	June 28, 1870	104, 837
Hose-coupling	J. W. Douglas	Middletown, Conn	Apr. 30, 1867	64, 292
Hose-coupling	J. Edson	Boston, Mass	Mar. 9, 1869	87, 554
Hose-coupling	J. Edson	Boston, Mass	Apr. 12, 1870	101, 841
Hose-coupling	J. Edson	Boston, Mass	Dec. 23, 1873	145, 731
Hose-coupling	R. J. Falconer	Washington, D. C	June 7, 1853	9, 768
Hose-coupling	L. M. Ferry	Chicopee, Mass	Oct. 7, 1856	15, 846
Hose-coupling	H. Feyh	Columbus, Ohio	Oct. 8, 1861	32, 430
Hose-coupling	J. E. Ford	Memphis, Tenn	Feb. 27, 1866	52, 839
Hose-coupling	L. B. Forester	Clyde, Mich	Aug. 11, 1868	80, 820
Hose-coupling	R. J. Gaines	Portland, Conn	Apr. 14, 1868	76, 622
Hose-coupling	A. M. George	Nashua, N. H	Oct. 25, 1864	44, 794
Hose-coupling	J. H. George	Newark, N. J	July 27, 1869	92, 954
Hose-coupling	G. S. Goble	Saint Louis, Mo	Sept. 2, 1873	142, 388
Hose-coupling	S. Groom	Troy, N. Y	July 12, 1853	9, 841
Hose-coupling	S. Groom	Troy, N. Y	Apr. 3, 1855	12, 626
Hose-coupling	S. Groom	Troy, N. Y	Mar. 29, 1859	23, 368
Hose-coupling	W. Hamilton	Chicopee, Mass	June 30, 1868	79, 343
Hose-coupling	G. Hancock	Providence, R. I	Aug. 7, 1860	29, 548
Hose-coupling	L. W. Hanson and S. Bush	Springfield, Mass	July 28, 1868	80, 407
Hose-coupling	C. F. Hartwig	West Haven, Conn	Feb. 11, 1868	74, 219
Hose-coupling	R. Heneage	Buffalo, N. Y	Aug. 16, 1859	25, 117
Hose-coupling	R. Heneage	Buffalo, N. Y	July 10, 1866	56, 223
Hose coupling	L. E. Hicks	Boston, Mass	May 23, 1855	12, 937
Hose-coupling	B. Holly	Lockport, N. Y	Nov. 20, 1860	30, 678
Hose-coupling	R. Hoskin	Dutch Flat, Cal	Apr. 8, 1862	34, 888
Hose-coupling	S. Ingersoll	Stamford, Conn	July 15, 1873	140, 830
Hose coupling	T. L. Johnson and J. Fitzgerald	Rochester, N. Y	Jan. 11, 1870	98, 774
Hose-coupling	E. B. Jucket	New Haven, Conn	Sept. 9, 1862	36, 410
Hose-coupling	E. B. Jucket	New Haven, Conn	Feb. 17, 1863	37, 721
Hose-coupling	E. S. Kennedy, J. Putman, and H. Smith.	Birmingham, Pa	Jan. 2, 1872	122, 388
Hose-coupling	J. W. Kennedy	Saint Louis, Mo	July 1, 1873	140, 374
Hose-coupling	J. W. Kennedy	Saint Louis, Mo	Aug. 12, 1873	141, 718
Hose-coupling	W. A. Kenyon, jr., and F. S. Wagenhals.	Lancaster, Ohio	Feb. 20, 1872	123, 914
Hose-coupling	J. Kerns	New York, N.Y	Sept. 17, 1867	68, 993
Hose-coupling	W. Kilburn	Napa, Cal	Oct. 7, 1873	143, 514
Hose-coupling	W. Knowles	Boston, Mass	Aug. 18, 1863	39, 580
Hose-coupling	R. B. Lawton and W. H. Bliss	Newport, R. I	Feb. 22, 1859	23, 033
Hose-coupling	H. Lewis	Bennington, Vt	Feb. 25, 1873	136, 328
Hose-coupling	A. S. Libby and L. H. Downing	Lawrence and North Andover, Mass.	Oct. 29, 1872	132, 589
Hose-coupling	C. Locker	Orville, Cal	Dec. 20, 1870	110, 253
Hose-coupling	S. H. Loring	Lawrence, Mass	Oct. 16, 1866	58, 850
Hose-coupling	S. H. Loring	Lawrence, Mass	June 11, 1867	65, 758
Hose-coupling	J. J. Lovell	New York, N. Y	May 9, 1871	114, 575
Hose coupling	A. H. Lowell	Manchester, N. H	July 19, 1859	24, 811
Hose-coupling	J. W. Magill and M. D. Herbert	Little Falls, N. Y	Aug. 13, 1872	130, 436
Hose-coupling	J. Mahony	Newport, R. I	Sept. 3, 1872	131, 013
Hose-coupling	T. J. Mayall	Boston, Mass	Nov. 25, 1873	144, 997
Hose-coupling	N. N. McLeod	Saint Louis, Mo	May 24, 1859	24, 179
Hose-coupling	B. Mee	Troy, N. Y	May 7, 1867	64, 437
Hose-coupling	J. B. Mitchell	Portland, Me	July 10, 1866	56, 324
Hose-coupling	A. J. Morse	Boston, Mass	Nov. 5, 1867	70, 456
Hose-coupling	P. H. Niles	Boston, Mass	Sept. 3, 1867	68, 380
Hose-coupling	P. H. Niles	Boston, Mass	Apr. 7, 1868	76, 503
Hose-coupling	W. J. Osbourne	New York, N. Y	Dec. 21, 1869	98, 184
Hose-coupling	W. J. Osbourne and G. B. Massey.	New York, N. Y	Mar. 5, 1867	62, 675
Hose-coupling	J. C. Osgood	Troy, N. Y	June 5, 1866	55, 354
Hose-coupling	D. T. Perkins and C. F. Hovey	Springfield, Mass	Sept. 10, 1867	68, 650
Hose-coupling	D. T. Perkins and C. F. Hovey	Springfield, Mass	Dec. 31, 1867	72, 892
Hose-coupling	C. Perley	New York, N. Y	Feb. 24, 1863	37, 767
Hose-coupling	L. D. Phillips	Chicago, Ill	Jan. 20, 1857	16, 450
Hose-coupling	C. Powell	Birmingham, England	Apr. 19, 1870	102, 038
Hose-coupling	L. J. Roberts	Corry, Pa	Oct. 8, 1867	69, 706
Hose-coupling	O. Salgee	New York, N. Y	Sept. 10, 1867	68, 656
Hose-coupling	G. Sewell	Brooklyn, N. Y	Feb. 8, 1870	99, 715
Hose-coupling	G. Sewell	Brooklyn, N. Y	May 31, 1870	103, 785
Hose-coupling	J. Singer	Cleveland, Ohio	Jan. 3, 1860	26, 713
Hose-coupling	W. H. Smith	Newport, R. I	Aug. 30, 1859	25, 283
Hose-coupling	C. F. Spencer	Rochester, N. Y	Oct. 2, 1860	30, 261
Hose-coupling	J. Steger	New York, N. Y	Feb. 25, 1868	74, 950
Hose-coupling	N. Thompson	Brooklyn, N. Y	Dec. 24, 1867	72, 565
Hose-coupling	T. J. Trapp	Williamsport, Pa	June 3, 1873	139, 632
Hose-coupling	C. Vander Woerd	Cambridge, Mass	June 8, 1858	20, 532
Hose-coupling	G. H. Van Vleck and H. Tupper	Buffalo, N. Y	Aug. 9, 1859	25, 065
Hose-coupling	S. W. Warren	Brooklyn, N. Y	May 8, 1860	28, 221
Hose-coupling	A. M. Waterhouse	New York, N. Y	June 19, 1855	13, 112
Hose-coupling	T. W. Welsh	Pittsburgh, Pa	Jan. 16, 1872	122, 873
Hose-coupling	G. Westinghouse, jr	Pittsburgh, Pa	Mar. 4, 1873	136, 397
Hose-coupling	L. Wharton	Salem, Ohio	June 13, 1871	115, 917
Hose-coupling	J. Williston	Vallejo, Cal	Nov. 25, 1862	37, 021
Hose-coupling	T. Wilson and L. B. Kendall	Kalamazoo, Mich	May 20, 1873	139, 223
Hose-coupling	M. A. Winham	North San Juan, Cal	Sept. 9, 1862	36, 433
Hose-coupling, Escape-valve	A. F. Allen	Providence, R. I	Sept. 3, 1867	68, 478
Hose-coupling, Metallic	E. S. Elmer	Hudson, N. Y	June 3, 1873	139, 459
Hose-coupling spanner	L. Pond	Foxborough, Mass	Nov. 25, 1873	144, 862
Hose couplings, Device for connecting dissimilar	W. H. Paige	Springfield, Mass	July 7, 1868	79, 592
Hose-couplings, Ring for securing	A. Schrader	New York, N. Y	July 5, 1870	105, 003
Hose-covering	T. McAuley and M. L. Cheney	San Francisco and Illinoistown, Cal.	Mar. 13, 1866	53, 161
Hose, Covering the ends of rubber	J. P. Rider and J. R. Bird	Brooklyn, N. Y	Mar. 1, 1870	100, 448
Hose, Device for carrying	P. Frentz	New Albany, Ind	Sept. 11, 1866	57, 886

Index of patents issued from the United States Patent Office from 1790 to 1873, inclusive—Continued.

Invention.	Inventor.	Residence.	Date.	No.
Hose, Device for handling fire-engine	J. Lowe	Washington, D. C	June 18, 1872	128, 155
Hose, Elbow-support for flexible	A. O. Bourn	Cranston, R. I	Aug. 11, 1868	80, 901
Hose, Engine	E. W. French	South Scituate, Mass	Sept. 3, 1867	68, 432
Hose, Engine	C. Lenzmann	Brooklyn, N. Y	Dec. 14, 1858	22, 296
Hose, Fire-engine	J. Boyd	Boston, Mass	May 30, 1821	....
Hose, Flexible	J. Davis	Pawtucket, R. I	Feb. 9, 1869	86, 652
Hose, Flexible play-pipe for	J. Greacen, jr	New York, N.Y	Dec. 9, 1873	145, 415
Hose for water, &c., Wooden	C. H. Proessdorf and E. Bauch	Boston Highland, Mass	Aug. 10, 1869	93, 476
Hose from cloth and gum-elastic, Manufacturing	S. D. Reed	Philadelphia, Pa	June 29, 1833	....
Hose-guard	D. P. Lewis	Huntsville, Ala	Nov. 19, 1867	71, 026
Hose holder, Garden	H. C. Smith	Cleveland, Ohio	Apr. 15, 1873	137, 802
Hose, Hydraulic	L. R. Blake	Brooklyn, N. Y	May 13, 1873	138, 845
Hose, Hydraulic	L. R. Blake	Brooklyn, N. Y	Oct. 14, 1873	143, 661
Hose, Hydraulic	L. H. Downing	North Andover, Mass	May 13, 1873	138, 867
Hose, Hydraulic	A. S. Libby	Lawrence, Mass	Dec. 10, 1872	133, 785
Hose, India-rubber	E. M. Chaffee	Roxbury, Mass	Jan. 9, 1834	....
Hose, India-rubber	H. H. Day and R. McMullin	Jersey City, N. J., and Great Barrington, Mass.	Sept. 17, 1850	7, 650
Hose, India-rubber and combination	W. A. Torry	Mont Clair, N. J	Dec. 13, 1870	110, 172
Hose-ladder strap	A. F. Allen	Providence, R. I	Nov. 16, 1869	96, 863
Hose-leak stopper	W. C. Davol, jr	Fall River, Mass	June 6, 1871	115, 715
Hose, Leather	C. F. W. Meyer	Oconomowoc, Wis	Sept. 10, 1867	68, 641
Hose, &c., Machine for cleaning	J. B. Alden, jr., and E. L. Gates	Worcester, Mass	Aug. 31, 1858	21, 307
Hose, Machine for making rubber	T. J. Mayall	Roxbury, Mass	May 15, 1860	28, 288
Hose, Machine for making rubber	J. Murphy	New York, N. Y	Apr. 9, 1872	125, 479
Hose, &c., Machine for making rubber	J. Murphy and A. H. Hook	New York, N. Y	June 15, 1869	91, 476
Hose, Machine for manufacture of rubber	J. Quinn	Leyland, Great Britain	Jan. 9, 1872	122, 649
Hose, Machine for manufacturing	T. J. Mayall	Boston, Mass	July 15, 1873	140, 934
Hose, &c., Machine for repairing	A. Shedlock	New York, N. Y	Mar. 28, 1871	113, 213
Hose, Machinery for making India-rubber	T. J. Mayall	Roxbury, Mass	Nov, 13, 1860	30, 637
Hose, Machinery for manufacturing India-rubber	T. B. De Forest	Birmingham, Conn	Apr. 14, 1857	17, 029
Hose, Making India-rubber	J. H. Howell	Ansonia, Conn	Oct. 21, 1856	15, 947
Hose, Manufacture of	G. H. Bonnaffon	Allegheny City, Pa	Sept. 3, 1867	68, 344
Hose, Manufacture of flexible rubber	C. McBurney	Boston, Mass	Apr. 23, 1872	126, 069
Hose, Manufacture of hydraulic	E. A. Street	South Orange, N. J	Aug. 19, 1873	142, 054
Hose, Manufacture of India-rubber	I. B. Harris	Newtown, Conn	Oct. 8, 1872	132, 006
Hose, Manufacture of leather	J. and J. A. Punderford	New Haven, Conn	Dec. 18, 1866	60, 555
Hose, Manufacture of rubber	T. J. Mayall	Roxbury, Mass	May 22, 1860	28, 388
Hose, Manufacture of rubber	E. L. Perry and W. A. Torrey	New York, N. Y., and Mont Clair, N. J.	July 9, 1867	66, 518
Hose, Manufacture of rubber or gutta-percha	E. L. Perry and C. Manheim	New York, N. Y	July 6, 1869	92, 353
Hose, Manufacture of textile	L. B. Cooley and J. C. Cooke	Middletown, Conn	Mar. 16, 1858	19, 625
Hose, Manufacture of water-proof	E. M. Chaffee	Providence, R. I	Oct. 2, 1866	58, 377
Hose, Manufacture of water-proof	T. J. Mayall	Roxbury, Mass	Nov. 29, 1859	26, 276
Hose, Manufacture of water-proof	T. S. Reed	Providence, R. I	Apr. 2, 1872	125, 331
Hose, Mending fire-engine	J. S. Mackay	Brooklyn, N. Y	Jan. 15, 1861	31, 143
Hose, Method of lining	H. Hartley	Pittsburgh, Pa	Dec. 24, 1867	72, 488
Hose, Method of stopping accidental breaches in fire.	R. Bulkley	New York, N. Y	May 8, 1840	1, 598
Hose, Mode of protecting the ends of vulcanized-rubber.	H. A. Alden	Matteawan, N. Y	Dec. 14, 1869	97, 853
Hose or pipe coupling, Compound	J. W. Osgood	Columbus, Ohio	May 20, 1851	8, 096
Hose or tubing, Manufacture of rubber	T. J. Mayall	Roxbury, Mass	Apr. 13, 1869	88, 888
Hose or tubing, Rubber	T. J. Mayall	Roxbury, Mass	Apr. 13, 1869	88, 887
Hose-pipe	J. Edson	Boston, Mass	Nov. 9, 1869	96, 682
Hose-pipe	R. Hollings	Boston, Mass	Jan. 4, 1853	9, 520
Hose-pipe	A. R. Reese	Phillipsburgh, N. J	June 13, 1871	115, 895
Hose-pipe	G. Smith	Macon, Ga	June 12, 1860	28, 694
Hose-pipe and sprinkler combined	W. W. Ransom	Buffalo, N. Y	Mar. 25, 1873	137, 095
Hose-pipe, Compound	M. Cronin	Washington, D. C	June 3, 1873	139, 550
Hose-pipe coupling	L. Button and R. Blake	Waterford, N. Y	Dec. 18, 1860	30, 909
Hose-pipe, India-rubber	C. McBurney	Roxbury, Mass	Jan. 11, 1859	22, 566
Hose, pipes, and tubes, Closing leaks in	R. Street	Albany, N. Y	Jan. 18, 1870	98, 893
Hose-pipes, Clamp for stopping leaks in	C. Rubsam	Newark, N. J	May 10, 1864	42, 735
Hose-pipes, Coating	G. H. Hinckley	Stonington, Conn	Sept. 29, 1857	18, 281
Hose-pipes, Manufacture of rubber	J. H. Cheever	Boston, Mass	Feb. 8, 1859	22, 854
Hose pipes, Self-packing patch for	J. B. Button	Milwaukee, Wis	Jan. 2, 1872	122, 435
Hose, Process of making suction rubber	G. C. Smith	Matteawan, N. Y	July 6, 1869	92, 381
Hose, Process of vulcanizing rubber	J. H. Cheever	New York, N. Y	Feb. 6, 1872	123, 454
Hose, Protecting the ends of vulcanized India-rubber or combination.	J. Schenck	Brooklyn, N. Y	Feb. 1, 1870	99, 485
Hose-protector	W. C. Bridges and D. P. Dieterich.	Philadelphia, Pa	Oct. 2, 1860	30, 199
Hose-protector	C. C. Converse	Brooklyn, N. Y	Feb. 2, 1869	86, 508
Hose-protector	D. Demarest	New York, N. Y	Nov. 1, 1853	10, 183
Hose-protector	I. H. Stone	Saint Louis, Mo	June 12, 1866	55, 550
Hose-protector, Railway	J. H. Bellamy	Charlestown, Mass	May 25, 1869	90, 336
Hose-reel	B. F. Lee and H. A. Alden	New York and Fishkill, N. Y.	Sept. 9, 1862	36, 438
Hose-reel	T. J. Mayall	Boston, Mass	July 15, 1873	140, 784
Hose-rest, Garden	C. Ryder	Cambridgeport, Mass	Nov. 11, 1873	144, 415
Hose-ring	G. Sewell	Brooklyn, N. Y	Feb. 21, 1871	111, 981
Hose, Rubber	E. L. Perry	New York, N. Y	Nov. 16, 1869	96, 932
Hose, Rubber	G. C. Smith	Fishkill, N. Y	Jan. 18, 1870	99, 018
Hose, Rubber and gutta-percha	E. M. Chaffee	Providence, R. I	May 26, 1868	78, 260
Hose, Rubber play-pipe for	E. L. Perry	New York, N. Y	May 2, 1871	114, 337
Hose-shield	J. A. Haase	Philadelphia, Pa	Dec. 10, 1867	72, 028
Hose, Sleeve or patch for stopping leaks in	J. F. Bellemere and F. R. Fleer	Reading, Pa	Aug. 5, 1873	141, 536
Hose-sprinkler	W. Anderson	San Francisco, Cal	Mar. 7, 1871	112, 309
Hose, Suction	A. F. Allen	Providence, R. I	Mar. 1, 1870	100, 244
Hose, Suction	W. C. Downs	Providence, R. I	Jan. 7, 1873	134, 654
Hose, Suction	C. McBurney	Roxbury, Mass	May 31, 1859	24, 222
Hose, Suction	J. Riley	Boston, Mass	Feb. 6, 1833	....
Hose, Suction	T. D. Stetson and W. Brandon	New York, N. Y	Mar. 19, 1872	124, 770
Hose-supporter	A. Johnson	Cairo, N. Y	Apr. 27, 1858	20, 069
Hose, syringe, &c., Distributing-apparatus for garden.	S. Barton, jr	London, England	Aug. 12, 1873	141, 620
Hose-tender	H. A. Gilbertson	New York, N. Y	Aug. 11, 1868	80, 941

Index of patents issued from the United States Patent Office from 1790 *to* 1873, *inclusive*—Continued.

Invention.	Inventor.	Residence.	Date.	No.
Hose to couplings, Mode of fastening	C. McBurney	Roxbury, Mass	Feb. 16, 1864	41, 630
Hose-tongs, clamp, wrench, and pick, Combined	H. Bitter	Philadelphia, Pa	Oct. 15, 1867	69, 896
Hose-tubing	H. A. Alden	Matteawan, N. Y	May 8, 1860	28, 231
Hose-tubing	T. J. Mayall	Roxbury, Mass	Mar. 20, 1860	27, 553
Hose-tubing	C. McBurney	Roxbury, Mass	Apr. 10, 1860	27, 819
Hose, tubing, and other rubber fabrics, Manufacture of.	J. Schenck	Brooklyn, N. Y	May 25, 1869	90, 397
Hose-tubing, Elastic	J. C. Boyd	Boston, Mass	Aug. 30, 1859	25, 239
Hose-tubing, Flexible	H. A. Alden	Matteawan, N. Y	May 22, 1860	28, 432
Hose-tubing, Rubber	T. J. Mayall	Roxbury, Mass	May 22, 1860	28, 389
Hose tubes or pipes for conveying fluids under pressure, Manufacture of flexible.	J. P. Ryder	Brooklyn, N. Y	May 25, 1869	90, 393
Hose-tunnel for railways	L. Myers	Philadelphia, Pa	Mar. 14, 1871	112, 731
Hose, Vulcanized India-rubber	D. C. Gately	Newton, Conn	Nov. 19, 1872	133, 219
Hose, Vulcanized rubber	H. A. Alden	Matteawan, N. Y	Feb. 22, 1870	100, 096
Hose, Water-proof	H. A. Alden	Matteawan, N. Y	Sept. 18, 1860	30, 032
Hose, Water-proof	T. J. Mayall	Roxbury, Mass	Mar. 5, 1861	31, 614
Hose, Water-proof	T. L. Reed	Providence, R. I	Mar. 26, 1872	124, 914
Hose, Water-proof leather	J. Penderford	New Haven, Conn	Apr. 26, 1859	23, 781
Hose-wrench	E. Bucklin, jr	Pawtucket, R. I	Nov. 25, 1873	144, 827
Hosiery goods, Uniting edges of	E. E. Killbourn	New Brunswick, N. J	Aug. 27, 1867	68, 087
Hosiery, Manufacture of	W. H. McNary	Brooklyn, N. Y	Dec. 23, 1856	16, 285
Hosiery, Manufacturing seamless	W. Goddard	New York, N. Y	Aug. 26, 1856	15, 607
Hosiery, Pressing	O. Osborne	Philadelphia, Pa	June 17, 1873	140, 068
Hosiery, Stretcher for	S. J. Wright	Ellsworth, N. Y	Sept. 24, 1867	69, 293
Hospital-table	S. J. A. Hussey	Cornwall, N. Y	May 23, 1865	47, 831
Hot-air and cupola furnace	L. V. Badger	Portsmouth, N. H	Dec. 2, 1835	....
Hot-air conductor, Portable	J. B. Oldershaw	Baltimore, Md	Dec. 10, 1867	72, 074
Hot-air engine	C. W. Baldwin	Charlestown, Mass	Feb. 14, 1865	46, 320
Hot-air engine	C. W. Baldwin	Boston, Mass	July 11, 1865	48, 639
Hot-air engine	C. W. Baldwin	Boston, Mass	June 19, 1866	55, 601
Hot-air engine	B. F. Craig	Washington, D. C	Jan. 17, 1860	26, 837
Hot-air engine	M. G. Crane	Chelsea, Mass	Jan. 31, 1865	46, 084
Hot-air engine	W. D. Grimshaw	Newark, N. J	June 19, 1860	28, 748
Hot-air engine	H. Kilbourne	Waterloo, Iowa	Nov. 7, 1865	50, 875
Hot-air engine	H. J. Kritzer	Albion, Mich	July 29, 1862	36, 008
Hot-air engine	C. P. Leavitt	New York, N. Y	Oct. 4, 1870	107, 928
Hot-air engine	C. P. Leavitt	New York, N. Y	Dec. 3, 1872	133, 649
Hot-air engine	W. Lehmann	Nuremberg, Germany	June 15, 1869	91, 239
Hot-air engine	T. McDonough	Middletown, Conn	Jan. 14, 1862	34, 155
Hot-air engine	T. McDonough	Middletown, Conn	Nov. 8, 1864	44, 966
Hot-air engine	T. McDonough	Middletown, Conn	Jan. 30, 1866	52, 305
Hot-air engine	T. McDonough	Newburgh, N. Y	Nov. 2, 1869	96, 335
Hot-air engine	H. Messer	Roxbury, Mass	Apr. 21, 1863	38, 237
Hot-air engine	H. Messer	Roxbury, Mass	July 21, 1863	39, 321
Hot-air engine	H. Messer	Roxbury, Mass	Jan. 5, 1864	41, 084
Hot-air engine	H. Messer	Roxbury, Mass	Mar. 7, 1865	46, 689
Hot-air engine	W. Reinlein	Barcelona, Spain	June 26, 1866	55, 971
Hot-air engine	A. K. Rider	New York, N. Y	Jan. 17, 1871	111, 087
Hot-air engine	A. K. Rider	New York, N. Y	July 16, 1872	128, 979
Hot-air engine	S. H. Roper	Boston, Mass	Feb. 4, 1862	34, 333
Hot-air engine	S. H. Roper	Roxbury, Mass	Mar. 18, 1862	34, 723
Hot-air engine	S. H. Roper	Boston, Mass	June 9, 1863	38, 866
Hot-air engine	S. H. Roper	Roxbury, Mass	June 5, 1866	55, 428
Hot-air engine	P. Shaw	Boston, Mass	May 28, 1861	32, 455
Hot-air engine	P. Shaw	Boston, Mass	Nov. 26, 1861	33, 799
Hot-air engine	P. Shaw	Boston, Mass	May 18, 1869	90, 128
Hot-air engine	L. A. L. Soderstrom	Paris, France	June 29, 1869	92, 117
Hot-air engine	C. Stevens	Boston, Mass	Oct. 17, 1865	50, 506
Hot-air engine	F. H. Wenham	London, England	Sept. 1, 1868	81, 853
Hot-air engine	M. D. Whipple	Cambridge, Mass	Sept. 30, 1862	36, 581
Hot-air engine	S. Wilcox, jr	Westerly, R. I	Feb. 14, 1860	27, 180
Hot-air engine	S. Wilcox, jr	Westerly, R. I	Apr. 2, 1861	31, 924
Hot-air engine	S. Wilcox, jr	Westerly, R. I	May 16, 1865	47, 759
Hot-air engine	S. Wilcox, jr	Westerly, R. I	Sept. 19, 1865	50, 062
Hot-air engine	A. O. Willcox	Philadelphia, Pa	Aug. 2, 1853	9, 909
Hot-air furnace	W. N. Abbott	New York, N. Y	Oct. 29, 1872	132, 554
Hot-air furnace	J. Albee	Boston, Mass	Sept. 10, 1867	68, 589
Hot-air furnace	J. Albee	Boston, Mass	June 30, 1868	79, 294
Hot-air furnace	C. Allen	Hartford, Conn	Apr. 27, 1869	89, 458
Hot-air furnace	C. Allen	Hartford, Conn	Jan. 17, 1871	111, 027
Hot-air furnace	J. Amory	West Roxbury, Mass	Mar. 31, 1868	76, 136
Hot-air furnace	H. Arden	Cincinnati, Ohio	Feb. 11, 1868	74, 270
Hot-air furnace	B. Arthurs	Pittsburgh, Pa	Nov. 12, 1867	70, 679
Hot-air furnace	J. R. Barker	Chicago, Ill	May 28, 1872	127, 140
Hot-air furnace	J. C. Barnes	Albany, N. Y	Oct. 3, 1871	119, 554
Hot-air furnace	A. H. Bartlett	King's Bridge, N. Y	Jan. 30, 1855	12, 305
Hot-air furnace	A. H. Bartlett	Spuyten Duyvil, N. Y	Sept. 18, 1860	30, 039
Hot-air furnace	W. D. Bartlett	Amesbury, Mass	Jan. 10, 1871	110, 889
Hot-air furnace	P. D. Beckwith	Dowagiac, Mich	June 1, 1869	90, 630
Hot-air furnace	O. Bellman	Hagerstown, Md	Nov. 26, 1872	133, 293
Hot-air furnace	O. Bellman and J. W. Garver	Hagerstown, Md	Feb. 22, 1870	100, 107
Hot-air furnace	J. M. Blackman	Decorah, Iowa	May 30, 1871	115, 270
Hot-air furnace	V. W. Blanchard	Bridport, Vt	Jan. 8, 1867	61, 043
Hot-air furnace	J. Bolton	Richmond, Va	Apr. 26, 1853	9, 686
Hot-air furnace	J. Bolton	Richmond, Va	Dec. 20, 1853	10, 333
Hot-air furnace	J. A. Bolton	Leicester, England	Dec. 3, 1861	33, 819
Hot-air furnace	W. H. Bond	Syracuse, N. Y	Apr. 2, 1872	125, 259
Hot-air furnace	L. Bonnell	Milwaukee, Wis	Apr. 5, 1870	101, 422
Hot-air furnace	L. Bonnell	Milwaukee, Wis	Jan. 3, 1871	110, 731
Hot-air furnace	J. Bouis	Baltimore, Md	Mar. 29, 1834	....
Hot-air furnace	R. Boyd and J. C. Hart	Rochester, N. Y	Nov. 30, 1869	97, 347
Hot-air furnace	N. A. Boynton	Boston, Mass	Feb. 8, 1853	9, 573
Hot-air furnace	N. A. Boynton	New York, N. Y	Aug. 22, 1854	11, 545
Hot-air furnace	N. A. Boynton	New York, N. Y	May 4, 1869	89, 549
Hot-air furnace	E. Brady and J. Sloan	Philadelphia, Pa	Apr. 14, 1868	76, 590
Hot-air furnace	S. A. Briggs	Providence, R. I	July 31, 1855	13, 374
Hot-air furnace	L. W. Brown	Cleveland, Ohio	Apr. 19, 1870	102, 086

Index of patents issued from the United States Patent Office from 1790 *to* 1873, *inclusive*—Continued.

Invention.	Inventor.	Residence.	Date.	No.
Hot-air furnace	L. W. Brown and I. L. Frankem	Indianapolis, Ind	Feb. 20, 1866	52, 671
Hot-air furnace	R. F. Brown	Chicago, Ill	Nov. 1, 1870	108, 754
Hot-air furnace	B. Brownell	Chicago, Ill	Sept. 18, 1866	58, 057
Hot-air furnace	W. Bryent	Boston, Mass	Sept. 25, 1847	5, 309
Hot-air furnace	J. T. Budd	New York, N. Y	Oct. 2, 1866	58, 373
Hot-air furnace	H. G. Burr	Minneapolis, Minn	Apr. 7, 1868	76, 397
Hot-air furnace	J. H. Burtis	Brooklyn, N. Y	June 6, 1871	115, 699
Hot-air furnace	J. H. Cahill	Philadelphia, Pa	Apr. 14, 1857	17, 022
Hot-air furnace	E. H. Camp	Jackson, Mich	Aug. 6, 1861	32, 974
Hot-air furnace	E. H. Camp	Jackson, Mich	May 7, 1867	64, 485
Hot-air furnace	B. F. Campbell	Boston, Mass	Feb. 28, 1871	112, 122
Hot-air furnace	J. Carton and J. Briggs	Utica, N. Y	Aug. 1, 1854	11, 414
Hot-air furnace	J. E. Chapman	Cannon Falls, Minn	Aug. 17, 1869	93, 675
Hot-air furnace	F. E. Chatard, jr	Baltimore, Md	Aug. 22, 1871	118, 195
Hot-air furnace	T. W. Chatfield	Utica, N. Y	Mar. 28, 1854	10, 700
Hot-air furnace	T. W. Chatfield	Utica, N. Y	Mar. 22, 1870	100, 979
Hot-air furnace	J. Chilcott	Brooklyn, N. Y	Jan. 9, 1866	51, 921
Hot-air furnace	J. Child	Elyria, Ohio	Mar. 23, 1858	19, 683
Hot-air furnace	G. Chilson	Boston, Mass	Aug. 4, 1845	4, 133
Hot-air furnace	G. Chilson	Boston, Mass	May 2, 1848	5, 550
Hot-air furnace	E. Clark	Lancaster, Pa	May 9, 1871	114, 528
Hot-air furnace	T. E. Coles	Troy, Ohio	Jan. 5, 1869	85, 564
Hot-air furnace	M. A. Cushing	Aurora, Ill	July 6, 1869	92, 282
Hot-air furnace	M. A. Cushing	Aurora, Ill	May 30, 1871	115, 444
Hot-air furnace	G. Darby	Augusta, Me	Feb. 2, 1858	19, 239
Hot-air furnace	G. B. Davis	Chicago, Ill	Aug. 13, 1867	67, 732
Hot-air furnace	G. W. Day	Haverhill, Mass	Dec. 16, 1873	145, 489
Hot-air furnace	H. G. Dayton	Maysville, Ky	July 31, 1866	56, 727
Hot-air furnace	W. W. Dodge	Boston, Mass	Aug. 27, 1872	130, 847
Hot-air furnace	C. M. Drennan	Boston, Mass	May 28, 1872	127, 321
Hot-air furnace	M. B. Dyott	Philadelphia, Pa	Aug. 30, 1853	9, 966
Hot-air furnace	H. A. and C. H. Engels and J. Wieland.	San Francisco, Cal	Apr. 2, 1867	63, 490
Hot-air furnace	W. Ennis	New York, N. Y	Mar. 29, 1853	9, 633
Hot-air furnace	A. Ernst and C. Shepard	Milwaukee, Wis	July 17, 1860	29, 150
Hot-air furnace	M. G. Fagan	Troy, N. Y	June 28, 1870	104, 842
Hot-air furnace	J. R. Ferguson	Brooklyn, N. Y	Nov. 30, 1858	22, 173
Hot-air furnace	R. R. Finch	Peekskill, N. Y	Oct. 24, 1871	120, 257
Hot-air furnace	J. Fridley, jr	Carlisle, Pa	Feb. 4, 1873	135, 422
Hot-air furnace	J. R. Gaston	Normal, Ill	Aug. 27, 1872	130, 913
Hot-air furnace	S. Gates	Albion, N. Y	Dec. 7, 1852	9, 444
Hot-air furnace	W. B. Geddes	Rochester, N. Y	June 13, 1871	115, 841
Hot-air furnace	W. B. Geddes	Rochester, N. Y	Jan. 14, 1873	134, 879
Hot-air furnace	S. F. Gold	Englewood, N. J	Apr. 8, 1873	137, 546
Hot-air furnace	B. Gommenginger	Rochester, N. Y	Mar. 1, 1870	100, 281
Hot-air furnace	B. Gommenginger	Rochester, N. Y	July 4, 1871	116, 701
Hot-air furnace	B. Gommenginger	Rochester, N. Y	June 4, 1872	127, 476
Hot-air furnace	B. Gommenginger and C. W. Trotter.	Rochester, N. Y	Oct. 5, 1869	95, 461
Hot-air furnace	J. E. Grant	Charlestown, Mass	Oct. 3, 1854	11, 770
Hot-air furnace	M. Greenebaum	Chicago, Ill	Jan. 23, 1855	12, 277
Hot-air furnace	C. B. Gregory	Beverly, N. J	Aug. 25, 1868	81, 492
Hot-air furnace	W. O. Grover	Boston, Mass	June 3, 1873	139, 505
Hot-air furnace	D. Gusweiler	Cincinnati, Ohio	June 15, 1869	91, 436
Hot-air furnace	J. Gwynn	Tiffin, Ohio	May 18, 1869	90, 095
Hot-air furnace	J. D. Hall	Brooklyn, N. Y	Apr. 15, 1873	137, 838
Hot-air furnace	S. J. Hare	Louisville, Ky	Nov. 24, 1868	84, 356
Hot-air furnace	W. H. Harris	Buffalo, N. Y	Nov. 19, 1872	133, 096
Hot-air furnace	O. N. Hart	Winona, Minn	July 20, 1869	92, 822
Hot-air furnace	J. B. Haupt	Philadelphia, Pa	July 9, 1872	128, 799
Hot-air furnace	J. P. Hayes	Philadelphia, Pa	May 23, 1854	10, 962
Hot-air furnace	J. I. Hess	Philadelphia, Pa	Oct. 13, 1868	82, 948
Hot-air furnace	F. C. Hesse	Cincinnati, Ohio	June 25, 1867	66, 083
Hot-air furnace	S. E. Hewes	Albany, N. Y	June 13, 1871	115, 956
Hot-air furnace	R. Hillson	Albany, N. Y	Feb. 29, 1848	5, 459
Hot-air furnace	I. H. Hobbs, A. W. Rand, and G. H. Sellers.	Philadelphia, Pa	Oct. 11, 1859	25, 735
Hot-air furnace	H. Holcomb	Painesville, Ohio	Dec. 22, 1863	41, 007
Hot-air furnace	B. Holly	Lockport, N. Y	Jan. 31, 1865	46, 107
Hot-air furnace	W. and W. James, jr	Montreal, Canada	Nov. 1, 1870	108, 790
Hot-air furnace	O. S. Kelsey	Hartford, Conn	Oct. 7, 1873	143, 453
Hot-air furnace	J. L. Kite	Philadelphia, Pa	June 19, 1855	13, 092
Hot-air furnace	A. Knobel	Monroe, Wis	Mar. 31, 1868	76, 203
Hot-air furnace	A. Kohler	Boston, Mass	Oct. 15, 1867	69, 820
Hot-air furnace	B. S. Koll	Pittsburgh, Pa	July 5, 1870	104, 962
Hot-air furnace	T. Kruse	La Fayette, Ind	July 16, 1872	129, 145
Hot-air furnace	J. A. Lawson	Troy, N. Y	Nov. 5, 1867	70, 445
Hot-air furnace	J. A. Lawson	Troy, N. Y	July 4, 1871	116, 723
Hot-air furnace	J. A. Lawson	Troy, N. Y	July 4, 1871	116, 724
Hot-air furnace	W. H. Lee and C. M. Hardenbergh.	Minneapolis, Minn	June 23, 1868	79, 075
Hot-air furnace	W. H. Lee and C. M. Hardenbergh.	Minneapolis, Minn	Nov. 2, 1869	96, 328
Hot-air furnace	J. Leeds	Philadelphia, Pa	Nov. 27, 1860	30, 738
Hot-air furnace	M. W. Lester	Chicago, Ill	Aug. 31, 1869	94, 224
Hot-air furnace	R. Z. Liddle	Brooklyn, N. Y	Nov. 26, 1867	71, 517
Hot-air furnace	H. N. Longfellow	Lawrence, Mass	Jan. 28, 1873	135, 227
Hot-air furnace	S. Macferran	Philadelphia, Pa	Feb. 5, 1856	14, 196
Hot-air furnace	J. Magee	Chelsea, Mass	Dec. 16, 1873	145, 512
Hot-air furnace	J. Magee	Chelsea, Mass	Dec. 16, 1873	145, 513
Hot-air furnace	W. R. Manning	Chicago, Ill	Nov. 26, 1872	133, 375
Hot-air furnace	J. Martin	Florence, Ala	Jan. 26, 1869	86, 239
Hot-air furnace	P. Martin	Cincinnati, Ohio	Jan. 15, 1867	61, 223
Hot-air furnace	P. Martin	Cincinnati, Ohio	Jan. 3, 1871	110, 664
Hot-air furnace	J. McCoy	Burlington, N. J	Aug. 23, 1870	106, 708
Hot-air furnace	S. Mead	New Haven, Conn	Dec. 24, 1861	34, 003
Hot-air furnace	J. Mealey	Ogdensburgh, N. Y	July 2, 1872	128, 555
Hot-air furnace	J. H. Mearns	Philadelphia, Pa	Mar. 19, 1872	124, 842

Index of patents issued from the United States Patent Office from 1790 *to* 1873, *inclusive*—Continued.

Invention.	Inventor.	Residence.	Date.	No.
Hot-air furnace	G. F. Merklee	New York, N. Y	Dec. 24, 1867	72, 520
Hot-air furnace	M. Metcalf	Grand Rapids, Mich	Sept. 1, 1868	81, 809
Hot-air furnace	G. H. Miller	Leavenworth, Kans	May 10, 1870	102, 956
Hot-air furnace	G. H. Miller	Leavenworth, Kans	July 1, 1873	140, 523
Hot-air furnace	J. A. Miller	New York, N. Y	Apr. 25, 1865	47, 443
Hot-air furnace	D. B. Montague	Springfield, Mass	Jan. 21, 1873	135, 003
Hot-air furnace	D. B. Morris	Pittsburgh, Pa	Dec. 30, 1873	146, 089
Hot-air furnace	J. R. Morris	New Haven, Conn	Aug. 12, 1846	4, 689
Hot-air furnace	H. Movers	Chicago, Ill	June 27, 1871	116, 339
Hot-air furnace	J. R. Nichols	Boston, Mass	Mar. 28, 1871	113, 197
Hot-air furnace	E. D. Norcross	Augusta, Me	Feb. 8, 1870	99, 585
Hot-air furnace	E. D. Norcross	Boston, Mass	June 11, 1872	127, 913
Hot-air furnace	S. W. Norton	Lemont, Ill	Feb. 20, 1866	52, 738
Hot-air furnace	A. L. Otis	Normal, Ill	Sept. 28, 1869	95, 256
Hot-air furnace	O. Paddock	Watertown, N. Y	Nov. 27, 1860	30, 752
Hot-air furnace	H. L. Palmer	New York, N. Y	Feb. 11, 1873	135, 838
Hot-air furnace	J. H. H. Perkins	Utica, N. Y	Dec. 23, 1856	16, 287
Hot-air furnace	J. S. Perry	Albany, N. Y	June 22, 1869	91, 560
Hot-air furnace	J. L. Pfau, jr	Quincy, Ill	Apr. 25, 1871	114, 035
Hot-air furnace	A. Pfund	New York, N. Y	Nov. 4, 1873	144, 355
Hot-air furnace	W. M. Phelps	Marshall, Mich	Aug. 20, 1872	130, 740
Hot-air furnace	C. L. Pierce	Natick, Mass	Apr. 22, 1873	138, 188
Hot-air furnace	C. L. Pierce	Natick, Mass	Apr. 22, 1873	138, 189
Hot-air furnace	S. Pierce	Troy, N. Y	May 20, 1851	8, 104
Hot-air furnace	W. T. Powell and R. F. Brown	Chicago, Ill	Aug. 20, 1872	130, 595
Hot-air furnace	H. Randall	Quincy, Ill	Aug. 8, 1871	117, 814
Hot-air furnace	A. M. Rice	Boston, Mass	Oct. 26, 1852	9, 358
Hot-air furnace	A. Richmond	Providence, R. I	Oct. 26, 1852	9, 360
Hot-air furnace	S. I. Russell	Chicago, Ill	May 15, 1855	12, 875
Hot-air furnace	S. C. Salisbury	New York, N. Y	Aug. 3, 1869	93, 351
Hot-air furnace	W. Sanford	Brooklyn, N. Y	May 6, 1862	35, 180
Hot-air furnace	W. Sanford	Brooklyn, N. Y	June 22, 1869	91, 569
Hot-air furnace	C. B. Sawyer	Fitchburgh, Mass	Mar. 18, 1862	34, 699
Hot-air furnace	B. C. Sayre	Montrose, Pa	Apr. 15, 1873	137, 800
Hot-air furnace	P. I. Schopp	Louisville, Ky	Oct. 12, 1869	95, 734
Hot-air furnace	J. H. Schwein	Cincinnati, Ohio	Feb. 13, 1866	52, 648
Hot-air furnace	M. D. Seward	Normal, Ill	Apr. 8, 1873	137, 727
Hot-air furnace	S. B. Sexton	Baltimore, Md	Feb. 21, 1871	112, 081
Hot-air furnace	S. B. Sexton and G. W. Beard	Baltimore, Md	Oct. 25, 1870	108, 733
Hot-air furnace	J. H. Shedd and B. Worcester	Waltham, Mass	Oct. 31, 1865	50, 739
Hot-air furnace	C. J. Shepard	Brooklyn, N. Y	Oct. 21, 1873	143, 851
Hot-air furnace	J. Siddons	Rochester, N. Y	Dec. 29, 1868	85, 335
Hot-air furnace	E. Slater	Philadelphia, Pa	Jan. 18, 1870	99, 015
Hot-air furnace	E. Slater and A. H. Platt	Philadelphia, Pa	Oct. 27, 1868	83, 415
Hot-air furnace	J. B. Smith and O. L. Giddings	Manchester and Exeter, N. H.	July 16, 1872	129, 180
Hot-air furnace	S. Smith	Worcester, Mass	June 16, 1868	79, 020
Hot-air furnace	W. Smith	Cicero, Ind	Feb. 15, 1870	99, 963
Hot-air furnace	G. S. G. Spence	Boston, Mass	Aug. 17, 1852	9, 204
Hot-air furnace	G. S. G. Spence	Boston, Mass	Feb. 20, 1855	12, 448
Hot-air furnace	J. Stuber	Utica, N. Y	May 1, 1860	28, 112
Hot-air furnace	J. Stuber and F. Frank	Utica, N. Y	June 1, 1858	20, 454
Hot-air furnace	B. F. Sturtevant	Jamaica Plains, Mass	Sept. 28, 1869	95, 281
Hot-air furnace	P. Sweeney	Buffalo, N. Y	July 11, 1854	11, 278
Hot-air furnace	J. M. Thatcher	Bergen, N. J	July 9, 1872	128, 926
Hot-air furnace	M. A. Thayer	Chicago, Ill	May 24, 1870	103, 390
Hot-air furnace	G. G. Thomas	Saint Louis, Mo	July 26, 1870	105, 864
Hot-air furnace	G. G. Thomas	Saint Louis, Mo	Aug. 22, 1871	118, 299
Hot-air furnace	W. D. Titus	Brooklyn, N. Y	May 4, 1869	89, 610
Hot-air furnace	J. C. Treat	East Hartford, Conn	Aug. 5, 1851	8, 276
Hot-air furnace	C. W. Trotter	Rochester, N. Y	Mar. 5, 1867	62, 577
Hot-air furnace	L. B. Tupper	New York, N. Y	Jan. 9, 1872	122, 684
Hot-air furnace	W. H. Turner	Indianapolis, Ind	Sept. 12, 1871	118, 988
Hot-air furnace	W. Twitchell	Syracuse, N. Y	Sept. 24, 1872	131, 723
Hot-air furnace	J. Van	Cincinnati, Ohio	Aug. 23, 1870	106, 746
Hot-air furnace	J. S. Van Buren	Green Island, N. Y	Apr. 5, 1870	101, 683
Hot-air furnace	J. J. Vogelgesang	Columbus, Ohio	July 18, 1871	117, 225
Hot-air furnace	G. W. Walker	Boston, Mass	Oct. 31, 1871	120, 406
Hot-air furnace	T. Wallace	Chicago, Ill	May 15, 1866	54, 795
Hot-air furnace	G. E. Waring	Stamford, Conn	Dec. 24, 1850	7, 863
Hot-air furnace	E. Webster	Hartford, Conn	June 12, 1866	55, 564
Hot-air furnace	E. Webster	Hartford, Conn	July 5, 1870	105, 017
Hot-air furnace	D. P. Weeks	Malden, Mass	Dec. 12, 1854	12, 080
Hot-air furnace	D. P. Weeks	Boston, Mass	Aug. 11, 1857	18, 002
Hot-air furnace	J. G. Weldon	Pittsburgh, Pa	Oct. 17, 1871	119, 955
Hot-air furnace	J. W. Wentworth	Minneapolis, Minn	Jan. 12, 1869	85, 774
Hot-air furnace	S. Wethered	Baltimore, Md	Jan. 3, 1860	26, 724
Hot-air furnace	T. Whitaker and J. Constantine.	Bolton and Manchester, England.	Nov. 26, 1867	71, 348
Hot-air furnace	C. White	West Roxbury, Mass	Dec. 17, 1872	134, 118
Hot-air furnace	J. Whitehill	Frederick, Md	Nov. 8, 1859	26, 064
Hot-air furnace	J. Whitehill	Frederick, Md	Mar. 24, 1868	76, 018
Hot-air furnace	H. Whittingham	New York, N. Y	Aug. 7, 1866	57, 026
Hot-air furnace	H. Whittingham	New York, N. Y	Aug. 7, 1866	57, 027
Hot-air furnace	H. Whittingham	New York, N. Y	Aug. 7, 1866	57, 028
Hot-air furnace	C. H. Whorf	Saint Louis, Mo	Mar. 3, 1868	75, 097
Hot-air furnace	I. T. Winchester	Boston, Mass	Mar. 9, 1869	87, 610
Hot-air furnace	H. G. Wing	New Bedford, Mass	May 23, 1846	4, 532
Hot-air furnace	C. Wood	Worcester, Mass	July 11, 1871	116, 908
Hot-air furnace	T. Yates	Milwaukee, Wis	Mar. 25, 1873	137, 277
Hot-air furnace and fire-grate for heating apartments, &c.	W. H. Whiteley	Charlestown, Mass	May 11, 1841	2, 085
Hot-air furnace, bake-oven, and heating room	J. Stahl	Baltimore, Md	May 29, 1832	....
Hot-air furnace, Base-burning	J. Gray	Albany, N. Y	Jan. 7, 1868	73, 091
Hot-air furnace, Base-burning	J. G. Porter	New York, N. Y	Apr. 12, 1870	101, 911
Hot-air furnace for heaters	J. B. Driscole	New York, N. Y	Oct. 6, 1868	82, 811
Hot-air furnace, Portable	J. P. Hayes	Boston, Mass	Mar. 20, 1849	6, 201
Hot-air furnace, Portable	H. L. B. Lewis	New York, N. Y	Sept. 19, 1845	4, 198

Index of patents issued from the United States Patent Office from 1790 to 1873, inclusive—Continued.

Invention.	Inventor.	Residence.	Date.	No.
Hot-air-furnace register	D. Culver	New York, N. Y	Aug. 10, 1848	5, 698
Hot-air-furnace register	J. W. Geddes	Baltimore, Md	Mar. 2, 1858	19, 562
Hot-air-furnace register	C. F. Tuttle	Williamsburgh, N. Y	Jan. 23, 1849	6, 060
Hot-air-furnace register	C. F. Tuttle	Williamsburgh, N. Y	Sept. 11, 1849	6, 708
Hot-air furnaces and for vapor-draft of grate-bars, Water-vessel for.	W. Moultrie	New York, N. Y	Aug. 25, 1857	18, 054
Hot-air furnaces, Apparatus to prevent an over-supply of coal in fire-boxes of.	L. W. Leeds	Germantown, Pa	Sept. 4, 1855	13, 525
Hot-air furnaces, Dome for	B. Gomminginger	Rochester, N. Y	Mar. 19, 1872	124, 807
Hot-air furnaces, Heat-regulator for	I. Hayes	PhiladelphiaPa.	July 2, 1872	128, 485
Hot-air furnaces, Heat-regulator for	A. H. Tingley	Providence, R. I	Oct. 30, 1866	59, 294
Hot-air furnaces, Radiating-attachment for	J. H. Keyser	New York, N. Y	Mar. 19, 1867	62, 962
Hot-air furnaces, Setting for	C. J. Shepard	Brooklyn, N. Y	June 18, 1872	128, 179
Hot-air furnaces, Setting for	C. J. Shepard	Brooklyn, N. Y	June 18, 1872	128, 180
Hot-air furnaces, Shield for feed-air pipes for	B. C. Bibb	Baltimore, Md	May 7, 1872	126, 373
Hot-air-pipe elbow	J. M. Thacher	Jersey City, N. J	June 6, 1871	115, 785
Hot-air-pipe evaporator	G. F. J. Colburn	Newark, N. J	Jan. 22, 1861	31, 152
Hot-air-pipe evaporator	J. L. Hutchinson	Baltimore, Md	Feb. 13, 1866	52, 572
Hot-air register	W. S. Bronson	Hartford, Conn	Apr. 26, 1870	102, 217
Hot-air register	T. W. Brown	Reading, Pa	Oct. 27, 1868	83, 359
Hot-air register	B. Bunce and R. Salt	New York, N. Y	Mar. 19, 1872	124, 663
Hot-air register	W. G. Creamer	Brooklyn, N. Y	Sept. 14, 1869	94, 872
Hot-air register	T. Dowling	Salem, Mass	Jan. 6, 1863	37, 338
Hot-air register	O. A. Ebert	Baltimore, Md	Feb. 15, 1870	99, 864
Hot-air register	D. T. Gale	Fort Wayne, Ind	June 27, 1871	116, 294
Hot-air register	W. Highton	Malden, Mass	June 23, 1868	79, 226
Hot-air register	W. Highton	Malden, Mass	Nov. 18, 1873	144, 616
Hot-air register	W. B. Kehew and C. H. Fifield	Salem, Mass	July 22, 1862	35, 938
Hot-air register	J. W. McGlashan	Montreal, Canada	June 27, 1871	116, 337
Hot-air register	T. E. McNeill	Philadelphia, Pa	Mar. 13, 1860	27, 461
Hot-air register	H. M. Phinney	Cambridge, Mass	Oct. 5, 1869	95, 510
Hot-air register	H. M. Phinney	Boston, Mass	Apr. 4, 1871	113, 447
Hot-air register	H. M. Phinney	Cambridge, Mass	Sept. 17, 1872	131, 459
Hot-air register	G. Pollock	Roxbury, Mass	Aug. 20, 1850	7, 582
Hot-air register	C. B. Sawyer	Fitchburgh, Mass	Mar. 25, 1862	34, 783
Hot-air register	S. B. Sexton	Baltimore, Md	July 23, 1861	32, 897
Hot-air register	S. J. Sherman	New York, N. Y.	Oct. 6, 1857	18, 356
Hot-air register	J. H. Simonds	New York, N. Y	Feb. 19, 1861	31, 485
Hot-air register	J. H. Simonds	New York, N. Y	Mar. 19, 1861	31, 748
Hot-air register	J. V. Tibbets	New York, N. Y	Aug. 18, 1857	18, 026
Hot-air register	W. H. Towers	Philadelphia, Pa	Nov. 15, 1853	10, 240
Hot-air register	W. Turton	Bushwick, N. Y	Mar. 16, 1852	8, 811
Hot-air register	W. Turton	Brooklyn, N. Y	Oct. 13, 1863	40, 294
Hot-air register	E. A. Tuttle	Williamsburgh, N. Y	Apr. 12, 1853	9, 664
Hot-air register	E. A. Tuttle	Williamsburgh, N. Y	Jan. 3, 1854	10, 371
Hot-air register	E. A. Tuttle	Brooklyn, N. Y	Sept. 6, 1859	25, 355
Hot-air register	E. A. Tuttle	Brooklyn, N. Y	Mar. 11, 1873	136, 682
Hot-air register	A. Watson	Jersey City, N. J	June 28, 1870	104, 802
Hot-air register	W. Young	Easton, Pa	Apr. 25, 1871	114, 247
Hot-air-register attachment	J. D. McBride	Mansfield, Ohio	Feb. 25, 1868	74, 968
Hot-air-register attachment	H. Sinclair	New York, N. Y	Sept. 1, 1868	81, 695
Hot-air-register frame, &c	C. R. Harvy	New York, N. Y	Jan. 29, 1867	61, 538
Hot-air registers, Composition-stone frame for	J. S. Elliott and J. F. Wood	Chelsea and Everett, Mass	Apr. 23, 1872	126, 042
Hot-air registers, Deflector for	S. H. Caughey	Baltimore, Md	July 30, 1867	67, 265
Hot-air, steam, and water gage	J. C. C. Walker	Waco Village, Tex	Oct. 8, 1867	69, 729
Hot-bed box or case	W. Wells	Salem, Mass	Aug. 9, 1870	106, 298
Hot-bed shutter	J. Weed	Muscatine, Iowa	Oct. 10, 1871	119, 805
Hot-blast apparatus	L. Blair	Painesville, Ohio	Mar. 11, 1856	14, 386
Hot-blast apparatus	R. Denholm	Newburgh, Ohio	Sept. 12, 1865	49, 862
Hot-blast apparatus, Furnace	J. Froggett	Youngstown, Ohio	Mar. 26, 1867	63, 237
Hot-blast-furnace lamp	J. H. Wilhelm	Chicago, Ill	Nov. 7, 1865	50, 865
Hot-blast furnaces, Collar for pipes in	H. McCullaugh	Marietta, Pa	Dec. 29, 1868	85, 392
Hot-blast oven	R. Long	Pittsburgh, Pa	Apr. 2, 1872	125, 138
Hot-blast pipe	C. Glidden	Milwaukee, Wis	Feb. 28, 1865	46, 559
Hot-blast pipe	W. B. Pollock	Youngstown, Ohio	Mar. 7, 1865	46, 698
Hot-house, Parlor	P. Griffith	Brooklyn, N. Y	Mar. 1, 1870	100, 287
Hot-water and hot-air heaters, Apparatus for automatic control of combustion in.	J. C. Wrightman and J. H. Mills.	Newton and Boston, Mass.	May 6, 1873	138, 547
Hot-water apparatus	J. Brown	New York, N. Y	May 30, 1854	10, 982
Hot-water apparatus	H. Humphreville, jr	Strasburgh, Pa	July 24, 1860	29, 267
Hot-water apparatus	J. F. Hunter	New York, N. Y	Mar. 27, 1860	27, 670
Hot-water apparatus, Boiler for	E. B. Cherevoy	New York, N. Y	Apr. 10, 1860	27, 858
Hot-water apparatus, Boiler for	E. Horelck	New York, N. Y	Apr. 9, 1861	31, 976
Hot-water apparatus for warming buildings	E. L. Miller	Brooklyn, N. Y	July 7, 1846	4, 625
Hot water boiler	P. Lesson	Newark, N. J.	Jan. 30, 1872	123, 269
Hot-water boiler	A. Müller	Brooklyn, N. Y	June 15, 1869	91, 475
Hot-water boiler	H. Steeger	New York, N. Y	Dec. 31, 1867	72, 760
Hot-water boiler	J. Tregeser	New York, N. Y	July 27, 1869	93, 139
Hot-water-furnace boiler	C. R. Ellis	Brooklyn, N. Y	Sept. 6, 1864	44, 083
Hot-water heater	E. C. Clay	Malden, Mass	Jan. 30, 1872	123, 153
Hot-water heater	H. Howard	Springfield, Mass	Jan. 1, 1867	60, 732
Hot-water heater	A. Janes	New York, N. Y	Jan. 29, 1850	7, 054
Hot-water heater	G. Nixon	Philadelphia, Pa	May 28, 1872	127, 262
Hot-water heater	A. C. Pentland	Philadelphia, Pa	May 9, 1871	114, 595
Hot-water heater	G. H. Sellers	Phœnixville, Pa	Oct. 27, 1868	83, 553
Hot-water heater for apartments	J. F. Hunter and F. W. Geissenhainer.	New York, N. Y	Sept. 18, 1860	30, 065
Hot-water heater for buildings	G. M. Dexter	Boston, Mass	Dec. 1, 1840	1, 875
Hot-water heater for buildings	A. E. Hitchings	New York, N. Y	Jan. 25, 1848	5, 418
Hot-water heater for buildings	A. Marriott	Saint Louis, Mo	Sept. 12, 1871	118, 953
Hot-water heaters, Automatic air-damper for	W. C. Baker	New York, N. Y	Aug. 30, 1864	44, 038
Hot-water heaters, Boiler for	C. R. Ellis	Brooklyn, N. Y	July 16, 1872	129, 542
Hot-water or steam heater	S. Williams	Cambridge, Mass	June 11, 1872	127, 817
Hotel and burglar alarm	C. S. Noé	Bergen Point, N. J	Mar. 15, 1870	100, 792
Hotel communicating-apparatus	N. A. Patterson	Nashville, Tenn	Jan. 11, 1870	98, 794
Hotel register	C. L. Hawes	Titusville, Pa	Apr. 16, 1867	63, 889
Hotel-register	J. L. Mitchell	Buffalo, N. Y	Apr. 16, 1867	63, 924
Hotels, &c., Elevator or hoisting-apparatus for	O. Tufts	Boston, Mass	Aug. 9, 1859	25, 061

Index of patents issued from the United States Patent Office from 1790 *to* 1873, *inclusive*—Continued.

Invention.	Inventor.	Residence.	Date.	No.
Hound and fifth-wheel combined	G. Archer	Massillon, Ohio	June 24, 1862	35, 662
House: *See* Bath-house. Bee-house. Bird-house. Brick-drying house. Corn-house. Dairy-house. Dry-house. Drying-house. Fire-engine house. Fire-proof house. Floating house. Grain store-house. Hot-house. Ice-house. Ice-preserving house. Iron house. Light-house. Locomotive-engine house. Malt-house. Metallic house. Milk-house. Navigable mercantile house. Portable house. Preserving-house. Refrigerating-house. Sheep-house. Signal-house. Slaughter-house. Smoke-house. Sugar-house. Switch-tender's house. Tobacco-drying house.				
House	O. C. Campbell	Omaha, Nebr	Mar. 11, 1873	136, 583
House	G. W. Clayton	Cleveland, Ohio	July 2, 1872	128, 464
House	D. L. Emerson	Oakland, Cal	Apr. 15, 1873	137, 833
House	H. M. Irwin	Charlotte, N. C	Aug. 24, 1869	94, 116
House	T. W. H. Moseley	Hyde Park, Mass	July 15, 1873	140, 941
House	T. W. H. Moseley	Hyde Park, Mass	July 15, 1873	140, 942
House	W. Ortwine	Baltimore, Md	Feb. 2, 1864	41, 445
House	J. Park	Joliet, Ill	Mar. 5, 1867	62, 676
House	A. Sidle	Minneapolis, Minn	Nov. 20, 1866	59, 929
House	R. B. Varden	Uniontown, Md	Mar. 5, 1872	124, 396
House	F. Walton	Staines, England	June 15, 1869	91, 289
House	W. Ward	New York, N. Y	Oct. 21, 1873	143, 946
House-alarm, Electro-magnetic	W. Whiting	Roxbury, Mass	July 20, 1858	20, 970
House, bridge, boat, and wagon-body, Combined	J. C. Adams	Baltimore, Md	Feb. 18, 1862	34, 399
House-furnaces, Regulating the draft of	S. L. Hay and H. B. Osgood	Reading and Dorchester, Mass.	Nov. 11, 1856	16, 055
House-hole cover	R. Liston	Albany, N. Y	May 31, 1870	103, 630
House-trimmings	J. R. Webber	Morris, Ill	Feb. 15 1870	99, 798
House-ventilation	J. H. Griscom	New York, N. Y	Jan. 18, 1859	22, 646
House-ventilator	R. Boyd	Evansville, Iowa	Nov. 12, 1867	70, 792
House-ventilator	B. J. Burnett	Mount Vernon, N. Y	Aug. 15, 1865	49, 373
House-ventilator	S. W. Williams	Centreville, N. Y	Sept. 17, 1861	33, 325
House-warmer and smoke-driver, Atmospherical	R. Annesly	Philadelphia, Pa	Apr. 27, 1814	....
House-warming furnace	P. Martin	Cincinnati, Ohio	Sept. 20, 1864	44, 329
Houses, &c., Composition for covering	A. Rogers	Hudson, N. Y	Mar. 14, 1817	....
Houses, Composition for covering and flooring	L. De Niroth	Baltimore, Md	Feb. 3, 1807	....
Houses, halls, &c., Ventilating	W. C. Grimes	Philadelphia, Pa	Feb. 9, 1869	86, 663
Houses, Ventilating	R. Mayo and R. Mills	Washington, D. C	Oct. 24, 1836	67
Houses with tin-plate, sheet-iron, or zinc, Covering	R. S. Tilden	Lynchburgh, Va	Sept. 10, 1829	....
Household and culinary operations, Machine for facilitating.	H. S. Shepardson	Shelburne Falls, Mass	Oct. 31, 1865	50, 779
Household-implement	T. Garrick	Providence, R. I	June 29, 1869	91, 839
Household-implement	T. Garrick	Providence, R. I	May 9, 1871	114, 665
Household-implement	A. Guinzburg	Boston, Mass	June 15, 1869	91, 227
Household-implements	R. A. McCauley	Baltimore, Md	July 11, 1871	116, 849
Household-utensil	H. R. Halsey	La Fayette, Ill	Aug. 15, 1871	118, 125
Household-utensil	W. C. McGill	Cincinnati, Ohio	Sept. 24, 1867	69, 111
Howel and croze	J. B. Siegfried	Pittsburgh, Pa	Sept. 15, 1868	82, 163
Hub	A. Huston	Cincinnati, Ohio	Oct. 9, 1866	58, 647
Hub and axle	J. W. and A. W. Beer	Myers' Dale and Rural Valley, Pa.	Aug. 6, 1872	130, 102
Hub and axle	C. R. Donner	Sonora, Cal	Feb. 14, 1871	111, 822
Hub and axle	N. Maxham	Hancock, Vt	Feb. 25, 1868	74, 841
Hub and axle, Carriage	L. D. Cook	West Liberty, Ohio	Jan. 2, 1872	122, 363
Hub and axle fastening	J. Henderson	Horseheads, N. Y	May 29, 1855	12, 946
Hub and axle lubricator	T. Wilson	Garton, England	Aug. 25, 1868	81, 449
Hub and axle, Vehicle	D. M. Buckhout	Mount Kisco, N. Y	Oct. 15, 1867	69, 760
Hub and axle, Vehicle	M. Chapin	Erie, Pa	Feb. 11, 1873	135, 689
Hub and axle-tree for wheels and pulleys	P. Slayton	Lockport, N. Y	Oct. 27, 1828	....
Hub and bearing, Car-wheel	D. Worstell and H. Pirrung	Lawrence County, Ohio	June 25, 1872	128, 348
Hub and box, Wheel	S. Mosher	Winchester, Ill	Oct. 13, 1868	83, 083
Hub and felly boring machine	D. Sperry	Colchester, Conn	May 12, 1812	....
Hub and journal, Carriage-wheel	R. W. McClelland	Springfield, Ill	June 10, 1862	35, 531
Hub and spoke, Carriage-wheel	J. Maris	Marietta, Ohio	Apr. 27, 1869	89, 322
Hub, axle, and box, Wagon	J. W. Pollock	Cross Bridges, Tenn	Mar. 17, 1868	75, 574
Hub band, Carriage	J. Jenkins and J. R. Cooke	Winsted, Conn	Dec. 12, 1854	12, 062
Hub band, Carriage	S. C. Talcott	Ashtabula, Ohio	June 21, 1864	43, 238
Hub-band, Metallic	S. D. Forbes	Wilmington, Del	Mar. 5, 1872	124, 202
Hub band, Vehicle-wheel	J. Ives	Mount Carmel, Conn	July 16, 1872	128, 961
Hub band, Vehicle-wheel	G. H. Johnson	Bridgeport, Conn	May 27, 1873	139, 317
Hub band, Wheel	S. H. Miller	Elizabethtown, N. J	Oct. 7, 1830	....
Hub-bands, Fastening	J. Ives	Mount Carmel, Conn	Feb. 7, 1871	111, 547
Hub-bands for carriage-wheels, Manufacture of	S. E. Farrand	Newark, N. J	June 11, 1839	1, 171
Hub-bands for wagon-wheels, Making	G. W. Beers	Bridgeport, Conn	Nov. 29, 1859	26, 242
Hub-bands for wagon-wheels, Making	J. A. Boughton	Poughkeepsie, N. Y	Nov. 29, 1859	26, 246
Hub-bands	E. B. Butler	New Britain, Conn	Dec. 31, 1867	72, 719
Hub bands, Roll for rolling wagon	J. Glass	Pittsburgh, Pa	Aug. 12, 1873	141, 641
Hub-blocks for the lathe, Machine for preparing	L. Eames	Kalamazoo, Mich	June 9, 1857	17, 500
Hub-borer	W. I. Casselman	Vernon, N. Y	May 9, 1854	10, 882
Hub-borer	J. Shaerer	Reading, Pa	Jan. 13, 1857	16, 402
Hub-borer	C. B. Wiley	Adrian, Mich	Apr. 5, 1859	23, 535
Hub boring and mortising machine	G. M. Atherton	Friendsville, Ill	Dec. 6, 1859	26, 330
Hub boring and mortising machine	C. H. Guard	Brownville, N. Y	July 24, 1855	13, 307
Hub boring and mortising machine	J. Tompkins	Conesville, N. Y	Oct. 6, 1837	419
Hub boring and mortising machine, Carriage	R. I. R. Stone	Berlin, Ohio	Mar. 7, 1854	10, 603
Hub boring and mortising machinery	J. J. Greenough	Washington, D. C	June 13, 1846	4, 565
Hub-boring and spoke-tenoning machine	J. Gardner	Bigler, Pa	Apr. 16, 1872	125, 803

Index of patents issued from the United States Patent Office from 1790 to 1873, inclusive—Continued.

Invention.	Inventor.	Residence.	Date.	No.
Hub-boring apparatus, Wheel	J. Hinds	Troy, N. Y	Nov. 25, 1837	488
Hub-boring machine	A. Bascomb	Seneca, Mich	May 28, 1872	127, 296
Hub-boring machine	S. L. Bond	Greenwood, S. C	Nov. 29, 1859	26, 245
Hub-boring machine	D. C. Breed	Lyndonville, N. Y	Mar. 27, 1866	53, 400
Hub-boring machine	E. Caswell	Lyons, N. Y	Aug. 5, 1873	141, 425
Hub-boring machine	F. W. Dexter	Randolph, N. Y	Jan. 19, 1869	86, 006
Hub-boring machine	L. A. Dole	Salem, Ohio	Mar. 13, 1866	53, 216
Hub-boring machine	Z. Doolittle	Perry, Ga	Dec. 8, 1857	18, 808
Hub-boring machine	J. Duncan and W. H. Arnold	Buchanan, Mich	July 29, 1873	141, 264
Hub-boring machine	J. Eisenmann	Chicago, Ill	May 13, 1873	138, 742
Hub-boring machine	N. Johnson	Berlin, Wis	Mar. 3, 1868	75, 167
Hub-boring machine	F. Jonas	Freeport, Ill	Mar. 16, 1869	87, 852
Hub-boring machine	F. Jonas	Burlington, Iowa	Nov. 18, 1873	144, 618
Hub-boring machine	D. Murphy	Dubuque, Iowa	Apr. 6, 1869	88, 579
Hub-boring machine	J. A. Newell	Kalamazoo, Mich	Dec. 16, 1873	145, 521
Hub-boring machine	D. J. Owen	Springville, Pa	Dec. 31, 1867	72, 887
Hub-boring machine	W. S. Owen	Oskaloosa, Iowa	Dec. 23, 1873	145, 894
Hub-boring machine	G. T. Pearsall and S. A. Garrison	Apalachin and Union, N. Y.	Sept. 30, 1862	36, 573
Hub-boring machine	D. Quimby	Littleton, N. H	July 12, 1859	24, 756
Hub-boring machine	J. L. Roberts and R. K. Daily	Waverly, Iowa	Apr. 9, 1872	125, 616
Hub-boring machine	P. Schuttler	Chicago, Ill	May 2, 1865	47, 576
Hub-boring machine	A. R. Silver	Salem, Ohio	Aug. 11, 1868	80, 837
Hub-boring machine	A. R. Silver	Salem, Ohio	July 12, 1870	105, 265
Hub-boring machine	A. Troup	Lewisburgh, Pa	May 3, 1870	102, 625
Hub-boring machine	J. W. Walters	Riceville, Iowa	Apr. 21, 1868	77, 137
Hub-boring machine	D. B. Wright	South Amesbury, Mass	Dec. 2, 1873	145, 142
Hub-boring-machine clamp	G. V. Brecht	Saint Louis, Mo	June 16, 1868	78, 922
Hub boring machine, Wagon	F. Bremerman	Indianapolis, Ind	Dec. 26, 1865	51, 688
Hub-boring machine, Wagon	E. Caswell	Newport, Mich	Mar. 29, 1864	42, 075
Hub-boring machine, Wagon	J. W. Emerson	Rochester, Minn	Feb. 4, 1868	74, 062
Hub-boring machine, Wagon	T. Harper	West Manchester, Pa	Aug. 21, 1866	57, 319
Hub-boring machine, Wagon	J. F. Kendall	Concord, N. H	Apr. 7, 1868	76, 467
Hub-boring machine, Wagon	J. Kirtch	Rochester, N. Y	Oct. 25, 1865	44, 807
Hub-boring machine, Wagon	S. Lavenue	Alton, Ill	Nov. 6, 1866	59, 417
Hub-boring machine, Wagon	L. Mason	Turin, N. Y	May 15, 1866	54, 750
Hub-boring machine, Wagon	J. R. McAlister	Richville, N. Y	Oct. 30, 1866	59, 245
Hub-boring machine, Wagon	W. Snodgrass	Macomb, Ill	July 2, 1867	66, 405
Hub-boring machine, Wagon	J. Thompson	Clifton, N. Y	Aug. 6, 1861	33, 012
Hub-boring machine, Wagon-wheel	J. Wharff	Bangor, Me	May 26, 1868	78, 246
Hub-boring tool	H. C. Garvin and J. H. King	Hagerstown, Md	Apr. 10, 1855	12, 677
Hub-boring tool	H. L. Mooney and W. B. Carter	Astoria, Ill	May 27, 1856	14, 968
Hub box, Carriage	J. A. Cramer	Brooklyn, N. Y	Apr. 2, 1861	31, 873
Hub-box, Cast-iron	J. Witt	Chambersburgh, Pa	Dec. 28, 1825	......
Hub-box molds, Preparing	J. G. Holt	Chicago, Ill	June 5, 1866	55, 296
Hub box, Wheel	I. Cooper	Baltimore, Md	Feb. 7, 1831	......
Hub box, Wheel	E. G. Woodside	San Francisco, Cal	Nov. 5, 1867	70, 670
Hub, Carriage	S. Allaire	New York, N. Y	Mar. 3, 1868	75, 105
Hub, Carriage	S. S. Barry	Brownhelm, Ohio	Jan. 6, 1852	8, 623
Hub, Carriage	S. W. Beach	Chicago, Ill	May 26, 1857	17, 360
Hub, Carriage	H. V. Belding	Oppenheim, N. Y	June 15, 1869	91, 302
Hub, Carriage	N. Bryan	Thomaston, Ga	Oct. 18, 1870	108, 441
Hub, Carriage	J. P. Chandler	Wilton, Me	Feb. 1, 1870	99, 405
Hub, Carriage	W. T. Dole	Peabody, Mass	Dec. 12, 1871	121, 857
Hub, Carriage	H. R. Fry	Wabash, Ind	Feb. 6, 1872	123, 472
Hub, Carriage	G. Kenny	Nashua, N. H	Apr. 6, 1869	88, 720
Hub, Carriage	C. Leavitt	Cleveland, Ohio	Jan. 20, 1863	37, 451
Hub, Carriage	J. A. Maynard	Newtonville, Mass	Oct. 25, 1870	108, 612
Hub, Carriage	J. W. Minor and D. P. Ward	New Bedford, Mass	Mar. 24, 1868	75, 948
Hub, Carriage	H. Nycum	Uniontown, Pa	Mar. 11, 1856	14, 407
Hub, Carriage	J. O'Connor	Jackson, Mo	May 3, 1870	102, 580
Hub, Carriage	J. Pruette	Aurora, Ill	Aug. 2, 1859	24, 955
Hub, Carriage	S. W. Reed	Berkshire, N. Y	Dec. 11, 1855	13, 919
Hub, Carriage	J. Smith	Sunbury, Ohio	Feb. 19, 1856	14, 294
Hub, Carriage	H. W. Stow	New Haven, Conn	May 2, 1871	114, 364
Hub, Carriage	J. W. Weston	New York, N. Y	July 6, 1869	92, 238
Hub, Carriage	M. Whelan	New Haven, Conn	Apr. 21, 1868	77, 147
Hub, Carriage	J. M. Whiting	Providence, R. I	Jan. 15, 1867	61, 293
Hub, Carriage	J. M. Whiting	Providence, R. I	Aug. 3, 1869	93, 259
Hub, Carriage	A. S. Woodward	Pepperell, Mass	Aug. 31, 1869	94, 372
Hub, Carriage-wheel	A. Allcott	Haverhill, Mass	June 11, 1872	127, 823
Hub, Carriage-wheel	E. A. Archibald	Methuen, Mass	Nov. 14, 1871	120, 845
Hub, Carriage-wheel	J. D. Arnold	Murray, N. Y	Oct. 1, 1872	131, 838
Hub, Carriage-wheel	W. J. Arrington	Jefferson County, Ga	Mar. 29, 1870	101, 208
Hub, Carriage-wheel	S. Atha	West Liberty, Ohio	Feb. 15, 1870	99, 809
Hub, Carriage-wheel	S. Atha	West Liberty, Ohio	Nov. 29, 1870	109, 707
Hub Carriage-wheel	J. B. Bauman	Shepherdstown, Pa	Sept. 24, 1872	131, 650
Hub, Carriage-wheel	J. F. Beckwith	South Alabama, N. Y	Mar. 27, 1860	27, 611
Hub, Carriage-wheel	J. E. Bowers	Bainbridge, Ga	Apr. 2, 1872	125, 164
Hub, Carriage-wheel	R. Brown and W. Aldrich	Dayton, Ohio	Oct. 15, 1872	132, 241
Hub, Carriage-wheel	J. Y. Burwell	Worthington, Pa	Oct. 24, 1871	120, 150
Hub, Carriage-wheel	B. Burr	Batavia, Ill	Mar. 22, 1870	100, 973
Hub, Carriage-wheel	A. F. Cooper	San Francisco, Cal	May 3, 1870	102, 657
Hub, Carriage-wheel	H. Delano	Skaneateles, N. Y	Oct. 28, 1837	448
Hub, Carriage-wheel	L. Dorman	Worcester, Mass	May 21, 1867	64, 955
Hub, Carriage-wheel	J. Dump	Kingston, Ohio	Mar. 15, 1870	100, 873
Hub, Carriage-wheel	J. Dump and A. Moore	Kingston, Ohio	Apr. 16, 1872	125, 727
Hub, Carriage-wheel	J. M. Emmerich	New Haven, Conn	Aug. 23, 1870	106, 678
Hub, Carriage-wheel	J. F. Fowler	Alliance, Ohio	Mar. 19, 1872	124, 731
Hub, Carriage-wheel	J. C. Garretson	Marshland, N. Y	May 7, 1872	126, 389
Hub, Carriage-wheel	H. D. Haraden	Hartford, Vt	May 19, 1868	77, 977
Hub, Carriage-wheel	L. T. Hazen	Coventry, N. Y	May 24, 1859	24, 120
Hub, Carriage-wheel	C. C. Holt	Lawrence, Mass	May 14, 1872	126, 810
Hub, Carriage-wheel	J. W. Jackson and L. W. Burchinal.	Smithfield, Pa	Sept. 1, 1857	18, 097
Hub, Carriage-wheel	J. Kritch	Cleveland, Ohio	Jan. 2, 1872	122, 322
Hub, Carriage-wheel	J. Kritch	Cleveland, Ohio	Jan. 2, 1872	122, 323
Hub, Carriage-wheel	J. Locke and J. Clark	Lewisburgh and Milroy, Pa.	Oct. 1, 1872	131, 887
Hub, Carriage-wheel	C. Merry	New York, N. Y	Dec. 15, 1857	18, 855

Index of patents issued from the United States Patent Office from 1790 *to* 1873, *inclusive*—Continued.

Invention.	Inventor.	Residence.	Date.	No.
Hub, Carriage-wheel	P. Murphy	South Amesbury, Mass	Aug. 12, 1873	141,659
Hub, Carriage-wheel	W. C. Pearsall	McMinnville, Tenn	Feb. 15, 1870	99,942
Hub, Carriage-wheel	N. Platt	Jackson, Miss	Aug. 3, 1858	21,083
Hub, Carriage-wheel	J. Ridge	Richmond, Ind	Nov. 12, 1872	132,924
Hub, Carriage-wheel	S. T. F. Sterrick	Washington, D. C	June 29, 1869	92,121
Hub, Carriage-wheel	H. Thomas	Medway, Mass	Feb. 8, 1828	....
Hub, Carriage-wheel	H. Thomas	Middleborough, Mass	Sept. 11, 1829	....
Hub, Carriage-wheel	H. E. Vick	Alliance, Ohio	Nov. 14, 1871	121,025
Hub, Carriage-wheel	H. E. Vick	Alliance, Ohio	Dec. 5, 1871	121,694
Hub, Carriage-wheel	A. Warner	Hamden, Conn	Feb. 5, 1867	61,900
Hub, Carriage-wheel	J. M. Whiting	New Bedford, Mass	Mar. 30, 1858	19,820
Hub, Carriage-wheel	M. Young	Frederick, Md	Dec. 26, 1865	51,773
Hub, Cast-iron	C. Washburn	Bridgewater, Mass	Jan. 18, 1833	....
Hub-centering machine and spoke-guide	K. P. Allen	Homer, Mich	July 11, 1871	116,790
Hub-clamp	J. McClelland	Geneva, N. Y	Sept. 12, 1871	118,868
Hub dividing and boring machine	J. B. Cleveland	Bergen County, N. J	June 7, 1809	....
Hub, Expanding wheel	A. I. Judge	Baltimore, Md	Jan. 14, 1868	73,342
Hub-fastening for eccentrics	G. W. Miller and J. D. Stevens	Scranton, Pa	May 12, 1868	77,747
Hub for carriage-wheels, Metallic	E. A. Archibald	Methuen, Mass	Apr. 9, 1872	125,514
Hub for carriage-wheels, Metallic	N. T. Edson	New Orleans, La	July 13, 1858	20,869
Hub for carriage-wheels, Metallic	N. T. Edson	New Orleans, La	Oct. 16, 1860	30,395
Hub for carriage-wheels, Metallic	J. Johnson	Garysburgh, N. C	Nov. 20, 1860	30,681
Hub for carriage-wheels, Metallic	S. I. Russell	Chicago, Ill	June 15, 1858	20,586
Hub for coaches, wagons, &c., Cast	W. Dickinson	Batavia, N. Y	Oct. 11, 1828	....
Hub for heavy-wheeled vehicles	J. J. Anderson	Indianapolis, Ind	Apr. 15, 1873	137,874
Hub for pulleys, &c., Self-oiling	D. M. Weston	Boston, Mass	Mar. 10, 1868	75,323
Hub for vehicle-wheels, Metallic	S. T. F. Sterick	Georgetown, D. C	Mar. 26, 1872	125,095
Hub for vehicles, Metallic	J. Abbott	Washington, Ind	Jan. 30, 1866	52,255
Hub for wheels, Cast-iron	B. Lyman	Manchester, Conn	Oct. 29, 1825	....
Hub for wheels, Cast-iron	B. Lyman	Manchester, Conn	Nov. 6, 1827	....
Hub for wheels of vehicles	E. D. Ives	Philadelphia, Pa	Feb. 6, 1872	123,480
Hub for wheels of vehicles	J. Monk	Norwich, Conn	Apr. 1, 1873	137,314
Hub for wheels of vehicles, Metallic	J. B. Stuart	Bunker Hill, Ill	Feb. 19, 1867	62,161
Hub-machine	L. Eames	Kalamazoo, Mich	May 11, 1858	20,197
Hub-machine	I. N. Felch	Hollis, Me	Jan. 22, 1861	31,161
Hub-machine	J. B. Ripsom	East Kendall, N. Y	Aug. 25, 1863	39,675
Hub, Metallic	W. W. Ball	Edinburgh, Ind	May 5, 1868	77,568
Hub, Metallic	H. B. Buch	Litiz, Pa	Sept. 11, 1866	57,855
Hub, Metallic	J. H. Harper	Pittsburgh, Pa	Mar. 7, 1871	112,452
Hub, Metallic	J. B. Hayden	Easton, N. Y	Feb. 7, 1854	10,506
Hub, Metallic	J. Oliphant	Springhill Furnace, Pa	June 30, 1868	79,496
Hub, Metallic carriage	H. Boardman	Lancaster, Pa	Feb. 14, 1860	27,097
Hub, Metallic wheel	J. Monk	Norwich, Conn	Dec. 5, 1871	121,462
Hub-mortising machine	E. Joslin and D. L. Gibbs	Keene, N. H., and Norwich, Conn.	May 27, 1862	35,420
Hub-mortising machine	P. Snyder	Grand Rapids, Mich	Oct. 17, 1871	120,117
Hub-mortising machinery	A. Thompson	Ridgeville, Md	Apr. 25, 1843	3,054
Hub or gig band, Carriage	C. Cornelius	Philadelphia, Pa	June 18, 1824	....
Hub-reamer	S. A. Garrison	Union, N. Y	Aug. 23, 1859	25,190
Hub-reamer	J. G. Robinson	Springfield, Ill	Mar. 22, 1870	101,164
Hub, Road-carriage	J. Raddin	Lynn, Mass	May 3, 1870	102,709
Hub-shaping machine	A. Goodyear	Albion, Mich	Aug. 31, 1869	94,201
Hub shell, Carriage	J. O'Connor	Jackson, Mo	Oct. 19, 1869	96,028
Hub turning and mortising machine	E. M. Scott	Auburn, N. Y	Mar. 4, 1862	34,595
Hub-turning machine	W. G. Beach	New Haven, Conn	Nov. 14, 1871	120,931
Hub-turning machine	A. Rickart	Schoharie, N. Y	Sept. 7, 1858	21,443
Hub-turning machine	A. Rickurt	Schoharie, N. Y	June 28, 1859	24,579
Hub-turning machine	J. J. Zufelt and R. Craig	Sheboygan Falls, Wis	Jan. 5, 1869	85,629
Hub, Vehicle	H. M. Du Bois	Philadelphia, Pa	Feb. 4, 1873	135,470
Hub, Vehicle	C. W. Fillmore	Marengo, Ill	Apr. 30, 1872	126,278
Hub, Vehicle	C. W. Fillmore	Chicago, Ill	Aug. 13, 1872	130,416
Hub, Vehicle	W. I. Lyman	East Hampton, Mass	Jan. 17, 1871	111,070
Hub, Vehicle	S. McGee	Madison, N. J	Feb. 4, 1873	135,570
Hub, Vehicle	A. Moffitt	Brownsville, Pa	Sept. 2, 1862	36,355
Hub, Vehicle	G. L. Rouse	Cincinnati, Ohio	June 10, 1873	139,691
Hub, Vehicle	T. Royer and G. L. Rouse	Cincinnati, Ohio	Mar. 4, 1873	136,459
Hub, Vehicle	S. T. F. Sterick	Georgetown, D. C	Jan. 14, 1873	134,945
Hub, Vehicle	J. P. Zeller	South Bend, Ind	Sept. 7, 1869	94,685
Hub, Vehicle-wheel	J. S. Ball	Charleston, Ill	June 25, 1872	128,354
Hub, Vehicle-wheel	D. Davis	New York, N. Y	Aug. 29, 1871	118,589
Hub, Vehicle-wheel	D. Davis	New York, N. Y	July 2, 1872	128,600
Hub, Vehicle-wheel	C. G. Gray	Ansonia, Conn	July 9, 1872	128,794
Hub, Vehicle-wheel	C. J. Harris	Louisville, Ky	Oct. 8, 1872	132,071
Hub, Vehicle-wheel	S. B. Hitt and R. W. Chapman	Waterloo, Iowa	Apr. 23, 1872	125,955
Hub, Vehicle-wheel	J. B. Hubbell	Naugatuck, Conn	July 23, 1872	129,666
Hub, Vehicle-wheel	E. D. Ives	Philadelphia, Pa	Aug. 22, 1871	118,369
Hub, Vehicle-wheel	P. Jones	Newark, N. J	July 2, 1872	128,546
Hub, Vehicle-wheel	P. Jones	Newark, N. J	July 2, 1872	128,547
Hub, Vehicle-wheel	W. A. Lewis	Chicago, Ill	Dec. 10, 1872	133,867
Hub, Vehicle-wheel	E. B. Lowe	Bellefontaine, Ohio	Oct. 8, 1872	132,091
Hub, Vehicle-wheel	W. F. Morton	New Haven, Conn	Dec. 24, 1872	134,155
Hub, Vehicle-wheel	A. M. Ocobock	Toledo, Ohio	Oct. 29, 1872	132,596
Hub, Vehicle-wheel	J. Ridge	Richmond, Ind	July 30, 1872	130,075
Hub, Vehicle-wheel	H. Taylor, A. P. Taylor, and C. Palm.	Milton, Mineral Ridge, and Jackson, Ohio.	Jan. 16, 1872	122,742
Hub, Vehicle-wheel	O. Vanorman	Fond du Lac, Wis	Oct. 10, 1871	119,803
Hub, Wagon	E. R. Baker	Fairhaven, Mass	Aug. 18, 1868	81,127
Hub, Wagon	J. H. Gaines	Durhamville, Tenn	Aug. 8, 1871	117,878
Hub, Wagon	J. D. Ham	Bethany, Ga	July 26, 1870	105,674
Hub, Wagon	J. B. Hubbell	Naugatuck, Conn	Sept. 20, 1870	104,497
Hub, Wagon	T. M. Jones and C. W. Fillmore.	Chicago, Ill	Jan. 16, 1872	122,835
Hub, Wagon	R. W. McClelland	Springfield, Ill	July 24, 1866	56,586
Hub, Wagon	N. Rixford	Mansfield Centre, Conn	Feb. 2, 1869	86,451
Hub, Wagon	S. W. Slocumb	Albany, Ill	Dec. 3, 1867	71,801
Hub, Wagon	A. E. Smith	Bronxville, N. Y	July 7, 1863	39,177
Hub, Wagon	W. C. Tucker	Richmond Switch, R. I	Sept. 10, 1867	68,808
Hub, Wagon	A. S. Woodward	Pepperell, Mass	Dec. 1, 1868	84,601

Index of patents issued from the United States Patent Office from 1790 to 1873, inclusive—Continued.

Invention.	Inventor.	Residence.	Date.	No.
Hub, Wagon-wheel	S. Altha	West Liberty, Ohio	Sept. 26, 1871	119, 295
Hub, Wagon-wheel	F. Nichols	Newport, Ky	Sept. 29, 1868	82, 630
Hub, Wagon-wheel	W. H. Rodeheaver	Miamisburgh, Ohio	Apr. 14, 1868	76, 818
Hub, Wheel	J. Atherton	Philadelphia, Pa	June 2, 1836	....
Hub, Wheel	W. F. Ehlers	Pottsville, Pa	Aug. 2, 1870	105, 926
Hub, Wheel	J. B. Hards	Chicago, Ill	Aug. 16, 1870	106, 361
Hub, Wheel	W. C. Johnson	Philadelphia, Pa	Mar. 28, 1871	113, 171
Hub, Wheel	O. B. Little	Wheeling, W. Va	Jan. 23, 1872	123, 030
Hub, Wheel	J. Summers	Raleigh Court-House, Va	Apr. 15, 1856	14, 678
Hub, Wheel	B F. Taft	Groton Junction, Mass	Sept. 3, 1867	68, 397
Hub, Wheel	J. L. Van West	Tolland, Mass	Feb. 7, 1871	111, 703
Hub, Wheel	C. P. Whitman	Charlemont, Mass	May 10, 1870	102, 988
Hub, Wheel	G. E. Whitmore	Housatonic, Mass	June 4, 1867	65, 521
Hub, Wheel	J. V. Woolsey	Sandusky, Ohio	Aug. 8, 1871	117, 849
Hubs and axles, Applying friction-roller to	J. B. and S. Wilson	Townsend's Inlet, N. J., and Kensington, Pa.	Feb. 25, 1851	7, 948
Hubs and axles, Attaching and detaching	R. D. Munson	Williston, Vt	Jan. 2, 1849	5, 995
Hubs and axles, Connecting	C. Chinnock	New York, N. Y	Jan. 9, 1849	6, 004
Hubs and axles, Connecting and disconnecting	A. M. Billings	Claremont, N. H	Jan. 14, 1851	7, 899
Hubs and axles, Connecting and disconnecting	A. M. Billings and T. A. Ambrose.	Claremont, N. H	June 25, 1850	7, 450
Hubs and axles, Manufacture of	S. R. Hunter and M. Merrill	Cortlandville, N. Y	Feb. 27, 1849	6, 147
Hubs and axles of vehicles, Connecting	D. Beard	Shippensburgh, Pa	Mar. 8, 1859	23, 148
Hubs and ships' blocks, Box for wheel	I. Cooper	Baltimore, Md	Jan. 27, 1830	....
Hubs and spokes of wrought-iron wheels, Process for forming the.	T. Ryan	Scott Bar, Cal	Aug. 10, 1869	93, 483
Hubs, Apparatus for setting box in carriage	C. Weidig	New Haven, Conn	Dec. 26, 1871	122, 295
Hubs, Arrangement of cutter for turning	G. Cooper	Berlin, Wis	Nov. 30, 1858	22, 167
Hubs, Attaching caps to	P. Heoter and R. Victor	Grand Rapids, Mich	May 30, 1871	115, 316
Hubs, Boring and mortising	H. Hayes	Quincy, Ill	Sept. 2, 1856	15, 652
Hubs, Box for carriage	A. C. Garratt	Roxbury, Mass	Feb. 26, 1856	14, 310
Hubs, Box-setter for boring in	E. Badlam, jr	Chester, Vt	Sept. 18, 1835	....
Hubs, Box-setter for wheel	F. W. Dexter	Randolph, N. Y	Apr. 22, 1862	35, 015
Hubs, Casting box for wheel	T. Ellis	Philadelphia, Pa	Dec. 6, 1859	26, 394
Hubs, Cutter for boring wheel	L. S. Maring	Westport, Mass	Oct. 4, 1853	10, 087
Hubs, Device for securing the boxing in	M. Turley	Council Bluffs, Iowa	Aug. 22, 1871	118, 406
Hubs for boring, Clamp for centering	J. Thrasher	Avon, N. Y	Nov. 10, 1857	18, 612
Hubs for boxes, Apparatus for boring	H. Sidle	Dillsburgh, Pa	Feb. 24, 1852	8, 761
Hubs for boxes, Boring	S. H. Yocum	Shelbyville, Ind	Sept. 2, 1856	15, 678
Hubs for boxes, Machine for preparing	W. R. Jones	Granville, N. Y	July 29, 1851	8, 267
Hubs for boxes, Machinery for preparing	I. Munden	Allegheny City, Pa	Dec. 11, 1849	6, 941
Hubs for boxes, Tool for preparing	S. Fahrney	Near Boonsborough, Md	Mar. 19, 1850	7, 185
Hubs for containing oil, Constructing carriage and wagon wheel.	A. Randel	Vernon, N. Y	Sept. 8, 1837	374
Hubs for linch-cups, Boring out the inside and the boxes of.	J. R. Morrison	Springfield, Ohio	May 13, 1834	....
Hubs for reception of boxes, Machine for preparing	H. Moore	Seneca County, Ohio	July 29, 1851	8, 265
Hubs for setting boxes, Machine for boring wheel	J. C. Hendry	Manchester, N. H	July 18, 1871	117, 170
Hubs, Greasing and keeping dust, &c., out of the boxes of carriage-wheel.	N. C. Day	Lunenburgh, Mass	July 10, 1840	1, 680
Hubs, Inserting India rubber into carriage	G. F. Wilson	East Providence, R. I	Nov. 24, 1868	84, 329
Hubs, Lathe for turning	J. Z. Wagner and H. R. Fry	Wabash, Ind	Apr. 29, 1873	138, 303
Hubs, Machine for forming	W. Patterson	Constantine, Mich	Dec. 20, 1859	26, 520
Hubs, Machine for hewing out	G. W. Miles and P. P. Lane	Michigan City, Ind., and Cincinnati, Ohio.	Mar. 8, 1859	23, 220
Hubs, Machine for turning the band portion of carriage.	Z. Doolittle	Perry, Ga	Dec. 22, 1857	18, 893
Hubs, Mortising wagon	P. Paullin	Dantown, Ohio	Apr. 17, 1833	....
Hubs of carriage-wheels, Lining metallic boxes for	M. Palmer	Baltimore, Md	Mar. 9, 1844	3, 463
Hubs of carriage-wheels, &c., Making metallic cases for chilling cast-iron-pipe boxes for.	J. Huntington	Zanesville, Ohio	Feb. 24, 1845	3, 929
Hubs of cars and other wheels, Hardening or chilling.	H. Thomas	Beaver Meadow, Pa	Oct. 13, 1838	980
Hubs of propellers, Securing the arms to the	H. O. Perry	Buffalo, N. Y	Dec. 7, 1858	22, 266
Hubs of wagon-wheels and the tenons of spokes to fit in the hubs, Mortising.	D. B. Goewey	Birmingham, Pa	Apr. 9, 1867	63, 628
Hubs of wheels, Boring and mortising	R. McCarty	Martinsburgh, N. Y	Mar. 31, 1834	....
Hubs of wheels, Machine for making	G. Lyon	New York	Dec. 28, 1826	....
Hubs on axles, Securing	A. E. Smith	Bronxville, N. Y	Feb. 17, 1857	16, 661
Hubs on axles, Securing carriage-wheel	J. Scheeper	New York, N. Y	Dec. 24, 1861	34, 015
Hubs, &c., Pattern for metal	J. Johnson	Geneseo, N. Y	July 6, 1852	9, 095
Hubs, Sand-cap for vehicle	G. H. Nevens	Livermore, Cal	May 2, 1871	114, 324
Hubs, Seasoning wagon	A. Bennighofen	Hamilton, Ohio	May 23, 1871	115, 017
Hubs, Securing box-metal in carriage	A. H. Ahlborn	Lawrenceville, Pa	Feb. 4, 1868	74, 029
Hubs, Securing boxes in	W. Greenleaf	Terre Haute, Ind	May 8, 1866	54, 529
Hubs, Securing boxes in metallic	J. B. Stuart	Bunker Hill, Ill	Feb. 5, 1867	61, 890
Hubs, &c., Securing boxes to wheel	J. Kritsch	Binghamton, N. Y	Sept. 9, 1862	36, 412
Hubs to axle-boxes, Attaching	T. C. Maris	Athens, Ohio	Mar. 5, 1867	62, 551
Hubs to axles, Attaching	P. Bargion	Stockton, Cal	Sept. 3, 1872	131, 077
Hubs to axles, Attaching	C. H. Denison	Brattleborough, Vt	Oct. 29, 1861	33, 577
Hubs to axles, Attaching	C. H. Guard	New York, N. Y	Sept. 25, 1866	58, 247
Hubs to axles, Attaching	J. Gunn	Salem Township, Ill	June 15, 1869	91, 228
Hubs to axles, Attaching	J. M. Perkins	New York, N. Y	May 15, 1855	12, 871
Hubs to axles, Attaching	J. M. Riley	Newark, N. J	Sept. 30, 1856	15, 818
Hubs to axles, Attaching	E. Sampson	Claremont, N. H	Feb. 21, 1854	10, 552
Hubs to axles, Attaching	E. S. Scripture	Green Point, N. Y	Dec. 18, 1855	13, 962
Hubs to axles, Attaching	H. B. Simonds	West Hartford, Vt	Mar. 4, 1856	14, 371
Hubs to axles, Attaching	J. Weathers	Greensburgh, Ind	May 12, 1868	77, 785
Hubs to axles, Attaching	J. M. White	Xenia, Ohio	Mar. 24, 1857	16, 891
Hubs to axles, Attaching	W. L. Williams	New York, N. Y	Jan. 2, 1872	122, 506
Hubs to axles, Attaching	L. Winslow	Rochester, N. Y	Apr. 14, 1857	17, 063
Hubs to axles, Attaching carriage	L. Adams	Amherst, Mass	Mar. 16, 1869	87, 815
Hubs to axles, Connecting	J. Kellogg	Madison, Ohio	Nov. 13, 1849	6, 870
Hubs to axles, Device for attaching	O. F. Shepard	Cincinnati, Ohio	July 1, 1873	140, 389
Hubs to axles of carriage-wheels, Attaching	J. London, jr	Auburn, N. Y	Oct. 12, 1839	1, 365
Hubs to axles, Securing	G. E. Clow	Jeffersonville, Ind	Sept. 20, 1870	107, 452
Hubs to axles, Securing	C. Darling	Utica, N. Y	Aug. 15, 1854	11, 521
Hubs to axles, Securing	L. J. Worden	Utica, N. Y	May 5, 1857	17, 247

Index of patents issued from the United States Patent Office from 1790 to 1873, inclusive—Continued.

Invention.	Inventor.	Residence.	Date.	No.
Hubs to axles, Securing the boxes of carriage-wheel.	L. Smith and G. R. Waring	Derby, Conn	June 17, 1840	1,640
Hubs to axles, Securing wheel	I. Osgood	Utica, N. Y	Nov. 6, 1866	59,445
Hubs to let in boxes, Cutting and boring the ends of	J. B. Francis	Loudoun County, Va	Mar. 2, 1832	....
Hubs to receive boxes, Tool for boring	U. Kimble	Penfield, N. Y	Mar. 27, 1855	12,598
Hubs to shafts, Securing wheel	E. Sanford	Hartford, Conn	Oct. 21, 1873	143,849
Hubs to the arms of axle-trees, Attaching carriage-wheel.	G. Hunt	Prattsville, N. Y	July 8, 1839	1,223
Hubs, Tool for fitting bands on	C. E. Stone and A. Herbert	Amesbury and Salisbury, Mass.	July 14, 1868	79,925
Hubs, tool-handles, &c., Machine for turning	S. Carpenter	Flushing, N. Y	Oct. 10, 1854	11,777
Hubs, &c., Turning	S. Beers	Naugatuck, Conn	June 6, 1854	11,016
Hubs, Turning carriage	A. Rickart	Schoharie, N. Y	July 21, 1857	17,846
Hubs, Turning wagon	W. W. Cleaveland	Cold Water, Mich	Sept. 29, 1868	82,600
Hubs with axles, Connecting	J. Foster	Bridgeport, Conn	July 24, 1849	6,608
Hubs with axles, Connecting	J. Foster	Bridgeport, Conn	Feb. 26, 1850	7,117
Hubs with axles, Connecting	E. Sampson and A. M. Billings	Claremont, N. H	Nov. 20, 1849	6,887
Hubs with axles, Connecting	E. S. Scripture	Green Point, N. Y	July 30, 1850	7,535
Hubs while being bored, Centering and holding	A. Moore	Honeoye Falls, N. Y	Mar. 31, 1857	16,932
Hubs while being bored, Machine for holding	P. Schuttler	Chicago, Ill	Feb. 21, 1865	46,501
Hull for steamboats	W. Hagerty	Monongahela, Pa	June 23, 1868	79,114
Hull of vessels	D. De Haven	New Orleans, La	July 19, 1870	105,437
Hull of vessels	L. P. Rider	Pittsburgh, Pa	Oct. 4, 1870	107,961
Hull of vessels	J. Van Pelt	Perry, Ill	Oct. 29, 1867	70,296
Huller: *See* Barley-huller. Buckwheat-huller. Clover-huller. Clover-seed huller. Corn-huller. Cotton-seed huller. Grain-huller. Grass-seed huller. Rice-huller. Strawberry-huller.				
Huller and screen	G. Stevenson and J. J. Crider	Zionsville and Greenfield, Ind.	Mar. 22, 1864	42,027
Hulling-machine	A. Angell	Newburgh, N. Y	Mar. 2, 1869	87,388
Hulling-machine	S. Bentz	Carroll County, Md	Feb. 11, 1862	34,346
Hulling-machine	A. Bertelson	West Salem, Wis	June 10, 1873	139,758
Hulling-machine	G. A. Buchholz	Shepherd's Bush, England	Nov. 23, 1869	97,039
Hulling-machine	G. A. Buchholz	Regent's Park, England	May 10, 1870	102,764
Hulling-machine	J. E. Carver	Bridgewater, Mass	July 26, 1870	105,643
Hulling-machine	J. M. Hendricks	Philadelphia, Pa	Sept. 2, 1862	36,350
Hulling-machine	S. R. Hockman	Urbana, Ohio	June 1, 1869	90,752
Hulling-machine	C. Jordan	East Bridgewater, Mass	Nov. 2, 1869	96,439
Hulling-machine	D. Kahnweiler	New York, N. Y	Mar. 29, 1870	101,271
Hulling-machine	C. Keniston and J. H. Sawyer	Somerville and Boston, Mass.	Oct. 5, 1869	95,485
Hulling-machine	J. McGregor	Saratoga, N. Y	May 12, 1840	1,600
Hulling-machine	C. Miller	Carroll Township, Pa	Dec. 25, 1855	13,992
Hulling-machine	D. Pease, jr	Floyd, N. Y	Apr. 3, 1849	6,271
Hulling-machine	D. Pease, jr	Floyd, N. Y	Apr. 10, 1849	6,289
Hulling-machine	W. Porter	Brooklyn, N. Y	Apr. 19, 1870	102,152
Hulling-machine	W. Seck	Frankfort, Prussia	May 7, 1872	126,416
Hulling machine, Barley, &c	J. J. Johnston and J. E. Weaver	Allegheny City and Temperanceville, Pa.	June 30, 1863	39,051
Hulling-machine, Cylindrical	S. Houpt	Philadelphia, Pa	Nov. 16, 1812	....
Hulling-machine feed-device	D. Kahnweiler	New York, N. Y	Dec. 19, 1871	122,028
Hulling machine, Grain	W. Zimmermann	Quincy, Ill	June 14, 1859	24,423
Hulling-mill	C. S. Bailey	New York, N. Y	Sept. 7, 1869	94,540
Hulling-mill, Centrifugal	C. S. Bailey	New York, N. Y	July 20, 1869	92,690
Hulling-mills, Dress of stones for	D. Collins	Jersey City, N. J	Mar. 9, 1858	19,605
Human power, Economizing	W. C. Moores	Bloomfield, Wis	May 10, 1864	42,678
Husk-hackling machines, Separating-attachment for.	G. B. Stacy	Richmond, Va	Oct. 18, 1870	108,532
Husk-splitter	B. Hollis	Randolph, Ill	Dec. 19, 1871	122,017
Husking-machine	E. Farnum and G. W. Scott	Blackstone, Mass	Feb. 5, 1867	61,823
Husking-machine	J. Russell	Brooklyn, N. Y	Nov. 26, 1867	71,539
Husking-palm	D. E. Shaw	Ross County, Ohio	Nov. 10, 1857	18,607
Husking-pin	F. Brown	Russell, Ohio	June 7, 1864	43,004
Husking-pin	W. L. Corson	Hennepin, Ill	Feb. 13, 1872	123,615
Husking-pin	W. Sherwin	Shelburne Falls, Mass	Sept. 24, 1861	33,369
Husking pin, Corn	E. Blair	Bucyrus, Ohio	Oct. 13, 1868	82,915
Husking-thimble	J. H. Gould	Smith, Ohio	May 13, 1856	14,864
Husking-thimble	H. J. Kintz	Greece, N. Y	Mar. 16, 1869	87,856
Hut, Portable	A. Derrom	Paterson, N. J	July 9, 1861	32,757
Hydrant	J. Allison	Cincinnati, Ohio	Nov. 10, 1868	83,899
Hydrant	G. C. Bailey	Pittsburgh, Pa	Dec. 21, 1869	98,011
Hydrant	G. C. Bailey	Pittsburgh, Pa	Mar. 19, 1872	124,777
Hydrant	H. J. Bailey	Pittsburgh, Pa	June 16, 1868	78,917
Hydrant	H. J. Bailey	Pittsburgh, Pa	Dec. 29, 1868	85,267
Hydrant	T. R. Bailey, jr	Lockport, N. Y	Mar. 10, 1868	75,344
Hydrant	W. Bailey	Troy, N. Y	July 4, 1865	48,504
Hydrant	W. Bailey	Troy, N. Y	Sept. 20, 1870	107,434
Hydrant	J. W. Baker	Parkersburgh, W. Va	Sept. 24, 1867	69,155
Hydrant	F. H. Bartholomew	New York, N. Y	Mar. 7, 1846	4,410
Hydrant	F. H. Bartholomew	New York	Aug. 12, 1846	4,692
Hydrant	F. H. Bartholomew	New York, N. Y	Feb. 15, 1859	22,927
Hydrant	F. Bauschtliker	Washington, D. C	Apr. 20, 1869	89,117
Hydrant	H. C. Biggs and A. C. Flinn	Lancaster, Pa	Apr. 17, 1866	53,944
Hydrant	W. W. Birny	Seneca Falls, N. Y	June 2, 1857	17,415
Hydrant	B. P. Bower	Cleveland, Ohio	May 22, 1866	54,845
Hydrant	G. N. Bowman	Pottsville, Pa	July 20, 1869	92,696
Hydrant	T. C. Bride	Quincy, Ill	Feb. 1, 1870	99,396
Hydrant	S. H. Brown	Troy, N. Y	May 13, 1862	35,213
Hydrant	T. Brown	Allegheny City, Pa	June 23, 1868	79,200
Hydrant	J. G. Bryan	Philadelphia, Pa	Apr. 20, 1869	89,123
Hydrant	J. Bryant	Brooklyn, N. Y	June 16, 1857	17,552
Hydrant	J. Bryant	Brooklyn, N. Y	Apr. 5, 1859	23,444
Hydrant	J. H. Buckley	New Haven, Conn	Mar. 19, 1867	62,933
Hydrant	S. G. Cabell	Quincy, Ill	Apr. 6, 1869	88,542

Index of patents issued from the United States Patent Office from 1790 *to* 1873, *inclusive*—Continued.

Invention.	Inventor.	Residence.	Date.	No.
Hydrant	S. G. Cabell and A. Q. Ross	Quincy, Ill., and Cincinnati, Ohio.	Mar. 1, 1870	100, 255
Hydrant	J. L. Chapman	Baltimore, Md	Oct. 12, 1842	2, 812
Hydrant	E. Clampitt	Baltimore, Md	Nov. 29, 1870	109, 587
Hydrant	S. P. Clark	Baltimore, Md	Sept. 3, 1861	33, 185
Hydrant	Z. E. Coffin	Boston, Mass	July 21, 1868	80, 143
Hydrant	R. H. Colvin	Columbia, Pa	July 20, 1843	3, 185
Hydrant	C. J. Cowperthwaite	Philadelphia, Pa	Jan. 15, 1856	14, 090
Hydrant	C. J. Cowperthwaite	Philadelphia, Pa	May 6, 1856	14, 805
Hydrant	D. C. Cregier	Chicago, Ill	Sept. 10, 1861	33, 239
Hydrant	T. A. Davies	New York	May 12, 1843	3, 082
Hydrant	R. De Charms	Philadelphia, Pa	Jan. 5, 1858	19, 022
Hydrant	W. P. Dickinson, D. S. Witman, and G. W. Rabold.	Reading, Pa	July 2, 1867	66, 225
Hydrant	W. H. Duffett	Rochester, N. Y	Aug. 15, 1871	118, 116
Hydrant	H. English	Baltimore, Md	Apr. 1, 1856	14, 557
Hydrant	J. Ewing	Philadelphia, Pa	Dec. 19, 1823	....
Hydrant	J. Farman	Cleveland, Ohio	Apr. 4, 1871	113, 507
Hydrant	W. Fields and S. Gerhard	Wilmington, Del	Feb. 3, 1857	16, 536
Hydrant	G. W. Fisher	Saint Louis, Mo	June 4, 1872	127, 587
Hydrant	L. Fitton	Wheeling, W. Va	Feb. 2, 1869	86, 384
Hydrant	A. S. Fort	Cincinnati, Ohio	Mar. 1, 1870	100, 389
Hydrant	S. P. Francisco and W. P. Dickinson.	Reading, Pa	Dec. 21, 1858	22, 357
Hydrant	C. E. Frazier	Baltimore, Md	Nov. 17, 1868	84, 104
Hydrant	J. Fricker	Cincinnati, Ohio	Dec. 12, 1871	121, 771
Hydrant	J. Fricker and A. Warden	Cincinnati, Ohio	June 7, 1870	104, 012
Hydrant	A. Fuller	Cincinnati, Ohio	Sept. 4, 1860	29, 872
Hydrant	B. G. Fuller	Baltimore, Md	Apr. 21, 1868	77, 027
Hydrant	J. P. Gallagher	Saint Louis, Mo	Jan. 5, 1864	41, 144
Hydrant	J. P. Gallagher	Saint Louis, Mo	May 2, 1871	114, 283
Hydrant	J. Gibson and M. Heberger	Cincinnati, Ohio	June 10, 1862	35, 515
Hydrant	T. Gibson	Philadelphia, Pa	Dec. 13, 1823	....
Hydrant	K. Goddard	Philadelphia, Pa	Feb. 9, 1858	19, 330
Hydrant	D. K. Gould	Chestertown, N. Y	Apr. 30, 1867	64, 307
Hydrant	N. B. Gousha	Baltimore, Md	Aug. 7, 1866	56, 927
Hydrant	W. H. Graham	Saint Louis, Mo	Dec. 9, 1873	145, 294
Hydrant	P. H. Griffin	Albany, N. Y	Dec. 12, 1871	121, 716
Hydrant	J. V. Hayes	Buffalo, N. Y	May 6, 1873	138, 562
Hydrant	N. Hayman	New York, N. Y	Sept. 8, 1863	39, 812
Hydrant	S. C. Higbie	Oppenheim, N. Y	Sept. 1, 1843	3, 240
Hydrant	J. R. Higgs	Utica, N. Y	Aug. 31, 1858	21, 338
Hydrant	A. Hoagland	Jersey City, N. J	Apr. 21, 1857	17, 093
Hydrant	B. Holly	Lockport, N. Y	Sept. 14, 1869	94, 749
Hydrant	D. Horne	Baltimore, Md	Mar. 31, 1836	....
Hydrant	J. Hyde	New York, N. Y	Jan. 5, 1858	19, 029
Hydrant	J. P. Hyde	New York, N. Y	Apr. 19, 1870	102, 126
Hydrant	W. James	Baltimore, Md	Nov. 1, 1859	25, 969
Hydrant	E. T. Jenkins	Ravenswood, N. Y	Jan. 19, 1869	86, 076
Hydrant	A. Johnson	Philadelphia, Pa	June 12, 1860	28, 674
Hydrant	W. Kearney	Union Township, N. J	Sept. 22, 1868	82, 326
Hydrant	J. P. Kenyon	Brooklyn, N. Y	Feb. 19, 1861	31, 464
Hydrant	J. R. Manny	Chicago, Ill	Jan. 25, 1870	99, 097
Hydrant	N. B. Marsh	Cincinnati, Ohio	Oct. 4, 1859	25, 660
Hydrant	J. W. Marshall	Williamsburgh, N. Y	Nov. 17, 1868	84, 202
Hydrant	J. McCann	Albany, N. Y	Mar. 21, 1871	112, 825
Hydrant	R. A. McCauley	Baltimore, Md	May 29, 1866	55, 136
Hydrant	S. McElroy	Brooklyn, N. Y	Oct. 14, 1862	36, 663
Hydrant	A. J. and T. McKenna	Pittsburgh, Pa	July 5, 1870	105, 106
Hydrant	J. McLelland	Washington, D. C	Mar. 3, 1863	37, 820
Hydrant	B. B. Moore	Detroit, Mich	Aug. 20, 1861	33, 095
Hydrant	J. G. Morgan	Brooklyn, N. Y	Jan. 20, 1857	16, 448
Hydrant	J. G. Murdock	Cincinnati, Ohio	May 26, 1863	38, 694
Hydrant	J. G. Murdock	Cincinnati, Ohio	Dec. 10, 1867	72, 071
Hydrant	J. W. Murphy	Baltimore, Md	Aug. 6, 1872	130, 233
Hydrant	J. Myers	Cincinnati, Ohio	May 3, 1870	102, 698
Hydrant	J. Neumann	Philadelphia, Pa	May 14, 1861	32, 338
Hydrant	W. P. Newhall	New York, N. Y	Dec. 4, 1866	60, 230
Hydrant	T. W. Newton and J. H. Laning	Philadelphia, Pa	July 30, 1833	....
Hydrant	J. Old	Pittsburgh, Pa	June 26, 1866	55, 893
Hydrant	J. and S. P. Parham	Philadelphia, Pa., and Trenton, N. J.	Mar. 2, 1858	19, 511
Hydrant	R. D. Patterson	Cincinnati, Ohio	Sept. 26, 1865	50, 155
Hydrant	J. Pandler, jr., and F. Bauschtliker.	Washington, D. C	June 5, 1866	55, 357
Hydrant	F. J. Paul	Cincinnati, Ohio	Apr. 1, 1873	137, 476
Hydrant	G. P. Perrine and J. E. Boyle	Richmond, Va	June 23, 1857	17, 632
Hydrant	J. L. Pillsbury	Columbus, Ohio	June 6, 1871	115, 637
Hydrant	A. Pixley and J. Robertson	Brooklyn, N. Y	June 19, 1866	55, 782
Hydrant	J. Powell	Cincinnati, Ohio	Mar. 2, 1858	19, 513
Hydrant	J. Pringle	Jersey City, N. J	Dec. 8, 1863	40, 852
Hydrant	D. J. Pruner	Bellefonte, Pa	Dec. 28, 1869	98, 411
Hydrant	W. Race and S. R. C. Mathews	Seneca Falls, N. Y	Jan. 26, 1858	19, 206
Hydrant	W. Race and S. R. C. Mathews	Seneca Falls, N. Y	Apr. 19, 1859	23, 706
Hydrant	W. Race and S. R. C. Mathews	Lockport, N. Y	Nov. 16, 1869	96, 959
Hydrant	T. Ragan	Philadelphia, Pa	Mar. 4, 1873	136, 545
Hydrant	J. Regester	Baltimore, Md	June 9, 1863	38, 841
Hydrant	J. Regester	Baltimore, Md	June 19, 1866	55, 711
Hydrant	J. Regester	Baltimore, Md	June 19, 1866	55, 712
Hydrant	J. Regester	Baltimore, Md	Oct. 22, 1867	70, 119
Hydrant	R. Reilly	Baltimore, Md	Nov. 10, 1868	83, 882
Hydrant	A. Richmond, jr	Dayton, Ohio	Nov. 22, 1870	109, 547
Hydrant	J. P. Riley	Philadelphia, Pa	Dec. 11, 1866	60, 423
Hydrant	G. W. Robertson	Philadelphia, Pa	Jan. 24, 1860	26, 931
Hydrant	G. H. Rodgers	Baltimore, Md	Sept. 10, 1861	33, 282
Hydrant	G. J. and H. W. Ross	New York, N. Y	June 3, 1862	35, 473
Hydrant	N. Schlesser	Chicago, Ill	Aug. 15, 1871	118, 059
Hydrant	G. M. Selden	Troy, N. Y	Aug. 13, 1861	33, 054

Index of patents issued from the United States Patent Office from 1790 *to* 1873, *inclusive*—Continued.

Invention.	Inventor.	Residence.	Date.	No.
Hydrant	J. Small	Washington, D. C	Nov. 28, 1871	121, 429
Hydrant	J. J. Smith and S. Wood	Cleveland, Ohio	Feb. 2, 1869	86, 466
Hydrant	J. N. Smith	Jersey City, N. J	Nov. 13, 1866	59, 670
Hydrant	J. N. Smith	Jersey City, N. J	Nov. 13, 1866	59, 671
Hydrant	J. N. Smith	Jersey City, N. J	Jan. 22, 1867	61, 366
Hydrant	J. N. Smith	Jersey City, N. J	July 9, 1872	128, 763
Hydrant	C. L. Stacy	Cincinnati, Ohio	Oct. 4, 1859	25, 683
Hydrant	C. L. Stacy	Cincinnati, Ohio	Sept. 10, 1872	131, 191
Hydrant	A. Stephenson	Saint Louis, Mo	Sept. 5, 1865	49, 801
Hydrant	R. Stileman	Philadelphia, Pa	Jan. 20, 1863	37, 466
Hydrant	E. Stocker	Lancaster, Pa	Feb. 20, 1866	52, 797
Hydrant	E. Stocker and A. Henpel	Lancaster, Pa	Oct. 1, 1872	131, 796
Hydrant	J. Stroop	Pittsburgh, Pa	Dec. 16, 1833	....
Hydrant	J. Stroop	Pittsburgh, Pa	Nov. 11, 1834	....
Hydrant	J. Swan	Brooklyn, N. Y	Oct. 19, 1858	21, 858
Hydrant	T. T. Tasker	Philadelphia, Pa	July 11, 1842	2, 714
Hydrant	S. Tice	Cincinnati, Ohio	Dec. 8, 1868	84, 848
Hydrant	T. Van Kannel	Cincinnati, Ohio	Dec. 14, 1869	97, 835
Hydrant	T. Van Kannel and G. Elvenhoefer.	Cincinnati, Ohio	Dec. 23, 1873	145, 768
Hydrant	S. T. Walker	Baltimore, Md	Nov. 26, 1835	....
Hydrant	S. T. Walker	Baltimore, Md	July 1, 1836	....
Hydrant	S. T. Walker	Baltimore, Md	July 2, 1836	....
Hydrant	J. Walsh	Philadelphia, Pa	Jan. 18, 1870	99, 037
Hydrant	J. Walsh	Philadelphia, Pa	May 16, 1871	114, 996
Hydrant	J. Walsh	Philadelphia Pa	June 18, 1872	128, 192
Hydrant	J. M. Ward	New York, N. Y	Aug. 21, 1866	57, 449
Hydrant	L. W. Werner	Saint Louis, Mo	Jan. 12, 1869	85, 879
Hydrant	M. Zwiebel	Pottsville, Pa	Aug. 10, 1869	93, 658
Hydrant and fire-plug	S. H. Davies	Cincinnati, Ohio	July 18, 1840	1, 702
Hydrant and fire-plug	J. I. Healey	Brooklyn, N. Y	June 17, 1873	139, 890
Hydrant and fire-plug, Combined	F. Bauschtliker	Washington, D. C	Nov. 1, 1870	108, 751
Hydrant and gas-pipe stop	F. Shickle	Saint Louis, Mo	Nov. 17, 1868	84, 222
Hydrant and street-washer	J. Farnan	Cleveland, Ohio	Sept. 24, 1872	131, 512
Hydrant, Anti-freezing	J. W. Slocum	Philadelphia, Pa	Apr. 6, 1869	88, 589
Hydrant-cap	N. W. Speers	Cincinnati, Ohio	June 6, 1854	11, 019
Hydrant-curb, Self-adjusting	J. A. Finnegan	Charlestown, Mass	Aug. 18, 1868	81, 266
Hydrant for aqueducts	D. Burt	Schenectady, N. Y	June 22, 1808	....
Hydrant for filtration	J. H. Carter	Cincinnati, Ohio	Oct. 18, 1859	25, 805
Hydrant-fountain	L. Gouley,	Philadelphia, Pa	Aug. 1, 1828	....
Hydrant, Non-freezing	J. E. Carter	Portland, Me	Nov. 3, 1868	83, 602
Hydrant, &c., Non-freezing	J. H. Hall	Peoria, Ill	Apr. 30, 1872	126, 202
Hydrant, Non-freezing	H. Merrie	Cincinnati, Ohio	June 13, 1871	115, 882
Hydrant, Non-freezing	J. H. Printz	Zanesville, Ohio	July 30, 1872	130, 069
Hydrant, Non-freezing	J. Walsh	Philadelphia, Pa	Aug. 6, 1872	130, 166
Hydrant stop-cock rod	H. Rausch	Brooklyn, N. Y	Oct. 12, 1869	95. 724
Hydrant waste-device	J. Culver	Baltimore, Md	Apr. 22, 1856	14, 712
Hydrants and fire-plugs, Thawing-device for	J. C. Moore	Philadelphia, Pa	July 26, 1870	105, 830
Hydrants, Incasing	W. Bramwell	New York, N. Y	June 9, 1857	17, 538
Hydrants, Stop-plate for	J. C. Eastman	Nashua, N. H	Aug. 2, 1870	105, 925
Hydrants, Waste-attachment to	E. J. Baker	Baltimore, Md	Apr. 8, 1856	14, 592
Hydraulic and pneumatic elevator for railways	H. H. Day	Bloomingdale, N. J	Sept. 5, 1871	118, 696
Hydraulic and steam engine, Cylinder and piston of.	J. Underwood	Lowell, Mass	Dec. 9, 1856	16, 202
Hydraulic apparatus	A. Desgoffe and A. Ollivier	Paris, France	Feb. 7, 1865	46, 315
Hydraulic apparatus	L. Eames	Kalamazoo, Mich	Sept. 9, 1862	36, 397
Hydraulic apparatus	G. H. Herring	Durand, Ill	Feb. 7, 1871	111, 536
Hydraulic apparatus	N. Nolan	New York, N. Y	Nov. 8, 1870	109, 042
Hydraulic apparatus	P. F. Powers	Genoa, Nev	May 2, 1871	114, 339
Hydraulic apparatus	J. F. Thompson	Greensborough, Pa	Dec. 8, 1868	84, 719
Hydraulic apparatus for producing blast	R. Cook	Saratoga Springs, N. Y	June 11, 1850	7, 423
Hydraulic apparatus for transmitting power	D. Hinman	Brunswick, Ohio	Oct. 31, 1848	5, 899
Hydraulic boat and mode of propelling the same	W. Shultz	Philadelphia, Pa	May 28, 1812	
Hydraulic cement, Manufacture of artificial	W. B. Tucker	Baltimore, Md	May 13, 1873	138, 770
Hydraulic disk	T. Shaw	Philadelphia, Pa	May 30, 1871	115, 365
Hydraulic elevator	C. W. Baldwin	Boston, Mass	May 28, 1872	127, 139
Hydraulic elevator	D. Bickford	Boston, Mass	Nov. 28, 1865	51, 129
Hydraulic elevator	T. Chambers	Saint Louis, Mo	Dec. 17, 1867	72, 165
Hydraulic elevator	D. Corey	New York, N. Y	Aug. 31, 1827	....
Hydraulic elevator	H. L. Ensign	Chicago, Ill	Jan. 30, 1872	123, 161
Hydraulic elevator	H. Flad	Saint Louis, Mo	Sept. 14, 1869	94, 732
Hydraulic elevator	H. Flad and G. P. Herthel, jr	Saint Louis, Mo	May 28, 1867	65, 200
Hydraulic elevator	T. Stebins	San Francisco, Cal	Oct. 8, 1872	132, 111
Hydraulic elevator	T. Stebins	San Francisco, Cal	June 3, 1873	139, 624
Hydraulic engine	W. E. B. Ainsworth and G. W. Scott.	Blackstone, Mass	Apr. 5, 1870	101, 445
Hydraulic engine	J. F. Burgin and A. Koch	Williamsport, Pa	Feb. 14, 1860	27, 104
Hydraulic engine	A. C. Carey and J. Smith	Ipswich, Mass	Mar. 21, 1854	10, 665
Hydraulic engine	R. F. Dobson	Belmont, Wis	Jan. 7, 1873	134, 651
Hydraulic engine	J. Dreisörner	New York, N. Y	Aug. 21, 1866	57, 298
Hydraulic engine	M. Everett	Kalamazoo, Mich	Sept. 17, 1867	68, 863
Hydraulic engine	J. S. Groynne	New York, N. Y	Jan. 26, 1858	19, 224
Hydraulic engine	J. D. Heaton	Dixon, Ill	Sept. 29, 1857	18, 2[illegible]0
Hydraulic engine	J. C. Hodges	Morristown, Tenn	Jan. 14, 1873	13[illegible], 806
Hydraulic engine	K. Horde	Philadelphia, Pa	Aug. 13, 1861	33, 039
Hydraulic engine	H. Hubbell	New Haven, Conn	Aug. 28, 1866	57, 572
Hydraulic engine	H. J. King and B. L. Beebe	Middletown, N. Y	Dec. 14, 1869	97, 779
Hydraulic engine	A. D. Laws	Bridgeport, Conn	May 10, 1870	102, 948
Hydraulic engine	T. G. McLaughlin	Philadelphia, Pa	Aug. 7, 1847	5, 214
Hydraulic engine	A. Miller	Grafton, Va	May 20, 1856	14, 922
Hydraulic engine	F. Ransom	Buffalo, N. Y	Apr. 28, 1868	77, 214
Hydraulic engine	L. M. Sabin	Saint Louis, Mo	Oct. 8, 1867	69, 591
Hydraulic engine	W. Snyder	Bullskin Township, Pa	Jan. 3, 1871	11[illegible], 796
Hydraulic engine	J. Strong	....	Mar. 24, 1801	....
Hydraulic engine	H. H. Stuart	Jamaica, N. Y	July 3, 1866	56, 120
Hydraulic engine	W. J. Tate	Philadelphia, Pa	Oct. 31, 1871	120, 548
Hydraulic engine	E. Thomas and W. K. Hall	Beverly and Philippi, Va	Dec. 4, 1860	30, 844
Hydraulic engine	J. H. and R. W. Welch	Georgetown, D. C	Dec. 17, 1872	134, 115

Index of patents issued from the United States Patent Office from 1790 to 1873, inclusive—Continued.

Invention.	Inventor.	Residence.	Date.	No.
Hydraulic engine and meter	J. S. Barden	Providence, R. I	Oct. 24, 1865	50, 651
Hydraulic engine, Atmospheric and condensing	A. J. Reynolds	Chicago, Ill	Dec. 7, 1869	97, 698
Hydraulic engine, Revolving	A. Hubbard	Windsor, Vt	Apr. 22, 1828	....
Hydraulic heater	W. H. Churchman	Philadelphia, Pa	May 2, 1854	10, 855
Hydraulic heater	L. W. Leeds and R. M. Smith	Philadelphia, Pa	May 16, 1854	10, 942
Hydraulic indicator	R. D. Bradley	Preston, Md	Nov. 8, 1864	44, 931
Hydraulic jack	R. Blackwood	Philadelphia, Pa	June 25, 1861	32, 612
Hydraulic jack	R. Dudgeon	New York, N. Y	Aug. 1, 1865	49, 097
Hydraulic jack	R. Dudgeon	New York, N. Y	Apr. 15, 1873	137, 765
Hydraulic jack	G. Lindsay	New York, N. Y	Mar. 10, 1857	16, 801
Hydraulic jack	A. V. Ojeda	San Francisco, Cal	July 29, 1873	141, 232
Hydraulic jack	J. Ryan	Saint Louis, Mo	Jan. 23, 1866	52, 210
Hydraulic jack	J. Ryan	Saint Louis, Mo	Jan. 30, 1866	52, 324
Hydraulic jack	E. J. Shaw and G. F. Eisenhardt.	Philadelphia, Pa	Jan. 9, 1872	122, 539
Hydraulic jack	T. H. Watson	New York, N. Y	Sept. 20, 1864	44, 358
Hydraulic lifting-jack	J. Tangye	Birmingham, England	Apr. 14, 1863	38, 187
Hydraulic machine	L. Brunier	New York, N. Y	July 8, 1842	2, 711
Hydraulic machine	C. Darmand	Georgetown, D. C	Aug. 7, 1817	....
Hydraulic machine	T. Hutchins	Reading, Pa	May 7, 1834	....
Hydraulic machine	J. James	Lynganore, Md	Nov. 18, 1818	....
Hydraulic machine, Balance-lever	D. Fisk	Springfield, N. Y	July 18, 1820	....
Hydraulic machine for raising water	B. Langdon	....	Feb. 20, 1801	....
Hydraulic machine for raising water	W. Shultz	Philadelphia, Pa	May 26, 1812	....
Hydraulic machinery	R. McCormick	Rockbridge County, Va	Oct. 1, 1830	....
Hydraulic meter	J. S. Barden	New Haven, Conn	Feb. 26, 1856	14, 335
Hydraulic motor	A. Barbarin and J. Albrecht	New Orleans, La	Aug. 1, 1871	117, 504
Hydraulic motor	A. Barbarin and J. Albrecht	New Orleans, La	Aug. 1, 1871	117, 505
Hydraulic motor	B. A. Bloch	Paris, France	Mar. 25, 1873	137, 171
Hydraulic motor	U. A. Boyden	Boston, Mass	Sept. 20, 1853	10, 027
Hydraulic motor	J. Harris	Boston, Mass	July 19, 1870	105, 450
Hydraulic motor	J. H. Jennings	New Bedford, Mass	Nov. 14, 1871	120, 975
Hydraulic motor	M. Keely and G. W. Cressman	Barren Hill, Pa	Oct. 11, 1859	25, 741
Hydraulic motor	W. Keunish	London, England	May 15, 1860	28, 282
Hydraulic motor	V. Kromer	Grand Rapids, Wis	June 13, 1871	115, 869
Hydraulic motor	C. Mesler	Almond, N. Y	Aug. 30, 1864	44, 010
Hydraulic motor	M. Millard	Franklin, Ohio	Mar. 19, 1872	124, 843
Hydraulic motor	M. A. Shepard	Orio, Ill	July 19, 1859	24, 826
Hydraulic motor	W. Walter	Arkada, Wash	Aug. 26, 1873	142, 183
Hydraulic motor	T. Welham	Brownsville, Nebr	May 31, 1864	42, 980
Hydraulic motor, Organ-lever coupling for	H. F. Wheeler	Boston, Mass	May 6, 1873	138, 546
Hydraulic power	Z. and A. Parker	Coshocton County, Ohio	Oct. 19, 1829	....
Hydraulic-power elevator	P. J. Borger	Cincinnati, Ohio	Aug. 13, 1872	130, 468
Hydraulic press	T. Baxter	Petersburgh, Va	Apr. 19, 1859	23, 741
Hydraulic press	J. T. Burr	Brooklyn, N. Y	Aug. 13, 1867	67, 719
Hydraulic press	M. H. Clark	Danville, Va	Sept. 20, 1859	25, 488
Hydraulic press	J. A. Desgoffe	Paris, France	Dec. 20, 1870	110, 346
Hydraulic press	R. Dudgeon	New York, N. Y	Jan. 25, 1859	22, 713
Hydraulic press	A. H. Emery	New York, N. Y	Apr. 16, 1867	63, 875
Hydraulic press	W. Ettenger and H. P. Edmund.	Richmond, Va	Apr. 14, 1868	76, 732
Hydraulic press	C. W. Flippen	Laurel Grove, Va	May 8, 1860	28, 167
Hydraulic press	R. Grant	New York, N. Y	Aug. 1, 1854	11, 455
Hydraulic press	T. Harbottle	Brooklyn, N. Y	May 4, 1869	89, 656
Hydraulic press	M. Hittinger	Charlestown, Mass	Aug. 23, 1864	43, 912
Hydraulic press	C. W. Holbrook	New York, N. Y	Mar. 3, 1868	75, 015
Hydraulic press	A. Judson and S. A. Farrington	Newark, N. J	June 29, 1869	92, 057
Hydraulic press	G. W. Rawson	Cambridgeport, Mass	Aug. 11, 1868	80, 877
Hydraulic press	C. E. Rymes	Charlestown, Mass	May 19, 1863	38, 606
Hydraulic press	J. B. Tunstall	Boydton, Va	Oct. 27, 1868	83, 424
Hydraulic press, Copper-lined cylinder for	C. Sellers	Philadelphia, Pa	May 28, 1872	127, 191
Hydraulic press for cotton, &c	C. Wilson	Williamsburgh, N. Y	Oct. 9, 1849	6, 776
Hydraulic press for peat, brick, &c	A. H. Emery	New York, N. Y	Aug. 21, 1866	57, 303
Hydraulic press, Portable	R. Dudgeon	New York, N. Y	July 8, 1851	8, 203
Hydraulic press, Pressure-accumulator for	W. D. Grimshaw	Ansonia, Conn	Jan. 30, 1872	123, 169
Hydraulic press, Retainer for	M. Madden	Saint Louis, Mo	July 15, 1862	35, 881
Hydraulic press, Retainer for	C. E. Rymes	Charlestown, Mass	Dec. 27, 1859	26, 622
Hydraulic press to prevent corrosion	T. P. Shaffner	Louisville, Ky	Dec. 18, 1866	60, 570
Hydraulic-pressure regulator	I. Judson	New Haven, Conn	June 25, 1867	66, 028
Hydraulic-pressure regulator	J. S. McDonald	New Orleans, La	June 25, 1872	128, 235
Hydraulic ram	B. S. Benson	Harford County, Md	Dec. 26, 1845	4, 328
Hydraulic ram	J. Cerneau and S. S. Hallet	New York, N. Y	Aug. 3, 1809	....
Hydraulic ram	W. Fields, jr	Providence, R. I	Jan. 28, 1851	7, 918
Hydraulic ram	J. L. Gatchel	Chester, Pa	Apr. 10, 1847	5, 052
Hydraulic ram	D. A. Leighton	Middleburgh, N. Y	Aug. 22, 1848	5, 725
Hydraulic ram	J. C. Strode	East Bradford, Pa	Mar. 27, 1847	5, 037
Hydraulic ram	J. C. Strode	East Bradford, Pa	Oct. 4, 1853	10, 079
Hydraulic ram	J. C. Strode	West Chester, Pa	May 23, 1854	10, 969
Hydraulic ram	J. F. Warner	Philadelphia, Pa	June 8, 1858	20, 523
Hydraulic ram	E. Webb	Parkersville, Pa	Dec. 5, 1854	12, 042
Hydraulic ram	J. D. West	New York, N. Y	Aug. 29, 1854	11, 626
Hydraulic ram, Tidal or current	J. W. Middleton	Philadelphia, Pa	Sept. 19, 1854	11, 697
Hydraulic ram, Apparatus for directing water to and maintaining continuous pressure upon.	T. Harbottle and M. F. Fowler	New York, N. Y	Oct. 29, 1861	33, 585
Hydraulic rams, Operating	C. Polly	May's Landing, N. J	Mar. 7, 1854	10, 610
Hydraulic rams, Operating the waste-gate in	J. Osborn	Hamden, Conn	Feb. 11, 1851	7, 932
Hydraulic regulator for machinery	L. B. Pitcher	Syracuse, N. Y	Apr. 9, 1850	7, 272
Hydraulic seal	J. H. Sutton	Honesdale, Pa	July 12, 1870	105, 277
Hydraulic steam-engine, Planetary	J. Black	Philadelphia, Pa	Sept. 20, 1853	10, 024
Hydraulic structures, Obtaining permanent foundations for.	I. W. P. Lewis	Boston, Mass	May 1, 1845	4, 030
Hydraulic wheel for raising water	P. D. Henry	New Orleans, La	Jan. 9, 1841	1, 930
Hydraulic-wheel press	G. A. Gray, jr	Hamilton, Ohio	Apr. 29, 1873	138, 393
Hydraulics	J. Hatheway	....	Oct. 29, 1794	....
Hydraulics	J. Roup	Kanawha County, Va	Oct. 6, 1827	....
Hydraulics	T. Somerby	Nantucket, Mass	Oct. 1, 1830	....
Hydro-atmospheric elevator	C. W. Baldwin	Boston, Mass	Jan. 17, 1871	111, 030
Hydrocarbon and other liquids, Package for	M. W. Brown	New York, N. Y	Apr. 4, 1871	113, 394

Index of patents issued from the United States Patent Office from 1790 to 1873, inclusive—**Continued.**

Invention.	Inventor.	Residence.	Date.	No.
Hydrocarbon and other liquids, Vessel and tank for holding.	J. G. Moody	New York, N. Y	Feb. 5, 1867	61,753
Hydrocarbon-apparatus, Self-ventilating safety-can for filling and discharging.	L. D. Towsley	Newark, N. J	Dec. 21, 1869	98,208
Hydrocarbon as fuel, Apparatus for burning	W. H. Russell	Brooklyn, N. Y	July 9, 1872	128,914
Hydrocarbon-heater	O. F. Morrill	Chelsea, Mass	Dec. 4, 1866	60,224
Hydrocarbon-liquid can	I. W. Shaler	Brooklyn, N. Y	Oct. 3, 1871	119,657
Hydrocarbon liquids, Apparatus for heating	P. B. Kitchen	Philadelphia, Pa	Nov. 8, 1859	26,070
Hydrocarbon liquids, Apparatus for making light from.	E. H. Covel	New York, N. Y	June 15, 1869	91,423
Hydrocarbon liquids, Apparatus for vaporizing	W. H. Burr	Philadelphia, Pa	Nov. 5, 1872	132,800
Hydrocarbon liquids as fuel, Burning	A. J. Works	Fair Haven, Conn	Jan. 8, 1867	61,131
Hydrocarbon liquids, Burning	C. Fisher	Trenton, N. J	July 23, 1867	67,108
Hydrocarbon liquids for illumination, Apparatus for vaporizing.	J. Griffin	Meriden, Conn	Apr. 26, 1864	42,469
Hydrocarbon liquids for light, Burning	T. J. Barrow	Brooklyn, N. Y	Feb. 28, 1871	112,206
Hydrocarbon liquids for transmitting heat, Use of	W. C. Baker	New York, N. Y	Oct. 16, 1866	58,755
Hydrocarbon liquids in the presence of steam, Process of vaporizing and decomposing.	F. Cook and J. A. Bassett	New York, N. Y., and Salem, Mass.	Sept. 10, 1867	68,708
Hydrocarbon liquids, Process of vaporizing	J. A. Bassett	Salem, Mass	Sept. 10, 1867	68,686
Hydrocarbon liquids, &c., Storing	P. C. P. L. Prefontaine	Paris, France	July 17, 1866	56,508
Hydrocarbon oils as fuel, Burning	F. Cook	New York, N. Y	Sept. 10, 1867	68,707
Hydrocarbon oils, Generating and burning vapor of.	H. K. Foote	Oil City, Pa	Sept. 18, 1866	58,087
Hydrocarbon on locomotives, Burning	A. C. Rand	Chicago, Ill	Oct. 10, 1871	119,720
Hydrocarbon-vapor apparatus	G. H. Bronson	Cincinnati, Ohio	Dec. 13, 1859	26,458
Hydrocarbon-vapor apparatus	E. H. Covel	New York, N. Y	May 31, 1859	24,200
Hydrocarbon-vapor apparatus	A. and A. Davis and C. Cunningham.	Lowell, Mass., and Nashua, N. H.	Jan. 15, 1856	14,129
Hydrocarbon-vapor apparatus	L. L. Hill	Greenport, N. Y	Dec. 20, 1859	26,497
Hydrocarbon-vapor apparatus	A. A. Moss	Philadelphia, Pa	Aug. 9, 1859	25,032
Hydrocarbon-vapor apparatus	F. S. Pease	Buffalo, N. Y	Mar. 13, 1860	27,470
Hydrocarbon vapor for illumination	W. H. Laubach	Philadelphia, Pa	Feb. 14, 1860	27,190
Hydrocarbon-vapor machine	J. F. Spence	Brooklyn, N. Y	Sept. 17, 1867	69,037
Hydrocarbon vapors, Apparatus for burning	F. C. Ambler	New York, N. Y	June 4, 1872	127,402
Hydrocarbon vapors, Apparatus for generating	W. R. Roberts	Philadelphia, Pa	May 9, 1871	114,603
Hydrocarbon vapors, Combustion-chamber for burning.	C. J. Eames	New York, N. Y	Mar. 25, 1873	137,132
Hydrocarbon vapors, Device for generating and burning.	C. Carpenter, jr	Seekonk, Mass	Sept. 23, 1873	143,061
Hydrocarbons, Apparatus for burning	M. Carbonel	Havana, Cuba	June 17, 1873	140,009
Hydrocarbons, Apparatus for burning	C. H. Cushing	Tidioute, Pa	Nov. 18, 1873	144,660
Hydrocarbons, Apparatus for burning	I. Kendrick	Philadelphia, Pa	Jan. 30, 1872	123,109
Hydrocarbons, Apparatus for burning	I. Kendrick	Philadelphia, Pa	Nov. 5, 1872	132,839
Hydrocarbons, Apparatus for burning	A. C. Rand	Aurora, Ill	Oct. 22, 1872	132,491
Hydrocarbons, Apparatus for burning	E. F. Rogers	Chelsea, Mass	Jan. 21, 1873	135,011
Hydrocarbons, Apparatus for burning	A. Schpakofsky and N. Stange.	St. Petersburg, Russia	Apr. 3, 1866	53,763
Hydrocarbons, Apparatus for burning	W. Sim and A. Barff	Glasgow, North Britain	Aug. 8, 1865	49,357
Hydrocarbons, Apparatus for burning	F. T. Suhr	Titusville, Pa	Apr. 21, 1868	77,124
Hydrocarbons, Apparatus for burning liquid	A. J. Works	Fair Haven, Conn	Aug. 21, 1866	57,429
Hydrocarbons, Apparatus for generating and burning vapor from.	J. Kidd	New York, N. Y	Jan. 10, 1871	110,857
Hydrocarbons, Apparatus for separating the products of distillation of.	L. N. Wilcox	Pittsburgh, Pa	July 25, 1865	49,020
Hydrocarbons, Apparatus for vaporizing	J. K. Caldwell	Philadelphia, Pa	Sept. 12, 1871	118,903
Hydrocarbons, Apparatus for vaporizing and aërating volatile.	L. Stevens	Fitchburgh, Mass	Dec. 20, 1864	45,568
Hydrocarbons, Apparatus for vaporizing and burning.	S. J. Whiting	Philadelphia, Pa	Mar. 26, 1872	124,993
Hydrocarbons, Apparatus for vaporizing and burning.	S. J. Whiting	Philadelphia, Pa	June 11, 1872	127,723
Hydrocarbons, Apparatus for vaporizing and burning liquid.	F. Cook	New York, N. Y	Sept. 10, 1867	68,702
Hydrocarbons, Apparatus for vaporizing and burning liquid.	O. F. Morrill	Chelsea, Mass	May 27, 1862	35,383
Hydrocarbons, Apparatus for vaporizing volatile	J. Butler	New York, N. Y	Apr. 25, 1871	113,978
Hydrocarbons as fuel, Using liquid	F. Cook	New York, N. Y	Sept. 10, 1867	68,705
Hydrocarbons, Burning	J. Chandler and D. A. Wray	Pioneer, Pa	Oct. 22, 1872	132,440
Hydrocarbons, Burning	I. Kendrick	Philadelphia, Pa	Nov. 8, 1870	109,131
Hydrocarbons for heating, &c., Vaporizing	H. H. Eames and C. J. Eames	Philadelphia, Pa., and New York, N. Y.	Oct. 15, 1872	132,266
Hydrocarbons, Producing light by the combination of solid and liquid.	J. Phillips	Cologne, Germany	Jan. 18, 1870	98,883
Hydrocarbons, Production of light from heavy	R. S. Merrill	Cambridge, Mass	Mar. 15, 1870	100,915
Hydrocarbons, Separating	R. A. Chesebrough	New York, N. Y	June 5, 1866	55,240
Hydrocarbons, Treating solid and liquid	F. Lambe	London, England	June 13, 1871	115,871
Hydrocarbons, Vaporizing and burning liquid	F. Cook	New York, N. Y	Sept. 10, 1866	68,703
Hydrocarbons with oxygen, Apparatus for supplying.	A. H. Webster	Hudson, N. Y	Feb. 8, 1859	22,907
Hydrodynamic engine	L. C. St. John	Buffalo, N. Y	May 2, 1854	10,869
Hydrodynamic machine for testing strength of materials.	F. C. Lowthorp	Trenton, N. J	May 26, 1857	17,383
Hydro-mechanical press	J. Beverly		Dec. 26, 1803	
Hydro-oxygen light, Apparatus for the production of Hare's.	G. H. Smith	Rochester, N. Y	Sept. 27, 1859	25,611
Hydropneumatic apparatus	J. Gregg	Rochester, N. Y	Jan. 24, 1842	2,433
Hydropneumatic engine	R. Crosbie	Newark, N. J	Jan. 6, 1812	
Hydropneumatic engine	J. A. Huss	Saint Louis, Mo	Aug. 15, 1865	49,410
Hydropneumatic engine for extinguishing fires, Platuary.	W. Loughbridge	Weverton, Md	Apr. 10, 1855	12,686
Hydropneumatic machine for exhausting and sealing vessels.	W. H. Elliot	Plattsburgh, N. Y	July 24, 1855	13,303
Hydropneumatic motor	C. L. Stevens	Galesburgh, Ill	Oct. 31, 1871	120,545
Hydrofuze fabric	J. Wansbrough	Southwark, England	Jan. 11, 1859	22,596
Hydrogen and air, Apparatus for mixing	D. Ashworth	Wappinger's Falls, N. Y	Apr. 26, 1870	102,203
Hydrogen generator and carbureter	B. Sloper	Saint Louis, Mo	July 20, 1869	92,892
Hydrogen-lighter, Döbereiner	J. Taylor and R. W. Brown	Westerly, R. I	Nov. 18, 1862	36,971
Hydrolator	Z. Swope	Lancaster, Pa	Feb. 26, 1850	7,128
Hydrometer	J. Adams	New York, N. Y	Jan. 15, 1861	31,100

Index of patents issued from the United States Patent Office from 1790 to 1873, inclusive—Continued.

Invention.	Inventor.	Residence.	Date.	No.
Hydrometer	P. Hogg	Brooklyn, N. Y	May 6, 1862	35, 152
Hydrometer	H. Petrie	Chicago, Ill	June 13, 1865	48, 203
Hydrometer	J. Saxton	Washington, D. C	Sept. 27, 1864	44, 460
Hydrometer	E. Southworth	New York, N. Y	Feb. 28, 1821	....
Hydrometer	G. Tagliabue	New York, N. Y	Oct. 16, 1866	58, 913
Hydrometer	J. C. Tucker and E. Dwelle	Boston, Mass	Dec. 23, 1822	....
Hydrometer	T. Welham	Brownsville, Nebr	May 31, 1864	42, 979
Hydrometric apparatus	L. Brauer	Memphis, Tenn	Sept. 26, 1865	50, 092
Hydrostatic and aërostatic machinery	E. C. Genet	Albany, N. Y	Oct. 31, 1825	....
Hydrostatic and hydraulic elevator	J. McCreary	Nobletown, Pa	Jan. 5, 1828	....
Hydrostatic and steam engine	S. H. Long and G. F. A. Hauto	Germantown, Pa	Sept. 3, 1813	....
Hydrostatic engine	T. Hansbron and B. B. Redding	Sacramento, Cal	Mar. 14, 1865	46, 795
Hydrostatic paradox to govern fluid-pressure, Application of.	C. D. Colden	New York, N. Y	June 2, 1815	....
Hydrostatic paradox to move machinery, Application of.	C. D. Colden	New York, N. Y	May 19, 1815	....
Hydrostatic press	T. Baxter	Petersburgh, Va	Apr. 13, 1836	....
Hydrostatic press	J. T. Burr	Brooklyn, N. Y	July 18, 1871	117, 043
Hydrostatic press	C. Graham	Kingston, Pa	Aug. 6, 1867	67, 531
Hydrostatic press	D. H. Mason and M. W. Baldwin.	Philadelphia, Pa	Dec. 2, 1829	....
Hydrostatic press	E. Merrill	New Bedford, Mass	Mar. 28, 1838	662
Hydrostatic press for cotton, &c	J. Houpt	Forkland, Ala	Aug. 21, 1841	2, 221
Hygienic basin and steam-bath	E. O. Murden	Hightstown, N. J	Jan. 21, 1873	135, 004
Hygieometer	B. M. Lawrence	Galesburgh, Ill	Mar. 23, 1869	88, 047
Hygrometer	R. Baeklen and W. Staehlen	Brooklyn, N. Y	Aug. 8, 1865	49, 221
Hygrometer	A. H. and C. R. Black	Indianapolis, Ind	Mar. 26, 1861	31, 773
Hygrometer	W. Edson	Boston, Mass	July 4, 1865	48, 620
Hygrometer	A. E. Lazell	West Meriden, Conn	Mar. 10, 1868	75, 432
Hygrometer	L. S. Ullman	Nashville, Tenn	Sept. 13, 1859	25, 457
Hygrometers, Device for actuating the index of	C. L. Clark	Rochester, N. Y	May 25, 1858	20, 326
Hygrometric regulator for hot-water apparatus	J. H. Ross	New York, N. Y	Dec. 12, 1854	12, 071
I.				
Ice, Accumulating	J. Dutton	Aston, Pa	May 8, 1843	3, 080
Ice and coal box	R. Hagen	Saint Louis, Mo	Aug. 21, 1866	57, 317
Ice and cooling air, liquids, &c., Manufacture of	C. Tellier	Paris, France	Mar. 8, 1870	100, 689
Ice and cooling, Apparatus for making	F. Windhausen	Brunswick, Germany	Jan. 24, 1871	111, 292
Ice and for other purposes, Manufacture of liquids for.	P. H. Vander Weyde	Philadelphia, Pa	Dec. 17, 1867	72, 431
Ice and in cooling air, &c., Manufacture of	D. E. Somes	Washington, D. C	Sept. 17, 1867	68, 908
Ice and refrigerating, Apparatus for making	A. B. Tripler	New Orleans, La	Mar. 14, 1871	112, 654
Ice and refrigerating machine, Manufacture of	S. B. Martin and J. M. Beath	San Francisco, Cal	May 28, 1872	127, 180
Ice and snow fender for roofs of buildings	L. Howe	Worcester, Mass	May 31, 1864	42, 992
Ice and the refrigerating of air and liquids, Manufacture of.	C. Tellier	Passy, near Paris, France	Jan. 5, 1869	85, 719
Ice, Apparatus for and manner of cutting	N. J. Wyeth	Cambridge, Mass	Mar. 18, 1829	....
Ice, Apparatus for loading	P. F. Whitney	Saugerties, N. Y	Mar. 2, 1869	87, 452
Ice, Apparatus for making	J. H. Bunnell	Massillon, Ohio	Nov. 10, 1863	40, 551
Ice, Apparatus for making	C. P. Leavitt	New York, N. Y	July 18, 1871	117, 087
Ice, Apparatus for making	A. C. Twining	New Haven, Conn	Apr. 15, 1862	34, 993
Ice, Apparatus for making	E. C. Weld	New York, N. Y	Aug. 22, 1871	118, 411
Ice, Apparatus for making	G. Willard	New York, N. Y	Apr. 1, 1873	137, 523
Ice, Apparatus for purifying water for the manufacture of.	S. F. Peterson	New Orleans, La	Mar. 22, 1870	101, 035
Ice, Apparatus for the manufacture of	S. Bennett	Jefferson Parish, La	Nov. 1, 1870	108, 868
Ice, Apparatus for the manufacture of	D. L. Holden	New Orleans, La	Apr. 12, 1870	101, 876
Ice, Apparatus for the manufacture of	T. S. C. H. Lowe	New York, N. Y	Apr. 2, 1867	63, 404
Ice, Apparatus for the manufacture of	A. Mühl	San Antonio, Tex	Nov. 28, 1871	121, 402
Ice, Apparatus for the manufacture of	J. E. Sears	Waco, Tex	Aug. 29, 1871	118, 649
Ice-bag	S. Stroock	New York, N. Y	July 11, 1871	117, 013
Ice, Bag for preserving	W. B. Coates	Philadelphia, Pa	Apr. 30, 1867	64, 283
Ice-boat	T. B. Kelley	Dundee, Ill	Sept. 8, 1868	81, 909
Ice-boat	J. B. Stoddard	Baltimore, Md	Apr. 2, 1867	63, 577
Ice-boat, Arrangement of means attached to	D. Large	Philadelphia, Pa	June 24, 1856	15, 187
Ice-boat for breaking ice	W. Hunt and J. Townsend	New York, N. Y	Oct. 3, 1838	958
Ice, Boat for traveling on	T. Chapin	Canandaigua, N. Y	Sept. 14, 1838	919
Ice, Boat for traveling on	J. R. Halsey	Newark, N. J	Oct. 16, 1866	58, 915
Ice-boat, Velocipede	F. G. Johnson	Brooklyn, N. Y	Feb. 11, 1862	34, 369
Ice-box or cooler	J. W. Campbell	New York, N. Y	Feb. 26, 1867	62, 392
Ice-breaker	C. W. Chapman	Hartford, Conn	July 3, 1860	28, 966
Ice-breaker	C. W. Dunlap	Brooklyn, N. Y	Sept. 18, 1866	58, 078
Ice-breaker	A. Flannigan	Trappe, Ind	Apr. 28, 1868	77, 366
Ice-breaker	M. Freytag	Philadelphia, Pa	July 2, 1836	....
Ice breaker and fender	F. Ford	Point Pleasant, W. Va	Sept. 23, 1873	143, 003
Ice-breaker and propeller in boats, Paddle to be used as an.	W. Van Dusen	Kensington, Pa	July 29, 1837	314
Ice-breaker for boats, &c	S. Nicolson	Boston, Mass	July 16, 1844	3, 668
Ice-breaker for navigating frozen waters	B. S. Gillespie	New York, N. Y	Mar. 8, 1837	138
Ice-breaking boat	H. and W. Brown	Philadelphia, Pa	Aug. 5, 1856	15, 472
Ice-breaking boat	D. Compton	Newport, N. J	Sept. 28, 1869	95, 326
Ice-breaking boat	J. D. Foster, H. C. Foster, and J. Q. Miller.	Cincinnati and Springfield, Ohio.	Dec. 8, 1857	18, 832
Ice-breaking boat	Z. Oram	Camden, N. J	Apr. 14, 1857	17, 052
Ice-breaking instrument	I. H. Griffing	New York, N. Y	Aug. 5, 1856	15, 483
Ice-breaking machine	J. I. Williams	Georgetown, D. C	Mar. 2, 1813	....
Ice-calk	G. W. Farley	Manchester, N. H	May 26, 1868	78, 270
Ice-calk	D. Krauser	Pottsville, Pa	Feb. 24, 1863	37, 788
Ice, Car and receiving-platform for receiving blocks of.	N. J. Wyeth	Cambridge, Mass	Oct. 28, 1843	3, 323
Ice-carriage	H. C. Moore	Springfield, Mass	Mar. 24, 1868	75, 953
Ice-carrier	L. Townsend	Terre Haute, Ind	Apr. 23, 1872	126, 108
Ice-chair	F. Ashley	New York, N. Y	Feb. 5, 1861	31, 290
Ice-chest for soda-apparatus and refrigerator	E. Bigelow	Springfield, Mass	Apr. 20, 1869	89, 010
Ice-chute, Automatic	J. A. Wolfer	Rondout, N. Y	Dec. 22, 1868	85, 156
Ice-clog	J. J. Starr	Cincinnati, Ohio	Oct. 28, 1862	36, 866
Ice, Cooling air and liquids and making	D. E. Somes	Washington, D. C	May 5, 1868	77, 669

Index of patents issued from the United States Patent Office from 1790 to 1873, inclusive—Continued.

Invention.	Inventor.	Residence.	Date.	No.
Ice, Cooling air in the manufacture of	O. Lugo and J. B. McPherson	Baltimore, Md	Mar. 14, 1871	112, 726
Ice, Cooling and producing	C. E. Haynes	Boston, Mass	June 21, 1870	104, 588
Ice, cooling buildings, &c., Apparatus for the manufacture of.	A. H. Tait	Jersey City, N. J	Nov. 11, 1873	144, 577
Ice-cream freezer	A. H. Austin	Baltimore, Md	Sept. 19, 1848	5, 775
Ice-cream freezer	A. S. Ballard	Mount Pleasant, Iowa	Dec. 29, 1868	85, 420
Ice-cream freezer	R. P. Beil	Peoria, Ill	Oct. 28, 1873	144, 053
Ice-cream freezer	G. W. Brown	New York, N. Y	May 29, 1860	28, 449
Ice-cream freezer	G. W. Brown	New York, N. Y	May 17, 1864	42, 744
Ice-cream freezer	H. B. Campbell	Racine, Wis	June 3, 1873	139, 453
Ice-cream freezer	A. Cavileer, W. McCuddy, and P. N. Woliston.	Springfield, Ohio	Jan. 1, 1867	60, 689
Ice-cream freezer	J. R. Champlin	Laconia, N. H	June 26, 1866	55, 820
Ice-cream freezer	J. R. Champlin	Laconia, N. H	Mar. 16, 1869	87, 909
Ice-cream freezer	G. Coffeen, jr	Warren County, Ohio	Nov. 13, 1849	6, 865
Ice-cream freezer	J. W. Condon	Logansport, Ind	July 9, 1872	128, 856
Ice-cream freezer	M. G. Crane	Chelsea, Mass	Nov. 6, 1866	59, 366
Ice-cream freezer	G. W. Davis	New Orleans, La	May 22, 1860	28, 354
Ice-cream freezer	J. Decker	Bel Air, Md	Aug. 21, 1849	6, 661
Ice-cream freezer	J. Dooling	Boston, Mass	Oct. 20, 1868	83, 265
Ice-cream freezer	J. Dooling	Boston, Mass	July 16, 1872	129, 326
Ice-cream freezer	J. Dooling	Boston, Mass	July 29, 1873	141, 209
Ice-cream freezer	J. W. Dougherty and F. W. Gerecke.	Newburgh, N. Y	July 16, 1867	66, 685
Ice-cream freezer	F. H. Duc	Charleston, S. C	Sept. 17, 1867	66, 968
Ice-cream freezer	A. W. Edwards	Mendota, Ill	Nov. 28, 1865	51, 160
Ice-cream freezer	S. S. Fitch	New York, N. Y	Mar. 4, 1873	136, 496
Ice-cream freezer	W. A. Garloch and W. D. Richards.	Belpre, Ohio	Sept. 29, 1868	82, 509
Ice-cream freezer	C. Gooch	Cincinnati, Ohio	June 11, 1867	65, 559
Ice-cream freezer	C. Gooch	Cincinnati, Ohio	Jan. 21, 1868	73, 596
Ice-cream freezer	C. Gooch	Cincinnati, Ohio	July 4, 1871	116, 584
Ice-cream freezer	M. F. Graves	Sunbury, Pa	Oct. 21, 1873	143, 822
Ice-cream freezer	J. Gray	Milwaukee, Wis	Sept. 10, 1867	68, 734
Ice-cream freezer	G. Guinot	Milton, Fla	Feb. 4, 1873	135, 474
Ice-cream freezer	E. Halloway	Belvidere, Ill	Sept. 24, 1872	131, 611
Ice-cream freezer	S. H. Hamilton and C. A. Ashton	Jacksonville, Ill	Nov. 11, 1862	36, 904
Ice-cream freezer	W. Hawkins	Oregon, Mo	Aug. 31, 1869	94, 313
Ice-cream freezer	B. H. Hilliar	New London, Conn	May 17, 1870	103, 182
Ice-cream freezer	L. A. Lipp	Coatesville, Pa	Feb. 26, 1867	62, 429
Ice-cream freezer	A. Lucetti	New York, N. Y	July 22, 1873	141, 060
Ice-cream freezer	B. G. Martin	Williamsburgh, N. Y	Mar. 1, 1870	100, 305
Ice-cream freezer	H. B. Masser	Sunbury, Pa	Dec. 12, 1848	5, 960
Ice-cream freezer	H. B. Masser	Sunbury, Pa	Jan. 19, 1858	19, 147
Ice-cream freezer	H. B. Masser	Sunbury, Pa	Apr. 23, 1861	32, 138
Ice-cream freezer	H. B. Masser	Sunbury, Pa	Mar. 19, 1867	63, 068
Ice-cream freezer	T. and G. M. Mills	Philadelphia, Pa	Mar. 15, 1870	100, 918
Ice-cream freezer	T. and G. M. Mills	Philadelphia, Pa	June 25, 1872	128, 414
Ice-cream freezer	C. W. Packer	Philadelphia, Pa	May 22, 1860	28, 402
Ice-cream freezer	C. W. Packer	Philadelphia, Pa	July 23, 1867	67, 133
Ice-cream freezer	O. Paddock	Watertown, N. Y	Dec. 3, 1861	33, 850
Ice-cream freezer	O. Paddock	Watertown, N. Y	Oct. 21, 1862	36, 725
Ice-cream freezer	O. Paddock	Watertown, N. Y	Oct. 15, 1867	69, 833
Ice-cream freezer	J. Parisette	Indianapolis, Ind	June 10, 1856	15, 083
Ice-cream freezer	A. L. Platt	Bloomington, Ill	Nov. 19, 1872	133, 251
Ice-cream freezer	A. L. Platt and J. W. Wilmeth	Bloomington, Ill	Aug. 15, 1871	118, 149
Ice-cream freezer	T. M. Powell	Baltimore, Md	Sept. 5, 1854	11, 651
Ice-cream freezer	J. H. Rae	Syracuse, N. Y	July 2, 1867	66, 389
Ice-cream freezer	J. E. Robinson	Boston, Mass	Jan. 22, 1867	61, 360
Ice-cream freezer	M. Rosenstein	Boston, Mass	May 30, 1871	115, 361
Ice-cream freezer	J. F. Rote	Reading, Pa	Apr. 4, 1871	113, 450
Ice-cream freezer	T. Sands	Laconia, N. H	Feb. 27, 1872	124, 015
Ice cream freezer	P. Schumacher	New York, N. Y	Nov. 2, 1869	96, 489
Ice-cream freezer	J. S. Shattuck	Medford, Mass	Sept. 5, 1865	49, 797
Ice-cream freezer	F. G. Siemers	Winona, Minn	Jan. 21, 1868	73, 658
Ice-cream freezer	J. S. Silva	Savannah, Ga	Oct. 2, 1860	30, 256
Ice-cream freezer	C. E. Simmons	Monroe, Wis	Sept. 16, 1873	142, 951
Ice-cream freezer	J. Sissons	Horncastle, England	May 25, 1869	90, 594
Ice-cream freezer	W. H. Skerrett	Cincinnati, Ohio	Aug. 6, 1867	67, 599
Ice-cream freezer	E. B. Tilden	Prairie City, Iowa	Aug. 12, 1873	141, 738
Ice-cream freezer	J. Tingley	Philadelphia, Pa	Feb. 11, 1868	74, 259
Ice-cream freezer	J. Tingley	Philadelphia, Pa	June 6, 1871	115, 657
Ice-cream freezer	E. P. Torrey	Jersey City, N. J	Jan. 17, 1860	26, 867
Ice-cream freezer	T. Weaver	Harrisburgh, Pa	Mar. 26, 1872	124, 992
Ice-cream freezer	D. Wiggins	Greenport, N. Y	Apr. 20, 1869	89, 259
Ice-cream freezer	W. G. Young	Baltimore, Md	May 30, 1848	5, 601
Ice-cream freezer, Bottom for	E. M. Manigle	Philadelphia, Pa	Dec. 12, 1865	51, 468
Ice-cream-freezer scraper	F. G. Siemers	Winona, Minn	Mar. 3, 1868	75, 209
Ice-cream holder and filler	H. R. Robbins	Baltimore, Md	Apr. 2, 1872	125, 333
Ice-cream, jelly, &c., Mold for	E. M. May	Fond du Lac, Wis	Aug. 19, 1873	141, 882
Ice-cream machine	J. R. Champlin	Laconia, N. H	Nov. 1, 1864	44, 851
Ice-cream, Machine for manufacturing	C. Gooch	Cincinnati, Ohio	May 14, 1872	126, 691
Ice-cream mold	P. Field	Boston, Mass	Feb. 2, 1869	86, 519
Ice-cream pail	G. A. Nash	Niles, Mich	Oct. 4, 1870	107, 949
Ice-cream receptacle	E. A. G. Roulstone	Boston, Mass	Mar. 9, 1869	87, 710
Ice-cream refrigerator	C. Vignal	New York, N. Y	July 19, 1870	105, 521
Ice-cream server	J. Oyarzabal	Malaga, Spain	Oct. 19, 1869	95, 929
Ice-cream, water-ice, &c., Manner of packing and conveying.	I. Allegretti	Philadelphia, Pa	Apr. 4, 1871	113, 239
Ice-creeper	G. L. Bailey	Portland, Me	Apr. 8, 1862	34, 869
Ice-creeper	W. B. Coates	Philadelphia, Pa	Jan. 11, 1870	98, 668
Ice-creeper	E. B. Colby	Franklin, N. H	Feb. 11, 1873	135, 629
Ice-creeper	D. Cumming	Philadelphia, Pa	Feb. 9, 1869	86, 894
Ice-creeper	E. F. Dieterichs	Philadelphia, Pa	June 25, 1872	128, 370
Ice-creeper	R. H. Earle	Saint John's, Newfoundland.	Dec. 9, 1873	145, 340
Ice-creeper	E. S. Ellis	Trenton, N. J	July 4, 1871	116, 575
Ice-creeper	W. Field	Providence, R. I	Jan. 27, 1863	37, 558

Index of patents issued from the United States Patent Office from 1790 *to* 1873, *inclusive*—Continued.

Invention.	Inventor.	Residence.	Date.	No.
Ice-creeper	E. M. Gardner	Nantucket, Mass	Jan. 23, 1866	52, 157
Ice-creeper	D. Greene	Troy, N. Y	Oct. 10, 1865	50, 353
Ice-creeper	A. S. Hadaway	Plymouth, Mass	June 12, 1866	55, 584
Ice-creeper	A. Heulings	Philadelphia, Pa	June 3, 1873	139, 506
Ice-creeper	C. Hoëller	Cincinnati, Ohio	Sept. 14, 1869	94, 888
Ice-creeper	L. G. Hoffman	Albany, N. Y	July 7, 1868	79, 760
Ice-creeper	I. S. and J. W. Hyatt, jr	Chicago, Ill	Aug. 18, 1863	39, 575
Ice-creeper	C. H. Johnson and W. Axford	White Hall, Ill	Dec. 16, 1873	145, 656
Ice-creeper	W. P. Patton	Harrisburgh, Pa	Dec. 24, 1867	72, 534
Ice-creeper	H. L. Schell	Philadelphia, Pa	Jan. 24, 1860	26, 961
Ice-creeper	T. Symonds	Portland, Me	Oct. 17, 1865	50, 512
Ice-creeper	W. A. Towers	Philadelphia, Pa	Mar. 25, 1856	14, 527
Ice-creeper	E. M. Turner	Pittsburgh, Pa	Aug. 8, 1871	117, 945
Ice-creeper	M. Warne	Philadelphia, Pa	May 16, 1871	114, 998
Ice-creeper	W. C. Wells	Philadelphia, Pa	June 2, 1868	78, 560
Ice-creeper	H. H. White	Leominster, Mass	Oct. 1, 1872	131, 836
Ice-creeper	A. Wilke	Brunswick, Germany	Dec. 7, 1869	97, 738
Ice-crusher	W. W. Armington	Lowell, Mass	Sept. 26, 1865	50, 198
Ice-crusher	M. J. Dixon and W. Tunstill	New York, and Brooklyn, N. Y.	June 17, 1873	140, 019
Ice-crusher	F. F. Fowler	Upper Sandusky, Ohio	Mar. 5, 1867	62, 539
Ice-crusher	J. W. Hollingsworth and H. D. Weaver.	Mount Vernon, Ind	Mar. 24, 1868	75, 913
Ice-crusher	F. Hudner and J. R. Little	New York, N. Y	Feb. 4, 1873	135, 556
Ice-crusher	O. B. Kinsey	Newark, N. J	Feb. 7, 1871	111, 652
Ice-crusher	J. Middleton	New York, N. Y	Jan. 1, 1861	31, 027
Ice-crusher	L. H. Moseley	Poughkeepsie, N. Y	Jan. 26, 1864	41, 388
Ice-crusher	E. F. Pryor	Dayton, Ohio	June 4, 1867	65, 504
Ice-crusher	N. Richardson	Gloucester, Mass	June 9, 1868	78, 830
Ice-crusher	J. L. Rowe	New York, N. Y	Mar. 19, 1861	31, 766
Ice-crushing machine	A. C. Hobbs and J. Brown	New York, N. Y	Sept. 4, 1849	6, 690
Ice-crushing table	S. H. and D. W. Davis	Detroit, Mich	Jan. 18, 1870	98, 852
Ice-cutter	E. Bacher	Findley, Ohio	Jan. 30, 1872	123, 076
Ice-cutter	G. L. Cummings	New York, N. Y	Dec. 22, 1868	85, 217
Ice-cutter	C. W. Flint	Washington, D. C	Dec. 22, 1868	85, 080
Ice-cutter	V. H. Hallock	Queens, N. Y	Apr. 5, 1870	101, 613
Ice-cutter	G. R. Marvin	Keokuk, Iowa	Feb. 11, 1868	74, 236
Ice-cutter	F. G. Siemers	Winona, Minn	Aug. 18, 1868	81, 303
Ice-cutter	L. Townsend	Terre Haute, Ind	Oct. 8, 1872	132, 035
Ice-cutter	J. B. Wilson	Malden, Mass	Nov. 6, 1847	5, 362
Ice cutter and marker	N. J. Wyeth	Cambridge, Mass	Oct. 25, 1843	3, 317
Ice, Cutting and removing	W. Blount and J. Cammett	Osterville, Mass	Aug. 26, 1846	4, 713
Ice-cutting apparatus	C. and G. Beatty	Portsmouth, Va	July 21, 1868	80, 119
Ice-cutting apparatus	J. Fielemeyer	Philadelphia, Pa	Feb. 5, 1861	31, 307
Ice-cutting attachment to vessels	T. Estlack	Philadelphia, Pa	May 5, 1857	17, 209
Ice-cutting machine	J. Baker	Philadelphia, Pa	Feb. 11, 1868	74, 278
Ice-cutting machine	L. Barnum	Bridgeport, Conn	Mar. 7, 1871	112, 408
Ice-cutting machine	A. Deutschel	Cleveland, Ohio	July 12, 1870	105, 183
Ice-cutting machine	W. F. Gordon	Detroit, Mich	Nov. 19, 1867	70, 994
Ice-cutting machine	S. Trask	Hallowell, Me	Apr. 21, 1836	....
Ice-cutting machine	N. J. Wyeth	Cambridge, Mass	Dec. 1, 1840	1, 877
Ice, Device for cutting and shaving	S. E. Blake	Worcester, Mass	Feb. 14, 1865	46, 331
Ice, Device to prevent slipping on	W. R. Landfear	Hartford, Conn	Nov. 8, 1864	44, 961
Ice-elevating device	H. Little	Middletown, N. Y	Aug. 27, 1867	68, 218
Ice-elevating machine	N. J. Wyeth	Cambridge, Mass	Nov. 6, 1843	3, 327
Ice-elevating track	A. Pfund	New York, N. Y	Oct. 28, 1873	144, 133
Ice-elevator	E. E. Conklin	New York, N. Y	June 24, 1873	140, 187
Ice-elevator	W. Eighmie	Beekman, N. Y	Nov. 5, 1872	132, 758
Ice-elevator	H. Little	Middletown, N. Y	May 21, 1867	64, 886
Ice-elevator	J. J. Neuman	Middletown, Ohio	Aug. 8, 1871	117, 804
Ice-elevator	W. F. B. Read	Chicago, Ill	Aug. 11, 1868	81, 004
Ice-elevator	L. Townsend	Terre Haute, Ind	Apr. 23, 1872	126, 106
Ice-elevator	T. C. Wolking	Covington, Ky	May 21, 1872	127, 131
Ice-elevator	N. J. Wyeth	Cambridge, Mass	Dec. 10, 1840	1, 886
Ice-elevator	R. S. Zilar	Cincinnati, Ohio	Nov. 3, 1868	83, 752
Ice-float	L. Townsend	Terre Haute, Ind	Apr. 23, 1872	126, 107
Ice for shipping and storing, Machine for preparing	N. J. Wyeth	Cambridge, Mass	Dec. 1, 1837	497
Ice for storage, Cutting	V. H. Hallock	Milton, N. Y	Oct. 4, 1864	44, 529
Ice for transportation, Machinery and car for elevating and depositing.	N. J. Wyeth	Cambridge, Mass	Dec. 10, 1841	2, 381
Ice from ponds, Machine for raising	W. W. Cowling	Richmond, Va	June 5, 1847	5, 145
Ice from rivers, Machinery for raising	C. R. Wortendyke	New York, N. Y	Nov. 13, 1855	13, 809
Ice from the water and depositing on sleds, Machinery for raising blocks of.	N. J. Wyeth	Cambridge, Mass	Dec. 1, 1840	1, 878
Ice, Heel-spur to prevent slipping on	H. Pollard	Boston, Mass	Jan. 26, 1858	19, 205
Ice-house	A. Baierle	Chicago, Ill	Dec. 22, 1868	85, 161
Ice-house	C. Liebmann	Brooklyn, E. D., N. Y	Nov. 7, 1871	120, 754
Ice-house	C. Liebmann	Brooklyn, N. Y	Apr. 16, 1872	125, 820
Ice-house	W. D. Parker	New York, N. Y	June 19, 1855	13, 098
Ice-house	W. Velte and J. Fagan	Pittsburgh, Pa	Mar. 29, 1870	101, 405
Ice house	A. Wilbur	Cedar City, Mo	Nov. 18, 1873	144, 720
Ice-house, Filling for walls of	C. Stoll	Brooklyn, N. Y	Dec. 3, 1872	133, 680
Ice-house floor	J. Barbian	Chicago, Ill	Jan. 7, 1868	72, 960
Ice-house for brewers and butchers	A. Baierle, F. Hartmann, and F. Reese.	Chicago, Ill	June 2, 1868	78, 507
Ice-house for packing and curing meat	H. A. Roberts	Boston, Mass	Aug. 22, 1871	118, 277
Ice-house, Upper floor of	P. Kephart	Baltimore, Md	Sept. 26, 1848	5, 798
Ice, Housing and shipping	B. Wise	Cincinnati, Ohio	Aug. 5, 1862	36, 121
Ice in blocks, Sleds for transportation of	N. J. Wyeth	Cambridge, Mass	Dec. 1, 1840	1, 876
Ice in rivers and harbors, Method of obstructing	P. Voorhis	New York, N. Y	Apr. 16, 1867	63, 968
Ice in rivers, Machine for planing away	R. W. Heywood	Baltimore, Md	Jan. 26, 1858	19, 195
Ice-increaser	L. Townsend	Terre Haute, Ind	Apr. 23, 1872	126, 109
Ice into blocks for storing, Machine for cutting	G. B. Gruman	Ridgefield, Conn	Oct. 1, 1867	69, 338
Ice leveling and smoothing machine	W. Wharton, jr	Philadelphia, Pa	Apr. 11, 1865	47, 237
Ice-lowering trap	C. Egnor	Catskill, N. Y	Aug. 30, 1870	106, 798
Ice-machine	D. Boyle	San Francisco, Cal	June 25, 1872	128, 488
Ice-machine	W. Bray	Lambertsville, N. J	Dec. 23, 1873	145, 838
Ice-machine	D. L. Holden	New Orleans, La	Sept. 28, 1869	95, 347

Index of patents issued from the United States Patent Office from 1790 to 1873, inclusive—Continued.

Invention.	Inventor.	Residence.	Date.	No.
Ice-machine	W. R. Johnston and W. Whitelaw.	Sedalia, Mo., and Memphis, Tenn.	Apr. 30, 1872	126, 305
Ice-machine	A. Mühl	Waco, Tex	Dec. 12, 1871	121, 888
Ice-machine	H. Norman and C. F. Dietrich	New Orleans, La	Feb. 4, 1873	135, 576
Ice-machine	C. Plagge	New York, N. Y	Aug. 23, 1870	106, 722
Ice-machine	R. Reece	Llandilo, Wales	Oct. 1, 1872	131, 783
Ice-machine	A. Rock	Boston, Mass	Feb. 28, 1871	112, 284
Ice-machine	C. A. Seely	New York, N. Y	Oct. 10, 1871	119, 795
Ice-machine	D. K. Tuttle and O. Lugo	Baltimore, Md	Apr. 5, 1870	101, 682
Ice-machine	D. K. Tuttle and O. Lugo	Baltimore, Md	Jan. 24, 1871	111, 280
Ice-machine	F. Windhausen	Brunswick, Germany	Mar. 22, 1870	101, 198
Ice-machine	F. Windhausen	Brunswick, Germany	Jan. 24, 1871	111, 293
Ice-machine, Carré	A. Vaass and F. Littmann	Halle, Prussia	May 2, 1871	114, 495
Ice-machine using ammonia	A. C. Twining	New Haven, Conn	Dec. 19, 1871	121, 975
Ice, Machinery for raising and depositing	N. J. Wyeth	Cambridge, Mass	Dec. 10, 1841	2, 380
Ice, Machinery for the manufacture of	W. T. Duval	Georgetown, D. C	Aug. 2, 1870	105, 924
Ice-making and air and liquid cooling apparatus	D. E. Somes	Washington, D. C	Jan. 28, 1868	73, 936
Ice, Manufacture of	A. Albertson	Jersey City, N. J	May 30, 1871	115, 409
Ice, Manufacture of	T. G. Bigger	Kansas City, Mo	Mar. 7, 1871	112, 409
Ice, Manufacture of	J. Dutton	Aston, Pa	Aug. 18, 1846	4, 697
Ice, Manufacture of	R. S. Egbert	Colfax, Cal	July 21, 1868	80, 063
Ice, Manufacture of	J. F. Gesner	West Farms, N. Y	Oct 4, 1870	107, 898
Ice, Manufacture of	C. M. Keller and J. Henderson	New York and Brooklyn, N. Y.	Sept. 12, 1865	49, 887
Ice, Manufacture of	S. J. Newsham, W. H. Haines, and W. S. Henson.	Mont Clair and Newark, N. J.	Nov. 1, 1870	108, 816
Ice, Manufacture of	M. Rosenstein	Boston, Mass	Oct. 4, 1870	107, 962
Ice, Manufacture of	P. H. Vander Weyde	New York, N. Y	Nov. 1, 1870	108, 851
Ice, Manufacturing	A. C. Twining	Hudson, Ohio	Nov. 8, 1853	10, 221
Ice-manufacturing machine	P. H. Vander Weyde	Philadelphia, Pa	Feb. 16, 1869	87, 084
Ice, Manufacturing or forming	T. B. Smith	Saint Louis, Mo	Jan. 23, 1841	1, 941
Ice, Mode of manufacturing	T. S. C. Lowe	New York, N. Y	Apr. 2, 1867	63, 413
Ice, Mode of obtaining and securing	J. E. Manuel	Philadelphia, Pa	July 12, 1839	1, 240
Ice, Packing and stowing	F. Tredor	Boston, Mass	May 4, 1838	726
Ice-pick	S. S. Boynton and D. Keefe	Peoria, Ill	Oct. 28, 1873	143, 957
Ice-pick	J. H. Bridgins	Astoria, N. Y	July 9, 1867	66, 557
Ice-pick	E. Brown	Green Point, N. Y	Apr. 5, 1870	101, 424
Ice-pick	E. G. Burnham	Bridgeport, Conn	May 22, 1866	54, 852
Ice-pick	W. T. Eames	New York, N. Y	Aug. 8, 1871	117, 754
Ice-pick	S. G. Hoyt	New York, N. Y	Jan. 28, 1868	73, 809
Ice-pick	J. L. Rowe	New York, N. Y	Dec. 21, 1858	22, 403
Ice-pick	T. A. Taylor	Bangor, Me	Aug. 15, 1871	118, 070
Ice-pick	J. T. Van Kirk	Philadelphia, Pa	July 17, 1860	29, 220
Ice-pick	M., H., H. T., and J. White	Philadelphia, Pa	May 10, 1859	23, 969
Ice pick and hook	E. H. Schmultz and J. Baker	New York, N. Y	Jan. 24, 1871	111, 260
Ice-pick and meat-maul combined	T. A. Conklin	New Britain, Conn	June 18, 1872	128, 020
Ice-pick, nut-cracker, &c	G. L. Witsil	Philadelphia, Pa	May 8, 1866	54, 653
Ice-planer	S. Lewis	Williamsburgh, N. Y	July 6, 1869	92, 325
Ice-planer	W. F. Pough	Esopus, N. Y	July 13, 1869	92, 542
Ice planer or cutter	J. Serrill	Philadelphia, Pa	Feb. 12, 1867	62, 075
Ice-planing machine	S. Lewis	Brooklyn, N. Y	Mar. 3, 1868	75, 029
Ice planing or shaving machine	H. D. J. Pratt	Washington, D. C	May 17, 1859	24, 052
Ice-preserver	J. Dunning	Bangor, Me	Sept. 19, 1871	119, 016
Ice-preserver	J. W. D. Patten	New York, N. Y	Oct. 6, 1868	82, 869
Ice-preserver	J. E. Pilkington	Baltimore, Md	Sept. 13, 1870	107, 405
Ice-preserver and water-cooler	R. Heneage	Buffalo, N. Y	May 25, 1869	90, 360
Ice-preserver, Portable	E. Powers	Hartford, Conn	Aug. 28, 1821	
Ice-preserving jar	W. L. Faxon	Quincy, Mass	Jan. 16, 1872	122, 821
Ice, Process for the artificial production of	J. Gorrie	New Orleans, La	May 6, 1851	8, 080
Ice-run	G. Burhans	Rondout, N. Y.	July 11, 1871	116, 802
Ice-sandal	T. J. Linton	Providence, R. I	Mar. 7, 1865	46, 682
Ice-shaver	R. Gilliland	Hudson, Mich	Feb. 2, 1869	86, 527
Ice-shaver, Rotary	C. C. Clawson	Raleigh, N. C	Sept. 5, 1871	118, 689
Ice-shaving machine	W. H. Collins	Boston, Mass	Sept. 26, 1871	119, 324
Ice-shaving machine	J. D. Freeman	Abbeville, Ala	Aug. 5, 1873	141, 553
Ice shoe or calk	D. Bennet	Stratford, Conn	Apr. 1, 1862	34, 809
Ice-spur	C. Monnin	Buffalo, N. Y	Feb. 16, 1858	19, 376
Ice-spur	M. V. B. White	Ballston, N. Y	Jan. 1, 1867	60, 811
Ice-spur	C. F. Wieland	Darmstadt, Ill	Nov. 24, 1868	84, 400
Ice-stand	H. A. Roberts	Hartford, Conn	Aug. 24, 1858	21, 275
Ice therefrom to platforms, Railway-car for discharging blocks of.	N. J. Wyeth	Cambridge, Mass	Dec. 10, 1841	2, 382
Ice to a uniform thickness, Machine for reducing blocks of.	N. J. Wyeth	Cambridge, Mass	Dec. 1, 1840	1, 879
Ice, Vessel for the formation of	B. T. Babbitt	New York, N. Y	Jan. 12, 1869	85, 781
Ice-water receptacle	W. H. Hart	Medfield, Mass	Feb. 19, 1867	62, 133
Iceland moss and carrageen, Manufacture from	W. J. Rand	Brooklyn, N. Y	Aug. 10, 1869	93, 478
Ichthyocolla, Manufacture of	E. Rowe	Rockport, Mass	Dec. 19, 1848	5, 978
Ichthyocolla or fish-glue, Extr cting	D. Waldron	New York, N. Y	Mar. 4, 1812	
Identifying-box	L. T. McNeiley	Danville, Mo	June 18, 1867	65, 928
Illuminating-apparatus	G. Cuppers	Brooklyn, E. D., N. Y	Nov. 28, 1871	121, 234
Illuminating-apparatus	C. Geisse	Taycheedah, Wis	July 10, 1866	56, 203
Illuminating-apparatus	J. H. Irwin	Chicago, Ill	May 8, 1866	54, 553
Illuminating-apparatus	J. A. Thompson	Brooklyn, N. Y	Dec. 30, 1873	146, 109
Illuminating-compound	J. Scott	Washington, D. C	Feb. 5, 1867	61, 768
Illuminating-plate, Flexible sheet-metal	T. Hyatt	New York, N. Y	Dec. 2, 1873	145, 185
Illuminating-signals, Igniting	J. J. Detwiller	Greenville, N. J	Dec. 18, 1866	60, 491
Illumination	G. H. Smith	Rochester, N. Y	Oct. 14, 1862	36, 689
Illusory decapitation	J. C. Withington	Brookline, Mass	Apr. 23, 1867	64, 179
Illustrating waves, Apparatus for	C. S. Lyman	New Haven, Conn	Nov. 19, 1867	71, 190
Image, Sectional	B. Day	West Hoboken, N. J	Dec. 20, 1870	110, 213
Immersion and steam bath combined	J. W. Caldwell	Cincinnati, Ohio	Dec. 27, 1870	110, 545
Implement	B. F. Bean	Schuylkill, Pa	June 23, 1868	79, 100
Implement	E. S. Bennett	New York, N. Y	Mar. 16, 1869	87, 754
Implement	J. W. Calep	Boston, Mass	Dec. 20, 1870	110, 199
Implement	T. C. Comstock	Harrodsburgh, Ky	Mar. 3, 1868	74, 992
Implement	E. R. Mecker	Elizabeth, N. J	Feb. 18, 1868	74, 568
Implement	W. K. Rairigh	Rural Valley, Pa	Nov. 10, 1868	83, 997

Index of patents issued from the United States Patent Office from 1790 *to* 1873, *inclusive*—Continued.

Invention.	Inventor.	Residence.	Date.	No.
Implement	W. A. Sharp	Tama City, Iowa	Apr. 12, 1870	101, 773
Implement	A. Thayer	Albany, N. Y	June 16, 1868	79, 027
Implement	C. Wetzler	Chicago, Ill	Jan. 19, 1869	86, 048
Implement	H. B. Whitehead	Holly Springs, Miss	June 10, 1873	139, 748
Implement	E. F. Wilder	Lowell, Mass	May 26, 1868	78, 248
Implement	W. Worley	Newark, W. Va	July 20, 1869	92, 768
Implement, Compound	B. G. Devoe, T. Rodgers, and J. C. Beals.	Fredericktown, Ohio	Jan. 9, 1872	122, 577
Implement, Compound	E. A. Edwards	San Buenaventura, Cal	July 16, 1872	129, 116
Implement, Compound	A. Iske	Lancaster, Pa	Mar. 11, 1873	136, 601
Implement, Compound	J. G. Powell	Philadelphia, Pa	Oct. 1, 1872	131, 904
Implement, Compound	C. W. Russell	Milford, Mass	Apr. 23, 1872	126, 094
Implement, Compound	J. C. Schlarbaum	San José, Cal	Apr. 2, 1872	125, 336
Incendiary compound	H. W. Libbey	Cleveland, Ohio	June 13, 1865	48, 187
Inclined-plane elevator	T. B. Simonton	Williamsburgh, N. Y	Mar. 15, 1870	100, 811
Inclined plane for drawing up vessels	E. Clark	New York, N. Y	June 27, 1825	....
Inclined-plane statical wheels, Machine to facilitate passage of boats in canals, removing earth, &c., by.	J. Williams	....	Mar. 23, 1804	....
Inclined planes and hoists, Brake for	L. Klee	Pittsburgh, Pa	Mar. 12, 1872	124, 592
Inclined planes, Device for assisting railway-car to ascend.	T. Agudio	Turin, Italy	June 9, 1863	38, 800
Inclined planes, Machinery for ascending	R. F. Stevens and S. B. Pitcher	Syracuse, N. Y	Dec. 28, 1846	4, 908
Inclined planes, Machinery for ascending and descending.	G. E. Sellers	Cincinnati, Ohio	Nov. 13, 1847	5, 367
Inclined planes, Method of ascending	J. S. French	Old Point Comfort, Va	June 22, 1852	9, 052
Inclined wheel	G. Lyman	Charles Parish, La	Mar. 4, 1822	....
Inclined wheel	J. Palmer	Cincinnati, Ohio	Dec. 29, 1818	....
Inclined wheel moved by animal weight, Horizontal	M. Isaacs and J. Wilbank	Philadelphia, Pa	Oct. 14, 1817	....
Inclined wheels, Method of propelling	N. H. Shaw	Charlotte Hall, Md	Oct. 9, 1824	....
Inclinometer	A. Chase, jr	Somerville, Mass	Aug. 29, 1865	49, 675
Incombustible, Rendering articles	A. G. Fell	Brooklyn, N. Y	Dec. 31, 1867	72, 830
Incubation, Artificial	W. J. Canteto	England	July 24, 1847	5, 204
Incubation, Artificial	L. G. Hoffman	Albany, N. Y	Feb. 20, 1847	4, 978
Incubator	J. and H. Graves	Boston, Mass	Dec. 27, 1870	110, 454
Incubator	H. J. Haight	New York, N. Y	Dec. 26, 1871	122, 249
Incubator	E. Woodward and N. J. Millet	Charlestown, Mass	Dec. 19, 1871	121, 977
Incubators, Electro-magnetic regulator for	J. Graves	Reading, Mass	June 18, 1872	127, 972
Index	O. F. Bullard	Media, Pa	Mar. 28, 1871	113, 015
Index	H. H. Hall	Boston, Mass	Oct. 27, 1868	83, 378
Index, Alphabetical	W. C. Huston	Eaton, Ohio	Apr. 26, 1870	102, 404
Index gage and calipers	D. F. Elmer	Springfield, Mass	Oct. 1, 1867	69, 418
Index, Ledger	C. A. Fitch and W. O. St. John	San Francisco and Oakland, Cal.	Mar. 22, 1870	100, 994
Index-lettering, Tool for	E. Crawley	Cincinnati, Ohio	July 8, 1856	15, 283
Index or book-marker	J. Johnson	New York, N. Y	Oct. 12, 1858	21, 759
Index, Universal	M. L. Powell	Newcastle, Ind	May 8, 1866	54, 593
Indexes, Case for containing	C. E. Fellows	Philadelphia, Pa	June 3, 1873	139, 503
Indexing books and ledgers	J. H. Swindell	Camden, N. J	Apr. 21, 1868	77, 126
Indexing for records, System of	A. Campbell	Frederick, Md	Feb 11, 1868	74, 299
India-rubber: *See* Gum-elastic.				
India-rubber and gutta-percha compound, Colored	J. Malcolm	New York, N. Y	Jan. 2, 1866	51, 846
India-rubber and gutta-percha compounds, Manufacture of colored.	J. Malcolm	New York, N. Y	Jan. 2, 1866	51, 847
India-rubber and gutta-percha compounds, Manufacture of colored.	J. Malcolm	New York, N. Y	Jan. 2, 1866	51, 848
India-rubber and gutta-percha compounds, Manufacture of colored vulcanized.	J. Malcolm	New York, N. Y	Jan. 2, 1866	51, 849
India-rubber and gutta-percha goods, Manufacture of.	J. Murphy	New York, N. Y	Feb. 15, 1870	99, 934
India-rubber and pneumatic spring	F. M. Ray	New York, N. Y	Aug. 1, 1848	5, 696
India-rubber and steel springs, Combined	E. T. Bussell	Shelbyville, Ind	Nov. 29, 1853	10, 280
India rubber, Apparatus for curing	J. B. Forsyth	Roxbury, Mass	Nov. 21, 1865	51, 036
India rubber, Apparatus for pressing and vulcanizing.	J. Banigan	Woonsocket, R. I	Aug. 31, 1869	94, 273
India-rubber bat-cloth, Mode of making	C. Goodyear	New Haven, Conn	Oct. 12, 1852	9, 319
India-rubber boat, Portable	H. H. Day	New York, N. Y	Jan. 15, 1846	4, 356
India rubber, Cleaning	A. G. Day	Seymour, Conn	June 10, 1856	15, 067
India-rubber-cloth goods, Manufacture of	W. Cable	Boston, Mass	Mar. 12, 1872	124, 542
India-rubber cloth, Preparing	G. B. Millerd	Colchester, Conn	Feb. 10, 1857	16, 601
India-rubber cloth, Preparing elastic	C. Winslow	Lynn, Mass	June 23, 1857	17, 649
India-rubber cloth, Process for making	H. G. Tyer and J. Helm	New Brunswick, N. J	Jan. 2, 1855	12, 144
India-rubber cloth, Process for making	H. G. Tyer and J. Helm	New Brunswick, N. J	Jan. 30, 1855	12, 334
India-rubber cloth, Process of preparing elastic	N. Hayward	Colchester, Conn	May 6, 1856	14, 811
India-rubber cloth, Process of rolling	F. D. Hayward and J. C. Bickford.	Colchester, Conn	Mar. 19, 1850	7, 189
India rubber, Coating metal articles with	C. Hingher	New Brunswick, N. J	Aug. 23, 1870	106, 585
India rubber, Covering metal articles with	C. Hingher	New Brunswick, N. J	Aug. 30, 1870	106, 819
India-rubber fabric	C. J. Gilbert and G. Gay	New York, N. Y	July 17, 1847	5, 196
India-rubber fabric	C. Goodyear	New York, N. Y	Mar. 9, 1844	3, 461
India-rubber fabric	C. Goodyear	New York, N. Y	Mar. 9, 1844	3, 462
India-rubber fabric	C. Goodyear	New York, N. Y	June 15, 1844	3, 633
India-rubber fabric	C. Goodyear	New Haven, Conn	July 5, 1845	4, 099
India-rubber fabric	C. Goodyear	New Haven, Conn	Aug. 16, 1859	25, 111
India-rubber fabric	E. E. Marcy	New York, N. Y	Dec. 6, 1859	26, 358
India-rubber fabric	E. E. Marcy	New York, N. Y	Dec. 6, 1859	26, 359
India-rubber fabric	E. E. Marcy	New York, N. Y	Dec. 6, 1859	26, 360
India-rubber fabrics, Manufacture of	N. Goodyear	Newtown, Conn	May 13, 1845	4, 047
India-rubber fabrics, Mode of making binding for	C. A. Ensign	Naugatuck, Conn	Nov. 28, 1865	51, 161
India rubber for the manufacture of hose, belting, packing, &c., Preparation of.	E. L. Simpson	Bridgeport, Conn	Feb. 28, 1865	46, 608
India-rubber goods, Finishing the surface of	J. T. Trotter and I. F. Williams	New York, N. Y	May 22, 1860	28, 420
India-rubber goods, Flocking	F. R. Taylor	New York, N. Y	Feb. 2, 1864	41, 453
India-rubber goods, Machine for manufacturing corrugated or shirred.	H. H. Day	Jersey City, N. J	Oct. 12, 1844	3, 788
India-rubber goods, Manufacture of	H. Hutchinson	Newark, N. J	Feb. 12, 1861	31, 391
India-rubber goods, Manufacture of	J. W. Sutton	Akron, Ohio	Oct. 15, 1872	132, 185

Index of patents issued from the United States Patent Office from 1790 *to* 1873, *inclusive*—Continued.

Invention.	Inventor.	Residence.	Date.	No.
India-rubber goods, Manufacture of	J. W. Sutton	Akron, Ohio	Oct. 15, 1872	132, 186
India rubber, gutta-percha, &c., Manufacture of	E. L. Simpson	Bridgeport, Conn	Oct. 16, 1866	58, 902
India rubber, gutta-percha, &c., Process of manufacturing.	E. L. Simpson	Bridgeport, Conn	Feb. 28, 1865	46, 610
India-rubber hollow-molded articles, Manufacture of.	D. D. Parmelee	Salem, Mass	Dec. 20, 1859	26, 551
India rubber in the liquid state, Preserving	H. L. Norris	New York, N. Y	July 26, 1853	9, 891
India rubber into shreds, Machine for cutting	J. Bogardus	New York, N. Y	Nov. 21, 1845	4, 280
India rubber into threads, Machine for cutting	J. W. Cox	Malden, Mass	June 14, 1859	24, 426
India rubber into threads, Machine for cutting	H. H. Day, H. G. Tyer, and J. Helen.	Jersey City and New Brunswick, N. J.	June 7, 1845	4, 073
India rubber into threads, Machine for cutting	H. Messer	Roxbury, Mass	May 31, 1859	24, 265
India rubber, Machine for cutting	H. G. Tyer and J. Helm	New Brunswick, N. J	Oct. 9, 1844	3, 782
India rubber, Machine for cutting sheets of	C. A. Ensign	Naugatuck, Conn	Nov. 29, 1870	109, 725
India rubber, Machine for joining irregular seams in	C. A. Ensign	Naugatuck, Conn	Nov. 29, 1870	109, 726
India rubber, Making hollow articles of	C. Goodyear	New Haven, Conn	Apr. 25, 1848	5, 536
India rubber, Manufacture of	N. Goodyear	New York, N. Y	May 6, 1851	8, 075
India rubber, Manufacture of	D. Hayward	Providence, R. I	Aug. 29, 1854	11, 608
India rubber, Manufacture of	N. Hayward	Colchester, Conn	Apr. 15, 1856	14, 657
India rubber, Manufacture of	D. McCurdy	Newark, N. J	Apr. 1, 1851	8, 016
India rubber, Manufacture of	R. Solis	New Brunswick, N. J	Feb. 1, 1853	9, 570
India rubber, Manufacture of	J. T. Trotter	New York, N. Y	Jan. 1, 1851	7, 880
India rubber, Manufacture of	H. G. Tyer and J. Helme	New Brunswick, N. J	Jan. 30, 1849	6, 066
India rubber, Mode of treating	S. Bourne	Headstone, Drive Harrow, England.	Feb. 12, 1867	61, 992
India-rubber-packing former	W. Webster	Middletown, Ohio	Apr. 25, 1865	47, 477
India rubber, Preparing	W. F. Ely	New York, N. Y	Apr. 17, 1847	5, 969
India rubber, Preparing	J. Thomas	New York, N. Y	Sept. 4, 1847	5, 271
India rubber, Preserving	F. Bronner	Vera Cruz, Mexico	Sept. 7, 1852	9, 246
India rubber, Printing and ornamenting	G. H. Lewis	Providence, R. I	May 26, 1863	38, 686
India rubber, Process for cleaning	T. Sault	Seymour, Conn	Nov. 11, 1856	16, 069
India rubber, Process for treating	E. D. S. Goodyear	Stapleton, N. Y	Mar. 28, 1854	10, 689
India rubber, Process of obtaining gum resembling	E. H. Rogers	Tuscaloosa, Ala	Mar. 11, 1873	136, 765
India rubber, Process of preparing	W. Mullee	Franklin, Pa	Feb. 12, 1867	62, 055
India-rubber rings, Manufacture of	W. H. Collins and C. E. Longden.	Prospect, Conn	July 1, 1873	140, 472
India-rubber roller, Manufacture of	J. B. Forsyth	Roxbury, Mass	Nov. 13, 1866	59, 580
India-rubber spring	W. F. Converse	Harrison, Ohio	Apr. 17, 1855	12, 738
India-rubber spring	A. Magowan	Trenton, N. J	Feb. 18, 1873	136, 079
India-rubber spring for cars, Manufacture of	F. M. Ray	New York, N. Y	Apr. 2, 1850	7, 251
India-rubber spring for upholstery purposes	F. Colton	New York, N. Y	Nov. 17, 1857	18, 630
India-rubber spring, Vulcanized	F. M. Ray	New York, N. Y	Oct. 8, 1850	7, 706
India-rubber springs, Operating	J. W. Wilder	New York, N. Y	Aug. 28, 1866	57, 647
India rubber, Treating	W. F. Shaw	Boston, Mass	Aug. 12, 1856	15, 531
India rubber, Treatment of	A. K. Eaton	New York, N. Y	June 19, 1860	28, 744
India rubber, Treatment of	H. W. Joslin	Trenton, N. J	Jan. 12, 1859	22, 560
India rubber, Treatment of	H. W. Joslin and A. K. Eaton	Trenton, N. J., and New York, N. Y.	Nov. 22, 1859	26, 233
India rubber, Treatment of	A. Shannon	New York, N. Y	Apr. 19, 1859	23, 717
India-rubber tube, Manufacture of	C. McBurney and J. B. Forsyth.	Roxbury, Mass	Apr. 17, 1866	53, 999
India-rubber tube or strip, Machine for making	B. M. Hotchkiss	Naugatuck, Conn	July 12, 1870	105, 335
India-rubber tubing	F. R. Taylor	New York, N. Y	May 31, 1864	42, 972
India rubber, Use of the oxide of tin in the manufacture of.	J. Pridham	New Brunswick, N. J	Mar. 19, 1850	7, 196
India rubber with leather, Mode of uniting	A. C. Andrews	New Haven, Conn	Apr. 16, 1867	63, 779
Indicating-tubes for ascertaining draft of boats	J. E. Vansant	Louisville, Ky	July 25, 1854	11, 393

Indicator:

See Billiard-indicator.
Canal-lock indicator.
Car-indicator.
Car-seat indicator.
Check-valve indicator.
Crank-indicator.
Door-indicator.
Electrical indicator.
Electro-magnetic indicator.
Fire-indicator.
Fluid-indicator.
Fuel-indicator.
Gas light and pressure indicator.
Gas-meter indicator.
Hand-indicator.
Heat-indicator.
High and low water indicator.
High-water indicator.
Knitting-machine indicator.
Knitting-machine-knot indicator.
Laundry-indicator.
Leeway-indicator.
Loom-indicator.
Low-water indicator.
Magnetic indicator.
Navigator's bearing-indicator.
Office-indicator.
Padlock-indicator.
Plumb and level indicator.
Power-indicator.
Pressure-indicator.
Privy-indicator.
Railway-train indicator.
Speed-indicator.
Station-indicator.
Station or street-registering indicator.
Steam-boiler indicator.
Steam-engine indicator.
Steam-generator indicator.
Steam-pressure indicator.
Street and station indicator.
Telegraph-indicator.
Telegraph-circuit indicator.
Telephonic indicator.
Time-indicator.
Time-table indicator.
Valve-indicator.
Wages-indicator.
Water-indicator.
Water-closet indicator.
Water-gage indicator.
Water-level indicator.

Invention.	Inventor.	Residence.	Date.	No.
Indicator and low-water alarm for boilers	M. Ellwood	Girard, Ohio	Mar. 11, 1873	136, 646
Indicator-lock	E. A. Cooper	Lancaster, N. Y	Mar. 25, 1873	137, 181
Indicator-lock	T. Lalor	Toronto, Canada	Aug. 4, 1868	80, 637
Indicator-lock	E. M. and J. E. Mix	Westfield, N. Y	Jan. 4, 1870	98, 513
Indigo-blue vat for coloring wool and cotton	G. Molt	Millbury, Mass	Jan. 7, 1873	134, 694

Index of patents issued from the United States Patent Office from 1790 *to* 1873, *inclusive*—Continued.

Invention.	Inventor.	Residence.	Date.	No.
Indigo, Compounding a blue color as a substitute for	S. De Riemer	Poughkeepsie, N. Y.	May 15, 1809	....
Indigo-dye	T. Weber	Alton, Ill	Jan. 19, 1869	86, 047
Indigo for dyeing and printing, Preparation of	P. Schutzenberger and F. De Lalande.	Paris, France	Mar. 11, 1873	136, 770
Indigo, Manufacture of	C. W. Smith	Highfield, near Stroud, England.	Aug. 20, 1872	130, 666
Indigo, Manufacture of	T. T. Woodruff	Philadelphia, Pa	May 14, 1872	126, 665
Indigo, Process and apparatus for the manufacture of.	T. T. Woodruff	Philadelphia, Pa	May 14, 1872	126, 663
Indigo, Process and apparatus for the manufacture of.	T. T. Woodruff	Philadelphia, Pa	May 14, 1872	126, 664
Indigo, Process of restoring waste	F. A. Sawyer, jr	Boston, Mass	Oct. 24, 1871	120, 215
Induction-coil	D. M. Cook	Mansfield, Ohio	Oct. 10, 1871	119, 825
Induction-coil apparatus and circuit-breaker	C. G. Page	Washington, D. C	Apr. 14, 1868	76, 654
Infant's chair	J. Hayes	Philadelphia, Pa	Aug. 22, 1871	118, 240
Inflammable liquid so as to prevent accidents, Mode of preparing.	T. J. Barron	Brooklyn, N. Y	Mar. 28, 1865	46, 987
Ingot-casting mold	H. Dickenson	Jersey City, N. J	July 30, 1867	67, 171
Ingot-mold	G. Abel and J. Pedder	Temperanceville, Pa	Aug. 31, 1869	94, 170
Ingot-mold	N. Churchman	Pioche, Nev	Dec. 9, 1873	145, 278
Ingot-mold	Z. S. Durfee	Troy, N. Y	Sept. 27, 1870	107, 766
Ingot-mold	Z. S. Durfee	New York, N. Y	Sept. 17, 1872	131, 332
Ingot-mold	A. L. Holley	Harrisburgh, Pa	July 30, 1872	129, 957
Ingot-mold	M. T. Mooney	Johnstown, Pa	Dec. 16, 1873	145, 677
Ingot-mold, Adjustable	M. S. Drake	Newark, N. J	Mar. 18, 1873	136, 821
Ingot mold, Combined steel and iron	P. Doyle	Newark, N. J	Feb. 1, 1870	99, 299
Ingot-mold for gold or silver	J. Feix	San Francisco, Cal	Nov. 18, 1873	144, 605
Ingot molds, Stopper for steel	S. T. Wellman	Nashua, N. H	July 23, 1872	129, 699
Ingots, Casting	Z. S. Durfee	New York, N. Y	Jan. 2, 1872	122, 310
Ingots, Casting brass	L. T. Wooster	Ansonia, Conn	Aug. 17, 1871	118, 174
Ingots, Casting steel	H. Disston and J. Marsden	Philadelphia, Pa	Aug. 17, 1869	93, 863
Ingots, Casting steel	J. A. Holmes	Philadelphia, Pa	Aug. 5, 1873	141, 506
Ingots for alloying gold and silver, Manufacture of copper.	J. Feix	San Francisco, Cal	Nov. 25, 1873	144, 896
Ingots for corrugated metal beams, Form of	R. Montgomery	New York, N. Y	Aug. 23, 1870	106, 713
Ingots in groups, Casting	F. J. Slade	Trenton, N. J	Mar. 29, 1870	101, 389
Ingots, Manufacture of steel	J. B. Tarr	Fairhaven, Mass	Oct. 20, 1868	83, 222
Ingots, Mold for casting steel	J. E. Fry	Johnstown, Pa	Nov. 12, 1867	70, 710
Ingots of iron and steel, Apparatus for casting	A. L. Holley	Brooklyn, N. Y	Nov. 12, 1872	133, 030
Ingots of iron and steel, Manufacture of refined	C. Shunk	Armstrong County, Pa	Nov. 5, 1867	70, 476
Ingots, Working scrap-iron or other metals into	D. D. Parmelee	New York, N. Y	Dec. 30, 1873	146, 092
Inhalation, Apparatus for generating and washing gas for.	A. W. Sprague	Boston, Mass	June 12, 1866	55, 548
Inhalation, Oxygenized compound for	J. L. Martin	Baltimore, Md	Feb. 13, 1872	123, 714
Inhalator	M. Vergnes	New York, N. Y	July 12, 1864	43, 536
Inhaler	G. Bastian and B. Segnitz	New York, N. Y	Sept. 1, 1863	39, 716
Inhaler	G. L. Chapin	Chicago, Ill	Oct. 15, 1872	132, 197
Inhaler	W. R. Crumb	Buffalo, N. Y	Jan. 14, 1873	134, 858
Inhaler	W. H. Cutler	Buffalo, N. Y	Feb. 4, 1873	135, 411
Inhaler	D. H. Goodwillie	New York, N. Y	Nov. 6, 1866	59, 386
Inhaler	R. B. Heintzelman	New York, N. Y	Nov. 14, 1871	120, 816
Inhaler	I. Holmes	Moscow, N. Y	Jan. 8, 1867	61, 008
Inhaler	C. D. Hunter and E. S. Woods	Marlborough, Mass	Apr. 29, 1873	138, 253
Inhaler	C. D. Hunter and E. S. Woods	Marlborough, Mass	July 15, 1873	140, 828
Inhaler	A. Prentiss	Prentiss Vale, Pa	May 14, 1867	64, 793
Inhaler	S. J. Shaw	Marlborough, Mass	July 22, 1873	141, 175
Inhaler, Air	Z. Rogers	Chicago, Ill	May 11, 1869	90, 051
Inhaler and fumigator	E. C. Kirkwood	Washington, D. C	June 4, 1872	127, 420
Inhaler and nasal douche combined	E. Schofield	Worcester, Mass	Jan. 28, 1873	135, 372
Inhaler and remedy for throat disease	G. H. Tichenor	Canton, Miss	Mar. 9, 1869	87, 603
Inhaler and vaporizer for administering anæsthetics.	E. E. Duncanson	Chicago, Ill	Oct. 10, 1871	119, 748
Inhaler, Chloroform	C. B. Porter	Ann Arbor, Mich	May 13, 1862	35, 257
Inhaler, Ether	W. F. G. Morton and A. A. Gould.	Boston, Mass	Nov. 13, 1847	5, 365
Inhaler for medical purposes	J. P. Brower	Syracuse, N. Y	Feb. 9, 1869	86, 638
Inhaler, Gas	S. W. Albee	Charlestown, N. H	May 10, 1864	42, 721
Inhaler, Gas	T. F. Frank	Buffalo, N. Y	Oct. 23, 1866	59, 000
Inhaler, Gas	C. L. Munns	Philadelphia, Pa	June 12, 1866	55, 527
Inhaler, Medical	N. I. Donaldson	Worcester, Mass	Jan. 14, 1873	134, 861
Inhaler, Medical	J. C. Parkinson	Vineland, N. J	Dec. 16, 1873	145, 679
Inhaler or lung-protector	L. P. Haslett	Louisville, Ky	June 12, 1849	6, 529
Inhaler, Vapor	M. M. Mills	Aurora, Ill	Jan. 28, 1868	73, 913
Inhaler, Vapor	D. Russell	Milford, Mass	Oct. 31, 1865	50, 735
Inhaling-apparatus	J. Montgomery	New York, N. Y	Nov. 23, 1869	97, 214
Inhaling-apparatus	E. W. Owen	Brooklyn, N. Y	May 4, 1869	89, 594
Inhaling-apparatus	A. E. Pursell	Indianapolis, Ind	Oct. 1, 1872	131, 780
Inhaling-apparatus	E. Schofield	Worcester, Mass	Dec. 23, 1873	145, 758
Inhaling-apparatus	J. C. Schooley	Cincinnati, Ohio	Aug. 18, 1857	18, 020
Inhaling-apparatus	S. H. T. Tilghman	Snow Hill, Md	Nov. 21, 1854	11, 976
Inhaling-apparatus	C. Warren	Milford, Mass	Sept. 13, 1864	44, 244
Inhaling apparatus, Vapor	A. P. Lighthill	Boston, Mass	Apr. 25, 1865	47, 434
Inhaling-fluid for cure of consumption and other diseases.	N. W. Abbott	Centralia, Ill	July 16, 1867	66, 667
Inhaling gases	W. Z. W. Chapman	New York, N. Y	Apr. 28, 1868	77, 168
Inhaling gases, Apparatus for	W. Z. W. Chapman	New York, N. Y	Sept. 6, 1864	44, 075
Inhaling gases, Apparatus for	P. H. Vander Weyde	New York, N. Y	June 5, 1866	55, 398
Inhaling medicinal agents, Means for	A. G. Hull	New York, N. Y	Apr. 21, 1857	17, 095
Inhaling powders, Instrument for	I. Warren	Boston, Mass	Mar. 16, 1852	8, 813
Inhaling pure air, Apparatus for	B. I. Lane	South Framingham, Mass	Feb. 21, 1865	46, 477
Inhaling-tube	C. Bullock	Cambridge, Mass	July 18, 1865	48, 789
Inhaling-tube	A. P. Messer	Boston, Mass	May 4, 1869	89, 591
Inhaling-tube	D. Minthorn	New York, N. Y	June 27, 1854	11, 171
Inhaling-tube	S. W. Sine	Easton, Pa	Dec. 17, 1867	72, 234
Injecting and douching instrument	J. Singer	Chicago, Ill	Oct. 6, 1863	40, 192
Injections, Apparatus for	C. A. Jozansi	Saint Romain, France	July 28, 1863	39, 348
Injector	A. J. Blakslee and G. C. Williams.	Du Quoin, Ill	June 15, 1869	91, 205

Index of patents issued from the United States Patent Office from 1790 *to* 1873, *inclusive*—Continued.

Invention.	Inventor.	Residence.	Date.	No.
Injector	P. C. Heinz	Funkville, Pa	Nov. 13, 1866	59, 602
Injector	C. Hughes	Niles, Ohio	July 6, 1869	92, 313
Injector	A. Morton	Glasgow, Scotland	Jan. 19, 1869	86, 030
Injector	C. F. Root	West Springfield, Mass	Dec. 20, 1870	110, 289
Injector	S. Rue	Philadelphia, Pa	Dec. 30, 1873	145, 969
Injector, Air	J. A. Bassett and O. C. Smith	Salem, Mass	May 30, 1865	48, 041
Injector and ejector	W. B. Mack	Boston, Mass	Dec. 17, 1872	133, 997
Injector, Boiler	J. Gresham	Manchester, Great Britain	Aug. 7, 1866	57, 057
Injector, Boiler	J. Gresham	Manchester, Great Britain	Feb. 11, 1868	74, 345
Injector, Boiler	J. Gresham	Manchester, Great Britain	Feb. 11, 1868	74, 346
Injector, Boiler	J. T. Hancock	Jamaica Plain, Mass	Jan. 26, 1869	86, 152
Injector, Feed-water	J. Robinson and J. Gresham	Manchester, Great Britain	May 29, 1866	55, 218
Injector, feed-water heater, and condenser, Combined.	E. Korting	Vienna, Austria	Jan. 30, 1872	123, 264
Injector for feeding boilers	W. Sellers	Philadelphia, Pa	Mar. 3, 1868	75, 059
Injector for feeding boilers	W. and C. Sellers	Philadelphia, Pa	July 22, 1873	141, 174
Injector for feeding boilers, Self-adjusting	W. Sellers	Philadelphia, Pa	July 22, 1873	141, 173
Injector for heating liquids	P. Hogg	Brooklyn, N. Y	Aug. 2, 1864	43, 691
Injector for insect-powder	C. Chinnock	Brooklyn, N. Y	Aug. 10, 1869	93, 597
Injector for insect-powder	M. Koppe	New York, N. Y	Sept. 17, 1867	68, 995
Injector for steam and other enginery, Oil	R. Brayton	Fremont, Ohio	Dec. 15, 1868	84, 853
Injector for steam-engines, Oil	G. B. Brayton	Boston, Mass	Aug. 14, 1866	57, 079
Injector for steam-generators	A. Barclay	Kilmarnock, North Britain	Mar. 19, 1867	62, 993
Injector for steam-generators	E. Ferguson	New Berne, N. C	Feb. 4, 1868	74, 067
Injector for steam-generators	S. Rue, jr	Chester County, Pa	Jan. 28, 1868	73, 757
Injector for steam-generators	S. Rue, jr	Paoli, Pa	Sept. 1, 1868	81, 822
Injector for steam-generators	S. Rue, jr	Philadelphia, Pa	Apr. 22, 1873	138, 198
Injector for steam-generators	S. Rue, jr	Philadelphia, Pa	Apr. 22, 1873	138, 199
Injector for steam-generators	E. Ware	Southford, Conn	Aug. 24, 1869	94, 152
Injector for the hair	A. A. Smith	Seneca Falls, N. Y	Oct. 25, 1864	44, 843
Injector, Gaseous liquid	C. Schultz and T. Warker	New York, N. Y	Oct. 17, 1865	50, 500
Injector, Giffard	N. Cope	Cincinnati, Ohio	Jan. 27, 1863	37, 542
Injector, Giffard	J. Gresham	Manchester, England	Apr. 12, 1870	101, 858
Injector, Giffard	J. Millholland	Reading, Pa	June 10, 1862	35, 575
Injector, Giffard	W. Sellers	Philadelphia, Pa	July 21, 1863	39, 312
Injector, Giffard	W. Sellers	Philadelphia, Pa	July 21, 1863	39, 313
Injector, Giffard	W. Sellers	Philadelphia, Pa	Aug. 15, 1865	49, 445
Injector, Metal	J. W. Hollingsworth	Mount Vernon, Ind	May 17, 1870	103, 188
Injector, Mode of operating Giffard's	J. Millholland	Reading, Pa	Dec. 17, 1861	33, 957
Injector, Pile-ointment	L. Heins	Brunswick, Ga	May 14, 1872	126, 700
Injector, Powder	D. Russell	Milford, Mass	Aug. 25, 1863	39, 678
Injector, Steam	S. and J. Benson	Centralia, Ill	Apr. 8, 1873	137, 651
Injector, Steam and water	S. S. Jamison, jr	Saltsburgh, Pa	Nov. 28, 1871	121, 376
Injector, Steam-boiler	H. C. Crowell	Morgan, Ohio	May 24, 1870	103, 304
Injector, Steam-boiler	J. P. F. Datichy	Brooklyn, N. Y	July 12, 1870	105, 181
Injector, Steam-boiler	W. B. Mack	Manchester, England	Sept. 5, 1871	118, 734
Injector, Steam-boiler	S. Maltby and C. Osborn	Dayton, Ohio	Sept. 26, 1865	50, 205
Injector, Steam-boiler	T. O'Rorke	Pittsburgh, Pa	July 16, 1872	129, 491
Injector, Vagina	G. W. King	Saratoga Springs, N. Y	Aug. 18, 1868	81, 278
Injector, Veterinary narcotic	M. Crohn	Saint Louis, Mo	Nov. 5, 1867	70, 418
Injector, Water	J. B. Collen	Philadelphia, Pa	Mar. 22, 1864	42, 044
Injectors, Feed-regulator for	J. A. Marden and T. S. Smith	Boston and Somerville, Mass.	Oct. 28, 1873	143, 988
Ink	C. F. Panknin	Charleston, S. C	Aug. 9, 1870	106, 198
Ink and preparation for indelible marking	E. W. Briggs	Brooklyn, N. Y	Nov. 19, 1872	133, 197
Ink, Black	D. A. Dougherty	Kittaning, Pa	Feb. 23, 1869	87, 094
Ink, &c., Black pigment or mineral black for printing.	J. Cist	Wilkesbarre, Pa	Oct. 28, 1808	......
Ink box, Indelible	C. L. Lochman	Carlisle, Pa	July 16, 1867	66, 858
Ink, coloring-matter, mordant, or other liquids, Supplying or steeping, &c.	J. Rennie	Lodi, N. J	June 9, 1834	....
Ink-composition	B. Owens	Saint Louis, Mo	May 21, 1867	65, 002
Ink, Composition for	J. Shaw	Bridgeport, Conn	Oct. 15, 1867	69, 846
Ink, Composition for bank-note and other	T. S. Hunt	Montreal, Canada	Dec. 8, 1863	40, 839
Ink, Composition for printing and copying	J. Underwood and F. V. Burt	London, England	Mar. 24, 1863	38, 008
Ink, Compound for making writing	A. D. Bowman	New York, N. Y	Nov. 13, 1866	59, 549
Ink, Compound for printers'	H. Loewenberg	New York, N. Y	Apr. 9, 1867	63, 733
Ink, Compound for printers'	H. Lowenberg	New York, N. Y	June 28, 1870	104, 862
Ink-compound for telegraphic and other purposes	G. Little	Rutherford Park, N. J	Oct. 1, 1872	131, 886
Ink, Copying	A. G. Buzby	Philadelphia, Pa	May 26, 1868	78, 258
Ink-cup	P. K. Holbrook	Malden, Mass	Oct. 30, 1866	59, 223
Ink-distributer	J. Prince	Philadelphia, Pa	Nov. 19, 1833	....
Ink-distributer, Self-moving	J. Prince	New York, N. Y	Apr. 23, 1830	....
Ink-distributer, Self-operating	J. Maxson	Schenectady, N. Y	Jan. 9, 1835	....
Ink for hand-stamp, &c	R. H. Rogers	New York, N. Y	Apr. 19, 1864	42, 405
Ink for paper-ruling	L. Francis	New York, N. Y	Dec. 24, 1867	72, 621
Ink for printing, &c	L. Francis	New York, N. Y	Apr. 26, 1870	102, 243
Ink for printing bank-notes, &c	A. K. Eaton	New York, N. Y	Apr. 28, 1863	38, 298
Ink for printing, Copying	C. McIlvaine	Philadelphia, Pa	Oct. 25, 1870	108, 615
Ink for printing postage-stamps, Manufacture of	J. Macdonough	New York, N. Y	Feb. 27, 1866	52, 869
Ink for printing stamps, drafts, and checks	J. P. Simonds	New York, N. Y	Mar. 22, 1870	101, 170
Ink for ruling and printing, Copying	C. McIlvaine	Philadelphia, Pa	July 25, 1871	117, 314
Ink for stamping-purposes	J. B. F. Jud	New York, N. Y	July 15, 1873	140, 782
Ink-fountain	E. Jordan	West Cummington, Mass	Nov. 20, 1849	6, 883
Ink-fountain	H. H. Shorp	Cleveland, Ohio	Feb. 13, 1872	123, 651
Ink from grahamite, Mode of compounding printers'	H. Wurtz	New York, N. Y	Feb. 4, 1868	74, 188
Ink from type, Composition for removing	A. M. Bonton	Newark, N. J	Oct. 2, 1866	58, 370
Ink, Green	G. Smillie	New York, N. Y	Mar. 24, 1863	37, 984
Ink-holder	D. Harrington	Philadelphia, Pa	Apr. 22, 1831	....
Ink, Indelible	J. M. and G. W. Caldwell	Burlington, N. J	May 10, 1870	102, 915
Ink, Making black	P. Ferris	Greenwich, Conn	Dec. 5, 1842	2, 870
Ink, Making printers'	B. Chase	New London, Conn	June 28, 1810	....
Ink, Making printers'	E. Clark	Brooklyn, N. Y	July 25, 1845	4, 102
Ink, Making printers'	C. A. Thompson	Adrian, Mich	Apr. 17, 1855	12, 733
Ink, Manufacture of	J. A. W. and H. H. Sangster	Buffalo, N. Y	Mar. 20, 1866	53, 379
Ink, Manufacture of indelible writing	T. J. Spear	New Orleans, La	July 16, 1841	2, 176
Ink, Manufacture of printing	G. Duryee	New York, N. Y	June 27, 1865	48, 385
Ink, Manufacture of writing	J. W. Carter	Boston, Mass	Mar. 12, 1872	124, 544

Index of patents issued from the United States Patent Office from 1790 *to* 1873, *inclusive*—Continued.

Invention.	Inventor.	Residence.	Date.	No.
Ink, &c., Mixing and grinding apparatus for	J. Martin	New York, N. Y	Dec. 31, 1872	134, 434
Ink or writing-fluid	C. Hebel	Louisville, Ky	Aug. 23, 1870	106, 582
Ink-pad	H. R. Towne and W. H. Taylor	Stamford, Conn	Feb. 18, 1873	135, 949
Ink-powder and dye from aniline colors	J. Zengeler	Chicago, Ill	May 25, 1869	90, 417
Ink-press, Composition for	H. G. Horton	Troy, N. Y	Sept. 17, 1872	131, 444
Ink, Printers'	C. Krĕci	Scranton, Pa	Aug. 31, 1869	94, 220
Ink, Printers'	C. McIlvaine	Philadelphia, Pa	July 25, 1871	117, 313
Ink, Printers'	M. Turly	Council Bluffs, Iowa	Apr. 18, 1871	113, 947
Ink, Printers'	M. Turly and B. F. Thomas	Council Bluffs, Iowa	July 5, 1870	105, 014
Ink, Printers'	S. H. Turner	Brooklyn, N. Y	Sept. 6, 1853	10, 006
Ink, Printers'	M. Weissberger	Saint Paul, Minn	June 18, 1867	65, 973
Ink, Printers'	C. Wulsten	La Fayette, Ind	June 16, 1868	79, 045
Ink, Printing	G. W. Casilear	Washington, D. C	June 21, 1870	104, 554
Ink, Printing	A. A. Hulot	Paris, France	May 23, 1865	47, 909
Ink, Printing	J. Kircher	Cannstadt, near Stuttgart, Würtemberg.	Dec. 13, 1870	110, 048
Ink, Printing	J. Kircher	New York, N. Y	Sept. 19, 1871	119, 154
Ink, Printing	J. Kircher	New York, N. Y	Nov. 5, 1872	132, 840
Ink, Printing	G. Matthews	Montreal, Canada	June 30, 1857	17, 688
Ink, Printing	J. C. White	Quincy, Ill	May 7, 1872	126, 601
Ink, Red	T. J. Summus	Lynn, Mass	Mar. 7, 1865	46, 684
Ink-roller	A. A. Hanscom	Saco, Me	June 29, 1858	20, 710
Ink-roller	A. Schimmelfennig and J. Ende.	Washington, D. C	June 8, 1858	20, 512
Ink-roller, Printer's	L. Francis and F. W. Letmate	New York, N. Y	June 21, 1864	43, 492
Ink slab, India	I. Speyer	Terre Haute, Ind	Nov. 19, 1872	133, 178
Ink-stains, &c., Compound for removing	V. G. Bloede	Brooklyn, N. Y	Feb. 23, 1869	87, 088
Ink, Sympathetic	D. C. McNeil	Osceola, Mo	June 30, 1868	79, 374
Ink, Use of resin oil in printer's	M. M. Mathews	Rochester, N. Y	Oct. 1, 1850	7, 686
Ink, Vessel for making	A. Harrison	Philadelphia, Pa	Feb. 24, 1852	8, 755
Ink-well	T. Bell	New York, N. Y	Dec. 3, 1867	71, 686
Ink-well	F. C. Brownell	Brooklyn, N. Y	Jan. 5, 1864	41, 136
Ink-well	F. C. Brownell	East Orange, N. J	May 9, 1865	47, 616
Ink-well	J. H. Kidder	Lawrence, Mass	Feb. 18, 1873	136, 069
Ink-well	J. W. Ross	Boston, Mass	Sept. 1, 1863	39, 754
Ink-well cover	J. A. Blake	New Haven, Conn	Mar. 19, 1867	62, 925
Ink-well cover	G. Munger	New York, N. Y	Oct. 16, 1866	58, 952
Ink-wells, Fastening the covers of	G. and H. M. Sherwood	Chicago, Ill	Dec. 9, 1862	37, 113
Ink-wells, Fastening the covers of	H. M. Sherwood	Chicago, Ill	Oct. 18, 1864	44, 753
Ink, Writing	R. G. Loftus	Chelsea, Mass	Nov. 15, 1870	109, 327
Ink, Writing	J. D. Myers	New York, N. Y	Sept. 26, 1835	....
Ink, Writing	J. Popper	Phillipsburgh, N. J	Oct. 8, 1872	132, 102
Ink, Writing	E. F. Walsh and S. Duryea	Brooklyn, N. Y	June 13, 1871	115, 915
Inking-apparatus	F. L. Bailey	Boston, Mass	Oct. 7, 1873	143, 489
Inking-apparatus	H. Barth	Cincinnati, Ohio	Apr. 12, 1870	101, 703
Inking-apparatus	G. P. Gordon	Rahway, N. J	Sept. 2, 1873	142, 457
Inking-apparatus	I. Hart	Cincinnati, Ohio	Nov. 18, 1873	144, 673
Inking-apparatus	G. W. Wood	Richmond, Ind	Apr. 16, 1867	63, 975
Inking-apparatus for printing in colors	T. L. Baylies and G. W. Wood	Richmond, Ind	Aug. 6, 1867	67, 400
Inking-pad	J. A. Miles	Charleston, Ill	Jan. 21, 1873	135, 043
Inking-pad	W. A. Whitney and C. G. Seaver.	Adrian, Mich	July 9, 1872	128, 834
Inking-roller	R. M. Hoe	New York, N. Y	Apr. 17, 1844	3, 550
Inking-roller	W. J. Stone	Washington, D. C	Oct. 20, 1829	....
Inking-roller, Composition for printer's	A. Van Bibber	Cincinnati, Ohio	Mar. 28, 1871	113, 224
Inking-roller, Lithographic	S. D. Tucker	New York, N. Y	June 1, 1869	90, 702
Inking-rollers, Regulating temperature of	E. W. Arnold	Boston, Mass	Sept. 21, 1837	394
Inkstand	J. L. Adam	London, England	July 9, 1872	128, 695
Inkstand	H. L. Andrews	Chicago, Ill	Mar. 31, 1868	76, 138
Inkstand	H. P. Andrews and M. E. Rawson.	Cleveland, Ohio	Oct. 20, 1868	83, 126
Inkstand	J. Axtman	East Cambridge, Mass	Aug. 8, 1865	49, 211
Inkstand	S. W. Baldwin	Yonkers, N. Y	Jan. 14, 1873	134, 836
Inkstand	O. Barber	Hartford, Conn	May 13, 1814	....
Inkstand	N. G. Bartlett	Keokuk, Iowa	Jan. 14, 1868	73, 284
Inkstand	J. Barwick	North Woolwich, England	Feb. 4, 1868	73, 946
Inkstand	J. M. Batchelder	Cambridge, Mass	Apr. 27, 1858	20, 028
Inkstand	E. O. Bennett	Mount Pleasant, Iowa	Nov. 26, 1867	71, 443
Inkstand	A. Bingham	Boston, Mass	Aug. 28, 1855	13, 515
Inkstand	J. A. Bowen	Boston, Mass	June 6, 1871	115, 686
Inkstand	H. Bradford	New York, N. Y	Apr. 3, 1866	53, 729
Inkstand	M. Braun	Brooklyn, N. Y	Feb. 21, 1860	27, 200
Inkstand	A. W. Brinkerhoff	Upper Sandusky, Ohio	May 7, 1872	126, 514
Inkstand	W. M. Brooke	Eaton, Ohio	Jan. 7, 1873	134, 587
Inkstand	B. Brower	New York, N. Y	Nov. 25, 1873	144, 825
Inkstand	W. Burnet	New York, N. Y	Aug. 23, 1859	25, 175
Inkstand	G. Burnham	Philadelphia, Pa	Dec. 30, 1841	2, 409
Inkstand	W. E. Carlile	New York, N. Y	Nov. 9, 1869	96, 673
Inkstand	S. C. Catlin	Cleveland, Ohio	May 2, 1871	114, 262
Inkstand	C. T. Chase	Albany, N. Y	June 30, 1868	79, 444
Inkstand	S. D. Clark	Minneapolis, Minn	Aug. 8, 1871	117, 743
Inkstand	D. Cumming, jr	New York, N. Y	Dec. 13, 1864	45, 391
Inkstand	S. Darling	Bangor, Me	Sept. 14, 1858	21, 554
Inkstand	S. Darling	Bangor, Me	Aug. 1, 1865	49, 093
Inkstand	S. Darling	Bangor, Me	Jan. 9, 1866	51, 931
Inkstand	S. Darling	Bangor, Me	Sept. 3, 1867	68, 588
Inkstand	S. Darling	Bangor, Me	Apr. 7, 1868	76, 409
Inkstand	S. Darling	Providence, R. I	Jan. 31, 1871	111, 435
Inkstand	S. Darling and J. E. Hall	Bangor, Me	Apr. 14, 1868	76, 719
Inkstand	O. Dean	Richmond, Va	Feb. 18, 1868	74, 671
Inkstand	T. Duncan	Brookville, Md	May 23, 1871	115, 041
Inkstand	J. R. Ender	Trenton, La	Jan. 3, 1860	26, 661
Inkstand	H. Evans	Philadelphia, Pa	Jan. 10, 1860	26, 755
Inkstand	A. Fife	Philadelphia, Pa	Aug. 21, 1849	6, 664
Inkstand	B. S. Fletcher	Cornish, N. H	Aug. 29, 1865	49, 617
Inkstand	V. Fogerty	Cambridgeport, Mass	Aug. 31, 1858	21, 395
Inkstand	R. Gleason, jr	Dorchester, Mass	Mar. 18, 1856	14, 451
Inkstand	K. Goddard	Philadelphia, Pa	Apr. 28, 1857	17, 147

Index of patents issued from the United States Patent Office from 1790 *to* 1873, *inclusive*—Continued.

Invention.	Inventor.	Residence.	Date.	No.
Inkstand	A. P. Griffing	East Cambridge, Mass	Feb. 4, 1862	34, 299
Inkstand	F. T. Grimes	Liberty, Mo	June 14, 1870	104, 300
Inkstand	W. O. Haskell	Boston, Mass	June 6, 1871	115, 733
Inkstand	S. Hawson and T. Sweeney	Jersey City, N. J., and Brooklyn, N. Y.	Apr. 8, 1873	137, 548
Inkstand	A. Haywood	Easton, Mass	Nov. 23, 1818	....
Inkstand	B. J. Heywood	London, England	May 26, 1857	17, 373
Inkstand	G. G. Hickman	Coatesville, Pa	Feb. 9, 1869	86, 670
Inkstand	F. L. Hicks	New York, N. Y	Jan. 3, 1865	45, 794
Inkstand	L. E. Hicks	Boston, Mass	Mar. 9, 1858	19, 613
Inkstand	L. E. Hicks	New York, N. Y	June 9, 1863	38, 863
Inkstand	P. K. Holbrook	Malden, Mass	Dec. 20, 1864	45, 498
Inkstand	T. P. How	New York, N. Y	Jan. 31, 1860	26, 992
Inkstand	T. S. Hudson	East Cambridge, Mass	Dec. 28, 1858	22, 429
Inkstand	T. S. Hudson	East Cambridge, Mass	June 3, 1862	35, 449
Inkstand	W. Hunt	New York, N. Y	May 29, 1845	4, 062
Inkstand	W. Hunt	New York, N. Y	Oct. 7, 1845	4, 221
Inkstand	W. Hunt	New York, N. Y	Dec. 11, 1845	4, 306
Inkstand	O. H. Jadwin	Carbondale, Pa	Nov. 23, 1858	22, 123
Inkstand	L. P. Jenks	Boston, Mass	Dec. 8, 1863	40, 841
Inkstand	G. R. G. Jones	Memphis, Tenn	Dec. 24, 1872	134, 144
Inkstand	W. F. Jones	Circleville, Ohio	July 2, 1872	128, 630
Inkstand	A. D. Judd	New Haven, Conn	Dec. 9, 1873	145, 352
Inkstand	J. H. Kidder and G. E. Hood	Lawrence, Mass	Feb. 18, 1873	136, 070
Inkstand	R. Lapham	New York, N. Y	Dec. 3, 1867	71, 766
Inkstand	J. G. Lucas	Newark, N. J	Nov. 26, 1872	133, 465
Inkstand	T. J. Mayall	Boston, Mass	Apr. 23, 1872	125, 973
Inkstand	G. Merritt	Brooklyn, N. Y	Apr. 14, 1868	76, 792
Inkstand	E. Morgan	Springfield, Mass	June 18, 1872	128, 163
Inkstand	F. E. Oliver	New York, N. Y	June 9, 1863	38, 836
Inkstand	J. Peard	New York, N. Y	Feb. 27, 1872	124, 152
Inkstand	S. and J. J. Perry	London, England	Nov. 12, 1867	70, 893
Inkstand	G. M. Prentiss	Worcester, Mass	Jan. 11, 1859	22, 582
Inkstand	J. S. Rankin	Minneapolis, Minn	Jan. 23, 1872	122, 909
Inkstand	W. Read, jr	Cambridge, Mass	May 27, 1873	139, 261
Inkstand	T. Robjohn	New York, N. Y	Aug. 25, 1857	18, 060
Inkstand	T. Robjohn	New York, N. Y	Aug. 16, 1859	25, 217
Inkstand	J. W. Ross	Boston, Mass	Apr. 30, 1861	32, 207
Inkstand	W. G. Shattuck	Boston, Mass	Oct. 20, 1868	83, 328
Inkstand	T. S. Shenston	Brantford, Canada	Nov. 25, 1873	144, 929
Inkstand	S. Slocomb	Cambridge, Mass	Oct. 2, 1860	30, 257
Inkstand	S. Slocomb	East Cambridge, Mass	June 3, 1862	35, 478
Inkstand	E. Southworth	Saybrook, Conn	June 29, 1833	....
Inkstand	J. C. Sparr	Irondequoit, N. Y	Jan. 30, 1872	123, 205
Inkstand	J. G. Thompson	Carbondale, Pa	Nov. 3, 1863	40, 517
Inkstand	J. B. Thurston and F. M. West	New York, N. Y	Jan. 28, 1873	135, 296
Inkstand	W. C. Tilden	Washington, D. C	Apr. 30, 1872	126, 348
Inkstand	W. H. Towers	New York, N. Y	Oct. 2, 1860	30, 266
Inkstand	H. Whitney, jr	Cambridge, Mass	July 1, 1851	8, 188
Inkstand	H. Whitney, jr	Cambridge, Mass	May 8, 1855	12, 841
Inkstand	C. H. Wright	Baltimore, Md	Apr. 12, 1870	101, 953
Inkstand and bell combined	A. J. Lyon	Mount Vernon, N. Y	Sept. 24, 1872	131, 692
Inkstand and calendar combined	G. G. Percival	Brooklyn, N. Y	July 24, 1866	56, 603
Inkstand and mucilage-holder combined	W. W. Beach	New York, N. Y	July 9, 1867	66, 448
Inkstand and pen-rack	S. Walker	Boston, Mass	Jan. 6, 1863	37, 313
Inkstand, Attaching pen-rack to	C. A. Roberts	West Meriden, Conn	June 8, 1869	91, 166
Inkstand, Barometer	T. S. Hudson	East Cambridge, Mass	June 4, 1861	32, 498
Inkstand-filler	M. C. Stebbins	Springfield, Mass	Feb. 25, 1873	136, 186
Inkstand, Fountain	C. T. Close	New York, N. Y	Dec. 11, 1855	13, 902
Inkstand, Fountain	F. Draper	East Cambridge, Mass	Jan. 7, 1851	7, 885
Inkstand from composition stone	J. S. Elliott and J. F. Wood	Chelsea and Everett, Mass	June 25, 1872	128, 218
Inkstand, pen-cleaner, &c., Arrangement of	F. Bailey	New York, N. Y	Sept. 27, 1864	44, 381
Inkstand, Revolving	J. M. Kennedy	Vicksburgh, Miss	Dec. 15, 1868	84, 950
Inkstand, wafer or sand box, calendar, letter and envelope holder, and pen-rack combined	G. Schmidt	New York, N. Y	July 3, 1866	56, 104
Inkstand with air-tight stopper	J. Farley	Washington, D. C	Jan. 30, 1841	1, 957
Inkstands, Calendar-attachment to	S. R. Dummer	New York, N. Y	Dec. 10, 1867	71, 861
Inkstands, Construction of	R. T. Fry	Spring Garden, Pa	July 11, 1854	11, 270
Inkstands, Construction of	D. Harrington	Philadelphia, Pa	Nov. 8, 1845	4, 258
Inkstands, Manufacture of	T. S. Hudson	East Cambridge, Mass	Aug. 24, 1869	94, 113
Inkstands, Manufacture of	D. J. Mandell	Springfield, Mass	Feb. 21, 1842	2, 460
Inkstands, &c., Method of making	H. Whitney	East Cambridge, Mass	Mar. 30, 1869	88, 354
Inkstands of stone, Manufacturing	A. H. Quincy	Boston, Mass	Sept. 3, 1811	....
Inkstands, Operating-parts of fountain	W. A. Wheeler	New York, N. Y	Mar. 28, 1865	47, 060
Inkstands to desks, Mode of attaching	L. R. Satterlee	Rochester, N. Y	Aug. 12, 1856	15, 527
Inkstands, Tool for molding	A. W. Brinkerhoff	Sandusky, Ohio	Dec. 16, 1873	145, 619
Inlaying	J. W. Hyatt, jr	Albany, N. Y	Oct. 10, 1871	119, 710
Inlaying gold in tortoise-shell	U. Bailey	West Newbury, Mass	Feb. 22, 1827	....
Inlaying, Machine for preparing wood for	C. F. Ritchel	Newark, N. J	Mar. 15, 1870	100, 927
Inlaying metallic surfaces	E. G. Wright	Boston, Mass	Mar. 8, 1870	100, 580
Inlaying wood, Method of	T. W. and H. K. Porter	Boston, Mass	Sept. 6, 1870	107, 097
Inoculating-apparatus	A. Stauch	Philadelphia, Pa	June 12, 1860	28, 697
Insect and reptile destroying apparatus	C. Russell	Pittsburgh, Pa	July 2, 1861	32, 720
Insect-bar	J. V. Hirley	Cincinnati, Ohio	July 16, 1872	129, 476
Insect-destroyer	W. Armistead	Prince William County, Va	Nov. 13, 1810	....
Insect-destroyer	J. M. Bennett	Janesville, Iowa	Nov. 28, 1871	121, 272
Insect-destroyer	T. Byrne and D. Strunk	New York, N. Y., and Lavaca County, Tex.	Dec. 6, 1870	109, 869
Insect-destroyer	S. Creighton	Lithopolis, Ohio	Aug. 29, 1871	118, 517
Insect-destroyer	C. R. Dudley	Canton, Miss	Dec. 24, 1872	134, 130
Insect-destroyer	J. A. Finney	Nashville, Ohio	July 15, 1873	140, 818
Insect-destroyer	J. G. G. Garrett	Port Gibson, Miss	Nov. 12, 1872	133, 023
Insect-destroyer	C. J. Hauck	Brooklyn, N. Y	Nov. 26, 1872	133, 444
Insect-destroyer	W. H. Lewis	New York, N. Y	Jan. 25, 1870	99, 214
Insect-destroyer	A. McKenzie	Henry, Ill	Mar. 17, 1868	75, 560
Insect-destroyer	D. G. Mosher	Mosherville, Mich	Sept. 26, 1871	119, 389
Insect-destroyer	D. G. Mosher	Mosherville, Mich	Apr. 1, 1873	137, 469
Insect-destroyer	J. Orin	Dayton, Ohio	Sept. 3, 1872	131, 116

Index of patents issued from the United States Patent Office from 1790 *to* 1873, *inclusive*—Continued.

Invention.	Inventor.	Residence.	Date.	No.
Insect-destroyer	E. D. Pugh	Fort Plain, Iowa	Aug. 13, 1872	130, 390
Insect-destroyer	P. Reynard	Saint Louis, Mo	July 16, 1872	129, 167
Insect-destroyer	R. V. Shockey	Mount Pleasant, Iowa	Jan. 9, 1872	122, 541
Insect-destroyer	W. B. Stewart	Brooklyn, N. Y	Oct. 8, 1872	132, 116
Insect-destroyer	S. W. Thomas	North Royalton, Ohio	May 27, 1873	139, 277
Insect-destroyer	P. S. Von Wagner	Saltfleet Township, Canada	Oct. 3, 1871	119, 547
Insect-destroying apparatus	H. S. Danziger	New York, N. Y	Dec. 31, 1872	134, 468
Insect-destroying composition	J. Ahearn	Baltimore, Md	Nov. 16, 1869	96, 861
Insect-destroying composition	G. M. Jaques	Boston, Mass	Oct. 2, 1866	58, 423
Insect-destroying composition	D. R. Prindle	East Bethany, N. Y	Mar. 26, 1867	63, 298
Insect-destroying composition	A. M. Turney	Butternuts, N. Y	Jan. 30, 1866	52, 345
Insect-destroying composition	W. Weaver	Phœnixville, Pa	Apr. 2, 1867	63, 589
Insect-destroying composition for trees	B. Best	Dayton, Ohio	June 2, 1868	78, 569
Insect-destroying compound	J. G. Barker	Watertown, Mass	May 28, 1872	127, 141
Insect-destroying compound	B. Best	Dayton, Ohio	July 27, 1869	93, 044
Insect-destroying compound	W. C. Bibb	Madison, Ga	Sept. 23, 1873	142, 985
Insect-destroying compound	W. R. Fairbairn	Ridotte Township, Ill	Dec. 15, 1868	84, 867
Insect-destroying compound	S. Galbraith	Pine Grove Plantation, La	Sept. 17, 1867	68, 867
Insect-destroying compound	M. Haas	New York, N. Y	Oct. 20, 1868	83, 280
Insect-destroying compound	J. Hinds	Hindsburgh, N. Y	Apr. 13, 1869	88, 873
Insect-destroying compound	J. B. Lunbeck	Leon, Iowa	Aug. 5, 1873	141, 512
Insect-destroying compound	A. McDougall	Manchester, England	July 16, 1867	66, 725
Insect-destroying compound	N. T. P. Robertson and T. Niles	Fairbury, Ill	Dec. 7, 1869	97, 555
Insect-destroying compound	W. B. Royal	Brenham, Tex	June 17, 1873	140, 079
Insect-destroying compound	P. B. Sheldon	Bath, N. Y	Feb. 9, 1869	86, 873
Insect-destroying compound for trees	D. Daniels	Fitchburgh, Mass	Feb. 11, 1868	74, 317
Insect-destroying compound for trees, &c	Z. F. De Moss	Pleasanton, Kans	Sept. 24, 1872	131, 603
Insect-destroying varnish	T. J. Elliott	New York, N. Y	Nov. 14, 1871	120, 812
Insect-exterminator	A. B. Ewing	Lewisburgh, Tenn	Apr. 9, 1872	125, 557
Insect-exterminator	E. Gilliam	Allegheny City, Pa	May 8, 1866	54, 528
Insect-guard	J. W. Brook	Lynchburgh, Pa	July 25, 1871	117, 379
Insect-powder	H. S. Danziger	New York, N. Y	Dec. 3, 1872	133, 634
Insect powder blower	P. Reynard and V. Varin	New York and Brooklyn, N. Y.	Sept. 20, 1859	25, 526
Insect-powder blower	P. Reynard and V. Varin	New York and Brooklyn, N. Y.	Nov. 8, 1859	26, 055
Insect-powder gun	W. H. Ball	Brooklyn, N. Y	Aug. 5, 1873	141, 481
Insect-shield for the head	J. Haven	Boston, Mass	Dec. 4, 1860	30, 818
Insect-trap	J. W. Anderson	Portland, Ind	Apr. 8, 1873	137, 588
Insect trap	J. J. Armstrong	Kings County, N. Y	Oct. 17, 1871	120, 020
Insect-trap	A. D. Chesebro	Grand Rapids, Mich	Dec. 31, 1872	134, 419
Insect-trap	C. W. Curtis	Osborn, Mo	Dec. 10, 1872	133, 763
Insect-trap	L. Endslow	Blair, Pa	Sept. 26, 1871	119, 337
Insect-trap	F. G. Fowler	Bridgeport, Conn	Oct. 8, 1872	132, 067
Insect-trap	P. Funk and J. N. Baader	Buffalo, N. Y	May 17, 1870	103, 037
Insect-trap	W. Ogden	Philadelphia, Pa	Jan. 16, 1866	52, 067
Insect-trap	C. E. Penny	Fort Wayne, Ind	Nov. 7, 1871	120, 595
Insect-trap	M. Rigell	Newton, Ala	Jan. 28, 1873	135, 366
Insect-trap	J. M. Smith	Fort Wayne, Ind	Mar. 25, 1873	137, 255
Insect-trap	J. W. Stell	Gonzales, Tex	Dec. 31, 1872	134, 444
Insect-trap	L. I. Way	Annawan, Ill	July 4, 1871	116, 652
Insect-trap	W. Weaver	Phœnixville, Pa	Nov. 26, 1867	71, 346
Insect trap	J. H. Welch and J. Baker	Fort Wayne, Ind	Feb. 27, 1872	124, 099
Insect-trap	T. Wier	Lacon, Ill	Oct. 10, 1871	119, 905
Insect-trap for protecting fruit while growing	B. M. Quint	Saint Joseph, Mich	May 18, 1869	90, 194
Insects and vermin, Device for destroying	P. Reynard and V. Varin	New York, N. Y	Sept. 20, 1864	44, 341
Insects, Apparatus for killing	H. Hill and L. E. P. Bush	Lexington, Ky	Nov. 26, 1867	71, 486
Insects, Cleansing hair and feathers from	W. Wisdom	Cleveland, Ohio	Dec. 20, 1853	10, 347
Insects, Composition to prevent depredations of	F. G. Johnson	Bellwood, Sag Harbor, N.Y	Mar. 27, 1860	27, 637
Insects, &c., Compound for destroying	E. B. Sears	San Francisco, Cal	Feb. 25, 1873	136, 185
Insects, Construction of apparatus for destroying	A. Isaacsen	New York, N. Y	Aug. 21, 1860	29, 699
Insects injurious to fruit-trees, Composition for destroying.	P. B. Sheldon	Prattsburgh, N. Y	Aug. 30, 1859	25, 281
Insects in trees and plants, Method of destroying	H. A. Graef	Brooklyn, N. Y	Oct. 20, 1868	83, 279
Insects in wheat, Composition for destroying	J. Newcomer	Baltimore, Md	Aug. 4, 1868	80, 655
Insects, Lotion for the destruction of	D. Leibert	Washington, D. C	Aug. 31, 1869	94, 221
Insects on hop-vines and other plants, Process of fumigation for destroying.	J. Deane	Conneaut, Ohio	Dec. 24, 1867	72, 613
Insects on plants, Compound for destroying	W. A. Phillips	Perry Centre, N. Y	Aug. 4, 1868	80, 660
Insects on plants, Mode of destroying	J. Nicolas	La Fourche, La	June 13, 1831	....
Insects on potato-plants, Composition for destroying	J. P. Wilson	Elmwood, Ill	Sept. 22, 1868	82, 468
Insects on trees and plants, Composition for destroying.	J. A. Elias	Le Roy, Ohio	Nov. 27, 1866	59, 986
Insects on trees, Apparatus for destroying	C. Baudouin and A. Fteley	New York, N. Y	Mar. 2, 1869	87, 324
Insects on trees, Apparatus for destroying	J. Hatch	Lynn, Mass	May 14, 1867	64, 667
Insects on trees, Compound for killing	H. D. Flower	Chicago, Ill	Nov. 3, 1868	83, 615
Insects on trees, plants, &c., Compound for destroying.	G. W. Spots	Jacksonville, Ill	Nov. 3, 1868	83, 737
Insects, Trap for destroying	T. C. Silliman	Chester, Conn	Nov. 3, 1868	83, 666
Insole	B. S. Bryant	Hanson, Mass	Feb. 13, 1872	123, 675
Instep-stretcher	C. C. Pease	Jamestown, N. Y	Jan. 21, 1868	73, 546
Instructor, Self	C. Varlé	Baltimore, Md	May 20, 1830	....
Insubmersible boat	A. Du Buc Marentille		Dec. 23, 1802	....
Insulating and finishing compound for conducting-wires.	T. L. Reed and E. F. Phillips	Providence, R. I	Nov. 18, 1873	144, 794
Insulating-compound	A. H. Castle	Ann Arbor, Mich	Nov. 23, 1869	97, 045
Insulating-wires for helices	J. J. Clark and H. Splitdorf	East Chester and New York, N. Y.	Sept. 18, 1866	58, 217
Insulation of telegraph line and apparatus, Total	G. W. Nichols	Chicago, Ill	Sept. 10, 1867	68, 779
Insulator	A. H. Castle	Ann Arbor, Mich	Mar. 10, 1868	75, 365
Insulator	A. H. Castle	Ann Arbor, Mich	July 14, 1868	79, 951
Insulator	G. Floyd	Cincinnati, Ohio	May 14, 1867	64, 654
Insulator	F. Scott	Brooklyn, N. Y	Aug. 24, 1869	94, 037
Insulator and bracket, Telegraphic	J. Robertson	Carbondale, Pa	Jan. 30, 1872	123, 198
Insulator, Battery	A. G. Davis	Baltimore, Md	July 16, 1872	129, 465
Insulator, Battery	O. W. Robertson	Milwaukee, Wis	July 12, 1870	105, 252
Insulator, Cramp-hook for telegraphic-wire	W. H. Dechaut	Philadelphia, Pa	July 26, 1870	105, 656
Insulator for battery-cups	J. H. Thomas	Baltimore, Md	Feb. 25, 1873	136, 191

Index of patents issued from the United States Patent Office from 1790 *to* 1873, *inclusive*—Continued.

Invention.	Inventor.	Residence.	Date.	No.
Insulator for electrodes	L. Engler and E. F. Krauss	Paris, France	May 15, 1860	28, 263
Insulator for lightning-conductors	E. Eagles	Mamaroneck, N. Y	Aug. 18, 1863	39, 613
Insulator for lightning-rods	J. Burnham	Batavia, Ill	Aug. 10, 1869	93, 410
Insulator for lightning-rods	R. Hickok	Fort Edward, N. Y	Mar. 29, 1859	23, 373
Insulator for lightning-rods	T. J. Lockhart and J. Locke	Pittsburgh, Pa	Nov. 12, 1867	70, 727
Insulator for lightning-rods	G. W. Otis	Lynn, Mass	Aug. 26, 1851	8, 316
Insulator for lightning-rods	D. Stebbins	Wallingford, Conn	June 25, 1867	66, 054
Insulator for lightning-rods	T. U. Webb	Jersey City, N. J	Aug. 15, 1854	11, 539
Insulator for lightning-rods, Supporting	N. N. McLeod	Saint Louis, Mo	Nov. 30, 1858	22, 188
Insulator for telegraph and other electric conductors, Composition.	J. M. Merrick, jr	Boston, Mass	Feb. 23, 1869	87, 187
Insulator for telegraph-wires	R. B. Baker	Philadelphia, Pa	May 17, 1870	103, 122
Insulator for telegraph-wires	J. M. Batchelder	Cambridge, Mass	Oct. 14, 1851	8, 418
Insulator for telegraph-wires	D. Brooks	Philadelphia, Pa	Nov. 29, 1864	45, 221
Insulator for telegraph-wires	D. Brooks	Philadelphia, Pa	Mar. 26, 1867	63, 206
Insulator for telegraph-wires	C. A. Bush	New London, Conn	Oct. 26, 1869	96, 198
Insulator for telegraph-wires	L. A. Cauvet	New York, N. Y	July 25, 1865	48, 906
Insulator for telegraph-wires	J. I. Conklin, jr	New York, N. Y	Apr. 23, 1872	126, 027
Insulator for telegraph-wires	A. B. Day	Oak Creek, Wis	June 18, 1867	65, 886
Insulator for telegraph-wires	M. De Montufar	New York, N. Y	Dec. 6, 1870	109, 879
Insulator for telegraph-wires	W. D. Guseman and E. C. Bright.	Morgantown, W. Va	Nov. 30, 1869	97, 392
Insulator for telegraph-wires	J. Holland	Conshohocken, Pa	Jan. 5, 1864	41, 157
Insulator for telegraph-wires	H. Read	Jersey City, N. J	May 30, 1871	115, 521
Insulator for telegraph-wires	Z. C. Robbins	Washington, D. C	Oct. 14, 1851	8, 419
Insulator for telegraph-wires	Z. C. Robbins	Washington, D. C	Feb. 28, 1871	112, 281
Insulator for telegraph-wires	A. G. Smith and W. Pettingell	Painesville, Ohio	Nov. 22, 1870	109, 461
Insulator for telegraph-wires	H. H. Ward	New York, N. Y	Aug. 15, 1865	49, 482
Insulator, Glass telegraph	J. Garity	East Birmingham, Pa	Jan. 3, 1871	110, 645
Insulator-holder	R. Calrow	Mamaroneck, N. Y	July 2, 1867	66, 215
Insulator, Lamp	A. Bliss	New Rochelle, N. Y	Dec. 23, 1862	37, 212
Insulator, Telegraph	W. W. Baldwin	Cleveland, Ohio	July 26, 1870	105, 625
Insulator, Telegraph	J. F. Boynton	Syracuse, N. Y	July 9, 1867	66, 453
Insulator, Telegraph	D. Brooks	Philadelphia, Pa	Oct. 8, 1867	69, 622
Insulator, Telegraph	W. H. Burnap and J. D. Brassington.	New York, N. Y	Oct. 9, 1866	58, 594
Insulator, Telegraph	S. Chester	Elizabeth, N. J	May 16, 1871	114, 924
Insulator, Telegraph	A. B. Ely	Boston, Mass	Jan. 9, 1866	51, 934
Insulator, Telegraph	A. B. Ely	Newton, Mass	Mar. 27, 1866	53, 426
Insulator, Telegraph	A. B. Ely	Newton, Mass	Mar. 27, 1866	53, 427
Insulator, Telegraph	A. B. Ely	Boston, Mass	July 24, 1866	56, 542
Insulator, Telegraph	D. R. P. Emminger	Harrisburgh, Pa	Feb. 20, 1872	123, 878
Insulator, Telegraph	M. G. Farmer	Salem, Mass	Mar. 5, 1872	124, 199
Insulator, Telegraph	M. G. Farmer	Salem, Mass	Mar. 5, 1872	124, 200
Insulator, Telegraph	M. G. Farmer	Salem, Mass	July 23, 1872	129, 724
Insulator, Telegraph	M. G. Farmer and J. M. Batchelder.	Salem and Cambridge, Mass.	Sept. 14, 1858	21, 492
Insulator, Telegraph	S. L. Finley	Morrisania, N. Y	Feb. 15, 1870	99, 873
Insulator, Telegraph	J. L. Finn	Elyria, Ohio	Apr. 24, 1868	75, 889
Insulator, Telegraph	C. Fox and E. G. Heston	Gap, Pa	Dec. 30, 1873	145, 997
Insulator, Telegraph	J. W. Fry	Elizabeth, N. J	July 6, 1869	92, 299
Insulator, Telegraph	N. Hendricks	Jersey City, N. J	Jan. 28, 1873	135, 234
Insulator, Telegraph	S. C. Hendrickson	New York, N. Y	Apr. 30, 1867	64, 315
Insulator, Telegraph	M. Y. Holley	Washington, D. C	Nov. 28, 1871	121, 368
Insulator, Telegraph	G. W. Kidwell	Elwood, Ind	Nov. 14, 1871	120, 884
Insulator, Telegraph	S. McKee	Pittsburgh, Pa	Sept. 6, 1870	107, 075
Insulator, Telegraph	S. Oakman	Boston, Mass	July 26, 1870	105, 834
Insulator, Telegraph	C. H. Pond	Cleveland, Ohio	Jan. 23, 1872	122, 961
Insulator, Telegraph	A. G. Safford	Saint Albans, Vt	Nov. 30, 1869	97, 318
Insulator, Telegraph	W. E. Simonds	Hartford, Conn	Nov. 26, 1867	71, 418
Insulator, Telegraph	W. E. Simonds	Hartford, Conn	June 29, 1869	92, 111
Insulator, Telegraph	W. W. Smith	Cincinnati, Ohio	Feb. 20, 1866	52, 761
Insulator, Telegraph	W. W. Smith	Cincinnati, Ohio	Mar. 13, 1866	53, 246
Insulator, Telegraph	D. W. Teller and W. L. Savage	North Greenwich, Conn	Oct. 22, 1867	70, 132
Insulator, Telegraph	S. F. Van Choate	New York, N. Y	Apr. 4, 1865	47, 141
Insulator, Telegraph	C. F. Varley	New York, N. Y	Mar. 24, 1868	75, 815
Insulator, Telegraph	C. F. Varley	New York, N. Y	Apr. 21, 1868	76, 960
Insulator, Telegraph	W. W. Waddell	Hillsborough, Ohio	Mar. 13, 1866	53, 201
Insulator, Telegraph	J. L. Waite	Burlington, Iowa	Oct. 22, 1867	70, 052
Insulator, Telegraph	M. L. Wood	Ithaca, N. Y	Nov. 26, 1867	71, 564
Insulator, Telegraph	J. Yandell	Saint Louis, Mo	Oct. 14, 1851	8, 438
Insulator, Telegraph-line	P. Ely and M. B. Fenninger	Kinzer's Post-Office, Pa	May 6, 1873	138, 489
Insulator, Telegraph-line	E. A. Merrell	East Lyme, Conn	May 6, 1873	138, 512
Insulator, Telegraph-line	I. I. Tatom	West Point, Ga	May 27, 1873	139, 436
Insulators, &c., Compound for telegraph	J. F. Boynton	Syracuse, N. Y	Jan. 22, 1867	61, 390
Insulators for telegraph-poles, Manufacture of	H. Brooke	New York, N. Y	Jan. 25, 1870	99, 145
Insulators, Former for segmentally screw-threaded	S. Oakman	Boston, Mass	Oct. 15, 1872	132, 214
Insulators, Former for segmentally screw-threaded	S. Oakman	Boston, Mass	Oct. 15, 1872	132, 215
Insulators, Mode of forming	J. M. Brookfield	Brooklyn, N. Y	May 31, 1870	103, 555
Insulators, Mode of making telegraph	W. Brookfield	New York, N. Y	Apr. 4, 1871	113, 3[illegible]3
Insulators, Molding telegraph	R. Hemingray	Covington, Ky	Dec. 19, 1871	122, 015
Insulators, Press for making glass	H. C. Deemer and J. W. Eufinger.	Brooklyn, N. Y	Sept. 10, 1872	131, 153
Insulators to telegraph-poles, Attaching	J. B. Stearns	Boston, Mass	Sept. 13, 1870	107, 303
Intagliotype-plate	E. B. Larcher	New York, N. Y	June 20, 1865	48, 2[illegible]0
Interest-table, Circular	J. Jelleff	Butternuts, N. Y	Sept. 8, 1812	......
Interest-table, Circular or revolving	G. D. Lowe	Hampfield, Pa	Aug. 12, 1815	......
Interfering-attachment	F. B. Doughty	New York, N. Y	May 7, 1867	64, 506
Interfering-bands in place, Device for holding	J. W. Currier	Springfield, Mass	June 1, 1869	90, 731
Interfering-pad	G. W. Fry	Uxbridge, Mass	Aug. 24, 1869	94, 0[illegible]8
Interfering-pad	W. A. Lovelace	Richmond, Mass	Sept. 14, 1869	94, 902
Interfering-pad for horses	W. Somerville	Buffalo, N. Y	June 14, 1870	104, 359
Interfering-strap for horses	E. Chesterman	Boston, Mass	Sept. 1, 1868	81, 601
Intestines, Composition for preserving animal	T. L. Reed	Providence, R. I	Dec. 3, 1867	71, 787
Intestines, Machine for cleaning animal	C. F. Dortenbach	Cleveland, Ohio	June 16, 1863	38, 888
Invalid back-rest	W. F. Brown	New Bedford, Mass	Aug. 11, 1863	39, 462
Invalid-chair	T. Armstrong	Sandwich, Ill	July 15, 1873	140, 868
Invalid-chair	T. J. Blackburn and E. P. Terrel	Spring Hills, Ohio	Sept. 12, 1865	49, 849

Index of patents issued from the United States Patent Office from 1790 to 1873, inclusive—Continued.

Invention.	Inventor.	Residence.	Date.	No.
Invalid-chair	A. P. Blunt and J. S. Smith	Washington, D. C	Feb. 16, 1869	86, 899
Invalid-chair	G. T. Fowler	East Somerville, Mass	May 2, 1871	114, 283
Invalid-chair	J. G. Holmes	Charleston, S. C	Sept. 24, 1844	3, 761
Invalid-chair	J. G. Holmes	Charleston, S. C	June 16, 1857	17, 567
Invalid-chair	D. S. James	New Market, Va	May 13, 1856	14, 872
Invalid-chair	G. A. Mansfield	Melrose, Mass	Aug. 25, 1863	39, 663
Invalid-chair	J. N. McMullen	West Liberty, Ohio	Mar. 5, 1867	62, 663
Invalid-chair	C. Messenger	Warren, Ohio	Sept. 25, 1860	30, 149
Invalid-chair	H. F. Siebold	New York, N. Y	Sept. 10, 1872	131, 230
Invalid-chair	C. L. Taillant	New York, N. Y	Sept. 2, 1856	15, 673
Invalid-chair	J. B. Wallace	Franklin, Ohio	Jan. 8, 1867	61, 126
Invalid-chair	L. M. Whitman	Sterling, Ill	Nov. 3, 1868	83, 746
Invalid-chair	R. Witherell	Huntington, Mass	Apr. 7, 1857	17, 008
Invalid chair and lounge	W. Y. Eastes	Summitville, Ind	Aug. 16, 1870	106, 343
Invalid-chair, Conduit for	G. Wells	Bethel, Conn	Mar. 23, 1869	88, 104
Invalid-chair, Traveling	C. L. Bander	Cleveland, Ohio	Nov. 10, 1863	40, 547
Invalid-chair, Traveling	C. L. Bauder	Cleveland, Ohio	Dec. 18, 1866	60, 464
Invalid locomotive-chair	T. S. Minniss	Meadville, Pa	May 10, 1853	9, 708
Invalid-rest	T. S. Kennard	Exeter, N. H	Sept. 15, 1868	82, 125
Invalid-supporter	J. T. Alston	Raleigh, N. C	Aug. 12, 1856	15, 504
Invalid-table	J. M. Allen	Worcester, Mass	Mar. 1, 1859	23, 068
Invalid-table	S. Ustick	Philadelphia, Pa	July 18, 1865	48, 854
Invalid-table	T. N. Webb	Baltimore, Md	June 27, 1871	116, 516
Invalid table and apparatus	S. Ustick	Philadelphia, Pa	Sept. 5, 1865	49, 807
Invalid-table and book-holder	P. J. Probeck and J. B. Corlett	Newburgh, Ohio	May 18, 1869	90, 124
Invalids, Apparatus for removing	J. Ruth	Philadelphia, Pa	Oct. 22, 1861	33, 545
Iodine from sea-water, Extracting	R. Cupper	New York, N. Y	Jan. 22, 1867	61, 404
Iodine, Manufacture of	J. Fongerat	New York, N. Y	May 19, 1868	78, 078
Iodine, Process for the manufacture of	J. Fongerat and L. A. Tartierè	Quogue, N. Y	Oct. 27, 1868	83, 372
Iron: *See* Andiron. Axle-iron. Boot and shoe iron. Branding-iron. Carriage-iron. Chafe-iron. Clinching-iron. Clothes-iron. Curling-iron. Draw-bar iron. Edge-iron. Flat-iron. Fluting-iron. Fluting and puffing iron. Fluting and sad iron. Fore-iron. Gas and fluting iron. Glossing and fluting iron. Grappling-iron. Hat-shell iron. Hod-iron. Holdback-iron. Lasting-iron. Laundry-iron. Mop-iron. Plane-iron. Plow-iron. Polishing-iron. Puffing-iron. Rub-iron. Sad-iron. Sad and embossing iron. Sad and fluting iron. Shaft-iron. Sheet-iron. Smoothing-iron. Soldering-iron. Tew-iron. Tuyere-iron. Wafer-iron. Waffle-iron.				
Iron against corrosion, Protecting	C. Godfrey and R. Lighthall	New York, N. Y	Nov. 22, 1870	109, 509
Iron and apparatus therefor, Manufacture of	J. D. Whelpley and J. J. Storer	Boston, Mass	Mar. 22, 1870	101, 067
Iron and castings, Manufacturing sheet, boiler, slit, and bar.	N. Callender	Cumberland County, Pa	Dec. 14, 1832	....
Iron and copper, Process of purifying	K. W. Zenger	Prague, Austria	June 18, 1872	128, 088
Iron and copper with tin, &c., Coating	E. P. Morewood	New York, N. Y	Sept. 17, 1844	3, 746
Iron and cutting nails, Machine for rolling and plating.	S. Couch and A. Towne	Boston, Mass	Dec. 10, 1808	....
Iron and granulating the same, Manufacture of	W. H. Perry	Sharon, Pa	Feb. 8, 1870	99, 588
Iron and making castings, Method of refining cast	Z. S. Durfee	New York, N. Y	Aug. 29, 1871	118, 597
Iron and other material to give greater strength for building, &c., Treating.	W. Haggett	Watford, England	Sept. 3, 1872	130, 992
Iron and other metals, Composition of	W. M. Arnold	New York, N. Y	June 12, 1866	55, 452
Iron and other metals, Construction of converter and furnace for treating.	J. Absterdam	New York, N. Y	Feb. 9, 1869	86, 796
Iron and other metals from potter's clay, Process of separating.	W. J. Lynd	Golden City, Colo	May 25, 1869	90, 565
Iron and other metals with protecting alloys, Coating:	C. Marshall	Philadelphia, Pa	Nov. 11, 1873	144, 403
Iron and other ores, Furnace and process for treating.	J. Y. Smith	Pittsburgh, Pa	Nov. 15, 1870	109, 355
Iron and other ores, Reduction of	H. S. Lucas	Chester, Mass	Nov. 1, 1864	44, 872
Iron and other oxides from clay, porcelain-earth, &c., Method of extracting.	W. J. Lynd	Golden City, Colo	July 27, 1869	92, 981
Iron and removing scale from iron for coating with other metals, Manufacture of sheet.	C. Marshall	Philadelphia, Pa	May 16, 1871	114, 956
Iron and steel and in the deoxidizing of iron-ores, Process of manufacturing malleable cast.	A. K. Eaton	New York, N. Y	June 25, 1861	32, 621
Iron and steel, Apparatus for refining	T. S. C. Lowe	Morristown, Pa	Aug. 13, 1872	130, 380
Iron and steel, Apparatus for the manufacture of	C. Adams	Philadelphia, Pa	Dec. 21, 1869	98, 139
Iron and steel, Apparatus for the manufacture of	W. Gerhardt	Pittsburgh, Pa	Feb. 13, 1866	52, 559
Iron and steel, Apparatus for the manufacture of	A. L. Holley and J. B. Pease	Swatara Township, Pa	Jan. 26, 1869	86, 304
Iron and steel bar, Combined	E. Wheeler	Hudson, Mass	June 14, 1870	104, 237
Iron and steel by applying anthracite coal, Manufacturing.	F. W. Geissenhainer	New York	Dec. 19, 1833	....
Iron and steel by means of blasts of air, Purifying	G. W. Swett	Troy, N. Y	June 30, 1863	39, 078
Iron and steel by rolling, Manufacturing	R. W. Bangs and S. W. Walbridge.	Bennington, Vt	Dec. 31, 1833	....
Iron and steel by the Bessemer or pneumatic process, Manufacture of.	A. L. Holley	Swatara Township, Pa	Jan. 26, 1869	86, 303
Iron and steel, Carbonizing and hardening	F. E. Sessions	Worcester, Mass	Mar. 24, 1868	75, 986
Iron and steel castings, Manufacture of	R. Yeilding	New York, N. Y	Oct. 3, 1871	119, 682
Iron and steel combined, Manufacture of bars and articles of.	W. M. Pickslay	Philadelphia, Pa	Nov. 13, 1866	59, 644
Iron and steel, Composition-flux for manufacture of	J. Jameson	Philadelphia, Pa	Sept. 7, 1869	94, 605
Iron and steel, Compound for hardening	J. McDonald	Kankakee, Ill	Dec. 5, 1871	121, 645

Index of patents issued from the United States Patent Office from 1790 *to* 1873, *inclusive*—Continued.

Invention.	Inventor.	Residence.	Date.	No.
Iron and steel, Compound for hardening	C. Pauvert	Targé, France	Mar. 23, 1858	19,710
Iron and steel direct from the ore, Manufacture of wrought.	W. Henderson	Glasgow, Scotland	Nov. 27, 1866	60,002
Iron and steel direct from the ore, Manufacture of wrought.	G. H. Smith	New York, N. Y	July 20, 1869	92,894
Iron and steel directly from the ore, Manufacture of	C. M. Dupuy	New York, N. Y	Feb. 28, 1865	46,549
Iron and steel, Electro-deposition of copper and brass in.	W. H. Walenn	London, England	June 7, 1870	103,947
Iron and steel, Flux for refining	E. T. Atwood	Minerva, Ohio	Apr. 25, 1871	114,090
Iron and steel, Flux for the manufacture of	L. Sibert	Staunton, Va	Aug. 1, 1871	117,693
Iron and steel, Flux for welding, puddling, and brazing.	A. J. Hindermeyer	Rohrerstown, Pa	Dec. 18, 1866	60,516
Iron and steel from granulated iron, Manufacture of.	C. Wood	Middlesborough-on-Tees, England.	Oct. 28, 1873	144,009
Iron and steel from iron sponge, Manufacture of wrought.	T. S. Blair	Pittsburgh, Pa	May 21, 1872	126,923
Iron and steel from rust, Preserving	M. Sorel	Paris, France	Dec. 7, 1837	510
Iron and steel furnace	T. J. Chubb	Williamsburgh, N. Y	June 8, 1869	90,924
Iron and steel furnace	G. E. Harding	New York, N. Y	Sept. 23, 1873	143,145
Iron and steel furnace	J. W. Nystrom	Philadelphia, Pa	Oct. 8, 1861	33,446
Iron and steel furnace	A. Pousard	Paris, France	Oct. 27, 1868	83,542
Iron and steel furnace	J. M. Roberts	Burlington, N. J	Nov. 28, 1871	121,422
Iron and steel furnace	H. Ross and D. F. Agnew	Pittsburgh, Pa	July 13, 1869	92,478
Iron and steel furnace	H. Ross and J. H. Clemens	Pittsburgh, Pa	Aug. 22, 1871	118,279
Iron and steel furnace	J. G. Trotter	Newark, N. J	Mar. 16, 1869	87,889
Iron and steel in furnace, and apparatus therefor, Manufacture of.	G. J. and T. C. Hinde	Wolverhampton, England, and Ynispenllweh, Wales.	Mar. 30, 1869	88,480
Iron and steel furnace and manufacture	T. C. Hinde	Fownhope, England	Dec. 12, 1871	121,872
Iron and steel, Furnace and process for the manufacture of.	F. Ellershausen	Montreal, Canada	May 12, 1868	77,722
Iron and steel, Furnace and tool for treating	W. Yates	Westminster, England	Feb. 23, 1869	87,231
Iron and steel, Furnace for converting cast-iron into.	W. Fields	Wilmington, Del	May 24, 1870	103,437
Iron and steel, Hardening	G. J. Farmer	Birmingham, England	Apr. 6, 1858	19,836
Iron and steel in run-ways, Refining	S. M. Wickersham	Pittsburgh, Pa	Dec. 3, 1872	133,690
Iron and steel, &c., Increasing the strength of wrought.	W. R. Johnson	Philadelphia, Pa	June 30, 1838	814
Iron and steel, Increasing the strength of wrought	W. R. Johnson	Philadelphia, Pa	July 9, 1838	826
Iron and steel, Machinery and buildings for manufacture of.	H. Bessemer	London, England	Aug. 15, 1871	117,968
Iron and steel, Machinery for the manufacture of	H. Bessemer	London, England	July 25, 1865	49,055
Iron and steel, Machinery to aid in puddling	H. Bennett	Wombridge, England	Feb. 23, 1864	41,671
Iron and steel, Making	W. P. Boyden	New York, N. Y	June 11, 1836	......
Iron and steel, Making	H. G. Spafford	Albany, N. Y	Oct. 30, 1822	......
Iron and steel, Manufacture of	G. F. Ansell	London, England	June 28, 1870	104,686
Iron and steel, Manufacture of	H. Bessemer	London, England	Nov. 11, 1856	16,082
Iron and steel, Manufacture of	H. Bessemer	London, England	July 25, 1865	49,051
Iron and steel, Manufacture of	H. Bessemer	London, England	July 25, 1865	49,052
Iron and steel, Manufacture of	H. Bessemer	London, England	July 25, 1865	49,053
Iron and steel, Manufacture of	H. Bessemer	London, England	Dec. 5, 1865	51,397
Iron and steel, Manufacture of	H. Bessemer	London, England	Dec. 5, 1865	51,398
Iron and steel, Manufacture of	H. Bessemer	London, England	Sept. 21, 1869	94,994
Iron and steel, Manufacture of	H. Bessemer	London, England	Sept. 21, 1869	94,995
Iron and steel, Manufacture of	H. Bessemer	London, England	Sept. 21, 1869	94,996
Iron and steel, Manufacture of	H. Bessemer	London, England	Sept. 21, 1869	94,997
Iron and steel, Manufacture of	O. Bolton, jr., and J. Pedder	Pittsburgh, Pa	July 15, 1873	140,761
Iron and steel, Manufacture of	J. B. Bradley and E. De Camp	Morristown, N. J	Aug. 27, 1872	130,787
Iron and steel, Manufacture of	J. P. Budd	Ystalyfera, near Swasea, Wales.	May 10, 1870	102,912
Iron and steel, Manufacture of	C. Carpenter, jr	Seekonk, Mass	Dec. 23, 1873	145,843
Iron and steel, Manufacture of	N. Cutter and E. Savage	Cincinnati, Ohio, and West Meriden, Conn.	Oct. 5, 1869	95,568
Iron and steel, Manufacture of	Z. S. Durfee	New York, N. Y	May 2, 1871	114,277
Iron and steel, Manufacture of	Z. S. Durfee	New York, N. Y	Jan. 2, 1872	122,312
Iron and steel, Manufacture of	F. Ellershausen, A. E. Stayner, and A. Guzman.	Ellershouse, Halifax, Nova Scotia, and New York, N. Y.	Nov. 17, 1868	84,053
Iron and steel, Manufacture of	W. Ennis	Philadelphia, Pa	Dec. 14, 1869	97,897
Iron and steel, Manufacture of	F. P. Fletcher and V. W. Blanchard.	Bridport, Vt	Nov. 16, 1869	96,905
Iron and steel, Manufacture of	A. L. Fleury and C. Adams	Philadelphia and Pittsburgh, Pa.	May 13, 1862	35,276
Iron and steel, Manufacture of	J. L. Floyd	Philadelphia, Pa	June 15, 1869	91,324
Iron and steel, Manufacture of	W. Gerhardt	Pittsburgh, Pa	Feb. 13, 1866	52,560
Iron and steel, Manufacture of	J. Henderson	Brooklyn, N. Y	Oct. 17, 1865	50,474
Iron and steel, Manufacture of	J. Henderson	New York, N. Y	Aug. 17, 1869	93,713
Iron and steel, Manufacture of	J. Henderson	New York, N. Y	Mar. 29, 1870	101,263
Iron and steel, Manufacture of	J. Henderson	New York, N. Y	Aug. 16, 1870	106,365
Iron and steel, Manufacture of	J. Jameson	Philadelphia, Pa	Mar. 23, 1869	88,173
Iron and steel, Manufacture of	J. J. Johnston	Allegheny City, Pa	July 27, 1869	93,155
Iron and steel, Manufacture of	J. A. Jones	Middlesborough, England	June 9, 1868	78,806
Iron and steel, Manufacture of	J. A. Jones	Middlesborough, England	Sept. 8, 1868	81,908
Iron and steel, Manufacture of	C. Low	England	May 28, 1846	4,540
Iron and steel, Manufacture of	J. G. Martien	Newark, N. J	Feb. 24, 1857	16,690
Iron and steel, Manufacture of	J. W. Middleton	Philadelphia, Pa	Mar. 21, 1871	112,828
Iron and steel, Manufacture of	R. Mushet	Coleford, England	May 26, 1857	17,389
Iron and steel, Manufacture of	C. M. Nes	York, Pa	Nov. 29, 1870	109,752
Iron and steel, Manufacture of	C. M. Nes	York, Pa	Jan. 30, 1872	123,191
Iron and steel, Manufacture of	G. Parry	Ebro Vale Iron-Works, England.	Apr. 25, 1865	47,506
Iron and steel, Manufacture of	E. Peckham	Antwerp, N. Y	Oct. 14, 1873	143,637
Iron and steel, Manufacture of	J. Pedder and G. Abel	Pittsburgh, Pa	Apr. 8, 1873	137,621
Iron and steel, Manufacture of	O. M. Phillips	New York, N. Y	Feb. 9, 1869	86,859
Iron and steel, Manufacture of	J. Player	Philadelphia, Pa	Oct. 19, 1869	95,933
Iron and steel, Manufacture of	J. Ralston, A. L. Thomas, and W. Parkenson.	Tamaqua, Pa	Mar. 23, 1869	88,208
Iron and steel, Manufacture of	L. Sibert	Mount Solon, Va	May 21, 1867	64,916
Iron and steel, Manufacture of	L. Sibert	Mount Solon, Va	June 23, 1868	79,152

Index of patents issued from the United States Patent Office from 1790 *to* 1873, *inclusive*—Continued.

Invention.	Inventor.	Residence.	Date.	No.
Iron and steel, Manufacture of	L. Sibert	Staunton, Va	Oct. 4, 1870	107, 058
Iron and steel, Manufacture of	C. W. Siemens	Westminster, England	Oct. 12, 1869	95, 843
Iron and steel, Manufacture of	G. H. Smith	New York, N. Y	July 12, 1870	105, 267
Iron and steel, Manufacture of	T. Southall and C. Crudgington	Kidderminster, England	Sept. 14, 1844	3, 742
Iron and steel, Manufacture of	H. Spencer and L. K. Saylor	Philadelphia, Pa	Dec. 7, 1869	97, 718
Iron and steel, Manufacture of	C. M. Tessié du Motay	Paris, France	Mar. 2, 1869	87, 479
Iron and steel, Manufacture of	C. Usher	Iowa Falls, Iowa	Mar. 5, 1867	62, 706
Iron and steel, Manufacture of	R. Yeilding	Detroit, Mich	Oct. 13, 1868	83, 119
Iron and steel, Manufacture of	H. K. York	Cardiff, Great Britain	Dec. 24, 1867	72, 579
Iron and steel, Manufacture of malleable	H. Bessemer	London, England	Dec. 5, 1865	51, 399
Iron and steel, Manufacture of malleable	H. Bessemer	London, England	Dec. 5, 1865	51, 400
Iron and steel, Manufacture of malleable	H. Bessemer	London, England	Dec. 5, 1865	51, 401
Iron and steel, Manufacture of malleable	J. C. Ridley	Newcastle-upon-Tyne, England.	Aug. 1, 1871	117, 684
Iron and steel, Manufacture of malleable	E. B. Wilson	Westminster, England	July 28, 1863	39, 364
Iron and steel, Manufacture of pure	H. Larkin, A. Leighton, and W. White.	Theydon Germon, Liverpool, and Hampstead, England.	Apr. 9, 1872	125, 464
Iron and steel, Manufacturing	O. Bolton, jr., and J. Pedder	Pittsburgh, Pa	Feb. 4, 1873	135, 512
Iron and steel, Metal plates of	J. Myers, jr	Williamsburgh, N. Y	Oct. 4, 1870	108, 042
Iron and steel, Metallic solution for coating	A. A. Lothrop	Neponset, Mass	Nov. 30, 1869	97, 417
Iron and steel, Method of uniting	W. and W. H. Terwilliger and J. S. Lockwood.	New York, N. Y	July 24, 1866	56, 680
Iron and steel, Mode of making bars, shafts, &c., composed of.	C. Sanderson	Sheffield, England	Sept. 19, 1865	50, 084
Iron and steel, Mode of plating, coating, and ornamenting articles of.	K. Frazer	Syracuse, N. Y	Oct. 26, 1869	96, 217
Iron and steel, Mode of utilizing tin-plate cuttings in the manufacture of.	D. D. Parmelee	New York, N. Y	June 29, 1869	91, 962
Iron and steel, Preventing corrosion of	R. A. Fisher	San Francisco, Cal	Dec. 16, 1873	145, 496
Iron and steel, Process and apparatus for making	J. Jameson	Philadelphia, Pa	Mar. 30, 1869	88, 299
Iron and steel, Process and apparatus for the manufacture of.	H. Bessemer	London, Great Britain	Feb. 22, 1870	100, 003
Iron and steel, Process and apparatus for the manufacture of.	J. Jameson	Philadelphia, Pa	Mar. 31, 1868	76, 196
Iron and steel, Process and apparatus for the manufacture of.	J. W. Middleton	Philadelphia, Pa	July 16, 1872	129, 243
Iron and steel, Process for refining	J. Absterdam	New York, N. Y	Jan. 23, 1866	52, 121
Iron and steel, Process of making	M. Lane	Washington, D. C	Dec. 17, 1861	33, 949
Iron and steel, Process of manufacturing axles, &c., from.	H. Bessemer	London, England	July 25, 1865	49, 054
Iron and steel, Process of mixing	J. Cartwright	Youngstown, Ohio	Jan. 7, 1868	73, 163
Iron and steel, Process of purifying	J. F. Bennett	Pittsburgh, Pa	Mar. 10, 1868	75, 240
Iron and steel, Process of treating	E. Savage	West Meriden, Conn	Feb. 16, 1869	86, 944
Iron and steel puddling machinery	T. Harrison	Tudhoe, England	June 7, 1864	43, 067
Iron and steel, &c., Purifying	E. Brady	Philadelphia, Pa	Oct. 5, 1869	95, 419
Iron and steel, Purifying	W. Gerhardt	New York, N. Y	Nov. 3, 1863	40, 472
Iron and steel, Refining	J. E. Atwood	Pittsburgh, Pa	Sept. 13, 1870	107, 322
Iron and steel, Refining	N. B. Hatch	Pittsburgh, Pa	Dec. 17, 1872	133, 937
Iron and steel Refining	W. H. Kimball	Boston, Mass	June 20, 1871	116, 065
Iron and steel, Refining	I. M. Phelps	Philadelphia, Pa	Dec. 16, 1873	145, 680
Iron and steel, Refining	A. H. Siegfried	South Bend, Ind	Apr. 4, 1871	113, 583
Iron and steel shavings, turnings, &c., Melting	G. Whitney	Philadelphia, Pa	Nov. 5, 1872	132, 743
Iron and steel to anneal and toughen them, Treating	A. F. Andrews	New Haven, Conn	Sept. 23, 1873	142, 977
Iron and steel, Welding	J. Popping	New York, N. Y	Sept. 16, 1873	142, 939
Iron and steel wire, Annealing	G. I. Washburn	Worcester, Mass	Oct. 7, 1862	36, 628
Iron and steel with copper, brass, &c., Process of uniting.	R. Savary	Pittsburgh, Pa	Aug. 11, 1863	39, 531
Iron and steel with gold, silver, &c., Coating	J. B. Thompson	Middlesex County, England	Sept. 11, 1866	58, 037
Iron and steel with molten iron, Coating	R. B. Fowler and D. F. Brandon	Chicago, Ill	Mar. 28, 1871	113, 040
Iron and the separation of copper, silver, and other metals from their solutions, Preparing finely divided.	G. Bishop, jr	Swansea, Great Britain	Feb. 25, 1868	74, 791
Iron, Annealing and polishing sheet	J. W. Ells	Pittsburgh, Pa	July 25, 1865	48, 918
Iron, Annealing sheet	W. D. Wood	Borough of McKeesport, Pa	July 9, 1867	66, 546
Iron, Apparatus and machinery for puddling	P. A. Dormoy	Vienna, Austria	May 18, 1869	90, 158
Iron, Apparatus for heating the converters and other vessels used in the Bessemer process of treating.	C. G. Larson	Stockholm, Sweden	June 20, 1871	116, 201
Iron, Apparatus for making pig-blooms in the manufacture of.	J. A. Burden	Troy, N. Y	Aug. 3, 1869	93, 170
Iron, Apparatus for puddling	J. Davies	Knoxville, Tenn	Oct. 21, 1873	143, 811
Iron, Apparatus for puddling	J. Griffiths	Litchurch, England	June 27, 1865	48, 485
Iron, Apparatus for puddling	J. McCarty	Reading, Pa	Oct. 5, 1852	9, 303
Iron, Apparatus for puddling and melting	H. A. V. Post	New York, N. Y	Aug. 6, 1872	130, 241
Iron, Apparatus for purifying	S. M. Wrickersham	Allegheny, Pa	Dec. 14, 1869	97, 844
Iron, Apparatus for refining	J. W. Niptrom	Philadelphia, Pa	Apr. 3, 1866	53, 658
Iron, Apparatus for the manufacture of	J. Yates	Mott Haven, N. Y	Mar. 1, 1864	41, 806
Iron, Apparatus for the manufacture of	J. Yates	Mott Haven, N. Y	Mar. 1, 1864	41, 808
Iron, Apparatus for the manufacture of malleable	A. Dickerson	Newark, N. J	Mar. 13, 1847	5, 013
Iron, Apparatus for the manufacture of pig-bloom from cast.	J. A. Burden	Troy, N. Y	Aug. 17, 1869	93, 857
Iron articles into steel, Process for converting	R. A. Jackson	Pittsburgh, Pa	Feb. 11, 1873	135, 646
Iron at the rolls, Machine for moving	C. Hewitt	Trenton, N. J	June 7, 1859	24, 304
Iron axles, Gage for	W. C. Bamberger	Washington, D. C	Nov. 15, 1859	26, 079
Iron, Balanced "heave-up" for rolling	J. Copley, jr	Allegheny City, Pa	Sept. 24, 1867	69, 077
Iron balls, blooms, and slabs, Manufacture of malleable.	J. Player	Norton, England	Apr. 28 1868	77, 210
Iron bands, Machine for contracting the circumference of wrought.	W. Massey	Greene County, Ill	July 3, 1849	6, 573
Iron bars and rods, Polishing	B. Lauth	Pittsburgh, Pa	Feb. 26, 1861	31, 546
Iron bars, Constructing cast or wrought	C. McCammon	Albany, N. Y	Feb. 14, 1860	27, 1[illegible]
Iron bars, &c., Furnace for heating	J. Pardoe	Worcester, Mass	Mar. 11, 1873	136, 667
Iron bars, Hollowing out	T. Rowand	Gloucester County, Pa	Dec. 24, 1819	
Iron bars, Machine for straightening	G. H. Sellers	Phœnixville, Pa	Mar. 22, 1864	42, 051
Iron bars, Rolling	B. Lauth	Pittsburgh, Pa	Aug. 21, 1860	29, 702
Iron beam, Wrought	L. Kirkup	East New York, N. Y	Apr. 22, 1873	138, 029

Index of patents issued from the United States Patent Office from 1790 to 1873, inclusive—**Continued.**

Invention.	Inventor.	Residence.	Date.	No.
Iron beams, Machine for forming flanges on wrought.	J. H. Kroehl	New York, N. Y	Jan. 2, 1855	12, 133
Iron boat for navigating rivers, &c	T. J. Bond	Baltimore, Md	Dec. 21, 1820	....
Iron boiler, Cast	J. A. Miller	New York, N. Y	Oct. 24, 1865	50, 611
Iron boilers, Machine for bending sheet	C. Post	Springport, N. Y	Mar. 11, 1828	....
Iron boiling and puddling furnace	J. Westerman	Sharon, Pa	May 24, 1870	103, 401
Iron bolt, Malleable cast	P. Frost	Springfield, Vt	Jan. 26, 1869	86, 220
Iron borings, drillings, and filings rendered available in furnaces.	M. Maison	Georgetown, D. C	Nov. 8, 1831	....
Iron, brass, &c., Mold for casting	J. Leonard, jr., J. S. Brainard, and A. Sizer.	Wallingford, Conn	Jan. 28, 1830	....
Iron by means of fluxes introduced by suction, Refining and purifying.	J. Absterdam	New York, N. Y	Jan. 19, 1869	86, 050
Iron by the "Ellershausen process," Manufacture of.	H. Davies	Newport, Ky	July 5, 1870	104, 939
Iron, Carbonating and smelting	A. Whitney	Southfield, N. Y	July 17, 1837	269
Iron, Casting	W. Hancock		Feb. 26, 1799	....
Iron chills, molds, pig-beds, &c., Cold-fix for lining	H. A. Laughlin	Pittsburgh, Pa	Jan. 5, 1869	85, 529
Iron chips for remelting, Process for preparing	A. Povey	Portland, Me	July 1, 1873	140, 383
Iron chips, shavings, &c., for melting, Preparing	E. C. Haserick	Lake Village, N. H	Nov. 28, 1871	121, 245
Iron chips, turnings, &c., Mode of melting and aggregating.	E. C. Haserick	Lake Village, N. H	Mar. 13, 1866	53, 142
Iron-cleaning machine, Sheet	E. A. Harvey	Wilmington, Del	Apr. 4, 1865	47, 103
Iron, Coating	G. A. Mariner and F. Dorsett	Chicago, Ill	Aug. 15, 1871	118, 032
Iron, Coating and bronzing	L. Dockstader	West Meriden, Conn	Apr. 26, 1870	102, 378
Iron, Composition	W. M. Arnold	New York, N. Y	Aug. 23, 1864	43, 891
Iron, Composition and manufacture of	A. H. Everitt	New York, N. Y	Apr. 11, 1865	47, 198
Iron, Composition for carbonizing	C. V. Wilson	Newark, N. J	Aug. 2, 1864	43, 732
Iron, Composition for cementing	J. Johnson	East Brooklyn, N. Y	June 14, 1859	24, 429
Iron, Composition for coating	D. Read and J. F. Galley	New York, N. Y	Mar. 20, 1866	53, 336
Iron, Composition for covering and protecting	Y. Bey	New York, N. Y	Mar. 31, 1863	38, 022
Iron, Composition for refining and toughening	J. H. Wolf	Philadelphia, Pa	May 12, 1868	77, 904
Iron, Composition for treating	S. Weaver	Pottstown, Pa	Mar. 28, 1871	113, 228
Iron, Composition to be used in puddling	I. E. Richards	Columbia, Pa	Oct. 29, 1867	70, 361
Iron, Compound for coating	J. A. Sewall	Normal, Ill	Aug. 22, 1871	118, 397
Iron, Compound for hardening bar	F. E. Blake	Mattoon, Ill	Mar. 26, 1872	125, 009
Iron, Compound for hardening cast	B. W. Nichols	Canton, Ohio	May 11, 1869	90, 014
Iron, Compound for improving the quality of	T. Jones and W. Morgan	Pittsburgh, Pa	Apr. 30, 1867	64, 338
Iron, Corrugating	J. L. Vernam	Rochester, N. Y	Jan. 1, 1861	31, 008
Iron cutting and punching machine	A. J. Peavey	South Montville, Me	Jan. 4, 1859	22, 510
Iron cutting and punching machine	A. Shogren	Mission, Ill	Sept. 1, 1863	39, 757
Iron, Cutting, punching, and slitting	J. V. Buskark	Norwalk, Ohio	Apr. 6, 1833	....
Iron-cutting shears	J. Nichol	New York, N. Y	May 18, 1869	90, 184
Iron, Decarbonizing	J. F. Allen	Tremont, N. Y	Apr. 14, 1868	76, 581
Iron direct from the ore, Apparatus for making wrought.	J. Renton	Newark, N. J	Dec. 23, 1851	8, 613
Iron direct from the ore, Making wrought	J. Benton	Cleveland, Ohio	Oct. 24, 1854	11, 838
Iron direct from the ore, Process for making malleable.	M. S. Salter	Newark, N. J	Nov. 20, 1849	6, 886
Iron directly from the ore, Combination of furnaces for manufacturing wrought.	C. S. Quilliard	Rondout, N. Y	Dec. 23, 1841	2, 394
Iron directly from the ore, Manufacture of wrought	C. M. Dupuy	New York, N. Y	Jan. 23, 1866	52, 149
Iron directly from the ore, Manufacturing malleable	G. A. Whipple	Newark, N. J	May 10, 1853	9, 715
Iron directly from the ore, Method of making wrought.	A. Dickerson	Newark, N. J	July 22, 1850	7, 519
Iron directly from the ore, Mode of manufacturing malleable.	J. F. Winslow	Troy, N. Y	May 16, 1846	4, 526
Iron directly from the ore, Process of obtaining wrought.	J. D. Whelpley and J. J. Storer	Boston, Mass	Sept. 28, 1869	95, 295
Iron-drilling machine	J. H. Currier and W. H. Taber	Fairhaven, Mass	July 8, 1839	1, 224
Iron-drilling machine	I. S. Lauback	New York, N. Y	Feb. 9, 1864	41, 516
Iron, Finery-fire for the manufacture of	T. Ash and Z. Walker	Trenton, N. J	Nov. 25, 1873	144, 821
Iron-finishing furnace, Sheet	W. D. Wood	McKeesport, Pa	Mar. 14, 1865	46, 841
Iron, Fire-plating	W. H. Thoss	San Francisco, Cal	Aug. 30, 1859	25, 291
Iron, Flux for puddling	J. Burnish, J. Talbot, and T. W. Yardley.	Pottsville, Pa	Feb. 1, 1859	22, 779
Iron for car-wheels, &c., Manufacture of cast	C. Burgess	Portsmouth, Ohio	Oct. 21, 1873	143, 874
Iron for fronts of buildings, &c., Coating	J. Alexander	Green Point, N. Y	Sept. 21, 1869	95, 064
Iron for making steel and malleable cast-iron, Mode of treating pig.	J. R. Speer	Pittsburgh, Pa	July 6, 1869	92, 221
Iron for malleable-iron castings, Process of melting and refining.	T. J. Chubb	Williamsburgh, N. Y	June 8, 1869	90, 927
Iron for strap-joint	W. D. Rinehart	Pittsburgh, Pa	Aug. 29, 1869	49, 683
Iron for the manufacture of stoves, pipes, &c., Machine for punching sheet.	S. Davis	Mifflin, Pa	Feb. 21, 1840	1, 495
Iron forging and crushing apparatus	R. Morrison	Newcastle-upon-Tyne, Great Britain.	Mar. 4, 1862	34, 587
Iron, &c., Forming oblique catches, &c., on plates and other pieces of cast.	J. S. Mott	New York, N. Y	Dec. 1, 1840	1, 872
Iron from blast-furnace slag, Manufacture of	J. J. Vinton	Sharon, Pa	Oct. 14, 1873	143, 600
Iron from brass turnings and filings, Magnetic machine for separating.	J. Jouson	Baltimore, Md	Aug. 27, 1867	68, 205
Iron from corrosion, Preserving	C. De Bussey	Paris, France	July 19, 1864	43, 630
Iron from corrosion, Preserving	G. W. Holley	Niagara, N. Y	Dec. 1, 1863	40, 752
Iron from furnace-cinders, Machinery for separating.	O. Walroth and L. Evans	Chittenango and Manlius, N. Y.	Dec. 21, 1852	9, 493
Iron from gas-purifiers, Treating and utilizing oxides of.	W. Cleland	Liverpool, England	May 24, 1864	42, 898
Iron from mill-cinders, Manufacture of	C. M. Nes	York, Pa	Apr. 2, 1872	125, 212
Iron from ore-cinders or slag, Manufacture of wrought.	J. D. Whelpley and J. J. Storer	Boston, Mass	May 3, 1870	102, 740
Iron from oxidation, Method of protecting	E. G. Pomroy	New York, N. Y	June 28, 1859	24, 604
Iron from oxidation, Process of protecting	P. Sumner and P. Naylor	New York, N. Y	Oct. 18, 1839	1, 374
Iron from pig-metal, Manufacturing malleable	T. C. Lewis	Pittsburgh, Pa	Oct. 1, 1830	....
Iron from sand, Machine for separating	G. H. Sanborn	Boston, Mass	Jan. 22, 1867	61, 471
Iron from the fire, Apparatus for drawing	D. N. Williams	Chicago, Ill	Mar. 19, 1867	63, 125
Iron from the ore, Furnace for making	M. Bell and E. B. Isett	Sabbath Rest and Tyrone City, Pa.	Nov. 14, 1854	11, 927

Index of patents issued from the United States Patent Office from 1790 to 1873, inclusive—Continued.

Invention.	Inventor.	Residence.	Date.	No.
Iron from the ore, Furnace for manufacturing wrought.	T. W. Harvey	New York, N. Y	Aug. 22, 1854	11, 584
Iron from the ore, Manufacture of cast	P. E. Jay	Saint Jean Baptiste, Canada.	Sept. 30, 1873	143, 350
Iron from the ore, Manufacture of purified	J. W. Middleton	Philadelphia, Pa	Oct. 4, 1870	107, 942
Iron from the ore, Manufacture of wrought	H. Boardman	New York, N. Y	June 27, 1865	48, 478
Iron from the ore, Manufacturing malleable	W. N. Clay	Flimby, England	July 25, 1845	4, 103
Iron from the ore, Mode of obtaining wrought	S. Broadmeadow	New York, N. Y	May 30, 1844	3, 605
Iron from the ore, Process of making cast	P. E. Jay	Saint Jean Baptiste, Canada.	Dec. 24, 1872	134, 289
Iron from the slag of blast-furnaces, Extracting	E. J. Bird	Frostburgh, Md	Oct. 11, 1870	108, 235
Iron from the water of salt wells and springs, Mode of separating compounds of.	G. H. Cook	New Brunswick, N. J	Oct. 29, 1861	33, 574
Iron from titaniferous iron-ore, Manufacture of	J. Player	New York, N. Y	July 7, 1868	79, 681
Iron-furnace	A. Hamar	New York, N. Y	May 12, 1868	77, 730
Iron-furnace	B. B. Howell	Philadelphia, Pa	Oct. 11, 1828	....
Iron, Furnace and oven for the manufacture of	J. Yates	Mott Haven, N. Y	Mar. 1, 1864	41, 807
Iron, Furnace for boiling	C. D. Baker	Wheeling, W. Va	June 27, 1865	48, 355
Iron, Furnace for making	D. W. Hendrickson and J. P. McLean.	New York, N. Y	Sept. 3, 1867	68, 566
Iron, Furnace for making malleable	M. S. Salter	Saltersville, N. J	Nov. 24, 1863	40, 710
Iron, Furnace for manufacture of	A. Reese	Pittsburgh, Pa	Jan. 9, 1872	122, 651
Iron, Furnace for manufacturing	I. C. Bryant	Philadelphia, Pa	Dec. 31, 1838	1, 057
Iron, Furnace for producing malleable	D. R. Nash	Brooklyn, N. Y	Sept. 9, 1873	142, 716
Iron-furnaces, Machine for charging	E. G. Scovil	Saint John, Canada	May 2, 1871	114, 346
Iron-galvanizing furnace	W. Blake	Boston, Mass	Mar. 19, 1861	31, 696
Iron gate or fence post	T. E., A., and E. King	Cherry Valley, Ohio	May 25, 1858	20, 352
Iron girder	P. H. Jackson	New York, N. Y	Nov. 28, 1871	121, 374
Iron, Hardening	W. C. Dunn	La Porte, Ind	July 24, 1866	56, 539
Iron, Hardening	H. Paddock	Saint Johnsbury, Vt	Feb. 12, 1867	61, 949
Iron, Hardening cast	T. Allin	New York, N. Y	Apr. 26, 1864	42, 453
Iron-hardening process	T. H. Jenkins	New York, N. Y	Dec. 26, 1865	51, 723
Iron, Hardening the surface of	R. T. King	Pana, Ill	Dec. 2, 1873	145, 110
Iron, Hearth for working and refining	R. S. Stewart, J. Christopher, and R. Forward.	Ligonier and Somerset, Pa	Mar. 1, 1859	23, 123
Iron-heating furnace, Bar	H. Burden	Troy, N. Y	Oct. 14, 1834	....
Iron-heating furnaces, Charging-apparatus for	W. F. Maharg	Saint Louis, Mo	Dec. 24, 1872	134, 151
Iron-holder	W. B. Coates	Philadelphia, Pa	Jan. 23, 1866	52, 140
Iron hoops, Straightening	B. Seymour		June 26, 1797	....
Iron house	D. D. Badger	New York, N. Y	Aug. 7, 1855	13, 379
Iron houses, Construction of	C. Mettam	New York, N. Y	July 11, 1854	11, 290
Iron in casting chilled rolls, Method of giving a rotary motion to the melted.	J. C. Parry	Pittsburgh, Pa	Oct. 16, 1849	6, 805
Iron in sea-water, Mode of preventing corrosion in pipes, bolts, and similar articles of.	R. Lighthall	Brooklyn, N. Y	Dec. 7, 1869	97, 657
Iron in the hearth of a blast-furnace, Refining	C. Shunk	Canton, Ohio	May 17, 1859	24, 060
Iron into bar iron and steel, Process of converting cast.	J. Heaton	Langley Mill, England	Aug. 13, 1867	67, 762
Iron into malleable iron, Converting cast	A. Hamer and G. H. Sellers	Philadelphia, Pa., and Wilmington, Del.	Feb. 2, 1869	86, 537
Iron into steel and malleable iron, Process of converting cast.	F. P. Fletcher and V. W. Blanchard.	Bridgeport, Vt	Aug. 6, 1867	67, 426
Iron into steel, Apparatus for converting	J. Webster	Birmingham, England	Apr. 8, 1873	137, 742
Iron into steel by cementation, Process of converting.	E. R. Weston	East Corinth, Me	June 11, 1861	32, 546
Iron into steel by means of hydrocarbon vapor, Apparatus for converting.	T. R. Scowden	Cincinnati, Ohio	Aug. 13, 1872	130, 540
Iron into steel, Composition for converting	T. Sheehan	Dunkirk, N. Y	June 4, 1867	65, 512
Iron into steel, Converting	J. F. Boynton	Syracuse, N. Y	July 16, 1867	66, 785
Iron into steel, Converting	E. F. Houghton	Philadelphia, Pa	June 18, 1872	128, 042
Iron into steel, Converting	S. C. Salisbury	New York, N. Y	Aug. 27, 1867	68, 118
Iron into steel, Converting articles made of wrought	L. La Breche-Viger	Montreal, Canada	Nov. 28, 1871	121, 263
Iron into steel, Converting articles of cast	B. W. Nichols	Canton, Ohio	July 20, 1869	92, 876
Iron into steel, Converting cast	W. Harris and A. Woolever	Allentown, Pa	Apr. 12, 1870	101, 963
Iron into steel, Converting pig	E. Leonard	Canton, Mass	Jan. 6, 1812	....
Iron into steel, Method of converting	E. G. Pomeroy	New York, N. Y	Sept. 4, 1860	29, 909
Iron into steel, Method of converting	T. Sheehan	Dunkirk, N. Y	Sept. 4, 1860	29, 919
Iron into steel, Method of converting cast	A. Hamar	Philadelphia, Pa	Feb. 2, 1869	86, 536
Iron into steel, Mode of converting articles of	R. A. Jackson	Lawrenceville, Pa	Dec. 3, 1867	71, 620
Iron into steel, Mode of converting articles of	B. W. Nichols	Canton, Ohio	Apr. 27, 1869	89, 329
Iron into steel, Mode of converting cast	T. Clark	Louisville, Ky	June 25 1872	128, 362
Iron into steel partially, Converting	E. Jenks	Colebrook, Conn	Jan. 28, 1818	....
Iron into steel, Process and apparatus for converting cast.	A. Thoma	New York, N. Y	Nov. 9, 1869	96, 633
Iron into steel, Process for converting	T. J. Barron	Brooklyn, N. Y	Jan. 1, 1867	60, 823
Iron into wrought iron and steel, Converting cast	J. R. Bradley	Ironton, Ohio	Apr. 15, 1862	34, 937
Iron into wrought iron and steel, Converting cast	A. C. Rand	Aurora, Ill	Mar. 18, 1873	137, 025
Iron, Machine for bending sheet or plate	J. Watchman	Baltimore, Md	June 1, 1843	3, 116
Iron, Machine for cutting rings from	C. H. Bassett	Birmingham, Conn	Aug. 1, 1865	49, 191
Iron, Machine for dressing	J. G. Tibbets	New York, N. Y	June 21, 1839	1, 183
Iron, &c., Machine for drilling	A. Morgan	Wooster, Ohio	July 20, 1843	3, 187
Iron, Machine for molding and rolling	J. White	Philadelphia, Pa	Mar. 14, 1810	....
Iron, Machine for rolling	H. B. Comer	Temperanceville, Pa	Mar. 29, 1859	23, 425
Iron, Machine for rolling bar	J. S. Haitupee and A. Alexander.	Pittsburgh, Pa	Apr. 19, 1853	9, 673
Iron, Machine for scraping	C. Zug	Pittsburgh, Pa	Apr. 4, 1871	113, 611
Iron, Machine for squeezing puddled balls of	S. Gissinger	Lawrenceville, Pa	Dec. 17, 1867	72, 387
Iron, Machine for squeezing puddled balls of	S. Gissinger	Lawrenceville, Pa	Dec. 17, 1867	72, 388
Iron, Machine for working	J. Reese	Pittsburgh, Pa	Oct. 6, 1868	82, 876
Iron, Machinery for bending sheet	H. A. Roe	Erie, Pa	Sept. 11, 1845	4, 187
Iron, Machinery for drawing out and compressing heated.	H. Burden	Troy, N. Y	Oct. 16, 1849	6, 792
Iron, &c., Machinery for puddling	V. S. Bloomhall	Conshohocken, Pa	Feb. 20, 1872	123, 860
Iron. Machinery for twisting and rolling	H. Ames	Falls Village, Conn	Oct. 9, 1847	5, 333
Iron, Machinery for working sheet	A. W. Whitney	Woodstock, Vt	Dec. 11, 1847	5, 389
Iron, &c., Machinery or apparatus operated by steam for forging, stamping, and cutting.	J. Nasmyth	Great Britain	Apr. 10, 1843	3, 042
Iron, Making	A. Hamar	New York, N. Y	Sept. 1, 1868	81, 775

Index of patents issued from the United States Patent Office from 1790 *to* 1873, *inclusive*—Continued.

Invention.	Inventor.	Residence.	Date.	No.
Iron malleable, Making cast	A. K. Eaton	New York, N. Y	June 16, 1857	17, 561
Iron malleable, Process of rendering cast	A. H. Siegfried	Lewisburgh, Pa	Sept. 24, 1872	131, 634
Iron, Manufacture of	J. Absterdam	New York, N. Y	Apr. 2, 1872	125, 245
Iron, Manufacture of	E. Brady	Philadelphia, Pa	Aug. 10, 1869	93, 406
Iron, Manufacture of	W. G. Brown and F. McKee	Birmingham, Pa	Apr. 3, 1860	27, 688
Iron, Manufacture of	J. Burt	Detroit, Mich	Jan. 19, 1869	86, 056
Iron, Manufacture of	J. Burt	Detroit, Mich	May 25, 1869	90, 494
Iron, Manufacture of	W. Bushnell	New York, N. Y	Feb. 18, 1873	135, 886
Iron, Manufacture of	A. G. Cook	Burlington, Vt	Mar. 12, 1867	62, 819
Iron, Manufacture of	G. Crane	London, England	Nov. 29, 1838	1, 024
Iron, Manufacture of	W. H. Dawes	West Bromwich, England	May 24, 1864	42, 899
Iron, Manufacture of	A. H. Everett	New York, N. Y	June 27, 1865	48, 483
Iron, Manufacture of	W. Fields	Wilmington, Del	May 10, 1870	102, 797
Iron, Manufacture of	A. L. Fleury	Philadelphia, Pa	July 16, 1861	32, 826
Iron, Manufacture of	A. L. Fleury	Troy, N. Y	Sept. 15, 1863	39, 991
Iron, Manufacture of	L. S. Goodrich	Waverly, Tenn	Nov. 14, 1871	120, 871
Iron, Manufacture of	D. W. Hendrickson	New York, N. Y	Dec. 3, 1867	71, 754
Iron, Manufacture of	J. Jameson	Philadelphia, Pa	May 7, 1867	64, 425
Iron, Manufacture of	J. J. Johnston	Allegheny City, Pa	Feb. 5, 1867	61, 743
Iron, Manufacture of	J. J. Johnston	Allegheny, Pa	Jan. 9, 1872	122, 524
Iron, Manufacture of	J. J. Johnston	Columbiana, Ohio	May 14, 1872	126, 709
Iron, Manufacture of	S. T. Jones	New York, N. Y	Sept. 16, 1851	8, 357
Iron, Manufacture of	W. Kelly	Lyon County, Ky	June 23, 1857	17, 628
Iron, Manufacture of	A. K. Kerpely	Oravitza, Austria	Feb. 13, 1866	52, 656
Iron, Manufacture of	B. Lauth	Pittsburgh, Pa	Aug. 23, 1859	25, 235
Iron, Manufacture of	W. M. Lyon	Pittsburgh, Pa	May 25, 1869	90, 372
Iron, Manufacture of	W. H. Perry	Saint Louis, Mo	Aug. 6, 1861	33, 006
Iron, Manufacture of	W. H. Richardson	Glasgow, North Britain	July 30, 1867	67, 350
Iron, Manufacture of	D. Stewart	Kittaning, Pa	Dec. 17, 1867	72, 335
Iron, Manufacture of	D. Stewart	Kittaning, Pa	July 13, 1869	92, 667
Iron, Manufacture of	D. Stewart	Kittaning, Pa	Feb. 1, 1870	99, 367
Iron, Manufacture of	D. Stewart	Kittaning, Pa	Feb. 1, 1870	99, 368
Iron, Manufacture of	F. D. Taylor	Brady's Bend Township, Pa.	Aug. 1, 1871	117, 576
Iron, Manufacture of	W. J. Taylor	High Bridge, N. J	Dec. 9, 1873	145, 461
Iron, Manufacture of	W. J. Taylor	High Bridge, N. J	Dec. 9, 1873	145, 462
Iron, Manufacture of	A. Thomas	Howard, Pa	Apr. 5, 1859	23, 512
Iron, Manufacture of	J. D. Williams	Allegheny City, Pa	Aug. 8, 1865	49, 331
Iron, Manufacture of	J. D. Williams	Allegheny, Pa	Dec. 26, 1865	51, 771
Iron, Manufacture of	J. D. Williams	Philadelphia, Pa	June 7, 1870	104, 091
Iron, Manufacture of	J. Yates	Mott Haven, N. Y	Mar. 1, 1864	41, 805
Iron, Manufacture of galvanized	J. D. Grey	Pittsburgh, Pa	Sept. 21, 1869	95, 017
Iron, Manufacture of malleable	J. J. G. Collins	Philadelphia, Pa	May 3, 1870	102, 501
Iron, Manufacture of malleable	A. Manvel	Elizabethport, N. J	July 19, 1864	43, 595
Iron, Manufacture of metal-coated sheet	B. Morison	Philadelphia, Pa	Feb. 20, 1872	123, 787
Iron, Manufacture of pig	J. Burt	Detroit, Mich	Sept. 7, 1869	94, 471
Iron, Manufacture of pig	C. M. Nes	York, Pa	Nov. 16, 1869	96, 927
Iron, Manufacture of refined cast	C. Burgess	Portsmouth, Ohio	July 29, 1873	141, 319
Iron, Manufacture of sheet	R. N. Allen and C. C. Hinsdale	Cleveland, Ohio	Mar. 20, 1866	53, 253
Iron, Manufacture of sheet	G. Atkins	Sharon, Pa	Mar. 23, 1869	88, 002
Iron, Manufacture of sheet	S. Barker	Hartford, Conn	Oct. 5, 1869	95, 554
Iron, Manufacture of sheet	J. Chandler	Attica, Ohio	Oct. 19, 1858	21, 817
Iron, Manufacture of sheet	I. E. Craig	Camden, Ohio	Feb. 11, 1868	74, 314
Iron, Manufacture of sheet	I. E. Craig	Camden, Ohio	May 31, 1870	103, 577
Iron, Manufacture of sheet	J. Dickson	Newcastle, Pa	Feb. 4, 1862	34, 294
Iron, Manufacture of sheet	W. Fields	Wilmington, Del	Dec. 28, 1869	98, 364
Iron, Manufacture of sheet	I. and T. Gray	Pittsburgh, Pa	Sept. 26, 1865	50, 203
Iron, Manufacture of sheet	J. D. Grey	Pittsburgh, Pa	May 24, 1870	103, 323
Iron, Manufacture of sheet	C. C. Hinsdale	Cleveland, Ohio	Apr. 27, 1869	89, 312
Iron, Manufacture of sheet	H. McCarty	Pittsburgh, Pa	Sept. 27, 1863	10, 047
Iron, Manufacture of sheet	D. A. Morris	Pittsburgh, Pa	Oct. 12, 1858	21, 772
Iron Manufacture of sheet	D. A. Morris	Pittsburgh, Pa	Jan. 22, 1861	31, 184
Iron, Manufacture of sheet	E. B. Nock	Cleveland, Ohio	Aug. 25, 1868	81, 528
Iron, Manufacture of sheet	C. H. Perkins	Providence, R. I	Mar. 27, 1866	53, 476
Iron, Manufacture of sheet	E. G. Pomeroy	Pittsburgh, Pa	Jan. 31, 1854	10, 482
Iron, Manufacture of sheet	D. L. Pratt	Bridgeport, Ohio	Feb. 14, 1865	46, 384
Iron, Manufacture of sheet	D. L. Pratt	Bethesda, Ohio	Mar. 26, 1872	125, 079
Iron, Manufacture of sheet	W. Rogers	Apollo, Pa	Jan. 23, 1872	122, 912
Iron, Manufacture of sheet	W. D. Wood	Wilmington, Del	May 14, 1861	32, 341
Iron, Manufacture of sheet and bar	J. M. Jones, B. Spaulding, and S. Parkins.	East Taunton, Mass., Port Richmond, N. Y., and Providence, R. I.	July 31, 1866	56, 759
Iron, Manufacture of sheet and plate	N. C. Gridley	Milwaukee, Wis	July 27, 1869	93, 083
Iron, Manufacture of sheet and plate	C. C. Hinsdale	Cleveland, Ohio	Sept. 8, 1868	81, 903
Iron, Manufacture of wrought	T. C. Coleman	Louisville, Ky	Jan. 19, 1869	86, 004
Iron, Manufacture of wrought	W. M. Lyon	Pittsburgh, Pa	May 25, 1869	90, 373
Iron, Manufacture of wrought	C. Sacre, S. Perkins, and W. Smellie.	Manchester and Gorton, Great Britain.	Oct. 18, 1870	108, 521
Iron, Manufacture of wrought	R. Thomas and G. Edwards	Columbiana, Ala	Aug. 1, 1865	49, 175
Iron, Manufacture of wrought and puddled	W. Ennis	Philadelphia, Pa	Aug. 16, 1870	106, 347
Iron, Manufacturing	J. Butland	Philadelphia, Pa	Nov. 12, 1807	....
Iron, Manufacturing and purifying	J. Webster	Birmingham, England	May 17, 1870	103, 109
Iron-manufacturing furnace	E. Peckham	Antwerp, N. Y	Oct. 24, 1871	120, 318
Iron-manufacturing furnace	H. Ross	Pittsburgh, Pa	Oct. 27, 1868	83, 551
Iron, Manufacturing hoop and sheet	J. H. Pierson	Ramapo Works, Rockland, N. Y.	Dec. 24, 1807	....
Iron, Manufacturing malleable cast	S. Boyden	Newark, N. J	Mar. 9, 1831	....
Iron, Manufacturing malleable cast	S. Boyden	Newark, N. J	Apr. 6, 1831	....
Iron, Manufacturing, puddling, and heating	W. Jones	Haversham, N. Y	Dec. 16, 1833	....
Iron, Manufacturing sheet	W. D. Wood	McKeesport, Pa	Apr. 8, 1873	137, 585
Iron, Manufacturing sheet or boiler	J. McNary	Brooklyn, N. Y	Dec. 17, 1834	....
Iron measuring and cutting machine	L. B. Griffith	Honeybrook, Pa	Nov. 4, 1851	8, 485
Iron, Melting and decarbonizing	C. Peters	Trenton, N. J	Nov. 2, 1869	96, 479
Iron, Melting and refining	G. P. Miller and H. Dougherty	Lancaster, Pa	Oct. 6, 1857	18, 347
Iron-melting furnace	W. McFarland	Saint Louis, Mo	Dec. 7, 1858	22, 257
Iron-melting furnace	G. H. Sellers	Wilmington, Del	Apr. 2, 1872	125, 147
Iron melting and refining furnace	A. F. Du Faur	New York, N. Y	Feb. 1, 1870	99, 415
Iron-mill	J. W. and C. Post	Washington, D. C	Apr. 11, 1829	....

Index of patents issued from the United States Patent Office from 1790 *to* 1873, *inclusive* —Continued.

Invention.	Inventor.	Residence.	Date.	No.
Iron, Mode of purifying	E. Brady	Philadelphia, Pa	Apr. 13, 1869	88, 941
Iron, Mode of treating conglomerates of cast	T. S. Blair	Pittsburgh, Pa	Dec. 14, 1869	97, 754
Iron, Mode of treating slag and cinders for the manufacture of.	R. Keck	Clintonville, N. Y	Oct. 1, 1867	69, 348
Iron or ores to the metallic state by coating them with certain fluxes, Process of reducing.	J. Tower	Madison, Ohio	Dec. 7, 1844	3, 850
Iron or steel bars or rods, Manufacture of	B. Lauth	Reading, Pa	Mar. 12, 1867	62, 758
Iron or steel with copper or brass or other alloys of copper, Coating.	G. J. Hinde	Wolverhampton, England	Sept. 7, 1869	94, 492
Iron-ore for smelting, Preparations of	C. Cochrane	Upper Gornal, England	July 9, 1872	128, 712
Iron-ore from other substances, Machine for separating magnetic.	J. Y. Smith	Pittsburgh, Pa	Nov. 15, 1870	109, 354
Iron-ore, Furnace for producing sponge from	G. Nock	New Monmouth, N. J	May 24, 1870	103, 489
Iron-ore in the manufacture of steel, Preparing and treating.	E. L. Seymour	New York, N. Y	May 23, 1871	115, 243
Iron-ore, Mode of desulphurizing	J. Little	Newburgh, N. Y	May 21, 1867	64, 887
Iron-ore, Mode of smelting	J. Lyon	Pottsville, Pa	July 31, 1837	332
Iron-ore, Mode of smelting	J. Richard	Philadelphia, Pa	Dec. 10, 1838	1, 032
Iron-ore, Reduction of	W. S. Cooke	East Fairfield, Ohio	May 29, 1847	5, 131
Iron-ore, Roasting and treating	H. Aitken	Falkirk, Scotland	Aug. 31, 1869	94, 374
Iron-ore, Smelting	H. Bessemer	London, England	Nov. 18, 1856	16, 083
Iron-ore, Smelting	T. S. Ridgway	Pottsville, Pa	Dec. 17, 1834	....
Iron-ore, &c., Treating	H. Aitken	Falkirk, Scotland	Sept. 29, 1868	82, 576
Iron-ore, Treatment and reduction of titaniferous.	C. Martin, W. Barrett, and T. S. Webb.	Chancery Lane and Norton, England.	Sept. 29, 1868	82, 539
Iron pig-bloom, Apparatus for the manufacture of	T. S. Blair	Pittsburgh, Pa	June 29, 1869	91, 901
Iron pig-bloom, Apparatus for the manufacture of	J. Coyne	Allegheny City, Pa	Aug. 31, 1869	94, 286
Iron pins, Making coated	D. and T. Fowler	North Branford, Mass	Jan. 17, 1860	26, 874
Iron pipe, Cast	C. Pomroy	Pottsville, Pa	May 4, 1858	20, 171
Iron-pipe joint	E. G. Blakeslee	Sing Sing, N. Y	July 16, 1867	66, 782
Iron pipe, Tin-lined	C. M. Platt	Waterbury, Conn	Dec. 26, 1871	122, 131
Iron-pipe-welding machinery	J. C. Vaughn and J. F. Winslow.	Greenbush and Troy, N. Y	Aug. 1, 1848	5, 695
Iron pipes, Cleansing galvanized	W. Blake	Boston, Mass	Aug. 21, 1860	29, 663
Iron pipes, Flask for cast	J. Firth and J. Ingham	Phillipsburgh, N. J	Dec. 2, 1862	37, 037
Iron pipes for welding, Machine for bending	E. J. Woodworth, J. Brandy, and F. Merithew.	Birmingham, Pa	Nov. 8, 1870	109, 094
Iron pipes, Machine for cleansing cast	W. Smith	Allegheny City, Pa	June 7, 1870	104, 072
Iron pipes, Method of employing centrifugal force in casting.	T. J. Lovegrove	Baltimore, Md	Dec. 26, 1848	5, 988
Iron pipes, &c., Mold for casting	J. D. Morris	Kensington, Pa	Dec. 2, 1835	....
Iron pipes, Molding	W. Doyle	Auburn, N. Y	May 15, 1860	28, 261
Iron plate, Method of treating sheet	C. H. Perkins	Providence, R. I	Feb. 13, 1866	52, 647
Iron-plating furnace	C. Wray	San Francisco, Cal	Feb. 21, 1860	27, 254
Iron ponton	H. Grundt	Berlin, Prussia	Feb. 4, 1862	34, 329
Iron post for wire fences	D. Kaufman	Boiling Springs, Pa	June 3, 1873	139, 469
Iron, Preserving the surface of cast or wrought	C. F. L. Oudry	Paris, France	Nov. 23, 1858	22, 132
Iron, Process and apparatus for making sheet	G. W. Francis	Hartford, Conn	June 22, 1869	91, 619
Iron, Process and apparatus for treating crude or pig.	H. Bessemer	London, England	Aug. 1, 1871	117, 507
Iron, Process for smelting	S. Macferran	Philadelphia, Pa	July 24, 1855	13, 316
Iron, Process for steelifying	T. Sheehan	Dunkirk, N. Y	Mar. 31, 1868	76, 256
Iron, Process for the manufacture of	C. Adams	Philadelphia, Pa	Nov. 28, 1871	121, 226
Iron, Process of coating	E. G. Pomeroy	Philadelphia, Pa	Oct. 13, 1857	18, 409
Iron, Process of hardening	T. Chappell and W. A. Jones	Winona, Minn	Aug. 9, 1864	43, 813
Iron, Process of manufacturing rolled	G. W. Billings	Chicago, Ill	May 27, 1873	139, 359
Iron, Process of manufacturing sheet	S. H. Kimball	Cleveland, Ohio	Mar. 23, 1869	88, 182
Iron, Process of manufacturing wrought	P. E. Jay	Saint Jean Baptiste, Canada	Dec. 24, 1872	134, 288
Iron, Process of preserving	W. H. Sterling	New York, N. Y	May 21, 1872	126, 909
Iron, Process of refining cast	H. M. Baker	Harlem, N. Y	Sept. 15, 1868	82, 194
Iron, Process of treating cleaned or scaled	W. D. Wood	Keysport, Pa	Jan. 8, 1867	61, 034
Iron, Protecting surfaces of articles of	T. Sellick	Greenwich, Conn	July 5, 1859	24, 665
Iron, Puddling	J. J. Johnston	Columbiana, Ohio	May 14, 1872	126, 710
Iron, Puddling	R. Savary	Steubenville, Ohio	Mar. 11, 1856	14, 412
Iron, Puddling	W. Sellers	Philadelphia, Pa	July 16, 1872	129, 430
Iron, Puddling	J. G. Willans	Bayswater, England	Mar. 12, 1867	62, 795
Iron-puddling apparatus	W. Sellers	Philadelphia, Pa	Mar. 5, 1872	124, 224
Iron, Puddling process for making wrought	P. Merryman and R. McCombs	West Fairview, Pa	Mar. 15, 1870	100, 916
Iron, Puddling process for the manufacture of wrought.	C. Hewitt	Hamilton Township, N. J	Mar. 30, 1869	88, 479
Iron, Purification of cast	J. W. Middleton	Philadelphia, Pa	Jan. 3, 1871	110, 666
Iron, Purifying cast	S. W. Kirk	Coatesville, Pa	Mar. 24, 1863	38, 003
Iron, reducing ore, &c., Refining	J. F. Allen	Tremont, N. Y	Feb. 21, 1871	112, 003
Iron, Refining	W. Kelly	Lyon, Ky	Dec. 22, 1857	18, 910
Iron, Refining	C. Shunk	Canton, Ohio	July 12, 1859	24, 766
Iron refining and carbonizing composition	J. E. Atwood	Trenton, N. J	Feb. 9, 1869	86, 801
Iron, Refining cast	H. S. Osborn	Easton, Pa	Sept. 22, 1868	82, 435
Iron, &c., remelting furnace	W. J. Keep	Troy, N. Y	Oct. 29, 1872	132, 667
Iron, Removing acid from the surface of	D. McDaniel and E. A. Harney	New Castle County and Wilmington, Del.	Dec. 3, 1861	33, 844
Iron, Restoring burned	G. W. Morris and W. Quann	Philadelphia, Pa	May 1, 1860	28, 103
Iron, Restoring tinned sheet	W. E. Brockway	New York, N. Y	July 22, 1873	141, 109
Iron roller for metals, Cast	J. Wood	Philadelphia, Pa	July 20, 1831	
Iron, Roller for rolling	A. B. Clemons	Derby, Conn	Apr. 19, 1864	42, 433
Iron roller for rolling iron, &c., Mode of casting large.	C. Alger	Boston, Mass	Mar. 30, 1811	....
Iron, Roller for slitting and other mills for rolling	B. Seymour		June 26, 1797	....
Iron, Rolling	A. S. Valentine	Bellefonte, Pa	Jan. 3, 1827	....
Iron, Rolling angle	J. L. Lewis	Pittsburgh, Pa	Apr. 26, 1864	42, 495
Iron-rolling machine	H. Burden	Troy, N. Y	Dec. 10, 1840	1, 890
Iron-rolling machinery	J. F. Lauth	Reading, Pa	Mar. 6, 1866	53, 012
Iron scraps, Remelting	A. Pevey	Lowell, Mass	Jan. 15, 1856	14, 114
Iron-shearing machine	A. Hardy	Boston, Mass	Sept. 22, 1863	40, 034
Iron-shearing machine, Angle	T. Reaney	Chester, Pa	Nov. 16, 1869	96, 960
Iron, Sheet	J. Miller	Cuba, N. Y	Apr. 16, 1867	63, 805
Iron-slitting machine	J. Reed	Marshfield, Mass	Apr. 21, 1825	....
Iron, Smelting and refining	W. W. Hubbell	Philadelphia, Pa	Feb. 8, 1870	99, 677
Iron, Smelting and refining	E. G. Pomeroy	New York, N. Y	Jan. 24, 1860	26, 923

Index of patents issued from the United States Patent Office from 1790 *to* 1873, *inclusive*—Continued.

Invention.	Inventor.	Residence.	Date.	No.
Iron-smelting furnace	S. M. Fales	Baltimore, Md	Feb. 8, 1859	22, 861
Iron-smelting furnace, Scrap	A. Ott	New York, N. Y	Sept. 27, 1870	107, 712
Iron-smelting process	F. Lang and C. A. Frey	Vienna, Austria, and Storré, Styria.	Dec. 12, 1865	51, 531
Iron sockets, Machine for making	L. Morse and H. Pettee	Foxborough, Mass	Dec. 28, 1838	1, 046
Iron, Softening cast or crude	A. Scott, jr., and A. Selden	Winchester, N. H	Nov. 8, 1814	....
Iron spindle and cast-iron driver, Cast or wrought	J. Barton	Milford, Pa	Apr. 23, 1831	....
Iron splaying or flaring machinery	N. Sutton	New York, N. Y	Apr. 28, 1825	....
Iron sponge	T. S. Blair	Pittsburgh, Pa	May 21, 1872	126, 924
Iron, steel, and malleable iron, Producing refined cast.	J. W. Middleton	Philadelphia, Pa	Jan. 17, 1871	110, 990
Iron, steel, and other material, Tool and wheel for cutting and polishing.	A. K. Eaton	Piermont, N. Y	May 24, 1870	103, 312
Iron, steel, and other metals, Processes for purifying.	J. F. Bennett	Pittsburgh, Pa	June 18, 1872	127, 953
Iron, steel, &c., Compound for treating	W. A. Skinner and J. E. Goodson.	Callao, Mo	May 27, 1873	139, 335
Iron, steel, &c., Manufacture of	A. T. Hay	Burlington, Iowa	Nov. 19, 1872	133, 098
Iron, steel, &c., Process of refining	J. Reese	Pittsburgh, Pa	June 18, 1867	65, 830
Iron structures, Clamp for	J. M. Moorehead	Brooklyn, N. Y	Oct. 6, 1868	82, 738
Iron, Surfacing cast	H. Tucker	Newton, Mass	May 24, 1870	103, 527
Iron-swaging machine	J. Foster	Brooklyn, N. Y	Mar. 3, 1857	16, 775
Iron-swaging machine	J. T. Willmarth	Worcester, Mass	Aug. 19, 1856	15, 588
Iron, Tool for cutting, punching, and upsetting	J. J. Rose	Elmwood, Ill	Nov. 12, 1867	70, 902
Iron, Tool for turning or planing	T. Cooper	Warwick, R. I	July 3, 1866	56, 141
Iron, Tools used in the manufacture of	A. L. Fleury	Philadelphia, Pa	Feb. 19, 1861	31, 450
Iron-tooth rake	N. C. Sanford and E. P. Parmlee	Meriden, Conn	June 8, 1830	....
Iron tubes, Heating skelps for the manufacture of wrought.	J. McCarty	Reading, Pa	Apr. 4, 1854	10, 747
Iron tubes, Welding	T. H. Russell	Wednesbury, England	Mar. 6, 1847	4, 992
Iron tubing, Manufacture of	J. H. Cotton	Boston, Mass	Aug. 12, 1862	36, 141
Iron tubings, &c., Mode of utilizing	C. S. and J. A. Lynch and C. E. Coffin.	Boston, Mass., and Muirkirk, Md.	Apr. 20, 1869	89, 228
Iron vessels, Coppering	W. B. Barnard	Waterbury, Conn	Mar. 24, 1863	37, 998
Iron vessels, Preventing the corrosion of cast	C. O. Greene	Troy, N. Y	Apr. 28, 1868	77, 186
Iron-ware, Coating articles of	H. Tucker	Newton, Mass	June 1, 1869	90, 891
Iron, Welding wrought	J. C. Cooke	Middletown, Conn	Apr. 3, 1860	27, 697
Iron wheels, Molding cast	G. S. Bosworth	Troy, N. Y	Oct. 30, 1860	30, 525
Iron with a harder metal, Mode of coating wrought or cast.	J. Rigg	Iowa Falls, Iowa	July 9, 1867	66, 630
Iron with brass or copper, Method of coating	H. Burgess	Kentish Town, England	July 18, 1854	11, 319
Iron with cast iron, Method of combining wrought	W. M. Arnold	New York, N. Y	May 22, 1866	54, 838
Iron with cast steel, Mode of coating wrought	J. W. Ells	Pittsburgh, Pa	Oct. 15, 1867	69, 904
Iron with copper, Coating	T. G. Bucklin	Troy, N. Y	Sept. 21, 1852	9, 270
Iron with copper or its alloy, Coating	E. G. Pomeroy	Saint Louis, Mo	Jan. 8, 1850	7, 005
Iron with other metals, Coating sheet	J. H. Whitling	Philadelphia, Pa	Nov. 20, 1866	59, 787
Iron with steel surface, Process for the manufacture of.	J. Reese	Pittsburgh, Pa	Apr. 30, 1867	64, 253
Iron with tin and other metals, Coating	E. Morewood	London, England	Sept. 4, 1866	57, 832
Iron with tin, Apparatus for coating sheet	W. O. Davies	Pittsburgh, Pa	June 4, 1872	127, 580
Iron with zinc, Process for coating	J. D. Grey and J. Lippincott	Baltimore, Md	June 4, 1872	127, 412
Iron, wood, and other material, Compound for coating.	H. E. F. De Brion	London, England	Aug. 13, 1867	67, 733
Iron-working furnace	J. Heatley	Etna, Pa	Sept. 22, 1868	82, 313
Iron, Working scrap and other	G. W. Jones	Ormsby, Pa	Feb. 8, 1870	99, 574
Iron, zinc, and copper, Composition metal of	A. Birkholz	Hartford, Conn	Mar. 11, 1862	34, 621
Ironing and kitchen table combined	J. N. Valley	North East, Pa	Sept. 12, 1871	118, 990
Ironing and pressing garments, Apparatus for	R. B. Sanson	London, England	Sept. 2, 1873	142, 516
Ironing and stiffening machine for linen and other fabrics.	J. Decoudun	Paris, France	Dec. 7, 1869	97, 614
Ironing and stretching board	J. W. Davis	Reno, Nev	June 27, 1871	116, 277
Ironing-apparatus	E. Sumption	South Bend, Ind	Feb. 4, 1873	135, 606
Ironing-apparatus	R. A. Tyler	Haverhill, Mass	Mar. 4, 1873	136, 471
Ironing-board	E. Bassett	Topeka, Kans	July 15, 1873	140, 758
Ironing-board	A. G. Beek	Philadelphia, Pa	May 9, 1871	114, 515
Ironing-board	J. H. Beidler	Adrian, Mich	Nov. 29, 1870	109, 712
Ironing-board	C. H. Bennett and W. H. Daggett.	South Vineland, N. J	Nov. 2, 1869	96, 384
Ironing-board	G. I. Birch	New York, N. Y	Apr. 14, 1868	76, 701
Ironing-board	B. D. Brown and D. Moyer	Philadelphia, Pa	Oct. 17, 1871	120, 030
Ironing-board	J. M. Childers	Corry, Pa	July 9, 1872	128, 709
Ironing-board	D. E. Crosby	South Vineland, N. J	Sept. 7, 1869	94, 572
Ironing-board	J. Fischer	Pittsburgh, Pa	Sept. 20, 1870	107, 469
Ironing-board	H. Hamill	New York, N. Y	Oct. 1, 1872	131, 876
Ironing-board	R. N. Herring	Chew's Landing, N. J	Apr. 29, 1873	138, 331
Ironing-board	G. M. Lane	Battle Creek, Mich	Mar. 5, 1872	124, 275
Ironing-board	A. Watson	Elizabeth, N. J	June 8, 1869	91, 033
Ironing-board	J. C. Miller	Lancaster, Ohio	Apr. 20, 1869	89, 162
Ironing-board	S. A. Mort	Dayton, Ohio	Aug. 14, 1866	57, 170
Ironing-board	L. N. Vallett	Providence, R. I	July 2, 1872	128, 569
Ironing-board and closet	J. N. Brewster	Brooklyn, N. Y	May 7, 1867	64, 480
Ironing clothes, Device for	C. S. Whipple	Waterford, Conn	May 21, 1872	126, 925
Ironing-machine	W. H. Bovell and C. Partinsky	San Francisco, Cal	Oct. 18, 1864	44, 765
Ironing-machine	G. Boxley	Troy, N. Y	Apr. 5, 1870	101, 573
Ironing-machine	L. Berenger	Paris, France	Mar. 30, 1869	88, 437
Ironing-machine	G. W. H. Calver	Burlington, N. J	Sept. 3, 1872	131, 052
Ironing-machine	G. W. Cottingham	Saint Mary's, Tex	Nov. 18, 1873	144, 743
Ironing-machine	P. J. Flanedy	San Francisco, Cal	Sept. 10, 1867	68, 617
Ironing-machine	A. G. Gardner	Troy, N. Y	Feb. 20, 1872	123, 819
Ironing-machine	G. Gilbert and A. N. Allen	New Haven, Conn	Aug. 13, 1867	67, 644
Ironing-machine	W. Jones	Oshkosh, Wis	June 28, 1870	104, 740
Ironing-machine	W. T. Nichols	Rutland, Vt	Apr. 22, 1862	35, 034
Ironing-machine	P. O'Thayne	New York, N. Y	Mar. 26, 1867	63, 301
Ironing-machine	C. F. Parker	Greenfield, Ohio	Sept. 27, 1870	107, 714
Ironing-machine	G. B. Perkins	Bridgeport, Conn	Mar. 12, 1867	62, 773
Ironing-machine	L. Ringler	New York, N. Y	July 13, 1869	92, 544
Ironing-machine	A. C. Sawyer	Canton, N. Y	June 28, 1870	104, 776
Ironing-machine	G. F. Taylor	New York, N. Y	Dec. 12, 1871	121, 908

Index of patents issued from the United States Patent Office from 1790 *to* 1873, *inclusive*—Continued.

Invention.	Inventor.	Residence.	Date.	No.
Ironing-machine	C. C. Thomas	Natchez, Miss	Jan. 9, 1872	122, 680
Ironing-machine	J. W. Thorp	Sanbornton Bridge, N. H	Mar. 12, 1867	62, 906
Ironing-machine	W. M. Tobey	New London, Conn	Mar. 26, 1867	63, 186
Ironing-machine	J. T. Walker	Albany, N. Y	Nov. 25, 1873	145, 034
Ironing-machine	A. B. Walters	Philadelphia, Pa	Aug. 6, 1872	130, 257
Ironing machine, Clothes	J. Shaefer	Lancaster, Pa	Sept. 7, 1858	21, 450
Ironing machine, Clothes	S. Swett, jr	Readfield, Me	Dec. 30, 1835	....
Ironing machine, Clothes	S. W. Woodward	Buffalo, N. Y	Mar. 24, 1863	37, 994
Ironing machine, Hosiery	W. Aiken	Franklin, N. H	Aug. 28, 1866	57, 455
Ironing or fluting machine	M. P. Carpenter	Buffalo, N. Y	May 6, 1862	35, 138
Ironing-stand	J. P. and M. L. Bever	Springfield, Ill	Mar. 3, 1863	37, 804
Ironing-stand and clothes-drier	B. S. Boydston	Richmond, Ind	Mar. 3, 1868	75, 120
Ironing-table	C. Blom, jr	Grand Haven, Mich	Dec. 16, 1873	145, 614
Ironing-table	P. Bostrom	Galesburgh, Ill	Jan. 9, 1872	122, 554
Ironing-table	F. A. Byram	Germantown, Pa	Aug. 8, 1871	117, 737
Ironing-table	P. P. Carroll	Washington, D. C	Oct. 27, 1868	83, 460
Ironing-table	J. R. Carzier	North East, Pa	Apr. 23, 1872	126, 016
Ironing-table	A. A. Chittenden	Boston, Mass	Nov. 19, 1867	70, 958
Ironing-table	J. M. Couthamal	Philadelphia, Pa	Nov. 26, 1872	133, 304
Ironing-table	H. T. De Montigny	West Troy, N. Y	Mar. 30, 1869	88, 281
Ironing-table	E. S. Farson	Philadelphia, Pa	Dec. 2, 1873	145, 162
Ironing-table	J. H. Frank	Ilion, N. Y	Aug. 12, 1873	141, 639
Ironing-table	J. Frankenfield and W. E. Kichline.	Bethlehem, Pa	July 1, 1873	140, 356
Ironing-table	J. F. Galley	New York, N. Y	Aug. 2, 1859	24, 929
Ironing-table	W. B. Grosh and S. H. Foreman	Reading, Pa	Sept. 23, 1873	143, 071
Ironing-table	J. R. Groves	Hightstown, N. J	Feb. 18, 1873	136, 056
Ironing-table	C. C. Hardy	Rutland, N. Y	Oct. 10, 1871	119, 703
Ironing-table	L. Harrington	Saugatuck, Mich	Oct. 27, 1868	83, 379
Ironing-table	O. C. Hotchkiss	Cortlandville, N. Y	Dec. 1, 1868	84, 549
Ironing-table	J. Johnson	New York, N. Y	June 24, 1862	35, 735
Ironing-table	F. Liller	Baltimore, Md	July 16, 1872	129, 042
Ironing-table	J. H. Mallory	La Porte, Ind	Dec. 6, 1870	109, 918
Ironing-table	J. H. Mallory	La Porte, Ind	Oct. 24, 1871	120, 296
Ironing-table	J. J. Märki	Chicago, Ill	Dec. 27, 1870	110, 577
Ironing-table	D. W. Marshall	Pawtucket, R. I	Jan. 3, 1871	110, 769
Ironing-table	J. F. Martin and W. A. Schaffner.	Harrisburgh, Pa	Sept. 13, 1870	107, 271
Ironing-table	H. McChesney	Buffalo, N. Y	Oct. 4, 1870	107, 940
Ironing-table	S. Merritt	Erie, Pa	Jan. 25, 1870	99, 219
Ironing-table	P. O'Thayne	New York, N. Y	Sept. 12, 1871	118, 878
Ironing-table	A. J. Palmberg	Boston, Mass	Sept. 9, 1873	142, 645
Ironing-table	H. Park	Columbus, Ohio	Mar. 31, 1868	76, 098
Ironing-table	W. P. Patten	Harrisburgh, Pa	July 6, 1869	92, 207
Ironing-table	J. T. Piercy	Martinsburgh, Ohio	Oct. 20, 1868	83, 310
Ironing-table	J. T. Plowman, sr	Baltimore, Md	June 11, 1872	127, 792
Ironing-table	T. Reed	Plainwell, Mich	Aug. 29, 1871	118, 643
Ironing-table	T. Reed	Plainwell, Mich	Oct. 17, 1871	120, 101
Ironing-table	T. M. Richards	Philadelphia, Pa	May 11, 1869	89, 943
Ironing-table	J. H. Ruff	Thomas' Run Post-Office, Md	Aug. 16, 1870	106, 509
Ironing-table	M. D. Safford	Boston, Mass	Jan. 29, 1867	61, 5[illegible]9
Ironing-table	B. Schoonmaker	Plainwell, Mich	Nov. 19, 1872	133, 122
Ironing-table	L. Scofield	Grand Haven, Mich	Nov. 25, 1873	144, 869
Ironing-table	H. Soggs	Columbus, Pa	June 8, 1869	90, [illegible]66
Ironing-table	W. H. Sparks	Absecom, N. J	Nov. 11, 1873	144, 418
Ironing-table	W. Vandenburg	New York, N. Y	Apr. 6, 1858	19, 883
Ironing-table	W. Vandenburg	New York, N. Y	May 11, 1858	20, 231
Ironing-table	W. Vandenburg and J. Harvey	New York, N. Y	Feb. 16, 1858	19, 390
Ironing-table	B. Van Gaasbeek	Mount Vernon, N. Y	Apr. 23, 1867	64, 170
Ironing-table	A. Wechsler	New York, N. Y	Jan. 28, 1873	135, 303
Ironing-table	A. T. Woolsey	Sandusky, Ohio	Oct. 22, 1867	70, 146
Ironing-table	G. F. Zimmann	Philadelphia, Pa	Jan. 3, 1860	26, 728
Ironing-table and bureau	M. White	Saratoga Springs, N. Y	June 14, 1870	104, 385
Ironing-table and clothes-drier	W. P. Adams	Brooklyn, N. Y	Dec. 14, 1869	97, 851
Ironing-table and clothes-drier	J. L. Devol	Parkersburgh, W. Va	June 29, 1869	91, 921
Ironing-table and skirt-board, Combined	A. S. Robinson and R. A. Harris	Albany, N. Y	Aug. 20, 1872	130, 749
Ironing-table bureau	M. White	Saratoga Springs, N. Y	Dec. 20, 1870	110, 410
Ironing-table, Folding	W. Edwards	Cleveland, Ohio	Apr. 8, 1873	137, 602
Ironing-table, lap-board, and step-ladder, Combined	H. E. Smith	Cleveland, Ohio	Mar. 4, 1873	136, 390
Ironing table, Skirt	A. S. Phillips	Boston, Mass	Dec. 17, 1867	72, 228
Irrigating-machine	W. W. Hull	Ashland, N. Y	Sept. 13, 1870	107, 375
Irrigation, Subterranean	W. H. Pugh	Chicago, Ill	Sept. 2, 1873	142, 413
Isinglass	W. Hall	Boston, Mass	Mar. 23, 1822	....
Isinglass, Apparatus for cooling the rollers used in the manufacture of.	E. Rowe	Rockport, Mass	Aug. 9, 1870	106, 212
Isinglass, Manufacture of	J. Manning	Rockport, Mass	Jan. 7, 1873	134, 690
Isinglass, Manufacturing	J. Hastings	Cambridge, Mass	Aug. 14, 1822	....
Isinglass, or ichthyocolla	W. Norwood, J. Burns, J. Rowe, and J. Haskell.	Gloucester, Mass	Jan. 21, 1834	....
Isinglass, Purifying	J. Lewis and I. Stanwood	Gloucester, Mass	Apr. 3, 1866	53, 639
Ivory and similar material, Elastic composition to imitate.	W. M. Welling	New York, N. Y	Apr. 20, 1869	89, 100
Ivory, Artificial	C. F. Dupper	Bridgeport, Conn	Nov. 21, 1865	51, 109
Ivory, Artificial	J. Hackert	New York, N. Y	Feb. 19, 1867	62, 128
Ivory, Artificial	A. Starr and W. M. Welling	New York, N. Y	June 9, 1868	78, 842
Ivory, Artificial	W. M. Welling	New York, N. Y	May 12, 1868	77, 938
Ivory-bleaching apparatus	J. Phyfe	New York, N. Y	Oct. 28, 1856	15, 983
Ivory-bleaching device	W. M. Welling	Brooklyn, N. Y	Dec. 11, 1855	13, 928
Ivory, bone, and wood combs, Method of making	J. Pratt	Meriden, Conn	Dec. 28, 1830	....
Ivory by steam, Machine for junking	L. Pratt and F. Bush	Meriden, Conn	Feb. 9, 1828	....
Ivory combs, Machine for cutting teeth in	P. Pratt	New York	Sept. 9, 1825	....
Ivory combs, Machine for manufacturing	D. Williams, 3d	Saybrook, Conn	Mar. 28, 1810	....
Ivory, Composition for artificial	A. Starr	New York, N. Y	Mar. 3, 1868	75, 067
Ivory, Composition for artificial	W. M. Welling	New York, N. Y	Apr. 27, 1869	89, 531
Ivory, Composition of artificial	J. Hackett	Bridgeport, Conn	May 31, 1864	42, 942
Ivory, Composition resembling	W. M. Welling	New York, N. Y	Apr. 27, 1869	89, 532
Ivory-cutting machine	C. B. Rogers	Deep River, Conn	May 2, 1865	47, 572

Index of patents issued from the United States Patent Office from 1790 *to* 1873, *inclusive*—Continued.

Invention.	Inventor.	Residence.	Date.	No.
Ivory-dust and other materials, Compound of	J. W. Hyatt, jr., and D. Blake	Albany and Spencertown, N. Y.	May 4, 1869	89, 582
Ivory, Factitious	L. Held	Brooklyn, N. Y	Aug. 4, 1857	17, 931
Ivory, Factitious	W. M. Welling	New York, N. Y	Aug. 4, 1857	17, 949
Ivory for piano-forte keys, Machine for gaging and toothing.	E. Savage	West Meriden, Conn	July 30, 1861	32, 967
Ivory-frame composition	I. M. M. Legaré	Aiken, S. C	June 15, 1858	20, 569
Ivory, Frame for bleaching	A. C. Breckenridge	Meriden, Conn	Aug. 19, 1856	15, 590
Ivory key-board, Machine for smoothing	M. Pratt	Meriden, Conn	July 31, 1866	56, 862
Ivory, Molding composition for	J. W. Hyatt, jr	Albany, N. Y	Apr. 6, 1869	88, 633
Ivory or other comb plates, Machine for planing and forming.	W. M. Fowler	North Branford, Conn	Oct. 28, 1840	1, 837
Ivory planing and dressing machine	U. Pratt	Deep River, Conn	Mar. 24, 1863	37, 976
Ivory, Process of bleaching	U. Pratt	Deep River, Conn	Jan. 6, 1852	8, 639
Ivory-slitting machine	P. Pratt	Meriden, Conn	Feb. 12, 1828	....
Ivory, wood, horn, &c., Composition for making imitation.	S. Gradenwitz	New York, N. Y	May 25, 1869	90, 443
J.				
Jack:				
See Aërostatic jack.				
Billiard-table-leveling jack.				
Boom-jack.				
Boot-jack.				
Builder's jack.				
Car-pushing jack.				
Car-replacing jack.				
Carriage-jack.				
Coupling-jack.				
Cross-head jack.				
Fence-jack.				
Fishing-jack.				
Hoisting-jack.				
Hoop-jack.				
Hop-pole jack.				
Horse-power jack.				
Horseshoeing-jack.				
Hydraulic jack.				
Knitting-needle jack.				
Lasting-jack.				
Lever-jack.				
Lifting-jack.				
Mechanical jack.				
Painter's jack.				
Pegging-jack.				
Railway-jack.				
Railway pushing-jack.				
Replacing car-jack.				
Saw-jack.				
Screw-jack.				
Self-acting jack.				
Shackle-jack.				
Shaft-turning jack.				
Shoe-jack.				
Shoe-finishing jack.				
Shoe-holding jack.				
Shoemaker's jack.				
Spinning-jack.				
Supporting-jack.				
Thill-shifting jack.				
Trimming-jack.				
Wagon-jack.				
Whirling-jack.				
Jack	A. N. Breneman	Lancaster, Pa	Jan. 29, 1867	61, 603
Jack	T. Wiles	Indianapolis, Ind	Dec. 24, 1867	72, 577
Jack	R. Wood	Grand Ledge, Mich	Mar. 13, 1860	27, 494
Jack-head	N. Thelen	Schenectady, N. Y	Apr. 30, 1872	126, 346
Jacket, Hunting	J. Garand	New York, N. Y	June 17, 1873	139, 950
Jack-screw	S. Vail	Morris, N. J	May 7, 1839	1, 146
Jacquard	W. K. Greene, jr	Schenectady, N. Y	Nov. 28, 1848	5, 939
Jacquard	H. Kelly	Manayunk, Pa	Nov. 28, 1848	5, 937
Jacquard-machine	A. Babbett	Auburn, N. Y	Oct. 18, 1859	25, 796
Jacquard-machine	J. Scott and J. Tannahell	Philadelphia, Pa	Mar. 18, 1851	7, 990
Jail, Iron-plate	E. Jacobs	Cincinnati, Ohio	Dec. 20, 1859	26, 500
Japan applied to leather	W. Gates	Hanover, N. Y	Nov. 14, 1835	....
Japan, &c., over fabrics, Machine for spreading	F. Sautermeister	Newark, N. J	Feb. 3, 1863	37, 592
Japanned metal surfaces, Decorating in gilt and upon	H. Petrie	Chicago, Ill	Nov. 28, 1871	121, 410
Japanning-apparatus	G. W. Hubbell	Derby, Conn	May 23, 1865	47, 826
Japanning eyelets, buttons, &c	S. N. Smith	Providence, R. I	Feb. 14, 1871	111, 882
Japanning metals	G. J. Sturdy and S. W. Young	Providence, R. I	July 21, 1868	80, 236
Japanning pins	J. I. Howe and T. Piper	Derby, Conn	June 10, 1856	15, 111
Jar:				
See Candy-jar.				
Confectionery-jar.				
Drill-jar.				
Drilling-jar.				
Fish-jar.				
Flower-jar.				
Fruit-jar.				
Ice-jar.				
Ice-preserving jar.				
Jelly-jar.				
Leech-jar.				
Oil-tool jar.				
Preserve-jar.				
Slop-jar.				
Soda-fountain - sirup jar.				
Jar and bottle cover	S. S. Butler	Petaluma, Cal	July 9, 1872	128, 849
Jar and bottle stopper	H. W. Ketcham	New York, N. Y	Apr. 19, 1870	102, 129
Jar and bottle stopper	A. Kline	Philadelphia, Pa	Oct. 27, 1863	40, 415
Jar and bottle stopper	T. O. Oliver	New York, N. Y	Apr. 3, 1866	53, 659
Jar, bottle, &c., stopper	N. Thompson	Saint Johnswood, England	May 16, 1865	47, 779
Jar, can, &c., cap	J. K. Chase	New York, N. Y	June 9, 1863	38, 810
Jar-cover fastening	J. F. Johnson	New York, N. Y	Jan. 7, 1868	73, 014
Jar-mold	G. Scott	Cincinnati, Ohio	June 5, 1860	28, 611
Jar-stopper	G. F. J. Colburn	Newark, N. J	Nov. 1, 1864	44, 852
Jars and bottles, Apparatus for stopping	N. Thompson	Saint Johnswood, England	Aug. 30, 1864	44, 054
Jars and cans, Instrument for lifting	J. E. Higby	West Meriden, Conn	Oct. 17, 1865	50, 476
Jars, cans, &c., Device for closing the mouths of	E. T. Jenkins	Ravenswood, N. Y	Apr. 14, 1868	76, 772
Jars for transportation, Device for securing cover to	M. H. Nichols	Hancock, N. Y	July 6, 1869	92, 343
Jars, Metallic screw-cap for	J. K. Chase	New York, N. Y	Oct. 27, 1857	18, 498
Jars, Sealing-ring for preserve	J. Adams	Pittsburgh, Pa	Jan. 2, 1866	51, 785
Jars, &c., Stopping	W. D. Ludlow	New York, N. Y	Aug. 6, 1861	33, 002
Jaw wrench, Revolving	N. Colver	Boston, Mass	May 14, 1850	7, 362
Jelly-glass	J. and T. B. Atterbury	Pittsburgh, Pa	May 31, 1870	103, 702
Jelly-glass	W. M. Kirchner	Pittsburgh, Pa	Dec. 6, 1870	109, 824
Jelly-glass	W. M. Kirchner	Pittsburgh, Pa	July 16, 1872	129, 037
Jelly-glass	D. C. Ripley	Pittsburgh, Pa	May 20, 1873	139, 081
Jelly-glass	T. A. Zellers	East Birmingham, Pa	Apr. 15, 1873	137, 995
Jelly-glass lid	J. Adams	Pittsburgh, Pa	May 13, 1873	138, 833
Jelly-glass lid	D. Challinor	Pittsburgh, Pa	May 13, 1873	138, 735
Jelly, Hand-press for	J. M. Newton	Dayton, Ky	Aug. 12, 1873	141, 809
Jelly-jar	W. C. King	Pittsburgh, Pa	May 6, 1873	138, 502
Jelly-tumbler	W. Doyle	Birmingham, Pa	Mar. 4, 1873	136, 492
Jet, Adjustable spray	A. Nickerson	East Somerville, Mass	Sept. 9, 1873	142, 719
Jewel-case	C. Beck	**Providence, R. I**	**Nov. 11, 1873**	**144, 431**

Index of patents issued from the United States Patent Office from 1790 *to* 1873, *inclusive*—Continued.

Invention.	Inventor.	Residence.	Date.	No.
Jewel-case	G. F. Kolb	Philadelphia, Pa	Apr. 25, 1865	47, 430
Jewel-case	G. F. Kolb	Philadelphia, Pa	Dec. 17, 1867	72, 206
Jewel-setting instrument	A. A. F. and A. Thoma	Piqua, Ohio	Aug. 6, 1867	67, 462
Jeweler's drilling-device	J. K. Laudermilch	Lebanon, Pa	Sept. 23, 1873	143, 084
Jeweler's scraps, Process for refining	S. B. Darling	Providence, R. I	Mar. 27, 1855	12, 585
Jeweler's stock, Manufacture of	T. Diebold, jr	New York, N. Y	Dec. 10, 1872	133, 835
Jeweler's tool	D. M. Bissell	Shelburne Falls, Mass	Sept. 6, 1870	107, 153
Jeweler's use, Manufacture of catches for	G. H. Fuller	Pawtucket, R. I	Dec. 30, 1873	145, 938
Jewelry	J. J. Huber	Geneva, Switzerland	Jan. 3, 1860	26, 678
Jewelry and silver-ware case	S. Bodwell and C. Beck	Providence, R. I	Mar. 19, 1872	124, 783
Jewelry, Apparatus for ornamenting gum	N. Lanphear	Monmouth, Ill	Sept. 22, 1863	40, 053
Jewelry-bases, Device for manufacturing	S. Cottle	New York, N. Y	July 8, 1873	140, 616
Jewelry-bases, Manufacture of	S. Cottle	New York, N. Y	Jan. 21, 1873	135, 087
Jewelry-bases, Manufacture of	S. Cottle	New York, N. Y	Jan. 21, 1873	135, 088
Jewelry-box	L. L. Burdon	Providence, R. I	Nov. 9, 1869	96, 669
Jewelry-boxes, Fastening for	H. Hoefer	Brooklyn, N. Y	May 28, 1872	127, 238
Jewelry-case	J. Smith	Philadelphia, Pa	June 27, 1871	116, 506
Jewelry, diamonds, &c., Chemical compound for cleaning.	A. H. Longley and E. R. Stickney.	Springfield, Mass	Dec. 2, 1873	145, 065
Jewelry-fastening	R. J. Pond	New York, N. Y	Dec. 26, 1871	122, 192
Jewelry, Fastening for	J. T. Folwell	Philadelphia, Pa	Sept. 29, 1857	18, 277
Jewelry, Loop-chain for	C. W. Dickinson	Newark, N. J	Mar. 2, 1858	19, 497
Jewelry, Making joint-wire or stock for	A. J. Wiley	South Attleborough, Mass	Nov. 26, 1861	33, 807
Jewelry, Manufacture of	S. Cottle	New York, N. Y	Dec. 10, 1872	133, 762
Jewelry, Manufacture of	J. S. Palmer	Providence, R. I	Dec. 17, 1867	72, 224
Jewelry, Method of fastening	J. B. Coppinger	New York, N. Y	Oct. 28, 1856	15, 969
Jewelry, Mode of connecting strung pearl	H. Dubosq	Philadelphia, Pa	Apr. 26, 1859	23, 760
Jewelry, Mode of imitating cluster	W. O. Draper, A. C. Sweetland, and G. H. Draper.	North Attleborough, Mass	Mar. 10, 1868	75, 393
Jewelry, &c., Ornament for	C. E. Richards	North Attleborough, Mass	Feb. 9, 1869	86, 864
Jewelry-ornaments, Pin for	J. P. Courtney	Brooklyn, N. Y	Dec. 30, 1873	145, 990
Jewelry-pin	I. Farjeon and W. H. Horton	New York, N. Y., and Jersey City, N. J.	Oct. 25, 1870	108, 578
Jewelry, plate, &c., Machine for ornamenting	O. S. Parmenter	Providence, R. I	May 23, 1865	47, 853
Jewelry, &c., Production of ornamental surfaces for	W. B. Woodbury	London, England	Apr. 28, 1868	77, 231
Jewelry, Rolling metal for	J. S. Palmer	Providence, R. I	June 14, 1859	24, 432
Jewelry, Setting for	S. J. Smith	New York, N. Y	Apr. 14, 1863	38, 184
Jewelry, Setting stones in	F. Stefani	New York, N. Y	Oct. 2, 1866	58, 501
Jib and stay connection	J. E. Seavey	Kennebunkport, Me	Jan. 20, 1863	37, 476
Jib-boom for vessels	C. L. Linnell	Truro, Mass	May 10, 1859	23, 926
Jib, &c., Hank for	W. A. Pratt	Deep River, Conn	Dec. 24, 1872	134, 164
Jib-hank for vessels	H. K. Eldridge	Cambridge, Mass	Feb. 22, 1870	100, 019
Jib-stays, Backer for	T. Lynch	Peacedale, R. I	Nov. 12, 1872	132, 971
Joiner, General	D. R. Williams, sr	Paris, Ky	Aug. 20, 1872	130, 681
Joiner, Universal	E. Passé	Fort Wayne, Ind	Sept. 2, 1873	142, 348
Joiner, Universal	J. H. Rayner and B. W. and W. P. Hatch.	Boston, Mass	Apr. 29, 1873	138, 284
Joiner's bench	J. E. Cryer	Peoria, Ill	May 24, 1859	24, 102
Joiner's bench	J. W. Mahan	Lexington, Ill	Oct. 6, 1857	18, 345
Joiner's bench-strip	C. T. Pearson	Chelsea, Mass	May 12, 1857	17, 292
Joiner's clamp	J. Clackson	Milford, Pa	May 31, 1859	24, 194
Joiner's clamp	W. H. Goodchild	Centreville, N. J	May 16, 1871	114, 800
Joiner's clamp	W. H. Goodchild	Centreville, N. J	May 16, 1871	114, 801
Joiner's clamp	W. H. Goodchild and S. F. Hay	Bayonne, N. J., and Brooklyn, N. Y.	Feb. 25, 1873	136, 152
Joiner's clamp	J. Kuneman	Canton, Ohio	Aug. 25, 1868	81, 381
Joiner's clamp	W. S. Todd	Mechanicsville, Iowa	Sept. 6, 1859	25, 353
Joiner's clamp	E. A. Walker	New Haven, Conn	July 8, 1873	140, 747
Joiner's gage	M. C. Ames	Hartford, Conn	Nov. 19, 1867	70, 937
Joiner's gage	S. H. Jennings and W. F. Arnold.	Deep River and Winthrop, Conn.	Dec. 30, 1873	146, 075
Joiner's gage	C. Sholl	Mount Joy, Pa	Mar. 8, 1864	41, 867
Joining and planing machine	J. S. Mott	Alburgh, Saint Albans, Vt	Mar. 8, 1806	....

Joint:
See Air-tight joint.
Ankle-joint.
Ankle-brace joint.
Ball-joint.
Ball-and-socket joint.
Box-joint.
Brace-joint.
Bracelet-joint.
Buggy-top joint.
Car-seat link-joint.
Carriage-prop joint.
Carriage-top joint.
Cement-pipe joint.
Chimney-joint.
Clapboard-joint.
Compass-joint.
Core-spindle joint.
Desk-joint.
Doll-joint.
Dovetailing-joint.
Felly-joint.
Fishing-rod joint.
Flue-joint.
Frame-joint.
Framing-joint.
Friction-joint.
Gas and water pipe joint.
Gas-pipe joint.
Globe-joint.
Iron-pipe joint.
Knee-joint.
Knuckle-joint.
Ladder-joint.
Lap-joint.
Lightning-rod joint.
Miter-joint.
Movable joint.
Pipe-joint.
Pivoted stump-joint.
Rail-joint.
Railway-rail joint.
Railway-track joint.
Rule-joint.
Seam-joint.
Seat and desk joint.
Shackle-joint.
Sheathing-joint.
Sheet-metal joint.
Sink-joint.
Skirt-hook joint.
Spectacle-bow joint.
Stationary joint.
Stave-joint.
Stove-pipe joint.
Stump-joint.
Sucker-rod joint.
Toggle-joint.
Toilet-glass joint.
Tongs-joint.
Tongue and groove joint.
Top-joint.
Tube-joint.
Universal joint.
Universal globe-joint.
Water-wheel joint.
Yielding joint.

Index of patents issued from the United States Patent Office from 1790 *to* 1873, *inclusive*—Continued.

Invention.	Inventor.	Residence.	Date.	No.
Joint of railway-rail	G. Palmer	Littlestown, Pa	Sept. 24, 1867	69, 241
Joint bit and check	R. D. Sterling	New York, N. Y	Oct. 22, 1867	70, 129
Joint-bolt	S. E. Jewett	Haverhill, Mass	Oct. 29, 1867	70, 223
Joint-clamp	D. E. Hall	Detroit, Mich	June 23, 1868	79, 116
Joint mold or flask	W. Culliss	Philadelphia, Pa	Jan. 4, 1870	98, 567
Joint tube, Free	R. Prosser	Birmingham, England	Sept. 21, 1852	9, 278
Jointed molds, Construction of	M. B. Stafford	New York, N. Y	Sept. 4, 1866	57, 791
Jointed-pipe connection	H. Smith	Norwalk, Ohio	Nov. 16, 1869	96, 982
Jointed pipe for steam, &c	L. Kirk	Reading, Pa	Oct. 9, 1847	5, 325
Jointer and gage-machine	J. Wadsworth	Milton, Mass	Feb. 10, 1810	....
Joist-hook	J. S. Miller	Terre Haute, Ind	Aug. 8, 1871	117, 798
Joist-protector	H. Galmann and C. Ruhe	Buchanan, Pa	Feb. 1, 1870	99, 305
Joists, Machine for dressing	J. H. Story	Cincinnati, Ohio	Aug. 21, 1860	29, 749
Joss-stick, Composition to be used instead of	F. Eckstein and J. McIlhenney	Philadelphia, Pa	Oct. 12, 1814	
Journal	D. R. Quick	New York, N. Y	Aug. 31, 1869	94, 439
Journal, Adjustable	M. Jackson	Bavington, Pa	Feb. 4, 1873	135, 477
Journal and axle box	P. S. Devlan	Jersey City, N. J	July 9, 1867	66, 471
Journal and bearing	J. Whitaker	Woonsocket, R. I	Dec. 9, 1873	145, 467
Journal and bearing composition	M. J. Marcotte	Saint Louis, Mo	Aug. 1, 1871	117, 652
Journal and bearing material	E. D. Murfey	New York, N. Y	Oct. 25, 1870	108, 722
Journal and box	T. Hopper and T. Garrison	New Brunswick, N. J	Jan. 2, 1849	5, 996
Journal and box	W. S. Mead	New York, N. Y	Mar. 6, 1866	53, 023
Journal and box for railway-cars	A. Jackman	Elizabethport, N. J	May 1, 1860	28, 129
Journal and journal-box	J. Stimpson	Baldwinsville, Mass	Sept. 16, 1862	36, 487
Journal and shaft bearing, Frictionless	J. Eccles	Philadelphia, Pa	July 22, 1873	141, 129
Journal, Anti-friction	M. E. Dayton	Syracuse, N. Y	Apr. 22, 1873	138, 011
Journal, Anti-friction	A. W. Hall and M. E. Dayton	Adrian, Ohio, and Syracuse, N. Y.	Mar. 12, 1872	124, 573
Journal, Anti-friction	P. W. Yarrell	Littleton, N. C	Mar. 29, 1870	101, 401
Journal, Anti-friction	P. W. Yarrell	Littleton, N. C	Dec. 20, 1870	110, 323
Journal axle, Carriage	H. McIlroy	Poplar Ridge, N. Y	Sept. 8, 1868	81, 924
Journal, Axle or shaft	C. D. Smith	Baldwinsville, Mass	Jan. 7, 1868	73, 055
Journal bearing	S. P. and H. B. Cook	San Francisco, Cal	Aug. 8, 1871	117, 868
Journal-bearing	D. A. Hopkins	Jersey City, N. J	Nov. 15, 1870	109, 319
Journal-bearing	A. F. Jones	New York, N. Y	June 27, 1871	116, 317
Journal bearing	J. H. Lindsay	Allegheny, Pa	Jan. 23, 1872	122, 898
Journal-bearing	E. D. Murfey	New York, N. Y	Oct. 4, 1870	108, 080
Journal-bearing	E. D. Murfey	New York, N. Y	Dec. 12, 1871	121, 804
Journal-bearing	I. P. Wendell	Philadelphia, Pa	Jan. 11, 1870	98, 825
Journal-bearing	R. Yeilding	Ypsilanti, Mich	Oct. 6, 1863	40, 220
Journal-bearing and lubricating-device	D. N. M. Peregoy	Johnson City, Tenn	Dec. 26, 1871	122, 277
Journal-bearing, Anti-friction	R. G. Hatfield	New York, N. Y	Feb. 8, 1870	99, 566
Journal-bearing, Anti-friction-roller	E. Turner	Chicago, Ill	Nov. 3, 1863	40, 471
Journal-bearing for hoisting-machines, &c., Anti-friction.	R. G. Hatfield	New York, N. Y	Dec. 28, 1869	98, 260
Journal-bearing, Railway-car	I. P. Wendell	Philadelphia, Pa	Nov. 24, 1868	84, 326
Journal-bearings, Composition for	S. Croll	Philadelphia, Pa	Mar. 12, 1872	124, 485
Journal-bearings, &c., Graphitized materials for	A. Hitchcock	New York, N. Y	Nov. 19, 1872	133, 101
Journal-bearings, Lining for	H. F. Boody and E. P. Merrill	Deering, Me	Sept. 24, 1872	131, 653
Journal-bearings, Lubricating	I. P. Wendell	Philadelphia, Pa	June 28, 1870	104, 805
Journal-bearings, &c., Metalline for	S. Gwynn	New York, N. Y	Apr. 12, 1870	101, 861
Journal-bearings, &c., Metalline for	S. Gwynn	New York, N. Y	Apr. 12, 1870	101, 862
Journal-bearings, &c., Metalline for	S. Gwynn	New York, N. Y	Apr. 12, 1870	101, 863
Journal-bearings, &c., Metalline for	S. Gwynn	New York, N. Y	Apr. 12, 1870	101, 864
Journal-bearings, &c., Metalline for	S. Gwynn	New York, N. Y	Apr. 12, 1870	101, 865
Journal-bearings, &c., Metalline for	S. Gwynn	New York, N. Y	Apr. 12, 1870	101, 866
Journal-bearings, &c., Metalline for	S. Gwynn	New York, N. Y	Apr. 12, 1870	101, 867
Journal-bearings, &c., Metalline for	S. Gwynn	New York, N. Y	Apr. 12, 1870	101, 868
Journal-bearings, steps, and other articles liable to friction, Process of forming metalline for.	S. Gwynn	New York, N. Y	Apr. 12, 1870	101, 869
Journal-box	S. Aland	Rome, N. Y	Feb. 25, 1873	136, 198
Journal-box	J. A. Althouse	New Harmony, Ind	May 27, 1873	139, 287
Journal-box	J. E. Atwood	Mansfield, Conn	Sept. 15, 1868	82, 066
Journal-box	J. D. Beers	Philadelphia, Pa	Nov. 21, 1871	121, 040
Journal-box	J. M. Brower	Syracuse, N. Y	June 9, 1863	38, 808
Journal-box	A. G. Brown	Wareham, Mass	Apr. 4, 1871	113, 490
Journal-box	J. T. Bruen	New York, N. Y	Oct. 17, 1865	50, 445
Journal-box	J. Bryant	Brooklyn, N. Y	Jan. 10, 1860	26, 747
Journal-box	T. F. Burgess	Lowell, Mass	Dec. 10, 1867	71, 849
Journal-box	A. B. Caldwell	Syracuse, N. Y	Sept. 1, 1868	81, 748
Journal-box	W. T. Carroll	Medway, Mass	Apr. 4, 1871	113, 494
Journal-box	N. W. Clark	Clarkston, Mich	Feb. 5, 1861	31, 297
Journal-box	R. Colburn and G. W. Gould	Norwich, Conn	July 14, 1868	79, 900
Journal-box	J. Connolly	Boston, Mass	Jan. 1, 1867	60, 696
Journal-box	G. H. Corliss	Providence, R. I	July 25, 1871	117, 386
Journal-box	R. Daniels	Almena, Mich	Apr. 12, 1859	23, 664
Journal-box	P. S. Devlan	Elizabethport, N. J	Sept. 25, 1860	30, 128
Journal-box	E. S. Dickinson and W. D. Nelson.	New York, N. Y	Sept. 27, 1864	44, 407
Journal-box	W. H. Doane	Cincinnati, Ohio	Aug. 22, 1865	49, 584
Journal-box	D. H. Dotterer	Memphis, Tenn	May 7, 1861	32, 243
Journal-box	S. C. Ellis	Jersey City, N. J	Feb. 1, 1870	99, 422
Journal-box	S. C. Ellis	Jersey City, N. J	May 9, 1871	114, 542
Journal-box	J. H. Hayward	New York, N. Y	Dec. 10, 1872	133, 776
Journal-box	O. T. L. Heine and E. Prussing	Chicago, Ill	Dec. 24, 1861	33, 992
Journal-box	G. H. Henfield	San Francisco, Cal	Nov. 10, 1868	83, 961
Journal-box	D. R. Hickok	Morrisville, Vt	Apr. 24, 1866	54, 156
Journal-box	S. B. Higgins	South Boston, Mass	Dec. 19, 1865	51, 654
Journal-box	T. Hill	Saint Louis, Mo	Oct. 17, 1865	50, 477
Journal-box	D. A. Hopkins	Paterson, N. J	Feb. 23, 1858	19, 424
Journal-box	J. Hughes	New Berne, N. C	Aug. 16, 1870	106, 486
Journal-box	G. G. Hunt	Bridgeport, Conn	Jan. 12, 1864	41, 221
Journal-box	J. P. Kenyon	Brooklyn, N. Y	Sept. 15, 1863	39, 932
Journal-box	H. A. Lee	Worcester, Mass	Aug. 1, 1865	49, 124
Journal-box	E. F. Light	Worcester, Mass	Feb. 20, 1866	52, 720
Journal-box	E. F. Light	Worcester, Mass	Aug. 21, 1866	57, 343
Journal-box	E. F. Light	Worcester, Mass	Oct. 15, 1867	69, 822
Journal-box	I. D. Mathews	Worcester, Mass	Mar. 31, 1868	76, 092

Index of patents issued from the United States Patent Office from 1790 *to* 1873, *inclusive*—Continued.

Invention.	Inventor.	Residence.	Date.	No.
Journal-box	J. S. McClure	Gold Hill, Nev	Oct. 1, 1872	131, 889
Journal-box	J. McIlvain	Churchville, Md	Jan. 18, 1870	98, 987
Journal-box	G. R. Meneely	West Troy, N. Y	June 11, 1872	127, 784
Journal-box	G. R. Meneely	West Troy, N. Y	Sept. 17, 1872	131, 454
Journal-box	G. R. Meneely	West Troy, N. Y	Nov. 26, 1872	133, 472
Journal-box	J. Millholland and J. J. Lahaye	Reading, Pa	Mar. 18, 1862	34, 690
Journal-box	J. A. Montgomery	Williamsport, Pa	Aug. 2, 1859	24, 947
Journal-box	J. R. Morris	Houston, Tex	May 13, 1873	138, 916
Journal-box	W. T. Morrow	Chicago, Ill	Feb. 3, 1863	37, 586
Journal-box	E. D. Murfey	New York, N. Y	Oct. 4, 1870	108, 079
Journal-box	E. D. Murfey	New York, N. Y	Oct. 4, 1870	108, 081
Journal-box	J. A. Norris	Philadelphia, Pa	May 25, 1858	20, 363
Journal-box	J. Old	Pittsburgh, Pa	Mar. 20, 1847	5, 027
Journal-box	W. S. Pratt	Brooklyn, N. Y	Apr. 19, 1859	23, 704
Journal-box	H. M. Preston	Unionville, Conn	Aug. 11, 1868	81, 002
Journal-box	M. Queen	Brooklyn, N. Y	Aug. 28, 1860	29, 817
Journal-box	M. J. Rice and W. H. Millen	Boston, Mass	Mar. 14, 1865	46, 823
Journal-box	M. J. Rice and W. H. Millen	Boston, Mass	Aug. 22, 1865	49, 591
Journal-box	J. T. Robinett	Petersburgh, Va	June 9, 1868	78, 832
Journal-box	J. T. Robinett	Petersburgh, Va	Feb. 8, 1870	99, 711
Journal-box	J. B. Sargent	New Haven, Conn	Aug. 29, 1871	118, 647
Journal-box	A. H. Sassaman	Scranton, Pa	Dec. 14, 1869	97, 968
Journal-box	J. Scheider	New York, N. Y	Jan. 31, 1871	111, 479
Journal-box	L. Schmid	Cleveland, Ohio	Sept. 17, 1872	131, 376
Journal-box	W. Sherburne	Charlestown, Mass	July 14, 1868	79, 867
Journal-box	R. Sibley	Greenville, Conn	May 4, 1869	89, 839
Journal-box	E. W. Skinner	Madison, Wis	Mar. 30, 1869	88, 333
Journal-box	E. H. Stearns	Erie, Pa	Dec. 13, 1870	110, 086
Journal-box	L. M. Taylor and W. D. Fowler	Washington, D. C	Oct. 1, 1867	69, 373
Journal-box	H. H. Thayer	Sandwich, Mass	Sept. 7, 1858	21, 474
Journal-box	W. O. Thoms and A. M. Miller	Fond du Lac, Wis	Oct. 14, 1862	36, 682
Journal-box	H. Turner	Charlestown, N. H	Apr. 27, 1852	8, 915
Journal-box	G. L. Weaver	Hartford, Conn	Sept. 29, 1868	82, 665
Journal-box	J. Webster	Brooklyn, N. Y	June 18, 1861	32, 604
Journal-box	I. P. Wendell	Philadelphia, Pa	Apr. 17, 1860	27, 940
Journal-box	D. Whitlock	Newark, N. J	June 18, 1872	128, 002
Journal-box and axle-bearing	S. T. F. Sterick	Georgetown, D. C	July 18, 1871	117, 217
Journal box and bearing	J. B. Caryl	Candor, N. Y	Apr. 9, 1867	63, 609
Journal box and bearing, Self-lubricating	C. Purdy	Bedford, Ohio	Feb. 26, 1867	62, 500
Journal box and bearing, Self-oiling	B. D. Stevens	Lawrence, Mass	June 23, 1863	39, 018
Journal-box, Anti-friction	B. C. Baker	Toledo, Ohio	July 11, 1871	116, 795
Journal-box, Anti-friction	H. M. Le Duc	Washington, D. C	Sept. 18, 1866	58, 109
Journal-box, Anti-friction	W. S. Pratt	Williamsburgh, N. Y	May 8, 1860	28, 195
Journal-box, Anti-friction	R. W. Smith	Toledo, Ohio	July 16, 1872	129, 256
Journal-box, Anti-friction	H. F. Stith	Stanton, Kans	Sept. 10, 1867	68, 667
Journal-box, Anti-friction	H. P. Westcott	Seneca Falls, N. Y	Mar. 14, 1871	112, 660
Journal-box, Car	G. H. Clemens	Baltimore, Md	Nov. 12, 1867	70, 697
Journal-box, Car	J. B. Collin	Altoona, Pa	Sept. 28, 1869	95, 196
Journal-box, Car	E. Von Jeinsen	Omaha, Nebr	Nov. 9, 1870	109, 159
Journal-box composition	F. W. Armstrong	New York, N. Y	May 27, 1862	35, 409
Journal-box for car-wheels	R. J. Hamilton	Chicago, Ill	Mar. 3, 1863	37, 814
Journal-box for carriages	S. B. Batchelor	Lowville, N. Y	Jan. 2, 1855	12, 154
Journal-box for land-carriages	E. P. Palmer	Milton, Del	Nov. 14, 1865	50, 948
Journal-box for lubricating axles	J. Schinneller	Temperanceville, Pa	Jan. 9, 1872	122, 662
Journal-box for railway-car axles	W. B. Gage	Saratoga Springs, N. Y	Apr. 1, 1856	14, 560
Journal-box for saw-mill carriages	W. M. Ferry, jr	Ferrysburgh, Mich	Nov. 29, 1859	26, 262
Journal-box for saw-mill carriages	C. R. Fox	Chicago, Ill	May 9, 1854	10, 882
Journal box, Glass	E. Campbell	Columbus, Ohio	Aug. 21, 1855	13, 454
Journal-box, Grindstone	T. W. Brown	New York, N. Y	Oct. 16, 1866	58, 768
Journal-box lining	J. Corduan	Brooklyn, N. Y	Mar. 12, 1861	31, 655
Journal-box lining	P. S. Devlan	Jersey City, N. J	Nov. 13, 1866	59, 564
Journal-box lining	S. Gwynn	New York, N. Y	May 6, 1873	138, 641
Journal-box lining	S. Gwynn	New York, N. Y	May 6, 1873	138, 642
Journal-box lining	S. Gwynn	New York, N. Y	May 6, 1873	138, 643
Journal-box lining	S. Gwynn	New York, N. Y	May 6, 1873	138, 644
Journal-box lining	S. Gwynn	New York, N. Y	May 6, 1873	138, 645
Journal-box lining	S. Gwynn	New York, N. Y	May 6, 1873	138, 646
Journal-box lining	S. Gwynn	New York, N. Y	July 15, 1873	140, 774
Journal-box lining	S. Gwynn	New York, N. Y	July 15, 1873	140, 775
Journal-box lubricator	A. and F. Brown	New York, N. Y	Sept. 28, 1869	95, 313
Journal-box lubricator, Railway-car	J. C. Geisendorff	Cincinnati, Ohio	Feb. 9, 1858	19, 291
Journal-box-oiling device	G. M. Morris	Cohoes, N. Y	Apr. 23, 1867	64, 130
Journal-box, Railway	O. Beecher and R. E. Rogers	Philadelphia, Pa	Dec. 8, 1863	40, 810
Journal-box, Railway	I. Dripps	Altoona, Pa	Aug. 8, 1871	117, 753
Journal-box, Railway	W. Groat	Green Island, N. Y	Feb. 16, 1864	41, 616
Journal-box, Railway	A. B. Nimbs	Buffalo, N. Y	Oct. 31, 1865	50, 771
Journal-box, Railway	R. C. Wright	Meadville, Pa	July 25, 1865	49, 022
Journal-box, Railway	R. C. Wright	Mead Township, Pa	Nov. 14, 1865	50, 982
Journal-box, Railway-car	W. B. Aitken	Philadelphia, Pa	Dec. 1, 1863	40, 729
Journal-box, Railway-car	R. N. Allen	Pittsford, Vt	July 8, 1873	140, 568
Journal-box, Railway-car	J. Ashenfelder	Philadelphia, Pa	July 31, 1860	29, 349
Journal-box, Railway-car	R. Brewer	Pontiac, Ill	Dec. 26, 1871	122, 219
Journal-box, Railway-car	D. A. Hopkins	Brooklyn, N. Y	Mar. 11, 1862	34, 636
Journal-box, Railway-car	C. Ihrig	Jersey City, N. J	July 26, 1870	105, 689
Journal-box, Railway-car	R. McWilliams	Philadelphia, Pa	Dec. 15, 1857	18, 884
Journal-box, Railway-car	R. McWilliams	Philadelphia, Pa	July 19, 1859	24, 846
Journal-box, Railway-car, &c	J. A. Norris	Philadelphia, Pa	Dec. 29, 1857	18, 934
Journal-box, Railway-car	A. L. Raplee	Edwin, N. Y	Sept. 13, 1870	107, 408
Journal-box, Railway-car	H. Rice	Concord, N. H	Aug. 7, 1860	29, 551
Journal-box, Railway-car	H. B. Rowley	Buffalo, N. Y	Apr. 27, 1869	89, 349
Journal-box, Railway-car	J. O. Scott	New York, N. Y	May 24, 1864	42, 881
Journal-box, Railway-car	M. Thornton	Macon, Ga	May 26, 1868	78, 244
Journal-box, Railway-car	P. Umholtz	Tremont, Pa	Mar. 15, 1859	23, 279
Journal-box, Roller	E. Wadhams	Yorkville, N. Y	Dec. 31, 1867	72, 942
Journal-box, Self-lubricating	S. F. Bond	Worcester, Mass	Apr. 23, 1867	64, 004
Journal-box, Self-lubricating	P. P. Lane	Cincinnati, Ohio	Nov. 26, 1867	71, 313
Journal-box, Self-oiling	J. Sault	South Manchester, Conn	Sept. 26, 1871	119, 412
Journal-box, Self-oiling	J. Whitlock	Birmingham, Conn	Apr. 23, 1867	64, 054

Index of patents issued from the United States Patent Office from 1790 *to* 1873, *inclusive*—Continued.

Index of patents issued from the United States Patent Office from 1790 *to* 1873, *inclusive*—Continued.

Invention.	Inventor.	Residence.	Date.	No.
Jug-top	P. Lauster	Allegheny, Pa	Sept. 8, 1868	82, 010
Jug-top	H. Wright	Pittsburgh, Pa	May 10, 1864	42, 712
Jug-top	H. Wright	Pittsburgh, Pa	Mar. 6, 1866	53, 076
Jug-top	H. Wright	Pittsburgh, Pa	Sept. 24, 1867	69, 151
Jug-top	H. Wright	Pittsburgh, Pa	Apr. 13, 1869	88, 829
Jug-top	H. Wright	Pittsburgh, Pa	July 16, 1872	129, 303
Jug-top, Casting	F. B. Richardson	Boston, Mass	June 22, 1869	91, 669
Jug-tops, Making	G. P. Lang and P. Lauster	Allegheny, Pa	Jan. 1, 1867	60, 750
Jump-seat	N. Warren and T. Underwood	Wilmington, Del	Dec. 5, 1871	121, 563
Jump-seat, Stump-joint	J. D. McAuliff	Saint Louis, Mo	July 2, 1872	128, 645
Jumping-hoop	C. L. Browne	Brooklyn, N. Y	July 23, 1867	67, 101
Jute fiber, Treating	J. Monach	Philadelphia, Pa	Feb. 4, 1868	74, 113
Jute twine, Preparing	J. B. Travers	New York, N. Y	Nov. 3, 1863	40, 520
Jute yarn, Preparing	J. Monach	Philadelphia, Pa	Apr. 21, 1868	77, 073
K.				
Kaleidoscope	J. Earnshaw	East Greenwich, R. I	June 25, 1867	66, 134
Kaleidoscope	M. Keller	Boston, Mass	July 12, 1870	105, 218
Kaleidoscope	A. C. McNulty and D. Lyman, jr.	New York, N. Y	Oct. 2, 1860	30, 238
Kaleidoscope-object	C. G. Bush	Boston, Mass	Sept. 30, 1873	143, 271
Kaleidoscope, Telescopic	J. Pool	Newark, N. J	Nov. 12, 1872	132, 978
Kalsomining wall, &c., Composition for	N. A. Frank	Chicago, Ill	June 30, 1868	79, 337
Kaolin, Apparatus for washing	D. Hull	Newburgh, N. Y	May 1, 1866	54, 355
Keel and bilge block	J. T. Parlour	Brooklyn, N. Y	Jan. 14, 1868	73, 376
Keel block or rest	J. T. Parlour	Brooklyn, N. Y	June 11, 1867	65, 691
Keel for boats, Ballast	H. A. Dirkes	New York, N. Y	Apr. 21, 1868	76, 896
Keel for ships and other navigable vessels	J. B. Tarr	Chicago, Ill	May 23, 1865	47, 879
Keel or bottom of vessels, Apparatus for ascertaining the curvature of.	H. E. Towle	Exeter, N. H	Apr. 23, 1861	32, 157
Keel, Swinging	J. Beale	Alexandria, Va	Jan. 15, 1812	....
Keel, Swinging false	P. Hall		Apr. 2, 1816	....
Keels, Apparatus for examining vessels	J. E. Simpson	East Boston, Mass	May 26, 1857	17, 396
Keels for vessels, Combination of center and bilge	S. P. Willeby	Philadelphia, Pa	May 20, 1873	139, 222
Keels of vessels, Construction of	G. Livingston		Oct. 22, 1803	....
Keels to vessels, Attaching	T. F. Griffith	New Market, Md	Apr. 26, 1845	4, 016
Keg and barrel for paint and other material	A. L. Freeman	Manchester, England	June 4, 1867	65, 368
Keg, Gunpowder	J. Wilson, C. Green, and W. Wilson, jr.	Brandywine, Del	Mar. 31, 1857	16, 944
Keg, Metallic	W. Hill	Pottsville, Pa	July 27, 1869	93, 085
Kegs, &c., Attaching the heads of metallic powder	J. Wilson, C. Green, and W. Wilson, jr.	Wilmington, Del	July 12, 1859	24, 772
Kegs, Machine for securing heads to metallic	W. Hill	Pottsville, Pa	Oct. 25, 1870	10-, 594
Kegs, Manufacture of oyster	G. Frauenberger	Rochester, N. Y	Aug. 17, 1869	93, 870
Kegs or casks, Machine for chamfering and crozing	J. A. Seaman	Saint Louis, Mo	Nov. 22, 1859	26, 206
Kerosene burning-fluid	A. Gesner	Williamsburgh, N. Y	June 27, 1854	11, 203
Kerosene burning-fluid	A. Gesner	Williamsburgh, N. Y	June 27, 1854	11, 204
Kerosene burning-fluid	A. Gesner	Williamsburgh, N. Y	June 27, 1854	11, 205
Kerosene-heater	Z. B. and C. E. Grandy	Stafford Springs, Conn	July 22, 1873	141, 046
Kerosene, Process for making	A. Gesner	Williamsburgh, N. Y	Mar. 27, 1855	12, 612
Kettle	W. J. Evans and W. Perry	New York, N. Y	June 3, 1873	139, 502
Kettle	H. Matthews	South Yarmouth, Mass	Dec. 31, 1867	72, 746
Kettle	A. N. Merrill	Batavia, Ill	May 14, 1867	64, 685
Kettle	A. Sower	New York, N. Y	May 28, 1867	65, 129
Kettle	D. Stuart	Philadelphia, Pa	Oct. 3, 1871	119, 666
Kettle and boiler for heating water	P. N. Griffith	Orange, N. J	Sept. 26, 1825	....
Kettle and vessel cover	E. K. Dean	Bangor, Me	May 3, 1870	102, 507
Kettle-bail	J. Britton	Williamsburgh, N. Y	Aug. 9, 1870	106, 115
Kettle, &c., bail	A. D. Crocker	Boston, Mass	Sept. 18, 1866	58, 224
Kettle-bail	J. T. Darnell	Florence, N. J	Oct. 22, 1872	132, 450
Kettle-bail	T. H. Dodge	Nashua, N. H	May 24, 1853	9, 744
Kettle-bail	W. Hailes	Albany, N. Y	Dec. 17, 1867	72, 195
Kettle, Boiling	H. H. Pember	New York, N. Y	June 9, 1868	78, 823
Kettle, Boiling	S. Spoor	Phelps, N. Y	Feb. 23, 1869	87, 306
Kettle-bottom	C. A. Moore	West Brook, Conn	May 29, 1866	55, 145
Kettle, Camp	C. Avery	Ashtabula, Ohio	May 3, 1864	42, 547
Kettle, Camp	J. C. Mulligan	Elizabeth City, N. J	Jan. 20, 1863	37, 456
Kettle, Cooking	B. W. Dunning	Brooklyn, N. Y	Apr. 16, 1867	63, 871
Kettle-cover	V. R. Holmes	Grand Rapids, Mich	Jan. 12, 1869	85, 827
Kettle, Dinner	J. Wagner	Cumberland, Md	Mar. 31, 1868	76, 121
Kettle-ears, Making	M. Wells	Brooklyn, N. Y	May 8, 1860	28, 222
Kettle for boiling by steam	J. O. Morse and G. D. Hiscox	Englewood, N. J., and Brooklyn, N. Y.	Sept. 14, 1869	94, 907
Kettle for boiling milk	H. Friedländer	New York, N. Y	Feb. 2, 1869	86, 385
Kettle for boiling salt, &c	J. Kendall	Fayette County, Pa	Mar. 18, 1818	....
Kettle for calcining plaster of Paris	J. B. King	New York, N. Y	Nov. 21, 1854	11, 966
Kettle for culinary purposes	J. Burns	New York, N. Y	Dec. 29, 1868	85, 429
Kettle for culinary purposes	D. H. Tuxworth, sr	Baltimore, Md	Jan. 27, 1863	37, 533
Kettle for manufacturing comfits	W. H. Holt	Hartford, Conn	Oct. 8, 1850	7, 702
Kettle for melting, mixing, and casting metals	W. S. Deeds	Baltimore, Md	July 19, 1870	105, 553
Kettle for trying oil	J. L. Alberger	Buffalo, N. Y	Nov. 30, 1858	22, 152
Kettle-furnace, Portable	J. Murdock	South Carver, Mass	June 20, 1865	48, 302
Kettle-handle	H. T. Robbins	Hyde Park, Mass	May 16, 1871	114, 860
Kettle-handle	J. Warner	New Britain, Conn	Jan. 22, 1861	31, 198
Kettle, Hatter's	R. Wildman	Hartford, Conn	Nov. 8, 1845	4, 256
Kettle, Hatter's planking	E. Joyce	Brooklyn, N. Y	June 6, 1825	....
Kettle, Lard-rendering	J. J. Bate	Brooklyn, N. Y	Oct. 21, 1856	15, 942
Kettle, Lard-rendering	J. J. Bate	Brooklyn, N. Y	Sept. 29, 1857	18, 271
Kettle, Lard-rendering	J. J. Bate	Brooklyn, N. Y	July 13, 1858	20, 856
Kettle, Lard-rendering	A. Lapham	Brooklyn, N. Y	Nov. 10, 1857	18, 622
Kettle-lifter	E. A. Thissell	Lowell, Mass	July 22, 1873	141, 020
Kettle machine, Brass	J. Cannon	Warren, R. I	Feb. 3, 1857	16, 564
Kettle machine, Brass	O. W. Minard	Waterbury, Conn	July 1, 1856	15, 247
Kettle-molding flask	G. Walworth	Peekskill, N. Y	Mar. 30, 1869	88, 425
Kettle or boiler used in salt manufacture, Separating and removing bitterings from.	D. Dear	Salina, N. Y	Oct. 18, 1837	428
Kettle, pail, &c., lid	S. B. Cox	Buffalo, N. Y	Aug. 20, 1867	67, 847

Index of patents issued from the United States Patent Office from 1790 *to* 1873, *inclusive*—Continued.

Invention.	Inventor.	Residence.	Date.	No.
Kettle, Shielding-arch for evaporating	J. English	Syracuse, N. Y	Nov. 17, 1868	84, 101
Kettle-spout attachment	M. Saulson	Troy, N. Y	Aug. 3, 1869	93, 354
Kettle, Stove or range	L. D. Lothrop	Dover, N. H	Feb. 13, 1872	123, 569
Kettle, Wash	J. Reist	Philadelphia, Pa	Nov. 29, 1864	45, 272
Kettles and other hollow ware, Machine for making brass.	F. J. Seymour	Wolcottville, Conn	Jan. 14, 1873	134, 938
Kettles and other vessels, Apparatus for discharging liquids from.	A. Brear	Saugatuck, Conn	Apr. 15, 1862	35, 007
Kettles and other vessels, Mode of discharging the contents of sugar.	A. Brear	Saugatuck, Conn	Apr. 1, 1862	34, 811
Kettles, boilers, &c., Bail for	W. M. Stratton	West Troy, N. Y	Feb. 13, 1872	123, 648
Kettles, boilers, stoves, &c., Attaching covers to	W. Hailes	Albany, N. Y	Dec. 18, 1866	60, 508
Kettles, Casting iron	C. McGinnes	Pittsburgh, Pa	July 20, 1858	20, 951
Kettles, Constructing and setting salt	C. C. P. Crosby	Brooklyn, N. Y	Jan. 11, 1840	1, 471
Kettles, Constructing hatters'	W. Porter and A. Sanger	Waltham, Mass	July 8, 1834	....
Kettles, Construction of salt	O. W. Seely	Albany, N. Y	Apr. 9, 1861	32, 005
Kettles, Construction of stamped sheet-metal	F. G. and W. F. Niedringhaus	Saint Louis, Mo	Dec. 10, 1867	71, 900
Kettles, Cover and lifting-device for	A. Selkirk	Albany, N. Y	Jan. 29, 1867	61, 636
Kettles, Die for manufacturing brass	O. Newton	Pittsburgh, Pa	Feb. 11, 1862	34, 374
Kettles, &c., for boilers, Setting	G. W. Robinson	Attleborough, Mass	July 17, 1818	
Kettles, &c., from disks of metal, Machinery for making.	H. W. Hayden	Waterbury, Conn	Dec. 16, 1851	8, 589
Kettles from metal disks, Machine for forming	L. C. Camp	Berlin, Conn	Jan. 9, 1855	12, 227
Kettles, Grinding the inner surfaces of cast-iron	C. C. Bradley, jr	Syracuse, N. Y	Feb. 24, 1857	16, 679
Kettles, Machine for making brass	E. C. Blakesley, E. Platt, jr., and E. Jordan.	Waterbury, Conn	Oct. 28, 1856	15, 961
Kettles, Making brass	O. W. Minard	Waterbury, Conn	Apr. 15, 1856	14, 696
Kettles, Making brass	O. W. Minard	Waterbury, Conn	Sept. 23, 1856	15, 772
Kettles, Making brass	F. J. Seymour	Waterbury, Conn	May 13, 1856	14, 887
Kettles, Metallic flask for casting large	W. Kelly	Eddyville, Ky	Dec. 3, 1850	7, 810
Kettles, Setting and arranging sugar	M. White	New Orleans, La	Sept. 17, 1839	1, 326
Kettles, Setting evaporating	W. S. Worthington	Newtown, N. Y	Sept. 8, 1863	39, 859
Kettles, Setting potash	D. B. Turner	Florence, Ohio	Sept. 18, 1841	2, 263
Kettles, Setting sugar	J. Malory	New Orleans, La	Sept. 20, 1838	936
Kettles, Setting sugar	H. Roth	Iberville Parish, La	Mar. 2, 1858	19, 515
Kettles with spouts, Molding	W. H. Pease	Dayton, Ohio	Oct. 28, 1851	8, 471
Key: *See* Bed-key. Bedstead-key. Clock-key. Cock-key. Coffin-key. Door-key. Edge-key. Fence-key. Lever-key. Lock-key. Organ-key. Piano-key. Sewing-machine-lock key. Sheet-metal key. Telegram-key. Telegraph-key. Telegraph transmitting-key. Tuning-key. Watch-key. Watch-pendant key. Wire-fence key.				
Key	T. S. Bowman	Saint Louis, Mo	May 15, 1866	54, 678
Key	J. Deally	Louisville, Ky	May 29, 1860	28, 457
Key	H. H. Elwell	South Norwalk, Conn	July 12, 1870	105, 322
Key	J. Linden	Seneca Falls, N. Y	Feb. 8, 1870	99, 577
Key	C. E. Lombard	Springfield, Mass	Mar. 26, 1867	63, 271
Key	E. Parker	New Britain, Conn	May 11, 1869	89, 886
Key	E. Parker	New Britain, Conn	May 18, 1869	90, 121
Key	S. Perry	Newport, N. Y	Aug. 4, 1857	17, 939
Key	A. Vang	Chicago, Ill	Jan. 19, 1869	86, 114
Key-blanks, Die for making	E. Parker	New Britain, Conn	July 2, 1872	128, 650
Key-board	R. Humphreys	Jonesborough, Tenn	June 14, 1859	24, 392
Key-board	J. P. Perry	Yarmouth Port, Mass	May 23, 1871	115, 096
Key-board, Chromatic	H. Downes	New York, N. Y	Jan. 9, 1872	122, 584
Key-board instruments, Key-cap for	J. P. Lord	Manchester, N. H	Nov. 11, 1873	144, 399
Key-board instruments, Melody-attachment to	C. Fogelberg	New York, N. Y	Jan. 26, 1869	86, 219
Key-board, Transposing	A. D. B. Wolff	Paris, France	Sept. 16, 1873	142, 974
Key-body-milling machine	F. Caffrey and J. L. Nettleton	West Cheshire, Conn	June 8, 1869	91, 082
Key bush, Lock	A. F. Whiting	Crestline, Ohio	June 25, 1872	128, 266
Key-check	J. J. De Barry	Chicago, Ill	Jan. 21, 1873	135, 092
Key-fastener	E. H. Bailey	Philadelphia, Pa	Apr. 9, 1861	31, 939
Key-fastener	R. S. Dunning	Fall River, Mass	May 4, 1869	89, 641
Key-fastener	H. E. Fowler	Northford, Conn	Aug. 2, 1864	43, 676
Key-fastener	M. P. Hall	Manchester, N. H	Aug. 7, 1866	56, 933
Key-fastener	J. E. Hills	Orange, Mass	June 8, 1869	90, 945
Key-fastener	H. Hungerford	Brooklyn, N. Y	Feb. 7, 1865	46, 239
Key-fastener	W. Neracker	Cleveland, Ohio	Apr. 29, 1873	138, 272
Key-fastener	L. Tobey	Naples, N. Y	May 16, 1871	114, 992
Key-fastening, Metallic	C. L. Rehn	Philadelphia, Pa	Oct. 9, 1860	30, 350
Key-guard	M. E. Berolzheimer	New York, N. Y	Nov. 23, 1869	97, 024
Key-guard	M. E. Berolzheimer	New York, N. Y	Jan. 11, 1870	98, 736
Key-guard	E. P. Furlong	Portland, Me	Nov. 6, 1866	59, 383
Key-guard	B. R. Hathaway	Mormon Island, Cal	Oct. 12, 1869	95, 682
Key-guard	P. G. Smith	Brooklyn, N. Y	Dec. 7, 1869	97, 713
Key-guard for door-locks	J. Wiard	New Britain, Conn	Nov. 26, 1867	71, 350
Key-guard for locks	J. M. Rix	Boston, Mass	Dec. 19, 1865	51, 621
Key-hole escutcheon	A. L. Crooker	Charlestown, Mass	Oct. 15, 1872	132, 202
Key-hole escutcheon	S. W. Drowne	Norwich, Conn	June 28, 1870	104, 838
Key-hole escutcheon	E. Field	Greenwich, Conn	Mar. 10, 1857	16, 792
Key-hole guard	D. P. Bird	Richwood, Ohio	Sept. 14, 1869	94, 862
Key-hole guard	J. F. Bodtker	Chicago, Ill	May 24, 1870	103, 288
Key-hole guard	G. C. Bovey	Cincinnati, Ohio	Apr. 4, 1871	113, 621
Key-hole guard	C. B. Davies	Dayton, Ohio	Nov. 10, 1868	83, 837
Key-hole guard	W. N. Hall	Mexia, Tex	Dec. 23, 1873	145, 865
Key-hole guard	T. G. Harold	Brooklyn, N. Y	May 29, 1860	28, 535
Key-hole guard	A. Huffnagle	Philadelphia, Pa	Aug. 11, 1868	80, 959
Key-hole guard	F. Jenny	Parkersburgh, W. Va	Oct. 10, 1871	119, 851
Key-hole guard	J. H. Jones	Binghamton, N. Y	Aug. 16, 1870	106, 368
Key-hole guard	C. Kurz	Newark, N. J	Aug. 20, 1872	130, 583
Key-hole guard	M. MacKenzie	Malden, Mass	May 18, 1869	90, 283

Index of patents issued from the United States Patent Office from 1790 *to* 1873, *inclusive*—Continued.

Invention.	Inventor.	Residence.	Date.	No.
Key-hole guard	C. Read	Jersey City, N. J	Sept. 29, 1868	82, 551
Key-hole guard	J. L. Russell	Prairie City, Iowa	Jan. 18, 1870	99, 005
Key-hole guard	E. E. Shepardson	Providence, R. I.	Mar. 9, 1869	87, 714
Key-hole guard	W. H. Taylor	Stamford, Conn	Aug. 12, 1873	141, 833
Key-hole guard	G. Wheeler	New York, N. Y	Oct. 9, 1860	30, 368
Key-hole guard and bolt-fastener for locks, Combined.	C. H. Townsend and A. F. Potter.	Oakland, Cal	Apr. 23, 1872	126, 104
Key-hole guard and door-securer, Combined	G. Wheeler	New Springville, N. Y	Apr. 30, 1872	126, 360
Key-hole guard and protector	H. M. Flanagin	Penn's Grove, N. J	Jan. 25, 1870	99, 178
Key-hole guard for door-locks	A. T. Brooks	New Britain, Conn	Apr. 28, 1868	77, 165
Key-hole guard for door-locks	W. Johnson, 2d	Haverhill, Mass	May 7, 1867	64, 426
Key-hole guard for door-locks	D. A. Pratt	Sing Sing, N. Y	Oct. 8, 1867	69, 700
Key-hole guard for door-locks	S. Shloss, F. Veerkamp, and C. F. Leopold.	Philadelphia, Pa	Nov. 19, 1867	71, 235
Key-hole guard for locks	J. M. Alden	New Rochelle, N. Y	Mar. 26, 1872	124, 874
Key-hole guard for locks	J. B. Whitney	New York, N. Y	June 18, 1872	128, 082
Key-hole-guard lock	H. R Towne	Stamford, Conn	May 17, 1870	103, 256
Key-hole guard, Portable	W. E. Dante	Washington, D. C	Mar. 9, 1869	87, 645
Key-hole guide	W. R. Pomeroy	Millersburgh, Ohio	July 25, 1865	49, 000
Key-hole protector	E. Kershaw	Boston, Mass	May 22, 1849	6, 467
Key-hole protector	J. H. Parker	Waltham, Mass	May 23, 1871	115, 093
Key-hole stop	J. Moulson	Philadelphia, Pa	Sept. 28, 1858	21, 616
Key-ring	W. T. Jenks	Toledo, Ohio	Feb. 11, 1873	135, 712
Key-ring	G. A. Libbey	Milwaukee, Wis	Feb. 2, 1869	86, 424
Key-ring and check	C. A. Wentworth	Boston, Mass	Aug. 10, 1869	93, 506
Key-ring and door-fastener, Combined	B. H. Melendy	Manchester, N. H	July 19, 1870	105, 474
Key-ring and door-fastener, Combined	B. H. Melendy	Manchester, N. H	Feb. 14, 1871	111, 762
Key-seat cutter	L. M. Mooney	Tecumseh, Mich	Sept. 10, 1872	131, 293
Key-seat cutter for wheels and pulleys	J. Barton	Cleveland, Ohio	Nov. 30, 1858	22, 155
Key-seat-cutting machine	T. R. Bailey	Lockport, N. Y	Oct. 15, 1872	132, 239
Key-seat, &c., cutting machine	W. B. Bement	Philadelphia, Pa	June 5, 1860	28, 627
Key-seat-cutting machine	J. C. Morgan	Alliance, Ohio	Mar. 7, 1865	46, 758
Key-seat-cutting machine	W. H. Smith and R. S. Eddy	La Crosse, Wis	Apr. 28, 1868	77, 225
Key-seat-cutting machinery	J. K. Dirner	Honesdale, Pa	Sept. 18, 1866	58, 076
Key-turning machine	J. Terry, jr., and S. M. Hunter	Terrysville, Conn	July 25, 1871	117, 348
Key-way gage	J. Donaldson	Carondelet, Mo	Nov. 19, 1872	133, 086
Keys, Attaching bows to	E. L. Gaylord	Terrysville, Conn	May 29, 1860	28, 466
Keys for locks, Manufacture of	W. Wilcox	Middletown, Conn	Jan. 7, 1873	134, 577
Keys in locks, Fastening	J. H. Desalusse	Philadelphia, Pa	Aug. 8, 1865	49, 338
Keys, Safety-drop for	R. K. Lee	Brooklyn, N. Y	May 18, 1858	20, 280
Keyed instruments, Producing the swell in	C. Pommer	Philadelphia, Pa	Aug. 30, 1819	
Kiln: *See* Bone-black kiln. Brick-kiln. Cement-kiln. Charcoal-kiln. Circular kiln. Desiccating-kiln. Drying-kiln. Drying and burning kiln. Earthenware-kiln. Glass-kiln. Grain-kiln. Hop-kiln. Lime-kiln. Malt-kiln. Ore-roasting kiln. Pipe-kiln. Pottery-kiln. Progressive kiln. Stone-kiln. Tile-kiln. Wood-charring kiln.				
Kiln	H. Aiken	Philadelphia, Pa	May 21, 1872	127, 007
Kiln	A. Batchelar	Brockham, Great Britain	Apr. 25, 1871	114, 095
Kiln	R. Connable	Jackson, Mich	Mar. 25, 1873	137, 180
Kiln	B. R. Hawley	Normal, Ill	Sept. 20, 1870	107, 612
Kiln	H. Howard	Brazil, Ind	June 25, 1872	128, 228
Kiln	P. C. Taylor	San Antonio, Tex	Sept. 27, 1870	107, 735
Kiln	G. A. Wedekind and H. Dueberg.	Baltimore, Md	May 2, 1871	114, 499
Kiln-drying machine	A. B. Winegar	San Francisco, Cal	June 30, 1868	79, 425
Kiln-furnace	O. Bennett	Boston, Mass	June 11, 1854	127, 836
Kindling-block	W. Loft	Bergen, N. J	Feb. 2, 1869	86, 427
Kindling-material	T. Schwartz	New York, N. Y	Apr. 4, 1871	113, 578
King-bolt	E. A. Keasey	Ligonier, Ind	Oct. 1, 1867	69, 446
King-bolt	F. B. Morse	New Haven, Conn	Feb. 23, 1869	87, 280
King-bolt	F. B. Morse	Plantsville, Conn	Oct. 11, 1870	108, 282
King-bolt	W. Vandersaar	Indianapolis, Ind	Mar. 7, 1871	112, 517
King-bolt and plate, Clip	H. M. Beecher	Plantsville, Conn	June 17, 1873	139, 855
King-bolt and whiffletree-plate for vehicles	L. Adams	Amherst, Mass	Oct. 20, 1868	83, 234
King-bolt blanks, Die for making	F. Van Patten	Auburn, N. Y	Aug. 12, 1873	141, 679
King-bolt blanks, Die for swaging clip	F. Van Patten	Auburn, N. Y	Aug. 12, 1873	141, 681
King-bolt clips, Die for making	A. Dillenbeck	Fort Plain, N. Y	Feb. 25, 1873	136, 143
King-bolt dies, Manufacturing clip	F. B. Prindle	Southington, Conn	Oct. 3, 1871	119, 643
King-bolt dies, Series of	F. B. Morse	Plantsville, Conn	Oct. 26, 1869	96, 251
King-bolt for car-trucks	S. W. Murray and B. P. Lamason.	Milton, Pa	Apr. 25, 1871	114, 028
King-bolt for carriages	W. Clark	Johnsonville, N. Y	Mar. 29, 1870	101, 227
King-bolt for carriages	D. Hyde and E. H. Andrews	Bridgeport, N. Y	Apr. 4, 1871	113, 521
King-bolt for carriages	J. L. H. Mosier	New York, N. Y	Feb. 25, 1873	136, 255
King-bolt for carriages	J. Phelps	Red Creek, N. Y	Nov. 13, 1866	59, 642
King-bolt for carriages	J. Reiber and J. Schrader	Bridgeport, Ill	May 5, 1868	77, 528
King-bolt for perch-carriages	J. L. H. Mosier	New York, N. Y	Apr. 29, 1873	138, 342
King-bolt for vehicles	J. Deeble	Plantsville, Conn	Nov. 4, 1873	144, 323
King-bolt for wagons	L. T. Swartwout	Locke, N. Y	Apr. 4, 1871	113, 595
King-bolt for wagons	J. J. Waldron	East Durham, N. Y	May 12, 1868	77, 782
King-bolt-forging die	F. B. Morse	Plantsville, Conn	June 13, 1871	115, 976
King-bolt-forming die	F. B. Prindle	Southington, Conn	May 3, 1870	102, 707
King-bolt-head die	F. B. Morse	Plantsville, Conn	Feb. 22, 1870	100, 178
King-bolt-head dies, Series of	R. R. Miller	Plantsville, Conn	Feb. 22, 1870	100, 053
King-bolt protector and anti-rattling attachment to wagon.	H. L. Taylor	Fredonia, N. Y	Jan. 2, 1872	122, 415
King-bolt socket, Carriage	F. B. Morse	Plantsville, Conn	May 17, 1870	103, 274
King-bolt-trimming die	F. B. Prindle	Southington, Conn	May 3, 1870	102, 708
King-bolts, Die for manufacture of clip	F. Van Patten	Auburn, N. Y	Aug. 12, 1873	141, 680
Kisses, &c., Machine for wrapping	C. C. Wilson	Baltimore, Md	Jan. 3, 1871	110, 812
Kit, Mess	E. Blakeslee	New Haven, Conn	Feb. 7, 1865	46, 210

Index of patents issued from the United States Patent Office from 1790 *to* 1873, *inclusive*—Continued.

Invention.	Inventor.	Residence.	Date.	No.
Kitchen and ironing table	J. Good	Lancaster, Pa	Oct. 28, 1873	143, 977
Kitchen-boiler	J. and W. C. Bailey	Farmington, Me	May 22, 1835	....
Kitchen-boiler	W. B. Scaife	Pittsburgh, Pa	Sept. 27, 1870	107, 817
Kitchen-boiler	W. B. Scaife	Pittsburgh, Pa	Oct. 18, 1870	108, 524
Kitchen-boiler	W. B. Scaife	Pittsburgh, Pa	Dec. 3, 1872	133, 668
Kitchen-boiler	W. B. Scaife	Pittsburgh, Pa	Dec. 3, 1872	133, 669
Kitchen-cabinet	N. B. Smith	Waukegan, Ill	Sept. 10, 1872	131, 186
Kitchen-implement	J. Frisch	Albany, N. Y	Sept. 29, 1868	82, 615
Kitchen-implement	C. S. Westland and J. B. Allen	Providence, R. I	Sept. 15, 1868	82, 187
Kitchen-knife	J. W. Androvall and T. W. Joline.	Tottenville, N. Y	May 2, 1871	114, 388
Kitchen, Movable	G. Youle	New York, N. Y	July 1, 1814	....
Kitchen or stove-caboose, Portable	J. Truman	Philadelphia, Pa	Dec. 3, 1811	....
Kitchen-table	H. Closterman, jr	Cincinnati, Ohio	Dec. 26, 1871	122, 225
Kitchen-table	F. Ehrhardt	Washington, D. C	Jan. 11, 1870	98, 751
Kitchen-table	J. Shaw	Whitby, Canada	Feb. 21, 1871	112, 083
Kitchen-table	B. Welteck	New York, N. Y	Sept. 19, 1871	119, 208
Kitchen, Tin	G. Richardson	South Reading, Mass	June 14, 1832	....
Kitchen, Traveling	M. Pinner	New York, N. Y	July 7, 1863	39, 170
Kitchen-utensil	J. J. Diehl	York, Pa	Aug. 17, 1869	93, 862
Kite	S. Clark	New York, N. Y	Nov. 9, 1869	96, 550
Kite	I. Ferris	Cincinnati, Ohio	July 25, 1871	117, 270
Kite	C. J. Hardekopf	Philadelphia, Pa	Feb. 18, 1873	135, 987
Kite	E. J. Hughes	Pittsburgh, Pa	Dec. 31, 1867	72, 855
Kite	O. Maddans	Brooklyn, N. Y	Nov. 21, 1871	121, 056
Kite	T. Perrins	Philadelphia, Pa	Jan. 2, 1866	51, 860
Kite	A. Horns and E. J. Smith	New York, N. Y	Mar. 24, 1868	76, 028
Kite-frame	J. Shepard	New Britain, Conn	Nov. 19, 1867	71, 232
Kite-string holder	G. G. Sheldon	Chicago, Ill	May 31, 1870	103, 668
Knapsack	G. D. Arrington	Charlestown, Mass	Sept. 24, 1861	33, 329
Knapsack	J. Bondy	New York, N. Y	Mar. 4, 1862	34, 560
Knapsack	R. C. Buchanan	United States Army	Oct. 4, 1859	25, 625
Knapsack	A. N. Clark	Boston, Mass	Mar. 17, 1863	37, 936
Knapsack	J. W. Frazier	Newark, N. J	Feb. 20, 1872	123, 885
Knapsack	T. Garrick	Providence, R. I	Sept. 24, 1861	33, 343
Knapsack	W. Griffiths	Philadelphia, Pa	May 10, 1859	23, 979
Knapsack	W. B. Johns	United States Army	Apr. 12, 1859	23, 582
Knapsack	J. K. Mizner	Detroit, Mich	Nov. 27, 1866	60, 034
Knapsack	G. H. Palmer	Monmouth, Ill	June 10, 1873	139, 731
Knapsack	A. Perrin	Paris, France	June 13, 1865	48, 246
Knapsack	J. Rush	Philadelphia, Pa	Mar. 25, 1862	34, 778
Knapsack	C. W. Schaefer	Philadelphia, Pa	July 11, 1871	116, 874
Knapsack	J. Short	New York, N. Y	Nov. 12, 1861	33, 726
Knapsack	J. Short	Boston, Mass	Jan. 28, 1862	34, 272
Knapsack	J. Short	Boston, Mass	Dec. 16, 1862	37, 203
Knapsack	E. F. Southward	Boston, Mass	May 26, 1863	38, 701
Knapsack	A. W. Süs	New York, N. Y	May 17, 1864	42, 805
Knapsack	C. E. Sweney	Charlestown, Mass	Feb. 4, 1862	34, 336
Knapsack	J. Weber	New York, N. Y	Jan. 31, 1865	46, 195
Knapsack	O. E. Woods	Philadelphia, Pa	May 24, 1864	42, 895
Knapsack	O. E. Woods	Philadelphia, Pa	Nov. 8, 1864	44, 993
Knapsack and bed combined	B. Frodsham and M. Levett	New York, N. Y	Oct. 1, 1861	33, 385
Knapsack-attachment	J. F. Fog	Philadelphia, Pa	Apr. 2, 1872	125, 188
Knapsack-attachment	J. Sherlock	New York, N. Y	July 27, 1869	93, 130
Knapsack-collar	J. E. Atwood	Washington, D. C	Dec. 16, 1862	37, 146
Knapsack-engine	J. W. Douglas	Middletown, Conn	Aug. 24, 1869	93, 974
Knapsack, Folio extension	J. Boyd	Boston, Mass	Nov. 19, 1833	....
Knapsack, Hospital	J. McEvoy	New York, N. Y	Jan. 7, 1862	34, 117
Knapsack, Military	J. P. Lherbette	Washington, D. C	Oct. 7, 1808	....
Knapsack, overcoat, and tent	J. Short, 2d	New York, N. Y	Oct. 8, 1861	33, 468
Knapsack-sling	D. C. Baxter	Philadelphia, Pa	Mar. 17, 1863	37, 893
Knapsack-sling	J. T. Warren	Stafford, N. Y	Feb. 14, 1865	46, 410
Knapsack-supporter	A. Dickey	Cincinnati, Ohio	Mar. 21, 1865	46, 886
Knapsack, tent, and litter combined	L. Joubert	Paris, France	July 7, 1863	39, 150
Knapsacks, Mode of carrying	W. Clifford	Holly Springs, Miss	Nov. 3, 1868	83, 762
Knapsacks, Mode of carrying	T. S. Truss	London, England	Sept. 20, 1864	44, 354
Knapsacks, Slinging	W. oftHman	United States Army	Jan. 9, 1872	122, 522
Knapsacks, Slinging	T. Miles	Philadelphia, Pa	Jan. 28, 1862	34, 260
Kneading-board	H. P. Jones	Davenport, Iowa	Nov. 18, 1873	144, 768
Kneading-machine	E. McAllister	Rochester, N. Y	Dec. 26, 1871	122, 184
Kneading-machine	W. S. Reinert	Philadelphia, Pa	Sept. 7, 1858	21, 442
Knee-brace lacer	J. Monroe	New York, N. Y	July 16, 1867	66, 728
Knee-joint, Artificial	E. Cotly	Washington, D. C	Dec. 9, 1862	37, 087
Kneeling-case for churches	L. Mooney	Baltimore, Md	Oct. 15, 1867	69, 831

Knife:

See Apple and vegetable knife.
Band-knife.
Band-cutting knife.
Belt-knife.
Blubber-knife.
Budding-knife.
Buffing-knife.
Butcher's knife.
Butter-knife.
Can-opening knife.
Carving-knife.
Chopping-knife.
Conical-edged knife.
Corn-knife.
Corn-stalk knife.
Counter-knife.
Crozing-knife.
Currier's knife.
Currying-knife.
Cylindrical cutting knife.
Dirk-knife.
Drawing-knife.
Folding-knife.
Fruit-knife.
Fruit-coring knife.
Gage-knife.
Harness round-knife.
Harvester-knife.
Hay-knife.
Hay and pruning knife.
Hide-knife.
Honey-comb knife.
Kitchen-knife.
Leather-splitting knife.
Mincing-knife.
Mowing-machine knife.
Nurseryman's knife.
Orange-knife.
Paper-sack knife.
Paring-knife.
Penknife.

Index of patents issued from the United States Patent Office from 1790 *to* 1873, *inclusive*—Continued.

Invention.	Inventor.	Residence.	Date.	No.
Knife—Continued.				
See Planer-knife. Pocket-knife. Polishing-knife. Pruning-knife. Putty-knife. Reaper-knife. Reaper and mower knife. Reaping-knife. Round knife. Saw-knife. Sheaf-band knife. Shoe-knife. Skinning-knife. Sole-trimming knife. Spiral knife. Splint-knife. Splitting-knife. Table-knife. Tanner's and currier's knife. Tobacco-knife. Trimming guard-knife. Vegetable-knife. Welt-knife.				
Knife	N. E. Babcock and G. D. Goodsell.	Rockford, Ill	Mar. 29, 1870	101, 209
Knife	N. W. Caughy	Baltimore, Md	Dec. 17, 1867	72, 365
Knife	L. Eddy	Taunton, Mass	July 22, 1873	140, 997
Knife	J. O. Ely	Philadelphia, Pa	Feb. 9, 1869	86, 657
Knife	R. H. Fisher	Beaver Falls, Pa	Dec. 29, 1868	85, 296
Knife	N. P. Mulloy	Waltham, Mass	Jan. 29, 1867	61, 556
Knife	W. Sausser	Hannibal, Mo	Aug. 7, 1866	56, 998
Knife and cane-stripper, Combined	J. Leffel	Springfield, Ohio	Dec. 5, 1865	51, 326
Knife and fork	J. Ball	Brooklyn, N. Y	Nov. 23, 1869	97, 152
Knife and fork	F. C. Beach and A. A. C. Klaucke.	Stratford, Conn., and Washington, D. C.	Sept. 4, 1866	57, 662
Knife and fork	W. D. Smith	Homerville, Ga	Dec. 17, 1872	134, 011
Knife and fork cleaner	S. Brackett	Fall River, Mass	Apr. 10, 1860	27, 773
Knife and fork cleaner	J. L. Hannum	Berea, Ohio	Nov. 26, 1872	133, 367
Knife and fork cleaner	B. F. Hughson	Cold Spring, N. Y	Aug. 27, 1867	68, 079
Knife and fork cleaner	E. Marwedel	Washington, D. C	July 15, 1873	140, 932
Knife and fork cleaner	J. Merritt	New York, N. Y	Dec. 10, 1867	72, 067
Knife and fork cleaner	J. Protz	Easton, Pa	Mar. 13, 1860	27, 472
Knife and fork cleaner	H. and F. Wegner	West Troy, N. Y	Aug. 30, 1870	106, 897
Knife and fork cleaner, knife-sharpener, and can-opener combined.	F. W. Echternach and M. J. Welch.	Philadelphia, Pa	Feb. 6, 1872	123, 386
Knife and fork cleaning and polishing machine	M. N. Armstrong	New York, N. Y	Sept. 14, 1843	3, 268
Knife and fork, Combined	A. W. Cox	Malden, Mass	Mar. 30, 1869	87, 370
Knife and fork, Combined	J. S. Jennings	Buffalo, N. Y	Sept. 11, 1866	57, 918
Knife and fork, Combined	J. McMorries	Columbus, Miss	Oct. 26, 1869	96, 134
Knife and fork, Combined	T. B. Thorpe	New York, N. Y	Mar. 14, 1865	46, 832
Knife and fork handle	C. A. Moore	Westbrook, Conn	Aug. 25, 1868	81, 399
Knife and fork handle	W. Sanderson and J. J. De Barry.	New York, N. Y	Jan. 25, 1870	99, 242
Knife and fork handle, Detachable	L. Boynton and W. F. Sweet	Jackson Township, Pa	Mar. 15, 1870	100, 717
Knife and fork holder	J. W. Patten	North Greenbush, N. Y	Oct. 27, 1868	83, 538
Knife and fork rack	J. B. Hawley	New Haven, Conn	May 8, 1866	54, 541
Knife and fork rest	L. J. Cherrington	Boston, Mass	July 11, 1871	116, 807
Knife and fork scourer	E. J. Marsh	Leominster, Mass	May 9, 1871	114, 578
Knife and fork scourer	D. S. Marvin	Watertown, N. Y	May 5, 1868	77, 632
Knife and fork scourer	M. Merrill	Marengo, Ill	Feb. 25, 1868	74, 763
Knife and fork scourer	R. R. Pattison	Springfield, Ill	Apr. 10, 1866	53, 863
Knife and fork scouring machine	C. Aumock	Columbus, Ohio	Jan. 13, 1852	8, 646
Knife and pencil-case, Combined	R. Cross	Attleborough, Mass	Feb. 26, 1856	14, 306
Knife and pencil-sharpener, Combined	J. McClure	Nashua, N. H	July 4, 1871	116, 732
Knife and scissors combined	C. W. Sykes	Suffield, Conn	Mar. 5, 1867	62, 702
Knife and scissors sharpener	W. H. Alcorn	New York, N. Y	Aug. 9, 1864	43, 752
Knife and scissors sharpener	M. T. Chapman	Galesburgh, Ill	Feb. 26, 1867	62, 472
Knife and scissors sharpener	A. Herthal	Bridgeport, Conn	July 16, 1867	66, 837
Knife and scissors sharpener	M. V. B. Howe	Ashburnham, Mass	June 26, 1866	55, 867
Knife and scissors sharpener	I. S. and J. W. Hyatt, jr	Chicago, Ill	June 17, 1862	35, 652
Knife and scissors sharpener	J. N. Jacobs	Worcester, Mass	Nov. 26, 1867	71, 491
Knife and scissors sharpener	T. T. Prosser and J. Lawson	Chicago, Ill	Aug. 28, 1866	57, 564
Knife and scissors sharpener	J. J. Russ	Worcester, Mass	July 24, 1866	56, 615
Knife and spoon cleaner	J. Macnish	Berlin, Wis	June 29, 1858	20, 724
Knife and spoon polisher	L. J. Williams	Rolla, Mo	Oct. 15, 1872	132, 343
Knife and watch-key combined	E. D. Norton	Bangor, Me	Mar. 26, 1867	63, 171
Knife-blade holder	S. A. Cummings	Middleton, Mass	Dec. 5, 1865	51, 297
Knife-blades, Machine for rolling	W. P. Lathrop	Winchester, Conn	Mar. 22, 1864	42, 005
Knife-blades, Machine for straightening	H. Pierce	Claremont, N. H	June 23, 1857	17, 634
Knife-carrier	P. E. Cummings	Sanford, Me	Oct. 23, 1866	58, 992
Knife-cleaner	J. J. Banta	Jersey City, N. J	July 20, 1858	20, 929
Knife-cleaner	J. T. Barnstead	Peabody, Mass	Apr. 30, 1872	126, 172
Knife-cleaner	W. S. Beebe, J. T. Baynes, and A. A. King.	West Troy, N. Y	Dec. 12, 1871	121, 747
Knife-cleaner	W. J. Burge	Atchison, Kans	June 25, 1867	66, 126
Knife-cleaner	W. Christian and J. H. Morrow	New York, N. Y	Sept. 11, 1866	57, 865
Knife-cleaner	A. B. Clarke	Wyandotte, Mich	Jan. 30, 1872	123, 238
Knife-cleaner	C. F. Dean	Saint Johnsbury, Vt	July 30, 1867	67, 169
Knife-cleaner	I. Detheridge, jr	New York, N. Y	Oct. 23, 1860	30, 468
Knife-cleaner	J. A. Eivins	South Boston, Mass	Oct. 15, 1867	69, 905
Knife-cleaner	T. M. Fell	Brooklyn, N. Y	June 2, 1863	38, 786
Knife-cleaner	H. C. Gilbert	New London, Conn	Mar. 18, 1873	136, 990
Knife-cleaner	T. and L. Gringras	Buffalo, N. Y	May 20, 1873	139, 057
Knife-cleaner	L. Goulding	Medfield, Mass	Dec. 10, 1867	72, 021
Knife-cleaner	J. F. Hammond	Providence, R. I	Aug. 6, 1867	67, 534
Knife-cleaner	C. H. Hardy	Bath, Me	Apr. 20, 1869	89, 218
Knife-cleaner	W. W. Hopkins	Chesterfield Factory, N. H	Aug. 19, 1856	15, 562
Knife-cleaner	I. Jewett	Boston, Mass	Aug. 6, 1867	67, 434
Knife-cleaner	A. C. Kaiser	Vienna, Mo	Dec. 1, 1868	84, 552
Knife-cleaner	W. T. Kosinski	New York, N. Y	Sept. 12, 1865	49, 892
Knife-cleaner	J. McNamee	Easton, Pa	Nov. 8, 1859	26, 040
Knife-cleaner	W. Miller	Waltham, Mass	May 25, 1858	20, 391
Knife-cleaner	W. Miller	Chicopee, Mass	Dec. 15, 1868	84, 960
Knife-cleaner	C.C. Morgan	Auburn, N. Y	July 30, 1867	67, 336
Knife-cleaner	R. R. Pattison	Chicago, Ill	Jan. 15, 1867	61, 245
Knife-cleaner	J. Seeberger	West Troy, N. Y	July 4, 1871	116, 635
Knife-cleaner	A. J. Sellon	Boston, Mass	May 8, 1866	54, 612

Index of patents issued from the United States Patent Office from 1790 *to* 1873, *inclusive*—Continued.

Invention.	Inventor.	Residence.	Date.	No.
Knife-cleaner	C. B. Sheldon	New York, N. Y	Dec. 23, 1873	145, 909
Knife-cleaner	G. Smith	New York, N. Y	July 23, 1861	32, 902
Knife-cleaner	O. Sweeney	Norwich, Conn	Jan. 22, 1861	31, 193
Knife-cleaner	I. F. Thompson	Providence, R. I	Jan. 21, 1868	73, 673
Knife-cleaner	J. H. Van Riper	New York, N. Y	Mar. 29, 1864	42, 133
Knife-cleaner	W. Vine	Norwalk, Conn	Nov. 9, 1869	96, 642
Knife-cleaner	H. Woodward	London, England	Oct. 8, 1867	69, 601
Knife cleaner and sharpener	S. A. W. Houmann and H. Nielson.	Saint Louis, Mo	Jan. 26, 1872	125, 050
Knife cleaner and sharpener	L. J. Johnson	Montville, Conn	Nov. 5, 1867	70, 574
Knife-cleaning box	C. D. Copeland	Fall River, Mass	Mar. 20, 1866	53, 275
Knife-cleaning box	F. A. Salisbury	Greene, N. Y	May 20, 1862	35, 328
Knife-cleaning machine	A. C. Ketchum	New York, N. Y	Apr. 29, 1856	14, 788
Knife-cleaning machine	T. Roberts	Lynn, Mass	Feb. 1, 1870	99, 480
Knife-cleaning machine	G. Weedon	New York, N. Y	Jan. 27, 1863	37, 537
Knife-die for cutting leather-straps for whips	C. Baeder	New York, N. Y	Oct. 31, 1854	11, 854
Knife, fork, and pie crimper	G. D. Bayley	Boston, Mass	Oct. 18, 1864	44, 699
Knife, fork, and spoon cleaning machine	E. and A. Buckman	East Greenbush, N. Y	Mar. 25, 1862	34, 736
Knife, fork, and spoon, Combination of	J. H. Cables	Plymouth Hollow, Conn	Mar. 18, 1862	34, 712
Knife, fork, and spoon, Combination of	J. W. Hardie and A. S. Hayward.	New York, N. Y., and Boston, Mass.	Jan. 7, 1862	34, 098
Knife, fork, and spoon, Combined	N. Ames	Saugus Centre, Mass	Sept. 17, 1861	33, 285
Knife, fork, and spoon Combined	A. Neill	Boston, Mass	Jan. 7, 1862	34, 069
Knife, fork, and spoon Combined	W. H. Richards	Newton, Mass	July 23, 1861	32, 916
Knife, fork, and spoon holder	G. L. Morse and L. M. Herrick	Harrison, N. J	Mar. 7, 1865	46, 692
Knife-grinding machine	E. S. M. Fernald	Saco, Me	Oct. 4, 1870	107, 894
Knife-grinding machine	A. Hankey and F. Stiles, jr	Leicester, Mass	Aug. 4, 1857	17, 952
Knife-grinding machinery	W. Fosket	Meriden, Conn	Dec. 26, 1865	51, 779
Knife-guard	G. K. Dearborn	Pawtucket, R. I	Mar. 29, 1870	101, 235
Knife-guard	E. A. Goodes	Philadelphia, Pa	Oct. 5, 1869	95, 463
Knife-guard	D. Perry	North Providence, R. I	Feb. 8, 1870	99, 701
Knife-guard	T. T. Woodward	Ansonia, Conn	Nov. 9, 1869	96, 758
Knife-handle	J. B. Bailey	New York, N. Y	May 24, 1870	103, 278
Knife-handle	L. Carrier	East Douglass, Mass	Mar. 6, 1860	27, 349
Knife, &c., handle	M. Chapman	Greenfield, Mass	Aug. 9, 1870	106, 254
Knife-handle	S. B. Everett	Plymouth Hollow, Conn	Aug. 19, 1862	36, 255
Knife-handle	J. L. Frary	New Britain, Conn	Dec. 7, 1869	97, 624
Knife-handle	D. H. Merriam	Fitchburgh, Mass	Apr. 24, 1866	54, 189
Knife-handle	A. L. Taylor	Springfield, Vt	Dec. 14, 1869	97, 991
Knife-handle	W. D. Woods	Bennington, N. H	Mar. 22, 1870	101, 201
Knife-handle and other table-cutlery	Z. K. Murdock	Meriden, Conn	Apr. 16, 1841	2, 052
Knife-handle bolster	W. P. Lathrop	West Winsted, Conn	July 6, 1869	92, 196
Knife, nut cracker and picker	G. A. Fairfield	Hartford, Conn	Apr. 5, 1870	101, 448
Knife-polisher	J. W. Battelle	Worcester, Mass	Dec. 19, 1865	51, 542
Knife-polisher	W. H. Cummings	Oxford, Mass	Nov. 28, 1871	121, 339
Knife-polisher	H. T. Field	New Braintree, Mass	May 25, 1858	20, 340
Knife-polisher	C. H. Lithgow	Chicago, Ill	May 17, 1870	103, 210
Knife-polisher	J. Palmer	Cleveland, Ohio	June 20, 1865	48, 305
Knife-polisher	E. H. Rockwood and C. D. Shepard.	Boston, Mass	July 16, 1872	129, 594
Knife-polisher	R. Shaler	Madison, Conn	Nov. 28, 1848	5, 941
Knife-polisher	F. W. Tarble and H. W. Sawyer.	Charlestown and Chelsea, Mass.	Feb. 25, 1873	136, 188
Knife-polisher	H. J. Wattles	Rockford, Ill	Oct. 15, 1872	131, 339
Knife-polisher and knife and scissors sharpener	J. Wheeler	Athol, Mass	June 5, 1866	55, 404
Knife polisher and sharpener	T. Bootsmann	Tompkinsville, N. Y	Oct. 21, 1873	143, 871
Knife-polishing machine	A. Munger and R. C. Taylor	Auburn, N. Y	Sept. 11, 1849	6, 707
Knife-rest	D. Sherwood	Lowell, Mass	Dec. 28, 1869	98, 431
Knife-ring	C. B. and W. A. N. Long	Worcester, Mass	Aug. 4, 1868	80, 553
Knife-ring	J. C. Wilmarth and A. Fobes	Saint Louis, Mo	Oct. 6, 1868	82, 912
Knife, saw, and spring-scale combined, Butcher's	J. Baggs	Easton, England	Apr. 29, 1873	138, 362
Knife-scourer	J. Q. Adams and S. R. Goodsell	Brooklyn, N. Y	Apr. 12, 1870	101, 698
Knife-scourer	H. E. French	Unity, N. H	Sept. 27, 1870	107, 674
Knife-scourer	S. R. Goodsell and J. Q. Adams	Brooklyn, N. Y	June 8, 1869	91, 011
Knife-scourer	F. O. Harvey	Kansas City, Mo	Dec. 3, 1872	133, 645
Knife-scourer	E. W. Haven	Brandon, Vt	Mar. 11, 1873	136, 598
Knife-scourer	H. B. Hutchins and W. Horter	Philadelphia, Pa	July 24, 1866	56, 564
Knife-scourer	D. S. Neal	Lynn, Mass	June 13, 1871	115, 885
Knife scourer and sharpener	D. Cromwell, jr	Yarmouth Port, Mass	June 20, 1871	116, 029
Knife scourer and sharpener	D. Hodgkins	Newburyport, Mass	Mar. 23, 1869	88, 167
Knife-scouring board	E. Dening	Seneca Falls, N. Y	Apr. 2, 1872	125, 124
Knife-scouring machine	G. M. Morris and J. Newton	Watertown, Conn	Dec. 4, 1855	13, 883
Knife, scythe, &c., sharpener and renovator	I. Jennings	New York	Mar. 17, 1829	....
Knife-sharpener	A. Anan	New York, N. Y	Nov. 16, 1858	22, 055
Knife-sharpener	J. J. and A. T. Armstrong	Brooklyn, N. Y	Aug. 3, 1858	21, 058
Knife-sharpener	A. B. Bean	New Haven, Conn	Nov. 26, 1867	71, 442
Knife-sharpener	C. A. Bogert	Bay City, Mich	Apr. 13, 1869	88, 839
Knife-sharpener	W. J. Dunn	New York	Aug. 18, 1829	....
Knife-sharpener	S. Gissinger	Pittsburgh, Pa	Jan. 9, 1872	122, 597
Knife-sharpener	T. Haynes	Saint Louis, Mo	Oct. 22, 1867	69, 993
Knife-sharpener	G. Hinman	New Haven, Conn	Feb. 8, 1859	22, 919
Knife-sharpener	W. H. Howland	San Francisco, Cal	Nov. 23, 1869	97, 090
Knife-sharpener	W. Hunt	New York	Feb. 19, 1829	....
Knife-sharpener	J. W., jr., and I. S. Hyatt	Chicago, Ill	Feb. 19, 1861	31, 461
Knife-sharpener	T. K. Knapp	Worcester, Mass	July 30, 1867	67, 200
Knife-sharpener	T. K. Knapp	Dansville, N. Y	May 28, 1872	127, 170
Knife-sharpener	G. B. Markham	Plymouth, Mich	Apr. 30, 1867	64, 342
Knife-sharpener	J. Meyer, jr	New York, N. Y	May 14, 1867	64, 687
Knife-sharpener	W. Nash	New Britain, Conn	Feb. 13, 1866	52, 587
Knife-sharpener	D. Porter	Cleveland, Ohio	Nov. 14, 1865	50, 951
Knife-sharpener	J. B. Pratt	Corning, N. Y	July 10, 1866	56, 264
Knife-sharpener	J. Quipp	Buffalo, N. Y	Aug. 20, 1872	130, 660
Knife-sharpener	C. H. Reynolds	New York, N. Y	May 19, 1868	78, 070
Knife-sharpener	J. C. Reynolds	Taunton, Mass	Oct. 29, 1867	70, 360
Knife-sharpener	Z. C. and H. A. Robbins	Washington, D. C	Dec. 7, 1869	97, 701
Knife-sharpener	J. J. Russ	Worcester, Mass	June 27, 1871	116, 495
Knife-sharpener	S. C. Stokes	Manchester, N. H	Mar. 6, 1860	27, 413
Knife-sharpener	S. C. Stokes	Manchester, N. H	May 3, 1870	102, 727

Index of patents issued from the United States Patent Office from 1790 *to* 1873, *inclusive*—Continued.

Invention.	Inventor.	Residence.	Date.	No.
Knife-sharpener	T. Vickery	Providence, R. I	Mar. 1, 1870	100,470
Knife-sharpener	T. H. White	Orange, Mass	Sept. 4, 1866	57,804
Knife-sharpener	P. M. Withington	Stoughton, Mass	May 17, 1870	103,113
Knife-sharpener	M. Young	Frederick, Md	July 23, 1867	67,011
Knife sharpener and cleaner	J. M. Farnam	Hartford, Conn	Apr. 10, 1860	27,787
Knife sharpener and cleaner	J. J. Mason	Drummondville, Canada	Dec. 10, 1872	133,711
Knife-sharpener and glass-cutter combined	S. C. Stokes	Manchester, N. H	Oct. 15, 1872	132,219
Knife sharpener and polisher	E. Beeman	Owego, N. Y	Jan. 2, 1866	51,789
Knife sharpener and scourer	A. T. Boon	Galesburgh, Ill	Sept. 15, 1863	39,877
Knife-sharpener, Domestic	P. Cornell	Brutus, N. Y	Jan. 15, 1830	......
Knife to handle, Attaching	W. Clayton	Bristol, Conn	Sept. 4, 1866	57,677
Knife, tweezers, and ear-spoon, Combined	B. C. English	Springfield, Mass	Sept. 26, 1865	50,106
Knives and forks, Construction of	J. W. Hardie	New York, N. Y	Nov. 12, 1861	33,703
Knives and forks, Construction of	P. Ulmer	Philadelphia, Pa	Feb. 4, 1862	34,337
Knives and forks, Manufacture of	J. D. Frary	New Britain, Conn	Sept. 25, 1866	58,242
Knives and forks, Manufacture of	J. B. H. Leonard	Meriden, Conn	Sept. 25, 1866	58,269
Knives and forks, Manufacture of	C. A. Moore	Westbrook, Conn	Dec. 15, 1868	85,020
Knives and forks, Manufacturing	J. H. Nichols	New Britain, Conn	Oct. 29, 1867	70,352
Knives, Attaching bolsters to	H. Barber	Greenfield, Mass	Nov. 12, 1867	70,778
Knives, Attaching handles to	M. Chapman	Greenfield, Mass	Dec. 16, 1862	37,150
Knives, spoons, &c., Impressed roller for making	J. Owings	Baltimore, Md	Jan. 25, 1814	......
Knit fabric	C. J. Appleton	Cohoes, N. Y	Sept. 27, 1870	107,749
Knit fabric	H. Boot	Philadelphia, Pa	Aug. 13, 1872	130,467
Knit fabric	B. J. McGee	Watertown, Mass	Nov. 25, 1873	144,915
Knit fabric	S. Thacker	Snenton, England	Oct. 17, 1871	120,126
Knit fabric and method of knitting	D. Bickford	New York, N. Y	Sept. 17, 1872	131,386
Knit fabric and method of knitting	D. Bickford	New York, N. Y	Sept. 17, 1872	131,387
Knit fabric and method of knitting	D. Bickford	New York, N. Y	Mar. 11, 1873	136,639
Knit fabrics, Machine for uniting the selvages of	T. Beven	Loughborough, England	Apr. 22, 1873	137,997
Knit fabrics, Machinery for drying and finishing tubular.	W. K. Greene, jr., and W. M. Pawling.	Amsterdam and Hagaman's Mills, N. Y.	Jan. 6, 1863	37,288
Knit fringe or border	J. Phipps	Philadelphia, Pa	June 10, 1873	139,810
Knit mitten	O. F. Tripp	Battle Creek, Mich	Aug. 26, 1873	142,301
Knit or woven fabric, Producing designs for	I. Rehn	Washington, D. C	Nov. 21, 1871	121,128
Knitted fabric	H. Boot	Philadelphia, Pa	May 2, 1871	114,397
Knitted fabric	G. S. Harwood	Newton, Mass	Jan. 12, 1864	41,218
Knitted fabric	G. Jelley	Roxbury, Mass	July 21, 1868	80,183
Knitted fabric	M. Landenberger, jr	Philadelphia, Pa	Aug. 3, 1869	93,314
Knitted fabric	J. Phipps	Philadelphia, Pa	Aug. 6, 1872	130,312
Knitted fabric	J. Vickerstaff	Philadelphia, Pa	June 16, 1857	17,612
Knitted fabric, Manufacturing	J. K. Kilbourn and E. E. Kilbourn.	Pittsfield, Mass., and Litchfield, Conn.	June 7, 1859	24,308
Knitted fabrics, Machine for drying and finishing tubular.	N. P. Akin	Philmont, N. Y	Mar. 2, 1869	87,317
Knitted fabrics, Machine for napping, brushing, &c., tubular.	C. and I. Tompkins	Troy, N. Y	June 24, 1873	140,320
Knitted goods, Mode of uniting edges of	S. Arnold	Claverack, N. Y	June 21, 1870	104,532
Knitted hosiery, Machine	B. F. Shaw	South Danvers, Mass	Apr. 23, 1867	64,154
Knitted mitten, Machine-made	A. C. Carey	Malden, Mass	Apr. 24, 1866	54,108
Knitted stocking, Machine	A. C. Carey	Malden, Mass	July 30, 1867	67,264
Knitted stocking, Machine-made	A. C. Carey	Malden, Mass	Apr. 24, 1866	54,109
Knitted stockings, Apparatus for cutting out	W. Martin	Philadelphia, Pa	Aug. 9, 1864	43,783
Knitting	C. Tompkins and J. Johnson	Troy, N. Y	Sept. 18, 1855	13,586
Knitting-burrs, Tool for manufacturing	H. Fisher	Waterford, N. Y	Nov. 10, 1863	40,587
Knitting-holder	H. P. Joy	South Berwick Junction, Me.	Aug. 12, 1873	141,796
Knitting-machine	W. H. Abel	Greenville, R. I	Feb. 11, 1868	74,266
Knitting-machine	W. H. Abel	Greenville, R. I	Nov. 3, 1868	83,583
Knitting-machine	W. H. Abel	Greenville, R. I	Nov. 3, 1868	83,584
Knitting-machine	W. H. Abel	Bennington, Vt	July 19, 1870	105,537
Knitting-machine	W. H. Abel	Lowell, Mass	July 19, 1870	105,538
Knitting-machine	W. H. Abel	Lowell, Mass	July 19, 1870	105,539
Knitting-machine	W. H. Abel	Bennington, Vt	Aug. 23, 1870	106,531
Knitting-machine	J. B. Aiken	Franklin, N. H	May 22, 1855	12,933
Knitting-machine	J. B. Aiken	Manchester, N. H	Aug. 2, 1859	24,916
Knitting-machine	J. B. Aiken	New York, N. Y	Feb. 5, 1861	31,287
Knitting-machine	J. B. and W. Aiken	Franklin, N. H	July 8, 1856	15,314
Knitting-machine	N. P. Aikin	Troy, N. Y	July 13, 1858	20,854
Knitting-machine	W. Aiken	Franklin, N. H	Dec. 1, 1857	18,725
Knitting-machine	W. Aiken	Franklin, N. H	Nov. 9, 1858	22,004
Knitting-machine	W. Aiken	Franklin, N. H	Aug. 7, 1860	29,450
Knitting-machine	W. Aiken	Franklin, N. H	July 2, 1861	32,674
Knitting-machine	W. Aiken	Franklin, N. H	June 21, 1864	43,170
Knitting-machine	W. Aiken	Franklin, N. H	Dec. 31, 1867	72,771
Knitting-machine	W. and J. B. Aiken	Franklin, N. H	Sept. 11, 1855	13,561
Knitting-machine	C. J. Appleton	Philadelphia, Pa	Feb. 14, 1860	27,183
Knitting-machine	C. J. Appleton	Cohoes, N. Y	Sept. 27, 1870	107,750
Knitting-machine	C. J. Appleton	Hamilton, Canada	Feb. 13, 1872	123,545
Knitting-machine	J. M. Armour	Craftsbury, Vt	May 15, 1866	54,812
Knitting-machine	J. M. Armour	Syracuse, N. Y	Apr. 25, 1871	113,965
Knitting-machine	J. M. Armour	Syracuse, N. Y	Mar. 4, 1873	136,480
Knitting-machine	H. L. Arnold	Ottawa, Ill	Sept. 10, 1872	131,144
Knitting-machine	V. G. Arnold	Providence, R. I	Aug. 2, 1870	106,025
Knitting-machine	T. Bailey	Ballston Spa, N. Y	Feb. 24, 1852	8,750
Knitting-machine	T. Bailey	Ballston Spa, N. Y	Nov. 20, 1855	13,811
Knitting-machine	J. H. Barsantee	Portsmouth, N. H	July 29, 1851	8,262
Knitting-machine	J. H. Barsantee	Philadelphia, Pa	May 30, 1854	10,980
Knitting-machine	C. E. Bean	Oswego, N. Y	Aug. 20, 1872	130,558
Knitting-machine	D. Bickford	Boston, Mass	Sept. 10, 1867	68,595
Knitting-machine	D. Bickford	Boston, Mass	July 21, 1868	80,121
Knitting-machine	D. Bickford	Boston, Mass	July 6, 1869	92,146
Knitting-machine	W. Binkley	Manchester, N. H	Oct. 4, 1859	25,698
Knitting-machine	C. W. Blakeslee	Northfield, Conn	Sept. 20, 1864	44,365
Knitting-machine	C. W. Blakeslee, E. B. Beecher, and A. G. Davis.	Watertown and Westville, Conn.	July 14, 1868	79,897
Knitting-machine	C. W. Blakeslee, A. G. Davis, and E. B. Beecher.	Watertown and Westville, Conn.	Aug. 11, 1868	80,835
Knitting-machine	H. Bogel	Watertown, Wis	Mar. 10, 1868	75,353

Index of patents issued from the United States Patent Office from 1790 *to* 1873, *inclusive*—Continued.

Invention.	Inventor.	Residence.	Date.	No.
Knitting-machine	B. Bollinger and G. G. Nodle	New Berlin, Ohio	Sept. 22, 1868	82, 281
Knitting-machine	J. Bradley	Lowell, Mass	May 14, 1872	126, 621
Knitting-machine	J. Bradley	Lowell, Mass	Sept. 24, 1872	131, 595
Knitting-machine	J. L. Branson	Chicago, Ill	June 18, 1872	127, 954
Knitting-machine	F. L. Buell	Manchester, Conn	Aug. 23, 1859	25, 230
Knitting-machine	J. Bullock	Cohoes, N. Y	Nov. 15, 1859	26, 085
Knitting-machine	F. Burns	Upper Gilmanton, N. H	Aug. 24, 1869	94, 071
Knitting-machine	W. W. Burson and J. Nelson	Rockford, Ill	June 23, 1868	79, 202
Knitting-machine	W. W. Burson and J. Nelson	Rockford, Ill	Oct. 4, 1870	108, 003
Knitting-machine	H. Burt	Newark, N. J	May 16, 1854	10, 915
Knitting-machine	H. Burt	Newark, N. J	July 4, 1854	11, 238
Knitting-machine	C. Callahan	Chelsea, Mass	July 15, 1873	140, 809
Knitting-machine	W. and W. Campion	Nottingham and Sneinton, England.	May 20, 1873	139, 114
Knitting-machine	A. C. Carey	Malden, Mass	Apr. 18, 1865	47, 354
Knitting-machine	A. C. Carey	Malden, Mass	July 30, 1867	67, 263
Knitting-machine	J. Chantrell	Bristol, Conn	Mar. 13, 1860	27, 430
Knitting-machine	J. Chantrell	Bristol, Conn	May 15, 1860	28, 255
Knitting-machine	J. Chantrell	Bristol, Conn	Jan. 7, 1868	73, 164
Knitting-machine	M. Christoffers	Hanover, Prussia	July 16, 1872	129, 529
Knitting-machine	W. W. Clay	Nottingham, England	Dec 22, 1863	40, 993
Knitting-machine	W. W. Clay	Philadelphia, Pa	Sept. 19, 1865	49, 980
Knitting-machine	W. W Clay	Philadelphia, Pa	June 5, 1866	55, 241
Knitting-machine	E. Colvin	Poultney, Vt	Jan. 13, 1857	16, 375
Knitting-machine	E. Colvin	Poultney, Vt	June 7, 1859	24, 284
Knitting-machine	F. M. Comstock	Cleveland, Ohio	Sept. 20, 1870	107, 454
Knitting-machine	F. M. Comstock	Cleveland, Ohio	Apr. 9, 1872	125, 543
Knitting-machine	J. A. Corwin	Newark, N. J	Sept. 26, 1854	11, 720
Knitting-machine	W. Cotton	Loughborough, England	Nov. 20, 1866	59, 892
Knitting-machine	W. Cotton	Loughborough, England	Aug. 13, 1867	67, 727
Knitting-machine	T. Crane	Fort Atkinson, Wis	Oct. 15, 1867	69, 775
Knitting-machine	T. Crane	Fort Atkinson, Wis	Oct. 15, 1867	69, 776
Knitting-machine	T. Crane	Fort Atkinson, Wis	Jan. 28, 1868	73, 697
Knitting-machine	T. Crane	Fort Atkinson, Wis	June 15, 1869	91, 214
Knitting-machine	J. Dalton	Brooklyn, N. Y	May 7, 1861	32, 241
Knitting-machine	J. Dalton	Williamsburgh, N. Y	Oct. 9, 1866	58, 610
Knitting-machine	J. Dalton	Brooklyn, N. Y	Apr. 28, 1868	77, 363
Knitting-machine	A. G. Davis, C. W. Blakeslee, and A. N. Allen.	Westville, Conn	Mar. 22, 1870	101, 106
Knitting-machine	O. Davis	New Lebanon, Ind	Sept. 13, 1870	107, 230
Knitting-machine	J. P. Delahunty	Cohoes, N. Y	July 27, 1858	21, 045
Knitting-machine	J. H. Doolittle	Waterbury, Conn	Oct. 16, 1855	13, 693
Knitting-machine	R. Ellis	Northampton, Mass	June 17, 1851	8, 163
Knitting-machine	E. S. Ells	Troy, N. Y	Aug. 23, 1859	25, 185
Knitting-machine	S. V. Essick	Mansfield, Ohio	Sept. 3, 1867	68, 425
Knitting-machine	S. V. Essick	Alliance, Ohio	July 8, 1873	140, 691
Knitting-machine	W. B. Evans	Holderness, N. H	Dec. 16, 1862	37, 161
Knitting-machine	S. D. Fairbanks	Cohoes, N. Y	Dec. 1, 1857	18, 792
Knitting-machine	G. W. Folts and J. L. Branson	Boston, Mass., and Chicago, Ill.	Feb. 21, 1871	112, 027
Knitting-machine	T. Fowler	Cohoes, N. Y	July 3, 1855	13, 165
Knitting-machine	W. Franz and W. Pope	Crestline, Ohio	Mar. 23, 1869	88, 027
Knitting-machine	W. Franz and W. Pope	Bucyrus and Crestline, Ohio.	July 12, 1870	105, 187
Knitting-machine	A. French	Waterbury, Conn	Nov. 6, 1855	13, 750
Knitting-machine	F. Gardner	Hamilton, Canada	July 6, 1869	92, 300
Knitting-machine	A. J. and D. Goffe	Cohoes, N. Y	Aug. 5, 1856	15, 484
Knitting-machine	A. J. and D. Goffe	Cohoes, N. Y	Nov. 22, 1859	26, 231
Knitting-machine	J. Goodman	Pawtucket, R. I	Mar. 27, 1866	53, 530
Knitting-machine	H. Guenther	New York, N. Y	Mar. 26, 1872	124, 950
Knitting-machine	H. Günther	New York, N. Y	Jan. 14, 1873	134, 881
Knitting-machine	H. V. Hartz	Cleveland, Ohio	Jan. 30, 1872	123, 100
Knitting-machine	H. V. Hartz and J. Feiss	Cleveland, Ohio	Feb. 8, 1870	99, 670
Knitting-machine	W. S. Hill	Manchester, N. H	May 24, 1870	103, 332
Knitting-machine	J. Hiller and J. Bullock	Cohoes, N. Y	Jan. 25, 1859	22, 769
Knitting-machine	J. Hinkley	Norwalk, Ohio	May 29, 1866	55, 103
Knitting-machine	J. Hinkley	Norwalk, Ohio	Apr. 12, 1870	101, 875
Knitting-machine	J. Hollen	White Township, Pa	July 16, 1850	7, 509
Knitting-machine	J. Hollen	White Township, Pa	Nov. 28, 1854	11, 995
Knitting-machine	J. Hollen	Fostoria, Pa	Oct. 18, 1859	25, 827
Knitting-machine	J. Hollen	Fostoria, Pa	Jan. 1, 1861	31, 020
Knitting-machine	J. Hollen	Fostoria, Pa	Sept. 2, 1862	36, 351
Knitting-machine	J. Hollen	Fostoria, Pa	July 19, 1870	105, 454
Knitting-machine	W. H. H. Hollen	Fostoria, Pa	Dec. 17, 1867	72, 296
Knitting-machine	W. H. H. Hollen	Fostoria, Pa	Jan. 3, 1871	110, 656
Knitting-machine	W. H. H. Hollen	Fostoria, Pa	Mar. 5, 1872	124, 357
Knitting-machine	I. M. Hopkins	Pascoag, R. I	May 16, 1854	10, 910
Knitting-machine	J. A. and H. A. House	Bridgeport, Conn	Mar. 13, 1866	53, 224
Knitting-machine	H. A. House	Bridgeport, Conn	Jan. 4, 1870	98, 500
Knitting-machine	H. A. House	Bridgeport, Conn	Apr. 12, 1870	101, 878
Knitting-machine	H. A. House	Bridgeport, Conn	Nov. 19, 1872	133, 103
Knitting-machine	J. M. Howe	Rochester, N. Y	Mar. 15, 1870	100, 765
Knitting-machine	J. M. Howe	Rochester, N. Y	Aug. 8, 1871	117, 886
Knitting-machine	S. W. Howland	Adams, Mass	May 14, 1861	32, 295
Knitting-machine	L. B. Hunt	Hyde Park, Mass	July 25, 1871	117, 293
Knitting-machine	L. B. Hunt	Hyde Park, Mass	Oct. 8, 1872	131, 955
Knitting-machine	G. Jacksen	Cohoes, N. Y	Aug. 22, 1854	11, 554
Knitting-machine	G. Jenson	Brooklyn, N. Y	Dec. 20, 1864	45, 501
Knitting-machine	G. Johnstone	Philadelphia, Pa	Aug. 11, 1868	80, 965
Knitting-machine	G. Johnstone	Philadelphia, Pa	Sept. 13, 1870	107, 381
Knitting-machine	L. Kavanaugh	Waterford, N. Y	Jan. 26, 1864	41, 381
Knitting-machine	C. G. Keeny	Manchester, Conn	Jan. 25, 1859	22, 735
Knitting-machine	J. Kent	New York, N. Y	Dec. 28, 1869	98, 272
Knitting-machine	D. Kidder	Franklin, N. H	Oct. 27, 1868	83, 390
Knitting-machine	E. E. Kilbourn	New Brunswick, N. J	May 23, 1865	47, 829
Knitting-machine	J. K. and E. E. Kilbourn	Norfolk, Conn	Feb. 16, 1858	19, 370
Knitting-machine	J. K. and E. E. Kilbourn	Norfolk, Conn	Oct. 12, 1858	21, 762
Knitting-machine	J. K. and E. E. Kilbourn	Norfolk, Conn	Apr. 9, 1861	31, 984
Knitting-machine	I. W. Lamb	Detroit, Mich	Sept. 15, 1863	39, 934

Index of patents issued from the United States Patent Office from 1790 *to* 1873, *inclusive*—Continued.

Invention.	Inventor.	Residence.	Date.	No.
Knitting-machine	I. W. Lamb	Rochester, N. Y	Oct. 10, 1865	50, 369
Knitting-machine	M. T. Lamb	Valparaiso, Ind	June 5, 1866	55, 420
Knitting-machine	T. Langham	Philadelphia, Pa	Jan. 21, 1862	34, 210
Knitting-machine	T. Langham	Philadelphia, Pa	Mar. 6, 1866	53, 013
Knitting-machine	S. Larkin	Bridgeport, Conn	Aug. 11, 1868	80, 866
Knitting-machine	J. and M. Lee and W. Carter	Needham, Mass	Dec. 27, 1870	110, 479
Knitting-machine	J. Leonard	Attica, Ohio	Feb. 11, 1873	135, 823
Knitting-machine	J. Y. Leslie	Cincinnati, Ohio	Oct. 3, 1854	11, 751
Knitting-machine	T. Lovelidge	Germantown, Pa	Aug. 31, 1858	21, 396
Knitting-machine	W. Mansfield	Dracut, Mass	Mar. 22, 1853	9, 626
Knitting-machine	M. Marshall	Lowell, Mass	Mar. 15, 1853	9, 621
Knitting-machine	J. Maxwell	Galesville, N. Y	Mar. 29, 1853	9, 637
Knitting-machine	J. McCune	Auburn, Ind	Sept. 14, 1869	94, 904
Knitting-machine	J. McMullen and J. Hollen, jr	Sinking Valley and Logan's Valley, Pa.	Feb. 11, 1837	126½
Knitting-machine	W. H. McNary	Brooklyn, N. Y	May 15, 1860	28, 290
Knitting-machine	W. H. McNary	Brooklyn, N. Y	Apr. 30, 1867	64, 241
Knitting-machine	G. Merrill	East Orange, N. J	July 8, 1873	140, 635
Knitting-machine	E. Morse	Winchendon, Mass	Aug. 16, 1870	106, 499
Knitting-machine	J. Nesmith	Lowell, Mass	July 29, 1856	15, 435
Knitting-machine	A. Paget	Loughborough, Great Britain.	May 21, 1867	64, 900
Knitting-machine	O. Parkhurst	Cohoes, N. Y	Mar. 5, 1861	31, 647
Knitting-machine	G. M. Patten	Bath, Me	Jan. 19, 1869	86, 033
Knitting-machine	H. Pease	Brockport, N. Y	June 25, 1872	128, 421
Knitting-machine	R. Peberdy	Philadelphia, Pa	Mar. 2, 1869	87, 508
Knitting-machine	J. Pepper	Portsmouth, N. H	Feb. 25, 1851	7, 945
Knitting-machine	J. Pepper	Portsmouth, N. H	June 24, 1851	8, 172
Knitting-machine	J. Pepper	Franklin, N. H	July 17, 1855	13, 289
Knitting-machine	J. Pepper	Lowell, Mass	Dec. 19, 1865	51, 618
Knitting-machine	J. Pepper	Lake Village, N. H	Aug. 27, 1867	68, 107
Knitting-machine	J. Pepper	Lake Village, N. H	Apr. 26, 1870	102, 313
Knitting-machine	J. Pepper, jr	Portsmouth, N. H	Dec. 5, 1854	12, 046
Knitting-machine	D. C. Philips and C. and I. Tompkins.	Philmont and Troy, N. Y	June 14, 1870	104, 346
Knitting-machine	F. Philip	Stockport, N. Y	Mar. 15, 1870	100, 798
Knitting-machine	C. H. Platt and G. A. Stanbery	Norwalk, Ohio	Apr. 8, 1873	137, 568
Knitting-machine	J. Powell	Waterbury, Conn	Oct. 2, 1855	13, 621
Knitting-machine	E. R. Pray	Holderness, N. H	Sept. 1, 1868	81, 683
Knitting-machine	A. B. Prouty	Worcester, Mass	Dec. 28, 1869	98, 410
Knitting-machine	H. Pudder and T. W. Hurlbut	Dayton, Ky., and Cincinnati, Ohio.	June 13, 1871	115, 983
Knitting-machine	W. H. Ramsdell	Lowell, Mass	Aug. 27, 1872	130, 866
Knitting-machine	E. M. Ray	Providence, R. I	May 30, 1854	10, 993
Knitting-machine	J. D. Reiff	Skippackville, Pa	Jan. 12, 1869	85, 765
Knitting-machine	J. W. Rist	Rochester, N. Y	Sept. 22, 1868	82, 348
Knitting-machine	M. L. Roberts	Mount Union, Ohio	Feb. 12, 1861	31, 404
Knitting-machine	M. L. Roberts	New Brunswick, N. J	July 27, 1869	93, 008
Knitting-machine	S. H. Roper	Boston, Mass	Aug. 24, 1869	94, 135
Knitting-machine	S. R. Roper	Roxbury, Mass	Aug. 8, 1871	117, 931
Knitting-machine	J. Rose	Nottingham, England	Jan. 9, 1872	122, 535
Knitting-machine	L. D. Sanborn	Cohoes, N. Y	Oct. 7, 1873	143, 468
Knitting-machine	F. Schott	Brooklyn, N. Y	Nov. 23, 1858	22, 135
Knitting-machine	H. B. Scudder	Needham, Mass	Mar. 15, 1870	100, 810
Knitting-machine	A. Sessions, jr	Springfield, Mass	Sept. 5, 1865	49, 835
Knitting-machine	E. Shore	Conshohocken, Pa	Sept. 2, 1862	36, 373
Knitting-machine	E. Shore	Conshohocken, Pa	Feb. 9, 1864	41, 540
Knitting-machine	J. and J. Smith	Philadelphia, Pa	June 17, 1873	140, 086
Knitting-machine	A. B. Stillings	Springfield, Mass	June 19, 1866	55, 734
Knitting-machine	E. Taélbouis and A. Renevey	St. Just-en-Chaussee, France.	Feb. 18, 1873	136, 012
Knitting-machine	W. A. Tangeman	Lockland, Ohio	Jan. 5, 1869	85, 546
Knitting-machine	J. Teachout	Waterford, N. Y	Dec. 10, 1867	72, 116
Knitting-machine	J. Teachout	Waterford, N. Y	Dec. 10, 1867	72, 117
Knitting-machine	J. Terrell	Philadelphia, Pa	Jan. 1, 1861	31, 042
Knitting-machine	S. T. Thomas	Laconia, N. H	Apr. 1, 1862	34, 852
Knitting-machine	A. Thornton	Nottingham, England	Sept. 27, 1864	44, 494
Knitting-machine	H. Thornton	Nottingham, England	Mar. 20, 1866	53, 368
Knitting-machine	J. and W. Thornton	Pease Hill Rise, Nottingham, England.	June 5, 1866	55, 444
Knitting-machine	E. Tiffany	Thompsonville, Conn	May 1, 1860	28, 133
Knitting-machine	C. Tompkins	Troy, N. Y	Dec. 23, 1856	16, 297
Knitting-machine	O. F. Tripp	Battle Creek, Mich	July 15, 1873	140, 800
Knitting-machine	O. Twombly	Ashland, N. H	Aug. 12, 1873	141, 836
Knitting-machine	O. Twombly and W. Noyes, jr	Holderness, N. H., and Newburyport, Mass.	July 7, 1868	79, 789
Knitting-machine	J. Vickerstaff	Philadelphia, Pa	Mar. 23, 1858	19, 740
Knitting-machine	J. Waldie	Ipswich, Mass	Aug. 11, 1868	80, 887
Knitting-machine	A. J. Wale	Philadelphia, Pa	July 21, 1868	80, 099
Knitting-machine	S. Wallis	Townsend, Mass	Oct. 4, 1864	44, 568
Knitting-machine	S. Wallis	Lowell, Mass	Dec. 19, 1865	51, 666
Knitting-machine	B. Ward	Troy, N. Y	Apr. 14, 1868	76, 854
Knitting-machine	J. F. Waterhouse	Germantown, Pa	Aug. 9, 1859	25, 067
Knitting-machine	J. C. Welsch	Edgerton, Ohio	Feb. 11, 1868	74, 460
Knitting-machine	J. Whittle	Philadelphia, Pa	Apr. 11, 1865	47, 239
Knitting-machine	H. L. Williams	Seneca Falls, N. Y	June 26, 1866	55, 968
Knitting-machine	J. G. Wilson	New York, N. Y	Aug. 12, 1862	36, 199
Knitting-machine	J. G. Wilson	New York, N. Y	May 22, 1866	55, 027
Knitting-machine	J. G. Wilson	New York, N. Y	June 5, 1866	55, 434
Knitting-machine	B. S. Wood	Burrillville, R. I	June 6, 1854	10, 998
Knitting-machine	L. Woodward	Nottingham, England	Dec. 2, 1873	145, 079
Knitting-machine	H. C. Work	Philadelphia, Pa	Nov. 2, 1869	96, 531
Knitting-machine	C. H. Young	Lake Village, N. H	Sept. 23, 1873	143, 051
Knitting-machine and knitted fabric	T. Crane	Fort Atkinson, Wis	June 15, 1869	91, 215
Knitting machine and needle	A. W. Allen	Indianapolis, Ind	Dec. 6, 1870	109, 793
Knitting-machine and process of knitting	N. Clark	Malden, Mass	Dec. 12, 1871	121, 924
Knitting-machine burr	C. Allardice	Cohoes, N. Y	May 3, 1864	42, 545
Knitting-machine burr	W. H. Carr	Troy, N. Y	July 26, 1864	43, 636
Knitting-machine burr	J. Clute	Cohoes, N. Y	May 9, 1865	47, 620

Index of patents issued from the United States Patent Office from 1790 *to* 1873, *inclusive*—Continued.

Invention.	Inventor.	Residence.	Date.	No.
Knitting-machine burr	H. Fisher	Waterford, N. Y	Nov. 24, 1863	40, 679
Knitting-machine burr	G. Jackson and G. Campbell	Cohoes, N. Y	July 28, 1863	39, 347
Knitting-machine burr	L. Kavanaugh	Waterford, N. Y	June 10, 1862	35, 565
Knitting-machine burr	L. Kavanaugh	Waterford, N. Y	July 25, 1871	117, 299
Knitting-machine burr	M. Van Auken	Amsterdam, N. Y	June 23, 1863	38, 997
Knitting-machine burr, Self-lubricating	C. Allardice	Cohoes, N. Y	May 29, 1866	55, 034
Knitting-machine burrs, Means for oiling	J. Maxwell	Amsterdam, N. Y	Sept. 27, 1864	44, 437
Knitting machine, Circular	G. Campbell	Waterford, N. Y	July 22, 1862	35, 968
Knitting machine, Circular	C. G. Cole	Bennington, Vt	July 6, 1869	92, 167
Knitting machine, Circular	J. Dalton	Brooklyn, N. Y	Sept. 27, 1864	44, 403
Knitting machine, Circular	W. Franz and W. Pope	Crestline, Ohio	Feb. 1, 1870	99, 425
Knitting machine, Circular	W. Franz and W. Pope	Crestline, Ohio	Feb. 1, 1870	99, 426
Knitting machine, Circular	T. Hawthorne	Philadelphia, Pa	Mar. 14, 1865	46, 848
Knitting machine, Circular	S. L. Otis	Manchester, Conn	Nov. 1, 1864	44, 885
Knitting machine, Circular	D. Scattergood	Nottingham, England	Aug. 11, 1863	39, 499
Knitting machine, Circular	C. Shirtcliff	Philadelphia, Pa	May 2, 1865	47, 579
Knitting-machine indicator	G. P. Fuller	Adrian, Mich	Mar. 10, 1868	75, 260
Knitting-machine indicator	J. C. Potter	Alfred, N. Y	Mar. 3, 1868	75, 194
Knitting-machine indicator	J. W. Rist	Rochester, N. Y	Sept. 29, 1868	82, 554
Knitting-machine knot-indicator	J. F. Steward	Plano, Ill	Apr. 23, 1867	64, 046
Knitting machine, Lace	H. Williamson	Williamsburgh, N. Y	Oct. 31, 1871	120, 474
Knitting-machine needle	W. Aiken	Franklin, N. H	May 24, 1864	42, 824
Knitting-machine needle	W. Aiken	Franklin, N. H	June 2, 1868	78, 506
Knitting-machine needle	R. Allen	Salem, Mich	July 28, 1868	80, 264
Knitting-machine needle	J. M. Armour	Craftsbury, Vt	Jan. 30, 1866	52, 256
Knitting-machine needle	N. H. Baldwin	Laconia, N. H	June 25, 1872	128, 274
Knitting-machine needle	D. Bickford	Boston, Mass	Dec. 1, 1868	84, 472
Knitting-machine needle	D. Bickford	New York, N. Y	Oct. 22, 1872	132, 382
Knitting-machine needle	A. C. Carey	Lynn, Mass	Apr. 25, 1865	47, 488
Knitting-machine needle	J. F. Daniels	Lake Village, N. H	Feb. 8, 1870	99, 647
Knitting-machine needle	R. Ellis	Boston, Mass	June 3, 1856	15, 006
Knitting-machine needle	L. W. Fifield	New Hampton, N. H	Dec. 19, 1865	51, 577
Knitting-machine needle	L. W. Fifield	Holderness, N. H	July 24, 1866	56, 544
Knitting-machine needle	L. W. Fifield	Melrose, Mass	Jan. 22, 1867	61, 413
Knitting-machine needle	J. K. and E. E. Kilbourn	Pittsfield, Mass., and Norfolk, Conn.	Sept. 21, 1858	21, 566
Knitting-machine needle	J. W. Lamb	Rochester, N. Y	Nov. 21, 1865	51, 115
Knitting-machine needle	J. H. Lane and C. D. House	Lake Village, N. H	Dec. 7, 1869	97, 526
Knitting-machine needle	J. Miller	Warren, R. I	Nov. 19, 1867	71, 039
Knitting-machine needle	J. Miller and J. A. Bidwell	Warren, R. I., and East Boston, Mass.	Jan. 29, 1867	61, 630
Knitting-machine needle	J. L. and S. L. Otis	Florence, Mass., and Manchester, Conn.	May 23, 1865	47, 852
Knitting-machine needle	B. F. Peaslee	Lake Village, N. H	Apr. 9, 1867	63, 746
Knitting-machine needle	S. Peberdy	Philadelphia, Pa	May 13, 1862	35, 254
Knitting-machine needle	C. H. Platt and G. A. Stanberg	Norwalk, Ohio	Feb. 25, 1873	136, 262
Knitting-machine needle	S. H. Roper	Boston, Mass	Oct. 3, 1871	119, 651
Knitting-machine needle	C. P. S. Wardwell	Lake Village, N. H	Feb. 26, 1867	62, 382
Knitting-machine-needle shanks, Machine for making.	J. S. Perkins	Lake Village, N. H	July 28, 1868	80, 304
Knitting-machine needles, Manufacture of	J. H. Bullard	Chicopee Falls, Mass	July 8, 1873	140, 611
Knitting-machine register	B. B. Bollinger	Louisville, Ohio	Sept. 8, 1868	81, 870
Knitting-machine register	W. V. Dubois	Covington, Ind	May 25, 1869	90, 435
Knitting-machine register	J. W. Rist	Rochester, N. Y	Sept. 29, 1868	82, 555
Knitting machine register	B. F. Wyman and B. H. Hartshorn.	Lancaster and Ashland, Mass.	Sept. 17, 1867	68, 928
Knitting-machine, Rotary	J. Hollen	Blair County, Pa	Mar. 6, 1866	53, 000
Knitting-machine, Rotary	S. W. Park and E. S. Ells	Troy, N. Y	Aug. 5, 1856	15, 492
Knitting-machine, Rotary	M. L. Roberts	Chatsworth, Ill	May 7, 1867	64, 572
Knitting-machine, Rotary	H. G. Sanford	Worcester, Mass	Nov. 30, 1852	9, 434
Knitting-machine, Rotary	D. Tainter	Worcester, Mass	Nov. 30, 1852	9, 435
Knitting-machine, Rotary	C. Tompkins and J. Johnson	Troy, N. Y	May 27, 1856	14, 975
Knitting-machine sinkers, Blade for	L. Kavanaugh	Waterford, N. Y	Jan. 6, 1863	37, 347
Knitting-machine, Weft-thread	C. Callahan	Chelsea, Mass	Feb. 11, 1873	135, 625
Knitting-machine-weight hook	C. H. Koegel	Buffalo, N. Y	Apr. 15, 1873	137, 778
Knitting-machines, Construction of cylinders of	C. H. Young	Lake Village, N. H	Sept. 13, 1870	107, 318
Knitting-machines, Dividing-wheel of weft-thread	H. Woodman	Saco, Me	Apr. 25, 1871	114, 077
Knitting-machines, looms, &c., Electro-magnetic stop-motion, &c., for.	H. E. Wells and D. H. Moran	Van Wert, Ohio	Jan. 30, 1872	123, 315
Knitting-machines, &c., Machine for producing reciprocating motion in.	S. Haslam, jr	New Britain, Conn	July 28, 1868	80, 283
Knitting-machines, Manufacture of latch-needles for.	S. Woodward	Manchester, N. H	June 27, 1871	116, 386
Knitting machines, Register for tubular	D. A. Lehman	Wakarusa, Ind	Dec. 16, 1873	145, 509
Knitting-machines, Rotary take-up for	J. Teachout	Waterford, N. Y	Dec. 10, 1867	72, 115
Knitting-machines, Set-up device for	H. L. Arnold	Ottawa, Ill	Sept. 2, 1873	142, 429
Knitting-machines, Setting-up and tension comb for.	W. Gaskill	Cincinnati, Ohio	June 7, 1870	104, 114
Knitting machines, Setting-up device for circular	W. Franz and W. Pope	Crestline, Ohio	May 3, 1870	102, 529
Knitting-machines, Setting-up work in	I. W. Lamb	Rochester, N. Y	Sept. 12, 1865	49, 895
Knitting-machines, Sinker-wheel for	H. C. Bradford	Providence, R. I	May 30, 1871	115, 426
Knitting-machines, Stop-motion for	R. Cushman	Pawtucket, R. I	May 22, 1855	12, 896
Knitting-machines, Stop-motion for	J. Dalton	Brooklyn, N. Y	June 28, 1864	43, 294
Knitting-machines, Stop-motion for	E. Kay	Philadelphia, Pa	June 5, 1866	55, 308
Knitting-machines, Stop-motion for	M. Lee	Needham, Mass	Sept. 19, 1865	50, 012
Knitting-machines, Stop-motion for	B. L. Mack	Essex, Conn	May 10, 1864	42, 672
Knitting-machines, Stop-motion for	T. F. Wynn	Atlanta, Ga	June 4, 1872	127, 539
Knitting-machines, Stop-motion and creel-stand for	H. C. Bradford	Providence, R. I	July 4, 1871	116, 677
Knitting-machines, Stop-motion and indicator for	E. Wilder	Chicopee Falls, Mass	Oct. 13, 1868	83, 116
Knitting-machines, Stop-motion for circular	P. W. Hart	Stamford, N. Y	Jan. 31, 1865	46, 186
Knitting-machines, Stopping-mechanism for	R. Cook	New Hartford, N. Y	Oct. 3, 1871	119, 448
Knitting-machines, Stopping-mechanism for	J. Kennedy	Claverack, N. Y	Aug 31, 1869	94, 218
Knitting-machines, Take-up device for	N. H. Baldwin	Laconia, N. H	July 16, 1872	129, 387
Knitting machines, Take-up for circular	S. Ward	Amsterdam, N. Y	Jan. 26, 1864	41, 415
Knitting-machines, Take-up mechanism for	T. Crane	Fort Atkinson, Wis	Jan. 29, 1867	61, 608
Knitting-machines, Take-up mechanism for	S. Ward	Amsterdam, N. Y	Feb. 19, 1867	62, 239
Knitting machines, Take-up of circular	H. Brockway	Cohoes, N. Y	Nov. 8, 1864	44, 932
Knitting machines, Taking-up mechanism for circular.	J. Lessels	Troy, N. Y	June 5, 1866	55, 422

Index of patents issued from the United States Patent Office from 1790 *to* 1873, *inclusive*—Continued.

Invention.	Inventor.	Residence.	Date.	No.
Knitting-machines, Transferring and loop-retaining comb for.	J. N. Kendall	Nashua, N. H	May 24, 1870	103, 471
Knitting-machines, Yarn-take-up and tension-device for.	H. L. Arnold	Ottawa, Ill	Aug. 5, 1873	141, 534
Knitting-machines, Yarn-tension device for	J. Crandell	Chicopee Falls, Mass	Sept. 28, 1869	95, 200
Knitting-machines, Yarn-tension device for	W. Franz and W. Pope	Bucyrus and Crestline, Ohio	Feb. 13, 1872	123, 687
Knitting machinery	J. Whitworth	Manchester, England	Feb. 1, 1848	5, 432
Knitting-needle	J. Hibbert	Providence, R. I	Jan. 9, 1849	6, 025
Knitting-needle jack	H. J. Wickham	Manchester, Conn	Aug. 18, 1868	81, 235
Knitting-needles, Machine	T. Sands	Gilford, N. H	June 23, 1863	38, 988
Knitting-needles, Machine for making	C. P. S. Wardwell	Lake Village, N. H	Sept. 26, 1865	50, 188
Knitting-needles, Machine for making the tongues of machine.	W. Aiken	Franklin, N. H	Nov. 12, 1867	70, 678
Knitting-needles, Machine for notching	W. Aiken	Franklin, N. H	Dec. 10, 1867	71, 836
Knitting-needles, Making	T. Sands	Gilford, N. H	June 23, 1863	38, 987
Knitting needles, Method of riveting latch	C. P. S. Wardwell	Lake Village, N. H	Dec. 5, 1871	121, 562
Knitting pile fabric	D. Bickford	Boston, Mass	Dec. 1, 1868	84, 473
Knitting lacings, &c., Machine for	W. W. Westcott and H. L. Walcott.	Providence, R. I., and Charles River Village, Mass.	July 18, 1865	48, 878
Knitting stockings and joining selvages, Method of	D. Bickford	New York, N. Y	Sept. 17, 1872	131, 388
Knitting stockings and stockinets, Machine for	A. Porter, J. Mead, and J. Stedwall.	Queensbury, N. Y	July 31, 1815	....
Knitting stockings, Machine for	H. Burt	Boston, Mass	Sept. 23, 1843	3, 275
Knitting stockings, &c., Machine for	A. French	Springfield, Mass	Mar. 18, 1842	2, 493
Knitting stockings, Machine for	B. Hutchinson	Springfield, Mass	Oct. 22, 1840	1, 834
Knitting stockings, &c., Machine for	J. McMullen and J. Hollen, jr.	Huntingdon County, Pa	Mar. 5, 1831	....
Knitting stockings, Method of	H. A. House	Bridgeport, Conn	Mar. 7, 1871	112, 346
Knob and door fastener, Combined base	H. H. Heskett and M. E. Ferguson.	McLean County, Ill	Jan. 25, 1870	99, 086
Knob, Atmospheric	O. H. Needham	New York, N. Y	Sept. 29, 1868	82, 629
Knob bolt, Door	L. Lillie	Troy, N. Y	July 5, 1859	24, 647
Knob-bolts, Operating and locking	O. Ellsworth	Hartford, Conn	June 7, 1853	9, 767
Knob, Carriage	J. Barclay	Attleborough, Mass	Oct. 22, 1867	69, 959
Knob, Carriage	R. D. Case and J. Barclay	New York, N. Y., and Attleborough Falls, Mass.	Oct. 22, 1867	69, 965
Knob, Commode	E. Skinner	Sandwich, N. H	June 11, 1829	....
Knob, Curtain	C. Z. Kroh	Tiffin, Ohio	Aug. 29, 1865	49, 634
Knob, Door	J. W. Bliss	Hartford, Conn	July 22, 1856	15, 367
Knob, Door	G. W. Cady	Providence, R. I	June 21, 1870	104, 420
Knob, Door	E. Day	Chicago, Ill	Jan. 12, 1869	85, 799
Knob, Door	H. H. Elwell	Meriden, Conn	July 15, 1856	15, 332
Knob, Door	H. H. Elwell	South Norwalk, Conn	Sept. 2, 1873	142, 451
Knob, Door	J. W. Grogan	Brooklyn, N. Y	Feb. 22, 1870	100, 026
Knob, Door	G. Jones	Peekskill, N. Y	May 31, 1870	103, 749
Knob, Door	J. J. King	New York, N. Y	Sept. 21, 1869	95, 024
Knob, Door	W. Leighton	Cambridge, Mass	Jan. 16, 1855	12, 265
Knob, Door	A. Longstreet	Chicago, Ill	Nov. 26, 1867	71, 315
Knob, Door	G. B. Lothrop	Boston, Mass	July 20, 1869	92, 733
Knob, Door	E. Parker	Meriden, Conn	May 5, 1863	38, 406
Knob, Door	T. J. Sloan	Bronxville, N. Y	Nov. 9, 1869	96, 627
Knob, Door	L. L. Smith	New York, N. Y	July 9, 1872	128, 764
Knob, Door	A. E. Young	Dorchester, Mass	Aug. 21, 1855	13, 473
Knob, Door, commode, &c	E. Robinson, F. Draper, and J. H. Lord.	Cambridge and Boston, Mass.	Dec. 2, 1836	98
Knob, Drawer	N. Thompson	Brooklyn, N. Y	Sept. 26, 1871	119, 429
Knob, Drawer, commode, &c	D. Hottman	Baltimore, Md	July 31, 1837	325
Knob, Extension door	W. T. Munger	Branford, Conn	July 18, 1865	48, 874
Knob for doors, &c., Ferrule	E. Robinson, F. Draper, and J. H. Lord.	Cambridge and Boston, Mass.	Oct. 20, 1836	65
Knob for doors, &c., Glass	F. Draper	East Cambridge, Mass	Sept. 10, 1840	1, 784
Knob for doors, Glass	C. D. Kellogg and W. L. Coan	Boston, Mass	Dec. 22, 1857	18, 911
Knob for fastening curtains, &c	W. Z. M. and J. W. Chapman	New York, N. Y	Mar. 20, 1855	12, 540
Knob for permutation-locks	H. Isham	New Britain, Conn	Jan. 18, 1870	98, 971
Knob for safe-doors	J. G. Kittredge	San Francisco, Cal	Apr. 19, 1870	102, 132
Knob, Furniture	J. H. Shelton	Waterbury, Conn	Nov. 22, 1870	109, 459
Knob, Glass	T. T. Abbot	Canton, Mass	Nov. 19, 1833	....
Knob, Glass	S. Richards	Cambridge, Mass	Oct. 31, 1831	....
Knob, Glass	W. S. Thompson and A. C. Hobbs.	Cambridge and Boston, Mass.	May 20, 1842	2, 628
Knob, Glass door	A. E. Young	Boston, Mass	Apr. 23, 1872	126, 004
Knob guard, Door	H. Ahrend	Newark, N. J	Apr. 1, 1873	137, 400
Knob guard, Door	J. W. Mellish	Walpole, N. H	June 8, 1869	91, 116
Knob label, Drawer	F. Hale and W. Manley	Philadelphia, Pa	Aug. 3, 1869	93, 259
Knob latch and lock, Combined	J. H. Vickers	Norwich, Conn	Aug. 10, 1869	93, 504
Knob-latch, Reversible	W. H. Andrews	New Haven, Conn	Jan. 17, 1871	111, 028
Knob-lock	J. Adt	Waterbury, Conn	Dec. 11, 1860	30, 902
Knob-lock	F. G. Johnson	Brooklyn, N. Y	June 12, 1866	55, 506
Knob-lock	J. Patterson	Pittsburgh, Pa	Oct. 19, 1829	....
Knob machine, Porcelain	T. J. Sloan	New York, N. Y	Sept. 14, 1869	94, 849
Knob machine, Porcelain	T. J. Sloan	Bronxville, N. Y	Apr. 25, 1871	114, 210
Knob-molding machine	C. H. Palmer	Brooklyn, E. D., N. Y	Jan. 31, 1871	111, 470
Knob or handle, Glass furniture	I. and J. P. Bakewell	Pittsburgh, Pa	May 14, 1828	....
Knob, Picture	E. D. Ives	New Haven, Conn	Feb. 18, 1868	74, 692
Knob, Picture	H. C. Luther and C. E. Richards.	Providence, R. I., and North Attleborough, Mass	June 21, 1870	104, 609
Knob rose, Door	W. H. Andrews	New Haven, Conn	July 14, 1868	79, 936
Knob rose, Door	C. L. Bates	New York, N. Y	May 25, 1869	90, 420
Knob rose, Door	E. A. Crew	Waterbury, Conn	Mar. 26, 1872	124, 936
Knob rose, Door	S. S. Day	New York, N. Y	Nov. 3, 1857	18, 537
Knob rose, Door	H. H. Elwell	South Norwalk, Conn	Dec. 10, 1872	133, 842
Knob rose, Door	W. Hall	Boston, Mass	Oct. 19, 1869	95, 899
Knob rose, Door	J. J. Henderson	New York, N. Y	Nov. 22, 1870	109, 409
Knob rose, Door	W. H. Mattson	Philadelphia, Pa	May 14, 1872	126, 642
Knob rose, Door	W. T. Munger	New Britain, Conn	Mar. 1, 1870	100, 314
Knob rose, Door	M. V. Nobles	Rochester, N. Y	May 30, 1867	48, 023
Knob roses, Bush-nut for door	C. Hood	Hartford, Conn	Aug. 6, 1872	130, 131

Index of patents issued from the United States Patent Office from 1790 *to* 1873, *inclusive*—Continued.

Invention.	Inventor.	Residence.	Date.	No.
Knob-roses, Fastening for door	J. M. Adolphus	Philadelphia, Pa	Jan. 31, 1871	111, 297
Knob-roses to doors, Attaching	L. P. Waterman and C. H. Porter.	Bridgeport, Conn	May 18, 1869	90, 209
Knob shank, Door	L. R. Livingston, J. J. Roggen, and C. Adams.	Pittsburgh, Pa	July 7, 1846	4, 620
Knob, Silvered-glass door	J. W. Haines	Cambridgeport, Mass	Aug. 5, 1873	141, 504
Knob spindle	W. Varrah	New Haven, Conn	Mar. 11, 1873	136, 795
Knob spindle	O. Newton	Pittsburgh, Pa	July 28, 1857	17, 887
Knob-spindles, Fastening door	A. Cooley	Hartford, Conn	Nov. 11, 1856	16, 047
Knob with glass shanks or screw, Glass	D. Jarves	Boston, Mass	June 13, 1829	....
Knobs and other molded articles, Plastic composition for the manufacture of.	H. E. Shepard	New Haven, Conn	July 8, 1873	140, 735
Knobs, Attaching door	D. B. Cobb	Jersey City, N. J	Sept. 17, 1867	68, 956
Knobs, Attaching screws to	W. E. Doolittle	West Haven, Conn	July 20, 1869	92, 801
Knobs, &c., Composition for door	J. Harrison	Stillwater, N. Y	Oct. 2, 1847	5, 315
Knobs, Die for making	J. Wise	Branford, Conn	Feb. 12, 1867	61, 977
Knobs, &c., Fastening commode	W. Price	Pitt, Pa	Nov. 14, 1826	....
Knobs, Fastening door	N. Benham	Hartford, Conn	Apr. 8, 1856	14, 595
Knobs for doors, &c., Making glass	H. Whitney and E. Robinson	Cambridge, Mass	Nov. 4, 1826	....
Knobs for furniture, Fastening for	L. B. Myers	Elmore, Ohio	Oct. 30, 1866	59, 250
Knobs for locks, &c., Attaching necks, shanks, screws, &c., to glass.	J. G. Hotchkiss, J. A. Davenport, and J. W. Quincy.	New Haven, Conn., and New York.	Nov. 16, 1841	2, 361
Knobs, Glass screw for	O. Newton	Pittsburgh, Pa	Oct. 17, 1835	....
Knobs, Grinding off the shanks of glass	J. W. Haines	Cambridge, Mass	Apr. 9, 1872	125, 565
Knobs, Instrument for manufacturing door	A. Rogers	Painesville, Ohio	Aug. 22, 1854	11, 562
Knobs, Machine for making porcelain	T. J. Sloan	Bronxville, N. Y	Dec. 19, 1871	122, 072
Knobs, Machine for stamping clay door	G. Lawton	Trenton, N. J	May 28, 1867	65, 244
Knobs, Making pressed glass	J. Robinson	Pittsburgh, Pa	Oct. 6, 1827	....
Knobs, Manufacture of door	O. Newton	Pittsburgh, Pa	Nov. 25, 1851	8, 547
Knobs, Manufacture of door	B. Nott	Bethlehem, N. Y	Mar. 9, 1852	8, 795
Knobs, Manufacture of door	G. O. Russell	Middletown, Conn	Oct. 7, 1845	4, 220
Knobs, Manufacture of glass	E. D. Ives	New Haven, Conn	Apr. 14, 1868	76, 766
Knobs, Manufacture of iron door	C. Carpenter	Hamilton, Canada	July 16, 1872	129, 458
Knobs, Manufacture of mineral	J. Jones	Brooklyn, N. Y	Sept. 17, 1867	68, 885
Knobs, Manufacturing glass	D. Jarves	Boston, Mass	Oct. 19, 1830	....
Knobs, Mold for casting metal	C. Rebstock and J. Ottner	Bridgeport, Conn	Jan. 14, 1873	134, 931
Knobs, Mold for casting metal door	C. Rebstock and A. Hart	Bridgeport and Meriden, Conn.	July 16, 1872	129, 056
Knobs of clay used in pottery and porcelain, Making door.	J. G. Hotchkiss, J. A. Davenport, and J. W. Quincy.	New Haven, Conn., and New York.	July 29, 1841	2, 197
Knobs of plate-metal, Die for making door and other	L. E. Hicks	Middletown, Conn	June 27, 1840	1, 654
Knobs, Securing the necks to door	T. Kennedy	Branford, Conn	Jan. 10, 1865	45, 836
Knobs to collars, Attaching door	L. R. Livingston, J. J. Roggen, C. Adams, and W. and R. Phillips.	Pittsburgh, Pa	Dec. 10, 1846	4, 883
Knobs to door-latches, Attaching	W. H. Andrews	New Haven, Conn	Apr. 23, 1867	63, 980
Knobs to doors, Attaching	J. A. Crever	Pittsburgh, Pa	Oct. 16, 1849	6, 800
Knobs to doors, Attaching	C. F. Langford	Brooklyn, N. Y	Dec. 7, 1869	97, 654
Knobs to doors, &c., Attaching roses for	N. Matthews	Pittsburgh, Pa	Apr. 6, 1852	8, 857
Knobs to doors, drawers, &c., Securing	C. Ward	Salem, Mass	Nov. 19, 1861	33, 759
Knobs to drawers, &c., Attaching	S. A. Brackett	Boston, Mass	Nov. 18, 1873	144, 597
Knobs to latches, Attaching	L. Michaels	Covington, Ky	Sept. 25, 1873	143, 174
Knobs to metallic sockets, Attaching glass	E. and G. W. Robinson	Boston, Mass	Oct. 20, 1837	434
Knobs to roses, Attaching	C. F. Rebstock	Meriden, Conn	Dec. 10, 1872	133, 889
Knobs to screws, Mode of attaching	C. H. Thurston	Marlborough, N. H	May 5, 1868	77, 550
Knobs to shanks, Attaching	J. Hoeflinger	Saint Joseph, Mo	Jan. 30, 1872	123, 102
Knobs to shanks, Attaching door	W. T. Munger	Branford, Conn	Mar. 19, 1867	62, 967
Knobs to shanks, Clasp for securing	J. W. Haines	Cambridge, Mass	May 14, 1872	126, 805
Knobs to shanks, Fastening door	G. Jones and B. E. Mead	Peekskill, N. Y	Nov. 13, 1866	59, 605
Knobs to shanks, Fastening door	M. V. Nobles	Rochester, N. Y	May 30, 1867	48, 024
Knobs to shanks, Fastening door	M. V. Nobles	Rochester, N. Y	May 30, 1867	48, 025
Knobs to shanks, Securing	T. J. Sloan	New York, N. Y	Apr. 13, 1869	88, 918
Knobs to shanks, Securing	T. J. Sloan	New York, N. Y	Apr. 13, 1869	88, 919
Knobs to spindles, Adjusting	A. Dawes	Hudson, Mass	Dec. 31, 1867	72, 815
Knobs to spindles, Adjusting	T. Kennedy	Branford, Conn	May 29, 1866	55, 115
Knobs to spindles, Attaching	M. Andrew	Melbourne, Colony of Victoria.	Apr. 12, 1870	101, 808
Knobs to spindles, Attaching	M. Andrew	Melbourne, Victoria	June 6, 1871	115, 675
Knobs to spindles, Attaching	W. H. Andrews	New Haven, Conn	Apr. 28, 1868	77, 344
Knobs to spindles, Attaching	M. W. Barse	Olean, N. Y	Oct. 31, 1871	120, 362
Knobs to spindles, Attaching	G. N. Cummings	Meriden, Conn	July 22, 1862	35, 924
Knobs to spindles, Attaching	W. A. Fenn	Wolcott, N. Y	Dec. 15, 1868	84, 868
Knobs to spindles, Attaching	W. A. Fenn	Rochester, N. Y	Feb. 28, 1871	112, 134
Knobs to spindles, Attaching	W. A. Fenn	Rochester, N. Y	July 11, 1871	116, 941
Knobs to spindles, Attaching	J. R. Gill and W. R. Baker	Hamilton and Wellington Square, Canada.	May 21, 1872	127, 046
Knobs to spindles, Attaching	A. G. Gray	Saint John, Canada	May 14, 1872	126, 694
Knobs to spindles, Attaching	A. M. Hill	Pittsburgh, Pa	July 19, 1864	43, 582
Knobs to spindles, Attaching	J. F. Keller and N. Schner	Hagerstown, Md	Mar. 7, 1871	112, 354
Knobs to spindles, Attaching	F. M. and J. B. Merriam	West Meriden, Conn	July 29, 1873	141, 285
Knobs to spindles, Attaching	E. M. and J. E. Mix	Westfield, N. Y	May 21, 1872	127, 089
Knobs to spindles, Attaching	C. Morrill	New York, N. Y	Jan. 17, 1871	110, 992
Knobs to spindles, Attaching	W. T. Munger	Branford, Conn	Feb. 9, 1869	86, 683
Knobs to spindles, Attaching	W. T. Munger	Branford, Conn	Feb. 16, 1869	86, 998
Knobs to spindles, Attaching	S. Selden	Erie, Pa	Jan. 7, 1873	134, 708
Knobs to spindles, Attaching	O. L. Smith	Providence, R. I	June 6, 1871	115, 650
Knobs to spindles, Attaching	H. B. Tuthill	New York, N. Y	May 4, 1869	89, 708
Knobs to spindles, Attaching	H. J. B. Whipple	Meriden, Conn	July 29, 1873	141, 408
Knobs to spindles, Attaching door	W. Boch	Newtown, N. Y	Mar. 24, 1868	75, 847
Knobs to spindles, Attaching door	C. B. Bristol	New Haven, Conn	Mar. 5, 1867	62, 599
Knobs to spindles, Attaching door	C. B. Bristol	New Haven, Conn	Dec. 24, 1867	72, 597
Knobs to spindles, Attaching door	J. Evans	New Haven, Conn	May 31, 1870	103, 580
Knobs to spindles, Attaching door	J. N. Karr	Buffalo, N. Y	Jan. 17, 1871	110, 977
Knobs to spindles, Attaching door	T. Marvin	Cambridge, Mass	May 12, 1868	77, 745
Knobs to spindles Attaching door	E. Parker	West Meriden, Conn	June 28, 1864	43, 332
Knobs to spindles, Attaching door	S. S. Putnam	Dorchester, Mass	July 7, 1868	79, 597
Knobs to spindles, Attaching door	E. Robinson	Boston, Mass	Jan. 10, 1843	2, 904
Knobs to spindles, Fastening	D. Skidmore	Seneca Falls, N. Y	July 15, 1862	35, 899

Index of patents issued from the United States Patent Office from 1790 *to* 1873, *inclusive*—Continued.

Invention.	Inventor.	Residence.	Date.	No.
Knobs to spindles, &c., Fastening door locks and latch.	A. O. Downer	Utica, N. Y	Dec. 21, 1842	2, 891
Knobs to spindles of door-locks, Attaching	W. T. Munger	Branford, Conn	Nov. 26, 1867	71, 502
Knobs, umbrella-handles, and other molded articles, Manufacture of door.	W. Sanderson and J. J. De Barry.	New York, N. Y	Jan. 25, 1870	99, 241
Knobs upon collars, Casting door	W. Heiggs	Utica, N. Y	Dec. 10, 1846	4, 884
Knock-down chair	C. R. Long	Louisville, Ky	Apr. 8, 1873	137, 613
Knock-down chair	V. Stockton	Williamsburgh, Ohio	July 11, 1871	117, 012
Knuckle-joint	J. K. Collins	Huntsville, Ala	Mar. 6, 1866	52, 969
Knuckle-joint	G. B. Garlinghouse	North Madison, Ind	May 14, 1867	64, 752
Knuckle-joint	J. H. Mears and C. W. Yale	Oshkosh, Wis	Dec. 11, 1866	60, 403
Knuckle-joint	S. N. Taylor	Horicon, Wis	Feb. 27, 1866	52, 908
Krout-cutting machine	W. K. Baylor and C. Rapp	Batesville, Ind	Apr. 13, 1869	88, 769
L.				
Label, Apothecary's	G. G. Percival	Philadelphia, Pa	Sept. 8, 1868	82, 028
Label-attaching machine	J. F. Zacharias	Leesburgh, Va	Aug. 13, 1867	67, 833
Label, Baggage	J. M. March	Washington, D. C	Sept. 10, 1867	68, 632
Label, Bottle	H. Frank	Pittsburgh, Pa	Jan. 21, 1873	135, 034
Label-card	J. Sharp	Roxbury, Mass	June 15, 1852	9, 039
Label, Cask	E. A. Locke	Boston, Mass	Mar. 24, 1868	75, 934
Label, Commercial	C. N. Morris	Cincinnati, Ohio	May 26, 1863	38, 692
Label, Direction	G. E. Pevey	Lowell, Mass	Mar. 5, 1867	62, 561
Label, Direction	M. J. Procter	Lowell, Mass	Mar. 13, 1866	53, 233
Label-fastener	C. R. Doane	Brooklyn, N. Y	Aug. 24, 1869	94, 088
Label for hat-lining, Ornamental	T. W Bracher	New York, N. Y	Dec. 9, 1873	145, 391
Label for trees, &c	W. W. Wade and F. T. Cordis	Long Meadow, Mass	Dec. 21, 1850	22, 390
Label, Fruit-can	J. Dunlap	Pittsburgh, Pa	June 29, 1869	92, 023
Label-gumming apparatus	J. H. Hudson	Winona, Minn	Feb. 20, 1872	123, 906
Label-holder	F. J. Collier	Philadelphia, Pa	Oct. 2, 1860	30, 205
Label-holder	C. A. Dickerman	New Haven, Conn	Dec. 8, 1868	84, 804
Label-holder	G. S. True	Leavenworth, Kans	Aug. 4, 1860	80, 683
Label holder and handle combined	H. Manieck	New York, N. Y	Aug. 16, 1870	106, 494
Label holder, Drawer-pull	J. A. Evarts and P. Cinquin	West Meriden, Conn	Apr. 4, 1871	113, 506
Label-holder for railway-car	J. H. Parsons	Quincy, Mich	Nov. 19, 1867	71, 051
Label holder, Lock	J. S. Ramsay	Baltimore, Md	July 21, 1868	80, 218
Label intended for second use	H. Loewenberg	New York, N. Y	Apr. 5, 1864	42, 208
Label, Mail-bag	S. Andrews	Perth Amboy, N. J	July 25, 1854	11, 348
Label, Mail-bag	O. S. North	New Britain, Conn	Mar. 13, 1844	3, 486
Label, Mail-bag, &c	T. P. Trott	Washington, D. C	Apr. 24, 1860	28, 023
Label or tag	T. B. and L. De Forest	Birmingham, Conn	Nov. 15, 1864	45, 029
Label or tag	E. W. Dennison	Boston, Mass	June 9, 1863	38, 871
Label-pasting machine	W. T. Slocum	Philadelphia, Pa	July 16, 1872	129, 606
Label, Sheep	C. H. Dana	West Lebanon, N. H	June 6, 1865	48, 055
Label, Tag	D. D. Foley	Washington, D. C	Apr. 26, 1870	102, 461
Label, Trunk	S. W. Downey	Piedmont, W. Va	Feb. 11, 1868	74, 322
Labels, Apparatus for cutting and attaching	C. M. Spencer	Manchester, Conn	Oct. 11, 1859	25, 770
Labels, Apparatus for damping and gumming	J. Benn and G. O. Luckman	Goat and Manchester, England.	Oct. 15, 1867	69, 895
Labels, Apparatus for pasting	W. E. Booraem	New York, N. Y	Nov. 10, 1868	83, 825
Labels, Apparatus for softening the gum of adhesive	B. Wilder	North Scituate, Mass	Feb. 21, 1865	46, 514
Labels for periodicals, &c., Accountant	R. Dick	Toronto, Canada	Oct. 4, 1859	25, 635
Labels, hat-linings, &c., Ornamenting	T. W. Bracker	New York, N. Y	Oct. 3, 1871	119, 563
Labels, Machine for making sheep's	C. H. Dana	West Lebanon, N. H	Aug. 29, 1865	49, 609
Labels, Metallic hook for	S. B. Fay	New York, N. Y	July 1, 1856	15, 226
Labels on glass ware, Mode of securing	E. M. Davis	Pittsburgh, Pa	Feb. 25, 1868	74, 804
Labels or tags, Machine for cutting and punching	C. S. Moseley	Boston, Mass	July 5, 1864	43, 458
Labels, tags, &c., Process for preparing wood for the manufacture of.	J. Melling	Rochester, N. Y	July 9, 1867	66, 512
Labels to bottles, Attaching	W. N. Walton	New York, N. Y	Sept. 23, 1862	36, 542
Labels to bottles, Mode of applying	J. S. and T. B. Atterbury	Pittsburgh, Pa	May 15, 1866	54, 665
Labels to newspapers, Machine for attaching	J. F. Zacharias	Leesburgh, Va	Apr. 28, 1868	77, 430
Labeling-machine	C. W. Greene	Jackson, Tenn	Feb. 25, 1873	136, 154
Labeling machine, Spool	Q. S. Backus	Winchendon, Mass	Feb. 9, 1869	86, 627
Labeling machine, Spool	H. Conant	Willimantic, Conn	June 10, 1862	35, 562
Lace and ribbon case	M. Deitzler	Ashland, Pa	Apr. 1, 1873	137, 297
Lace-drying frame	A. Chatain	Washington, D. C	Sept. 26, 1871	119, 318
Lace, &c., Fabric for making imitation	H. Loewenberg	New York, N. Y	Nov. 17, 1863	40, 633
Lace-making machine	G. Osborn	Brooklyn, N. Y	July 27, 1869	92, 995
Lacing and shoe-fastener, Slide	R. Adams	Cincinnati, Ohio	Oct. 2, 1866	58, 364
Lacing-device	E. C. C. Kellogg	Hartford, Conn	Feb. 13, 1866	52, 645
Lacing-device	J. Nealey, jr	Bangor, Me	Mar. 31, 1868	76, 234
Lacing-device	A. Young	Philadelphia, Pa	Dec. 31, 1867	72, 955
Lacing-eye	A. G. Mead	Boston, Mass	Oct. 12, 1869	95, 708
Lacings, Apparatus for tagging	F. J. Seymour	Wolcottville, Conn	May 17, 1864	42, 798
Lacings, Machine for cutting	H. B. Harvey	Northbridge, Mass	Sept. 13, 1864	44, 187
Lacteal instrument	C. H. Davidson	Charlestown, Mass	Nov. 9, 1858	22, 018
Ladder	P. M. Ackerman	Webster, N. Y	June 11, 1867	65, 525
Ladder	F. Aldrich, jr	Augusta, Mich	Sept. 15, 1863	39, 871
Ladder	G. Aldrich	Armada, Mich	Apr. 30, 1861	32, 170
Ladder	H. O. Baker	New York, N. Y	Feb. 26, 1867	62, 305
Ladder	M. D. Boyd	Buffalo, N. Y	Sept. 17, 1867	68, 836
Ladder	W. L. Burlingame	Leslie, Mich	July 7, 1868	79, 727
Ladder	M. Burt	Norton, Mass	Feb. 26, 1867	62, 314
Ladder	E. P. H. Capron	Springfield, Ohio	Sept. 29, 1868	82, 594
Ladder	G. Ckertizza	New York, N. Y	Oct. 23, 1866	58, 984
Ladder	J. H. Conley	Philadelphia, Pa	Mar. 14, 1871	112, 552
Ladder	C. Croley	Dayton, Ohio	Oct. 16, 1866	58, 781
Ladder	C. Croley	Dayton, Ohio	Aug. 13, 1867	67, 636
Ladder	R. L. Dodge	Portland, Me	Apr. 6, 1869	88, 616
Ladder	J. Eagon	Batesville, Ohio	July 22, 1873	141, 126
Ladder	F. Ellis	Walton, N. Y	Dec. 26, 1871	122, 238
Ladder	G. H. Ellis	London, England	Mar. 14, 1871	112, 607
Ladder	D. Fitzgerald	New York, N. Y	Apr. 12, 1859	23, 669
Ladder	F. W. Hovey	Boston, Mass	Jan. 1, 1867	60, 731
Ladder	J. Hughes	New Berne, N. C	Oct. 18, 1870	108, 483

Index of patents issued from the United States Patent Office from 1790 *to* 1873, *inclusive*—Continued.

Invention.	Inventor.	Residence.	Date.	No.
Ladder	E. J. Knowlton	Lyon, Mich	May 19, 1863	38, 586
Ladder	T. B. Luzier and G. A. Haas	Philadelphia, Pa	Feb. 4, 1868	74, 102
Ladder	J. Maxwell and T. Gray	Philadelphia, Pa	July 18, 1871	117, 186
Ladder	W. H. McHench	Cobleskill, N. Y	May 10, 1870	102, 845
Ladder	A. C. McKendree	Conneaut, Ohio	June 16, 1868	78, 988
Ladder	E. T. Olds	Lyon, Mich	Apr. 5, 1864	42, 219
Ladder	P. M. Papin	Saint Louis, Mo	Oct. 13, 1868	83, 084
Ladder	W. G. Philips	Newport, Del	Nov. 3, 1863	40, 500
Ladder	B. Pickering	Dayton, Ohio	May 15, 1866	54, 767
Ladder	B. Schoonmaker	Plainwell, Mich	July 20, 1869	92, 755
Ladder	A. Simmerman and J. S. Simmerman.	Glassborough and Millville, N. J.	Jan. 21, 1868	73, 468
Ladder	A. P. Smith	Sacramento, Cal	Dec. 30, 1873	146, 029
Ladder	M. L. Smith	Battle Creek, Mich	Nov. 2, 1869	96, 495
Ladder	J. Stiles	Salem, Mich	Sept. 15, 1863	39, 970
Ladder	D. B. Taylor	Avon, Mich	Feb. 19, 1867	62, 234
Ladder	C. C. Thomas and F. A. S. Raymond.	Beverly, Mass	June 30, 1868	79, 520
Ladder	J. A. Thompson	Auburn, N. Y	Apr. 12, 1870	101, 943
Ladder	B. F. Turner	Bridgeton, N. J	July 9, 1867	66, 655
Ladder	G. W. Willis	Atchinson, Kans	Mar. 12, 1872	124, 526
Ladder, Adjustable fruit	S. Wright	Hillsborough, Mo	Aug. 23, 1870	106, 645
Ladder, Adjustable lifting	G. Claflin	Miller's Corners, N. Y	Sept. 13, 1870	107, 225
Ladder, Adjustable step	R. R. Croasdale and P. Rink	Reaville, N. J	July 20, 1869	92, 708
Ladder, Adjustable step	J. J. Hardenbrook and S. Belford.	Columbus Grove, Ohio	May 20, 1873	139, 004
Ladder and chair	E. Kohn and J. L. Natcher	Sidney, Ohio	Oct. 1, 1867	69, 448
Ladder and ironing-board, Combined step	H. H. Kendrick	Fulton, N. Y	Dec. 23, 1873	145, 738
Ladder and scaffold combined	R. L. Upchurch	Pana, Ill	Nov. 26, 1872	133, 505
Ladder and staging, Artisan's	C. Monson	New Haven, Conn	Feb. 4, 1862	34, 313
Ladder apparatus, Fire	D. Fobes and H. M. Hartshorn	Boston and Malden, Mass	June 17, 1862	35, 601
Ladder basket-rest	W. E. Ludlow	Cincinnati, Ohio	Aug. 31, 1869	94, 422
Ladder, bench, and clothes-frame, Combined step	H. H. Barker	Hartford, Conn	Dec. 23, 1873	145, 716
Ladder, Convertible	H. B. Malbone	Geneva, N. Y	Aug. 3, 1869	93, 323
Ladder, Elevating	A. Miller	New York, N. Y	May 20, 1873	139, 175
Ladder, Elevating or scaling	S. D. Wollison	Pittsfield, Mass	Sept. 8, 1863	39, 857
Ladder, Extension	H. Barns	Somers, Wis	Dec. 24, 1867	72, 444
Ladder, Extension	C. R. Bryant	Frankfort, N. Y	Mar. 19, 1867	62, 932
Ladder, Extension	M. T. Burbank	Lawrence, Mass	Apr. 13, 1869	88, 842
Ladder, Extension	R. F. Delmot	Flemington, Pa	Apr. 4, 1871	113, 273
Ladder, Extension	C. Eaton	Webster, N. Y	Mar. 7, 1865	46, 648
Ladder, Extension	J. C. Hearne and D. Adams	Pleasant Hill, Mo	Apr. 8, 1873	137, 679
Ladder, Extension	G. L. Johnson	Fairfield, N. Y	Mar. 26, 1867	63, 163
Ladder, Extension	F. and C. Kavemann and B. Hoerstmann.	Cincinnati, Ohio	Apr. 24, 1860	28, 039
Ladder, Extension	J. Kearns	Louisville, Ohio	Feb. 18, 1868	74, 548
Ladder, Extension	M. M. Knowles	Elmira, N. Y	Aug. 18, 1868	81, 279
Ladder, Extension	T. F. Mantey	New Orleans, La	May 26, 1868	78, 302
Ladder, Extension	L. N. Millener	Adams' Basin, N. Y	Feb. 20, 1872	123, 834
Ladder, Extension	W. Morehead	Parkersburgh, W. Va	Mar. 14, 1865	46, 812
Ladder, Extension	W. Morehead	Parkersburgh, W. Va	Dec. 8, 1868	84, 839
Ladder, Extension	J. F. Morse	Montgomery, Ill	May 5, 1868	77, 640
Ladder, Extension	J. Moulton	Boston, Mass	May 17, 1859	24, 044
Ladder, Extension	M. D. Myers	Ilion, N. Y	Feb. 2, 1864	41, 443
Ladder, Extension	G. B. Nickle and J. M. Carville	New York, N. Y	Apr. 10, 1860	27, 821
Ladder, Extension	W. O'Connor	Berville, Mich	Apr. 15, 1873	137, 948
Ladder, Extension	B. Pine	New York, N. Y	Mar. 3, 1868	75, 052
Ladder, Extension	J. Pine	Brooklyn, N. Y	June 10, 1873	139, 689
Ladder, Extension	N. Pullman	New Oregon, Iowa	July 5, 1870	105, 126
Ladder, Extension	H. M. Quackenbush	Herkimer, N. Y	Oct. 22, 1867	70, 016
Ladder, Extension	J. L. Ripley	Fremont, Ohio	June 13, 1865	48, 210
Ladder, Extension	A. Rogers	Painesville, Ill	Aug. 10, 1869	93, 642
Ladder, Extension	J. W. Scott	Philadelphia, Pa	Sept. 27, 1870	107, 822
Ladder, Extension	E. O. Shepardson	Saint Louis, Mo	July 18, 1871	117, 117
Ladder, Extension	J. A. Smith	Lacon, Ill	July 28, 1868	80, 367
Ladder, Extension	W. F. Trautman	Llewellyn, Pa	July 13, 1869	92, 492
Ladder, Extension	C. G. Udell	Chicago, Ill	Mar. 26, 1867	63, 188
Ladder, Extension	C. G. Udell	Chicago, Ill	Aug. 10, 1869	93, 502
Ladder, Extension	L. F. Ward	Marathon, N. Y	Apr. 28, 1863	38, 345
Ladder, Extension	T. Watson and C. Perry	Brooklyn, N. Y	Oct. 30, 1866	59, 297
Ladder, Extension	T. Watson and C. Perry	Brooklyn, N. Y	Oct. 15, 1867	69, 875
Ladder, Extension	T. Watson and C. Perry	Brooklyn, N. Y	Dec. 13, 1870	110, 094
Ladder, Extension	F. Willis	Marathon, N. Y	Jan. 20, 1863	37, 470
Ladder, Extension	T. C. Wood	Augusta, Mich	Jan. 24, 1865	46, 042
Ladder, Extension and step	H. J. Hancock	New York, N. Y	Mar. 9, 1869	87, 666
Ladder, Extension and step	S. E. Hewes	Albany, N. Y	Mar. 24, 1868	75, 907
Ladder, Extension fruit	A. Rogers	Painesville, Ohio	May 15, 1866	54, 778
Ladder, Extension fruit	J. E. Treat	Oxford, Mich	July 2, 1867	66, 420
Ladder, Extension platform	G. A. Schachtel	Newark, N. J	Nov. 15, 1870	109, 256
Ladder, Extension step	H. D. Chance	Llewellyn, Pa	Aug. 23, 1870	106, 659
Ladder, Extension step	L. B. Covert	New York, N. Y	May 5, 1868	77, 463
Ladder, Extension step	L. B. Covert	Detroit, Mich	Apr. 29, 1873	138, 234
Ladder, Extension step	G. W. Packer	Toulon, Ill	June 2, 1868	78, 476
Ladder, Extension step	A. W. and J. E. Walker	Detroit, Mich	Mar. 18, 1873	137, 040
Ladder, Fire	J. Blomgren	Galesburgh, Ill	Dec. 10, 1867	71, 962
Ladder, Fire	G. K. Foster	San Francisco, Cal	Apr. 30, 1867	64, 212
Ladder, Fire	D. F. Haasz	Philadelphia, Pa	Aug. 17, 1869	93, 877
Ladder, Fire	J. L. Hannah	New York	Aug. 19, 1834	......
Ladder, Fire	M. B. Hill	Fairfield, Ohio	Sept. 1, 1843	3, 245
Ladder, Fire	S. Lehman	Philadelphia, Pa	Aug. 9, 1826	......
Ladder, Fire	B. Lies	New York	June 20, 1815	......
Ladder, Fire	A. Lotz	Franklin, Tenn	Nov. 9, 1869	96, 710
Ladder, Fire	J. Welte	Buffalo, N. Y	June 29, 1858	20, 752
Ladder, Fire and escape	J. Van Amringe	Cincinnati, Ohio	May 13, 1856	14, 891
Ladder, Fire-escape	C. G. Buttkereit	Toledo, Iowa	Mar. 19, 1872	124, 541
Ladder, Fire-escape	J. H. Grimsley	New Lexington, Ohio	July 13, 1858	20, 875
Ladder, Fire-escape	H. T. Hartman	Norwood, Va	July 13, 1869	92, 527
Ladder, Fire-escape	D. Hayes and W. Free	San Francisco, Cal	Feb. 25, 1868	74, 821

Index of patents issued from the United States Patent Office from 1790 *to* 1873, *inclusive*—Continued.

Invention.	Inventor.	Residence.	Date.	No.
Ladder, Fire-escape	I. Henderson	Philadelphia, Pa	Feb. 26, 1867	62, 418
Ladder, Fire-escape	A. Iske	Lancaster, Pa	Nov. 26, 1861	33, 785
Ladder, Fire-escape	H. Johnston and W. I. Matthews.	Collinsville, Ill	July 26, 1859	24, 909
Ladder, Fire-escape	H. Loewenberg	New York, N. Y	Sept. 22, 1857	18, 262
Ladder, Fire-escape	R. Loomis	Saratoga Springs, N. Y	Apr. 9, 1867	63, 735
Ladder, Fire-escape	H. and G. Luckenbach	Philadelphia, Pa	Sept. 19, 1871	119, 161
Ladder, Fire-escape	H. Lyles	Washington, D. C	Mar. 18, 1873	136, 844
Ladder, Fire-escape	W. H. Paige	Springfield, Mass	July 5, 1864	43, 425
Ladder, Fire-escape	W. W. Parsons	Stanstead, Canada	Sept. 2, 1873	142, 409
Ladder, Fire-escape	W. B. Peregoy	Baltimore, Md	Feb. 28, 1871	112, 276
Ladder, Fire-escape	G. Pfluger	Deerfield, N. Y	Oct. 7, 1873	143, 462
Ladder, Fire-escape	J. S. Pierson	Brooklyn, N. Y	Sept. 23, 1873	143, 182
Ladder, Fire-escape	R. G. Pike	Middletown, Conn	June 21, 1864	43, 261
Ladder, Fire-escape	W. P. Prewitt	Elkton, Ky	Mar. 5, 1867	62, 564
Ladder, Fire-escape	S. R. Roscoe	Carlisle, N. Y	Apr. 3, 1855	12, 643
Ladder, Fire-escape	J. C. Salomon, jr	Georgetown, D. C	June 1, 1852	8, 987
Ladder, Fire-escape	W. A. Shannon	Washington, D. C	Dec. 17, 1861	33, 963
Ladder, Fire-escape	G. Skinner	Brooklyn, N. Y	Sept. 22, 1868	82, 358
Ladder, Fire-escape	C. Weidling	New York, N. Y	June 8, 1869	90, 975
Ladder, Fire-escape	C. H. White	White's Station, Mich	Apr. 29, 1873	138, 304
Ladder, Fire-escape	J. Withers	Collinsville, Ill	Dec. 14, 1858	22, 324
Ladder, Fire-escape	T. Witmer	Buffalo, N. Y	Apr. 18, 1871	113, 958
Ladder, Fireman's	M. Cronin	Washington, D. C	Nov. 25, 1873	144, 958
Ladder, Fireman's	D. Fitzgerald	New York, N. Y	July 12, 1859	24, 728
Ladder, Fireman's	D. Giambastiani	Washington, D. C	Oct. 14, 1856	15, 883
Ladder, Fireman's	P. Porta	Milan, Italy	Sept. 2, 1873	142, 349
Ladder, Fireman's extension	G. W. Harris	Lancaster, N. Y	Dec. 23, 1873	145, 867
Ladder, Fireman's extension	R. H. Jones	San Francisco, Cal	July 7, 1868	79, 763
Ladder, Folding	J. M. White and A. J. Kee	Hannibal, Mo	Apr. 8, 1873	137, 582
Ladder, Folding step	M. Mattern	Poughkeepsie, N. Y	Aug. 27, 1872	130, 929
Ladder, Fruit	C. S. and C. D. Cannon	Chicago, Ill	Sept. 6, 1870	106, 996
Ladder, Fruit	J. Hannan	Lyon, Mich	Dec. 15, 1863	40, 926
Ladder, Fruit	J. Hannan	South Lyon, Mich	Oct. 25, 1864	44, 799
Ladder, Fruit	S. Hudson	Milford, Mich	Jan. 23, 1866	52, 172
Ladder, Fruit	A. and J. B. Longcor	Paw Paw, Mich	June 24, 1873	140, 286
Ladder, Fruit	D. McMaster	Bath, N. Y	Nov. 28, 1865	51, 205
Ladder, Fruit	A. W. Olds	Green Oak, Mich	Apr. 26, 1864	42, 500
Ladder, Fruit	A. W. Olds	Green Oak, Mich	Mar. 7, 1865	46, 694
Ladder, Fruit	E. F. Olds	South Lyon, Mich	Dec. 13, 1864	45, 430
Ladder, Fruit	D. C. Smith	Adrian, Mich	June 18, 1867	65, 959
Ladder, Fruit	B. F. Turner	Bridgeton, N. J	Sept. 17, 1867	69, 049
Ladder, Fruit	J. H. Verity	Lansing, Mich	Feb. 22, 1870	100, 219
Ladder, Fruit	S. S. Whaley	Tidioute, Pa	Aug. 20, 1867	67, 933
Ladder, Fruit and extension	M. Tuell	Penn Yan, N. Y	Apr. 12, 1870	101, 789
Ladder, Fruit extension	E. Slater	Girard, Pa	Nov. 13, 1866	59, 666
Ladder, Fruit or step	W. E. Bond	Cleveland, Ohio	Nov. 14, 1865	50, 894
Ladder, Fruit step	J. F. Winchell	Springfield, Ohio	Jan. 8, 1867	61, 129
Ladder-hook	S. D. Fish	Schuyler Falls, N. Y	Sept. 26, 1871	119, 226
Ladder-hook	J. G. Rockwell	Cortland, N. Y	Oct. 9, 1866	58, 734
Ladder-hook, Adjustable	W. T. Farrar	Concord, Mass	Apr. 9, 1861	31, 963
Ladder joint, Step	S. E. Hewes	Albany, N. Y	Sept. 15, 1868	82, 221
Ladder, Library step	C. C. Schmitt	New York, N. Y	Jan. 23, 1866	52, 211
Ladder, Life, escape, and fire	J. Johnson	Baltimore, Md	Apr. 18, 1831	......
Ladder or fire-escape	T. Armitage	Philadelphia, Pa	Apr. 18, 1854	10, 797
Ladder or step for street-lamp lighter	M. M. Smith	Nashville, Tenn	Sept. 8, 1868	81, 954
Ladder, Orchard	C. Hays	Madison, Wis	Jan. 31, 1865	46, 105
Ladder, Orchard	W. H. Heffley	Rochester, Ind	Sept. 17, 1872	131, 442
Ladder, Platform	J. Sheetz	Sunbury, Pa	Nov. 16, 1869	96, 974
Ladder-platform	C. Wolf	Wilmore, Pa	Sept. 3, 1872	131, 042
Ladder-splice and safety-hook combined	J. Edmunds	South Adams, Mass	Mar. 26, 1872	124, 940
Ladder-stand	D. R. Burkholder	Plainfield, Pa	June 6, 1871	115, 696
Ladder, Step	M. E. Abbott	Bethlehem, Pa	Nov. 19, 1867	70, 933
Ladder, Step	E. R. Austin	Elmira, N. Y	Apr. 13, 1869	88, 831
Ladder, Step	E. R. Austin	Elmira, N. Y	Nov. 23, 1869	97, 148
Ladder, Step	J. H. Balsley	Dayton, Ohio	Jan. 7, 1862	34, 100
Ladder, Step	J. H. Balsley	Dayton, Ohio	Feb. 8, 1870	99, 621
Ladder, Step	J. Barnett	Dayton, Ohio	Sept. 26, 1865	50, 091
Ladder, Step	W. W. Berntheisel	West Hempfield Township, Pa.	Aug. 13, 1867	67, 710
Ladder, Step	C. E. Boman	San Francisco, Cal	June 22, 1869	91, 706
Ladder, Step	W. E. Bond	Cleveland, Ohio	Apr. 5, 1864	42, 162
Ladder, Step	J. Booher	Dayton, Ohio	Sept. 27, 1864	44, 391
Ladder, Step	C. W. Brown	Newark, N. J	Apr. 3, 1866	53, 566
Ladder, Step	E. P. H. Capron	Springfield, Ohio	June 26, 1866	55, 817
Ladder, Step	J. Charleville	Saint Louis County, Mo	Nov. 24, 1868	84, 260
Ladder, Step	T. Curtis	New Hudson, Mich	May 3, 1864	42, 565
Ladder, Step	W. J. Emens	Saint Louis, Mo	Apr. 16, 1872	125, 728
Ladder, Step	V. Fountain, jr	Factoryville, N. Y	Sept. 1, 1863	39, 726
Ladder, Step	C. Frizell	Boston, Mass	July 1, 1873	140, 359
Ladder, Step	C. Frizell	Boston, Mass	July 8, 1873	140, 578
Ladder, Step	M. B. Geary	New York, N. Y	Mar. 7, 1871	112, 442
Ladder, Step	H. P. Hammond and W. A. Hathaway.	North Kingston, R. I	June 5, 1866	55, 288
Ladder, Step	D. B. Hedden	Newark, N. J	July 30, 1867	67, 193
Ladder, Step	W. Huey	Galena, Md	Oct. 24, 1871	120, 158
Ladder, Step	J. S. Lash	Philadelphia, Pa	Aug. 7, 1866	56, 957
Ladder, Step	M. C. Longacre	Cleveland, Ohio	Oct. 20, 1868	83, 291
Ladder, Step	M. N. Lovell	Erie, Pa	Apr. 16, 1872	125, 683
Ladder, Step	M. N. Lovell	Erie, Pa	Mar. 11, 1873	136, 659
Ladder, Step	P. Martin	Newark, N. J	Jan. 23, 1872	123, 032
Ladder, Step	A. Moore, jr	Hillsborough, Ohio	May 1, 1866	54, 387
Ladder, Step	E. M. Norton	Bridgeport, Conn	Jan. 30, 1872	123, 287
Ladder, Step	J. S. Oakley	Passaic, N. J	Nov. 21, 1871	121, 190
Ladder, Step	J. S. Oakley and B. M. Post	Passaic, N. J	May 20, 1873	139, 183
Ladder, Step	A. W. O'Blenus	New York, N. Y	Dec. 16, 1873	145, 522
Ladder, Step	W. G. Philips	Newport, Del	June 8, 1869	91, 039
Ladder, Step	W. G. Philips	Newport, Del	Jan. 3, 1871	110, 785

Index of patents issued from the United States Patent Office from 1790 *to* 1873, *inclusive*—Continued.

Invention.	Inventor.	Residence.	Date.	No.
Ladder, Step	C. S. Rouse	Dowagiac, Mich	Nov. 3, 1868	83, 795
Ladder, Step	A. F. Saunders	Boston, Mass	Dec 5, 1865	51, 394
Ladder, Step	H. T. Smith	Brooklyn, N. Y	June 25, 1867	66, 179
Ladder, Step	D. I. Stagg	New York, N. Y	Feb. 3, 1863	37, 594
Ladder, Step	O. M. Sweet	Forestville, N. Y	Mar. 12, 1872	124, 518
Ladder, Step	C. G. Udell	Chicago, Ill	June 8, 1869	90, 973
Ladder, Step	T. Vogelmann	Hamilton, Ohio	July 30, 1867	67, 235
Ladder, Step	S. Wright	Hillsborough, Mo	May 13, 1873	138, 977
Ladder, Step and extension	H. W. Covert	Rochester, N. Y	Oct 1, 1867	69, 321
Ladder, Step and extension	C. J. Komar	Willoughby, Ohio	Aug. 27, 1867	68, 090
Ladder, Step and extension	W. E. Ludlow	Cincinnati, Ohio	Aug. 31, 1869	94, 423
Ladder, Step and extension	G. S. Walker	Erie, Pa	Feb. 4, 1868	74, 177
Ladders, Construction of step	J. P. McCandless	Cincinnati, Ohio	Oct. 15, 1872	132, 212
Ladders, Construction of step	C. G. Udell	Chicago, Ill	Nov. 12, 1872	132, 938
Ladies' dress-spring	S. J. Sherman	Brooklyn, N. Y	Feb. 25, 1862	34, 531
Ladies' figure-stand	J. R. Palmenberg	New York, N. Y	Aug. 1, 1865	49, 145
Ladies' supporting-brace	M. Harris	New York, N. Y	Feb. 28, 1871	112, 244
Ladies' work-stand	J. B. Atwater	Chicago, Ill	July 25, 1865	48, 888
Ladies' work-table	C. R. P. Foster	Canandaigua, N. Y	Apr. 8, 1851	8, 030
Ladle and fork	W. B. Dunbar	Waterbury, Conn	Dec. 6, 1859	26, 393
Ladle, Bent wood handle and socket for	S. Stedman	Hartford, Conn	July 25, 1818	
Ladle, Bullet	M. Babcock	Charlestown, Mass	June 14, 1864	43, 151
Ladle, Bullet	G. Rugg	Potsdam, N. Y	Apr. 24, 1860	28, 013
Ladle, Culinary	J. C. Haines	Dublin, Ind	Feb. 23, 1858	19, 419
Ladle for pouring metal	H. W. Benton	Lebanon, N. H	July 2, 1867	66, 285
Ladle-stopper	F. A. Ostrander	Troy, N. Y	Aug. 8, 1871	117, 918
Ladle with fork attached	N. S. Warner and H. S. Benedict.	Bridgeport, Conn	Oct. 16, 1860	30, 437
See, also, Culinary ladle.				
Fruit-canning ladle.				
Pouring-metal ladle.				
Wax-ladle.				
Lagging, Machine for cutting	B. B. Slade	Smithfield, R. I	Jan. 27, 1838	586
Lambrequins	H. M. Johnston	New York, N. Y	Jan. 18, 1870	98, 974
Lambrequins	H. M. Johnston	New York, N. Y	July 26, 1870	105, 692
Lambrequins	H. M. Johnston	New York, N. Y	July 26, 1870	105, 693
Lambrequins, Wooden	H. Weber	Detroit, Mich	Apr. 2, 1872	125, 356
Lamp	J. Adair	Pittsburgh, Pa	July 31, 1860	29, 347
Lamp	J. Adair	Pittsburgh, Pa	Oct. 13, 1863	40, 222
Lamp	J. Adair	Pittsburgh, Pa	Jan. 10, 1865	45, 805
Lamp	S. Adam, jr., and J. R. Fogg	Portland, Me	Oct. 14, 1862	36, 692
Lamp	H. W. Adams	Brooklyn, N. Y	Sept. 6, 1859	25, 310
Lamp	F. Adt	Wolcottville, Conn	July 14, 1868	79, 887
Lamp	A. Albertson	Jersey City, N. J	Jan. 10, 1871	110, 816
Lamp	J. B. Alexander	Washington, D. C	Dec. 3, 1867	71, 566
Lamp	J. B. Alexander	Washington, D. C	Apr. 12, 1870	101, 961
Lamp	J. Allen	New York, N. Y	Jan. 14, 1868	73, 278
Lamp	J. E. Ambrose	Batavia, Ill	Oct. 16, 1860	30, 381
Lamp	J. E. Ambrose	Lena, Ill	Apr. 23, 1861	32, 110
Lamp	J. K. Andrews	Antrim, Ohio	July 11, 1865	48, 760
Lamp	J. K. Andrews	Antrim, Ohio	July 2, 1867	66, 278
Lamp	A. C. Arnold and E. Blackman	Norwalk, Conn	Feb. 11, 1868	74, 271
Lamp	J. S. and T. B. Atterbury	Pittsburgh, Pa	Apr. 7, 1863	38, 087
Lamp	J. S. and T. B. Atterbury	Pittsburgh, Pa	Sept. 29, 1868	82, 579
Lamp	J. S. and T. B. Atterbury	Pittsburgh, Pa	Apr. 26, 1870	102, 204
Lamp	J. S. and T. B. Atterbury	Pittsburgh, Pa	Nov. 22, 1870	109, 370
Lamp	T. B. Atterbury	Pittsburgh, Pa	July 22, 1873	140, 988
Lamp	L. J. Atwood	Waterbury, Conn	Sept. 16, 1862	36, 493
Lamp	L. J. Atwood	Waterbury, Conn	Oct. 13, 1863	40, 226
Lamp	L. J. Atwood	Waterbury, Conn	Oct. 13, 1863	40, 227
Lamp	L. J. Atwood	Waterbury, Conn	Mar. 1, 1864	41, 751
Lamp	L. J. Atwood	Waterbury, Conn	Aug. 1, 1865	49, 064
Lamp	L. J. Atwood	Waterbury, Conn	Jan. 21, 1868	73, 488
Lamp	L. J. Atwood	Waterbury, Conn	July 21, 1868	80, 111
Lamp	L. J. Atwood	Waterbury, Conn	June 22, 1869	91, 590
Lamp	L. J. Atwood	Waterbury, Conn	Aug. 29, 1871	118, 421
Lamp	L. J. Atwood	Waterbury, Conn	July 16, 1872	129, 509
Lamp	L. J. Atwood	Waterbury, Conn	Feb. 11, 1873	135, 749
Lamp	S. K. Ayers	Delton, Wis	Mar. 3, 1868	74, 972
Lamp	C. H. Bagley	Elgin, Ill	Aug. 14, 1866	57, 064
Lamp	L. Bailey and R. Thayer	Charlestown and Boston, Mass.	May 4, 1858	20, 134
Lamp	J. R. Baker	Kendallville, Ind	Apr. 12, 1864	42, 266
Lamp	T. J. Barron	Brooklyn, N. Y	Feb. 11, 1862	34, 390
Lamp	J. Barson, E. Daniels, and J. Farrell.	New York, N. Y	Aug. 11, 1868	80, 851
Lamp	H. J. Batchelder	Marlborough, Mass	Sept. 18, 1860	30, 104
Lamp	W. W. Batchelder	New York, N. Y	Dec. 28, 1858	22, 409
Lamp	W. W. Batchelder	New York, N. Y	Feb. 28, 1865	46, 538
Lamp	J. M. Batchelor	Foxcroft, Me	July 12, 1859	24, 711
Lamp	J. T. Beale	London, England	Nov. 29, 1838	1, 023
Lamp	G. A. Beidler	Chicago, Ill	Nov. 26, 1867	71, 268
Lamp	H. M. Beidler	Chicago, Ill	May 7, 1867	64, 474
Lamp	J. H. Beidler	Lincoln, Ill	June 26, 1866	55, 806
Lamp	J. Bellerjean	Philadelphia, Pa	Sept. 15, 1868	82, 196
Lamp	N. Benedict	Washington, D. C	Nov. 9, 1869	96, 664
Lamp	L. Berns	Middletown, N. Y	Sept. 9, 1873	142, 609
Lamp	A. Black	Barnstable, Mass	Oct. 3, 1817	
Lamp	S. G. Blackman	Waterbury, Conn	Jan. 7, 1862	34, 048
Lamp	W. S. Blaisdell and C. K. Young	Factory Point, Vt	Sept. 2, 1873	142, 432
Lamp	E. C. Blakeslee	Waterbury, Conn	Dec. 2, 1862	37, 068
Lamp	G. Blanchard	New York, N. Y	Aug. 7, 1860	29, 456
Lamp	A. Bliss	New York, N. Y	Jan. 14, 1862	34, 125
Lamp	E. Boesch	San Francisco, Cal	Dec. 26, 1871	122, 217
Lamp	W. G. A. Boneville	Dover, Del	Apr. 12, 1864	42, 269
Lamp	C. Boschan, J. Bindtner, and W. Caffon.	Vienna, Austria	July 4, 1865	48, 635
Lamp	H. H. Boucher	Doylestown, Pa	Aug. 4, 1868	80, 705

Index of patents issued from the United States Patent Office from 1790 *to* 1873, *inclusive*—Continued.

Invention.	Inventor.	Residence.	Date.	No.
Lamp	B. S. Boydston	Richmond, Ind	Sept. 29, 1868	82, 590
Lamp	O. Boynton	Hinesburgh, Vt	Jan. 21, 1868	73, 495
Lamp	S. C. Brockington	Groton, Conn	Aug. 4, 1868	80, 590
Lamp	E. Brown	Burlington, Vt	Apr. 17, 1866	53, 945
Lamp	W. Brown	Newburyport, Mass	Mar. 5, 1872	124, 249
Lamp	A. W. Browne	Brooklyn, N. Y	Jan. 28, 1868	73, 870
Lamp	G. Brownlee	Princeton, Ind	Apr. 22, 1873	138, 070
Lamp	A. Burbank	Rochester, N. Y	May 17, 1870	103, 011
Lamp	M. Burnett	South Boston, Mass	Mar. 22, 1864	41, 972
Lamp	F. Burrows	Peoria, Ill	Jan. 15, 1867	61, 152
Lamp	H. E. Burton	Boston, Mass	Nov. 26, 1867	71, 449
Lamp	C. W. Cahoon	Portland, Me	Feb. 19, 1861	31, 511
Lamp	C. W. Cahoon	Portland, Me	Dec. 3, 1861	33, 824
Lamp	C. W. Cahoon	Portland, Me	Dec. 3, 1861	33, 825
Lamp	C. W. Cahoon	Portland, Me	Sept. 2, 1862	36, 386
Lamp	C. W. Cahoon	Portland, Me	Sept. 16, 1862	36, 451
Lamp	C. W. Cahoon	Portland, Me	Oct. 13, 1863	40, 240
Lamp	C. W. Cahoon	Portland, Me	Oct. 13, 1863	40, 241
Lamp	C. W. Cahoon	Portland, Me	July 21, 1868	80, 133
Lamp	J. Calkins	New York, N. Y	Sept. 1, 1868	81, 749
Lamp	M. L. Callender	New York, N. Y	Apr. 23, 1861	32, 118
Lamp	M. L. Callender	New York, N. Y	Apr. 28, 1863	38, 284
Lamp	M. L. Callender	New York, N. Y	June 6, 1865	48, 128
Lamp	D. Challinor	Birmingham, Pa	July 26, 1870	105, 644
Lamp	O. M. Chamberlain	New York, N. Y	Oct. 11, 1870	108, 106
Lamp	L. Chandor	St. Petersburg, Russia	July 21, 1868	80, 137
Lamp	R. S. Chapin	New York, N. Y	Mar. 4, 1862	34, 562
Lamp	S. Cheney	Cleveland, Ohio	Jan. 25, 1859	22, 703
Lamp	R. Christy	Mansfield Valley, Pa	Aug. 12, 1873	141, 763
Lamp	E. Clark	Philadelphia, Pa	Apr. 27, 1814	....
Lamp	H. M. Clark	Cleveland, Ohio	Mar. 24, 1868	75, 731
Lamp	H. M. Clark	Meriden, Conn	Apr. 5, 1870	101, 584
Lamp	I. Clark	Brooklyn, N. Y	Jan. 16, 1866	52, 031
Lamp	J. M. Clark	Lancaster, Pa	July 15, 1873	140, 764
Lamp	P. J. Clark	Meriden, Conn	June 7, 1839	1, 164
Lamp	J. F. Clegg	Philadelphia, Pa	Apr. 2, 1861	31, 870
Lamp	T. Clough	Dobb's Ferry, N. Y	Nov. 26, 1867	71, 280
Lamp	T. Clough	Dobb's Ferry, N. Y	Nov. 26, 1867	71, 281
Lamp	T. Clough	Dobb's Ferry, N. Y	Oct. 27, 1868	83, 462
Lamp	G. F. J. Colburn	Newark, N. J	Apr. 14, 1863	38, 149
Lamp	M. H. Collins	Chelsea, Mass	Sept. 19, 1865	49, 984
Lamp	M. H. Collins	Chelsea, Mass	Feb. 4, 1868	74, 049
Lamp	A. Combs	Burlingame, Kans	Apr. 23, 1872	126, 025
Lamp	C. J. Conway	New York, N. Y	July 19, 1853	9, 857
Lamp	R. Cornelius	Philadelphia, Pa	Dec. 9, 1862	37, 086
Lamp	E. T. Covell	New Bedford, Mass	Feb. 18, 1862	34, 408
Lamp	N. Cradit	Ripley, Ohio	Dec. 14, 1858	22, 327
Lamp	A. Crook	New York, N. Y	May 25, 1869	90, 428
Lamp	E. E. Dailey, W. H. Johnson, and C. C. Du Bois.	Brooklyn, E. D., N. Y	Feb. 18, 1868	74, 670
Lamp	J. Davidson	Baltimore, Md	July 2, 1846	4, 617
Lamp	J. Davis, jr	Templeton, Mass	Apr. 23, 1867	63, 999
Lamp	C. T. Day	Newark, N. J	Oct. 20, 1863	40, 334
Lamp	F. B. De Keravenan	New York, N. Y	Oct. 23, 1860	30, 466
Lamp	J. M. A. Dew	Chicago, Ill	Aug. 31, 1869	94, 295
Lamp	M. A. Dietz	Brooklyn, N. Y	Mar. 8, 1859	23, 160
Lamp	M. A. Dietz	Brooklyn, N. Y	May 3, 1859	23, 832
Lamp	R. E. ietz	New York, N. Y	Oct. 28, 1873	144, 070
Lamp	J. Dodin	New York, N. Y	Feb. 18, 1862	34, 415
Lamp	C. H. Dolbeare	Boston, Mass	Jan. 1, 1861	31, 012
Lamp	H. W. Dopp	Buffalo, N. Y	Oct. 16, 1860	30, 444
Lamp	J. L. Drake	Cincinnati, Ohio	May 17, 1859	24, 015
Lamp	J. L. Drake	Cincinnati, Ohio	Sept. 20, 1859	25, 493
Lamp	L. F. Drake and E. Egginton	Portland, Me	Dec. 17, 1867	72, 376
Lamp	J. P. Driver	Marengo, Ind	Aug. 29, 1865	49, 613
Lamp	J. Duke	New York, N. Y	Aug. 15, 1821	....
Lamp	J. Dunn	New York, N. Y	Oct. 18, 1870	108, 339
Lamp	M. B. Dyott	Philadelphia, Pa	Sept. 29, 1863	40, 094
Lamp	M. B. Dyott	Philadelphia, Pa	May 29, 1866	55, 075
Lamp	M. B. Dyott	Philadelphia, Pa	May 7, 1867	64, 508
Lamp	R. N. Eagle	Washington, D. C	Sept. 6, 1864	44, 081
Lamp	O. C. Evans	New York, N. Y	Jan. 22, 1861	31, 160
Lamp	H. B. Fernald	Boston, Mass	May 17, 1844	3, 591
Lamp	J. S. Fish	Cleveland, Ohio	Apr. 26, 1870	102, 240
Lamp	F. A. Flanegin	Fagundus City, Pa	Mar. 4, 1873	136, 427
Lamp	J. C. Fletcher	Springfield, Ohio	July 12, 1838	834
Lamp	A. L. Fleury	Baltimore, Md	July 5, 1859	24, 622
Lamp	W. Foster, jr., and L. Foster	Boston, Mass	Aug. 20, 1813	....
Lamp	S. W. Fowler	Brooklyn, N. Y	Jan. 21, 1868	73, 443
Lamp	S. W. Fowler	Brooklyn, N. Y	Apr. 12, 1870	101, 723
Lamp	W. Freeman	Mount Carmel, Conn	May 21, 1861	32, 358
Lamp	J. A. Frey	Washington, D. C	Jan. 28, 1868	73, 710
Lamp	G. P. Fuller	Humphrey, N. Y	July 5, 1870	105, 063
Lamp	J. B. Fuller	Norwich, Conn	June 15, 1869	91, 325
Lamp	J. B. Fuller	Norwich, Conn	May 24, 1870	103, 442
Lamp	W. Fulton	Cranberry, N. J	Aug. 3, 1858	21, 069
Lamp	W. Fulton	Cranberry, N. J	July 24, 1860	29, 260
Lamp	J. Funck	Tompkinsville, N. Y	Aug. 17, 1869	93, 871
Lamp	B. Garvey	New York, N. Y	Jan. 20, 1863	37, 442
Lamp	J. Gibbs	Brooklyn, N. Y	Dec. 1, 1868	84, 623
Lamp	D. Gilbert	Brattleborough, Vt	Mar. 29, 1834	....
Lamp	H. J. Goff	Dubuque, Iowa	Mar. 14, 1871	112, 586
Lamp	S. J. Gold	Cornwall, Conn	May 8, 1843	3, 073
Lamp	C. Goodwin	Chicago, Ill	Apr. 26, 1864	42, 534
Lamp	R. A. Goodyear	New Haven, Conn	June 5, 1866	55, 280
Lamp	W. J. Gordon and M. W. House	Cleveland, Ohio	July 23, 1872	129, 728
Lamp	W. J. Gordon and M. W. House	Cleveland, Ohio	Oct. 29, 1872	132, 655
Lamp	E. H. Green	Baltimore, Md	Sept. 5, 1865	49, 752

Index of patents issued from the United States Patent Office from 1790 *to* 1873, *inclusive*—Continued.

Invention.	Inventor.	Residence.	Date.	No.
Lamp	B. F. Greenough	Boston, Mass	Apr. 10, 1841	2,039
Lamp	F. T. Grimes	Liberty, Mo	Jan. 3, 1871	110,648
Lamp	F. T. Grimes	Liberty, Mo	Feb. 7, 1871	111,531
Lamp	F. T. Grimes	Liberty, Mo	Oct. 10, 1871	119,842
Lamp	P. C. Guion and P. K. Wombaugh.	Cincinnati, Ohio	Aug. 26, 1856	15,636
Lamp	S. Guthrie	New Orleans, La	Feb. 14, 1860	27,124
Lamp	J. S. Gwynn	Plainfield, N. J	Aug. 27, 1861	33,140
Lamp	M. P. Hadley	Bluffton, Wis	Oct. 15, 1872	132,283
Lamp	E. J. Hale	Foxcroft, Me	June 12, 1860	28,666
Lamp	E. J. Hale	Foxcroft, Me	June 19, 1860	28,749
Lamp	E. J. Hale	Foxcroft, Me	June 26, 1860	28,855
Lamp	E. J. Hale and C. H. Chandler	Foxcroft, Me	Mar. 1, 1859	23,085
Lamp	H. Halvorson	Cambridge, Mass	Sept. 20, 1859	25,506
Lamp	H. L. Hanson	Portland, Me	Sept. 24, 1867	69,091
Lamp	E. Harris	Boston, Mass	Mar. 20, 1855	12,550
Lamp	J. Harris, jr	Boston, Mass	Aug. 15, 1854	11,524
Lamp	A. Harroun, jr	Onondaga, N. Y	Feb. 20, 1866	52,711
Lamp	H. Hasseupflug	Huntingdon, Pa	Aug. 20, 1861	33,084
Lamp	H. W. Hayden	Waterbury, Conn	Apr. 14, 1863	38,162
Lamp	H. W. Hayden	Waterbury, Conn	May 9, 1865	47,680
Lamp	H. W. Hayden	Waterbury, Conn	Jan. 21, 1868	73,599
Lamp	H. W. Hayden	Waterbury, Conn	Jan. 21, 1868	73,600
Lamp	H. W. Hayden	Waterbury, Conn	Dec. 14, 1869	97,773
Lamp	H. W. Hayden	Waterbury, Conn	July 23, 1872	129,821
Lamp	P. Hayden	Pittsburgh, Pa	Oct. 14, 1862	36,649
Lamp	E. K. Haynes	Boston, Mass	June 29, 1869	91,933
Lamp	B. F. Hebard	Dorchester, Mass	Aug. 23, 1864	43,911
Lamp	F. Heidrick	Philadelphia, Pa	Apr. 2, 1861	31,887
Lamp	F. Heidrick	Philadelphia, Pa	May 7, 1861	32,248
Lamp	A. A. Henderson	Buffalo, N. Y	Feb. 17, 1863	37,690
Lamp	J. E. Hendricks	Waterbury, Conn	Aug. 25, 1868	81,370
Lamp	A. Hicks	Factoryville, N. Y	Apr. 25, 1865	47,493
Lamp	J. Higgins	Cambridge, Mass	Feb. 23, 1864	41,702
Lamp	S. A. Hill and D. Alter	Freeport, Pa	Apr. 12, 1859	23,579
Lamp	C. F. A. Hinrichs	New York, N. Y	Dec. 27, 1870	110,464
Lamp	R. Hitchcock	Watertown, N. Y	Jan. 7, 1873	134,547
Lamp	R. Hitchcock	Watertown, N. Y	Aug. 26, 1873	142,103
Lamp	R. Hitchcock	Watertown, N. Y	Dec. 2, 1873	145,176
Lamp	J. H. Hobbs	Wheeling, W. Va	May 24, 1870	103,460
Lamp	B. J. Hoffacker	New York, N. Y	July 17, 1866	56,497
Lamp	P. Hoffmann	Constableville, N. Y	Oct. 1, 1867	69,436
Lamp	E. B. Horn	Boston, Mass	Jan. 8, 1842	2,413
Lamp	J. Horton	New York, N. Y	Dec. 28, 1869	98,264
Lamp	T. Houghton	Philadelphia, Pa	Jan. 10, 1860	26,767
Lamp	M. W. House	Cleveland, Ohio	Apr. 14, 1868	76,764
Lamp	M. W. House	Cleveland, Ohio	Dec. 5, 1871	121,521
Lamp	D. Howarth	Portland, Me	June 19, 1866	55,771
Lamp	J. J. Hoyt	Chelmsford, Mass	July 23, 1872	129,828
Lamp	D. Hughes	Rochester, N. Y	Dec. 17, 1861	33,945
Lamp	J. S. Hull	Cincinnati, Ohio	May 14, 1867	64,768
Lamp	J. S. Hull	Cincinnati, Ohio	Mar. 19, 1872	124,823
Lamp	W. Hunt	New York, N. Y	May 21, 1861	32,402
Lamp	T. A. Hunter and J. Blewitt	New York, N. Y	Nov. 3, 1868	83,711
Lamp	H. C. Hutchinson	Cayuga, N. Y	Apr. 15, 1862	34,964
Lamp	H. C. Hutchinson	Cayuga, N. Y	Jan. 3, 1865	45,719
Lamp	J. Ingersoll	Cleveland, Ohio	Feb. 18, 1868	74,542
Lamp	J. H. Irwin	Beardstown, Ill	Oct. 29, 1861	33,588
Lamp	J. H. Irwin	Chicago, Ill	May 28, 1867	65,230
Lamp	J. H. Irwin	Philadelphia, Pa	May 6, 1873	138,653
Lamp	J. H. Irwin	Philadelphia, Pa	May 6, 1873	138,654
Lamp	J. Ives	Mount Carmel, Ill	Feb. 21, 1865	46,471
Lamp	J. Ives	Mount Carmel, Conn	July 18, 1865	48,816
Lamp	W. W. Jacobs	Hagerstown, Md	Dec. 22, 1868	85,099
Lamp	R. Jenkins	Covington, Ky	Jan. 25, 1859	22,729
Lamp	I. Jennings	New York, N. Y	Sept. 22, 1836	31
Lamp	I. Jennings	New York, N. Y	Jan. 19, 1847	4,935
Lamp	J. C. Jennison and A. Hale	Foxcroft, Me	June 12, 1860	28,717
Lamp	M. Jincks	Dansville, N. Y	Nov. 12, 1867	70,856
Lamp	E. F. Jones	Boston, Mass	May 4, 1858	20,159
Lamp	G. A. Jones	New York, N. Y	Nov. 10, 1863	40,566
Lamp	A. Judson	Brooklyn, N. Y	Apr. 9, 1861	31,983
Lamp	A. Judson	Brooklyn, N. Y	Oct. 15, 1867	69,815
Lamp	A. Judson	Brooklyn, N. Y	Sept. 1, 1868	81,645
Lamp	A. Kaestner	New York, N. Y	Nov. 5, 1861	33,649
Lamp	F. Kampfe	New York, N. Y	Oct. 7, 1873	143,452
Lamp	J. Keim	Philadelphia, Pa	Mar. 29, 1834	....
Lamp	E. G. Kelley	New York, N. Y	Aug. 17, 1869	93,719
Lamp	S. J. Kelly	Pemberton, N. J	Feb. 3, 1863	37,581
Lamp	J. P. and E. Kenyon	Brooklyn, N. Y	Aug. 31, 1858	21,344
Lamp	M. R. Kenyon	Providence, R. I	Apr. 26, 1864	42,492
Lamp	W. M. Kimball	Rochester, N. Y	Feb. 26, 1856	14,317
Lamp	J. Kirby, jr	Adrian, Mich	Aug. 19, 1873	141,933
Lamp	C. A. Kleeman	Erfurt, Prussia	Mar. 10, 1863	37,867
Lamp	A. Kleinsteiber	Milwaukee, Wis	Oct. 9, 1860	30,377
Lamp	A. H. Knapp	Newton Centre, Mass	Nov. 8, 1859	26,071
Lamp	H. Knowles	New London, Conn	Jan. 25, 1859	22,771
Lamp	W. Kuebler and H. Beierlein	Philadelphia, Pa	Sept. 9, 1862	86,413
Lamp	E. L. Lambie	Washington, D. C	Mar. 26, 1872	125,058
Lamp	E. M. Lang and J. Gilman	Westbrook and Portland, Me.	May 9, 1865	47,681
Lamp	C. B. Lashar	New York, N. Y	Apr. 14, 1863	38,170
Lamp	W. Lassell	Boston, Mass	Sept. 5, 1865	49,769
Lamp	A. B. Latta	Cincinnati, Ohio	July 7, 1863	39,154
Lamp	H. H. Laughlin	Philadelphia, Pa	May 30, 1871	115,329
Lamp	F. Leclair	New York, N. Y	Dec. 8, 1857	18,818
Lamp	J. Lee	New York, N. Y	Nov. 12, 1867	70,867
Lamp	J. K. Leedy	Woodstock, Va	Jan. 24, 1860	26,910

Index of patents issued from the United States Patent Office from 1790 *to* 1873, *inclusive*—Continued.

Invention.	Inventor.	Residence.	Date.	No.
Lamp	H. Leibert	Norristown, Pa	Jan. 1, 1861	31, 024
Lamp	J. E. Leighton	Lowell, Mass	May 6, 1873	138, 509
Lamp	W. Lewis	Boston, Mass	Jan. 23, 1818	....
Lamp	J. Lindsley	Pawtucket, R. I	Aug. 27, 1872	130, 924
Lamp	E. A. Locke and W. N. Weeden	Boston, Mass	Jan. 21, 1868	73, 538
Lamp	L. Loeffler	East Cambridge, Mass	Aug. 11, 1863	39, 488
Lamp	G. H. Lomax	Somerville, Mass	Sept. 20, 1870	107, 514
Lamp	J. R. Loomis	Winsted, Conn	Sept. 6, 1859	25, 342
Lamp	H. Long	Kittaning, Pa	Sept. 21, 1869	95, 027
Lamp	D. Lubin	New York, N. Y	June 4, 1872	127, 621
Lamp	C. E. Lyon	Worcester, Mass	Apr. 12, 1870	101, 748
Lamp	A. M. Mace	Needham, Mass	July 30, 1872	129, 973
Lamp	C. D. Macqueen	Philadelphia, Pa	Jan. 31, 1871	111, 360
Lamp	J. Mallory	New York, N. Y	Nov. 20, 1812	....
Lamp	L. Mangeon	New York, N. Y	Apr. 13, 1869	88, 968
Lamp	C. B. and S. S. Mann	Baltimore, Md	May 16, 1871	114, 954
Lamp	G. Marlow and M. Ralphe	Cincinnati, Ohio	Aug. 30, 1859	25, 304
Lamp	E. Marsh	Newark, N. J	Jan. 24, 1865	46, 053
Lamp	R. Marsh	Flushing, N. Y	June 17, 1873	139, 964
Lamp	S. Marshall	Wilmington, Del	Aug. 19, 1862	36, 226
Lamp	W. M. Marshall	Philadelphia, Pa	Nov. 19, 1872	133, 107
Lamp	A. J. Martin	Catskill, N. Y	Nov. 12, 1872	132, 972
Lamp	C. F. Martine	Dorchester, Mass	May 20, 1862	35, 349
Lamp	C. F. Martine	Dorchester, Mass	Sept. 19, 1865	50, 019
Lamp	C. Mayer	Baltimore, Md	May 13, 1873	138, 812
Lamp	T. Mayhew	Poughkeepsie, N. Y	Jan. 7, 1862	34, 065
Lamp	F. McDaniels	Philadelphia, Pa	Sept. 13, 1870	107, 276
Lamp	S. T. McDougall	Brooklyn, N. Y	Feb. 18, 1868	74, 563
Lamp	J. K. Mentzer	New Holland, Pa	Dec. 21, 1869	98, 083
Lamp	W. Mentzel and A. and J. W. Geddes.	Baltimore, Md	Mar. 11, 1862	34, 638
Lamp	R. S. Merrill	Lynn, Mass	June 14, 1859	24, 397
Lamp	R. S. Merrill	Boston, Mass	May 22, 1866	54, 935
Lamp	R. S. Merrill	Boston, Mass	Aug. 7, 1866	56, 974
Lamp	R. S. Merrill	Hyde Park, Mass	June 21, 1870	104, 481
Lamp	R. S. Merrill	Hyde Park, Mass	June 21, 1870	104, 623
Lamp	R. S. Merrill	Boston, Mass	Sept. 13, 1870	107, 277
Lamp	R. S. Merrill	Boston, Mass	Sept. 13, 1870	107, 278
Lamp	R. S. Merrill	Boston, Mass	Jan. 17, 1871	111, 072
Lamp	R. S. Merrill and W. Carleton	Boston, Mass	Jan. 28, 1868	73, 912
Lamp	L. Michaels	Covington, Ky	Mar. 18, 1873	136, 849
Lamp	C. Miller	Saint Louis, Mo	June 12, 1860	28, 683
Lamp	J. J. Miller	Chicago, Ill	Mar. 10, 1863	37, 887
Lamp	J. J. Miller	Chicago, Ill	June 30, 1863	39, 101
Lamp	G. W. Mitchell	Saint Louis, Mo	Jan. 2, 1866	51, 852
Lamp	D. A. Moore	Syracuse, N. Y	Dec. 10, 1861	33, 898
Lamp	S. C. Moore	Boston, Mass	Nov. 10, 1868	83, 874
Lamp	L. E. C. Moore and J. S. Hamilton.	Pittston, Pa	Apr. 12, 1870	101, 902
Lamp	W. Morehouse	Buffalo, N. Y	Nov. 19, 1861	33, 749
Lamp	S. G. Morrison	Williamsport, Pa	Jan. 6, 1863	37, 301
Lamp	T. H. Mott	New York, N. Y	Apr. 23, 1872	126, 075
Lamp	C. Mulchahey	Springfield, Mass	Oct. 2, 1866	58, 457
Lamp	W. Mulholland	Brooklyn, N. Y	Sept. 28, 1858	21, 617
Lamp	W. Mullally	Boston, Mass	Oct. 29, 1867	70, 247
Lamp	W. Mullally	Boston, Mass	Feb. 11, 1868	74, 403
Lamp	J. Mulvany	New York, N. Y	Nov. 26, 1861	33, 791
Lamp	G. Neilson	Boston, Mass	Jan. 24, 1860	26, 952
Lamp	G. Neilson	Boston, Mass	Sept. 18, 1860	30, 082
Lamp	G. Neilson	Boston, Mass	Dec. 17, 1867	72, 219
Lamp	G. Neilson	Boston, Mass	Jan. 14, 1868	73, 372
Lamp	C. Newman	Pittsburgh, Pa	Aug. 13, 1861	33, 047
Lamp	A. H. North	Hartford, Conn	Apr. 5, 1859	23, 483
Lamp	E. D. Norton	Bradford, Pa	June 18, 1867	65, 935
Lamp	J. E. Noyes	New Albany, Ind	Sept. 15, 1868	82, 150
Lamp	P. Noyes	Lowell, Mass	July 28, 1868	80, 302
Lamp	R. Nutting	Randolph, Vt	Jan 24, 1871	111, 237
Lamp	A. Odel and W. A. Burrows	New York, N. Y	Oct. 8, 1861	33, 447
Lamp	W. J. Palmer	Flushing, N. Y	Feb. 25, 1862	34, 521
Lamp	J. M. Parker	La Grange, Mo	Feb. 6, 1872	123, 415
Lamp	G. T. Parkhurst	Baltimore, Md	Sept. 13, 1859	25, 438
Lamp	W. P. Patton	Harrisburgh, Pa	Aug. 20, 1861	33, 102
Lamp	J. A. Pease	Boston, Mass	Apr. 22, 1873	138, 185
Lamp	J. A. Pease	Boston, Mass	Oct. 28, 1873	144, 130
Lamp	O. N. Perkins	Meriden, Conn	Aug. 13, 1872	130, 530
Lamp	O. N. Perkins	Meriden, Conn	Oct. 28, 1873	144, 132
Lamp	J. M. Perkins and M. W. House	Cleveland, Ohio	Mar. 14, 1865	46, 819
Lamp	J. M. Perkins and M. W. House	Cleveland, Ohio	Dec. 11, 1866	60, 416
Lamp	J. Phelps	Owego, N. Y	Aug. 3, 1869	93, 224
Lamp	D. L. Pickard	Rochester, N. Y	Jan. 30, 1866	52, 316
Lamp	W. H. Pierce	Somerville, Mass	May 5, 1863	38, 453
Lamp	W. Pitt	Ithaca, N. Y	July 23, 1861	32, 883
Lamp	P. Plant	Washington, D. C	Apr. 6, 1858	19, 896
Lamp	A. H. Platt and W. S. Rosecrans	Cincinnati, Ohio	Oct. 1, 1861	33, 402
Lamp	W. Porter	Belleville Township, N. J	July 23, 1867	67, 070
Lamp	S. S. and A. J. Post	Jersey City, N. J	July 16, 1861	32, 851
Lamp	G. Pugh	Cleveland, Ohio	June 15, 1869	91, 263
Lamp	R. B. Pullam	Cincinnati, Ohio	Dec. 24, 1861	34, 029
Lamp	W. H. Racey	Saint Augustine, Fla	Sept. 28, 1858	21, 627
Lamp	W. H. Racey	Saint Augustine, Fla	Oct. 30, 1860	30, 546
Lamp	M. Rae	Manchester, England	Aug. 12, 1862	36, 167
Lamp	T. Raymond	Franklinville, N. Y	Apr. 8, 1862	34, 904
Lamp	O. Redmond	Rochester, N. Y	July 26, 1853	9, 878
Lamp	C. Reichmand	Philadelphia, Pa	Sept. 21, 1858	21, 576
Lamp	E. B. Requa	Jersey City, N. J	May 6, 1862	35, 175
Lamp	F. Rhind	Brooklyn, N. Y	Sept. 9, 1873	142, 730
Lamp	F. C. Richer	Gilmer, Tex	Aug. 3, 1869	93, 228
Lamp	C. and C. Richman	Philadelphia, Pa	June 11, 1841	2, 129

Index of patents issued from the United States Patent Office from 1790 *to* 1873, *inclusive*—Continued.

Invention.	Inventor.	Residence.	Date.	No.
Lamp	C. W. Richter, sr	Madison, Ga	Feb. 22, 1859	23, 044
Lamp	C. W. Richter	Madison, Ga	Nov. 29, 1859	26, 295
Lamp	F. C. Rider	Providence, R. I	Mar. 13, 1855	12, 520
Lamp	K. Riedel	Guttenberg, N. J	Oct. 2, 1866	58, 481
Lamp	D. C. Ripley	Pittsburgh, Pa	Mar. 2, 1869	87, 367
Lamp	A. J. Ritter	Rahway, N. J	May 31, 1864	42, 998
Lamp	C. H. Robinson	Boston, Mass	July 22, 1862	35, 955
Lamp	S. Rodman	New Bedford, Mass	Oct. 17, 1848	5, 870
Lamp	E. F. Rogers	Boston, Mass	Jan. 30, 1866	52, 369
Lamp	E. D. Rosencrantz	Cincinnati, Ohio	Dec. 24, 1861	34, 030
Lamp	J. Russell	Troy, N. Y	July 2, 1861	32, 721
Lamp	J. F. Russell	Washington, D. C	Feb. 1, 1870	99, 482
Lamp	S. Russell	Waterbury, Conn	July 16, 1872	129, 596
Lamp	S. Russell	Waterbury, Conn	July 16, 1872	129, 597
Lamp	C. W. Russell and N. Clifford	New York, N. Y	Dec. 24, 1867	72, 683
Lamp	S. Rust	New York, N. Y	Apr. 20, 1837	176
Lamp	S. Rust	New York, N. Y	June 30, 1837	246
Lamp	S. Rust	New York, N. Y	Sept. 25, 1837	406
Lamp	S. Rust	New York, N. Y	Jan. 9, 1838	564
Lamp	S. Rust	New York, N. Y	June 7, 1838	772
Lamp	S. Rust	New York, N. Y	Sept. 14, 1843	3, 260
Lamp	L. L. Sagendorph	Boston, Mass	Mar. 10, 1868	75, 467
Lamp	M. Samuels	San Francisco, Cal	May 30, 1871	115, 528
Lamp	J. F. Sanford	Keokuk, Iowa	Feb. 9, 1869	86, 867
Lamp	J. F. Sanford	Keokuk, Iowa	May 4, 1869	89, 600
Lamp	H. Sangster	Buffalo, N. Y	Aug. 30, 1864	44, 017
Lamp	J. Sangster	Buffalo, N. Y	Mar. 25, 1862	34, 782
Lamp	W. H. Sangster	Chicago, Ill	Nov. 27, 1866	60, 068
Lamp	S. Sargent	Watertown, Mass	July 24, 1860	29, 321
Lamp	O. Sanlay	New Orleans, La	May 8, 1860	28, 199
Lamp	W. Scarlett	Aurora, Ill	May 9, 1871	114, 609
Lamp	W. Scarlett	Aurora, Ill	Nov. 14, 1871	121, 008
Lamp	J. Schelly	Savannah, Ga	July 24, 1860	29, 324
Lamp	B. B. Schneider	New York, N. Y	Apr. 19, 1870	102, 163
Lamp	H. A. Schottky and T. Simendinger.	New York, N. Y	Jan. 31, 1871	111, 481
Lamp	C. Seidhof	Lancaster, Mass	July 27, 1852	9, 157
Lamp	I. W. Shaler	Brooklyn, N. Y	Oct. 10, 1871	119, 887
Lamp	I. W. Shaler	Brooklyn, N. Y	Feb. 20, 1872	123, 940
Lamp	W. F. Shaw	Boston, Mass	Jan. 4, 1859	22, 516
Lamp	S. Shea and E. W. Gillman	Long Island City, N. Y	Mar. 17, 1868	75, 587
Lamp	J. W. Shehan	San Francisco, Cal	July 13, 1869	92, 661
Lamp	W. F. Shaw	Boston, Mass	Dec. 31, 1842	2, 893
Lamp	J. W. Shulz and J. Trull	Medford, Mass	Mar. 19, 1831	....
Lamp	A. M. Silber and F. White	London, England	May 21, 1872	127, 108
Lamp	G. H. Simmons	Bennington, Vt	Mar. 4, 1873	136, 553
Lamp	J. B. Slawson	New Orleans, La	Nov. 5, 1867	70, 635
Lamp	F. V. Sleeth	Keokuk, Iowa	Apr. 19, 1870	102, 168
Lamp	E. F. Slocum	Chicago, Ill	June 11, 1861	32, 537
Lamp	A. G. Smith	Jersey City, N. J	Aug. 28, 1866	57, 587
Lamp	J. H. Smith	Brewster Station, N. Y	Apr. 14, 1868	76, 836
Lamp	R. H. Smith	Pittsburgh, Pa	Nov. 16, 1869	96, 986
Lamp	S. P. Smith	Waterford, N. Y	May 4, 1869	89, 696
Lamp	T. B. Smith	Marietta, Ohio	Aug. 28, 1860	29, 831
Lamp	W. H. Smith	New York, N. Y	Jan. 19, 1869	8, 118
Lamp	W. H. Smith	New York, N. Y	Mar. 10, 1868	75, 479
Lamp	D. M. Smyth	Orange, N. J	July 14, 1868	79, 924
Lamp	O. and H. S. Snow	West Meriden, Conn	June 12, 1860	28, 695
Lamp	T. E. Sparks	Norwich, Conn	Aug. 28, 1866	57, 589
Lamp	C. F. Spencer	Rochester, N. Y	Jan. 21, 1868	73, 663
Lamp	S. Spoor	Phelps, N. Y	Jan. 21, 1868	73, 664
Lamp	W. Staehlen	Williamsburgh, N. Y	Jan. 16, 1862	122, 864
Lamp	W. Staehlen	Brooklyn, N. Y	Apr. 16, 1872	125, 852
Lamp	C. C. Stansell	Middleborough, Mass	June 17, 1862	35, 655
Lamp	C. C. Stansell	Middleborough, Mass	July 22, 1862	35, 979
Lamp	W. H. Starr	New York, N. Y	June 6, 1846	4, 561
Lamp	R. Steinmann	Boston, Mass	Apr. 6, 1858	19, 898
Lamp	W. G. Sterling	Bridgeport, Conn	Mar. 17, 1863	37, 928
Lamp	W. G. Sterling	Bridgeport, Conn	Oct. 20, 1863	40, 368
Lamp	L. Sterling and T. W. Willson	New York, N. Y	June 14, 1870	104, 372
Lamp	C. St. John	Charlestown, Mass	Nov. 19, 1867	71, 242
Lamp	G. Stobwasser	Berlin, Prussia	Sept. 25, 1866	58, 361
Lamp	J. Street	Philadelphia, Pa	May 2, 1846	4, 489
Lamp	T. F. Strong	London, England	July 7, 1846	4, 621
Lamp	T. F. Strong	New York, N. Y	July 20, 1846	4, 639
Lamp	J. Stuber	Utica, N. Y	May 1, 1860	28, 132
Lamp	J. Stuber and F. Frank	Utica, N. Y	Apr. 23, 1861	32, 155
Lamp	J. Stuber and T. Harden	Utica, N. Y	Oct. 30, 1855	13, 723
Lamp	J. Stuber and R. Hughes	Utica, N. Y	Feb. 2, 1858	19, 266
Lamp	J. Sutton	New York, N. Y	Apr. 11, 1848	5, 506
Lamp	Z. Swope	Lancaster, Pa	Mar. 13, 1860	27, 500
Lamp	D. Symonds	Lowell, Mass	Mar. 7, 1865	46, 730
Lamp	I. W. Taber	New Bedford, Mass	July 17, 1860	29, 203
Lamp	A. Taplin	Providence, R. I	Oct. 14, 1862	36, 680
Lamp	A. Taplin	Providence, R. I	Mar. 1, 1864	41, 794
Lamp	A. Taplin	Somerville, Mass	Feb. 25, 1868	74, 863
Lamp	J. Thomas	New York, N. Y	Apr. 23, 1861	32, 156
Lamp	J. Thomas	New York, N. Y	July 23, 1861	32, 906
Lamp	W. S. Thompson	Rochester, N. Y	Jan. 7, 1862	34, 080
Lamp	G. W. Thomson	Buffalo, N. Y	July 26, 1870	105, 867
Lamp	G. W. Thompson	New York, N. Y	Oct. 24, 1871	120, 172
Lamp	H. Tilden	Boston, Mass	Jan. 14, 1868	73, 409
Lamp	W. H. Topham	New Bedford, Mass	Feb. 19, 1861	31, 496
Lamp	W. Totheroh	Reading, Pa	July 28, 1868	80, 519
Lamp	G. A. Tremeschini	Vicenza, Austria	Jan. 24, 1865	46, 059
Lamp	C. Tribby	Washington, D. C	Feb. 13, 1872	123, 749
Lamp	E. Tritten	Phila elphia, Pa	Jan. 29, 1861	31, 271
Lamp	E. Tritten	Philadelphia, Pa	Dec. 3, 1861	33, 259

Index of patents issued from the United States Patent Office from 1790 *to* 1873, *inclusive*—Continued.

Invention.	Inventor.	Residence.	Date.	No.
Lamp	A. Turnbull	West Meriden, Conn	Jan. 28, 1862	34, 277
Lamp	A. P. Tyler	Cleveland, Ohio	Dec. 31, 1867	72, 764
Lamp	C. N. Tyler	Buffalo, N. Y	Mar. 27, 1866	53, 506
Lamp	C. N. Tyler	Buffalo, N. Y	Nov. 19, 1867	71, 248
Lamp	I. Van Bunschoten	New York, N. Y	Nov. 21, 1854	11, 979
Lamp	J. T. Van Kirk	Philadelphia, Pa	July 17, 1860	29, 221
Lamp	J. T. Vankirk and W. M. Fulton.	Frankfort, Pa., and Cranberry N. J.	Nov. 29, 1859	26, 310
Lamp	U. B. Vidal	Philadelphia, Pa	July 30, 1861	32, 930
Lamp	C. Von Bonhorst	Hancock, Md	Feb. 21, 1860	27, 248
Lamp	A. V. H. Webb	New York, N. Y	Feb. 19, 1839	1, 083
Lamp	W. Webb	Waterbury, Conn	Apr. 5, 1864	42, 262
Lamp	C. Weber and H. Rimann	West Meriden, Conn	Sept. 22, 1868	82, 458
Lamp	H. B. Wellman	Allegheny City, Pa	Jan. 7, 1868	73, 145
Lamp	H. Weston	Towanda, Pa	Aug. 27, 1867	68, 136
Lamp	J. C. Wharton	Nashville, Tenn	Sept. 2, 1873	142, 425
Lamp	L. White	Hartford, Conn	Sept. 13, 1859	25, 475
Lamp	T. H. White and E. Knight	Cleveland, Ohio	Mar. 4, 1873	136, 573
Lamp	A. Whitlock	Danbury, Conn	Sept. 8, 1868	82, 051
Lamp	C. Wilhelm	Philadelphia, Pa	Feb. 6, 1866	52, 480
Lamp	J. T. Williams	York Borough, Pa	Aug. 28, 1860	29, 838
Lamp	T. S. Williams	Boston, Mass	Dec. 27, 1870	110, 526
Lamp	J. S. Wood	Plainfield, N. J	Apr. 27, 1869	89, 536
Lamp	O. D. Woodbury	Derby, Conn	Mar. 6, 1866	53, 089
Lamp	L. J. Worden and A. Leach	Utica, N. Y	July 23, 1861	32, 918
Lamp	J. C. Wright	Minersville, Pa	July 16, 1861	32, 847
Lamp	F. Yeiser and S. W. Hedger	Lancaster, Ky	May 23, 1871	115, 262
Lamp	H. Young	Cincinnati, Ohio	Nov. 26, 1867	71, 354
Lamp	C. Yaiser	Newark, N. J	June 8, 1869	91, 195
Lamp, Advertising	F. R. Warner	Paris, France	Aug. 19, 1873	141, 971
Lamp, Alcohol and turpentine	I. Jennings	New York, N. Y	Sept. 22, 1836	29
Lamp and blow-pipe, Combined soldering	A. C. Frieseke	Owasso, Mich	Aug. 22, 1871	118, 229
Lamp and burner	I. W. Shaler	Brooklyn, N. Y	Feb. 27, 1872	124, 017
Lamp and burning ingredients for	I. Jennings	New York, N. Y	Mar. 3, 1829	....
Lamp and candle stand and holder	G. Uttley	Chapel Hill, N. C	Aug. 7, 1866	57, 017
Lamp and candle wick	S. R. Weeden	Providence, R. I	Jan. 1, 1861	31, 045
Lamp and gas shade holder	C. W. Emerson	Hartford, Conn	Nov. 19, 1867	71, 150
Lamp and gas stove, Coal-oil	J. E. Ambrose	Middletown, N. Y	Jan. 24, 1865	46, 045
Lamp and lantern	W. Harvie	Glasgow, North Britain	Sept. 13, 1870	107, 366
Lamp and lantern, Combination of chamber	C. H. Peters	Cincinnati, Ohio	May 12, 1863	38, 500
Lamp and lantern, Street	J. N. Aronson	New York, N. Y	May 9, 1871	114, 513
Lamp and liquid to be burned therein	J. Hawkins	Baltimore, Md	Apr. 21, 1810	....
Lamp and producing light	I. Jennings	New York, N. Y	Sept. 10, 1829	....
Lamp and reflector combined	H. J. Goff	Dubuque, Iowa	Apr. 4, 1871	113, 650
Lamp and reflector, Light-house	A. Folger	Nantucket, Mass	Mar. 28, 1842	2, 520
Lamp and stove, Combined	C. B. Guy	Lybrand, Iowa	July 11, 1865	48, 678
Lamp and torch	J. Martin	Boston, Mass	June 30, 1837	248
Lamp, Argand	L. F. Gallup	Woodstock, Vt	Mar. 9, 1831	....
Lamp, Argand	H. W. Hayden	Waterbury, Conn	Nov. 5, 1872	132, 831
Lamp, Argand	B. Hemmenway	Roxbury, Mass	Jan. 20, 1841	1, 934
Lamp, Argand	C. F. A. Hinrichs and R. Knopp	Brooklyn and New York, N. Y.	Dec. 24, 1872	134, 281
Lamp, Argand	A. B. Howland	Titusville, Pa	Sept. 21, 1869	94, 961
Lamp, Argand	D. Melville	Newport, R. I	Nov. 19, 1818	....
Lamp, Argand	J. L. Tough	Baltimore, Md	May 11, 1841	2, 091
Lamp, Argand	J. G. Webb	New York, N. Y	Oct. 9, 1855	13, 674
Lamp-attachment	W. W. Wade and C. Burnham	Long Meadow and Springfield, Mass.	Apr. 6, 1858	19, 885
Lamp-attachment for preventing smoke, &c	R. Thomas	Hoboken, N. J	May 4, 1858	20, 178
Lamp baker, Coal-oil	W. B. Billings	New York, N. Y	Oct. 18, 1864	44, 700
Lamp-base	J. Kintz	Meriden, Conn	June 18, 1872	128, 048
Lamp-basket	P. J. Clark	West Meriden, Conn	Feb. 6, 1872	123, 379
Lamp, Billiard-table	R. Kleeman	Chicago, Ill	Apr. 12, 1870	101, 888
Lamp, Binnacle	W. Lawrence	Meriden, Conn	Nov. 14, 1832	....
Lamp-black and colophane, Manufacture of	E. Clark	Brooklyn, N. Y	Jan. 2, 1849	6, 001
Lamp-black, Apparatus for the manufacture of	M. Matlack	Philadelphia, Pa	Mar. 24, 1868	75, 943
Lamp-black, Making	J. Hastings	Cambridge, Mass	Aug. 15, 1826	....
Lamp-black, Making	G. Mini	Philadelphia, Pa	Nov. 13, 1844	3, 824
Lamp-black, Making	J. A. Roth	Philadelphia, Pa	June 9, 1857	17, 519
Lamp-black, Manufacture of	W. G. W. Iaeger	Baltimore, Md	July 18, 1854	11, 331
Lamp-black, Manufacture of	A. Millochau	New York, N. Y	Dec. 10, 1867	72, 068
Lamp-black, Manufacture of	A. Millochau	New York, N. Y	Nov. 17, 1868	84, 131
Lamp-black, Manufacture of	R. N. Perlee	Jersey City, N. J	Dec. 10, 1867	72, 078
Lamp-black, Manufacture of	A. Prenate	Elizabeth, N. J	Oct. 17, 1865	50, 493
Lamp-boiler	P. H. Sims	Waterloo, Canada	Sept. 24, 1872	131, 712
Lamp boiler, Kerosene	W. T. Rossman	Hudson, N. Y	July 28, 1868	80, 508
Lamp-bowls, &c., to their stands, Means of attaching.	S. S. Barrie	New York, N. Y	Mar. 26, 1872	125, 002
Lamp-boxes, Draft-attachment for	C. M. Bromwick	Boston, Mass	Sept. 2, 1862	36, 337
Lamp-bracket	W. Ascough	Buffalo, N. Y	Aug. 25, 1868	81, 324
Lamp-bracket	T. W. Brown	New York, N. Y	July 17, 1866	56, 361
Lamp-bracket	R. Devereux	Buffalo, N. Y	Nov. 12, 1867	70, 701
Lamp-bracket	E. L. Ferguson	Buffalo, N. Y	Mar. 10, 1868	75, 256
Lamp-bracket	L. Hull	Charlestown, Mass	Aug. 26, 1873	142, 106
Lamp-bracket	T. A. Hunter and J. Blewitt	New York, N. Y	June 29, 1869	92, 053
Lamp-bracket	G. Jones	Peekskill, N. Y	June 3, 1873	139, 582
Lamp-bracket	R. Marsh	New York, N. Y	Sept. 27, 1870	107, 699
Lamp-bracket	W. W. McDonald	Bucyrus, Ohio	Feb. 6, 1872	123, 498
Lamp-bracket	R. S. Merrill	Boston, Mass	Sept. 13, 1870	107, 279
Lamp-bracket	C. H. Miller	Buffalo, N. Y	Mar. 16, 1869	87, 957
Lamp-bracket	R. S. Tay and F. T. Fracker	Milford and Cambridge, Mass.	Nov. 27, 1866	60, 086
Lamp-bracket	A. F. Young and J. F. Foote	Mystic Bridge, Conn	Nov. 5, 1872	132, 785
Lamp, Bracket and chandelier	J. Ives	Mount Carmel, Conn	Oct. 25, 1864	44, 803
Lamp bracket and reflector	E. Boesch	San Francisco, Cal	Apr. 16, 1872	125, 785
Lamp bracket, Coach	S. Boudren	Bridgeport, Conn	Jan. 4, 1870	98, 468
Lamp bracket or support	H. Campbell	San Francisco, Cal	Dec. 5, 1871	121, 583

Index of patents issued from the United States Patent Office from 1790 to 1873, inclusive—Continued.

Invention.	Inventor.	Residence.	Date.	No.
Lamp-burner cones, Mode of forming	C. St. John and C. E. Marston	Charlestown, Mass	Dec. 14, 1869	97, 825
Lamp, Burning-fluid	M. B. Dyott	Philadelphia, Pa	Jan. 13, 1857	16, 379
Lamp, Burning-fluid	H. Todd	Columbus, Ohio	June 29, 1858	20, 746
Lamp, Camphene	R. V. De Guinon	Williamsburgh, N. Y	Jan. 6, 1852	8, 630
Lamp, Camphene	M. B. Dyott	Philadelphia, Pa	Aug. 25, 1840	1, 742
Lamp, Camphene, &c	S. J. Gold	Cornwall, Conn	July 16, 1841	2, 168
Lamp, Camphene	E. J. Hale	Foxcroft, Me	June 22, 1858	20, 641
Lamp, Camphene	E. B. Horn	Boston, Mass	Feb. 6, 1849	6, 103
Lamp, Camphene	J. Newell	Boston, Mass	Oct. 4, 1853	10, 099
Lamp, Camphene	I. Van Bunchoten	New York, N. Y	Mar. 2, 1852	8, 784
Lamp, Candle	J. A. Pease	Boston, Mass	May 13, 1873	138, 925
Lamp, Candle	T. S. Williams and F. A. Taber	Boston, Mass	May 31, 1870	103, 810
Lamp, Candle	T. S. Williams and F. A. Taber	Boston, Mass	June 21, 1870	104, 525
Lamp, Candle	T. S. Williams and F. A. Taber	Boston, Mass	July 5, 1870	105, 020
Lamp-cap	W. Bell	Boston, Mass	Nov. 14, 1854	11, 928
Lamp-cap	F. Draper	East Cambridge, Mass	Mar. 20, 1844	3, 500
Lamp-cap	R. H. Eddy	Boston, Mass	May 10, 1844	3, 582
Lamp, Car	J. E. Cross	Adrian, Mich	Aug. 19, 1873	141, 856
Lamp, Car	W. Westlake	Chicago, Ill	July 16, 1872	129, 259
Lamp, Car	W. Westlake	Chicago, Ill	July 16, 1872	129, 260
Lamp, Carbon-oil	O. V. Flora	Cincinnati, Ohio	Apr. 28, 1863	38, 303
Lamp, Carbureted-air	J. D. Jenkins	Charlestown, Mass	Jan. 28, 1868	73, 900
Lamp case, Locomotive	S. Bidwell	Rochester, N. Y	Sept. 4, 1855	13, 537
Lamp case, Locomotive	I. A. Williams	Utica, N. Y	Nov. 16, 1858	22, 099
Lamp, Cast-metal	T. F. Hammer	Branford, Conn	Mar. 1, 1870	100, 398
Lamp-chandelier	J. Aulbach	West Meriden, Conn	Aug. 26, 1873	142, 189
Lamp-chandelier	R. S. Mains	New York, N. Y	Dec. 3, 1872	133, 589
Lamp-chimney	E. R. Addison	Wheeling, W. Va	Nov. 19, 1867	70, 935
Lamp-chimney	A. Albertson	New York, N. Y	Dec. 8, 1863	40, 890
Lamp-chimney	H. Andes	Wilkesbarre, Pa	Nov. 5, 1872	132, 786
Lamp-chimney	H. C. Appleby	Conneaut, Ohio	June 12, 1866	55, 450
Lamp-chimney	E. D. Ashe	Brompton, England	Aug. 1, 1871	117, 500
Lamp-chimney	L. J. Atwood	Waterbury, Conn	Oct. 13, 1863	40, 228
Lamp-chimney	J. F. Bartlett	Winsted, Conn	May 25, 1869	90, 484
Lamp-chimney	J. Bellerjean	Philadelphia, Pa	June 4, 1867	65, 467
Lamp-chimney	E. Blackman	New York, N. Y	Feb. 6, 1872	123, 325
Lamp-chimney	E. S. Blake	Pittsburgh, Pa	July 15, 1862	35, 866
Lamp-chimney	E. S. Blake	Pittsburgh, Pa	Feb. 9, 1864	41, 473
Lamp-chimney	E. S. Blake	Pittsburgh, Pa	Oct. 4, 1864	44, 508
Lamp-chimney	H. Booth, jr	New York, N. Y	June 16, 1863	38, 881
Lamp-chimney	H. Brooke	New York, N. Y	July 6, 1869	92, 256
Lamp-chimney	H. Brown	New York, N. Y	Mar. 25, 1862	34, 733
Lamp-chimney	V. W. Brown and D. Frankish	Camden, N. J., and Philadelphia, Pa.	Jan. 16, 1866	52, 025
Lamp-chimney	V. W. Brown and D. Frankish	Camden, N. J., and Philadelphia, Pa.	May 29, 1866	55, 054
Lamp-chimney	E. S. Chase	Eau Claire, Wis	Nov. 30, 1869	97, 358
Lamp-chimney	G. F. J. Colburn	Newark, N. J	Apr. 14, 1863	38, 150
Lamp-chimney	R. R. Crosby	Boston, Mass	July 14, 1868	79, 958
Lamp-chimney	W. H. Culp	Hammondsville, Ohio	Aug. 1, 1865	49, 089
Lamp-chimney	T. A. Davies	New York, N. Y	Apr. 25, 1871	114, 113
Lamp-chimney	E. P. Dickie	Fishkill Landing, N. Y	Feb. 4, 1862	34, 293
Lamp-chimney	E. Dithridge	Pittsburgh, Pa	Oct. 12, 1869	95, 667
Lamp-chimney	E. and E. D. Dithridge	Pittsburgh, Pa	Oct. 8, 1861	32, 428
Lamp-chimney	C. Dorflinger	Brooklyn, N. Y	Aug. 5, 1862	36, 078
Lamp-chimney	R. N. Eagle	Washington, D. C	Jan. 26, 1869	86, 291
Lamp-chimney	R. N. Eagle	Washington, D. C	Jan. 18, 1870	98, 936
Lamp-chimney	R. N. Eagle	Washington, D. C	Jan. 18, 1870	98, 937
Lamp-chimney	J. G. Floyd	Keokuk, Iowa	Sept. 19, 1865	49, 993
Lamp-chimney	S. W. Fowler	Brooklyn, N. Y	Dec. 10, 1867	72, 010
Lamp-chimney	S. W. Fowler	Brooklyn, N. Y	Jan. 4, 1870	98, 484
Lamp-chimney	S. W. Fowler	Brooklyn, N. Y	May 30, 1871	115, 458
Lamp-chimney	W. T. Gillinder	Philadelphia, Pa	Feb. 9, 1869	86, 660
Lamp-chimney	J. Gracie and R. H. Boyd	Pittsburgh and Hulton Station, Pa.	July 28, 1868	80, 282
Lamp-chimney	G. W. Griswold	Factoryville, Pa	Mar. 7, 1871	112, 446
Lamp-chimney	P. Hagen	Elizabeth City, N. J	Nov. 6, 1866	59, 389
Lamp-chimney	E. J. Hale	Foxcroft, Me	Sept. 13, 1859	25, 406
Lamp-chimney	M. W. House	Cleveland, Ohio	Mar. 7, 1871	112, 460
Lamp-chimney	E. Jones	South Boston, Mass	Mar. 8, 1870	100, 637
Lamp-chimney	A. Judson	Brooklyn, N. Y	Aug. 7, 1866	56, 952
Lamp-chimney	W. L. Jukes	Covington, Ky	May 12, 1863	38, 492
Lamp-chimney	J. W. Larimore	Chicago, Ill	Jan. 5, 1869	85, 677
Lamp-chimney	J. Lewis	Mohawk, N. Y	July 19, 1864	43, 592
Lamp-chimney	R. S. Merrill	Boston, Mass	Jan. 17, 1871	111, 073
Lamp-chimney	W. O. B. Merrill	Philadelphia, Pa	June 10, 1862	35, 535
Lamp-chimney	H. W. Millar	Utica, N. Y	July 21, 1863	39, 301
Lamp-chimney	W. Onions and H. Roberts	Saint Louis, Mo	Mar. 31, 1868	76, 239
Lamp-chimney	J. M. Rankin, jr	Brooklyn, N. Y	Feb. 28, 1871	112, 182
Lamp-chimney	C. H. Reichmann	New York, N. Y	Jan. 1, 1867	60, 937
Lamp-chimney	E. B. Requa	Jersey City, N. J	Dec. 3, 1867	71, 789
Lamp-chimney	J. Robinson	Funkville, Pa	Apr. 21, 1868	77, 100
Lamp-chimney	E. O. Schartau	Philadelphia, Pa	Aug. 7, 1866	56, 999
Lamp-chimney	E. O. Schartau	Philadelphia, Pa	Dec. 3, 1867	71, 651
Lamp-chimney	W. Sedgwick	Poughkeepsie, N. Y	Mar. 26, 1872	125, 090
Lamp-chimney	D. E. Somes	Washington, D. C	July 25, 1871	117, 341
Lamp-chimney	W. Starkey	Bridgeport, N. J	Mar. 1, 1864	41, 792
Lamp-chimney	M. Sweeney	Wheeling, W. Va	June 23, 1868	79, 277
Lamp-chimney	G. W. Taylor and W. P. Jenney	Fairhaven, Mass	Dec. 20, 1870	110, 308
Lamp-chimney	T. J. Townsend	Baltimore, Md	Mar. 1, 1864	41, 796
Lamp-chimney	S. W. Warren	Boston, Mass	May 10, 1870	102, 892
Lamp-chimney	A. Whitlock	Danbury, Conn	July 20, 1869	92, 766
Lamp-chimney	J. D. Willoughby	Shippensburgh, Pa	Dec. 17, 1867	72, 253
Lamp-chimney	S. R. Wilmot	Bridgeport, Conn	Dec. 15, 1868	85, 047
Lamp-chimney	G. L. Witsil	Philadelphia, Pa	June 5, 1866	55, 436
Lamp-chimney	C. H. Wolcott	Randolph, N. Y	Apr. 27, 1869	89, 367
Lamp-chimney adjuster	J. R. Fogg	Portland, Me	May 19, 1863	38, 628
Lamp chimney and burner	M. H. Collins	Chelsea, Mass	Apr. 30, 1872	126, 184

Index of patents issued from the United States Patent Office from 1790 to 1873, inclusive—Continued.

Invention.	Inventor.	Residence.	Date.	No.
Lamp-chimney and dish-washer	C. S. and S. A. Moore	Worcester, Mass	Jan. 31, 1871	111, 368
Lamp chimney and shade combined	J. H. Connelly	Wheeling, W. Va	July 3, 1866	56, 010
Lamp-chimney attachment	W. L. Fish	Newark, N. J	June 17, 1862	35, 598
Lamp-chimney attachment	F. N. Gisborne and H. Allman	London, England	July 14, 1868	79, 823
Lamp-chimney attachment	J. Hughes	New Berne, N. C	Feb. 15, 1870	99, 901
Lamp-chimney attachment	D. M. Shea	Chicopee, Mass	July 21, 1868	80, 093
Lamp-chimney attachment	F. J. Tinker	Cincinnati, Ohio	Oct. 16, 1866	58, 915
Lamp-chimney attachment	J. Zengeler	Chicago, Ill	Jan. 6, 1863	37, 316
Lamp-chimney, bottle, or can cleaner	C. N. Pond	Oberlin, Ohio	July 3, 1866	56, 092
Lamp-chimney cap	J. G. Dreyfus	New York, N. Y	May 4, 1869	89, 743
Lamp-chimney cleaner	F. Ashley	New York, N. Y	Nov. 20, 1866	59, 808
Lamp-chimney cleaner	L. J. Baker	Machias, Me	May 5, 1868	77, 567
Lamp-chimney cleaner	W. B. Barnard	Waterbury, Conn	Aug. 28, 1866	57, 460
Lamp-chimney cleaner	J. S. Black	Oakland, Ill	Mar. 10, 1868	75, 351
Lamp-chimney cleaner	C. Boeckh	Strasburg, France	Oct. 7, 1862	36, 600
Lamp-chimney cleaner	O. B. Brown	Malden, Mass	Nov. 5, 1867	70, 403
Lamp-chimney cleaner	B. E. Chollar and J. G. Johannes.	Leavenworth, Kans	Mar. 11, 1873	136, 704
Lamp-chimney cleaner	C. B. Clark	Armada, Mich	Aug. 8, 1871	117, 741
Lamp-chimney cleaner	H. Fellows	Bloomington, Ind	May 28, 1872	127, 230
Lamp-chimney cleaner	A. A. Ford	North Abington, Mass	June 11, 1872	127, 754
Lamp-chimney cleaner	R. G. Forsyth	Oakland, Ill	Apr. 8, 1873	137, 670
Lamp-chimney cleaner	L. Granger	Armada, Mich	May 14, 1872	126, 693
Lamp-chimney cleaner	J. R. Hamilton	Portland, Oreg	Jan. 14, 1868	73, 324
Lamp-chimney cleaner	E. T. Hazeltine	Warren, Pa	Mar. 22, 1864	41, 994
Lamp-chimney cleaner	E. T. Hazeltine	Warren, Pa	June 5, 1866	55, 891
Lamp-chimney cleaner	J. P. Howell	New York, N. Y	Oct. 1, 1867	69, 439
Lamp-chimney cleaner	R. R. Howell	Fostoria, Ohio	July 9, 1872	128, 879
Lamp-chimney cleaner	F. Imhorst	New York, N. Y	Dec. 20, 1864	45, 500
Lamp-chimney cleaner	F. Imhorst	New York, N. Y	Jan. 2, 1866	51, 835
Lamp-chimney cleaner	W. J. Johnson	Newton, Mass	Aug. 14, 1866	57, 149
Lamp-chimney cleaner	L. Keiler	Catawissa, Pa	Sept. 25, 1866	58, 265
Lamp-chimney cleaner	M. R. Kenyon	Providence, R. I	July 22, 1862	35, 939
Lamp-chimney cleaner	G. Leas	Shirleysburgh, Pa	Dec. 17, 1867	72, 306
Lamp-chimney cleaner	J. Lee	New York, N. Y	June 2, 1868	78, 528
Lamp-chimney cleaner	J. H. Lightner	Shirleysburgh, Pa	Nov. 26, 1867	71, 396
Lamp-chimney cleaner	M. J. Lourrentz	Leavenworth, Kans	Oct. 27, 1868	83, 391
Lamp-chimney cleaner	M. N. Lovell	Erie, Pa	Sept. 8, 1868	81, 917
Lamp-chimney cleaner	E. P. Morse	Batavia, N. Y	Aug. 6, 1872	130, 307
Lamp-chimney cleaner	R. B. Musson	Champaign, Ill	Dec. 10, 1867	72, 072
Lamp-chimney cleaner	A. G. Newkirk	Warren, Pa	Jan. 28, 1868	73, 745
Lamp-chimney cleaner	R. Pattin	Marietta, Ohio	Jan. 8, 1867	61, 092
Lamp-chimney cleaner	R. Pearson	Allegheny City, Pa	July 25, 1871	117, 452
Lamp-chimney cleaner	H. W. Prouty	Boston, Mass	Apr. 23, 1872	126, 084
Lamp-chimney cleaner	H. W. Prouty	Boston, Mass	Feb. 11, 1873	135, 665
Lamp-chimney cleaner	J. Ryan	Brooklyn, N. Y	July 12, 1870	105, 256
Lamp-chimney cleaner	A. D. Smith	Grafton, Ohio	Oct. 18, 1870	108, 527
Lamp-chimney cleaner	I. Smith	New York, N. Y	Oct. 10, 1871	119, 890
Lamp-chimney cleaner	N. A. Vurgason	Brooklyn, N. Y	Sept. 1, 1868	81, 710
Lamp-chimney cleaner	J. J. Wait	Oreana, Nev	May 19, 1868	78, 160
Lamp-chimney cleaner	L. Ward	Poughkeepsie, N. Y	June 20, 1871	116, 121
Lamp chimney, Coal-oil	W. Howard	Flushing, N. Y	June 3, 1862	35, 447
Lamp-chimney fastening	J. A. Allen and E. F. Lewis	Washington, D. C	Jan. 7, 1868	73, 068
Lamp-chimney fastening	W. Fulton	Elizabeth City, N. J	Apr. 8, 1862	34, 884
Lamp-chimney fastening	E. Russell	Naugatuck, Conn	Nov. 26, 1867	71, 538
Lamp-chimney fastening	J. and A. W. Sangster	Buffalo, N. Y	Oct. 28, 1862	36, 802
Lamp-chimney fastening	W. S. Thompson	Rochester, N. Y	Dec. 16, 1862	37, 183
Lamp-chimney fastening	L. C. White	Waterbury, Conn	Dec. 9, 1862	37, 119
Lamp-chimney, Glass	G. F. J. Colburn	Newark, N. J	June 23, 1863	38, 947
Lamp-chimney handle	M. W. Brown	New York, N. Y	Feb. 7, 1865	46, 212
Lamp-chimney heater	J. B. Greene	Providence, R. I	Sept. 30, 1862	36, 584
Lamp-chimney holder	J. F. Hechtle	Waterbury, Conn	Dec. 21, 1869	98, 060
Lamp-chimney holder	J. H. Irwin	Philadelphia, Pa	Dec. 17, 1872	134, 063
Lamp-chimney holder	O. Newton	Pittsburgh, Pa	Apr. 8, 1862	34, 900
Lamp-chimney holder	T. Rider	Providence, R. I	Apr. 23, 1867	64, 039
Lamp-chimney holder, Sectional	C. T. Haughey	Wabash, Ind	June 4, 1872	127, 603
Lamp-chimney, Lens	S. Constant	Brooklyn, N. Y	June 13, 1854	11, 060
Lamp-chimney lifter	J. H. Wilhelm and J. W. Larimore.	Chicago, Ill	Apr. 24, 1866	54, 239
Lamp-chimney, Metal-top	F. I. Seymour	Wolcottville, Conn	Mar. 10, 1868	75, 302
Lamp-chimney, Mica	J. Baird and W. Fisher	Hamilton, Ohio	Aug. 9, 1870	106, 247
Lamp-chimney, Mica	G. M. Bull	New Baltimore, N. Y	Jan. 9, 1872	122, 560
Lamp-chimney, Mica	E. C. Gaylord	Hartford, Conn	June 24, 1862	35, 675
Lamp-chimney, Mica	B. J. C. Howe	Syracuse, N. Y	Jan. 27, 1863	37, 506
Lamp-chimney, Mica	J. Y. Humphrey	Philadelphia, Pa	July 17, 1860	29, 172
Lamp-chimney, Mica	J. W. Schreiber	New York, N. Y	Aug. 19, 1862	36, 237
Lamp-chimney, Mica	L. Van Order	Ithaca, N. Y	Apr. 8, 1862	34, 917
Lamp-chimney, Mica	W. P. Ware	New York, N. Y	Mar. 3, 1863	37, 837
Lamp-chimney protector	E. Stern and S. Blair	New York, N. Y	Oct. 29, 1872	132, 606
Lamp-chimney, Reflecting	H. L. Hervey	Philadelphia, Pa	Aug. 17, 1869	93, 882
Lamp-chimney, Reflecting	A. Kunkle	Birmingham, Pa	Dec. 31, 1872	134, 432
Lamp-chimney spring	O. Snow	West Meriden, Conn	Apr. 8, 1862	34, 908
Lamp-chimney spring-catch	M. Fowler	Meriden, Conn	June 17, 1862	35, 649
Lamp-chimney supporter	W. Mears and H. Davies	Newport, Ky	June 24, 1873	140, 214
Lamp-chimneys, Apparatus for preventing the breakage of.	B. Vitalis	Athens, Greece	Apr. 23, 1872	125, 915
Lamp-chimeys, bottles, &c., Device for cleaning	J. T. Walker	Palmyra, N. Y	Aug. 15, 1865	49, 458
Lamp-chimneys, Combined holding and supporting spring for.	W. N. Weeden	Waterbury, Conn	Oct. 1, 1872	131, 918
Lamp-chimneys, Construction of glass	C. L. Daboll	New London, Conn	May 26, 1863	38, 657
Lamp-chimneys, Device for attaching	J. B. Alexander	Washington, D. C	Aug. 27, 1867	68, 142
Lamp-chimneys, Device for removing	G. Asmus	Houghton, Mich	June 27, 1865	48, 354
Lamp-chimneys, Device for securing	J. J. Marcy	Meriden, Conn	Dec. 4, 1860	30, 856
Lamp-chimneys, Forming glass	M. H. Collins	Chelsea, Mass	Sept. 3, 1867	68, 416
Lamp-chimneys, Glass mold for forming	M. H. Collins	Chelsea, Mass	Mar. 18, 1873	136, 971
Lamp-chimneys, Implement for handling	A. H. Merrill	Boston, Mass	Aug. 27, 1861	33, 170
Lamp-chimneys, Instrument for cleaning	T. B. De Forest	Birmingham, Conn	Jan. 15, 1861	31, 111
Lamp-chimneys, Machine for forming glass	M. H. Collins	Chelea, Mass	Sept. 26, 1871	119, 222

Index of patents issued from the United States Patent Office from 1790 *to* 1873, *inclusive*—Continued.

Invention.	Inventor.	Residence.	Date.	No.
Lamp chimneys, Mode of affixing argand	D. Jarves	Boston, Mass	July 20, 1829	....
Lamp-chimneys, Mode of cleaning	C. D. Blinn	Port Huron, Mich	Sept. 8, 1863	39, 792
Lamp-chimneys, Mode of elevating	J. G. Leffingwell	Newark, N. J	Mar. 17, 1863	37, 917
Lamp-chimneys, Mode of securing	E. Bowen	Meriden, Conn	Dec. 10, 1861	33, 875
Lamp-chimneys, Mode of securing	T. T. Jacobs	Mount Carroll, Ill	May 13, 1862	35, 239
Lamp-chimneys, Mold for making	E. Dithridge	Pittsburgh, Pa	Mar. 5, 1872	124, 343
Lamp-chimneys, Molding	E. Eltinge	Kingston, N. Y	Sept. 26, 1865	50, 105
Lamp-chimneys, Tool for sizing	L. J. Atwood	Waterbury, Conn	Dec. 10, 1867	71, 838
Lamp-cleaner	B. F. Horn	Boston, Mass	June 26, 1866	55, 862
Lamp-cleaner	E. F. Sutton	New York, N. Y	Apr. 10, 1866	53, 898
Lamp-cleaner	R. White	Princeton, Canada	Oct. 3, 1865	50, 310
Lamp, Coach	C. B. Brown and E. Andrews	Placerville, Cal	Oct. 4, 1864	44, 511
Lamp, Coach	W. Lawrence	Wallingford, Conn	Apr. 7, 1838	683
Lamp, Coal-oil	J. E. Ambrose	Jersey City, N. J	Apr. 26, 1864	42, 455
Lamp, Coal-oil	L. Bader	Philadelphia, Pa	July 7, 1863	39, 195
Lamp, Coal-oil	H. Behn	New York, N. Y	June 17, 1862	35, 579
Lamp, Coal-oil	W. B. Billings	New York, N. Y	June 16, 1863	38, 879
Lamp, Coal-oil	C. E. Corbett	Corbettsville, N. Y	Aug. 18, 1863	39, 551
Lamp, Coal-oil	G. Y. Custer	Morristown, Pa	Aug. 19, 1862	36, 206
Lamp, Coal-oil	B. Garvey	New York, N. Y	Apr. 28, 1863	38, 305
Lamp, Coal-oil	D. E. Hall	Brooklyn, N. Y	Dec. 16, 1862	37, 197
Lamp, Coal-oil	J. M. Hancock	Philadelphia, Pa	Dec. 2, 1862	37, 080
Lamp, Coal-oil	A. J. Harwood	Utica, N. Y	Oct. 25, 1864	44, 800
Lamp, Coal-oil	J. G. Hunt	Cincinnati, Ohio	Jan. 19, 1864	41, 300
Lamp, Coal-oil	J. H. Irwin	Beardstown, Ill	May 6, 1862	35, 158
Lamp, Coal-oil	A. Judson	Brooklyn, N. Y	Apr. 21, 1863	38, 263
Lamp, Coal-oil	A. C. Ketchum	New York, N. Y	Sept. 22, 1863	40, 080
Lamp, Coal-oil	A. Leach	Utica, N. Y	Jan. 6, 1863	37, 349
Lamp, Coal-oil	R. S. Merrill	Lynn, Mass	June 19, 1860	28, 762
Lamp, Coal-oil, &c	R. S. Merrill	Lynn, Mass	June 3, 1862	35, 460
Lamp, Coal-oil	W. Morehouse	Buffalo, N. Y	Apr. 1, 1862	34, 841
Lamp, Coal-oil	J. Mowery	Philadelphia, Pa	Aug. 23, 1864	43, 953
Lamp, Coal-oil	A. H. Platt	Yellow Springs, Ohio	Apr. 25, 1865	47, 451
Lamp, Coal-oil	J. Ridge	Richmond, Ind	Apr. 15, 1862	34, 984
Lamp, Coal-oil	J. Ridge	Richmond, Ind	Jan. 5, 1864	41, 093
Lamp, Coal-oil	J. Ridge	Richmond, Ind	Jan. 5, 1864	41, 094
Lamp, Coal-oil, &c	G. Rimmington	South Brooklyn, N. Y	May 25, 1858	20, 373
Lamp, Coal-oil	J. N. Smith	New York, N. Y	June 10, 1862	35, 549
Lamp, Coal-oil	C. F. Spencer	Rochester, N. Y	Mar. 24, 1863	37, 986
Lamp, Coal-oil	W. D. Taylor	Fort Madison, Iowa	Mar. 22, 1864	42, 028
Lamp, Coal-oil	J. White and A. Agnew	Philadelphia, Pa	Sept. 2, 1862	36, 380
Lamp, Coal-oil	G. L. Witsil	Philadelphia, Pa	Oct. 21, 1862	36, 755
Lamp, Coal-oil hand	E. Roberts	Moorestown, N. J	Sept. 12, 1865	49, 924
Lamp, Coal-oil-stove	J. Bowles	Augusta, Ga	Feb. 7, 1871	111, 512
Lamp, coffee-pot, and boiler combined	L. A. Plumb	Biddeford, Me	Jan. 22, 1867	61, 454
Lamp-collar	J. C. Beers	Brooklyn, N. Y	Jan. 12, 1864	41, 255
Lamp-collar	H. Kelly and W. H. Locke	Boston, Mass	May 2, 1871	114, 302
Lamp-collar	A. Taplin	Providence, R. I	Mar. 1, 1864	41, 795
Lamp-collar and safety-tube	C. B. Mann	Baltimore, Md	Apr. 9, 1872	125, 588
Lamp-cone	C. H. Buckelew	Jersey City, N. J	Apr. 4, 1865	47, 087
Lamp-cone	M. B. Dyott	Philadelphia, Pa	Mar. 24, 1863	37, 956
Lamp-cone	E. B. Requa	Jersey City, N. J	Jan. 5, 1864	41, 092
Lamp cone, Coal-oil	W. Fulton	Elizabeth City, N. J	May 6, 1862	35, 151
Lamp-crane	A. C. Black	Kaukauna, Wis	July 13, 1869	92, 570
Lamp cup, Kerosene	N. L. Bradley	Meriden, Conn	Jan. 31, 1871	111, 430
Lamp, Electric	H. M. Collier and H. N. Baker	Binghamton and New York, N. Y.	May 18, 1858	20, 255
Lamp, Electric	G. G. Percival	Brooklyn, N. Y	Apr. 3, 1866	53, 669
Lamp, Essential-oil	M. B. Dyott	Philadelphia, Pa	May 30, 1842	2, 658
Lamp extension-supporter	L. Hull	Charleston, Mass	Jan. 16, 1872	122, 767
Lamp-extinguisher	C. E. Abbott	Boston, Mass	July 3, 1866	55, 973
Lamp-extinguisher	C. E. Abbott	Malden, Mass	Nov. 13, 1866	59, 534
Lamp-extinguisher	C. E. Abbott	Malden, Mass	Nov. 19, 1867	71, 112
Lamp-extinguisher	C. E. Abbott	Malden, Mass	Aug. 11, 1868	80, 796
Lamp-extinguisher	J. M. Aubrey	Portage City, Wis	June 29, 1869	91, 812
Lamp-extinguisher	S. J. Batchelder	Manchester, N. J	June 23, 1868	79, 187
Lamp-extinguisher	W. P. Bennett	East Pepperell, Mass	Nov. 5, 1867	70, 392
Lamp-extinguisher	G. V. Bunker	Yankton, Dak	Nov. 30, 1869	97, 271
Lamp-extinguisher	W. I. Bunker	Yankton, Dak	Dec. 14, 1869	97, 757
Lamp-extinguisher	S. C. Catlin and A. Corbin	Cleveland, Ohio, and Ellenville, N. Y.	Feb. 1, 1870	99, 289
Lamp-extinguisher	W. C. Ebert	Hannibal, Mo	May 10, 1870	102, 926
Lamp-extinguisher	J. W. Fowle	Boston, Mass	Apr. 7, 1868	76, 319
Lamp-extinguisher	J. M. Goodridge	Norfolk, Va	Mar. 12, 1872	124, 567
Lamp-extinguisher	W. Grayson and D. C. Hyndman.	Odell, Ill	Aug. 24, 1869	93, 986
Lamp-extinguisher	M. C. Heptinstall	Enfield, N. C	June 29, 1869	91, 934
Lamp-extinguisher	F. B. Hill and W. H. McCoy	Cleveland, Ohio, and Des Moines, Iowa.	Nov. 5, 1867	70, 567
Lamp-extinguisher	F. Hille	Lyons, Iowa	Dec. 23, 1873	145, 870
Lamp-extinguisher	J. N. Howe	Franklin, N. H	Nov. 13, 1866	59, 607
Lamp-extinguisher	J. Hughes	New Berne, N. C	May 17, 1870	103, 048
Lamp-extinguisher	R. Lange	Baltimore, Md	May 27, 1873	139, 398
Lamp-extinguisher	C. Lanzendorfer	Chicago, Ill	Jan. 19, 1869	86, 084
Lamp-extinguisher	W. D. Lindsley	Wathena, Kans	Aug. 12, 1873	141, 801
Lamp-extinguisher	C. Newman	San Francisco, Cal	May 25, 1869	90, 572
Lamp-extinguisher	J. H. Noyes	Abington, Mass	Feb. 13, 1855	12, 394
Lamp-extinguisher	S. W. Perkins	Geneseo, Ill	Dec. 28, 1869	98, 405
Lamp-extinguisher	D. Pike and O. B. Graham	Bridgeport, Conn	Sept. 2, 1873	142, 508
Lamp-extinguisher	J. Pons	Baltimore, Md	Mar. 2, 1869	87, 511
Lamp-extinguisher	D. J. Powers	Chicago, Ill	May 31, 1870	103, 652
Lamp-extinguisher	J. H. Prentice	Ashtabula, Ohio	July 21, 1868	80, 083
Lamp-extinguisher	W. A. Richardson and H. D. Ward.	Worcester, Mass	Apr. 3, 1866	53, 679
Lamp-extinguisher	W. A. Richardson and H. D. Ward.	Worcester, Mass	Sept. 4, 1866	57, 768
Lamp-extinguisher	E. Richmond	Abington, Mass	Dec. 18, 1855	13, 972
Lamp-extinguisher	F. Rohrer	San Francisco, Cal	Nov. 19, 1867	71, 221

Index of patents issued from the United States Patent Office from 1790 *to* 1873, *inclusive*—Continued.

Invention.	Inventor.	Residence.	Date.	No.
Lamp-extinguisher	W. G. Ruge	Holstein, Mo	Aug. 22, 1871	118, 280
Lamp-extinguisher	M. Safford	Boston, Mass	May 7, 1867	64, 578
Lamp-extinguisher	L. M. Sargent	Worcester, Mass	Oct. 22, 1867	70, 125
Lamp-extinguisher	E. Shaw	Portland, Me	June 11, 1867	65, 696
Lamp-extinguisher	G. Simpson and W. H. Edmunds.	Waterbury, Vt	July 9, 1867	66, 526
Lamp-extinguisher	W. H. and C. W. Terpening	Geneseo, Ill	Aug 10, 1869	93, 565
Lamp-extinguisher	E. J. Toof	Fort Madison, Iowa	Mar. 10, 1868	75, 317
Lamp-extinguisher	F. C. Wireman	New York, N. Y	Aug. 19, 1873	142, 067
Lamp extinguisher and regulator	C. E. Lyon	Worcester, Mass	July 31, 1866	56, 770
Lamp, Extinguisher for fluid	W. B. Carpenter	Brooklyn, N. Y	Sept. 9, 1856	15, 686
Lamp extinguisher, Street	G. S. Dunbar	Pittsfield, Mass	Jan. 16, 1872	122, 762
Lamp-deflectors, Buffer for inserting coiled wire around the edges of.	M. H. Mosman	Waterbury, Conn	Nov. 23, 1869	97, 215
Lamp, Döbereiner self-lighting	G. Müller	Newark, N. J	June 9, 1868	78, 755
Lamp-fastening	L. B. Carpenter	Buffalo, N. Y	July 25, 1854	11, 354
Lamp fastening, Coach	M. De Voursney	Newark, N. J	Nov. 5, 1867	70, 422
Lamp-feed regulator, Automatic	H. S. Hudson	Selma, Ala	May 4, 1869	89, 580
Lamp-feed trap	M. W. House	Cleveland, Ohio	May 24, 1870	103, 334
Lamp-feeder	T. P. Gibbons	Baltimore, Md	Sept. 15, 1868	82, 219
Lamp-feeder	T. Mayhew	Poughkeepsie, N. Y	Jan. 13, 1863	37, 404
Lamp-filler	C. R. Landmann	New York, N. Y	Aug. 29, 1854	11, 612
Lamp-filler	H. H. V. Lilley	Milford, Mass	Apr. 29, 1873	138, 415
Lamp-filler	H. W. Stables	Saco, Me	Oct. 19, 1869	95, 948
Lamp, Fire-kindling	S. J. Lowe	Quincy, Ill	Nov. 26, 1867	71, 515
Lamp, Fluid	S. F Allen	New York, N. Y	Nov. 29, 1853	10, 270
Lamp, Fluid	W. Bennett	Brooklyn, N. Y	Nov. 27, 1855	13, 860
Lamp, Fluid	S. Bidwell	Rochester, N. Y	Sept. 9, 1856	15, 724
Lamp, Fluid	D. H. Chamberlain	West Roxbury, Mass	May 8, 1855	12, 814
Lamp, Fluid	A. Von Schuttenbach	St. Petersburg, Russia	June 21, 1859	24, 509
Lamp, Fluid	T. Ward	Columbus, Ohio	Mar. 22, 1870	101, 192
Lamp for boiling water, &c	T. G. Fessenden	Boston, Mass	Jan. 31, 1827	....
Lamp for burning compound, spirits, or oil	W. Magee	New York, N. Y	June 27, 1832	....
Lamp for burning compound, tallow, lard, beeswax, &c.	N. Rublee	Montpelier, Vt	Dec. 4, 1832	....
Lamp for burning evaporable ingredients	I. Jennings	New York, N. Y	Aug. 1, 1831	....
Lamp for burning fat, &c	J. Warren	Cincinnati, Ohio	Feb. 20, 1821	....
Lamp for burning heavy oil	R. Hitchcock	Watertown, N. Y	Apr. 23, 1872	125, 954
Lamp for burning lard and other concrete substances.	E. T. Williams and L. T. Tew	Newport, R. I	June 26, 1841	2, 140
Lamp for burning off paint	W. W. Wakeman, jr	New York, N. Y	June 11, 1867	65, 621
Lamp for burning tallow	M. S. Underwood	Marshallton, Pa	Sept. 18, 1841	2, 257
Lamp for burning tallow and other fatty substances.	I. Jennings	New York, N. Y	May 20, 1830	....
Lamp for burning turpentine, camphene, &c., Argand.	C. Carr	Philadelphia, Pa	Mar. 28, 1842	2, 514
Lamp for burning vapor of benzole, &c	C. Warner	Washington, D. C	Oct. 14, 1851	8, 433
Lamp for burning volatile ingredients	I. Jennings	New York, N. Y	Sept. 11, 1841	2, 254
Lamp for burning volatile ingredients	I. Jennings	New York, N. Y	Oct. 12, 1844	3, 793
Lamp for burning volatile liquids	E. N. Horsford and J. R. Nichols.	Cambridge and Haverhill, Mass.	Oct. 30, 1855	13, 729
Lamp for burning volatile materials	G. Egles	Boston, Mass	May 22, 1835	....
Lamp for burning volatile materials, Argand	D. Pettibone	Philadelphia, Pa	Mar. 28, 1842	2, 516
Lamp for cooking-purposes	M. E. Hatch	Beloit, Wis	Apr. 13, 1869	88, 784
Lamp for cooking-purposes, Coal-oil	W. B. Billings	Brooklyn, N. Y	Nov. 14, 1865	50, 892
Lamp for heating curling-irons	D. T. Burrell	Bridgewater, Mass	Mar. 14, 1865	46, 775
Lamp for lighting gas	C. McIntosh	Jersey City, N. J	June 15, 1858	20, 573
Lamp for medicated-vapor bath, Alcohol	L. E. Hicks	Middletown, Conn	Mar. 16, 1844	3, 488
Lamp for vessels, Head	H. Langster	Buffalo, N. Y	June 2, 1863	38, 764
Lamp for volatile materials	A. V. H. Webb	New York, N. Y	Nov. 4, 1842	2, 843
Lamp-fount to bracket, chandelier, &c	G. Bohner	Chicago, Ill	Dec. 6, 1870	109, 864
Lamp, Fountain	H. W. Adams	New York, N. Y	June 30, 1857	17, 658
Lamp, Fountain	J. P. Driver	Marengo, Iowa	Aug. 21, 1866	57, 299
Lamp, Fountain	G. W. Harrold	Rochester, N. Y	July 15, 1862	35, 875
Lamp, Fountain	N. Linden	Jersey City, N. J	June 24, 1856	15, 198
Lamp, Fountain	M. Samuels	New York, N. Y	Dec. 15, 1868	85, 029
Lamp, Fountain	J. H. Seaman	Brooklyn, N. Y	June 9, 1863	38, 846
Lamp-fountains, Mold for glass	J. Wing	Boston, Mass	Oct. 21, 1873	143, 864
Lamp, &c., frame, Street	J. Bradston	San Francisco, Cal	May 17, 1870	103, 238
Lamp, Gas	S. Andrews	Perth Amboy, N. J	May 10, 1832	....
Lamp, Gas	J. H. Brown	New York, N. Y	Dec. 13, 1870	110, 005
Lamp, Gas	J. Gray	New York, N. Y	Oct. 17, 1871	120, 058
Lamp, Gas	D. Melville	Newport, R. I	Mar. 18, 1813	....
Lamp, Gas	D. Melville	Newport, R. I	Mar. 24, 1810	....
Lamp, Gas	H. G. Sickel	Philadelphia, Pa	Aug. 7, 1849	6, 624
Lamp, Gas-burning	S. Andrews	Perth Amboy, N. J	June 3, 1856	14, 994
Lamp, Gas-conducting	R. Cornelius	Philadelphia, Pa	Mar. 18, 1841	2, 000
Lamp, Gas hand	H. J. Rice	Cincinnati, Ohio	July 25, 1871	117, 331
Lamp generator, Vapor	S. D. Baldwin	Milwaukee, Wis	May 8, 1860	28, 143
Lamp generator, Vapor	J. S. Hull	Cincinnati, Ohio	May 14, 1867	64, 769
Lamp, Glass	J. Adams	Birmingham, Pa	Jan. 10, 1871	110, 815
Lamp, Glass	J. S. and T. B. Atterbury	Pittsburgh, Pa	Nov. 22, 1870	109, 369
Lamp, Glass	J. Reighard	Birmingham, Pa	July 2, 1861	32, 739
Lamp, Glass	D. C. Ripley	Pittsburgh, Pa	Jan. 7, 1868	73, 122
Lamp, Glass	D. C. Ripley, jr	Birmingham, Pa	Apr. 5, 1870	101, 512
Lamp, Glass	D. C. Ripley	Pittsburgh, Pa	Sept. 20, 1870	107, 544
Lamp globe-shade	H. Whitney	East Cambridge, Mass	Feb. 23, 1869	87, 128
Lamp, Hall, garden, &c	A. W. Wehrhan	Columbia, S. C	May 10, 1870	102, 985
Lamp-hanger	T. S. Hudson	East Cambridge, Mass	Feb. 7, 1865	46, 236
Lamp, Hanging	G. Bohner	Chicago, Ill	Apr. 30, 1872	126, 254
Lamp, Hanging	A. Platt	Middletown, Conn	Oct. 8, 1836	42
Lamp, Hanging	A. Platt	Middletown, Conn	Oct. 20, 1836	....
Lamp-heater	W. T. Eddy	West Hoboken, N. J	Apr. 7, 1863	38, 102
Lamp-heater	W. Friel	San Francisco, Cal	May 20, 1873	139, 140
Lamp-heater	C. J. Hauck	Brooklyn, N. Y	July 16, 1872	129, 555
Lamp-heater	L. E. Truesdell	Warren, Mass	Sept. 24, 1872	131, 578
Lamp-heater for nursery-cups	M. Fox and F. Reinhard	West Eau Claire, Wis	July 16, 1872	129, 275
Lamp-heater for nursery-flasks	S. Hughes	Jersey City, N. J	Aug. 13, 1872	130, 429

Index of patents issued from the United States Patent Office from 1790 *to* 1873, *inclusive*—Continued.

Invention.	Inventor.	Residence.	Date.	No.
Lamp heater, Kerosene	W. L. Brown	Shelburne Falls, Mass	Aug. 24, 1869	93, 961
Lamp heater, Vehicle	E. H. Reynolds	Rising Sun, Md	May 21, 1867	65, 010
Lamp heating and shaving apparatus	C. S. Bourne	Springfield, Mass	Sept. 13, 1864	44, 252
Lamp-heating apparatus	A. Rittenhouse	Philadelphia, Pa	Nov. 8, 1870	109, 054
Lamp-heating apparatus	W. N. White	Winchendon, Mass	Nov. 1, 1870	108, 863
Lamp-holder	C. Monson	New Haven, Conn	Jan. 18, 1859	22, 664
Lamp holder and shade	W. H. Hawkins	Cleveland, Ohio	Jan. 23, 1872	122, 886
Lamp, Hunting	J. T. Staples	Bladen Springs, Ala	Apr. 2, 1872	125, 150
Lamp, Hydrocarbon-vapor	J. Cook	New York, N. Y	May 14, 1872	126, 625
Lamp, Hydrocarbon-vapor	R. R. Crosby	Boston, Mass	Feb. 9, 1858	19, 287
Lamp, Hydrocarbon-vapor	A. M. Mace	Springfield, Mass	Apr. 22, 1856	14, 727
Lamp, Hydrocarbon-vapor	I. Suggitt	Providence, R. I	Sept. 29, 1857	18, 307
Lamp, Hydrocarbon-vapor	T. Varney	San Francisco, Cal	Sept. 30, 1856	15, 829
Lamp, Hydrogen	C. Hagen and F. Aurnhammer	New York, N. Y	Mar. 28, 1865	47, 012
Lamp, Hydrogen	J. Pusey	Philadelphia, Pa	July 23, 1872	129, 681
Lamp, Hydropneumatic	S. S. Lee	Providence, R. I	Apr. 21, 1842	2, 570
Lamp, Hydrostatic	E. D. Kendall	New York, N. Y	May 21, 1872	127, 067
Lamp, Hydrostatic safety	H. S. Whitfield	Tuscaloosa, Ala	May 21, 1872	127, 004
Lamp, Insect-destroying	F. M. Briggs	Livonia, Mich	July 28, 1868	80, 446
Lamp, Insect-destroying	G. C. Cranston	South Bend, Ind	Mar. 23, 1869	88, 140
Lamp, Insect-destroying	J. Zimmerman	Royalton Centre, N. Y	July 30, 1867	67, 244
Lamp, Kerosene	S. Adlaur, jr., and J. R. Fogg	Portland, Me	Apr. 21, 1863	28, 208
Lamp, Kerosene	J. S. and T. B. Atterbury	Pittsburgh, Pa	Jan. 6, 1863	37, 267
Lamp, Kerosene	J. H. Beidler and A. R. Crihfield.	Lincoln, Ill	Mar. 13, 1866	53, 103
Lamp, Kerosene	H. H. Dickinson	Hartford, Conn	Dec. 10, 1861	33, 879
Lamp, Kerosene	A. Meucci	Clifton, N. Y	Aug. 12, 1862	36, 192
Lamp, Kerosene or coal-oil	J. Ridge	Richmond, Ind	Dec. 9, 1862	37, 144
Lamp, Lantern	R. M. Merrill	Chicago, Ill	Sept. 30, 1862	36, 570
Lamp, Lantern	J. Silvins and W. T. Hain	Sunbury, Pa	June 11, 1867	65, 698
Lamp, lantern, &c., Street	J. Binney	Boston, Mass	Mar. 7, 1865	46, 631
Lamp, Lard	E. S. Archer	Philadelphia, Pa	June 18, 1842	2, 671
Lamp, Lard	A. H. Baird	New York, N. Y	May 26, 1842	2, 644
Lamp, Lard	H. P. Benton	Norwalk, Ohio	Apr. 10, 1843	3, 040
Lamp, Lard	J. S. Brown	Washington, D. C	Oct. 9, 1855	13, 675
Lamp, Lard	J. S. Brown	Washington, D. C	Feb. 3, 1857	16, 524
Lamp, Lard	D. H. Chamberlain	Boston, Mass	Aug. 29, 1854	11, 633
Lamp, Lard	P. J. Clark	Meriden, Conn	Mar. 21, 1843	3, 009
Lamp, Lard	I. N. Coffin	Washington, D. C	Mar. 17, 1857	16, 825
Lamp, Lard	R. Cornelius	Philadelphia, Pa	Apr. 6, 1843	3, 030
Lamp, Lard	J. T. Creighton	Alexandria, D. C	Aug. 11, 1842	2, 751
Lamp, Lard	S. Davis	New Holland, Pa	May 6, 1856	14, 806
Lamp, Lard	J. Grannis	Oberlin, Ohio	Aug. 25, 1842	2, 763
Lamp, Lard	J. D. Hays	Mount Morris, Ill	July 3, 1855	13, 170
Lamp, Lard	B. H. Horn	Boston, Mass	May 26, 1842	2, 641
Lamp, Lard	T. Houghton and J. F. Wallace	Philadelphia, Pa	Nov. 15, 1843	3, 333
Lamp, Lard	H. Howard	Wooster, Ohio	Feb. 10, 1843	2, 946
Lamp, Lard	L. Jones	Utica, N. Y	May 26, 1843	3, 101
Lamp, Lard	A. Kayser	Fulton, Mo	Aug. 20, 1845	4, 154
Lamp, Lard	D. Kinnear	Circleville, Ohio	Feb. 4, 1851	7, 921
Lamp, Lard	J. Lee	Wellsville, Ohio	Nov. 21, 1842	2, 854
Lamp, Lard	B. K. Maltby and J. Neal	Rootstown and Middlebury, Ohio.	May 4, 1842	2, 604
Lamp, Lard	W. S. Moore	Springfield, Ill	Feb. 10, 1843	2, 950
Lamp, Lard	S. P. Moorehouse	Ludlowville, N. Y	Apr. 26, 1830	....
Lamp, Lard	C. F. Reese	Millersville, Pa	June 12, 1866	55, 535
Lamp, Lard	P. Robinson	Chillicothe, Ohio	Dec. 5, 1842	2, 873
Lamp, Lard	H. Rose	Rushville, Ill	Aug. 4, 1843	3, 206
Lamp, Lard	J. S. Senseny	Chambersburgh, Pa	July 15, 1856	15, 364
Lamp, Lard	T. Sewell	New York, N. Y	Oct. 2, 1847	5, 311
Lamp, Lard	I. Smith and J. Stonesifer	Boonsborough, Md	Aug. 8, 1854	11, 497
Lamp, Lard	L. A. Stockwell	Batavia, N. Y	Sept. 13, 1853	10, 022
Lamp, Lard	F. H. Southworth	Washington, D. C	July 2, 1842	2, 703
Lamp, Lard	F. H. Southworth	Washington, D. C	Oct. 22, 1842	2, 827
Lamp, Lard	T. Terrel	Spring Hill, Ohio	Aug. 14, 1866	57, 215
Lamp, Lard	S. B. Terry	Plymouth, Conn	Feb. 24, 1843	2, 976
Lamp, Lard	J. Tobin	Bloomfield, N. J	Mar. 26, 1844	3, 513
Lamp, Lard	H. Tomlinson	Geneva, N. Y	Sept. 1, 1843	3, 243
Lamp, Lard	W. Thompson	Cincinnati, Ohio	Apr. 10, 1843	3, 044
Lamp, Lard	C. Wilhelm	Philadelphia, Pa	Apr. 6, 1843	3, 032
Lamp, Lard	Z. Worrell	Chester Hill, Ohio	Feb. 7, 1842	2, 448
Lamp, Lard or oil solar	J. G. Webb	Williamsburgh, N. Y	Oct. 14, 1851	8, 436
Lamp, Lard, tallow, &c	G. Carr	Buffalo, N. Y	Sept. 4, 1841	2, 238
Lamp, Lard, tallow, &c	N. S. Cate and J. H. Putnam	Charlestown and Malden, Mass.	Nov. 16, 1841	2, 358
Lamp-light	I. Jennings	New York, N. Y	Oct. 16, 1830	....
Lamp, Light-house, &c	C. Cornelius	Philadelphia, Pa	Dec. 28, 1822	....
Lamp, Light-house	S. K. Jackson	Norfolk, Va	Dec. 3, 1872	133, 647
Lamp, Light-house	W. Lewis, sr., and B. Hemmenway.	Boston and Roxbury, Mass	Aug. 7, 1844	3, 692
Lamp, Light-house	C. Wheeler	Rockport, Mass	Apr. 11, 1846	4, 454
Lamp-lighter	A. Assmann	Rahway, N. J	Feb. 4, 1868	74, 033
Lamp-lighter	C. M. Clinton	Ithaca, N. Y	Dec. 1, 1863	40, 734
Lamp-lighter	J. Rigby	Fort Howard, Wis	Aug. 24, 1869	94, 031
Lamp-lighter	S. and J. Thomas	Brooklyn, N. Y	Mar. 8, 1859	23, 203
Lamp-lighter and burglar and fire alarm	J. B. Irwin	Newark, Ohio	May 24, 1870	103, 338
Lamp lighter, Street	M. McGill and J. E. Tynan	Paterson, N. J	May 14, 1867	64, 684
Lamp-lighting apparatus	M. A. Lynch	Boston, Mass	May 27, 1873	139, 255
Lamp-lighting apparatus	W. Wuerz	New York, N. Y	July 30, 1872	130, 000
Lamp-lighting device	N. Allen	West Meriden, Conn	Jan. 13, 1863	37, 378
Lamp-lighting device	N. L. Cole	Norwich, Conn	Mar. 17, 1868	75, 529
Lamp-lighting device	G. K. Proctor	Beverly, Mass	Jan. 24, 1860	26, 953
Lamp, Locomotive	S. E. and H. B. Cleveland	Buffalo, N. Y	June 24, 1856	15, 172
Lamp, Locomotive	L. A. Hamblen	Chicago, Ill	Jan. 13, 1857	16, 384
Lamp, Locomotive	N. J. Knapp	Chicago, Ill	Sept. 13, 1859	25, 421
Lamp, Locomotive	J. Radley	New York, N. Y	July 31, 1860	29, 402
Lamp, Locomotive	J. Radley	New York, N. Y	June 26, 1866	55, 903
Lamp, Locomotive	T. Snook and S. Hill	Rochester, N. Y	Dec. 21, 1852	9, 490

Index of patents issued from the United States Patent Office from 1790 *to* 1873, *inclusive*—Continued.

Invention.	Inventor.	Residence.	Date.	No.
Lamp, Locomotive	S. H. Trimmons	Memphis, Tenn	Apr. 8, 1862	34, 914
Lamp, Locomotive	I. A. Williams	Utica, N. Y	Oct. 10, 1854	11, 799
Lamp, Locomotive	I. A. Williams	Utica, N. Y	Apr. 29, 1862	35, 122
Lamp, Locomotive and railway	J. Stuber	Utica, N. Y	May 20, 1856	14, 942
Lamp, Locomotive coal-oil	P. Budenbach	New York, N. Y	Mar. 24, 1863	38, 001
Lamp, Locomotive head	P. Budenback	New York, N. Y	Nov. 28, 1865	51, 135
Lamp, Locomotive reflector	F. J. Seymour	Waterbury, Conn	July 8, 1856	15, 305
Lamp, Magnesium	I. S. Russell and H. R. Russell	New Market, Md., and Woodbury, N. J.	Apr. 29, 1873	138, 346
Lamp, Magnesium	R. H. Thurston	Providence, R. I	May 1, 1866	54, 442
Lamp, Magnesium	R. H. Thurston	Providence, R. I	June 5, 1866	55, 393
Lamp, Miner's	W. G. Dowd	Scranton, Pa	May 25, 1869	90, 434
Lamp, Miner's	W. McClare	Hyde Park, Pa	Aug. 15, 1865	49, 477
Lamp, Miner's	W. Seybold	McKeesport, Pa	May 13, 1862	35, 264
Lamp, Miner's	W. Seybold	McKeesport, Pa	June 8, 1869	90, 963
Lamp, Miner's	J. C. Sommerville	Snow Shoe, Pa	June 8, 1869	90, 967
Lamp, Miner's	G. W. Trimble	Bloomsburgh, Pa	Mar. 17, 1868	75, 603
Lamp, Miner's	W. C. Winfield	Hubbard, Ohio	May 23, 1871	115, 143
Lamp, Miner's	W. C. Winfield	Hubbard, Ohio	May 7, 1872	126, 606
Lamp mold, Glass	J. Bridges	Allentown, Pa	July 23, 1872	129, 781
Lamp mold, Glass	H. Dillaway	Sandwich, Mass	July 18, 1871	117, 157
Lamp mold, Glass	J. Jenkins	Boston, Mass	Sept. 18, 1860	30, 066
Lamp, Naval store-room or coach	J. J. Walton	New York, N. Y	July 1, 1873	140, 448
Lamp, Night	T. W. Howchin	Morrisania, N. Y	May 6, 1862	35, 154
Lamp, Nurse	W. Howe	Boston, Mass	Dec. 31, 1812	....
Lamp, Oil	J. Calkins	Long Branch, N. J	July 23, 1872	129, 786
Lamp, Oil	T. S. Speakman	Camden, N. J	Apr. 4, 1865	47, 173
Lamp, Oil or spirit gas	J. M. and G. Truman	Penn Township, Pa	Nov. 19, 1833	....
Lamp or candle stand	F. A. Marshall	Marlborough, Mass	Jan. 29, 1861	31, 251
Lamp or candlestick and match-box combined	T. Shanks	Baltimore, Md	Jan. 19, 1858	19, 158
Lamp or gas-light bracket	C. Monson	New Haven, Conn	Sept. 15, 1863	39, 945
Lamp or gas shade	M. J. Wellman	New York, N. Y	Aug. 4, 1863	39, 438
Lamp or holder, Candle	M. Weis	Boston, Mass	Feb. 20, 1872	123, 803
Lamp, Paraffine	H. Ryder	Somerville, Mass	Apr. 29, 1873	138, 441
Lamp, Petroleum	W. B. Billings	Brooklyn, N. Y	Oct. 11, 1864	44, 592
Lamp, Petroleum	J. and J. Hinks	Birmingham, England	Dec. 10, 1867	72, 040
Lamp, Petroleum argand	A. K. Murray and A. B. Howland.	Titusville, Pa	Jan. 21, 1868	73, 455
Lamp, Plumbers' and painters'	G. Wanier	New York, N. Y	Dec. 3, 1867	71, 826
Lamp, Pocket	J. Espelding and W. E. Huttmann.	Chicago, Ill	June 29, 1869	91, 833
Lamp, Pocket	A. P. Foster and D. B. H. Bartlett.	Lowell, Mass	Aug. 31, 1869	94, 300
Lamp, Pocket	H. L. Hervey	Quincy, Ill	Dec. 9, 1856	16, 180
Lamp, Portable condensing	T. Philipps	New York, N. Y	June 25, 1823	....
Lamp, Portable cooking-apparatus by	C. A. Harper	Little Rock, Ark	May 1, 1866	54, 337
Lamp-post	L. A. Gouch	Yonkers, N. Y	Nov. 21, 1871	121, 165
Lamp-post	L. A. Gouch	Yonkers, N. Y	June 3, 1873	139, 504
Lamp-post	J. W. Graham	Chillicothe, Ohio	May 25, 1869	90, 356
Lamp-post	E. A. Heath	New York, N. Y	Jan. 23, 1872	122, 887
Lamp-post attachment for supporting electrical wires.	J. P. Tirrell	Charlestown, Mass	June 3, 1873	139, 630
Lamp post, Street	P. H. Branson	Saint Louis, Mo	Feb. 21, 1865	46, 444
Lamp post, Street gas	S. W. France	Brooklyn, N. Y	Dec. 5, 1871	121, 456
Lamp post, Street gas	J. T. P. Hunt	Manchester, N. H	Jan. 3, 1865	45, 717
Lamp-post, Sediment-chamber for	G. C. Bovey	Cincinnati, Ohio	Feb. 8, 1870	99, 527
Lamp-post, Sediment-chamber for	J. W. Graham	Chillicothe, Ohio	Sept. 26, 1871	119, 347
Lamp, Railway	J. Carton	Utica, N. Y	Aug. 26, 1862	36, 274
Lamp, Railway	J. Carton	Utica, N. Y	Oct. 7, 1862	36, 602
Lamp, Railway	J. M. A. Dew	Chicago, Ill	Aug. 10, 1869	93, 602
Lamp, Railway	L. Hover	Chicago, Ill	May 24, 1864	42, 853
Lamp, Railway	T. J. Newland	Utica, N. Y	Dec. 23, 1862	37, 239
Lamp, Railway	L. S. White	Chicopee, Mass	Sept. 5, 1854	11, 657
Lamp, Railway and marine signal	J. F. Veronee	Charleston, S. C	July 16, 1872	129, 502
Lamp, Railway-car	J. L. Howard	Hartford, Conn	Oct. 7, 1873	143, 411
Lamp, Railway-car	W. C. Marshall	Hartford, Conn	Aug. 20, 1872	130, 586
Lamp, Railway-car	R. S. Merrill	Boston, Mass	Nov. 22, 1870	109, 537
Lamp, Railway-car	W. H. Paige	Springfield, Mass	Jan. 16, 1872	122, 734
Lamp, Railway-car	T. S. Williams and P. S. Page	Boston, Mass	May 19, 1863	38, 620
Lamp, Railway-car coal-oil	W. O. B. Merrill	Philadelphia, Pa	June 10, 1862	35, 534
Lamp, Reflecting	J. C. Fletcher	Springfield, Ohio	Mar. 30, 1836	....
Lamp-reflector	J. W. Haines	Cambridge, Mass	July 11, 1871	116, 953
Lamp-reflector	F. Judson	Castleton, N. Y	Feb. 15, 1870	99, 908
Lamp, Reflector	J. H. Pease	Reading, Pa	June 8, 1852	9, 001
Lamp-reflector	J. T. Williams	Philadelphia, Pa	June 17, 1862	35, 644
Lamp-reflector, Adjustable	A. Haye	New York, N. Y	July 25, 1871	117, 282
Lamp reflector and chimney protector	G. F. J. Colburn	Newark, N. J	June 10, 1862	35, 507
Lamp-reservoirs, Placing	C. N. Orpen	New York, N. Y	July 29, 1862	36, 060
Lamp, Resin-oil	F. Blake	Needham, Mass	July 17, 1855	13, 259
Lamp, Resin-oil	S. Constant	Brooklyn, N. Y	Jan. 24, 1854	10, 443
Lamp, Resin-oil	S. Constant	Brooklyn, N. Y	Aug. 8, 1854	11, 474
Lamp, Resin-oil	I. Pitman	Reading, Mass	Sept. 19, 1854	11, 701
Lamp, Resin-oil	P. Sargent	Newburyport, Mass	Mar. 4, 1856	14, 369
Lamp, Resin-oil argand	I. Van Bunschoten	New York, N. Y	Mar. 18, 1856	14, 478
Lamp, Safety	J. Defossez	Paris, France	Sept. 2, 1862	36, 341
Lamp, Safety	A. H. Emery	New York, N. Y	Dec. 29, 1868	85, 373
Lamp, Safety	J. W. Hoffman	Southwark District, Pa	Apr. 23, 1850	7, 303
Lamp, Safety	C. R. Laudmann	New York, N. Y	Sept. 26, 1854	11, 730
Lamp, Safety	W. Pratt	Baltimore, Md	Nov. 24, 1857	18, 704
Lamp, Safety	F. Saunders	Aberdeen, Miss	May 4, 1869	89, 795
Lamp, Safety	C. Smith	Hermon, Me	Dec. 7, 1869	97, 560
Lamp safety-apparatus	C. J. Brown	Plymouth, N. H	Jan. 21, 1873	135, 074
Lamp safety-attachment	J. F. Sanford	Keokuk, Iowa	Apr. 27, 1869	89, 508
Lamp, Safety compass	J. Feinour, jr	Philadelphia, Pa	Apr. 5, 1828	....
Lamp, Safety derrick	J. Dillen	Petroleum Centre, Pa	May 3, 1870	102, 663
Lamp, Safety derrick	H. Freeman	Petroleum Centre, Pa	May 14, 1872	126, 688
Lamp safety-tube	G. M. Hopkins and J. A. Straight	Albion, N. Y	Nov. 29, 1870	109, 735
Lamp safety-tube	F. Stewart	Philadelphia, Pa	July 2, 1850	7, 484

Index of patents issued from the United States Patent Office from 1790 *to* 1873, *inclusive*—Continued.

Invention.	Inventor.	Residence.	Date.	No.
Lamp screw-collar	W. H. Russell	Brooklyn, N. Y	June 7, 1870	104, 066
Lamp, Self-acting blow-pipe	D. W. C. McCloskey	New York, N. Y	Aug. 26, 1851	8, 324
Lamp, Self-extinguishing	J. J. Cuthbert	Duck Port, La	June 18, 1872	127, 963
Lamp, Self-generating gas	S. Clayton and Y. Bailey	West Chester, Pa	Aug. 27, 1850	7, 594
Lamp, Self-lighting	W. W. Batchelder	Boston, Mass	May 2, 1871	114, 389
Lamp, Self-lighting	W. W. Batchelder	Boston, Mass	Oct. 3, 1871	119, 497
Lamp, Self-lighting	A. Bennett	New York, N. Y	Mar. 27, 1849	6, 244
Lamp, Self-lighting	T. W. Carroll	Baltimore, Md	Jan. 24, 1860	26, 884
Lamp, Self-lighting	W. A. Leonard	Boston, Mass	May 14, 1872	126, 640
Lamp, Self-supplying	E. B. Horn	Boston, Mass	Sept. 11, 1844	3, 735
Lamp-shade	J. B. Alexander	Washington, D. C	Mar. 30, 1869	88, 430
Lamp-shade	B. Arnold	East Greenwich, R. I	July 9, 1861	32, 740
Lamp-shade	D. W. Bashore	Palmyra, Pa	Aug. 27, 1867	68, 278
Lamp-shade	D. W. Bashore	Palmyra, Pa	June 2, 1868	78, 413
Lamp-shade	R. W. Churchill	Bridgeport, Conn	Sept. 21, 1869	95, 000
Lamp-shade	M. H. Collins	Chelsea, Mass	Apr. 25, 1871	113, 982
Lamp-shade	M. H. Collins	Chelsea, Mass	May 16, 1871	114, 766
Lamp-shade	M. H. Collins	Chelsea, Mass	July 25, 1871	117, 260
Lamp-shade	A. Combs	Burlingame, Kans	Aug. 27, 1872	130, 899
Lamp-shade	G. H. Dimond and G. Doolittle	Bridgeport, Conn	Feb. 9, 1869	86, 825
Lamp-shade	T. B. Doolittle	Bridgeport, Conn	Mar. 10, 1868	75, 390
Lamp-shade	J. V. Dunlap	Hartford, Conn	Nov. 13, 1866	59, 566
Lamp-shade	M. B. Dyott	Philadelphia, Pa	Jan. 16, 1855	12, 239
Lamp-shade	J. Emery	Bucksport, Me	July 9, 1867	66, 576
Lamp-shade	J. Emery	Bucksport, Me	Mar. 3, 1868	75, 136
Lamp-shade	F. E. Foster	Cressona, Pa	Jan. 19, 1869	85, 921
Lamp-shade	H. M. Hartshorn	Malden, Pa	Mar. 10, 1868	75, 418
Lamp-shade	A. W. Hasskell	Boston, Mass	Aug. 2, 1870	105, 940
Lamp-shade	E. K. Haynes	Hanover, N. H	Nov. 19, 1867	71, 168
Lamp-shade	E. K. Haynes	Hanover, N. H	May 19, 1868	78, 084
Lamp-shade	E. K. Haynes	Boston, Mass	Nov. 30, 1869	97, 294
Lamp-shade	A. B. Hendryx	Ansonia, Conn	June 8, 1869	91, 130
Lamp-shade	A. M. Hills	Hockanum, Conn	May 9, 1865	47, 638
Lamp-shade	C. E. L. Holmes	Waterbury, Conn	Apr. 1, 1862	34, 860
Lamp-shade	W. H. Horton	Chelsea, Mass	Feb. 18, 1868	74, 537
Lamp-shade	S. W. Huntington	Augusta, Me	Apr. 7, 1868	76, 462
Lamp-shade	S. W. Huntington	Augusta, Me	Apr. 21, 1868	76, 921
Lamp-shade	W. H. Johnson	Springfield, Mass	Mar. 8, 1870	100, 533
Lamp-shade	W. Kemble and W. H. C. Bartlett.	New York and West Point, N. Y.	Oct. 20, 1857	18, 456
Lamp-shade	A. D. Laws	Bridgeport, Conn	Feb. 16, 1869	86, 987
Lamp-shade	N. F. Leete	Middletown, Conn	Mar. 24, 1868	75, 932
Lamp-shade	E. A. Locke and W. N. Needen	Boston, Mass	Sept. 8, 1868	81, 915
Lamp-shade	G. Macbridge	Carthage, Ill	Sept. 10, 1872	131, 173
Lamp-shade	J. McFarland	Clinton, Ill	Mar. 20, 1866	53, 320
Lamp-shade	J. F. Phelps	Havana, N. Y	Mar. 2, 1869	87, 428
Lamp-shade	W. Reed, jr	Boston, Mass	June 7, 1870	103, 928
Lamp-shade	W. Robinson	Spring Valley, N. Y	Mar. 1, 1870	100, 450
Lamp-shade	S. Rust	New York, N. Y	Jan. 9, 1838	565
Lamp-shade	W. F. Shaw	Boston, Mass	Nov. 26, 1867	71, 333
Lamp-shade	A. Shepard	Ashland, Mass	Dec. 18, 1866	60, 577
Lamp-shade	W. Simons	Charleston, S. C	Mar. 18, 1873	137, 034
Lamp-shade	A. A. Smith	Seneca Falls, N. Y	Apr. 21, 1868	77, 115
Lamp-shade	M. Stewart	Philadelphia, Pa	Feb. 20, 1866	52, 768
Lamp-shade	C. St. John	Boston, Mass	July 4, 1865	48, 632
Lamp-shade	C. St. John	Boston, Mass	Oct. 17, 1865	50, 509
Lamp-shade	J. F. Travis	New York, N. Y	May 18, 1869	90, 135
Lamp-shade	G. W. Tucker	Waterbury, Conn	May 4, 1869	89, 707
Lamp-shade	J. H. Webber	Charlestown, Mass	July 24, 1866	56, 647
Lamp-shade	G. Wedekind	Philadelphia, Pa	May 12, 1868	77, 936
Lamp-shade	G. Wedekind	Philadelphia, Pa	June 30, 1868	79, 526
Lamp-shade	G. Wedekind	Philadelphia, Pa	Nov. 30, 1869	97, 463
Lamp-shade	A. M. Weeks	New York, N. Y	June 16, 1868	79, 039
Lamp-shade	M. J. Williams and J. J. Greenough.	New York, N. Y	Feb. 28, 1865	46, 600
Lamp-shade	C. and A. C. Wilhelm	Philadelphia, Pa	May 3, 1859	23, 875
Lamp-shade	H. Zahn	New York, N. Y	Mar. 7, 1865	46, 748
Lamp shade and drip-cup, Combined	J. S. and T. B. Atterbury	Pittsburgh, Pa	Dec. 16, 1873	145, 548
Lamp shade and reflector	J. M. Kenerson	Newport, N. H	Aug. 9, 1870	106, 173
Lamp-shade clasp	J. J. Marcy	West Meriden, Conn	July 2, 1867	66, 366
Lamp-shade clasp	C. Reichman	Philadelphia, Pa	Feb. 4, 1862	34, 332
Lamp-shade clasp	C. Reichman	Philadelphia, Pa	Sept. 12, 1865	49, 921
Lamp-shade, Folding	H. M. Hartshorn	Malden, Mass	Jan. 3, 1871	110, 651
Lamp-shade holder	L. J. Atwood	Waterbury, Conn	May 30, 1865	47, 913
Lamp-shade holder	E. E. Conrad	Philadelphia, Pa	Apr. 8, 1862	34, 921
Lamp shade holder	E. E. Conrad	Philadelphia, Pa	Feb. 11, 1873	135, 778
Lamp-shade holder	J. Fallows	Philadelphia, Pa	July 5, 1864	43, 397
Lamp-shade holder	J. Foller	Washington, D. C	Sept. 6, 1870	107, 026
Lamp-shade holder	E. P. Gleason	New York, N. Y	Nov. 16, 1869	96, 909
Lamp-shade holder	J. Hanley	New York, N. Y	Apr. 25, 1865	47, 418
Lamp-shade holder	M. W. House	Cleveland, Ohio	Mar. 7, 1871	112, 459
Lamp-shade holder	M. W. House	Cleveland, Ohio	July 29, 1873	141, 223
Lamp-shade holder	J. W. Lyon	Brooklyn, N. Y	May 3, 1870	102, 691
Lamp-shade holder	J. W. Lyon	Brooklyn, N. Y	Mar. 28, 1871	113, 181
Lamp-shade holder	W. C. Matthews	Chelsea, Mass	Mar. 25, 1862	34, 805
Lamp-shade holder	C. M. Mitchell	Waterbury, Conn	May 18, 1869	90, 287
Lamp-shade holder	T. B. Peacock	Dresden, Ohio	Aug. 30, 1870	106, 866
Lamp-shade holder	O. N. Perkins	Meriden, Conn	Dec. 10, 1872	133, 885
Lamp-shade holder	J. T. Pope	New York, N. Y	Mar. 23, 1869	88, 206
Lamp-shade holder	J. T. Pope	New York, N. Y	May 11, 1869	90, 021
Lamp-shade holder	C. H. Reichmann	New York, N. Y	Sept. 30, 1862	36, 574
Lamp-shade holder	P. A. Stecher	New York, N. Y	Aug. 26, 1862	36, 308
Lamp-shade holder	C. St. John	Boston, Mass	Feb. 21, 1865	46, 534
Lamp-shade holder	C. St. John	New York, N. Y	Jan. 28, 1870	104, 898
Lamp shade or globe holder	L. J. Atwood	Waterbury, Conn	July 29, 1873	141, 251
Lamp-shade panel	G. W. Bruiker	Philadelphia, Pa	Feb. 12, 1867	61, 994
Lamp-shade protector	C. and A. C. Wilhelm	Philadelphia, Pa	Apr. 3, 1855	12, 646
Lamp-shade support	J. B. Alexander	Washington, D. C	Apr. 5, 1870	101, 409

Index of patents issued from the United States Patent Office from 1790 *to* 1873, *inclusive*—Continued.

Invention.	Inventor.	Residence.	Date.	No.
Lamp-shade support	J. H. Connelly	Wheeling, W. Va	July 3, 1866	56, 011
Lamp-shade support	E. E. Conrad	Philadelphia, Pa	June 21, 1870	104, 558
Lamp-shade support	C. W. Emerson	Hartford, Conn	Oct. 5, 1869	95, 575
Lamp-shade support	G. W. Hatch	Brooklyn, N. Y	June 25, 1872	128, 308
Lamp-shade support	E. W. Holt	Corinna, Me	June 7, 1870	103, 885
Lamp-shade support	J. W. Lyon	Brooklyn, N. Y	Apr. 21, 1868	77, 059
Lamp-shade support	G. C. James	Cincinnati, Ohio	Oct. 22, 1867	70, 094
Lamp-shads support	C. Reichmann	Philadelphia, Pa	Oct. 20, 1863	40, 361
Lamp-shade support	E. Russell	Waterbury, Conn	May 30, 1871	115, 527
Lamp-shade support	W. F. Shaw	Boston, Mass	Dec. 14, 1858	22, 311
Lamp-shade supports, Process of making	W. N. Weeden	Waterbury, Conn	Dec. 16, 1873	145, 702
Lamp-shades, Incombustible paper	G. Wedekind	Philadelphia, Pa	May 12, 1863	38, 523
Lamp-shades, Manufacture of	J. H. Hobbs, C. W. Brockemier, and W. Leighton, jr.	Wheeling, W. Va	Apr. 25, 1871	114, 140
Lamp-shades, Mode of packing	E. A. Locke and W. N. Weeden	Boston, Mass	Nov. 24, 1868	84, 289
Lamp, Shoemaker's	J. M. Sparer	Philadelphia, Pa	Dec. 3, 1867	71, 658
Lamp sign, Street	H. D. Coleman	New Orleans, La	Oct. 1, 1872	131, 741
Lamp, Signal	R. P. Bailey	Niagara, N. Y	May 12, 1857	17, 253
Lamp, Signal	J. and J. Feinour, jr	Philadelphia, Pa	Aug. 1, 1828	....
Lamp, Signal	J. Wall	New York, N. Y	June 25, 1867	66, 191
Lamp, Spirit	A. J. Walker	New York, N. Y	May 24, 1853	9, 751
Lamp-snuffer	R. N. Eagle	New York, N. Y	Sept. 30, 1862	36, 530
Lamp-snuffer	K. Schow	La Fayette, Ind	Apr. 10, 1866	53, 890
Lamp snuffer and extinguisher	M. L. Battle	Bainbridge, Ga	May 7, 1872	126, 437
Lamp, Solar	J. Hassell	Brooklyn, N. Y	June 9, 1857	17, 507
Lamp-stand and clothes-dryer, Combined	J. Donaldson	Rockford, Ill	Aug. 8, 1865	49, 244
Lamp-stands, &c., Mode of ornamenting	H. W. Hayden	Waterbury, Conn	June 9, 1863	38, 823
Lamp, Street	T. P. Austin	New York, N. Y	Aug. 30, 1870	106, 764
Lamp, Street	J. Benson	Yonkers, N. Y	May 9, 1871	114, 518
Lamp, Street	E. Boesch	San Francisco, Cal	Sept. 6, 1870	106, 990
Lamp, Street	C. J. Bowell	Cleveland, Ohio	June 25, 1872	128, 355
Lamp, Street	G. Brandon	New York, N. Y	Mar. 21, 1871	112, 774
Lamp, Street	A. Burger	New York, N. Y	May 14, 1872	126, 783
Lamp, Street	O. Case and B. D. Evans	Columbus, Ohio	Dec. 1, 1868	84, 535
Lamp, Street	C. Cuppers	New York, N. Y	Oct. 4, 1870	107, 884
Lamp, Street	M. B. Dyott	Philadelphia, Pa	Mar. 19, 1872	124, 673
Lamp, Street	M. B. Dyott	Philadelphia, Pa	Apr. 8, 1873	137, 538
Lamp, Street	J. H. Eyrse	Pekin, Ill	Sept. 23, 1873	143, 000
Lamp, Street	J. S. Fish	Cleveland, Ohio	June 11, 1872	127, 685
Lamp, Street	B. Giroux	Chicago, Ill	July 16, 1872	129, 221
Lamp, Street	J. S. Hagerty	Baltimore, Md	Sept. 9, 1873	142, 693
Lamp, Street	J. F. Harly	Kipton Station, Ohio	Feb. 1, 1870	99, 314
Lamp, Street	F. Hartman	Chicago, Ill	Mar. 4, 1873	136, 373
Lamp, Street	C. F. Jacobsen and A. Burger	New York, N. Y	May 27, 1873	139, 394
Lamp, Street	J. F. Kerno	Baltimore, Md	Jan. 10, 1871	110, 922
Lamp, Street	J. H. Kramer and A. Burger	New York, N. Y	Nov. 9, 1869	96, 707
Lamp, Street	F. Lange	Chicago, Ill	Dec. 15, 1868	84, 952
Lamp, Street	T. T. Markland	Philadelphia, Pa	June 19, 1866	55, 686
Lamp, Street	T. T. Markland, jr	Philadelphia, Pa	Apr. 7, 1868	76, 335
Lamp, Street	J. F. Marsh	Dubuque, Iowa	June 24, 1873	140, 149
Lamp, Street	A. J. McDowell and D. T. Bates	Richmond, Ind	July 23, 1872	129, 846
Lamp, Street	J. G. Miner	Morrisania, N. Y	May 29, 1866	55, 143
Lamp, Street	J. G. Miner	Morrisania, N. Y	June 17, 1873	140, 062
Lamp, Street	A. Nahe	Brooklyn, N. Y	Sept. 2, 1873	142, 500
Lamp, Street	T. North	Cincinnati, Ohio	June 27, 1871	116, 476
Lamp, Street	J. H. Robinson	Washington, D. C	May 7, 1872	126, 577
Lamp, Street	W. G. Schmidlen and J. W. Driscoll.	New York, N. Y	Apr. 12, 1870	101, 769
Lamp, Street	W. G. Schmidlin and J. W. Driscoll.	Hoboken, N. J., and New York, N. Y.	July 25, 1871	117, 470
Lamp, Street	W. G. Schmidlin and J. W. Driscoll.	New York, N. Y	Feb. 6, 1872	123, 516
Lamp, Street	F. Schumann	Washington, D. C	Dec. 3, 1872	133, 672
Lamp, Street	T. A. Skelton	Middlesex County, England.	Jan. 2, 1872	122, 409
Lamp, Street	R. H. Smith	Pittsburgh, Pa	June 20, 1871	116, 234
Lamp, Street	J. Stratton	Brooklyn, N. Y	June 27, 1865	48, 461
Lamp, Street	W. O. Wheeler	Deposit, N. Y	June 23, 1868	79, 1 9
Lamp, Street gas	W. A. Shaw and G. Parker	Boston, Mass	Apr. 11, 1854	10, 760
Lamp, Submarine	L. H. Hasse	New York, N. Y	Oct. 9, 1860	30, 314
Lamp-supporter	E. Russell and F. W. Platt	Waterbury, Conn	June 20, 1871	116, 224
Lamp, Suspended bunker	H. J. Van Thiel	Stapleton, N. Y	Jan. 5, 1864	41, 108
Lamp, Suspension barrel	I. Couch and J. A. Frary	Meriden, Conn	Mar. 13, 1833	....
Lamp, Tallow	J. Love		June 11, 1798	....
Lamp, Torch	H. Brewer	East Boston, Mass	Jan. 10, 1854	10, 419
Lamp-top	C. Dorflinger	Brooklyn, N. Y	Aug. 5, 1862	36, 079
Lamp-top	F. Hewitt	Newark, N. J	June 9, 1863	38, 826
Lamp-top	S. E. Winslow	Kensington, Pa	Oct. 24, 1848	5, 880
Lamp tops, rivets, &c., Method of making	L. C. White	Meriden, Conn	Sept. 7, 1852	9, 256
Lamp-train gearing	R. Hitchcock and G. A. Jones	New York, N. Y	Feb. 25, 1868	74, 914
Lamp trimmer and extinguisher	C. P. Snow	Freeport, Ill	Nov. 30, 1869	97, 453
Lamp-trimmer's shears	W. B. Barnard	Waterbury, Conn	Dec. 27, 1864	45, 574
Lamp-tube	I. Jennings	New York, N. Y	July 9, 1850	7, 492
Lamp-tube-forming machine	W. Wallace	Ansonia, Conn	June 16, 1868	79, 036
Lamp-tubes, Device for adjusting wicks in	H. A. De Graw	Newburgh, N. Y	Jan. 27, 1863	37, 495
Lamp-tubes, Method of forming	T. B. Doolittle	Bridgeport, Conn	July 28, 1868	80, 464
Lamp, Turpentine	L. Jones	New York, N. Y	Nov 25, 1838	1, 022
Lamp, Vapor	C. E. Atherton	Paterson, N. J	Nov. 6, 1860	30, 556
Lamp, Vapor	H. Bateman	Boston, Mass	May 25, 1858	20, 386
Lamp, Vapor	W. B. Billings	New York, N. Y	Oct. 2, 1860	30, 194
Lamp, Vapor	M. V. B. Buell	Buffalo, N. Y	May 1, 1860	28, 063
Lamp, Vapor	M. L. Callender	New York, N. Y	Nov. 26, 1861	33, 773
Lamp, Vapor	J. Clarke	Syracuse, N. Y	June 12, 1860	28, 652
Lamp, Vapor	T. G. Clayton	Washington, D. C	May 15, 1860	28, 258
Lamp, Vapor	N. Clifford	Brooklyn, N. Y	July 9, 1861	32, 755
Lamp, Vapor	C. C. Coe	Rome, N. Y	July 9, 1861	32, 808
Lamp, Vapor	L. F. Conover	Philadelphia, Pa	Feb. 28, 1860	27, 272
Lamp, Vapor	E. Crooker	Buffalo, N. Y	July 10, 1860	29, 124

Index of patents issued from the United States Patent Office from 1790 *to* 1873, *inclusive*—Continued.

Invention.	Inventor.	Residence.	Date.	No.
Lamp, Vapor	M. W. Dillingham	Charlestown, Mass	Nov. 6, 1860	30,568
Lamp, Vapor	H. W. Dopp	Buffalo, N. Y	Nov. 13, 1860	30,621
Lamp, Vapor	T. Drake	Windsor, Conn	Mar. 3, 1863	37,811
Lamp, Vapor	D. A. Fiske	Milwaukee, Wis	Jan. 7, 1862	34,053
Lamp, Vapor	A. Geiger	Dayton, Ohio	May 29, 1860	28,467
Lamp, Vapor	J. G. Gilbert	New York, N. Y	Sept. 8, 1859	18,142
Lamp, Vapor	J. S. Gray	New York, N. Y	Dec. 18, 1860	30,965
Lamp, Vapor	H. Johnson	Washington, D. C	Feb. 28, 1860	27,293
Lamp, Vapor	H. Johnson	Washington, D. C	Apr. 17, 1860	27,912
Lamp, Vapor	C. B. Loveless	Tom's River, N. J	Jan. 22, 1861	31,179
Lamp, Vapor	S. W. Lowe	Philadelphia, Pa	May 29, 1860	28,536
Lamp, Vapor	A. M. Mace	Springfield, Mass	June 22, 1858	20,649
Lamp, Vapor	H. N. Macomber	Lynn, Mass	May 18, 1858	20,283
Lamp, Vapor	J. K. O'Neil	Kingston, N. Y	Nov. 22, 1859	26,200
Lamp, Vapor	I. W. Pettibone	Norfolk, Conn	May 15, 1860	28,299
Lamp, Vapor	W. H. Racey	Saint Augustine, Fla	June 29, 1858	20,729
Lamp, Vapor	T. S. Ray and A. C. Rand	Buffalo, N. Y	May 29, 1860	28,503
Lamp, Vapor	C. W. Richter	Madison, Ga	June 12, 1860	28,691
Lamp, Vapor	J. J. Riddle	Cincinnati, Ohio	Feb. 7, 1865	46,266
Lamp, Vapor	J. H. Rollins	Wapello, Iowa	May 15, 1860	28,306
Lamp, Vapor	L. Short	Buffalo, N. Y	Jan. 29, 1861	31,282
Lamp, Vapor	I. Van Bunschoten	New York, N. Y	July 31, 1860	29,421
Lamp, Vapor	G. Walker	Philadelphia, Pa	June 26, 1860	28,924
Lamp, Vapor	F. W. Willard	New York, N. Y	Dec. 11, 1860	30,899
Lamp, Vapor-burning	D. H. Chamberlain	West Roxbury, Mass	Nov. 24, 1857	18,719
Lamp, Vapor-burning	N. Mason	Chelsea, Mass	July 20, 1858	20,952
Lamp, Vapor-burning	S. Whitmarsh	Northampton, Mass	Aug. 12, 1856	15,547
Lamp, Vapor-burning street	C. E. Smith and H. J. Rice	Columbus, Ohio	Aug. 23, 1870	106,630
Lamp, Vehicle	S. P. Dodge	Boston, Mass	June 5, 1866	55,255
Lamp, Vehicle	C. F. Waldron	New York, N. Y	Apr. 14, 1868	76,652
Lamp ventilator and shade	C. Recht	New York, N. Y	July 31, 1866	56,799
Lamp ventilator, Petroleum-oil	S. D. Pinkham	Fond du Lac, Wis	Apr. 22, 1862	35,039
Lamp, Vulcanizing	B. W. Franklin	New York, N. Y	Feb. 3, 1863	37,575
Lamp-wick	P. Backer	Chicago, Ill	Dec. 20, 1870	110,188
Lamp-wick	T. Bingham	Newburgh, N. Y	Dec. 13, 1864	45,384
Lamp-wick	E. D. Boyd	Helena, Ark	Sept. 14, 1869	94,707
Lamp-wick	J. H. Connelly	Wheeling, W. Va	Jan. 6, 1863	37,276
Lamp-wick	J. H. Connelly and J. Cook	Wheeling, Va	Aug. 26, 1862	36,277
Lamp-wick	J. Farrell	Brooklyn, N. Y	May 23, 1871	115,045
Lamp-wick	W. A. Gensch	New York, N. Y	Aug. 25, 1868	81,359
Lamp-wick	I. L. Hoard	Bristol, R. I	Nov. 20, 1866	59,839
Lamp-wick	C. P. Ladd	Bloomfield, N. J	Feb. 20, 1872	123,917
Lamp-wick	C. B. Larchar	New York, N. Y	July 7, 1863	39,153
Lamp-wick	C. W. Le Count	Norwalk, Conn	Oct. 30, 1866	59,237
Lamp-wick	C. W. Le Count and S. Chard	Norwalk, Conn	Feb. 27, 1866	52,862
Lamp-wick	J. Y. Leslie	Brooklyn, N. Y	Oct. 26, 1858	21,890
Lamp-wick	P. Martin	Medford, Mass	Dec. 16, 1873	145,667
Lamp-wick	C. F. Martine	Boston, Mass	Mar. 5, 1867	62,657
Lamp-wick	F. McKee	Pittsburgh, Pa	Dec. 23, 1862	37,237
Lamp-wick	P. Noyes	Lowell, Mass	Dec. 19, 1865	51,611
Lamp-wick	H. T. Robbins	Hyde Park, Mass	May 30, 1871	115,524
Lamp-wick	D. T. Robinson	Boston, Mass	Dec. 27, 1870	110,591
Lamp-wick	S. Rust	New York, N. Y	May 10, 1845	4,040
Lamp-wick	E. O. Schartau	Philadelphia, Pa	May 12, 1868	77,767
Lamp-wick	P. J. Skinner	Oswego, N. Y	Sept. 13, 1870	107,299
Lamp-wick	C. L. Topliff	New York, N. Y	June 19, 1866	55,745
Lamp-wick	B. F. Walton	Philadelphia, Pa	Jan. 23, 1866	52,231
Lamp-wick	A. J. White	Brooklyn, N. Y	Apr. 28, 1863	38,349
Lamp-wick	J. B. Wortendyke	Godwinsville, N. J	Apr. 26, 1859	23,801
Lamp-wick, Adjustable	G. Finley	Collins Township, Pa	Apr. 28, 1863	38,302
Lamp-wick adjuster	L. B. Lathrop	San José, Cal	Sept. 26, 1871	119,275
Lamp-wick adjuster	C. C. Stansell	Middleborough, Mass	Feb. 13, 1866	52,618
Lamp-wick adjuster	H. Woodbury	Huntsville, Mo	Apr. 23, 1872	126,002
Lamp-wick, Automatic lighting	W. H. Weeks	New York, N. Y	May 17, 1870	103,110
Lamp-wicks, Feed-roller for	C. H. Bagley	Elgin, Ill	June 18, 1867	65,861
Lamp-wick feeder	G. Cade	Long Branch, N. J	Apr. 5, 1870	101,427
Lamp-wick gage	J. A. Paddock	Chandlerville, Ill	Dec. 24, 1872	134,161
Lamp wick-holder, Argand	C. Moeller	Newark, N. J	Feb. 12, 1856	14,248
Lamp-wick inserter	W. Y. A. Boardman	New Haven, Conn	Dec. 4, 1866	60,130
Lamp-wick raiser	A. Combs	Philadelphia, Pa	Feb. 18, 1873	136,034
Lamp-wick regulator	H. Beebe	Jersey City Heights, N. J	May 16, 1871	114,906
Lamp-wick regulator	L. W. Leary	Norfolk, Va	Sept. 20, 1870	107,617
Lamp-wick regulator	H. Mund and E. Hoffmann	Chicago, Ill	July 10, 1866	56,250
Lamp-wick regulator	J. Pomeroy	Derby, Conn	June 16, 1863	38,933
Lamp-wick trimmer	H. Halvorsen	Cambridge, Mass	July 19, 1859	24,804
Lamp-wick trimmer	E. C. Jenkins, jr	Worcester, Mass	Dec. 14, 1869	97,776
Lamp-wick trimmer	S. Naylor and A. Fairchild	Independence, Iowa	Aug. 26, 1873	142,265
Lamp-wick trimmer	J. F. Sanford	Keokuk, Iowa	Feb. 9, 1869	86,868
Lamp-wick trimmer	D. Warner	Boston, Mass	Aug. 4, 1868	80,577
Lamp-wick trimmer, chimney cleaner and lifter, Combined.	C. M. Tyler	Indianapolis, Ind	July 20, 1869	92,908
Lamp-wick tube	F. H. Fuller and O. S. Severance	Boston, Mass	Dec. 1, 1868	84,542
Lamp-wick tube	J. H. Gray	Boston, Mass	Mar. 1, 1870	100,285
Lamp-wick tube	E. K. Haynes	Boston, Mass	June 6, 1871	115,607
Lamp-wick tube	A. D. Laws	Bridgeport, Conn	Mar. 10, 1868	75,281
Lamp-wick tube	E. W. Perry	Boston, Mass	Oct. 12, 1842	2,811
Lamp-wick tube	S. Rust	New York, N. Y	Mar. 21, 1846	4,428
Lamp wick-tubes, Camphene	J. Maclean	Philadelphia, Pa	Dec. 26, 1845	4,326
Lamp-wicks, Chemically-prepared	A. M. Daniels	Hartford, Conn	Apr. 4, 1871	113,269
Lamp-wicks, Device for raising and adjusting	J. B. Alexander	Washington, D. C	Nov. 17, 1868	84,045
Lamp-wicks, Device for trimming	H. F. Bond	Waltham, Mass	Jan. 3, 1865	45,692
Lamp-wicks, Elevator-tube for	R. Cornelius and C. Wilhelm	Philadelphia, Pa	July 24, 1849	6,6[illegible]3
Lamp-wicks, Machine for making	J. H. Connelly	Wheeling, W. Va	Jan. 14, 1864	43,097
Lamp-wicks, Process for treating	S. F. Sutherland	Baltimore, Md	Jan. 28, 1873	135,252
Lamp-wicks, Raising	S. Rust	New York, N. Y	Mar. 9, 1844	3,467
Lamp-wicks, Raising and lowering	S. Rust	New York, N. Y	Dec. 16, 1835	....
Lamp-wicks, Tube for raising	R. C. Overton	New York, N. Y	Nov. 7, 1848	5,912
Lamp with argand burner, Lard	R. Cornelius	Philadelphia, Pa	Apr. 6, 1843	3,028

Index of patents issued from the United States Patent Office from 1790 to 1873, inclusive—Continued.

Invention.	Inventor.	Residence.	Date.	No.
Lamp without chimney, Coal-oil	J. W. Schreiber	New York, N. Y	Mar. 24, 1863	37, 983
Lamp, Wooden	J. H. Mather	Saybrook, Conn	July 20, 1831	....
Lamps and economy of light, Construction of	I. Jennings	New York, N. Y	June 11, 1829	....
Lamps and other lights, Chimney and shade for	M. J. Wellman and J. J. Greenough.	New York, N. Y	Dec. 1, 1863	40, 785
Lamps, Apparatus for extinguishing street gas	G. S. Dunbar	Pittsfield, Mass	Oct. 3, 1871	119, 455
Lamps, Apparatus for supplying naphtha to vapor-burning street.	F. M. Randell	Greenburgh, N. Y	Jan. 10, 1871	110, 866
Lamps, Attachment for converting other fluid into coal-oil.	A. G. Tisdel and W. Nash	Watertown, N. Y	Mar. 24, 1863	37, 987
Lamps, binnacles of ships, &c., Lighting street	J. E. Smith	Baltimore, Md	June 1, 1825	....
Lamps by electricity, Method of lighting street	C. W. Smith	Evans, N. Y	Oct. 12, 1858	21, 781
Lamps, Cap of glass	F. Draper	East Cambridge, Mass	Jan. 17, 1842	2, 424
Lamps, Case for ratchet-wheel for	L. Hover	Chicago, Ill	Aug. 18, 1863	39, 574
Lamps, Coating	G. W. Thomson	Buffalo, N. Y	Apr. 4, 1871	113, 365
Lamps, Combination of globe and chimney for	E. B. Regna	Jersey City, N. J	Feb. 24, 1863	37, 773
Lamps, Composition for burning in	I. Jennings	New York, N. Y	Dec. 31, 1839	1, 453
Lamps, Composition for supplying	H. Porter	Bangor, Me	Apr. 8, 1835	....
Lamps, Construction of	J. J. Hoyt and J. E. Crane	Chelmsford and Lowell, Mass.	Mar. 14, 1871	112, 598
Lamps, Construction of	C. West	Baltimore, Md	Oct. 7, 1844	3, 781
Lamps, Construction of burning-fluid	D. F. Randall	Chicopee, Mass	Jan. 13, 1857	16, 398
Lamps, Detachable fluid-reservoir for street	L. A. Gouch	Yonkers, N. Y	May 9, 1871	114, 550
Lamps, Device for securing collars to	E. S. Kennedy	Birmingham, (Buchanan Post-Office,) Pa.	May 9, 1871	114, 689
Lamps, Device for snuffing	J. Ives	Brooklyn, N. Y	July 30, 1867	67, 309
Lamps, Device for suspending	D. F. Hubbell	Bethel, Conn	Apr. 30, 1867	64, 319
Lamps, Electric fan for	C. T. Mason	Sumter, S. C	June 9, 1868	78, 674
Lamps, Fixture for ratchet-wheel for	H. Topping and M. Gally	Marion and Auburn, N. Y.	July 3, 1866	56, 126
Lamps, Flame-contracting cap for night	J. Christison	New York, N. Y	Apr. 12, 1864	42, 328
Lamps, Flame regulator and extinguisher for	W. H. H. Hinds	Groton, Mass	May 14, 1867	64, 671
Lamps, gas-burners, &c., Apparatus for lighting	P. B. Tyler, W. M. Chandler, and L. F. Standish.	Chicopee and Springfield, Mass.	Jan. 15, 1867	61, 284
Lamps, Generator for vapor	W. S. Mead	Buffalo, N. Y	Feb. 7, 1860	27, 061
Lamps, Glass deflector for	W. H. Matthews	Chelsea, Mass	Apr. 22, 1862	35, 061
Lamps, Glass globe for	G. M. Irwin	Birmingham, Pa	Nov. 7, 1871	120, 747
Lamps, Glass globe or reservoir for	A. Otto	Braunfels, Tex	Apr. 16, 1872	125, 754
Lamps, Heating-attachment for oil	E. O. Schartau	Philadelphia, Pa	May 28, 1867	65, 286
Lamps, Hinged collar for	G. F. J. Colburn	Newark, N. J	July 14, 1863	39, 253
Lamps, &c., Indicating filler for	J. A. Ryan	Cleveland, Ohio	Jan. 14, 1873	134, 819
Lamps, lanterns, &c., Glass lights or panes for	B. B. Schneider	New York, N. Y	Feb. 27, 1872	124, 016
Lamps, Lantern globe or shade for street	J. N. Aronson	New York, N. Y	Sept. 24, 1872	131, 649
Lamps, Lenticular globe for	G. W. Morrison	New Albany, Ind	May 9, 1871	114, 701
Lamps, Lighting	E. Hubbard and W. L. Cheney	Chester, Mass	Mar. 14, 1834	....
Lamps, Lighting and trimming	J. Gallagher	Brooklyn, N. Y	June 24, 1862	35, 733
Lamps, Lighting street	H. Elliot	New York, N. Y	Aug. 11, 1863	39, 471
Lamps, Lock-fastener for	J. Harding	Warrington, Great Britain.	Dec. 17, 1867	73, 196
Lamps, Making glass	P. F. Slave and J. Golding	East Cambridge, Mass	Jan. 23, 1845	3, 892
Lamps, Making reservoir of metallic	P. J. Clark	Meriden, Conn	May 28, 1850	7, 393
Lamps, Manufacture of	C. T. Close	New York, N. Y	May 27, 1862	35, 363
Lamps, Mechanical movement for	F. B. De Keravenan	New York, N. Y	Feb. 10, 1863	37, 659
Lamps, Mechanical movement for	F. B. De Keravenan	New York, N. Y	June 9, 1863	38, 859
Lamps, Mode of attaching chimneys to	W. Morehouse	Buffalo, N. Y	Apr. 1, 1862	34, 842
Lamps, Mode of attaching chimneys to	H. H. Swift	Hart's Village, N. Y	Jan. 6, 1863	37, 366
Lamps, Mode of attaching extinguishers to	F. A. Jewett	Abington, Mass	Dec. 11, 1855	13, 910
Lamps, Mode of hanging torch	C. H. Cooper	New York, N. Y	Sept. 18, 1860	30, 048
Lamps, Mode of lighting street	H. Lenox	Trenton, N. J	Apr. 26, 1870	102, 284
Lamps, Mode of manufacturing	H. Whitney and T. Leighton	Cambridge, Mass	Jan. 11, 1839	1, 068
Lamps, Mode of removing chimney and filling	J. R. Baker	Kendallville, Ind	June 24, 1862	35, 664
Lamps, Mold for pressing glass fountain	H. W. Adams	New York, N. Y	Aug. 19, 1856	15, 548
Lamps, Mode of preventing explosion of	C. P. Grosvenor	McGrawville, N. Y	Dec. 10, 1867	71, 872
Lamps, Mode of preventing the explosion of	E. Howard	Redhill, England	Nov. 12, 1867	70, 721
Lamps, Name-plate for street	T. T. Markland, jr	Philadelphia, Pa	Apr. 7, 1868	76, 334
Lamps, Name-plate for street	C. J. O'Hara	New Orleans, La	Sept 29, 1868	82, 544
Lamps or torches, Frame for swinging	L. T. Pitkin	Hartford, Conn	Sept. 18, 1860	30, 089
Lamps, Packing-tube for vapor	W. E. Jervey	New Orleans, La	Apr. 4, 1871	113, 524
Lamps, Raising and lowering signal	T. G. Crosby	Buffalo, N. Y	June 20, 1865	48, 334
Lamps, Reflector and elastic tube for	J. and J. Flagg, jr	Boston, Mass	May 17, 1810	....
Lamps, Regulating the flame of	J. S. Trough	Baltimore, Md	July 17, 1839	1, 246
Lamps, Regulating the flow of oil to wick in carcel	A. Coates	New York, N. Y	Mar. 25, 1856	14, 492
Lamps, Regulating the interior draft of	E. Whelan	Philadelphia, Pa	Mar. 26, 1845	3, 966
Lamps, Safety-apparatus for	C. Applebee	Lyndon, Vt	Nov. 17, 1868	84, 078
Lamps, Safety-device for	H. W. M. Washington	Green Plains, Va	Oct. 18, 1870	108, 416
Lamps, Sign for street	W. H. Bell and J. G. Jory	Baltimore, Md	Feb. 6, 1872	123, 324
Lamps, Sign for street	J. T. Foley	New York, N. Y	Dec. 19, 1871	121, 938
Lamps, Sign for street	W. Graham, W. Snyder, and P. O'Brien.	Pittsburgh, Pa	Dec. 5, 1871	121, 614
Lamps, Sign for street	W. P. Ware	New York, N. Y	Mar. 3, 1868	75, 086
Lamps, Slide-pendant for hanging	J. A. Evarts	Meriden, Conn	Dec. 16, 1873	145, 637
Lamps, Spring-catch for	D. A. Draper	Cambridge, Mass	July 7, 1863	39, 131
Lamps, Supplying oil to street	R. F. De Guinon	Jersey City, N. J	July 15, 1873	140, 815
Lamps, &c., Suspension-device for	H. Weed	New Haven, Conn	Nov. 13, 1866	59, 689
Lamps, Suspension-spring for	J. A. Evarts	West Meriden, Conn	Apr. 7, 1868	76, 317
Lamps, Thick tube for	H. W. Hayden	Waterbury, Conn	Mar. 31, 1863	38, 079
Lamps to lanterns, Fastening	E. Sirret	Buffalo, N. Y	Oct. 24, 1854	11, 843
Lamps to lanterns, Method of attaching	J. Fleming	Pittsburgh, Pa	July 6, 1858	20, 785
Lamps to lanterns, Method of fastening	E. Sirret and W. H. Scott	Buffalo, N. Y	Apr. 22, 1856	14, 741
Lamps to lanterns, Securing	W. Porter	Williamsburgh, N. Y	Oct. 24, 1854	11, 849
Lamps, Train for	G. A. Jones	New York, N. Y	Feb. 18, 1868	74, 695
Lamps, Tube for glass	D. Jarves	Cambridge, Mass	Feb. 2, 1822	....
Lamps, Wind-guard and air-heater for	J. B. Capewell	Gloucester, N. J	Oct. 24, 1865	50, 560
Lance, Bomb	O. Allen	San Francisco, Cal	Oct. 27, 1863	40, 387
Lance, Bomb	A. F. and J. H. Andrews	Avon, Conn	Nov. 16, 1858	22, 054
Lance, Bomb	C. C. Brand	Ledyard, Conn	June 22, 1852	9, 047
Lance, Bomb	I. Goodspeed	Norwich, Conn	Aug. 9, 1859	25, 080
Lance, Bomb	J. Grudchos and S. Eggers	New Bedford, Mass	May 26, 1857	17, 370
Lance, Bomb	Z. Kelley	New Bedford, Mass	June 9, 1868	78, 673
Lance, Bomb	E. Pierce	Hallowell, Me	Aug. 22, 1865	49, 548

Index of patents issued from the United States Patent Office from 1790 *to* 1873, *inclusive*—Continued.

Invention.	Inventor.	Residence.	Date.	No.
Lance, Bomb	E. Pierce	Hallowell, Me	June 1, 1869	90, 868
Lance, Bomb	R. Sibley	Greenville, Conn	Apr. 28, 1857	17, 173
Lance, Bomb	R. Sibley	Greenville, Conn	Aug. 17, 1858	21, 219
Lance, Harpoon	N. Scholfield	Norwich, Conn	Aug. 24, 1858	21, 278
Lance, Whaling	O. Allen	Norwich, Conn	Sept. 19, 1846	4, 764
Lances, Cushion for wings of	N. Scholfield	Norwich, Conn	Dec. 8, 1857	18, 824
Lancet	H. Mellish	Walpole, N. H	Apr. 3, 1855	12, 636
Lancet	S. Wilmot	Bridgeport, Conn	Oct. 25, 1832	....
Lancet, Retreating-spring	T. R. Williams	Newport, R. I	July 16, 1824	....
Lancet, Revolving	T. C. Harrison	New Egypt, N. J	June 20, 1836	....
Lancet, Spring	G. J. Capewell	West Cheshire, Conn	June 19, 1866	55, 620
Lancet, Spring	J. Dewey	Chelsea, Vt	Apr. 2, 1824	....
Lancet, Spring	J. H. Genrig	Philadelphia, Pa	Apr. 11, 1846	4, 450
Lancet, Spring	J. W. W. Gordon	Catonsville, Md	Jan. 27, 1857	16, 479
Lancet, Spring	J. Ives	Bristol, Conn	Mar. 27, 1849	6, 240
Lancet, Spring	J. H. Johnson	Saint Louis, Mo	Apr. 10, 1849	6, 288
Lancet, Spring	W. Parkinson	Marshall County, Va	Aug. 11, 1857	17, 994
Lancet, Spring	J. M. Van Osdel	Chicago, Ill	Apr. 24, 1841	2, 061
Land and sea, Ascertaining position and direction on.	B. Garvey	Ashland, N. Y	Feb. 4, 1862	34, 298
Land-clearing machine	T. Oxley	Norfolk, Va	Apr. 18, 1821	....
Land-conveyance	L. F. Hake	Salem, Ohio	Feb. 19, 1867	62, 264
Land-conveyance	R. J. Nunn	Savannah, Ga	Apr. 23, 1867	64, 134
Land-conveyance brake	W. T. Kosinski	Philadelphia, Pa	Oct. 18, 1870	108, 366
Land-conveyance brake	W. T. Koskinski	Philadelphia, Pa	Dec. 3, 1872	133, 587
Land-conveyance, Propeller for	P. Dickson	Utica, Minn	Nov. 25, 1862	36, 995
Land, Device for cutting marshy	A. McFarlane	South Genesee, Wis	Apr. 15, 1862	34, 974
Land-leveler	F. Monroe	Bruce, Mich	Apr. 17, 1866	54, 003
Land-leveler	R. Shepard	Shaker Village, N. H	Apr. 1, 1862	34, 851
Land, Machine for fallowing	R. J. Gatling	Indianapolis, Ind	Jan. 27, 1857	16, 476
Land, Machine for marking and furrowing	J. R. Dikeman and J. J. Hewlett	Hempstead, N. Y	Apr. 29, 1862	35, 087
Land-roller	W. W. Andrew	La Porte, Ind	Sept. 26, 1871	119, 294
Land-roller	E. P. H. Capron	Springfield, Ohio	Jan. 5, 1869	85, 641
Land-roller	J. Cole	Fredericksburgh, Iowa	Nov. 21, 1871	121, 157
Land-roller	H. L. Currier	Oregon, Ill	Sept. 10, 1867	68, 610
Land-roller	W. W. Ballard	Davisburgh, Mich	Oct. 8, 1867	69, 611
Land-roller	G. R. Burt	Perry, N. Y	Sept. 24, 1867	69, 175
Land-roller	J. W. Dilley	Macomb, Ill	Dec. 27, 1870	110, 554
Land-roller	G. C. Dolph	West Andover, Ohio	Dec. 13, 1870	110, 019
Land-roller	D. Fuller and D. Swain	Oakwood, Mich	Dec. 1, 1868	84, 541
Land-roller	J. F. Glidden	De Kalb, Ill	June 11, 1872	127, 687
Land-roller	W. H. Crow and C. M. Sloan	Darien, Kans	Nov. 4, 1873	144, 272
Land-roller	J. T. Hudnet and H. W. Mathews	Reaville and Frenchtown, N. J.	Mar. 28, 1871	113, 169
Land-roller	E. F. Hutchinson	Auburn, Me	Apr. 21, 1868	77, 044
Land-roller	E. O. Jones	Brandon, Mich	Apr. 21, 1868	77, 048
Land-roller	E. O. Jones	Brandon, Mich	Mar. 30, 1869	88, 391
Land-roller	E. J. Knowlton	South Lyon, Mich	May 1, 1866	54, 368
Land-roller	J. Lanyon	Mineral Point, Wis	Sept. 10, 1872	131, 222
Land-roller	H. P. Manley	Ellsworth, N. Y	Jan. 19, 1869	86, 087
Land-roller	S. F. Mann	Indianapolis, Ind	Dec. 24, 1867	72, 653
Land-roller	H. W. Mathews	Frenchtown, N. J	May 7, 1872	126, 471
Land-roller	H. W. Mathews	Frenchtown, N. J	Sept. 17, 1872	131, 364
Land-roller	H. W. Mathews	Frenchtown, N. J	Oct. 29, 1872	132, 678
Land-roller	J. D. McAnally	Waterloo, Ind	Nov. 26, 1872	133, 468
Land-roller	N. S. McLay	Olathe, Kans	Jan. 12, 1869	85, 841
Land-roller	M. Miller	East Gaines, N. Y	Apr. 30, 1867	64, 242
Land-roller	E. F. Olds	Lyon, Mich	Nov. 12, 1867	70, 739
Land-roller	R. Provost	Raritan, N. J	Dec. 24, 1872	134, 311
Land-roller	H. Retzlaff	Saint Louis, Mo	Dec. 7, 1869	97, 697
Land-roller	R. Sandiford	Joliet, Ill	Dec. 29, 1868	85, 333
Land-roller	I. W. Searles	Tiffin, Ohio	Oct. 4, 1870	107, 057
Land-roller	J. S. Shafer	Plymouth, Mich	Apr. 9, 1867	63, 657
Land-roller	A. S. Skiff	Trenton Falls, N. J	May 1, 1866	54, 425
Land-roller	W. H. Staats, A. C. Schwanke, and L. Stadler.	La Prairie and Bowen, Ill	May 18, 1869	90, 131
Land-roller	E. A. Uehling	Richwood, Wis	Nov. 2, 1869	96, 513
Land-roller	E. Whitcomb and D. A. Gunn	Waterville, Ohio	July 26, 1870	105, 749
Land-roller	W. Williams	New Berlin, Ill	Nov. 11, 1873	144, 489
Land-roller	J. Winans	Plymouth, Mich	Apr. 2, 1867	63, 597
Land-roller	W. Winer	Freeport, Ill	July 25, 1871	117, 497
Land-roller	J. Woolridge	Dean's Corners, Ill	Jan. 21, 1873	135, 191
Land-roller	W. S. Worley	Tuscola, Ill	Dec. 17, 1867	72, 438
Land-roller and clod-pulverizer, Combined	J. Breiver	New Vienna, Ohio	Oct. 6, 1868	82, 793
Land-roller and harrow	L. French	Crawfordsville, Iowa	Aug. 20, 1872	130, 710
Land roller and marker	A. Mains	Olena, Ill	July 16, 1867	66, 861
Land roller and marker, Combined	H. F. Baker	Centreville, Ind	Dec. 20, 1870	110, 333
Land-roller and plaster-sower, Combined	G. F. Brock and E. Brondige	Davisburgh, Mich	Oct. 22, 1867	70, 070
Land-roller, fertilizer, and seed-sower combined	L. D. Taylor	Granville Centre, Pa	June 23, 1868	79, 280
Land roller, marker, and harrow, Combined	R. Sandiford	Joliet, Ill	June 8, 1869	90, 961
Land-roller, Pulverizing	F. Post	Plano, Ill	Oct. 20, 1868	83, 311
Land-roller, Sulky	P. Schmitt	Stewartsville, Mo	July 20, 1869	92, 888
Land-spading machine	W. E. Ward	Port Chester, N. Y	Oct. 20, 1857	18, 479
Land-tiller	G. W. Zeigler	Maumee, Ohio	Apr. 14, 1868	76, 687
Lantern	J. E. Ambrose	Lombard, Ill	Aug. 10, 1869	93, 577
Lantern	E. S. Archer	New York, N. Y	Mar. 17, 1868	75, 510
Lantern	J. S. and T. B. Atterbury	Pittsburgh, Pa	May 12, 1863	38, 457
Lantern	J. S. and T. B. Atterbury	Pittsburgh, Pa	Nov. 17, 1863	40, 594
Lantern	J. S. and T. B. Atterbury	Pittsburgh, Pa	Apr. 18, 1865	47, 267
Lantern	F. A. Balch	Hingham, Wis	Aug. 29, 1871	118, 576
Lantern	C. S. S. Baron	Bellaire, Ohio	Jan. 23, 1872	122, 931
Lantern	C. S. S. and A. L. Baron	Bellaire, Ohio	Oct. 18, 1870	108, 430
Lantern	H. Beebe	Hudson, N. J	Dec. 24, 1867	72, 589
Lantern	H. Beebe	Jersey City Heights, N. J	Jan. 7, 1873	134, 631
Lantern	J. Bellerjean	Philadelphia, Pa	Dec. 4, 1866	60, 123
Lantern	W. J. Berry	Brooklyn, N. Y	Mar. 15, 1870	100, 845
Lantern	W. J. Berry	Brooklyn, N. Y	Sept. 23, 1873	142, 984
Lantern	L. F. Betts	Saint Louis, Mo	Nov. 21, 1865	51, 005

Index of patents issued from the United States Patent Office from 1790 *to* 1873, *inclusive*—Continued.

Invention.	Inventor.	Residence.	Date.	No.
Lantern	L. F. Betts	New York, N. Y	Nov. 13, 1866	59, 705
Lantern	L. F. Betts	New York, N. Y	Nov. 26, 1867	71, 444
Lantern	L. F. Betts	Chicago, Ill	Sept. 1, 1868	81, 584
Lantern	L. F. Betts	Philadelphia, Pa	July 29, 1873	141, 311
Lantern	P. Blake	New Haven, Conn	Jan. 13, 1852	8, 650
Lantern	W. H. Bonnell	Buffalo, N. Y	Mar. 10, 1868	75, 354
Lantern	W. H. Bonnell	Buffalo, N. Y	Nov. 16, 1869	96, 772
Lantern	E. Boorse	Philadelphia, Pa	Apr. 2, 1867	63, 359
Lantern	G. R. Boynton	Chicago, Ill	July 1, 1862	35, 800
Lantern	T. H. Brady	New Britain, Conn	May 26, 1868	78, 180
Lantern	J. H. Breckenridge	West Meriden, Conn	Mar. 17, 1863	37, 899
Lantern	J. D. Brown	Cincinnati, Ohio	May 29, 1860	28, 450
Lantern	T. B. Burgert	Crestline, Ohio	Oct. 26, 1869	96, 082
Lantern	W. Burns	Chicago, Ill	Jan. 2, 1866	51, 798
Lantern	W. Burns	Chicago, Ill	Mar. 19, 1867	63, 010
Lantern	C. H. Butterfield	South Lancaster, Mass	July 17, 1855	13, 250
Lantern	J. Caldwell	Providence, R. I	Mar. 3, 1868	74, 988
Lantern	D. Challinor	Birmingham, Pa	July 12, 1870	105, 171
Lantern	R. Chester	Chicago, Ill	Dec. 4, 1866	60, 139
Lantern	R. Chester	Chicago, Ill	Dec. 4, 1866	60, 140
Lantern	P. J. Clark	West Meriden, Conn	Feb. 17, 1863	37, 718
Lantern	P. J. Clark	West Meriden, Conn	May 19, 1868	78, 058
Lantern	P. J. Clark	West Meriden, Conn	July 13, 1869	92, 515
Lantern	P. J. Clark and J. Kintz	Meriden, Conn	Apr. 23, 1867	64, 075
Lantern	E. B. Coffin	Johnstown, R. I	Apr. 23, 1861	32, 161
Lantern	J. P. Connor and M. T. Hynes	Boston, Mass	Apr. 29, 1873	138, 232
Lantern	E. Conrad	Philadelphia, Pa	Apr. 17, 1866	54, 059
Lantern	J. A. Cowles	Chicago, Ill	May 31, 1864	42, 928
Lantern	A. R. Cribfield	Lincoln, Ill	Apr. 2, 1867	63, 480
Lantern	J. E. Cross	Chicago, Ill	Aug. 6, 1867	67, 510
Lantern	T. A. Davies	New York, N. Y	Feb. 25, 1873	136, 218
Lantern	C. Dean	Buffalo, N. Y	Nov. 16, 1869	96, 897
Lantern	C. Davies	New York, N. Y	Apr. 28, 1863	38, 355
Lantern	C. Deaves and E. S. Archer	New York, N. Y	June 21, 1864	43, 257
Lantern	T. B. De Forest	New York, N. Y	Feb. 14, 1860	27, 186
Lantern	T. B. De Forest	New York, N. Y	Mar. 27, 1860	27, 666
Lantern	T. B. De Forest	New York, N. Y	Apr. 17, 1860	27, 892
Lantern	T. B. De Forest	New York, N. Y	Aug. 7, 1860	29, 472
Lantern	A. M. Duburn	Chicago, Ill	June 26, 1866	55, 839
Lantern	A. M. Duburn	Chicago, Ill	Oct. 2, 1866	58, 392
Lantern	A. M. Duburn	Chicago, Ill	Sept. 22, 1868	82, 297
Lantern	R. Dunham	Portland, Me	Oct. 24, 1865	50, 567
Lantern	M. B. Dyott	Philadelphia, Pa	Oct. 17, 1871	119, 920
Lantern	C. Engelskirken	Buffalo, N. Y	Feb. 7, 1865	46, 227
Lantern	H. Evans, jr	Newark, N. J	Sept. 26, 1848	5, 806
Lantern	A. French	Philadelphia, Pa	Sept. 25, 1866	58, 244
Lantern	A. French	Philadelphia, Pa	Nov. 28, 1871	121, 241
Lantern	A. French	Philadelphia, Pa	Nov. 12, 1872	132, 906
Lantern	F. M. Ford	Sebec, Me	July 25, 1871	117, 399
Lantern	M. H. Fowler	New York, N. Y	Oct. 27, 1863	40, 401
Lantern	H. A. Fox	Cincinnati, Ohio	June 28, 1864	43, 303
Lantern	N. Gear	Newark, Ohio	May 2, 1871	114, 432
Lantern	C. Gersten	Brooklyn, N. Y	Jan. 25, 1859	22, 723
Lantern	A. H. Golden	La Fayette, Ind	Apr. 6, 1858	19, 845
Lantern	O. L. Gridley and C. Engelskerhen.	Buffalo, N. Y	June 20, 1871	116, 177
Lantern	E. J. Hale	Foxcroft, Me	Mar. 19, 1861	31, 713
Lantern	E. J. Hale	Foxcroft, Me	July 7, 1868	79, 569
Lantern	J. Hale and J. S. Atterbury	Pittsburgh, Pa	Mar. 19, 1861	31, 714
Lantern	H. W. Harkness	New Britain, Conn	Nov. 30, 1869	97, 396
Lantern	J. O. Harris	Reading, Pa	Apr. 10, 1866	53, 819
Lantern	J. O. Harris	Reading, Pa	Oct. 16, 1866	58, 942
Lantern	J. O. Harris	Reading, Pa	Nov. 13, 1866	59, 717
Lantern	C. Hart	Wakefield, Mass	Mar. 1, 1870	100, 289
Lantern	E. F. Haskell	Sherman, Me	Jan. 18, 1870	98, 962
Lantern	E. K. Haynes	Boston, Mass	Mar. 18, 1873	136, 994
Lantern	E. K. Haynes	Boston, Mass	Nov. 4, 1873	144, 200
Lantern	W. H. H. Hinds	Groton, Mass	May 10, 1864	42, 658
Lantern	T. Houghton	Philadelphia, Pa	Apr. 6, 1869	88, 713
Lantern	L. Hover	Jersey City, N. J	Jan. 23, 1855	12, 291
Lantern	J. Hughes	Buchanan, Pa	Aug. 10, 1869	93, 535
Lantern	J. Hughes	Buchanan, Pa	Dec. 28, 1869	98, 383
Lantern	S. Hughes	Jersey City Heights, N. J	Feb. 11, 1873	135, 711
Lantern	J. J. Hull and J. Haufman	Brooklyn, N. Y	Oct. 3, 1871	119, 518
Lantern	J. H. Irwin	Chicago, Ill	Oct. 28, 1862	36, 841
Lantern	J. H. Irwin	Chicago, Ill	Nov. 17, 1863	40, 624
Lantern	J. H. Irwin	Chicago, Ill	May 24, 1864	42, 856
Lantern	J. H. Irwin	Chicago, Ill	May 2, 1865	47, 551
Lantern	J. H. Irwin	Chicago, Ill	Oct. 24, 1865	50, 591
Lantern	J. H. Irwin	Chicago, Ill	Sept. 11, 1866	57, 914
Lantern	J. H. Irwin	Chicago, Ill	May 28, 1867	65, 229
Lantern	J. H. Irwin	Chicago, Ill	Jan. 7, 1868	73, 012
Lantern	J. H. Irwin	Chicago, Ill	Feb. 2, 1869	86, 548
Lantern	J. H. Irwin	Chicago, Ill	Feb. 2, 1869	86, 549
Lantern	J. H. Irwin	Chicago, Ill	May 4, 1869	89, 770
Lantern	J. H. Irwin	Philadelphia, Pa	Feb. 1, 1870	99, 442
Lantern	J. H. Irwin	New York, N. Y	June 14, 1870	104, 318
Lantern	J. H. Irwin	New York, N. Y	July 5, 1870	105, 083
Lantern	J. H. Irwin	New York, N. Y	July 12, 1870	105, 339
Lantern	G. B. Isham	Burlington, Vt	May 13, 1873	138, 805
Lantern	J. Ives	Mount Carmel, Conn	Aug. 8, 1865	49, 274
Lantern	A. S. Jackson	Kokomo, Ind	Apr. 14, 1868	76, 767
Lantern	E. N. Jenkins	Chicago, Ill	July 24, 1866	56, 567
Lantern	J. M. Jenness	Boston, Mass	May 3, 1870	102, 551
Lantern	C. Jones	Brooklyn, N. Y	June 28, 1864	43, 312
Lantern	J. B. Jones	Williamsburgh, N. Y	Feb. 28, 1860	27, 332
Lantern	W. M. Kimball and K. Hartmann.	Cleveland, Ohio	Mar. 15, 1859	23, 253

Index of patents issued from the United States Patent Office from 1790 to 1873, inclusive—Continued.

Invention.	Inventor.	Residence.	Date.	No.
Lantern	J. Kintz	West Meriden, Conn	Aug. 15, 1865	49, 417
Lantern	J. Kintz	West Meriden, Conn	Dec. 30, 1873	146, 079
Lantern	A. Lanergan	Boston, Mass	July 11, 1854	11, 289
Lantern	E. M. Lang	Portland, Me	May 11, 1869	89, 871
Lantern	T. Langston	Meriden, Conn	May 26, 1868	78, 378
Lantern	T. Langston	Brooklyn, N. Y	Dec. 1, 1868	84, 556
Lantern	T. Langston	Brooklyn, N. Y	Mar. 16, 1869	87, 857
Lantern	T. Langston	Brooklyn, N. Y	June 29, 1869	92, 063
Lantern	T. Langston	Brooklyn, N. Y	June 29, 1869	92, 064
Lantern	T. Langston	Meriden, Conn	Jan. 4, 1870	98, 602
Lantern	L. W. Leary	Norfolk, Va	May 31, 1870	103, 628
Lantern	L. W. Leary	Norfolk, Va	Jan. 28, 1870	104, 745
Lantern	F. Leclére	Watertown, N. Y	Mar. 12, 1867	62, 861
Lantern	J. G. Leffingwell	Newark, N. J	Apr. 11, 1865	47, 212
Lantern	A. Loeffelholz and A. Prier	Milwaukee, Wis	Mar. 25, 1873	137, 223
Lantern	G. H. Magersuppe	New York, N. Y	Apr. 2, 1861	31, 896
Lantern	I. C. Mayo	Gloucester, Mass	Dec. 21, 1869	98, 080
Lantern	W. McKay	Newburyport, Mass	July 22, 1873	141, 008
Lantern	G. C. Merrill	Chicago, Ill	Jan. 24, 1865	46, 010
Lantern	R. M. Merrill	Chicago, Ill	Oct. 31, 1865	50, 725
Lantern	F. Meyrose	Saint Louis, Mo	Sept. 18, 1866	58, 122
Lantern	J. H. Miltimore	Milwaukee, Wis	Aug. 8, 1865	49, 290
Lantern	C. F. Moller	Newark, N. J	Aug. 14, 1866	57, 167
Lantern	R. Mood and H. M. Britton	Cincinnati, Ohio	Feb. 23, 1869	87, 189
Lantern	J. C. Moore	Philadelphia, Pa	Mar. 6, 1866	53, 027
Lantern	F. Morandi	Boston, Mass	Feb. 5, 1856	14, 201
Lantern	P. A. Morley	Brooklyn, N. Y	Apr. 17, 1860	27, 924
Lantern	G. Mortimer	Jersey City Heights, N. J	Mar. 5, 1872	124, 372
Lantern	W. Mullin	Steubenville, Ohio	Sept. 22, 1863	40, 057
Lantern	S. Naylor	Independence, Iowa	July 22, 1873	141, 162
Lantern	L. C. Ober	Boston, Mass	July 8, 1862	35, 835
Lantern	J. Orphy	Buffalo, N. Y	May 2, 1871	114, 471
Lantern	S. Peters	Crescent, N. Y	May 10, 1870	102, 858
Lantern	G. Peugeot	Buffalo, N. Y	Jan. 9, 1867	51, 966
Lantern	E. J. Pohl and J. A. Higgins	Philadelphia, Pa	Nov. 26, 1872	133, 479
Lantern	W. Porter	New York, N. Y	Aug. 18, 1863	39, 586
Lantern	W. Porter, jr	New York, N. Y	June 13, 1871	115, 891
Lantern	W. Porter, sr., and W. Porter, jr.	New York, N. Y	Apr. 23, 1867	64, 141
Lantern	W. Porter, sr., and W. Porter, jr.	New York, N. Y	Aug. 3, 1869	93, 226
Lantern	W. Porter and E. A. Tuttle	Williamsburgh, N. Y	July 5, 1853	9, 833
Lantern	E. T. Prindle and J. Wellfare	Aurora, Ill	Dec. 31, 1867	72, 900
Lantern	G. W. Putnam	Boston, Mass	Oct. 6, 1868	82, 750
Lantern	J. H. Reighard	Birmingham, Pa	Jan. 26, 1858	19, 207
Lantern	J. H. Reighard	Birmingham, Pa	Apr. 6, 1858	19, 897
Lantern	J. H. Richardson	Philadelphia, Pa	Dec. 24, 1867	72, 542
Lantern	S. Roebuck	New York, N. Y	Sept. 12, 1865	49, 953
Lantern	J. H. Rohrman	Philadelphia, Pa	Sept. 1, 1857	18, 105
Lantern	H. Sangster	Buffalo, N. Y	May 28, 1867	65, 285
Lantern	H. and J. Sangster	Buffalo, N. Y	June 10, 1851	8, 154
Lantern	S. Shannon	Buffalo, N. Y	Sept. 23, 1856	15, 782
Lantern	A. G. Smith	Jersey City, N. J	Nov. 20, 1866	59, 865
Lantern	A. G. Smith	Jersey City, N. J	Jan. 28, 1868	73, 934
Lantern	J. O. Smith	New York, N. Y	Dec. 31, 1867	72, 759
Lantern	T. Smith	Cleveland, Ohio	May 20, 1873	139, 030
Lantern	W. P. Smith	Louisville, Ky	Sept. 11, 1866	57, 988
Lantern	S. C. Spaulding	Rutland, Vt	Sept. 14, 1858	21, 521
Lantern	C. F. Spencer	Rochester, N. Y	Mar. 26, 1867	63, 321
Lantern	C. F. Spencer	Rochester, N. Y	July 16, 1867	66, 902
Lantern	J. Staniford, jr., and A. D. Allen.		May 10, 1804	
Lantern	W. G. and C. Sterling	New York, N. Y	Apr. 30, 1867	64, 377
Lantern	J. Straszer	Saint Louis, Mo	Jan. 27, 1863	37, 529
Lantern	P. Sweeny	New York, N. Y	Oct. 4, 1870	108, 064
Lantern	C. J. Sykes	Chicago, Ill	Jan. 30, 1872	123, 208
Lantern	C. J. Sykes	Chicago, Ill	Apr. 30, 1872	126, 345
Lantern	C. Tabor	Craftsbury, Vt	Sept. 27, 1870	107, 835
Lantern	A. E. Taylor	New Britain, Conn	May 30, 1871	115, 385
Lantern	N. Thompson	Brooklyn, N. Y	Nov. 3, 1868	83, 799
Lantern	N. Thompson	Brooklyn, N. Y	Apr. 27, 1869	89, 520
Lantern	N. Thompson	Brooklyn, N. Y	Nov. 23, 1869	97, 246
Lantern	W. S. Thompson	Rochester, N. Y	Apr. 2, 1867	63, 441
Lantern	W. D. Titus	Brooklyn, N. Y	Dec. 5, 1854	12, 041
Lantern	F. G. Tucker and C. Crawford	Albany, N. Y	Apr. 14, 1863	38, 189
Lantern	A. Tufts	Malden, Mass	Mar. 13, 1860	27, 485
Lantern	A. Tufts	Malden, Mass	Aug. 29, 1865	49, 665
Lantern	G. Wallingford	Charlestown, Mass	Nov. 12, 1872	132, 995
Lantern	C. Waters	Brooklyn, N. Y	July 17, 1855	13, 286
Lantern	H. Wentworth and V. B. Clark	Ripon, Wis	Jan. 21, 1868	73, 481
Lantern	W. Westlake	Milwaukee, Wis	Aug. 2, 1864	43, 730
Lantern	W. Westlake	Chicago, Ill	July 18, 1865	48, 858
Lantern	W. Westlake	Chicago, Ill	Sept. 26, 1865	50, 192
Lantern	W. Westlake	Chicago, Ill	Dec. 12, 1865	51, 526
Lantern	W. Westlake	Brooklyn, N. Y	Jan. 1, 1867	60, 810
Lantern	W. Westlake	Brooklyn, N. Y	Apr. 7, 1868	76, 366
Lantern	W. Westlake	Chicago, Ill	Sept. 7, 1869	94, 535
Lantern	W. Westlake	Chicago, Ill	Sept. 7, 1869	94, 536
Lantern	W. Westlake	Chicago, Ill	May 10, 1870	102, 895
Lantern	W. Westlake	Chicago, Ill	May 10, 1870	102, 896
Lantern	W. Westlake	Chicago, Ill	Sept. 6, 1870	107, 140
Lantern	W. Westlake	Chicago, Ill	Oct. 11, 1870	108, 221
Lantern	W. Westlake	Chicago, Ill	Oct. 11, 1870	108, 222
Lantern	W. Westlake	Chicago, Ill	Oct. 3, 1871	119, 549
Lantern	W. Westlake	Chicago, Ill	Aug. 26, 1873	142, 134
Lantern	G. Wheeler	Chicago, Ill	Jan. 14, 1868	73, 415
Lantern	A. Whelden	South Dennis, Mass	July 14, 1868	80, 040
Lantern	H. J. White	Boston, Mass	Dec. 6, 1870	109, 856

Index of patents issued from the United States Patent Office from 1790 *to* 1873, *inclusive*—Continued.

Invention.	Inventor.	Residence.	Date.	No.
Lantern	S. R. Wilmot	Bridgeport, Conn	Dec. 11, 1866	60, 452
Lantern	A. Withmar	Saint Louis, Mo	Mar. 8, 1870	100, 578
Lantern	G. W. Woodward	New York, N. Y	Apr. 5, 1864	42, 249
Lantern	M. B. Wright	Meriden, Conn	Nov. 12, 1867	70, 931
Lantern	H. C. Yerby	Leslie, Mich	June 8, 1869	91, 192
Lantern	M. Young	Frederick, Md	July 26, 1870	105, 762
Lantern, Advertising	S. Kuh	Jefferson, Iowa	Nov. 19, 1872	133, 158
Lantern, Advertising	S. Kuh	Jefferson, Iowa	Aug. 12, 1873	141, 719
Lantern and axle lubricator, Combined carriage	J. Scheeper	New York, N. Y	Dec. 3, 1861	33, 853
Lantern and candlestick, Combined	M. A. Shepard	Fort Branch, Ind	Sept. 20, 1870	107, 551
Lantern and foot-warmer, Combined	W. F. Bartlett	Hillsdale, Mich	Nov. 20, 1866	59, 810
Lantern and foot-warmer, Combined	S. M. Wirts and F. Swift	Hudson, Mich	Aug. 6, 1867	67, 622
Lantern and lamp, Combining	J. S. Grimes	Lansingburgh, N. Y	Apr. 18, 1848	5, 516
Lantern and lamp frame	G. F. J. Coburn	Newark, N. J	May 24, 1864	42, 839
Lantern and oil-can combined	W. G. Russell	New York, N. Y	May 19, 1857	17, 351
Lantern and reflector, Attachment of	W. C. Owen	Brooklyn, N. Y	May 12, 1863	38, 498
Lantern, Astronomical	J. F. Clarke	West Roxbury, Mass	Dec. 27, 1870	110, 435
Lantern-attachment to caps	J. C. Cary	New York, N. Y	Sept. 14, 1858	21, 485
Lantern, Carriage	T. Wigley	Bridgeport, Conn	Dec. 10, 1872	133, 735
Lantern-carrier	S. J. Shaw and H. J. Batchelder.	Marlborough, Mass	Feb. 28, 1860	27, 315
Lantern, Coal-oil	P. J. Clark	West Meriden, Conn	Apr. 14, 1863	38, 199
Lantern, Coal-oil	M. H. Fowler	New York, N. Y	Jan. 13, 1863	37, 394
Lantern, Coal-oil	M. Miller	Brooklyn, N. Y	Aug. 17, 1858	21, 209
Lantern, Coal-oil	S. Sargent	Watertown, Mass	Sept. 17, 1861	33, 312
Lantern, Coal-oil	J. W. Schreiber	New York, N. Y	Sept. 2, 1862	36, 371
Lantern, Dock	T. Langston	Meriden, Conn	June 18, 1872	128, 153
Lantern, Decorative	C. C. E. Schwartz	Philadelphia, Pa	Sept. 2, 1873	142, 520
Lantern-fastening	C. Mounin and W. M. Booth	Buffalo, N. Y	Aug. 1, 1854	11, 437
Lantern, Fishing	J. Goodrich	Muscoda, Wis	Feb. 10, 1863	37, 621
Lantern, foot-warmer, and water-heater, Combined	G. A. Wells	Oskaloosa, Iowa	Jan. 29, 1867	61, 587
Lantern for lighting street-gas	J. Reese and C. N. Tyler	Washington, D. C	June 23, 1857	17, 637
Lantern for marine-telegraph	J. W. Moore and W. H. Elliot	Plattsburgh, N. Y	May 20, 1862	35, 322
Lantern for railway-car, Coal-oil	S. B. H. Vance	New York, N. Y	Jan. 27, 1863	37, 553
Lantern for railway-switch, Signal	S. N. Lennon	Deposit, N. Y	July 27, 1858	21, 006
Lantern-frame	R. S. Laird	Sandwich, Ill	Jan. 31, 1865	46, 114
Lantern-frame	E. F. Parker	Proctorsville, Vt	Feb. 1, 1853	9, 566
Lantern-frame	E. F. Parker	Proctorsville, Vt	Jan. 30, 1855	12, 324
Lantern, Gas-lighting	A. Wilson	Philadelphia, Pa	Dec. 1, 1857	18, 784
Lantern, Glass	P. A. Morley	Brooklyn, N. Y	Aug. 29, 1854	11, 632
Lantern, Globe	J. S. and T. B. Atterbury	Pittsburgh, Pa	Apr. 18, 1865	47, 268
Lantern, Globe	C. P. Lindley	Waterbury, Conn	June 2, 1863	38, 754
Lantern-globe	I. C. Mayo	Gloucester, Mass	May 3, 1870	102, 692
Lantern-globe	M. Sweeney	Wheeling, W. Va	Oct. 19, 1869	96, 053
Lantern-globe	W. Westlake	Brooklyn, N. Y	Feb. 4, 1868	74, 020
Lantern-globe holder	J. S. Dennis and W. F. Kistler	Chicago, Ill	May 23, 1871	115, 035
Lantern-globe holder	J. H. Irwin	Philadelphia, Pa	May 23, 1871	115, 059
Lantern-guard	T. Brown, jr., and J. L. Lowry	Pittsburgh, Pa	Oct. 24, 1865	50, 653
Lantern-guard	C. H. Butterfield	South Lancaster, Mass	Sept. 11, 1855	13, 539
Lantern-guard	C. H. Butterfield	Nashua, N. H	Dec. 25, 1855	14, 006
Lantern-guard	W. R. P. Cross	Portland, Me	Oct. 4, 1864	44, 577
Lantern-guard	J. Kintz	West Meriden, Conn	June 28, 1870	104, 857
Lantern-guard	W. H. Pierce	East Cambridge, Mass	Feb. 18, 1862	34, 447
Lantern-guard	T. Smith	Cleveland, Ohio	Mar. 11, 1873	136, 782
Lantern-guard	W. Westlake	Milwaukee, Wis	Apr. 26, 1864	42, 520
Lantern-guard attachment	T. Brown, jr	Allegheny County, Pa	May 24, 1864	42, 906
Lantern, Hand	C. Engelskircher	Buffalo, N. Y	July 24, 1866	56, 543
Lantern, Hand	G. Peugeot	Buffalo, N. Y	July 24, 1866	56, 604
Lantern handle, Signal	A. N. Towne	Chicago, Ill	Apr. 16, 1867	63, 820
Lantern holder, Body	S. E. Fessenden	Stamford, Conn	Mar. 7, 1871	112, 436
Lantern, Insect-destroying	S. C. Wilt	Hartleton, Pa	Oct. 7, 1846	4, 808
Lantern, Insect-trap	R. W. Pitman	West Point, Iowa	Mar. 5, 1867	62, 563
Lantern, Kerosene	E. G. Tobey	Portland, Me	Aug. 12, 1862	36, 198
Lantern-lamp	R. M. Merrill	Chicago, Ill	May 27, 1862	35, 382
Lantern, Locomotive	J. H. Kelley	Rochester, N. Y	Sept. 18, 1855	13, 577
Lantern, Magic	A. G. Buzby	Philadelphia, Pa	June 17, 1873	139, 865
Lantern, Magic	A. G. Buzby	Philadelphia, Pa	Nov. 4, 1873	144, 314
Lantern, Magic	P. Diehl	Chicago, Ill	Feb. 25, 1873	136, 142
Lantern, Magic	A. Krüss	Hamburg, Germany	Jan. 25, 1870	99, 211
Lantern, Magic	L. J. Marcy	Newport, R. I	Apr. 28, 1868	77, 300
Lantern, Magic	L. J. Marcy	Newport, R. I	July 6, 1869	92, 330
Lantern, Magic	G. Sibbald	Philadelphia, Pa	Apr. 19, 1864	42, 412
Lantern, Miners' safety	M. L. Beaufils and J. Rexroth	Paris, France	Sept. 28, 1869	95, 184
Lantern, Omnibus	F. O. Dechamps	Philadelphia, Pa	July 26, 1853	9, 885
Lantern, Pocket	C. Mackh	Chicago, Ill	May 11, 1869	90, 007
Lantern, Pocket	J. A. Minor	Middletown, Conn	Jan. 24, 1865	46, 011
Lantern, Pocket	G. W. Putnam	Peterborough, N. Y	Nov. 17, 1868	84, 213
Lantern, Pocket	A. Ralston	West Middletown, Pa	Mar. 24, 1857	16, 880
Lantern, Portable	C. Deavs	New York, N. Y	Jan. 31, 1865	46, 184
Lantern, Portable	H. L. Kassebaum	New York, N. Y	May 16, 1865	47, 733
Lantern, Portable	N. Waterman	Boston, Mass	Dec. 25, 1849	6, 978
Lantern, Portable and stationary	L. W. Leary	Norfolk, Va	Dec. 15, 1868	85, 014
Lantern, Railway-signal	S. N. Lennon	Deposit, N. Y	Feb. 15, 1859	22, 960
Lantern, Railway-signal	N. A. Menaar	Buffalo, N. Y	May 21, 1861	32, 377
Lantern, Railway-signal	J. L. Wager	Deposit, N. Y	Feb. 15, 1859	22, 992
Lantern, Railway-switch	C. Byrne	Kingsville, Ohio	Jan. 27, 1863	37, 493
Lantern, Railway-switch	E. E. and A. B. Dickerson	Oshkosh, Wis	Aug. 21, 1866	57, 295
Lantern, Reflecting	E. Boesch	San Francisco, Cal	Oct. 22, 1872	132, 433
Lantern, Reflecting	A. E. Young	Dorchester, Mass	May 6, 1862	35, 194
Lantern, Reflecting and magnifying	W. Lewis	Boston, Mass	June 8, 1810	
Lantern, Self lighting and extinguishing	A. Roesler and C. Frey	Warsaw, Ill	May 18, 1858	20, 302
Lantern, Ship's	A. Saunders	South Kingston, R. I	Jan. 28, 1873	135, 371
Lantern, Ship's	H. Saunders	South Kingston, R. I	May 6, 1873	138, 532
Lantern, Signal	D. Ammen	United States Navy	Feb. 16, 1858	19, 332
Lantern, Signal	L. V. Badger	Chicago, Ill	Dec. 17, 1867	72, 262
Lantern, Signal	G. Callard	Buffalo, N. Y	July 31, 1849	6, 617
Lantern, Signal	R. Chester	Chicago, Ill	Dec. 2, 1862	37, 032

Index of patents issued from the United States Patent Office from 1790 to 1873, inclusive—Continued.

Invention.	Inventor.	Residence.	Date.	No.
Lantern, Signal	S. W. Clark and W. R. Sykes	London, England	June 17, 1873	140, 015
Lantern, Signal	T. A. Davis	New York, N. Y	Feb. 11, 1873	135, 783
Lantern, Signal	H. C. Felthousen	Buffalo, N. Y	Dec. 3, 1861	33, 831
Lantern, Signal	H. B. Fernald	Washington, D. C	Aug. 26, 1873	142, 223
Lantern, Signal	J. Graham	Grafton, W. Va	Dec. 15, 1868	85, 004
Lantern, Signal	L. Hover	Flushing, N. Y	Oct. 4, 1859	25, 645
Lantern, Signal	W. Howard	Flushing, N. Y	June 1, 1858	20, 431
Lantern, Signal	S. H. Miller	Brooklyn, N. Y	Aug. 26, 1873	142, 257
Lantern, Signal	S. H. Miller	Brooklyn, N. Y	Nov. 11, 1873	144, 554
Lantern, Signal	C. G. Moeller	Newark, N. J	Sept. 17, 1872	131, 368
Lantern, Signal	J. W. Moffitt	Harrisburgh, Pa	June 18, 1872	128, 058
Lantern, Signal	J. R. Pierce and L. B. Austin	Oswego, N. Y	Nov. 10, 1857	18, 602
Lantern, Signal	W. S. Roberts and E. H. Fiske	East Greenwich, R. I	Aug. 22, 1871	118, 278
Lantern, Signal	H. Sangster	Buffalo, N. Y	Dec. 18, 1849	6, 959
Lantern, Signal	I. W. Shaler	Brooklyn, N. Y	Oct. 7, 1873	143, 425
Lantern, Signal	D. Todd	Detroit, Mich	Nov. 9, 1869	96, 637
Lantern, Signal	E. G. Turner	New Bedford, Mass	July 13, 1869	92, 493
Lantern, Signal	T. F. Woodward	South Reading, Mass	July 10, 1860	29, 118
Lantern-spring	J. A. Cowles	Chicago, Ill	May 17, 1864	42, 751
Lantern, Steam-boat	J. M. Read	Louisville, Ky	Dec. 28, 1838	1, 049
Lantern, Street	J. P. Avery and W. L. Nichols	Norwich, Conn	June 18, 1867	65, 860
Lantern, Street	J. W. Bartlett	New York, N. Y	June 7, 1870	103, 828
Lantern, Street	J. W. Ba tlett	New York, N. Y	June 27, 1871	116, 399
Lantern, Street	A. Burger	New York, N. Y	Nov. 28, 1871	121, 327
Lantern, Street	J. Cook	New York, N. Y	Mar. 19, 1872	124, 665
Lantern, Street	J. F. Harley	Kipton Station, Ohio	Sept. 26, 1871	119, 355
Lantern, Street	M. A. Heath	Providence, R. I	Oct. 10, 1871	119, 762
Lantern, Street	C. F. Hollis	Boston, Mass	Mar. 19, 1872	124, 819
Lantern, Street	A. R. and E. A. Hunt	Newark, N. J	Dec. 11, 1866	60, 377
Lantern, Street	B. A. Johnson	Jeffersonville, Ind	Aug. 14, 1866	57, 148
Lantern, Street	J. Neumann	Philadelphia, Pa	Apr. 26, 1870	102, 305
Lantern, Street	J. Reese and C. N. Tyler	Washington, D. C	June 30, 1857	17, 696
Lantern, Street	A. Tufts	Malden. Mass	Mar. 7, 1871	112, 396
Lantern, Street-car	A. A. Young	Boston, Mass	Aug. 18, 1868	81, 124
Lantern, Stove	D. L. Jacques	Hudson, Mich	Mar. 14, 1865	46, 802
Lantern, Submarine	G. W. Fuller	Cambridgeport, Mass	Mar. 24, 1863	37, 959
Lantern, Submarine	C. M. Gould and C. B. Lamb	Worcester, Mass	Apr. 8, 1856	14, 608
Lantern, Submarine telescopic	H. Thompson	Mobile, Ala	Feb. 16, 1869	87, 012
Lantern, Tubular	J. S. Dennis	Chicago, Ill	Feb. 27, 1872	123, 980
Lantern, Tubular	J. S. Dennis	Chicago, Ill	Feb. 27, 1872	123, 981
Lantern, Tubular	J. S. Dennis	Chicago, Ill	Feb. 27, 1872	123, 982
Lantern, Wax	J. Headford	New York, N. Y	July 16, 1872	129, 280
Lanterns and lamps, Glass for	M. V. Shaver	Adrian, Mich	Apr. 16, 1872	125, 849
Lanterns, Attachment for lighting	A. C. Richard	Newtown, Conn	Jan. 5, 1858	19, 044
Lanterns, Lenses for	E. Barrett	New York, N. Y	Nov. 26, 1867	71, 441
Lanterns, Machine for making	W. Westlake	Chicago, Ill	Apr. 25, 1865	47, 478
Lanterns, Machinery for manufacturing	W. Westlake	Chicago, Ill	Oct. 17, 1865	50, 538
Lanterns, Regulator for the wicks of	H. W. Bleyer	Buffalo, N. Y	June 20, 1865	48, 254
Lanterns, Removable flanch-bar for securing glass of.	H. Crout	Baltimore, Md	Jan. 15, 1856	14, 087
Lanterns, Securing glass in	H. Sangster	Buffalo, N. Y	Aug. 22, 1854	11, 570
Lanterns, Transparent slide for magic	S. Solomons	London, England	Apr. 2, 1867	63, 437
Lanterns, Wind-brake for	A. Davis	Chicago, Ill	Aug. 5, 1862	36, 077
Lanyard, Elastic	J. E. Jones	Waretown, N. J	Dec. 29, 1868	85, 388
Lanyard, Elastic	J. E. Jones	Waretown, N. J	Feb. 1, 1870	99, 320
Lanyard, Elastic	J. E. Jones	Watertown, N. J	Jan. 30, 1872	123, 178
Lanyard-scores in dead-eyes, Machine for cutting and boring holes in.	T. Blanchard	New York, N. Y	Aug. 10, 1836	....
Lap-board	M. Church	Chicago, Ill	Nov. 26, 1872	133, 302
Lap-board	D. P. Cook	Hartford, Conn	Feb. 28, 1871	112, 126
Lap-board	W. F. Gammel	Elizabeth, N. J	May 3, 1870	102, 530
Lap-board	S. Mahan	Cleveland, Ohio	July 9, 1872	128, 739
Lap-board	S. Mahan	Cleveland, Ohio	Jan. 21, 1873	135, 040
Lap-board	G. K. Proctor	Salem, Mass	June 17, 1873	140, 074
Lap-board	E. J. Sprague	Youngstown, Ohio	June 24, 1873	140, 313
Lap-board	C. Trefethen	Manchester, N. H	Aug. 22, 1871	118, 303
Lap-board	W. B. White and B. S. Bryant	Abington and Hanson, Mass.	July 30, 1872	130, 094
Lap-board, Dress-maker's	R. L. Woodbury	Lexington, Mass	Aug. 2, 1870	106, 016
Lap-joint for belting	H. Underwood	New York, N. Y	Sept. 11, 1866	58, 004
Lap-shaver and leather-splitter	J. Harvey and F. Herkstroeter	Saint Louis, Mo	Apr. 18, 1865	47, 300
Lap-welded tubes, Apparatus for manufacturing	J. Nicholson	Philadelphia, Pa	Apr. 30, 1867	64, 246
Lap-welded tubes, Machine for finishing	P. L. Weimer	Lebanon, Pa	Oct. 31, 1865	50, 781
Lapping-machine	S. Campbell	New York Mills, N. Y	Oct. 9, 1849	6, 785
Lapping-machine feed-regulater, Cotton	R. Kitson	Lowell, Mass	July 11, 1871	116, 840
Lard and butter box	W. Pratt	New York, N. Y	Feb. 21, 1871	111, 970
Lard and butter box	C. L. Tucker	Chicago, Ill	Dec. 1, 1868	84, 657
Lard and butter cutter	W. M. Bleakley	Verplanck, N. Y	July 30, 1872	129, 921
Lard and tallow, Drying	A. Smith	Cincinnati, Ohio	Sept. 9, 1873	142, 743
Lard, Apparatus for refining	G. C. Napheys	Philadelphia, Pa	Nov. 21, 1865	51, 075
Lard, Apparatus for stirring and cooling	E. F. Ring	Saint Louis Mo	Nov. 22, 1870	109, 548
Lard, Apparatus for stirring and cooling	G. B. Williams	New York, N. Y	Feb. 5, 1867	61, 907
Lard boiling, &c., Apparatus for condensation of vapors in.	C. C. and G. F. Peirson	Philadelphia, Pa	July 14, 1868	80, 006
Lard, butter, &c., Package for	G. M. Huntley	Grand Rapids, Mich	Nov. 29, 1870	109, 739
Lard-boiler	W. Branagan	Burlington, Iowa	Nov. 6, 1866	59, 350
Lard, Can or bucket for transporting	J. A. Curtis	Baltimore, Md	Oct. 22, 1872	132, 356
Lard-cooler	A. E. Camp and C. L. Reids	Louisville, Ky	Oct. 10, 1871	119, 818
Lard-cooler	G. C. Cassard	Baltimore, Md	June 16, 1868	78, 925
Lard-cooler	G. C. Cassard	Baltimore, Md	Oct. 1, 1872	131, 847
Lard-cooler	J. Ring	Saint Louis, Mo	Aug. 13, 1872	130, 534
Lard-cooler	V. E. Rusco	Chicago, Ill	Nov. 17, 1863	40, 644
Lard-cooler	W. J. Wilcox	New York, N. Y	Dec. 9, 1862	37, 120
Lard, Cooling	W. J. Wilcox	New York, N. Y	Sept. 4, 1866	57, 805
Lard-expresser	C. Bixler	Rogersville, Ohio	Jan. 24, 1860	26, 879
Lard, Machine for clarifying, mixing, and bleaching.	O. J. Backus	San Francisco, Cal	Apr. 2, 1872	125, 162
Lard or tallow, Apparatus for drying and cooling	A. Smith	Cincinnati, Ohio	Sept. 9, 1873	142, 744

Index of patents issued from the United States Patent Office from 1790 *to* 1873, *inclusive*—Continued.

Invention.	Inventor.	Residence.	Date.	No.
Lard or tallow into separate substances, Process of converting.	B. Lapham	Lexington, Ky	Mar. 1, 1833	....
Lard-package	J. King	Saint Louis, Mo	May 16, 1871	114, 971
Lard, Package for holding and shipping	C. L. Tucker	Chicago, Ill	July 2, 1867	66, 268
Lard-packing boxes, Cementing and strengthening	C. L. Tucker	Chicago, Ill	Aug. 18, 1868	81, 229
Lard, Preparing	H. A. Amelung	Alton, Ill	Nov. 13, 1844	3, 827
Lard-press	S. S. Avis	Penn's Grove, N. J	Apr. 28, 1868	77, 239
Lard-press	W. F. Pagett and C. F. Rohrer	Fremont, Ohio	Oct. 22, 1872	132, 486
Lard-press	J. Rayner	Piqua, Ohio	Oct. 17, 1865	50, 494
Lard-press and sausage-stuffer	J. B. Cassell	Worcester Township, Pa	Sept. 1, 1868	81, 751
Lard-press and sausage-stuffer	A. J. Truxell	Salem, Va	July 14, 1868	80, 035
Lard-press, Rotary	B. Hubbe	New York, N. Y	Jan. 29, 1867	61, 541
Lard, Process for refining	H. S. Lewis	Chicago, Ill	June 25, 1861	32, 633
Lard-rendering, &c., Mode of condensing noxious vapors from.	S. Davis	New York, N. Y	June 18, 1867	65, 884
Lard-scraps drying and pressing apparatus	A. K. Howe	New York, N. Y	Nov. 1, 1870	108, 788
Lard stirring and cooling apparatus	A. R. Judson	New York, N. Y	Apr. 18, 1865	47, 361
Lard stirring and cooling apparatus	J. N. Meriam	Cambridgeport, Mass	Dec. 22, 1868	85, 117
Lard-stirring machine	W. J. Wilcox	New York, N. Y	Jan. 20, 1863	37, 469
Lard, tallow, and grease from the refuse of rendering-tanks, Machine for separating.	P. Andrew	Cincinnati, Ohio	Feb. 7, 1865	46, 204
Lard, tallow, &c., Mode of treating	G. B. Turrell	New York, N. Y	June 28, 1864	43, 352
Larder, Ventilated	J. Sloan	Philadelphia, Pa	Aug. 28, 1866	57, 642
Last	J. Auzer	Ashtabula, Ohio	Oct. 29, 1872	132, 556
Last	W. L. Beardsley	Binghamton, N. Y	Aug. 7, 1866	56, 882
Last	T. Bullivant	Newark, N. J	Mar. 30, 1869	88, 269
Last	A. W. Cheever	Lynn, Mass	Aug. 6, 1867	67, 414
Last	P. C. Clapp	Dorchester, Mass	Nov. 6, 1866	59, 359
Last	C. O. Crosby	New Haven, Conn	June 22, 1869	91, 720
Last	C. S. Dunbrack	Swampscott, Mass	Nov. 22, 1870	109, 399
Last	E. T. Green	Stoneham, Mass	Oct. 30, 1866	59, 209
Last	B. Hitchings	Lynn, Mass	Oct. 13, 1868	82, 951
Last	S. T. Hutchins	North Anson, Me	July 24, 1866	56, 565
Last	P. Jackson	Saugus, Mass	Aug. 8, 1865	49, 275
Last	N. Jones	Syracuse, N. Y	Aug. 23, 1870	106, 591
Last	G. Marshall	Brooklyn, N. Y	Oct. 3, 1865	50, 252
Last	S. Mawhinney	Worcester, Mass	Apr. 20, 1869	89, 229
Last	A. W. Merritt	Scituate, Mass	Mar. 16, 1869	87, 788
Last	W. J. B. Mills	Philadelphia, Pa	Feb. 20, 1872	123, 835
Last	W. J. B. Mills	Philadelphia, Pa	Feb. 27, 1872	124, 075
Last	W. J. B. Mills	Philadelphia, Pa	Apr. 8, 1873	137, 616
Last	M. H. Pool	East Abington, Mass	Mar. 9, 1869	87, 590
Last	N. M. Rosinsky	New York, N. Y	July 16, 1872	129, 174
Last	I. N. C. Sands	Stoughton, Mass	May 17, 1864	42, 796
Last	I. N. C. Saville	Worcester, Mass	June 22, 1869	91, 774
Last	I. N. C. Saville	Worcester, Mass	May 20, 1873	139, 196
Last	W. C. Shipherd	Saratoga Springs, N. Y	Dec. 9, 1862	37, 114
Last	O. S. Squire	North Haven, Conn	July 19, 1859	24, 830
Last	G. H. Taylor	Shelburne, Mass	Jan. 4, 1859	22, 534
Last	A. Taylor	Osawatomie, Kans	Nov. 5, 1867	70, 648
Last	J. C. Toot	Clearfield, Pa	Jan. 2, 1872	122, 338
Last	D. M. True	Rockland, Me	Feb. 8, 1859	22, 904
Last	S. S. Turner	Westborough, Mass	May 11, 1858	20, 228
Last	G. M. Wells	London, England	Feb. 4, 1868	74, 180
Last	G. M. Wells	London, England	Feb. 4, 1868	74, 181
Last	W. Wells	South Danvers, Mass	Mar. 27, 1866	53, 510
Last	H. M. Whitmarsh	Abington, Mass	Sept. 13, 1864	44, 266
Last	G. B. Whitney	Natick, Mass	July 2, 1872	128, 688
Last	H. Wight	Malden, Mass	Aug. 29, 1871	118, 663
Last and shoe holder	A. J. Rock	Union Village, Va	May 12, 1868	77, 764
Last and shoe holder	B. J. Tayman	Philadelphia, Pa	May 27, 1873	139, 343
Last-block elevator and instep-stretcher	S. Richmond	Annapolis, Md	Dec. 1, 1868	84, 647
Last-block fastener	D. Huard	Ashland, Wis	July 15, 1873	140, 827
Last-block fastener	N. R. Streeter	Groton, N. Y	Nov. 25, 1873	145, 028
Last-block fastener	L. S. Wright	Groton, N. Y	Sept. 9, 1873	142, 603
Last-block fastener	N. R. Streeter	Groton, N. Y	Nov. 4, 1873	144, 368
Last-block fastening	J. Bullivant	Newark, N. J	Nov. 22, 1870	109, 492
Last-block fastening	J. A. Heckenbach and A. Haertle.	Mayville, Wis	June 17, 1873	140, 038
Last-block fastening	D. Lynahan and H. H. Koch	Buffalo, N. Y	Apr. 18, 1865	47, 314
Last-block fastening	L. R. Lockwood	Upton, Mass	Sept. 16, 1851	8, 370
Last-block fastening	A. J. Tewksbury	Haverhill, Mass	Sept. 20, 1864	44, 352
Last-block holder	D. Daniels	Park's Corners, Ill	July 22, 1873	141, 123
Last, Boot and shoe	C. F. Carr and G. F. Holbrook	Norwich, N. Y	Feb. 25, 1868	74, 890
Last, Boot and shoe	J. H. Noyes	Abington, Mass	Mar. 19, 1861	31, 733
Last, Boot and shoe	J. C. Plumer	Portland, Me	Jan. 3, 1865	45, 748
Last, Boot and shoe	W. H. Rounds	Campello, Mass	Apr. 15, 1873	137, 799
Last, Boot and shoe	S. M. Shorey	Chicago, Ill	Dec. 6, 1870	109, 845
Last, Boot and shoe	J. H. Swain	San Francisco, Cal	May 12, 1868	77, 931
Last, Boot and shoe	P. Ware, jr	Boston, Mass	Sept. 2, 1873	142, 539
Last, Boot and shoe	B. L. White	Westford, Mass	June 30, 1863	39, 104
Last, Boot and shoe	W. Young	Philadelphia, Pa	May 20, 1807	....
Last, Boot and shoe	W. Young	Baltimore, Md	June 10, 1817	....
Last, Boot and shoe nailing	J. A. Safford	Winchester, Mass	Sept. 24, 1872	131, 565
Last, Darning	G. H. Babcock	Providence, R. I	Sept. 25, 1866	58, 199
Last, Darning	D. E. Holden	Cleveland, Ohio	Nov. 21, 1865	51, 047
Last-fastener	H. Brown	Burton, Ohio	Dec. 8, 1868	84, 677
Last-fastening	W. S. Huntington	New York, N. Y	July 30, 1867	67, 307
Last-finishing machine	J. McComber	Herkimer, N. Y	Jan. 5, 1864	41, 121
Last-finishing machine	J. D. Spiller	Malden, Mass	Aug. 6, 1872	130, 251
Last for machine-sewed turned shoe	L. R. Blake	Boston, Mass	Aug. 16, 1870	106, 459
Last-head block	C. F. Pollard	Lynn, Mass	Dec. 16, 1862	37, 179
Last-holder	J. Dickinson	Painesville, Ohio	Nov. 13, 1860	30, 620
Last-holder	A. G. Mack	Rochester, N. Y	Feb. 21, 1860	27, 229
Last-holder	G. H. Smith	Lowell, Mass	Sept. 8, 1863	39, 839
Last-holder	A. J. Tewksbury	Haverhill, Mass	Nov. 10, 1857	18, 614
Last holder or jack	G. M. Wells	Chicago, Ill	Apr. 28, 1868	77, 425
Last-holder, Revolving	H. G. De Witt	Napanock, N. Y	June 29, 1852	9, 067

Index of patents issued from the United States Patent Office from 1790 *to* 1873, *inclusive*—Continued.

Invention.	Inventor.	Residence.	Date.	No.
Last-holder, Revolving	B. Marshall	Philadelphia, Pa	Apr. 28, 1857	17, 160
Last-holder, Revolving	J. Mumford	Clarksburgh, Ohio	June 3, 1856	15, 045
Last-holder, Rotary	D. Philbrick	Manchester, N. H	May 25, 1858	20, 393
Last-holder, Shoemaker's	H. Calder and G. Burgess	Richmond, Va	Oct. 26, 1869	96, 297
Last-holder, Shoemaker's	M. J. Stein	New York, N. Y	Mar. 23, 1869	88, 227
Last, Hollow metallic	S. H. Whorf	Malden, Mass	Sept. 29, 1857	18, 310
Last, Iron	J. Godfrey	New York, N. Y	May 18, 1869	90, 165
Last-lock	D. Goodyear	Ithaca, N. Y	June 11, 1867	65, 560
Last-machine	R. R. Frohock	Boston, Mass	Mar. 23, 1869	88, 029
Last-machine	J. W. Town	South Woodbury, Vt	Sept. 8, 1863	39, 847
Last, Metal	G. G. Townsend	Rochester, N. Y	Dec. 8, 1868	84, 720
Last-shaping machine	E. R. Stilwell	Dayton, Ohio	Sept. 17, 1861	33, 313
Last, Shoe	S. K. Abbott	Salem, N. H	Jan. 31, 1865	46, 174
Last, Shoe	H. R. Bean	Marlborough, Mass	Apr. 20, 1869	89, 009
Last, Shoe	T. Daugherty	Erie, Pa	Mar. 21, 1854	10, 679
Last, Shoe	J. C. F. Deecken	New York, N. Y	Feb. 15, 1859	22, 942
Last, Shoe	N. Jones	Homer, N. Y	Sept. 16, 1862	36, 495
Last, Shoe	F. Maynard	Cambridge, Mass	Sept. 25, 1860	30, 148
Last, Shoe	W. J. B. Mills	Philadelphia, Pa	Jan. 30, 1872	123, 116
Last, Shoe	W. J. B. Mills	Philadelphia, Pa	Mar. 5, 1872	124, 280
Last, Shoe	W. J. B. Mills	Philadelphia, Pa	May 14, 1872	126, 824
Last, Shoe	W. J. B. Mills	Philadelphia, Pa	May 21, 1872	126, 896
Last, Shoe	P. P. Paul	Brooklyn, N. Y	Dec. 2, 1873	145, 124
Last, Shoe	J. Schmitt	New Albany, Ind	Aug. 11, 1868	81, 014
Last, Shoe	W. C. Shipherd	New York, N. Y	May 17, 1864	42, 800
Last, Shoe	W. C. Shipherd	Willoughby, Ohio	Oct. 26, 1869	96, 152
Last, Shoe	J. Whistler	Carlisle, Pa	Apr. 24, 1849	6, 381
Last, Shoemaker's	S. Carsley, 2d	Harrison, Me	Apr. 2, 1830	....
Last, Shoemaker's	J. C. Plumer	Portland, Me	July 17, 1860	29, 225
Last, Slide-block	P. A. Newhall	Lynn, Mass	Feb. 18, 1868	74, 580
Last, yarn and needle holder combined, Stocking	D. B. Keith	East Boston, Mass	Nov 22, 1870	109, 519
Lasts, Form-plate for bottoms of shoe	J. G. Ross	Philadelphia, Pa	June 17, 1873	140, 078
Lasts, &c., from the same pattern, Machinery for turning right and left.	S. Huntington	Middlefield, N. Y	Feb. 20, 1849	6, 131
Lasts, Machine for manufacturing shoe	A. and W. P. Haskell	North Brookfield, Mass	Mar. 31, 1857	16, 917
Lasts, Machine for mounting the uppers of boots and shoes on.	J. P. Molliere	Lyons, France	Dec. 18, 1855	13, 951
Lasts, &c., Machinery for turning	E. Webber and C. Hartshorn	Gardiner, Me	Apr. 3, 1849	6, 253
Lasts, Machinery for turning right and left	C. Hartshorn and W. B. Shaw	Gardiner, Me	Nov. 13, 1849	6, 869
Lasts, Securing and releasing blocks of	A. J. Bramhart	Hartford, N. Y	Apr. 15, 1856	14, 647
Laster and binder	J. M. Hirlinger	Red Rock, Pa	Dec. 10, 1867	72, 041
Laster and punch, Shank	F. Henderson	Marietta, Ohio	June 28, 1870	104, 733
Laster, Boot	L. Barnett	Leechburgh, Pa	Mar. 8, 1870	100, 490
Laster, Boot and shoe shank	O. R. Clark	La Fayette, Ind	Oct. 10, 1871	119, 741
Laster, Boot-shank	C. W. Bliss and O. M. Adams	Milford, Mass	Sept. 24, 1867	69, 066
Laster, Shank	J. H. Bean	Cincinnati, Ohio	Mar. 18, 1873	136, 896
Laster, Shank	J. and A. B. Cain	Dubuque, Iowa	June 6, 1865	48, 052
Laster, Shank	D. G. Chase	Boston, Mass	Nov. 1, 1859	25, 995
Laster, Shank	O. R. Clark and F. H. Slyter	Marengo, Ill	Feb. 6, 1866	52, 389
Laster, Shank	W. N. Flagg	Boylston, Mass	Jan. 30, 1866	52, 286
Laster, Shank	W. H. Hanna	Chico, Cal	Nov. 25, 1873	144, 903
Laster, Shank	F. Henderson	Marietta, Ohio	Mar. 16, 1869	87, 773
Laster, Shank	R. B. Perkins	Almond, N. Y	May 13, 1873	138, 926
Laster, Shank	L. Rastetter and A. Simcox	Fort Wayne, Ind	May 28, 1867	65, 117
Laster, Shank	W. Steele and F. Henderson	Sistersville, W. Va	Oct. 22, 1867	70, 040
Laster, Shank	D. Witt	Hubbardstown, Mass	July 16, 1872	129, 301
Laster, Shoe	P. Thompson	Sardis, Ohio	June 23, 1868	79, 087
Lasting and nailing boots and shoes, Device for	L. R. Blake	Fort Wayne, Ind	Nov. 5, 1872	132, 794
Lasting boots and shoes	I. N. Beals	North Bridgewater, Mass	Apr. 20, 1869	89, 007
Lasting boots and shoes	J. C. Wightman	Boston, Mass	Mar. 4, 1873	136, 476
Lasting boots and shoes	J. Brackett	Lynn, Mass	Aug. 5, 1873	141, 420
Lasting boots and shoes	R. C. Lambart	Quincy, Mass	Jan. 21, 1873	135, 133
Lasting boots and shoes, Device for	J. G. Rust	Xenia, Ohio	Mar. 17, 1868	75, 580
Lasting boots and shoes, Machine for	C. W. Glidden	Lynn, Mass	Feb. 4, 1873	135, 539
Lasting boots and shoes, Machine for	C. W. Glidden	Lynn, Mass	Feb. 4, 1873	135, 540
Lasting boots and shoes, Machine for	C. W. Glidden	Lynn, Mass	Feb. 25, 1873	136, 317
Lasting boots and shoes, Machine for	S. Hart	New Haven, N. Y	June 20, 1846	4, 587
Lasting boots and shoes, Machine for	J. Kimball	Boston, Mass	Sept. 8, 1857	18, 152
Lasting boots and shoes, Machine for	J. Purinton, jr	Lynn, Mass	Oct. 4, 1859	25, 673
Lasting boots and shoes, Machinery for	C. H. Trask and H. F. Wheeler	Lynn and Boston, Mass	Sept. 9, 1873	142, 657
Lasting boots and shoes, Method of	W. Chambers	Lynn, Mass	June 28, 1870	104, 828
Lasting boots and shoes, Mode of	I. R. Rogers	Lynn, Mass	Jan. 7, 1868	73, 048
Lasting boots and shoes, Nailer for	G. McKay	Cambridge, Mass	Dec. 17, 1872	134, 083
Lasting boots and shoes, Nailing machine for	H. F. Wheeler	Boston, Mass	Dec. 24, 1872	134, 231
Lasting boots, Implement for	H. Conant	Worcester, Mass	Aug. 24, 1852	9, 213
Lasting boots, Instrument for	B. Livermore	Hartland, Vt	Aug. 24, 1852	9, 216
Lasting boots, &c., Machine for	J. Eells, jr	Unadilla, N. Y	Mar. 25, 1825	....
Lasting implement, Boot and shoe	F. O. Claflin	Brooklyn, N. Y	Sept. 29, 1868	82, 599
Lasting-instrument	T. Daugherty	Erie, Pa	Aug. 1, 1854	11, 453
Lasting-iron	A. J. Smith	Canal Dover, Ohio	June 22, 1869	91, 781
Lasting-jack	G. F. Seaver	Haverhill, Mass	Oct. 8, 1872	131, 976
Lasting-jack	J. C. Wightman	Newton, Mass	Oct. 22, 1872	132, 510
Lasting-jack for boots, &c	J. C. Drew	Auburn, Me	Dec. 9, 1873	145, 407
Lasting-knee	C. H. Haskell	Lynn, Mass	May 28, 1872	127, 343
Lasting-machine	W. E. Fischer	Boston, Mass	Nov. 1, 1864	44, 916
Lasting-machine	C. W. Glidden	Lynn, Mass	Jan. 21, 1873	135, 111
Lasting-machine	C. W. Glidden	Lynn, Mass	June 24, 1873	140, 265
Lasting-machine	K. Grassau	Virginia City, Nev	Apr. 5, 1870	101, 456
Lasting-machine	H. N. Moyon and J. E. Demercier.	Paris, France	Mar. 22, 1870	101, 151
Lasting-machine	W. Wells	Boston, Mass	May 1, 1860	28, 120
Lasting-machine	W. Wells	Middleton, Mass	Mar. 15, 1864	41, 967
Lasting-machine	T. Wolcott	Stowe, Mass	Sept. 8, 1863	39, 870
Lasting machine, Boot	D. Harrington	German Flats, N. Y	Aug. 26, 1845	4, 163
Lasting machine, Boot	B. Livermore and N. F. English.	Hartland, Vt	Apr. 17, 1847	5, 063
Lasting machine, Shoe	C. H. Trask	Lynn, Mass	Oct. 4, 1870	107, 981
Lasting-machines, Toe-piece for	W. E. Fischer	Boston, Mass	June 1, 1869	90, 651

Index of patents issued from the United States Patent Office from 1790 to 1873, inclusive—Continued.

Invention.	Inventor.	Residence.	Date.	No.
Lasting-machines, Toe-piece for	A. S. McIntire and N. S. Thompson.	Stoneham, Mass	Feb. 14, 1865	46, 375
Lasting-mechanism	L. R. Blake	Brooklyn, N. Y	Sept. 30, 1873	143, 322
Lasting shoes	W. H. Lovejoy	Lowell, Mass	Feb. 21, 1871	112, 157
Lasting-tool	J. H. Bean	Cincinnati, Ohio	July 29, 1873	141, 255
Lasting-tool	L. R. Blake	Brooklyn, N. Y	Feb. 25, 1873	136, 300
Lasting-tool	C. W. Glidden	Lynn, Mass	Feb. 25, 1873	136, 318
Lasting-tool	F. Henderson	Marietta, Ohio	Apr. 13, 1869	88, 958
Latch	H. and S. W. Budd	Philadelphia, Pa	Feb. 9, 1869	86, 641
Latch	J. A. Clarke	New York, N. Y	Feb. 15, 1870	99, 845
Latch	T. Dolan	Albany, N. Y	May 31, 1870	103, 586
Latch	S. W. Drowne	Norwich, Conn	Mar. 15, 1870	100, 872
Latch	H. H. Elwell	South Norwalk, Conn	Mar. 7, 1865	46, 650
Latch	B. Erbe	Snowden, Pa	Dec. 8, 1863	40, 829
Latch	M. P. Favor	East Northwood, N. H	May 24, 1870	103, 314
Latch	J. H. Fisher	Chicago, Ill	Feb. 20, 1872	123, 817
Latch	F. W. Gammell	Spring Valley, Iowa	Jan. 11, 1870	98, 684
Latch	R. Geselbracht and F. Frey	Galena, Ill	Dec. 7, 1869	97, 498
Latch	H. Isbell	Meriden, Conn	Nov. 8, 1834	....
Latch	R. Kinsly	Springfield, Mass	Aug. 7, 1847	5, 225
Latch	G. W. Large	Yellow Springs, Ohio	Dec. 14, 1869	97, 934
Latch	N. Petré	New York, N. Y	Oct. 5, 1869	95, 508
Latch	S. M. Richardson	New York, N. Y	Nov. 8, 1864	44, 976
Latch	J. C. Robie	Binghamton, N. Y	Nov. 15, 1864	45, 078
Latch	G. A. Seaver	New York, N. Y	Nov. 9, 1869	96, 624
Latch	J. Shepard and J. Sigourney	Bristol, Conn	Aug. 13, 1867	67, 811
Latch	A. Spangler	Philadelphia, Pa	Nov. 17, 1868	84, 227
Latch	A. A. Stuart	Plainfield, Iowa	Nov. 8, 1870	109, 074
Latch	E. A. Tuttle	Brooklyn, N. Y	May 16, 1865	47, 758
Latch	M. J. Woodruff	New Britain, Conn	Nov. 30, 1869	97, 332
Latch	G. C. Worth	Upper Sandusky, Ohio	June 7, 1864	43, 063
Latch and bolt, Combined	P. P. Child	Saint Louis, Mo	Jan. 7, 1868	73, 078
Latch and bolt, Combined	M. J. Meyer	Washington, D. C	Oct. 3, 1865	50, 260
Latch and catch	J. B. Farmer	Indianapolis, Ind	Aug. 6, 1867	67, 518
Latch and door-check	R. L. Omensetter	Philadelphia, Pa	Dec. 16, 1873	145, 523
Latch and door-lock, Combined knob	H. W. Busse	Chicago, Ill	Mar. 10, 1868	75, 360
Latch and lock, Case for reversible	J. Kinzer	Pittsburgh, Pa	Feb. 13, 1872	123, 567
Latch and lock combined	J. Adair	Pittsburgh, Pa	Jan. 16, 1872	122, 794
Latch and lock, Combined	A. M. Adams	Sacramento, Cal	July 29, 1873	141, 193
Latch and lock, Combined	J. H. Allen and J. Schwab	Louisville, Ky	Aug. 31, 1869	94, 172
Latch and lock, Combined	W. N. Bailey	Duplain, Mich	Jan. 18, 1870	98, 907
Latch and lock, Combined	J. Brady	Norwich, Conn	Jan. 26, 1869	86, 276
Latch and lock, Combined	G. W. Cilley	Norwich, Conn	May 11, 1869	89, 853
Latch and lock, Combined	W. H. Cloud	Fremont, Ohio	Jan. 5, 1869	85, 644
Latch and lock, Combined	C. L. Dean	Newark, N. J	Mar. 15, 1870	100, 734
Latch and lock, Combined	N. Edwards	Newark, Ohio	Sept. 8, 1868	81, 990
Latch and lock, Combined	V. Frazee	San Francisco, Cal	Dec. 12, 1871	121, 769
Latch and lock, Combined	V. Frazee	San Francisco, Cal	Sept. 3, 1872	131, 092
Latch and lock, Combined	S. A. Green	Lexington, Ind	Sept. 1, 1868	81, 625
Latch and lock, Combined	T. Hahn	New York, N. Y	July 12, 1870	105, 197
Latch and lock, Combined	A. J. Hollenback	North Bangor, N. Y	Mar. 11, 1873	136, 723
Latch and lock, Combined	F. Hoppe	Stuttgart, Germany	May 10, 1870	102, 820
Latch and lock, Combined	F. L. Johnson	Wallingford, Conn	Sept. 22, 1868	82, 418
Latch and lock, Combined	G. Müllar	New York, N. Y	Mar. 4, 1873	136, 448
Latch and lock, Combined	J. H. Morse	Peoria, Ill	Nov. 28, 1871	121, 253
Latch and lock, Combined	A. Ochsner	New Haven, Conn	June 15, 1869	91, 479
Latch and lock, Combined	H. G. Pein	Peoria, Ill	June 29, 1869	91, 865
Latch and lock, Combined	N. Petré	New York, N. Y	Oct. 5, 1869	95, 506
Latch and lock, Combined	N. Petré	New York, N. Y	Oct. 5, 1869	95, 507
Latch and lock, Combined	F. P. Pfleghar	New Haven, Conn	Mar. 8, 1870	100, 661
Latch and lock, Combined	A. Sprague	Poland, N. Y	Nov. 24, 1868	84, 391
Latch and lock, Combined	M. P. Warner and E. W. Payne	Morrison, Ill	Oct. 3, 1871	119, 673
Latch and lock, Combined	T. Weaver	Harrisburgh, Pa	Oct. 26, 1869	96, 175
Latch and lock, Combined knob	C. Brown	Goshen, N. Y	Jan. 21, 1868	73, 498
Latch and lock, Combined knob	M. B. Foote	New England Village, Mass	Oct. 22, 1867	70, 081
Latch and lock, Combined knob	J. Imray	London, England	Apr. 23, 1872	125, 957
Latch and lock, Combined knob	J. B. Kelley	Brandon, Vt	Mar. 16, 1869	88, 791
Latch and lock, Combined knob	J. McLeod	San Francisco, Cal	Apr. 13, 1869	88, 974
Latch and lock, Combined knob	J. Peck	New Haven, Conn	July 16, 1872	129, 422
Latch and lock, Combined knob	F. Raymond	Woodhaven, N. Y	Apr. 6, 1869	88, 737
Latch and lock, Combined knob	G. Schumacher	New York, N. Y	Mar. 3, 1868	75, 204
Latch and lock, Combined knob	B. Seegmuller	New York, N. Y	July 23, 1867	67, 140
Latch and lock, Combined knob	W. H. Sullenberger	Harrisburgh, Pa	Sept. 7, 1869	94, 524
Latch and lock, Combined knob	J. H. Vickers	Norwich, Conn	May 5, 1868	77, 552
Latch and lock, Door	A. Bingham	Unity, Me	July 7, 1835	....
Latch and lock, Door	A. Iske and J. Teufel	Lancaster, Pa	Oct. 25, 1859	25, 900
Latch and lock for doors, Gravitating combined	G. E. Sellers	Cincinnati, Ohio	May 12, 1842	2, 621
Latch and lock for gates, Combined	H. R. Van Eps	Peoria, Ill	Aug. 16, 1870	106, 519
Latch and lock for sliding doors, Combined	C. W. Chappell	Watertown, Wis	Apr. 2, 1872	125, 168
Latch and lock for sliding doors, Combined	A. F. Whiting	Bath, Me	Feb. 6, 1872	123, 533
Latch and lock, Knob	S. H. Wheeler	Dowagiac, Mich	July 11, 1871	117, 023
Latch and lock, Spring	W. A. Ives	New Haven, Conn	Nov. 18, 1856	16, 089
Latch, Automatic	E. Voigt	Philadelphia, Pa	Feb. 22, 1870	100, 220
Latch-bolt	J. Adt	Waterbury, Conn	May 14, 1861	32, 270
Latch-bolt, Constructing	S. Oppenheimer	Peru, Ind	Jan. 8, 1867	61, 090
Latch, bolt, and lock, Spiral-spring	A. Blanvelt	New Brunswick, N. J	Feb. 9, 1825	....
Latch-bolt, Reversible	B. G. Hosmer	Nashua, N. H	Jan. 26, 1864	41, 379
Latch-bolt, Spring	E. M. Ray	Norfolk County, Mass	Oct. 23, 1849	6, 808
Latch, Carriage-door	N. Belvallette	Paris, France	Nov. 19, 1867	71, 123
Latch, Coach-door	J. Thielemann	Newark, N. J	June 3, 1873	139, 485
Latch, Cupboard	E. Doen	New Britain, Conn	June 14, 1864	43, 105
Latch, Cupboard	B. P. Grover	Holyoke, Mass	Aug. 15, 1865	49, 398
Latch, Cupboard	A. D. Judd	New Haven, Conn	July 9, 1867	66, 595
Latch, Cupboard	A. D. Judd	New Haven, Conn	Apr. 25, 1871	114, 152
Latch, Cupboard	W. E. Sparks	New Haven, Conn	June 14, 1870	104, 370
Latch, Cupboard	W. E. Sparks	New Haven, Conn	Sept. 13, 1870	107, 418
Latch, Cupboard	W. E. Sparks and P. Bradford	New Haven, Conn	Mar. 26, 1872	124, 981
Latch, Cupboard	W. B. Wadsworth	Cleveland, Ohio	June 9, 1863	38, 852

Index of patents issued from the United States Patent Office from 1790 to 1873, inclusive—Continued.

Invention.	Inventor.	Residence.	Date.	No.
Latch, Door	J. Alrichs	Wilmington, Del	May 24, 1845	4,060
Latch, Door	F. and F. Babcock	Middletown, Conn	Sept. 29, 1868	82,581
Latch, Door	T. C. Ball	Keene, N. H	Mar. 16, 1858	19,614
Latch, Door	E. V. Bitner	Lock Haven, Pa	Apr. 18, 1871	113,729
Latch, Door	B. M. Bosworth	Warren, R. I	Mar. 6, 1840	1,507
Latch, Door	P. Bradford	New Haven, Conn	Dec. 10, 1867	71,965
Latch, Door	E. K. Breckenridge	West Meriden, Conn	May 16, 1871	114,910
Latch, Door	E. W. Brettell	Elizabeth, N. J	Apr. 20, 1869	89,197
Latch, Door	E. W. Brettell	Elizabeth, N. J	Oct. 19, 1869	95,978
Latch, Door	F. Clymer	Galion, Ohio	Dec. 10, 1867	71,982
Latch, Door	J. H. Cooper	Philadelphia, Pa	Feb. 3, 1863	37,567
Latch, Door	F. M. Crossett	Piermont, N. Y	Jan. 13, 1863	37,386
Latch, Door	J. C. Curryer and W. C. Young	Thorntown, Ind	July 6, 1869	92,280
Latch, Door	F. W. Dean	Tremont, Ill	Aug. 17, 1869	93,685
Latch, Door	J. Dearborn	Seabrook, N. H	Aug. 19, 1862	36,207
Latch, Door	H. L. De Zeng	Geneva, N. Y	Apr. 3, 1866	53,585
Latch, Door	H. Duntze	New Haven, Conn	July 6, 1839	1,220
Latch, Door	B. Erbe	Pittsburgh, Pa	Nov. 6, 1866	59,373
Latch, Door	J. B. Evans	Millville, N. J	Nov. 26, 1867	71,474
Latch, Door	S. W. Fosdick and A. C. Dakin	Clinton, Mass	Mar. 7, 1865	46,653
Latch, Door	H. Hackman, jr	Paque, Pa	Dec. 7, 1858	22,234
Latch, Door	C. F. Herrick	Independence, Iowa	June 29, 1869	92,047
Latch, Door	J. M. Hoggan	New Haven, Conn	Nov. 25, 1841	2,376
Latch, Door	M. Howland	Waterbury, Conn	May 31, 1859	24,214
Latch, Door	O. Judd	Cherry Valley, N. Y	Dec. 10, 1840	1,888
Latch, Door	E. King	Taunton, Mass	Nov. 12, 1867	70,860
Latch, Door	E. Manley	Marion, N. Y	Jan. 9, 1866	51,954
Latch, Door	E. Manley	Marion, N. Y	Aug. 3, 1869	93,324
Latch, Door	J. Moser	Mendota, Ill	July 27, 1869	93,112
Latch, Door	O. B. Rand	Kalamazoo, Mich	June 24, 1873	140,307
Latch, Door	O. B. Rand	Kalamazoo, Mich	Nov. 4, 1873	144,288
Latch, Door	A. Rix	San Francisco, Cal	Oct. 21, 1862	36,732
Latch, Door	A. H. Rowe	Hartford, Conn	Feb. 20, 1866	52,750
Latch, Door	H. Sanders	Utica, N. Y	May 9, 1871	114,608
Latch, Door	S. W. Skinner	Lyons, N. Y	May 21, 1872	126,907
Latch, Door	T. Slaight	Newark, N. J	Nov. 6, 1860	30,594
Latch, Door	F. Stowe	Huron, Ohio	Nov. 25, 1873	144,931
Latch, Door	R. S. Underhill	Bath, N. Y	May 22, 1860	28,425
Latch, Door	H. Underwood	Tolland, Conn	Sept. 27, 1864	44,476
Latch, Door	J. Walton	Brooklyn, N. Y	Sept. 11, 1860	30,016
Latch, Door and gate	W. Broomhall	Circleville, Ohio	Jan. 29, 1867	61,604
Latch, Door and gate	L. Brumbach	Reading, Pa	Mar. 16, 1869	87,907
Latch, Door and gate	J. A. Park	Lansing, Mich	Feb. 12, 1867	62,061
Latch, Door and window	S. W. Gear	Whitestone, N. Y	June 2, 1868	78,521
Latch fastening, Door	E. Morris	Burlington, N. J	May 30, 1848	5,606
Latch, Door knob	A. Bingham	Newtonville, Mass	Feb. 25, 1868	74,746
Latch, Door knob	L. and W. Hotchkiss	Derby, Conn	Mar. 6, 1840	1,508
Latch, Door knob	A. W. Upton	Lowell, Mass	Mar. 12, 1867	62,908
Latch, door-lock, &c	E. Robinson and W. Hall	Boston, Mass	Mar. 3, 1841	1,995
Latch, Door-lock	J. P. Sherwood	Sandy Hill, N. Y	May 6, 1841	2,082
Latch, Door locking	E. Field	Greenwich, Conn	July 3, 1855	13,163
Latch, Door locking	E. Halsey	San José, Cal	Oct. 14, 1873	143,692
Latch, Door thumb	E. Parker	Meriden, Conn	Mar. 28, 1842	2,518
Latch, Double-door	J. W. Still	San Francisco, Cal	Dec. 14, 1869	97,985
Latch-fastening	H. N. Avery	New York, N. Y	Jan. 2, 1866	51,787
Latch-fastening	H. M. Jones	West Meriden, Conn	Aug. 14, 1866	57,150
Latch-fastening	L. R. Livingston, J. J. Roggen, and C. Adams.	Pittsburgh, Pa	Nov. 24, 1846	4,862
Latch fastening, Door	J. F. Tozer	Binghamton, N. Y	Feb. 3, 1863	37,598
Latch for blinds or shutters	B. S. Huntington	New York, N. Y	Nov. 28, 1865	51,186
Latch for doors, Locking knob	J. Wertsbaugher	La Grange, Ind	Dec. 17, 1867	72,347
Latch for farm-gates, Catch	J. Summers	Raleigh, Va	Apr. 20, 1858	20,008
Latch, Gate	H. D. Alderfer	Grater's Ford, Pa	Oct. 3, 1871	119,491
Latch, Gate and door	J. H. Allison	Charlottesville, Ind	Oct. 3, 1871	119,552
Latch, Gate	E. Barks and L. H. Emmons	Noblesville, Ind	Sept. 20, 1870	107,589
Latch, Gate	C. Bartlow	Richmond, Ind	Dec. 31, 1872	134,454
Latch, Gate	M. J. Briar	Oxford, Ind	Nov. 12, 1867	70,794
Latch, Gate	J. T. Bryan	Lebanon, Ind	Apr. 2, 1867	63,466
Latch, Gate	C. B. Clark	Buffalo, N. Y	Sept. 26, 1871	119,263
Latch, Gate	H. Clomo	Galena, Ill	May 17, 1870	103,143
Latch, Gate	C. Cole	Ithaca, N. Y	Nov. 8, 1870	108,971
Latch, Gate	C. Cole	Dayton, Ohio	Jan. 17, 1871	110,957
Latch, Gate	J. P. Curry	Vincennes, Ind	July 16, 1872	129,324
Latch, Gate	A. K. Davis	Carey, Ohio	Nov. 5, 1867	70,537
Latch, Gate	B. F. Dickey	Marshall, Mich	Apr. 30, 1872	126,191
Latch, Gate	E. E. Earll and J. B. Hunter	Brooklyn and New York, N. Y.	June 14, 1870	104,287
Latch, Gate	J. Ellis	Detroit, Mich	Aug. 6, 1861	32,990
Latch, Gate	J. Ellis	Detroit, Mich	Nov. 25, 1862	37,028
Latch, Gate	J. R. Finley	Delphi, Ind	Jan. 14, 1873	134,877
Latch, Gate	W. Fosket	Meriden, Conn	Feb. 7, 1871	111,626
Latch, Gate	C. W. Fox	Saint Louis, Mo	June 20, 1871	116,042
Latch, Gate	C. H. Frietog	Springfield, Ill	July 25, 1871	117,401
Latch, Gate	E. O. Frink	Indianapolis, Ind	Dec. 26, 1865	51,710
Latch, Gate	W. R. Goodrich	Whitestone, N. Y	June 21, 1870	104,447
Latch, Gate	B. Hendrickson	Huntington, N. Y	Dec. 15, 1868	84,945
Latch, Gate	G. B. Howland	Gardner, Ill	Apr. 21, 1868	76,918
Latch, Gate	S. Ingersoll	Mianus, Conn	Feb. 25, 1868	74,915
Latch, Gate	J. Johnson and S. Ingersoll	Brooklyn, N. Y	June 14, 1870	104,320
Latch, Gate	W. H. Kellogg	Duquoin, Ill	Mar. 5, 1867	62,637
Latch, Gate	F. Ketcham	Elizabeth Township, Pa	Aug. 14, 1866	57,152
Latch, Gate	J. Kindel	Wilmington, Ohio	Feb. 22, 1870	100,044
Latch, Gate	J. E. Klein	Oskaloosa, Iowa	Feb. 5, 1867	61,839
Latch, Gate	G. W. Large	Yellow Springs, Ohio	Feb. 5, 1867	61,744
Latch, Gate	G. W. Large	Yellow Springs, Ohio	Jan. 30, 1872	123,267
Latch, Gate	J. H. Lee	Marshall, Tex	Sept. 26, 1871	119,374
Latch, Gate	J. Leonard	Wilmington, Ohio	June 20, 1865	48,292
Latch, Gate	E. Manross	Bristol, Conn	Jan. 10, 1860	26,774

Index of patents issued from the United States Patent Office from 1790 *to* 1873, *inclusive*—Continued.

Invention.	Inventor.	Residence.	Date.	No.
Latch, Gate	J. A. Martin	Strasburgh, Pa	Jan. 12, 1869	85, 839
Latch, Gate	J. W. McLean	Indianapolis, Ind	June 5, 1866	55, 326
Latch, Gate	C. E. Merrifield	Indianapolis, Ind	Jan. 21, 1868	73, 624
Latch, Gate	A. E. Morgan	Poughkeepsie, N. Y	July 28, 1857	17, 908
Latch, Gate	E. Nicholson	Rockport, Ohio	July 9, 1867	66, 515
Latch, Gate, &c	W. Patton	Towanda, Pa	Aug. 30, 1870	106, 955
Latch, Gate	G. M. D. Pomeroy	Ithaca, Ind	Apr. 23, 1867	64, 140
Latch, Gate	F. M. Ranons	Little Shasta Valley, Cal	Oct. 26, 1869	96, 147
Latch, Gate	P. Rasar and D. J. Mayes	Illiopolis, Ill	Aug. 18, 1868	81, 209
Latch, Gate	W. Redhead	Des Moines, Iowa	Feb. 16, 1869	87, 065
Latch, Gate	C. W. Rhoads	Indianapolis, Ind	Apr. 2, 1867	63, 562
Latch, Gate	C. Roberts	Winchester, Ill	Feb. 5, 1867	61, 763
Latch, Gate, &c	C. Roberts	Winchester, Ill	Feb. 25, 1873	136, 183
Latch, Gate	H. W. Robie	Binghamton, N. Y	Dec. 19, 1871	122, 061
Latch, Gate	J. C. Rogers	Alden, N. Y	Dec. 8, 1868	84, 711
Latch, Gate, &c	G. N. Sharp	La Plata, Mo	June 18, 1872	128, 075
Latch, Gate, &c	G. N. Sharp	La Plata, Mo	Dec. 9, 1873	145, 368
Latch, Gate	D. Sheets	Suisun, Cal	Oct. 19, 1869	96, 041
Latch, Gate	W. Shumard	Richmond, Ind	Mar 22, 1870	101, 055
Latch, Gate	J. W. Still	San Francisco, Cal	Feb. 2, 1869	86, 601
Latch, Gate	W. T. Wells	Decatur, Ill	Feb. 12, 1867	62, 102
Latch, Gate	W. T. Wells	Decatur, Ill	Apr. 9, 1867	63, 769
Latch, Gate	S. B. Williams	Leavenworth, Kans	Dec. 8, 1863	40, 875
Latch, Gate	L. Woodruff	Ann Arbor, Mich	Jan. 29, 1867	61, 594
Latch, Gate and door	C. C. Holcomb	Madison, Wis	Apr. 21, 1868	77, 040
Latch, Gate and door	B. D. Shaw	Beverly, Ohio	Dec. 18, 1866	60, 644
Latch, Gate locking	J. L. Devol	Parkersburgh, W. Va	May 30, 1871	115, 448
Latch, Guard-bolt	E. H. Kent	New York, N. Y	Oct. 18, 1870	108, 361
Latch handle, Door	J. Ambrun and E. Olfermann	Leavenworth, Kans	Apr. 22, 1873	138, 112
Latch keeper, Door	W. Harvey	Albany, N. Y	Dec. 19, 1871	122, 011
Latch, Knob	T. C. Ball	Bellows Falls, Vt	Sept. 15, 1868	82, 070
Latch, Knob	J. H. Barnes	Brooklyn, N. Y	Nov. 22, 1864	45, 129
Latch, Knob	A. T. Brooks	New Britain, Conn	July 6, 1869	92, 151
Latch, Knob	A. T. Brooks	New Britain, Conn	July 6, 1869	92, 152
Latch, Knob	A. T. Brooks	New Britain, Conn	July 6, 1869	92, 153
Latch, Knob	A. B. Clemons	Ansonia, Conn	Mar. 23, 1869	88, 133
Latch, Knob	C. J. Colby	Waterbury, Vt	July 25, 1865	48, 909
Latch, Knob	J. Davis	Terre Haute, Ind	June 28, 1870	104, 712
Latch, Knob	J. M. A. Dew	Chicago, Ill	Feb. 25, 1873	136, 141
Latch, Knob	H. H. Elwell	Norwalk, Conn	Mar. 27, 1866	63, 526
Latch, Knob	A. Heyse	Terre Haute, Ind	Aug. 12, 1873	141, 714
Latch, Knob	C. M. Jordan	Stillwater, Minn	June 11, 1872	127, 773
Latch, Knob	R. Lee	Cincinnati, Ohio	Sept. 11, 1866	57, 930
Latch, Knob	W. T. Munger	Branford, Conn	Apr. 3, 1866	53, 745
Latch, Knob	W. T. Munger	Branford, Conn	Nov. 3, 1868	83, 652
Latch, Knob	E. Parker	New Britain, Conn	Jan. 21, 1873	135, 147
Latch, Knob	F. P. Pfleghar	New Haven, Conn	Jan. 11, 1870	98, 795
Latch, Knob	F. B. Pye	Trenton, N. J	Jan. 23, 1866	52, 201
Latch, Knob	O. B. Rand	Kalamazoo, Mich	Feb. 2, 1869	86, 448
Latch, Knob	S. M. Richardson	New York, N. Y	Nov. 8, 1864	44, 977
Latch, Knob	H. Richmond and A. Cloude	West Meriden, Conn	Oct. 3, 1865	50, 273
Latch, Knob	T. Slaight	Newark, N. J	June 8, 1869	91, 174
Latch, Knob	A. W. Upton	Lowell, Mass	Mar. 20, 1866	53, 364
Latch, Knob	R. L. Webb	West Meriden, Conn	Nov. 15, 1864	45, 100
Latch, Knob	A. F. Whiting	Greenville, Conn	June 15, 1869	91, 394
Latch, Knob	A. Williams	Norwich, Conn	June 6, 1865	48, 123
Latch, Knob	J. T. Williams	Chicago, Ill	Sept. 20, 1870	107, 578
Latch-lever	C. C. Lewis	Gainesville, Ala	Dec. 3, 1872	133, 588
Latch, Lever spring	J. N. Houston	Decatur, Ind	July 5, 1864	43, 410
Latch, Lock	N. Petre	New York, N. Y	July 30, 1867	67, 213
Latch lock	N. Petré	New York, N. Y	Feb. 7, 1871	111, 567
Latch lock, Door	H. Bosh	Mount Vernon, N. Y	Oct. 8, 1867	69, 619
Latch lock, Door	J. C. Kline	Pittsburgh, Pa	Jan. 16, 1855	12, 240
Latch, Locking	J. Hardy, 2d, and B. B. Floyd	Andover and Lawrence, Mass.	Sept. 1, 1868	81, 779
Latch, Locking knob	S. W. Drowne	Norwich, Conn	Mar. 24, 1868	75, 879
Latch, Locking knob	C. H. Eiffler	New York, N. Y	Dec. 3, 1867	71, 721
Latch, Locking knob	F. Hegner	Boston, Mass	Apr. 16, 1872	125, 734
Latch, Locking knob	W. L. Imlay	Philadelphia, Pa	Jan. 14, 1868	73, 334
Latch, Locking knob	H. C. Storrs	New York, N. Y	Jan. 21, 1868	73, 666
Latch, Locking-spindle door	W. H. McNamee	Philadelphia, Pa	Jan. 30, 1855	12, 320
Latch-mortise	W. Coover	Erie, Pa	Feb. 17, 1836	....
Latch, Mortise door	L. Foster	Boston, Mass	June 27, 1838	811
Latch, Mortise door	C. S. Gay	Nashua, N. H	Sept. 28, 1837	413
Latch, Mortise door	R. Kinsley	Springfield, Mass	Mar. 7, 1846	4, 402
Latch, Mortise door	W. T. Munger	New Britain, Conn	May 6, 1873	138, 680
Latch, Mortise door	D. N. Ropes	Portland, Me	Sept. 28, 1837	414
Latch, Mortise door	W. Wilson	Northampton, Mass	Nov. 26, 1844	3, 839
Latch, Mortise knob	G. H. Palmer	New Bedford, Mass	Aug. 18, 1867	67, 795
Latch-needles, Manufacture of	T. T. Mayne	Philadelphia, Pa	June 24, 1873	140, 150
Latch, Night	E. W. Brettell	Newark, N. J	July 23, 1867	66, 940
Latch, Night	C. C. Dickerman	Boston, Mass	Feb. 14, 1871	111, 732
Latch, Night	R. S. Foster	Madison, N. J	Jan. 25, 1870	99, 076
Latch, Oblique door	R. T. Fry	Spring Garden, Pa	Dec. 26, 1848	5, 984
Latch or fastener for doors	J. R. Webber	Morris, Ill	June 21, 1870	104, 672
Latch or fastener for window shutters and blinds	A. Hill	Steubenville, Ohio	Feb. 24, 1843	2, 977
Latch, Railway-car-door	J. Stephenson	New York, N. Y	Dec. 13, 1864	45, 442
Latch, Reversible	W. H. Andrews	New Haven, Conn	July 5, 1870	105, 026
Latch, Reversible	W. E. and H. H. Clark	New Haven, Conn	Jan. 26, 1869	86, 326
Latch, Reversible	H. Dotzenroth	Pittsburgh, Pa	Jan. 3, 1871	110, 637
Latch, Reversible	C. R. Fisher	Chelsea, Mass	Aug. 4, 1868	80, 539
Latch, Reversible	E. Halley	Branford, Conn	Nov. 14, 1865	50, 931
Latch, Reversible	J. Hamill	Allegheny, Pa	June 11, 1872	127, 870
Latch, Reversible	C. F. Herrick	Independence, Iowa	July 5, 1870	105, 075
Latch, Reversible	A. D. Judd	New Haven, Conn	Mar. 26, 1872	124, 960
Latch, Reversible	J. Kinzer	Pittsburgh, Pa	Aug. 11, 1868	80, 974
Latch, Reversible	J. Kinzer	Pittsburgh, Pa	May 17, 1870	103, 053
Latch, Reversible	J. Kinzer	Pittsburgh, Pa	May 17, 1870	103, 054

Index of patents issued from the United States Patent Office from 1790 *to* 1873, *inclusive*—Continued.

Invention.	Inventor.	Residence.	Date.	No.
Latch, Reversible	W. T. Munger	Branford, Conn	Sept. 8, 1868	81, 931
Latch, Reversible	W. T. Munger	Branford, Conn	Dec. 8, 1868	84, 704
Latch, Reversible	W. T. Munger	Branford, Conn	Jan. 12, 1869	85, 753
Latch, Reversible	W. T. Munger	Branford, Conn	Jan. 12, 1869	85, 754
Latch, Reversible	W. T. Munger	New Britain, Conn	May 18, 1869	90, 288
Latch, Reversible	W. T. Munger	New Britain, Conn	Mar. 1, 1870	100, 313
Latch, Reversible	W. T. Munger	New Britain, Conn	May 13, 1873	138, 919
Latch, Reversible	G. A. Seaver	New York, N. Y	Nov. 24, 1868	84, 382
Latch, Reversible	W. R. Smith	Branford, Conn	Apr. 5, 1870	101, 673
Latch, Reversible	W. E. Sparks	New Haven, Conn	Oct. 19, 1869	96, 049
Latch, Reversible	W. E. Sparks	New Haven, Conn	May 3, 1870	102, 723
Latch, Reversible	R. L. Webb	New Britain, Conn	Dec. 31, 1867	72, 946
Latch, Reversible	E. Whitney	New Haven, Conn	Dec. 29, 1868	85, 497
Latch, Reversible door	T. Lyons	Hartford, Conn	Mar. 16, 1869	87, 784
Latch, Reversible knob	W. H. Andrews	New Haven, Conn	Mar. 25, 1873	137, 166
Latch, Reversible knob	H. P. Appleton	Norwich, Conn	Sept. 20, 1870	107, 387
Latch, Reversible knob	A. Aston	New Britain, Conn	Mar. 30, 1869	88, 261
Latch, Reversible knob	H. H. Elwell	South Norwalk, Conn	Apr. 28, 1868	77, 264
Latch, Reversible knob	H. H. Elwell	South Norwalk, Conn	July 4, 1861	116, 697
Latch, Reversible knob	B. Erbe	Birmingham, Pa	May 19, 1868	77, 966
Latch, Reversible knob	B. Erbe	East Birmingham, Pa	Feb. 23, 1869	87, 096
Latch, Reversible knob	B. Erbe	Pittsburgh, Pa	Oct. 1, 1872	131, 860
Latch, Reversible knob	B. Erbe	Pittsburgh, Pa	Aug. 5, 1873	141, 430
Latch, Reversible knob	A. M. Hill	New Haven, Conn	Oct. 11, 1870	108, 261
Latch, Reversible knob	W. A. Hopkins	New York, N. Y	July 12, 1870	105, 334
Latch, Reversible knob	C. S. Jennings	Meriden, Conn	Apr. 8, 1873	137, 685
Latch, Reversible knob	C. Moody	Toronto, Canada	Aug. 19, 1873	141, 883
Latch, Reversible knob	W. T. Munger	Branford, Conn	Mar. 3, 1868	75, 184
Latch, Reversible knob	W. T. Munger	New Britain, Conn	Apr. 18, 1871	113, 909
Latch, Reversible knob	W. A. C. Oaks	Reading, Pa	Aug. 25, 1868	81, 529
Latch, Reversible knob	J. E. Parker	Meriden, Conn	May 5, 1868	77, 650
Latch, Reversible knob	J. E. Parker	Meriden, Conn	May 12, 1868	77, 909
Latch, Reversible knob	F. P. Pfleghar	New Haven, Conn	Mar. 2, 1869	87, 510
Latch, Reversible knob	F. P. Pfleghar	New Haven, Conn	July 6, 1869	92, 354
Latch, Reversible knob	H. M. Ritter	Cincinnati, Ohio	May 7, 1867	64, 570
Latch, Reversible knob	H. M. Ritter	Cincinnati, Ohio	May 7, 1867	64, 571
Latch, Reversible knob	H. M. Ritter	Cincinnati, Ohio	Mar. 28, 1871	113, 096
Latch, Reversible knob	H. M. Ritter	Cincinnati, Ohio	Mar. 28, 1871	113, 097
Laich, Reversible knob	R. S. Webb	New Britain, Conn	Mar. 25, 1873	137, 159
Latch, Reversible knob	A. F. Whiting	Crestline, Ohio	Mar. 25, 1873	137, 271
Latch, Reversible knob	A. F. Whiting	Crestline, Ohio	Mar. 25, 1873	137, 272
Latch, Reversible knob	J. Whittingham	Pittsburgh, Pa	Mar. 15, 1870	100, 830
Latch, Right and left door	G. E. Sellers	Cincinnati, Ohio	May 7, 1842	2, 610
Latch, Safety	J. W. Fifield	Franklin, N. H	Oct. 4, 1870	107, 895
Latch, Sliding-door	H. Belfield	Philadelphia, Pa	Jan. 10, 1860	26, 809
Latch, Stop mortise	F. A. Richardson	Poultney, Vt	Oct. 23, 1866	59, 076
Latch, Thumb	P. Eli and W. and J. A. Blake	New Haven, Conn	July 21, 1840	1, 704
Latch, Thumb	J. Range	Meriden, Conn	May 14, 1861	32, 339
Latch, Window	A. Bingham	Newtonville, Mass	Apr. 23, 1867	63, 988
Latch, Window	E. F. Hofmann	Pougkeepsie, N. Y	Jan. 1, 1867	60, 727
Latches, Adjustable escutcheon for night	W. T. Munger	Branford, Conn	Mar. 19, 1867	62, 968
Latches, Catch for gate and door	G. C. Lawton	Algona, Iowa	May 30, 1871	115, 330
Latches, Guard-attachment for door	W. Miller	Boston, Mass	Nov. 5, 1861	33, 682
Latches, Handle-plate for door	F. Corbin	New Britain, Conn	May 13, 1873	138, 862
Latches, Strike for door	J. E. Terry	Philadelphia, Pa	Apr. 14, 1863	38, 203
Lath and billets for rolling, Metallic	J. Reese	Pittsburgh, Pa	July 5, 1870	104, 997
Lath and clapboard sawing-machine	E. C. Gilman	Canaan, N. H	Aug. 23, 1844	3, 715
Lath-bolts, Machine for cutting	T. Crispin	Bay City, Mich	Aug. 22, 1871	118, 205
Lath-bolting machine	J. C. McIntyre	Fort Edward, N. Y	Sept. 2, 1873	142, 489
Lath-bolting machine	J. M. Stowell	Milwaukee, Wis	Apr. 30, 1872	126, 162
Lath-cutting machine	W. Bullock	Philadelphia, Pa	July 9, 1850	7, 487
Lath-cutting machine	I. Carpenter	Cincinnati, Ohio	May 2, 1832	....
Lath-cutting machine	D. M. Cradit	Ithaca, N. Y	May 22, 1835	....
Lath-cutting machine	O. H. Dibble	Evans, N. Y	Jan. 22, 1833	....
Lath-cutting machine	S. F. Finch and J. Wheeler	Rootstown, Ohio	Aug. 16, 1845	4, 147
Lath-cutting machine	R. Haynes	Oberlin, Ohio	Oct. 5, 1858	21, 675
Lath-cutting machine	D. Hubba	Wheeling, Va	July 18, 1834	....
Lath-cutting machine	C. Learned	Indianapolis, Ind	Apr. 24, 1866	54, 180
Lath-cutting machine	W. W. Lesner	Urbana, Ohio	Oct. 8, 1834	....
Lath-cutting machine	J. N. Lynch	Dillsburgh, Pa	Feb. 16, 1830	....
Lath-cutting machine	C. F. Packard	Greenwich, Conn	Feb. 21, 1854	10, 556
Lath-cutting machine	W. Pierson	Lebanon, Ohio	Dec. 31, 1833	....
Lath-cutting machine	S. Willard	Cincinnati, Ohio	Mar. 16, 1832	....
Lath-cutting machine	J. M. Winslow	Campbell's Port, Ohio	Oct. 3, 1846	4, 787
Lath cutting and splitting machine	B. Langdon	Troy, N. Y	Dec. 15, 1835	....
Lath for building	J. L. Brabyn	New York, N. Y	June 16, 1857	17, 550
Lath for building	D. Phillips	Shaftsbury, Vt	Jan. 17, 1865	45, 936
Lath for fire-proof ceilings of houses, Metallic	P. Sumner	New York, N. Y	Apr. 25, 1844	3, 566
Lath for plastering	E. D. Averell	New York, N. Y	May 28, 1867	65, 153
Lath-frame	A. Reed	Mankato, Minn.	July 23, 1867	67, 073
Lath from boards, Cutting	J. Jameson	Springfield, Ohio	Aug. 24, 1831	....
Lath-machine	J. Black	Memphis, Tenn	Nov. 16, 1858	22, 058
Lath-machine	A. Blaikie and W. Clark	Saint Clair, Mich	Nov. 6, 1855	13, 746
Lath-machine	S. P. Brower and W. H. Knox	Glen's Falls, N. Y	Sept. 30, 1873	143, 325
Lath-machine	J. L. Brown	Indianapolis, Ind	Jan. 15, 1856	14, 126
Lath-machine	T. Bruno	Saginaw, Mich	Dec. 16, 1873	145, 486
Lath-machine	J. H. Butler	Hampden, Me	Feb. 27, 1872	124, 034
Lath-machine	J. Dobbins	Litchfield, Mich	Sept. 28, 1869	95, 207
Lath-machine	H. Frisbee	Olmstead, Ohio	May 16, 1854	10, 905
Lath-machine	R. J. Gatling	Indianapolis, Ind	July 10, 1860	29, 072
Lath-machine	J. Gilman	Nashua, N. H	Mar. 25, 1856	14, 499
Lath-machine	A. Gummer	Indianapolis, Ind	Apr. 16, 1861	32, 103
Lath-machine	J. T. Hall	Alma, Mich	Nov. 8, 1870	109, 005
Lath-machine	S. and C. Howard	Easton and Bristol, Mass	Dec. 17, 1834	....
Lath-machine	J. L. Knowlton	Philadelphia, Pa	Aug. 16, 1870	106, 374
Lath-machine	J. C. Mackay	Ionia, Mich	Feb. 28, 1871	112, 261
Lath-machine	O. C. Macklett	Saint Paul, Minn	June 30, 1868	79, 366
Lath-machine	J. W. McLean and A. Gummer	Indianapolis, Ind	June 26, 1860	28, 883

Index of patents issued from the United States Patent Office from 1790 *to* 1873, *inclusive*—Continued.

Invention.	Inventor.	Residence.	Date.	No.
Lath-machine	O. C. Meigs	Dubuque, Iowa	Aug. 20, 1872	130, 650
Lath-machine	W. Merrell	Randolph, Ohio	Sept. 23, 1851	8, 376
Lath-machine	W. B. Noyes	Manchester, N. H	Jan. 18, 1870	98, 996
Lath-machine	J. B. Okey	Indianapolis, Ind	Jan. 27, 1857	16, 493
Lath-machine	J. Oothoudt	Minneapolis, Minn	Apr. 5, 1870	101, 653
Lath-machine	S. M. Palmer	Glen's Falls, N. Y	July 16, 1872	129, 492
Lath-machine	S. M. and G. E. Palmer	Sandy Hill, N. Y	Jan. 21, 1870	104, 637
Lath-machine	J. Pefley	Bainbridge, Ind	Dec. 28, 1858	22, 449
Lath-machine	J. H. and A. E. Redstone	Indianapolis, Ind	Nov. 15, 1859	26, 125
Lath-machine	S. B. Rittenhouse	Larwill, Ind	Feb. 25, 1873	136, 182
Lath-machine	C. Schleicher	Louisville, Ky	Jan. 24, 1871	111, 259
Lath-machine	I. R. Shank	Buffalo, Va	June 21, 1853	9, 804
Lath-machine	I. R. Shank	Buffalo, Va	June 6, 1854	11, 035
Lath-machine	E. Smith	Ithaca, N. Y	Apr. 28, 1836	....
Lath-machine	H. C. Smith	Cleveland, Ohio	Sept. 28, 1852	9, 286
Lath-machine	G. W. Tolhurst	Cleveland, Ohio	Dec. 9, 1851	8, 584
Lath-machine	T. Wright and A. P. Howell	Cincinnati, Ohio	Oct. 9, 1827	....
Lath, Machine for making siding	L. Rice	Lockport, N. Y	July 23, 1828	....
Lath machine, Sheet-metal	E. May	Indianapolis, Ind	Apr. 13, 1869	88, 972
Lath-machines, Method of feeding the bolts in	J. Nevison	Morgan, Ohio	May 18, 1858	20, 292
Lath, Metallic	J. B. Cornell	New York, N. Y	June 22, 1858	20, 629
Lath, Metallic	I. V. Holmes	New York, N. Y	Dec. 15, 1868	84, 881
Lath, Metallic	I. V. Holmes	New York, N. Y	Jan. 26, 1869	86, 305
Lath, Metallic	I. V. Holmes	New York, N. Y	Dec. 26, 1871	122, 171
Lath, Metallic	J. M. Manterstock	New York, N. Y	Apr. 12, 1859	23, 597
Lath, Metallic	K. Nirison	Chicago, Ill	Mar. 12, 1872	124, 611
Lath, Metallic	T. O'Callahan	Boston, Mass	Aug. 19, 1873	141, 945
Lath, Metallic	J. F. Walter, jr	New York, N. Y	Jan. 28, 1868	73, 765
Lath, Metallic	C. Welding	New York, N. Y	Apr. 8, 1873	137, 743
Lath, Metallic	W. E. Worthen	New York, N. Y	Feb. 24, 1857	16, 702
Lath, Metallic	W. E. Worthen	New York, N. Y	May 3, 1859	23, 877
Lath-mill	J. S. Hyde	Pent Water, Mich	Nov. 16, 1869	96, 811
Lath-sawing machine	G. P. Appleton	Suncook, N. H	July 8, 1873	140, 570
Lath-sawing machine	S. P. Brower	Glen's Falls, N. Y	Jan. 14, 1873	134, 846
Lath-sawing machine	T. Bruno	Saginaw, Mich	May 5, 1868	77, 449
Lath-sawing machine	W. G. Bulgin	Vienna, N. J	Aug. 25, 1868	81, 475
Lath-sawing machine	J. H. Butler	Hampden, Me	Mar. 15, 1870	100, 722
Lath-sawing machine	E. H. Ewell	Saint Louis, Mich	Oct. 19, 1869	95, 888
Lath-sawing machine	W. P. Hale	Ionia, Mich	Oct. 5, 1869	95, 582
Lath-sawing machine	E. H. Hancock	Augusta, Ga	Jan. 4, 1859	22, 499
Lath-sawing machine	I. A. Hedges and J. M. Story	Cincinnati, Ohio	May 5, 1868	77, 484
Lath-sawing machine	A. Rodgers	Muskegon, Mich	July 8, 1873	140, 647
Lath-sawing machine	W. Tuxworth	Sheridan, Mich	Jan. 18, 1870	98, 898
Lath-sawing machine	E. T. Wheeler and W. H. Vaughn.	Cannelton, Ind	Jan. 7, 1868	73, 214
Lath-sawing machine, Bed for	T. R. Markillie	Winchester, Ill	Jan. 1, 1856	14, 027
Lath, &c., slitting gage	B. K. Crandell	Lockport, N. Y	Aug. 25, 1828	....
Lath surface, Continuous sheet-metal	J. B. Cornell	New York, N. Y	May 13, 1856	14, 854
Lath surface, Metallic	J. B. Cornell	New York, N. Y	Aug. 10, 1858	21, 118
Lath with circular saw, Sawing	C. Madison	Miamisburgh, Ohio	Apr. 4, 1831	....
Laths from the block, Machine for riving	J. L. Brown	Indianapolis, Ind	Feb. 8, 1859	22, 853
Laths, Machine for bundling	J. S. Anderson	Flintville, Wis	Aug. 20, 1872	130, 687
Laths, Machine for sawing bevels on	G. W. Corson	Corson's Post-Office, Pa	July 10, 1860	29, 062
Laths, Machinery for cutting	S. Cheney	Cleveland, Ohio	Sept. 27, 1845	4, 206
Laths, palings, &c., Machine for sawing	S. Head	Millersburgh, Pa	Nov. 24, 1863	40, 690
Lathing and plastering, Mode of	J. G. Vaughn	Middleborough, Mass	Jan. 13, 1857	16, 425
Lathing-apparatus	W. L. Beardsley	Binghamton, N. Y	May 28, 1867	65, 048
Lathing buildings	B. F. Gold	New Haven, Conn	Aug. 1, 1854	11, 423
Lathing, Continuous metallic	B. Cornell	New York, N. Y	Mar. 2, 1858	19, 487
Lathing-machine	G. N. Creamer	Trenton, N. J	Oct. 5, 1869	95, 435
Lathing-machine	G. N. Creamer	Trenton, N. J	Nov. 22, 1870	109, 391
Lathing-machine	T. Hill	New Centreville, Wis	Jan. 21, 1868	73, 604
Lathing-machine	W. S. Wright	Ithaca, N. Y	Apr. 20, 1869	89, 112
Lathe	W. Aldrich	Lowell, Mass	Apr. 10, 1855	12, 662
Lathe	J. P. Allen	Springfield, Ohio	May 9, 1871	114, 512
Lathe	E. and N. Avord	Westfield, Mass	Oct. 12, 1836	....
Lathe	T. R. Bailey	Lockport, N. Y	July 1, 1851	8, 196
Lathe	W. B. Bement	Philadelphia, Pa	May 21, 1867	64, 938
Lathe	A. H. Brown	Albany, N. Y	Apr. 29, 1856	14, 787
Lathe	J. T. Bunce	East Haddam, Conn	May 25, 1858	28, 323
Lathe	H. Chavons	Union City, Ind	Sept. 19, 1871	119, 012
Lathe	C. S. Clark	Providence, R. I	June 29, 1869	91, 910
Lathe	H. O. Clark	Worcester, Mass	June 6, 1854	11, 037
Lathe	T. Collins and A. Castiel	Saint Louis, Mo	Mar. 30, 1869	88, 452
Lathe	J. Cook	Buffalo, N. Y	Apr. 3, 1860	27, 696
Lathe	L. L. Crane	Cleveland, Ohio	June 5, 1866	55, 412
Lathe	M. Gore	Shelby, Mich	June 29, 1869	91, 930
Lathe	O. M. Grimes	Newark, N. J	June 29, 1869	92, 041
Lathe	L. Griscom	Mahanoy Plane, Pa	May 12, 1868	77, 812
Lathe	A. Hatch	New Haven, Conn	Mar. 20, 1866	53, 374
Lathe	D. Heer	Philadelphia, Pa	July 29, 1873	141, 347
Lathe	J. Hess	Niagara Falls, N. Y	July 19, 1859	24, 805
Lathe	H. R. Hill and N. W. Twiss	New Haven, Conn	Sept. 12, 1871	118, 938
Lathe	J. Kievlan	Chicago, Ill	Oct. 12, 1869	95, 694
Lathe	J. Kievlan	Chicago, Ill	Feb. 21, 1871	112, 051
Lathe	S. U. King	Windsor, Vt	Dec. 6, 1870	109, 912
Lathe	B. Lawrence	Lowell, Mass	Dec. 12, 1871	121, 878
Lathe	B. M. Lewy	Montgomery, Ala	Oct. 8, 1867	69, 685
Lathe	F. B. Mattson	Rockford, Ill	June 28, 1870	104, 751
Lathe	D. W. Maurice	Springfield, Ohio	Apr. 24, 1866	54, 187
Lathe	L. F. Munger	Rochester, N. Y	Jan. 7, 1868	73, 027
Lathe	H. E. Nickerson	Atchison, Kans	Dec. 3, 1872	133, 593
Lathe	L. Pastlewait	Russellville, Ohio	Oct. 16, 1860	30, 420
Lathe	P. H. Pitts	Waverly, Mo	Aug. 16, 1870	106, 404
Lathe	E. Putman	Gardner, Mass	Dec. 9, 1829	....
Lathe	S. W. Putnam and J. Q. Wright	Fitchburgh, Mass	July 21, 1863	39, 326
Lathe	E. Rhodes	Kingsbury, N. Y	Nov. 11, 1830	....

Index of patents issued from the United States Patent Office from 1790 *to* 1873, *inclusive*—Continued.

Invention.	Inventor.	Residence.	Date.	No.
Lathe	J. F. C. Rider and E. P. Brownell.	South New Market, N. H., and Providence, R. I.	Apr. 18, 1871	113, 798
Lathe	M. Roberts	Worcester, Mass	July 10, 1860	29, 103
Lathe	J. M. Scribner	Middleburgh, N. Y	May 15, 1860	28, 308
Lathe	W. Sellers	Philadelphia, Pa	Dec. 28, 1869	98, 423
Lathe	H. C. Spalding	New York, N. Y	Apr. 1, 1856	14, 578
Lathe	J. Stark	Waltham, Mass	May 30, 1865	47, 997
Lathe	T. Sullivan	Belleville, Canada	Jan. 23, 1872	123, 058
Lathe	J. Watson	Philadelphia, Pa	Mar. 1, 1870	100, 473
Lathe	B. D. Whitney	Winchendon, Mass	Aug. 7, 1860	29, 534
Lathe and chuck, Watchmaker's	C. Hopkins	Philadelphia, Pa	Nov. 5, 1872	132, 763
Lathe and gear cutter combined	U. Opperman	Washington, D. C	Aug. 22, 1871	118, 262
Lathe-arbors, Machine for grinding	C. Coester, jr., and A. B. Lawther.	Bridgeport, Conn	Sept. 10, 1867	68, 698
Lathe-attachment	C. Kilburn	Burlington, Vt	Nov. 22, 1859	26, 192
Lathe-attachment for chasing designs on molds	J. E. Miller	Birmingham, Pa	Nov. 7, 1871	120, 593
Lathe-attachment for cutting veneers	B. F. Sturtevant	Boston, Mass	Dec. 27, 1859	26, 627
Lathe-attachment for turning irregular forms	M. Roberts	Belfast, Me	May 20, 1856	14, 941
Lathe-attachment for turning tapers on bars	C. Jillson	Worcester, Mass	June 28, 1864	43, 311
Lathe, Automatic	A. Edmonds	Mount Pulaski, Ill	July 14, 1857	17, 782
Lathe, Automatic	J. McNary	Brooklyn, N. Y	May 4, 1858	20, 166
Lathe-back center	A. Lefever and G. C. Barnes	Battle Creek, Mich	Jan. 31, 1860	26, 998
Lathe-back rest	W. H. Hendrick and J. Jacobs	Mount Vernon, Ohio	Jan. 31, 1860	26, 988
Lathe, Block-making gang	E. Whiting		Apr. 1, 1802	....
Lathe boring-attachment	C. E. McBeth	Hamilton, Ohio	Sept. 25, 1866	58, 273
Lathe boring-attachment	F. B. Williams	Sterling, Ill	Apr. 29, 1873	138, 460
Lathe box and journal	A. Wood	Worcester, Mass	Dec. 10, 1867	72, 147
Lathe burnishing-attachment	J. F. Ray	East Haddam, Conn	Dec. 28, 1858	22, 452
Lathe-carriage	G. W. Lewin	Worcester, Mass	Mar. 24, 1868	75, 933
Lathe-carrier	F. Ferrin	Camden, N. J	Apr. 1, 1873	137, 298
Lathe-carrier	J. Zook	Harrisburgh, Pa	Feb. 28, 1854	10, 585
Lathe-carrier, Extension	W. A. Lorenz	Newark, N. J	Apr. 2, 1872	125, 203
Lathe carrier or dog	L. P. Wilcox	Brooklyn, N. Y	Dec. 11, 1866	60, 451
Lathe-center	H. K. Porter	Boston, Mass	July 5, 1870	104, 995
Lathe-centers, Apparatus for grinding	W. S. Jarboe	New York, N. Y	Dec. 24, 1872	134, 287
Lathe-centers, Tool for grinding	D. Slate	Hartford, Conn	Apr. 30, 1872	126, 160
Lathe, Centering	B. F. Bee	Harwich, Mass	July 28, 1868	80, 327
Lathe centering-device	H. Gray	Bristol, Conn	Aug. 19, 1873	142, 014
Lathe centering-device	J. A. Talpey	Somerville, Mass	June 16, 1863	38, 922
Lathe centering-machine	D. Kelly	Philadelphia, Pa	Mar. 15, 1870	100, 772
Lathe-chuck	J. C. Miller	Danville, Ky	May 9, 1871	114, 700
Lathe, Cooper's	I. Hoover	Miamisburgh, Ohio	July 2, 1835	....
Lathe cutter-head	M. Roberts	Belfast, Me	May 13, 1856	14, 899
Lathe cutter, Wood-turning	J. Shannon	Cohoes, N. Y	Aug. 1, 1865	49, 161
Lathe-dog	C. Buss	Marlborough, N. H	Oct. 18, 1870	108, 325
Lathe-dog	E. Cliff and O. A. Haynes	South Saint Louis, Mo	Nov. 5, 1872	132, 804
Lathe-dog	W. Emmett	Paterson, N. J	Aug. 18, 1868	81, 264
Lathe-dog	R. H. Lecky	Allegheny City, Pa	June 23, 1868	79, 134
Lathe-dog	C. W. Le Count	Norwalk, Conn	Oct. 24, 1865	50, 604
Lathe-dog	C. W. Le Count	Norwalk, Conn	Jan. 1, 1867	60, 751
Lathe-dog	N. M. Phillips	New York, N. Y	May 18, 1858	20, 298
Lathe-dog	W. Pimbott	Syracuse, N. Y	Dec. 4, 1866	60, 238
Lathe-dog	J. W. Russell	Springfield, Mass	Aug. 4, 1868	80, 770
Lathe-dog	J. S. Skinner	Lebanon, N. H	Aug. 9, 1870	106, 224
Lathe-dog	D. M. Smith	Springfield, Vt	May 16, 1854	10, 939
Lathe-dog	W. Vine	Norwalk, Conn	Aug. 1, 1865	49, 180
Lathe-dog	J. B. West	Geneseo, N. Y	Jan. 25, 1870	99, 269
Lathe-dog	H. K. White	Chelsea, Mass	Mar. 14, 1871	112, 759
Lathe-dog	L. P. Whiting	Poughkeepsie, N. Y	Nov. 5, 1872	132, 882
Lathe-dog	C. L. Wickham	Utica, N. Y	Apr. 29, 1873	138, 305
Lathe-dog and bench-vise jaw	N. Wilton	Groton, N. H	Oct. 15, 1867	69, 883
Lathe-dog, Automatic	A. E. Wenzel	New York, N. Y	Dec. 18, 1860	30, 997
Lathe, Engine	A. B. Couch	Worcester, Mass	Apr. 27, 1869	89, 468
Lathe, Engine	J. L. Johnson	Ashburnham, Mass	Mar. 7, 1865	46, 677
Lathe, Engine	S. Teal	Rochester, N. Y	Aug. 18, 1868	81, 309
Lathe, Engine	A. E. Whitmore	Boston, Mass	Jan. 5, 1869	85, 552
Lathe, Engine-turning	C. W. Dickinson	Newark, N. J	Dec. 13, 1864	45, 455
Lathe-fastening	J. M. Stone	North Andover, Mass	May 23, 1865	47, 876
Lathe-fastening	J. G. Stowe	Providence, R. I	Dec. 14, 1869	97, 986
Lathe for chasing and backing down taps	W. X. Stevens	Worcester, Mass	Mar. 19, 1867	62, 977
Lathe for cutting fluted moldings	J. Anderson, J. McLaren, and J. Bryant.	New York, N. Y	Nov. 25, 1856	16, 108
Lathe for cutting grooves in metal rollers	H. Waters	Boston, Mass	May 16, 1871	115, 000
Lathe for cutting irregular forms	P. Barker	Battle Creek, Mich	Apr. 28, 1868	77, 158
Lathe for cutting screws from wire	G. W. Daniels	Waltham, Mass	Oct. 19, 1858	21, 864
Lathe for cutting tenons for clock-movement	R. Peck	Bristol, Conn	Feb. 23, 1858	19, 472
Lathe for finishing the driving-wheels of locomotives.	H. D. Stover	New York, N. Y	June 22, 1869	91, 788
Lathe for forming shafting	A. Wood	Worcester, Mass	Dec. 10, 1867	72, 148
Lathe for grinding and polishing pivots of watch-work.	C. V. Woerd	Waltham, Mass	May 25, 1869	90, 620
Lathe for making awl-handles	F. Reed	Canton, Mass	June 21, 1870	104, 645
Lathe for manufacture of clothes-pins, &c	J. Humphrey	Keene, N. H	Sept. 22, 1857	18, 268
Lathe for setting jewels in time-pieces	H. H. Heskett	McLean County, Ill	Mar. 15, 1870	100, 891
Lathe for squaring nuts	J. Flower	Detroit, Mich	Mar. 29, 1870	101, 249
Lathe for turning a peculiar species of curves	H. G. Thompson	New York, N. Y	Apr. 23, 1850	7, 311
Lathe for turning and polishing soft-metal and composition buttons.	L. Meriam	New Haven, Conn	Jan. 4, 1815	....
Lathe for turning axles	A. J. Bower	Albion, Ill	May 30, 1871	115, 424
Lathe for turning balls	J. B. West	Geneseo, N. Y	Aug. 18, 1868	81, 315
Lathe for turning beaded work	F. Baldwin	South Wardsborough, Vt	Aug. 24, 1858	21, 240
Lathe for turning bent sticks	T. Ott	Green Township, Pa	Aug. 6, 1872	130, 237
Lathe for turning billiard-balls	L. A. Johnson	San Francisco, Cal	Sept. 8, 1863	39, 817
Lathe for turning broom-handles	P. Prescott	Boonville, N. Y	Oct. 8, 1861	33, 466
Lathe for turning buttons	D. C. Robie	Springfield, Mass	Mar. 10, 1868	75, 462
Lathe for turning eccentrics	J. B. Gayle	Portsmouth, Va	July 30, 1867	67, 288
Lathe for turning fancy handles	L. Wentworth	Burlington, Iowa	Jan. 9, 1855	12, 224
Lathe for turning gun-barrels	D. Dana and A. Olney	Canton, Mass	Aug. 24, 1818	....

Index of patents issued from the United States Patent Office from 1790 *to* 1873, *inclusive*—Continued.

Invention.	Inventor.	Residence.	Date.	No.
Lathe for turning heads of nails, tacks, &c	W. H. Nichols and H. H. Abbe.	Chatham, Conn	Aug. 1, 1865	49, 130
Lathe for turning hoe-handles, &c	R. D. Bartlett	Harmony, Me	Dec. 26, 1848	5, 983
Lathe for turning hubs	A. Goodyear	Albion, Mich	Aug. 31, 1869	94, 200
Lathe for turning interior and exterior surfaces	N. Chapin	New York, N. Y	Jan. 11, 1853	9, 529
Lathe for turning interior surfaces of hollow-ware	P. Teal and C. Tyler	Philadelphia, Pa	May 23, 1854	10, 946
Lathe for turning iron	W. Sellers	Philadelphia, Pa	Feb. 18, 1868	74, 609
Lathe for turning irregular curved forms	W. Payton	London, England	Jan. 28, 1868	73, 830
Lathe for turning irregular forms	F. Baker	Pepperell, Mass	May 16, 1854	10, 922
Lathe for turning irregular forms	F. T. Grant	Augusta, Me	Mar. 24, 1863	37, 960
Lathe for turning irregular forms	B. F. Jenkins and L. L. Knight.	Barre, Mass	Jan. 4, 1853	9, 521
Lathe for turning irregular forms	W. P. Lathrop	West Winsted, Conn	Aug. 20, 1872	130, 725
Lathe for turning irregular forms	E. Lumley	Elizabeth City, N. J	May 5, 1863	38, 399
Lathe for turning irregular forms	L. Smith	Plymouth, Conn	Dec. 9, 1856	16, 192
Lathe for turning irregular forms	C. and A. Spring	Boston, Mass	May 10, 1859	23, 957
Lathe for turning irregular forms	D. Werst and A. Puderbaugh	Waltz Township, Ind	Feb. 14, 1860	27, 178
Lathe for turning irregular forms	C. and M. T. Whipple and J. Sprague.	Douglass, Mass	Apr. 3, 1835	....
Lathe for turning irregular forms	E. K. Wisell	Warren, Ohio	May 28, 1872	127, 286
Lathe for turning irregular forms, Automatic	S. N. Baker	New Haven, Conn	July 28, 1857	17, 866
Lathe for turning irregular forms, Automatic	W. D. Sloan	New York, N. Y	Mar. 31, 1857	16, 936
Lathe for turning irregular forms, Wood	A. R. Stewart	Douglass Harbor, Canning, New Brunswick.	Apr. 2, 1867	63, 574
Lathe for turning knobs, Wood	J. Stever and J. A. Way	Bristol, Conn	Jan. 8, 1867	61, 111
Lathe for turning locomotive-crank pins	S. S. and D. Cheney	Galesburgh, Ill	Aug. 11, 1863	39, 465
Lathe for turning locomotive-drivers	J. M. Stone	Manchester, N. H	Apr. 10, 1855	12, 708
Lathe for turning masts, &c	P. H. Niles	Boston, Mass	Dec. 28, 1858	22, 447
Lathe for turning metal shafting	W. Sellers	Philadelphia, Pa	June 1, 1858	20, 446
Lathe for turning oval frames	J., W., and G. Gardner	New York, N. Y	June 29, 1858	20, 705
Lathe for turning ovals	R. Eickemeyer	Yonkers, N. Y	Apr. 5, 1870	101, 447
Lathe for turning ovals	R. Lawson	Shelburne Falls, Mass	Nov. 23, 1869	97, 096
Lathe for turning ovals	F. Smith	Allegheny City, Pa	Feb. 27, 1866	52, 900
Lathe for turning pen-handles	J. Keith	Worcester, Mass	Nov. 6, 1866	59, 407
Lathe for turning rake-handles, &c	J. Haven	Newport, N. H	Oct. 14, 1835	....
Lathe for turning regular forms	J. Moessinger	New York, N. Y	Nov. 19, 1872	133, 241
Lathe for turning scythe-snaths	S. Hinton	Jackson, Mich	Sept. 4, 1866	57, 716
Lathe for turning shafting	J. J. Conley	New York, N. Y	Oct. 9, 1866	58, 605
Lathe for turning shafting, Engine	A. Wood	Worcester, Mass	Mar. 31, 1868	76, 130
Lathe for turning spherical shapes	R. B. Carsley	New Bedford, Mass	Jan. 19, 1864	41, 277
Lathe for turning spherical shot and shell	C. Foster	Pittsburgh, Pa	Jan. 23, 1866	52, 244
Lathe for turning spokes	T. Derington	Duquoin, Ill	Jan. 26, 1864	41, 368
Lathe for turning spools	L. W. Boynton	Hartford, Conn	Nov. 26, 1872	133, 403
Lathe for turning spools	J. T. Hawkins	Annapolis, Md	Mar. 16, 1869	87, 929
Lathe for turning spools	A. T. Wing	Old Town, Me	Aug. 26, 1873	142, 314
Lathe for turning tobacco-pipes	C. Kapf	Philadelphia, Pa	July 29, 1873	141, 279
Lathe for turning tool-handles	H. K. Jones	Kensington, Conn	June 27, 1865	48, 409
Lathe for turning wagon-hubs	H. Inman	Hagaman's Mills, N. Y	Aug. 14, 1866	57, 145
Lathe for turning wagon-hubs	H. C. Sisco	Indianapolis, Ind	Mar. 20, 1866	53, 382
Lathe for turning whip-stocks	G. Dayton and A. Mallory	Westfield, Mass	Apr. 27, 1832	....
Lathe for turning whip-stocks	L. Hull	Charlestown, Mass	July 17, 1866	56, 419
Lathe for turning wire-eyed buttons, Old pin	W. Lawrence	Meriden, Conn	Apr. 12, 1815	....
Lathe-gearing, Adjustable	J. Humphreys	Chicopee, Mass	Nov. 3, 1868	83, 774
Lathe, Gun-stock	A. Town	Woodbury, Vt	Feb. 28, 1836	....
Lathe-head	A. F. Shaw	West Roxbury, Mass	Sept. 8, 1868	81, 952
Lathe, Metal-planing	W. W. Hubbard	Boston, Mass	Nov. 25, 1856	16, 118
Lathe, Metal-turning	E. F. and G. A. Barnes	New Haven, Conn	Jan. 21, 1873	135, 064
Lathe, Metal-turning	G. Henderson and J. Steelle	Allegheny, Pa	July 6, 1858	20, 794
Lathe, Metal-turning	J. W. Hyatt, jr	Albany, N. Y	Apr. 9, 1872	125, 571
Lathe, Metal-turning	W. Kitson	Lowell, Mass	Jan. 16, 1872	122, 769
Lathe, Metal-turning	F. W. Pomroy	Sandusky, Ohio	Aug. 13, 1872	130, 387
Lathe, Metal-turning	S. W. Putnam	Fitchburgh, Mass	Nov. 26, 1872	133, 483
Lathe, Metal-turning	W. Sellers	Philadelphia, Pa	May 28, 1872	127, 195
Lathe, Metal-turning	M. G. Stolp	Ithaca, N. Y	May 14, 1872	126, 758
Lathe-mill, Vertical	E. Skinner and J. Webster	Sandwich, N. H	Aug. 26, 1831	....
Lathe-pattern for turning lasts, &c	H. Mellish	Walpole, N. H	Dec. 23, 1834	....
Lathe, Pivot	R. Lissenbee	Greenville, Tenn	Apr. 17, 1866	53, 994
Lathe polishing-apparatus, Watch-maker's	J. M. Bottum	New York, N. Y	Mar. 13, 1855	12, 502
Lathe, Potters'	W. Meek	Anderson, Ind	July 22, 1873	141, 157
Lathe, Potters'	H. B. Morris	Burlington, N. J	Apr. 30, 1872	126, 318
Lathe, Prism	A. Kelsey	Charlestown, Mass	Dec. 5, 1865	51, 388
Lathe, Railway-car-axle	G. A. Gray, jr	Cincinnati, Ohio	May 9, 1871	114, 670
Lathe-rest	C. A. Bauer	Hamilton, Ohio	Mar. 11, 1873	136, 579
Lathe-rest	H. K. Smith	Norwich, Conn	Jan. 1, 1867	60, 950
Lathe-rest	H. K. Smith	Norwich, Conn	Feb. 18, 1868	74, 621
Lathe-rest	E. Stahlbrodt	Rochester, N. Y	May 5, 1868	77, 672
Lathe-rest	W. Thompson	Worcester, Mass	July 14, 1868	79, 877
Lathe-rest, Adjustable	J. E. Burdge	Cincinnati, Ohio	Aug. 13, 1867	67, 630
Lathe-rest attachment	D. White, jr	Lowell, Mass	Feb. 16, 1858	19, 395
Lathe-rest engine	H. Morrison	Paterson, N. J	Oct. 1, 1861	33, 399
Lathe-rest for turning balls	W. F. Vose	Newton, Mass	June 19, 1866	55, 749
Lathe-rest for turning spheres	P. Wenzel	Mayence-on-the-Rhine, Germany.	Sept. 14, 1869	94, 930
Lathe-rest, Sliding	E. S. Gardner	Philadelphia, Pa	Sept. 1, 1857	18, 120
Lathe rest, Turning	W. H. Barney	Columbus, Ohio	July 5, 1864	43, 386
Lathe, Rose-engine	T. Lippiatt	New York, N. Y	Aug. 14, 1866	57, 158
Lathe, Rose-engine	T. Lippiatt	New York, N. Y	May 28, 1867	65, 245
Lathe, Rose-engine	T. Lippiatt	New York, N. Y	July 28, 1868	80, 292
Lathe, Rotary	G. Tugnot	New York, N. Y	Nov. 28, 1854	12, 008
Lathe, Screw-cutting	T. D. Bassett	Charlestown, Mass	July 7, 1868	79, 721
Lathe, Screw-cutting	C. Ellis	Canton, Mass	Apr. 14, 1868	76, 612
Lathe, Screw-cutting	J. P. T. Lang	Washington, D. C	July 7, 1868	79, 767
Lathe, Screw-cutting	H. F. Shaw	West Roxbury, Mass	Oct. 27, 1868	83, 554
Lathe, Screw-cutting	J. Tangye	Birmingham, England	July 10, 1866	56, 339
Lathe-slide	E. K. Root	Hartford, Conn	May 15, 1855	12, 874
Lathe slide-rest	C. A. Noyes	Pittsfield, Mass	Apr. 17, 1855	12, 742
Lathe slide-rest	J. G. Rominger	Philadelphia, Pa	Sept. 10, 1867	68, 654
Lathe slide-rest	W. Stephens	Richmond, Ind	Feb. 20, 1855	12, 423
Lathe slide-rest	C. Van Haagen	Philadelphia, Pa	Mar. 11, 1873	136, 794
Lathe slide-rest	C. Van Horn	Springfield, Mass	Apr. 17, 1855	12, 747

Index of patents issued from the United States Patent Office from 1790 *to* 1873, *inclusive*—Continued.

Invention.	Inventor.	Residence.	Date.	No.
Lathe-spindle	J. E. Boutelle	Fisherville, N. H	Apr. 19, 1870	101, 975
Lathe-spindles, Hold-fast for	L. R. Faught	Philadelphia, Pa	May 30, 1871	115, 290
Lathe, Spoke	G. F. Almy	Toledo, Ohio	Mar. 15, 1870	100, 836
Lathe, Spoke	I. S. Roland	Reading, Pa	June 1, 1869	90, 689
Lathe, Spoke	C. B. Conant and H. Thompson.	Worcester, Mass	Mar. 1, 1870	100, 261
Lathe, Spoke	J. Pierce and A. B. Curtis	New Haven, Conn	Aug. 15, 1871	118, 148
Lathe-stand	H. D. Stover	New York, N. Y	Oct. 17, 1865	50, 510
Lathe taper-turning attachment	M. Neckermann	Pittsburgh, Pa	Sept. 30, 1873	143, 296
Lathe to prevent counterfeiting, Figure	D. H. Mason	Philadelphia, Pa	June 7, 1832	....
Lathe-tool	J. C. Shackleton	Lawrence, Mass	July 9, 1867	66, 641
Lathe-tool	T. Shaw	Philadelphia, Pa	Mar. 3, 1868	75, 060
Lathe tool and rest holder	J. Walfenden	Jersey City, N. J	July 31, 1866	56, 840
Lathe tool-elevator	W. Hamilton	Chicopee, Mass	Nov. 29, 1864	45, 297
Lathe tool-holder	C. E. Albro	Syracuse, N. Y	Dec. 31, 1872	134, 343
Lathe tool-holder	J. Baillie	Salem, Ohio	Feb. 25, 1868	74, 877
Lathe tool-holder	W. Coulter	Birmingham, Conn	Dec. 10, 1872	133, 833
Lathe tool-holder	J. Edson	Boston, Mass	Dec. 24, 1867	72, 468
Lathe tool-holder	C. H. Fowler	Roxbury, Mass	June 2, 1868	78, 520
Lathe tool-holder	W. O. Hickok and G. W. Reisinger.	Harrisburgh, Pa	Nov. 10, 1868	83, 963
Lathe tool-holder	C. Peck	New Haven, Conn	June 28, 1859	24, 574
Lathe tool-holder	L. Reder	Wilmington, Del	Dec. 16, 1873	145, 684
Lathe tool-holder	I. P. Richards	Providence, R. I	Dec. 24, 1872	134, 219
Lathe tool-rest	T. J. Currier and A. M. Black	Worcester, Mass	Jan. 15, 1867	61, 166
Lathe tool-rest	C. F. Hadley	Chicopee, Mass	Oct. 29, 1872	132, 573
Lathe tool-rest	J. Service	Greenville, Conn	Dec. 4, 1866	60, 266
Lathe tool-rest	J. G. Stowe	Providence, R. I	Dec. 7, 1869	97, 723
Lathe tool-rest, Engine	C. Newhall	Hinsdale, N. H	June 30, 1868	79, 381
Lathe tool-supporter	I. F. Brown	New London, Conn	July 4, 1871	116, 545
Lathe, Turning	W. Aldrich	Lowell, Mass	Mar. 15, 1853	9, 616
Lathe, Turning	E. and N. Alvord	Westfield, Mass	Oct. 11, 1836	43
Lathe, Turning	F. A. Armbruster	New York, N. Y	Jan. 15, 1867	61, 135
Lathe, Turning	J. Ball	New York, N. Y	Oct. 4, 1864	44, 503
Lathe, Turning	J. G. Baker	Philadelphia, Pa	Nov. 1, 1864	44, 915
Lathe, Turning	E. Bancroft and W. Sellers	Philadelphia, Pa	Feb. 7, 1854	10, 491
Lathe, Turning	J. Bisbee	Plainfield, Mass	Nov. 8, 1832	....
Lathe, Turning	C. G. Bloomer	Wickford, R. I	Dec. 20, 1864	45, 555
Lathe, Turning	J. Burt	Sturgis, Mich	Nov. 13, 1866	59, 552
Lathe, Turning	R. J. Cole	Poultney, Vt	Dec. 22, 1863	40, 994
Lathe, Turning	F. W. Coy	Boston, Mass	Oct. 8, 1867	69, 544
Lathe, Turning	H. Doane and A. Goodman	Dana, Mass	Nov. 6, 1849	6, 843
Lathe, Turning	R. W. Drew	Lowell, Mass	Dec. 4, 1866	60, 155
Lathe, Turning	L. R. Faught	Philadelphia, Pa	Dec. 4, 1866	60, 163
Lathe, Turning	M. J. Gardner	Yorkborough, Pa	Dec. 14, 1830	....
Lathe, Turning	A. Goodyear	Springport, Mich	Aug. 23, 1864	43, 908
Lathe, Turning	N. Green	Bedford County, Tenn	Apr. 27, 1832	....
Lathe, Turning	P. J. Hardy	New York, N. Y	Mar. 28, 1871	113, 048
Lathe, Turning	N. Harper	Newark, N. J	Feb. 14, 1865	46, 352
Lathe, Turning	J. Hunt	Shutesbury, Mass	June 8, 1826	....
Lathe, Turning	W. Johnson	Lambertville, N. J	May 28, 1867	65, 089
Lathe, Turning	F. Kirst	Westfield, Mass	July 12, 1864	43, 512
Lathe, Turning	S. R. Lewis	Rockford, Ill	Oct. 20, 1868	83, 181
Lathe, Turning	H. Locke	Grand Rapids, Mich	Sept. 20, 1870	107, 513
Lathe, Turning	J. P. Luther	Berlin, Wis	Aug. 19, 1873	141, 940
Lathe, Turning	G. Meldrum	West Philadelphia, Pa	Apr. 18, 1854	10, 783
Lathe, Turning	F. B. Miles	Philadelphia, Pa	Feb. 14, 1871	111, 859
Lathe, Turning	J. Moore	Leverett, Mass	Jan. 19, 1828	....
Lathe, Turning	H. L. Morse	New Bedford, Mass	Apr. 16, 1867	63, 928
Lathe, Turning	W. Patrick	Leverett, Mass	Apr. 24, 1827	....
Lathe, Turning	J. Richards	Cincinnati, Ohio	Aug. 13, 1867	67, 676
Lathe, Turning	W. Ross	Chester County, Pa	July 10, 1809	....
Lathe, Turning	W. Sellers	Philadelphia, Pa	Mar. 13, 1860	27, 478
Lathe, Turning	W. Sellers	Philadelphia, Pa	June 11, 1872	127, 929
Lathe, Turning	F. Shaller	Hudson, N. Y	Aug. 13, 1867	67, 810
Lathe, Turning	F. G. Sheldon	Hudson, Mich	Mar. 19, 1872	124, 861
Lathe, Turning	S. D. Sheldon	Fitchburgh, Mass	Mar. 21, 1871	112, 971
Lathe, Turning	W. Simpson, Z. Oman, and S. Foreman.	Nottawa, Mich	Sept. 11, 1866	57, 984
Lathe, Turning	D. Slate	Hartford, Conn	Jan. 31, 1865	46, 152
Lathe, Turning	W. D. Sloan	New York, N. Y	Jan. 5, 1858	19, 051
Lathe, Turning	Q. H. Spencer	North Providence, R. I	Aug. 10, 1869	93, 492
Lathe, Turning	A. Warth	New York, N. Y	Oct. 10, 1854	11, 798
Lathe, Turning	J. D. White	Hartford, Conn	May 21, 1850	7, 390
Lathe, Turning	C. White and A. Hill	Hadley, Mass	Mar. 30, 1824	....
Lathe, Turning	J. C. Wybell	West Meriden, Conn	Aug. 7, 1866	57, 054
Lathe, Turning	L. D. Wynhoop	Owassa, Mich	Apr. 5, 1864	42, 250
Lathe, Turning	F. Zibell	Saint Louis, Mo	Mar. 24, 1868	75, 822
Lathe, Uniform-turning	A. Haggett	Middlesex County, Mass	June 10, 1809	....
Lathe, Variety-frame	A. C. Wicker and L. W. Williams.	Fair Haven, Vt	Aug. 6, 1867	67, 473
Lathe, Watch-maker's	E. J. Chapin	Ottawa, Ill	Nov. 18, 1862	36, 942
Lathe, Watch-maker's	R. Cowles	Cleveland, Ohio	Jan. 14, 1873	134, 787
Lathe, Watch-maker's	E. Harris	Princeton, Ill	Mar. 29, 1859	23, 370
Lathe, Watch-maker's	G. Hunziker	Summit, Miss	Jan. 16, 1872	122, 768
Lathe, Watch-maker's	E. H. Kelly	Jackson, Tenn	Nov. 11, 1873	144, 547
Lathe, Watch-maker's	G. Jackson	Cincinnati, Ohio	Oct. 28, 1862	36, 842
Lathe, Watch-maker's	J. M. Oram	Gosport, Ind	Oct. 7, 1873	143, 458
Lathe, Watch-maker's	C. C. Powell and J. Belknap	Saint Johnsbury, Vt	July 13, 1869	92, 545
Lathe, Watch-maker's	F. Shaller	Hudson, N. Y	Sept. 4, 1866	57, 775
Lathe, Watch-maker's	R. H. St. John	Bellefontaine, Ohio	Sept. 15, 1857	18, 213
Lathe, Watch-maker's	R. H. St. John	Bellefontaine, Ohio	Mar. 29, 1859	23, 406
Lathe, Watch-maker's	C. Tribby	Winchester, Va	Feb. 5, 1861	31, 340
Lathe, Watch-maker's	A. Wild	New York, N. Y	July 17, 1860	29, 227
Lathe, Watch-maker's	C. V. Woerd	Waltham, Mass	Oct. 28, 1873	144, 178
Lathe way-smoother	A. Thomas	Worcester, Mass	Nov. 5, 1867	70, 483
Lathe-wheel	B. Hoadley	New Haven, Conn	Dec. 16, 1814	....
Lathe, Wheelwright's felly	J. Sitton and J. A. Black	Columbia, S. C	May 5, 1828	....
Lathe, Wood	H. T. Clay	Philadelphia, Pa	Oct. 8, 1867	69, 543
Lathe, Wood	L. H. Dwelley	Dorchester, Mass	May 17, 1870	103, 033

Index of patents issued from the United States Patent Office from 1790 to 1873, inclusive—Continued.

Invention.	Inventor.	Residence.	Date.	No.
Lathe, Wood	G. W. Feltes	Carbondale, Ill	Apr. 14, 1868	76,733
Lathe, Wood	A. Pries and H. Arnd	Saint Louis, Mo	Nov. 28, 1871	121,417
Lathe, Wood	W. H. Wussow	Aurora, Ill	Dec. 4, 1866	60,317
Lathe, Wood-turning	L. Atwood and E. Ripley	Rutland, Vt	Feb. 15, 1870	99,810
Lathe, Wood-turning	F. Baldwin	Brattleborough, Vt	June 18, 1867	65,864
Lathe, Wood-turning	F. Baldwin	Janesville, Wis	Aug. 30, 1870	106,767
Lathe, Wood-turning	A. J. and N. M. Barnes	Tiffin, Ohio	Nov. 9, 1869	96,659
Lathe, Wood-turning	A. Basse	Quincy, Ill	Oct. 16, 1866	58,758
Lathe, Wood-turning	H. C. Berry	Wauseon, Ohio	Apr. 16, 1867	63,836
Lathe, Wood-turning	C. A. Blessing	Philadelphia, Pa	Apr. 22, 1873	137,999
Lathe, Wood-turning	A. P. C. Bonte	Cincinnati, Ohio	June 16, 1868	78,919
Lathe, Wood-turning	J. C. Brown	Washington, D. C	July 12, 1870	105,166
Lathe, Wood-turning	W. W. Carey	Lowell, Mass	Dec. 19, 1865	51,554
Lathe, Wood-turning	J. Chase	Rochester, N. Y	Mar. 26, 1867	63,213
Lathe, Wood-turning	J. Chase	Rochester, N. Y	Mar. 15, 1870	100,855
Lathe, Wood-turning	J. Chase	Rochester, N. Y	Sept. 30, 1873	143,221
Lathe, Wood-turning	J. Coleman	Argyle, N. Y	July 12, 1864	43,477
Lathe, Wood-turning	M. F. Connett	Ladoga, Ind	Mar. 24, 1868	75,736
Lathe, Wood-turning	D. Dick	Corning, N. Y	Dec. 17, 1867	72,273
Lathe, Wood-turning	D. Dick	Corning, N. Y	May 5, 1868	77,596
Lathe, Wood-turning	L. H. Dwelley	Dorchester, Mass	Mar. 5, 1867	62,618
Lathe, Wood-turning	R. W. George	Boston, Mass	Aug. 22, 1865	49,516
Lathe, Wood-turning	A. Goodyear	Springport, Mich	June 25, 1867	66,146
Lathe, Wood-turning	T. Gray	Lowell, Mass	July 12, 1864	43,492
Lathe, Wood-turning	S. L. Hart	Milwaukee, Wis	Aug. 13, 1867	67,757
Lathe, Wood-turning	H. R. Hawkins	Akron, Ohio	June 11, 1867	65,568
Lathe, Wood-turning	L. Hull	Charlestown, Mass	Mar. 6, 1866	53,003
Lathe, Wood-turning	S. U. King	Windsor, Vt	May 5, 1868	77,495
Lathe, Wood-turning	J. E. F. Leland	New York, N. Y	Oct. 16, 1866	58,947
Lathe, Wood-turning	G. Lewis	Westfield, Ohio	Feb. 4, 1868	73,985
Lathe, Wood-turning	H. Locke	Boston, Mass	July 25, 1865	49,035
Lathe, Wood-turning	J. McMicheal	Philadelphia, Pa	Jan. 22, 1867	61,442
Lathe, Wood-turning	N. Norris and C. S. and H. S. Black.	Buchanan, Mich	Feb. 8, 1870	99,696
Lathe, Wood-turning	G. H. Ober	Newburgh, Ohio	June 27, 1865	48,428
Lathe, Wood-turning	A. R. Park	Columbia, Tex	May 20, 1873	139,076
Lathe, Wood-turning	O. H. Perry	Golconda, Ill	Apr. 5, 1870	101,502
Lathe, Wood-turning	J. Phillips, jr	Chicago, Ill	Mar. 19, 1867	63,089
Lathe, Wood-turning	J. Richards	Cincinnati, Ohio	July 16, 1867	66,885
Lathe, Wood-turning	I. Rood	Elyria, Ohio	Jan. 10, 1871	110,871
Lathe, Wood-turning	O. S. Smiley and C. O. Small	Augusta, Me	May 20, 1873	139,028
Lathe, Wood-turning	M. Spenli	Detroit, Mich	Dec. 18, 1866	60,647
Lathe, Wood-turning	M. Spenli	Detroit, Mich	Apr. 2, 1867	63,571
Lathe, Wood-turning	J. J. Urmston	Rahway, N. J	Apr. 5, 1870	101,548
Lathe, Wood-turning	A. J. Van Ornum	Hartford, Vt	May 26, 1868	78,341
Lathe, Wood-turning	A. Warth	Stapleton, N. Y	Nov. 9, 1858	22,043
Lathe, Wood-turning	W. J. Watson	Benton Centre, N. Y	Apr. 7, 1868	76,363
Lathe, Wood-turning	J. B. West	Lakeville, N. Y	Apr. 17, 1866	54,047
Lathe, Wood-turning	A. N. Wilcox	Watervliet, N. Y	Jan. 5, 1858	19,056
Lathe, Wood-turning	E. Williams	Rowlesburgh, W. Va	Mar. 24, 1868	76,022
Lathe, Wood-turning	J. Wilson and R. Hughes	Boston, Mass	Aug. 20, 1867	68,020
Lathe, Wood-turning	E. K. Wisell	Warren, Ohio	Jan. 14, 1868	73,423
Lathes, Adjustable tool-rest for	J. S., A. N., and O. Wheeler	Worcester, Mass	June 27, 1871	116,517
Lathes and other machinery, Mechanism for transmitting power to.	J. F. Stewart, F. Klinkerman, and J. Lamb.	Aurora, Ind	Apr. 26, 1870	102,445
Lathes, Centering heavy articles in	J. S. French	Boston, Mass	Jan. 19, 1864	41,289
Lathes, Cutting-tool for turning	C. Rahsskopff	San Francisco, Cal	Dec. 24, 1872	134,313
Lathes, Device for securing the tail-stocks of	A. Thomas	Worcester, Mass	Mar. 6, 1866	53,058
Lathes, Drill-rest for chucking	G. E. Johnson and W. Boston	Biddeford, Me	Aug. 29, 1871	118,614
Lathes, Drill-rest for turning	H. H. Folsom	Lowell, Mass	Feb. 14, 1871	111,736
Lathes, Drilling-attachment for turning	J. McCrum	Locust Grove, Ohio	Aug. 28, 1866	57,536
Lathes, Expanding and contracting or universal chuck for.	S. Fairman	Waterford, N. Y	July 18, 1840	1,692
Lathes, Eye-protector or chip-arrester for	C. T. Lamphere	Greenfield, Mass	July 7, 1868	79,766
Lathes, Feed-stop for	A. P. Brown	Worcester, Mass	July 10, 1866	56,172
Lathes for beaded work, Mode of operating radial cutters in.	G. W. Walton and H. Edgarton	Wilmington, Del	July 7, 1857	17,762
Lathes, Gear-cutting attachment to	W. P. Hopkins	Lawrence, Mass	Sept. 30, 1873	143,242
Lathes, Gear-cutting attachment to	T. O. Mills	Rockford, Ill	Nov. 4, 1873	144,214
Lathes, Gearing for	W. Fitzgerald	Boston, Mass	May 8, 1866	54,519
Lathes, Method of centering in watch-maker's	P. Leffel and J. H. Mulholland	Springfield, Ohio	Sept. 6, 1859	25,339
Lathes, Method of feeding the tool-carriage in turning.	A. Rennie	Binghamton, N. Y	July 20, 1858	20,956
Lathes, Method of operating feed-nuts in	W. A. Patrick	Ludlow, Vt	Feb. 14, 1860	27,149
Lathes or planing-machines, Tool-post for	W. H. Leach	Uxbridge, Mass	May 12, 1868	77,742
Lathes, Revolving center-rest for wood	H. J. Durgin	Rochester, N. Y	Apr. 5, 1870	101,596
Lathes, Screw-cutting attachment for	W. Gleason	Rochester, N Y	Oct. 21, 1873	143,898
Lathes, Securing buttons in	C. W. Robinson	Providence, R. I	June 27, 1809	....
Lathes, Securing pinions, &c., of watches in	J. M. Bottum	New York, N. Y	July 15, 1851	8,216
Lathes, Socket-coupling for	G. N. Trowbridge	Lowell, Mass	Sept. 1, 1857	18,112
Lathes, Steady-rest for	J. Brodie	San Francisco, Cal	Aug. 17, 1869	93,801
Lathes, Tail-stock for turning	L. B. Tyng	Lowell, Mass	Mar. 7, 1854	10,602
Lathes to turn tapering shafts, Tool for adjusting	W. N. Berkeley	Cedar Rapids, Iowa	July 10, 1866	56,164
Lathes, Tool-adjuster for	L. J. Parsons	New Haven, Conn	Apr. 27, 1869	89,335
Lathes, Tool-elevator for	C. Knox	Chicopee, Mass	Apr. 26, 1870	102,278
Lathes, Tool for turning	A. A. Burr	Rockdale, N. Y	Feb. 9, 1864	41,478
Lathes, Tool-grinding attachment to	J. Schaughnessy	Cincinnati, Ohio	May 17, 1870	103,242
Lathes, Tool-holder for turning	S. Gissinger	Lawrenceville, Pa	Sept. 17, 1867	68,870
Lathes, Tool-rest for turning	A. Hathaway	Chicopee, Mass	Mar. 8, 1864	41,889
Lathes, Tool-rest for turning	M. H. Merriam and W. W. Nichols.	Chelsea and Boston, Mass	Aug. 22, 1854	11,586
Lathes, Tool-rest for turning	F. A. Pratt	Hartford, Conn	Sept. 16, 1862	36,479
Lathes, Tool-rest for turning	G. A. Rollins	Nashua, N. H	Feb. 28, 1854	10,582
Lathes, Tool-holder for turning	W. Sellers	Philadelphia, Pa	Aug. 5, 1862	36,111
Lathes, Universal slide-rest for	E. G. Chormann	Philadelphia, Pa	Aug. 1, 1871	117,513
Lathes, Varying the speed of the mandrel in	W. A. Chapin, jr	Saint Johnsbury, Vt	Oct. 2, 1849	6,755
Lathes, Work-holder for	W. P. Hopkins	Lawrence, Mass	Oct. 22, 1872	132,400
Lathes, Work-holder for	W. P. Hopkins	Lawrence, Mass	Dec. 31, 1872	134,381
Latitude and longitude, Instrument for determining	J. Stinson	Danville, N. J	Sept. 18, 1855	13,584

Index of patents issued from the United States Patent Office from 1790 *to* 1873, *inclusive*—Continued.

Invention.	Inventor.	Residence.	Date.	No.
Latitude at sea, Method of determining approximate.	E. Cavendy	New York, N. Y	Nov. 10, 1857	18, 572
Latitude, longitude, and difference of time, Apparatus for showing.	S. D. Tillman	Jersey City, N. J	July 30, 1872	129, 909
Launching vessels, Apparatus for	T. F. Rowland	Green Point, N. Y	May 5, 1863	38, 418
Launching vessels, Greasing ways for	W. Mead	Greenwich, Conn	Oct. 25, 1832	....
Laundry and tailor's press	G. W. Jennings	Boston, Mass	Feb. 14, 1860	27, 131
Laundry-bench	A. H. Spencer	Providence, R. I	Feb. 8, 1870	99, 608
Laundry-box	M. A. H. Saurman	Philadelphia, Pa	May 23, 1871	115, 109
Laundry, Family	H. E. Smith	Cincinnati, Ohio	Nov. 14, 1865	50, 964
Laundry-heater	C. H. Goss	Troy, N. Y	June 15, 1869	91, 433
Laundry-indicator	R. Clarke	Macon, Ga	Mar. 8, 1870	100, 601
Laundry-iron heater	R. Diven	New York, N. Y	June 7, 1870	103, 997
Laundry, Portable	J. Morrison	Indianapolis, Ind	Feb. 14, 1871	111, 864
Laundry, Portable	J. Van	Cincinnati, Ohio	May 4, 1869	89, 815
Lavatory-apparatus	D. Wellington	Boston, Mass	May 3, 1870	102, 737
Law-blank case	C. I. Walker	Charleston, S. C	May 20, 1873	139, 094
Lawn or garden sprinkler	H. E. Cooke	Cleveland, Ohio	Aug. 12, 1873	141, 632
Lawn-seat	H. H. Gratz	Lexington, Ky	Nov. 11, 1873	144, 533
Lawn-seat	W. C. Medcalf	Rochester, N. Y	Sept. 16, 1873	142, 927
Lawn-sprinkler	J. Lessler	Buffalo, N. Y	Dec. 19, 1871	121, 949
Lawns, Machinery for dressing	R. L. Hawes	Worcester, Mass	Apr. 24, 1855	12, 771
Lay figures, dolls, &c	C. L. Parent	Havre, France	July 30, 1872	130, 068
Leach, Tanner's	S. Snyder	Cincinnati, Ohio	Feb. 23, 1869	87, 119
Leach-tub	W. Bauzelt	Brooklyn, N. Y	Oct. 13, 1868	83, 020
Leaching and tanning apparatus	C. Korn	Brooklyn, N. Y	Mar. 18, 1873	137, 004
Leaching-vat	T. H. Baker	Springfield, Ohio	Feb. 23, 1869	87, 235
Lead, acetate of copper, and acetate of iron, Manufacture of carbonate of.	O. Jacobi	Philadelphia, Pa	Aug. 31, 1869	94, 214
Lead and acetic acid, Manufacture of sugar of	L. D. Gale and I. M. Gattman	Washington, D. C., and New York, N. Y.	Aug. 17, 1869	93, 817
Lead and flake-white, Manufacturing white	A. J. Hamilton	New York	Nov. 8, 1813	....
Lead, and in the purification of the products of combustion for the same, Manufacture of white.	H. Hannen	Philadelphia, Pa	Feb. 9, 1869	86, 835
Lead and packing can, White	C. J. Fortin and D. H. Drake	Cincinnati, Ohio	Aug. 3, 1869	93, 191
Lead and Paris-white, Grinding white	S. Watrous	Groton, Conn	Jan. 26, 1833	....
Lead and pulpy materials, Apparatus for drying white.	J. B. Pollock	Port Richmond, N. Y	Jan. 2, 1872	122, 404
Lead and other metallic pipe, Manufacture of	R. M. Seydle and L. Ward	Milton, Pa	Aug. 1, 1838	862
Lead and saltpeter, Manufacture of white	C. Delafield	Factoryville, N. Y	June 5, 1866	55, 249
Lead and zinc for the manufacture of pigments, Treating the ores of.	E. O. Bartlett	Birmingham, Pa	Oct. 18, 1870	108, 433
Lead and zinc, Refining and whitening the products from ores of.	G. S. Lewis	Philadelphia, Pa	Mar. 14, 1871	112, 607
Lead and zinc, Refining and whitening the product from sulphurets of.	G. S. Lewis	Philadelphia, Pa	Mar. 14, 1871	112, 606
Lead and zinc, Sheet-metal from	C. C. C. Morgan	Auburn, N. Y	May 11, 1869	89, 881
Lead, Apparatus for making white	T. J. Coggeshall	New York, N. Y	Dec. 27, 1864	45, 587
Lead, Apparatus for manufacture of white	C. L. Wheeler	Pittsburgh, Pa	May 28, 1872	127, 395
Lead, Apparatus for manufacturing white	H. Hannen	Dubuque, Iowa	Mar. 30, 1858	19, 771
Lead, Apparatus for manufacturing white	R. Rowland	New York, N. Y	June 29, 1858	20, 731
Lead, Apparatus for refining and desilvering	G. Luce	Marseilles, France	Nov. 25, 1873	144, 993
Lead apparatus, White	D. R. Erdmann	Philadelphia, Pa	Aug. 16, 1859	25, 106
Lead by precipitation, Manufacture of white	R. Baker	Newark, N. Y	Apr. 3, 1855	12, 616
Lead-cutting machine	C. Koehler	Evansville, Ind	Apr. 9, 1872	125, 462
Lead, Die for making sheet	N. York	Boston, Mass	July 1, 1873	140, 567
Lead-drier	A. D. Armstrong	Allegheny City, Pa	Dec. 23, 1873	145, 713
Lead-foil coated with tin, Manufacture of	D. S. Hines	Brooklyn, N. Y	Mar. 26, 1867	63, 249
Lead for pigments, Manufacture of white	E. Milner	Springfield, England	July 8, 1873	140, 721
Lead from dross, Method of obtaining	C. Pickering	Saint Louis, Mo	Oct. 23, 1866	59, 063
Lead from precious metals, Separating	S. W. Kirk	Philadelphia, Pa	Nov. 29, 1870	109, 742
Lead, &c., from mixing-tubs to millstones, Apparatus for feeding.	J. B. Pollock	Port Richmond, N. Y	Nov. 28, 1871	121, 416
Lead-holder for pencil	P. Gabriel	Seymour, Conn	Aug. 6, 1867	67, 528
Lead, Machine for cutting sheet	S. E. Chubbuck	Roxbury, Mass	May 28, 1867	65, 173
Lead, Machine for making sheet	D. G. Batchelder	Salem, Mass	Feb. 25, 1873	136, 124
Lead, Machine for making sheet	H. Hannen	Philadelphia, Pa	June 6, 1871	115, 730
Lead, Machine for making sheet	J. Millingar	Philadelphia, Pa	July 5, 1870	105, 112
Lead, Machine for manufacturing sheet	J. Robertson	Brooklyn, N. Y	Sept. 19, 1865	50, 036
Lead, Machine for preparing bar	R. J. Bailey	Pittsburgh, Pa	June 29, 1869	91, 814
Lead machines, Combination of dies for sheet	J. Robertson	Brooklyn, N. Y	May 27, 1851	8, 112
Lead, &c., Machinery for making pipes continuously from.	G. E. Sellers	Upper Darby, Pa	Dec. 17, 1840	1, 908
Lead, Machinery for manufacture of carbonate of	C. Ripley	Saugerties, N. Y	July 11, 1837	264
Lead, Making sulphate, carbonate, and other pigments of.	H. Holland	Westfield, Mass	Nov. 3, 1838	994
Lead, Making tinned sheet	T. Ervbank	New York	May 30, 1823	....
Lead, Making white	E. Clark	Philadelphia, Pa	Mar. 28, 1818	....
Lead, Making white	E. De Butts	Baltimore, Md	May 5, 1814	....
Lead, Manufacture of acetate of	J. E. Fournier	Courville, France	July 17, 1866	56, 565
Lead, Manufacture of carbonate of	H. Hannen	Philadelphia, Pa	May 14, 1867	64, 763
Lead, Manufacture of carbonate of	G. S. Lewis	Philadelphia, Pa	Mar. 14, 1871	112, 608
Lead, Manufacture of carbonate of	H. Hannen	Philadelphia, Pa	Feb. 12, 1867	62, 130
Lead, Manufacture of chloride of	F. F. Meyer	New York, N. Y	Oct. 23, 1860	30, 521
Lead, Manufacture of dry white	E. O. Bartlett	Birmingham, Pa	Mar. 1, 1870	100, 353
Lead, Manufacture of oxychloride of	L. Brumlen	Hoboken, N. J	Jan. 29, 1861	31, 224
Lead, Manufacture of red	S. Witherell, jr	Philadelphia, Pa	Nov. 1, 1811	....
Lead, Manufacture of the subacetate of	L. Brumlen	Hoboken, N. J	Aug. 13, 1872	130, 410
Lead, Manufacture of white	F. Albert	Brooklyn, N. Y	May 3, 1859	23, 815
Lead, Manufacture of white	W. Archer and C. Rice	New York, N. Y	Mar. 7, 1865	46, 706
Lead, Manufacture of white	W. Baker	Sheffield, England	June 13, 1865	48, 243
Lead, Manufacture of white	S. R. Bradley	New York, N. Y	Sept. 21, 1869	95, 075
Lead, Manufacture of white	L. Brumlen	Hoboken, N. J	Mar. 28, 1871	113, 014
Lead, Manufacture of white	T. H. Burridge	Saint Louis, Mo	May 2, 1871	114, 405
Lead, Manufacture of white	T. Cobley	London, England	Apr. 28, 1863	38, 283
Lead, Manufacture of white	H. Cory	England	Oct. 8, 1840	1, 804
Lead, Manufacture of white	J. Cuddy	Pittsburgh, Pa	Sept. 28, 1869	95, 201

Index of patents issued from the United States Patent Office from 1790 *to* 1873, *inclusive*—Continued.

Invention.	Inventor.	Residence.	Date.	No.
Lead, Manufacture of white	J. Cuddy	Pittsburgh, Pa	July 19, 1870	105, 431
Lead, Manufacture of white	J. Cuddy and G. S. Selden	Pittsburgh and Philadelphia, Pa.	June 21, 1870	104, 434
Lead, Manufacture of white	C. Delafield	Staten Island, N. Y	Jan. 23, 1866	52, 144
Lead, Manufacture of white	C. Delafield	Factoryville, N. Y	Apr. 3, 1866	53, 583
Lead, Manufacture of white	C. W. Dwelle	Saint Louis, Mo	Sept. 21, 1869	95, 097
Lead, Manufacture of white	C. W. Dwelle	Saint Louis, Mo	Oct. 25, 1870	108, 571
Lead, Manufacture of white	T. M. and A. G. Fell	New York, N. Y	July 24, 1866	56, 685
Lead, Manufacture of white	T. M. and A. G. Fell	Brooklyn, N. Y	Nov. 20, 1866	59, 901
Lead, Manufacture of white	T. M. and A. G. Fell	Brooklyn, N. Y	Nov. 20, 1866	59, 902
Lead, Manufacture of white	T. M. and A. G. Fell	Brooklyn, N. Y	June 25, 1867	66, 137
Lead, Manufacture of white	T. M. and A. G. Fell	Brooklyn, N. Y	June 25, 1867	66, 138
Lead, Manufacture of white	T. M. and A. G. Fell	Brooklyn, N. Y	June 25, 1867	66, 139
Lead, Manufacture of white	S. Gardner	New York	Aug. 26, 1843	3, 232
Lead, Manufacture of white	I. M. Gattman	New York, N. Y	Nov. 19, 1867	70, 990
Lead, Manufacture of white	I. M. Gattman	New York, N. Y	July 20, 1869	92, 816
Lead, Manufacture of white	H. S. Hannen	Philadelphia, Pa	July 21, 1868	80, 168
Lead, Manufacture of white	R. F. Hatfield	New York, N. Y	Nov. 8, 1870	109, 125
Lead, Manufacture of white	O. Jacobi	Philadelphia, Pa	May 12, 1868	77, 818
Lead, Manufacture of white	G. T. Lewis and E. O. Bartlett	Philadelphia and Birmingham, Pa.	Dec. 14, 1869	97, 936
Lead, Manufacture of white	F. F. Mayer	New York, N. Y	June 15, 1869	91, 466
Lead, Manufacture of white	A. P. Meylert	New Britain, Conn	Mar. 4, 1873	136, 446
Lead, Manufacture of white	H. J. Overman	New York, N. Y	Oct. 23, 1866	59, 135
Lead, Manufacture of white	T. Reakish	Philadelphia, Pa	Apr. 20, 1869	89, 074
Lead, Manufacture of white	J. Richards	Philadelphia, Pa	Dec. 2, 1836	95
Lead, Manufacture of white	R. Rowland	New York, N. Y	Apr. 19, 1864	42, 407
Lead, Manufacture of white	B. F. Smith	Few York, N. Y	Nov. 9, 1858	22, 036
Lead, Manufacture of white	B. F. Smith	New York, N. Y	Jan. 18, 1859	22, 679
Lead, Manufacture of white	P. Spence	Newton Heath, Great Britain.	Mar. 6, 1866	53, 093
Lead, Manufacture of white	P. H. Vander Weyde	Philadelphia, Pa	Feb. 12, 1867	62, 097
Lead, Manufacture of white	D. Wadsworth	Brooklyn, N. Y	Nov. 14, 1871	120, 916
Lead, Manufacture of white	J. Welsh and T. Evans	Philadelphia, Pa	Aug. 15, 1814	....
Lead, Manufacturing a white pigment to be used as a substitute for white.	W. Cumberland	New York	June 7, 1838	767
Lead, Manufacturing white	E. A. Boehne	Saint Louis, Mo	Aug. 26, 1873	142, 199
Lead, Manufacturing white	G. F. Hagner	Philadelphia, Pa	Oct. 13, 1817	....
Lead, Manufacturing white	G. A. Harrison	New York, N. Y	July 20, 1831	....
Lead, Manufacturing white	P. Phillips	Campbell County, Ky	Apr. 17, 1837	160
Lead, Manufacturing white	J. Richards	Philadelphia, Pa	May 28, 1818	....
Lead, Method of manufacturing sheet	J. Robertson	Brooklyn, N. Y	Oct. 3, 1848	5, 820
Lead, Method of producing white pigments from	J. G. Dale and E. Milner	Warrington, England	Jan. 12, 1869	85, 796
Lead mixing and feeding machine, White	J. Haslett, jr	Allegheny City, Pa	Dec. 3, 1867	71, 612
Lead, Mode of making oxychloride of	L. Brumlen	Hoboken, N. J	Aug. 21, 1860	29, 665
Lead, Mode of producing white	J. G. Dale and E. Milner	Warrington, England	Nov. 30, 1869	97, 365
Lead, Mode of separating the oxide from the metal in the manufacture of white.	E. Clark	Saugerties, N. Y	July 11, 1839	1, 231
Lead, Mode of separating the oxide from the metal during the manufacture of white.	E. Clark	Saugerties, N. Y	Dec. 5, 1839	1, 424
Lead, Mode of treating lead salts for the manufacture of white.	T. M. and A. G. Fell	Brooklyn, N. Y	June 25, 1867	66, 140
Lead, Mode of washing white	S. Witherlee, jr	Philadelphia, Pa	Oct. 29, 1811	....
Lead-ore, Treating galena, or	J. E. Schwabe	New York, N. Y	Dec. 18, 1855	13, 961
Lead or other pipes from frost, &c., Method of securing.	D. Lownes		July 21, 1803	....
Lead-ores and collecting precious metals, Reducing	J. V. Z. Blaney and R. F. Adams	Chicago, Ill	Dec. 31, 1872	134, 457
Lead, Packing bar	Z. Kinsey	Dubuque, Iowa	Feb. 22, 1859	23, 031
Lead-pipe against the action of water, Mode of protecting.	L. Brandeis	Brooklyn, N. Y	Feb. 23, 1864	41, 675
Lead pipe against the action of water, Protecting	L. Brandeis	Brooklyn, N. Y	Dec. 15, 1863	40, 904
Lead pipe, Apparatus for making	W. Spillman	Columbus, Miss	Feb. 26, 1867	62, 450
Lead pipe, Coating	R. P. Berry	Newport, R. I	Nov. 5, 1872	132, 746
Lead pipe, &c., Composition for lining	T. Hodgson	Brooklyn, N. Y	Aug. 18, 1863	39, 615
Lead-pipe connection	W. D. Richardson	Springfield, Ill	Aug. 4, 1868	80, 667
Lead pipe, Die for manufacturing	B. Tatham	New York, N. Y	Oct. 12, 1869	95, 850
Lead-pipe die, Tin-lined	W. A. Shaw	New York, N. Y	July 24, 1866	56, 622
Lead pipe for conveying water, Making	A. Todd	Oxford, N. H	Apr. 18, 1814	....
Lead pipe for stills, &c., Tin-plated	T. Ewbank	New York	May 9, 1823	....
Lead-pipe machine	S. Parks, jr	Brooklyn, N. Y	Oct. 17, 1848	5, 852
Lead-pipe machine	C. E. Rockwell	New York, N. Y	Feb. 9, 1858	19, 313
Lead pipe, Machine for forming tin-lined	A. Hamon	Paris, France	Nov. 17, 1868	84, 114
Lead pipe, Machine for making	J. Laing	Bordentown, N. J	Mar. 30, 1843	3, 023
Lead pipe, Machine for making tin-lined	H. J. Bailey	Pittsburgh, Pa	June 29, 1869	91, 815
Lead pipe, Machine for making tin-lined	S. E. Chubbuck	Boston, Mass	July 7, 1868	79, 548
Lead pipe, Machine for making tin-lined	S. E. Chubbuck and J. H. Chadwick.	Boston, Mass	July 7, 1868	79, 549
Lead pipe, Machine for making tin-lined	J. E. Granniss	New York, N. Y	Mar. 12, 1872	124, 568
Lead pipe, Machine for making tin-lined	H. Merrie	Cincinnati, Ohio	Apr. 27, 1868	89, 326
Lead pipe, Machine for making tin-lined	J. Robertson	Brooklyn, N. Y	June 1, 1869	90, 872
Lead pipe, Machine for pressing	R. Layng	New York, N. Y	June 5, 1866	55, 315
Lead-pipe machinery	I. Adams	Boston, Mass	Mar. 8, 1848	5, 466
Lead-pipe machinery	N. Buttrick, jr	Chelmsford, Mass	Nov. 8, 1845	4, 259
Lead-pipe machinery	S. G. Cornell	Greenwich, Conn	Aug. 21, 1847	5, 253
Lead-pipe machinery	C. and G. E. Sellers	Cincinnati, Ohio	Mar. 9, 1844	3, 475
Lead-pipe machinery	B. Tatham	New York, N. Y	May 11, 1852	8, 943
Lead pipe, Making	J. Robertson	Brooklyn, N. Y	Aug. 26, 1856	15, 620
Lead pipe, Manufacture and coating of	F. Bennett	Watford, England	Jan. 1, 1867	60, 851
Lead pipe, Manufacture of	W. P. Tatham	Philadelphia, Pa	Sept. 3, 1850	7, 624
Lead pipe, Manufacture of tin-coated	D. Turner	Cambridgeport, Mass	Sept. 3, 1872	131, 072
Lead pipe, Manufacture of tin-lined	H. J. Bailey	Pittsburgh, Pa	Feb. 2, 1869	86, 347
Lead pipe, Manufacture of tin-lined	J. Farrell	Pittsburgh, Pa	Feb. 25, 1868	74, 755
Lead pipe, Manufacture of tin-lined	E. W. Newton	Franklin Grove, Ill	Mar. 26, 1872	124, 911
Lead pipe, Manufacture of tin-lined	W. A. Shaw	New York, N. Y	Feb. 18, 1868	74, 611
Lead pipe, Manufacture of tin-lined	W. A. Shaw	New York, N. Y	Feb. 18, 1868	74, 612
Lead pipe, Manufacture of tin-lined	W. A. Shaw	New York, N. Y	Feb. 18, 1868	74, 613
Lead pipe, Manufacture of tinned	W. A. Shaw and G. Willard	New York, N. Y	Jan. 26, 1864	41, 401

Index of patents issued from the United States Patent Office from 1790 *to* 1873, *inclusive*—Continued.

Index of patents issued from the United States Patent Office from 1790 *to* 1873, *inclusive*—Continued.

Invention.	Inventor.	Residence.	Date.	No.
Leather cloth, imitation leather, &c., Manufacture of.	N. C. Szerelmey	Clapham Common, England	Nov. 1, 1864	44, 910
Leather, cloth, &c., Machine for cutting objects from	W. H. Johnson	Springfield, Mass	Nov. 17, 1863	40, 626
Leather, cloth, &c., Manufacture of enameled	F. Longhurst and A. L. Murdock.	Boston, Mass	May 3, 1864	42, 584
Leather cloth. Ornamenting	A. Pellet	Paris, France	Oct. 2, 1860	30, 242
Leather, cloth, &c., to render water and fire proof, Mode of treating.	R. O. Lowrey	Salem, N. Y	May 19, 1868	77, 990
Leather, Coloring	J. Koppitz and F. B. Mayer	Boston and Cambridgeport, Mass.	Sept. 16, 1873	142, 797
Leather, Coloring and finishing	H. Hibbard	Attica, N. Y	Dec. 15, 1837	511
Leather-coloring composition	L. C. May	Cochituate, Mass	Mar. 26, 1872	124, 965
Leather-coloring compound	G. Jäger	Indianapolis, Ind	Nov. 28, 1871	121, 375
Leather, Coloring tanned	B. H. Lightfoot	Philadelphia, Pa	Dec. 22, 1863	41, 011
Leather, Composition for artificial	S. Whitmarsh	Northampton, Mass	May 25, 1858	20, 383
Leather, Composition for blacking	J. Hayward	Cleveland, Ohio	May 9, 1846	4, 498
Leather, Composition for bleaching and stuffing	L. W. Fiske	Louisville, Ky	Feb. 6, 1855	12, 368
Leather, Composition for coating	E. Brown	Indianapolis, Ind	Apr. 16, 1867	63, 847
Leather, Composition for currying	I. S. Hershey	Hagerstown, Md	June 12, 1847	5, 151
Leather, Composition for dressing and coloring	Z. A. Taylor	Haverhill, Mass	July 11, 1871	116, 886
Leather, Composition for finishing wax	P. Maguire	New Boston, Mass	May 3, 1870	102, 565
Leather, Composition for preserving and coloring	O. F. Battey	Worcester, Mass	July 18, 1871	117, 144
Leather, Composition for preserving and waterproofing.	R. K. Wright	Philadelphia, Pa	Apr. 28, 1863	38, 353
Leather, Composition for stuffing	R. Andrews	Milwaukee, Wis	May 25, 1869	90, 333
Leather, Composition for stuffing	J. Haseltine	Warren, N. H	Mar. 10, 1868	75, 264
Leather, Composition for stuffing	J. Rose	Newark, N. J	May 6, 1856	14, 832
Leather, Composition for stuffing and finishing	T. D. Williams	Chicago, Ill	Sept. 20, 1870	107, 579
Leather, Composition for the manufacture and preserving of.	R. Andrews	Milwaukee, Wis	Jan. 22, 1867	61, 379
Leather, Composition for treating	A. Taw	Philadelphia, Pa	Aug. 30, 1864	44, 025
Leather, Composition for water-proofing	R. Goldenblum and F. Steiner.	East Hampton, Mass	Feb. 19, 1861	31, 453
Leather, Composition to be applied to	W. Lehman	Newville, Pa	May 26, 1868	78, 295
Leather, Compound for oiling, polishing, and blacking.	G. F. Whitney	Boston, Mass	May 24, 1870	103, 402
Leather, Compound for stuffing	J. Merril	Boston, Mass	Mar. 8, 1870	100, 652
Leather, Compound for stuffing	A. Doepp	Newark, N. J	Nov. 17, 1868	84, 096
Leather, Compound for treating	S. S. Walbank	Superior City, Wis	Dec. 27, 1870	110, 610
Leather-compressing machine, Sole	J. Elliot	Sharon, Conn	Jan. 18, 1810	....
Leather cord or rope	O. I. Harrington and E. Weaver	Wood's Run, Pa	Sept. 7, 1869	94, 690
Leather-creasing machine	W. S. Bullen	Indianapolis, Ind	Nov. 27, 1860	30, 785
Leather-creasing machine	B. R. Hamilton and S. Swan	South Deerfield and Conway, Mass.	May 30, 1871	115, 312
Leather creasing or ornamenting	J. M. Bent	Wayland, Mass	Oct. 16, 1866	58, 761
Leather-creasing roller	S. B. Randall	Cincinnati, Ohio	Dec. 2, 1873	145, 240
Leather creasing, slicking, and skiving	C. C. Bellows	New Ipswich, N. H	Oct. 23, 1866	58, 978
Leather crimping and folding machine	G. Platts and J. Walden	Newark, N. J	Oct. 21, 1873	143, 783
Leather-crimping machine	G. and M. Algar	Greenport, N. Y	Nov. 25, 1838	1, 016
Leather-crimping machine	L. Upham	Putney, Vt	Jan. 9, 1838	541
Leather-crimping machine	H. Wing	Buffalo, N. Y	July 31, 1860	29, 429
Leather-crimping machine for boots, &c	J. Adams	Fair Haven, Vt	Mar. 26, 1838	656
Leather-crimping machine for boots, &c	C. H. Jaquith	Keene, N. H	Mar. 21, 1838	642
Leather-crimping machine, Mode of regulating the jaws of.	J. Goodwin, jr	Sterling, Mass	May 30, 1839	1, 162
Leather cushioning and ventilating machine	E. Thomas	Philadelphia, Pa	Aug. 2, 1870	105, 998
Leather-cutter	C. S. Stearns	Marlborough, Mass	June 9, 1868	78, 696
Leather, Cutter for cutting washers from	E. R. Brown and J. Long	Mauch Chunk and Packerton, Pa.	May 23, 1871	115, 157
Leather cutting and embossing machine	S. D. Tripp	Lynn, Mass	June 7, 1864	43, 050
Leather cutting and punching machine	I. Genung	Newark, N. J	July 30, 1831	....
Leather cutting and splitting machine	J. Edson	Boston, Mass	Jan. 31, 1860	26, 977
Leather-cutting machine	H. F. Baker	Pawtucket, R. I	Feb. 21, 1871	111, 900
Leather-cutting machine	H. H. Bigelow	Worcester, Mass	Apr. 4, 1871	113, 618
Leather-cutting machine	J. S. Bishop	Wayne, Me	June 29, 1833	....
Leather-cutting machine	R. Crocker	Marshalltown, Iowa	Jan. 28, 1873	135, 266
Leather-cutting machine	G. Domett	Boston, Mass	July 20, 1831	....
Leather-cutting machine	H. Eichler	New Lisbon, Wis	Mar. 2, 1869	87, 481
Leather-cutting machine	E. A. Holbrook	East Randolph, Mass	Apr. 9, 1872	125, 457
Leather-cutting machine	A. Keith	North Bridgewater, Mass	Mar. 7, 1871	112, 353
Leather-cutting machine	S. H. King	Tunbridge, Vt	Dec. 22, 1868	85, 102
Leather-cutting machine	L. N. Leland	Grafton, Mass	Sept. 28, 1837	416
Leather-cutting machine	N. Paquette	Summerfield, Mich	Apr. 15, 1873	137, 952
Leather-cutting machine	J. W. Richardson	South Braintree, Mass	Nov. 27, 1860	30, 761
Leather-cutting machine	J. F. Severance	East Bridgewater, Mass	May 23, 1865	47, 865
Leather-cutting machine	C. S. Stearns	Marlborough, Mass	Sept. 25, 1866	58, 315
Leather-cutting machine	A. L. Zent	Roanoke, Ind	Oct. 11, 1870	108, 229
Leather, Cutting out	A. B. Keith and T. K. Reed	North Bridgewater and East Bridgewater, Mass.	Apr. 30, 1867	64, 228
Leather-cutting press	N. J. Simonds	Woburn, Mass	Apr. 12, 1870	101, 931
Leather-cutting tool	J. Demarest	Belleville, N. J	May 27, 1873	139, 240
Leather-cutting tool	J. Sweesy	Elizabethville, Pa	Oct. 29, 1872	132, 609
Leather-cutting tool for soles	L. Marble	Henrietta, N. Y	Mar. 28, 1828	....
Leather, Cutting up and punching sole	L. D. Hawkins	Stoneham, Mass	Nov. 7, 1871	120, 642
Leather, Currying and finishing	S. Parker	Bellerica, Mass	July 9, 1808	....
Leather, Currying and finishing	S. Parker		Apr. 26, 1809	....
Leather, Device for stretching	W. Strevell	Jersey City, N. J	Oct. 30, 1866	59, 292
Leather, Device for trimming	W. F. Foley	Albany, N. Y	Dec. 30, 1873	145, 937
Leather, Dicing and polishing	R. and H. Brackett	Boston and Woburn, Mass	Mar. 15, 1845	3, 957
Leather dicing and polishing machine	L. Hedge	Windsor, Vt	July 22, 1819	....
Leather-dicing machine	A. Shidloch	New York, N. Y	Dec. 26, 1871	122, 136
Leather-dresser's table	C. T. Woodman	Boston, Mass	Feb. 2, 1864	41, 458
Leather-dressing	J. N. Baratta	Ayer, Mass	Nov. 26, 1872	133, 400
Leather dressing and scouring machine	E. Fitzhenry	Boston, Mass	Apr. 14, 1868	76, 619
Leather-dressing composition	M. Shaw	Abington, Mass	Feb. 25, 1862	34, 530
Leather-dressing composition	C. Werner	Philadelphia, Pa	Jan. 2, 1855	12, 171
Leather-dressing compound	T. G. Bell	New York, N. Y	Jan. 28, 1873	135, 310
Leather-dressing machine	J. A. Enos	Peabody, Mass	May 13, 1873	138, 874
Leather-dressing machine	E. Fitzhenry	Boston, Mass	Feb. 26, 1867	62, 324

Index of patents issued from the United States Patent Office from 1790 *to* 1873, *inclusive*—Continued.

Invention.	Inventor.	Residence.	Date.	No.
Leather-dressing machine	E. Fitzhenry	Boston, Mass	Mar. 1, 1870	100, 387
Leather-dressing machine	E. Fitzhenry	Boston, Mass	Aug. 15, 1871	118, 003
Leather-dressing machine	C. Korn	Meriden, Conn	June 25, 1861	32, 629
Leather-dressing machine	C. Korn	Wurtsborough, N. Y	June 18, 1867	65, 919
Leather-dressing machine	R. Lee	Newark, N. J	June 13, 1865	48, 186
Leather-dressing machine	C. A. McDonald	Woburn, Mass	Jan. 2, 1872	122, 395
Leather-dressing machine	H. P. Reed	Peabody, Mass	July 2, 1872	128, 658
Leather-dressing machine	C. Slawson	Norwich, N. Y	Nov. 13, 1849	6, 875
Leather, Drop for hammering sole	J. H. Walker	Worcester, Mass	May 10, 1864	42, 703
Leather drying and pressing machine	G. Harvey	New York, N. Y	Aug. 15, 1865	49, 402
Leather drying and polishing machine	S. Couillard, jr	Boston, Mass	June 27, 1827	......
Leather-edge-trimming machine	W. F. Foley	Albany, N. Y	Apr. 23, 1872	125, 946
Leather, Enameling	J. Rose	Newark, N. J	Mar. 9, 1858	19, 583
Leather, Fabric from waste	E. Waite	Franklin, Mass	May 2, 1871	114, 373
Leather, Fabrics prepared to imitate	H. A. Clark	Boston, Mass	Sept. 23, 1873	143, 122
Leather, Factitious enameled	J. W. Munroe	Fall River, Mass	May 10, 1859	23, 987
Leather fastener, Metallic	H. Beals	East Stoughton, Mass	Mar. 26, 1872	125, 007
Leather fastening	H. Beals	East Stoughton, Mass	Aug. 6, 1872	130, 181
Leather filling, &c., Machine for cutting	R. C. Turner	Mendon, Mich	July 21, 1868	80, 098
Leather-finisher	C. Bassett	Boston, Mass	Mar. 25, 1835	......
Leather-finishing	A. Bertram	New Albany, Ind	May 12, 1868	77, 863
Leather, Finishing and dressing sheep-skins for	R. Andrews	Milwaukee, Wis	July 13, 1869	92, 416
Leather-finishing machine	J. Burr	Johnstown, N. Y	Nov. 24, 1820	......
Leather-finishing machine	S. P. Cobb	South Danvers, Mass	June 5, 1860	28, 562
Leather-finishing machine	S. P. Cobb	South Danvers, Mass	Dec. 1, 1863	40, 735
Leather-finishing machine	W. Ellard	Woburn, Mass	Apr. 2, 1861	31, 879
Leather-finishing machine	E. Fitzhenry and J. Ball	Portland, Oreg	Jan. 15, 1867	61, 182
Leather-finishing machine	J. P. Friend	Peabody, Mass	Aug. 8, 1871	117, 877
Leather-finishing machine	W. P. Martin	Salem, Mass	Feb. 28, 1860	27, 300
Leather-finishing machine	W. P. Martin	Salem, Mass	May 1, 1860	28, 108
Leather-finishing machine	J. Pyle	Wilmington, Del	Sept. 30, 1856	15, 816
Leather-finishing machine	R. L. and C. Smith	Stockport, N. Y	Jan. 10, 1860	26, 792
Leather-finishing machine	T. F. Weston	Salem, Mass	June 7, 1859	24, 344
Leather-finishing machine	C., T. F., and J. W. Weston	Salem, Mass	Sept. 25, 1855	13, 605
Leather-finishing, Method of	L. B. Fox	Williamsport, Pa	Apr. 21, 1868	77, 025
Leather, Finishing split	J. Putnam	Danvers, Mass	July 27, 1869	93, 002
Leather fluting and glazing machine	G. Crossley	Philadelphia, Pa	Feb. 13, 1872	123, 681
Leather-folding machine	J. Lombard	Springfield, Me	Oct. 1, 1867	69, 454
Leather-folding machine	E. B. Stimpson	New York, N. Y	July 29, 1873	141, 398
Leather for boots and shoes, Method of preparing sole.	D. M. Ayer	Lewiston, Me	Jan. 1, 1867	60, 810
Leather for boots, &c., Machine for crimping	N. Woodbury	Calais, Me	Aug. 16, 1838	885
Leather for cards, Twilling-machine for pricking	P. Earle	......	Dec. 6, 1803	......
Leather for cotton or woolen cards, Machine for finishing.	G. S. Quick	Auburn, N. Y	Apr. 19, 1864	42, 399
Leather for enameling, Treating	T. P. Howell and N. F. Blanchard.	Newark, N. J	Nov. 20, 1855	13, 819
Leather for floor-covering, Artificial	S. M. Allen	Woburn, Mass	Oct. 15, 1867	69, 742
Leather for gloves, wafers, &c., Machine for cutting.	E. M. Ely	Paris, France	Jan. 28, 1829	......
Leather for harness, Machine for creasing and blacking.	A. Stempel	Oquawka, Ill	Nov. 2, 1858	21, 989
Leather for shoe-binding, Mode of striping alum-dressed.	S. Frothingham and G. Harriss	Middletown, Conn	Apr. 20, 1808	......
Leather for soling, Process of preparing the flanking of.	D. Kern	York, Pa	June 25, 1861	32, 628
Leather for suspender-straps, Mode of preparing	O. Webb	New York, N. Y	Oct. 14, 1807	......
Leather for the manufacture of boots and shoes, Machine for preparing.	C. Rice and S. H. Whorf	Boston and Roxbury, Mass	Nov. 20, 1855	13, 827
Leather for the soles and heels of boots and shoes, Machine for hammering.	J. P. Molliere	Lyons, France	Jan. 22, 1856	14, 140
Leather for wear, Preparing	G. V. Sheffield and J. F. Coburn	Hopkinton, Mass	May 7, 1867	64, 587
Leather from curriers' shavings or buffings, Making sheets of.	C. F. Crockett	Newark, N. J	June 17, 1856	15, 121
Leather from scrap-leather, &c., Preparation of artificial.	S. Sorenson	Ebeltoft, Denmark	May 24, 1870	103, 517
Leather from vats, Apparatus for raising	C. E. Robinson and L. D. Sanborn.	Concord, N. H	Apr. 27, 1858	20, 093
Leather-gage	L. W. Beecher	Avon, N. Y	May 4, 1852	8, 917
Leather, Gilding and silvering	A. Allen	New York, N. Y	Dec. 26, 1817	......
Leather, Glazed	E. G. Adams	Decatur, Ga	June 30, 1836	......
Leather goods, Water-proof	S. La Forge	Cleveland, Ohio	Feb. 28, 1860	27, 337
Leather-graining composition	O. P. Whitman	Lynn, Mass	Feb. 27, 1866	52, 920
Leather, Gripe for holding	B. Rowe	Albany, N. Y	July 19, 1853	9, 862
Leather-handling machine	W. G. Waterman	Sullivan, N. Y	Sept. 16, 1834	......
Leather, harness, &c., Blacking for	T. James	Medford, Mass	May 29, 1866	55, 203
Leather, hides, and skins, Composition for dyeing and coloring.	C. Bond	New York, N. Y	Jan. 4, 1870	98, 466
Leather-holder	C. Bates	Conestoga, Pa	Sept. 14, 1869	94, 697
Leather-holding clamp	R. Emans	Mansfield, N. J	Oct. 6, 1837	420
Leather-holding clamp	A. Rittenhouse	Philadelphia, Pa	Aug. 18, 1868	81, 293
Leather impermeable to water, Rendering	J. Board	New York, N. Y	Nov. 4, 1813	......
Leather, &c., impervious to gas, Process of rendering.	E. Osgood	Boston, Mass	July 30, 1872	129, 901
Leather impervious to hydrocarbon liquids, Rendering.	A. Warth	Stapleton, N. Y	Oct. 4, 1870	108, 072
Leather, &c., Infusing oil into	H. Guest	......	July 16, 1801	......
Leather into bales, Machine for rolling	N. Burk	Fulton, N. Y	Aug. 10, 1858	21, 114
Leather into counters, Machine for cutting	A Keith	North Bridgewater, Mass	Sept. 20, 1864	44, 318
Leather into hollow-ware forms, Machine for cutting.	F. Durand and O. Recqueur	Paris, France	June 11, 1850	7, 424
Leather into round bands, Machine for cutting	C. G. Burnham	East Hartford, Conn	Aug. 17, 1869	93, 803
Leather into strips for boot and shoe soles and heels, Machine for cutting.	J. P. Molliere	Lyons, France	June 19, 1855	13, 095
Leather, Japanned	F. S. Merritt	Boston, Mass	May 9, 1871	114, 586
Leather lacings, Instrument for cutting	C. A. Woodbury	Woodstock, Vt	July 2, 1872	128, 519
Leather, Lap cutting and leveling	H. Underwood	Tolland, Conn	Oct. 16, 1847	5, 329
Leather, Machine for creasing straps of	D. H. Hovey	Kilborn, Ohio	Nov. 15, 1853	10, 231

Index of patents issued from the United States Patent Office from 1790 *to* 1873, *inclusive*—Continued.

Invention.	Inventor.	Residence.	Date.	No.
Leather, Machine for cutting articles from	C. Rice and S H. Whorf	Boston and Roxbury, Mass	Dec. 11, 1855	13, 920
Leather, Machine for forming staple seams in	S. W. Shorey	Boston, Mass	Feb. 25, 1873	136, 340
Leather, Machine for removing grease from	W. A. Perkins	Salem, Mass	Feb. 13, 1872	123, 643
Leather, Machine for removing grease from	J. Perkins and G. L. Newcomb	Peabody and Salem, Mass	Dec. 26, 1871	122, 130
Leather, Machine for removing grease from	J. Starratt	Salem, Mass	July 19, 1870	105, 506
Leather, Machinery for striking out	S. Hutchinson	Leeds, England	Apr. 26, 1870	102, 270
Leather, Making	R. Downey	New Albany, Ind	June 15, 1844	3, 632
Leather, Making scabbards, sheaths, &c., of	F. Durand and O. Pecquer	France	Feb. 15, 1847	4, 986
Leather, Making sod-oil for	G. Poyzer	Philadelphia, Pa	May 10, 1805	....
Leather, Manufacture of	W. Adamson	Philadelphia, Pa	June 23, 1868	79, 177
Leather, Manufacture of	F. A. Holcomb	Grand Rapids, Mich	Dec. 27, 1870	110, 465
Leather, Manufacture of	S. B. Jenks and F. A. Holcomb	Grand Rapids, Mich	Jan. 24, 1871	111, 214
Leather, Manufacture of	W. Pyle	Wilmington, Del	July 12, 1870	105, 247
Leather, Manufacture of	G. Rawle and W. N. Evans	Bristol and Bedminster, England.	Aug. 5, 1873	141, 459
Leather, Manufacture of	C. J. Tinnerholm	Quincy, Ill	Feb. 13, 1872	123, 748
Leather, Manufacture of	A. H. Van Gieson	Newark, N. J	Nov. 24, 1863	40, 718
Leather, Manufacture of artificial	H. B. Meech	Fort Edward, N. Y	Mar. 17, 1868	75, 690
Leather, Manufacture of artificial	C. Saffray	New York, N. Y	Feb. 26, 1867	62, 503
Leather, Manufacture of artificial	W. W. Waite	South Natick, Mass	Dec. 26, 1865	51, 781
Leather, Manufacture of artificial	F. Walton	Staines, England	Jan. 17, 1871	111, 100
Leather, Manufacture of enameled and japanned	I. W. Dawson	Newark, N. J	Oct. 29, 1867	70, 176
Leather, Manufacture of japanned	F. S. Merritt	Boston, Mass	May 23, 1871	115, 083
Leather, Manufacture of japanned	J. L. Newton	Boston, Mass	Oct. 9, 1866	58, 733
Leather, Manufacture of water-proof	R. Spoonse	Jersey Shore, Pa	May 7, 1867	64, 589
Leather, Manufacturing artificial	E. and J. K. Cushman	Amherst, Mass	Apr. 5, 1859	23, 454
Leather, Method of laying off patterns for stitching on.	W. P. Wolfington	Louisville, Ky	Mar. 8, 1870	100, 579
Leather, Method of oiling	L. Holcomb	Granby, Conn	Oct. 9, 1860	30, 320
Leather, Method of stuffing	F. A. White	Roxbury, Mass	Aug. 5, 1856	15, 499
Leather, Mode of covering rounded articles with	J. H. Osgood	Boston, Mass	May 21, 1872	127, 098
Leather, Mode of economizing the manufacture of articles of.	W. Adamson	Philadelphia, Pa	Jan. 31, 1865	46, 061
Leather, Mode of embossing	G. W. Pratt	Salem, Mass	July 18, 1865	48, 876
Leather, Mode of manufacturing	A. Hickman and E. S. Davenport.	Abingdon, Va	Aug. 1, 1838	866
Leather, Mode of manufacturing fair	J. L. Turner	Philadelphia, Pa	July 15, 1840	1, 688
Leather, Mode of polishing	A. Bayrd	South Reading, Mass	Dec. 29, 1828	....
Leather, Mode of softening	H. Cunningham	Albany, N. Y	Oct. 12, 1869	95, 779
Leather, Mode of treating tanned	B. H. Lightfoot	Philadelphia, Pa	June 14, 1864	43, 120
Leather, Mode of whitening	J. C. Booth	Philadelphia, Pa	Dec. 5, 1840	1, 881
Leather more durable and flexible, Mode of rendering.	L. G. Sourzar and L. Bombail	Bordeaux, France	Nov. 13, 1866	59, 736
Leather, Oil-blacking for	J. L. Baumer	Columbus, Ohio	Dec. 14, 1869	97, 857
Leather oil or polish	S. M. Farnham	Tully, N. Y	Apr. 21, 1868	77, 021
Leather-oiling machine, Tanner's	G. Huttelmaier	Allegheny, Pa	Sept. 22, 1863	40, 079
Leather, Ornamenting	L. Wolfson	Boston, Mass	Sept. 26, 1871	119, 291
Leather or skins, Machine for softening or dressing.	F. J. Burcham	Racine, Wis	June 25, 1867	66, 125
Leather-ornamenting machine	C. T. Woodman	Boston, Mass	Mar. 29, 1864	42, 136
Leather, paper, &c., Drying	S. M. Allen	Woburn, Mass	Jan. 19, 1864	41, 348
Leather, paper, &c., finishing machine	P. Farrell	Albany, N. Y	June 15, 1869	91, 219
Leather, paste-board, and paper, Manufacture of	A. N. Mathiew	Paris, France	Apr. 20, 1858	20, 020
Leather pebbling and embossing machine	B. Merritt, jr	Chelsea, Mass	Aug. 19, 1862	36, 228
Leather-perforating machinery for harness-makers	L. D. Woodmansee	Hamilton, Ohio	Dec. 16, 1873	145, 543
Leather pieces, Mode of connecting	B. F. Sturtevant	Boston, Mass	June 21, 1864	43, 236
Leather-piercing machine	A. Eggleston	Fall River, Mass	Apr. 18, 1871	113, 863
Leather-piercing machine	G. V. Sheffield and J. F. Coburn.	Hopkinton, Mass	Sept. 4, 1866	57, 780
Leather-polish	H. A. and R. G. Sawyer	Milwaukee, Wis	May 2, 1871	114, 344
Leather, Polish for	J. Herold and M. Brown	Frederick, Md	May 26, 1868	78, 372
Leather polishing and graining machine, Morocco	J. Perkins	Boston, Mass	June 26, 1809	....
Leather-polishing machine	G. T. Adler	Philadelphia, Pa	Feb. 7, 1860	27, 028
Leather-polishing machine	W. P. Gamble	Philadelphia, Pa	Apr. 8, 1856	14, 606
Leather-polishing machine	J. W. Hildreth	Boston, Mass	Apr. 21, 1868	76, 914
Leather-polishing machine	J. M. Poole	Wilmington, Del	Sept. 28, 1852	9, 292
Leather-polishing machine	F. Seibert	Williamsburgh, N. Y	Nov. 29, 1853	10, 287
Leather-polishing machine	R. A. Stratton	Philadelphia, Pa	Jan. 24, 1860	26, 932
Leather-polishing machine	P. P. Tapley	Lynn, Mass	Jan. 3, 1854	10, 379
Leather, Preparation of	M. W. Page	Franklin, N. H	May 8, 1866	54, 587
Leather, Prepared	S. Dyar	Charlestown, Mass	Feb. 19, 1867	62, 120
Leather, Preparing cloth and paper, &c., to imitate	H. A. Clark	Boston, Mass	July 22, 1873	141, 118
Leather, Preservative blacking for	W. B. Moore	Winchester, Mo	Nov. 12, 1867	70, 737
Leather-preserving composition	J. Lamplugh	Philadelphia, Pa	Dec. 30, 1873	146, 080
Leather-preserving composition	T. P. Merriam	New Bedford, Mass	Nov. 4, 1842	2, 844
Leather-pressing machine	G. F. Dressing	Buffalo, N. Y	Oct. 11, 1864	44, 616
Leather-pricking machine	G. C. Converse	Spencer, Mass	Feb. 18, 1873	135, 969
Leather-pricking machine	R. U. and M. Richards	Norfolk, Conn	Apr. 24, 1810	....
Leather-pricking machine	S. Sheldon	Cincinnati, Ohio	Oct. 10, 1840	1, 817
Leather-pricking machine	S. Sheldon	Cincinnati, Ohio	Oct. 10, 1840	1, 818
Leather-pricking machine	J. H. Walker	Worcester, Mass	Feb. 25, 1862	34, 537
Leather-pricking machine for harness and coach makers.	J. W. Briggs, and L. C. and J. S. Carmer.	Painesville, Ohio	Mar. 26, 1838	658
Leather, Process and apparatus for stuffing	W. A. Perkins and J. A. Enos	Salem and Peabody, Mass	Oct. 1, 1872	131, 777
Leather, Process for making japanned	H. L. Hall	Beverly, Mass	Jan. 9, 1855	12, 226
Leather, Process for the manufacture of artificial	H. A. Clark	Boston, Mass	July 22, 1873	141, 116
Leather, Process of and machinery for stamping and staining.	R. Robinson	Newburyport, Mass	Mar. 19, 1810	
Leather, Process of finishing	H. C. Williams	Lancaster, Pa	May 12, 1863	38, 525
Leather, Process of oiling, preserving, and strengthening.	L. S. Robbins	New York, N. Y	Mar. 24, 1868	75, 795
Leather, Punching	J. M. Bent	Wayland, Mass	Oct. 16, 1866	58, 762
Leather punching and cutting machine	E. E. Hardy and N. Dubrul	Joliet, Ill	Feb. 7, 1871	111, 533
Leather punching and feeding machine	W. Barry	Carthage, N. Y	Feb. 14, 1871	111, 719
Leather-punching machine	T. Burkitt and J. K. Krieg	Brooklyn and New York, N. Y.	Sept. 17, 1872	131, 330
Leather-punching machine	J. Matheis	Ottawa, Ill	Feb. 21, 1871	111, 957
Leather-punching machine	L. H. Wood	Marlborough, Mass	June 27, 1865	48, 474

Index of patents issued from the United States Patent Office from 1790 to 1873, inclusive—Continued.

Invention.	Inventor.	Residence.	Date.	No.
Leather-quilting machine	A. G. Brewer	Washington, D. C	Oct. 29, 1867	70, 158
Leather raising, creasing, and slicking machine	C. W. Guest	Dexter, Mich	Nov. 10, 1863	40, 561
Leather-reducing machine	W. C. Joslin	Putnam, Conn	Dec. 1, 1868	84, 631
Leather remnants, Mode of treating	B. C. Young	Boston, Mass	Feb. 4, 1873	135, 615
Leather, Removing color from	E. S. Frye	Salem, Mass	Mar. 22, 1870	100, 997
Leather, Removing grease from waste	J. P. Rush	Peabody, Mass	Oct. 11, 1870	108, 191
Leather, Restoring old	B. Schmidt	Hoboken, N. J	Apr. 14, 1868	76, 824
Leather, &c., Rigid hoop-knife for splitting	D. H. Chamberlian	West Roxbury, Mass	June 16, 1857	17, 554
Leather-roller	J. Gardiner	Petersburgh, Pa	Apr. 25, 1843	3, 055
Leather-roller	J. T. Harris	Swampscott, Mass	June 23, 1868	79, 070
Leather-roller	D. H. Priest	Watertown, Mass	Sept. 19, 1865	50, 079
Leather, Rolling	J. Whitney	Manchester, Mass	Sept. 22, 1863	40, 069
Leather-rolling machine	J. G. Busfield	Feltonville, Mass	Jan. 26, 1864	41, 363
Leather-rolling machine	W. Coburn	Lancaster, Pa	Oct. 31, 1839	1, 388
Leather-rolling machine	J. G. Curtis	Emporium, Pa	Mar. 29, 1870	101, 234
Leather-rolling machine	J. G. Curtis	Warren, Pa	May 30, 1871	115, 443
Leather-rolling machine	W. H. Leach	Uxbridge, Mass	June 30, 1868	79, 360
Leather-rolling machine	N. Linsley	Lena, Ill	Nov. 28, 1871	121, 389
Leather-rolling machine	W. P. Martin	Salem, Mass	Feb. 20, 1866	52, 728
Leather-rolling machine	J. McLaughlin and H. Hill	Sunderland, Vt	Apr. 28, 1836	....
Leather-rolling machine	J. E. Merritt	Winn, Me	Feb. 27, 1866	52, 871
Leather-rolling machine	C. W. Monson	Upton, Iowa	Aug. 10, 1869	93, 465
Leather-rolling machine	H. Paign	Alexander, N. Y	Dec. 23, 1828	....
Leather-rolling machine	J. A. Safford	Winchester, Mass	Jan. 18, 1870	98, 889
Leather-rolling machine	Q. Stoddard	Jackson, Mich	Dec. 4, 1866	60, 277
Leather-rolling machine	J. H. Walker	Worcester, Mass	Dec. 10, 1867	71, 929
Leather-rolling machine	H. J. Weston	Buffalo, N. Y	Mar. 22, 1870	101, 197
Leather-rolling machine	J. Whitney	Winchester, Mass	Mar. 19, 1872	124, 709
Leather rolling, shaving, softening, and finishing machine.	T. Chace	New York, N. Y	Sept. 12, 1838	914
Leather-rounding machine	P. Beckman	Napierville, Ill	July 28, 1863	39, 335
Leather rolling machine, Green or wet	J. Whitney	Winchester, Mass	Mar. 24, 1863	37, 991
Leather-rounding machine	J. H. Tizzard	Eaton, Ohio	Apr. 4, 1871	113, 707
Leather-rounding machine	J. Yeager	Berrysburgh, Pa	Jan. 8, 1867	61, 132
Leather-rounding tool	L. A. Sweatt	San Francisco, Cal	Aug. 1, 1871	117, 575
Leather, Rubber-coated	H. W. Joslin	Brooklyn, N. Y	Nov. 6, 1866	59, 402
Leather rubbing and polishing machine	J. T. Flanders	Newburyport, Mass	Oct. 4, 1853	10, 067
Leather-sammier	D. C. Guttridge	Pittsburgh, Pa	Feb. 1, 1870	99, 431
Leather-scalloping machine	A. Goodyear	Albion, Mich	Jan. 14, 1868	73, 243
Leather-scarfing machine	C. H. Helms	Poughkeepsie, N. Y	Aug. 11, 1868	80, 952
Leather-scarfing machine	C. H. Helms	Poughkeepsie, N. Y	Mar. 15, 1870	100, 890
Leather scouring and setting machine	P. E. Hummel	Pulaski, N. Y	June 16, 1857	17, 576
Leather scouring and setting-out machine	C. Holmes	Boston, Mass	June 1, 1869	90, 664
Leather scouring and setting-out machine	A. W. Reid	Schenectady, N. Y	Apr. 5, 1870	101, 508
Leather scouring and stretching machine	J. C. Parmerlee	Bean Blossom, Ind	Aug. 8, 1871	117, 921
Leather scouring, blacking, and finishing machine	F. W. Rust	Umatilla, Oreg	Nov. 10, 1868	84, 001
Leather, Scouring, cleansing, &c	R. Emes	Boston, Mass	Nov. 21, 1831	....
Leather scouring, glassing, or setting machine	F. A. Lockwood	Winthrop, Mass	Oct. 21, 1873	143, 829
Leather-scouring machine	D. P. Burdon	New York, N. Y	July 19, 1870	105, 419
Leather-scouring machine	W. M. Clark	Butternuts, N. Y	Aug. 29, 1865	49, 606
Leather-scouring machine	F. Davis	Lawrence, Kans	Dec 4, 1866	60, 149
Leather-scouring machine	H. C. Havemeyer and D. P. Burdon.	New York, N. Y	Nov. 1, 1870	108, 782
Leather-scouring machine	H. C. Havemeyer and D. P. Burdon.	New York, N. Y	Nov. 1, 1870	108, 783
Leather-scouring machine	I. W. Pray and E. Fitzhenry	Portland, Oreg	Jan. 15, 1867	61, 250
Leather-scouring machine	A. W. Roberts	Hartford, Conn	Mar. 26, 1867	63, 307
Leather-scouring machine	A. W. Reid	Schenectady, N. Y	July 16, 1872	129, 251
Leather-scouring machine	N. F. Snow	Salem, Mass	Oct. 1, 1872	131, 831
Leather scouring, polishing, and glassing machine	D. Harrington	Boston, Mass	May 16, 1871	114, 809
Leather-scraping machine	J. T. Barnstead	Peabody, Mass	Mar. 22, 1870	101, 081
Leather scraps, Method of treating	H. B. Farwell	Boston, Mass	Apr. 10, 1866	53, 800
Leather, Seam for	E. Shaw	Milwaukee, Wis	Dec. 10, 1872	133, 897
Leather, Shaving	H. Johnson	Brooklyn, Conn	June 16, 1836	....
Leather, Shaving and splitting	S. Parker	Billerica, Mass	Apr 5, 1813	....
Leather, Shaving and splitting	M. Tullar	Royalton, Vt	July 29, 1808	....
Leather-shaving knife	J. B. Wentworth	Lynn, Mass	July 13, 1858	20, 911
Leather-shaving machine	D. Y. Haas	Hepler, Pa	Nov. 4, 1873	144, 273
Leather-shaving machine	J. P. Houston and O. Batchelder.	Bedford, N. H	Apr. 11, 1825	....
Leather-shaving machine	J. Reilly	Waynesborough, Pa	Feb. 28, 1833	....
Leather, Silvering and gilding	L. Kenton	Philadelphia, Pa	June 21, 1830	....
Leather, skins, &c., Machine for pressing and finishing.	N. D. Morey	Saratoga Springs, N. Y	Dec. 12, 1871	121, 727
Leather, Skiving	B. S. Mathews	Stamford, Conn	Apr. 10, 1849	6, 290
Leather skiving and splitting machine	P. D. Cummings	Portland, Me	Mar. 25, 1873	137, 183
Leather skiving and whitening machine	S. Graham	Roxbury, Mass	May 10, 1838	732
Leather-skiving machine	G. Andrews	Oxford, Me	Oct. 7, 1873	143, 398
Leather-skiving machine	J. W. Chandler	East Corinth, Me	June 3, 1873	139, 495
Leather-skiving machine	J. P. Crooks	Hopkinton, Mass	Oct. 25, 1870	108, 687
Leather-skiving machine	E. T. Ingalls	Haverhill, Mass	June 5, 1860	28, 580
Leather-skiving machine	S. Keen	East Bridgewater, Mass	Nov. 27, 1860	30, 736
Leather-skiving machine	E. B. Stimpson	New York, N. Y	May 13, 1873	138, 951
Leather-skiving machine	W. S. Williams	Lynn, Mass	July 31, 1860	29, 428
Leather-slicker	G. B. Fowle	Boston, Mass	June 11, 1872	127, 756
Leather-slicker	H. L. Sultzbach	Marietta, Pa	Apr. 27, 1858	20, 098
Leather so as to render it suitable for the manufacture of gloves, &c., Process of treating.	N. C. Russell	Gloversville, N. Y	Aug. 17, 1869	93, 910
Leather-softening machine	H. Cunningham	Albany, N. Y	Nov. 12, 1872	132, 902
Leather-softening machine	J. Greenleaf	Lowell, Mass	Sept. 30, 1856	15, 807
Leather-softening machine	J. B. Wentworth	Lynn, Mass	Feb. 5, 1856	14, 211
Leather, Splitting	E. Howard and J. Butters	Boston, Mass	May 3, 1820	....
Leather, Splitting	C. S. Stearns	Marlborough, Mass	June 14, 1864	43, 159
Leather splitting and fleshing machine	J. A. Safford	Winchester, Mass	Jan. 18, 1870	98, 888
Leather splitting and paring machine	H. Aiken	Middlesex, Mass	June 12, 1835	....
Leather splitting and rolling machine	C. S. Stearns	Marlborough, Mass	June 9, 1868	78, 697
Leather splitting and skiving machine	A. Cochran	Athens, Ohio	Apr. 4, 1871	113, 631
Leather splitting and skiving machine	S. H. Schenck	Zionsville, Ind	Oct. 9, 1866	58, 681

Index of patents issued from the United States Patent Office from 1790 to 1873, inclusive—Continued.

Invention.	Inventor.	Residence.	Date.	No.
Leather splitting and stretching apparatus	B. Rowe	Albany, N. Y	Apr. 30, 1850	7, 327
Leather splitting and stripping machine	C. S. Stearns	Marlborough, Mass	Feb 9, 1864	41, 583
Leather-splitting gage	J. N. Baird	Bloomfield, Iowa	May 14, 1872	126, 668
Leather-splitting knife	H. Aiken	Brighton, Mass	Mar. 12, 1823	....
Leather-splitting knife	A. H. Van Gieson	Newark, N. J	June 9, 1868	78, 701
Leather-splitting machine	J. H. Abbott and J. A. Marden	Malden and Boston, Mass	Mar. 24, 1868	75, 823
Leather-splitting machine	C. W. Baldwin and L. D. Hawkins.	Charlestown and Stoneham, Mass.	Apr. 10, 1866	53, 771
Leather-splitting machine	J. Butters	New York, N. Y	May 31, 1822	....
Leather-splitting machine	D. H. Chamberlain	West Roxbury, Mass	May 10, 1859	23, 900
Leather-splitting machine	H. E. Chapman	Albany, N. Y	Nov. 23, 1858	22, 108
Leather-splitting machine	H. Cunningham	Albany, N. Y	Oct. 12, 1869	95, 780
Leather-splitting machine	A. Dawes	Hudson Mass	Oct. 29, 1867	70, 175
Leather-splitting machine	J. P. Fairlamb	Wilmington, Del	Feb. 22, 1848	5, 456
Leather-splitting machine	J. F. Flanders	Boston, Mass	Aug. 14, 1860	29, 649
Leather-splitting machine	J. F. Flanders and J. A. Marden.	Newburyport, Mass	Aug. 29, 1854	11, 604
Leather-splitting machine	J. Goebel and J. Preis	Caledonia, Wis	Nov. 25, 1873	144, 893
Leather-splitting machine	C. Keniston	Somerville, Mass	Dec. 21, 1869	98, 068
Leather-splitting machine	J. A. Marden and H. A. Butters	Newburyport and Haverhill, Mass.	Nov. 6, 1855	13, 756
Leather-splitting machine	J. A. Marden	Newburyport, Mass	Apr. 3, 1866	53, 741
Leather-splitting machine	J. A. Marden	Newburyport, Mass	May 8, 1866	54, 571
Leather-splitting machine	M. H. Merriam and J. B. Crosby.	Chelsea and Stoneham, Mass.	Feb. 13, 1855	12, 392
Leather-splitting machine	W. Panton	Milton, Mass	July 15, 1851	8, 227
Leather-splitting machine	E. Pratt	Salem, Mass	Dec. 19, 1854	12, 114
Leather-splitting machine	E. Pratt	Salem, Mass	Mar. 11, 1856	14, 430
Leather-splitting machine	A. Richardson	Boston, Mass	Apr. 23, 1831	....
Leather-splitting machine	A. Richardson	Boston, Mass	Feb. 9, 1841	1, 967
Leather-splitting machine	A. Richardson	Boston, Mass	Apr. 17, 1844	3, 541
Leather-splitting machine	A. Richardson	North Enfield, N. H	Sept. 16, 1851	8, 369
Leather-splitting machine	B. Rowe	Albany, N. Y	June 2, 1863	38, 763
Leather-splitting machine	J. A. Safford	Boston, Mass	Mar. 19, 1861	31, 746
Leather-splitting machine	J. A. Safford	Boston, Mass	Feb. 2, 1864	41, 448
Leather-splitting machine	J. P. Shaw and J. C. Briggs	Boston, Mass	Dec. 31, 1833	....
Leather-splitting machine	C. S. Stearns	Marlborough, Mass	Feb. 22, 1870	100, 082
Leather-splitting machine	A. F. Stowe	Worcester, Mass	June 10, 1873	139, 744
Leather-splitting machine	J. Taggart	Boston, Mass	Nov. 10, 1868	83, 888
Leather-splitting machine	S. S. Turner	Westborough, Mass	Oct. 30, 1860	30, 553
Leather-splitting machine	A. H. Van Gieson	Newark, N. J	July 8, 1862	35, 850
Leather-splitting machine	A. H. Van Gieson	Newark, N. J	Apr. 17, 1866	54, 043
Leather-splitting machine	F. J. Vittum	Newburyport, Mass	Feb. 18, 1868	74, 734
Leather-splitting machine	C. Weston	Salem, Mass	Aug. 30, 1853	9, 980
Leather-splitting machine	H. White	Binghamton, N. Y	Aug. 2, 1839	1, 272
Leather-splitting machine	H. Wing	Buffalo, N. Y	Aug. 25, 1863	39, 695
Leather-splitting machine, Bed-spring of	J. B. Tay	North Woburn, Mass	Aug. 7, 1855	13, 407
Leather-splitting machine, Feed of	D. H. Chamberlain	West Roxbury, Mass	June 5, 1860	28, 559
Leather-splitting machine, Roll for	C. S. Stearns	Marlborough, Mass	Feb. 13, 1872	123, 589
Leather splitting machine, Sole, &c	E. Putnam	Danvers, Mass	Nov. 20, 1838	1, 010
Leather-splitting mill, Cylindrical	P. Dow	Boston, Mass	July 12, 1810	....
Leather stamping and shaping machine	B. B. Harris	Lockport, Ill	Nov. 5, 1867	70, 558
Leather-stitching machine	S. Sheldon	Cincinnati, Ohio	Aug. 3, 1839	1, 277
Leather stock, &c., Machine for cutting up	F. L. Walker	Boston, Mass	Feb. 2, 1869	86, 611
Leather strap and tube, Seamless	A. Escheulohr	Munich, Bavaria	May 12, 1868	77, 724
Leather-strap clasp	A. B. Conde	Albany, N. Y	Jan. 9, 1866	51, 926
Leather straps, Clasp for securing end of	O. O. Storle	Milwaukee, Wis	June 3, 1873	139, 626
Leather straps, Creasing-roll for	J. F. Moloney	Cincinnati, Ohio	Dec. 3, 1872	133, 660
Leather straps for harness, &c., Process for forming.	G. Earle	Dover, Ohio	Sept. 30, 1862	36, 562
Leather straps, Machine for creasing	S. Cronce	Flemington, N. J	Oct. 24, 1848	5, 882
Leather straps, Machine for creasing the edges of	W. M. Thornton	Pottsville, Pa	May 15, 1855	12, 878
Leather straps, Machine for creasing and finishing.	W. M. K. Thornton	Niles, Mich	Jan. 8, 1861	31, 094
Leather straps, &c., Machine for cutting	J. Haberbush, J. Bentz, and V. S. Vogel.	Lancaster, Pa	Feb. 23, 1869	87, 167
Leather straps, Machine for cutting and punching.	J. F. Hollister	Plano, Ill	May 27, 1873	139, 388
Leather straps, Machine for raising and creasing	G. W. Pruyne	Mexico, N. Y	Apr. 15, 1856	14, 698
Leather straps, Machine for shecoring	C. Morris	New Haven, Conn	Feb. 20, 1855	12, 444
Leather straps, Machine for waving or embossing	D. H. Unger	Philadelphia, Pa	Oct. 7, 1873	143, 477
Leather straps, Tool for chamfering	J. Bridger	Richland, Iowa	Nov. 2, 1858	21, 937
Leather-stretching frame	C. P. Oliver	Newark, N. J	Feb. 11, 1873	135, 836
Leather-stretching frame, Wet	R. W. Dawson	Newark, N. J	June 25, 1867	66, 131
Leather-stretching machine	W. R. Andrews and R. Dingwell.	Newark, N. J	Sept. 15, 1868	82, 063
Leather-stretching machine	J. F. Connelly and W. B. Hughes	Newark, N. J	Oct. 8, 1867	69, 633
Leather-stretching machine	J. H. Haskell	Baltimore, Md	May 15, 1860	28, 271
Leather-stretching machine	A. W. Roberts	Hartford, Conn	Feb. 8, 1859	22, 893
Leather-stretching machine	B. Rowe	Albany, N. Y	Apr. 22, 1851	8, 052
Leather-stretching machine	W. Strevel and D. Brown	Albany, N. Y	Apr. 22, 1851	8, 054
Leather-stretching pole	H. Dawson	Baltimore, Md	Apr 6, 1869	88, 697
Leather stripping and cutting machine	J. E. Coffin, C. E. Morrill, and G. F. Hall.	Portland and Westbrook, Me.	Mar. 28, 1871	113, 189
Leather-stripping machine	A. R. E. Falck and P. Stoerger	Newark, N. J	Oct. 20, 1857	18, 441
Leather-stripping machine	D. Ridgway	....	Oct. 26, 1809	....
Leather, Stuffing	J. Armstrong	Woburn Centre, Mass	Feb. 10, 1857	16, 573
Leather, Stuffing and currying	T. McDonald	Roxbury, Mass	Aug. 14, 1866	57, 165
Leather stuffing and currying machine	F. Carl	Charlestown, Mass	Apr. 16, 1867	63, 856
Leather-stuffing machine	F. Fischer	Cambridge, Mass	Mar. 24, 1868	75, 896
Leather-stuffing machine	H. Muller	North Cambridge, Mass	June 9, 1868	78, 815
Leather-stuffing machine	J. W. Schayer	Boston, Mass	June 9, 1868	78, 835
Leather, Substitute for	P. Wenzel	Mayence, Hesse-Darmstadt	Mar. 27, 1866	53, 549
Leather-tapering machine	W. Mannheim	New York, N. Y	Sept. 24, 1867	69, 229
Leather-tearing machine	E. S. Hidden	Millburn, N. Y	June 20, 1871	116, 055
Leather to a uniform thickness, Machine for reducing roller.	W. C. Joslin	West Thompson, Conn	Sept. 24, 1867	69, 219
Leather-trimmer	L. P. Curtis	Marlborough, Mass	Apr. 14, 1868	76, 608
Leather trimmer, Cue	L. A. Johnson, H. W. Collinder, and J. E. Boyle.	New York, N. Y	Aug. 31, 1869	94, 317

Index of patents issued from the United States Patent Office from 1790 to 1873, inclusive—Continued.

Invention.	Inventor.	Residence.	Date.	No.
Leather trimming, chamfering, and skiving machine.	W. H. Rounds	Campello, Mass	Feb. 7, 1860	27,073
Leather-trimming machine	S. H. Jones	East Albany, N. Y	July 8, 1873	140,631
Leather-trimming machine	S. P. Pope	Burlington, N. Y	Sept. 10, 1861	33,264
Leather-trimming machine, Cue	J. E. Boyle	New York, N. Y	May 4, 1869	89,624
Leather, &c., Treating	J. Burrill	Lynn, Mass	May 3, 1864	42,619
Leather, Treating tanned	B. K. Lightfoot	Philadelphia, Pa	Nov. 17, 1863	40,669
Leather, Treatment of tanned	A. Dietz	New York, N. Y	Oct. 16, 1860	30,393
Leather-varnish	C. Brumby	Rochester, N. Y	Sept. 12, 1871	118,842
Leather-varnishing composition	O. S. Boyden and M. C. Fredericks.	Newark, N. J	May 25, 1858	28,320
Leather-varnishing machine	L. L. Allen	Hallowell, Me	Sept. 3, 1872	130,969
Leather-washer	G. Miller and C. M. Andrews	Providence, R. I	Dec. 20, 1859	26,510
Leather-washer, Annular	P. L. Gibbs	Chicago, Ill	Dec. 30, 1873	146,062
Leather-washing machine	A. Howard and G. F. Howard	Wellsville, Md., and Chicago, Ill.	May 28, 1867	65,224
Leather-washing machine	A. Howard and G. F. Howard	Wellsville, N. Y., and Chicago, Ill.	July 14, 1868	79,832
Leather-washing machine	I. Sipes	Boston, Mass	Dec. 12, 1865	51,487
Leather water-proof and mode of applying same, Composition for rendering.	C. F. Miller	Baltimore, Md	Aug. 28, 1840	1,749
Leather, Water-proof composition for	W. J. Roome	New York, N. Y	Jan. 6, 1844	3,401
Leather water-proof, Composition for rendering	W. A. Ronald and H. Miller	Rowan County, N. Y	Jan. 24, 1842	2,431
Leather water-proof, Making	A. Straub	Milton, Pa	June 30, 1828	
Leather water-proof, Making	D. Kizer	New York, N. Y	Nov. 19, 1827	
Leather water-proof, Method of dressing and making.	O. A. Goold	Portland, Me	July 9, 1872	128,873
Leather water-proof, Process for rendering	G. Conkling	Conklingsville, N. Y	Sept. 5, 1865	49,826
Leather-work, Metal fastening material for	C. Keniston	Somerville, Mass	Dec. 16, 1873	145,658
Leather-work, Nail for	S. W. Baldwin	Baldwinsville, N. Y	Sept. 27, 1864	44,382
Leather-work ornament	L. E. Sallee	Peoria, Ill	June 23, 1868	79,148
Leather-work, Seam for	C. Keniston	Somerville, Mass	Aug. 19, 1873	142,026
Leather-work, Seam for	S. W. Shorey	Boston, Mass	Sept. 10, 1872	131,308
Leather-work, Seam for	S. W. Shorey	Boston, Mass	May 27, 1873	139,429
Leather-working tool	J. S. Alexander	Pittsburgh, Pa	July 1, 1873	140,336
Leathern tubes, Machine for making	N. Wyllys	South Glastonbury, Conn	Dec. 23, 1851	8,604
Leathern tubes or hose, &c., Making	A. L. Pennock and J. Sellers	Philadelphia, Pa	July 6, 1818	
Ledger account, Keeping	A. J. Folger	Nantucket, Mass	Oct. 30, 1849	6,825
Lee-board	J. H. and J. Swain	Cape May Court-House, N.J	Apr. 10, 1811	
Lee-board for vessels	A. Jouan	San Francisco, Cal	Nov. 18, 1856	16,090
Lee-way indicator	A. A. Wilder	Detroit, Mich	Jan. 21, 1851	7,912
Lee-way indicator for vessels	T. Byrne	New York, N. Y	July 3, 1866	56,003
Leech, Artificial	F. Wolff	New York, N. Y	Oct. 30, 1866	59,306
Leech, Mechanical	M. Delluc	New York, N. Y	May 7, 1850	7,341
Leech, Mechanical	F. A. Stohlmann and A. H. Smith.	Brooklyn and New York, N. Y.	Feb. 22, 1870	100,210
Leeches, Bleeding with and breeding	J. Kunitz	Philadelphia, Pa	May 7, 1805	
Leg and foot, Artificial	M. H. Hawking	New Haven, Conn	May 18, 1869	90,099
Leg and foot rest	E. Withall	Rochester, N. Y	Oct. 24, 1871	120,176
Leg, Artificial	J. J. Austin	New York, N. Y	June 20, 1865	48,251
Leg, Artificial	D. Bly	Rochester, N. Y	Apr. 19, 1859	23,656
Leg, Artificial	D. Bly	Rochester, N. Y	May 17, 1859	24,062
Leg, Artificial	D. Bly	Rochester, N. Y	Aug. 30, 1859	25,238
Leg, Artificial	D. Bly	Rochester, N. Y	Feb. 19, 1861	31,438
Leg, Artificial	D. Bly	Rochester, N. Y	May 19, 1863	38,549
Leg, Artificial	D. Bly	Rochester, N. Y	May 19, 1863	38,550
Leg, Artificial	D. Bly	New York, N. Y	Sept. 4, 1866	57,666
Leg, Artificial	D. Bly	Macon, Ga	Mar. 9, 1869	87,264
Leg, Artificial	J. Bringhurst	Philadelphia, Pa	June 12, 1866	55,459
Leg, Artificial	T. Burr	Battle Creek, Mich	Apr. 18, 1865	47,353
Leg, Artificial	H. L. Byrd	Augusta, Ga	Mar. 6, 1866	52,964
Leg, Artificial	T. J. Cain	Cleveland, Ohio	Oct. 18, 1864	44,766
Leg, Artificial	E. Carlton and E. Goss	Portland, Me	Mar. 12, 1867	62,731
Leg, Artificial	R. Clement	Philadelphia, Pa	Apr. 18, 1865	47,281
Leg, Artificial	J. Condell	Morristown, N. Y	July 11, 1865	48,660
Leg, Artificial	J. Condell	Morristown, N. Y	July 18, 1865	48,792
Leg, Artificial	P. Daniels	Le Roy, N. Y	Jan. 19, 1864	41,282
Leg, Artificial	D. D. Douglass	Springfield, Mass	Jan. 10, 1860	26,753
Leg, Artificial	J. S. Drake	New York, N. Y	Aug. 31, 1852	9,232
Leg, Artificial	J. S. Drake	New York, N. Y	Sept. 4, 1866	57,691
Leg, Artificial	E. Elliott and C. E. Newton	Lowell, Mass	Jan. 28, 1868	73,787
Leg, Artificial	J. Emery	Cedar Falls, Iowa	May 28, 1867	65,187
Leg, Artificial	T. F. Englebrecht, R. Baeklen, and W. Staehlen.	New York and Brooklyn, N. Y.	Jan. 6, 1863	37,282
Leg, Artificial	J. A. Foster	West Stockholm, N. Y	Aug. 8, 1865	49,253
Leg, Artificial	J. A. Foster	Detroit, Mich	June 29, 1869	92,031
Leg, Artificial	J. Travel	Saint Louis, Mo	Sept. 15, 1863	39,912
Leg, Artificial	D. Gilson	Nashua, N. H	June 19, 1866	55,645
Leg, Artificial	S. G. Gregory	Albany, N. Y	Aug. 17, 1869	93,876
Leg, Artificial	G. W. Hall	Lyndonville, N. Y	June 2, 1863	38,739
Leg, Artificial	J. E. Hanger	Staunton, Va	Feb. 14, 1871	111,741
Leg, Artificial	W. M. Hawkins	Elba, Ala	July 16, 1872	129,340
Leg, Artificial	G. B. Head	Albany, N. Y	Oct. 27, 1868	83,496
Leg, Artificial	W. Hinds	Little Falls, N. Y	Oct. 18, 1864	44,726
Leg, Artificial	B. W. Jewett	Gilford, N. H	Jan. 6, 1857	16,360
Leg, Artificial	B. W. Jewett	Gilford, N. H	Aug. 7, 1860	29,494
Leg, Artificial	G. B. Jewett	Salem, Mass	June 24, 1862	35,686
Leg, Artificial	G. B. Jewett	Salem, Mass	July 22, 1862	35,937
Leg, Artificial	G. B. Jewett	Salem, Mass	Oct. 4, 1864	44,534
Leg, Artificial	G. B. Jewett	Salem, Mass	Aug. 22, 1865	49,528
Leg, Artificial	G. B. Jewett	Salem, Mass	Aug. 22, 1865	49,529
Leg, Artificial	G. B. Jewett	Salem, Mass	Dec. 19, 1865	51,593
Leg, Artificial	S. B. Jewett	Laconia, N. H	Dec. 7, 1869	97,647
Leg, Artificial	H. A. Kimball and A. J. Lawrence.	Philadelphia, Pa	May 1, 1866	54,364
Leg, Artificial	L. Legran	Allegheny City, Pa	Jan. 16, 1866	52,057
Leg, Artificial	L. Legran	Allegheny City, Pa	Sept. 3, 1867	68,758

Index of patents issued from the United States Patent Office from 1790 to 1873, inclusive—Continued.

Invention.	Inventor.	Residence.	Date.	No.
Leg, Artificial	P. Lockie	Rochester, N. Y	Oct. 27, 1863	40,417
Leg, Artificial	R. G. Lockwood	Battle Creek, Mich	Oct. 31, 1865	50,770
Leg, Artificial	J. Madden	Cleveland, Ohio	Jan. 9, 1866	51,953
Leg, Artificial	A. A. Marks	New York, N. Y	Mar. 7, 1865	46,687
Leg, Artificial	D. B. Marks	New York, N. Y	Mar. 7, 1854	10,611
Leg, Artificial	A. McOmber	Schenectady, N. Y	Nov. 19, 1867	71,197
Leg, Artificial	A. Mennel	New York, N. Y	Aug. 29, 1865	49,645
Leg, Artificial	H. L. Mills	Saint Paul, Minn	Oct. 15, 1867	69,829
Leg, Artificial	J. Monroe	New York, N. Y	Mar. 15, 1864	41,934
Leg, Artificial	J. Monroe	New York, N. Y	Oct. 11, 1864	44,644
Leg, Artificial	J. Monroe	New York, N. Y	July 25, 1865	49,038
Leg, Artificial	K. Moore	Oswego, N. Y	July 20, 1869	92,739
Leg, Artificial	F. W. Neubert	Pittsburgh, Pa	Nov. 22, 1864	45,169
Leg, Artificial	R. H. Nicholas	Chicago, Ill	Aug. 7, 1866	56,983
Leg, Artificial	R. H. Nicholas and D. Bly	Rochester, N. Y	July 28, 1857	17,888
Leg, Artificial	B. F. Palmer	Meredith, N. H	Nov. 4, 1846	4,834
Leg, Artificial	B. F. Palmer	Meredith, N. H	Feb. 20, 1849	6,122
Leg, Artificial	B. F. Balmer	Philadelphia, Pa	Aug. 17, 1852	9,200
Leg, Artificial	B. F. Palmer	Philadelphia, Pa	Apr. 8, 1873	137,711
Leg, Artificial	D. D. Parmelee	New York, N. Y	Feb. 10, 1863	37,637
Leg, Artificial	L. F. Pingree	Portland, Me	July 5, 1870	104,994
Leg, Artificial	H. D. Reinhardt	Baltimore, Md	Feb. 9, 1864	41,535
Leg, Artificial	J. Reichenbach	Pittsburgh, Pa	Jan. 12, 1864	41,237
Leg, Artificial	J. Reichenbach	Pittsburgh, Pa	Jan. 12, 1864	41,238
Leg, Artificial	W. H. Rhodes	Berlin, N. Y	July 31, 1855	13,360
Leg, Artificial	J. Russell	Philadelphia, Pa	Aug. 17, 1852	9,202
Leg, Artificial	F. Schmitt	Springfield, Ill	July 16, 1867	66,744
Leg, Artificial	J. Schneider	Cincinnatti, Ohio	Aug. 15, 1865	49,443
Leg, Artificial	W. Selpho	New York, N. Y	May 6, 1856	14,836
Leg, Artificial	G. L. Shephard	New York, N. Y	May 17, 1864	42,799
Leg, Artificial	S. P. Sleppy	Wilkesbarre, Pa	May 22, 1866	54,970
Leg, Artificial	I. D. Small	North Fairfield, Ohio	Dec. 15, 1863	40,956
Leg, Artificial	U. Smith	Battle Creek, Mich	July 28, 1863	39,361
Leg, Artificial	A. Spaulding	Lowell, Mass	Aug. 7, 1855	13,494
Leg, Artificial	W. C. Stone	Boston, Mass	Dec. 17, 1850	7,847
Leg, Artificial	A. Strasser	Montgomery, Ala	Mar. 31, 1868	76,267
Leg, Artificial	C. Sweet	Vicksburgh, Miss	Nov. 26, 1867	71,424
Leg, Artificial	J. Taggart	Roxbury, Mass	Sept. 25, 1855	13,611
Leg, Artificial	L. Tassius	Norwalk, Ohio	Aug. 11, 1868	81,033
Leg, Artificial	T. Uren	San Francisco, Cal	Aug. 18, 1863	39,599
Leg, Artificial	R. M. Vaughan	Glasgow, Mo	Dec. 22, 1863	41,033
Leg, Artificial	J. Walber	New York, N. Y	Sept. 12, 1865	49,936
Leg, Artificial	A. T. Watson	New York, N. Y	Feb. 5, 1867	61,780
Leg, Artificial	J. W. Weston, F. Buchner, and R. Bocklen.	New York and Buffalo, N. Y.	Apr. 10, 1866	53,931
Leg Artificial	J. W. Weston and T. B. Stanley	New York, N. Y	June 6, 1865	48,138
Leg, Artificial	T. E. M. White	New Bedford, Mass	June 24, 1862	35,723
Leg Artificial	O. D. Wilcox	Easton, Pa	Jan. 13, 1857	16,420
Leg, Artificial	O. D. Wilcox	Easton, Pa	Sept. 30, 1856	15,831
Legs Artificial	G. W. Yerger	Philadelphia, Pa	Mar. 19, 1850	7,204
Legs, Apparatus for relief of debility or weakness in.	S. P. W. Douglass	Palmyra, N. Y	May 15, 1841	2,092
Legs, Attachment to artificial	O. D. Wilcox	Elmira, N. Y	Aug. 24, 1858	21,289
Legs, Boring-tool for making wooden	E. Osborne	Philadelphia, Pa	Sept. 4, 1866	57,756
Legs, Support for artificial	P. Daniels	Le Roy, N. Y	Feb. 24, 1863	37,738
Legs, Support for artificial	J. W. Weston	New York, N. Y	Dec. 27, 1864	45,662
Legs, Tool for making wooden	E. Osborne	Philadelphia, Pa	Sept. 4, 1866	57,755
Legging	A. L. Munson	New Haven, Conn	Mar. 13, 1866	53,231
Legging	W. G. Rule	New York, N. Y	Jan. 15, 1867	61,265
Legging	A. P. Smith	Rock Falls, Ill	June 4, 1872	127,652
Legging, Ladies'	E. F. and S. S. Bedford	Johnstown, N. Y	Apr. 5, 1864	42,157
Legging, Saddle	W. B. Johns	Georgetown, D. C	Dec. 24, 1861	33,996
Leggotyping	W. A. Leggo	Montreal, Canada	Mar. 21, 1871	112,933
Leghorn, straw, &c., Mode of whitening	H. Cooper	Franklin County, Pa	Nov. 12, 1828	....
Lemonade, Composition for	A. Warner	Cleveland, Ohio	Jan. 6, 1863	37,314
Lemonade, Preparing materials for	J. Warren, jr	New York, N. Y	Sept. 11, 1847	5,292
Lemon-squeezer	A. Barbarin	New Orleans, La	Oct. 8, 1867	69,531
Lemon-squeezer	A. Bassford	Fort Wayne, Ind	Jan. 21, 1873	135,067
Lemon-squeezer	A. Bigelow	Killingworth, Conn	Feb. 18, 1873	136,023
Lemon-squeezer	L. S. Chichester	New York, N. Y	July 3, 1860	28,967
Lemon-squeezer	V. Fogerty	Boston, Mass	Feb. 27, 1866	52,928
Lemon-squeezer	O. Hesselbacher and H. Moesta	Detroit, Mich	June 18, 1867	65,809
Lemon-squeezer	V. Himmer	New York, N. Y	Mar. 22, 1870	101,128
Lemon-squeezer	J. L. Jensen	Brooklyn, N. Y	June 14, 1870	104,159
Lemon-squeezer	J. Klepzig	San Francisco, Cal	Nov. 17, 1868	84,125
Lemon-squeezer	J. H. Mead	New York, N. Y	July 15, 1873	140,785
Lemon-squeezer	T. Reece and A. Clarke	Philadelphia, Pa	Jan. 15, 1867	61,251
Lemon-squeezer	A. J. T. Reuter	Boston, Mass	Mar. 26, 1867	63,304
Lemon-squeezer	E. M. Sammis	Babylon, N. Y	Sept. 2, 1873	142,414
Lemon-squeezer	T. C. Smith	New York, N. Y	Apr. 7, 1868	76,539
Lemon-squeezer	G. M. Thomas	New York, N. Y	June 10, 1862	35,554
Lemon-squeezer	W. A. and J. W. Whitney	Brooklyn, N. Y	Nov. 26, 1872	133,511
Lenitive, Pain	E. Audler	New York, N. Y	Oct. 26, 1818	....
Lens	C. B. Boyle	New York, N. Y	Feb. 20, 1866	52,672
Lens, Achromatic	H. Fitz	New York, N. Y	Sept. 27, 1864	44,483
Lens, Double-globe	J. Schnitzer	New York, N. Y	May 24, 1864	42,880
Lens, Fluid	S. Kakeles	New York, N. Y	Apr. 24, 1860	27,983
Lens, Fluid	O. Warden	Cincinnati, Ohio	Aug. 28, 1866	57,602
Lens, Fluid	D. A. Woodward	Baltimore, Md	Nov. 27, 1866	60,109
Lens, Glass	H. M. Paine	North Oxford, Mass	Oct. 3, 1846	4,786
Lenses, Adjusting	W. and W. H. Lewis	New York, N. Y	Dec. 16, 1851	8,590
Lenses, &c., Machine for grinding and polishing	H. D. Knight	Elgin, Ill	Nov. 15, 1870	109,325
Lenses, Machine for grinding and polishing	M. Witmer	South Pekin, N. Y	Oct. 13, 1863	40,303
Lenses, Manufacture of glass	J. L. Gilliland	New York, N. Y	Aug. 10, 1852	9,184
Lenses, Mounting fluid	A. M. Cole	Windham, Me	Mar. 16, 1858	19,624
Let-off mechanism	G. Bailey	Putnam, Conn	June 13, 1871	115,806
Let-off mechanism	A. J. Woodman	Indian Orchard, Mass	July 30, 1872	129,915
Letter and envelope combined	E. B. Gleason	Boston, Mass	Mar. 15, 1859	23,242

Index of patents issued from the United States Patent Office from 1790 *to* 1873, *inclusive*—Continued.

Invention.	Inventor.	Residence.	Date.	No.
Letter and figure, Metallic projecting	L. Child	New York	Oct. 4, 1817	....
Letter-balance	B. Chambers, jr	Washington, D. C	Mar. 10, 1868	75, 244
Letter-balance and pen-holder, Combined	R. B. Kepner	Philadelphia, Pa	Aug. 17, 1869	93, 720
Letter-board for object-teaching	J. H. Palm	Mansfield, Ohio	May 30, 1871	115, 349
Letter-box	C. R. Bancroft	Boston, Mass	June 7, 1870	103, 826
Letter-box	W. H. Bramble	Decatur, Ill	Apr. 3, 1866	53, 562
Letter-box	J. W. Brown	New York, N. Y	Apr. 15, 1862	34, 940
Letter-box	L. De Mets	Elizabeth, N. J	Nov. 12, 1872	132, 954
Letter-box	J. A. Farrington	Brooklyn, N. Y	Nov. 7, 1871	120, 578
Letter-box	C. P. Gorely	Boston, Mass	Dec. 17, 1867	72, 190
Letter-box	C. P. Gorely	Boston, Mass	June 23, 1868	79, 222
Letter-box	B. B. Gurley	Minneapolis, Minn	Nov. 25, 1873	144, 976
Letter-box	G. C. Jenks	New York, N. Y	Oct. 9, 1860	30, 324
Letter-box	D. P. Jordan	Chicago, Ill	June 30, 1868	79, 355
Letter-box	C. Lewis	Philadelphia, Pa	Jan. 31, 1871	111, 461
Letter-box	S. Maltby	Washington, D. C	July 19, 1864	43, 594
Letter-box	E. T. Marsh	Rochester, N. Y	June 7, 1870	104, 043
Letter-box	E. Mill and S. L. Chubbuck	Cleveland, Ohio	Apr. 30, 1872	126, 151
Letter-box	A. Potts	Philadelphia, Pa	Dec. 6, 1870	109, 939
Letter-box	A. Rosenstjerna	Chicago, Ill	July 29, 1873	141, 237
Letter-box	J. A. Ryan	Cleveland, Ohio	Aug. 15, 1871	118, 057
Letter-box	A. T. Sinclaire	New York, N. Y	Nov. 5, 1872	132, 865
Letter-box	J. W. and J. D. Smith	Washington, D. C	Feb. 4, 1868	74, 015
Letter-box	J. H. Springer	Philadelphia, Pa	Mar. 22, 1864	42, 052
Letter-box	S. Strong	Washington, D. C	Aug. 31, 1869	94, 449
Letter-box	E. C. Weld	New York, N. Y	Oct. 4, 1870	108, 073
Letter-box	S. S. Williams	Chicago, Ill	Dec. 3, 1872	133, 691
Letter-box	F. Wittram	San Francisco, Cal	June 22, 1869	91, 583
Letter-box	J. E. Woodruff	Buffalo, N. Y	Mar. 24, 1868	76, 026
Letter-box, Alarm	S. H. Morris	Charlestown, Mass	May 7, 1872	126, 477
Letter-box alarm	E. H. Ripley	Boston Highlands, Mass	Nov. 16, 1869	96, 965
Letter-box alarm	A. Taylor	Brooklyn, N. Y	Nov. 21, 1871	121, 218
Letter-box alarm	A. Taylor	Brooklyn, N. Y	June 10, 1873	139, 833
Letter-box alarm and door-plate, Combined	Z. E. Fobes	Troy, N. Y	Feb. 20, 1872	123, 880
Letter-box attachment for door	E. K. Mull	Reading, Pa	Nov. 26, 1872	133, 474
Letter box, Drop	J. North	Middletown, Conn	Mar. 13, 1860	27, 466
Letter-box, Postal	D. D. Foley	Washington, D. C	June 19, 1866	55, 636
Letter-box, Post-office	W. H. Taylor	Stamford, Conn	Aug. 12, 1873	141, 834
Letter-box, Post-office	L. Yale, jr	Shelburne Falls, Mass	Sept. 19, 1871	119, 212
Letter-box, Street	C. L. Evert and A. Johnson	Washington, D. C., and New York, N. Y.	Feb. 25, 1873	136, 313
Letter-box, Street	S. Strong	Washington, D. C	Mar. 30, 1869	88, 528
Letter-box, Street	D. and G. H. White	Chicago, Ill	Apr. 14, 1868	76, 864
Letter-box to lamp-posts, Mode of attaching metallic.	A. Potts	Philadelphia, Pa	Mar. 9, 1858	19, 578
Letter-clip	L. E. Osborne	New Haven, Conn	May 10, 1864	42, 683
Letter-clip and paper-binder, Combined	G. W. McGill	New York, N. Y	Dec. 23, 1873	145, 809
Letter-copying apparatus	A. L. Adams	Philadelphia, Pa	Oct. 9, 1860	30, 283
Letter-lock	H. L. Arnold	Chicago, Ill	May 4, 1869	89, 826
Letter-opener	H. Gross	Tiffin, Ohio	Jan. 26, 1869	86, 300
Letter-opener	R. Johnson	Urbana, N. Y	May 17, 1864	42, 777
Letter-package	J. W. Burns	Medway, Ohio	Nov. 3, 1868	83, 600
Letter, parcel, &c., Method of receiving and delivering.	A. E. Beach	Stratford, Conn	Nov. 13, 1866	59, 739
Letter-pouch	P. Davis	Newport News, Va	June 30, 1868	79, 324
Letter-press for proof, &c	A. Ramage	Philadelphia, Pa	May 19, 1823	....
Letter sheet and envelope combined	T. Orton	Chicago, Ill	Mar. 11, 1873	136, 666
Letter-sheet blank	A. C. Fletcher	New York, N. Y	May 28, 1872	127, 330
Letter writer, Manifold	J. K. Park	New York, N. Y	Nov. 13, 1844	3, 825
Letters and parcels, Method of transporting	A. E. Beach	Stratford, Conn	Oct. 26, 1869	96, 187
Letters, &c., Device for folding	G. W. R. Lewin	Rochester, N. Y	Aug. 24, 1869	94, 012
Letters for signs, &c., Composition	A. M. and J. P. Todd	Des Moines, Iowa	Aug. 17, 1869	93, 924
Letters for signs, &c., Mode of constructing	T. Motley	Brooklyn, N. Y	Apr. 28, 1857	17, 167
Letters, Machinery for post-marking	E. N. Moore	Boston, Mass	Jan. 16, 1849	6, 036
Letters, Making block	L. Katen	New York, N. Y	Sept. 20, 1844	3, 750
Letters, Making shaded	G. Bruce	New York	Apr. 4, 1820	....
Letters, &c., Mode of sealing	J. Saxton	Washington, D. C	Feb. 15, 1859	22, 982
Letters, &c., Mold for casting	G. F. Sack	New York, N. Y	Aug. 4, 1868	80, 567
Letters on metal plates, Device for raising	R. R. Foote	Chicago, Ill	Aug. 15, 1871	118, 005
Letters on the circumference of metal disks, Device for forming.	S. M. Ott	Newark, N. J	Mar. 26, 1872	124, 970
Letters, packages, &c., Mode of transporting and delivering.	A. E. Beach	Stratford, Conn	Sept. 5, 1865	49, 699
Letters without discovery, Device to prevent opening.	C. S. Westcott	New York, N. Y	May 20, 1862	35, 342
Lettering-block, Adjustable	E. J. Griffin	Hart's Falls, N. Y	Jan. 11, 1870	98, 762
Levee	A. C. Brawner	Frankfort, Ky	July 13, 1869	92, 575
Levee	W. P. Craig	Milton, Ky	Sept. 21, 1869	95, 090
Levees, Inclined way for building	J. Bradford	Wilmington, Del	July 1, 1873	140, 458
Levees, Machine for making	E. J. Brown	Carroll Parish, La	July 13, 1869	92, 421
Levees, Machine for making	E. Comeaux	Bayou Goula, La	Dec. 10, 1867	71, 854
Level	W. P. Cutter	Chelsea, Mass	July 28, 1868	80, 337
Level	H. G. Loomis	Hartford, Conn	July 7, 1868	79, 582
Level	D. Masten	Bingham, N. Y	Sept. 25, 1866	58, 271
Level	W. L. Richardson	Reading, Mass	Jan. 1, 1867	60, 788
Level, Adjustable spirit	P. Clifford	Holyoke, Mass	Nov. 26, 1867	71, 279
Level, Adjustable spirit	L. L. Davis	Springfield, Mass	Sept. 17, 1867	68, 961
Level, Adjustable spirit	L. L. Davis	Springfield, Mass	Mar. 17, 1868	75, 534
Level, Adjustable spirit	J. Morss and F. B. Abel	Philadelphia, Pa	Apr. 25, 1871	114, 027
Level, Adjustable spirit	E. A. and C. M. Stratton	Greenfield, Mass	July 16, 1872	129, 183
Level, Adjustable spirit	W. J. Tate	New Haven, Conn	Jan. 21, 1868	73, 670
Level, Adjusting spirit	S. N. Chapin and A. Stanley	New Britain, Conn	Sept. 10, 1867	68, 603
Level, Adjusting spirit	J. A. Traut	New Britain, Conn	Oct. 6, 1868	82, 769
Level and clinometer, Pendulum	W. Johnson	Edisto Island, S. C	Apr. 25, 1871	114, 014
Level and clinometer, Plumb	N. Barnum	La Porte, Ind	May 7, 1872	126, 372
Level and measure, Land	F. A. Archibald	Concord, N. H	July 20, 1869	92, 773
Level and plumb, Balance and pendulum	L. Lewis	Newfield, N. Y	Oct. 3, 1838	956
Level and plumb, Combined	P. Clifford	Holyoke, Mass	June 11, 1867	65, 726

Index of patents issued from the United States Patent Office from 1790 *to* 1873, *inclusive*—Continued.

Invention.	Inventor.	Residence.	Date.	No.
Level and plumb instrument	J. E. Eldred	Rochester, N. Y	Aug. 25, 1831	....
Level and square, Plumb	D. G. Davison, E. Pullen, and S. Davison.	Prospect Plain and Cranberry, N. J.	July 7, 1863	39,124
Level and square, Reflecting spirit	F. Wilbar	Roxbury, Mass	Apr. 20, 1852	8,807
Level, Collimating	J. Locke	Cincinnati, Ohio	July 2, 1850	7,477
Level, Combined spirit	A. Cahoon, jr	Harwich, Mass	May 20, 1862	35,298
Level, Combined square and	O. R. Chaplin	Salem, Mass	May 8, 1866	54,503
Level, Fluid	W. G. Ladd, jr	Cambridge, Mass	Apr. 9, 1850	7,263
Level, Grading	J. Thornley	Charlottesville, Va	Oct. 21, 1873	143,942
Level, Graduating	N. Hollingsworth	Rosetta, Ill	July 2, 1867	66,342
Level or inclinometer	T. A. Chandler	Rockford, Ill	Apr. 14, 1857	17,023
Level, Pendulum	T. A. Chandler	Rockford, Ill	May 17, 1853	9,722
Level, Pendulum	A. C. L. Delsarte	Paris, France	Aug. 19, 1873	142,005
Level, Pendulum	A. Munger	Oberlin, Ohio	Aug. 17, 1835	....
Level, Plumb	B. F. St. John	Shelbyville, Ind	Sept. 5, 1865	49,802
Level, Plumb	G. Cuppers	New York, N. Y	Dec. 19, 1865	51,564
Level, Plumb	M. W. E. Doran	Indianapolis, Ind	Apr. 10, 1866	53,795
Level, Plumb	C. Ensminger and A. W. Elmer	Springfield, Mass	July 16, 1867	66,695
Level, Plumb	A. Girard	Mobile, Ala	July 2, 1846	4,611
Level, Plumb	W. H. and D. D. Polleys	La Crosse, Wis	Oct. 27, 1863	40,428
Level, Spirit	J. T. Zimmerman and H. Baker	Lancaster, Pa	Oct. 8, 1867	69,602
Level, Spirit	L. Brooks	Great Falls, N. H	Aug. 29, 1854	11,596
Level, Spirit	R. T. Burnett	New York, N. Y	Sept. 5, 1865	49,710
Level, Spirit	L. L. Davis	Springfield, Mass	Mar. 17, 1868	75,533
Level, Spirit	L. L. Davis	Springfield, Mass	Nov. 21, 1871	121,088
Level, Spirit	H. W. Evans	Philadelphia, Pa	Feb. 13, 1855	12,384
Level, Spirit	C. F. Hill	Hamilton, Ohio	Dec. 15, 1868	84,880
Level, Spirit	T. N. Hosmer	Todd's Valley, Cal	Nov. 11, 1862	36,906
Level, Spirit	R. Ledig	Newark, N. J	Feb. 13, 1866	52,579
Level, Spirit	H. Lewis	Bennington, Vt	June 30, 1868	79,363
Level, Spirit	W. T. Nicholson	Providence, R. I	May 1, 1860	28,104
Level, Spirit	F. W. Marston	Philadelphia, Pa	June 17, 1873	140,055
Level, Spirit	T. S. Scoville	Rochester, N. Y	Sept. 13, 1859	25,446
Level, Spirit	G. A. Shelley	Madison, Conn	Nov. 7, 1871	120,675
Level, Spirit	H. S. Shepardson	Shelburne Falls, Mass	Sept. 13, 1864	44,225
Level, Spirit	J. D. Sibley	Middletown, Conn	June 25, 1868	79,266
Level, Spirit	E. A. and C. M. Stratton	Greenfield, Mass	Mar. 1, 1870	100,463
Level, Spirit	J. A. Traut	New Britain, Conn	July 2, 1872	128,513
Level, Spirit	A. J. Vandegrift	Cincinnati, Ohio	May 8, 1866	54,626
Level, square, compass, and plumb-staff, Combined	J. R. Abbott	Midway, Ind	Dec. 26, 1865	51,675
Level, Surveying	C. Becker	Brooklyn, N. Y	Dec. 1, 1857	18,728
Level, Surveyor's	S. D. Hailey	Jackson, Tenn	Jan. 10, 1860	26,762
Levels, Adjustments applied to pendulum	C. Cole	Tarrytown, N. Y	July 28, 1857	17,870
Levels, Mounting spirit	S. I. Sherman	New York, N. Y	July 19, 1853	9,867
Levels to squares, Attaching spirit	J. Steger	Matteawan, N. Y	Jan. 24, 1860	26,956
Leveler, Rut	C. Marshall	North Easton, Mass	Jan. 3, 1871	110,663
Leveling and grading instrument	H. E. Towle	New York, N. Y	May 10, 1870	102,882
Leveling-instrument	J. Rohrer	South Bend, Ind	Mar. 5, 1872	124,389
Leveling-instrument for ditching, &c	J. Gray	Raymond, Miss	Nov. 24, 1857	18,689
Leveling-instrumegt, Self-adjustable	J. Redhead	Woodville, Miss	Dec. 21, 1858	22,378
Leveling-instrument, Surveyor's	L. Lea	Jackson, Tenn	Aug. 21, 1860	29,703
Lever, Alternate propelling	P. Smith	Port Gibson, Miss	June 9, 1825	....
Lever and pendulum power engine	S. Woods	Freeport, Me	Jan. 7, 1835	....
Lever and pulley power applied to a standing press	G. W. Grater	Boston, Mass	Oct. 1, 1830	....
Lever and sling, Truss	S. Wells	Boston, Mass	July 24, 1822	....
Lever and stump-puller	J. G. Fox	Oregon, Wis	Jan. 25, 1870	99,179
Lever, Angle	J. Barron	Norfolk, Va	May 14, 1822	....
Lever balance, Bent	R. G. Kimball	Albany, N. Y	Sept. 14, 1869	94,755
Lever, Balance spring-catch	L. Kendall	New York, N. Y	May 8, 1809	....
Lever-bottom press, Combined	E. Venear	Halifax, N. C	May 10, 1826	....
Lever, Circular propelling	J. Turner	Marcellus, N. Y	Mar. 12, 1825	....
Lever-clasp, Spring	G. D. Pike	Chatham, Conn	Apr. 3, 1866	53,671
Lever, Compound	J. Simpson	Marietta, Ga	Jan. 12, 1869	85,864
Lever, Compound pendulum	C. Berry	Poughkeepsie, N. Y	Nov. 1, 1825	....
Lever-crane	E. Smith	Paris, N. Y	Dec. 31, 1818	....
Lever, crank, weight, and balance wheel, Combination of.	E. T. Merrel	Parkman, Me	Apr. 22, 1835	....
Lever, Differential	G. I. Washburn	Worcester, Mass	Mar. 14, 1865	46,838
Lever, Engine	T. W. Goodwin	Portsmouth, Va	Apr. 28, 1863	38,364
Lever, Engine	T. W. Goodwin	Portsmouth, Va	Aug. 18, 1863	39,564
Lever escapement	P. Humbert	Boston, Mass	Jan. 1, 1861	31,022
Lever escapement	C. E. Jacob	New York, N. Y	July 28, 1846	4,664
Lever escapement, Detached	S. B. Terry	Waterbury, Conn	July 1, 1873	140,560
Lever for drawing framed timber together	J. N. Eames	Wilmington, Mass	Sept. 9, 1873	142,618
Lever for presses	J. P. Gates	Red River County, Tex	Mar. 4, 1873	136,431
Lever for supporting ships, Bilge	J. Thomas	New York, N. Y	Nov. 6, 1826	....
Lever for turning wheels, rowing boats, &c	R. Hopkins	Hopkinton, N. Y	Mar. 25, 1808	....
Lever-grapnel	E. B. Dewey	Pontiac, Mich	Dec. 8, 1868	84,683
Lever, Hand	E. J. Durant	Lebanon, N. H	May 31, 1859	24,202
Lever-jack	N. Badgley	New York, N. Y	June 23, 1863	38,936
Lever-jack	J. Bernheisel, sr	Green Park, Pa	Aug. 25, 1868	81,467
Lever-jack	L. H. Davis	Kennett's Square, Pa	May 4, 1852	8,919
Lever-jack	T. M. Kane	Goshen, N. Y	Dec. 15, 1863	40,981
Lever-jack	T. J. Kindleberger	Eaton, Ohio	Nov. 19, 1867	71,183
Lever-jack	J. Leffel	Springfield, Ohio	Dec. 10, 1850	7,820
Lever-jack	J. Leffel	Springfield, Ohio	Nov. 15, 1864	45,056
Lever-jack	F. Stamm	Lampter, Pa	Mar. 15, 1859	23,273
Lever-jack, Extension	W. H. Hartman	Fostoria, Ohio	Sept. 19, 1865	49,999
Lever key and lock	A. Prutyman	Philadelphia, Pa	Mar. 4, 1836	....
Lever, Lifting	R. Sealy	New York, N. Y	July 1, 1822	....
Lever-paddle	J. Harris	London, England	June 21, 1864	43,271
Lever-power	T. W. Lafetra	New York, N. Y	Jan. 23, 1834	....
Lever-power	H. A. Woodman and M. Person	Charlotte, Mich	Sept. 3, 1872	131,142
Lever-power, Accumulating	J. Cochran	South Hadley, Mass	Mar. 26, 1834	....
Lever-power and inclined-wheel	P. Wykoff	Westport, Ky	July 20, 1831	....
Lever-power, Crossing or collapsing	W. Scarbrough	Darien, Ga	May 19, 1824	....
Lever-power, Double-ratchet	J. S. Williams	Chicago, Ill	Sept. 8, 1868	81,967
Lever-power for pressing	R. Sanderson	Athens, Ohio	Feb. 20, 1844	3,446

Index of patents issued from the United States Patent Office from 1790 *to* 1873, *inclusive*—Continued.

Invention.	Inventor.	Residence.	Date.	No.
Lever-power for sewing and knitting machines	G. Stackpole	New York, N. Y	Nov. 17, 1868	84, 144
Lever, Power gained	E. G. Fitch	Blakely, Ala	Oct. 5, 1827	....
Lever-power, Mode of applying	G. E. Clay	Stillwater, Minn	June 22, 1858	20, 624
Lever-power, Mode of applying	E. Harris	Princeton, Ill	May 17, 1859	24, 025
Lever-power, Projectile	G. Wood	Vernon, Ind	June 9, 1830	....
Lever-power, Vibrating	N. Tripp	Niagara Falls, N. Y	Feb. 18, 1868	74, 732
Lever-press	T. Blake	Turner, Me	Feb. 20, 1821	....
Lever-press	H. G. Guyon	New York, N. Y	July 2, 1836	....
Lever-press	W. Linn	Danville, Va	Apr. 25, 1828	....
Lever-press	B. Morris	Binghamton, N. Y	Feb. 2, 1830	....
Lever-press	J. Outland	Barnesville, Ohio	Aug. 8, 1871	117, 919
Lever-press	J. Payne	Russellville, Ky	Sept. 9, 1835	....
Lever-press	H. Sharman	Scriba, N. Y	Nov 1, 1830	....
Lever-press, Compound	J. Rodgers	Maury County, Tenn	Jan. 21, 1829	....
Lever-press, Progressive	J. Perkins	Philadelphia, Pa	Feb. 27, 1819	....
Lever-press, Spiral	A. O. Stansbury	New York, N. Y	Mar. 21, 1821	....
Lever-presses, Windlass tackle for	J. B. Carpenter	Henderson, N. Y	Jan. 30, 1830	....
Lever, Pumping	J. S. Appel	Kulpsville, Pa	Jan. 1, 1869	90, 718
Lever-purchase	J. B. Case	Fletcher, Vt	Mar. 31, 1868	76, 050
Lever-purchase, rivet, and bolt-cutter	H. Seely	Unadilla, N. Y	May 12, 1825	....
Lever, Ratchet	H. Baldwin	Nashua, N. H	Mar. 23, 1854	10, 955
Lever, Ratchet	H. W. Millar	Utica, N. Y	May 31, 1870	103, 763
Lever shears	J. J. Sandgren	Lyons, Iowa	Apr. 16, 1867	63, 945
Lever-spring, Compensating	W. S. Pratt	Williamsburgh, N. Y	July 24, 1860	29, 304
Lever-truss and sling for yards of vessels	S. A. Wells	Boston, Mass	Dec. 23, 1824	....
Levers, &c., Adjusting attachment to reversing	G. W. Jordan	Passaic, N. J	Apr. 1, 1873	137, 374
Levers, Combination of	C. Tomkins	Montgomery, Ala	Nov. 14, 1834	....
Levers, Hanging the griping-jaw of spike machines in weighted.	J. H. Swett	Pittsburgh, Pa	Mar. 14, 1854	10, 645
Leverage	W. W. Wills	Janesville, Wis	Oct. 10, 1865	50, 409
Lewis-lever	T. Lidgerwood	Brooklyn, N. Y	Mar. 12, 1850	7, 172
Library-register	W. T. Ray	Philadelphia, Pa	Sept. 5, 1865	49, 788
Library-shears	L. Prang	Boston, Mass	July 5, 1870	104, 996
Library-system	J. M. W. Geist	Lancaster, Pa	Jan. 2, 1866	51, 823
Lid and plate lifter and pot-hook	T. Simpson	Newark, Ohio	May 17, 1870	103, 247
Lid-lifter	W. Van Gaasbeck	Hudson, N. Y	Dec. 30, 1873	145, 035
Lid-supporter	J. C. Barlow	Brimfield, Mass	Oct. 16, 1866	58, 936
Life and anchor boat	J. Francis	New York, N. Y	Jan. 11, 1839	1, 067
Life and property saving vessels, Arrangement of means for balancing and propelling.	J. Minifie	Baltimore, Md	July 8, 1856	15, 298
Life and treasure buoy	F. D. Lee	Charleston, S. C	Apr. 27, 1858	20, 072
Life and treasure buoy for vessel, Arrangement of	F. D. Lee	Charleston, S. C	Dec. 8, 1857	18, 819
Life-berth for vessel	J. P. McLean	New York, N. Y	Dec. 7, 1858	22, 258
Life-boat	J. Adams	Gloucester, N. J	Sept. 16, 1873	142, 883
Life-boat	J. Allen	New York, N. Y	Mar. 6, 1855	12, 473
Life-boat	L. Ball	Auburn, N. Y	Apr. 20, 1858	19, 977
Life-boat	E. A. Barrett	New York, N. Y	Apr. 29, 1873	138, 225
Life-boat	P. K. Beaupré	Metropolis City, Ill	Aug. 29, 1871	118, 580
Life-boat	H. Berdan	New York, N. Y	Feb. 13, 1855	12, 375
Life-boat	M. E. D. Brown	Utica, N. Y	Apr. 5, 1859	23, 442
Life-boat	M. M. Camp	New Haven, Conn	Sept. 22, 1857	18, 233
Life-boat	D. Dodge and P. Burgess	New York, N. Y., and East Boston, Mass.	Aug. 9, 1853	9, 915
Life-boat	C. D. Flynt	Philadelphia, Pa	June 1, 1869	90, 833
Life-boat	Y. Foreman	New York, N. Y	Oct. 11, 1853	10, 120
Life-boat, &c	J. Francis	New York, N. Y	Mar. 26, 1841	2, 018
Life-boat	L. F. Frazee	New Brunswick, N. J	Nov. 22, 1853	10, 266
Life-boat	L. F. Frazee	Jersey City, N. J	June 23, 1868	79, 111
Life-boat	J. R. Grace	Brooklyn, N. Y	Sept. 1, 1868	81, 623
Life-boat	P. Henrich	Allegheny City, Pa	Apr. 26, 1870	102, 261
Life-boat	H. Hensel	Carver, Minn	Aug. 28, 1866	57, 633
Life-boat	W. Hughes	Waupun, Wis	July 31, 1866	56, 855
Life-boat	R. Humble	Milwaukee, Wis	Oct. 4, 1870	107, 912
Life-boat	O. R. Ingersoll	Brooklyn, N. Y	Feb. 20, 1872	123, 908
Life-boat	G. W. La Baw	Jersey City, N. J	Apr. 1, 1856	14, 565
Life-boat	G. W. La Baw	Jersey City, N. J	Nov. 1, 1859	25, 971
Life-boat	M. Ludlum	Essex, N. Y	Jan. 27, 1857	16, 491
Life-boat	M. Ludlum	Fair Haven, Vt	Mar. 29, 1859	23, 380
Life-boat	T. Masac	Good Hope Plantation, La	May 31, 1870	103, 637
Life-boat	S. Mills	New York, N. Y	Aug. 7, 1860	29, 508
Life-boat	M. V. Nobles	Elmira, N. Y	June 18, 1867	65, 824
Life-boat	L. Raymond	New York, N. Y	Feb. 4, 1843	2, 938
Life-boat	R. J. Robeson	Oskaloosa, Iowa	Aug. 8, 1871	117, 817
Life-boat	J. T. Scholl	Port Washington, Wis	Feb. 26, 1861	31, 562
Life-boat	A. L. Shears	Omro, Wis	May 25, 1858	20, 374
Life-boat	A. L. Shears	Flint, Mich	Aug. 1, 1865	49, 197
Life-boat	J. M. Starr	Fond du Lac, Wis	Dec. 27, 1870	110, 600
Life-boat	H. Thompson	Mobile, Ala	Oct. 5, 1869	95, 538
Life-boat	G. H. Tier	Fremont, Ohio	Aug. 7, 1866	57, 012
Life-boat	W. A. Vail	Clinton, Conn	Aug. 5, 1873	141, 609
Life-boat	F. Vié	Havre, France	May 4, 1869	89, 816
Life-boat	W. H. Wylly	Savannah, Ga	Jan. 22, 1867	61, 500
Life-boat and bed-bottom, Combined	C. Bütgenbach	Louisville, Ky	Aug. 15, 1871	117, 977
Life-boat constructed of mattresses	J. M. Woodward	New York, N. Y	Sept. 7, 1858	21, 462
Life-boat, Folding	E. S. Berthon	Fareham, England	Mar. 20, 1855	12, 537
Life-boat, Folding	C. Locher	New York, N. Y	Jan. 2, 1855	12, 184
Life-boat, Folding	H. Martin	Louisville, Ky	Apr. 12, 1859	23, 595
Life-boat, Reversible	G. P. Tewksbury	Boston, Mass	Aug. 7, 1849	6, 640
Life-boat, Reversible	N. Thompson, jr	Williamsburgh, N. Y	Apr. 11, 1854	10, 766
Life-boat, Sectional	C. Pond	Philadelphia, Pa	July 9, 1872	128, 750
Life-boat, Self-inflating and folding	W. and T. Schnebly	New York, N. Y	Jan. 23, 1849	6, 053
Life-boat, Sofa	P. Van Zile, S. M. Griffin, and W. S. Dey.	New York, N. Y	Nov. 6, 1855	13, 771
Life-boat, Sporting	J. M. Cayce	Franklin, Tenn	Aug. 20, 1867	67, 846
Life-boat, Surf	W. M. Camp	New Haven, Conn	Oct. 25, 1859	25, 882
Life-boat, Surf	J. R. Grace	Brooklyn, N. Y	Mar. 6, 1860	27, 362
Life-boat, Water-cask	W. N. Clark	Chester, Conn	May 3, 1859	23, 824
Life-boats, Apparatus for supplying air to	P. F. Schenck	Riceville, N. J	June 18, 1867	65, 953

Index of patents issued from the United States Patent Office from 1790 to 1873, inclusive—Continued.

Invention.	Inventor.	Residence.	Date.	No.
Life-boats, Expansible float for	C. Segros	New York, N. Y	Sept. 21, 1858	21, 570
Life-boats, Form of air-chambers of	J. D. Greene	Cambridge, Mass	Sept. 25, 1849	6, 737
Life-boats, Mode of constructing	G. James	Philadelphia, Pa	Jan. 21, 1839	1, 071
Life-buoy	W. Flower	Philadelphia, Pa	May 14, 1808	....
Life-pole and fire-engine assistant	W. W. Van Loan	Catskill, N. Y	Feb. 2, 1822	....
Life-preserver	S. Albro	Buffalo, N. Y	Mar. 2, 1852	8, 767
Life-preserver	G. M. Allerton	New York, N. Y	July 14, 1868	79, 934
Life-preserver	E. S. Bennett	New York, N. Y	Jan. 26, 1869	86, 127
Life-preserver	J. M. Billhofer	Irvington, N. Y	May 21, 1867	64, 939
Life-preserver	J. Bond	Norfolk, Va	Apr. 6, 1869	88, 692
Life-preserver	H. B. Bourne	New York, N. Y	May 8, 1840	1, 596
Life-preserver	T. B. Boyd	Saint Louis, Mo	June 21, 1870	104, 546
Life-preserver	R. Bulkley	New York, N. Y	May 8, 1840	1, 595
Life-preserver	C. J. Bunker	New York, N. Y	Sept. 29, 1857	18, 274
Life-preserver	A. Chandler	New York, N. Y	Oct. 3, 1844	3, 774
Life-preserver	E. M. Crandal	Marshalltown, Iowa	Oct. 4, 1870	108, 006
Life-preserver	E. G. Fitch	New Orleans, La	Aug. 18, 1846	4, 699
Life-preserver	A. J. Gibson	Worcester, Mass	Dec. 8, 1857	18, 809
Life-preserver	A. Gregory	New York, N. Y	Apr. 27, 1869	89, 402
Life-preserver	N. E. Guérin	New York	Nov. 16, 1841	2, 359
Life-preserver	D. H. Heyen	New York, N. Y	Dec. 3, 1867	71, 755
Life-preserver	L. Jacobs	Albion, Mich	May 10, 1870	102, 823
Life-preserver	J. Knight	New York, N. Y	June 2, 1857	17, 434
Life-preserver	C. Krejci	Scranton, Pa	Mar. 15, 1870	100, 906
Life-preserver	M. Ormsbee	Brooklyn, N. Y	Oct. 17, 1871	120, 089
Life-preserver	H. Palmer	Augusta, Mich	Oct. 12, 1858	21, 776
Life-preserver	N. Platt	Jackson, Miss	Feb. 7, 1860	27, 066
Life-preserver	R. Robinson	San Francisco, Cal	Oct. 22, 1867	70, 121
Life-preserver	E. Rousseil	West Newark, N. J	Jan. 9, 1866	51, 971
Life-preserver	S. Scholfield	Norwich, Conn	July 7, 1863	39, 175
Life-preserver	T. R. Scott	New York, N. Y	Nov. 21, 1871	121, 132
Life-preserver	J. E. Serrell and W. Davies	New York, N. Y	Dec. 15, 1857	18, 869
Life-preserver	J. Shaw	Bridgeport, Conn	May 28, 1867	65, 287
Life-preserver	W. A. Simonds	Boston, Mass	Feb. 3, 1857	16, 555
Life-preserver	B. W. Taylor	Henderson, Ky	May 30, 1871	115, 386
Life-preserver	J. E. Thomson and E. Clark	Buffalo, N. Y., and Louisville, Ky.	Aug. 15, 1871	118, 168
Life preserver, buoy, &c	C. A. De Liancourt	Paris, France	May 10, 1844	3, 584
Life-preserver, Carriage	H. G. De Grandval	Portsmouth, N. H	Oct. 3, 1817	....
Life-preserver for railway-cars, Brakesman's	J. Worsley	Providence, R. I	Sept. 6, 1864	44, 134
Life-preserver for steamboats, &c	M. Pearson	Newburyport, Mass	Jan. 27, 1843	2, 937
Life-preserver, Marine	J. B. Verdier	Paris, France	Feb. 16, 1864	41, 654
Life-preserver or buoyant dress	R. Porter	Billerica, Mass	May 25, 1840	1, 619
Life-preserver, Safety	J. J. White	Philadelphia, Pa	Apr. 7, 1838	679
Life-preservers, &c., Float for	J. W. Weston	New York, N. Y	Jan. 26, 1869	86, 193
Life-preservers to vests, Mode of attaching	R. L. Nelson	Ocala, Fla	Nov. 21, 1854	11, 969
Life-preserving and swimming apparatus	H. Olsen	San Francisco, Cal	Feb. 25, 1868	74, 931
Life-preserving apparatus	J. B. Stoner	New York, N. Y	Feb. 4, 1868	74, 168
Life-preserving bed for ship	G. K. Hooper	Boston, Mass	Sept. 11, 1855	13, 546
Life-preserving bedstead and sofa	J. T. Garlick	New York, N. Y	Mar. 31, 1857	16, 946
Life-preserving berth	J. J. Clyde	Williamsburgh, N. Y	Mar. 5, 1867	62, 609
Life-preserving berth for steam and other vessels	E. Foster	Hartford, Conn	Sept. 1, 1857	18, 090
Life-preserving boat, Automatic	A. Carson	Memphis, Tenn	July 16, 1867	66, 933
Life-preserving bucket	N. Thompson, jr	Williamsburgh, N. Y	Oct. 18, 1853	10, 140
Life-preserving bucket-raft	C. French	Jersey City, N. J	Apr. 20, 1858	19, 989
Life-preserving buoy	B. Burling	Buffalo, N. Y	Mar. 16, 1858	19, 618
Life-preserving buoy	O. E. Woods	Philadelphia, Pa	Oct. 11, 1859	25, 781
Life-preserving cape	G. and C. Palmer	Morris Run, Pa	Nov. 11, 1873	144, 561
Life-preserving chair, stool, &c	W. H. Shecut	New York	Mar. 9, 1842	2, 482
Life-preserving door	J. T. Pheatt	Toledo, Ohio	Apr. 17, 1855	12, 743
Life-preserving dress	C. S. Merriman	Vallisca, Iowa	July 16, 1872	128, 971
Life-preserving float	G. W. Hamilton	Watkins, N. Y	Mar. 16, 1858	19, 632
Life-preserving float	T. Hosmer	Sandusky, Ohio	Mar. 4, 1873	136, 436
Life-preserving garment	E. R. Cogswell	New York, N. Y	Aug. 12, 1873	141, 631
Life-preserving garment	W. Morris	Philadelphia, Pa	Apr. 26, 1870	102, 419
Life-preserving hammock, Arrangement of the sections in a.	S. I. Seely	New York, N. Y	July 10, 1849	6, 581
Life-preserving hat	S. W. White	England	Oct. 14, 1840	1, 826
Life-preserving mattress	J. Adams	Gloucester, N. J	Apr. 16, 1872	125, 779
Life-preserving mattress	L. Bauhoefer	Philadelphia, Pa	Dec. 4, 1860	30, 794
Life-preserving mattress	C. P. Crossman and E. M. Quimby.	Warren, Mass	Feb. 16, 1858	19, 350
Life-preserving mattress	J. Golding	New York, N. Y	June 18, 1867	65, 901
Life-preserving mattress	J. D. Greene	Cambridge, Mass	Aug. 17, 1869	93, 819
Life-preserving mattress	J. D. Greene	New York, N. Y	Apr. 4, 1871	113, 652
Life-preserving mattress	D. E. Hall	Detroit, Mich	Aug. 3, 1869	93, 195
Life-preserving mattress	J. Hunt	Providence, R. I	Sept. 20, 1870	104, 499
Life-preserving mattress	H. B. Mountain	New York, N. Y	Mar. 11, 1873	136, 749
Life-preserving mattress	W. H. Pack and J. S. Vanhorn	Jersey City, N. J	Oct. 29, 1872	132, 686
Life-preserving mattress	J. F. Woodfine	Hoboken, N. J	Sept. 30, 1873	143, 313
Life-preserving mattress and raft	J. Golding	New York, N. Y	July 23, 1867	67, 039
Life-preserving raft	A. Baker	Appleton, Wis	Jan. 17, 1860	26, 825
Life-preserving raft	A. G. Mack	Rochester, N. Y	July 12, 1859	24, 747
Life-preserving raft	F. Z. Tucker	Brooklyn, N. Y	Jan. 2, 1855	12, 143
Life-preserving raft	C. W. Wailey	New Orleans, La	Mar. 12, 1867	62, 909
Life-preserving raft of bouyant mattresses	W. Urquhart	New York, N. Y	Mar. 9, 1858	19, 593
Life-preserving rafts, Canvas sheet connected with	L. Taggart	Philadelphia, Pa	Jan. 26, 1858	19, 216
Life-preserving seat	H. Matthews	Brooklyn, N. Y	Feb. 26, 1867	62, 433
Life-preserving seat	G. B. Tewksbury	Boston, Mass	Oct. 19, 1852	9, 349
Life-preserving seat	N. Thompson, jr	Williamsburgh, N. Y	Oct. 18, 1853	10, 141
Life-preserving seat	N. Thompson, jr	Williamsburgh, N. Y	Dec. 19, 1854	12, 108
Life-preserving seat	N. Thompson, jr	Williamsburgh, N. Y	Jan. 9, 1855	12, 220
Life-preserving skirt	S. E. Saul	Brooklyn, N. Y	July 26, 1870	105, 730
Life-preserving state-room for navigable vessel, Bouyant.	H. Hallock	Brookhaven, N. Y	June 1, 1858	20, 426
Life-preserving stool	H. T. Pratt	New York, N. Y	Mar. 15, 1870	100, 801
Life-preserving trunk	L. Robstock	Hollidaysburgh, Pa	May 2, 1871	114, 475
Life-preserving trunk	O. E. Woods	Philadelphia, Pa	Dec. 28, 1858	22, 467

Index of patents issued from the United States Patent Office from 1790 *to* 1873, *inclusive*—Continued.

Invention.	Inventor.	Residence.	Date.	No.
Life-preserving vessel	J. Macintosh	New York, N. Y	Nov. 11, 1837	462
Life-preserving vest	T. A. Delano	New York, N. Y	Nov. 9, 1858	22, 021
Life-raft	H. C. Calkins	New York, N. Y	Nov. 28, 1871	121, 275
Life-raft	W. B. Davies	New York, N. Y	Feb. 11, 1873	135, 784
Life-raft	L. F. Frazee	South Amboy, N. J	July 10, 1866	56, 201
Life-raft	D. McFarland	New York, N. Y	Dec. 13, 1870	110, 059
Life-raft	J. Murtaugh	New York, N. Y	Dec. 18, 1866	60, 640
Life-raft	E. L. Perry	New York, N. Y	Nov. 15, 1864	45, 073
Life-raft	J. Rider	New York, N. Y	Feb. 22, 1870	100, 191
Life-raft, Extensible	C. Furbush	Kittery, Me	Nov. 30, 1858	22, 175
Life-rafts, Float for	J. Mason and G. W. Rogers	Detroit, Mich., and Brooklyn, N. Y.	Sept. 16, 1873	142, 801
Life-saving apparatus	J. B. Stoner	New York, N. Y	Sept. 24, 1872	131, 719
Life-saving raft	G. Blanchard	Washington, D. C	Feb. 20, 1855	12, 405
Life-saving raft	S. B. Broad	New York, N. Y	May 8, 1860	28, 150
Lifter:				
See Barrel-lifter. Plate-lifter.				
Box-lifter. Shoe-lifter.				
Exercising-lifter. Skirt-lifter.				
Griddle-lifter. Stone-lifter.				
Kettle-lifter. Stove-lifter.				
Lamp-chimney lifter. Stove-cover lifter.				
Stove-lid lifter.				
Lid lifter. Tie-lifter.				
Log-lifter. Track-lifter.				
Pan-lifter. Transom-lifter.				
Pattern-lifter. Wagon-lifter.				
Pitcher-lid lifter. Wagon-body lifter.				
Lifter and tongs, Combined	J. Hyslop, jr., and C. E. Phillips.	Abington, Mass	Mar. 10, 1868	75, 427
Lifter for kitchen use	T. S. Coffin	Harrington, Me	Jan. 25, 1870	99, 061
Lifting-apparatus	D. P. Butler	Boston, Mass	Mar. 2, 1869	87, 465
Lifting-apparatus	C. H. Douglas	Hartford, Conn	June 13, 1871	115, 942
Lifting-apparatus	W. L. Jones	Baldwin City, Kans	Apr. 2, 1867	63, 526
Lifting-apparatus	F. W. Reilly	Chicago, Ill	Apr. 21, 1868	76, 944
Lifting-apparatus	G. B. Windship	Boston, Mass	Sept. 12, 1865	49, 945
Lifting-apparatus	G. B. Windship	Boston, Mass	Feb. 20, 1872	123, 804
Lifting-apparatus, Spring	D. P. Butler	Boston, Mass	July 20, 1869	92, 793
Lifting-bar	D. P. Butler	Boston, Mass	June 19, 1866	55, 618
Lifting-bar	D. P. Butler	Boston, Mass	July 20, 1869	92, 792
Lifting-handle	P. Bradford	New Haven, Conn	Mar. 12, 1867	62, 729
Lifting-handle	J. Ottner	New Britain, Conn	June 26, 1860	28, 940
Lifting-handle	J. B. Sargent	New Britain, Conn	Nov. 2, 1858	21, 981
Lifting-jack	G. H. Alger	Ames, N. Y	Apr. 21, 1868	76, 967
Lifting-jack	W. C. Anderson	Saint Louis, Mo	May 15, 1860	28, 245
Lifting-jack	O. A. Anthony	Mayfield, N. Y	Jan. 10, 1871	110, 818
Lifting-jack	W. F. Arnold	New Britain, Conn	Dec. 29, 1868	85, 264
Lifting-jack	W. F. Arnold	New Britain, Conn	Jan. 7, 1873	134, 625
Lifting-jack	J. B. Ausbourne	Milwaukee, Wis	Aug. 10, 1869	93, 512
Lifting-jack	H. Ballentine	Philadelphia, Pa	July 4, 1871	116, 534
Lifting-jack	J. H. Bean	Marietta, Ohio	May 21, 1867	64, 937
Lifting-jack	J. W. Bemis	Fall River, Mass	Feb. 4, 1868	74, 040
Lifting-jack	G. Benjamin	Avoca, N. Y	Sept. 18, 1860	30, 043
Lifting-jack	L. J. Blades and J. Mahoney	Harrington and Wilmington, Del.	Mar. 15, 1870	100, 846
Lifting-jack	J. W. Bliss	Hartford, Conn	Mar. 21, 1854	10, 663
Lifting-jack	J. S. Bodge	La Porte, Ind	Apr. 16, 1872	125, 784
Lifting-jack	W. A. Bowyer	Helen Furnace, Pa	Aug. 10, 1869	93, 405
Lifting-jack	W. and C. H. Brady	Mount Joy, Pa	July 6, 1869	92, 255
Lifting-jack	A. C. Brinfer	Middletown, Pa	Sept. 17, 1867	68, 837
Lifting-jack	W. A. Brunker	Farmersburgh, Ind	May 20, 1873	139, 041
Lifting-jack	W. S. Burgin	Washington, Vt	Nov. 21, 1871	121, 081
Lifting-jack	C. Butterworth	Miamisburgh, Ohio	May 21, 1867	64, 946
Lifting-jack	J. Callahan	Stroud Glades, Va	Dec. 29, 1857	18, 955
Lifting-jack	J. Camp	Olney, Ill	Sept. 22, 1868	82, 381
Lifting-jack	J. S. Chesnut	Philadelphia, Pa	Apr. 14, 1857	17, 024
Lifting-jack	S. J. Clark	Detroit, Mich	July 23, 1867	67, 103
Lifting-jack	W. Clark	Decatur, Ill	May 3, 1870	102, 493
Lifting-jack	H. Clement	Rising Sun, Ind	Mar. 19, 1872	124, 721
Lifting-jack	C. T. Close	Brooklyn, N. Y	Feb. 12, 1867	62, 009
Lifting-jack	C. B. Conant	Hardwick, Mass	Feb. 25, 1862	34, 482
Lifting-jack	J. Cook	Palmersville, Pa	May 26, 1863	38, 654
Lifting-jack	J. Coulter, jr	Xenia, Ohio	Feb. 4, 1868	73, 954
Lifting-jack	I. A. Crippen	Philadelphia, Pa	Sept. 9, 1873	142, 680
Lifting-jack	J. Q. Crosby	Northborough, Mass	Dec. 1, 1868	84, 536
Lifting-jack	C. Crow	Onargo, Ill	May 21, 1867	64, 949
Lifting-jack	A. M. Culver	Bedford, Ohio	Feb. 19, 1867	62, 117
Lifting-jack	A. M. Culver	Bedford, Ohio	Feb. 18, 1868	74, 512
Lifting-jack	H. Culver	Richfield, Ohio	Jan. 2, 1866	51, 807
Lifting-jack	J. Dampman	Lebanon, Pa	Mar. 23, 1869	88, 015
Lifting-jack	A. A. Davis	Clark's Green, Pa	Nov. 14, 1871	120, 864
Lifting-jack	R. W. and D. Davis	Yellow Springs, Ohio	Apr. 14, 1857	17, 027
Lifting-jack	D. Diver	Boone, Iowa	Mar. 5, 1867	62, 616
Lifting-jack	A. Dom	Mount Healthy, Ohio	Nov. 5, 1872	132, 753
Lifting-jack	W. M. Doty	New York, N. Y	Apr. 22, 1873	138, 078
Lifting-jack	F. Drew	South Boston, Mass	Nov. 20, 1855	13, 835
Lifting-jack	S. S. Eccleston	South New Berlin, N. Y	Jan. 14, 1873	134, 865
Lifting-jack	W. C. Ellis	Patriot, Ind	June 18, 1872	127, 968
Lifting jack	J. H. Elward	Polo, Ill	Nov. 2, 1869	96, 408
Lifting-jack	D. Fasig	Rowsburgh, Ohio	Aug. 12, 1862	36, 144
Lifting-jack	D. Fasig	Rowsburgh, Ohio	May 15, 1866	54, 706
Lifting-jack	L. W. Fifield	Worcester, Mass	Apr. 5, 1870	101, 450
Lifting-jack	L. P. Garcin	San Francisco, Cal	Apr. 5, 1870	101, 454
Lifting-jack	E. R. Gard	Chicago, Ill	June 27, 1871	116, 296
Lifting-jack	E. R. Gard	Chicago, Ill	Jan. 23, 1872	123, 010
Lifting-jack	R. W. Genung	Blooming Grove, N. Y	July 11, 1854	11, 298
Lifting-jack	A. M. Gilmore	Greensburgh, Ind	May 14, 1872	125, 800
Lifting-jack	A. E. Goddard	Essex, Conn	Dec. 12, 1871	121, 774

Index of patents issued from the United States Patent Office from 1790 to 1873, inclusive—Continued.

Invention.	Inventor.	Residence.	Date.	No.
Lifting-jack	G. F. Graves	Mount Upton, N. Y	Jan. 1, 1867	60, 881
Lifting-jack	H. Gray	Bristol, Conn	Dec. 16, 1856	16, 233
Lifting-jack	W. Green	Holly, Mich	Oct. 1, 1867	69, 400
Lifting-jack	W. Green	Holly, Mich	Sept. 15, 1868	82, 111
Lifting-jack	W. H. Greenwalt	Strickersville, Pa	Oct. 31, 1871	120, 511
Lifting-jack	W. H. Greenwalt	Strickersville, Pa	Apr. 30, 1872	126, 286
Lifting-jack	S. Gulick	Kline's Grove, Pa	Feb. 19, 1867	62, 193
Lifting-jack	J. S. Haldeman	Kansas City, Mo	Apr. 1, 1873	137, 363
Lifting-jack	J. T. Hamilton	Greensburgh, Ind	Mar. 19, 1872	124, 814
Lifting-jack	J. T. Hamilton and E. F. Conner.	Boone County and Decatur, Ind.	Nov. 25, 1873	144, 977
Lifting-jack	J. F. Hammond	Providence, R. I	Aug. 21, 1866	57, 439
Lifting-jack	A. Healy	Kalamazoo, Mich	June 11, 1867	65, 569
Lifting-jack	W. G. Hermance	Albany, N. Y	Feb. 19, 1867	62, 268
Lifting-jack	A. Higley	South Bend, Ind	May 29, 1866	55, 102
Lifting-jack	J. H. Hinton	Wayne County, Ohio	Feb. 4, 1873	135, 551
Lifting-jack	C. Holmes	Washington, Ohio	Oct. 29, 1867	70, 214
Lifting-jack	E. Hoyt	Stamford, Conn	Feb. 6, 1872	123, 347
Lifting-jack	A. Jackson	Clifton Springs, N. Y	Sept. 10, 1867	68, 750
Lifting-jack	J. C. Jackson	Rochester, N. Y	Aug. 31, 1858	21, 342
Lifting-jack	B. F. Johnson	Glasgow, Mo	Feb. 21, 1871	111, 942
Lifting-jack	I. D. Johnson	Kennett's Square, Pa	Dec. 12, 1871	121, 721
Lifting-jack	S. G. Jones	Fitzwatertown, Pa	Jan. 23, 1855	12, 280
Lifting-jack	S. D. Jones	Blue Island, Ill	Aug. 15, 1871	118, 020
Lifting-jack	L. J. Knowles	Warren, Mass	Nov. 10, 1857	18, 592
Lifting-jack	J. Kohler	New York, N. Y	Apr. 21, 1868	76, 927
Lifting-jack	F. Krick	Fidelity, Ohio	Nov. 24, 1868	84, 285
Lifting-jack	I. L. Landis	Manheim Township, Lancaster County, Pa.	Jan. 7, 1862	34, 063
Lifting-jack	W. J. Lane	Chappaqua, N. Y	Oct. 23, 1860	30, 484
Lifting-jack	S. C. Leonard	Oberlin, Ohio	Nov. 29, 1870	109, 635
Lifting-jack	S. M. Lusk	Centreton, Ohio	Jan. 30, 1872	123, 270
Lifting-jack	E. L. Marsh	Greenwich, Ohio	Sept. 17, 1867	68, 891
Lifting-jack	C. A. Masterson	Decatur, Ill	Feb. 20, 1872	123, 784
Lifting-jack	T. Maxon	Springfield, Ohio	Sept. 12, 1871	118, 955
Lifting-jack	T. Maxon	Springfield, Ohio	Aug. 20, 1872	130, 733
Lifting-jack	S. McGuffin	Rising Sun, Ind	Mar. 26, 1872	125, 065
Lifting-jack	W. McMille	Milan, Ohio	Nov. 5, 1867	70, 591
Lifting-jack	W. A. Middleton	Newville, Pa	Dec. 17, 1872	134, 086
Lifting-jack	D. L. Miller	Madison, N. J	Dec. 1, 1857	18, 760
Lifting-jack	D. L. H. Mitchell	Forest, Miss	June 17, 1873	140, 063
Lifting-jack	H. T. Morrison	Lawrenceville, Va	May 23, 1871	115, 086
Lifting-jack	T. W. H. Moseley	Hyde Park, Mass	Mar. 4, 1873	136, 533
Lifting-jack	C. W. Mosher	East Leon, N. Y	Jan. 18, 1870	98, 992
Lifting-jack	D. Mulligan and J. C. Imlay	Greensburgh, Ind	Apr. 23, 1872	126, 076
Lifting-jack	B. Newbury	Catskill, N. Y	May 27, 1851	8, 110
Lifting-jack	J. D. Otstot	Springfield, Ohio	Dec. 2, 1862	37, 051
Lifting-jack	J. D. Otstot	Springfield, Ohio	Dec. 17, 1872	133, 949
Lifting-jack	J. N. Parker	Darlington, Wis	Aug. 27, 1867	68, 230
Lifting-jack	S. J. Parmele	Killingworth, Conn	Jan. 9, 1867	51, 963
Lifting-jack	S. Perry	Seaford, Del	May 12, 1868	77, 756
Lifting-jack	J. W. Pettingill	Rockford, Ill	Feb. 13, 1866	52, 596
Lifting-jack	O. L. Pinney	Brunswick, Ohio	Oct. 22, 1867	70, 115
Lifting-jack	L. B. Prindle	Litchfield, Conn	Jan. 28, 1868	73, 833
Lifting-jack	D. Putnam	Stockton, N. Y	Sept. 2, 1873	142, 510
Lifting-jack	G. Race	Norwich, N. Y	Feb. 26, 1867	62, 442
Lifting-jack	C. E. and F. A. Randall	Boston, Mass	Nov. 26, 1872	133, 484
Lifting-jack	W. S. Rayment	Union City, Mich	Nov. 14, 1871	120, 002
Lifting-jack	R. M. Reynolds	Oakville, Mich	Nov 30, 1869	97, 444
Lifting-jack	A. C. Richard	Newtown, Conn	May 25, 1858	20, 372
Lifting-jack	J. Riddlesberger	Waynesborough, Pa	Feb. 23, 1869	87, 292
Lifting-jack	E. S. Robertson and A. B. Collins.	Mount Liberty, Ohio	Apr. 28, 1868	77 406
Lifting-jack	J. Rosecrans	Berkshire, Ohio	Apr. 12, 1870	101, 920
Lifting-jack	W. F. Rundell	East Genoa, N. Y	Feb. 17, 1863	37, 707
Lifting-jack	B. and C. Schopf	Pittsburgh, Ind	Jan. 25, 1870	99, 244
Lifting-jack	H. G. Seekins and C. H. Goss	Elyria, Ohio	July 7, 1857	17, 757
Lifting-jack	J. W. Shankland	Summerfield, Ohio	Jan. 7, 1873	134, 709
Lifting-jack	A. Shaver	Warnerville, N. Y	Feb. 8, 1870	99, 717
Lifting-jack	T. Sheldon	Enfield, Conn	June 5, 1866	55, 430
Lifting-jack	H. S. Shepardson	Shelburne Falls, Mass	Oct. 3, 1865	50, 306
Lifting-jack	T. Shiver	Newburgh, Ind	June 4, 1867	65, 514
Lifting-jack	S. Shoemaker	Smithville, Ohio	Oct. 29, 1861	33, 608
Lifting-jack	E. Shopbell	Ashland, Ohio	Oct. 22, 1867	70, 034
Lifting-jack	E. Shopbell	Ashland, Ohio	Jan. 21, 1868	73, 657
Lifting-jack	D. Shumusk	Castile, N. Y	Aug. 13, 1872	130, 542
Lifting-jack	A. S. Skinner	Windsor, Mich	Oct. 29, 1872	132, 697
Lifting-jack	J. Sloneker	Jefferson, Pa	May 7, 1872	126, 496
Lifting-jack	R. T. and R. T. Smart, jr	Troy, N. Y	Sept. 17, 1872	131, 472
Lifting-jack	C. A. Smith	New York, N. Y	Aug. 25, 1868	81, 424
Lifting-jack	F. B. Smith	Craigsville, N. Y	July 11, 1854	11, 303
Lifting-jack	F. S. Smith	Geneva, Ohio	Dec. 10, 1872	133, 728
Lifting-jack	J. U. Smith	Orion, Mich	June 20, 1871	116, 232
Lifting-jack	L. Smith	Rochester, Minn	July 3, 1866	56, 111
Lifting-jack	L. P. Smith	Middletown, Pa	Sept. 21, 1869	95, 157
Lifting-jack	L. P. Smith	Middletown, Pa	Oct. 31, 1871	120, 402
Lifting-jack	T. Stebins	San Francisco, Cal	A r. 13, 1869	88, 923
Lifting-jack	J. St. John	New York, N. Y	July 8, 1851	8, 214
Lifting-jack	J. Stoody	Ripley, Ohio	Nov. 5, 1867	70, 644
Lifting-jack	J. J. Stewart	New York, N. Y	Aug. 13, 1872	130, 449
Lifting-jack	G. P. Sweezy	Riverhead, N. Y	Nov. 26, 1872	133, 499
Lifting-jack	S. Taber	Saint Joseph, Mo	June 4, 1867	65, 447
Lifting-jack	S. Taylor	Cold Water, Mich	Feb. 25, 1873	136, 281
Lifting-jack	J. Terrell	Royal Centre, Ind	Feb. 13, 1872	123, 593
Lifting-jack	W. Thomas	Hingham, Mass	May 19, 1857	17, 344
Lifting-jack	J. M. and R. M. Thompson	Coshocton, Ohio	Sept. 23, 1862	36, 540
Lifting-jack	R. M. Thompson	Coshocton, Ohio	May 5, 1868	77, 678
Lifting-jack	W. M. K. Thornton	Clinton Junction, Wis	Feb. 28, 1865	46, 599

Index of patents issued from the United States Patent Office from 1790 *to* 1873, *inclusive*—Continued.

Invention.	Inventor.	Residence.	Date.	No.
Lifting-jack	W. Thurber	Oleon, N. Y	Dec 8, 1863	40, 867
Lifting-jack	B. B. Tomlinson	Mount Carroll, Ill	Mar. 12, 1872	124, 519
Lifting-jack	I. Vaughan	Harrisburgh, Pa	Oct. 15, 1872	132, 189
Lifting-jack	A. F. Wagner	Ilion, N. Y	Jan. 16, 1866	52, 036
Lifting-jack	R. Walker	Batavia, N. Y	July 30, 1867	67, 386
Lifting-jack	H. H. Warren	Buffalo, N. Y	Apr. 22, 1873	138, 059
Lifting-jack	L. D. Warren	Havana, Ill	Oct. 18, 1870	108, 414
Lifting-jack	A. M. Waters	Cuyahoga Falls, Ohio	Aug. 23, 1870	106, 639
Lifting-jack	F. C. White	Euclid, Ohio	Mar. 15, 1870	100, 828
Lifting-jack	J. Wilkinson	Bowling Green, Mo	May 19, 1868	78, 167
Lifting-jack	H. J. Wilson	Mason, Mich	Apr. 4, 1871	113, 381
Lifting-jack	G. W. Windsor	Allegheny City, Pa	July 19, 1870	105, 584
Lifting-jack	E. Young	Camden Centre, Mich	Feb. 23, 1864	47, 738
Lifting-jack	E. Zimerman	Pamelia Four Corners, N. Y	Feb. 19, 1867	62, 175
Lifting-jack	J. Zimmerman	Bloomfield, Pa	June 17, 1862	35, 646
Lifting-jack and cant-hook	D. Fasig	Rowsburgh, Ohio	Nov. 10, 1868	83, 947
Lifting-jack and derrick, Combined	H. Senseman and W. F. Pagett	Tremont, Ohio	July 30, 1872	130, 079
Lifting-jack and spike-extractor	J. Douglass	McConnellstown, Pa	Mar. 3, 1868	75, 034
Lifting-jack and wrench, Combined	R. A. York	Reading, Mich	May 4, 1869	89, 723
Lifting-jack. Carriage	J. Jenkins	Monroe, N. Y	Nov. 7, 1854	11, 920
Lifting-jack for dumping-cart	R. B. Little	Providence, R. I	Dec. 19, 1871	122, 036
Lifting-jack, Hydraulic	J. Ryan	Saint Louis, Mo	Mar. 2, 1869	87, 435
Lifting-jack. Hydrostatic	J. Robertson	Brooklyn, N. Y	July 12, 1859	24, 759
Lifting-jack, Suspended	H. C. Havemeyer	New York, N. Y	Oct. 24, 1871	120, 275
Lifting-jacks. Trammel for	E. H. Hake	Canfield, Ohio	Feb. 25, 1873	136, 156
Lifting-machine	T. S. Crane	Newark, N. J	Mar. 26, 1872	124, 885
Lifting-machine	A. Kriebel	Hereford, Pa	Sept. 15, 1868	82, 128
Lifting-machine	C. H. Mann	Orange, N. J	July 5, 1870	104, 973
Lifting-machine	F. B. O'Conner	New York, N. Y	Sept. 5, 1871	118, 740
Lifting-machine	C. A. Simmons	Waldo, Fla	July 25, 1871	117, 339
Lifting-shovel	W. E. Davis	Brooklyn, N. Y	Oct. 27, 1863	40, 398
Light: *See* Artificial light. Billiard-light. Buoy-light. Calcium light. Deck-light. Drop-light. Electric-light. Entry-light. Fire-light. Foot-light. Gas-light. Head-light. Hydro-oxygen light. Locomotive signal-light. Port-light. Ship-light. Signal-light. Sky-light. Torch-light. Vault-light. Vessel-light.				
Light and air through steps, &c., Admitting	J. B. Cornell	New York, N. Y	June 8, 1858	20, 484
Light and heat, and applying the same, Producing	S. S. Hill	New York, N. Y	Apr. 7, 1863	38, 137
Light and heat, Generating	H. L. Barnum	New York, N. Y	June 2, 1836	....
Light and heat, Method of using explosive liquids for the production of.	W. Beschke	Philadelphia, Pa	Aug. 14, 1866	57, 245
Light from fuel, Mode of obtaining	B. Henfrey		Apr. 16, 1802	....
Light-house. Floating	J. B. Stoner	New York, N. Y	Apr. 29, 1873	138, 292
Light-house, Floating	J. B. Stoner	New York, N. Y	Apr. 29, 1873	138, 293
Light-houses, piers, &c., Construction of foundation for.	L. Bail	New Haven, Conn	Feb. 25, 1862	34, 474
Light. Method of producing	P. Carlevaris	Turin, Italy	July 10, 1866	56, 336
Light produced by combination of liquids	I. Jennings	New York	June 13, 1831	....
Light the same as daylight, Means for rendering artificial.	N. H. Gillet	New York, N. Y	July 19, 1864	43, 581
Lighter and alarm	T. N. Howell	Circleville, Ohio	Dec. 31, 1867	72, 854
Lighters, Tool for making	A. Kleinschmidt and F. Schlater.	Philadelphia, Pa	Jan. 23, 1866	52, 173
Lighting and heating apparatus	A. M. Silber and F. White	London, England	Sept. 12, 1871	118, 979
Lighting-apparatus, Catoptric	W. Hodgins	Albany, N. Y	Feb. 2, 1864	41, 432
Lighting-apparatus, Electric	A. P. Berlioz	Paris, France	May 23, 1871	115, 154
Lighting-apparatus, &c., System of flambeau-like	E. A. L. d'Argy	Paris, France	Feb. 18, 1868	74, 513
Lighting factories and other buildings, Mode of	A. J. White	Ballston Spa, N. Y	Sept. 17, 1867	68, 923
Lighting, heating, vaporizing, and drying apparatus.	J. Kidd	New York, N. Y	Oct. 22, 1872	132, 403
Lighting rooms, Method of	D. G. Haskins and J. Winlock	Cambridge; Mass	Nov. 5, 1867	70, 433
Lighting up picture-galleries	E. M. Smith	New York, N. Y	Aug. 4, 1868	80, 570
Lightning-arrester	A. Barbarin	New Orleans, La	Sept. 3, 1867	68, 407
Lightning-conductor	R. L'Anglois	Assumption, La	Oct. 7, 1846	4, 807
Lightning-conductor	N. Brittan	Chicago, Ill	July 19, 1864	43, 565
Lightning-conductor	I. Johnson	Lodi Station, Ill.	Apr. 18, 1865	47, 310
Lightning-conductor	J. A. Kissell and N. Blickensdufer.	Chicago, Ill	May 14, 1867	64, 774
Lightning-conductor	D. Munson	Indianapolis, Ind	Nov. 1, 1864	44, 880
Lightning-conductor	W. G. Pike	Philadelphia, Pa	Nov. 12, 1867	70, 741
Lightning-conductor	O. Preston	South Dansville, N. Y	Oct. 31, 1871	120, 457
Lightning-conductor	O. White	Racine, Wis	Mar. 30, 1858	19, 819
Lightning-conductors, Attachment for	J. Spratt	Cincinnati, Ohio	Feb. 5, 1850	7, 076
Lightning-conductors, Machine for making tubular	B. F. Housel and S. O. Thayer	Winona, Minn	Sept. 6, 1870	107, 031
Lightning-conductors, Making	C. Stearns	Lowell, Mass	Sept. 20, 1859	25, 554
Lightning-conductors, Mode of attaching the receiving and discharging points of.	W. A. Orcutt	Boston, Mass	Oct. 9, 1841	2, 284
Lightning-rod	J. S. Barber	Gloucester, Mass	Mar. 5, 1839	1, 026
Lightning-rod	N. Brittan	Lockport, N. Y	Nov. 6, 1860	30, 560
Lightning-rod	E. Calender	Boston, Mass	Oct. 3, 1808	....
Lightning-rod	H. B. Conner and J. Denton	Pittsburgh, Pa	Oct. 2, 1866	58, 382
Lightning-rod	S. D. Cushman	New Lisbon, Ohio	Feb. 13, 1866	52, 541
Lightning-rod	S. D. Cushman	New Lisbon, Ohio	Mar. 7, 1871	112, 426
Lightning-rod	J. Drew	Wenona, Ill	Dec. 23, 1873	145, 851
Lightning-rod	H. W. Farley	Oswego, Ill	June 22, 1869	91, 530
Lightning-rod	D. A. Foot and A. Chadwick	Winona, Minn	Aug. 10, 1869	93, 609
Lightning-rod	D. A. Foot and G. S. Knapp	Winona, Minn	Mar. 28, 1870	113, 151
Lightning-rod	J. R. Fricke	Pittsburgh, Pa	Jan. 17, 1871	111, 052
Lightning-rod	J. R. Fricke	Pittsburgh, Pa	Feb. 28, 1871	112, 137
Lightning-rod	A. A. Gaylord	East Cleveland, Ohio	Apr. 18, 1871	113, 869
Lightning-rod	F. O. Goodwin	Philadelphia, Pa	Mar. 5, 1872	124, 264

Index of patents issued from the United States Patent Office from 1790 *to* 1873, *inclusive*—Continued.

Invention.	Inventor.	Residence.	Date.	No.
Lightning-rod	W. Hall	Indianapolis, Ind	Oct. 18, 1859	25, 823
Lightning-rod	W. Hall	Dubuque, Iowa	May 12, 1868	77, 729
Lightning-rod	J. W. Hankenson	Minneapolis, Minn	July 2, 1872	128, 617
Lightning-rod	L. J. Hawley	Baltimore, Md	Feb. 6, 1866	52, 411
Lightning-rod	H. H. Homan	Cincinnati, Ohio	Sept. 14, 1852	9, 260
Lightning-rod	C. Jillson	Worcester, Mass	Feb. 5, 1867	61, 741
Lightning-rod	L. King	East Cleveland, Ohio	Apr. 4, 1871	113, 530
Lightning-rod	T. T. Kinsey	Philadelphia, Pa	June 8, 1869	90, 949
Lightning-rod	G. Kirtland	New Haven, Conn	June 2, 1868	78, 459
Lightning-rod	A. Lyon	Worcester, Mass	July 11, 1854	11, 261
Lightning-rod	S. H. Miner	Winona, Minn	Dec. 12, 1871	121, 884
Lightning-rod	S. J. Mitchell	Saint Louis, Mo	May 23, 1865	47, 846
Lightning-rod	J. M. Mott	Chicago, Ill	May 21, 1872	127, 094
Lightning-rod	W. B. Munn	Cleveland, Ohio	Feb. 4, 1873	135, 574
Lightning-rod	D. Munson	Indianapolis, Ind	Aug. 5, 1856	15, 491
Lightning-rod	D. Munson	Indianapolis, Ind	Feb. 11, 1868	74, 406
Lightning-rod	D. Munson	Indianapolis, Ind	Nov. 17, 1868	84, 210
Lightning-rod	D. Munson	Indianapolis, Ind	Feb. 1, 1870	99, 461
Lightning-rod	D. Munson	Indianapolis, Ind	Mar. 8, 1870	100, 549
Lightning-rod	D. Munson	Indianapolis, Ind	Jan. 3, 1871	110, 778
Lightning-rod	D. Munson	Indianapolis, Ind	Sept. 19, 1871	119, 043
Lightning-rod	D. Munson	Indianapolis, Ind	July 23, 1872	129, 675
Lightning-rod	D. Munson	Indianapolis, Ind	July 23, 1872	129, 676
Lightning-rod	D. Munson	Indianapolis, Ind	July 23, 1872	129, 677
Lightning-rod	G. W. Otis	Lynn, Mass	July 21, 1868	80, 205
Lightning-rod	N. Otis	Charles City, Iowa	July 13, 1869	92, 638
Lightning-rod	M. D. Phelps	Bristolville, Ohio	Dec. 19, 1871	122, 055
Lightning-rod	J. Pratt	Chicago, Ill	Dec. 27, 1864	45, 632
Lightning-rod	W. S. Reyburn and F. J. Martin.	Philadelphia, Pa	Sept. 14, 1869	94, 773
Lightning-rod	W. S. Reyburn and F. J. Martin.	Philadelphia, Pa	Oct. 26, 1869	96, 268
Lightning-rod	J. Robertson	New York, N. Y	July 23, 1872	129, 683
Lightning-rod	G. Row	Indiana, Pa	Nov. 22, 1870	109, 455
Lightning-rod	A. S. Sherwood	Detroit, Mich	Jan. 30, 1866	52, 329
Lightning-rod	J. C. Schoonmaker	Winona, Minn	July 9, 1872	128, 818
Lightning-rod	C. P. Snow	Freeport, Ill	Dec. 26, 1871	122, 290
Lightning-rod	J. Spratt	Cincinnati, Ohio	May 4, 1852	8, 930
Lightning-rod	C. Stearns	Lowell, Mass	June 11, 1867	65, 775
Lightning-rod	G. A. Stephenson and O. L. Sutliff.	Wooster, Ohio	Aug. 15, 1871	118, 166
Lightning-rod	L. D. Vermilye, W. S. Reyburn, and E. A. W. Hunter.	Dayton, Ohio, and Philadelphia, Pa.	July 13, 1869	92, 551
Lightning-rod	W. A. Wells and G. House	Saint Paul, Minn., and Gallipolis, Ohio.	Feb. 13, 1872	123, 600
Lightning-rod	D. F. Welsh	Nevada, Ohio	Feb. 20, 1872	123, 958
Lightning-rod	J. D. West	New York, N. Y	Feb. 1, 1870	99, 381
Lightning-rod	J. J. White	Juliustown, N. J	Oct. 21, 1873	143, 862
Lightning-rod	D. Wooster	Seymour, Conn	Sept. 4, 1860	29, 933
Lightning-rod	E. L. Yancey	Batavia, N. Y	Oct. 29, 1872	132, 615
Lightning rod and conductor	T. Garlick	Cleveland, Ohio	Dec. 7, 1869	97, 496
Lightning rod and conductor	J. W. Hankenson	Minneapolis, Minn	Jan. 25, 1870	99, 184
Lightning-rod attachment	R. D. Dwyer	Richmond, Va	July 3, 1855	13, 162
Lightning rod bracket, Pointed	M. Wells	New York, N. Y	July 7, 1868	79, 618
Lightning-rod clasp	R. Street	Albany, N. Y	Aug. 16, 1870	106, 426
Lightning-rod connection	A. Codington	Bound Brook, N. J	Sept. 26, 1871	119, 323
Lightning-rod, Corrugated	J. A. Kissell and N. Blickensderfer.	Chicago, Ill	July 16, 1867	66, 854
Lightning-rod coupling	D. W. Demorest	Newark, N. J	Oct. 24, 1871	120, 251
Lightning-rod coupling	W. S. Reyburn and E. A. W. Hunter.	Philadelphia, Pa	May 25, 1869	90, 578
Lightning-rod coupling, Tubular	J. W. Fritsch	Cincinnati, Ohio	Dec. 16, 1873	145, 641
Lightning-rod coupling, Tubular	J. H. Weston	Cincinnati, Ohio	June 10, 1873	139, 841
Lightning-rod for vessels	R. B. Forbes	Boston, Mass	July 4, 1854	11, 217
Lightning-rod joint	L. King	East Cleveland, Ohio	Aug. 29, 1865	49, 633
Lightning-rod joint	J. B. Lyon	Cleveland, Ohio	Oct. 10, 1865	50, 372
Lightning-rod joint	N. E. Smith	Cleveland, Ohio	Oct. 10, 1865	50, 398
Lightning-rod point	J. F. Boynton	Syracuse, N. Y	Oct. 26, 1869	96, 194
Lightning-rods, Connecting	J. E. Strong	Boston, Mass	Apr. 19, 1841	2, 056
Lightning-rods, Construction of	L. S. Baldwin and L. Parks	Le Roy, N. Y	Aug. 9, 1859	25, 077
Lightning-rods, Construction of	J. M. Patterson	Woodbury, N. J	July 31, 1860	29, 398
Lightning-rods, Device for attaching	J. B. Elliott	Philadelphia, Pa	Dec. 29, 1857	18, 963
Lightning-rods, Device for securing	J. A. Enggren	Brooklyn, N. Y	May 24, 1859	24, 110
Lightning-rods, Device for securing	V. Schrage	Cincinnati, Ohio	July 13, 1858	20, 916
Lightning-rods, Machine for joining tubular	G. S. Knapp	Winona, Minn	June 27, 1871	116, 453
Lightning-rods, Machine for making	G. S. Knapp	Winona, Minn	Oct. 17, 1871	119, 987
Lightning-rods, Method of insulating and supporting.	M. Fox	Stamford, Conn	Apr. 24, 1860	27, 977
Lightning-rods, Method of insulating and supporting.	E. C. Rogers	Boston, Mass	Oct. 26, 1858	21, 905
Lightning-rods, Mode of constructing	G. B. Hamilton and J. S. Stevenes.	Wellington and Cleveland, Ohio.	Jan. 26, 1869	86, 151
Lightning-rods to houses, Mode of securing	J. Brown and G. W. Robinson	Providence, R. I	Sept. 24, 1825	....
Lightning-rods with sheet metal, Machine for covering.	W. S. Reyburn and F. J. Martin.	Philadelphia, Pa	Aug. 17, 1869	93, 834
Limb, Artificial	R. Briody	Detroit, Mich	May 19, 1868	78, 048
Limb, Artificial	J. Coombs	Greenfield, Mass	Aug. 8, 1865	49, 234
Limb, Artificial	R. H. Dutton	Philadelphia, Pa	Apr. 12, 1859	23, 559
Limb, Artificial	S. G. Gregory	Albany, N. Y	Feb. 2, 1869	86, 530
Limb, Artificial	S. B. Kepperling and T. B. Kreiter.	Neffsville, Pa	May 23, 1871	115, 065
Limb, Artificial	H. A. Kimball	Philadelphia, Pa	Aug. 18, 1863	39, 578
Limb, Artificial	G. C. Kirschmann	New York, N. Y	June 7, 1864	43, 031
Limb, Artificial	A. A. Marks	New York, N. Y	Dec. 1, 1863	40, 763
Limb, Artificial	A. McOmber	Schenectady, N. Y	July 10, 1866	56, 243
Limb, Artificial	A. Mennel	New York, N. Y	Dec. 20, 1864	45, 511
Limb, Artificial	J. Monroe	New York, N. Y	Sept. 25, 1866	58, 351
Limb, Artificial	J. Reichenbach	Pittsburgh, Pa	July 9, 1872	128, 907
Limb, Artificial	J. W. Weston	New York, N. Y	Feb. 2, 1864	41, 456

Index of patents issued from the United States Patent Office from 1790 *to* 1873, *inclusive*—Continued.

Invention.	Inventor.	Residence.	Date.	No.
Limb-extension, Counter	S. H. Whittlesey	Central Mine, Mich	Oct. 22, 1872	132, 509
Limb-supporter	D. H. B. Allen	Chelsea, Vt	Oct. 4, 1864	44, 499
Limbs, Apparatus for fractured	J. M. Pitts	Sumpter, S. C	Nov. 6, 1860	30, 588
Limbs, Apparatus for supporting and ventilating wounded.	G. S. Towles	New Castle, Me	Oct. 6, 1863	40, 164
Limbs, Clasp for artificial	S. F. Burd and J. F. Denniston	Mercer and Pittsburgh, Pa	Mar. 14, 1871	112, 683
Limbs, Lining for artificial	J. W. Weston	New York, N. Y	Mar. 13, 1866	53, 206
Limbs, Screw for extending crooked or setting fractured.	J. Merrill	Glasgow, Ky	June 22, 1832	....
Limbs, Ventilating apparatus for wounded	T. C. Ball	Springfield, Vt	Aug. 4, 1863	39, 453
Lime, Acetate of	C. J. T. Burcey	Black Rock, Conn	Sept. 12, 1871	118, 787
Lime, and apparatus therefor, Protecting unslacked	H. Bisbing	Bridgeton, N. J	Sept. 5, 1871	118, 680
Lime and brick and boiling-kettles, Burning	S. Hill	New Milford, Conn	Feb. 12, 1827	....
Lime and guano spreader	W. Croasdale	Hartsville, Pa	June 24, 1856	15, 171
Lime and manure, Spreading	L. Cooper	Coopersville, Pa	Oct. 19, 1852	9, 339
Lime and marl ashes on land and exchanging soils, Machine for spreading.	D. T. Hill	Plainfield, N. J	Feb. 10, 1838	598
Lime and mortar, Preparation of	L. B. Pitcher	Salina, N. Y	Aug. 9, 1870	106, 282
Lime and other fertilizers, Machine for spreading	P. Seymour	East Bloomfield, N. Y	Dec. 1, 1857	18, 774
Lime and sand for mortar, Machine for mixing	H. W. Hunt and J. Sands	Peekskill, N. Y., and Greenwich, Conn.	Apr. 8, 1856	14, 613
Lime and yeast-powder, Manufacture of phosphate of.	E. N. Horsford	Cambridge, Mass	Aug. 6, 1872	130, 298
Lime, Apparatus for making acid sulphite of	L. Gamotis and S. Martin	New Orleans, La	July 21, 1857	17, 830
Lime, Apparatus for preparing the oxymuriate of	J. Kendall	Lincoln, Mass	Jan. 28, 1820	....
Lime, ashes, &c., Machine for spreading	F. H. Smith	Baltimore, Md	July 5, 1837	258
Lime-bin	J. Smith	Fairton, N. J	Dec. 10, 1872	133, 729
Lime, Burning	S. Garber and H. Swartzengrover.	Norristown, Pa	Mar. 25, 1837	152
Lime, Burning	P. Lossing		Aug. 4, 1800	....
Lime, Burning	E. A. Smith	Saint Albans, Vt	Apr. 19, 1864	42, 413
Lime-burning and ore-smelting furnace	J. Owings	Adams County, Pa	May 29, 1835	....
Lime-burning furnace	T. R. Hartell	Philadelphia, Pa	Dec. 28, 1858	22, 424
Lime, dirt, and sediment from boiling liquid, Device for removing.	A. J. Simmons	Indianapolis, Ind	May 30, 1871	115, 367
Lime from pyroligneous acid, Manufacture of acetate of.	J. Bell	Dover, N. H	May 9, 1871	114, 517
Lime from water, Apparatus for heating and separating.	S. F. Jackson and J. A. Davis	Eureka, Ill	Mar. 30, 1869	88, 484
Lime-kiln	A. G. Anderson	Quincy, Ill	Nov. 3, 1857	18, 531
Lime-kiln	G. Atkins	Sharon, Pa	May 21, 1867	64, 827
Lime-kiln	G. Atkins	Sharon, Pa	Feb. 23, 1869	87, 131
Lime-kiln	L. Averill	Elmira, N. Y	Aug. 19, 1856	15, 549
Lime-kiln	D. T. Barrett	Port Byron, Ill	Jan. 9, 1872	122, 550
Lime-kiln	L. Bomeisler	Philadelphia, Pa	Sept. 10, 1829	....
Lime-kiln	S. Brown	Berwick, Pa	Aug. 26, 1851	8, 319
Lime-kiln	G. W. Calkins and H. White	Cleveland, Ohio	June 15, 1858	20, 549
Lime-kiln	J. Clarkson and G. W. Decker	Washington, D. C	May 3, 1870	102, 496
Lime-kiln	C. B. Corey	Cleveland, Ohio	Dec. 17, 1872	133, 925
Lime-kiln	J. Coryeon	Bay City, Mich	Dec. 31, 1872	134, 421
Lime-kiln	U. Cummings	Buffalo, N. Y	Nov. 20, 1866	59, 760
Lime-kiln	N. Davis	Ida, Mich	July 16, 1872	129, 539
Lime-kiln	R. Donaldson	Mount Nebo, Pa	Feb. 19, 1861	31, 445
Lime-kiln	J. M. Downing	Bristol, Pa	Aug. 25, 1831	....
Lime-kiln	E. B. English	Philadelphia, Pa	Feb. 23, 1864	41, 691
Lime-kiln	T. Ennett	Rockford, Ill	June 14, 1870	104, 291
Lime-kiln	H. R. Fell	Texas, Md	Jan. 5, 1858	19, 023
Lime-kiln	P. J. Gerbault Guichard	Saint Berthevin-le-Laval, France.	Aug. 2, 1870	106, 054
Lime-kiln	L. Gibbs	Fremont, Ohio	Aug. 27, 1867	68, 063
Lime-kiln	P. Griscom and C. S. Denn	Baltimore, Md	Nov. 17, 1857	18, 635
Lime-kiln	P. Griscom and J. H. Miller	Baltimore, Md	May 15, 1866	54, 717
Lime-kiln	M. Groh and J. V. Weitz	Cleveland, Ohio	Oct. 31, 1871	120, 378
Lime-kiln	G. Harper	Jenkerstown, Pa	Jan. 22, 1817	....
Lime kiln	G. Hensler	Kankakee, Ill	Oct. 26, 1869	96, 109
Lime-kiln	D. Hills	Richville, N. Y	Aug. 30, 1870	106, 818
Lime-kiln	C. Hinkley	Williamsville, N. Y	June 25, 1867	66, 024
Lime-kiln	S. C. Hotchkiss	Sylvania, Ohio	July 23, 1867	66, 962
Lime-kiln	A. Jeffries	Allegheny County, Pa	Apr. 21, 1857	17, 098
Lime-kiln	D. Keeports	Logansport, Ind	Dec. 24, 1872	134, 292
Lime-kiln	T. A. Kirk	Kansas City, Mo	Mar. 7, 1871	112, 468
Lime-kiln	J. L. Livingston	Mount Carroll, Ill	Jan. 16, 1866	52, 061
Lime-kiln	J. L. Livingston	Mount Carroll, Ill	Oct. 29, 1867	70, 234
Lime-kiln	J. McAdams	Mount Pleasant, Pa	Sept. 10, 1872	131, 290
Lime-kiln	W. McCoy	Fannettsburgh, Pa	Nov. 5, 1850	7, 757
Lime-kiln	J. McGregor	Selma, Ala	July 28, 1857	17, 885
Lime-kiln	J. Q. Merriam and A. J. Dietrick.	Fort Scott, Kans	Oct. 24, 1871	120, 307
Lime-kiln	J. L. Messer	Troy, N. Y	Sept. 16, 1873	142, 805
Lime-kiln	L. Montgomery	Newstead, N. Y	Aug. 13, 1867	67, 667
Lime-kiln	L. Montgomery	Akron, N. Y	Nov. 11, 1873	144, 555
Lime-kiln	R. Neisch	New York, N. Y	Aug. 15, 1854	11, 529
Lime-kiln	J. Newkirk	Factoryville, N. Y	Dec. 1, 1857	18, 764
Lime-kiln	E. Orcutt	Bennington, Vt	Sept. 26, 1848	5, 808
Lime-kiln	C. D. Page	Rochester, N. Y	Feb. 7, 1854	10, 510
Lime-kiln	C. D. Page	Rochester, N. Y	July 28, 1857	17, 889
Lime-kiln	C. D. Page	Rochester, N. Y	Dec. 7, 1858	22, 239
Lime-kiln	C. D. Page	Grand Rapids, Mich	Apr. 26, 1864	42, 501
Lime-kiln	C. D. Page	Rochester, N. Y	June 19, 1866	55, 699
Lime-kiln	C. D. Page	Rochester, N. Y	Sept. 8, 1868	82, 023
Lime-kiln	W. C. Pettijohn	Saint Louis, Mo	Sept. 15, 1868	82, 242
Lime-kiln	H. Petton	Erie, Pa	May 22, 1866	54, 943
Lime-kiln	D. H. Pequea	Lancaster, Pa	Aug. 28, 1855	13, 494
Lime-kiln	L. Phleger	Philadelphia, Pa	July 14, 1857	17, 807
Lime-kiln	W. W. Potts	Bridgeport, Pa	Feb. 4, 1868	73, 998
Lime-kiln	T. Power		July 12, 1804	....
Lime-kiln	T. Power	Hudson, N. Y	Jan. 7, 1817	....
Lime-kiln	T. H. Powers	Sodus, N. Y	Apr. 17, 1833	....

Index of patents issued from the United States Patent Office from 1790 *to* 1873, *inclusive*—Continued.

Invention.	Inventor.	Residence.	Date.	No.
Lime-kiln	F. W. Raeder	Saint Louis, Mo	Jan. 21, 1873	135, 045
Lime-kiln	E. Randall	Mason City, Iowa	July 30, 1872	129, 985
Lime-kiln	S. H. Robinson	Baltimore, Md	Jan. 16, 1855	12, 242
Lime-kiln	W. Robinson	Baltimore, Md	Apr. 14, 1857	17, 056
Lime-kiln	J. Sands	Sands' Mills, N. Y	Mar. 11, 1856	14, 411
Lime-kiln	R. E. Schroeder	Rochester, N. Y	May 6, 1851	8, 079
Lime-kiln	S. I. Seely	New York, N. Y	May 17, 1853	9, 736
Lime-kiln	F. Shelly	Alton, Ill	Jan. 18, 1870	99, 012
Lime-kiln	R. A. Smith	Atchison, Kans	Mar. 22, 1870	101, 173
Lime-kiln	J. B. Speed	Louisville, Ky	Nov. 23, 1869	97, 243
Lime-kiln	D. Stephens	Elmira, N. Y	Aug. 11, 1857	17, 986
Lime-kiln	W. P. Stuart	Hannibal, Mo	July 16, 1872	129, 617
Lime-kiln	J. Swope	Lancaster, Pa	Mar. 14, 1817	....
Lime-kiln	D. H. Turbett	New Bloomfield, Pa	July 18, 1871	117, 222
Lime-kiln	A. H. Tyson	Baltimore, Md	Sept. 28, 1839	1, 343
Lime-kiln	M. Verhoeven	Rochester, N. Y	Apr. 7, 1868	76, 563
Lime-kiln	A. B. Weeks	Rockland, Me	Mar. 2, 1858	19, 525
Lime-kiln	G. W. White	Greensburgh, Ind	Jan. 22, 1867	61, 495
Lime-kiln	B. Zwart	Keokuk, Iowa	Apr. 20, 1858	20, 015
Lime-kiln arch	A. McBride	Lowell, Mich	Nov. 21, 1871	121, 183
Lime-kiln blast-attachment	W. Hollenbaugh	New Germantown, Pa	Oct. 24, 1871	120, 278
Lime-kiln chimney	W. Rennyson	Morristown, Pa	Feb. 14, 1871	111, 874
Lime-kiln dome	D. Wentzel	New Bloomfield, Pa	Dec. 31, 1872	134, 501
Lime-kiln for anthracite coal	S. Griscom	Reading, Pa	June 26, 1832	....
Lime-kiln, Perpetual	W. Gorsuch	Baltimore, Md	Aug. 17, 1811	....
Lime-kiln, Perpetual	A. Jeans	Mill Creek Hundred, Del	Feb. 15, 1827	....
Lime-kilns, Construction of	J. H. Bower	Walnut, Pa	Sept. 4, 1874	5, 270
Lime-kilns, Mode of regulating draft of	I. Richardson	Paoli, Pa	Feb. 21, 1840	1, 496
Lime, Manufacture and application of bisulphate of.	W. Marr	New York, N. Y	Nov. 5, 1867	70, 588
Lime, Manufacture and use of neutral sulphate of.	E. N. Horsford	Cambridge, Mass	Sept. 15, 1863	39, 922
Lime, Manufacture of acid sulphite of	P. L. Bernard and J. Albrecht	New Orleans, La	Oct. 9, 1855	13, 632
Lime, Manufacture of superphosphate of	A. De Figanière	Philadelphia, Pa	Nov. 27, 1866	59, 978
Lime, Manufacture of superphosphate of	C. Morfit	Sudbrook Park, England	Jan. 31, 1871	111, 370
Lime, Manufacture of superphosphate of	B. Tanner	New Brighton, England	Feb. 13, 1872	123, 744
Lime, Manufacture of superphosphate of	B. Tanner	New Brighton, England	July 1, 1873	140, 559
Lime, Manufacturing marl into	J. Peasely		Feb. 1, 1803	....
Lime, Mode of manufacturing superphosphate of	E. P. Baugh	Philadelphia, Pa	May 9, 1865	47, 610
Lime on land, Spreading	M. L. Wilson	Quakertown, Pa	June 22, 1832	....
Lime-oven	A. Califf	Danville, Ill	Apr. 25, 1871	114, 104
Lime, plaster, &c., spreading and sowing machine	L. Rice	West Chester, Pa	Aug. 31, 1837	372
Lime, Preparing tannate of	O. Rich	Cambridge, Mass	Dec. 9, 1856	16, 189
Lime, Reducing calcareous matter into	C. Varle		May 2, 1806	....
Lime-spreader	W. C. Burnett	Burns' Mills, Pa	June 28, 1870	104, 694
Lime-spreader	H. Ludington and S. R. Supton	Addison, Pa	Mar. 13, 1855	12, 508
Limestone, Hardening magnesian	J. McMurty	Mount Sterling, Ky	June 21, 1870	104, 619
Limestone or asphalt, Machine for cutting or screening bituminous.	Q. A. Gillmore	New York, N. Y	Aug. 30, 1859	25, 254
Linch-pin	T. Laly	Philadelphia, Pa	Sept. 20, 1864	44, 321
Linch-pin	G. Wright	Washington, D. C	May 9, 1865	47, 674
Linch-pin and washer	W. Mansfield and H. S. Thistle	South Braintree, Mass., and Washington, D. C.	May 8, 1847	5, 109
Linch-pin, Securing	C. M. Risley	Woodbury, N. J	Aug. 29, 1865	49, 651
Line and strap holder	F. W. Jay	Canton, Ohio	Apr. 2, 1872	125, 136
Line and wire tightener	T. Crosby	Manchester, Iowa	Nov. 1, 1870	108, 766
Line-clamp	W. A. Ford	Greensburgh, Ind	Mar. 26, 1872	125, 038
Line-connector	C. Jarvis	New York, N. Y	Dec. 19, 1871	122, 021
Line-fastener	W. P. W. Dana	Newport, R. I	Feb. 11, 1868	74, 203
Line-holder	J. Bryant	Akron, Ind	Apr. 6, 1869	88, 606
Line-holder	S. J. Clark	Detroit, Mich	Dec. 3, 1867	71, 702
Line-holder	C. S. Hutchins	Canton, Conn	June 15, 1869	91, 340
Line-holder	D. W. C. McMaster	Southborough, Mass	Oct. 13, 1868	83, 075
Line-holder	W. Morse	Boston, Mass	Aug. 27, 1867	68, 304
Line-holder	D. Peters and R. F. Williams	Keokuk, Iowa	Oct. 8, 1867	69, 696
Line-holder	J. D. Sater	Greensburgh, Ind	Nov. 14, 1871	121, 006
Line-holder	J. H. Zinn	Harrisburgh, Pa	June 23, 1868	79, 288
Line-holder and wardrobe-hook, Combined	C. Goddard	Alliance, Ohio	Aug. 22, 1871	118, 361
Line holder for mason-work, &c	L. A. Green	Rocky Hill, Conn	Mar. 11, 1862	34, 621
Line holder, Mackerel	E. L. Decker	Southport, Me	Mar. 7, 1871	112, 336
Line-loop	D. McMillan	Osage, Iowa	Aug. 12, 1873	141, 804
Line-machine, Self-winding	W. A. Coventry	Paterson, N. J	Sept. 21, 1869	95, 002
Line-reel	L. Martin	Indianapolis, Ind	Jan. 7, 1868	73, 019
Lines, Device for rounding	L. B. Gates	Bane Centre, N. Y	Sept. 17, 1867	68, 972
Linen bosoms, Apparatus for folding	H. Beebe	New Haven, Conn	Feb. 27, 1866	52, 817
Linen, &c., Compound for removing mildew from	M. Faurot	Scranton, Pa	Feb. 16, 1869	87, 036
Linen, &c., Glazing and polishing	W. Smith	New York, N. Y	May 9, 1805	....
Linen, Machine for smoothing	W. P. Smith and J. Odell	Durham, N. H	July 17, 1810	....
Linen-smother	H. Rodd	Chestnut Hill, Pa	Apr. 9, 1861	32, 001
Liniment	M. P. Munder	Baltimore, Md	Sept. 2, 1873	142, 499
Liniment	W. H. Burr	Monterey, Ky	May 24, 1870	103, 411
Liniment	J. M. Barrett	Plymouth, N. C	Aug. 31, 1869	94, 379
Liniment	P. Baumann, jr	New Athens, Ill	Aug. 13, 1867	67, 706
Liniment	J. Benda	Yatton, Iowa	Apr. 29, 1873	138, 313
Liniment	J. C. Branch and H. P. Quinn	Washington, Ga	Nov. 8, 1870	108, 963
Liniment	J. W. Burnham	Winterport, Me	Feb. 19, 1867	62, 111
Liniment	E. N. Carpenter	Elkhart, Ind	May 28, 1867	65, 170
Liniment	E. Conway	Dayton, Ohio	Mar. 17, 1863	37, 901
Liniment	L. B. Cram	Weathersfield, Vt	Apr. 23, 1867	63, 995
Liniment	A. J. Creel	Hopkinton, Iowa	May 19, 1868	78, 068
Liniment	J. C. Dustan	Mount Vernon, N. J	May 14, 1867	64, 649
Liniment	D. Edwards	Saint Anthony, Minn	Feb. 20, 1866	52, 697
Liniment	E. C. Evens	Forest Hill, Ind	Nov. 23, 1869	97, 180
Liniment	H. Fedder	Lancaster, N. Y	Sept. 15, 1868	82, 212
Liniment	J. Gifford	Smithport, Pa	Feb. 26, 1867	62, 403
Liniment	D. J. Gish	Hopkinsville, Ky	July 8, 1873	140, 698
Liniment	J. E. Gould	Jamestown, N. Y	Aug. 16, 1864	43, 847
Liniment	J. R. Grows	Brunswick, Me	July 5, 1864	43, 403
Liniment	W. P. Hamlin	Exira, Iowa	May 12, 1868	77, 814

Index of patents issued from the United States Patent Office from 1790 *to* 1873, *inclusive*—Continued.

Invention.	Inventor.	Residence.	Date.	No.
Liniment	O. P. Hare	Petersburgh, Va	June 14, 1870	104, 146
Liniment	H. J. Kintz	Greece, N. Y	Jan. 14, 1873	134, 808
Liniment	J. Knights	Biddeford, Me	Aug. 20, 1872	130, 644
Liniment	J. S. Lightner	Westford, Wis	Mar. 26, 1867	63, 269
Liniment	J. D. Love	Harrisburgh, Oreg	Oct. 18, 1870	108, 371
Liniment	B. Marstella	Wolf Creek, Va	May 9, 1865	47, 653
Liniment	J. H. M. Morris	Reading, Ill	June 13, 1865	48, 198
Liniment	R. M. Moyle	South Manchester, Conn	June 5, 1866	55, 342
Liniment	P. R. Myers	Oxford, Ohio	Mar. 26, 1872	124, 910
Liniment	R. Newcomb	South Brooklyn, N. Y	June 11, 1867	65, 688
Liniment	T. B. Randell	New York, N. Y	Sept. 19, 1871	119, 179
Liniment	W. E. Rose	Hornellsville, N. Y	Apr. 8, 1873	137, 624
Liniment	G. W. Smith	North Whitehall Township, Pa.	Mar. 7, 1865	46, 715
Liniment	H. Stonebraker	Baltimore, Md	Feb. 19, 1867	62, 298
Liniment	W. H. Wagoner	Hurd Post-Office, Pa	May 28, 1872	127, 283
Liniment	W. H. Hallack	Corunna, Ind	July 27, 1869	93, 143
Liniment	W. W. Wells	Freehold, N. J	Apr. 13, 1869	88, 826
Liniment	G. S. Wood	Vassalborough, Me	Feb. 28, 1871	112, 201
Liniment	J. W. Woodring	Greensburgh, Ky	Feb. 20, 1866	52, 778
Liniment and medicine	J. King	Warren County, Ohio	Oct. 12, 1869	95, 695
Liniment, Anti-rheumatic	J. Galette	New York, N. Y	July 9, 1867	66, 482
Liniment, Compound for	M. C. Ross	New York, N. Y	Nov. 17, 1863	40, 642
Liniment for burns, scalds, &c	L. Wolff	Chicago, Ill	Mar. 29, 1870	101, 340
Liniment for cure of foot-rot in sheep	W. H. Lawes	Somerville, N. J	Mar. 5, 1867	62, 644
Liniment for horses, &c	D. O'Halloran	New York, N. Y	June 22, 1869	91, 659
Liniment for horses and cattle	H. H. Hanmer	Nashville, Tenn	Nov. 15, 1870	109, 203
Liniment for horses and other animals	W. Christman	Philadelphia, Pa	Mar. 24, 1868	75, 729
Liniment for neuralgia, rheumatism, &c	B. Bissell	New London, N. Y	Nov. 30, 1869	97, 345
Liniment for treating neuralgia, &c	G. L. Fearis	Connersville, Ind	June 28, 1870	104, 721
Liniment for treating nervous diseases	J. F. Smith	Forest Hill, Cal	Jan. 11, 1870	98, 809
Liniment, Mammalial	S. Rueger	Kansas City, Mo	Dec. 1, 1868	84, 580
Liniment, Medical vegetable	T. L. Upton	Farmington, W. Va	Apr. 16, 1867	63, 965
Liniment or medical compound	G. H. Fairchild	Saint Louis, Mo	Sept. 5, 1871	118, 706
Liniment or medical compound	D. C. Randall	Niles City, Mich	Dec. 22, 1868	85, 242
Liniment, Rheumatic	L. E. Anderson	Saint Louis, Mo	Nov. 17, 1863	40, 668
Liniment, Rheumatic	A. Brown		Oct. 17, 1804	......
Liniment for rheumatism	J. M. Cantrell	Polk County, Oreg	Mar. 19, 1872	124, 718
Liniment for rheumatism	A. M. Dennen	Folsom City, Cal	Oct. 6, 1868	82, 696
Liniment for rheumatism	S. Galland	Jefferson City, Mo	Nov. 24, 1863	40, 686
Liniment for rheumatism	J. W. Helmes	Bainbridge, Ga	Sept. 19, 1871	119, 144
Liniment for rheumatism	A. A. Riddick	Franklin, Va	Dec. 20, 1870	110, 288
Liniment for rheumatism, &c	G. Shepherd	Philadelphia, Pa	Mar. 29, 1864	42, 115
Liniment for rheumatism	J. A. Schuetz	Saint Louis, Mo	Mar. 6, 1860	27, 389
Liniment, Surgical and flesh-wound	J. S. Stratton	Newfane, Vt	Apr. 9, 1872	125, 628
Link	G. W. Dyer	Crestline, Ohio	Mar. 12, 1872	124, 556
Link	A. Goodhart	Newville, Pa	May 26, 1868	78, 275
Link	J. H. McIntire	Crestline, Ohio	Dec. 12, 1871	121, 725
Link	R. C. Schenck, jr	Dayton, Ohio	Mar. 4, 1873	136, 550
Link, Adjustable	A. J. Dexter	North Foster, R. I	Aug. 2, 1870	105, 921
Link, Adjustable	J. Hinkley	Norwalk, Ohio	Dec. 23, 1862	37, 229
Link, Adjustable	F. Pfeiffer	Giesborough, D. C	Dec. 19, 1865	51, 661
Link, Adjustable	C. J. E. Thompson	Providence, R. I	Aug. 5, 1862	36, 118
Link and hook, Connecting	R. Creuzbaur	New York, N. Y	Apr. 23, 1867	63, 996
Link-boring apparatus	C. Kellogg	Detroit, Mich	Feb. 11, 1868	74, 376
Link, Connecting	R. Creuzbaur	Newark, N. J	July 2, 1867	66, 303
Link, Connecting	R. Creuzbaur	New York, N. Y	Oct. 29, 1867	70, 172
Link, Connecting	W. N. Pelton	New London, Conn	Apr. 7, 1868	76, 517
Link-motion engines, Reversing	T. C. Crowley	Parker Township, Pa	Feb. 4, 1873	135, 410
Link-motion for operating valves	W. Knowles	Boston, Mass	Apr. 12, 1864	42, 297
Link-motion for steam-engines	J. F. Allen	New York, N. Y	Apr. 29, 1862	35, 070
Link-motion for steam-engines	N. V. Stevens	Lowell, Mass	Feb. 20, 1866	52, 766
Link, Snap	C. W. Saladee	Newark, Ohio	Sept. 26, 1865	50, 167
Link, Snap	C. W. Saladee	Newark, Ohio	Nov. 21, 1865	51, 088
Link-welding die	F. Van Patten and O. A. Anthony.	Ilion, N. Y	June 12, 1866	55, 570
Linseed, &c., Apparatus for triturating and heating	T. Rowe	Brooklyn, N. Y	Mar. 21, 1865	56, 945
Linseed, &c., Machine for crushing	T. Rowe	New York, N. Y	Oct. 28, 1862	36, 799
Lint and basket holder	W. Richardson	Baltimore, Md	Mar. 31, 1868	76, 248
Lint for manufacture, Preparing back	A. K. Smede	Lexington, Ky	Oct. 11, 1828	......
Lint, Manufacture of	R. D. Dwyer	New York, N. Y	Aug. 2, 1859	24, 970
Lint-room and cotton-press, Combined	I. H. Albertson	Bedias, Tex	Dec. 20, 1870	110, 327
Lip-shield	E. C. Philbrick	Bath, Mo	Nov. 16, 1869	96, 953
Liquid and fluid meter, and in the method of operating valves.	J. C. Horton and S. K. Hawkins	New York and Lansingburgh, N. Y.	Sept. 10, 1867	68, 746
Liquid and gas meter	J. Mason	Paterson, N. J	Sept. 17, 1867	69, 007
Liquid and gas meter	D. B. Spooner	Syracuse, N. Y	Apr. 4, 1871	113, 591
Liquid and gas meter	D. B. Spooner	Syracuse, N. Y	Oct. 22, 1872	132, 497
Liquid and spirit meter	E. S. Hutchinson	Baltimore, Md	Aug. 27, 1867	68, 081
Liquid and spirit meter	W. Murphy	Cork, Ireland	July 3, 1866	56, 156
Liquid-can	A. V. Smith	San Francisco, Cal	July 2, 1872	128, 668
Liquid-concentrating apparatus	C. A. Wood	Dorchester, Mass	Apr. 4, 1865	47, 168
Liquid-cooler	J. L. Baudelot	Harancourt, France	Nov. 1, 1859	25, 992
Liquid-cooler	J. W. Campbell, sr	New York, N. Y	May 16, 1871	114, 916
Liquid-cooler	J. W. Campbell, sr	New York, N. Y	Apr. 2, 1872	125, 268
Liquid-cooler	W. A. Colston	Great Bend, Pa	Dec. 15, 1868	84, 994
Liquid-cooler	A. Fischer	New York, N. Y	May 15, 1866	54, 708
Liquid-cooler	H. C. Fowler	Great Bend, Pa	Mar. 31, 1868	76, 181
Liquid-cooler	M. Gould	New York, N. Y	Oct. 15, 1867	69, 797
Liquid-cooler	C. Greenlee and W. H. Redfield	Belvidere, Ill	May 18, 1869	90, 094
Liquid-cooler	E. and M. A. F. Haas	Mendota, Ill	Aug. 11, 1868	80, 945
Liquid-cooler	V. Hall	New York, N. Y	Sept. 20, 1859	25, 505
Liquid-cooler	P. and F. Kinkel	New York, N. Y	July 25, 1865	48, 941
Liquid-cooler	B. G. Martin	Philadelphia, Pa	Mar. 6, 1866	53, 082
Liquid-cooler	H. Meidinger	Carlsruhe, Germany	Mar. 4, 1873	136, 445
Liquid-cooler	R. Morton	Stockton-on-Tees, England	Mar. 3, 1868	75, 044
Liquid-cooler	J. R. Neil	Walcottville, Conn	July 1, 1873	140, 529
Liquid-cooler	H. Pietsch	New York, N. Y	Dec. 22, 1868	85, 125

Index of patents issued from the United States Patent Office from 1790 *to* 1873, *inclusive*—Continued.

Invention.	Inventor.	Residence.	Date.	No.
Liquid-cooler	P. Reutlinger and A. Herbert	Williamsburgh, N. Y	Apr. 17, 1866	54, 016
Liquid-cooler	J. R. B. Schwarze	Washington, D. C	June 26, 1866	55, 962
Liquid-cooler	A. Smith	New York, N. Y	Dec. 22, 1863	41, 024
Liquid-cooler	E. L. Strohecker and A. Iverson, jr.	Macon, Ga	Aug. 6, 1872	130, 331
Liquid-cooler	C. P. Zimmerman	Newark, N. J	Nov. 28, 1865	51, 275
Liquid cooler and filter	D. E. Somes	Washington, D. C	Sept. 29, 1868	82, 651
Liquid-cooler and ice-water stand	G. Giebrich	Burlington, Iowa	Dec. 17, 1872	133, 979
Liquid-cooler for liquids under pressure	E. Bigelow	Springfield, Mass	Oct. 18, 1870	108, 318
Liquid cooling and aerating apparatus	F. Haeck	Brussels, Belgium	Nov. 8, 1864	45, 004
Liquid-cooling coil	W. Gee	New York, N. Y	Oct. 14, 1873	143, 689
Liquid-cooling vessel	H. Waldstein and M. Fausky	New York, N. Y	Nov. 13, 1866	59, 687
Liquid-decanting apparatus	T. Molinier	New Orleans, La	Oct. 19, 1869	95, 924
Liquid-defecating apparatus	J. E. Wilson	Baltimore, Md	June 17, 1862	35, 645
Liquid-diffusing apparatus	C. H. Hudson	Roxbury, Mass	Apr. 28, 1868	77, 286
Liquid drawing and measuring apparatus	R. F. Stevens	Syracuse, N. Y	Mar. 25, 1851	7, 999
Liquid drawing and weighing apparatus	E. S. Harris and S. S. Robinson	Morrison, Ill	June 25, 1867	66, 149
Liquid-drawing apparatus	A. Bain	New York, N. Y	Nov. 15, 1864	45, 013
Liquid-drawing apparatus, Effervescent	T. Warker	New York, N. Y	Nov. 18, 1873	144, 809
Liquid-effervescing bottle	H. Codd	Camberwell, England	Apr. 29, 1873	138, 230
Liquid-elevator	R. Porter	Melrose, Mass	Sept. 3, 1861	33, 226
Liquid-elevator, Pneumatic	J. P. Gruber	New York, N. Y	June 14, 1870	104, 301
Liquid, Evaporating	C. Sholes	Phelps, N. J	July 26, 1826	....
Liquid evaporating and decomposing apparatus	H. Eckstein	Vienna, Austria	Oct. 5, 1869	95, 445
Liquid-evaporating apparatus and process	J. J. Sherman	Albany, N. Y	Feb. 16, 1869	86, 948
Liquid-evaporator	E. Constantine	New York, N. Y	May 8, 1860	28, 158
Liquid-evaporator	J. P. Dake	Salem, Ohio	July 21, 1868	80, 149
Liquid-evaporator	L. C. England	Philadelphia, Pa	May 28, 1867	65, 065
Liquid-evaporator	G. F. Gray	Brooklyn, N. Y	May 31, 1870	103, 602
Liquid-evaporator	H. Hathaway and B. Lathrop	Detroit, Mich., and Tolland, Conn.	Jan. 1, 1861	31, 018
Liquid-evaporator	J. Howarth	Salem, Mass	May 2, 1871	114, 295
Liquid-evaporator	J. J. Johnston	Allegheny City, Pa	Nov. 14, 1865	50, 935
Liquid-evaporator	T. Oxnard	Marseilles, France	May 17, 1864	42, 789
Liquid-evaporator	A. Peek	Manchester, England	Jan. 5, 1864	41, 089
Liquid-evaporator	G. Stephenson	Zionsville, Ind	May 8, 1860	28, 211
Liquid-evaporator	J. Trageser	New York, N. Y	June 11, 1861	32, 542
Liquid-evaporator, Air-blast	G. Clark	Buffalo, N. Y	June 6, 1871	115, 572
Liquid evaporator and concentrator	J. Howarth	Salem, Mass	Mar. 7, 1871	112, 347
Liquid evaporator and concentrator	J. Howarth	Salem, Mass	Mar. 7, 1871	112, 348
Liquid-evaporator for boiling food for stock	A. J. Andrews	Linden, Mich	July 18, 1871	117, 030
Liquid-flask	H. W. Dee	London, England	May 30, 1871	115, 283
Liquid, Forcing	J. Van Norman	Easton, Pa	Mar. 28, 1871	113, 115
Liquid-forcing apparatus	M. E. Ogden	New York, N. Y	Nov. 14, 1871	120, 895
Liquid-freezer	F. P. E. Carré	Paris, France	Oct. 2, 1860	30, 201
Liquid-freezer	S. S. Fitch	New York, N. Y	Oct. 7, 1873	143, 446
Liquid-freezer	F. Sagno	Milan, Italy	May 13, 1873	138, 941
Liquid-gage	J. J. Greenough	Cincinnati, Ohio	Oct. 19, 1858	21, 814
Liquid gage and hydrometer, Combined	E. T. Bussell and J. Smith	New York, N. Y	June 7, 1870	103, 976
Liquid, gas, &c., meter	S. R. P. Camp	Syracuse, N. Y	Sept. 2, 1873	142, 336
Liquids heater and evaporator, Steam	B. T. Babbitt	New York, N. Y	Mar. 12, 1872	124, 530
Liquid heater, Portable	W. N. Hancock	Salem, N. J	Mar. 7, 1865	46, 665
Liquid-heating apparatus	D. H. Burrell	Little Falls, N. Y	Feb. 28, 1871	112, 120
Liquid-heating boiler	W. Nims	Fort Ann, N. Y	Oct. 24, 1846	4, 826
Liquid-heating machine	S. Bolton	Philadelphia, Pa	Jan. 16, 1809	....
Liquid-heating machine	R. Volentine	Canaan, N. Y	June 13, 1812	....
Liquid-measuring apparatus	E. K. Dutton	Manchester, England	Nov. 20, 1866	59, 761
Liquid-meter	R. H. Atwell	Baltimore, Md	Sept. 17, 1867	68, 830
Liquid-meter	R. H. Atwell	Baltimore, Md	Mar. 24, 1868	75, 721
Liquid-meter	P. Ball and B. Fitts	Worcester, Mass	Nov. 22, 1870	109, 372
Liquid-meter	J. S. Barden	Providence, R. I	Apr. 2, 1872	125, 114
Liquid-meter	N. L. Blanchard	Spuyten Duyvil, N. Y	May 3, 1870	102, 650
Liquid-meter	J. A. Bradshaw and W. H. Brown.	Lowell, Mass	Sept. 28, 1869	95, 189
Liquid-meter	F. Buttner	New York, N. Y	Mar. 28, 1871	113, 017
Liquid-meter	H. Chandler	Buffalo, N. Y	Sept. 19, 1871	119, 076
Liquid-meter	H. Chandler	Buffalo, N. Y	May 14, 1872	126, 623
Liquid-meter	W. F. Class and W. Napp	Cleveland, Ohio	Jan. 11, 1870	98, 667
Liquid-meter	I. Cook	Saint Louis, Mo	June 4, 1872	127, 575
Liquid-meter	T. Cook	New York, N. Y	May 2, 1871	114, 411
Liquid-meter	J. C. Cooke	Middletown, Conn	Nov. 5, 1861	33, 675
Liquid-meter	J. W. Crenim	New York, N. Y	Dec. 26, 1871	122, 231
Liquid-meter	R. Creuzbaur	Brooklyn, E. D., N. Y	June 7, 1870	103, 989
Liquid-meter	R. Creuzbaur	Brooklyn, E. D., N. Y	June 7, 1870	103, 990
Liquid-meter	T. A. Curtis	Springfield, Mass	Jan. 16, 1872	122, 759
Liquid-meter	T. A. Curtis and J. S. Curtis	Springfield, Mass., and Hartford, Conn.	Dec. 2, 1873	145, 096
Liquid-meter	J. F. De Navarro and H. C. Sergeant.	New York, N. Y., and Newark, N. J.	May 2, 1871	114, 420
Liquid-meter	J. F. De Navarro and H. C. Sergeant.	New York, N. Y	Apr. 16, 1872	125, 794
Liquid-meter	G. W. Devoe	New York, N. Y	Feb. 9, 1869	86, 823
Liquid-meter	E. M. Du Boys	Paris, France	Oct. 6, 1868	82, 812
Liquid-meter	N. Finck	Elizabeth, N. J	Oct. 10, 1871	119, 699
Liquid-meter	W. Fischer	Essen, Prussia	Sept. 27, 1870	107, 769
Liquid-meter	W. H. Fruen	Boston, Mass	Sept. 20, 1870	107, 607
Liquid-meter	C. A. and G. W. Geissenhainer	Pittsburgh, Pa	Sept. 15, 1868	82, 105
Liquid-meter	O. Gilmore	Raynham, Mass	Mar. 16, 1869	87, 837
Liquid-meter	T. C. Hargrave	Boston, Mass	Mar. 22, 1870	101, 125
Liquid-meter	D. W. Huntington and W. A. Hempstead.	South Coventry, Conn	Dec. 26, 1871	122, 174
Liquid-meter	O. G. Leopold	Cincinnati, Ohio	Apr. 7, 1857	16, 983
Liquid-meter	J. Mason	Paterson, N. J	Sept. 15, 1868	82, 138
Liquid-meter	G. B. Massey	New York, N. Y	Feb. 23, 1869	87, 183
Liquid-meter	G. B. Massey	New York, N. Y	Nov. 1, 1870	108, 804
Liquid-meter	H. W. Mather	New York, N. Y	Sept. 12, 1871	118, 867

Index of patents issued from the United States Patent Office from 1790 *to* 1873, *inclusive*—Continued.

Invention.	Inventor.	Residence.	Date.	No.
Liquid-meter	H. S. Maxim	Brooklyn, N. Y	June 11, 1872	127, 907
Liquid-meter	J. Mead	Charlestown, Mass	July 18, 1871	117, 094
Liquid-meter	J. Minor and N. W. and G. W. Nesmith.	Peoria and Metamora, Ill	Mar. 16, 1869	87, 959
Liquid-meter	C. Moore	New York, N. Y	Dec. 13, 1870	110, 062
Liquid-meter	C. Moore	New York, N. Y	Jan. 24, 1871	111, 234
Liquid-meter	F. A. Norley	Syracuse, N. Y	Apr. 8, 1873	137, 559
Liquid-meter	H. B. Nickerson	Boston, Mass	July 11, 1871	116, 859
Liquid-meter	C. Nida	Greenville, N. J	Aug. 31, 1869	93, 235
Liquid-meter	A. K. Rider	Elizabeth City, N. J	Jan. 4, 1870	98, 634
Liquid-meter	C. H. Riggs	Warwick, N. Y	Aug. 25, 1868	81, 408
Liquid-meter	W. Robjohn	New York, N. Y	July 2, 1867	66, 395
Liquid-meter	T. A. Searle	Providence, R. I	Mar. 2, 1869	87, 516
Liquid-meter	H. C. Sergeant	Newark, N. J	Aug. 30, 1870	106, 878
Liquid-meter	H. C. Sergeant	Newark, N. J	Jan. 24, 1871	111, 262
Liquid-meter	H. C. Sergeant	Newark, N. J	May 2, 1871	114, 480
Liquid-meter	H. C. Sergeant	New York, N. Y	Apr. 16, 1872	125, 848
Liquid-meter	J. H. Shedd	Waltham, Mass	Mar. 23, 1869	88, 259
Liquid-meter	G. Sickels	Boston, Mass	Sept. 24, 1872	131, 633
Liquid-meter	J. P. Smith	Cleveland, Ohio	June 15, 1869	91, 374
Liquid-meter	J. P. Smith	Buffalo, N. Y	July 12, 1870	105, 379
Liquid-meter	W. E. Snediker	New York, N. Y	Nov. 1, 1870	108, 842
Liquid-meter	W. E. Snediker	New York, N. Y	Jan. 24, 1871	111, 268
Liquid-meter	W. E. Snediker	New York, N. Y	Feb. 6, 1872	123, 521
Liquid-meter	D. B. Spooner	Syracuse, N. Y	Nov. 21, 1871	121, 211
Liquid-meter	W. M. Storm	New York, N. Y	May 7, 1867	64, 457
Liquid-meter	W. G. Stuart	Chicopee, Mass	Feb. 21, 1871	112, 091
Liquid-meter	J. Sutherland	Brooklyn, N. Y	June 2, 1868	78, 551
Liquid-meter	I. P. Tice	New York, N. Y	Apr. 2, 1872	125, 152
Liquid-meter	S. C. Treat	Tabor, Iowa	May 10, 1870	102, 883
Liquid-meter	F. Wagner	New York, N. Y	Jan. 25, 1870	99, 266
Liquid-meter	A. Werckmeister	Berlin, Prussia	Nov. 22, 1870	109, 564
Liquid-meter	A. M. White	Bridgeport, Conn	Mar. 11, 1873	136, 799
Liquid-meter	D. Williamson	New York, N. Y	July 18, 1871	117, 229
Liquid-meter	J. Winsborrow	Dalston, England	Nov. 16, 1869	96, 858
Liquid-meter	R. W. Young	Richmond, Va	Aug. 20, 1867	68, 025
Liquid-meter, Piston	R. Creuzbaur	Brooklyn, N. Y	Dec. 28, 1869	98, 353
Liquid-meters, Valve arrangement for	J. Johnson	New York, N. Y	Sept. 26, 1871	119, 364
Liquid purifying and rectifying apparatus	T. H. Sinclaire	New York, N. Y	Apr. 27, 1869	89, 353
Liquid receptacle and funnel	W. H. Mumber	Boston, Mass	Apr. 18, 1871	113, 786
Liquid-removing press	J. Waters	West Sutton, Mass	Oct. 25, 1870	108, 657
Liquid-sampler	A. Barbarin	New Orleans, La	Oct. 13, 1868	83, 024
Liquid-separator	J. J. Miller	Chicago, Ill	Aug. 23, 1864	43, 951
Liquid-still separator	G. W. Shaw	Buffalo, N. Y	Feb. 22, 1870	100, 196
Liquids and other substances, Mode of cooling and freezing.	A. H. Tait	New York, N. Y	Aug. 31, 1869	94, 450
Liquids, Boiler for evaporating	J. Moffat		Feb. 1, 1803	
Liquids by compressed air, Device for drawing	A. L. Webster	New York, N. Y	Aug. 2, 1870	106, 008
Liquids by steam, Heating	J. B. Armstrong	Wilkes County, Ga	May 29, 1833	
Liquids by the expansion and condensation of gas, Apparatus for forcing.	J. S. Baldwin	Newark, N. J	Dec. 5, 1871	121, 482
Liquids, Can for holding	L. Jennings	Erving, Mass	Oct. 10, 1854	11, 789
Liquids, Conveying	T. Donelly and J. H. and A. Anchors.	Pittsburgh and Bear Creek Station, Pa.	June 14, 1870	104, 284
Liquids, Decalcifying	F. Garcia	New Orleans, La	Mar. 5, 1861	31, 650
Liquids, Device for mixing	G. Watkins	Brooklyn, N. Y	Sept. 17, 1867	69, 054
Liquids, Evaporating	W. Henry	La Porte, Ind	Jan. 8, 1839	1, 061
Liquids for aiding digestion, Process for preparing	J. J. Sherman	Albany, N. Y	May 8, 1855	12, 834
Liquids for heating by means of pipes, &c	L. Grimm and J. Corvin	Magdeburg, Prussia	Apr. 22, 1873	138, 082
Liquids from boiling over the sides of vessels, Device for preventing.	J. Liblong	Waterbury, Conn	Aug. 26, 1856	15, 633
Liquids from paints and other solid substances, Machine for separating.	D. M. Weston	Boston, Mass	Feb. 4, 1868	74, 021
Liquids from solids, Method of expressing	D. A. James	Cincinnati, Ohio	Apr. 28, 1868	77, 194
Liquids on draft, Apparatus for cooling	J. Link	United States Army	Dec. 15, 1868	84, 888
Liquids on draft, Apparatus for cooling	E. Whyte	Philadelphia, Pa	May 28, 1867	65, 320
Liquids on draft, Apparatus for cooling and dispensing.	B. M. Bastian	New York, N. Y	Oct. 22, 1872	132, 430
Liquids, Pneumatic device for forcing	C. F. Bowman and S. Slyker	Wilkesbarre, Pa	June 22, 1869	91, 707
Liquids under pressure, Regulator and controller for the flow of.	J. S. Baldwin	Newark, N. J	May 4, 1869	89, 617
Liquids, Vessel for holding	J. Stimpson	Baltimore, Md	Oct. 17, 1854	11, 819
Liquids with carbonic acid, Apparatus for charging	J. Matthews	New York, N. Y	Aug. 5, 1873	141, 571
Liquids with nitric acid, Apparatus for treatment of	C. W. Volney	Boston, Mass	Apr. 9, 1872	125, 635
Liquor, Apparatus for improving	C. Walters and I. Rudigier	Pittsburgh, Pa	May 13, 1873	138, 776
Liquor-cooler	J. W. Collier	New York, N. Y	May 6, 1873	138, 478
Liquor-cooler	T. H. Miller	Allentown, Pa	Aug. 7, 1866	56, 978
Liquor-cooler	D. Lownes		July 21, 1803	
Liquor-cooler	E. Wright	Philadelphia, Pa	Aug. 1, 1871	117, 712
Liquor-discharge, Automatic	A. C. Rowe	Chicago, Ill	May 9, 1871	114, 607
Liquor drawing and measuring apparatus	J. H. Hecker	Hinkletown, Pa	Dec. 19, 1848	5, 970
Liquor-drawing apparatus, Compressed-air	J. Johnson	Genesee, N. Y	May 30, 1837	217
Liquor-flask	R. Heneage	Buffalo, N. Y	Jan. 14, 1864	43, 154
Liquor-forcing apparatus	J. F. Bennett	Pittsburgh, Pa	Apr. 23, 1872	126, 010
Liquor-fount	J. Douglass	New York	June 25, 1821	
Liquor-gaging device	A. and A. S. Wolcott	East Bloomfield, N. Y	Mar. 3, 1857	16, 766
Liquor-gate	O. H. and C. H. Bush	Fall River, Mass	Mar. 30, 1843	3, 002
Liquor on tap, Device to prevent injury to	J. G. Cullmann	Cincinnati, Ohio	June 8, 1869	91, 092
Liquor purifying and ageing apparatus	A. J. Gibson	Cincinnati, Ohio	Dec. 22, 1868	85, 225
Liquor-ripening process	W. M. Storm	New York, N. Y	Aug. 7, 1866	57, 009
Liquor-thief	J. F. Collins	New York, N. Y	June 22, 1869	91, 522
Liquor thief and measure	S. C. Catlin	Cleveland, Ohio	Dec. 20, 1870	110, 202
Liquor thief and test	N. Stafford	Brooklyn, N. Y	Jan. 2, 1866	51, 879
Liquors and for distillation, Boiling	J. Morris	New Haven, Conn	Mar. 3, 1813	
Liquors, Apparatus for treating and ageing spirituous.	G. Goewey	Philadelphia, Pa	June 29, 1869	91, 840
Liquors, Compound for fining, purifying, and mellowing spirituous.	W. Thompson	Dublin, Ireland	Feb. 27, 1866	52, 943

Index of patents issued from the United States Patent Office from 1790 *to* 1873, *inclusive*—Continued.

Invention.	Inventor.	Residence.	Date.	No.
Liquors, Cooling and discharging fermented	F. Brunon	Philadelphia, Pa	Dec. 15, 1863	40, 977
Liquors, Drawing and corking sparkling	S. A. Bille	New York	Nov. 8, 1828	....
Liquors, Elevating	J. A. Smith	New London, Conn	Aug. 7, 1826	....
Liquors, Extracting specimen of	J. M. Na_lee	Philadelphia, Pa	July 24, 1866	56, 596
Liquors, Method of determining the strength of	W. Cornell	Brooklyn, N. Y	Aug. 20, 1827	....
Liquors on draft, Cooler for	J. Gateley	Philadelphia, Pa	Aug. 20, 1867	67, 869
Liquors, Refining spirituous	G. W. Ferris	Quincy, Ill	July 10, 1866	56, 197
Litharge, paint, &c., Machine for separating light from heavy particles of.	W. Atwood	Cape Elizabeth, Me	July 14, 1868	79, 802
Lithographic and other impressions, Mode of preparing cloth for receiving.	C. Volkmar, jr	New York, N. Y	Nov. 12, 1867	70, 920
Lithographic copy from photographic negatives, Mode of producing.	E. Rye	Copenhagen, Denmark	Nov. 22, 1870	109, 551
Lithographic power-press	B. Ackermann	New York, N. Y	July 2, 1861	32, 672
Lithographic power-press	E. Reynolds	Mansfield, Conn	Aug. 9, 1864	43, 796
Lithographic press	G. Gooper	New York, N. Y	June 16, 1868	78, 930
Lithographic press	J. Crawley	Brooklyn, N. Y	July 6, 1869	92, 276
Lithographic press	J. Donlevy	New York, N. Y	Sept. 11, 1847	5, 283
Lithographic press	G. Dunlop	Brooklyn, N. Y	Apr. 2, 1867	63, 489
Lithographic press	A. Hoen	Baltimore, Md	May 11, 1869	89, 997
Lithographic press	B. Huber	Boston, Mass	Dec. 9, 1873	145, 420
Lithographic press	T. Hunter	Philadelphia, Pa	Feb. 14, 1871	111, 749
Lithographic press	C. C. Maurice	New York, N. Y	Feb. 25, 1868	74, 840
Lithographic press	C. C. Maurice	New York, N. Y	June 27, 1871	116, 335
Lithographic press	C. C. Maurice	New York, N. Y	May 20, 1873	139, 171
Lithographic press	R. McNie	New York, N. Y	Mar. 26, 1861	31, 818
Lithographic press	J. Taggart	Roxbury, Mass	Sept. 8, 1863	39, 867
Lithographic press	J. Walton	Brooklyn, N. Y	July 20, 1869	92, 765
Lithographic press	J. Koehler	New York, N. Y	Mar. 20, 1866	53, 309
Lithographic printing	O. Dubois	Fall River, Mass	Apr. 4, 1871	113, 276
Lithographic printing	R. C. Manners	Boston, Mass	Feb. 15, 1833	....
Lithographic printing-form	I. Reynolds	Dayton, Ohio	Nov. 18, 1873	144, 796
Lithographic printing-press	E. S. Boynton	Brooklyn, N. Y	June 27, 1871	116, 406
Lithographic printing-press	A. H. Marinoni	Paris, France	Mar. 16, 1869	87, 950
Lithographic printing-press	E. Reynolds	Mansfield, Conn	Mar. 8, 1864	41, 862
Lithographic printing-press	E. Reynolds	Mansfield, Conn	Feb. 14, 1865	46, 390
Lithographic printing-press	G. H. Reynolds	New York, N. Y	Feb. 17, 1863	37, 727
Lithographic printing-press	J. W. H. Stübee	Boston, Mass	Mar. 29, 1864	42, 125
Lithographic printing-press	W. H. Stubbe	Boston, Mass	Jan. 4, 1859	22, 519
Lithographic prints to glass, Process of transferring.	O. P. Wolff	Cincinnati, Ohio	Dec. 9, 1873	145, 472
Lithographic stone	G. H. Moore and L. Bagger	Norwich, Conn., and Washington, D. C.	Aug. 19, 1873	141, 943
Lithographic stone, Mounting	G. H. Reynolds	New York, N. Y	Jan. 1, 1861	31, 033
Lithographic transferred print	O. P. Wolff	Cincinnati, Ohio	Dec. 16, 1873	145, 606
Lithography	D. Henderson	Jersey City, N. J	Sept. 17, 1824	....
Lithontriptor	I. Lukens	Philadelphia, Pa	Dec. 30, 1826	....
Live spindle and flier	W. MacLardy and J. Lewis	Manchester, England	July 3, 1849	6, 572
Live spindles, Mode of steadying	A. Anderson	Paterson, N. J	Nov. 21, 1845	4, 281
Loader, Car	D. F. Halliburton	Rutherford Station, Tenn	July 16, 1872	129, 549
Loader for locomotive-tenders	A. D. Smith	Grafton, Ohio	May 19, 1868	78, 022
Loader for locomotive-tenders	A. H. Spencer	Providence, R. I	Feb. 8, 1870	99, 728
Loading and dumping machine	M. Phillips	Springfield, Ill	July 25, 1871	117, 322
Loading and dumping machine, Portable	W. D. Dorsey	Decatur, Ill	Dec. 20, 1870	110, 216
Loading and unloading vessels, Method of	G. A. Des Corats	Paris, France	July 23, 1867	67, 031
Loading and unloading vessels, Portable machine for.	J. E. Walsh	New York, N. Y	Oct. 8, 1872	131, 982
Loading and unloading wagons	R. Thelfall	Centreville, Cal	Dec. 24, 1872	134, 330
Loading apparatus, Cart	J. Byerly	Frederick, Md	July 15, 1873	140, 884
Loading apparatus, Cart	B. G. Fitzhugh	Frederick, Md	Aug. 6, 1872	130, 121
Loading apparatus for railway-cars, Wood	J. Nicholson	Monticello, Ind	Aug. 9, 1870	106, 274
Loading apparatus, Stone	H. S. Perkins	Rutland, Vt	Mar. 19, 1872	124, 852
Loading cars, Adjusting-rolls for	W. Dame	Woonsocket, R. I	Oct. 22, 1872	132, 449
Loading carts, wagons, &c	N. Osborn	Fairfield, Conn	May 23, 1826	....
Loading coal, Apparatus for	S. S. Roberts	Elizabeth, Pa	July 13, 1869	92, 476
Loading coal, Device for	J. F. Logan	Coal Bluff, Pa	Jan. 11, 1870	98, 782
Loading logs, stone, and hay, Wagon for	J. Sutherland	Morris, Ill	Feb. 11, 1868	74, 253
Loading lumber, Apparatus for	J. S. Preston	Winona, Minn	Mar. 5, 1872	124, 385
Loading of boats, &c., Ascertaining the weight of goods in.	T. Cohoon	Troy, N. Y	June 18, 1829	....
Loading vessels, Swinging railway for	J. L. Cheeseman	Gardiner, Me	Nov. 28, 1871	121, 333
Loading vessels, Winch and gin for	E. Allen	New York	Feb. 29, 1832	....
Loam-luting, Mode of revivifying	J. Chilcott	Brooklyn, N. Y	Oct. 3, 1865	50, 220

Lock:

See Alarm-lock.
Bag-lock.
Bank-lock.
Barrel-hoop lock.
Berth-lock.
Bottle-lock.
Brake-lock.
Burglar-alarm lock.
Can-cover lock.
Canal-lock.
Car-lock.
Car-seat lock.
Cell-lock.
Chain-lock.
Chronometric lock.
Combination lock.
Cross-bar lock.
Door-lock.
Drawer-lock.
Egyptian lock.
Faucet-lock.
Fire-arm lock.
Freight-car lock.
Fruit-box lock.
Furniture-lock.
Fuse-lock.
Gun-lock.
Gun and pistol lock.
Hand-cuff lock.
Hasp-lock.
Hoop-lock.
Indicator-lock.
Knob-lock.
Last-lock.
Magnetic lock.
Mail-bag lock.
Master-key lock.
Maze-lock.
Mortise-lock.
Mortise and latch lock.
Nut-lock.
Oar-lock.
Padlock.
Percussion-lock.
Permutation-lock.
Perplexing unpickable lock.

Index of patents issued from the United States Patent Office from 1790 to 1873, inclusive—Continued.

Invention.	Inventor.	Residence.	Date.	No.
Lock—Continued. *See* Piano-lock. Pin-lock. Pipe-coupling lock. Powder-proof lock. Prison-lock. Pump-lock. Railway-switch lock. Reversible lock. Ring-lock. Row-lock. Safe-lock. Sash-lock. Seal-lock. Sewing-machine lock. Ship-lock. Ship's pump lock. Sleigh-lock. Spool-holder lock. Spring-lock. Switch-lock. Till-lock. Till alarm-lock. Time-lock. Trace-lock. Trick-lock. Trunk-lock. Tumbler-lock. Umbrella-lock. Umbrella-stand lock. Valise-lock. Vault-cover lock. Vehicle-seat lock. Vehicle-wheel lock. Vest-chain lock. Vine-lock. Wagon-lock. Wagon-brake lock. Wagon-seat lock. Watch-chain lock.				
Lock	C. Ackerman	Newark, N. J	Sept. 21, 1858	21, 543
Lock	J. Adt	Waterbury, Conn	June 25, 1861	32, 605
Lock	J. Adt	Waterbury, Conn	Dec. 23, 1862	37, 209
Lock	W. H. Aikins	Berkshire, N. Y	May 13, 1856	14, 848
Lock	H. Allman	London, England	Jan. 30, 1866	52, 372
Lock	E. G. F. Arndt and A. Hühne	Rondout, N. Y	Nov. 11, 1862	36, 886
Lock	T. A. Auberlin	San Francisco, Cal	Apr. 17, 1866	53, 941
Lock	S. T. Bacon	Boston, Mass	July 12, 1859	24, 709
Lock	L. Baier	Pittsburgh, Pa	June 30, 1857	17, 714
Lock	L. Baier	Cincinnati, Ohio	Apr. 27, 1858	20, 027
Lock	H. H. Baker	New Market, N. J	Nov. 15, 1864	45, 014
Lock	M. O. Baker	New York, N. Y	May 31, 1870	103, 542
Lock	G. Bayer	New York, N. Y	Feb. 6, 1872	123, 444
Lock	L. Beer	New York, N. Y	Jan. 4, 1870	98, 545
Lock	E. S. Bennett	Brooklyn, N. Y	Feb. 18, 1862	34, 460
Lock	A. Betteley	Boston, Mass	Apr. 6, 1852	8, 851
Lock	O. Billings	La Grange, Ohio	May 17, 1859	24, 075
Lock	I. H. A. Bleckman	Ronsdor, Prussia	Nov. 20, 1855	13, 814
Lock	C. Bodmer	Fayetteville, Ill	Feb. 11, 1873	135, 702
Lock	W. Bohannan	New York, N. Y	May 29, 1866	55, 047
Lock	W. Bohannan and F. G. Johnson.	New York and Brooklyn, N. Y.	May 29, 1866	55, 048
Lock	T. Bowles	New York, N. Y	Dec. 18, 1855	13, 968
Lock	J. A. Braden	La Grange, Ga	June 8, 1858	20, 476
Lock	J. Brady	Branford, Conn	June 7, 1870	103, 838
Lock	J. Brady	Branford, Conn	Oct. 11, 1870	108, 098
Lock	J. Brady	Branford, Conn	Oct. 11, 1870	108, 099
Lock	J. B. Brennan	Mount Vernon, N. Y	Nov. 7, 1854	11, 885
Lock	E. W. Brettell	Newark, N. J	June 26, 1860	28, 828
Lock	E. W. Brettell	Newark, N. J	Mar. 8, 1864	41, 827
Lock	E. W. Brettell	Newark, N. J	Aug. 8, 1865	49, 223
Lock	B. V. M. Bronse	Kokomo, Ind	June 12, 1866	55, 460
Lock	C. F. Brown	Bridgeport, Conn	July 10, 1860	29, 120
Lock	H. and S. W. Budd	Philadelphia, Pa	June 26, 1866	55, 814
Lock	R. B. Burchell	Brooklyn, N. Y	Sept. 11, 1866	57, 857
Lock	I. D. Bush	Detroit, Mich	June 19, 1866	55, 616
Lock	W. C. Bussey	Jackson, Cal	Sept. 5, 1865	49, 712
Lock	W. H. Butler	New York, N. Y	Oct. 28, 1856	15, 962
Lock	A. Clabaugh	Altoona, Pa	July 7, 1863	39, 117
Lock	C. Claude	Newark, N. J	Oct. 15, 1861	33, 476
Lock	C. Claude	New York, N. Y	May 1, 1866	54, 297
Lock	G. Clay	New York, N. Y	Apr. 12, 1859	23, 662
Lock	J. M. Cook	Hinsdale, N. Y	June 2, 1857	17, 424
Lock	G. W. Coppernoll	Ohio, N. Y	Sept. 30, 1856	15, 800
Lock	W. Corbett and W. Burns	Saint Louis, Mo	June 5, 1866	55, 244
Lock	H. W. Covert	Rochester, N. Y	Sept. 15, 1857	18, 228
Lock	G. W. Dana	Durand, Ill	Dec. 27, 1859	26, 569
Lock	V. R. David	Newark, Ill	Mar. 31, 1857	16, 908
Lock	H. and J. Day	New York, N. Y	Apr. 4, 1820	......
Lock	L. P. Decker	Williamsburgh, N. Y	Jan. 8, 1867	61, 055
Lock	A. E. Deitz	Brooklyn, N. Y	Nov. 28, 1865	51, 152
Lock	W. Denney	Philadelphia, Pa	Mar. 16, 1858	19, 628
Lock	L. Derby	New York, N. Y	May 8, 1860	28, 162
Lock	C. C. Dickerman	Boston, Mass	Mar. 29, 1870	101, 355
Lock	C. C. Dickerman	Boston, Mass	Nov. 1, 1870	108, 770
Lock	W. Dickson	Albany, N. Y	Mar. 26, 1867	63, 230
Lock	L. Diss	Ilion, N. Y	Apr. 28, 1857	17, 139
Lock	L. Diss	Oriskany, N. Y	Jan. 3, 1860	26, 659
Lock	W. B. Dodds	Cincinnati, Ohio	Mar. 14, 1865	46, 858
Lock	W. B. Dodds	Cincinnati, Ohio	Apr. 17, 1866	53, 959
Lock	T. Dougherty	Macon, Ga	June 7, 1859	24, 287
Lock	M. Ducharme	Cohoes, N. Y	May 14, 1861	32, 334
Lock	C. Duckworth	Hartford, Conn	Nov. 1, 1859	25, 959
Lock	J. Ebling	New York, N. Y	May 4, 1869	89, 746
Lock	R. S. Eddy and A. O. Miles	Nashua, N. H	Dec. 17, 1861	33, 931
Lock	G. and G. F. Elliott	Manchester, Conn	Feb. 21, 1860	27, 262
Lock	H. H. Elwell	South Norwalk, Conn	Oct. 14, 1862	36, 645
Lock	H. H. Elwell	South Norwalk, Conn	Jan. 13, 1863	37, 390
Lock	H. H. Elwell	South Norwalk, Conn	July 21, 1863	39, 289
Lock	H. H. Elwell	South Norwalk, Conn	Feb. 9, 1864	41, 573
Lock	J. Euteneuer	Peoria, Ill	Sept. 19, 1865	49, 991
Lock	M. Erb and F. C. Goffin	Newark, N. J	June 17, 1856	15, 124
Lock	J. Farrel	New York, N. Y	Nov. 18, 1862	36, 946
Lock	E. Fay	Washington, D. C	Jan. 4, 1870	98, 577
Lock	P. S. Felter	Cincinnatus, N. Y	Jan. 12, 1864	41, 211
Lock	P. S. Felter	Cincinnatus, N. Y	Nov. 1, 1864	44, 859
Lock	P. S. Felter	Cincinnatus, N. Y	July 17, 1866	56, 394

Index of patents issued from the United States Patent Office from 1790 *to* 1873, *inclusive*—Continued.

Invention.	Inventor.	Residence.	Date.	No.
Lock	R. S. Foster	Sing Sing, N. Y	June 23, 1863	39, 002
Lock	R. S. Foster	Sing Sing, N. Y	June 23, 1863	39, 003
Lock	R. S. Foster	Sing Sing, N. Y	June 23, 1863	39, 004
Lock	R. S. Foster	Sing Sing, N. Y	June 23, 1863	39, 005
Lock	R. S. Foster	Sing Sing, N. Y	June 23, 1863	39, 006
Lock	R. S. Foster	Sing Sing, N. Y	June 23, 1863	39, 007
Lock	R. S. Foster	Sing Sing, N. Y	June 23, 1863	39, 008
Lock	R. S. Foster	Sing Sing, N. Y	June 23, 1863	39, 009
Lock	R. S. Foster	Sing Sing, N. Y	June 23, 1863	39, 010
Lock	R. S. Foster	Sing Sing, N. Y	Sept. 22, 1863	40, 077
Lock	J. B. Frost	Jackson, Miss	Feb. 27, 1872	124, 125
Lock	F. Garachon	New York, N. Y	June 29, 1852	9, 070
Lock	J. Gerard	Trenton, N. J	June 7, 1870	103, 868
Lock	C. T. Gibson	Baltimore, Md	Aug. 1, 1865	49, 101
Lock	F. C. Goffin	New York, N. Y	Oct. 26, 1852	9, 361
Lock	F. Gorris	Cleveland, Ohio	Sept. 2, 1873	142, 335
Lock	F. Gould	Huntington, N. Y	Aug. 17, 1858	21, 193
Lock	W. H. Greenwood	Galva, Ill	Aug. 14, 1860	29, 588
Lock	J. T. Guthrie	Leesburgh, Ohio	Sept. 4, 1866	57, 704
Lock	J. L. Hall	Cincinnati, Ohio	Sept. 22, 1857	18, 243
Lock	J. L. Hall	Cincinnati, Ohio	Sept. 25, 1860	30, 140
Lock	J. L. Hall	Cincinnati, Ohio	May 5, 1863	38, 384
Lock	W. Hall	Brookline, Mass	Jan. 6, 1863	37, 290
Lock	W. Hall	Brookline, Mass	May 23, 1865	47, 817
Lock	G. Hopson	Bridgeport, Conn	July 12, 1864	43, 501
Lock	W. W. Harder	New York, N. Y	Aug. 28, 1866	57, 503
Lock	A. Hardy and F. L. and G. A. Walker.	Boston, Mass	Mar. 6, 1866	52, 995
Lock	T. G. Harold	Brooklyn, N. Y	Aug. 5, 1862	36, 124
Lock	T. G. Harold	Brooklyn, N. Y	Nov. 24, 1863	40, 724
Lock	T. G. Harold	Brooklyn, N. Y	Dec. 8, 1863	40, 882
Lock	T. G. Harold	Brooklyn, N. Y	Jan. 5, 1864	41, 177
Lock	T. G. Harold	Brooklyn, N. Y	Jan. 5, 1864	41, 178
Lock	H. Hartwig	New York, N. Y	Apr. 30, 1861	32, 225
Lock	H. Hartwig	New York, N. Y	Apr. 30, 1861	32, 226
Lock	S. Hiatt	Indianapolis, Ind	Dec. 28, 1858	22, 425
Lock	A. M. Hill	Branford, Conn	July 29, 1862	36, 053
Lock	A. M. Hill	Philadelphia, Pa	Mar. 31, 1863	38, 080
Lock	J. J. Hirshbuhl	Louisville, Ky	May 14, 1861	32, 289
Lock	A. Hoagland	Jersey City, N. J	Mar. 9, 1858	19, 564
Lock	E. D. Hodgson	London, England	Nov. 7, 1871	120, 743
Lock	L. G. Hoffman	Albany, N. Y	Apr 24, 1866	54, 162
Lock	R. G. Holmes and W. H. Butler	New York, N. Y	Oct. 30, 1855	13, 722
Lock	E. R. Hopkins	New York, N. Y	Aug. 28, 1866	57, 508
Lock	D. W. G. Humphrey	Gray, Me	Nov. 6, 1855	13, 753
Lock	J. Hutson	Janesville, Wis	Dec. 27, 1864	45, 611
Lock	A. Inglis	Indianapolis, Ind	Nov. 23, 1869	97, 196
Lock	H. Isham	New Britain, Conn	July 1, 1856	15, 239
Lock	H. Isham	New Britain, Conn	July 7, 1857	17, 740
Lock	H. Isham	New Britain, Conn	Feb. 28, 1860	27, 294
Lock	H. Jackson	Brooklyn, N. Y	Feb. 7, 1865	46, 241
Lock	C. Jagy	New York, N. Y	May 27, 1862	35, 376
Lock	C. Jagy and F. Denzler	New York, N. Y	Mar. 1, 1864	41, 817
Lock	C. F. and C. F., jr., and W. W. Johnson.	Owego, N. Y	Sept. 11, 1860	29, 975
Lock	F. G. Johnson	Brooklyn, N. Y	Feb. 5, 1861	31, 320
Lock	W. Jones	Brooklyn, N. Y	Nov. 1, 1864	44, 869
Lock	H. W. Kahlke	Brooklyn, N. Y	May 17, 1864	42, 778
Lock	J. Keath	Ash Grove, Ill	Dec. 17, 1872	133, 991
Lock	W. M. Keefer	Chicago, Ill	June 21, 1870	104, 598
Lock	E. Kershaw	Boston, Mass	Dec. 4, 1855	13, 880
Lock	R. Ketchum	Seneca Castle, N. Y	Dec. 7, 1852	9, 451
Lock	W. S. Kirkham	Brantford, Conn	Nov. 6, 1860	30, 580
Lock	E. Lawshé	Atlanta, Ga	Nov. 13, 1866	59, 615
Lock	L. Layman	Westfield, N. Y	Nov. 6, 1860	30, 606
Lock	W. C. Leach and M. J. Knox	Knox Corners, N. Y	Aug. 19, 1862	36, 221
Lock	E. Leman, jr	Boston, Mass	Nov. 28, 1822	....
Lock	L. Lillie	Troy, N. Y	Aug. 8, 1865	49, 282
Lock	M. T. Lincoln	Washington, D. C	May 31, 1864	42, 954
Lock	J. M. Lippincott	Pittsburgh, Pa	May 13, 1856	14, 896
Lock	J. M. Lippincott	Pittsburgh, Pa	Aug. 5, 1856	15, 489
Lock	J. P. Lipps	Newark, N. J	July 6, 1858	20, 850
Lock	J. Loch and G. Bayer	New York, N. Y	Feb. 27, 1866	52, 864
Lock	S. N. Long	South Chatham, Mass	Dec. 9, 1862	37, 135
Lock	S. N. Long and M. E. Hathaway.	Chatham and Wareham, Mass.	June 30, 1863	39, 056
Lock	J. P. Lord	Manchester, N. H	Aug. 31, 1858	21, 346
Lock	W. Lorenz	Lebanon, Pa	Aug. 30, 1864	44, 007
Lock	J. W. Lyon	Brooklyn, N. Y	Aug. 19, 1862	36, 224
Lock	N. Macneale	Cincinnati, Ohio	July 5, 1864	43, 457
Lock	N. Macneale and W. B. Dodds	Cincinnati, Ohio	Nov. 11, 1862	36, 908
Lock	L. H. Magott	Springfield, Mass	Dec. 12, 1865	51, 467
Lock	W. K. Marvin	New York, N. Y	Sept. 13, 1864	44, 207
Lock	W. K. Marvin	New York, N. Y	May 23, 1865	47, 842
Lock	W. Maurer	New York, N. Y	Apr. 8, 1856	14, 616
Lock	F. McCollum	Rockville, Conn	Feb. 9, 1864	41, 518
Lock	W. C. McGill	Cincinnati, Ohio	May 1, 1866	54, 383
Lock	M. McGonnigle	Allegheny City, Pa	Jan. 23, 1866	52, 183
Lock	A. O. Miles	Nashua, N. H	Oct. 23, 1866	59, 048
Lock	J. W. Miller	Waynesborough, Pa	June 7, 1864	43, 078
Lock	J. C. Mix	Terryville, Conn	Jan. 27, 1863	37, 515
Lock	W. Moore	Brooklyn, N. Y	Nov. 23, 1858	22, 146
Lock	A. B. Mullett	Washington, D. C	Nov. 13, 1866	59, 628
Lock	L. F. Munger	Le Roy, N. Y	July 14, 1857	17, 804
Lock	L. F. Munger	Rochester, N. Y	Feb. 22, 1859	23, 040
Lock	L. F. Munger	Rochester, N. Y	Feb. 21, 1865	46, 531
Lock	J. T. Mygatt	Binghamton, N. Y	Feb 17, 1863	37, 700
Lock	R. Newill	New York	June 14, 1843	3, 135

Index of patents issued from the United States Patent Office from 1790 *to* 1873, *inclusive*—Continued.

Invention.	Inventor.	Residence.	Date.	No.
Lock	H. Oaks	Waynesborough, Pa	Apr. 18, 1865	47,325
Lock	J. M. Perkins	New York, N. Y	Mar. 2, 1858	19,533
Lock	J. M. Perkins	Cleveland, Ohio	Aug. 5, 1862	36,106
Lock	J. M. Perkins	Cleveland, Ohio	Dec. 23, 1862	37,240
Lock	S. Perry	Newport, N. Y	May 12, 1857	17,293
Lock	N. Petré	New York, N. Y	Oct. 5, 1869	95,509
Lock	N. Petré	New York, N. Y	Dec. 28, 1869	98,406
Lock	N. Petré	New York, N. Y	Feb. 1, 1870	99,344
Lock	S. Pettes	Bellows Falls, Vt	July 12, 1834	....
Lock	W. Pie	New York	Aug. 5, 1813	....
Lock	J. A. Pimentel and W. H. Shute	New York, N. Y	July 29, 1862	36,024
Lock	J. H. Pomeroy	Bloomington, Ill	Dec. 4, 1855	13,884
Lock	J. Post	Newark, N. J	Aug. 8, 1865	49,297
Lock	J. Post	Newark, N. J	Aug. 8, 1865	49,298
Lock	D. Powers	Philadelphia, Pa	Mar. 1, 1859	23,113
Lock	T. Powers	Philadelphia, Pa	Jan. 29, 1861	31,258
Lock	T. Powers	New York, N. Y	Sept. 6, 1870	107,100
Lock	D. A. Pratt	New York, N. Y	May 22, 1866	54,952
Lock	F. Prockert	New York, N. Y	Nov. 23, 1869	97,227
Lock	T. L. Pye	New York, N. Y	Sept. 28, 1858	21,636
Lock	P. B. Quimby	Belfast, Me	May 23, 1836	....
Lock	A. Rankin	Philadelphia, Pa	May 22, 1860	28,406
Lock	D. B. Read and J. H. Clapp	Providence, R. I	Mar. 8, 1870	100,556
Lock	E. S. Renwick	New York, N. Y	May 17, 1864	42,793
Lock	A. A. Richards	Urbana, Ohio	Feb. 15, 1859	22,980
Lock	H. D. Richardson	Florence, Mass	June 13, 1865	48,209
Lock	H. D. Richardson	Northampton, Mass	May 7, 1867	64,568
Lock	J. C. Riethmüller	Pittsburgh, Pa	Mar. 3, 1857	16,749
Lock	G. A. Rogers	Augusta, Me	Jan. 20, 1830	....
Lock	G. Rossner	Rochester, N. Y	Sept. 18, 1860	30,092
Lock	E. A. G. Roulstone	Roxbury, Mass	Dec. 4, 1866	60,257
Lock	F. Rudolph	New York, N. Y	July 25, 1865	48,980
Lock	H. D. Russell	Naugatuck, Conn	Sept. 9, 1856	15,708
Lock	J. Sargent	Rochester, N. Y	Aug 28, 1866	57,574
Lock	J. Sargent and H. W. Covert	Rochester, N. Y	May 2, 1865	47,575
Lock	J. Sargent and H. W. Covert	Rochester, N. Y	Jan. 9, 1866	51,973
Lock	Z. Sargent	Rochester, N. H	Feb. 22, 1870	100,195
Lock	W. Sellers	New York, N. Y	Oct. 16, 1866	58,898
Lock	F. G. Shalling	Somerset, Mass	Nov. 11, 1862	36,919
Lock	E. M. Shaw	Baltimore, Md	Apr. 6, 1858	19,879
Lock	H. L. Shepardson	Shelburne Falls, Mass	Jan. 18, 1870	99,013
Lock	G. A. Sherlock	Boston, Mass	July 31, 1866	56,808
Lock	J. P. Sherwood	Fort Edward, N. Y	Sept. 8, 1857	18,162
Lock	A. W. Snow	Norwich, Conn	Oct. 9, 1866	58,688
Lock	M. Stachelin and H. Young	Port Chester, N. Y	June 12, 1866	55,549
Lock	N. Stafford	Brooklyn, N. Y	Mar. 7, 1865	46,721
Lock	H. Stein	New York, N. Y	Dec. 9, 1873	145,370
Lock	T. Stewart	Philadelphia, Pa	Dec. 5, 1871	121,680
Lock	S. C. St. John	Edmeston, N. Y	Jan. 29, 1861	31,267
Lock	O. B. Thompson	Hudson, Ohio	Nov. 2, 1858	21,994
Lock	C. F. Toll	Boston, Mass	Aug. 9, 1864	43,806
Lock	C. F. Toll	Boston, Mass	Jan. 23, 1866	52,226
Lock	H. R. Towne	Stamford, Conn	Sept. 12, 1871	118,986
Lock	H. B. Tyler	Norwich, Conn	Aug. 15, 1865	49,481
Lock	B. M. Van Der Veer	Clyde, N. Y	Apr. 25, 1865	47,470
Lock	B. M. Van Der Veer	Clyde, N. Y	Dec. 26, 1865	51,763
Lock	E. Voigh and P. P. Mathes	Philadelphia, Pa	Apr. 5, 1870	101,684
Lock	R. Vollschwitz	New York, N. Y	Dec. 12, 1865	51,498
Lock	H. Underwood	Tolland, Conn	Sept. 27, 1864	44,475
Lock	C. P. Wagner	New York, N. Y	May 2, 1865	47,588
Lock	H. B. Weaver	Hartford, Conn	Nov. 2, 1869	96,369
Lock	R. L. Webb	West Meriden, Conn	Jan. 12, 1864	41,247
Lock	T. K. Webster	Lawrence, Mass	May 10, 1859	23,967
Lock	C. S. Westcott	New York, N. Y	Dec. 14, 1858	22,319
Lock	W. Whiting and H. Pickford	Roxbury and Boston, Mass	Mar. 24, 1857	16,892
Lock	W. Whiting and H. Pickford	Roxbury and Boston, Mass.	July 14, 1857	17,815
Lock	E. P. Whitney	Westfield, N. Y	Sept. 4, 1860	29,929
Lock	T. Wilbur	Woodbury, Conn	Dec. 23, 1834	....
Lock	H. Willard	Evansville, Ind	Dec. 20, 1870	110,412
Lock	H. Willard and P. Trunz	Evansville, Ind	Sept. 6, 1870	107,145
Lock	A. Williams and E. P. Cummings.	Philadelphia, Pa	May 5, 1857	17,245
Lock	C. Wilson	Springfield, Mass	Nov. 7, 1854	11,916
Lock	G. M. Wood	Decatur, Ill	Aug. 21, 1866	57,423
Lock	L. Woodruff	New Britain, Conn	July 15, 1862	35,912
Lock	T. B. Worrell and T. Walker	Philadelphia, Pa	Dec. 7, 1869	97,581
Lock	G. C. Worth	Upper Sandusky, Ohio	Sept. 23, 1862	36,547
Lock	W. E. Worthen	New York, N. Y	Apr. 10, 1860	27,854
Lock	H. Wynblad	New York, N. Y	Aug. 19, 1856	15,589
Lock	H. Wynblad	West Hoboken, N. J	Aug. 24, 1858	21,293
Lock	L. Yale, jr	Newport, N. Y	June 3, 1856	15,031
Lock	L. Yale, jr	Philadelphia, Pa	Nov. 9, 1858	22,048
Lock	L. Yale, jr	Philadelphia, Pa	June 12, 1860	28,710
Lock	L. Yale, jr	Philadelphia, Pa	Jan. 29, 1861	31,278
Lock	L. Yale, jr	Philadelphia, Pa	May 14, 1861	32,330
Lock	L. Yale, jr	Philadelphia, Pa	May 14, 1861	32,331
Lock	L. Yale, jr	Shelburne Falls, Mass	June 27, 1865	48,475
Lock	L. Yale, jr	Bernardston, Mass	Oct. 12, 1869	95,865
Lock and bolt	E. Fink and C. E. McDonald	Indianapolis, Ind	July 14, 1863	39,223
Lock and bolt	H. C. Howells	Putnam, Ohio	Feb. 10, 1834	....
Lock and detector	J. H. Lyons	New York, N. Y	Sept. 13, 1859	25,428
Lock and escutcheon	R. Kinsley	Springfield, Mass	Sept. 19, 1848	5,791
Lock and key	J. R. Bugbee	Boston, Mass	Apr. 22, 1851	8,060
Lock and key	F. A. H. Gaebel	New York, N. Y	May 12, 1863	38,476
Lock and key	E. M. Hendrickson	Brooklyn, N. Y	May 27, 1856	14,958
Lock and key	J. Hoffacker	New York, N. Y	Nov. 2, 1858	21,962
Lock and key	H. B. Weaver	Hartford, Conn	Jan. 1, 1867	60,974
Lock and key	W. M. Williams	England	Dec. 1, 1840	1,874

Index of patents issued from the United States Patent Office from 1790 *to* 1873, *inclusive*—Continued.

Invention.	Inventor.	Residence.	Date.	No.
Lock and key	W. Wilson	Northampton, Mass	Jan. 27, 1843	2, 931
Lock and key	L. Yale, jr	Newport, N. Y	May 6, 1851	8, 071
Lock and knob-latch	A. M. Hill	Branford, Conn	Jan. 11, 1861	32, 551
Lock and knob-latch	W. S. Kirkham	Branford, Conn	Jan. 11, 1861	32, 521
Lock and knob-latch combined	E. G. F. Arndt and C. E. L. Molbius.	New York, N. Y	Aug. 20, 1867	67, 938
Lock and label-sheath, Compound	S. W. Marsh	Washington, D. C	Apr. 3, 1860	27, 728
Lock and latch	J. Chittenden	Burlington, Vt	Aug. 12, 1820	....
Lock and latch	H. H. Elwell	South Norwalk, Conn	Feb. 23, 1864	41, 741
Lock and latch	W. S. Kirkham	Branford, Conn	Mar. 15, 1859	23, 254
Lock and latch	B. Mallory	New Haven, Conn	May 5, 1863	38, 400
Lock and latch	B. Mallory	New Haven, Conn	June 7, 1864	43, 035
Lock and latch	J. F. Thayer	Boston, Mass	May 18, 1822	....
Lock and latch, Combined	G. B. Allen	Norwalk, Conn	Mar. 15, 1870	100, 708
Lock and latch, Combined	G. W. Cilley	Norwich, Conn	May 11, 1869	89, 852
Lock and latch, Combined	R. Kinsley	Springfield, Mass	Apr. 16, 1845	4, 060
Lock and latch, Combined	F. M. Ranous	Yreka City, Cal	Dec. 13, 1870	110, 071
Lock and latch, Combined	S. C. Weddington	Jonesborough, Ind	Sept. 6, 1870	107, 207
Lock and latch, Combined	S. H. Wheeler	Dowagiac, Mich	Apr. 4, 1871	113, 377
Lock-attachment	P. P. Stephan	Newark, N. J	Feb. 5, 1861	31, 337
Lock-attachment	J. M. Wilson	Philadelphia, Pa	June 7, 1859	24, 346
Lock-bolt	T. Kromer	Neustadt, Baden	Sept. 19, 1871	119, 036
Lock-bolt	G. Washburn	New York, N. Y	July 14, 1868	79, 879
Lock-bolt spindle	J. Sargent	Rochester, N. Y	Jan. 4, 1870	98, 623
Lock-bolts, Cam for throwing	H. W. Covert	Rochester, N. Y	Oct. 5, 1858	21, 655
Lock-bolts, Method of operating	T. Slaight	Newark, N. J	Mar. 12, 1850	7, 176
Lock-case, Machine for riveting	T. W. Baxter and E. Brown	Chicago, Ill., and Green Point, N. Y.	June 25, 1872	128, 205
Lock catch or stopper	E. P. McCeney	Washington County, D. C	Jan. 14, 1868	73, 358
Lock, Combination	W. C. McGill	Cincinnati, Ohio	Mar. 11, 1873	136, 606
Lock-cutting machine	I. L. Penney	Minneapolis, Minn	May 13, 1873	138, 815
Lock-fastener	H. S. Wilcox	West Meriden, Conn	June 14, 1864	43, 147
Lock-gate wicket	J. Jack	Nunda, N. Y	May 21, 1850	7, 379
Lock-gates by water-power, Method of working	J. D. Willoughby	Susquehanna Post-Office, Pa.	Oct. 24, 1848	5, 885
Lock-gates, Manner of suspending, opening, and closing.	H. McCarty	Pittsburgh, Pa	Mar. 16, 1844	3, 493
Lock-guard	L. Schroder	Cincinnati, Ohio	June 21, 1859	24, 523
Lock-guard attachment	P. S. Felter	Cincinnatus, N. Y	Dec. 17, 1861	33, 933
Lock guard or tumbler	L. Kellogg	Charleston, Ohio	Aug. 15, 1848	5, 706
Lock-hasp	M. Newman, 2d	Oak Hill, N. Y	Apr. 15, 1856	14, 675
Lock-joints in sheet-metal, Machine for forming	J. Shaw	Hinckley, Ohio	Sept. 11, 1866	57, 980
Lock-keeper, Adjustable	T. Strobridge	Pittsburgh, Pa	July 31, 1866	56, 828
Lock-key	A. G. Batchelder	Lowell, Mass	Sept. 3, 1867	68, 348
Lock-key	P. A. Gladwin	Boston, Mass	June 12, 1866	55, 582
Lock-key	W. E. Hawkins	Wallingford, Conn	Mar. 5, 1872	124, 266
Lock-key	J. Koenig	Erie, Pa	Feb. 27, 1872	124, 066
Lock-key	W. Park	Norwich, Conn	Dec. 27, 1870	110, 491
Lock-key	E. Reynolds	Mansfield, Conn	Mar. 7, 1865	46, 704
Lock-key	F. W. Smith, jr	Bridgeport, Conn	Mar. 2, 1869	87, 518
Lock-key	A. A. Stuart	Cedar Rapids, Iowa	Dec. 24, 1872	134, 325
Lock-key	P. Van Antwerp	New York, N. Y	Oct. 25, 1859	25, 923
Lock-key	A. F. Whiting	Bath, Me	Feb. 6, 1872	123, 532
Lock-key fastener	J. B. Ayer	Malden, Mass	Sept. 12, 1865	51, 410
Lock-key, Device for fastening	J. R. Tempest	Philadelphia, Pa	July 4, 1865	48, 602
Lock-key guard	C. Leavitt	Cleveland, Ohio	Mar. 21, 1865	46, 914
Lock-key, Sheet-metal	T. Powers	New York, N. Y	July 18, 1871	117, 110
Lock label and seal, Combined	E. J. Brooks	New York, N. Y	June 10, 1873	139, 764
Lock-nut	L. Arnold and J. B. Atwood	Belchertown and Palmer, Mass.	July 23, 1872	129, 707
Lock-nut	A. G. Binns	Pittsburgh, Pa	Apr. 5, 1870	101, 421
Lock-nut	C. F. Brush	Cleveland, Ohio	Jan. 30, 1872	123, 148
Lock-nut	A. W. Bunnell	Linesville, Pa	Sept. 5, 1871	118, 684
Lock-nut	A. W. Bunnell	Linesville, Pa	Sept. 5, 1871	118, 685
Lock-nut	J. Dinsmore	Dinsmore, Pa	May 28, 1872	127, 329
Lock-nut	P. Dyer, jr., A. Parker, and W. B. Way.	Pontiac, Mich	Nov. 8, 1870	108, 957
Lock-nut	W. P. Ewing and I. S. De Ford.	Elkton, Md	Sept. 7, 1869	94, 585
Lock-nut	P. L. Gibbs	Dunleith, Ill	Oct. 26, 1869	96, 218
Lock-nut	S. Barber	Beaver, Pa	July 7, 1868	79, 649
Lock-nut	M. W. Griswold	New York, N. Y	Oct. 19, 1869	95, 897
Lock-nut	T. Hagan	Rochester, Pa	July 20, 1869	92, 721
Lock-nut	R. W. Hamilton	New York, N. Y	Aug. 8, 1871	117, 881
Lock-nut	W. Hamilton	Toronto, Canada	Dec. 29, 1868	85, 381
Lock-nut	J. Harvey	Painesville, Ohio	Aug. 22, 1871	118, 238
Lock-nut	G. G. Hermance	Claverack, N. Y	Feb. 1, 1870	99, 315
Lock-nut	J. W. Hilton	Bradford, Pa	Mar. 30, 1869	88, 385
Lock-nut	S. W. Kirk	Coatesville, Pa	Mar. 31, 1868	76, 201
Lock-nut	J. S. Kirkpatrick	Punta Arenas, Cal	Aug. 13, 1872	130, 377
Lock-nut	S. T. Lamb	New Albany, Ind	Nov. 21, 1871	121, 175
Lock-nut	S. T. Lamb	New Albany, Ind	Nov. 21, 1871	121, 176
Lock-nut	M. Langhorne	Washington, D. C	Feb. 15, 1870	99, 919
Lock-nut	E. T. Ligon	Demopolis, Ala	May 24, 1870	103, 348
Lock-nut	W. C. Mason	Beaver Falls, Pa	Oct. 12, 1869	95, 764
Lock-nut	J. Miller, jr	Marshalltown, Iowa	Nov. 1, 1870	108, 927
Lock-nut	A. T. Morris	Nevada, Ohio	Jan. 16, 1872	122, 845
Lock-nut	J. A. Morrison	Parker's Landing, Pa	Apr. 30, 1872	126, 223
Lock-nut	W. H. Nichols	East Hampton, Conn	Jan. 28, 1873	135, 242
Lock-nut	F. Oakley	Toronto, Canada	Aug. 17, 1869	93, 943
Lock-nut	G. Palmer	Littlestown, Pa	Feb. 2, 1869	86, 582
Lock-nut	J. A. Peabody and A. F. Champlin.	Westerly, R. I	May 7, 1872	126, 482
Lock-nut	J. L. Randolph	Berkeley Springs, W. Va	Dec. 27, 1870	110, 499
Lock-nut	A. Roff	Southport, Conn	Apr. 13, 1869	88, 908
Lock-nut	A. Roff	Southport, Conn	Aug. 3, 1869	93, [illegible]
Lock-nut	J. Rogers	Sterling, Ill	June 30, 1868	79, 397
Lock-nut	H. Rosamyer	Rochester, Pa	Aug. 13, 1869	93, 230
Lock-nut	G. P. Rose	Milwaukee, Wis	Mar. 5, 1872	124, 390

Index of patents issued from the United States Patent Office from 1790 *to* 1873, *inclusive*—Continued.

Index of patents issued from the United States Patent Office from 1790 *to* 1873, *inclusive*—Continued.

Invention.	Inventor.	Residence.	Date.	No.
Locomotive-axle bearings	J. P. Laird	Altoona, Pa	Mar. 6, 1866	53, 009
Locomotive-axle bearings	D. Matthew	Philadelphia, Pa	Dec. 28, 1858	22, 439
Locomotive-boiler	J. Briggs	Louisville, Ky	Oct. 27, 1863	40, 389
Locomotive-boiler	B. Crawford	Pittsburgh, Pa	July 7, 1863	39, 123
Locomotive-boiler	S. Crawford	New York, N. Y	Aug. 8, 1865	49, 236
Locomotive-boiler	J. J. Dutcher	New Haven, Conn	Apr. 10, 1855	12, 673
Locomotive-boiler	B. L. Griffith	Hazleton, Pa	July 26, 1859	24, 862
Locomotive-boiler	S. J. Hayes and E. S. Jeffrey	Chicago, Ill	July 11, 1871	116, 836
Locomotive-boiler	G. F. Johnson	Philadelphia, Pa	May 19, 1863	38, 584
Locomotive-boiler	J. E. McConnell	Wolverton, England	June 2, 1857	17, 436
Locomotive-boiler	J. Millholland	Reading, Pa	Jan. 19, 1864	41, 316
Locomotive-boiler	J. Thompson	East Boston, Mass	May 29, 1860	28, 520
Locomotive-boiler furnace	P. W. Mackenzie	Blauveltville, N. Y	Jan. 16, 1872	122, 841
Locomotive-boiler furnace	A. J. Stevens	San Francisco, Cal	Feb. 14, 1871	111, 884
Locomotive-boiler furnace	R. Winans	Baltimore, Md	Apr. 27, 1859	20, 117
Locomotive-boiler furnace	J. Wood, jr	Conshohocken, Pa	Jan. 23, 1872	123, 069
Locomotive-boilers, Attaching steam-gage to	J. L. Eastman	Boston, Mass	Nov. 10, 1857	18, 578
Locomotive-boilers, Fire-box of	R. Winans	Baltimore, Md	Apr. 27, 1859	20, 115
Locomotive-boilers, Process for using anthracite-coal dust or slack as fuel in.	A. Berney	Jersey City, N. J	Mar. 18, 1873	136, 960
Locomotive-brake	A. De Bergue	Paris, France	Mar. 24, 1868	75, 873
Locomotive, Carriage	M. W. Baldwin	Philadelphia, Pa	May 16, 1846	4, 530
Locomotive, Carriage	W. Heston	Philadelphia, Pa	Oct. 11, 1830	......
Locomotive, Carriage	W. Howard	Baltimore, Md	Dec. 10, 1828	......
Locomotive-carriage and rail therefor	E. Kimber	Kimberton, Pa	Mar. 5, 1831	......
Locomotive-carriage, Machinery of a	S. Olcott	Harsimus, N. J	Sept. 11, 1829	......
Locomotive carrying its own railway, Traction	C. F. Mann	Troy, N. Y	Nov. 22, 1859	26, 195
Locomotive center-bearing	B. W. Healey	Providence, R. I	June 8, 1869	91, 015
Locomotive cow-catcher	J. Mitchell	Osceola, Iowa	Oct. 6, 1857	18, 348
Locomotive cow-catcher	J. Noble	Pittsburgh, Pa	Apr. 21, 1868	76, 938
Locomotive cow-catcher	J. Wisner	Aurora, N. Y	Feb. 10, 1857	16, 623
Locomotive crank-pins from true center, Apparatus for determining deviation of.	C. J. Clifford	New Hampton, N. J	Jan. 14, 1868	73, 233
Locomotive-cross-head jack	J. S. Funk	Marysville, Pa	Mar. 1, 1870	100, 390
Locomotive cross-heads, Rest for	S. A. Alexander and E. Dunn	Sunbury, Pa	Aug. 31, 1869	94, 171
Locomotive draft-pipe	J. M. Foss	Concord, N. H	June 12, 1866	55, 479
Locomotive draft-pipe	A. Pearsall	Atlanta, Ga	Mar. 5, 1867	62, 680
Locomotive draw-bar	D. C. Canneil	La Fayette, Ind	Apr. 23, 1867	64, 071
Locomotive driving-box	J. E. Wootton	Philadelphia, Pa	Dec. 1, 1857	18, 786
Locomotive driving-power	W. D. Arnett	Denver, Colo	Feb. 7, 1871	111, 603
Locomotive driving-wheel	J. Birkenhead	Canton, Mass	July 28, 1868	80, 384
Locomotive driving-wheel	W. J. Johnson	New Orleans, La	Mar. 23, 1869	88, 178
Locomotive driving-wheels, Apparatus for distributing sand to produce adhesion of.	E. Tolles	New York	Oct. 9, 1841	2, 283
Locomotive driving-wheels, Apparatus for turning crank-pins on.	C. J. Clifford	New Hampton, N. J	Sept. 3, 1867	68, 415
Locomotive driving-wheels, Increasing adhesion of.	E. L. Miller	Charleston, S. C	June 19, 1834	......
Locomotive driving-wheels, Mode of increasing adhesion of.	J. L. Mott	New York	Aug. 28, 1841	2, 228
Locomotive-engine	H. W. Adams	Philadelphia, Pa	Aug. 20, 1872	130, 685
Locomotive-engine	W. S. G. Baker	Baltimore, Md	Nov. 6, 1866	59, 340
Locomotive-engine	M. W. Baldwin	Philadelphia, Pa	Dec. 31, 1840	1, 921
Locomotive-engine	F. Barker	Baltimore, Md	Sept. —, 1836	......
Locomotive-engine	C. W. Cahoon	Portland, Me	Aug. 31, 1869	94, 396
Locomotive-engine	H. R. Campbell	Philadelphia, Pa	Feb. 5, 1836	......
Locomotive-engine	J. Clark	London, England	Nov. 7, 1871	120, 713
Locomotive-engine	J. M. Coale	Baltimore, Md	May 21, 1867	64, 842
Locomotive engine	I. H. Congdon	Springfield, Ill	Aug. 23, 1864	43, 898
Locomotive-engine	E. Crane	Dorchester, Mass	Nov. 22, 1859	26, 165
Locomotive-engine	R. F. Fairlie	London, England	Jan. 16, 1866	52, 117
Locomotive-engine	R. F. Fairlie	Westminster, England	Oct. 3, 1871	119, 591
Locomotive-engine	A. J. Graham	Portland, Oreg	Sept. 4, 1860	29, 879
Locomotive-engine	H. Gray	Boston, Mass	May 5, 1857	17, 213
Locomotive-engine	W. S. Hudson	Paterson, N. J	Dec. 10, 1872	133, 859
Locomotive-engine	S. L. Langdon	New Orleans, La	Mar. 10, 1868	75, 280
Locomotive-engine	J. Lewis	Manchester, England	Nov. 26, 1872	133, 321
Locomotive-engine	Z. H. Mann and L. B. Thyng	Lowell, Mass	Mar. 10, 1838	628
Locomotive-engine	L. Rarchaert	Paris, France	May 21, 1867	65, 008
Locomotive-engine	H. R. Remsen and P. M. Hutton	Troy, N. Y	June 29, 1852	9, 078
Locomotive-engine	S. Samuels and W. J. Brassington.	Mott Haven and Brooklyn, N. Y.	June 18, 1867	65, 951
Locomotive-engine	C. and G. Sellers	Philadelphia, Pa	May 22, 1835	......
Locomotive-engine	S. Skillman	Jersey City, N. J	Feb. 6, 1872	123, 367
Locomotive-engine	S. Skillman	Jersey City, N. J	Aug. 27, 1872	130, 875
Locomotive-engine	R. Winans	Baltimore, Md	Apr. 6, 1858	19, 889
Locomotive-engine	R. Winans	Baltimore, Md	Apr. 13, 1858	19, 962
Locomotive-engine	R. Winans	Baltimore, Md	Aug. 24, 1858	21, 290
Locomotive-engine	R. Winans	Baltimore, Md	Jan. 11, 1859	22, 597
Locomotive-engine	D. Winder	Xenia, Ohio	May 10, 1853	9, 716
Locomotive-engine	S. Wright	Philadelphia, Pa	Dec. 26, 1837	540
Locomotive-engine alarm	R. A. Filkins	North Adams, Mass	Jan. 3, 1871	110, 643
Locomotive-engine and car, Combining	M. McDowell and N. W. Wheeler.	New York and Brooklyn, N. Y.	Jan. 31, 1860	27, 002
Locomotive-engine and railway-car	M. W. Baldwin	Philadelphia, Pa	Sept. 10, 1834	......
Locomotive-engine boiler	J. Penniman	Baltimore, Md	Apr. 24, 1840	1, 568
Locomotive-engine boiler	R. Winans	Baltimore, Md	Apr. 27, 1859	20, 116
Locomotive-engine boilers, Fire-box of	R. Winans	Baltimore, Md	Apr. 27, 1859	20, 114
Locomotive-engine, car, and carriage	A. C. Jones	Philadelphia, Pa	June 14, 1834	......
Locomotive-engine, Condensing double-end	J. P. Woodbury	West Roxbury, Mass	Feb. 21, 1871	112, 001
Locomotive-engine driving-wheel	A. F. Cooper	San Francisco, Cal	Sept. 19, 1871	119, 014
Locomotive-engine for ascending inclined planes	L. Burnell	Elyria, Ohio	June 26, 1834	......
Locomotive-engine for ascending inclined planes	D. H. Dotterer and T. Jackson	Reading, Pa	Mar. 25, 1839	1, 108
Locomotive-engine for ascending inclined planes	S. Marsh	West Roxbury, Mass	Sept. 10, 1861	33, 255
Locomotive-engine for inclined planes	J. Ruggles	Thomaston, Me	July 13, 1836	......
Locomotive-engine for inclined planes, Railway	E. F. Aldrich	New York	Oct. 12, 1837	426
Locomotive-engine for producing increased adhesion to the rails.	F. Windhausen	Duderstadt, Hanover	Dec. 29, 1857	19, 000
Locomotive-engine for railways, Constructing	H. Waterman	Hudson, N. Y	Feb. 10, 1841	1, 969

Index of patents issued from the United States Patent Office from 1790 *to* 1873, *inclusive*—Continued.

Invention.	Inventor.	Residence.	Date.	No.
Locomotive-engine furnace	R. Gill and G. W. Grier	Altoona, Pa	July 26, 1859	24, 867
Locomotive-engine furnace	D. W. Wyman	New York, N. Y	Apr. 20, 1869	89, 261
Locomotive-engine furnaces, Apparatus for increasing draft of.	J. Sweeney	Chicago, Ill	Aug. 6, 1861	33, 011
Locomotive-engine houses, Smoke-stack for	H. Clayton	Tamaqua, Pa	May 3, 1859	23, 825
Locomotive-engine lubricator	E. C. Hamlin	Pavilion, N. Y	June 23, 1863	38, 962
Locomotive-engine running-gear	S. Norris	Philadelphia, Pa	Sept. 26, 1854	11, 733
Locomotive-engine running-gear	J. L. Whetstone	Cincinnati, Ohio	Apr. 10, 1860	27, 850
Locomotive-engine smoke-stack	R. Frazer	Camden County, N. J	Sept. 30, 1873	143, 339
Locomotive-engine smoke-stack	L. Fulton	Cincinnati, Ohio	Apr. 13, 1869	88, 864
Locomotive-engine smoke-stack	C. P. Noble	Chicago, Ill	Aug. 11, 1863	39, 493
Locomotive-engine wheel	G. S. Griggs	Roxbury, Mass	Dec. 29, 1857	18, 966
Locomotive-engine wheel and boiler-tube	M. W. Baldwin	Philadelphia, Pa	Apr. 3, 1835	....
Locomotive-engine wheel and boiler-tube for railways.	M. W. Baldwin	Philadelphia, Pa	Aug. 17, 1835	....
Locomotive-engine wheel, axle, and railway	R. Berrian	New York	Nov. 13, 1832	....
Locomotive-engines and railway-cars, Making springs for.	I. Oberhausser	Charleston, S. C	Feb. 15, 1838	605
Locomotive-engines, Apparatus for reversing or stopping.	J. Cunningham	Reading, Pa	Oct. 22, 1850	7, 730
Locomotive-engines, Apparatus for supplying pure water to.	H. V. P. Draper and N. Covington.	Hannibal, Mo	Dec. 24, 1872	134, 263
Locomotive-engines, Arrangement for disposing of the sparks from.	L. P. Teed	White Deer Mills, Pa	Aug. 7, 1860	29, 531
Locomotive-engines, Arrangement of cylinders and their connection for.	A. Smethurst	Philadelphia, Pa	Nov. 24, 1857	18, 712
Locomotive-engines, Arrangement of exhaust-pipes in.	J. Williams	Dunkirk, N. Y	Mar. 6, 1855	12, 495
Locomotive-engines, Ash-pan for	G. Davis	Roxbury, Mass	Oct. 11, 1853	10, 104
Locomotive-engines, Boiler and water-heater of	T. Perkins	Baltimore, Md	June 26, 1849	6, 561
Locomotive-engines by hand-power, Moving	J. E. Wootten	Philadelphia, Pa	Nov. 29, 1859	26, 318
Locomotive-engines, Device for discharging the smoke, cinders, &c., from.	J. Greacen, jr	New York, N. Y	Sept. 4, 1860	29, 880
Locomotive-engines, Exhaust-device for	J. Densmore, sr	Erie, Pa	Jan. 5, 1869	85, 571
Locomotive-engines, Exhaust-pipe of	G. W. Lathrop	Weedsport, N. Y	Dec. 18, 1860	30, 976
Locomotive-engines, Fire-box for	L. Crossman and S. Atkinson	Elizabeth City, N. J	July 20, 1858	20, 937
Locomotive-engines, Fire-box of	R. Greenwood	Altoona, Pa	Aug. 9, 1859	25, 009
Locomotive-engines for removing objects from the track, Attachment to.	C. H. Eisenbrandt	Baltimore, Md	June 14, 1859	24, 383
Locomotive-engines for working heavy grades, Boiler and gearing of.	G. E. Sellers	Cincinnati, Ohio	July 9, 1850	7, 498
Locomotive-engines, Managing and supplying fire for generating steam in.	M. W. Baldwin	Philadelphia, Pa	Oct. 15, 1836	54
Locomotive-engines on railways, Braking	M. W. Verdin	Baltimore, Md	July 5, 1859	24, 680
Locomotive-engines on railways, Propelling	M. W. Verdin	Baltimore, Md	July 5, 1859	24, 679
Locomotive-engines, Sand-box for	T. Price and J. H. Wilson	Covington, Ky	Mar. 13, 1866	53, 182
Locomotive-engines, Signal-mechanism for	J. M. Cook	Taunton, Mass	Apr. 29, 1862	35, 082
Locomotive-engines, Spark-arrester and chimney of.	W. A. Peaslee and J. O. D. Lilly.	Indianapolis, Ind	Dec. 6, 1859	26, 373
Locomotive-engines, Superheating steam for	J. Martin	Toronto, Canada	Dec. 4, 1860	30, 857
Locomotive-engines with water, Supplying	A. D. Merritt and C. Kemplen	Woodstock, Ill	Dec. 4, 1860	30, 829
Locomotive-engines, Traction for	T. Selleck	Greenwich, Conn	Sept. 16, 1862	36, 486
Locomotive exhaust-pipe	A. S. Smith	Boston, Mass	Apr. 9, 1872	125, 414
Locomotive, Farm	D. F. Leach	Forsyth, Ill	Aug. 3, 1869	93, 316
Locomotive-fender with fuel, Apparatus for supplying.	A. C. Land	Garlandville, Miss	May 7, 1872	126, 466
Locomotive fire-box	C. Dean	Saint Catharine's, Canada	Dec. 1, 1863	40, 743
Locomotive fire-box	C. H. Farley	Portland, Me	Jan. 14, 1873	134, 741
Locomotive fire-box	J. P. Laird	Altoona, Pa	Nov. 17, 1863	40, 630
Locomotive fire-box	W. R. Thomas	Catasauqua, Pa	Mar. 15, 1859	23, 278
Locomotive fire-box	J. L. Vauclain	La Fayette, Ind	Aug. 13, 1861	33, 057
Locomotive fire-box	R. and T. Winans	Baltimore, Md	May 9, 1854	10, 901
Locomotive-fire-box crown	H. C. Darby	Wyandotte, Kans	June 24, 1873	140, 251
Locomotive fire-box, Removable	J. I. De Haven	Reading, Pa	Apr. 24, 1849	6, 376
Locomotive foot-board	G. F. Morse	Portland, Me	Apr. 26, 1870	102, 299
Locomotive for ascending and descending inclined planes.	E. Coleman	Philadelphia, Pa	Dec. 31, 1845	4, 342
Locomotive for ascending inclined planes	A. Cathcart	Madison, Iowa	Oct. 23, 1849	6, 818
Locomotive for ascending inclined planes	N. Riggenbach	Olten, Switzerland	Feb. 13, 1872	123, 729
Locomotive for consuming sparks	F. B. Longmire and H. J. Brooke.	Philadelphia, Pa	July 10, 1840	1, 681
Locomotive for plowing, &c	T. S. Minniss	Meadville, Pa	Jan. 15, 1867	61, 231
Locomotive, Friction	T. S. Minniss	Meadville, Pa	Sept. 27, 1870	107, 702
Locomotive-furnace	O. W. Bayley	Manchester, N. H	Apr. 17, 1855	12, 749
Locomotive-furnace	O. W. Bayley	Boston, Mass	Feb. 2, 1858	19, 277
Locomotive-furnace	T. Davis	Cleveland, Ohio	Dec. 31, 1872	134, 422
Locomotive-furnace	G. S. Griggs	Roxbury, Mass	Dec. 15, 1857	18, 882
Locomotive-furnace	A. J. Stevens	Sacramento, Cal	Dec. 23, 1873	145, 819
Locomotive guide-wheel	S. Norris	Philadelphia, Pa	Apr. 24, 1860	28, 007
Locomotive, Horse-power	J. C. Miller	Union Township, Ohio	Oct. 11, 1859	25, 752
Locomotive, Horse-power	J. R. Sleeper	Philadelphia, Pa	Oct. 30, 1834	....
Locomotive-links, Machine for slotting	W. H. Denny	Philadelphia, Pa	June 24, 1873	140, 121
Locomotive, Marine	S. Lee	New York, N. Y	May 16, 1871	114, 832
Locomotive-pilot	B. F. Partridge, jr	Columbus, Ky	May 28, 1867	65, 264
Locomotive, portable, and other steam-engines	L. Perkins	London, England	Dec. 16, 1873	145, 525
Locomotive, Portable traction	J. Barrans	Peckham, England	Nov. 8, 1859	26, 074
Locomotive-power machine for removing houses	S. Compton, jr	Elmira, N. Y	July 31, 1837	331
Locomotive recording-instrument	J. D. Richardson	Houston, Tex	Apr. 29, 1873	138, 437
Locomotive, Road	E. C. James	Baltimore, Md	Dec. 17, 1867	72, 397
Locomotive, Road, &c	J. Robingson	New Brighton, Pa	Sept. 30, 1856	15, 820
Locomotive running-gear	G. S. Griggs	Roxbury, Mass	June 17, 1851	8, 166
Locomotive running-gear	J. H. Murrill	Manchester, Va	Oct. 7, 1851	8, 410
Locomotive running-gear	T. H. Neal	Pittsburgh, Pa	June 30, 1863	39, 060
Locomotive running-gear	J. C. Story	Cincinnati, Ohio	Nov. 21, 1865	51, 098
Locomotive running-gear	R. Winans	Baltimore, Md	Dec. 2, 1851	8, 571
Locomotive sanding-device	M. V. Nobles	Elmira, N. Y	Apr. 1, 1873	137, 472
Locomotive-signal	A. E. Turnbull	Springfield, Ohio	June 15, 1858	20, 590
Locomotive signal-light	A. Dick	Hamilton, Canada	July 23, 1872	129, 797

Index of patents issued from the United States Patent Office from 1790 *to* 1873, *inclusive*—Continued.

Invention.	Inventor.	Residence.	Date.	No.
Locomotive smoke and cinder conveyer	A. M. Rodgers	Brooklyn, N. Y	May 9, 1871	114, 605
Locomotive smoke and spark conveyer	E. A. C. Fox	Frederick City, Md	Sept. 16, 1873	142, 906
Locomotive smoke-conductor	A. Storm	Matteawan, N. Y	Sept. 24, 1872	131, 639
Locomotive smoke-stack	J. Atkins	Washington, D. C	July 28, 1868	80, 437
Locomotive smoke-stack	L. Bell	Washington, D. C	May 23, 1871	115, 153
Locomotive smoke-stack	J. V. Bishop	Atlanta, Ga	Nov. 25, 1873	144, 884
Locomotive smoke-stack	H. Brooks and J. Ball	Zanesville, Ohio	Oct. 30, 1866	59, 175
Locomotive smoke-stack	E. A. Castellaw	Savannah, Ga	Apr. 25, 1871	113, 980
Locomotive smoke-stack	E. Fontaine	Fort Wayne, Ind	Nov. 1, 1870	108, 899
Locomotive smoke-stack	S. Ham	Philadelphia, Pa	Nov. 22, 1864	45, 204
Locomotive smoke-stack	J. Hughes	Scranton, Pa	Sept. 16, 1873	142, 851
Locomotive smoke-stack	J. A. W. Justi	Savannah, Ga	Oct. 27, 1868	83, 506
Locomotive smoke-stack	W. J. Mehary	Philadelphia, Pa	July 5, 1870	105, 107
Locomotive smoke-stack	A. S. Sweet, jr	Detroit, Mich	June 23, 1863	38, 992
Locomotive smoke-stack	J. L. Vauclain	La Fayette, Ind	Aug. 20, 1861	33, 114
Locomotive smoke-stack	E. H. Winchell	New York, N. Y	Oct. 10, 1871	119, 803
Locomotive smoke-stack, Self-cleaning	S. M. Cummings and H. Isreal	Allegheny, Pa	Dec. 27, 1870	110, 441
Locomotive smoke-stack, Arrangement of vertical-tube feed-water heaters in.	M. W. Baldwin and D. Clark	Philadelphia, Pa	Feb. 14, 1854	10, 514
Locomotive smoke-stack, Base-piece of	P. S. Ebbert	Chicago, Ill	July 1, 1856	15, 225
Locomotive smoke-stack, Cone for	D. B. Strope	Fort Wayne, Ind	Nov. 14, 1871	120, 835
Locomotive smoke-stack, Exhaust-device for	R. H. Lecky	Allegheny City, Pa	July 13, 1869	92, 617
Locomotive spark and smoke conductor	J. Viall	Somerville, Mass	June 1, 1869	90, 899
Locomotive spark-arrester	M. Brassill	Hartford, Conn	Oct. 14, 1873	143, 664
Locomotive spark-extinguisher	P. H. Corlett	Manchester, Pa	Aug. 18, 1863	39, 552
Locomotive steam-boiler	E. Ayer	Norwich, Conn	May 2, 1848	5, 546
Locomotive steam-engine	M. W. Baldwin	Philadelphia, Pa	Aug. 25, 1842	2, 759
Locomotive steam-engine	A. Haeupel and J. Reinhardt	Philadelphia, Pa	June 30, 1868	79, 341
Locomotive steam-engine	R F. Fairlie	London, England	Dec. 22, 1868	85, 076
Locomotive steam-engine	E. F. Johnson	Middletown, Conn	Dec. 31, 1844	3, 866
Locomotive steam-engine	G. G. Jones	Rushsylvania, Ohio	July 6, 1869	92, 193
Locomotive steam-engine, &c	S. H. Long	Philadelphia, Pa	Dec. 28, 1832	....
Locomotive steam-engine	J. S. Stuart	Philadelphia, Pa	May 4, 1869	89, 809
Locomotive steam-engine	A. Whitney	Rotterdam, N. Y	June 27, 1840	1, 653
Locomotive steam-engine	R. Winans	Baltimore, Md	July 29, 1837	308
Locomotive steam-engine	R. Winans	Baltimore, Md	July 29, 1837	311
Locomotive steam-engine	R. Winans	Baltimore, Md	July 28, 1843	3, 201
Locomotive steam-engine carriage	S. Norris and W. Knight	Philadelphia, Pa	Feb. 10, 1843	2, 951
Locomotive steam-engine for rail and other roads	J. Ruggles	United States	July 28, 1836	1
Locomotive steam-engines, Framing of	R. Winans	Baltimore, Md	July 29, 1837	305
Locomotive steam-engines, Method of connecting driving-wheels of.	R. S. Stevens	New York, N. Y	Sept. 3, 1842	2, 773
Locomotive-tender	W. Ross and W. E. Rutter	Providence, R. I	May 16, 1848	5, 584
Locomotive-tender	A. Sturrock	Doncaster, England	Nov. 24, 1863	40, 715
Locomotive-tender	R. and T. Winans	Baltimore, Md	May 23, 1854	10, 971
Locomotive-tender with water, Device for supplying.	W. E. Prall	Washington, D. C	July 4, 1871	116, 752
Locomotive-tender with water, Method of supplying.	W. J. Brassington and W. B. Burtnett.	Brooklyn and New York, N. Y.	June 18, 1867	65, 869
Locomotive-tenders, Machine for loading	J. N. Jackson	Brookhaven, Miss	June 15, 1869	91, 236
Locomotive, Traction	T. S. Winniss	Meadville, Pa	Jan. 28, 1873	135, 240
Locomotive vehicle for running on ice or in water	N. Wiard	Janesville, Wis	Jan. 24, 1860	26, 960
Locomotive, Water	G. A. Fall	Hoboken, N. J	Oct. 12, 1869	95, 785
Locomotive water-supply pipe	F. Gerrard	Kansas City, Mo	Jan. 11, 1870	98, 685
Locomotive-wheel	M. J. Montgomery	New York, N. Y	May 31, 1864	42, 958
Locomotive-wheel	J. R. Richardson	Max Meadows, Va	July 19, 1870	105, 598
Locomotive-wheel for ascending inclined planes on railways.	E. Town	Montpelier, Vt	July 31, 1837	339
Locomotive-wheels, Apparatus for applying chalk to.	N. Sehner	Hagerstown, Md	Feb. 27, 1872	124, 165
Locomotive-wheels, Counterpoise to cast	H. A. Chase	Boston, Mass	May 23, 1854	10, 958
Locomotive-wheels, Mode of chilling rims for	H. W. Moore	Jersey City, N. J	Apr. 26, 1859	23, 775
Locomotive-window	J. H. Dinsmore	Boston, Mass	Dec. 23, 1873	145, 790
Locomotive-window	H. Skinner	Fulton, N. Y	Sept. 29, 1857	18, 304
Locomotive with driving-axle above the boiler	R. H. Emerson	Portland, Me	May 1, 1849	6, 401
Locomotive with spading apparatus, Agricultural	C. Locher	Oroville, Cal	Oct. 13, 1868	82, 963
Locomotives, Adjusting cross-heads of	J. D. Akley and R. English	Oil City, Pa	May 24, 1870	103, 407
Locomotives, Apparatus for burning soft coal in	A. C. Rand	Aurora, Ill	Dec. 24, 1872	134, 314
Locomotives, Arrangement of deflecting-plates and spark-receivers in.	W. H. Bullock	Boston, Mass	Dec. 29, 1857	18, 953
Locomotives, Arrangement of means for regulating the discharge of exhaust steam in.	J. E. Wootten	Philadelphia, Pa	May 1, 1855	12, 805
Locomotives, Arrangement of means for regulating the fire of coal-burning.	J. M. Hartnett	Waukegan, Ill	July 21, 1857	17, 834
Locomotives, Automatic apparatus for operating valves of exhaust-pipes of.	R. C. Huse, jr	Georgetown, Mass	Apr. 4, 1871	113, 299
Locomotives by stationary power, Propelling	J. A. Etzler	Philadelphia, Pa	Dec. 23, 1841	2, 396
Locomotives, Conveyer of smoke and cinders for	A. M. Rodgers	Brooklyn, N. Y	Oct. 10, 1871	119, 790
Locomotives, Device for preventing collision of	H. Payne, sr	Mount Vernon, Ohio	Jan. 22, 1867	61, 354
Locomotives for ascending inclined planes, Cog-gearing of.	W. Hoyt	Dupont, Ind	Apr. 17, 1849	6, 321
Locomotives for railways, Constructing and operating.	R. P. Morgan	Chicago, Ill	Nov. 8, 1864	44, 997
Locomotives, Heating feed-water apparatus of	P. S. Ebbert	Chicago, Ill	May 5, 1857	17, 208
Locomotives in engine-houses, Arrangement for carrying off smoke from.	J. O. D. Lilly, J. S. Vauclain, and J. W. Lilly.	La Fayette, Ind	Feb. 23, 1858	19, 469
Locomotives, Machinery for oiling the journals of	S. Scotton	Richmond, Ind	Jan. 11, 1859	22, 586
Locomotives, Manner of attaching legs to walking	S. G. Hoge	Bellefontaine, Ohio	Feb. 23, 1858	19, 468
Locomotives, Means for increasing draft in	C. F. Thomas	Taunton, Mass	July 10, 1855	13, 238
Locomotives, Method of increasing traction in	C. W. T. Krausch	Philadelphia, Pa	Jan. 29, 1867	61, 546
Locomotives, &c., Mode of applying steam-power to	J. Ericsson	Sweden	Nov. 5, 1840	1, 847
Locomotives, Mode of supplying water to	S. Vail	Morristown, N. J	July 12, 1839	1, 234
Locomotives, Mode of turning	G. M. Cole	Folsom City, Cal	Sept. 17, 1861	33, 294
Locomotives, Petticoat-pipe for	W. G. Freeman	Richmond, Va	June 9, 1868	78, 658
Locomotives, Storing power in pneumatic	H. F. C. Krumme	Ridgway, Pa	Apr. 5, 1870	101, 631
Locomotives with compressed air from fixed stations, Mode of charging the receivers of.	S. Carson	New York, N. Y	Dec. 9, 1856	16, 220

Index of patents issued from the United States Patent Office from 1790 *to* 1873, *inclusive*—Continued.

Invention.	Inventor.	Residence.	Date.	No.
Log and leeway indicator, Marine	A. E. Lozier	New York, N. Y	Mar. 15, 1864	41, 932
Log and sounding line, Combined	A. Pécoul	Marseilles, France	Dec. 25, 1855	13, 994
Log-binder	W. L. Dean	Dayton, N. Y	Sept. 5, 1871	118, 697
Log-boat	A. Olmsted	Windsor, Mich	Apr. 27, 1869	89, 332
Log-canting apparatus	B. R. Stevens	Grand Rapids, Mich	Jan. 5, 1869	85, 543
Log-canting machine	A. Rodgers	Muskegon, Mich	Aug. 26, 1873	142, 280
Log, Electro-magnetic	W. I. Reid	Brooklyn, N. Y	Apr. 7, 1868	76, 521
Log, Hand	J. R. St. John	New York, N. Y	May 6, 1851	8, 074
Log-lifter	G. B. Sims	Elizabeth, Ind	Mar. 5, 1872	124, 294
Log-loader	J. H. Harvey	Chanticleer, Ohio	May 31, 1870	103, 608
Log loader and piler	J. G. Brady	Forest Hill, Mich	June 21, 1870	104, 548
Log-moving machine	G. M. Hinckley and A. Rodgers	Milwaukee, Wis., and Muskegon, Mich.	June 3, 1873	139, 578
Log, Nautical	T. Hotchkiss	Stratford, Conn	Nov. 15, 1864	45, 042
Log, Registering marine	A. Gordan	New York, N. Y	Dec. 8, 1863	40, 834
Log-rolling device	T. Emery	Peshtigo, Wis	Feb. 25, 1873	136, 145
Log-rolling machine	J. Torrent	Muskegon, Mich	Oct. 24, 1871	120, 220
Log-rolling machine	J. Torrent	Muskegon, Mich	July 11, 1871	117, 017
Log-rossing machine	G. W. Nichols	Clinton, Iowa	June 18, 1872	128, 060
Log-turner	H. T. Hunter	Spring Lake, Mich	Oct. 8, 1872	131, 956
Log-turner	J. C. Moore	Ypsilanti, Mich	Apr. 29, 1873	138, 270
Log-turner	I. W. Pool	Eau Claire, Wis	July 1, 1873	140, 386
Log-turner	E. Tarrant	Muskegon, Mich	Aug. 12, 1873	141, 736
Log-turner	J. Torrent	Muskegon, Mich	Aug. 12, 1873	141, 739
Log-turner	C. Van Vleck	Sidney, Mich	Aug. 19, 1873	141, 969
Log turning and canting device	D. S. Griffes	Flint, Mich	May 23, 1871	115, 194
Log turning and loading device	S. Snyder	Delaware, Ohio	July 6, 1869	92, 388
Log-turning apparatus	L. J. Stannard and M. L. Perry	Newark, N. Y	Mar. 23, 1869	88, 093
Log-turning device for mills	I. H. Newton	Grand Rapids, Mich	Mar. 23, 1869	88, 063
Log-turning machine	G. W. Baker	Elizabeth City, N. C	Nov. 19, 1872	133, 185
Log-turning machine	E. C. Dicey	Whitehall, Mich	Aug. 27, 1872	130, 904
Log-turning machine	W. E. Hill	Erie, Pa	Aug. 9, 1870	106, 160
Log-turning machine	E. H. Stearns	Erie, Pa	Dec. 20, 1870	110, 398
Log-turning machine	E. H. Stearns	Erie, Pa	Aug. 8, 1871	117, 828
Logs, Building flumes for floating	J. Du Bois	Williamsport, Pa	Jan. 19, 1864	41, 285
Logs, Hauling up	S. H. Richardson	Bangor, Me	May 24, 1870	103, 372
Logs, Implement for rolling and piling	W. Todd	Cherryfield, Me	Aug. 31, 1858	21, 385
Logs on wagons, Apparatus for loading	P. Gilbert	Alexandria, Ohio	Sept. 22, 1857	18, 242
Logging-skid	G. W. Nichols	River Falls, Wis	Jan. 1, 1867	60, 926
Logotrope	C. Richardson and J. Graeme, jr	New York, N. Y	Oct. 1, 1867	69, 482
Logotype	W. H. Wilkinson	Southwick, Mass	Aug. 23, 1870	106, 641
Longimeter	A. Bayrd	South Reading, Mass	July 16, 1828	....
Longitude, Mode of finding	J. Henry	Maysville, Ky	Sept. 13, 1820	....
Looking-glass	G. H. Chinnock	Brooklyn, N. Y	Sept. 24, 1872	131, 665
Looking-glass	H. Willard	Vergennes, Vt	May 20, 1862	35, 344
Looking-glass	U. A. Woodbury	Morrisville, Vt	Apr. 28, 1868	77, 229
Looking-glass and match-holder	D. Cumming, jr	New York, N. Y	July 21, 1863	39, 277
Looking-glass attachment	C. J. Hartmann	London, England	June 1, 1869	90, 746
Looking-glass bracket	F. C. Brock	Oriskany Falls, N. Y	Aug. 20, 1872	130, 563
Looking-glass frames to bureaus, Attaching	C. Kilburn	Philadelphia, Pa	Aug. 2, 1870	106, 065
Looking-glass or mirror	S. F. Brooks	Weston, Mass	Apr. 24, 1860	27, 965
Looking-glass support	W. H. Gray	New York, N. Y	Dec. 15, 1868	84, 944
Looking-glasses, Silvering	T. Drayton	Brighton, England	Aug. 12, 1844	3, 702
Looking-glasses, Silvering	J. Webster	Brooklyn, N. Y	May 8, 1855	12, 840
Looking-glasses and pictures, Device for hanging	D. Crowell, jr	Yarmouth Port, Mass	June 6, 1871	115, 713
Looking-glasses in dressing-cases, Device for the Adjustment of.	A. A. Gray and W. C. Hyde	Detroit, Mich	Nov. 12, 1867	70, 713
Looking-glasses, Process for making	L. P. Augenard	New York, N. Y	Jan. 31, 1865	46, 062
Loom	R. W. Andrews	Staffordville, Conn	Oct. 3, 1865	50, 215
Loom	H. Bachofner	Springfield, Mass	Oct. 30, 1849	6, 823
Loom	C. F. Bartling	Greenville, Ohio	Apr. 17, 1866	53, 943
Loom	C. Bergen	New York	June 14, 1826	....
Loom	A. Bigelow and J. Butler	Granville, Ohio	Jan. 16, 1849	6, 035
Loom	E. B. Bigelow	Boston, Mass	Apr. 8, 1856	14, 590
Loom	E. B. Bigelow	Boston, Mass	Apr. 20, 1869	89, 011
Loom	E. B. Bigelow	Boston, Mass	Mar. 22, 1870	101, 088
Loom	S. Boorn	Lowell, Mass	July 22, 1862	35, 918
Loom	S. Boorn	Lowell, Mass	May 16, 1871	114, 756
Loom	W. H. Boozer	Potter's Mills, Pa	Mar. 14, 1871	112, 679
Loom	A. H. Boyd	Saco, Me	Oct. 15, 1850	7, 717
Loom	J. L. Branson	Pittsburgh, Pa	Nov. 23, 1869	97, 158
Loom	J. Braun	Philadelphia, Pa	Aug. 15, 1865	49, 369
Loom	J. Braun	Philadelphia, Pa	Mar. 20, 1866	53, 265
Loom	D. Briggs	Canaan, Conn	Feb. 19, 1814	....
Loom	S. Briggs	Canaan, Conn	May 27, 1813	....
Loom	A. Brigham	Manchester, N. H	Jan. 2, 1855	12, 120
Loom	W. Breitenstein	New York, N. Y	Apr. 14, 1863	38, 195
Loom	W. Breitenstein	New York, N. Y	Jan. 24, 1865	45, 969
Loom	J. Broadbent	Oak Grove, Ky	Aug. 7, 1855	13, 382
Loom	S. Bronson	Kent, Conn	Nov. 25, 1814	....
Loom	E. W. Brown	Fall River Mass	July 25, 1854	11, 352
Loom	O. C. Burr	Millbury, Mass	July 17, 1835	....
Loom	A. Campbell	Kingston, Pa	Mar. 28, 1812	....
Loom	A. Canis and F. Higgins	Manchester, N. H	Feb. 25, 1873	136, 133
Loom	N. B. Carney	New York, N. Y	May 19, 1857	17, 353
Loom	S. Chapman	Otsego, N. Y	Aug. 30, 1816	....
Loom	S. Chidester	Windham, N. Y	Sept. 2, 1826	....
Loom	T. Clarkson	South Adams, Mass	May 1, 1860	28, 125
Loom	G. Cliff	Memphis, Mich	Apr. 28, 1868	77, 169
Loom	J. H. Clifton	Newcastle, Pa	May 29, 1860	28, 453
Loom	R. Collins	Cabotsville, Mass	Oct. 2, 1847	5, 318
Loom	A. Conant	Willimantic, Conn	Apr. 17, 1860	27, 889
Loom	J. C. Cooke	Middletown, Conn	Apr. 10, 1860	27, 860
Loom	C. Cooper	Lebanon, Pa	Nov. 4, 1808	....
Loom	C. Cooper and G. Shaake	Lebanon, Pa	Aug. 28, 1812	....
Loom	C. Cooper and G. Shalk	Lebanon, Pa	May 31, 1815	....
Loom	H. M. Cooper	Lindley, Mo	Nov. 19, 1867	70, 964
Loom	G. Copeland	Lewiston, Me	Jan. 23, 1855	12, 293

Index of patents issued from the United States Patent Office from 1790 *to* 1873, *inclusive*—Continued.

Invention.	Inventor.	Residence.	Date.	No.
Loom	G. Copeland	North Gray, Me	Apr. 9, 1861	31, 956
Loom	J. D. Cottrell	Milford, Mass	Oct. 27, 1863	40, 442
Loom	J. D. Cottrell and G. Draper	Milford, Mass	Oct. 1, 1867	69, 320
Loom	J. Coulter and S. Gano	Berkeley County, Va	Apr. 18, 1814	....
Loom	S. Craige	Philadelphia, Pa	Oct. 18, 1814	....
Loom	G. Crompton	Worcester, Mass	Apr. 27, 1858	20, 044
Loom	G. Crompton	Worcester, Mass	Apr. 23, 1861	32, 123
Loom	G. Crompton	Worcester, Mass	Oct. 22, 1861	33, 520
Loom	G. Crompton	Worcester, Mass	July 16, 1867	66, 682
Loom	G. Crompton	Worcester, Mass	Mar. 17, 1868	75, 530
Loom	G. Crompton	Worcester, Mass	Apr. 28, 1868	77, 361
Loom	G. Crompton	Worcester, Mass	Aug. 4, 1868	80, 608
Loom	G. Crompton	Worcester, Mass	Aug. 11, 1868	80, 810
Loom	G. Crompton	Worcester, Mass	Aug. 18, 1868	81, 070
Loom	G. Crompton	Worcester, Mass	Aug. 25, 1868	81, 347
Loom	G. Crompton	Worcester, Mass	Dec. 29, 1868	85, 432
Loom	G. Crompton	Worcester, Mass	Aug. 31, 1869	94, 401
Loom	G. Crompton	Worcester, Mass	Sept. 7, 1869	94, 571
Loom	G. Crompton	Worcester, Mass	Sept. 14, 1869	94, 873
Loom	G. Crompton	Worcester, Mass	Sept. 21, 1869	95, 092
Loom	G. Crompton	Worcester, Mass	July 15, 1873	140, 894
Loom	R. Crosbie	Newark, N. J	Jan. 6, 1812	....
Loom	C. Grossley	Ellington, Conn	June 14, 1859	24, 378
Loom	J. F. Crowley	Philadelphia, Pa	May 16, 1871	114, 929
Loom	B. Cummings	Townsend, Mass	Oct. 3, 1817	....
Loom	P. C. Curtis	Paris, N. Y	Nov. 17, 1810	....
Loom	H. Dale	Philadelphia, Pa	Oct. 9, 1866	58, 609
Loom, Circular	W. Darker, jr	Philadelphia, Pa	July 7, 1863	39, 197
Loom	J. F. Davies and W. E. Yates	Manchester, Great Britain	June 13, 1871	115, 938
Loom	H. D. Davis	North Andover, Mass	Aug. 16, 1870	106, 334
Loom	H. D. Davis	North Andover, Mass	Jan. 10, 1871	110, 904
Loom	H. G. Davis	Worcester, Mass	Nov. 21, 1871	121, 161
Loom	H. D. Davis	North Andover, Mass	July 29, 1873	141, 262
Loom	B. G. Dawley	North Providence, R. I	Dec. 23, 1856	16, 306
Loom	W. Day	Newark, N. J	July 16, 1872	129, 216
Loom	J. Deakin	Gloucester, N. J	Jan. 14, 1868	73, 237
Loom	J. Detweiler	West Liberty, Ohio	Aug. 15, 1871	118, 114
Loom	R. M. Dill	Holyoke, Mass	July 10, 1855	13, 217
Loom	W. F. Draper	Hopedale, Mass	June 16, 1868	78, 941
Loom	J. W. Drummond	New York, N. Y	Nov. 24, 1863	40, 685
Loom	G. W., J., and J. C. Duckworth	Pittsfield, Mass	Nov. 23, 1869	97, 175
Loom	J. C. Duckworth	Mount Carmel, Conn	Apr. 15, 1873	137, 898
Loom	F. Durand	Paris, France	Jan. 6, 1857	16, 354
Loom	W. Dutcher	Bennington, Vt	June 27, 1846	4, 606
Loom	J. Earnshaw	East Greenwich, R. I	July 9, 1867	66, 574
Loom	J. Eccles	Philadelphia, Pa	Jan. 23, 1855	12, 274
Loom	R. Elliott	Chester, Pa	Jan. 3, 1871	110, 640
Loom	H. A. Ellis	Mystic River, Conn	Nov. 27, 1866	59, 987
Loom	D. Farrior and D. P. J. Murphy	Aberfoil, Ala	Dec. 4, 1843	3, 364
Loom	A. Faulkner	Walpole, N. H	Oct. 23, 1849	6, 813
Loom	G. S. Faulkner	Indianapolis, Ind	Jan. 25, 1870	99, 177
Loom	A. R. Field	Centre Falls, R. I	Apr. 19, 1870	101, 989
Loom	W. T. Flinn	Bridesburgh, Pa	Dec. 14, 1868	84, 872
Loom	J. F. Fosdick	Lowell, Mass	Dec. 23, 1863	37, 223
Loom	D. K. Fretz	Cono, Iowa	Oct. 1, 1867	69, 422
Loom	A. Frey	New York, N. Y	May 7, 1861	32, 245
Loom	J. G. Frick	Pottsville, Pa	Aug. 1, 1871	117, 616
Loom	A. L. Fuller	Clinton, Mass	Dec. 23, 1856	16, 271
Loom	M. A. Furbush	Philadelphia, Pa	Nov. 24, 1863	40, 683
Loom	W. Gadd and J. Moore	Manchester, Great Britain	July 13, 1869	92, 523
Loom	J. G. Garretson	Cincinnati, Ohio	Nov. 3, 1868	83, 621
Loom	J. C. and A. P. Garretson	Jackson Township, Iowa	May 13, 1862	35, 229
Loom	W. V. Gee	New Haven, Conn	Feb. 27, 1855	12, 457
Loom	W. V. Gee	Philadelphia, Pa	Oct. 31, 1871	120, 510
Loom	W. R., W. R., jr., and J. A. Giffard.	Milo, Me	Apr. 4, 1871	113, 288
Loom	C. W. Gilbert	Worcester, Mass	May 31, 1870	103, 600
Loom	L. M. Gilbert	Cow Run, Ohio	Aug. 30, 1870	106, 809
Loom	R. B. Goodyear	Wilmington, Del	Aug. 1, 1871	117, 533
Loom	O. W. Gordon and N. T. Frame	Salem, Iowa	Jan. 23, 1866	52, 161
Loom	J. Graham	New York, N. Y	May 7, 1867	64, 525
Loom	W. Graichen and C. H. Hoffman	Clinton, Mass	May 7, 1861	32, 269
Loom	D. Grieve		June 8, 1797	....
Loom	J. M. M. Guiramond	Baltimore, Md	Jan. 7, 1814	....
Loom	F. Haigh	Methuen, Mass	Nov. 26, 1867	71, 480
Loom	W. Hainsworth	Philadelphia, Pa	Nov. 26, 1867	71, 299
Loom	H. Halvorson	Northampton, Mass	Oct. 1, 1850	7, 682
Loom	G. Hancock	Holyoke, Mass	Aug. 16, 1864	43, 848
Loom	M. M. Hankins	Vandalia, Ill	Nov. 12, 1867	70, 837
Loom	D. S. Harris	Coventry, R. I	Mar. 27, 1855	12, 593
Loom	E. B. Hastings	Palmer, Mass	Jan. 31, 1871	111, 343
Loom	J. Haworth	Frankford, Pa	Mar. 20, 1847	5, 021
Loom	J. and J. Haworth	Frankford, Pa	Oct. 31, 1848	5, 891
Loom	J. Heavin	Montgomery County, Va	Oct. 16, 1812	....
Loom	J. Heavin	Montgomery County, Va	Feb. 18, 1814	....
Loom	J. G. Henderson	Keokuk, Iowa	Nov. 2, 1869	96, 320
Loom	B. W. and H. Hendrick	Woonsocket Falls, R. I	May 12, 1842	2, 616
Loom	J. J. Herbert	Philadelphia, Pa	June 7, 1870	103, 880
Loom	J. Hillsley	West Manayunk, Pa	May 21, 1872	127, 055
Loom	I. N. Hodson	Mount Pleasant, Iowa	Sept. 4, 1866	57, 717
Loom	S. Holdsworth	Durham, England	June 3, 1862	35, 445
Loom	W. J. Horstman	Philadelphia, Pa	July 8, 1856	15, 295
Loom	W. H. Howard	Philadelphia, Pa	May 26, 1857	17, 375
Loom	J. Hubbard	Jamestown, N. C	Nov. 6, 1826	....
Loom	O. B. Hubbard	Lowell, Mass	Mar. 21, 1865	46, 971
Loom	J. P. Humaston	New York, N. Y	July 14, 1868	79, 908
Loom	F. W. Huppelsberg	New York, N. Y	July 16, 1867	66, 844
Loom	W. S. Irish	Middlebury, N. Y	May 27, 1855	12, 596

Index of patents issued from the United States Patent Office from 1790 *to* 1873, *inclusive*—Continued.

Invention.	Inventor.	Residence.	Date.	No.
Loom	R. Jackson	Attleborough, Pa	July 23, 1816	....
Loom	W. H. Jackson and G. Merrill	Brooklyn, N. Y., and Newburyport, Mass.	May 5, 1868	77, 619
Loom	W. Janes	Ashford, Conn	Sept. 1, 1810	....
Loom	J. Jelleffs	Butternuts, N. Y	May 24, 1815	....
Loom	B. H. Jenks	Bridesburgh, Pa	Oct. 24, 1854	11, 833
Loom	B. H. Jenks	Bridesburgh, Pa	Apr. 3, 1855	12, 630
Loom	B. H. Jenks	Bridesburgh, Pa	May 12, 1863	38, 489
Loom	B. H. Jenks	Bridesburgh, Pa	Nov. 24, 1868	84, 423
Loom	B. H. Jenks	Bridesburgh, Pa	Mar. 2, 1869	87, 341
Loom	B. H. Jenks and R. B. Goodyear	Philadelphia, Pa	May 22, 1866	55, 011
Loom	C. J. Kane	Milford, Conn	Mar. 25, 1873	137, 077
Loom	J. J. Kendall	Corinth, Miss	Feb. 28, 1860	27, 296
Loom	O. A. Kelly	Woonsocket, R. I	Nov. 22, 1853	10, 252
Loom	J. B. Kershaw	Indianapolis, Ind	Sept. 11, 1866	58, 024
Loom	J. Knowles	Cohoes, N. Y	Nov. 8, 1853	10, 209
Loom	L. J. Knowles	Warren, Mass	June 24, 1856	15, 186
Loom	L. K. Knowles	Warren, Mass	Nov. 4, 1856	16, 015
Loom	L. J. Knowles	Worcester, Mass	Jan. 21, 1873	134, 992
Loom	W. S. Laycock	Sheffield, England	Dec. 13, 1870	110, 050
Loom	J. O. Leach	Ballston, N. Y	Oct. 30, 1855	13, 724
Loom	T. Lovelidge	Philadelphia, Pa	Feb. 14, 1860	27, 136
Loom	F. C. Lowell and P. T. Jackson	Boston, Mass	Feb. 23, 1815	....
Loom	J. Lyall	New York, N. Y	Aug. 11, 1868	80, 982
Loom	J. Lyall	New York, N. Y	Dec. 10, 1872	133, 868
Loom	J. A. Marden	Newburyport, Mass	Oct. 1, 1861	33, 397
Loom	B. A. Mann	West Meriden, Conn	Nov. 24, 1863	40, 698
Loom	J. G. Melville and W. Brayshaw	Wetheredville, Md	Apr. 24, 1855	12, 762
Loom	S. C. Mendenhall	Richmond, Ind	Oct. 10, 1854	11, 790
Loom	S. C. Menhenhall	Richmond, Ind	Jan. 4, 1859	22, 533
Loom	S. C. Mendenhall	Richmond, Ind	Aug. 29, 1865	49, 644
Loom	C. Miller	Saint Louis, Mo	Apr. 23, 1867	64, 127
Loom	J. Miller	Eldridge, Ill	Oct. 11, 1870	108, 281
Loom	R. Mueller	New York, N. Y	Apr. 30, 1872	126, 319
Loom	W. Murkland	Lowell, Mass	Feb. 5, 1861	31, 324
Loom	W. and J. W. Murkland	Lowell, Mass	Jan. 23, 1872	123, 037
Loom	J. Nields	Taunton, Mass	Nov. 27, 1847	5, 379
Loom	A. Nimmo	Philadelphia, Pa	May 30, 1871	115, 508
Loom	J. R. Norfolk	Salem, Mass	Feb. 8, 1870	99, 586
Loom	E. L. Norfolk, S. S. Stanley, and J. A. Marden.	Salem and Newburyport, Mass.	Feb. 22, 1848	5, 450
Loom	B. Oldfield	Williamsburgh, N. Y	Jan. 23, 1866	52, 192
Loom	B. and E. Oldfield	Newark, N. J	Jan. 14, 1868	73, 374
Loom	B. and E. Oldfield	Norwich, Conn	Mar. 30, 1869	88, 503
Loom	E. Oldfield	Norwich, Conn	Aug. 1, 1871	117, 672
Loom	I. Orndorff	Russellville, Ky	Apr. 10, 1866	53, 857
Loom	I. Orndorff	Russellville, Ky	Apr. 10, 1866	53, 858
Loom	F. Painter	East Hampton, Mass	May 26, 1857	17, 404
Loom	A. Parkinson	Norwich, Conn	Dec. 18, 1860	30, 948
Loom	F. Peabody	Salem, Mass	Mar. 5, 1861	31, 617
Loom	N. Perry	Boston, Mass	Sept. 25, 1811	....
Loom	J. Phelps	Manlius, N. Y	Aug. 27, 1812	....
Loom	R. Pilson and S. P. Heath	Laurel, Md	May 27, 1856	14, 971
Loom	W. J. Porter and W. Cross	New York, N. Y., and Jersey City, N. J.	Oct. 11, 1870	108, 292
Loom	W. J. Quinn	Philadelphia, Pa	Feb. 25, 1868	74, 938
Loom	P. F. Ramel and J. Dogat	Lyons, France	Apr. 19, 1870	102, 039
Loom	R. Reynolds	Stockport, N. Y	Sept. 23, 1862	36, 529
Loom	W. Reynolds	Manchester, N. H	July 22, 1862	35, 981
Loom	E. Robbins	Milford, Mass	May 27, 1862	35, 394
Loom	E. Robinson	Augusta, Mass	July 24, 1813	....
Loom	T. Robjohn	New York, N. Y	May 7, 1867	64, 573
Loom	R. C. Rogers	Maine, Mass	Nov. 29, 1811	....
Loom	R. C. Rogers	District of Maine, Mass	June 23, 1812	....
Loom	W. Rossetter	Accrington, England	May 4, 1869	89, 692
Loom	G. Roth	New York, N. Y	Dec. 12, 1854	12, 072
Loom	J. Salsbury	Central Falls, R. I	Aug. 18, 1868	81, 113
Loom	J. Schofield	Worcester, Mass	Feb. 5, 1867	61, 767
Loom	E. A. Scholfield	Westerly, R. I	Aug. 25, 1857	18, 061
Loom	J. Scholfield	Stonington, Conn	Sept. 9, 1825	....
Loom	C. W. Schonherr	Saxony	May 8, 1837	187
Loom	L. Scofield	Farmington, Wis	Oct. 8, 1867	69, 711
Loom	L. Schofield	Farmington, Wis	Aug. 21, 1866	57, 444
Loom	E. M. Scott	Auburn, N. Y	Sept. 7, 1858	21, 448
Loom	J. Shaw	Ballardvale, Mass	Mar. 10, 1868	75, 305
Loom	S. Shepard	Taunton, Mass	July 29, 1862	36, 029
Loom	J. Shinn	Leverington, Pa	Apr. 30, 1861	30, 236
Loom	J. Short	New Brunswick, N. J	July 16, 1872	129, 602
Loom	T. Siddall	Bristol Township, Pa	July 9, 1814	....
Loom	J. Silbermann and G. Unger	New York, N. Y	Mar. 3, 1868	75, 062
Loom	A. W. Silvis	Birmingham, Iowa	Oct. 27, 1868	83, 557
Loom	J. Smith and P. McMahon	Somerville and Charlestown, Mass.	Feb. 13, 1872	123, 647
Loom	O. C. Smith	Salem, Mass	Apr. 21, 1863	38, 265
Loom	D. W. Snell	Woonsocket, R. I	May 29, 1855	12, 970
Loom	D. W. Snell and S. S. Bartlett	Woonsocket, R. I	Jan. 13, 1857	16, 405
Loom	J. Sprinkle	Wythe County, Va	Apr. 27, 1814	....
Loom	J. Standish	Providence, R. I	Nov. 11, 1830	....
Loom	T. Stibbs	Wooster, Ohio	Aug. 12, 1862	36, 178
Loom	A. Stockwell	Providence, R. I	June 1, 1869	90, 888
Loom	A. Stockwell and B. D. Humes	Millbury, Mass	Feb. 18, 1862	34, 451
Loom	L. Stone	Nelson, N. H	Dec. 14, 1869	97, 826
Loom	R. Sugden	Boston, Mass	Dec. 2, 1813	....
Loom	J. S. Templeton	Glasgow, Scotland	July 2, 1872	128, 675
Loom	S. T. Thomas	Lawrence, Mass	July 3, 1855	13, 186
Loom	S. T. Thomas	Lawrence, Mass	July 3, 1855	13, 187
Loom	S. T. Thomas	Gilford, N. H	Jan. 7, 1873	134, 572
Loom	S. T. Thomas	Gilford, N. H	Dec. 9, 1873	145, 316

Index of patents issued from the United States Patent Office from 1790 *to* 1873, *inclusive*—Continued.

Invention.	Inventor.	Residence.	Date.	No.
Loom	S. T. Thomas and J. H. Dolley	Gilford, N. H	Dec. 10, 1867	72, 119
Loom	T. G. Thompson and B. Ballard	Richmond, Ind	Sept. 3, 1867	68, 583
Loom	W. Tongue	Philadelphia, Pa	Jan. 9, 1855	12, 229
Loom	H. E. Towle	Newark, N. J	Jan. 24, 1871	111, 278
Loom	W. Townsend	Seneca Falls, N. Y	Aug. 10, 1869	93, 500
Loom	W. Townshend	Hinsdale, Mass	Mar. 1, 1853	9, 603
Loom	W. Townshend	Hinsdale, Mass	Nov. 15, 1853	10, 241
Loom	W. Tunstill	New York, N. Y	May 9, 1865	47, 687
Loom	A. Urbahn	Paterson, N. J	Jan. 30, 1872	123, 210
Loom	L. Van Rifer	Spring Valley, N. H	Mar. 20, 1855	12, 565
Loom	A. O. Very	Andover, N. Y	Oct. 1, 1867	69, 515
Loom	J. Wade	Palmer, Mass	Jan. 23, 1872	122, 976
Loom	W. B. Walker	Salem, Iowa	Oct. 1, 1867	69, 516
Loom	A. Ware, jr	Franklin, Mass	June 18, 1813	....
Loom	E. Warren	New York	Aug. 27, 1818	....
Loom	E. Warren	New York	Dec. 11, 1821	....
Loom	E. Warren	New York	May 1, 1822	....
Loom	L. R. Wattles	Newton, Mass	July 31, 1860	29, 427
Loom	J. R. Weber	Bourbon, Ind	Mar. 12, 1867	62, 791
Loom	J. Welsh	Philadelphia, Pa	Oct. 3, 1854	11, 768
Loom	J. Welsh	Philadelphia, Pa	Jan. 9, 1855	12, 225
Loom	J. Welsh	Philadelphia, Pa	May 29, 1855	12, 981
Loom	J. Welsh	Philadelphia, Pa	July 20, 1858	20, 969
Loom	J. Welsh	Philadelphia, Pa	Aug. 3, 1858	21, 098
Loom	J. Welsh	Philadelphia, Pa	Sept. 5, 1865	49, 814
Loom	J. Welsh	Philadelphia, Pa	July 17, 1866	56, 474
Loom	G. L. White	Woonsocket, R. I	Nov. 27, 1866	60, 101
Loom	R. Whitehill	New York, N. Y	Dec. 27, 1870	110, 524
Loom	W. Whiteside and J. Shinn	Philadelphia, Pa	June 5, 1855	13, 022
Loom	B. Wilcox	Franklin County, Ohio	Apr. 21, 1826	....
Loom	T. R. Williams	Newport, R. I	July 3, 1813	....
Loom	J. Wilson	Abbeville District, S. C	May 29, 1849	6, 487
Loom	F. M. Wolf	Glanchan, Saxony	Sept. 13, 1864	44, 271
Loom	E. Wood	Philadelphia, Pa	July 17, 1855	13, 284
Loom	J. C. Wood	Conshohocken, Pa	Jan. 19, 1869	85, 981
Loom	E. Wright and B. Fitts	Worcester, Mass	May 17, 1864	42, 812
Loom	W. Wright	Brookline, N. H	Oct. 12, 1814	....
Loom	H. Wyman	Worcester, Mass	Oct. 29, 1867	70, 309
Loom	H. Wyman	Worcester, Mass	July 2, 1872	128, 694
Loom	G. Yates and E. Clayton	Lancaster, Pa	June 13, 1854	11, 100
Loom	J. Zürcher	Lancaster, Pa	Aug. 19, 1873	141, 907
Loom	C. Zwicki	Pittsburgh, Pa	June 4, 1861	32, 501
Loom	C. Zwicki	Pittsburgh, Pa	Jan. 12, 1864	41, 254
Loom	C. Zwicki	Chicago, Ill	Mar. 26, 1867	63, 353
Loom, Bag	L. B. Jillson and G. Sparhawk	Lewiston, Me	Nov. 27, 1855	13, 848
Loom-beam	S. T. Thomas and E. Everett	Lawrence, Mass	Nov. 7, 1854	11, 919
Loom beam and roller	G. L. Garsed	Wilmington, Del	Apr. 8, 1873	137, 672
Loom-box-operating mechanism	H. Wyman	Worcester, Mass	Jan. 31, 1871	111, 417
Loom, Broad power	J. Leland	Milbury, Mass	Mar. 9, 1833	....
Loom, Brussels	E. B. Bigelow	Boston, Mass	Mar. 20, 1847	5, 020
Loom-cam	G. S. Faulkner	Staffordsville, Conn	July 16, 1867	66, 818
Loom-cam	F. E. Howe and L. Washburn	Stafford, Conn	Oct. 1, 1867	69, 438
Loom-cam	G. O. Wickers and T. J. McClary	North Andover, Mass	Mar. 8, 1870	100, 575
Loom, Carpet	J. R. Clark	South Coventry, Conn	May 7, 1830	....
Loom, Carpet	J. Haight	Harsimus, N. J	Mar. 17, 1834	....
Loom, Carpet	W. Sherwood	Ridgefield, Conn	July 20, 1846	4, 644
Loom, Carpet	J. A. Van Riper	New York, N. Y	Nov. 16, 1852	9, 401
Loom, Carpet and rug	W. Bacon	Philadelphia, Pa	Apr. 7, 1830	....
Loom, Cassinet	J. Hammond and J. McClelland	Williamsport, Pa	July 3, 1839	....
Loom, Carpet, &c., power	T. Flint	Boston, Mass	June 27, 1842	2, 696
Loom, Check and plaid power	H. Burt and O. D. and A. H. Boyd	Manchester, Conn	Aug. 19, 1828	....
Loom, Circular	M. C. Bryant	Lowell, Mass	Jan. 11, 1870	98, 738
Loom, Circular	J. J. Greenough	New York, N. Y	Nov. 21, 1865	51, 040
Loom, Circular	J. A. Grunwald	New York, N. Y	Dec. 27, 1859	26, 585
Loom, Circular	W. H. Walton and A. Nandam	New York and West Farms, N. Y.	Nov. 1, 1864	44, 902
Loom, Circular hat-weaving	J. V. Reed	New York, N. Y	Mar. 5, 1872	124, 288
Loom, Circular-weaving	J. Buser	New York, N. Y	Dec. 17, 1867	72, 362
Loom, Circular-weaving	A. Wagner	New York, N. Y	Aug. 25, 1868	81, 438
Loom cloth-beam	J. P. Hillard	Fall River, Mass	Aug. 1, 1871	117, 630
Loom-combs, Machine for shaping the elastic dents of.	M. Robinson	Accrington, England	Feb. 4, 1873	135, 444
Loom, Common and power	B. Lapham	Waterford, N. Y	Jan. 20, 1838	575
Loom, Cross-weaving	C. Roder	Philadelphia, Pa	Nov. 14, 1865	50, 990
Loom, Crank	E. Cutter	Walpole, N. H	Sept. 27, 1811	....
Loom, Crank-power	E. Burt	Manchester, Conn	Aug. 8, 1837	348
Loom, Damask	Tompkins and Gilroy	North Providence, R. I	May 9, 1835	....
Loom, Domestic	B. Maltby	New York, N. Y	Apr. 13, 1822	....
Loom, Domestic and factory	W. Janes	Ashford, Conn	Mar. 20, 1813	....
Loom, Family	C. Hathaway	Walton, N. Y	Oct. 11, 1814	....
Loom, Fancy	G. Crompton	Worcester, Mass	Jan. 9, 1866	51, 928
Loom, Fancy	L. J. Knowles	Warren, Mass	Feb. 24, 1863	37, 760
Loom, Fancy	C. Roder	Ceralvo, Ky	Oct. 18, 1859	25, 869
Loom, Fancy	S. T. Thomas	Laconia, N. H	Feb. 11, 1862	34, 381
Loom, Fancy check	E. Burt	Manchester, Conn	Jan. 24, 1854	10, 442
Loom, Fancy check power	E. Burt	Manchester, Conn	Feb. 4, 1851	7, 925
Loom, Figure power	W. Crompton	Taunton, Mass	Nov. 25, 1837	491
Loom, Figure power	J. M. Hoggan	New Haven, Conn	Oct. 14, 1834	....
Loom filling-fork	W. G. Duce and A. C. Eddy	Baltic, Conn., and Providence, R. I.	Oct. 20, 1868	83, 267
Loom filling-fork	J. H. Knowles	Lawrence, Mass	May 27, 1873	139, 251
Loom for figured fabrics	S. Eccles	Kensington, Pa	Mar. 5, 1850	7, 137
Loom for figured fabrics	J. Reynolds	Providence, R. I	Oct. 16, 1849	6, 797
Loom for making fishing and other nets	B. F. Jouannin and F. M. and A. Baudouin.	Paris, France	Aug. 16, 1864	43, 888
Loom for making fringe	G. Roth	New York, N. Y	Sept. 21, 1869	95, 145
Loom for making weavers' harness	K. Vogel	Chelsea, Mass	Oct. 25, 1853	10, 173

Index of patents issued from the United States Patent Office from 1790 *to* 1873, *inclusive*—Continued.

Invention.	Inventor.	Residence.	Date.	No.
Loom for operating shuttle-boxes	J. Ashworth	North Andover, Mass	Mar. 9, 1869	87, 616
Loom for piled fabric	J., jr., and J. Turnbull	Simsbury, Conn	Jan. 29, 1850	7, 061
Loom for regulating the delivery of the warp from the warp-beam.	W. H. Brayton	Warren, R. I	Jan. 6, 1844	3, 397
Loom for warping	C. Whitney	Dedham, Mass	Feb. 1, 1814	....
Loom for weaving	A. Faulkner	Walpole, N. H	Apr. 17, 1849	6, 316
Loom for weaving	C. Whiting, J. Goulding, and W. Janes.	Dedham, Mass	June 25, 1824	....
Loom for weaving bags	C. Baldwin	Manchester, N. H	Dec. 2, 1851	8, 553
Loom for weaving bags	W. Talbot	Sanford, Me	Nov. 28, 1854	12, 005
Loom for weaving bags	S. T. Thomas	Lawrence, Mass	Apr. 22, 1856	14, 746
Loom for weaving blinds, &c	B. R. Murphy	Parkersburgh, W. Va	Nov. 26, 1872	133, 332
Loom for weaving Brussels carpets, &c	E. B. Bigelow	Boston, Mass	Mar. 10, 1849	6, 153
Loom for weaving Brussels carpets, &c	E. B. Bigelow	Boston, Mass	Mar. 13, 1849	6, 186
Loom for weaving carpets	H. Baker	North Salem, N. Y	Apr. 14, 1826	....
Loom for weaving carpets, &c	E. B. Bigelow	Lancaster, Mass	May 16, 1842	2, 625
Loom for weaving carpets, &c	E. B. Bigelow	Lancaster, Mass	May 26, 1842	2, 639
Loom for weaving carpets	J. Marsden	Halifax, England	July 20, 1869	92, 856
Loom for weaving carpets, &c., Power	H. Skinner	Yonkers, N. Y	May 4, 1869	88, 694
Loom for weaving chip	P. Bennet	Rochester, Mass	Feb. 8, 1806	....
Loom for weaving cloth	A. Whittemore	....	Nov. 17, 1796	....
Loom for weaving cloth for stocks	C. Kile	Philadelphia, Pa	Oct. 11, 1836	47
Loom for weaving cloth with swells or gores	M. Opper	New York, N. Y	Aug. 14, 1866	57, 177
Loom for weaving coach-lace	J. H. Murrill	Richmond, Va	Oct. 4, 1853	10, 096
Loom for weaving coach-lace, &c., Power	E. B. Bigelow	West Boylston, Mass	Apr. 20, 1837	169
Loom for weaving concave and convex surfaces, Power.	W. Breitenstein	New York, N. Y	Oct. 9, 1866	58, 589
Loom for weaving corsets	W. Breitenstein	New York, N. Y	Jan. 27, 1863	37, 556
Loom for weaving corsets, &c	W. P. Brown	New York, N. Y	May 31, 1864	42, 986
Loom for weaving corsets	B. J. Goullioud	Paris, France	Jan. 27, 1863	37, 547
Loom for weaving counterpanes, &c., Power	E. B. Bigelow	Lancaster, Mass	July 28, 1842	2, 741
Loom for weaving counterpanes, &c., Power	E. B. Bigelow	Lancaster, Mass	Aug. 2, 1842	2, 744
Loom for weaving the covering of cord, &c., Circular	J. E. Palmer	Middletown, Conn	Dec. 27, 1864	45, 629
Loom for weaving coverlet	J. Meily, jr., and S. Mellinger	Lebanon, Pa	June 26, 1835	....
Loom for weaving cut-pile fabric	E. B. Bigelow	Boston, Mass	Oct. 17, 1854	11, 803
Loom for weaving cut-pile fabric	M. C. Bryant	Lowell, Mass	June 25, 1850	7, 452
Loom for weaving cut-pile fabric	M. C. Bryant	Lowell, Mass	Aug. 5, 1851	8, 283
Loom for weaving cut-pile fabrics, Jacquard	E. B. Bigelow	Clinton, Mass	Mar. 18, 1851	7, 983
Loom for weaving double-faced pile-fabrics	P. Joyot, jr	Paris, France	Feb. 14, 1865	46, 433
Loom for weaving embroidered fabrics	J. G. Spitzli	Millville, Mass	Nov. 21, 1865	51, 095
Loom for weaving fancy goods	B. F. Rice	Clinton, Mass	Oct. 18, 1853	10, 138
Loom for weaving figured cloth	W. Levally	Canterbury, Conn	Mar. 20, 1829	....
Loom for weaving figured fabrics	C. W. Blanchard	Clinton, Mass	Aug. 3, 1852	9, 162
Loom for weaving figured fabrics	G. Crompton	Worcester, Mass	Nov. 14, 1854	11, 933
Loom for weaving figured fabrics	S. and J. Eccles	Kensington, Pa	Aug. 3, 1852	9, 168
Loom for weaving figured fabric	R. Garsed	Frankford, Pa	Nov. 6, 1849	6, 845
Loom for weaving figured fabrics, &c	C. G. Gilroy	Great Britain	Mar. 12, 1842	2, 486
Loom for weaving figured fabric	C. G. Gilroy	New York	Apr. 15, 1843	3, 047
Loom for weaving figured fabrics	B. H. Jenks and R. B. Goodyer.	Bridesburgh, and Philadelphia, Pa.	Apr. 13, 1852	8, 874
Loom for weaving figured fabrics	M. Marshall	Lowell, Mass	Dec. 11, 1849	6, 939
Loom for weaving figured fabrics	S. T. Thomas and E. Everett	Lowell and Lawrence, Mass	Dec. 24, 1850	7, 861
Loom for weaving figured fabrics, Hand	W. Townshend	Rochester, N. H	Apr. 26, 1845	4, 015
Loom for weaving figured goods	A. Babbett	Auburn, N. Y	Oct. 8, 1850	7, 714
Loom for weaving figured goods	H. Baker	North Salem, N. Y	Aug. 30, 1827	....
Loom for weaving figured work	J. Smith	Shafferstown, Pa	Apr. 22, 1835	....
Loom for weaving fringe	L. D. Valletton	Philadelphia, Pa	Aug. 4, 1868	80, 786
Loom for weaving fringe, Hand	E. A. B. Judkins	Portland, Me	Feb. 2, 1839	1, 075
Loom for weaving fringed counterpanes, Power	E. B. Bigelow	Lancaster, Mass	Apr. 24, 1840	1, 561
Loom for weaving hair-cloth	I. Angell	Pawtucket, R. I	Nov. 13, 1860	30, 652
Loom for weaving hair-cloth	S. B. and S. M. Chaffee	Providence, R. I	Oct. 12, 1858	21, 793
Loom for weaving hair-cloth	H. Halvorson	Hartford, Conn	Sept. 27, 1853	10, 044
Loom for weaving hair-cloth	W. S. Laycock	Sheffield, England	Oct. 8, 1872	132, 014
Loom for weaving hair-cloth	I. Lindsley	Providence, R. I	June 25, 1861	32, 634
Loom for weaving hair-cloth	I. Lindsley	North Providence, R. I	Nov. 15, 1864	45, 107
Loom for weaving hair-cloth	I. Lindsley	Pawtucket, R. I	Sept. 26, 1871	119, 277
Loom for weaving hair-cloth	I. Lindsley	Pawtucket, R. I	Sept. 26, 1871	119, 278
Loom for weaving hair-cloth	J. Turpie	New York, N. Y	June 17, 1873	139, 981
Loom for weaving hair-cloth	J. S. Winsor	Providence, R. I	May 31, 1864	42, 982
Loom for weaving hair-cloth and fabrics produced thereon.	I. Lindsley	Pawtucket, R. I	Sept. 26, 1871	119, 276
Loom for weaving hair-cloth, Power	I. Lindsley	Pawtucket, R. I	Oct. 25, 1864	44, 808
Loom for weaving hats	P. Brooks	New Haven, Conn	June 15, 1869	91, 305
Loom for weaving hats, &c	P. L. Slayton	New York, N. Y	Nov. 22, 1864	45, 208
Loom for weaving hats, Circular	L. Bonard	New York, N. Y	June 4, 1861	32, 461
Loom for weaving hats, &c., Circular	P. L. Slayton	New York, N. Y	Feb. 2, 1864	41, 466
Loom for weaving hats, Circular	P. L. Slayton and C. I. Kane	New York, N. Y	Oct. 23, 1866	59, 138
Loom for weaving ingrain carpets	E. B. Bigelow	Boston, Mass	Feb. 9, 1869	86, 806
Loom for weaving ingrain carpets, Power	W. and J. W. Murkland	Lowell, Mass	Nov. 23, 1869	97, 106
Loom for weaving ingrain carpets, Power	A. Murray	Lowell, Mass	Apr. 20, 1869	89, 065
Loom for weaving irregular fabrics	A. Goullioud	Barcelona, Spain	Mar. 22, 1870	101, 006
Loom for weaving irregular fabrics	C. Heptonstall	Providence, R. I	July 20, 1869	92, 722
Loom for weaving knotted counterpanes, &c	E. B. Bigelow	West Boylston, Mass	Jan. 6, 1838	546
Loom for weaving palm-leaf, &c	G. W. Chandler	Fitchburgh, Mass	Dec. 10, 1867	71, 852
Loom for weaving palm-leaf	J. C. Smith	Chicopee, Mass	July 14, 1868	79, 923
Loom for weaving palm-leaf, straw, &c	J. M. Baker	Providence, R. I	Nov. 15, 1864	45, 115
Loom for weaving pile-fabrics	E. B. Bigelow	Clinton, Mass	Jan. 14, 1851	7, 898
Loom for weaving pile-fabrics	E. B. Bigelow	Boston, Mass	Nov. 15, 1853	10, 222
Loom for weaving pile-fabrics	E. B. Bigelow	Boston, Mass	Dec. 18, 1855	13, 936
Loom for weaving pile-fabrics	E. B. Bigelow	Boston, Mass	May 5, 1857	17, 189
Loom for weaving pile-fabrics	M. C. Bryant	Lowell, Mass	Mar. 19, 1850	7, 180
Loom for weaving pile-fabrics	E. K. Davis	New York, N. Y	Feb. 9, 1869	86, 651
Loom for weaving pile-fabrics	J. C. Ellison	Philadelphia, Pa	Oct. 28, 1873	144, 077
Loom for weaving pile-fabrics	L. Ferguson	Lowell, Mass	Aug. 25, 1868	81, 487
Loom for weaving pile-fabrics	L. Ferguson	Lowell, Mass	Nov. 9, 1869	96, 570
Loom for weaving pile-fabrics	W. G. Hartley	Saxonville, Mass	Apr. 9, 1867	63, 631
Loom for weaving pile-fabrics	J. Johnson	Troy, N. Y	Mar. 12, 1850	7, 168
Loom for weaving pile-fabrics	J. Johnson	Troy, N. Y	Aug. 5, 1851	8, 281

Index of patents issued from the United States Patent Office from 1790 to 1873, inclusive—Continued.

Invention.	Inventor.	Residence.	Date.	No.
Loom for weaving pile-fabrics	C. A. Maxfield	Troy, N. Y	Jan. 13, 1852	8,656
Loom for weaving pile-fabrics	E. Pickford	New Brunswick, N. J	May 27, 1873	139,328
Loom for weaving pile-fabrics	S. Richardson	Claremont, N. H	Aug. 10, 1852	9,188
Loom for weaving pile-fabrics	H. Skinner	Yonkers, N. Y	Dec. 30, 1873	146,101
Loom for weaving pile-fabrics	W. Webster	Morrisania, N. Y	Oct. 6, 1868	82,773
Loom for weaving pile-fabrics	W. Webster	Morrisania, N. Y	Aug. 27, 1872	130,960
Loom for weaving pile-fabrics	W. Webster	Morrisania, N. Y	Aug. 27, 1872	130,961
Loom for weaving pile-fabrics	W. Webster	Morrisania, N. Y	Sept. 10, 1872	131,197
Loom for weaving pile-fabrics double	E. B. Bigelow	Boston, Mass	Jan. 13, 1857	16,370
Loom for weaving pile-fabrics, Power	E. B. Bigelow	Boston, Mass	Aug. 17, 1869	93,800
Loom for weaving pile-fabrics without the figuring-wires.	R. W. Sievier	Cavendish Square, England.	June 1, 1852	8,988
Loom for weaving plaids, &c	M. A. Furbush and G. Crompton.	Worcester, Mass	May 31, 1859	24,206
Loom for weaving plaids, &c., Power	E. B. Bigelow	Boston, Mass	Apr. 10, 1846	3,987
Loom for weaving plain cloth, Power	F. Downing	Enfield, Mass	Jan. 27, 1843	2,935
Loom for weaving plush or pile fabric	S. Holt	Newark, N. J	Mar. 7, 1865	46,754
Loom for weaving narrow goods, Silk	C. Bergen	Brooklyn, N. Y	Dec. 18, 1839	1,432
Loom for weaving narrow ware	B. Oldfield	Newark, N. J	Oct. 2, 1866	58,464
Loom for weaving ribbon, &c	J. Rushworth	New York, N. Y	Oct. 8, 1867	69,708
Loom for weaving satinet, kerseymere, and other cloth.	J. D. Seagrave	Uxbridge, Mass	May 17, 1838	740
Loom for weaving seamless bags	S. Northrop	New Milford, Conn	Jan. 1, 1851	7,876
Loom for weaving skirt-fringe	J. Beck	New York, N. Y	Aug. 31, 1858	21,312
Loom for weaving skirts, Warp-beam for	F. X. Loughery	Kelleysville, Pa	May 11, 1869	89,935
Loom for weaving slat-blinds	J. L. Devol	Parkersburgh, W. Va	May 28, 1872	127,318
Loom for weaving slat-blinds	M. Free	Philadelphia, Pa	July 17, 1866	56,493
Loom for weaving slats, window-shades, &c	E. B. Hastings, W. L. Merritt, and W. E. Nichols.	Palmer and Templeton, Mass.	Feb. 4, 1873	135,427
Loom for weaving slatted window-shades	G. Hasecoster	Richmond, Va	Sept. 11, 1866	57,898
Loom for weaving stock-frames, Power	F. Goodell and F. W. Harvey	Ramapo, N. Y	Dec. 2, 1835	....
Loom for weaving stocks	C. Kile	Erie, Pa	Sept. 18, 1835	
Loom for weaving suspender-webbing	W. V. Gee	New Haven, Conn	Sept. 18, 1855	13,571
Loom for weaving tape, &c	J. Duckworth	Pittsfield, Mass	Nov. 9, 1869	96,564
Loom for weaving tape, ribbon, &c	L. J. Knowles	Warren, Mass	May 15, 1866	54,742
Loom for weaving tapestry and Brussels carpets	E. B. Bigelow	Clinton, Mass	Sept. 24, 1850	7,660
Loom for weaving tapestry carpets with parti-colored warp.	E. B. Bigelow	Clintonville, Mass	Jan. 7, 1851	7,884
Loom for weaving trimmings	L. D. Valetton	Philadelphia, Pa	Apr. 12, 1864	42,335
Loom for weaving tufted pile-fabrics, Power	H. Skinner	West Farms, N. Y	Sept. 1, 1863	39,759
Loom for weaving, Water, steam, or other power	S. Blydenburgh and H. Healy	Worcester, Mass	Feb. 20, 1815	....
Loom for weaving webbing, tape, &c	A. G. Bill and G. Spalding	Middletown, Conn	Mar. 28, 1831	
Loom for weaving wire	S. Holdsworth	Maspeth, N. Y	Apr. 22, 1873	138,090
Loom for weaving wire	G. W. Smith	Mauch Chunk, Pa	Dec. 25, 1855	14,000
Loom for weaving wire-cloth	L. Kittinger	Massillon, Ohio	Jan. 26, 1869	86,233
Loom for weaving wire-cloth	C. H. Waters	Groton, Mass	Sept. 2, 1862	36,377
Loom for weaving wire-cloth	G. F. Wright and C. K. Sawyer	Clinton, Mass	Sept. 24, 1872	131,584
Loom for weaving wire-cloth, Power	E. B. Bigelow	Boston, Mass	Oct. 6, 1857	18,320
Loom for weaving wire-gauze	M. Bretzger	Pittsburgh, Pa	Dec. 5, 1846	4,873
Loom for weaving wire, Power	J. S. Gustin	New York	Feb. 23, 1826	....
Loom for working any number of heddles, Weaver's	G. McCrae	Baltimore, Md	Oct. 18, 1843	3,309
Loom, Fringe	S. Walker	Roxbury, Mass	Nov. 9, 1858	22,042
Loom, Garment-weaving	S. Leather	Dalton, England	May 5, 1868	77,498
Loom, Hair-cloth	J. Noblit	Philadelphia, Pa	Apr. 9, 1861	31,992
Loom, Hand	J. Albertson and S. C. Byers	Richmond, Ind	July 21, 1868	80,047
Loom, Hand	J. L. Branson	Cincinnati, Ohio	Mar. 27, 1866	53,398
Loom, Hand	J. L. Branson	Cincinnati, Ohio	July 14, 1868	79,946
Loom, Hand	J. D. Browne	Cincinnati, Ohio	June 25, 1867	65,997
Loom, Hand	A. Carter and R. Spake	Salem, Iowa	May 8, 1866	54,501
Loom, Hand	T. A. Dugdale	Richmond, Ind	Sept. 17, 1867	68,969
Loom, Hand	G. W. Firestone	Fredericksburgh, Ohio	Oct. 29, 1867	70,186
Loom, Hand	W. S. Freeman	West Union, Ohio	Aug. 4, 1868	80,717
Loom, Hand	I. H. Garretson	Clay, Iowa	Feb. 18, 1851	7,936
Loom, Hand	J. G. Garretson	Salem, Iowa	May 21, 1850	7,378
Loom, Hand	J. G. Garretson	Salem, Iowa	Jan. 19, 1864	41,290
Loom, Hand	G. Harsin and T. M. Kirkpatrick.	Kirkville, Iowa	Feb. 11, 1868	74,351
Loom, Hand	J. W. Hayse	Salem, Iowa	Aug. 22, 1865	49,589
Loom, Hand	J. G. Henderson	Palmyra, Mo	Jan. 1, 1861	31,019
Loom, Hand	H. D. Hunt	Danville, Ill	Oct. 8, 1867	69,673
Loom, Hand	J. G. and H. T. Henderson	Salem, Iowa	Mar. 14, 1865	46,798
Loom, Hand	A. Jones	Clinton, Ill	Oct. 26, 1869	96,238
Loom, Hand	R. Lloyd	Philadelphia, Pa	Feb. 8, 1810	....
Loom, Hand	E. Lowe	Burrows, Ind	Sept. 1, 1868	81,656
Loom, Hand	C. L. McDowell	Wassonville, Iowa	Jan. 30, 1866	52,306
Loom, Hand	D. Mendenhall	Fairfield, Iowa	Oct. 16, 1866	58,865
Loom, Hand	S. C. Mendenhall	Richmond, Ind	Jan. 13, 1857	16,392
Loom, Hand	S. C. Mendenhall	Richmond, Ind	Jan. 19, 1864	41,313
Loom, Hand	S. C. Mendenhall	Richmond, Ind	Nov. 12, 1867	70,877
Loom, Hand	S. C. Mendenhall and O. and E. King.	Richmond, Ind., and Salem, Iowa.	Nov. 9, 1852	9,388
Loom, Hand	S. C. Mendenhall and S. Sparks	Richmond, Ind	Oct. 23, 1866	59,132
Loom, Hand	A. and P. P. Meredith	Maxintuckee, Ind	Nov. 19, 1867	71,035
Loom, Hand	H. H. Mitchell	Mineral Point, Wis	Dec. 31, 1867	72,879
Loom, Hand	J. A. Mitchell	Ringgold, Ga	Dec. 20, 1853	10,340
Loom, Hand	A. R. Nixon	Rhea Springs, Tenn	Oct. 11, 1859	25,756
Loom, Hand	J. E. Nute	Lincoln, Me	Sept. 6, 1870	107,094
Loom, Hand	J. E. Nute and G. H. Hathorn	Lincoln, Me	Jan. 14, 1868	73,256
Loom, Hand	J. Pelsor	Brooklyn, Ill	Mar. 20, 1866	53,330
Loom, Hand	C. Roder	Ceralvo, Ky	Mar. 8, 1864	41,863
Loom, Hand	A. Rosenberger	Brandonville, W. Va	Apr. 9, 1868	63,752
Loom, Hand	J. Seaman and W. Y. Henderson.	Andover, N. Y	Sept 19, 1865	50,041
Loom, Hand	O. Strong	Green Centre, Ind	Jan. 14, 1868	73,266
Loom, Hand	T. G. Thompson and A. F. Fox.	Richmond and Greensborough, Ind.	Sept. 17, 1867	69,047
Loom, Hand	T. H. Tibbles	Kansas City, Mo	Nov. 19, 1867	71,087
Loom, Hand	C. Unverzagt	Richmond, Ind	Feb. 13, 1866	52,626

Index of patents issued from the United States Patent Office from 1790 to 1873, inclusive—Continued.

Invention.	Inventor.	Residence.	Date.	No.
Loom, Hand	C. Unverzagt	Terre Haute, Ind	Jan. 28, 1868	73, 852
Loom, Hand	W. B. Walker and N. D. Hartley	Salem, Iowa	Aug. 14, 1866	57, 226
Loom, Hand	C. Wandel	Milton, Iowa	Sept. 17, 1867	68, 918
Loom, Hand	J. Whitehead	Oskaloosa, Iowa	Dec. 4, 1866	60, 305
Loom, Hand and power	J. Thorp	Providence, R. I	Mar. 28, 1812	....
Loom, Hand-power	G. A. Brown and J. A. Hunt	Oakalla, Ill	Jan. 19, 1869	86, 055
Loom, Hand-power	J. M. Deen, W. B. Bolding, and H. Perry.	Dayton, Iowa	Mar. 26, 1867	63, 143
Loom-harness	J. Blackmar	Brooklyn, Conn	Oct. 20, 1836	64
Loom, Harness	D. C. Brown	Lowell, Mass	Jan. 2, 1866	51, 794
Loom-harness	D. C. Brown	Lowell, Mass	Jan. 7, 1868	72, 967
Loom-harness	A. B. Corey	Providence, R. I	Sept. 28, 1869	95, 198
Loom-harness	G. Matoon	Chicopee Falls, Mass	Sept. 15, 1857	18, 208
Loom, Harness	H. Parsons	Waterloo, N. Y	Aug. 12, 1862	36, 166
Loom-harness-actuating mechanism	J. F. Foss	Lowell, Mass	July 2, 1872	128, 480
Loom-harness board, Jacquard	E. Everett and S. T. Thomas	Lawrence and Lowell, Mass	Jan. 18, 1853	9, 545
Loom-harness clasp	G. Copeland	Danville, Me	Sept. 26, 1854	11, 719
Loom-harness connection	J. T. Holden	Elmwood, R. I	Mar. 16, 1869	87, 848
Loom-harness frame	D. C. Brown	Lowell, Mass	May 5, 1868	77, 446
Loom-harness frame	J. Greenhalgh, sr	Pascoag, R. I	May 15, 1860	28, 270
Loom-harness, Heddle-eye for	J. L. Cheney	Lowell, Mass	Jan 9, 1866	51, 920
Loom-harness, Machine for making	J. Blackmar	Brooklyn, Conn	Oct. 20, 1836	61
Loom-harness, Machine for making	J. Senneff	Philadelphia, Pa	June 26, 1855	13, 152
Loom-harness, Machine for making	J. Sladdin	Lawrence, Mass	Aug. 4, 1868	80, 774
Loom-harness, Machine for making	J. Sladdin	Lawrence, Mass	Nov. 28, 1871	121, 258
Loom-harness, Machine for making wire heddles for.	D. C. Brown and J. Ashworth	Lowell and North Andover, Mass.	May 12, 1868	77, 713
Loom-harness, Machine for making wire heddles for.	A. Howe and S. S. Granniss	Morrisville, N. Y	Sept. 30, 1841	2, 273
Loom-harness, Machine for manufacturing	D. C. Brown	Lowell, Mass	Feb. 15, 1853	9, 584
Loom-harness, Machine for manufacturing	J. H. Crowell	Providence, R. I	July 25, 1871	117, 389
Loom-harness, Machine for varnishing and dressing	E. J. Ellis	Lewiston, Me	Aug. 23, 1870	106, 676
Loom-harness, Machine for weaving	G. W. and W. R. Harris	Lowell, Mass., and Manchester, England.	Sept. 13, 1864	44, 186
Loom-harness, Machine for weaving	S. Holton, jr., and W. R. Harris	Middlebury, Vt	Sept. 4, 1849	6, 691
Loom-harness, Manufacture of wire heddles for	A. Howe and S. S. Grannis	Morrisville, N. Y	Oct. 11, 1841	2, 299
Loom-harness mechanism	J. Booth	Pottstown, Pa	Dec. 12, 1871	121, 842
Loom-harness mechanism	A. F. Gibboney	Belleville, Pa	Sept. 28, 1869	95, 221
Loom-harness mechanism	J. Greenhalgh	Woonsocket, R. I	Feb. 25, 1868	74, 819
Loom-harness, Mechanism for operating	O. Plummer and J. Schofield	Worcester, Mass	Aug. 11, 1868	80, 876
Loom-harness, Mode of counterbalancing	J. Greenhalgh	Waterford, Mass	Nov. 2, 1852	9, 377
Loom-harness, Mode of hanging and guiding	L. S. Fisher	Broadhead, Wis	Apr. 16, 1867	63, 877
Loom-harness needle, Machinery for filling	L. S. Reynolds	Manchester, N. H	Mar. 29, 1859	23, 393
Loom-harness, Operating	F. Traub	Philadelphia, Pa	Apr. 28, 1868	77, 335
Loom-harness-operating mechanism	W. R. Andrews	Mystic River, Conn	Oct. 23, 1866	58, 969
Loom-harness-operating mechanism	J. Ashworth	North Andover, Mass	Aug. 16, 1870	106, 306
Loom-harness-operating mechanism	J. Bachelder and W. H. Bliss, 2d.	Norwich, Conn	Mar. 31, 1868	76, 036
Loom-harness-operating mechanism	E. B. Bigelow	Boston, Mass	Aug. 17, 1869	93, 799
Loom-harness-operating mechanism	G. Crompton	Worcester, Mass	Apr. 7, 1868	76, 406
Loom-harness-operating mechanism	G. Crompton	Worcester, Mass	Jan. 31, 1871	111, 324
Loom-harness-operating mechanism	J. C. Duckworth	Pittsfield, Mass	Dec. 6, 1870	109, 884
Loom-harness-operating mechanism	G. S. Faulkner	Indianapolis, Ind	Jan. 25, 1870	99, 176
Loom-harness operating mechanism	A. R. Field	Central Falls, R. I	Aug. 23, 1870	106, 571
Loom-harness-operating mechanism	J. F. Gebhart	New Albany, Ind	June 15, 1869	91, 431
Loom-harness-operating mechanism	R. B. Goodyear	Wilmington, Del	Oct. 3, 1871	119, 459
Loom-harness-operating mechanism	O. Plummer	Worcester, Mass	Sept. 12, 1871	118, 969
Loom-harness, Shuttles for machines for knitting	D. C. Brown and J. Ashworth	Lowell, Mass	June 21, 1864	43, 252
Loom-harness, Wire	E. Brown	Cazenovia, N. Y	Oct. 30, 1828	....
Loom-harness, Wire heddle for	M. Finkle	New York, N. Y	Aug. 8, 1865	49, 251
Loom harness, Warp-eyes of wire heddle for	D. C. Brown	Lowell, Mass	May 21, 1867	64, 944
Loom-heddle	J. Ashworth	Andover, Mass	Jan. 16, 1872	122, 795
Loom-heddle	E. G. Jelley	Pawtucket, R. I	May 25, 1869	90, 364
Loom-heddle	C. Kennedy	Philadelphia, Pa	Sept. 3, 1867	68, 444
Loom-heddle	J. Senneff	Philadelphia, Pa	Dec. 31, 1867	72, 917
Loom-heddle	C. Whipple	Providence, R. I	Feb. 13, 1872	123, 601
Loom heddle and harness	B. Hartford and W. B. Tilton	Enfield, N. H	Dec. 29, 1837	544
Loom-heddle frame	M. Finkle	New York, N. Y	June 6, 1865	48, 057
Loom-heddle, Metallic	W. Janes and D. Bolles	Ashford, Conn	Apr. 26, 1815	....
Loom-heddle, Metallic	C. Strong	Hartford, Vt	Apr. 24, 1840	1, 563
Loom-heddle motion	B. H. Jenks and R. B. Goodyear	Philadelphia, Pa	May 22, 1866	55, 010
Loom-heddles, Apparatus for varnishing	J. L. Lairdieson	Troy, N. Y	Dec. 17, 1861	33, 948
Loom index-chain	B. H. Jenks	Bridesburgh, Pa	Feb. 19, 1867	62, 204
Loom-indicator	A. Fellows, T. B. Harrison, and H. Dyer.	Marquoketa, Iowa	Feb. 4, 1868	73, 962
Loom, Ingrain carpet	W. Sherwood	Worcester, Mass	Jan. 28, 1830	....
Loom-jack-operating mechanism	C. H. Knowlton	Camden, N. J	Nov. 16, 1869	96, 925
Loom, Jacquard	C. B. Bigelow	Clintonville, Mass	Oct. 23, 1849	6, 806
Loom, Jacquard	A. Calderhead	Philadelphia, Pa	Feb. 3, 1841	1, 964
Loom, Jacquard	J. C. Cooke	Waterbury, Conn	Sept. 9, 1856	15, 717
Loom, Jacquard	J. Goulding	Worcester, Mass	Aug. 3, 1852	9, 171
Loom, Jacquard	J. Goulding	Worcester, Mass	July 8, 1856	15, 291
Loom, Jacquard	H. W. Hensel and L. D. Valetton.	Philadelphia, Pa	June 16, 1863	38, 929
Loom, Jacquard	L. Holms	Andover, Mass	Mar. 27, 1847	5, 033
Loom, Jacquard	J. Perrins	Philadelphia, Pa	May 28, 1846	4, 537
Loom, Jacquard	M. Wolf	Philadelphia, Pa	June 9, 1868	78, 779
Loom jaw-temple	E. B. Bigelow	Boston, Mass	Feb. 24, 1845	3, 925
Loom jaw-temple	G. Draper	Ware, Mass	Feb. 27, 1849	6, 144
Loom jaw-temple	B. Peck	Rehoboth, Mass	Sept. 26, 1848	5, 797
Loom jaw-temples, Opening and closing	E. Williams and D. L. Huntington.	Norwich, Conn	Nov. 16, 1841	2, 363
Loom, Knitting	W. Henson	Newark, N. J	Mar. 2, 1852	8, 773
Loom, Knitting	P. E. Ladrange	Vignory, France	Oct. 16, 1844	3, 798
Loom, Knitting	J. Mee	Lowell, Mass	May 10, 1853	9, 718
Loom, Knitting	R. Walker and J. McIntire	Portsmouth, N. H	Feb. 12, 1844	3, 436
Loom, Lappet	W. Aspinall	Philadelphia, Pa	June 28, 1870	104, 687
Loom, Lappet	J. Clough and J. Crompton	Chicopee, Mass	Mar. 21, 1871	112, 785

Index of patents issued from the United States Patent Office from 1790 *to* 1873, *inclusive*—Continued.

Invention.	Inventor.	Residence.	Date.	No.
Loom, Lappet or embroidering	F. W. Newton	South Orange, N. J	Dec. 8, 1868	84,753
Loom, Lappet-weaving	W. Aspinall	Manayunk, Pa	Oct. 31, 1865	50,764
Loom let-off	L. C. Briggs	Boston, Mass	Aug. 18, 1868	81,133
Loom let-off	G. Crompton	Worcester, Mass	Apr. 26, 1864	42,461
Loom let-off	G. Draper	Hopedale, Mass	Aug. 4, 1868	80,534
Loom let-off	S. Estes	Newburyport, Mass	Apr. 17, 1866	54,066
Loom let-off	H. Fiske	Farnumsville, Mass	Aug. 15, 1865	49,472
Loom let-off	W. H. Gray	Dover, N. H	Apr. 9, 1861	31,969
Loom let-off	D. Hussey	Nashua, N. H	Oct. 23, 1866	59,024
Loom let-off	J. A. Marden	Chelsea, Mass	Jan. 14, 1868	73,353
Loom let-off	J. Phillips	Pawtucket, R. I	Sept. 26, 1865	50,156
Loom let-off	W. W. Pomeroy	East Hampton, Mass	Aug. 15, 1865	49,479
Loom let-off	G. Richardson	Lowell, Mass	Apr. 23, 1867	64,147
Loom let-off	G. Richardson	Lowell, Mass	June 11, 1867	65,606
Loom let-off	J. Remick	Newburyport, Mass	Oct. 29, 1867	70,265
Loom let-off	T. S. Smith	Boston, Mass	Jan. 21, 1868	73,470
Loom let-off	E. Wright	Worcester, Mass	Feb. 20, 1866	52,780
Loom let-off and take-up	J. Pender	Worcester, Mass	Feb. 4, 1868	73,993
Loom let-off and take-up	J. F. Kirkwood	Thistle, Md	Oct. 2, 1866	58,433
Loom let-off and take-up mechanism	J. A. Marden	Newburyport, Mass	Apr. 30, 1867	64,235
Loom let-off and take-up motion	D. Bassett	Killingly, Conn	Nov. 21, 1865	51,003
Loom let-off and tension mechanism, Power	E. B. Bigelow	Boston, Mass	Dec. 13, 1870	110,000
Loom let-off mechanism	E. B. Bigelow	Boston, Mass	May 24, 1870	103,415
Loom let-off mechanism	E. B. Bigelow	Boston, Mass	Apr. 4, 1871	113,483
Loom let-off mechanism	T. Booth and C. C. Sanderson	Norway, Me	Nov. 24, 1868	84,253
Loom let-off mechanism	L. C. Briggs and A. Howard	Boston, Mass	June 15, 1869	91,207
Loom let-off mechanism	M. Brookfield	Brookfield, Iowa	Apr. 9, 1867	63,699
Loom let-off mechanism	M. C. Burleigh	Somersworth, N. H	Aug. 23, 1870	106,543
Loom let-off mechanism	M. C. Burleigh	Somersworth, N. H	Aug. 6, 1872	130,107
Loom let-off mechanism	B. F. Carter	Manville, R. I	Aug. 25, 1868	81,342
Loom let-off mechanism	B. F. Carter	Woonsocket, R. I	Jan. 19, 1869	85,905
Loom let-off mechanism	W. R. Clark	North Adams, Mass	June 8, 1869	90,996
Loom let-off mechanism	J. Clegg	Warwick, R. I	Feb. 14, 1871	111,816
Loom let-off mechanism	D. M. Collins	Lowell, Mass	Sept. 28, 1869	95,197
Loom let-off mechanism	J. Day	Paterson, N. J	May 3, 1870	102,506
Loom let-off mechanism	G. Draper	Milford, Mass	Aug. 11, 1863	39,469
Loom let-off mechanism	G. Draper	Milford, Mass	Jan. 7, 1868	72,987
Loom let-off mechanism	L. N. Fletcher and J. M. Page	Lowell, Mass	July 5, 1870	104,949
Loom let-off mechanism	W. T. Flinn	Philadelphia, Pa	Apr. 4, 1871	113,415
Loom let-off mechanism	C. Gahren and L. Langlotz	New York, N. Y	Apr. 29, 1873	138,391
Loom let-off mechanism	T. Gorrell	Warren, R. I	May 9, 1871	114,667
Loom let-off mechanism	F. W. Graichen	Olneyville, R. I	Aug. 8, 1871	117,768
Loom let-off mechanism	W. Hall	North Adams, Mass	Aug. 4, 1868	80,625
Loom let-off mechanism	W. Hall	North Adams, Mass	July 6, 1869	92,305
Loom let-off mechanism	W. H. Howard	Philadelphia, Pa	Aug. 8, 1871	117,779
Loom let-off mechanism	D. Hussey	Nashua, N. H	Apr. 14, 1868	76,636
Loom let-off mechanism	S. M. Linscott	Buxton, Me	Aug. 12, 1873	141,720
Loom let-off mechanism	D. Long and J. Preston	Fairview, Pa	Jan. 18, 1870	99,044
Loom let-off mechanism	J. Magee	Usquepaugh, R. I	June 14, 1870	104,328
Loom let-off mechanism	E. McDaniel	Lowell, Mass	Dec. 21, 1869	98,082
Loom let-off mechanism	P. McGee	North Providence, R. I	Sept. 29, 1868	82,625
Loom let-off mechanism	W. Potter and L. J. Labounty	Lowell, Mass	Sept. 14, 1869	94,769
Loom let-off mechanism	H. A. Remington	Anthony, R. I	Jan. 17, 1871	111,002
Loom let-off mechanism	R. Reynolds	Stockport, N. Y	Apr. 13, 1869	88,906
Loom let-off mechanism	C. R. Saatweber	New York, N. Y	Mar. 12, 1872	124,632
Loom let-off mechanism	C. Schilling	Auburn, N. Y	Nov. 21, 1871	121,131
Loom let-off mechanism	T. S. Smith	Charlestown, Mass	Dec. 1, 1868	84,515
Loom let-off mechanism	T. S. Smith and A. B. Ely	Boston and Newton, Mass	Jan. 14, 1868	73,399
Loom let-off mechanism	B. Tweedle	Philadelphia, Pa	July 16, 1872	129,628
Loom let-off mechanism	A. J. Woodman	Indian Orchard, Mass	Nov. 14, 1871	120,843
Loom let-off mechanism	E. Wright	Worcester, Mass	Dec. 31, 1867	72,769
Loom let-off mechanism	C. H. Young	River Point, R. I	Aug. 25, 1868	81,453
Loom let-off mechanism, Narrow-ware	J. N. Leavenworth	Hamden, Conn	Jan. 15, 1867	61,218
Loom let-off mechanism, Power	P. Boylan	Gloucester, N. J	May 10, 1864	42,638
Loom let-off motion	S. Estes	Newburyport, Mass	Sept. 12, 1865	49,950
Loom let-off motion	F. W. Graichen	Providence, R. I	June 27, 1871	116,437
Loom let-off motion	W. H. Gray	Dover, N. H	June 28, 1859	24,553
Loom let-off motion	W. H. Gray	Dover, N. H	June 28, 1859	24,602
Loom let-off motion	J. Knowles	Buffalo, N. Y	Apr. 30, 1850	7,324
Loom let-off motion	J. Myers	Biddeford, Me	Mar. 10, 1849	6,159
Loom let-off motion	R. Reynolds	Stockport, N. H	May 21, 1861	32,382
Loom let-off motion	G. Smith	Cumberland, R. I	Dec. 17, 1867	72,235
Loom let-off motion	A. B. Taylor and S. Wilcox, jr	Mystic, Conn., and Westerly, R. I.	Feb. 22, 1853	9,596
Loom let-off motion	R. Walker	Milford, Mass	Feb. 19, 1867	62,168
Loom let-off motion	R. Walker	Milford, Mass	Oct. 29, 1867	70,297
Loom let-off motion, Power	S. O. Colvin	Coventry, R. I	Jan. 12, 1858	19,073
Loom let-off motion, Power	W. H. Gray	Dover, N. H	Aug. 4, 1857	17,926
Loom let-off motion, Power	R. Reynolds	Stockport, N. Y	June 28, 1864	43,338
Loom let-off motion, Power	N. Wyllys	South Glastenbury, Conn	Mar. 16, 1858	19,664
Loom, Lint	F. Hall	Charlestown, Mass	Nov. 28, 1817	......
Loom measuring-apparatus	J. A. Bradshaw and S. L. Crockett	Lowell, Mass	Nov. 1, 1870	108,871
Loom, Narrow-ware	R. B. Fowler	Worcester, Mass	Nov. 18, 1873	144,610
Loom, Narrow-ware	L. J. Knowles	Warren, Mass	Nov. 13, 1866	59,613
Loom, Narrow-ware	L. J. Knowles	Warren, Mass	Dec. 11, 1866	60,388
Loom, Narrow-ware	S. Walker	Boston, Mass	Jan. 29, 1867	61,583
Loom, Narrow-ware	A. Williamson	New York, N. Y	May 1, 1860	28,135
Loom or lathe for weaving	W. Harris		Mar. 15, 1800	......
Loom-picker	J. Bardsley and A. W. Gordon	Willimantic, Conn	Apr. 26, 1864	42,456
Loom-picker	S. Boorn	Lowell, Mass	Mar. 26, 1861	31,776
Loom-picker	S. Boorn	Lowell, Mass	Apr. 9, 1861	31,948
Loom-picker	S. Boorn	Lowell, Mass	Sept. 30, 1873	143,218
Loom-picker	J. Cady	Staffordville, Conn	Sept. 1, 1863	39,714
Loom-picker	W. E. Card and P. Andrews	Phœnix, R. I	Oct. 29, 1867	70,164
Loom-picker	A. H. Carroll	Baltimore, Md	Apr. 13, 1869	88,845
Loom-picker	I. M. Colburn	Derby, Conn	Sept. 26, 1848	5,796
Loom-picker	B. F. Day and C. H. Nelson	Biddeford, Me	Oct. 3, 1865	50,227

Index of patents issued from the United States Patent Office from 1790 *to* 1873, *inclusive*—Continued.

Invention.	Inventor.	Residence.	Date.	No.
Loom-picker	J. Hunt and A. Stockwell	Providence, R. I	Apr. 12, 1870	101, 734
Loom-picker	R. Joslin	Pawtucket, R. I	Feb. 18, 1868	74, 545
Loom-picker	R. Leach	Linwood Station, Pa	Jan. 28, 1868	73, 818
Loom-picker	Z. Lyford	Lowell, Mass	Feb. 23, 1858	19, 428
Loom-picker	T. J. Mayall	Roxbury, Mass	June 2, 1857	17, 468
Loom-picker	G. P. Medbery	Sprague, Conn	Jan. 5, 1864	41, 165
Loom-picker	B. W. Nichols	Phœnix Village, R. I	Oct. 1, 1867	69, 472
Loom-picker	J. M. Parker	Leicester, Mass	Jan. 31, 1871	111, 378
Loom-picker	O. A. Sawyer	Lowell, Mass	Dec. 23, 1873	145, 905
Loom-picker	O. B. Smith	Palmer, Mass	July 23, 1867	66, 996
Loom-picker	W. Taylor	Maynard, Mass	Jan. 28, 1873	135, 253
Loom-picker	E. Wright	Worcester, Mass	Oct. 1, 1867	69, 380
Loom-picker arm	G. W. Paterson	Newburyport, Mass	June 26, 1866	55, 959
Loom-picker check	L. J. Labounty	Lowell, Mass	Sept. 5, 1865	49, 830
Loom-picker check	W. Montgomery	Northampton, Mass	Apr. 19, 1870	102, 145
Loom-picker cushion	G. L. Richardson and J. C. Moody	Brunswick, Me	July 13, 1869	92, 471
Loom-picker cushion	W. J. Thorn	Westbrook, Me	Feb. 12, 1867	62, 087
Loom-picker motion	S. Boorn	Lowell, Mass	May 5, 1857	17, 193
Loom-picker motion	J. Cady	Staffordville, Conn	May 2, 1865	47, 517
Loom-picker motion	J. C. Fisher	Providence, R. I	June 20, 1871	116, 251
Loom-picker motion	W. Nugent	North Providence, R. I	Apr. 30, 1867	64, 354
Loom-picker motion	J. Pilkington	Frankford, Pa	Jan. 7, 1868	73, 037
Loom-picker motion	J. Robinson	North Andover, Mass	June 18, 1861	32, 586
Loom-picker motion	E. Robbins	Milford, Mass	Mar. 11, 1862	34, 647
Loom-picker motion	J. Stever	Bristol, Conn	Sept. 24, 1872	131, 575
Loom-picker motion, Power	H. Elliott	Globe Village, Mass	Aug. 7, 1866	56, 913
Loom-picker motion, Power	W. Stearns	Manchester, N. H	July 5, 1859	24, 668
Loom-picker shoe	E. J. Hall	Cambridge, Mass	June 1, 1869	90, 838
Loom-picker staff	R. Collins	Chicopee, Mass	Oct. 1, 1867	69, 319
Loom-picker staff	S. Estes	Newburyport, Mass	Nov. 23, 1858	22, 114
Loom-picker staff	E. Robbins	Hopedale, Mass	May 27, 1862	35, 395
Loom-picker-staff check	C. A. Hooper	Lewiston, Me	Oct. 17, 1871	120, 068
Loom-picker-staff check	E. and H. C. Phillips	Blackstone, Mass	Feb. 4, 1868	74, 124
Loom-picker-staff connection	W. Weiland	Dedham, Mass	Aug. 1, 1865	49, 182
Loom-picker-operating mechanism	E. A. Wareham and W. W. Waggoner	Kirkville, Iowa	Sept. 13, 1870	107, 313
Loom-picker-staff-actuating mechanism	M. C. Burleigh	Somersworth, N. H	Aug. 9, 1870	106, 116
Loom-picker-staff arrester	E. and H. C. Phillips	Blackstone, Mass	Jan. 30, 1866	52, 367
Loom-picker-staff-operating mechanism	S. Mortimer	Leicester, Mass	Feb. 5, 1867	61, 754
Loom-picker-staff-operating mechanism	W. W. and V. J. Phillips	Wellsville, N. Y	Apr. 2, 1867	63, 421
Loom-picker-staff-operating mechanism	E. P. Terrel	West Liberty, Ohio	July 20, 1869	92, 900
Loom-picker staves, Operating	W. Rouse	Taunton, Mass	Sept. 13, 1870	107, 412
Loom-picker staves, Operating	P. A. Sherman and F. Baxter	Pawtucket, R. I	Apr. 10, 1866	53, 892
Loom-picker staves, Operating	E. Wright	Worcester, Mass	Oct. 29, 1867	70, 308
Loom-picker staves, Operating	J. F. Wicks	Millville, Mass	Apr. 27, 1869	89, 365
Loom picking-mechanism	M. C. Burleigh	Somersworth, N. H	Aug. 6, 1872	130, 108
Loom picking-mechanism	J. C. Fisher	Providence, R. I	Jan. 30, 1872	123, 094
Loom picking-mechanism	H. R. Fry	Wabash, Ind	Aug. 29, 1871	118, 446
Loom picking-mechanism	E. D. Gove	Holyoke, Mass	Nov. 28, 1871	121, 359
Loom picking-mechanism	H. T. Henderson	Keokuk, Iowa	Aug. 20, 1852	130, 636
Loom picking-mechanism	O. D. Lombard	Lowell, Mass	Sept. 14, 1869	94, 760
Loom picking-mechanism	E. W. Moffatt	Hyde Park, Mass	Dec. 24, 1872	134, 153
Loom picking-mechanism	L. E. Ross	Providence, R. I	Apr. 15, 1873	137, 798
Loom picking-mechanism	L. E. Ross	Providence, R. I	Sept. 30, 1873	143, 257
Loom picking-mechanism	C. E. Smith	Lowell, Mass	Nov. 26, 1872	133, 493
Loom picking-mechanism	W. Stearns	Manchester, N. Y	Sept. 9, 1873	142, 592
Loom picking-mechanism	J. Stever	Bristol, Conn	June 17, 1873	140, 091
Loom picking-mechanism	A. Stockwell	Woonsocket, R. I	Mar. 11, 1873	136, 786
Loom picking-mechanism	E. P. Terrel	West Liberty, Ohio	Feb. 20, 1872	123, 952
Loom picking-mechanism	W. Townsend	Auburn, N. Y	July 22, 1873	141, 187
Loom picking-mechanism	H. A. Whitten	New Bedford, Mass	Mar. 26, 1872	124, 995
Loom picking-motion	H. A. Whitten and E. D. Gove	Holyoke, Mass	Feb. 7, 1871	111, 595
Loom picking-stick check	R. Davis	Lawrence, Mass	Aug. 5, 1873	151, 496
Loom, Pneumatic	C. Richardson	London, England	Nov. 26, 1872	133, 386
Loom, Power	A. Allen	Wilmington, Del	Apr. 15, 1856	14, 644
Loom, Power	A. Allen	Wilmington, Del	Aug. 4, 1857	17, 912
Loom, Power	W. Baird	Philadelphia, Pa	Nov. 29, 1853	10, 290
Loom, Power	J. T. Barnes	Manayunk, Pa	Oct. 17, 1854	11, 804
Loom, Power	D. Barnum	Bridgeport, Conn	Mar. 26, 1845	3, 968
Loom, Power	E. B. Bigelow	Boston, Mass	Aug. 18, 1846	4, 696
Loom, Power	E. B. Bigelow	Boston, Mass	Feb. 12, 1856	14, 222
Loom, Power	E. B. Bigelow	Boston, Mass	Feb. 9, 1869	86, 805
Loom, Power	J. Bullough	Accrington, England	Jan. 18, 1870	98, 919
Loom, Power	W. H. Cheetham, jr	New York, N. Y	Apr. 5, 1859	23, 446
Loom, Power	J. L. Cheney	Lowell, Mass	Mar. 17, 1857	16, 824
Loom, Power	J. Crawshaw	Rochester, N. Y	Nov. 16, 1858	22, 065
Loom, Power	J. C. Duckworth	Pittsfield, Mass	May 18, 1869	90, 085
Loom, Power	C. Duxbury and J. Nield	Taunton, Mass	Apr. 21, 1842	2, 574
Loom, Power	J. Earnshaw	Providence, R. I	Apr. 9, 1867	63, 622
Loom, Power	E. Fairman	Stafford, Conn	Feb. 6, 1838	595
Loom, Power	M. A. Furbush and G. Crompton	Worcester, Mass	Sept. 4, 1860	29, 873
Loom, Power	W. G. Gavit	Washington County, R. I	Apr. 8, 1835	....
Loom, Power	W. Gilmore	Smithfield, R. I	Oct. 28, 1820	
Loom, Power	J. Gledhill	New York, N. Y	Nov. 15, 1853	10, 223
Loom, Power	W. Graichen and C. Hoffman	Clinton, Mass	Jan. 14, 1862	34, 141
Loom, Power	W. H. Gray	Dover, N. H	Jan. 17, 1860	26, 875
Loom, Power	J. Greenhalgh, jr	Waterford, Mass	Nov. 8, 1853	10, 205
Loom, Power	J. Greenhalgh, sr	Waterford, Mass	Mar. 4, 1856	14, 358
Loom, Power	E. Hall	Rochester, N. Y	Feb. 12, 1856	14, 237
Loom, Power	P. W. Hart	Stamford, Conn	Sept. 4, 1866	57, 712
Loom, Power	A. Holbrook and B. F. French	Dunstable, N. H	June 17, 1834	....
Loom, Power	H. Holcroft and C. S. Smith	Media and Chester Valley, Pa.	July 19, 1864	43, 583
Loom, Power	T. H. and H. James	Stockport, N. Y	Jan. 28, 1862	34, 255
Loom, Power	B. H. Jenks and J. Shinn	Bridesburgh, Pa	May 12, 1863	38, 534
Loom, Power	J. Johnson	Troy, N. Y	Feb. 19, 1856	14, 285

Index of patents issued from the United States Patent Office from 1790 *to* 1873, *inclusive*—Continued.

Invention.	Inventor.	Residence.	Date.	No.
Loom, Power	J. C. Kempton	Manayunk, Pa	June 14, 1834	....
Loom, Power	T. King	West Farms, N. Y	Apr. 16, 1861	32, 068
Loom, Power	B. Lapham	Waterford, N. Y	June 28, 1836	....
Loom, Power	W. B. Leonard	Fishkill, N. Y	May 23, 1827	....
Loom, Power	W. B. Leonard	Fishkill, N. Y	May 23, 1827	....
Loom, Power	F. C. Lewis	Grafton, Mass	Mar. 30, 1836	....
Loom, Power	R. Lightbown	Eaton, N. Y	Oct. 30, 1849	6, 832
Loom, Power	W. Mason	Taunton, Mass	Oct. 18, 1853	10, 135
Loom, Power	T. H. Müller	New York, N. Y	Jan. 25, 1870	99, 223
Loom, Power	J. Nield	Taunton, Mass	May 25, 1844	3, 599
Loom, Power	J. Nield	Taunton, Mass	Mar. 15, 1845	3, 954
Loom, Power	J. Nield	Taunton Mass	Mar. 15, 1845	3, 955
Loom, Power	J. Pender	Worcester, Mass	Oct. 18, 1853	10, 137
Loom, Power	W. B. Pender and N. C. Horn	Wolfsborough, N. H	Aug. 15, 1838	884
Loom, Power	R. Reynolds	Valatia Village, N. Y	June 1, 1852	8, 984
Loom, Power	C. Roder	Ceralvo, Ky	June 21, 1864	43, 263
Loom, Power	G. B. Sanford	Putney, Vt	Mar. 17, 1843	3, 005
Loom, Power	J. Shinn	Philadelphia, Pa	Mar. 25, 1873	137, 252
Loom, Power	J. Shuttleworth	Frankford, Pa	May 21, 1850	7, 387
Loom, Power	J. Shuttleworth	Frankford, Pa	Jan. 3, 1854	10, 382
Loom, Power	A. Smith and H. Skinner	West Farms, N. Y	Nov. 14, 1856	16, 037
Loom, Power	A. Stone	Providence, R. I	Apr. 30, 1829	....
Loom, Power	W. Weild	Manchester, Great Britain	Jan. 13, 1857	16, 415
Loom-power	D. Whitman	Windham, Conn	Mar. 20, 1835	....
Loom power, Silk	G. Gay	Poughkeepsie, N. Y	Sept. 26, 1835	....
Loom, Rattan-weaving	S. L. Fitts	Ashburnham, Mass	Aug. 16, 1870	106, 477
Loom-reed	H. Holt	Rutland Township, Ohio	Oct. 17, 1831	....
Loom-reed	J. Senereff	Philadelphia, Pa	Oct. 1, 1830	....
Loom-reed	B. F. Whitcomb	Claremont, N. H	June 29, 1869	91, 888
Loom-reed	J. H. Williams	Pleasant Hill, Ohio	June 28, 1870	104, 808
Loom, Revolving weaving	J. Hess	Hempfield, Pa	Oct. 2, 1813	....
Loom, Rib-knitting	B. Wainwright	East Boston, Mass	Mar. 5, 1876	62, 579
Loom, Ribbon	C. V. Ganahl	Innspruck, Austria	Oct. 3, 1846	4, 782
Loom, Ribbon	W. J. Horstmann	Philadelphia, Pa	Mar. 23, 1858	19, 698
Loom, Rotary-power stocking	R. Walker	Portsmouth, N. H	Dec. 5, 1859	1, 421
Loom, Satinet power	R. Mitchell and N. Butterworth.	Troy, Mass	Mar. 22, 1828	....
Loom shedding-mechanism	G. Crompton	Worcester, Mass	July 1, 1873	140, 476
Loom shedding-mechanism	G. Crompton	Worcester, Mass	July 8, 1873	140, 682
Loom shedding-mechanism	G. Crompton	Worcester, Mass	Aug. 12, 1873	141, 768
Loom shedding-mechanism	C. H. Landenberger and H. Atkinson.	Philadelphia, Pa	Dec. 2, 1873	145, 216
Loom shedding-mechanism	J. Lomas	Kellyville, Pa	Apr. 15, 1873	137, 937
Loom shedding-mechanism	E. Redfearn	Philadelphia, Pa	Aug. 5, 1873	141, 589
Loom shedding-mechanism	I. L. Wilber	Taunton, Mass	Oct. 28, 1873	144, 176
Loom-shuttle	E. Baggett	Fall River, Mass	Sept. 1, 1868	81, 578
Loom-shuttle	E. P. Ball	Chicopee, Mass	Dec. 23, 1873	145, 831
Loom-shuttle	F. and A. J. Blanding	Lowell, Mass	Jan. 21, 1873	134, 971
Loom-shuttle	J. Brown	Brooklyn, N. Y	Sept. 2, 1873	142, 373
Loom-shuttle	H. H. Bryant	Boston, Mass	Jan. 14, 1873	134, 848
Loom-shuttle	W. H. Burns	Grafton, Mass	Jan. 25, 1870	99, 152
Loom-shuttle	E. F. Burrows	Mystic River, Conn	Apr. 26, 1870	102, 368
Loom-shuttle	H. Carstaedt	New York, N. Y	Sept. 13, 1870	107, 220
Loom-shuttle	H. Carstaedt	New York, N. Y	Feb. 27, 1872	123, 978
Loom-shuttle	N. D. Chapman	Rollingsford, N. H	Nov. 18, 1873	144, 657
Loom-shuttle	D. G. Chase	Boston, Mass	Mar. 17, 1868	75, 522
Loom-shuttle	S. E. Chase	Boston, Mass	Apr. 14, 1868	76, 604
Loom-shuttle	A. D. Clark	Wilkinsonville, Mass	June 21, 1864	43, 182
Loom-shuttle	A. D. Clark	Wilkinsonville, Mass	Aug. 25, 1868	81, 344
Loom-shuttle	T. Clark	Ware, Mass	Oct. 11, 1864	44, 604
Loom-shuttle	J. H. Coburn	Lowell, Mass	Mar. 12, 1842	2, 489
Loom-shuttle	J. H. Coburn	Lowell, Mass	June 13, 1865	48, 154
Loom-shuttle	J. H. Coburn	Lowell, Mass	Dec. 14, 1869	97, 882
Loom-shuttle	G. Crompton	Worcester, Mass	Nov. 9, 1869	96, 676
Loom-shuttle	G. Crompton	Worcester, Mass	Nov. 9, 1869	96, 677
Loom-shuttle	E. Cross	Southbridge, Mass	Feb 4, 1868	73, 955
Loom-shuttle	A. M. Damon and J. Whitaker	Lowell, Mass	July 7, 1868	79, 557
Loom-shuttle	D. C. G. Field	Lowell, Mass	Mar. 11, 1873	136, 715
Loom-shuttle	M. F. Fields	Lawrence, Mass	Oct. 28, 1873	143, 976
Loom-shuttle	C. H. Fiske	Lowell, Mass	Aug. 10, 1869	93, 608
Loom-shuttle	E. H. Graham	Biddeford, Me	June 27, 1871	116, 435
Loom-shuttle	T. Isherwood	Stonington, Conn	June 6, 1871	115, 614
Loom-shuttle	J. Kuttner	New York, N. Y	Apr. 25, 1871	114, 155
Loom-shuttle	J. E. Ladd and H. T. Henderson.	Bristol, Conn	Dec. 17, 1872	134, 074
Loom-shuttle	P. A. Lambert	Ansonia, Conn	Dec. 17, 1872	134, 075
Loom-shuttle	L. Litchfield	Southbridge, Mass	May 1, 1855	12, 780
Loom-shuttle	J. Lofvendahl	Woonsocket, R. I	May 31, 1870	103, 757
Loom-shuttle	J. Lofvendahl	Woonsocket, R. I	July 18, 1871	117, 183
Loom-shuttle	J. Lyall	New York, N. Y	July 4, 1871	116, 609
Loom-shuttle	E. P. Marble	New Worcester, Mass	May 1, 1855	12, 781
Loom-shuttle	E. W. Marble	Sutton, Mass	Apr. 25, 1871	114, 167
Loom-shuttle	J. Martin	Lowell, Mass	Jan. 30, 1872	123, 184
Loom-shuttle	T. Martin	Chelsea, Mass	Feb. 11, 1873	135, 652
Loom-shuttle	T. K. McIntyre	Salem, Mass	Aug. 12, 1873	141, 803
Loom-shuttle	F. Miller	New York, N. Y	Mar. 26, 1872	124, 967
Loom-shuttle	L. L. Northup	Olneyville, R. I	May 7, 1872	126, 569
Loom-shuttle	E. A. Paine	Grafton, Mass	Mar. 1, 1870	100, 317
Loom-shuttle	C. L. F. Pfefferkorn and W. Aust.	Manchester, N. H	Oct. 21, 1873	143, 927
Loom-shuttle	C. R. Sawyer and G. F. Wright	Clinton, Mass	Feb. 4, 1873	135, 446
Loom-shuttle	F. W. Sawyer	Grafton, Mass	Oct. 22, 1867	70, 026
Loom-shuttle	L. Scofield	Farmington, Wis	Feb. 4, 1868	74, 013
Loom-shuttle	A. Simpson	Cumberland, R. I	Jan. 19, 1869	85, 969
Loom-shuttle	C. E. Smith	Lowell, Mass	Mar. 14, 1871	112, 643
Loom-shuttle	C. E. Smith and F. T. Jaques	Lowell, Mass	Oct. 27, 1868	83, 559
Loom-shuttle	E. A. Thissell	Lowell, Mass	July 18, 1871	117, 220

Index of patents issued from the United States Patent Office from 1790 *to* 1873, *inclusive*—Continued.

Invention.	Inventor.	Residence.	Date.	No.
Loom-shuttle	E. A. Thissell	Lowell, Mass	July 23, 1872	129, 692
Loom-shuttle	F. O. Tucker	Stonington, Conn	Aug. 22, 1871	118, 405
Loom-shuttle	F. O. Tucker	Stonington, Conn	Sept. 19, 1871	119, 201
Loom-shuttle	F. O. Tucker	Stonington, Conn	Nov. 28, 1871	121, 262
Loom-shuttle	W. Tucker	Blackstone, Mass	Dec. 28, 1852	9, 507
Loom-shuttle	W. Tucker	Fiskedale, Mass	June 17, 1873	140, 098
Loom-shuttle	W. Tunstill	New York, N. Y	Dec. 27, 1864	45, 682
Loom-shuttle	C. H. Waters and W. Orr, jr	Groton and Clinton, Mass	Nov. 12, 1871	121, 830
Loom-shuttle	R. Webendörfer	New York, N. Y	Jan. 11, 1870	98, 822
Loom-shuttle	H. Wells	Hopkinton, R. I	May 24, 1870	103, 400
Loom-shuttle	R. Whitehill	New York, N. Y	Apr. 23, 1872	126, 119
Loom-shuttle	W. Wilder	Wilkinsonville, Mass	Aug. 16, 1864	43, 877
Loom-shuttle	W. Wilder	Wilkinsonville, Mass	Aug. 23, 1864	43, 940
Loom-shuttle	W. Wilder	Wilkinsonville, Mass	Jan. 24, 1865	46, 040
Loom-shuttle	N. A. Williams	Utica, N. Y	Feb. 2, 1869	86, 483
Loom-shuttle	D. Wright	Waltham, Mass	May 16, 1871	115, 006
Loom-shuttle-acting mechanism	T. Martin	Chelsea, Mass	Mar. 12, 1872	124, 501
Loom-shuttle-actuating mechanism	J. R. Norfolk	Salem, Mass	Nov. 4, 1873	144, 218
Loom-shuttle-actuating mechanism	G. V. Sheffield and W. S. Horton.	Providence, R. I	Apr. 9, 1872	125, 493
Loom-shuttle and bobbin	C. Unverzagt	Richmond, Ind	May 1, 1866	54, 445
Loom-shuttle binder	C. Duckworth	Mount Carmel, Conn	Aug. 21, 1866	57, 300
Loom-shuttle binder	G. B. Hammond and J. A. Tucker.	Putnam, Conn., and Attleborough, Mass.	Apr. 22, 1873	138, 020
Loom-shuttle binder	M. E. Haskell	Lowell, Mass	Apr. 20, 1869	89, 043
Loom-shuttle binder	J. C. Poland, jr., and B. R. Cotton.	Auburn and Lewiston, Me	Jan. 22, 1867	61, 357
Loom-shuttle binder	E. Prentice	Wauregan, Conn	Feb. 9, 1869	86, 688
Loom-shuttle box	A. F. Gibboney	Union Township, Pa	Dec. 28, 1858	22, 420
Loom-shuttle-box mechanism	J. Ashworth	North Andover, Mass	Dec. 12, 1871	121, 838
Loom-shuttle-box mechanism	W. Atkinson and C. C. Klein	Philadelphia, Pa	Feb. 11, 1873	135, 748
Loom-shuttle-box mechanism	R. B. Goodyear	Wilmington, Del	May 27, 1873	139, 384
Loom-shuttle-box mechanism	J. M. Stone	North Andover, Mass	Dec. 12, 1871	121, 907
Loom-shuttle-box motion	C. Duckworth	Thompsonville, Conn	June 28, 1853	9, 815
Loom-shuttle-box-operating mechanism	R. B. Goodyear	Elkton, Md	Mar. 2, 1869	87, 337
Loom-shuttle-box-operating mechanism	L. J. Knowles	Warren, Mass	Dec. 13, 1870	110, 146
Loom-shuttle-box-operating mechanism	A. Nimmo	Philadelphia, Pa	Aug. 10, 1869	93, 637
Loom-shuttle-box-operating mechanism	A. Nimmo	Philadelphia, Pa	Dec. 7, 1869	97, 679
Loom-shuttle-box-operating mechanism	W. Whiteside	Philadelphia, Pa	May 18, 1869	90, 212
Loom-shuttle box, Actuating	M. Rice	Upland, Pa	Aug. 11, 1868	81, 005
Loom-shuttle boxes, Apparatus for operating	A. Allen	Wilmington, Del	Sept. 4, 1849	6, 693
Loom-shuttle boxes, Apparatus for operating	R. B. Goodyear	Philadelphia, Pa	Mar. 13, 1849	6, 170
Loom-shuttle boxes, Operating	J. Ashworth	North Andover, Mass	Sept. 8, 1868	81, 865
Loom-shuttle boxes, Operating	J. Brierley	Worcester, Mass	July 28, 1868	80, 385
Loom-shuttle boxes, Operating	T. Cleary	Millbury, Mass	Feb. 2, 1869	86, 365
Loom-shuttle boxes, Operating	R. B. Goodyear and B. Hirst	Philadelphia and Manayunk, Pa.	Jan. 1, 1850	6, 986
Loom-shuttle check	E. W. Bray	Mooresville, Ind	Dec. 3, 1872	133, 624
Loom-shuttle check	J. N. Kingston	Bristol County, Mass	Oct. 1, 1872	131, 761
Loom-shuttle check	D. Pickman	Lowell, Mass	Dec. 21, 1869	98, 188
Loom-shuttle driver, Power	A. S. Lyon	Lawrence, Mass	Jan. 27, 1863	37, 513
Loom-shuttle-driving mechanism	J. Stever	Bristol, Conn	Jan. 18, 1870	99, 024
Loom-shuttle guard	C. H. Hudson	Clinton, Mass	Apr. 8, 1873	137, 551
Loom-shuttle guard	J. Rycroft and L. A. White	Millbury, Mass	June 27, 1871	116, 499
Loom-shuttle guard	E. M. Stevens	Boston, Mass	Apr. 30, 1867	64, 261
Loom-shuttle guard	E. M. Stevens	Boston, Mass	Dec. 23, 1823	145, 763
Loom-shuttle guard, Power	P. Miggett	Hoosick Falls, N. Y	Aug. 8, 1854	11, 505
Loom-shuttle guide	J. B. Bancroft	Hopedale, Mass	Jan. 14, 1873	134, 837
Loom-shuttle guide	J. B. Bancroft	Hopedale, Mass	Dec. 30, 1873	145, 925
Loom-shuttle guide	W. A. Hastings	Palmer, Mass	Feb. 9, 1869	86, 668
Loom-shuttle guide	H. T. Robbins	Lowell, Mass	Sept. 14, 1852	9, 263
Loom-shuttle guide	H. T. Robbins	Lowell, Mass	July 24, 1855	13, 322
Loom-shuttle guide or guard	S. R. Bixby and J. B. Bancroft	Worcester and Milford, Mass.	Aug. 16, 1864	43, 828
Loom-shuttle motion	J. Avery	Lowell, Mass	May 29, 1855	12, 941
Loom-shuttle motion	L. Ferguson	Lowell, Mass	May 19, 1857	17, 323
Loom-shuttle motion	A. Gartenmann	New York, N. Y	Mar. 4, 1873	136, 501
Loom-shuttle motion	O. A. Kelly	Woonsocket, R. I	June 4, 1850	7, 417
Loom-shuttle motion	G. W. Perry	Thompson, Conn	Nov. 11, 1851	8, 506
Loom-shuttle motion	G. J. Wardwell	Hanover, Me	Aug. 5, 1851	8, 278
Loom-shuttle motion	T. T. Willcox	Norwich, Conn	Nov 26, 1850	7, 803
Loom-shuttle motion, Power	W. Crighton	Fall River, Mass	Oct. 18, 1853	10, 126
Loom-shuttle-operating device	J. J. and E. Harrison	Manchester, England	Nov. 20, 1866	59, 767
Loom-shuttle-operating machinery	R. P. Cunningham	Abington, Conn	Dec. 16, 1845	4, 308
Loom-shuttle-operating mechanism	D. Brickford	Boston, Mass	Nov. 21, 1865	51, 006
Loom-shuttle-operating mechanism	E. P. Terrel	West Liberty, Ohio	Dec. 13, 1870	110, 171
Loom-shuttle, Narrow-ware	J. Y. Hamilton	Clinton, Mass	Jan. 29, 1867	61, 535
Loom-shuttle relief-mechanism	W. C. Macomber	Mystic River, Conn	Dec. 23, 1873	145, 882
Loom-shuttle spindle	N. I. Allen and J. C. Moody	Brunswick, Me	Nov. 23, 1869	97, 147
Loom-shuttle spindle	A. Morton	Salmon Falls, N. H	Aug. 1, 1871	117, 556
Loom-shuttle stopping-mechanism	S. Boorn	Lowell, Mass	Apr 12, 1870	101, 817
Loom-shuttle stopping-device, Power	D. S. Esten	Hinsdale, Mass	Sept. 8, 1863	39, 803
Loom-shuttle tongue	C. B. Thorp	Smithfield, R. I	Apr. 17, 1837	162
Loom-shuttles, Operating	J. Schottenfels	New York, N. Y	Mar. 20, 1866	53, 350
Loom, Stocking	J. Bazier, jr	Canton, Mass	Oct. 28, 1814	....
Loom, Stocking	E. Herrick	Stockbridge, Mass	Oct. 22, 1813	....
Loom, Stocking	A. Porter	Queensbury, N. Y	Aug. 22, 1816	....
Loom, Stocking	J. Vickerstaff	Philadelphia, Pa	Mar. 20, 1847	5, 023
Loom stop-apparatus	E. Burt	Manchester, Conn	June 20, 1845	4, 088
Loom-stop-actuating mechanism	A. J. Woodman	Springfield, Mass	July 9, 1872	128, 772
Loom-stop, Electrical	G. F. Lufbery	Montargis, France	Sept. 16, 1873	142, 925
Loom stop-mechanism	C. E. Barnes	Boston, Mass	May 20, 1873	138, 987
Loom stop-mechanism	J. Cowngill	Philadelphia, Pa	Dec. 20, 1870	110 208
Loom stop-mechanism	G. Crompton	Worcester, Mass	Feb. 9, 1869	86, 735
Loom stop-mechanism	G. K. Dearborn	Smithfield, R. I	Aug. 24, 1869	93, 971
Loom stop-mechanism	W. Edson	Boston, Mass	June 3, 1873	139, 560
Loom stop-mechanism	T. Isherwood and W. Nuttall	Westerly, R. I	May 13, 1873	138, 893
Loom stop-mechanism	L. J. Knowles	Worcester, Mass	Sept. 2, 1873	142, 401

Index of patents issued from the United States Patent Office from 1790 to 1873, inclusive—Continued.

Invention.	Inventor.	Residence.	Date.	No.
Loom stop-mechanism	J. J. Switzer	Chelsea, Mass	June 8, 1869	91,050
Loom stop-mechanism	J. J. Switzer	Boston, Mass	Mar. 18, 1873	136,880
Loom stop-mechanism	H. F. Wheeler and W. O. Wakefield.	Boston, Mass	Oct. 7, 1873	143,552
Loom stop-motion	A. A. Barker	Lewiston, Me	June 1, 1869	90,627
Loom stop-mechanism, Power	E. B. Bigelow	Boston, Mass	Apr. 20, 1869	89,012
Loom stop-motion	A. Carmichel and W. Barney	Westerly, R. I	Nov. 27, 1866	59,962
Loom stop-motion	G. Draper	Milford, Mass	Sept. 1, 1863	39,723
Loom stop-motion	C. Duckworth	Mount Carmel, Conn	Sept. 6, 1864	44,080
Loom stop-motion	E. Hall	Cabotville, Mass	Sept. 24, 1850	7,665
Loom stop-motion	L. B. Hoit	Millbury, Mass	Feb. 17, 1852	8,740
Loom stop-motion	A. J. Loisean	New York, N. Y	Jan. 22, 1867	61,439
Loom stop-motion	A. Marchington	Upland, Pa	Aug. 21, 1866	57,350
Loom stop-motion	G. E. Milroy	Lowell, Mass	Mar. 30, 1869	88,322
Loom stop-motion	W. and D. Pilkington	Frankford and Chester, Pa	July 16, 1867	66,734
Loom stop-motion	J. J. Switzer	Roxbury, Mass	Aug. 11, 1868	80,885
Loom stop-motion	J. Taylor and J. Woodhead	Middletown, Pa	Apr. 21, 1831	....
Loom stop-motion and shuttle	W. W. Tucker	Hartford, Conn	Jan. 16, 1872	122,745
Loom stop-motion, Hair-cloth	R. J. Stafford	Smithfield, R. I	Mar. 23, 1858	19,719
Loom stop-motion, Power	T. Foden	Holyoke, Mass	Oct. 1, 1851	33,383
Loom take-up	C. Gabren	New York, N. Y	Dec. 10, 1872	133,845
Loom take-up	G. H. Holmes	New Brunswick, N. J	Aug. 10, 1869	93,441
Loom take-up	W. Rouse	Taunton, Mass	May 9, 1871	114,606
Loom take-up, Corset	J. Kuttner	New York, N. Y	Mar. 7, 1871	112,356
Loom-take-up mechanism	S. Estes	Newburyport, Mass	Mar. 25, 1873	137,188
Loom-take-up mechanism	C. Gabren	New York, N. Y	Nov. 18, 1873	144,756
Loom-take-up mechanism	J. Michna and G. Fischer	New York, N. Y	Mar. 7, 1871	112,478
Loom-take-up mechanism	A. Mitchell and H. S. Combs	Providence, R. I., and Warren, Mass.	Sept. 17, 1872	131,455
Loom-take-up mechanism	J. Sparks	Concord, Ky	Apr. 27, 1869	89,445
Loom-take-up motion	L. L. Shaw	Lewiston, Me	Aug. 16, 1864	43,868
Loom-take-up motion	A. Stone	Johnston, R. I	Aug. 17, 1835	....
Loom-take-up motion, Power	J. P. Comins	Killingly, Conn	Oct. 20, 1836	....
Loom-take-up motion, Power	J. Whitaker	Philadelphia, Pa	Oct. 28, 1862	36,875
Loom take-up, Narrow-ware	M. Welf	New York, N. Y	Oct. 9, 1866	58,708
Loom take-up, Trimming	S. Walker	Roxbury, Mass	Sept. 20, 1859	25,538
Loom tappet-chain	J. Nuttall	Walmersly, England	Mar. 15, 1870	100,794
Loom-temple	J. H. Allen	Biddeford, Me	Jan 16, 1855	12,236
Loom-temple	N. J. Allen and J. C. Moody	Auburn and Brunswick, Me.	Sept. 24, 1872	131,491
Loom-temple	N. J. Allen and E. S. Stimpson	Milford, Mass	Nov. 18, 1873	144,647
Loom-temple	E. A. Angell	Killingly, Conn	Oct. 19, 1838	987
Loom-temple	J. Beard and A. Whitney	Waltham, Mass	Apr. 6, 1842	2,536
Loom-temple	L. Briggs	Grosvenor Dale, Conn	Oct. 6, 1868	82,684
Loom-temple	N. Chapman	Hopedale, Mass	Apr. 22, 1873	138,004
Loom-temple	N. Chapman	Hopedale, Mass	June 17, 1873	140,013
Loom-temple	N. Chapman	Hopedale, Mass	Nov. 11, 1873	144,507
Loom-temple	L. K. and P. Day	Sacarappa, Me	Feb. 27, 1849	6,143
Loom-temple	E. and W. W. Dutcher	North Bennington, Vt	Dec. 16, 1851	8,586
Loom-temple	E. and W. W. Dutcher	North Bennington, Vt	Dec. 28, 1852	9,502
Loom-temple	F. J. Dutcher	Hopedale, Mass	Mar. 11, 1873	136,644
Loom-temple	W. W. Dutcher	Milford, Mass	Apr. 2, 1867	63,372
Loom-temple	W. W. Dutcher	Hopedale, Mass	June 2, 1868	78,441
Loom-temple	W. W. Dutcher	Hopedale, Mass	Dec. 14, 1869	97,895
Loom-temple	W. W. Dutcher	Hopedale, Mass	Aug. 2, 1870	106,038
Loom-temple	W. W. Dutcher	Hopedale, Mass	Sept. 6, 1870	107,168
Loom-temple	W. W. Dutcher	Hopedale, Mass	June 17, 1873	139,884
Loom-temple	W. W. Dutcher and G. Draper	Milford, Mass	Apr. 16, 1867	63,872
Loom-temple	A. H. Gilman	Boston, Mass	June 14, 1870	104,137
Loom-temple	J. B. Greene	Worcester, Mass	Nov. 8, 1863	10,206
Loom-temple	W. H. Howard	Media, Pa	May 31, 1870	103,616
Loom-temple	A. Jillson	Woonsocket, R. I	Sept. 16, 1851	8,368
Loom-temple	P. Newell	Waterford, N. Y	Jan. 5, 1833	....
Loom-temple	P. Newell	Waterford, N. Y	Dec. 2, 1836	96
Loom-temple	W. and J. W. Murkland	Lowell, Mass	Feb. 27, 1872	124,004
Loom-temple	A. Palmer	Lebanon, N. Y	Feb. 10, 1846	4,379
Loom-temple	R. Pilson	Laurel, Md	Sept. 14, 1858	21,515
Loom-temple	R. Reynolds	Stockport, N. Y	Feb. 19, 1856	14,292
Loom-temple	B. Rogers	Beaver Creek, Ohio	Nov. 24, 1814	....
Loom-temple	C. T. K. Rollins	Lowell, Mass	Aug. 30, 1833	....
Loom-temple	J. A. Scholfield	Westerly, R. I	Aug. 2, 1853	9,900
Loom-temple	E. F. Shaw	Boston, Mass	Nov. 26, 1867	71,541
Loom-temple	J. Simpson	Millbury, Mass	July 20, 1869	92,890
Loom-temple	J. Smith	Laurel, Md	Aug. 7, 1855	13,413
Loom-temple	E. S. Stimpson	Hopedale, Mass	June 17, 1873	139,929
Loom-temple	J. C. Tilton	Sanbornton Bridge, N. H	May 15, 1855	12,879
Loom-temple	J. C. Tilton	Sanbornton Bridge, N. H	Jan. 11, 1859	22,594
Loom-temple	G. L. White	Cumberland, R. I	Aug. 28, 1866	57,646
Loom temple, Cloth	J. Bazin, jr	Canton, Mass	Aug. 8, 1823	....
Loom temple, Double-cloth	S. Everett	Biddeford, Me	Apr. 2, 1850	7,237
Loom temple, Perpetual	D. A. Smith	Townsend, Mass	Feb. 23, 1816	....
Loom temple, Power	J. Standish	Providence, R. I	Nov. 11, 1830	....
Loom temple, Revolving-burr	E. B. Harris and A. R. Arnold	Woodstock, Conn	Apr. 13, 1830	....
Loom-temple, Rotary	I. Draper	Saugus, Mass	Apr. 1, 1829	....
Loom-temple, Rotary	I. C. Lane	Waltham, Mass	Mar. 26, 1844	3,511
Loom-temple, Self-acting rotary	W. Craig and J. Cochrane	England	Nov. 25, 1841	2,377
Loom-temple, Self-adjusting	S. P. Mason	Newport, R. I	July 22, 1837	291
Loom-temple, Self-operating	E. R. Otis	River Head, Conn	Apr. 2, 1830	....
Loom-temples, Making	O. M. Stillman	Brookfield, N. Y	Jan. 30, 1826	....
Loom-tension-regulating device	M. Sigler	Paterson, N. J	Feb. 2, 1869	86,465
Loom-tension regulator	E. B. Bigelow	Boston, Mass	Mar. 12, 1845	3,948
Loom-tension regulator	F. Painter	East Hampton, Mass	Oct. 29, 1867	70,253
Loom-treadles and harness-shafts, Operating	O. W. Gordon	Mount Pleasant, Iowa	Apr. 30, 1867	64,216
Loom-treadles, Operating	R. W. Andrews	Stafford, Conn	Jan. 18, 1853	9,540
Loom, Upright power	S. Shepard and J. Thorp	Taunton, Mass	July 25, 1816	....
Loom, Vibrating-cam	W. H. Howard	Worcester, Mass	Feb. 21, 1830	....

Index of patents issued from the United States Patent Office from 1790 *to* 1873, *inclusive*—Continued.

Invention.	Inventor.	Residence.	Date.	No.
Loom warp-feeding mechanism	A. B. Ely	Newton, Mass	Dec. 17, 1867	72, 379
Loom warp-shedding mechanism	J. Holding and J. Eccles	Manchester, Great Britain	Oct. 10, 1871	119, 848
Loom warp-stand	P. D. Tifft	Eagleville, Ohio	May 10, 1870	102, 989
Loom warp-suspender	W. Squire	New York, N. Y	Jan. 17, 1815	....
Loom warp-tension mechanism	J. F. Randall	Warren, Ohio	Nov. 1, 1870	108, 826
Loom, Water	B. Brown	Albemarle, Va	Mar. 21, 1814	....
Loom, Water	P. C. Curtis	Paris, N. Y	May 28, 1816	....
Loom, Water	T. Mussy	Philadelphia, Pa	Mar. 4, 1811	....
Loom, Water	T. Mussey	New London, Conn	June 10, 1817	....
Loom, Water	S. Shepard	Philadelphia, Pa	Apr. 27, 1812	....
Loom, Weaving	B. Slingerland	Paterson, N. J	Apr. 1, 1845	3, 977
Loom, Weaving	R. Wilcox	Columbus, Ohio	Mar. 21, 1825	....
Loom yarn-beam	B. A. Bailey	Lewiston, Me	Aug. 4, 1868	80, 529
Loom yarn-beam	A. W. Cole	West Killingly, Conn	Sept. 12, 1871	118, 910
Loom yarn-beam	S. Pettee	Foxborough, Mass	June 15, 1813	....
Loom yarn-beams, Friction-device for	A. Beebe and F. L. Bestor	Enfield, Mass	Aug. 23, 1864	43, 895
Looms, Belt-shifting mechanism for	L. J. Knowles	Warren, Mass	Aug. 9, 1870	106, 267
Looms, Belt-shipper for	L. J. Knowles	Warren, Mass	July 14, 1868	79, 986
Looms, Belt-shipping mechanism for	L. J. Knowles	Warren, Mass	Feb. 2, 1869	86, 417
Looms, Brake for power	R. and G. B. Reynolds	Stockport, N. Y	June 21, 1859	24, 488
Looms, Brake for yarn-beam of	J. J. and E. Harrison	Broughton, England	Sept. 15, 1868	82, 220
Looms by electricity, Operating	G. Bonelli	Turin, Sardinia	Dec. 12, 1854	12, 050
Looms, Cloth-registering attachment for	C. C. Temple	Saco, Me	June 20, 1865	48, 325
Looms, Cloth-winding mechanism for	W. A. Arnold	Rockport, Mass	Dec. 6, 1870	109, 794
Looms, Cylinder for figuring	E. M. Hastings and J. Shepherdson.	Jamestown, N. Y	Mar. 18, 1851	7, 988
Looms, Delivery and take-up motion of	A. H. Boyd	Saco, Me	Mar. 10, 1849	6, 157
Looms, Device for supplying weft to the shuttles in	T. Ingram	Brantford, England	July 12, 1864	43, 556
Looms, Eye of wire heddles for	J. Ashworth	North Andover, Mass	Jan. 29, 1867	61, 501
Looms, Finger for shuttle-stop rod in	E. S. Laney	Waterloo, N. Y	May 12, 1868	77, 820
Looms for twilling, Operating treadle-cams in	R. Garsed	Frankford, Pa	July 20, 1846	4, 645
Looms for weaving bolting-cloth, Mounting and using harness in.	R. White	Williamstown, Vt	Apr. 1, 1842	2, 528
Looms for weaving counterpanes, &c., Mounting	E. B. Bigelow	Lancaster, Mass	May 30, 1842	2, 653
Looms for weaving figured fabrics, Shuttle-box motion in.	C. Duckworth	Mount Carmel, Conn	Jan. 9, 1866	51, 932
Looms for weaving goods with elastic strands, Tension-mechanism for.	L. Hull	Charlestown, Mass	Aug. 8, 1865	49, 271
Looms for weaving irregular fabrics, Take-up mechanism for.	H. Carstaedt	New York, N. Y	Mar. 30, 1869	88, 365
Looms, Friction-mechanism for warp-beams of	O. Kenison and A. J. McClary	Lawrence, Mass	Feb. 21, 1865	46, 475
Looms, Graduating take-up motion for power	H. Hendrick	Killingly, Conn	Sept. 22, 1836	30
Looms, Harness-motion for	A. L. Anderson	Ware, Mass	May 15, 1866	54, 661
Looms, Harness-motion for	R. Collins	Chicopee, Mass	Apr. 17, 1866	54, 058
Looms, Harness-motion for	G. Crompton	Worcester, Mass	Nov. 27, 1866	59, 972
Looms, Harness-motion for	J. F. Gebhart	New Albany, Ind	July 16, 1867	66, 827
Looms, Harness-motion for	J. F. Gebhart	New Albany, Ind	July 16, 1867	66, 828
Looms, Harness-motion for	B. F. Knowles	Providence, R. I	Jan. 22, 1861	31, 176
Looms, Harness-motion for	L. J. Knowles	Warren, Mass	Aug. 27, 1867	68, 303
Looms, Harness-motion for	W. W. McGregor	Dedham, Mass	July 3, 1866	56, 077
Looms, Harness-motion for	C. Miller	Saint Louis, Mo	Aug. 7, 1866	56, 976
Looms, Harness-motion for	C. Schilling	Auburn, N. Y	Sept. 10, 1867	68, 795
Looms, Harness-motion for	H. Yount	Dayton, Ohio	Aug. 13, 1867	67, 699
Looms, Harness-motion for power	D. M. Ayer	Lewiston, Me	Oct. 21, 1862	36, 697
Looms, Harness-motion for power	C. Duckworth	Mount Carmel, Conn	Aug. 1, 1865	49, 096
Looms, Harness-motion for power	J. Greenhalgh	Pascoag, R. I	May 3, 1864	42, 575
Looms, Heddle-connection for	T. K. McIntyre	Warner, N. H	July 9, 1872	128, 807
Looms, Heddle-eye for	T. Clegg	North Andover, Mass	Mar. 7, 1854	10, 623
Looms, Hook-temple for	W. W. Dutcher and G. Draper.	Milford, Mass	May 12, 1857	17, 267
Looms, Jacquard apparatus for	J. A. Elder	Westbrook, Me	June 21, 1853	9, 795
Looms, Jacquard apparatus for	L. D. Valetton	Philadelphia, Pa	Nov. 14, 1865	50, 993
Looms, Jacquard mechanism for	E. K. Davis	New York, N. Y	Nov. 9, 1869	96, 553
Looms, Lay for power	B. Brooks	New Market, N. H	Dec. 17, 1834	....
Looms, Lubricating temple for	E. S. Stimpson	Milford, Mass	May 18, 1869	90, 133
Looms, Lubricator for picker-spindles in	T. Parker	Shelby, N. C	Sept. 19, 1871	119, 171
Looms, Machine for casting metallic eyes or mail of heddles for.	J. Senneff	Philadelphia, Pa	Aug. 22, 1854	11, 589
Looms, Making and using	J. Goulding	Dedham, Mass	July 7, 1830	....
Looms, Measuring cloth on	J. G. Webster	Middlesex County, Mass	Feb. 19, 1850	7, 112
Looms, Mechanism for operating pile-wire	W. Weild	Manchester, England	Aug. 25, 1868	81, 443
Looms, Mechanism for presenting palm-leaf to	I. Angell	Malden, Mass	Dec. 10, 1867	71, 945
Looms, Mode of applying friction to the yarn-beam of power.	S. Kimball	Putney, Vt	May 30, 1838	758
Looms, Mode of delivering warp in	J. Coulter	Xenia, Ohio	Apr. 25, 1843	3, 057
Looms, Mode of operating the treadle in power	E. Horton	Stafford, Conn	Feb. 22, 1838	617
Looms, Mode of throwing shuttle in	S. C. Mendenhall	Richmond, Ind	Nov. 9, 1852	9, 387
Looms, Mode of throwing shuttle in power	R. P. Cunningham	Pomfret, Conn	Apr. 25, 1843	3, 062
Looms, Motion of the lay in	J. Goulding	Worcester, Mass	June 15, 1852	9, 024
Looms, Mounting lease-rod for	W. A. Hastings	Thorndike, Mass	Sept. 21, 1869	94, 957
Looms, Needle of knitting	W. Bamford	Ipswich, Mass	Sept. 12, 1848	5, 761
Looms, Pattern-card for Jacquard	S. T. Thomas and E. Everett	Lowell and Lawrence, Mass	Mar. 16, 1852	8, 810
Looms, Pattern-chain for	H. D. Davis	North Andover, Mass	Apr. 19, 1870	102, 094
Looms, Pattern-chain for	B. H. Jenks	Philadelphia, Pa	July 24, 1860	29, 283
Looms, Reeds, heddles, or harness for	J. A. Wilkinson	Providence, R. I	July 17, 1835	....
Looms, Regulating motion of yarn-beam in power	W. A. Potter	Cranston, R. I	Nov. 23, 1837	475
Looms, Roller for pattern-chain for	G. Crompton and M. A. Furbush.	Worcester, Mass	Nov. 14, 1854	11, 956
Looms, Roller jaw-temple for	W. H. Burns	Grafton, Mass	Sept. 28, 1869	95, 314
Looms, Roller jaw-temple for	W. H. Burns	Grafton, Mass	Feb. 1, 1870	99, 287
Looms, Roller-temple for	W. H. Burns	Grafton, Mass	Nov. 29, 1870	109, 583
Looms, Roller-temple for	N. Chapman	Hopedale, Mass	Jan. 8, 1867	61, 050
Looms, Roller-temple for	W. W. Dutcher	Milford, Mass	June 16, 1857	17, 559
Looms, Roller-temple for	W. W. Dutcher	Milford, Mass	Mar. 24, 1863	37, 954
Looms, Roller-temple for	W. W. Dutcher	Milford, Mass	Mar. 27, 1866	53, 423
Looms, Roller-temple for	W. W. Dutcher	Hopedale, Mass	Aug. 30, 1870	106, 797
Looms, Roller-temple for	J. Mathis	Dornbein, Austria	Apr. 24, 1866	54, 269
Looms, Rotary cam for	R. Sargent	Norwich, Conn	Nov. 5, 1867	70, 623
Looms, Rotary temple for power	G. Draper	Saugus, Mass	Feb. 21, 1842	2, 464

Index of patents issued from the United States Patent Office from 1790 *to* 1873, *inclusive*—Continued.

Invention.	Inventor.	Residence.	Date.	No.
Looms, Rotary temple for weaving	G. Draper	Palmer, Mass	Oct. 28, 1840	1, 838
Looms, Sectional take-up for corset	S. Ottenheimer	New York, N. Y	May 28, 1867	65, 112
Looms, Self-acting temple for	K. Gibbs	South Berwick, Me	Feb. 13, 1839	1, 080
Looms, Shipper-lever for	G. W. Hathaway	Hinsdale, Mass	Feb. 3, 1863	37, 577
Looms, Shuttle-box-operating lever for	B. H. Jenks	Bridesburgh, Pa	July 13, 1869	92, 612
Looms, Spooling-machine for tape	I. Gibbs	Warren, Mass	Nov. 21, 1865	51, 038
Looms, Stopping power	O. M. Stillman	Stonington, Conn	Nov. 10, 1841	2, 333
Looms, Treadle-cam for	B. H. Jenks	Bridesburgh, Pa	Mar. 19, 1867	63, 055
Looms, Weft-feeding device for hair cloth	J. Blanchard	Pawtucket, R. I	Feb. 21, 1865	46, 442
Looms, Weft-fork for	J. H. Knowles	Lawrence, Mass	May 2, 1871	114, 307
Looms, Weft-stop mechanism for	W. Comey and S. S. Turner	Westborough, Mass	July 16, 1872	129, 212
Looms, Weft-stop mechanism for	J. D. Cottrell	Hopedale, Mass	May 4, 1869	89, 561
Looms, Weft-stop mechanism for	T. Nayler	Manchester, Iowa	May 7, 1872	126, 410
Looms, Weft-stop mechanism for	J. J. Switzer	Boston, Mass	Oct. 10, 1871	119, 798
Looms, Weft-stop motion for	G. Crompton	Worcester, Mass	Oct. 22, 1867	69, 974
Looms, Weft-stop motion for	J. C. Fifield	Lowell, Mass	Mar. 24, 1868	75, 888
Looms, Wire heddle for	D. C. Brown	Lowell, Mass	May 3, 1864	42, 553
Looms, Yarn-delivering mechanism for	G. Draper	Milford, Mass	Mar. 8, 1864	41, 834
Loop-banding machine	G. W. Weeks	Clinton, Mass	Nov. 14, 1871	121, 031
Loop, Body	R. R. Miller	Plantsville, Conn	June 27, 1871	116, 472
Loop-making machine	A. Babbett	Auburn, N. Y	Feb. 9, 1864	41, 470
Loop, Metallic	C. H. Littlefield	Turner, Me	July 30, 1867	67, 322
Lotteries, and mode of drawing, System of	F. W. Dana	Boston, Mass	Dec. 28, 1825	....
Lotteries, Calculating and drawing	W. Seger	Bergen, N. J	May 25, 1825	....
Lotteries, Constructing and drawing	J. J. Cohen	Baltimore, Md	Aug. 28, 1828	....
Lotteries, Constructing and drawing	M. Davis	Cumberland City, Me	May 9, 1825	....
Lotteries, Drawing	J. I. Cohen	Baltimore, Md	Mar. 3, 1825	....
Lotteries, Drawing	J. C. Rives	Washington, D. C	Dec. 22, 1826	....
Lotteries, Instituting and drawing	J. K. Casey	New York, N. Y	Mar. 28, 1831	....
Lotteries, Method of drawing	W. C. Conine	Baltimore, Md	Nov. 14, 1826	....
Lotteries, Method of drawing	J. Howard and A. Hayward	Providence, R. I	Nov. 8, 1825	....
Lotteries, Mode of drawing	D. Malcomb	Orleans, La	July 30, 1829	....
Lotteries, Mode of drawing	W. E. Spalding	Brooklyn, Conn	Feb. 15, 1828	....
Lottery	J. Vannini	Washington, D. C	Dec. 4, 1815	....
Lottery, American	E. Nott	Union College, N. Y	Sept. 9, 1825	....
Lottery-scheme	E. Gratton	Providence, R. I	May 16, 1826	....
Lottery-scheme	J. Vannini	Philadelphia, Pa	Oct. 18, 1823	....
Lottery-schemes, Construction of	J. W. Tagno	Philadelphia, Pa	May 16, 1825	....
Lottery-schemes, Mathematical operation of drawing.	J. Vannini	New York, N. Y	July 18, 1840	1, 700
Lounge	L. H. Baker	Tarrytown, N. Y	June 23, 1868	79, 050
Lounge	J. G. Broemser	Saint Louis, Mo	Jan. 25, 1859	22, 699
Lounge	W. H. Colley	Leavenworth, Kans	May 24, 1870	103, 301
Lounge	A. Eliaers	Boston, Mass	Oct. 25, 1853	10, 150
Lounge	W. C. Hart	Nantucket, Mass	May 26, 1868	78, 204
Lounge	W. H. Hart	Lowell, Mass	Sept. 30, 1873	143, 283
Lounge	F. Hayek	New York, N. Y	Mar. 2, 1869	87, 411
Lounge	A. Hayer	Philadelphia, Pa	Dec. 9, 1873	145, 418
Lounge	F. Krater	Pittsburgh, Pa	Jan. 14, 1873	134, 752
Lounge	O. Kubitschky	Chicago, Ill	Jan. 10, 1871	110, 923
Lounge	C. H. Levy and S. T. Herschmann.	New York, N. Y	Oct. 15, 1872	132, 162
Lounge	A. and D. Short	West Liberty, Ohio	July 31, 1866	56, 809
Lounge and bath-tub, Combined	C. Wendel	New York, N. Y	Apr. 29, 1873	138, 457
Lounge and bed	J. C. Hall	Cincinnati, Ohio	Feb. 21, 1871	112, 031
Lounge and bed	A. Hansen	San Francisco, Cal	Jan. 4, 1870	98, 589
Lounge and bed, Combined	P. Frank	New York, N. Y	Aug. 6, 1872	130, 290
Lounge and bedstead	A. R. Harper and C. B. Dake	Hobart, Ind	Dec. 21, 1869	98, 056
Lounge and bedstead, Combined	J. C. Hall	Cincinnati, Ohio	Oct. 24, 1871	120, 270
Lounge and bedstead, Combined	S. Higgins	Portland, Me	June 17, 1873	139, 957
Lounge and camp-stool, Combined	O. P. Pell	Flushing, N. Y	Sept. 24, 1861	33, 362
Lounge and chair, Combined	S. Herschmann	New York, N. Y	May 9, 1871	114, 679
Lounge and chair, Combined	A. T. Linikin	Roxbury, Mass	July 17, 1849	6, 594
Lounge and chair, Combined	M. P. Robinson	Fayette County, Ky	Dec. 9, 1873	145, 452
Lounge and chair seat	M. Le Page and F. Raymond	Woodhaven, N. Y	Apr. 23, 1867	64, 115
Lounge and sofa-bed	R. M. Reeby	Worcester, Mass	Oct. 26, 1869	96, 266
Lounge, bureau, and table	A. Failer	New York, N. Y	July 21, 1868	80, 159
Lounge, Extension	V. O. Hesh and J. Benjamin	Naples, N. Y	Aug. 8, 1871	117, 775
Lounge, Extension	N. H. Hill	Cincinnati, Ohio	Sept. 20, 1870	107, 494
Lounge, Extension	J. S. Paine	Cambridge, Mass	Apr. 4, 1871	113, 333
Lounge, Extension	C. Streit	Indianapolis, Ind	Nov. 17, 1868	84, 147
Lounge, Extension	C. Streit	Cincinnati, Ohio	Feb. 13, 1872	123, 649
Lounge, Extension	G. H. Thomas	Quincy Point, Mass	July 23, 1872	129, 766
Lounge, Extension bed	C. F. Vollmer	Harrisburgh, Pa	Dec. 24, 1867	72, 704
Lounge, Folding	J. Beiersdorf	Chicago, Ill	Apr. 26, 1870	102, 360
Lounge, Folding	W. B. Coats	Philadelphia, Pa	Apr. 30, 1872	126, 266
Lounge, Folding	D. Forbes	Chicago, Ill	Nov. 10, 1868	83, 949
Lounge, Folding	W. H. Lotz	Chicago, Ill	Dec. 29, 1868	85, 391
Lounge, Folding	A. Lundberg	Chicago, Ill	Mar. 18, 1873	136, 843
Lounge, Folding	M. Marso	Chicago, Ill	Dec. 27, 1870	110, 578
Lounge, Folding	J. W. McDonough	Chicago, Ill	Dec. 11, 1866	60, 400
Lounge, Folding	W. Prafrock	Saint Louis, Mo	June 27, 1871	116, 350
Lounge, Hammock	J. C. Craft	Baltimore, Md	Sept. 2, 1873	142, 327
Lounge, Invalid	R. Leaman	Hillsborough, Ohio	Jan. 1, 1867	60, 911
Lounge, Nursery	S. Buttenheim	New York, N. Y	Aug. 27, 1867	68, 164
Lounge, Ottoman	G. H. C. Williams	Chicago, Ill	July 28, 1868	80, 322
Lounge, Reversible	S. Lloyd	Washington, D. C	Feb. 26, 1867	62, 349
Lounges, Leg for folding	W. Seng	Chicago, Ill	Sept. 3, 1872	131, 031
Lounges, Leg for folding	J. F. Tobey	Medford, Mass	May 6, 1873	138, 716
Lowering and detaching boats from davits, Device for.	C. H. Ramsten	Carlskrona, Sweden	Oct. 23, 1866	59, 153
Low land, Device for filling	G. Howell	Philadelphia, Pa	Dec. 27, 1870	110, 468
Low-water alarm	J. C. Leistner and A. Kayser	Cincinnati, Ohio	Sept. 19, 1871	119, 037
Low-water alarm	J. Stanton	Philadelphia, Pa	Apr. 25, 1871	114, 216
Low-water alarm and fire-extinguisher for steam-boiler.	B. R. Singleton	Saint Louis, Mo	Feb. 20, 1872	123, 795
Low-water alarm and indicator	J. Ross	Greenville, Mich	Mar. 25, 1873	136, 097
Low-water-alarm apparatus for steam-boilers	C. H Brown	Fitchburgh, Mass	Aug. 7, 1860	29, 455

Index of patents issued from the United States Patent-Office from 1790 *to* 1873, *inclusive*—Continued.

Invention.	Inventor.	Residence.	Date.	No.
Low-water alarm for boilers	J. Atkins	Washington, D. C	June 30, 1868	79, 431
Low-water alarm for boilers	F. S. Davenport	Jerseyville, Ill	Jan. 7, 1868	73, 168
Low-water alarm for boilers	E. W. Mandeville	Ithaca, N. Y	Nov. 17, 1868	84, 201
Low-water alarm in boilers	A. S. Lyman	Upper Alton, Ill	Dec. 18, 1849	6, 955
Low-water alarm for steam-boilers	J. R. Brown and W. A. Fosket	New Haven, Conn	Feb. 4, 1873	135, 519
Low-water alarm for steam-boilers	S. Dustin	Detroit, Mich	Apr. 26, 1859	23, 761
Low-water alarm for steam-boilers	A. Carr	Paterson, N. J	Apr. 16, 1861	32, 045
Low-water alarm for steam-boilers	J. C. Leistner and A. Kayser	Cincinnati, Ohio	June 25, 1872	128, 318
Low-water alarm for steam-boilers	W. H. Miller	Philadelphia, Pa	May 14, 1861	32, 305
Low-water alarm for steam-boilers	L. Savage	Ashtabula, Ohio	Jan. 23, 1872	122, 970
Low-water alarm for steam-boilers, Electro-magnetic.	W. Duryea	Glen Cove, N. Y	Aug. 9, 1870	106, 141
Low-water alarm for steam-generators	R. F. Crane	Chicago, Ill	Aug. 6, 1867	67, 506
Low-water alarm for steam-generators	S. B. Palmer	Syracuse, N. Y	Feb. 19, 1867	62, 150
Low-water alarm for steam-generators	S. B. Palmer	Syracuse, N. Y	Sept. 8, 1868	82, 025
Low-water alarm for steam-generators	J. H. Springer and W. M. Bartram.	Philadelphia, Pa	Feb. 19, 1867	62, 230
Low-water and high-pressure alarm	G. B. Massey	New York, N. Y	Aug. 23, 1870	106, 598
Low-water and high-pressure indicator	G. M. Hopkins and J. A. Straight	Albion, N. Y	Aug. 30, 1870	106, 821
Low-water and high-steam indicator	L. F. Smith	Philadelphia, Pa	Nov. 24, 1868	84, 444
Low-water and high-steam indicator	L. F. Smith	Philadelphia, Pa	Sept. 6, 1870	107, 114
Low-water detector	J. Ashcroft	Brooklyn, N. Y	June 5, 1866	55, 221
Low-water detector	W. C. Baker	New York, N. Y	Apr. 21, 1863	28, 212
Low-water detector	B. H. Bartol	Philadelphia, Pa	Dec. 18, 1866	60, 463
Low-water detector	C. F. Cosfeldt, jr	Philadelphia, Pa	Nov. 12, 1867	70, 807
Low-water detector	J. Cosfeldt	Philadelphia, Pa	June 13, 1865	48, 158
Low-water detector	J. Delery	Saint Bernard Parish, La	July 3, 1866	56, 017
Low-water detector	T. Firth	Cincinati, Ohio	Nov. 13, 1866	59, 574
Low-water detector	J. E. Gillespie	Boston, Mass	Mar. 12, 1867	62, 837
Low-water detector	J. Harrison, jr	Philadelphia, Pa	Mar. 24, 1868	75, 754
Low-water detector	G. W. Hewitt and J. B. Haley	Cincinnati, Ohio	May 8, 1866	54, 544
Low-water detector	L. L. Lee	Milwaukee, Wis	June 27, 1871	116, 460
Low-water detector	G. B. Massey	New York, N. Y	Sept. 5, 1871	118, 737
Low-water detector	M. Peck	New Haven, Conn	Jan. 23, 1866	52, 194
Low-water detector	J. R. Shupplee and B. K. Wright	Philadelphia, Pa	Dec. 26, 1865	51, 759
Low-water detector	S. W. Warren	Brooklyn, N. Y	Oct. 21, 1862	36, 741
Low-water detector	J. Yates	Mott Haven, N. Y	Jan. 31, 1865	46, 173
Low-water detector and safety-valve, Combined	D. Burns	Bay City, Mich	May 5, 1868	77, 451
Low-water detector and steam-pressure alarm	D. McFarland	New York, N. Y	July 7, 1868	79, 673
Low-water detector for boilers	J. Aschcroft	New York, N. Y	Aug. 4, 1868	80, 700
Low-water detector for boilers	C. F. Cosfeldt, jr	Philadelphia, Pa	Jan. 14, 1868	73, 301
Low-water detector for boilers	T. Dutton	Port Jervis, N. Y	Sept. 8, 1868	81, 989
Low-water detector for boilers	G. B. Massey	New York, N. Y	Sept. 28, 1869	95, 245
Low-water detector for boiler-feeds	J. N. B. Bond	New York, N. Y	Feb. 25, 1868	74, 883
Low-water detector for steam-boilers	E. S. Bennett	New York, N. Y	Jan. 26, 1869	86, 125
Low-water detector for steam-boilers	G. W. Blake	New York, N. Y	Aug. 20, 1861	33, 067
Low-water detector for steam-boilers	J. W. Hopper	New York, N. Y	July 24, 1860	29, 277
Low-water detector for steam-boilers	B. Schaffer	Buckau, Magdeburg, Prussia.	Nov. 22, 1864	45, 207
Low-water detector for steam-boilers	M. S. Vosburgh	Attica, N. Y	Dec. 31, 1872	134, 499
Low-water detector for steam-boilers	L. Youmans	Fulton, N. Y	Sept. 3, 1861	33, 224
Low-water detector for steam-boilers, Electro-magnetic.	W. Duryea	Glen Cove, N. Y	Aug. 9, 1870	106, 142
Low-water detector for steam-generators	W. W. Bailey	New York, N. Y	Mar. 17, 1868	75, 618
Low-water detector for steam-generators	G. B. Massey	New York, N. Y	Feb. 18, 1868	74, 706
Low-water detector for steam-generators	G. B. Massey	New York, N. Y	Feb. 23, 1869	87, 181
Low-water detector for steam-generators	D. C. Mead and C. Maggi	Pittsburgh, Pa	Oct. 24, 1865	50, 660
Low-water detector for steam-generators	T. Savill	Philadelphia, Pa	Sept. 18, 1866	58, 179
Low-water detector for steam-generators	J. V. Weitz	Cleveland, Ohio	Sept. 3, 1867	68, 473
Low-water indicator	J. O. Alter	Saint Louis, Mo	Aug. 13, 1867	67, 702
Low-water indicator	F. S. Barns	New York, N. Y	July 13, 1869	92, 564
Low-water indicator	R. Berryman	Philadelphia, Pa	Feb. 15, 1870	99, 996
Low-water indicator	J. W. Bishop	New Haven, Conn	Apr. 11, 1865	47, 183
Low-water indicator	W. A. Bradford	Cincinnati, Ohio	Dec. 21, 1869	98, 019
Low-water indicator	W. A. Bradford	Cincinnati, Ohio	Apr. 18, 1871	113, 843
Low-water indicator	J. Brahn	Jersey City, N. J	June 22, 1869	91, 599
Low-water indicator	R. R. Carpenter	Tippecanoe, Ohio	July 19, 1870	105, 422
Low-water indicator	T. G. Eiswald	Providence, R. I	July 28, 1868	80, 399
Low-water indicator	T. G. Eiswald	Providence, R. I	July 28, 1868	80, 400
Low-water indicator	T. G. Eiswald and J. Barbour	Providence, R. I	Aug. 25, 1868	81, 483
Low-water indicator	R. A. Filkins	North Adams, Mass	Dec. 3, 1867	71, 729
Low-water indicator	L. Fulton	Edinburgh, Ind	Apr. 23, 1867	64, 007
Low-water indicator	T. C. Hargrave and W. B. Charlton.	Boston, Mass	May 19, 1868	77, 978
Low-water indicator	G. M. Hopkins	Albion, N. Y	Sept. 15, 1868	82, 222
Low-water indicator	G. M. Hopkins	Albion, N. Y	Feb. 23, 1869	87, 102
Low-water indicator	G. M. Hopkins and J. A. Straight.	Albion, N. Y	Aug. 31, 1869	94, 209
Low-water indicator	G. M. Hopkins and J. A. Straight.	Albion, N. Y	Aug. 30, 1870	106, 820
Low-water indicator	E. Joyce	New York, N. Y	Nov. 3, 1868	83, 638
Low-water indicator	H. Kimball	Randolph, Vt	Feb. 28, 1871	112, 256
Low-water indicator	W. H. Laubach	Philadelphia, Pa	Mar. 19, 1867	63, 061
Low-water indicator	L. L. Lee	Milwaukee, Wis	Oct. 5, 1869	95, 595
Low-water indicator	G. McAllister	San Francisco, Cal	May 14, 1867	64, 781
Low-water indicator	W. Moore	Kokomo, Ind	Oct. 13, 1868	82, 974
Low-water indicator	A. F. W. Neynaber	Philadelphia, Pa	Oct. 25, 1870	108, 618
Low-water indicator	E. D. Pritchard	New York, N. Y	June 22, 1869	91, 769
Low-water indicator	J. C. Raymond and F. T. Allyn	Brooklyn, N. Y	June 2, 1868	78, 481
Low-water indicator	F. Raynsford	Grand Rapids, Mich	Nov. 1, 1870	108, 828
Low-water indicator	P. W. Reinshagen	Cincinnati, Ohio	Apr. 26, 1870	102, 433
Low-water indicator	H. Reynolds	New Haven, Conn	Nov. 23, 1869	97, 228
Low-water indicator	J. W. Richards	Newark, N. J	Dec. 1, 1868	84, 646
Low-water indicator	L. F. Smith	Philadelphia, Pa	Nov. 12, 1867	70, 758
Low-water indicator	L. F. Smith	Philadelphia, Pa	Dec. 3, 1867	71, 803
Low-water indicator	F. Steele	Brooklyn, N. Y	Dec. 9, 1873	145, 369
Low-water indicator	W. W. Virdin	Baltimore, Md	May 4, 1869	89, 714
Low-water indicator	C. S. Watson	Philadelphia, Pa	July 7, 1868	79, 708

Index of patents issued from the United States Patent Office from 1790 *to* 1873, *inclusive*—Continued.

Invention.	Inventor.	Residence.	Date.	No.
Low-water indicator	D. Wiehl	Cincinnati, Ohio	July 5, 1870	105, 155
Low-water indicator and alarm	W. Moore	Kokomo, Ind	July 16, 1872	129, 159
Low-water indicator and alarm	F. Strange	Detroit, Mich	Aug. 26, 1873	142, 181
Low-water indicator and safety-valve, Combined	C. Burley	Cincinnati, Ohio	Dec. 17, 1867	72, 361
Low-water indicator and steam-pressure alarm	D. McFarland	New York, N. Y	July 7, 1868	79, 672
Low-water indicator and feed-water regulator	R. Berryman	Philadelphia, Pa	Jan. 25, 1870	99, 234
Low-water indicator for boilers	J. S. Griffith	Saint Louis, Mo	Feb. 2, 1869	86, 531
Low-water indicator for boilers	H. Kimball	Randolph, Vt	June 1, 1869	90, 853
Low-water indicator for boilers	C. S. Watson	Philadelphia, Pa	May 31, 1870	103, 630
Low-water indicator for boilers	P. D. Wesson	Providence, R. I	July 6, 1869	92, 237
Low-water indicator for steam-boilers	C. Brooke	Norfolk, Va	Aug. 6, 1872	130, 185
Low-water indicator for steam-boilers	G. A. Riedel	Philadelphia, Pa	Apr. 18, 1865	47, 365
Low-water indicator for steam-boilers	L. Fulton	Edinburgh, Ind	July 21, 1868	80, 162
Low-water indicator for steam-generators	L. H. Colborn	Chicago, Ill	Apr. 21, 1868	76, 893
Low-water register, Steam-boiler	W. S. Belt	Cincinnati, Ohio	Aug. 27, 1872	130, 786
Low-water safety-apparatus for steam-boilers	J. Ashcroft	Lynn, Mass	Mar. 20, 1860	27, 508
Low-water signal	T. Shaw	Philadelphia, Pa	July 11, 1865	48, 730
Low-water signal for steam-boilers	T. Shaw	Philadelphia, Pa	Dec. 27, 1864	45, 680
Lozenge	F. Ernst	New York, N. Y	Nov. 27 1866	59, 988
Lozenge and cracker machine	C. A. Oehl	Portsmouth, N. H	Jan. 19, 1869	86, 096
Lozenge and other paste, Machine for rolling	C. A. Oehl	New York, N. Y	May 2, 1871	114, 468
Lozenge, Cathartic	W. M. Du Bois	Poughkeepsie, N. Y	July 7, 1868	79, 641
Lozenge-cutting machine	G. H. Copping	Toronto, Canada	Nov. 4, 1873	144, 258
Lozenge, &c., cutting machine	W. E. Damant	West Hoboken, N. J	Apr. 4, 1871	113, 635
Lozenge-cutting machine	W. P. Eayrs	Nashua, N. H	Feb. 28, 1871	112, 231
Lozenge-cutting machine	W. J. McClelland	New York, N. Y	May 8, 1860	28, 239
Lozenge-cutting machine	J. W. Pepper	Salem, Mass	July 2, 1850	7, 482
Lozenge-machine	E. Belling	New York, N. Y	Aug. 30, 1859	25, 236
Lozenge-machine	O. B. and S. E. Chase	Boston and Charlestown, Mass.	May 12, 1857	17, 262
Lozenge-machine	O. R. Chase	Birmingham, England	July 7, 1863	39, 196
Lozenge-machine	E. Greenfield	New York, N. Y	July 16, 1872	129, 277
Lozenge-machine	E. Günther	New York, N. Y	Jan. 9, 1872	122, 602
Lozenge-machine	J. S. Howell and C. W. Carter	Portsmouth, N. H	June 11, 1867	65, 575
Lozenge-machine	G. Kober	New York, N. Y	May 22, 1860	28, 381
Lozenge-machine	C. A. Oehl	Portsmouth, N. H	Feb. 15, 1870	99, 939
Lozenge-machine	D. Sowle	Fall River, Mass	June 19, 1860	28, 815
Lozenge-package	H. W. Booth	Toronto, Canada	June 18, 1872	128, 011
Lozenges and confectionery, Manufacture of	E. Greenfield	New York, N. Y	Feb. 15, 1870	99, 379
Lubricating-apparatus	J. C. Eggleston	Waterbury, Conn	Sept. 25, 1866	58, 339
Lubricating-axle	M. F. Kellogg	Pittsfield, Ohio	Apr. 12, 1870	101, 885
Lubricating-box	C. Andrews	Providence, R. I	Aug. 30, 1870	106, 762
Lubricating-box	T. J. Rowley and W. Roland	Chillicothe, Ohio	Mar. 10, 1868	75, 465
Lubricating-box and bearing for upright shafts	J. P. Grosvenor	Lowell, Mass	Feb. 13, 1872	123, 562
Lubricating-composition	A. Bauer	Philadelphia, Pa	May 12, 1863	38, 459
Lubricating-composition	C. Chitterling	Dunkirk, N. Y	July 24, 1860	29, 243
Lubricating-composition	J. Cordnan	Brooklyn, N. Y	Mar. 31, 1868	76, 165
Lubricating-composition	G. W. Goodhue	Cincinnati, Ohio	Jan. 5, 1864	41, 151
Lubricating-composition	C. A. Grandey	Rutland, Vt	Nov. 26, 1867	71, 296
Lubricating-composition	C. Grath	Saint Louis, Mo	Dec. 8, 1863	40, 835
Lubricating-composition	A. Hamill	Baltimore, Md	June 9, 1863	38, 822
Lubricating-composition	E. E. Hendrick	New York, N. Y	July 1, 1862	35, 753
Lubricating-composition	W. Hilton	Belvidere, N. J	Jan. 5, 1864	41, 068
Lubricating-composition	J. Turner	New York, N. Y	July 7, 1863	39, 185
Lubricating-composition for machinery	E. E. Hendrick	New York, N. Y	July 1, 1862	35, 754
Lubricating-composition, Metallic	C. B. Lawrence	Munda, N. Y	Oct. 8, 1861	33, 438
Lubricating-compound	W. N. Abbott	Boston, Mass	Nov. 30, 1869	97, 335
Lubricating-compound	F. T. Allyn	New York, N. Y	Nov. 26, 1867	71, 356
Lubricating-compound	E. Brown, jr	Binghamton, N. Y	July 10, 1855	13, 211
Lubricating-compound	R. R. Brown	Buffalo, N. Y	Jan. 18, 1859	22, 621
Lubricating-compound	G. A. Colehamer	Reading, Pa	July 25, 1854	11, 359
Lubricating-compound	J. and W. W. Cumberland	Mobile, Ala., and New Albany, Ind.	Apr. 3, 1849	6, 263
Lubricating-compound	T. E. Curtiss	Titusville, Pa	May 26, 1868	78, 189
Lubricating-compound	P. S. Devlan	Reading, Pa	Jan. 16, 1849	6, 028
Lubricating-compound	N. Dresser	Rochester, N. Y	Apr. 17, 1855	12, 716
Lubricating-compound	E. Dyer, jr	Providence, R. I	Jan. 31, 1871	111, 441
Lubricating-compound	B. French	Rochester, N. Y	Apr. 30, 1872	126, 282
Lubricating-compound	B. French	Rochester, N. Y	Sept. 10, 1872	131, 264
Lubricating-compound	E. F. Gerdom and C. W. Schindler.	Albany, N. Y	Jan. 7, 1868	73, 089
Lubricating-compound	E. F. Gerdom and C. W. Schindler.	Albany, N. Y	July 28, 1868	80, 471
Lubricating-compound	G. W. Gladden	Cincinnati, Ohio	Mar. 8, 1864	41, 840
Lubricating-compound	A. S. Grenville	Westborough, Mass	Jan. 30, 1849	6, 071
Lubricating-compound	H. Grogan	Flatbush, N. Y	Aug. 2, 1870	106, 053
Lubricating-compound	H. Grogan	Flatbush, N. Y	July 4, 1871	116, 704
Lubricating-compound	T. Hull and A. H. Vail	Poughkeepsie, N. Y	Feb. 11, 1868	74, 366
Lubricating-compound	H. R. Hutchinson	Empire Iron Works, Ky	Jan. 14, 1873	134, 890
Lubricating-compound	L. Kirk and J. Dodsworth	Reading, Pa	Oct. 16, 1847	5, 333
Lubricating-compound	A. Lebküchler	Belleville, Ill	June 25, 1861	32, 631
Lubricating-compound	I. Lossiel	Philadelphia, Pa	Sept. 19, 1865	50, 015
Lubricating-compound	S. Y. Love	Philadelphia, Pa	Sept. 30, 1873	143, 362
Lubricating-compound	J. Marshall	Reading, Pa	May 22, 1855	12, 918
Lubricating-compound	J. B. McNum	Port Jervis, N. Y	Jan. 3, 1860	26, 693
Lubricating-compound	A. Millochau	New York, N. Y	May 30, 1871	115, 342
Lubricating-compound	C. S. Moore	Erie, Pa	Nov. 23, 1869	97, 262
Lubricating-compound	J. B. Norris	Richmond, Va	Dec. 23, 1873	145, 812
Lubricating-compound	V. Radspinner and W. H. Moss	New Richmond, Ohio	July 3, 1860	29, 005
Lubricating-compound	C. Keck	Floraville, Ill	June 4, 1872	127, 429
Lubricating-compound	E. Roat	Buffalo, N. Y	Mar. 18, 1873	136, 868
Lubricating-compound	E. F. Prentiss	Philadelphia, Pa	May 29, 1855	12, 964
Lubricating-compound	V. Purdy	Poughkeepsie, N. Y	Mar. 7, 1871	112, 489
Lubricating-compound	J. Selgrath	Pottsville, Pa	June 24, 1851	8, 176
Lubricating-compound	E. B. S. Shoemaker	Towsontown, Md	July 19, 1870	105, 499
Lubricating-compound	M. D. Smith	Prairie City, Iowa	Nov. 25, 1873	144, 875
Lubricating-compound	J. H. Smyser	Pittsburgh, Pa	Mar. 29, 1870	101, 325
Lubricating-compound	J. H. Smyser	Pittsburgh, Pa	Jan. 3, 1871	110, 690

Index of patents issued from the United States Patent Office from 1790 *to* 1873, *inclusive*—Continued.

Invention.	Inventor.	Residence.	Date.	No.
Lubricating-compound	D. C. Taylor	Goshen, N. Y	Sept. 19, 1865	50, 049
Lubricating-compound	E. A. Tonner	Saint Louis, Mo	Mar. 18, 1873	136, 945
Lubricating-compound	W. Turner	Phœnixville, Pa	Feb. 19, 1861	31, 519
Lubricating-compound	H. Vaughn and W. Hutton	Providence, R. I., and Baltimore, Md.	Aug. 2, 1859	24, 965
Lubricating-compound	J. Williams	Sharpsburgh, Pa	Oct. 22, 1872	132, 379
Lubricating-compounds, Solution for thinning	R. Patterson	Philadelphia, Pa	Oct. 11, 1859	25, 792
Lubricating-cup	M. F. Carson	Cleveland, Ohio	Dec. 22, 1868	85, 164
Lubricating-cup	J. B. Collin	Altoona, Pa	Dec. 17, 1867	72, 168
Lubricating-cup	T. Cooper	Warwick, R. I	May 6, 1873	138, 558
Lubricating-cup	J. Songster	Buffalo, N. Y	July 11, 1865	48, 776
Lubricating-cup	C. S. Westlandt	Providence, R. I	Oct. 22, 1861	33, 562
Lubricating-cup, Automatic	J. A. Bryan and W. Stamfield	Kent, Ohio	Jan. 24, 1871	111, 105
Lubricating-cup for steam and other enginery	I. Dreyfus	New York, N. Y	Jan. 19, 1869	86, 009
Lubricating-device	T. S. Brown	Poughkeepsie, N. Y	Mar. 5, 1867	62, 601
Lubricating-device	F. Colligon	Buffalo, N. Y	July 28, 1868	80, 457
Lubricating-device	C. Hirsch	Buffalo, N. Y	Mar. 26, 1872	124, 954
Lubricating-device	G. M. Morris	Cohoes, N. Y	Jan. 1, 1867	60, 925
Lubricating-device	B. H. Reynolds and J. Bachelder.	Canterbury and Norwich, Conn.	June 16, 1868	78, 895
Lubricating-device	M. Senior	Frankford, Pa	Mar. 17, 1868	75, 584
Lubricating machinery, Composition for	S. Balsden	Brooklyn, N. Y	Oct. 13, 1863	40, 229
Lubricating machinery, Composition for	E. Q. Henderson	Charlotte, N. C	Mar. 7, 1871	112, 453
Lubricating machinery, Composition for	P. Zieber, P. S. Devlan, and J. Hancock.	Reading and Philadelphia, Pa.	Sept. 11, 1847	5, 288
Lubricating machinery, Method of	P. E. Proust	Paris, France	Mar. 14, 1865	46, 856
Lubricating machinery, Mode of	A. P. Gross	Vallejo, Cal	Oct. 22, 1872	132, 395
Lubricating-material	C. Carpenter, jr	Astoria, N. Y	Sept. 15, 1868	82, 083
Lubricating-material	W. P. Ewing	Rochester, N. Y	Sept. 11, 1866	58, 020
Lubricating-material	W. Little	Middlesex County, England	Aug. 1, 1854	11, 406
Lubricating-sleeve	G. A. Lloyd	San Francisco, Cal	Dec. 21, 1869	98, 172
Lubricating-spindle	A. H. Gilman	Hopedale, Mass	Dec. 17, 1867	72, 185
Lubricating substances, Mode of applying	J. Dougall	Sterlingshire, Scotland	May 10, 1864	42, 714
Lubricating under pressure, Method of	J. D. Custer	Norristown, Pa	Mar. 24, 1857	16, 871
Lubricator: *See* Atmospheric lubricator. Automatic lubricator. Axle-lubricator. Axle-box lubricator. Car-wheel lubricator. Cylinder-lubricator. Hammer-lubricator. Hub and axle lubricator. Journal-lubricator. Machinery-lubricator. Nail-machine lubricator. Piston-lubricator. Portable lubricator. Pulley-lubricator. Shaft-lubricator. Shaft-bearing lubricator. Ship's-block lubricator. Spindle-lubricator. Spindle-step lubricator. Spinning-machine lubricator. Steam lubricator. Steam-engine lubricator. Steam-valve lubricator. Valve lubricator. Veloci-trot lubricator. Ventilator-lubricator. Wheel-lubricator. Wood lubricator.				
Lubricator	D. Adamson	New York, N. Y	Dec. 7, 1869	97, 470
Lubricator	D. Adamson	Bremen, Germany	June 21, 1870	104, 400
Lubricator	J. F. A. Aerts	Antwerp, Belgium	Nov. 28, 1865	51, 276
Lubricator	J. F. A. and P. F. Aerts	Antwerp and Brussels, Belgium.	Jan. 31, 1865	46, 196
Lubricator	H. V. Aiken	Gibsonburgh, Pa	May 27, 1873	139, 286
Lubricator	C. D. Austin	Newcastle-upon-Tyne, England.	Dec. 6, 1870	109, 795
Lubricator	E. J. Baker	Baltimore, Md	Apr. 1, 1856	14, 549
Lubricator	I. Barnett and P. Eckford	Cincinnati, Ohio	Jan. 2, 1872	122, 427
Lubricator	C. A. Baumgart	Allegheny City, Pa	Oct. 12, 1869	95, 635
Lubricator	M. Bergner and O. Netzow	Baltimore, Md	May 18, 1869	90, 223
Lubricator	A. Breaer	Saugatuck, Conn	May 6, 1856	14, 797
Lubricator	F. Bresson	Paris, France	Apr. 18, 1865	47, 367
Lubricator	J. Broughton	New York, N. Y	Jan. 3, 1865	45, 694
Lubricator	J. Broughton	New York, N. Y	Dec. 12, 1865	51, 419
Lubricator	J. Broughton	New York, N. Y	Nov. 2, 1869	96, 389
Lubricator	P. G. Brown	Schenectady, N. Y	Apr. 12, 1859	23, 661
Lubricator	H. A. Burr and L. E. Rockwell.	New York, N. Y	Dec. 9, 1862	37, 084
Lubricator	A. S. Cameron	New York, N. Y	Nov. 30, 1869	97, 354
Lubricator	H. Campbell	Newton, Conn	Nov. 24, 1863	40, 673
Lubricator	E. Clampitt	Baltimore, Md	Oct. 19, 1858	21, 816
Lubricator	W. A. Clarke	Westville, Conn	Jan. 2, 1872	122, 361
Lubricator	B. F. Clemenshaw	Troy, N. Y	Feb. 11, 1873	135, 777
Lubricator	J. L. Courcier	Paris, France	Feb. 12, 1867	62, 013
Lubricator	H. Crossley	Brooklyn, N. Y	Aug. 11, 1868	80, 923
Lubricator	D. Currie	Saint Louis, Mo	June 14, 1870	104, 279
Lubricator	H. A. Daniels	Thomaston, Conn	July 6, 1869	92, 283
Lubricator	A. C. Dewies	Crefeld, Prussia	Oct. 21, 1862	36, 703
Lubricator	J. L. Dickinson	Dubuque, Iowa	Dec. 20, 1870	110, 215
Lubricator	I. Dreyfus	New York, N. Y	Aug. 9, 1870	106, 261
Lubricator	I. Dreyfus	New York, N. Y	Nov. 8, 1870	109, 118
Lubricator	E. Ehlin	San Francisco, Cal	Oct. 10, 1871	119, 751
Lubricator	W. E. Everett	New York, N. Y	Mar. 4, 1856	14, 352
Lubricator	J. Fanyon	Bridgeport, Conn	June 11, 1867	65, 734
Lubricator	E. Faull	Maldon, Australia	Aug. 13, 1867	67, 743
Lubricator	J. H. Ferguson	Springfield, Mass	Oct. 3, 1865	50, 233
Lubricator	F. Ficht	Marinette, Wis	May 28, 1872	127, 161
Lubricator	W. T. Garratt	San Francisco, Cal	Dec. 5, 1871	121, 611
Lubricator	W. T. Garratt	San Francisco, Cal	May 14, 1872	126, 799
Lubricator	W. T. Garratt	San Francisco, Cal	Nov. 19, 1872	133, 217
Lubricator	J. Gates	Portland, Ohio	Sept. 20, 1870	107, 478
Lubricator	J. Gates	Portland, Ohio	Apr. 29, 1873	138, 243

Index of patents issued from the United States Patent Office from 1790 *to* 1873, *inclusive*—Continued.

Invention.	Inventor.	Residence.	Date.	No.
Lubricator	S. F. Gates	Cambridge, Mass	Feb. 21, 1871	112, 029
Lubricator	W. Gee	New York, N. Y	Feb. 12, 1856	14, 236
Lubricator	W. Gee	New York, N. Y	Aug. 9, 1870	106, 150
Lubricator	G. Gerdom	New York, N. Y	Apr. 26, 1870	102, 391
Lubricator	G. Gerdom	Albany, N. Y	May 17, 1870	103, 170
Lubricator	E. F. Gerdom and C. W. Schindler.	Albany, N. Y	May 19, 1868	77, 974
Lubricator	F. R. Glascock	Hillsborough, Ohio	Dec. 3, 1872	133, 581
Lubricator	T. W. Godwin	Portsmouth, Va	Nov. 3, 1863	40, 475
Lubricator	T. W. Godwin	Portsmouth, Va	Nov. 10, 1863	40, 560
Lubricator	T. W. Godwin	Portsmouth, Va	June 28, 1864	43, 305
Lubricator	T. W. Godwin	Portsmouth, Va	Feb. 7, 1865	46, 231
Lubricator	D. A. Greene	New York, N. Y	June 14, 1870	104, 139
Lubricator	W. Hamilton	Chicopee, Mass	Oct. 2, 1866	58, 409
Lubricator	J. Harlin	New York, N. Y	Sept. 29, 1868	82, 522
Lubricator	A. W. Harris	Providence, R. I	July 6, 1869	92, 307
Lubricator	H. J. Hawkins	Mobile, Ala	July 14, 1857	17, 822
Lubricator	J. E. Hendrick	Waterbury, Conn	Sept. 6, 1870	107, 044
Lubricator	J. W. and R. Hewitt	Allegheny City, Pa	May 11, 1869	89, 865
Lubricator	R. S. Hildreth	South Adams, Mass	Jan. 4, 1870	98, 655
Lubricator	M. Hinmann	West Stockbridge, Mass	Dec. 13, 1870	110, 040
Lubricator	J. Hodge	Harrison, N. J	June 27, 1871	116, 442
Lubricator	T. Holland	New York, N. Y	Nov. 10, 1868	83, 965
Lubricator	T. Holland	New York, N. Y	Feb. 15, 1870	99, 894
Lubricator	T. Holland and J. T. Cody	Cincinnati, Ohio	June 2, 1868	78, 594
Lubricator	J. J. Hoyt	Chelmsford, Mass	Nov. 23, 1869	97, 091
Lubricator	S. Hutchinson, jr	Salem, Mass	Nov. 26, 1872	133, 316
Lubricator	E. Johnson	Wilkins, Pa	Apr. 14, 1868	76, 774
Lubricator	A. J. Judge	Baltimore, Md	Feb. 10, 1863	37, 627
Lubricator	A. J. Judge	Baltimore, Md	Jan. 19, 1864	41, 302
Lubricator	S. E. Kleinschmidt	Cleveland, Ohio	Oct. 31, 1865	50, 719
Lubricator	F. A. Lane	Swanzey, N. H	Feb. 2, 1869	86, 421
Lubricator	L. S. Lapham	Providence, R. I	Mar. 20, 1860	27, 548
Lubricator	L. Leigh	Seymour, Conn	Aug. 30, 1864	44, 005
Lubricator	S. Lemon, jr	Hoboken, N. J	Aug. 6, 1867	67, 438
Lubricator	J. E. Lonergan	Sacramento, Cal	Oct. 22, 1872	132, 477
Lubricator	F. Lunkenheimer and J. Meehan.	Cincinnati, Ohio, and Covington, Ky.	Apr. 22, 1873	138, 169
Lubricator	C. Mather	Steubenville, Ohio	May 2, 1871	114, 317
Lubricator	E. McCoy	Ypsilanti, Mich	May 27, 1873	139, 407
Lubricator	F. P. McCullon and W. Woodcock.	Philadelphia and Scranton, Pa.	Jan. 26, 1869	86, 174
Lubricator	W. McCully	Paterson, N. J	Mar. 2, 1869	87, 423
Lubricator	H. McGraw	Detroit, Mich	Dec. 31, 1872	134, 436
Lubricator	J. Meehan	Covington, Ky	Aug. 23, 1870	106, 712
Lubricator	J. F. Monroe	Fitchburgh, Mass	Nov. 4, 1856	16, 018
Lubricator	T. J. Moors	Blossburgh, Pa	Sept. 6, 1870	107, 192
Lubricator	W. Morris	Dayton, Ohio	Apr. 29, 1873	138, 341
Lubricator	D. F. Mosman	Chelsea, Mass	Nov. 7, 1871	120, 660
Lubricator	C. Nelson	Brooklyn, N. Y	Aug. 30, 1870	106, 857
Lubricator	H. T. Neuss	Williamsburgh, N. Y	Aug. 16, 1864	43, 859
Lubricator	T. J. Nottingham	Cincinnati, Ohio	Mar. 1, 1870	100, 437
Lubricator	T. J. Nottingham	Cincinnati, Ohio	Dec. 13, 1870	110, 156
Lubricator	T. J. Nottingham	Cincinnati, Ohio	Mar. 28, 1871	113, 200
Lubricator	L. H. Olmsted	Newark, N. J	May 2, 1865	47, 561
Lubricator	J. A. Osenbrück	Hemelingen, Germany	Apr. 29, 1873	138, 274
Lubricator	E. Painter	East Hampton, Mass	Feb. 9, 1864	41, 531
Lubricator	E., B., and T. S. Parker	Schenectady, N. Y	Aug. 7, 1860	29, 515
Lubricator	C. H. Parshall	Detroit, Mich	June 18, 1872	128, 167
Lubricator	T. G. Pelton and J. Brewer	Lyons, Iowa, and Albany, Ill.	Aug. 1, 1865	49, 148
Lubricator	W. E. Phillips	Silver City, Idaho	Apr. 25, 1871	114, 194
Lubricator	N. W. Pomeroy	Meriden, Conn	Sept. 23, 1856	15, 775
Lubricator	N. W. Pomeroy	Meriden, Conn	Feb. 3, 1857	16, 553
Lubricator	J. M. Power	Port Discovery, Wash. Ty	Jan. 7, 1873	134, 611
Lubricator	J. M. Power	Port Discovery, Wash. Ty	Oct. 14, 1873	143, 639
Lubricator	W. A. Pratt	Baltimore, Md	Oct. 7, 1873	143, 530
Lubricator	W. Pratt and N. B. Williams	Providence, R. I., and New York, N. Y.	Jan. 3, 1871	110, 677
Lubricator	D. T. Pray	Boston, Mass	July 22, 1873	141, 168
Lubricator	H. Pringle	Green Point, N. Y	Aug. 30, 1870	106, 959
Lubricator	T. Reeves	Little River, Cal	Mar. 21, 1871	112, 847
Lubricator	H. W. Regan	Renovo, Pa	Apr. 29, 1873	138, 436
Lubricator	J. Regester	Baltimore, Md	Dec. 5, 1854	12, 047
Lubricator	W. M. Rennyson	Pottsville, Pa	Nov. 17, 1868	84, 135
Lubricator	D. M. Reynolds	Port Deposit, Md	Oct. 6, 1868	82, 878
Lubricator	J. Richey	Cincinnati, Ohio	June 30, 1868	79, 501
Lubricator	J. Richter	Cincinnati, Ohio	Jan. 10, 1871	110, 868
Lubricator	T. R. Robinson and R. E. Jones	Providence, R. I	Dec. 4, 1866	60, 251
Lubricator	J. Ross	Charlestown, Mass	Oct. 7, 1873	143, 422
Lubricator	R. Ross	Bethlehem, Pa	Mar. 1, 1864	41, 785
Lubricator	R. Ross and B. E. Lehman	Bethlehem, Pa	June 11, 1867	65, 609
Lubricator	M. G. Ryan	Frostburgh, Md	May 9, 1871	114, 716
Lubricator	C. Schott	Nashville, Tenn	Aug. 14, 1866	57, 196
Lubricator	G. Scott	Kensington, Pa	Mar. 13, 1866	53, 190
Lubricator	J. R. Sees	Philadelphia, Pa	Oct. 27, 1863	40, 433
Lubricator	N. Seibert	Nevada, Cal	Sept. 14, 1869	94, 780
Lubricator	N. Seibert	San Francisco, Cal	Feb. 14, 1871	111, 881
Lubricator	A. Shafer	Wellsville, N. Y	Apr. 20, 1869	89, 175
Lubricator	W. R. Shaw	Buffalo, N. Y	Sept. 24, 1872	131, 570
Lubricator	L. F. Smith	Philadelphia, Pa	Sept. 27, 1870	107, 826
Lubricator	W. K. Stevens	Alexandria, La	Feb. 16, 1858	19, 385
Lubricator	H. Strait	Covington, Ky	Apr. 21, 1857	17, 118
Lubricator	H. Taylor	Cincinnati, Ohio	May 12, 1868	77, 851
Lubricator	H. Taylor	Cincinnati, Ohio	June 23, 1868	79, 279
Lubricator	J. Fenwick	Grantham, England	Jan. 31, 1871	111, 401
Lubricator	H. Thomas	New York, N. Y	Feb. 16, 1869	87, 079
Lubricator	H. Thomas	New York, N. Y	July 6, 1869	92, 399

Index of patents issued from the United States Patent Office from 1790 *to* 1873, *inclusive*—Continued.

Invention.	Inventor.	Residence.	Date.	No.
Lubricator	R. H. Tradenick	Pittsburgh, Pa	Sept. 15, 1868	82, 182
Lubricator	J. E. Uhl	Renovo, Pa	Mar. 7, 1871	112, 515
Lubricator	W. Van Anden	Poughkeepsie, N. Y	Dec. 6, 1864	45, 365
Lubricator	E. Von Jeinsen	San Francisco, Cal	Sept. 27, 1870	107, 739
Lubricator	R. M. Wade	Wadesville, Va	June 6, 1854	11, 024
Lubricator	R. M. Wade	Wadesville, Va	May 1, 1855	12, 803
Lubricator	G. Waters	Cincinnati, Ohio	Aug. 6, 1867	67, 469
Lubricator	G. Waters	Cincinnati, Ohio	June 23, 1868	79, 167
Lubricator	G. Waters	Cincinnati, Ohio	Sept. 22, 1868	82, 368
Lubricator	G. Waters	Cincinnati, Ohio	June 29, 1869	92, 131
Lubricator	J. Webster	New York, N. Y	Apr. 11, 1854	10, 767
Lubricator	D. M. Weston	Boston, Mass	Jan. 19, 1864	41, 335
Lubricator	J. B. Wickersham	Philadelphia, Pa	Oct. 22, 1867	70, 058
Lubricator	J. B. Wickersham	Philadelphia, Pa	Sept. 29, 1868	82, 667
Lubricator	J. B. Wickersham	Philadelphia, Pa	July 26, 1870	105, 750
Lubricator	J. L. Wiggin and E. Folsom	South New Market, N. H	July 10, 1866	56, 306
Lubricator	J. H. Wilkinson	South New Market, N. H	Apr. 12, 1870	101, 954
Lubricator	W. W. W. Wood	Philadelphia, Pa	May 12, 1863	38, 526
Lubricator and anti-friction bearing	T. Roddick and J. Lockhead	Stranraer and Glasgow, Scotland.	Oct. 24, 1871	120, 213
Lubricator and water-conductor, Combined	T. A. Shock	Boston, Mass	Jan. 14, 1868	73, 262
Lubricators, Oil-hole cover for	S. A. Skinner	Hoosick Falls, N. Y	Nov. 7, 1871	120, 677
Lumber, Apparatus for washing and elevating	S. E. Worrell	Quincy, Ill	Mar. 15, 1870	100, 831
Lumber, Arrangement of devices for dressing pieces of.	H. Brown	New York, N. Y	May 19, 1857	17, 313
Lumber by superheated steam, Apparatus for drying and seasoning.	C. F. Allen and L. W. Campbell	Aurora, Ill	May 7, 1867	64, 398
Lumber, Device for canting or turning logs during the process of sawing them into.	C. Foster	Oshkosh, Wis	Jan. 6, 1863	37, 341
Lumber, Device for tonguing and grooving	N. G. Norcross	Lowell, Mass	May 2, 1854	10, 844
Lumber-dryer	J. Allonas	Mansfield, Ohio	Jan. 4, 1870	98, 541
Lumber-dryer	J. Brakeley	Bordentown, N. J	Apr. 14, 1868	76, 591
Lumber-dryer	J. Du Bois	Williamsport, Pa	Aug. 18, 1868	81, 074
Lumber-dryer	A. Edwards	New Haven, Conn	Jan. 7, 1873	134, 529
Lumber-dryer	R. E. Ferguson	Chicago, Ill	Oct. 4, 1870	107, 892
Lumber-dryer	J. W. Hanna	Wabash, Ind	Nov. 27, 1866	59, 998
Lumber-dryer	R. P. Johnson	Wabash, Ind	July 9, 1867	66, 594
Lumber-dryer	J. B. and T. E. Johnson	Indianapolis, Ind	Dec. 14, 1869	97, 777
Lumber-dryer	R. P. Johnson and E. J. Sumner	Wabash, Ind	July 7, 1868	79, 661
Lumber-dryer	F. J. Norton	Fremont, Ohio	Nov. 2, 1869	96, 471
Lumber-dryer	J. H. Osgood, jr	Boston, Mass	Nov. 27, 1866	60, 043
Lumber-dryer	J. H. Rae	Syracuse, N. Y	June 9, 1868	78, 692
Lumber-dryer	W. C. Scott	Richmond, Ind	May 11, 1869	90, 627
Lumber-dryer	E. J. Sumner	Chicago, Ill	Mar. 26, 1872	125, 098
Lumber, Drying and seasoning	E. C. Bender and W. Steffe	York and Philadelphia, Pa	Dec. 17, 1867	72, 157
Lumber-drying kiln	S. R. Kirby	Detroit, Mich	Oct. 21, 1873	143, 912
Lumber-drying kiln	H. W. Oliver	New Haven, Conn	June 24, 1862	35, 700
Lumber-edging machine	W. G. Caldwell	Three Rivers, Mich	Aug. 19, 1873	141, 852
Lumber-edging machine	L. A. Ensworth and B. Barker	Williamsport, Pa	Jan. 27, 1863	37, 496
Lumber, Feed-roll for	W. P. Hale	Lock Haven, Pa	Dec. 10, 1872	133, 775
Lumber-feeding rollers, Device for governing the parallel yielding of.	J. B. Pomroy	Chicago, Ill	Nov. 18, 1856	16, 105
Lumber for building purposes, Machine for preparing.	J. F. Gyles	Chicago, Ill	June 25, 1872	128, 388
Lumber for joiner's use, Prepared	J. Mullay	Somerville, Mass	Feb. 25, 1873	136, 332
Lumber from logs, Machine for dressing	A. Wolcott	Detroit, Mich	July 24, 1855	13, 342
Lumber-grooving machine	M. E. Carter	Syracuse, N. Y	Aug. 28, 1866	57, 474
Lumber-grooving machine	B. Holly and J. W. Wheeler	Seneca Falls, N. Y	July 8, 1851	8, 206
Lumber into irregular forms, Machinery for working.	R. Powers	Prescott, Mass	May 8, 1849	6, 436
Lumber-jointing machines, Clamp and mouth-piece for.	C. F. Bauersfeld	Cincinnati, Ohio	Mar. 27, 1855	12, 580
Lumber, Machine for cutting thin	B. F. Betts	Tonawanda, N. Y	July 7, 1863	39, 113
Lumber, Machine for resawing	E. H. Titus and J. Sharp	Wilkesbarre and Phillipsburgh, Pa.	June 29, 1858	20, 745
Lumber, &c., Machine for tallying	G. R. Lewis	Ashtabula, Ohio	June 6, 1865	48, 078
Lumber, Manufacturing	E. B. Rowe	Chicago, Ill	Apr. 20, 1869	89, 172
Lumber-marker	W. Merritt	Weymouth, Mass	Oct. 21, 1873	143, 773
Lumber, Process of seasoning	H. H. Beach	Rome, N. Y	Oct. 4, 1870	107, 854
Lumber-rack	W. H. Powers	Grand Rapids, Mich	June 11, 1872	127, 921
Lumber-rack	C. Sach	Grand Rapids, Mich	Jan. 18, 1870	99, 006
Lumber-rack	R. B. Woodcock	Grand Rapids, Mich	May 16, 1871	114, 897
Lumber-register	F. Nehanus	Ellenburgh Centre, N. Y	June 5, 1866	55, 327
Lumber preserving and drying	J. Oliver	Toronto, Canada	Sept. 2, 1873	142, 347
Lumber-scribing machine	J. Shellenberger	Indianapolis, Ind	Sept. 10, 1850	7, 636
Lumber-seasoning apparatus	M. R. Moore	Philadelphia, Pa	May 31, 1859	24, 230
Lumber-slitting gage	F. P. Hart	Chandlersville, Pa	May 15, 1855	12, 861
Lumber to rotary planers, Feeding tapering	P. H. Woolsey	Andes, N. Y	Feb. 19, 1861	31, 508
Lumber tonguing and grooving cutter	D. Perrin	McGregor, Iowa	Apr. 15, 1873	137, 861
Lumber-trimming machine	C. Lamb and T. J. Frazier	Clinton, Iowa	May 6, 1873	138, 505
Lunch-bag, Traveling	J. H. Noyes	Oneida, N. Y	Apr. 18, 1865	47, 324
Lunch-box	R. Cook	Saratoga Springs, N. Y	Jan. 28, 1862	34, 243
Lunch-box	J. Elson	Northampton, Mass	Nov. 8, 1870	108, 989
Lunch-box	J. Erpelding	Chicago, Ill	Jan. 12, 1869	85, 807
Lunch-box	J. Erpelding	Chicago, Ill	Mar. 22, 1870	101, 113
Lunch-box	M. G. Fagan	Troy, N. Y	Sept. 14, 1869	94, 729
Lunch-box	E. C. Lockwood	Bridgeport, Conn	Feb. 11, 1873	135, 651
Lunch-box	D. Miller	Allegheny City, Pa	Jan. 17, 1871	110, 991
Lunch-box	J. F. Morgan	Boston, Mass	May 2, 1865	47, 559
Lunch-box	J. F. Morgan	Boston, Mass	Nov. 5, 1867	70, 598
Lunch-box	P. H. Niles	Boston, Mass	Mar. 10, 1868	75, 451
Lunch-box	A. Norman	Rochester, N. Y	Aug. 3, 1869	93, 334
Lunch-box	J. S. Tibbals	Milford, Conn	Sept. 24, 1867	69, 272
Lunch-box	D. Troxell	Newark, N. J	May 2, 1871	114, 370
Lunch-box	W. H. Van Allen	Albany, N. Y	Nov. 19, 1872	133, 129
Lunch-box and shopping-bag, Combined	C. C. Cobleigh	Brighton, Mass	Nov. 11, 1873	144, 385
Lunch-box, Folding	C. S. Hurlburt	Springfield, Mass	Jan. 30, 1866	52, 291
Lunch-box, Folding	J. A. Minor	Middletown, Conn	Jan. 9, 1867	51, 957

Index of patents issued from the United States Patent Office from 1790 *to* 1873, *inclusive*—Continued.

Invention.	Inventor.	Residence.	Date.	No.
Lunch-box, Folding	F. B. Parks	Cambridgeport, Mass	July 16, 1867	66, 881
Lunch-heater	H. M. Kinsley	Chicago, Ill	Jan. 1, 1867	60, 743
Lye and melting potash, Cleansing	J. Jones	Fabius, N. Y	Feb. 11, 1809	....
Linch-pin	L. B. Gusman	Philadelphia, Pa	Mar. 13, 1866	53, 220
M.				
Macaroni-server	A. L. Lincoln	Boston, Mass	July 1, 1856	15, 266
Macaroni, vermicelli, Italian pastes, &c., Making	J. B. Sartori	Philadelphia, Pa	July 13, 1807	....
Mace, Policeman's	M. Warne	Philadelphia, Pa	Oct. 20, 1868	83, 228
Machine-drill	N. P. Eddy	Fall River, Mass	Apr. 24, 1866	54, 248
Machine for propelling, Hydrostatic and pneumatic.	R. Mills and H. B. Fernald	Washington, D. C., and Portland, Me.	Oct. 10, 1835	....
Machine, Horizontal double-rotary	W. C. Sullivan	Jefferson County, Ind	June 5, 1824	....
Machine-stand	W. B. Bement	Philadelphia, Pa	June 3, 1862	35, 433
Machinery, Apparatus for propelling	W. Z. W. Chapman	New York, N. Y	Sept. 14, 1869	94, 809
Machinery, Apparatus for propelling	E. J. Leyburn	Lexington, Va	Sept. 7, 1869	94, 618
Machinery, Apparatus to prevent rapping noises in	G. Ames	Rochester, N. Y	May 27, 1873	139, 227
Machinery, Application of the screw in propelling	W. H. Godfrey	Rochester, N. Y	Nov. 27, 1826	....
Machinery, Applying water in propelling	G. M. Gibbs	Prince William's Park, S. C	Mar. 10, 1828	....
Machinery-brake	D. S. Baker	West Bloomfield, N. Y	Sept. 1, 1868	81, 729
Machinery-brake	R. D. Napier	Birkenhead, England	Aug. 10, 1869	93, 469
Machinery-brake	R. D. Napier	Limehouse, England	Oct. 31, 1871	120, 527
Machinery by balance-lever, Propelling	S. H. Emmons and G. Upham	Massillon, Ohio	May 16, 1833	....
Machinery by combining air, water, and animal power, Propelling.	T. P. Codman	Boston, Mass	Apr. 20, 1833	....
Machinery by helping-power, Propelling	W. Loomis	Ashford, N. Y	Apr. 29, 1833	....
Machinery by means of cams and inclined planes, Propelling.	P. C. Curtis	Utica, N. Y	May 29, 1835	....
Machinery by weights, Propelling	C. Broyles	Tellico, Tenn	Oct. 19, 1827	....
Machinery by weights, Propelling	E. Turner	North Pownal, Me	June 12, 1835	....
Machinery, Device for connecting parts of	E. S. Pierce	Hartford, Conn	Apr. 13, 1869	88, 902
Machinery, Driving	W. S. Wells and S. B. Wells	New York and Middleburgh, N. Y.	June 27, 1865	48, 467
Machinery-lubricator	B. Hilbert	Cincinnati, Ohio	Apr. 23, 1867	64, 105
Machinery-lubricator	G. E. Smith	Fitchburgh, Mass	May 23, 1871	115, 247
Machinery-register	S. F. Payne	Chicopee, Mass	Nov. 25, 1873	144, 861
Machinery, Regulating	N. Schoolfield	Norwich, Conn	May 17, 1836	....
Machinery-regulator	T. C. Van Wyck and W. Kent	Poughkeepsie, N. Y	Feb. 1, 1870	99, 376
Machinery, Propelling	L. M. Edwards	Trenton, Tenn	Mar. 2, 1835	....
Machinery, Propelling	T. D. Newsom	Nashville, Tenn	Jan. 22, 1831	....
Machinery, Propelling	T. D. Newsom and J. C. Shule	Nashville, Tenn	Dec. 14, 1830	....
Machinery, Propelling	D. Russel	Tuscumbia, Ala	Oct. 27, 1835	....
Machinery, Propelling	W. Stanton	Centre Township, Ind	Apr. 23, 1827	....
Machinery with the foot, Driving light	A. Clarke	Greenwich, Conn	Jan. 20, 1841	1, 936
Machinists' gage	S. H. Bellows	Middletown, Conn	Sept. 16, 1873	142, 761
Mackerel-latch	C. S. H. Foster	Deer Isle, Me	Mar. 23, 1869	88, 026
Mackerel, Machine for splitting	S. S. Turner	Lewiston, Me	Oct. 21, 1856	15, 941
Madder and other plants, Process of separating coloring-matter from.	A. Paraf	Boston, Mass	Feb. 25, 1868	74, 935
Madder colors, Dyeing and printing	A. Paraf	New York, N. Y	Oct. 31, 1871	120, 393
Madder-extractor	J. Hunter	Philadelphia, Pa	Apr. 12, 1870	101, 735
Madder extracts, Obtaining	J. Hunter	Philadelphia, Pa	Feb. 15, 1870	99, 904
Madder for dyeing and printing, Extract of	A. Paraf	New York, N. Y	Sept. 21, 1869	95, 039
Madder for dyeing, Composition of	A. Paraf	New York, N. Y	Oct. 31, 1871	120, 392
Madder for dyeing, Preparation of coloring-matter from.	T. Briston	Cranston, R. I	June 14, 1870	104, 259
Madder for printing cloth, &c., Composition of	A. Paraf	New York, N. Y	Apr. 18, 1871	113, 919
Madder, Material for dyeing and printing obtained from.	A. Paraf	New York, N. Y	Aug. 17, 1869	93, 900
Madder, Obtaining extract of	A. Paraf	New York, N. Y	Jan. 24, 1871	111, 142
Madder, Process of extracting the coloring-matter of.	A. Paraf	New York, N. Y	Feb. 16, 1869	86, 939
Madder, Process of extracting the coloring-matter of.	A. Paraf	New York, N. Y	Jan. 17, 1871	110, 995
Madder, Products from	A. Paraf	New York, N. Y	Apr. 18, 1871	113, 918
Magazines, Flooding and entering powder	C. W. Copeland	Brooklyn, N. Y	Nov. 6, 1849	6, 842
Magic lantern: *See* Lantern.				
Magnesia compounds in chemical manufacture, Employment of.	C. G. Clemm	Dresden, Germany	Jan. 19, 1864	41, 349
Magnesia, Making hydrate or milk of	C. H. Phillips and L. Reid	New York, N. Y	July 22, 1873	141, 167
Magnesia, Manufacture of hydrate of	C. Wandel	Waldau, Germany	July 6, 1869	92, 405
Magnesia, Preparing	W. Dunn	Boston, Mass	Apr. 1, 1812	....
Magnesia, Preparing	W. Dunn	Boston, Mass	Apr. 8, 1813	....
Magnesium for burning, Preparing	C. H. Wing	Newton Corner, Mass	Apr. 24, 1866	54, 266
Magnesium, Manufacture and purification of	E. Sonstadt	Loughborough, Great Britain.	Dec. 27, 1864	45, 684
Magnesium, Manufacture of	J. O. Christian and J. and H. Charlton.	Manchester and Strangeways, England.	Sept. 17, 1867	68, 955
Magnesium Manufacture of	E. Sonstadt	Loughborough, Great Britain.	Dec. 6, 1864	45, 370
Magnet, Electro	T. A. Edison	Newark, N. J	Aug. 27, 1872	130, 795
Magnet, Receiving	N. Parks	Rome, N. Y	Feb. 16, 1858	19, 379
Magnet, Receiving	S. F. Van Choate	New York, N. Y	Sept. 29, 1863	40, 133
Magnet, Relay	W. G. Brownson	Wellsville, Ohio	July 30, 1867	67, 160
Magnet, Relay	L. G. D'Arlincourt	Paris, France	July 26, 1870	105, 653
Magnet, Relay	C. Durant	Jersey City, N. J	May 19, 1868	78, 076
Magnet, Relay	C. Durant	Jersey City, N. J	June 30, 1868	79, 330
Magnet, Relay	C. Durant	Jersey City, N. J	June 30, 1868	79, 331
Magnet, Relay	T. A. Edison	Newark, N. J	Aug. 12, 1873	141, 771
Magnet, Relay	S. Marcus	Vienna, Austria	Sept. 10, 1861	33, 254
Magnet, Reversible	J. L. Churchill	Nokomis, Ill	Feb. 11, 1873	135, 690
Magnet, Self-adjusting relay	J. M. Fairchild	New Haven, Conn	Dec. 10, 1867	71, 863
Magnet, Sounder	J. J. Clark and H. Splitdorf	New York, N. Y	Sept. 12, 1865	49, 857
Magnet, Sounder	S. F. Day	Ballston, N. Y	June 19, 1866	55, 627
Magnet spring, Relay	W. N. McInnis	Northumberland, Pa	Feb. 25, 1868	74, 925
Magnet, Telegraph	S. F. Van Choate	New York, N. Y	June 2, 1863	38, 774

Index of patents issued from the United States Patent Office from 1790 *to* 1873, *inclusive*—Continued.

Invention.	Inventor.	Residence.	Date.	No.
Magnets, Helix for	L. Bradley	New York, N. Y	Aug. 1, 1865	49, 074
Magnetic cylinder	L. Browning	Franconia, N. H	Oct. 13, 1812	....
Magnetic electric machine	J. Kidder	New York, N. Y	Mar. 15, 1864	41, 927
Magnetic engine	H. M. Paine	Newark, N. J	May 17, 1870	103, 229
Magnetic iron non-magnetic substances, Process and apparatus for separating.	F. A. H. La Rue and O. Audet	Quebec, Canada	Jan. 26, 1869	86, 167
Magnetic lock	L. C. Springer	Chicago, Ill	Jan. 1, 1867	60, 953
Magnetic needle on ships, Mode of compensating the local attraction of the.	C. Kline	New York, N. Y	Mar. 17, 1857	16, 845
Magnetic needles, Correcting	C. Guiteau	Syracuse, N. Y	Mar. 26, 1850	7, 216
Magnetic signals on railway, Transmitting	H. Maule	Philadelphia, Pa	Oct. 5, 1858	21, 688
Magnetism and electro-magnetism, Propelling machinery by.	T. Davenport	Brandon, Vt	Feb. 25, 1837	132
Magneto-electric apparatus	J. Kidder	New York, N. Y	Sept. 18, 1866	58, 105
Magneto-electric battery for firing fuses, &c	B. G. Noble	Brooklyn, N. Y	Mar. 5, 1872	124, 216
Magneto-electric machine	A. N. Allen	Pittsfield, Mass	Feb. 6, 1872	123, 438
Magneto-electric machine	G. W. Beardslee	Flushing, N. Y	Dec. 27, 1859	26, 557
Magneto-electric machine	G. W. Beardslee	Flushing, N. Y	Dec. 27, 1859	26, 558
Magneto-electric machine	C. Carpenter, jr	Pawtucket, Mass	Nov. 1, 1853	10, 175
Magneto-electric machine	C. Carpenter, jr	Providence, R. I	Apr. 8, 1856	14, 598
Magneto-electric machine	A. Davis	New York, N. Y	Aug. 1, 1854	11, 415
Magneto-electric machine	L. T. Gramme and E. L. C. D'Ivernois.	Paris, France	Oct. 17, 1871	120, 057
Magneto-electric machine	D. F. J. Lontin and E. L. C. de Ivernois.	Paris, France	Aug. 24, 1869	94, 014
Magneto-electric machine	E. C. Shepard	New York, N. Y	Aug. 19, 1856	15, 596
Magneto-electric machine	H. Wilde	Manchester, England	Nov. 13, 1866	59, 738
Magneto-electric machine for giving shocks	G. H. and B. H. Horn	Boston, Mass., and New York, N. Y.	Apr. 11, 1848	5, 507
Mail and packages from railway-cars while in motion, Method of delivering.	C. D. Everett	Cleveland, Ohio	Nov. 28, 1865	51, 162
Mail and packages on railway-cars, Mode of receiving and delivering.	W. J. Ketcham	Washington, D. C	July 25, 1865	48, 954
Mail and traveling bag, India-rubber	E. M. Chaffee	Boston, Mass	Dec. 31, 1833	....
Mail-axle	W. H. Saunders	Hastings-upon-Hudson, N. Y	Mar. 5, 1850	7, 150
Mail axle and hub	A. E. Smith	Bronxville, N. Y	Jan. 13, 1857	16, 404
Mail-bag	F. C. Brest and P. Wonsey	Spencerport, N. Y	June 8, 1869	91, 075
Mail-bag	D. F. Dodge	Lowville, N. Y	June 23, 1868	79, 214
Mail-bag	J. Fye	Hamilton, Ohio	June 5, 1866	55, 416
Mail-bag	R. Gornall	Baltimore, Md	May 10, 1859	23, 916
Mail-bag	T. J. Hardaway	Macon, Ga	May 20, 1873	139, 061
Mail-bag	T. J. Lamdin	Baltimore, Md	May 10, 1859	23, 924
Mail-bag	T. J. Lamdin	Baltimore, Md	Jan. 30, 1872	123, 113
Mail-bag	P. Laporte	Richmond, Va	May 5, 1824	....
Mail-bag	R. O. Lowrey	Salem, N. Y	Oct. 12, 1869	95, 817
Mail-bag	W. Ruddach	Baltimore, Md	May 3, 1859	23, 863
Mail-bag	L. M. Sims	Lincoln, Ill	Nov. 12, 1867	70, 907
Mail-bag	E. Stevens and J. A. Knight	Saint Louis, Mo	Nov. 5, 1867	70, 480
Mail-bag	Z. T. Sweet	Davisville, Cal	Sept. 7, 1869	94, 526
Mail-bag, Air-tight	C. A. Robbins and H. Allen	Iowa City, Iowa, and Allen Grove, Wis.	Sept. 7, 1852	9, 253
Mail-bag and carpet-sack lock	J. B. Logan	Thorntown, Ind	July 3, 1866	56, 069
Mail-bag bolt	J. Atkins	Hanover, N. H	Feb. 17, 1836	....
Mail-bag catcher for railway-cars	L. F. Ward	Elyria, Ohio	Jan. 29, 1867	61, 584
Mail-bag clasp	L. Fox	New York, N. Y	Sept. 18, 1866	58, 091
Mail-bag clasp and lock	H. C. Jones	Newark, N. Y	Oct. 23, 1837	440
Mail-bag delivering and receiving device	G. W. Hildreth	Lockport, N. Y	Sept. 12, 1871	118, 937
Mail-bag fastener	S. Denison	Portlandville, N. Y	Dec. 10, 1867	71, 997
Mail-bag fastener	D. G. Gay	Eugene City, Oreg	Oct. 11, 1870	108, 251
Mail-bag fastener	R. O. Lowrey	Salem, N. Y	Oct. 12, 1869	95, 818
Mail-bag fastener	H. M. Stephenson and J. B. Tyer.	Wabash, Ind	Apr. 4, 1871	113, 359
Mail-bag fastening	B. X. Blair	Huntingdon, Pa	June 3, 1873	139, 536
Mail-bag fastening	D. F. Dodge	Lowville, N. Y	Feb. 7, 1871	111, 521
Mail-bag fastening	J. C. Garland	Chicago, Ill	Feb. 8, 1859	22, 866
Mail-bag fastening	T. McGrane	New York, N. Y	Dec. 5, 1871	121, 646
Mail-bag fastening	J. A. Paul	Huntingdon, Pa	July 8, 1873	140, 639
Mail-bag fastening	A. D. Perry	New York	May 23, 1848	5, 596
Mail-bag fastening	W. J. Stowell	Baltimore, Md	Jan. 7, 1873	134, 571
Mail-bag fastening	J. A. Truitt	Oakland, Pa	Apr. 27, 1869	89, 447
Mail-bag fastening, &c	M. V. B. White	Fort Edwards, N. Y	Apr. 25, 1871	114, 239
Mail-bag lock	W. W. Gingrich and L. S. Coates.	Mexico, Pa	Dec. 16, 1862	37, 164
Mail-bag lock	R. O. Lowrey	Salem, N. Y	Oct. 12, 1869	95, 816
Mail-bag receiver, Railway	C. D. Everett	Cleveland, Ohio	Oct. 31, 1865	50, 698
Mail-bag tag	C. A. Snyder	Richmond, Va	Dec. 28, 1869	98, 441
Mail-bag, Water-proof	J. M. Jarrett	Brooklyn, N. Y	Oct. 30, 1866	59, 228
Mail bag, Way	H. Johnson	Washington, D. C	Apr. 25, 1846	4, 475
Mail-bags, Adjustable label-holder for	G. A. Lamb	Jeffersonville, N. Y	Oct. 29, 1867	70, 228
Mail-bags, Apparatus for receiving and delivering	J. J. Cooper	Olmsted, Ohio	Mar. 27, 1866	53, 417
Mail-bags, &c., by nailing, Making	A. L. Pennock and J. Seilers	Philadelphia, Pa	July 6, 1818	....
Mail-bags, Device for closing	G. M. Rhoades	East Hamilton, N. Y	Feb. 10, 1863	37, 641
Mail-bags, &c., Mouth-piece of	J. Sellers and A. L. Pennock	Philadelphia, Pa	June 12, 1840	1, 633
Mail-bags on railway-cars, Receiving and delivering	P. N. Maine	Olmsted Falls, Ohio	Sept. 11, 1866	57, 936
Mail-bags to and from railway trains and stations, Apparatus for receiving and delivering.	A. Chavanne	Paris, France	July 25, 1865	49, 056
Mail-bags to cars, Device for delivering	J. B. McLain	Newark, Ohio	Nov. 15, 1870	109, 233
Mail-bags upon railway-cars, Receiving and delivering.	J. Fredenburgh and G. A. Davidson.	Green, N. Y	May 15, 1866	54, 710
Mail-box	A. W. Allen and C. Reitz	Indianapolis, Ind	Sept. 20, 1870	107, 586
Mail-lock	T. Ascherfeld	Elkton, Md	Mar. 13, 1866	53, 096
Mail-pouch	M. Smith	Saint Louis, Mo	Apr. 28, 1863	38, 362
Mail-pouch	M. Smith	Saint Louis, Mo	May 17, 1864	42, 802
Mail-pouch holder and catcher, Railway	B. F. Bean	Pawling, Pa	Oct. 14, 1873	143, 657
Mail pouch or box	M. Smith	Saint Louis, Mo	May 15, 1866	54, 787
Mails, Device for receiving and delivering	F. K. Sibley and L. C. Wade	Auburndale and Newton Upper Falls, Mass.	Dec. 22, 1868	85, 183
Main: *See* Gas-main. Hydraulic main. Water-main.				

Index of patents issued from the United States Patent Office from 1790 *to* 1873, *inclusive*—Continued.

Invention.	Inventor.	Residence.	Date.	No.
Maize-husking machine	T. C. Hargreaves	Schenectady, N. Y	Oct. 4, 1853	10, 070
Maize, Process of mashing	F. Seitz	Easton, Pa	Jan. 20, 1852	8, 665
Maize suitable for grinding, Preparation to render	S. P. McCroskey	Monroe, Iowa	May 28, 1861	32, 448
Malleable-iron and steel furnace	H. Bessemer	London, England	July 25, 1871	117, 247
Malleable metal pipes, Connection for	W. H. Harrison	Philadelphia, Pa	Jan. 18, 1870	98, 961
Mallet	L. W. Blanchard	Whitingham, Vt	Mar. 29, 1859	23, 346
Mallet	J. B. Davids	New York, N Y	Aug. 1, 1871	117, 609
Mallet	A. Holbrook	Providence, R. I	Aug. 19, 1873	141, 930
Mallet	W. Lance	Olney, Ill	July 12, 1859	24, 744
Mallet	A. Partridge	Medway, Mass	Mar. 21, 1865	46, 972
Mallet	A. Partridge	Medway, Mass	May 28, 1872	127, 363
Mallet	S. M. Willis	Athol, Mass	July 1, 1873	140, 395
Mallet, Calker's	S. C. Searles	Wilmington, Del	Mar. 8, 1870	100, 562
Mallet, Elastic	A. C. Eddy	Providence, R. I	Feb. 20, 1866	52, 696
Mallet for driving tubes, &c	C. Stewart	Worcester, Mass	Dec. 16, 1873	145, 537
Mallet, Serving	T. Batty	New York, N. Y	Aug. 20, 1850	7, 571
Mallet, Serving	D. H. Southworth	New York, N. Y	Nov. 16, 1852	9, 412
Mallet, Serving and worming	J. B. Petitval	Charleston, S. C	July 11, 1837	266
Malt and other liquors, Apparatus for fermenting	G. Wallace	Cincinnati, Ohio	Mar. 5, 1867	62, 581
Malt and other liquors, Treating	G. Storey	Wheeling, W. Va	Mar. 31, 1868	76, 266
Malt, Apparatus for sprouting	J. Gecmen	Chicago, Ill	Mar. 13, 1866	53, 135
Malt, Apparatus for sprouting	J. Gecmen	Chicago, Ill	Apr. 14, 1868	76, 624
Malt, Condensed extract of	T. Hawks	Rochester, N. Y	Dec. 11, 1866	60, 370
Malt-degerminating machine	C. Sauter	New York, N. Y	Feb. 7, 1871	111, 686
Malt-dryer	H. A. and C. H. Engels and J. Wieland.	San Francisco, Cal	Apr. 2, 1867	63, 374
Malt-dryer	W. L. Phillips	Normal, Ill	June 14, 1870	104, 347
Malt-dryer	J. G. Schiffer	San Francisco, Cal	Dec. 30, 1873	146, 025
Malt-dryer	P. Schuff	New York, N. Y	Sept. 16, 1873	142, 946
Malt, Drying	W. W. Hughes	Philadelphia, Pa	May 25, 1869	90, 545
Malt-drying kiln	J. Gecmen	New York, N. Y	Mar. 25, 1873	137, 194
Malt-drying kiln	C. Oefinger and S. Grupp	San Francisco, Cal	Apr. 7, 1868	76, 508
Malt-drying kiln	J. A. Remer	New York, N. Y	Feb. 7, 1871	111, 569
Malt-extract	L. Hoff	New York, N. Y	Dec. 18, 1866	60, 631
Malt-extract, Concentrated	T. Hawks	Rochester, N. Y	June 16, 1868	78, 875
Malt, &c., for making beer, ale, and porter, Extract of	T. Hawks	Rochester, N. Y	Feb. 3, 1863	37, 578
Malt-house	T. Krausch	New York, N. Y	Nov. 28, 1871	121, 387
Malt house or kiln	W. Appleton	Albany, N. Y	Nov. 19, 1867	71, 117
Malt in mash-tubs, Self-acting sparger for distributing water upon.	H. J. Brooks and F. B. Longmire.	Philadelphia, Pa	May 8, 1840	1, 597
Malt in the kiln, Apparatus for stirring	W. Toepfer	Milwaukee, Wis	May 30, 1871	115, 390
Malt-kiln	C. P. Berckhemer	Cincinnati, Ohio	Nov. 8, 1870	109, 105
Malt-kiln	W. Blakey	Baltimore, Md	Oct. 8, 1867	69, 534
Malt-kiln	J. Gecmen	Chicago, Ill	Nov. 28, 1865	51, 169
Malt-kiln	J. Gecmen	Chicago, Ill	Nov. 13, 1866	59, 583
Malt-kiln	J. Gecmen	Chicago, Ill	Sept. 17, 1867	68, 869
Malt-kiln	J. G. White	Albany, N. Y	Feb. 22, 1870	100, 091
Malt-kiln and malt-house	W. W. Hughes	Philadelphia, Pa	Aug. 24, 1869	94, 114
Malt-kiln floor	H. Franz	Philadelphia, Pa	Sept. 25, 1860	30, 137
Malt-kiln floor	W. Gerhard, jr	Florence, Mass	Aug. 17, 1869	93, 701
Malt-kiln floor	T. S. Smith	Cincinnati, Ohio	Feb. 21, 1860	27, 237
Malt-kiln floor	G. Taylor	Camden, N. J	Sept. 11, 1866	57, 997
Malt-kiln-floor plate	M. Stewart	Philadelphia, Pa	Jan. 3, 1854	10, 370
Malt-kiln flooring	W. W. Hughes and J. C. Adams	Philadelphia, Pa	Jan. 15, 1867	61, 202
Malt-kiln, Steam-drying	D. S. Davis	Hibernia, Tenn	Aug. 30, 1822	....
Malt-kilns, Device to operate the tilting doors of	J. C. Rhodes	Chicago, Ill	June 29, 1869	92, 096
Malt-kilns, Material for floor of	R. Speidel	New York, N. Y	Sept. 10, 1867	68, 664
Malt-kilns, Stirrer for	E. Schmidt	Pittsburgh, Pa	Apr. 29, 1873	138, 288
Malt liquor cooler	O. Hœpfner and C. Schnepf	Philadelphia, Pa	Oct. 13, 1863	40, 262
Malt-liquor cooler	W. Partington	Philadelphia, Pa	Oct. 15, 1867	69, 930
Malt-liquor preserving and discharging apparatus	M. Reeder	Philadelphia, Pa	Jan. 24, 1860	26, 930
Malt-liquors and cider, Ripening and keeping	R. Hare	Philadelphia, Pa	Aug. 22, 1811	....
Malt-liquors, Apparatus for boiling, cooling, and fermenting.	T. Haigh and R. A. Robertson	Liverpool, England, and Philadelphia, Pa.	Apr. 23, 1867	64, 011
Malt-liquors, Apparatus for cooling	W. Allenderff	Philadelphia, Pa	Nov. 6, 1866	59, 336
Malt-liquors, Apparatus for cooling	F. L. Wissmann	Philadelphia, Pa	Oct. 3, 1865	50, 293
Malt-liquors, Apparatus for drawing and preserving	H. Mittendorf	York, Pa	Sept. 7, 1867	69, 013
Malt-liquors, &c., Apparatus for fermenting	A. Hammer	New York, N. Y	Dec. 13, 1864	45, 406
Malt-liquors, Apparatus for preserving and discharging.	L. Dungan	Philadelphia, Pa	Nov. 5, 1861	33, 637
Malt-liquors, Compound for treating	W. Zinsser	New York, N. Y	Dec. 6, 1870	109, 990
Malt-liquors, Decolorizing	C. R. M Wall	Brooklyn, N. Y	Oct. 17, 1865	50, 523
Malt-liquors from becoming flat, Apparatus for preventing.	F. M. Ruschhaupt	New York, N. Y	Mar. 18, 1862	34, 695
Malt, Machine for bruising	J. F. Tibble	Detroit, Mich	May 16, 1871	114, 882
Malt-polishing apparatus	W. W. Stoll	Brooklyn, N. Y	Nov. 25, 1873	145, 027
Malt-polishing machine	C. Stoll	Brooklyn, N. Y	June 24, 1873	140, 316
Malt-reservoir	F. C. Spiess and A. Dobler	New York, and Brooklyn, N. Y.	June 6, 1871	115, 652
Malt-reservoir	C. Stoll	Brooklyn, N. Y	Dec. 27, 1870	110, 601
Malt-shovel	W. Beach	Albany, N. Y	Aug. 14, 1866	57, 069
Malt-steep	W. W. Stoll	Brooklyn, N. Y	Oct. 21, 1873	143, 938
Malt, Steeping, growing, and drying	A. Kreusler	New Lebanon, N. Y	Sept. 5, 1865	49, 768
Malt-sirup, Manufacture of	T. Hawks	Rochester, N. Y	June 27, 1865	48, 396
Malting and brewing vessel	C. Berckhemer	Cincinnati, Ohio	May 14, 1867	64, 622
Malting and drying kiln	H. Parsons	Colebrook, N. H	July 9, 1812	....
Malting-apparatus	M. Riely	Morrow, Ohio	Nov. 14, 1865	50, 958
Man-hole covers, Device for securing	J. J. Craven	Jersey City, N. J	July 15, 1873	140, 893
Man-power, Machine for application of	T. H. Hoskings	Detroit, Mich	May 21, 1861	32, 369
Man's assistant, One-armed	G. W. Dalbey	Carrollton, Miss	May 17, 1870	103, 023
Manacle	A. Delestatins	Philadelphia, Pa	Dec. 10, 1861	33, 917
Manacle	A. Rankin	Philadelphia, Pa	Feb. 20, 1866	52, 745
Mandrel, Adjustable	C. W. Le Count	Norwalk, Conn	June 19, 1866	55, 681
Mandrel, Buffing	G. B. Dunham	Lynn, Mass	Mar. 25, 1873	137, 187
Mandrel, Buffing	N. S Thompson	Augusta, Me	Aug. 5, 1873	141, 523
Mandrel, Circular-saw	T. H. Keeney	Newport, Ky	July 31, 1855	13, 357
Mandrel-cutters in turning tapering sticks, Device to operate.	P. H. Niles	Boston, Mass	Feb. 24, 1857	16, 705

Index of patents issued from the United States Patent Office from 1790 *to* 1873, *inclusive*—Continued.

Invention.	Inventor.	Residence.	Date.	No.
Mandrel, Expanding	D. L. Allen	Williamsport, Pa	Apr. 7, 1868	76, 372
Mandrel, Expanding	D. L. Allen	Williamsport, Pa	Sept. 21, 1869	95, 065
Mandrel, Expanding	I. Beetison	New Britain, Conn	Nov. 27, 1866	59, 948
Mandrel, Expanding	J. Brewer	Philadelphia, Pa	June 23, 1868	79, 198
Mandrel, Expanding	M. Gardner, sr	Carlisle, Pa	Apr. 28, 1868	77, 369
Mandrel, Expanding	J. B. Mallalieu	Chicago, Ill	Apr. 30, 1867	64, 234
Mandrel, Expanding	A. F. Nagle	Providence, R. I	Aug. 4, 1868	80, 560
Mandrel, Expanding	W. Sherrod	Providence, R. I	Dec. 2, 1851	8, 562
Mandrel, Expanding	J. P. Simons	Houston, Tex	Jan. 23, 1872	122, 920
Mandrel, Expanding	J. D. Smith	New York, N. Y	May 13, 1873	138, 947
Mandrel for cutting tapering sticks	G. Turner	Edinborough, Pa	May 22, 1855	12, 926
Mandrel for holding carriage-hubs, &c	N. Hawley	Rome, N. Y	Feb. 27, 1855	12, 455
Mandrel for holding tapered cast rings	G. E. Brettell	Rochester, N. Y	June 24, 1873	140, 113
Mandrel for lining cylinders with metal, Revolving	G. Potts	Cincinnati, Ohio	Sept. 13, 1853	10, 013
Mandrel for loading case-shot, &c	S. Sawyer	Fitchburgh, Mass	Dec. 24, 1861	34, 041
Mandrel for making gutta-percha tubing	J. Reynolds	New York, N. Y	June 10, 1856	15, 086
Mandrel for turning-machinery, Expanding	W. Sherrod	Providence, R. I	June 21, 1853	9, 805
Mandrel or boring-tool	J. C. Millerd	River Point, R. I	Sept. 1, 1868	81, 664
Mandrel, Riveting	J. Berry	Buffalo, N. Y	Nov. 29, 1870	109, 575
Mandrel, Universal	J. Gallatin, jr	New York, N. Y	Mar. 14, 1871	112, 580
Mandrel, Wheelwright's guide	J. Sykes	Mercer, Pa	June 26, 1855	13, 142
Mane-turner, Elastic	C. Phelps	Clayton, N. Y	Dec. 13, 1870	110, 068
Magnates and permagnates, Preparations of	B. Schmidt	Hoboken, N. J	Apr. 14, 1868	76, 825
Manganese, ferro-manganese, and spie geleisen, Manufacfacture of.	O. Bolton, jr	Pittsburgh, Pa	Aug. 5, 1873	141, 419
Manganese, Manufacture of oxide of	C. T. Dunlop	Glasgow, North Britain	Aug. 7, 1860	29, 474
Manganese ores, Reducing	C. Adams	Philadelphia, Pa	Nov. 26, 1867	71, 355
Manganese, Process for preparing blue peroxide of	J. W. Hobbs	Boston, Mass	Apr. 6, 1869	88, 570
Manger	C. E. Albright	Muncy, Pa	June 7, 1870	103, 956
Manger	E. Carlin	United States Army	June 4, 1872	127, 563
Manger	A. Chambers	Unionville, N. Y	Aug. 17, 1869	93, 673
Manger	F. Denzler and J. Miller	Brooklyn, N. Y	Sept. 17, 1867	68, 965
Manger	J. Johnson	Kent, Ind	Sept. 25, 1866	58, 263
Manger	J. E. Kelly	New York, N. Y	Apr. 10, 1860	27, 812
Manger	D. Sager	Albany, N. Y	Nov. 15, 1864	45, 081
Manger	W. F. Stanley	Cazenovia, N. Y	Nov. 3, 1868	83, 738
Manger, Automatic	D. A. Townsend	Unionville, S. C	Feb. 25, 1873	136, 192
Manger for horses	J. C. Higgins	Millstone, N. J	Jan. 7, 1873	134, 601
Manger, Hay	J. Packer	Philadelphia, Pa	Mar. 29, 1859	23, 386
Manger, Portable	T. Fisler	Camden, N. J	Sept. 10, 1861	33, 242
Mangle	J. Beaumont	Chambersburgh, Pa	June 23, 1868	79, 188
Mangle	C. C. Converse	Elmira, N. Y	May 20, 1862	35, 301
Mangle	C. C. Converse	Dubuque, Iowa	Sept. 22, 1863	40, 022
Mangle	G. Coombs	West Falls, N. Y	Apr. 17, 1860	27, 890
Mangle	J. T. Coxell	Brooklyn, N. Y	Nov. 22, 1859	26, 227
Mangle	D. Cumming, jr	Mobile, Ala	July 27, 1858	21, 044
Mangle	R. A. Duncan	Brooklyn, N. Y	Sept. 23, 1873	142, 997
Mangle	T. Farnsworth	Cleveland, Ohio	May 16, 1865	47, 708
Mangle	R. Gage	Kingston, Canada	June 18, 1872	127, 971
Mangle	H. Gransden	Dubuque, Iowa	July 30, 1867	67, 293
Mangle	W. D. Grimshaw	New York, N. Y	Dec. 4, 1860	30, 812
Mangle	E. Gundlach	Hackensack, N. J	Nov. 18, 1873	144, 671
Mangle	P. H. Hink and H. Kaack	Moline, Ill	Aug. 31, 1869	94, 206
Mangle	W. W. Hollman	Eddyville, Ky	Sept. 13, 1859	25, 413
Mangle	J. Johansen	Springfield, Ill	Oct. 15, 1867	69, 813
Mangle	J. Johnson	New York, N. Y	Apr. 4, 1865	47, 152
Mangle	W. T. Littlejohn	Kalamazoo, Mich	Oct. 4, 1859	25, 656
Mangle	S. W., J. F., and N. Palmer	Auburn, N. Y	Jan. 8, 1867	61, 018
Mangle	W. Price	Cincinnati, Ohio	Jan. 17, 1865	45, 938
Mangle	H. W. Putnam	Cleveland, Ohio	Aug. 26, 1862	36, 303
Mangle	W. Radbourne	Rahway, N. J	Feb. 7, 1865	46, 263
Mangle	N. Soderstrom	Chicago, Ill	Aug. 10, 1869	93, 491
Mangle	R. A. Stratton	Philadelphia, Pa	Mar. 24, 1857	16, 887
Mangle	R. A. Stratton	Philadelphia, Pa	May 28, 1867	65, 134
Mangle	H. Tregellas	Calumet, Mich	Aug. 29, 1871	118, 654
Mangle	H. Wehdeking	Edgerton, Ohio	Nov. 12, 1867	70, 766
Mangle	J. B. Westwick	Galena, Ill	Nov. 9, 1869	96, 646
Mangle and ironing-machine	J. Seaman	Chicago, Ill	July 14, 1868	80, 019
Mangle and wringer	D. Cumming, jr	New York, N. Y	Apr. 3, 1866	53, 580
Mangle, Clothes	J. B. Greenhut	Chicago, Ill	Dec. 12, 1865	51, 452
Mangle, Clothes	H. E. Smith	New York, N. Y	Nov. 2, 1869	96, 361
Mangle, Domestic	S. Nowlan	New York, N. Y	Apr. 20, 1858	20, 002
Mangle, Domestic-calender	I. Doolittle	Bennington, Vt	Feb. 1, 1833	....
Mangle or rolling-press	A. Hermann	New Haven, Conn	May 8, 1866	54, 543
Mangle, Steam	P. Rundquist	New York, N. Y	May 16, 1871	114, 863
Mangles, washing-machines, &c., Composition bowl for.	T. Hardcastle	Bolton, England	May 25, 1869	90, 447
Mangling and ironing machine	S. Williams	Philadelphia, Pa	Nov. 29, 1870	109, 699
Mangling and wringing machine	A. Hollings	Providence, R. I	Sept. 3, 1872	131, 001
Mangling or wringing machine	T. Hall and J. Newton	Lawrence, Mass	June 3, 1873	139, 574
Manila and other fibrous substances, Machine for braiding.	D., J., and E. Fitzgerald	New York, N. Y	Apr. 24, 1840	1, 566
Mantel-bar	W. P. Chadwick	Edgartown, Mass	June 1, 1858	20, 404
Mantel, False	J. T. Fleehearty	Baltimore, Md	May 20, 1873	139, 137
Mantel for fire-place	E. Skinner	Sandwich, N. Y	July 20, 1831	....
Mantel-piece	H. Tucker	Cambridge, Mass	Apr. 2, 1850	7, 253
Mantel-pieces of cast iron or other soft metals, Making.	I. Deaves	Philadelphia, Pa	Feb. 21, 1821	....
Manual-power	I. E. Overpeck	Overpeck's Station, Ohio	Apr. 25, 1865	47, 447
Manual-power	J. H. Yager	Trenton, Ohio	Oct. 23, 1866	59, 117
Manual-power, Driving machinery by	A. West	Green, Me	Aug. 16, 1839	1, 289
Manual-power machine	T. L. Kenworthy and A. Silvers	Collinsville, Ohio	July 30, 1867	67, 315
Manure	W. D. Hall	Hamden, Conn	Mar. 14, 1865	46, 847
Manure	P. G. Renny	Rahway, N. J	July 2, 1867	66, 357
Manure and fertilizer, Composting	D. Ruggles	Fredericksburgh, Va	May 17, 1870	103, 085
Manure and hay fork	G. B. Flint	Sing Sing, N. Y	May 18, 1869	90, 087
Manure and lime spreader	J. W. Fawkes	Christiana, Pa	Aug. 29, 1854	11, 602
Manure and oil, Treating fish for	R. C. Demolon and G. A. C. Thurneyssen.	Paris, France	Mar. 6, 1855	12, 480

Index of patents issued from the United States Patent Office from 1790 *to* 1873, *inclusive*—Continued.

Invention.	Inventor.	Residence.	Date.	No.
Manure and sand loader	H. G. Marchant	Annisquam, Mass	May 9, 1854	10, 892
Manure, Apparatus for distributing liquid	J. W. Clark	Iowa City, Iowa	Feb. 12, 1867	62, 006
Manure, Artificial	W. S. Amies	Guernsey, Great Britain	July 2, 1872	128, 578
Manure, Artificial	D. Bruce	Paspebiac, Canada	Jan. 11, 1859	22, 544
Manure, Artificial	P. S. and W. H. Chappell	Baltimore, Md	Mar. 27, 1849	6, 234
Manure, Artificial	E. Von Nordhausen	New York, N. Y	Jan. 19, 1864	41, 331
Manure-beds, Preparing	C. F. Spieker	New York, N. Y	Apr. 13, 1858	19, 974
Manure crusher and sower	T. T. Nelson	Clarke County, Va	Dec. 20, 1853	10, 325
Manure, Device for spreading	J. H. Stevens	East Durham, N. Y	Jan. 3, 1865	45, 767
Manure-distributer	J. W. Barnes	Murfreesborough, N. C	Oct. 14, 1856	15, 876
Manure-distributer	J. Cadwell	Dexter, Mich	Nov. 17, 1863	40, 605
Manure-distributer	D. E. Cripe	Pyrmont, Ind	Apr. 30, 1872	126, 187
Manure-distributer	J. B. Crowell	Greencastle, Pa	June 23, 1863	38, 949
Manure-distributer	H. S. Palmer	Norvell, Mich	May 14, 1867	64, 697
Manure-distributer and seed-sower	S. L. Fraser	West Town, N. Y	Oct. 18, 1870	108, 469
Manure-drag	W. Geahr	East Earl Township, Pa	Jan. 4, 1870	98, 487
Manure-drag	C. H. and J. H. Harnley	Penn Township, Pa	May 21, 1867	64, 861
Manure-drag	J. D. Heebner	Norrittonville, Pa	Aug. 9, 1870	106, 158
Manure-drag	U. Hummer	White Oak, Pa	Jan. 28, 1868	73, 894
Manure-drag	A. H. Shock	Piqua, Pa	July 9, 1867	66, 525
Manure-drag	A. H. Shock and H. R. Shirk	Lancaster, Pa	Oct. 5, 1869	95, 523
Manure-drag	J. S. and H. D. Spangler and D. Madlem.	Ephratah, Pa	Feb. 16, 1869	86, 951
Manure-drag	D. Wingenroth	Ephratah, Pa	Mar. 16, 1869	87, 812
Manure-drag, Stop-jointed	S. B. Minnich	Landisville, Pa	Apr. 9, 1867	63, 650
Manure-drill	G. B. Singeltary	Greenville, N. C	Dec. 13, 1859	26, 448
Manure-excavator	A. R. Hurst	Harrisburgh, Pa	Aug. 29, 1854	11, 610
Manure for transportation, Preparing	S. W. Boynton	Hartford, Conn	Nov. 26, 1872	133, 404
Manure for transportation, storage, or market, Preparing.	H. C. Babcock	Hartford, Conn	Aug. 20, 1872	130, 616
Manure-fork	A. S. Brinser and H. Bricker	Falmouth, Pa	Jan. 3, 1871	110, 734
Manure-fork	B. H. Franklin	Worcester, Mass	Dec. 20, 1853	10, 322
Manure-fork	J. W. Horst	Annville, Pa	July 14, 1868	79, 831
Manure-fork	W. Truby	Brush Valley, Pa	Oct. 22, 1867	70, 135
Manure-fork	P. Yengst	Union Deposit, Pa	Nov. 26, 1867	71, 353
Manure fork and hook, Combined	C. C. Cole	Phelps, N. Y	Feb. 25, 1868	74, 895
Manure fork and hook, Combined	R. Dean	Coboctah, Mich	Feb. 23, 1869	87, 249
Manure fork and scraper, Combined	J. W. Clark	Kingston, Wis	Apr. 19, 1870	102, 092
Manure from fish, Manufacture of	J. B. Hyde	Newark, N. J	Nov. 12, 1861	33, 706
Manure-hook	S. B. Minnich	Landisville, Pa	Mar. 1, 1870	100, 311
Manure-hook	M. Stoll and H. Gross	Middletown, Pa	Dec. 8, 1868	84, 691
Manure hook or drag	H. Gross	Middletown, Pa	May 25, 1869	90, 526
Manure hook or drag	H. Gross	Middletown, Pa	May 25, 1869	90, 527
Manure hook or drag	H. Gross	Middletown, Pa	Nov. 9, 1869	96, 583
Manure, Machine for sowing pulverulent	J. G. Stegall	Thomasville, Ga	Jan. 5, 1869	85, 619
Manure, Machine for sowing pulverulent	N. G. Swift	Hart's Village, N. Y	July 9, 1861	32, 800
Manure, Making	C. Baer and J. Gouliart	Baltimore, Md	June 24, 1843	3, 139
Manure, Manufacture of	P. Eley	Philadelphia, Pa	Aug. 11, 1863	39, 525
Manure, Manufacture of	A. F. Mosselman	Paris, France	Sept. 6, 1864	44, 147
Manure, Preparation of animal and other	R. Hare	Philadelphia, Pa	Jan. 29, 1850	7, 053
Manure, Preparing night-soil for	R. B. Fitts	Philadelphia, Pa	Feb. 17, 1863	37, 685
Manure, Process of treating feldspar for	C. Bickell	Baltimore, Md	Nov. 25, 1856	16, 111
Manure-spreader	P. Eley	New York, N. Y	May 17, 1864	42, 758
Manure-spreader	S. A. Hedges	Lancaster, Ohio	Jan. 11, 1853	9, 535
Manure-spreader	D. Hill	New Vienna, Ohio	Mar. 15, 1870	100, 762
Manure-spreader	E. Marshall	Clinton, N. J	Aug. 22, 1854	11, 557
Manure-spreading machine	J. H. Stevens	East Durham, N. Y	Mar. 25, 1862	34, 790
Manure, Treating	E. P. Baugh	Philadelphia, Pa	May 9, 1865	47, 611
Map and chart holder	E. A. and A. C. Apgar	Trenton, N. J	Mar. 10, 1868	75, 343
Map and chart rack	F. G. Johnson	Brooklyn, N. Y	Apr. 23, 1872	125, 969
Map, Dissected	C. J. Higgins	Indianapolis, Ind	Sept. 2, 1873	142, 338
Map-drawing apparatus	G. S. Ormsby	Xenia, Ohio	Nov. 30, 1869	97, 432
Map exhibiter and cabinet	W. A. and G. Rice	Framingham, Mass	Oct. 14, 1873	143, 717
Map, Geographical	E. A. and A. C. Apgar	Philadelphia, Pa	Oct. 16, 1866	58, 748
Map-holder	J. Highfield	Centreville, Ind	July 30, 1872	130, 050
Map-holder, Portable	G. Rice	Framingham, Mass	Mar. 9, 1869	87, 591
Map, Ribbon	M. Coloney and S. B. Fairchild	Saint Louis, Mo	July 9, 1867	66, 463
Map-stand	W. F. Phelps	Winona, Minn	Oct. 25, 1870	108, 625
Map-suspending frame	W. F. Phelps	Winona, Minn	Nov. 5, 1867	70, 467
Map to teach geography, &c., Outline	E. F. Anderson	Mansfield, Conn	Oct. 20, 1868	83, 236
Maps and charts, Roller-case for	J. S. Ostrander	Albany, N. Y	June 22, 1869	91, 766
Maps, charts, &c., Apparatus for exhibiting	N. K. Lombard	Boston, Mass	Oct. 27, 1836	....
Maps, charts, &c., Apparatus for preserving and exhibiting.	N. K. Lombard, jr	Boston, Mass	Oct. 27, 1836	70
Maps, charts, &c., Instrument for delineating	A. Bailey	Newberry, S. C	Sept. 4, 1840	1, 770
Maps, Coloring	L. Stebbins	Hartford, Conn	Mar. 12, 1840	1, 510
Maps, Constructing block or plate for printing	G. W. Bacon	London, Great Britain	Aug. 7, 1866	57, 056
Maps, Making dissected	S. McCleary and J. Pierce	Hoosick, N. Y	Sept. 25, 1849	6, 747
Maps, Process and apparatus for coloring	L. Stebbins	Hartford, Conn	July 11, 1837	263
Maps, songs, curtains, &c., Apparatus for winding	J. S. Ostrander	Albany, N. Y	June 22, 1869	91, 660
Mapping lands, &c., Instrument for	J. Deneale	Dumfries, Va	Aug. 3, 1820	....
Marble and freestone scouring machine	R. P. Henry	Akron, Ohio	Oct. 14, 1862	36, 650
Marble and other stone or wood, Sawing	C. B. Austin	Kensington, Pa	Dec. 24, 1833	....
Marble and other stone, Sawing, molding, and polishing.	I. D. Kirk	Philadelphia, Pa	Feb. 26, 1833	....
Marble and other substances, Composition for grinding and polishing.	J. C. McAfee	West Alexander, Pa	Aug. 25, 1868	81, 520
Marble and other substances, Process for imitating	W. Bonney	New York, N. Y	Aug. 8, 1854	11, 468
Marble and stone sawing machine	G. J. Wardwell	Hatley, Canada	Nov. 4, 1856	16, 033
Marble and veining the same, Process of manufacturing artificial.	F. H. Hall	Hackensack, N. J	Dec. 9, 1873	145, 345
Marble and wood polishing machine	J. C. Mateer	Kankakee, Ill	Oct. 22, 1872	132, 404
Marble, Artificial	B. Auguste	Chicago, Ill	July 8, 1873	140, 668
Marble, Artificial	G. H. Mellen	New York, N. Y	Dec. 24, 1872	134, 300
Marble, Artificial	C. Vlockmann	Columbus, Ohio	Nov. 16, 1869	97, 004
Marble, Artificial variegated	A. and I. Straub	Milton, Pa., and Cincinnati, Ohio.	Nov. 17, 1863	40, 650
Marble, &c., Carving machine	W. A. Pease	Goshen, Ind	Apr. 10, 1860	27, 827
Marble, Coloring	S. Gardner	New York, N. Y	Feb. 1, 1870	99, 306

Index of patents issued from the United States Patent Office from 1790 *to* 1873, *inclusive*—Continued.

Invention.	Inventor.	Residence.	Date.	No.
Marble, Coloring and staining	J. Zengeler	Chicago, Ill	Mar. 28, 1871	113, 127
Marble, &c., Composition for cleaning	W. Ginnaugh	Niles, Mich	Apr. 18, 1865	47, 295
Marble, Compound for polishing and cleaning	P. Doyle and F. P. Colton	Hartford, Conn	June 14, 1870	104, 285
Marble, Compound for the manufacture of artificial	W. Meyer	Canton, Ohio	Feb. 16, 1869	86, 962
Marble, &c., Compounding China clay to imitate	D. Kaempfe and C. List	Nevohaus, Schwarzburg-Rudolstadt.	July 5, 1870	105, 089
Marble cutting and finishing machine	J. W. Maloy	Boston, Mass	Apr. 10, 1866	53, 845
Marble, Cutting and polishing	J. F. Mangin		Feb. 16, 1797	
Marble cutting and polishing machine	S. Mulliken		Mar. 11, 1791	
Marble cutting, grooving, and beading machine	J. D. Buzzell	Cape Elizabeth, Me	Aug. 13, 1838	880
Marble-cutting machine	G. W. Wheeler and H. I. Stevens	New Fairfield and Bethel, Conn.	Dec. 17, 1867	72, 250
Marble-cutting machine	W. Wickersham	Boston, Mass	Mar. 3, 1868	75, 230
Marble dressing and carving machine	G. V. Black	Jacksonville, Ill	Jan. 16, 1872	122, 754
Marble-dressing machine	R. P. Bailey	Niagara Falls, N. Y	June 25, 1867	65, 990
Marble-finishing machine	H. W. Kent	Battle Creek, Mich	Nov. 7, 1865	50, 829
Marble for painting on, Preparing	A. Robertson		Mar. 5, 1804	
Marble-graining tool	S. Wiggins	Bridgeport, Conn	June 28, 1864	43, 359
Marble, granite, &c., Composition for making	L. Matthey	Brooklyn, N. Y	Mar. 7, 1827	
Marble, granite, &c., Process for cleaning	J. Sawyer	Northfield, Vt	Dec. 30, 1873	145, 971
Marble grinding and polishing machine	J. H. Volk	Chicago, Ill	Apr. 19, 1870	102, 184
Marble, Grinding, smoothing, and polishing	S. Risley	Philadelphia, Pa	June 14, 1834	
Marble, Grooving and beading	A. J. Torrey	New York	Nov. 1, 1825	
Marble, Imitating	B. Trembley	New York	Nov. 13, 1827	
Marble, Imitating	H. Tucker	Cambridge, Mass	Oct. 14, 1851	8, 420
Marble, Imitation	S. W. Davis	Cincinnati, Ohio	May 22, 1849	6, 471
Marble, Imitation	A. and W. Fisher	New York, N. Y	Mar. 25, 1862	34, 746
Marble in kerfs of varying angles, Machine for sawing.	S. Nickelson	Pulaski, Tenn	May 6, 1856	14, 823
Marble in obelisk forms, Machine for sawing	J. A. Bailey	Detroit, Mich	Mar. 25, 1856	14, 532
Marble in obelisk forms, Machine for sawing	J. E. Haviland	Galveston, Tex	Apr. 15, 1856	14, 658
Marble in obelisk forms, Machine for sawing	I. A. Heald	Springfield, Mass	Mar. 25, 1856	14, 536
Marble in obelisk forms, Machine for sawing	J. Miller	Buffalo, N. Y	Apr. 22, 1856	14, 729
Marble in obelisk forms, Machine for sawing	P. Schrag and W. J. Von Kammerhueber.	Washington, D. C	Feb. 19, 1856	14, 296
Marble in taper forms, Machine for sawing	C. Amazeen	New Castle, N. H	Mar. 4, 1856	14, 342
Marble in taper forms, Machine for sawing	M. M. Manly	South Dorset, Vt	Sept. 30, 1856	15, 814
Marble in taper forms, Machine for sawing	G. T. Pearsall	Apalachin, N. Y	Dec. 11, 1855	13, 916
Marble in taper forms, Machine for sawing	C. A. Schultz	Chicago, Ill	Mar. 18, 1856	14, 471
Marble, Lettering	J. Tusel	Kendallville, Ind	May 12, 1868	77, 781
Marble, Machine for cutting, chiseling and polishing.	A. Clark	West Stockbridge, Mass	Apr. 10, 1810	
Marble, Machine for sawing out tapering blocks of	J. A. Cole	Washington, D. C	Dec. 4, 1855	13, 866
Marble, Machine for scouring	H. C. Lull	Montpelier, Vt	Feb. 19, 1867	62, 142
Marble, Making artificial	C. F. Barker	Boston, Mass	Aug. 12, 1873	141, 689
Marble, Manufacture of artificial	R. Lamb	New York, N. Y	Jan. 31, 1860	27, 022
Marble, Manufacture of artificial	W. Myer	Canton, Ohio	Feb. 16, 1869	86, 963
Marble, Manufacture of artificial	G. A. Trear	Chicago, Ill	Aug. 9, 1870	106, 263
Marble, Manufacture of imitation	G. Davey	London, England	Apr. 2, 1872	125, 122
Marble, Manufacture of imitation	J. S. Elliott and J. F. Wood	Chelsea and Everett, Mass	Mar. 12, 1872	124, 557
Marble, Manufacture of imitation	A. Fischer	New York, N. Y	Mar. 12, 1872	124, 562
Marble-molding machine	J. Finn	Boston, Mass	Jan. 7, 1873	134, 532
Marble-molding machine	J. Finn	Boston, Mass	July 29, 1873	141, 213
Marble obelisks, Adjusting angles in machines for sawing.	L. Brooks	Great Falls, N. H	Apr. 15, 1856	14, 688
Marble obelisks, Machine for sawing	A. Straub	Milton, Pa	Jan. 8, 1856	14, 072
Marble or pipe-stone turner	D. Clark	West Stockbridge, Mass	Apr. 9, 1810	
Marble or plastic material, Artificial	H. A. Garvey	Memphis, Tenn	Sept. 14, 1869	94, 736
Marble or stone, Machine for grinding the surface of	J. A. Bachman	Lambertville, N. J	Apr. 23, 1867	63, 982
Marble, Ornamenting	S. Gardner	New York, N. Y	Oct. 22, 1867	69, 985
Marble, Ornamenting	A. Hill	Norwalk, Conn	Nov. 27, 1866	60, 007
Marble, Ornamenting	H. Hoffman and C. F. Hill	New York, N. Y	Apr. 15, 1851	8, 046
Marble planing and carving machine	G. A. Haley	Chicago, Ill	Dec. 10, 1872	133, 704
Marble-polishing machine	W. Emmet	Galveston, Tex	Mar. 20, 1860	27, 538
Marble-polishing machine	M. Mallon	Rahway, N. J	Jan. 23, 1872	122, 951
Marble-polishing machine	J. W. Maloy	Boston, Mass	Oct. 16, 1866	58, 852
Marble-polishing machine	J. W. Maloy	Boston, Mass	Nov. 20, 1866	59, 916
Marble-polishing machine	E. T. Nichols	Joliet, Ill	July 18, 1865	48, 832
Marble-polishing machine	E. and E. B. Price	Norwalk, Conn	Feb. 1, 1870	99, 348
Marble-polishing machine	A. M. Tomb	Lyons, N. Y	Feb. 14, 1865	46, 407
Marble-polishing machine	J. Ziegler	Philadelphia, Pa	Mar. 12, 1845	3, 937
Marble, Producing mosaics upon	C. C. Fitzgerald	New York, N. Y	June 7, 1870	104, 008
Marble-running machinery	E. Ferris	West Chester, N. Y	Mar. 7, 1826	
Marble, Sawing	I. D. Kirk	Philadelphia, Pa	Dec. 28, 1832	
Marble, Sawing and polishing	L. Bissel	Otsego, N. Y	June 7, 1811	
Marble, Sawing and polishing	J. Dickerson		Jan. 9, 1796	
Marble, Sawing and polishing	J. F. Mangin		July 2, 1796	
Marble sawing and polishing machine	A. McAllister	Salem, N. Y	Feb. 18, 1828	
Marble-sawing machine	J. Ashenfelder	Philadelphia, Pa	June 3, 1856	14, 995
Marble-sawing machine	C. Avery	Tunkhannock, Pa	June 17, 1856	15, 115
Marble-sawing machine	G. W. Bishup	Brooklyn, N. Y	Nov. 13, 1855	13, 773
Marble-sawing machine	W. and G. Bull	Tonawanda, Pa	Feb. 19, 1856	14, 277
Marble-sawing machine	H. Burt	Newark, N. J	Nov. 6, 1855	13, 742
Marble-sawing machine	D. M. Campbell and J. Stevens	Oswego, N. Y	Dec. 29, 1868	85, 280
Marble-sawing machine	I. Carter	Malone, N. Y	July 15, 1856	15, 328
Marble-sawing machine	W. C. Chipman	Sandwich, Mass	Nov. 13, 1855	13, 777
Marble-sawing machine	J. Cochrane	Baltimore, Md	Sept. 11, 1855	13 540
Marble-sawing machine	R. S. Craig and A. H. Woodward.	Dover, N. Y	Mar. 1, 1870	100, 378
Marble-sawing machine	L. B. Fisher	Branch County, Mich	Nov. 13, 1855	13, 784
Marble-sawing machine	L. S. Fisher	Waynesborough, Pa	July 29, 1856	15, 419
Marble-sawing machine	J. E. French and J. M. Stephenson.	Pendleton, Ind	Sept. 27, 1870	107, 675
Marble-sawing machine	W. D. Gallaher	Bensalem Township, Pa	Nov. 18, 1856	16, 086
Marble-sawing machine	G. W. Hubbard	Middletown, Conn	Nov. 13, 1855	13, 788
Marble-sawing machine	E. W. Judd	Middlebury, Vt	Aug. 14, 1822	
Marble-sawing machine	W. B. Kimball	Peterborough, N. H	Nov. 13, 1855	13, 791
Marble-sawing machine	H. Lawrence	New York, N. Y	July 1, 1856	15, 242

Index of patents issued from the United States Patent Office from 1790 *to* 1873, *inclusive*—Continued.

Invention.	Inventor.	Residence.	Date.	No.
Marble-sawing machine	J. Lyon	New York, N. Y	Jan. 11, 1859	22, 605
Marble-sawing machine	J. M. Mott, jr	Lansingburgh, N. Y	July 22, 1856	15, 363
Marble-sawing machine	R. Myers	Factory Point, Vt	June 3, 1856	15, 015
Marble-sawing machine	F. Noette and A. Schmidt	Brooklyn, N. Y	Nov. 20, 1855	13, 829
Marble-sawing machine	C. H. G. Pease	Danbury, Conn	Mar. 9, 1869	87, 701
Marble-sawing machine	R. G. Pine	Newark, N. J	Nov. 6, 1855	13, 762
Marble-sawing machine	A. H. Tingley	Providence, R. I	Mar. 19, 1850	7, 201
Marble-sawing machine	J. Toll	Locust Grove, Ohio	Sept. 9, 1856	15, 713
Marble-sawing machine	J. A. Toll	Sugar Ridge, Ohio	June 3, 1856	15, 024
Marble-sawing machine	P. J. Torney	Washington, D. C	Dec. 1, 1868	84, 518
Marble-sawing machine	A. F. Ward	Louisville, Ky	May 6, 1856	14, 839
Marble-sawing machine	C. T. Warren	Malden, Mass	Nov. 13, 1855	13, 805
Marble-sawing machine	A. Webster and D. K. Bennett	Montpelier, Vt	Sept. 23, 1856	15, 792
Marble-scouring device	W. Weaver	Nashua, N. H	Nov. 5, 1867	70, 656
Marble-shooter	W. F. Falls	Boston, Mass	Nov. 12, 1867	70, 822
Marble slabs, Machine for molding the edges of	J. Finn	Boston, Mass	Apr. 22, 1873	138, 079
Marble, Smoothing and polishing	H. A. Torrey	New York	Nov. 1, 1825	....
Marble, stone, &c., Composition for cleaning	A. C. Ford	Lynn, Mass	May 25, 1869	90, 438
Marble, stone, &c., Machine for carving	W. I. Casselman	Vernon Village, N. Y	July 11, 1854	11, 286
Marble, stone, &c., Machine for sawing	J. Norman and A. R. McLean	West Dresden, N. Y	Sept. 28, 1858	21, 622
Marble to preserve it, Treating	C. C. Fitzgerald	New York, N. Y	June 7, 1870	104, 007
Marble with a revolving saw, Sawing	I. D. Kirk	Philadelphia, Pa	July 3, 1832	....
Marble-working machine	A. Saffer	New York, N. Y	Apr. 23, 1872	125, 990
Marble-working machine	C. Warner	Washington, D. C	June 1, 1858	20, 458
Marine alarm and fog-signal	L. Lewenberg	New York, N. Y	May 25, 1858	20, 354
Marine boilers, Arrangement of flues in	R. F. Soper	Philadelphia, Pa	Apr. 17, 1849	6, 367
Marine camel	M. Toulmin	New Orleans, La	Apr. 2, 1872	125, 352
Marine camel	W. P. Walker	Memphis, Tenn	Oct. 22, 1872	132, 506
Marine camel	S. Woolston	Vincent Town, N. J	Oct. 21, 1862	36, 745
Marine camel	S. Woolston	Vincent Town, N. J	May 12, 1863	38, 527
Marine drag	J. C. Beals	Searsport, Me	Sept. 30, 1873	143, 318
Marine drag	O. R. Ingersoll	New York, N. Y	Aug. 13, 1872	130, 503
Marine drag or floating anchor	T. Wilson and A. Crawford	Detroit, Mich	July 23, 1872	129, 878
Marine engine	J. Ericsson	New York, N. Y	Aug. 30, 1870	106, 800
Marine engine	W. B. Reaney	Chester, Pa	Dec. 6, 1870	109, 944
Marine-engine indicator and register	A. R. Harris	Saint Louis, Mo	June 13, 1871	115, 846
Marine engines, Apparatus for signaling the directions of motions of.	A. R. Harris	Saint Louis, Mo	July 5, 1870	105, 071
Marine engines, Refrigerator for	J. Merrill and G. Patten	Boston, Mass	Aug. 8, 1854	11, 485
Marine foundation	E. Manico	London, Great Britain	Jan. 21, 1868	73, 617
Marine leak-signal	W. Huston	Wilmington, Del	Mar. 1, 1864	41, 777
Marine motor	W. P. Kirkland	San Francisco, Cal	Jan. 15, 1867	61, 211
Marine signal	T. H. Dodge	Nashua, N. H	Mar. 23, 1852	8, 822
Marine signal	T. H. Dodge	Nashua, N. H	Nov. 9, 1852	9, 384
Marine steam-engine	W. Kennish, jr	New York, N. Y	Jan. 25, 1859	22, 736
Marine steam-engine	W. A. Lighthall	Albany, N. Y	Apr. 11, 1842	2, 545
Marine steam-engine	J. Maudslay and J. Field	Lambeth, England	June 11, 1842	2, 668
Marker and furrower, Ground	G. W. Martin, W. G. Parish, and J. A. Petrie.	Elizabeth, N. J	Oct. 3, 1871	119, 628
Marker and planter, Combined	R. A. Green	Martinsville, Ohio	Mar. 30, 1869	88, 382
Marker, Carpenter's	G. N. Nickason	Ellenville, N. Y	Jan. 4, 1870	98, 617
Marker, Corn	J. H. Beam	Woodside, Ill	Apr. 20, 1869	89, 008
Marker, Corn	J. Bearden	Bath, Ill	Sept. 14, 1869	94, 698
Marker, Corn	A. G. Brassfield	Henry, Ill	Aug. 29, 1871	118, 584
Marker, Corn	J. T. Corbitt	Des Moines, Iowa	Dec. 28, 1869	98, 350
Marker, Corn	J. W. Eardly	Grand Rapids, Mich	Nov. 3, 1868	83, 612
Marker, Corn	J. P. Olin	Westfield, Ohio	Mar. 3, 1868	75, 190
Marker, Corn	W. E. Phelps	Elmwood, Ill	June 16, 1868	78, 997
Marker, Corn	J. B. Thomas	Centreville, Ind	July 26, 1870	105, 865
Marker, Corn	J. W. Tucker	Eugene, Ill	Mar. 23, 1869	88, 234
Marker, Corn	S. J. Woland	Lincoln, Ill	June 1, 1869	90, 908
Marker, Corn-ground	O. H. Catey	Williamsburgh, Ind	Aug. 31, 1869	94, 185
Marker, Corn-ground	G. Johnson	Decorah, Iowa	Dec. 26, 1865	51, 725
Marker, Corn-ground	F. B. Kendall	Monmouth, Ill	July 11, 1871	116, 838
Marker, Corn-ground	J. H. Rynerson	Clayton, Ind	Oct. 28, 1873	144, 146
Marker, Corn-row	A. Cooke and D. W. Cross	Brooklandville, Md	June 3, 1873	139, 456
Marker, Corn-row	J. Roberts	Greenfield, Ind	Aug. 24, 1869	94, 032
Marker, Field	J. M. Canterbury	Mexico, Mo	Feb. 11, 1868	74, 302
Marker for corn-planting, Land	A. C. Smith	Joyner's Depot, N. C	Oct. 25, 1870	108, 644
Marker for planting corn	J. Burnham	La Salle, Ill	Nov. 19, 1867	70, 953
Marker for planting corn	D. A. Freeman	Belleville, Mich	Aug. 27, 1867	68, 297
Marker for planting, Field	W. Goltry	La Grange, Iowa	July 4, 1865	48, 551
Marker for seeding-machines	G. Armstrong	Elmira, Ill	Sept. 21, 1869	95, 067
Marker, Ground	S. P. Lionberger	Saint Mary's Township, Ill	May 18, 1869	90, 277
Marker, Land	J. H. Anderson	Hillsborough, Ohio	Sept. 3, 1872	131, 045
Marker, Land	G. W. Betts	Shadeville, Ohio	Aug. 19, 1873	141, 913
Marker, Land	W. L. Bower	Joliet, Ill	Sept. 22, 1868	82, 282
Marker, Land	J. Cuff	Emery, Ohio	Feb. 11, 1873	135, 695
Marker, Land	J. V. Gray	Washington, Ind	Feb. 6, 1872	123, 474
Marker, planter, and cultivator, Combined corn	W. Stirk	Fort Wayne, Ind	Mar. 15, 1870	100, 815
Marker, planter, and cultivator, Corn	E. Barto	Tiffin, Ohio	May 11, 1869	89, 843
Marker, Ready	T. W. Wisner	Oceola, Mich	July 22, 1862	35, 964
Marker, Seed	S. D. Fisher	Normal, Ill	Aug. 2, 1870	106, 042
Marker, seed-planter, and cultivator	C. H. and F. Logan	Pawling Post-Office, Pa	Dec. 2, 1873	145, 220
Market-box	F. Gearing	Pittsburgh, Pa	Sept. 24, 1867	69, 204
Marking and cutting gage, Dressmaker's	M. Blauvelt	Ithica, N. Y	Aug. 19, 1873	141, 849
Marking-brand	B. B. Hill	Chicopee, Mass	Dec. 3, 1861	33, 868
Marking-brand	G. H. Strong	Buffalo, N. Y	July 14, 1863	39, 261
Marking-can	F. W. Marvin	Sacramento, Cal	July 7, 1868	79, 666
Marking-compound	H. Loewenberg	New York, N. Y	Apr. 9, 1867	63, 734
Marking corn-furrows, Device for	G. G. Crose	Schoolcraft, Mich	Oct. 15, 1861	33, 478
Marking corn-ground, Machine for	G. Sprague	Spring Hill, Kans	June 4, 1867	65, 443
Marking corn-ground, with rake-attachment, Machine for.	M. Dickle and E. P. Cowan	Ottumwa, Iowa	Mar. 9, 1869	87, 552
Marking ground for planting, Device for	W. W. Potts	Rushville, Ill	Nov. 22, 1864	45, 174
Marking ground for planting, Machine for	W. R. McKinley	Lucas County, Iowa	Mar. 10, 1868	75, 288
Marking-machine	E. Spencer	Saint Louis, Mo	Aug. 19, 1862	36, 243
Marking-pot	J. F. W. Dorman	Baltimore, Md	Aug. 6, 1872	130, 111

Index of patents issued from the United States Patent Office from 1790 *to* 1873, *inclusive*—**Continued.**

Invention.	Inventor.	Residence.	Date.	No.
Marking-pot	W. H. Green	New York, N. Y	Sept. 13, 1870	107, 250
Marking-pot	J. L. Tarbox	New York, N. Y	May 7, 1872	126, 501
Marking-roller, Elastic	L. R. Witherell	Galesburgh, Ill	Feb. 12, 1867	62, 104
Marking-roller for planting	F. Roth and B. Fürst	Lacon, Ill	June 13, 1871	115, 900
Marking stock	G. W. Devin	Ottumwa, Iowa	Apr. 24, 1866	54, 127
Marking-wheel	H. Holt	Brooklyn, N. Y	Jan. 23, 1866	52, 169
Marl, dirt, &c., Machine for raising	T. F. Christman	Wilson, N. C	June 22, 1858	20, 623
Marl, Process for treating and compounding	R. B. Fitts	Philadelphia, Pa	May 30, 1865	47, 941
Marquetry	J. Dill	Grand Rapids, Mich	Aug. 29, 1871	118, 595
Marquetry	L. F. Groebl	Philadelphia, Pa	Jan. 23, 1855	12, 276
Marsh and swamp lands, Mode of reclaiming	S. B. Driggs	New York, N. Y	June 27, 1865	48, 382
Marsh-drainer	J. Blanc	New Orleans, La	July 2, 1836	....
Marsh-filling device	G. Howell and W. Smith	Philadelphia, Pa	Sept. 15, 1868	82, 224
Marsh-filling machine	G. Howell	Philadelphia, Pa	Nov. 12, 1867	70, 849
Marshes, Filling	G. Howell	Philadelphia, Pa	Mar. 31, 1868	76, 194
Marshes, Filling	S. B. Woods and J. T. Chapman.	Jersey City, N. J., and Brooklyn, N. Y.	Nov. 10, 1868	83, 895
Martingale	P. J. McGuiness	New York, N. Y	Dec. 1, 1868	84, 643
Martingale	R. R. Miller	Southington, Conn	Feb. 11, 1873	135, 657
Martingale	F. G. Moody	North Monmouth, Me	Aug. 26, 1873	142, 261
Martingale	W. B. Perrie	Horse Head, Md	Sept. 8, 1868	82, 029
Martingale for preventing horses or mules from throwing or breaking fences, Pricking.	D. and J. Pfouts	Holmes County, Ohio	Sept. 2, 1862	36, 363
Martingale-ring	G. T. Bushnell	Birmingham, Conn	Nov. 15, 1859	26, 086
Martingale-ring	W. F. Gilbert	Birmingham, Conn	Nov. 9, 1869	96, 690
Martingale-ring	D. C. Lockwood	Derby, Conn	Oct. 2, 1860	30, 233
Martingale-ring	W. M. Welling	New York, N. Y	Mar. 17, 1863	37, 941
Martingale-ring, Composition	C. B. Bristol	New Haven, Conn	Feb. 23, 1869	87, 240
Mash and beer cooler	C. Schenck	Manheim, Baden	Aug. 27, 1867	68, 242
Mash, Apparatus for ascertaining the proof-spirits in fermented.	G. Tagliabue	New York, N. Y	Feb. 14, 1871	111, 885
Mash-cooler	F. P. Bush	Cincinnati, Ohio	Nov. 5, 1867	70, 518
Mash-cooling apparatus	A. Naegeli	Wegeleben, Germany	Nov. 5, 1872	132, 730
Mash-cooling machine	J. Shilling	Troy, Ohio	Aug. 11, 1857	17, 984
Mash for distillation, Preparing	G. Seitz	Easton, Pa	Jan. 26, 1858	19, 210
Mash from grain, Saccharifying	C. H. Frings	Centreton, Mo	Dec. 6, 1870	109, 887
Mash-machine	M. Brand and C. P. Hoffman	Chicago, Ill	Dec. 11, 1866	60, 330
Mash-machine	J. T. Forrer	Peoria, Ill	Apr. 3, 1866	53, 597
Mash-machine	A. Hammer	Philadelphia, Pa	Jan. 9, 1855	12, 205
Mash of beer, &c., Mode of cooling	A. Smith	New York, N. Y	Dec. 22, 1863	41, 025
Mash, Process and apparatus for forming	F. W. Wolf	Chicago, Ill	Apr. 9, 1872	125, 645
Mash-tub	L. Klee	Pittsburgh, Pa	June 16, 1868	78, 973
Mash-tub	A. Miller and J. R. Stauffer	Pennsville, Pa	Sept. 1, 1863	39, 739
Mash-tub	B. Roop	Pekin, Ohio	May 6, 1844	3, 575
Mash-tub and vapor-cooler	A. W. Cram	Saint Louis, Mo	Mar. 9, 1869	87, 640
Mash-tub, Distiller's	D. P. Patterson	Fayette County, Pa	Aug. 21, 1860	29, 713
Mash-tubs, Apparatus for heating	A. Hammer	Reading, Pa	Aug. 17, 1858	21, 195
Mash-tubs, Heating and cooling coil for	M. J. Allen	New York, N. Y	Apr. 27, 1869	89, 374
Mash-tun	J. Chilcott	Brooklyn, N. Y	May 30, 1865	47, 931
Mash-tun	C. Stoll	Brooklyn, N. Y	Dec. 3, 1872	133, 679
Mash-tun	R. Wicks and J. Faulkner, jr	Williamsburgh, N. Y	May 11, 1852	8, 946
Mash-tun pipes, Preventing coating of	A. Adams	Springfield, Mass	Aug. 28, 1866	57, 454
Mashing	N. G. Thorn	Dayton, Ohio	Mar. 1, 1859	23, 125
Mashing-apparatus	E. Haeckel	Cincinnati, Ohio	Sept. 6, 1859	25, 329
Mashing-machine	L. C. Field	Galesburgh, Ill	Sept. 14, 1869	94, 730
Mashing-machine	A. Harris	New York, N. Y	July 12, 1870	105, 331
Mashing, Treating cereal grains for	C. H Frings	Central, Mo	Apr. 30, 1872	126, 283
Mashing-tub	J. C. Birket	Peoria, Ill	Sept. 19, 1871	119, 072
Mashing-tub	J. Wright	Waterloo, N. Y	June 24, 1851	8, 183
Mask and respirator, Fireman's	I. P. Nelson	Cambridge, Mass	Mar. 17, 1857	16, 863
Mask, Safety	D. W. Flora	Newaygo, Mich	May 28, 1872	127, 331
Masonic emblems, Apparatus for display of	R. H. Lyon and G. B. Rand	Dubuque, Iowa	Nov. 19, 1872	133, 236
Mast and rigging	G. T. May	Tompkinsville, N. Y	July 15, 1862	35, 882
Mast and spar for vessels	W. T. Griffenberg	Wilmington, Del	May 21, 1872	127, 047
Mast and truss hoops, Bending	J. Mulford	Northern Liberties, Pa	Jan. 16, 1835	....
Mast coat	A. J. Gore	San Francisco, Cal	July 11, 1865	48, 767
Mast-coat	G. Perkins and J. A. Stetson	Lynnfield and Gloucester, Mass.	Nov. 5, 1872	132, 774
Mast for navigable vessels	C. P. Coles	Southsea, England	Apr. 14, 1863	38, 151
Mast for vessels	W. T. Griffenberg	Wilmington, Del	July 30, 1872	130, 035
Mast for vessels	E. E. Holley	New York, N. Y	Feb. 22, 1870	100, 151
Mast for vessels, Hinged	J. E. Hammond	New Bedford, Mass	Dec. 16, 1873	145, 648
Mast-head	S. T. Nickerson	South Orrington, Me	Feb. 27, 1872	124, 006
Mast-hoop	E. F. Burrows	Mystic River, Conn	July 13, 1869	92, 577
Mast-hoop	F. B. Dunton	Centre Lincolnville, Me	Apr. 20, 1869	89, 135
Mast-hoop	W. T. Maddocks	Northport, Me	Feb. 23, 1869	87, 105
Mast-hoop	W. McKay	Newburyport, Mass	Dec. 22, 1868	85, 234
Mast-hoop	D. R. Procter	Gloucester, Mass	Mar. 10, 1863	37, 872
Mast-hoop, Anti-friction	D. H. Hussey	Portsmouth, N. H	Aug. 3, 1869	93, 303
Mast-hoop balance-line	J. Conway	Harrison, Md	May 4, 1869	89, 632
Mast-hoops, Machine for manufacturing	E. W. Scott	Lowell, Mass	Oct. 31, 1854	11, 875
Mast-hoops, Sail-link to	J. E. Seavey	Kennebunkport, Me	June 17, 1862	35, 654
Mast, steeple, &c., Iron	E. S. Boynton	Alexandria, Va	Mar. 26, 1861	31, 777
Masts and spars, Connection of telescopic	C. F. Brown	Warren, R. I	June 17, 1851	8, 164
Masts, Apparatus for lowering ship's	G. Dowling	Fair Haven, Conn	Mar. 17, 1868	75, 671
Masts, Attaching topmasts and top-gallant	L. Brown	New York, N. Y	Apr. 18, 1871	113, 738
Masts, Ball for	E. C. Seely	Port Medway, Canada	Aug. 27, 1872	130, 825
Masts, Cheek-piece for ships'	J. Baymore	Philadelphia, Pa	Apr. 4, 1871	113, 386
Masts in the decks of vessels, Supporting	T. J. Woodworth	Salem, N. J	Jan. 27, 1857	16, 510
Masts of navigable vessels, Connection of the gaff to the	C. R. Fisher	Chelsea, Mass	Mar. 28, 1865	47, 004
Masts of ships, Fid for the	B. Rotch	Bristol, Great Britain	Dec. 10, 1823	....
Masts of vessels, Cap or withe for	J. W. Sikes	Plymouth, N. C	Aug. 1, 1854	11, 461
Masts of vessels, Purchase for hoisting and lowering top-.	W. Winchester	Portland, Me	Sept. 14, 1869	94, 804
Masts of vessels, Securing	D. S. Stevens and L. Snedecor.	Red Bank, N. J	Aug. 11, 1868	80, 884
Masts, Supporting ship's top	T. Batty	Brooklyn, N. Y	Sept. 4, 1855	13, 518
Master-key lock	A. Rankin	Philadelphia, Pa	July 29, 1873	141, 381

Index of patents issued from the United States Patent Office from 1790 *to* 1873, *inclusive*—Continued.

Invention.	Inventor.	Residence.	Date.	No.
Master-wheel, Double-lever	D. E. Mitchell	Rural Dale, Ohio	June 21, 1870	104, 484
Mastic, Composition	A. Thiele	Yickfaw, La	Sept. 9, 1873	142, 595
Mastic composition	J. Thompson	North Wrentham, Mass	Mar. 30, 1858	19, 802
Masting and rigging vessels	W. Webster	Jefferson County, Wash. T.	June 22, 1858	20, 673
Mat: *See* Bird-cage mat. Cage-mat. Door-mat. Metallic mat. Oil-press mat. Rubber mat. Straw mat. Table-mat. Toilet-mat. Wooden mat.				
Mat	J. M. Groh	Benevola, Md	Sept. 22, 1868	82, 308
Mat	D. M. Harman	Baltimore, Md	Sept. 17, 1872	131, 440
Mat	S. Lewis	Williamsburgh, N. Y	Oct. 11, 1870	108, 160
Mat	P. W. Neefus	New York, N. Y	Feb. 7, 1871	111, 559
Mat and scraper	W. Farrington	Morrisania, N. Y	June 29, 1869	92, 026
Mat, Husk	J. P. Smucker	Ashland, Ohio	Aug. 31, 1869	94, 252
Mats, &c., Machinery for making	D. Hodgman and A. D. Wyckoff	New York, N. Y	May 14, 1849	6, 412
Mats, Manufacture of Manila	S. S. Williams	Roxbury, Mass	Aug. 22, 1828	....
Match and cigar box	G. Graetz	Alexandria, Va	Dec. 22, 1868	85, 170
Match-blocks, Machine for making	E. Andrews and W. Tucker	Portland, Me., and Philadelphia, Pa.	Nov. 16, 1869	96, 764
Match-blocks, Machine for splitting	V. R. Powell	Troy, N. Y	June 16, 1863	38, 911
Match-box	E. B. and L. W. Beecher	New Haven, Conn	Feb. 23, 1869	87, 236
Match-box	L. W. Beecher	Westville, Conn	June 22, 1869	91, 703
Match-box	J. Bonsfield	Cleveland, Ohio	Nov. 2, 1869	96, 303
Match-box	D. Burhans	Burlington, Iowa	May 30, 1871	115, 430
Match-box	T. Crommelin	Brooklyn, N. Y	Mar. 26, 1872	124, 886
Match-box	G. Dowler	Birmingham, England	Mar. 22, 1864	41, 976
Match-box	S. W. Francis	New York, N. Y	Apr. 30, 1861	32, 187
Match-box	J. H. Goodfellow and M. Russell, jr.	Troy, N. Y	May 10, 1870	102, 807
Match-box	J. Holtz	Baltimore, Md	Aug. 12, 1873	141, 793
Match-box	H. Howson	Philadelphia, Pa	Jan. 21, 1862	34, 230
Match-box	H. Howson	Philadelphia, Pa	Sept. 15, 1863	39, 994
Match-box	A. D. Judd	Hew Haven, Conn	Dec. 12, 1871	121, 788
Match-box	A. D. Judd	Hew Haven, Conn	Jan. 16, 1872	122, 836
Match-box	J. Kirchfeld and F. Heyl	Riegelsville, Pa	Mar. 31, 1868	76, 084
Match-box	A. Martin	Basel, Switzerland	Apr. 14, 1868	76, 786
Match-box	J. Matthias	New York, N. Y	Nov. 11, 1873	144, 404
Match-box	L. O. P. Meyer	Newtown, Conn	Aug. 29, 1871	118, 627
Match-box	L. O. P. Meyer	Newtown, Conn	Nov. 14, 1871	120, 889
Match-box	A. Meyersberg	New York, N. Y	Aug. 5, 1873	141, 581
Match-box	J. Monaghan	Tuckahoe, N. Y	Nov. 15, 1870	109, 235
Match-box	M. L. Orum	Philadelphia, Pa	Nov. 4, 1873	144, 283
Match-box	O. Paddock	Watertown, N. Y	Jan. 14, 1873	134, 764
Match-box	C. Prahl	Stapleton, N. Y	July 15, 1873	140, 949
Match-box	L. Schoerken	Kohn, Prussia	Mar. 19, 1867	63, 104
Match-box	J. H. Wheeler	Addison, Vt	Jan. 28, 1866	73, 855
Match-box	H. J. Wickham	Manchester, Conn	May 4, 1869	89, 719
Match-box and candle-stick combined	C. R. Stickney	Hartford, Conn	June 4, 1867	65, 445
Match-box and taper-holder combined	J. A. Whipple	Cambridge, Mass	May 28, 1867	65, 142
Match-box, Pocket	S. W. Francis	New York, N. Y	Jan. 28, 1862	34, 249
Match-box, Pocket	A. Roesler and C. Frey	Warsaw, Ill	Feb. 15, 1859	22, 979
Match-boxing machine	N. B. Forrest	Auburn, N. Y	June 15, 1869	91, 429
Match-card	M. G. Crane	Boston, Mass	Sept. 5, 1865	49, 727
Match cards, Rack for holding comb	E. G. Byam and B. E. Parkhurst.	Boston, Mass., and Brunswick, Me.	Oct. 12, 1858	21, 794
Match-case	F. B. Coleman	Southampton, Mass	May 3, 1870	102, 499
Match-case	H. Ellig	Bridgeport, Conn	Feb. 19, 1867	62, 260
Match-case, Pocket	L. O. P. Meyer	Newtown, Conn	Nov. 14, 1871	120, 892
Match-case, Portable	D. Ellis and P. Hine	Waterbury, Conn	Apr. 24, 1860	27, 974
Match-composition	J. F. Babcock, W. A. Leonard, and E. B. Crane.	Boston, Mass	Mar. 18, 1873	136, 953
Match-composition	W. H. Rogers	New York, N. Y	Oct. 27, 1868	83, 412
Match composition, Friction	N. B. and D. Shaw	Sanbornton, N. H	Aug. 29, 1865	49, 659
Match-compound	L. Lanszweert	San Francisco, Cal	Mar. 27, 1866	53, 454
Match cord, Pocket-safe for friction	W. H. Rogers	New York, N. Y	Oct. 13, 1868	83, 097
Match-cutting machine	P. D. Cummings, M. Smith, and J. C. Jordan.	Portland, Me	June 3, 1873	139, 457
Match-cutting machine	T. G. Murphy	Akron, Ohio	Jan. 7, 1873	134, 565
Match for lighting cigars, &c	G. Graetz	Alexandria, Va	Dec. 8, 1868	84, 821
Match for lighting cigars, &c	J. Howe	Allegheny, Pa	July 2, 1872	128, 626
Match for retaining fire, Friction	J. H. Stevens	New York, N. Y	Nov. 16, 1839	1, 414
Match-frames, Machine for filling	W. Gates, jr	Frankfort, N. Y	Oct. 31, 1854	11, 860
Match-frames, Machine for filling	A. Sohn	Monroeville, Ohio	Apr. 18, 1854	10, 809
Match, Friction	W. K. Ashard	New York	June 24, 1843	3, 146
Match, Friction	G. W. Carleton	Bath, Me	May 20, 1842	2, 635
Match, Friction	J. R. Dey	Hudson City, N. J	Apr. 7, 1863	38, 096
Match, Friction	S. C. Moore	Boston, Mass	May 23, 1865	47, 848
Match, Friction	A. D. Phillips	Springfield, Mass	Oct. 24, 1836	68
Match, Friction	W. H. Rogers	New York, N. Y	Oct. 12, 1869	95, 730
Match, Friction	C. W. Smith	New York, N. Y	Mar. 11, 1862	34, 649
Match, Friction	E. Smith	Erving, Mass	Oct. 3, 1844	3, 773
Match, Friction	B. F. Woodside	McDonald, Tenn	Jan. 14, 1868	73, 275
Match-holder	L. J. Johnson	Philadelphia, Pa	July 2, 1861	32, 735
Match-holder	L. O. P. Meyer	Newtown, Conn	Aug. 29, 1871	118, 628
Match-holder	L. O. P. Meyer	Newtown, Conn	Nov. 14, 1871	120, 890
Match-holder	L. O. P. Meyer	Newtown, Conn	Nov. 14, 1871	120, 891
Match-hook	J. D. Leach	Penobscot, Me	May 2, 1871	114, 451
Match-igniter	J. H. Merrill	Norwalk, Conn	Aug. 22, 1865	49, 542
Match light and box	R. and N. S. Allison	Burlington, N. J	Mar. 26, 1814	......
Match-lighter	W. Adamson	Philadelphia, Pa	July 16, 1867	66, 770
Match-machine	E. Andrews and W. Tucker	Portland, Me	Dec. 1, 1868	84, 464
Match-machine	E. Andrews and W. Tucker	Portland, Me	July 13, 1869	92, 415
Match-machine	W. B. Eltonhead	Philadelphia, Pa	Aug. 10, 1869	93, 525
Match-machine	L. Holms	Paterson, N. J	July 1, 1856	15, 238
Match-machine	L. T. Luther	Oak Grove, Pa	Feb. 15, 1870	99, 780
Match-machine	S. Miller and W. Gates, jr	Hammond and Frankfort, N. Y.	Mar. 9, 1858	19, 608

Index of patents issued from the United States Patent Office from 1790 to 1873, inclusive—Continued.

Invention.	Inventor.	Residence.	Date.	No.
Match-machine	L. and J. Thomas	Allegheny City, Pa	Jan. 23, 1855	12, 297
Match machine, Friction	C. D. Smith and H. Patterson	Baldwinville, Mass	July 22, 1856	15, 393
Match-making machine	M. Young	Frederick, Md	Aug. 29, 1871	118, 502
Match, Parlor	F. Zaiss	Philadelphia, Pa	Apr. 16, 1872	125, 874
Match, Percussion	W. Servant	Providence, R. I	Apr. 18, 1871	113, 937
Match-plates, Method and apparatus for molding	J. R. Davies	Racine, Wis	July 25, 1871	117, 391
Match-safe	W. H. Andrews	New Haven, Conn	Dec. 20, 1864	45, 554
Match-safe	J. E. Auld	Buffalo, N. Y	Oct. 2, 1866	58, 536
Match-safe	C. A. Babcock	Frankfort, N. Y	May 28, 1867	65, 041
Match-safe	C. Bünger	New Haven, Conn	June 11, 1867	65, 722
Match-safe	L. Burnell	Milwaukee, Wis	Jan. 3, 1860	26, 648
Match-safe	P. D. M. Carmichael	Le Roy, N. Y	Jan. 7, 1868	72, 973
Match-safe	L. Darozir	Worcester, Mass	Jan. 7, 1873	134, 523
Match-safe	S. De Mackiewicz	New York, N. Y	June 21, 1864	43, 186
Match-safe	J. W. Durham	Ripley, Tenn	July 6, 1869	92, 290
Match-safe	J. A. Evarts	West Meriden, Conn	Jan. 28, 1868	73, 706
Match-safe	G. Geer	Meriden, Conn	May 3, 1870	102, 676
Match-safe	J. Gibbs	Brooklyn, N. Y	July 27, 1869	92, 955
Match-safe	C. Goldthwait	South Weymouth, Mass	Oct. 18, 1870	108, 346
Match-safe	T. S. Hathaway	Decatur, Ill	Sept. 25, 1866	58, 253
Match safe	J. and A. Helm	Washington, D. C	Nov. 18, 1873	144, 615
Match-safe	A. Hoyt	New York, N. Y	June 30, 1868	79, 353
Match-safe	R. W. Jenks	Providence, R. I	Sept. 5, 1865	49, 760
Match-safe	M. Jincks	Dansville, N. Y	May 19, 1868	77, 984
Match-safe	A. D. Judd	New Haven, Conn	May 10, 1870	102, 828
Match-safe	P. Killin and H. C. Yates	Decatur, Ill	July 30, 1867	67, 317
Match-safe	C. H. Leggett	Elizabeth, N. J	Apr. 29, 1873	138, 261
Match-safe	H. Richmond	West Meriden, Conn	Oct. 6, 1868	82, 752
Match-safe	J. Roebuck	New York, N. Y	Apr. 16, 1867	63, 943
Match-safe	A. Roesler	Warsaw, Ill	Oct. 29, 1867	70, 362
Match-safe	G. H. Snow	New Haven, Conn	Apr. 19, 1864	42, 414
Match-safe	A. Steinbök	New York, N. Y	Oct. 26, 1869	96, 163
Match-safe	W. Stine	Elmore, Ohio	Apr. 4, 1871	113, 360
Match-safe	H. M. Woodford	Kensington, Conn	June 30, 1868	79, 426
Match-safe, Pocket	C. E. Fowler	Carmel, N. Y	Nov. 20, 1866	59, 830
Match-safe, Pocket	A. M. Smith	New York, N. Y	Jan. 17, 1860	26, 865
Match safe, Portable and water-proof friction	P. Merrill	Port Sanilac, Mich	Oct. 12, 1858	21, 770
Match-safe, Sheet-metal	J. Musgrove	Newark, N. J	May 2, 1871	114, 323
Match-safes, Manufacture of metallic	J. Fallows	Philadelphia, Pa	Feb. 27, 1872	124, 043
Match, Safety	H. House	London, Great Britain	Feb. 20, 1872	123, 905
Match-slitting machine	M. Smith and J. C. Jordan	Portland, Me	Apr. 2, 1872	125, 149
Match-splint card	B. Hotchkiss	New Haven, Conn	June 13, 1865	48, 176
Match-splint-cutting machine	T. Cook	New York, N. Y	May 12, 1857	17, 263
Match-splint-cutting machine	E. Fitzgerald	New York	July 18, 1840	1, 695
Match-splint-cutting machine	H. Law	Wilmington, N. C	Aug. 28, 1844	3, 719
Match-splint-cutting machine	C. E. Warner	New York	July 2, 1842	2, 700
Match-splint-framing machine	J. Bentz	Brooklyn, N. Y	Sept. 3, 1867	68, 342
Match-splint machine	R. F. Gustine	Chicago, Ill	Apr. 12, 1853	9, 660
Match-splint-splitting machinery	L. Smith	New York, N. Y	Mar. 28, 1848	5, 496
Match splints and arranging them in the dipping-frame, Machine for making.	A. Fessenden and L. L. Knight	Templeton and Barre, Mass	Apr. 26, 1845	4, 013
Match splints for dipping, Machine for framing lucifer.	A. and E. B. Beecher	New Haven, Conn	Feb. 3, 1863	37, 562
Match-splints, Machine for making	E. Andrews and W. Tucker	Springfield, Mass	Dec. 20, 1864	45, 465
Match-splints, Machine for making	E. Andrews and W. Tucker	Portland, Me	Aug. 27, 1867	68, 144
Match-splints, Machine for making	D. Burhans	Burlington, Iowa	Aug. 16, 1870	106, 319
Match splints, Machine for making	C. L. Zeidler	Cincinnati, Ohio	June 18, 1867	65, 853
Match splints or sticks, Manufacture of friction and other.	N. T. Winans and T. Hyatt	New York	Nov. 26, 1840	1, 867
Match-splints, Splitting	M. D. Murphy and O. C. Barber	Middlebury, Ohio	Feb. 2, 1869	86, 576
Match-stand	G. G. Taylor	Charlestown, Mass	Apr. 7, 1868	76, 549
Match stand, Friction	N. Waterman	Boston, Mass	June 16, 1863	38, 923
Match-stick	J. K. Robinson	Middlebury, Ohio	July 26, 1870	105, 727
Match-sticks, Machine for making	F. D. Bowens	Philadelphia, Pa	Apr. 27, 1869	89, 472
Match-sticks, Machine for making	F. Zaiss	Philadelphia, Pa	Mar. 18, 1873	136, 893
Match-sticks, Machinery for cutting	F. De Bowens	Philadelphia, Pa	Aug. 31, 1869	94, 189
Match-sticks, Manufacture of	S. C. Ellis	Jersey City, N. J	Sept. 22, 1863	40, 027
Match, Water-proof friction	L. T. Luther	Oak Grove, Pa	Nov. 19, 1867	71, 189
Matches, Apparatus for dipping lucifer	S. A. Bell and T. Higgins	Stratford and Bow, England.	June 16, 1863	38, 878
Matches, Composition for	E. K. Smith	Philadelphia, Pa	June 18, 1867	65, 838
Matches, Composition for friction	W. P. Allen	Dubuque, Iowa	Jan. 11, 1859	22, 538
Matches, Composition for friction	L. Langszweert	San Francisco, Cal	Oct. 10, 1865	50, 379
Matches, Composition for the manufacture of friction.	J. J. Karlen	Eilenbach, Switzerland	Apr. 5, 1870	101, 625
Matches, Composition for the manufacture of friction.	N. T. Winans and T. and T. Hyatt.	New York	Dec. 23, 1841	2, 402
Matches, Composition for manufacture of friction	N. T. Winans and T. and T. Hyatt.	New York	Dec. 23, 1841	2, 403
Matches, Composition for the manufacture of safety and other friction.	W. Austin	London, England	Nov. 3, 1868	83, 683
Matches, Composition-matter of friction	J. H. Stevens	New York, N. Y	Nov. 16, 1839	1, 413
Matches, Compound for making friction	O. C. Green	Copenhagen, Denmark	June 8, 1869	91, 114
Matches, Dipping-frame for the manufacture of	D. Helmer	Mohawk, N. Y	Mar. 15, 1864	41, 918
Matches, Dipping loco-foco	J. Hatfield	Stillwater, N. Y	June 3, 1837	219
Matches, Drying	M. Young	Frederick, Md	Aug. 27, 1872	130, 834
Matches for preserving from accidental ignition, Manufacture of friction.	J. H. Stevens	New York, N. Y	Nov. 16, 1839	1, 412
Matches, Friction-mat for lighting	J. B. Colt	Brooklyn, N. Y	July 23, 1872	129, 716
Matches, Ignitible compound for friction	S. Blaisdell	Brunswick, Me	Mar. 18, 1842	2, 494
Matches in frames for dipping, Machine for placing friction.	G. Sebold	Durlach, Germany	June 9, 1868	78, 837
Matches, Machine for cutting	A. Welsch	Chicago, Ill	June 28, 1864	43, 358
Matches, Machine for framing	J. Bentz	Brooklyn, N. Y	June 26, 1866	55, 808
Matches, Machine for gathering and depositing dipped.	T. Cook	New York, N. Y	May 19, 1857	17, 316
Matches, Machine for making and dipping	J. F. Cranston	Springfield, Mass	Jan. 21, 1868	73, 507
Matches, Machine for making friction	W. Gates, jr., and H. J. Harwood.	Franklin and Utica, N. Y	Apr. 4, 1854	10, 737

Index of patents issued from the United States Patent Office from 1790 *to* 1873, *inclusive*—Continued.

Invention.	Inventor.	Residence.	Date.	No.
Matches, Machine for manufacturing friction	A. Underwood	German Flats, N. Y	Apr. 29, 1856	14, 782
Matches, Machine for matting the ends of blocks in making.	H. E. Pierce	Charlemont, Mass	Jan. 10, 1854	10, 413
Matches, Machinery for making	E. Adler	New York, N. Y	Aug. 1, 1854	11, 405
Matches, Machinery for making	I. H. Smith	Wolcott, Conn	Apr. 29, 1851	8, 066
Matches, Manufacture and packing of friction	E. Precht and V. Joepken	New York, N. Y	Nov. 18, 1862	36, 967
Matches, Manufacture of	E. Andrews	Portland, Me	Oct. 15, 1867	69, 891
Matches, Manufacture of	E. Andrews and W. Tucker	Fiskedale, Mass	Mar. 26, 1867	63, 197
Matches, Manufacture of	G. Graham	New Haven, Conn	Jan. 26, 1869	86, 299
Matches, Manufacture of	E. J. Hill	Milwaukee, Wis	Dec. 24, 1867	72, 637
Matches, Manufacture of friction	W. Baustian	Davenport, Iowa	Mar. 19, 1867	62, 922
Matches, Manufacture of friction	C. F. Bonhack	New York, N. Y	Aug. 5, 1873	141, 483
Matches, Manufacture of friction	G. G. Dennis	Dover, N. H	July 25, 1865	48, 913
Matches, Manufacture of friction	J. W. Hjerpe	Stockholm, Sweden	Oct. 13, 1863	40, 259
Matches, Manufacture of friction	S. Krackowizer	New York, N. Y	Apr. 18, 1865	47, 311
Matches, Manufacture of friction	V. Powell	Troy, N. Y	July 25, 1865	49, 002
Matches, Manufacture of friction	V. Powell	Troy, N. Y	Aug. 22, 1865	49, 549
Matches, Manufacture of friction	P. B. Tyler, W. M. Chandler, and L. F. Standish.	Springfield and Chicopee Falls, Mass.	Nov. 7, 1865	50, 869
Matches, Manufacture of safety	L. O. P. Meyer	Bethlehem, Pa	June 25, 1867	66, 101
Matches, Manufacture of safety	L. O. P. Meyer	Newtown, Conn	Jan. 17, 1871	111, 075
Matches, Manufacturing	O. H. Hicks	Baltimore, Md	Apr. 15, 1873	137, 840
Matches, Manufacturing splints for friction	S. Dole	Concord, N. H	May 30, 1837	214
Matches, Pocket-safe for friction	J. S. Morton	Hartford, Conn	Jan. 25, 1870	99, 224
Matches, Putting up	E. J. Hill	Milwaukee, Wis	July 23, 1867	67, 052
Matches, tapers, &c., Machine for filling dipping-clamps in the manufacture of.	T. Higgins	Bon, England	Sept. 16, 1862	36, 463
Matches water-proof, Rendering friction	L. J. Henry	New York, N. Y	Apr. 5, 1859	23, 465
Matrices, &c., Mode of constructing	J. J. C. Smith	Covington, Ky	Sept. 20, 1859	25, 533
Matter, Composition of	G. F. J. Colburn	Newark, N. J	Jan. 28, 1868	73, 694
Matter, Composition of	L. Francis	New York, N. Y	June 26, 1866	55, 951
Matter, Composition of	W. K. Wyckoff	Ripon, Wis	Oct. 2, 1866	58, 532
Matting-apparatus	J. H. Reilly	Brooklyn, N. Y	Jan. 17, 1871	111, 086
Matting for carpet-linings, &c	H. C. Heermance	Claverack, N. Y	Feb. 5, 1867	61, 831
Matting for cars, Slat	W. Barton	Troy, N. Y	Oct. 6, 1868	82, 678
Matting for floors	J. Michell	West Farms, N. Y	Apr. 16, 1867	63, 921
Matting, India-rubber binding for	F. S. Beardsley	Bridgeport, Conn	Jan. 23, 1872	122, 933
Matting, Slat	W. Barton	Troy, N. Y	Mar. 3, 1868	74, 975
Mattock, hoe, &c., swaging die	C. Konald	Snowden, Pa	Sept. 3, 1867	68, 446
Mattocks, picks, &c., Machine for forming the eyes of.	J. Donald	Cohoes, N. Y	Nov. 26, 1872	133, 360
Mattress	E. F. Bassett	Seymour, Conn	Apr. 15, 1862	34, 935
Mattress	T. G. Clinton	Cincinnati, Ohio	Apr. 20, 1852	8, 883
Mattress	C. Enscoe	New York, N. Y	Oct. 15, 1872	132, 268
Mattress	W. Foehlich	Harburg, Hanover	June 24, 1862	35, 732
Mattress	W. P. Ford	Cheneyville, La	July 14, 1857	17, 785
Mattress	J. J. Haley	Newton, Mass	Feb. 13, 1872	123, 628
Mattress	H. W. Henley	New York, N. Y	June 7, 1859	24, 302
Mattress	R. Krause	New York, N. Y	Apr. 4, 1865	47, 112
Mattress	J. Maas	Westfield, Mass	Dec. 12, 1871	121, 723
Mattress	A. G. Morey	Chicago, Ill	Oct. 13, 1868	82, 975
Mattress	W. H. Pack and J. S. Vanhorn	Jersey City, N. J	Feb. 28, 1871	112, 175
Mattress	W. H. Robertson	New London, Conn	Dec. 17, 1846	4, 896
Mattress	T. Roller	Allegheny City, Pa	Mar. 18, 1873	137, 030
Mattress	G. Schott	New York, N. Y	Jan. 21, 1868	73, 466
Mattress	H. E. Smith	Fitchburgh, Mass	May 20, 1873	139, 202
Mattress	B. F. Walton	Philadelphia, Pa	Oct. 1, 1867	69, 376
Mattress	W. F. Wehrmann	Cincinnati, Ohio	Sept. 24, 1872	131, 727
Mattress and bed	F. Eider	Winnsborough, S. C	July 10, 1860	29, 066
Mattress and chair spring	W. Hersee	Buffalo, N. Y	Nov. 10, 1857	18, 585
Mattress and cushion tacker	T. A. Watson and A. H. Phillips.	Brenham, Tex	July 2, 1872	128, 571
Mattress and life-preserving float	L. Bauhoefer	Philadelphia, Pa	Apr. 16, 1867	63, 831
Mattress, bolster, &c	A. Salisbury and J. Uram	Troy, N. Y	July 1, 1836	....
Mattress, Columnar	H. E. Smith	New York, N. Y	May 17, 1870	103, 096
Mattress, Earth	T. W. Phinney	Newport, R. I	Apr. 18, 1871	113, 931
Mattress, Floating	W. Williams	Saint Louis, Mo	Jan. 1, 1861	31, 048
Mattress, Folding	M. Sulzbacher	New York, N. Y	July 6, 1869	92, 396
Mattress, Folding	W. Wells	Harrisburgh, Pa	Apr. 27, 1859	20, 112
Mattress, Folding-box spring	S. P. Kittle	Brooklyn, N. Y	Dec. 7, 1869	97, 522
Mattress for invalids, Spring	D. Baird	New York, N. Y	Jan. 7, 1851	7, 882
Mattress frame, Elastic-fabric	J. E. Whittlesey and J. W. C. Peters.	Chicago, Ill	Sept. 16, 1873	142, 827
Mattress frame, Wire	G. V. and W. I. Bunker	Yankton, Dak	Dec. 16, 1873	145, 553
Mattress frame, Wire	G. C. Perkins	Hartford, Conn	Apr. 4, 1871	113, 559
Mattress frame, Wire	J. G. Smith	Saint Louis, Mo	Feb. 25, 1873	136, 271
Mattress frame, Woven-wire	G. C. Perkins	Hartford, Conn	Oct. 8, 1872	131, 968
Mattress frame, Woven-wire	D. J. Powers	Chicago, Ill	Feb. 25, 1873	136, 189
Mattress, Hair	W. F. Phyfe	New York, N. Y	Oct. 1, 1830	....
Mattress or seaman's bed	I. Martin	Boston, Mass	Sept. 17, 1833	....
Mattress, Palm-leaf	E. Howe	Cambridge, Mass	Sept. 13, 1834	....
Mattress, sofa, &c	R. Cromelon	New York, N. Y	Nov. 7, 1831	....
Mattress, Spring	E. L. Bushnell	Poughkeepsie, N. Y	Apr. 12, 1853	9, 658
Mattress, Spring	E. L. Bushnell	Poughkeepsie, N. Y	Oct. 15, 1872	132, 248
Mattress, Spring	C. and G. H. Chinnock	New York, N. Y	Aug. 20, 1872	130, 698
Mattress, Spring	R. T. Cornel	Poughkeepsie, N. Y	May 9, 1871	114, 648
Mattress, Spring	F. Forst	Philadelphia, Pa	June 16, 1846	4, 578
Mattress, Spring	C. Fulton	New York, N. Y	Oct. 11, 1870	108, 128
Mattress, Spring	H. Gardiner, R. Lowe, J. and J. Wood, and J. Pickering.	Manchester, Great Britain	Oct. 24, 1871	120, 154
Mattress, Spring	H. A. Gaston	San José, Cal	Apr. 8, 1873	137, 673
Mattress, Spring	A. Gebhard	Indianapolis, Ind	Feb. 18, 1868	74, 528
Mattress, Spring	J. G. F. Grote	Cincinnati, Ohio	Apr. 19, 1864	42, 368
Mattress, Spring	J. G. F. Grote	Cincinnati, Ohio	Apr. 19, 1864	42, 369
Mattress, Spring	W. B. Judson	Poughkeepsie, N. Y	Sept. 13, 1870	107, 383
Mattress, Spring	H. Jury	New York, N. Y	Aug. 14, 1860	29, 600
Mattress, Spring	S. P. Kittle	Brooklyn, N. Y	Nov. 8, 1864	44, 960
Mattress, Spring	J. V. McElwee	Philadelphia, Pa	Apr. 2, 1850	7, 248

Index of patents issued from the United States Patent Office from 1790 *to* 1873, *inclusive*—Continued.

Invention.	Inventor.	Residence.	Date.	No.
Mattress, Spring	G. W. Mitchell	Saint Louis, Mo	Feb. 28, 1865	46, 573
Mattress, Spring	P. O'Neil	Philadelphia, Pa	Sept. 4, 1849	6, 686
Mattress, Spring	W. F. Ressegine	Cincinnati, Ohio	June 11, 1850	7, 429
Mattress, Spring	J. W. H. Scott	Rochester, N. Y	Apr. 14, 1868	76, 663
Mattress, Spring	D. N. Sellig	Newburgh, N. Y	Dec. 16, 1873	145, 597
Mattress, Spring	R. Stillwell	New York, N. Y	Sept. 20, 1864	44, 375
Mattress, Spring	C. Van Dyeck	Nashville, Tenn	Aug. 14, 1866	57, 223
Mattress, Spring	H. H. Vere	New York, N. Y	Aug. 13, 1867	67, 822
Mattress, Spring	J. Waters	Southwark, Pa	Jan. 20, 1852	8, 669
Mattress, Spring	L. Whitehead and S. P. Kittle	New York, N. Y	Aug. 28, 1860	29, 846
Mattress-spring	E. L. Bushnell	Poughkeepsie, N. Y	Oct. 19, 1869	95, 984
Mattress-spring	W. B. Judson	Poughkeepsie, N. Y	Sept. 19, 1871	119, 153
Mattress-stuffer	T. A. Watson	Brenham, Tex	Apr. 2, 1872	125, 233
Mattress-stuffing	F. A. Lane	Swanzey, N. H	May 25, 1869	90, 368
Mattress-stuffing machine	E. L. Wright	Sterling, Ill	Sept. 12, 1871	118, 892
Mattress-stuffing, Maching for cutting wood into shreds and crimping them for.	E. K. Browning	Utica, N. Y	July 15, 1851	8, 217
Mattress-stuffing, &c., Process of preparing moss for.	A. Rock	New Orleans, La	Jan. 7, 1873	134, 562
Mattress, Ventilating	T. Tolman	West Townsend, Mass	Dec. 15, 1857	18, 886
Mattress, Ventilating spring	R. W. Geraghty	Newark, N. J	Mar. 27, 1860	27, 667
Mattress, Wire	E. Parker	New Britain, Conn	May 20, 1873	139, 077
Mattress, Wire	D. J. Powers	Chicago, Ill	Nov. 11, 1873	144, 564
Mattress, Wire	J. G. Smith	Saint Louis, Mo	Dec. 2, 1873	145, 249
Mattress, Wire spring	F. W. Hoffmann	Morrisania, N. Y	Dec. 3, 1872	133, 533
Mattress, Wire spring	F. R. Wegman	Saxony	June 16, 1868	79, 040
Mattress, Woven-wire	G. C. Perkins	Hartford, Conn	Nov. 22, 1870	109, 446
Mattress, Woven-wire	J. W. C. Peters	Chicago, Ill	Oct. 15, 1872	132, 175
Mattresses and cusions, Elastic material for	T. B. Smith	Marietta, Ohio	Nov. 9, 1858	22, 037
Mattresses and making paper, Material for filling	F. C. Cone	San Francisco, Cal	Feb. 20, 1872	123, 810
Mattresses, Bracing and fastening spiral springs for.	O. Fuller	San Francisco, Cal	Oct. 25, 1864	44, 793
Mattresses, cushions, and other articles of upholstery, Material for filling.	W. H. Sowers	Boston, Mass	July 11, 1871	117, 018
Mattresses, cushions, &c., Making	C. L. Fleischmann	Washington, D. C	Sept. 4, 1847	5, 275
Mattresses, cushions, &c., Material for	B. Frodsham	New York, N. Y	Nov. 26, 1861	33, 777
Mattresses, Frame for wire	G. C. Perkins	Hartford, Conn	July 2, 1872	128, 652
Mattresses, Machine for cutting shavings for	F. Skinner	New Haven, Conn	Jan. 10, 1860	26, 791
Mattresses, Machine for hackling shucks for	D. A. Cole	Nashville, Tenn	Feb. 9, 1869	86, 734
Mattresses, Machine for preparing husks for	W. Nash and M. Kenfield	Malden, Mich	Oct. 27, 1868	83, 402
Mattresses, Machinery for preparing husks for	A. Olcott	Newark, N. J	Nov. 20, 1847	5, 373
Mattresses, &c., Material for filling	W. J. Woodley	San Francisco, Cal	Aug. 6, 1872	130, 171
Mattresses, Material for filling	B. Harrington	China, Me	Jan. 23, 1872	123, 016
Mattresses, &c., Material for stuffing	H. R. Hildreth and W. H. Smith.	Dutch Flat, Cal	Jan. 15, 1867	61, 194
Mattresses, &c., Material for stuffing	R. J. Kellett	San Francisco, Cal	Aug. 6, 1872	130, 141
Mattresses, pillows, &c., Stuffing for	A. C. Crondal	New York, N. Y	Sept. 22, 1863	40, 024
Mattresses, Process for treating moss for	S. Barker	New York, N. Y	Apr. 7, 1857	16, 961
Mattresses, seats, backs of sofas, &c., from paper pulp, sponge, &c., Manufacture of.	J. L. Kendall	Foxborough, Mass	Aug. 6, 1872	130, 223
Mattresses, seats, &c., Flexible support for	C. Pullinger	Philadelphia, Pa	Apr. 8, 1873	137, 718
Mattresses, seats, &c., Spring-coupling for	W. B. Judson	New York, N. Y	Jan. 14, 1873	134, 893
Mattresses, sofas, &c., from paper pulp, sponge, &c., Manufacture of.	J. L. Kendall	Foxborough, Mass	Aug. 6, 1872	130, 222
Mattresses, sofas, seats, &c., Stuffing for	J. M. Gilbert	Troy, N. Y	Feb. 11, 1868	74, 340
Mattresses, Stuffing for	H. A. Alden	Matteawan, N. Y	Feb. 16, 1864	41, 589
Mattresses, Tightening-device for woven-wire	D. J. Powers	Chicago, Ill	Sept. 16, 1873	142, 940
Mattresses, &c., Treating cork for	L. Bauhoefer	Philadelphia, Pa	Nov. 6, 1866	59, 342
Mattresses, &c., Treating moss for	C. G. Augeroth, jr	Philadelphia, Pa	Aug. 30, 1864	43, 962
Mattresses, &c., Treating sponge for stuffing	R. O. Doremus	New York, N. Y	Nov. 13, 1866	59, 714
Maxiliary compress	C. E. Davis	Saint Helena, Cal	Sept. 6, 1870	107, 011
Maze-lock	T. Nicholson	Falmouth, Va	Sept. 30, 1851	8, 396
Mead, Carbonated	J. Heaton	New York	July 28, 1819	....
Mead, Composition of	T. T. Kimball and A. H. White	Dedham, Mass	Sept. 26, 1835	....
Mead, Making carbonated sarsaparilla	J. S. Schieffelin	New York	July 19, 1824	....
Mead, Portable sarsaparilla	J. C. Brigham	Methuen, Mass	July 25, 1833	....
Meal and flour dryer	H. Cutter	Ashland, Mass	Sept. 7, 1869	94, 574
Meal and flour making machine	G. W. Holman	Beloit, Wis	Feb. 1, 1859	22, 806
Meal and flour mill	J. C. Smith	Lacon, Ill	Sept. 29, 1837	396
Meal, Cooling and drying	J. Denchfield	Oswego, N. Y	Apr. 20, 1858	19, 984
Meal-separator	W. Hawkins	Brooklyn, N. Y	June 8, 1869	91, 128
Measure, Adjustable	H. Kraut	Saint Louis, Mo	Oct. 16, 1866	58, 838
Measure and funnel combined	H. L. Dickenson	East Berlin, Conn	May 14, 1867	64, 752
Measure and funnel, Combined	S. R. Dummer	New York, N. Y	Apr. 5, 1864	42, 174
Measure and funnel, Combined	J. Fanyon	Providence, R. I	May 4, 1869	89, 569
Measure and funnel, Combined	E. Grattem	Williamstown, Mich	Nov. 13, 1866	59, 590
Measure and funnel, Liquid	J. Huff	Ironton, Mo	Aug. 23, 1870	106, 588
Measure and weigher, Combined	A. B. Hurd	Watkins, N. Y	Oct. 6, 1868	82, 842
Measure, Beer	E. Bagot	New York, N. Y	Jan. 10, 1860	26, 807
Measure, Beer	C. Chinnock	Brooklyn, N. Y	Apr. 15, 1862	34, 943
Measure, Board	A. D. Hoffman	Belleville, Mich	Feb. 24, 1863	37, 755
Measure, Board	G. S. Tiffany	Palmyra, Mich	July 25, 1865	49, 011
Measure, Cloth	C. V. Hemenway	New London, Ohio	Feb. 11, 1873	135, 804
Measure, Dry	J. H. Teahl	Eberly's Mills, Pa	Jan. 5, 1869	85, 706
Measure, Foot	J. C. F. Bremser	Saint Louis, Mo	Aug. 16, 1870	106, 314
Measure for cutting dresses	S. R. Windle	Chillicothe, Ohio	Mar. 31, 1868	76, 128
Measure for cutting garments	S. Severson	Baltimore, Md	Feb. 15, 1826	....
Measure for gaging the liquid contents of vessels, Tape.	H. E. Wrigley	Titusville, Pa	July 16, 1872	129, 639
Measure for laying out garments, Tailor's	F. A. Fairchild	Columbus, Ga	Oct. 31, 1835	....
Measure for stock, Feed	C. E. Spaulding	Theresa, N. Y	Sept. 28, 1869	95, 280
Measure for the human body	G. Beard	Salineville, Ohio	July 11, 1865	48, 644
Measure, Funnel	J. J. Hillman	Boston, Mass	July 28, 1863	39, 346
Measure, funnel, and faucet, Combined	G. W. McCann	Springfield, Ohio	July 2, 1867	66, 371
Measure, funnel, and faucet, Combined	H. Mitchell	Richmond, Ind	Mar. 7, 1865	46, 690
Measure, Grain	J. Brown	Lawn Ridge, Ill	Dec. 28, 1858	22, 411
Measure, Grain	N. M. Lurton	Newbern, Ill	Sept. 3, 1867	68, 761
Measure, Grain	E. O. Melvin	Brooklyn, Wis	Jan. 7, 1868	73, 112
Measure, Grain	J. A. Rosbeck	Hermon, N. Y	Aug. 3, 1869	93, 231

Index of patents issued from the United States Patent Office from 1790 *to* 1873, *inclusive*—Continued.

Invention.	Inventor.	Residence.	Date.	No.
Measure, Grain	H. C. Smith	Chicago, Ill	June 25, 1867	66, 050
Measure, Hat	J. Wehle	New York, N. Y	Aug. 16, 1859	25, 154
Measure, Hatter's head	H. Bonham	Philadelphia, Pa	Aug. 2, 1870	105, 895
Measure, Hatter's head	F. A. Fenton	Camden, N. J	Nov. 18, 1873	144, 606
Measure, Heel	J. T. Siegert	Washington, D. C	Aug. 6, 1867	67, 598
Measure-holder	G. W. Burwell	Zanesville, Ohio	July 20, 1869	92, 791
Measure, Liquid	J. L. Abbott	North Providence, R. I	Sept. 17, 1867	68, 931
Measure, Liquid	P. Brinkerhoff	Chillicothe, Mo	June 4, 1867	65, 335
Measure, Liquid	P. Brinkerhoff	Chillicothe, Mo	Aug. 20, 1867	67, 948
Measure, Liquid	J. Dailey	Albany, N. Y	Aug. 14, 1866	57, 097
Measure, Liquid	T. W. Ellis	Macon, Ga	June 14, 1870	104, 129
Measure, Liquid	A. Hicks	Flushing, N. Y	Feb. 16, 1864	41, 620
Measure, Liquid	S. Mainster and J. F. Kirkwood	Thistle, Md	Mar. 12, 1867	62, 762
Measure, Liquid	M. McDevitt	Hampton, Va	Mar. 1, 1870	100, 307
Measure, Liquid	T. Rogers	Montgomery Square, Pa	Feb. 25, 1862	34, 526
Measure, Liquid	W. Sprague	Sandy Creek, N. Y	July 28, 1868	80, 370
Measure, Liquid	J. S. Tough	Baltimore, Md	July 23, 1841	2, 187
Measure, Liquid	J. Trent	Millerton, N. Y	Jan. 1, 1867	60, 806
Measure, Liquid	W. L. Wever and A. W. Johnson	Plattsburgh, N. Y	Dec. 9, 1873	145, 377
Measure, Liquid	W. M. Wright	Chambersburgh, Pa	Mar. 28, 1871	113, 237
Measure, Lumber	F. S. Baldwin	Saint Louis, Mo	Apr. 29, 1873	138, 310
Measure, Lumber	A. M. Olds	Chicago, Ill	Aug. 9, 1864	43, 789
Measure, Lumber	A. M. Olds	Chicago, Ill	Jan. 3, 1865	45, 741
Measure, Lumber	W. S. Poulson	Cadiz, Ohio	Feb. 20, 1872	123, 931
Measure, Lumber	W. E. Walson	Chester, Pa	Nov. 7, 1871	120, 685
Measure, Pantaloon	E. Grimston	Danvers, Mass	Sept. 21, 1837	388
Measure, Rotary disk	E. V. Lawrence	Brooklyn, N. Y	Nov. 25, 1862	37, 002
Measure, Self-counting	E. Dexter	Holmes' Hole, Mass	May 13, 1856	14, 857
Measure, Self-registering lumber	D. E. Pease and C. Richards	Richland Centre, Wis	Mar. 23, 1869	88, 203
Measure, Shoemaker's	M. Lane	Washington, D. C	Sept. 17, 1861	33, 307
Measure, Shoemaker's foot	C. Cross	Louisville, Ky	May 11, 1869	89, 919
Measure, Shoemaker's foot	J. McNichol	Pontiac, Ill	June 6, 1871	115, 761
Measure, Shoemaker's foot	W. Wilson	Boston, Mass	Aug. 22, 1871	118, 414
Measure, Spring tape	W. H. Bangs, jr	West Meriden, Conn	Dec. 6, 1864	45, 372
Measure, Surveyor's	W. H. Paine	Sheboygan, Wis	Aug. 7, 1860	29, 514
Measure, Tailor's	W. R. Acton	Richmond, Va	Sept. 5, 1846	4, 742
Measure, Tailor's	W. W. Allen	Bordentown, N. J	Sept. 10, 1850	7, 641
Measure, Tailor's	G. Beard	Salineville, Ohio	Mar. 6, 1866	52, 950
Measure, Tailor's	G. Beard	Philadelphia, Pa	Mar. 6, 1866	52, 951
Measure, Tailor's	J. Beaudry	Montreal, Canada	Oct. 14, 1873	143, 556
Measure, Tailor's	A. A. Bogardus	Newburgh, N. Y	Nov. 29, 1845	4, 294
Measure, Tailor's	J. Carpenter	Uniontown, Pa	Apr. 10, 1849	6, 286
Measure, Tailor's	J. P. Combs	Trenton, N. J	Nov. 9, 1844	3, 820
Measure, Tailor's	R. Damo	Hanover, N. H	May 8, 1840	1, 584
Measure, Tailor's	L. Derby	New York, N. Y	Jan. 27, 1857	16, 472
Measure, Tailor's	H. Donges	Newport, Pa	Apr. 25, 1846	4, 477
Measure, Tailor's	I. Du Bois	Brooklyn, N. Y	May 23, 1871	115, 180
Measure, Tailor's	G. Ecklee and S. X. Ball	Flint Creek, N. Y	Sept. 28, 1843	3, 286
Measure, Tailor's	G. W. T. Harley	Frederick, Md	Feb. 13, 1866	52, 566
Measure, Tailor's	G. W. T. Harley	Frederick, Md	Jan. 30, 1872	123, 170
Measure, Tailor's	H. Isham	Montpelier, Vt	May 30, 1844	3, 603
Measure, Tailor's	J. A. Johnston	Montreal, Canada	July 1, 1873	140, 507
Measure, Tailor's	C. Kile	Nashville, Ohio	June 27, 1846	4, 596
Measure, Tailor's	L. P. Lemont	Bath, Me	Jan. 9, 1829	....
Measure, Tailor's	W. and C. F. Lillibridge	Zanesville, Ohio	Oct. 31, 1854	11, 868
Measure, Tailor's	J. Maginnis	Lockport, N. Y	Dec. 2, 1851	8, 566
Measure, Tailor's	B. G. Martin	Richmond, Va	Oct. 29, 1846	4, 831
Measure, Tailor's	L. B. and E. Miller	Wallkill, N. Y	May 29, 1841	2, 106
Measure, Tailor's	C. Morey and D. Hummer	McArthurstown, Ohio	July 8, 1843	3, 161
Measure, Tailor's	I. Moses	New York, N. Y	May 24, 1870	103, 487
Measure, Tailor's	T. Oliver	New York, N. Y	Mar. 30, 1843	3, 024
Measure, Tailor's	M. T. Rowlands	Pittston Ferry, Pa	Apr. 18, 1854	10, 779
Measure, Tailor's	M. G. Simril	Chesterville, S. C	Jan. 7, 1847	4, 923
Measure, Tailor's	W. Sinnott and J. McNaughton	Brooklyn, N. Y	June 23, 1868	79, 083
Measure, Tailor's	W. R. Stace	Rochester, N. Y	July 6, 1858	20, 826
Measure, Tailor's	A. Stocker	Ogdensburgh, N. Y	May 28, 1850	7, 402
Measure, Tailor's	A. Stocker	Rome, N. Y	Sept. 30, 1856	15, 824
Measure, Tailor's	E. Virtue	Philadelphia, Pa	Dec. 16, 1851	8, 600
Measure, Tailor's	A. Ward	Camden, N. J	Dec. 26, 1845	4, 327
Measure, Tailor's	T. Watt	Hubbard Township, Ohio	Feb. 20, 1847	4, 975
Measure, Tailor's	W. T. Wells	Shelbyville, Tenn	Apr. 20, 1852	8, 895
Measure, Tailor's	J. W. Whitham	Washington, D. C	Nov. 6, 1849	6, 856
Measure, Tape	L. P. Bradley	New Haven, Conn	July 13, 1869	92, 573
Measure, Tape	J. A. Evarts	West Meriden, Conn	Feb. 13, 1872	123, 624
Measure, Tape	A. J. Fellows	New Haven, Conn	July 14, 1868	79, 965
Measure, Tape	H. B. Tyler	West Meriden, Conn	July 8, 1873	140, 745
Measure, Variable	L. Coates	Collamer, Pa	Oct. 30, 1866	59, 186
Measure, Yard	J. Douglass	McConnellstown, Pa	Dec. 3, 1867	71, 719
Measures, Apparatus for tapering	A. Hatch	New Haven, Conn	Feb. 19, 1867	62, 267
Measures, Construction of funnel-attachments for liquid.	C. C. Judwin	Honesdale, Pa	Mar. 12, 1872	124, 492
Measures, Forming sheet-metal	J. Coover	Chambersburgh, Pa	Oct. 22, 1872	132, 386
Measures, Machine for forming shell for dry	J. Gage	Henniker, N. H	Sept. 7, 1869	94, 591
Measures, Machine for graduating linear	S. C. Hubbard	Middletown, Conn	June 16, 1857	17, 606
Measures, Manufacture of graduated glass	W. Hodgson, jr	Philadelphia, Pa	Feb. 18, 1862	34, 424
Measurements and laying out garments, Pattern for applying	S. C. Ewing	Indianapolis, Ind	Feb. 21, 1872	112, 024
Measurements, Apparatus for electric	L. Bradley	Jersey City, N. J	Jan. 7, 1873	134, 636
Measurements, Device for taking tailor's	J. Smith	Burton, Ohio	Aug. 6, 1872	130, 161
Measurer and register, Grain	R. B. Clark	Lyle, Minn	Mar. 11, 1873	136, 643
Measurer and register, Grain	J. A. Cleuxton	Bentonville, Ohio	Feb. 26, 1861	31, 530
Measurer, Board	A. Lafever	Battle Creek, Mich	Sept. 13, 1859	25, 423
Measurer, Fluid	J. L. Perry and M. Burt	Mansfield and Morton, Mass.	Apr 19, 1859	23, 701
Measurer, Grain	J. W. Smith and H. Griesa	North Cohocton, N. Y	Sept. 7, 1869	94, 659
Measurer, Self-regulating grain	G. W. Atkins	Milton, Del	May 11, 1858	20, 186
Measurer, Ship's model	A. S. Hosley	New York, N. Y	Aug. 19, 1851	8, 307
Measurer, Standard coat	C. Barber	Boston, Mass	Dec. 26, 1837	539

Index of patents issued from the United States Patent Office from 1790 *to* 1873, *inclusive*—Continued.

Invention.	Inventor.	Residence.	Date.	No.
Measurer, Tailor's	J. S. Rockafellow	Flemington, N. J	Sept. 18, 1835	....
Measuring altitudes, &c., Instrument for	G. C. Ayling	Boston, Mass	Dec. 21, 1858	22, 396
Measuring and cutting garments	J. Henderson and C. K. Watson	Zanesville, Ohio	Oct. 22, 1831	....
Measuring and cutting garments	W. and C. Kahler	Bloomsburgh, Pa	Jan. 20, 1838	574
Measuring and cutting garments	D. L. Pendill	Gilboa, N. Y	June 14, 1843	3, 130
Measuring and cutting garments	J. Pudney	Waterford, N. Y	Sept. 3, 1831	....
Measuring and cutting garments	H. Seger	Macon, Ga	Apr. 29, 1842	2, 590
Measuring and cutting garments	J. G. Wilson	New York	May 3, 1833	....
Measuring and cutting out dress-waists, Pattern for.	J. M. Lent	Schuyler's Lake, N. Y	June 22, 1869	91, 642
Measuring and cutting out dresses, System of	J. Malnight	Grass Lake, Mich	Nov. 19, 1867	71, 192
Measuring and drafting garments	I. J. Hendryx	Troy, N. Y	Apr. 18, 1840	1, 557
Measuring and folding cloth	L. Hillman	Newton, N. J	Aug. 10, 1869	93, 619
Measuring and folding machine, Cloth	L. Harvey	Newcastle, Ind	July 16, 1872	129, 554
Measuring and laying out garments	J. Lenley, jr	Newtown, Va	Nov. 26, 1867	71, 520
Measuring and laying out garments	H. Wengel	Philadelphia, Pa	May 5, 1868	77, 704
Measuring and laying out garments, Device for	I. J. Ordway	West Edmeston, N. Y	Jan. 4, 1870	98, 618
Measuring and laying out garments, Pattern for	F. Wetmore	Chicago, Ill	Dec. 13, 1870	110, 097
Measuring and marking, Combination-tool for	C. M. Lane	Cincinnati, Ohio	June 16, 1868	78, 974
Measuring and marking out garments	A. Ward	Philadelphia, Pa	Jan. 7, 1835	....
Measuring and recording by the tape	E. A. Preston	Battle Creek, Mich	Dec. 7, 1858	22, 241
Measuring and registering device, Liquid	A. Werrkmeister	Charlottenburg, Prussia	Mar. 1, 1870	100, 475
Measuring and registering grain	W. H. Pfrimmer	Lanesville, Ind	May 5, 1868	77, 521
Measuring and sacking grain	D. M. Cochvan	Richmond, Ind	Sept. 3, 1861	33, 186
Measuring and transferring liquid	T. W. Whitley	Paterson, N. J	Feb. 1, 1833	....
Measuring and weighing apparatus	N. Smith	Lansing, Iowa	May 12, 1863	38, 511
Measuring and weighing grain, Machine for	C. A. Postley	Philadelphia, Pa	Aug. 7, 1855	13, 397
Measuring and winding machine, Cloth	J. E. Race and H. Whitney	Chicago, Ill	June 15, 1869	91, 264
Measuring apparatus, Silk and thread	L. Dimock	Leeds, Mass	Apr. 25, 1871	114, 114
Measuring-apparatus, Tailor's	F. Mueller and H. Koeller	New York, N. Y	July 20, 1869	92, 873
Measuring apparatus, Thread	L. F. Dunn	Oneida, N. Y	Dec. 26, 1871	122, 236
Measuring-attachment for packaged fabrics	E. Morgan	Washington, D. C	Jan. 24, 1871	111, 235
Measuring boards, Apparatus for	I. J. W. Adams	Salisbury, Md	Apr. 19, 1870	101, 967
Measuring boards, Implement for	M. Tuells	Penn Yan, N. Y	May 10, 1870	102, 885
Measuring boards, Instrument for	J. Jones	Rochester, N. Y	Jan. 6, 1857	16, 359
Measuring boards, Machine for	W. S. Finley	Fond-du-Lac, Wis	Aug. 6, 1872	130, 120
Measuring-can	D. H. Sumner	South Boston, Mass	Sept. 17, 1867	68, 913
Measuring-can	A. W. Tice and N. F. Smalley	Boston, Mass	Nov. 18, 1873	144, 713
Measuring-can	A. G. Wilson	Chicago, Ill	Nov. 15, 1870	109, 283
Measuring-can, Automatic	T. D. Arkle and H. C. Greer	Bridgeport, Ohio	Sept. 3, 1867	68, 480
Measuring-can for liquids	W. Barry	Carthage, N. Y	June 8, 1869	90, 916
Measuring-can for liquids	C. M. Bridge	Leon, Iowa	Sept. 19, 1871	119, 114
Measuring-can for liquids	J. S. Gold	Springfield, Ill	May 11, 1869	90, 044
Measuring-can for oils, &c	W. B. Sherman	Clinton, N. Y	Feb. 18, 1873	136, 103
Measuring cloth, &c	A. Smith	Perrysburgh, Ohio	June 19, 1847	5, 167
Measuring cloth, Apparatus for	T. Bisbing	Buckstown, Pa	Feb. 23, 1869	87, 134
Measuring cloth, Apparatus for	T. M. Brintnall	Medina, Ohio	Apr. 19, 1870	101, 979
Measuring cloth, Apparatus for	R. H. Kent	Middlebury, Ohio	Nov. 24, 1868	84, 281
Measuring cloth, Apparatus for	S. B. Luckett	Corydon, Ind	Nov. 24, 1868	84, 290
Measuring cloth, Apparatus for	S. B. Luckett	Corydon, Ind	Mar. 1, 1870	100, 425
Measuring cloth, Apparatus for	G. R. McIntire	Houghton, Mich	Apr. 28, 1868	77, 305
Measuring cloth, Apparatus for	J. E. Race and A. Smith	Chicago, Ill	Dec. 15, 1868	84, 903
Measuring cloth, Apparatus for	C. L. Shotnell	Allamuchy, N. J	May 26, 1868	78, 399
Measuring cloth, Device for	O. R. Holbrook	Staffordville, Conn	Sept. 16, 1873	142, 787
Measuring cloth in the piece or roll, Device for	W. Beaton	Grinnell, Iowa	Nov. 22, 1864	45, 131
Measuring cloth, Machine for	J. I. Benham	New London, Conn	Feb. 6, 1866	52, 378
Measuring cloth, Machine for	W. F. Boggs	Petersburgh, Ill	Mar. 24, 1868	75, 848
Measuring cloth, Machine for	J. M. Brintnall	Medina, Ohio	Sept. 26, 1871	119, 303
Measuring cloth, Machine for	J. W. Drummond	New York, N. Y	May 31, 1859	24, 260
Measuring cloth, Machine for	W. Hebdon	New York, N. Y	May 18, 1869	90, 262
Measuring cloth, Machine for	D. Max	Newton, Ill	Sept. 8, 1868	81, 921
Measuring cloth, Machine for	C. M. Swany	Philadelphia, Pa	Dec. 15, 1863	40, 962
Measuring cloth, Machine for	T. Weeden and T. Tribe	Hillsdale, Mich	Sept. 6, 1870	107, 205
Measuring cloth, Machine for	W. W. Wythes	Saint Clair, Pa	Sept. 29, 1857	18, 313
Measuring cloth on looms, Mechanism for	H. Halvorson	Boston, Mass	Aug. 1, 1854	11, 435
Measuring cloth in the cloth-beam, Method of	W. H. Woodworth	Salmon Falls, N. H	Dec. 21, 1852	9, 496
Measuring coats	T. E. Tilden	Baltimore, M. D	Dec. 5, 1840	1, 880
Measuring-device for horses' feet	H. B. Ferren	Batavia, N. Y	Sept. 15, 1868	82, 213
Measuring-device for horses' feet	H. B. Ferren	Batavia, N. Y	Sept. 15, 1868	82, 214
Measuring device, Skirt	L. G. Rice	Montague, Mass	Dec. 1, 1868	84, 579
Measuring distances	R. Porter	Billerica, Mass	June 11, 1836	....
Measuring distances from a single station, Instrument for.	E. A. Crandall	Friendship, N. Y	Aug. 26, 1856	15, 602
Measuring distances, Instrument for	G. Achelis and H. Poppenhusen	Westchester, Pa., and New York, N. Y.	Mar. 24, 1868	75, 824
Measuring distances, Instrument for	W. H. Chaffee	Flint, Mich	Apr. 8, 1862	34, 879
Measuring distances, Instrument for	J. J. Daly	New Orleans, La	Mar. 27, 1866	53, 420
Measuring distances, Instrument for	J. Frankle	Amesbury, Mass	Sept. 25, 1866	58, 241
Measuring distances, Parallactic instrument for	H. L. Hervey	Quincy, Ill	June 3, 1856	15, 040
Measuring distances, Parallactic instrument for	W. Würdemann	Washington, D. C	Sept. 11, 1849	6, 709
Measuring, drafting, and cutting garments	W. W. Allen	Philadelphia, Pa	Oct. 23, 1837	435
Measuring dry goods, Instrument for	T. A. Shinn	Baden, Pa	Aug. 27, 1867	68, 245
Measuring fluids, Apparatus for	J. C. Horton and S. K. Hawkins	New York and Lansingburgh, N. Y.	Apr. 2, 1867	63, 388
Measuring fluids, Apparatus for	E. G. Shortt	Carthage, N. Y	June 29, 1869	92, 110
Measuring fluids, Machine for	J. Bogardus	New York, N. Y	Oct. 12, 1837	425
Measuring fluids while drawing, Method of	S. Krauser	Reading, Pa	Oct. 7, 1856	15, 853
Measuring for garments	G. W. M. Bacon	New York, N. Y	Apr. 20, 1833	....
Measuring for garments	J. Mendenhall	Westchester, Pa	Apr. 16, 1833	....
Measuring for garments	D. Williams	New York, N. Y	May 22, 1833	....
Measuring garments	W. C. Bishop	Ovid, N. Y	Apr. 25, 1837	179
Measuring garments	J. B. West	New York, N. Y	Jan. 3, 1865	45, 780
Measuring garments, Instrument for	H. C. Brundage	Middletown, N. Y	Feb. 7, 1842	2, 450
Measuring garments, Instrument for	J. and J. F. Knowland	Brownsborough, Ky	July 20, 1842	2, 730
Measuring garments, Instrument for	P. F. L. Veret	Warrenton, Ga	May 26, 1842	2, 640
Measuring garments, Instrument for	D. Williams	New York, N. Y	Apr. 26, 1839	1, 136
Measuring grain	D. Murray	Fairfield, N. C	Aug. 23, 1859	25, 212
Measuring grain and seed	P. H. Schenck	New York, N. Y	Feb. 15, 1810	....

Index of patents issued from the United States Patent Office from 1790 *to* 1873, *inclusive*—Continued.

Invention.	Inventor.	Residence.	Date.	No.
Measuring grain, Machine for	C. Ross	Hartland, Mich	Sept. 15, 1863	39, 959
Measuring grain, Machine for	T. Wilson	Matamora, Ill	Sept. 15, 1863	39, 980
Measuring grain, Tally-box for	A. J. Burke	Grandy Centre, Iowa	Oct. 9, 1866	58, 593
Measuring heights and distances	B. Wilmer	Baltimore, Md	Aug. 20, 1819	....
Measuring heights and distances	C. Wilson	Clinton, Pa	Sept. 29, 1868	82, 669
Measuring-instrument, Tailor's	J. B. Barnett and F. Story	Catskill, N. Y	Nov. 12, 1839	1, 406
Measuring-instrument, Tailor's	J. M. Krider	Newtown Stephensburgh, Va.	Oct. 31, 1854	11, 866
Measuring-instrument, Tailor's	J. M. Krider	Madison Court-House, Va	Aug. 13, 1867	67, 774
Measuring-instrument, Tailor's	W. J. Lemmond	Lancaster, S. C	Apr. 18, 1840	1, 556
Measuring, laying out, and cutting garments, Device for.	G. P. Sweezy	Riverhead, N. Y	Apr. 20, 1869	89, 091
Measuring liquids	G. W. Devoe	New York, N. Y	Dec. 11, 1866	60, 344
Measuring liquids, Apparatus for	F. H. Houghton	Brooklyn, N. Y	May 8, 1866	54, 548
Measuring liquids, Apparatus for	H. James	Barclay, Ill	June 5, 1860	28, 581
Measuring liquids, Apparatus for	P. Noyes	Lowell, Mass	Nov. 2, 1869	96, 340
Measuring liquids, Apparatus for	J. C. Rankin	Mount Vernon, N. Y	June 12, 1860	28, 690
Measuring liquids, Apparatus for	J. Schmidt	Vienna, Austria	Dec. 26, 1871	122, 195
Measuring liquids, Apparatus for	E. F. Wilder	Lowell, Mass	Mar. 22, 1870	101, 069
Measuring liquids, Apparatus for	M. H. Wiley and T. Miller	Boston, Mass	May 24, 1870	103, 404
Measuring liquids, Device for	A. Fickett and J. C. Ware	Titusville, Pa	June 18, 1867	65, 897
Measuring liquids, Device for	R. F. Fisher and S. L. Bell	Boston and Wellfleet, Mass	June 4, 1872	127, 410
Measuring lumber, Instrument for	C. Fleming	Ypsilanti, Mich	Feb. 5, 1861	31, 308
Measuring lumber, Instrument for	C. Littlefield	Kennebunk, Me	July 21, 1868	80, 077
Measuring lumber, Instrument for	T. B. Stevenson	Dayton, Ohio	Sept. 17, 1867	68, 911
Measuring packages, Rod for	D. Stanbury	Laurel Hill, N. J	Oct. 26, 1818	....
Measuring register, Cloth	S. Crocker	Port Allen, Iowa	Apr. 15, 1873	137, 829
Measuring register, Land	W. Frasius	Chatsworth, Ill	May 23, 1871	115, 186
Measuring register, Liquid	M. Springwater	Evansville, Ind	Mar. 25, 1873	137, 155
Measuring-rod	A. D. King	Granville Corners, Mass	Mar. 12, 1867	62, 756
Measuring-rod and dividers, Combined extension	G. H. Disher	Mobile, Ala	Nov. 4, 1873	144, 264
Measuring sails, Instrument for	J. Dominis	Sandwich Islands	Sept. 30, 1842	2, 790
Measuring-scale	J. J. Hendryx	Bennington, Vt	May 29, 1822	....
Measuring scale, Log	L. Smith	Lowell, Mass	June 14, 1870	104, 368
Measuring-tape	O. W. Minard	Waterbury, Conn	May 1, 1860	28, 101
Measuring the body of garments, Pattern for	M. Palmer, jr., and E. W. Anderson.	Boston, Mass., and Washington, D. C.	July 19, 1870	105, 486
Measuring the foot and leg and cutting leather for boots.	T. Howe	Worcester, Mass	Apr. 18, 1820	....
Measuring the length of braces in carpentry, Instrument for.	H. Whipple	South Shaftsbury, Vt	Feb. 26, 1856	14, 329
Measuring the person and cutting out ladies' dresses, Guide for.	L. L. Jackson	Richmond, Ind	May 25, 1869	90, 363
Measuring the superfices of boards, Machine for	S. C. Kennard	South Newmarket, N. H	Jan 1, 1858	19, 031
Measuring-wheel	T. R. Way	Springfield, Ohio	Jan. 21, 1873	135, 185
Measuring-wheel for wagon-tires	T. Whiting	Niles, Mich	May 13, 1873	138, 829
Measuring-wheel, Revolving	L. Young	New York, N. Y	Nov. 20, 1855	13, 833
Meat and brine, Device for keeping	J. Burgum	Concord, N. H	June 12, 1866	55, 463
Meat, and for other purposes, Portable apparatus for curing.	H. J. Noyes and S. C. Talcott	Ashtabula, Ohio	Jan. 21, 1870	104, 632
Meat and other articles, Machine for cutting	P. Vande Sande	Rochester, N. Y	Nov. 3, 1868	83, 676
Meat and other biscuit, Manufacture of	J. Carr and C. Lucop	Clapham and Drummond Road, England.	Apr. 21, 1868	76, 889
Meat and vegetable chopper	L. S. Chichester	New York, N. Y	Aug. 28, 1860	29, 766
Meat and vegetable chopper	S. F. Emerson	Seville, Ohio	Feb. 1, 1870	99, 301
Meat and vegetable chopper	W. H. Peirce	Bangor, Me	May 9, 1871	114, 594
Meat and vegetable chopper	E. Smith	Claremont, N. H	Apr. 2, 1867	63, 566
Meat and vegetable chopping machine	H. W. Russell	Stoughton, Mass	Feb. 6, 1866	52, 449
Meat and vegetable cutter	P. Bonfils and J. Grossman	West Hoboken, N. J	Sept. 20, 1870	107, 592
Meat and vegetable cutter	G. E. Bringman	Philadelphia, Pa	Dec. 13, 1870	110, 111
Meat and vegetable cutter	E. Covert	Farmers' Village, N. Y	June 6, 1871	115, 579
Meat and vegetable cutter	M. L. Edwards	Salem, Ohio	June 3, 1873	139, 501
Meat and vegetable cutter	J. R. Hall and C. H. Ellis	Salem, Ohio	July 15, 1873	140, 824
Meat and vegetable cutter	J. Shepard	Bristol, Conn	Apr. 24, 1866	54, 223
Meat and vegetable cutter	A. R. Silver	Salem, Ohio	July 11, 1871	117, 003
Meat and vegetable cutter	J. Stepp	Somerville, Mass	Aug. 13, 1872	130, 544
Meat and vegetable cutter	E. P. Whitney	Stamford, Conn	July 10, 1866	56, 305
Meat and vegetable cutter and grater	E. Culver	Shelburne, Mass	Aug. 16, 1864	43, 837
Meat and vegetable cutting machine	S. Blim	Salem, Ohio	Dec. 31, 1872	134, 458
Meat and vegetable masher	G. A. Anderson and C. J. Baker	Albany, N. Y	Oct. 11, 1870	108, 083
Meat, Apparatus for slaughtering and curing	G. W. Fulton	Fulton, Tex	June 29, 1869	92, 035
Meat-baker	W. K. Wyckoff	Ripon, Wis	Sept. 17, 1872	131, 383
Meat-biscuit	A. Kirkwood	Toronto, Canada	Mar. 28, 1871	113, 063
Meat-block	L. H. Ives	Syracuse, N. Y	May 18, 1869	90, 104
Meat-broiler	G. B. Ransom	Chester, Conn	Jan. 20, 1863	37, 460
Meat, Building for packing	D. E. Somes	Biddeford, Me	Nov. 27, 1860	30, 770
Meat by steam, Apparatus for cooking	G. H. Munroe	New York, N. Y	Apr. 26, 1870	102, 302
Meat chest or bin	W. B. Smith	Oskaloosa, Iowa	Oct. 18, 1870	108, 403
Meat-chopper	F. Ashley	New York, N. Y	May 1, 1866	54, 274
Meat-chopper	D. W. Baker	Harwich, Mass	May 15, 1866	54, 667
Meat-chopper	J. Battey	Manchester, N. J	Jan. 28, 1873	135, 258
Meat-chopper	C. B. Becker	Lancaster County, Pa	May 8, 1860	28, 145
Meat-chopper	J. H. Blake	Westbrook, Me	Sept. 28, 1869	95, 309
Meat-chopper	C. N. Brumm	Minersville, Pa	Mar. 23, 1869	88, 126
Meat-chopper	P. H. Chesley	Lynn, Mass	Mar. 23, 1858	19, 681
Meat-chopper	P. Clareton	New York, N. Y	June 8, 1869	90, 994
Meat-chopper	A. J. Curtis	Monroe, Me	Aug. 11, 1868	80, 811
Meat-chopper	J. H. De Poe	Boonton, N. J	July 5, 1870	104, 940
Meat-chopper	A. F. Doebert	Lancaster, N. Y	Aug. 20, 1867	67, 854
Meat-chopper	L. A. Dole	Salem, Ohio	May 18, 1858	20, 258
Meat-chopper	J. A. Eberly	Reamstown, Pa	Sept. 20, 1870	107, 466
Meat-chopper	M. L. Edwards	Salem, Ohio	Aug. 15, 1871	117, 999
Meat-chopper	C. H. Finson	Bangor, Me	June 1, 1869	90, 737
Meat-chopper	C. A. Foster	Winchendon, Mass	June 5, 1866	55, 414
Meat-chopper	O. Gardner	New York Mills, N. Y	July 8, 1873	140, 696
Meat-chopper	C. L. Gilpatric	Boston, Mass	June 30, 1868	79, 465
Meat-chopper	J. L. Good	Elizabethtown, Pa	May 16, 1871	114, 799
Meat-chopper	J. A. Hard	Lawrence, Kans	Mar. 25, 1873	137, 074

Index of patents issued from the United States Patent Office from 1790 to 1873, inclusive—Continued.

Invention.	Inventor.	Residence.	Date.	No.
Meat-chopper	N. L. Hatch	Cape Elizabeth, Me	Mar. 31, 1868	76, 187
Meat-chopper	F. C. Heberhart	Madison, Ind	Sept. 23, 1862	36, 518
Meat-chopper	J. G. Hirzel	Wilmington, Del	Aug. 11, 1868	80, 955
Meat-chopper	J. H. Landis	Eden, Pa	Mar. 12, 1861	31, 669
Meat-chopper	J. Massey	New York, N. Y	July 25, 1865	48, 965
Meat-chopper	A. McCarter	Salem, Ohio	Dec. 14, 1869	97, 946
Meat-chopper	A. McCarter	Salem, Ohio	June 7, 1870	104, 049
Meat-chopper	G. H. Mellen	New York, N. Y	Nov. 1, 1870	108, 809
Meat-chopper	H. D. Musselman	Lancaster, Pa	Feb. 28, 1860	27, 334
Meat-chopper	H. Obrecht	Mahanoy City, Pa	July 23, 1867	66, 989
Meat-chopper	W. P. Patton	Harrisburgh, Pa	Dec. 19, 1871	122, 051
Meat-chopper	T. Payne	Grand Rapids, Mich	Sept. 8, 1868	81, 939
Meat-chopper	D. Peters	Fort Madison, Iowa	Feb. 13, 1872	123, 725
Meat-chopper	H. P. Rankin	Allegheny, Pa	July 15, 1873	140, 950
Meat-chopper	W. Reitz	Cleveland, Ohio	Aug. 20, 1872	130, 745
Meat-chopper	N. E. Russell	China, Me	June 15, 1869	91, 273
Meat-chopper	S. Rutschman	Philadelphia, Pa	Sept. 1, 1868	81, 824
Meat-chopper	F. G. Siemers	Winona, Minn	Feb. 9, 1869	86, 783
Meat-chopper	A. R. Silver	Salem, Ohio	Aug. 30, 1870	106, 879
Meat-chopper	A. R. Silver	Salem, Ohio	Aug. 12, 1873	141, 731
Meat-chopper	A. F. Spaulding and S. M. Scott	Winchester, Mass	July 11, 1865	48, 734
Meat-chopper	L. Steigert	Cincinnati, Ohio	Jan. 11, 1870	98, 812
Meat-chopper	W. H. H. Walker	Bangor, Me	July 16, 1867	66, 916
Meat-chopper	N. T. Worthley	Brunswick, Me	Oct. 24, 1871	120, 359
Meat-chopper and vegetable-slicer	J. M. Taft	Arcadia, Wis	Jan. 3, 1871	110, 801
Meat-chopper, Rotating	G. W. Dechant	Berrysburgh, Pa	Apr. 22, 1862	35, 014
Meat, Chopping	J. Masser and S. Smith	Megentown, Pa	June 2, 1836	......
Meat, &c., chopping machine	W. H. Allen	Lowell, Mass	Feb. 13, 1855	12, 373
Meat-chopping machine	H. Chamberlain	Calais, Me	Sept. 24, 1867	69, 073
Meat-chopping machine	J. L. Good	Elizabethtown, Pa	May 14, 1872	126, 632
Meat-chopping machine	G. W. Sargent and P. H. Chesley	Chelsea, Mass	June 27, 1865	48, 451
Meat-chopping machine	H. Smith	Philadelphia, Pa	July 19, 1811	......
Meat-chopping machine	A. F. Spaulding and S. M. Scott	Winchendon, Mass	Jan. 31, 1865	46, 153
Meat-chopping machine	G. Spiehlman	Strasburgh, Pa	Mar. 11, 1873	136, 624
Meat-cleaver	E. Pollard	Albany, N. Y	Feb. 1, 1859	22, 839
Meat, Composition for covering	J. J. Bate and F. S. Lowe	Brooklyn, N. Y., and Jersey City, N. J.	Sept. 29, 1857	18, 270
Meat-compound	B. M. Fowler	Hackensack, N. J	Nov. 28, 1865	51, 259
Meat, Cooling and packing	D. E. Somes	Washington, D. C	Jan. 29, 1867	61, 638
Meat-crusher	A. J. Crane	Waterbury, Vt	Feb. 11, 1873	135, 781
Meat-crusher	O. P. Dennison	Mulberry, Ohio	June 20, 1871	116, 166
Meat-crusher	R. V. Jones	Canton, Ohio	June 6, 1865	48, 072
Meat, &c., Curing	A. C. Allen	Houston, Tex	Sept. 12, 1846	4, 754
Meat, Curing	G. A. Scherpf	New York, N. Y	May 7, 1845	4, 035
Meat, Curing	J. C. Schooley	Cincinnati, Ohio	Mar. 13, 1855	12, 530
Meat, Curing	W. H. Silberhorn	New York, N. Y	June 22, 1869	91, 678
Meat, Curing	G. Starkweather	Hartford, Conn	Nov. 5, 1850	7, 766
Meat, Curing and packing	I. Atkinson	Hamilton, Canada	Jan. 21, 1873	134, 965
Meat, &c., Curing and preserving	A. S. Lyman	New York, N. Y	June 22, 1869	91, 552
Meat, &c., Curing and preserving	A. Rock	New Orleans, La	July 26, 1870	105, 851
Meat, Curing and preserving	M. B. Sherwood	Buffalo, N. Y	Dec. 29, 1868	85, 485
Meat-curing apparatus	W. G. Bell	Boston, Mass	Dec. 19, 1871	121, 925
Meat-curing apparatus	M. B. Sherwood, jr	Buffalo, N. Y	Nov. 16, 1869	96, 978
Meat, Curing ham, beef, and other	O. M. Martin	Ann Arbor, Mich	Dec. 15, 1868	84, 893
Meat-cutter	C. Adams	Pittsburgh, Pa	July 23, 1861	32, 852
Meat-cutter	J. S. Artley	Danville, Pa	Jan. 3, 1871	110, 721
Meat-cutter	D. Bearly	Newcastle, Ind	Nov. 14, 1865	50, 984
Meat-cutter	W. G. Bell	Boston, Mass	Jan. 3, 1865	45, 788
Meat-cutter	L. Bonnet	New York, N. Y	July 24, 1860	29, 239
Meat-cutter	A. Burdick	Saratoga County, N. Y	Aug. 21, 1849	6, 652
Meat-cutter	W. Burns	Rome, Ohio	May 11, 1852	8, 936
Meat-cutter	A. Clark	Doe Run Post-Office, Pa	May 9, 1846	4, 507
Meat-cutter	A. B. Davenport	Petersham, Mass	Mar. 9, 1858	19, 547
Meat-cutter	P. Demeure	Brooklyn, N. Y	Apr. 20, 1858	20, 019
Meat-cutter	J. Derr	Oley, Pa	June 5, 1866	55, 251
Meat-cutter	D. Deshon, 2d	New London, Conn	June 13, 1848	5, 629
Meat-cutter	B. Dewalt and C. E. Schrader	Reading, Pa	Feb. 22, 1859	23, 016
Meat-cutter	J. G. Divoll	Sonora, Cal	June 22, 1869	91, 526
Meat-cutter	D. S. Early	Hummelstown, Pa	June 11, 1867	65, 551
Meat-cutter	A. J. Eddy	New Britain, Conn	Apr. 21, 1868	77, 014
Meat-cutter	H. H. Everts	Mount Morris, N. Y	Oct. 12, 1843	3, 303
Meat-cutter	C. A. Foster	Fitchburgh, Mass	July 14, 1868	79, 902
Meat-cutter	C. L. Gilpatric	Lewiston, Me	Oct. 4, 1864	44, 527
Meat-cutter	J. C. Hasfele	New York, N. Y	Aug. 27, 1867	68, 068
Meat-cutter	S. Hartley	Bridgeport, Ohio	Sept. 13, 1870	107, 364
Meat-cutter	H. Havell	Newark, N. J	May 31, 1859	24, 210
Meat-cutter	J. L. Haven	Cincinnati, Ohio	Jan. 22, 1867	61, 424
Meat-cutter	W. B. Heintze	Williamsburgh, N. Y	Feb. 22, 1870	100, 035
Meat-cutter	J. Hetzel and S. H. Hager	Miamisburgh, Ohio	Feb. 9, 1869	86, 669
Meat-cutter	J. C. Howe	Worcester, Mass	Aug. 18, 1868	81, 088
Meat-cutter	J. K. Hoyer	Reading, Pa	Sept. 7, 1858	21, 421
Meat-cutter	R. V. Jones	Canton, Ohio	Sept. 5, 1865	49, 762
Meat-cutter	D. R. Kenyon	Raritan, N. J	Mar. 19, 1872	124, 748
Meat-cutter	A. Klein	New York, N. Y	July 27, 1869	93, 094
Meat-cutter	P. Miles	New Haven, Conn	May 22, 1860	28, 393
Meat-cutter	P. Miles	New Haven, Conn	Jan. 8, 1861	31, 097
Meat-cutter	W. M. Miller	Tulpehocken, Pa	June 25, 1867	66, 033
Meat-cutter	S. Millet	New York, N. Y	May 24, 1853	9, 748
Meat-cutter	J. Morris	Derby, Conn	Feb. 25, 1835	......
Meat-cutter	G. Mosman	Boston, Mass	Mar. 16, 1869	87, 961
Meat-cutter	D. H. Mundy and H. W. Hoffman	Camden, N. J	May 4, 1869	89, 787
Meat-cutter	H. B. Murphy	Allegheny City, Pa	Feb. 7, 1871	111, 670
Meat-cutter	J. Nacher	La Crosse, Wis	Feb. 11, 1868	74, 409
Meat-cutter	M. Newman	Oak Hill, N. Y	Sept. 14, 1858	21, 514
Meat-cutter	A. Nittinger, jr	Philadelphia, Pa	Aug. 11, 1863	39, 492
Meat-cutter	A. Nittinger, jr	Philadelphia, Pa	Sept. 21, 1869	95, 038
Meat-cutter	A. Nittinger, jr	Philadelphia, Pa	May 24, 1870	103, 360

Index of patents issued from the United States Patent Office from 1790 *to* 1873, *inclusive*—Continued.

Invention.	Inventor.	Residence.	Date.	No.
Meat-cutter	H. Obrecht	Mahanoy City, Pa	July 7, 1868	79, 678
Meat-cutter	J. G. Perry	Kingston, R. I	Mar. 15, 1859	23, 262
Meat-cutter	J. G. Perry	Kingston, R. I	Aug. 2, 1859	24, 953
Meat-cutter	J. G. Perry	South Kingston, R. I	July 31, 1860	29, 400
Meat-cutter	J. G. Perry	South Kingston, R. I	Sept. 18, 1860	30, 088
Meat-cutter	J. G. Perry	South Kingston, R. I	Dec. 2, 1862	37, 053
Meat-cutter	J. G. Perry	South Kingston, R. I	July 14, 1863	39, 243
Meat-cutter	J. G. Perry	South Kingston, R. I	July 5, 1864	43, 427
Meat-cutter	J. G. Perry	South Kingston, R. I	July 5, 1864	43, 428
Meat-cutter	J. G. Perry	South Kingston, R. I	Jan. 3, 1865	45, 744
Meat-cutter	J. G. Perry	Kingston, R. I	Oct. 20, 1868	83, 308
Meat-cutter	J. G. Perry	Kingston, R. I	Apr. 6, 1869	88, 666
Meat-cutter	J. G. Perry	Kingston, R. I	Apr. 6, 1869	88, 667
Meat-cutter	J. Potts	Yocumtown, Pa	June 15, 1852	9, 035
Meat-cutter	F. S. Rutschman	Philadelphia, Pa	May 25, 1869	90, 588
Meat-cutter	C. Schiller	Baltimore, Md	Jan. 5, 1869	85, 615
Meat-cutter	H. Seib	New York, N. Y	June 14, 1870	104, 363
Meat-cutter	I. Siegrist	Steinsburgh, Pa	Apr. 16, 1872	125, 699
Meat-cutter	F. G. Siemers	Winona, Minn	Aug. 25, 1868	81, 421
Meat-cutter	F. G. Siemers	Winona, Minn	June 13, 1871	115, 987
Meat-cutter	D. Slaughter	West Hempfield Township, Pa.	Aug. 4, 1868	80, 675
Meat-cutter	J. E. Smith	Buffalo, N. Y	Feb. 25, 1868	74, 947
Meat-cutter	J. E. Smith	Buffalo, N. Y	June 21, 1870	104, 506
Meat-cutter	L. S. Starrett	Newburyport, Mass	May 23, 1865	47, 875
Meat-cutter	S. L. Stockstill and H. H. Dille	Medway, Ohio	June 23, 1868	79, 085
Meat-cutter	E. D. Tippett	Georgetown, D. C	Dec. 28, 1846	4, 909
Meat-cutter	G. P. Trenlieb	Baltimore, Md	Dec. 28, 1869	98, 318
Meat-cutter	C. Welte	Frankford, Pa	Apr. 21, 1868	77, 143
Meat-cutter	J. M. Wilder	Peterborough, N. H	May 10, 1845	4, 042
Meat-cutter	O. D. Woodruff	Southington, Conn	Jan. 10, 1860	26, 806
Meat-cutter	O. D. Woodruff	Southington, Conn	Mar. 9, 1869	87, 611
Meat-cutter	J. Zeitler and J. Zintl	Dayton, Ohio	Aug. 13, 1861	33, 060
Meat-cutters, Casting cylinders for	S. R. Plumb	Southington, Conn	July 31, 1860	29, 440
Meat-cutting apparatus	J. G. Perry	South Kingston, R. I	Feb. 26, 1850	7, 123
Meat-cutting machine	W. Bliesner	Saint Louis, Mo	July 30, 1867	67, 254
Meat-cutting machine	G. V. Brecht	Saint Louis, Mo	May 20, 1856	14, 901
Meat-cutting machine	A. Kirn	Buffalo, N. Y	Nov. 2, 1869	96, 326
Meat-cutting machine	J. Morris	Derby, Conn	Nov. 20, 1837	459
Meat-cutting machine	J. Morris	Derby, Conn	Feb. 28, 1842	2, 476
Meat-cutting machine	W. E. Richardson	Chicago, Ill	Mar. 21, 1865	46, 942
Meat-cutting machine	T. Vanderslice	Valley Forge, Pa	May 6, 1851	8, 072
Meat-cutting machine	F. Wobfersberger	Salem Station, Ohio	Mar. 23, 1858	19, 728
Meat-holder	S. Beissel	Shamokin, Pa	Dec. 9, 1873	145, 329
Meat-hook	S. Weaver	Pottstown, Pa	Sept. 27, 1870	107, 839
Meat in market-places, Apparatus for preserving and holding butcher's.	A. Seltzer	Baltimore, Md	Sept. 5, 1840	1, 774
Meat-mangle	G. A. Cover	Macomb, Ill	June 18, 1867	65, 878
Meat-mangle	S. D. Ingham	Ripley, Ohio	Sept. 23, 1873	143, 158
Meat-mangler	A. T. Adams	Indianapolis, Ind	Nov. 12, 1867	70, 675
Meat-mangler	J. T. Harvey and W. Dixon	Marysville, Pa	Oct. 4, 1870	108, 021
Meat-mangler	G. H. Tift	Morrisville, Vt	Sept. 11, 1866	58, 000
Meat, Manufacture of fluid	S. Darby	London, England	Jan. 9, 1872	122, 574
Meat-masher	J. Lefeber	Cambridge City, Ind	Aug. 20, 1867	67, 991
Meat-masher	G. W. Putnam	Peterborough, township of Smithfield, N. Y.	Dec. 27, 1864	45, 633
Meat-masher	G. Storer	New Britain, Conn	July 19, 1859	24, 831
Meat-mill	J. Gardiner	Philadelphia, Pa	Sept. 8, 1868	81, 904
Meat-mincer	J. M. Dronyer	Carondelet, Mo	June 16, 1868	78, 942
Meat-mincer	A. W. Hale	New Britain, Conn	Mar. 15, 1859	23, 246
Meat-mincer	L. Holzwarth, H. Froelsch, and T. Gerhards.	Philadelphia, Pa	Feb. 20, 1872	123, 899
Meat-mincing machine	A. Lightheiser	Reading, Pa	Dec. 25, 1855	13, 990
Meat-mincing machine	A. McInturff	Liberty, Va	Jan. 17, 1854	10, 430
Meat-mincing machine	P. Miles	New Haven, Conn	Aug. 19, 1862	36, 229
Meat-mincing machine	P. Miles	New York, N. Y	July 12, 1864	43, 520
Meat-mincing machine	O. Moses	Malone, N. Y	July 1, 1856	15, 248
Meat or fruit press	L. S. Starrett	Athol Depot, Mass	Apr. 15, 1873	137, 803
Meat, paint, &c., Metallic can for	E. M. Wood	Needham, Mass	Mar. 18, 1873	136, 890
Meat-pounder and ice-pick	G. Murray, jr	Cambridgeport, Mass	May 12, 1868	77, 752
Meat-pounder and potato-masher	J. A. McNeil	Grand Rapids, Mich	Oct. 31, 1865	50, 724
Meat pounder, block, and chopping bowl	G. B. Mill	Buffalo, N. Y	Mar. 8, 1870	100, 545
Meat, Preparing and preserving	H. Endemann	New York, N. Y	May 31, 1870	103, 728
Meat, Preparing extract of	M. S. Valentine	Richmond, Va	Apr. 25, 1871	114, 234
Meat-press	J. I. Danforth	Newburyport, Mass	Nov. 4, 1873	144, 322
Meat, &c., Process and apparatus for curing and packing.	D. E. Somes	Washington, D. C	Feb. 26, 1867	62, 449
Meat, pulverizing spice, &c., Chopping	B. P. Coston	Philadelphia, Pa	June 13, 1831	....
Meat-roaster	S. Pierce	New York, N. Y	July 12, 1838	835
Meat-roasting apparatus	J. G. Brown and J. P. Derby	South Reading, Mass	May 12, 1857	17, 259
Meat-roasting spit	E. W. Bigelow	Worcester, Mass	Apr. 28, 1868	77, 349
Meat, Salting	T. Davison	New York, N. Y	Aug. 7, 1849	6, 623
Meat, Salting	T. Spencer	Syracuse, N. Y	Oct. 21, 1862	36, 735
Meat, &c., Salting and preserving	E. Darnell	Fox, Ill	Apr. 2, 1867	63, 484
Meat, Salting and seasoning	W. Sadler	Lockport, N. Y	Apr. 2, 1867	63, 432
Meat-smoking apparatus	E. Bickford	Ogden, N. Y	June 17, 1862	35, 580
Meat-smoking apparatus	R. T. Burnett	New York, N. Y	Jan. 19, 1869	85, 998
Meat-smoking apparatus	J. Wright	Wilmington, Del	July 29, 1856	15, 452
Meat-spit	P. Fisher	Williamsburgh, N. Y	Nov. 19, 1867	70, 984
Meat-stuffer	A. R. Silver	Salem, Ohio	Nov. 26, 1872	133, 342
Meat-tenderer	W. Beach	Philadelphia, Pa	Aug. 16, 1853	9, 932
Meat-tenderer	M. M. Pettes	Worcester, Mass	Mar. 5, 1872	124, 383
Meat-tenderer	L. B. Tarbox	Collinsville, N. Y	Feb. 21, 1871	112, 092
Meat tendering or chopping device	F. W. Codding	West Rutland, Mass	Jan. 28, 1870	104, 706
Meat, Utensil for compressing	E. Mingay	Boston, Mass	Dec. 10, 1872	133, 791
Meat with brine, Apparatus for injecting	J. C. Adams	Baltimore, Md	Sept. 10, 1861	33, 228
Mechanical adjustment	W. F. Parker	Meriden, Conn	Jan. 19, 1869	86, 098
Mechanical adjustment	R. B. Perkins	Meriden, Conn	Apr. 6, 1869	88, 733

Index of patents issued from the United States Patent Office from 1790 to 1873, inclusive—Continued.

Invention.	Inventor.	Residence.	Date.	No.
Mechanical adjustment	R. B. Perkins	Meriden, Conn	Apr. 6, 1869	88, 734
Mechanical device	J. M. Folsom	Louisville, Ky	Feb. 13, 1872	123, 686
Mechanical escapement	E. C. Hopping	Madison, N. J	Nov. 5, 1872	132, 833
Mechanical jack	A. Jones	Lebanon, N. H	Aug. 3, 1858	21, 107
Mechanical power	N. S. Baker	Covington, Ky	Mar. 12, 1872	124, 415
Mechanical power	F. S. Bremer	Montgomery, Ill	Aug. 20, 1872	130, 696
Mechanical power	C. P. Carter	Poughkeepsie, N. Y	Apr. 28, 1868	77, 167
Mechanical power	E. Matteson	Troy, N. Y	Sept. 28, 1858	21, 611
Mechanical power	E. W. Morton	Brooklyn, N. Y	Feb. 18, 1873	136, 088
Mechanical power	C. Van Derzee	Albany, N. Y	Aug. 12, 1873	141, 741
Mechanical power	A. B. Wood	Hamburg, Ark	June 18, 1867	65, 981
Mechanical power, Mode of applying	L. Butts	Canterbury, Conn	July 9, 1823	....
Mechanical power, Rotary	A. G. Waterhouse	San Francisco, Cal	Sept. 21, 1869	95, 057
Medallion case	J. Powell	Cincinnati, Ohio	Nov. 13, 1866	59, 648
Medals, breast-pins, &c., Pin-fastening for	G. O. Monroe	New York, N. Y	Nov. 15, 1864	45, 121
Medals, &c., Machine for producing reduced copies of.	C. J. Hill	London, England	Apr. 14, 1868	76, 631
Medical apparatus, Electro	L. Drescher	New York, N. Y	July 4, 1871	116, 695
Medical apparatus, Electro-magnetic	H. Soltmann	New York, N. Y	Mar. 15, 1859	23, 272
Medical apparatus, Electro-magnetic	A. J. Steel	Brooklyn, N. Y	June 20, 1871	116, 110
Medical apparatus for treating diseases by vacuum	G. Hadfield	Cincinnati, Ohio	May 8, 1866	54, 530
Medical apparatus, Sponge-holder for electro	J. Kidder	New York, N. Y	June 20, 1871	116, 197
[Medical.] Burns and scalds, Method of curing	L. Maxwell	Mitchell, Ind	Feb. 28, 1865	46, 570
[Medical] Catarrhal douche	R. M. Moyle	New York, N. Y	Mar. 28, 1871	113, 080
Medical extract	H. S. Draper	Rochester, N. Y	Sept. 28, 1869	95, 209
Medical extract from hemlock-bark	S. H. Kennedy	Johnstown, N. Y	Dec. 5, 1871	121, 631
Medical extracts, Mode of evaporating solutions, decoctions, &c., and preparing.	J. W. W. Gordon	Baltimore, Md	Oct. 8, 1840	1, 805
[Medical.] Means of applying heat and cold in the treatment of diseases.	J. Chapman	Portman Square, England	Feb. 21, 1865	46, 535
Medical or aromatic substances, Apparatus for diffusing the vapors of.	A. L. Fleury	New York, N. Y	Nov. 13, 1866	59, 576
Medical purposes, Apparatus for forming vapor for.	J. Gardette and H. Rance	New Orleans, La	Jan. 3, 1860	26, 667
Medical purposes, Citrates of iron and manganese for.	A. K. Clapp	Boston, Mass	May 2, 1871	114, 264
Medical purposes, Electrical instrument for	A. Page	Springfield, Mass	Aug. 19, 1862	36, 231
Medical purposes, Electro-magnetic machine for	L. Drescher	New York, N. Y	Feb. 1, 1870	99, 414
Medical purposes, Voltaic pile for	A. C. Garratt	Boston, Mass	Dec. 29, 1868	85, 300
Medical rubbing-apparatus	G. H. Taylor	New York, N. Y	Jan. 2, 1872	122, 500
Medical steam-bath	J. Deverin and J. B. Isabelle	Lexington, Ky	July 6, 1818	....
Medical topical application, Apparatus for	G. E. B. French	Washington, D. C	Jan. 3, 1860	26, 663
Medical use, Electro-magnetic apparatus for	C. Reitz	Indianapolis, Ind	Sept. 20, 1870	107, 626
Medical use, Oscillating rubbing-machine for	G. H. Taylor	New York, N. Y	May 12, 1868	77, 933
Medical vacuum-apparatus	G. Hadfield	Cincinnati, Ohio	Jan. 1, 1867	60, 883
Medical vacuum-apparatus	J. G. Hadfield	Cincinnati, Ohio	Dec. 24, 1867	72, 631
Medical vibrating and kneading machine	G. H. Taylor	New York, N. Y	Feb. 2, 1869	86, 604
Medicated balsam	L. F. Griffin	New York, N. Y	Dec. 3, 1867	71, 743
Medicated bath	B. Swett	Mount Morris, N. Y	Dec. 31, 1845	4, 347
Medicated-bath apparatus	J. Davenport	Philadelphia, Pa	May 21, 1872	126, 878
Medicated bath, Compound for saline	C. Lennig	Philadelphia, Pa	Jan. 1, 1867	60, 753
Medicated candy	B. H. Bener and M. H. Burgess	Erie, Pa	July 11, 1865	48, 645
Medicated candy	S. B Colton	Bloomfield, Conn	Jan. 21, 1873	134, 979
Medicated candy	L. Morgenthau	Manheim, Baden	Apr. 25, 1865	47, 504
Medicated confections	N. Saltabassi	New York, N. Y	Nov. 5, 1872	132, 861
Medicated cracker	J. L. Halliman	Grand Rapids, Mich	May 4, 1869	89, 654
Medicated fabric	H. Glynn	Baltimore, Md	Dec. 21, 1858	22, 362
Medicated liquid magnesia	J. Cullen	Philadelphia, Pa	May 4, 1818	....
Medicated liquid magnesia	R. Jordan and M. Anderson	Philadelphia, Pa	June 25, 1834	....
Medicated lotion	J. G. Popp	Vienna, Austria	Dec. 4, 1860	30, 834
Medicated malt-liquor, Manufacture of	H. H. Kessler	Detroit, Mich	May 10, 1870	102, 833
Medicated or pure steam to brutes, Apparatus for administering.	L. Quetsch	Urbana, Ill	Oct. 19, 1869	96, 036
Medicated pad	W. D. Titus	Brooklyn, N. Y	Jan. 3, 1860	26, 719
Medicated pad or belt	A. F. Cooper	Cambridge, Mass	Nov. 4, 1873	144, 315
Medicated paper	C. J. Eames	New York, N. Y	Jan. 21, 1873	134, 983
Medicated plaster	A. D. Richards	Somerville, Mass	Feb. 7, 1871	111, 682
Medicated powders, Instrument for depositing	A. M. A. Laforgue	France	Mar. 16, 1869	87, 946
Medicated shampoo-bath	R. D. Mott	Boston, Mass	Dec. 16, 1833	....
Medicated-steam bath, Machine used for	B. Marshall	New York, N. Y	Dec. 7, 1821	....
Medicated towel	L. Stewart	San Francisco, Cal	Mar. 4, 1873	136, 466
Medicated troches	C. H. Needles	Philadelphia, Pa	Nov. 27, 1866	60, 039
Medicated-vapor apparatus	A. F. Rose	Brooklyn, N. Y	July 13, 1858	20, 896
Medicated vapors, Apparatus for administering	J. C. Cook	New Haven, Conn	Feb. 1, 1870	99, 408
Medicated vapors, Chair for disseminating	J. W. Smith	Iowa Point, Kans	May 21, 1861	32, 384
Medicated voltaic plaster	W. C. Collins	Bucksport, Me	July 4, 1871	116, 562
Medicator, Dermatic	I. R. Weisiger	Danville, Ky	May 11, 1869	90, 058
Medicator, Mechanical	P. A. Royce	Buffalo, N. Y	Sept. 25, 1866	58, 301
Medicinal beverage	L. F. Lastreto	San Francisco, Cal	Mar. 22, 1870	101, 027
Medicinal beverage	A. S. Taylor	San Francisco, Cal	Aug. 8, 1871	117, 942
Medicinal beverage, Aerated	A. F. Drickwitz	New York, N. Y	Oct. 26, 1869	96, 089
Medicinal extracts, Preparation of	A. J. Despinoy	Lille, France	Sept. 8, 1860	30, 050
Medicinal preparation of iron	W. A. Aspinall	Brooklyn, N. Y	June 13, 1871	115, 805
Medicinal purposes, &c., Solution of balsamic gum for.	O. Oldberg	Washington, D. C	Sept. 12, 1871	118, 813
Medicine	H. Adolph	Clinton, Kans	June 1, 1869	90, 715
Medicine	A. Alexander	Eugene City, Oreg	Apr. 28, 1868	77, 343
Medicine	J. S. Anders	North Wales, Pa	June 12, 1866	55, 449
Medicine	J. Baker	Jefferson Township, Ohio	May 5, 1831	....
Medicine	J. P. Barnett	Navasota, Tex	Oct. 7, 1873	143, 399
Medicine	T. A. Barry and B. A. Patten	San Francisco, Cal	Mar. 24, 1868	75, 837
Medicine	S. Bass	Savannah, Ga	Nov. 3, 1868	83, 755
Medicine	J. Bates	Salineville, Ohio	Feb. 26, 1867	62, 389
Medicine	P. Becker	South Bethlehem, Pa	May 31, 1870	103, 545
Medicine	D. D. T. Benedict	Havana, N. Y	Nov. 20, 1866	59, 756
Medicine	R. S. Bernard	Norfolk, Va	Aug. 17, 1835	....
Medicine	E. M. and L. M. Berry	Saltillo, Ind.	Nov. 10, 1868	83, 823
Medicine	G. B. Bieler	Cincinnati, Ohio	Jan. 9, 1866	51, 914

Index of patents issued from the United States Patent Office from 1790 to 1873, inclusive—Continued.

Invention.	Inventor.	Residence.	Date.	No.
Medicine	F. Bird	Hancock, Ga	Apr. 16, 1828	....
Medicine	O. W. Blanchard	Delavan, Wis	Aug. 27, 1867	68, 159
Medicine	G. Bode	Milwaukee, Wis	June 5, 1866	55, 234
Medicine	A. Brown	Fond du Lac, Wis	May 6, 1873	138, 473
Medicine	A. R. Brown	Litchfield, Mich	Feb. 5, 1867	61, 802
Medicine	E. D. Burrill	Providence, R. I	Mar. 22, 1870	101, 094
Medicine	T. J. Butcher	Wenona Station, Ill	July 20, 1869	92, 937
Medicine	L. J. Buttrill	Jackson, Ga	July 12, 1870	105, 168
Medicine	J. H. Butts	Stroudsburgh, Pa	Sept. 8, 1868	81, 876
Medicine	J. Cahn	Selma, Ala	June 18, 1872	128, 198
Medicine	W. R. Call and T. F. Griffin	Gloucester, Mass	Jan. 4, 1870	98, 554
Medicine	M. H. Campbell	Syracuse, N. Y	June 18, 1872	128, 110
Medicine	J. Carnrick	New York, N. Y	Mar. 2, 1869	87, 393
Medicine	F. J. Carrall	Millville, N. J	Apr. 17, 1866	53, 946
Medicine	M. Cary	Racine, Wis	Dec. 8, 1868	84, 796
Medicine	W. Cawein	Louisville, Ky	Feb. 13, 1866	52, 530
Medicine	G. W. Chambers	Talladega, Ala	Sept. 5, 1871	118, 686
Medicine	H. Clark	Northampton, Mass	Mar. 24, 1868	75, 732
Medicine	S. P. Clayton	South Amboy, N. J	Jan. 16, 1866	52, 032
Medicine	H. W. Cloud	Evansville, Ind	Oct. 4, 1870	107, 877
Medicine	J. H. Conway	Richmond, Va	May 29, 1866	55, 214
Medicine	J. Conzelman	Saint Louis, Mo	May 20, 1873	139, 045
Medicine	T. Copland	Hamilton, Canada	Dec. 17, 1872	133, 924
Medicine	A. P. Coryell	Janesville, Wis	Feb. 21, 1865	46, 449
Medicine	A. M. Cox	Elizabeth, N. J	July 2, 1867	66, 301
Medicine	J. L. A. Creuse	Brooklyn, N. Y	Nov. 28, 1871	121, 338
Medicine	J. D. Curl and J. G. Bartlett	Mokena, Ill	May 4, 1869	89, 562
Medicine	W. Davidson	Binghamton, N. Y	Dec. 3, 1867	71, 712
Medicine	D. Davis	Sheakleyville, Pa	Dec. 24, 1872	134, 129
Medicine	J. Dean	Freeport, Ill	Aug. 27, 1867	68, 051
Medicine	G. Déclat	Paris, France	Dec. 23, 1873	145, 850
Medicine	H. De Lapaturelle	New York, N. Y	May 1, 1866	54, 304
Medicine	J. Dent	Augusta, Ga	July 2, 1828	....
Medicine	P. M. Devos	New York, N. Y	July 24, 1866	56, 533
Medicine	C. A. F. Dietz	New York, N. Y	Feb. 11, 1868	74, 205
Medicine	O. G. Ditmars	New York, N. Y	Mar. 17, 1868	75, 536
Medicine	O. G. Ditmars	New York, N. Y	Mar. 17, 1868	75, 537
Medicine	B. W. Donaldson	Dixfield, Me	Apr. 7, 1868	76, 308
Medicine	L. Dow	Hebron, Conn	Nov. 24, 1820	....
Medicine	J. D. Doyle	Rochester, N. Y	June 25, 1872	128, 373
Medicine	N. T. Drake	High Point, N. C	Sept. 28, 1869	95, 330
Medicine	C. J. Eames	New York, N. Y	May 10, 1870	102, 788
Medicine	M. C. Edey	New York, N. Y	Mar. 31, 1868	76, 176
Medicine	J. B. Edwards	Knightstown, Ind	Sept. 11, 1866	57, 884
Medicine	J. P. Eisenhut	Monroe, Mich	June 12, 1866	55, 474
Medicine	W. H. Farnham	Sparta, Wis	Apr. 3, 1866	53, 595
Medicine	J. Fay	Lacon, Ill	May 9, 1871	114, 544
Medicine	S. Field	Galva, Ill	Dec. 16, 1873	145, 560
Medicine	J. Fischer	Saint Louis, Mo	July 2, 1867	66, 315
Medicine	W. B. Foster	Ridgeville, Ohio	Mar. 5, 1867	62, 622
Medicine	E. Frese	San Francisco, Cal	Aug. 4, 1868	80, 718
Medicine	A. Fuqua	Woodville, Ky	July 1, 1873	140, 408
Medicine	G. C. Furber	Yreka, Cal	June 28, 1870	104, 727
Medicine	H. Gahn	Upsala, Sweden	Apr. 8, 1873	137, 542
Medicine	G. W. C. Gamble	Millersburgh, Iowa	Feb. 18, 1868	74, 680
Medicine	O. V. Garnett	Versailles, Ky	Apr. 9, 1867	63, 627
Medicine	L. L. Gebhart	New Providence, Ind	Aug. 23, 1870	106, 573
Medicine	R. A. Gettings	Marion, Ky	Oct. 15, 1872	132, 275
Medicine	F. Goetsch	Bloomington, Ill	Apr. 2, 1872	125, 286
Medicine	C. Goffinet	Leopold, Ind	June 25, 1872	128, 385
Medicine	G. H. Goltry and F. W. Hogarth	Port Allegheny, Pa	May 30, 1871	115, 460
Medicine	D. C. Gould	Sterling, Ill	Oct. 15, 1867	69, 796
Medicine	J. W. Gray	Riceville, Pa	Aug. 13, 1872	130, 368
Medicine	J. Greenwald	Cincinnati, Ohio	Dec. 17, 1867	72, 193
Medicine	J. Gregory	Marion, Ohio	Jan. 8, 1867	61, 063
Medicine	H. H. Grigg	Philadelphia, Pa	Jan. 11, 1870	98, 690
Medicine	B. Gutierrez	Santa Barbara, Cal	Sept. 5, 1871	118, 714
Medicine	J. M. F. Hall	Davenport, Iowa	Sept. 25, 1866	58, 249
Medicine	C. L. Hammond	Java, N. Y	Sept. 14, 1869	94, 886
Medicine	M. J. Hanson	Mauston, Wis	Jan. 18, 1870	98, 959
Medicine	J. Harrigan	East Boston, Mass	Nov. 12, 1867	70, 838
Medicine	O. F. Harris	Norwich, Conn	Dec. 6, 1870	109, 895
Medicine	J. B. Haskins	Brooklyn, N. Y	July 5, 1870	105, 073
Medicine	G. M. Hay	Americus, Ga	Jan. 5, 1869	85, 659
Medicine	W. Habbert	Manchester, England	Apr. 27, 1869	89, 481
Medicine	M. A. Hilt	Syracuse, N. Y	Oct. 22, 1867	69, 995
Medicine	A. J. Hobbs	Van Wirt, Ga	June 30, 1868	79, 352
Medicine	W. Horner	Washington, D. C	Dec. 13, 1864	45, 410
Medicine	J. M. Hughes	Menomonee, Wis	Dec. 29, 1868	85, 385
Medicine	J. P. Hulse	Acra, N. Y	May 8, 1866	54, 550
Medicine	J. P. Humes	Winnebago City, Minn	May 19, 1868	78, 096
Medicine	S. W. Ingraham	Wooster, Ohio	Feb. 2, 1869	86, 404
Medicine	J. G. Jeffrey	South New Berlin, N. Y	Apr. 16, 1867	63, 902
Medicine	H. D. and I. D. Jewett	Saint Omer, Ind	May 21, 1872	127, 060
Medicine	A. Johnson	Brockport, N. Y	Mar. 2, 1869	87, 343
Medicine	N. Joly	Paris, France	Aug. 6, 1867	67, 555
Medicine	W. N. Jordan	Cambridge, Mass	June 22, 1869	91, 751
Medicine	C. Judson	Omro, Wis	Aug. 13, 1867	67, 657
Medicine	A. W. Kidder	South Norridgewock, Me	Apr. 6, 1869	88, 644
Medicine	N. Kieffer	New Orleans, La	Sept. 18, 1866	58, 106
Medicine	J. W. M. Kirkpatrick	Hamburgh, Ark	Feb. 16, 1869	86, 986
Medicine	J. B. Knoebel	Shoal Creek, Ill	June 14, 1864	43, 118
Medicine	M. Knotts	Carondelet, Mo	May 28, 1867	65, 235
Medicine	J. Kornitzer	New York, N. Y	May 14, 1872	126, 637
Medicine	D. Laugell	Apple Creek, Ohio	Sept. 7, 1869	94, 616
Medicine	P. Lee and L. Matthews	Lebanon, Oreg	Mar. 29, 1870	101, 279
Medicine	C. L. Lege	San Antonio, Tex	June 11, 1867	65, 580
Medicine	J. Leich	Brooklyn, N. Y	Sept. 20, 1864	44, 323

Index of patents issued from the United States Patent Office from 1790 *to* 1873, *inclusive*—Continued.

Invention.	Inventor.	Residence.	Date.	No.
Medicine	G. H. Leithead	East Birmingham, Pa	June 4, 1867	65, 401
Medicine	J. Levy	Chicago, Ill	Aug. 14, 1866	57, 155
Medicine	N. C. Lincoln	Brunswick, Me	Oct. 9, 1866	58, 652
Medicine	H. Lister	Houston, Tex	Feb. 7, 1871	111, 655
Medicine	H. H. Lockwood	Madison, Wis	May 29, 1866	55, 123
Medicine	A. Loosley	Philadelphia, Pa	Mar. 5, 1867	62, 650
Medicine	E. A. Lowdermill	Grand Junction, Tenn	Mar. 24, 1868	75, 776
Medicine	W. S. Lyon	Tranquility, Ohio	Sept. 10, 1867	68, 632
Medicine	L. Marmaduke	Shelbyville, Mo	July 3, 1866	56, 072
Medicine	A. A. Marsh	Frankfort, Mich	Aug. 30, 1870	106, 846
Medicine	Z. Marshall	Andersonville, Ind	Nov. 20, 1866	59, 776
Medicine	P. and P. Mays	Clearfield, Pa	Jan. 18, 1870	98, 875
Medicine	W. L. McCord	Abbeville, S. C	June 29, 1869	91, 954
Medicine	G. Mershon	Brookville, Iowa	Apr. 10, 1866	53, 852
Medicine	A. Messex	Waynesborough, Ga	Feb. 19, 1867	62, 282
Medicine	L. B. Millard	San José, Cal	Nov. 26, 1872	133, 328
Medicine	E. Miller	New York, N. Y	Oct. 18, 1870	108, 504
Medicine	W. J. Miller	Linesville, Pa	Jan. 29, 1867	61, 631
Medicine	C. H. Mitchell	Bristow Station, Ky	May 14, 1867	64, 787
Medicine	G. Mohler	Yates City, Ill	Sept. 29, 1868	82, 541
Medicine	G. Montgomery	Canton, Ill	Aug. 21, 1866	57, 357
Medicine	L. O. Moreno	New York, N. Y	Oct. 23, 1866	59, 052
Medicine	T. Morley	Cambridge, Mass	Jan. 14, 1873	134, 762
Medicine	C. A. Morse	Williamsport, Ohio	Sept. 6, 1864	44, 106
Medicine	G. F. Munro	Leon, Iowa	Nov. 25, 1873	145, 002
Medicine	E. Myers	Davis, Ill	June 28, 1870	104, 758
Medicine	W. B. Myres	Frenchtown, N. J	June 29, 1869	92, 086
Medicine	F. H. Norton	New York, N. Y	June 14, 1864	43, 128
Medicine	W. W. Oglesby	Benton County, Oreg	Oct. 18, 1870	108, 508
Medicine	H. C. S. Otto	New York, N. Y	Mar. 29, 1870	101, 305
Medicine	S. E. Paddock	Delaware, Ohio	Sept. 2, 1873	142, 408
Medicine	B. Paine	McMinnville, Tenn	Nov. 5, 1867	70, 602
Medicine	S. Payne	Louisville, Ky	Nov. 26, 1867	71, 407
Medicine	E. Penfield	Oberlin, Ohio	Mar. 28, 1865	47, 035
Medicine	J. Penoyer	Massillon, Ohio	Jan. 12, 1869	85, 761
Medicine	M. Perl	New Orleans, La	Dec. 10, 1867	71, 907
Medicine	E. G. Perry	Knoxville, Ill	May 22, 1866	54, 944
Medicine	W. H. H. Peters	Tuskegee, Ala	June 1, 1869	90, 779
Medicine	N. Petre	New York, N. Y	Aug. 28, 1866	57, 561
Medicine	W. W. Pettengill	Georgia, Vt	Oct. 1, 1872	131, 778
Medicine	S. Pitcher	Barnstable, Mass	May 12, 1868	77, 758
Medicine	H. Pool	Montgomery County, Tenn	June 22, 1869	91, 768
Medicine	H. G. Pope and H. F. Herrick	New Berlin, N. Y	June 11, 1867	65, 693
Medicine	P. Poucin	Minneapolis, Minn	June 12, 1866	55, 534
Medicine	J. Prentiss	New London, Conn	Mar. 27, 1822	....
Medicine	J. L. Putegnat	Brownsville, Tex	Mar. 18, 1873	136, 937
Medicine	M. H. Ramsaur	Lincolnton, N. C	July 6, 1869	92, 209
Medicine	J. Ramsburgh, sr	New Madrid, Mo	Oct. 6, 1868	82, 873
Medicine	W. Ranson	Portage, Wis	Oct. 23, 1866	59, 070
Medicine	J. J. Reeves	Sulphur Springs, Tex	June 26, 1860	28, 904
Medicine	M. Richardson	Norristown, Pa	May 29, 1866	55, 160
Medicine	J. A. Robbins	Medford, Mass	Apr. 4, 1871	113, 569
Medicine	E. Robinson	Fairport, N. Y	May 18, 1869	90, 308
Medicine	L. Rogers	Morehouse Parish, La	May 19, 1868	78, 017
Medicine	L. Rogers	Morehouse Parish, La	May 19, 1868	78, 018
Medicine	T. Rosbrugh	Bellefontaine, Ohio	Feb. 4, 1868	74, 147
Medicine	P. G. Rosenblatt	Greenville, Tenn	June 12, 1866	55, 540
Medicine	A. Rullmann	New York, N. Y	Sept. 24, 1867	69, 252
Medicine	A. J. Runyan	Ashland, Ind	Mar. 20, 1866	53, 348
Medicine	W. T. Salie	Bowdoinham, Me	June 26, 1866	55, 911
Medicine	B. Samson	Cortlandville, N. Y	Nov. 20, 1866	59, 750
Medicine	L. Schultz	Buffalo, N. Y	May 16, 1865	47, 750
Medicine	I. W. Scranton	West Liberty, Iowa	Sept. 8, 1868	82, 038
Medicine	S. P. Sedgwick	Wheaton, Ill	Apr. 14, 1868	76, 832
Medicine	A. F. Shannon	Quincy, Ill	July 11, 1871	116, 875
Medicine	G. V. and J. A. Sheffield	Northbridge Centre, Mass	Feb. 15, 1870	99, 959
Medicine	O. Skidmore	Albany, N. Y	June 8, 1869	90, 962
Medicine	G. H. Smith	New Orleans, La	Apr. 2, 1867	63, 567
Medicine	G. H. Smith	New Orleans, La	May 10, 1870	102, 980
Medicine	J. Smith	Schodack Centre, N. Y	Feb. 18, 1868	74, 622
Medicine	W. C. Smith	Lafargeville, N. Y	Feb. 20, 1866	52, 760
Medicine	F. W. Speissegger	Charleston, S. C	June 19, 1866	55, 730
Medicine	J. H. Sperling	Peru, Ind	June 18, 1867	65, 840
Medicine	J. T. Stewart	Peoria, Ill	Jan. 21, 1868	73, 552
Medicine	J. T. Stewart	Peoria, Ill	May 31, 1870	103, 675
Medicine	L. Stroever	Philadelphia, Pa	Aug. 20, 1867	67, 926
Medicine	P. O. R. Stromski	Boston, Mass	Mar. 16, 1869	87, 802
Medicine	D. W. Stutsman	Upshur, Ohio	Nov. 20, 1866	59, 803
Medicine	W. C. Tait	Alexandria, La	June 28, 1870	104, 791
Medicine	A. S. Talbert	Lexington, Ky	Jan. 16, 1866	52, 091
Medicine	W. H. Tate	Orleans, Ind	Sept. 3, 1867	68, 582
Medicine	D. W. Taylor	Bell County, Tex	Jan. 7, 1868	73, 135
Medicine	T. H. Taylor	Saratoga Springs, N. Y	Nov. 26, 1867	71, 549
Medicine	C. K. Tayntor	Cuyler, N. Y	Mar. 31, 1868	76, 272
Medicine	H. Themel	Esconawba, Mich	Nov. 5, 1872	132, 876
Medicine	H. Themel	Esconawba, Mich	Mar. 11, 1873	136, 680
Medicine	J. M. Thompson	Saltilloville, Ind	Jan. 28, 1868	73, 939
Medicine	S. Thompson	Boston, Mass	Jan. 28, 1823	....
Medicine	A. A. Thomson	Newburgh, Pa	Nov. 30, 1869	97, 459
Medicine	W. P. Thurber	Chicago, Ill	Apr. 7, 1868	76, 555
Medicine	J. Thurmon	Pike County, Mo	Apr. 4, 1865	47, 140
Medicine	J. P. and L. Thurmon	Warrenton, Mo	May 28, 1867	65, 302
Medicine	H. A. Tilden	New Lebanon, N. Y	July 16, 1872	129, 626
Medicine	T. S. Tuggle	Columbus, Ga	Apr. 6, 1869	88, 680
Medicine	T. S. Tuggle	Columbus, Ga	Apr. 6, 1869	88, 681
Medicine	T. H. W. Upshur	Norfolk, Va	Sept. 8, 1868	82, 046
Medicine	J. B. Vann	Elyton, Ala	Jan. 26, 1869	86, 331
Medicine	P. W. Vaughn	Columbia, Ky	Mar. 9, 1869	87, 605

Index of patents issued from the United States Patent Office from 1790 *to* 1873, *inclusive*—Continued.

Invention.	Inventor.	Residence.	Date.	No.
Medicine	J. Ward	East Hardwick, Vt	May 10, 1864	42, 704
Medicine	J. Ward	Evansville, Ind	Sept. 21, 1869	95, 173
Medicine	J. Weaver	Knightstown, Ind	Jan. 5, 1864	41, 111
Medicine	W. Weber	Cincinnati, Ohio	June 29, 1869	91, 886
Medicine	R. B. Weese	Charlottesville, Ind	May 24, 1864	42, 890
Medicine	C. H. Whittemore	Lewiston, Me	Aug. 18, 1868	81, 122
Medicine	A. D. Willis	Crawfordsville, Ind	Jan. 4, 1870	98, 535
Medicine	J. A. Willis	Cherry Valley, N. Y	May 7, 1867	64, 611
Medicine	E. Wills	Winslow, N. J	Sept. 28, 1869	95, 298
Medicine	J. T. Wilson	Brooklyn, N. Y	Dec. 10, 1867	72, 143
Medicine	W. M. Wilson and D. S. Denton	Polk County, Mo	Aug. 30, 1870	106, 905
Medicine	L. W. Wollenweber	Jeffersonville, Ind	Oct. 22, 1872	132, 424
Medicine	T. T. Wright	Memphis, Tenn	July 15, 1873	140, 866
Medicine	J. Yount	Gettysburgh, Pa	July 23, 1867	67, 012
Medicine	H. Zoeger	New York, N. Y	Feb. 26, 1867	62, 516
Medicine, Ague	T. M. Daniel	Athens, Ga	Oct. 1, 1867	69, 414
Medicine, Ague	J. Monfort	Jessamine County, Ky	Dec. 18, 1866	60, 538
Medicine and apparatus for making the same	P. Fahrney	Chicago, Ill	Dec. 13, 1870	110, 025
Medicine and apparatus for the cure of asthma	J. Pinchard	San Francisco, Cal	Mar. 25, 1873	137, 093
Medicine and disinfectant	J. Walton	Newark, Ohio	June 11, 1872	127, 814
Medicine and liniment	W. S. Crooker	Shamburgh, Pa	Aug. 9, 1870	106, 127
Medicine and medicated food	J. M. O. Tamin	New York, N. Y	Mar. 11, 1873	136, 678
Medicine and remedial agents, and apparatus therefor, Mode of applying.	J. Allen	New York, N. Y	Mar. 12, 1867	62, 916
Medicine and surgery, Application of carbonated water in.	A. D. Puffer	Somerville, Mass	Aug. 24, 1869	94, 029
[Medicine,] Anodyne and alterative sirup	R. Thompson	Rome, N. Y	Nov. 14, 1835	....
Medicine, Anti-bilious	J. J. Giraud	Baltimore, Md	Sept. 9, 1817	....
Medicine, Anti-rheumatic	J. Schmoll	New York, N. Y	Mar. 5, 1867	62, 692
Medicine, Anti-spasmodic tincture	H. Howard	Columbus, Ohio	Aug. 25, 1832	....
Medicine, Anti-typhus	J. P. Fertig	Saint Louis, Mo	Sept 8, 1863	39, 861
Medicine, Apparatus for administering	S. Curtiss, jr	Williamsburgh, N. Y	Apr. 10, 1866	53, 790
Medicine, Apparatus for administering pulverulent	J. Moore and D. P. Adams	Marietta, Ohio	July 24, 1855	13, 318
Medicine, Astringent	E. W. Ferris	Macon, Miss	Oct. 16, 1860	30, 396
Medicine, Astringent tonic	H. Howard	Columbus, Ohio	Aug. 25, 1832	....
Medicine, Bitter tonic	H. Howard	Columbus, Ohio	Aug. 25, 1832	....
[Medicine.] Bitters	T. Beck	Omaha, Nebr	Dec. 27, 1870	110, 423
[Medicine.] Bitters	J. Bender	Lonaconing, Md	Apr. 28, 1868	77, 245
[Medicine.] Bitters	G. W. Brown	Portland, Oreg	Aug. 13, 1872	130, 409
[Medicine.] Bitters	E. E. Crady	Sioux City, Iowa	Oct. 17, 1871	120, 042
[Medicine.] Bitters	A. Desaulniers	Oswego, N. Y	Mar. 14, 1871	112, 558
[Medicine.] Bitters	J. Frechette	Chicago, Ill	May 21, 1872	126, 948
[Medicine.] Bitters	W. P. and J. Haubert	Canton, Ohio	Feb. 23, 1869	87, 169
[Medicine.] Bitters	M. Holst	Memphis, Tenn	June 27, 1871	116, 310
[Medicine.] Bitters	J. D. Holtzermann	Piqua, Ohio	May 7, 1867	64, 421
[Medicine.] Bitters	A. T. Hyde	Rochester, Minn	Aug. 25, 1868	81, 508
[Medicine.] Bitters	L. Kappler	Woodstock, Ill	Dec. 2, 1873	145, 062
[Medicine.] Bitters	J. Lladó	New Orleans, La	Aug. 3, 1869	93, 209
[Medicine.] Bitters	A. M. Loryea	East Portland, Oreg	July 11, 1871	116, 846
[Medicine.] Bitters	E. Phillips	Peoria, Ill	June 18, 1872	128, 169
[Medicine.] Bitters	G. V. Rambaut	Petersburgh, Va	Feb. 25, 1868	74, 940
[Medicine.] Bitters	W. T. Sherwood	Ripon, Wis	Apr. 18, 1871	113, 938
[Medicine.] Bitters	R. G. Turner	Columbia, Tex	Jan. 23, 1872	122, 975
[Medicine.] Bitters	S. R. Whitlow	Limestone Township, Ill	Apr. 4, 1871	113, 713
[Medicine.] Bitters, Anti-dyspeptic	S. M. Barnes	La Fayette, Va	Oct. 15, 1872	132, 233
[Medicine.] Bitters, Anti-dyspeptic	C. T. Provost	New York, N. Y	Aug. 20, 1867	68, 004
[Medicine.] Bitters for chills and fevers	T. B. Owens	Gatesville, Tex	May 20, 1873	139, 185
[Medicine.] Bitters, Jaundice	J. Wheaton		Jan. 17, 1801	....
[Medicine.] Bitters, Stomach	L. W. Harris	Chicago, Ill	Jan. 5, 1869	85, 657
[Medicine.] Bitters, Stomach	D. Rinkle	Indianapolis, Ind	Mar. 22, 1870	101, 042
[Medicine.] Bitters, Stomach	F. Stutzel	Wabashaw, Minn	Oct. 7, 1873	143, 427
[Medicine.] Bitters, Tonic	R. Blacklidge	Bridgeport, Conn	Apr. 6, 1869	88, 537
[Medicine.] Bitters, Tonic	F. Fullerton	Williamsport, Pa	Oct. 20, 1868	83, 273
[Medicine.] Bitters, Tonic	I. Hellman	Saint Louis, Mo	Apr. 11, 1865	47, 204
[Medicine.] Bitters, Tonic	R. P. Jenkins	Nashville, Tenn	Feb. 2, 1869	86, 551
[Medicine.] Bitters, Tonic	J. H. McCartney	Dansville, N. Y	May 11, 1869	90, 048
Medicine, Botanic	S. Thompson	Boston, Mass	May 6, 1836	....
[Medicine.] Bronchial troche	S. C. Henszey, jr	Westchester, Pa	June 5, 1866	55, 292
Medicine by means of steam, Administering	C. Whitlaw	New York	Feb. 16, 1825	....
Medicine called bilious cordial	S. Chamberlaine		Dec. 31, 1804	....
Medicine called canker-drops	D. Albrook	Onondaga County, N. Y	Mar. 28, 1814	....
Medicine called the essence of tansy	I. Newton	Norwich, Vt	Feb. 28, 1806	....
[Medicine.] Carbolated cod-liver oil	J. H. Willson	Brooklyn, N. Y	May 21, 1872	127, 005
Medicine-case	M. O. Baldwin	Wamego, Kans	Sept. 16, 1873	142, 887
Medicine-case	A. Button	Dunkirk, N. Y	May 17, 1870	103, 012
Medicine-case	W. H. Cutler	Buffalo, N. Y	Mar. 18, 1873	136, 973
Medicine-case	E. H. Hance	Philadelphia, Pa	May 31, 1864	42, 943
Medicine-case	A. Harrington	Otsego County, N. Y	July 16, 1811	....
Medicine-case	J. R. Haynes and A. F. Worthington.	Newport, Ky., and Cincinnati, Ohio.	Aug. 27, 1867	68, 073
Medicine-case	T. M. Perot	Philadelphia, Pa	Sept. 15, 1863	39, 952
Medicine-case	H. M. Smith	New York, N. Y	May 23, 1871	115, 121
Medicine, Chemical anti-dysenteric	J. G. Vought	Rochester, N. Y	Dec. 4, 1821	....
Medicine-chest	R. B. Parkinson and J. M. Maris	Philadelphia, Pa	Nov. 29, 1864	45, 266
Medicine-chest	E. Sander	Saint Louis, Mo	Apr. 20, 1869	89, 173
Medicine-chest	E. M. Skinner	Boston, Mass	Oct. 23, 1866	59, 086
Medicine-chest for veterinary surgeon	J. W. Gladsden	Philadelphia, Pa	Mar. 12, 1872	124, 432
[Medicine.] Chlorine, Cosmetic	D. West	Hudson, N. Y	Jan. 11, 1833	....
[Medicine.] Compound tincture of myrrh	H. Howard	Columbus, Ohio	Aug. 25, 1832	....
[Medicine.] Cough-lozenges	E. Gauvreau	Quebec, Canada	July 30, 1872	129, 942
[Medicine.] Cough-mixture	M. Connell	Jersey City, N. Y	Apr. 16, 1872	125, 723
[Medicine.] Cough-mixture	P. M. Huffman	Harvard, Ill	June 18, 1867	65, 813
[Medicine.] Cough-mixture	S. R. Whitlow	Rosefield, Ill	Aug. 25, 1868	81, 446
[Medicine.] Cough-mixture for candies	L. Violet	New Lebanon, N. Y	June 6, 1871	115, 790
[Medicine.] Cough-sirup	C. Benedict, jr	Omro, Wis	Apr. 24, 1866	54, 098
[Medicine.] Cough-sirup	B. F. Burroughs	West Perry Township, Pa	June 6, 1871	115, 598
[Medicine.] Cough-sirup	G. W. Chatfield	New Haven, Conn	Mar. 22, 1870	101, 096
[Medicine.] Cough-sirup	J. Conzelman	Saint Louis, Mo	Aug. 2, 1870	105, 916

Index of patents issued from the United States Patent Office from 1790 *to* 1873, *inclusive*—Continued.

Invention.	Inventor.	Residence.	Date.	No.
[Medicine.] Curing sores	R. Rood	Centre Lisle, N. Y	Feb. 20, 1836	....
Medicine-dropper	P. J. McElroy	Cambridge, Mass	June 30, 1868	79, 487
[Medicine.] Elixir for the cure of bronchitis, &c., Vegetable.	P. Faulkner	Rockville, Pa	Sept. 23, 1843	3, 282
Medicine-evaporator	E. O. Schartau	Philadelphia, Pa	Mar. 23, 1869	88, 218
Medicine, Eye	D. R. Morgan	San Francisco, Cal	Jan. 14, 1868	73, 369
Medicine, Eye	H. M. Sanderson	Noble, Ill	Feb. 4, 1868	74, 151
Medicine, Eye	A. Schuster	Saint Louis, Mo	Jan. 2, 1866	51, 868
[Medicine.] Eye-balsam	W. Klingbeil	Champaign City, Ill	June 20, 1871	116, 199
[Medicine.] Eye-wash	P. Paul	Black Earth, Wis	Dec. 5, 1871	121, 467
[Medicine.] Eye-wash	F. Pawlewski and J. Schulz	Milwaukee, Wis	Dec. 23, 1873	145, 749
Medicine, Febrifuge	A. Johnson	Northumberland County, Pa.	Jan. 11, 1812	....
[Medicine.] Felon, &c., curing compound	R. Feibelman	Columbus, Ind	Aug. 18, 1868	81, 152
Medicine, Fever	C. F. Holing	Hudson City, N. J	Aug. 13, 1867	67, 763
Medicine, Fever	S. Thompson	Surry, N. H	Mar. 2, 1813	....
Medicine, Fever and ague	W. H. Barker	West Farms, N. Y	May 21, 1872	127, 014
Medicine, Fever and ague	G. Behrens	Brunswick, Mo	Sept. 3, 1872	131, 078
Medicine, Fever and ague	L. Bodenheimer	Paducah, Ky	Dec. 19, 1871	121, 984
Medicine, Fever and ague	W. Campbell	Camden, N. J	Oct. 29, 1872	132, 629
Medicine, Fever and ague	W. Campbell	Camden, N. J	Jan. 7, 1873	134, 639.
Medicine, Fever and ague	M. Cannon	New Orleans, La	Oct. 1, 1830	....
Medicine, Fever and ague	J. M. Ferguson	Summit, Miss	Oct. 26, 1869	96, 215
Medicine, Fever and ague	M. Ferro	Iuka, Miss	Sept. 6, 1870	107, 024
Medicine, Fever and ague	G. W. Gay	Bath, Me	May 21, 1872	126, 882
Medicine, Fever and ague	S. Laighton	Newark, N. J	July 12, 1870	105, 224
Medicine, Fever and ague	A. V. Lee	Clayton, Ala	Aug. 18, 1868	81, 181
Medicine, Fever and ague	J. Mabrey	Jefferson City, Mo	June 23, 1868	79, 241
Medicine, Fever and ague	J. H. Morris	Eaton, Ohio	Feb. 8, 1870	99, 583
Medicine, Fever and ague	O. Scidmore	Albany, N. Y	Sept. 19, 1871	119, 053
Medicine, Fever and ague	O. M. Spiller	Akron, Ohio	July 5, 1870	105, 011
Medicine, Fever and ague	G. E. Swan	Mount Vernon, Ohio	Mar. 7, 1871	112, 509
Medicine for asthma	C. B. Hurst	Rochester, Pa	Sept. 10, 1872	131, 167
Medicine for asthma and other diseases	J. C. Lewis	Belfast, Me	May 18, 1869	90, 179
Medicine for bronchial and lung diseases	S. W. Helm	Bachelor Valley, Cal	Mar. 14, 1871	112, 595
Medicine for cancer	J. Andrus	Hillsborough, N. H	July 30, 1816	....
Medicine for cancer, &c	F. Baker	New York, N. Y	June 27, 1871	116, 530
Medicine for cancer	J. Harrigan	Boston, Mass	Nov. 26, 1872	133, 443
Medicine for cancer	J. Harrison	Castalia, N. C	May 16, 1871	114, 810
Medicine for cancer	N. McKelfresh	Elizabeth, Ind	Dec. 5, 1871	121, 534
Medicine for cancer	L. Roy	Plattsburgh, Mo	Oct. 26, 1869	96, 269
Medicine for cancer	M. Williams	Middletown, Ohio	Nov. 2, 1869	96, 371
Medicine for cancer and other diseases	P. A. Cobb	Lynchburgh, Va	Apr. 19, 1870	102, 093
Medicine for catarrh	J. A. Carpenter	Waverly, N. Y	Jan. 21, 1873	135, 077
Medicine for catarrh	F. P. Gardiner	New Haven, Conn	Feb. 5, 1867	61, 730
Medicine for catarrh	W. V. Green and J. M. Kintzel	Wellsville, Ohio	June 17, 1873	139, 951
Medicine for catarrh	J. Snow	Grand Rapids, Mich	Nov. 2, 1869	96, 496
Medicine for catarrh and asthma	E. Field	Ostrander, Ohio	Mar. 7, 1871	112, 329
Medicine for cattle and other animals	O. E. Hornidy	Chauncey, Ill	Apr. 27, 1869	89, 314
Medicine for cattle and other animals	D. P. Mathews	Winthrop, Mass	Sept. 8, 1868	81, 920
Medicine for chills and fever	R. Bevil	Bowie County, Tex	Oct. 10, 1871	119, 812
Medicine for chills and fever	I. Brown	Philadelphia, Pa	Dec. 19, 1871	121, 989
Medicine for chills and fever	W. P. Fennell	Louisville, Ky	May 16, 1871	114, 784
Medicine for cholera	J. Houck	Baltimore, Md	Oct. 25, 1832	....
Medicine for cholera	A. Hunn, sr	Lancaster, Ky	Aug. 12, 1833	....
Medicine for cholera	A. E. Parrott	Norfolk, Va	June 26, 1866	55, 898
Medicine for cholera	A. Racicot	Troy, N. Y	July 8, 1873	140, 642
Medicine for cholera-infantum, &c	A. Haws	Battle Mountain Station, Nev.	July 16, 1872	129, 343
Medicine for cholera-morbus	J. Pool	Hermon, N. Y	Sept. 28, 1869	95, 376
Medicine for colds	W. E. Chilson	Troy, Pa	Jan. 17, 1871	111, 044
Medicine for colic	W. W. Gardner	Russellville, Ky	July 30, 1872	129, 886
Medicine for colic	J. A. McKinnon	Selma, Ala	Mar. 22, 1870	101, 145
Medicine for constipation	C. A. Simmons	Waldo, Fla	Dec. 12, 1871	121, 904
Medicine for consumption	T. Bell	New York, N. Y	Apr. 24, 1866	54, 097
Medicine for consumption	T. W. Bethel	Brooklyn, N. Y	Sept. 26, 1871	119, 301
Medicine for consumption	N. C. Jarrell	High Point, N. C	May 7, 1872	126, 397
Medicine for consumption	J. E. Larkin	Newark, N. J	Oct. 3, 1871	119, 620
Medicine for consumption, &c	C. S. Long	Philadelphia, Pa	May 27, 1812	....
Medicine for consumption, &c	W. H. H. White	San Francisco, Cal	June 24, 1873	140, 320
Medicine for corns and bunions	J. L. S. Hall	Wheeling, W. Va	Sept. 14, 1869	94, 819
Medicine for coughs, &c	J. Cushings	Wellington, Ohio	Aug. 16, 1870	106, 333
Medicine for coughs and colds	D. E. Smith	Cornwall, Conn	June 30, 1828	....
Medicine for coughs and colds	A. Young	Pittsborough, N. C	Apr. 19, 1870	102, 194
Medicine for coughs, colds, &c	C. P. Crossman	West Warren, Mass	May 23, 1871	115, 032
Medicine for coughs, colds, &c	W. H. H. Irvin	Philadelphia, Pa	Feb. 22, 1870	100, 039
Medicine for coughs, colds, &c	J. D. Richmond, jr	New Liberty, Ky	Aug. 8, 1871	117, 930
Medicine for coughs, colds, &c	J. Willey	Oakland, Cal	July 2, 1872	128, 692
Medicine for croup	C. Sanborn	South Berwick, Me	Apr. 8, 1862	34, 905
Medicine for cure of diseases in cattle	E. A. Wilder	Dennysville, Me	June 21, 1870	104, 677
Medicine for cure of gravel	B. W. Deal	New Market, Md	Apr. 6, 1869	88, 554
Medicine for cure of gravel	T. G. Eiswald	Providence, R. I	Aug. 16, 1870	106, 474
Medicine for cure of foot-rot in sheep	G. Eckert	Charlestown, Ohio	Jan. 5, 1869	85, 650
Medicine for cure of foot-rot in sheep	W. T. Sherman	Marengo, Ohio	Apr. 4, 1871	113, 459
Medicine for curing foot-rot in horses, &c	E. W. Wakefield	San Francisco, Cal	Oct. 13, 1863	40, 297
Medicine for cutaneous diseases	E. A. Shewell	Boston, Mass	Nov. 21, 1871	121, 205
Medicine for diarrhœa, &c	A. B. Dorman	Cape Girardeau, Mo	Nov. 19, 1872	133, 213
Medicine for diarrhœa	R. A. Walton	Shawneetown, Ill	May 30, 1871	115, 547
Medicine for diptheria	A. J. Denison	Paris, Mich	May 25, 1869	90, 347
Medicine for diptheria, &c	T. J. Glines	Hebron, N. Y	Sept. 20, 1864	44, 299
Medicine for diptheria	H. Pruden	Erie, Pa	June 24, 1873	140, 163
Medicine for diptheria	H. L. Sheffer	Marston, Wis	May 24, 1870	103, 512
Medicine for dropsy	C. Brown	New Albany, Ind	Oct. 20, 1868	83, 249
Medicine for dropsy	J. Newton and I. Barfield	Richland Parish, La	Feb. 1, 1870	99, 464
Medicine for dropsy	J. R. Strickland	Sayville, N. Y	July 6, 1869	92, 394
Medicine for dropsy and epilepsy	J. S. Fall	Rattlesnake Springs, Ga	Nov. 25, 1831	....
Medicine for dysentery and dyspepsia	T. Powell	Burlington, Vt	Feb. 2, 1828	....
Medicine for dyspepsia	C. Brightman	Crittenden County, Ky	Sept. 3, 1872	131, 048

Index of patents issued from the United States Patent Office from 1790 to 1873, inclusive—Continued.

Invention.	Inventor.	Residence.	Date.	No.
Medicine for dyspepsia	N. H. Chesebrough	New York, N. Y	Apr. 17, 1866	54, 056
Medicine for dyspepsia, &c	D. Mayon and E. Champlain	Cloverdale, Cal	Aug. 13, 1872	130, 518
Medicine for erysipelas	H. A. Lamb	Portland, Me	July 4, 1865	48, 567
Medicine for fowls	S. H. Kipp	Millerstown, Pa	Oct. 25, 1870	108, 603
Medicine for gonorrhœa, &c	S. F. Alford	Goshen Springs, Miss	Oct. 22, 1872	132, 425
Medicine for gout	E. Smith	New York	Dec. 15, 1828	....
Medicine for gout and rheumatism	W. A. Parker	Accomac County, Va	Nov. 4, 1831	....
Medicine for healing wounds, &c	N. W. Gaddy	Nichols, S. C	July 16, 1872	129, 469
Medicine for heart-disease, &c	M. D. Britten	Eaton, Mich	July 2, 1872	128, 528
Medicine for kidney-diseases	R. Hawkins and A. A. Hill	Beallsville, Pa	Aug. 1, 1871	117, 537
Medicine for kidney-diseases	W. A. Ruth	Wyoming, Del	Apr. 1, 1873	137, 488
Medicine for liver and other diseases	C. A. Simmons	Waldo, Fla	July 25, 1871	117, 338
Medicine for liver-disease	J. M. Cunningham	Mount Morris, Ill	June 10, 1873	139, 707
Medicine for lung and other diseases	W. Trinder	Philadelphia, Pa	Oct. 8, 1872	132, 119
Medicine for lung-diseases	W. A. Smith	Concord, N. C	Aug. 13, 1872	130, 392
Medicine for miasmatic diseases	J. C. Hascall	Corunna, Mich	Apr. 22, 1862	35, 022
Medicine for neutralizing the taste of cod-liver oil	C. D. Bradley	Taunton, Mass	July 22, 1873	141, 030
Medicine for organs of the voice	F. Frisiani	New York, N. Y	Aug. 24, 1869	93, 982
Medicine for piles	L. E. Brady	Gatesville, N. C	July 9, 1872	128, 784
Medicine for piles	W. Carr	Bath, Me	May 12, 1863	38, 466
Medicine for piles	S. S. Davis	Edgerton, Wis	July 13, 1869	92, 587
Medicine for piles	L. Heins	New York, N. Y	Jan. 9, 1872	122, 608
Medicine for piles	L. L. Moore and R. B. S. Whayre	Calhoun, Ky	June 27, 1871	116, 474
Medicine for piles	M. Rinehart	Lockport, Ill	Nov. 24, 1868	84, 441
Medicine for piles	P. Roskopf	Brooklyn, E. D., N. Y	Feb. 20, 1872	123, 934
Medicine for piles	J. W. Ward	Lowell, Ohio	June 11, 1872	127, 719
Medicine for purifying the blood	R. Guinn	Baltimore, Md	Apr. 20, 1869	89, 039
Medicine for purifying the blood	P. H. Schmid	New York, N. Y	Oct. 24, 1871	120, 333
Medicine for rheumatism, &c	H. A. Chase	Tully, N. Y	July 19, 1870	105, 425
Medicine for rheumatism	G. Conroy	Mendocino County, Cal	Oct. 12, 1869	95, 771
Medicine for rheumatism, &c	W. Curless	Truckee, Cal	Aug. 8, 1871	117, 870
Medicine for rheumatism, &c	W. H. Farrar	Richmond, Va	Apr. 25, 1871	113, 995
Medicine for rheumatism	R. Gilkinson	New York, N. Y	Jan. 24, 1871	111, 195
Medicine for rheumatism, &c	T. B. Hick	New Haven, Conn	Aug. 13, 1872	130, 497
Medicine for rheumatism	A. J. Jenkins	Virginia City, Nev	Nov. 5, 1870	109, 216
Medicine for rheumatism	N. Jenkins	New Orleans, La	Dec. 27, 1870	110, 470
Medicine for rheumatism	J. Keiser	York, Pa	Oct. 17, 1871	120, 074
Medicine for rheumatism	W. Landert and J. Deggeller	Chicago, Ill	May 3, 1870	102, 686
Medicine for rheumatism	R. Lighthall	Brooklyn, N. Y	June 29, 1869	92, 065
Medicine for rheumatism, &c	G. Lucy	Mobile, Ala	Apr. 29, 1873	138, 335
Medicine for rheumatism	H. H. Munroe	Louisville, Ky	Dec. 14, 1869	97, 794
Medicine for rheumatism	H. Sawyer	Wilson, N. C	May 9, 1871	114, 717
Medicine for rheumatism	C. F. Washburn	San Francisco, Cal	Nov. 7, 1871	120, 802
Medicine for rheumatism	A. Whitney	Boston, Mass	Sept. 5, 1871	118, 766
Medicine for rheumatism and other diseases	J. B. Rodgers	Saint Louis, Mo	Apr. 19, 1870	102, 161
Medicine for ringbone, spavin, &c	W. A. Cleveland	Waterville, N. Y	Mar. 23, 1869	88, 134
Medicine for ringbone, spavin, &c	A. M. French	Burton, Ohio	May 28, 1867	65, 203
Medicine for ringbone, spavin, &c	H. H. Vance	New Corydon, Ind	Nov. 24, 1868	84, 322
Medicine for ringbone, spavin, splint, &c	W. A. Cleveland	Waterville, N. Y	Oct. 22, 1867	69, 970
Medicine for scarlatina, sore-throat, &c	M. A. Ragsdale	Grass Valley, Cal	Aug. 16, 1864	43, 866
Medicine for scrofula and cancer	E. Willard	Albany, N. Y	Oct. 8, 1810	....
Medicine for scrofula, liver-complaint, &c	P. Fay, sr	Lacon, Ill	Mar. 22, 1870	113, 151
Medicine for scrofula, rheumatism, &c	G. Jaques	Wilmington, Del	July 16, 1824	
Medicine for skin-diseases	F. W. A. Bergengren	Stockholm, Sweden	Mar. 19, 1872	124, 660
Medicine for skin-diseases	A. B. Simpson	Camden, Ohio	Mar. 4, 1873	136, 554
Medicine for small-pox	A. Haws	Battle Mountain Station, Nev.	July 16, 1872	129, 342
Medicine for sore-throat	F. M. Moore	Chico, Cal	Oct. 10, 1871	119, 782
Medicine for spavin	W. J. Andrews	East Machias, Me	June 20, 1871	116, 135
Medicine for the teeth	J. C. Hassell	Nevada City, Cal	July 2, 1872	128, 620
Medicine for toothache	F. J. Oswald	New York, N. Y	Dec. 24, 1872	134, 217
Medicine for toothache	S. Pennington	Mount Pleasant, Ohio	July 30, 1829	....
Medicine for toothache	W. P. Sigsby	Delta, Ohio	Oct. 5, 1869	95, 610
Medicine for toothache	J. Utley	Hartford, Conn	Nov. 5, 1817	....
Medicine for toothache	T. White	Saint Clairsville, Ohio	July 29, 1829	....
Medicine for treating catarrh, &c., by inhalation	D. Slade	Chicago, Ill	Mar. 21, 1871	112, 858
Medicine for treating diseases of the lungs, chest, &c.	M. E. Perrin	Montreal, Canada	July 11, 1871	116, 863
Medicine for treating horses' hoofs, &c	J. C. Burroughs	Clinton, Mich	Dec. 2, 1873	145, 047
Medicine for treating ringworm, &c	P. Roskopf	Brooklyn, N. Y	June 11, 1872	127, 925
Medicine for wounds, inflammation, &c	O. Troemel	Manitowoc, Wis	Mar. 8, 1864	41, 873
Medicine for yellow-fever, &c	T. H. Thompson	Grenada, Miss	Mar. 12, 1872	124, 458
Medicine from chloral	T. H. Hazard	Richmond, Va	Mar. 15, 1870	100, 760
Medicine from globe-flowers	J. S. Pemberton	Atlanta, Ga	Mar. 1, 1870	100, 439
[Medicine.] Glanders in horses, Compound for cure of.	J. Althouse	East Cocalico Township, Pa.	Mar. 26, 1867	63, 195
Medicine-glass	T. G. Boggs	Philadelphia, Pa	Dec. 22, 1868	85, 203
Medicine, Hog	J. Shannon	Palmyra, Mo	Apr. 9, 1872	125, 492
Medicine, Hog	D. W. Stow	Thorntown, Ind	Dec. 18, 1866	60, 591
Medicine, Hog and chicken cholera	M. N. George	Evansville, Ind	July 18, 1871	117, 065
Medicine, Hog and chicken cholera	A. C. McMahan	Lincoln, Ill	Aug. 31, 1869	94, 428
Medicine, Hog-cholera	J. B. Ball	Lebanon, Ind	June 23, 1868	79, 098
Medicine, Hog-cholera	S. S. Barger	Golconda, Ill	July 6, 1869	92, 248
Medicine, Hog-cholera	G. H. Baugh	Oskaloosa, Iowa	Nov. 27, 1866	59, 946
Medicine, Hog-cholera	M. Boyes	Pocahontas, Ill	Oct. 2, 1866	58, 371
Medicine, Hog-cholera	A. J. Carver and E. P. Horn	Greenhill, Tenn	Sept. 15, 1868	82, 205
Medicine, Hog-cholera	N. H. Cass	Henryville, Ind	Aug. 4, 1868	80, 597
Medicine, Hog-cholera	T. L. Cotten	Madison County, Miss	Oct. 18, 1870	108, 453
Medicine, Hog-cholera	E. Fabra	Covington, Ky	Oct. 15, 1872	132, 204
Medicine, Hog-cholera	H. C. Fahlbush	Marshall, Ill	June 25, 1872	128, 380
Medicine, Hog-cholera	G. W. Gish and W. H. Ferguson.	Daviess County, Ky	Jan. 7, 1868	72, 996
Medicine, Hog-cholera	J. and M. J. Holton	Mahaska County, Iowa	May 21, 1872	126, 961
Medicine, Hog-cholera	A. M. Johnston and H. H. Avrit.	Clarksville, Tenn	Aug. 3, 1869	93, 206
Medicine, Hog-cholera	C. L. Jones	Pedler Township, Va	Oct. 5, 1871	121, 523
Medicine, Hog-cholera	F. La Rew	Hamilton, Ohio	Apr. 14, 1863	38, 169
Medicine, Hog-cholera	J. Lighter	Clay Village, Ky	Mar. 19, 1861	31, 723

Index of patents issued from the United States Patent Office from 1790 *to* 1873, *inclusive*—Continued.

Invention.	Inventor.	Residence.	Date.	No.
Medicine, Hog-cholera	J. S. Mason	Jefferson County, Ky	July 3, 1866	56, 075
Medicine, Hog-cholera	J. H. Mesler	Symmes' Corners, Ohio	May 28, 1867	65, 256
Medicine, Hog-cholera	W. B. Robuck	Oxford, Miss	Jan. 12, 1869	85, 855
Medicine, Hog-cholera	W. M. Runyon, R. H. Haller, and D. B. Morris.	Oskaloosa, Iowa	Oct. 16, 1866	58, 890
Medicine, Horse	J. Andrews	Hardin County, Ohio	Jan. 14, 1873	134, 721
Medicine, Horse	T. Burns	Jefferson County, Ohio	Apr. 23, 1872	125, 933
Medicine, Horse	P. Cope	Perryopolis, Pa	Apr. 9, 1872	125, 441
Medicine, Horse	C. L. Hammond	North Java, N. Y	Oct. 30, 1866	59, 216
Medicine, Horse	V. Vance	Havana, N. Y	July 25, 1865	49, 015
Medicine, Horse and cattle powder	A. Rudisill and M. Sell	Sell Station, Pa	Feb. 14, 1871	111, 876
Medicine, Horse, cattle, &c	G. Van Wagenen	Racine, Wis	Sept. 1, 1868	81, 711
[Medicine.] Horse-powder	T. H. Russell	Lebanon, N. H	Aug. 15, 1871	118, 056
[Medicine.] Lotion	J. Hopkins	Philadelphia, Pa	Aug. 24, 1809	....
Medicine, Nutritive	S. C. Upham	Philadelphia, Pa	Feb. 12, 1867	62, 091
Medicine of vegetable alkaloids	G. W. Scollay	New York, N. Y	Dec. 19, 1871	122, 065
Medicine or apple ginger	D. Nixon	Philadelphia, Pa	Sept. 12, 1871	118, 812
Medicine or blood-purifier	D. Parlow	Philadelphia, Pa	May 13, 1873	138, 759
Medicine or capsicum plaster	J. and I. Coddington	New York, N. Y	Apr. 2, 1872	125, 173
Medicine or coffee-antidote	E. D. D. St. Cyr	Lowell, Mass	Dec. 5, 1871	121, 665
Medicine or cordial	J. Ambrose	Nashville, Tenn	Oct. 5, 1869	95, 407
Medicine or cordial	J. Shapley and A. D. Hutchinson.	Rosefield, Ill	June 29, 1869	92, 107
Medicine or ethereal oil for treating rheumatism, &c	L. J. Bell	College Corner, Ind	Feb. 13, 1872	123, 666
Medicine or infant's powders	J. Fehr	Hoboken, N. J	Sept. 23, 1873	143, 133
Medicine or liver-invigorator	W. S. Simmons	Weatherford, Tex	Jan. 9, 1872	122, 672
Medicine or liver-lozenge	S. Gilbert	Portsmouth, N. H	Dec. 16, 1873	145, 498
Medicine or lotion for the skin	A. Field	Springfield, Mass	June 24, 1873	140, 126
Medicine or milk of magnesia	C. H. Phillips and L. Reid	New York and Brooklyn, N. Y.	Apr. 29, 1873	138, 282
Medicine or mosquito-lotion	A. D. Breazeale	Selma, Ala	Sept. 24, 1872	131, 655
Medicine from oak bark	S. J. Miller	Economy, Ind	Nov. 16, 1869	96, 943
Medicine or opium-cure	U. H. Kellogg	Jamestown, N. Y	Sept. 23, 1873	143, 164
Medicine or pain-remedy	J. J. Mills and W. Perry	Frederick's Hall, Va	Nov. 1, 1870	108, 811
Medicine or pectoral sirup	G. La Montague	Muskegon, Mich	Sept. 6, 1870	107, 066
Medicine or piles-liniment	P. H. Steenbergen	Chico, Cal	Aug. 19, 1873	141, 961
Medicine or pill	J. P. Waddell	Brookhaven, Miss	Aug. 9, 1870	106, 239
Medicine or powders for horses, &c	J. Norcross	West Newton, Pa	Mar. 28, 1871	113, 198
Medicine or powders for treating cattle and hogs	R. H. Pullin	McDowell, Va	Aug. 15, 1871	118, 052
Medicine or restorative balsam	G. H. Brecht	Burton, Ill	Nov. 7, 1871	120, 705
Medicine or worm-candy	J. C. Wills	Owingsville, Ky	Jan. 2, 1872	122, 507
[Medicine.] Pain-destroying compound	S. T. Bond	Edenton, N. C	July 21, 1868	80, 123
[Medicine.] Pain-killer	H. Stonebraker	Baltimore, Md	Feb. 19, 1867	62, 297
Medicine, Preparation of aerated drink for	P. H. Vander Weyde and J. Matthews, jr.	New York, N. Y	Mar. 30, 1869	88, 348
Medicine, Preparation of cod-liver and other oils for	T. Hyatt	Atchison, Kans	Jan. 5, 1869	85, 668
Medicine, Preparation of iron for	J. E. Siebel	Chicago, Ill	July 27, 1869	93, 131
Medicine, Process of preparing homeopathic	B. Fincke	Brooklyn, N. Y	Aug. 24, 1869	93, 980
[Medicine.] Prophylactic remedy	C. H. Thomas	New Orleans, La	Oct. 16, 1860	30, 434
Medicine, Putting up	F. Kraus	Cincinnati, Ohio	Oct. 4, 1870	108, 034
[Medicine.] Salt-rheums, Remedy for	W. B. Trufant	Bath, Me	Feb. 10, 1838	597
Medicine-spoon and bottle-stopper combined	S. C. Currie	New York, N. Y	July 6, 1869	92, 278
Medicine, Sugar tablet for containing	H. F. Wiesecke	New York, N. Y	Oct. 28, 1862	36, 816
[Medicine.] Sweating-powders	H. Howard	Columbus, Ohio	Aug. 25, 1832	....
[Medicine.] Syphilis, Cure of	N. W. Badeau	New York	Mar. 8, 1832	....
[Medicine.] Syphilis, Cure of	J. Moser	Philadelphia, Pa	Jan. 26, 1816	....
Medicine to animals, Instrument for giving fluid	H. A. Brandes	Newark, N. J	June 21, 1870	104, 416
Medicine to diseased parts, Instrument for conveying.	P. J. Frank	Ashford, N. Y	Nov. 2, 1869	96, 415
Medicine to prevent or cure the effects of the bite of a mad-dog.	W. Stoy	Lebanon, Pa	June 9, 1809	....
Medicine to prevent scurvy	J. W. Armour	Fredericktown, Md	Sept. 28, 1827	....
Medicine to the ear, Apparatus for applying	S. Van Etten	Corning, N. Y	May 21, 1872	127, 123
[Medicine.] Tonic or bitters	J. Heisler	Schuylkill Haven, Pa	Mar. 3, 1868	75, 161
Medicine, Worm	A. E. Wright	Hamlin, N. Y	May 30, 1871	115, 407
Medicine, Worm-destroying	J. J. Oellig	Waynesborough, Pa	Oct. 28, 1837	445
Medicine: *See* Liniment, Ointment, Salve.				
Melodeon	C. Austin	Concord, N. H	June 19, 1849	6, 543
Melodeon	C. Austin	Concord, N. H	Jan. 22, 1867	61, 305
Melodeon	J. Berger	Knoxville, Ill	May 14, 1867	64, 741
Melodeon	W. H. Bigelow	South Framingham, Mass	Sept. 3, 1861	33, 180
Melodeon	T. Brett	Geneva, Ohio	Oct. 25, 1870	108, 560
Melodeon	J. C. Briggs	Woodbury, Conn	Feb. 10, 1857	16, 574
Melodeon	J. C. Briggs	Ansonia, Conn	Sept. 27, 1870	107, 655
Melodeon	R. Burditt	Brattleborough, Vt	Sept. 10, 1861	33, 278
Melodeon	R. Burditt and H. P. Green	Brattleborough, Vt	Mar. 10, 1857	16, 786
Melodeon	C. G. Burke	Utica, N. Y	Dec. 6, 1859	26, 344
Melodeon	P. Corbach	Cleveland, Ohio	Aug. 28, 1866	57, 472
Melodeon	J. Carhart	New York, N. Y	Apr. 17, 1855	12, 713
Melodeon	J. Carhart	New York, N. Y	July 1, 1856	15, 218
Melodeon	J. R. Cluxton	Russellville, Ohio	Sept. 14, 1869	94, 713
Melodeon	W. Cooper	Deposit, N. Y	Jan. 7, 1868	73, 080
Melodeon	K. Ebermeyer	Ellwangen, Germany	Dec. 3, 1867	71, 593
Melodeon	P. Engers	Jefferson Furnace, Pa	Apr. 25, 1871	113, 994
Melodeon	W. Evans	Lockport, Ill	June 9, 1857	17, 501
Melodeon	O. N. Frary	Ansonia, Conn	Aug. 8, 1854	11, 479
Melodeon	H. N. Goodman	New Haven, Conn	June 28, 1853	9, 816
Melodeon	R. Goodrich	Pittsfield, Mass	Mar. 15, 1864	41, 915
Melodeon	E. Hamlin	Winchester, Mass	July 21, 1868	80, 167
Melodeon	F. Hoddick	Buffalo, N. Y	Oct. 23, 1866	59, 128
Melodeon	G. G. Hunt	Wolcottville, Conn	Oct. 23, 1855	13, 704
Melodeon	S. A. Jewett	Cleveland, Ohio	Oct. 13, 1857	18, 399
Melodeon	S. A. Jewett	Cleveland, Ohio	May 28, 1867	65, 088
Melodeon	S. H. Jones	Jamaica Plains, Mass	July 3, 1860	29, 037
Melodeon	W. J. Kinnard and B. Dreher	Cleveland, Ohio	Apr. 4, 1865	47, 110

Index of patents issued from the United States Patent Office from 1790 *to* 1873, *inclusive*—Continued.

Invention.	Inventor.	Residence.	Date.	No.
Melodeon	L. Louis	Boston, Mass	Nov. 18, 1856	16, 094
Melodeon	L. Louis	Buffalo, N. Y	June 24, 1862	35, 737
Melodeon	L. Louis	Providence, R. I	June 18, 1867	65, 822
Melodeon	W. N. Manning	Rockport, Mass	June 3, 1856	15, 014
Melodeon	A. Nudd	Waupun, Wis	June 18, 1867	65, 936
Melodeon	P. J. Puetz	Milwaukee, Wis	June 27, 1865	48, 437
Melodeon	I. Rehn	Philadelphia, Pa	Nov. 16, 1858	22, 089
Melodeon	J. A. Rollins	Buffalo, N. Y	June 3, 1856	15, 019
Melodeon	W. T. Sheldon	East Mendon, N. Y	June 25, 1861	32, 654
Melodeon	G. O. Spence	Elmira, N. Y	Dec. 23, 1856	16, 296
Melodeon	A. L. Swan	Cherry Valley, N. Y	May 7, 1850	7, 352
Melodeon	A. L. Swan	Cherry Valley, N. Y	Mar. 9, 1852	8, 792
Melodeon	T. T. Thornton	Buffalo, N. Y	Mar. 3, 1857	16, 760
Melodeon	C. J. Van Oeckelen	New York, N. Y	July 3, 1860	29, 023
Melodeon	W. C. Whipple and W. C. Bowe	Westville, Conn	June 5, 1855	13, 021
Melodeon	A. W. Wilcox	Worcester, Mass	Apr. 9, 1867	63, 683
Melodeon	G. Wood	Detroit, Mich	July 23, 1867	67, 150
Melodeon-action	J. C. Briggs	Ansonia, Conn	Feb. 22, 1870	100, 009
Melodeon and other wind-instruments	R. W. Carpenter	Brattleborough, Vt	May 28, 1867	65, 171
Melodeon-attachment	D. L. Sprague and R. Burditt	Townsend and Brattleborough, Vt.	July 7, 1857	17, 755
Melodeon-case	G. S. Randall	Providence, R. I	Mar. 23, 1869	88, 073
Melodeon-coupling	E. B. Carpenter and E. N. Merriam.	Brattleborough and East Poultney, Vt.	Nov. 24, 1857	18, 676
Melodeon, Flutina	J. Müller and F. Kaiser	Philadelphia, Pa	Nov. 19, 1867	71, 204
Melodeon or organ swells, Knee-lever for	O. C. Whitney	Meadville, Pa	July 12, 1870	105, 284
Melodeon, Reed for	R. A. Abbott	Worcester, Mass	Sept. 22, 1868	82, 268
Melodeon-stops and other articles, Manufacture of	J. Gardner	New Haven, Conn	Apr. 14, 1868	76, 623
Melodeons and other reed-instruments, Tuning	E. E. Shepardson and E. Lucas	New Bedford, Mass	June 28, 1853	9, 821
Melodeons, &c., Attaching keys to	A. G. Burton	Shelburne Falls, Mass	July 5, 1864	43, 389
Melodeons, &c., Base-damper for	R. Burditt	Brattleborough, Vt	July 8, 1856	15, 316
Melodeons, &c., Basso-tenuto for	A. Zawadski	Syracuse, N. Y	Apr. 26, 1870	102, 353
Melodeons, Machine for manufacturing reed-board for.	J. Carhart	New York, N. Y	June 10, 1856	15, 061
Melodeons, Mode of operating swell in	C. E. Bacon	Buffalo, N. Y	Dec. 10, 1867	71, 947
Melodeons, Mode of operating the swell of	D. Clark and T. Stephenson	Buffalo, N. Y	Feb. 14, 1865	46, 335
Melodeons, Octave-coupling for	B. W. Nichols	Fair Haven, Conn	Mar. 4, 1862	34, 588
Melodeons, &c., Pedal-base for	G. W. Lane and W. N. Manning.	Rockport, Mass	Dec. 1, 1857	18, 751
Melodeons, Reed-plate for	G. W. Ingalls	Concord, N. H	Oct. 23, 1866	59, 027
Melodeons, &c., Swell for	J. Carhart	New York, N. Y	May 5, 1857	17, 196
Melodeons, &c., Swell for	T. F. Thornton	Buffalo, N. Y	May 22, 1855	12, 938
Melodeons, &c., Tremulo-attachment for	R. W. Carpenter	Brattleborough, Vt	June 4, 1867	65, 472
Melodeons, &c., Valve and valve-spring for	A. L. Swan	Cherry Valley, N. Y	Apr. 13, 1869	88, 925
Melodeons, Valve for	E. A. Jones and J. A. Bidwell	Sturgis, Mich	Nov. 10, 1868	83, 971
Melody-attachment	C. Fogelberg	Boston, Mass	Apr. 5, 1870	101, 452
Melon-shaped mold	E. Hodgetts	Brooklyn, N. Y	Sept. 19, 1871	119, 146
Melting and casting metal under pressure, Apparatus for.	H. Bessemer	London, England	July 25, 1871	117, 249
Melting and smelting furnace	J. Green	Norristown, Pa	Mar. 26, 1867	63, 240
Melting and smelting furnace	W. Quann and W. T. Smith	Philadelphia, Pa	Nov. 28, 1865	51, 266
Melting and smelting furnace	J. F. Stileman and Z. Ellis	Philadelphia, Pa	May 12, 1863	38, 513
Melting-furnace	S. A. Whitney	Glassborough, N. J	Dec. 4, 1866	60, 307
Melting-pot covering	A. C. Lewis	New York, N. Y	Apr. 4, 1871	113, 312
Menstrual-receiver	T. A. Gamage	Boston, Mass	Dec. 15, 1868	84, 874
Menstrual-receiver	S. L. Hockert	Chicago, Ill	Nov. 12, 1867	70, 843
Mensuration, Apparatus for teaching	I. Harrington	Huntington, Conn	Mar. 25, 1873	137, 075
Merchandise-elevator	W. McNaughton	Utica, Ohio	Dec. 20, 1870	110, 382
Mercurial heater	T. C. Hargrave and K. W. King	Boston, Mass	June 20, 1865	48, 276
Mercurial regulator for vulcanizing and other heaters.	J. B. Coolidge	Boston, Mass	Nov. 23, 1869	97, 052
Mercury from its ores, Apparatus for obtaining	H. Johnson	San Francisco, Cal	Apr. 9, 1872	125, 398
Mercury from its ores, Extracting	A. Patera	Vienna, Austria	Aug. 15, 1871	118, 147
Mercury from the washings of gold and silver ores, Apparatus for saving.	W. Spicer	Summit County, Colo	July 5, 1870	105, 010
Meridian-finder	W. Stackpole	Brooklyn, N. Y	Sept. 26, 1865	50, 182
Meridian, &c., Finding true	W. A. Burt	Mount Vernon, Mich	Feb. 25, 1836	....
Meridian-instrument	F. Yeiser	Danville, Ky	Dec. 17, 1861	33, 970
Metal, Anti-friction	A. B. Cook	Manchester, Tenn	June 1, 1869	90, 729
Metal, Anti-friction	G. S. Hunt	Sacramento, Cal	Feb. 25, 1873	136, 163
Metal, Apparatus for bundling scrap	L. J. Atwood	Waterbury, Conn	Jan. 15, 1867	61, 136
Metal-backed comb	E. E. Warner, J. W. Walton, and R. Vine.	Philadelphia, Pa	July 5, 1870	105, 148
Metal bands, Apparatus for shrinking	W. Carr	Logansport, Ind	Aug. 19, 1873	141, 992
Metal bars, Apparatus for bending	G. J. Neveil	Philadelphia, Pa	June 7, 1864	43, 037
Metal bars, Drawing or forging	W. Beach	Philadelphia, Pa	Apr. 7, 1863	38, 090
Metal bars, Fagot for	E. Wheeler	Philadelphia, Pa	Dec. 19, 1871	122, 090
Metal bars, Machine for bending	O. Long	Joliet, Ill	Apr. 2, 1867	63, 403
Metal bars, Machine for bending and twisting	S. Holmes	Philadelphia, Pa	Aug. 19, 1873	142, 018
Metal bars, Machine for breaking	W. Richter	Eintrachtshuette, Germany	Jan. 28, 1873	135, 364
Metal bars, Machine for cutting	D. R. Knowles	Centre Groton, Conn	Sept. 21, 1858	21, 567
Metal bars, Machine for shaping and heading	H. Thompson	Worcester, Mass	Apr. 9, 1867	63, 766
Metal bars, Machine for straightening	S. P. M. Tasker	Philadelphia, Pa	June 25, 1872	128, 335
Metal bars, Machine for straightening heavy	I. B. Howe	Northfield, Vt	Dec. 19, 1854	12, 098
Metal bars, Machine for twisting	A. D. Williams	London, England	Oct. 3, 1871	119, 486
Metal bars, Machinery for crimping	G. Slocum and M. T. Sayles	Lansingburgh, N. Y	Nov. 9, 1852	9, 392
Metal bars, Machinery for reducing	D. H. Chamberlain	Boston, Mass	Jan. 18, 1853	9, 542
Metal bars, Method of rolling	J. Horner	New York, N. Y	July 5, 1870	104, 957
Metal bars, rails, tubes, &c., Machine for straightening.	B. Lauth	Buchanan, Pa	Jan. 19, 1864	41, 306
Metal bars, Tender for heating and delivering	S. A. Darrack	Newburgh, N. Y	Apr. 21, 1868	77, 006
Metal box	J. L. Kite	Philadelphia, Pa	May 19, 1863	38, 585
Metal box for wood or iron to turn in	W. Shotwell	....	Feb. 7, 1800	....
Metal boxes, Machine for making	C. L. Rehn	Philadelphia, Pa	Oct. 20, 1868	83, 209
Metal boxes, Machine for making	J. Wild	Philadelphia, Pa	Nov. 26, 1867	71, 351
Metal boxes, Manufacture of	W. T. Slocum	Philadelphia, Pa	May 2, 1865	47, 606
Metal bracket, Cast	P. Bradford	New Haven, Conn	June 9, 1868	78, 784
Metal burnishing and spinning machine	F. J. Seymour	Wolcottville, Conn	Aug. 11, 1868	80, 836

Index of patents issued from the United States Patent Office from 1790 *to* 1873, *inclusive*—Continued.

Invention.	Inventor.	Residence.	Date.	No.
Metal can	J. Pollock and T. J. Diederick	Philadelphia, Pa	Nov. 3, 1868	83, 660
Metal can and box for paints, &c	F. W. Devoe	New York, N. Y	May 28, 1867	65, 181
Metal can, case, box, &c., for preserving food, gunpowder, liquids, paints, &c.	J. Bouvet	La Rochelle, France	June 28, 1864	43, 378
Metal can or box	H. Everett	Philadelphia, Pa	May 30, 1865	47, 939
Metal can or case for putting up alkalies	J. McCoy	Philadelphia, Pa	May 12, 1868	77, 899
Metal cans, Machine for cutting and pressing the heads of	R. Brady	New York, N. Y	Jan. 4, 1870	98, 550
Metal cans, Machine for manufacturing	J. Mays and E. W. Bliss	Brooklyn, N. Y	June 15, 1869	91, 248
Metal cans, Securing caps to	D. W. Pepper	Philadelphia, Pa	Jan. 29, 1867	61, 563
Metal caps for boxes, Machine for making	C. Diederick and W. T. Slocum	Philadelphia, Pa	Apr. 3, 1860	27, 756
Metal-casting mold	C. B. Cotter	Milford, Pa	Nov. 18, 1862	36, 943
Metal-casting mold	H. Davies	Covington, Ky	Jan. 11, 1870	98, 673
Metal castings, Machine for grinding	G. H. Spencer	York, Pa	Mar. 25, 1873	137, 104
Metal cleaning and silvering composition	C. P. Brockett, E. Todd, and J. Brockett.	New Haven, Conn	Mar. 13, 1860	27, 425
Metal-cleaning powder	J. H. Musgrave and J. M. Biedel.	Chambersburgh, Pa	May 7, 1872	126, 409
Metal-coated ware	G. Booth	Philadelphia, Pa	Feb. 18, 1873	136, 025
Metal-coating composition	J. Weisman	Philadelphia, Pa	July 5, 1864	43, 444
Metals, Coloring	J. Kintz	West Meriden, Conn	May 27, 1873	139, 319
Metal comb	H. Durell	New York, N. Y	June 20, 1836	....
Metal comb	R. A. Ives	Bristol, Conn	July 1, 1836	....
Metal combs, Making	R. A. Ives	Bristol, Conn	Sept. 25, 1838	945
Metal, Composition	J. Hackert	Bridgeport, Conn	May 17, 1864	42, 762
Metal-compressing and wire-pointing machine	A. B. Hendryx and A. W. Webster.	Ansonia, Conn	Nov. 19, 1872	133, 224
Metal-corrugating machine	J. B. Williams	Glastenbury, Conn	June 29, 1869	91, 889
Metal cupping and raising die	S. R. Wilmot	Bridgeport, Conn	June 26, 1866	55, 944
Metal cutter and shearer	I. Dubois	Boonesborough, Iowa	Nov. 15, 1870	109, 192
Metal cutter and shearer	T. C. Robinson	Boston, Mass	July 6, 1869	92, 368
Metal-cutter, Rotary	A. P. Stephens	Brooklyn, N. Y	Mar. 16, 1869	87, 985
Metal cutting and planing machine	D. Saunders	New York, N. Y	Oct. 28, 1862	36, 803
Metal cutting and punching machine	N. De Telescheff	St. Petersburg, Russia	Apl. 26, 1864	42, 531
Metal cutting and punching machine	A. L. Hastings	Horton, Iowa	May 7, 1872	126, 543
Metal cutting and shaping machine	C. Van Haagen	Philadelphia, Pa	Dec. 27, 1870	110, 609
Metal-cutting machine	A. S. Bunker	Lawrence, Mass	Apl. 30, 1872	126, 179
Metal-cutting machine	J. Tetlon	Salem, Mass	July 14, 1857	17, 813
Metal cutting, punching, and upsetting machine	H. B. Sevey	Vienna, Me	July 9, 1872	128, 758
Metal-cutting shears	H. Barth	Cincinnati, Ohio	Oct. 10, 1865	50, 325
Metal-cutting shears	G. Christian	Mentz, N. Y	Sept. 20, 1834	....
Metal-cutting shears	J. Hornig	Oswego, N. Y	Feb. 7, 1865	46, 237
Metal-cutting shears	G. M. Marshall	New Haven, Wis	Dec. 26, 1871	122, 270
Metal-cutting shears	T. F. Taft	Worcester, Mass	Aug. 18, 1857	18, 025
Metal-cutting tool	L. F. Goodyear	New Haven, Conn	Mar. 29, 1859	23, 364
Metal-cutting tool	J. Mooney	Providence, R. I	June 24, 1856	15, 190
Metal-die	B. Mead and G. Richards	Boston, Mass	July 2, 1814	....
Metal, Die for forming articles of	E. J. Manville	Waterbury, Conn	June 5, 1866	55, 425
Metal-drill	W. C. Burch	Gloucester, N. J	Dec. 31, 1872	134, 355
Metal-drill	W. R. Jones	Granville, N. Y	Jan. 11, 1836	....
Metal-drill	J. A. Kirkpatrick and G. W. Hornby.	Evansville, Ind	Aug. 10, 1869	93, 451
Metal-drill	W. Lyon	New York, N. Y	Sept. 20, 1853	10, 035
Metal-drill	J. Murphy	Half Moon, Pa	May 17, 1859	24, 045
Metal-drill	A. Pease	Enfield, Conn	Apr. 4, 1854	10, 744
Metal-drill	W. Wakeley	Homer, N. Y	Nov. 28, 1854	12, 010
Metal drill or borer, Governing the feed of a	J. R. Grout	Birmingham, Mich	Apr. 25, 1844	3, 564
Metal drilling and boring machine	C. Sellers	Philadelphia, Pa	Mar. 1, 1864	41, 789
Metal-drilling machine	R. Wilson	Milton, Pa	Dec. 14, 1858	22, 323
Metal-drilling machine	A. Woodworth	Cambridge Village, N. Y	Dec. 23, 1873	145, 923
Metal-drilling tool	E. Davies and R. H. Taunton	Birmingham, England	Nov. 26, 1867	71, 463
Metal expansion and contraction to generate power	H. Knowles	Hartford, Conn	Nov. 8, 1836	....
Metal-folding machine	D. Newton	Southington, Conn	Jan. 23, 1855	12, 282
Metal-forging machine	E. Wheeler	Feltonville, Mass	Jan. 27, 1857	16, 505
Metal, glass, and pottery, Manufacturing	E. W. Nohl	Ripon, Wis	Apr. 30, 1867	64, 353
Metal, glass, &c., Furnace for melting	W. P. Prickett	Philadelphia, Pa	May 19, 1868	78, 007
Metal-heating apparatus	J. Roy	Boston, Mass	Feb. 2, 1869	86, 456
Metal-heating furnace	I. J. Conklin	Philadelphia, Pa	Oct. 15, 1872	132, 139
Metal hoop	S. Lewis	Rochester, N. Y	Aug. 22, 1871	118, 252
Metal hoop, Flaring	H. D. Barnes	New Haven, Conn	Nov. 15, 1864	45, 015
Metal, Joining plates of	E. Jacobs	Cincinnati, Ohio	Feb. 21, 1860	27, 222
Metal melting and refining furnace	T. and J. Barnhurst and W. Walker.	Philadelphia, Pa	May 4, 1814	....
Metal-melting furnace	J. Harrison	East Hampton, Conn	July 22, 1873	141, 139
Metal-melting furnace	W. Shea and L. D. Harvey	Harvey, Mich	July 14, 1868	79, 866
Metal-melting furnace	W. A. Sweet	Syracuse, N. Y	July 11, 1865	48, 739
Metal-melting furnace	J. Thomson and T. Widdowfield.	New York and Brooklyn, N. Y.	June 7, 1864	43, 048
Metal-melting furnace and pot	S. Grilley	Waterbury, Conn	Mar. 17, 1838	644
Metal-mold	J. W. Hollingsworth	Mount Vernon, Ind	Apr. 12, 1870	101, 877
Metal or wood into the ground, Apparatus for driving.	W. W. Winter	Cortlandville, N. Y	Feb. 20, 1866	52, 799
Metal pipe and tube for conveying water, gas, &c	D. D. Parmelee	New York, N. Y	June 18, 1872	128, 166
Metal-pipe-bending machine	J. Perkins and W. H. Burnet	Newark, N. J	Oct. 14, 1856	15, 899
Metal pipe or tube, Composition	B. Titus	Ulysses, N. Y	Apr. 19, 1831	....
Metal pipes, Casting	B. S. Benson	Baltimore, Md	Dec. 3, 1867	71, 687
Metal pipes, Casting	J. J. C. Smith	Philadelphia, Pa	May 18, 1869	90, 201
Metal pipes, Composition for lining	W. Johnston and H. Forbes	Brooklyn, N. Y	Jan. 18, 1859	22, 654
Metal pipes, Connection for soft	I. Davis	Brooklyn, N. Y	June 23, 1868	79, 055
Metal pipes, Determining the thickness of	H. Allen	New York, N. Y	July 8, 1843	3, 145
Metal pipes, Machine for coiling	P. L. Weimer	Lebanon, Pa	Nov. 9, 1858	22, 044
Metal pipes, Machine for coiling	P. L. Weimer	Lebanon, Pa	Aug. 30, 1859	25, 294
Metal pipes, Molding	H. Parmelee	Philadelphia, Pa	Sept. 4, 1860	29, 908
Metal pipes, roofing, &c., Composition for coating	J. Specht	Memphis, Tenn	Jan. 14, 1873	134, 822
Metal pipes with gutta-percha lining	A. D. Puffer	Somerville, Mass	May 20, 1856	14, 929
Metal-planing gage	H. H. Jennings	New Haven, Conn	Aug. 4, 1863	39, 404
Metal-planing machine	F. A. Pratt	Hartford, Conn	Aug. 17, 1869	93, 903
Metal-planing machine	A. Whitcomb	Worcester, Mass	Aug. 17, 1869	93, 847

Index of patents issued from the United States Patent Office from 1790 *to* 1873, *inclusive*—Continued.

Invention.	Inventor.	Residence.	Date.	No.
Metal-plate-bending machine	J. W. Easby	Washington, D. C	July 4, 1865	48, 538
Metal-plate-bending machine	R. Tippett	Harrisburgh, Pa	Feb. 23, 1869	87, 224
Metal-plate-heating and ore-calcining furnace	H. Chess	Pittsburgh, Pa	July 25, 1871	117, 382
Metal-plate-polishing machine	E. C. Atkins	Indianapolis, Ind	June 16, 1868	78, 915
Metal-plates, Apparatus for cutting tined implement from.	A. Clark	Saint Johnsville, N. Y	Nov. 25, 1862	36, 993
Metal plates, Beveling the edges of	W. H. Singer	Pittsburgh, Pa	Dec. 6, 1870	109, 953
Metal plates, Device for grinding	T. Handy	Decatur, Ill	Dec. 4, 1866	60, 181
Metal plates, Machine for manufacturing waved and corrugated.	R. Montgomery	New York, N. Y	July 26, 1859	24, 882
Metal plates, Machine for manufacturing waved and corrugated.	R. Montgomery	New York, N. Y	July 26, 1859	24, 883
Metal plates, Machine for straightening rolled	R. Crooker	Boston, Mass	June 25, 1872	128, 287
Metal plates of unequal thickness, Machine for uniting.	J. Carhart	New York, N. Y	Aug. 1, 1854	11, 413
Metal plates or sheets, Machine for cutting	M. Gregg	Wilmington, Del	Apr. 1, 1842	2, 532
Metal-pointing machine	A. Moltz	New York, N. Y	Dec. 22, 1863	41, 014
Metal-polishing machine	R. Cave	Louisville, Ky	Feb. 1, 1859	22, 781
Metal-pouring ladle	A. S. Wells	New Britain, Conn	Mar. 15, 1870	100, 826
Metal-press	C. H. Schubens	Newark, N. J	May 7, 1867	64, 584
Metal punching and shearing machine	E. A. Sloat	Stone Mill, N. Y	Jan. 16, 1872	122, 860
Metal-punching machine	M. Burnett and C. Vander Woerd.	Boston, Mass	Apr. 17, 1855	12, 712
Metal-punching machine	D. and G. Fowler	Wallingford, Conn	Apr. 17, 1855	12, 723
Metal raising and trimming die	S. Simpson	Wallingford, Conn	May 25, 1869	90, 596
Metal-rod-bending machine	G. J. Neveil	Philadelphia, Pa	Aug. 8, 1865	49, 234
Metal rods and tubes, Making	J. Newman	Birmingham, England	Nov. 28, 1854	12, 000
Metal rods, Machine for forming eyes on	C. Kellogg	Detroit, Mich	Feb. 11, 1868	74, 377
Metal, Rolling: *See* Rolling metal.				
Metal scouring and burnishing machine	J. B. Lyons	Milton, Conn	Apr. 5, 1870	101, 479
Metal-scraping machine	J. Stever	Bristol, Conn	Dec. 12, 1854	12, 076
Metal screen	J. Hopkins	New York, N. Y	Nov. 18, 1862	36, 957
Metal screens, Machine for punching	J. W. Nesmith	Black Hawk, Colo	Aug. 31, 1869	94, 333
Metal-separator	E. Borlase	Bristol, Conn	July 7, 1857	17, 721
Metal-separator	T. J. Chubb	New York, N. Y	Dec. 19, 1854	12, 090
Metal-separator	H. W. Lathrop	Baltimore, Md	June 25, 1872	128, 404
Metal-separator	S. J. Peet	Boston, Mass	Apr. 29, 1873	138, 276
Metal-separator, Magnetic	P. Righter	Newark, N. J	Sept. 24, 1867	69, 128
Metal-shaping die	L. Dodge	Waterford, N. Y	July 19, 1859	24, 842
Metal-shaping machine	W. E. Cass	Newark, N. J	July 30, 1872	130, 016
Metal-shaping machine	M. Wells	Williamsburgh, N. Y	Apr. 23, 1867	64, 176
Metal-shearing apparatus	I. Lamplugh	Springfield, Ill	Mar. 22, 1864	42, 003
Metal-shearing device	J. E. Heath	Niles, Mich	June 11, 1867	65, 745
Metal-shearing machine	W. J. Adams	Grand Rapids, Mich	May 2, 1871	114, 385
Metal-shearing machine	R. Briggs	Philadelphia, Pa	June 22, 1869	91, 512
Metal-shearing machine	E. Doty and H. Richardson	Janesville, Wis	Mar. 29, 1870	101, 356
Metal-shearing machine	A. A. Kent	Lyons, Iowa	Sept. 6, 1870	107, 179
Metal-shearing machine	N. B. Reynolds	Auburn, N. Y	Feb. 19, 1867	62, 154
Metal shearing, punching, and bending apparatus	I. Lamplugh	Springfield, Ill	Mar. 22, 1864	42, 004
Metal-sheathing composition	G. F. Muntz	Birmingham, England	Aug. 1, 1848	5, 694
Metal springs, Machine for polishing	A. B. Doolittle	Hartford, Conn	May 7, 1867	64, 505
Metal-striking-up press	J. North	Brooklyn, N. Y	Jan. 9, 1867	51, 960
Metal strips, Machine for connecting	H. A. Bartlett	Philadelphia, Pa	Sept. 3, 1867	68, 547
Metal surfaces, Apparatus for stippling	E. Dimes	New York, N. Y	June 25, 1872	128, 290
Metal surfaces, Mode of uniting	W. B. Barnard	Waterbury, Conn	Jan. 27, 1863	37, 555
Metal surfaces, Tool for matting	D. Mosman	West Meriden, Conn	May 7, 1872	126, 408
Metal tools, Setting diamond or carbon points in the faces of	W. McKenna	Boston, Mass	Sept. 23, 1873	143, 021
Metal-trimming machine	J. H. Ferguson and H. W. Lovejoy.	New York, N. Y	Apr. 2, 1867	63, 491
Metal tubes and columns, Manufacture of	F. H. Smith	Baltimore, Md	Feb. 21, 1871	112, 884
Metal tubes, Apparatus for brazing	T. G. Gorman and N. W. Peregoy.	Springfield, Ill	Mar. 25, 1873	137, 068
Metal tubes, Arrangement of dies and stocks for ornamenting.	S. M. Cate and E. Jordan	Waterbury, Conn	Apr. 10, 1855	12, 669
Metal tubes, Arrangement of rollers for making	M. R. Griswold	Watertown, Conn	Apr. 17, 1855	12, 739
Metal tubes, Die for making seamless	T. D. Jackson	New York, N. Y	Feb. 28, 1854	10, 569
Metal tubes from forming-mandrels, Liberating	J. Cutler	Birmingham, England	July 29, 1851	8, 250
Metal tubes, Implement for cutting	T. J. Lloyd	Pottsville, Pa	Dec. 22, 1857	18, 918
Metal tubes, Machine for forming	E. Ells	Ansonia, Conn	July 17, 1855	13, 255
Metal tubes, Machine for making	H. Hotchkiss	Plainfield, N. J	July 24, 1866	56, 561
Metal tubes, Machine for making	C. G. Smith	Chelsea, Mass	Sept. 4, 1866	57, 783
Metal tubes, Machine for making	P. L. Weimer	Lebanon, Pa	July 3, 1866	56, 152
Metal tubes, Machine for polishing	J. A. Thomas	Buffalo, N. Y	Feb. 5, 1867	61, 894
Metal tubes, Making	S. Vanstone	Providence, R. I	May 18, 1869	90, 322
Metal tubes, Making seamless	W. F. Brooks	New York, N. Y	Apr. 1, 1856	14, 551
Metal tubes, Making seamless	J. Pratt	Taunton, Mass	June 6, 1854	11, 009
Metal tubes, Making seamless	J. J. Speed, jr., and J. A. Bailey	Detroit, Mich	July 15, 1856	15, 348
Metal tubes, Manufacture of	G. F. Muntz, jr	Birmingham, England	June 14, 1853	9, 782
Metal tubes, Manufacturing	W. Beasley	Smithwick, England	Jan. 9, 1855	12, 228
Metal tubes to glass cups or vessels, Fastening	G. Gerdom	Albany, N. Y	Aug. 31, 1869	94, 408
Metal-twisting machine	T. Smith	Green Island, N. Y	June 4, 1872	127, 436
Metal vessels, Pressing and polishing	J. Newmann	New York, N. Y	Mar. 3, 1863	37, 822
Metal wheel	A. Haines and J. Kirkman	Pekin and Kickapoo, Ill	July 3, 1866	56, 042
Metal with metal, Process of coating	J. Poleux	New York, N. Y	Oct. 21, 1856	15, 953
Metal with rubber, Inlaying	B. L. Rowley	New Britain, Conn	Nov. 19, 1872	133, 259
Metal-working machine, Compound	G. L. Jones	Vanville, Wis	Sept. 2, 1873	142, 398
Metal-working machine, Compound	H. B. Sevey	Vienna, Me	Sept. 9, 1873	142, 588
Metal-working machine, Compound	H. B. Sevey	Vienna, Me	Oct. 14, 1873	143, 722
Metals and minerals, Treating	Z. A. Willard and W. G. Adams	Franklin, Mass	May 19, 1868	78, 168
Metals, Apparatus for amalgamating	O. G. Warren	New York, N. Y	Jan. 17, 1865	45, 944
Metals, Apparatus for collecting precious	J. T. McDougall	San Francisco, Cal	Jan. 7, 1868	73, 021
Metals, Apparatus for collecting precious	J. T. McDougall	San Francisco, Cal	Nov. 10, 1868	83, 868
Metals, Apparatus for collecting precious	J. T. McDougall	San Francisco, Cal	Nov. 22, 1870	109, 533
Metals, Apparatus for melting	J. A. Thayer	Cambridgeport, Mass	Aug. 20, 1872	130, 601
Metals, Apparatus for refining	A. Millochau	New York, N. Y	Mar. 21, 1871	112, 831
Metals, Apparatus for saving precious	W. C. Knight	Yankee Jim's, Cal	Dec. 17, 1867	72, 205

Index of patents issued from the United States Patent Office from 1790 *to* 1873, *inclusive*—Continued.

Invention.	Inventor.	Residence.	Date.	No.
Metals, Apparatus for testing the tensile strength of	A. B. Davis	Philadelphia, Pa	Aug. 6, 1872	130, 284
Metals, Apparatus for testing the tensile strength of.	T. Olsen	Philadelphia, Pa	Sept. 16, 1873	142, 937
Metals, Bath for coating metals with other	G. Rogers	Enfield, England	May 16, 1854	10, 914
Metals, Casting molten	J. D. Rosthorn	Vienna, Austria	Jan. 17, 1865	45, 947
Metals, Coating	J. D. Güneberg	Camden, N. J	Apr. 26, 1870	102, 254
Metals, Coating	S. Hiler	Haverstraw, N. Y	Oct. 12, 1858	21, 797
Metals, Coating and bronzing	F. Weil	Paris, France	July 12, 1864	43, 557
Metals, Combining wrought and cast	E. L. Brown	Philadelphia, Pa	June 9, 1868	78, 786
Metals, &c., Composition for coating	B. Oertly and H. Fendrick	Washington, D. C	July 21, 1868	80, 086
Metals, Composition for covering	J. H. Green	Christiansburgh, Iowa	Nov. 29, 1859	26, 267
Metals, &c., Compound for cleaning	W. Z. Moore	Vallejo, Cal	Nov. 4, 1873	144, 215
Metals, Compound for cleaning polished	M. McGlenn	Aurora, Ill	July 22, 1873	141, 156
Metals, Compressing, condensing, and extending	J. F. Shearman	Brooklyn, N. Y	July 31, 1866	56, 807
Metals, Core for casting	O. D. Hunter	Plymouth, Conn	Sept. 16, 1873	142, 790
Metals, Cutting	R. Anderson and A. Vancleve	United States Army, and Trenton, N. J.	Nov. 4, 1856	16, 001
Metals from other substances, Apparatus for separating the precious.	V. B. Ryerson	New York, N. Y	Apr. 10, 1866	53, 885
Metals from the beds of rivers, Process of obtaining.	J. Johnson	Saco, Me	Jan. 1, 1867	60, 898
Metals, &c., Furnace for reheating	H. Chisholm	Cleveland, Ohio	Mar. 4, 1873	136, 413
Metals, Granulating	J. Feix	San Francisco, Cal	Sept. 16, 1856	15, 733
Metals, Hardening	A. Wheeler	Warwick, Mass	July 31, 1849	6, 615
Metals in liquid state, Apparatus for compressing cast.	H. W. Barnum	Omaha, Nebr	Dec. 9, 1873	145, 325
Metals, ivory, &c., Method of forming designs upon.	T. Skinner	Pittsburgh, Pa	Dec. 24, 1867	72, 553
Metals, Machine for cleaning and brightening particles of precious.	J. D. Whelpley and J. J. Storer	Boston, Mass	Sept. 11, 1866	58, 011
Metals, Machine for forging	E. Wheeler	Feltonville, Mass	Sept. 1, 1857	18, 115
Metals, Machine for routing	H. Cottrell	Newark, N. J	July 2, 1872	128, 596
Metals, Machinery for hammering	J. Ramsbottom	Crewe, England	July 25, 1865	49, 058
Metals, Manufacture and preservation of	J. Corson	Washington, D. C	Oct. 19, 1869	95, 832
Metals, Manufacture of	A. Wall	Poplar Blackwell, England	Aug. 10, 1844	3, 698
Metals, &c., Melting	J. Walworth	New York	Oct. 17, 1809	....
Metals or ores, Composition to be used in the fusion, refining, or working of	H. B. Chew and E. V. Freeman	Baltimore, Md	Apr. 17, 1829	....
Metals, Quality and ornamentation of	W. Rose	Halesowen, England	July 7, 1863	39, 174
Metals, Refining	J. Ramdohr	Virginia City, Nev	June 27, 1865	48, 438
Metals, Separating	W. Elmer	New York, N. Y	Mar. 19, 1867	63, 026
Metals, Separating and refining	C. S. Eyster	Denver, Colo	June 14, 1870	104, 130
Metals, Shaping and punching	J. C. Dickey	Saratoga Springs, N. Y	Mar. 2, 1858	19, 498
Metals, Treating	M. Lane	Washington, D. C	Aug. 20, 1861	33, 090
Metals while in the molten state, Treating	H. W. Woodruff	Watertown, N. Y	Oct. 11, 1853	10, 115
Metals with anthracite coal, Manufacturing	R. Willcox	Paterson, N. J	Apr. 5, 1831	....
Metals with bituminous coal, charcoal, &c., Manufacturing.	R. Willcox	Paterson, N. J	Apr. 5, 1831	....
Metals with metal, Coating	I. Adams, jr	Boston, Mass	Aug. 21, 1866	57, 271
Metals, wood, &c., Composition for cleaning	C. S. Toms	Utica, N. Y	May 28, 1867	65, 139
Metallic and alkaline sulphates, Separating metal from a mixture of.	A. Monnier	Philadelphia, Pa	Dec. 12, 1871	121, 798
Metallic and alkaline sulphates to separate copper, &c., Treating.	A. Monnier	Philadelphia, Pa	Dec. 12, 1871	121, 799
Metallic and non-metallic surfaces, Plastic compound for protecting.	T. L. Roux	Toulon, France	Apr. 9, 1867	63, 654
Metallic and other hard substances, Machine for cutting.	J. H. Hall	Harper's Ferry, Va	Mar. 7, 1827	....
Metallic axles and other hollow metallic articles, Process of making hollow.	W. A. Lewis	Chicago, Ill	Oct. 10, 1871	119, 864
Metallic axles, Process of making hollow	W. A. Lewis	Chicago, Ill	Mar. 19, 1872	124, 833
Metallic bar	S. Stone	North Manchester, Conn	Oct. 12, 1869	95, 746
Metallic-bar cutter	T. Snell and W. Tucker	Philadelphia, Pa	Oct. 6, 1868	82, 886
Metallic bar, Triple-ribbed	J. Y. Smith	Alexandria, Va	May 15, 1866	54, 786
Metallic bars and rods, Cutter for	J. Gallagher	New York, N. Y	May 2, 1854	10, 858
Metallic bars, Machine for cutting	S. Hall	New York, N. Y	Dec. 8, 1857	18, 811
Metallic-bath furnace	J. S. Brown	Pawtucket, R. I	June 7, 1864	43, 006
Metallic boat	J. Francis	New York, N. Y	Mar. 23, 1858	19, 693
Metallic boat	L. Raymond	New York, N. Y	June 23, 1863	38, 984
Metallic bodies, Machine for squeezing and compressing.	E. A. Lester	Boston, Mass	Jan. 10, 1854	10, 410
Metallic box	I. C. Mayo	Gloucester, Mass	June 11, 1872	127, 782
Metallic box for presses, &c	J. Foster, jr., and P. Evens, jr	Cincinnati, Ohio	June 28, 1853	9, 823
Metallic bracket	A. D. Judd	New Haven, Conn	June 21, 1870	104, 463
Metallic bracket	A. D. Judd	New Haven, Conn	Dec. 9, 1873	145, 351
Metallic bracket	A. D. and E. M. Judd	New Haven, Conn	June 21, 1870	104, 464
Metallic-can bottom	H. W. Shepard	Mannsville, N. Y	Jan. 4, 1870	98, 526
Metallic-can cap	G. H. Perkins	Brooklyn, N. Y	Feb. 22, 1870	100, 186
Metallic-can cover	H. W. Shepard	Mannsville, N. Y	Apr. 12, 1870	101, 929
Metallic case or stove for kettles, boilers, stills, &c.	C. Gaston	Williamsport, Pa	Jan. 31, 1812	....
Metallic chair	A. W. Hopkins	New York, N. Y	Oct. 15, 1867	69, 808
Metallic chair-bottom	V. Stockton	Williamsburgh, Ohio	Aug. 7, 1860	29, 553
Metallic clasp	C. Marshall	Lockport, N. Y	June 24, 1873	140, 207
Metallic combs, Making	R. A. Ives	Bristol, Conn	Mar. 28, 1838	661
Metallic compound or alloy	J. Hackert	New York, N. Y	June 11, 1867	65, 563
Metallic cup and stand	S. J. Ladd	Providence, R. I	May 8, 1860	28, 238
Metallic fabric for grain scouring and percolating purposes.	A. J. Vandegrift	Covington, Ky	Feb. 4, 1873	135, 608
Metallic fabric for scouring and percolating purposes.	A. J. Vandegrift	Covington, Ky	Jan. 7, 1873	134, 619
Metallic handle	J. Hatton	New York, N. Y	Jan. 28, 1868	73, 804
Metallic hoop	J. L. Alberger	Buffalo, N. Y	Apr. 14, 1863	38, 139
Metallic hoop	E. C. Hamlin	Pavilion, N. Y	Oct. 10, 1871	119, 844
Metallic-hoop clasp	J. R. Speer	Pittsburgh, Pa	Dec. 1, 1857	18, 779
Metallic house	S. J. Seely	Brooklyn, N. Y	Dec. 23, 1862	37, 248
Metallic mat	F. G. Johnson	Brooklyn, N. Y	Aug. 27, 1872	130, 808
Metallic mill	J. C. Gentry	Dayton, Ohio	Mar. 18, 1836	....
Metallic mills, Dressing	S. G. Reynolds	Providence, R. I	Mar. 18, 1835	....
Metallic molds for casting metals, Preparing	J. Revere	Boston, Mass	Feb. 25, 1862	34, 524
Metallic molds, Process in casting in	J. Williams	Philadelphia, Pa	May 13, 1873	138, 775

Index of patents issued from the United States Patent Office from 1790 *to* 1873, *inclusive*—Continued.

Invention.	Inventor.	Residence.	Date.	No.
Metallic molds, Wash or coating for	J. Kinniburgh	Motherwell, Scotland	Mar. 18, 1862	34,719
Metallic moldings, Machine for making	F. M. Campbell	Cleveland, Ohio	July 16, 1872	129,001
Metallic oxides and in refining the metal resulting therefrom, Reducing.	J. Reese	Pittsburgh, Pa	Sept. 11, 1866	57,969
Metallic pipe	W. S. Mayo	New York, N. Y	May 24, 1859	24,132
Metallic pipe or tubing	P. Naylor	New York, N. Y	Jan. 2, 1872	122,482
Metallic pipes and tubes, Machine for making	J. and C. Hanson	Huddersfield, England	Mar. 29, 1841	2,021
Metallic pipes, Machinery and process of manufacturing.	J. E. Serrell	New York	Jan. 20, 1843	2,918
Metallic pipes, Machinery for making	G. N. and B. Tathem, jr	Philadelphia, Pa	Oct. 11, 1841	2,296
Metallic plate, Corrugated	R. Montgomery	New York, N. Y	July 31, 1866	56,779
Metallic plates, Apparatus for bending	J. W. Bean	Henry, Ill	Feb. 25, 1868	74,880
Metallic plates, Joining and riveting	W. Beschke	Alexandria, Va	Nov. 22, 1853	10,247
Metallic plates, Machine for flanging and wiring	O. W. Stow	Plantsville, Conn	Nov. 12, 1867	70,917
Metallic plates, Press for bending	J. Cochrane	New York, N. Y	Sept. 15, 1863	39,886
Metallic plates to each other, Method of uniting	S. Pratt	Cohasset, Mass	Aug. 14, 1849	6,647
Metallic points, Removing pains, &c., by	E. Perkins		Feb. 19, 1796	....
Metallic rods, Apparatus for rolling tapered	W. Clay	Clifton Lodge, England	Apr. 22, 1851	8,055
Metallic rods, &c., Device for cutting tenons on	C. Gillson	Worcester, Mass	Aug. 9, 1864	43,776
Metallic rods or bars, Machine for upsetting and forming articles from.	M. Seward	New Haven, Conn	Nov. 6, 1866	59,465
Metallic spring	L. Bissell	New York, N. Y	Feb. 20, 1855	12,406
Metallic spring	J. Harrison, jr	New York, N. Y	Aug. 24, 1858	21,255
Metallic spring	G. W. McMinn	Covington, Ky	Feb. 19, 1851	31,517
Metallic substances, Apparatus for concentrating and condensing volatile.	J. C. Coult	San Francisco, Cal	Apr. 2, 1867	63,365
Metallic surfaces, Coating	W. and W. A. Butcher	Philadelphia, Pa	June 29, 1858	20,697
Metallic surfaces, Composition for coating	G. Brown, E. E. Burnham, and J. Morrise.	Gloucester, Mass	June 11, 1867	65,639
Metallic surfaces, Device for producing pearl finish on.	B. D. Beiderhase and C. Witteck.	New York, N. Y	Jan. 7, 1873	134,581
Metallic surfaces, Device for producing satin finish on.	B. D. Beiderhase and C. Witteck.	New York, N. Y	Jan. 7, 1873	134,582
Metallic surfaces, Machine for finishing	G. Cowing	Seneca Falls, N. Y	May 19, 1863	38,565
Metallic surfaces, Mode of uniting	W. B. Barnard	Waterbury, Conn	Nov. 8, 1864	44,928
Metallic surfaces to prevent oxidation, Compound for coating.	A. Wall	Shadwell, England	June 22, 1841	2,136
Metallic tip	B. F. Sparrow	Boston, Mass	Aug. 20, 1867	67,920
Metallic tubes, Casting	J. Smith, jr	Norton, Mass	Dec. 2, 1856	16,166
Metallic tubes, Device for cutting off	S. P. M. Tasker	Philadelphia, Pa	May 31, 1870	103,685
Metallic tubes, Galvano-plastic coating for the interior of.	J. Matthews, jr	New York, N. Y	June 5, 1860	28,590
Metallic tubes, Machine for making	E. Valentine and M. T. Ridout.	Milwaukee, Wis	Feb. 7, 1865	46,311
Metallic tubes, Making	J. B. Root	New York, N. Y	Oct. 19, 1869	96,037
Metallic tubes or spouts, Machine for making	E. Valentine and M. T. Ridout.	Milwaukee, Wis	July 25, 1865	49,045
Metallic tubes, Punching	B. Mackerley	New Petersburgh, Ohio	May 4, 1858	20,165
Metallic washer	M. G. Hubbard	Syracuse, N. Y	Feb. 11, 1873	135,809
Metallurgic and other furnaces	T. Weiss	Dresden, Saxony	Nov. 1, 1870	108,862
Metallurgic and other furnaces, Draft-apparatus for.	G. Wingate	Boston, Mass	Nov. 11, 1873	144,585
Metallurgic and other furnaces, Feeding fuel to	J. D. Whelpley and J. J. Storer.	Boston, Mass	May 31, 1870	103,695
Metallurgic and other furnaces, Feeding pulverized fuel to.	J. D. Whelpley and J. J. Storer	Boston, Mass	May 31, 1870	103,804
Metallurgic and other furnaces, Heating	J. D. Whelpley and J. J. Storer.	Boston, Mass	Nov. 29, 1870	109,795
Metallurgic and other furnaces, Lining for	J. C. Hill	Alexandria, Va	Nov. 26, 1872	133,446
Metallurgic furnace	A. G. Bevin	East Hampton, Conn	May 4, 1869	89,620
Metallurgic furnace	H. Boetius	Hanover, Prussia	Oct. 22, 1867	69,963
Metallurgic furnace	Z. S. Durfee	New York, N. Y	Aug. 6, 1872	130,200
Metallurgic furnace	J. L. Hamilton and H. McDonald.	Pittsburgh, Pa	Jan. 2, 1872	122,380
Metallurgic furnace	J. Montgomery	Sing Sing, N. Y	Feb. 1, 1870	99,456
Metallurgic furnace	G. Perry and J. Webb	Chicago, Ill	May 6, 1873	138,523
Metallurgic furnace	J. Y. Smith	Pittsburgh, Pa	Oct. 17, 1871	120,005
Metallurgic furnace for iron and steel	J. Y. Smith	Pittsburgh, Pa	June 14, 1870	104,219
Metallurgic-furnace lining	S. Lansdowne	Sharon, Pa	Apr. 8, 1873	137,554
Metallurgic-furnace lining	J. Y. Smith	Pittsburgh, Pa	Oct. 17, 1871	120,006
Metallurgic-furnace regenerator	H. Frank	Pittsburgh, Pa	Dec. 31, 1872	134,373
Metallurgic-furnace regenerator	A. Pousard	Paris, France	Aug. 6, 1872	130,313
Metallurgic furnaces, Apparatus for condensing and purifying the smoke of.	W. J. Johnson	Allendale, England	Feb. 18, 1873	136,066
Metallurgic furnaces, Apparatus for feeding the charge to.	G. Edwards	Tannehill, Ala	Aug. 27, 1872	130,849
Metallurgic furnaces, Apparatus for generating and burning gas in.	J. W. Ells	Pittsburgh, Pa	July 19, 1870	105,557
Metallurgic furnaces, Constructing and heating	C. E. Detmold	New York, N. Y	July 15, 1843	3,176
Metallurgic furnace, Construction and operation of.	H. Bessemer	London, England	July 25, 1871	117,250
Metallurgic furnaces, Gas and air heating apparatus for.	J. W. Ells	Pittsburgh, Pa	July 19, 1870	105,558
Metallurgic furnaces so as to use the waste heat under steam-boiler, Arrangement of.	G. Nimmo	Jersey City, N. J	Nov. 19, 1872	133,114
Metallurgic operations by the introduction of chemical agents.	R. Willcox	Paterson, N. J	Apr. 5, 1831	....
Metallurgical process and furnace	C. W. Siemens	Westminster, England	Apr. 27, 1869	89,441
Metamorphoscope	E. A. Goodes	Philadelphia, Pa	Nov. 12, 1867	70,831
Meteorological changes, Electro-magnetic apparatus for noting.	S. Chester	Elizabeth, N. J	Jan. 2, 1872	122,437
Meter: *See* Cable-meter. Dynamometer. Electro-magnetic meter. Fluid-meter. Gas-meter. Gas and water meter. Grain-meter. Heliometer. Horometer. Hydraulic meter. Hydrometer. Hygicometer. Hygrometer. Liquid-meter. Power-meter. Proof-meter. Pyrometer. Rotary meter. Sounding-meter. Spirit-meter. Water-meter. Way-meter.				

Index of patents issued from the United States Patent Office from 1790 *to* 1873, *inclusive*—Continued.

Index of patents issued from the United States Patent Office from 1790 *to* 1873, *inclusive*—Continued.

Invention.	Inventor.	Residence.	Date.	No.
Milk, Apparatus for testing	C. S. Brown	New York, N. Y	June 20, 1865	48, 256
Milk, Apparatus for transporting	J. H. Whitney	Brooklyn, N. Y	Dec. 17, 1872	134, 020
Milk, Apparatus for treating	G. D. Greenleaf and D. C. Larkins.	Depauville, N. Y	Apr. 14, 1868	76, 749
Milk, Apparatus for treating	J. A. Otis and T. Barber	Watertown, N. Y	Oct. 6, 1868	82, 866
Milk-boiler	O. Zwicker	Philadelphia, Pa	Nov. 18, 1873	144, 819
Milk-boiler alarm	S. Mangold	New York, N. Y	May 20, 1873	139, 072
Milk-box	C. W. Eastwood	New York, N. Y	Feb. 1, 1870	99, 417
Milk-can	E. Abbott	Chagrin Falls, Ohio	Sept. 25, 1866	58, 194
Milk-can	E. Abbott	Willoughby, Ohio	Jan. 28, 1868	73, 687
Milk-can	T. W. Akin	Patterson, N. Y	Nov. 10, 1868	83, 898
Milk-can	S. O. Avery	Brewster's Station, N. Y	Oct. 29, 1867	70, 150
Milk-can	N. P. Barnes	Carmel, N. Y	Mar. 20, 1866	53, 256
Milk-can	T. M. Bell	New York, N. Y	Jan. 31, 1871	111, 302
Milk-can	T. M. Bell	New York, N. Y	Oct. 3, 1871	119, 559
Milk-can	J. A. Bennett	Millerton, N. Y	Aug. 11, 1868	80, 800
Milk-can	N. C. Burnap	rgusville, N. Y	Nov. 5, 1867	70, 516
Milk-can	D. Burnett	Bedford Station, N. Y	Nov. 21, 1871	121, 154
Milk-can	C. L. Camp and J. H. Chappel.	Brooklyn, N. Y	May 17, 1870	103, 137
Milk-can	J. F. Cass	L'Original, Canada	Dec. 9, 1873	145, 334
Milk-can	J. Cochran	Purdy's Station, N. Y	Sept. 28, 1869	95, 324
Milk-can	J. Cochran	Purdy's Station, N. Y	Mar. 1, 1870	100, 376
Milk-can	A. P. Cook	Collins Centre, N. Y	Aug. 25, 1868	81, 345
Milk-can	A. Crittenden	West Turin, N. Y	July 14, 1832	....
Milk-can	A. P. Curry	Chagrin Falls, Ohio	June 11, 1867	65, 548
Milk-can	J. L. Davis	Salem, N. J	Feb. 20, 1866	52, 691
Milk-can	J. E. Dean	Canaan, Conn	May 19, 1868	78, 072
Milk-can	E. R. Denniston	Middletown, N. Y	June 14, 1859	24, 381
Milk-can	S. J. Dwyer	Albany, N. Y	Apr. 27, 1869	89, 392
Milk-can	J. H. Farley	Lowell, Mass	Mar. 19, 1867	62, 949
Milk-can	J. L. Finch	Warwick, N. Y	Mar. 19, 1867	63, 029
Milk-can	W. Frost	Amenia, N. Y	Nov. 15, 1859	26, 098
Milk-can	E. A. Grant	Louisville, Ky	Sept. 17, 1872	131, 436
Milk-can	G. A. Huggins	Mannsville, N. Y	July 14, 1868	79, 907
Milk-can	H. L. McAvoy and E. Mills	Baltimore, Md	Nov. 5, 1867	70, 450
Milk-can	J. C. Milligan	Brooklyn, N. Y	Sept. 20, 1870	107, 521
Milk-can	O. J. Nutting	Warwick, N. Y	Nov. 3, 1868	83, 724
Milk-can	H. Preston and J. Mahood	Philadelphia, Pa	Mar. 24, 1863	37, 977
Milk-can	W. Ralph	Utica, N. Y	June 18, 1867	65, 828
Milk-can	C. A. Reight	Middleton, N. Y	Apr. 21, 1868	77, 092
Milk-can	H. Langster	Buffalo, N. Y	Mar. 1, 1870	100, 454
Milk-can	S. P. Sauer	Wheaton, Ill	Aug. 20, 1872	130, 752
Milk-can	D. W. Shaw	Baltimore, Md	July 7, 1868	79, 693
Milk-can	A. C. Shorte and H. B. Todd	Plymouth, Conn	Jan. 19, 1869	86, 039
Milk-can	S. Stroock	New York, N. Y	Sept. 12, 1871	118, 887
Milk-can	L. A. Sunderland	Chagrin Falls, Ohio	Dec. 15, 1868	84, 973
Milk-can	L. A. Sunderland	Madison, Ohio	Apr. 12, 1870	101, 784
Milk-can	P. Teets	New York, N. Y	May 28, 1861	32, 439
Milk-can	I. Vanderslice	Philadelphia, Pa	Aug. 18, 1868	81, 119
Milk-can	I. F. Van Duzer	Middletown, N. Y	June 19, 1866	55, 791
Milk-can	H. M. Viet	Carlisle, Ohio	Dec. 15, 1868	84, 921
Milk-can	R. C. Wickham	Pawlet, Vt	Dec. 18, 1866	60, 603
Milk can and cooler	S. Shattuc	Kipton, Ohio	Sept. 6, 1870	107, 107
Milk-can bottom, &c	T. M. Bell	New York, N. Y	July 5, 1870	104, 924
Milk-can bottom	A. Burnham	Arkwright, N. Y	Sept. 5, 1865	49, 711
Milk-can bottom	M. Wiles and J. C. Wock	Fort Plain, N. Y	Aug. 6, 1867	67, 474
Milk-can cover	A. Brightman	New Bedford, Mass	Nov. 1, 1864	44, 848
Milk-can cover	J. Fandel	Boston, Mass	Apr. 21, 1868	77, 020
Milk-can covers, Locking	D. Burnett	Bedford Station, N. Y	June 6, 1871	115, 697
Milk-can covers, Securing	N. C. Burnap	Argusville, N. Y	Mar. 12, 1872	124, 539
Milk-can fastener	D. D. Edgerton	Ava, N. Y	Nov. 22, 1870	109, 410
Milk, Can for preserving and transporting	M. M. Clark	Monroe, N. Y	Mar. 7, 1865	46, 640
Milk-can handle	A. C. Beckwith and G. H. Graham.	Oriskany, N. Y	Oct. 4, 1870	107, 855
Milk-can handle	J. W. Hannan	Elyria, Ohio	Nov. 21, 1871	121, 168
Milk-can, &c., handle	C. Millar	Utica, N. Y	Apr. 3, 1866	53, 649
Milk-can handle	H. W. Shepard	Mannsville, N. Y	June 7, 1870	103, 933
Milk-can holder, Condensed	R. J. Tanner	Lake View, N. J	Dec. 17, 1872	134, 015
Milk-can, Refrigerating	W. W. White and M. King	Lowville, N. Y	June 9, 1868	78, 777
Milk-can stopper	J. M. Burghartt	Great Barrington, Mass	Aug. 3, 1869	93, 171
Milk-can, Ventilating	L. B. Arnold	Lansing, N. Y	Mar. 14, 1871	112, 674
Milk-cans, Apparatus for stamping	W. M. Storm	New York, N. Y	Sept. 7, 1858	21, 473
Milk-cans, &c., Clasp for securing covers to	G. D. C. Ransom and C. Smith	Brooklyn, N. Y	June 11, 1872	127, 794
Milk-cans, Metallic bottom for	T. M. Bell	New York, N. Y	Jan. 14, 1873	134, 723
Milk-carrier	L. Morris	Havre de Grace, Md	Dec. 13, 1870	110, 152
Milk-concentrating apparatus	A. S. Lyman	New York, N. Y	Sept. 2, 1862	36, 354
Milk, Concentration of	G. Borden, jr	Brooklyn, N. Y	Aug. 19, 1856	15, 553
Milk, Condensed	G. Borden and J. G. Borden	White Plains and South East, N. Y.	Nov. 4, 1873	144, 311
Milk, Condensing	J. G. Borden	South East, N. Y	Oct. 29, 1872	132, 621
Milk, Condensing	J. R. Pond	New Hartford, Conn	Jan. 5, 1864	41, 090
Milk, Condensing	J. R. Pond	New Hartford, Conn	Nov. 14, 1865	50, 950
Milk, Condensing	J. R. Pond	New Hartford, Conn	Dec. 26, 1865	51, 749
Milk-cooler	L. B. Arnold	Lansing, N. Y	Dec. 22, 1868	85, 160
Milk-cooler	A. E. Baldwin	Newark, N. J	Aug. 10, 1869	93, 583
Milk-cooler	L. Bauer	McHenry County, Ill	Sept. 17, 1872	131, 325
Milk-cooler	A. Beeman	Potter's Corners, Pa	June 18, 1872	127, 952
Milk-cooler	H. Blake	Panama, N. Y	Oct. 8, 1872	131, 993
Milk-cooler	A. M. Blanchard	Ellington, N. Y	Apr. 9, 1872	125, 531
Milk-cooler	B. C. Bort and T. Bryant	Chateaugay, N. Y	June 18, 1872	128, 102
Milk-cooler	B. C. Bort and T. Bryant	Chateaugay, N. Y	Nov. 5, 1872	132, 795
Milk-cooler	N. C. Burnap	Argusville, N. Y	Aug. 13, 1867	67, 717
Milk-cooler	A. P. Bussey	Westernville, N. Y	Nov. 2, 1869	96, 390
Milk-cooler	W. O. Campbell	Montgomery Centre, Vt	Mar. 11, 1873	136, 584
Milk-cooler	S. F. Cowles	Coventry, Vt	Feb. 25, 1873	136, 215
Milk-cooler	A. J. Crane and D. J. Hadley	Cambridge, Vt	Nov. 25, 1873	144, 835
Milk-cooler	J. Dingee	Downington, Pa	Aug. 10, 1869	93, 424
Milk-cooler	C. A. Douglas	Franklin, N. Y	Feb. 20, 1872	123, 813

Index of patents issued from the United States Patent Office from 1790 *to* 1873, *inclusive*—Continued.

Invention.	Inventor.	Residence.	Date.	No.
Milk-cooler	C. A. Douglas	Franklin, N. Y	Nov. 11, 1873	144, 448
Milk-cooler	W. Eaton	Norwich, N. Y	Oct. 22, 1872	132, 454
Milk-cooler	W. Eaton	Norwich, N. Y	Dec. 17, 1872	134, 043
Milk-cooler	A. Foot	Warren, Conn	Jan. 19, 1869	85, 920
Milk-cooler	R. D. Gardner	Watertown, N. Y	June 29, 1869	91, 928
Milk-cooler	W. H. Gould	Manchester, N. H	Aug. 30, 1870	106, 811
Milk-cooler	C. Graves, G. C. Powers, and C. O. Graves.	Waterbury, Vt	Dec. 23, 1873	145, 732
Milk-cooler	J. F. Harly	Cleveland, Ohio	June 6, 1871	115, 605
Milk-cooler	J. F. Hawkins	Hannibal, Mo	July 15, 1873	140, 919
Milk-cooler	L. T. Hawley	Salina, N. Y	Mar. 30, 1869	88, 476
Milk-cooler	W. Hodgdon	North Craftsbury, Vt	Apr. 22, 1873	138, 088
Milk-cooler	I. Houghtling	Houghton, Mich	Mar. 16, 1869	87, 851
Milk-cooler	N. and R. E. Johnson	Jasper, N. Y	Apr. 4, 1871	113, 302
Milk-cooler	J. Jones, C. D. Faulkner, F. L. Jones, and H. K. Faulkner.	Utica, N. Y	June 4, 1872	127, 608
Milk-cooler	W. M. King	Morrison, Ill	Feb. 23, 1869	87, 265
Milk-cooler	A. J. Kuster and X. Marzolf	Strykersville, N. Y	June 11, 1872	127, 898
Milk-cooler	C. H. Latham	Randolph, N. Y	Jan. 9, 1872	122, 616
Milk-cooler	C. H. Latham	Randolph, N. Y	Oct. 22, 1872	132, 473
Milk-cooler	R. P. Main	Canton, N. Y	Dec. 30, 1873	146, 084
Milk-cooler	J. Mansfield	Jefferson, Wis	Jan. 18, 1859	22, 660
Milk-cooler	E. Martin	Potsdam, N. Y	Nov. 4, 1873	144, 343
Milk-cooler	E. L. Matteson	Randolph, N. Y	May 21, 1872	127, 081
Milk-cooler	W. D. McDonald	Randolph, N. Y	Apr. 8, 1873	137, 557
Milk-cooler	J. R. McKay	Rockton, Ill	Apr. 4, 1871	113, 438
Milk-cooler	J. W. McMillan	Granger, Ohio	June 14, 1870	104, 180
Milk-cooler	H. Messenger	Massena, N. Y	Mar. 5, 1872	124, 369
Milk-cooler	N. W. Miller	East Randolph, N. Y	Jan. 30, 1872	123, 278
Milk-cooler	N. W. Miller	East Randolph, N. Y	May 7, 1872	126, 563
Milk-cooler	J. O. Moore	Washingtonville, N. Y	July 31, 1866	56, 780
Milk-cooler	A. Northorp	Elyria, Ohio	Apr. 4, 1871	113, 690
Milk-cooler	F. S. Oakes	Cattaraugus, N. Y	Jan. 28, 1873	135, 284
Milk-cooler	J. Pearl	Lawrenceville, N. Y	Oct. 21, 1873	143, 840
Milk-cooler	J. R. Pond	New Hartford, Conn	Oct. 12, 1869	95, 722
Milk-cooler	M. F. Potter	Kaneville, Ill	June 15, 1869	91, 481
Milk-cooler	C. G. Riggs and H. C. Markham	Turin and West Turin, N. Y	Dec. 29, 1868	85, 330
Milk-cooler	C. L. Sabin	Barnard, Vt	Mar. 30, 1869	88, 332
Milk-cooler	C. L. Sheldon	Lowville, N. Y	Aug. 27, 1867	68, 316
Milk-cooler	C. L. Sheldon	Lowville, N. Y	Dec. 31, 1867	72, 920
Milk-cooler	J. C. Sherwood	West Cornwall, Conn	June 15, 1869	91, 372
Milk-cooler	D. Smith	Jamestown, N. Y	Nov. 11, 1873	144, 572
Milk-cooler	R. Smith	Richford, Vt	July 16, 1872	129, 610
Milk-cooler	E. P. Spencer	Scott, N. Y	Mar. 2, 1869	87, 521
Milk-cooler	S. Spencer	Kiantone, N. Y	Jan. 21, 1873	135, 169
Milk-cooler	J. L. Tallman and J. V. De Puy	Tecumseh, Mich	Apr. 6, 1869	88, 677
Milk-cooler	J. C. Thayer	Dunton, Ill	July 7, 1868	79, 613
Milk-cooler	J. C. Thayer	Dunton, Ill	Mar. 23, 1869	88, 096
Milk-cooler	A. Thompson	Hudson, Ohio	May 11, 1869	89, 900
Milk-cooler	A. Thompson and J. Darling	Hudson and Northfleet Township, Ohio.	May 11, 1869	89, 901
Milk-cooler	I. Wheeler	Massena, N. Y	Nov. 5, 1872	132, 880
Milk-cooler	G. A. Whitcomb	Morrison, Ill	Aug. 24, 1869	94, 160
Milk-cooler	O. H. Willard	Randolph, N. Y	Mar. 19, 1872	124, 870
Milk-cooler	D. F. Woodward	Jasper, N. Y	June 14, 1870	104, 392
Milk cooler and deodorizer	A. Osborn	Eagleville, Ohio	Oct 12, 1869	95, 719
Milk-cooler, Portable	J. M. Sanborn	Hardwick, Vt	June 10, 1862	35, 569
Milk, Cooling	W. Peck	York, Ill	Apr. 23, 1867	64, 028
Milk-cooling device	L. T. Hawley and A. Westcott	Salina and Syracuse, N. Y	Mar. 3, 1868	75, 160
Milk, Dietetic compound from	V. A. Jagrilski	London, England	Aug. 8, 1871	117, 889
Milk, Equalizing the temperature of	S. Parry	New York, N. Y	Aug. 16, 1864	43, 863
Milk, &c., evaporator	A. F. Dalson	New York, N. Y	June 27, 1854	11, 193
Milk for cattle, Substitute for	R. K. Blodgett	Fulton, Ill	Dec. 10, 1867	71, 961
Milk-house	H. Peregoy	Mount Carmel, Md	Jan. 3, 1871	110, 784
Milk-house	J. R. Pond	New Hartford, Conn	Oct. 12, 1869	95, 723
Milk-house	J. A. Price	Beckleysville, Md	Oct. 31, 1871	120, 534
Milk-house	F. Schaller	Mattoon, Ill	Sept. 7, 1869	94, 655
Milk-house	H. Yerty	Covington, Ohio	Aug. 27, 1867	68, 276
Milk keelers or pans, Making	A. Miner	Camillus, N. Y	July 20, 1825	....
Milk-man's signal	E. B. Blake	Tarrytown, N. Y	June 20, 1871	116, 014
Milk pail and strainer	J. L. Drake	Boston, N. Y	July 16, 1867	66, 814
Milk-pan	N. C. Burnap	Argusville, N. Y	Jan. 25, 1870	99, 150
Milk-pan	B. F. Jewett	North Bangor, N. Y	Apr. 13, 1869	88, 878
Milk-pan	F. J. May	New York, N. Y	Oct. 3, 1865	50, 256
Milk-pan	E. L. Pratt	Philadelphia, Pa	Mar. 29, 1859	23, 429
Milk-pan	C. F. Spaulding	Saint Johnsbury, Vt	July 23, 1861	32, 903
Milk-pan	W. Templeton	Rockville, Pa	July 30, 1867	67, 228
Milk-pan of stone-ware	S. S. Perry	Troy, N. Y	Sept. 9, 1825	....
Milk-pan rack	G. B. Lewis	Moreau Station, N. Y	Jan. 3, 1860	26, 688
Milk-rack	W. Boswell	Pontiac, Mich	Aug. 15, 1871	118, 098
Milk-rack	R. Cruikshank	Salem, N. Y	May 19, 1863	38, 625
Milk-rack	A. Jackson	Clifton Springs, N. Y	July 3, 1866	56, 056
Milk-rack	I. P. Tice	New York, N. Y	Feb. 18, 1868	74, 633
Milk rack and cooler	L. Shipman	Barker, N. Y	Mar. 22, 1870	101, 054
Milk-rack and fruit-drier, Combined	C. B. and G. W. Hart	Victor, N. Y	June 18, 1867	65, 907
Milk-rack and table combined	J. McCoun	Lockport, N. Y	Jan. 29, 1867	61, 675
Milk-rack, Revolving	H. Blake	Panama, N. Y	Apr. 23, 1872	125, 928
Milk-receptacle	G. M. Fillmore	Saint Paul, Minn	Dec. 20, 1870	110, 348
Milk-separator	A. E. Baldwin	Newark, N. J	Aug. 10, 1869	93, 582
Milk-shelf	W. Veber, jr	Shingle Creek, N. Y	Mar. 31, 1868	76, 120
Milk-skimmer	S. Martin	Milwaukee, Wis	July 30, 1872	129, 975
Milk-stand	Z. Sanders	West Windsor, Vt	Oct. 31, 1865	50, 736
Milk-stand	A. R. Titus	Warren, Pa	Aug. 8, 1865	49, 321
Milk-stirrer	J. H. Soule	Georgetown, N. Y	Mar. 31, 1868	76, 260
Milk tempering and cooling apparatus	I. B. Tuttle	East Randolph, N. Y	Feb. 1, 1870	99, 502
Milk to obtain useful products, Process of treating	A. E. Baldwin	Newark, N. J	June 9, 1868	78, 640
Milk-vat	J. A. Edwards	Waterford, Pa	Oct. 6, 1868	82, 698
Milk-vat	E. C. Hale	Steamburgh, N. Y	Oct. 1, 1872	131, 874

Index of patents issued from the United States Patent Office from 1790 *to* 1873, *inclusive*—Continued.

Invention.	Inventor.	Residence.	Date.	No.
Milk-vat	L. C. and J. M. Schermerhorn	North Gage, N. Y	Apr. 9, 1867	63, 755
Milk-vat	O. H. Willard and A. Pope	Randolph and Conewago, N. Y.	Jan. 30, 1872	123, 318
Milk-vessel	M. L. Shade	Fair Grove, Mo	Sept. 3, 1872	131, 069
Milker, Cow	L. O. Colvin	Cincinnatus, N. Y	May 22, 1860	28, 351
Milker, Cow	L. O. Colvin	Cincinnatus, N. Y	May 29, 1860	28, 455
Milker, Cow	L. O. Colvin	Philadelphia, Pa	Feb. 17, 1863	37, 676
Milker, Cow	L. O. Colvin	Philadelphia, Pa	Jan. 12, 1864	41, 196
Milker, Cow	L. O. Colvin	Philadelphia, Pa	Mar. 28, 1865	46, 994
Milker, Cow	L. O. Colvin	New York, N. Y	May 22, 1866	54, 865
Milker, Cow	L. O. Colvin	New York, N. Y	Apr. 5, 1870	101, 589
Milker, Cow	L. C. Colvin	Humphreyville, Pa	Nov. 1, 1870	108, 881
Milker, Cow	G. H. Gardner	Philadelphia, Pa	Dec. 20, 1864	45, 486
Milker, Cow	G. H. Gardner	Philadelphia, Pa	Sept. 3, 1867	68, 258
Milker, Cow	E. A. Hewitt	Groton, Conn	Feb. 21, 1871	111, 932
Milker, Cow	C. Knapp	New York, N. Y	Nov. 27, 1849	6, 904
Milker, Cow	W. Reading	Offutt's Cross-Roads, Md	Oct. 17, 1871	120, 100
Milker's protector	J. M. Weare	Seabrook, N. H	May 2, 1854	10, 870
Milkers, Tail-clasp for	H. H. Dickinson	West Northfield, Mass	June 16, 1868	78, 936
Milking-apparatus	A. C. Black	Kaukauna, Wis	Jan. 25, 1870	99, 053
Milking-apparatus	E. Spedden	Astoria, Oreg	Oct. 19, 1869	95, 947
Milking cows	L. T. Blake	New Haven, Conn	Jan. 29, 1867	61, 599
Milking cows, Apparatus for	M. L. Baker	Milford, N. Y	May 21, 1861	32, 343
Milking cows, Device for	W. D. Nichols	Davenport, Iowa	May 21, 1861	32, 379
Milking implement, Cow	W. H. Whitman	Bailey Hollow, Pa	Aug. 26, 1856	15, 629
Milking-machine	H. V. Belding	Oppenheimer, N. Y	Aug. 9, 1864	43, 754
Milking-machine	B. F. Graves	Groton, Mass	Mar. 10, 1868	75, 261
Milking machine, Cow	L. O. Colvin	New York, N. Y	June 30, 1868	79, 317
Milking machine, Cow	I. Cook	Philadelphia, Pa	Dec. 3, 1867	71, 582
Milking machine, Cow	J. W. Kingman	Dover, N. H	Aug. 9, 1859	25, 019
Milking machine, Cow	T. H. Lindley	Taunton, Mass	Nov. 3, 1868	83, 777
Milking machine, Cow	S. W. Lowe	Philadelphia, Pa	Aug. 9, 1859	25, 022
Milking machinery, Cow	L. O. Colvin	New York, N. Y	Feb. 18, 1868	74, 507
Milking-seat, Folding	S. Snow	Somers, Conn	Apr. 13, 1869	88, 921
Milking-shield	O. H. Needham	New York, N. Y	Jan. 6, 1857	16, 361
Milking, Teat-cup for	B. F. Graves	Groton, Mass	July 7, 1868	79, 568

Mill:

See Apple-mill.
Apple-grinding mill.
Bark-mill.
Bone-grinding mill.
Boring-mill.
Cane-mill.
Cider-mill.
Clamp-mill.
Clay-mill.
Coffee-mill.
Corn-mill.
Corn and cob mill.
Cotton-seed mill.
Cracking-mill.
Crushing-mill.
Curd-mill.
Cylinder-mill.
Family mill.
Fanning-mill.
Feed-mill.
Feed-grinding mill.
Felly-mill.
Fish-bait mill.
Flax and hemp mill.
Floating-mill.
Flour-mill.
Flouring-mill.
Flouring and grist mill.
Foot-mill.
Founder's cleansing-mill.
Fulling-mill.
Gig-mill.
Grain-mill.
Grape-mill.
Grinding-mill.
Grinding and bolting mill.
Grinding and chopping mill.
Grinding and crushing mill.
Grinding and hulling mill.
Grinding and pounding mill.
Grist-mill.
Grist and coffee mill.
Grist and flour mill.
Grist and saw mill.
Gunpowder-mill.
Hide and fulling mill.
Hominy-mill.
Hominy and pearling mill.
Hominy and smut mill.
Horizontal mill.
Horse-mill.
Hulling-mill.
Iron mill.
Meat-mill.
Metallic mill.
Mixing-mill.
Mortar-mill.
Mortar-tub mill.
Mud and ore mill.
Ore-crushing mill.
Ore-stamp mill.
Paint-mill.
Paper-mill.
Pendulum-mill.
Percussion-mill.
Platting-mill.
Pug-mill.
Quartz-mill.
Rolling-mill.
Saw-mill.
Screw-mill.
Shingle-mill.
Slitting-mill.
Smut-mill.
Smut-fanning mill.
Spice-mill.
Stamp-mill.
Stamping-mill.
Starch-mill.
Steam rice-mill.
Steel mill.
Sugar-mill.
Sugar-cane mill.
Tide-mill.
Tide or current mill.
Turning and boring mill.
Universal mill.
Warping-mill.
Water-mill.
Windmill.
Winnowing-mill.

Invention.	Inventor.	Residence.	Date.	No.
Mill	S. Brewster	Middlesex, N. J	Mar. 15, 1816	-------
Mill	E. Bryant	New London, Conn	Feb. 8, 1814	-------
Mill	H. Chase	-------	Dec. 16, 1799	-------
Mill	E. D. Clark	Earlville, N. Y	June 26, 1860	28, 834
Mill	S. Coleman	New Orleans, La	June 12, 1860	28, 655
Mill	E. D. Curtis	Boston, Mass	Nov. 23, 1837	477
Mill	J. W. Curtiss	-------	Aug. 24, 1797	-------
Mill	J. Frisbie	Pennsylvania	June 30, 1810	-------
Mill	J. Goulding	Leicester, Mass	Mar. 1, 1811	-------
Mill	C. Hamaker	Lancaster	May 21, 1807	-------
Mill	J. Harman, jr	New York	Mar. 30, 1836	-------

Index of patents issued from the United States Patent Office from 1790 *to* 1873, *inclusive*—Continued.

Invention.	Inventor.	Residence.	Date.	No.
Mill	F. B. Hunt	Cincinnati, Ohio	Jan. 10, 1860	26, 818
Mill	G. D. Jones	New York, N. Y	Feb. 21, 1860	27, 224
Mill	W. Joslin	Cleveland, Ohio	July 3, 1860	28, 989
Mill	D. L. Latourette	Saint Louis, Mo	June 13, 1854	11, 075
Mill	A. E. Pirkey	Bradford, Ill	May 31, 1859	24, 235
Mill	J. Quinby	Wilmington, N. C	Feb. 3, 1812	....
Mill	W. Stewart	Philadelphia, Pa	Feb. 19, 1861	31, 492
Mill	H. C. Velie	Poughkeepsie, N. Y	June 26, 1860	28, 923
Mill	A. H. Wagner	Staunton, Va	July 31, 1860	29, 422
Mill	W. Warwick	Birmingham, Pa	Oct. 24, 1854	11, 845
Mill	J. Woodward	Philadelphia, Pa	Mar. 30, 1858	19, 807
Mill and grain-drill hopper	R. P. Johnson	Griffin, Ga	July 26, 1870	105, 807
Mill and machinery	L. Valcourt	Lewisville, Ky	June 25, 1805	....
Mill, and means of communicating a rotary motion by an auger, spindle, &c.	S. Mans and J. Black	Pennsylvania	Feb. 18, 1809	....
Mill-attachment	N. Bryan	Thomaston, Ga	Oct. 18, 1870	108, 442
Mill-bolt	W. H. Berdan	Mooreville, Mich	June 7, 1870	103, 833
Mill-bolt	B. C. White	Des Moines, Iowa	Oct. 17, 1871	120, 137
Mill-buhr dresser	W. P. Stalcup	Brookville, Ind	Oct. 29, 1867	70, 373
Mill-buhr feeder	A. P. Lawshaw	Harper's Ferry, W. Va	Apr. 26, 1870	102, 279
Mill-buhrs, Device for cooling	E. Embry and T. J. Blackburn	West Liberty, Ohio	June 18, 1872	128, 132
Mill-buhrs, Preventing the heating of	M. Bradley	Peru, Ind	Dec. 3, 1867	71, 690
Mill-buhrs, Stopping and starting	W. Crawford and S. Clarke	Attica, Ind	Mar. 22, 1870	101, 105
Mill-bush	J. Aulabaugh	East Berlin, Pa	July 18, 1840	1, 701
Mill-bush	E. Casner	Penn Yan, N. Y	Oct. 9, 1860	30, 296
Mill-bush	G. L. Dulaney	Long Meadow, Va	Oct. 24, 1854	11, 828
Mill-bush	G. L. Dulaney	Mount Jackson, Va	June 12, 1855	13, 029
Mill-bush	J. M. Evril	Centre, Pa	Apr. 28, 1868	77, 265
Mill-bush	H. F. Frisbie	Danville, Ill	Jan. 10, 1871	110, 843
Mill-bush	J. Heck	Boonsborough, Md	Mar. 26, 1844	3, 505
Mill-bush	H. Knowles	Washington, D. C	Jan. 16, 1849	6, 039
Mill-bush	J. G. Shafer	Fulton County, Pa	July 3, 1860	29, 013
Mill-bush	G. Strause	Boonsborough, Md	Feb. 16, 1858	19, 386
Mill-bush	R. M. Wade	Summit Point, Va	May 25, 1844	3, 601
Mill-bush	J. Wells	Baltimore, Md	Mar. 23, 1858	19, 727
Mill-bush and other spindles	C. T. Weston	Scranton, Pa	Nov. 22, 1870	110, 565
Mill-bush and spindle	J. Williams	Sullivan, Ill	May 4, 1869	89, 721
Mill-bush, Self-regulating and self-oiling	S. Moore	Chambersburgh, Pa	Apr. 8, 1840	1, 541
Mill-can shoe	J. Dennis	Providence, R. I	Sept. 8, 1834	....
Mill-dress	G. L. Dulaney	Mount Jackson, Va	June 19, 1855	13, 115
Mill-dress	J. W. Kane	New Carlisle, Ohio	Aug. 10, 1852	9, 192
Mill-driver	F. Walters	Covington, Ky	May 31, 1859	24, 268
Mill-feed, Adjustable	W. T. Duvall	Georgetown, D. C	Mar. 21, 1871	112, 909
Mill-feed apparatus	J. Sherlock and W. Brackbill	Portugal, Pa	June 18, 1850	7, 445
Mill-feed regulator	J. Ross	Brooklyn, N. Y	June 2, 1868	78, 541
Mill-feeder	J. M. Clark	Lancaster, Pa	Jan. 28, 1862	34, 239
Mill for mashing vegetables and mixing clay	C. Alvord	Geddes, N. Y	Aug. 31, 1852	9, 226
Mill for rolling irregular shapes by means of a cam-pattern.	J. S. Hall	Columbus, Ohio	Jan. 30, 1849	6, 079
Mill for tempering oleagenous seeds	W. Wilber	New Orleans, La	Jan. 27, 1857	16, 508
Mill-gudgeon	M. Withers	Strasburgh, Pa	Aug. 24, 1813	....
Mill or machine wheel, and applying water, wind, or steam thereto.	L. Copley	New Lebanon, N. Y	May 15, 1834	....
Mill-pick	A. Babcock and J. B. Martin	Silver Creek, N. Y	Aug. 6, 1872	130, 178
Mill-pick	C. Crossley	Philadelphia, Pa	Feb. 20, 1866	52, 688
Mill-pick	J. Cummings	West Charleston, Vt	July 11, 1871	116, 817
Mill-pick	J. Cummings	West Charleston, Vt	June 24, 1873	140, 118
Mill-pick	I. Faver	New York, N. Y	Apr. 9, 1861	31, 964
Mill-pick	E. S. Forgy	Dayton, Ohio	Apr. 22, 1873	138, 080
Mill-pick	H. H. Gillett	Warsaw, Mo	Sept. 15, 1868	82, 107
Mill-pick	B. Hostler	Brookfield, N. Y	Apr. 30, 1861	32, 190
Mill-pick	F. Kortick	Mendota, Ill	June 11, 1872	127, 774
Mill-pick	C. McNeal	Silver Creek, N. Y	Apr. 23, 1872	126, 072
Mill-pick	C. Metzger and G. R. Roraback	De Soto, Mo	Sept. 28, 1869	95, 252
Mill-pick	L. M. Osborn	Hamilton, N. Y	Oct. 6, 1863	40, 183
Mill-pick	C. Pennoyer	Lockport, N. Y	Oct. 29, 1872	132, 598
Mill-pick	A. Rasner	Dayton, Ohio	Oct. 31, 1871	120, 536
Mill-pick	H. N. Relyea	Warsaw, N. Y	Jan. 15, 1867	61, 252
Mill-pick	G. M. Rhoades	East Hamilton, N. Y	May 19, 1863	38, 603
Mill-pick	T. Sheehan	Dunkirk, N. Y	May 24, 1864	42, 882
Mill-pick	W. B. Stephens	Stephens' Mills, N. Y	Sept. 3, 1867	68, 394
Mill-pick	U. Stewart	Berlin, Wis	Dec. 3, 1867	71, 811
Mill-pick	A. Van Doren	Fredericksburgh, Va	Oct. 8, 1872	132, 121
Mill-pick	T. R. Way	Springfield, Ohio	Mar. 25, 1873	137, 116
Mill-pick	J. A. Wheeler	Cleveland, Ohio	Oct. 4, 1864	44, 570
Mill-pick	S. G. Williams	Trumansburgh, N. Y	Sept. 27, 1870	107, 745
Mill-pick	M. N. Young	Kelseyville, Cal	Aug. 26, 1873	142, 321
Mill-pick, Combination	J. W. Nance and D. Lillard	Cornersville, Tenn	June 10, 1873	139, 806
Mill pick handle	N. Rose	Belmont, N. Y	July 6, 1869	92, 369
Mill-pick holder	J. P. Brady	Mount Joy, Pa	Aug. 17, 1858	21, 184
Mill-pick holder	J. P. Sinclair	Mottville, N. Y	Sept. 5, 1871	118, 753
Mill power, Floating	A. G. Heitmann	Brooklyn, N. Y	Mar. 15, 1870	100, 761
Mill power, Increasing	R. Burns	Mifflin County, Pa	June 20, 1822	....
Mill-races, Floating sluice for	M. A. Shepard	Parkersburgh, Ill	May 8, 1860	28, 204
Mill-roller	W. S. Seymour	Ravenna, Ohio	Oct. 3, 1865	50, 281
Mill-shaft coupling	W. Kean	Chicago, Ill	Feb. 7, 1871	111, 550
Mill-spindle	E. T. Butler	Buffalo, N. Y	Jan. 27, 1852	8, 682
Mill-spindle	J. C. Gentry	Dayton, Ohio	Sept. 25, 1839	1, 340
Mill-spindle	J. A. Hafner	Pittsburgh, Pa	Dec. 30, 1873	146, 063
Mill-spindle	S. Hoyt	Wilmington, Del	Jan. 31, 1860	26, 993
Mill-spindle	J. H. McMinn	Logansport, Ind	Dec. 24, 1867	72, 661
Mill-spindle, &c	E. G. Patter	Lebanon, Ill	July 18, 1840	1, 703
Mill-spindle	S. P. Ruff	Weaver's Old Stand, Pa	Mar. 6, 1860	27, 386
Mill-spindle, &c., bush	R. E. and F. A. Howe	Somonauk, Ill	Aug. 13, 1872	130, 373
Mill-spindle bush	H. W. Vett	Union, Mo	Aug. 15, 1871	118, 077
Mill-spindle driver	J. M. Laogan	Springfield, Ill	May 17, 1870	103, 211
Mill-spindle driver	J. McConnell	Beaver, Pa	Apr. 17, 1847	5, 074
Mill-spindle spring	T. Alsop	Elkhart, Ill	July 7, 1868	79, 537

Index of patents issued from the United States Patent Office from 1790 *to* 1873, *inclusive*—Continued.

Invention.	Inventor.	Residence.	Date.	No.
Mill-spindle springs, Attaching	T. Alsop	Elkhart City, Ill	Aug. 25, 1868	81, 456
Mill-spindle step	J. N. Parker	Lewiston, Me	Dec. 18, 1855	13, 954
Mill-spindle step	G. S. Young	Clearfield, Pa	Jan. 3, 1871	110, 814
Mill-spindle step and tram-box	W. G. Norment	Dyersburgh, Tenn	Feb. 14, 1871	111, 768
Mill-spindle-step support	J. Russell	Prairie, Mo	Feb. 1, 1870	99, 353
Mill-spindles from dirt, Oiling and protecting	J. Hubbard	Watertown, Conn	Apr. 8, 1840	1, 538
Mill-spindles, &c., Gudgeon or pivot and step of	J. Staub	Georgetown, D. C	May 4, 1841	2, 075
Mill-spindles, Hanging	W. H. Naracon	Auburn, N. Y	June 15, 1852	9, 033
Mill-spindles, Hanging step of	G. Hotchkiss	Windsor, N. Y	June 29, 1852	9, 072
Mill-spindles, Securing and adjusting the steps of	G. Hotchkiss	Windsor, N. Y	Aug. 17, 1858	21, 199
Mill-spindles, Self-tightening bush for	H. Flinchbaugh	Lampeter Township, Pa	Oct. 19, 1838	984
Mill-spindles, Spring equalizer for	T. L. Clark	Mount Vernon, Ohio	Mar. 3, 1868	75, 124
Mill-spindles, Step and bearing of	T. S. Minniss	Meadville, Pa	June 22, 1852	9, 057
Mill-spindles, Tightening followers to	G. W. Wilson	Galesburgh, Ill	Nov. 18, 1862	36, 975
Mill-staff	A. Eshelman	Earlville, Pa	Feb. 4, 1869	86, 518
Mill-step	C. Corbit	Christiana, Pa	Mar. 24, 1868	75, 865
Mili-step	J. C. Dickey	Saratoga Springs, N. Y	July 3, 1855	13, 161
Mill-wheel	J. Cowen		Dec. 14, 1802	
Mill-wheel	S. H. Freeman	Cecilton, Md	May 17, 1836	
Mill-wheel	R. Lathrop	West Springfield, Mass	Nov. 12, 1812	
Mill-wheel	E. Rew	Canandaigua, N. Y	Feb. 22, 1809	
Mill wheel dresser	J. York	Columbus, Pa	July 1, 1836	
Mill-wheel, Tide	D. Wheeler	Bristol, Mass	Mar. 2, 1811	
Mill-wheel, Trailing	J. Read	Lime, Conn	Mar. 6, 1818	
Mill-wheels, Applying water to	W. Kendall	Fairfield, Me	Jan. 12, 1831	
Mill-wheels, Applying water to	J. Mi henor	Clinton, Ohio	Mar. 9, 1831	
Mills and machinery, Impelling-power for	B. S. Ridgway	Charleston, S. C	Feb. 6, 1829	
Mills, Application of the pendulum to	I. Cobb	Walpole, N. H	Mar. 1, 1809	
Mills, Application of water-power to	Dart and Wood	Fabius, N. Y	Sept. 9, 1835	
Mills, boats, &c., by supporting-chains, Propelling	B. Todd	Marietta, Ohio	Aug. 24, 1831	
Mills, boats, cars, &c., Propelling machinery for	A. Parkhurst and S. Bacon	Scriba, N. Y	Apr. 11, 1831	
Mills, boats, land-vehicles, &c., by wind, Propelling	B. Dugdale	Trenton, N. J	May 3, 1832	
Mills, boats, &c., Lever-power machinery for propelling.	W. Rhodes	Trenton, Tenn	Apr. 27, 1832	
Mills, Bushing and regulating	E. W. Welsh	Paris, Va	Jan. 28, 1840	1, 481
Mills by concentric machinery, Propelling	H. Buell	Randolph, Tenn	May 3, 1833	
Mills, &c., by rolling and balance wheel power, Propelling.	J. Estes	Brownsville, Tenn	Aug. 15, 1831	
Mills, &c., by weights, Propelling	O. R. Marston	Java, N. Y	Jan. 9, 1835	
Mills, Chain-band, with buckets or float-boards, to give motion to the machinery of.	S. Belknap	New York	Dec. 28, 1809	
Mills, Constructing and securing wing-gudgeons for	D. Philips	Georgetown, Pa	Dec. 28, 1840	1, 917
Mills, Cooling and feeding material to	B. F. Harrington and U. B. Burriss.	Missouri City, Mo	Apr. 19, 1859	23, 682
Mills, cotton-gins, &c., Construction of driving-shaft for.	J. Massey	Thomasville, Ga	June 21, 1859	24, 475
Mills, Feed-regulator for	M. Weaver	East Earl Township, Pa	May 11, 1869	89, 904
Mills, Feeding device and tell-tale for	J. D. Mines	Moffatt's Creek, Va	July 16, 1872	129, 577
Mills, Flume for	A. Messer	Providence, R. I	Nov. 19, 1822	
Mills, Frame-work of	W. S. Parham	Sussex County, Va	June 14, 1834	
Mills, Grinding-surface of	O. N. Stanford	Cincinnati, Ohio	Jan. 11, 1859	22, 588
Mills, Hanging runner-stone in	N. Oakley	Babylon, N. Y	Oct. 31, 1848	5, 893
Mills, Machinery for	M. B. Poiteaux	Richmond, Va	May 20, 1830	
Mills, Metallic grinding-rings for	G. and H. O'Connor	Mishawaka, Ind	Nov. 26, 1872	133, 333
Mills, Plate to	S. Hull	Poughkeepsie, N. Y	May 22, 1860	28, 373
Mills, Raising water for	A. Brookfield		Oct. 24, 1800	
Mills, Spindle, bush, ring, and ball for	W. P. Wing	Greenwich, Mass	Feb. 20, 1835	
Mills, thrashing-machines, &c., Machinery for	A. Sawyer	Hopewell, N. Y	Oct. 1, 1830	
Mills, Tram and level for	J. M. Seldomridge	Spring Valley, Ohio	May 12, 1863	38, 508
Mills, vessels, &c., by manual or other power, Propelling.	J. G. Aldebert	Opelousas, La	Dec. 29, 1831	
Mills, Working	J. Rumsay		Aug. 26, 1791	
Miller's stone-staff	S. Barnes	Rochester, Mich	Sept. 16, 1862	36, 444
Milling	J. Reeder	Lebanon, Ohio	Oct. 31, 1829	
Milling and cutting metals, Machine for	M. G. Wilder	West Meriden, Conn	Oct. 21, 1862	36, 744
Milling-machine	A. H. Brainard	Hyde Park, Mass	Oct. 1, 1872	131, 733
Milling-machine	J. R. Brown	Providence, R. I	Feb. 21, 1865	46, 521
Milling-machine	L. Chapman	Collinsville, Conn	Jan. 25, 1870	99, 059
Milling-machine	W. Hawkins	San Francisco, Cal	Oct. 31, 1871	120, 433
Milling-machine	B. Lawrence	Lowell, Mass	Apr. 29, 1873	138, 260
Milling-machine	W. A. N. Long	Worcester, Mass	Sept. 5, 1871	118, 732
Milling-machine	G. T. Pillings	Philadelphia, Pa	July 22, 1873	141, 012
Milling-machine	W. H. Robertson	Hartford, Conn	Oct. 5, 1852	9, 307
Milling-machine	E. S. Stiles	Portland, Me	Feb. 14, 1865	46, 401
Milling-machine	J. Watson	Philadelphia, Pa	Jan. 11, 1870	98, 821
Milling-machine	J. Wheelock	Worcester, Mass	July 23, 1872	129, 876
Milling-machine head-stock	A. H. Brainard	Hyde Park, Mass	Jan. 31, 1871	111, 311
Milling-machines, Back-center for	A. H. Brainard	Hyde Park, Mass	Jan. 17, 1871	110, 951
Milling-tool	J. Buckingham and J. H. Baird	Watertown, Conn	Apr. 15, 1851	8, 036
Milling-tool	A. W. Gifford	Worcester, Mass	Aug. 27, 1867	68, 064
Milling-tool	L. Griswold	Branford, Conn	Feb. 21, 1871	111, 927
Milling-tool	A. J. Lutz and H. Reiss	New York, N. Y	Aug. 3, 1869	93, 212
Milling-tool	J. T. Smith	Middletown, Conn	Mar. 27, 1866	53, 496
Milling-tools, Machine for cutting	B. F. Bee	Harwick, Mass	July 13, 1869	92, 565
Millstone	W. Bahme	New Media, Pa	Dec. 24, 1867	72, 587
Millstone	T. Barnett	Beverly, England	Oct. 12, 1852	9, 313
Millstone	J. H. Burrows	Cincinnati, Ohio	Apr. 23, 1842	2, 581
Millstone	D. Drawbaugh	Eberly's Mills, Pa	Apr. 28, 1863	38, 296
Millstone	E. T. Hanon-Valcke	Paris, France	Apr. 15, 1851	8, 037
Millstone	F. Kelsey	New York, N. Y	Aug. 29, 1848	5, 738
Millstone	J. W. Masury	Brooklyn, N. Y	Feb. 21, 1871	111, 956
Millstone	W. Miller	Lampeter, Pa	Aug. 15, 1811	
Millstone	D. Philips	Jefferson County, Miss	Apr. 3, 1829	
Millstone	J. Preslow	Auburn, N. Y	Feb. 12, 1833	
Millstone-adjusting device	A. W. Winall	Cincinnati, Ohio	Apr. 8, 1873	137, 748
Millstone-adjustment	J. G. Siemers	Saint Louis, Mo	July 15, 1856	15, 346
Millstone, Artificial	E. Cutter and S. Blanchard	Cincinnati, Ohio	Sept. 10, 1846	4, 753
Millstone, Artificial	C. Rands	Englewood, N. J	Nov. 8, 1864	44, 973

Index of patents issued from the United States Patent Office from 1790 *to* 1873, *inclusive*—Continued.

Invention.	Inventor.	Residence.	Date.	No.
Millstone bail and driver	A. G. Waldo	Milwaukee, Wis	Feb. 7, 1871	111, 704
Millstone-balance	J. A. Althouse	New Harmony, Ind	May 2, 1871	114, 249
Millstone-balance	W. C. Benn	San Francisco, Cal	Sept. 22, 1868	82, 278
Millstone-balance	D. Collins	Zanesfield, Ohio	Feb. 13, 1872	123, 678
Millstone-balance	Z. Dawson	Cole Creek, Ind	Aug. 15, 1871	118, 113
Millstone-balance	A. Frederick	Toledo, Ohio	Mar. 9, 1869	87, 656
Millstone-balance	J. M. Schramm	Pontoosuc, Ill	Dec. 31, 1872	134, 491
Millstone-balance	W. A. Vance	Kalamazoo, Mich	May 16, 1871	114, 890
Millstone-balance	J. Walsh	Galena, Ill	Jan. 24, 1871	111, 282
Millstone-balance	G. W. Wilson	Tolono, Ill	Aug. 31, 1869	94, 370
Millstone-balance	G. W. Wilson	Chebanse, Ill	Feb. 11, 1873	135, 740
Millstone balance and cooler	W. B. Dolsen	Waterloo, Iowa	June 14, 1870	104, 283
Millstone-bearing	P. Plamondon	Atchison, Kans	Aug. 9, 1870	106, 283
Millstone, Buhr	O. Evans		May 28, 1796	....
Millstone-bush	J. Brown	Fonda, N. Y	Aug. 15, 1871	118, 101
Millstone-bush	C. Custer	Philadelphia, Pa	Dec. 31, 1867	72, 812
Millstone-bush	M. De Camp	South Bend, Ind	May 24, 1859	24, 106
Millstone-bush	M. French	Leesville, Ind	Apr. 3, 1860	27, 709
Millstone-bush	L. S. Ives	Brooklyn, N. Y	Aug. 16, 1859	25, 120
Millstone-bush	G. W. Landon	Graham, Ind	Mar. 15, 1864	41, 928
Millstone-bush	J. F. McKray	Harmonsburgh, Pa	June 23, 1863	38, 971
Millstone-bush	N. Taylor	Urbana, N. Y	July 23, 1827	....
Millstone bush and bearing	A. Ortlip	East Vincent, Pa	Oct. 4, 1870	108, 045
Millstone bush and spindle	R. S. Cathcart	Cincinnati, Ohio	Apr. 1, 1872	137, 418
Millstone-cleaning composition	G. M. Pettee	Detroit, Mich	Nov. 30, 1869	97, 439
Millstone, Concave and convex	J. Sawyer and E. Clark	Royalston, Mass	Feb. 18, 1826	....
Millstone condenser and ventilator	D. Heffner	Independence, Iowa	May 9, 1871	114, 678
Millstone-cooling device	J. J. Roth	Philadelphia, Pa	Sept. 30, 1873	143, 258
Millstone-cooling device	E. D. Shanton and S. Shaver	Minneapolis, Minn	Apr. 22, 1873	138, 050
Millstone-curb	G. W. Eggleston	Monroe County, N. Y	July 18, 1871	117, 160
Millstone-curb	O. L. Richardson	Athens, Ga	Aug. 14, 1860	29, 617
Millstone-curb	A. Van Vleck and T. Phillips	Jordan, N. Y	July 19, 1870	105, 610
Millstone-dress	W. Ager	Rohrsburgh, Pa	June 29, 1852	9, 063
Millstone-dress	F. Bellinger	Lockport, N. Y	Apr. 27, 1858	20, 029
Millstone-dress	D. Bowman	Knoxville, Tenn	Apr. 2, 1867	63, 360
Millstone-dress	D. Bowman	Johnson City, Tenn	June 14, 1870	104, 107
Millstone-dress	J. Broughton	New York, N. Y	May 22, 1860	28, 344
Millstone-dress	W. P. Coleman	New Orleans, La	Oct. 7, 1856	15, 868
Millstone-dress	E. Deer	Annapolis, Ind	Jan. 2, 1872	122, 369
Millstone-dress	P. Dickson	Woodcock Township, Pa	Sept. 12, 1854	11, 665
Millstone-dress	W. G. Dunnivay and H. Osborn	New Cumberland, Ind	July 12, 1870	105, 186
Millstone-dress	J. Fairclough	Saint Joseph, Mo	Nov. 30, 1869	97, 373
Millstone-dress	W. Finkle	Cole Creek, Ind	July 4, 1854	11, 216
Millstone-dress	M. Fries	Philadelphia, Pa	Sept. 19, 1871	119, 134
Millstone-dress	E. P. Gaines	Melrose, Tex	July 25, 1854	11, 372
Millstone-dress	J. W. Gaines	Melrose, Tex	June 26, 1860	28, 850
Millstone-dress	J. W. Gaines	Clarksville, Tex	Nov. 24, 1868	84, 349
Millstone-dress	A. N. Garland	West Charleston, Vt	Dec. 3, 1867	71, 733
Millstone-dress	J. P. Harris	Greensborough, Ga	Apr. 22, 1873	138, 021
Millstone-dress	A. C. Hartsock	Douglas, Ill	July 21, 1868	80, 171
Millstone-dress	N. Haywood	Cleveland, Ohio	Dec. 15, 1857	18, 848
Millstone-dress	C. V. Littlepage	Austin, Tex	Mar. 20, 1860	27, 551
Millstone-dress	G. W. Loy	Jefferson, Tex	July 20, 1858	20, 950
Millstone-dress	G. W. Loy	Nacogdoches, Tex	Mar. 8, 1870	100, 537
Millstone-dress	G. Natcher	Indianapolis, Ind	Apr. 27, 1858	20, 083
Millstone-dress	G. Natcher	Indianapolis, Ind	Apr. 27, 1858	20, 084
Millstone-dress	G. Natcher	Sidney, Ohio	July 18, 1865	48, 831
Millstone-dress	G. and H. O'Conner	Mishawaka, Ind	Nov. 22, 1870	109, 542
Millstone-dress	H. O'Conner	Mishawaka, Ind	Oct. 22, 1872	132, 514
Millstone-dress	D. N. M. Peregoy	Elizabethtown, Tenn	May 16, 1871	114, 967
Millstone-dress	J. Sedgebeer	Cincinnati, Ohio	July 8, 1862	35, 858
Millstone-dress	B. C. Stephens	Houston, Mo	Dec. 15, 1868	84, 971
Millstone-dress	T. B. Stout	Keyport, N. J	Nov. 11, 1856	16, 074
Millstone-dress	H. L. Spencer	Social Circle, Ga	June 9, 1868	78, 768
Millstone-dress	D. W. Thompson	Somerset, Pa	Sept. 27, 1864	44, 471
Millstone-dress	J. Wilson	Davis' Mills, Va	May 16, 1871	114, 896
Millstone-dress	J. P. H. Wohlenberg	Lyons, Iowa	Oct. 18, 1870	108, 422
Millstone-dress	S. Wolff	Vicksburgh, Miss	June 1, 1858	20, 462
Millstone-dress	N. W. Wortham	Union Point, Ga	Dec. 17, 1867	72, 256
Millstone-dress	P. W. Yarrell	Garysburgh, N. C	Apr. 2, 1872	125, 157
Millstone-dress for cleaning grain	W. Ager	Rohrsburgh, Pa	Sept. 5, 1854	11, 636
Millstone-dresser	Z. G. Hurd	El Dorado, Iowa	Dec. 9, 1862	37, 097
Millstone-dresser	J. E. Karelsen	New York, N. Y	Oct. 21, 1862	36, 716
Millstone-dressing	J. Black	Helena, Ark	Aug. 10, 1844	3, 699
Millstone-dressing	C. Carlisle and E. Estabrook	Norwich, Vt	Aug. 28, 1846	4, 724
Millstone-dressing	H. Dolmetch	Canton, Pa	Oct. 31, 1871	120, 505
Millstone-dressing	I. W. Elmore	Lyons, N. Y	Sept. 25, 1834	....
Millstone-dressing	E. P. Gaines	Nacogdoches County, Tex	Mar. 4, 1851	7, 980
Millstone-dressing	M. P. Gardner	Huntington County, Ind	Jan. 7, 1862	34, 054
Millstone-dressing	S. Golay	Noyon, Switzerland	Jan. 21, 1868	73, 524
Millstone-dressing	M. Holden	Lawrenceburgh, Ind	Aug. 5, 1851	8, 272
Millstone-dressing	N. Jacobs	Newark, Ohio	Mar. 11, 1837	145
Millstone-dressing	J. Pennabecker	Eagle Mills, Durlach, Pa	Aug. 4, 1868	80, 762
Millstone-dressing	T. Rogers	Leesville, Ohio	Apr. 16, 1872	125, 760
Millstone-dressing	A. Sheek	Smith Grove, N. C	Feb. 1, 1859	22, 831
Millstone-dressing	R. M. Smith	Rutherford County, Pa	Feb. 8, 1840	1, 488
Millstone-dressing	T. W. Trussell	Winchester, Va	Nov. 21, 1854	11, 978
Millstone-dressing	I. Whissen	Mount Jackson, Va	Feb. 2, 1858	19, 273
Millstone-dressing	A. Wing	Mayville, N. Y	Mar. 21, 1865	46, 963
Millstone-dressing	J. Yorbrough	Milton, N. C	Jan. 24, 1860	26, 943
Millstone dressing and cracking machine	J. J. Faulkner	McMinnville, Tenn	Oct. 29, 1872	132, 651
Millstone dressing and furrowing apparatus	J. L. Norton	London, England	Dec. 6, 1870	109, 928
Millstone dressing and furrowing apparatus	J. L. Norton	London, England	Dec. 13, 1870	110, 064
Millstone-dressing apparatus	J. Dickinson	New York, N. Y	Oct. 28, 1862	36, 770
Millstone-dressing apparatus	H. Robinson and J. Smith	Lewishan, Carshalton, England.	Nov. 29, 1870	109, 760
Millstone-dressing gage	R. Ruston	Bockville, Ind	June 22, 1869	91, 773
Millstone-dressing guide	J. North	New York, N. Y	May 2, 1871	114, 326

Index of patents issued from the United States Patent Office from 1790 *to* 1873, *inclusive*—Continued.

Invention.	Inventor.	Residence.	Date.	No.
Millstone-dressing machine	M. H. Bacon	Mystic, Conn	July 10, 1860	29, 047
Millstone-dressing machine	M. H. Bacon	Mystic, Conn	July 31, 1860	29, 350
Millstone-dressing machine	E. C. Badger	Warner, N. H	Sept. 19, 1854	11, 686
Millstone-dressing machine	W. Bold	Sheboygan Falls, Wis	Oct. 20, 1868	83, 243
Millstone-dressing machine	J. Bowman	Somerset, Ohio	June 5, 1860	28, 553
Millstone-dressing machine	C. D. Brewer	Lewisburgh, Pa	Sept. 4, 1860	29, 857
Millstone-dressing machine	J. S. Carr	Alliance, Ohio	Dec. 14, 1869	97, 878
Millstone-dressing machine	Z. Chesebrough	Alden, N. Y	Dec. 10, 1838	1, 029
Millstone-dressing machine	W. B. Cummings and N. P. Dadman.	Tyngsborough and Chelmsford, Mass.	Nov. 22, 1853	10, 269
Millstone-dressing machine	S. W. Draper and R. M. Draper	South Dedham and Roxborough, Mass.	May 13, 1856	14, 859
Millstone-dressing machine	W. A. Dryden and J. H. Montgomery.	Monmouth, Ill	Oct. 24, 1865	50, 565
Millstone-dressing machine	J. East	Romeo, Mich	Aug. 17, 1869	93, 867
Millstone-dressing machine	S. East	Memphis, Mich	May 2, 1871	114, 425
Millstone-dressing machine	W. Farrow	Alexandria, Ill	Aug. 19, 1873	142, 009
Millstone-dressing machine	H. B. Gill	Ogden, N. Y	June 14, 1859	24, 385
Millstone-dressing machine	J. T. Gilmore and L. Anderson	Painesville, Ohio	Jan. 26, 1869	86, 297
Millstone-dressing machine	J. T. Gilmore and J. S. Crane	Lake Village, N. H	June 6, 1871	115, 600
Millstone-dressing machine	J. B. Harris	Ottawa, Ill	Sept. 21, 1869	94, 956
Millstone-dressing machine	J. B. Harris	Ottawa, Ill	Apr. 1, 1873	137, 364
Millstone-dressing machine	E. W. Hazard and C. H. Jenner	Binghamton and Rochester, N. Y.	Sept. 16, 1851	8, 364
Millstone-dressing machine	J. Hine	Cockermouth, England	Oct. 19, 1869	95, 901
Millstone-dressing machine	G. Z. Hockenburry	Pittsburgh, Pa	July 24, 1860	29, 275
Millstone-dressing machine	F. A. Hoyt	La Crosse, Wis	Mar. 4, 1873	136, 517
Millstone-dressing machine	J. W. Kennedy and J. T. Plummer.	Plainfield, Conn	July 12, 1859	24, 742
Millstone-dressing machine	S. K. Landes	West Cocalico, Pa	Aug. 21, 1860	29, 701
Millstone-dressing machine	A. Lane	Addison, N. Y	June 30, 1868	79, 359
Millstone-dressing machine	W. A. McGlaughlin	Greenland, Pa	Jan. 19, 1869	85, 948
Millstone-dressing machine	R. D. Nesmith	Lake Village, N. Y	Jan. 15, 1856	14, 109
Millstone-dressing machine	L. Norcross	Livermore, Me	May 6, 1823	....
Millstone-dressing machine	J. Norman	Glasgow, Scotland	Apr. 18, 1871	113, 914
Millstone-dressing machine	F. Ogden and F. A. Gilbert	Mansfield, Ohio	June 25, 1872	128, 419
Millstone-dressing machine	J. W. Parish	McFarland's, Va	July 29, 1873	141, 288
Millstone-dressing machine	J. Pepler	Bath, England	Nov. 8, 1870	109, 046
Millstone-dressing machine	W. Pickens and P. Dalrymple	Chicago, Ill	May 10, 1870	102, 968
Millstone-dressing machine	S. Prettyman and J. Wallis	Mumford and Greenwich, England.	May 25, 1869	90, 461
Millstone-dressing machine	L. Randolph	Jerseyville, Ill	Aug. 19, 1873	141, 949
Millstone-dressing machine	J. Richmond	Lockport, N. Y	July 15, 1862	35, 894
Millstone-dressing machine	J. B. Sacket	Lawton, Mich	Mar. 25, 1862	34, 780
Millstone-dressing machine	I. G. Shands	Saint Louis, Mo	Mar. 21, 1854	10, 674
Millstone-dressing machine	S. Shine and D. Savage	Forks and Orangeville, Pa.	July 29, 1873	141, 392
Millstone-dressing machine	M. Shirk	Lancaster County, Pa	July 24, 1860	29, 345
Millstone-dressing machine	F. Simmons	New Orleans, La	Nov. 30, 1869	97, 320
Millstone-dressing machine	S. Teague	Newton, Ohio	Apr. 5, 1859	23, 511
Millstone-dressing machine	H. B. Weaver	South Windham, Conn	Jan. 15, 1861	31, 142
Millstone-dressing machine	C. Woolverton and B. Ridgeway, jr.	Morris Villa, Pa	May 24, 1811	....
Millstone-dressing machine	J. I. Yount	Tippecanoe, Ohio	Oct. 3, 1871	119, 551
Millstone-dressing machine, Diamond	F. S. Davenport	Jerseyville, Ill	Aug. 26, 1873	142, 153
Millstone-dressing machine, Diamond	D. Larer	Pottsville, Pa	Nov. 25, 1873	144, 851
Millstone-dressing machinery	A. Fisk, jr., L. D. Ramsay, and O. S. Gregory.	Sullivan, Pa	Feb. 20, 1847	4, 973
Millstone-dressing machinery	A. Woodworth	Cambridge, N. Y	Aug. 6, 1872	130, 265
Millstone dressing or pecking machine	S. Trumbull	Suffield, Conn	Aug. 2, 1839	1, 269
Millstone-dressing, Staff or gage for	L. B. Woolever	Mechanicsville, Ohio	Feb. 18, 1873	135, 956
Millstone-dressing, Tool for laying off furrows for	J. C. Hunt and J. Temple	Terre Haute, Ind	Sept. 1, 1868	81, 639
Millstone-driver	S. Darkess	Deerfield, Ind	Nov. 2, 1869	96, 312
Millstone-driver	W. A. Gustine	Ipava, Ill	Apr. 30, 1872	126, 143
Millstone-driver	A. Miller	New Berne, N. C	Aug. 9, 1859	25, 028
Millstone-driver	J. Plance	Constantine, Mich	Mar. 18, 1873	136, [illegible]63
Millstone driver	D. B. Ritter	Glasgow, Ky	Sept. 7, 1869	94, 650
Millstone-driver	J. J. Tomlinson	Bazeman City, Mont	Jan. 24, 1871	111, 277
Millstone-drivers, Bearing for	E. Clark	Lancaster, Pa	Sept. 22, 1857	18, 234
Millstone-elevator	J. B. Hall	Lebanon, N. H	June 1[illegible], 1871	115, 844
Millstone-exhaust	D. Baird	Bloody Run, Pa	Dec. 3, 1867	71, 677
Millstone-exhaust	D. Baird	Bloody Run, Pa	Apr. 21, 1868	76, 9[illegible]2
Millstone-exhaust	J. Lingenfelter	Bloody Run, Pa	Apr. 4, 1871	113, 676
Millstone-exhaust	J. Lingenfelter	Bloody Run, Pa	June 6, 1871	115, 747
Millstone-face tester	J. Kuhn	Mount Pleasant, Pa	Sept. 8, 1868	81, 912
Millstone-feed	J. W. Anderson	Portland, Ind	Nov. 26, 1872	133, 288
Millstone-feed	M. De Camp	South Bend, Ind	Apr. 16, 1867	63, 786
Millstone-feed apparatus	A. Dewey	Princeton, Ky	July 25, 1871	117, 267
Millstone-feed regulator	G. W. Clapper	Martinsville, Ind	May 18, 1869	90, 079
Millstone feeder and grain scourer	P. H. Massey	South Bend, Ind	Feb. 14, 1871	111, 760
Millstone for dressing and cooling flour	J. B. Mirick	Manlius, N. Y	Aug. 10, 1833	....
Millstone for flour-mills	J. Clyde	Crawford County, Pa	July 27, 1831	....
Millstone, Grist	D. Stern	Vanderburgh County, Ind	Jan. 5, 1832	....
Millstone leveling	D. A. Balmer	Lexington, Ind	Oct. 23, 1860	30, 454
Millstone-leveling apparatus	B. H. Meitzler and P. L. Hoos	Fort Seneca, Ohio	Apr. 9, 1872	125, 472
Millstone-leveling device	A. Dray	Portland, Oreg	Jan. 18, 1861	32, 596
Millstone-leveling device	W. Ring	Owen County, Ind	June 3, 1873	139, 479
Millstone-machine	E. C. and R. A. Henderson	Albia, Iowa	Aug. 11, 1868	80, 954
Millstone-pick	H. J. Brunner	Nazareth, Pa	Feb. 14, 1871	111, 810
Millstone-pick	C. R. Elmer	Bridgeton, N. J	Nov. 21, 1865	51, 030
Millstone-pick	S. Etheridge	Tecumseh, Mich	Mar. 2, 1863	....
Millstone-pick	A. S. Jones	Joliet, Ill	Apr. 15, 1862	34, 966
Millstone-pick	D. C. Stone	Kingston, N. Y	Jan. 24, 1865	46, 035
Millstone-pick	H. P. Straub	Cincinnati, Ohio	June 15, 1869	91, 284
Millstone-pick	T. R. Way	Springfield, Ohio	Aug. 5, 1873	141, 613
Millstone-pick	C. Whitehouse	Bridgetown, England	Mar. 2, 1869	87, 530
Millstone-pick, Handle for	A. Rowe	Atalissa, Iowa	Aug. 12, 1862	36, 170
Millstone-pick, Stock for	F. Trump	Springfield, Ohio	Apr. 15, 1873	137, 979
Millstone-picking machine	S. A. Bell	Newtown, Ohio	Oct. 24, 1871	120, 181

Index of patents issued from the United States Patent Office from 1790 *to* 1873, *inclusive*—Continued.

Invention.	Inventor.	Residence.	Date.	No.
Millstone-picking machine	E. W. Daniels	Springfield, Mass	Jan. 24, 1860	26, 897
Millstone-picking machine	J. H. Gray	Boston, Mass	Dec. 4, 1866	60, 176
Millstone-picking machine	J. Kelly	West Milton, Ohio	Aug. 21, 1860	29, 748
Millstone-picking machine	R. D. Nesmith	Franklin, N. H	Aug. 30, 1859	25, 274
Millstone-propelling machinery	J. Case	Sodus, N. Y	June 13, 1831	......
Millstone-redressing machine	W. Y. Gill	Henderson, Ky	May 19, 1857	17, 326
Millstone-regulating device	S. Benson	Centralia, Ill	June 30, 1868	79, 528
Millstone-spindles, Oiling	J. J. Chubb	Decatur, Ind	Apr. 22, 1873	138, 074
Millstone to grind grain and hull clover-seed	B. Myers	Chambersburgh, Pa	Sept. 28, 1831	....
Millstone tram and level combined	J. Hutchins	Portsmouth, Ohio	Feb. 11, 1873	135, 812
Millstone tramming, staffing, and fine-dressing machine.	J. T. Gilmore	Burton, Ohio	Oct. 28, 1862	36, 776
Millstone trimming and dressing machine	J. I. Gilmore	Painesville, Ohio	Jan. 12, 1869	85, 737
Millstone-ventilating device	R. Symes	Saint Charles, Mo	June 30, 1868	79, 517
Millstone-ventilator	W. K. Fuller	Modena, Ill	Dec. 24, 1867	72, 474
Millstone-ventilator	A. Grube	West Hempfield Township, Pa.	Feb. 8, 1870	99, 564
Millstone-ventilator	J. Metherell	Rockford, Ill	Sept. 8, 1868	81, 927
Millstone, with ventilator for cooling the flour, Dressing.	P. Cheek	Flat Rock, Ga	July 26, 1845	2, 222
Millstones and curbs, Cooling	J. Mallin	Bedford, Ohio	Dec. 31, 1867	72, 744
Millstones, Balance-iron for	J. H. Glover	Skreggs Creek, Ky	Dec. 1, 1857	18, 741
Millstones, Balancing	S. Culver	Sandwich, Ill	Jan. 2, 1872	122, 367
Millstones, Balancing	E. and Z. Dawson and B. Hilton.	Brumersburgh, Ohio	Sept. 17, 1867	68, 963
Millstones, Balancing	J. Fairclough	Louisville, Ky	Dec. 21, 1858	22, 356
Millstones, Balancing	D. Fellenbaum	Lancaster, Pa	Sept. 4, 1860	29, 869
Millstones, Balancing	J. Foley	Cleveland, Ohio	Sept. 17, 1867	68, 865
Millstones, Balancing	C. V. Foreman	Mechanicstown, Md	Sept. 6, 1870	107, 029
Millstones, Balancing	W. J. Lemuth	Greencastle, Ind	July 29, 1862	36, 009
Millstones, Balancing	G. S. Thompson	Philadelphia, Pa	Nov. 23, 1869	97, 134
Millstones, Balancing	R. Van Ormer and W. J. Bell	McAllisterville, Pa	May 6, 1862	35, 187
Millstones, Balancing and ventilating	S. N. Page	Salona, Pa	July 7, 1863	39, 167
Millstones, Balancing or adjusting	M. L. Chase	Baltimore, Md	June 13, 1831	....
Millstones, Blacking-staff for facing	A. Heartsill	Louisville, Tenn	Mar. 9, 1869	87, 669
Millstones, Casting the face of	G. Mauter	Londonderry, N. H	Dec. 8, 1818	....
Millstones, Cementing	S. Hoyt	Wilmington, Del	May 1, 1860	28, 083
Millstones, Composition for cleaning	D. Kindig	Newville, Pa	Feb. 25, 1868	74, 833
Millstones, Composition for dressing	W. H. Hayes	Rock Island, Ill	Sept. 13, 1864	44, 189
Millstones, Composition for taking off the glaze from.	J. Noble	Derby, Conn	June 10, 1813	....
Millstones, Cooling	F. Freleigh	Stowe, Ohio	Mar. 26, 1845	3, 969
Millstones, Cooling	A. Taylor	Littleton, N. H	Mar. 31, 1836	....
Millstones, Crane, screw, and bale for raising	T. Reynolds	West Fallowfield, Pa	Nov. 8, 1814	....
Millstones, Device for adjusting	A. W. Winall	Cincinnati, Ohio	Jan. 28, 1873	135, 393
Millstones, Device for guiding diamonds for dressing.	B. D. Tripp	Moravia, N. Y	Dec. 18, 1860	30, 995
Millstones, Diamond-protector for dressing	J. Dickinson	Brooklyn, N. Y	Sept. 3, 1861	33, 187
Millstones, Elevating	A. M. Bruckart	Brunnersville, Pa	Feb. 24, 1863	37, 732
Millstones, Eye for	E. Munson	Utica, N. Y	July 19, 1853	9, 859
Millstones, Feeding	M. Smith	Princeton, Iowa	Jan. 19, 1858	19, 156
Millstones, Finishing and balancing	G. Todd	Saint Louis, Mo	Nov. 8, 1851	8, 538
Millstones for scouring and hulling buckwheat, Dressing.	B. I. Harris	Auburn, Pa	Dec. 25, 1855	13, 985
Millstones, Forming and balancing	E. Munson	Utica, N. Y	Aug. 7, 1849	6, 639
Millstones, Grain and middlings feeder for	B. B. Broyles	Freedom, Tenn	July 9, 1872	128, 702
Millstones, Hand-mill for picking	S. Hodge and J. Dorr		Oct. 23, 1792	....
Millstones, Hanging	E. Clark	Lancaster, Pa	Sept. 15, 1857	18, 191
Millstones, Hanging	W. A. Clark, S. D. Porter, and W. D. Simpson.	Saint Louis, Mo	June 2, 1857	17, 421
Millstones, Hanging	R. Cochran	Cincinnati, Ohio	Aug. 7, 1855	13, 384
Millstones, Hanging	G. P. Dance	Columbia, Tex	May 29, 1860	28, 534
Millstones, Hanging	S. Fagin	Cincinnati, Ohio	Oct. 30, 1866	59, 201
Millstones, Hanging	J. A. Forsman	Cincinnati, Ohio	Aug. 31, 1858	21, 330
Millstones, Hanging	J. H. Glover	Glasgow, Ky	May 22, 1860	28, 362
Millstones, Hanging	D. Marsh	Bridgeport, Conn	Jan. 15, 1856	14, 128
Millstones, Hanging	Z. McDaniel	Bowling Green, Ky	Sept. 11, 1860	30, 025
Millstones, Hanging	C. Schneider	Galion, Ohio	July 19, 1870	105, 496
Millstones, Hanging	H. P. Straub	Cincinnati, Ohio	Oct. 18, 1870	108, 405
Millstones, Hanging	A. W. Winall	Cincinnati, Ohio	June 15, 1869	91, 295
Millstones, Hanging	A. W. Winall	Cincinnati, Ohio	June 3, 1873	139, 642
Millstones, Hanging and using	J. McDuffee	Bradford, Vt	Feb. 21, 1833	....
Millstones, Hanging-device for	P. Zimmerman	Delaware Water Gap, Pa	July 7, 1868	79, 717
Millstones, Instrument for facilitating the facing of	B. D. Sanders	Holliday's Cove, Va	Feb. 10, 1857	16, 615
Millstones, Laying off and dressing the runners of	C. Vest	Stokes County, N. C	Oct. 2, 1838	954
Millstones, Machine for facing and polishing	E. Munson	Utica, N. Y	Jan. 29, 1861	31, 255
Millstones, Machine for furrowing	J. J. Zinn	Albion, Pa	July 2, 1867	66, 439
Millstones, Machine for leveling the faces of	D. Drawbaugh	Eberle's Mills, Pa	May 12, 1863	38, 472
Millstones, Manufacture of	J. S. Elliott and J. F. Wood	Chelsea and Everett, Mass	Mar. 12, 1872	124, 558
Millstones, Method of putting face-dress on	J. S. Braley and J. A. and P. L. Schmitt.	Utica, Mo	Aug. 23, 1870	106, 540
Millstones, Process of repairing	J. S. Elliott and J. F. Wood	Chelsea and Everett, Mass	Mar. 12, 1872	124, 559
Millstones, Steadying the motion of	J. J. Doughty	Lake City, Minn	July 13, 1869	92, 433
Millstones, Stopping	A. Reeder	Buckingham, Pa	June 2, 1863	38, 761
Millstones, Swinging spout for feeding	M. and C. Painter	Owing's Mills, Md	July 1, 1856	15, 250
Millstones, Tram for gaging	T. R. James	Saint Louis, Mo	Oct. 27, 1868	83, 503
Millstones, Ventilating	J. Campbell	Peoria, Ill	Jan. 28, 1868	73, 776
Millstones, Ventilating	C. Cerisier	Mung sur Loire, France	Jan. 4, 1870	98, 555
Millstones, Ventilating	J. Gray	Dubuque, Iowa	Oct. 29, 1867	70, 198
Millstones, Ventilating	A. D. Lagoguey	Paris, France	Apr. 24, 1866	54, 268
Millstones, Ventilating	H. McEldowney	Dixon, Ill	Oct. 1, 1867	69, 459
Millstones, Ventilating	S. G. Morrison	Williamsport, Pa	June 18, 1861	32, 579
Millstones, Ventilating	G. P. Plant and J. Raith	Saint Louis, Mo	Oct. 23, 1860	30, 498
Millstones, Ventilating	L. Racine	Joliet, Ill	July 6, 1858	20, 818
Mincing-cleaver	S. J. Tongue	Philadelphia, Pa	July 21, 1868	80, 242
Mincing-knife	S. C. Ketchum	Winchendon, Mass	May 28, 1867	65, 233
Mincing-knife	H. M. Remington	Springfield, Mass	Feb. 12, 1867	61, 954
Mincing-machine	W. H. Peirce	Bangor, Me	Aug. 31, 1869	93, 237

Index of patents issued from the United States Patent Office from 1790 *to* 1873, *inclusive*—Continued.

Invention.	Inventor.	Residence.	Date.	No.
Mine-ventilator	J. B. Jones	Laurel Run, Pa	May 3, 1870	102, 681
Mine-water, Compound for purifying	F. J. Delker	Ashland, Pa	Dec. 20, 1870	110, 345
Mines, Device for raising tailings from	W. A. Rogers	Folsom, Cal	Jan. 9, 1872	122, 657
Mines, Door for	N. Jensen	Carbon Station, Wyo	June 25, 1872	128, 312
Mines, Purifying water from	H. Burgess	Royer's Ford, Pa	Nov. 18, 1873	144, 737
Mines, Safety-cage for	J. Evans	Virginia City, Nev	Aug. 7, 1866	56, 915
Mines, Safety dumping-cage for	W. Z. Hatcher	Plymouth, Pa	June 28, 1870	104, 848
Mines, Ventilating	J. S. Fisk and J. Westerman	Meadville and Sharon, Pa	Jan. 17, 1865	45, 918
Mines, Ventilation of	J. L. Beadle	Ashland, Pa	May 16, 1865	47, 694
Mines, Ventilation of	H. Haupt	Cambridge, Mass	June 6, 1865	48, 065
Mines, Winding-machinery for	E. M. Ivens	Tamaqua, Pa	Apr. 14, 1857	17, 065
Mineral and fossil substances, Machine for separating.	R. George	Denver City, Colo	Aug. 2, 1870	106, 047
Mineral and fossil substances with bitumenized clay to form a plastic material, Mixing.	P. Lea	Goliad, Tex	Nov. 11, 1873	144, 397
Mineral and metallic substances, Apparatus for collecting and separating.	W. T. Duvall	Georgetown, D. C	July 28, 1868	80, 398
Mineral and other hard substances, Machine for crushing.	D. C. Ebaugh	Charleston, S. C	July 26, 1870	105, 659
Mineral bath to imitate mineral-water, Preparation of.	O. Gavron	New York, N. Y	June 21, 1870	104, 446
Mineral composition resembling jasper	J. P. Pepper	New Britain, Conn	Dec. 16, 1851	8, 592
Minerals, &c., crushing and grinding machine	T. O. Cutler	New York, N. Y	Dec. 12, 1854	12, 054
Mineral-fountain, Family	G. E. Pendergast	Philadelphia, Pa	Aug. 18, 1815	....
Mineral-washing machine	P. G. Gardiner	New York, N. Y	Oct. 4, 1864	44, 526
Mineral-water	C. E. Michel	Saint Louis, Mo	Dec. 3, 1867	71, 777
Mineral-water, Apparatus for making	J. S. Ewing	Philadelphia, Pa	Feb. 10, 1817	....
Mineral-water, Apparatus for making	S. Fahnestock	Lancaster, Pa	Apr. 23, 1819	....
Mineral-water, Giving temperature to artificial	J. Hawkins	Philadelphia, Pa	Apr. 5, 1823	....
Mineral-water, Imitating natural	J. Hawkins	Baltimore, Md	Sept. 7, 1809	....
Mineral-water, Manufacture of	J. Matthews	New York, N. Y	July 16, 1872	129, 044
Mineral-water, Manufacture of bottled	T. Maloney	Des Moines, Iowa	Mar. 14, 1871	112, 610
Mineral-water, Manufacturing	B. Marshall	Albany, N. Y	Dec. 5, 1815	....
Mineral-water, Manufacturing	W. Mead	Newburgh, N. Y	Apr. 12, 1831	....
Miners' bar	R. B. Platt	Hazleton, Pa	Apr. 22, 1873	138, 098
Miners' pick	S. H. Wallace	Parnassus, Pa	Oct. 29, 1867	70, 298
Miners' pick	R. K. Walton	Clarington, Ohio	Nov. 18, 1873	144, 808
Miners' squib	J. Holmes	Saint Clair, Pa	Dec. 12, 1871	121, 783
Miniature-case	H. Halvorson	Boston, Mass	Mar. 25, 1856	14, 501
Miniature-cases, &c., Composition for	M. Tomlinson	Birmingham, Conn	Aug. 24, 1858	21, 285
Miniature-cases, Press for making	H. T. Anthony	New York, N. Y	Jan. 31, 1854	10, 465
Minie-balls, Apparatus for casting	C. B. Tatham	Brooklyn, N. Y	May 20, 1862	35, 334
Minie-balls, Device for making	P. Naylor	New York, N. Y	May 17, 1864	42, 785
Minie-balls, Grooving and sizing	P. Naylor	New York, N. Y	May 17, 1864	42, 786
Mining and tunneling	H. Haupt and J. Y. Smith	Cambridge, Mass., and Alexandria, Va.	Apr. 4, 1865	47, 168
Mining and tunneling machine	R. C. M. Lovell	Covington, Ky	July 30, 1867	67, 323
Mining-apparatus, Coal	G. E. Donisthorpe, W. Firth, and R. Ridley.	Leeds, England	May 10, 1864	42, 724
Mining apparatus, Coal	R. H. Lamborn	Altoona, Pa	May 10, 1864	42, 669
Mining apparatus, Hydraulic	F. H. Fisher	Nevada City, Cal	Dec. 20, 1870	110, 222
Mining apparatus, Hydraulic	T. Watson	Nevada City, Cal	Oct. 25, 1870	108, 658
Mining-drill	H. Burk	Mineral Point, Ohio	Apr. 15, 1862	34, 941
Mining-machine	D. Morris	Bartlett, Ohio	July 20, 1869	92, 871
Mining-machine	D. Williamson	Elizabeth, Pa	Mar. 2, 1869	87, 453
Mining-pan	J. A. Brock	Chicago, Ill	Apr. 23, 1861	32, 115
Mining-pick	G. Hofman	Scott Bar, Cal	May 12, 1863	38, 481
Mining-pick	H. L. Lowman	Virginia City, Nev	June 6, 1865	48, 080
Mining, Placer	C. H. Smith	Rock Island, Ill	Mar. 28, 1865	47, 046
Mining-shaft and building ventilator	T. Boyd	Cambridgeport, Mass	Jan. 1, 1867	60, 677
Mining-shaft safety-guard	E. O. Leermo	Gold Hill, Nev	June 23, 1868	79, 076
Mining sluice and riffle	C. J. Garland	Gwin Mine, Cal	Oct. 17, 1871	119, 978
Mining-sluices, Apparatus for collecting the precious metals in.	J. B. Beers	San Francisco, Cal	July 23, 1872	129, 644
Mirror	P. A. Dailey	New York, N. Y	July 31, 1866	56, 723
Mirror	S. R. Scottron	Springfield, Mass	Mar. 31, 1866	76, 253
Mirror, Adjustable	L. F. Neagle	Philadelphia, Pa	Nov. 26, 1867	71, 320
Mirror, Adjustable	G. S. Roberts	Meredith Village, N. H	June 13, 1871	115, 898
Mirror, Adjustable duplex	G. S. Roberts	Meredith Village, N. H	Apr. 23, 1872	126, 091
Mirror and frame hanger	J. S. Thomson	Camptown, Pa	Apr. 23, 1872	125, 913
Mirror and picture-frame	W. McConnell	Clarksville, N. J	June 21, 1870	104, 617
Mirror and picture-frame support	D. Hartmann	Mansfield, Ohio	July 4, 1871	116, 589
Mirror and stand, Adjustable	H. W. Southworth	Mittineague, Mass	Feb. 2, 1869	86, 599
Mirror, Back-reflecting	R. Mason	Newark, N. J	Mar. 7, 1871	112, 474
Mirror for attachment to windows	G. P. Bertrand	Easton, Pa	July 22, 1862	35, 917
Mirror-frame	O. L. Gardner	New York, N. Y	Sept. 10, 1867	68, 730
Mirror-frame, Hand-toilet	E. P. Williams and G. H. Chinnock.	Elizabeth, N. J., and New York, N. Y.	May 31, 1870	103, 807
Mirror frame, Hand toilet	E. P. Williams and G. H. Chinnock.	Elizabeth, N. J., and New York, N. Y.	May 31, 1870	103, 808
Mirror frame, Hand toilet	E. P. Williams and G. H. Chinnock.	Elizabeth, N. J., and New York, N. Y.	May 31, 1870	103, 809
Mirror-frame, Swivel	C. W. Hafermaly	Providence, R. I	Jan. 12, 1869	85, 815
Mirror, Hand	J. F. Dohan	Binghampton, N. Y	Dec. 23, 1873	145, 729
Mirror, Hand	W. U. Dudley	New York, N. Y	July 20, 1869	92, 942
Mirror, Hand	A. C. Estabrook	Northamton, Mass	Jan. 23, 1872	123, 008
Mirror, Hand	J. E. Faris	Baltimore, Md	Feb. 13, 1866	52, 551
Mirror, Hand	W. M. Welling	New York, N. Y	Apr. 16, 1872	125, 867
Mirror-holder	W. Simpson	Berlin, Canada	July 15, 1873	140, 851
Mirror, &c., Ornamental	W. M. Davis	Brook Haven, N. Y	Apr. 16, 1872	125, 726
Mirror, Suspension toilet	R. T. Sargent	Norwich, Vt	Mar. 28, 1871	113, 100
Mirror, Toilet	G. H. Chinnock	New York, N. Y	Sept. 14, 1869	94, 711
Mirror, Toilet	G. H. Chinnock	New York, N. Y	Sept. 14, 1869	94, 712
Mirror, Toilet	G. H. Chinnock	New York, N. Y	Mar. 22, 1870	100, 980
Mirror, Toilet	G. H. Chinnock and E. P. Williams.	Elizabeth N. J	June 7, 1870	103, 980
Mirror, Toilet	J. Johnson	Saco, Me	Oct. 24, 1865	50, 659
Mirror, Toilet	W. H. King	Newark, N. J	Mar. 22, 1870	101, 022

Index of patents issued from the United States Patent Office from 1790 *to* 1873, *inclusive*—Continued.

Invention.	Inventor.	Residence.	Date.	No.
Mirror, Toilet	W. F. Patterson	Boston, Mass	July 2, 1872	128, 503
Mirror, Toilet	W. F. Patterson	Boston, Mass	July 2, 1872	128, 651
Mirror, Toilet	L. H. Rogers	Boston, Mass	Mar. 23, 1869	88, 077
Mirror, Toilet	G. Wattis	New York, N. Y	Sept. 14, 1869	94, 798
Mirror, Window	A. and A. Olander	Glen Gardner, N. J., and New York, N. Y.	Apr. 1, 1873	137, 383
Mirror, Window	H. Voss and H. P. Brandt	Davenport, Iowa	Aug. 5, 1873	141, 476
Mirrors, &c., Adjustable hanger for	J. Wright	New York, N. Y	Dec. 23, 1873	145, 825
Mirrors, Coating and protecting the silvering of	D. Briansky	St. Petersburg, Russia	Apr. 12, 1864	42, 276
Mirrors, Composition for silvering	H. Poissonnier	New York, N. Y	June 19, 1860	28, 773
Mirrors, Device for adjusting	O. L. and W. Gardner	Glen Gardners, N. J	Apr. 18, 1871	113, 868
Mirrors, Hanging	F. Brown	Detroit, Mich	July 9, 1867	66, 458
Mirrors, Hanging	W. C. Cumming	Peekskill, N. Y	Jan. 1, 1867	60, 699
Mirrors, Hanging	J. S. and H. F. Gray	Chelsea, Mass	June 19, 1866	55, 647
Mirrors in traps, Arrangement of	J. Stevens	Middleton, Md	June 11, 1850	7, 431
Mirrors, Manufacture of	R. Kleck	New York, N. Y	Nov. 7, 1865	50, 874
Mirrors, Manufacture of silvered	D. Durand	New York, N. Y	Jan. 30, 1872	123, 247
Mirrors, Metallizing	E. Dodé	Paris, France	Nov. 13, 1866	59, 734
Mirrors, Ornamenting	W. M. Davis	Mount Sinai, N. Y	Feb. 14, 1871	111, 731
Mirrors, Ornamenting	T. C. March	London, England	Mar. 5, 1867	62, 654
Mirrors, Silvering	R. Keck	Chicago, Ill	July 17, 1866	56, 507
Mirrors, Silvering	T. Petitjean	Tottenham Court Road, England.	Oct. 21, 1856	15, 950
Mirrors to furniture, Suspending	L. Pierce	Charleston, Mass	Mar. 8, 1870	100, 663
Missiles, Means of checking and resisting	A. C. Twining	New Haven, Conn	July 28, 1863	39, 363
Miter and bevel square, Carpenters'	J. S. Halsted and C. J. Ackerman.	New York, N. Y	June 26, 1855	13, 127
Miter and beveling machine	L. Holliday	Rogersville, N. Y	May 15, 1855	12, 862
Miter and other joints, Machine for cutting	F. A. Gleason	Rome, N. Y	Mar. 6, 1855	12, 482
Miter-box	J. W. Bower and A. D. Mohler	Greencastle and Huntington, Ind.	May 14, 1872	126, 669
Miter-box	S. D. Bowker	Kansas City, Mo	Aug. 10, 1869	93, 589
Miter-box	D. Bull	Amboy, Ill	June 12, 1866	55, 576
Miter-box	J. Bullard	Hyde Park, Vt	Feb. 1, 1870	99, 398
Miter-box	R. and R. T. Burchell	Trenton, N. J	Apr. 20, 1869	89, 124
Miter-box	H. N. Burr	Mansfield, Ohio	July 19, 1870	105, 421
Miter-box	C. Caton	Coshocton, Ohio	May 21, 1872	126, 873
Miter-box	C. Caton	Coshocton, Ohio	June 17, 1873	139, 867
Miter-box	G. L. Chapin	Perrysburgh, N. Y	June 9, 1857	17, 494
Miter-box	E. C. Cheek	Placerville, Cal	July 2, 1867	66, 217
Miter-box	J. Devoge	Meadville, Pa	Nov. 15, 1870	109, 190
Miter-box	T. B. Dooley	Boston, Mass	Mar. 19, 1872	124, 797
Miter-box	A. C. Hall	Carbondale, Pa	July 9, 1872	128, 795
Miter-box	J. and T. Hanks	New Brunswick and Jersey City, N. J.	July 16, 1872	129, 557
Miter-box	G. E. Hedges	Ashland, Nebr	Aug. 1, 1871	117, 538
Miter-box	W. H. Herbert	Blissfield, Mich	Sept. 1, 1868	81, 782
Miter-box	D. Howell, jr., and M. K. Kellam.	New York, N. Y	May 3, 1864	42, 582
Miter-box	G. Keating	Thomaston, Me	Aug. 21, 1866	57, 334
Miter-box	J. King	Northampton, Mass	July 16, 1872	128, 963
Miter-box	L. W. Langdon	Northampton, Mass	Nov. 15, 1864	45, 055
Miter-box	C. F. Linscott	Chicago, Ill	Dec. 6, 1870	109, 915
Miter-box	H. Markle	Spencer, Ind	June 6, 1871	115, 752
Miter-box	J. A. McKinstry	Monson, Mass	Oct. 23, 1866	59, 046
Miter-box	J. A. McKinstry and W. Walden.	Springfield, Mass	Feb. 11, 1873	135, 832
Miter-box	H. B. Nash	Sandy Hill, N. Y	Sept. 11, 1860	30, 027
Miter-box	J. Pons	Baltimore, Md	Mar. 9, 1869	87, 703
Miter-box	J. W. Richard son	Boston, Mass	Dec. 31, 1867	72, 906
Miter-box	C. Robinson	Fox Lake, Wis	Apr. 28, 1868	77, 217
Miter-box	C. Robinson	Fox Lake, Wis	Sept. 22, 1868	82, 351
Miter-box	E. Root	Parma, Mich	June 8, 1869	90, 960
Miter-box	M. O. Royce	Boston, Mass	July 20, 1869	92, 887
Miter-box	M. O. Royce	East Cambridge, Mass	Sept. 5, 1871	118, 748
Miter-box	S. D. Skiff	Columbia, Ohio	Aug. 20, 1872	130, 665
Miter-box	M. Spear	Bowdoinham, Me	Sept. 30, 1851	8, 401
Miter-bok	M. Spear	Bowdoinham, Me	May 16, 1854	10, 936
Miter-box	M. Spear	Bowdoinham, Me	May 1, 1855	12, 796
Miter-box	G. M. Stevens	Portland, Me	July 15, 1873	140, 853
Miter-box	A. F. Tarr	Rockport, Mass	Dec. 21, 1858	22, 386
Miter-box	J. P. Tierney	Sacramento, Cal	Sept. 16, 1873	142, 878
Miter-box	W. Tinsley	Glen's Falls, N. Y	Nov. 28, 1854	12, 006
Miter-box	C. H. Underwood	Dorchester, Mass	Mar. 23, 1869	88, 236
Miter-box	W. S. Wheeler and S. E. Brickford.	Franklin, N. H	Mar. 12, 1867	62, 910
Miter-box	E. M. Wilcox	Bloomer, Wis	July 26, 1870	105, 752
Miter-box	W. P. Wood	Washington, D. C	June 3, 1856	15, 053
Miter-box and miter-saw	D. McAllaster	Malden, Mass	July 5, 1870	104, 975
Miter-box for cutting printers' rules	T. H. Mead	New York, N. Y	Nov. 11, 1873	144, 406
Miter-cutting machine	F. D. Green	Williamsport, Pa	Oct. 8, 1872	131, 949
Miter-cutting machine	S. W. Hall	Williamsport, Pa	Aug. 17, 1858	21, 194
Miter-cutting machine	Z. Kelley	Easton, Md	Apr. 22, 1812	......
Miter-gage	G. W. La Baw, A. T. Hutchinson, and I. W. Fleming.	Jersey City, N. J., Harlem, N. Y., and Jersey City, N. J.	Apr. 21, 1868	77, 053
Miter-joints, Machine for trimming	A. Johnson	Putnam, Conn	June 12, 1866	55, 499
Miter-machine	J. H. Brown	Brockport, N. Y	May 4, 1869	89, 627
Miter-machine	D. Bull	Amboy, Ill	July 15, 1873	140, 763
Miter-machine	J. H. Estes	Boston, Mass	Sept. 29, 1868	82, 504
Miter-machine	L. Henkle	Rochester, N. Y	July 2, 1872	128, 486
Miter-machine	J. R. Howell	Buffalo, N. Y	June 21, 1870	104, 458
Miter-machine	F. C. Jones	Ouachita Parish, La	May 24, 1870	103, 341
Miter-machine	T. E. King	Boston, Mass	Mar. 7, 1871	112, 355
Miter-machine	G. W. La Baw	Jersey City, N. J	June 27, 1854	11, 164
Miter-machine	G. W. La Baw	Jersey City, N. J	May 29, 1855	12, 956
Miter-machine	J. Nonnenbacher	New York, N. Y	Apr. 5, 1870	101, 497
Miter-machine	T. Pooley	Blackberry Station, Ill	Dec. 23, 1873	145, 751

Index of patents issued from the United States Patent Office from 1790 *to* 1873, *inclusive*—Continued.

Invention.	Inventor.	Residence.	Date.	No.
Miter-machine	L. W. Rosecrans	Marshalltown, Iowa	Oct. 11, 1870	108, 296
Miter-machine	D. B. Ware	Athens, Mich	Aug. 30, 1870	106, 894
Miter-planing machine	J. Mannebach	New York, N. Y	Dec. 3, 1872	133, 653
Miter-point borer	B. Hall	Canaan, N. Y	Mar. 8, 1810	....
Miters, Templet for cutting	A. H. Bartlett	New Haven, Conn	July 16, 1872	129, 516
Mitering-bench	J. W. Mahan	Lexington, Ill	Apr. 1, 1856	14, 569
Mitering-machine	F. D. Green	Williamsport, Pa	June 24, 1873	140, 267
Mitering-machine	J. Holzberger	Newark, N. J	July 26, 1870	105, 686
Mitering-machine	F. A. Howard	Belfast, Me	Aug. 21, 1866	57, 325
Mitering-machine	F. A. Howard	Belfast, Me	Aug. 25, 1868	81, 373
Mitering-machine	C. Loetscher	Dubuque, Iowa	Dec. 9, 1873	145, 354
Mitering-machine	H., G. W., and A. D. Malin	Pleasantville, Pa	Apr. 16, 1872	125, 745
Mitering-machine	J. J. Landers, jr	New York, N. Y	May 26, 1868	78, 330
Mitering-machine	J. M. Seymour	Newark, N. J	July 7, 1868	79, 783
Mitering-machine	E. Shaw	Tarr Farm, Pa	Aug. 9, 1870	106, 218
Mitering-machine	G. M. Stevens	Portland, Me	May 7, 1872	126, 588
Mitering-machine	R. F. Tompkins and H. T. Williams.	New York, N. Y	Nov. 10, 1868	83, 891
Mitering-machine	G. Trimble	Philadelphia, Pa	Dec. 19, 1865	51, 633
Mitering-machine	J. H. Van Ness	Charlotte, N. C	Sept. 30, 1873	143, 265
Mitering machine, Wood	R. F. Tompkins	New York, N. Y	Oct. 1, 1867	69, 510
Mitten	C. Neumann	Boston, Mass	July 8, 1862	35, 833
Mitten	F. L. Oakley	New York, N. Y	Apr. 29, 1873	138, 273
Mitten	A. P. Smith	Sterling, Ill	Jan. 1, 1867	60, 949
Mitten	G. Topping	Chicago, Ill	Nov. 3, 1868	83, 800
Mitten	E. V. Whitaker	Gloversville, N. Y	Dec. 3, 1872	133, 689
Mitten	J. L. Whitten	Essex Junction, Vt	Nov. 19, 1872	133, 183
Mixer:				
See Batter-mixer. Mortar-mixer.				
Clay-mixer. Paint-mixer.				
Mixing, heating, and cooling substances, Apparatus for.	M. Nolden	Frankfort-on-the-Main, Prussia.	Sept. 20, 1870	107, 621
Mixing-machine	J. W. Stockwell	Portland, Me	June 24, 1873	140, 171
Mixing-mill	W. Martien	Baltimore, Md	Jan. 2, 1872	122, 393
Moccasin	C. Elliott	New York, N. Y	Jan. 26, 1825	....
Moccasin	T. Hersey	Bangor, Me	Feb. 4, 1873	135, 548
Moccasin	A. Margesson	Bangor, Me	Aug. 26, 1873	142, 252
Moccasin	A. Tillotson	New York, N. Y	Nov. 22, 1823	....
Moccasin and sock	W. Bolton	Northampton, Mass	Oct. 21, 1814	....
Moccasin, Boot	J. J. Drown	Plattsburgh, N. Y	Dec. 17, 1872	134, 041
Moccasin, sock, &c	D. Williams	New York, N. Y	Oct. 25, 1832	....
Moccasin-sock or overshoe	W. Brower	Rensselaer County, N. Y	Mar. 6, 1822	....
Moccasin, Water-proof	J. Syms	New York, N. Y	Nov. 10, 1827	....
Moccasins, Application of hat-trimmings in making	J. Jackson	Burlington, N. J	Apr. 1, 1815	....
Moccasins, &c., in molds, Making	J. J. Staples and W. H. Cornell	New York, N. Y	Dec. 15, 1824	....
Model-stand, Artist's	G. A. Gilbert	New York, N. Y	May 14, 1872	126, 690
Molasses-cup	E. R. Cook	Trenton, N. J	May 17, 1864	42, 750
Molasses-cup	D. W. Messer	Boston, Mass	Apr. 14, 1857	17, 048
Molasses-cup	W. S. Rooney	Albany, N. Y	Sept. 3, 1867	68, 532
Molasses-cup, Sheet-metal	G. B. Halsted	New York, N. Y	Feb. 4, 1868	74, 081
Molasses-gate	J. Hegarty	Jersey City, N. J	Jan. 5, 1869	85, 660
Molasses-gate	G. W. Hubbard	Meriden, Conn	Jan. 10, 1860	26, 769
Molasses-gate	J. D. Kellogg and J. Wright	Northampton, Mass	Aug. 16, 1839	1, 290
Molasses-gate	T. Mace	New York, N. Y	Mar. 5, 1867	62, 652
Molasses-gate	E. Stebbins	Chicopee, Mass	Oct. 15, 1850	7, 724
Molasses-gate	A. West	Greene, Me	Aug. 28, 1840	1, 750
Molasses-gate	J. Whittemore	Boston, Mass	Oct. 9, 1839	1, 359
Molasses gate and cock for drawing liquors	L. Lincoln	Hartford, Conn	Nov. 10, 1841	2, 342
Molasses-gate, Automatic	P. H. Kimball	Prophetstown, Ill	July 30, 1867	67, 319
Molasses gate or faucet	E. Stebbins	Chicopee, Mass	July 15, 1851	8, 228
Molasses-gates, Casting	M. R. Chace	Fall River, Mass	Apr. 24, 1860	27, 969
Molasses, Manufacture of	G. W. Sayre	Pisgah, Ohio	Apr. 10, 1866	53, 887
Molasses, Process for improving the color of	J. Von Bohm	Melbourne, Australia	Feb. 14, 1865	46, 431
Molasses-vessel	W. Briggs	Bristol, R. I	May 28, 1867	65, 159
Mold:				
See Baking-mold. Electrotype-mold.				
Bonnet-front mold. Gig-mold.				
Boot and shoe sole mold. Glass-mold.				
Bottle-mold. Glass-blowers mold.				
Brick-mold. Glass-jar mold.				
Building-block mold. Glass-lamp mold.				
Bullet-mold. Glass-makers' mold.				
Burial-case mold. Glass-pressing mold.				
Butter-mold. Glass-ware mold.				
Button-mold. Hard-rubber mold.				
Candle-mold. Hat-forming mold.				
Car-wheel mold. Hub-box mold.				
Casting-mold. Ice-cream mold.				
Cement-mold. Ingot-mold.				
Chill-mold. Ingot-casting mold.				
Cigar-mold. Jar-mold.				
Cigar-bunch mold. Joint-mold.				
Cigar-drying mold. Jointed mold.				
Clay-mold. Lamp-mold.				
Clay and cement mold. Lamp-chimney mold.				
Clay and cement pipe mold. Melon-shaped mold.				
Collar-mold. Metal-mold.				
Concrete-block mold. Metallic mold.				
Confectioners' mold. Needle-threader mold.				
Corn-cake mold. Oil-cake mold.				
Cornice-mold. Pencil-tip mold.				
Cylindrical mold. Pipe-mold.				
Die-mold. Pipe-casting mold.				
Doughnut-mold. Plaster-mold.				
Drain-pipe mold. Plow-point mold.				
Elastic-mold. Post-mold.				
Printing-type mold.				
Projectile-mold.				

Index of patents issued from the United States Patent Office from 1790 *to* 1873, *inclusive*—**Continued.**

Invention.	Inventor.	Residence.	Date.	No.
Mold—Continued.				
See Pudding-mold. Rifle-ball mold. Saddle-tree mold. Sealing-wax mold. Sectional mold. Sewer-pipe mold. Sheaf-mold. Soap-mold. Sole-mold. Spoon-mold. Stereotype-mold. Sugar-mold. Suppository-mold. Syringe-mold. Tile-mold. Tooth-mold. Type-mold. Vulcanizable-gum mold. Water-trap mold. Wrought-iron mold.				
Mold	T. J. Barron	Brooklyn, N. Y	May 16, 1871	114, 965
Mold	D. N. Spogle	Annapolis, Md	July 25, 1871	11 , 343
Mold and core for metallic castings	W. Hainsworth	Allegheny, Pa	Dec. 6, 1870	109, 894
Mold-board and colter, Adjustable	G. D. Rowell	Menomonee Falls, Wis	July 27, 1869	93, 011
Mold-board blank	W. M. Watson	Tonica, Ill	Nov. 5, 1861	33, 672
Mold-board press	A. Thompson	Ottawa, Ill	June 7, 1870	103, 942
Mold-board, Revolving	J. S. Godfrey	Leslie, Mich	Apr. 28, 1868	77, 184
Mold-boards, Casting	J. Oliver	South Bend, Ind	Apr. 21, 1868	76, 939
Mold-boards, Chill for casting	J. Oliver	South Bend, Ind	Feb. 2, 1869	86, 579
Mold-facing material	J. B. Hyde	Newark, N. J	June 5, 1860	28, 579
Mold for casting molds	J. S. Elliott	Chelsea, Mass	Apr. 15, 1873	137, 901
Mold for composition articles	I. Smith	New York, N. Y	July 18, 1871	117, 119
Mold-forming machine	S. J. Peet and L. Sawyer	Boston, Mass	Mar. 22, 1870	101, 033
Molds, Composition for facing	R. B. Smith	Mount Pleasant, Ohio	July 31, 1866	56, 817
Molds, Device for blackwashing	J. B. Aston	Pittsburgh, Pa	Sept. 9, 1873	142, 661
Molds, Facing	J. and A. Hursh	Philadelphia, Pa	Aug. 8, 1865	49, 272
Molds for casting off alto-relievo figures, &c., Process of making.	N. Heintzelman	New York, N. Y	June 7, 1870	103, 879
Molds, Heating	H. Mooers	Toledo, Ohio	Dec. 3, 1861	33, 846
Molds, Machine for producing	S. J. Peet	New York, N. Y	Mar. 1, 1870	100, 319
Molds, Powder for facing	W. Batty	Cincinnati, Ohio	Jan. 1, 1867	60, 824
Molds, Powder for facing	J. Nichols and W. Batty	Cincinnati, Ohio	Feb. 28, 1865	46, 578
Molds, &c., with metallic surface, Covering	J. A. Adams	Brooklyn, N. Y	Feb. 15, 1870	99, 806
Molded and coated articles, Composition for forming	S. Whitmarsh	Northampton, Mass	July 7, 1868	79, 794
Molded articles from plaster materials, Manufacture of.	A. C. Estabrook	Northampton, Mass	Jan. 30, 1872	123, 254
Molded articles, Manufacture of	W. B. Gleason	Boston, Mass	Mar. 3, 1868	75, 149
Molder's bench	C. L. Bishop	Meriden, Conn	Feb. 7, 1865	46, 209
Molder's clamp	J. B. Crowley	Cincinnati, Ohio	Jan. 9, 1866	51, 995
Molder's clamp	W. W. Tice	California, Ohio	Nov. 27, 1866	60, 091
Molder's clamp	C. Trusdale	Cincinnati, Ohio	Jan. 2, 1867	51, 903
Molder's flask	J. A. Blake	Providence, R. I	Apr. 23, 1872	126, 012
Molder's flask	M. M. Donnelly	Cincinnati, Ohio	Aug. 20, 1807	67, 855
Molder's flask	M. M. Donnelly	Cincinnati, Ohio	Aug. 20, 1867	67, 856
Molder's flask	J. G. Holliday	Wheeling, W. Va	Sept. 11, 1866	57, 903
Molder's flask	E. F. Jones	Foxborough, Mass	May 23, 1871	115, 214
Molder's flask	E. C. Little	Saint Louis, Mo	Sept. 18, 1866	58, 173
Molder's flask	L. K. Livingston, J. J. Roggen, and C. Adams.	Pittsburgh, Pa	Oct. 10, 1848	5, 841
Molder's flask	A. Merriman	Plantsville, Conn	June 13, 1871	115, 973
Molder's flask	C. Truesdale	Cincinnati, Ohio	May 13, 1873	138, 960
Molder's flask	A. Wilcox	Green Island, N. Y	Jan. 11, 1870	98, 728
Molder's flask	S. Williamson	Cincinnati, Ohio	May 28, 1872	127, 398
Molder's flasks, Packer for packing sand in	B. S. Benson	Baltimore, Md	Oct. 13, 1868	83, 029
Molder's gate or sprue	J. M. Killin	Pittsburgh, Pa	May 28, 1872	127, 351
Molder's match-board	W. H. Jefts	Lowell, Mass	July 16, 1872	129, 033
Molder's match-plates, Composition for	G. P. Darrow	Cincinnati, Ohio	July 10, 1866	55, 187
Molder's match-plates, Manufacture of	C. Truesdale and A. J. Sennett	Cincinnati, Ohio	Dec. 5, 1865	51, 366
Molder's pot or ladle	J. McCleary	Martin's Ferry, Ohio	Mar. 18, 1873	136, 927
Molder's slicker	W. J. Fryer, jr	New York, N. Y	May 23, 1871	115, 049
Molder's sprue	M. R. Howell	Elizabethport, N. J	Jan. 3, 1865	45, 716
Molder's sprue	P. W. Lamb	Albany, N. Y	Jan. 1, 1867	60, 909
Molder's table	J. Lebeau	Cincinnati, Ohio	Mar. 14, 1865	46, 859
Molding	P. Conver	Farmington, Ill	Aug. 27, 1872	130, 900
Molding	J. Dougherty	Cold Spring, N. Y	May 15, 1860	28, 260
Molding	A. King	Utica, N. Y	Sept. 4, 1866	57, 822
Molding and carving machine	A. S. Gear	Boston, Mass	Apr. 30, 1872	126, 198
Molding and casting apparatus	G. Ross	Newport, Ky	Apr. 10, 1866	53, 883
Molding and glazing forms made of plastic material	S. Hart	Marietta, Ohio	Jan. 25, 1870	99, 083
Molding and pressing bricks by hydraulic pressure, Machine for.	E. Rogers	Cleveland, Ohio	Sept. 23, 1873	143, 188
Molding and venting machine	W. Newsham	Philadelphia, Pa	Aug. 9, 1870	106, 273
Molding-apparatus	W. C. Amish	Hainesport, N. J	Dec. 23, 1873	145, 828
Molding-apparatus	S. Ashford	Philadelphia, Pa	Sept. 13, 1864	44, 149
Molding beads or hollow ware	C. Neal	Philadelphia, Pa	July 5, 1859	24, 701
Molding-box	T. L. Luders	Olney, Ill	July 9, 1867	66, 508
Molding circular and undercut work	W. Sellers and J. Walker	Cincinnati, Ohio	Oct. 2, 1855	13, 625
Molding-cutter	N. Jenkins	New York, N. Y	Oct. 4, 1870	108, 026
Molding-cutter	T. J. Shannon	Lawrenceburgh, Ind	Feb. 8, 1870	99, 716
Molding-cutter head	E. H. Hinners	New York, N. Y	July 8, 1873	140, 626
Molding-cutter head	D. Strauley	Cincinnati, Chio	Apr. 16, 1872	125, 765
Molding-cutter head	J. K. Wood	Allegheny, Pa	Aug. 20, 1872	130, 684
Molding earthenware, Machine for	T. G. Green	Church Gresley Pottery, England.	Nov. 1, 1870	108, 779
Molding-facing machine	R. Howdon	Cincinnati, Ohio	Nov. 12, 1867	70, 847
Molding-flask	J. J. Johnston	Allegheny, Pa	Apr. 22, 1856	14, 724
Molding-flask	T. G. Lucas	Middletown, Conn	Nov. 16, 1869	96, 819
Molding-flask	J. McLaughlin	Philadelphia, Pa	Aug. 30, 1870	106, 850
Molding-flask	J. Yocom, jr	Philadelphia Pa	Jan. 15, 1867	61, 303
Molding for cast-iron plates with dovetailed recesses.	T. A. Smith	Albany, N. Y	Mar. 8, 1853	9, 613
Molding from plaster of Paris, Composition for	G. Schlueter	Brooklyn, N. Y	Feb. 1, 1870	99, 355
Molding hollow ware	J. J. Johnston and J. V. Cunningham.	Allegheny and Pittsburgh, Pa.	June 6, 1854	11, 044
Molding hollow ware	W. Rogers	Philadelphia, Pa	Feb. 8, 1848	5, 442
Molding in flasks, Apparatus for	E. Satterlee	Albany, N. Y	Jan. 13, 1852	8, 661

Index of patents issued from the United States Patent Office from 1790 *to* 1873, *inclusive*—Continued.

Invention.	Inventor.	Residence.	Date.	No.
Molding in flasks, Machine for	L. A. Orcutt	Albany, N. Y	Mar. 8, 1853	9, 612
Molding-machine	J. Albright	Midway, Tenn	May 16, 1871	114, 900
Molding-machine	W. J. Boda	Dayton, Ohio	Sept. 16, 1873	142, 890
Molding-machine	M. W. Clark	Worcester, Mass	Sept. 24, 1872	131, 666
Molding-machine	J. H. Culver	San Francisco, Cal	Mar. 14, 1871	112, 691
Molding-machine	J. Demarest	Mott Haven, N. Y	July 25, 1871	117, 266
Molding-machine	J. S. Dewing	Orange, Mass	June 25, 1872	128, 289
Molding-machine	A. S. Gear	New Haven, Conn	Nov. 23, 1869	97, 188
Molding-machine	J. W. Gheen	Wilmington, Del	Apr. 22, 1873	138, 015
Molding-machine	T. Glover	Philadelphia, Pa	Aug. 9, 1870	106, 264
Molding-machine	J. P. Grosvenor	Lowell, Mass	Sept. 8, 1868	81, 997
Molding-machine	G. S. Hudson	Ellisburgh, N. Y	May 28, 1867	65, 226
Molding-machine	N. Jenkins	Madison, Conn	Dec. 23, 1873	145, 804
Molding-machine	I. L. Lamb	Darien, Wis	July 18, 1871	117, 083
Molding-machine	H. A. Lee	Worcester, Mass	Oct. 13, 1863	40, 269
Molding-machine	H. A. Lee	Worcester, Mass	July 7, 1868	79, 663
Molding-machine	W. McConnell	Clarksville, N. J	Nov. 23, 1869	97, 210
Molding-machine	C. H. Mellor	Philadelphia, Pa	Aug. 4, 1868	80, 648
Molding-machine	C. B. Morse	Rhinebeck, N. Y	Jan. 16, 1855	12, 248
Molding-machine	O. H. Perry	Cincinnati, Ohio	Nov. 26, 1867	71, 531
Molding-machine	G. M. Ramsay	New York, N. Y	Jan. 16, 1855	12, 243
Molding-machine	C. Schilling	Pekin, Ill	June 27, 1871	116, 503
Molding-machine	H. I. Seymour	Troy, N. Y	Oct. 13, 1863	40, 286
Molding-machine	E. M. Smith	Indianapolis, Ind	Feb. 21, 1860	27, 261
Molding-machine	I. P. Tice	New York, N. Y	Feb. 2, 1869	86, 605
Molding-machine	C. Warner	New York, N. Y	May 24, 1859	24, 168
Molding-machine	J. A. Woodbury	Boston, Mass	Aug. 1, 1871	117, 586
Molding-machine cutter	E. Benjamin	Chicago, Ill	Dec. 5, 1871	121, 447
Molding-machine-cutter guide	D. Jordan	Charlestown, Mass	Apr. 30, 1867	64, 227
Molding-machine-cutter head	J. Alloways	Decatur, Ill	Sept. 9, 1873	142, 548
Molding-machine-cutter head	M. W. Clark	Worcester, Mass	Sept. 12, 1871	118, 907
Molding-machine-cutter head	J. M. Smith	New York, N. Y	Sept. 13, 1853	10, 017
Molding-machine cutter, Wood	D. Jordan	Charlestown, Mass	June 7, 1870	103, 890
Molding-machine cutters, Machine for shaping	W. H. Brown	Worcester, Mass	Dec. 3, 1872	133, 627
Molding-machine feed	L. Gould	Norwich Conn	July 14, 1863	39, 226
Molding machine, Power	H. A. Billings	Providence, R. I	Sept. 17, 1861	33, 291
Molding machine, Variety	J. Hatch	San Francisco, Cal	July 13, 1869	92, 609
Molding machine, Variety	N. Jenkins	New York, N. Y	Jan. 25, 1870	99, 202
Molding machine, Wood	N. Jenkins	New Haven, Conn	Jan. 2, 1872	122, 321
Molding-machines, Adjusting the planes in	W. D. Jones	Dayton, Ohio	Apr. 24, 1860	27, 986
Molding-machinery	W. T. Horrobin	Bennington, Vt	Nov. 24, 1868	84, 279
Molding materials admitting of cohesion, Machine for.	J. Greacen, jr., A. Foster, and J. H. Cooper.	New York, N. Y., and Philadelphia, Pa.	July 17, 1866	56, 496
Molding-pit for casting cylinders and pipes	R. Cartwright	Chicago, Ill	May 31, 1870	103, 560
Molding plaster cornices, Device for	S. Ferris	New York, N. Y	Sept. 13, 1870	107, 246
Molding pots and crucibles, Apparatus for	R. Taylor and F. Strow	Philadelphia, Pa	Dec. 1, 1868	84, 593
Molding-sand, &c., Apparatus for crushing, grinding, and mixing.	J. O. Hanctin	Paris, France	Mar. 11, 1873	136, 721
Molding-tool	A. P. Odholm	Bridgeport, Conn	Feb. 23, 1869	87, 282
Molding, Wood	L. Bushnell	New Bedford, Mass	Nov. 11, 1873	144, 438
Moldings and beads, Machine for turning	E. Coddington	Thompson, N. Y	July 31, 1837	321
Moldings and frames, Machine for making	G. Henze	New York, N. Y	Sept. 5, 1865	49, 829
Moldings and other purposes, Composition for manufacture of.	H. Lowenberg	Boston, Mass	Mar. 25, 1862	34, 757
Moldings, &c., Apparatus for applying metal leaf to.	R. J. Marcher	New York, N. Y	Sept. 22, 1863	40, 055
Moldings, Apparatus for filling, polishing, and varnishing.	M. Hamburger, I. J. Siskind, and A. Klein.	New York, N. Y	Jan. 7, 1873	134, 599
Moldings, &c., Apparatus for laying metal leaf on	R. Marcher	New York, N. Y	May 10, 1859	23, 930
Moldings, &c., Apparatus for laying metal leaf on.	R. Marcher	New York, N. Y	July 30, 1861	32, 950
Moldings, Arrangement of device for planing	H. B. Smith	Lowell, Mass	July 6, 1858	20, 824
Moldings, Burnishing	R. Marcher	New York, N. Y	June 7, 1859	24, 319
Moldings, Composition for	J. and W. Thiem	Lawrenceburgh, Ind	Feb. 9, 1869	86, 710
Moldings, Composition for ornamental	C. E. Bonnet	Philadelphia, Pa	Dec. 22, 1868	85, 055
Moldings, Cutter for planing	R. M. Evans	Gilford, N. H	Dec. 6, 1853	10, 304
Moldings, Cutting curved	J. J. Westerfield	New Brunswick, N. J	Dec. 9, 1856	16, 197
Moldings, Forming	J. Dill and E. Rice	Grand Rapids, Mich	Oct. 5, 1869	95, 441
Moldings, Machine for cutting	S. Kennedy	New York, N. Y	Apr. 10, 1830	....
Moldings, Machine for cutting	C. B. Rogers	Norwich, Conn	Jan. 10, 1860	26, 787
Moldings, Machine for cutting	J. B. Winslow	Charlestown, Mass	May 29, 1860	28, 527
Moldings, Machine for cutting ornamental	H. and R. S. Schevenell	Athens, Ga	Sept. 25, 1855	13, 602
Moldings, Machine for cutting twist	H. Newhouse	Chicago, Ill	Feb. 18, 1862	34, 467
Moldings, Machine for cutting wood	W. A. McDonald	Morrisania, N. Y	May 11, 1869	89, 937
Moldings, Machine for cutting wooden	A. H. Brown	Albany, N. Y	Jan. 22, 1861	31, 207
Moldings, Machine for cutting wooden curved	I. P. Tice	Baltimore, Md	May 3, 1859	23, 872
Moldings, Machine for fluting	R. Van Riper	Brooklyn, N. Y	July 5, 1870	105, 145
Moldings, Machine for forming sheet-metal	V. Fischer	New York, N. Y	Feb. 4, 1868	74, 068
Moldings, Machine for forming sheet-metal	H. G. Fiske	San Francisco, Cal	Aug. 30, 1870	106, 803
Moldings, Machine for making	T. J. Close	Philadelphia, Pa	Oct. 9, 1866	58, 599
Moldings, Machine for making	N. Jenkins	New York, N. Y	Aug. 6, 1867	67, 553
Moldings, Machine for making metallic	H. Adler and C. Rogers	Pittsburgh, Pa	Sept. 10, 1872	131, 239
Moldings, Machine for making metallic	J. Parkin and J. H. Smith	Cleveland, Ohio	Sept. 10, 1872	131, 179
Moldings, Machine for making metallic	B. K. Price	Cleveland, Ohio	Apr. 16, 1872	125, 691
Moldings, Machine for manufacture of	J. C. Cooke and H. A. Whiteley	Preston, Conn	Feb. 14, 1871	111, 817
Moldings, Machine for ornamenting	J. W. Campbell	New York, N. Y	Nov. 20, 1866	59, 817
Moldings, Machine for ornamenting	G. F. Goetze	New York, N. Y	Dec. 4, 1866	60, 173
Moldings, Machine for planing	J. B. H. Chatain	New York, N. Y	Mar. 19, 1850	7, 181
Moldings, Machine for planing oval	F. Brandon	Albany, N. Y	May 12, 1863	38, 461
Moldings, Machine for planing oval	G. Hammer	Cincinnati, Ohio	Mar. 11, 1862	34, 663
Moldings, Machine for polishing and varnishing	C. and J. Gschwind	Union Hill, N. J	Jan. 23, 1872	122, 945
Moldings, Machine for preparing	G. Henze	New York, N. Y	Nov. 8, 1864	44, 952
Moldings, Machine for sand-papering	J. Barker	Chicago, Ill	Aug. 23, 1870	106, 536
Moldings, Machine for sand-papering	F. G. Chapman	Chicago, Ill	May 31, 1870	103, 717
Moldings, Machine for smoothing	J. S. Loomis	Brooklyn, N. Y	Aug. 29, 1871	118, 466
Moldings, Machine for turning spiral	P. P. Ruger	New York, N. Y	Oct. 4, 1853	10, 089
Moldings, Machine for turning spiral	E. A. Stockton	San Francisco, Cal	July 25, 1871	117, 477
Moldings, Machinery for making	A. T. Serrell	New York	May 16, 1848	5, 575

Index of patents issued from the United States Patent Office from 1790 *to* 1873, *inclusive*—Continued.

Invention.	Inventor.	Residence.	Date.	No.
Moldings, Manufacture of imitation gold	H. W. Ladd	New York, N. Y	May 19, 1863	38, 588
Moldings, Metal	A. Destouy	New York, N. Y	Aug. 30, 1864	43, 979
Moldings on marble, Machine for cutting	H. L. Houghton	Springfield, Vt	Jan. 29, 1856	14, 177
Moldings on planks, piles, &c., Sticking	J. Judge	Washington, D. C	June 13, 1831	....
Moldings on wood, Machine for planing	A. Church, jr	Canandaigua, N. Y	July 29, 1837	296
Moldings, Pattern for	J. Alexander	Brooklyn, N. Y	Aug. 16, 1859	25, 157
Moldings, Planing-machine for	E. H. Ripley	North Chelmsford, Mass	June 16, 1868	79, 006
Moldings, Rest for finishing	D. A. Stevens	Toledo, Ohio	Oct. 8, 1872	132, 115
Moldings, Rolled-iron	M. M. Manly and S. Sellers	Philadelphia, Pa	Jan. 7, 1873	134, 553
Moldings, Tool for cutting	E. C. Austin	Monroe Village, Wis	Jan. 22, 1867	61, 306
Moldings, Tool for cutting	D. W. Perry	Wilkesbarre, Pa	Dec. 15, 1868	84, 964
Moldings, Tool for cutting wood	C. E. Boynton	San Francisco, Cal	Sept. 27, 1870	107, 653
Moldings, Tool for grooving	R. I. Marcher	Salisbury Mills, N. Y	May 22, 1855	12, 916
Moldings, Tool for grooving	R. I. Marcher	Salisbury Mills, N. Y	May 22, 1855	12, 917
Moldings, Tool for turning	W. W. Carey	Lowell, Mass	Aug. 25, 1868	81, 338
Moldings, Wood	W. J. Miller and J. W. Campbell.	New York, N. Y	Oct. 5, 1869	95, 598
Mole-killer	J. Wilson	Little Falls, N. Y	Dec. 7, 1869	97, 742
Mole-trap	J. Adams	Greencastle, Ind	Jan. 3, 1871	110, 618
Mole-trap	W. C. Akers	Petersburgh, Va	Jan. 4, 1870	98, 539
Mole-trap	L. Andrews	Plantsville, Conn	Nov. 4, 1862	36, 827
Mole-trap	D. Boswell	Greencastle, Ind	Aug. 22, 1871	118, 189
Mole-trap	T. G. Brown	Tarrytown, N. Y	Nov. 26, 1872	133, 407
Mole-trap	G. W. Hardwick	Wyandotte, Ind	Sept. 26, 1871	119, 354
Mole-trap	R. I. Huggins	Bethel, Ohio	Dec. 30, 1873	146, 003
Mole-trap	J. A. Myers	Lovely Dale, Ind	Aug. 20, 1872	130, 737
Mole-trap	C. Polley	Sinking Spring, Ohio	May 19, 1868	78, 128
Mole-trap	C. Polley	McMinnville, Tenn	May 7, 1872	126, 573
Money-box	G. W. McDaniel	Georgetown, D. C	Mar. 22, 1870	101, 144
Money-box for stages, &c	T. W. Gibbons	Franklin, N. J	June 7, 1859	24, 295
Money-counting apparatus	J. W. Meaker	Chicago, Ill	Jan. 28, 1868	73, 911
Money-drawer	C. E. Hays	Lancaster, Pa	Feb. 19, 1867	62, 195
Money-drawer	C. L. Page	Cambridge, Mass	Mar. 23, 1869	88, 068
Money-drawer	A. G. and P. M. Shults	Avoca, N. Y	Oct. 2, 1866	58, 490
Money-drawer	H. Unna	San Francisco, Cal	June 25, 1872	128, 439
Money-drawer alarm	R. M. Campbell	East Cambridge, Mass	July 12, 1859	24, 775
Money-drawer alarm	J. M. Case	Lansing, Mich	May 14, 1872	126, 785
Money-drawer alarm	I. Robbins	Hughesville, Pa	Sept. 3, 1867	68, 531
Money-drawer alarm	J. F. Winchell	Springfield, Ohio	Jan. 15, 1867	61, 297
Money-drawer catch	S. Hubbell, jr	Urbana, Ohio	Apr. 10, 1866	53, 827
Money-drawer lock and alarm attachment	J. H. Weaver	Columbus, Ohio	Sept. 10, 1867	68, 816
Money-table	W. Painter	Wilmington, Del	Aug. 3, 1858	21, 082
Monkey-wrench	L. Coes	Worcester, Mass	Aug. 10, 1869	93, 521
Monkey-wrench	S. Zarley	Niantic, Ill	Dec. 14, 1869	97, 849
Monument	F. M. Jones	Independence, Miss	Nov. 28, 1871	121, 380
Monument	J. M. Martin	Cleveland, Ohio	May 13, 1862	35, 278
Monument	J. C. Taylor	Roanoke, Ala	Dec. 2, 1873	145, 254
Monuments, &c., Casting	H. Siebenan	New York, N. Y	Nov. 17, 1830	....
Monuments, Fixing likenesses in	W. Boyd	Garrettsville, Ohio	Jan. 31, 1854	10, 468
Monuments, Protecting likenesses in	I. H. Wells	Pagetown, Ohio	Dec. 17, 1867	72, 345
Mop	C. L. W. Baker	Hartford, Conn	Sept. 6, 1870	106, 986
Mop	A. T. Boon and W. W. Spaulding	Galesburgh, Ill	Mar. 21, 1865	46, 872
Mop	P. Cook, jr	Sioux City, Iowa	Dec. 14, 1869	97, 885
Mop	R. H. Ewing	Elizabethtown, Ohio	June 2, 1863	38, 734
Mop	A. Hemmenway	Townsend, Ohio	Oct. 1, 1861	33, 390
Mop	M. Jincks	Danville, N. Y	May 16, 1865	47, 730
Mop	M. W. Kirkwood and S. H. Riley	Iowa City, Iowa	Apr. 18, 1871	113, 777
Mop	J. Law	New York, N. Y	May 25, 1869	90, 562
Mop	J. and H. R. Lee	Galesburgh, Ill	Mar. 28, 1865	47, 066
Mop	W. Mallerd	Bridgeport, Conn	July 13, 1869	92, 624
Mop	T. F. Rooney	Chicago, Ill	June 13, 1871	115, 899
Mop	J. Sangster and M. Boyd	Buffalo, N. Y	June 5, 1866	55, 429
Mop	M. H. Wiley	Bucksport, Me	May 1, 1866	54, 473
Mop	G. W. Tolhurst	Circleville, Ohio	Feb. 14, 1865	46, 406
Mop	J. A. Wright	Keene, N. H	Sept. 8, 1868	82, 055
Mop and brush	A. C. Bacon	Cleveland, Ohio	June 11, 1867	65, 528
Mop and brush combined	H. McClay	Niles, Mich	Oct. 19, 1858	21, 842
Mop and brush holder	D. Edward	Montreal, Canada	July 4, 1871	116, 574
Mop and clothes wringer	E. Youngs	Tuscarora, N. Y	Aug. 4, 1868	80, 794
Mop and mop-wringer combined	J. L. Boyden	Bridgeport, Conn	Feb. 4, 1873	135, 400
Mop and rubber scrubber	B. C. Davis	Binghamton, N. Y	Feb. 8, 1870	99, 648
Mop and scourer	W. T. Grant	Jacksonville, Ill	Feb. 20, 1866	52, 705
Mop and scrubber	W. T. Grant	Jacksonville, Ill	Sept. 4, 1866	57, 703
Mop and scrubber, Combination of	R. Price	New York, N. Y	Mar. 6, 1860	27, 383
Mop and scrubber combined	E. S. Bennett	New York, N. Y	Feb. 2, 1869	86, 351
Mop and scrubber head	W. H. Kline	Jersey Shore, Pa	Oct. 18, 1870	108, 364
Mop and scrubber head	E. M. Shoemaker and S. McCleery.	Franklin County, Ohio	Feb. 8, 1870	99, 603
Mop and scrubbing-brush holder	P. O'Brian	Philadelphia, Pa	May 19, 1868	78, 121
Mop and wringer	J. M. Enos	Saint Joseph, Mich	June 5, 1866	55, 413
Mop and wringer combined	J. A. Wright	Keene, N. H	June 30, 1868	79, 428
Mop-cloth	W. B. Link	Taberg, N. Y	Feb. 26, 1867	62, 347
Mop for crockery and lamp-chimnies	C. S. Moore and H. P. Boyd	Worcester, Mass	Dec. 19, 1871	121, 955
Mop-handle	H. B. Malbone	Middletown, Conn	May 29, 1866	55, 135
Mop-handle	H. and J. S. B. Norton	Farmington, Me	May 17, 1859	24, 049
Mop-head	F. Allen	Worcester, Mass	Dec. 2, 1856	16, 137
Mop-head	J. W. Barbour	Winooski Falls, Vt	Feb. 11, 1873	135, 753
Mop-head	A. Barns	Ashtabula, Ohio	Nov. 20, 1855	13, 812
Mop-head	W. H. Beatty	Columbus, Ohio	May 24, 1870	103, 413
Mop-head	G. E. Brettell	Rochester, N. Y	May 18, 1869	90, 153
Mop-head	J. Brizee	Alvarado, Cal	June 13, 1871	115, 815
Mop-head	J. D. Browne	Cincinnati, Ohio	Sept. 22, 1868	82, 286
Mop-head	C. B. Clark	Buffalo, N. Y	Aug. 13, 1867	67, 722
Mop-head	C. B. Clark	Buffalo, N. Y	July 9, 1872	128 711
Mop-head	C. B. Clark	Buffalo, N. Y	Dec. 9, 1873	145, 397
Mop-head	C. B. Clark and E. L. Ferguson	Buffalo, N. Y	Sept. 22, 1868	82, 383
Mop-head	C. B. Clark and O. S. Garretson	Buffalo, N. Y	May 22, 1866	54, 869

Index of patents issued from the United States Patent Office from 1790 *to* 1873, *inclusive*—Continued.

Invention.	Inventor.	Residence.	Date.	No.
Mop-head	S. Cooper	Antwerp, N. Y	Apr. 25, 1865	47, 398
Mop-head	W. S. Crinklaw	Lanark, Ill	Mar. 22, 1870	100, 987
Mop-head	J. Davis	Chicago, Ill	Mar. 14, 1871	112, 556
Mop-head	J. Davis	Chicago, Ill	Sept. 2, 1873	142, 328
Mop-head	H. Dodge	Albany, N. Y	Apr. 12, 1870	101, 716
Mop-head	J. Doty, jr	West Falls, N. Y	July 17, 1860	29, 148
Mop-head	W. Duncan	Lebanon, N. H	Jan. 14, 1873	134, 739
Mop-head	L. H. Emmons	Noblesville, Ind	June 2, 1868	78, 444
Mop-head	R. W. English	Buffalo, N. Y	Mar. 10, 1868	75, 253
Mop-head	J. Fahrney	Boonsborough, Md	Mar. 23, 1869	88, 150
Mop-head	J. Fazig	West Salem, Ohio	Sept. 13, 1859	25, 400
Mop-head	A. Field	Jericho, Vt	May 21, 1872	127, 037
Mop-head	B. French	Rochester, N. Y	June 20, 1871	116, 174
Mop-head	S. Gantz	Beaver Creek, Md	May 25, 1869	90, 520
Mop-head	O. S. Garretson	Buffalo, N. Y	June 11, 1867	65, 662
Mop-head	O. S. Garretson	Cincinnati, Ohio	Aug. 13, 1867	67, 643
Mop-head	O. S. and J. G. Garretson	Buffalo, N. Y	Oct. 4, 1870	107, 897
Mop-head	H. M. Guild	Springfield, Mass	Aug. 20, 1867	67, 873
Mop-head	G. H. Hammer	Newville, Pa	Apr. 24, 1866	54, 146
Mop-head	F. M. Hardison and J. A. Hooper.	South Berwick, Me	Oct. 16, 1866	58, 817
Mop-head	J. S. Harris	East Poultney, Vt	July 28, 1857	17, 877
Mop-head	J. S. Harris	Poultney, Vt	Sept. 1, 1863	39, 730
Mop-head	P. R. Higley	Buffalo, N. Y	Dec. 12, 1865	51, 514
Mop-head	N. Homes	Laona, N. Y	May 9, 1865	47, 642
Mop-head	J. H. Hood	Dansville, N. Y	Oct. 29, 1872	132, 580
Mop-head	M. L. Horton	Claremont, N. H	Mar. 18, 1862	34, 687
Mop-head	C. Karr	Buffalo, N. Y	Sept. 26, 1865	50, 137
Mop-head	F. C. Kendall	Hartford, Wis	Nov. 22, 1870	109, 421
Mop-head	S. C. Ketchum	Winchendon, Mass	Apr. 23, 1872	125, 895
Mop-head	W. A. Lewis	Springfield, Vt	Apr. 9, 1867	63, 731
Mop-head	D. W. Marston and J. S. Skinner.	Lebanon, N. H	Dec. 30, 1873	145, 958
Mop-head	H. H. Mason and J. Messinger	Springfield, Vt	Feb. 19, 1867	62, 214
Mop-head	H. H. Mason and J. Messinger	Springfield, Vt	May 14, 1867	64, 683
Mop-head	H. H. Mason and J. Messinger	Springfield, Vt	July 16, 1867	66, 863
Mop-head	H. H. Mason and S. W. Shelden	Springfield, Vt	Jan. 5, 1864	41, 081
Mop-head	G. Meader	Prairie Centre, Ill	June 11, 1867	65, 584
Mop-head	J. Messinger	Springfield, Vt	Mar. 14, 1871	112, 615
Mop-head	H. Murch	Lebanon, N. H	June 14, 1853	9, 781
Mop-head	E. D. Murphy	Claremont, N. H	July 21, 1863	39, 305
Mop-head	H. C. Parsons	Dexter, Me	Sept. 28, 1869	95, 259
Mop-head	T. Randlett	Endfield, N. H	Nov. 15, 1853	10, 237
Mop-head	E. R. and N. F. Reed	Hyde Park, Vt	July 22, 1862	35, 951
Mop-head	W. A. Robinson	Grand Rapids, Mich	Mar. 24, 1868	75, 796
Mop-head	L. C. Rodier	Springfield, Mass	June 26, 1866	55, 961
Mop-head	O. Root	Wendell, Mass	Nov. 10, 1868	84, 000
Mop-head	G. W. Sanders	Springfield, Vt	Oct. 1, 1867	69, 368
Mop-head	S. Seldon and M. Griswold, jr	Erie, Pa	Apr. 4, 1871	113, 457
Mop-head	J. W. Shaw	Concord, N. H	Feb. 26, 1867	62, 373
Mop-head	E. and E. G. Sirret	Buffalo, N. Y	Apr. 13, 1869	88, 917
Mop-head	W. W. Spaulding	Galesburgh, Ill	Aug. 22, 1865	49, 566
Mop-head	J. A. Taylor	Alden, N. Y	Feb. 6, 1855	12, 365
Mop-head	L. Taylor	North Springfield, Vt	Feb. 15, 1859	22, 990
Mop-head	L. W. Taylor	Upper Falls Post-Office, Vt	Aug. 8, 1871	117, 943
Mop-head	L. H. Thomas	Waterbury, Vt	Apr. 3, 1866	53, 704
Mop-head	E. P. Thompson	Worcester, Mass	June 9, 1857	17, 527
Mop-head	H. Thompson and R. Q. Tuson	Lebanon, N. H	July 15, 1856	15, 352
Mop-head	J. Todd	Webster, N. Y	June 27, 1871	116, 512
Mop-head	J. Troxel	Reedsburgh, Ohio	Oct. 1, 1867	69, 512
Mop-head	W. P. Valentine	Buffalo, N. Y	Oct. 11, 1870	108, 304
Mop-head	I. E. Weston	Winchendon, Mass	Oct. 30, 1866	59, 299
Mop-head	H. H. Wetmore	Barre, Vt	Jan. 24, 1871	111, 161
Mop-head	G. W. Wheat	Philipsburgh, Pa	Feb. 11, 1873	135, 679
Mop-head	R. W. Whitney and A. C. Stockin.	South Berwick, Me	Oct. 9, 1866	58, 703
Mop-head	L. Williams	Arlington, Vt	Dec. 13, 1870	110, 180
Mop-head	J. A. Wilson	Spencer, Mass	May 7, 1867	64, 612
Mop-head and scrubbing-brush, Combined	W. S. Van Hoesen	Saugerties, N. Y	Nov. 3, 1868	83, 806
Mop-head, and securing same to handles	J. Howe	Worcester, Mass	June 15, 1837	241
Mop head and wringer	G. Gullman	Poughkeepsie, N. Y	Aug. 18, 1868	81, 272
Mop head and wringer, Combined	W. Gage	Buffalo, N. Y	Aug. 23, 1870	106, 684
Mop-holder	W. R. Axe	Beloit, Wis	Mar. 5, 1861	31, 586
Mop-holder	T. C. Ball	Springfield, Vt	Feb. 7, 1865	46, 290
Mop-holder	J. Brizee	Alvarado, Cal	Feb. 21, 1871	112, 014
Mop-holder	B. W. Bruel	Beloit, Wis	Aug. 14, 1860	29, 566
Mop-holder	G. Fliedner	Portland, Oreg	July 16, 1872	129, 019
Mop-holder	C. L. Haines	Carmel, Me	Mar. 18, 1873	136, 912
Mop-holder	W. Morehouse	Buffalo, N. Y	Oct. 16, 1866	58, 872
Mop-holder	E. M. Naramore	Underhill, Vt	Oct. 14, 1873	143, 633
Mop-holder	G. Stackpole	New York, N. Y	May 4, 1869	89, 803
Mop-holder	S. S. Whitman	Little Falls, N. Y	Jan. 24, 1842	2, 436
Mop-iron	S. Johnson	Lee, N. Y	Oct. 18, 1843	3, 306
Mop, mat, and wiping-cushion	H. E. Smith	Newark, N. J	Apr. 12, 1870	101, 777
Mop-press	L. S. Covey and J. Duffy	Saint Croix County, Wis	Mar. 10, 1868	75, 247
Mop-squeezer	E. N. Porter	Morrisville, Vt	Mar. 26, 1867	63, 294
Mop-squeezer	E. S. Wilkins and J. Straw	Stowe, Vt	Apr. 16, 1867	63, 973
Mop, Tar	J. W. Midwinter	Port Washington, N. J	Oct. 27, 1863	40, 420
Mop-wringer	J. Adams	Janesville, Wis	Oct. 29, 1867	70, 310
Mop-wringer	M. J. Barcalo	Mount Morris, N. Y	Sept. 1, 1868	81, 583
Mop-wringer	O. C. Barnes	Stowe, Vt	June 4, 1867	65, 330
Mop-wringer	O. D. Barrett	Fulton, N. Y	July 31, 1855	13, 347
Mop-wringer	C. Bradway	Maquoketa, Iowa	May 10, 1870	102, 759
Mop-wringer	E. E. Brewster	Holly, Mich	Jan. 16, 1872	122, 804
Mop-wringer	O. M. Brooks and E. J. Matteson	Janesville, Wis	Nov. 12, 1867	70, 692
Mop-wringer	A. W. Bunnell	Linesville, Pa	Nov. 25, 1873	144, 828
Mop-wringer	M. P. Carpenter	Buffalo, N. Y	Jan. 23, 1866	52, 136

Index of patents issued from the United States Patent Office from 1790 *to* 1873, *inclusive*—Continued.

Invention.	Inventor.	Residence.	Date.	No.
Mop-wringer	B. B. Choate	Springfield, Vt	Apr. 28, 1868	77, 255
Mop-wringer	O. Choate	Morrisville, Vt	May 15, 1860	28, 227
Mop-wringer	A. J. Davis	Hartford, Mich	May 19, 1868	78, 071
Mop-wringer	H. L. Ennes	Birmingham, Ohio	Nov. 15, 1870	109, 307
Mop-wringer	J. Filkins	Sandwich, Ill	June 9, 1868	78, 657
Mop-wringer	W. W. Finch	Mishawaka, Ind	June 11, 1867	65, 737
Mop-wringer	M. M. Follett	Westborough, Mass	Apr. 7, 1868	76, 430
Mop-wringer	C. W. Gage and J. Northrup	Homer, N. Y	June 11, 1867	65, 558
Mop-wringer	W. and W. S. Gillett	Stowe, Vt	Nov. 13, 1866	59, 586
Mop-wringer	J. H. Gridley	Washington, D. C	Mar. 17, 1868	75, 679
Mop-wringer	W. Hall	North Adams, Mass	Apr. 20, 1869	89, 041
Mop-wringer	Z. Howe	Lowell, Mass	Dec. 10, 1867	72, 043
Mop-wringer	A. S. Lesner	Boston, Mass	July 16, 1867	66, 720
Mop wringer	J. M. May and W. M. Colton	Jonesville and Stoughton, Wis.	Apr. 28, 1868	77, 391
Mop-wringer	J. H. Mears	Oshkosh, Wis	Feb 11, 1868	74, 400
Mop-wringer	W. R. Mills	Hartford, Mich	May 5, 1868	77, 639
Mop-wringer	J. H. Newton	Paxton, Ill	June 24, 1873	140, 216
Mop-wringer	D. Peck	Rochelle, Ill	Oct. 8, 1867	69, 694
Mop-wringer	R. T. Reed	Binghamton, N. Y	Oct. 22, 1867	70, 017
Mop-wringer	F. Rhines	Watertown, N. Y	Mar. 27, 1866	53, 489
Mop-wringer	A. J. Robinson	Troy, N. Y	Jan. 1, 1867	60, 790
Mop-wringer	L. F. Rollins	Bangor, Me	Mar. 5, 1867	62, 567
Mop-wringer	H. B. Rorke	California, Mo	May 12, 1868	77, 919
Mop-wringer	H. Russell	New Richmond, Wis	July 9, 1867	66, 639
Mop-wringer	C. S. Schmidt	New York, N. Y	Sept. 25, 1860	30, 159
Mop-wringer	R. W. Soper	Janesville, Wis	Oct. 15, 1867	69, 942
Mop-wringer	A. G. Starkweather	Burlington, Vt	July 9, 1867	66, 647
Mop-wringer	D. J. Stone	Norwich, R. I	June 2, 1868	78, 618
Mop-wringer	A. D. Tate	Peekskill, N. Y	Jan. 26, 1869	86, 328
Mop-wringer	B. Van Alstine	Clannahan Township, Ill	Feb. 19, 1867	62, 237
Mop-wringer	C. E. Wareham	Sedalia, Mo	July 30, 1867	67, 388
Mop-wringer	G. Wells and S. A. Haynes	Island Pond, Vt	Nov. 10, 1868	84, 034
Mop-wringer	J. F. White	Brattleborough, Vt	Nov. 27, 1866	60, 102
Mop-wringer	H. B. Willoughby	Ottawa, Ill	July 7, 1868	79, 795
Mop-wringer and brush	W. S. Simpson	Berea, Ohio	May 5, 1868	77, 543
Mop-wringer and scrubber, Combined	W. R. Winter	Cedar Falls, Iowa	July 1, 1873	140, 566
Mop-wringer and scrubbing-brush	L. L. Gordon	Detroit, Mich	Feb. 15, 1870	99, 772
Mops, Wringing	R. Holgate and C. Hart	Wyoming and Farmington, Ill	June 4, 1872	127, 416
Mordant	E. F. Prentiss	Philadelphia, Pa	Apr. 24, 1866	54, 203
Mordant for fixing aniline colors	A. Schultz	Lyons, France	Jan. 13, 1863	37, 426
Morocco and leather, Polishing and graining	R. Emes	Boston, Mass	Apr. 20, 1831	....
Morocco, Machine for figuring and polishing	E. L. Norton	Charlestown, Mass	May 6, 1856	14, 821
Morocco, Machine for finishing	E. Bookhout and H. Cochen, jr	Williamsburgh, N. Y	June 18, 1850	7, 433
Morocco, Machine for finishing and dicing	M. Hall	Charlestown, Mass	Dec. 31, 1817	....
Morocco, Machine for graining	G. R. Johnson	Wilmington, Del	May 1, 1866	54, 360
Morocco, Polishing, graining, and dicing	R. Emes	Boston, Mass	Nov. 21, 1831	....
Morocco, Tool for figuring	S. Green	Lynn, Mass	Mar. 25, 1856	14, 534
Mortar	J. A. Fremon	Montgomery, Ala	Dec. 22, 1868	85, 223
Mortar and brick elevator	P. Anderson and J. A. Wait	Chicago, Ill	Aug. 27, 1872	130, 889
Mortar and cement, and material therefor, Manufacture of.	E. A. Ellsworth	Washington, D. C	Jan. 28, 1868	73, 882
Mortar and making bricks, Tempering	A. Kinsley	....	Dec. 20, 1794	....
Mortar, cement, paint, &c., Composition for	C. B. Hutchins	Ann Arbor, Mich	Feb. 2, 1869	86, 546
Mortar, cement, &c., Preparation of	H. Y. D. Scott	Ealing, England	Nov. 7, 1871	120, 672
Mortar for bricks, Preparing	O. W. Seely	Williamson, N. Y	July 20, 1831	....
Mortar for making building-blocks, &c., Apparatus for mixing.	J. Shelley	Harlem, N. Y	Nov. 19, 1867	71, 231
Mortar, Hominy	J. Keezer	Chillicothe, Ohio	Mar. 2, 1858	19, 507
Mortar-machine	S. H. Hinsdell	Camillus, N. Y	Aug. 1, 1871	117, 540
Mortar-machine and mason's tender	S. Whitman	Danville, Ill	Mar. 11, 1835	....
Mortar-making machine	D. Hunt	Stamford, Conn	Mar. 22, 1810	....
Mortar-mill	A. A. Anderson	Galesburgh, Ill	Dec. 24, 1867	72, 440
Mortar-mill	H. Hill	Chicago, Ill	Apr. 2, 1872	125, 295
Mortar-mill	J. McIntyre	Syracuse, N. Y	June 28, 1870	104, 870
Mortar-mill	L. B. Pitcher	Salina, N. Y	Nov. 27, 1866	60, 055
Mortar-mill	L. B. Pitcher	Salina, N. Y	July 26, 1870	105, 840
Mortar-mixer	A. A. Hoagland and W. Nickel.	Oneonta, N. Y	Oct. 14, 1873	143, 575
Mortar-mixer	C. Pierce	Portland, Mich	Aug. 19, 1873	141, 947
Mortar-mixer	C. W. Wasson	Friendship, N. Y	Mar. 4, 1873	136, 571
Mortar-mixer	S. Wetmore	Wellsborough, Pa	June 21, 1870	104, 519
Mortar mixer and grinder	E. Spaulding	Keene, N. H	May 27, 1873	139, 338
Mortar-mixer, Portable	H. Rosen	Elkhart, Ind	Aug. 13, 1872	130, 536
Mortar, Mixing	J. Peck	Buffalo, N. Y	June 29, 1852	9, 077
Mortar-mixing machine	A. C. Ellithorpe	Chicago, Ill	June 20, 1871	116, 172
Mortar-mixing machine, &c	Everhart and Swimley	Washington, D. C	June 20, 1835	....
Mortar-mixing machine	B. F. Field	Beloit, Wis	Oct. 7, 1856	15, 847
Mortar-mixing machine	A. Kirkpatrick	Urbana, Ohio	Nov. 12, 1831	....
Mortar and molding and pressing brick, Machine for mixing.	J. M. Brookings	Wiscasset, Me	Jan. 23, 1826	....
Mortar or grinding apples, Preparing	T. Streeter and J. S. Wibert	Chili, N. Y	June 12, 1827	....
Mortar-tub mill	J. Parker	Gardiner, Me	Mar. 28, 1826	....
Mortars, Mounting hand	W. F. Goodwin	New York, N. Y	Jan. 31, 1865	46, 101
Mortars, Process and machinery for making	L. B. Pitcher	Salina, N. Y	Sept. 20, 1870	187, 535
Mortise and common door-lock	T. Whitehouse	Boston, Mass	Sept. 29, 1825	....
Mortise and draw-bore delineator	R. Bloss	Rochester, N. Y	May 3, 1864	42, 550
Mortise and latch lock	J. G. Hotchkiss	New Haven, Conn	June 19, 1835	....
Mortise-lock	P. and E. W. Blake	New Haven, Conn	Feb. 5, 1836	....
Mortises, Cutting straight or curved	E. Q. Smith	Cincinnati, Ohio	July 3, 1855	13, 184
Mortises, Machine for boring dovetailed	H. J. Betjemann	Cincinnati, Ohio	Dec. 3, 1850	7, 807
Mortises, Machine for cutting square	C. H. Thompson	Cleveland, Ohio	Mar. 5, 1872	124, 230
Mortises, Machine for laying out	J. M. Rodkey	Indiana, Pa	July 25, 1871	117, 334
Mortises, Tool for dressing	C. E. Littlefield	Carver's Harbor, Me	June 18, 1872	128, 050
Mortising	J. B. Stark	Fisherville, N. H	July 31, 1866	56, 820
Mortising and boring	G. Page	Keene, N. H	Mar. 13, 1836	....
Mortising and boring machine	J. M. Jay	Canton, Ohio	July 28, 1857	17, 880
Mortising and boring machine	M. Sands	Franklin, N. Y	Oct. 1, 1830	....
Mortising and boring machine	G. N. Stearns	Syracuse, N. Y	Mar. 11, 1856	14, 416

Index of patents issued from the United States Patent Office from 1790 *to* 1873, *inclusive*—Continued.

Invention.	Inventor.	Residence.	Date.	No.
Mortising and dovetailing machine	I. Brainerd	Aurora, Ohio	Jan. 9, 1838	570
Mortising and making tenons on the ends of spokes of wheels, panel-work, carpetery, &c.	L. Davis	Northampton, Mass	Apr. 14, 1829	....
Mortising and tenoning	J. H. Darby	Leominster, Mass	June 15, 1835	....
Mortising and tenoning machine	A. Bailey	Jefferson, Ohio	June 20, 1840	1,649
Mortising and tenoning machine	S. Baldwin	Montpelier, Vt	Mar. 17, 1810	....
Mortising and tenoning machine	T. A. Chandler	Rockford, Ill	Dec. 9, 1846	4,876
Mortising and tenoning machine	E. Mudge	Genesee County, N. Y	Oct. 1, 1830	....
Mortising and tenoning machine	W. L. Peters	Frankfort, Pa	Dec. 12, 1842	2,878
Mortising and tenoning machine	E. M. Shaw	Brooklyn, N. Y	Oct. 20, 1834	....
Mortising and tenoning machine	E. M. Shaw	Brooklyn, N. Y	Apr. 8, 1835	....
Mortising and tenoning machine	W. H. Sible	Harrisburgh, N. Y	Aug. 24, 1869	94,140
Mortising and tenoning machine	L. Sprague, M. Noyes, T. Lund, jr., and J. Ayres.	Dunstable, N. H	Nov. 19, 1833	....
Mortising and tenoning machine	H. D. Stover	New York, N. Y	Nov. 3, 1868	83,671
Mortising and tenoning machine	E. Street	Montville, Conn	Mar. 20, 1855	12,563
Mortising and tenoning timber	J. McClintic	Chambersburgh, Pa	Oct. 22, 1835	....
Mortising and tenoning timber, Machine for	H. Barnes	Munson, Ohio	Apr. 28, 1838	716
Mortising and tenoning timber, Machine for	W. Jackson and J. J. Speed, jr	Speedville, N. Y	Sept. 10, 1828	....
Mortising and tenoning timber, Machine for	J. McClintic	Chambersburgh, Pa	Oct. 8, 1827	....
Mortising in wood	G. Page	Keene, N. H	Mar. 13, 1833	....
Mortising-machine	W. R. Axe	Beloit, Wis	Sept. 20, 1859	25,479
Mortising-machine	S. E. Babcock	Olstead, N. H	June 16, 1836	....
Mortising-machine	C. Baggerman and J. Green	Saint Louis, Mo	Jan. 21, 1868	73,428
Mortising-machine	T. R. Bailey	Lockport, N. Y	Aug. 5, 1856	15,467
Mortising-machine	S. S. Bartlett	Woonsocket, R. I	Sept. 24, 1861	33,330
Mortising-machine	C. Bennett	Pepperell, Mass	Sept. 17, 1845	4,195
Mortising-machine	N. Bickford	Cincinnati, Ohio	Sept. 19, 1871	119,007
Mortising-machine	G. Bradley	Auburn, N. Y	Feb. 25, 1835	....
Mortising-machine	J. E. Brown and S. S. Bartlett	Woonsocket, R. I	Jan. 24, 1854	17,464
Mortising-machine	S. C. Brown	Richmond, Ind	June 25, 1867	65,996
Mortising-machine	M. E. Campfield	Newark, N. J	Jan. 25, 1870	99,155
Mortising-machine	C. Carter	Auburn, N. Y	Oct. 6, 1868	82,692
Mortising-machine	J. B. Chambers	Pittsburgh, Pa	May 25, 1852	8,966
Mortising-machine	J. H. Chandler	Worcester, Mass	Dec. 17, 1834	....
Mortising-machine	J. C. Channell	Dunstable, N. H	June 16, 1836	....
Mortising-machine	F. G. Chapman	Chicago, Ill	Mar. 8, 1870	100,502
Mortising-machine	D. Clark	Brooklyn, Conn	Sept. 14, 1836	24
Mortising-machine	J. G. Clifton	Middletown, Ohio	Feb. 26, 1867	62,394
Mortising-machine	E. Coffin	Indianapolis, Ind	July 17, 1860	29,141
Mortising-machine	J. Cox	Portland, Oreg	Jan. 25, 1870	99,064
Mortising-machine	E. Dayton	Meriden, Conn	Apr. 2, 1872	125,123
Mortising-machine	F. A. Deland and L. Phillips	Memphis, Mich	Oct. 13, 1868	82,926
Mortising-machine	J. Dennis and I. J. Richardson	Palmyra, N. Y	Nov. 19, 1833	....
Mortising-machine	W. Downing and W. H. Soley	Philadelphia, Pa	Feb. 28, 1871	112,228
Mortising-machine	J. Driver	Marysville, Cal	June 24, 1873	140,257
Mortising-machine	L. Eames	Kalamazoo, Mich	Apr. 17, 1860	27,895
Mortising-machine	W. L. Epperson	Louisville, Ky	Dec. 1, 1868	84,620
Mortising-machine	H. K. Forbis	Danville, Ky	Jan. 26, 1869	86,293
Mortising-machine	A. Foster	Phillipstown, Mass	Apr. 29, 1830	....
Mortising-machine	J. W. Fowle	Boston, Mass	Feb. 14, 1871	111,834
Mortising-machine	C. Gates	Centre Antrim, N. H	Sept. 26, 1835	....
Mortising-machine	D. L. Gibbs	Worcester, Mass	Aug. 27, 1867	68,298
Mortising-machine	D. L. Gibbs	Worcester, Mass	May 12, 1868	77,877
Mortising-machine	D. L. Gibbs	Worcester, Mass	Dec. 8, 1868	84,817
Mortising-machine	D. L. Gibbs	Worcester, Mass	Feb. 2, 1869	86,389
Mortising-machine	D. L. Gibbs	Worcester, Mass	Nov. 7, 1871	120,737
Mortising-machine	S. Glover and D. Parmelee	Newtown, Conn	June 8, 1807	....
Mortising-machine	C. Green	Bethel, Ohio	Jan. 6, 1857	16,332
Mortising-machine	A. Greenleaf and H. Amidons	Mexico, N. Y	Dec. 28, 1827	....
Mortising-machine	J. Guild	Cincinnati, Ohio	Nov. 30, 1852	9,431
Mortising-machine	E. Hammond	Cumberland, Md	May 6, 1873	138,649
Mortising-machine	F. H. Harwood	Rushville, N. Y	Jan. 31, 1860	26,986
Mortising-machine	J. Hawkins	Stockbridge, Mass	June 20, 1836	....
Mortising-machine	B. Holly	Seneca Falls, N. Y	July 25, 1854	11,403
Mortising-machine	E. J. Houston	Nacoochee, Ga	Feb. 18, 1873	135,917
Mortising-machine	L. Houston	Smithville, N. J.	Jan. 28, 1873	135,337
Mortising-machine	J. W. Ingle	Upperville, Va	Jan. 21, 1847	4,936
Mortising-machine	J. M. Johnson and J. Herig	Cleveland, Ohio	Sept. 8, 1868	82,004
Mortising-machine	E. Joslyn	Keene, N. H	Apr. 1, 1856	14,564
Mortising-machine	W. Kegg	Lasellsville, N. Y	Apr. 12, 1859	23,586
Mortising-machine	J. J. Kellogg	Richmond, N. Y	Jan. 24, 1828	....
Mortising-machine	J. King	Morristown, N. J	Mar. 18, 1841	2,010
Mortising-machine	J. Lemman	Cincinnati, Ohio	May 20, 1862	35,318
Mortising-machine	S. Leroy	Mexico, N. Y	July 10, 1827	....
Mortising-machine	H. Marsh	Morristown, N. J	June 13, 1831	....
Mortising-machine	W. E. Marsh	Westfield Township, Essex County, N. J.	Apr. 18, 1828	....
Mortising-machine	J. McBride	Richmond, Ind	Nov. 26, 1835	....
Mortising-machine	N. Mead and J. Cutel	Walpole, N. H	Nov. 3, 1817	....
Mortising-machine	H. Mellish	Walpole, N. H	Nov. 19, 1833	....
Mortising-machine	L. G. Merrill	Angels, Cal	Oct. 18, 1870	108,503
Mortising-machine	J. A. Merriman	Hinsdale, Mass	Jan. 15, 1856	14,106
Mortising-machine	J. Moreland	Adrian, Mich	Feb. 22, 1853	9,595
Mortising-machine	W. H. Morrison and M. W. E. Doran.	Indianapolis, Ind	Aug. 1, 1854	11,430
Mortising-machine	P. A. Mowers	Cleversburgh, Pa	Jan. 12, 1869	85,752
Mortising-machine	J. Munsell	Painted Post, N. Y	Aug. 12, 1846	4,695
Mortising-machine	C. Muschar	Pittsburgh, Pa	Jan. 14, 1873	134,815
Mortising-machine	E. Myers and S. R. Smith	Cincinnati, Ohio	Nov. 14, 1871	120,993
Mortising-machine	W. Nangel	Philadelphia, Pa	Feb. 14, 1860	27,145
Mortising-machine	W. Nangel	Philadelphia, Pa	Oct. 16, 1866	58,878
Mortising-machine	A. O. Neal	Hyde Park, Mass	Sept. 21, 1869	95,129
Mortising-machine	G. V. Orton and J. Richards	Cincinnati, Ohio	June 23, 1868	79,249
Mortising-machine	B. H. Otis	Dedham, Mass	Feb. 20, 1846	4,387
Mortising-machine	G. F. Outten	Norfolk, Va	Apr. 15, 1873	137,950
Mortising-machine	G. Page	Keene, N. H	Mar. 18, 1836	....
Mortising-machine	J. Page	Henniker, N. H	June 12, 1835	....

Index of patents issued from the United States Patent Office from 1790 *to* 1873, *inclusive*—Continued.

Invention.	Inventor.	Residence.	Date.	No.
Mortising-machine	F. Parker	Petaluma, Cal	Oct. 20, 1868	83,306
Mortising-machine	H. Parmele	Killingworth, Conn	Feb. 15, 1810	....
Mortising-machine	L. E. Payne and O. Pier	Stowe, Vt	Nov. 6, 1855	13,759
Mortising-machine	J. A. Peabody	Philadelphia, Pa	Aug. 18, 1868	81,286
Mortising-machine	G. B. Phillips	Poughkeepsie, N. Y	Apr. 14, 1868	76,808
Mortising-machine	H. and S. H. Plumb	Honesdale, Pa	June 27, 1854	11,177
Mortising-machine	F. Purden	Baltimore, Md	June 14, 1853	9,784
Mortising-machine	J. Richards	Cincinnati, Ohio	Aug. 25, 1868	81,407
Mortising-machine	J. Richards and W. H. Doane	Cincinnati, Ohio	Sept. 10, 1867	68,791
Mortising-machine	I. J. Richardson	Palmyra, N. Y	July 7, 1835	....
Mortising-machine	E. Ripley	New York, N. Y	Dec. 22, 1827	....
Mortising-machine	E. J. Rowe	Eureka, Cal	Dec. 5, 1871	121,551
Mortising-machine	G. T. Savary	Newburyport, Mass	Aug. 16, 1870	106,511
Mortising-machine	A. Schmackers	Cincinnati, Ohio	Mar. 8, 1870	100,672
Mortising-machine	H. Selick	Lewistown, Pa	Mar. 16, 1869	87,879
Mortising-machine	E. M. Shaw	Wilbraham, Mass	Mar. 23, 1836	....
Mortising-machine	W. C. Shaw	Madison, Ind	July 27, 1852	9,156
Mortising-machine	D. L. Shearer	Waupun, Wis	Sept. 13, 1870	107,413
Mortising-machine	W. H. Sible	Harrisburgh, Pa	Dec. 13, 1870	110,079
Mortising-machine	J. Skipp	Newark, N. J	Mar. 15, 1870	100,813
Mortising-machine	H. B. Smith	Manchester, N. H	Apr. 17, 1849	6,343
Mortising-machine	H. B. Smith	Lowell, Mass	Jan. 10, 1854	10,422
Mortising-machine	H. B. Smith	Lowell, Mass	Oct. 9, 1855	13,663
Mortising-machine	H. B. Smith	Lowell, Mass	June 30, 1857	17,701
Mortising-machine	H. B. Smith	Lowell, Mass	Aug. 16, 1859	25,221
Mortising-machine	H. C. Smith	Clarksville, Ohio	Sept. 4, 1860	29,920
Mortising-machine	H. C. Smith	Clarksville, Ohio	Aug. 11, 1863	39,502
Mortising-machine	H. C. Smith	Lawrence, Kans	July 19, 1870	105,603
Mortising-machine	A. Spencer, jr	Southport, N. Y	Aug. 9, 1859	25,055
Mortising-machine	S. Spencer	Angelica, N. Y	Sept. 10, 1850	7,635
Mortising-machine	J. H. Stahl	Stanton's Mills, Pa	Feb. 4, 1873	135,604
Mortising-machine	W. Stoddard	Lowell, Mass	Jan. 8, 1856	14,071
Mortising-machine	J. Stufflebern	Milwaukee, Wis	Nov. 21, 1865	51,100
Mortising-machine	J. Swan	Seymour, Conn	Oct. 29, 1872	132,608
Mortising-machine	C. R. Tompkins	Rochester, N. Y	Oct. 22, 1867	70,051
Mortising-machine	A. C. Vaughn	Rainsburgh, Pa	Mar. 6, 1860	27,397
Mortising-machine	E. Wallace	Huntsville, Pa	Sept. 10, 1872	131,236
Mortising-machine	J. J. Weeks	Buckram, N. Y	Apr. 17, 1849	6,350
Mortising machine	L. W. Wolfe	Jacksonville, Ill	Feb. 18, 1868	74,739
Mortising-machine	I. Wright	Centre Antrim, N. H	July 17, 1835	....
Mortising-machine	C. L. Zeidler	Cincinnati, Ohio	July 18, 1865	48,863
Mortising-machine	C. L. Zeidler	Cincinnati, Ohio	Nov. 12, 1867	70,771
Mortising-machine, Bearded-chisel	S. Metcalf	Wilmington, Vt	Jan. 17, 1827	....
Mortising-machine clamp	A. Schmackers	Cincinnati, Ohio	Sept. 26, 1871	119,414
Mortising-machine gage	T. Beach	Freeport, Pa	July 4, 1871	116,538
Mortising-machine, Lever-power	T. Green	Manlius, N. Y	Apr. 12, 1826	....
Mortising-machine, Power	G. W. Gould	Norwich, Conn	June 16, 1863	38,898
Mortising machine, Timber	J. Andrews	Sudbury, Mass	May 30, 1838	752
Mortising machine, Timber	F. and T. Burdick	Brooklyn, N. Y	Apr. 7, 1838	684
Mortising machine, Timber	J. A. Fay	Keene, N. H	Jan. 17, 1842	2,425
Mortising machine, Timber	T. H. Haskings	Springfield, Ohio	June 30, 1837	247
Mortising machine, Timber	I. McLaughlin	Sunderland, Vt	Apr. 7, 1838	685
Mortising machine, Timber	E. M. Shaw	Baltimore, Md	Sept. 22, 1838	937
Mortising-machines, Device for throwing into and out of gear the tools of.	L. Kittinger	East Greenville, Ohio	Dec. 29, 1857	18,977
Mortising-machines, Feeding	R. P. Benton	Rochester, N. Y.	Jan. 23, 1855	12,287
Mortising-machines, Regulating length of stroke in.	E. Gould	Newark, N. J	Sept. 25, 1855	13,594
Mortising-machines, Reversing the chisels in	M. Marshall	Lowell, Mass	Apr. 28, 1857	17,183
Mortising-machines, Reversing the chisels in	F. Stamm	Lancaster, Pa	Oct. 12, 1858	21,783
Mortising, sawing, and boring machine	R. Medley	Bloomfield, Ky	July 7, 1829	....
Mortising, slotting, and dovetailing machine	J. Felber	Saint Louis, Mo	Oct. 6, 1868	82,816
Mortising, tenoning, and boring machine	B. Richtmyre and J. H. Martin	Conesville, N. Y	July 16, 1839	1,241
Mortising timber	J. W. Bliss	Somers, Conn	Mar. 20, 1834	....
Mortising timber	G. Page	Keene, N. H	Mar. 21, 1836	....
Mortising-tool	T. Board and C. N. Austin	Jackson Court-House, Va	Jan. 10, 1860	26,745
Mortising-tool	T. and J. H. Burdick	Albany, N. Y	Jan. 26, 1869	86,209
Mortising-tool	L. Eames	Kalamazoo, Mich	Apr. 10, 1860	27,782
Mortising-tool	H. K. Forbis	Danville, Ky	Dec. 30, 1873	146,058
Mortising-tool	A. C. Hitchcock and C. H. Amidon.	Greenfield, Mass	Mar. 18, 1856	14,454
Mortising-tool	H. Knowles	New York, N. Y	Jan. 29, 1856	14,160
Mortising-tool, Revolving	W. Zimmerman	Quincy, Ill	July 11, 1865	48,759
Mortising wood	C. Thompson	Poughkeepsie, N. Y	Jan. 30, 1834	....
Mosaics, Manufacture of wooden	L. De Forest	Derby, Conn	Jan. 24, 1860	26,898
Mosquito and fly net	E. O. Carrington	Philadelphia, Pa	Dec. 24, 1867	72,452
Mosquito and fly net	J. W. Cran and A. S. Randolph.	Norwalk, Conn., and Plainfield, N. J.	Aug. 13, 1867	67,729
Mosquito and fly net	A. M. Rodgers	Brooklyn, N. Y	May 11, 1869	90,050
Mosquito and fly net	W. Weston, jr	West Newton, Mass	Aug. 20, 1867	68,018
Mosquito and fly screen	J. W. Boughton	Philadelphia, Pa	July 2, 1872	128,582
Mosquito-bar	A. L. Carrier	Washington, D. C	July 7, 1863	39,116
Mosquito-bar	W. Field	Providence, R. I	Nov. 15, 1864	45,032
Mosquito-bar	R. F. S. Heath	Philadelphia, Pa	May 9, 1871	114,556
Mosquito-bar	J. S. Hunter	Hartford, Conn	May 28, 1867	65,085
Mosquito-bar	S. Kemper	Berger, Mo	May 10, 1870	102,830
Mosquito-bar	T. Masac	Good Hope Plantation, La	Oct. 25, 1870	108,611
Mosquito-bar	S. Roebuck	Brooklyn, N. Y	Sept. 24, 1861	33,365
Mosquito-bar	E. A. G. Roulstone	Roxbury, Mass	Aug. 7, 1866	56,994
Mosquito-bar	L. Sawyer	Springfield, Vt	Apr. 24, 1866	54,259
Mosquito-bar	T. S. Scoville	New York, N. Y	Feb. 21, 1860	27,236
Mosquito-bar	E. Steinel	Amsterdam, N. Y	May 19, 1868	78,024
Mosquito-bar	C. L. S. Walker	Newark, N. J	July 3, 1866	56,128
Mosquito-bar	T. S. Williams	Enterprise, Miss	May 31, 1859	24,255
Mosquito-bar and window-screen	A. J. Whittier	Roxbury, Mass	Aug. 27, 1867	68,326
Mosquito-bar, Canopy or	J. B. Platt	Augusta, Ga	Mar. 9, 1869	87,589
Mosquito-bar for windows, &c	V. Barker	Otisfield, Me	Feb. 12, 1867	61,986
Mosquito-bar for windows	A. C. Flint	Boston, Mass	June 16, 1868	78,948

Index of patents issued from the United States Patent Office from 1790 *to* 1873, *inclusive*—Continued.

Invention.	Inventor.	Residence.	Date.	No.
Mosquito-bar for windows	C. T. Warren	Linden, N. J	June 30, 1868	79, 419
Mosquito-bar frame	J. Higgins	Chicago, Ill	July 30, 1861	32, 943
Mosquito-bar frame	L. A. P. Jacques	Cincinnati, Ohio	Apr. 24, 1849	6, 390
Mosquito-bar frame	D. McHugh	Mainville, Ohio	Aug. 13, 1867	67, 784
Mosquito-bar frame	H. Searle	Washington, D. C	May 11, 1869	90, 054
Mosquito-bar frame, Adjustable	L. J. Parsons	New Bedford, Mass	Oct. 20, 1868	83, 202
Mosquito-bar netting, Elastic frame for	R. M. Holland and A. J. Hibbs	Philadelphia, Pa	June 11, 1867	65, 673
Mosquito bar or tent	A. W. Price	Adrian, Mich	June 20, 1865	48, 308
Mosquito-bars, Adjusting	F. C. Payne	Hebron, Conn	July 27, 1858	21, 019
Mosquito-bars to window-blinds, doors, &c., Attaching.	J. Hebron	Buffalo, N. Y	Dec. 14, 1869	97, 916
Mosquito-bars to windows, Fastening	A. F. Tarr	Rockport, Mass	Feb. 16, 1864	41, 650
Mosquito-canopy	M. Bliss	Grinnell, Iowa	May 15, 1866	54, 674
Mosquito-canopy	W. Bourquignon	Providence, R. I	Sept. 23, 1873	143, 119
Mosquito-canopy	L. J. Henry	New York, N. Y	Aug. 19, 1856	15, 592
Mosquito-canopy	I. E. Palmer	Hackensack, N. J	Feb. 25, 1868	74, 933
Mosquito-canopy	S. Roebuck	Brooklyn, N. Y	Apr. 28, 1863	38, 339
Mosquito-canopy	S. and J. Roebuck	New York, N. Y	Mar. 30, 1869	88, 514
Mosquito-canopy frame	I. E. Palmer	Hackensack, N. J	Sept. 25, 1866	58, 283
Mosquito-curtain	J. S. Martin	Boston, Mass	Dec. 11, 1855	13, 931
Mosquito-curtain	B. B. Webster	Boston, Mass	Oct. 3, 1854	11, 764
Mosquito-frame	L. S. Thompson	Brooklyn, N. Y	Feb. 24, 1863	37, 779
Mosquito-frame, Folding	S. C. Maine	Boston, Mass	Aug. 11, 1868	80, 830
Mosquito-frame for windows	A. E. Horton	North Leominster, Mass	Oct. 1, 1867	69, 343
Mosquito-guard	H. Harris	Newark, N. J	Sept. 10, 1867	68, 759
Mosquito-guard	A. D. Puffer	Somerville, Mass	Mar. 20, 1866	53, 334
Mosquito-guard	R. Themar	Sheboygan, Wis	June 22, 1869	91, 793
Mosquito-guard for bedsteads	A. D. Puffer	Somerville, Mass	May 8, 1866	54, 595
Mosquito-killer	H. D. Forbes	Cambridge, Mass	June 16, 1868	78, 950
Mosquito-net	J. D. De Coursey	Philadelphia, Pa	June 12, 1866	55, 473
Mosquito-net	J. J. A. Ergenzinger	Atlanta, Ga	Feb. 21, 1871	112, 023
Mosquito-net	J. B. Holmes	Philadelphia, Pa	Mar. 16, 1869	87, 849
Mosquito-net	T. G. Voorhis and W. B. Whiteman.	New York, N. Y	Apr. 9, 1861	32, 018
Mosquito-net	T. S. Winslow	New York, N. Y	Mar. 5, 1872	124, 239
Mosquito-net	J. Zengler	Chicago, Ill	Mar. 15, 1864	41, 957
Mosquito-net, Adjustable frame for	E. Howard	New York, N. Y	July 23, 1872	129, 665
Mosquito-net and shades for windows	R. B. Burchell	Brooklyn, N. Y	Feb. 14, 1860	27, 103
Mosquito-net frame	U. W. Armstrong	Evansville, Ind	Feb. 16, 1869	86, 966
Mosquito-net frame	W. A. Griffith	Boston, Mass	Aug. 6, 1867	67, 532
Mosquito-net frame	S. E. Hartwell	New York, N. Y	Apr. 15, 1856	14, 655
Mosquito-net frame	S. Hughes	Jersey City, N. J	July 16, 1872	129, 030
Mosquito-net frame	G. T. Palmer	Brooklyn, N. Y	Apr. 28, 1868	77, 207
Mosquito-net frame	N. Peterson and R. Roecher	Memphis, Tenn	Mar. 4, 1873	136, 456
Mosquito-net frame	M. L. Treadwell	New York, N. Y	Aug. 6, 1867	67, 612
Mosquito-net frame	W. M. White	Canton, Conn	Nov. 19, 1872	133, 279
Mosquito-net frames, Metallic connection for	U. W. Armstrong and I. Keeney	Evansville, Ind	June 14, 1870	104, 245
Mosquito-net frames, Rubber connection for	U. W. Armstrong and I. Keeney	Evansville, Ind	Mar. 22, 1870	100, 966
Mosquito-net holder	C. Messenger	Cleveland, Ohio	Aug. 16, 1870	106, 386
Mosquito-net in window-blinds	G. W. Miles	Philadelphia, Pa	Dec. 3, 1867	71, 778
Mosquito-net stand	A. Strasser and B. M. Lewy	Montgomery, Ala	Aug. 4, 1868	80, 782
Mosquito-net support	B. M. Lewy and A. Strasser	Montgomery, Ala	June 8, 1869	91, 030
Mosquito-netting, Machine for stretching and folding.	J. A. and L. Van Riper	Spring Valley, N. Y	Sept. 8, 1863	39, 868
Mosquito-screen	J. M. Griswold	Auburndale, Mass	July 13, 1869	92, 441
Mosquito-screen	E. Spaulding	Brooklyn, N. Y	June 9, 1868	78, 767
Mosquito-screen for windows	G. A. Keene	Newburyport, Mass	June 4, 1872	127, 609
Mosquito-screen for windows	T. Stover	Cambridgeport, Mass	July 9, 1867	66, 534
Mosquito-screen frame	W. W. Wooley	New York, N. Y	June 28, 1870	104, 809
Mosquito tent or frame	J. Stewart	New York, N. Y	Mar. 15, 1864	41, 946
Moss, Cleaning and carding	L. Boudreaux	Thibodeaux, La	Aug. 4, 1857	17, 954
Moss-cleaning machine	H. Hull	Petersonville, La	June 25, 1867	66, 026
Moth-destroying apparatus	C. F. Worch	New York, N. Y	June 4, 1867	65, 462
Moth-killer	W. A. Flanders	Sharon, Vt	June 6, 1854	11, 013
Moth-proof box	R. M. Seldis	New York, N. Y	Nov. 14, 1871	120, 904
Moth-proof case	J. W. Aiken and J. H. Stone	Philadelphia, Pa	Nov. 12, 1867	70, 676
Moth-proof clothing-chest	R. S. Jennings	Philadelphia, Pa	Nov. 14, 1871	120, 976
Moth-proof lining	J. R. Smith	Chicago, Ill	Aug. 31, 1869	94, 357
Moth-protector	F. F. Voigt	New Orleans, La	Oct. 3, 1871	119, 481
Moth-trap	J. Frew	Meadville, Pa	Feb. 10, 1863	37, 618
Moth-trap	J. M. Heard	Aberdeen, Miss	Feb. 7, 1860	27, 048
Motion and multiplying power, Extending	W. Kendall, jr	Waterville, Me	Jan. 16, 1826	....
Motion and preventing reaction, Transmitting rotary.	C. Sellers	Philadelphia, Pa	May 28, 1872	127, 192
Motion, Apparatus for converting	J. J. Chenal	Génissiat, France	Nov. 7, 1871	120, 622
Motion, Apparatus for converting	L. S. Fithian	Brooklyn, N. Y	June 6, 1871	115, 589
Motion, Apparatus for converting	C. L. Spencer	Providence, R. I	Feb. 23, 1869	87, 218
Motion, Apparatus for converting	K. J. Winslow	Twickenham, England	June 23, 1868	79, 092
Motion, Apparatus for converting reciprocating into rotary.	R. M. Fryer	Nashville, Tenn	Mar. 19, 1872	124, 805
Motion, Apparatus for converting reciprocating into rotary.	C. L. Spencer	Providence, R. I	Feb. 23, 1869	87, 217
Motion, Apparatus for converting reciprocating into rotary.	F. Wagner	New York, N. Y	Mar. 21, 1871	112, 991
Motion, Apparatus for converting rotary into reciprocating.	D. Morrison	Portland, Me	Nov. 26, 1867	71, 506
Motion, Apparatus for preventing back	A. H. and J. H. Race	Brooklyn, N. Y	Sept. 30, 1873	143, 252
Motion, Apparatus for producing reciprocating	J. J. P. Lyon	Ypsilanti, Mich	Oct. 13, 1868	82, 964
Motion, Apparatus for reproducing	C. T. Chester	Englewood, N. J	May 27, 1873	139, 294
Motion, Apparatus for transmitting	E. H. Bancroft	Syracuse, N. Y	Aug. 1, 1871	117, 503
Motion, Apparatus for transmitting	F. M. La Boiteaux	Cincinnati, Ohio	Oct. 29, 1872	132, 671
Motion, Apparatus for transmitting	P. Merhof and F. Healy	Croton, N. Y., and Barrington, R. I.	Jan. 2, 1872	122, 398
Motion, Apparatus for transmitting	S. S. Rembert	Memphis, Tenn	Aug. 29, 1871	118, 554
Motion, Apparatus for transmitting rotary	M. Clemens	Springfield, Mass	Nov. 2, 1869	96, 395
Motion, Apparatus for transmitting rotary	S. R. Morgan	Philadelphia, Pa	Mar. 23, 1869	88, 061
Motion, Application of steam to produce rotary	J. Mount	Nashville, Tenn	July 16, 1829	....
Motion by belts, Transmitting	J. Rand	Roxbury, Mass	Aug. 30, 1864	44, 014

Index of patents issued from the United States Patent Office from 1790 *to* 1873, *inclusive*—Continued.

Invention.	Inventor.	Residence.	Date.	No.
Motion by means of friction, Apparatus for converting rotary into reciprocating.	R. Sammer	Vineland, N. J	Apr. 18, 1871	113, 801
Motion by means of friction, Device for obtaining	S. Marden	Newton, Mass	Feb. 11, 1868	74, 234
Motion by varying and irregular power, Machinery to produce uniform.	J. Kello	New York, N. Y	Apr. 1, 1815	....
Motion, Changing	R. G. Turner and H. Stone	Dedham, Mass	Dec. 15, 1863	40, 965
Motion, Changing reciprocating, eccentric, or irregular to rotary.	J. Woodhull	Rochester, N. Y	Apr. 6, 1829	....
Motion, Changing reciprocating into rotary	A. M. Bouton and A. Perry	Newark, N. J	Aug. 18, 1842	2, 753
Motion, Changing reciprocating into rotary	J. Harris, jr	Boston, Mass	Jan. 14, 1851	7, 902
Motion, Changing reciprocating into rotary	T. J. Morton	Racine, Wis	June 25, 1872	128, 240
Motion, Changing reciprocating into rotary	J. V. Strait	Litchfield, Ohio	July 22, 1851	8, 233
Motion, Changing reciprocating into rotary	P. Yates	Milwaukee, Wis	Apr. 23, 1850	7, 316
Motion, Changing rotary into reciprocating	A. Dean	Penn Yan, N. Y	Nov. 29, 1864	45, 230
Motion, Changing rotary into reciprocating	I. D. Garlick	Lyons, N. Y	Aug. 20, 1850	7, 577
Motion, Changing rotary into reciprocating	J. H. and A. E. Redstone	Indianapolis, Ind	Apr. 29, 1862	35, 131
Motion, Changing rotary into reciprocating	S. L. Wiegand	Philadelphia, Pa	May 25, 1858	20, 384
Motion, Combined lever and crank	E. Nolt	Earl Township, Pa	Apr 14, 1863	38, 176
Motion, Communicating	W. Bicknell	Hartford, Me	Dec. 4, 1866	60, 126
Motion, Compensation for loss of	T. Shaw	Philadelphia, Pa	Feb. 7, 1865	46, 309
Motion, Converting	W. H. Akins	Dryden, N. Y	July 19, 1864	43, 558
Motion, Converting	H. Burk	Mineral Point, Ohio	May 28, 1867	65, 166
Motion, Converting	P. Dickson	Utica, Minn	Mar. 24, 1863	37, 953
Motion, Converting	J. W. Drummond	New York, N. Y	Mar. 8, 1864	41, 835
Motion, Converting	A. Eckert	Trenton, Ohio	Jan. 15, 1867	61, 179
Motion, Converting	W. H. Hurlbut	Elgin, Ill	Mar. 26, 1867	63, 254
Motion, Converting	S. C. Ketchum	Winchendon, Mass	Apr. 28, 1863	38, 311
Motion, Converting	A. E. Kline	Woodville, Pa	Nov. 29, 1864	45, 252
Motion, Converting	T. D. Lakin	Greenfield, N. H	Dec. 17, 1872	133, 994
Motion, Converting	T. A. Macaulay	New York, N. Y	Mar. 1, 1864	41, 780
Motion, Converting	W. H. McNary	Brooklyn, N. Y	July 15, 1862	35, 883
Motion, Converting	L. Planer	New York, N. Y	May 22, 1860	28, 403
Motion, Converting	C. L. Spencer	Providence, R. I	Mar. 4, 1863	34, 597
Motion, Converting alternate rectilinear into rotary	P. Cooper	New York, N. Y	Apr. 28, 1828	....
Motion, Converting circular into reciprocating	T. and C. Plumleigh	Dundee, Ill	Oct. 8, 1867	69, 584
Motion-converting mechanism	E. S. Pierce	Hartford, Conn	June 29, 1869	92, 091
Motion, Converting reciprocal into rotary	D. Morrison	Portland, Me	Jan. 21, 1868	73, 631
Motion, Converting reciprocating into rotary	J. C. Butterworth and B. Arnold.	Providence and East Greenwich, R. I.	July 9, 1861	32, 749
Motion, Converting reciprocating into rotary	I. Chapman	New York, N. Y	Aug. 3, 1858	21, 065
Motion, Converting reciprocating into rotary	J. Cupps and A. R. Harper	Chicago, Ill., and Grandville, Mich.	Jan. 1, 1867	60, 837
Motion, Converting reciprocating into rotary	J. F. Foss	Lowell, Mass	Aug. 29, 1865	49, 618
Motion, Converting reciprocating into rotary	C. P. Gallagher	San Francisco, Cal	Nov. 28, 1854	11, 993
Motion, Converting reciprocating into rotary	S. Gissinger and J. W. Kellberg	Allegheny, Pa	Dec. 23, 1856	16, 303
Motion, Converting reciprocating into rotary	C. A. Harper	Fort Worth, Tex	Nov. 1, 1859	25, 966
Motion, Converting reciprocating into rotary	J. F. Hartmann	Richmond, Ind	Apr. 26, 1870	102, 260
Motion, Converting reciprocating into rotary	F. H. Harwood	Rushville, N. Y	Apr. 6, 1858	19, 818
Motion, Converting reciprocating into rotary	C. Howard	Alton, Ill	Aug. 10, 1852	9, 185
Motion, Converting reciprocating into rotary	R. Maffett	Bradford, Pa	Apr. 1, 1856	14, 568
Motion, Converting reciprocating into rotary	E. Matteson	Brooklyn, N. Y	Aug. 28, 1860	29, 802
Motion, Converting reciprocating into rotary	C. B. Parsons	Burr Oak, Mich	Feb. 28, 1860	27, 307
Motion, Converting reciprocating into rotary	E. A. Smead	Tioga, Pa	Feb. 15, 1859	22, 984
Motion, Converting reciprocating into rotary	A. T. Underhill	New York, N. Y	Sept. 20, 1859	25, 550
Motion, Converting reciprocating into rotary	C. A. Watson	Green River, Ky	Aug. 1, 1838	864
Motion, Converting reciprocating into rotary	T. Williams	Providence, R. I	Mar. 5, 1861	31, 649
Motion, Converting reciprocating rectilinear into continuous circular.	B. Babbett	Bangor, Me	Oct. 11, 1836	....
Motion, Converting reciprocating rotary into reciprocating rectilinear.	A Carson	New York, N. Y	July 20, 1852	9, 131
Motion, Converting rectilinear into circular	R. Dennis	Rahway, N. J	Nov. 3, 1825	....
Motion, Converting rectilinear into rotary	J. B. Eads	Saint Louis, Mo	Jan. 2, 1866	51, 815
Motion, Converting rectilinear into rotary	J. A. Ehle	Greenbush, Wis	May 21, 1867	64, 852
Motion, Converting rectilinear into rotary	J. G. Harroun	Sag Harbor, N. Y	July 25, 1871	117, 280
Motion, Converting rectilinear into rotary	J. and T. G. McLaughlin	Pennsylvania	Aug. 29, 1848	5, 741
Motion, Converting rectilinear into rotary	G. W. Richardson and R. Glover	Grayville, Ill	Feb. 15, 1859	23, 002
Motion, Converting rectilinear into rotary	J. T. Savage	Raleigh, N. C	Aug. 9, 1823	....
Motion, Converting rotary into double-acting reciprocating.	C. Willson	Bedford, N. Y	Oct. 5, 1839	1, 356
Motion, Converting rotary into reciprocating	S. F. Ames	Stamford, Conn	July 4, 1865	48, 502
Motion, Converting rotary into reciprocating	H. Baker	Catskill, N. Y	June 7, 1853	9, 761
Motion, Converting rotary into reciprocating	A. E. Kline	Goodville, Pa	Sept. 20, 1864	44, 319
Motion, Converting rotary into reciprocating	P. C. Van Brocklin	Buffalo, N. Y	July 15, 1862	35, 903
Motion, Converting rotary into reciprocating	A. Warth	New York, N. Y	Dec. 2, 1856	16, 155
Motion, Converting rotary into reciprocating	J. J. Weeks	Locust Valley, N. Y	July 20, 1858	20, 980
Motion, Converting rotary into reciprocating	P. Werni	Manchester, Mich	Jan. 3, 1865	45, 779
Motion, Converting rotary into reciprocating rectilinear.	A. Broughton	Malone, N. Y	Sept. 13, 1859	25, 465
Motion, Converting rotary into reciprocating rectilinear.	W. S. Lazelle	New York, N. Y	Feb. 7, 1860	27, 057
Motion, Crank	E. P. Brownell	East Haddam, Conn	Mar. 25, 1862	34, 735
Motion, Crank	M. A. Rowe	Martinsville, Ill	Apr. 26, 1870	102, 323
Motion, Crank	T. Shaw	Philadelphia, Pa	May 19, 1863	38, 608
Motion, Device for arresting	J. Everding	Philadelphia, Pa	Nov. 7, 1871	120, 730
Motion, Device for changing rotary into reciprocal	D. Shively	Bayard, Ohio	May 28, 1872	127, 196
Motion, Device for converting	W. H. Abel	Greenville, R. I	Feb. 11, 1868	74, 265
Motion, Device for converting	H. H. Bishop	Bristol, Conn	Mar. 3, 1863	37, 806
Motion, Device for converting	F. Brewer	Collinsville, Ill	June 27, 1865	48, 360
Motion, Device for converting	W. M. Cox	Blanket Hill, Pa	Dec. 30, 1873	145, 931
Motion, Device for converting	B. Eybel	New York, N. Y	May 7, 1867	64, 511
Motion, Device for converting	M. M. Follett	Westborough, Mass	Apr. 7, 1868	76, 429
Motion, Device for converting	J. Hawthorn	Coshocton, Ohio	Feb. 5, 1867	61, 829
Motion, Device for converting	T. A. Mitchell	Washington, D. C	June 22, 1869	91, 555
Motion, Device for converting	J. P. Taylor	Hudson City, N. J	Dec. 16, 1873	145, 599
Motion, Device for converting	P. Werni	Elizabeth, N. J	July 23, 1872	129, 771
Motion, Device for converting	H. Wheeler	Racine, Wis	Dec. 16, 1873	145, 540
Motion, Device for converting alternate circular motion into direct circular.	A. Bartholf	New York, N. Y	Apr. 19, 1859	23, 651

Index of patents issued from the United States Patent Office from 1790 *to* 1873, *inclusive*—Continued.

Invention.	Inventor.	Residence.	Date.	No.
Motion, Device for convertng circular into reciprocating.	L. Dame	Newburyport, Mass	Feb. 11, 1873	135,697
Motion, Device for converting reciprocating into alternate circular.	H. Ehrenfeld	New York, N. Y	June 21, 1859	24,448
Motion, Device for converting reciprocating into rotary.	G. L. Gavett	Sandstone, Mich	Nov. 28, 1871	131,354
Motion, Device for converting reciprocating into rotary.	J. Hathaway	Marietta, Ga	Apr. 3, 1860	27,715
Motion, Device for converting reciprocating into rotary.	G. S. Nutter	Bunker Hill, Ill	July 16, 1872	129,582
Motion, Device for converting rotary into reciprocating.	E. Walker	Dundee, Ill	Mar. 10, 1868	75,495
Motion, Device for convert ingrotary into reciprocating rectilinear.	C. F. Hadley	Chicopee, Mass	Mar. 23, 1869	88,163
Motion, Device for reversing	C. F. Hadley	Chicopee, Mass	Mar. 1, 1870	100,395
Motion, Device for stopping and changing	F. A. Pratt	Hartford, Conn	Sept. 4, 1860	29,939
Motion, Device for transferring	W. H. Benson	Waynesborough, Va	July 15, 1873	140,878
Motion, Device for transmitting	J. Brizee	Alvarado, Cal	July 10, 1866	56,171
Motion, Device for transmitting	J. W. Browning	Mattoon, Ill	Oct. 11, 1864	44,594
Motion, Device for transmitting	T. C. Entwistle	New York, N. Y	Sept. 4, 1866	57,693
Motion, Device for transmitting	E. Gallagher	Brooklyn, N. Y	Feb. 11, 1868	74,337
Motion, Device for transmitting	P. Palmlund	Brooklyn, N. Y	July 4, 1871	116,745
Motion, Device for transmitting	N. Thompson	Farmington, Mich	Dec. 3, 1867	71,820
Motion, Device for transmitting	L. Tilton	Brooklyn, E. D., N. Y	May 28, 1867	65,304
Motion, Device for transmitting and reversing rotary.	W. H. Merrick	Philadelphia, Pa	Feb. 28, 1871	112,269
Motion, Device for transmitting rotary	L. B. Flanders	Philadelphia, Pa	Oct. 15, 1867	69,787
Motion, Device for transmitting rotary	H. J. Hancock	New York, N. Y	Aug. 2, 1870	105,937
Motion, Device for transmitting rotary	H. Morris	West Philadelphia, Pa	Dec. 28, 1858	22,445
Motion, Directing	A. Buchanan	Jersey City, N. J	Sept. 22, 1363	40,075
Motion, Feed	A. J. Shipley	Waterbury, Conn	Dec. 22, 1868	85,249
Motion for machinery, Electrical stop	E. Maertens	Philadelphia, Pa	June 24, 1873	140,287
Motion for machinery, Pendulum	J. Jacob	Montreal, Canada	Dec. 29, 1826	....
Motion for preserving rolling contact, &c	G. P. Gordon and F. O. Degener	New York, N. Y	June 16, 1857	17,565
Motion for propelling machinery, Alternate	B. Elliott	Danby, N. Y	Oct. 4, 1834	....
Motion for stamping and other machines, Progressive reciprocating.	R. L. Barclay	Brooklyn, N. Y	Nov. 23, 1869	97,153
Motion from reciprocating rectilinear motion, Mechanism for obtaining rotary.	A. R. Gilmore	Bath, Me	Apr. 5, 1859	23,461
Motion in car-brakes and other machinery, Transmitting.	J. Steger	New York, N. Y	Apr. 13, 1869	88,990
Motion, Instrument for registering reciprocating and rotary.	F. B. Hall	Hartford, Conn	Apr. 23, 1861	32,133
Motion, Machine for converting oscillating motion into direct circular.	L. Planer	New York, N. Y	June 7, 1859	24,359
Motion, Machine for converting reciprocating into intermittent rotary.	H. Ehrenfeld	New York, N. Y	Nov. 29, 1859	26,261
Motion, Machine for converting reciprocating into rotary.	J. G. Deshler	Allentown, Pa	Sept. 22, 1868	82,296
Motion, Machine for converting reciprocating into rotary.	J. B. Page	Chicago, Ill	Aug. 14, 1866	57,178
Motion, Machine for converting rotary into reciprocating.	C. C. Welsh	Pleasant Valley, Pa	Oct. 6, 1868	82,907
Motion, Machine for producing rotary	D. F. Mosman	Chelsea, Mass	Dec. 17, 1872	133,999
Motion, Machine for transmitting	C. B. Grout	New York, N. Y	Apr. 16, 1872	125,809
Motion, Machinery for changing	C. L. Pyron and R. Bruce	Manchester, Tenn	Jan. 3, 1860	26,709
Motion, Machinery for transmitting	M. Kaefer	New York, N. Y	May 31, 1859	24,218
Motion, Mechanical	A. Kloman	Pittsburgh, Pa	Aug. 2, 1870	105,951
Motion, Mechanical	D. Lynahan	Buffalo, N. Y	Dec. 6, 1864	45,330
Motion, Mechanism for converting reciprocating into rotary.	W. Simpson and A. Gardner	Ilford, England	Nov. 23, 1869	97,240
Motion, Mechanism for converting rotary into reciprocating.	W. N. Brown	New York, N. Y	Oct. 25, 1859	25,880
Motion, Mechanism for obtaining intermitted rotary.	L. J. Knowles	Warren, Mass	Mar. 20, 1866	53,308
Motion, Mechanism for producing rotary	C. E. De Lorierè	London, England	June 11, 1872	127,682
Motion, Mechanism for transmitting	C. C. Hull	Williamsburgh, N. Y	Apr. 13, 1869	88,876
Motion, Mechanism for transmitting rotary	G. Bancker and A. Campbell	New York, N. Y	Nov. 2, 1858	21,934
Motion, Metallic band for communicating	J. Eve	Augusta, Ga	May 1, 1828	....
Motion obtained from rectilinear, Rotary	A. Smith	Rhinebeck, N. Y	Aug. 30, 1823	....
Motion of machinery, Changing the	O. Pettee	Newton, Mass	Mar. 15, 1825	....
Motion of machinery, Registering	S. L. Wiegand	Philadelphia, Pa	Aug. 3, 1858	21,101
Motion of shafts into reciprocating motion, Converting rotary.	J. B. Siner	Lawrence, Mass	Mar. 11, 1873	136,675
Motion, Producing alternate lateral	N. Moody	Hallowell, Mass	Feb. 19, 1819	....
Motion, Producing continuous circular from reciprocating rectilinear.	C. S. Harris	Holyoke, Mass	June 20, 1854	11,123
Motion, Producing reciprocating and lateral	M. Ingalls	Burlington, Pa	Mar. 13, 1847	5,015
Motion, Producing vertical and horizontal reciprocating.	M. Steigers	Saint Louis, Mo	Mar. 9, 1858	19,586
Motion, Reciprocating and rotary	I. Van Doren	Somerville, N. J	Mar. 23, 1858	19,726
Motion, Reciprocating crank	B. K. Dorwart	Lancaster, Pa	Oct. 31, 1865	50,695
Motion, Stop	B. Stott	Westerly, R. I	Aug. 30, 1864	44,024
Motion, Stopping and changing	F. A. Pratt	Hartford, Conn	Nov. 26, 1861	33,816
Motion to machinery, Apparatus for communicating	S. Kilburn	Sterling, Mass	Apr. 16, 1831	....
Motion to machinery, Communicating	C. H. Miller	Buffalo, N. Y	Aug. 2, 1870	105,964
Motion to machinery, Transmitting	W. Phelps	Sycamore, Ill	July 3, 1860	29,000
Motion to machines, Device for communicating	L. Cheetham	Lewiston, Me	July 16, 1872	129,316
Motion, Transmitting	M. M. Ammidown	Boston, Mass	Oct. 6, 1868	82,674
Motion, Transmitting	S. Andress	Chesaning, Mich	Feb. 26, 1861	31,522
Motion, Transmitting	J. W. Howlett	Greensborough, N. C	Apr. 9, 1861	31,978
Motion, Transmitting	M. Kaefer	Alexandria, Pa	May 5, 1857	17,222
Motion, Transmitting	M. Kaefer	Factoryville, N. Y	Mar. 5, 1861	31,607
Motion, Transmitting	N. Read	Jewett City, Conn	Nov. 9, 1869	96,615
Motion, Transmitting	W. Rowell	New York, N. Y	May 7, 1867	64,450
Motion, Transmitting	W. Rowell	New York, N. Y	May 7, 1867	64,451
Motion, Transmitting	C. Sellers	Philadelphia, Pa	Sept. 10, 1861	33,283
Motion, Transmitting	H. F. Shaw	West Roxbury, Mass	Oct. 3, 1865	50,283

Index of patents issued from the United States Patent Office from 1790 *to* 1873, *inclusive*—Continued.

Invention.	Inventor.	Residence.	Date.	No.
Motion, Transmitting	E. Wadhaws	Hamilton, Canada	July 11, 1865	48, 780
Motion, Transmitting	H. F. Wheeler	Boston, Mass	Sept. 26, 1871	119, 437
Motion, Transmitting	F. Yeiser	Danville, Ky	Jan. 30, 1866	52, 354
Motion, Transmitting and arresting	W. Sellers	Philadelphia, Pa	Jan. 21, 1862	34, 217
Motion, Traverse	D. Walker	East Hampton, Mass	Jan. 16, 1872	122, 790
Motion, Weft-stop	F. O. Tucker	Westerly, R. I	Aug. 1, 1871	117, 698
Motive-power	J. B. Atwater	Chicago, Ill	Apr. 13, 1869	88, 767
Motive-power	A. M. Bacon	Boston, Mass	Nov. 9, 1869	96, 657
Motive-power	C. Batcheller	Polk County, Iowa	Aug. 23, 1870	106, 537
Motive-power	J. N. Bethune	Warrenton, Va	Sept. 26, 1871	119, 260
Motive-power	A. Bouchard	New Orleans, La	Jan. 16, 1872	122, 802
Motive-power	I. W. Brown	New Haven, Conn	Feb. 11, 1868	74, 295
Motive-power	J. Bruce	Brooklyn, N. Y	Sept. 23, 1862	36, 506
Motive-power	J. M. Cayce	Franklin, Tenn	Sept. 13, 1870	107, 221
Motive-power	A. Girandat	New York, N. Y	Aug. 27, 1861	33, 139
Motive-power	J. K. Glenn	New York, N. Y	Oct. 22, 1867	69, 987
Motive-power	J. Green	Providence, R. I	Sept. 13, 1864	44, 181
Motive-power	W. H. Hartman	Fostoria, Ohio	May 16, 1865	47, 717
Motive-power	J. H. Haven	Lewiston, N. Y	Jan. 30, 1866	52, 287
Motive-power	J. B. Hunter	Ashley, Ill	Sept. 26, 1871	119, 362
Motive-power	A. Johnson and W. H. Elliott	Bloomington, Ill	May 14, 1867	64, 770
Motive-power	W. B. Jones	Franklin, Ky	May 21, 1867	64, 881
Motive-power	F. Kettler	Milwaukee, Wis	Feb. 3, 1863	37, 582
Motive-power	C. Mans	Danville, Pa	Mar. 22, 1859	23, 305
Motive-power	J. McKnight and W. S. Deisher	Reading, Pa	June 5, 1866	55, 325
Motive-power	G. Meyer	New Richmond, Mich	Apr. 15, 1873	137, 856
Motive-power	J. G. Mitchell	Collington, Md	Mar. 1, 1859	23, 104
Motive-power	W. T. Nichols	Rutland, Vt	Dec. 2, 1862	37, 050
Motive-power	T. J. Polson	Kilmichael, Miss	Aug. 29, 1871	118, 641
Motive-power	W. S. Reeder	Saint Louis, Mo	May 28, 1872	127, 368
Motive-power	B. J. Sage	New Orleans, La	May 28, 1872	127, 189
Motive-power	J. A. Schulo	New York, N. Y	Apr. 24, 1866	54, 220
Motive-power	J. S. Swan	Kanawha County, W. Va	Sept. 12, 1871	118, 982
Motive-power	L. Tilliers	New York, N. Y	Aug. 20, 1861	33, 112
Motive-power	J. Toll	Locust Grove, Ohio	Jan. 26, 1864	41, 406
Motive-power	L. D. Towsley and E. Matteson	New York and Brooklyn, N. Y.	July 24, 1860	29, 329
Motive-power	W. M. Watson	Tonica, Ill	Aug. 16, 1870	106, 523
Motive-power and balance-car	J. Bayma	San Francisco, Cal	Nov. 28, 1871	121, 316
Motive-power apparatus	T. Davis	Detroit, Mich	Dec. 20, 1870	110, 212
Motive-power apparatus	J. T. Gilbert	Asbury, Ill	Apr. 4, 1871	113, 417
Motive-power apparatus	A. Jackson	Lebanon, Tenn	June 21, 1870	104, 595
Motive-power apparatus	L. W. Lathrop	Poughkeepsie, N. Y	Nov. 1, 1870	108, 915
Motive-power apparatus	J. B. Martel	Philadelphia, Pa	June 25, 1872	128, 408
Motive-power apparatus	C. A. Mills	Bridgeport, Conn	Jan. 3, 1871	110, 667
Motive-power apparatus	S. G. Monce	Marathon, Ohio	June 23, 1868	79, 289
Motive-power apparatus	H. Wilkins and W. H. Sangster	Paris, Ky	Feb. 15, 1870	99, 992
Motive-power, Apparatus for obtaining	R. N. Thomson	Edinburgh, Scotland	May 22, 1866	55, 030
Motive-power, Apparatus for obtaining and applying.	W. Huston	Wilmington, Del	Jan. 22, 1867	61, 339
Motive-power, Apparatus for producing	E. Stockton and W. O. St. John	Folsom, Cal	May 5, 1868	77, 674
Motive-power, Apparatus for the application of electro-magnetism as a.	S. Stimpson	New York	Sept. 12, 1838	910
Motive-power, Applying	J. Schley	Savannah, Ga	May 16, 1871	114, 866
Motive-power for the vertical rise and fall of the tide, Device for producing.	A. W. Scharit	Saint Louis, Mo	Jan. 10, 1865	45, 867
Motive-power, Combining springs for	G. Terry	New York, N. Y	Dec. 8, 1863	40, 866
Motive-power, Controlling	F. E. Sickels	New York, N. Y	Sept. 12, 1848	5, 765
Motive-power, Device for transmitting	R. T. Smith	Nashua, N. H	Jan. 28, 1873	135, 293
Motive-power engine	J. S. Johnstone	Ayr, North Britain	June 13, 1871	115, 963
Motive-power engine	J. Robertson	Glasgow, Scotland	Aug. 2, 1870	106, 078
Motive-power, Engine for employing steam, &c., under pressure to obtain.	J. W. Durham	Durhamville, Tenn	July 17, 1860	29, 149
Motive-power engines, Apparatus for forming an explosive mixture of air and hydrocarbon-vapor for use in.	J. Kidd	London, England	Mar. 12, 1867	62, 856
Motive-power engines, Recovering and regenerating steam in.	F. M. A. Motard	Paris, France	Aug. 19, 1873	142, 038
Motive-power for driving street-cars	P. E. McDonnell	Lyons, Ill	Aug. 23, 1870	106, 600
Motive-power for locomotion, &c	D. A. Pratt	Tremont, N. Y	Oct. 8, 1867	69, 699
Motive-power from petroleum and other oils, Obtaining.	R. D. McCreary	Oil City, Pa	Nov. 12, 1867	70, 732
Motive-power, Liquid used as a	J. C. Salomon, jr	Baltimore, Md	July 22, 1856	15, 391
Motive-power machine	H. Wickham, jr	Chicago, Ill	July 26, 1870	105, 870
Motive-power, Machine for obtaining	R. Lide	Union Street Borough, England.	Aug. 18, 1868	81, 302
Motive-power, Machinery for transmitting and distributing.	J. Richmond	Lockport, N. Y	Jan. 24, 1871	111, 251
Motive-power, Obtaining	P. Daniel	Franklin County, Ky	June 29, 1858	20, 701
Motive-power, Obtaining	R. M. Marchant	London, England	June 13, 1871	115, 877
Motive-power, Obtaining	J. Pringle	Summerhill, Pa	July 24, 1860	29, 303
Motive-power, Obtaining	W. M. Storm	Troy, N. Y	Feb. 4, 1851	7, 922
Motive-power, Pipe for transmitting pneumatic currents for.	R. Spear	New Haven, Conn	May 17, 1870	103, 252
Motive-power regulator	J. E. Gillespie	Boston, Mass	Nov. 19, 1867	71, 160
Motive power, Self	J. J. Girand	Baltimore, Md	Mar. 31, 1836	....
Motive-power, Spring	T. B. Fogarty	New York, N. Y	Jan. 7, 1873	134, 533
Motive-power, Spring	J. B. Howell	Wilkesbarre, Pa	Sept. 24, 1872	131, 614
Motive-power, Steam-enginery	A. C. Pilliner and J. C. Hill	Newport, England	Oct. 26, 1869	96, 262
Motive-power, Tidal	W. R. Close	Bangor, Me	Dec. 27, 1870	110, 436
Motive-power to machinery, Applying	R. R. Moffatt	Brooklyn, N. Y	Dec. 20, 1870	110, 268
Motive-power, Transmitting	T. Hanson	New York, N. Y	June 5, 1866	55, 289
Motive-power, Transmitting	J. Richmond	Lockport, N. Y	June 20, 1871	116, 221
Motive-power, Transmitting	R. S. Smith	Nashua, N. H	Oct. 23, 1866	59, 089
Motive-powers, Adaptation of substances as	F. M. Ruschhaupt	New York, N. Y	Mar. 20, 1860	27, 569

Index of patents issued from the United States Patent Office from 1790 to 1873, inclusive—Continued.

Invention.	Inventor.	Residence.	Date.	No.
Motor: *See* Air-motor. Atmospheric motor. Ball-motor. Churn-motor. Compressed-air motor. Electro-motor. Electro-magnetic motor. Fluid-motor. Grinding-motor. Hand-motor. Hydraulic motor. Hydro-pneumatic motor. Marine motor. Rotary motor. Rotary fluid-motor. Sewing-machine motor. Signal-alarm motor. Spring-motor. Steam-motor. Thermal motor. Tremolo-motor. Tydal motor. Water-motor.				
Motor	M. Comstock	Chicago, Ill	May 13, 1873	138, 737
Motor	L. H. Dean	Tecumseh, Mich	Dec. 16, 1873	145, 490
Motor	P. Guzman	Paris, France	Oct. 31, 1871	120, 379
Motor	C. L. Langdon	New Orleans, La	Mar. 19, 1872	124, 829
Motor	C. J. Schumacher	Portland, Me	Sept. 24, 1872	131, 631
Motor	L. E. Truesdell	Warren, Mass	Feb. 25, 1873	136, 286
Motor-engine	T. D. Richardson	North Providence, R. I	Apr. 15, 1873	137, 865
Motor for driving machinery	A. N. Proctor	Boston, Mass	Aug. 19, 1873	142, 041
Motor-machine	I. G. Hubbard	Nokomis, Ill	May 23, 1871	115, 058
Mouse and animal trap	G. L. Hart	New Britain, Conn	Mar. 21, 1871	112, 807
Mouse-trap	W. K. Bachman	Columbia, S. C	Sept. 27, 1870	107, 647
Mouse-trap	E. and A. Buckman	Brooklyn, N. Y	Oct. 11, 1870	108, 104
Mouse-trap	J. N. Bunnell	Unionville, Conn	May 4, 1869	89, 551
Mouse-trap	J. N. Bunnell	Unionville, Conn	Nov. 1, 1870	108, 876
Mouse-trap	A. G. Davis	Watertown, Conn	Mar. 30, 1869	88, 456
Mouse-trap	A. G. Davis	Watertown, Conn	June 22, 1869	91, 722
Mouse-trap	A. G. Davis and H. S. Frost	Watertown, Conn	July 5, 1870	105, 048
Mouse-trap	G. L. Hart	New Britain, Conn	Jan. 2, 1872	122, 318
Mouse-trap	H. C. Hart	Unionville, Conn	Dec. 20, 1870	110, 357
Mouse-trap	H. C. Hart	Unionville, Conn	Mar. 28, 1871	113, 051
Mouse-trap	A. A. Hotchkiss	Sharon, Conn	Apr. 5, 1870	101, 620
Mouse-trap	C. A. Hotchkiss	Bridgeport, Conn	Sept. 5, 1871	118, 721
Mouse-trap	A. Ovaitt	Unionville, Conn	July 4, 1871	116, 743
Mouse-trap	H. S. Weller	Watertown, Conn	May 9, 1871	114, 629
Mousing-hook	S. G. Coleman	Providence, R. I	May 1, 1860	28, 126
Mousing-hook	J. North	Middletown, Conn	Nov. 20, 1860	30, 687
Mousing-hook	H. Wynblad	West Hoboken, N. J	July 3, 1860	29, 029
Movable axle	G. N. Reynolds	Charleston, S. C	Apr. 29, 1819	....
Movable joint for tables, &c	H. N. House	Washington, D. C	May 16, 1865	47, 726
Movement and gas-burner combined, Mechanical	E. Omensetter	Philadelphia, Pa	Jan. 23, 1872	123, 042
Movement, Apparatus for producing intermittent rotative.	R. C. Reynolds	Lawrence, Mass	July 15, 1873	140, 951
Movement, Apparatus for producing plane or parallel.	A. G. Barrett	Barrett, Kans	Dec. 26, 1871	122, 145
Movement, Apparatus for treating diseases by mechanical.	A. L. Wood	New York, N. Y	Dec. 7, 1869	97, 744
Movement, Application of wind-power to produce a reciprocating.	C. C. Moore	New York, N. Y	Aug. 5, 1862	36, 101
Movement, Cam	J. E. Goodwin	Brewerton, N. Y	Sept. 30, 1873	143, 235
Movement-cure apparatus	D. Wark	Montreal, Canada	Aug. 1, 1871	117, 580
Movement, Electro-magnetic	F. L. Pope	Elizabeth, N. J	May 17, 1870	103, 077
Movement for actuating presses, Mechanical	P. C. Ingersoll	Green Point, N. Y	Dec. 28, 1869	98, 385
Movement for converting motion, Mechanical	A. B. Hendryx	Ansonia, Conn	Aug. 13, 1872	130, 371
Movement for converting oscillating into rotary motion, Mechanical.	J. Peabody	Bangor, Me	Jan. 28, 1873	135, 288
Movement for converting power into speed, Mechanical.	W. F. Goodwin	East New York, N. Y	Dec. 31, 1867	72, 842
Movement, Mechanical	E. Allen	Norwich, Conn	Oct. 22, 1867	70, 063
Movement, Mechanical	J. Armstrong	Toledo, Ohio	Sept. 2, 1873	142, 365
Movement, Mechanical	H. H. Baker	New Market, N. J	May 28, 1872	127, 294
Movement, Mechanical	W. H. Baker	Tamaqua, Pa	Nov. 15, 1859	26, 078
Movement, Mechanical	J. S. Barden	Providence, R. I	Aug. 25, 1868	81, 329
Movement, Mechanical	J. and W. F. Barnes	Rockford, Ill	May 9, 1871	114, 514
Movement, Mechanical	W. B. Bartram	Danbury, Conn	Aug. 27, 1872	130, 784
Movement, Mechanical	A. Benneckendorf	Hoboken, N. J	Oct. 10, 1871	119, 735
Movement, Mechanical	P. Bloomsburg, jr., and J. Molyneux.	Bordentown, N. J	May 28, 1867	65, 051
Movement, Mechanical	M. Bockman	Brooklyn, N. Y	Mar. 3, 1868	74, 979
Movement, Mechanical	N. Bradford	Addison, Me	June 7, 1870	103, 971
Movement, Mechanical	W. Brant	Paris, Ill	July 17, 1866	56, 354
Movement, Mechanical	A. W. Browne	Brooklyn, N. Y	Dec. 31, 1867	72, 792
Movement, Mechanical	A. W. Browne	Brooklyn, N. Y	Aug. 24, 1869	94, 068
Movement, Mechanical	A. W. Browne and W. F. Goodwin.	Brooklyn and New York, N. Y.	Aug. 18, 1868	81, 248
Movement, Mechanical	A. R. Buffington	United States Army	Dec. 8, 1868	84, 678
Movement, Mechanical	H. C. Burk	Mineral Point, Ohio	Jan. 7, 1868	72, 972
Movement, Mechanical	A. C. Burner	Green Bank, W. Va	May 28, 1872	127, 219
Movement, Mechanical	C. W. Carr	Paola, Kans	Jan. 28, 1873	135, 263
Movement, Mechanical	H. J. Case	Auburn, N. Y	Apr. 20, 1869	89, 026
Movement, Mechanical	E. Chapman	Rochester, Minn	Jan. 30, 1872	123, 237
Movement, Mechanical	A. Clark	Albany, Ill	Oct. 10, 1871	119, 821
Movement, Mechanical	T. J. and G. M. Clark	Higganum, Conn	Aug. 10, 1869	93, 598
Movement, Mechanical	I. A. Clippinger	Newton, Iowa	Oct. 31, 1865	50, 687
Movement, Mechanical	R. F. Cochran	Barry, Md	Feb. 27, 1872	124, 036
Movement, Mechanical	J. H. Cooper	Philadelphia, Pa	Oct. 3, 1871	119, 576
Movement, Mechanical	J. Corley	Near Coalfield, Kans	Apr. 25, 1871	114, 111
Movement, Mechanical	E. Courtright	Partello, Mich	Oct. 8, 1872	131, 940
Movement, Mechanical	J. H. Cranston	Norwich, Conn	Dec. 9, 1873	145, 402
Movement, Mechanical	F. Creamer	Brooklyn, N. Y	Jan. 19, 1869	86, 060
Movement, Mechanical	C. M. Currey	Pontiac, Mich	Dec. 24, 1867	72, 608
Movement, Mechanical	J. Daguier	Cincinnati, Ohio	July 8, 1873	140, 683
Movement, Mechanical	R. B. Davidson	Greenville, Ky	Sept. 15, 1863	39, 891
Movement, Mechanical	J. H. Davis	Woburn, Mass	Nov. 2, 1858	22, 003

Index of patents issued from the United States Patent Office from 1790 *to* 1873, *inclusive*—Continued.

Invention.	Inventor.	Residence.	Date.	No.
Movement, Mechanical	J. P. Davis	Stiles, Wis	May 5, 1868	77, 464
Movement, Mechanical	M. S. Davis	Vincennes, Ind	June 11, 1872	127, 855
Movement, Mechanical	J. B. Doolittle	Wallingford, Conn	June 3, 1873	139, 499
Movement, Mechanical	A. Duvall	Baltimore, Md	Feb. 9, 1869	86, 826
Movement, Mechanical	A. Eckert	Trenton, Ohio	Oct. 18, 1870	108, 340
Movement, Mechanical	R. Eickemeyer	Yonkers, N. Y	May 27, 1873	139, 379
Movement, Mechanical	W. H. Fenn	Rochester, N. Y	Mar. 14, 1871	112, 699
Movement, Mechanical	P. Ferguson and F. G. Bates	New Haven, Conn., and Springfield, Mass.	Nov. 14, 1871	120, 955
Movement, Mechanical	E. E. Furney	Chicopee, Mass	May 18, 1869	90, 254
Movement, Mechanical	D. T. Gale	Poughkeepsie, N. Y	Nov. 2, 1869	96, 419
Movement, Mechanical	W. Galladay	Sheboygan Falls, Wis	May 21, 1867	64, 858
Movement, Mechanical	W. Garrison	Clarkstown, N. Y	Mar. 15, 1870	100, 747
Movement, Mechanical	L. Goodall	Deering, N. H	Apr. 30, 1872	126, 284
Movement, Mechanical	W. F. Goodwin	East New York, N. Y	Mar. 17, 1868	75, 676
Movement, Mechanical	W. F. Goodwin	New York, N. Y	Aug. 18, 1868	81, 271
Movement, Mechanical	R. B. Hamel and J. B. Holden	Jersey City, N. J	May 17, 1870	103, 175
Movement, Mechanical	W. Hammill	Parma, Mich	June 21, 1870	104, 451
Movement, Mechanical	M. A. Hardy	Cambridge, Mass	Nov. 2, 1869	96, 316
Movement, Mechanical	W. M. Henderson	Philadelphia, Pa	Oct. 5, 1869	95, 586
Movement, Mechanical	A. B Hendryx	Ansonia, Conn	Aug. 13, 1872	130, 372
Movement, Mechanical	E. C. Hopping	Madison, N. J	Apr. 15, 1873	137, 924
Movement, Mechanical	H. R. Huling	Boston, Mass	Mar. 23, 1869	88, 040
Movement, Mechanical	C. W. Hurd	Comstock's Landing, N. Y.	Aug. 15, 1871	118, 016
Movement, Mechanical	J. James	Ogdensburgh, N. Y	Mar. 5, 1861	31, 606
Movement, Mechanical	J. M. Johnson	Northcutt's Store, N. Y.	July 6, 1869	92, 192
Movement, Mechanical	W. F. Jones	Easton, Kans	Nov. 14, 1871	120, 883
Movement, Mechanical	D. E. Keating	Oswego Falls, N. Y	Oct. 3, 1871	119, 617
Movement, Mechanical	J. J. Kimball	Naperville, Ill	Sept. 5, 1871	118, 723
Movement, Mechanical	M. Laemmel	Bay Ridge, N. Y	Dec. 1, 1868	84, 632
Movement, Mechanical	S. H. Lancaster	Le Claire, Iowa	Dec. 27, 1864	45, 616
Movement, Mechanical	C. B. Lewis	Clifton, Ohio	Mar. 20, 1866	53, 312
Movement, Mechanical	L. V. Lewis	Sun Prairie, Wis	Aug. 12, 1873	141, 800
Movement, Mechanical	E. Matteson	South Brooklyn, N. Y	June 20, 1365	48, 294
Movement, Mechanical	S. C. Matteson	Osceola, Wis	Nov. 14, 1865	50, 943
Movement, Mechanical	J. H. McCanny	Wytheville, Va	July 4, 1871	116, 617
Movement, Mechanical	J. H. McCanny	Wytheville, Va	Jan. 9, 1872	122, 530
Movement, Mechanical	W. C. McGill	Cincinnati, Ohio	Apr. 18, 1865	47, 315
Movement, Mechanical	W. C. McGill	Cincinnati, Ohio	Sept. 12, 1865	49, 905
Movement, Mechanical	W. S. Mead	New York, N. Y	Sept. 17, 1867	69, 009
Movement, Mechanical	C. Meiners	Indianapolis, Ind	Apr. 29, 1873	138, 424
Movement, Mechanical	E. Melton	Flemingsburgh, Ky	July 5, 1870	114, 977
Movement, Mechanical	H. Merriman	Bloomington, Ill	June 15, 1869	91, 251
Movement, Mechanical	D. S. Merritt	Mount Morris, Mich	Aug. 25, 1868	81, 393
Movement, Mechanical	G. R. Metten	Cleveland, Ohio	Oct. 6, 1868	82, 860
Movement, Mechanical	G. R. Metten	Cleveland, Ohio	Mar. 2, 1869	87, 353
Movement, Mechanical	J. Niebergall	New York, N. Y	Oct. 15, 1867	69, 929
Movement, Mechanical	A. Nimmo	Philadelphia, Pa	Apr. 4, 1871	113, 687
Movement, Mechanical	W. Oliver	Haynesville, La	Sept. 9, 1873	142, 578
Movement, Mechanical	I. E. Palmer	Hackensack, N. J	Sept. 22, 1868	82, 436
Movement, Mechanical	W. Park	Norwich, Conn	Jan. 14, 1873	134, 766
Movement, Mechanical	J. H. Pelton	Cleveland, Tenn	July 9, 1867	66, 623
Movement, Mechanical	T. H. Percival	Harper's Ferry, W. Va	Dec. 12, 1871	121, 893
Movement, Mechanical	O. Plummer	Worcester, Mass	Dec. 14, 1869	97, 959
Movement, Mechanical	N. Read	Winchendon, Mass	July 27, 1869	93, 004
Movement, Mechanical	G. Rich	Smithville, Mo	Apr. 28, 1868	77, 322
Movement, Mechanical	G. N. Rhodes	Hamilton, N. Y	Jan. 4, 1870	98, 632
Movement, Mechanical	E. O. Rood	Lodi, Ill	Sept. 1, 1868	81, 819
Movement, Mechanical	E. O. Rood	Lodi, Ill	Feb. 9, 1869	86, 692
Movement, Mechanical	E. G. Russell	Ravenna, Ohio	Mar. 8, 1871	112, 498
Movement, Mechanical	C. W. Saladee and W. Veach	Newark, Ohio	Sept. 25, 1866	58, 304
Movement, Mechanical	E. W. Sargent	Lowell, Mass	June 16, 1868	79, 013
Movement, Mechanical	C. B. Sawyer	Fitchburgh, Mass	Sept. 24, 1872	131, 566
Movement, Mechanical	L. Scofield	Watertown, Wis	Aug. 6, 1872	130, 158
Movement, Mechanical	J. M. Scott	Kinsman, Ohio	Dec. 31, 1867	72, 915
Movement, Mechanical	J. See	Mitchell, Ind	June 30, 1868	79, 503
Movement, Mechanical	J. A. Shanner	Plain View, Ill	Nov. 10, 1868	84, 007
Movement, Mechanical	H. F. Shaw	West Roxbury, Mass	Aug. 10, 1869	93, 485
Movement, Mechanical	A. Shear	Plymouth, Mich	May 16, 1871	114, 870
Movement, Mechanical	H. Shutts	Oregon, Mo	June 8, 1869	91, 048
Movement, Mechanical	H. Smith	Salem, Mass	Jan. 30, 1866	52, 334
Movement, Mechanical	P. J. Smith	Philadelphia, Pa	May 22, 1866	54, 972
Movement, Mechanical	C. D. Snell and J. W. Renney	Mechanic Falls, Me	Oct. 8, 1867	69, 719
Movement, Mechanical	E. Soper	New York, N. Y	Jan. 14, 1868	73, 401
Movement, Mechanical	M. F. Spore	Preble, N. Y	Mar. 17, 1868	75, 591
Movement, Mechanical	C. R. Squier	Cleveland, Ohio	May 7, 1872	126, 421
Movement, Mechanical	W. R. Swinnerton	Peoria, Ill	Feb. 25, 1868	74, 952
Movement, Mechanical	H. Thomas and R. Wallace	New York, N. Y	Sept. 21, 1869	95, 165
Movement, Mechanical	J. C. Vancleave	Hamburgh, Ark	Feb. 8, 1870	99, 611
Movement, Mechanical	A. Van Guysling	Albany, N. Y	Nov. 9, 1869	96, 746
Movement, Mechanical	J. H. Wait	Portsmouth, Ohio	May 15, 1860	28, 318
Movement, Mechanical	W. Walker	Odin, Ill	Dec. 13, 1870	110, 175
Movement, Mechanical	A. Wallace	Jacksonville, Fla	Jan. 2, 1872	122, 502
Movement, Mechanical	M. Wappich	Sacramento, Cal	Oct. 11, 1864	44, 681
Movement, Mechanical	A. G. Waterhouse	Portage, Wis	Mar. 25, 1873	137, 115
Movement, Mechanical	W. Weaver	Greenwich, N. Y	Oct. 3, 1871	119, 674
Movement, Mechanical	M. Webb	Chenango Forks, N. Y	Nov. 24, 1868	84, 451
Movement, Mechanical	J. H. Whitney	Rochester, Minn	Nov. 8, 1870	109, 163
Movement, Mechanical	J. Woolf	Burr Oak, Mich	Oct. 18, 1870	108, 547
Movement, Mechanical	J. Woolf	Burr Oak, Mich	Nov. 29, 1870	109, 790
Movement, Mechanical	J. Woolf	Burr Oak, Mich	Dec. 13, 1870	110, 185
Movement, Mechanical	J. Woolf	Burr Oak, Mich	Jan. 31, 1871	111, 503
Movement, Mechanical	J. Woolf	Burr Oak, Mich	May 30, 1871	115, 406
Movement, Mechanical	E. E. Young	York, Me	Aug. 29, 1871	118, 572
Movement, Mechanical	D. Zeigler	Lewistown, Pa	Sept. 13, 1870	107, 432
Movement, Mechanical	C. Zeitler	Benton's Port, Iowa	Mar. 19, 1872	124, 712
Movement or substitute for cog-wheels, Mechanical	G. H. Knight	Cincinnati, Ohio	Sept. 10, 1867	68, 754
Movement, Parallel	A. J. Vandegrift	Cincinnati, Ohio	Nov. 26, 1867	71, 342

Index of patents issued from the United States Patent Office from 1790 *to* 1873, *inclusive*—Continued.

Invention.	Inventor.	Residence.	Date.	No.
Movement, Pneumatic reciprocating	D. V. Wood	Ann Arbor, Mich	Jan. 18, 1870	98, 901
Movement, Rotary	W. H. Mitchell	San Francisco, Cal	Aug. 16, 1859	25, 132
Movement, Rotary mechanical	A. G. Waterhouse	San Francisco, Cal	Apr. 26, 1870	102, 454
Moving heavy bodies, Apparatus for	C. Whittaker	Milwaukee, Wis	July 14, 1868	79, 883
Moving heavy bodies, Device for	J. A. Woodworth	Hickory Corners, Mich	Apr. 20, 1869	89, 111
Moving-machine for the sick	J. C. Jenckes	Providence, R. I	May 29, 1823	....
Mower and hedge-trimmer	J. O. Taber	Salem, Ohio	Mar. 22, 1870	101, 660
Mower and reaper knife sharpener	W. B. Deuel	Ithaca, N. Y	Nov. 3, 1868	83, 767
Mower and reaper knives, Machine for grinding	H. Millard	New York, N. Y	Sept. 21, 1869	95, 033
Mower and reaper sickle sharpener	E. Dowd	Oshkosh, Wis	July 12, 1870	105, 184
Mower-cutters, Machine for grinding	H. C. Fisk	Wellsville, N. Y	Mar. 7, 1871	112, 330
Mower, Lawn	W. Allen	Worcester, Mass	July 11, 1871	116, 791
Mower, Lawn	W. Allen	Worcester, Mass	Nov. 18, 1873	144, 725
Mower, Lawn	J. Arbeiter	East Hartford, Conn	Apr. 20, 1869	89, 004
Mower, Lawn	J. Arbeiter	East Hartford, Conn	Mar. 15, 1870	100, 840
Mower, Lawn	D. H. Chamberlain	West Roxbury, Mass	Dec. 12, 1871	121, 705
Mower, Lawn	L. Chapman	Collinsville, Conn	Dec. 28, 1869	98, 348
Mower, Lawn	L. Chapman	Collinsville, Conn	July 16, 1872	129, 099
Mower, Lawn	C. M. Clinton	Ithaca, N. Y	Apr. 5, 1870	101, 585
Mower, Lawn	T. Coldwell and G. L. Chadborn	Newburgh, N. Y	July 26, 1870	105, 781
Mower, Lawn	T. Coldwell and G. L. Chadborn	Newburgh, N. Y	Mar. 18, 1873	136, 969
Mower, Lawn	W. H. Drake	Musconetcong, N. J	Nov. 22, 1870	109, 398
Mower, Lawn	N. Eaton	Woburn, Mass	Apr. 19, 1870	102, 101
Mower, Lawn	J. C. Field	Chicago, Ill	Mar. 15, 1870	100, 742
Mower, Lawn	G. W. Fischer	Dayton, Ohio	Apr. 23, 1872	126, 102
Mower, Lawn	E. R. Gard	Chicago, Ill	June 27, 1871	116, 295
Mower, Lawn	T. Garrick	Providence, R. I	May 11, 1869	89, 981
Mower, Lawn	H. W. Harkness	New Britain, Conn	Jan. 4, 1870	98, 590
Mower, Lawn	B. and D. H. Harnish	Lancaster and Pequea, Pa	Mar. 7, 1871	112, 338
Mower, Lawn	H. C. Hart	Unionville, Conn	Dec. 27, 1870	110, 457
Mower, Lawn	A. M. Hills	Hockanum, Conn	Jan. 28, 1868	73, 807
Mower, Lawn	A. M. Hills	Hockanum, Conn	June 20, 1871	116, 188
Mower, Lawn	A. Ingrham	Philadelphia, Pa	Feb. 28, 1871	112, 146
Mower, Lawn	S. D. King	Middletown, N. Y	Feb 4, 1873	135, 563
Mower, Lawn	S. D. King	Middletown, N. Y	Apr. 8, 1873	137, 691
Mower, Lawn	W. J. Lane	Millbrook, N. Y	Oct. 15, 1872	132, 160
Mower, Lawn	B. Merritt, jr	Newton, Mass	Apr. 20, 1869	89, 230
Mower, Lawn	J. S. Oakley and O. D. Wood	Passaic, N. J., and Newburgh, N. Y.	Nov. 14, 1871	120, 994
Mower, Lawn	A. J. Ohmer	Hamilton, Ohio	Dec. 20, 1870	110, 388
Mower, Lawn	E. G. Passmore	Philadelphia, Pa	Feb. 23, 1869	87, 286
Mower, Lawn	E. G. Passmore	Philadelphia, Pa	Apr. 4, 1871	113, 336
Mower, Lawn	L. Ross	Worcester, Mass	Dec. 27, 1870	110, 562
Mower, Lawn	S. W. Sears	New York, N. Y	Apr. 14, 1868	76, 831
Mower, Lawn	W. Sellers	Haverhill, Mass	Oct. 21, 1873	143, 787
Mower, Lawn	J. Shaw	Brooklyn, N. Y	Oct. 13, 1868	83, 101
Mower, Lawn	T. Loetbeer	Irvington, N. Y	June 10, 1873	139, 741
Mower, Lawn	B. L. Walker	Sing Sing, N. Y	Sept. 24, 1872	131, 726
Mower, Lawn	M. S. G. Wilde	Somerville, Mass	Feb. 16, 1869	86, 960
Mower, Lawn	A. W. C. Williams	London, England	May 2, 1871	114, 501
Mowers and reapers, Frame for combined	J. A. Moore and A. H. Patch	Louisville, Ky	June 30, 1857	17, 693
Mowers and reapers, Grinding the teeth of	D. Hinman	Berea, Ohio	Sept. 13, 1859	25, 412
Mowing and reaping machine	A. Amsden	Rochester, N. Y	May 26, 1857	17, 357
Mowing and reaping machine	W. Burgess	London, England	Sept. 18, 1855	13, 565
Mowing and reaping machine	J. Butter	Buffalo, N. Y	Feb. 7, 1860	27, 034
Mowing and reaping machine	V. M. Chafee	Xenia, Ohio	Mar. 20, 1860	27, 523
Mowing and reaping machine	R. Ketcham	South Dansville, N. Y	May 15, 1860	28, 283
Mowing and reaping machine	J. W. Mulley	Amsterdam, N. Y	Dec. 16, 1856	16, 247
Mowing and reaping machine	A. M. Wilson	Rhinebeck, N. Y	June 10, 1837	197
Mowing and reaping machine frame	M. G. Hubbard	Penn Yan, N. Y	July 15, 1856	15, 338
Mowing and reaping machine, Grain, grass, &c	A. M. Wilson	Rhinebeck, N. Y	May 15, 1837	196
Mowing and reaping machines, Cutting-apparatus for.	C. H. McCormick	Chicago, Ill	Nov. 5, 1861	33, 656
Mowing and reaping machines, Driver's seat for	F. Getz	Amherst, N. Y	Nov. 6, 1860	30, 572
Mowing and thrashing	E. Briggs and G. G. Carpenter	Fort Covington, N. Y	Feb. 5, 1836	....
Mowing grass and cutting grain, Machine for	A. S. Hathaway	Columbia, Me	July 1, 1856	15, 265
Mowing-machine	H. Adkins	Round Prairie, Ill	Jan. 15, 1850	7, 012
Mowing-machine	A. B. Allen	New York, N. Y	July 24, 1860	29, 228
Mowing-machine	H. Allen	Fayetteville, Tenn	June 2, 1836	....
Mowing-machine	W. Allen	Worcester, Mass	Mar. 24, 1868	75, 720
Mowing-machine	T. A. Anderson	McMinn County, Tenn	June 29, 1833	....
Mowing-machine	W. Anderson	Newton, N. J	Apr. 9, 1872	125, 429
Mowing-machine	B. Atwood	Stanstead, Canada	June 11, 1872	127, 730
Mowing-machine	C. Aultman and L. Miller	Canton, Ohio	June 17, 1856	15, 160
Mowing-machine	W. Bacheller	West Newbury, Mass	Aug. 11, 1857	17, 956
Mowing-machine	J. Baily	Chester County, Pa	Feb. 13, 1822	....
Mowing-machine	J. C. Baker	Mechanicsburgh, Ohio	Sept. 5, 1871	118, 673
Mowing-machine	E. Ball	Canton, Ohio	Aug. 12, 1856	15, 507
Mowing-machine	E. Ball	Canton, Ohio	Dec. 1, 1857	18, 788
Mowing-machine	A. Bank	Salem, Ohio	June 26, 1866	55, 905
Mowing-machine	J. F. Barrett	North Granville, N. Y	May 13, 1856	14, 898
Mowing-machine	S. S. Bartlett	Providence, R. I	May 16, 1865	47, 692
Mowing-machine	L. M. Batty	Canton, Ohio	Sept. 19, 1865	49, 962
Mowing-machine	L. D. Bidwell	Birmingham, Conn	Dec. 8, 1868	84, 790
Mowing-machine	A. Bolander	Akron, Ohio	Mar. 21, 1871	112, 892
Mowing-machine	S. P. Briggs	Saratoga Springs, N. Y	June 2, 1857	17, 417
Mowing-machine	A. Brown, L. G. Kniffen, and T. H. Dodge.	Worcester, Mass	Nov. 8, 1864	44, 935
Mowing-machine	C. B. Brown	Alton, Ill	Sept. 4, 1855	13, 517
Mowing-machine	J. J. and F. Bulfinch	Freeport and Waldoborough, Me.	Feb. 20, 1872	123, 864
Mowing-machine	C. Bullock	Jamestown, N. Y	June 5, 1860	28, 628
Mowing-machine	H. M. Burdick and T. V. Le Roy	Ilion, N. Y	Nov. 12, 1872	133, 013
Mowing-machine	O. H. Burdick and O. F. Dagget.	Auburn, N. Y	Apr. 2, 1872	125, 264
Mowing-machine	T. D. Burrall	Geneva, N. Y	Apr. 27, 1858	20, 035
Mowing-machine	G. E. Burt	Harvard, Mass	Mar. 3, 1868	74, 986
Mowing-machine	G. E. Burt	Harvard, Mass	Mar. 9, 1869	87, 539

Index of patents issued from the United States Patent Office from 1790 *to* 1873, *inclusive*—Continued.

Invention.	Inventor.	Residence.	Date.	No.
Mowing-machine	J. Butler	Buffalo, N. Y	June 8, 1858	20, 479
Mowing-machine	A. H. Caryl	Sandusky, Ohio	Sept. 15, 1857	18, 187
Mowing-machine	O. R. Chaplin	Saint Johnsbury, Vt	Jan. 10, 1860	26, 814
Mowing-machine	L. S. Clark	Bethel, Conn	Aug. 8, 1871	117, 864
Mowing-machine	T. J. and G. M. Clark	Higganum, Conn	Oct. 12, 1869	95, 769
Mowing-machine	S. Clarridge	Mount Sterling, Ohio	May 28, 1872	127, 229
Mowing-machine	C. Colahan	Cleveland, Ohio	July 15, 1873	140, 890
Mowing-machine	S. Comfort, jr	Morrisville, Pa	Mar. 18, 1856	14, 445
Mowing-machine	A. J. Cook	Delphi, Ind	Nov. 20, 1846	4, 861
Mowing-machine	E. Cope and T. Hoopes, jr	Chester County, Pa	May 18, 1825	......
Mowing-machine	W. Crook	New Hope, Pa	Apr. 13, 1858	19, 913
Mowing-machine	M. V. Cumming	Winthrop, Me	May 12, 1868	77, 719
Mowing-machine	A. D. Darby	White Creek, N. Y	Sept. 17, 1861	33, 296
Mowing-machine	T. H. Dodge	Washington, D. C	Nov. 15, 1859	26, 095
Mowing-machine	T. H. Dodge	Washington, D. C	Feb. 19, 1861	31, 520
Mowing-machine	G. C. Dolph	West Andover, Ohio	Sept. 8, 1859	18, 141
Mowing-machine	J. Drummond	Waterford, N. Y	June 30, 1836	......
Mowing-machine	J. G. Dunham	Raritan, N. J	Dec. 18, 1860	30, 961
Mowing-machine	R. Dutton	Dayton, Ohio	Mar. 19, 1861	31, 705
Mowing-machine	R. Dutton	Brooklyn, N. Y	Feb. 2, 1864	41, 460
Mowing-machine	R. Dutton	Brooklyn, N. Y	Mar. 15, 1864	41, 958
Mowing-machine	R. Emerson, jr., and F. Graham	Rockford, Ill	Jan. 14, 1862	34, 180
Mowing-machine	H. Fisher	Canton, Ohio	Jan. 12, 1858	19, 083
Mowing-machine	H. Fisher	Alliance, Ohio	May 7, 1861	32, 244
Mowing-machine	H. L. Frailey	Lancaster, Pa	Dec. 5, 1865	51, 301
Mowing-machine	A. Franklin, W. J. Hastings, and A. Gates.	Florence, Ind	Feb. 2, 1869	86, 523
Mowing-machine	S. Fuller	Somers, Conn	Mar. 14, 1871	112, 579
Mowing-machine	P. Gaillard	Lancaster, Pa	Dec. 4, 1812	......
Mowing-machine	A. Gale	Poughkeepsie, N. Y	July 25, 1854	11, 367
Mowing-machine	A. Gale	Poughkeepsie, N. Y	Aug. 11, 1857	17, 964
Mowing-machine	F. Gardiner	Gardiner, Me	July 31, 1860	29, 371
Mowing-machine	F. A. Geisler	Bristol, R. I	Dec. 29, 1868	85, 378
Mowing-machine	A. M. George	Nashua, N. H	May 1, 1860	28, 072
Mowing-machine	L. Gorden	Vanlue, Ohio	Mar. 19, 1872	124, 678
Mowing-machine	J. P. Greeley and L. W. Buxton	Nashua, N. H	July 21, 1863	39, 286
Mowing-machine	W. C. Greenleaf	Andover, Me	July 1, 1836	......
Mowing-machine	A. M. Hall	West Falmouth, Me	Dec. 23, 1856	16, 274
Mowing-machine	T. Harding	Springfield, Ohio	May 19, 1857	17, 350
Mowing-machine	D. K. and J. K. Harris	Allensville, Ind	Nov. 6, 1849	6, 846
Mowing-machine	R. Heath	West Newbury, Mass	Apr. 26, 1833	......
Mowing-machine	E. F. and J. Hennington	West Hoosick, N. Y	Apr. 9, 1861	31, 973
Mowing-machine	W. Heston	Bedford, Ohio	Feb. 6, 1872	123, 399
Mowing-machine	E. and D. Hinchley	Worcester, Mass	June 19, 1866	55, 654
Mowing-machine	E. C. Hopping	Madison, N. J	Dec. 23, 1873	145, 872
Mowing-machine	C. Howell	Cleveland, Ohio	Mar. 2, 1858	19, 504
Mowing-machine	M. G. Hubbard	Syracuse, N. Y	Oct. 11, 1864	44, 628
Mowing-machine	M. G. Hubbard	Syracuse, N. Y	Oct. 11, 1864	44, 629
Mowing-machine	M. G. Hubbard	Syracuse, N. Y	Nov. 22, 1864	45, 158
Mowing-machine	M. G. Hubbard	Syracuse, N. Y	Nov. 29, 1864	45, 246
Mowing-machine	O. Hussey	Baltimore, Md	July 5, 1859	24, 641
Mowing-machine	B. Illingworth	Le Roy, Minn	Nov. 15, 1870	109, 322
Mowing-machine	E. Ingersoll	Farmington, Mich	May 7, 1830	......
Mowing-machine	S. E. and M. P. Jackson	Booneville, N. Y	Dec. 29, 1857	18, 975
Mowing-machine	S. E. and M. P. Jackson	Booneville, N. Y	Dec. 29, 1857	18, 976
Mowing-machine	J. Jann	New Windsor, Md	Feb. 28, 1865	46, 566
Mowing-machine	G. W. Jennings	Boston, Mass	Mar. 12, 1861	31, 668
Mowing-machine	G. T. and M. Jerome	Mineola, N. Y	Sept. 28, 1858	21, 607
Mowing-machine	S. Johnston	Brockport, N. Y	June 27, 1871	116, 315
Mowing-machine	W. G. Kenyon	Wakefield, R. I	Oct. 12, 1869	95, 807
Mowing-machine	W. G. Kenyon	Wakefield, R. I	Mar. 12, 1872	124, 440
Mowing-machine	W. F. Ketchum	Buffalo, N. Y	May 27, 1856	14, 961
Mowing-machine	W. A. Kirby	Auburn, N. Y	Aug. 26, 1873	142, 111
Mowing-machine	L. G. Kniffen	Worcester, Mass	Dec. 24, 1861	33, 998
Mowing-machine	L. G. Kniffen	Worcester, Mass	Dec. 13, 1864	45, 416
Mowing-machine	W. H. Knight	East Machias, Me	Apr. 27, 1869	89, 412
Mowing-machine	S. Lamb	New York, N. Y	June 20, 1840	1, 645
Mowing-machine	I. Lard	Ashley, Mo	Nov. 20, 1846	4, 859
Mowing-machine	H. F. Lovejoy	Nineveh, N. Y	Nov. 7, 1865	50, 830
Mowing-machine	C. M. Lufkin	Ackworth, N. H	May 13, 1856	14, 874
Mowing-machine	J. J. Mann	Westville, Ind	Mar. 11, 1856	14, 404
Mowing-machine	J. S. Manning	Philadelphia, Pa	Jan. 22, 1856	14, 138
Mowing-machine	J. P. Manny	Rockford, Ill	Oct. 27, 1857	18, 510
Mowing-machine	J. P. Manny	Rockford, Ill	Mar. 25, 1862	34, 763
Mowing-machine	C. D. Mansfield	Lynn, Mass	Apr. 6, 1869	88, 650
Mowing-machine	H. Marcellus	Amsterdam, N. Y	May 4, 1858	20, 164
Mowing-machine	H. Marcellus	Amsterdam, N. Y	Sept. 4, 1860	29, 895
Mowing-machine	H. C. and D. C. Markham	Collinsville, N. Y	Mar. 7, 1871	112, 365
Mowing-machine	N. F. Mathewson	Barrington, R. I	May 5, 1868	77, 634
Mowing-machine	N. F. Mathewson	Barrington, R. I	Apr. 6, 1869	88, 653
Mowing-machine	H. Michaux	Hockanum, Conn	Mar. 4, 1873	136, 380
Mowing-machine	H. Michaux	Higganum, Conn	Sept. 23, 1873	143, 025
Mowing-machine	S. M. Moore	Beloit, Wis	Oct. 28, 1862	36, 850
Mowing-machine	A. Palmer	Brockport, N. Y	Dec. 26, 1865	51, 746
Mowing-machine	E. R. Pease	Poughkeepsie, N. Y	Mar. 12, 1861	31, 692
Mowing-machine	H. Pease	Brockport, N. Y	Jan. 8, 1856	14, 078
Mowing-machine	J. G. Perry	South Kingston, R. I	Mar. 21, 1865	46, 933
Mowing-machine	J. G. Perry	South Kingston, R. I	Mar. 6, 1866	53, 039
Mowing-machine	J. W. Pierce	Milbury, Mass	Mar. 17, 1868	75, 573
Mowing-machine	R. G. Pike	Middletown, Conn	Jan. 30, 1866	52, 317
Mowing-machine	J. Pine	Troy, N. Y	Oct. 11, 1864	44, 655
Mowing-machine	S. Ray and M. R. Shalters	Alliance, Ohio	Apr. 10, 1860	27, 869
Mowing-machine	C. Reed	Springfield, Ohio	July 3, 1866	56, 093
Mowing-machine	J. H. Rible	Dayton, Ohio	Apr. 2, 1861	31, 907
Mowing-machine	E. S. Ritchie	Brookline, Mass	May 13, 1862	35, 263
Mowing-machine	S. Rockafellow	Coatsville, Pa	July 3, 1855	13, 181
Mowing-machine	H. J. Ruggles	Poultney, Vt	Apr. 28, 1868	77, 407
Mowing-machine	F. Russell	Boston, Mass	Feb. 27, 1855	12, 463

Index of patents issued from the United States Patent Office from 1790 *to* 1873, *inclusive*—Continued.

Invention.	Inventor.	Residence.	Date.	No.
Mowing-machine	F. Russell	Boston, Mass	Mar. 20, 1855	12, 559
Mowing-machine	F. Russell	South Boston, Mass	Oct. 12, 1858	21, 777
Mowing-machine	G. T. Savery	Newburyport, Mass	Jan. 24, 1871	111, 258
Mowing-machine	W. and T. Schnebly	Hagerstown, Md	Aug. 22, 1833	....
Mowing-machine	W. H. Seymour	Brockport, N. Y	Nov. 23, 1869	97, 126
Mowing-machine	M. R. Shalters and S. Ray	Alliance, Ohio	Aug. 22, 1871	118, 284
Mowing-machine	G. F. Shaw	West Roxbury, Mass	Sept. 6, 1870	107, 108
Mowing-machine	H. F. Shaw	West Roxbury, Mass	Sept. 13, 1870	107, 296
Mowing-machine	J. W. Shipman	Springfield Centre, N. Y	July 24, 1860	29, 325
Mowing-machine	A. Soule	Yarmouth, Me	June 1, 1869	90, 697
Mowing-machine	W. S. Stone	New Philadelphia, Ohio	Aug. 19, 1873	142, 053
Mowing-machine	Z. Swope	Lancaster, Pa	Jan. 3, 1871	110, 799
Mowing-machine	J. Taggart	Roxbury, Mass	Mar. 17, 1857	16, 855
Mowing-machine	J. W. Thompson	Greenfield, Mass	July 15, 1856	15, 354
Mowing-machine	J. B. Tinker	Buffalo, N. Y	Dec. 5, 1865	51, 364
Mowing-machine	A. P. Trask and D. Aldrich	Ellington, N. Y	Oct. 16, 1839	1, 370
Mowing-machine	W. Van Auden	Poughkeepsie, N. Y	Sept. 27, 1864	44, 477
Mowing-machine	W. V. Walker	Dresserville, N. Y	Aug. 5, 1873	141, 611
Mowing-machine	J. B. Wardwell	Methuen, Mass	June 16, 1857	17, 597
Mowing-machine	F. W. Warner	East Haddam, Conn	July 31, 1860	29, 425
Mowing-machine	J. D. L. Watkins and R. Bryson.	Schenectady, N. Y	July 23, 1861	32, 908
Mowing-machine	A. Wemple	Chicago, Ill	Jan. 29, 1867	61, 644
Mowing-machine	R. W. Whitney and F. M. Hardison.	South Berwick, Me	Dec. 5, 1865	51, 396
Mowing-machine	J. D. Wilber	Poughkeepsie, N. Y	Feb. 10, 1863	37, 656
Mowing-machine	J. D. Wilber	Poughkeepsie, N. Y	Apr. 5, 1870	101, 554
Mowing-machine	A. M. Wilson	New York, N. Y	Sept. 3, 1846	4, 734
Mowing-machine	C. Wilson and W. Moore, jr	Yardleyville, Pa	May 29, 1855	12, 980
Mowing-machine	T. Windell	New Albany, Ind	Jan. 11, 1859	22, 608
Mowing-machine	T. H. Witherby	Worcester, Mass	Mar. 29, 1870	101, 338
Mowing-machine	J. E. Wood	Worcester, Mass	Mar. 17, 1863	37, 932
Mowing-machine	W. A. Wood	Hoosick Falls, N. Y	Feb. 22, 1859	23, 057
Mowing-machine	G. W. N. Yost	Corry, Pa	June 9, 1868	78, 712
Mowing-machine	G. W. N. Yost	Corry, Pa	Apr. 15, 1873	137, 817
Mowing-machine and harvester	J. H. Manny	Waddam's Grove, Ill	Sept. 23, 1851	8, 385
Mowing-machine and hay-spreader, Combined	J. G. Perry	Kingston, R. I	Apr. 13, 1869	88, 899
Mowing-machine and hay-tedder, Combined	J. H. Dater	Eagle Mills, N. Y	Feb. 22, 1870	100, 124
Mowing-machine blade	G. Stone	Beloit, Wis	Jan. 8, 1856	14, 070
Mowing-machine cutter	F. Russell	Manchester, N. H	Feb. 14, 1860	27, 159
Mowing-machine-cutter grinder	C. B. Curtis	Jordan, N. Y	Oct. 6, 1868	82, 808
Mowing-machine-cutter grinder	A. French	Philadelphia, Pa	June 2, 1868	78, 448
Mowing-machine-cutter grinder	D. W. Jameson	Warren, Ohio	Oct. 13, 1868	83, 067
Mowing-machine-cutter grinder	D. W. Jameson	Warren, Ohio	Aug. 31, 1869	94, 415
Mowing-machine-cutter grinder	S. W. and J. C. Palmer	Auburn, N. Y	Aug. 4, 1868	80, 761
Mowing-machine-cutter grinder	B. B. Snow and T. J. Dickerson.	Auburn, N. Y	Aug. 4, 1868	80, 571
Mowing-machine-cutter grinder	A. P. Thayer	Syracuse, N. Y	May 26, 1868	78, 401
Mowing-machine-cutter grinder	G. P. Yorke and W. H. Wilson	Westfield, N. Y	June 16, 1868	78, 909
Mowing-machine cutters, Clamp-bar for holding	M. C. Cronk	Auburn, N. Y	Mar. 30, 1869	88, 279
Mowing-machine cutters, Machine for sharpening	G. Sanford	Bergen Point, N. J	Oct. 20, 1868	83, 322
Mowing-machine cutters while being ground, Machine for holding.	D. W. Jameson	Warren, Ohio	Nov. 17, 1868	84, 191
Mowing-machine draft-pole	C. W. Cardot	Fredonia, N. Y	May 14, 1867	64, 746
Mowing-machine gearing	J. V. Strait	Litchfield, Ohio	Oct. 19, 1869	95, 950
Mowing machine, Grass, &c	P. Baker	Long Island, N. Y	Feb. 19, 1814	....
Mowing machine, Grass and grain	D. Lewis, jr	Bern, N. Y	Apr. 14, 1838	692
Mowing mach'ne, Grass, grain, &c	I. Wheeler	Salem, N. H	May 30, 1838	754
Mowing-machine guards, Manufacture of	A. Simonds	Fitchburgh, Mass	May 8, 1866	54, 652
Mowing machine, Hand	L. M. Dondwa	Amherst, N. H	Oct. 29, 1861	33, 578
Mowing machine, Hand	H. Fisher	Alliance, Ohio	May 5, 1863	38, 381
Mowing machine, Hand	G. W. Jennings	Boston, Mass	Mar. 28, 1865	47, 022
Mowing machine, Hand	J. M. Spencer	Enfield, Me	Sept. 4, 1860	29, 922
Mowing-machine-knife grinder	M. C. Cronk	Auburn, N. Y	June 16, 1868	78, 863
Mowing-machine-knife grinder	M. W. Knox	Sheridan, N. Y	Jan. 14, 1873	134, 809
Mowing-machine-knife grinder	D. F. Welsh	Nevada, Ohio	Nov. 8, 1870	109, 162
Mowing-machine knives, Machine for grinding	J. A. Thompson	Auburn, N. Y	June 16, 1868	79, 030
Mowing-machine knives while being ground, Clamp for.	H. J. Case	Auburn, N. Y	Aug. 4, 1868	80, 596
Mowing machine, Lawn	J. A. and H. A. House	Bridgeport, Conn	July 25, 1865	48, 947
Mowing-machine thill	O. A. Hillman	Providence, R. I	Dec. 10, 1872	133, 706
Mowing-machine track-cleaner	W. J. Andrews	Columbia, Tenn	July 2, 1872	128, 453
Mowing-machine track-cleaner	B. F. Power	McConnellsville, Ohio	Jan. 17, 1871	111, 083
Mowing-machine track-clearer	J. M. Davis	Saint Louis, Mo	Apr. 24, 1866	54, 124
Mowing-machine track-clearer	A. Marcellus	Amsterdam, N. Y	Dec. 29, 1857	18, 983
Mowing-machine track-clearer	A. Marcellus	Amsterdam, N. Y	Apr. 13, 1858	19, 936
Mowing-machine track-clearer	A. W. Morse	Eaton, N. Y	Jan. 21, 1862	34, 213
Mowing-machine track-clearer	P. A. Spicer	Marshall, Mich	Dec. 7, 1869	97, 563
Mowing-machine track-clearer	J. Timms	Malta, Ohio	Nov. 28, 1871	121, 300
Mowing-machines, Attaching draft-pole to	C. W. Cardot	Fredonia, N. Y	May 14, 1867	64, 745
Mowing-machines, Cutting-apparatus for	C. Bullock	Jamestown, N. Y	Dec. 8, 1857	18, 800
Mowing-machines, Cutting-apparatus for	M. G. Hubbard	Syracuse, N. Y	Sept. 13, 1864	44, 193
Mowing-machines, Cutting-apparatus for	H. F. Shaw	West Roxbury, Mass	June 14, 1870	104, 364
Mowing-machines, Device for sharpening the cutters of.	W. Tanner	Chicago, Ill	May 25, 1869	90, 407
Mowing-machines, Dividing-shoe for	W. A. Wood	Hoosick Falls, N. Y	June 24, 1856	15, 203
Mowing-machines, Friction-clutch for	G. S. Reynolds	Lebanon, N. H	Apr. 9, 1872	125, 409
Mowing-machines, Machine for grinding cutters of	H. Dodge	Washington Mills, N. Y	July 13, 1869	92, 431
Mowing-machines, Machine for grinding cutters of	M. Fowks	Leeds, N. Y	Jan. 19, 1869	86, 010
Mowing-machines, &c., Machine for grinding cutters of.	W. H. Laubach	Philadelphia, Pa	Feb. 18, 1868	74, 701
Mowing-machines, Machine for grinding cutters of	H. T. Phillips and H. W Leonard.	Auburn, N. Y	Aug. 18, 1868	81, 105
Mowing-machines, Machine for grinding cutters of	A. P. Thayer	Syracuse, N. Y	Nov. 16, 1869	97, 000
Mowing-machines, Machine for grinding cutters of	S. D. Wackman	Auburn, N. Y	June 2, 1868	78, 623
Mowing-machines, Mounting the cutters of	G. Hart	Dillsborough, Ind	Oct. 8, 1850	7, 700
Mucilage	A. Klozcewskie and D. Mindeleff.	Washington, D. C	Mar. 25, 1873	137, 213

Index of patents issued from the United States Patent Office from 1790 *to* 1873, *inclusive*—Continued.

Invention.	Inventor.	Residence.	Date.	No.
Mucilage and varnish bottle top	W. Burnet	New York, N. Y	July 2, 1867	66,213
Mucilage-bottle	H. O. Baker	New York, N. Y	Apr. 7, 1868	76,379
Mucilage-bottle	W. W. Beach	New York, N. Y	July 9, 1867	66,447
Mucilage-bottle	D. W. De Forrest	Brooklyn, N. Y	Apr. 21, 1868	77,008
Mucilage-bottle	B. De Long	Washington, D. C	Jan. 19, 1869	85,915
Mucilage-bottle	H. H. Ebaugh	Hereford, M. D	Mar. 17, 1868	75,672
Mucilage-bottle	A. M. Olds	New York, N. Y	June 25, 1867	66,036
Mucilage-bottle	D. A. Robinson, jr	Union Springs, N. Y	Feb. 2, 1869	86,453
Mucilage-bottle top	J. W. Boughton	Appleton, Wis	July 4, 1865	48,508
Mucilage-bottles, Attachment to	D. Bly	Macon, Ga	Apr. 16, 1867	63,841
Mucilage-bottles, Brush and top for	W. Burnet	New York, N. Y	Aug. 27, 1867	68,163
Mucilage-bottles, Cap for	T. N. Hickcox	Brooklyn, N. Y	Oct. 2, 1866	58,417
Mucilage-brush	G. D. Wilcox	Providence, R. I	Apr. 28, 1868	77,429
Mucilage-cup	J. E. Crosby	New York, N. Y	Dec. 2, 1873	145,155
Mucilage-holder	W. W. Beach	New York, N. Y	July 9, 1867	66,449
Mucilage-holder	C. C. Converse	Brooklyn, N. Y	Aug. 23, 1870	106,664
Mucilage-holder	A. H. Fatzinger	New York, N. Y	Mar. 12, 1872	124,429
Mucilage-holder	J. M. Keep	New York, N. Y	Mar. 8, 1870	100,640
Mucilage-pot	O. Dean	Richmond, Va	Oct. 1, 1867	69,415
Mucilage-stand	E. Morgan	Springfield, Mass	July 16, 1867	66,868
Mucilaginous compound	V. G. Bloede	Brooklyn, N. Y	Feb. 12, 1867	61,991
Mud and ore mill	J. Kellet	Elizabethport, N. J	Sept. 20, 1870	104,503
Mud-boat	A. D. Manley	Washington, Mich	Feb. 12, 1867	62,045
Mud, Floating machine for removing	S. Davis	Baltimore, Md	Jan. 22, 1817	....
Mud-machine	S. Savis	Baltimore, Md	Mar. 28, 1818	....
Mud-machine	J. Eveleth	Georgetown, D. C	Aug. 9, 1821	....
Mud-machine, Common	J. Hart	Middletown, Conn	Nov. 20, 1838	1,011
Mud-machine for beds of streams, rivers, &c	E. H. Holmes	Norwich, Conn	Nov. 18, 1830	....
Mud, Machine for raising	J. Eveleth	....	Feb. 21, 1806	....
Muff	L. Bauhoefer	Philadelphia, Pa	Aug. 20, 1867	67,841
Muff	J. W. Gay	Brooklyn, N. Y	Jan. 5, 1864	41,146
Muff	J. W. Gay	Brooklyn, N. Y	Mar. 1, 1864	41,766
Muff	C. Hollwede and J. Brzezinsky	New York, N. Y	Jan. 22, 1867	61,430
Muff	M. A. King	New York, N. Y	Dec. 10, 1867	72,049
Muff	M. A. King	New York, N. Y	Apr. 21, 1868	77,051
Muff	B. Levy	Boston, Mass	Sept. 28, 1869	95,240
Muff	M Mahler	Montreal, Canada	Apr. 3, 1866	53,760
Muff	J. Phillips	New York, N. Y	Sept. 27, 1859	25,582
Muff	R. M. Seldis	New York, N. Y	June 28, 1870	104,780
Muff	R. M. Seldis	New York, N. Y	Dec. 3, 1872	133,674
Muff	R. M. Seldis	New York, N. Y	Jan. 21, 1873	135,163
Muff	M. Stern and R. M. Seldis	New York, N. Y	June 15, 1869	91,382
Muff and collar box	H. Fowler	Detroit, Mich	Mar. 8, 1870	100,703
Muff and collar box	G. L. Jaeger	New York, N. Y	Nov. 15, 1870	109,323
Muff and collar box	J. P. Jones	Bloomfield, N. J	Oct. 26, 1869	96,118
Muff and collar box	H. Manneck	New York, N. Y	Sept. 12, 1871	118,866
Muff and porte-monnaie	H. Manneck and C. Strohmenger	New York, N. Y	Nov. 22, 1870	109,435
Muff-block, Expanding	C. F. Butterworth	Troy, N. Y	Dec. 21, 1869	98,025
Muff-box	R. S. Jennings	Philadelphia, Pa	May 14, 1872	126,636
Muff, collar, and cuff box	J. C. Budd	New York, N. Y	Nov. 1, 1870	108,755
Muff, Ear, cheek, and chin	W. P. Ware	Cincinnati, Ohio	July 6, 1858	20,834
Muff, Foot	W. M. Marshall	Philadelph a, Pa	Aug. 4, 1868	80,753
Muff-forming block	E. W. L. Bellander	Jersey City, N. J	Feb. 11, 1868	74,286
Muff, Ladies'	T. Eckert	New York, N. Y	June 5, 1866	55,260
Muff-lining	C. F. Butterworth	Troy, N. Y	June 17, 1873	139,942
Muff-stand	L. Bergtold	Buffalo, N. Y	Dec. 9, 1873	145,330
Mug, Shaving	F. B. Clock	Boston, Mass	Sept. 20, 1870	107,450
Mug, Shaving	A. J. Furr and W. C. Knaus	Boonsborough, Mo	Nov. 18, 1873	144,667
Mule-drums, Preparing groove for	J. Butterworth	Northern Liberties, Pa	Dec. 30, 1828	....
Mule-head	H. Hopkins	Providence, R. I	Feb. 22, 1833	....
Mule, Self-acting	H. Crossman	Providence, R. I	Sept. 24, 1867	69,188
Mule, Self-acting	A. G. Cunnock	Lowell, Mass	Sept. 12, 1865	49,860
Mule, Self-acting	L. C. Dam and J. Wigley	Lowell, Mass., and Milford, N. H.	Jan. 9, 1866	51,930
Mule, Self-acting	J. Harris	Lawrence, Mass	Nov. 6, 1855	13,752
Mule, Self-acting	R. Lakin and J. Wain	Ardwick and Manchester, England.	Mar. 22, 1864	42,048
Mule, Self-acting	W. Mason	Taunton, Mass	Oct. 3, 1846	4,779
Mule, Self-acting	R. Roberts	Manchester, England	Oct. 11, 1841	2,305
Mule, Self-acting	W. Rouse	Taunton Mass	Nov. 2, 1852	9,378
Mule, Self-acting	J. Sutherland	East Hampton, Mass	July 31, 1866	56,829
Mule, Self-acting	W. W. Wier and W. Grover	Chicopee and Holyoke, Mass.	May 27, 1856	14,992
Mules, Belt-shipping device for self-acting	H. Macon	Providence, R. I	May 2, 1871	114,310
Mules, Regulator for self-acting	E. C. Sanger	Salem, Mass	July 3, 1849	6,570
Mules, Roller-catch for self-acting	M. R. Pearson	Georgetown, Mass	Oct. 24, 1854	11,850
Mules, Winding-mechanism for self-acting	G. O. Wickers	Lawrence, Mass	Dec. 24, 1872	134,180
Multiple-tool	H. J. Harris	Shreveport, La	Sept. 13, 1870	107,255
Multiplying-camera	W. Southworth	Newcastle, Me	June 17, 1862	35,635
Multiplying motion on a single shaft, Device for	L. S. Fithian	New York, N. Y	Apr. 21, 1868	76,902
Multiplying numbers, Machine for	J. D. Smith	Brantingham, N. Y	Nov. 24, 1857	18,711
Multiplying power, Apparatus for	J. R. Dubois	Virginia City, Nev	Sept. 10, 1872	131,215
Multiplying revolutions around an axis, Device for	L. S. Fithian	New York, N. Y	Apr. 21, 1868	76,901
Music	N. A. Lete	New York, N. Y	June 11, 1818	....
Music, Apparatus for holding	A. A. Gaget	Paris, France	Nov. 10, 1857	18,581
Music, &c., Apparatus for teaching	L. A. Seward	New Orleans, La	Oct. 26, 1869	96,151
Music-binder	L. B. Covert	Brooklyn, N. Y	Mar. 30, 1869	88,369
Music-blackboard	O. B. Brown and H. G. Caroy	Malden, Mass	Dec. 16, 1873	145,622
Music-book	I. Ickes	Massillon, Ohio	Dec. 29, 1868	85,448
Music-books, Apparatus for turning leaves of	I. Gallup	Mystic Bridge, Conn	Apr. 17, 1855	12,724
Music, books, &c., Apparatus for turning leaves of	R. J. Stuart	Yonkers, N. Y	Feb. 27, 1872	124,172
Music-books, &c., Instrument for turning the leaves of	C. B. Thayer	Boston, Mass	Jan. 18, 1859	22,690
Music-box cylinders, Manufacture of	A. Frankfeld	New York, N. Y	Oct. 28, 1873	144,084
Music by means of a dumb instrument, Teaching	G. Geib	New York, N. Y	Aug. 30, 1819	....
Music by steam or compressed air, Apparatus for producing	J. E. Stoddard	Worcester, Mass	Oct. 9, 1855	13,668
Music, Composing	W. Fischer	Philadelphia, Pa	June 5, 1855	13,001

Index of patents issued from the United States Patent Office from 1790 *to* 1873, *inclusive*—Continued.

Invention.	Inventor.	Residence.	Date.	No.
Music, Keyed instrument of	F. Peabody	Salem, Mass	Nov. 10, 1863	40, 573
Music-leaf holder	A. Extein and J. C. Mills	Springfield, Mass	July 4, 1871	116, 577
Music-leaf turner	C. I. Adkins	Pataskala, Ohio	Apr. 2, 1872	125, 113
Music-leaf turner	A. Altenburg and G. J. Lambrix	Buffalo, N. Y	Oct. 10, 1871	119, 810
Music-leaf turner	C. P. Brown	Spring Lake, Mich	May 7, 1872	126, 516
Music-leaf turner	A. W. Bush and M. McComb	Saint Cloud, Minn	Aug. 1, 1871	117, 511
Music-leaf turner	A. B. Clark	Richmond, Ind	Jan. 1, 1867	60, 833
Music-leaf turner	G. C. A. Class	Chicago, Ill	Oct. 17, 1871	120, 038
Music-leaf turner	G. Downing	Weymouth, England	Jan. 28, 1873	135, 322
Music-leaf turner	A. M. Ernsberger	Indianapolis, Ind	Dec. 31, 1872	134, 367
Music-leaf turner	T. Folks	Melrose, N. Y	Dec. 31, 1872	134, 369
Music-leaf turner	J. H. Gerry	Boston, Mass	May 20, 1873	139, 141
Music-leaf turner	J. B. Geyser	Pittsburgh, Pa	May 6, 1873	138, 631
Music-leaf turner	T. Goodman	Providence, R. I	Feb. 4, 1868	73, 970
Music-leaf turner	N. P. Hall	Janesville, Wis	Aug. 12, 1873	141, 644
Music-leaf turner	C. Heyer	Racine, Wis	Sept. 12, 1871	118, 935
Music-leaf turner	G. N. Hopkins	Albion, N. Y	Jan. 9, 1866	51, 999
Music-leaf turner	J. W. Mellor	Philadelphia, Pa	Sept. 20, 1870	107, 519
Music-leaf turner	L. F. Morawetz	Baltimore, Md	Jan. 7, 1873	134, 557
Music-leaf turner	S. Nickelson	Gallatin, Tenn	Aug. 20, 1872	130, 588
Music-leaf turner	A. K. Noyes	Lynn, Mass	Dec. 15, 1868	84, 962
Music-leaf turner	E. B. Robinson	Portland, Me	July 26, 1870	105, 726
Music-leaf turner	G. Robb	McDonald, Pa	Dec. 24, 1872	134, 221
Music-leaf turner	A. Rosenfield	San Francisco, Cal	Apr. 15, 1873	137, 797
Music-leaf turner	W. Weaver	Phœnixville, Pa	Mar. 26, 1872	124, 924
Music-leaf turner	G. W. White	Brooklyn, N. Y	July 8, 1873	140, 664
Music, Machine for registering	H. F. Bond	Hudson, Wis	Nov. 29, 1859	26, 244
Music-notator	S. R. Brooks	Saint Louis, Mo	June 13, 1871	115, 930
Music-note block	C. J. and J. Costello	Kingston and New York, N. Y.	June 28, 1870	104, 833
Music, Notes in	W. C. Phillips	Lunenburgh Court-House, Va.	June 18, 1829	....
Music, Printing	A. Law		May 12, 1802	....
Music-portfolio	J. C. Koch	Brooklyn, N. Y	Jan. 21, 1873	135, 130
Music-portfolio	F. C. Schumann	New York, N. Y	Jan. 7, 1873	134, 613
Music-rack	C. S. Ely	Branford, Conn	Jan. 7, 1873	134, 530
Music-rack	J. R. Hill	Hyde Park, Mass	May 17, 1870	103, 046
Music-rack	A. Schäffer	Dubuque, Iowa	Apr. 20, 1869	89, 084
Music-rack	G. Tefft	Salem, N. Y	Apr. 7, 1868	76, 551
Music-rack	T. Ward	Birmingham, Pa	May 20, 1856	14, 935
Music rack or desk	F. X. Rizy	Monroe, Mich	Apr. 23, 1867	64, 148
Music-registering machine	H. B. Horton	Akron, Ohio	Dec. 18, 1855	13, 946
Music, Scale for instrumental	A. S. Smith	Rochester, N. Y	Oct. 21, 1856	15, 937
Music-sheets, Medium for turning	A. Krafft	Berlin, Prussia	Nov. 8, 1870	109, 023
Music-staff	V. C. Taylor	Des Moines, Iowa	Nov. 26, 1867	71, 550
Music-stand	O. Behr	Saint Joseph, Mo	July 16, 1872	129, 388
Music-stand	L. V. Brown	Salisbury, N. C	Sept. 13, 1870	107, 219
Music-stand	L. V. Brown	Salisbury, N. C	Aug. 6, 1872	130, 186
Music-stand	E. Conley	Cincinnati, Ohio	Nov. 30, 1869	97, 360
Music-stand	J. David	Brooklyn, N. Y	Oct. 17, 1865	50, 460
Music-stand	H. W. Holly	Stamford, Conn	Feb. 6, 1849	6, 094
Music-stand	W. C. James	Fishersville, N. H	Nov. 14, 1871	120, 880
Music-stand	C. Lehnert	West Roxbury, Mass	Sept. 27, 1870	107, 791
Music-stand	A. R. Nettleton	Unionville, Conn	Oct. 19, 1869	96, 026
Music-stand	J. Newton	Hanover, N. H	Dec. 20, 1870	110, 272
Music-stand	O. Schulze	Saint Louis, Mo	July 1, 1873	140, 547
Music-stand	A. W. White	Boston, Mass	Oct. 29, 1872	132, 704
Music-stand clamp	I. H. K. Downes	Charlestown, Mass	Nov. 19, 1872	133, 214
Music-stand, Portable	A. Iske	Lancaster, Pa	June 18, 1872	127, 978
Music-stand, Portable	D. M. White	Malden, Mass	May 12, 1868	77, 940
Music-stand standard	J. Thornton	Oswego, N. Y	Feb. 13, 1872	123, 594
Music-stands, Metallic frame for	W. M. Greenwood and B. Roux	Cincinnati, Ohio	Nov. 2, 1869	96, 424
Music, Teaching	J. C. Climo	Philadelphia, Pa	Sept. 24, 1867	69, 179
Music, &c., Temporary binder for	J. C. Clarke	Jersey City, N. J	Dec. 3, 1867	71, 703
Music to printed books, Arranging	M. Elliot	Boscawen, N. H	June 9, 1818	....
Music, Transposing scales for teaching	P. G. Bryan	Louisville, Ky	Nov. 8, 1870	109, 109
Music turner and holder, Sheet	F. J. Herpers and M. M. Sommer	Newark, N. J	Feb. 8, 1870	99, 568
Music, Turning the leaves of	I. M. Ross	Petersburgh, Ind	Nov. 2, 1869	96, 483
Music-type	E. L. Balch	Boston, Mass	June 16, 1868	78, 855
Music-type	U. K. Hill	New York, N. Y	Feb. 7, 1812	....
Music-type by combining printer's type	G. Bruce	New York	Nov. 27, 1830	....
Music, Writing	T. Harrison	Springfield, Ohio	Oct. 26, 1839	1, 383
Musical blackboard	J. M. Thomson and J. Cordley	Adrian, Mich	May 6, 1873	138, 714
Musical box	A. Paillard	St. Croix, Switzerland	Aug. 2, 1870	105, 972
Musical cage for animals	A. Suppernio	Millburn, N. J	Dec. 26, 1871	122, 138
[Musical.] Cornet and other wind instruments	W. C. Fritz	Boston, Mass	July 5, 1870	105, 059
Musical demonstrating-board	W. H. and G. W. Bowlsby	Monroe, Mich	July 12, 1864	43, 473
Musical dial	N. Petersen	Columbus, Miss	Apr. 30, 1867	64, 360
Musical instrument	J. D. Akin	Spartansburgh, Pa	June 1, 1858	20, 397
Musical instrument	T. Atkins and H. Drewer	Cincinnati, Ohio	July 11, 1871	116, 793
Musical instrument	M. Cannon	Salt Lake City, Utah	May 27, 1873	139, 293
Musical instrument	R. W. Carpenter	Brattleborough, Vt	Sept. 11, 1866	57, 861
Musical instrument	M. H. Collins	Chelsea, Mass	July 23, 1872	129, 653
Musical instrument	W. David	Tamaqua, Pa	Oct. 20, 1863	40, 333
Musical instrument	J. and E. Foster	Keene, N. H	Apr. 10, 1866	53, 803
Musical instrument	J. J. Hawkins		Oct. 24, 1800	....
Musical instrument	G. Herrick	Waverly, N. Y	Aug. 24, 1869	94, 108
Musical instrument	C. F. Hill	New York, N. Y	Dec. 21, 1869	98, 163
Musical instrument	U. C. and C. F. Hill	New York, N. Y	June 19, 1847	5, 164
Musical instrument	U. C. and C. F. Hill	Jersey City, N. J., and New York, N. Y.	Feb. 9, 1858	19, 296
Musical instrument	U. C. Hill and H. J. Newton	Jersey City, N. J., and New York, N. Y.	Feb. 28, 1860	27, 288
Musical instrument	A. F. Hunt and J. S. Bradish	Warren, Ohio	Jan. 9, 1849	6, 006
Musical instrument	W. Leigh	Bridgeport, Conn	Mar. 25, 1873	137, 219
Musical instrument	L. Marques	New York, N. Y	Mar. 20, 1866	53, 375
Musical instrument	H. T. Merrill	Galena, Ill	June 14, 1859	24, 396
Musical instrument	A. L. Mora	New York, N. Y	July 23, 1872	129, 746

Index of patents issued from the United States Patent Office from 1790 *to* 1873, *inclusive*—Continued.

Invention.	Inventor.	Residence.	Date.	No.
Musical instrument	J. F. Munson	Washington, D. C	May 23, 1865	47, 851
Musical instrument	F. Peabody	Salem, Mass	Oct. 10, 1865	50, 381
Musical instrument	J. W. Prescott	Concord, N. H	Apr. 17, 1849	6, 356
Musical instrument	A. Schoenhut	Philadelphia, Pa	Mar. 25, 1873	137, 243
Musical instrument	D. Schuyler	Buffalo, N. Y	July 3, 1866	56, 106
Musical instrument	K. Smith	Lowell, Mass	July 11, 1871	116, 879
Musical instrument	W. Standing, sr	Duquoin, Ill	July 2, 1872	128, 566
Musical instrument	F. Suter	Williamsburgh, N. Y	June 1, 1869	90, 793
Musical instrument	G. W. Van Dusen	Williamsburgh, N. Y	Dec. 10, 1867	72, 129
Musical instrument	W. Vogel	Norwich, Conn	Nov. 10, 1868	84, 027
Musical instrument	A. White	New Haven, Conn	May 8, 1860	28, 224
Musical instrument	J. and H. W. Whitney	Boston, Mass	Apr. 8, 1873	137, 643
Musical instrument	G. Woods	Cambridge, Mass	Jan. 3, 1865	45, 800
Musical instrument	G. Woods	Cambridge, Mass	Aug. 15, 1865	49, 483
Musical instrument	G. Woods	Cambridge, Mass	Sept. 11, 1866	58, 032
Musical instrument	A. Young	Norfolk, Va	Feb. 20, 1872	123, 969
Musical instrument, Automatic	H. Groves	New York, N. Y	Nov. 4, 1856	16, 009
Musical instrument, Electro-magnetic	L. Wesson	Chillicothe, Ohio	Apr. 4, 1865	47, 144
Musical instrument, Keyed	J. Alley and H. W. Poole	Newburyport and Worcester, Mass.	July 3, 1849	6, 565
Musical instrument, Keyed	H. C. Baudet	Paris, France	Oct. 16, 1866	58, 959
Musical instrument, Keyed	E. Hamlin	Winchester, Mass	July 11, 1871	116, 834
Musical instrument keys	E. C. Riley	New York	Aug. 22, 1833	....
Musical instrument, Military-band	R. H. Gates	Lancaster, Ohio	June 4, 1872	127, 591
Musical instrument, Military brass	B. F. Quinby	Boston, Mass	Apr. 9, 1872	125, 614
Musical instrument, Reed	C. Austin	Concord, N. H	Apr. 30, 1850	7, 317
Musical instrument, Reed	J. A. Bazin	Canton, Mass	Aug. 2, 1853	9, 892
Musical instrument, Reed	B. T. Blodget and H. B. Horton	Akron, Ohio	June 19, 1849	6, 531
Musical instrument, Reed	J. C. Briggs	Ansonia, Conn	Mar. 19, 1867	63, 006
Musical instrument, Reed	R. Burdett	Chicago, Ill	Nov. 6, 1866	59, 354
Musical instrument, Reed	R. Burditt	Brattleborough, Vt	Jan. 9, 1866	51, 994
Musical instrument, Reed	J. Foster	Keene, N. H	Aug. 29, 1871	118, 445
Musical instrument, Reed	L. K. Fuller and H. K. White	Brattleborough, Vt	Dec. 1, 1868	84, 486
Musical instrument, Reed	W. H. Gerrish	Boston, Mass	Aug. 2, 1870	105, 936
Musical instrument, Reed	F. A. Gleason	Rome, N. Y	June 20, 1854	11, 121
Musical instrument, Reed	E. Hamlin	Boston, Mass	May 27, 1856	14, 955
Musical instrument, Reed	E. P. Needham	New York, N. Y	Apr. 12, 1859	23, 601
Musical instrument, Reed	E. P. Needham	New York, N. Y	June 2, 1868	78, 536
Musical instrument, Reed	I. T. Packard	Chicago, Ill	May 1, 1866	54, 395
Musical instrument, Reed	I. T. Packard	Chicago, Ill	July 26, 1870	105, 718
Musical instrument, Reed	G. S. Shepard	Canaan, N. H	July 31, 1855	13, 365
Musical instrument, Reed	J. P. Sleeper	Worcester, Mass	Oct. 29, 1850	7, 747
Musical instrument, Reed	G. Woods	Boston, Mass	May 8, 1860	28, 230
Musical instrument, Reed and pipe	I. T. Packard	Chicago, Ill	Jan. 15, 1867	61, 241
Musical instrument, Rotary-valve wind	T. D. Paine	Smithfield, R. I	Nov. 14, 1848	5, 919
Musical-instrument sounding-board	C. Bogart	Charlestown, Mass	Dec. 9, 1851	8, 575
Musical-instrument sounding-board	J. H. Schucht	London, England	Dec. 26, 1871	122, 196
Musical instrument, Stringed	G. Schleicher	Mount Vernon, N. Y	Aug. 11, 1868	81, 012
Musical instrument, Stringed	G. L. Wild	Baltimore, Md	Sept. 5, 1854	11, 655
Musical instrument, Wind	C. H. Eisenbrandt	Baltimore, Md	Jan. 26, 1858	19, 187
Musical instrument, Wind	W. Fischer	New York, N. Y	Dec. 22, 1868	85, 222
Musical instrument, Wind	I. Fiske	Worcester, Mass	Apr. 29, 1873	138, 389
Musical instrument, Wind	I. Fiske	Worcester, Mass	Sept. 23, 1873	143, 134
Musical instrument, Wind	G. E. Parker	Newark, N. J	June 17, 1873	140, 069
Musical instrument, Wind	C. J. Van Oeckelen	New York, N. Y	Nov. 23, 1858	22, 139
Musical instruments, Automatic attachment for keyed.	E. D. Bootman	Janesville, Wis	Dec. 18, 1860	30, 955
Musical instruments, Automatic attachment for keyed.	H. Downes	New York, N. Y	May 30, 1871	115, 286
Musical instruments, Bellows for	J. Carhart	Buffalo, N. Y	Dec. 28, 1846	4, 912
Musical instruments, Bellows for	O. Follett	Pittsfield, Mass	Jan. 21, 1868	73, 592
Musical instruments, Bellows for reed	A. W. Wilcox	Providence, R. I	Apr. 28, 1868	77, 428
Musical instruments by means of bellows, Applying wind to.	P. L. and G. Grosh	Lancaster, Pa	Apr. 8, 1833	....
Musical instruments, Crook for	I. Fiske	Worcester, Mass	Nov. 12, 1867	70, 824
Musical instruments, Device for carrying off water from.	G. F. Dallon	Flatbush, N. Y	Apr. 21, 1863	38, 218
Musical instruments, Electro-pneumatic action for	W. F. and H. Schmoele, jr	Philadelphia, Pa	Dec. 16, 1873	145, 532
Musical instruments, Key-board attachment for	E. Tourjee	Providence, R. I	Nov. 26, 1867	71, 427
Musical instruments, Key-board for	W. D. Edgar	Ottawa, Kans	Sept. 26, 1871	119, 335
Musical instruments, Key-board for	E. Hamlin	Winchester, Mass	Dec. 12, 1871	121, 778
Musical instruments, Key-board for	G. B. Kirkham	New York, N. Y	Jan. 1, 1867	60, 744
Musical instruments, Key-coupling for	J. Berger	Knoxville, Ill	Jan. 7, 1868	73, 072
Musical instruments, Machine for cutting key-boards, &c., for.	D. E. Butler	Chesterfield, Ohio	Feb. 16, 1858	19, 345
Musical instruments, Making pivot-holes in keys of.	S. H. Wilder	Deep River, Conn	Apr. 2, 1872	125, 239
Musical instruments, Material for	H. Huebscher	Klingenthal, Germany	Aug. 20, 1872	130, 641
Musical instruments, Metallic reed for	C. N. Cutter	Worcester, Mass	June 30, 1868	79, 321
Musical instruments, Metallic reed for	C. N. Cutter	Worcester, Mass	June 30, 1868	79, 322
Musical instruments, Metallic reed for	A. H. Hammond	Worcester, Mass	Aug. 27, 1861	33, 143
Musical instruments, Metallic reed for	A. H. Hammond	Worcester, Mass	Feb. 2, 1869	86, 394
Musical instruments, Mute for	J. F. Stratton	New York, N. Y	Dec. 5, 1865	51, 363
Musical instruments, Operating swell in	G. Woods	Boston, Mass	Oct. 21, 1862	36, 756
Musical instruments, Pedal-attachment for	S. Schoenbrum	New York, N. Y	Feb. 27, 1872	124, 163
Musical instruments, Pedal for	H. Haas	New York, N. Y	Oct. 7, 1873	143, 509
Musical instruments, Preparing wood for	W. H. May	Bridgeport, Conn	Dec. 31, 1867	72, 877
Musical instruments, Reed-box for	M. F. Landfeur	Manchester, Conn	May 2, 1854	10, 848
Musical instruments, Reed for	J. C. Briggs	Woodbury, Conn	Oct. 21, 1856	15, 921
Musical instruments, Reed for	R. R. Martino	New York, N. Y	Jan. 19, 1869	86, 022
Musical instruments, Reed-stop for	A. P. Hughes	Philadelphia, Pa	Sept. 29, 1857	18, 284
Musical instruments, Sounding-board bridge for	G. Woods	Cambridgeport, Mass	May 6, 1873	138, 725
Musical instruments, Sounding-plate for	A. Schoenhut	Philadelphia, Pa	Oct. 28, 1873	144, 148
Musical instruments, String for	W. Randle	Florida, N. Y	Apr. 7, 1857	16, 995
Musical instruments, Swell for	T. Loud	Philadelphia, Pa	Aug. 1, 1865	49, 127
Musical instruments, Swell for	H. Woodbury	Buffalo, N. Y	Sept. 17, 1861	33, 322
Musical instruments, Tremolo-motor for	J. R. Lomas	New Haven, Conn	Nov. 28, 1871	121, 390
Musical instruments, Tremolo-regulator for reed	I. T. Packard	Chicago, Ill	Mar. 30, 1869	88, 328

Index of patents issued from the United States Patent Office from 1790 *to* 1873, *inclusive*—Continued.

Invention.	Inventor.	Residence.	Date.	No.
Musical instruments, Tuning-pin for	S. Clarke	New York, N. Y	Mar. 5, 1861	31, 589
Musical instruments, Valve for wind	C. H. Eisenbrandt	Baltimore, Md	July 4, 1854	11, 215
Musical instruments, Valve for wind	G. Hammer	Cincinnati, Ohio	Apr. 3, 1855	12, 628
Musical notation	F. W. Acee	Columbus, Ga	Dec. 26, 1871	122, 096
Musical notation	P. Phillips	Middletown Point, N. J	May 27, 1856	14, 970
Musical notation	P. Phillips	Matawan, N. J	Mar. 17, 1868	75, 572
Musical notation	J. P. Powell	Cincinnati, Ohio	July 22, 1873	141, 013
Musical notation	E. P. Stewart	Cotton Plant, Miss	Apr. 22, 1873	138, 104
Musical notation	E. Van Heeringen	Pickensville, Ala	June 12, 1849	6, 528
Musical picture-case	F. F. Kullrich	Berlin, Prussia	Oct. 19, 1869	95, 915
Musical pitch-pipe	W. H. Clark	Dayton, Ohio	Jan. 23, 1872	122, 879
Musical reed	J. C. Briggs	Woodbury, Conn	May 1, 1860	28, 060
Musical reed	N. B. Jewett	Worcester, Mass	Dec. 31, 1845	4, 339
Musical return-ball	J. Burke	Brooklyn, N. Y	Apr. 7, 1868	76, 396
Musical scale	W. E. Catlin	Wayne Township, Pa	Aug. 27, 1867	68, 287
Musical scale, Revolving	S. D. Tillman	Seneca Falls, N. Y	Nov. 8, 1853	10, 217
Musical scales, Representing	W. B. Billings	Eastport, Me	Sept. 10, 1850	7, 628
Musical staff	J. C. Cline	Philadelphia, Pa	Feb. 13, 1866	52, 534
Musical staff	H. Wright	Akron, Ohio	June 14, 1870	104, 393
Musical tablet	J. Branique	New York, N. Y	Nov. 26, 1867	71, 359
Musical-tone index	W. H. Clarke	Dayton, Ohio	July 2, 1872	128, 591
Musical transposing-board	J. M. Bruner	Roanoke, Ohio	May 30, 1871	115, 428
Musical transposing-index	J. L. Ober	Beverly Farms, Mass	Dec. 17, 1872	134, 001
Musician's hand exerciser	J. Monestier	Saint Denis, near Paris, France.	Mar. 30, 1858	19, 814
Musket and gun barrels, Altering the caliber of	C. E. Bailey	Springfield, Mass	Dec. 31, 1867	72, 777
Musket-balls, Apparatus for compressing	C. S. Bulkley	New York, N. Y	May 27, 1862	35, 411
Musket-balls, Machine for compressing	P. Naylor	New York, N. Y	Apr. 1, 1862	34, 844
Mustache-guard	M. C. Heptinstall	Enfield, N. C	June 29, 1869	91, 936
Mustache-guard	C. E. Mitchell and M. Moriarty	Bangor, Me	July 23, 1867	66, 985
Mustache-guard	E. J. F. Randolph	New York, N. Y	Feb. 20, 1872	123, 839
Mustache-guard for drinking-vessels	E. W. H. Bass	Quincy, Mass	Mar. 8, 1870	100, 586
Mustache-shield for cups	G. P. Cutler	Lawrence, Mass	June 28, 1870	104, 711
Mustard-box, &c	R. W. Betts	New York, N. Y	May 3, 1864	42, 549
Mustard-pot	T. M. Schleier	Knoxville, Tenn	Feb. 4, 1873	135, 595
Mustard, Preparation of flour of	S. Hopkins	Rahway, N. J	Aug. 19, 1813	....
Mustard, &c., Preparation of flour of	S. Hopkins	Philadelphia, Pa	May 31, 1817	....
Muzzle, Dog	H. Belmer and C. H. S. Schultz	Cincinnati, Ohio	July 23, 1867	67, 016
Muzzle, Dog	C. D. Quillfelat	New York, N. Y	Aug. 27, 1872	130, 846
Muzzle, Dog	H. Kaempf	Newark, N. J	July 6, 1869	92, 315
Muzzle for dogs and other animals	C. F. Schmidt	Williamsburgh, N. Y	Apr. 2, 1861	31, 911

N.

Nail:
See Barbed nail. Brass, copper, and composition nail. Button-nail. Capped nail. Carpet-nail. Coffin-nail. Copper sheathing-nail. Cut nail. Galvanized nail. Harness-nail. Horseshoe nail. Leather-work nail. Picture-nail. Screw-nail. Sheathing-nail. Shoe-nail. Siding-nail. Slating-nail. Tree nail. Trunk-nail. Upholstery-nail. Weather-tiling nail. Weather tiling and siding nail. Wrought nail.

Invention.	Inventor.	Residence.	Date.	No.
Nail	W. N. Hall	Mexia, Tex	Jan. 21, 1873	135, 114
Nail	H. A. Harvey	Orange, N. J	July 2, 1867	66, 331
Nail	W. E. Lockwood	Philadelphia, Pa	July 9, 1867	66, 601
Nail	B. T. Nichols	Roselle, N. J	Aug. 12, 1873	141, 810
Nail	M. Otis	East Bridgewater, Mass	Mar. 20, 1834	....
Nail	A. Patterson	Worcester, Mass	Aug. 16, 1870	106, 501
Nail and nail machine	W. Hunt	New York, N. Y	Nov. 12, 1839	1, 407
Nail and peg driver	F. O. Claflin	New York, N. Y	Dec. 21, 1869	98, 030
Nail and screw head	W. H. Nichols and F. D. Strong	East Hampton, Conn	June 11, 1861	32, 527
Nail and spike machine	H. Burden	Troy, N. Y	Dec. 2, 1834	....
Nail and spike, Grooved and flanged	T. W. Harvey	Jamestown, N. Y	Sept. 13, 1833	....
Nail and spike making machine	B. Allison and R. French		June 8, 1804	....
Nail and spike making machine	T. Wood	Matteawan, N. Y	Feb. 22, 1870	100, 230
Nail and spike pointing machine	W. Spink	Providence, R. I	July 5, 1859	24, 705
Nail and tack	R. Speer	Passaic, N. J	July 3, 1866	56, 114
Nail and tack cutting and heading machine	J. Reed	Marshfield, Mass	Apr. 22, 1825	....
Nail and tack feeding apparatus	J. Reed	Marshfield, Mass	Apr. 22, 1825	....
Nail and tack heading machine	L. Bolles	New York, N. Y	Sept. 19, 1822	....
Nail and tack machine	D. J. Farmer	Wheeling, W. Va	Aug. 22, 1871	118, 357
Nail and tack making machinery	O. L. Bassett, T. R. Bearse, and W. B. Wilber.	Taunton, Mass	Oct. 9, 1866	58, 577
Nail and tack machine feeder	W. Hunt	New York, N. Y	Oct. 12, 1843	3, 305
Nail and tack plate feeder	D. J. Farmer	Wheeling, W. Va	Apr. 4, 1871	113, 644
Nail-assorting apparatus	J. Coyne	Pittsburgh, Pa	June 3, 1873	139, 548
Nail-assorting machine	B. Berelander	Boston, Mass	May 13, 1873	138, 783
Nail-brad-cutting machine, Cylindrical vibrating	E. A. Lester	Baltimore, Md	Dec. 27, 1815	....
Nail-button	E. Kolben	New York, N. Y	Jan. 2, 1872	122, 468
Nail-clincher	J. Koyl	Rockford, Ill	Sept. 3, 1867	68, 567
Nail-clincher	J. Kunz	Napoleon, Ohio	Jan. 7, 1873	134, 680
Nail-cutter, Cylindrical	J. Perkins	Newburyport, Mass	Nov. 1, 1815	....
Nail cutters, Manufacture of barbed	E. Riley	Cleveland, Ohio	Dec. 19, 1871	121, 964
Nail cutting and heading machine	H. Hayes	Philadelphia, Pa	Feb. 24, 1816	....
Nail cutting and heading machine	J. Jelleff	Butternuts, N. Y	June 14, 1808	....
Nail cutting and heading machine	R. C. Lambert	Quincy, Mass	Feb. 27, 1872	124, 067
Nail cutting and heading machine	W. Leslie	....	Nov. 5, 1861	....
Nail cutting and heading machine	W. Miller	New York, N. Y	Sept. 26, 1810	....
Nail cutting and heading machine	T. Morgan	Hempstead, N. Y	Sept. 5, 1821	....
Nail cutting and heading machine	M. Otis	Boston, Mass	Dec. 31, 1817	....
Nail cutting and heading machine	N. Read	....	Jan. 8, 1798	....
Nail cutting and heading machine	J. Reed	Boston, Mass	Feb. 22, 1807	....

Index of patents issued from the United States Patent Office from 1790 *to* 1873, *inclusive*—Continued.

Invention.	Inventor.	Residence.	Date.	No.
Nail cutting and heading machine	J. Reed	Kingston, Mass	Sept. 16, 1810	....
Nail cutting and heading machine	M. and R. Reeve	Philadelphia, Pa	Nov. 14, 1811	....
Nail cutting and heading machine	B. R. Reid	Salem, Mass	Aug. 18, 1810	....
Nail cutting and heading machine	G. Thompson	Philadelphia, Pa	Sept. 15, 1821	....
Nail cutting and heading machine	F. Young	....	Aug. 23, 1800	....
Nail-cutting machine	F. J. Ayers	Roxbury, Mass	July 2, 1842	2, 698
Nail-cutting machine	W. H. Battelle	Youngstown, Ohio	Oct. 6, 1868	82, 679
Nail-cutting machine	E. Bigelow	Baltimore, Md	Apr. 1, 1807	....
Nail-cutting machine	L. Bissel	Hartwick, N. Y	June 27, 1809	....
Nail-cutting machine	E. Bless	Indianapolis, Ind	Jan. 3, 1871	110, 624
Nail-cutting machine	N. Boureau	....	Aug. 21, 1804	....
Nail-cutting machine	J. Coyne	Allegheny City, Pa	Aug. 31, 1869	94, 285
Nail-cutting machine	D. J. Farmer	Wheeling, W. Va	Nov. 1, 1870	108, 895
Nail-cutting machine	F. W. Geyssenhayner	....	Dec. 22, 1804	....
Nail-cutting machine	C. W. Glidden	Lynn, Mass	Feb. 11, 1873	135, 797
Nail-cutting machine	J. C. Gould	Oxford, N. J	Mar. 16, 1869	87, 924
Nail-cutting machine	M. and D. Hain	Hermann, Mo	Dec. 17, 1872	133, 935
Nail-cutting machine	J. Hicks	Hadley, Mass	Jan. 28, 1808	....
Nail-cutting machine	S. P. Hollis and H. B. Chest	Pittsburgh, Pa	Feb. 18, 1873	135, 915
Nail-cutting machine	C. D. Hunt	Fairhaven, Mass	Nov. 17, 1868	84, 189
Nail-cutting machine	C. D. Hunt	Fairhaven, Mass	Jan. 5, 1869	85, 665
Nail-cutting machine	C. D. Hunt	Fairhaven, Mass	Oct. 12, 1869	95, 802
Nail-cutting machine	T. M. Lawrence	Piqua, Ohio	Jan. 2, 1872	122, 324
Nail-cutting machine	H. Lutgen	Marietta, Pa	Sept. 8, 1813	....
Nail-cutting machine	F. B. Miles and J. Lawrence	Philadelphia, Pa	July 16, 1872	129, 156
Nail-cutting machine	R. Moore	Lycoming County, Pa	Jan. 13, 1814	....
Nail-cutting machine	J. G. Peerson	....	Mar. 23, 1794	....
Nail-cutting machine	J. Perkins	....	Jan. 16, 1795	....
Nail-cutting machine	B. R. Reid	Danvers, Mass	Mar. 8, 1810	....
Nail-cutting machine	E. Reynolds	Mansfield, Conn	May 2, 1865	47, 605
Nail-cutting machine	M. J. Rice	Birmingham, England	Feb. 16, 1869	87, 067
Nail-cutting machine	P. Richards	Philadelphia, Pa	Mar. 1, 1870	100, 447
Nail-cutting machine	W. A. Sweet	Syracuse, N. Y	Nov. 3, 1863	40, 514
Nail-cutting machine	M. Thibault	Hull, Canada	Dec. 16, 1873	145, 696
Nail-cutting machine	E. West	....	July 6, 1802	....
Nail-cutting machine	A. Whittemore	....	Nov. 19, 1796	....
Nail-cutting machine	W. Wickersham	Boston, Mass	June 26, 1860	28, 928
Nail-cutting machine, Vertical balance-wheel	E. A. Lester	Herkimer, N. Y	May 8, 1811	....
Nail-cutting machine, Vibrating	E. A. Lester	Herkimer, N. Y.	Apr. 2, 1814	....
Nail-cutting machines, Feeder for	C. Isbister	Allegheny, Pa	Dec. 31, 1844	3, 868
Nail-delivering machine	S. S. Putnam	Boston, Mass	Dec. 23, 1873	145, 755
Nail-distributer	A. Morrison	Harrisburgh, Pa	Nov. 11, 1873	144, 407
Nail-distributing machine	C. M. Chase	Boston, Mass	Jan. 14, 1873	134, 852
Nail-distributing machine	A. Knowlton	Boston, Mass	Aug. 22, 1871	118, 250
Nail-drawer	S. Marden	Newton, Mass	Dec. 24, 1867	72, 655
Nail-drawing implement	H. W. Holly	Norwich, Conn	Jan. 14, 1868	73, 330
Nail-drawing instrument	E. P. Whitney	Stamford, Conn	Apr. 28, 1868	77, 427
Nail-drawing tool	D. T. Robinson	Boston, Mass	Nov. 13, 1866	59, 657
Nail-driving instrument	S. P. Carpenter	Milford, Mass	July 13, 1852	9, 112
Nail-driving machine	D. L. Falardo	New York, N. Y	June 4, 1867	65, 360
Nail-driving machine	S. W. Phelps	Sandusky, Ohio	Oct. 25, 1870	108, 624
Nail-driving machine	A. Smith	Perrysburgh, Ohio	Dec. 23, 1873	145, 818
Nail-extracting implement	D. Morris	Bartlett, Ohio	Feb. 11, 1868	74, 401
Nail-extractor	J. B. Breathitt	Cooper County, Mo	June 30, 1868	79, 307
Nail-extractor	G. J. Capewell	Cheshire, Conn	July 16, 1872	129, 210
Nail-extractor	G. J. Capewell	Cheshire, Conn	Oct. 7, 1873	143, 496
Nail-extractor	I. F. Davis	Rockford, Ill	Aug. 10, 1869	93, 522
Nail-extractor	J. Edson	Boston, Mass	Jan. 21, 1868	73, 587
Nail-extractor	E. A. Franklin	Brenham, Tex	May 14, 1872	126, 686
Nail-extractor	J. H. Hogan	Saint Louis, Mo	May 29, 1866	55, 105
Nail-extractor	H. Jeffrey	Saint Charles, Mo	Aug. 6, 1867	67, 552
Nail-extractor	W. Knaus	Otterville, Mo	June 15, 1869	91, 452
Nail-extractor	A. Marden and A. H. Burgess	Philadelphia, Pa	June 25, 1867	66, 032
Nail-extractor	R. McConnell	Lawrenceville, Pa	Nov. 7, 1865	50, 834
Nail-extractor	D. R. Morland and C. Blocher	Broad Ford, Pa	Apr. 15, 1873	137, 945
Nail-extractor	I. A. Pinnell	Boonville, Mo	June 2, 1868	78, 479
Nail-extractor	G. C. Taft	Worcester, Mass	Sept. 6, 1870	107, 121
Nail-extractor	G. C. Taft	Worcester, Mass	Dec. 30, 1873	145, 975
Nail-extractor	J. Tyzick	Saint John, New Brunswick	June 9, 1868	78, 699
Nail-extractor	J. Tyzick and H. W. Eskildson	New York, N. Y	Sept. 29, 1868	82, 568
Nail-extractor	W. G. Ward	Fayette, N. Y	Dec. 13, 1870	110, 176
Nail-forging machine	S. S. Putnam	Boston, Mass	Aug. 17, 1858	21, 213
Nail-forging machine	S. J. Seely	New York, N. Y	Aug. 4, 1857	17, 941
Nail-forging machine, Wrought	H. W. Mather	Deep River, Conn	Apr. 6. 1869	88, 652
Nail-head-plating machine	W. H. Van Gilson	Newark, N. J	Nov. 30, 1858	22, 211
Nail heads, Machine for cutting metal caps for	Z. Walsh	Newark, N. J	Oct. 6, 1857	18, 372
Nail heading and cutting machine	I. Garrettson	....	Nov. 16, 1796	....
Nail-heading machine	E. Bigelow	Massachusetts	Jan. 29, 1817	....
Nail-heading machine	A. Chandler and S. Shepard	Taunton, Mass	Sept. 16, 1808	....
Nail-heading machine	S. Rogers	Plymouth, Mass	Dec. 15, 1807	....
Nail-heading machine	S. Rogers	Bridgewater, Mass	Jan. 26, 1814	....
Nail-heading machine	H. Sedam	Troy, N. Y	Sept. 6, 1815	....
Nail-heading machine	B. S. Walcott	....	Sept. 4, 1802	....
Nail-machine	J. M. Allen	Cambridge, Mass	Nov. 5, 1867	70, 496
Nail-machine	E. Bartholomew and W. Church	Boston, Mass	Aug. 18, 1815	....
Nail-machine	S., sr., and S. Briggs, jr	....	Aug. 2, 1791	....
Nail-machine	W. Caruthers	....	Dec. 13, 1802	....
Nail-machine	F. Davison	Richmond, Va	Apr. 28, 1868	77, 262
Nail-machine	F. Davison	Richmond, Va	Aug. 31, 1869	94, 290
Nail-machine	F. Davison	Richmond, Va	July 21, 1868	80, 150
Nail-machine	D. Dodge	Keesville, N. Y	June 3, 1856	15, 054
Nail-machine	D. Dodge	Keesville, N. Y	Aug. 23, 1859	25, 183
Nail-machine	D. Dodge	Keesville, N. Y	Aug. 30, 1859	25, 309
Nail-machine	J. Elgar	York, Pa	June 21, 1815	....
Nail-machine	N. Elgar	York, Pa	June 19, 1815	....
Nail-machine	J., sr., and J. Elgar	Yorktown, Pa	Sept. 1, 1809	....
Nail-machine	D. J. Farmer	Wheeling, W. Va	May 31, 1870	103, 730
Nail-machine	W. H. Field	Taunton, Mass	May 26, 1863	38, 665

Index of patents issued from the United States Patent Office from 1790 *to* 1873, *inclusive*—Continued.

Invention.	Inventor.	Residence.	Date.	No.
Nail-machine	N. Fobes		Aug. 2, 1802	....
Nail-machine	F. A. Gleason	Brooklyn, N. Y	Jan. 5, 1869	85, 522
Nail-machine	S. L. Gould	Skowhegan, Me	Oct. 9, 1866	58, 637
Nail-machine	R. C. Grant	Middleport, Ohio	May 5, 1868	77, 604
Nail-machine	H. Greene and W. J. Jordon	Philadelphia, Pa	Apr. 27, 1858	20, 126
Nail-machine	E. G. Hall	New York, N. Y	May 7, 1861	32, 263
Nail-machine	S. Hart		Jan. 4, 1799	....
Nail-machine	J. Kello and J. Carter	Philadelphia, Pa	June 10, 1817	....
Nail-machine	J. S. King	Raynham, Mass	Oct. 20, 1857	18, 457
Nail-machine	J. B. Kingham	East Bridgewater, Mass	Oct. 11, 1864	44, 637
Nail-machine	J. Krauser	Philadelphia, Pa	Sept. 23, 1826	....
Nail-machine	J. L. Krauser	Reading, Pa	May 18, 1858	20, 312
Nail-machine	J. L. Krauser	Philadelphia, Pa	Aug. 6, 1861	33, 018
Nail-machine	R. Lario	Reading, Pa	May 31, 1864	42, 993
Nail-machine	H. McCluer	Homer, N. Y	Apr. 18, 1817	....
Nail-machine	M. McManus	Brooklyn, N. Y	June 28, 1864	43, 324
Nail-machine	C. H. Merrick	Pittsburgh, Pa	Sept. 12, 1865	49, 906
Nail-machine	A. V. B. Orr	Lancaster, Pa	Oct. 25, 1859	25, 910
Nail-machine	G. Osborn	Lakeville, Mass	July 27, 1869	92, 994
Nail-machine	A. Owen	Buffalo, N. Y	Feb. 12, 1861	31, 397
Nail-machine	A. W. Paull and J. Morgan, jr	Wheeling, W. Va	July 4, 1871	116, 747
Nail-machine	C. Philips, jr	Sackett's Harbor, N. Y	Oct. 21, 1815	....
Nail-machine	A. M. Polsey	Boston, Mass	Nov. 27, 1866	60, 056
Nail-machine	A. M. Polsey	Boston, Mass	Mar. 5, 1867	62, 682
Nail-machine	H. Reese	Baltimore, Md	Feb. 28, 1871	112, 280
Nail-machine	H. Reese	Baltimore, Md	Sept. 5, 1871	118, 744
Nail-machine	M. Reeve	Philadelphia, Pa	June 27, 1816	....
Nail-machine	M. and R. Reeve	Philadelphia, Pa	Sept. 28, 1813	....
Nail-machine	S. G. Reynolds	Worcester, Mass	Jan. 20, 1852	8, 677
Nail-machine	S. G. Reynolds	Bristol, R. I	Sept. 18, 1866	58, 137
Nail-machine	S. G. Reynolds	Bristol, R. I	Apr. 23, 1867	64, 035
Nail-machine	B. Robinson	Boston, Mass	Dec. 17, 1867	72, 230
Nail-machine	M. T. Rounds	Taunton, Mass	Oct. 11, 1864	44, 661
Nail-machine	J. Russell	Brooklyn, N. Y	Aug. 29, 1865	49, 653
Nail-machine	J. Russell	Brooklyn, N. Y	Apr. 9, 1867	63, 655
Nail-machine	G. W. Sargent	New York, N. Y	Mar. 31, 1868	76, 104
Nail-machine	H. Scheuerle	New York, N. Y	May 24, 1870	103, 376
Nail-machine	E. W. Scott and A. M. George	Lowell, Mass., and Nashua, N. H.	June 9, 1857	17, 523
Nail-machine	H. W. Taylor	Birmingham, Pa	July 6, 1858	20, 829
Nail-machine	R. Ward	Alder Creek, N. Y	May 22, 1866	54, 981
Nail-machine	W. Waring and C. Philips, jr	Jefferson, N. Y	Oct. 21, 1815	....
Nail-machine	W. Wickersham	Boston, Mass	Feb. 6, 1866	52, 479
Nail-machine	G. B. Wiggin and J. W. Hoard	Providence, R. I	Oct. 7, 1862	36, 629
Nail-machine	P. A. Wilbur	Newcastle, Pa	Oct. 14, 1856	15, 910
Nail-machine	A. P. Winslow	Cleveland, Ohio	Oct. 11, 1870	108, 225
Nail-machine	J. Wooton	Boonton, N. J	Aug. 29, 1854	11, 628
Nail-machine, Compound-lever and parallel-shaft	R. Turner	Boston, Mass	Nov. 25, 1811	....
Nail machine, Cut	T. H. Barlow	Lexington, Ky	June 20, 1854	11, 114
Nail machine, Cut	J. Berry	New York, N. Y	Feb. 14, 1860	27, 096
Nail machine, Cut	F. Davison	Liberty, Va	Oct. 31, 1871	120, 501
Nail machine, Cut	D. J. Farmer	Wheeling, W. Va	Aug. 22, 1871	118, 355
Nail machine, Cut	D. J. Farmer	Wheeling, W. Va	Aug. 22, 1871	118, 356
Nail machine, Cut	S. K. Jones and G. H. Snow	New Haven, Conn	May 11, 1869	90, 003
Nail machine, Cut	J. B. Kingham	Dorchester, Mass	May 18, 1869	90, 108
Nail machine, Cut	G. C. Goodhaus	Jamestown, Ohio	Mar. 16, 1858	19, 631
Nail machine, Cut	A. W. Paull and J. Morgan, jr	Wheeling, W. Va	June 20, 1871	116, 091
Nail machine, Cut	L. Richards	Providence, R. I	Dec. 21, 1869	98, 109
Nail machine, Cut	J. P. Sherwood	Fort Edward, N. Y	Dec. 19, 1854	12, 103
Nail machine, Cut	J. E. Sweet and J. B. Elliott	Syracuse and New York, N. Y.	Dec. 22, 1868	85, 145
Nail machine, Cylindrical wrought	C. I. Richards	New York, N. Y	Oct. 17, 1848	5, 853
Nail-machine, Eccentric-roller	J. Fairbanks	Boston, Mass	Aug. 14, 1810	....
Nail-machine feeder	H. B. Landers	Williamsburgh, N. Y	Apr. 4, 1871	113, 310
Nail-machine-feeding apparatus	J. Nolan	Oxford, N. J	Apr. 5, 1870	101, 496
Nail-machine-feeding apparatus	U. S. Wolff	Burrel Township, Pa	Apr. 19, 1870	102, 073
Nail-machine-feeding machinery	T. Odiorne	Malden, Mass	Apr. 3, 1829	....
Nail-machine-grinding die	G. B. Wiggin and J. W. Hoard	Providence, R. I	Feb. 10, 1863	37, 657
Nail-machine lubricator	L. A. Dodge	Keeseville, N. Y	June 30, 1868	79, 326
Nail-machine, Self-feeding	T. Morgan	Paterson, N. J	June 17, 1824	....
Nail-machine toggle-pin	W. H. Ridley	Pittsburgh, Pa	May 16, 1871	114, 857
Nail machine, Wrought	D. Armstrong	Chicago, Ill	May 30, 1871	115, 267
Nail machine, Wrought	C. Clareni	New York, N. Y	July 5, 1859	24, 616
Nail machine, Wrought	J. Nichols	Rhode Island	July 11, 1807	....
Nail machine, Wrought	L. Norcross	Livermore, Me	May 18, 1824	....
Nail machine, Wrought	A. V. B. Orr and G. Bantz	Frederick, Md	Dec. 7, 1858	22, 238
Nail machine, Wrought	H. Reese	Baltimore, Md	Nov. 26, 1872	133, 486
Nail machine, Wrought	J. P. Sherwood	Fort Edward, N. Y	May 10, 1853	9, 712
Nail machine, Wrought	T. Somerby	Wells, Me	Jan. 22, 1840	1, 477
Nail machine, Wrought	J. Taggart	Boston, Mass	Apr. 6, 1869	88, 753
Nail-machines, Grinding the upper cutters of	G. B. Wiggin and J. W. Hoard	Providence, R. I	May 5, 1863	38, 439
Nail-machines, Nippers for	S. Chubbuck	Wareham, Mass	Jan. 13, 1835	....
Nail-machines, Swedge for forming cutters for	E. Bless	Indianapolis, Ind	June 6, 1871	115, 563
Nail-machinery	T. Fowler	Seymour, Conn	May 21, 1867	64, 963
Nail-machinery	J. Holcomb	Brandon, Vt	July 14, 1846	4, 634
Nail machinery, Wrought	D. Dodge	Keeseville, N. Y	June 22, 1852	9, 051
Nail machinery, Wrought	W. M. Gooding	Newark, N. J	Mar. 28, 1848	5, 489
Nail machinery, Wrought	A. W. Gray	Middletown, Vt	May 16, 1846	4, 525
Nail-plate-cutting machine	T. Searls	Pottstown, Pa	Dec. 3, 1872	133, 673
Nail-plate feeder	P. S. Bradford	Bridgewater, Mass	May 17, 1864	42, 742
Nail-plate feeder	W. Diehl	Norristown, Pa	Apr. 10, 1849	6, 291
Nail-plate feeder	D. Drawbaugh	Eberle's Mills, Pa	Dec. 12, 1865	51, 435
Nail-plate feeder	D. Drawbaugh	Eberle's Mill, Pa	Nov. 19, 1867	71, 148
Nail-plate feeder	J. Ferguson and J. Turner	Bridgewater, Mass	Aug. 22, 1871	118, 224
Nail-plate feeder	J. S. Fisk	Youngstown, Ohio	June 30, 1863	39, 105
Nail-plate feeder	J. C. Gould	Boonton, N. J	May 12, 1857	17, 273
Nail-plate feeder	A. Heddaeus	Pittsburgh, Pa	Aug. 12, 1856	15, 515
Nail-plate feeder	J. W. Hoard and T. A. Searle	Providence, R. I	Aug. 17, 1858	21, 198

Index of patents issued from the United States Patent Office from 1790 *to* 1873, *inclusive*—Continued.

Invention.	Inventor.	Residence.	Date.	No.
Nail-plate feeder	J. W. Hoard and T. A. Searle	Providence, R. I	Feb. 28, 1860	27, 289
Nail-plate feeder	C. D. Hunt	Fairhaven, Mass	Apr. 10, 1866	53, 924
Nail-plate feeder	C. D. Hunt	Fairhaven, Mass	Feb. 4, 1868	74, 094
Nail-plate feeder	J. Iler and W. Fitzpatrick	Troy, N. Y	June 13, 1854	11, 053
Nail-plate feeder	C. Isbister	Allegheny, Pa	Jan. 27, 1852	8, 687
Nail-plate feeder	E. B. Lake	Bridgeport, N. J	Nov. 26, 1867	71, 521
Nail-plate feeder	J. McNeely	Pittsburgh, Pa	Sept. 23, 1873	143, 022
Nail-plate feeder	S. K. Paden	Pulaski, Pa	Apr. 8, 1873	137, 710
Nail-plate feeder	H. B. and E. C. Perkins	Oxford, N. J	Jan. 30, 1872	123, 290
Nail-plate feeder	J. Poulson	Frenchtown, N. J	Dec. 16, 1873	145, 681
Nail-plate feeder	J. Newell	Lowell, Mass	Dec. 20, 1859	26, 513
Nail-plate feeder	D. Reed	Birmingham, Conn	Dec. 29, 1868	85, 478
Nail-plate feeder	J. R. Richardson	Newcastle, Pa	Apr. 13, 1869	88, 809
Nail-plate feeder	W. Riley, jr	Reading, Pa	Aug. 14, 1860	29, 613
Nail-plate feeder	T. A. Searle	Providence, R. I	June 19, 1866	55, 722
Nail-plate feeder	W. Sherman and J. Russell	New York and Brooklyn, N. Y.	Sept. 1, 1868	81, 830
Nail-plate feeder	J. P. Sherwood	Fort Edward, N. Y	Aug. 9, 1859	25, 051
Nail-plate feeder	J. H. Swett	Pittsburgh, Pa	Aug. 17, 1858	21, 222
Nail-plate feeder	J. L. Wiggin	South New Market, N. H	June 19, 1866	55, 755
Nail-plate feeder and turner	M. Otis	East Bridgewater, Mass	Dec. 3, 1850	7, 812
Nail-plate-feeding machine	F. J. Ayers	St. John, New Brunswick	Nov. 26, 1850	7, 792
Nail-plate-feeding machine	J. C. Gould	Oxford, N. J	Sept. 26, 1871	119, 228
Nail-plate-feeding machine	J. C. Gould	Oxford, N. J	Oct. 24, 1871	120, 190
Nail-plate-feeding machine	F. Hain	Gasconade County, Mo	Dec. 17, 1867	72, 392
Nail-plate-feeding machine	L. Richards	Providence, R. I	Sept. 10, 1872	131, 181
Nail-plate-feeding machine	J. P. Sherwood	Fort Edward, N. Y	Mar. 18, 1856	14, 474
Nail-plate-feeding machine	W. Wallick	Philadelphia, Pa	June 25, 1867	66, 192
Nail-plate holder	W. H. Battelle	Newcastle, Pa	July 14, 1857	17, 778
Nail-plate holder	J. C. Rhodes	South Abington, Mass	Nov. 19, 1867	71, 064
Nail-plate nippers	G. B. Johnson	Wheeling, W. Va	Apr. 2, 1872	125, 196
Nail-plates and sheet-strips, Process of manufacturing.	J. Trowen, E. Hemmings, and J. Sheldon.	Niles, Ohio	June 20, 1871	116, 046
Nail-plates, Feeding	F. E. Boyd	Boston, Mass	Mar. 17, 1868	75, 620
Nail-plates, Feeding	A. L. Reed	Pittsburgh, Pa	July 24, 1846	4, 657
Nail-plates, Feeding	P. A. Wilbur	Newcastle, Pa	Oct. 21, 1856	15, 938
Nail-plates, Mechanism for feeding	R. C. Grant	Middleport, Ohio	Feb. 25, 1873	136, 235
Nail-plates, Rolling	O. H. Dewey, J. Dudley, and F. Lindeman.	Wheeling, W. Va	Mar. 28, 1871	113, 029
Nail-plates while cutting, Vibrating	J. Bernard	Troy, N. Y	Apr. 16, 1811	....
Nail-separating apparatus	A. Morrison	Harrisburgh, Pa	Sept. 30, 1873	143, 246
Nail-separating device	H. B. Chess	Pittsburgh, Pa	Dec. 23, 1873	145, 720
Nail-separating machine	B. Bevelander	Boston, Mass	Jan. 28, 1873	135, 312
Nail slitting and heading machine	M. Garber	....	Apr. 17, 1804	....
Nail, spike, &c	W. Osgood and E. Hunt	Troy, N. Y	Mar. 14, 1834	....
Nail, spike, tack, horseshoe and nail, screw-bolt, &c., Wrought-iron.	W. and T. Schnebly	Hagerstown, Md	Mar. 3, 1834	....
Nails and brads, Machine for cutting	W. Hunt	New York, N. Y	Nov. 13, 1840	1, 853
Nails and brads, Machine for cutting	I. Kimball	New Hampshire	May 1, 1805	....
Nails and brads, Machine for cutting	P. Zacharie	....	May 4, 1796	....
Nails and making wrought nails, Machine for cutting and heading.	M. and R. Reeve	Burlington, N. J	Feb. 3, 1812	....
Nails and making wrought nails, Machine for cutting and heading.	M. and R. Reeve	Philadelphia, Pa	Apr. 16, 1813	....
Nails by rolling, Method of making	E. H. Collier	Scituate, Mass	June 18, 1850	7, 437
Nails, Casting iron, brass, and composition	J. Walworth	New York	Nov. 22, 1809	....
Nails, Claw for drawing	J. S. Allen and O. Gillmore	Norwich, Conn	May 14, 1867	64, 733
Nails, Covering heads of	W. H. Van Gieson	Waterbury, Conn	Dec. 29, 1857	18, 007
Nails, Cutting and heading	G. Chandlee	....	Dec. 12, 1796	....
Nails, Cutting and heading	N. and D. Chickering	Dover, Mass	Mar. 9, 1812	....
Nails, Cutting and heading	N. Elgar	Yorktown, Pa	July 25, 1815	....
Nails, Cutting and heading	J. Kersey	....	Feb. 24, 1797	....
Nails, Cutting and heading	E. Mason	Milton, Vt	June 6, 1832	....
Nails, Cutting and heading	J. Nevill	....	Aug. 12, 1797	....
Nails, Cutting and heading	T. Newhall, jr	Essex County, Mass	Sept. 7, 1816	....
Nails, Cutting and heading	I. Northrop	Oneida County, N. Y	Jan. 7, 1817	....
Nails, Cutting and heading	J. Reed	Kingston, Mass	Aug. 14, 1811	....
Nails, Cutting and heading	J. Reed	Hanover, Mass	Dec. 16, 1814	....
Nails, Cutting and heading	M. Reeve	Philadelphia, Pa	June 26, 1816	....
Nails, Cutting and heading	M. and R. Reeve	Burlington, N. J	Oct. 31, 1812	....
Nails, Cutting and heading	S. Rogers	Bridgewater, Mass	Oct. 21, 1814	....
Nails, Cutting and heading	F. A. Russel	Georgetown, D. C	Mar. 14, 1816	....
Nails, Cutting and heading	J. Spence	....	Feb. 16, 1797	....
Nails, Cutting and heading	J. White	Philadelphia, Pa	Jan. 11, 1812	....
Nails, Cutting and heading	S. Willard	Hudson, N. Y	Nov. 30, 1807	....
Nails, Cutting, griping, and heading	J. Reed	Hanover, Mass	Oct. 22, 1814	....
Nails from felt, Process of separating	W. Jones	New York, N. Y	Aug. 9, 1870	106, 171
Nails, Heading	J. Byington	....	Dec. 23, 1796	....
Nails, Heading	J. Frost	....	Dec. 23, 1796	....
Nails, Heading and cutting	E. West	....	July 6, 1802	....
Nails, Horizontal wheel for cutting and heading	J. Reed	Kingston, Mass	Apr. 15, 1809	....
Nails, Machine for cutting off	M. Gormley	Wilna, N. Y	July 21, 1868	80, 068
Nails, Machine for making and heading	M. Garber	....	Feb. 20, 1801	....
Nails, Machine for selecting and delivering	A. Knowlton	Boston, Mass	Jan. 28, 1873	135, 345
Nails, Machine for selecting and delivering	A. Knowlton	Boston, Mass	Feb. 18, 1873	136, 074
Nails, Making	J. Byington	....	Jan. 15, 1796	....
Nails, Making	J. M. Cooper	Philadelphia, Pa	Mar. 30, 1869	88, 454
Nails, Making	L. Fling	....	Dec. 19, 1797	....
Nails, Making	J. Perkins	....	Feb. 14, 1799	....
Nails, Manufacturing	J. Bigelow	....	Nov. 19, 1796	....
Nails, Manufacturing	J. Brest and E. Gilmore	Roxbury, Mass	June 1, 1830	....
Nails, Manufacturing	J. Hearsey	Wareham, Mass	May 13, 1830	....
Nails, Manufacturing	J. L. Krauser	Reading, Pa	July 27, 1858	21, 005
Nails, Manufacturing	F. Palmer	Buffalo, N. Y	Oct. 25, 1832	....
Nails, Manufacturing	F. Palmer	Buffalo, N. Y	Aug. 7, 1834	....
Nails, Manufacturing	T. Perkins	....	Feb. 7, 1794	....
Nails, Method of rolling iron for	J. Reed	....	July 15, 1802	....
Nails, Mode of pointing	H. A. Wills	Vergennes, Vt	Dec. 6, 1870	109, 987

Index of patents issued from the United States Patent Office from 1790 *to* 1873, *inclusive*—Continued.

Invention.	Inventor.	Residence.	Date.	No.
Nails milled out of heated rods	J. Reed		June 9, 1801	
Nails or brads, Machine for cutting	J. P. Sawin and J. Skinner	Roxbury, Mass	Feb. 8, 1811	
Nails, screws, &c., Attaching ornamental head to	T. C. Richards	New York, N. Y	Dec. 31, 1867	72, 905
Nails, Tool for clinching	D. J. Hendrickson	Otego, N. Y	Feb. 16, 1858	19, 364
Nailing and pegging machine	J. F. Sargent	Boston, Mass	Apr. 21, 1868	77, 104
Nailing-machine	L. R. Blake and A. S. Libby	Boston and Lawrence, Mass.	Mar. 31, 1868	76, 150
Nailing or pegging machine	J. B. Crosby	Boston, Mass	Mar. 2, 1869	87, 473
Name-plate and letter-chute, Combination of	A. M. Damon and S. G. Lyford	Lowell, Mass	Dec. 22, 1868	85, 218
Name-plate and letter-slide, Combined	J. M. Coombs	Boston, Mass	Mar. 26, 1867	63, 141
Naphtha storing and discharging apparatus	T. Forstall	New Orleans, La	June 27, 1871	116, 423
Napkin-holder	W. L. Dewey	Bridgeport, Conn	June 11, 1867	65, 653
Napkin holder and clasp, Child's table	D. Essex	New York, N. Y	Feb. 25, 1873	136, 146
Napkin-ring	C. E. Green and J. H. Bodwell	Nowark, N. J	Aug. 20, 1872	130, 576
Napkin-ring and salt-cup combined	G. Pine	Trenton, N. J	Mar. 24, 1868	75, 968
Napkin, Table	A. L. Munson	New Haven, Conn	Mar. 27, 1866	53, 468
Napkins to the person, Device for attaching	P. H. Raiford	Houston, Tex	May 23, 1871	115, 238
Napping-machine	C. P. Ladd	Bloomfield, N. J	July 30, 1872	129, 895
Napping-machine	S. Marsh	Jericho, Vt	Jan. 11, 1836	
Napping-machine, Metallic	R. Daniels	Woodstock, Vt	June 26, 1835	
Nasal irrigator	M. Mattson	New York, N. Y	Mar. 10, 1868	75, 285
Nasal plug, Elastic	J. J. Essex	Newport, R. I	Sept. 13, 1870	107, 238
Nautical alarm	S. G. Cabell	Quincy, Ill	June 20, 1871	116, 152
Nautical alarm	J. F. Haskins	Fitchburgh, Mass	Dec. 13, 1870	110, 036
Nautical alarm	E. L. Seymour	New York, N. Y	Mar. 3, 1857	16, 776
Nautical ventilator	B. Wynkoop		June 19, 1795	
Naval architecture	J. Thomas	Washington, D. C	Apr. 14, 1825	
Naval architecture	B. F. Wells	Georgetown, D. C	Oct. 18, 1859	25, 857
Navigable mercantile house	R. Haskell	Geneva, N. Y	July 7, 1821	
Navigating and propelling vessels by action of wind and waves.	J. A. Etzler	Philadelphia, Pa	Apr. 1, 1842	2, 533
Navigating-machine	J. J. Giraud	Baltimore, Md	Jan. 31, 1827	
Navigating rivers and shoal water	M. Williams	Albany, N. Y	Feb. 24, 1809	
Navigation	W. W. Van Loan	New York	Jan. 15, 1831	
Navigation, Chute for river	I. A. Putman	Mexico, Me	Apr. 28, 1868	77, 212
Navigation, Internal	J. Renwick	New York	Nov. 7, 1823	
Navigation, Machine for the improvement of	A. Hunn		Mar. 24, 1804	
Navigation, Machine to be used in	A. Heilbronn	New York	Mar. 16, 1829	
Navigator's bearing-indicator	J. D. Leach	Penobscot, Me	July 6, 1869	92, 197
Navigator's bearing-indicator	J. D. Leach	Penobscot, Me	Apr. 26, 1870	102, 281
Neck-scarf and collar supporter	E. A. Bushee	East Boston, Mass	Feb. 9, 1864	41, 568
Neck-scarf, Reversible	A. S. Saroni	New York, N. Y	Dec. 23, 1862	37, 247
Neck-stock	C. Fanshaw	New York	Oct. 3, 1846	4, 788
Neck-stock	W. Watson	Lowell, Ill	June 12, 1860	28, 705
Neck-stock and cravat-stiffener, Metallic frame for	J. Johnson	New York	Dec. 31, 1838	1, 058
Neck-tie	M. Altman	Baltimore, Md	Aug. 30, 1870	106, 761
Neck-tie	J. Bachelder	Norwich, Conn	June 15, 1869	91, 405
Neck-tie	J. Bachelder	Norwich, Conn	Feb. 8, 1870	99, 619
Neck-tie	H. Bendix and J. H. Fleisch	New York, N. Y	Mar. 13, 1866	53, 102
Neck-tie	J. R. Cluxton	Russellville, Ohio	Mar. 20, 1866	53, 272
Neck-tie	A. M. Dexter	Philadelphia, Pa	Dec. 5, 1865	51, 299
Neck-tie	F. Field	Troy, N. Y	Sept. 21, 1869	94, 951
Neck-tie	H. G. Fisk and T. J. Flagg	New York, N. Y	Jan. 28, 1868	73, 790
Neck-tie	F. D. Ford	New Bedford, Mass	Oct. 29, 1867	70, 331
Neck-tie	J. B. Gardiner	Springfield, Mass	Feb. 13, 1866	52, 557
Neck-tie	L. A. Grill	New York, N. Y	Oct. 11, 1870	108, 133
Neck-tie	I. W. Hamlet	Nashua, N. H	Sept. 18, 1866	58, 168
Neck-tie	W. H. Hart, jr	Philadelphia, Pa	May 8, 1866	54, 536
Neck-tie	W. H. Hart, jr	Philadelphia, Pa	Dec. 10, 1867	72, 034
Neck-tie	J. G. Hitchcock	New York, N. Y	Oct. 4, 1870	107, 907
Neck-tie	A. Hoffstadt	New York, N. Y	Oct. 31, 1871	120, 437
Neck-tie	A. Johnson	Brooklyn, N. Y	Sept. 22, 1868	82, 323
Neck-tie	G. A. Keene	Boston, Mass	Mar. 26, 1867	63, 164
Neck-tie	G. Kennedy	Philadelphia, Pa	Aug. 10, 1869	93, 624
Neck-tie	O. Kueppers	Union Hill, N. J	Nov. 26, 1872	133, 457
Neck-tie	G. Lane	New York, N. Y	Feb. 15, 1870	99, 918
Neck-tie	W. A. Laverty	Philadelphia, Pa	Feb. 19, 1867	62, 210
Neck-tie	W. A. Laverty	Philadelphia, Pa	Oct. 1, 1872	131, 884
Neck-tie	J. Silberman and G. Unger	New York, N. Y	Feb. 11, 1868	74, 436
Neck-tie	J. H. Newman	Boston, Mass	July 10, 1866	56, 253
Neck-tie	R. R. Parker	Indianapolis, Ind	Apr. 22, 1873	138, 097
Neck-tie	J. K. P. Pine	Troy, N. Y	Sept. 4, 1866	57, 762
Neck-tie	P. F. Smith	New York, N. Y	Jan. 29, 1861	31, 265
Neck-tie	W. S. Smoot	Washington, D. C	Dec. 15, 1868	85, 036
Neck-tie	G. B. Taylor	New York, N. Y	Mar. 13, 1866	53, 197
Neck-tie	E. Thomas	Philadelphia, Pa	May 19, 1868	78, 027
Neck-tie	E. C. Townsend	New York, N. Y	Nov. 19, 1861	33, 766
Neck-tie	H. P. Wetmore and J. G. Hitchcock.	Elizabeth, N. J., and New York, N. Y.	Oct. 11, 1870	108, 309
Neck-tie	J. R. Wilber and O. W. Peirce	Providence, R. I	Apr. 18, 1871	113, 957
Neck-tie	T. S. Wiles	Troy, N. Y	Mar. 21, 1865	46, 962
Neck-tie and bow	E. Woodward	Brooklyn, N. Y	July 20, 1869	92, 921
Neck-tie and collar combined	J. Varley	Hudson, N. J	Dec. 21, 1869	98, 129
Neck-tie and watch-guard combined	T. J. Flagg	New York, N. Y	June 23, 1868	79, 063
Neck-tie clasp	W. S. Barnes	Watertown, N. Y	June 27, 1865	48, 356
Neck-tie clasp	C. M. Hyatt	Albany, N. Y	Sept. 4, 1866	57, 821
Neck-tie, collar, and bosom combined	G. V. Woods	Belchertown, Mass	Oct. 2, 1866	58, 561
Neck-tie fastener	W. B. Dean	New York, N. Y	Oct. 8, 1872	131, 999
Neck-tie fastener	J. A. Hard	Lawrence, Kans	July 1, 1873	140, 412
Neck-tie fastener	E. Side	Brooklyn, N. Y	Nov. 7, 1871	120, 600
Neck-tie fastener	P. S. White	Providence, R. I	Nov. 30, 1869	97, 331
Neck-tie fastening	J. Bachelder	Norwich, Conn	June 15, 1869	91, 406
Neck-tie fastening	J. Bachelder	Norwich, Conn	Mar. 1, 1870	100, 250
Neck-tie fastening	G. W. Bishop	Baltimore, Md	Nov. 8, 1870	109, 106
Neck-tie fastening	W. J. Cowing	Washington, D. C	Dec. 27, 1870	110, 439
Neck-tie fastening	R. A. Goodyear	New Haven, Conn	Jan. 14, 1868	73, 244
Neck-tie fastening	H. M. Heineman	Williamsburgh, N. Y	Oct. 2, 1866	58, 414
Neck-tie fastening	H. M. Heineman	San Francisco, Cal	Dec. 7, 1869	97, 637

Index of patents issued from the United States Patent Office from 1790 to 1873, inclusive—Continued.

Invention.	Inventor.	Residence.	Date.	No.
Neck-tie fastening	J. G. Hitchcock	New York, N. Y	June 26, 1866	55, 859
Neck-tie fastening	A. Komp	New York, N. Y	Mar. 28, 1871	113, 177
Neck-tie fastening	P. G. Shirts	Highland Falls, N. Y	Sept. 20, 1870	107, 634
Neck-tie fastening	W. J. Terry	Walla Walla, Wash	June 11, 1867	65, 776
Neck-tie fastening	D. H. Tierney	Waterbury, Conn	Dec. 15, 1868	84, 974
Neck-tie fastening	W. A. Wicks	Baltimore, Md	Sept. 27, 1870	107, 744
Neck-tie fastening	C. Winterbottom	Philadelphia, Pa	May 14, 1872	126, 766
Neck-tie fastening	E. Woodward	Brooklyn, N. Y	July 7, 1868	79, 796
Neck-tie frame	S. W. Roof	New York, N. Y	May 12, 1868	77, 765
Neck-tie frame	P. Welch	New York, N. Y	May 12, 1868	77, 788
Neck-tie holder	A. F. Bixby	Decatur, Ill	Apr. 2, 1872	125, 116
Neck-tie holder	F. H. Brown	Stamford, Conn	Feb. 11, 1868	74, 294
Neck-tie holder	J. A. Eshleman	Philadelphia, Pa	Jan. 31, 1865	46, 096
Neck-tie holder	J. A. Eshleman	Philadelphia, Pa	July 10, 1866	56, 196
Neck-tie holder	W. H. Hart, jr	Philadelphia, Pa	Apr. 26, 1870	102, 259
Neck-tie holder	W. H. Hart, jr., and H. H. Thayer.	Philadelphia, Pa	July 5, 1870	105, 072
Neck-tie holder	J. Hayden and W. H. Hart, jr	Philadelphia, Pa	Oct. 3, 1871	119, 608
Neck-tie holder	J. O. Schoener	Saint Louis, Mo	Feb. 27, 1872	124, 164
Neck-tie holder	H. H. Thayer	Philadelphia, Pa	Sept. 19, 1871	119, 197
Neck-tie, India-rubber	W. H. Halsey	Hoboken, N. J	Sept. 25, 1866	58, 344
Neck-tie or scarf, Reversible	L. Kahn	New York, N. Y	Apr. 2, 1872	125, 305
Neck-tie retainer	J. B. Bishop	New Haven, Conn	Jan. 11, 1870	98, 660
Neck-tie retainer	S. A. Fite	Philadelphia, Pa	July 13, 1869	92, 522
Neck-tie retainer	E. B. Gibbud	Waterbury, Conn	Feb. 28, 1871	112, 237
Neck-tie retainer	W. J. Ketcham	Washington, D. C	Apr. 7, 1868	76, 468
Neck-tie retainer	H. Laurence	New Orleans, La	Nov. 8, 1870	109, 133
Neck-tie retainer	H. Laurence	New Orleans, La	Oct. 17, 1871	120, 077
Neck-tie retainer	C. P. Lawton	Webster, Mass	Apr. 18, 1871	113, 780
Neck-tie retainer	J. McGee	Lancaster, N. H	July 18, 1871	117, 189
Neck-tie retainer	J. C. Merritt	New York, N. Y	Mar. 5, 1867	62, 665
Neck-tie retainer	P. C. Moulton	New Haven, Conn	June 8, 1869	91, 153
Neck-tie retainer	P. C. Moulton	New Haven, Conn	Oct. 12, 1869	95, 713
Neck-tie retainer	J. Varley	Hudson City, N. J	Mar. 23, 1869	88, 239
Neck-tie supporter	B. F. Bean	Lynn, Mass	June 7, 1864	43, 068
Neck-tie supporter	S. Bruhl	New York, N. Y	Apr. 27, 1869	89, 382
Neck-tie supporter	G. R. Eager	East Boston, Mass	Apr. 7, 1868	76, 312
Neck-tie supporter	J. A. Eshleman	Philadelphia, Pa	Sept. 4, 1866	57, 694
Neck-tie supporter	W. H. Hart, jr	Philadelphia, Pa	Aug. 14, 1866	57, 129
Neck-tie supporter	A. Komp	New York, N. Y	June 17, 1873	140, 049
Neck-tie supporter	E. Peltz	New York, N. Y	Mar. 4, 1873	136, 538
Neck-tie supporter	T. Rosenthal	New York, N. Y	Aug. 7, 1866	57, 048
Neck-tie supporter	G. K. Snow	Watertown, Mass	Feb. 14, 1865	46, 426
Neck-tie supporter	W. A. Wicks	Baltimore, Md	Dec. 13, 1870	110, 178
Neck-tie supporter, Collar	C. L. Lockwood	New York, N. Y	June 5, 1866	55, 317
Neck-ties, Device for adjusting and buttoning	O. Wilson	Sandusky City, Ohio	Aug. 10, 1869	93, 656
Neck-ties, Device for securing fastening-loop to	B. F. Weishampel	Baltimore, Md	Jan. 12, 1869	85, 773
Neck-ties, Mode of packing	W. H. Hart, jr	Philadelphia, Pa	Mar. 31, 1868	76, 186
Necklace-fastening	E. S. Dodge	Providence, R. I	July 16, 1872	129, 111
Needle	A. I. Ambler	Milwaukee, Wis	Mar. 24, 1863	37, 996
Needle	G. Cooper	Thompsonville, Conn	Jan. 22, 1861	31, 153
Needle	J. W. Donaldson and D. Sheets	Suisun, Cal	Apr. 26, 1870	102, 380
Needle	S. Ivers	New Bedford, Mass	Oct. 27, 1868	83, 501
Needle	W. H. Marriott	Baltimore, Md	Feb. 9, 1869	86, 769
Needle	H. G. Suplee	San Francisco, Cal	Oct. 25, 1870	108, 737
Needle and tool holder for work-table	H. H. Wescott	Portland, Me	July 25, 1871	117, 354
Needle-arranging machine	C. O. Crosby	New Haven, Conn	Apr. 12, 1870	101, 829
Needle-blanks, Machine for reducing wire for	T. Fowler	Seymour, Conn	Sept. 3, 1867	68, 430
Needle, Belt-lacing	J. W. Smith	Bryan, Ohio	May 6, 1873	138, 540
Needle-book	J. F. Umpleby	Albany, N. Y	May 5, 1868	77, 551
Needle-case	W. Avery and A. Fenton	Redditch, England	Jan. 18, 1870	98, 904
Needle-case	W. Avery and A. Fenton	Redditch, England	May 3, 1870	102, 471
Needle-case	W. B. Field and B. Beach	Rockville and Meriden, Conn.	Mar. 15, 1870	100, 879
Needle-case	T. G. Harold	Brooklyn, N. Y	Apr. 24, 1860	28, 037
Needle-case	W. J. Marchant and H. B. Hart	Philadelphia, Pa	Sept. 9, 1873	142, 639
Needle-case	J. F. Newhall	Waltham, Mass	June 25, 1867	66, 166
Needle-case	G. L. Turney	London, England	Nov. 13, 1866	59, 737
Needle-case	G. Wempe	San Francisco, Cal	Jan. 14, 1873	134, 829
Needle case and index	C. D. Wheeler	New York, N. Y	May 17, 1859	24, 070
Needle case, Sewing	J. O. Whitcomb	Brooklyn, N. Y	Apr. 15, 1862	35, 004
Needle, Crochet	J. M. Hoadley	Derby, Conn	Aug. 18, 1863	39, 572
Needle, Crochet	J. M. Hoadley	Birmingham, Conn	Apr. 21, 1868	76, 916
Needle-cushion	W. H. Bixler	Easton, Pa	July 8, 1873	140, 609
Needle-electrodes for electro-surgical uses	J. Kidder	New York, N. Y	Dec. 19, 1871	122, 031
Needle-envelope	A. James	Redditch, England	Oct. 20, 1868	83, 284
Needle for caning chairs	M. E. Hurley	Baltimore, Md	July 17, 1866	56, 421
Needle-grinding machine	C. Marsh	Bridgeport, Conn	Dec. 7, 1869	97, 664
Needle-guard	J. C. Howells	Madison, Wis	Apr. 10, 1860	27, 805
Needle-machine	W. Aiken	Franklin, N. H	Nov. 12, 1867	70, 677
Needle-making machine	R. T. Barton	New Haven, Conn	Sept. 10, 1867	68, 593
Needle-making machine	C. O. Crosby	New Haven, Conn	Nov. 28, 1865	51, 150
Needle-making machine	C. O. Crosby	New Haven, Conn	June 14, 1870	104, 278
Needle-making machinery	C. Kaiser	New York, N. Y	Nov. 12, 1861	33, 707
Needle-making machine	F. W. Mallett	New Haven, Conn	Apr. 29, 1873	138, 419
Needle-making machine	F. Plant	New York, N. Y	June 19, 1860	28, 772
Needle-making machine	C. P. S. Wardwell	Lake Village, N. H	Sept. 24, 1867	69, 280
Needle-making machinery	C. P. S. Wardwell	Lake Village, N. H	Sept. 24, 1867	69, 281
Needle-manufacturing machine	J. B. Blanchard	Boston, Mass	Aug. 2, 1870	106, 026
Needle, Neutralizing local attraction of the	C. Kline	Brooklyn, N. Y	Nov. 23, 1858	22, 125
Needle-paper wrapper	M. E. Baylis	Redditch, England	Aug. 15, 1871	117, 967
Needle-papering machine	C. O. Crosby	New Haven, Conn	July 12, 1870	105, 313
Needle-papering machine	G. P. Farmer	Philadelphia, Pa	July 1, 1862	35, 748
Needle-papering machine	G. P. Farmer	Philadelphia, Pa	Apr. 10, 1866	53, 799
Needle, Plated	R. J. Roberts	New York, N. Y	Feb. 15, 1870	99, 952
Needle-points, Machine for grinding	C. Kaiser	New York, N. Y	Aug. 13, 1861	33, 042
Needle polisher, sharpener, and lubricator	H. W. Little	Muncie, Ind	Nov. 16, 1869	96, 931
Needle-polishing machine	C. O. Crosby	New Haven, Conn	Mar. 16, 1869	87, 826

Index of patents issued from the United States Patent Office from 1790 to 1873, inclusive—Continued.

Invention.	Inventor.	Residence.	Date.	No.
Needle-polishing machine	C. O. Crosby	New Haven, Conn	July 12, 1870	105, 312
Needle-polishing machine	F. W. Mallett	New Haven, Conn	Oct. 3, 1871	119, 525
Needle scouring machine	C. O. Crosby	New Haven, Conn	Feb. 9, 1869	86, 819
Needle-scouring machine	F. W. Mallett	New Haven, Conn	Mar. 28, 1871	113, 182
Needle-setter and thread-pinchers	C. T. Barber and B. T. Loomis	New York, N. Y	May 4, 1869	89, 618
Needle setter and threader	R. W. Brown	Providence, R. I	June 14, 1870	104, 260
Needle setter and threader	C. F. Martine	Boston, Mass	Apr. 4, 1871	113, 542
Needle-setter, needle-sharpener, needle-case, and ripper.	E. E. Hendrick	Carbondale, Pa	May 16, 1871	114, 815
Needle setter, threader, and cutter	A. Johnston	Ottumwa, Iowa	Feb. 4, 1873	135, 479
Needle, Sewing	J. Cottrill	Studley, England	June 1, 1858	20, 409
Needle, Sewing	B. Garvey	New York, N. Y	May 12, 1857	17, 272
Needle, Sewing	G. A. Lloyd and S. Fetlow	San Francisco, Cal	Oct. 1, 1867	69, 453
Needle, Sewing	A. Morrall	Studley, England	May 25, 1869	90, 460
Needle, Sewing	J. Wilcox and S. H. Whitridge	Philadelphia, Pa	Oct. 3, 1854	11, 769
Needle-sharpener	A. S. Dinsmore	New York, N. Y	May 25, 1869	90, 433
Needle-sharpener	T. Harris	Cote Saint Paul, Canada	Dec. 19, 1871	121, 940
Needle-sharpener	E. K. Haynes	Boston, Mass	June 28, 1870	104, 732
Needle-sharpener	J. M. Mason	Chelsea, Mass	July 23, 1872	129, 673
Needle-sharpening device	R. S. Barnum and H. C. Goodrich.	Chicago, Ill	Sept. 9, 1873	142, 552
Needle shuttle, threader, and knife	J. Slack	Brooklyn, N. Y	Oct. 14, 1873	143, 726
Needle, Spiral fissure	S. Hodgins	Saint Louis, Mo	Aug. 6, 1867	67, 545
Needle, Tag	O. Low	Chelsea, Mass	Apr. 3, 1866	53, 638
Needle-threader	N. Barnum	La Porte, Ind	July 26, 1870	105, 628
Needle-threader	O. Cox	Alexandria, Va	Feb. 9, 1869	86, 647
Needle-threader	F. R. Gaspary	Berlin, Prussia	Nov. 25, 1873	144, 974
Needle-threader	A. Hunter	Buffalo, N. Y	Apr. 6, 1869	88, 717
Needle-threader	A. Huston	Bristol, Me	Oct. 22, 1867	69, 997
Needle-threader	S. L. Mercer	Washington, D. C	Jan. 30, 1872	123, 277
Needle-threader	J. O'Kane	New York, N. Y	Apr. 19, 1864	42, 394
Needle-threader	C. W. Simpson	Bangor, Me	Sept. 21, 1869	94, 980
Needle-threader	C. L. Spencer	Providence, R. I	Sept. 25, 1866	58, 314
Needle-threader	E. J. Stanley	Montezuma, Iowa	Oct. 8, 1872	132, 110
Needle-threader and spool combined	G. P. Farmer	Brooklyn, N. Y	Dec. 26, 1871	122, 241
Needle-threader, Magnet	O. Cox	Washington, D. C	June 14, 1864	43, 100
Needle-threaders, Mold for casting	S. S. Burlingame	Warwick, R. I	July 24, 1860	29, 230
Needle-threading apparatus	J. O'Kane	Philadelphia, Pa	Feb. 3, 1863	37, 587
Needle-threading hook	H. Wells	Woburn, Mass	Dec. 30, 1873	145, 978
Needle-threading instrument	S. S. Burlingame	Warwick, R. I	Feb. 15, 1859	22, 934
Needle-wrapper	R. Bennett	Redditch, England	Apr. 19, 1859	23, 742
Needle-wrapper	O. H. Blood and F. C. Treadwell, jr.	New York, N. Y	Oct. 13, 1863	40, 305
Needle-wrapper	W. Byfield	New York, N. Y	May 22, 1866	54, 856
Needle-wrapper	J. Clark	Redditch, England	May 19, 1868	78, 059
Needle-wrapper	R. Crowley	New York, N. Y	Dec. 13, 1859	26, 418
Needle-wrapper	D. Evans	Studley, England	Mar. 19, 1867	63, 028
Needle-wrapper	G. Ewart	New York, N. Y	May 11, 1869	89, 921
Needle-wrapper	J. Hill	Redditch, England	Nov. 12, 1867	70, 716
Needle-wrapper	C. B. James	Redditch, England	Feb. 9, 1869	86, 757
Needle-wrapper	V. Milward	Ipsley, England	July 5, 1870	105, 113
Needle-wrapper	J. Pitzler	New York, N. Y	July 25, 1871	117, 323
Needle-wrapper	J. Pitzler	New York, N. Y	July 25, 1871	117, 324
Needle-wrapper	C. Schleicher	Schoenthal, Rhenish Prussia.	May 30, 1871	115, 529
Needle-wrapper	A. Shrimpton	Redditch, England	Feb. 8, 1870	99, 605
Needle-wrapper	S. E. Tolten	Brooklyn, N. Y	Oct. 23, 1866	59, 095
Needle-wrapper	J. W. Whitfield	New York, N. Y	Apr. 23, 1867	64, 053
Needles, Machine for forming the hooks of machine.	N. Paine	Milford, Mass	Sept. 16, 1873	142, 863
Needles, Machine for polishing the eyes of	W. H. Dayton and J. Aldis	Wolcottville, Conn	June 4, 1872	127, 582
Needles, Machine for polishing the eyes of	C. Kaiser	New York, N. Y	Dec. 17, 1861	33, 947
Needles, Machine for polishing the eyes of	R. Thompson	Waterbury, Conn	June 4, 1872	127, 529
Needles, Machine for scouring the eyes of	H. A. Nettleton and E. R. Lawton.	West Cheshire, Conn	Nov. 8, 1870	109, 040
Needles, Manufacture of	A. Morral	Studley, Great Britain	Dec. 21, 1839	1, 437
Needles, Manufacture of sewing	H. Walker	London, England	Nov. 23, 1858	22, 140
Needles, Manufacture of split	C. H. Palmer	New York, N. Y	May 2, 1871	114, 332
Needles, Manufacture of split	C. H. Palmer	New York, N. Y	Oct. 10, 1871	119, 876
Needles, Papering or putting up	D. Evans	Studley, England	Dec. 12, 1871	121, 860
Needles, Polishing	C. O. Crosby	New Haven, Conn	Feb. 9, 1869	86, 817
Needles, Polishing	C. O. Crosby	New Haven, Conn	Feb. 9, 1869	86, 818
Negative plate for "Smee" battery	L. L. Smith	Brooklyn, N. Y	Feb. 22, 1870	100, 202
Ne plus ultra, Machine called the	J. Baily	Philadelphia, Pa	Mar. 21, 1825	....
Nest, Hen's	S. S. Bent	Port Chester, N. Y	June 16, 1868	78, 857
Nest, Hen's	C. W. Blackman	Bridgeport, Conn	Oct. 1, 1867	69, 392
Nest, Hen's	C. Campbell	Yellow Head, Ill	Feb. 20, 1866	52, 675
Nest, Hen's	B. F. Hayward	Nebraska City, Nebr	Sept. 22, 1868	82, 312
Nest, Hen's	D. P. Leach	Franklin, Ind	Oct. 5, 1869	95, 594
Nest, Hen's	N. Stevens	New Haven, Conn	May 22, 1866	54, 974
Nest, Wooden bird's	J. A. Deknatel	New York, N. Y	July 2, 1872	128, 538
Nests, Device for preserving eggs in hens'	C. V. Ament	Dansville, N. Y	Mar. 21, 1854	10, 675
Net, Adjustment of fishing	W. Randolph	Bloomington, Ill	Aug. 25, 1863	39, 676
Net and sash, Combined window	I. Wiswell	Springfield, Vt	Oct. 9, 1860	30, 370
Net, Fish	W. and J. Carr and J. Shannon	Sunbury, Pa	Sept. 14, 1844	3, 741
Net, Fish	T. Cartwright	Davenport, Iowa	Sept. 29, 1868	82, 490
Net, Fish	C. Livandais	New Orleans, La	Oct. 22, 1872	132, 476
Net, Fish	P. G. Sabins	Westport, Mass	Feb. 8, 1870	99, 713
Net, Fishing	L. H. Alexander	Gloucester, Mass	Aug. 15, 1871	117, 957
Net, Fishing, &c	B. Arnold	East Greenwich, R. I	July 16, 1867	66, 669
Net, Fishing	B. Arnold	East Greenwich, R. I	June 9, 1868	78, 716
Net, Fishing	E. A. Field	Sidney, Me	June 19, 1866	55, 635
Net, Fishing	T. Hall	Gloucester, Mass	Apr. 27, 1858	20, 125
Net, Fishing	S. Harper	Leipersville, Pa	Oct. 27, 1868	83, 493
Net, Fishing	W. Maxwell	Washington, N. J	Nov. 6, 1866	59, 429
Net, Fishing	F. A. Werdmüller	New York, N. Y	Mar. 9, 1869	87, 740
Net, Fly	C. K. Burkholder	York Springs, Pa	Feb. 18, 1868	74, 664
Net, Fly	J. Frymire	Orangeville, Pa	Nov. 9, 1869	96, 574

Index of patents issued from the United States Patent Office from 1790 *to* 1873, *inclusive*—Continued.

Invention.	Inventor.	Residence.	Date.	No.
Net, Fly	L. B. and G. W. Lee	Jerusalem, N. Y	Sept. 3, 1867	68, 518
Net, Fly	W. Sohier	Boston, Mass	May 30, 1871	115, 374
Net, Fly	R. Wilson	Milton, Pa	May 11, 1858	20, 235
Net for anglers, Landing-net	C. De Saxe	New York, N. Y	Apr. 18, 1854	10, 794
Net for catching fish at sea	B. Merritt, jr	Charlestown, Mass	June 29, 1858	20, 725
Net for catching mackerel, &c., in deep water	B. W. Hale	Newbury, Mass	June 4, 1838	763
Net for fishing, Pound	P. E. Tierman	Waukegan, Ill	Apr. 18, 1871	113, 817
Net for horses	H. Korn	Philadelphia, Pa	June 22, 1832	....
Net for horses	H Korn	Philadelphia, Pa	Apr. 19, 1834	....
Net for horses	H. Korn	Philadelphia, Pa	Apr. 19, 1834	....
Net for horses, Fly	J. Cantner	Millheim, Pa	Mar. 9, 1869	87, 631
Net for horses, Fly	J. Cantner and D. W. Zeigler	Millheim, Pa	Apr. 16, 1872	125, 659
Net for horses, Fly	J. Graham	New York, N. Y	Apr. 19, 1870	102, 113
Net for horses, Fly	J. M. Harman	Orangeville, Pa	Feb. 1, 1870	99, 434
Net for horses, Fly	W. H. Horstman	Philadelphia, Pa	July 28, 1831	....
Net for horses, Fly	D. W. Hurst	Petersburgh, Pa	July 21, 1868	80, 182
Net for horses, Fly	J. S. Huston	Mechanicsburgh, Pa	May 24, 1870	103, 337
Net for horses, Fly	S. B. Kline	Mechanicsburgh, Pa	Mar. 2, 1869	87, 499
Net for horses, Fly	H. Korn	Philadelphia, Pa	Sept. 12, 1829	....
Net for horses, Fly	H. Korn	Philadelphia, Pa	Dec. 8, 1831	....
Net for horses, Fly	H. Korn	Philadelphia, Pa	July 2, 1836	....
Net for horses, Fly	P. Mintzer	Philadelphia, Pa	July 20, 1831	....
Net for horses, Tail	F. Parson and E. W. Parker	Saint Louis, Mo., and Penn Yan, N. Y.	June 11, 1872	127, 918
Net for window, Fly	W. C. McGowan and J. M. Hale.	Georgia Plains, Vt	Aug. 18, 1868	81, 099
Net frame, Insect	H. Stoltz	Evansville, Ind	Aug. 6, 1872	130, 163
Net gear, Fishing	C. C. Crosman	Portland, Me	Feb. 26, 1867	62, 481
Net, Gill	H. Cook	Haddam, Mass	Mar. 17, 1843	3, 004
Net, Gill	J. Drummond	New York	Aug. 27, 1818	....
Net, Gill	D. Wills	Camden, N. J	Mar. 31, 1868	76, 284
Net, Insect	C. B. Seaman	Honesdale, Pa	Nov. 24, 1868	84, 381
Net, Lady's hair	J. Dalton	New York, N. Y	Mar. 5, 1872	124, 340
Net, Lady's head	D. M. Smyth	New York, N. Y	Feb. 2, 1864	41, 465
Net-making machine	B. Arnold	East Greenwich, R. I	Jan. 4, 1870	98, 461
Net, Metallic fish	W. Beck	New York, N. Y	Oct. 25, 1870	108, 553
Net or seine, Fish	S. May	Columbia, Pa	Jan. 11, 1812	....
Net, Portable fly and mosquito	T. M. Prentiss	Boston, Mass	Nov. 18, 1873	144, 792
Net straps, Machine for cutting and punching fly	J. Yeager	Berrysburgh, Pa	Nov. 19, 1867	71, 259
Net straps, Machine for rounding fly	C. K. Burkholder and H. Lerew	York Springs, Pa	Sept. 17, 1867	68, 949
Net supporter, Fish	B. Ryder, jr	South Orrington, Me	Apr. 4, 1871	113, 572
Net, Trap	W. S. Wilcox	Wellington, Ohio	Oct. 27, 1868	83, 429
Nets, Machine for cutting fly	A. D. Hoffman	Detroit, Mich	June 26, 1866	55, 860
Nets, Machine for cutting fly	H. D. Martin	Ypsilanti, Mich	Jan. 1, 1867	60, 764
Nets, Machine for cutting leather fly	A. Worden	Ypsilanti, Mich	Feb. 2, 1861	41, 459
Nets, Machine for making fly	A. Prutzmann	Canton, Ohio	Dec. 14, 1869	97, 809
Nets, Machine for making seine	B. Arnold	East Greenwich, R. I	Sept. 23, 1862	36, 499
Nets, Machine for punching leather straps for fly	J. Matheis	Ottawa, Ill	July 21, 1868	80, 081
Nets, Machine for punching leather straps for fly	J. Matheis	Ottawa, Ill	May 18, 1869	90, 281
Nets, Machine for weaving fish	L. Van Hoesen	New Haven, Conn	Oct. 7, 1842	2, 798
Nets, Securing lash in fly	M. Heilman	Lebanon, Pa	Jan. 4, 1870	98, 493
Nets, Securing lash in fly	J. S. Huston	Mechanicsburgh, Pa	Dec. 21, 1869	98, 164
Netted or laced fabric, Machine for making	H. A. Oesterle	Philadelphia, Pa	Dec. 26, 1865	51, 744
Netting, Horse	J. Cantner and M. Ulrich	Millheim, Pa	Aug. 7, 1866	56, 900
Netting-machine	J. McMullen	Baltimore, Md	June 27, 1846	4, 608
Netting-machine	J. McMullen	Baltimore, Md	July 1, 1856	15, 245
Netting-machine	P. Moulton	New Rochelle, N. Y	Aug. 22, 1846	4, 707
Netting, Machine for cutting leather fly	T. Tulley	Springfield, Ill	June 11, 1872	127, 716
Netting, Machine for making	A. Vivarttas	New York, N. Y	June 14, 1870	104, 381
Netting, Machine for making wire	J. Nesmith	Lowell, Mass	Apr. 4, 1854	10, 743
Netting, Window	J. R. Wharry	Moundsville, W. Va	Mar. 17, 1868	75, 612
News-distributer	J. H. Pratt	New York, N. Y	Feb. 3, 1863	37, 590
Newspaper-clamp	J. D. Hall	Philadelphia, Pa	Aug. 2, 1864	43, 685
Newspaper-wrapper	L. P. Mara	New York, N. Y	Feb. 19, 1861	31, 516
Newspapers, &c., Machine for feeding up, cutting, and pasting directions on.	R. W. Wright	New Haven, Conn	Dec. 20, 1859	26, 543
Newspapers, music, &c., Ready-binder for	E. Ripley	Troy, N. Y	Dec. 26, 1837	536
Nickel, Electro-deposition of	I. Adams, jr	Boston, Mass	Aug. 3, 1869	93, 157
Nickel, Electro-deposition of	I. Adams, jr	Boston, Mass	May 10, 1870	102, 748
Nickel, Melting, casting, and hardening	I. Adams, jr	Boston, Mass	May 25, 1869	90, 476
Nickel-plating	N. S. Keith	New York, N. Y	Nov. 28, 1871	121, 383
Nickel-plating	W. J. Kuhns	Brooklyn, N. Y	May 17, 1870	103, 201
Night-chair	J. Macferran	Philadelphia, Pa	Mar. 24, 1868	75, 778
Night-lock, Traveler's	A. V. Thomas	Frederick, Md	Oct. 31, 1865	50, 751
Night-signals, Mode of firing	I. Edge and C. C. Hyde	Jersey City, N. J., and Stonington, Conn.	Apr. 29, 1862	35, 089
Night-soil, Boat for transporting	J. Paterson	Hoboken, N. J	July 30, 1872	129, 982
Night-soil for agricultural purposes, Treating	R. B. Fitts	Philadelphia, Pa	Aug. 11, 1863	39, 472
Night-soil, Treating	R. B. Fitts	Philadelphia, Pa	Sept. 9, 1862	36, 399
Night-soil, Vessel for reception and transportation of.	R. A. Smith	Philadelphia, Pa	July 18, 1865	48, 847
Nine-pin ball	J. Taggart	Roxbury, Mass	July 26, 1859	24, 893
Nipper-block, Self-acting	W. Waley	New London, Conn	May 30, 1854	10, 997
Nipper-block, Self-acting	J. Whipple, jr	Milford, Mass	May 22, 1855	12, 928
Nippers	T. G. Hall	New York, N. Y	May 14, 1867	64, 664
Nippers	E. F. Stacy	Gloucester, Mass	June 2, 1868	78, 546
Nippers and pinchers, Combined	D. Sweetman	Homer, N. Y	Oct. 13, 1863	40, 291
Nippers, Clinching	J. B. Wilder	Mannsville, N. Y	July 14, 1868	79, 928
Nippers, Cutting	F. A. Adams	Shelburne, Mass	Oct. 15, 1872	132, 191
Nippers, &c., Cutting	P. Broadbooks	Batavia, N. Y	Nov. 18, 1873	144, 734
Nippers, Cutting	W. A. Stevens	East Brookfield, Mass	May 7, 1872	126, 423
Nippers, Cutting	N. Thompson	Brooklyn, N. Y	July 14, 1868	80, 031
Nippers, Friction	D. Thomas	Hingham, Mass	Aug. 4, 1868	80, 575
Nippers, Police	J. B. Craig and M. Haughey	Saint Louis, Mo	Mar. 4, 1873	136, 419
Nippers, Police	W. G. Phillips	Brooklyn, N. Y	Aug. 10, 1869	93, 474
Nippers, Policeman's	G. Lutz, D. H. Royce, M. Trenor, and R. Chadwick.	Columbus, Ohio	May 3, 1870	102, 563
Nipple, Artificial	H. D. Lockwood	Charlestown, Mass	Dec. 7, 1869	97, 659

Index of patents issued from the United States Patent Office from 1790 *to* 1873, *inclusive*—Continued.

Invention.	Inventor.	Residence.	Date.	No.
Nipple, Artificial	E. Pratt	New York, N. Y	Aug. 4, 1855	4, 131
Nipple, Nursing	F. H. Holton	Brooklyn, N. Y	June 9, 1868	78, 741
Nipple, Rubber	G. Stevenson	Zionsville, Ind	Sept. 3, 1872	131, 130
Nipple-shield	W. Baxton	Woburn, Mass	Apr. 2, 1835	....
Nipple-shield	S. C. Foster	New York, N. Y	May 2, 1871	114, 281
Nipple-shield	T. McLaughlin	Millville, N. J	Apr. 28, 1868	77, 393
Nipple-shield	J. Parker	Boston, Mass	Jan. 13, 1857	16, 396
Nipple-shield	C. H. and J. M. Wilder	New York, N. Y	June 18, 1867	65, 978
Nipple-shield, Artificial metallic	J. J. Heintzleman	Philadelphia, Pa	Mar. 16, 1833	....
Nitric acid, Method of purifying	G. M. Mowbray	Titusville, Pa	Sept. 21, 1869	94, 969
Nitro-glycerine	A. Nobel	Hamburg, Germany	Oct. 24, 1865	50, 617
Nitro-glycerine	A. Nobel	New York, N. Y	Aug. 14, 1866	57, 175
Nitro-glycerine	A. Nobel	Hamburg, Germany	May 26, 1868	78, 317
Nitro-glycerine	A. Nobel	Hamburg, Germany	Aug. 5, 1873	141, 455
Nitro-glycerine, Apparatus for the manufacture of	G. M. Mowbray	North Adams, Mass	Aug. 23, 1870	106, 606
Nitro-glycerine can	J. Taylor	Petroleum Centre, Pa	Aug. 6, 1871	117, 577
Nitro-glycerine, &c., Can for transporting	A. Hamar	Philadelphia, Pa	Apr. 1, 1873	137, 439
Nitro-glycerine, Manufacture of	A. Hamar	Philadelphia, Pa	Apr. 1, 1873	137, 440
Nitro-glycerine, Manufacture of	G. M. Mowbray	Titusville, Pa	Apr. 7, 1868	76, 499
Nitro-glycerine, Manufacture of	E. A. L. Roberts	Titusville, Pa	Mar. 21, 1871	112, 848
Nitro-glycerine, Manufacture of	E. A. L. Roberts	Titusville, Pa	Mar. 21, 1871	112, 849
Nitro-glycerine, Manufacture of	E. A. L. Roberts	Titusville, Pa	Dec. 12, 1871	121, 898
Nitro-glycerine, Manufacture of	T. P. Shaffner	Louisville, Ky	Aug. 17, 1869	93, 756
Nitro-glycerine, Method of exploding	G. M. Mowbray	Titusville, Pa	July 27, 1869	93, 113
Nitro-glycerine, Method of packing	T. P. Shaffner	Louisville, Ky	July 24, 1866	56, 620
Nitro-glycerine, Method of preparing	S. Chester and O. Bürstenbinder	New York, N. Y	Jan. 19, 1869	85, 906
Nitro-gylcerine, nitro-benzole, &c., Apparatus for the manufacture of.	G. M. Mowbray	North Adams, Mass	Aug. 23, 1870	106, 607
Nitro-glycerine, Preserving	T. P. Shaffner	Louisville, Ky	Feb. 9, 1869	86, 701
Nitroleum and other explosive liquids, Process of preserving.	T. P. Shaffner	Louisville, Ky	Dec. 28, 1869	98, 426
Nitroleum, Method of blasting with	T. P. Shaffner	Louisville, Ky	Dec. 18, 1866	60, 573
Nitrous oxide and other gases, Apparatus for washing.	E. E. Pursell	Indianapolis, Ind	July 9, 1872	128, 753
Nitrous oxide, &c., Apparatus for making	M. B. Renslow	Springfield, Mass	Aug. 20, 1867	67, 907
Nitrous oxide as an anaesthetic agent, Use of	W. P. Barker	Grand Rapids, Mich	Mar. 2, 1869	87, 319
Non-freezing battery	E. H. Ashcroft	Boston, Mass	Dec. 2, 1873	145, 143
Noodle-machine	D. Obergfell	Wheeling, W. Va	May 10, 1870	102, 962
Nose-bag, Automatic	G. A. Hankinson	Manahocking, N. J	Mar. 17, 1863	37, 914
Notations, Phonetic and diacritical	J. W. Shearer	Madison, N. C	July 16, 1872	129, 600
Notes, checks, &c., Device for canceling	E. M. Scott	Auburn, N. Y	Apr. 1, 1862	34, 850
Notes, checks, &c., from alteration, Protecting	E. Watson	Albany, N. Y	Aug. 3, 1838	871
Notes, checks, &c., To prevent alteration in	C. Folsom	New York, N. Y	Dec. 13, 1870	110, 129
Nozzle	R. R. and J. Craig	Nevada City, Cal	Apr. 23, 1872	125, 884
Nozzle	C. Oyston	Little Falls, N. Y	Aug. 25, 1863	39, 674
Nozzle	T. Watson	Nevada, Cal	Nov. 28, 1871	121, 265
Nozzle and cap, Oil-can	E. T. Covell	Brooklyn, N. Y	June 21, 1870	104, 430
Nozzle and cap, Oil-can	J. A. Bostwick	New York, N. Y	Nov. 7, 1871	120, 702
Nozzle and sealing-cap for sheet-metal cans	J. A. Bostwick	New York, N. Y	June 21, 1870	104, 413
Nozzle and stopper for cans, &c	M. Bray	Boston, Mass	May 24, 1870	103, 420
Nozzle and turn-pipe, Hydraulic	D. L. Gorman	Michigan Bluff, Cal	Aug. 13, 1872	130, 366
Nozzle and ventilating oil-can, Extension	W. Bonner	Saint Louis, Mo	Sept. 24, 1867	69, 067
Nozzle-bending die, Coffee-pot	W. H. Miller	Brandenburgh, Ky	Jan. 29, 1867	61, 678
Nozzle, Can	F. W. Devol	New York, N. Y	Oct. 20, 1868	83, 140
Nozzle, Can	C. Pratt	New York, N. Y	Apr. 20, 1869	89, 167
Nozzle, Elastic hose	J. Greacen, jr	New York, N. Y	Jan. 17, 1872	122, 719
Nozzle, Exhaust	A. F. Du Faur	New York, N. Y	Sept. 20, 1870	107, 603
Nozzle, Exhaust	C. H. Frisbie	Chicago, Ill	May 31, 1870	103, 596
Nozzle, Exhaust	J. Lewis	Chicago, Ill	July 4, 1871	116, 605
Nozzle, Expansible hose	C. Crook	Yonkers, N. Y	Sept. 10, 1867	68, 712
Nozzle, Fire-engine	C. Burchardt	New York, N. Y	Mar. 13, 1866	53, 109
Nozzle, Fire-engine	L. Button and R. Blake	Waterford, N. Y	Nov. 15, 1859	26, 088
Nozzle, Fire-engine, &c	J. J. Hofer	New Orleans, La	Dec. 3, 1867	71, 756
Nozzle, Fire-engine	J. C. Howells	Madison, Wis	June 26, 1860	28, 865
Nozzle, Fire-engine	C. Oyston	Little Falls, N. Y	Mar. 13, 1866	53, 175
Nozzle for cans and casks	A. L. Webster	Cleveland, Ohio	July 5, 1870	105, 151
Nozzle for cans, Cut-off	J. M. Murphy	New York, N. Y	Oct. 19, 1869	95, 927
Nozzle for cans, Screw	F. W. Devoe	New York, N. Y	Mar. 16, 1869	87, 915
Nozzle for cans, Tap	H. Miller	New York, N. Y	July 26, 1870	105, 713
Nozzle for drawing liquor, Cut-off	F. C. Edwards	Chicago, Ill	Sept. 2, 1873	142, 383
Nozzle for exhaust-pipes of locomotives	W. E. Cooper	Dunkirk, N. Y	Dec. 18, 1855	13, 939
Nozzle for exhaust-pipes of locomotives, Movable tapering.	F. Espenschade	Mifflintown, Pa	Mar. 14, 1854	10, 634
Nozzle for fire-engines, Water-spreading	W. Gurley	Troy, N. Y	June 5, 1866	55, 285
Nozzle for hose-pipes, hydrants, &c	I. O. Endicott	Manchester, N. H	Nov. 19, 1872	133, 215
Nozzle for hose-pipes, Stop	J. N. Allen	Providence, R. I	Nov. 25, 1873	144, 942
Nozzle for lead-pipe machine	J. B. Collan	Reading, Pa	Jan. 1, 1851	7, 867
Nozzle for liquid-jars, Measuring	J. H. Pein	New York, N. Y	Aug. 2, 1864	43, 703
Nozzle for liquid-package	H. H. Rogers	Brooklyn, N. Y	Aug. 15, 1871	118, 154
Nozzle for liquid-package, Screw	H. H. Rogers and G. F. Walter	Brooklyn, N. Y	Aug. 15, 1871	118, 155
Nozzle for locomotives, Exhaust	G. W. Richardson	Troy, N. Y	Mar. 11, 1873	136, 619
Nozzle for steam-engines, Exhaust	W. A. Foster	Fitchburgh, Mass	Sept. 20, 1870	107, 605
Nozzle for steam-engines, Exhaust	C. H. Frisbie	Chicago, Ill	Aug. 3, 1869	93, 294
Nozzle, Hose	M. S. Curtis and G. W. Harris	New York, N. Y	Dec. 17, 1867	72, 372
Nozzle, Hose	E. A. Day	Oberlin, Ohio	Feb. 25, 1873	136, 309
Nozzle, Hose	W. Jeffers	Pawtucket, R. I	Sept. 3, 1861	33, 200
Nozzle, Hose	C. F. Macy and S. Martin	Little York, Cal	Dec. 8, 1863	40, 847
Nozzle, Hose	J. B. Mitchell	Portland, Me	Aug. 25, 1868	81, 396
Nozzle, Hose	J. Trees	Greenburgh, Pa	Aug. 6, 1867	67, 613
Nozzle, Hose	G. O. Wickers	Lawrence, Mass	Apr. 9, 1872	125, 510
Nozzle, Hose and pipe	H. B. Morrison	Mount Morris, N. Y	Oct. 14, 1862	36, 688
Nozzle, Hose-pipe	A. F. Allen	Providence, R. I	Apr. 27, 1869	89, 456
Nozzle, Hose-pipe	A. F. Allen	Providence, R. I	Nov. 16, 1869	96, 862
Nozzle, Hose-pipe	A. F. Allen	Providence, R. I	Oct. 29, 1872	132, 617
Nozzle, Hose-pipe	O. J. Backus	San Francisco, Cal	Oct. 6, 1868	82, 676
Nozzle, Hose-pipe	B. A. Berryman and A. B. Wakefield.	Saint Louis, Mo	Apr. 15, 1873	137, 881
Nozzle, Hose-pipe	J. A. Cushman	Seneca Falls, N. Y	Aug. 18, 1868	81, 257

Index of patents issued from the United States Patent Office from 1790 *to* 1873, *inclusive*—Continued.

Invention.	Inventor.	Residence.	Date.	No.
Nozzle, Hose-pipe	F. E. Hall	Boston, Mass	Aug. 12, 1873	141,787
Nozzle, Hose-pipe	N. Hotz	Trenton, N. J	Jan. 31, 1860	26,994
Nozzle, Hose-pipe	R. F. Moore	Manchester, N. H	Nov. 27, 1866	60,036
Nozzle, Hose-pipe	A. W. Roberts	Hartford, Conn	May 8, 1855	12,831
Nozzle, Hose-pipe	A. M. White	New York, N. Y	Apr. 9, 1867	63,680
Nozzle, Hose-pipe	A. Williscroft	Wilmington, Del	Aug. 31, 1869	94,369
Nozzle, Hose-pipe	E. M. Wright	Marysville, Cal	Oct. 15, 1861	33,504
Nozzle, Hydraulic	A. Harris	La Porte, Cal	Dec. 28, 1869	98,257
Nozzle, Hydraulic	H. Shaw	Nevada City, Cal	Mar. 21, 1871	112,969
Nozzle of oil-cans, Sealing	J. A. Bostwick	New York, N. Y	Jan. 9, 1872	122,512
Nozzle, Oil-can	J. A. Boswick	New York, N. Y	Mar. 28, 1871	113,137
Nozzle, Oil-can	S. P. Doane	San Francisco, Cal	Sept. 19, 1871	119,081
Nozzle, Oil-can	G. H. Perkins	Philadelphia, Pa	Dec. 23, 1873	145,898
Nozzle, Oil-can	C. Pratt	New York, N. Y	Dec. 28, 1869	98,408
Nozzle, Pipe	F. S. Babbitt	Taunton, Mass	Aug. 18, 1868	81,242
Nozzle, Pump	C. Wilson	Columbia, Ohio	Mar. 4, 1873	136,478
Nozzle, Revolving hose	H. B. Morrison	Le Roy, N. Y	June 30, 1868	79,493
Nozzle, Revolving hose	I. Smith and H. D. Tewksbury	New York, N. Y	July 10, 1866	56,284
Nozzle, Steam	W. Ebbitt	New York, N. Y	July 15, 1873	140,902
Nozzle-stopper for oil-cans	J. H. Noyes	Abington, Mass	May 2, 1871	114,467
Nozzle, Valved oil-can	E. J. Durant	Lebanon, N. H	July 2, 1872	128,473
Nozzle, Variable exhaust	M. E. Brown	Buffalo, N. Y	Feb. 21, 1865	46,445
Nozzles and screw-caps for oil-can, Manufacture of	E. T. Covell	Brooklyn, N. Y	Nov. 16, 1869	96,890
Nozzles, Attaching revolving tip to hose	H. B. Morrison	Le Roy, N. Y	Aug. 25, 1863	39,700
Nozzles, Blast-protector for exhaust	J. H. Setchel	Cincinnati, Ohio	July 11, 1871	116,873
Nozzles, Elastic clasp for elastic	E. A. Day	Oberlin, Ohio	Dec. 31, 1872	134,469
Nubia	T. Dolan	Philadelphia, Pa	July 18, 1871	117,158
Nubia and veil, Combined	J. W. Tuttle	Watertown, Mass	Dec. 30, 1873	145,977
Numbering and paging machine	G. J. Hill	Buffalo, N. Y	June 27, 1865	48,488
Numbering and paging machine	A. C. Sine	Cincinnati, Ohio	Feb. 28, 1871	112,292
Numbering-head	C. W. Dickinson	Belleville, N. J	July 15, 1873	140,899
Numbering-machine	J. C. Clapp	South Boston, Mass	Apr. 12, 1864	42,274
Numbering-machine	W. H. Forbush	Buffalo, N. Y	Feb. 4, 1868	73,963
Numbering-machine	C. L. Sholes	Milwaukee, Wis	Sept. 12, 1871	118,978
Numbering-machine	J. D. Smith	Washington, D. C	Sept. 6, 1870	107,202
Numbering-machine	J. D. Smith	Washington, D. C	Dec. 16, 1873	145,598
Numbering-machine	S. W. Soulé	Milwaukee, Wis	June 18, 1867	65,839
Numbering-machine	S. W. Soulé and C. L. Sholes	Milwaukee, Wis	Nov. 13, 1866	59,675
Numbering-machine	H. Sutcliffe	Brooklyn, N. Y	Sept. 28, 1869	95,282
Numeral-frame	J. H. R. Reffelt	Hoboken, N. J	Mar. 3, 1863	37,825
Nursery-chair	A. B. Anderson, jr	Brooklyn, N. Y	June 30, 1863	39,026
Nursery-chair	C. M. Loring and E. Averell	Charlestown, Mass	Nov. 26, 1867	71,397
Nursery-chair	S. S. May	Sterling, Mass	June 4, 1850	7,418
Nursery-chair	S. Rainey	New Orleans, La	Nov. 28, 1865	51,215
Nursery-cup	J. F. Leslie and E. A. Tibbetts	Woburn, Mass	July 28, 1868	80,355
Nursery-gate	J. W. Boughton	Philadelphia, Pa	May 27, 1873	139,232
Nursery-gate	E. A. Tuttle	Brooklyn, N. Y	Aug. 12, 1873	141,677
Nursery gate, Child's	E. Howard	Milford, Mass	July 11, 1871	116,960
Nursery gate, Child's	E. Howard	Milford, Mass	Feb. 20, 1872	123,904
Nurseryman's knife	S. S. Jackson	Cincinnati, Ohio	Aug. 30, 1864	43,996
Nursing and other chairs	A. Woods	Charlestown, Va	Mar. 13, 1822	
Nursing-bottle	L. P. Dodge	New York, N. Y	Nov. 7, 1871	120,575
Nursing-cup	S. W. Talbot	Dedham, Mass	Apr. 19, 1833	
Nursing-table	J. Larkin	Unionville, S. C	Apr. 13, 1869	88,881
Nut	F. L. Delfer	Burlington, Iowa	Jan. 7, 1873	134,649
Nut	W. Harris and C. Browning	Rush Run, Ohio	Jan. 29, 1867	61,537
Nut	B. D. Sanders	Wellsburgh, W. Va	Oct 20, 1868	83,213
Nut	B. D. Sanders	Wellsburgh, W. Va	Oct. 20, 1868	83,214
Nut	L. Till	Sandusky, Ohio	July 3, 1866	56,124
Nut	W. Van Anden	Poughkeepsie, N. Y	Apr. 1, 1873	137,395
Nut, Adjustable	J. Russell	Springfield, Mass	Feb. 25, 1873	136,184
Nut and bolt dressing machine	J. Kohler	Cincinnati, Ohio	Jan. 21, 1873	135,131
Nut and bolt fastening	J. T. Antill and W. J. Sloan	Smith's Ferry, Pa., and Liverpool Township, Ohio.	Aug. 19, 1873	141,980
Nut and bolt fastening	L. L. Dunlap	Pontiac, Mich	Oct. 8, 1872	131,945
Nut and bolt fastening	P. Hayden	Allegheny, Pa	Oct. 1, 1872	131,755
Nut and bolt fastening	P. F. King	Saint Louis, Mo	Aug. 19, 1873	141,877
Nut and bolt fastening	M. F. McIntyer	Girard, Pa	Dec. 10, 1872	133,873
Nut and bolt fastening	A. C. Smith	Fort Madison, Iowa	Nov. 5, 1872	132,867
Nut and bolt fastening	J. L. Williams	Jackson, Tenn	Oct. 1, 1872	131,922
Nut and bolt fastening	E. B. Wingate	Friendship, N. Y	Jan. 14, 1873	134,963
Nut and coffee roaster	D. A. T. Gale	Poughkeepsie, N. Y	June 8, 1869	91,009
Nut and coffee roaster	D. T. Gale	Fort Wayne, Ind	May 20, 1873	139,054
Nut and washer	D. B. Hart	Mentor, Ohio	Aug. 6, 1867	67,539
Nut and washer machine	H. Carter and J. Rees	Pittsburgh, Pa	Aug. 26, 1851	8,322
Nut and washer machine	W. R. Wilbur	Cleveland, Ohio	June 10, 1873	139,696
Nut and washer machine	J. T. Wood and E. C. Smith	Pittsburgh, Pa	Jan. 12, 1864	41,252
Nut, Axle	E. A. Stanley	Brewer, Me	May 10, 1870	102,876
Nut-bar	S. Vanstone	Providence, R. I	June 1, 1869	90,897
Nut blank, Pressed	F. Washbourne	Brooklyn, N. Y	Feb. 21, 1871	111,996
Nut-blanks, Making	R. H. Cole	Saint Louis, Mo	Sept. 28, 1858	21,599
Nut-blanks, Rolled bar for	H. W. Oliver, jr	Pittsburgh, Pa	Feb. 14, 1871	111,772
Nut-blanks, tubes, &c., Piles for	J. Ostrander	Manchester, Va	Apr. 25, 1871	114,187
Nut-blanks, tubes, &c., Piles for	J. Ostrander	Manchester, Va	Apr. 25, 1871	114,188
Nut-box	R. H. Cole	Saint Louis, Mo	Jan. 1, 1856	14,011
Nut-box, Adjustable	J. Turner	Richmond, Va	Nov. 13, 1866	59,731
Nut, Cast	G. P. Darrow	Cincinnati, Ohio	Aug. 25, 1868	81,480
Nut, Clamp	W. Pearson	Windsor Locks, Conn	Aug. 11, 1868	80,997
Nut-cracker	P., E. W., and J. A. Blake	New Haven, Conn	Sept. 6, 1853	9,985
Nut-cracker	P. Ceredo	Dusseldorf, Prussia	Nov. 22, 1870	109,495
Nut-cracker	L. A. Clark	Bridgeport, Conn	Jan. 24, 1860	26,885
Nut-cracker	T. Earle	Providence, R. I	Dec. 8, 1863	40,825
Nut-cracker	R. Frisbie	Middletown, Conn	May 17, 1859	24,018
Nut-cracker	C. Hayden	Collinsville, Conn	Nov. 10, 1868	83,959
Nut-cracker	E. L. Pratt	Boston, Mass	Mar. 31, 1868	76,247
Nut-cracker	J. Pusey	Philadelphia, Pa	Nov. 8, 1870	119,144
Nut-cracker	E. Ripley	Troy, N. Y	May 31, 1859	24,238

Index of patents issued from the United States Patent Office from 1790 *to* 1873, *inclusive*—Continued.

Invention.	Inventor.	Residence.	Date.	No.
Nut-cracker	S. J. Smith	New York, N. Y	May 15, 1860	28, 311
Nut-fastener	C. Buckley, jr	Rochester, N. Y	Sept. 24, 1867	69, 173
Nut-fastener	J. Christy	Philadelphia, Pa	Apr. 21, 1868	76, 995
Nut-fastener	J. Davis	New Bedford, Mass	Apr. 20, 1869	89, 206
Nut-fastener	M. Grover	Clyde, Ohio	Sept. 14, 1869	94, 885
Nut-fastener	W. Harris and C. Browning	Rush Run, Ohio	Jan. 29, 1867	61, 620
Nut-fastening	W. Harris	Rush Run, Ohio	Dec. 24, 1867	72, 486
Nut-fastening	F. Myers	New York, N. Y	Oct. 18, 1870	108, 381
Nut-fastening	F. Myers	New York, N. Y	Oct. 18, 1870	108, 382
Nut-fastening	A. M. Rouse	Saint Louis, Mo	Dec. 31, 1872	134, 489
Nut-fastening	S. A. Todd	Wellsville, Ohio	Dec. 10, 1872	133, 812
Nut-fastening device	D. Sawyer	Topeka, Kans	Sept. 24, 1872	131, 630
Nut-fastenings, Washer for	H. P. Hood	Indianapolis, Ind	Jan. 21, 1873	134, 990
Nut-finishing machine	F. P. Pleghar and W. Schöllhorn	New Haven, Conn	Oct. 25, 1864	44, 818
Nut for joint-bolts	J. Baldwin	Manchester, N. H	Apr. 7, 1868	76, 380
Nut for screw-bolts	N. Thompson	Brooklyn, N. Y	Feb. 8, 1870	99, 729
Nut for securing piston-heads to piston-rods	J. Wheelock	Worcester, Mass	June 24, 1873	140, 328
Nut-forging machine	E. Paye and S. Hall	New York, N. Y	Sept. 8, 1859	18, 156
Nut-forging machine	A. P. Plant and A. Shepard	Plantsville, Conn	June 23, 1868	79, 255
Nut, Jam	B. W. Nichols	Canton, Ohio	Apr. 20, 1869	89, 066
Nut, Jam	W. H. Van Cleve	Ypsilanti, Mich	Nov. 7, 1871	120, 798
Nut, Jam	J. M. Winslow	Rochester, N. Y	June 4, 1872	127, 445
Nut-lock	S. C. Adams	Buffalo, N. Y	June 15, 1869	91, 197
Nut-lock	J. M. Allison	Salina, Pa	Apr. 8, 1873	137, 646
Nut-lock	W. E. Ball	Belmont, Ohio	Apr. 27, 1869	89, 460
Nut-lock	W. E. Ball	Bethesda, Ohio	Aug. 20, 1872	130, 689
Nut-lock	M. L. Ballard	Canton, Ohio	Apr. 1, 1873	137, 401
Nut-lock	H. Beagle, jr	Philadelphia, Pa	May 31, 1870	103, 544
Nut-lock	J. Bell	New York, N. Y	Oct. 4, 1870	107, 998
Nut-lock	F. A. Bishop	Shingle Springs, Cal	Dec. 10, 1872	133, 694
Nut-lock	W. H. Bowman	London, Ohio	Nov. 25, 1873	144, 886
Nut-lock	F. H. Bradley	Mystic River, Conn	July 23, 1872	129, 710
Nut-lock	K. Brown	Gardner, Ill	Apr. 27, 1869	89, 283
Nut-lock	J. M. Connel	Newark, Ohio	Aug. 1, 1871	117, 517
Nut-lock	E. A. Cooper	Lancaster, N. Y	Nov. 11, 1873	144, 511
Nut-lock	J. R. Cribbs	Gardner, Ill	Mar. 16, 1869	87, 761
Nut-lock	D. Cumming, jr	Brooklyn, N. Y	Aug. 23, 1870	106, 668
Nut-lock	M. A. Cushing and O. R. Glover	Ottawa, Ill	Dec. 6, 1870	109, 809
Nut-lock	E. Czarniecki	Allegheny, Pa	June 24, 1873	140, 119
Nut-lock	B. W. Davis	Fort Madison, Iowa	Mar. 18, 1873	136, 906
Nut-lock	J. Dennis	Churchville, N. Y	Aug. 9, 1870	106, 138
Nut-lock	L. L. Deweese	Canton, Ohio	June 22, 1869	91, 724
Nut-lock	C. Dittman	Leacock, Pa	May 24, 1870	103, 308
Nut-lock	C. Dittman	Leacock Post-Office, Pa	Nov. 25, 1873	144, 964
Nut-lock	E. H. Dooley	New York, N. Y	Jan. 16, 1872	122, 761
Nut-lock	H. W. Dopp	Buffalo, N. Y	Aug. 26, 1873	142, 215
Nut-lock	J. R. Eichenberger and C. G. Binkly.	Burton City, Ohio	Apr. 1, 1873	137, 426
Nut-lock	L. Fay	Worcester, Mass	Mar. 14, 1871	112, 574
Nut-lock	J. H. Fisher	Chicago, Ill	Oct. 19, 1869	95, 890
Nut-lock	O. S. Freeland	Newport, R. I	Feb. 22, 1870	100, 023
Nut-lock	B. Furst and P. Oettinger	Lacon, Ill	Dec. 31, 1872	134, 426
Nut-lock	R. Gilliland	Hudson, Mich	Sept. 20, 1870	107, 609
Nut-lock	G. Hart	Columbiana, Ohio	June 25, 1872	128, 391
Nut-lock	F. C. Hays	Uhricksville, Ohio	July 23, 1872	129, 822
Nut-lock	M. Hays	Binghamton, N. Y	Apr. 15, 1873	137, 919
Nut-lock	D. Hoffman and W. Johnston	Noblestown and Havelock, Pa.	June 7, 1870	104, 033
Nut-lock	H. P. Hood and W. Coombs	Indianapolis, Ind., and Bangor, Me.	Feb. 4, 1873	135, 553
Nut-lock	W. P. Horton	Milwaukee, Wis	Dec. 24, 1872	134, 204
Nut-lock	C. A. Howard	Pontiac, Mich	Nov. 25, 1873	144, 905
Nut-lock	A. C. Hull and J. S. Brown	Saint Louis, Mo., and Mount Pleasant, D. C.	June 3, 1873	139, 467
Nut-lock	E. Kaylor	Pittsburgh, Pa	Dec. 9, 1873	145, 427
Nut-lock	G. W. Keen	Zanesville, Ohio	Apr. 1, 1873	137, 453
Nut-lock	L. Leeds	New London, Conn	Dec. 30, 1873	146, 081
Nut-lock	K. H. Loomis	New York, N. Y	Nov. 26, 1872	133, 324
Nut-lock	K. H. Loomis	Cleveland, Ohio	Sept. 9, 1873	142, 709
Nut-lock	H. C. Lowe	North East, Md	Sept. 16, 1873	142, 858
Nut-lock	H. W. McAuley	Sterling, Wis	Oct. 25, 1870	108, 613
Nut-lock	T. B. McConaughey and J. Adams.	Newark, Del	May 17, 1870	103, 220
Nut-lock	H. McCown	Enon Valley, Pa	May 25, 1869	90, 567
Nut-lock	J. Miller, jr	Richmond, Ind	Aug. 20, 1872	130, 736
Nut-lock	L. J. Miller	Orrville, Ohio	Apr. 29, 1873	138, 268
Nut-lock	J. Moorcroft	Newport, R. I	Sept. 27, 1870	107, 704
Nut-lock	A. Morley	New Troy, Mich	May 28, 1872	127, 359
Nut-lock	I. E. Nagle	New Orleans, La	Apr. 1, 1873	137, 382
Nut-lock	G. Palmer	Littletown, Pa	June 1, 1869	90, 867
Nut-lock	M. Payne	Cardington, Ohio	Dec. 21, 1869	98, 187
Nut-lock	S. Peatfield	Ipswich, Mass	Oct. 28, 1873	144, 131
Nut-lock	R. G. Peterson and J. Coulter	Perryville, Ohio	June 4, 1872	127, 509
Nut-lock	W. H. Phipps	Southborough, Mass	June 13, 1871	115, 889
Nut-lock	A. Porter	Chicago, Ill	May 27, 1873	139, 269
Nut-lock	W. P. Porter	Pittsburgh, Pa	June 1, 1869	90, 685
Nut-lock	P. F. Randolph	Jerseyville, Ill	May 2, 1871	114, 474
Nut-lock	J. A. Reed	Dunellen, N. J	July 8, 1873	140, 644
Nut-lock	T. E. Rhine	Quaker City, Ohio	Nov. 19, 1872	133, 253
Nut-lock	J. F. Saiger	Shelby, Ohio	July 9, 1872	128, 815
Nut-lock	D. Sawyer	Washington, Ind	Sept. 23, 1873	143, 097
Nut-lock	A. C. Smith	Fort Madison, Iowa	Dec. 30, 1873	146, 102
Nut-lock	C. S. Southwick and D. H. Baker.	Newport, R. I	Apr. 5, 1870	101, 674
Nut-lock	G. C. Stamper	Osceola, Iowa	May 18, 1869	90, 132
Nut-lock	J. B. Sweetland	Pontiac, Mich	Dec. 30, 1873	146, 106
Nut-lock	C. H. Taylor	New York, N. Y	Sept. 16, 1873	142, 877
Nut-lock	D. O. Thompson and B. A. Rice	Pontiac, Mich	Feb. 11, 1873	135, 864

Index of patents issued from the United States Patent Office from 1790 to 1873, inclusive—Continued.

Invention.	Inventor.	Residence.	Date.	No.
Nut-lock	F. P. Thompson	Frederickton, Canada	July 29, 1873	141, 403
Nut-lock	S. I. Thompson	New Waterford, Ohio	May 16, 1871	114, 991
Nut-lock	E. Turner	Greensburgh, Pa	Apr. 1, 1873	137, 513
Nut-lock	W. B. Wait	New York, N. Y	Apr. 22, 1873	138, 214
Nut-lock	C. R. Watrous	Mystic, Conn	Dec. 30, 1873	146, 113
Nut-lock	S. H. Wheeler	Dowagiac, Mich	Oct. 11, 1870	108, 223
Nut-lock	G. S. Willard	Salesville, Ohio	Oct. 8, 1872	132, 079
Nut-lock	A. Williams	Burrell, Pa	Jan. 14, 1873	134, 781
Nut-lock and washer	E. A. Ellsworth	Washington, D. C	Dec. 3, 1867	71, 722
Nut-lock bolt	T. T. Prosser	Chicago, Ill	May 31, 1870	103, 775
Nut-lock for fish-plates	J. W. Hazelton, A. A. Southard, and O. Merwin.	Drayton Plains and Elba, Mich.	May 4, 1869	89, 659
Nut-lock plate	E. D. Taylor	Hornellsville, N. Y	Apr. 5, 1870	101, 541
Nut-lock washer	M. F. McIntyer	Girard, Pa	Mar. 18, 1873	136, 848
Nut-locking device	I. Allen	Taunton, Mass	July 30, 1872	130, 003
Nut-locking device	J. L. Estel	Salem, Ohio	Nov. 28, 1871	121, 348
Nut-locking device	A. B. Davis	Philadelphia, Pa	Jan. 23, 1872	122, 999
Nut-locking device	J. M. Horton	Milwaukee, Wis	Sept. 26, 1871	119, 360
Nut-locking device	P. L. Gibbs	Dunleith, Ill	May 28, 1872	127, 336
Nut-locking device	T. W. Kirkwood	McKeesport, Pa	Apr. 2, 1872	125, 309
Nut-locking device	K. H. Loomis	New York, N. Y	Jan. 23, 1872	122, 900
Nut-locking device	S. B. Lowe	Chattanooga, Tenn	Feb. 20, 1872	123, 831
Nut-locking device	J. Maitland	Newburgh, Ohio	Dec. 12, 1871	121, 723
Nut-locking device	H. McCown	Enon Valley, Pa	Mar. 16, 1869	87, 953
Nut-locking device	A. McKenney	Maumee, Ohio	Aug. 22, 1871	118, 256
Nut-locking device	W. Moorehouse	Buffalo, N. Y	Oct. 13, 1868	83, 081
Nut-locking device	P. Philippi	Beardstown, Ill	Apr. 6, 1869	88, 668
Nut-locking device	H. L. Purdie	Buffalo, N. Y	Dec. 5, 1871	121, 546
Nut-locking device	H. L. Purdie	Buffalo, N. Y	Dec. 5, 1871	121, 547
Nut-locking device	H. L. Purdie	Buffalo, N. Y	Dec. 5, 1871	121, 548
Nut-locking device	H. L. Purdie	Buffalo, N. Y	Dec. 5, 1871	121, 549
Nut-locking device	J. Wetmore	Salem, Ohio	Jan. 21, 1873	135, 052
Nut-locking washer	J. H. Gridley	Washington, D. C	Feb. 26, 1867	62, 483
Nut-locking washer	R. Rutter	Vallejo, Cal	May 31, 1870	103, 780
Nut-locking washer	W. H. Williams	Canton, Ohio	July 6, 1869	92, 241
Nut-locking washer	L. Winslow	Rochester, N. Y	July 30, 1872	130, 097
Nut-machine	A. B. Bean	New Haven, Conn	Apr. 28, 1868	77, 347
Nut-machine	J. R. Blakeslee	Youngstown, Ohio	Oct. 8, 1872	132, 047
Nut-machine	J. Braun	Philadelphia, Pa	Apr. 22, 1873	138, 066
Nut-machine	R. Brayton	Buffalo, N. Y	Aug. 4, 1857	17, 914
Nut-machine	O. C. Burdict	New Haven, Conn	Apr. 10, 1866	53, 782
Nut-machine	O. C. Burdict	New Haven, Conn	Sept. 3, 1867	68, 556
Nut-machine	H. Carter and J. Rees	Pittsburgh, Pa	Nov. 22, 1853	10, 249
Nut-machine	R. H. Cole	Saint Louis, Mo	July 17, 1855	13, 252
Nut-machine	R. H. Cole	Saint Louis, Mo	June 3, 1856	15, 001
Nut-machine	R. H. Cole	Saint Louis, Mo	May 5, 1857	17, 197
Nut-machine	R. H. Cole	Saint Louis, Mo	Oct 27, 1857	18, 499
Nut-machine	R. H. Cole	Saint Louis, Mo	May 4, 1858	20, 145
Nut-machine	R. H. Cole	Saint Louis, Mo	Sept. 21, 1858	21, 551
Nut-machine	J. C. Day	Jersey City, N. J	Dec. 22, 1857	18, 892
Nut-machine	G. Dunham	Unionville, Conn	June 27, 1865	48, 383
Nut-machine	G. Dunham	Unionville, Conn	Nov. 14, 1865	50, 923
Nut-machine	G. H. Fuller	Unionville, Conn	May 9, 1871	114, 664
Nut-machine	A. B. Glover	Birmingham, Conn	July 7, 1868	79, 749
Nut-machine	R. Griffiths	Allegheny, Pa	Oct. 30, 1855	13, 720
Nut-machine	R. Griffiths	Allegheny City, Pa	Mar. 18, 1856	14, 452
Nut-machine	R. Griffiths	Philadelphia, Pa	Dec. 2, 1856	16, 142
Nut-machine	J. S. Hall	Pittsburgh, Pa	Aug. 11, 1868	80, 946
Nut-machine	J. Harlem	Philadelphia, Pa	June 4, 1867	65, 380
Nut-machine	D. Howell	Louisville, Ky	July 2, 1867	66, 238
Nut-machine	E. Hubner and C. Hall	New York, N. Y	Dec. 5, 1865	51, 316
Nut-machine	P. Koch	Manchester, England	May 9, 1871	114, 691
Nut-machine	P. Miles	New Haven, Conn	Jan. 1, 1861	31, 056
Nut-machine	C. Ratcliff	Cincinnati, Ohio	Dec. 9, 1856	16, 188
Nut-machine	J. B. Savage	Southington, Conn	Dec. 14, 1858	22, 310
Nut-machine	A. Schwebel	New Haven, Conn	Sept. 24, 1867	69, 256
Nut-machine	A. J. Shepard	Buffalo, N. Y	Dec. 13, 1859	26, 446
Nut-machine	L. Thomas	Allegheny City, Pa	Aug. 11, 1863	39, 504
Nut-machine	W. E. Ward	Port Chester, N. Y	Oct. 7, 1856	15, 861
Nut-machine	F. Watkins	Birmingham, England	Dec. 8, 1868	84, 781
Nut-machine	S. H. Whitaker	Cincinnati, Ohio	Jan. 27, 1857	16, 507
Nut-machine	S. H. Whitaker	Cincinnati, Ohio	June 9, 1857	17, 534
Nut-machine	S. H. Whitaker	Cincinnati, Ohio	Sept. 22, 1857	18, 259
Nut-machine	S. H. Whitaker	Cincinnati, Ohio	Oct. 19, 1858	21, 860
Nut-machine die	O. C. Burdict	New Haven, Conn	Jan. 5, 1864	41, 137
Nut-machine die	O. C. Burdict	New Haven, Conn	Apr. 10, 1866	53, 780
Nut-machines, Reversible die-box for	W. Anderson	Pittsburgh, Pa	Nov. 16, 1869	96, 763
Nut-making machine	E. R. Addison	Wheeling, W. Va	July 21, 1868	80, 108
Nut-making machine	A. B. Bean	New Haven, Conn	Mar. 19, 1867	62, 923
Nut-making machine	J. R. Blakeslee	Youngstown, Ohio	July 16, 1872	129, 520
Nut-making machine	B. H. Bradley	New York, N. Y	July 5, 1870	105, 033
Nut-making machine	J. R. Bridges	Pittsburgh, Pa	Aug. 6, 1867	67, 403
Nut-making machine	O. C. Burdict	New Haven, Conn	Dec. 8, 1863	40, 815
Nut-making machine	W. Chisholm	Cleveland, Ohio	Nov. 17, 1863	40, 606
Nut-making machine	P. Coleman	Philadelphia, Pa	Feb. 14, 1865	46, 337
Nut-making machine	G. Dunham	Unionville, Conn	Aug. 6, 1867	67, 421
Nut-making machine	A. Emerson	New York, N. Y	May 7, 1867	64, 510
Nut-making machine	M. H. Foster and H. C. Hart	Unionville, Conn	May 5, 1868	77, 476
Nut-making machine	G. H. Fuller	Unionville, Conn	Feb. 16, 1869	86, 914
Nut-making machine	A. B. Glover	Yonkers, N. Y	May 10, 1864	42, 654
Nut-making machine	W. Horsfall	Sing Sing, N. Y	Feb. 16, 1869	87, 046
Nut-making machine	W. W. Hubbard	Philadelphia, Pa	Sept. 18, 1866	58, 104
Nut-making machine	E. Kaylor	ittsburgh, Pa	July 3, 1866	56, 062
Nut-making machine	E. Kaylor	Perrysville, Pa	Dec. 1, 1868	84, 494
Nut-making machine	P. Koch	New Haven, Conn	May 5, 1863	38, 397
Nut-making machine	L. A. Livingood	Womelsdorf, Pa	Sept. 3, 1872	131, 105
Nut-making machine	J. Paton	Newburgh, Ohio	Nov. 29, 1864	45, 268
Nut-making machine	J. Reese	Sharon, Pa	Feb. 7, 1854	10, 493

Index of patents issued from the United States Patent Office from 1790 *to* 1873, *inclusive*—Continued.

Invention.	Inventor.	Residence.	Date.	No.
Nut-making machine	T. R. Taylor	Cleveland, Ohio	Oct. 21, 1862	36, 737
Nut-making machine	S. Vanstone	Providence, R. I	Nov. 3, 1868	83, 745
Nut-making machine	L. Weckesser	New Haven, Conn	Dec. 13, 1864	45, 446
Nut-mandrel	G. A. Gray, jr	Cincinnati, Ohio	June 29, 1869	91, 841
Nut-roaster	D. A. T. Gale	Poughkeepsie, N. Y	Aug. 18, 1868	81, 159
Nut, Rubber jam	W. Todd	Allegheny City, Pa	Jan. 14, 1873	134, 823
Nut, Screw	L. Derby	New York, N. Y	Jan. 20, 1863	37, 435
Nut, Self-locking	J. H. Raymond and W. J. Brassington.	Brooklyn, N. Y	Feb. 14, 1865	46, 389
Nut, Spring	D. R. Pratt	Worcester, Mass	Jan. 25, 1870	99, 232
Nut, Take-up	J. Ross and W. W. Graham	Rutland, Vt	Nov. 15, 1870	109, 348
Nut-tapping and screw-cutting machine	G. Bennett and R. Dalzell	Waddington, N. Y	Jan. 22, 1861	31, 148
Nut-tapping machine	J. W. Durrell	Dunkirk, N. Y	May 30, 1871	115, 451
Nut-tapping machine	J. L. Gill, jr	Columbus, Ohio	Apr. 4, 1871	113, 512
Nut-tapping machine	A. B. Glover	Burmingham, Conn	July 7, 1857	17, 734
Nut-tapping machine	T. Hull and N. Thomas	Chicago, Ill	Aug. 13, 1872	130, 502
Nut-tapping machine	J. Kirkley	Chicago, Ill	Nov. 12, 1867	70, 862
Nut-tapping machine	S. W. Putnam, jr	Fitchburgh, Mass	Sept. 2, 1873	142, 511
Nut-tapping machine	S. W. Putnam, jr	Fitchburgh, Mass	July 8, 1873	147, 727
Nut-tapping machine	W. Scully	Detroit, Mich	Aug. 9, 1870	106, 215
Nut-tapping machine	T. Votteler	Cincinnati, Ohio	June 26, 1866	55, 965
Nut-tapping machine	F. Watkins	Birmingham, England	Nov. 15, 1864	45, 110
Nut-tapping machinery	D. Reese	Newburgh, Ohio	July 3, 1866	56, 095
Nut-threading machine	O. P. and L. W. Briggs	Chicago, Ill	Apr. 15, 1873	137, 757
Nut-threading machine	W. W. Hubbard	Philadelphia, Pa	Nov. 8, 1864	44, 954
Nut-turning device	B. Boardman	Norwich, Conn	Apr. 21, 1868	76, 983
Nut-wrench and drill	T. A. Chandler	Rockford, Ill	Sept. 26, 1871	119, 317
Nuts and bolt-heads, Machine for dressing	J. King	Bordentown, N. J	Feb. 27, 1849	6, 140
Nuts and bolts, Clamp for holding	W. J. Lewis	Pittsburgh, Pa	Apr. 12, 1864	42, 299
Nuts and washers, Machine for making	R. Brayton	Buffalo, N. Y	Jan. 9, 1855	12, 194
Nuts and washers, Machine for making	C. P. Geissenhainer	Pittsburgh, Pa	Aug. 7, 1866	57, 042
Nuts and washers, &c., Machine for making	W. Kenyon	Steubenville, Ohio	Oct. 14, 1851	8, 427
Nuts and washers, Machine for making	I. Scoville	Chicago, Ill	Aug. 18, 1863	39, 590
Nuts, bolts, &c., Manufacturing	J. Marsden	Orrell, England	Apr. 7, 1863	38, 116
Nuts, Device for dressing the sides of	I. Doeg	New Market, N. H	Nov. 4, 1873	144, 265
Nuts, Device for locking screw	W. F. Vernier	Philadelphia, Pa	Nov. 10, 1863	40, 583
Nuts from unscrewing, Mode of preventing	S. Noblet	Halifax, Pa	Sept. 21, 1858	21, 574
Nuts, Locking	D. Cumming, jr	New York, N. Y	June 16, 1868	78, 864
Nuts, Locking	W. Mullins	Pittsburgh, Pa	Apr. 7, 1868	76, 500
Nuts, Machine for making hot-pressed	L. Thierry and G. B. Hill	Detroit, Mich	July 2, 1867	66, 414
Nuts, Machine for making metallic	A. S. Upson	Unionville, Conn	June 15, 1869	91, 288
Nuts, Machine for manufacturing iron	B. Hobbs	Newburgh, Ohio	Jan. 26, 1869	86, 302
Nuts, Machine for polishing metallic	R. H. and J. C. Cole	Saint Louis, Mo	June 3, 1856	15, 004
Nuts, Machine for rolling hollow bars for	J. Ostrander	Manchester, Va	Apr. 4, 1871	113, 332
Nuts, Making	R. H. Cole	Saint Louis, Mo	June 3, 1856	15, 003
Nuts, Making	I. H. Steer	Winchester, Va	June 19, 1855	13, 118
Nuts, Making	J. H. Sternbergh	Reading, Pa	Aug. 18, 1868	81, 224
Nuts, Making	O. W. Yale	Hartford, Conn	Sept. 22, 1868	82, 470
Nuts, Making metallic	S. W. Wood	Washington, D. C	Apr. 27, 1859	23, 118
Nuts, Making screw	F. Rheydt	Chicago, Ill	Dec. 29, 1868	85, 399
Nuts, Manufacture of	J. W. Gaskill and J. Christie	Phillipsburgh, N. J	Dec. 14, 1869	97, 901
Nuts, Manufacture of	L. Thomas	Allegheny City, Pa	Apr. 12, 1864	42, 320
Nuts, Manufacture of screw-threaded	C. H. Guard	Toronto, Canada	Feb. 20, 1872	123, 823
Nuts on axles, Securing	T. W. Williams	Philadelphia, Pa	May 26, 1857	17, 410
Nuts on bolts, Locking	G. J. Harris	Brooklyn, N. Y	June 25, 1872	128, 389
Nuts on bolts, Machine for putting	B. S. Walker	Sing Sing, N. Y	Jan. 23, 1872	122, 927
Nuts, Stop-washer for	H. N. Armstrong	Erie, Pa	June 27, 1865	48, 353
Nuts, Substitute for jam	T. B. Young	Louisville, Ky	May 3, 1870	102, 644
Nuts, Tapping	H. C. Hart and J. B. Blakeslee	Unionville, Conn	Dec. 24, 1867	72, 487
Nuts to bolts, Device for applying rubber	J. Minetree	Petersburgh, Va	Dec. 10, 1872	133, 716
Nutritive and curative preparation	S. C. Upham	Philadelphia, Pa	Feb. 26, 1867	62, 512
O.				
Oakum by machinery, Picking	E. Waterhouse	Gardiner, Me	Apr. 24, 1826	....
Oakum, Combination of machinery for picking	J. Stansbury and W. Ridgaway	Baltimore, Md	Jan. 17, 1842	2, 428
Oakum, curled hair, &c., Machine for picking	H. Burnham	Boston, Mass	Oct. 5, 1838	962
Oakum from junk, Manufacturing	E. Cook	Haddam, Conn	July 9, 1832	....
Oakum-machine	J. Tibbals	Haddam, Conn	Feb. 8, 1839	1, 077
Oakum, Machine for balling	S. G. Archibald	Edinburgh, North Britain	Apr. 18, 1871	113, 831
Oakum, Machine for picking	O. Allen	Tewksbury, Mass	July 16, 1842	2, 723
Oakum, Machine for picking	M. Morrison	Newburyport, Mass	Feb. 14, 1806	....
Oakum, Machine for tarring	R. Mansley	Philadelphia, Pa	Jan. 18, 1859	22, 662
Oakum, Machine for twisting	L. and C. Howard	Watkins, N. Y	Jan. 28, 1873	135, 338
Oakum, Machinery for the manufacture of	T. Henry and D. Tibbal	Chatham, Conn	July 8, 1843	3, 163
Oakum, Making	D. Rider	New York	Apr. 26, 1808	....
Oakum, Manufacture of	T. H. Dunham	Boston, Mass	Oct. 1, 1872	131, 859
Oakum, Manufacture of	L. S. Robbins and J. A. Southmayd.	New York, N. Y., and Elizabeth, N. J.	Mar. 2, 1869	87, 434
Oakum, Manufacturing	E. Cook and S. Usher	Haddam, Conn	June 7, 1834	....
Oakum, Manufacturing old junk or old rigging into.	D. French	Middletown, Conn	Oct. 9, 1805	....
Oakum, Process for preparing	J. A. and G. Cormack	New York, N. Y	June 8, 1852	8, 995
Oakum with tar, Coating	J. F. Stairs	Halifax, Canada	Dec. 30, 1873	146, 105
Oar	H. W. Connor	Troy, N. Y	Mar. 23, 1869	88, 013
Oar	N. Davenport	Troy, N. Y	June 6, 1871	115, 584
Oar	S. W. Francis	New York, N. Y	Oct. 22, 1867	69, 983
Oar	R. E. Gleason	Libertyville, Ill	June 15, 1869	91, 226
Oar	W. J. Hough	Martinez, Cal	Jan. 18, 1870	98, 868
Oar	A. S. Jacobs	Saint Louis, Mo	July 16, 1867	66, 847
Oar	B. J. Kellam	Tremont, N. Y	Mar. 12, 1867	62, 755
Oar	T. Lindsley	Moline, Ill	Aug. 6, 1861	33, 001
Oar, Boat	R. Rode	Manchester Township, Pa	Sept. 23, 1856	15, 794
Oar, Boat	R. Rode	Manchester, Pa	Feb. 17, 1857	16, 656
Oar-collar	J. Robison	Curwinsville, Pa	Aug. 13, 1867	67, 804
Oar for steering, Head	C. Seaman	Washington County, Md	Aug. 15, 1817	....
Oar, Jointed	C. Dann	La Crosse, Wis	Sept. 27, 1870	107, 669
Oar-lock	W. E. Beman	Portland, Me	June 29, 1869	91, 817
Oar-lock	W. P. Glading	New York, N. Y	Feb. 21, 1854	10, 543

Index of patents issued from the United States Patent Office from 1790 *to* 1873, *inclusive*—Continued.

Invention.	Inventor.	Residence.	Date.	No.
Oar-lock	H. Hempstead	Greenport, N. Y	Aug. 24, 1869	93, 992
Oar-lock	M. A. Lanagan	Brooklyn, N. Y	Dec. 31, 1867	72, 865
Oar or boat-fin, Rigged	R. Smith	Brooklyn, N. Y	Sept. 12, 1865	49, 930
Oar, Sculling	F. T. Angers	Canastota, N. Y	Oct. 12, 1869	95, 754
Oar, Steel-plated	W. H. McMillan	New York, N. Y	Apr. 19, 1864	42, 389
Oar-swivel	M. Fryer	Greenbush, N. Y	Oct. 2, 1866	58, 402
Oars for boats, Machine for sawing boards into	J. Benson, E. Page, and R. T. Hough.	Boston, Mass., and West Leyden, N. Y.	May 26, 1842	2, 643
Oars, Machine for turning boat	B. and A. F. Potter	Hubbardston, Mass	Jan. 20, 1844	3, 408
Oars, &c., Machinery for turning	E. Page	Barcelona, N. Y	Dec. 26, 1845	4, 325
Oat-cleaner	S. Dickens, jr	Milwaukee, Wis	Dec. 13, 1870	110, 122
Oat-cleaning machine	W. D. Freeman	Tomales, Cal	June 8, 1869	91, 110
Oat-dusting machine	R. Exelby and G. W. Marshall	Buffalo, N. Y	Oct. 13, 1868	83, 054
Oat-separator	L. Patrie	Victor, N. Y	May 19, 1863	38, 630
Oats, barley, &c., Machine for hulling	J. Andrews and E. Piper	Camden, Me	Dec. 10, 1840	1, 894
Oats from wheat, Screen for separating	A. C. Ferren	Decorah, Iowa	July 29, 1862	36, 052
Observatory	L. B. Sawyer	Charlestown, Mass	Oct. 9, 1866	58, 680
Observatory and signal-tower, Portable	E. Tanner	Bowmansville, N. Y	Aug. 25, 1863	39, 689
Observatory, Military	T. Welham	Nemaha County, Nev	Dec. 16, 1862	37, 207
Obstetric chair and supporter	A. Blood	Janesville, Wis	Aug. 27, 1850	7, 590
Obstetrics, Manakin, with fœtus, &c., for illustrating the practice of.	B. H. Aylworth	Oxford, N. Y	Mar. 30, 1869	88, 432
Obstetrical bandage	J. O. Hamilton	Jerseyville, Ill	Oct. 8, 1867	69, 659
Obstetrical chair	C. C. Wingo	Newport, Va	Aug. 24, 1858	21, 291
Obstetrical extractor	A. C. Buffum	Chicago, Ill	Aug. 21, 1855	13, 453
Obstetrical supporter	F. H. Chase, A. Weston, and L. Babbit.	Clintonville and Au Sable, N. Y.	Aug. 20, 1850	7, 573
Obstetrical supporter	W. S. Daniels	Panama, N. Y	Mar. 14, 1854	10, 649
Obstetrical supporter	W. W. Finch, J. Blaisdell, and L. Babbit.	Essex County, N. Y	Jan. 15, 1850	7, 019
Obstetrical supporter	S. B. Manly	Corry, Pa	Sept. 3, 1867	68, 521
Obstetrical supporter	A. Pollard and S. Minkler	Clinton County, N. Y	Apr. 24, 1849	6, 385
Octave-coupling for reed-instruments	B. O. Church and H. Smith	Brattleborough, Vt	July 9, 1867	66, 460
Odometer	S. Beers	Naugatuck, Conn	Aug. 12, 1856	15, 509
Odometer	D. L. Branning	Tampa, Fla	Feb. 28, 1871	112, 116
Odometer	A. Carter	Forestville, Conn	Nov. 4, 1856	16, 003
Odometer	J. Clarke	Powhatan County, Va	Nov. 30, 1818	....
Odometer	N. R. Coburn	Lowell, Mass	Oct. 17, 1865	50, 456
Odometer	A. D. Hoffman	Wayne, Mich	Nov. 22, 1864	45, 157
Odometer	A. F. Howard	Hartford, Conn	July 3, 1860	28, 985
Odometer	L. W. Nicholls	North Brookfield, N. Y	Nov. 13, 1860	30, 640
Odometer	W. Oldroyd	Mount Vernon, Ohio	Nov. 14, 1848	5, 914
Odometer	W. H. Prescott and W. Judson	Galesburgh, Ill	Feb. 19, 1867	62, 152
Odometer	W. H. Prescott and W. Judson	Galesburgh, Ill	Mar. 24, 1868	75, 789
Odometer	J. C. Spencer	Phelps, N. Y	Sept. 17, 1867	69, 038
Odometer	M. W. Stevens and E. H. Drake	Stoughton, Mass	Oct. 1, 1867	69, 504
Odometer	J. Thompson	Middleborough, Mass	Oct. 31, 1854	11, 878
Odometer	S. R. Thorp	Batavia, N. Y	Oct. 31, 1854	11, 877
Odometer	H. Walker	Hartford, Vt	Mar. 22, 1859	23, 329
Odometer	J. M. Whitney	Bolton, Mass	Mar. 20, 1860	27, 589
Odometer	T. K. Work	Hartford, Conn	Feb. 8, 1859	22, 912
Odometer and counting-machine	J. L. Martin	Baltimore, Md	June 17, 1856	15, 140
Odometer-case	J. P. Seipel and C. B. Alsover	Easton, Pa	Jan. 17, 1871	111, 008
Odometer for carriage-wheels	S. Beers	Waterbury, Conn	Sept. 14, 1839	1, 325
Odometer for carriages	W. A. Turner	Plymouth, N. C	Nov. 19, 1833	....
Odometer, Railway	M. F. Potter	Charlemont, Mass	June 13, 1854	11, 082
Odometer-register	H. F. Hart	New York, N. Y	Dec. 10, 1867	72, 033
Odometer-register	W. Yorke	Portland, Me	May 17, 1870	103, 272
Odor-trap for sinks, &c	W. E. Hatfield	Newark, N. J	June 17, 1862	35, 606
Odors, Compound for cleansing the human body from offensive.	H. D. Bird	Petersburgh, Va	Apr. 30, 1867	64, 189
Offal in the manufacture of fertilizers, Treating	J. J. Storer	Boston, Mass	Mar. 18, 1873	136, 943
Offal and manufacturing gas, Treating	J. Turner	Chicago, Ill	July 1, 1873	140, 391
Offal, Apparatus for treating	J. N. B. Bond	New York, N. Y	June 10, 1873	139, 759
Offal, Apparatus for treating	W. L. Bradley	Boston, Mass	Jan. 14, 1873	134, 844
Offal, Apparatus for treating	J. B. Chenoweth and E. P. Baugh.	Philadelphia, Pa	May 26, 1868	78, 261
Offal, Apparatus for treating	A. Edwards	New Haven, Conn	Jan. 28, 1873	135, 324
Offal, Apparatus for treating	A. W. Louth	Philadelphia, Pa	Mar. 28, 1865	47, 027
Offal, Apparatus for treating	J. A. Slater	New Haven, Conn	Feb. 11, 1873	135, 733
Offal-drier	M. Anderson	Chicago, Ill	June 3, 1873	139, 528
Offal, Treating	W. Adamson	Philadelphia, Pa	Feb. 14, 1865	46, 318
Office-chair	R. Fitts, jr	Fitchburgh, Mass	July 23, 1867	67, 034
Office-indicator	L. Burger	Chicago, Ill	June 20, 1871	116, 018
Office indicator and register	J. M. Keep	New York, N. Y	June 22, 1869	91, 543
Office register and directory	H. Rentchler	Bellville, Ill	May 3, 1870	102, 592
Office table and bed combined	D. Walker	Newark, N. J	Jan. 21, 1873	135, 021
Oil and burning-fluid by pneumatic pressure, Raising	J. H. Smith	Allegheny City, Pa	Dec. 10, 1867	72, 101
Oil and cider press	W. H. Hoeg	Ashtabula County, Ohio	May 20, 1826	....
Oil and naphtha from paraffine wax, &c., Process of separating.	R. B. and W. W. Lucas	Cleveland, Ohio	June 6, 1871	115, 622
Oil and other cans, Packing-case for	T. Scantlin	Evansville, Ind	May 14, 1872	126, 750
Oil and other liquids, Vessel for	G. W. Banker	New York, N. Y	June 27, 1871	116, 393
Oil and other mixed liquids, Apparatus for testing coal.	H. J. Smith and W. Jones	Philadelphia, Pa	May 6, 1862	35, 184
Oil and paint box and can	H. B. Everest and A. P. Ross	Rochester, N. Y	Nov. 12, 1872	132, 955
Oil and paint can	J. T. Williams	Philadelphia, Pa	Mar. 4, 1873	136, 575
Oil and paint, Can for transporting	H. Everett	Philadelphia, Pa	Oct. 1, 1872	131, 863
Oil and spirits of turpentine for paint, Preparing	J. G. Pendergast	Palmyra, N. Y	July 9, 1830	....
Oil and spirits of turpentine from pine wood, Producing.	S. L. Cole	Burlington, Vt	Dec. 1, 1863	40, 737
Oil and tallow cup	R. A. Copeland	Brooklyn, N. Y	July 6, 1869	92, 170
Oil and varnish can	E. T. Woodward	Charlestown, Mass	Aug. 12, 1862	36, 188
Oil, Anti-friction	J. F. Boynton	Syracuse, N. Y	Jan. 1, 1867	60, 829
Oil, Apparatus for coal	R. Shroder	Darlington, Pa	Dec. 16, 1856	16, 255
Oil, Apparatus for collecting floating	A. Ralston	West Middletown, Pa	Apr. 17, 1866	54, 014
Oil, Apparatus for collecting floating	J. J. Serrell	Hudson County, N. J	Feb. 5, 1867	61, 880
Oil, Apparatus for containing and measuring	E. F. Wilder	Lowell, Mass	Aug. 27, 1872	130, 833

Index of patents issued from the United States Patent Office from 1790 *to* 1873, *inclusive*—Continued.

Invention.	Inventor.	Residence.	Date.	No.
Oil, Apparatus for drawing and measuring	P. Noyes	Lowell, Mass	Oct. 4, 1870	107, 951
Oil, &c., Apparatus for extracting	L. Smith	Erie, Pa	Jan. 24, 1865	46, 033
Oil, Apparatus for filtering and refining	M. H. Krüger	New York, N. Y	Jan. 1, 1867	60, 747
Oil, Apparatus for manufacture of coal	H. K. Symmes	Newton, Mass	Nov. 1, 1859	26, 000
Oil, Apparatus for purifying	S. Gwynn and S. M. Clark	New York, N. Y., and Washington, D. C.	Aug. 25, 1868	81, 496
Oil, Apparatus for purifying and filtering	D. M. Lamb	Strathroy, Canada	Nov. 18, 1873	144, 621
Oil, Apparatus for purifying coal	B. Crawford	Allegheny, Pa	Mar. 28, 1871	113, 023
Oil, Apparatus for purifying mineral	W. Adamson	Philadelphia, Pa	Nov. 15, 1864	45, 007
Oil, Apparatus for redistillation of coal	L. Atwood	New York, N. Y	May 15, 1860	28, 246
Oil, Apparatus for testing coal	J. Tagliabue	New York, N. Y	Sept. 16, 1862	36, 488
Oil, Apparatus for testing coal	G. Tagliabue	New York, N. Y	Oct. 28, 1862	36, 826
Oil, Apparatus for treating	T. H. Burridge	Saint Louis, Mo	July 5, 1870	105, 038
Oil, Apparatus for treating linseed	D. D. Templeton	New York, N. Y	Mar. 18, 1873	136, 881
Oil as a substitute for linseed-oil, Process of preparing.	A. Millochau	New York, N. Y	May 19, 1863	38, 641
Oil, Ascertaining the quality of lamp	J. W. Harris	Dorchester, Mass	Sept. 3, 1840	1, 764
Oil-barrels, Process for lining	C. B. Hutchinson	Auburn, N. Y	July 4, 1865	48, 625
Oil, Bleaching and refining	O. Loew	New York, N. Y	Mar. 29, 1870	101, 284
Oil-boring apparatus	L. Atwood	Norwich, Conn	May 9, 1865	47, 609
Oil-box	J. W. Rhoades	Clyde, Ohio	Nov. 17, 1868	84, 215
Oil, Box and axle for saving	B. Kraft	Reading, Pa	July 1, 1851	8, 187
Oil-box for axles with conical journals	W. D. Titus	Brooklyn, N. Y	Feb. 12, 1856	14, 261
Oil, Burning	H. C. Dewitt	Saint Louis, Mo	Mar. 26, 1867	63, 229
Oil-burning apparatus,	A. J. White	Balston Spa, N. Y	Feb. 18, 1868	74, 648
Oil-bush and cap-neck spindle	J. Hinman	Clinton Township, Pa	Mar. 2, 1835	....
Oil-cabinet	T. Miller	East Boston, Mass	Nov. 25, 1873	144, 856
Oil-cabinet	M. H. Wiley	East Boston, Mass	Mar. 22, 1870	101, 070
Oil-cabinet	M. H. Wiley	East Boston, Mass	June 14, 1870	104, 389
Oil-cake mold	R. Macdonald	Hull, England	Nov. 11, 1873	144, 400
Oil-cake-packing apparatus	W. Hawes	Port Richmond, N. Y	Nov. 15, 1870	109, 2[illegible]6
Oil-cake stripper	W. Hawes	Port Richmond, N. Y	Dec. 31, 1872	134, 429
Oil-cake trimmer	W. Hawes	Port Richmond, N. Y	Nov. 12, 1872	132, 962
Oil-cake trimmer	A. Judson	Newark, N. J	June 6, 1871	115, 739
Oil, Camphene	A. V. H. Webb	New York, N. Y	Feb. 19, 1839	1, 082
Oil-can	T. K. Anderson	Addison, N. Y	Oct. 15, 1861	33, 471
Oil-can	W. C. Arthur	Baltimore, Md	Apr. 10, 1860	27, 876
Oil-can	J. Ashton	Fall River, Mass	July 3, 1866	55, 975
Oil-can	N. Aubin	Montreal, Canada	Dec. 28, 1869	98, 332
Oil-can	J. E. Auld	Buffalo, N. Y	Aug. 12, 1873	141, 619
Oil-can	B. F. Barnes	Boston, Mass	Nov. 24, 1868	84, 405
Oil-can	J. F. Bérendorf	Paris, France	Dec. 29, 1857	18, 949
Oil-can	J. Born	New Orleans, La	Sept. 24, 1872	131, 495
Oil-can	J. A. Bostwick	New York, N. Y	July 9, 1872	128, 699
Oil-can	J. Broughton	New York, N. Y	Mar. 7, 1865	46, 635
Oil-can	C. J. Brown	Plymouth, N. H	July 23, 1872	129, 783
Oil-can	J. Burson	Yates City, Ill	Aug. 22, 1871	118, 194
Oil-can	P. Childs	Northampton, Mass	July 11, 1871	116, 808
Oil-can	C. Chinnock	Brooklyn, N. Y	Sept. 7, 1869	94, 564
Oil, &c., can	G. H. Chinnock	Brooklyn, N. Y	Sept. 9, 1873	142, 613
Oil-can	B. Clark	New York, N. Y	Sept. 19, 1865	49, 979
Oil-can	P. C. Clark	Philadelphia, Pa	Feb. 1, 1870	99, 407
Oil-can	J. R. Cole	Morrisania, N. Y	Aug. 19, 1873	141, 920
Oil-can	J. R. Compton and C. M. Toms	Rahway, N. J	May 24, 1870	103, 428
Oil-can	W. G. Cowell	Wallingford, Conn	Oct. 21, 1873	143, 810
Oil-can	P. J. Dwyer	Elizabethport, N. J	Apr 12, 1870	101, 839
Oil-can	L. S. Enos	Olean, N. Y	Feb. 12, 1856	14, 229
Oil-can	J. G. Evans	Saint Louis, Mo	Aug. 24, 1869	94, 093
Oil-can	W. A. Fenn	Rochester, N. Y	Apr. 27, 1869	89, 299
Oil-can	S. Field and C. W. Heald	Barre, Mass	Aug. 31, 1852	9, 233
Oil-can	J. L. Folsom	East Boston, Mass	Mar. 1, 1870	100, 279
Oil-can	W. A. Foster	Fitchburgh, Mass	Oct. 14, 1873	143, 567
Oil-can	O. H. Gardner	Fulton, N. Y	Jan. 5, 1869	85, 521
Oil-can	J. B. Gale	Raleigh, N. C	Dec. 17, 1867	72, 384
Oil-can	J. D. Gray	Cincinnati, Ohio	Mar. 1, 1870	100, 393
Oil-can	W. R. Hallock	Westfield, N. J	Dec. 30, 1873	145, 999
Oil-can	G. Hatch	Pomeroy, Ohio	Sept. 3, 1867	68, 503
Oil-can	C. J. Hauck	Williamsburgh, N. Y	Mar. 24, 1868	75, 904
Oil-can	L. W. Hemp and J. Z. Skinner	Saint Louis, Mo	June 18, 1872	128, 144
Oil-can	L. W. Hemp and J. Z. Skinner	Saint Louis, Mo	June 25, 1872	128, 394
Oil-can	L. W. Hemp and J. Z. Skinner	Saint Louis, Mo	Sept. 3, 1872	130, 999
Oil-can	G. L. Holt	Springfield, Mass	Apr. 24, 1866	54, 250
Oil, &c., can	M. W. House	Cleveland, Ohio	July 29, 1873	141, 224
Oil-can	G. P. Hunt	New York, N. Y	Oct. 18, 1859	25, 831
Oil-can	E. D. Hurst	Lancaster, Pa	Dec. 27, 1864	45, 610
Oil-can	J. Jackson, jr	Westerly, R. I	July 31, 1860	29, 377
Oil-can	W. E. Jenkins	Auburn, N. Y	July 27, 1869	93, 092
Oil-can	M. S. Kavanagh	Detroit, Mich	Aug. 1, 1871	117, 544
Oil-can	H. Keller	Sauk Centre, Minn	May 27, 1873	139, 318
Oil-can	L. W. Kent	Cleveland, Ohio	Sept. 12, 1871	118, 860
Oil-can	G. A. Knowlton	Natick, Mass	Dec. 10, 1867	71, 887
Oil-can	D. D. Mackay and C. Butler	Whitestone and New York, N. Y.	Nov. 7, 1871	120, 653
Oil-can	J. J. Marcy	West Meriden, Conn	Apr. 7, 1868	76, 483
Oil-can	J. Mayher	East Hampton, Mass	July 7, 1863	39, 157
Oil-can	J. Mayher	East Hampton, Mass	Sept. 5, 1865	49, 776
Oil-can	J. S. McIntire	Chicago, Ill	May 7, 1867	64, 436
Oil-can	M. McNamara	Bridgeport, Conn	Oct. 29, 1872	132, 680
Oil-can	W. C. Newkirk	Piqua, Ohio	Dec. 11, 1866	60, 409
Oil-can	J. Ogdin	Philadelphia, Pa	May 12, 1868	77, 755
Oil-can	L. H. Olmsted	Newark, N. J	May 3, 1864	42, 593
Oil-can	G. C. Overshiser	Binghamton, N. Y	Sept. 27, 1864	44, 485
Oil-can	J. M. Perkins and M. W. House	Cleveland, Ohio	July 4, 1865	48, 585
Oil-can	A. H. Phillippi	Reading, Pa	Feb. 12, 1867	62, 065
Oil-can	W. Polyblank	Cleveland, Ohio	Apr. 13, 1869	88, 903
Oil-can	W. T. Prall	Pomeroy, Ohio	Oct. 23, 1866	59, 066
Oil-can	C. Pratt	New York, N. Y	Nov. 16, 1869	96, 958
Oil-can	T. Priestley	Saxonville, Mass	Apr. 8, 1856	14, 641

Index of patents issued from the United States Patent Office from 1790 *to* 1873, *inclusive*—Continued.

Invention.	Inventor.	Residence.	Date.	No.
Oil-can	G. S. Prior	Boston, Mass	Dec. 5, 1871	121, 545
Oil-can	A. P. Quinby	Newark, N. J	Mar. 4, 1873	136, 544
Oil-can	F. W. Read	Marquette, Mich	Dec. 12, 1871	121, 813
Oil-can	M. Robbins	Cincinnati, Ohio	Oct. 1, 1867	69, 485
Oil-can	S. Sargent	Lowell, Mass	Dec. 11, 1866	60, 429
Oil-can	P. Scanlan	Indianapolis, Ind	Feb. 15, 1870	99, 789
Oil-can	E. S. Scripture	New York, N. Y	Feb. 25, 1862	34, 529
Oil-can	E. S. Scripture	Brooklyn, N. Y	Feb. 9, 1864	41, 537
Oil-can	E. S. Scripture	Brooklyn, N. Y	June 28, 1864	43, 345
Oil-can	S. Short and E. S. Scripture	Brooklyn, N. Y	Oct. 31, 1865	50, 742
Oil-can	G. W. and G. H. Simmons	Bennington, Vt	July 14, 1857	17, 810
Oil-can	F. Skinner	Cleveland, Ohio	Nov. 9, 1869	96, 626
Oil-can	H. M. Smith	Chicago, Ill	June 17, 1873	139, 927
Oil-can	D. G. Starkey	New York, N. Y	Apr. 9, 1850	7, 276
Oil-can	A. C. Stoessiger	Chicago, Ill	Nov. 5, 1872	132, 875
Oil-can	J. M. Thompson	Holyoke, Mass	Oct. 28, 1856	15, 986
Oil-can	B. W. Tuttle	Galena, Ill	Dec. 3, 1872	133, 683
Oil-can	C. O. Twining	Saint Louis, Mo	Sept. 12, 1871	118, 989
Oil-can	R. Wallace	Wallingford, Conn	July 2, 1872	128, 570
Oil-can	H. C. Warfel	Philipsburgh, Pa	Nov. 29, 1870	109, 693
Oil-can	O. H. Warren	Baldwinsville, N. Y	Sept. 23, 1873	143, 107
Oil-can	H. B. Wellman	Indianapolis, Ind	June 16, 1868	78, 906
Oil-can	H. Wells	Florence, Mass	Apr. 21, 1857	17, 124
Oil-can	W. Westlake	Chicago, Ill	Sept. 6, 1870	107, 141
Oil-can	J. A. Whitman	Auburn, Me	Apr. 2, 1867	63, 591
Oil-can	F. C. Wilson	Chicago, Ill	Aug. 5, 1873	141, 615
Oil-can	G. D. Winchell	Cincinnati, Ohio	Sept. 26, 1871	119, 441
Oil-can and lamp, Lubricating	E. Ashcroft	South Boston, Mass	Dec. 4, 1866	60, 117
Oil-can and oiler	J. E. Weaver	Temperanceville, Pa	Sept. 18, 1866	58, 157
Oil-can and powder-canister	E. W. Woodruff	Washington, D. C	July 26, 1864	43, 652
Oil-can and torch combined	W. Kelley	Baltimore, Md	Mar. 11, 1873	136, 738
Oil-can cap	J. H. Breckenridge	Meriden, Conn	Apr. 2, 1861	31, 866
Oil-can cap	J. H. Noyes	Centre Abington, Mass	Apr. 14, 1868	76, 802
Oil-can cap and nozzle	J. R. Hathway	Cortland, N. Y	July 23, 1872	129, 820
Oil-can cover	E. A. More	Saint Louis, Mo	Feb. 26, 1867	62, 357
Oil-can, Dripping	L. R. Boyd	New Brunswick, N. J	Jan. 21, 1873	135, 072
Oil, Can for transporting	J. E. Pimley	Newark, N. J	Dec. 30, 1873	146, 020
Oil-can from paper-stuff	E. G. Kelley	New York, N. Y	July 6, 1869	92, 346
Oil-can holder	W. S. Brick	Fitchburgh, Mass	Mar. 14, 1871	112, 537
Oil-can holder	J. N. Wilkins	Chicago, Ill	July 2, 1872	128, 517
Oil-can, Lubricating	F. Fildes	Media, Pa	Apr. 12, 1859	23, 560
Oil-can, Metallic	G. W. Banker	New York, N. Y	May 20, 1873	138, 986
Oil can or tank, Coal	C. H. Phelps	New York, N. Y	Sept. 16, 1862	36, 478
Oil can, Plumbago	D. D. Mackay and C. Butler	Whitestone and New York, N. Y.	Apr. 18, 1871	113, 783
Oil-can, Portable	C. J. Hauck	Brooklyn, N. Y	Apr. 16, 1872	125, 813
Oil-can, Railway	S. W. Murray and B. P. Lamason.	Milton, Pa	Feb. 22, 1870	100, 058
Oil-can, Safety	J. M. Perkins and M. W. House	Cleveland, Ohio	Dec. 6, 1864	45, 340
Oil-can spout	S. Moyle, jr	Bridgeport, Conn	May 31, 1870	103, 644
Oil-can spout	J. Robinson	Baltimore, Md	Oct. 15, 1872	132, 323
Oil-can stopper	L. R. Boyd	New Brunswick, N. J	Jan. 14, 1873	134, 842
Oil-can stopper	E. C. Godwin	Norfolk, Va	Nov. 12, 1872	133, 024
Oil-can stopper	C. Seimel	Brooklyn, N. Y	Oct. 26, 1869	96, 274
Oil-can top	M. Fowler	Meriden, Conn	Oct. 1, 1872	131, 748
Oil-can, Transportation	J. C. Moore	Philadelphia, Pa	May 21, 1872	127, 092
Oil-can, Transportation	J. C. Moore	Philadelphia, Pa	Nov. 19, 1872	133, 242
Oil-can tube	E. B. Beach	West Meriden, Conn	July 7, 1868	79, 723
Oil-can vent-spout	J. J. Marcy	Meriden, Conn	Jan. 18, 1870	98, 986
Oil-cans, Apparatus for filling	J. A. Bostwick	New York, N. Y	June 21, 1870	104, 414
Oil-cans, Apparatus for filling	E. F. Wilder	Lowell, Mass	Feb. 11, 1873	135, 680
Oil-cans, Boxing	D. Saunderson	Saint Louis, Mo	June 18, 1867	65, 836
Oil-cans, Packing-case for	J. M. Murphy	New York, N. Y	Dec. 14, 1869	97, 954
Oil-cans, Reversible cap and spout for	E. T. Covell	Brooklyn, N. Y	June 14, 1870	104, 275
Oil-cans, Screw-cap for	W. Rigg	London, England	Feb. 25, 1868	74, 850
Oil-cans, Ventilating decanting-apparatus for	W. Cleveland	Orange, N. J	Aug. 15, 1871	117, 984
Oil-case	I. Van Hagen	New York, N. Y	Jan. 27, 1857	16, 501
Oil, &c., case or box	S. Selden	Erie, Pa	June 30, 1863	39, 071
Oil, Cask for transporting	W. Jenkins	Abergavenny, England	Oct. 28, 1873	144, 103
Oil-casks, Sink for	T. Miller	Boston, Mass	Mar. 5, 1872	124, 279
Oil, Clarifying castor	J. Hallock	Burlington County, N. J	Nov. 5, 1818	......
Oil, Cleansing fish	P. Fitz Simmons	West Troy, N. Y	Apr. 4, 1846	4, 439
Oil, Cleansing whale	B. Folger	......	Jan. 2, 1792	......
Oil-cloth: *See* Floor-cloth.				
Oil-cloth and similar fabrics, Drying	O. S. Washburn	Philadelphia, Pa	Jan. 3, 1871	110, 702
Oil-cloth by machinery, Manufacturing	W. Powers	Lansingburgh, N. Y	Feb. 18, 1832	......
Oil-cloth carpeting, Printing	J. Albro, jr	Elizabethtown, N. J	Sept. 18, 1847	5, 299
Oil-cloth, carpets, &c., Metal binding for	J. Piper	Utica, N. Y	May 4, 1869	89, 599
Oil-cloth, Composition for	R. Hoskin	Brooklyn, N. Y	Oct. 27, 1863	40, 411
Oil-cloth, Composition for making	O. C. Washburn	Philadelphia, Pa	Feb. 18, 1862	34, 453
Oil-cloth, &c., Compound for coating	J. L. Tenney and J. W. Bailey	Skowhegan, Me	May 28, 1867	65, 301
Oil-cloth cutter	J. Rauch	Selin's Grove, Pa	Feb. 1, 1870	99, 474
Oil-cloth-cutting instrument	M. A. Sunderland	Utica, N. Y	Jan. 23, 1866	52, 123
Oil-cloth-drying machine	D. Sampson	Winthrop, Me	May 30, 1837	206
Oil-cloth, Machine for coating	M. Sawyer	Cincinnati, Ohio	Oct. 20, 1863	40, 364
Oil-cloth, Machine for painting floor	C. W. Strout and A. Wilder	Hallowell, Me	Sept. 21, 1869	94, 985
Oil-cloth, Machine for preparing floor	S. W. Herrick and C. G. Gilbert, jr.	Salem, N. J	July 30, 1867	67, 195
Oil-cloth, Machine for varnishing floor	C. W. Strout	Hallowell, Me	Mar. 16, 1869	87, 803
Oil-cloth, Machine for varnishing floor	C. W. Strout and A. Wilder	Hallowell, Me	July 27, 1869	93, 135
Oil-cloth, Making furniture	I. Macauley	Philadelphia, Pa	Apr. 4, 1825	......
Oil-cloth, Manufacture of	W. Berri, jr	Brooklyn, N. Y	Oct. 18, 1870	108, 437
Oil-cloth, Manufacture of	G. Sampson	Manchester, Me	June 20, 1865	48, 314
Oil-cloth, Manufacture of	J. B. Stevenson, jr	Philadelphia, Pa	Dec. 31, 1867	72, 932
Oil-cloth, Manufacture of floor	J. and W. Weems	Johnstone, Great Britain	Aug. 17, 1869	93, 776
Oil-cloth, Marbleized	T. Carson	Brooklyn, N. Y	Aug. 1, 1871	117, 512
Oil-cloth, Metal binding for	D. M. Knowles	New Bedford, Mass	Sept. 19, 1871	119, 035
Oil cloth, Oil-paint for coating	T. Potter	Philadelphia, Pa	Jan. 17, 1871	110, 999

Index of patents issued from the United States Patent Office from 1790 *to* 1873, *inclusive*—Continued.

Invention.	Inventor.	Residence.	Date.	No.
Oil-cloth, Printing	J. Albro	Elizabeth, N. J	June 7, 1859	24, 270
Oil-cloth-printing apparatus	J. Marchbank	Lansingburgh, N. Y	Jan. 2, 1867	51, 898
Oil-cloth-printing block	J. Jenkins	Elizabeth, N. J	May 11, 1852	8, 941
Oil-cloth, Printing floor	L. Moore	Ballston Spa, N. Y	Mar. 26, 1850	7, 221
Oil-cloth, Printing floor	N. B. Powers	Lansingburgh, N. Y	Mar. 26, 1850	7, 228
Oil-cloth-printing machine	C. Rommel	Elizabeth, N. J	Apr. 4, 1871	113, 570
Oil-cloth-printing machine	C. Rommel	Elizabeth, N. J	Apr. 15, 1873	137, 962
Oil-cloth-printing machine	W. E. Worth	San Francisco, Cal	Dec. 9, 1873	145, 379
Oil-cloth-printing machinery	A. F. Buchanan	East Newark, N. J	Oct. 25, 1870	108, 563
Oil-cloth-printing machinery	E. A. Goodes	Philadelphia, Pa	Mar. 7, 1871	112, 443
Oil-cloth-printing machinery	W. H. Halsey	Philadelphia, Pa	Feb. 21, 1871	112, 032
Oil-cloth-printing machinery	J. Kraft	Newark, N. J	Aug. 30, 1870	106, 834
Oil-cloth, Registering-block for printing	J. Albro	Elizabethtown, N. J	June 5, 1855	12, 989
Oil-cloth, Surfacing floor	W. Berry	Bedford, N. Y	Oct. 23, 1849	6, 816
Oil, Coloring lamp	E. Bourne	New York, N. Y	Mar. 4, 1834	....
Oil, Composition for paint	J. D. Baldwin	Columbus, Ga	Oct. 11, 1859	25, 712
Oil, Composition paint	J. McCafferty	New York, N. Y	Apr. 15, 1873	137, 854
Oil, Composition to prevent the absorption of animal and fish.	N. Hathaway	Fairhaven, Mass	May 22, 1835	....
Oil, Compound	A. Dart	Dartford, Wis	Nov. 15, 1864	45, 028
Oil, Compound for the manufacture of lubricating	E. E. Hendrick	Carbondale, Pa	May 18, 1869	90, 100
Oil, Compound paint	Z. S. Doty	Janesville, Wis	May 17, 1864	42, 757
Oil, Converting the residuum of petroleum into	C. C. Mengel and A. P. Von Pohrnhoff.	Brooklyn, N. Y., and Saint Catharine's, Canada.	July 11, 1871	116, 852
Oil cool in lamps or burners, Keeping	J. Allen	Washington, D. C	Sept. 26, 1865	50, 087
Oil-cup	A. C. Ancona	Evansville, Ind	Apr. 12, 1870	101, 807
Oil-cup	A. T. Ballantine	Titusville, Pa	Sept. 30, 1873	143, 216
Oil-cup	C. W. Bancroft	Columbus, Ohio	June 7, 1864	43, 000
Oil-cup	S. Charnley	Portage City, Wis	July 28, 1868	80, 333
Oil-cup	C. F. Davis	New Market, N. H	Nov. 27, 1866	59, 976
Oil-cup	C. Ferger	New York, N. Y	Apr. 14, 1868	76, 616
Oil-cup	J. Fogle	Putnam, Ohio	Sept. 20, 1864	44, 297
Oil-cup	J. H. Gomer	New York, N. Y	Dec. 10, 1867	71, 869
Oil-cup	J. P. Haines	New York, N. Y	July 20, 1869	92, 820
Oil-cup	E. Hurd	Virginia City, Nev	Sept. 22, 1868	82, 321
Oil-cup	T. C. Longton	Trenton, N. J	Mar. 24, 1868	75, 937
Oil-cup	H. McGraw	Detroit, Mich	Apr. 9, 1872	125, 602
Oil-cup	C. H. Parshall	Detroit, Mich	Apr. 9, 1872	125, 405
Oil-cup	W. P. Patton and J. R. Miller	Harrisburgh, Pa	Oct. 8, 1867	69, 582
Oil-cup	R. Poole	Baltimore, Md	Aug. 8, 1865	49, 352
Oil-cup	W. Rider	Almont, Mich	Aug. 22, 1871	118, 276
Oil-cup	J. Ross	North Cambridge, Mass	May 10, 1870	102, 973
Oil-cup	H. K. Sears and S. L. Holt	Hartford, Conn	May 28, 1867	65, 126
Oil cup	H. Stanley	Saint Louis, Mo	Oct. 6, 1868	82, 888
Oil-cup	E. Weissenborn	Hudson City N. J	Jan. 17, 1865	45, 945
Oil-cup	C. Williams	Vineland, N. J	Oct. 15, 1867	69, 882
Oil-cup	N. B. Williams	Providence, R. I	Mar. 3, 1868	75, 100
Oil-cup	N. B. Williams	Providence, R. I	Aug. 11, 1868	81, 051
Oil-cup and bearing for spindles	E. Convers	Oswego, N. Y	June 14, 1870	104, 115
Oil-cup, Automatic	J. H. Wilkinson	South New Market, N. H	May 28, 1872	127, 205
Oil-cup for adjustable boxes	C. C. Klein	Philadelphia, Pa	Feb. 9, 1869	86, 761
Oil-cup for journal-boxes	A. Richardson	Bellows Falls, Vt	July 29, 1851	8, 251
Oil-cup for lubricating engines	S. H. Whitmore	Cincinnati, Ohio	June 22, 1858	20, 674
Oil-cup for lubricating shafts	E. J. Gerdom and C. W. Schindler.	Albany, N. Y	Jan. 28, 1868	73, 795
Oil-cup for machinery	W. Douglas and H. M. Ingler	Bellaire, Ohio	May 14, 1867	64, 646
Oil-cup for machinery	T. W. Godwin	Portsmouth, Va	Sept. 22, 1863	40, 032
Oil-cup for machinery	R. Ross and W. Holland	Philadelphia, Pa	June 22, 1858	20, 665
Oil-cup for movable bearings	D. Harrigan	Somerville, Mass	Nov. 2, 1869	96, 427
Oil-cup for steam-engines	E. H. Ashcroft	Lynn, Mass	Nov. 3, 1868	83, 587
Oil-cup for steam-engines	J. C. Carroll	Litchfield, Ill	Aug. 11, 1868	80, 911
Oil-cup for steam-engines	T. Chatterton	Cleveland, Ohio	Aug. 6, 1867	67, 496
Oil-cup for steam-engines	D. Clark	Philadelphia, Pa	Jan. 10, 1854	10, 399
Oil-cup for steam-engines	F. H. Furniss	Cleveland, Ohio	Aug. 20, 1867	67, 867
Oil-cup for steam-engines	M. Hawkins	Birmingham, Conn	Mar. 19, 1867	63, 045
Oil-cup for steam-engines	F. Lunkenheimer	Cincinnati, Ohio	June 25, 1867	66, 157
Oil-cup for steam-engines	G. Trott	Pittsburgh, Pa	Mar. 28, 1854	10, 712
Oil-cup for steam-engines	N. J. White	Lawrence, Mass	Aug. 21, 1866	57, 418
Oil-cup for steam-engines	J. Wildhack	Pekin, Ill	Feb. 12, 1867	62, 103
Oil-cup for steam-pressure	F. P. McCullon and W. Woodcock.	Philadelphia, Pa	May 12, 1868	77, 900
Oil-cup, Globe	V. Giroud	New York, N. Y	Aug. 29, 1865	49, 624
Oil-cup, Lubricating	A. Higley	Cleveland, Ohio	Apr. 2, 1872	125, 133
Oil-cup, Lubricating	J. H. Palmer	Greenbush, N. Y	Apr. 17, 1866	54, 070
Oil-cup, Lubricating	E. N. Roland	Baltimore, Md	Dec. 15, 1857	18, 863
Oil-cup of connecting-rods or movable bearings of machinery.	D. Harrigan	Somerville, Mass	Feb. 2, 1869	86, 540
Oil-cup or lubricator	R. Ross	Pittsburgh, Pa	Oct. 15, 1861	33, 508
Oil-cups and inkstands, Spring-closing cap for	H. H. Heskett	Le Roy, Ill	June 7, 1870	104, 032
Oil-cups, Tip for	F. P. Pfleghar and W. Shölborn	New Haven, Conn	Feb. 28, 1865	46, 586
Oil, &c., Device for drawing off and skimming	I. Peck and W. H. H. Glover	Southold and New York, N. Y.	July 7, 1863	39, 169
Oil, Discoloring and drying linseed	N. S. Parmantier	Philadelphia, Pa	Jan. 30, 1809	....
Oil-drill	L. H. Bowman	Morristown, Pa	Sept. 5, 1865	49, 704
Oil-dripper	J. M. Thompson	Holyoke, Mass	Aug. 7, 1855	13, 406
Oil, Factitious	H. W. Adams	New York, N. Y	Apr. 3, 1855	12, 614
Oil, Factitious	J. W. Harmon	Elizabethtown, N. J	Sept. 29, 1857	18, 279
Oil-feeder	J. Benson	Boston, Mass	Feb. 28, 1844	3, 454
Oil-feeder	R. Cornelius	Philadelphia, Pa	Apr. 6, 1843	3, 031
Oil-feeder	E. S. Scripture	Syracuse, N. Y	Mar. 15, 1845	3, 952
Oil-feeder	J. Turner	Cambridgeport, Mass	July 31, 1860	29, 446
Oil-filter	T. Antisell	New York, N. Y	Nov. 12, 1850	7, 767
Oil fire-tester, Carbon	G. E. Shaw	Pittsburgh, Pa	July 3, 1866	56, 107
Oil, Fish	W. D. Hall	Hamden, Conn	July 17, 1860	29, 164
Oil for burning and lubricating, Compound	R. A. Gilman	Woodland, Wis	Dec. 15, 1863	40, 924
Oil for burning and lubricating, Compound	W. G. Hermance	West Sandlake, N. Y	June 9, 1863	38, 825
Oil for coating leather and metal, Compound	W. K. Wyckoff	Ripon, Wis	July 20, 1869	92, 924
Oil for curriers' use	J. B. Kendall	Boston, Mass	June 15, 1869	91, 446

Index of patents issued from the United States Patent Office from 1790 *to* 1873, *inclusive*—Continued.

Invention.	Inventor.	Residence.	Date.	No.
Oil for lubricating, Compound for the treatment of	C. J. Eames and C. A. Seely	New York, N. Y	July 9, 1867	66, 573
Oil for lubricating from petroleum	J. K. Truax	Pittsburgh, Pa	Mar. 25, 1873	137, 157
Oil for lubricating machinery, &c	J. H. Lester	Brooklyn, N. Y	Jan. 30, 1866	52, 301
Oil for lubricating, paint, &c., Compound	W. H. Spooner	Bristol, R. I	Nov. 15, 1864	45, 090
Oil for manufacture of varnish and other purposes, Treating and drying.	F. Walton	Denton, England	Nov. 12, 1861	33, 722
Oil for mixing paint, Compound	P. M. Wallover	Smith's Ferry, Pa	Dec. 21, 1869	98, 211
Oil for mixing paint, Compound	E. K. Wood and R. W. Henry	De Witt, Iowa	Jan. 29, 1867	61, 696
Oil for paint, Manufacture of drying	F. Huot	New York, N. Y	Sept. 3, 1867	68, 367
Oil for paintings, Preparing linseed or other	D. E. Breinig	New York, N. Y	Jan. 29, 1867	61, 653
Oil for purposes of illumination, Mode of using volatile.	B. F. Greenough	Boston, Mass	Nov. 15, 1843	3, 339
Oil for wool	W. H. Moss	New Richmond, Ohio	Aug. 4, 1868	80, 559
Oil from acid residuum of petroleum, Process of obtaining useful.	A. Farrar	Longwood, Mass	Oct. 26, 1869	96, 097
Oil from animal and vegetable substances, Apparatus for extracting.	C. O. Heyl	Berlin, Prussia	Sept. 3, 1867	68, 506
Oil from animal and vegetable substances, Apparatus for extracting.	H. A. Sterns	Smithfield, R. I.	May 24, 1870	103, 519
Oil from bitumens, Preparing	L. and W. Atwood	Waltham, Mass	Aug. 12, 1856	15, 506
Oil from bone-dust, Apparatus for extracting	W. W. Lucas	Cleveland, Ohio	Feb. 7, 1871	111, 658
Oil from cannel-coal, Production of	L. and W. Atwood	Waltham, Mass	Aug. 12, 1856	15, 505
Oil from cans, Apparatus for drawing	J. G. Evenden	Chicago, Ill	June 17, 1873	140, 022
Oil from castor beans, Making warm-pressed	T. Pharo	Tuckerton, N. J	Feb. 5, 1822	....
Oil from coal, Extracting	E. N. Horner	New Brighton, Pa	Jan. 25, 1859	22, 727
Oil, &c., from coal, Extraction of volatile	L. Atwood	Brooklyn, N. Y	Oct. 19, 1858	21, 805
Oil from cotton-seed, Extracting	C. L. Fleischmann	Washington, D. C	May 4, 1869	89, 720
Oil from cotton-seed, Extracting	A. A. Hayes	Boston, Mass	Apr. 8, 1856	14, 610
Oil from cotton-seed, Extracting	G. Palmer	Montville, Pa	Dec. 14, 1830	....
Oil from cotton-seed, Extracting	C. Whiting	....	Mar. 2, 1799	....
Oil from cotton-seed, Manufacture of	G. G. Henry	Mobile, Ala	Jan. 17, 1860	26, 845
Oil from cotton-seed, Process of making	W. M. Force	Newark, N. J	Aug. 20, 1872	130, 573
Oil from cotton-waste, Apparatus for separating	A. N. Cole	Brockville, Canada	July 2, 1872	128, 594
Oil from cotton-waste, Separating	A. N. Cole	Brockville, Canada	June 6, 1871	115, 708
Oil from fish, Cylinder-press for extracting	J. G. Stoddard and B. F. Gallup	Groton, Conn	June 18, 1867	65, 962
Oil from fish, Extracting	W. H. H. Glover	Southold, N. Y	Nov. 2, 1869	96, 315
Oil from fish, Extracting	A. M. Millochau	New York, N. Y	Mar. 19, 1861	31, 765
Oil from flax-seed, Extracting	D. Dodge	Hamilton, Mass	May 14, 1827	....
Oil from flax-seed, Pressing	C. Howe, F. Brewster, and J. Johnson.	Burlington, Vt	May 24, 1828	....
Oil from grain and other materials, Apparatus for separating.	E. T. Hutchinson	Baltimore, Md	Jan. 17, 1871	111, 064
Oil from grain and seed, Apparatus and process for removing.	E. T. Hutchinson	Baltimore, Md	Mar. 7, 1871	112, 350
Oil from herbs, and for other purposes, Apparatus for extracting.	L. Daubert	Louisville, Ky	May 7, 1867	64, 499
Oil from leather, &c., Extracting	J. A. Dean	Easton, Mass	May 1, 1866	54, 303
Oil, &c., from minerals, Apparatus for extracting	H. P. Gengembre	Pittsburgh, Pa	Jan. 30, 1866	52, 284
Oil, &c., from minerals, Process for extracting	H. P. Gengembre	Pittsburgh, Pa	Jan. 30, 1866	52, 283
Oil from oils extracted from bituminous minerals, Preparation of drying.	C. Cherry	Pittsburgh, Pa	Sept. 2, 1856	15, 644
Oil from palma christi, Extracting	J. G. Gebhard	....	Feb. 4, 1800	....
Oil from petroleum, Lubricating	T. E. Merrick	Cleveland, Ohio	June 22, 1869	91, 654
Oil from petroleum, Manufacturing lubricating	S. Gibbons	Freedom, Pa	Mar. 2, 1869	87, 485
Oil from petroleum residuum, Preparing paint	A. Millochau	New York, N. Y	Mar. 17, 1863	37, 918
Oil from pigs' feet, Extracting	C. Wahl	Chicago, Ill	Aug. 12, 1862	36, 184
Oil from poppy-seed, Obtaining	D. Markham	....	Dec. 2, 1824	....
Oil from resin, Lubricating	L. S. Robbins	New York, N. Y	Nov. 4, 1851	8, 489
Oil from resin, Paint	L. S. Robbins	New York, N. Y	Nov. 4, 1851	8, 491
Oil from resin, Tanners'	L. S. Robbins	New York, N. Y	Nov. 4, 1851	8, 488
Oil from running streams, Apparatus for obtaining	T. S. Scoville	Williamsport, Pa	July 18, 1865	48, 841
Oil from seed, Machinery for expressing	J. Hallock	Little Egg Harbor, N. J	Jan. 22, 1822	....
Oil from seed, Extracting	J. Robertson	Brooklyn, N. Y	Jan. 22, 1867	61, 463
Oils from seed, Extracting fat	J. Preston	Dorchester, Mass	Apr. 13, 1858	19, 948
Oil from seed, meal, &c., Apparatus for removing	E. S. Hutchinson	Baltimore, Md	July 26, 1870	105, 688
Oil from streams, Apparatus for collecting floating	E. C. Summers	Huntingdon, Pa	June 5, 1866	55, 388
Oil from streams, Apparatus for collecting floating	H. E. Towle	New York, N. Y	May 29, 1866	55, 178
Oil from sunflowers, Manufacturing	C. A. Barnitz	Spring Garden, Pa	Oct. 20, 1830	....
Oil from tanks, Device for discharging	W. J. Brundred	Oil City, Pa	Jan. 17, 1871	110, 953
Oil from vegetable and animal substances, Extracting.	T. Richardson, J. J. Lundy, and R. Irvine.	Newcastle-upon-Tyne, Leith, and Midlothian, Great Britain.	May 31, 1864	42, 987
Oil from vegetable and other matter, Apparatus and process for extracting.	T. Sim and E. S. Hutchinson	Baltimore, Md	Mar. 23, 1869	88, 222
Oil from vegetable and other matter, Apparatus for extracting.	E. S. Hutchinson	Baltimore, Md	Apr. 30, 1872	126, 300
Oil from vegetable and other matter, Apparatus for removing.	E. S. Hutchinson	Baltimore, Md	Mar. 7, 1871	112, 349
Oil from vegetable and other matter, Apparatus for removing.	T. Sim	Baltimore, Md	Mar. 7, 1871	112, 389
Oil from vegetable and other matter, Process and apparatus for extracting.	T. Sim	Baltimore, Md	Aug. 10, 1869	93, 645
Oil from water and other impurities, Apparatus for separating fish.	T. L. Robinson	Boston, Mass	Apr. 18, 1865	47, 333
Oil from wells, Apparatus for inducing the flow of.	J. B. Christian	Mount Carroll, Ill	May 22, 1866	54, 859
Oil, &c., from wells, Apparatus for raising	H. W. Faucett	Titusville, Pa	July 8, 1873	140, 692
Oil from wells, Obtaining	L. Phleger and G. G. Lobdell	Philadelphia, Pa., and Wilmington, Del.	Aug. 7, 1866	56, 989
Oil, &c., Furnace for the combustion of coal	J. Calkins, P. A. Hover, and C. W. Grannis.	Hudson and Gowanda, N. Y.	May 8, 1860	28, 154
Oil, Gelatinizing	E. Quern	New York, N. Y	Aug. 30, 1859	25, 277
Oil-globe for steam-chest	C. A. Wilson	Cincinnati, Ohio	Sept. 22, 1868	82, 467
Oil, grease, gums, &c., from cotton and woolen waste, Apparatus for removing and recovering.	J. W. Krepps	Chicago, Ill	Apr. 19, 1870	102, 018
Oil, Grinding cotton and other seed for	W. Wilber	New Orleans, La	Sept. 11, 1855	13, 558
Oil, Illuminating	A. W. Burrows	Cleveland, Ohio	Nov. 5, 1867	70, 405
Oil, Illuminating	D. W. Fowler	Goshen, Ind	Mar. 3, 1868	75, 147
Oil, Illuminating	J. E. Noyes	New Albany, Ind	Sept. 15, 1868	82, 151

Index of patents issued from the United States Patent Office from 1790 to 1873, inclusive—Continued.

Invention.	Inventor.	Residence.	Date.	No.
Oil in making paints, Manufacturing a mucilagenous compound as a substitute for.	L. J. Vankleek	Poughkeepsie, N. Y	Nov. 23, 1829	....
Oil, Instrument for testing the inflammability of illuminating.	H. M. Hartshorn	Malden, Mass	June 29, 1869	91, 843
Oil, Lubricating	E. L. Brady	New Orleans, La	Sept. 17, 1867	68, 942
Oil, Lubricating	R. A. Chesebrough	New York, N. Y	May 19, 1868	77, 959
Oil, Lubricating	H. and C. Fink	Baltimore, Md	May 28, 1872	127, 328
Oil, Lubricating	G. W. Gladden	Cincinnati, Ohio	Nov. 5, 1872	132, 759
Oil, Lubricating	D. R. P. Hill	Fredericktown, Pa	Dec. 20, 1870	110, 361
Oil, Lubricating	J. A. Kestler	Chicago, Ill	Aug. 18, 1868	81, 093
Oil, Lubricating	J. M. Lippincott	Pittsburgh, Pa	Sept. 4, 1866	57, 737
Oil, Lubricating	W. H. Mason	Boston, Mass	May 25, 1852	8, 971
Oil, Lubricating	C. Moore	East Saginaw, Mich	Nov. 5, 1867	70, 454
Oil, Lubricating	G. O. Spence	Titusville, Pa	Apr. 14, 1868	76, 838
Oil, Lubricating	P. H. Vander Weyde	Philadelphia, Pa	Feb. 12, 1867	62, 092
Oil, Making benne	W. Few	New York	Mar. 1, 1809	....
Oil, Making paraffine	J. Young	Manchester, England	Mar. 23, 1852	8, 833
Oil, Manufacture of coal	L. Atwood	New York, N. Y	May 29, 1860	28, 448
Oil, Manufacture of coal	H. P. Gengembre	Allegheny, Pa	Aug. 16, 1859	25, 109
Oil, Manufacture of compound	F. L. De Gerbeth	Dalston, England	Aug. 18, 1868	81, 071
Oil, Manufacture of illuminating	C. C. Menger and A. P. Von Pohrnhoff.	Brooklyn, N. Y., and St. Catharine's, Canada.	Mar. 26, 1872	125, 067
Oil, Manufacture of lubricating	L. M. Mott	Chicago, Ill	Apr. 24, 1866	54, 192
Oil, Manufacture of lubricating	S. M. Mott	Wellsville, N. Y	July 23, 1861	32, 885
Oil, Manufacture of lubricating	H. T. Slemmer	Norristown, Pa	Feb. 27, 1866	52, 897
Oil, Manufacture of pyrogenic	L. Atwood	Brooklyn, N. Y	Dec. 28, 1858	22, 406
Oil, Manufacture of resin	W. T. Clough	Jersey City, N. J	Apr. 22, 1845	4, 008
Oil, Manufacturing resin	S. W. Hawes	Boston, Mass	Apr. 12, 1853	9, 662
Oil, Manufacture of resin	J. Merrill	Boston, Mass	Apr. 27, 1869	89, 493
Oil, Manufacture of resin	J. Treat	New York, N. Y	Mar. 15, 1870	100, 953
Oil, Manufacturing fish, whale, and sperm	D. C. and C. H. Knapp and A. H. Howland.	Oswego, N. Y	May 29, 1832	....
Oil obtained from mineral coal, Apparatus for purifying.	C. Cherry	Pittsburgh, Pa	Sept. 2, 1856	15, 642
Oil, Obtaining flaxseed	E. Horner	East Brook, Pa	Feb. 9, 1847	4, 961
Oil, Odorized kerosene	S. Jarden	Baltimore, Md	Mar. 4, 1862	34, 577
Oil of cotton-seed for all purposes of linseed-oil, Application of.	G. P. Digges	Albemarle County, Va	Dec. 16, 1820	....
Oil of hazze, Preparation of	Preswick and Fisher	New York	Aug. 17, 1835	....
Oil of vitriol, Concentrating	W. H. Belmain	Saint Helen's, Great Britain	Apr. 26, 1870	102, 205
Oil of vitriol, Storing and transporting	W. H. Belmain	Saint Helen's, Great Britain	Apr. 26, 1870	102, 206
Oil on surface of rivers, Mode of collecting	J. Cannon	New Richmond, Pa	Mar. 29, 1864	42, 073
Oil or blubber press	W. P. Chadwick	Edgartown, Mass	July 11, 1854	11, 299
Oil or fluid can	J. R. Nichols	Haverhill, Mass	Sept. 20, 1853	10, 035
Oil or suet cup	R. Ross	Bethlehem, Pa	Aug. 25, 1868	81, 542
Oil, Paint	R. Bartholow	Cincinnati, Ohio	Apr. 4, 1865	47, 083
Oil, Paint	V. Cordier	Paris, France	Sept. 29, 1868	82, 604
Oil, Paint	W. Johnson	Allegheny City, Pa	Sept. 20, 1864	44, 316
Oil, Paint	S. Melsom	Erie, Pa	Apr. 23, 1867	64, 125
Oil, Paint	A. Millochau	New York, N. Y	May 19, 1863	38, 640
Oil, Paint	W. W. Nichols	Lockport, N. Y	Nov. 21, 1865	51, 118
Oil, Paint, &c	D. S. Robinson	Oswego, N. Y	Dec. 18, 1866	60, 557
Oil, Paint, &c	W. Ward	Cleveland, Ohio	Mar. 10, 1868	75, 497
Oil-paintings from one surface to another, Method of transferring.	J. F. Wachsmuth	Highland, Ill	Apr. 20, 1869	89, 188
Oil, &c., Portable tank for	J. F. Keeler	Pittsburgh, Pa	Mar. 27, 1866	53, 451
Oil, Preparation of cod-liver	M. O'Reilly	New York, N. Y	Dec. 20, 1870	110, 275
Oil, Preparation of wool	T. Barrows	Dedham, Mass	Aug. 28, 1855	13, 486
Oil, Preparing	S. Gwynn	New York, N. Y	Aug. 25, 1868	81, 495
Oil, Preparing lubricating	L. Atwood	Boston, Mass	Mar. 29, 1853	9, 630
Oil, Preparing paraffine	W. Brown	Glasgow, Scotland	Sept. 27, 1853	10, 055
Oil-press	W. P. Chadwick	Edgartown, Mass	Nov. 9, 1852	9, 382
Oil-press	I. Clowes	Hampton, Va	Nov. 3, 1825	....
Oil-press	W. M. Force	Newark, N. J	Mar. 11, 1873	136, 716
Oil-press	E. Hills	Cincinnati, Ohio	Mar. 12, 1850	7, 166
Oil-press	A. Judson	Newark, N. J	Aug. 1, 1871	117, 641
Oil-press	D. L. Latourette	Saint Louis, Mo	Oct. 28, 1851	8, 469
Oil-press	W. W. Marsh	Jacksonville, Ill	Jan. 13, 1857	16, 391
Oil-press	J. Marshall	Pentonville Road, England	June 6, 1865	48, 140
Oil-press	W. V. McKenzie	Jersey City, N. J	May 6, 1862	35, 168
Oil-press	W. V. McKenzie	Jersey City, N. J	Apr. 25, 1865	47, 441
Oil-press	I. S. Schuyler	New York, N. Y	Aug. 27, 1861	33, 173
Oil-press	J. White	....	Nov. 27, 1804	....
Oil-press case and wrapping	W. M. Force	Newark, N. J	July 16, 1872	129, 021
Oil-press, Double hydrostatic	O. Badger and O. Lull	Otsego, N. Y	Aug. 9, 1832	....
Oil-press, Double hydrostatic	O. Badger and O. Lull	Otsego and Waterloo, N. Y	Oct. 5, 1836	40
Oil, Press for packing pulp for linseed	C. Moore	Trenton, N. J	Jan. 19, 1858	19, 149
Oil-press, Hydraulic	W. R. Fee	Cincinnati, Ohio	Sept. 13, 1859	25, 402
Oil-press, Hydraulic	W. Wilber	New Orleans, La	Dec. 25, 1855	14, 005
Oil press, Linseed	C. Moore	Trenton, N. J	Mar. 23, 1858	19, 708
Oil-press mat	H. Nutt	Brooklyn, N. Y	Aug. 25, 1863	39, 671
Oil-presses, Press-board for	J. Shinn	Philadelphia, Pa	Apr. 16, 1867	63, 951
Oil-presses, Securing and guiding the boxes of	W. W. Marsh	Jacksonville, Ill	Apr. 28, 1857	17, 161
Oil-presses, Steam-heating seed for	J. Criswell	Pittsburgh, Pa	July 31, 1837	334
Oil-pressing machinery	W. Wilber	New Orleans, La	Jan. 13, 1857	16, 422
Oil-pressing machinery	W. Wilber	New York, N. Y	Oct. 6, 1857	18, 367
Oil, &c., Process of extracting	C. A. Seely	New York, N. Y	Aug. 31, 1869	94, 246
Oil, Producing Columbian	T. Paul	Baltimore, Md	Feb. 23, 1811	....
Oil, Purifying	T. Drayton	Brooklyn, N. Y	July 4, 1854	11, 239
Oil, Purifying	H. Halvorson	Cambridge, Mass	Apr. 28, 1857	17, 181
Oil, Purifying and deodorizing	D. Symonds	Lowell, Mass	May 28, 1867	65, 137
Oil, Purifying and refining	W. M. Sloane	New York, N. Y	Nov. 29, 1870	109, 772
Oil, Purifying animal	A. Traband	New York	Nov. 21, 1842	2, 855
Oil, &c., Purifying coal	R. A. Chesebrough	New York, N. Y	Aug. 22, 1865	49, 502
Oil, Purifying cotton-seed	J. J. Gerrard	New York	Aug. 14, 1833	....
Oil, Purifying linseed	R. T. Clarke	Cincinnati, Ohio	May 17, 1870	103, 141
Oil, Purifying resin	S. L. Dana	Lowell, Mass	Apr. 19, 1853	9, 680
Oil, Purifying sperm and whale	J. L. Embree	New York	June 13, 1831	....

Index of patents issued from the United States Patent Office from 1790 *to* 1873, *inclusive*—Continued.

Invention.	Inventor.	Residence.	Date.	No.
Oil, Purifying spermaceti	R. Robotham		Aug. 14, 1799	....
Oil, Pyrogenous lubricating	S. Downer and J. Merrill	Boston, Mass	July 29, 1856	15, 418
Oil-refineries, Process of treating acid residuum from.	A. Farrar	Brookline, Mass	Mar. 15, 1870	100, 876
Oil, Solidifying	C. A. Jordery	Paris, France	May 7, 1872	126, 552
Oil, Steam apparatus for extracting vegetable	W. Wilber	New Orleans, La	Sept. 11, 1855	13, 557
Oil-still: *See* Still.				
Oil-stone box or case	G. O. Wichers	Lawrence, Mass	Feb. 9, 1864	41, 561
Oil-stone holder	H. Brown	Hamilton, Ill	Apr. 26, 1870	102, 218
Oil-storing tank	P. Noyes	Lowell, Mass	Apr. 12, 1870	101, 761
Oil, Substitute for linseed	Todd and Peabody	Washington, D. C	July 7, 1835	
Oil, tallow, &c., Extracting	A. Randel	New York, N. Y	Mar. 18, 1862	34, 722
Oil-tank	G. W. Glass	Pittsburgh, Pa	Oct. 24, 1871	120, 263
Oil-tank	G. W. Howard	Pontiac, Mich	Feb. 18, 1862	34, 426
Oil-tank	J. K. Ingall	New York, N. Y	Apr. 10, 1866	53, 829
Oil-tank	S. H. Ingalls	New Bedford, Mass	Aug. 4, 1863	39, 446
Oil-tank	T. C. Joy	Titusville, Pa	Oct. 9, 1866	58, 648
Oil-tank	J. Lessler	Buffalo, N. Y	Nov. 5, 1872	132, 843
Oil-tank	H. Pierce and J. B. Button	Cleveland, Ohio	Jan. 22, 1867	61, 453
Oil-tank	J. Reese	Pittsburgh, Pa	Feb. 11, 1862	34, 373
Oil-tank	H. F., G. S., and A. Snyder	Williamsport and Blairsville, Pa.	Aug. 17, 1869	93, 918
Oil-tank	H. F., G. S., and A. Snyder	Williamsport and Freeport, Pa.	May 2, 1871	114, 361
Oil-tank	H. F., G. S., and A. Snyder	Williamsport and Freeport, Pa.	May 23, 1871	115, 127
Oil-tank	H. F., G. S., and A. Snyder	Williamsport and Freeport, Pa.	June 27, 1871	116, 366
Oil-tank	W. C. Strickler	Chicago, Ill	Aug. 20, 1872	130, 762
Oil-tank	A. G. Wilson	New York, N. Y	Sept. 8, 1868	81, 968
Oil-tank fitting	H. F. Snyder	Williamsport, Pa	Sept. 10, 1872	131, 190
Oil-tank, Incased	J. Shea and T. Campbell	Boston, Mass	Nov. 19, 1872	133, 123
Oil-tank safety-chamber	E. H. Knight	Washington, D. C	Jan. 15, 1867	61, 213
Oil-tank, Self-measuring	J. Schalk, jr	Guttenbergh, N. J	May 20, 1873	139, 084
Oil-tanks, Apparatus for filling and emptying	D. Birdsall	Brooklyn, N. Y	Feb. 20, 1872	123, 859
Oil-tanks, grain-cars, &c., Discharge-apparatus for.	W. J. Brundred	Oil City, Pa	June 13, 1871	115, 817
Oil-tanks, Machine for turning the flanges on the heads of.	G. Selden	Erie, Pa	July 4, 1871	116, 636
Oil-tanks, Man-hole cover for	H. F. Snyder	Williamsport, Pa	Mar. 19, 1872	124, 768
Oil, tobacco, and cotton press	G. Palmer	Montville, Conn	Dec. 14, 1830	....
Oil, tobacco, and other presses	E. Thomas	Craigsville, Va	July 14, 1868	79, 875
Oil to render it drying, Treating cotton-seed	R. Farley	Cincinnati, Ohio	Jan. 31, 1871	111, 444
Oil to render it drying, Treating cotton-seed	H. Goldmann	New York, N. Y	Aug. 5, 1873	141, 503
Oil-tool extractor	S. Buckley and P. H. Lawrence	Lawrenceburgh, Pa	July 1, 1873	140, 345
Oil-tool jar	R. H. Lecky	Allegheny City, Pa	May 30, 1865	47, 961
Oil-tools, Socket-joint for	R. Gracey	Pittsburgh, Pa	May 29, 1866	55, 086
Oil, Transporting	W. W. Horton	Freeport, Ill	Mar. 21, 1865	46, 906
Oil, Treating	P. Marsh	South Adams, Mass	Jan. 1, 1856	14, 042
Oil, &c., Treating	H. L. Smith	Gambier, Ohio	Nov. 27, 1866	60, 076
Oil, Treating	H. K. Taylor and D. M. Graham	Cleveland, Ohio	May 22, 1866	54, 978
Oil, Treating cod-liver and castor	G. W. Fox	Manchester, Great Britain	Sept. 13, 1870	107, 244
Oil, Treating vegetable	R. T. Clark	Covington, Ky	Jan. 2, 1872	122, 439
Oil-trying apparatus	J. M. Hunter	New York, N. Y	Nov. 6, 1860	30, 575
Oil-vessel	G. W. Banker	Saint Louis, Mo	Oct. 13, 1863	40, 230
Oil, Vessel for holding	A. T. Woodward	New York, N. Y	Feb. 7, 1871	111, 599
Oil, Vessel for storing and transporting	T. J. McGarry	Cleveland, Ohio	Apr. 23, 1867	64, 123
Oil-vessels, Coating for	S. Gwynn	New York, N. Y	July 4, 1865	48, 552
Oil which has been used in lubricating machinery, Purifying.	C. Houlker	Frankford, Pa	Dec. 20, 1870	110, 364
Oil which wool contains, Composition to start the animal.	J. Goulding	Dedham, Mass	Aug. 24, 1827	....
Oils and fats by acids, Purification of	R. G. Loftus	Chelsea, Mass	Apr. 18, 1871	113, 782
Oils and fats for burning in lamps, lubricating machinery, &c., Treating.	S. Lewis	Rochester, N. Y	June 10, 1862	35, 527
Oils and fats, Process of chilling	J. E. Richardson	New York, N. Y	May 28, 1867	65, 275
Oils and fats, Purifying	R. C. Barton	Brooklyn, N. Y	Apr. 21, 1868	76, 974
Oils and fats, Treating and refining	J. W. Burton	Leeds, England	Aug. 6, 1872	130, 277
Oils and fats to form composition for illuminating and other purposes, Treating.	S. Lewis	Rochester, N. Y	June 14, 1864	43, 156
Oils, Apparatus for burning hydrocarbon	M. T. Gosnell	Baltimore, Md	Nov. 12, 1867	70, 712
Oils, Apparatus for burning hydrocarbon	J. R. Lee	Grass Valley, Cal	Mar. 7, 1871	112, 358
Oils, Apparatus for distilling heavy	H. Ryder	Somerville, Mass	Sept. 2, 1873	142, 515
Oils, Apparatus for extracting essential	G. G. Percival	Waterville, Me	Jan. 17, 1871	110, 998
Oils, Apparatus for gasifying and burning carbon	O. F. Morrill	Chelsea, Mass	Aug. 5, 1862	36, 102
Oils, Apparatus for testing hydrocarbon	J. B. Blair	Philadelphia, Pa	June 10, 1873	139, 654
Oils, Apparatus for treating hydrocarbon	J. J. Johnston	Allegheny, Pa	July 25, 1871	117, 426
Oils, Apparatus for treating hydrocarbon	L. M. Rice and S. E. Adams	Hartford, Conn., and Charlestown, Mass.	May 25, 1869	90, 392
Oils as fuel, Using hydrocarbon	G. B. Hill	New York, N. Y	Sept. 15, 1863	39, 918
Oils, Bleaching vegetable	T. Leonhard	Paterson, N. J	Dec. 3, 1867	71, 768
Oils, Burning hydrocarbon	D. D. Van Norman, L. B. Brown, and E. R. Morrison.	Petroleum Centre, Pa	Aug. 1, 1865	49, 179
Oils, Burning petroleum and other hydrocarbon	L. E. Truesdale	Warren, Mass	Apr. 12, 1870	101, 964
Oils, Clarifying sperm, whale, cotton-seed, and other.	E. C. Moss	New York	Dec. 28, 1832	....
Oils, Drier for petroleum and heavy	S. F. Rogers	Malden, Mass	Dec. 18, 1866	60, 558
Oils for use, Preparing hydrocarbon	E. Dennis	Sing Sing, N. Y	Feb. 13, 1872	123, 683
Oil in casks, Putting up	P. G. Finn	Erie, Pa	May 21, 1867	64, 855
Oils, Machine for extracting essential	V. Squarza	San Francisco, Cal	Feb. 20, 1866	52, 764
Oils, Machine for testing lubricating	R. H. Thurston	Hoboken, N. J	Dec. 24, 1872	134, 229
Oils, Manufacture of deodorized heavy hydrocarbon	J. Merrill	Boston, Mass	May 18, 1869	90, 284
Oils, Manufacture of hydrocarbon	L. Atwood	New York, N. Y	Mar. 26, 1861	31, 858
Oils, Manufacture of hydrocarbon	J. Merrill	Boston, Mass	July 2, 1861	32, 705
Oils, Manufacture of hydrocarbon	W. Spears	Jamestown, N. Y	Sept. 27, 1870	107, 734
Oils, Manufacture of siccative	J. Roux	New York, N. Y	Jan. 24, 1860	26, 929
Oils, &c., Pressing and straining	J. Cunningham and S. S. Edmonston.	New York, N. Y	June 22, 1812	....
Oils, Process and apparatus for extracting essential.	G. G. Percival	Waterville, Me	Nov. 7, 1871	120, 596

Index of patents issued from the United States Patent Office from 1790 *to* 1873, *inclusive*—Continued.

Invention.	Inventor.	Residence.	Date.	No.
Oils, Process and apparatus for extracting essential.	G. G. Percival	Waterville, Me	Dec. 10, 1872	133, 719
Oils, &c., Process and apparatus for producing	J. D. Stanley	Baltimore, Md	Sept. 10, 1872	131, 312
Oils, Process of preparing linseed for pressing in extracting.	C. Moore	Trenton, N. J	Aug. 19, 1856	15, 594
Oils, Purifying fish and other	L. Lecesne	New York, N. Y	June 23, 1812	....
Oils, Purifying hydrocarbon	J. Fordred	Blackheath, England	Apr. 24, 1866	54, 267
Oils, Purifying hydrocarbon	J. Merrill	Boston, Mass	June 28, 1864	43, 325
Oils, Storing	H. P. Genembre	Tarentum, Pa	Feb. 18, 1862	34, 462
Oils, Treating hydrocarbon	A. M. Burke and S. Wright	Cleveland, Ohio	June 25, 1867	65, 999
Oils, Treating hydrocarbon	H. K. Taylor and D. M. Graham	Cleveland, Ohio	Nov. 20, 1866	59, 751
Oiled fabric	G. Streat	New York, N. Y	Jan. 21, 1868	73, 473
Oiled silk and linen, Manufacturing	R. Hodgson		Feb. 1, 1793	....
Oiler	J. Broughton	New York, N. Y	Mar. 6, 1866	52, 959
Oiler	G. J. Capewell	West Cheshire, Conn	July 3, 1866	56, 004
Oiler	J. H. Godwin	Scotland Neck, N. C	Aug. 20, 1867	67, 974
Oiler	C. E. and Z. B. Grandy	Stafford Springs, Conn	Apr. 1, 1873	137, 361
Oiler	W. H. Hart	Meriden, Conn	May 23, 1865	47, 896
Oiler	C. J. Hauck	Williamsburgh, N. Y	May 8, 1866	54, 540
Oiler	C. J. Hauck	Brooklyn, N. Y	Sept. 17, 1872	131, 441
Oiler	G. L. Holt	Springfield, Mass	Apr. 23, 1867	64, 014
Oiler	J. King	Ansonia, Conn	July 10, 1866	56, 233
Oiler	J. Kips and W. Allmendinger	Melrose, N. Y	Feb. 12, 1867	61, 039
Oiler	P. Laflin	Warren, Mass	Sept. 18, 1866	58, 107
Oiler	A. D. Laws	Bridgeport, Conn	Jan. 17, 1871	111, 068
Oiler	E. McDuff and E. D. Forron	Warwick, R. I	Mar. 7, 1871	112, 367
Oiler	J. A. Metcalf	Lawrence, Mass	Aug. 31, 1869	94, 430
Oiler	L. H. Olmsted	Binghamton, N. Y	Nov. 19, 1861	33, 751
Oiler	L. H. Olmsted	Stamford, Conn	May 1, 1866	54, 394
Oiler	F. P. Pfleghar and W. Shellhorn.	New Haven, Conn	Mar. 5, 1867	62, 562
Oiler	F. Stone	New York, N. Y	July 30, 1867	67, 225
Oiler	C. B. Wood	Mansfield, Mass	Dec. 24, 1872	134, 342
Oiler and filler	A. Millar	Roxbury, Mass	Aug. 20, 1867	67, 896
Oiler for looms, Automatic	J. Crook	Ludlow, Vt	Dec. 26, 1871	122, 159
Oiler for loose pulleys	E. L. Conkey	Boston, Mass	Mar. 5, 1872	124, 333
Oiler for loose pulleys	C. A. King	Springfield, Mass	June 14, 1870	104, 322
Oiler for loose pulleys	S. R. Norris	New York, N. Y	Mar. 28, 1871	113, 199
Oiler for machinery	C. A. Morton	Biddeford, Me	July 20, 1869	92, 872
Oiler for machinery	D. C. Smiley	New York, N. Y	Apr. 25, 1854	10, 836
Oiler for machinery	N. Tallman	West New Brighton, N. Y	Sept. 1, 1868	81, 705
Oiler for steam-engines, Automatic	T. Sims	Philadelphia, Pa	May 14, 1872	126, 651
Oiler for the slide of steam-engines	C. C. Tracy	New York, N. Y	Aug. 17, 1869	93, 772
Oiler, Journal	J. A. Weaver	Temperanceville, Pa	June 26, 1866	55, 938
Oiler, Pneumatic	C. E. Decker	Chicago, Ill	Aug. 12, 1873	141, 703
Oiler, Portable	G. W. Crossley	Cleveland, Ohio	Jan. 23, 1872	122, 995
Oiler, Portable	J. A. Mathewson	Johnston, R. I	Aug. 20, 1872	130, 732
Oiler, Steam-cylinder	S. F. Stanton and O. Ripley	Manchester, N. H., and Charlestown, Mass.	Aug. 24, 1869	94, 040
Oiler, Steam-engine	A. Worden	Ypsilanti, Mich	July 29, 1873	141, 304
Oiler-tube	F. J. Seymour	Meriden, Conn	Dec. 29, 1868	85, 484
Oilers, Attaching tubes to	S. V. Beckwith	Hamden, Conn	July 10, 1866	56, 163
Oiling-apparatus for crank-pins	J. F. Allen	Mott Haven, N. Y	Dec. 12, 1871	121, 744
Oiling-device	A. M. and J. M. Schoenfelt	Waterside, Pa	July 6, 1869	92, 378
Oiling spindles, top-rolls, &c., of spinning and other machinery, Device for.	S. H. Barber	East Greenwich, R. I	Feb. 12, 1867	61, 915
Oily matter from water, Apparatus for separating	J. Naughten	Cincinnati, Ohio	Sept. 29, 1857	18, 293
Oily particles held in suspension by steam, Apparatus for separating the.	R. Hale	Roxbury, Mass	June 30, 1857	17, 675
Ointment	J. Bogan and J. B. McCrary	Clarksville, Ohio	Aug. 11, 1868	80, 854
Ointment	T. Combs	Yonkers, N. Y	Apr. 26, 1870	102, 224
Ointment	M. A. Curtis	San Francisco, Cal	Feb. 20, 1873	136, 308
Ointment	W. Dony	Honesdale, Pa	Nov. 28, 1865	51, 154
Ointment	G. D. Field	New Orleans, La	Mar. 8, 1870	100, 516
Ointment	W. W. Gray	Richmond, Va	Mar. 18, 1835	....
Ointment	J. G. Hucks	San Francisco, Cal	May 27, 1873	139, 315
Ointment	W. H. Jones	Henry County, Ala	May 7, 1872	126, 551
Ointment	L. H. Moseley	Franklin, Tenn	Feb. 4, 1868	74, 118
Ointment	N. Shepherd	Belmont County, Ohio	July 9, 1830	....
Ointment	L. Strober	Jersey City, N. J	Nov. 21, 1865	51, 099
Ointment	D. Wigg	Hyde Park, N. Y	Jan. 5, 1869	85, 714
Ointment	H. F. Wood	San Francisco, Cal	May 26, 1868	78, 409
Ointment, Cancer	E. Gilman	Licking, Ohio	Mar. 31, 1836	....
Ointment for bruises, burns, &c	J. T. Mulkey	Walton's Ford, Ga	Feb. 22, 1870	100, 057
Ointment for curing spavin, splint, &c., in horses	B. Lehmann	Piqua, Ohio	Mar. 12, 1867	62, 759
Ointment for horses	G. P. Barnum	Marion, Iowa	Aug. 13, 1867	67, 627
Ointment for horses, cattle, &c	R. Jones	New York, N. Y	July 28, 1868	80, 353
Ointment for horses' hoofs	A. McDougall	Goffstown, N. H	Aug. 6, 1872	130, 229
Ointment for piles	A. M. Irwin	Brooklyn, N. Y	Nov. 14, 1871	120, 879
Ointment for piles	W. W. Riley	Mansfield, Ohio	Jan. 31, 1844	3, 417
Ointment for treating diseases in horses and other animals.	J. S. Williams	Warsaw, Ohio	Mar. 5, 1867	62, 717
Ointment for various uses	W. Judkins	Smithfield, Ohio	June 26, 1816	....
Ointment, Machine for making mercurial	J. W. W. Gordon	Baltimore, Md	June 5, 1844	3, 619
Ointment, Nitrated mercurial	C. Larned	Columbus, Ohio	Feb. 17, 1863	37, 697
Ointment or salve, Vegetable	W. C. and J. H. Roney	Gallupville, N. Y	June 28, 1864	43, 341
Ointment, Rheumatic	J. Gecmen	New York, N. Y	Feb. 18, 1873	135, 983
Oleine from fatty matter, Extracting	C. F. A. Simonin	Philadelphia, Pa	Oct. 28, 1873	144, 000
Omnibus	D. C. Macomber	New York, N. Y	Sept. 23, 1856	15, 769
Omnibus	H. M. Stow	San Francisco, Cal	Jan. 19, 1869	86, 042
Omnibus and car register	R. Cromelien and W. R. Crisp	Washington, D. C	May 26, 1863	38, 655
Omnibus and car register	S. R. Stinard	Canandaigua, N. Y	May 26, 1863	38, 704
Omnibus-brake	H. Bothe	Saint Louis, Mo	Aug. 15, 1871	117, 972
Omnibus-coffer	J. T. Curtis	New York, N. Y	May 5, 1857	17, 206
Omnibus-register	L. Brauer	Washington, D. C	July 27, 1858	20, 986
Omnibus-register	F. Brunon	Philadelphia, Pa	Mar. 26, 1861	31, 778
Omnibus-register	L. Cromwell	Baltimore, Md	Apr. 15, 1856	14, 652
Omnibus-register	E. S. Dawson and A. Weeks	Syracuse, N. Y	Mar. 26, 1861	31, 788

Index of patents issued from the United States Patent Office from 1790 *to* 1873, *inclusive*—Continued.

Invention.	Inventor.	Residence.	Date.	No.
Omnibus-register	F. O. Deschamps	Philadelphia, Pa	Mar. 2, 1852	8,770
Omnibus-register	F. O. Deschamps	Philadelphia, Pa	Feb. 7, 1854	10,503
Omnibus-register	R. E. House	Binghamton, N. Y	May 25, 1858	20,349
Omnibus-register	H. C. Howells and J. C. Howells	New York, N. Y., and Madison, Wis.	June 14, 1859	24,390
Omnibus-register	W. M. Keague	Brooklyn, N. Y	Oct. 11, 1859	25,740
Omnibus-register	L. W. Mallory	Philadelphia, Pa	June 20, 1854	11,148
Omnibus-register	W. Morris	Philadelphia, Pa	Nov. 14, 1854	11,939
Omnibus-register	I. B. Person and J. L. Brockett	Baltimore, Md	Aug. 19, 1851	8,304
Omnibus-register	J. Rodgers	New York, N. Y	Mar. 18, 1856	14,470
Omnibus-register	P. Taltavull	Washington, D. C	Jan. 25, 1853	9,563
Omnibus-register	I. Z. A. Wagner	Philadelphia, Pa	Sept. 28, 1852	9,289
Omnibus-register	R. F. White	New York, N. Y	June 7, 1859	24,345
Omnibus-registers, Self-locking device for	M. Offley	Baltimore, Md	Aug. 7, 1860	29,549
Omnibus-spring	J. H. Dennis	Louisville, Ky	Jan. 14, 1862	34,136
Omnibus-step	J. Ashenfelder	Philadelphia, Pa	Mar. 23, 1852	8,816
Omnibus-step	W. H. Hoyt	New York, N. Y	May 27, 1851	8,119
Omnibus-step protector	T. Coles	New York, N. Y	July 25, 1854	11,360
Omnibus-steps, Turning up	S. Burdett	New York, N. Y	Aug. 27, 1850	7,591
Onion-seed from grain, Machine for separating	L. Ruggles and D. Tomlinson	Brookfield, Conn	Mar. 16, 1810	....
Open or middle ring	B. F. Zinn	Mount Rock, Pa	Nov. 17, 1868	84,155
Open-ring	A. H. Bixler	Carlisle, Pa	May 5, 1868	77,441
Opera-chair	A. Abel	New York, N. Y	Jan. 30, 1872	123,140
Opera-chair	B. Koechling	New York, N. Y	Feb. 8, 1870	99,576
Opera-chairs, Perfuming	S. Fredrick	New York, N. Y	Aug. 13, 1872	130,421
Opera-glass, Folding	J. Saxton	Washington, D. C	June 29, 1869	91,969
Operating-chair	J. B. Morrison	Saint Louis, Mo	Sept. 29, 1868	82,542
Operating-drill	T. D. Keith	Mayville, Wis	May 11, 1869	89,932
Operating-drill	E. G. Lamson	Shelburne Falls, Mass	Sept. 17, 1867	68,888
Opium from the poppy-plant, Pressing	L. Cook	Shrewsbury, N. J	Apr. 20, 1832	....
Ophthalmic vapor-apparatus	T. F. Frank	Ischua, N. Y	Aug. 30, 1859	25,252
Optical instrument	O. B. Brown	Malden, Mass	Aug. 10, 1869	93,594
Optical instrument	L. Heath	Boston, Mass	May 8, 1866	54,542
Optical instrument	R. B. Tolles	Canastota, N. Y	Sept. 25, 1855	13,603
Optical instruments, Adjustment for	C. B. Richard	Hartford, Conn	May 23, 1865	47,860
Optical instruments, Binocular eye-piece for	R. B. Tolles	Canastota, N. Y	July 3, 1866	56,125
Optical instruments, Double eye-piece for	A. Beckers	New York, N. Y	Dec. 13, 1859	26,407
Oracular wheel or center-table	W. O. George	Richmond, Va	May 20, 1856	14,910
Orange-knife	E. L. Cooke	Hartford, Conn	Mar. 15, 1870	100,862
Ordnance	B. T. Babbitt	New York, N. Y	Feb. 5, 1861	31,291
Ordnance	B. T. Babbitt	New York, N. Y	Feb. 25, 1862	34,472
Ordnance	E. D. Baker	Claremont, N. H	Feb. 4, 1862	34,287
Ordnance	T. A. Blakly	London, England	Feb. 16, 1864	41,662
Ordnance	B. Britten	Red Hill, England	June 27, 1871	116,408
Ordnance	L. W. Broadwell	New Orleans, La	June 19, 1866	55,609
Ordnance	J. B. Eads	Saint Louis, Mo	Sept. 4, 1866	57,692
Ordnance	J. Ericsson	New York, N. Y	Jan. 12, 1864	41,208
Ordnance	G. H. Ferriss	Utica, N. Y	May 3, 1864	42,571
Ordnance	T. Fowlds	Trevorton, Pa	May 6, 1862	35,148
Ordnance	E. Heaton	New Haven, Conn	June 3, 1862	35,443
Ordnance	J. C. C. Holenshade	Cincinnati, Ohio	Apr. 7, 1863	38,110
Ordnance	W. W. Hubbell	Philadelphia, Pa	Sept. 13, 1864	44,194
Ordnance	J. L. Lowry	Pittsburgh, Pa	May 16, 1865	47,740
Ordnance	T. J. Mayall	Roxbury, Mass	Nov. 27, 1860	30,742
Ordnance	T. J. Mayall	Roxbury, Mass	May 21, 1861	32,376
Ordnance	R. Montgomery	New York, N. Y	Mar. 11, 1862	34,666
Ordnance	W. Ostrander, I. D. Reeder, and J. Cordwan.	Brooklyn, N. Y	Apr. 15, 1862	34,977
Ordnance	W. Palliser	Pall Mall, England	June 29, 1869	91,864
Ordnance	J. B. Read	Tuscaloosa, Ala	Oct. 5, 1869	95,604
Ordnance	C. W. Stafford	New York, N. Y	Sept. 6, 1864	44,119
Ordnance	C. W. Stafford	Old Saybrook, Conn	Feb. 21, 1865	46,506
Ordnance	N. Wiard	New York, N. Y	May 26, 1863	38,709
Ordnance	N. Wiard	New York, N. Y	Aug. 18, 1863	39,604
Ordnance	N. Wiard	New York, N. Y	Apr. 12, 1864	42,315
Ordnance	C. Walter	Bridgeport, Conn	Mar. 13, 1855	12,522
Ordnance	W. E. Woodbridge	Little Falls, N. Y	Jan. 1, 1867	60,979
Ordnance	L. W. Wright	Thorndike, Mass	May 29, 1866	55,193
Ordnance	D. F. Yeakel	La Fayette, Ind	Apr. 29, 1862	35,124
Ordnance, Adjustable front sight for	J. B. Learock	Boston, Mass	Mar. 29, 1864	42,091
Ordnance, Adjustable sight for	C. B. Long	Worcester, Mass	Dec. 2, 1862	37,076
Ordnance and fire-arm	R. White	Bridgeport, Conn	Aug. 28, 1866	57,607
Ordnance and other fire-arms	O. Abbruzzo	New York, N. Y	Dec. 1, 1868	84,525
Ordnance and other fire-arms, Chamber of	J. P. Schenkl	Worcester, Mass	Apr. 19, 1859	23,746
Ordnance and other fire-arms, Wad for	E. D. Williams	Philadelphia, Pa	May 13, 1862	35,273
Ordnance and projectile	B. F. Bates and C. R. Macy	New York, N. Y	Mar. 7, 1865	46,625
Ordnance and projectile	C. Heaton	New York, N. Y	June 25, 1872	128,392
Ordnance and projectile therefor	B. C. Pole	Washington, D. C	Sept. 30, 1873	143,251
Ordnance, Apparatus for casting	J. Millholland	Reading, Pa	July 15, 1862	35,884
Ordnance, Arranging water-tubes for cooling the breech of.	J. Hursh	Philadelphia, Pa	July 15, 1862	35,877
Ordnance, Attaching trunnions to	R. Gibbons	Oakland, Cal	Aug. 2, 1864	43,681
Ordnance, Automatic revolving	T. J. Campbell	Lincoln, Ill	June 10, 1862	35,504
Ordnance, Breech-loading	C. Alger	Hudson, N. Y	Dec. 24, 1861	33,976
Ordnance, Breech-loading	W. Bacon	Monticello, Kans	May 11, 1869	89,965
Ordnance, Breech-loading	P. R. Beaupre	Metropolis, Ill	Apr. 14, 1868	76,586
Ordnance, Breech-loading	G. W. Bishup	Brooklyn, N. Y	Sept. 9, 1856	15,682
Ordnance, Breech-loading	R. C. Bristol	Chicago, Ill	Mar. 25, 1862	34,730
Ordnance, Breech-loading	L. W. Broadwell	St. Petersburg, Russia	Dec. 10, 1861	33,876
Ordnance, Breech-loading	L. W. Broadwell	London, Great Britain	July 12, 1864	43,553
Ordnance, Breech-loading	L. W. Broadwell	New Orleans, La	June 19, 1866	55,762
Ordnance, Breech-loading	E. B. Butterfield	New York, N. Y	Nov. 5, 1861	33,635
Ordnance, Breech-loading	J. S. Butterfield	Philadelphia, Pa	Apr. 30, 1861	30,238
Ordnance, Breech-loading	J. A. De Brame	New York, N. Y	Dec. 24, 1861	34,025
Ordnance, Breech-loading	L. Evans	Morgantown, Va	Oct. 9, 1860	30,305
Ordnance, Breech-loading	G. H. Ferris	Utica, N. Y	Mar. 22, 1864	41,984
Ordnance, Breech-loading	T. L. Fortune	Weston, Mo	July 26, 1864	43,655
Ordnance, Breech-loading	J. A. France	Cobleskill, N. Y	Sept. 10, 1861	33,244

Index of patents issued from the United States Patent Office from 1790 *to* 1873, *inclusive*—Continued.

Invention.	Inventor.	Residence.	Date.	No.
Ordnance, Breech-loading	G. P. Ganster	New York, N. Y	Sept. 9, 1862	36, 401
Ordnance, Breech-loading	W. F. Goodwin	Powhatan, Ohio	May 20, 1862	35, 311
Ordnance, Breech-loading	W. F. Goodwin	Powhatan, Ohio	June 14, 1864	43, 110
Ordnance, Breech-loading	W. F. Goodwin	Powhatan, Ohio	June 14, 1864	43, 111
Ordnance, Breech-loading	W. F. Goodwin	Powhatan, Ohio	June 28, 1864	43, 306
Ordnance, Breech-loading	C. Gordon	Goswell Road, England	Apr. 22, 1873	138, 146
Ordnance, Breech-loading	C. Grimshaw	Milwaukee, Wis	Oct. 29, 1861	33, 582
Ordnance, Breech-loading	J. S. Hall	Pittsburgh, Pa	Apr. 29, 1862	35, 095
Ordnance, Breech-loading	E. Hamilton	Chicago, Ill	July 9, 1861	32, 768
Ordnance, Breech-loading	W. S. Henson	Newark, N. J	Nov. 5, 1861	33, 646
Ordnance, Breech-loading	W. W. Hubbell	Philadelphia, Pa	May 29, 1860	28, 486
Ordnance, Breech-loading	W. W. Hubbell	Philadelphia, Pa	July 5, 1864	43, 412
Ordnance, Breech-loading	J. Lee	Bolivar, Ohio	May 19, 1863	38, 638
Ordnance, Breech-loading	C. F. Leisen	Philadelphia, Pa	Oct. 22, 1861	33, 530
Ordnance, Breech-loading	L. M. Lull and J. T. Starr	Walnut Grove, Ill	Feb. 18, 1868	74, 557
Ordnance, Breech-loading	H. F. Mann	La Porte, Ind	Nov. 26, 1861	33, 787
Ordnance, Breech-loading	T. J. Mayall	Roxbury, Mass	Oct. 9, 1860	30, 335
Ordnance, Breech-loading	E. R. McCabe	Rochester, Iowa	May 27, 1862	35, 380
Ordnance, Breech-loading	J. A. Miller	Paducah, Ky	Feb. 7, 1865	46, 259
Ordnance, Breech-loading	R. L. Mills	Columbus, Ohio	Feb. 16, 1864	41, 631
Ordnance, Breech-loading	R. R. Moffatt	La Crosse, Wis	July 8, 1862	35, 863
Ordnance, Breech-loading	J. B. Moody	Cincinnati, Ohio	Mar. 6, 1866	53, 026
Ordnance, Breech-loading	C. Morrill	New York, N. Y	June 17, 1862	35, 619
Ordnance, Breech-loading	C. C. W. Müller	New Orleans, La	Aug. 13, 1867	67, 792
Ordnance, Breech-loading	M. J. Murphy	Cork, Ireland	Dec. 18, 1866	60, 540
Ordnance, Breech-loading	A. F. Potter	San Francisco, Cal	May 17, 1870	103, 078
Ordnance, Breech-loading	R. B. Reynolds	United States Navy	Sept. 29, 1863	40, 121
Ordnance, Breech-loading	T. C. Rice	Cambridgeport, Mass	Jan. 28, 1862	34, 266
Ordnance, Breech-loading	W. Schmoele, jr	Philadelphia, Pa	Jan. 19, 1864	41, 325
Ordnance, Breech-loading	W. G. Sherwin, J. McFarland, and C. Thieme.	Cincinnati, Ohio	Nov. 19, 1861	33, 756
Ordnance, Breech-loading	F. Shilling	Fort Delaware	Oct. 13, 1863	40, 295
Ordnance, Breech-loading	A. B. Smith	Clinton, Pa	July 28, 1863	39, 359
Ordnance, Breech-loading	H. Stanhope	Philadelphia, Pa	July 23, 1861	32, 921
Ordnance, Breech-loading	W. Stow	Utica, N. Y	July 29, 1862	36, 036
Ordnance, Breech-loading	E. A. Sutcliffe	New York, N. Y	Aug. 18, 1863	39, 596
Ordnance, Breech-loading	N. Thompson	Brooklyn, N. Y	May 28, 1872	127, 202
Ordnance, Breech-loading	N. Thompson	Brooklyn, N. Y	Aug. 20, 1872	130, 673
Ordnance, Breech-loading	N. Thompson	Brooklyn, N. Y	Oct. 1, 1872	131, 833
Ordnance, Breech-loading	N. Thompson	Brooklyn, N. Y	Oct. 15, 1872	132, 220
Ordnance, Breech-loading	E. H. Tobey	Chicago, Ill	Dec. 28, 1869	98, 316
Ordnance, Breech-loading	W. B. Treadwell	Albany, N. Y	June 17, 1862	35, 637
Ordnance, Breech-loading	T. Tully	Waukegan, Ill	June 28, 1864	43, 351
Ordnance, Breech-loading	E. A. Turpin	New York, N. Y	Aug. 26, 1862	36, 325
Ordnance, Breech-loading	W. Wallace	Ansonia, Conn	Sept. 20, 1864	44, 356
Ordnance, Breech-loading	G. I. Washburn	Worcester, Mass	Oct. 6, 1863	40, 205
Ordnance, Breech-loading	L. C. T. Weber	Rochester, N. Y	Jan. 1, 1861	31, 044
Ordnance, Breech-loading	A. Weeks	Minneapolis, Minn	Oct. 1, 1867	69, 519
Ordnance, Breech-loading	J. Whitworth	Manchester, England	Nov. 14, 1871	120, 842
Ordnance, Breech-loading	W. Williams	Saint Louis, Mo	Nov. 1, 1864	44, 905
Ordnance, Breech-loading	T. Yates	Milwaukee, Wis	Feb. 14, 1865	46, 417
Ordnance, Breech-loading	D. F. Yeakel	La Fayette, Ind	Feb. 11, 1862	34, 388
Ordnance, Bronze	S. B. Dean	Boston, Mass	May 18, 1869	90, 244
Ordnance, Carriage for raising and transporting	C. Bates	Kingston, Mass	Sept. 27, 1864	44, 390
Ordnance, Cast-iron	J. A. Dahlgren	Philadelphia, Pa	Aug. 6, 1861	32, 983
Ordnance, Cast-iron	J. A. Dahlgren	Philadelphia, Pa	Aug. 6, 1861	32, 984
Ordnance, Cast-iron	J. A. Dahlgren	Philadelphia, Pa	Aug. 6, 1861	32, 985
Ordnance, &c., Casting	T. J. Rodman	Pittsburgh, Pa	Aug. 14, 1847	5, 236
Ordnance, Chambered trunnion for disabling	P. B. Lawson and A. Berney	Cold Spring, N. Y., and Jersey City, N. J.	Mar. 10, 1863	37, 870
Ordnance, Counterpoise-platform for	A. T. Brewer	Brighton, Mass	Apr. 26, 1870	102, 213
Ordnance, Device for indicating the elevation of	C. B. Long	Worcester, Mass	July 29, 1862	36, 054
Ordnance, Discharging	W. Johnson	Milwaukee, Wis	May 26, 1863	38, 683
Ordnance-discharging apparatus	H. Arden	Brooklyn, N. Y	Oct. 24, 1871	120, 179
Ordnance, Elastic breech for	J. F. Clew	New York, N. Y	Dec. 27, 1864	45, 586
Ordnance, Elastic breech for	H. H. Day	New York, N. Y	Dec. 2, 1862	37, 035
Ordnance exhausted of air, Tampion-cap for the muzzles of.	J. F. Clew	New York, N. Y	Nov. 15, 1864	45, 020
Ordnance, Firing-mechanism for	N. Thompson	Brooklyn, N. Y	Aug. 13, 1872	130, 452
Ordnance for use under water	J. C. Tilton	Geneseo, Ill	Apr. 22, 1862	35, 048
Ordnance gas-check	J. W. Pearson	Watertown, Mass	Oct. 12, 1869	95, 720
Ordnance, Hooped	R. P. Parrott	Cold Spring, N. Y	May 6, 1862	35, 171
Ordnance, Implement for disabling	E. E. Bean	Boston, Mass	Jan. 5, 1864	41, 115
Ordnance, Implement for disabling	A. Bonzano	Detroit, Mich	Mar. 24, 1863	37, 946
Ordnance in boring-mills, Apparatus for adjusting	S. P. Dean	Boston, Mass	Dec. 8, 1863	40, 824
Ordnance lighting and firing device	G. K. Farrington	Alcatraz Island, Cal	Dec. 27, 1870	110, 448
Ordnance, Loading	J. B. Eads	Saint Louis, Mo	Dec. 3, 1867	71, 592
Ordnance, Loading	W. E. Moore	Crawfordsville, Ind	Mar. 15, 1859	23, 258
Ordnance, Loading	D. E. Rice	Detroit, Mich	June 28, 1864	43, 339
Ordnance-lock	N. Thompson	Brooklyn, N. Y	Sept. 10, 1872	131, 194
Ordnance-lock	J. Vavasseur	Southwark, England	May 18, 1869	90, 323
Ordnance, Machinery for finishing rim-bases of	E. F. Howard	Boston, Mass	Sept. 22, 1863	40, 039
Ordnance, Manufacture of	H. Ames	Falls Village, Conn	Apr. 11, 1865	47, 177
Ordnance, Manufacture of	G. P. Harding	Chiswick, England	Feb. 19, 1867	62, 266
Ordnance, Manufacture of	R. P. Parrott	Cold Spring, N. Y	Oct. 1, 1861	33, 401
Ordnance, Manufacture of	T. E. Vickers	Sheffield, Great Britain	Mar. 21, 1865	46, 978
Ordnance, Mounting	B. H. Bartol	Philadelphia, Pa	Apr. 7, 1863	38, 089
Ordnance, Mounting	C. F. Brown	Warren, R. I	July 17, 1855	13, 249
Ordnance, Mounting	C. F. Brown	Warren, R. I	Aug. 26, 1862	36, 273
Ordnance, Mounting	M. Stoddard	Buffalo, N. Y	Aug. 25, 1863	39, 687
Ordnance, Mounting and operating	J. B. Eads	Saint Louis, Mo	Jan. 3, 1865	45, 704
Ordnance, Mounting and operating	J. Taggert	Roxbury, Mass	Apr. 21, 1863	38, 266
Ordnance, Mounting and working	A. Moncrieff	Culfargie, Scotland	May 30, 1871	115, 502
Ordnance, Mounting field	F. A. De Mey	New York, N. Y	Sept. 16, 1862	36, 457
Ordnance, Mounting field	W. F. Goodwin	Powhatan, Ohio	Sept. 15, 1863	39, 914
Ordnance, Muzzle-loading	F. W. Alexander	Baltimore, Md	Feb. 18, 1868	74, 478

Index of patents issued from the United States Patent Office from 1790 *to* 1873, *inclusive*—Continued.

Invention.	Inventor.	Residence.	Date.	No.
Ordnance on gun-boats, &c., Operating	W. L. Winans	Baltimore, Md., now residing in St. Petersburg, Russia.	Feb. 21, 1865	46, 516
Ordnance on war-vessels, Operating	W. P. Miller	Marysville, Cal	Apr. 7, 1863	38, 118
Ordnance, Operating	M. L. Callender and N. W. Northrup.	New York and Greene, N. Y.	Mar. 17, 1863	37, 935
Ordnance, Operating	J. B. Eads	Saint Louis, Mo	Feb. 16, 1864	41, 611
Ordnance, Operating	J. B. Eads	Saint Louis, Mo	May 3, 1864	42, 566
Ordnance, Operating	J. B. Eads	Saint Louis, Mo	Feb. 14, 1865	46, 342
Ordnance, Operating	J. B. Eads	Saint Louis, Mo	July 3, 1866	56, 021
Ordnance, Operating	J. B. Eads	Saint Louis, Mo	Feb. 28, 1871	112, 230
Ordnance, Operating	J. H. Field	Saugerties, N. Y	July 17, 1866	56, 395
Ordnance, Operating	C. Perley	New York, N. Y	Dec. 12, 1865	51, 475
Ordnance, Operating	J. Ridgway	Boston, Mass	Oct. 21, 1862	36, 730
Ordnance, Operating	E. A. Stevens	Hoboken, N. J	Jan. 6, 1863	37, 364
Ordnance, Operating	O. Tufts	Boston, Mass	Dec. 23, 1862	37, 250
Ordnance, Operating	M. Wappich	Sacramento, Cal	Sept. 22, 1863	40, 067
Ordnance, Operating heavy	J. B. Eads	Saint Louis, Mo	Jan. 12, 1864	41, 206
Ordnance, Pendulum-sight for	W. F. Goodwin	New York, N. Y	Nov. 8, 1864	44, 947
Ordnance, Provision for the recoil of	J. B. Atwater	Chicago, Ill	July 5, 1864	43, 382
Ordnance, Recoil-obviator for	S. F. Hawley	Constableville, N. Y	Sept. 12, 1871	118, 933
Ordnance, Repeating	E. C. C. Kellogg	Hartford, Conn	Nov. 3, 1863	40, 540
Ordnance, Repeating	J. A. Matthews	Saint Louis, Mo	July 31, 1860	29, 437
Ordnance, &c., Repeating	W. McCord and E. Maher	Sing Sing and New York, N. Y.	Nov. 26, 1861	33, 813
Ordnance Repeating	J. P. Taylor	Elizabethton, Tenn	July 4, 1871	116, 775
Ordnance Repeating	A. H. Townsend	Georgetown, Colo	June 6, 1871	115, 659
Ordnance, Repeating	G. Utley	Louisburgh, N. C	May 11, 1858	20, 229
Ordnance, Reversible	J. D. Murphy	Baltimore, Md	June 30, 1868	79, 377
Ordnance, Revolving	J. L. Bonham	Hellen, Pa	Sept. 30, 1862	36, 355
Ordnance, Revolving	J. A. De Brame	New York, N. Y	Dec. 24, 1861	34, 024
Ordnance, Revolving	B. Franke	New York, N. Y	July 29, 1862	35, 998
Ordnance, Revolving	M. F. Hardy	Seward, N. Y	Aug. 12, 1862	36, 148
Ordnance, Revolving	N. L. Milburn	Saint Louis, Mo	Sept. 4, 1866	57, 751
Ordnance, Revolving	N. Taliaferro	Augusta, Ky	Jan. 14, 1862	34, 171
Ordnance, Rifling	H. E. Dimick	Saint Louis, Mo	Jan. 13, 1857	16, 377
Ordnance, Rifling	B. B. Hotchkiss	New York, N. Y	Jan. 21, 1868	73, 447
Ordnance, Rifling	J. Seipel	Washington, D. C	Oct. 17, 1865	50, 502
Ordnance, Rifling	T. Steece	United States Navy	Mar. 17, 1863	37, 924
Ordnance, Serving	J. H. Coon	Deposit, N. Y	Jan. 6, 1863	37, 279
Ordnance, Shield for	P. Andrew	Cincinnati, Ohio	Nov. 3, 1863	40, 453
Ordnance, Sight for	J. Brady	Philadelphia, Pa	Feb. 14, 1865	46, 329
Ordnance, Strengthening	P. M. Parsons	Blackheath, England	July 19, 1864	43, 629
Ordnance, Submarine	D. Fitzgerald	New York, N. Y	Dec. 5, 1871	121, 455
Ordnance to facilitate unspiking, Vent-bushing for	B. G. Martin	New York, N. Y	Dec. 23, 1862	37, 257
Ordnance, Vent-hole of	W. W. Gould	Skowhegan, Me	Apr. 22, 1862	35, 019
Ordnance-vent stopper	J. J. Hirschbuhl	Louisville, Ky	May 28, 1861	32, 422
Ore after being pulverized, Machine for washing and separating.	H. Trumbull	Jersey City, N. J	Aug. 9, 1859	25, 060
Ore and bone crusher	A. Newell	Red Wing, Minn	Sept. 8, 1868	81, 932
Ore and coal, Desulphurizing	W. H. Letterman	Philadelphia, Pa	June 26, 1860	28, 875
Ore and extracting gold and silver, Sulphurizing	W. W. Hubbell	Philadelphia, Pa	June 4, 1867	65, 387
Ore and grain separator	S. Duvall	Philadelphia, Pa	Apr. 16, 1872	125, 797
Ore and metal furnace	W. S. Wood	Newtown, N. Y	June 20, 1871	116, 129
Ore and mineral separator	S. R. Krom	New York, N. Y	Apr. 7, 1868	76, 331
Ore and mineral washer	E. Platt	Charleston, S. C	Oct. 4, 1870	107, 957
Ore and mineral washer	J. B. Platt	Augusta, Ga	Oct. 4, 1870	107, 958
Ore and mineral washing pans, Arrangement of	S. Porter	Hartford, Conn	Dec. 9, 1851	8, 582
Ore and rock, Machine for pulverizing	V. Ryerson	New York, N. Y	Nov. 30, 1869	97, 445
Ore and stone crusher	R. Learmouth	Buffalo, N. Y	Feb. 6, 1872	123, 406
Ore by means of fuel, chemicals, and fluxes, Treatment of.	J. B. Whelpley and J. J. Storer	Boston, Mass	July 11, 1871	116, 903
Ore-calcining apparatus	G. W. White	New York, N. Y	Feb. 7, 1865	46, 287
Ore-calcining furnace	J. B. Britton	Philadelphia, Pa	June 7, 1864	43, 003
Ore-calcining furnace	J. H. Dunstan	Belleville, Canada	Apr. 9, 1872	125, 384
Ore-chloridizing furnace	I. M. Phelps	Chicago, Ill	Feb. 25, 1873	136, 261
Ore, Chlorodizing silver	E. N. Riotte	San Francisco, Cal	Mar. 28, 1871	113, 208
Ore-cleaner	D. Pollock	Lancaster, Pa	Jan. 27, 1857	16, 495
Ore cleaner and separator]	J. H. Hillman	Trigg Furnace, Ky	Oct. 29, 1872	132, 577
Ore cleaning and separating machine	N. Carpenter	Ellenville, N. Y	May 3, 1864	42, 558
Ore cleaning machine	B. O. Bryan	Marietta, Pa	June 26, 1855	13, 136
Ore, coal, and similar material, Apparatus for the discharge and preparation of granular.	P. Osterspey	Mechernich, Prussia	Nov. 2, 1869	96, 472
Ore, Concentrating and cleaning	V. Baron	France	May 6, 1862	35, 197
Ore-concentrating apparatus	J. Hepburn	Mokelumne Hill, Cal	Dec. 15, 1863	40, 932
Ore-concentrating machine	T. Varney	San Francisco, Cal	Mar. 19, 1867	62, 983
Ore-concentrating table	G. Kustel	Dayton, Nev	Mar. 14, 1865	46, 806
Ore-concentrating table	H. A. Thompson	Grant, Gipps Land, Victoria.	Nov. 27, 1866	60, 089
Ore-concentrator	H. Donnelly	Virginia City, Nev	Jan. 14, 1868	73, 306
Ore-concentrator	L. Goodwin and S. A. West	San Francisco, Cal	Sept. 17, 1867	68, 978
Ore-concentrator	M. Hungerford	San Francisco, Cal	Mar. 12, 1867	62, 749
Ore-concentrator	M. Hungerford	San Francisco, Cal	July 2, 1872	128, 628
Ore-concentrator	J. A. Peer	Grass Valley, Cal	Sept. 9, 1873	142, 646
Ore-concentrator	J. W. V. Rawlins and S. Stephens.	Houghton, Mich	June 29, 1869	91, 966
Ore-concentrator	E. L. Seymour	New York, N. Y	Aug. 30, 1859	25, 280
Ore-concentrator	W. C. Stiles	Nevada City, Cal	May 2, 1871	114, 486
Ore-concentrator	G. W. and W. L. Strong	San Francisco, Cal	May 18, 1869	90, 205
Ore concentrator and amalgamator	J. S. Bradford	New York, N. Y	Oct. 12, 1869	95, 640
Ore concentrator and amalgamator	S. Fountain	Silver City, Nev	Dec. 17, 1867	72, 182
Ore concentrator and endless sluice-blanket	C. D. Smith	Drytown, Cal	July 20, 1869	92, 893
Ore concentrator and separator	J. Edgar	New York, N. Y	Dec. 7, 1869	97, 488
Ore, Concentrator for dressing	H. Weston and G. C. Langtry	Dayton, Nev	Oct. 20, 1868	83, 231
Ore-concentrator, Self-discharging blanket	J. M. Bryan	Lincoln, Cal	May 4, 1869	89, 734
Ore-crusher	J. Burns	New York, N. Y	June 4, 1872	127, 561
Ore-crusher	J. W. Cummings	Georgetown, Colo	Apr. 23, 1872	126, 034
Ore-crusher	M. B. Dodge	New York, N. Y	Dec. 4, 1866	60, 154

Index of patents issued from the United States Patent Office from 1790 *to* 1873, *inclusive*—**Continued.**

Invention.	Inventor.	Residence.	Date.	No.
Ore-crusher	M. B. Dodge	Brooklyn, N. Y	May 25, 1869	90, 510
Ore-crusher	J. Fowler	Rahway, N. J	May 22, 1866	54, 884
Ore-crusher	A. W. Hall	New York, N. Y	Oct. 31, 1865	50, 703
Ore-crusher	J. Hamilton, L. E. Hanson, and G. W. and J. Hamilton.	Wheeling, W. Va	May 25, 1869	90, 532
Ore crusher	W. P. Hammond	Napa City, Cal	Sept. 24, 1872	131, 612
Ore-crusher	W. P. Hammond	Napa City, Cal	Dec. 24, 1872	134, 201
Ore crusher	B. Hotchkiss and S. C. Goodsell	New Haven, Conn	Mar. 26, 1867	63, 160
Ore-crusher	S. Ingersoll	Stamford, Conn	July 25, 1865	49, 032
Ore-crusher	S. R. Krom	New York, N. Y	June 22, 1869	91, 641
Ore-crusher	S. R. Krom	New York, N. Y	July 16, 1872	129, 238
Ore-crusher	G. Mitchell	Philadelphia, Pa	May 28, 1872	127, 260
Ore-crusher	G. Mitchell	Philadelphia, Pa	Nov. 19, 1872	133, 111
Ore-crusher	W. R. Parrott and J. J. Boardman.	Boston, Mass	Jan. 8, 1867	61, 019
Ore-crusher	A. F. W. Partz	Wurtsborough, N. Y	Sept. 27, 1864	44, 450
Ore-crusher	J. V. Pomeroy	Utica, N. Y	July 25, 1865	48, 999
Ore-crusher	J. Reese	Pittsburgh, Pa	Mar. 31, 1868	76, 100
Ore-crusher	C. W. Stafford	Saybrook, Conn	Nov. 13, 1866	59, 676
Ore-crusher	E. and W. Thornton	Sharon, Pa	July 30, 1872	130, 084
Ore-crusher	J. Webster and J. G. Morgan	Brooklyn, N. Y	Nov. 21, 1865	51, 104
Ore-crusher	Z. Wheeler	San Francisco, Cal	Mar. 6, 1866	53, 066
Ore-crusher	J. T. Whelpley	Boston, Mass	Nov. 8, 1864	44, 989
Ore crusher and amalgamator	L. Griswold	Denver, Colo	Mar. 21, 1871	112, 804
Ore crusher and amalgamator	J. Hart	Melbourne, Victoria	May 7, 1867	64, 416
Ore crusher and grinder	S. Hughes	San Francisco, Cal	Apr. 7, 1868	76, 461
Ore crusher, grinder, and amalgamator	J. A. Collins	Virginia City, Nev	July 14, 1868	79, 954
Ore-crushers, Chilled plate for	J. L. Agnew and C. E. Wright	Negaunee, Mich	Mar. 16, 1869	87, 750
Ore crushing and pulverizing machine	A. Buffum	Perth Amboy, N. J	Jan. 9, 1855	12, 191
Ore crushing and pulverizing machine	S. Gardiner, jr	New York, N. Y	July 25, 1854	11, 368
Ore-crushing machine	N. Conkling	Brooklyn, N. Y	Mar. 16, 1858	19, 670
Ore, &c., crushing machine	A. W. Hall	New York, N. Y	Apr. 18, 1865	47, 297
Ore-crushing machine	A. W. Hall	New York, N. Y	Oct. 17, 1865	50, 467
Ore-crushing machine	S. F. Hodge	Detroit, Mich	May 26, 1857	17, 374
Ore-crushing machine	S. Hughes	Charleston, S. C	Nov. 14, 1871	120, 972
Ore-crushing machine	J. T. Plass	New York, N. Y	Oct. 25, 1864	44, 819
Ore-crushing machine	W. H. Plumb	New York, N. Y	Dec. 5, 1854	12, 031
Ore, Machine for crushing mineral	W. P. Parrott	Boston, Mass	May 29, 1860	28, 499
Ore-crushing mill	S. F. Mackie	New York, N. Y	Apr. 30, 1867	64, 339
Ore-crushing mill	J. Moore	San Francisco, Cal	May 6, 1862	35, 202
Ore-crushing roller	W. H. Plumb	New York N. Y	Nov. 14, 1856	16, 039
Ore-crushing stamp	G. D. Crocker	San Francisco, Cal	Aug. 5, 1873	141, 493
Ore-crushing stamp	J. M. McFarland	Golden City, Colo	July 1, 1873	140, 425
Ore, Method of decomposing and desulphurizing	R. Spencer	New York, N. Y	Nov. 8, 1864	44, 986
Ore, Deodorizing	I. Rogers	North Haverstraw, N. Y	Aug. 21, 1860	29, 746
Ore, Deoxidizing and carbonizing iron	C. Adams	Philadelphia, Pa	Feb. 22, 1870	100, 095
Ore-deoxidizing furnace, Iron	A. H. Brainerd	Rome, N. Y	Nov. 15, 1870	109, 173
Ore, Desulphurizing	H. Bradford	New York, N. Y	June 28, 1864	43, 363
Ore, Desulphurizing	E. A. Burns	Meadow Lake, Cal	Mar. 8, 1870	100, 497
Ore, Desulphurizing	M. B. Dodge	New York, N. Y	Apr. 4, 1865	47, 094
Ore, Desulpurizing	J. A. Hitchings	Denver City, Colo	Sept. 18, 1866	58, 100
Ore, Desulphurizing	J. F. Osgood	Boston, Mass	Apr. 5, 1870	101, 654
Ore, Desulphurizing	V. Ryerson	New York, N. Y	Aug. 14, 1866	57, 194
Ore desulphurizing and amalgamating apparatus	E. P. Gardiner	New York, N. Y	Jan. 24, 1865	45, 991
Ore, Desulphurizing and disintegrating	J. C. Ayer	Lowell, Mass	Mar. 7, 1865	46, 619
Ore, Desulphurizing and disintegrating	J. C. Ayer	Lowell, Mass	Mar. 7, 1865	46, 620
Ore, Desulphurizing and disintegrating	D. Minthorn	New York, N. Y	Feb. 27, 1866	52, 875
Ore, Desulphurizing and disintegrating	G. Vining	Boston, Mass	Aug. 8, 1865	49, 322
Ore desulphurizing and oxidizing apparatus	J. M. Rohrer	Pine Grove, Pa	Apr. 6, 1869	88, 741
Ore, Desulphurizing and oxidizing metallic	M. B. Mason	New York, N. Y	Jan. 3, 1865	45, 803
Ore, Desulphurizing and treating	J. W. Bailey	Hamilton, Nev	July 25, 1871	117, 363
Ore-desulphurizing apparatus	J. S. Briggs	Chicago, Ill	Mar. 20, 1866	53, 266
Ore-desulphurizing apparatus	F. W. Crosby and W. Helen	New York, N. Y	Aug. 21, 1866	57, 293
Ore-desulphurizing apparatus	B. G. Noble	New York, N. Y	Jan. 23, 1866	52, 191
Ore-desulphurizing apparatus	I. N. Stanley	Brooklyn, N. Y	Jan. 29, 1867	61, 577
Ore-desulphurizing apparatus	A. H. Tait and J. W. Avis	New York, N. Y	Jan. 16, 1866	52, 090
Ore-desulphurizing furnace	E. C. Hegeler and F. W. Matthiesen.	La Salle, Ill	Nov. 25, 1873	144, 904
Ore-desulphurizing furnace	M. B. Mason	New York, N. Y	Jan. 3, 1865	45, 804
Ore-desulphurizing furnace	J. G. Trotter	Newark, N. J	Nov. 22, 1870	109, 559
Ore-desulphurizing furnace, Isomeric diaphragm	D. Menthorn	New York, N. Y	Nov. 7, 1865	50, 836
Ore, Desulphurizing gold and silver	A. L. Fleury	Pittsburgh, Pa	July 3, 1866	56, 027
Ore-desulphurizing machine, Metallic	J. S. Sullivan	New York, N. Y	Aug. 18, 1834	....
Ore-desulphurizing process	J. Absterdam	New York, N. Y	Jan. 23, 1866	52, 120
Ore-desulphurizing process	J. F. Alexander	Shelby, N. C	Aug. 11, 1868	80, 798
Ore-desulphurizing process	M. B. Mason	New York, N. Y	Apr. 3, 1866	53, 742
Ore-desulphurizing process	J. S. Oliver	New York, N. Y	Feb. 27, 1872	124, 077
Ore-disintegrating apparatus	J. B. Cox	San Francisco, Cal	Apr. 28, 1868	77, 259
Ore-disintegrating apparatus	A. B. Paul	San Francisco, Cal	Apr. 5, 1870	101, 500
Ore, Disintegrating, desulphurizing, and oxidizing	J. C. Ayer	Lowell, Mass	Mar. 7, 1865	46, 621
Ore-dressing apparatus	C. F. Collom	Calstock, Great Britain	Aug. 29, 1871	118, 437
Ore-feeder for grinding and crushing mills	C. P. Stanford	San Francisco, Cal	May 20, 1873	139, 204
Ore furnace and apparatus, Pyritous	J. Fretz	Angel, Cal	Sept. 20, 1859	25, 500
Ore, Furnace and process for treating and reducing	V. W. Blanchard	Bridport, Vt	Nov. 16, 1869	96, 872
Ore, Furnace for making iron direct from the	J. M. Quinby, A. H. Brown, G. H. Penton, and J. Criswell.	Newark, N. J	Aug. 9, 1859	25, 044
Ore, Furnace for producing iron direct from the	W. Griffith, jr	Pottsville, Pa	Aug. 31, 1869	94, 306
Ore-grinder	J. W. V. Rawlins and S. Stephens.	Houghton, Mich	Sept. 14, 1869	94, 771
Ore grinding and amalgamating apparatus	M. B. Dodge	New York, N. Y	May 23, 1865	47, 808
Ore grinding and amalgamating machine	P. Hinkle	San Francisco, Cal	Apr. 16, 1867	63, 979
Ore-grinding mill	W. Ball	Chicopee, Mass	Mar. 30, 1852	8, 835
Ore-grinding mill	W. Ball	Chicopee, Mass	Apr. 11, 1854	10, 754
Ore-grinding mill	J. S. Niswander	Oakland, Cal	Mar. 28, 1871	113, 085
Ore-heater	A. C. Vandyke	Greenupsburgh, Ky	Apr. 24, 1860	28, 025
Ore-jigger	J. F. Utsch	Iserlohn, Germany	Dec. 3, 1872	133, 606
Ore, Jigging-machine for dressing	W. W. Spalding	Greenland, Mich	June 11, 1867	65, 703

Index of patents issued from the United States Patent Office from 1790 *to* 1873, *inclusive*—Continued.

Invention.	Inventor.	Residence.	Date.	No.
Ore of copper and other metals to obtain metal and other products therefrom, Treating	W. Henderson	Glasgow, Scotland	Dec. 18, 1866	60, 514
Ore or quartz crusher	J. Fowler	Rahway, N. J	Sept. 18, 1866	58, 089
Ore or quartz crusher	E. P. McCarthy	San Francisco, Cal	Sept. 18, 1866	58, 116
Ore-oxidizing furnace	T. J. Chubb	Brooklyn, N. Y	Aug. 6, 1867	67, 497
Ore-oxidizing furnace	I. M. Phelps	Chicago, Ill	June 6, 1871	115, 769
Ore, Process and apparatus for reducing iron	T. S. Blair	Pittsburgh, Pa	May 21, 1872	126, 922
Ore-pulverizer	S. Gardner	New York, N. Y	Nov. 4, 1873	144, 330
Ore-pulverizer	F. C. Morse	Buckskin, Colo	Oct. 25, 1870	108, 617
Ore-pulverizing machine	A. K. Eaton	New York, N. Y	Feb. 28 1854	10, 567
Ore, Purifying and preparing glass	C. Carter	Newburgh, N. Y	Apr. 9, 1867	63, 702
Ore reducing and smelting furnace	L. G. Marshall	Philadelphia, Pa	June 16, 1863	38, 906
Ore-reducing apparatus	G. B. Simpson	Washington, D. C	Aug. 21, 1866	57, 394
Ore-reducing furnace	J. E. Ware	Saint Louis, Mo	May 9, 1871	114, 734
Ore-reducing furnace for refractory metals	W. Quann	Philadelphia, Pa	Sept. 14, 1869	94, 911
Ore-reducing furnace, Iron	A. Thorna	New York, N. Y	Oct. 22, 1867	70, 047
Ore-reducing furnace, Iron	J. Wilson	Dover, N. J	July 16, 1872	128, 993
Ore-reducing furnace, Iron	J. Wilson	Dover, N. J	Dec. 9, 1873	145, 471
Ore-reducing furnace, Metallic	E. W. Nohl	Ripon, Wis	June 11, 1867	65, 586
Ore-reducing furnaces, Apparatus for feeding fuel to.	J. H. Boyd	Chicago, Ill	Nov. 4, 1873	144, 185
Ore, Reducing sulphur	C. W. Moore	San Francisco, Cal	June 1, 1869	90, 677
Ore, Reduction of iron	W. Quann	Philadelphia, Pa	Aug. 6, 1861	33, 020
Ore, Reduction of refractory iron	C. Martin	Chancery Lane, England	Feb. 11, 1868	74, 397
Ore-roaster	F. W. Crosby	Toledo, Iowa	Jan. 7, 1868	72, 980
Ore-roaster	C. A. Griffin	Chelsea, Mass	Oct. 18, 1864	44, 769
Ore, Roasting	G. Kustel and F. W. Smith	San Francisco, Cal., and Ellsworth, Nev.	Oct. 3, 1872	131, 882
Ore roasting and calcining furnace	E. Westman	Stockholm, Sweden	Jan. 12, 1869	85, 881
Ore roasting and chloridizing furnace	H. Tindall	Chicago, Ill	Oct. 13, 1868	83, 122
Ore roasting and chloridizing, Process of	H. Tindall	Chicago, Ill	Oct. 13, 1868	83, 121
Ore, Roasting and desulphurizing	G. W. Baker	New York, N. Y	Mar. 28, 1865	46, 984
Ore, Roasting and desulphurizing	H. Bradford	New York, N. Y	Feb. 21, 1865	46, 520
Ore, Roasting and desulphurizing	R. P. Wilson	New York, N. Y	Nov. 1, 1864	44, 906
Ore, Roasting and desulphurizing	R. P. Wilson	New York, N. Y	Mar. 7, 1865	46, 745
Ore roasting and reducing apparatus	G. W. White	New York, N. Y	Sept. 6, 1864	44, 145
Ore roasting and reducing furnace	J. Maunton	New York, N. Y	Dec. 3, 1867	71, 776
Ore roasting and smelting furnace	W. Quann	Philadelphia, Pa	Oct. 24, 1871	120, 165
Ore roasting and smelting furnace	L. Stevens	Washington, D. C	July 22, 1873	141, 181
Ore, Roasting and treating	J. W. Bailey	San Francisco, Cal	Nov. 7, 1871	120, 695
Ore roasting and treating furnace	R. George	Mineral Point, Wis	July 21, 1868	80, 065
Ore roasting and treating furnace	C. Stetefeldt	Austin, Nev	Dec. 31, 1867	72, 931
Ore roasting and treating furnace	Z. A. Willard	Boston, Mass	Jan. 14, 1873	134, 782
Ore roasting by waste gas, Iron	A. Hamar	New York, N. Y	Nov. 17, 1868	84, 113
Ore, Roasting, desulphurizing, and smelting	J. D. Whelpley and J. J. Storer	Boston, Mass	Sept. 11, 1866	58, 012
Ore-roasting furnace	J. Agrell and J. Klepzig	San Francisco, Cal	Sept. 3, 1867	68, 406
Ore-roasting furnace	S. F. Ambler	Monitor, Cal	July 25, 1871	117, 360
Ore-roasting furnace	J. P. Arey	Georgetown, Colo	June 6, 1871	115, 559
Ore-roasting furnace	C. N. Atkins and A. Govan	Pottsville, Pa	May 24, 1870	103, 409
Ore-roasting furnace	J. W. Bailey	San Francisco, Cal	Nov. 7, 1871	120, 696
Ore-roasting furnace	N. Bartlett	Centreville, N. J	May 19, 1868	77, 950
Ore-roasting furnace	M. P. Boss	Eureka, Nev	Nov. 12, 1872	132, 892
Ore-roasting furnace	E. Brady and J. Sloan	Philadelphia, Pa	Nov. 26, 1867	71, 271
Ore-roasting furnace	W. Buckner	Central City, Colo	Nov. 26, 1867	71, 448
Ore-roasting furnace	S. W. Bullock	Elizabeth, N. J	Mar. 2, 1869	87, 227
Ore-roasting furnace	D. C. Collier, S. Cushman, and N. E. Farrell.	Central City, Colo	June 16, 1868	78, 928
Ore-roasting furnace	J. Collom	Empire City, Colo	Aug. 23, 1870	106, 553
Ore-roasting furnace	G. B. Field	New York, N. Y	Aug. 20, 1867	67, 862
Ore-roasting furnace	G. B. Field	New York, N. Y	Aug. 27, 1867	68, 058
Ore-roasting furnace	H. Goulding	Silver City, Nev	Nov. 19, 1867	70, 995
Ore-roasting furnace	C. B. Grubb	Lancaster, Pa	Jan. 26, 1864	41, 376
Ore-roasting furnace	E. P. Hudson	New York, N. Y	June 2, 1868	78, 456
Ore-roasting furnace	J. Jenkins and H. Weissenborn	Elizabeth and Newark, N. J.	Nov. 18, 1862	36, 954
Ore-roasting furnace	F. Kesseler	San Francisco, Cal	June 27, 1871	116, 324
Ore-roasting furnace	R. F. Knox and J. Osborn	San Francisco, Cal	July 25, 1871	117, 430
Ore-roasting furnace	R. F. Knox and J. Osborn	San Francisco, Cal	July 2, 1872	128, 637
Ore-roasting furnace	C. Millinger	Cornwall, Pa	May 19, 1868	78, 113
Ore-roasting furnace	R. B. Norman	Sacramento, Cal	Apr. 8, 1862	34, 901
Ore-roasting furnace	D. J. O'Harra and C. B. Thompson.	Empire City, Nev	Aug. 6, 1867	67, 445
Ore-roasting furnace for precious metals	J. S. Akin	Rye Patch, Nev	June 6, 1871	115, 673
Ore-roasting furnace, Metallic	F. W. Crosby	Washington, D. C	May 23, 1871	115, 031
Ore-roasting furnace, Revolving	F. Ernst	San Francisco, Cal	Sept. 1, 1868	81, 762
Ore-roasting furnace, Feeder for	J. P. Arey	Georgetown, Colo	Aug. 8, 1871	117, 718
Ore-roasting furnace, Feeder for	C. Stetefeldt	Austin, Nev	Mar. 1, 1870	100, 337
Ore, Roasting iron	H. Aitken	Falkirk, Scotland	Apr. 13, 1860	88, 939
Ore-roasting kiln	J. M. Rohrer and J. H. Bassler	Pine Grove, Pa	Aug. 10, 1869	93, 643
Ore-roasting kiln, Iron	A. Thoma	New York, N. Y	Oct. 22, 1867	70, 045
Ore roasting, oxidizing, and chloridizing furnace	J. P. Arey	Georgetown, Colo	May 17, 1870	103, 006
Ore, Separating	T. J. Chubb	New York, N. Y	Aug. 25, 1857	18, 038
Ore, Separating	A. Huet and A. Geyler	Paris, France	Mar. 12, 1872	124, 578
Ore separating and concentrating apparatus	A. Hunter	Solano County, Cal	May 23, 1865	47, 828
Ore separating and concentrating apparatus	J. Jenkins	South Bethlehem, Pa	June 6, 1871	115, 737
Ore separating and concentrating apparatus	E. L. Seymour	New York, N. Y	Dec. 19, 1865	51, 662
Ore separating and dressing machine	H. Trumbull	Jersey City, N. J	June 2, 1863	38, 773
Ore separating and dressing process	N. Troughton	Swansea, England	July 22, 1844	3, 674
Ore, Separating and sorting	E. Lawson	Island Pond, Vt	Feb. 23, 1864	41, 710
Ore separating and treating machine	S. R. Krom	New York, N. Y	Dec. 5, 1871	121, 526
Ore-separating apparatus	H. Bradford and E. Fitzgerald	New York, N. Y	Feb. 22, 1853	9, 590
Ore separating, concentrating, and amalgamating apparatus.	H. H. Eames	Gold Hill, N. C	June 3, 1873	139, 556
Ore-separating jigger	W. H. Plumb	Paterson, N. J	July 1, 1873	140, 535
Ore, Separating precious metals from	S. W. Kirk	Philadelphia, Pa	Nov. 18, 1873	144, 772
Ore-separator	S. F. Ambler	Brooklyn, N. Y	Dec. 26, 1865	51, 678
Ore-separator	H. Ball	New York, N. Y	June 7, 1870	103, 825
Ore-separator	Y. Bates	Pinos Altos, N. Mex	May 3, 1870	102, 476

Index of patents issued from the United States Patent Office from 1790 *to* 1873, *inclusive*—Continued.

Invention.	Inventor.	Residence.	Date.	No.
Ore-separator	W. O. Bourne	New York, N. Y	Nov. 24, 1857	18, 672
Ore-separator	W. O. Bourne	New York, N. Y	June 14, 1859	24, 367
Ore-separator	W. O. Bourne	New York, N. Y	Oct. 9, 1860	30, 290
Ore-separator	H. Bradford	New York, N. Y	June 29, 1858	20, 756
Ore separator	H. Bradford	Reading, Pa	Oct. 7, 1873	143, 492
Ore-separator	F. Cazin	Frumet, Mo	July 2, 1872	128, 536
Ore-separator	T. J. Chubb	New York, N. Y	Sept. 1, 1857	18, 085
Ore-separator	T. J. Chubb	New York, N. Y	Oct. 13, 1857	18, 388
Ore-separator	T. J. Chubb	Williamsburgh, N. Y	Aug. 13, 1872	130, 478
Ore-separator	G. Copeland	Denver, Colo	May 31, 1870	103, 574
Ore-separator	A. Goodhart	Newville, Pa	June 10, 1873	139, 782
Ore-separator	D. Gross	Maxatawny, Pa	Aug. 13, 1872	130, 425
Ore-separator	C. D. Hicks	Denver, Colo	Apr. 19, 1864	42, 372
Ore-separator	W. Hooper	Ticonderoga, N. Y	Mar. 22, 1870	101, 132
Ore-separator	W. Hooper	Ticonderoga, N. Y	Mar. 21, 1871	112, 918
Ore-separator	W. Hooper	Ticonderoga, N. Y	Mar. 21, 1871	112, 919
Ore-separator	S. R. Krom	New York, N. Y	Sept. 1, 1868	81, 794
Ore-separator	S. Man	Chicago, Ill	July 31, 1860	29, 438
Ore-separator	J. J. Müller	New York, N. Y	Aug. 27, 1861	33, 171
Ore-separator	H. P. Minot	Chicago, Ill	Dec. 9, 1873	145, 441
Ore-separator	F. A. Morley	Sodus Point, N. Y	Apr. 19, 1864	42, 392
Ore-separator	R. C. Morton	West Lubec, Me	June 2, 1868	78, 468
Ore-separator	D. Nevin	Georgetown, Colo	July 19, 1870	105, 480
Ore-separator	D. Nevin	Georgetown, Colo	Sept. 3, 1872	131, 064
Ore-separator	P. P. Parkhurst	Princeton, Mass	Nov. 22, 1859	26, 236
Ore-separator	H. P. Russ	Russville, Cal	June 22, 1858	20, 666
Ore-separator	E. L. Seymour	New York, N. Y	July 17, 1860	29, 198
Ore-separator	E. L. Seymour	New York, N. Y	Jan. 3, 1865	45, 757
Ore-separator	R. Shaler	Madison, Conn	Mar. 20, 1855	12, 576
Ore-separator	L. Stadtmuller	Bristol, Conn	Nov. 23, 1858	22, 138
Ore-separator	J. F. Utsch	Iserholn, Germany	Oct. 15, 1872	131, 336
Ore-separator	J. Watson	Cliff Mine, Mich	Apr. 25, 1865	47, 476
Ore-separator	J. D. Whelpley and J. J. Storer	Boston, Mass	Apr. 3, 1866	53, 718
Ore separator and amalgamator	H. Pietsch	New York, N. Y	May 3, 1864	42, 596
Ore separator and concentrator	R. George	Denver City, Colo	Aug. 2, 1870	106, 049
Ore separator and concentrator	T. N. Paine and S. Stephens	Grass Valley, Cal	Nov. 5, 1867	70, 603
Ore separator and concentrator	R. D. Symons, J. T. Harvy, and S. Stephens.	Grass Valley, Cal	Oct. 20, 1868	83, 221
Ore separator and washer	J. J. Müller	New York, N. Y	Mar. 18, 1862	34, 721
Ore-separator, Centrifugal	S. T. Pearce	New York, N. Y	Mar. 2, 1869	87, 360
Ore-separator, Electro-magnetic	R. Cook	Plattsburgh, N. Y	Feb. 20, 1849	6, 121
Ore separator, Gold	H. Merrill	Brooklyn, N. Y	Aug. 22, 1871	118, 379
Ore separator, Gold	T. Rhoads	Ottawa, Ill	Jan. 7, 1868	73, 041
Ore separator, Iron	J. Goulding	Reesville, N. Y	July 18, 1832	......
Ore-separator or jigging machine	T. Davey	Houghton, Mich	May 15, 1866	54, 697
Ores, Smelting and desulphurizing iron	A. Hamar	New York, N. Y	Feb. 11, 1868	74, 217
Ore, Smelting copper	W. L. Faber	New York, N. Y	Apr. 25, 1865	47, 407
Ore, Smelting copper	S. F. Tracy	New York, N. Y	Dec. 16, 1851	8, 599
Ore-smelting flux	S. D. Young	Elizabethtown, Ill	Aug. 26, 1873	142, 186
Ore-smelting furnace	E. Balbach, jr	Newark, N. J	Aug. 19, 1873	141, 912
Ore-smelting furnace	E. J. Hall	San Francisco, Cal	Mar. 22, 1864	41, 989
Ore-smelting furnace	J. Neville	Jersey City, N. J	Apr. 29, 1873	138, 428
Ore smelting furnace	G. W. Swett	Troy, N. Y	Apr. 14, 1863	38, 186
Ore-smelting furnace	J. Winchester	New York, N. Y	June 7, 1864	43, 061
Ore smelting furnace, Gold, silver, &c	E. W. Nohl	Chicago, Ill	Sept. 1, 1868	81, 671
Ore-smelting furnace, Iron	P. Kerr	New Bethlehem, Pa	May 28, 1861	32, 426
Ore-smelting furnace, Iron	A. Thoma	New York, N. Y	Oct. 22, 1867	70, 046
Ore-smelting furnace, Lead	D. P. Webster	New York, N. Y	Apr. 7, 1868	76, 364
Ore, Smelting iron	C. Cochrane	The Ellowes, Upper Gornal, England.	July 4, 1871	116, 558
Ore, Smelting lead	A. H. Everett	New York, N. Y	May 3, 1864	42, 570
Ore, Smelting lead	D. P. Webster	New York, N. Y	Apr. 14, 1868	76, 679
Ore-stamp	L. D. Webb	Greenville, Cal	Oct. 7, 1873	143, 480
Ore-stamp feeder	J. D. Cusenbary and J. A. Mars.	San Francisco, Cal	June 24, 1873	140, 250
Ore-stamp feeder	J. Tullock	Jamestown, Cal	Nov. 18, 1873	144, 714
Ore, Stamp-head for crushing	F. Murray	Baltimore, Md	Aug. 26, 1862	36, 302
Ore-stamp mill	G. S. Olin	Deer Lodge, Mont	Aug. 26, 1873	142, 172
Ore-stamps, Sectional cam for	J. M. Thompson	San Francisco, Cal	June 24, 1873	140, 318
Ore-stamper	T. Reaney	Philadelphia, Pa	June 15, 1852	9, 036
Ore-stamping machine	W. Ball	Chicopee, Mass	Mar. 23, 1852	8, 818
Ore-stamping machine	W. Ball	Chicopee, Mass	Dec. 31, 1867	72, 715
Ore-stamping machine	J. F. Laird	Philadelphia, Pa	July 4, 1854	11, 229
Ore-stamping machine	V. Woodcock	Swanzey, N. H	June 15, 1852	9, 042
Ore, Stamping-machine for crushing	W. Murray	Baltimore, Md	Aug. 23, 1859	25, 213
Ore, Sweeping and washing	W. Davis	New York, N. Y	Apr. 28, 1836	......
Ore to obtain precious metals, Process of desulphurizing.	H. H. Eames	Saint Paul, Minn	Feb. 2, 1869	86, 513
Ore, Treating	G. W. Baker	New York, N. Y	Mar. 28, 1865	46, 983
Ore, Treating	C. F. Carpenter	Louisville, Ky	June 26, 1866	55, 818
Ore, Treating	M. Laflin	Chicago, Ill	July 22, 1873	141, 147
Ore, Treating	S. F. Mackie	New York, N. Y	Aug. 29, 1865	49, 637
Ore, Treating	H. B. Seidel	Wilmington, Del	Aug. 29, 1865	49, 658
Ore-treating and iron and steel manufacturing apparatus.	E. Peckham	Antwerp, N. Y	June 24, 1873	140, 158
Ore-treating apparatus	J. C. Coult and J. Roach	San Francisco, Cal	July 17, 1866	56, 378
Ore-treating apparatus	H. Halvorson	North Cambridge, Mass	Oct. 17, 1865	50, 533
Ore-treating apparatus	R. Spencer	New York, N. Y	May 23, 1865	47, 873
Ore-treating apparatus	R. Spencer	New York, N. Y	May 23, 1865	47, 874
Ore, Treating copper	H. J. M. Priestienne	Paris, France	Mar. 4, 1862	34, 591
Ore-treating furnace	W. Ennis	Philadelphia, Pa	Mar. 14, 1871	112, 698
Ore-treating furnace	S. H. Folsom	Winchester, Mass	July 28, 1868	80, 279
Ore-treating furnace	W. Hendrick	New York, N. Y	Oct. 31, 1865	50, 717
Ore-treating furnace	L. Stevens	Washington, D. C	Oct. 15, 1872	132, 331
Ore-treating furnace, Auriferous	A. E. Griffiths	Philadelphia, Pa	Mar. 27, 1866	53, 440
Ore-treating furnace, Iron	I. Rogers	Haverstraw, N. Y	Feb. 26, 1861	31, 580
Ore, Treating gold	H. Halvorson	North Cambridge, Mass	Feb. 6, 1866	52, 490
Ore, Treating metalliferous	W. L. Raht	Baltimore, Md	Aug. 21, 1866	57 376

Index of patents issued from the United States Patent Office from 1790 *to* 1873, *inclusive*—Continued.

Invention.	Inventor.	Residence.	Date.	No.
Ore, Treatment of	C. M. Du Motay	Paris, France	May 24, 1870	103, 434
Ore, Treatment of sulphureted	I. Gattman	Philadelphia, Pa	Apr. 20, 1858	19, 991
Ore, Utilizing the sulphur fumes from copper	A. Bigelow	Newark, N. J	Sept. 28, 1869	95, 307
Ore-washer	W. Ball	Chicopee, Mass	Feb. 5, 1856	14, 182
Ore-washer	H. Barnard	Morristown, N. J	Feb. 16, 1858	19, 337
Ore-washer	K. L. Bean	Chicago, Ill	Nov. 2, 1869	96, 327
Ore-washer	R. Boehm	Chicago, Ill	May 18, 1869	90, 227
Ore-washer	H. Bradford	New York, N. Y	Aug. 12, 1856	15, 554
Ore-washer	H. Bradford	Reading, Pa	Nov. 8, 1870	108, 962
Ore-washer	A. Buffam	Brooklyn, N. Y	Oct. 21, 1851	8, 443
Ore-washer	W. L. Carter	Marietta, Pa	Mar. 11, 1856	14, 388
Ore-washer	W. L. Carter	Marietta, Pa	Nov. 15, 1859	26, 089
Ore-washer	W. L. Carter	Marietta, Pa	Aug. 24, 1869	93, 965
Ore-washer	S. T. Dorland	Beekman, N. Y	Oct. 29, 1872	132, 642
Ore-washer	R. Gidley	Moore's Mills, N. Y	June 3, 1873	139, 570
Ore-washer	L. M. Gochnauer	Marietta, Pa	June 11, 1872	127, 688
Ore-washer	I. T. Halstead	Fredonia, N. Y	Sept. 16, 1873	142, 848
Ore-washer	G. B. Hamilton	Washington, D. C	Mar. 13, 1866	53, 221
Ore-washer	W. Hooper	Ticonderoga, N. Y	May 27, 1873	139, 390
Ore-washer	W. M. Hughes	Howard County, Mo	Jan. 8, 1850	7, 002
Ore-washer	P. P. Martin	Paris, France	May 26, 1857	17, 385
Ore-washer	G. E. Mills	New York, N. Y	Mar. 13, 1860	27, 463
Ore-washer	B. O'Bryan	Lancaster, Pa	Sept. 8, 1860	30, 086
Ore-washer	C. B. Parson and G. W. Fisher	San Francisco and Saint Louis, Mo.	July 16, 1872	128, 973
Ore-washer	J. Paull	Clifton, Mich	Oct. 13, 1857	18, 406
Ore-washer	M. Peckham and L. O. Palmer	Utica, N. Y	Jan. 4, 1853	9, 522
Ore-washer	J. A. Peer	Grass Valley, Cal	Sept. 9, 1873	142, 647
Ore-washer	J. Pritchett	Philadelphia, Pa	Oct. 9, 1849	6, 781
Ore-washer	G. H. Reynolds	New York, N. Y	Sept. 17, 1872	131, 373
Ore-washer	G. H. Reynolds	New York, N. Y	Sept. 17, 1872	131, 374
Ore-washer	W. T. Rickard	Monitor, Cal	May 14, 1872	126, 744
Ore-washer	N. S. Ryder	Greenland, Mich	Sept. 17, 1867	69, 030
Ore-washer	R. Solliday	Allentown, Pa	May 13, 1873	138, 948
Ore-washer	R. Solliday	Allentown, Pa	Dec. 9, 1873	145, 455
Ore-washer	S. Thomas	Allentown, Pa	Sept. 30, 1856	15, 827
Ore-washer	P. Von Schmidt	New York, N. Y	Oct. 16, 1849	6, 791
Ore-washer	S. Wheeler	Albany, N. Y	Dec. 23, 1873	145, 771
Ore-washer	J. Wicks	Greenland, Mich	Apr. 23, 1867	64, 178
Ore-washer	T. Wise	Boston, Mass	Mar. 24, 1863	38, 016
Ore-washer	T. Wren	Hamilton, Nev	Apr. 29, 1873	138, 307
Ore washer and amalgamator	J. N. Wyckoff and T. M. Fell	Brooklyn, N. Y., and Orange Mines, Va.	July 26, 1859	24, 901
Ore washer and concentrator	W. G. Heslep and T. H. Cochrane.	Jamestown, Cal	June 22, 1869	91, 627
Ore washer and separator	J. H. Rae	Syracuse, N. Y	Jan. 26, 1869	86, 249
Ore-washer, Gold	S. L. Beckwith	San Francisco, Cal	Feb. 11, 1868	74, 281
Ore-washer gudgeon or bearing	S. Thomas	Catasauqua, Pa	Sept. 2, 1873	142, 530
Ore-washing apparatus	P. Scheuerman	Hancock, Mich	Dec. 15, 1863	40, 950
Ore-washing apparatus	R. Uren	Houghton, Mich	Jan. 8, 1867	61, 124
Ore-washing apparatus	J. Watson	Cliff Mine, Mich	May 23, 1865	47, 884
Ore-washing machine	J. Collom	Houghton, Mich	May 12, 1863	38, 467
Ore-washing machine	R. Edwards	Eagle River, Mich	Nov. 29, 1853	10, 282
Ore-washing machine	F. Fredly	Logan Township, Centre County, Pa.	Apr. 20, 1837	171
Ore-washing machine	N. H. Lebby	Charleston, S. C	Apr. 12, 1870	101, 890
Ore, &c., washing machine	R. R. Osgood	Green Island, N. Y	July 23, 1872	129, 678
Ore-washing machine	M. A. Woodside	Georgetown, Cal	Oct. 16, 1866	58, 931
Ore with chlorine, Apparatus for treating	E. Gaussoin	Baltimore, Md	July 24, 1866	56, 549
Ore with copper amalgam by means of electric currents, Process of treating.	A. L. Nolf and F. L. A. Pioche	San Francisco, Cal	Aug. 17, 1869	93, 899
Ore with spongy iron, Treating metallic	J. J. A. De Bronac and A. J. M. Deherrypon.	Paris, France	Sept. 6, 1859	25, 320
Ore with superheated steam, Treating	L. E. Rivot	Paris, France	Jan. 28, 1868	73, 840
Ore, Working	V. and J. King	Bel Air, Md	May 30, 1811	....
Ore, Working Franklinite	T. Selleck	Greenwich, Conn	Jan. 30, 1855	12, 329
Ore, Working silver	W. L. Faber	New York, N. Y	Apr. 18, 1865	47, 286
Ore, Working silver	H. Janin	Virginia City, Nev	Mar. 2, 1869	87, 340
Ores and amalgamating precious metals, Apparatus for grinding.	C. H. Griffin	Lynn, Mass	July 22, 1862	35, 980
Ores and amalgamating precious metals, Machine for grinding.	J. G. Randall	Canyon City, Colo	Nov. 3, 1863	40, 501
Ores and extracting precious metals, Furnace for reducing.	J. H. Boyd	Chicago, Ill	Nov. 4, 1873	144, 186
Ores and fertilizers, Machine for washing and screening.	A. Duvall	Baltimore, Md	Sept. 6, 1870	107, 169
Ores and metals, Apparatus for assaying and testing.	J. S. Phillips	San Francisco, Cal	Sept. 7, 1869	94, 508
Ores and metals, Flux for treating	P. N. Mackay	San Francisco, Cal	Dec. 16, 1873	145, 580
Ores and metals, Melting	D. Walworth	Middletown, Conn	May 27, 1814	....
Ores and minerals, Apparatus for concentrating	S. K. Krom	New York, N. Y	Aug. 4, 1868	80, 747
Ores and minerals, Calcining	B. Keith, A. Behr, and N. S. Keith.	New York, N. Y	Sept. 9, 1862	36, 437
Ores and minerals, Disintegrating and desulphurizing.	W. F. Goodwin and C. R. Squires.	East New York and New York, N. Y.	Sept. 3, 1867	68, 561
Ores and minerals, Machine for concentrating and separating.	R. George	Denver City, Colo	Apr. 27, 1869	89, 476
Ores and minerals, Process and machinery for obtaining metals and other products from.	J. D. Whelpley and J. J. Storer	Boston, Mass	Nov. 13, 1866	59, 696
Ores and minerals, Roasting and desulphurizing	S. Stevens	New York, N. Y	June 7, 1864	43, 046
Ores and other granular substances, Machine for separating.	S. T. Pearce	New York, N. Y	Aug. 4, 1868	80, 764
Ores and other granular substances, Machine for separating.	S. T. Pearce	New York, N. Y	Mar. 2, 1869	87, 361
Ores and other materials, Apparatus for roasting and drying.	A. Duvall	Baltimore, Md	Sept. 20, 1870	107, 600
Ores and other materials, Machine for separating	S. T. Pearce	New York, N. Y	Aug. 4, 1868	80, 763

Index of patents issued from the United States Patent Office from 1790 *to* 1873, *inclusive*—Continued.

Invention.	Inventor.	Residence.	Date.	No.
Ores a d other materials of different specific gravities, Machine for separating and concentrating.	R. George	Denver City, Colo	Aug. 2, 1870	106, 048
Ores and oxides, Reducing and refining metallic	S. C. Salisbury	New York, N. Y	May 28, 1867	65, 122
Ores and refining metals, Flux for reducing	W. Quann	Philadelphia, Pa	Oct. 17, 1871	120, 099
Ores, Apparatus and process for separating, concentrating, and amalgamating	W. C. Shaw	Philadelphia, Pa	July 4, 1871	116, 763
Ores, Apparatus for chlorinating and leaching	M. H. Stowe	Washington, D. C	May 7, 1872	126, 424
Ores, Apparatus for cleaning sulphurets and other	C. C. Coleman	San Francisco, Cal	Feb. 7, 1871	111, 612
Ores, Apparatus for concentrating auriferous	J. A. Bertola	New York, N. Y	Mar. 29, 1864	42, 065
Ores, Apparatus for extracting precious metals from.	A. Ott	New York, N. Y	Mar. 10, 1868	75, 294
Ores, Apparatus for reducing quicksilver	J. C. Coult	San Francisco, Cal	Oct. 29, 1867	70, 321
Ores, Apparatus for separating metals from	J. H. Elward	Polo, Ill	June 5, 1866	55, 261
Ores, Apparatus for separating metals from	S. R. Krom	New York, N. Y	Oct. 16, 1866	58, 839
Ores, Apparatus for separating metals from	S. R. Krom	New York, N. Y	Nov. 6, 1866	59, 509
Ores, Apparatus for separating metals from	S. R. Krom	New York, N. Y	Nov. 6, 1866	59, 510
Ores, Apparatus for separating metals from	S. R Krom	New York, N. Y	Nov. 20, 1866	59, 773
Ores, Apparatus for separating metals from	J. D. Whelpley and J. J. Storer	Boston, Mass	June 13, 1865	48, 226
Ores, Apparatus for separating volatile metals from.	J. C. Coult and J. Roach	San Francisco, Cal	Apr. 17, 1866	53, 951
Ores, Apparatus for smelting zinc	A. Borgnet	Swansea, England	Feb. 5, 1867	61, 706
Ores, Burning, roasting, and smelting	J. D. Whelpley and J. J. Storer	Boston, Mass	Jan. 12, 1864	41, 250
Ores, "Chilian mill" for pulverizing metallic	J. A. Bertola	New York, N. Y	Apr. 19, 1864	42, 341
Ores, Collecting gold, silver, &c., from	J. Corson	Washington, D. C	May 5, 1868	77, 591
Ores containing precious metals, Furnace for reducing.	J. H. Boyd	Chicago, Ill	Nov. 4, 1873	144, 184
Ores containing sulphur, &c., Stall for roasting	J. T. Reese	Baltimore, Md	Mar. 19, 1867	63, 094
Ores, Desulphurizing and volatilizing lead and silver.	G. T. Lewis	Philadelphia, Pa	July 4, 1871	116, 604
Ores, Desulphurizing auriferous pyrites and other sulphuret.	J. H. Rae and T. T. Davis	Syracuse, N. Y	Aug. 10, 1869	93, 640
Ores, Desulphurizing auropyrites and other	L. Sibert	Staunton, Va	Sept. 20, 1870	107, 553
Ores, Disintegrating and desulphurizing gold, silver, and copper.	F. F. Brower and G. C. Campbell.	Ottawa, Ill	Jan. 23, 1866	52, 132
Ores, Extracting gold and silver from	C. F. Carpenter	Louisville, Ky	Apr. 3, 1866	53, 569
Ores, Extracting metals from	W. P. McConnell	Washington, D. C	Apr. 25, 1871	114, 175
Ores, Extracting precious metals from	G. P. B. Hill	Virginia City, Nev	Sept. 22, 1868	82, 315
Ores, Extracting precious metals from	A. Ott	New York, N. Y	Mar. 10, 1868	75, 293
Ores, Extracting precious metals from	V. Ryerson	New York, N. Y	May 1, 1866	54, 412
Ores for smelting-furnaces, Device for heating	A. Royer	Reed's Mills, Ohio	May 5, 1863	38, 419
Ores for the extraction of precious metal, Desulphurizing and treating.	H. H. Eames	Saint Paul, Minn	Feb. 2, 1869	86, 514
Ores, Flux for extracting precious metals from their.	W. W. Hubbell	Philadelphia, Pa	May 4, 1869	89, 579
Ores, Furnace and condenser for reducing cinnabar and other volatile.	R. F. Knox and J. Osborn	San Francisco, Cal	June 14, 1870	104, 323
Ores, Hydraulic machine for washing	J. M. Allenwood	Timbuctoo, Cal	July 12, 1864	43, 468
Ores into powder and preparing it for various uses, Grinding and reducing.	H. Alexander	Baltimore, Md	Oct. 27, 1813	....
Ores, Lifting-stamp for crushing	D. E. Rice	Detroit, Mich	June 28, 1859	24, 578
Ores, limestones, &c., Kiln for calcining	C. W. Siemens	Westminster, England	Jan. 21, 1870	104, 655
Ores, Machine for separating and concentrating magnetic and other.	E. L. Seymour	New York, N. Y	Jan. 5, 1869	85, 700
Ores, Machinery for separating metal from	J. A. Hitchings	Denver City, Colo	May 15, 1866	54, 726
Ores, metals, and minerals, Treating	C. A. Stevens	New York, N. Y	July 7, 1868	79, 701
Ores, mills, &c., Machine for concentrating the dust from.	S. H. Rounds and W. L. Strong	San Francisco, Cal	July 26, 1864	43, 646
Ores of copper and silver, Reducing	J. T. McDougall	San Francisco, Cal	Mar. 29, 1864	42, 100
Ores of copper, &c., Treating	C. M. T. Du Motay and W. Hillegeirt.	Paris, France, and Clausthal, Hanover.	June 18, 1872	128, 026
Ores of copper, Treating sulphurous	J. D. Whelpley and J. J. Storer	Boston, Mass	Nov. 13, 1866	59, 693
Ores of gold and silver, Treatment of	J. A. Bertola	New York, N. Y	Dec. 1, 1857	18, 789
Ores of gold and silver with vapor of mercury, Machine for treating.	W. A. Thompson	New York, N. Y	Jan. 7, 1868	73, 137
Ores of gold or silver, Triturating and amalgamating apparatus for treating.	L. Wray	Ramsgate, England	Mar. 9, 1869	87, 748
Ores of gold, silver, and copper, Treating	J. McCullock	San Francisco, Cal	June 26, 1860	28, 882
Ores of gold, silver, and other metals, Flux for smelting.	C. W. Moore	San Francisco, Cal	June 29, 1869	91, 862
Ores of gold, silver, copper, &c., Smelting	W. Quann	Philadelphia, Pa	Dec. 16, 1862	37, 198
Ores of gold, silver, &c., Roasting, desulphurizing, and disintegrating.	A. K. Johnson	New York, N. Y	June 30, 1865	48, 284
Ores of iron, Magnetic machine for cleaning and separating.	J. Y. Smith	Pittsburgh, Pa	June 14, 1870	104, 221
Ores of iron, Purifying and reducing magnetic	J. Y. Smith	Pittsburgh, Pa	June 14, 1870	104, 220
Ores of precious metal, Reducing	A. I. Frick and J. B. Le Clerc	San Francisco, Cal	Nov. 16, 1869	96, 790
Ores of precious metals, Treating the	J. S. Wethered and S. E. Woodworth.	San Francisco, Cal	Feb. 19, 1861	31, 499
Ores of zinc, lead, iron, &c., Treating certain	R. George	Mineral Point, Wis	Apr. 3, 1860	27, 710
Ores, Preparing	E. N. Kent	New York, N. Y	Nov. 15, 1864	45, 048
Ores, Reducing	J. H. Boyd	Chicago, Ill	Oct. 14, 1873	143, 622
Ores, Reducing	T. J. Chubb	Williamsburgh, N. Y	Dec. 14, 1869	97, 761
Ores, Reducing	C. D. Williams and W. H. Nobles.	Saint Croix, Minn	Nov. 9, 1869	96, 755
Ores, Reducing lead	C. F. Carpenter	Louisville, Ky	Jan. 28, 1868	73, 874
Ores, Reducing refractory gold, silver, and copper	C. W. Harvey	Buffalo, N. Y	Mar. 27, 1866	53, 444
Ores, &c., Reduction of	A. T. Hay	Burlington, Iowa	Nov. 19, 1872	133, 099
Ores, Reduction of	A. Parkes	Birmingham, Great Britain	Jan. 30, 1849	6, 062
Ores, Refining	J. L. Constable	New York, N. Y	July 14, 1863	39, 257
Ores, Refining and smelting	J. P. McLean	New York, N. Y	July 14, 1868	79, 848
Ores, Roasting sulphurets and other	A. F. W. Partz	Wurtsborough, N. Y	June 14, 1864	43, 129
Ores, Separating cobalt and nickel from other	A. Monnier	Philadelphia, Pa	May 19, 1868	78, 001
Ores, Separating gold and silver from	W. Crooks	Wine-Office Court, England	May 15, 1866	54, 829
Ores, Separating gold and silver from	J. H. Elward and J. L. Hayes	Polo, Ill., and Boston, Mass	Feb. 27, 1866	52, 834
Ores, Separating gold and silver from	M. Laflin	Chicago, Ill	Sept. 20, 1864	44, 320
Ores, Separating metals from	S. W. Kirk	Philadelphia, Pa	Feb. 28, 1871	112, 153
Ores, Shaft-furnace for roasting	T. McGlew	Austin, Nev	July 15, 1873	140, 837
Ores, Shaking-machine for separating	A. W. Schell	Clausthal, Hanover	Mar. 10, 1863	37, 888

Index of patents issued from the United States Patent Office from 1790 *to* 1873, *inclusive*—Continued.

Invention.	Inventor.	Residence.	Date.	No.
Ores, Shaking-table for concentrating	P. S. Buckminster	Gold Hill, Nev	Mar. 5, 1867	62, 602
Ores, Shaking-table for separating	C. A. Stetefeldt	New York, N. Y	May 31, 1864	42, 970
Ores, Smelting iron and other	R. D. and W. F. Cox	Philadelphia, Pa	Apr. 4, 1871	113, 401
Ores, Submerged rocker for separating	O. Edes	Plymouth, Mass	Apr. 23, 1850	7, 302
Ores, sulphurets, animal matter, &c., Apparatus for treating.	H. H. and C. J. Eames	New York, N. Y	June 10, 1873	139, 664
Ores, Treating auriferous	W. A. Ott	Brooklyn, N. Y	Apr. 25, 1865	47, 497
Ores, Treating auriferous and argentiferous	J. H. Rae	Syracuse, N. Y	Feb. 5, 1867	61, 866
Ores, Treating aurjferous and argentiferous	L. E. Rivot	Paris, France	May 31, 1864	42, 997
Ores, Treating gold and silver	J. Hastings and L. P. Yantier	San Francisco, Cal	Mar. 5, 1861	31, 598
Ores, Treating pyritous and other sulphur	M. W. Sinding	Norway	Aug. 25, 1863	39, 684
Ores, Working gold and silver	G. Kustel	San Francisco, Cal	Dec. 1, 1868	84, 497
Organ	R. Burdett and B. O. Church	Chicago, Ill	Aug. 8, 1871	117, 859
Organ, &c	B. O. Church and H. Smith	Brattleborough, Vt	Dec. 17, 1867	72, 369
Organ	C. H. Davie and W. Jackson	Chicago, Ill	May 31, 1870	103, 723
Organ	A. and G. Gemünder	Springfield, Mass	June 15, 1852	9, 022
Organ	H. N. Goodman	Syracuse, N. Y	May 24, 1870	103, 448
Organ	G. B. Kirkham	New York, N. Y	May 21, 1867	64, 986
Organ	J. Meads	Albany, N. Y	Nov. 25, 1838	1, 021
Organ	A. M. Peasley	Boston, Mass	Nov. 11, 1818	....
Organ	W. Robj hn	New York, N. Y	May 8, 1866	54, 603
Organ	W. Robjohn	New York, N. Y	Aug. 3, 1869	93, 349
Organ	I. Roush and J. W. Truby	Otto, N. Y	June 23, 1868	79, 082
Organ	D. Schuyler	Buffalo, N. Y	Feb. 27, 1866	52, 892
Organ	W. Sumner	Worcester, Mass	Feb. 28, 1854	10, 580
Organ	S. R. Warren	Montreal, Canada	Oct. 8, 1872	131, 984
Organ	G. W. Woodruff	Hartford, Conn	Aug. 1, 1871	117, 587
Organ	G. Woods	Cambridgeport, Mass	Apr. 16, 1872	125, 778
Organ-action	W. N. Manning	Rockport, Mass	Aug. 27, 1872	130, 928
Organ-action	J. H. Odell	New York, N. Y	Jan. 16, 1872	122, 777
Organ-action	A. Storck	Cincinnati, Ohio	July 20, 1869	92, 760
Organ action, Electric	H. L. Roosevelt	New York, N. Y	Apr. 13, 1869	88, 909
Organ action, Reed	W. N. Manning	Rockport, Mass	Nov. 18, 1873	144, 781
Organ and book-case, Combination	S. Taylor	Worcester, Mass	June 22, 1869	91, 687
Organ and melodeon	S. Taylor	Worcester, Mass	Feb. 13, 1866	52, 623
Organ and melodeon	S. Taylor	Worcester, Mass	June 22, 1869	91, 686
Organ and melodeon	S. Taylor	Worcester, Mass	Oct. 5, 1869	95, 536
Organ and melodeon coupling	C. Ling and G. S. Chandler	Detroit, Mich	Nov. 26, 1867	71, 314
Organ and melodeon coupling	H. W. Smith	Boston, Mass	June 12, 1855	13, 048
Organ and melodeon reed	A. Newell	Chicago, Ill	Mar. 7, 1871	112, 483
Organ and melodeon tremolo	J. R. Lomas	New Haven, Conn	Aug. 1, 1871	117, 550
Organ and melodeon valves, Machine for clothing	C. D. Goodman	Cleveland, Ohio	Oct. 3, 1871	119, 458
Organ and piano-forte, Combining	R. M. Ferris	New York, N. Y	Dec. 16, 1851	8, 587
Organ attachment, Reed	E. P. Needham	New York, N. Y	Sept. 3, 1872	131, 112
Organ-bellows	J. R. Perry	Wilkesbarre, Pa	Dec. 14, 1869	97, 803
Organ-bellows	J. R. Perry	Wilkesbarre, Pa	June 7, 1870	104, 060
Organ-bellows	J. R. Perry	Wilkesbarre, Pa	Aug. 9, 1870	106, 275
Organ-bellows	E. H. Thomas, jr	Brattleborough, Vt	Nov. 19, 1867	71, 245
Organ bellows action, Reed	S. Taylor	Brighton, Mass	Sept. 6, 1870	107, 122
Organ bellows, Reed	S. H. Jones	Boston, Mass	June 29, 1869	91, 852
Organ bellows, Reed	J. R. and S. R. Perry	Wilkesbarre, Pa	June 6, 1871	115, 635
Organ bellows, Reed	J. Schatz	New Haven, Conn	May 5, 1868	77, 661
Organ bellows, Reed	H. W. Smith	West Newton, Mass	Dec. 17, 1872	134, 104
Organ, Cabinet	R. W. Blake	Worcester, Mass	Sept. 10, 1872	131, 209
Organ, &c., Cabinet	H. A. Clifford	Rockville, Conn	Aug. 2, 1870	105, 909
Organ, Cabinet	H. N. Goodman	Syracuse, N. Y	Dec. 10, 1872	133, 851
Organ, Cabinet	G. Woods	Cambridge, Mass	Aug. 8, 1865	49, 355
Organ, Cabinet	G. Woods	Cambridgeport, Mass	June 6, 1871	115, 671
Organ-case	G. Woods	Cambridgeport, Mass	Dec. 17, 1872	134, 023
Organ case, Cabinet	A. J. Sorenson	Erie, Pa	Jan. 7, 1873	134, 569
Organ case, Reed	C. E. Bacon	Buffalo, N. Y	Jan. 10, 1871	110, 819
Organ, Church	W. Jackson	Albany, N. Y	May 10, 1870	102, 822
Organ, &c., coupler	R. W. Carpenter	Chicago, Ill	Mar. 31, 1868	76, 159
Organ-coupler	R. E. Letton	Quincy, Ill	Aug. 5, 1873	141, 444
Organ coupler, Sub-base reed	G. Cook	New Haven, Conn	May 17, 1870	103, 019
Organ free-reed	H. E. Hodgson	Knoxville, Tenn	Aug. 12, 1873	141, 792
Organ, harmonium, &c	W. Dawes	Leeds, England	Feb. 27, 1866	52, 940
Organ-keys, Device for holding down	A. Fowler	Mount Vernon, N. Y	Sept. 3, 1872	131, 090
Organ, melodeon, &c	H. A. Clark	Philadelphia, Pa	July 26, 1870	105, 780
Organ-melodeon	T. F. Thornton	Buffalo, N. Y	Dec. 18, 1855	13, 959
Organ, Musical instrument entitled the vocal	J. W. Campbell	Attica, Ind	Nov. 26, 1840	1, 859
Organ or harmonium, Cabinet	T. Atkins	Cincinnati, Ohio	Apr. 4, 1865	47, 081
Organ, Parlor	J. Cordley	Adrian, Mich	Jan. 10, 1871	110, 900
Organ, Parlor	E. Hamlin	Winchester, Mass	Aug. 23, 1870	106, 692
Organ, Parlor	T. Sands	Chelsea, Mass	May 27, 1856	14, 992
Organ-pipe	E. B. Andrews	Osborn Hollow, N. Y	May 7, 1867	64, 470
Organ-pipe	D. Boardman	Mount Vernon, N. H	Apr. 18, 1848	5, 520
Organ-pipe	G. H. Brock	Huntington, N. Y	Sept. 1, 1868	81, 591
Organ-pipe	J. Gorham	Bairdstown, Ga	Mar. 13, 1860	27, 443
Organ-pipe	J. Lorenz	Cincinnati, Ohio	May 19, 1868	78, 107
Organ pipe, Reed	C. W. Small	Worcester, Mass	Aug. 23, 1870	106, 628
Organ-pipes, Wind-regulator for	D. George	Nazareth, Pa	Oct. 9, 1855	13, 642
Organ, Portable pipe	A. B. Felgemaker and S. L. Derrick.	Buffalo, N. Y	May 5, 1868	77, 703
Organ, Reed	C. Austin	Concord, N. H	Feb. 18, 1873	135, 875
Organ, Reed	R. Burdett	Chicago, Ill	Feb. 23, 1869	87, 241
Organ, Reed	R. Burdett	Chicago, Ill	Dec. 28, 1869	98, 345
Organ, Reed	R. Burdett	Erie, Pa	Feb. 11, 1873	135, 687
Organ, Reed	R. Burdett	Erie, Pa	Nov. 11, 1873	144, 382
Organ, Reed	R. W. Carpenter	Chicago, Ill	Mar. 2, 1869	87, 394
Organ, Reed	S. J. Crocket	Boston, Mass	June 17, 1873	139, 874
Organ, Reed	A. Dreher	Cleveland, Ohio	Apr. 1, 1873	137, 350
Organ, Reed	L. K. Fuller	Brattleborough, Vt	July 23, 1872	129, 810
Organ, Reed	L. K. Fuller	Brattleborough, Vt	June 10, 1873	139, 666
Organ, Reed	L. K. Fuller	Brattleborough, Vt	Sept. 9, 1873	142, 690
Organ, Reed	W. H. Gerrish	Boston, Mass	Jan. 7, 1873	134, 536
Organ, Reed	H. N. Goodman	Syracuse, N. Y	Nov. 8, 1870	109, 003
Organ, Reed	A. K. Hebard	Boston, Mass	Nov. 28, 1871	121, 246

Index of patents issued from the United States Patent Office from 1790 *to* 1873, *inclusive*—Continued.

Invention.	Inventor.	Residence.	Date.	No.
Organ, Reed	F. Hoddick	Buffalo, N. Y	Apr. 14, 1868	76, 759
Organ, Reed	W. J. Kent	Buffalo, N. Y	Sept. 2, 1873	142, 399
Organ, Reed	W. J. Kent	Buffalo, N. Y	Nov. 11, 1873	144, 549
Organ, Reed	R. E. Letton	Quincy, Ill	Jan. 14, 1873	134, 754
Organ, Reed	R. E. Letton	Quincy, Ill	Aug. 5, 1873	141, 510
Organ, Reed	E. Oakes	Danville, Ill	Mar. 11, 1873	136, 614
Organ, Reed	C. W. Palmer	Cleveland, Ohio	June 17, 1873	139, 913
Organ, Reed	B. E. Riggs	Homer, Mich	Nov. 19, 1872	133, 255
Organ, Reed	T. A. Rousseau	Belleville, near Paris, France.	May 3, 1859	23, 884
Organ, Reed	G. W. Scribner	Detroit, Mich	June 21, 1870	104, 653
Organ, Reed	J. A. Smith	Erie, Pa	Aug. 5, 1873	141, 468
Organ, Reed	A. J. Sorensen	Erie, Pa	June 3, 1873	139, 483
Organ, Reed	D. Tripp	Owego, N. Y	Aug. 27, 1872	130, 956
Organ, Reed	F. H. Wells	Brattleborough, Vt	Mar. 24, 1868	76, 013
Organ, Reed	H. K. White and E. A. Foster	Brattleborough, Vt	Aug. 20, 1872	130, 677
Organ, Reed	A. W. Wilcox	New Haven, Conn	Jan. 26, 1869	86, 335
Organ, Reed	G. W. Woodruff	Hartford, Conn	Apr. 4, 1871	113, 714
Organ, Reed	G. W. Woodruff	Hartford, Conn	Dec. 5, 1871	121, 700
Organ, Reed	G. Woods	Cambridgeport, Mass	Sept. 13, 1870	107, 316
Organ, Reed	G. Woods	Cambridgeport, Mass	Jan. 23, 1872	122, 979
Organ, Reed	G. Woods	Cambridgeport, Mass	Jan. 14, 1873	134, 830
Organ, Reed	G. Woods	Cambridgeport, Mass	Dec. 2, 1873	145, 078
Organ-reed	A. M. Brush	Clayton, N. Y	Sept. 4, 1866	57, 671
Organ-reed	M. C. Morgan	Boston, Mass	June 17, 1873	139, 912
Organ reed, Reed	M. O. Nichols	Clyde, Ohio	Nov. 11, 1873	144, 409
Organ reed, Reed	J. R. Perry	Wilkesbarre, Pa	Feb. 4, 1873	135, 581
Organ-reeds, Adjustable tongue for	M. Procopé	Stockholm, Sweden	Sept. 23, 1873	143, 093
Organ, Seraphina or harmonicon	L. Zwahlen	New York	May 5, 1832	....
Organ-stop	A. R. Hebard	Boston, Mass	July 19, 1870	105, 571
Organ-stop action	J. A. Smith	Erie, Pa	Aug. 5, 1873	141, 469
Organ-stop action, Reed	I. T. Packard	Fort Wayne, Ind	July 1, 1873	140, 381
Organ-stop action, Reed	J. A. Smith	Erie, Pa	Oct. 22, 1872	132, 419
Organ-stop action, Reed	J. A. Smith	Erie, Pa	Mar. 25, 1873	137, 102
Organ-stop handle	W. Boyrer	New York, N. Y	Nov. 9, 1869	96, 543
Organ-stops, Name-plate for	C. E. Bacon and W. J. Kent	Buffalo, N. Y	May 23, 1871	115, 146
Organ swell, Mechanism for reed	M. J. Kerigan	Boston, Mass	Dec. 26, 1871	122, 257
Organ swell, Reed	J. R. Lomas	New Haven, Conn	July 29, 1873	141, 282
Organ-treadle	J. A. Smith	Erie, Pa	Mar. 25, 1873	137, 101
Organ-valve arrangement	M. Baumgarten, jr	New Haven, Conn	Oct. 20, 1868	83, 241
Organs and melodeons, Knee-swell for	S. Taylor	Worcester, Mass	Jan. 30, 1866	52, 339
Organs, Blowing-apparatus for	J. F. Barker	Springfield, Mass	Nov. 28, 1871	121, 269
Organs, &c., Blowing apparatus for	M. J. Kerigan	Boston, Mass	Feb. 1, 1870	99, 322
Organs, &c., Chime for reed and pipe	C. Lehnert	Boston, Mass	Dec. 12, 1871	121, 790
Organs, &c., Enharmonic key-board for	H. W. Poole	South Danvers, Mass	Jan. 28, 1868	73, 753
Organs, Fall for parlor	W. O. Trowbridge	Newton, Mass	Oct. 28, 1873	144, 167
Organs, &c., Key-board for	M. Baumgarten	New Haven, Conn	Apr. 3, 1866	53, 728
Organs, Key-board for reed and pipe	C. Fogelberg	Boston, Mass	Dec. 12, 1871	121, 768
Organs, Mechanism for operating swell of reed	A. E. Thompson	Boston, Mass	Feb. 20, 1866	52, 771
Organs, &c., Melody-attachment for	C. Fogelberg	Boston, Mass	Apr. 5, 1870	101, 605
Organs, Motor-regulator and register-attachment for	W. H. Topham	New Haven, Conn	June 4, 1867	65, 452
Organs, Operating fan-tremolo for	A. C. Bradley	Washington, D. C	July 9, 1872	128, 845
Organs, Operating the stop for	H. R. Moore	Erie, Pa	Mar. 11, 1873	136, 664
Organs, Operating tremolo in	R. W. Carpenter	Chicago, Ill	Mar. 2, 1869	87, 395
Organs or melodeons, Reverberating reed-cells for	C. W. Small	Worcester, Mass	Apr. 18, 1871	113, 940
Organs, &c., Pedal for	T. Robjohn	New York, N. Y	Feb. 9, 1858	19, 312
Organs, Pneumatic action for	T. Winans	Baltimore, Md	Oct. 14, 1873	143, 602
Organs, Producing a vibrating swell in	A. Hitchcock	New York, N. Y	Jan. 28, 1868	73, 719
Organs, Reed-board for	R. Burdett	Chicago, Ill	Sept. 27, 1870	107, 755
Organs, Reed-board for	W. Munroe	Worcester, Mass	Oct. 28, 1873	144, 121
Organs, Reed-board for	J. R. Perry	Wilkesbarre, Pa	Aug. 6, 1872	130, 239
Organs, Reed-board for	A. W. Wilcox	New Haven, Conn	July 8, 1873	140, 750
Organs, &c., Reed for	W. Munroe	Cambridge, Mass	Jan. 7, 1868	73, 114
Organs, Swell for pipe	F. H. Hastings	Boston, Mass	Dec. 3, 1872	133, 531
Organs, Transposing-mechanism for	W. G. Day	Baltimore, Md	Sept. 13, 1870	107, 342
Organs, Tremolo-attachment for reed or pipe	M. J. Kerigan	Boston, Mass	Apr. 12, 1870	101, 742
Organs, Tremolo for	J. R. Lomas	New Haven, Conn	Feb. 22, 1870	100, 048
Organs, Tremolo for reed	L. K. Fuller	Brattleborough, Vt	Dec. 5, 1871	121, 609
Organs, Tremolo for reed	L. K. Fuller	Brattleborough, Vt	Dec. 5, 1871	121, 610
Organs, Tremolo for reed	J. R. Lomas	New Haven, Conn	Aug. 8, 1871	117, 788
Organs, Tremulant for	T. P. Sanborn	Boston, Mass	Sept. 20, 1870	107, 549
Organs, &c., Valve for	J. R. Mortimore	New York, N. Y	Jan. 11, 1870	98, 790
Organs, Valve for pneumatic draw in	A. Stein	Westfield, Mass	Aug. 30, 1870	106, 884
Ornament manufacture, Metal	F. J. Emery	Chicago, Ill	May 4, 1869	89, 642
Ornaments, &c., Composition-matter for forming	E. A. Hildreth	Wheeling, W. Va	Feb. 11, 1868	74, 225
Ornaments from anthracite or other coal, Making	J. W. Kirk	Schuylkill County, Pa	June 13, 1831	....
Ornaments of sheet-metal, Constructing architectural.	H. Baeuerle	New York, N. Y	June 29, 1869	91, 813
Ornaments, Producing metal	W. Henigst	Columbus, Ohio	Nov. 12, 1872	133, 029
Ornaments used for picture and mirror frames and for architectural purposes, Device for cutting up composition.	R. J. Marcher	New York, N. Y	Sept. 9, 1862	36, 416
Ornamental bars or rods of metal, Manufacture of	S. Tuddenham	London, England	Jan. 17, 1871	111, 097
Ornamental compounds, Cementing-material for	C. L. Grau	Hamburg, Germany	Apr. 26, 1853	9, 693
Ornamental purposes, Composition of matter for	A. H. Wright	Camden, N. J	Aug. 9, 1859	25, 074
Ornamental transfer	P. Kneipp	Philadelphia, Pa	Sept. 28, 1869	95, 356
Ornamentation, Composition for articles of	W. E. Houston	New Haven, Conn	Oct. 1, 1861	33, 391
Ornamentation of metal, glass, &c	B. G. George	London, England	Apr. 18, 1871	113, 758
Ornamenting	H. Harrop	Greenwich, N. Y	Feb. 14, 1865	46, 354
Ornamenting and dressing glass and metal surfaces, &c.	G. F. Morse	New York, N. Y	Nov. 21, 1871	121, 119
Ornamenting and lettering glass	T. R. Hartell	Philadelphia, Pa	Oct. 2, 1866	58, 410
Ornamenting and lettering hard and uneven surfaces	W. H. McElcheran	Hamilton, Canada	Nov. 7, 1871	120, 656
Ornamenting and lettering looking-glasses, signs, &c., Machine for.	W. M. Davis	Brook Haven, N. Y	Oct. 18, 1870	108, 458
Ornamenting columns, chairs, ivory, &c	R. Thompson	Washington, Ohio	Oct. 1, 1830	....
Ornamenting hollow articles of metal	D. R. Pruden	West Meriden, Conn	July 16, 1861	32, 850
Ornamenting metal	N. A. Batchelor	New York, N. Y	Jan. 5, 1864	41, 132

Index of patents issued from the United States Patent Office from 1790 *to* 1873, *inclusive*—Continued.

Invention.	Inventor.	Residence.	Date.	No.
Ornamenting metallic surfaces	H. Tucker	Newton, Mass	Apr. 27, 1869	89, 523
Ornamenting textile fabrics, Process and material for.	C. Günther	Berlin, Prussia	Feb. 15, 1870	99, 773
Ornamenting tin, &c	L. Fitzmaier	New York, N. Y	July 30, 1867	67, 184
Ornamenting surface of metals by electro-depositions from solutions.	R. O'Neil	New York, N. Y	Apr. 4, 1871	113, 331
Ornamenting surface of wood, &c., Machine for	T. T. Ponsonby	Nottingham, England	Sept. 27, 1870	107, 718
Ornamenting wood	A. Beckers	New York, N. Y	June 19, 1866	55, 604
Ornamenting wood, Composition for and mode of	T. H. Davis	Philadelphia, Pa	May 3, 1870	102, 660
Ornamenting wood, metals, &c	R. Parke	New York, N. Y	Mar. 12, 1872	124, 616
Orrery	J. G. Moore	Philadelphia, Pa	Nov. 21, 1865	51, 072
Orthography, System of pronouncing	E. Leigh	Saint Louis, Mo	May 26, 1868	78, 296
Oscillating chair	W. T. Doremus	New York, N. Y	Dec. 31, 1872	134, 423
Oscillating chair	W. T. Doremus	New York, N. Y	Feb. 18, 1873	135, 974
Oscillating engine	W. E. Bird	New York, N. Y	May 21, 1867	64, 940
Oscillating engine	M. E. Bollinger	Littlestown, Pa	Mar. 26, 1861	31, 775
Oscillating engine	F. Brown	New York, N. Y	Jan. 16, 1866	52, 024
Oscillating engine	F. Brown	New York, N. Y	June 12, 1866	55, 575
Oscillating engine	H. Chavons	Union City, Ind	Oct. 15, 1872	132, 251
Oscillating engine	W. Craig	New York, N. Y	Sept. 12, 1854	11, 664
Oscillating engine	M. V. Cummings	Winthrop, Me	Aug. 6, 1867	67, 511
Oscillating engine	T. W. Godwin	Norfolk, Va	Nov. 28, 1871	131, 355
Oscillating engine	A. Goulding	Worcester, Mass	Dec. 5, 1871	121, 457
Oscillating engine	J. Goulding	Worcester, Mass	June 13, 1871	115, 952
Oscillating engine	L. Griscom	Port Carbon, Pa	Sept. 3, 1872	130, 988
Oscillating engine	T. Hill	Vallejo, Cal	July 12, 1870	105, 204
Oscillating engine	M. C. Kilgore and W. Eberhard	Washington, Iowa	Nov. 22, 1864	45, 159
Oscillating engine	W. H. King	Philadelphia, Pa	July 4, 1865	48, 563
Oscillating engine	A. B. Latta	Cincinnati, Ohio	Oct. 11, 1853	10, 119
Oscillating engine	H. B. Martin	San Francisco, Cal	Mar. 2, 1869	87, 419
Oscillating engine	C. R. Otis	Yonkers, N. Y	Oct. 2, 1860	30, 240
Oscillating engine	E. G. Otis	Yonkers, N. Y	Oct. 2, 1860	30, 241
Oscillating engine	C. Reed	Georgetown, D. C	Sept. 17, 1872	131, 461
Oscillating engine	J. A. Reed	New York, N. Y	Jan. 9, 1855	12, 216
Oscillating engine	M. Runkel	New York, N. Y	Apr. 12, 1859	23, 612
Oscillating engine	A. Schmid	Zurich, Switzerland	Feb. 27, 1872	124, 162
Oscillating engine	W. Sellers	Philadelphia, Pa	June 11, 1872	127, 928
Oscillating engine	H. Shlarbaum	New York, N. Y	Sept. 1, 1863	39, 756
Oscillating engine	J. W. Van Sant	Perth Amboy, N. Y	Apr. 16, 1872	125, 773
Oscillating engine	H. Wightman and W. Warden	Allegheny, Pa	Jan. 15, 1856	14, 124
Oscillating engine	G. F. Wood	Ulysses, N. Y	Jan. 23, 1855	12, 299
Oscillating engines and pumps, Arrangement of	G. Sprenkel and T. W. Basford	Harrisonburgh, Pa	Dec. 22, 1857	18, 933
Oscillating engines, Mechanism for	H. Broomell	Christiana, Pa	June 7, 1870	103, 840
Oscillating engines, Trunnion-box lining for	J. A. Reed	Jersey City, N. Y	July 5, 1859	24, 663
Oscillating piston-engine	W. M. Elrod	Saint Louis, Mo	Aug. 20, 1867	67, 966
Oscillating steam-engine	W. D. Andrews	New York, N. Y	Nov. 11, 1862	36, 885
Oscillating steam-engine	J. S. Barden	New Haven, Conn	Feb. 23, 1858	19, 464
Oscillating steam-engine	M. Cridge and S. Wadsworth	Pittsburgh, Pa	Dec. 12, 1854	12, 052
Oscillating steam-engine	I. M. Forrester	Bridgeport, Conn	Dec. 20, 1870	110, 352
Oscillating steam-engine	M. J. Gardner	York, Pa	Aug. 23, 1853	9, 954
Oscillating steam-engine	R. J. King	Lancaster City, Pa	Sept. 15, 1868	82, 126
Oscillating steam-engine	G. F. Lowry	Lake Mississippi, Miss	Mar. 19, 1872	124, 754
Oscillating steam-engine	F. Millward	Cincinnati, Ohio	Nov. 17, 1868	84, 206
Oscillating steam-engine	T. Murgatroyd	Cleveland, Ohio	Jan. 14, 1862	34, 161
Oscillating steam-engine	D. H. Nation and T. B. Hall	Saint Louis, Mo	Aug. 8, 1865	49, 293
Oscillating steam-engine	A. Nittinger, jr	Philadelphia, Pa	Aug. 30, 1871	106, 869
Oscillating steam-engine	J. Pattison	Brooklyn, N. Y	Feb. 5, 1856	14, 204
Oscillating steam-engine	F. C. Ricker	Gilmer, Tex	July 6, 1869	92, 363
Oscillating steam-engine	J. Robertson	Brooklyn, N. Y	May 7, 1872	126, 576
Oscillating steam-engine	M. Runkel	New York, N. Y	Mar. 6, 1860	27, 387
Oscillating steam-engine	E. Russell	Brooklyn, N. Y	July 9, 1861	32, 787
Oscillating steam-engine	E. P. Ryder	Brooklyn, N. Y	Aug. 6, 1872	130, 157
Oscillating steam-engine	J. B. Sweetland	Pontiac, Mich	July 9, 1872	128, 766
Oscillating steam-engine	J. Wallace	Pittsburgh, Pa	July 28, 1857	17, 903
Ottoman	A. O. and E. G. Ganiard	New York, N. Y	Feb. 9, 1869	86, 744
Ottoman and hassock	G. B. Green	Staffordshire, England	Feb. 20, 1872	123, 821
Ottoman and hassock filler	E. G. Ganiard	New York, N. Y	Aug. 4, 1868	80, 619
Oval-cutting machine	C. W. Packer	Philadelphia, Pa	Aug. 16, 1864	43, 883
Oval frames, Machine for jointing	D. Garrison	Philadelphia, Pa	May 22, 1866	54, 890
Oval frames, Machine for jointing	J. E. Rogers	Chelsea, Mass	Feb. 7, 1865	46, 268
Oval frames, Machine for turning	I. P. Tice	Baltimore, Md	Mar. 26, 1861	31, 837
Oval-turning machine	A. H. Brown	Albany, N. Y	Sept. 24, 1861	33, 334
Oval-turning machine	J. Irving	New York, N. Y	July 5, 1859	24, 642
Oval-turning machine	P. Pryibil	New York, N. Y	Aug. 24, 1869	94, 130
Oven	W. H. Akins	Berkshire, N. Y	June 2, 1836	....
Oven	J. Allen and E. Pick	Brooklyn, N. Y	June 24, 1862	35, 661
Oven	H. Ball	New York, N. Y	Sept. 23, 1856	15, 753
Oven	H. Ball	New York, N. Y	July 19, 1870	105, 541
Oven	J. Barnitt	Upper Providence, Pa	Mar. 3, 1809	....
Oven	B. Blaney	Boston, Mass	Dec. 17, 1834	....
Oven	J. S. Brown	Washington, D. C	Apr. 13, 1858	19, 965
Oven	D. P. Burdon	Brooklyn, N. Y	Feb. 1, 1859	22, 778
Oven	E. Bussey	Troy, N. Y	May 5, 1868	77, 581
Oven	J. C. Carlisle	Chesterville, Me	Apr. 19, 1833	....
Oven	D. S. Coburn	Amboy, Ill	Jan. 30, 1872	123, 086
Oven	A. Crumbie	Jersey City, N. J	Feb. 8, 1870	99, 645
Oven	J. S. Dunham and J. Green	Saint Louis, Mo	Oct. 12, 1869	95, 671
Oven	M. Dunn	Bloomington, Ill	Mar. 29, 1870	101, 357
Oven	S. S. Eaton	Mansfield, Conn	Dec. 17, 1834	....
Oven	D. J. George	Arcade, Ind	Aug. 16, 1870	106, 352
Oven	P. J. Gindre and J. Doerler	Cincinnati, Ohio	June 30, 1863	39, 042
Oven	G. Y. Gray	Niles, Mich	Mar. 30, 1869	88, 472
Oven	J. P. Hayes	Philadelphia, Pa	July 31, 1855	13, 375
Oven	J. P. Hayes	Philadelphia, Pa	Dec. 2, 1856	16, 143
Oven	J. F. Hoffmeister	Alton, Ill	May 24, 1859	24, 122
Oven	G. C. Jennison	Ware, Mass	Feb. 1, 1859	22, 809
Oven	D. A. Kennedy	Beloit, Wis	June 15, 1869	91, 451
Oven	D. McKenzie	Brooklyn, N. Y	May 1, 1860	28, 1[illegible]0

Index of patents issued from the United States Patent Office from 1790 *to* 1873, *inclusive*—Continued.

Invention.	Inventor.	Residence.	Date.	No.
Oven	D. McKenzie	Brooklyn, N. Y	June 6, 1871	115, 626
Oven	D. Moore	Davenport, Iowa	Feb. 8, 1870	99, 694
Oven	G. R. Moore	Philadelphia, Pa	Nov. 2, 1869	96, 338
Oven	J. Ohmert	Mount Morris, Ill	May 6, 1856	14, 825
Oven	E. G. Otis	Yonkers, N. Y	Aug. 24, 1858	21, 271
Oven	S. Pollard	Orono, Me	Feb. 3, 1836	....
Oven	C. Preston	Wheeling, W. Va	Sept. 7, 1869	94, 643
Oven	J. Rainey	Brooklyn, N. Y	Oct. 17, 1870	108, 389
Oven	J. M. Read	Boston, Mass	Aug. 28, 1860	29, 818
Oven	T. N. Reid	Baltimore, Md	May 18, 1852	8, 960
Oven	T. Russell	New York, N. Y	Feb. 1, 1859	22, 828
Oven	W. Sellers	Philadelphia, Pa	Jan. 22, 1861	31, 192
Oven	I. H. Shaver	Cedar Rapids, Iowa	July 4, 1871	116, 762
Oven	C. Swain	Laconia, N. H	May 3, 1870	102, 614
Oven	R. R. Tongue	Fryburgh, Me	Aug. 15, 1834	....
Oven	E. Treadwell	New York, N. Y	July 19, 1853	9, 864
Oven	F. C. Treadwell, jr	New York, N. Y	Aug. 22, 1854	11, 578
Oven	J. R. Treadwell	Brooklyn, N. Y	Oct. 12, 1869	95, 750
Oven	J. Vatter	Philipsburgh, Ohio	May 4, 1869	89, 713
Oven	W. C. Wedge	Chicopee, Mass	July 27, 1869	93, 145
Oven	J. West	Stafford County, Va	Oct. 27, 1808	....
Oven	A. Whittock	Danbury, Conn	Oct. 2, 1866	58, 522
Oven	A. Wilson	Buffalo, N. Y	Oct. 3, 1871	119, 680
Oven and bath, Combined	J. Perry	Brooklyn, N. Y	May 26, 1868	78, 231
Oven and drum, Combined	J. J. Stout, S. J. Russell, and N. Mendenhall.	Greensburgh, Ind	Sept. 21, 1869	95, 161
Oven and drum, Detachable	C. A. Harper and I. A. Crane	Rahway, N. J	Nov. 26, 1867	71, 481
Oven and range, Combined baker's	J. Williams	Topeka, Kans	June 24, 1873	140, 330
Oven and stove-pipe drum, Combined	J. W. Clough	Montville, Me	Feb. 16, 1869	86, 907
Oven, Bake	A. A. and J. A. Aull	Bellefontaine, Ohio	Oct. 11, 1870	108, 086
Oven, Bake	H. Ball	Philadelphia, Pa	Nov. 19, 1850	7, 778
Oven, Bake	J. Bonis	Baltimore, Md	Dec. 30, 1812	....
Oven, Bake	P. Grouvell and L. N. and E. Mouchat.	Paris, France	Mar. 20, 1847	5, 028
Oven, Bake	J. Hall	Cincinnati, Ohio	Oct. 11, 1870	108, 258
Oven, Bake	I. A. Hammer	Newton, Iowa	Sept. 21, 1869	95, 106
Oven, Bake	J. P. Hayes	Philadelphia, Pa	July 29, 1856	15, 422
Oven, Bake	D. Moore	Davenport, Iowa	June 1, 1869	90, 771
Oven, Bake	W. Pettet	New York, N. Y	Aug. 10, 1858	21, 147
Oven, Bake	J. J. Piggott	Saint Louis, Mo	Apr. 22, 1873	138, 042
Oven, Bake	H. Reip	Baltimore, Md	Sept. 27, 1825	....
Oven, Bake	J. Vale	Beloit, Wis	July 7, 1868	79, 615
Oven, Baker's	G. E. Bailey	Mansfield, Mass	Dec. 5, 1871	121, 573
Oven, Baker's	G. E. Bailey	Mansfield, Mass	July 16, 1872	129, 080
Oven, Baker's	H. Berdan	New York, N. Y	Oct. 20, 1857	18, 429
Oven, Baker's, &c	H. Chatain	Washington, D. C	Apr. 9, 1872	125, 439
Oven, Baker's	J. Chilcott	Brooklyn, N. Y	June 9, 1857	17, 495
Oven, Baker's	A. Crumbie	Jersey City, N. J	Apr. 9, 1872	125, 546
Oven, Baker's	D. Donald	Williamsburgh, N. Y	July 8, 1873	140, 686
Oven, Baker's	C. F. Holmes	Saint Louis, Mo	Apr. 16, 1872	125, 735
Oven, Baker's	W. Johnson and H. Davies	Brooklyn, N. Y	Feb. 25, 1862	34, 502
Oven, Baker's	J. G. Klinge	New York, N. Y	Oct. 12, 1869	95, 811
Oven, Baker's	I. W. Knapp	New York, N. Y	Feb. 18, 1862	34, 432
Oven, Baker's	T. P. Mahon	New York, N. Y	Apr. 26, 1870	102, 415
Oven, Baker's	A. Martin	Allegheny, Pa	Feb. 14, 1871	111, 759
Oven, Baker's	D. McKenzie	Brooklyn, N. Y	Oct. 15, 1872	132, 305
Oven, Baker's	D. McKenzie	Brooklyn, N. Y	June 10, 1873	139, 802
Oven, Baker's	D. McKenzie	Brooklyn, N. Y	Oct. 7, 1873	143, 522
Oven, Baker's	W. R. Nevins and J. J. Yates	New York, N. Y	Sept. 28, 1858	21, 620
Oven, Baker's	J. Rayney	Brooklyn, N. Y	Oct. 18, 1870	108, 517
Oven, Baker's	J. Rayney and W. Cairns	Brooklyn, N. Y., and Jersey City, N. J.	Mar. 4, 1873	136, 546
Oven, Baker's	N. F. Rice	New Orleans, La	Nov. 15, 1859	26, 126
Oven, Baker's	C. Robbins	Charlestown, Mass	Apr. 15, 1873	137, 961
Oven, Baker's	R. Sanderson	Cleveland, Ohio	Nov. 14, 1871	120, 901
Oven, Baker's	A. B. Smith	Geneva, N. Y	July 30, 1872	129, 993
Oven, Baker's	G. H. Smith	Galesburgh, Ill	June 25, 1872	128, 433
Oven, Baker's	E. B. Strong	Buffalo, N. Y	Apr. 11, 1836	....
Oven, Baker's	J. Taylor	Galesburgh, Ill	Oct. 29, 1872	132, 548
Oven, Baker's	J. and C. L. Vale	Chicago, Ill	June 7, 1870	104, 083
Oven, Baker's	J. H. Vance and F. B. Hall	Ann Arbor, Mich	Sept. 10, 1872	131, 315
Oven, Baking	A. A. and J. A. Aull	Bellefontaine, Ohio	Dec. 12, 1871	121, 839
Oven, Baking	G. S. Blodgett and P. T. Sweet	Burlington, Vt	Dec. 5, 1854	12, 018
Oven, Baking	E. Moody	Northfield, Mass	Apr. 5, 1828	....
Oven, Baking and cooking	S. J. Gold	Cornwall, Conn	Mar. 29, 1834	....
Oven, Chimney	E. Smith	Ithaca, N. Y	July 1, 1836	....
Oven, Circular coke	F. J. F. Laumonier	Angers, France	June 18, 1867	65, 820
Oven, Coke	L. F. and A. Beckwith	New York, N. Y	July 19, 1870	105, 413
Oven, Coke	J. Bowers	Connellsville, Pa	Oct. 24, 1865	50, 552
Oven, Coke	J. Erichsen and J. G. Maardt	London, England	July 23, 1872	129, 803
Oven, Coke	T. G. Kenny	Prospect, Pa	June 18, 1872	128, 151
Oven, Coke	G. Lambert	Mons, Belgium	Jan. 9, 1855	12, 209
Oven, Coke	T. G. Meier	Saint Louis, Mo	Sept. 19, 1871	119, 092
Oven, Coke	L. Schantl	Saint Louis, Mo	May 24, 1870	103, 507
Oven, Combined furnace and tempering	R. L. and A. F. Hewett	Worcester and Milbury, Mass.	May 7, 1872	126, 393
Oven, Cooking	J. Chilcott	Brooklyn, N. Y	Aug. 22, 1865	49, 503
Oven, Desulphurizing	J. C. Brewster	New York, N. Y	Jan. 25, 1870	99, 143
Oven, Detachable	A. B. Nott	Fairhaven, Mass	June 13, 1865	48, 200
Oven-door	C. H. Finn	Syracuse, N. Y	Jan. 12, 1869	85, 809
Oven-door flue	B. Simonds, jr	Bedford, Mass	Oct. 15, 1821	....
Oven-door, Hollow	G. W. Eddy	Waterford, N. Y	Feb. 10, 1846	4, 377
Oven-doors, Attachment for	S. Morton	Valparaiso, Ind	June 9, 1868	78, 754
Oven, Dutch	C. Schneidt	Chicago, Ill	Aug. 13, 1873	130, 444
Oven, Elevated	J. C. Kennedy	Albany, N. Y	July 11, 1854	11, 295
Oven, Enameling	V. Keller	Allegheny, Pa	Jan. 11, 1870	98, 777
Oven, Fire-heating	C. G. Nye	Onondaga, N. Y	June 1, 1869	90, 775
Oven for annealing iron	S. Reynolds	Pittsburgh, Pa	Feb. 13, 1866	52, 605

Index of patents issued from the United States Patent Office from 1790 *to* 1873, *inclusive*—Continued.

Invention.	Inventor.	Residence.	Date.	No.
Oven for bakers, Reel	J. A. Groffet	Saint Louis, Mo	Dec. 17, 1867	72, 194
Oven for baking and heating houses, &c	J. Baldwin	New York, N. Y	Oct. 27, 1835	....
Oven for baking bread and other aliments	J. L. Rolland	Paris, France	Nov. 27, 1855	13, 855
Oven for baking bread, Railway	S. Short	Nantucket, Mass	Oct. 6, 1837	421
Oven for baking fire-brick	L. A. Boisson	Lyons, France	Jan. 22, 1861	31, 150
Oven for baking, &c., Revolving	S. Harris	Norfolk, Va	Dec. 5, 1825	....
Oven for converting iron into steel	W. A. Sweet	Syracuse, N. Y	Jan. 10, 1865	45, 878
Oven for cooling and annealing glass	J. Franck	Millville, N. J	May 12, 1843	3, 081
Oven for cooling castings	P. F. Geisse	Wellsville, Ohio	Apr. 12, 1859	23, 671
Oven for cooling window-glass	D. Bievez	Haine St. Pierre, Belgium	Dec. 3, 1867	71, 571
Oven for drying fruit, Portable	G. Diffenderfer	Lewisburgh, Pa	July 30, 1867	67, 172
Oven for plate-glass, Flattening	J. B. Boulicault	New Albany, Ind	Nov. 25, 1873	144, 946
Oven, Fruit-drying	D. Lippy and S. Linn	Mansfield, Ohio	Sept. 20, 1864	44, 324
Oven, Gas and air-heating	G. F. Wilson	Providence, R. I	June 18, 1872	128, 004
Oven-heating furnace	A. H. Quincy	Boston, Mass	Feb. 17, 1812	....
Oven in chimney or large bakery, Double	R. Bacon and W. E. Marshall	Walpole, N. H	Feb. 9, 1831	....
Oven, Peanut	D. Dill	New Haven, Conn	June 8, 1869	91, 098
Oven, Perpetual	J. Deneale	Dumfries, Va	Dec. 4, 1806	....
Oven, Perpetual	W. T. Hunter	New York	Jan. 8, 1818	....
Oven, Perpetual rotary	J. L. D. Mathies	Rochester, N. Y	Dec. 21, 1831	....
Oven, Portable	G. W. Ayres	Rahway, N. J	Mar. 18, 1861	34, 726
Oven, Portable	T. B. Curtis	New Haven, Conn	June 27, 1839	1, 201
Oven, Portable	C. Doane	Wareham, Mass	Oct. 9, 1849	6, 772
Oven, Portable	J. A. Frey	Washington, D. C	Mar. 15, 1864	41, 914
Oven, Portable	W. Goddard	Portsmouth, N. H	Oct. 12, 1831	....
Oven, Portable	F. L. Hedenberg	New York	Apr. 26, 1828	....
Oven, Portable	M. H. Leland	Millbury, Mass	July 30, 1867	67, 321
Oven, Portable	F. C. Miller	Evans Centre, N. Y	May 26, 1868	78, 226
Oven, Portable	Z. N. Morrell	Cameron, Tex	July 12, 1859	24, 752
Oven, Portable	W. S. Robinson	Oskaloosa, Iowa	July 25, 1871	117, 464
Oven, Portable	E. Wassenich	Cincinnati, Ohio	Feb. 25, 1862	34, 540
Oven, Portable bake	D. Hull	Utica, N. Y	Mar. 28, 1834	....
Oven, Portable elevated	P. Killin	Mount Healthy, Ohio	Oct. 7, 1851	8, 415
Oven, Portable rotary	R. Wilcox	New York, N. Y	Apr. 20, 1820	....
Oven-reel	A. Crumbie	Brooklyn, N. Y	Aug. 19, 1873	142, 003
Oven, Reflecting	B. Ames	Ithaca, N. Y	June 25, 1836	....
Oven, Reflecting	C. D. Van Allen	Penn Yan, N. Y	Mar. 31, 1836	....
Oven, Revolving	J. A. Kinkele	Sacramento City, Cal	Dec. 24, 1867	72, 645
Oven, Rotary	D. A. Kennedy	Darien, Wis	July 20, 1869	92, 840
Oven, Rotary bake	J. Vale	Beloit, Wis	Oct. 4, 1870	107, 984
Oven, Steam	E. Brown	Burlington, Vt	Jan. 30, 1872	123, 230
Oven, Steam	J. G. Whitlock	New York, N. Y	Jan. 23, 1866	52, 235
Oven, Tin bake	W. Lewis	Franklin, N. Y	Jan. 6, 1832	....
Oven, Union	J. S. Gold	Norwich, Conn	Dec. 16, 1853	....
Oven with reflector	J. Hurd, jr	Boston, Mass	Nov. 10, 1829	....
Ovens, Admission of steam to baking	J. Y. Betts	Coventry, England	Mar. 2, 1869	87, 390
Ovens and boilers for generation of steam, Combining cooking.	R. McMillen	Middleburgh, Ohio	Dec. 14, 1841	2, 388
Ovens and other like furnaces, Heating pottery	H. Speeler	Trenton, N. J	Dec. 24, 1867	72, 694
Ovens and rooms, Heating	M. B. Poitiaux	Richmond, Va	Jan. 17, 1827	....
Ovens by anthracite, Heating	F. C. Treadwell	Brooklyn, N. Y	June 16, 1836	....
Ovens by steam, Heating	H. Lyon	Cincinnati, Ohio	Sept. 28, 1858	21, 610
Ovens, Construction and application of	S. Pollard	Bucksport, Me	June 12, 1835	....
Ovens, Heating elevated	C. W. Grannis	Collins, N. Y	Feb. 1, 1847	4, 950
Ovens, Heating elevated	P. A. Palmer	Le Roy, N. Y	Sept. 24, 1850	7, 672
Ovens to be combined with cooking-stoves, Method of constructing flues of elevated.	R. D. Granger	Albany, N. Y	Oct. 11, 1841	2, 308
Ovens, Turn-table for baker's	R. Shaler	Madison, Conn	Oct. 2, 1866	58, 489
Ovens, Valve for elevated	H. Stanley	West Poultney, Vt	June 11, 1842	2, 664
Ovens with cooking-stoves so as to render them movable, Combining elevated.	E. C. Robinson	Troy, N. Y	Dec. 30, 1841	2, 404
Ovens with stoves, Combining elevated	S. B. Spaulding	Brandon, Vt	Aug. 28, 1841	2, 235
Overall	E. F. and J. H. Stacy	Gloucester, Mass	Apr. 23, 1872	125, 994
Overall	E. Weil	New York, N. Y	Oct. 21, 1873	143, 947
Overcoat and tent, Convertible	H. J. Phillips	New York, N. Y	Nov. 19, 1861	33, 752
Overflow-alarm	T. Mayes	Albany, N. Y	Nov. 18, 1873	144, 687
Overshoe	D. H. Bond	Canterbury, Conn	June 30, 1836	....
Overshoe	A. O. Bourn	Cranston, R. I	Feb. 23, 1869	87, 137
Overshoe	A. O. Bourn	Providence, R. I	Nov. 28, 1871	121, 446
Overshoe	P. Dorn	Philadelphia, Pa	May 7, 1850	7, 343
Overshoe	J. Evans	New Brunswick, N. J	Jan. 30, 1872	123, 162
Overshoe	D. Hodgman	New York, N. Y	Mar. 23, 1842	2, 504
Overshoe	H. L. Hotchkiss	New Haven, Conn	June 2, 1868	78, 595
Overshoe	J. J. Mellon	Pana, Ill	Oct. 29, 1872	132, 681
Overshoe	W. W. Swann	Richmond, Va	May 24, 1870	103, 522
Overshoe	H. G. Tyer	Andover, Mass	Sept. 3, 1867	68, 398
Overshoe	A. Vetter	Philadelphia, Pa	May 2, 1846	4, 497
Overshoe	A. G. Waterhouse	San Francisco, Cal	Sept. 7, 1869	94, 531
Overshoe	G. Watkinson	New Haven, Conn	Nov. 18, 1873	144, 810
Overshoe	P. S. Whitman	Providence, R. I	Aug. 29, 1871	118, 568
Overshoe	J. Wild	Woonsocket, R. I	Mar. 12, 1872	124, 525
Overshoe	I. F. Williams	Bristol, R. I	Sept. 10, 1872	131, 199
Overshoe	I. F. Williams	Bristol, R. I	Sept. 10, 1872	131, 200
Overshoe	I. F. Williams	Bristol, R. I	Sept. 10, 1872	131, 201
Overshoe and boot and shoe	W. S. McEwen and N. A. Patterson.	Kingston, Tenn	Apr. 17, 1860	27, 918
Overshoe-fastening	H. E. Starrett	Lawrence, Kans	Feb. 22, 1870	100, 207
Overshoe, India-rubber	E. F. Bickford	Malden, Mass	July 16, 1872	129, 389
Overshoe-spike	G. W. Bradley	Weston, Conn	Mar. 30, 1869	88, 441
Overshoe spur-attachment	H. Schwandt	New Orleans, La	Feb. 14, 1871	111, 880
Overshoe, Straw and wood	F. W. Mitchel, W. C. Wilcox, and H. T. Miller.	Utica, N. Y	June 1, 1858	20, 439
Overshot-bucket wheel, used in hydraulic works	D. S. Howard	Lyonsdale, N. Y	Feb. 16, 1831	....
Ox-bow	A. L. D. Moore	La Grange, Tex	May 24, 1870	103, 355
Ox-bow fastening	E. Kenney	Livermore, Me	May 5, 1863	38, 395
Ox-bow pin	I. Glover	Newtown, Conn	Aug. 1, 1871	117, 618
Ox-bow pin	E. N. Hartwell	Sharon, Conn	July 30, 1872	129, 888

Index of patents issued from the United States Patent Office from 1790 *to* 1873, *inclusive*—Continued.

Invention.	Inventor.	Residence.	Date.	No.
Ox-bow pin	B. B. Hotchkiss	New York, N. Y	Apr. 16, 1867	63, 894
Ox-bow pin	O. D. Hunter	Terrysville, Conn	Apr. 16, 1867	63, 796
Ox-bow pin	H. P. Judson	Bethlehem, Conn	Dec. 8, 1868	84, 831
Ox-bow pin	J. Low	New Britain, Conn	Apr. 9, 1867	63, 736
Ox-bow pin	E. S. Woodford	Winchester, Conn	Feb. 27, 1866	52, 923
Ox-bows, Shaping wood for	H. S. Denison	Colraine, Mass	Oct. 13, 1863	40, 249
Ox-shoe	H. Lake	Shelburne, Vt	Oct. 29, 1867	70, 227
Ox-shoe	I. Merrill and A. Maxwell	Shelburne, Mass	July 19, 1864	43, 623
Ox-shoe	A. Van Valkenburgh	Griffin's Corners, N. Y	May 8, 1860	28, 218
Ox-shoe-forging die	H. Colburn	Stafford, Conn	Apr. 26, 1870	102, 372
Ox-shoe machine, with movable dies, Roller	P. P. Read	Bowdoin, Me	Jan. 23, 1849	6, 048
Ox-shoes, Die for forming	J. Deeble	Plantsville, Conn	Dec. 16, 1873	145, 633
Ox-shoes, Die for manufacturing	B. S. Parker	Greenfield, Mass	Feb. 20, 1872	123, 927
Ox-shoes, Machine for making	R. French	Seymour, Conn	Dec. 16, 1873	145, 561
Ox-shoes, Machine for making	W. Hamilton	Neversink, N. Y	Mar. 18, 1873	136, 915
Ox-shoes, Machine for making	A. L. Wooding	Bristol, Conn	June 7, 1870	103, 954
Oxides, Purifying metallic	A. Monnier	Philadelphia, Pa	Mar. 21, 1865	46, 924
Oxides, Reducing metallic	J. Reese	Pittsburgh, Pa	June 19, 1866	55, 710
Oxidizing metal, Apparatus for	W. Atwood	Cape Elizabeth, Me	Mar. 7, 1865	46, 618
Oxygen, and applying the same to useful purposes, Mode of preparing.	H. A. Archereau	Paris, France	July 2, 1867	66, 279
Oxygen and chlorine, Process for producing	J. T. A. Mallet	Paris, France	Jan. 21, 1868	73, 540
Oxygenating-furnace	G. Stamm	Pittsburgh, Pa	June 29, 1869	92, 120
Oyster and other dredger	J. Walmer	Ludlow, Ky	Oct. 28, 1873	144, 169
Oyster-basket	C. D. Martin	Port Chester, N. Y	Apr. 9, 1872	125, 470
Oyster-lock	W. Rankin	New York, N. Y	Feb. 6, 1872	123, 509
Oyster-boats, Dredge-roller for	C. T. Belbin	Baltimore, Md	June 5, 1866	55, 228
Oyster-bucket, Shell	E. H. Frazier	Baltimore, Md	Aug. 20, 1872	130, 631
Oyster-can	C. H. Dexter	Baltimore, Md	July 22, 1873	141, 124
Oyster-can	J. A. Tillery	Baltimore, Md	May 28, 1872	127, 388
Oyster-dredge	W. C. Baker	Baltimore, Md	Nov. 28, 1871	121, 227
Oyster-dredge	C. T. Belbin	Baltimore, Md	June 2, 1868	78, 509
Oyster-dredge	W. Belbin	Baltimore, Md	Jan 17, 1865	45, 904
Oyster-dredge	E. Fairbanks	Baltimore, Md	Aug. 8, 1865	49, 340
Oyster-dredge	W. L. Force	Keyport, N. J	Feb. 21, 1860	27, 213
Oyster-dredge	W. C. Hornfager and M. Fitz Gibbons.	New York, N. Y	Aug. 5, 1873	141, 439
Oyster-dredge	I. A. Ketcham	Breslau, N. Y	Oct. 29, 1872	132, 668
Oyster-dredge	I. A. Ketcham	Breslau, N. Y	Apr. 22, 1873	138, 164
Oyster-dredge	E. B. Lake	Mauricetown, N. J	Nov. 28, 1871	121, 249
Oyster-dredge	T. W. Landon	Fairmount, Md	Apr. 23, 1872	125, 964
Oyster-dredge	T. F. Mayhew	Port Morris, N. J	Apr. 27, 1869	89, 323
Oyster-dredge	T. F. Mayhew	Port Morris, N. J	Nov. 30, 1869	97, 420
Oyster-dredge	T. P. Sink	Fairton, N. J	Oct. 4, 1859	25, 680
Oyster-dredge	T. P. Sink	Fairton, N. J	Aug. 18, 1868	81, 304
Oyster-dredge	T. P. Sink	Fairton, N. J	Oct. 31, 1871	120, 463
Oyster-dredge winding-apparatus	T. F. Mayhew	Port Morris, N. J	Jan. 16, 1872	122, 843
Oyster-dredge windlass	W. T. Howard and O. Reeder	Baltimore, Md	Feb. 15, 1870	99, 900
Oyster-dredges, Hoisting	J. Whitecar	Philadelphia, Pa	May 5, 1863	38, 436
Oyster-dredges, Machine for raising	W. T. Howard	Baltimore, Md	Apr. 8, 1873	137, 550
Oyster-dredges, Roller for boarding	T. P. Sink	Fairton, N. J	June 4, 1867	65, 442
Oyster-dredges, Winder for	C. T. Belbin	Baltimore, Md	Nov. 8, 1870	109, 104
Oyster-dredges, Winder for	H. S. Lawson	Baltimore, Md	Mar. 2, 1865	46, 912
Oyster-dredging apparatus	W. Belbin	Baltimore, Md	Nov. 20, 1866	59, 812
Oyster-dredging machinery	T. P. Sink	Fairton, N. J	Feb. 25, 1868	74, 857
Oyster-keg stoppers, Device for removing	S. Swartz	Buffalo, N. Y	Apr. 17, 1866	54, 039
Oyster-knife	P. Blake	New Haven, Conn	Apr. 18, 1854	10, 798
Oyster-knife	G. W. Huffnagle	New Hope, Pa	Mar. 24, 1868	75, 917
Oyster-knife and ice-pick, Combined	W. Pattberger	Philadelphia, Pa	Dec. 21, 1869	98, 102
Oyster-nursery	B. F. Syford	San Francisco, Cal	June 11, 1872	127, 903
Oyster-opener	J. E. Alger	New York, N. Y	July 30, 1867	67, 246
Oyster-opener	J. Seipel and W. Rupp	Washington, D. C	Jan. 5, 1858	19, 050
Oyster-opener	J. H. Starin, jr	New York, N. Y	Nov. 19, 1872	133, 267
Oyster-opening device	M. C. Boyer	Norristown, Pa	Mar. 21, 1871	112, 895
Oyster-opening device	D. M. Cleary	Baltimore, Md	Oct. 28, 1873	144, 063
Oyster-opening machine	W. H. Towers	Philadelphia, Pa	Apr. 18, 1854	10, 810
Oyster-packing can	L. R. Comstock	Baltimore, Md	Jan. 21, 1873	135, 083
Oyster-platform	J. Vreeland	New York	Apr. 29, 1828	......
Oyster-rake	A. Barrett	Baltimore, Md	Apr. 14, 1868	76, 697
Oyster-rake	J. Johnson	Brooklyn, N. Y	Oct. 2, 1866	58, 426
Oyster-shucker	G. Holtzman	Baltimore, Md	May 30, 1871	115, 474
Oyster-tongs	J. Johnson	Brooklyn, N. Y	Mar. 17, 1868	75, 550
Oyster-tongs	J. W. Sands	Annapolis, Md	July 19, 1870	105, 495
Oyster-tongs	J. Smith	Bruceport, Wash	Jan. 21, 1873	135, 167
Oyster-tongs	E. Ward	Smyrna, Del	Sept. 27, 1870	107, 740
Oyster-tongs, Working	G. Jury	Baltimore, Md	Oct. 11, 1864	44, 634
Oyster-winder	W. H. Kanne	Baltimore, Md	Mar. 11, 1873	136, 737
Oyster-winder	S. S. Shaw	Newport, N. J	Apr. 21, 1868	77, 110
Oyster, Apparatus for opening	W. Beach	Baltimore, Md	Sept. 29, 1857	18, 273
Oysters, clams, &c., Package for	M. W. Brown	New York, N. Y	Apr. 4, 1871	113, 395
Oysters, &c., Cooking	L. P. Keach	Baltimore, Md	Apr. 18, 1854	10, 806
Ozone, and treating liquids with, Generation of	T. A. Hoffman	Beardstown, Ill	June 25, 1872	128, 227
Ozone, Apparatus for making	C. H. Johnson	New Brighton, Pa	Jan. 28, 1873	135, 226
Ozone, Apparatus for the manufacture of	C. F. Dunderdale	New York, N. Y	Nov. 29, 1870	109, 601
Ozone, Apparatus for the production of	C. F. Dunderdale	New York, N. Y	Mar. 15, 1870	100, 736
Ozone, Generating	P. A. Royce	Suspension Bridge, N. Y	Sept. 12, 1871	118, 976
Ozone-generator	W. Elmer and H. G. Hubert	New York, N. Y	Nov. 5, 1867	70, 426
Ozone-generator	R. Heneage	Buffalo, N. Y	Mar. 4, 1873	136, 511
Ozone-generator	R. Heneage	Buffalo, N. Y	Mar. 4, 1873	136, 512
Ozone-generator	O. Loew	New York, N. Y	May 28, 1872	127, 175
Ozone or ozonized air, Process of obtaining	O. Loew	New York, N. Y	Sept. 6, 1870	107, 071
P.				
Package band or fastener	F. R. Hunt	Leavenworth, Kans	Mar. 15, 1870	100, 897
Package-clamp	D. Dick	New York, N. Y	May 9, 1871	114, 651
Package-register	G. W. Moore	Philadelphia, Pa	Mar. 25, 1873	137, 231
Package-strap	J. C. Carey	New York, N. Y	Apr. 26, 1870	102, 370

Index of patents issued from the United States Patent Office from 1790 *to* 1873, *inclusive*—Continued.

Invention.	Inventor.	Residence.	Date.	No.
Packages for dry goods	A. Robertson	Upper Holloway, England	Aug. 24, 1858	21,274
Packages, Marking and lettering	W. Francis and W. Johnston	Wayneville, N. C	Oct. 3, 1844	3,771
Packages, Sling for	F. Hohorst	New York, N. Y	Oct. 10, 1871	119,765
Packages, Spring frame for	H. B. Osgood	Dorchester, Mass	Nov. 4, 1856	16,019
Packer, Flour	S. Taggart	Indianapolis, Ind	Aug. 2, 1859	24,963
Packing	J. F. Barrett	Granville, N. Y	Jan. 23, 1836	....
Packing	L. Katzenstein	New York, N. Y	July 19, 1870	105,462
Packing and atomizing can for insect-powder	F. L. Palmer, sr	New York, N. Y	Dec. 21, 1869	98,101
Packing and bearing	E. D. Murfey	New York, N. Y	July 12, 1870	105,361
Packing and bearing	E. D. Murfey	New York, N. Y	Oct. 11, 1870	108,286
Packing and bearing	E. D. Murfey	New York, N. Y	Dec. 12, 1871	121,805
Packing and tubing, Manufacture of	T. J. Mayall	Roxbury, Mass	Nov. 29, 1859	26,278
Packing-apparatus	C. E. Haynes	Boston, Mass	Mar. 15, 1870	100,889
Packing-bag	E. A. Merrill	New York, N. Y	Aug. 29, 1871	118,626
Packing, belting, and hose, India-rubber	J. Murphy	New York, N. Y	Apr. 12, 1870	101,905
Packing-box	D. L. Bartlett	Ogdensburgh, N. Y	May 20, 1873	139,106
Packing-box	J. H. Foster	Chicago, Ill	July 18, 1871	117,061
Packing-box	L. A. Fullgraff	New York, N. Y	Jan. 21, 1873	135,108
Packing-box	A. Gregg	Watertown, Mass	May 30, 1871	115,306
Packing-box and show-case, Combined	J. Nelson	Bridgeport, Conn	Dec. 31, 1872	134,480
Packing-box for bottles	J. J. Soloman	Philadelphia, Pa	Dec. 6, 1870	109,961
Packing-box for fruit, provisions, &c	G. M. Huston	Putnam, Ohio	Dec. 6, 1870	109,821
Packing-box, Metallic	H. L. Hopkins	San Francisco, Cal	July 11, 1865	48,686
Packing-box, Paper	J. W. Tuttle	Newton, Mass	July 25, 1871	117,350
Packing-box, Pasteboard	J. W. Tuttle	Newton, Mass	July 25, 1871	117,349
Packing-box, Rotary-steam-cylinder	S. Deacon and J. Russell	Lawrence, Mass	Nov. 29, 1870	109,722
Packing-boxes, Construction of	E. L. Perkins	Roxbury, Mass	Sept. 27, 1859	25,581
Packing-boxes, Metallic strap for	H. C. Dixon	Brooklyn, N. Y	Sept. 24, 1872	131,605
Packing-can	N. P. Lindergreen	Boston, Mass	Oct. 13, 1868	83,070
Packing-case	W. W. Cadwell and J. M. Case	Lansing, Mich	July 23, 1872	129,713
Packing-case	E. A. Pharo	Philadelphia, Pa	Feb. 4, 1873	135,583
Packing-case	H. D. Stover	New York, N. Y	Feb. 26, 1861	31,569
Packing case, Bottle and jar	W. Thompson	Dublin, Ireland	June 29, 1869	91,884
Packing-case, Folding	L. Selling	Detroit, Mich	Dec. 17, 1872	134,103
Packing-case for perfumery, &c	T. P. Spencer	Jersey City, N. J	July 12, 1870	105,269
Packing-cases, Safety-strip for	S. Baker	Newark, N. J	Feb. 2, 1869	86,492
Packing, Composition for steam	W. R. Bunnell	Jersey City, N. J	June 6, 1871	115,694
Packing, Compound hard and soft metal	A. Fulton	Pittsburgh, Ind	Mar. 26, 1850	7,214
Packing-device	W. Riddle	10 Larkhall Lane, England	Aug. 10, 1869	93,555
Packing for artesian wells	P. C. Heinz	Pioneer, Pa	Mar. 9, 1869	87,670
Packing for artesian wells	S. Swartz	Buffalo, N. Y	July 4, 1865	48,599
Packing for axle oil-boxes, Adjusting the	W. Groat	Troy, N. Y	June 11, 1850	7,426
Packing for barometers	J. P. Simmons	Fulton, N. Y	Mar. 12, 1861	31,679
Packing for car axles and boxes	S. R. Dummer	New York, N. Y	July 28, 1868	80,465
Packing for car-wheels, Elastic	J. N. Farrar	Trumansburgh, N. Y	Aug. 6, 1872	130,118
Packing for carriage-shackles	F. B. Morse	New Haven, Conn	Apr. 28, 1868	77,399
Packing for deep-well tubes	J. J. Parker	Marietta, Ohio	Oct. 2, 1866	58,468
Packing for deep wells	A. D. Griffin	Titusville, Pa	Aug. 13, 1867	67,749
Packing for deep wells	F. Martin	New York, N. Y	Sept. 12, 1865	49,903
Packing for deep wells	J. R. Robinson and D. A. Strong	Washington, D. C	Feb. 6, 1866	52,448
Packing for deep wells, Piston	C. H. Jackson	Angola, N. Y	Sept. 5, 1865	49,759
Packing for doors and windows	S. Strook	New York, N. Y	Aug. 8, 1871	117,939
Packing for engines, &c	H. C. White	New York, N. Y	Feb. 27, 1872	124,183
Packing for engines, pumps, &c	W. H. Miller	Philadelphia, Pa	Apr. 28, 1868	77,204
Packing for exterior of pumps in deep oil-wells, &c., Elastic	J. Moulton	Boston, Mass	Mar. 14, 1865	46,860
Packing for hose-couplings	R. J. Gould	New York, N. Y	Mar. 3, 1868	75,151
Packing for hydrocarbon-burners	J. Waterman	New York, N. Y	Sept. 6, 1864	44,144
Packing for joints	J. E. Wooten	Reading, Pa	Mar. 31, 1868	76,286
Packing for joints of doors, lids, &c., Elastic	D. E. Somes	Washington, D. C	Nov. 12, 1867	70,910
Packing for joints of steam and water pipes	H. Brocard	Paris, France	May 19, 1868	77,954
Packing for joints of steam-engines	W. J. Towne	Newtonville, Mass	Nov. 10, 1868	84,022
Packing for joints of steam-engines, Paper	J. Ross	Somerville, Mass	Nov. 6, 1866	59,456
Packing for joints, valves, &c	N. Jenkins	Boston, Mass	Oct. 15, 1867	69,811
Packing for journal-boxes	J. Conk	Red Bank, N. J	Jan. 12, 1864	41,197
Packing for journal-boxes	E. S. Hanna	Pittsburgh, Pa	Feb. 27, 1872	124,133
Packing for journal-boxes, joints, &c., Composition for	J. T. Davis and W. C. Selden	Jersey City, N. J., and Brooklyn, N. Y.	Nov. 5, 1867	70,421
Packing for journal-boxes of railway-cars	N. Monroe	Boston, Mass	May 17, 1870	103,071
Packing for journals and bearings, Fibrous mineral	C. A. Stevens	New York, N. Y	Mar. 14, 1871	112,648
Packing for journals, Lubricating	J. T. Robinson	New York, N. Y	May 21, 1872	126,987
Packing for journals, shafting, &c., Lubricating	S. Baxendale	Boston, Mass	Dec. 9, 1873	145,384
Packing for machinery and linings for journal-boxes and bearing-surfaces	C. A. Stevens and C. Butler	New York, N. Y	Aug. 22, 1871	118,295
Packing for man-holes of steam-generators	J. P. McLean	Brooklyn, N. Y	Mar. 19, 1867	63,073
Packing for oil-well tubes	G. E. Mills	New York, N. Y	Sept. 5, 1865	49,778
Packing for oil-well tubes	C. L. Noe	Bergen Point, N. J	Aug. 22, 1865	49,544
Packing for oil-well tubes	J. Parham, jr	Philadelphia, Pa	Sept. 5, 1865	49,783
Packing for oil-wells	H. S. Cate	Deerfield, Pa	Aug. 15, 1871	117,980
Packing for oil-wells	J. R. Cross	Chicago, Ill	Feb. 7, 1865	46,217
Packing for oil-wells	J. R. Cross	Chicago, Ill	Nov. 14, 1865	50,910
Packing for oil-wells	O. Redmond	Rochester, N. Y	Oct. 30, 1866	59,319
Packing for oil-wells	W. S. Wilkinson	Baltimore, Md	Mar. 6, 1866	53,073
Packing for piston and other rods	J. Johnson	Roxbury, Mass	Feb. 10, 1863	37,663
Packing for piston and valve rods	R. Martin	Brooklyn, N. Y	June 24, 1862	35,698
Packing for piston-heads	J. W. Adams	Richfield, Mich	June 16, 1868	78,911
Packing for piston-rods and other enginery, Wooden	C. N. Peterson	Chicago, Ill	Jan. 26, 1869	86,316
Packing for piston-rods and valve-stems	W. Heston	Akron, Ohio	July 22, 1873	141,052
Packing for piston stuffing-boxes	C. McBurney	Roxbury, Mass	June 28, 1859	24,569
Packing for pistons and stuffing-boxes, Tubular	W. C. Moat	London, England	Dec. 25, 1849	6,974
Packing for pistons, &c., Lubricating	D. J. Browne and C. W. Baldwin	Boston, Mass	Nov. 22, 1864	45,136
Packing for pistons, &c., Manufacture of	I. B. and W. H. Miller	Philadelphia, Pa	Apr. 4, 1865	47,119
Packing for printing-presses, Piston	C. B. Cottrell	Westerly, R. I	Apr. 2, 1872	125,177
Packing for pumps, Adjustable	C. Hood	Seneca Falls, N. Y	Jan. 2, 1866	51,831
Packing for pumps, hydrants, &c	J. W. Murphy	Baltimore, Md	Aug. 26, 1873	142,264
Packing for pumps, &c., Self-regulating water	J. Smart	Philadelphia, Pa	July 10, 1855	13,233
Packing for rifled projectiles	J. P. Schenkl	Boston, Mass	Jan. 17, 1865	45,951

Index of patents issued from the United States Patent Office from 1790 *to* 1873, *inclusive*—Continued.

Invention.	Inventor.	Residence.	Date.	No.
Packing for stationary joints, Metallic	H. T. Lee	Marysville, Cal	Apr. 26, 1870	102, 282
Packing for stationary joints, Metallic	I. J. Saunders	Davisville, Cal	July 2, 1872	128, 664
Packing for steam-boiler pistons	T. J. Hudson	New Berne, N. C	Oct. 18, 1859	25, 829
Packing for steam-engine-piston rods	D. Neahr	Fort Yuma, Cal	Jan. 5, 1869	85, 666
Packing for steam-engine-piston rods, &c	C. F. Pike	Providence, R. I	Aug. 12, 1840	1, 715
Packing for steam-engine pistons	J. Barwick and S. Tindall	Silvertown, England	Nov. 19, 1867	71, 121
Packing for steam-engine pistons	F. A. Brown	Ithaca, N. Y	Jan. 5, 1869	85, 561
Packing for steam-engine pistons	L. P. Garner	Ashland, Pa	Aug. 17, 1869	93, 697
Packing for steam-engine pistons	J. Gates	Portland, Oreg	June 8, 1869	91, 010
Packing for steam-engine pistons	H. Horton	New York, N. Y	Oct. 5, 1858	21, 678
Packing for steam-engine pistons	J. C. Merriam	Olneyville, R. I	Apr. 4, 1871	113, 547
Packing for steam-engine pistons	T. R. Morgan	Pittsburgh, Pa	Mar. 16, 1869	87, 789
Packing for steam-engine pistons	J. W. Pettis	Hillsdale, Mich	Dec. 11, 1855	13, 917
Packing for steam-engine pistons	E. Sullivan	Pittsburgh, Pa	Oct. 26, 1869	96, 284
Packing for steam-engine pistons	J. Williamson	New York, N. Y	Aug. 31, 1837	368
Packing for steam-engine pistons	J. Young	Franklin Furnace, Ohio	Mar. 26, 1861	31, 847
Packing for steam-engine pistons and stuffing-boxes	P. Clark	Rahway, N. J	June 2, 1857	17, 422
Packing for steam-engine pistons, Metallic	G. H. Corliss	Providence, R. I	Sept. 8, 1859	18, 136
Packing for steam and other enginery, Gasket	B. Densmore	New York, N. Y	Dec. 15, 1868	84, 864
Packing for steam and other engines	W. Peters	Baltimore, Md	Jan. 28, 1862	34, 283
Packing for steam and water joints	H. T. Lee	Chicago, Ill	July 9, 1872	128, 737
Packing for steam-cylinder pistons	W. Johnson	Lambertville, N. J	Aug. 17, 1869	93, 889
Packing for steam-engines	J. L. Bates	Providence, R. I	Nov. 25, 1862	36, 987
Packing for steam-engines	J. H. Gould	Philadelphia, Pa	Feb. 19, 1861	31, 512
Packing for steam-engines	W. V. Grinnel	Middletown, Conn	Apr. 5, 1832	....
Packing for steam-engines	F. Raymond, H. Dunn, and C. Colt.	Nashville, Tenn	July 9, 1872	128, 754
Packing for steam-engines, Fibrous	J. L. Husband	Philadelphia, Pa	July 3, 1866	56, 054
Packing for steam-engines, Fibrous	W. H. Miller	Philadelphia, Pa	Nov. 29, 1864	45, 261
Packing for steam-engines, Fibrous	W. H. Miller	Philadelphia, Pa	May 14, 1867	64, 688
Packing for steam-engines, gas-pipes, water-pipes, &c.	A. J. Simmons	Indianapolis, Ind	Oct. 17, 1871	120, 003
Packing for steam-engines, Manufacture of	J. Glanding	Philadelphia, Pa	Apr. 28, 1868	77, 275
Packing for steam-engines, &c., Manufacture of	W. H. Miller	Philadelphia, Pa	May 12, 1868	77, 902
Packing for steam-engines, Metallic	G. S. Cox	Barbour County, Ala	Oct. 2, 1849	6, 757
Packing for steam-engines, Pasteboard	J. Ross	Somerville, Mass	Oct. 8, 1867	69, 590
Packing for steam-engines, pumps, &c	W. Beschke	Philadelphia, Pa	Feb. 8, 1870	99, 627
Packing for steam-piston and other rods	H. Bertrand and F. Velut	Paris, France	Jan. 19, 1869	85, 988
Packing for steam-pistons	T. Hanson	New York, N. Y	July 18, 1871	117, 071
Packing for steam-pistons	T. Thurber	Auburn, N. Y	July 9, 1867	66, 537
Packing for steam-pistons	J. Wheelock	Worcester, Mass	Feb. 15, 1870	99, 990
Packing for steam-pistons	W. Wilson	Galesburgh, Ill	Aug. 11, 1868	81, 052
Packing for steam-pistons, Metallic	H. D. Dunbar	Springfield, Mass	Oct. 31, 1865	50, 697
Packing for steam-pistons, Metallic	D. Lasher	Brooklyn, N. Y	June 30, 1857	17, 683
Packing for stuffing-boxes	W. W. Allmand	East Boston, Mass	Dec. 7, 1869	97, 585
Packing for stuffing-boxes	F. J. Roth	Schenectady, N. Y	May 5, 1868	77, 534
Packing for stuffing-boxes	C. M. Templeton	Concord, N. H	Dec. 3, 1867	71, 664
Packing for stuffing-boxes, &c	J. H. Tuck	Pall Mall, England	June 26, 1855	13, 145
Packing for stuffing-boxes	R. and T. Winans	Baltimore, Md	Feb. 5, 1861	31, 346
Packing for stuffing-boxes, Metallic	J. F. Chuse	Litchfield, Ill	Mar. 5, 1867	62, 527
Packing for stuffing-boxes, Metallic	A. H. Hall and T. Locher	Sacramento, Cal	May 30, 1871	115, 463
Packing for stuffing-boxes, Metallic	C. F. Jauriet	Aurora, Ill	Feb. 13, 1866	52, 575
Packing for stuffing-box of steam and other engines.	M. Botticher	Washington, D. C	Oct. 4, 1864	44, 575
Packing for stuffing-boxes of steam-engines, pumps, &c., Manufacture of.	W. H. Miller	Philadelphia, Pa	May 21, 1867	64, 995
Packing for tubes of boilers or condensers	J. Newkirk	Factoryville, N. Y	Aug. 8, 1865	49, 295
Packing from asbestus and other fibrous materials, Machinery for forming.	C. A. Stevens and I. Lindsley.	New York, N. Y., and Pawtucket, R. I.	Mar. 14, 1871	112, 651
Packing from asbestus and other fibrous minerals.	C. A. Stevens	New York, N. Y	Mar. 14, 1871	112, 647
Packing from stuffing boxes, Instrument for extracting.	H. N. Smade	Manistee, Mich	Sept. 12, 1871	118, 823
Packing, Gasket	G. W. Coffee	San Francisco, Cal	May 14, 1872	126, 624
Packing, Gasket	W. W. Girdwood	Barking Road, Bromley, Great Britain.	Sept. 19, 1871	119, 027
Packing, Impregnating fibrous materials for	E. D. Murfey	New York, N. Y	Oct. 11, 1870	108, 284
Packing in steam-engines, Metallic stuffing-box for	E. Winship	New York, N. Y	Aug. 31, 1852	9, 239
Packing, India-rubber	A. C. Andrews	New Haven, Conn	Nov. 9, 1869	96, 654
Packing journals and bearings, Material for	E. D. Murfey	New York, N. Y	May 10, 1870	102, 958
Packing, Lubricating-compound for steam and other.	W. M. Canfield	Philadelphia, Pa	Dec. 27, 1870	110, 432
Packing-machine	A. Ralph	Philadelphia, Pa	Sept. 9, 1873	142, 651
Packing, Making gaskets for steam	M. C. Gardiner	Rochester, N. Y	Aug. 2, 1870	105, 934
Packing, Manufacture of elastic	W. Jenkins	Boston, Mass	May 8, 1866	54, 554
Packing, Manufacturing Ind a-rubber piston	I. B. Harris	Edinburgh, Scotland	May 21, 1872	126, 953
Packing, Metallic piston	A. G. Bill	Cuyahoga Falls, Ohio	Apr. 5, 1859	23, 439
Packing, Metallic piston	W. A. Boyden	Harrisburgh, Pa	Oct. 28, 1873	143, 956
Packing, Metallic piston	D. H. Fairbanks	South Norwalk, Conn	Oct. 1, 1872	131, 817
Packing, Metallic piston	J. Massey	Chester, Pa	Sept. 9, 1873	142, 640
Packing, Metallic piston	H. L. Russell	Hudson, Mich	Oct. 25, 1853	10, 159
Packing, Metallic piston	W. Wright	Providence, R. I	Feb. 27, 1849	6, 146
Packing, Metallic piston and valve rod	G. M. Cruickshank	Providence, R. I	Mar. 7, 1871	112, 423
Packing, Metallic piston-rod	C. S. Barry	Providence, R. I	June 11, 1872	127, 834
Packing, Metallic piston-rod	D. Devore	Philadelphia, Pa	Jan. 9, 1872	122, 578
Packing, Metallic piston-rod	J. C. Furness	Boston, Mass	Oct. 22, 1872	132, 361
Packing, Metallic piston-rod	J. C. Furness	Boston, Mass	May 6, 1873	138, 627
Packing, Metallic piston-rod	F. W. Pettee	Boston, Mass	Feb. 11, 1873	135, 663
Packing, Metallic piston-rod	P. W. Richards	Boston, Mass	Oct. 8, 1872	132, 024
Packing, Metallic piston-rod	J. H. Webster	Saint Louis, Mo	July 16, 1867	66, 758
Packing of engine and pump pistons, Lightening the.	J. Crabtree and J. Hopkinson	Philadelphia, Pa	Nov. 29, 1853	10, 283
Packing of rotary engines, Adjusting the	H. G. Thompson	New York, N. Y	Feb. 4, 1851	7, 926
Packing of pistons for steam-engines, Adjusting the.	J. Crabtree	Philadelphia, Pa	May 2, 1854	10, 856
Packing of pistons for steam-engines, Setting out the.	G. H. Hoagland	Port Jervis, N. Y	July 14, 1857	17, 788
Packing of pistons in deep wells, Adjusting the	E. D. Brown	Buffalo, N. Y	Nov. 14, 1865	50, 898
Packing of stuffing-boxes, &c., Lubricating the	I. B. and W. H. Miller	Philadelphia, Pa	Apr. 4, 1865	47, 170
Packing, Oil-proof rubbers for steam	J. M. Flagg	Providence, R. I	Aug. 15, 1871	118, 004

Index of patents issued from the United States Patent Office from 1790 *to* 1873, *inclusive*—Continued.

Invention.	Inventor.	Residence.	Date.	No.
Packing or bearing for journals, &c	E. D. Murfey	New York, N. Y	Dec. 12, 1871	121,806
Packing or show case	G. G. Bates	New York, N. Y	Aug. 26, 1873	142,073
Packing, Paper	A. L. Jones	New York, N. Y	Dec. 19, 1871	122,023
Packing, Piston	J. R. Abbe	Providence, R. I	June 7, 1864	42,999
Packing, Piston	E. B. Allen	Portland, Me	Aug. 6, 1867	67,476
Packing, Piston	J. Anthony and T. B. Purves	Greenbush, N. Y	Aug. 10, 1869	93,578
Packing, Piston	J. Askwith	Chicago, Ill	Mar. 5, 1867	62,590
Packing, Piston	G. R. Babbitt	Providence, R. I	Feb. 27, 1872	123,971
Packing, Piston	G. R. Babbitt and W. A. Harris	Providence, R. I	Mar. 4, 1873	136,405
Packing, Piston	M. W. Bailey	West Chester, Pa	Oct. 15, 1867	69,747
Packing, Piston	C. S. Barry	Providence, R. I	June 11, 1872	127,833
Packing, Piston	C. S. Barry	Providence, R. I	July 8, 1873	140,608
Packing, Piston	T. A. Birbee	Dayton, Ohio	Apr. 7, 1868	76,295
Packing, Piston	G. F. Blake	Boston, Mass	Sept. 12, 1871	118,838
Packing, Piston	G. F. Blake	Boston, Mass	Nov. 7, 1871	120,565
Packing, Piston	J. Broughton	Lambertsville, N. J	Oct. 16, 1866	58,766
Packing, Piston	W. Buchanan	New York, N. Y	Sept. 12, 1865	49,853
Packing, Piston	J. C. Budd	Union Township, N. J	Sept. 23, 1873	143,057
Packing, Piston	L. W. Campbell	Aurora, Ill	Nov. 5, 1867	70,521
Packing, Piston	L. W. Campbell	Aurora, Ill	Nov. 5, 1867	70,522
Packing, Piston	A. S. Cameron	New York, N. Y	July 31, 1866	56,699
Packing, Piston	A. S. Cameron	New York, N. Y	Mar. 14, 1871	112,685
Packing, Piston	J. W. Carey	Baltimore, Md	June 24, 1873	140,244
Packing, Piston	T. Carpenter	Providence, R. I	Apr. 8, 1862	34,878
Packing, Piston	C. H. Clark	Wilmington, Del	Apr. 17, 1866	53,949
Packing, Piston	C. H. Clark	Wilmington, Del	May 7, 1867	64,491
Packing, Piston	D. Clark	Hazleton, Pa	Aug. 15, 1865	49,379
Packing, Piston	J. Clark	Harrisburgh, Pa	Feb. 15, 1870	99,846
Packing, Piston	J. Clark	Harrisburgh, Pa	May 17, 1870	103,140
Packing, Piston	J. Clark	Harrisburgh, Pa	Sept. 12, 1871	118,843
Packing, Piston	J. J. Clause	Cleveland, Ohio	Jan. 9, 1872	122,570
Packing, Piston	J. M. Clay	Lancaster, Pa	Dec. 3, 1872	133,519
Packing, Piston	J. H. Cooper	Philadelphia, Pa	July 23, 1872	129,793
Packing, Piston	G. Dryden	Worcester, Mass	Jan. 29, 1867	61,614
Packing, Piston	G. Dryden	South Boston, Mass	July 23, 1872	129,799
Packing, Piston	P. Estes	Leavenworth, Kans	May 2, 1871	114,427
Packing, Piston	W. J. Ford	Chicago, Ill	Sept. 13, 1870	107,243
Packing, Piston	D. R. Fraser	Chicago, Ill	Mar. 25, 1862	34,749
Packing, Piston	D. R. Fraser	Chicago, Ill	June 3, 1862	35,439
Packing, Piston	A. Fulton	Pittsburgh, Pa	July 4, 1865	48,547
Packing, Piston	J. Gates	Portland, Oreg	Nov. 8, 1870	108,999
Packing, Piston	S. M. Goodale	Saint Louis, Mo	Aug. 6, 1872	130,214
Packing, Piston	S. Goodfellow	Troy, N. Y	Dec. 12, 1865	51,449
Packing, Piston	G. Gwynn	New York, N. Y	Feb. 20, 1872	123,891
Packing, Piston	A. W. Harris	Providence, R. I	Apr. 2, 1872	125,132
Packing, Piston	C. A. Hodge	Knoxville, Tenn	Sept. 16, 1873	142,914
Packing, Piston	J. W. Holloway	Akron, Ohio	July 25, 1865	48,945
Packing, Piston	W. S. Hudson	Paterson, N. J	July 23, 1867	66,964
Packing, Piston	J. Hughes	Saxton, Pa	June 21, 1870	104,459
Packing, Piston	A. W. Jackson	Centralia, Ill	Sept. 11, 1866	57,916
Packing, Piston	T. J. Jones	Summit, N. J	Feb. 13, 1866	52,644
Packing, Piston	J. Keesey	Chester, Pa	Sept. 6, 1870	107,062
Packing, Piston	O. Kelsey	Worcester, Mass	May 24, 1870	103,343
Packing, Piston	E. Kendall	New Lebanon, N. Y	July 11, 1865	48,692
Packing, Piston	J. King	Hoboken, N. J	Mar. 24, 1868	75,927
Packing, Piston	L. J. Knowles	Worcester, Mass	July 19, 1870	105,466
Packing, Piston	P. L. Kreuter	Bloomington, Ill	May 27, 1862	35,378
Packing, Piston	B. Lowe	Fall River, Mass	Dec. 11, 1866	60,395
Packing, Piston	T. J. Mayall	Boston, Mass	Oct. 14, 1873	143,705
Packing, Piston	J. McAlonan	New York, N. Y	June 27, 1871	116,336
Packing, Piston	H. P. McCarroll	Pittsburgh, Pa	Nov. 14, 1871	120,888
Packing, Piston	F. McConnell	Dowagiac, Mich	Sept. 13, 1870	107,274
Packing, Piston	J. R. McCormick	Cincinnati, Ohio	June 4, 1872	127,625
Packing, Piston	J. P. McLean	Brooklyn, N. Y	Mar. 19, 1867	63,071
Packing, Piston	A. McMullin	Paterson, N. J	June 15, 1869	91,250
Packing, Piston	J. J. Miller	Chicago, Ill	Aug. 23, 1864	43,950
Packing, Piston, &c	E. D. Murfey	New York, N. Y	Mar. 19, 1872	124,846
Packing, Piston	C. A. Murray	Pittsburgh, Pa	July 29, 1862	36,016
Packing, Piston	J. Myers	Allegheny City, Pa	Oct. 24, 1865	50,661
Packing, Piston	W. Ord	Brooklyn, Ohio	Nov. 15, 1870	109,242
Packing, Piston	C. E. Perkins	Malone, N. Y	May 6, 1873	138,690
Packing, Piston	W. A. and T. F. Powers	Brooklyn, N. Y	Sept. 18, 1866	58,176
Packing, Piston	E. T. Prindle	Aurora, Ill	Aug. 21, 1866	57,375
Packing, Piston	G. W. Reisinger	Harrisburgh, Pa	Nov. 12, 1872	133,054
Packing, Piston	G. Robinson	Detroit, Mich	July 16, 1867	66,888
Packing, Piston	F. J. Roth	Newark, Ohio	Nov. 27, 1866	60,062
Packing, Piston	F. J. Roth	Schenectady, N. Y	May 5, 1868	77,532
Packing, Piston	F. J. Roth	Schenectady, N. Y	May 5, 1868	77,533
Packing, Piston	O. C. Smith	Salem, Mass	May 13, 1862	35,267
Packing, Piston	W. G. Snook and O. C. Patchell	Corning, N. Y	Oct. 23, 1866	59,143
Packing, Piston	A. J. Stevens	San Francisco, Cal	Mar. 7, 1865	46,723
Packing, Piston	J. H. Strehli	Cincinnati, Ohio	Jan. 23, 1872	122,923
Packing, Piston	E. Sullivan	Pittsburgh, Pa	Apr. 9, 1867	63,672
Packing, Piston	E. Sullivan	Pittsburgh, Pa	Dec. 7, 1869	97,564
Packing, Piston	E. Sullivan	Mount Washington, Pa	Sept. 26, 1871	119,248
Packing, Piston	A. Tannock	Litchfield, Ill	Nov. 7, 1865	50,855
Packing, Piston	J. H. Teal	Memphis, Tenn	Apr. 16, 1872	125,857
Packing, Piston	W. R. Thomas	Catasauqua, Pa	Mar. 7, 1865	46,733
Packing, Piston	T. Thurber	Auburn, N. Y	Mar. 26, 1867	63,185
Packing, Piston	C. Tibbetts and D. L. Weaver	Riverton, Ky	July 16, 1872	129,070
Packing, Piston	S. Tucker	Worcester, Mass	Sept. 17, 1867	68,914
Packing, Piston	L. Turner	New Orleans, La	Oct. 25, 1870	108,653
Packing, Piston	E. A. Walker	Hyannis, Mass	Apr. 12, 1870	101,792
Packing, Piston	S. L. Weigand	Philadelphia, Pa	Sept. 19, 1871	119,100
Packing, Piston	G. Wells	Providence, R. I	Feb. 21, 1871	111,997
Packing, Piston	J. Wheelock	Worcester, Mass	Apr. 5, 1864	42,245
Packing, Piston	J. Wheelock	Worcester, Mass	Nov. 28, 1865	51,250
Packing, Piston	W. D. Whitmore	Bloomington, Ill	Mar. 1, 1870	100,477

Index of patents issued from the United States Patent Office from 1790 *to* 1873, *inclusive*—Continued.

Invention.	Inventor.	Residence.	Date.	No.
Packing, Piston	W. F. Williams	Schenley, Pa	Sept. 24, 1872	131, 645
Packing, Piston and stop-valve	P. Giffard	Paris, France	May 6, 1873	138, 633
Packing, Piston and valve	W. M. Stevenson and A. Pearce	Harmony, Pa	Nov. 16, 1869	96, 993
Packing, Piston and valve rod	G. Tetley and C. D. B. Fisk	Providence, R. I	May 21, 1872	126, 997
Packing, Piston, cylinder, valve, &c	J. Schenck	Brooklyn, N. Y	Aug. 22, 1871	118, 394
Packing, Piston-rod	C. P. Benoit	Detroit, Mich	Sept. 25, 1866	58, 204
Packing, Piston-rod	J. H. Blessing	Albany, N. Y	Mar. 21, 1871	112, 770
Packing, Piston-rod	O. Collier	Sacramento, Cal	Oct. 13, 1868	83, 123
Packing, Piston-rod	P. and J. Eckford	Cincinnati, Ohio	Apr. 12, 1870	101, 840
Packing, Piston-rod	J. M. Flagg	Providence, R. I	Aug. 13, 1872	130, 418
Packing, Piston-rod	T. R. Grant	Newark, Ohio	Nov. 20, 1866	59, 907
Packing, Piston-rod	W. Hartley	Rockford, Ill	Dec. 13, 1870	110, 035
Packing, Piston-rod	J. F. Hermance	New Haven, Conn	Apr. 1, 1873	137, 444
Packing, Piston-rod	O. H. Jewell	Chicago, Ill	Jan. 12, 1869	85, 890
Packing, Piston-rod, &c	S. Katzenstein	New York, N. Y	Jan. 21, 1873	135, 127
Packing, Piston-rod, &c	R. C. Koch and T. B. McLaughlin.	Pittston and South Easton, Pa.	Sept. 17, 1872	131, 355
Packing, Piston-rod	H. T. Lee	Marysville, Cal	Apr. 19, 1870	102, 137
Packing, Piston-rod	S. Lockard	La Grange, Ind	Sept. 22, 1868	82, 331
Packing, Piston-rod	T. Lord	Providence, R. I	Apr. 15, 1873	137, 783
Packing, Piston-rod	J. W. Lynch	Richmond, Va	May 28, 1872	127, 254
Packing, Piston-rod	R. Martin	Brooklyn, N. Y	Aug. 14, 1866	57, 163
Packing, Piston-rod	J. P. McLean	Brooklyn, N. Y	Mar. 19, 1867	63, 072
Packing, Piston-rod	J. P. McLean and J. Vandercar	Brooklyn, N. Y	Nov. 20, 1866	59, 918
Packing, Piston-rod	W. H. Miller	Philadelphia, Pa	Mar. 26, 1867	63, 285
Packing, Piston-rod	W. H. Miller	Philadelphia, Pa	Jan. 21, 1868	73, 454
Packing, Piston-rod	E. T. Prindle	Aurora, Ill	Mar. 27, 1866	53, 542
Packing, Piston-rod	G. S. Prindle	Aurora, Ill	Feb. 20, 1866	52, 742
Packing, Piston-rod	E. A. Richmond	San Francisco, Cal	Oct. 11, 1870	108, 188
Packing, Piston-rod	C. Thayer and D. McPherson	Rome, N. Y	Feb. 22, 1870	100, 213
Packing, Piston-rod	S. S. Turner	Westborough, Mass	Dec. 17, 1867	72, 340
Packing, Piston-rod	W. W. Vanderbilt	New York, N. Y	July 15, 1873	140, 973
Packing, Piston-rod	H. C. White	New York, N. Y	Feb. 27, 1872	124, 182
Packing, Piston-rod	W. P. Woodruff	New York, N. Y	Aug. 14, 1866	57, 243
Packing, Piston-rod	F. Wright	Galesburgh, Ill	July 24, 1866	56, 658
Packing, Piston-rod	J. Young	Brooklyn, N. Y	Feb. 5, 1867	61, 789
Packing-press	J. Y. Parce	Fairport, N. Y	Oct. 9, 1860	30, 342
Packing, Pump	W. Race	Seneca Falls, N. Y	Nov. 24, 1857	18, 705
Packing pump-joints	B. Fitts	Newark, N. J	June 18, 1867	65, 803
Packing, Pump-piston	M. N. Allen, J. Rouse, and J. Melcher.	Titusville, Pa., and Waterbury, Conn.	Jan. 9, 1866	51, 990
Packing, Pump-piston	M. J. Althouse	Waupun, Wis	Apr. 2, 1867	63, 454
Packing, Pump-piston	J. Carter	Philadelphia, Pa	May 1, 1866	54, 295
Packing, Pump-piston	E. T. Ford	Stillwater, N. Y	July 6, 1869	92, 180
Packing, Pump-piston	D. B. Fuller	Buffalo, N. Y	Jan. 24, 1865	45, 989
Packing, Pump-piston	E. A. Jeffery	Corning, N. Y	Dec. 11, 1849	6, 937
Packing ring, Piston	J. B. Root	New York, N. Y	Aug. 14, 1866	57, 189
Packing ring, Piston-rod	W. C. Conwell	Scranton, Pa	Sept. 5, 1865	49, 725
Packing-ring for pistons of steam-engines	J. F. Bogardus	Brooklyn, N. Y	Aug. 7, 1866	57, 037
Packing, Rotary-engine	J. Loader	London, England	June 7, 1870	103, 901
Packing, Rotary-pump	A. W. Cary	Rockport, N. Y	May 15, 1849	6, 456
Packing, Rotary-pump	O. Sweet and M. E. Hicks	Providence, R. I	Nov. 25, 1862	37, 015
Packing, Rotary steam-engine	G. Sickels	Brooklyn, N. Y	Aug. 25, 1857	18, 063
Packing, Rubber	C. L. Frink	Rockville, Conn	May 8, 1866	54, 523
Packing, Rubber	J. Johnson	Brooklyn, E. D., N. Y	Oct. 26, 1869	96, 237
Packing, &c., Rubber flanged tubing for	T. J. Mayall	Boston, Mass	Apr. 9, 1872	125, 594
Packing, Spring for piston	D. Maydole	Norwich, N. Y	Aug. 16, 1870	106, 495
Packing, Steam	A. O. Bourn	Providence, R. I	July 4, 1871	116, 671
Packing, Steam	A. O. Bourn	Providence, R. I	June 4, 1872	127, 555
Packing, Steam	W. Brown	Hoboken, N. J	July 5, 1870	105, 036
Packing, Steam	A. H. Hall and H. T. Lee	Marysville, Cal	June 8, 1869	91, 120
Packing, Steam and hydraulic	W. M. Canfield	Philadelphia, Pa	Apr. 4, 1871	113, 252
Packing, Steam and water	W. C. Selden	Brooklyn, N. Y	Mar. 10, 1868	75, 301
Packing to projectiles, Securing soft-metal	H. F. Mann	La Porte, Ind	Jan. 6, 1863	37, 352
Packing, Tube	S. L. Fox	Philadelphia, Pa	Jan. 10, 1865	45, 822
Packing, Valve and piston	J. Marks	Boston, Mass	Feb. 1, 1870	99, 333
Packing, Vulcanized-rubber	D. C. Gately	Newtown, Conn	Jan. 26, 1869	86, 296
Packing, Well	J. Calkins and J. Fraser	Buffalo, N. Y	Oct. 24, 1865	50, 558
Packing, Well	J. H. Luther	Petroleum Centre, Pa	Jan. 21, 1873	134, 995
Packing, Well	J. J. Parker	Marietta, Ohio	Oct. 2, 1866	58, 469
Packing, Well-tube	S. E. Hewes	Albany, N. Y	Jan. 9, 1866	51, 945
Packing, Well-tube	P. Sicourt	Saragossa, Spain	Aug. 22, 1865	49, 599
Pad:				
See Blotting-pad. Bosom-pad. Breast-pad. Bronzing-pad. Carpet-pad. Coach-pad. Collar-pad. Corset-pad. Electrotype-backing pad. Foot-pad. Furniture-pad. Harness-pad. Harness and saddle pad. Harness coach-pad. Harness-saddle pad. Harness-tree pad. Hernia-pad. Hernial pad. Hoof-pad. Horse foot-pad. Horse head-pad. Horse hoof-pad. Horse interfering-pad. Horse neck-pad. Horseshoe rubber pad. Ink-pad. Inking-pad. Interfering-pad. Medicated pad. Neck-pad. Paper pad. Rouge-pad. Saddle-pad. Scrubbing-pad. Shoulder-pad. Stair-pad. Truss-pad. Ventilating-pad.				
Pad and trunk lock	A. M. Adams	Washington, D. C	Nov. 8, 1870	108, 951
Pad-billet	L. Hays	Ames, Iowa	June 9, 1868	78, 736
Pad-crimp press	H. H. Beers	Toulon, Ill	Aug. 27, 1867	68, 156
Pad-crimp press	G. Kennedy	Clarksville, Iowa	Jan. 7, 1868	73, 186

Index of patents issued from the United States Patent Office from 1790 *to* 1873, *inclusive*—Continued.

Invention.	Inventor.	Residence.	Date.	No.
Pad-hook	G. Roos and M. White	Buffalo, N. Y	Sept. 25, 1866	58,352
Pad-hooks to pads, Attaching	G. D. Gillett	Meridian, N. Y	July 14, 1868	79,822
Pad-tree	A. J. Cronk	Peoria, Ill	Aug. 6, 1867	67,507
Paddle	C. C. Everson	Palmyra, N. Y	July 16, 1872	129,012
Paddle and water wheel, Bracing the arms of	W. F. Julian	Hatsville, Ind	June 7, 1841	2,117
Paddle, Boat	P. E. Barbour	Louisville, Ky	Jan. 27, 1835	....
Paddle, Boat	J. Cochran	Baltimore, Md	Apr. 28, 1836	....
Paddle-floats, Means of adjusting	J. B. Baptista	New York, N. Y	Sept. 16, 1873	142,835
Paddle, Folding-boat	E. Jenks	Colebrook, Conn	June 6, 1817	....
Paddle for propelling boats	R. Claiborne	....	Feb. 23, 1802	....
Paddle for propelling ships, &c	J. L. Sullivan	Boston, Mass	Apr. 20, 1818	....
Paddle for steam-vessels, Elastic-plate	A. Jouan	San Francisco, Cal	Oct. 7, 1856	15,850
Paddle for vessels	A. C. Semple	Cincinnati, Ohio	July 5, 1853	9,834
Paddle-gate	E. Trumbull	Little Falls, N. Y	Apr. 4, 1833	....
Paddle-mechanism for boats	C. Howard	New York, N. Y	Mar. 19, 1872	124,746
Paddle-mechanism for boats	C. Howard	New York, N. Y	Feb. 11, 1873	135,710
Paddle of paddle-wheels for vessels, Arranging	P. G. Gardiner	New York, N. Y	May 4, 1841	2,076
Paddle, Propelling	J. J. Giraud	Baltimore, Md	Sept. 18, 1827	....
Paddle, Pendulum	A. F. W. Neynaber	Philadelphia, Pa	Aug. 12, 1862	36,165
Paddle, Reciprocating	P. C. Clark	Reading, Pa	May 25, 1858	20,328
Paddle, Revolving	J. W. McDermott	New York, N. Y	Feb. 11, 1873	135,653
Paddle-wheel	L. Ames and M. Miles	Waubeck and Pepin, Wis	Oct. 20, 1863	40,377
Paddle-wheel	D. Anderson	Philadelphia, Pa	Feb. 15, 1870	99,807
Paddle-wheel	J. Atkins	Louisville, Ky	Sept. 2, 1873	142,322
Paddle-wheel	E. H. Bailey	Philadelphia, Pa	Dec. 15, 1863	40,896
Paddle-wheel	E. Banks	Millport, N. Y	Jan. 8, 1867	60,988
Paddle-wheel	S. Bateman	Asniere, near Paris, France	Nov. 15, 1870	109,286
Paddle-wheel	T. S. Bigelow	Lake Mills, Wis	Aug. 11, 1863	39,522
Paddle-wheel	A. T. Boon	Galesburgh, Ill	Nov. 3, 1863	40,455
Paddle-wheel	E. T. Bostrom	Newnan, Ga	Nov. 19, 1867	70,947
Paddle-wheel	A. Buchanan	New York, N. Y	Mar. 2, 1858	19,482
Paddle-wheel	J. Burson	Yates, Ill	June 2, 1868	78,574
Paddle-wheel	F. W. Capen	Newton, Mass	Dec. 18, 1855	13,940
Paddle-wheel	J. C. Chaffee	Titusville, Pa	May 13, 1873	138,856
Paddle-wheel	A. Chapman	Fairfax, Vt	Oct. 3, 1854	11,741
Paddle-wheel	A. R. Chase	Cincinnati, Ohio	Nov. 9, 1842	2,845
Paddle-wheel	W. Choate	Newburyport, Mass	Oct. 10, 1865	50,417
Paddle-wheel	E. T. Colburn	Boston, Mass	July 31, 1866	56,713
Paddle-wheel	M. H. Collins and W. H. Holland.	Chelsea, Mass	Oct. 9, 1866	58,603
Paddle-wheel	G. Collyer and A. H. Patterson	Philadelphia, Pa	Mar. 13, 1860	27,618
Paddle-wheel	A. M. Comstock	Old Lyme, Conn	May 10, 1864	42,640
Paddle-wheel	R. H. Connelly	Philadelphia, Pa	Aug. 16, 1870	106,331
Paddle-wheel	R. Cromelien	Washington, D. C	Jan. 28, 1868	73;783
Paddle-wheel	M. A. Crooker	New York, N. Y	Sept. 13, 1870	107,340
Paddle-wheel	G. H. Cushman	Plymouth County, Mass	Apr. 9, 1872	125,381
Paddle-wheel	D. S. Darling	Brooklyn, N. Y	Aug. 15, 1871	117,991
Paddle-wheel	D. De Haven	New Orleans, La	Sept. 6, 1870	107,014
Paddle-wheel	D. Deshon, 2d	New London, Conn	Apr. 4, 1846	4,446
Paddle-wheel	G. W. Dickinson	Charleston, Ill	Oct. 29, 1872	132,641
Paddle-wheel	N. T. Edson	New Orleans, La	Apr. 30, 1867	64,295
Paddle-wheel	N. T. Edson	New Orleans, La	June 25, 1872	128,217
Paddle-wheel	H. Ehrhart	Muscatine, Iowa	Oct. 19, 1858	21,826
Paddle-wheel	P. Emerson	Carondelet, Mo	Sept. 22, 1868	82,395
Paddle-wheel	H. B. Fay	New York, N. Y	Feb. 7, 1860	27,042
Paddle-wheel	A. C. Fletcher	New York, N. Y	Aug. 11, 1863	39,473
Paddle-wheel	A. C. Fletcher	New York, N. Y	June 15, 1869	91,221
Paddle-wheel	A. C. Fletcher	New York, N. Y	Dec. 23, 1873	145,857
Paddle-wheel	C. Fletcher	Cincinnati, Ohio	Mar. 25, 1856	14,497
Paddle-wheel	W. C. Ford	Brooklyn, N. Y	Feb. 3, 1863	37,572
Paddle-wheel	R. Germain	Buffalo, N. Y	Apr. 10, 1860	27,794
Paddle-wheel	R. Germain	Buffalo, N. Y	Sept. 8, 1863	39,807
Paddle-wheel	J. Goodier and J. F. Kilshaw	Chester and New Brighton, England.	June 5, 1866	55,439
Paddle-wheel	W. Goodwin	Boston, Mass	Oct. 22, 1867	70,084
Paddle-wheel	W. Gorman	New York, N. Y	July 5, 1859	24,627
Paddle-wheel	J. Granger	Zanesville, Ohio	Apr. 10, 1866	53,812
Paddle-wheel	E. Haight	Buffalo, N. Y	Jan. 31, 1860	26,984
Paddle-wheel	J. W. Harris	Durhamville, N. Y	June 7, 1859	24,300
Paddle-wheel	A. Heuston	San Francisco, Cal	Sept. 11, 1866	57,901
Paddle-wheel	B. Hill	Rochester, N. Y	Dec. 25, 1855	13,988
Paddle-wheel	J. C. Hite	Mound City, Ill	Aug. 29, 1871	118,532
Paddle-wheel	W. H. Holland	Boston, Mass	June 19, 1866	55,770
Paddle-wheel	W. H. Holland	Chelsea, Mass	Dec. 18, 1866	60,517
Paddle-wheel	W. H. Holland	Chelsea, Mass	Dec. 18, 1866	60,518
Paddle-wheel	W. H. Holland	Boston, Mass	Aug. 2, 1870	105,943
Paddle-wheel	A. Houseworth	New York, N. Y	Aug. 19, 1856	15,564
Paddle-wheel	W. Huffman	Oshkosh, Wis	Nov. 30, 1869	97,404
Paddle-wheel	W. Hunter	Detroit, Mich	Nov. 12, 1867	70,851
Paddle-wheel	B. Irving	Green Point, N. Y	Sept. 6, 1853	10,000
Paddle-wheel	E. Jones	San Francisco, Cal	Mar. 6, 1866	53,008
Paddle-wheel	G. A. Keene	Lynn, Mass	Oct. 8, 1872	132,013
Paddle-wheel	S. Kepner	Pottstown, Pa	Apr. 15, 1862	34,969
Paddle-wheel	C. A. Kirkpatrick	Somerville, Mass	July 25, 1865	48,956
Paddle-wheel	W. F. Knowlton	Saint Cloud, Minn	June 21, 1870	104,603
Paddle-wheel	R. B. Locke	Stapleton, N. Y	Oct. 26, 1858	21,892
Paddle-wheel	J. Mahony	Newport, R. I	Nov. 9, 1869	96,603
Paddle-wheel	W. R. Manley	New York, N. Y	May 26, 1868	78,301
Paddle-wheel	W. R. Manley	New York, N. Y	Mar. 23, 1869	88,052
Paddle-wheel	B. G. Martin	Philadelphia, Pa	July 11, 1865	48,771
Paddle-wheel	E. Mathers	Harrisville, W. Va	Apr. 25, 1871	114,022
Paddle-wheel	E. Matteson	South Brooklyn, N. Y	June 4, 1867	65,495
Paddle-wheel	J. May	Columbus, Ga	Feb. 8, 1859	22,884
Paddle-wheel	E. J. McCarthy	Saugerties, N. Y	Dec. 28, 1847	5,405
Paddle-wheel	A. McKenzie	Newport, Ky	May 28, 1867	65,252
Paddle-wheel	J. Merkel	Mount Pleasant, Iowa	Aug. 7, 1866	56,973
Paddle-wheel	W. H. Muntz	Norton, Mass	Nov. 15, 1853	10,235
Paddle-wheel	W. H. Muntz	Norton, Mass	July 4, 1854	11,231

Index of patents issued from the United States Patent Office from 1790 *to* 1873, *inclusive*—Continued.

Invention.	Inventor.	Residence.	Date.	No.
Paddle-wheel	L. F. Noe	New York, N. Y	Mar. 25, 1862	34, 767
Paddle-wheel	L. A. Norton	Healdsburgh, Cal	June 5, 1864	43, 424
Paddle-wheel	N. Orcutt	Binghamton, N. Y	Jan. 18, 1859	22, 688
Paddle-wheel	J. Perkins	London, England	Nov. 30, 1829	....
Paddle-wheel	J. W. Post	Castile, Pa	Oct. 25, 1870	108, 626
Paddle-wheel	E. Pratt	New York, N. Y	Apr. 4, 1871	113, 561
Paddle-wheel	J. Rees	Pittsburgh, Pa	Aug. 23, 1870	106, 620
Paddle-wheel	W. C. Rice	Oquawka, Ill	May 25, 1869	90, 465
Paddle-wheel	F. B. Scott	Buffalo, N. Y	July 14, 1863	39, 248
Paddle-wheel	T. Shaw	Philadelphia, Pa	Nov. 26, 1872	133, 341
Paddle-wheel	N. P. Sheldon	San Francisco, Cal	Aug. 8, 1871	117, 822
Paddle-wheel	E. C. Smith	Old Ripley, Ill	Nov. 5, 1867	70, 637
Paddle-wheel	N. Smith	Berwick, La	Apr. 27, 1858	20, 096
Paddle-wheel	J. Speers	West Manchester, Pa	Oct. 18, 1859	25, 871
Paddle-wheel	E. Spencer	Ottawa, Canada	May 21, 1867	65, 023
Paddle-wheel	J. S. Swann	Kanawha County, W. Va	May 14, 1872	126, 760
Paddle-wheel	J. Thompson and M. L. Doty	Chariton, Iowa	May 17, 1859	24, 067
Paddle-wheel	N. Thompson	Bridgeport, Conn	Mar. 22, 1859	23, 324
Paddle-wheel	W. Thompson	Madison, Wis	Dec. 5, 1871	121, 689
Paddle-wheel	N. Thorn	Philadelphia, Pa	Jan. 26, 1869	86, 258
Paddle-wheel	G. W. Tinsley	Minneapolis, Minn	Nov. 19, 1872	133, 179
Paddle-wheel	C. A. Todd	New York, N. Y	July 31, 1866	56, 834
Paddle-wheel	B. W. Tucker	Brooklyn, N. Y	Mar. 15, 1870	100, 820
Paddle-wheel	A. Van Antwerp	Albany, N. Y	July 25, 1854	11, 393
Paddle-wheel	T. G. Walker and H. H. Holbrooke.	New York, N. Y	Jan. 7, 1873	134, 620
Paddle-wheel	W. P. Walker	Memphis, Tenn	Dec. 10, 1872	133, 905
Paddle-wheel	J. U. Wallis	Dansville, N. Y	Jan. 23, 1855	12, 298
Paddle-wheel	J. U. Wallis	Dansville, N. Y	July 3, 1855	13, 190
Paddle-wheel	A. Wingard	San Francisco, Cal	Feb. 1, 1870	99, 510
Paddle-wheel	G. Wingate	Philadelphia, Pa	June 22, 1858	20, 676
Paddle-wheel	L. W. Wright	Brooklyn, N. Y	Jan. 21, 1870	104, 528
Paddle-wheel, Adjustable	L. M. Dehart	Reading, Pa	Apr. 3, 1855	12, 621
Paddle-wheel and mode of propelling canal-boat	B. M. Smith	Rochester, N. Y	May 22, 1835	....
Paddle-wheel bucket	A. M. Glover	Waterborough, S. C	June 12, 1855	13, 058
Paddle-wheel bucket	A. C. Loud	San Francisco, Cal	Aug. 23, 1870	106, 704
Paddle-wheel buckets, Arrangement of	M. A. Crooker	New York, N. Y	Oct. 28, 1856	15, 967
Paddle-wheel, Chain	A. Rogers	Middletown, Pa	Apr. 6, 1830	....
Paddle-wheel for and manner of connecting to vessels.	B. Phillips	Philadelphia, Pa	June 13, 1831	....
Paddle-wheel for propelling boats	J. J. Greenough	Boston, Mass	Aug. 18, 1837	360
Paddle-wheel for propelling boats	W. F. Kearsing	New York	Dec. 8, 1826	....
Paddle-wheel for propelling boats	I. McCord	Harrisburgh, Pa	Sept. 22, 1838	938
Paddle-wheel for propelling boats	J. Ong	North Huntingdon, Pa	May 22, 1837	199
Paddle wheel for propelling boats, &c., Inclined float.	F. W. Stevens	Chigwell, England	Sept. 5, 1840	1, 771
Paddle-wheel for propelling steam and other boats.	W. A. Douglas	Albany, N. Y	July 31, 1837	326
Paddle-wheel for propelling steamboats, &c	M. W. King	New York, N. Y	Jan. 15, 1840	1, 473
Paddle-wheel for propelling vessels, Constructing and arranging.	W. W. Van Loan	Catskill, N. Y	Mar. 29, 1841	2, 022
Paddle-wheel for steam-vessels	A. Connison	Newark, N. J	June 18, 1842	2, 672
Paddle-wheel for steamers, Feathering	A. H. Brown	Washington, D. C	July 19, 1853	9, 854
Paddle-wheel, Horizontal	P. Lear and E. Buck	Boston, Mass	Feb. 20, 1844	3, 448
Paddle-wheel, Oblique-bucket	G. S. Weeks	Oswego, N. Y	Apr. 13, 1852	8, 880
Paddle-wheel propulsion, Engine for	I. L. Thompson	Sun Fish, Ohio	Sept. 3, 1872	131, 135
Paddle-wheel, Steam and horse boat	J. P. Espy	Philadelphia, Pa	May 22, 1833	....
Paddle-wheel, Steamboat, &c	R. D. Chatterton	Derby, England	July 24, 1844	3, 679
Paddle-wheel, Steamboat	T. Hunt	Boston, Mass	Feb. 17, 1831	....
Paddle-wheel, Stern	D. K. Peoples	Philadelphia, Pa	Sept. 18, 1860	30, 087
Paddle-wheel, Submerged	S. B. Howd	Arcadia, N. Y	June 14, 1845	4, 081
Paddle-wheel, Submerged	W. F. Ketchum	Buffalo, N. Y	Aug. 1, 1854	11, 429
Paddle-wheels, Application of steam to	W. Thornton	Washington, D. C	Dec. 23, 1814	....
Paddle-wheels, Arrangement of feathering floats with.	J. P. Worthing	Binghamton, N. Y	May 27, 1862	35, 425
Paddle-wheels, Construction and location of	H. Randall	Philadelphia, Pa	Jan. 13, 1863	37, 421
Paddle-wheels, Driving	J. M. Story	Cincinnati, Ohio	July 3, 1866	56, 118
Paddle-wheels, Driving	J. M. Story	Cincinnati, Ohio	May 30, 1871	115, 380
Paddle-wheels, Fastening	E. Reid	Bazetta, Ohio	Aug. 28, 1860	29, 819
Paddle-wheels, Feathering	F. Barbaires	Solano County, Cal	June 21, 1864	43, 173
Paddle-wheels, Feathering	R. Bell	East Saginaw, Mich	June 2, 1868	78, 415
Paddle-wheels, Feathering	J. Burson	Yates City, Ill	Aug. 8, 1865	49, 226
Paddle-wheels, Feathering	A. Buttrick	Chelsea, Vt	Sept. 22, 1863	40, 009
Paddle-wheels, Feathering	S. and T. Champion	Washington, D. C	Jan. 3, 1854	10, 359
Paddle-wheels, Feathering	T. and S. Champion	Washington, D. C	June 6, 1854	11, 031
Paddle-wheels, Feathering	M. G. Collins	Cumberland, Md	Oct. 10, 1865	50, 338
Paddle-wheels, Feathering	T. Cooley	Brockport, N. Y	Oct. 22, 1861	33, 518
Paddle-wheels, Feathering	B. Densmore	Brockport, N. Y	Dec. 4, 1860	30, 804
Paddle-wheels, Feathering	J. V. Dinsmore	Milford, Mass	Dec. 11, 1866	60, 345
Paddle-wheels, Feathering	P. Emerson	Carondelet, Mo	Sept. 28, 1869	95, 333
Paddle-wheels, Feathering	F. Felter	Perth Amboy, N. J	Nov. 28, 1854	11, 992
Paddle-wheels, Feathering	S. F. Gates	Boston, Mass	Oct. 31, 1865	50, 702
Paddle-wheels, Feathering	A. Gilman	Charlestown, Mass	Aug. 1, 1865	49, 102
Paddle-wheels, Feathering	G. K. Glenn	Ashton, Ill	July 22, 1873	141, 135
Paddle-wheels, Feathering	E. Haight	Buffalo, N. Y	Nov. 7, 1865	50, 817
Paddle-wheels, Feathering	L. T. Howard	Smith's Mills, Miss	Feb. 10, 1857	16, 589
Paddle-wheels, Feathering	L. T. Howard	Smith's Mills, Miss	Sept. 15, 1857	18, 202
Paddle-wheels, Feathering	T. L. Jones	Poughkeepsie, N. Y	Jan. 10, 1854	10, 424
Paddle-wheels, Feathering	G. A. Keene	Newburyport, Mass	Jan. 10, 1865	45, 835
Paddle-wheels, Feathering	G. A. Keene	Newburyport, Mass	Aug. 6, 1867	67, 437
Paddle-wheels, Feathering	G. A. Keene	Lynn, Mass	Oct. 8, 1872	132, 012
Paddle-wheels, Feathering	A. Lefebvre	San Francisco, Cal	May 13, 1873	138, 900
Paddle-wheels, Feathering	H. Lull	Hoboken, N. J	May 20, 1856	14, 920
Paddle-wheels, Feathering	W. R. Manley	New York, N. Y	June 24, 1862	35, 697
Paddle-wheels, Feathering	W. R. Manley	New York, N. Y	Mar. 16, 1869	87, 860
Paddle-wheels, Feathering	W. R. Manley	New York, N. Y	Mar. 16, 1869	87, 861
Paddle-wheels, Feathering	W. L. R. Mattason	Rochester, N. Y	Mar. 13, 1860	27, 499
Paddle-wheels, Feathering	M. T. Miles	Chicago, Ill	Sept. 3, 1872	131, 019
Paddle-wheels, Feathering	J. G. Shands	Saint Louis, Mo	June 17, 1856	15, 149
Paddle-wheels, Feathering	J. C. H. : tut	San Francisco, Cal	Dec. 24, 1872	134, 326

Index of patents issued from the United States Patent Office from 1790 to 1873, inclusive—Continued.

Invention.	Inventor.	Residence.	Date.	No.
Paddle-wheels, Feathering	R. Williams and S. Wilson	Buffalo, N. Y	June 12, 1860	28,707
Paddle-wheels, Method of arranging and operating submerged horizontal.	P. Lear	Boston, Mass	Feb. 27, 1855	12,460
Paddle-wheels of steam and other boats, Machine for bending the stirrup for the.	A. B. Latta	Cincinnati, Ohio	Mar. 30, 1843	3,022
Paddle-wheels of vessels, Machinery for gearing and ungearing.	B. S. Doxey	United States Navy	Feb. 9, 1821	....
Paddle-wheels, Opening and closing bucket for	G. Tingle	New York, N. Y	Aug. 20, 1850	7,585
Paddle wheels, Valve or gate for oblique float	J. C. Carncross	Philadelphia, Pa	June 15, 1852	9,012
Padlock	A. M. Adams	Washington, D. C	Apr. 12, 1870	101,803
Padlock	W. H. Akins and H. E. Abell	Ithaca and Brooklyn, N. Y	May 2, 1871	114,386
Padlock	J. H. Ames	Stamford, Conn	Jan. 17, 1871	110,948
Padlock	S. Andrews	Perth Amboy, N. J	July 8, 1856	15,270
Padlock	W. Ball	Washington, D. C	Apr. 16, 1842	2,565
Padlock	T. Bernhard	Hartford, Conn	Dec 21, 1869	98,015
Padlock	A. S. Blake	Waterbury, Conn	Nov. 17, 1868	84,162
Padlock	H. D. Blake	New Britain, Conn	Aug. 15, 1865	49,468
Padlock	W. Bohannan	Baltimore, Md	Apr. 17, 1860	27,883
Padlock	W. Bohannan	New York, N. Y	Feb. 28, 1865	46,539
Padlock, &c	W. Bohannan	Brooklyn, N. Y	Aug. 6, 1867	67,401
Padlock	D. T. Brown	Newton, N. Y	Apr. 18, 1865	47,352
Padlock	D. T. Brown	Plainfield, N. J	July 4, 1871	116,544
Padlock	S. W. Budd	Philadelphia, Pa	Jan. 23, 1872	122,991
Padlock	I. D. Bush	Detroit, Mich	Apr. 2, 1867	63,467
Padlock	S. G. Cabell	Quincy, Ill	Dec. 14, 1869	97,875
Padlock	G. L. Chamberlin	Marietta, Ohio	June 11, 1872	127,678
Padlock	B. Chambers	Washington, D. C	Apr. 10, 1847	5,057
Padlock	C. J. Clements	New York, N. Y	Dec. 17, 1867	72,166
Padlock	J. Corbett	Brooklyn, N. Y	Nov. 29, 1870	109,717
Padlock	E. Coyle	Albany, N. Y	Mar. 7, 1865	46,643
Padlock	A. Crosby	Westfield, N. Y	June 22, 1869	91,610
Padlock	G. W. Dana	Racine, Wis	Oct. 6, 1868	82,694
Padlock	J. M. A. Dew	Chicago, Ill	May 3, 1870	102,661
Padlock	B. J., S. B., and A. H. Ebert	Frederick, Md	Oct. 8, 1867	69,645
Padlock	F. Egge	Bridgeport, Conn	Feb. 2, 1869	86,377
Padlock	F. Egge	Bridgeport, Conn	Dec. 23, 1873	145,853
Padlock	C. H. Eiffler	New York, N. Y	Jan. 12, 1869	85,805
Padlock	H. Eiffler	New York, N. Y	Dec. 17, 1867	72,378
Padlock	V. Enders	New York, N. Y	Dec. 19, 1865	51,576
Padlock	H. Essex	Meadville, Pa	Aug. 6, 1872	130,203
Padlock	H. Essex	Meadville, Pa	July 8, 1873	140,617
Padlock	E. L. Gaylord	Terryville, Conn	June 21, 1870	104,572
Padlock	C. F. Gerlach	Milwaukee, Wis	Mar. 19, 1872	124,677
Padlock	C. T. Gibson	Baltimore, Md	Apr. 5, 1870	101,607
Padlock	J. A. Grewey	Albany, N. Y	Feb. 8, 1859	22,869
Padlock	F. C. Goffin and C. Liebrick	Philadelphia, Pa	June 12, 1849	6,522
Padlock	J. B. Green	Darien, Conn	Dec. 31, 1867	72,845
Padlock	H. F. Haack	New York, N. Y	May 10, 1870	102,810
Padlock	E. P. Hall	Chicago, Ill	May 17, 1870	103,043
Padlock	J. Harrison, jr	Milwaukee, Wis	Jan. 8, 1856	14,059
Padlock	W. Harvey	Albany, N. Y	Feb. 11, 1868	74,220
Padlock	L. Hillebrand	Philadelphia, Pa	Mar. 1, 1870	100,402
Padlock	L. Hillebrand	Philadelphia, Pa	Mar. 1, 1870	100,403
Padlock	J. J. Hirschbuhl	Louisville, Ky	June 18, 1861	32,563
Padlock	A. Huffer and N. Sehner	Hagerstown, Md	July 4, 1865	48,558
Padlock	J. Ingels	Milton, Ind	July 26, 1870	105,691
Padlock	H. Jackson	New York, N. Y	Aug. 1, 1865	49,113
Padlock	H. Jackson	Brooklyn, N. Y	June 4, 1867	65,388
Padlock	B. Rensler	Hartford, Conn	June 11, 1872	127,695
Padlock	R. Kinsley	Springfield, Mass	Dec. 7, 1852	9,452
Padlock	J. E. Kohout	Brooklyn, N. Y	Apr. 14, 1868	76,780
Padlock	I. W. Lamb	Salem, Mich	Feb. 2, 1869	86,420
Padlock	F. F. Landis	Lancaster, Pa	May 12, 1868	77,740
Padlock	T. Lanston	Washington, D. C	Mar. 8, 1870	100,644
Padlock	C. Liebrick	Philadelphia, Pa	Aug. 11, 1863	39,486
Padlock	J. W. Lyon	Brooklyn, N. Y	Apr. 22, 1862	35,030
Padlock	O. D. Madge	Washington, D. C	July 11, 1871	116,972
Padlock	G. McGregor, R. Lee, and T. J. Clinton.	Cincinnati, Ohio	Aug. 26, 1851	8,314
Padlock	W. McIntyre	New York, N. Y	Dec. 6, 1870	109,922
Padlock	J. H. McWilliams	New York, N. Y	July 11, 1871	116,977
Padlock	G. Merkel and C. H. Meyer	New York, N. Y	Apr. 4, 1871	113,322
Padlock	C. H. Miller	Frederick, Pa	Feb. 26, 1867	62,353
Padlock	D. K. Miller	Reading, Pa	July 26, 1870	105,710
Padlock, &c	D. K. Miller	Philadelphia, Pa	Oct. 28, 1873	143,990
Padlock	E. M. and J. E. Mix	Ithaca, N. Y	Nov. 2, 1858	22,000
Padlock	E. M. and J. E. Mix	Ithaca, N. Y	Dec. 10, 1861	33,920
Padlock	E. M. and J. E. Mix	Westfield, N. Y	Sept. 25, 1866	58,278
Padlock, &c	W. H. Murphy	Versailles, Ohio	Sept. 3, 1867	68,527
Padlock	H. Nelson	Jerome, N. Y	Aug. 29, 1871	118,473
Padlock, &c	J. Nock	Philadelphia, Pa	July 16, 1839	1,243
Padlock	J. North	Middletown, Conn	Dec. 18, 1860	31,003
Padlock	I. J. Oldis	Wheeler, N. Y	Jan. 1, 1856	14,030
Padlock	D. F. Randall	Chicopee, Mass	July 14, 1868	79,859
Padlock	J. S. Rankin	Ann Arbor, Mich	Dec. 14, 1869	97,812
Padlock	M. F. Ridout	Milwaukee, Wis	Sept. 9, 1862	36,421
Padlock	H. Ritchie	Newark, N. J	Aug. 23, 1853	9,963
Padlock	L. C. Rodier	Springfield, Mass	Mar. 8, 1864	41,864
Padlock	E. H. Roper and W. Ball	Washington, D. C	Nov. 9, 1842	2,848
Padlock	F. Roos and F. Spoehr	New York, N. Y	July 31, 1860	29,406
Padlock	C. W. Saladee	Newark, Ohio	May 29, 1866	55,166
Padlock	C. W. Saladee	Newark, Ohio	May 29, 1866	55,167
Padlock	C. W. Saladee	Newark, Ohio	July 17, 1866	56,450
Padlock	C. W. Saladee	Newark, Ohio	July 24, 1866	56,616
Padlock	C. W. Saladee	Newark, Ohio	July 24, 1866	56,617
Padlock	C. W. Saladee and W. Armstrong.	Newark, Ohio	July 17, 1866	56,451
Padlock	J. Schneider	Chicago, Ill	Jan. 26, 1858	19,208

Index of patents issued from the United States Patent Office from 1790 *to* 1873, *inclusive*—Continued.

Invention.	Inventor.	Residence.	Date.	No.
Padlock	A. Seeger	New York, N. Y	Jan. 9, 1872	122, 666
Padlock	H. S. Shepardson	Shelburne Falls, Mass	Apr. 26, 1870	102, 440
Padlock	T. Slaight	Newark, N. J	Oct. 14, 1851	8, 431
Padlock	T. Slaight	Newark, N. J	Jan. 2, 1855	12, 186
Padlock	T. Slaight	Newark, N. J	Nov. 12, 1861	33, 715
Padlock	T. Slaight	Newark, N. J	Dec. 12, 1865	51, 488
Padlock	T. Slaight	Newark, N. J	Nov. 23, 1869	97, 127
Padlock	A. F. Smith and J. H. Vickers	Norwich, Conn	July 7, 1868	79, 694
Padlock	F. W. Smith, jr	Bridgeport, Conn	Jan. 25, 1870	99, 115
Padlock	F. W. Smith, jr	Bridgeport, Conn	Jan. 25, 1870	99, 116
Padlock	D. Snell	Little Falls, N. Y	Mar. 29, 1870	101, 391
Padlock	M. P. Thatcher	Pontiac, Mich	Jan. 7, 1868	73, 136
Padlock	J. E. Thompson	Buffalo, N. Y	Oct. 18, 1864	44, 757
Padlock	D. Tilton	Stoneham, Mass	Aug. 26, 1851	8, 318
Padlock	C. A. Wallin	Stockholm, Sweden	Sept. 10, 1872	131, 237
Padlock	S. White	Newark, N. J	June 20, 1854	11, 149
Padlock	W. Wilcox	Middletown, Conn	Oct. 31, 1871	120, 557
Padlock	W. Wilcox	Middletown, Conn	July 23, 1872	129, 879
Padlock	J. E. Wootten	Cressona, Pa	Dec. 11, 1866	60, 454
Padlock	L. Yale	Newport, N. Y	Sept. 8, 1857	18, 169
Padlock	D. Zeiler	Sumneytown, Pa	Feb. 27, 1872	124, 189
Padlock-case	S. Andrews	Perth Amboy, N. J	Dec. 16, 1856	16, 224
Padlock, Combination	G. K. Farrington	Alcatraz Island, Cal	Feb. 15, 1870	99, 872
Padlock, Combination	J. L. Hall	Cincinnati, Ohio	Sept. 1, 1868	81, 630
Padlock, Combination	W. N. Hall	Mexia, Tex	Dec. 23, 1873	145, 864
Padlock, Combination	A. S. Harding and N. Reed	Otisville, N. Y	Apr. 9, 1867	63, 722
Padlock, Combination	J. Kittle	East Bend, S. C	Feb. 18, 1873	135, 992
Padlock, Combination	W. C. Langenau	Cleveland, Ohio	Oct. 28, 1873	144, 114
Padlock, Combination	A. Leyden	Atlanta, Ga	Mar. 12, 1867	62, 862
Padlock, Combination	E. E. Stubbs	West Elkton, Ohio	Mar. 3, 1868	75, 216
Padlock, Combination	G. Thompson	Trenton, N. J	May 3, 1870	102, 620
Padlock for mail-bags, &c	P. A. Altmaier	Harrisburgh, Pa	Jan. 5, 1869	85, 630
Padlock for mail-bags, &c	S. Andrews	Perth Amboy, N. J	Dec. 5, 1840	1, 882
Padlock for mail-bags	H. C. Jones	Newark, N. J	Apr. 1, 1842	2, 525
Padlock, Hermetically closed and keyless	S. Bickerstaff	Cincinnati, Ohio	May 19, 1868	77, 953
Padlock, Indicator	H. W. Dopp	Buffalo, N. Y	Feb. 13, 1872	123, 558
Padlock, Indicator	F. J. Hoyt	New York, N. Y	Mar. 5, 1872	124, 270
Padlock-key	E. L. Gaylord	Bridgeport, Conn	July 18, 1871	117, 064
Padlock, Keyless	W. Bohannan	Brooklyn, N. Y	Feb. 28, 1871	112, 211
Padlock, Permutation	S. Beadle and W. P. Yates	Elmira, N. Y	Aug. 31, 1869	94, 274
Padlock, Permutation	L. H. Gano	New York, N. Y	Sept. 12, 1871	118, 798
Padlock, Permutation	S. Loyd	New York, N. Y	Nov. 14, 1871	120, 822
Padlock, Permutation	J. E. Treat	Oxford, Mich	Aug. 10, 1869	93, 501
Padlock, Register	J. L. Chambers	Brooklyn, N. Y	Nov. 5, 1867	70, 409
Padlock, Sealing	Z. Kelly	New Bedford, Mass	Mar. 5, 1867	62, 636
Padlock, Sealing	R. M. Webb and I. Hermann	New York, N. Y	July 23, 1867	67, 089
Padlock-shield	C. C. Gale	Indianapolis, Ind	June 21, 1870	104, 441
Paging and numbering machine	P. Koch and G. Schüle	New York, N. Y	Feb. 9, 1869	86, 763
Paging-machine	F. O. Degener	New York, N. Y	July 24, 1855	13, 302
Paging-machine	G. Hodgkinson	Cincinnati, Ohio	Nov. 14, 1854	11, 936
Paging-machine	C. L. Sholes	Milwaukee, Wis	June 15, 1869	91, 277
Paging-machine	S. W. Soule	Milwaukee, Wis	Nov. 16, 1869	96, 990
Paging-machine	S. W. Soule and C. L. Sholes	Milwaukee, Wis	Sept. 27, 1864	44, 488
Pail	C. W. Bartlett	New York, N. Y	Apr. 16, 1872	125, 715
Pail	J. J. Dutcher	Brooklyn, N. Y	Feb. 15, 1859	22, 999
Pail and can fastening	J. K. Chace	New York, N. Y	Apr. 5, 1870	101, 430
Pail and commode, Combined slop	G. Howland	Brunswick, N. Y	Mar. 18, 1873	136, 999
Pail and commode, Combined slop	D. Paterson	New York, N. Y	July 16, 1872	129, 294
Pail and lantern, Dinner	C. Britain	Saint Joseph, Mich	Jan. 19, 1864	41, 274
Pail and lantern, Dinner	N. C. Burnap	Argusville, N. Y	May 31, 1870	103, 558
Pail and night-stool combined, Chamber	S. Smith	Brooklyn, N. Y	May 28, 1872	127, 378
Pail and stool combined, Milking	M. A. Green	Fitchburgh, Mass	Nov. 5, 1872	132, 720
Pail and strainer, Milk	D. S. Curtiss	Washington, D. C	Oct. 12, 1869	95, 781
Pail and wringer, Mop	E. H. Lord and E. Hinman	Homer and Syracuse, N. Y.	Sept. 3, 1867	68, 447
Pail, Ash	G. W. Walker	Boston, Mass	Aug. 13, 1861	33, 058
Pail bail and handle	E. T. Covell	Brooklyn, N. Y	Nov. 16, 1869	96, 889
Pail-bail ear	W. J. McLea	Leroy, N. Y	Aug. 25, 1868	81, 390
Pail-bail ear	J. Walton	Brooklyn, N. Y	May 12, 1868	77, 853
Pail-bails, &c., Machinery for bending	R. Bunker	Rochester, N. Y	Dec. 7, 1852	9, 438
Pail, Butter	H. P. Westcott	Seneca Falls, N. Y	Sept. 8, 1868	82, 048
Pail, Butter	H. P. Westcott	Seneca Falls, N. Y	Sept. 6, 1870	107, 139
Pail, Chamber	G. A. Giggins	New York, N. Y	Feb. 1, 1870	99, 437
Pail, Chamber	J. S. Jennings	Brooklyn, N. Y	Dec. 7, 1869	97, 516
Pail, Chamber	J. H. Orr	Long Island City, N. Y	July 23, 1867	67, 067
Pail, Dinner	N. C. Burnap	Argusville, N. Y	Dec. 14, 1869	97, 873
Pail, Dinner	S. B. Cox	Buffalo, N. Y	Sept. 10, 1867	68, 608
Pail, Dinner	J. O. Fairbairn	Milwaukee, Wis	Nov. 23, 1869	97, 181
Pail, Dinner	J. H. Foote	Pittsfield, Mass	Apr. 21, 1868	76, 905
Pail, Dinner	D. Howarth	Portland, Me	May 15, 1866	54. 817
Pail, Dinner	D. Howarth	Portland, Me	Nov. 26, 1867	71, 305
Pail, Dinner	H. Joyce and A. Ernst	Troy, N. Y	Apr. 18, 1871	113, 893
Pail, Dinner	H. C. and W. W. Ketcham	Newark, N. J	Oct. 31, 1871	120, 442
Pail, Dinner	E. C. Luks	Massillon, Ohio	May 6, 1873	138, 569
Pail, Dinner	A. Macqueen, jr	Philadelphia, Pa	Mar. 9, 1869	87, 574
Pail, Dinner	P. Molan	New York, N. Y	Mar. 25, 1873	137, 230
Pail, Dinner	A. Nevin	Philadelphia, Pa	Nov. 19, 1872	133, 113
Pail, Dinner	M. Saulson	Troy, N. Y	Oct. 20, 1868	83, 323
Pail, Dinner	M. Saulson	Troy, N. Y	June 8, 1869	91, 168
Pail, Dinner	H. L. and A. R. Simonson	Graniteville, N. Y	Feb. 27, 1872	124, 019
Pail, Dinner	R. S. Squires	Cleveland, Ohio	Feb. 18, 1873	135, 946
Pail, Dinner	J. T. Sturtevant	Cleveland, Ohio	Dec. 24, 1872	134, 175
Pail, Dinner	G. Wetzler	Peoria, Ill	July 4, 1871	116, 780
Pail, Dinner	L. K. Williams	Hudson, Mich	Feb. 20, 1872	123, 963
Pail-ear	G. E. Eastman	Washington Mills, N. Y	July 28, 1868	80, 341
Pail-ear	J. A. Otis	Rutland, N. Y	June 21, 1870	104, 488
Pail ear, Water	J. G. Kritchbaum	Youngstown, Ohio	May 4, 1869	89, 666
Pail, Earth-chamber	H. Terry	Waterbury, Conn	July 19, 1870	105, 607

Index of patents issued from the United States Patent Office from 1790 *to* 1873, *inclusive*—Continued.

Invention.	Inventor.	Residence.	Date.	No.
Pail, Folding	A. Du Chateau and J. D. Williams.	Green Bay, Wis	Nov. 25, 1873	144, 893
Pail, Folding	R. Sabin	Benona, Mich	Sept. 10, 1872	131, 303
Pail, Garbage and slop	C. T. Voss	Stapleton, N. Y	Apr. 16, 1872	125, 860
Pail, Heat-retaining	J. C. Brain	Brooklyn, N. Y	Feb. 9, 1869	86, 727
Pail-hoop	E. Hill	Orange, Mass	June 25, 1872	128, 310
Pail lid, Slop	W. B. Sawyer	New York, N. Y	Feb. 9, 1869	86, 698
Pail-making machine	R. Hyde	Winchester, Mass	Apr. 19, 1822	....
Pail, Measuring	J. B. Wightman	Brooklyn, N. Y	May 16, 1871	114, 894
Pail, Milk	J. Carton and W. Ralph	Utica, N. Y	June 19, 1866	55, 621
Pail, Milk	H. Marshall	Atlanta, Ga	Jan. 12, 1869	85, 837
Pail, Milk	S. Oppenheimer	Peru, Ind	Dec. 1, 1868	84, 572
Pail, Milk	S. E. Oviatt	West Richfield, Ohio	Aug. 6, 1872	130, 147
Pail, Milk	C. Tabor	Craftsbury, Vt	Sept. 27, 1870	107, 834
Pail, Milk-straining	H. Buckins	Canton, Ohio	Sept. 18, 1860	30, 046
Pail, Milking	R. A. Fish	Worcester, Mass	Nov. 17, 1868	84, 179
Pail, Milking	A. J. Johnson	Vicksburgh, Mich	Sept. 16, 1873	142, 917
Pail, Milking	G. T. Lincoln	Leominster, Mass	June 10, 1873	139, 679
Pail, Milking	S. Oppenheimer	Peru, Ind	Mar. 16, 1858	19, 648
Pail, Milking	H. Springer	Brady Township, Mich	Apr. 8, 1873	137, 731
Pail, Paint	C. R. Otis	Chicago, Ill	Apr. 26, 1870	102, 308
Pail, Paper	A. and I. Jennings	Fairfield, Conn	Sept. 29, 1868	82, 526
Pail rest, Milk	G. C. Taft	Worcester, Mass	Mar. 26, 1872	124, 919
Pail, Scrubbing	A. P. Hawse and L. J. Adams	Morrisville, Vt	Feb. 1, 1859	22, 799
Pail, Self-sealing	C. A. Marshall	Cleveland, Ohio	Feb. 27, 1872	124, 145
Pail, Slop	J. S. Jennings	Brooklyn, N. Y	Nov. 26, 1872	133, 370
Pail, Slop	J. G. Roth	New York, N. Y	May 14, 1872	126, 648
Pail, Tin	H. Sangster	Buffalo, N. Y	Apr. 5, 1870	101, 516
Pail with strainer attached, Milk	L. M. Doddridge	New Mount Pleasant, Ind	May 17, 1870	103, 158
Pail, Wooden	L. A. Fleming	West Mount Vernon, N. Y	Jan. 4, 1870	98, 578
Pails, Apparatus for graining	J. R. and A. J. Cross	Chicago, Ill	Dec. 27, 1864	45, 590
Pails, Apparatus for painting or graining	J. Carter	Winchendon, Mass	Dec. 10, 1867	71, 978
Pails, Attachment of handle to tin	T. Evans	Watkins, N. Y	June 21, 1859	24, 451
Pails, buckets, &c., Mode of making and affixing the ears and bales of.	J. F. Phelps	Havana, N. Y	May 12, 1840	1, 605
Pails, firkins, &c., Cover for	T. F. Conklin	Fond du Lac, Wis	July 5, 1870	104, 933
Pails, &c., Machine for crozing and dressing the inside of.	R. W. How and C. E. Patterson	Brooklyn, N. Y	Oct. 22, 1872	132, 401
Pails, &c., Machine for driving hoops on	W. Raymond	Marlborough, N. H	June 28, 1859	24, 576
Pails, &c., Making	A. Miner	Camillus, N. Y	Apr. 25, 1823	....
Pails, &c., Manufacturing water	J. W. Jarboe	Green Point, N. Y	Aug. 20, 1867	67, 984
Pails, &c., Non-conducting filling for the walls of dinner.	O. M. Spiller	Cleveland, Ohio	Nov. 4, 1873	144, 232
Pails, Tool for painting	J. Carter	Winchendon, Mass	Jan. 31, 1865	46, 075
Paint	E. Battley	Mont Clair, N. J	Aug. 15, 1865	49, 467
Paint	H. W. Bradley	New Berlin, N. Y	Mar. 5, 1867	62, 598
Paint	E. S. Burns	Portland, Me	Aug. 17, 1869	93, 804
Paint	W. J. Byrne	Russellville, Ky	Nov. 28, 1871	121, 330
Paint	T. H. Currey	Cincinnati, Ohio	Feb. 15, 1870	99, 856
Paint	W. J. Dodge	Syracuse, N. Y	Dec. 24, 1867	72, 614
Paint	J. S. D'Orsey	New York, N. Y	July 27, 1858	20, 993
Paint	J. L. Geroldsek	Livingston, N. Y	July 10, 1866	56, 322
Paint	J. B. Hodgskin	New York, N. Y	Dec. 5, 1865	51, 317
Paint	W. N. Jordon	Cambridge, Mass	Apr. 25, 1871	114, 017
Paint	M. W. Kidder	Lowell, Mass	July 25, 1871	117, 301
Paint	B. F. King	Annapolis Junction, Md	Nov. 7, 1871	120, 751
Paint	T. C. Lamb	Chicago, Ill	Oct. 16, 1866	58, 840
Paint	H. A. and D. E. Longsdorf	Mechanicsburgh, Pa	June 12, 1866	55, 510
Paint	J. Lucas	Gibbsborough, N. J	Aug. 12, 1873	141, 802
Paint	F. C. Semelroth	Logansport, Ind	July 27, 1869	93, 129
Paint	R. B. Smith	Mount Pleasant, Ohio	July 31, 1866	56, 814
Paint	I. Tyson, jr	Baltimore, Md	June 4, 1861	32, 491
Paint	J. P. Vinsonheller	Urbana, Ohio	July 30, 1867	67, 385
Paint and artificial stone, Compound for	T. Hodgson	Brooklyn, N. Y	Jan. 12, 1869	85, 825
Paint and chemical mixing machine	R. Poole	Baltimore, Md	Jan. 12, 1869	85, 762
Paint and drug mill	M. Bishop	La Fayette, Ind	Mar. 6, 1866	52, 953
Paint and drugs, Mill for grinding	G. D. Jones	Jersey City, N. J	Apr. 1, 1851	8, 009
Paint, and in coating wood, stone, &c	W. Coggeshall	Springfield, Ohio	Dec. 18, 1866	60, 478
Paint and other compounds from bituminous slates, &c., Manufacture of.	W. F. Patterson	Vanceburgh, Ky	Mar. 5, 1867	62, 679
Paint and liquid can	H. Miller	New York, N. Y	Feb. 18, 1873	136, 085
Paint and package for	A. T. Stevens	New York, N. Y	Nov. 5, 1872	132, 874
Paint and pigment	J. Brenton	Pittston, Pa	Oct. 19, 1869	95, 977
Paint and pigment	S. W. Isham	Hinckley, Ohio	Apr. 12, 1870	101, 880
Paint and putty, Compounds for	R. M. Breinig	Brooklyn, N. Y	Apr. 2, 1872	125, 261
Paint and varnish, Manufacture of	G. F. Cornelius	Westminster, Great Britain	Aug. 30, 1870	102, 784
Paint, Apparatus for mixing and grinding oil	W. H. Dolson	New York, N. Y	Nov. 3, 1857	18, 528
Paint box or can	H. Tucker	Newton, Mass	May 20, 1873	139, 214
Paint-burner	W. W. Wakeman, jr., and R. Ross.	New York and Brooklyn, N. Y.	Jan. 15, 1867	61, 288
Paint-can	L. U. Bean and H. A. Stevenson	Philadelphia, Pa	May 14, 1872	126, 775
Paint-can	G. W. Bennett	Brooklyn, N. Y	Aug. 27, 1867	68, 158
Paint-can	G. M. Bligh	New York, N. Y	Feb. 28, 1860	27, 265
Paint-can	P. Brown	Brooklyn, N. Y	Nov. 29, 1859	26, 248
Paint-can	C. Burnham	Philadelphia, Pa	Aug. 13, 1867	67, 718
Paint, &c., can	G. H. Chinnock	Brooklyn, N. Y	Sept. 9, 1873	142, 612
Paint, &c., can	J. R. Cole	New York, N. Y	Dec. 9, 1873	145, 400
Paint-can	J. F. Drummond	New York, N. Y	May 21, 1861	32, 397
Paint-can	J. F. Drummond	New York, N. Y	May 21, 1872	127, 034
Paint-can	H. Everett	Philadelphia, Pa	July 30, 1867	67, 278
Paint-can	H. Everett	Philadelphia, Pa	Sept. 3, 1872	131, 029
Paint-can	W. L. Gilroy	Philadelphia, Pa	Oct. 9, 1860	30, 310
Paint-can	E. B. Hamlin	Saint Louis, Mo	Nov. 11, 1868	83, 849
Paint, &c., can	W. A. Hopkins	New York, N. Y	Jan. 22, 1867	61, 337
Paint-can	E. A. Leland	Brooklyn, N. Y	Aug. 14, 1860	29, 651
Paint-can, &c	J. W. Masury	Brooklyn, N. Y	July 12, 1859	24, 748
Paint-can	J. W. Masury	New York, N. Y	July 10, 1860	29, 090
Paint-can	J. W. Masury	Brooklyn, N. Y	Dec. 16, 1873	145, 670

Index of patents issued from the United States Patent Office from 1790 *to* 1873, *inclusive*—Continued.

Invention.	Inventor.	Residence.	Date.	No.
Paint-can	H. Miller	Hoboken, N. J	Mar. 26, 1867	63, 283
Paint-can	D. Miller	Allegheny City, Pa	Mar. 31, 1868	76, 229
Paint, &c., can	H. Miller	New York, N. Y	Dec. 9, 1873	145, 440
Paint-can	O. E. Walker	Cincinnati, Ohio	Sept. 16, 1873	142, 971
Paint can and box	F. W. Devoe	New York, N. Y	June 4, 1867	65, 476
Paint can and box	J. F. Drummond	New York, N. Y	June 4, 1867	65, 477
Paint-can and device for removing the contents	F. L. Miller	Brooklyn, N. Y	Dec. 9, 1873	145, 439
Paint can and pail	F. W. Devoe	New York, N. Y	June 25, 1867	66, 077
Paint-can ear	C. F. Brand	Philadelphia, Pa	Jan. 31, 1865	46, 176
Paint, Can for packing and transporting	H. Tucker	Newton, Mass	Jan. 14, 1873	134, 947
Paint-can, Metallic	G. H. Chinnock	Brooklyn, N. Y	Nov. 19, 1872	133, 203
Paint-can, Metallic	G. H. Chinnock	Brooklyn, N. Y	June 17, 1873	139, 868
Paint-can, Metallic	G. H. Chinnock	Brooklyn, N. Y	June 17, 1873	139, 869
Paint-can, Sealing	J. F. Walter, jr	Brooklyn, N. Y	July 8, 1873	140, 748
Paint-cans, Machine for filling	J. Wilcox	Springfield, Mass	Jan. 14, 1868	73, 272
Paint-canister, Metallic	J. W. Masury	Brooklyn, N. Y	Apr. 28, 1857	17, 163
Paint, cement, &c., Composition for	C. W. Westover	Saint Joseph, Mo	Apr. 5, 1870	101, 687
Paint, cement, hard and soft rubber, &c., Composition to be used in the manufacture of.	F. Dickenson, jr	Hartford, Conn	June 15, 1869	91, 216
Paint, chemicals, &c., Machine for rubbing and mixing.	R. Poole	Baltimore, Md	May 28, 1867	65, 268
Paint, Combination of ochers for	D. S. Wood	Tiskilwa, Ill	Aug. 23, 1870	106, 643
Paint-composition	J. Ball and J. P. Ford	Zanesville, Ohio	Aug. 1, 1865	49, 066
Paint-composition	E. F. Barnes	New York, N. Y	June 23, 1863	39, 000
Paint-composition	J. H. Beardsley	New York, N. Y	Oct. 19, 1858	21, 810
Paint-composition	G. Blondin	New York, N. Y	June 20, 1854	11, 112
Paint-composition	R. Brannon	Baltimore, Md	Feb. 5, 1847	4, 958
Paint-composition	A. B. W. Bullard	Worcester, Mass	Jan. 5, 1864	41, 119
Paint-composition	P. Caubet	Paris, France	June 23, 1863	39, 020
Paint-composition	W. F. Cronkhite	Syracuse, N. Y	Aug. 23, 1864	43, 944
Paint-composition	E. P. Emerson	Blairsville, Pa	Feb. 28, 1860	27, 280
Paint-composition	A. Ferris	Benville, Ind	Nov. 13, 1866	59, 573
Paint-composition	W. H. Foran	Boston, Mass	June 18, 1872	128, 136
Paint-composition	J. L. Geroldsex	Livingston, N. Y	Sept. 13, 1864	44, 178
Paint-composition	R. P. Hinds	Chicago, Ill	Dec. 29, 1868	85, 308
Paint-composition	W. C. Hurd	New York, N. Y	Oct. 2, 1866	58, 421
Paint-composition	P. A. Klipstine	New Baltimore, Va	July 20, 1831	....
Paint-composition	R. Love	New York, N. Y	Feb. 1, 1870	99, 332
Paint-composition	C. Metcalf	Hanover, N. H	Feb. 25, 1831	....
Paint-composition	J. Miller	Moore Township, Pa	July 7, 1863	39, 202
Paint-composition	E. Oberlander, H. Weiss, and E. Bagnicki.	New York, N. Y	Mar. 20, 1866	53, 387
Paint-composition	W. Ten Eyck	Pike Township, Pa	Feb. 23, 1864	47, 729
Paint-composition	J. Trippe	Orange, N. J	Dec. 6, 1864	45, 362
Paint-composition	J. C. Wendrem	Albany, N. Y	Aug. 1, 1865	49, 201
Paint, Composition bronze	A. Towne	Auburn, Me	June 21, 1870	104, 665
Paint, Composition for drying	N. Hemenway	West Springfield, Mass	Feb. 22, 1833	....
Paint, Composition for mixing with	U. Lee	Burlington, Mich	Jan. 3, 1860	26, 685
Paint, Composition for mixing with	G. W. Slagle	Washington, D. C	Oct. 25, 1859	25, 935
Paint, Composition in preparing	J. Seabury	Brooklyn, N. Y	July 7, 1863	39, 204
Paint, Composition of	H. Hibbard	Darien, N. Y	Sept. 20, 1836	....
Paint, Composition of	H. Hibbard	United States	Sept. 20, 1836	27
Paint, Composition of metallic oxide for white	F. Sheppard	Fredericksburgh, Va	Mar. 18, 1835	....
Paint, &c., Composition of water-proof	D. M. Cummings	Enfield, N. H	Dec. 22, 1868	85, 216
Paint, Composition stone	E. Skinner and J. Webster	Sandwich, N. H	May 28, 1833	....
Paint-compound	D. R. Averill	Newburgh, Ohio	July 16, 1867	66, 773
Paint-compound	E. Blakeley	Neponset, Ill	Mar. 8, 1870	100, 558
Paint compound	A. H. Bourne	Fort Scott, Kans	July 26, 1870	105, 634
Paint compound	H. W. Bradley	Binghamton, N. Y	June 23, 1868	79, 196
Paint-compound	H. W. Bradley	Binghamton, N. Y	May 18, 1869	90, 073
Paint-compound	W. T. Burrell	Weymouth, Mass	Apr. 9, 1872	125, 436
Paint-compound	C. Campbell	New York, N. Y	Aug. 5, 1873	141, 544
Paint-compound	D. P. Flinn	Geneva, N. Y	Sept. 19, 1865	50, 068
Paint-compound	I. Gattman	Philadelphia, Pa	Sept. 30, 1856	15, 806
Paint-compound	J. Gould	Lexington, Mass	Sept. 28, 1869	95, 403
Paint-compound	W. G. Huyett	Williamsburgh, Pa	May 11, 1858	20, 205
Paint-compound	W. N. Jordan	Cambridge, Mass	Sept. 6, 1870	107, 059
Paint-compound	S. F. Mathews	Harrisburgh, Pa	Apr. 23, 1872	125, 972
Paint-compound	J. M. Merrymon	Logansport, Ind	Dec. 29, 1857	18, 006
Paint-compound	J. D. Mills	Dayton, Ohio	Jan. 21, 1868	73, 627
Paint-compound	H. Nelson	New York, N. Y	June 10, 1873	139, 729
Paint-compound	J. Quarterman	Flushing, N. Y	Oct. 21, 1873	143, 841
Paint-compound	M. Schneider	Cleveland, Ohio	July 2, 1867	66, 257
Paint-compound	C. F. Sibbald	Philadelphia, Pa	May 10, 1853	9, 717
Paint-compound	C. A. Sitzler	Kitzingen, Germany	Dec. 2, 1873	145, 130
Paint-compound	F. Staling	Harrisonburgh, Va	Dec. 24, 1872	134, 323
Paint-compound	F. A. Stall	Boston, Mass	Aug. 19, 1873	141, 897
Paint-compound	J. S. Vosseller	Plainfield, N. J	Jan. 21, 1873	135, 049
Paint-compound	A. F. Willson	Fort Wayne, Ind	Sept. 1, 1868	81, 721
Paint, Compound for mixing	J. L. Newell	Port Huron, Mich	July 29, 1873	141, 231
Paint-compound, Vehicle for	F. Kuhlman	Lille, France	Aug. 12, 1856	15, 520
Paint, Construction of vessel or apparatus for preserving.	J. Rand	United States	Sept. 11, 1841	2, 252
Paint, Drying	A. Jones	New York, N. Y	Nov. 12, 1850	7, 773
Paint, drugs, &c., Machine for grinding	I. Eddy	Boston, Mass	Mar. 19, 1824	....
Paint, dye-stuffs, &c., Mill for grinding	W. G. Johnson	Philadelphia, Pa	May 3, 1832	....
Paint, Fire and water proof	T. Brinkmann	Greeneville, Tenn	Dec. 21, 1869	98, 022
Paint, Fire and water proof	J. Weisman	Philadelphia, Pa	Jan. 31, 1844	3, 420
Paint, Fire-proof	C. P. Crossman	West Warren, Mass	Jan. 7, 1873	134, 522
Paint, Fire-proof	E. Kunzendorf	New York, N. Y	Oct. 12, 1869	95, 696
Paint, Fire-proof	L. Paimbeouf	Washington, D. C	Nov. 11, 1837	449
Paint, Fire-proof	G. W. Powell	Chinese Camp, Cal	Oct. 27, 1863	40, 429
Paint fire-proof, Composition for rendering	J. B. Harris	Germantown, Ky	July 3, 1866	56, 044
Paint for bituminous roofing-fabrics	A. Robinson	New York, N. Y	Nov. 29, 1864	45, 275
Paint for buildings, roofs, &c	J. H. and G. O. Smith	Chicago, Ill	Nov. 10, 1868	84, 011
Paint for houses, Composition for	W. Cox	Dayton, Ohio	Oct. 23, 1837	438
Paint for liquid compasses	E. S. Ritchie	Brookline, Mass	May 12, 1868	77, 763
Paint for marbleizing wood	F. G. Pokorny	New York, N. Y	July 19, 1870	105, 596

Index of patents issued from the United States Patent Office from 1790 *to* 1873, *inclusive*—Continued.

Invention.	Inventor.	Residence.	Date.	No.
Paint for oil-barrels, &c	W. L.Elkins	Philadelphia, Pa	July 15, 1873	140, 817
Paint for plastered houses, Oil	A. Thompson	Bethany, N. Y	Feb. 2, 1828	....
Paint for protecting iron from rust, &c., Galvanic	L. Knapp	New York	July 1, 1840	1, 664
Paint for roofs	T. C. Rice	Worcester, Mass	June 21, 1870	104, 648
Paint for roofing, wood, metal, &c	J. C. and C. M. Bills	Ottumwa and Albia, Iowa	July 22, 1873	140, 991
Paint for ships' bottoms	G. B. Bair	Baltimore, Md	July 9, 1872	128, 777
Paint for ships' bottoms, &c	J. Bowker	Baltimore, Md	May 24, 1870	103, 290
Paint for ships' bottoms	J. G. Fuller	Brooklyn, N. Y	Feb. 28, 1860	27, 331
Paint for ships' bottoms	S. A. Gilman	Boston, Mass	Mar. 5, 1872	124, 204
Paint for ships' bottoms	L. A. Messinger	Philadelphia, Pa	Apr. 12, 1870	101, 899
Paint for ships' bottoms	G. W. Morse	New York, N. Y	Oct. 2, 1866	58, 458
Paint for ships' bottoms	D. Parkhurst	Gloucester, Mass	July 4, 1865	48, 583
Paint for ships' bottoms	H. Roundy	San Francisco, Cal	May 25, 1869	90, 395
Paint for ships' bottoms	J. G. Tarr and A. H. Wonson	Gloucester, Mass	Nov. 3, 1863	40, 515
Paint for ships' bottoms	J. G. Tarr and A. H. Wonson	Gloucester, Mass	June 13, 1865	48, 221
Paint for ships' bottoms	J. G. Tarr and A. H. Wonson	Gloucester, Mass	June 20, 1865	48, 323
Paint for ships' bottoms	T. D. Teal	Philadelphia, Pa	Feb. 1, 1870	99, 497
Paint for ships' bottoms	I. J. Wyman	New York, N. Y	Jan. 24, 1871	111, 295
Paint for ships' bottoms and other surfaces	W. Jesty	Gosport, England	Jan. 23, 1872	122, 949
Paint for ships' bottoms, Copper	T. F. Griffin and R. Tarr, jr	Gloucester, Mass	Oct. 18, 1870	108, 474
Paint for wood and other materials	J. K. Palmer	Cambridge, Mass	Oct. 19, 1869	96, 031
Paint for wood, metal, and woven fabrics	C. D. Smith	Chicago, Ill	Sept. 10, 1867	68, 661
Paint from bituminous coal, Process of making	C. Mortimer	Philadelphia, Pa	Apr. 9, 1850	7, 266
Paint from iron-ore, Manufacturing	R. Nicholson	Bairdstown, Ky	June 23, 1812	....
Paint from iron-ore, Manufacturing	B. Wilbor	Whitestown, N. Y	Feb. 10, 1814	....
Paint from magnesite	J. Briggs	Boston, Mass	Jan. 16, 1872	122, 755
Paint from zinc-ores, Manufacture of white	N. Bartlett and G. T. Lewis	Birmingham and Philadelphia, Pa.	Aug. 20, 1867	67, 839
Paint, fruit, &c., can	H. Miller	Hoboken, N. J	June 28, 1864	43, 326
Paint, Grinding	P. Breasted	Greene County, N. Y	Apr. 18, 1814	....
Paint-grinding machine	E. Skinner	Sandwich, N. H	Dec. 20, 1818	....
Paint-grinding mill	S. Cook	Mendon, Mass	Mar. 3, 1829	....
Paint-grinding mill	A. Holcomb	Butternuts, N. Y	May 14, 1827	....
Paint-grinding mill	T. R. Kingdon and L. P. Dodge	New York, N. Y	Mar. 18, 1873	136, 838
Paint-grinding mill	C. Thomas	West Newbury, Mass	Apr. 27, 1858	20, 102
Paint-grinding mill	J. W. Webb	Mount Morris, N. Y	Feb. 15, 1839	1, 081
Paint-guard and brush-holder	W. T. Bailey	Fairfax County, Va	Mar. 26, 1872	124, 928
Paint, India-rubber	W. and W. A. Butcher	Philadelphia, Pa	Sept. 15, 1857	18, 183
Paint, indigo, &c., Machine for grinding	O. C. Harris	Waterville, N. Y	Mar. 15, 1832	....
Paint, Hydraulic	D. R. Prindle	East Bethany, N. Y	Mar. 26, 1867	63, 299
Paint-keg, Metallic	J. C. Adams	Philadelphia, Pa	June 18, 1867	65, 855
Paint, Machine for coating cloth with	D. Cushing	Wheeling, Va	June 10, 1856	15, 065
Paint, Machine for grinding oil	H. W. Gear	New York, N. Y	Apr. 19, 1864	42, 364
Paint, Manufacture of	D. R. Averill	New Centreville, N. Y	Nov. 21, 1871	121, 146
Paint, Manufacture of	D. R. Averill	New Centreville, N. Y	Nov. 21, 1871	121, 147
Paint, Manufacture of	D. R. Averill	Centreville, N. Y	Dec. 5, 1871	121, 478
Paint, Manufacture of	H. Curtis	New York, N. Y	Dec. 3, 1867	71, 585
Paint, Manufacture of	G. F. De Douhet	Paris, France	Sept. 9, 1851	8, 356
Paint, Manufacture of	F. W. Gerdes	Allegheny City, Pa	Dec. 13, 1870	110, 027
Paint, Manufacture of	W. C. Hurd	New York, N. Y	Feb. 26, 1867	62, 490
Paint, Manufacture of	W. C. Hurd	New York, N. Y	Sept. 22, 1868	82, 416
Paint, Manufacture of	T. C. Rice	Worcester, Mass	Nov. 29, 1870	109, 757
Paint, Manufacture of brown metallic	P. Winter	Horicon, Wis	May 21, 1867	65, 032
Paint, Manufacture of pigment for	R. S. Stenton	New York, N. Y	Oct. 5, 1869	95, 534
Paint, Marine	J. J. Currier and L. Cook	Gloucester, Mass	Dec. 26, 1871	122, 110
Paint, Marine	G. Hook	West Harwich, Mass	Mar. 24, 1868	75, 915
Paint, Marine	W. H. Lewis	Boston, Mass	Oct. 5, 1869	95, 494
Paint-material	W. H. Hubbard	West Meriden, Conn	Apr. 6, 1869	88, 715
Paint, Metallic	H. E. Colton	New York, N. Y	Sept. 24, 1867	69, 185
Paint, Metallic	F. G. Holland	Washington, D. C	July 21, 1868	80, 177
Paint, Metallic alloy	C. Wetterstedt	Marseilles, France	Aug. 5, 1851	8, 275
Paint-mill	W. R. Axe	Rockton, Ill	July 13, 1869	92, 561
Paint-mill	W. R. Axe	Rockton, Ill	July 4, 1871	116, 662
Paint-mill	C. Belcher	Newark, N. J	Apr. 7, 1868	76, 386
Paint-mill	J. A. Berrill	Waterville, N. Y	Oct. 9, 1860	30, 288
Paint-mill	J. A. Berrill	Waterville, N. Y	Mar. 30, 1869	88, 360
Paint-mill	C. W. Brown	Boston, Mass	Feb. 20, 1855	12, 431
Paint-mill	R. Byrne	Brooklyn, N. Y	Nov. 18, 1873	144, 656
Paint-mill	C. Clifton	Jersey City, N. J	Jan. 1, 1867	60, 691
Paint-mill	W. W. Draper	Greenfield, Mass	Sept. 10, 1850	7, 630
Paint-mill	S. J. Goodwin	Rockton, Ill	Apr. 2, 1867	63, 502
Paint-mill	J. W. Masury	Brooklyn, N. Y	Oct. 4, 1870	107, 939
Paint-mill	R. J. McGrew	Evansville, Ind	Oct. 15, 1872	132, 304
Paint-mill	D. E. Paynter	Philadelphia, Pa	Apr. 10, 1855	12, 707
Paint-mill	G. P. Zindgraf	Philadelphia, Pa	June 12, 1866	55, 572
Paint-mill, Horizontal cast-iron	O. Packard	Wilmington, Vt	Feb. 12, 1827	....
Paint, Mineral	E. B. Hexel	Vincent, Pa	Apr. 19, 1870	102, 118
Paint, Mineral	J. S. Kenyon and L. Fox	Baldwinsville, N. Y	Nov. 1, 1870	108, 914
Paint, Mineral	J. R. Smith	Keyport, N. J	Oct. 15, 1867	69, 852
Paint, Mineral	W. H. Towers	Boston, Mass	Mar. 19, 1872	124, 772
Paint-mixing machine	C. W. Brown and G. W. Bauker	Boston, Mass	May 15, 1860	28, 324
Paint-mixing machine	S. G. Cheever	Boston, Mass	June 5, 1860	28, 560
Paint, Mode of making brown	T. Miller, jr	Madison County, New York	Jan. 25, 1815	....
Paint, Mode of mixing	J. Linnell, jr	Orleans, Mass	Mar. 26, 1831	....
Paint, Mode of preparing white-lead	J. N. Troville	Christiansburgh, Va	Mar. 31, 1840	1, 535
Paint-oil	D. R. P. Hill	Morgantown, W. Va	Sept. 29, 1868	82, 523
Paint-oil compound	W. E. Tascott	Cleveland, Ohio	June 22, 1869	91, 578
Paint-oils, varnish, &c., Vessel for	R. Mansure, jr	Philadelphia, Pa	June 24, 1873	140, 148
Paint or mastic, Machine for spreading	J. W. Wheeler	Cleveland, Ohio	Nov. 17, 1868	84, 074
Paint-pots to sides of buildings, Device for hanging	J. H. Flagg	Perkinsville, Vt	Jan. 8, 1867	61, 002
Paint, Preparing hydrocarbon-liquids to serve as vehicle for.	A. Meucci	Clifton, N. Y	May 26, 1863	38, 714
Paint, Preparing white	H. C. Otto	Philadelphia, Pa	Feb. 28, 1842	2, 470
Paint-preserving can	E. Clark	New York, N. Y	Oct. 12, 1858	21, 796
Paint, Process for making	W. F. Davis	New York, N. Y	Aug. 17, 1852	9, 208
Paint, Process for preparing	H. S. Lucas	Chester, Mass	Nov. 23, 1852	9, 422
Paint, Process for treating	G. Blondin	New York, N. Y	June 20, 1854	11, 111
Paint, Process of separating impalpable powder for	G. W. Griswold	Carbondale, Pa	July 18, 1854	11, 326

Index of patents issued from the United States Patent Office from 1790 *to* 1873, *inclusive*—Continued.

Invention.	Inventor.	Residence.	Date.	No.
Paint-restoring compound	W. O. Bradstreet	Gratis, Ohio	Nov. 19, 1872	133, 195
Paint, Roofing	C. W. Langworthy	Bergen, N. J	June 15, 1869	91, 455
Paint, Rubber	S Truscott	Cleveland, Ohio	Apr. 4, 1871	113, 368
Paint, soap, &c., Machine for mixing	J. Stainthorp and I. Cole	New York, N. Y	Apr. 26, 1870	102, 330
Paint, Stone	B. V. Betterton	Piqua, Ohio	Apr. 21, 1868	76, 882
Paint to stencil-plates, Apparatus for applying	C. Bates	Kingston, Mass	May 30, 1865	47, 917
Paint to surfaces, Applying	J. W. Kingman	North Bridgewater, Mass.	May 1, 1866	54, 365
Paiut, &c., Treating oil for	D. D. Cattanach	Providence, R. I	Dec. 30, 1873	146, 044
Paint, Varnish	W. B. Finch	Chicago, Ill	Dec. 24, 1867	72, 617
Paint, varnish, &c., Apparatus for removing	T. F. Moody	Toledo, Ohio	Mar. 21, 1871	112, 832
Paint, varnish, &c., Composition for removing	H. C. Kearney and J. W. Harrison.	Philadelphia, Pa	Apr. 12, 1864	42, 295
Paint-vehicle	A. C. Church	Union City, Mich	Jan. 5, 1858	19, 014
Paint-vehicle	I. Gattam	Philadelphia, Pa	Dec. 1, 1857	18, 794
Paint vessel and package	J. W. Masury	Brooklyn, N. Y	July 22, 1873	141, 061
Paint, Water-color	M. J. Green	Boston, Mass	Apr. 30, 1872	126, 142
Paint, Water-proof	E. E. and J. F. Ellery	New York, N. Y	July 5, 1859	24, 620
Paint, Water-proof	J A. Moffitt	Boston, Mass	Dec. 8, 1868	84, 702
Paint, Water-proof	T. E. Warren	Troy, N. Y	Mar. 15, 1845	3, 956
Paint, White	F. Shepherd	New Haven, Conn	July 19, 1837	284
Paint, White water-color	F. Shepherd	New Haven, Conn	July 17, 1837	285
Painting wire-cloth, Machine for	J. H. De Witt	Orange, N. J	Sept. 30, 1873	143, 227
Paint, Yellow	Baron Carrendeffez	New York	Sept. 2, 1806	....
Painted-cloth cover	W. D. Richardson	Pawtucket, R. I	Sept. 16, 1873	142, 868
Painted or varnished surfaces, Compound for filling or stuffing.	D. R. Averill	New Centreville, N. Y	Nov. 28, 1871	121, 311
Painted surfaces, Apparatus for sanding	E. May	Indianapolis, Ind	Dec. 6, 1859	26, 363
Painted surfaces, Composition for cleaning	B. F. Shaw	Peabody, Mass	Feb. 9, 1869	86, 702
Painted surfaces, Finish for	H. T. Payne and W. Ayres	San Francisco, Cal	Jan. 14, 1868	73, 377
Painters' canvas	C. Volkmar	Baltimore, Md	Jan. 7, 1868	73, 060
Painters' canvas-stretcher	J. Fairman	New York, N. Y	Mar. 7, 1871	112, 435
Painters' canvas-stretcher	M. K. Kellogg	Baltimore, Md	Oct. 1, 1867	69, 447
Painters' colors, Mill for grinding	C. Greene	....	July 23, 1801	....
Painters' colors, Mill for grinding	J. J. Wells	Hartford, Conn	June 13, 1809	....
Painters' composition, Coach	J. W. Tully	Delano, Pa	Feb. 8, 1870	99, 610
Painters' composition, Filling for	R. Sharp	Pittsburgh, Pa	Aug. 5, 1873	141, 518
Painters' holder	S. J. Newell	Dirigo, Me	Mar. 12, 1872	124, 610
Painters' hook	J. W. Pattee	Thornton, N. H	July 14, 1868	79, 853
Painters' jack, Carriage	A. G. Rykert	Attica, N. Y	Feb. 4, 1873	135, 593
Painters' palette	O. L. R. Andrews	Boston, Mass	July 2, 1872	128, 522
Painters' panel	A. G. Collins	Washington, D. C	Aug. 25, 1863	39, 632
Painter's pot	T. E. Costello	Brooklyn, N. Y	Apr. 1, 1872	137, 423
Painters' stand, Carriage	J. Burtch	Hall's Corners, N. Y	Feb. 11, 1873	135, 771
Painter stopper, Shank	C. Perley and J. Terry	New York, N. Y	June 5, 1849	6, 509
Painters' striping-implement	D. Lydon and C. Halsey	San Francisco, Cal	Nov. 19. 1872	133, 163
Painters' striping-instrument	J. J. McCormick and G. Crossingham.	New York and Croton Falls, N. Y.	Nov. 24, 1857	18, 699
Painters' trestle	D. Moritz	New York, N. Y	Dec. 12, 1871	121, 801
Painting and coating wood, Compound for	H. F. Snow and J. H. Davis	Dover, N. H	Oct. 7, 1873	143, 472
Painting and staining furniture	H. K. Wilson	Barbourville, Ky	Nov. 7, 1871	120, 605
Painting and varnishing, Composition for	S. Page	Chelsea, Mass	May 7, 1867	64, 443
Painting and varnishing machine	H. Thayer and L. L. Martin	Warsaw, N. Y	Feb. 9, 1858	19, 316
Painting and varnishing wood and metals	W. Piotrowski	New York, N. Y	Feb. 19, 1867	62, 222
Painting-apparatus	A. P. Merritt	Charlotte, Mich	May 31, 1870	103, 640
Painting barrels, casks, &c., Machine for	W. Burnet and G. B. Fisher	Chicago, Ill	Aug. 5, 1873	141, 542
Painting cloth, Machine for	O. Ferrin	Lansingburgh, N. Y	Nov. 28, 1831	....
Painting coach-bodies, &c., Trestle for	W. M. Knapp	Muncie, Ind	Nov. 4, 1873	144, 339
Painting, Composition for encaustic	W. H. F. Kehrwieder	Philadelphia, Pa	July 26, 1870	105, 809
Painting, Decorative oil	J. M. Lasché	Paris, France	Jan. 16, 1872	122, 771
Painting for ornamenting walls, floors, &c., Elastic-stamp.	T. Boynton	Windsor, Vt	July 12, 1832	....
Painting, graining, &c., Compound for ornamental	B. S. Harrington	Pontiac, Mich	Feb. 9, 1869	86, 836
Painting metallic roofs, Composition for	C. D. Smith	Chicago, Ill	May 1, 1866	54, 426
Painting, Oil	J. C. Walker	Washington, D. C	Jan. 14, 1873	134, 827
Painting on cloth	L. Jarosson	Jersey City, N. J	June 7, 1853	9, 771
Painting on glass, Ornamental	J. W. Bowers	Brookline, Mass	Jan. 13, 1852	8, 651
Painting on translucent surfaces	J. B. Hall	Philadelphia, Pa	Mar. 28, 1848	5, 495
Painting, Pigments for distemper	H. M. Johnston	New York, N. Y	Feb. 15, 1870	99, 907
Painting, Preparing pigments for water-color	E. R. Campbell	New York, N. Y	Oct. 17, 1871	119, 966
Painting, Process for decorating oil	P. Cousin and P. Oury	Paris, France	Dec. 7, 1869	97, 607
Painting rooms	J. Selby	....	Dec. 30, 1803	....
Painting signs on cloth, Process of	A. Stempel	Chicago, Ill	Nov. 11, 1873	144, 574
Palette and brush for graining	F. Cole	Kankakee, Ill	Oct. 26, 1869	96, 085
Palliasses, Apparatus for making	J. Pigot	Brooklyn, N. Y	May 23, 1854	10, 966
Palm-leaf fabrics, Manufacture of	F. Perrin	Cambridge, Mass	Dec. 9, 1862	37, 138
Palm leaf-splitting machine	C. McFarland	Barre, Mass	Mar. 31, 1841	2, 024
Palm-leaf-splitting machine	C. Wadsworth	Barre, Mass	Nov. 17, 1834	....
Palm-leaf-splitting machine	C. Wadsworth	Barre, Mass	Dec. 23, 1834	....
Palm-leaf warp and woof for weaving, Preparation of.	F. Perren	Cambridge, Mass	May 28, 1867	65, 266
Pamphlet-cover, Self-sealing	W. H. Nichols	East Hampton, Conn	Feb. 4, 1868	74, 121
Pamphlet-covering machine	W. H. Clague and R. B. Randall.	Rochester, N. Y	Feb. 13, 1872	123, 555

Pan:

See Amalgamating-pan. Gravel-pan.
Ash-pan. Grinding-pan.
Bake pan. Locomotive ash-pan.
Baking-pan. Milk-pan.
Bed-pan. Mining-pan.
Biscuit-pan. Nitro-glycerine pan.
Bread-pan. Refrigerator-pan.
Cake-pan. Rendering-pan.
Confection-pan. Saucepan.
Dish-pan. Scale-pan.
Drip-pan. Sheet-metal pan.
Dripping-pan. Stereotype-pan.
Dust-pan. Stew-pan.
Egg-poaching pan. Urinal-pan.
Evaporating-pan. Vacuum-pan.
Frying-pan. Water-closet pan.

Index of patents issued from the United States Patent Office from 1790 *to* 1873, *inclusive*—Continued.

Invention.	Inventor.	Residence.	Date.	No.
Pan-folding machine	W. Hamilton	Philadelphia, Pa	Aug. 25, 1868	81, 498
Pan-former	J. Finn	Decorah, Iowa	May 12, 1868	77, 726
Pan-forming machine	E. Kuessner	Cincinnati, Ohio	Dec. 16, 1873	145, 660
Pan-lifter	W. H. S. Lawrence and C. I. Collamore.	Bangor, Me	June 27, 1871	116, 330
Pan-lifter	M. A. Shepard	Bridgeport, Ill	Feb. 25, 1868	74, 855
Pan-making, Machine for	J. Dane, jr	Newark, N. J	Mar. 26, 1872	124, 938
Pan-wiring machine	J. Shephard	Bristol, Conn	May 19, 1868	78, 021
Pans and other cooking-utensils, Handle for	J. A. Lawson	Troy, N. Y	Oct. 26, 1869	96, 122
Panacea	J. Houck	Baltimore, Md	May 9, 1833	....
Panel-cutting machine	F. D. Green	Williamsport, Pa	Aug. 25, 1868	81, 361
Panel-molding machine	C. H. Williamson and J. H. Allyn.	Whitestown, N. Y	Mar. 11, 1873	136, 800
Panel-raising machine	F. D. Green	Williamsport, Pa	Aug. 19, 1873	142, 015
Panel-raising machine	J. H. Millspaugh	Williamsport, Pa	Jan. 16, 1872	122, 774
Panel-raising machine	D. F. Walker	Minneapolis, Minn	Oct. 31, 1871	120, 405
Panel-table	J. G. Greene	Port Henry, N. Y	Oct. 15, 1867	69, 799
Panel, Wooden	E. Maitre	Veuxaulles, France	Dec. 2, 1873	145, 221
Panels to wooden frames, Constructing and attaching iron.	J. Shaefer	Lancaster, Pa	Sept. 9, 1862	36, 427
Paneling-machine	S. Heyser	Jackson, Mich	Mar. 5, 1872	124, 267
Paneling-machine	N. Jenkins	New York, N. Y	Nov. 23, 1869	97, 092
Paneling-machine	N. Jenkins	New York, N. Y	Oct. 4, 1870	108, 027
Paneling-machine	D. F. Walker	Minneapolis, Minn	July 11, 1871	117, 020
Paneling-machine	H. P. Westcott	Seneca Falls, N. Y	Jan. 7, 1862	34, 083
Panier	J. S. Colby	New York, N. Y	Aug. 19, 1873	141, 854
Panoramas, Electro-magnetic apparatus for moving	L. Finger	Cambridge, Mass	July 26, 1870	105, 664
Pantaloons	G. R. Eager	Boston, Mass	June 13, 1871	115, 832
Pantaloons	G. R. Eager	Boston, Mass	June 11, 1872	127, 750
Pantaloons	J. D. Ehlers and J. P. Herron	Baltimore, Md., and Washington, D. C.	Mar. 11, 1862	34, 635
Pantaloons	W. Franklin	New Haven, Conn	Jan. 17, 1860	26, 842
Pantaloons	B. J. Greely	Springfield, Mass	June 29, 1858	20, 708
Pantaloons	B. J. Greely	New York, N. Y	Oct. 3, 1865	50, 242
Pantaloons	F. T. Hoyt	Brooklyn, N. Y	June 24, 1873	140, 197
Pantaloons	H. Osler	Philadelphia, Pa	Aug. 18, 1863	39, 584
Pantaloons	T. R. Sloan	Brooklyn, N. Y	July 26, 1870	105, 856
Pantaloons	E. T. Taylor	Philadelphia, Pa	Oct. 21, 1873	143, 792
Pantaloons, Attaching strap to	S. Heller	New York, N. Y	June 17, 1862	35, 608
Pantaloons, Device for preserving shape of	W. D. Sinclair	Trenton, N. J	Aug. 2, 1864	43, 714
Pantaloons, Drying and pressing apparatus for	J. Braun	Rochester, N. Y	Dec. 22, 1868	85, 059
Pantaloons, Fastening for	I. Stratton	Keene, N. H	Nov. 5, 1867	70, 645
Pantaloons, Guard for	S. B. Smith and E. Lindsley	Cleveland, Ohio	Apr. 23, 1867	64, 161
Pantaloons, Guide for cutting out	J. Ramsey and A. B. Smith	Clinton, Pa	Apr. 23, 1861	32, 145
Pantaloons, Presser-block for	L. Cohen and J. C. Walker	San Francisco, Cal	Apr. 1, 1873	137, 346
Pantaloons, Protecting-guard for	E. Lindsley	Cleveland, Ohio	Jan. 7, 1868	73, 106
Pantaloons, Protector for	C. W. S. Heaton	Belleville, Ill	July 10, 1866	56, 221
Pantaloons, Steaming and drying apparatus for shaping.	E. B. Viets	Boston, Mass	Nov. 18, 1873	144, 644
Pantaloons, Strap-clasp for	W. H. Miller	New York	July 8, 1842	2, 705
Pantaloons, Strap-fastening for	W. W. Riley	Columbus, Ohio	Oct. 31, 1848	5, 887
Pantaloons, Strapfor	S. Heller	New York, N. Y	Dec. 15, 1863	40, 931
Pantaloons, Strap for	M. H. Simpson	Boston, Mass	May 30, 1839	1, 161
Pantaloons, Strap for	C. F. Walker	Benford's Store Post-Office, Pa.	Aug. 2, 1864	43, 727
Pantaloons, Strap for	E. Wilder	Boston, Mass	Nov. 19, 1833	....
Pantaloons, Stretcher for	L. Kaltenbacher	Louisville, Ky	Aug. 19, 1873	142, 025
Pantaloons, Stretcher for	J. D. Ryan	New York, N. Y	Nov. 18, 1873	144, 703
Pantaloons, Stretching-apparatus for	J. Mottet	Philadelphia, Pa	July 19, 1864	43, 596
Pantaloons, Stretching-board for	E. F. Smith	Salem, Mass	Mar. 26, 1872	124, 916
Pantaloons, Stretching-device for	S. C. Wells	Le Roy, N. Y	Sept. 29, 1868	82, 666
Pantaloons, Stretching-frame for	J. H. Hulse	Baltimore, Md	Apr. 21, 1868	76, 919
Pantaloons, Tree for	M. Taine	South Boston, Mass	June 24, 1873	140, 172
Pantaloons, Tree for	E. B. Viets	Boston, Mass	May 27, 1873	139, 283
Pantaloons, &c., Waistband-fastening for	J. A. Baarvig	Chicago, Ill	Nov. 4, 1873	144, 334
Pantiles, Making	W. Harwood	Richmond, Va	July 3, 1806	....
Pantry	J. Shattuck	Brookline, Mass	July 4, 1865	48, 631
Pantograph	L. F. Bruce and N. J. Wolcott	Springfield, Mass	Sept. 12, 1871	118, 902
Pantograph	A. Judd	Chicopee, Mass	Aug. 13, 1850	7, 564
Pantograph	H. Neumeyer	Macungie, Pa	Dec. 2, 1856	16, 148
Pantograph and parallel ruler	J. J. Hawkins	....	May 17, 1803	....
Pantographic engraving-machine	B. L. Phillips	Providence, R. I	Dec. 16, 1862	37, 177
Pantographic machine	E. Oldham	Brooklyn, N. Y	May 15, 1866	54, 759
Pantographic machine	R. Whittman	Accrington, England	Oct. 1, 1861	33, 421
Pantographic reversing-instrument	C. L. Getz	Philadelphia, Pa	Aug. 19, 1862	36, 212
Paper	H. E. Wagner	Dresden, Germany	Apr. 29, 1873	138, 302
Paper, Adhesive composition for fly	G. N. Harcy	Lyme, Ohio	June 25, 1872	128, 223
Paper, Adhesive or gum	S. L. Weigand	Philadelphia, Pa	Aug. 2, 1864	43, 731
Paper and card-board to fix marks by metallic pencils. Treating.	H. M. Johnston	New York, N. Y	Nov. 18, 1873	144, 678
Paper and cloth collar	J. Barton	Battle Creek, Mich	Feb. 20, 1866	52, 659
Paper and cloth combined, Machine for forming shirt-bosoms from.	E. P. Furlong	Portland, Me	June 1, 1869	90, 739
Paper and cloth, Machine for uniting	G. K. Snow	Watertown, Mass	Dec. 17, 1872	134, 105
Paper and leather board, Manufacture of	C. B. Hutchins	Ann Arbor, Mich	Feb. 21, 1871	112, 045
Paper and other fabrics, Apparatus for sizing, stretching, and coloring.	P. Hild	New York, N. Y	June 11, 1872	127, 692
Paper and other fabrics, Friction-calender for	W. W. Harding and W. H. Soley	Philadelphia, Pa	Feb. 16, 1869	87, 044
Paper and other fabrics incorrodible, Rendering	T. G. Chase	Philadelphia, Pa	Nov. 9, 1858	22, 015
Paper and other fabrics. Machine for drying	F. Beck	New York, N. Y	Dec. 27, 1870	110, 540
Paper and other fabrics, Producing colored prints on	M. Laemmel	Bay Ridge, N. Y	July 4, 1871	116, 720
Paper and other materials with solutions, Process of coating sheets of.	J. C. Crosman	Boston, Mass	Feb. 19, 1867	62, 256
Paper and paper-boards, Machine for making	J. F. Jones	Rochester, N. Y	Oct. 13, 1863	40, 265
Paper and paper-pulp, Manufacture of	J. M. Allen	Marion, Mass	Aug. 22, 1871	118, 177
Paper and paper-pulp, Manufacture of	J. Brown	London, England	May 10, 1859	23, 897
Paper and paper-pulp, Treatment of	T. Taylor	Grove End Road, England	May 16, 1871	114, 880
Paper and print roller	H. J. Smith	Boston, Mass	June 1, 1869	90, 695

Index of patents issued from the United States Patent Office from 1790 *to* 1873, *inclusive*—Continued.

Invention.	Inventor.	Residence.	Date.	No.
Paper and trimming books, Machine for cutting	F. J. Austin	New York, N. Y	June 16, 1841	2, 141
Paper and vegetable fibrous substances, Treating	A. T. Schmidt	Pittsburgh, Pa	Apr. 4, 1871	113, 454
Paper and woven fabrics to produce water-proof sheets and slabs, Treating.	J. Scoffern	Finsbury, England	Jan. 19, 1869	86, 103
Paper, Apparatus for the manufacture of albumenized.	J. Klein	Hesse-Darmstadt, North German Confederation.	July 19, 1870	105, 464
Paper, Apparatus for pasting lining to	B. F. Fields	Beloit, Wis	July 18, 1871	117, 060
Paper, Apparatus for securing pulverized and other materials to.	W. Adamson	Philadelphia, Pa	May 28, 1867	65, 038
Paper, &c., Apparatus for separating wood fiber for.	A. Fickett	Rochester, N. Y	Apr. 26, 1870	102, 239
Paper articles, Apparatus for making	H. Grambo	Philadelphia, Pa	Dec. 3, 1867	71, 742
Paper, &c., Application of the hibiscus moscheutos to the manufacture of.	W. J. Cantello	Philadelphia, Pa	May 13, 1862	35, 215
Paper articles of apparel, Button-hole for	B. M. Smith	New York, N. Y	Mar. 3, 1868	75, 212
Paper articles of wearing-apparel, Button-hole for	W. C. Reeves and L. L. Solomon	Charlestown, Mass	Apr. 7, 1868	76, 347
Paper bag	A. Adams	Chagrin Falls, Ohio	Aug. 2, 1870	105, 877
Paper bag	J. Arkell and B. and A. Smith	Canajoharie, N. Y	Apr. 25, 1865	47, 376
Paper bag	J. Arkell and B. Smith	Canajoharie, N. Y	June 6, 1865	48, 036
Paper bag	L. D. Benner	Boston, Mass	Dec. 24, 1872	134, 244
Paper bag	L. D. Benner	Boston, Mass	Sept. 30, 1873	143, 321
Paper bag	G. H. Chinnock	Brooklyn, N. Y	June 17, 1873	139, 870
Paper bag	L. C. Crowell	West Dennis, Mass	May 28, 1867	65, 176
Paper bag	L. C. Crowell	Boston, Mass	Feb. 20, 1872	123, 811
Paper bag	L. C. Crowell	Boston, Mass	Apr. 8, 1873	137, 533
Paper bag	G. and D. Dare	Auburn, N. Y	Nov. 19, 1867	71, 143
Paper bag	D. and C. F. Manuel	Boston, Mass	Nov. 26, 1867	71, 398
Paper bag	O. W. Stow	Plantsville, Conn	Nov. 4, 1873	144, 238
Paper bag and box making machine	J. Miller, jr	Baltimore, Md	July 9, 1861	32, 777
Paper-bag and envelope making machine	J. A. Smith and S. E. Pettee	Clinton and Foxborough, Mass.	May 1, 1855	12, 786
Paper-bag and material therefor	A. Jameson	Trenton, N. J	June 22, 1869	91, 748
Paper-bag cutter	L. D. Barraud	New York, N. Y	Apr. 24, 1860	27, 959
Paper-bag engine	I. N. Crehore and F. Stiles, jr	Boston and Leicester, Mass	Jan. 25, 1859	22, 707
Paper-bag-folding machine	E. A. Hollingsworth	South Braintree, Mass	Jan. 24, 1865	45, 999
Paper-bag-folding machine	G. J. Moffatt	Brooklyn, N. Y	Mar. 29, 1870	101, 299
Paper-bag machine	A. Adams	Chagrin Falls, Ohio	May 16, 1871	114, 743
Paper-bag machine	C. Amazeen	New York, N. Y	Nov. 17, 1868	84, 076
Paper-bag machine	C. F. Annan	Boston, Mass	June 1, 1869	90, 624
Paper-bag machine	C. F. Annan	Boston, Mass	Dec. 27, 1870	110, 536
Paper-bag machine	C. F. Annan	Boston, Mass	Feb. 14, 1871	111, 802
Paper-bag machine	C. F. Annan	Boston, Mass	Feb. 14, 1871	111, 803
Paper-bag machine	C. F. Annan	Boston, Mass	Nov. 12, 1872	132, 890
Paper-bag machine	C. F. Annan	Boston, Mass	Jan. 7, 1873	134, 580
Paper-bag machine	H. G. Armstrong	Philadelphia, Pa	Oct. 2, 1860	30, 191
Paper-bag machine	H. G. Armstrong	Trenton, N. J	May 20, 1873	139, 104
Paper-bag machine	P. E. Armstrong	Philadelphia, Pa	Feb. 21, 1871	112, 005
Paper-bag machine	J. Arkell	Canajoharie, N. Y	Dec. 26, 1871	122, 099
Paper-bag machine	L. D. Benner	Boston, Mass	Jan. 9, 1872	122, 510
Paper-bag machine	L. D. Benner	Boston, Mass	May 13, 1873	138, 844
Paper-bag machine	T. and J. Bilby, J. Baron, and W. Baron.	Burmley and Rochdale, England.	Oct. 1, 1872	141, 841
Paper-bag machine	B. S. Binney	Somerville, Mass	Oct. 17, 1871	119, 915
Paper-bag machine	J. N., J. P., and S. H. Bryant	Temperanceville, Pa	Sept. 26, 1871	119, 307
Paper-bag machine	B. Cole	Brooklyn, N. Y	Dec. 17, 1872	134, 035
Paper-bag machine	L. C. Crowell	Boston, Mass	Feb. 20, 1872	123, 812
Paper-bag machine	L. C. Crowell	Boston, Mass	Oct. 14, 1873	143, 674
Paper-bag machine	G. Dunham	Unionville, Conn	Aug. 19, 1873	141, 862
Paper-bag machine	B. F. Ellis	Dayton, Ohio	Sept. 5, 1865	49, 736
Paper-bag machine	E. W. Goodale	Clinton, Mass	May 29, 1855	12, 945
Paper-bag machine	W. Goodale	Clinton, Mass	July 12, 1859	24, 734
Paper-bag machine	W. Goodale	Clinton, Mass	Aug. 23, 1859	25, 191
Paper-bag machine	E. W. Goodale	Clinton, Mass	Sept. 12, 1865	49, 951
Paper-bag machine	J. J. Greenough	New York, N. Y	Feb. 3, 1863	37, 573
Paper-bag machine	G. Guild	Saint Louis, Mo	Jan. 23, 1872	123, 013
Paper-bag machine	T. Hotchkiss	Stafford, Conn	Jan. 28, 1873	135, 275
Paper-bag machine	G. L. Jaeger	New York, N. Y	May 7, 1867	64, 537
Paper-bag machine	J. Keller	York County, Pa	Mar. 2, 1858	19, 506
Paper-bag machine	M. E. Knight	Boston, Mass	July 11, 1871	116, 842
Paper-bag machine	H. Law	Chatham, N. J	June 14, 1870	104, 169
Paper-bag machine	W. Liddell	Buffalo, N. Y	Sept. 30, 1873	143, 358
Paper-bag machine	H. C. Lockwood	Baltimore, Md	Mar. 9, 1869	87, 689
Paper-bag machine	N. Lorton and J. S. Davison	Cranberry, N. J	July 5, 1870	105, 099
Paper-bag machine	G. F. Lufberg	New York, N. Y	May 8, 1860	28, 188
Paper-bag machine	G. H. Mallary	New York, N. Y	July 28, 1868	80, 298
Paper-bag machine	G. H. Mallary	Poughkeepsie, N. Y	Nov. 3, 1868	83, 648
Paper-bag machine	C. H. Morgan	Philadelphia, Pa	Feb. 17, 1863	37, 726
Paper-bag machine	H. B. Morris	Burlington, N. J	Oct. 15, 1872	132, 312
Paper-bag machine	M. Murphy	Cincinnati, Ohio	Jan 21, 1873	135, 145
Paper-bag machine	E. B. Olmsted	Washington, D. C	Nov. 5, 1867	70, 601
Paper-bag machine	J. S. Ostrander	Dayton, Ohio	Oct. 21, 1873	143, 925
Paper-bag machine	S. E. Pettee	Philadelphia, Pa	May 29, 1860	28, 537
Paper-bag machine	S. E. Pettee	Philadelphia, Pa	May 5, 1863	38, 452
Paper-bag machine	J. P. Pultz	Plantsville, Conn	Sept. 7, 1869	94, 511
Paper-bag machine	J. P. Raymond	Cincinnati, Ohio	Dec. 2, 1873	145, 125
Paper-bag machine	B. F. Rice	Clinton, Mass	Apr. 28, 1857	17, 184
Paper-bag machine	J. Wells	Hoboken, N. J	Apr. 21, 1863	38, 253
Paper-bag machine	J. Wells	Hoboken, N. J	Sept. 15, 1863	40, 001
Paper-bag machine	J. Wells	Hoboken, N. J	Apr. 12, 1864	42, 313
Paper-bag machine	F. Wolle	Bethlehem, Pa	May 29, 1855	12, 982
Paper-bag machine	F. Wolle	Bethlehem, Pa	July 6, 1858	20, 838
Paper-bag machine	F. Wolle	Bethlehem, Pa	Oct. 26, 1852	9, 355
Paper-bag machines, Pasting-apparatus for	J. Arkell	Canajoharie, N. Y	Nov. 26, 1872	133 395
Paper bag machinery	J. Wells	Brooklyn, N. Y	Mar. 9, 1869	87, 608
Paper-bag-manufacturing tool	S. M. Kirk and E. J. Howlett	Philadelphia, Pa	Feb. 26, 1867	62, 342
Paper bags, Apparatus for making	B. B. Taggart	Watertown, N. Y	Aug. 15, 1865	49, 454
Paper bags, Closing	M. W. Brown	New York, N. Y	Sept. 7, 1869	94, 558
Paper bags, &c., Knife to cut	H. R. David	New York, N. Y	Oct. 5, 1858	21, 657

Index of patents issued from the United States Patent Office from 1790 *to* 1873, *inclusive*—Continued.

Invention.	Inventor.	Residence.	Date.	No.
Paper barrel	W. H. Bailey	New York, N. Y	Mar. 19, 1872	124, 714
Paper-binder, Temporary	F. M. Smith	Syracuse, N. Y	Oct. 4, 1870	107, 968
Paper-binding	W. P. Read	Long Meadow, Mass	Aug. 27, 1867	68, 236
Paper, Bleaching stock for	S. T. Merrill	Beloit, Wis	Nov. 12, 1867	70, 878
Paper, Blotting	N. Floyd	New York, N. Y	Sept. 3, 1872	131, 056
Paper, Bluing compound for the manufacture of	J. Hogben	Cleveland, Ohio	Nov. 2, 1869	96, 321
Paper board, Attachment to roll for calendering	E. A. Seeley	Scotch Plains, N. J	Oct. 19, 1869	96, 039
Paper board for building	C. B. Ayer	Beloit, Wis	Mar. 5, 1872	124, 315
Paper board for building	F. N. Davis	Beloit, Wis	Mar. 19, 1872	124, 795
Paper board for building, &c., Manufacture of	F. N. Davis	Beloit, Wis	July 9, 1872	128, 863
Paper board, Machine for making	J. F. Jones	Rochester, N. Y	Aug. 1, 1865	49, 119
Paper board, Machine for making	J. F. Jones	Rochester, N. Y	Sept. 12, 1865	49, 884
Paper-board manufacture	A. E. Seeley	Scotch Plains, N. J	May 16, 1871	114, 868
Paper board, pipe, &c., Manufacture of	N. C. Szerelmey	Middlesex, Great Britain	Apr. 17, 1866	54, 082
Paper boat	P. S. Tyler	Boston, Mass	Nov. 3, 1868	83, 804
Paper, books, &c., Cutting	J. Woodhouse	Otsego, N. Y	May 30, 1825	
Paper bosom	J. C. Arms	Northampton, Mass	Feb. 27, 1866	52, 812
Paper bosom	T. A. Curtis	Springfield, Mass	Feb. 20, 1866	52, 784
Paper bosom	E. P. Furlong	Portland, Me	Nov. 9, 1869	96, 687
Paper bosom	J. B. Gardiner	Springfield, Mass	Sept. 5, 1865	49, 747
Paper bosom	C. M. Lee	Springfield, Mass	Mar. 27, 1866	53, 455
Paper bosom	W. E. Lockwood	Philadelphia, Pa	Feb. 28, 1865	46, 567
Paper bosom	C. C. Taylor	Springfield, Mass	Feb. 27, 1866	52, 907
Paper bosom and collar	J. C. Arms	Northampton, Mass	June 12, 1866	55, 451
Paper-bosom machine	S. B. Hill	Chicopee, Mass	Apr. 3, 1866	53, 615
Paper box	W. Armour	Belfast, Ireland	June 9, 1868	78, 715
Paper box	E. B. Beecher and W. H. Swift	Westville, Conn., and Wilmington, Del.	Mar. 5, 1872	124, 319
Paper box	P. Beer	Detroit, Mich	May 16, 1871	114, 752
Paper box	H. H. Corbin	New Britain, Conn	Aug. 17, 1869	93, 809
Paper box	C. O. Crosby	New Haven, Conn	Mar. 15, 1870	100, 865
Paper box	H. A. Devendorf	Port Jackson, N. Y	June 8, 1869	91, 096
Paper box	M. W. Dillingham	Amsterdam, N. Y	Feb. 4, 1873	135, 469
Paper box	F. Field	Troy, N. Y	Jan. 18, 1870	98, 860
Paper box	G. M. Hendrickson	Albany, N. Y	Oct. 26, 1869	96, 108
Paper box	A. B. Hendryx	Ansonia, Conn	Mar. 23, 1869	88, 165
Paper box	G. L. Jaeger	New York, N. Y	July 20, 1869	92, 831
Paper box	G. L. Jaeger	New York, N. Y	July 25, 1871	117, 423
Paper box	B. J. Magee and J. F. Wall	Watertown, Mass	Apr. 4, 1871	113, 319
Paper box	H. Matier	Belfast, Ireland	May 9, 1871	114, 582
Paper box	J. W. Millet	Albany, N. Y	Mar. 28, 1865	47, 069
Paper box	B. Osborn	New York, N. Y	Mar. 8, 1870	100, 658
Paper box	B. Osborn	Newark, N. J	Dec. 31, 1872	134, 394
Paper box	C. L. Palmer	Norwich, Conn	Oct. 22, 1872	132, 368
Paper box	J. L. Reber	Philadelphia, Pa	June 22, 1869	91, 666
Paper box	J. Root	New Haven, Conn	Mar. 15, 1870	100, 806
Paper box	J. Root	New Haven, Conn	June 7, 1870	103, 932
Paper box	H. D. Scudder	Amsterdam, N. Y	Apr. 14, 1868	76, 830
Paper box	G. K. Snow	Watertown, Mass	July 8, 1873	140, 736
Paper box	R. N. Stewart	Philadelphia, Pa	Jan. 8, 1867	61, 028
Paper box	J. W. Tuttle	Newton Corner, Mass	Apr. 5, 1870	101, 546
Paper box	J. W. Wilcox	New York, N. Y	Feb. 28, 1871	112, 306
Paper box cutter, Tubular	J. and E. Spooner	New York, N. Y	Aug. 31, 1869	94, 358
Paper-box fastening	D. Crannell	Amsterdam, N. Y	Nov. 11, 1873	144, 513
Paper-box fastening	G. F. Wright	Clinton, Mass	Feb. 20, 1866	52, 800
Paper-box-finishing machine	R. Smith	Sherbrook, Canada	June 22, 1869	91, 780
Paper-box machine	G. R. Clarke	Pompton, N. J	May 7, 1872	126, 378
Paper-box machine	W. Gates, D. J. Lloyd, and S. Miller.	Frankfort, and South Hammond, N. Y.	Dec. 29, 1868	85, 301
Paper-box machine	C. R. Grimm	New York, N. Y	Oct. 1, 1872	131, 751
Paper-box machine	J. Hatfield	Troy, N. Y	Jan. 21, 1862	34, 229
Paper-box machine	C. B. Hatfield	Philadelphia, Pa	July 23, 1867	67, 050
Paper-box machine	C. B. Hatfield	Philadelphia, Pa	July 23, 1867	67, 051
Paper-box machine	C. B. Hatfield	Philadelphia, Pa	Mar. 8, 1870	100, 621
Paper-box machine	R. L. Hawes	Worcester, Mass	Jan. 16, 1855	12, 255
Paper-box machine	H. R. Heyl	Philadelphia, Pa	Oct. 31, 1871	120, 436
Paper-box machine	H. R. Heyl	Philadelphia, Pa	Oct. 8, 1872	132, 076
Paper-box machine	H. R. Heyl	Philadelphia, Pa	Oct. 8, 1872	132, 077
Paper-box machine	H. R. Heyl and A. Brehmer	Philadelphia, Pa	Oct. 8, 1872	132, 078
Paper-box machine	G. L. Jaeger	New York, N. Y	Feb. 25, 1873	136, 246
Paper-box machine	J. M. D. Keating and T. V. Waymouth.	New York, N. Y	June 8, 1869	91, 025
Paper-box machine	L. Koch	New York, N. Y	Mar. 13, 1855	12, 511
Paper-box machine	T. C. Luther	Waterbury, Conn	June 27, 1865	48, 490
Paper-box machine	C. A. Maxfield	New York, N. Y	Feb. 6, 1866	52, 432
Paper-box machine	C. A. Maxfield	New York, N. Y	June 21, 1870	104, 477
Paper-box machine	C. A. Maxfield	New York, N. Y	Oct. 3, 1871	119, 631
Paper-box machine	W. Orr, jr., and G. F. Wright	Clinton, Mass	Aug. 13, 1867	67, 669
Paper-box machine	B. F. Quinby	Boston, Mass	Apr. 27, 1869	89, 433
Paper-box machine	D. Simmons, jr	New York, N. Y	Feb. 22, 1870	100, 198
Paper-box machine	D. Simmons	New York, N. Y	Mar. 1, 1870	100, 459
Paper-box machine	R. Smith	Sherbrooke, Canada	Nov. 17, 1868	84, 246
Paper-box machine	J. and E. Spooner	New York, N. Y	June 29, 1869	91, 983
Paper-box machine	S. B. Terry	Terrysville, Conn	Sept. 6, 1859	25, 373
Paper-box machine	J. Worell	Philadelphia, Pa	Mar. 19, 1872	124, 872
Paper boxes, Apparatus for making	R. T. Barton	New Haven, Conn	Nov. 27, 1866	59, 944
Paper boxes, Apparatus for making	F. Leclére	Boston, Mass	Dec. 8, 1868	84, 835
Paper boxes, Finishing	S. Wheeler and E. Jerome	Albany, N. Y	Mar. 21, 1871	112, 874
Paper boxes from pulp, Manufacture of	S. Wheeler and E. Jerome	Albany, N. Y	Apr. 13, 1869	88, 827
Paper boxes, Machinery for making	C. O. Crosby	New Haven, Conn	Mar. 29, 1870	101, 354
Paper boxes, Machinery for making	F. Gates	Vineland, N. J	Apr. 5, 1870	101, 455
Paper boxes, Making blanks for	S. Wheeler and E. Jerome	Albany, N. Y	July 16, 1867	66, 920
Paper boxes, Manufacture of	S. Wheeler and E. Jerome	Albany, N. Y	Feb. 12, 1867	61, 968
Paper boxes, Manufacturing angular	E. Waters	Troy, N. Y	Feb. 2, 1858	19, 270
Paper braid	C. E. Richards	North Attleborough, Mass	June 18, 1867	65, 833
Paper, Burnisher for enameled	S. Shepard and A. M. George	Nashua, N. H	Oct. 6, 1868	82, 883
Paper by machinery, Making	T. Gilpin	Philadelphia, Pa	Dec. 24, 1816	
Paper by machinery, Manufacturing	R. Fairchild	Trumbull, Conn	May 4, 1829	

Index of patents issued from the United States Patent Office from 1790 *to* 1873, *inclusive*—Continued.

Invention.	Inventor.	Residence.	Date.	No.
Paper by single sheets, Separating	J. P. Comly	Dayton, Ohio	Mar. 22, 1853	9, 623
Paper case or envelope	J. W. Wilcox	New York, N. Y	Apr. 5, 1864	42, 247
Paper cask	J. W. Jarboe	Green Point, N. Y	Apr. 14, 1868	76, 771
Paper-calendering machine	S. Moore	Sudbury, Mass	Dec. 12, 1871	121, 800
Paper-calendering machine	T. B. De Forest	Birmingham, Conn	Apr. 9, 1872	125, 548
Paper-calendering roll	J. H. Garfield	Newton Lower Falls, Mass	Feb. 27, 1872	124, 048
Paper cap	M. G. Imbach and I. Weidenman.	Hartford, Conn	July 28, 1868	80, 352
Paper cap	N. Waterman and A. T. Perkins.	Toledo, Ohio	June 30, 1868	79, 420
Paper cap	P. S. White	Providence, R. I	Nov. 22, 1870	109, 566
Paper, Cap and other	D. D. Foley and J. J. Johnson	Washington, D. C	Oct. 24, 1871	120, 152
Paper, Carbonized	L. H. Rogers	New York, N. Y	Oct. 15, 1872	132, 324
Paper, Cigarette	F. X. Hazman and L. L. Arnold	New York, N. Y	July 25, 1865	48, 936
Paper-clamp	A. Faber	Washington, D. C	Feb. 6, 1866	52, 403
Paper-clamp	A. Palmer	Lee, Mass	Oct. 12, 1858	21, 775
Paper-clamp and inkstand combined	E. J. Toof	Fort Madison, Iowa	Apr. 21, 1868	77, 131
Paper-clasp	J. C. Arms	Northampton, Mass	Sept. 1, 1868	81, 727
Paper, Cleaning and preparing rags for the manufacture of.	W. E. Newton	London, England	May 10, 1870	102, 854
Paper-cleaning machine, Rag	W. Debit	Hartford, Conn	July 21, 1835	....
Paper-clip	W. A. Arnberg	Chicago, Ill	July 6, 1869	92, 141
Paper-clip	C. E. Candee	Jersey City, N. J	Feb. 11, 1868	74, 301
Paper-clip	J. M. Crull	Harrisburgh, Pa	Mar. 28, 1871	113, 026
Paper-clip	G. W. Emerson	Chicago, Ill	Jan. 4, 1870	98, 479
Paper-clip	A. H. Fatzinger	Washington, N. J	May 26, 1868	78, 271
Paper-clip	E. A. Franklin	Brenham, Tex	June 4, 1872	127, 472
Paper-clip	J. A. Harvey and C. A. Mills	Dubuque, Iowa	July 1, 1856	15, 232
Paper-clip	T. Orton	Chicago, Ill	May 20, 1873	139, 019
Paper-clip	J. H. Parsons	Quincy, Mich	Dec. 3, 1867	71, 635
Paper-clip	M. A. Wheeler	San Francisco, Cal	May 27, 1873	139, 285
Paper-clip, Suspension	W. Stokes	New York, N. Y	July 12, 1870	105, 274
Paper, cloth, and other materials, Coating and decorating.	F. Beck	New York, N. Y	Jan. 3, 1871	110, 727
Paper-cloth bosom	C. Arnold	Chicago, Ill	July 3, 1866	55, 974
Paper, cloth, &c., Coating and ornamenting the surface of.	F. Beck	New York, N. Y	Nov. 22, 1870	109, 486
Paper, cloth, &c., Staining and printing	B. Mesteyer	New York	June 28, 1815	....
Paper collar	W. S. Bell, jr	Boston, Mass	June 13, 1865	48, 148
Paper collar	G. F. Bigelow	Chicago, Ill	June 27, 1865	48, 359
Paper collar	C. A. Brown	Troy, N. Y	May 13, 1873	138, 849
Paper collar	C. K. Brown	Troy, N. Y	Jan. 20, 1863	37, 431
Paper collar	L. M. Crane	Ballston Spa, N. Y	Sept. 18, 1866	58, 223
Paper collar	W. L. Duff	Quincy, Ill	May 30, 1865	48, 016
Paper collar	A. A. Evans	Boston, Mass	May 26, 1863	38, 664
Paper collar	N. Evans, jr	Boston, Mass	Mar. 14, 1871	112, 571
Paper collar	G. A. Goldsmith	New York, N. Y	Apr. 12, 1870	101, 855
Paper collar	S. S. Gray	Boston, Mass	Apr. 14, 1863	38, 160
Paper collar	S. S. Gray	Boston, Mass	July 5, 1864	43, 400
Paper collar	S. S. Gray	Boston, Mass	Jan. 23, 1866	52, 163
Paper collar	S. S. Gray	Boston, Mass	July 11, 1871	116, 948
Paper collar	W. B. Harris	Springfield, Mass	May 15, 1866	54, 720
Paper collar	S. F. Hilton	Providence, R. I	Sept. 12, 1871	118, 803
Paper collar	S. B. Hutchinson	Nashua, N. H	Aug. 8, 1865	49, 347
Paper collar	A. Kaufmann	New York, N. Y	Jan. 29, 1867	61, 669
Paper collar	E. E. Mack	Albany, N. Y	Mar. 29, 1870	101, 286
Paper collar	G. W. Ray	Springfield, Mass	Jan. 8, 1867	61, 100
Paper collar	C. E. Richards	North Attleborough, Mass	Feb. 13, 1866	52, 607
Paper collar	S. J. Shaw	Marlborough, Mass	Oct. 24, 1865	50, 663
Paper collar	D. F. Smith	Sullivan, N. H	Apr. 12, 1870	101, 932
Paper collar	G. K. Snow	Watertown, Mass	Feb. 7, 1865	46, 310
Paper collar	C. Spofford and W. S. Bell, jr	Boston, Mass	Aug. 30, 1864	44, 022
Paper collar	C. Spofford and V. Fogerty	Boston, Mass	Dec. 27, 1864	45, 681
Paper collar	S. S. Stone	Troy, N. Y	July 31, 1866	56, 827
Paper collar and bosom	M. Pinner	New York, N. Y	Oct. 9, 1866	58, 673
Paper collar and cuff	H. Parmenter	New York, N. Y	Apr. 16, 1872	125, 842
Paper-collar band	G. K. Snow	Watertown, Mass	Jan. 30, 1866	52, 371
Paper-collar binding and folding apparatus	W. E. Lockwood	Philadelphia, Pa	Oct. 11, 1864	44, 640
Paper-collar binding and folding apparatus	J. F. Schuyler	Philadelphia, Pa	Feb. 24, 1863	37, 792
Paper-collar box	L. Churchill	Troy, N. Y	Feb. 7, 1871	111, 514
Paper-collar box	S. Kaufmann	New York, N. Y	June 28, 1870	104, 853
Paper-collar box	E. A. Locke and W. N. Weeden	Boston, Mass	July 20, 1869	92, 850
Paper-collar-boxing machine	D. Stoner and J. Ligwalt, jr	Chicago, Ill	Dec. 4, 1866	60, 278
Paper-collar button-hole	J. Merwin	New York, N. Y	Apr. 23, 1867	64, 022
Paper-collar button-hole	B. F. Porter	Nashua, N. H	Feb. 13, 1866	52, 599
Paper-collar button-hole	W. H. Robinson	Rochester, N. Y	Jan. 5, 1869	85, 535
Paper-collar button-hole	D. Stoner	Chicago, Ill	Apr. 10, 1866	53, 897
Paper-collar button-hole	N. F. Whiting	Lawrence, Mass	June 19, 1866	55, 753
Paper-collar button-hole patch, Self-sealing	T. Griffin	Roxbury, Mass	Aug. 21, 1866	57, 314
Paper-collar card, &c., from printing or embossing press, Removing.	T. Tebbetts	New York, N. Y	Oct. 17, 1865	50, 514
Paper collar, Colored	S. W. H. Ward	New York, N. Y	Aug. 2, 1864	43, 728
Paper-collar-creasing apparatus	T. Tebbetts	New York, N. Y	Nov. 13, 1866	59, 680
Paper-collar-cutting die	E. Jeffers	Boston, Mass	Jan. 26, 1869	86, 162
Paper-collar-cutting die	G. K. Snow	Watertown, Mass	Nov. 28, 1865	51, 235
Paper-collar-cutting machine	A. L. Elliott	Boston, Mass	Jan. 24, 1871	111, 113
Paper-collar-cutting machine	D. M. Smyth	Orange, N. J	Nov. 19, 1867	71, 074
Paper-collar die, Adjustable	C. Spofford	Boston, Mass	Nov. 19, 1867	71, 076
Paper-collar fastening	C. R. Everson	Palmyra, N. Y	June 18, 1867	65, 894
Paper-collar fastening	S. Hodgins	Saint Louis, Mo	July 31, 1866	56, 854
Paper-collar finishing and boxing apparatus	J. J. Currier and S. Wells, jr	Boston, Mass	Dec. 18, 1866	60, 484
Paper-collar-folding apparatus	A. H. Hook	New York, N. Y	Mar. 7, 1865	46, 669
Paper-collar-folding machine	J. J. Currier	Boston, Mass	May 22, 1866	55, 005
Paper-collar-folding machine	C. E. Moore	Roxbury, Mass	Oct. 23, 1866	59, 134
Paper-collar-folding machine	E. Vossnack	New York, N. Y	Oct. 10, 1865	50, 426
Paper collar, Ladies'	S. S. Gray	Boston, Mass	Feb. 7, 1865	46, 232
Paper-collar machine	C. K. Brown and W. Wright	Troy, N. Y	Apr. 24, 1866	54, 244

Index of patents issued from the United States Patent Office from 1790 to 1873, inclusive—Continued.

Invention.	Inventor.	Residence.	Date.	No.
Paper-collar machine	G. W. Cilley and G. H. Spaulding.	Norwich, Conn	Jan. 28, 1868	73, 693
Paper-collar machine	C. H. Dennison	New York, N. Y	Jan. 2, 1872	122, 444
Paper-collar machine	M. P. Dorsch	New York, N. Y	Feb. 14, 1865	46, 340
Paper-collar machine	J. C. Ford	Cambridge, Mass	Aug. 21, 1866	57, 308
Paper-collar machine	J. W. Griswold and J. Sigwalt, jr	Chicago, Ill	Oct. 16, 1866	58, 812
Paper-collar machine	H. F. Knapp	New York, N. Y	May 8, 1866	54, 561
Paper-collar machine	H. F. Knapp	New York, N. Y	Nov. 6, 1866	59, 410
Paper-collar machine	G. W. Livermore	Cambridge, Mass	Nov. 1, 1864	44, 871
Paper-collar machine	T. McSpedon	New York, N. Y	Dec. 27, 1864	45, 621
Paper-collar machine	C. E. Moore and M. L. Wyman	Boston and Melrose, Mass	Sept. 13, 1870	107, 399
Paper-collar machine	S. Shepherd	Nashua, N. H	Nov. 2, 1869	96, 491
Paper-collar machine	S. Shepherd and A. M. George	Nashua, N. H	Sept. 12, 1865	49, 927
Paper-collar machine	D. M. Smyth	New York, N. Y	Dec. 5, 1865	51, 361
Paper-collar machine	G. K. Snow	Watertown, Mass	Nov. 28, 1865	51, 233
Paper-collar machine	G. K. Snow	Watertown, Mass	Nov. 28, 1865	51, 234
Paper-collar machine	C. Spofford and C. H. Montague	Boston, Mass	Dec. 15, 1868	85, 037
Paper-collar machine	W. H. Walton	Brooklyn, N. Y	July 2, 1867	66, 429
Paper-collar machine	O. F. Washburn	Bridgewater, Vt	Dec. 24, 1867	72, 705
Paper-collar machinery, Die for	G. Harrington and G. D. Rollins.	Springfield, Mass	May 20, 1873	139, 146
Paper-collar-making apparatus	D. M. Smyth	New York, N. Y	Aug. 16, 1864	43, 870
Paper-collar-molding machine	S. S. Gray	Boston, Mass	Mar. 16, 1869	87, 841
Paper-collar press	J. H. Darlington	New York, N. Y	Feb. 20, 1866	52, 785
Paper-collar-shaping apparatus	W. E. Lockwood and H. Howson.	Philadelphia, Pa	Sept. 12, 1865	49, 952
Paper-collar-stretching machine	C. Spofford and W. S. Bell, jr	Boston, Mass	Feb. 14, 1865	46, 400
Paper-collar, Turn-down enameled	J. H. Hoffman	New York, N. Y	Jan. 24, 1865	45, 998
Paper collars and cuffs, Manufacture of	D. A. Alden	Roxbury, Mass	Nov. 27, 1866	59, 938
Paper collars, cuffs, &c., Fabric for the manufacture of.	J. Restein	Philadelphia, Pa	Dec. 29, 1868	85, 397
Paper-collars, Device for applying cloth patch to	J. P. Courtney and C. Redmayne.	Brooklyn, N. Y	Aug. 18, 1868	81, 255
Paper collars, Folding	W. S. Bell	Boston, Mass	Feb. 7, 1865	46, 291
Paper collars, Folding	J. H. Darlington	New York, N. Y	Aug. 2, 1864	43, 737
Paper collars, Folding	M. P. Dorsch	New York, N. Y	Oct. 11, 1864	44, 615
Paper collars, Machine for applying cloth patch to.	C. H. Denison	Springfield, Mass	June 30, 1868	79, 460
Paper collars, Machine for patching, punching, and embossing button-holes of.	C. Lang	Jersey City, N. J	Feb. 28, 1871	112, 155
Paper collars, Making	A. F. Gray	Boston, Mass	Nov. 8, 1864	44, 948
Paper collars, &c., Manufacture of	G. W. Ray	Springfield, Mass	May 1, 1866	54, 404
Paper collars, Manufacture of sweat-proof	J. H. Hoffman	New York, N. Y	Apr. 4, 1865	47, 107
Paper collars, Pressing	E. E. Ranous	New York, N. Y	Aug. 2, 1864	43, 741
Paper, Colored	G. La Monte, G. G. Saxe, and C. H. Clayton.	New York, N. Y	Dec. 19, 1871	121, 946
Paper, Coloring and drying	C. C. Fitzgerald	New York, N. Y	June 7, 1870	104, 006
Paper, &c., coloring apparatus	C. Williams	Philadelphia, Pa	Sept. 21, 1858	21, 584
Paper-coloring machine	C. K. Brown	Troy, N. Y	Dec. 29, 1868	85, 426
Paper, Composition for chemical-telegraph	G. Little	Rutherford Park, N. J	Aug. 27, 1872	130, 810
Paper, &c, Composition for saturating	A. T. Boon and J. Stafford	Galesburgh, Ill	Oct. 8, 1867	69, 618
Paper, Composition for treating vegetable	C. S. Dikeman	New York, N. Y	Oct. 21, 1862	36, 704
Paper, Compound for making water-proof	C. S. Rauh	Boston, Mass	Oct. 26, 1869	96, 148
Paper cop-tube	A. McCausland	Providence, R. I	July 7, 1857	17, 746
Paper, cordage, twine, &c., Manufacture of	J. H. McConnell	Springfield, Ill	Feb. 7, 1871	111, 661
Paper corset	J. H. Beal, E. J. Sawyer, and G. S. Webster.	Boston, Mass	July 30, 1867	67, 250
Paper cuff	S. A. Bemis, A. E. Foth, and J. W. Goodrich.	Springfield, Mass	Oct. 25, 1870	108, 676
Paper cuff	G. K. Snow	Watertown, Mass	May 25, 1869	90, 599
Paper cuff or wristband	A. A. Evans	Boston, Mass	July 31, 1866	56, 737
Paper-cutter	C. Cropper	Cincinnati, Ohio	Mar. 5, 1867	62, 530
Paper-cutter	St. C. Denny and I. L. G. Rice	Pittsburgh, Pa., and Cambridge, Mass.	Nov. 1, 1870	108, 891
Paper-cutter	J. Hatch	South Windham, Conn	June 22, 1869	91, 742
Paper-cutter	I. L. G. Rice	Cambridge, Mass	Nov. 15, 1870	109, 249
Paper-cutter	S. R. Timby	Saratoga, N. Y	Jan. 21, 1868	73, 674
Paper-cutter	J. Worrell	Philadelphia, Pa	June 13, 1871	115, 922
Paper-cutter and rule, Combined	F. E. Oliver	New York, N. Y	Feb. 26, 1861	31, 556
Paper-cutter and ruler	S. W. Wilcox	Mendon, Mass	Oct. 13, 1868	83, 115
Paper-cutter, Book-binder's	M. Riehl	Philadelphia, Pa	Oct. 2, 1866	58, 482
Paper-cutter, Desk	J. W. Wetmore	Erie, Pa	May 14, 1872	126, 659
Paper-cutter for continuous sheets	G. Keeble	Suffolk, Great Britain	Oct. 2, 1866	58, 563
Paper-cutter gage	C. C. Child	Boston, Mass	Feb. 15, 1870	99, 759
Paper-cutter, Rotary	H. S. Miller	Philadelphia, Pa	May 20, 1873	139, 176
Paper, Cutting	J. P. Farnum	Andover, Mass	Dec. 28, 1852	9, 503
Paper cutting and arranging machinery	J. C. Kneeland and G. M. Phelps	Troy, N. Y	Sept. 12, 1848	5, 767
Paper cutting and folding machine	C. Chambers, jr	Philadelphia, Pa	Apr. 4, 1871	113, 256
Paper cutting and folding machine	C. Chambers, jr., and W. Mendham.	Philadelphia, Pa	Apr. 4, 1871	113, 257
Paper cutting and folding machine	E. H. McArthur	Hillsdale, N. J	Sept. 3, 1867	68, 450
Paper cutting and folding machine	C. Moore	Hartford, Conn	Dec. 16, 1856	16, 256
Paper cutting and punching machine	J. T. S. Smith	New York, N. Y	Aug. 13, 1867	67, 813
Paper cutting and scoring machine	G. Bates	Philadelphia, Pa	Nov. 7, 1871	120, 697
Paper, Cutting and trimming edges of	J. Bateman	Harvard, Mass	Mar. 12, 1834	....
Paper cutting and twisting machine	I. P. Tice	New York, N. Y	Aug. 16, 1864	43, 874
Paper-cutting implement	J. Cook	Grand Manan, Canada	June 18, 1872	128, 021
Paper-cutting machine	J. L. Allen	New York, N. Y	Apr. 7, 1868	76, 373
Paper-cutting machine	J. Arkell	Canajoharie, N. Y	Aug. 22, 1871	118, 327
Paper-cutting machine	G. Bates	Philadelphia, Pa	Jan. 31, 1871	111, 423
Paper-cutting machine	E. P. Beckwith	Windham, Conn	Dec. 19, 1865	51, 669
Paper-cutting machine	M. B. Bigelow	Boston, Mass	July 13, 1858	20, 858
Paper-cutting machine	J. Bole	Grand Rapids, Mich	Aug. 2, 1870	105, 894
Paper-cutting machine	S. Brown	Philadelphia, Pa	Mar. 7, 1871	112, 414
Paper-cutting machine	E. Burroughs	Rochester, N. Y	Aug. 9, 1859	24, 989
Paper-cutting machine	S. M. Clark	Washington, D. C	July 16, 1872	129, 319
Paper-cutting machine	J. M. Clintic	Chambersburgh, Pa	Dec. 17, 1825	....
Paper-cutting machine	J. E. Coffin	Portland, Me	Jan. 3, 1871	110, 631
Paper-cutting machine	E. Cowles	Cleveland, Ohio	Oct. 10, 1871	119, 827

Index of patents issued from the United States Patent Office from 1790 *to* 1873, *inclusive*—Continued.

Invention.	Inventor.	Residence.	Date.	No.
Paper-cutting machine	A. W. Currier	Grand Rapids, Mich	Mar. 8, 1870	100, 506
Paper-cutting machine	A. T. De Puy	New York, N. Y	Oct. 29, 1872	132, 525
Paper-cutting machine	T. B. Dooley	Boston, Mass	May 28, 1872	127, 226
Paper-cutting machine	A. Dougherty	Brooklyn, N. Y	June 17, 1862	35, 592
Paper-cutting machine	S. Ellsworth	Lacon, Ill	Oct. 20, 1868	83, 143
Paper-cutting machine	F. A. Fletcher	Newark, Del	Nov. 28, 1871	121, 239
Paper-cutting machine	H. A. Gage	Manchester, N. H	Mar. 1, 1870	100, 391
Paper-cutting machine	H. Gilman	Troy, N. Y	Sept. 4, 1849	6, 692
Paper-cutting machine	J. L. Gregorie	Chicago, Ill	Mar. 5, 1872	124, 352
Paper-cutting machine	A. Hardy	Boston, Mass	Apr. 12, 1870	101, 873
Paper-cutting machine	F. Hesse	Bethlehem, Pa	July 19, 1853	9, 872
Paper-cutting machine	P. H. Hopkins	Brooklyn, N. Y	Feb. 20, 1872	123, 901
Paper-cutting machine	T. W. Houchin	Worcester, Mass	Dec. 20, 1859	26, 499
Paper-cutting machine	F. B. Howell	Lockport, Ohio	May 31, 1830	....
Paper-cutting machine	H. Law	New York, N. Y	Sept. 16, 1856	15, 738
Paper-cutting machine	H. Law	Chatham, N. J	Apr. 5, 1864	42, 200
Paper-cutting machine	H. Law	Chatham, N. J	Oct. 27, 1868	83, 513
Paper-cutting machine	J. Leviness	Brooklyn, N. Y	Aug. 1, 1871	117, 650
Paper-cutting machine	J. Leviness and P. Van Horn	Brooklyn, N. Y	Mar. 19, 1872	124, 690
Paper-cutting machine	J. E. Mallory	New York, N. Y	Sept. 21, 1852	9, 276
Paper-cutting machine	C. Montague	Hartford, Conn	Mar. 13, 1866	53, 169
Paper-cutting machine	C. Montague	Boston, Mass	Jan. 4, 1870	98, 611
Paper-cutting machine	C. W. L. Montague	Brooklyn, N. Y	Jan. 16, 1872	122, 844
Paper-cutting machine	C. Paine	Boston, Mass	Mar. 14, 1871	112, 623
Paper-cutting machine	E. Pine	Troy, N. Y	July 20, 1831	....
Paper-cutting machine	T. C. Robinson	Boston, Mass	Nov. 23, 1869	97, 120
Paper-cutting machine	T. C. Robinson	Boston, Mass	Dec. 14, 1869	97, 816
Paper-cutting machine	T. C. Robinson	Boston, Mass	May 2, 1871	114, 340
Paper-cutting machine	G. H. Sanborn	Boston, Mass	Dec. 3, 1867	71, 793
Paper-cutting machine	G. H. Sanborn	New York, N. Y	Nov. 24, 1868	84, 306
Paper-cutting machine	G. H. Sanborn	New York, N. Y	Nov. 24, 1868	84, 307
Paper-cutting machine	F. Schoettle	Philadelphia, Pa	July 9, 1872	128, 817
Paper-cutting machine	E. R. and T. W. Sheridan	New York, N. Y	Aug. 27, 1872	130, 873
Paper-cutting machine	E. R. and T. W. Sheridan	New York, N. Y	Oct. 15, 1872	132, 327
Paper-cutting machine	J. Shugert	Quincy, Pa	Dec. 14, 1830	....
Paper-cutting machine	H. Skidmore	Mount Vernon, N. Y	Aug. 11, 1868	81, 021
Paper-cutting machine	F. B. Sweetland	New Haven, Conn	Mar. 1, 1870	100, 465
Paper-cutting machine	J. F. and G. W. Tapley	Springfield, Mass	July 24, 1866	56, 679
Paper-cutting machine	W. H. Topham	New York, N. Y	Feb. 7, 1871	111, 702
Paper-cutting machine	F. L. Walker	Boston, Mass	Nov. 23, 1869	97, 140
Paper-cutting machine	G. A. Walker	Boston, Mass	Aug. 22, 1871	118, 308
Paper-cutting machine	G. A. Walker	Boston, Mass	Mar. 19, 1872	124, 774
Paper-cutting machine	G. A. Walker	Boston, Mass	Dec. 9, 1873	145, 464
Paper-cutting machine	B. Weaver	Springfield, Mass	Mar. 5, 1872	124, 236
Paper-cutting machine	C. Wells and H. Barth	Cincinnati, Ohio	July 25, 1865	49, 018
Paper-cutting machine	J. Worell	Philadelphia, Pa	Mar. 12, 1872	124, 650
Paper-cutting machine, Rotary	W. Bullock	Philadelphia, Pa	Mar. 1, 1870	100, 367
Paper-cutting machines, Clamp for	E. T. Hazeltine and C. Park	Warren, Pa	Apr. 15, 1873	137, 920
Paper-cutting machinery	Z. M. Crane	Dalton, Mass	July 18, 1848	5, 670
Paper-cutting machinery	N. Gavit	Philadelphia, Pa	May 9, 1854	10, 889
Paper-cutting machinery	M. Wilder	Peterborough, N. H	July 18, 1848	5, 672
Paper-cutting machinery	G. L. Wright	Springfield, Mass	July 18, 1848	5, 671
Paper-cutting shears	C. Brombacher	New York, N. Y	July 11, 1865	48, 650
Paper-cutting shears	W. P. Deane	Canton, Mass	May 27, 1873	139, 373
Paper, Cylinder-molds for making	J. F. Jones	Rochester, N. Y	May 26, 1863	38, 684
Paper-dampener	J. A. Lynch	Boston, Mass	Apr. 7, 1868	76, 479
Paper-damping apparatus	C. A. Waterbury	New York, N. Y	Apr. 27, 1859	20, 111
Paper damping and printing machine	A. Overend	Philadelphia, Pa	Mar. 14, 1854	10, 627
Paper-damping machine	A. Dougherty	New York, N. Y	July 9, 1861	32, 759
Paper-damping machine	R. Martin	Philadelphia, Pa	Nov. 27, 1860	30, 788
Paper, Designer's	A. Akeroyd	Manchester, N. H	Dec. 9, 1873	145, 269
Paper, Die for cutting out ornamental forms of	A. Delkescamp	Brooklyn, N. Y	Mar. 25, 1873	137, 063
Paper doll	J. T. Brown	Albion, N. Y	May 14, 1872	126, 622
Paper-dryer	A. E. Harding	Middletown, Ohio	Feb. 9, 1869	86, 666
Paper-dryer	E. A. Seeley	Scotch Plains, N. J	Oct. 28, 1873	143, 999
Paper-dryer	W. H. Soley	Philadelphia, Pa	May 30, 1871	115, 375
Paper dryer, Flock or velvet	T. A. Blanchard	New York, N. Y	Aug. 13, 1872	130, 406
Paper-dryer regulator	D. Crosby	Hampden, Me	Mar. 7, 1871	112, 422
Paper, Drying	J. Hartin	New York, N. Y	Aug. 9, 1853	9, 918
Paper, Drying	H. Howe	Shirley, Mass	Mar. 12, 1836	....
Paper drying and finishing apparatus	H. Dodge	Albany, N. Y	Nov. 14, 1871	120, 810
Paper drying and pressing machine	J. North	Middletown, Conn	Apr. 14, 1857	17, 050
Paper-drying cylinder	H. P. Howe	Shirley, Mass	Sept. 20, 1836	....
Paper-drying machine	J. Hoyt	Cleveland, Ohio	July 23, 1861	32, 870
Paper-drying machine	G. S. Rogers	Thetford Centre, Vt	Jan. 24, 1865	46, 026
Paper-drying machine	N. W. Taylor and J. W. Brightman.	Cleveland, Ohio	June 23, 1863	38, 993
Paper-drying machinery	E. L. Perkins	Roxbury, Mass	June 7, 1859	24, 328
Paper, Drying sized	J. Kingsland, jr., and N. White	Saugerties and New York, N. Y.	Aug. 10, 1852	9, 186
Paper, Drying sized	J. Norton, jr	Boston, Mass	Apr. 25, 1846	4, 480
Paper, Drying thick	E. and J. R. Cushman	Amherst, Mass	Aug. 1, 1854	11, 411
Paper during manufacture, Apparatus for smoothing and pressing.	D. Crosby	Hampden, Me	Feb. 14, 1871	111, 820
Paper-engine	W. Dickinson	Worcester, Mass	Sept. 3, 1840	1, 760
Paper-engine	E. Hawkins	Islip, N. Y	Jan. 11, 1870	98, 691
Paper-engine bed-plate	W. Clarke	Dayton, Ohio	Oct. 9, 1849	6, 784
Paper-engine grinding-cylinder	C. T. Frost	Medfield, Mass	Apr. 3, 1866	53, 598
Paper-engine washer	C. Rice	Watertown, N. Y	Apr. 26, 1834	....
Paper-fastener	C. L. Alexander	Washington, D. C	Oct. 2, 1866	58, 365
Paper-fastener	Z. W. Denham	Washington, D. C	May 2, 1865	47, 526
Paper-fastener	J. C. Jensen	Chicago, Ill	Apr. 16, 1872	125, 682
Paper-fastener	A. J. Kletzker	Saint Louis, Mo	Nov. 3, 1868	83, 640
Paper-fastener	J. W. McGill	Washington, D. C	Aug. 20, 1867	67, 895
Paper-fastener	G. G. W. Morgan	Washington, D. C	Oct. 9, 1866	58, 667
Paper-fastener	A. Smith	Perrysburgh, Ohio	July 5, 1864	43, 435
Paper-fastener	J. F. Tapley	Springfield, Mass	May 11, 1869	89, 898
Paper-fastener	W. M. Tileston	New York, N. Y	Sept. 15, 1868	82, 181

Index of patents issued from the United States Patent Office from 1790 to 1873, inclusive—Continued.

Invention.	Inventor.	Residence.	Date.	No.
Paper-fastener, Metallic	G. W. McGill	Washington, D. C	July 24, 1866	56, 587
Paper-fasteners, Press for attaching	G. W. McGill	Washington, D. C	Aug. 13, 1867	67, 665
Paper-fastening	J. M. Keep	New York, N. Y	Apr. 16, 1867	63, 905
Paper-fastening	M. H. N. Kendig	Washington, D. C	July 23, 1867	66, 968
Paper-feeder	E. Allen	Newark, N. J	Sept. 15, 1863	39, 872
Paper-feeder	J. T. Ashley	Brooklyn, E. D., N. Y	Sept. 13, 1870	107, 320
Paper-feeder	J. T. Ashley	Brooklyn, E. D., N. Y	Oct. 4, 1870	107, 851
Paper-feeder	J. A. Graves	Washington, D. C	Apr. 4, 1871	113, 291
Paper-feeder	J. H. Gray and W. B. Turner	Saint Anthony, Minn	Dec. 28, 1869	98, 253
Paper-feeder	O. Norelius	Minneapolis, Minn	Feb. 22, 1870	100, 059
Paper-feeding device	A. A. Dunk	Philadelphia, Pa	Nov. 12, 1872	132, 904
Paper-feeding machine	E. R. Andrews, R. B. Randall, and W. H. Clague.	Rochester, N. Y	Apr. 25, 1871	114, 087
Paper-feeding machine	J. T. Ashley	Williamsburgh, N. Y	May 20, 1873	138, 983
Paper-feeding machine	J. T. and F. Ashley	Brooklyn, N. Y	Oct. 21, 1873	143, 740
Paper-feeding machine	A. M. Crane	Staunton, Va	Mar. 12, 1872	124, 550
Paper-feeding machine	D. Dick	New York, N. Y	Jan. 30, 1872	123, 157
Paper-feeding machine	M. E. Knight	Boston, Mass	Nov. 15, 1870	109, 224
Paper-feeding machine	W. Mowry	Minneapolis, Minn	Feb. 2, 1869	86, 575
Paper-feeding machine	M. Piedra	Williamsburgh, N. Y	Mar. 11, 1873	136, 668
Paper-feeding machine	S. Scholfield and C. E. Baker	Providence, R. I., and Mont Clair, N. J.	Feb. 4, 1873	135, 597
Paper-feeding machine	S. Scholfield and C. E. Baker	Providence, R. I., and Mont Clair, N. J.	Feb. 4, 1873	135, 598
Paper-feeding machine	S. Scholfield and C. E. Baker	Providence, R. I., and Mont Clair, N. J.	Feb. 4, 1873	135, 599
Paper-feeding machine	C. M. Wielings	San Francisco, Cal	Dec. 16, 1873	145, 542
Paper felt or wadding	W. W. Glentworth and W. H. Gandey.	Philadelphia, Pa., and Lambertville, N. J.	Nov. 23, 1869	97, 189
Paper, Figure-tinted	G. La Monte and G. G. Saxe	New York, N. Y	Feb. 20, 1872	123, 782
Paper, Finishing	J. White and L. Gale	Newburgh, Vt	Feb. 28, 1827	....
Paper-finishing machine	T. Gilpin	Philadelphia, Pa	June 25, 1830	....
Paper flour-sack	J. M. Hurd	Auburn, N. Y	Nov. 12, 1867	70, 852
Paper, Fly	T. E. Peck	Bridgeport, Conn	Apr. 2, 1872	125, 326
Paper-folder	E. Gomez	New York, N. Y	Aug. 9, 1859	25, 007
Paper-folder	E. P. Tiffany	Hartford, Conn	Apr. 26, 1870	102, 334
Paper folding and pasting machine	G. K. Snow	Watertown, Mass	Mar. 6, 1860	27, 392
Paper folding and registering machine	A. F. Endriss	New York, N. Y	Mar. 8, 1859	23, 161
Paper, Folding and ruling	J. P. Herron	Washington, D. C	June 30, 1863	39, 045
Paper folding and stitching machine	J. H. Tanner	Franenfeld, Switzerland	Sept. 9, 1862	36, 428
Paper folding, cutting, and trimming machine	C. Chambers, jr	Philadelphia, Pa	Aug. 5, 1873	141, 487
Paper-folding machine	C. Chambers, jr	Kennett's Square, Pa	Oct. 7, 1856	15, 842
Paper-folding machine	C. Chambers, jr	Philadelphia, Pa	Nov. 3, 1857	18, 533
Paper-folding machine	C. Chambers, jr	Philadelphia, Pa	Apr. 5, 1859	23, 445
Paper-folding machine	C. Chambers, jr	Philadelphia, Pa	Nov. 15, 1859	26, 090
Paper-folding machine	C. Chambers, jr	Philadelphia, Pa	Nov. 27, 1860	30, 719
Paper-folding machine	C. Chambers, jr	Philadelphia, Pa	Dec. 18, 1860	30, 910
Paper-folding machine	C. Chambers, jr	Philadelphia, Pa	July 19, 1870	105, 424
Paper-folding machine	C. Chambers, jr., and W. Mendham.	Philadelphia, Pa	Aug. 5, 1873	141, 489
Paper-folding machine	C. Chambers, jr., and W. Mendham.	Philadelphia, Pa	Aug. 5, 1873	141, 490
Paper-folding machine	J. S. Gallaher, jr	Washington, D. C	Dec. 4, 1860	30, 853
Paper-folding machine	R. R. Gubbins	Troy, N. Y	Nov. 7, 1871	120, 639
Paper-folding machine	R. R. Gubbins	West Troy, N. Y	Dec. 5, 1871	121, 615
Paper-folding machine	J. Macnair	New York, N. Y	June 29, 1869	91, 948
Paper-folding machine	J. McAdams	Brooklyn, N. Y	Feb. 28, 1871	112, 264
Paper-folding machine	W. Mendham	Philadelphia, Pa	Feb. 1, 1870	99, 336
Paper-folding machine	W. Mendham	Philadelphia, Pa	June 21, 1870	104, 621
Paper-folding machine	W. H. and J. Milliken	Manchester, N. H	Feb. 26, 1861	31, 578
Paper-folding machine	O. Norelius	Minneapolis, Minn	Apr. 25, 1871	114, 031
Paper-folding machine	J. North	Middletown, Conn	Apr. 15, 1856	14, 697
Paper-folding machine	J. North	Middletown, Conn	June 24, 1862	35, 738
Paper-folding machine	J. North	Middletown, Conn	June 9, 1863	38, 874
Paper-folding machine	L. E. Osborn	New Haven, Conn	May 13, 1863	35, 249
Paper-folding machine	E. N. Smith	West Brookfield, Mass	Nov. 27, 1849	6, 896
Paper-folding machine	E. N. Smith	Springfield, Mass	May 19, 1857	17, 352
Paper-folding machine	G. K. Snow	Boston, Mass	Oct. 15, 1850	7, 722
Paper-folding machine	R. J. Stuart	Yonkers, N. Y	Oct. 28, 1873	144, 160
Paper-folding machine	T. Thompson	Niversville, N. Y	Feb. 12, 1856	14, 260
Paper-folding machine	G. W. D. Upton	Springfield, Mass	July 29, 1873	141, 405
Paper-folding machine	A. Washburn	Medina, Ohio	Nov. 26, 1872	133, 393
Paper-folding machine	A. Washburn	Medina, Ohio	Aug. 12, 1873	141, 742
Paper-folding machine	J. F. Weeks	Columbus, Ohio	June 9, 1857	17, 535
Paper-folding machine	C. P. Wiggins, A. H. Nordyke, and B. Strawbridge.	Richmond, Ind	Aug. 25, 1857	18, 075
Paper-folding machines, Packing and carrying device for.	W. Farrington	Brooklyn, N. Y	Aug. 5, 1873	141, 501
Paper-folding machines, Packing-box for	C. Chambers, jr	Philadelphia, Pa	Aug. 5, 1873	141, 486
Paper-folding machines, Packing-device for	C. Clark	New York, N. Y	Aug. 5, 1873	141, 491
Paper-folding machinery	C. O. Crosby	New Haven, Conn	Dec. 23, 1856	16, 266
Paper folding, pasting, and cutting machine	C. Chambers, jr	Philadelphia, Pa	Mar. 5, 1861	31, 588
Paper folding, pasting, and insetting machine	C. Chambers, jr	Philadelphia, Pa	Aug. 5, 1873	141, 488
Paper for automatic telegraphy and in recovering chemicals from waste-paper, Preparing.	G. Little	Rutherford Park, N. Y	July 23, 1872	129, 841
Paper for automatic transmitter, Apparatus for perforating.	G. Little	Hudson City, N. J	June 15, 1869	91, 241
Paper for bank-notes, bonds, &c	J. M. Wilcox	Glen Mills, Pa	May 16, 1871	115, 005
Paper for bank-notes, &c., Safety	J. Jameson	Gateshead, England	Apr. 15, 1873	137, 775
Paper for book-binders, Folding	G. K. Snow	Watertown, Mass	Feb. 14, 1860	27, 166
Paper for bonds, drafts, &c	R. Price	New York, N. Y	July 5, 1870	105, 125
Paper for boxes, Machine for cutting, cornering, and sewing.	A. Dennison	Brunswick, Me	Apr. 15, 1851	8, 044
Paper for boxes, Machine for cutting, scoring, and cornering.	J. M. Keer	Philadelphia, Pa	Nov. 29, 1870	109, 630
Paper for buildings, Preparing	C. B. Ayer	Beloit, Wis	Mar. 5, 1872	124, 314
Paper for buildings	F. N. Davis	Beloit, Wis	July 18, 1871	117, 155
Paper for checks, drafts, notes, &c	G. F. Thomas, jr	Brooklyn, N. Y	Apr. 12, 1870	101, 786

Index of patents issued from the United States Patent Office from 1790 *to* 1873, *inclusive*—Continued.

Invention.	Inventor.	Residence.	Date.	No.
Paper for chemical telegraph, &c	T. A. Edison	Newark, N. J	Oct. 22, 1872	132, 455
Paper for collars, &c., Manufacture of	S. M. Allen	Woburn, Mass	Mar. 31, 1863	38, 019
Paper for collars, Manufacture of	J. M. Willcox	Glen Mills, Pa	July 21, 1868	80, 105
Paper for collars, Surface-sized	W. F. Moseley	Brooklyn, N. Y	Nov. 7, 1871	120, 765
Paper for copying-presses, Damping	G. Burnham	Philadelphia, Pa	Dec. 24, 1850	7, 851
Paper for covering walls, Water and damp proof	C. Goessling	Jersey City, N. J	Aug. 4, 1868	80, 620
Paper for envelopes, &c., Apparatus for cutting and creasing.	M. Dimock	Newark, N. J	Oct. 11, 1864	44, 687
Paper for hangings and other purposes, Composite	W. Campbell	Millburn, N. J	May 25, 1869	90, 497
Paper for keeping off mosquitoes, Prepared	J. Bagnol	Baltimore, Md	Sept. 10, 1872	131, 242
Paper for manufacture of floor-covering, belting, window-shades, &c., Preparing.	J. J. Ott	Washington, D. C	Aug. 18, 1868	81, 199
Paper for manufacture of letter and invoice files	J. L. Rile	New York, N. Y	June 4, 1867	65, 436
Paper for manufacture of neck-ties, cravats, &c., Composition for coating.	M. W. Brown	New York, N. Y	Apr. 20, 1869	89, 198
Paper for manufacture of paper bags	W. E. Farrell	Philadelphia, Pa	June 15, 1869	91, 220
Paper for napkins, &c	E. Parrish	Philadelphia, Pa	May 30, 1871	115, 511
Paper for notes, checks, &c., Manufacture of	A. Varnham	London, England	Aug. 9, 1845	4, 143
Paper for postage-stamps	H. Loewenberg	New York, N. Y	Mar. 6, 1866	53, 081
Paper for printing postage and revenue stamps, Mode of preparing.	S. Lenher and H. H. Spencer	Philadelphia, Pa	Dec. 7, 1869	97, 528
Paper for printing, writing, and other purposes	J. Langtree	New York, N. Y	Sept. 6, 1870	107, 067
Paper for protecting goods from moths, &c., Manufacture of.	R. W. Russell	New York, N. Y	Mar. 30, 1869	88, 519
Paper for roofing and similar purposes, Preparing.	C. H. Smith	Baltimore, Md	Mar. 29, 1870	101, 406
Paper for transferring stamps, &c., Composition for preparing.	M. Rosenthal	Philadelphia, Pa	June 2, 1868	78, 610
Paper for twine, &c., Machine for cutting	J. B. Wortendyke	Goodwinville, N. J	Sept. 13, 1864	44, 249
Paper for various purposes, Process of treating	C. F. Crebose	Newton Lower Falls, Mass	Mar. 3, 1868	74, 996
Paper for water-closets, Medicated	G. W. Thompson	Brooklyn, N. Y	Dec. 22, 1868	85, 188
Paper for wrapping tobacco, snuff, soap, &c., Method of preparing.	M. W. Brown	West Farms, N. Y	Feb. 11, 1868	74, 195
Paper, &c., Forming metallic characters on	J. Lanza	New York, N. Y	June 19, 1866	55, 679
Paper fringe, Machine for cutting	P. and G. Pengeot	Buffalo, N. Y	Dec. 19, 1848	5, 972
Paper from alga or sea-weed, Mode of making	S. Green	New London, Conn	Feb. 15, 1809	....
Paper from corn-husks, Manufacturing	B. Allison and J. Hawkins	....	Dec. 30, 1802	....
Paper from curriers' shavings, Making	J. Condit, jr	....	Dec. 28, 1801	....
Paper from cutting-engines, Machinery for taking and laying.	J. M. Hollingsworth	Milton, Mass	Apr. 17, 1849	6, 337
Paper from grain, Process of manufacturing	C. V. and J. Stehlin and H. A. Haan.	New York and Brooklyn, N. Y.	Nov. 4, 1873	144, 294
Paper, &c., from husks of Indian corn, Manufacture of.	A. C. De Welsbach	Vienna, Austria	Apr. 21, 1863	38, 220
Paper from rags and straw and corn-husks, Mode of making white.	J. W. Cooper	Washington Township, Pa	Feb. 7, 1829	....
Paper from reeds	H. Lowe	Belleville, N. Y	Mar. 1, 1859	23, 099
Paper from resinous bark, Preparation of	C. C. Hall	Portland, Me	Feb. 20, 1855	12, 414
Paper from sand-grass, Manufacture of brown	I. Sanderson	Milton, Mass	Feb. 22, 1838	614
Paper from sorghum, Manufacture of	H. Pemberton	East Tarentum, Pa	Mar. 31, 1863	38, 059
Paper from Spanish grass, Manufacture of	W. B. Newbery	Dorchester, Mass	May 24, 1864	42, 866
Paper from straw, hay, &c., Making	W. Magaw	Meadville, Pa	Mar. 8, 1830	....
Paper from straw, Manufacture of	R. T., I. W., and A. I. Smart	Troy, N. Y	Mar. 27, 1860	27, 653
Paper from straw, Manufacture of fine white writing.	L. Bomeisler	Philadelphia, Pa	Sept. 10, 1829	....
Paper from straw, Process for making	W. Clarke	Dayton, Ohio	May 6, 1856	14, 804
Paper from the "ulvamarina," Making	E. H. Collier	Plymouth County, Mass	Apr. 15, 1828	....
Paper from vulcanized gum, Detaching	A. C. Richard	Newtown, Conn	Jan. 11, 1859	22, 584
Paper from wood, Manufacture of	S. M. Allen	Woburn, Mass	Mar. 31, 1863	38, 020
Paper from wood, Manufacture of	C. Watt and H. Burgess	London, England	July 18, 1854	11, 343
Paper from wood, Manufacturing	L. Wooster and J. B. Holmes	Meadville, Pa	Aug. 3, 1830	....
Paper-glazing	F. Beck	New York, N. Y	Jan. 1, 1867	60, 672
Paper goods, Manufacture of corded-edge	A. T. Denison	Poland, Me	Apr. 20, 1869	89, 132
Paper-hanger's apparatus	W. F. Trautman	Llewellyn, Pa	May 24, 1870	103, 393
Paper-hangings, Apparatus for hanging up and carrying off.	T. Van Deventer	New Brunswick, N. J	July 20, 1858	20, 965
Paper-hangings, Apparatus for making stamp-gilt	W. Bailey	New York, N. Y	Nov. 19, 1867	71, 120
Paper-hangings, Apparatus for trimming	E. Boothby	Saco, Me	Oct. 18, 1864	44, 764
Paper-hangings, Machine for coloring	J. Heist	New York, N. Y	Apr. 12, 1870	101, 732
Paper-hangings, Machine for marking	H. Lightbown	Pendleton, Great Britain	July 16, 1872	129, 573
Paper-hangings, Machine for trimming the edges of	J. Waugh	Elmira, N. Y	Oct. 5, 1858	21, 710
Paper-hangings, Paste for	J. Jones	New York, N. Y	Mar. 7, 1871	112, 464
Paper-hangings, Printing	R. L. Hawes	Worcester, Mass	Apr. 18, 1848	5, 525
Paper-hangings, Printing	W. M. Shaw and E. Gould	Newark, N. J	May 1, 1849	6, 404
Paper-hangings, Printing and embossing	E. S. Ormsby	New York, N. Y	Nov. 23, 1869	97, 221
Paper-hangings with light-colored satin ground, Manufacturing.	H. Steel	Hudson, N. Y	Sept. 8, 1813	....
Paper-holder	C. A. Barrows	Willimantic, Conn	Aug. 13, 1872	130, 464
Paper-holder	J. W. Foard	San Francisco, Cal	Sept. 5, 1865	49, 744
Paper-holder	Y. Orton	Chicago, Ill	Dec. 24, 1872	134, 160
Paper-holder	D. M. Smith	Springfield, Vt	Oct. 9, 1866	58, 687
Paper-holder	E. J. Toop	Fort Madison, Iowa	Jan. 30, 1866	52, 343
Paper-holder, Abrasive	G. A. Howe	Niles, Michigan	Dec. 19, 1871	122, 018
Paper-holder shelf	G. V. Hawkins	Clayton, Mich	Nov. 18, 1873	144, 763
Paper, Hot-pressing	F. Bailey	Salisbury, Pa	July 31, 1809	....
Paper, Implements for perforating and severing	A. H. Hook	New York, N. Y	Sept. 13, 1864	44, 190
Paper, &c., in making cards, pasteboard, &c., Machine for applying paste to sheets of.	D. H. Gilbert	Dorchester, Mass	Apr. 11, 1842	2, 553
Paper in paper-making machine, Machinery for drying.	O. Ellsworth	Boston, Mass	Feb. 12, 1867	61, 930
Paper, &c., in rolls, Apparatus for winding	G. D. Harrington	Columbus, Ohio	July 23, 1872	129, 660
Paper in the manufacture of boxes, Machine for cutting the corners of.	D. Whitlock	Newark, N. J	Sept. 18, 1866	58, 183
Paper in the manufacture of paper cop-tubes, Rolling or winding.	S. Burgess	Providence, R. I	Aug. 13, 1867	67, 716
Paper in the ream, Cutting and trimming	J. Ames	Springfield, Mass	Feb. 28, 1834	....
Paper in the ream, Cutting and trimming	J. Ames	Springfield, Mass	Feb. 28, 1834	....
Paper in the sheet, Machine for making	M. Haddock	New York	July 17, 1828	....

Index of patents issued from the United States Patent Office from 1790 *to* 1873, *inclusive*—Continued.

Invention.	Inventor.	Residence.	Date.	No.
Paper in writing and drawing, Clamp for holding	E. B. Forbush	Buffalo, N. Y	Sept. 3, 1850	7, 617
Paper into sheets, Machine for cutting	J. Hatch	South Windham, Conn	June 27, 1865	48, 395
Paper-knife handle	E. Kelsey	Centre Brook, Conn	July 25, 1865	48, 953
Paper, Leather	S. M. Allen	Woburn, Mass	Aug. 4, 1863	39, 371
Paper, Leather	E. F. and T. Blank	New York	Feb. 16, 1830	....
Paper, leather, and other thin substances, Machine for cutting,	H. A. Snowden	Brooklyn, N. Y	Sept. 17, 1872	131, 378
Paper, leather, &c , Apparatus for cutting ornaments in.	J. D. Mets	Dubuque, Iowa	May 5, 1863	38, 402
Paper, Letter	T. H. Dodge	Washington, D. C	Apr. 23, 1861	32, 167
Paper, Letter	R. Magee	Philadelphia, Pa	May 10, 1864	42, 673
Paper-machine	J. Ames	Springfield, Mass	Feb. 20, 1835	....
Paper-machine	M. J. Kearney	Philadelphia, Pa	Sept. 3, 1872	131, 103
Paper-machine	W. and A. L. Knight and E. F. Condit.	Whippany, N. J	Sept. 25, 1839	1, 336
Paper-machine	P. Scanlan	Indianapolis, Ind	Nov. 22, 1870	109, 552
Paper-machine	C. Whealen	Dayton, Ohio	Oct. 28, 1873	144, 172
Paper-machine attachment to prevent the straining and breaking of paper during the manufacture.	L. Dean	Fort Edward, N. Y	Aug. 9, 1870	106, 134
Paper-machine-belt guide	R. Hutton	Holyoke, Mass	July 1, 1873	140, 418
Paper-machine-crimping attachment	W. H. Bleasdale	Chagrin Falls, Ohio	May 10, 1870	102, 754
Paper-machine, Cylinder	I. Sanderson	Milton, Mass	Apr. 18, 1829	....
Paper-machine, Cylinder	J. P. Sherwood	Fort Edward, N. Y	Sept. 21, 1869	95, 153
Paper-machine deckel	C. McBurney and L. Hollingsworth.	Boston, Mass	Aug. 29, 1871	118, 624
Paper-machine-felt guide	P. A. Wait	Sandy Hill, N. Y	Apr. 8, 1856	14, 621
Paper-machine-felt washer	A. Anderson	Milwaukee, Wis	Sept. 6, 1864	44, 059
Paper, Machine for boiling and washing rags for manufacturing.	G. Spafford	Windham, Conn	Sept. 2, 1840	1, 753
Paper, Machine for cutting, printing, and folding	W. P. Corsa	Catskill, N. Y	May 31, 1870	103, 575
Paper, Machine for cutting rags for	M. Y. Beach	Springfield, Mass	Sept. 11, 1828	....
Paper, Machine for drying sized	N. W. Taylor and J. W. Brightman.	Cleveland, Ohio	Apr. 29, 1862	35, 117
Paper, Machine for folding and cutting the edges of	J. E. Coffin	Portland, Me	Nov. 17, 1868	84, 091
Paper, &c., Machine for grinding wood-pulp for	J. S. Elliott and J. F. Wood	Chelsea and Everett, Mass	Apr. 23, 1872	126, 041
Paper, Machine for making and sizing	G. W. Turner	London, Great Britain	Jan. 27, 1852	8, 698
Paper, Machine for making card	E. L. Perkins	Boston, Mass	July 10, 1840	1, 679
Paper, Machine for making lace	C. Lang	Worcester, Mass	Jan. 31, 1865	46, 115
Paper, Machine for polishing enameled	S. Shepherd and A. M. George	Nashua, N. H	July 17, 1866	56, 457
Paper, Machine for polishing enameled	W. F. Wright	Nashua, N. H	July 24, 1866	56, 682
Paper, Machine for preparing rags for making	B. Cox	Northampton, Mass	Mar. 28, 1818	....
Paper, Machine for preparing top and swingled tow for.	A. Frost	....	Mar. 19, 1804	....
Paper, Machine for printing both sides of a continuous sheet of.	T. French	Ithaca, N. Y	Nov. 20, 1837	468
Paper, Machine for printing house	M. D. Whipple	Lowell, Mass	Sept. 16, 1851	8, 372
Paper, &c., Machine for punching and perforating	G. C. Howard	Philadelphia, Pa	May 21, 1861	32, 370
Paper, Machine for receiving and piling	J. A. Wilkinson	Brooklyn, N. Y	Aug. 9, 1859	25, 068
Paper, &c., Machine for smoothing and polishing	W. Coolidge	Boston, Mass	Apr. 18, 1808	....
Paper, Machine for stamping lace	A. Gerandat	New York, N. Y	Mar. 15, 1870	100, 749
Paper, Machine for tearing and dusting rags, &c., used in the manufacture of.	H. Clark and W. Albertson	New London, Conn	Sept. 19, 1838	927
Paper, Machine for trimming edges of	J. McClintic	Chambersburgh, Pa	Mar. 31, 1827	....
Paper, Machine for washing rags in the manufacture of.	R. Carter	Elkton, Md	Feb. 22, 1838	615
Paper-machine, Fourdrinier	C. Hofmann	Elkton, Md	May 4, 1869	89, 766
Paper-machine suction-apparatus	W. McLaughlin	Lee, Mass	Apr. 22, 1873	138, 173
Paper-machine suction-box	F. Curtis	Foxborough, Mass	July 2, 1872	128, 469
Paper-machine vat, Cylinder	A. Howland	Sandy Hill, N. Y	Oct. 20, 1868	83, 165
Paper-machines, &c., Apparatus for controlling the motion of traveling webs in.	F. Thiry	Huy, Belgium	Dec. 21, 1867	72, 564
Paper-machines, Apparatus for removing and replacing the dandy-rolls of.	J. F. Marshall	Louisville, Ky	Feb. 13, 1872	123, 573
Paper-machines, Apparatus for repairing knotters, strainers, or screen-plates of.	J. Robertson	Lasswade, Great Britain	June 24, 1873	140, 166
Paper-machines, Bearings for the table-rolls of Fourdrinier.	G. S. Barton	Worcester, Mass	Mar. 29, 1870	101, 345
Paper-machines, Dandy-roller for	J. Whitehead	Manchester, England	Oct. 21, 1873	143, 801
Paper-machines, Elastic apron for	A. B. Lovell	Pomfret, N. Y	Oct. 6, 1868	82, 854
Paper-machines, Jacket-stretcher for couch-roller of	R. L. Howe	Westbrook, Me	May 24, 1864	42, 854
Paper-machines, Rubber roller for	R. A. Kelty	Lambertsville, N. J	Feb. 14, 1871	111, 751
Paper-machines, Stuff-regulator for	D. Hamel	Holyoke, Mass	Nov. 25, 1873	144, 902
Paper-machinery	H. Chapman	Catawissa, Pa	Dec. 5, 1865	51, 293
Paper, Machinery for cutting rags for making	A. S. Woodward and B. F. Bartlett.	Lowell and Pepperell, Mass	Oct. 31, 1854	11, 882
Paper, Machinery for measuring pulp in the manufacture of.	H. Pohl	Paterson, N. J	July 9, 1850	7, 497
Paper, Machinery for manufacturing	S. Rossman	Stuyvesant, N. Y	Jan. 5, 1858	19, 045
Paper, Machinery for manufacturing composite	W. Campbell	Millburn, N. J	July 6, 1869	92, 161
Paper, Machinery for piling	J. C. Kneeland	Northampton, Mass	July 27, 1858	21, 004
Paper, Machinery for plating or finishing sheets of.	C. T. Bainbridge	Brooklyn, N. Y	Sept. 5, 1865	49, 691
Paper, Machinery for separating sand, &c., from pulp in the manufacture of.	W. Bishop	Coventry, Conn	Dec. 31, 1845	4, 341
Paper-machinery, Screen-plate for	F. Curtis	Auburndale, Mass	Mar. 19, 1867	62, 942
Paper-machinery, Suction-box of	C. J. Bradbury	Charlestown, Mass	Oct. 1, 1872	131, 732
Paper-maker's felt, Washing and cleansing	S. E. Foster	Brattleborough, Vt	May 25, 1832	....
Paper-maker's molds, Twisting or bobbing machine for weaving the wire-work of.	P. Bernard	Whitestown, N. Y	July 7, 1809	....
Paper, Making	C. Forbes	East Hartford, Conn	Feb. 20, 1836	....
Paper, Making	J. M. and L. Hollingsworth	Boston, Mass	Dec. 4, 1843	3, 362
Paper, Making	J. Truman	Bridgeport, Pa	Feb. 28, 1834	....
Paper, Making	L. W. Wright	London, England	Mar. 27, 1847	5, 041
Paper, &c., Making cane-hemp for making	B. C. Smith	Burlington, N. J	Nov. 21, 1843	3, 349
Paper-making, Filtering for	T. Trench	Ithaca, N. Y	May 26, 1832	....
Paper-making machine	J. Ames	Springfield, Mass	May 14, 1822	....
Paper-making machine	E. B. Bingham	Brooklyn, N. Y	June 30, 1857	17, 663
Paper-making machine	J. Burns and J. Campbell	Bloomfield, N. J	Mar. 26, 1872	124, 881
Paper-making machine	P. Clark	Rahway, N. J	Aug. 4, 1857	17, 917

Index of patents issued from the United States Patent Office from 1790 *to* 1873, *inclusive*—Continued.

Invention.	Inventor.	Residence.	Date.	No.
Paper-making machine	E. N. Foote	Saratoga Springs, N. Y.	Nov. 22, 1864	45,149
Paper-making machine	E. T. Ford	Stillwater, N. Y.	Nov. 3, 1868	83,617
Paper-making machine	E. T. Ford	Stillwater, N. Y	July 13, 1869	92,596
Paper-making machine	R. C. Harris	Harrisville, N. J	June 29, 1869	91,842
Paper-making machine	I. Jennings	Fairfield, Conn	July 7, 1868	79,659
Paper-making machine	N. Keely	Philadelphia, Pa	July 29, 1873	141,358
Paper-making machine	C. Kinsey	Essex, N. J	May 8, 1807	....
Paper-making machine	T. Lindsay and W. Geddes	Westville and Seymour, Conn.	July 27, 1858	21,008
Paper-making machine	O. Marland	Boston, Mass	Dec. 5, 1854	12,028
Paper-making machine	O. Marland	Boston, Mass	Dec. 5, 1854	12,027
Paper-making machine	T. Nugent	Whippany, N. J	Mar. 12, 1872	124,612
Paper-making machine	J. B. Pignatelli	New York	Dec. 2, 1819	....
Paper-making machine	G. E. Rutledge	Dayton, Ohio	May 26, 1863	38,698
Paper-making machine	J. Scanlan	Lebanon, Pa	June 20, 1865	48,347
Paper-making machine	C. B. Van Valkenburgh	Valatie, N. Y	June 1, 1869	90,898
Paper-making machine	J. Viney	Manchester, N. H	Nov. 17, 1868	84,235
Paper-making machine	E. Wilmont	Laona, N. Y	Nov. 19, 1867	71,108
Paper-making machines, Drier-felts for	S. W. Baker	Providence, R. I	Oct. 10, 1865	50,323
Paper-making machines, Feeding pulp to	I. Kinsey	Hohokus, N. J	Oct. 7, 1856	15,852
Paper-making machines, Self-acting felt-guide for	T. Baker	Stillwater, N. Y	June 28, 1864	43,280
Paper-making machines, Vacuum-box of	J. L. Seaverns	Worcester, Mass	Aug. 11, 1863	39,500
Paper-making machinery	E. B. Bingham	Newark, N. J	Sept. 18, 1866	58,051
Paper-making machinery	J. S. Blake	Claremont, N. H	Jan. 20, 1857	16,430
Paper-making machinery	G. Burbank	Worcester, Mass	Apr. 12, 1826	....
Paper-making machinery	W. W. Harding	Philadelphia, Pa	Mar. 15, 1870	100,755
Paper-making machinery	J. Harper	East Haven, Conn	Mar. 11, 1862	34,633
Paper-making machinery	S. G. and G. S. Rogers	Thetford, Vt	Nov. 13, 1866	59,661
Paper-making machinery	G. J. Wheeler, G. W. Dunnell, and W. Sharp.	Bloomfield, N. J	Jan. 22, 1861	31,215
Paper-making machinery, Rag-engine of	T. Lindsay	Montville, Conn	May 16, 1865	47,739
Paper-making, Manufacture of suction-boxes for	F. Curtis	Newton, Mass	Nov. 5, 1867	70,534
Paper-making, Preparing resin size for use in	T. Gray	London, England	Nov. 3, 1868	83,707
Paper-making, Rag engine for	J. M. Shew	Paper Mills, Md	Dec. 18, 1866	60,645
Paper-making roll and drier	O. Marland	Boston, Mass	Feb. 27, 1855	12,442
Paper, Making, ruling, and cutting	J. Ames	Springfield, Mass	July 31, 1840	1,709
Paper, Making the leaves of books from a continuous sheet of.	T. T. Morrell	Farmington, N. H	Oct. 21, 1862	36,722
Paper, Making thick	S. G. Levis	Delaware County, Pa	Feb. 14, 1854	10,519
Paper-making, Washing rags for	J. Ames	Springfield, Mass	Apr. 6, 1831	....
Paper-making, Washing rags for	S. Eckstein	Philadelphia, Pa	June 13, 1831	....
Paper-manufacture and treatment of paper-pulp	A. T. Schmidt	Pittsburgh, Pa	Jan. 15, 1867	61,267
Paper, Manufacture of	J. M. Allen	Marion, Mass	Aug. 8, 1865	49,209
Paper, Manufacture of	S. D. Baldwin	Marysville, Cal	Jan. 9, 1872	122,548
Paper, Manufacture of	E. B. Bingham	Newark, N. J	Feb. 22, 1870	100,109
Paper, Manufacture of	J. W. Dixon	Philadelphia, Pa	Dec. 19, 1865	51,568
Paper, Manufacture of	L. Dodge	Waterford, N. Y	June 14, 1870	104,281
Paper, Manufacture of	L. Dodge	Waterford, N. Y	June 14, 1870	104,282
Paper, Manufacture of	E. T. Ford	Stillwater, N. Y	Nov. 3, 1868	83,616
Paper, Manufacture of	I. Hoffman	Oregon, N. Y	Apr. 26, 1870	102,265
Paper, Manufacture of	G. W. Hurlbut	Fair Haven, Vt	Oct. 16, 1866	58,944
Paper, Manufacture of	C. B. Hutchins	Ann Arbor, Mich	Nov. 29, 1870	109,621
Paper, Manufacture of	J. S. Kenyon and L. Fox	Baldwinsville, N. Y	Nov. 1, 1870	108,913
Paper, Manufacture of	G. E. Marshall	Laurel, N. Y	July 19, 1870	105,585
Paper, Manufacture of	S. Nowlan	New York, N. Y	Aug. 16, 1864	43,860
Paper, Manufacture of	C. E. O'Hara	New York, N. Y	Oct. 18, 1870	108,509
Paper, Manufacture of	H. Pemberton	Allegheny City, Pa	July 19, 1870	105,594
Paper, Manufacture of	H. Pemberton	Allegheny City, Pa	Oct. 11, 1870	108,177
Paper, Manufacture of	J. Pickles	Wigan, England	June 15, 1869	91,480
Paper, Manufacture of	J. B. Read	Tuscaloosa, Ala	Dec. 26, 1865	51,751
Paper, Manufacture of	J. B. Read	Tuscaloosa, Ala	Aug. 24, 1869	94,131
Paper, Manufacture of	A. E. Reed	Wookey, England	July 16, 1872	128,978
Paper, Manufacture of	W. A. Russell	Lawrence, Mass	Aug. 14, 1866	57,192
Paper, Manufacture of	T. P. Shaffner	Louisville, Ky	Oct. 30, 1866	59,281
Paper, Manufacture of	A. and N. A. Sprague	Fredonia, N. Y	Oct. 31, 1828	....
Paper, Manufacture of	J. H. Tiemann	New York, N. Y	Feb. 13, 1872	123,747
Paper, Manufacture of	Z. C. Warren	Brooklyn, N. Y	Mar. 23, 1869	88,102
Paper, Manufacture of	R. Waterman and G. W. Annis	Providence, R. I	Aug. 30, 1828	....
Paper, Manufacture of lace	C. Lang	Bergen, N. J	Apr. 17, 1866	53,991
Paper, Manufacture of parchment	A. J. Sheldon	Buffalo, N. Y	Jan. 25, 1870	99,248
Paper, Manufacture of printing and writing	W. W. Harding	Philadelphia, Pa	Jan. 19, 1869	86,012
Paper, Manufacture of tarred	H. F. Evans	Beloit, Wis	Feb. 2, 1869	86,380
Paper, Manufacture of tobacco	R. Antigüedad	New York, N. Y	July 12, 1870	105,160
Paper, Manufacture of tobacco	P. M. Consuegra and R. Antigüedad.	New York, N. Y	July 13, 1869	92,427
Paper, Manufacture of water and fire proof	T. Irving, J. McNeil, G. W. Rich, and C. J. Fay.	Elwood, N. J	Dec. 18, 1866	60,635
Paper, Manufacture of water-proof	S. M. Allen	Woburn, Mass	Mar. 13, 1866	53,094
Paper, Manufacturing	J. Ames	Springfield, Mass	May 14, 1832	....
Paper, Manufacturing	C. Austin		Dec. 14, 1798	....
Paper, Manufacturing	J. Biddis		Mar. 31, 1794	....
Paper, Manufacturing	I. Burbank	Worcester, Mass	Sept. 8, 1824	....
Paper, Manufacturing	S. S. Crocker and G. E. Marshall.	Lawrence, Mass	June 14, 1859	24,377
Paper, Manufacturing	R. R. Livingston		Oct. 28, 1799	....
Paper, Manufacturing	S. Stimpson	Newbury, Vt	Mar. 12, 1831	....
Paper, Manufacturing	S. Wheeler	Albany, N. Y	July 16, 1872	128,992
Paper, Manufacturing copying	W. Mann	Philadelphia, Pa	Jan. 11, 1853	9,536
Paper, Marbleizing	T. Carson	Brooklyn, N. Y	Mar. 14, 1871	112,544
Paper, Mark of opacity in	J. Reich and E. Star	Philadelphia, Pa	May 11, 1816	....
Paper, Marking and ornamenting	T. Mackenzie and A. Trochsler	Boston, Mass	Mar. 22, 1859	23,304
Paper, Material for making	J. J. Gillet-Damitte, H. D. Dubois, and A. Boissonneau.	Paris, France	Feb. 16, 1869	86,917
Paper, Material for making	W. Magaw	Meadville, Pa	Mar. 8, 1828	....
Paper-mill	T. Langstroth		May 1, 1804	....
Paper-mill engines, Bed-plate for	G. A. Corser	Leicester, Mass	June 7, 1864	43,070
Paper-mill engines, Bed-plate for	O. Morse	Needham Lower Falls, Mass.	May 32, 1865	47,849

Index of patents issued from the United States Patent Office from 1790 *to* 1873, *inclusive*—Continued.

Invention.	Inventor.	Residence.	Date.	No.
Paper-mills, Machine for cleaning rags for	W. Debit	East Hartford, Conn	Jan. 13, 1829	....
Paper-mills, Self-adjusting guide-roll for	R. L. Howe	Westbrook, Me	Mar. 19, 1867	62, 958
Paper-mold	W. Brewer and J. Smith	Surrey County, England	Mar. 4, 1851	7, 959
Paper-mold	J. Carnes, jr		Apr. 11, 1793	....
Paper-molding	W. W. Webster	Chelsea, Mass	Jan. 3, 1871	110, 704
Paper molding, cornice, &c	L. W. Kimball	Pittsford, Vt	Dec. 29, 1868	87, 312
Paper, moth-repellent	S. Crane	Dalton, Mass	Mar. 5, 1872	124, 336
Paper neck-tie	E. Child, jr	Springfield, Mass	May 22, 1866	54, 858
Paper neck-tie	H. Whitney	Watertown, Mass	July 23, 1867	67, 091
Paper neck-tie	J. Sangster and O. W. Seely	Buffalo, N. Y	May 8, 1866	54, 607
Paper on straw-board, &c., Machine for pasting	M. Fitzgibbons	New York, N. Y	June 20, 1871	116, 173
Paper or paper and cloth neck-ties	R. L. Walter	Washington, D. C	June 14, 1870	104, 382
Paper or linen, Machine for preparing and hackling tow for.	J. Tatterson	Southampton, Long Island, N. Y.	Dec. 7, 1805	....
Paper, Packing in cylinders for drying	W. B. Fowler	Lawrence, Mass	June 8, 1869	91, 007
Paper pad	S. Van Campen	New York, N. Y	Oct. 1, 1872	131, 915
Paper pail	P. C. Schuyler	New York, N. Y	Apr. 30, 1872	126, 335
Paper pan, bowl, box, and dish	J. W. Jarboe	Green Point, N. Y	Jan. 25, 1870	99, 260
Paper pantalets	E. P. Furlong	Portland, Me	Jan. 15, 1867	61, 187
Paper, pasteboard, &c., Machine for cutting	S. D. Tucker	New York, N. Y	Oct. 29, 1867	70, 292
Paper-perforating apparatus, Telegraph	L. Bradley	Jersey City, N. J	June 27, 1865	48, 479
Paper-perforating apparatus, Telegraph	T. A. Edison	Newark, N. J	Oct. 22, 1872	132, 456
Paper-perforating apparatus, Telegraph	G. Little	Hudson City, N. J	June 15, 1869	91, 240
Paper-perforating apparatus, Telegraph	G. Little	Rutherford Park, N. J	Nov. 2, 1869	96, 330
Paper-perforating apparatus, Telegraph	G. Little	Rutherford Park, N. J	Nov. 2, 1869	96, 331
Paper-perforating machine	G. C. Howard	Philadelphia, Pa	July 2, 1861	32, 693
Paper-perforator	J. C. Gaston	Cincinnati, Ohio	July 13, 1869	92, 600
Paper-planishing machine	J. F. Schuyler	Philadelphia, Pa	Feb. 24, 1863	37, 790
Paper-polisher, Composition	H. T. Cushman	North Bennington, Vt	June 1, 1869	90, 733
Paper polishing and calendering machine	P. Coburn	East Walpole, Mass	May 31, 1870	103, 569
Paper-polishing machine	E. L. Perkins	Roxbury, Mass	Oct. 24, 1854	11, 837
Paper-polishing machine	S. Shepard and A. M. George	Nashua, N. H	Oct. 6, 1868	82, 882
Paper-polishing machine	A. and G. F. Wright	Clinton, Mass	Jan. 10, 1871	110, 947
Paper, Preparation of	J. L. Jullion	Aberdeen, Great Britain	May 18, 1860	28, 182
Paper, Preparation of copying	C. Cowan	New York, N. Y	May 4, 1869	89, 738
Paper, Preparation of fiber for the manufacture of	J. E. Mallory	New York, N. Y	Mar. 26, 1861	31, 814
Paper, Preparation of paper-pulp and manufacture of.	J. Denis	London, England	Apr. 4, 1871	113, 592
Paper, Preparation of straw for the manufacture of	L. Dean	Fort Edward, N. Y	Nov. 29, 1870	109, 596
Paper, Preparing albumenized	H. F. Anthony	New York, N. Y	Mar. 1, 1864	41, 750
Paper, &c., Preparing corn-husks for the manufacture of.	H. Holland	Westfield, Mass	Aug. 13, 1838	878
Paper, &c., Preparing vegetable fiber for	J. B. Fuller	Claremont, N. H	Nov. 17, 1863	40, 659
Paper, Preparing vegetable substances for making	W. Magaw	Meadville, Pa	May 22, 1828	....
Paper-press	W. R. Dingman	Stuyvesant Falls, N. Y	Oct. 20, 1863	40, 336
Paper-press cylinder, Hot and cold	A. H. Jervis and T. Trench	Ithaca, N. Y	Nov. 6, 1832	....
Paper-press, Vertical-cutting	B. Morris	Oxford, N. Y	Feb. 15, 1835	....
Paper, Printing and cutting	J. F. Tapley	Springfield, Mass	Mar. 4, 1862	34, 601
Paper, Printing	J. E. Hover	Philadelphia, Pa	Apr. 16, 1867	63, 896
Paper, Producing designs on	W. B. Woodbury	London, England	Apr. 28, 1868	77, 230
Paper-product	J. L. Kendall	Foxborough, Mass	Nov. 11, 1873	144, 548
Paper-punching machine	S. M. Clark	Washington, D. C	Apr. 30, 1867	64, 197
Paper-puncturing device	T. F. Corry	Madison, Ind	Dec. 3, 1872	133, 567
Paper-pulp	J. W. Dixon	Philadelphia, Pa	Dec. 12, 1865	51, 433
Paper-pulp	J. W. Dixon	Philadelphia, Pa	Dec. 19, 1865	51, 572
Paper-pulp	C. C. Fitzgerald	Phœnix, N. Y	Dec. 22, 1868	85, 079
Paper-pulp	C. B. Sawyer	Fitchburgh, Mass	Nov. 29, 1870	109, 766
Paper-pulp	P. F. Schliecker	Baltimore, Md	July 4, 1871	116, 759
Paper-pulp and bleaching apparatus	J. W. Dixon	Philadelphia, Pa	June 26, 1866	55, 834
Paper pulp and fiber, Treating plants to produce	J. Dupont	Nimes, France	Aug. 6, 1872	130, 114
Paper-pulp, Apparatus for making	J. W. Dixon	Philadelphia, Pa	Feb. 13, 1866	52, 543
Paper-pulp, Apparatus for making	J. W. Dixon	Philadelphia, Pa	Feb. 13, 1866	52, 544
Paper-pulp, Apparatus for making	H. and F. Marx	Baltimore, Md	Oct. 23, 1866	59, 042
Paper-pulp, Apparatus for manufacture of	J. W. Dixon	Philadelphia, Pa	Dec. 12, 1865	51, 430
Paper-pulp, Apparatus for manufacture of	J. W. Dixon	Philadelphia, Pa	Dec. 12, 1865	51, 431
Paper-pulp, Apparatus for manufacture of	J. W. Dixon	Philadelphia, Pa	Feb. 20, 1866	52, 694
Paper-pulp, Apparatus for manufacture of	J. W. Dixon and G. Harding	Philadelphia, Pa	May 8, 1866	54, 510
Paper-pulp, Apparatus for manufacture of	J. Easton, jr., and F. Thiry	Grove Southwark, England, and Huy, Belgium.	Feb. 27, 1866	52, 941
Paper-pulp, Apparatus for manufacture of	M. L. Keen	Jersey City, N. J	May 2, 1871	114, 301
Paper-pulp, Apparatus for manufacture of	H. B. Meech	Troy, N. Y	July 11, 1871	116, 980
Paper-pulp, Apparatus for manufacture of	J. B. Palser and G. Howland	Fort Edward, N. Y	June 21, 1859	24, 484
Paper-pulp, Apparatus for molding hollow articles from.	J. L. Kendall and R. H. Trested	Foxborough, Mass., and New York, N. Y.	Apr. 16, 1872	125, 740
Paper-pulp, Apparatus for reducing wood to	H. Marx	Pikesville, Md	Dec. 1, 1868	84, 640
Paper-pulp, Apparatus for straining	L. Hollingsworth	Boston, Mass	July 2, 1872	128, 625
Paper-pulp-beating engine	G. Ames	New York, N. Y	Aug. 15, 1871	118, 092
Paper-pulp, Bleaching	J. Campbell	Chatham Village, N. Y	Apr. 16, 1872	125, 658
Paper-pulp, Bleaching	J. W. Dixon	Philadelphia, Pa	Dec. 19, 1865	51, 569
Paper-pulp, Bleaching	W. C. Joy and J. Campbell	Penn Yan, N. Y	July 2, 1867	66, 353
Paper-pulp, Bleaching	J. G. and J. H. Kendall	Leominster, Mass	July 2, 1846	4, 616
Paper-pulp-bleaching apparatus	H. L. Jones and D. S. Farquharson.	Rochester, N. Y	Mar. 13, 1866	53, 152
Paper-pulp, Boiler for making	M. L. Keen	Royer's Ford, Pa	June 16, 1863	38, 901
Paper-pulp, Boiler for preparing	L. Dean	Fort Edward, N. Y	Nov. 29, 1870	109, 595
Paper-pulp boiler or digester	W. F. Ladd	New York, N. Y	May 30, 1871	115, 327
Paper-pulp boilers, Device for filling and packing rotary.	A. Frickett	Rochester, N. Y	Dec. 3, 1867	71, 728
Paper-pulp, Disintegrating and bleaching wood, &c., to form.	J. B. Biron	Carpentras, France	Aug. 20, 1867	67, 941
Paper-pulp, Disintegrating fibrous materials for	A. H. F. Diminger	Berlin, Prussia	Oct. 11, 1870	108, 241
Paper-pulp, Disintegrating vegetable fiber for	W. Riddell	London, England	Aug. 1, 1871	117, 683
Paper-pulp, Disintegrating vegetable substances for.	Z. G. A. W. P. Oriolé, A. A. Fredet, and P. A. H. Matussiere.	Paris, France	Apr. 25, 1865	47, 505
Paper-pulp, Disintegrating woody fiber for	T. B. Armitage	New York, N. Y	Sept. 9, 1873	142, 550
Paper-pulp-dressing machine	N. Hebard	Dorchester, Mass	Dec. 27, 1839	1, 441
Paper-pulp-dressing machine	G. E. Sellers	Seller's Landing, Ill	July 2, 1867	66, 258

Index of patents issued from the United States Patent Office from 1790 to 1873, inclusive—Continued.

Invention.	Inventor.	Residence.	Date.	No.
Paper-pulp engine	S. L. Gould	Skowhegan, Me	Oct. 24, 1871	120, 265
Paper-pulp engine	J. Hatch	South Windham, Conn	Dec. 12, 1871	121, 780
Paper-pulp engine	J. Kingsland, jr	Franklin, N. J	Dec. 23, 1856	16, 278
Paper-pulp engine	H. B. Meech	Fort Edward, N. Y	July 11, 1871	116, 978
Paper-pulp engine	S. Moore and R. H. Hurlbut	Sudbury, Mass	Nov. 11, 1873	144, 557
Paper-pulp engine	T. Nugent	Whippany, N. J	July 30, 1872	130, 067
Paper-pulp engine	W. Parkison	Monongahela City, Pa	Feb. 9, 1869	86, 858
Paper-pulp engine	P. Rose	Norwich, Conn	Sept. 14, 1869	94, 843
Paper-pulp engine	C. Smith	South Windham, Conn	Dec. 19, 1871	121, 970
Paper-pulp engines, Bed-plate for	P. Frost	Medfield, Mass	June 20, 1871	116, 045
Paper-pulp engines, Grinding-plate for	P. Frost	Medfield, Mass	Sept. 14, 1869	94, 816
Paper-pulp for packing and transportation, Preparing	W. H. Merrick	Philadelphia, Pa	July 12, 1870	105, 354
Paper-pulp from beet and other refuse, Preparing	R. H. Collyer	Camden, N. J	Oct. 13, 1857	18, 389
Paper-pulp from coloring-matter, Cleansing	S. W. Wilder	Lawrence, Mass	Feb. 8, 1870	99, 735
Paper-pulp from corn-stalks, Making	J. W. Dixon	Philadelphia, Pa	Dec. 12, 1865	51, 432
Paper-pulp from ivory, Making	W. N. Clark	Chester, Conn	Sept. 15, 1857	18, 190
Paper-pulp from reeds, &c	R. W. Russell and T. Howland	Brooklyn and Stockport, N. Y.	Nov. 5, 1867	70, 474
Paper-pulp from reeds, Preparing	H. Lowe	Baltimore, Md	May 25, 1858	20, 355
Paper-pulp from straw, Manufacture of	C. M. Cresson	Philadelphia, Pa	July 11, 1871	116, 933
Paper-pulp from straw, &c., Manufacture of	A. K. Eaton	New York, N. Y	Mar. 22, 1864	41, 982
Paper-pulp from straw, Manufacture of	A. K. Haxstun	Fort Edward, N. Y	Mar. 6, 1866	52, 994
Paper-pulp from straw, &c., Manufacture of	J. A. Rothe	Philadelphia, Pa	Nov. 21, 1871	121, 130
Paper-pulp from straw, Preparing	J. Priestly and T. C. Bradbury	New York and Poughkeepsie N. Y.	June 19, 1866	55, 706
Paper-pulp from straw, &c., Preparing	J. Tiffany	Albany, N. Y	Aug. 27, 1867	68, 261
Paper-pulp from wood	H. Voelter	Heidenheim, Germany	May 22, 1866	55, 031
Paper-pulp from wood, Apparatus for making	J. Bridge	Augusta, Me	Jan. 2, 1872	122, 353
Paper-pulp from wood, Apparatus for making	H. W. Higley	Curtisville, Mass	Aug. 27, 1872	130, 803
Paper-pulp from wood, Boiler for making	M. L. Keen	Royer's Ford, Pa	Sept. 13, 1859	25, 418
Paper-pulp from wood, Making	A. Meucci	Clifton, N. Y	Mar. 13, 1866	53, 165
Paper-pulp from wood, Manufacture of	J. W. Dixon	Philadelphia, Pa	Dec. 20, 1864	45, 480
Paper-pulp from wood, Manufacture of	C. Marzoni	New York, N. Y	Dec. 21, 1858	22, 401
Paper-pulp from wood, Manufacture of	H. B. Meech	Fort Edward, N. Y	Aug. 23, 1870	106, 710
Paper-pulp from wood, straw, &c., Apparatus for making.	J. W. Dixon	Philadelphia, Pa	June 26, 1866	55, 835
Paper-pulp from wood, straw, &c., Making	J. W. Dixon	Philadelphia, Pa	Feb. 13, 1866	52, 545
Paper-pulp from wood, straw, &c., Making	J. W. Dixon	Philadelphia, Pa	Feb. 13, 1866	52, 546
Paper-pulp from wood, Manufacture of	G. Vining	Pittsfield, Mass	Dec. 21, 1869	98, 210
Paper-pulp grinder, Wood	M. S. and M. E. Otis	Rochester, N. Y	Nov. 4, 1873	144, 354
Paper-pulp, Grinding	J. Kingsland, jr	Franklin, N. J	Dec. 23, 1856	16, 316
Paper-pulp-grinding and sizing machine	J. Jordan, jr., and T. Eustice	East Hartford and Hartford, Conn.	May 18, 1858	20, 277
Paper-pulp-grinding machine	J. F. Jones	Rochester, N. Y	Apr. 25, 1865	47, 425
Paper-pulp-grinding machine, Wood	J. G. Moore	Lisbon, N. H	Nov. 19, 1872	133, 243
Paper-pulp-grinding machinery	J. Kingsland, jr	Franklin, N. J	Dec. 16, 1856	16, 239
Paper-pulp-grinding mill	J. Jordan, jr	East Hartford, Conn	Feb. 5, 1861	31, 321
Paper-pulp-grinding mill	G. Sanford	New York, N. Y	Feb. 12, 1861	31, 408
Paper-pulp in vats, Heating	J. Robeson	Montgomery, Pa	May 22, 1816	......
Paper-pulp machine	J. M. Burghardt and F. Burghardt.	Great Barrington and Curtisville, Mass.	July 9, 1872	128, 788
Paper-pulp, Machine for making	W. Miller	Herkimer, N. Y	May 12, 1868	77, 829
Paper-pulp, Machine for making	J. Taggart	Roxbury, Mass	Mar. 31, 1868	76, 270
Paper-pulp, Machine for disintegrating wood for	H. Dodge	Albany, N. Y	Jan. 25, 1870	99, 071
Paper-pulp, Machine for mixing coloring-matter with.	J. Wrinkle	Lee, Mass	Dec. 22, 1868	85, 157
Paper-pulp, Machine for preparing wood for the manufacture of.	F. Burghardt	Curtisville, Mass	Feb. 23, 1869	87, 139
Paper-pulp, Machine for reducing wood for the manufacture of.	J. Bridge	Augusta, Me	Apr. 4, 1871	113, 488
Paper-pulp machines, Water-regulator for	D. Hunter	North Bennington, Vt	Dec. 29, 1868	85, 386
Paper-pulp, Machinery for cleaning	S. S. Crocker	Lawrence, Mass	Apr. 15, 1862	34, 945
Paper-pulp, Machinery for making	L. Koch	New York, N. Y	Aug. 7, 1855	13, 412
Paper-pulp, Machinery for reducing wood to fiber for.	C. D. Negri	Hornsey Road, Great Britain.	Oct. 8, 1872	131, 944
Paper-pulp, Manufacture of	H. E. Bailliere	Hoboken, N. J	Jan. 21, 1868	73, 429
Paper-pulp, Manufacture of	L. Bardoux	Poitiers, France	Apr. 5, 1864	42, 155
Paper-pulp, Manufacture of	C. D. Brown and E. B. Denison	Portland, Me	Dec. 16, 1873	145, 620
Paper-pulp, Manufacture of	E. Clems	Toronto, Canada	July 10, 1860	29, 059
Paper-pulp, Manufacture of	F. De Campoloro	France	Aug. 7, 1860	29, 471
Paper-pulp, Manufacture of	G. Demailly	Argenteuil, France	Mar. 5, 1872	124, 196
Paper-pulp, Manufacture of	J. W. Dixon	Philadelphia, Pa	Dec. 6, 1864	45, 321
Paper-pulp, Macufacture of	J. W. Dixon	Philadelphia, Pa	Dec. 19, 1865	51, 570
Paper-pulp, Manufacture of	J. W. Dixon	Philadelphia, Pa	Dec. 19, 1865	51, 571
Paper-pulp, Manufacture of	J. W. Dixon	Philadelphia, Pa	Jan. 2, 1866	51, 813
Paper-pulp, Manufacture of	J. W. Dixon	Philadelphia, Pa	May 1, 1866	54, 308
Paper-pulp, Manufacture of	A. K. Eaton	Piermont, N. Y	Aug. 9, 1870	106, 143
Paper-pulp, Manufacture of	A. K. Eaton	Brooklyn, N. Y	Sept. 26, 1871	119, 224
Paper-pulp, Manufacture of	D. A. Fyfe	Manchester, England	Oct. 1, 1872	131, 749
Paper-pulp, Manufacture of	H. Glynn	Baltimore, Md	Feb. 6, 1855	12, 361
Paper-pulp, Manufacture of	M. L. Keen	Jersey City, N. J	Oct. 3, 1871	119, 464
Paper-pulp, Manufacture of	H. B. Meech	Fort Edward, N. Y	May 22, 1866	54, 932
Paper-pulp, Manufacture of	H. B. Meech	Troy, N. Y	July 11, 1871	116, 979
Paper-pulp, Manufacture of	M. A. C. Mellier	Paris, France	May 26, 1857	17, 387
Paper-pulp, Manufacture of	J. B. Palser and G. Howland	Fort Edward, N. Y	Nov. 22, 1859	26, 202
Paper-pulp, Manufacture of	C. A. Rose	Columbus, Ga	May 7, 1867	64, 449
Paper-pulp, Manufacture of	J. A. Roth	Philadelphia, Pa	Aug. 15, 1865	49, 480
Paper-pulp, Manufacture of	C. F. Sturgis	Dallas County, Ala	Mar. 31, 1857	16, 949
Paper-pulp, Manufacture of	A. H. Tait and W. H. Holbrook	Jersey City, N. J., and New York, N. Y.	Nov. 24, 1863	40, 728
Paper-pulp, Manufacture of	F. W. Zanders	Erfert, Germany	Feb. 13, 1872	123, 757
Paper-pulp, Manufacture of articles from	F. Curtis	Foxborough, Mass	Jan. 10, 1871	110, 833
Paper-pulp, Material for manufacture of	A. De Gogorza	New York, N. Y	Jan. 2, 1866	51, 810
Paper-pulp, Material for manufacture of	H. V. P. Draper	Hannibal, Mo	Sept. 19, 1871	119, 084
Paper-pulp, Molding articles from	E. H. Knight	Washington, D. C	Apr. 3, 1866	53, 631
Paper pulp or stock	H. Lowe	Baltimore, Md	Aug. 20, 1861	33, 092
Paper pulp or stock, Manufacture of	M. L Keen	Jersey City, N. J	Oct. 3, 1871	119, 465
Paper-pulp, Preparation of fibers for	M. Nixon	Philadelphia, Pa	May 18, 1858	20, 294

Index of patents issued from the United States Patent Office from 1790 *to* 1873, *inclusive*—Continued.

Invention.	Inventor.	Residence.	Date.	No.
Paper-pulp, Preparation of straw for	J. B. Palser and G. Howland	Fort Edward, N. Y	Mar. 20, 1860	27, 564
Paper-pulp, Preparing and bleaching	J. Campbell	Chatham Village, N. Y	June 20, 1871	116, 020
Paper-pulp, Preparing and bleaching	W. C. Joy and J. Campbell	Penn Yan, N. Y	Mar. 16, 1869	87, 779
Paper-pulp, Preparing	J. Mayrhofer	New York, N. Y	Nov. 1, 1859	25, 975
Paper-pulp, Preparing sawdust for	H. B. Meech	Fort Edward, N. Y	Dec. 20, 1864	45, 510
Paper-pulp, Preparing wood for	M. D. Whipple	Charlestown, Mass	May 29, 1855	12, 978
Paper-pulp, Recovering waste alkali used in treating.	C. M. T. Du Motay	Paris, France	Oct. 22, 1872	132, 452
Paper-pulp, Reducing straw and other fibrous substances for the manufacture of.	R. Sherwood	Fort Edward, N. Y	Dec. 13, 1864	45, 440
Paper-pulp, Reducing wood and woody fiber for	V. E. Keegan	Boston, Mass	Oct. 26, 1869	96, 239
Paper-pulp, Reducing wood fiber to	H. Voelter	Heidenheim, Würtemburg, Germany.	Aug. 10, 1858	21, 161
Paper-pulp, Rotary boiler for the manufacture of	H. B. Meech	Fort Edward, N. Y	Jan. 10, 1865	45, 845
Paper-pulp screen, Metallic	A. St. Clair	Boston, Mass	July 26, 1870	105, 755
Paper-pulp screen, Metallic	A. S. Winchester	Boston, Mass	Mar. 2, 1869	87, 385
Paper-pulp, Separating fibers of palm, palmetto, &c., for the manufacture of.	J. W. Dixon	Philadelphia, Pa	May 1, 1866	54, 309
Paper-pulp, Separating fibers of wood, &c., for the manufacture of.	A. S. Lyman	New York, N. Y	Mar. 4, 1862	34, 581
Paper-pulp, Treating fibrous substances for	A. Ungurer	Pforzheim, Germany	Oct. 7, 1873	143, 546
Paper-pulp, Treating hemp, flax, &c., for the manufacture of.	M. A. Cushing	Glen's Falls, N. Y	Oct. 10, 1865	50, 419
Paper-pulp, Treating straw and other materials for manufacture of.	H. B. Meech	Fort Edward, N. Y	Feb. 5, 1867	61, 848
Paper-pulp, Treating straw for	H. B. Meech	Fort Edward, N. Y	Nov. 7, 1865	50, 835
Paper-pulp, Treating straw for	F. A. Nixon	Philadelphia, Pa	Apr. 11, 1865	47, 217
Paper-pulp, Treating vegetable fiber for manufacture of.	J. W. Dixon	Philadelphia, Pa	Dec. 26, 1865	51, 706
Paper-pulp, &c., Treating vegetable substances for making.	H. B. Meech	Fort Edward, N. Y	Sept. 13, 1864	44, 209
Paper-pulp, Treating vegetable substances for making.	B. C. Tilghman	Philadelphia, Pa	Nov. 5, 1867	70, 485
Paper-pulp, Treating wood for manufacture of	V. E. Keegan	Boston, Mass	Apr. 13, 1869	88, 879
Paper-pulp, Treating wood, straw, &c., for manufacture of.	H. L. Jones and D. S. Farquharson.	Rochester, N. Y	June 5, 1866	55, 418
Paper-pulp washer	J. Piercy	Bloomfield, N. J	Jan. 21, 1862	34, 214
Paper-pulp washer	L. M. Wright	Fort Edward, N. Y	May 22, 1866	54, 993
Paper-pulp-washing machine	S. and J. Deacon	Lawrence, Mass	Dec. 2, 1873	145, 159
Paper, &c., Purifying and cleaning sizing for	N. J. Wells	Huntington, Mass	Jan. 22, 1867	61, 371
Paper-rag engine	N. W. Taylor and J. W. Brightman.	Cleveland, Ohio	Nov. 14, 1871	120, 837
Paper-rag engine	J. Storm	Woonsocket, R. I	Feb. 14, 1860	57, 167
Paper-rag engine, Regulator for	J. M. Hollingsworth	Braintree, Mass	Dec. 31, 1838	1, 059
Paper-stock, Recovering soda used in the manufacture of.	H. Lowe	Baltimore, Md	Dec. 17, 1861	33, 953
Paper, Recovering waste alkali used in the manufacture of.	T. F. Lehmann	Allegheny, Pa	Apr. 10, 1866	53, 839
Paper, Register for sheets of	S. T. Bacon	Boston, Mass	Mar. 8, 1859	23, 146
Paper, Register for sheets of	J. North	Middletown, Conn	Mar. 8, 1859	23, 221
Paper, Removing ink and colors from printed	J. A. Veazie	Boston, Mass	Feb. 11, 1868	74, 260
Paper-ribbon holder	W. Orr, jr., and G. F. Wright	Clinton, Mass	Oct. 2, 1866	58, 466
Paper, Roll for pressing, sizing, and calendering	F. Curtis	Malden, Mass	Jan. 9, 1866	51, 929
Paper-ruler	D. Munson	Indianapolis, Ind	Mar. 22, 1864	42, 012
Paper-ruling machine, Striker for	J. D. Connolly	Lincoln, Nebr	Oct. 14, 1873	143, 672
Paper-sack knife	L. H. Mealey	Alpha, Ohio	July 21, 1868	80, 197
Paper, Safety	L. M. Crane	Ballston, N. Y	Jan. 22, 1867	61, 321
Paper, Safety	H. Hayward	Chicago, Ill	Mar. 11, 1862	34, 634
Paper, Safety	M. A. Howell jr	Ottawa, Ill	May 22, 1860	28, 370
Paper, Safety	J. P. Oliver	Paris, France	June 9, 1863	38, 835
Paper, Safety	J. M. Willcox	Glen Mills, Pa	July 24, 1866	56, 650
Paper-"satining" machine	T. Christy	New York, N. Y	Jan. 19, 1869	86, 058
Paper, Shears for separating	J. A. Wilkinson	Brooklyn, N. Y	Aug. 30, 1859	25, 298
Paper shirt	H. M. Remington	Springfield, Mass	May 1, 1866	54, 408
Paper-sizing	J. Ames	Springfield, Mass	Sept. 1, 1822	....
Paper-sizing	E. Blake	Alstead, N. H	Nov. 19, 1833	....
Paper, &c., Sizing	J. M. Dorlan	East Brandywine Township, Pa.	June 11, 1872	127, 858
Paper-sizing	P. Grady	New York, N. Y	Jan. 14, 1873	134, 797
Paper-sizing	J. E. Hover	Philadelphia, Pa	Mar. 23, 1869	88, 169
Paper-sizing	G. E. Van Derburgh	Mamaroneck, N. Y	Apr. 19, 1864	42, 450
Paper-sizing	Z. C. Warren and H. C. Hulbert	Brooklyn, N. Y	Sept. 14, 1869	94, 797
Paper-sizing	W. W. Wilson and C. Dickerman	Westfield, Mass	Aug. 3, 1839	1, 286
Paper, &c., Sizing and water-proofing	J. N. Sigel	Alexandria, Va	Jan. 12, 1864	41, 241
Paper, Sizing for bank-note	J. M. Sturgeon	New York, N. Y	Dec. 3, 1867	71, 663
Paper, Sizing for colored	C. Williams	Philadelphia, Pa	Feb. 1, 1859	22, 836
Paper-sizing machine	J. Ames	Springfield, Mass	Dec. 1, 1837	495
Paper-sizing machine	L. D. Brown	Lee, Mass	Mar. 4, 1842	2, 479
Paper skirt, Lady's	J. H. Hayward	New York, N. Y	Oct. 9, 1866	58, 641
Paper sock	J. W. B. Covington	New York, N. Y	Feb. 13, 1866	52, 540
Paper-staining machine	C. Bartholomew	New York, N. Y	Dec. 3, 1867	71, 681
Paper stamper and feeder	J. O. Montignani and J. Gibson, jr.	Albany, N. Y	Mar. 26, 1872	125, 070
Paper, Stamping lace	A. Rhorbeck	New York, N. Y	May 11, 1869	89, 945
Paper-stock	M. L. Keen	Jersey City, N. J	Nov. 29, 1870	109, 742
Paper-stock	H. S. Lucas	Chester, Mass	Sept. 3, 1867	68, 370
Paper-stock and other fibers, Recovering waste alkalies from.	C. D. J. Seitz	Bury, England	Nov. 3, 1868	83, 733
Paper-stock, Apparatus for boiling and treating	G. Sinclair	Leith, Scotland	Jan. 10, 1871	110, 873
Paper-stock, Apparatus for manufacturing	M. L. Keen	Jersey City, N. J	July 9, 1872	128, 732
Paper-stock, Apparatus for preparing	J. Tiffany	Albany, N. Y	Jan. 7, 1868	73, 138
Paper-stock bleach	J. W. Rossman	Stockport, N. Y	Jan. 16, 1872	122, 783
Paper-stock, Bleaching	G. E. Marshall	Laurel, Ind	Sept. 28, 1869	95, 365
Paper-stock, Bleaching	S. T. Merrill	Beloit, Wis	Mar. 17, 1868	75, 691
Paper-stock, Bleaching	J. Tiffany and H. B. Meech	Albany and Fort Edward, N. Y.	July 31, 1866	56, 833
Paper-stock, Bleaching	F. Shelden	Fitchburgh, Mass	June 28, 1870	104, 781
Paper-stock, Boiler for treating	M. Nixon	Philadelphia, Pa	Nov. 22, 1859	26, 199

Index of patents issued from the United States Patent Office from 1790 *to* 1873, *inclusive*—Continued.

Invention.	Inventor.	Residence.	Date.	No.
Paper-stock, box-board, roofing-paper, &c	R. W. Russell	New York, N. Y	Mar. 30, 1869	88, 515
Paper-stock-cutting machine	A. L. Knight	Baltimore, Md	Nov. 12, 1867	70, 863
Paper-stock engine	V. O. Balcom and C. H. Hill	Bedford and Billerica, Mass	Dec. 2, 1856	16, 162
Paper-stock engine	J. G. Fuller	Brooklyn, N. Y	Mar. 21, 1865	46, 893
Paper-stock, &c., Forming, drying, and packing	G. S. Sellers	Seller's Landing, Ill	Jan. 5, 1864	41, 102
Paper-stock from reeds	H. Lowe	Baltimore, Md	July 13, 1858	20, 884
Paper-stock from wood, Boiler for preparation of	M. L. Keene	Jersey City, N. J	July 25, 1871	117, 427
Paper-stock from wood, Manufacture of	P. A. Chadbourne	Williamstown, Mass	Mar. 24, 1863	37, 951
Paper-stock from wood, Obtaining	W. Adamson	Philadelphia, Pa	July 18, 1871	117, 136
Paper-stock from wood, Preparing	J. H. Hawes	Stockbridge, Mass	Apr. 20, 1869	89, 220
Paper-stock, Machine for disintegrating wood for	J. Stutt	Fermanagh County, Ireland	Apr. 20, 1869	89, 255
Paper-stock, &c., Machine for pulping wood for	B. F. Barker	Curtisville, Mass	Sept. 19, 1871	119, 107
Paper-stock, Manufacture of	W. Adamson	Philadelphia, Pa	July 18, 1871	117, 134
Paper-stock, Manufacture of	W. P. Arnold	New York, N. Y	Aug. 13, 1872	130, 462
Paper-stock, Manufacture of	H. Betts	Norwalk, Conn	Aug. 1, 1865	49, 069
Paper-stock, Manufacture of	W. Deltour	New York, N. Y	Jan. 3, 1865	45, 791
Paper-stock, Manufacture of	M. L. Keen	Jersey City, N. J	Oct. 18, 1870	108, 487
Paper-stock, Manufacture of	A. T. Sturdevant	Mount Pleasant, N. Y	Oct. 21, 1873	143, 940
Paper-stock, Manufacture of	J. Tiffany	Albany, N. Y	July 31, 1866	56, 832
Paper stock, Manufacture of leather	E. and J. R. Cushman	Amherst, Mass	Aug. 21, 1860	29, 672
Paper stock, Manufacture of wood	G. H. Bliss and M. Rees	Stockbridge, Mass	June 7, 1870	103, 968
Paper-stock, Manufacturing	A. T. Sturdevant	Scipio Centre, N. Y	July 16, 1872	129, 185
Paper-stock, Preparation of	H. M. Baker	Washington, D. C	Aug. 2, 1870	105, 884
Paper-stock, Preparation of	G. E. Marshall	Louisville, Ky	May 25, 1869	90, 566
Paper-stock, Preparing	A. Randel	New York, N. Y	Apr. 30, 1861	32, 203
Paper-stock, Preparing vegetable fiber for	G. E. Sellers	Seller's Landing, Ill	Jan. 5, 1864	41, 101
Paper-stock, Preparing wood fiber for	G. E. Marshall	Laurel, Ind	Aug. 31, 1869	94, 228
Paper-stock, Preparing wood for	A. Fickett	Rochester, N. Y	June 29, 1869	92, 027
Paper-stock, Preparing woody fiber for	G. E. Sellers	Hardin County, Ill	Oct. 6, 1863	40, 217
Paper-stock, Pulp-washer for	G. E. Sellers	Seller's Landing, Ill	Jan. 24, 1865	46, 030
Paper-stock, Pulping and bleaching	J. W. Goodwyn	Petersburgh, Va	Mar. 8, 1870	100, 523
Paper-stock, &c., Rag whipper and duster for treating.	L. Brainard	Hartford, Conn	Mar. 15, 1870	100, 718
Paper-stock, Reducing and bleaching	M. B. Meech	Fort Edward, N. Y	Aug. 23, 1870	106, 711
Paper-stock to make pulp, Treating	J. Tiffany	Albany, N. Y	July 30, 1867	67, 229
Paper-stock, Treating fibrous substances for	T. Routledge	Sunderland, England	Apr. 1, 1873	137, 484
Paper-stock, Treating fibrous substances in the manufacture of.	T. Routledge	Sunderland, England	July 22, 1873	141, 016
Paper-stock, &c., Treating tarred rope, cordage, &c., for the manufacture of.	C. F. A. Simonin	Philadelphia, Pa	Sept. 19, 1871	119, 186
Paper-stock machine, Washing and bleaching	H. Monroe	Baldwinsville, N. Y	June 27, 1871	116, 338
Paper-stock-washing machine	J. E. Andrews	Coeymans Hollow, N. Y	July 14, 1868	79, 935
Paper-stock-washing machine	G. L. Lovett	Fitchburgh, Mass	Apr. 8, 1873	137, 696
Paper-stock-washing machine	H. W. Peaslee	Malden Bridge, N. Y	Jan. 23, 1855	12, 283
Paper-stuff, Apparatus for washing	S. Lenher and H. H. Spencer	Philadelphia, Pa	Mar. 21, 1865	46, 915
Paper-stuff, Boiler for preparing	C. S. Buchanan	Ballston Spa, N. Y	May 1, 1860	28, 062
Paper-stuff, Manufacture of	J. T. Coupier and M. A. C. Mellier.	Paris, France	Aug. 2, 1853	9, 910
Paper-stuff, Treating	J. A. Roth	Philadelphia, Pa	July 28, 1857	17, 895
Paper, Substitute for lining	G. Munger	New Haven, Conn	Oct. 17, 1865	50, 485
Paper-tins	E. H. Boswell	Philadelphia, Pa	May 15, 1866	54, 676
Paper to cutting-machines, Device for feeding	J. F. Schuyler	Philadelphia, Pa	Feb. 24, 1863	37, 791
Paper to printing-press, Feeding	O. Talcott	Chicago, Ill	Sept. 3, 1861	33, 217
Paper to remove ink and recover the pulp, Treatment of printed.	B. Lambert	Dover Road, Surrey, England.	July 23, 1861	32, 879
Paper to wood, Transferring impression from	A. C. Baker and M. Biddle	Albany, N. Y	Feb. 7, 1822	....
Paper, Tobacco	H. J. Hale	New York, N. Y	Feb. 7, 1865	46, 233
Paper together, Mode of fastening sheets of	E. L. Swartwout	Utica, N. Y	Mar. 22, 1859	23, 322
Paper, Top-press roller for making	M. Hunting	Watertown, Mass	Oct. 20, 1828	....
Paper, Transfer	W. Grüne	Berlin, Germany	Oct. 29, 1872	132, 659
Paper, Treating straw, wood, &c., for the manufacture of.	A. M. Hastings and S. Pettibone.	Rochester and Niagara Falls, N. Y.	Oct. 8, 1867	69, 663
Paper, Treating tracing	J. Moog	Karlsruhe, Germany	Apr. 19, 1870	102, 030
Paper, &c., Treating vegetables to obtain fiber for	J. A. Rothe	Philadelphia, Pa	Jan. 11, 1870	98, 712
Paper-trimmer	J. Hatch	South Windham, Conn	Feb. 26, 1867	62, 486
Paper-trimmer	J. Kotch	South Adams, Mass	Sept. 4, 1860	29, 891
Paper-trimmer	P. Shee	Philadelphia, Pa	Feb. 7, 1807	....
Paper-trimming machine	W. G. Ayres and S. L. Cole	Brooklyn, N. Y	Aug. 29, 1871	118, 423
Paper-trimming machine	W. G. Ayres and S. L. Cole	Brooklyn, N. Y	Feb. 13, 1872	123, 605
Paper-trimming machine	D. Pirie and A. Croom	Dundee, North Britain	Feb. 27, 1872	124, 155
Paper-trimming machine	J. F. Schuyler	Tiffin, Ohio	Nov. 29, 1870	109, 767
Paper-trimming machine	O. L. Starkey and E. W. Poston	Fort Wayne, Ind	Aug. 15, 1871	118, 066
Paper-trimming machine	H. L. Todd	Corning, N. Y	Sept. 12, 1871	118, 985
Paper-trimming machine	W. P. Yeoman	Waukegan, Ill	Feb. 14, 1871	111, 797
Paper tubes, &c., Machine for making	J. Arkell	Canajoharie, N. Y	Feb. 11, 1868	74, 190
Paper tubes, Machine for making	C. Hotz	Zurich, Switzerland	June 14, 1870	104, 312
Paper tubes, Machine for making	J. A. Olney	Providence, R. I	June 11, 1867	65, 590
Paper twine, Manufacturing	E. B. Bingham	Newark, N. J	Feb. 7, 1865	46, 208
Paper twine, Manufacturing	I. P. Till	New York, N. Y	Feb. 14, 1865	46, 405
Paper twine, Machine for making	R. V. De Guinon	South Bergen, N. J	Feb. 8, 1870	99, 654
Paper twine, Machine for making	J. B. Wortendyke	Goodwinville, N. J	May 24, 1864	42, 896
Paper under-sleeve, Ladies'	H. M. Remington	Springfield, Mass	May 15, 1866	54, 773
Paper, Using pelt for manufacturing	J. M. Thorndyke	New York, N. Y	Mar. 7, 1814	....
Paper, Utilizing the waste chloride of zinc in treating.	D. W. Harma	Pittsburgh, Pa	Oct. 31, 1871	120, 380
Paper veneers, Manufacture of	C. Walker and G. Willson	Windsor County, Vt	Mar. 20, 1849	6, 208
Paper vessel	E. M. Slayton	Conquest, N. Y	Nov. 18, 1873	144, 800
Paper vessels, Manufacture of	A. and I. Jennings	Fairfield, Conn	Nov. 26, 1867	71, 390
Paper, wadding, &c., Machine for drying	E. C. Wilson	Medway, Mass	Feb. 20, 1872	123, 850
Paper, wadding, &c., Machine for drying	E. C. Wilson	Medway, Mass	June 18, 1872	128, 085
Paper, Water-marks on ready-made	B. Westerberg	Saint Paul, Minn	Aug. 20, 1872	130, 676
Paper, Water-proof or damp-proof	S. C. Bishop	New York, N. Y	Jan. 14, 1868	73, 287
Paper washer	E. S. Hanna	Pittsburgh, Pa	Oct. 3, 1871	119, 604
Paper wearing-apparel	G. W. Day	Charlestown, Mass	June 11, 1867	65, 730
Paper wearing-apparel	A. T. Denison and E. P. Furlong.	Poland and Portland, Me	Dec. 18, 1866	60, 488
Paper wearing-apparel	T. W. Mann	Holyoke, Mass	Mar. 5, 1867	62, 550
Paper-weight	J. B. Wilson	Philadelphia, Pa	Dec. 10, 1872	133, 912

Index of patents issued from the United States Patent Office from 1790 *to* 1873, *inclusive*—Continued.

Invention.	Inventor.	Residence.	Date.	No.
Paper-weight and pen-wiper, Combined	D. W. Wright	New York, N. Y	July 9, 1867	66, 547
Paper-weight and sponge-holder, Combined	D. R. Manning	Brooklyn, N. Y	Dec. 23, 1873	145, 883
Paper-wetting apparatus	M. S. Beach	Brooklyn, N. Y	Nov. 9, 1858	22, 009
Paper-wetting machine	J. A. Lynch	Boston, Mass	Apr. 27, 1858	20, 077
Paper-wetting machine	W. Overend	Cincinnati, Ohio	Jan. 24, 1854	10, 458
Paper-wetting machine	A. Overend	Philadelphia, Pa	June 26, 1860	28, 895
Paper window-shade	L. A. Colbert	Baltimore, Md	Apr. 10, 1866	53, 786
Paper with mucilage, Machine for coating	H. E. Rile	New York, N. Y	June 18, 1867	65, 948
Paper, Wrapping	B. E. Hale	New York, N. Y	Oct. 10, 1871	119, 843
Paper, Wrapping	S. Wheeler	Albany, N. Y	July 25, 1871	117, 355
Paper, Writing	J. E. Hover	Philadelphia, Pa	Jan. 22, 1867	61, 338
Papers, Adhesive fastening for	G. R. Burdon	Waltham, Mass	Jan. 31, 1865	46, 071
Papers, Apparatus for filing and binding	P. Mac Vicar	Philadelphia, Pa	Jan. 14, 1873	134, 908
Papers for soda-powders, Machine for crimping package.	C. A. Cook	Lowell, Mass	Dec. 2, 1851	8, 556
Papers Machine for filling and folding medical powder.	M. S. Palmer	New Bedford, Mass	June 25, 1861	32, 642
Papering-machine	G. M. Lane	De Graff, Ohio	Sept. 13, 1870	107, 388
Papier-maché compound	G. F. Goetze	New York, N. Y	July 6, 1869	92, 303
Papier-maché, Making stereotype-molds of	A. Chase	Ithaca, N. Y	June 25, 1872	128, 285
Papier-maché, Mode of producing an extra surface on.	E. S. Judge	Baltimore, Md	Mar. 24, 1868	75, 766
Paraffine and obtaining it in crystals, Treating	F. Lambe	London, England	Apr. 19, 1870	102, 135
Paraffine and other hydrocarbon oils, Apparatus for burning.	H. H. Doty	London, England	Nov. 15, 1870	109, 303
Paraffine and paraffine oil, Manufacture of	H. W. C. Tweddle	Pittsburgh, Pa	Feb. 15, 1870	99, 975
Paraffine, &c., from oil, Apparatus for extracting	J. B. Meriam	Cleveland, Ohio	Feb. 12, 1867	61, 946
Paraffine, &c., Obtaining oil from	S. L. Wiegand	Philadelphia, Pa	Mar. 5, 1867	62, 583
Paraffine, Purifying	F. X. Byerley	Cleveland, Ohio	Oct. 22, 1872	132, 353
Paraffine, Purifying	S. H. Crocker	Pittsburgh, Pa	July 16, 1872	129, 463
Paraffine, Purifying	C. C. Parsons	New York, N. Y	Aug. 17, 1869	93, 739
Paraffine, Saturating wood, cloth, paper, &c., with	S. Gwynn	New York, N. Y	Feb. 20, 1866	52, 788
Paraffine, Treatment and purification of	R. M. Letchford and W. B. Nation.	Bethnal Green, England	Nov. 12, 1872	133, 042
Parallel or other rods	C. H. Clark	Wilmington, Del	Dec. 26, 1865	51, 694
Parasol	L. Battles	Gloucester, Mass	May 24, 1870	103, 412
Parasol	A. G. Davis and H. S. Frost	Watertown, Conn	Sept. 25, 1860	30, 127
Parasol	J. L. Jacquin	New York, N. Y	Dec. 5, 1871	121, 625
Parasol	C. St. John	Charlestown, Mass	Dec. 24, 1867	72, 695
Parasol and fan	J. T. Eichberg	New York, N. Y	Oct. 2, 1860	30, 213
Parasol and umbrella	E. Young	Philadelphia, Pa	Nov. 23, 1858	22, 142
Parasol, Feather-covered	G. Anton	Philadelphia, Pa	July 31, 1866	56, 691
Parasol, Reversible	J. Williams	New York, N. Y	June 29, 1869	91, 994
Parasols, Machine for retaining, adjusting, and sewing.	W. J. Tate	Philadelphia, Pa	July 26, 1870	105, 862
Parceling, Machine for manufacturing	J. H. Raynard	Lynn, Mass	Dec. 1, 1868	84, 578
Parchment, Manufacture of vegetable	E. P. Hudson	New York, N. Y	Mar. 16, 1869	87, 937
Parchment, Manufacture of vegetable	X. Karcheski	New York, N. Y	Dec. 18, 1860	30, 945
Parchment or water-proof paper, Apparatus for preparing.	E P. Hudson	New York, N. Y	Sept. 27, 1870	107, 686
Parchment-paper, Manufacture of rubber-coated	E. P. Hudson	New York, N. Y	Sept. 27, 1870	107, 687
Parchment, Solution for treating vegetable fiber for the manufacture of.	S. Gwynn	New York, N. Y	Jan. 14, 1868	73, 322
Parer and corer, Apple	S. Cruttenden	Guilford, Conn	Aug. 25, 1809	....
Parer and corer, Apple	H. S. Leonard	Attica, Ohio	Dec. 17, 1872	134, 078
Parer and corer, Apple	R. W. Mitchell	Springfield, Ohio	Apr. 13, 1838	686
Parer and corer, Apple	H. Selick	Lewiston, Pa	Aug. 21, 1866	57, 391
Parer and cutter, Peach	H. S. Hilsman	Madison, Ga	Dec. 4, 1860	30, 820
Parer and quarterer, Apple	C. S. Pratt	Paris, Me	Dec. 28, 1833	....
Parer and slicer, Apple	G. W. Brokaw	Lodi, N. Y	May 11, 1869	89, 971
Parer and slicer, Apple	C. A. Foster	Fitchburgh, Mass	July 11, 1871	116, 943
Parer and slicer, Apple	D. H. Whittemore	Worcester, Mass	Aug. 10, 1869	93, 574
Parer, Apple	A. G. Batchelder	Lowell, Mass	Mar. 2, 1869	87, 322
Parer, Apple	G. Bergner	Washington, Mo	Feb. 11, 1873	135, 622
Parer, Apple	J. D. Browne	Cincinnati, Ohio	Sept. 9, 1856	15, 683
Parer, Apple	J. Bullock, jr., and S. Benson	New York, N. Y	July 24, 1847	5, 197
Parer, Apple	C. P. Carter	Ware, Mass	Oct. 16, 1849	6, 789
Parer, Apple	C. P. Carter	Ware, Mass	Aug. 26, 1856	15, 603
Parer, Apple	C. P. Carter	Ware, Mass	Nov. 18, 1856	16, 104
Parer, Apple	R. P. Clark	Johnstown, N. Y	Sept. 20, 1859	25, 489
Pare , Apple	M. Coates	....	Feb. 14, 1803	....
Parer, Apple	H. A. Frost	Worcester, Mass	Aug. 6, 1861	33, 016
Parer, Apple	W. M. Griscom	Reading, Pa	May 14, 1872	126, 803
Parer, Apple	W. M. and C. W. Hardy	East Strong, Me	Jan. 29, 1861	31, 238
Parer, Apple	J. W. Hatch	Bedford County, Va	Feb. 3, 1836	....
Parer, Apple	S. S. Hersey	Farmington, Me	June 18, 1861	32, 561
Parer, Apple	S. S. Hersey	Farmington, Me	Aug. 30, 1864	43, 990
Parer, Apple	F. W. Hudson	Leominster, Mass	Dec. 2, 1862	37, 038
Parer, Apple	F. W. Hudson	Leominster, Mass	Mar. 5, 1872	124, 272
Parer, Apple	H. Keyes	Terre Haute, Ind	Nov. 20, 1866	59, 843
Parer, Apple	H. Keyes	Leominster, Mass	Dec. 16, 1856	16, 240
Parer, Apple	W. H. Lazelle	New York, N. Y	Jan. 25, 1853	9, 558
Parer, Apple	E. Manley	Marion, N. Y	Nov. 17, 1863	40, 640
Parer, Apple	S. N. Maxam	Shelburne Falls, Mass	Apr. 10, 1855	12, 689
Parer, Apple	J. F. and E. P. Monroe	Fitchburgh, Mass	Mar. 24, 1868	75, 951
Parer, Apple	R. and A. Mosher	Galway, N. Y	Dec. 28, 1829	....
Parer, Apple	W. A. C. Oakes	Reading, Pa	Dec. 10, 1872	133, 796
Parer, Apple	E. L. Pratt	Worcester, Mass	Oct. 4, 1853	10, 078
Parer, Apple	E. L. Pratt	Philadelphia, Pa	Apr. 29, 1856	14, 775
Parer, Apple	E. L. Pratt	Boston, Mass	Oct. 6, 1863	40, 185
Parer, Apple	E L. Pratt	Boston, Mass	Jan. 12, 1864	41, 266
Parer, Apple	E. L. Pratt	Boston, Mass	Aug. 23, 1864	43, 955
Parer, Apple	W. Robb, jr	South Stoddard, N. H	Nov. 22, 1870	109, 454
Parer, Apple	M. Smith	New Haven, Conn	Aug. 26, 1856	15, 625
Parer, Apple	J. Voak	Penn Yan, N. Y	July 12, 1864	43, 537
Parer, Apple	J. White	Antrim, N. H	June 3, 1862	35, 482
Parer, Apple	J. White	Antrim, N. H	Feb. 16, 1864	41, 658
Parer, Apple	D. H. Whittemore	Worcester, Mass	Nov. 20, 1866	59, 884

Index of patents issued from the United States Patent Office from 1790 *to* 1873, *inclusive*—Continued.

Invention.	Inventor.	Residence.	Date.	No.
Parer, Apple	C. A. Wiggin	North Sandwich, N. H	Aug. 4, 1868	80, 580
Parer, Apple	G. H. Wilde	Aurora, Ill	July 6, 1869	92, 240
Parer, Apple and potato	J. L. Clewell, jr., and W. F. Schatz.	Nazareth, Pa	Nov. 20, 1860	38, 667
Parer, corer, &c., Apple	G. W. Bennett	Smithville, Ind	Feb. 25, 1873	136, 206
Parer, corer, and cutter, Apple	W. A. Coe	Greensborough, N. C	Feb. 19, 1867	62, 184
Parer, corer, and slicer, Apple	J. I. Armfield	Jamestown, N. C	Dec. 20, 1859	26, 464
Parer, corer, and slicer, Apple	A. Clark	La Fayette, Ind	June 9, 1868	78, 721
Parer, corer, and slicer, Apple	J. J. Parker	Marietta, Ohio	Sept. 11, 1860	29, 988
Parer, corer, and slicer, Apple	J. Shobe	Upper Principio, Md	Nov. 15, 1870	109, 350
Parer, corer, and slicer combined, Apple	G. Borgner	Washington, Mo	Jan. 10, 1872	122, 553
Parer, cutter, and corer, Apple	G. W. Bennett	Harrodsburgh, Ind	May 24, 1870	103, 284
Parer, Cutter-head for apple	E. L. Pratt	Boston, Mass	Oct. 6, 1863	40, 216
Parer, Fork for peach	D. H. Goodell	Antrim, N. H	May 10, 1870	102, 934
Parer, Fork for peach	C. D. House	Lake Village, N. H	Aug. 17, 1869	93, 887
Parer, Fruit	A. G. Batchelder	Lowell, Mass	Jan. 3, 1872	122, 305
Parer, Fruit	L. D. Farwell and A. W. Goddard.	Lancaster and Clinton, Mass.	May 5, 1868	77, 600
Parer, Fruit	D. H. Goodell	Antrim, N. H	June 18, 1867	65, 804
Parer, Fruit	R. P. Scott	Cadiz, Ohio	May 16, 1871	114, 867
Parer, Fruit	A. Turnbull and R. L. Webb	New Britain, Conn	June 10, 1873	139, 837
Parer, Fruit	L. S. Woodbury	South Antrim, N. H	Dec. 20, 1870	110, 322
Parer, Fruit and vegetable	A. G. Batchelder	Lowell, Mass	June 7, 1870	103, 830
Parer, Fruit and vegetable	A. G. Batchelder	Lowell, Mass	Mar. 26, 1872	125, 004
Parer, Fruit and vegetable	T. Learing	South Norwalk, Conn	Dec. 5, 1871	121, 669
Parer, Fruit and vegetable	H. Soggs	Columbus, Pa	Dec. 13, 1870	110, 167
Parer, Peach	J. H. Brown	Mitchell, Ind	Oct. 6, 1868	82, 794
Parer, Peach	W. A. Coe	Greensborough, N. C	Oct. 30, 1860	30, 527
Parer, Peach	A. Hermans	Henderson, Tex	Dec. 27, 1859	26, 640
Parer, Peach	E. L. Pratt	Boston, Mass	Aug. 23, 1864	43, 956
Parer, Peach	M. Smith	New Haven, Conn	May 15, 1860	28, 310
Parer, Potato	W. B. Coates	Philadelphia, Pa	Nov. 29, 1859	26, 254
Parer, Potato	C. Disston	Philadelphia, Pa	July 3, 1860	28, 973
Parer, Potato	H. Underwood	Tolland, Conn	Aug. 2, 1864	43, 725
Parer, Potato	W. Zeiger	Elmore, Ohio	Apr. 20, 1869	89, 190
Parer, Potato and fruit	M. B. Atkinson	Georgetown, D. C	Sept. 13, 1870	107, 321
Parer, Vegetable	M. P. Smith	Baltimore, Md	Apr. 2, 1872	125, 225
Parer, Vegetable and fruit	E. D. Averell and J. Malan	Brooklyn, N. Y	Mar. 8, 1870	100, 583
Paring and chopping machine	S. H. Morse	North Jay, Me	June 27, 1871	116, 342
Paring and coring fruit	P. W. Hardwick	Wayne County, Ind	Sept. 25, 1849	6, 735
Paring and cutting machine, Peach	J. O. Ward	Pleasant Valley, N. Y	Oct. 21, 1851	8, 460
Paring and slicing knife	H. P. Brooks	Waterbury, Conn	Oct. 17, 1871	119, 963
Paring and slicing machine, Apple	G. H. Hubbard	Shelburne Falls, Mass	Jan. 27, 1857	16, 517
Paring and slicing machine, Apple	R. W. Thickins	Brasher Iron Works, N. Y	July 28, 1857	17, 901
Paring and slicing machine, Apple	L. Van Hoesen	New Haven, Conn	Dec. 4, 1855	13, 891
Paring and slicing machine, Apple	D. H. Whittemore	Chicopee Falls, Mass	Jan. 13, 1857	16, 417
Paring apples, Machine for	J. D. Browne	Cincinnati, Ohio	May 6, 1856	14, 800
Paring apples, Machine for	J. O. M. Ingersoll	Ithaca, N. Y	Jan. 20, 1857	16, 443
Paring apples, Machine for	B. F. Joslyn	Worcester, Mass	Mar. 17, 1857	16, 843
Paring apples, Machine for	H. Keyes	Leominster, Mass	June 17, 1856	15, 133
Paring apples, Machine for	J. F. and E. P. Monroe	Fitchburgh, Mass	Jan. 27, 1863	37, 516
Paring apples, Machine for	J. J. Parker	Marietta, Ohio	Apr. 7, 1857	16, 993
Paring apples, Machine for	J. D. Seagrave	Milford, Mass	Apr. 18, 1854	10, 785
Paring apples, Machine for	J. D. Seagrave	Worcester, Mass	June 17, 1856	15, 148
Paring apples, Machine for	D. H. Whittemore	Worcester, Mass	Feb. 17, 1857	16, 666
Paring apples, potatoes, &c., Machine for	E. L. Pratt	Philadelphia, Pa	Nov. 11, 1856	16, 080
Paring, coring, and quartering apples, Machine for	C. F. Bosworth	Petersham, Mass	June 9, 1857	17, 484
Paring, coring, and slicing apples	J. Weed	Painesville, Ohio	July 31, 1849	6, 619
Paring, coring, and slicing apples, Machine for	A. M. Blain	Deerfield, Va	June 11, 1867	65, 534
Paring, coring, and slicing apples, Machine for	J. Stewart and W. Campbell	Fowler, Ill	June 24, 1873	140, 315
Paring, cutting, and coring machine	W. Weaver	Phœnixville, Pa	Apr. 3, 1866	53, 754
Paring-knife	H. P. Brooks	Waterbury, Conn	Sept. 17, 1872	131, 329
Paring-knife	J. H. Bruen	Elmira, N. Y	Jan. 23, 1872	122, 934
Paring-knife	J. Lebeau	Cincinnati, Ohio	July 10, 1866	56, 235
Paring-knife	G. H. Vickroy	Washington, D. C	June 18, 1872	127, 998
Paring knife, Apple	A. Oot	Minetto, N. Y	Oct. 5, 1858	21, 695
Paring knife, Vegetable	W. Veber, jr	Shingle Creek, N. Y	May 25, 1869	90, 474
Paring knife, Vegetable and fruit	H. Soggs	Columbus, Pa	Jan. 25, 1870	99, 119
Paring machine, Fruit	W. M. Howland	Leominster, Mass	Aug. 30, 1864	44, 044
Paring, quartering, and coring apples, Machine for.	W. Badger	Boston, Mass	Feb. 16, 1809	......
Paring, slicing, and coring machine, Apple	J. J. Parker	Marietta, Ohio	July 6, 1858	20, 814
Park and garden seat	S. Macferran	Philadelphia, Pa	May 4, 1869	89, 779
Parlor-bath	J. R. Worster	New York, N. Y	Apr. 27, 1869	89, 370
Parlor-fountain for diffusing liquids	W. Altick	Dayton, Ohio	Mar. 1, 1870	100, 349
Parlor-heater	J. Carten	Utica, N. Y	Oct. 15, 1861	33, 474
Parlors and saloons, Covering for walls of	J. M. Soutrenon	New York, N. Y	Nov. 5, 1867	70, 640
Paris white and whiting	P. Ferris	Greenwich, Conn	Apr. 22, 1835	......
Parral and bow	J. Dame	Portsmouth, N. H	Sept. 26, 1854	11, 721
Partition, Fire-proof	C. F. Brand	Philadelphia, Pa	Nov. 11, 1873	144, 501
Partition, Movable	D. J. Williams	Birmingham, England	Apr. 4, 1871	113, 476
Passe-partout frame	P. Mignot	New York, N. Y	Apr. 26, 1870	102, 296
Passenger and freight elevator	S. Nash	Cambridge, Mass	Aug. 5, 1873	141, 513
Passenger and station register	J. C. Hackett	Sacramento, Cal	Aug. 10, 1869	93, 531
Passenger-boat	N. Cushman and I. Loomis	Whitehall, N. Y	Jan. 8, 1831	......
Passenger-elevator	J. S. Baldwin	Newark, N. J	Apr. 14, 1868	76, 694
Passenger-recorder	G. H. Aldrich	New York, N. Y	Apr. 25, 1871	114, 084
Passenger-recorder, Vehicle	J. T. Aldrich	New York, N. Y	June 13, 1871	115, 801
Passenger-register	S. L. Cooper	Lewistown, Pa	Aug. 5, 1873	141, 492
Passenger-register	J. Enright	Louisville, Ky	June 23, 1868	79, 060
Passenger-register	E. Hackett	New York, N. Y	Mar. 7, 1865	46, 663
Passenger-register	E. Hambujer	New York, N. Y	Jan. 16, 1866	52, 043
Passenger-register	W. Helffricht	Philadelphia, Pa	Jan. 8, 1867	61, 065
Passenger-register	W. Helffricht	Philadelphia, Pa	Mar. 26, 1867	63, 245
Passenger-register	T. S. Huntington and A. Fulton	Bellefontaine, Ohio	July 14, 1868	79, 911
Passenger-register	T. Jacobs	Philadelphia, Pa	Aug. 4, 1868	80, 741
Passenger-register	J. Melling	Rochester, N. Y	Oct. 8, 1867	69, 688
Passenger-register	A. F. Nagle	Providence, R. I	Apr. 7, 1868	76, 502

Index of patents issued from the United States Patent Office from 1790 *to* 1873, *inclusive*—Continued.

Invention.	Inventor.	Residence.	Date.	No.
Passenger-register	M. Springwoter	Louisville, Ky	Apr. 25, 1871	114, 060
Passenger-register, Automatic	H. H. Trenor	New York, N. Y	Oct. 5, 1869	95, 540
Passenger-register, Automatic vehicle	C. Ottinger	Philadelphia, Pa	Apr. 23, 1870	106, 610
Passenger-register, Car	J. Kurz	Philadelphia, Pa	Feb. 21, 1871	111, 948
Passenger-register, Horse-car	E. L. Fitch	West Eau Claire, Wis	June 21, 1870	104, 566
Passenger-register, Street-car	W. Willis	Chicago, Ill	Nov. 2, 1869	96, 526
Passenger-register, Street-car, &c., electro-magnetic.	W. H. Mumler	Boston, Mass	July 2, 1872	128, 500
Passenger-register, Vehicle	T. Ollis	Liverpool, England	Dec. 21, 1869	98, 098
Passenger-register, Vehicle	J. Rhoads	Harrisburgh, Pa	Aug. 2, 1870	105, 980
Passing-box	E. A. G. Roulstone	Roxbury, Mass	Mar. 25, 1862	34, 777
Paste	J. E. Hover	Philadelphia, Pa	Nov. 19, 1867	71, 174
Paste, Broom-corn-seed flour for manufacture of	J. W. Tallmadge	Boston, Mass	Dec. 30, 1873	145, 032
Paste-making apparatus	G. G. Noah	Charlestown, Mass	May 6, 1873	138, 519
Paste-making machine	G. G. Noah	Boston, Mass	Jan. 7, 1868	73, 115
Paste, Manufacture of	G. G. Noah	Boston, Mass	June 14, 1870	104, 190
Paste, Prepared	S. Merwin	Springfield, Mass	Oct. 2, 1866	58, 454
Paste, Preserving and making	G. G. Noah	Charlestown, Mass	May 21, 1872	127, 097
Pasteboard, Adhesive compound for preparing	G. L. Jaeger	New York, N. Y	July 5, 1870	105, 084
Pasteboard box	L. Carrier	Providence, R. I	July 23, 1867	67, 022
Pasteboard box	S. W. Millichamp	Rochester, N. Y	Nov. 26, 1872	133, 330
Pasteboard box	W. Pratt	New York, N. Y	Feb. 21, 1871	112, 075
Pasteboard boxes, Machine for forming	C. A. Maxfield	New York, N. Y	May 5, 1868	77, 502
Pasteboard boxes, Machine for making	J. T. Foster	Jersey City, N. J	Mar. 19, 1867	62, 952
Pasteboard, composition-board, panels, &c., from reed, Manufacture of.	W. P. Arnold	New York, N. Y	Aug. 13, 1872	130, 463
Pasteboard, &c., cutting machine	D. Burhaus	Burlington, Iowa	July 7, 1857	17, 723
Pasteboard cutting machine	E. E. Clarke	New Haven, Conn	Feb. 21, 1865	46, 448
Pasteboard-cutting machine	H. A. Gage	Manchester, N. H	Nov. 16, 1869	96, 791
Pasteboard-cutting machine	S. Orth	Philadelphia, Pa	Nov. 19, 1867	71, 048
Pasteboard cutting and scoring machine	S. Orth	Philadelphia, Pa	Sept. 12, 1865	49, 910
Pasteboard, Drying	J. H. Patterson	Schaghticoke, N. Y	Mar. 26, 1861	31, 822
Pasteboard for boxes, Machine for cutting	E. E. Clarke	New Haven, Conn	Mar. 3, 1857	16, 719
Pasteboard for boxes, Machine for cutting	E. E. Clarke	New Haven, Conn	Feb. 21, 1865	46, 522
Pasteboard for boxes, Machine for cutting	E. E. Clarke	New Haven, Conn	Feb. 28, 1865	46, 604
Pasteboard lining and drying machine	G. L. Jaeger	New York, N. Y	May 10, 1870	102, 942
Pasteboard-lining machine	G. H. Dickerman	Somerville, Mass	Mar. 5, 1872	124, 258
Pasteboard-lining machine	G. L. Jaeger	New York, N. Y	Feb. 14, 1871	111, 849
Pasteboard-lining machine	G. L. Jaeger	New York, N. Y	Apr. 30, 1872	126, 301
Pasteboard, Machine for making	L. Koch	New York, N. Y	Mar. 31, 1857	16, 927
Pasteboard, Machinery for making	O. W. Fiske	Dedham, Mass	July 11, 1854	11, 313
Pasteboard. Machinery for pressing water out of	L. Koch	New York, N. Y	Mar. 31, 1857	16, 928
Pasteboard, Manufacture of	W. E. Hale	Chicago, Ill	Mar. 2, 1869	87, 407
Pasteboard, Manufacture of	H. L. Palmer	Stillwater, N. Y	Mar. 2, 1869	87, 359
Pasteboard, Manufacture of stone	J. Smith	Mendon, N. Y	Mar. 13, 1855	12, 531
Pasteboard, &c., Manufacturing	F. A. Taft	Dedham, Mass	July 20, 1831	....
Pasteboard or paper, Machine for drying	P. Clark	Rahway, N. J	June 9, 1857	17, 489
Pasteboard paper for bandbox, &c	I. Sanderson	Milton, Mass	May 15, 1830	....
Pasteboard, sheet-iron, &c., Shears for cutting	A. Hardy	Boston, Mass	Aug. 12, 1831	....
Pastile, Disinfecting	A. Lanergan	Boston, Mass	July 1, 1856	15, 243
Pastry-board	L. E. Higby	Shelburne Falls, Mass	Jan. 31, 1860	26, 989
Pastry cutter and crimper	W. R. Marie	Boston, Mass	Apr. 13, 1869	88, 883
Pastry-cutter	J. Stephen	Womelsdorf, Pa	Aug. 6, 1867	67, 601
Pastry-jigger	J. Redding and J. B. Coe	Charlestown and Boston, Mass.	Apr. 14, 1868	76, 813
Pastry-roller	J. S. Foster	San Francisco, Cal	Nov. 20, 1866	59, 905
Pastry-roller	H. S. Maltby	Cincinnati, Ohio	Aug. 31, 1869	94, 227
Pastry-roller	W. C. Pierce	Philadelphia, Pa	Nov. 5, 1872	132, 736
Pastry-roller	A. L. Taylor	Springfield, Vt	July 16, 1867	66, 909
Patch-bolts, Machine for making	J. Kaylor	Reserve Township, Pa	Oct. 15, 1867	69, 816
Patient-elevator	C. Marx	New York, N. Y	May 19, 1863	38, 595
Pattern: *See* Car-seat-end pattern. Carpet-pattern. Casting-pattern. Coat-pattern. Dress-pattern. Embroidery-pattern. Garment-pattern. Hat-brim pattern. Heel-pattern. Oil-cloth pattern. Pipe-pattern. Pipe-elbow pattern. Plaster pattern. Shawl-pattern. Shirt-pattern. Shirt-bosom pattern. Waving pattern. Weaving-pattern.				
Pattern chart	C. Freetsha	Paterson, N. J	Jan. 29, 1867	61, 531
Pattern-cutting machine, Segment	J. Bean, jr	Pent Water, Mich	June 27, 1871	116, 258
Pattern-lifter	E. Card	North Providence, R. I	May 28, 1867	65, 169
Pattern-making machine	S. Huffman	Carthage Township, Ill	June 26, 1866	55, 869
Pattern, Transferable embroidery	C. H. Hudson	Aske Terrace, England	Mar. 1, 1864	41, 776
Patterns for flaring vessels, Drawing	O. B. Vandenburg	Findlay, Ohio	July 4, 1871	116, 647
Patterns for molding, Preparing	H. Wright	Milford, N. Y	Jan. 1, 1861	31, 051
Patterns, Measuring and laying out ladies' dress	H. M. Burrows	Norwich, N. Y	May 4, 1869	89, 735
Patterns to supporting-plates, Attaching	J. H. Harper	Pittsburgh, Pa	Jan. 7, 1873	134, 663
Pavement	W. W. Boyington	Chicago, Ill	May 19, 1868	78, 046
Pavement	G. H. Chinnock	New York, N. Y	Nov. 5, 1872	132, 801
Pavement	S. J. Davenport and J. Ward	Albany, N. Y	Oct. 28, 1873	143, 965
Pavement	W. H. De Valin	Sacramento, Cal	Nov. 18, 1873	144, 748
Pavement	A. Dilger	Saint Louis, Mo	July 5, 1870	104, 941
Pavement	L. S. Filbert	Philadelphia, Pa	Oct. 25, 1870	108, 696
Pavement	R. Foley	New York, N. Y	Sept. 15, 1868	82, 102
Pavement	E. Gomez	New York, N. Y	Dec. 17, 1872	134, 049
Pavement	H. A. Gunther	New York, N. Y	Oct. 24, 1871	120, 268
Pavement	I. H. Hobbs	Philadelphia, Pa	July 25, 1871	117, 286
Pavement	B. B. Hotchkiss	New York, N. Y	Mar. 26, 1867	63, 162
Pavement	W. W. Hubbell	Philadelphia, Pa	May 30, 1871	115, 475
Pavement	D. Huestis	Cold Spring, N. Y	July 24, 1866	56, 563
Pavement	J. S. Kelly	New York, N. Y	Aug. 30, 1870	106, 832
Pavement	J. S. Kelly	New York, N. Y	Dec. 5, 1871	121, 524
Pavement	W. Kilburn	Napa, Cal	July 15, 1873	140, 835
Pavement	H. A. Lacey	Detroit, Mich	Dec. 17, 1872	134, 073
Pavement	H. G. McGonegal	New York, N. Y	Feb. 4, 1868	74, 110

Index of patents issued from the United States Patent Office from 1790 *to* 1873, *inclusive*—Continued.

Invention.	Inventor.	Residence.	Date.	No.
Pavement	A. B. McKeon	Rutherford Park, N. J	May 10, 1870	102, 846
Pavement	J. Merlette, sr	Boundbrook, N. J	June 4, 1872	127, 628
Pavement	B. F. Miller	New York, N. Y	Apr. 23, 1867	64, 126
Pavement	B. F. Miller	New York, N. Y	July 7, 1868	79, 770
Pavement	S. C. Prescott	Jersey City, N. J	June 18, 1872	127, 986
Pavement	S. C. Prescott	Jersey City, N. J	July 23, 1872	129, 861
Pavement	W. H. Shurtliff	Providence, R. I	Feb. 12, 1867	61, 958
Pavement	C. W. Stafford	Saybrook, Conn	Mar. 26, 1867	63, 324
Pavement	C. W. Stafford	Saybrook, Conn	June 25, 1867	66, 201
Pavement	C. W. Stafford	Saybrook, Conn	Sept. 24, 1867	69, 297
Pavement	C. W. Stafford	Saybrook, Conn	Mar. 10, 1868	75, 309
Pavement	A. Stevens and L. A. Cauvet	New York, N. Y	July 5, 1870	105, 138
Pavement	A. Van Camp and M. M. Hodgman.	Washington, D. C., and Saint Louis, Mo.	May 17, 1870	103, 105
Pavement	C. G. Von Tagen	Philadelphia, Pa	Feb. 27, 1872	124, 022
Pavement	E. Whitehead	Cincinnati, Ohio	July 10, 1866	56, 303
Pavement	A. Wyckoff	Elmira, N. Y	June 2, 1868	78, 635
Pavement	C. Williams	New York, N. Y	Mar. 10, 1868	75, 504
Pavement	C. Williams	New York, N. Y	Mar. 24, 1868	76, 021
Pavement	C. Williams	New York, N. Y	July 21, 1868	80, 260
Pavement	C. Williams	New York, N. Y	July 21, 1868	80, 261
Pavement	M. E. Worrell	Monmouth, Ill	Jan. 7, 1873	134, 622
Pavement and for other purposes, Composition for street.	J. W. Brown	Washington, D. C	Feb. 2, 1869	86, 355
Pavement and foundation for the same, Composition	G. H. Moore	Norwich, Conn	Apr. 16, 1872	125, 831
Pavement and gutter, Cast-iron	M. Pennock	Kennett's Square, Pa	Jan. 26, 1864	41, 412
Pavement, and machinery for, Manufacture of concrete	W. H. Smith	New York, N. Y	Jan. 2, 1872	122, 498
Pavement and railway combined	B. C. Smith	Burlington, N. J	Jan. 15, 1861	31, 131
Pavement and railway track, Mode of constructing a combined street.	R. Montgomery	New York, N. Y	Jan. 18, 1859	22, 665
Pavement and sidewalk, Composition	N. B. Abbott	Brooklyn, N. Y	Feb. 13, 1872	123, 658
Pavement, Apparatus for laying concrete	G. H. Moore	Norwich, Conn	Feb. 27, 1872	124, 148
Pavement, Apparatus for the manufacture of asphalt.	A. D. Foote	Guilford, Conn	Aug. 6, 1872	130, 289
Pavement, Asphalt	C. P. Alsing	New York, N. Y	May 5, 1868	77, 565
Pavement, Asphalt	C. P. Alsing	New York, N. Y	Jan. 4, 1870	98, 460
Pavement, Asphalt	C. Burlew	Washington, D. C	Nov. 21, 1871	121, 082
Pavement, Asphalt	A. C. Campbell	New York, N. Y	Oct. 25, 1870	108, 566
Pavement, Asphalt	A. J. Crawford	Brooklyn, N. Y	Feb. 6, 1872	123, 458
Pavement, Asphalt	E. J. De Smedt	New York, N. Y	Nov. 29, 1870	109, 597
Pavement, Asphalt	S. R. Scharf	Baltimore, Md	Apr. 6, 1869	88, 746
Pavement, Asphalt	S. R. Scharf	Baltimore, Md	Jan. 24, 1871	111, 151
Pavement, Asphalt and crushed-rock	A. Van Camp	Washington, D. C	Nov. 14, 1871	120, 914
Pavement, Asphalt road and	E. J. De Smedt	New York, N. Y	May 31, 1870	103, 582
Pavement, Asphaltum or concrete	A. G. Day	Seymour, Conn	June 21, 1870	104, 562
Pavement, Belgian	C. Guidet	New York, N. Y	Oct. 2, 1866	58, 407
Pavement, Bituminous-rock	J. L. Fulton and J. Brace	Covington, Ky	May 17, 1870	103, 168
Pavement, Block-connecting device for wood	D. H. Mulford	Saratoga Springs, N. Y	Aug. 8, 1871	117, 802
Pavement-blocks, Forming	G. L. Eagan	San Francisco, Cal	Apr. 16, 1872	125, 798
Pavement-blocks from furnace-slag	T. A. Luckenbach	New York, N. Y	June 11, 1872	127, 699
Pavement, building-blocks, &c., Composition for	J. K. Griffin	Waterdown, Canada	Oct. 15, 1872	132, 153
Pavement, Cast-iron	A. Farran	Longwood, Mass	Jan. 4, 1870	98, 480
Pavement, Cast-iron	P. M. Hutton	Troy, N. Y	Mar. 4, 1856	14, 384
Pavement, Cast-iron	G. Neilson	Boston, Mass	Apr. 17, 1855	12, 755
Pavement, Cast-iron	G. M. Ramsay	New York, N. Y	Sept. 23, 1856	15, 776
Pavement, Cast-iron	A. P. Robinson	New York, N. Y	Apr. 22, 1856	14, 736
Pavement, Cast-iron	S. T. Savage	Albany, N. Y	Feb. 8, 1859	22, 896
Pavement, Cast-iron	C. J. Shepard	Brooklyn, N. Y	Mar. 3, 1857	16, 757
Pavement, Cast-iron	C. Warner	New York, N. Y	Jan. 2, 1855	12, 172
Pavement, Cast iron	C. Warner	New York, N. Y	Feb. 1, 1859	22, 835
Pavement, Cellular-iron	S. H. Titus and O. Des Granges	Saint Louis, Mo	Oct. 13, 1857	18, 417
Pavement, Cementing and water-proofing block	B. F. Camp	New York, N. Y	Oct. 29, 1872	132, 561
Pavement, Combination	L. S. Filbert and J. Taylor	Philadelphia, Pa., and New York, N. Y.	Jan. 2, 1872	122, 448
Pavement, Combined concrete and wood	M. Flanigan	Detroit, Mich	May 30, 1871	115, 457
Pavement, Combined wood and concrete	J. W. Brown	Washington, D. C	June 21, 1870	104, 551
Pavement, Combined wood and concrete	S. C. Prescott	Jersey City, N. J	June 20, 1871	116, 217
Pavement, Composition	N. B. Abbott	Brooklyn, N. Y	June 17, 1873	139, 848
Pavement, Composition	A. G. Anderson	Hoboken, N. J	June 13, 1871	115, 924
Pavement, Composition	H. L. Crawford	Brooklyn, N. Y	Aug. 3, 1869	93, 280
Pavement, Composition	J. P. Cranford	Brooklyn, N. Y	Mar. 23, 1869	88, 139
Pavement, Composition	J. R. Hayes	Philadelphia, Pa	Feb. 27, 1872	124, 059
Pavement, Composition	D. C. Heller	Reading, Pa	Dec. 12, 1865	51, 513
Pavement, Composition	D. C. Heller	Reading, Pa	Apr. 22, 1873	138, 023
Pavement, Composition	E. S Martin	Rahway, N. J	July 9, 1872	128, 805
Pavement, Composition	T. Price	Pittsburgh, Pa	Mar. 12, 1872	124, 509
Paving-composition	A. Thiele	Tickfaw, La	Sept. 9, 1873	142, 594
Pavement, Composition	Z. Waters and H. B. Bellamy.	Bloomington, Ill	Dec. 31, 1872	134, 500
Pavement, &c., Composition blocks for	C. W. M. Smith	San Francisco, Cal	Aug. 27, 1872	130, 949
Pavement, Composition cement for	A. B. McKeon	Rutherford Park, N. J	Aug. 10, 1869	93, 659
Pavement, Composition for	N. B. Abbott	Brooklyn, N. Y	Mar. 21, 1871	112, 764
Pavement, Composition for	R. Atkinson	Mount Vernon, N. Y	Oct. 11, 1864	44, 589
Pavement, Composition for	C. P. Burges and J. R. Stevenson.	Rochester, N. Y	Feb. 14, 1871	111, 724
Pavement, Composition for	G. W. Davis	Fitchburgh, Mass	Oct. 9, 1866	58, 718
Pavement, &c., Composition for	W. C. Dodge	Washington, D. C	Oct. 16, 1866	58, 789
Pavement, Composition for	G. Dubellé	Boston, Mass	Nov. 24, 1868	84, 272
Pavement, Composition for	G. L. Eagan	San Francisco, Cal	May 20, 1873	139, 127
Pavement, Composition for	L. S. Filbert	Philadelphia, Pa	Nov. 14, 1871	120, 956
Pavement, Composition for	W. P. Ford and A. A. Moore	Concord, N. H	June 11, 1867	65, 660
Pavement, Composition for	P. Harder	Danville, Pa	May 19, 1863	38, 582
Pavement, Composition for	C. B. Harris	New York, N. Y	Mar. 7, 1871	112, 339
Pavement, &c., Composition for	J. Hartlieb	Reading, Pa	Nov. 6, 1866	59, 391
Pavement, Composition for	J. R. Hayes	New York, N. Y	Sept. 20, 1870	107, 489
Pavement, Composition for	F. E. Matthews	Chicago, Ill	Apr. 25, 1871	114, 172
Pavement, Composition for	G. H. Moore	Norwich, Conn	Nov. 28, 1871	121, 397
Pavement, &c., Composition for	G. Scrimshaw	Milesburgh, Pa	Aug. 21, 1860	29, 722

Index of patents issued from the United States Patent Office from 1790 *to* 1873, *inclusive*—Continued.

Invention.	Inventor.	Residence.	Date.	No.
Pavement, Composition for	H. F. Snow and J. H. Davis	Dover, N. H	Sept. 1, 1868	81, 698
Pavement, Composition for	H. F. Snow and J. H. Davis	Dover, N. H	Nov. 16, 1869	96, 988
Pavement, Composition for	H. F. Snow and J. H. Davis	Dover, N. H	June 18, 1872	127, 992
Pavement, Composition for	A. Wanner	Chicago, Ill	Dec. 17, 1872	134, 112
Pavement, Composition for	E. Wenger	Richmond, Ind	Aug. 25, 1868	81, 564
Pavement, Composition for	S. J. Whiting	Philadelphia, Pa	Sept. 9, 1873	142, 601
Pavement, Composition for concrete	H. Burlew	Lockhaven, Pa	Sept. 2, 1862	36, 338
Pavement, Composition for concrete	R. Fisk	New York, N. Y	Oct. 8, 1867	69, 738
Pavement, Composition for concrete	E. W. Ranney	New York, N. Y	Jan. 4, 1870	98, 522
Pavement, Composition for concrete	E. Seeley	Scranton, Pa	Sept. 6, 1864	44, 117
Pavement, Composition for concrete	A. Van Camp and M. M. Hodgman.	Washington, D. C., and Saint Louis, Mo.	Mar. 15, 1870	100, 954
Pavement, Composition for forming blocks for	J. C. Tucker	New York, N. Y	July 2, 1872	128, 680
Pavement, Compound for	J. C. and M. V. Campbell	Syracuse, N. Y	Jan. 21, 1868	73, 434
Pavement, Compound for	R. Crawford, jr	Brooklyn, N. Y	Apr. 9, 1872	125, 545
Pavement, Compound for making concrete	H. M. Conklin	Carlstadt, N. J	Mar. 15, 1870	100, 730
Pavement, Concrete	D. W. Bailey	Chelsea, Mass	Nov. 23, 1869	97, 149
Pavement, Concrete	G. Bassett	Syracuse, N. Y	Oct. 29, 1872	132, 620
Pavement, Concrete	G. W. Dean	Stamford, Conn	Sept. 2, 1873	142, 444
Pavement, Concrete	N. H. Downs	Birmingham, Conn	Apr. 19, 1870	102, 097
Pavement, Concrete	S. Filbert	Philadelphia, Pa	Jan. 9, 1872	122, 591
Pavement, Concrete	W. Gilbert	Detroit, Mich	Apr. 2, 1872	125, 284
Pavement, Concrete	J. M. Hawes	Covington, Ky	Oct. 3, 1871	119, 607
Pavement, Concrete	R. A. Jackson and S. Gissinger	Pittsburgh, Pa	Oct. 10, 1871	119, 850
Pavement, Concrete	B. N. Lampman	Rutland, Vt	Apr. 16, 1867	63, 908
Pavement, Concrete	G. Leverich and A. H. Emery	New York, N. Y	June 14, 1870	104, 325
Pavement, Concrete	A. B. McKeon	Rutherford Park, N. J	Nov. 15, 1870	109, 333
Pavement, Concrete	H. Meyers	Hyde Park, Pa	May 3, 1864	42, 589
Pavement, Concrete	G. H. Moore	Norwich, Conn	Nov. 28, 1871	121, 294
Pavement, Concrete	G. H. Moore	Norwich, Conn	Jan. 30, 1872	123, 280
Pavement, Concrete	A. H. Perkins	Chicago, Ill	Mar. 12, 1872	124, 620
Pavement, Concrete	J. J. Schillinger	New York, N. Y	July 19, 1870	105, 599
Pavement, Concrete	J. J. Schillinger	New York, N. Y	Feb. 14, 1871	111, 879
Pavement, Concrete	C. Scriber	Washington, D. C	May 28, 1872	127, 190
Pavement, Concrete	R. Skinner and B. Bonnett	San Francisco, Cal	July 19, 1870	105, 502
Pavement, Concrete	T. Smith	Washington, D. C	Feb. 16, 1869	87, 007
Pavement, Concrete	W. H. Smith	New York, N. Y	Jan. 3, 1872	122, 497
Pavement, Concrete	H. B. Steele	Winchester, Conn	July 6, 1869	92, 390
Pavement, Concrete	A. Van Camp	Washington, D. C	July 27, 1869	93, 142
Pavement, Concrete	W. H. White	Troy, Ohio	Feb. 25, 1868	74, 963
Pavement, Concrete and stone	S. B. Brittan	Newark, N. J	Oct. 24, 1871	120, 236
Pavement, Concrete and tile	G. A. Aschback	New York, N. Y	Mar. 5, 1872	124, 312
Pavement, Concrete and wooden	W. B. Coates	Philadelphia, Pa	Mar. 15, 1870	100, 727
Pavement, Concrete asphaltic	W. B. Parisen	Newark, N. J	June 13, 1871	115, 887
Pavement, Concrete compound for	J. W. Smith	Washington, D. C	Nov. 16, 1869	96, 984
Pavement, Concrete for	G. H. S. Duffus	Baltimore, Md	Oct. 25, 1870	108, 693
Pavement, &c., Concrete for	A. Van Camp and M. M. Hodgman.	Washington, D. C., and Saint Louis, Mo.	Nov. 16, 1869	96, 850
Pavement, Connecting and disconnecting the blocks of iron or other.	B. C. Smith	Burlington, N. J	Sept. 22, 1857	18, 251
Pavement, &c., Constructing	J. F. Le Moulnier	New York, N. Y	Jan. 2, 1855	12, 136
Pavement, Construction of metallic side	P. H. Jackson	New York, N. Y	Oct. 19, 1858	21, 834
Pavement, Corrugated-iron	J. Montgomery	New York, N. Y	Feb. 22, 1859	23, 038
Pavement, Corrugated metallic	J. Wilkes	New York, N. Y	Oct. 4, 1870	107, 989
Pavement, Cutting blocks for wood	W. W. Ballard and B. B. Waddell.	Elmira, N. Y., and Memphis, Tenn.	Aug. 24, 1869	94, 068
Pavement-driver	H. Landhop	New York, N. Y	Mar. 22, 1864	42, 022
Pavement, floors, walks, &c., Cement composition for	B. Doud	Cortland, N. Y	Jan. 8, 1867	61, 056
Pavement for streets, Cast-iron	W. D. Ferry	Boston, Mass	July 12, 1843	3, 168
Pavement for streets, Cemented	J. L. Sullivan	New York, N. Y	Oct. 20, 1831	....
Pavement for streets, Cemented	J. L. Sullivan	New York, N. Y	Nov. 9, 1831	....
Pavement for streets, Composition	J. W. Byrnes	Washington, D. C	Mar. 24, 1868	75, 855
Pavement for streets, Iron	G. W. Bishup	Brooklyn, N. Y	June 30, 1857	17, 662
Pavement, Form for laying brick	S. C. Brewer	Water Valley, Miss	Oct. 29, 1872	132, 560
Pavement, &c., Forming blocks for wooden	A. C. De Lisle	Paris, France	July 10, 1840	1, 683
Pavement from rot, Process for preserving wooden	A. B. Tripler	Philadelphia, Pa	May 7, 1872	126, 592
Pavement-foundation	S. Reeve	Chicago, Ill	Dec. 20, 1870	110, 395
Pavement-foundation	J. B. Wickersham	New York, N. Y	July 11, 1854	11, 251
Pavement, Gutter in foot	J. Read	Philadelphia, Pa	Nov. 3, 1868	83, 663
Pavement, Heating and preparing stone for	C. Allen	Albany, N. Y	Oct. 17, 1871	120, 016
Pavement, Implement for laying wood	P. Hinkle	San Francisco, Cal	Oct. 26, 1869	96, 113
Pavement, Inlaying plastic	G. H. Moore	Norwich, Conn	Dec. 5, 1871	121, 651
Pavement, Iron	C. Mettam	New York, N. Y	Feb. 24, 1857	16, 692
Pavement, Iron	R. Montgomery	New York, N. Y	Dec. 28, 1858	22, 444
Pavement, Iron	J. Montgomery	New York, N. Y	Mar. 8, 1859	23, 188
Pavement, Iron	J. Montgomery	New York, N. Y	Mar. 8, 1859	23, 189
Pavement, Iron	B. C. Smith	Philadelphia, Pa	Jan. 8, 1861	31, 093
Pavement, Iron	L. Stebbins	New York, N. Y	Oct. 29, 1861	33, 610
Pavement, Iron	A. R. Tewkesbury	East Boston, Mass	Mar. 9, 1858	19, 592
Pavement, Laying cement	A. F. Shyrma	New York, N. Y	July 15, 1873	140, 956
Pavement, Laying concrete	J. Hopke and J. W. Heesch	Hoboken, N. J	Feb. 27, 1872	124, 135
Pavement, Laying concrete	J. W. and A. W. Krause, jr	Philadelphia, Pa	Apr. 22, 1873	138, 030
Pavement, Laying concrete	J. W. Snyder	Chicago, Ill	May 27, 1873	139, 272
Pavement, Laying street	G. Warren	Boston, Mass	Oct. 19, 1869	96, 062
Pavement, Laying wood	E. P. Morong	Boston, Mass	Oct. 29, 1872	132, 682
Pavement, Laying wooden-block	D. L. De Golyer	Chicago, Ill	Apr. 6, 1869	88, 765
Pavement, Machine for grooving blocks for wood	P. D. Cummings	Portland, Me	Dec. 27, 1870	110, 440
Pavement, Machine for making key-board for wood	P. D. Cummings	Portland, Me	Nov. 29, 1870	109, 591
Pavement, Machine for roughing	E. Hambujer	New York, N. Y	Aug. 2, 1864	43, 687
Pavement, Machinery for preparing wooden blocks for.	W. O. Robbins and C. W. Stafford.	New York, N. Y	Dec. 7, 1869	97, 554
Pavement, Manufacture of slabs for	R. E. Stephens	Owen Sound, Canada	July 2, 1872	128, 670
Pavement, Manufacturing, laying, and finishing artificial-stone.	G. L. Eagan	San Francisco, Cal	July 23, 1872	129, 800
Pavement, Metallic	S. B. Ellithorp	New York, N. Y	Aug. 5, 1856	15, 479
Pavement, Metallic	J. Dean	West Roxbury, Mass	Oct. 22, 1867	70, 076
Pavement, Metallic	J. B. Tarr	Chicago, Ill	Jan. 29, 1867	61, 580
Pavement, Metallic	S. D. Tillman	Jersey City, N. J	June 21, 1870	104, 513

Index of patents issued from the United States Patent Office from 1790 *to* 1873, *inclusive*—Continued.

Invention.	Inventor.	Residence.	Date.	No.
Pavement, Mixer and heater for preparing concrete for.	H. Franke	Brooklyn, N. Y	Nov. 28, 1871	131, 351
Pavement, Mode of making	S. D. Tillman	New York, N. Y	Mar. 27, 1860	27, 658
Pavement or road, Laying asphalt or concrete	E. J. De Smedt	New York, N. Y	May 31, 1870	103, 581
Pavement, Preparation of composition for	J. R. Hayes	New York, N. Y	Sept. 20, 1870	107, 490
Pavement, Preparing blocks for wood	H. M. Stow	San Francisco, Cal	Apr. 9, 1872	125, 418
Pavement, Preserving wooden	E. P. Morong	Boston, Mass	Dec. 31, 1872	134, 479
Pavement, Prismatic blocks or wooden	J. Abbott	Wilton, N. H	Sept. 25, 1841	2, 265
Pavement, Ramming-machine for wood and other	A. L. Lansing	Philadelphia, Pa	Nov. 29, 1870	109, 745
Pavement, Road	M. K. Couzens	Yonkers, N. Y	Mar. 19, 1872	124, 722
Pavement, roads, &c., Manufacture of asphaltic composition for.	A. Ruttkay	New York, N. Y	Sept. 27, 1870	107, 815
Pavement, Roadway	J. T. Parson	Washington, D. C	Apr. 28, 1868	77, 208
Pavement, roadways, &c., Compounds for	R. Fisk	New York, N. Y	Oct. 12, 1869	95, 673
Pavement-roller	E. M. Sealand	Cleveland, Ohio	Apr. 27, 1869	89, 351
Pavement, roofing, &c., Composition for	J. Clarke	Syracuse, N. Y	Nov. 8, 1864	44, 938
Pavement, roofing, &c., Composition for	J. Clarke and D. French	Syracuse, N. Y	Feb. 18, 1862	34, 404
Pavement, roofing, &c., Composition for	J. E. Dotch and E. Duempelman	Washington, D. C	May 18, 1869	90, 084
Pavement, roofing, &c., Composition for	E. Duempelman and J. E. Dotch	Washington, D. C	Feb. 16, 1869	87, 031
Pavement, roofing, &c., Composition for	H. Staples	Nashua, N. H	July 27, 1869	93, 018
Pavement, roofing, drain-pipes, &c., Composition for.	J. L. Kidwell	Washington, D. C	May 18, 1869	90, 106
Pavement, Securing ranges of short plank in	J. E. Ware	Saint Louis, Mo	Feb. 11, 1851	7, 933
Pavement, Sidewalk	J. B. Cornell	New York, N. Y	Apr. 28, 1857	17, 138
Pavement, sidewalk, &c., Composition for	A. G. Anderson	Hoboken, N. J	Feb. 9, 1869	86, 799
Pavement, Stone	J. Bolliger	San Francisco, Cal	Oct. 8, 1872	131, 931
Pavement, Stone	C. Guideh	New York, N. Y	Jan. 12, 1869	85, 814
Pavement, Stone	A. Meckert	Guttenberg, N. J	Mar. 11, 1873	136, 747
Pavement, Stone	T. D. Owens	Pittsburgh, Pa	Oct. 14, 1873	143, 587
Pavement, Stone	C. G. Waterbury	New York, N. Y	May 3, 1870	102, 629
Pavement, Street	J. C. Blake	Elizabeth, N. J	Apr. 26, 1870	102, 361
Pavement, Street	P. Cadac and W. H. De Valin	San Francisco and Sacramento, Cal.	Aug. 11, 1868	80, 856
Pavement, Street	D. C. Colby	Washington, D. C	Mar. 22, 1870	101, 100
Pavement, Street	W. H. De Valin	Sacramento, Cal	Aug. 29, 1871	118, 592
Pavement, Street	A. Dilger	Saint Louis, Mo	July 5, 1870	104, 942
Pavement, Street	E. M. Fowler	New York, N. Y	Feb. 22, 1870	100, 134
Pavement, Street	G. W. Grader and M. H. Baldwin.	Memphis, Tenn	Apr. 23, 1867	64, 094
Pavement, Street	A. Hamar	New York, N. Y	Dec. 3, 1867	71, 746
Pavement, Street	L. F. Noe	New York, N. Y	Apr. 5, 1864	42, 218
Pavement, Street	N. M. Shafer	New York, N. Y	Sept. 27, 1864	44, 464
Pavement, Street	W. E. Shaw	Portland, Me	Jan. 3, 1871	110, 794
Pavement, Street	H. M. Stow	San Francisco, Cal	Dec. 10, 1867	72, 110
Pavement, Street	H. M. Stow	San Francisco, Cal	Dec. 10, 1867	72, 111
Pavement, Street	H. M. Stow	San Francisco, Cal	Feb. 25, 1868	74, 862
Pavement, Street	W. H. Williams	Little Falls, N. Y	Apr. 5, 1870	101, 691
Pavement, Street and walk	A. Hoyt	Chicago, Ill	June 2, 1868	78, 455
Pavement, Substrata for	H. P. Russ	New York, N. Y	Mar. 14, 1848	5, 475
Pavement, &c., Treating bituminous substances for	J. O'Friel	Brooklyn, N. Y	Aug. 23, 1870	106, 717
Pavement, Treating brick for	W. H. DeValin	San Rafael, Cal	Nov. 18, 1873	144, 749
Pavement-washer	A. C. Neall	Philadelphia, Pa	July 25, 1871	117, 316
Pavement-washer, hose-hydrant, and hitching-post	C. M. Alburger	Philadelphia, Pa	Aug. 22, 1854	11, 563
Pavement, Wedge-driving machine for	W. J. Harris	Elizabeth, N. J	Oct. 24, 1871	120, 193
Pavement, Wood	A. M. Adams	Washington, D. C	Aug. 16, 1870	106, 447
Pavement, Wood	M. H. Alexander	Newark, N. J	Feb. 4, 1873	135, 460
Pavement, Wood	H. J. Alvord	Washington, D. C	Aug. 30, 1870	106, 910
Pavement, Wood	W. W. Ballard	Elmira, N. Y	Feb. 1, 1870	99, 391
Pavement, Wood	W. W. Ballard	Elmira, N. Y	July 12, 1870	105, 291
Pavement, Wood	W. W. Ballard	Elmira, N. Y	Mar. 26, 1872	125, 001
Pavement, Wood	W. W. Ballard	Elmira, N. Y	Apr. 30, 1872	126, 171
Pavement, Wood	W. W. Ballard and B. B. Waddell.	Elmira, N. Y., and Memphis, Tenn.	Aug. 24, 1869	94, 062
Pavement, Wood	W. O. Barton	Elizabeth, N. J	Oct. 31, 1871	120, 481
Pavement, Wood	G. A. Beidler	Philadelphia, Pa	Jan. 30, 1872	123, 219
Pavement, Wood	G. A. Beidler	Philadelphia, Pa	Feb. 6, 1872	123, 445
Pavement, Wood	G. A. Beidler	Philadelphia, Pa	May 14, 1872	126, 776
Pavement, Wood	H. M. Beidler	Philadelphia, Pa	Jan. 31, 1871	111, 425
Pavement, Wood	A. Betteley	Boston, Mass	Aug. 24, 1869	94, 066
Pavement, Wood	A. Betteley	Boston, Mass	Nov. 16, 1869	96, 870
Pavement, Wood	A. Betteley	Boston, Mass	Mar. 29, 1870	101, 346
Pavement, Wood	A. Betteley	Boston, Mass	Sept. 6, 1870	106, 989
Pavement, Wood	A. Betteley	Boston, Mass	Sept. 6, 1870	107, 152
Pavement, Wood	L. H. Boole	New York, N. Y	Mar. 1, 1870	100, 363
Pavement, Wood	J. W. Brocklebank and W. W. Tubbs.	New York, N. Y., and Elizabeth, N. J.	May 31, 1870	103, 713
Pavement, Wood	J. L. Brown	New York, N. Y	Nov. 5, 1867	70, 514
Pavement, Wood	C. P. Burgess	Rochester, N. Y	Sept. 19, 1871	119, 010
Pavement, Wood	J. J. Burrows	Washington, D. C	May 28, 1872	127, 304
Pavement, Wood	W. Bushnell	Elizabeth, N. J	Jan. 17, 1871	110, 954
Pavement, Wood	W. Bushnell	Elizabeth, N. J	Feb. 21, 1871	111, 904
Pavement, Wood	L. Caldwell	Elmira, N. Y	Jan. 2, 1872	122, 358
Pavement, Wood	L. Caldwell	Elmira, N. Y	Mar. 26, 1872	125, 019
Pavement, Wood	L. Cauvet	New York, N. Y	June 7, 1870	103, 979
Pavement, Wood	G. H. Chinnock	New York, N. Y	May 7, 1872	126, 521
Pavement, Wood	C. C. Converse	Brooklyn, N. Y	Aug. 31, 1869	94, 284
Pavement, Wood	T. Cowing	San Francisco, Cal	Apr. 5, 1870	101, 590
Pavement, Wood	J. Defoe	Detroit, Mich	July 2, 1872	128, 471
Pavement, Wood	D. L. De Golyer	Chicago, Ill	Nov. 30, 1869	97, 278
Pavement, Wood	H. Dowson	Springfield, Ill	July 26, 1870	105, 658
Pavement, Wood	G. W. Dyer	Washington, D. C	Mar. 19, 1872	124, 799
Pavement, Wood	J. Dyer	Rouseville, Pa	July 30, 1872	130, 027
Pavement, Wood	J. Dyer	Rouseville, Pa	July 30, 1872	130, 028
Pavement, Wood	J. Q. Felt	Detroit, Mich	Jan. 21, 1873	134, 985
Pavement, Wood	M. Fitzgibbons	New York	Sept. 13, 1870	107, 352
Pavement, Wood	M. Fitzgibbons	New York, N. Y	Oct. 10, 1871	119, 837
Pavement, Wood	M. Flanigan	Detroit, Mich	May 21, 1872	126, 881
Pavement, Wood	L. T. Follansbee	Washington, D. C	Oct. 25, 1870	108, 697

Index of patents issued from the United States Patent Office from 1790 *to* 1873, *inclusive*—Continued.

Invention.	Inventor.	Residence.	Date.	No.
Pavement, Wood	Z. E. Forbes	Troy, N. Y	May 7, 1872	126, 454
Pavement, Wood	C. Gaytes	Chicago, Ill	Oct. 25, 1870	108, 582
Pavement, Wood	J. F. Gyles	Chicago, Ill	Aug. 29, 1871	118, 528
Pavement, Wood	H. Haupt	Philadelphia, Pa	May 24, 1870	103, 328
Pavement, Wood	L. Hill	Alexandria, Va	June 13, 1871	115, 960
Pavement, Wood	D. W. Hunt	San Francisco, Cal	Sept. 22, 1868	82, 320
Pavement, Wood	S. H. Ingersoll	New York, N. Y	Mar. 29, 1870	101, 270
Pavement, Wood	S. H. Ingersoll	New York, N. Y	Oct. 3, 1871	119, 519
Pavement, Wood	S. H. Ingersoll	Brooklyn, N. Y	Feb. 27, 1872	124, 064
Pavement, Wood	J. Judge	New York, N. Y	July 18, 1871	117, 177
Pavement, Wood	G. A. May	Chicago, Ill	Mar. 21, 1871	112, 945
Pavement, Wood	H. G. McGonegal	New York, N. Y	Jan. 19, 1869	86, 025
Pavement, Wood	H. G. McGonegal	New York, N. Y	Nov. 12, 1872	133, 047
Pavement, Wood	D. McKenzie	Brooklyn, N. Y	July 26, 1870	105, 707
Pavement, Wood	D. McKenzie	Utica, N. Y	Nov. 8, 1870	109, 140
Pavement, Wood	E. McMullen	Montreal, Canada	July 4, 1871	116, 734
Pavement, Wood	A. R. McNair	New York, N. Y	Aug. 24, 1869	94, 019
Pavement, Wood	A. Miller	Chicago, Ill	June 29, 1869	91, 861
Pavement, Wood	A. Miller and C. Mason	Chicago, Ill	Mar. 31, 1868	76, 227
Pavement, Wood	W. S. Morse	Washington, D. C	Dec. 13, 1870	110, 153
Pavement, Wood	D. H. Mulford	Saratoga Springs, N. Y	Aug. 29, 1871	118, 471
Pavement, Wood	H. E. Paine	Milwaukee, Wis	Apr. 25, 1871	114, 190
Pavement, Wood	J. F. Paul	Boston, Mass	June 8, 1869	91, 158
Pavement, Wood	E. W. Perrin	Portland, Oreg	Dec. 30, 1873	146, 019
Pavement, Wood	H. E. Perry	San Francisco, Cal	Feb. 27, 1872	124, 081
Pavement, Wood	J. I. Peyton	Washington, D. C	Apr. 9, 1872	125, 482
Pavement, Wood	J. I. Peyton	Washington, D. C	June 4, 1872	127, 510
Pavement, Wood	R. C. Phillips	Cincinnati, Ohio	Dec. 5, 1871	121, 544
Pavement, Wood	A. W. Platt	New York, N. Y	Mar. 8, 1870	100, 554
Pavement, Wood	A. W. Platt	New York, N. Y	Oct. 11, 1870	108, 181
Pavement, Wood	A. Potts	Philadelphia, Pa	Apr. 25, 1871	114, 038
Pavement, Wood	D. C. Reeves	Washington, D. C	Feb. 13, 1872	123, 728
Pavement, Wood	P. H. Reinhard and E. T. M. Faehtz.	Washington, D. C	Apr. 18, 1871	113, 929
Pavement, Wood	W. O. Robbins	New York, N. Y	Aug. 17, 1869	93, 837
Pavement, Wood	J. J. Schroyer	Springfield, Ill	June 28, 1870	104, 778
Pavement, Wood	W. E. Shaw	Portland, Me	July 18, 1871	117, 212
Pavement, Wood	F. H. Smith	Washington, D. C	Dec. 31, 1872	134, 492
Pavement, Wood	H. M. Stow	San Francisco, Cal	Apr. 9, 1872	125, 416
Pavement, Wood	H. M. Stow	San Francisco, Cal	Apr. 9, 1872	125, 417
Pavement, Wood	H. M. Stowe	San Francisco, Cal	Dec. 31, 1872	134, 404
Pavement, Wood	H. M. Stow	San Francisco, Cal	Apr. 15, 1873	137, 973
Pavement, Wood	H. H. Thayer	Philadelphia, Pa	May 9, 1871	114, 727
Pavement, Wood	A. Van Camp and M. M. Hodgman.	Washington, D. C., and Saint Louis, Mo.	Nov. 16, 1869	96, 849
Pavement, Wood	C. Wheeler, jr	Auburn, N. Y	Mar. 4, 1873	136, 572
Pavement, Wood	J. H. Wilkinson	Concord, N. H	Apr. 12, 1870	101, 955
Pavement, Wood	R. H. Willet	Washington, D. C	May 16, 1871	114, 895
Pavement, Wood	A. Wyckoff	Elmira, N. Y	May 7, 1872	126, 617
Pavement, Wood	G. F. Ziegler	Jersey City, N. J	June 4, 1872	127, 669
Pavement, Wood and concrete	E. A. L. Roberts	Titusville, Pa	Aug. 1, 1871	117, 685
Pavement, Wooden	G. A. Beidler	Philadelphia, Pa	Aug. 23, 1870	106, 651
Pavement, Wooden	G. A. Beidler	Philadelphia, Pa	Apr. 18, 1871	113, 838
Pavement, Wooden	J. W. Brocklebank and C. Trainer.	New York, N. Y	Jan. 12, 1869	85, 786
Pavement, Wooden	W. H. Chappell	Saint Louis, Mo	Apr. 19, 1864	42, 347
Pavement, Wooden	J. R. Cushier	New York, N. Y	Mar. 30, 1869	88, 455
Pavement, Wooden	C. K. Deutsch	New York, N. Y	Aug. 22, 1871	118, 216
Pavement, Wooden	H. Fayette	Port Chester, N. Y	May 21, 1867	64, 959
Pavement, Wooden	H. Fayette	Port Chester, N. Y	May 21, 1867	64, 960
Pavement, Wooden	J. Grant	Milwaukee, Wis	Feb. 28, 1871	112, 239
Pavement, Wooden	L. Holms	Paterson, N. J	Nov. 2, 1869	96, 433
Pavement, Wooden	P. Howard	San Francisco, Cal	Nov. 16, 1869	96, 918
Pavement, Wooden	S. Kneass	Philadelphia, Pa	Apr. 26, 1870	102, 277
Pavement, Wooden	P. Koch	New Haven, Conn	Aug. 27, 1867	68, 089
Pavement, Wooden	H. McDougall	Chicago, Ill	Mar. 24, 1868	75, 780
Pavement, Wooden	D. McKenzie	New York, N. Y	July 7, 1868	79, 674
Pavement, Wooden	S. Nicolson	Boston, Mass	Aug. 8, 1854	11, 491
Pavement, Wooden	R. L. Ream	New York, N. Y	Oct. 8, 1867	69, 703
Pavement, Wooden	W. D. Richardson	Springfield, Ill	Aug. 25, 1868	81, 540
Pavement, Wooden	E. Shaw	Portland, Me	Mar. 7, 1871	112, 503
Pavement, Wooden	J. K. Thompson	Chicago, Ill	Mar. 1, 1870	100, 339
Pavement, Wooden	C. G. Waterbury	New York, N. Y	Oct. 1, 1867	69, 517
Pavement, Wooden-block	A. Trenaunay	Neuilly Sur Seine, near Paris, France.	May 10, 1870	102, 884
Pavement, Wooden street	C. G. Waterbury	New York, N. Y	May 10, 1870	102, 981
Paving and roofing, Composition	A. B. Vandermark	Jersey City, N. J	Aug. 8, 1871	117, 946
Paving and roofing, Concrete for	E. Duempelman	New York, N. Y	Nov. 29, 1870	109, 724
Paving-block	R. Shaler	Madison, Conn	Dec. 21, 1869	98, 197
Paving-block, Concrete	R. Norwood	Baltimore, Md	Jan. 11, 1870	98, 791
Paving-blocks, Device for keying	P. D. Cummings	Portland, Me	Nov. 8, 1870	108, 977
Paving-blocks, Frame for setting	W. Gilbert	Detroit, Mich	Dec. 31, 1872	134, 427
Paving-blocks, Machine for making	S. W. Brooks	Brownsville, Tex	Oct. 24, 1871	120, 237
Paving blocks, Machine for preparing	W. O. Robbins and C. W. Stafford.	New York, N. Y	Dec. 7, 1869	97, 553
Paving-blocks, Machine for sawing	W. W. Ballard	Elmira, N. Y	Aug. 24, 1869	94, 669
Paving-blocks of wood, Machine for sawing	A. Nash	Calais, Me	Nov. 13, 1840	1, 857
Paving-composition	H. Saunders	Chester, Pa	Oct. 3, 1871	119, 476
Paving-composition	J. W. Smith	Washington, D. C	Sept. 14, 1869	94, 785
Paving, &c., Concrete	J. E. Dotch	Washington, D. C	Dec. 14, 1869	97, 893
Paving, Concrete and tile	C. Burlew	Lockhaven, Pa	Aug. 27, 1867	68, 284
Paving, Concrete blocks for	J. J. Schillinger	New York, N. Y	Jan. 31, 1871	111, 480
Paving machine, Street	T. Davidson, jr	Kensington, Pa	Jan. 8, 1856	14, 054
Paving, Rammer for street	E. F. M. Faehtz	Washington, D. C	Feb. 27, 1872	123, 986
Paving-roller	E. W. Kittridge	Cincinnati, Ohio	Sept. 8, 1868	82, 009
Paving, roofing, and other purposes, Composition for.	J. E. Doth and E. Duempelman	Washington, D. C	June 1, 1869	90, 825
Paving, roofing, &c., Conglomerate for	F. N. Hopkins	Baltimore, Md	Nov. 23, 1869	97, 088

Index of patents issued from the United States Patent Office from 1790 *to* 1873, *inclusive*—Continued.

Invention.	Inventor.	Residence.	Date.	No.
Paving, &c., Stone and metal conglomerate for	G. H. Knight	Cincinnati, Ohio	Apr. 1, 1851	8, 020
Paving streets, &c., Forming blocks of wood for	S. Carey	New Orleans, La	Feb. 3, 1841	1, 966
Paving streets, &c., Forming blocks of wood for	J. H. Patterson	New York	Jan. 27, 1841	1, 953
Pawl and ratchet	G. G. Taylor	Worcester, Mass	Sept. 22, 1863	40, 086
Pawl and ratchet device	A. M. Hills	East Hartford, Conn	Apr. 9, 1872	125, 391
Pawl-attachment	N. A. Baker	Covington, Ky	Mar. 12, 1872	124, 416
Pawl-drill	S. Ingersoll	New York, N. Y	Oct. 10, 1854	11, 787
Pawl, Friction	J. Moore	San Francisco, Cal	May 7, 1867	64, 554
Pawl, Jointed	S. S. Walley	Philadelphia, Pa	Sept. 11, 1849	6, 705
Pawl or ratchet	E. P. Russell	Manlius, N. Y	July 24, 1860	29, 310
Pawl, Take-up-detaining	G. Richardson	Lowell, Mass	Nov. 15, 1870	109, 345
Pea and bean sheller	L. Cutting	San Francisco, Cal	Apr. 17, 1866	53, 956
Pea and bean sheller	W. K. Lewis	Boston, Mass	May 22, 1866	54, 927
Pea-dropper	A. J. Williams	Barnesville, Ga	May 4, 1869	89, 722
Pea-picker	A. Quinn	Wilmington, N. C	Mar. 9, 1869	87, 705
Pea-rake	E. W. Rowley, jr	Antwerp, N. Y	July 16, 1867	66, 891
Pea-rake	S. Skinner	Clayton, N. Y	Mar. 19, 1869	87, 718
Pea-sheller	W. H. Bridgens	New York, N. Y	Nov. 26, 1867	71, 273
Pea-sheller	G. B. Price	Watervliet, N. Y	Aug. 16, 1864	43, 864
Pea-sheller	G. T. Savary	Groveland, Mass	Sept. 25, 1866	58, 305
Pea-sheller	S. Ustick	Philadelphia, Pa	May 5, 1868	77, 683
Pea-sheller and cherry-stoner	G. Sanford	New York, N. Y	Oct. 3, 1865	50, 278
Pea-shelling machine	G. Clark, jr	Boston, Mass	July 17, 1866	56, 370
Pea-shelling machine	M. Gray and J. A. Talpey	Boston and Somerville, Mass.	Jan. 30, 1866	52, 356
Pea-shelling machine	W. J. Stevenson	New York, N. Y	Mar. 30, 1858	19, 800
Pea-vines, Machine for gathering	J. B. Stanley	Copiah County, Miss	Jan. 9, 1849	6, 009
Peach-cutter	W. J. Hill	Fayetteville, Tenn	June 10, 1873	139, 714
Peach-cutter	J. T. Vaughn	Griffin, Ga	Jan. 19, 1869	86, 045
Peach cutter and stoner	J. C. Kuhn	Booneville, Ark	Mar. 15, 1859	23, 255
Peach cutting and stoning machine	W. D. Mayfield	Ashley, Ill	Apr. 3, 1866	53, 642
Peach-stoner	W. D. Hatch	Antrim, N. J	Dec. 10, 1872	133, 855
Peach-stoner	J. S. Lester	Atlanta, Ga	May 28, 1872	127, 355
Peach-stone extractor	T. E. Johnson	Indianapolis, Ind	Aug. 8, 1871	117, 892
Peaches, Box for assorting	E. W. Lockwood	Middletown, Del	Feb. 25, 1873	136, 252
Pea-nut and bean sheller	J. H. McNash	Waverly Station, Va	Oct. 1, 1872	131, 768
Pea-nut and coffee polisher	B. F. Walters	Norfolk, Va	July 2, 1872	128, 515
Pea-nut cleaning and polishing machine	J. M. Keating	Norfolk, Va	Mar. 8, 1870	100, 639
Pea-nut digger	C. W. Nicholson	Southampton County, Va	Oct. 1, 1872	131, 891
Pea-nut heater	D. Kellogg	Ypsilanti, Mich	Oct. 28, 1873	144, 108
Pea-nut huller	J. Johnson	New York, N. Y	Apr. 18, 1871	113, 889
Pea-nut picker	W. A. Crocker	Norfolk, Va	Nov. 30, 1869	97, 364
Pea-nut picker	S. C. Newell and G. Baars	Beardstown, Tenn	July 16, 1872	129, 581
Pea-nut picking and cleaning machine	J. C. Underwood	Surry Court-House, Va	Aug. 25, 1868	81, 562
Pea-nut sheller	J. W. Sands and B. F. Walters	Norfolk, Va	Jan. 16, 1872	122, 856
Pea-nut warmer	W. Chrysler	Lockport, N. Y	July 9, 1872	128, 710
Pea-nuts, Bleaching	F. M. Ironmonger	Norfolk, Va	May 7, 1872	126, 550
Pea-nuts, Steaming and polishing	S. Shepherd	Nashua, N. H	Sept. 29, 1857	18, 302
Pearl on solid substances, Imitation of	C. Sticht	Paris, France	Feb. 26, 1867	62, 507
Pearl ornaments in handles of cast metal, Securing	C. Dickinson and W. Bellamy	Newark, N. J	July 8, 1856	15, 286
Pearl-shell, Process for cutting and shaping	E. Carrington	West Meriden, Conn	Jan. 14, 1873	134, 851
Pearl-ash	J. and N. Paree	Lincklean, N. Y	May 14, 1836	....
Pearl-ash in kettles without the use of ovens, Making	R. Ainsworth		May 14, 1808	....
Pearl-ash, Manufacture of	J. W. Brown	Washington, D. C	Feb. 26, 1867	62, 312
Pearl-ash, Manufacture of	W. A. Edwards	Clinton, Mich	Feb. 20, 1849	6, 117
Pearl-ash, Manufacture of	W. Wentworth and G. W. Cleveland.	Constable, N. Y	Aug. 20, 1872	130, 613
Peat and brick machine	G. B. Fisher	Chicago, Ill	Nov. 19, 1867	70, 983
Peat and brick machine	E. Hall	Dunnamora, N. Y	Dec. 31, 1867	72, 847
Peat and brick machine	C. D. Wrightington and B. P. Rider.	Fairhaven and Boston, Mass.	Dec. 10, 1867	71, 939
Peat and feeding it to furnaces, Method of preparing	J. Yates	Mott Haven, N. Y	Apr. 17, 1866	54, 052
Peat-carbonizing apparatus	J. Adams	Rochester, N. Y	Nov. 9, 1869	96, 649
Peat charcoal, Apparatus for making	J. B. Lyons	Litchfield, Conn	May 14, 1867	64, 777
Peat, Cistern for draining	E. H. Ashcroft	Lynn, Mass	Apr. 10, 1866	53, 769
Peat, clay, &c., Machine for pressing	E. M. Hamilton	Minneapolis, Minn	Nov. 19, 1867	71, 163
Peat-cleaning machine	E. H. Ashcroft	Lynn, Mass	Oct. 25, 1864	44, 776
Peat-compressing machine	J. Jones	Baltimore, Md	Oct. 8, 1867	69, 677
Peat-condensing machine	N. C. Sawyer	Boston, Mass	Oct. 4, 1864	44, 557
Peat-crib	M. S. Roberts	Racine, Wis	Feb. 4, 1868	74, 141
Peat-digger	J. B. Lyons	Litchfield, Conn	Dec. 24, 1867	72, 513
Peat-digging machine	A. Fitts	Worcester, Mass	Apr. 22, 1856	14, 716
Peat-disintegrating machine	A. Michelbacher	New York, N. Y	July 2, 1867	66, 376
Peat-dryer	H. Bradford	New York, N. Y	Aug. 28, 1866	57, 469
Peat-dryer	G. Weissenborn	New York, N. Y	Oct. 16, 1866	58, 926
Peat, Drying and charring	F. L. H. Danchell	London, England	Apr. 4, 1865	47, 162
Peat, Drying and preparing	T. G. Walker	New York, N. Y	July 2, 1867	66, 428
Peat-drying floor	J. B. Lyons	Milton, Conn	Mar. 1, 1870	100, 301
Peat-elevator	E. J. Hulbert and A. N. N. Aubin.	Portland, Conn	Aug. 1, 1871	117, 634
Peat, &c., Expelling volatile matter from	T. G. Walker	New York, N. Y	Feb. 8, 1870	99, 733
Peat, Expelling volatile matter from	T. G. Walker	New York, N. Y	Nov. 1, 1870	108, 853
Peat for composting, Treatment of	J. B. Hyde	Newark, N. J	Apr. 5, 1859	23, 467
Peat for fuel, Apparatus for preparing	J. Webster	Malden, Mass	Aug. 20, 1867	68, 016
Peat for fuel, Machine for preparing	S. Marden	Newton, Mass	Mar. 27, 1866	53, 538
Peat for fuel, Preparing	W. S. Tisdale	New York, N. Y	Sept. 24, 1872	131, 577
Peat for fuel, Treatment of	E. J. Hulbert and A. N. N. Aubin.	Portland, Conn	Dec. 26, 1871	122, 173
Peat fuel, Machine and process for manufacture of	T. H. Leavitt	Boston, Mass	Dec. 26, 1871	122, 181
Peat fuel, Manufacture of	D. Aikman	Montreal, Canada	Apr. 4, 1871	113, 478
Peat fuel, Manufacture of	R. A. Griffin	Montreal, Canada	Apr. 4, 1871	113, 423
Peat-gathering machine	A. Michelbacher	New York, N. Y	May 14, 1867	64, 784
Peat-grinder	J. B. Lyons	Cornwall, Conn	Dec. 11, 1866	60, 398
Peat-grinding machine	S. Mills	Madison, Wis	Aug. 27, 1867	68, 101
Peat into charcoal, Converting	J. B. Hyde	New York, N. Y	June 29, 1858	20, 758
Peat, Kiln for drying and preparing	S. Chapman	Newark, N. J	Aug. 27, 1867	68, 044
Peat-machine	E. Atkinson	Brookline, Mass	Mar. 5, 1867	62, 519

Index of patents issued from the United States Patent-Office from 1790 *to* 1873, *inclusive*—Continued.

Invention.	Inventor.	Residence.	Date.	No.
Peat-machine	A. N. N. Aubin	Portland, Conn	Aug. 2, 1870	105, 881
Peat-machine	N. Aubin	Plattsburgh, N. Y	Oct. 27, 1868	83, 441
Peat-machine	N. Aubin	Montreal, Canada	Dec. 28, 1869	98, 333
Peat-machine	N. A. Barbour	New York, N. Y	Jan. 1, 1867	60, 667
Peat-machine	A. Bridges	Newton, Mass	May 21, 1867	64, 832
Peat-machine	H. and H. Clayton, jr., and F. Howlett.	London, England	Oct. 14, 1873	143, 617
Peat-machine	J. T. Foster	Jersey City, N. J	Mar. 19, 1867	62, 951
Peat-machine	S. D. and S. Gilson	Oswego Falls, N. Y	July 10, 1866	56, 206
Peat-machine	G. D. Goodrich	Chicago, Ill	Dec. 10, 1867	71, 870
Peat-machine	S. B. Greacen	Norwich, Conn	Jan. 22, 1867	61, 418
Peat-machine	J. Hodges	Penny Hill, Bagshot, England.	Apr. 10, 1866	53, 935
Peat-machine	B. Hotchkiss	New Haven, Conn	Jan. 1, 1867	60, 730
Peat-machine	L. P. Jenks	Boston, Mass	Mar. 12, 1867	62, 752
Peat-machine	J. S. Kelly	New York, N. Y	Oct. 12, 1869	95, 806
Peat-machine	T. H. Leavitt	Boston, Mass	Mar. 6, 1866	53, 011
Peat-machine	Z. Ludington	Uniontown, Pa	Dec. 22, 1868	85, 178
Peat-machine	Z. Ludington	Fayette County, Pa	Mar. 28, 1871	113, 180
Peat-machine	C. Luxton	Hudson, N. J	May 14, 1867	64, 681
Peat-machine	C. Luxton	Hudson City, N. J	Apr. 5, 1870	101, 478
Peat-machine	J. B. Lyons	Cornwall, Conn	Dec. 11, 1866	60, 397
Peat-machine	J. B. Lyons	Milton, Conn	Mar. 1, 1870	100, 302
Peat-machine	J. B. Lyons	New Haven, Conn	Apr. 26, 1870	102, 414
Peat-machine	S. Marden	Newton, Mass	Apr. 24, 1866	54, 186
Peat-machine	S. Marden	Newton, Mass	Jan. 1, 1867	60, 918
Peat-machine	H. Mielisch	Racine, Wis	Mar. 21, 1871	112, 830
Peat-machine	A. Moffatt	Washington, D. C	June 5, 1866	55, 339
Peat-machine	H. C. Moore	Springfield, Mass	May 7, 1867	64, 553
Peat-machine	N. F. Potter	Providence, R. I	May 1, 1866	54, 402
Peat-machine	J. H. Rae	Syracuse, N. Y	May 22, 1866	55, 018
Peat-machine	M. S. Roberts	Lewiston, N. Y	Apr. 9, 1867	63, 751
Peat-machine	M. S. Roberts	Racine, Wis	July 28, 1868	80, 309
Peat-machine	A. Robinson	McLean, N. Y	Jan. 22, 1867	61, 464
Peat-machine	A. Root	New Haven, Conn	Feb. 19, 1867	62, 289
Peat-machine	M. B. Stafford	New York, N. Y	July 24, 1866	56, 626
Peat-machine	J. Tisdale	South Dedham, Mass	Feb. 18, 1868	74, 635
Peat-machine	W. S. Tisdale	New York, N. Y	Oct. 10, 1871	119, 801
Peat-machine	D. Wellington	Boston, Mass	Mar. 6, 1866	53, 065
Peat-machine	D. Wellington	Boston, Mass	Feb. 12, 1867	61, 965
Peat-machine	T. J. Wells	New York, N. Y	Feb. 6, 1866	52, 472
Peat-machine	T. J. Wells	New York, N. Y	Nov. 20, 1866	59, 882
Peat-machine	T. J. Wells	Saint Anthony, Minn	Mar. 26, 1867	63, 342
Peat, Machine for slicing and drying	A. H. Emery	New York, N. Y	Sept. 18, 1866	58, 190
Peat, Machine for tempering and preparing	W. F. Potter	Providence, R. I	Apr. 18, 1865	47, 331
Peat-machines, Material for dusting the mold of	T. H. Leavitt	Boston, Mass	Mar. 6, 1866	53, 014
Peat-mold	K. Goddard	Richmond, N. Y	Aug. 31, 1869	94, 302
Peat molding and drying apparatus	E. H. Ashcroft	Lynn, Mass	May 15, 1866	54, 663
Peat-molding apparatus	E. Weissenborn	Hudson City, N. J	June 12, 1866	55, 565
Peat-molding machine	N. Aubin	Montreal, Canada	June 25, 1870	99, 130
Peat-molding machine	J. B. Lyons	New Haven, Conn	Jan. 21, 1868	73, 616
Peat, Preparation of	C. E. L. Holmes	New York, N. Y	Nov. 8, 1870	109, 009
Peat, Preparing	W. H. Buckland	Glamorgan County, Great Britain.	Feb. 11, 1862	34, 391
Peat-preparing apparatus	E. H. Ashcroft	Lynn, Mass	May 22, 1866	54, 839
Peat-preparing apparatus	C. W. Baldwin	Boston, Mass	Dec. 17, 1867	72, 357
Peat-preparing apparatus	L. Elsberg	New York, N. Y	Aug. 28, 1866	57, 489
Peat-preparing apparatus	H. Y. Hind	Fredrickton, New Brunswick.	Mar. 27, 1866	53, 547
Peat-preparing apparatus	S. Marden	Newton, Mass	May 29, 1866	55, 205
Peat-preparing apparatus	M. S. Roberts	Lewiston, N. Y	Aug. 15, 1865	49, 438
Peat-preparing machine	E. H. Ashcroft	Lynn, Mass	Apr. 24, 1866	54, 088
Peat-preparing machine	C. W. Inglis	Paterson, N. J	Mar. 24, 1868	75, 763
Peat-pressing machine	F. Leach	Tioga, N. Y	June 25, 1867	66, 093
Peat, Process for solidifying	L. Elsberg	New York, N. Y	Oct. 4, 1864	44, 520
Peat, Process of treating	B. R. Hawley and A. R. Morgan.	Normal, Ill., and New York, N. Y.	June 4, 1872	127, 415
Peat, Tank for preparing	A. Betteley	Boston, Mass	Jan. 9, 1866	51, 913
Peat, Treating	H. Ball	New York, N. Y	Feb. 12, 1867	61, 985
Peat, Treating	J. H. Smith	New York, N. Y	Oct. 31, 1865	50, 743
Peat, Treating	J. H. Smith	New York, N. Y	Nov. 28, 1865	51, 231
Peat-wringer	A. N. N. Aubin	Portland, Conn	Aug. 1, 1871	117, 591
Pedecycle	G. Brownlee	Princeton, Ind	July 27, 1869	92, 936
Pedestal to column, Fastening	W. and W. H. Lewis	New York, N. Y	June 24, 1851	8, 181
Pedo-motive, Infant's	E. J. Gorham	Bangor, Me	Aug. 4, 1863	39, 392
Pedrometer, Sea	B. Lies	New York	June 20, 1815	....
Peg, Condensed-leather	C. and J. G. Rowland	Quincy, Ill	Aug. 20, 1867	68, 005
Peg-cutter	C. H. Bacon	Boston, Mass	Dec. 23, 1873	145, 714
Peg-cutter	H. M. Buell	Waterbury, Conn	Sept. 3, 1872	130, 974
Peg-cutter	C. Holmes	Lynn, Mass	June 18, 1872	128, 147
Peg-cutter	J. G. Klock	Mansfield, Ohio	Mar. 19, 1872	124, 826
Peg-cutter	E. R. Pease and R. R. Hayman	Poughkeepsie, N. Y	Jan. 11, 1859	22, 577
Peg-cutter	D. B. Sanderson	Lewiston, Me	Apr. 21, 1868	76, 949
Peg-cutter	L. S. Smith	Gorsuch's Mills, Md	Sept. 24, 1867	69, 265
Peg-cutter	W. M. Watt	Como, Miss	Aug. 8, 1871	117, 839
Peg-cutter	A. Whittemore	Cambridgeport, Mass	Dec. 9, 1873	145, 469
Peg cutter, Boot and shoe	H. E. Chapman	Albany, N. Y	Dec. 11, 1855	13, 901
Peg cutter, Boot and shoe	S. R. Jones	Baltimore, Md	Jan. 8, 1856	14, 060
Peg cutter, Boot and shoe	W. H. Sellers	Keokuk, Iowa	Apr. 6, 1869	88, 588
Peg cutter, Boot and shoe	N. S. Wakefield	Camden, N. J	Dec. 31, 1872	134, 449
Peg-cutter, Hand	P. Summiter	Marlborough, Mass	July 8, 1873	140, 599
Peg-cutting instrument, Boot and shoe	F. Osborne	South Hanson, Mass	July 13, 1869	92, 637
Peg-cutting machine	H. Thurber	Painted Post, N. Y	May 22, 1828	....
Peg-cutting machine	J. D. Vinnedge	Indianapolis, Ind	May 27, 1873	139, 441
Peg-cutting machine, Boot and shoe	J. Essex	Killingly, Conn	July 17, 1837	281, 000
Peg-cutting machine, Boot and shoe	A. Whittemore	Cambridgeport, Mass	July 15, 1873	140, 803
Peg-cutting machine, Shoe	S. K. Baldwin	Guilford, N. H	July 16, 1842	2, 275
Peg-cutting tool, Boot and shoe	D. D. Allen	Adams, Mass	Oct. 19, 1852	9, 340

Index of patents issued from the United States Patent Office from 1790 *to* 1873, *inclusive*—Continued.

Invention.	Inventor.	Residence.	Date.	No.
Peg-feeding apparatus	S. D. Tripp	Stoneham, Mass	Sept. 13, 1859	25, 472
Peg-float	W. Miller, J. J. Becker, and A. Simcox.	Fort Wayne, Ind	May 26, 1868	78, 227
Peg-float	C. H. Odell	Poughkeepsie, N. Y	Dec. 22, 1863	41, 015
Peg-float	P. A. Schoellhorn	Buffalo, N. Y	Mar. 4, 1873	136, 551
Peg float, Shoe	G. G. Townsend	Rochester, N. Y	Dec. 18, 1866	60, 596
Peg machine, Shoe	S., S., jr., and E. Beckwith	Westmoreland, N. Y	Jan. 27, 1829	....
Peg machine, Shoe	A. H. Boyd	Saco, Me	Mar. 23, 1858	19, 730
Peg machine, Shoe	A. Brown	West Waterford, Vt	Nov. 16, 1858	22, 061
Peg machine, Shoe	C. Cook	Hillsborough, N. H	Apr. 5, 1859	23, 451
Peg machine, Shoe	C. A. Priest	Winslow, Me	Aug. 7, 1860	29, 519
Peg machine, Shoe	E. T. Weeks	Franconia, N. H	Nov. 22, 1859	26, 238
Peg machine, Shoe	A. Woodward	Keene N. H	Feb. 23, 1858	19, 461
Peg pointing and splitting machine, Shoe	J. Reid	Marshfield, Mass	Apr. 5, 1859	23, 494
Peg-pointing machine	W. A. Greenwood	Palmer, N. Y	Feb. 19, 1831	....
Peg-pointing machine, Shoe	S. G. Brett	Newport, N. H	Aug. 14, 1860	29, 565
Peg-pointing machine, Shoe	J. Ladd	Holderness, N. H	Dec. 23, 1856	16, 280
Peg-rasp	J. W. Foard	San Francisco, Cal	Feb. 4, 1873	135, 419
Peg-rasp	J. Sawyer and L. Clark	South Royalston, Mass	Dec. 13, 1853	10, 316
Peg-rasper	E. Townsend	Boston, Mass	Mar. 13, 1866	53, 200
Peg-sharpener	A. Kehl	Indianapolis, Ind	July 18, 1871	117, 081
Peg, Shoe	O. G. Healy	Abington, Mass	Apr. 30, 1872	126, 206
Peg, Shoe	J. H. Oliver	Baltimore, Md	July 16, 1872	129, 490
Peg, Shoe	R. H. Thompson	Rochester, N. Y	Mar. 23, 1836	....
Peg-splitting machine, Shoe	N. H. Shaw	Tamworth, N. H	Feb. 17, 1857	16, 659
Peg-splitting machine, Shoe	W. D. Shaw	Tamworth, N. H	Apr. 12, 1859	23, 614
Peg-splitting machine, Shoe	M. Wilder	Peterborough, N. H	Aug. 15, 1835	....
Peg strip, Continuous leather	C. Rowland	Quincy, Ill	Feb. 4, 1868	74, 148
Peg tube and driver	J. P. Kemp	Charlestown, Mass	Dec. 6, 1859	26, 398
Pegs and cement into soles, &c., Machine for introducing.	E. Townsend	Boston, Mass	Aug. 30, 1864	44, 029
Pegs and grain, Apparatus for drying shoe	S. Kimball and W. Sawyer	Boxford, Mass	Apr. 19, 1859	23, 689
Pegs for boots and shoes, Machine for making wire for.	W. Wickersham	Boston, Mass	Aug. 22, 1871	118, 318
Pegs for boots and shoes, Screw	J. M. Estabrook	Milford, Mass	Dec. 29, 1868	85, 374
Pegs for boots and shoes, Wire for making	D. H. Campbell and E. Woodward.	Sunderland, Scotland, and Charlestown, Mass.	Dec. 27, 1870	110, 431
Pegs for boots and shoes, Wooden	T. Rowell	Hartford, Vt	Nov. 24, 1820	....
Pegs for shoes, &c., Pointing, splitting, and waxing	T. Rowell	Hartford, Vt	Feb. 11, 1825	....
Pegs in boots and shoes, Machine for inserting metallic.	T. T. Prosser	Chicago, Ill	Dec. 10, 1872	133, 798
Pegs, Machine for arranging	J. Ladd	Holderness, N. H	Sept. 13, 1859	25, 422
Pegs, Machine for making	J. S. L. Babbs	New Albany, Ind	Apr. 28, 1868	77, 241
Pegs, Machine for making condensed-leather	C. and J. G. Rowland	Quincy, Ill	Aug. 20, 1867	68, 006
Pegs, Machine for making metallic shoe	T. T. Prosser	Chicago, Ill	June 20, 1871	116, 219
Pegs, Machine for making wooden	T. A. Robertson	Georgetown, D. C	Aug. 16, 1845	4, 148
Pegs, Machine for making wooden	H. P. Westcott	Seneca Falls, N. Y	Nov. 28, 1848	5, 946
Pegs, Machine for manufacturing shoe	I. G. Worth	Vassalborough, Me	Aug. 3, 1858	21, 104
Pegs, Manufacture of shoe	G. W. Day	Clarlestown, Mass	Mar. 13, 1866	53, 124
Pegs or pins, Pointing	J. Hull	North Bridgewater, Mass	July 7, 1832	....
Pegs, Preparation for shoe	B. F. Sturtevant	Boston, Mass	July 15, 1862	35, 902
Pegs, Preparing shoe	C. Gifford	Braintree, Mass	Sept. 19, 1848	5, 780
Pegs, Putting leather together by wooden or metal	S. B. Browne	Westchester County, N. Y.	Apr. 2, 1814	....
Pegger, Hand	J. H. Brown	Boston, Mass	Oct. 3, 1865	50, 298
Pegging and nailing machine, Shoe	L. R. Blake	Boston, Mass	Jan. 19, 1869	86, 054
Pegging boots and shoes	N. A. Fisher	Westborough, Mass	Apr. 3, 1834	....
Pegging boots and shoes	A. Swingle	Boston, Mass	Feb. 12, 1856	14, 269
Pegging boots and shoes	A. Woodruff	Westford, Mass	Sept. 6, 1833	....
Pegging-jack	A. Bailey	Amesbury, Mass	July 29, 1856	15, 406
Pegging-jack	T. D. Bailey	Lowell, Mass	Dec. 21, 1858	22, 340
Pegging-jack	W. Billings	Brooklyn, N. Y	May 3, 1864	42, 618
Pegging-jack	E. Everson	Haverhill, Mass	July 24, 1860	29, 257
Pegging-jack	E. L. Harlow	Monmouth, Me	July 31, 1860	29, 434
Pegging-jack	J. Jenkins	Andover, Mass	Oct. 15, 1850	7, 721
Pegging-jack	G. A. Knowlton	Natick, Mass	May 1, 1866	54, 369
Pegging-jack	J. K. Krieg	New York, N. Y	Nov. 5, 1867	70, 577
Pegging-jack	W. R. Landfear	Hartford, Conn	Nov. 28, 1865	51, 262
Pegging-jack	D. Mead and E. W. Watson	Danversport, Mass	July 12, 1870	105, 353
Pegging-jack	H. Ricard	Haverhill, Mass	July 5, 1870	104, 999
Pegging-jack	W. H. Sweetland	Marblehead, Mass	Aug. 10, 1869	93, 497
Pegging-jack	A. Swingle	Boston, Mass	July 29, 1856	15, 462
Pegging-jack	C. Varney	East Brimfield, Mass	May 10, 1870	102, 888
Pegging-jack	A. K. Washburn	Bridgewater, Mass	Jan. 16, 1866	52, 095
Pegging-jack	J. G. Ziegler	Salt River, Mich	Oct. 29, 1872	132, 616
Pegging jack, Boot	A. W. Moore	Stafford, Conn	Feb. 21, 1865	46, 484
Pegging-jack or shoemaker's head-block	T. D. Bailey	Lowell, Mass	May 6, 1856	14, 799
Pegging-jack post	D. Bowker	Boston, Mass	June 7, 1870	103, 835
Pegging-machine	J. C. Briggs	Saratoga, N. Y	Oct. 9, 1845	4, 228
Pegging-machine	B. Q. Budding	Milford, Mass	May 7, 1867	64, 481
Pegging-machine	W. Fitzgerald	Salem, Mass	July 1, 1862	35, 749
Pegging-machine	W. Fitzgerald	Boston, Mass	Nov. 20, 1866	59, 903
Pegging-machine	J. J. Greenough	New York, N. Y	June 26, 1860	28, 852
Pegging-machine	L. Hall	Boston, Mass	Mar. 8, 1864	41, 888
Pegging-machine	O. G. Healy	Abington, Mass	Apr. 30, 1872	126, 205
Pegging-machine	H. Kuhlman	Glückstadt, Germany	Aug. 6, 1872	130, 301
Pegging-machine	W. R. Landfear	Hartford, Conn	Aug. 23, 1859	25, 204
Pegging-machine	G. W. Manson	Auburn, Me	Mar. 11, 1873	136, 743
Pegging-machine	M. Marshall	Lowell, Mass	Nov. 5, 1861	33, 679
Pegging-machine	W. B. Randall	Fayette, Me	Aug. 6, 1832	....
Pegging-machine	G. L. Roberts	Boston, Mass	Dec. 23, 1873	145, 900
Pegging-machine	J. and A. W. Sangster	Buffalo, N. Y	Apr. 5, 1859	23, 501
Pegging-machine	J. F. Sargent	Boston, Mass	Aug. 30, 1864	44, 048
Pegging machine	J. F. Sargent	Boston, Mass	Aug. 30, 1864	44, 049
Pegging-machine	J. F. Sargent	Boston, Mass	Oct. 11, 1864	44, 691
Pegging-machine	J. W. Soule	Boston, Mass	Sept. 22, 1868	82, 448
Pegging-machine	E. M. Stevens	Boston, Mass	Aug. 3, 1858	21, 091
Pegging-machine	E. M. Stevens	Boston, Mass	May 28, 1867	65, 294
Pegging-machine	H. C. Stone	Brookfield, Mass	Sept. 17, 1867	69, 040

Index of patents issued from the United States Patent Office from 1790 *to* 1873, *inclusive*—Continued.

Invention.	Inventor.	Residence.	Date.	No.
Pegging-machine	J. Taggart	Roxbury, Mass	Jan. 14, 1862	34,170
Pegging-machine	E. Townsend	Boston, Mass	Feb. 7, 1860	27,085
Pegging-machine	C. Varney	East Brimfield, Mass	June 21, 1870	104,668
Pegging-machine	P. Wells	Middleton, Mass	Dec. 18, 1860	30,950
Pegging-machine	L. D. Wheeler	Fitchburgh, Mass	Nov. 5, 1872	132,742
Pegging-machine	L. H. Wood	Marlborough, Mass	Nov. 1, 1859	25,989
Pegging-machine awl	E. P. Richardson	Lawrence, Mass	Feb. 20, 1872	123,841
Pegging machine, Boot and shoe	J. A. Bradshaw	Lowell, Mass	Aug. 15, 1854	11,514
Pegging machine, Boot and shoe	A. C. Gallahue	Metamoras, Ohio	Oct. 28, 1851	8,465
Pegging machine, Boot and shoe	A. C. Gallahue	Allegheny City, Pa	Aug. 16, 1853	9,947
Pegging machine, Boot and shoe	A. C. Gallahue	New York, N. Y	July 11, 1854	11,287
Pegging machine, Boot and shoe	A. C. Gallahue	North East Centre, N. Y	Mar. 29, 1859	23,361
Pegging machine, Boot and shoe	A. C. Gallahue	Katonah, N. Y	Aug. 26, 1862	36,292
Pegging machine, Boot and shoe	F. Gray	Rowley, Mass	Apr. 6, 1832	....
Pegging machine, Boot and shoe	J. J. Greenough	New York, N. Y	Jan. 17, 1854	10,427
Pegging machine, Boot and shoe	H. Halvorson	Hartford, Conn	Jan. 10, 1854	10,407
Pegging machine, Boot and shoe	W. N. Hawley	Hartford, Conn	Jan. 31, 1860	26,987
Pegging machine, Boot and shoe	W. B. Johnson	Sandwich, N. H	Jan. 1, 1856	14,020
Pegging machine, Boot and shoe	W. Kidder	Newburyport, Mass	Aug. 15, 1854	11,542
Pegging machine, Boot and shoe	L. Lackey	Sutton, Mass	May 16, 1854	10,917
Pegging machine, Boot and shoe	J. La Dow	Granville, Ohio	July 31, 1849	6,613
Pegging machine, Boot and shoe	W. R. Landfear	Hartford, Conn	Sept. 13, 1864	44,203
Pegging machine, Boot and shoe	W. R. Landfear	Hartford, Conn	Nov. 5, 1867	70,581
Pegging machine, Boot and shoe	N. Leonard	Merrimack, N. H	June 11, 1829	....
Pegging machine, Boot and shoe	M. Marshall	Lowell, Mass	Feb. 11, 1862	34,370
Pegging machine, Boot and shoe	M. Marshall	Lowell, Mass	Dec. 9, 1862	37,136
Pegging machine, Boot and shoe	W. Miller	Boston, Mass	Oct. 20, 1863	40,351
Pegging machine, Boot and shoe	D. D. Palmer	Waltham, Mass	Dec. 31, 1867	72,753
Pegging machine, Boot and shoe	T. T. Prosser	Chicago, Ill	Dec. 23, 1873	145,754
Pegging machine, Boot and shoe	J. F. Sargent	Boston, Mass	Feb. 4, 1862	34,335
Pegging machine, Boot and shoe	G. Schuh and P. L. Slayton	Madison, Ind	Mar. 4, 1856	14,370
Pegging machine, Boot and shoe	J. Standish	Cuyahoga Falls, Ohio	Feb. 14, 1854	10,521
Pegging machine, Boot and shoe	A. F. Strong	Northampton, Mass	Feb. 13, 1872	123,592
Pegging machine, Boot and shoe	B. F. Sturtevant	Boston, Mass	June 9, 1857	17,544
Pegging machine, Boot and shoe	B. F. Sturtevant	Boston, Mass	Aug. 11, 1857	17,998
Pegging machine, Boot and shoe	B. F. Sturtevant	Boston, Mass	Sept. 21, 1858	21,593
Pegging machine, Boot and shoe	S. D. Tripp	Rochester, Mass	Apr. 12, 1853	9,629
Pegging machine, Boot and shoe	S. D. Tripp	Winchester, Mass	Sept. 8, 1857	18,170
Pegging machine, Boot and shoe	G. J. Wardwell	Andover, Me	July 18, 1854	11,346
Pegging machine, Boot and shoe	W. Wells	Boston, Mass	Dec. 15, 1857	18,879
Pegging machine, Boot and shoe	D. Whittemore	North Bridgewater, Mass	Nov. 5, 1867	70,662
Pegging machine, Boot and shoe	E. Woodward	Charlestown, Mass	Feb. 11, 1873	135,681
Pegging machine, Boot and shoe hand	R. H. Thompson	Buffalo, N. Y	June 26, 1855	13,144
Pegging machine, Boot and shoe hand	F. J. Vittum	Chelsea, Mass	Dec. 4, 1860	40,846
Pegging-machine feeder	G. M. Cram	East Rindge, N. H	Mar. 5, 1872	124,335
Pegging-machine, Hand	W. W. Batchelder	New York, N. Y	June 10, 1856	15,055
Pegging-machine, Hand	C. H. Binger and W. E. Fischer	Boston, Mass	Mar. 8, 1864	41,884
Pegging-machine, Hand	J. H. Brown	Watertown, Mass	Jan. 1, 1867	60,681
Pegging-machine, Hand	J. H. Brown	Watertown, Mass	Mar. 5, 1867	62,525
Pegging-machine, Hand	L. Goddu	Lowell, Mass	Dec. 5, 1865	51,387
Pegging-machine, Hand	E. T. Ingalls	Haverhill, Mass	Aug. 12, 1862	36,157
Pegging-machine, Hand	E. M. Stevens	Medford, Mass	Aug. 6, 1861	33,022
Pegging-machine, Hand	A. Swingle	Boston, Mass	May 29, 1855	12,985
Pegging-machine, Hand	F. J. Vittum	Newburyport, Mass	Nov. 10, 1868	84,025
Pegging-machine jack	W. Fitzgerald	Boston, Mass	May 22, 1860	28,358
Pegging-machine peg-box	A. W. Moore	East Brimfield, Mass	Oct. 25, 1870	108,719
Pegging machine, Shoe	J. H. Brown	Watertown, Mass	Dec. 22, 1868	85,206
Pegging machine, Shoe	W. G. Budlong	Providence, R. I	May 12, 1863	38,463
Pegging machine, Shoe	A. C. Gallahue	Riverdale, N. Y	June 29, 1869	92,037
Pegging machine, Shoe	C. Keniston	West Cambridge, Mass	Aug. 27, 1861	33,151
Pegging machine, Shoe	L. Lackey	Sutton, Mass	July 27, 1858	21,051
Pegging machine, Shoe	J. Robinson	Methuen, Mass	Oct. 31, 1848	5,896
Pegging machines, Blank for shoe	B. F. Sturtevant	Boston, Mass	Aug. 16, 1859	25,149
Pegging-machinery, Peg-feed stop for	S. A. Holt and C. H. Williams	Hudson, Mass	Aug. 18, 1868	81,275
Pen	T. B. Briggs	Brooklyn, N. Y	July 16, 1872	129,455
Pen	R. M. and D. Cameron	Edinburgh, North Britain	Jan. 15, 1867	61,156
Pen	R. H. Chinn	Washington, D. C	Dec. 20, 1870	110,342
Pen	G. Crandell	Washington, D. C	Oct. 18, 1870	108,455
Pen	L. M. De Spraugh	New York	June 29, 1818	....
Pen	D. D. Foley	Washington, D. C	Jan. 18, 1870	98,945
Pen	W. R. Glover	Glasgow, Ky	Aug. 1, 1854	11,421
Pen	H. S. Goodspeed	New York, N. Y	Mar. 9, 1869	87,661
Pen	R. Hirst	Hudson, N. Y	Mar. 24, 1868	75,909
Pen	J. H. Holland	Hancock, Mich	Apr. 7, 1868	76,454
Pen	T. S. Hudson	East Cambridge, Mass	Apr. 3, 1866	53,622
Pen	W. A. Morse	Philadelphia, Pa	Jan. 14, 1868	73,255
Pen	W. A. Morse	Philadelphia, Pa	Dec. 7, 1869	97,676
Pen	E. L. Pratt	Boston, Mass	Feb. 18, 1868	74,720
Pen	E. L. Pratt	Beverly, Mass	May 11, 1869	89,940
Pen	J. T. Price	Arrow Rock, Mo	Jan. 19, 1869	85,959
Pen	E. P. Tiffany	Hartford, Conn	Mar. 1, 1870	100,340
Pen	J. W. Truman	Macon, Ga	June 8, 1869	91,056
Pen	M. Wagner	Cincinnati, Ohio	June 20, 1871	116,242
Pen	A. F. Warren	Brooklyn, N. Y	June 30, 1863	39,084
Pen	A. G. Waterhouse	San Francisco, Cal	Dec. 7, 1869	97,735
Pen	C. S. Westcott	Elizabeth, N. J	Aug. 13, 1872	130,458
Pen	G. W. Woolley	Washington, D. C	Dec. 13, 1870	110,186
Pen and eraser	W. F. Beaton	Philadelphia Pa	Aug. 27, 1867	68,153
Pen and eraser combined	J. Schott	Chicago, Ill	July 17, 1866	56,455
Pen and ink case	A. G. Buzby	Philadelphia, Pa	Jan. 15, 1867	61,154
Pen and pen-holder	J. T. Foster	Jersey City, N. J	Aug. 7, 1866	56,920
Pen and pencil calendar	T. B. Briggs	Brooklyn, N. Y	Dec. 10, 1872	133,695
Pen and pencil case	T. Addison	New York, N. Y	May 10, 1838	736
Pen and pencil case	A. G. Bagley	New York, N. Y	Jan. 1, 1850	6,981
Pen and pencil case	E. Baptis	Hoboken, N. J	Apr. 29, 1856	14,760
Pen and pencil case	E. Baptis	Hudson, N. J	July 14, 1857	17,776
Pen and pencil case	G. S. Clark	New York, N. Y	Nov. 1, 1853	10,177
Pen and pencil case	J. M. Clark	Jersey City, N. J	Apr. 25, 1871	114,110
Pen and pencil case	J. M. Clark	Jersey City, N. J	June 6, 1871	115,704

Index of patents issued from the United States Patent Office from 1790 *to* 1873, *inclusive*—Continued.

Invention.	Inventor.	Residence.	Date.	No.
Pen and pencil case	J. M. Clark	Jersey City, N. J	Aug. 29, 1871	118, 434
Pen and pencil case	J. Cockburn	New York, N. Y	Apr. 6, 1858	19, 831
Pen and pencil case	J. Cockburn	New York, N. Y	Apr. 10, 1860	27, 777
Pen and pencil case	J. Cockburn	New York, N. Y	Jan. 16, 1872	122, 708
Pen and pencil case	F. W. Cox	Brooklyn, N. Y	June 27, 1865	48, 374
Pen and pencil case	F. Edgell	Buffalo, N. Y	Apr. 15, 1873	137, 900
Pen and pencil case	G. E. Frew	Brooklyn, N. Y	June 12, 1860	28, 664
Pen and pencil case	G. E. Frew	Brooklyn, N. Y	July 31, 1860	29, 432
Pen and pencil case	J. J. Hatcher	Philadelphia, Pa	Nov. 6, 1846	4, 839
Pen and pencil case	J. J. Hatcher	Philadelphia, Pa	July 11, 1854	11, 314
Pen and pencil case	W. S. Hicks	New York, N. Y	Sept. 12, 1865	49, 878
Pen and pencil case	W. S. Hicks	New York, N. Y	Oct. 31, 1871	120, 520
Pen and pencil case	W. S. Hicks	New York, N. Y	Dec. 26, 1871	122, 169
Pen and pencil case	J. H. Knapp	New York, N. Y	Sept. 2, 1856	15, 660
Pen and pencil case	J. H. Knapp	New York, N. Y	Aug. 26, 1873	142, 243
Pen and pencil case	C. W. Lord	New York, N. Y	Dec. 10, 1861	33, 894
Pen and pencil case	J. J. Lownds	New York, N. Y	Oct. 3, 1854	11, 752
Pen and pencil case	J. J. Lownds	New York, N. Y	Dec. 8, 1863	40, 846
Pen and pencil case	J. Mabie	English Neighborhood, N. Y	Oct. 3, 1854	11, 762
Pen and pencil case	J. Monaghan and T. Flynn	New York and Brooklyn, N. Y.	Dec. 19, 1871	122, 047
Pen and pencil case	J. H. Rauch	New York, N. Y	Jan. 6, 1852	8, 640
Pen and pencil case	J. H. Rauch	New York, N. Y	Dec. 8, 1863	40, 855
Pen and pencil case	J. Richardson	New York, N. Y	Oct. 17, 1854	11, 820
Pen and pencil case	J. Richardson	New York, N. Y	Oct. 24, 1854	11, 840
Pen and pencil case	J. Richardson	New York, N. Y	Oct. 31, 1854	11, 873
Pen and pencil case	J. Richardson	New York, N. Y	July 12, 1859	24, 758
Pen and pencil case	J. Richardson	New York, N. Y	Feb. 28, 1860	27, 311
Pen and pencil case	R. H. Ryne	New York, N. Y	Dec. 24, 1867	72, 684
Pen and pencil case	L. F. Standish	Springfield, Mass	Aug. 4, 1868	80, 780
Pen and pencil case	C. H. Streightoff	New York, N. Y	Oct. 21, 1873	143, 859
Pen and pencil case	H. Withers	New York, N. Y	Mar. 19, 1836	....
Pen and pencil case, Sliding	J. H. Rauch	New York, N. Y	Oct. 24, 1854	11, 839
Pen and pencil holder	A. G. Bagley	New York, N. Y	June 6, 1846	4, 557
Pen and pencil holder	G. H. Byron	Governor's Island, N. Y	Mar. 3, 1857	16, 711
Pen and pencil holder	J. Jacobs	Columbus, Ohio	Oct. 14, 1862	36, 652
Pen and pencil holder	J. T. Price	Arrow Rock, Mo	Sept. 24, 1867	69, 126
Pen and pencil holder	A. Tower	New York, N. Y	June 9, 1868	78, 698
Pen and pencil holder	T. D. Richardson	New York, N. Y	Oct. 11, 1859	25, 793
Pen and pencil holder	R. Ryne	New York, N. Y	Sept. 13, 1864	44, 261
Pen and pencil holder	R. H. Ryne	New York, N. Y	Oct. 19, 1867	62, 227
Pen and pencil holder and knife, Combined	W. A. Morse	Philadelphia, Pa	Mar. 9, 1869	87, 583
Pen-barrels by drawing, Forming metallic	A. G. Bagley	New York, N. Y	Feb. 27, 1847	4, 991
Pen blanks, Machinery for punching steel	E. A. Marsh and J. P. Kelly	Chicopee, Mass	Jan. 1, 1867	60, 763
Pen, button, and jewelry press	J. M. Riley	Newark, N. J	Apr. 20, 1869	89, 246
Pen-cleaner	S. Darling	Providence, R. I	Jan. 17, 1871	110, 959
Pen-cleaner	J. Warren	Brooklyn, N. Y	Oct. 2, 1860	30, 270
Pen cleaner and holder	T. S. Hudson	Boston, Mass	Apr. 27, 1858	20, 065
Pen-distributer	S. A. Potter	Philadelphia, Pa	July 11, 1865	48, 717
Pen for weaning calves	J. B. Dow	Davenport, Iowa	Apr. 3, 1866	53, 589
Pen, Fountain	N. Bartlett	Canajoharie, N. Y	Sept. 9, 1843	3, 253
Pen, Fountain	G. A. Becker	Seymour, Conn	Jan. 25, 1870	99, 134
Pen, Fountain	G. A. Becker	Seymour, Conn	Sept. 6, 1870	106, 987
Pen, Fountain	H. A. Brown and J. Wiley	Brooklyn, N. Y	Feb. 19, 1856	14, 276
Pen, Fountain	A. S. Carleton	Providence, R. I	June 14, 1870	104, 109
Pen, Fountain	J. S. Charles	Omaha, Nebr	Aug. 13, 1867	67, 720
Pen, Fountain	R. H. Chinn	Washington, D. C	Sept. 29, 1868	82, 598
Pen, Fountain	R. H. Chinn	Washington, D. C	Feb. 8, 1870	99, 635
Pen, Fountain	P. C. Clark	Reading, Pa	Feb. 5, 1861	31, 298
Pen, Fountain	W. Cleveland	Orange, N. J	Jan. 31, 1854	10, 469
Pen, Fountain	F. C. Cone	San Francisco, Cal	Oct. 11, 1870	108, 110
Pen, Fountain	J. S. Cutts	Philadelphia, Pa	Nov. 9, 1858	22, 017
Pen, Fountain	T. M. Davis	Philadelphia, Pa	Feb. 6, 1872	123, 460
Pen, Fountain	A. G. Day	Seymour, Conn	July 29, 1856	15, 417
Pen, Fountain	J. Goodyear	Groton, N. Y	July 15, 1873	140, 771
Pen, Fountain	T. W. Grinter	Russellville, Ky	May 31, 1870	103, 737
Pen, Fountain	H. N. Hamilton	White Plains, N. Y	Dec. 2, 1873	145, 102
Pen, Fountain	G. F. Hawkes	New York, N. Y	Oct. 17, 1865	50, 470
Pen, Fountain	G. F. Hawkes	New York, N. Y	Apr. 2, 1872	125, 291
Pen, Fountain	M. F. Hoit	Livingston, Ala	Sept. 11, 1847	5, 286
Pen, Fountain	W. Hunt	New York, N. Y	Jan. 13, 1847	4, 927
Pen, Fountain	D. Hyde	Reading, Pa	May 20, 1830	....
Pen, Fountain	E. H. Hyde and R. Dawson	Haydenville, Mass	Mar. 26, 1850	7, 217
Pen, Fountain	J. Johnson	New York, N. Y	Oct. 26, 1858	21, 881
Pen, Fountain	J. G. Kenyon	Ferndale, Cal	Nov. 9, 1869	96, 598
Pen, Fountain	M. Klein and H. W. Wynne	Keokuk, Iowa	Sept. 3, 1867	68, 445
Pen, Fountain	G. Kneip	New York, N. Y	Oct. 6, 1868	82, 850
Pen, Fountain	L. H. Knisely	New Philadelphia, Ohio	Jan. 30, 1872	123, 263
Pen, Fountain	C. W. Krebs	Baltimore, Md	Nov. 26, 1850	7, 798
Pen, Fountain	B. J. La Mothe	New York, N. Y	Apr. 25, 1865	47, 432
Pen, Fountain	D. A. Macomber	New York, N. Y	Aug. 28, 1849	6, 672
Pen, Fountain	H. Madeheim	Brooklyn, N. Y	Aug. 14, 1866	57, 162
Pen, Fountain	E. Mathers	Fairmount, Va	Dec. 4, 1860	30, 828
Pen, Fountain	H. K. McClelland	Eldersville, Pa	Apr. 17, 1855	12, 727
Pen, Fountain	G. R. Metten	Saint Louis, Mo	Oct. 9, 1866	58, 732
Pen, Fountain	G. R. Metten	Saint Louis, Mo	Nov. 5, 1867	70, 453
Pen, Fountain	G. J. Nolty	New York, N. Y	Apr. 26, 1864	42, 536
Pen, Fountain	F. A. Odermatt and F. Ettlin	San Francisco, Cal	Apr. 6, 1869	88, 656
Pen, Fountain	N. A. Prince	New Gloucester, Me	Sept. 30, 1851	8, 399
Pen, Fountain	N. A. Prince	Brooklyn, N. Y	Jan. 23, 1855	12, 301
Pen, Fountain	N. A. Prince	Brooklyn, N. Y	Dec. 25, 1855	13, 995
Pen, Fountain	J. Reid	Fort Wayne, Ind	Feb. 9, 1864	41, 534
Pen, Fountain	C. A. Rosefield	Columbus, Ga	May 12, 1857	17, 298
Pen, Fountain	L. M. Sandford and J. P. Beebe	Clinton, Iowa, and Morris, Ill.	Nov. 21, 1865	51, 090
Pen, Fountain	F. Schifferle	Saint Louis, Mo	Nov. 15, 1870	109, 257
Pen, Fountain	J. C. Silvy	New Orleans, La	Jan. 27, 1857	16, 514
Pen, Fountain	S. A. Skinner	West Berkshire, Vt	Dec. 18, 1860	30, 935

Index of patents issued from the United States Patent Office from 1790 *to* 1873, *inclusive*—Continued.

Invention.	Inventor.	Residence.	Date.	No.
Pen, Fountain	N. B. Slayton	Madison, Ind	Aug. 26, 1856	15, 622
Pen, Fountain	S. E. Taylor	East Cambridge, Mass	June 29, 1858	20, 741
Pen, Fountain	M. Wagner	Cincinnati, Ohio	Nov. 12, 1867	70, 765
Pen, Fountain	W. R. Walker	Concord, N. H	July 27, 1869	93, 024
Pen, Fountain	A. F. Warren	Brooklyn, N. Y	Dec. 23, 1856	16, 299
Pen, Fountain	A. F. Warren	Brooklyn, N. Y	Oct. 6, 1857	18, 365
Pen, Fountain	A. F. and C. M. M. Warren	Brooklyn, N. Y	Mar. 11, 1856	14, 425
Pen, Fountain	J. Weller	Washington Court-House, Ohio.	Sept. 29, 1863	40, 135
Pen, Fountain	G. W. White	Mount Vernon, N. Y	Sept. 4, 1855	13, 534
Pen, Fountain	G. W. Woolley	Philadelphia, Pa	Dec. 4, 1860	30, 851
Pen, Fountain ruling	C. Ketcham	Penn Yan, N. Y	Sept. 2, 1856	15, 659
Pen, Gold	E. H. Bard and H. H. Wilson	Philadelphia, Pa	Dec. 20, 1853	10, 348
Pen, Gold	A. W. Rapp	Philadelphia, Pa	Jan. 6, 1852	8, 641
Pen-handle	W. A. Morse	Boston, Mass	Feb. 7, 1860	27, 063
Pen-handle tube, Machine for finishing the ends of	S. Wesson	Worcester, Mass	May 1, 1866	54, 455
Pen, Hog	R. M. Abbe	Thompsonville, Conn	Aug. 29, 1854	11, 592
Pen, Hog	B. Gifford	Pedee, Iowa	Jan. 15, 1867	61, 189
Pen-holder	A. Alden	Barre, Mass	Mar. 4, 1843	2, 981
Pen-holder	E. D. Babbitt	New York, N. Y	Oct. 9, 1866	58, 572
Pen-holder	F. Brackett	Calais, Me	Nov. 14, 1865	50, 897
Pen-holder	J. Bryant	Brooklyn, N. Y	Jan. 29, 1861	31, 225
Pen-holder	B. Charles	Akron, Ohio	Mar. 21, 1871	112, 781
Pen-holder	G. H. Christian	New York, N. Y	Sept. 23, 1862	36, 507
Ped-holder	B. Cole	Brooklyn, N. Y	Feb. 15, 1859	22, 938
Pen-holder	A. T. Cross	Providence, R. I	July 1, 1873	140, 478
Pen-holder	W. J. Dewey	New York, N. Y	Apr. 9, 1872	125, 443
Pen-holder	H. G. Eastman	Poughkeepsie, N. Y	June 9, 1868	78, 655
Pen-holder	W. Fife	Philadelphia, Pa	Sept. 28, 1839	1, 345
Pen-holder	D. D. Foley	Washington, D. C	Dec. 17, 1867	72, 382
Pen-holder	P. Gabriel	Seymour, Conn	Sept. 10, 1867	68, 727
Pen-holder	A. M. George	Sand Fly, Tex	Jan. 24, 1871	111, 194
Pen-holder	D. E. Hall	Detroit, Mich	Jan. 19, 1869	85, 928
Pen-holder	E. W. Hanson	Spring Garden, Pa	Dec. 6, 1853	10, 294
Pen-holder	D. Harrington	Philadelphia, Pa	Sept. 2, 1845	4, 174
Pen-holder	G. Harrison	New York, N. Y	May 18, 1869	90, 168
Pen-holder	J. Holland	Cincinnati, Ohio	Aug. 22, 1871	118, 367
Pen-holder	I. Jacobs	New York, N. Y	Feb. 14, 1871	111, 846
Pen-holder	J. Johnson	New York, N. Y	Oct. 12, 1858	21, 758
Pen-holder	R. B. Lawrence	Wheeling, W. Va	Aug. 23, 1870	106, 596
Pen-holder	R. B. Lawrence	Wheeling, W. Va	Nov. 1, 1870	108, 916
Pen-holder	T. K. Lyon	Richmond, Va	Aug. 5, 1856	15, 490
Pen-holder	A. Masson and P. H. Cary	Paris, France	Oct. 17, 1865	50, 543
Pen-holder	J. C. Mullett	Fredonia, N. Y	Dec. 24, 1872	134, 156
Pen-holder	J. S. Orndorff	Virginia City, Nev	July 29, 1873	141, 287
Pen-holder	O. A. Pennoyer	Washington, D. C	Apr. 4, 1871	113, 558
Pen-holder	M. Phineas	New York, N. Y	Oct. 24, 1854	11, 836
Pen-holder	G. G. Ray	Boston, Mass	Apr. 15, 1862	34, 983
Pen-holder	H. Roth	Virginia City, Nev	Feb. 7, 1871	111, 683
Pen-holder	H. Smith	Georgetown, D. C	Jan. 15, 1818	......
Pen-holder	D. M. Somers	Brooklyn, N. Y	Apr. 30, 1872	126, 339
Pen-holder	D. M. Somers	Brooklyn, N. Y	Aug. 20, 1872	130, 670
Pen-holder	H. A. Spencer and R. S. Cutting.	Cleveland, Ohio, and Providence, R. I.	Apr. 27, 1869	89, 354
Pen-holder	W. H. Towers	Philadelphia, Pa	Apr. 17, 1855	12, 734
Pen-holder	W. H. Towers	New York, N. Y	Nov. 18, 1862	36, 972
Pen-holder	A. R. Turner	Malden, Mass	Apr. 26, 1859	23, 800
Pen-holder	S. Walker	Boston, Mass	Jan. 10, 1865	45, 884
Pen-holder	A. F. Warren	Brooklyn, N. Y	June 19, 1860	28, 797
Pen-holder	C. M. H. Warren	Brooklyn, N. Y	Feb. 4, 1868	74, 179
Pen-holder	J. Warren	Brooklyn, N. Y	July 31, 1860	29, 426
Pen-holder	A. Warth	Stapleton, N. Y	Apr. 28, 1868	77, 422
Pen-holder	H. F. Wheeler	Boston, Mass	Dec. 17, 1867	72, 435
Pen holder	C. T. Widstrand	Washington, D. C	Oct. 1, 1872	131, 802
Pen-holder	C. G. Wilson	Brooklyn, N. Y	Aug. 17, 1869	93, 780
Pen-holder	O. O. Witherell	Plaistow, N. H	Mar. 12, 1867	62, 913
Pen-holder and eraser	W. A. Morse	Philadelphia, Pa	Aug. 20, 1867	67, 899
Pen-holder and ink-eraser	D. E. Holmes	Halifax, Mass	May 17, 1864	42, 769
Pen-holder and letter-balance, Combined	D. C. Lawrence	Cedar Falls, Iowa	Aug. 20, 1861	33, 091
Pen holder and nib, Fountain	A. S. Lyman and M. W. Baldwin.	Philadelphia, Pa	Sept. 19, 1848	5, 789
Pen-holder, Elastic	J. W. Pearson	Winchester, Mass	Aug. 27, 1861	33, 160
Pen holder, Flexible	F. J. Klein	New York, N. Y	Feb. 19, 1856	14, 286
Pen holder, Fountain	C. Cleveland	Middlebury, Vt	June 1, 1852	8, 977
Pen-holder, Metallic	A. Granger	New York, N. Y	Mar. 29, 1859	23, 366
Pen-holder or pencil, Weighing-attachment for	D. A. B. Savy	Paris, France	July 25, 1865	49, 059
Pen-holder, pen-case, and money-safe, Combination of.	W. E. Rose	Waukon, Iowa	May 3, 1864	42, 599
Pen-holder, Reservoir	J. Darling	Stane, North Britain	Sept. 3, 1867	68, 418
Pen-holder, Spring	D. Hall	Brooklyn, N. Y	Apr. 19, 1864	42, 436
Pen-holder springs, Machine for forming	J. Keith	Worcester, Mass	May 1, 1866	54, 362
Pen-holder tip	D. M. Somers	Brooklyn, N. Y	May 21, 1872	127, 113
Pen machine, Microscopic	J. Bennett	Lowell, Mass	Oct. 11, 1828	....
Pen, Marking	D. S. Holman	Philadelphia, Pa	Sept. 30, 1873	143, 347
Pen, Metallic	T. Alden	Barre, Mass	Dec. 12, 1842	2, 877
Pen, Metallic	A. Craytey	Brooklyn, N. Y	July 1, 1856	15, 223
Pen, Metallic	M. S. Fife	Philadelphia, Pa	Apr. 3, 1849	6, 278
Pen, Metallic	W. Fife	Philadelphia, Pa	Aug. 29, 1848	5, 737
Pen, Metallic	R. Griswold	Bainbridge, N. Y	May 20, 1862	35, 312
Pen, Metallic	M. Phineas	New York, N. Y	July 12, 1853	9, 843
Pen, Metallic	M. Phineas	New York, N. Y	Feb. 5, 1856	14, 203
Pen, Metallic	W. H. Towers	Philadelphia, Pa	Nov. 1, 1853	10, 192
Pen, Metallic	F. A. Wait	Philadelphia, Pa	July 7, 1857	17, 761
Pen, Metallic	J. Wilcox	Philadelphia, Pa	Oct. 28, 1856	15, 992
Pen, Metallic	P. Williamson	New York	Mar. 30, 1835	....
Pen, Metallic	T. Woodward	Brooklyn, N. Y	Dec. 1, 1842	2, 874
Pen, Metallic pointed	B. R. Norton	Syracuse, N. Y	June 21, 1853	9, 809
Pen, Metallic writing	P. Williamson	Baltimore, Md	Nov. 22, 1809	......

Index of patents issued from the United States Patent Office from 1790 *to* 1873, *inclusive*—Continued.

Invention.	Inventor.	Residence.	Date.	No.
Pen, pencil, knife, and tooth-pick case, Combined	J. F. Sturdy	Attleborough, Mass	June 21, 1859	24, 502
Pen-rack	C. J. Bouché	Louisville, Ky	Oct. 13, 1868	82, 916
Pen-rack	E. Clarke	Winchester, Conn	Oct. 8, 1872	131, 937
Pen-rack	W. E. Clarke	Attleborough, Mass	July 21, 1868	80, 140
Pen-rack	G. Jedamski	New York, N. Y	July 23, 1867	67, 120
Pen-rack	M. O'Reardon	New York, N. Y	July 12, 1870	105, 237
Pen-rack	O. P. Smith	New York, N. Y	July 1, 1862	35, 786
Pen-rack	J. Young	Boston, Mass	Jan. 3, 1860	26, 727
Pen-rack and bill-holder	C. P. Crossman	West Warren, Mass	Oct. 9, 1866	58, 716
Pen-rack, calendar, and letter-balance, Combination of.	H. N. Taft	Washington, D. C	Jan. 3, 1865	45, 770
Pen rack, cleaner, and pencil-sharpener	H. C. Haskell	Marshall, Mich	Feb. 28, 1860	27, 287
Pen-rack, Spring	J. Adair	Pittsburgh, Pa	Oct. 2, 1866	58, 363
Pen-rack, Spring	J. M. Keep	New York, N. Y	Jan. 21, 1868	73, 448
Pen-rest	R. B. Huguuin	Cleveland, Ohio	Mar. 3, 1868	75, 164
Pen, Ruling	A. Hathaway	Boston, Mass	Jan. 28, 1851	7, 920
Pen, Ruling	A. Hathaway	Charlestown, Mass	Jan. 5, 1869	85, 585
Pen, Ruling	W. O. Hickok	Harrisburgh, Pa	June 24, 1873	140, 272
Pen, Ruling	E. Ingram	Springfield, Mass	Feb. 20, 1872	123, 827
Pen, Ruling	R. McVeen	Cleveland, Ohio	July 6, 1869	92, 336
Pen, Self-supplying	C. Cleveland	Middleburgh, Vt	Dec. 16, 1833	....
Pen, Self-supplying fountain	M. T. C. Gould	Philadelphia, Pa	Oct. 1, 1830	....
Pen, Sheep	B. Gifford	Pedee, Iowa	Jan. 15, 1867	61, 190
Pen-stand	C. J. Rooney and D. Renshaw	New York, N. Y	Apr. 3, 1860	27, 739
Pen, Steel writing	D. Thomas	Hingham, Mass	July 3, 1840	1, 671
Pen-wiper	H. S. Ball	Spartanburgh, S. C	Dec. 23, 1873	145, 778
Pen-wiper	S. Darling	Bangor, Me	Feb. 4, 1868	73, 956
Pen-wiper	J. H. Kidder	Lawrence, Mass	Sept. 2, 1873	142, 480
Pen-wiper and paper-weight	J. L. Rowe	New York, N. Y	Mar. 8, 1859	23, 195
Pen, Writing	R. H. Chinn	Washington, D. C	Jan. 21, 1873	135, 080
Pen, Writing	J. Holland	Cincinnati, Ohio	Jan. 21, 1873	135, 118
Pen, Writing	J. F. Reeve	Richmond, Va	Jan. 27, 1857	16, 496
Pen, Writing	S. Warrington	Philadelphia, Pa	July 24, 1866	56, 645
Pens, Alloy for manufacture of	J. Holland	Cincinnati, Ohio	Oct. 8, 1872	132, 008
Pens, &c., Box, case, and card for	J. Mason	Birmingham, Great Britain.	May 6, 1862	35, 201
Pens for ruling paper, Mode of setting and holding.	S. V. Collins	Charlestown, Mass	June 20, 1854	11, 146
Pens from holders, Device for extracting	F. W. Mattern	Chicago, Ill	Aug. 22, 1871	118, 376
Pens, Ink-reservoir for	R. B. Fitts	Philadelphia, Pa	May 8, 1860	28, 235
Pens, Ink-retaining attachment for	A G. Brown	Hartford, Conn	Jan. 25, 1870	99, 147
Pens, Machine for rolling gold	A. Morton	New York, N. Y	Aug. 28, 1860	29, 809
Pens, Making metallic	H. C. Windle, J. Gillott, and S. Morris.	Great Britain	Mar. 21, 1838	648
Pens, Manufacture of	G. Stimpson, jr	New York, N. Y	Mar. 20, 1866	53, 356
Pens, Manufacture of	E. Wiley	Brooklyn, N. Y	Jan. 14, 1868	73, 419
Pens, Ruling-guide for fountain	J. H. Hobbs	Philadelphia, Pa	Feb. 19, 1861	31, 515
Pencil	O. Cleveland	Jersey City, N. J	July 5, 1864	43, 391
Pencil	H. T. Cushman	North Bennington, Vt	Aug. 8, 1871	117, 748
Pencil	H. T. Cushman	North Bennington, Vt	Nov. 5, 1872	132, 713
Pencil	J. L. Faber	Stein, Bavaria	Aug. 13, 1861	33, 034
Pencil	J. Reckendorfer	New York, N. Y	Oct. 28, 1862	36, 854
Pencil	F. Rowell and E. C. Loud	Springfield, Mass	Feb. 15, 1870	99, 955
Pencil and eraser	W. R. Evans and L. D. Benner	Thomaston, Me	Apr. 25, 1865	47, 406
Pencil and eraser, Combination of lead	H. L. Lipman	Philadelphia, Pa	Mar. 30, 1858	19, 783
Pencil and pen case, Ever-pointed	J. I. Lownds	Philadelphia, Pa	Sept. 22, 1836	32
Pencil and rubber eraser, Combined lead	J. Illfelder	New York, N. Y	Nov. 4, 1873	144, 337
Pencil and sponge holder for cleaning slates, &c	J. L. Rowe	New York, N. Y	Aug. 25, 1863	39, 704
Pencil-attachment	E. Weissenborn	Hudson City, N. J	Oct. 12, 1869	95, 861
Pencil blank, Machine for grooving lead	P. Hufeland	New York, N. Y	May 31, 1870	103, 746
Pencil-case	W. N. Bartholomew	Newton Centre, Mass	Mar. 29, 1870	101, 344
Pencil-case	A. T. Cross	Providence, R. I	July 1, 1873	140, 477
Pencil-case	W. H. Davis	Hartsville, Ind	May 7, 1872	126, 448
Pencil-case	C. H. Downes	Hudson City, N. J	Sept. 27, 1870	107, 672
Pencil-case	J. Durant	New York, N. Y	June 27, 1846	4, 602
Pencil-case	F. R. Goulding	Roswell, Ga	June 25, 1871	117, 273
Pencil-case	W. S. Hicks	New York, N. Y	Mar. 21, 1871	112, 917
Pencil-case	J. A. Kemmis	New Orleans, La	Nov. 9, 1869	96, 597
Pencil-case	J. H. Knapp	New York, N. Y	Aug. 6, 1867	67, 558
Pencil-case	J. H. Knapp	New York, N. Y	Feb. 6, 1872	123, 485
Pencil-case	J. H. Knapp	New York, N. Y	Feb. 6, 1872	123, 486
Pencil-case	W. A. Ludden	Brooklyn, N. Y	July 20, 1869	92, 853
Pencil-case	W. A. Ludden	Flushing, N. Y	Oct. 17, 1871	119, 935
Pencil-case	W. Maginn	New York, N. Y	Apr. 21, 1868	77, 062
Pencil-case	D. A. Peirce	East Greenwich, R. I	Apr. 5, 1859	23, 487
Pencil-case	J. H. Rauch	New York, N. Y	June 22, 1869	91, 665
Pencil-case	S. S. Rembert	Memphis, Tenn	July 15, 1873	140, 842
Pencil-case	D. T. Warren	New York, N. Y	June 26, 1866	55, 966
Pencil case, Ever-pointed	J. Hague	New York, N. Y	Aug. 16, 1839	1, 291
Pencil case, Ever-pointed	J. Saxton	Philadelphia, Pa	Apr. 11, 1829	....
Pencil case, Ever-pointed	G. W. Simons	Philadelphia, Pa	Oct. 12, 1839	1, 364
Pencil case, Ever-pointed	T. Woodward	Brooklyn, N. Y	Oct. 14, 1840	1, 823
Pencil case, Lead and slate	T. B. McCaughan	Moscow, Tenn	Apr. 7, 1868	76, 488
Pencil-cases, Ornamenting	E. S. Johnson	Jersey City, N. J	Dec. 5, 1871	121, 627
Pencil, &c., coloring and polishing machine, Lead	E. Weissenborn	Hudson City, N. J	Apr. 2, 1872	125, 360
Pencil, Composition steatite	R. Langtrom	Cincinnati, Ohio	July 18, 1871	117, 084
Pencil, Eraser and stamp	E. Faber	New York, N. Y	June 16, 1863	38, 892
Pencil, Ever-pointed	A. G. Bagley	New York	Nov. 6, 1846	4, 840
Pencil, Ever-pointed	J. Boss	Philadelphia, Pa	July 20, 1846	4, 651
Pencil, Ever-pointed	W. Jackson	Philadelphia, Pa	July 27, 1829	....
Pencil, Ever-pointed	T. Woodward	Brooklyn, N. Y	June 10, 1840	1, 625
Pencil, Ever-pointed lead	E. Meeds	Philadelphia, Pa	June 26, 1835	....
Pencil-finishing machine	T. H. Müller	Yonkers, N. Y	July 15, 1873	140, 946
Pencil-head mold, India-rubber	A. Neill	Boston, Mass	Jan 22, 1861	31, 187
Pencil-holder	J. M. Clark	Jersey City, N. J	July 3, 1866	56, 007
Pencil-holder	N. B. Cooper	Liberty, Ind	Nov. 26, 1867	71, 456
Pencil-holder	C. A. Eaton	Minneapolis, Minn	Oct. 25, 1870	108, 573
Pencil-holder	A. J. Hall	San Francisco, Cal	Apr. 5, 1870	101, 457
Pencil-holder	H. M. Hally	Norwich, Conn	Sept. 18, 1866	58, 102
Pencil-holder	J. L. Moore	Bridgeport, Conn	Mar. 23, 1869	88, 160

Index of patents issued from the United States Patent Office from 1790 *to* 1873, *inclusive*—Continued.

Invention.	Inventor.	Residence.	Date.	No.
Pencil-holder	L. B. Myers	Elmore, Ohio	July 10, 1866	56, 251
Pencil-holder	T. Orton	Chicago, Ill	Oct. 15, 1872	132, 174
Pencil-holder	G. A. Smith	Boston, Mass	Feb. 4, 1873	135, 603
Pencil-holder	J. Smith	Tiffin, Ohio	Dec. 31, 1867	72, 926
Pencil-holder	E. J. Toof	Fort Madison, Iowa	May 19, 1868	78, 158
Pencil-holder	H. J. Wickham	Manchester, Conn	May 16, 1871	114, 893
Pencil holder and adjuster, Lead	E. Weissenborn	Hudson City, N. J	Apr. 2, 1872	125, 359
Pencil-holder for compass	W. G. Hillegass	Philadelphia, Pa	Nov. 12, 1867	70, 840
Pencil holder, Slate	E. G. Ward	Hoboken, N. J	Oct. 12, 1869	95, 859
Pencil, Indelible	E. P. Clark	Northampton, Mass	July 10, 1866	56, 180
Pencil, Indelible	W. B. Hale	Northampton, Mass	Jan. 1, 1867	60, 885
Pencil, Lead	E. Faber	New York, N. Y	Jan. 28, 1868	73, 883
Pencil, Lead	J. Gray	Medford, Mass	Feb. 25, 1873	136, 319
Pencil, Lead	S. D. Hovey	Brooklyn, N. Y	Dec. 31, 1872	134, 382
Pencil, Lead	P. Hufeland	New York, N. Y	Feb. 18, 1873	136, 122
Pencil, Lead	W. A. Morse	Philadelphia, Pa	Dec. 24, 1872	134, 215
Pencil, Lead	T. H. Müller	Yonkers, N. Y	May 21, 1872	126, 980
Pencil, Lead	T. H. Müller	Yonkers, N. Y	June 4, 1872	127, 633
Pencil, Lead	T. H. Müller	Yonkers, N. Y	Mar. 25, 1873	137, 089
Pencil, Lead	J. Reckendorfer	New York, N. Y	Jan. 6, 1863	37, 360
Pencil, Lead	J. Reckendorfer	New York, N. Y	Jan. 19, 1869	86, 101
Pencil machine, Slate	D. J. Tittle	Albany, N. Y	June 4, 1872	127, 440
Pencil-point protector	G. Merritt	New York, N. Y	Mar. 5, 1867	62, 555
Pencil-point protector and mark-eraser	J. B. Hodgskin	New York, N. Y	Feb. 14, 1865	46, 358
Pencil-points, Composition of matter for, and making.	G. C. Baldwin	Ticonderoga, N. Y	Dec. 2, 1835	....
Pencil-points, Machine for making, and composition for.	G. C. Baldwin	Ticonderoga, N. Y	Dec. 2, 1835	....
Pencil-polishing machine	G. Braun	Hoboken, N. J	Oct. 29, 1872	132, 519
Pencil-scale	B. Worcester	Waltham, Mass	Dec. 4, 1866	60, 815
Pencil-sharpener	H. P. Andrews	Cleveland, Ohio	Aug. 8, 1865	~~49, 240~~
Pencil-sharpener	A. G. Batchelder	Lowell, Mass	Nov. 8, 1870	109, 100
Pencil-sharpener	H. Burgess	San Francisco, Cal	May 28, 1867	65, 165
Pencil-sharpener	H. Burgess	Oakland, Cal	May 10, 1870	102, 766
Pencil-sharpener	J. L. and D. H. Coles	New York, N. Y	July 21, 1868	80, 145
Pencil-sharpener	E. M. Crandal	Chicago, Ill	Mar. 11, 1873	136, 590
Pencil-sharpener	S. Darling	Providence, R. I	June 27, 1871	116, 276
Pencil-sharpener	S. W. Davis and C. P. Elliott	Norristown, Pa	Dec. 19, 1871	121, 996
Pencil-sharpener	M. W. Dillingham	Amsterdam, N. Y	Aug. 17, 1869	93, 687
Pencil-sharpener	M. W. Dillingham	Amsterdam, N. Y	July 19, 1870	105, 440
Pencil-sharpener	M. W. Dillingham	Amsterdam, N. Y	Aug. 30, 1870	106, 795
Pencil-sharpener	W. K. Foster	Bangor, Me	Apr. 27, 1858	20, 056
Pencil-sharpener	W. K. Foster	Bangor, Me	May 18, 1858	20, 262
Pencil-sharpener	D. Hoffman	Pottsville, Pa	July 30, 1872	129, 956
Pencil-sharpener	J. MacMullen	New York, N. Y	Jan. 2, 1866	51, 845
Pencil-sharpener	J. McClure	Nashua, N. H	Feb. 1, 1870	99, 335
Pencil-sharpener	L. B. Myers	Elmore, Ohio	May 18, 1869	90, 289
Pencil-sharpener	E. P. Needham	New York, N. Y	Mar. 21, 1871	112, 951
Pencil-sharpener	C. C. Plaisted	Hartford, Conn	Mar. 16, 1869	87, 967
Pencil-sharpener	A. E. Schatz	New York, N. Y	May 13, 1873	138, 943
Pencil-sharpener	E. Spencer	Lambertville, N. J	Aug. 20, 1867	67, 922
Pencil-sharpener	O. C. Squyer	Penn Yan, N. Y	Feb. 18, 1868	74, 728
Pencil-sharpener	J. W. Strange and S. Darling	Bangor, Me	Sept. 22, 1857	18, 265
Pencil-sharpener	W. N. Weeden	Boston, Mass	Nov. 9, 1869	96, 748
Pencil-sharpener	S. S. Woodcock	Somerville, Mass	Apr. 20, 1869	89, 109
Pencil-sharpener	R. Wright	New York, N. Y	Feb. 7, 1865	46, 289
Pencil-sharpener	H. L. D. Zeng	Geneva, N. Y	Feb. 6, 1872	123, 462
Pencil sharpener and holder, Slate	W. H. Alcorn	New York, N. Y	July 30, 1867	67, 245
Pencil sharpener, Slate	F. G. Bottner	Bridgeport, Conn	Aug. 6, 1867	67, 488
Pencil sharpener, Slate	W. Burnet	New York, N. Y	Oct. 5, 1858	21, 649
Pencil sharpener, Slate	A. C. Funston	Philadelphia, Pa	July 30, 1861	32, 940
Pencil sharpener, Slate	J. M. Hicks	Boston, Mass	Sept. 29, 1863	40, 102
Pencil sharpener, Slate and lead	J. Soumeillan	Philadelphia, Pa	Oct. 8, 1872	131, 977
Pencil-sharpeners, Making blade for	W. K. Foster	Bangor, Me	Jan. 26, 1858	19, 191
Pencil-sharpeners, Mold for casting	W. K. Foster	Bangor, Me	Apr. 17, 1855	12, 722
Pencil-sharpening instrument, Slate	G. Sickles	Brooklyn, N. Y	May 11, 1858	20, 219
Pencil-sheath	C. E. Abbott and R. S. Merrill	Malden and Hyde Park, Mass.	Aug. 8, 1871	117, 851
Pencil-sheath	S. Ayres	Danville, Ky	Sept. 22, 1868	82, 272
Pencil-sheath	J. Danner	Canton, Ohio	June 16, 1868	78, 931
Pencil-sleeve and eraser, Combination of	F. E. Oliver	New York, N. Y	June 3, 1862	35, 467
Pencil, Tailor's marking	W. Coover	Erie, Pa	Aug. 26, 1843	3, 236
Pencil-tip molds, Vulcanizing rubber	J. Banigan and G. W. Miller	Smithfield, R. I	Jan. 18, 1870	99, 045
Pencil-tips, Mold for making rubber	W. Weicker	Blackstone, Mass	May 10, 1870	102, 894
Pencil-varnishing machine	T. H. Müller and H. C. Benson	New York, N. Y	Nov. 29, 1870	109, 750
Pencil, &c., varnishing machine	A. Warth, P. Hufeland, and G. Braun.	Stapleton and New York, N. Y.	May 10, 1870	102, 893
Pencil varnishing or coloring machine, Lead	E. Weissenborn	Hudson City, N. J	Apr. 2, 1872	125, 362
Pencils, Attaching eraser to	G. L. Holt	Springfield, Mass	Feb. 9, 1869	86, 753
Pencils, Attaching India-rubber to	W. C. Vosburgh and W. A. Ludden.	Brooklyn, N. Y	May 20, 1862	35, 355
Pencils, Attaching rubber eraser to lead	T. H. Müller	Yonkers, N. Y	Apr. 30, 1872	126, 224
Pencils, Attaching rubber to	J. Reckendorfer	New York, N. Y	Jan. 19, 1869	85, 961
Pencils, Composition for	E. P. Clark	Holyoke, Mass	May 31, 1859	24, 195
Pencils, Composition for	W. Geller	New York, N. Y	Nov. 8, 1870	109, 000
Pencils, Composition for	S. C. Pruden	Athens, Ohio	May 14, 1867	64, 702
Pencils, Composition for making black lead	C. Osgood	Salem, Mass	Mar. 26, 1814	....
Pencils, Composition of slate	L. J. Cohen	New York, N. Y	Nov. 7, 1848	5, 903
Pencils, Cutter-head for wood of lead	F. G. Jenkins	New York, N. Y	May 12, 1863	38, 488
Pencils, Device for polishing	E. Weissenborn	Hudson City, N. J	Apr. 27, 1869	89, 530
Pencils for drawing-machines, Arrangement of	M. Nutting	Portland, Me	Nov. 21, 1854	11, 970
Pencils, Machine for cutting and smoothing ends of	T. H. Müller and H. C. Benson	New York, N. Y	Oct. 25, 1870	108, 721
Pencils, Machine for cutting ends of wooden	T. H. Müller	New York, N. Y	Jan. 3, 1871	110, 777
Pencils, Machine for cutting lead	A. Warth	Stapleton, N. Y	June 21, 1864	43, 267
Pencils, Machine for cutting off lead	P. Schrag and P. Hufeland	New York, N. Y	May 17, 1870	103, 241
Pencils, Machine for forming lead and other	E. Weissenborn	Hudson City, N. J	Apr. 2, 1872	125, 361
Pencils, Machine for making lead	E. Weissenborn	Hudson City, N. J	Sept. 10, 1867	68, 819

Index of patents issued from the United States Patent Office from 1790 *to* 1873, *inclusive*—Continued.

Invention.	Inventor.	Residence.	Date.	No.
Pencils, Machine for making slate	D. R. Satterlee	New Haven, Conn	Aug. 24, 1869	94, 136
Pencils, Machine for making wooden case for lead.	A. Weiller	New York, N. Y	June 23, 1863	39, 019
Pencils, Machine for manufacture of slate	D. R. Satterlee	New Haven, Conn	Apr. 21, 1868	77, 105
Pencils, Machine for sand-papering	P. Schrag	New York, N. Y	Dec. 27, 1864	45, 679
Pencils, Manufacture of lead	F. G. Jenkins	Brooklyn, E. D., N. Y	May 18, 1869	90, 269
Pencils, Manufacture of slate	N. C. Harris	Poultney, Vt	Apr. 24, 1855	12, 759
Pencils, Metallic slide and case for ever-pointed	J. Bogardus	New York, N. Y	Sept. 17, 1833	....
Pencils, Packing-board for	O. Cleveland	Jersey City, N. J	Oct. 29, 1872	132, 562
Pencils, Press for stamping lead	T. H. Müller	Yonkers, N. Y	Feb. 20, 1872	123, 925
Pencils, Producing slate	H. O. Brown	Castleton, Vt	July 16, 1872	129, 096
Pencils, Rubber head for	S. D. Hovey	Brooklyn, N. Y	July 28, 1868	80, 485
Pencils, Rubber head for lead	J. B. Blair	Philadelphia, Pa	July 23, 1867	66, 938
Pencils, Rubber head for lead	W. W. Shaw	Troy, N. Y	Apr. 5, 1859	23, 504
Pendulums and levers, Concentrating the power of.	A. and C. Berry	Poughkeepsie, N. Y	Mar. 25, 1826	....
Pendulum-balance	E. Sampson	Claremont, N. H	Nov. 6, 1849	6, 852
Pendulum, Balance	H. Twiss	Meriden, Conn	May 13, 1834	....
Pendulum, Compensating	M. Bemis	Ashburnham, Mass	Nov. 8, 1859	26, 008
Pendulum, Compensating	L. Bradley	Hartford, Conn	Feb. 26, 1861	31, 524
Pendulum, Compensating	G. Buchanan	Hickory, Pa	Dec. 4, 1860	30, 798
Pendulum, Compensation	H. B. James	Trenton, N. J	Oct. 31, 1871	120, 385
Pendulum, Compound	D. Bickford	Westerly, R. I	Mar. 2, 1858	19, 479
Pendulum, Compound	C. M. Rice and J. E. Harrington	Worcester and Millbury, Mass.	Jan. 19, 1858	19, 153
Pendulum, Double	R. Leslie	....	Jan. 30, 1793	....
Pendulum, Electro-magnetic	J. Hamblet, jr., and B. F. Edmands.	Boston, Mass	Jan. 12, 1864	41, 217
Pendulum for time-pieces, Torsion	A. D. Crane	Newark, N. J	Jan. 9, 1855	12, 196
Pendulum in calendar-clocks, Arrangement of	H. B. Horton	Ithaca, N. Y	Jan. 4, 1870	98, 498
Pendulum level and sight combined	C. Morrill	New York, N. Y	Apr. 16, 1867	63, 807
Pendulum-lock, Overbalanced	S. Goodwin	Baltimore, Md	July 7, 1809	....
Pendulum-machine	J. Lefever	Providence, R. I	June 13, 1810	....
Pendulum-machine for mechanical operations	C. L. Allen	Newark, N. J	Oct. 13, 1815	....
Pendulum-mill	W. Grandin	Hector, N. Y	Jan. 6, 1813	....
Pendulum-mill	S. Thayer	Braintree, Mass	Oct. 9, 1811	....
Pendulum-power	J. Tallman	Albany, N. Y	Mar. 31, 1808	....
Pendulum-power, Applying	A. Slevin	Ann Arbor, Mich	Mar. 30, 1858	19, 798
Pendulum-power for propelling machinery	A. Wade	Eagletown, N. Y	Mar. 25, 1835	....
Pendulum-power regulated by a fly-wheel	W. W. McIntosh and W. Barnheart.	Aurelius, Ohio	Apr. 13, 1825	....
Pendulum-screen	L. Dupré	Charleston, S. C	Apr. 1, 1807	....
Pendulum, Self-adjusting	J. S. Greig	Walden, N. Y	Mar. 13, 1847	5, 017
Pendulums upon the earth's surface, Apparatus for illustrating the motion of.	G. M. Dimmock	Springfield, Mass	July 12, 1873	9, 839
Penman's assistant	H. G. Eastman	Poughkeepsie, N. Y	Nov. 5, 1861	33, 638
Penman's assistant	H. G. Eastman	Poughkeepsie, N. Y	Feb. 9, 1864	41, 495
Penman's assistant	H. A. Hutson	Newbury, N. Y	Jan. 1, 1867	60, 896
Penman's assistant	W. King	Hopewell, New Brunswick	Dec. 12, 1865	51, 530
Penman's finger-shield	C. W. Brainerd	Hartford, Conn	Apr. 13, 1869	88, 942
Penmanship, Method of teaching	W. S. MacLawrin	New York, N. Y	Feb. 13, 1855	12, 391
Pepper and ink top	W. Markland	New York, N. Y	Oct. 1, 1830	....
Pepper and salt boxes, Cap for	A. A. Bingham and W. H. Ashfield.	Syracuse, N. Y	Dec. 31, 1872	134, 455
Pepper and spice box	B. Morahan	Brooklyn, N. Y	Nov. 28, 1871	121, 399
Pepper and spice box	A. E. Turnbull, jr	Clifton, Ohio	Dec. 27, 1870	110, 608
Pepper-box	J. Fallows	Philadelphia, Pa	May 8, 1866	54, 517
Pepper-box	A. H. Newton	Worcester, Mass	Oct. 31, 1865	50, 727
Pepper-box, Air-tight	E. Brown	Lynn, Mass	Jan. 26, 1858	19, 182
Pepper-box cover	H. E. Thomas	San Francisco, Cal	June 25, 1872	128, 261
Pepper-box top	J. S. Ewbank	New York, N. Y	May 31, 1864	42, 934
Pepper-box-top fastener	J. Bounds	Bridgeport, Conn	June 16, 1868	78, 858
Pepper-boxes, Manufacture of	J. H. Stone	Philadelphia, Pa	Feb. 26, 1867	62, 451
Pepper-cruet	H. T. Clawson	New Berne, N. C	Dec. 21, 1858	22, 349
Peppermint-dropper	H. H. Snow	New Haven, Conn	Mar. 4, 1851	7, 963
Perambulator	T. Galt	Rock Island, Ill	July 30, 1872	129, 941
Perambulator	A. Clifford	West Point, Mass	Oct. 27, 1829	....
Perambulator	C. Lyne	Padstow, England	Aug. 25, 1868	81, 387
Perambulator	A. W. Richards	Indianola, Iowa	Mar. 18, 1873	136, 867
Perambulator, Folding	A. Christian	New York, N. Y	Oct. 27, 1868	83, 363
Perambulator, Folding	C. Lyne	Padstow, England	Apr. 20, 1869	89, 058
Perambulator or child's carriage	J. A. Crandall	New York, N. Y	Jan. 15, 1861	31, 110
Perch-irons, Die for making	J. W. Sheppard	Plantsville, Conn	Feb. 20, 1872	123, 943
Perch-plate-forming die	R. R. Miller	Plantsville, Conn	Mar. 15, 1870	100, 783
Perch-spring clip	J. Deeble	Plantsville, Conn	Apr. 12, 1870	101, 834
Percolating apparatus	C. A. Smith	Cincinnati, Ohio	Nov. 20, 1847	5, 372
Percolator	P. H. Vander Weyde	Philadelphia, Pa	Apr. 10, 1866	53, 905
Percolator, Adjustable	J. Q. Hill	Worcester, Mass	Sept. 20, 1864	44, 311
Percolator and filtering-machine, Continuous	P. H. Vander Weyde	Philadelphia, Pa	Dec. 11, 1866	60, 445
Percolator for druggists and others	A. Merrell	Cincinnati, Ohio	June 14, 1870	104, 181
Percolator, Reversible	S. Woolston and W. Corfield	Vincentown, N. J., and Philadelphia, Pa.	Feb. 22, 1870	100, 232
Percussion-cap	A. C. Hobbs	Bridgeport, Conn	Sept. 14, 1869	94, 743
Percussion-cap	B. F. Woodside	McDonald Station, Tenn	July 26, 1870	105, 874
Percussion-cap box	T. Harvey	Baltimore, Md	July 12, 1864	43, 497
Percussion-cap, &c., box pocket	G. Gros	Bordeaux, France	Aug. 10, 1869	93, 432
Percussion-cap carrier	R. Paulson	Washington, D. C	Oct. 10, 1871	119, 879
Percussion-cap holder	T. B. Lamb	Hamilton, Mich	Nov. 3, 1863	40, 487
Percussion-cap holder	F. J. Seymour and O. N. Perkins.	Meriden, Conn	June 15, 1869	91, 276
Percussion-cap holder	C. Verniaud and W. Spencer	La Grange, Mo	Sept. 24, 1872	131, 642
Percussion-cap holder	J. F. Warren	Stafford, N. Y	Feb. 16, 1864	41, 655
Percussion-cap primer	C. R. Alsop	Middletown, Conn	Apr. 8, 1862	34, 919
Percussion-cap primer	G. W. Baker	Burlington, Vt	Sept. 1, 1857	18, 117
Percussion-cap primer	E. D. Seely	Brookline, Mass	Oct. 27, 1863	40, 432
Percussion-cap primer for fire-arms	H. R. Jones	Addison, N. Y	June 10, 1862	35, 524
Percussion-cap primer, Portable	J. K. Ely and R. Cook	Franklin, Ohio	Aug. 6, 1867	67, 424
Percussion-cap-trimming machine	J. H. Fowler and A. J. French	Waterbury, Conn	July 17, 1866	56, 492
Percussion-cap-trimming machine	D. N. Goff	Wolcottville, Conn	July 30, 1867	67, 190
Percussion-cap, Water-proof	J. Chattaway	Hampden County, Mass	June 10, 1856	15, 063

Index of patents issued from the United States Patent Office from 1790 *to* 1873, *inclusive*—Continued.

Invention.	Inventor.	Residence.	Date.	No.
Percussion-caps, &c., Composition for	A. Hockstätter	Langen, Hesse-Darmstadt	Jan. 12, 1864	41, 259
Percussion-caps, Facing ends of	J. Goldmark	New York, N. Y	Nov. 22, 1853	10, 262
Percussion-caps, Feeding	A. J. French	Bridgeport, Conn	Dec. 22, 1868	85, 224
Percussion-caps, Implement for feeding	L. Lelong and J. Decamp	Newark, N. J	Oct. 22, 1861	33, 531
Percussion-caps, Machine for forming and charging	G. Wright	Washington, D. C	Sept. 24, 1850	7, 675
Percussion-caps, Machine for lining	A. S. Blake	Waterbury, Conn	July 30, 1867	67, 253
Percussion-caps, Machine for lining	A. J. French	Waterbury, Conn	Sept. 10, 1867	68, 725
Percussion-caps, Machine for lining	A. J. French	Waterbury, Conn	Apr. 16, 1872	125, 875
Percussion-caps, Machine for lining	D. N. Goff	Wolcottville, Conn	July 30, 1867	67, 189
Percussion-caps, Machine for making	R. M. Bouton	West Troy, N. Y	Mar. 20, 1849	6, 196
Percussion-caps, Machine for ramming	C. Hicks	Haverstraw, N. Y	Feb. 10, 1857	16, 587
Percussion-caps, Machine for varnishing	C. Hicks	Haverstraw, N. Y	Feb. 17, 1857	16, 646
Percussion-caps, Machine for varnishing and lining	W. A. McIntire	Troy, N. Y	Feb. 6, 1866	52, 435
Percussion-caps, &c., Manufacture of water-proof	B. Burton	Brooklyn, N. Y	Aug. 11, 1868	81, 057
Percussion-caps, Self-feeding machine for charging	M. W. Fisher	Washington, D. C	Nov. 21, 1848	5, 928
Percussion-lock	M. Carleton	Haverhill, N. H	Dec. 23, 1830	....
Percussion-lock	D. G. Colburn	Port Byron, N. Y	Oct. 25, 1832	....
Percussion-lock	F. Dowler	Wayne Township, Ohio	Apr. 9, 1832	....
Percussion-lock	S. Forker	Meadville, Pa	Feb. 13, 1830	....
Percussion-lock	I. B. Richardson	Palmyra, N. Y	Feb. 17, 1832	....
Percussion-lock and walking-cane rifle and pistol	A. D. Cushing	Troy, N. Y	July 20, 1831	....
Percussion-lock, Concealed	J. Newbury	Poughkeepsie, N. Y	Apr. 27, 1830	....
Percussion-lock for discharging mining-blasts	E. Hughes	McCartysville, Cal	Oct. 11, 1864	44, 630
Percussion-mill for reducing or grinding grain	F. McCarthy	Demopolis, Ala	Feb. 24, 1845	3, 924
Percussion-pellet	C. Sharps	Hartford, Conn	June 28, 1853	9, 820
Percussion-powder	M. Kling	Reading, Pa	Aug. 18, 1857	18, 016
Percussion-powder	F. M. Ruschampt and J. Schutte	New York, N. Y	June 3, 1862	35, 477
Percussion-powder for discharging arms	S. Guthrie	Sackett's Harbor, N. Y	Aug. 21, 1834	....
Percussion-primer and gun-lock	E. Maynard	Washington, D. C	Sept. 22, 1845	4, 208
Percussion-tape primer	J. Chattaway	Springfield, Mass	July 22, 1856	15, 370
Perforating-machine	F. Anderson	Peekskill, N. Y	Dec. 26, 1871	122, 098
Perfume and volatile liquids, Device for nebulizing	E. P. Roche	Bath, Me	May 24, 1870	103, 374
Perfumery	G. Smith	Groton Junction, Mass	Nov. 29, 1870	109, 773
Perfuming and disinfecting apparatus	O. Boldeman	New York, N. Y	Apr. 4, 1871	113, 619
Permutation-lock	J. T. Adams	Washington, D. C	May 9, 1871	114, 510
Permutation-lock	G. B. Atwood	Philadelphia, Pa	Mar. 26, 1867	63, 133
Permutation-lock	G. B. Atwood	Philadelphia, Pa	Apr. 16, 1867	63, 824
Permutation-lock	T. A. Auberlin	Detroit, Mich	Sept. 28, 1869	95, 181
Permutation-lock	J. E. and L. P. Barnes	Davers Centre and Fitchburgh, Mass.	Sept. 20, 1870	107, 590
Permutation-lock	E. A. Barrows	Willimantic, Conn	Feb. 26, 1867	62, 388
Permutation-lock	E. W. Brettell	Newark, N. J	Aug. 27, 1867	68, 281
Permutation-lock	E. W. Brettell	Elizabeth, N. J	Oct. 20, 1868	83, 129
Permutation-lock	E. W. Brettell	Elizabeth, N. J	Oct. 25, 1870	108, 561
Permutation-lock	E. W. Brettell	New Haven, Conn	Mar. 18, 1873	136, 963
Permutation-lock	F. H. Brown	Chicago, Ill	Nov. 8, 1870	108, 964
Permutation-lock	W. C. Bussey	San Francisco, Cal	Mar. 12, 1872	124, 540
Permutation-lock	W. A. Carpenter	Elgin, Ill	Feb. 21, 1860	27, 202
Permutation-lock	S. L. Cole and W. G. Ayres	Brooklyn, N. Y	Feb. 16, 1869	86, 908
Permutation-lock	J. Connell	Rochester, N. Y	June 9, 1863	38, 858
Permutation-lock	J. Corbett	Salt Lake, Utah	Feb. 5, 1867	61, 810
Permutation-lock	G. L. Damon	Cambridge, Mass	May 30, 1871	115, 445
Permutation-lock	G. L. Damon	Boston, Mass	Apr. 2, 1872	125, 180
Permutation-lock	G. W. Dana	Racine, Wis	Jan. 19, 1869	86, 085
Permutation-lock	C. Diebold and J. Obernesser	Cincinnati, Ohio	Aug. 16, 1870	106, 472
Permutation-lock	W. F. Ensign	Troy, N. Y	Oct. 20, 1868	83, 144
Permutation-lock	W. F. Ensign	New York, N. Y	Nov. 17, 1868	84, 177
Permutation-lock	J. Farrel	New York, N. Y	Apr. 16, 1872	125, 669
Permutation-lock	C. Flesch	Rochester, N. Y	Feb. 19, 1867	62, 191
Permutation-lock	C. Flesch	Rochester, N. Y	Nov. 26, 1867	71, 373
Permutation-lock	C. Flesch	Rochester, N. Y	Dec. 27, 1870	110, 560
Permutation-lock	C. R. C. French	Berkley, Mass	Sept. 15, 1868	82, 104
Permutation-lock	L. H. Gano	New York, N. Y	Jan. 3, 1872	122, 452
Permutation-lock	J. B. Gray	Fredericksburgh, Va	Sept. 18, 1841	2, 261
Permutation-lock	H. Gross	Tiffin, Ohio	Mar. 29, 1870	101, 362
Permutation-lock	H. Gross	Tiffin, Ohio	May 24, 1870	103, 450
Permutation-lock	H. Gross	Cincinnati, Ohio	Oct. 11, 1870	108, 134
Permutation-lock	H. Gross	Cincinnati, Ohio	Nov. 18, 1873	144, 670
Permutation-lock	H. Gross	Cincinnati, Ohio	Dec. 2, 1873	145, 171
Permutation-lock	H. Gross and J. L. Hall	Cincinnati, Ohio	Sept. 6, 1870	107, 174
Permutation-lock	M. Hainque	San Francisco, Cal	Sept. 13, 1870	107, 251
Permutation-lock	P. W. Hall	Calvert, Tex	Sept. 9, 1873	142, 695
Permutation-lock	W. Hall	Boston, Mass	May 18, 1869	90, 096
Permutation-lock	W. N. Hall	Springfield, Tex	Sept. 27, 1870	107, 775
Permutation-lock	J. C. Hintz, jr	Cincinnati, Ohio	May 17, 1870	103, 183
Permutation-lock	J. C. Hintz, jr	Cincinnati, Ohio	Oct. 18, 1870	108, 481
Permutation-lock	H. Isham	Hartford, Conn	Jan. 18, 1870	98, 972
Permutation-lock	A. W. Johnson and G. Thompson.	New York, N. Y	May 21, 1867	64, 880
Permutation-lock	F. G. Johnson	Brooklyn, N. Y	June 30, 1857	17, 681
Permutation-lock	W. Johnson	Milwaukee, Wis	Nov. 17, 1868	84, 192
Permutation-lock	G. Kaiser	New York, N. Y	May 19, 1868	78, 097
Permutation-lock	W. F. Kistler	Chicago, Ill	July 9, 1867	66, 502
Permutation-lock	W. F. Kistler	Cincinnati, Ohio	Sept. 7, 1869	94, 613
Permutation-lock	W. F. Kistler	Chicago, Ill	July 5, 1870	105, 091
Permutation-lock	J. Klein	Williamsburgh, N. Y	May 5, 1868	77, 623
Permutation-lock	W. Kock	Cincinnati, Ohio	July 5, 1870	104, 961
Permutation-lock	I. W. Lamb	Salem, Mich	Dec. 17, 1867	72, 408
Permutation-lock	L. W. Langdon	Northampton, Mass	June 8, 1869	91, 139
Permutation-lock	J. H. Larry	Weston, Mass	Apr. 5, 1870	101, 632
Permutation-lock	R. A. Lee	Cleveland, Ohio	Oct. 15, 1872	132, 161
Permutation-lock	S. Loyd	New York, N. Y	Mar. 14, 1871	112, 724
Permutation-lock	C. L. Lucas	Plymouth, Mass	Aug. 27, 1867	68, 219
Permutation-lock	N. Macneale	Cincinnati, Ohio	July 5, 1870	105, 102
Permutation-lock	D. K. Miller	Reading, Pa	July 4, 1871	116, 737
Permutation-lock	D. K. Miller	Reading, Pa	May 26, 1868	78, 310
Permutation-lock	L. H. Miller	Baltimore, Md	Jan. 28, 1868	73, 742
Permutation-lock	L. H. Miller	Baltimore, Md	Sept. 6, 1870	107, 190

Index of patents issued from the United States Patent Office from 1790 *to* 1873, *inclusive*—Continued.

Invention.	Inventor.	Residence.	Date.	No.
Permutation-lock	S. Miller	Gratis, Ohio	Apr. 23, 1872	126, 074
Permutation-lock	S. Miller	Gratis, Ohio	Dec. 2, 1873	145, 223
Permutation-lock	J. H. Morse	Peoria, Ill	Mar. 30, 1858	19, 815
Permutation-lock	J. H. Morse	Peoria, Ill	June 15, 1869	91, 359
Permutation-lock	J. Obernesser	Cincinnati, Ohio	May 23, 1871	115, 230
Permutation-lock	J. Obernesser	Cincinnati, Ohio	May 23, 1871	115, 231
Permutation-lock	O. E. Pillard	New Britain, Conn	Jan. 28, 1868	73, 831
Permutation-lock	O. E. Pillard	New Britain, Conn	Mar. 24, 1868	75, 967
Permutation-lock	O. E. Pillard	New Britain, Conn	June 30, 1868	79, 388
Permutation-lock	O. E. Pillard	New Britain, Conn	June 1, 1869	90, 683
Permutation-lock	O. E. Pillard	New Britain, Conn	Dec. 7, 1869	97, 550
Permutation-lock	O. E. Pillard	New Britain, Conn	May 9, 1871	114, 706
Permutation-lock	O. E. Pillard	New Britain, Conn	Oct. 31, 1871	120, 396
Permutation-lock	O. E. Pillard	New Britain, Conn	Apr. 23, 1872	125, 905
Permutation-lock	F. B. Pye	New York, N. Y	Mar. 7, 1846	4, 406
Permutation-lock	J. Sargent	Rochester, N. Y	Jan. 4, 1870	98, 622
Permutation-lock	J. Sargent	Rochester, N. Y	Dec. 22, 1868	85, 245
Permutation-lock	J. P. Schmucker	Ashland, Ohio	Jan. 4, 1870	98, 523
Permutation-lock	D. Snell	Little Falls, N. Y	Oct. 12, 1869	95, 739
Permutation-lock	E. Stockwell	Stamford, Conn	July 25, 1871	117, 478
Permutation-lock	E. Stockwell	Stamford, Conn	Sept. 2, 1873	142, 529
Permutation-lock	W. Streeter	Rochester, N. Y	July 12, 1870	105, 383
Permutation-lock	T. J. Sullivan	Albany, N. Y	Aug. 25, 1868	81, 430
Permutation-lock	T. J. Sullivan	Albany, N. Y	Sept. 14, 1869	94, 789
Permutation-lock	T. J. Sullivan	Albany, N. Y	Nov. 12, 1872	132, 936
Permutation-lock	J. T. Taylor	Newman, Ga	May 21, 1872	127, 117
Permutation-lock	W. Terwilliger	New York, N. Y	Apr. 4, 1871	113, 465
Permutation-lock	H. C. Thrall	Springfield, Mass	Sept. 24, 1872	131, 576
Permutation-lock	D. L. Tower	New York, N. Y	Oct. 17, 1871	120, 128
Permutation-lock	H. R. Towne	Stamford, Conn	Feb. 7, 1871	111, 587
Permutation-lock	J. F. Vinton and G. A. Hines	Brattleborough, Vt	Nov. 14, 1871	120, 915
Permutation-lock	S. C. Weddington	Jonesborough, Ind	Oct. 31, 1871	120, 472
Permutation-lock	J. Weimar	Mount Vernon, Ohio	Aug. 24, 1869	94, 156
Permutation-lock	A. Wetzel	Cincinnati, Ohio	Feb. 25, 1868	74, 781
Permutation-lock	S. Wheeler	Albany, N. Y	May 7, 1867	64, 605
Permutation-lock	S. Wheeler	Albany, N. Y	Sept. 17, 1867	68, 922
Permutation-lock	T. B. Worrell	Philadelphia, Pa	Feb. 18, 1873	135, 957
Permutation-lock	L. Yale, jr	Shelburne Falls, Mass	Sept. 15, 1868	82, 192
Permutation-lock	L. Yale, jr	Bernardstown, Mass	Jan. 4, 1870	98, 536
Permutation-lock, Auricular	E. E. Stubbs	West Elkton, Ohio	July 2, 1872	128, 672
Permutation-lock for doors, &c	J. H. Morse	Peoria, Ill	Nov. 5, 1867	70, 599
Permutation-lock for doors, &c	T. J. Sullivan	Rochester, N. Y	Aug. 20, 1867	67, 927
Permutation-lock for doors, vaults, &c., Manifold	R. Newell	New York, N. Y	Sept. 25, 1838	944
Permutation-lock for vaults, safes, &c	D. W. Maples	Geneva, N. Y	Dec. 4, 1844	3, 842
Permutation-lock indicator	A. Hardy	Boston, Mass	Dec. 22, 1868	85, 227
Permutation-locks, Adjustable tumbler for	O. E. Pillard	New Britain, Conn	Dec. 3, 1867	71, 639
Permutation-locks, Adjustable tumbler for	O. E. Pillard	New Britain, Conn	Sept. 8, 1868	82, 039
Permutation-locks, Adjustable tumbler for	J. H. Porter	New York, N. Y	Dec. 3, 1867	71, 640
Permutation-locks, Tumbler for	E. W. Brettell	New Haven, Conn	Dec. 16, 1873	145, 618
Permutation-locks, Tumbler for	C. Flesch	Rochester, N. Y	Apr. 4, 1871	113, 414
Permutation-locks, Tumbler for	W. Kock	Cincinnati, Ohio	July 25, 1871	117, 302
Permutation plate-lock, Rotating	H. Ritchie	Newark, N. J	June 26, 1849	6, 555
Permutation-wheel for lock	E. W. Brettell	Elizabeth, N. J	June 13, 1871	115, 814
Perpetual register	J. S. and J. H. Hood	Washington, D. C	June 16, 1868	78, 967
Perplexing unpickable lock	J. C. S. Daubreville	Ramapo, N. Y	May 13, 1833	....
Perspective protracter	B. Otis	Philadelphia, Pa	Mar. 14, 1815	....
Pessary	J. B. Andrews	New York, N. Y	Nov. 13, 1849	6, 861
Pessary	B. Atkinson	Davenport, Iowa	Feb. 18, 1868	74, 482
Pessary	E. P. Banning, sr	New York, N. Y	July 29, 1873	141, 253
Pessary	L. Barker	New York, N. Y	Sept. 12, 1822	....
Pessary	E. T. Brigham	Philadelphia, Pa	Dec. 3, 1867	71, 692
Pessary	W. F. Chrisman	Trenton, Tenn	Sept. 8, 1868	81, 877
Pessary	H. H. Christie	Herkimer, N. Y	Nov. 13, 1866	59, 556
Pessary	W. Elmer	New York, N. Y	Aug. 17, 1858	21, 189
Pessary	D. Ewing	Indianapolis, Ind	Apr. 30, 1867	64, 211
Pessary	C. E. Flack	Jacksonville, Ill	Dec. 23, 1873	145, 854
Pessary	W. R. Gardner	Leonardsville, N. Y	July 12, 1870	105, 191
Pessary	L. A. Gescheidt	New York, N. Y	May 9, 1848	5, 556
Pessary	C. R. Gorgas	Brooklyn, N. Y	July 21, 1868	80, 163
Pessary	C. R. Gorgas	Roughsburgh, Ohio	May 18, 1869	90, 166
Pessary	W. G. Grant	Clyde, Ohio	June 18, 1867	65, 903
Pessary	E. B. Harding	Northampton, Mass	Nov. 27, 1866	59, 999
Pessary	J. M. Heard	Aberdeen, Wis	May 29, 1860	28, 480
Pessary	E. F. Hofman	Poughkeepsie, N. Y	June 4, 1867	65, 382
Pessary	E. F. Hofmann	New York, N. Y	Jan. 11, 1870	98, 769
Pessary	B. Joseph	Philadelphia, Pa	June 23, 1868	79, 231
Pessary	J. B. Merriman	Sheffield, Mass	Oct. 24, 1846	4, 825
Pessary	O. M. Muncaster	Washington, D. C	July 22, 1873	141, 069
Pessary	M. J. Rhees	Mount Holly, N. J	Nov. 5, 1867	70, 616
Pessary	J. H. Robinson	Charlestown, Mass	Aug. 7, 1849	6, 629
Pessary	J. H. Robinson	Charlestown, Mass	Nov. 19, 1850	7, 788
Pessary	J. R. Rowand	Philadelphia, Pa	Oct. 24, 1848	5, 877
Pessary	T. C. Sachse	Chicago, Ill	Oct. 22, 1867	70, 124
Pessary	H. V. Scattergood	Albany, N. Y	Dec. 15, 1863	40, 949
Pessary	C. Schmidt	Saint Louis, Mo	Mar. 29, 1870	101, 317
Pessary	I. Stealy	Crestline, Ohio	June 19, 1866	55, 733
Pessary	B. F. Taft	Blackstone, Mass	July 29, 1862	36, 040
Pessary	J. Warrington and W. Scattergood.	Philadelphia, Pa	Sept. 5, 1831	....
Pessary	F. F. Wells	Texana, Tex	Feb. 22, 1859	23, 052
Pessary	F. F. Wells	Texana, Tex	Mar. 13, 1860	27, 491
Pessary	J. P. Willms	Baltimore, Md	Dec. 4, 1866	60, 312
Pessary	D. D. Young	Dayton, Ohio	Aug. 3, 1869	93, 387
Pessary-adjuster	O. M. Muncaster	Washington, D. C	Mar. 4, 1873	136. 449
Pessary, Elastic	J. A. Wadsworth	Providence, R. I	Jan. 24, 1860	26, 941
Pessary for prolapsus uteri	S. K. Jennings	Baltimore, Md	Jan. 20, 1843	2, 921
Pessaries, Construction of	F. Roesler	New York, N. Y	Feb. 19, 1856	14, 293
Pestles with mortars, Arrangement of	P. C. Ingersoll	Elmira, N. Y	May 9, 1854	10, 890

Index of patents issued from the United States Patent Office from 1790 *to* 1873, *inclusive*—Continued.

Invention.	Inventor.	Residence.	Date.	No.
Petroleum and fluids made therefrom, Apparatus for burning.	F. Cook	New York, N. Y	Sept. 10, 1867	68, 704
Petroleum and other hydrocarbons, Apparatus for burning.	H. C. Van Tine	Pittsburgh, Pa	June 18, 1867	65, 846
Petroleum and other hydrocarbon liquids, Apparatus for combustion of.	H. Taylor	Montreal, Canada	May 3, 1870	102, 619
Petroleum and other hydrocarbon oils, Treating	J. Young	Kelly, North Britain	June 4, 1872	127, 446
Petroleum and other inflammable liquids, Apparatus for storing.	F. Bizard and P. Labarre	Marseilles, France	Jan. 15, 1867	61, 148
Petroleum and other liquid fuel, Process of burning.	D. Dick	Meadville, Pa	June 2, 1863	38, 732
Petroleum and other liquids from gas, Apparatus for freeing.	A. H. Hook	New York, N. Y	May 11, 1869	89, 998
Petroleum and other liquids to prevent loss by fire, Mode of storing.	S. Stevens	New York, N. Y	Jan. 1, 1867	60, 956
Petroleum and other liquids, Vessel for holding	E. Waters	Troy, N. Y	Aug. 25, 1868	81, 441
Petroleum and other oils to produce a vehicle for paint and varnish, Treating.	A. Meucci	Clifton, N. Y	Sept. 9, 1862	36, 419
Petroleum, Apparatus and process for treating	S. Van Syckel	Titusville, Pa	Oct. 21, 1873	143, 945
Petroleum, Apparatus for burning	E. McKinney	Middletown, Pa	July 25, 1865	48, 967
Petroleum, Apparatus for burning crude	H. Baldwin	Titusville, Pa	Mar. 10, 1868	75, 238
Petroleum, Apparatus for burning crude	C. Saffray	New York, N. Y	Sept. 24, 1867	69, 253
Petroleum, Apparatus for refining and distilling	J. Perkins and W. H. Burnet	Newark, N. J	Apr. 4, 1865	47, 125
Petroleum, Apparatus for treating	D. H. Burket and J. C. Gray	Half Moon and Putney-ville, Pa.	Aug. 21, 1866	57, 285
Petroleum, Apparatus for treating	A. Thirault	Williamsburgh, N. Y	Apr. 16, 1867	63, 963
Petroleum as fuel, Apparatus for burning	F. Cook	New York, N. Y	Sept. 10, 1867	68, 706
Petroleum, benzole, &c., Apparatus for deodorizing	J. Green	Rochester, N. Y	Mar. 14, 1865	46, 794
Petroleum, Burning	A. C. Rand	Aurora, Ill	July 2, 1872	128, 656
Petroleum, &c., Composition for coating vessel for	W. L. Dooling and D. Christman, jr.	Zanesville, Ohio	Mar. 20, 1866	53, 280
Petroleum, &c., Deodorizing	S. Lewis	Rochester, N. Y	May 10, 1864	42, 671
Petroleum, Deodorizing	O. Lugo	New York, N. Y	Jan. 1, 1867	60, 757
Petroleum, Deodorizing	R. Newall	Marietta, Ohio	Apr. 3, 1866	53, 656
Petroleum, Deodorizing	T. Restieaux	Boston, Mass	Apr. 9, 1867	63, 749
Petroleum, &c., Deodorizing	J. W. W. Tindall	Liverpool, England	Mar. 31, 1863	38, 069
Petroleum, Device for burning	F. A. Hull	Beloit, Wis	Feb. 4, 1873	135, 476
Petroleum, Device for heating and conveying	J. Casey	Washington, D. C	May 16, 1865	47, 701
Petroleum for lubricating, Preparing	W. H. Young	Athens, Ohio	Mar. 12, 1867	62, 798
Petroleum for manufacture of lubricating-oil, Treating.	G. O. Spence	Titusville, Pa	June 2, 1868	78, 545
Petroleum for manufacture of paints, &c., Preparing.	R. Bartholow	Cincinnati, Ohio	Apr. 4, 1865	47, 084
Petroleum-forge or blow-pipe	H. S. Saroni	Cincinnati, Ohio	Apr. 30, 1872	126, 158
Petroleum from the surface of water-courses, Device for collecting.	L. H. Cowley	Silver Creek, N. Y	Sept. 21, 1869	95, 089
Petroleum, &c., in conjunction with steam or heated air, Apparatus for burning.	G. O. Spence	Titusville, Pa	Apr. 30, 1867	64, 260
Petroleum, Manufacture of oils from	C. Toppan	Wakefield, Mass	Feb. 7, 1870	99, 500
Petroleum, Material for purifying and discoloring	J. Ellis	New York, N. Y	May 5, 1868	77, 470
Petroleum, naphtha, &c., Deodorizing	S. Lewis	Rochester, N. Y	Sept. 13, 1864	44, 258
Petroleum, naphtha, &c., Deodorizing	R. N. Warfield	Rochester, N. Y	Sept. 22, 1863	40, 068
Petroleum, Obtaining useful products from the "tarry residuum" of.	A. Millochaw	New York, N. Y	Jan. 5, 1864	41, 085
Petroleum, Packing-vessels for	J. W. Masury	Brooklyn, N. Y	Sept. 25, 1866	58, 272
Petroleum, Product from	R. A. Chesebrough	New York, N. Y	June 4, 1872	127, 568
Petroleum, &c., Purifying	R. A. Chesebrough	New York, N. Y	July 10, 1866	56, 179
Petroleum, Purifying	C. C. Parsons	New York, N. Y	Apr. 13, 1869	88, 978
Petroleum, Recovering waste acid for refining	O. W. Farrar	Pittsburgh, Pa	June 4, 1867	65, 361
Petroleum, Refined heavy-oil	H. B. Everest	Rochester, N. Y	Feb. 23, 1869	87, 157
Petroleum, Refining	S. Van Syckel	Titusville, Pa	July 15, 1873	140, 801
Petroleum, &c., Storing	E. Quinn	Washington, D. C	Oct. 2, 1866	58, 475
Petroleum-storing apparatus	I. Mathei	Antwerp, Belgium	Oct. 20, 1868	83, 192
Petroleum-tank	M. C. C. Church	Parkersburgh, W. Va	Aug. 28, 1866	57, 479
Petroleum, &c., tank	M. C. C. Church and E. H. Knight.	Parkersburgh, W. Va., and Washington, D. C.	Sept. 11, 1866	57, 866
Petroleum-tank	P. Jacovence	Bucharest, Wallachia	June 26, 1866	55, 970
Petroleum, Tank for storing	P. Andrew	Cincinnati, Ohio	May 28, 1867	65, 045
Petroleum, Tank for storing	J. Fraser and J. Calkins	Buffalo, N. Y	Oct. 10, 1865	50, 348
Petroleum to produce oil and gas, Apparatus for treating.	A. C. Rand and W. M. Sloane	New York, N. Y	Feb. 23, 1869	87, 199
Petroleum to remove the more volatile portions, Treating.	F. Huot	New York, N. Y	Dec. 3, 1867	71, 619
Petroleum to remove the more volatile portions, Treating.	R. G. Loftus	Chelsea, Mass	Sept. 1, 1868	81, 654
Petroleum, Transporting	J. H. Smith	Allegheny City, Pa	Dec. 10, 1867	72, 102
Petroleum, Transportation of	H. J. Lombaert	Philadelphia, Pa	Sept. 12, 1865	49, 901
Petroleum, Treating	J. C. Pedrick	Washington, D. C	Aug. 13, 1867	67, 796
Petroleum, Treating	C. S. Potter	Brooklyn, N. Y	July 5, 1864	43, 429
Petroleum, Treating	E. Schalk	New York, N. Y	Dec. 3, 1872	133, 598
Petroleum, Treating	J. A. Tatro	Hartford, Conn	Feb. 8, 1870	99, 728
Petroleum, Treating heavy	J. J. Looney	Saint Petersburgh, Pa	May 20, 1873	139, 009
Petroleum, &c., Vaporizing	H. R. Foote	Boston, Mass	Dec. 21, 1869	98, 046
Petroleum, &c., vessel	E. S. Allen	Fair Haven, Vt	May 8, 1866	54, 657
Petroleum-vessel	J. Brakeley	Philadelphia, Pa	Jan. 16, 1866	52, 023
Petroleum, Vessel for holding	J. W. Barnum and P. M. Mc-Noah.	Detroit, Mich	July 25, 1865	48, 891
Petroleum, Vessel for holding	J. M. Batchelder	Cambridge, Mass	Feb. 7, 1865	46, 206
Pew-shelf	D. Buchanan	Philadelphia, Pa	Dec. 20, 1870	110, 197
Phaeton, Pony carriage	J. C. Ham	New York, N. Y	May 31, 1870	103, 605
Phaeton, Shifting top for basket	C. P. Kimball	Portland, Me	Apr. 18, 1871	113, 775
Pharmaceutic composition	J. C. Bay	Mount Pleasant, Ohio	Jan. 26, 1825	....
Phial, Tubular essence	J. Staniford	Boston, Mass	Oct. 17, 1831	....
Phosphate and fertilizer, Manufacture of soluble	D. W. Prescott	Edinburgh, Va	Oct. 17, 1871	119, 994
Phosphate and for other purposes, Mixing liquid with dust or powder for the manufacture of.	A. Duvall	Baltimore, Md	Jan. 26, 1869	86, 289
Phosphate and in extracting phosphoric acid from bones, Manufacture of.	G. F. Wilson and E. N. Horsford.	East Providence, R. I., and Cambridge, Mass.	Mar. 10, 1868	75, 336

Index of patents issued from the United States Patent Office from 1790 *to* 1873, *inclusive*—Continued.

Invention.	Inventor.	Residence.	Date.	No.
Phosphate and other acid liquids, Apparatus for conveying acid.	G. F. Wilson	East Providence, R. I	Mar. 10, 1868	75,335
Phosphate-drying apparatus	F. J. Kimball	Philadelphia, Pa	May 2, 1871	114,304
Phosphate-drying apparatus	E. Frank and J. B. Adt	Baltimore, Md	Aug. 9, 1870	106,147
Phosphate, &c., Bag for	B. R. Croasdale	Philadelphia, Pa	Mar. 5, 1872	124,254
Phosphate, Burning bones for the manufacture of acid.	G. F. Wilson	East Providence, R. I	Mar. 10, 1868	75,332
Phosphate, Drying acid	G. F. Wilson	East Providence, R. I	Mar. 10, 1868	75,330
Phosphate for agricultural purposes, Manufacture of.	G. F. Wilson	East Providence, R. I	Mar. 10, 1868	75,327
Phosphate, Granulating acid	G. F. Wilson	East Providence, R. I	Mar. 10, 1868	75,334
Phosphate, Grinding and pulverizing acid	G. F. Wilson	East Providence, R. I	Mar. 10, 1868	75,333
Phosphate in the manufacture of yeast-powder, &c., Treating acid.	G. F. Wilson and E. N. Horsford.	East Providence, R. I., and Cambridge, Mass.	Mar. 10, 1868	75,338
Phosphate, Manufacture of a non-hygroscopic pulverulent acid.	E. N. Horsford	Cambridge, Mass	June 7, 1870	104,035
Phosphate, Manufacture of acid	H. Storck and F. M. Lyte	Asnieres, near Paris, France.	Apr. 8, 1873	137,635
Phosphate, Manufacture of acid	G. F. Wilson	East Providence, R. I	Mar. 10, 1868	75,328
Phosphate, Manufacturing fertilizing	G. A. Liebig and E. K. Cooper	Baltimore, Md	Jan. 17, 1865	45,961
Phosphate of lime and farinaceous matter in order to granulate them, Treating the mixture of acid.	G. F. Wilson and E. N. Horsford.	East Providence, R. I., and Cambridge, Mass.	Mar. 10, 1868	75,337
Phosphate of lime and soda for culinary and other purposes, Double.	E. N. Horsford	Cambridge, Mass	Mar. 29, 1864	42,140
Phosphate of lime, Apparatus for concentrating acid.	G. F. Wilson and E. N. Horsford.	East Providence, R. I., and Cambridge, Mass.	Mar. 10, 1868	75,339
Phosphate of lime, Apparatus for the manufacture of super.	R. B. Potts and F. Klett	Camden, N. J	Dec. 27, 1864	45,631
Phosphate of lime for culinary and other purposes, Preparation of.	E. N. Horsford	Cambridge, Mass	Mar. 1, 1864	41,815
Phosphate of lime, Preparation of acid	E. N. Horsford	Cambridge, Mass	Mar. 10, 1868	75,271
Phosphate, Preparation of bones for the manufacture of phosphoric acid.	G. F. Wilson	East Providence, R. I	Mar. 10, 1868	75,326
Phosphate, Prepared	A. A. Moses	Charleston, S. C	Feb. 2, 1869	86,574
Phosphate, Preparing bones for the manufacture of acid.	G. F. Wilson	East Providence, R. I	Mar. 10, 1868	75,329
Phosphate, Process and apparatus for the manufacture of acid.	G. F. Wilson	East Providence, R. I	Mar. 10, 1868	75,331
Phosphate to be used in food, Manufacture of acid.	E. N. Horsford	Cambridge, Mass	Mar. 10, 1868	75,272
Phosphate, Treating insoluble	H. Bower	Philadelphia, Pa	Jan. 26, 1869	86,275
Phosphate, Treating vitriolized	A. Duvall	Baltimore, Md	July 12, 1870	105,319
Phosphatic rock, Treating	N. A. Pratt and G. T. Lewis	Charleston, S. C., and Philadelphia, Pa.	July 9, 1872	128,752
Phosphatic substances, Grinding	G. T. Lewis	Philadelphia, Pa	May 9, 1871	114,693
Photo-electrotype	W. A. Leggo and G. E. Desbarats.	Quebec, Canada	May 30, 1865	48,035
Photo-engraving on metal	W. A. McGill and R. G. Pine	Memphis, Tenn	Mar. 26, 1872	124,905
Photo-galvanographic process for printing	P. Pretsch	Austria	Aug. 25, 1857	18,056
Photo-galvanography	P. E. Placet	Paris, France	Aug. 9, 1864	43,821
Photo-lithographic transfer	J. W. Osborne	Melbourne, Australia	Aug. 27, 1861	33,172
Photo-lithographic transfers, Producing	I. Rehn	Washington, D. C	Mar. 15, 1870	100,924
Photo-lithography	J. A. Cutting and L. H. Bradford.	Boston, Mass	Mar. 16, 1858	19,626
Photo-lithography	J. W. Osborne	Melbourne, Australia	June 25, 1861	32,668
Photo-sculpture	F. Willéme	Paris, France	Aug. 9, 1864	43,822
Photograph-camera	J. A. Scott	Lexington, Va	Dec. 22, 1868	85,247
Photograph-chair, Child's	M. H. Prescott, jr	La Crosse, Wis	May 27, 1873	139,329
Photograph, Compound	T. Miltenberger	Bellefontaine, Ohio	May 11, 1858	20,213
Photograph-cutter	S. W. Robinson	Champaign, Ill	Nov. 21, 1871	121,198
Photograph-drier	A. C. Platt	Oberlin, Ohio	Sept. 11, 1866	57,963
Photograph-exhibiting stand	J. S. Reid	Orange, Ind	Nov. 16, 1869	96,961
Photograph, Granulated	W. A. Leggo	Montreal, Canada	May 30, 1871	115,489
Photograph-holder	J. E. Treat	Boston, Mass	June 9, 1863	38,849
Photograph-mount	A. C. Partridge	Boston, Mass	June 11, 1872	127,705
Photograph on uneven surfaces, Apparatus to	J. H. Pein	Hoboken, N. J	Aug. 30, 1859	25,276
Photograph-press	R. Sibley	Greenville, Conn	June 10, 1862	35,570
Photograph-washing machine	L. V. Moulton	Beaver Dam, Wis	Aug. 12, 1873	141,658
Photographs and engravings, Mounting	J. L. Duffee	Washington, D. C	May 26, 1868	78,195
Photographs, Apparatus for cutting	L. A. Foulley	Boston, Mass	Jan. 31, 1865	46,099
Photographs, Apparatus for exhibiting	M. Dimock	Newark, N. J	Oct. 12, 1869	95,666
Photographs, Apparatus for exhibiting	W. H. Fay	Chester, Mass	June 28, 1864	43,299
Photographs, Apparatus for fixing	C. A. Gale	Piqua, Ohio	Oct. 10, 1871	119,753
Photographs, &c., Apparatus for pasting and mounting.	M. Ormsbee	New York, N. Y	July 7, 1863	39,166
Photographs, Apparatus for producing vignette	E. R. Scott	Philadelphia, Pa	July 8, 1862	35,857
Photographs, Apparatus for taking stereoscopic	J. H. Marston	Philadelphia, Pa	June 19, 1855	13,093
Photographs, Apparatus for washing and drying	A. Pearsch	New Orleans, La	Oct. 11, 1864	44,648
Photographs, Apparatus to preserve and exhibit	C. Robinson	Springfield, Mass	Apr. 11, 1865	47,222
Photographs, Applying tints to	H. Vander Weyde	New York, N. Y	Dec. 5, 1871	121,475
Photographs, Burnisher for	C. B. Conant	Lewiston, Me	Dec. 2, 1873	145,152
Photographs, Burnisher for	W. G. Entrekin	Philadelphia, Pa	Dec. 2, 1873	145,161
Photographs, Burnisher for	E. R. Weston	East Corinth, Me	Sept. 10, 1872	131,319
Photographs, Burnishing-apparatus for	E. R. Weston	East Corinth, Me	Sept. 10, 1872	131,320
Photographs, Burnishing-press for	W. G. Entrekin and J. Brambles, jr.	Philadelphia, Pa	May 20, 1873	139,132
Photographs, Coloring	J. F. Bodtker	Madison, Wis	Apr. 14, 1863	38,144
Photographs, Coloring	R. Winter	San Francisco, Cal	June 14, 1870	104,241
Photographs, &c., Composition for coloring and water-proofing.	W. F. Spieler	Philadelphia, Pa	June 9, 1863	38,847
Photographs, Cutting and pasting	J. Franke	Quincy, Ill	Feb. 11, 1868	74,336
Photographs, &c., Finishing	S. A. L. Hardinge	Brooklyn, N. Y	Jan. 5, 1869	85,584
Photographs for exhibition, Mounting	I. Rowell and F. E. Mills	San Francisco, Cal	Jan. 22, 1867	61,469
Photographs from glass to paper, Removing	E. Howell	Ashtabula, Ohio	May 19, 1857	17,330
Photographs, &c., Glazing	T. J. Denne and A. Hentschel	London, England	Oct. 29, 1872	132,640
Photographs, Guide for cutting out	J. Schofield	Philadelphia, Pa	Apr. 22, 1873	138,202
Photographs in the camera, Making positive and negative.	F. B. Gage	Saint Johnsbury, Vt	July 9, 1867	66,581
Photographs, Instrument for cutting	T. Bergner	Philadelphia, Pa	Jan. 31, 1865	46,066

Index of patents issued from the United States Patent Office from 1790 *to* 1873, *inclusive*—Continued.

Invention.	Inventor.	Residence.	Date.	No.
Photographs, Instrument for enlarging	D. Shire	Philadelphia, Pa	Mar. 22, 1859	23, 316
Photographs, Machine for pressing and smoothing	D. and J. Rupp	New York, N. Y	Oct. 24, 1865	50, 632
Photographs, Mounting	A. C. Platt	Sandusky, Ohio	June 11, 1872	127, 920
Photographs on cards, Apparatus for pasting	C. S. Lucas	Poughkeepsie, N. Y	Nov. 29, 1864	45, 253
Photographs on glass, Background for	J. W. Wykes	Wheeling, Va	June 23, 1857	17, 651
Photographs on paper, Varnishing	D. W. S. Rawson	Galena, Ill	Apr. 2, 1861	31, 906
Photographs, Printing	J. Albert	Munich, Bavaria	Nov. 30, 1869	97, 336
Photographs, Printing	A. S. Kilby	Huntington, Ind	Aug. 18, 1868	81, 277
Photographs, Printing	C. Meinerth	Newburyport, Mass	July 16, 1867	66, 726
Photographs, Printing	J. W. Swan	Newcastle - upon - Tyne, England.	Jan. 22, 1867	61, 368
Photographs, Printing	W. B. Woodbury	Manchester, England	Feb. 20, 1866	52, 803
Photographs, Producing surfaces for printing from	W. B. Woodbury	London, England	Apr. 28, 1868	77, 232
Photographs, Roller for finishing	W. J. Gordon	Philadelphia, Pa	Nov. 6, 1866	59, 507
Photographs, &c., Roller-press for finishing	D. Marshall and B. Marshall	Pittsburgh, Pa., and Marietta, Ohio.	July 12, 1864	43, 515
Photographs, Roller-press for polishing	N. S. Bowdish	Richfield Springs, N. Y	July 1, 1873	140, 456
Photographs, Solution for toning	J. C. Rutherford	Derby Line, Vt	Feb. 7, 1860	27, 076
Photographs, Tinting	C. Elveena	San Francisco, Cal	Aug. 7, 1866	56, 914
Photographs to tomb-stones, Securing	N. W. Langley, H. Jones, and A. S. Drake.	East Cambridge and Stoughton, Mass.	Dec. 6, 1859	26, 370
Photographs, &c., Treating	E. C. Hawkins	Cincinnati, Ohio	Dec. 22, 1857	18, 901
Photographers and others, Enameled flexible metallic plate for.	W. Hodge and J. R. Peck	Washington, Pa	June 6, 1871	115, 612
Photographer's clip or paper-holder	V. M. Griswold	Peekskill, N. Y	Nov. 23, 1869	97, 081
Photographer's curtain	L. G. Bigelow	Grand Rapids, Mich	Sept. 19, 1871	119, 111
Photographer's decanter	G. W. Doty and E. A. and W. F. Stein.	Ravenna, Ohio	July 11, 1865	48, 664
Photographer's dripping and drying rack	V. M. Griswold	Peekskill, N. Y	Nov. 23, 1869	97, 082
Photographer's movable screen	E. J. Foss	Cambridge, Mass	July 26, 1870	105, 665
Photographer's plate-vise	V. M. Griswold	Peekskill, N. Y	Oct. 19, 1869	95, 999
Photographer's posing-apparatus	G. Cramer and J. Gross	Saint Louis, Mo	Nov. 10, 1868	83, 835
Photographer's refrigerator	M. A. Thornton	Perrysburgh, Ohio	July 18, 1871	117, 124
Photographer's rest	E. L. Wilson	Philadelphia, Pa	Oct. 26, 1869	96, 292
Photographer's steeping-tank	J. W. Osborne	Brooklyn, N. Y	Apr. 25, 1871	114, 186
Photographic album	H. T. Anthony and F. Phoebus	New York, N. Y	May 28, 1861	32, 404
Photographic and other negatives, Producing pictures from.	W. B. Woodbury	Greenhitke, England	May 14, 1872	126, 861
Photographic apparatus	C. G. Anthoni	Paris, France	Sept. 10, 1861	33, 230
Photographic apparatus	N. F. English	Hartland, Vt	May 20, 1862	35, 308
Photographic apparatus	D. H. Houston	Cambria, Wis	Aug. 20, 1867	67, 981
Photographic apparatus	C. A. Leach	Philadelphia, Pa	June 19, 1866	55, 682
Photographic apparatus	W. A. Leggo	Montreal, Canada	May 6, 1873	138, 663
Photographic apparatus	R. H. Mims	Edgefield Court-House, S. C	Mar. 12, 1872	124, 607
Photographic apparatus	E. Müller	New York, N. Y	June 12, 1866	55, 525
Photographic apparatus	O. Nicour	Paris, France	Nov. 19, 1867	71, 205
Photographic apparatus	J. and J. Stock	New York, N. Y	July 4, 1871	116, 771
Photographic apparatus	J. and J. Stock	New York, N. Y	Feb. 6, 1872	123, 522
Photographic apparatus	N. Wright	New York, N. Y	Jan. 23, 1866	52, 239
Photographic background	D. and D. Bendann	Baltimore, Md	Apr. 9, 1872	125, 522
Photographic background	L. G. Bigelow	Grand Rapids, Mich	June 13, 1871	115, 812
Photographic background	J. Buchtel	Portland, Oreg	Dec. 16, 1873	145, 487
Photographic background	A. B. Costello	Jersey City, N. J	Dec. 16, 1873	145, 557
Photographic background	P. F. Finch	Lebanon, Ohio	July 16, 1872	129, 122
Photographic background	A. R. Gould	Carrollton, Ohio	Dec. 12, 1871	121, 715
Photographic background	D. W. Van Riper	Columbus, Ga	Sept. 6, 1870	107, 129
Photographic bank-note	L. Eidlitz	New York, N. Y	Feb. 14, 1860	27, 116
Photographic bath	A. Cobbs	Pittsburgh, Pa	Jan. 1, 1867	60, 834
Photographic bath	B. Hufnagel	New York, N. Y	Oct. 5, 1858	21, 679
Photographic bath	W. and W. H. Lewis	New York, N. Y	Dec. 16, 1856	16, 242
Photographic bath	W. and W. H. Lewis	New York, N. Y	June 26, 1860	28, 876
Photographic bath	J. H. Morrow	Baltimore, Md	Apr. 14, 1857	17, 066
Photographic bath	I. Rehn	Philadelphia, Pa	Dec. 4, 1855	13, 885
Photographic bath	R. E. Sisson	Wickford, R. I	June 12, 1866	55, 590
Photographic bath	N. Wright	New York, N. Y	Nov. 14, 1865	50, 981
Photographic bath and pans, Constructing	G. Mathiot	Washington, D. C	Apr. 28, 1857	17, 162
Photographic-camera embossing-press	E. B. Barker	Brooklyn, N. Y	Oct. 14, 1873	143, 608
Photographic camera	T. Barbour	Boston, Mass	Jan. 15, 1867	61, 139
Photographic camera	B. M. Clinedinst	Staunton, Va	July 16, 1872	129, 104
Photographic camera	E. M. Corbett	New York, N. Y	June 19, 1860	28, 739
Photographic camera	N. Flammang	New York, N. Y	Aug. 27, 1872	130, 800
Photographic camera	F. B. Gage	Saint Johnsbury, Vt	Dec. 24, 1867	72, 627
Photographic camera	W. A. Leggo	Montreal, Canada	Mar. 28, 1871	113, 067
Photographic camera	H. C. and C. L. Moore	Springfield, Mass	Feb. 16, 1864	41, 634
Photographic camera	G. W. Parker	Watkins, N. Y	July 18, 1871	117, 106
Photographic camera	A. Semmendinger	New York, N. Y	Feb. 21, 1860	27, 241
Photographic camera	A. Semmendinger	New York, N. Y	Aug. 7, 1860	29, 523
Photographic camera	E. P. Spahn	Newark, N. J	May 20, 1873	139, 090
Photographic camera	J. Stock	New York, N. Y	July 5, 1859	24, 671
Photographic camera	J. Stock	New York, N. Y	Aug. 4, 1863	39, 432
Photographic camera	J. Stock	New York, N. Y	May 31, 1864	42, 971
Photographic camera	J. Stock	New York, N. Y	Aug. 8, 1865	49, 315
Photographic camera	J. and J. Stock	New York, N. Y	June 6, 1871	115, 655
Photographic camera	J. and J. Stock	New York, N. Y	Dec. 5, 1871	121, 553
Photographic camera	J. and J. Stock	New York, N. Y	Jan. 16, 1872	122, 785
Photographic camera	J. and J. Stock	New York, N. Y	May 7, 1872	126, 498
Photographic camera	I. H. Stoddard	Ansonia, Conn	Jan. 18, 1870	99, 026
Photographic camera	H. W. Vaughan	San Francisco, Cal	Apr. 23, 1872	125, 914
Photographic camera	G. W. Venner	Charlestown, Mass	May 28, 1867	65, 314
Photographic camera	C. A. Waterbury	New York, N. Y	July 9, 1872	128, 832
Photographic camera	C. A. Waterbury	New York, N. Y	Nov. 26, 1872	133, 394
Photographic camera	J. H. and W. K. Wickoff	Ripon, Wis	Aug. 11, 1868	81, 055
Photographic camera	A. B. Wilson	Waterbury, Conn	Mar. 25, 1862	34, 800
Photographic camera	S. Wing	Waterville, Me	Dec. 4, 1860	30, 850
Photographic camera	S. Wing	Boston, Mass	May 26, 1868	78, 408
Photographic camera	T. E. Wood	Boston, Mass	Apr. 26, 1864	42, 524
Photographic-camera box	L. M. Boles and W. G. Smith	Cooperstown, N. Y	Feb. 17, 1857	16, 637
Photographic-camera box	M. Flammang	New York, N. Y	Mar. 19, 1872	124, 674

Index of patents issued from the United States Patent Office from 1790 *to* 1873, *inclusive*—Continued.

Invention.	Inventor.	Residence.	Date.	No.
Photographic-camera lens	C. C. Harrison and J. Schnitzer	New York, N. Y	June 17, 1862	35, 605
Photographic-camera plate-holder	A. D. Bollens	Newburgh, N. Y	June 1, 1858	20, 401
Photographic-camera plate-holder	W. and W. H. Lewis	New York, N. Y	Oct. 7, 1856	15, 854
Photographic-camera stand	J. Haworth	Frankford, Pa	Apr. 9, 1867	63, 633
Photographic-camera stand	H. Manger	Philadelphia, Pa	June 13, 1865	48, 193
Photographic-camera stand	G. S. Knapp	Winona, Minn	Aug. 24, 1869	94, 120
Photographic-camera stand	T. M. Schleier	Knoxville, Tenn	July 9, 1872	128, 915
Photographic-camera stand	J. Scouler	San Francisco, Cal	Feb. 14, 1865	46, 396
Photographic-camera stand	F. E. Wilke	Brooklyn, N. Y	Aug. 13, 1867	67, 830
Photographic cameras, &c., Achromatic object-glass for.	H. Roettger	Philadelphia, Pa	Apr. 18, 1865	47, 336
Photographic cameras, Cover for the lenses of	O. W. Noble	Darlington, Wis	July 25, 1871	117, 318
Photographic cameras, Diaphragm for	F. Miller and A. Wirsching	New York, N. Y	June 7, 1859	24, 356
Photographic cameras, Diaphragm for	C. C. Harrison and J. Schnitzer	New York, N. Y	Sept. 7, 1858	21, 470
Photographic cameras, Diaphragm for	J. R. Werner	New York, N. Y	Sept. 15, 1857	18, 218
Photographic cameras, Multiplying-reflector for	D. W. S. Rawson	Peru, Ill	Dec. 10, 1867	71, 911
Photographic cameras, Plate-frame for	W. and W. H. Lewis	New York, N. Y	Feb. 2, 1858	19, 252
Photographic-card case	W. Miller	New York, N. Y	Oct. 15, 1861	33, 507
Photographic-card holder	W. E. Eastman	Derby Line, Vt	June 22, 1869	91, 728
Photographic-card mount	T. Mayhew	Poughkeepsie, N. Y	Jan. 24, 1865	46, 008
Photographic-card mount	S. Wing	Boston, Mass	Oct. 13, 1863	40, 302
Photographic-card press	J. B. Blair	Philadelphia, Pa	June 14, 1864	43, 083
Photographic cards, Apparatus for mounting and printing.	G. W. Doty	Lockport, N. Y	Aug. 29, 1865	49, 612
Photographic chair	C. G. Pease	Charlestown, Mass	Mar. 6, 1866	53, 038
Photographic-copying board	W. D. Blackman	Defiance, Ohio	July 2, 1867	66, 288
Photographic dark-chambers, Portable	I. F. Woodward	McMinnville, Tenn	Oct. 31, 1871	120, 558
Photographic-embossing apparatus	H. Holt	Brooklyn, N. Y	May 28, 1872	127, 348
Photographic filtering-apparatus	R. M. Linn	Lookout Mountain, Tenn	Sept. 26, 1871	119, 375
Photographic gallery, Portable	S. Weaver	Gettysburgh, Pa	Dec. 15, 1863	40, 970
Photographic-glass holder	J. Longking	New Windsor Township, N. Y.	Feb. 24, 1857	16, 689
Photographic-glass rack	W. G. Smith	Carlisle, Pa	Dec. 27, 1864	45, 648
Photographic grounds for wood-engravers, Varnish to prepare.	R. Price	Worcester, Mass	May 5, 1857	17, 231
Photographic head-rest	G. Cramer and J. Gross	Saint Louis, Mo	Aug. 3, 1869	93, 279
Photographic head-rest	G. A. Emery	Boston, Mass	May 10, 1864	42, 650
Photographic head-rest	W. Kenyon	Crawfordsville, Ind	Sept. 14, 1869	94, 754
Photographic images, Apparatus for multiplying	D. W. S. Rawson	Peru, Ill	Apr. 27, 1869	89, 342
Photographic impressions, Preparation of oil ground to receive.	J. H. Tatum	Baltimore, Md	Apr. 15, 1856	14, 679
Photographic instrument	D. J. Kellogg	Rochester, N. Y	Sept. 30, 1856	15, 809
Photographic lens	C. B. Boyle	New York, N. Y	Oct. 31, 1865	50, 681
Photographic lens	J. H. Dallmeyer	London, England	Feb. 5, 1867	61, 812
Photographic lens	R. Morrison	Brooklyn, N. Y	May 21, 1872	126, 979
Photographic lens	J. Schnitzer	New York, N. Y	Aug. 1, 1865	49, 160
Photographic lenses, Combining	C. B. Boyle	New York, N. Y	Jan. 23, 1866	52, 129
Photographic medal	H. E. Coply	Waterbury, Conn	Apr. 2, 1861	31, 871
Photographic medal	D. F. Maltby	Waterbury, Conn	Aug. 14, 1860	29, 652
Photographic mercury-bath	J. Moulson	Philadelphia, Pa	Sept. 2, 1851	8, 335
Photographic monocular-glass	A. J. Dallemagne and L. Triboulet.	Paris, France	Dec. 30, 1873	146, 052
Photographic name-plate	T. E. Mackerley	Paint, Ohio	Apr. 25, 1865	47, 438
Photographic negative	J. Kirk	Newark, N. J	Mar. 4, 1873	136, 439
Photographic-negative baths, Heating	H. J. Sunter	Buffalo, N. Y	Mar. 25, 1873	137, 110
Photographic negatives, Apparatus for vignetting	F. Wolf	Philadelphia, Pa	Feb. 18, 1868	74, 654
Photographic negatives, Holder for retouching	N. Sarony	New York, N. Y	Feb. 5, 1867	61, 877
Photographic negatives, Retouching	D. H. Wright	Terre Haute, Ind	Nov. 18, 1873	144, 723
Photographic negatives, Varnish for coating	J. W. Morgeneier	Sheboygan, Wis	Dec. 6, 1870	109, 833
Photographic panoramic views, Apparatus for taking.	J. R. Johnson and J. A. Harrison.	London, England	Nov. 28, 1865	51, 279
Photographic-paper	H. M. Hedden	Worcester, Mass	Dec. 28, 1869	98, 375
Photographic-paper, &c., Apparatus for flowing and sensitizing.	W. S. Lukenbach	Newport, Pa	Apr. 27, 1869	89, 488
Photographic paper, &c., Apparatus for fumigating	J. L. Winner	Elizabeth City, N. C	May 30, 1871	115, 554
Photographic paper, Box for silvering and albumenizing.	E. C. Thompson and M. B. Wheaton.	New York, N. Y	July 31, 1860	29, 418
Photographic paper, Sensitized	J. G. Coffin	Portsmouth, Ohio	Jan. 30, 1872	123, 240
Photographic-paper-sensitizing frame	L. V. Moulton	Muskegon, Mich	July 12, 1870	105, 233
Photographic picture	W. H. F. Talbot	Lacock Abbey, England	June 26, 1847	5, 171
Photographic-picture holder	J. Buckett	Harlem, N. Y	May 2, 1865	47, 515
Photographic picture on glass	J. A. Cutting	Boston, Mass	July 11, 1854	11, 267
Photographic pictures, Apparatus for enlarging	J. A. Whitley	Owego, N. Y	Jan. 1, 1861	31, 060
Photographic pictures, Apparatus for obtaining	J. J. L. R. De Lafarge	Paris, France	June 21, 1864	43, 270
Photographic pictures, Apparatus for preserving and exhibiting.	C. Robinson	Springfield, Mass	Oct. 17, 1865	50, 498
Photographic pictures, Bath for toning	R. Anson	New York, N. Y	Feb. 19, 1861	31, 432
Photographic pictures, Bituminous ground for	V. M. Griswold	Lancaster, Ohio	Oct. 21, 1856	15, 924
Photographic pictures by artificial light, Taking	P. F. and W. S. Dodge	West Cambridge, Mass	Feb. 19, 1861	31, 444
Photographic pictures, Cabinet for exhibiting	J. A. Thompson	Auburn, N. Y	Sept. 6, 1864	44, 125
Photographic pictures, Collodion for	V. M. Griswold	Lancaster, Ohio	July 15, 1856	15, 336
Photographic pictures, Coloring	A. O. Dayton	Washington, D. C	Mar. 12, 1850	7, 160
Photographic pictures, Composition for coating	H. Happel	New York, N. Y	Apr. 4, 1871	113, 426
Photographic pictures, Composition for making	J. A. Cutting	Boston, Mass	July 11, 1854	11, 266
Photographic pictures, Device for exhibiting	A. G. Walton	San Francisco, Cal	Nov. 22, 1870	109, 472
Photographic pictures, engravings, &c., Preparation of.	S. D. Humphrey	New York, N. Y	Apr. 28, 1857	17, 152
Photographic pictures, Frame for printing	T. E. Sexton	Wilmington, Del	May 1, 1866	54, 416
Photographic pictures, Guide for cutting	L. T. Young	Philadelphia, Pa	Dec. 16, 1873	145, 711
Photographic pictures, &c., Machine for rolling	C. Sellers	Philadelphia, Pa	July 29, 1862	36, 028
Photographic pictures, Manufacture of	J. R. Johnson	London, England	July 20, 1869	92, 836
Photographic pictures on ceramic ware, Securing	J. B. Obernetter	Munich, Bavaria	Feb. 6, 1866	52, 503
Photographic pictures on glass	A. Bisbee and Y. Day	Columbus, Ohio, and Nashville, Tenn.	May 27, 1856	14, 946
Photographic pictures on glass, &c	F. Langenheim	Philadelphia, Pa	Nov. 19, 1850	7, 784
Photographic pictures on glass, Coloring	D. B. and H. B. Spooner	Springfield, Mass	Aug. 5, 1856	15, 497
Photographic pictures on japanned surfaces	H. L. Smith	Gambier, Ohio	Feb. 19, 1856	14, 300
Photographic pictures, Preparation of collodion for	J. A. Cutting	Boston, Mass	July 4, 1854	11, 213
Photographic pictures, Printing	J. Rehn	Philadelphia, Pa	Apr. 14, 1868	76, 660

Index of patents issued from the United States Patent Office from 1790 *to* 1873, *inclusive*—Continued.

Invention.	Inventor.	Residence.	Date.	No.
Photographic pictures, Process and composition for printing.	V. M. Griswold	Peekskill, N. Y	Apr. 10, 1866	53, 815
Photographic pictures, Revolving tablet for multiplying.	J. Lawrence	Ripon, Wis	May 28, 1867	65, 243
Photographic pictures, Tinting	G. Langdell and M. A. Root	Philadelphia, Pa	July 15, 1856	15, 341
Photographic pictures, Treating	J. B. Hall	New York, N. Y	Jan. 20, 1857	16, 438
Photographic pictures upon transparent media, Producing.	J. A. Whipple and W. B. Jones	Boston, Mass	June 25, 1850	7, 458
Photographic-plate dipper	D. H. Cross	Bennington, Vt	Mar. 15, 1870	100, 866
Photographic-plate holder	H. T. Anthony	New York, N. Y	Oct. 1, 1872	131, 837
Photographic-plate holder	H. T. Anthony	New York, N. Y	Apr. 29, 1873	138, 359
Photographic-plate holder	H. K. Averill, jr	Decorah, Iowa	May 13, 1862	35, 206
Photographic-plate holder	J. Buchtel	Portland, Oreg	Aug. 17, 1869	93, 669
Photographic-plate holder	J. Buchtel	Portland, Oreg	Jan. 7, 1873	134, 512
Photographic-plate holder	R. F. Channell and A. Bonzano	Phœnixville, Pa	Oct. 15, 1872	132, 250
Photographic-plate holder	W. L. Gill	Lancaster, Pa	Dec. 3, 1872	133, 526
Photographic-plate holder	W. Lewis	Brooklyn, N. Y	Apr. 29, 1873	138, 414
Photographic-plate holder	W. and W. H. Lewis	New York, N. Y	Mar. 3, 1857	16, 738
Photographic-plate holder	J. W. Moore	Bellefonte, Pa	June 7, 1870	104, 054
Photographic-plate holder	A. Semmendinger	Fort Lee, N. J	Nov. 25, 1873	145, 020
Photographic-plate holder	J. Stock	New York, N. Y	Dec. 1, 1857	18, 780
Photographic-plate holder	A. B. Wilson	Waterbury, Conn	Dec. 23, 1862	37, 252
Photographic-plate holder, Rotary	C. H. Shute	Edgartown, Mass	Feb. 21, 1865	46, 503
Photographic-plate holders, Catch for	H. C. Eddy	Jersey City, N. J	Sept. 13, 1864	44, 169
Photographic-plate holders, Corner for	J. L. Winner	Elizabeth City, N. C	Oct. 17, 1871	120, 141
Photographic-plate shield	H. Bryant and R. D. O. Smith	Washington, D. C	Nov. 30, 1858	22, 158
Photographic-plate shield	W. Campbell	Jersey City, N. J	Feb. 28, 1860	27, 269
Photographic-plate vise	L. Chapman	New York, N. Y	Feb. 5, 1856	14, 184
Photographic-plate vise	J. W. Jarboe	New York, N. Y	Mar. 17, 1857	16, 841
Photographic-plate vise	S. W. and W. L. Pearsall	New York, N. Y	Feb. 1, 1859	22, 822
Photographic-plate vise	M. H. Rison	Paris, Tenn	Jan. 25, 1859	22, 748
Photographic-plate vise	C. W. Stimson	Cleveland, Ohio	Jan. 17, 1854	10, 433
Photographic plates, Apparatus for drying	J. Thompson	Taunton, Mass	Oct. 11, 1864	44, 678
Photographic plates, Box for dry	J. and J. Stock	New York, N. Y	Oct. 15, 1861	33, 500
Photographic portraiture, Lens for	J. H. Dallmeyer	London, England	June 11, 1867	65, 729
Photographic posing-chair	N. S. Bowdish	Richfield Springs, N. Y	Oct. 31, 1871	120, 485
Photographic posing-chair	W. H. Lewis	New York, N. Y	Sept. 19, 1871	119, 090
Photographic posing-chair	C. A. Schindler	West Hoboken, N. J	Oct. 17, 1871	120, 110
Photographic posing-chair	I. O. Swett	Brighton, Mass	June 4, 1872	127, 439
Photographic-print cutter	J. Haworth	Philadelphia, Pa	Nov. 29, 1870	109, 615
Photographic-print cutter	T. M. Saurman	Norristown, Pa	Aug. 2, 1870	106, 080
Photographic-print-cutting apparatus	W. S. Burgess and G. A. Leuzi	Norristown, Pa	Mar. 29, 1870	101, 222
Photographic prints, Apparatus for gumming, cutting, and mounting.	D. H. Cross	Shaftsbury, Vt	June 21, 1864	43, 184
Photographic prints, Coloring	I. O. Beyse	Saint Louis, Mo	Nov. 7, 1865	50, 871
Photographic prints, Washing	M. Ormsbee	New York, N. Y	Mar. 21, 1865	46, 927
Photographic printing	E. Edwards	Firs Willesden, Great Britain.	May 25, 1869	90, 514
Photographic printing	E. G. Fowx	Baltimore, Md	Oct. 19, 1869	95, 892
Photographic-printing apparatus	E. and E. W. Brown	Roxbury and Boston, Mass	Feb. 4, 1868	73, 948
Photographic-printing apparatus	L. J. Marcy	Newport, R. I	Nov. 30, 1869	97, 419
Photographic-printing apparatus	J. Stehman	Lancaster, Pa	Mar. 10, 1868	75, 482
Photographic printing, Contact-pad for	J. Buchtel	Portland, Oreg	Nov. 17, 1868	84, 168
Photographic-printing frame	G. W. and W. Bowlsby	Monroe, Mich	May 10, 1864	42, 637
Photographic-printing frame	F. Burrows	Peoria, Ill	Aug. 16, 1864	43, 831
Photographic-printing frame	J. W. Campbell	New York, N. Y	Aug. 14, 1866	57, 083
Photographic-printing frame	J. G. Coffin	Portsmouth, Ohio	Oct. 24, 1871	120, 245
Photographic-printing frame	G. W. Cook	Saint Paul, Minn	May 12, 1863	38, 468
Photographic-printing frame	N. F. English	Hartland Four Corners, Vt	Mar. 28, 1865	47, 063
Photographic-printing frame	J. Heddon	Elkhart, Ind	Nov. 5, 1867	70, 564
Photographic-printing frame	W. H. Jacoby	Minneapolis, Minn	Oct. 14, 1873	143, 577
Photographic-printing frame	S. K. Jones	New Haven, Conn	Aug. 22, 1865	49, 531
Photographic-printing frame	C. L. Lochman	Carlisle, Pa	Nov. 20, 1866	59, 800
Photographic-printing frame	P. Murphy	New York, N. Y	Feb. 1, 1870	99, 462
Photographic-printing frame	B. Reed	Kokomo, Ind	Oct. 27, 1868	83, 548
Photographic-printing frame	D. Shive	Philadelphia, Pa	Feb. 27, 1866	52, 896
Photographic-printing frame	J. F. Shoemaker	Van Wert County, Ohio	Apr. 30, 1867	64, 374
Photographic-printing frame	M. Witt	Columbus, Ohio	July 7, 1863	39, 191
Photographic-printing frame and slide	M. Ormsbee	New York, N. Y	Apr. 28, 1863	38, 326
Photographic-printing frame for porcelain or glass pictures.	W. J. Kuhns	Brooklyn, N. Y	Dec. 5, 1865	51, 323
Photographic-printing press	C. Fontayne	Cincinnati, Ohio	Sept. 20, 1859	25, 540
Photographic-printing press	J. H. Hamilton	Sioux City, Iowa	Apr. 12, 1870	101, 728
Photographic printing, Pressure-frame for	L. E. Denison	Winthrop, Conn	Dec. 26, 1865	51, 699
Photographic process	M. M. Griswold	Columbus, Ohio	Nov. 5, 1867	70, 551
Photographic process	C. M. Du Motay and C. R. Maréchal.	Metz, France	Nov. 6, 1866	59, 520
Photographic process	J. Wothly	Paris, France	Aug. 15, 1865	49, 488
Photographic process for copying drawings, &c	W. Willis	Birmingham, England	June 12, 1866	55, 592
Photographic purposes, Arrangement of solar mirrors for.	W. Crane and W. H. Pease	Goshen, Ind	Apr. 24, 1866	54, 118
Photographic purposes, Combination of lenses for.	J. Zentmayer	Philadelphia, Pa	May 29, 1866	55, 195
Photographic purposes, Composition for	T. Cummings	Lancaster, Pa	July 5, 1870	104, 937
Photographic purposes, Curvette and tray for	H. T. Anthony	New York, N. Y	Mar. 29, 1864	42, 061
Photographic purposes, Duplicating-deflector for	D. Shive	Philadelphia, Pa	Oct. 3, 1865	50, 284
Photographic purposes, Preparing paper for	P. H. Murray	Portsmouth, Ohio	Apr. 4, 1871	113, 327
Photographic reflector	W. Kurtz	New York, N. Y	July 5, 1870	104, 963
Photographic rest	C. E. Krüger	New York, N. Y	Nov. 3, 1868	83, 714
Photographic rest	N. Sarony	New York, N. Y	Feb. 18, 1868	74, 604
Photographic rest	O. Sarony	Scarborough, England	June 5, 1866	55, 443
Photograhic roller-press	E. B. McCoy	Winsted, Conn	Feb. 25, 1862	34, 515
Photographic room	G. K. Proctor	Salem, Mass	Oct. 27, 1868	83, 545
Photographic screen	J. A. Anderson	Chicago, Ill	Feb. 22, 1870	100, 099
Photographic sensitizing-box	W., jr., and A. L. Hudson	Hingham, Mass	Jan. 2, 1866	51, 834
Photographic shield	E. Gordon	New York, N. Y	Oct. 19, 1858	21, 829
Photographic show-case	J. F. Adams	Buffalo, N. Y	Mar. 22, 1870	100, 962
Photographic sky-light	C. D. Mosher	Cook County, Ill	Apr. 15, 1873	137, 790
Photographic tent	I. F. Woodward	McMinnville, Tenn	Oct. 17, 1871	119, 957

Index of patents issued from the United States Patent Office from 1790 *to* 1873, *inclusive*—Continued.

Invention.	Inventor.	Residence.	Date.	No.
Photographic transfer	O. P. Howe	Augusta, Me	Apr. 7, 1868	76, 458
Photographic transfers, Making	A. G. Mowan	South Bergen, N. J	June 25, 1867	66, 102
Photographic tray	D. J. Kellogg	Rochester, N. Y	Apr. 7, 1857	16, 979
Photographic use, Compound lens for	J. H. Dallmeyer	Middlesex County, England.	June 30, 1868	79, 323
Photographic use, Preparation of paper, &c., for	W. Gibson	New York, N. Y	Dec. 18, 1866	60, 626
Photographic use, Preparing paper for	J. D. Brinckerhoff	New York, N. Y	Nov. 21, 1865	51, 010
Photographic vignettes, Apparatus for printing	J. E. Richard	Columbia, S. C	Aug. 17, 1869	93, 835
Photographing children, Apparatus for	W. W. Sloan	Jefferson, Tex	July 4, 1871	116, 765
Photographing engravers' blocks, Mode of	B. F. Taylor	Philadelphia, Pa	Feb. 19, 1867	62, 299
Photographing, Preparing articles of gold, silver, and glass for.	J. Wilder and A. B. Stow	Middletown, Conn	Mar. 16, 1869	87, 996
Photography	E. L. Bergstresser	Hublersburgh, Pa	June 17, 1873	140, 001
Photography	E. J. Foss	Boston, Mass	Aug. 15, 1871	118, 006
Photography	F. Glessner	Cincinnati, Ohio	Nov. 9, 1869	96, 691
Photography	C. A. Guilmette	Boston, Mass	Apr. 3, 1866	53, 608
Photography	W. H. Hill	Worcester, Mass	July 29, 1873	141, 351
Photography	H. A. Marchant	Philadelphia, Pa	July 21, 1857	17, 858
Photography	W. H. Smith	London, England	Oct. 17, 1865	50, 544
Photography	F. A. Wenderoth	Philadelphia, Pa	Oct. 17, 1871	120, 136
Photography on wood	C. B. Boyle	Albany, N. Y	Feb. 8, 1859	22, 852
Photography to printing, Application of	A. L. Poitevin	Paris, France	Oct. 28, 1862	36, 821
Photometer	S. G. Elliott	San Francisco, Cal	June 19, 1866	55, 797
Photometer	H. Vogel	Berlin, Prussia	Mar. 10, 1868	75, 493
Phrenology, Apparatus for teaching	G. P. Wilcox and W. Butler	Little Falls, N. Y	Apr. 15, 1856	14, 685
Physical culture, Apparatus for	G. B. Windship	Boston, Mass	Jan. 7, 1873	134, 578
Physician's prescription-case	A. S. Hamlin	Saint Louis, Mo	Oct. 21, 1862	36, 710
Physiognotrace	D. Atherton	Richmond, Va	May 4, 1805	......
Physiognotrace	I. Todd, A. Day, and W. Bache		June 15, 1803	......
Physiological battery	A. C. Garratt	Boston, Mass	July 6, 1869	92, 301
Piano	S. P. Brooks	Somerville, Mass	Jan. 9, 1872	122, 557
Piano	O. Bull	New York, N. Y	Nov. 15, 1870	109, 172
Piano	C. F. Chew	George street, Chalk Farm Road, England.	Aug. 23, 1870	106, 661
Piano	A. F. Dessau	Washington, D. C	Feb. 8, 1870	99, 655
Piano	A. Faas	Philadelphia, Pa	Aug. 22, 1871	118, 353
Piano	A. Gunther	Philadelphia, Pa	Aug. 15, 1865	49, 399
Piano	A. H. Hastings	New York, N. Y	Feb. 25, 1862	34, 491
Piano	C. Hattersley	Trenton, N. J	May 24, 1870	103, 456
Piano	C. Kurtzman	Buffalo, N. Y	May 24, 1870	103, 345
Piano	G. C. Manner	New York, N. Y	July 22, 1873	141, 152
Piano	F. Mathushek	New York, N. Y	Mar. 28, 1871	113, 073
Piano	F. B. McGregor	Pontiac, Mich	Sept. 22, 1868	82, 428
Piano	W. Nordhoff	Baltimore, Md	Aug. 14, 1866	57, 257
Piano	C. F. Oliver	Lynn, Mass	June 14, 1870	104, 341
Piano	J. W. Otto	Saint Louis, Mo	Aug. 28, 1866	57, 558
Piano	S. W. Parker	Somerville, Mass	Aug. 23, 1870	106, 612
Piano	R. Raven	New York, N. Y	Aug. 14, 1866	57, 186
Piano	C. Robbins	Chicago, Ill	Feb. 13, 1866	52, 610
Piano	G. Trayser	Indianapolis, Ind	July 9, 1867	66, 653
Piano	W. F. Ulman	Boston, Mass	Jan. 10, 1871	110, 940
Piano	T. Winans	Baltimore, Md	Dec. 24, 1872	134, 340
Piano-action	S. P. Brooks	Somerville, Mass	Nov. 5, 1872	132, 709
Piano-action	F. Koth	North New York, N. Y	Oct. 28, 1873	143, 986
Piano-action	R. Kreter	New York, N. Y	Dec. 23, 1873	145, 879
Piano-action	L. Kühner	New York, N. Y	Feb. 11, 1873	135, 820
Piano-action	L. Matt and B. Grueter	Boston, Mass	May 23, 1871	115, 077
Piano-action	F. B. McGregor	Pontiac, Mich	Dec. 6, 1870	109, 921
Piano-action	E. G. Newman and P. Anderson	New York, N. Y	July 1, 1873	140, 428
Piano-action	C. F. Oliver	Lynn, Mass	Aug. 16, 1870	106, 396
Piano-action	C. F. T. Steinway	New York, N. Y	June 6, 1871	115, 782
Piano-action	F. L. Trayser	Maysville, Ky	Oct. 14, 1873	143, 647
Piano action, Grand	D. F. Haaz	Philadelphia, Pa	Apr. 28, 1857	17, 148
Piano-action, Hammer-flange butt of	D. L. Bollermann	Mount Vernon, N. Y	Nov. 21, 1871	121, 077
Piano action, Upright	T. Atkins and H. Drewer	Cincinnati, Ohio	Oct. 29, 1872	132, 557
Piano action, Upright	R. Kreter	New York, N. Y	Mar. 18, 1873	137, 005
Piano action, Upright	G. C. Manner	Mott Haven, N. Y	Apr. 1, 1873	137, 377
Piano-agraffe	C. F. Chickering	New York, N. Y	Dec. 24, 1872	134, 194
Piano and organ attachment	L. J. Trémaux	New Orleans, La	Dec. 23, 1873	145, 793
Piano and organ key	C. F. L. Goffrie and J. H. Schucht.	London, England	Apr. 8, 1873	137, 544
Piano and organ printing-attachment	G. Heydrich	New Ulm, Minn	Dec. 17, 1872	133, 987
Piano, Bell	C. G. Buttkereit	Toledo, Iowa	Oct. 31, 1871	120, 415
Piano, Bell	C. G. Buttkereit	Toledo, Iowa	Sept. 16, 1873	142, 768
Piano, Bell	C. Williams and E. F. Falconnet.	Nashville, Tenn	Sept. 24, 1861	33, 372
Piano-bridge	C. W. Brewer	Racine, Wis	Apr. 27, 1869	89, 381
Piano-bridge	W. Compton	New York, N. Y	July 10, 1860	29, 061
Piano, couch, and bureau, Combined	C. Hess	Cincinnati, Ohio	July 17, 1866	56, 413
Piano-coupling attachment	J. Clark	Philadelphia, Pa	Oct. 25, 1870	108, 685
Piano-cover Water-proof	H. F. Herkner and J. W. Post	Brooklyn, N. Y	Feb. 7, 1871	111, 535
Piano, Double-action	J. J. Bender	New York, N. Y	Aug. 21, 1866	57, 434
Piano-forte hammers, Composition for paste for attaching the covering to.	J. Munson	Burlington, Vt	June 19, 1860	28, 768
Piano-forte	C. F. L. Albrecht	Philadelphia, Pa	Mar. 18, 1828	......
Piano-forte	A. Babcock	Boston, Mass	Dec. 17, 1825	......
Piano-forte	A. Babcock	Boston, Mass	Oct. 31, 1839	1, 389
Piano-forte	E. Badlam	Potsdam, N. Y	Oct. 25, 1845	4, 241
Piano-forte	W. W. Batchelder	New York, N. Y	June 21, 1864	43, 174
Piano-forte	W. W. Bennett	Jersey City, N. J	Aug. 26, 1873	142, 194
Piano-forte	E. Bloomfield and D. P. Otis	New York, N. Y	May 18, 1869	90, 150
Piano-forte	C. Bossert and J. Shomacker	Philadelphia, Pa	Apr. 29, 1842	2, 595
Piano-forte	W. Bourne	Boston, Mass	Feb. 17, 1863	37, 717
Piano-forte	W. Bourne	Boston, Mass	June 14, 1870	104, 256
Piano-forte	S. P. Brooks	Boston, Mass	Nov. 24, 1857	18, 673
Piano-forte	E. Brown	Boston, Mass	Nov. 20, 1838	1, 014
Piano-forte	E. Brown	Boston, Mass	Jan. 27, 1843	2, 934
Piano-forte	L. H. Browne	Boston, Mass	Sept. 23, 1851	8, 383

Index of patents issued from the United States Patent Office from 1790 *to* 1873, *inclusive*—Continued.

Invention.	Inventor.	Residence.	Date.	No.
Piano-forte	L. Caldera and L. Montu	Turin, Italy	Mar. 10, 1868	75, 362
Piano-forte	G. Charters	New York	July 8, 1816	....
Piano-forte	J. Chickering	Boston, Mass	Oct. 8, 1840	1, 802
Piano-forte	J. Chickering	Boston, Mass	Sept. 1, 1843	3, 238
Piano-forte	P. E. Chollet	New York, N. Y	May 7, 1867	64, 407
Piano-forte	I. Clark	Cincinnati, Ohio	Mar. 2, 1836	....
Piano-forte	O. M. Coleman	Philadelphia, Pa	Apr. 17, 1844	3, 548
Piano-forte	B. E. Colley	Cambridge, Mass	June 21, 1853	9, 793
Piano-forte	W. Compton	New York, N. Y	June 15, 1852	9, 014
Piano-forte	W. Compton	New York, N. Y	Sept. 6, 1853	9, 928
Piano-forte	W. Cumston	Boston, Mass	Aug. 3, 1839	1, 275
Piano-forte	D. Decker	New York, N. Y	June 2, 1863	38, 731
Piano-forte	D. Decker	New York, N. Y	Jan. 10, 1865	45, 818
Piano-forte	J. J. and D. Decker	New York, N. Y	Jan. 29, 1867	61, 612
Piano-forte	S. W. Draper	Boston, Mass	June 25, 1845	4, 082
Piano-forte	S. B. Driggs	Detroit, Mich	Dec. 18, 1855	13, 942
Piano-forte	S. B. Driggs	New York, N. Y	Jan. 12, 1858	19, 081
Piano-forte	S. B. Driggs	New York, N. Y	May 3, 1859	23, 834
Piano-forte	S. B. Driggs	New York, N. Y	Mar. 13, 1860	27, 435
Piano-forte	S. B. Driggs	New York, N. Y	Mar. 22, 1864	41, 977
Piano-forte	S. B. Driggs	New York, N. Y	Apr. 11, 1865	47, 196
Piano-forte	G. Ely	New York, N. Y	May 3, 1870	102, 668
Piano-forte	L. Farnesworth	Nashua, N. H	Aug. 9, 1870	106, 145
Piano-forte	J. U. Fischer	New York, N. Y	July 26, 1859	24, 905
Piano-forte	J. U. Fischer	New York, N. Y	July 10, 1860	29, 068
Piano-forte	L. Fissore	Baltimore, Md	July 22, 1833	....
Piano-forte	J. Geib	New York, N. Y	Oct. 3, 1817	....
Piano-forte	L. Gilbert	Boston, Mass	July 10, 1841	2, 167
Piano-forte	T. Gilbert	Boston, Mass	Sept. 30, 1851	8, 389
Piano-forte	C. Glassborow	Philadelphia, Pa	July 26, 1859	24, 805
Piano-forte	P. Gmehlin	New York, N. Y	July 23, 1872	129, 727
Piano-forte	H. Goldsmith	Philadelphia, Pa	Dec. 8, 1857	18, 810
Piano-forte	O. Gori and P. Ernstz	New York, N. Y	Mar. 26, 1844	3, 504
Piano-forte	J. A. Gray	Albany, N. Y	Mar. 27, 1849	6, 223
Piano-forte	J. A. Gray	Albany, N. Y	May 1, 1860	28, 137
Piano-forte	J. A. Guttwaldt	New York, N. Y	Aug. 27, 1818	....
Piano-forte	A. Hall	Lloydsville, Ohio	Jan. 24, 1854	10, 456
Piano-forte	A. Hall	Lloydsville, Ohio	Jan. 30, 1855	12, 315
Piano-forte	H. Hartye	Baltimore, Md	Mar. 12, 1836	....
Piano-forte	A. H. Hastings	New York, N. Y	Mar. 15, 1870	100, 888
Piano-forte	A. H. Hastings	New York, N. Y	Oct. 10, 1871	119, 760
Piano-forte	J. J. Hawkins	....	Feb. 12, 1800	....
Piano-forte	H. Herrick	New York, N. Y	Oct. 26, 1839	1, 379
Piano-forte	H. Herrick	Boston, Mass	July 21, 1868	80, 073
Piano-forte	G. Hews, R. C. Marsh, and N. M. Tileston.	Boston and Dorchester, Mass.	Apr. 10, 1843	3, 045
Piano-forte	C. Horst	New Orleans, La	Apr. 17, 1849	6, 342
Piano-forte	G. H. Hulskamp	Troy, N. Y	July 14, 1857	17, 789
Piano-forte	G. H. Hulskamp	Troy, N. Y	July 10, 1860	29, 081
Piano-forte	G. H. Hulskamp	New York, N. Y	Dec. 18, 1866	60, 519
Piano-forte	T., H. O., G. T., and W. F. Kearsing.	New York, N. Y	June 13, 1831	....
Piano-forte	J. G. Kunze	New York, N. Y	Dec. 20, 1859	26, 503
Piano-forte	F. C. Lighte	New York, N. Y	Sept. 13, 1859	25, 426
Piano-forte	F. C. Lighte	New York, N. Y	July 1, 1862	35, 766
Piano-forte	H. Lindeman	New York, N. Y	Aug. 7, 1860	29, 502
Piano-forte	G. C. Manner	New York, N. Y	Nov. 13, 1866	59, 619
Piano-forte	T. Marschall	New York, N. Y	Jan. 7, 1862	34, 114
Piano-forte	T. Marschall	New York, N. Y	Apr. 13, 1867	88, 970
Piano-forte	W. H. Mason	Boston, Mass	Oct. 16, 1866	58, 950
Piano-forte	F. Mathushek	New York, N. Y	Oct. 28, 1851	8, 470
Piano-forte	F. Mathushek	New York, N. Y	Oct. 2, 1860	30, 279
Piano-forte	L. Matt	Boston, Mass	Aug. 25, 1863	39, 664
Piano-forte	J. and J. McDonald	New York, N. Y	Oct. 5, 1852	9, 304
Piano-forte	W. H. McDonald	Brooklyn, N. Y	Apr. 2, 1867	63, 547
Piano-forte	J. S. McLean	....	May 27, 1796	....
Piano-forte	M. Miller	Rochester, N. Y	July 1, 1851	8, 194
Piano-forte	G. W. Neill	Boston, Mass	July 7, 1868	79, 591
Piano-forte	D. B. Newhall	Boston, Mass	Nov. 3, 1841	2, 330
Piano-forte	C. F. Oliver and G. M. Jackson	Lynn, Mass	May 1, 1845	4, 019
Piano-forte	S. T. Parmelee	New Haven, Conn	June 24, 1862	35, 703
Piano-forte	S. T. Parmelee	New Haven, Conn	June 24, 1862	35, 704
Piano-forte	S. T. Parmelee	New Haven, Conn	Mar. 7, 1865	46, 759
Piano-forte	D. T. Peck	New York, N. Y	Mar. 2, 1869	87, 509
Piano-forte	C. A. Peterson	New York, N. Y	Aug. 10, 1869	93, 473
Piano-forte	J. Pethick	Mount Morris, N. Y	Feb. 12, 1836	....
Piano-forte	L. Philleo	Utica, N. Y	July 2, 1846	4, 612
Piano-forte	J. Pirsson	New York, N. Y	Aug. 13, 1850	7, 568
Piano-forte	L. Reuckert	Baltimore, Md	Mar. 12, 1845	3, 940
Piano-forte	L. Ricketts	Baltimore, Md	June 24, 1844	3, 643
Piano-forte	T. J. V. Roz	Paris, France	Apr. 30, 1867	64, 371
Piano-forte	J. Ruck	New York, N. Y	Apr. 23, 1850	7, 308
Piano-forte	C. S. Sackmeister	New York	May 17, 1830	....
Piano-forte	J. H. Schomacker	Philadelphia, Pa	June 13, 1848	5, 631
Piano-forte	H. Schonacker	Detroit, Mich	Dec. 18, 1855	13, 960
Piano-forte	J. Schriber	New York	Oct. 29, 1846	4, 832
Piano-forte	P. Schuler	Philadelphia, Pa	Oct. 16, 1866	58, 896
Piano-forte	C. P. Seabury	New York	May 20, 1830	....
Piano-forte, &c	R. Shaw	Boston, Mass	July 2, 1807	....
Piano-forte	M. Stange	New York, N. Y	Jan. 21, 1862	34, 220
Piano-forte	C. F. T. Steinway	New York, N. Y	Dec. 14, 1869	97, 982
Piano-forte	C. F. T. Steinway	New York, N. Y	May 28, 1872	127, 383
Piano-forte	C. F. T. Steinway	New York, N. Y	May 28, 1872	127, 384
Piano-forte	H. Steinway, jr	New York, N. Y	Nov. 29, 1859	26, 300
Piano-forte	T. Steinway	New York, N. Y	Aug. 18, 1868	81, 306
Piano-forte	W. Steinway	New York, N. Y	June 5, 1866	55, 385
Piano-forte	D. Stirn	Milwaukee, Wis	May 4, 1869	89, 806
Piano-forte	W. F. Ulman	Boston, Mass	Aug. 22, 1871	118, 407

Index of patents issued from the United States Patent Office from 1790 *to* 1873, *inclusive*—Continued.

Invention.	Inventor.	Residence.	Date.	No.
Piano-forte	M. Vergnes	New York, N. Y	Oct. 4, 1864	44, 566
Piano-forte	M. Vergnes	New York, N. Y	Jan. 17, 1865	45, 943
Piano-forte	G. Vogt	Philadelphia, Pa	Mar. 1, 1859	23, 130
Piano-forte	E. L. Walker and G. W. Cherry	Carlisle, Pa., and Alexandria, D. C.	Jan. 16, 1845	3, 888
Piano-forte	S. R. Warren	Montreal, Canada	July 10, 1845	4, 109
Piano-forte action	J. Becker	New York, N. Y	June 3, 1856	14, 998
Piano-forte action	J. D. Boedicker	New York, N. Y	Nov. 5, 1861	33, 633
Piano-forte action	S. P. Brooks	Boston, Mass	Apr. 17, 1855	12, 737
Piano-forte action	S. P. Brooks	Somerville, Mass	May 30, 1865	47, 922
Piano-forte action	G. Brown	Boston, Mass	Jan. 27, 1852	8, 680
Piano-forte action	H. S. Calenberg	New York, N. Y	Oct. 8, 1861	32, 427
Piano-forte action	B. E. Colley	Boston, Mass	Oct. 26, 1869	96, 204
Piano-forte action	B. E. Colley	Boston, Mass	Oct. 15, 1872	132, 198
Piano-forte action	W. Compton	New York, N. Y	May 22, 1860	28, 352
Piano-forte action	W. G. Day	Baltimore, Md	Dec. 13, 1870	110, 120
Piano-forte action	D. Decker	New York, N. Y	Sept 13, 1859	25, 393
Piano-forte action	S. B. Driggs	New York, N. Y	May 19, 1857	17, 320
Piano-forte action	T. C. Faulder	Albany, N. Y	Mar. 10, 1863	37, 860
Piano-forte action	F. Frickinger	Shodack, N. Y	July 23, 1861	32, 863
Piano-forte action	T. Gilbert	Boston, Mass	Feb. 10, 1841	1, 970
Piano-forte action	T. Gilbert	Boston, Mass	Aug. 7, 1847	5, 216
Piano-forte action	J. A. Gray	Albany, N. Y	Sept. 9, 1851	8, 352
Piano-forte action	J. A. Gray	Albany, N. Y	Mar. 17, 1857	16, 832
Piano-forte action	N. J. Haines	New York, N. Y	May 24, 1859	24, 119
Piano-forte action	A. Hall	Lloydsville, Ohio	Apr. 11, 1854	10, 776
Piano-forte action	I. I. Harwood	Boston, Mass	Feb. 12, 1861	31, 385
Piano-forte action	A. K. Hebard	Cambridge, Mass	Dec. 9, 1873	145, 417
Piano-forte action	H. Herrick	Boston, Mass	Feb. 19, 1867	62, 134
Piano-forte action	M. Herter	New York, N. Y	Dec. 19, 1865	51, 588
Piano-forte action	G. Howe	Boston, Mass	Sept. 28, 1852	9, 282
Piano-forte action	G. Howe	Roxbury, Mass	Oct. 20, 1857	18, 453
Piano-forte action	G. H. Hulskamp	Troy, N. Y	May 22, 1860	28, 374
Piano-forte action	G. H. Hulskamp	Troy, N. Y	Oct. 21, 1862	36, 712
Piano-forte action	A. B. Irving	Indianapolis, Ind	Dec. 31, 1867	72, 738
Piano-forte action	C. L. and A. B. Irving	Fort Wayne, Ind	Oct. 23, 1866	59, 155
Piano-forte action	R. M. Kerrison	Philadelphia, Pa	Sept. 9, 1851	8, 353
Piano-forte action	R. M. Kerrison	Philadelphia, Pa	June 19, 1855	13, 091
Piano-forte action	J. Kohnle	New York, N. Y	Jan. 24, 1860	26, 909
Piano-forte action	F. Koth	New York, N. Y	July 4, 1865	48, 565
Piano forte action	R. Kreter	New York, N. Y	Sept. 9, 1851	8, 350
Piano-forte action	H. A. Leaman	New York, N. Y	Apr. 6, 1858	19, 857
Piano-forte action	E. A. Lee	Roxbury, Mass	May 23, 1854	10, 948
Piano-forte action, &c	T. Loud	Philadelphia, Pa	Dec. 7, 1837	504
Piano-forte action	T. Loud	Spring Garden, Pa	Apr. 24, 1847	5, 086
Piano-forte action	J. H. Low	Boston, Mass	Dec. 26, 1848	5, 985
Piano-forte action	N. M. Lowe	Boston, Mass	Mar. 25, 1856	14, 509
Piano-forte action	F. Marschall	New York, N. Y	Aug. 30, 1859	25, 305
Piano-forte action	J. V. Marshall	Albany, N. Y	June 8, 1858	20, 500
Piano-forte action	F. Mathushek	New York, N. Y	Dec. 20, 1859	26, 550
Piano-forte action	F. Mathushek	New York, N. Y	Mar. 28, 1871	113, 074
Piano-forte action	L. Matt	Boston, Mass	Dec. 19, 1871	121, 952
Piano-forte action	F. B. McGregor	Pontiac, Mich	Apr. 18, 1871	113, 785
Piano-forte action	J. S. Morton	New York, N. Y	Sept. 18, 1855	13, 580
Piano-forte action	W. Munroe	Boston, Mass	Apr. 24, 1855	12, 763
Piano-forte action	F. W. Niehaus	Boston, Mass	July 3, 1860	28, 997
Piano-forte action	J. F. Nunns	New York, N. Y	May 5, 1831	....
Piano-forte action	A. W. Perry	Saint Joseph, Mo	Dec. 7, 1869	97, 686
Piano-forte action	F. Pistorias	Chicago, Ill	July 12, 1864	43, 524
Piano-forte action	G. Pratt	West Roxbury, Mass	Dec. 27, 1864	45, 675
Piano-forte action	C. E. Rogers	Boston, Mass	Mar. 5, 1872	124, 289
Piano-forte action	J. Ruck	New York, N. Y	Apr. 16, 1850	7, 292
Piano-forte action	J. Ruck	New York, N. Y	Mar. 11, 1851	7, 976
Piano-forte action	T. S. Seabury	Stony Brook, N. Y	Oct. 25, 1859	25, 919
Piano-forte action	H. Seidel	Roxbury, Mass	May 12, 1868	77, 845
Piano-forte action	J. Shandelle	Huntsville, Ala	Jan. 14, 1873	134, 940
Piano-forte action	D. H. Shirley	Boston, Mass	Nov. 28, 1854	12, 003
Piano-forte action	D. H. Shirley	Boston, Mass	Mar. 30, 1869	88, 522
Piano-forte action	H. Steinway	New York, N. Y	May 5, 1857	17, 238
Piano-forte action	H. Steinway	New York, N. Y	June 15, 1858	20, 595
Piano-forte action	H. Steinway, jr	New York, N. Y	May 21, 1861	32, 386
Piano-forte action	H. Steinway, jr	New York, N. Y	May 21, 1861	32, 387
Piano-forte action	H. Steinway, jr	New York, N. Y	Apr. 8, 1862	34, 910
Piano-forte action	T. Steinway	New York, N. Y	Aug. 10, 1869	93, 647
Piano forte action	F. Taylor	New York, N. Y	Dec. 11, 1855	13, 924
Piano-forte action	J. Thompson	New York	Aug. 6, 1831	....
Piano-forte action	J. H. Tibbits	Omaha City, Nebr	July 11, 1865	48, 741
Piano-forte action	W. V. Wallace	New York, N. Y	Oct. 30, 1866	59, 295
Piano-forte action	J. J. Wise	Baltimore, Md	June 27, 1839	1, 205
Piano-forte action	J. J. Wise	Baltimore, Md	Dec. 26, 1848	5, 990
Piano-forte action, Upright	A. H. Hastings	Jersey City, N. J	June 20, 1871	116, 182
Piano-forte and cabinet, Combination	E. Cotter	Boston, Mass	Dec. 21, 1869	98, 038
Piano-forte and organ key	W. F. Furgang	Albany, N. Y	Apr. 20, 1852	8, 887
Piano-forte and organ key	E. P. Needham	New York, N. Y	July 1, 1873	140, 528
Piano-forte attachment	S. B. Driggs	Detroit, Mich	Jan. 24, 1854	10, 446
Piano-forte bottoms, Construction of	W. F. Senior	New York, N. Y	Nov. 21, 1845	4, 282
Piano-forte bridge	C. H. De Vine	Buffalo, N. Y	Oct. 13, 1868	82, 928
Piano-forte bridge	W. C. Ellis	Worcester, Mass	Apr. 22, 1873	138, 012
Piano-forte bridge	T. E. Power	Columbia, Mo	May 12, 1857	17, 296
Piano-forte bridge, Metallic	G. H. Hulskamp	Troy, N. Y	July 21, 1857	17, 838
Piano-forte cases, Machinery for making	S. Porter	Leominster, Mass	Dec. 7, 1869	97, 551
Piano-forte covers, &c., Ornamenting	J. F. Burgess	Brooklyn, N. Y	Nov. 27, 1866	59, 959
Piano-forte frame	H. S. Ackerly	New York, N. Y	Feb. 27, 1855	12, 432
Piano-forte frame	A. F. Dessau	Washington, D. C	Mar. 3, 1868	75, 132
Piano-forte frame	G. W. Neill	Boston, Mass	Apr. 6, 1869	88, 729
Piano-forte frame	J. Piffant	New Orleans, La	Jan. 25, 1853	9, 559
Piano-forte, Grand	F. C. Lighte	New York, N. Y	Feb. 21, 1860	27, 226
Piano-forte hammers, Covering	R. Kreter	New York, N. Y	Jan. 4, 1853	9, 526

Index of patents issued from the United States Patent Office from 1790 *to* 1873, *inclusive*—Continued.

Invention.	Inventor.	Residence.	Date.	No.
Piano-forte, Horizontal	E. R. Currier	Boston, Mass	Apr. 22, 1831	....
Piano-forte, Horizontal	F. C. Reichenbach	Philadelphia, Pa	May 19, 1841	2, 099
Piano-forte, Horizontal square	G. Bacon and R. Raven	New York, N. Y	Aug. 26, 1851	8, 320
Piano-forte, Instruments for teaching music with the.	E. Von Heeringen	Pickensville, Ala	June 26, 1849	6, 558
Piano-forte key	L. H. Browne	Boston, Mass	Aug. 6, 1861	32, 972
Piano-forte key	H. W. Hindsley	New Haven, Conn	Feb. 7, 1860	27, 050
Piano-forte key	J. Hoffacker and J. Richards	New York, N. Y	June 14, 1859	24, 404
Piano-forte key	C. J. Schoensmann	New York, N. Y	Aug. 7, 1860	29, 522
Piano-forte key-board	M. Philippi	Troy, N. Y	Oct. 11, 1859	25, 760
Piano-forte leg	A. Goodman and L. Hale	North Dana, Mass	Oct. 6, 1863	40, 166
Piano-forte lids, Raising	H. N. Oliver	Rahway, N. J	Feb. 18, 1873	136, 090
Piano-forte lock	P. F. Dodge	West Cambridge, Mass	May 3, 1859	23, 833
Piano-forte lock	S. Walker	New York, N. Y	Apr. 18, 1865	47, 349
Piano-forte pedal-attachment	G. A. Scott and W. B. Frisbee	San Francisco, Cal	July 5, 1870	105, 002
Piano-forte, Portable	M. Vergnes	New York, N. Y	Apr. 26, 1864	42, 518
Piano-forte, Portable	M. Vergnes	New York, N. Y	June 21, 1864	43, 245
Piano-forte soft-pedal attachment	J. Greener	Elmira, N. Y	Feb. 9, 1869	86, 747
Piano-forte sound-board	L. Matt	Boston, Mass	Dec. 31, 1867	72, 745
Piano-forte sound-board	J. Newman	Baltimore, Md	Apr. 7, 1857	16, 990
Piano-forte sound-board	C. F. Th. Steinway	New York, N. Y	Apr. 6, 1869	88, 749
Piano-forte sounding-board	J. A. Gray	Albany, N. Y	Feb. 6, 1855	12, 362
Piano-forte sounding-board	A. Speer and E. Marx	Aquackanonck, N. J	Sept. 28, 1852	9, 287
Piano-forte sounding-board	R. Swan, jr	New Bedford, Mass	Nov. 20, 1849	6, 889
Piano-forte, Square	C. F. Chickering	New York, N. Y	Apr. 23, 1861	32, 119
Piano-forte string	H. J. Newton	New York, N. Y	Oct. 21, 1851	8, 452
Piano-forte tuning-attachment	R. Beebe	West Springfield, Mass	Sept. 22, 1863	40, 015
Piano-forte, Upright	G. H. Davis	Boston, Mass	Oct. 28, 1873	143, 967
Piano-forte, Upright	L. Gilbert	Boston, Mass	June 18, 1850	7, 441
Piano-forte, Upright	H. Klepfer	Cincinnati, Ohio	Mar. 25, 1851	8, 002
Piano-forte, Upright	R. E. Letton	Quincy, Ill	Oct. 5, 1852	9, 301
Piano-forte, Upright	J. Thompson	New York	Oct. 1, 1830	....
Piano-fortes, Apparatus for augmenting sound in	M. Vergnes	New York, N. Y	Aug. 14, 1866	57, 225
Piano-fortes, Arranging the keys in	J. Dwight and D. B. Newhall	Boston, Mass	May 6, 1841	2, 081
Piano-fortes, Combination of reed instrument with	R. W. Carpenter	Brooklyn, N. Y	Dec. 23, 1862	37, 217
Piano-fortes, Combining metallic reeds with	M. Coburn	Savannah, Ga	Feb. 1, 1847	4, 948
Piano-fortes, Compensating tube for	T. Loud	Philadelphia, Pa	July 7, 1835	....
Piano-fortes, Composition finger-key for	L. Wolf	West Meriden, Conn	July 16, 1867	66, 928
Piano-fortes, Constructing hammer-heads in	E. Forbes	Boston, Mass	Feb. 10, 1841	1, 971
Piano-fortes, Constructing rest-pin for	D. Walker	New York, N. Y	June 19, 1838	790
Piano-fortes, Construction and action of	A. Babcock	Philadelphia, Pa	Dec. 31, 1833	....
Piano-fortes, Cross-stringing	A. Babcock	Philadelphia, Pa	May 24, 1830	....
Piano-fortes, Damper-action for upright	G. W. Neill	Boston, Mass	Mar. 8, 1870	100, 550
Piano-fortes, Duplex agraffe-scales for	C. F. T. Steinway	New York, N. Y	May 14, 1872	126, 848
Piano-fortes, Elevating the tops of	C. Meyer	Philadelphia, Pa	Apr. 10, 1849	6, 282
Piano-fortes, Glass string for	R. Bury	Albany, N. Y	Aug. 21, 1819	....
Piano-fortes, Harmonicon-attachment to	C. Peters	New London, Conn	July 16, 1872	128, 976
Piano-fortes, Iron frame for	A. Ludolff	New York, N. Y	Feb. 20, 1866	52, 725
Piano-fortes, Iron frame for upright	S. P. Brooks	Boston, Mass	Apr. 11, 1854	10, 770
Piano-fortes, Longitudinal bar in	J. Dwight	Boston, Mass	July 29, 1824	....
Piano-fortes, Melodeon-attachment for	L. Louis	New York, N. Y	Apr. 13, 1869	88, 967
Piano-fortes, Metallic frame for	T. Gilbert	Boston, Mass	July 24, 1847	5, 202
Piano-fortes, Mover for	D. I. Langworthy	Jamestown, N. Y	Apr. 2, 1867	63, 533
Piano-fortes, Prop-stick for	C. E. McDonald	Brooklyn, N. Y	Apr. 2, 1867	63, 546
Piano-fortes, Repeating-action for	J. Hura	Boston, Mass	Dec. 18, 1866	60, 520
Piano-fortes, Shifting-movement for square or horizontal.	T. Loud	Philadelphia, Pa	Apr. 1, 1842	2, 523
Piano-fortes, String-bearing for	R. A. Tooker	New York, N. Y	Oct. 20, 1863	40, 384
Piano-fortes, Stringing and tuning	W. J. Richards, F. M. Sofge, and J. H. Richards.	La Fayette, Ind	Feb. 8, 1870	99, 708
Piano-fortes, Swell-mute attachment to	A. G. Corliss	Portland, Me	Apr. 11, 1854	10, 773
Piano-fortes, Treble-attachment for	A. V. T. Barberie	Brooklyn, N. Y	Jan. 12, 1869	85, 889
Piano-fortes, &c., Working pedals of	R. H. Hooper	West Roxbury, Mass	July 14, 1868	79, 830
Piano frame, Upright	J. Whitney	Boston, Mass	Jan. 30, 1872	123, 212
Piano-frames, Brace for	D. Gibbons	Rochester, N. Y	Jan. 2, 1855	12, 188
Piano, Grand	G. H. Davis	Boston, Mass	Mar. 1, 1870	100, 266
Piano, Grand	D. F. Haasz	Philadelphia, Pa	Mar. 4, 1856	14, 383
Piano, Grand	G. Steck	New York, N. Y	June 20, 1871	116, 109
Piano, Grand	G. Steck	New York, N. Y	Oct. 21, 1873	143, 789
Piano, Grand	H. Steinway	New York, N. Y	Dec. 20, 1859	26, 532
Piano-hammer	C. W. Brewer	Racine, Wis	Sept. 8, 1868	81, 872
Piano-hammer	J. Percival	Auburn, N. Y	July 5, 1859	24, 659
Piano-hammers, Capping	J. Bullard	Readville, Mass	Apr. 22, 1873	138, 001
Piano, Horizontal	T. Loud, jr	Philadelphia, Pa	May 15, 1827	....
Piano-key	W. A. Reed	Deep River, Conn	Aug. 22, 1871	118, 389
Piano key-board	J. S. Allen and A. P. Wilkins	Allen's Grove, Wis	June 30, 1868	79, 295
Piano key-board arrangement	A. G. and C. Marsh	Seneca Falls, N. Y	May 17, 1859	24, 021
Piano-key-shaping machine	M. Pratt	Deep River, Conn	July 29, 1873	141, 380
Piano-keys, &c., Mode of balancing	F. J. Steinhauser	Lancaster, Pa	Aug. 3, 1869	93, 243
Piano-leg	F. and C. Gelin	New York, N. Y	June 7, 1859	24, 294
Piano-leg attachment	W. Clark	New York, N. Y	May 27, 1856	14, 948
Piano-legs, Making	H. Willgohs	New York, N. Y	Dec. 14, 1869	97, 845
Piano-lid prop-stick	M. Doherty	Cambridgeport, Mass	May 20, 1873	139, 125
Piano, &c., lock	H. Ahrend	Brooklyn, N. Y	June 1, 1869	90, 805
Piano-lock	H. Ahrend	Newark, N. J	Sept. 20, 1870	107, 585
Piano-lock	E. F. French	New York, N. Y	Mar. 16, 1869	87, 834
Piano-lock	E. L. Gaylord	Terryville, Conn	Aug. 1, 1865	49, 100
Piano-lock	E. L. Gaylord	Terryville, Conn	Sept. 3, 1867	68, 496
Piano-lock	E. L. Gaylord	Terryville, Conn	Sept. 3, 1867	68, 497
Piano-lock	A. Helmstaedter	Newark, N. J	Dec. 10, 1867	72, 038
Piano-lock	R. J. Jordan	New York, N. Y	Mar. 30, 1869	88, 488
Piano-lock	J. Murphy	Roslindale, Mass	Oct. 4, 1870	107, 947
Piano-lock	A. F. Pfeifer	Newark, N. J	Aug. 1, 1865	49, 149
Piano-lock	A. F. Pfeifer	Newark, N. J	Feb. 23, 1869	87, 196
Piano-lock	T. Powers	New York, N. Y	Mar. 26, 1872	124, 912
Piano-lock	J. Thielemann	Newark, N. J	Aug. 18, 1868	81, 226
Piano-lock	J. Thielemann and P. Meyer	Newark, N. J	May 24, 1870	103, 525
Piano-lock	E. D. Weyburn	Pittsburgh, Pa	Nov. 7, 1871	120, 804

Index of patents issued from the United States Patent Office from 1790 *to* 1873, *inclusive*—Continued.

Invention.	Inventor.	Residence.	Date.	No.
Piano-lock	N. Wilton	Boston, Mass	Mar. 2, 1858	19,529
Piano-lock, Eccentric	P. H. Niles	Boston, Mass	Aug. 7, 1849	6,636
Piano music-desk	M. Doherty	Cambridgeport, Mass	Jan. 14, 1873	134,860
Piano-orchestra	J. B. Schalkenbach	Triers, Prussia	June 3, 1862	35,490
Piano, Organ	R. Nutting, 2d	Romeo, Mich	Feb. 8, 1848	5,438
Piano, &c., pedal	A. A. Mixer	Hamilton, Ohio	Oct. 26, 1869	96,137
Piano, &c., pedal	E. Zachariae	Loehnberg, Prussia	Feb. 2, 1869	86,621
Piano-pedal attachment	G. C. A. Class	Philadelphia, Pa	May 13, 1873	138,857
Piano-pedal attachment	C.W. Held, jr., and A. M. Baugh	Brooklyn, N. Y	Jan. 14, 1873	134,886
Piano-pedal attachment	S. Schoenbrun	New York, N. Y	Sept. 19, 1871	119,184
Piano-pedal attachment	W. B. Stetson	Taylor, N. Y	Nov. 2, 1858	21,990
Piano-pedal attachment	A. A. Stinson	Herkimer, N. Y	Feb. 6, 1872	123,429
Piano pedal-stool	T. Springer	Hunter's Point, N. Y	Dec. 30, 1873	146,103
Piano pedal-stool	J. H. Wells	Chicago, Ill	Jan. 2, 1872	122,343
Piano-pins, Screw-clamp for	C. M. Lindsay	Foreston, Ill	Sept. 13, 1870	107,391
Piano-seat	L. Postawka	Boston, Mass	Oct. 30, 1866	59,261
Piano-seat	L. Postawka and A. Krasinski	Boston, Mass	May 1, 1866	54,401
Piano ound-board	J. Benz	Philadelphia, Pa	Oct. 16, 1866	58,763
Piano sound-board	P. Schuler	Philadelphia, Pa	Apr. 3, 1866	53,687
Piano sounding-board	A. De Kuhn	New York, N. Y	Sept. 11, 1866	57,912
Piano sounding-board	G. M. Guild	Boston, Mass	May 26, 1868	78,276
Piano sounding-board	T. Kater	Hamilton, Canada	Jan. 21, 1873	135,126
Piano sounding-board	W. Kells	Albany, N. Y	Oct. 1, 1872	131,760
Piano sounding-board	C. Meyer	Philadelphia, Pa	July 9, 1850	7,494
Piano sounding-board	F. E. Ramm	Philadelphia, Pa	Oct. 23, 1866	59,069
Piano sounding-board	C. F. T. Steinway	New York, N. Y	Feb. 11, 1873	135,857
Piano sounding-board	J. Stewart	Boston, Mass	Nov. 14, 1822	....
Piano sounding-board	F. Strothmann	Louisville, Ky	Jan. 22, 1867	61,483
Piano sounding-board	E. L. Taylor	Jersey City Heights, N. J.	Dec. 5, 1871	121,686
Piano-string bridge	W. C. Ellis	Worcester, Mass	Nov. 25, 1873	144,842
Piano, Upright	O. Altenburg	New York, N. Y	Mar. 12, 1872	124,470
Piano, Upright	C. F. Chickering	New York, N. Y	Nov. 28, 1871	121,334
Piano, Upright	G. C. Manner	New York, N. Y	Dec. 14, 1869	97,943
Piano, Vertical	E. Fobes	Boston, Mass	May 17, 1853	9,724
Piano, Vertical	G. Traeyser	Cincinnati, Ohio	Mar. 29, 1853	9,640
Piano with melodeon-attachment	L. Louis	Buffalo, N. Y	June 10, 1862	35,528
Pianos, Adjustable extension-pedal for	A. G. Brewer	Hopkinton, Mass	Jan. 19, 1869	85,901
Pianos, Auxiliary key-board for	T. J. V. Roz	New York, N. Y	Jan. 19, 1869	85,965
Pianos, Carriage for moving	H. C. Camp and R. C. Munger	Saint Paul, Minn	June 25, 1872	128,210
Pianos, Child's pedal for	C. A. Class	Saint Louis, Mo	Oct. 13, 1868	83,039
Pianos, Clamp for moving	D. Benson	Syracuse, N. Y	June 11, 1872	127,837
Pianos, desks, &c., Lid-support for	C. T. Faber	New York, N. Y	Oct. 1, 1867	69,329
Pianos, Device for moving	S. D. Reynolds	Rochelle, Ill	Jan. 23, 1872	122,966
Pianos, Fitting hammer-heads for	J. Mackay	Boston, Mass	Aug. 14, 1828	....
Pianos, Former for bending and gluing cases of grand.	D. Decker	New York, N. Y	July 5, 1870	105,049
Pianos, Front-fall for	A. H. Hastings	New York, N. Y	May 23, 1871	115,053
Pianos, Hand-guide for	M. Gether	Saint Louis, Mo	Nov. 23, 1869	97,074
Pianos, &c., Hand-supporter for	C. Sangalli	New York, N. Y	Dec. 8, 1868	84,844
Pianos, Instructing-scale for	S. Winner	Philadelphia, Pa	Dec. 22, 1863	41,043
Pianos, Iron frame for	G. Steck	New York, N. Y	Aug. 9, 1870	106,227
Pianos, Iron frame for	H. Worcester	New York, N. Y	June 3, 1862	35,484
Pianos, Iron frame for upright	G. Steck	New York, N. Y	Mar. 15, 1870	100,948
Pianos, Metal frame for	M. Martins	New York, N. Y	Sept. 4, 1866	57,743
Pianos, Metal spring for	I. B. Thompson	Philadelphia, Pa	Nov. 29, 1859	26,304
Pianos, Octave-coupler for	J. Clark	Philadelphia, Pa	July 16, 1872	129,530
Pianos, Prepared-wood sounding-board for	J. B. Dunham	East Chester, N. Y	July 25, 1871	117,393
Pianos, Repeating-action for	F. Farholtz	Louisville, Ky	Aug. 15, 1865	49,394
Pianos, Stringing	A. Choplain and P. E. Chollet	New York, N. Y	Dec. 19, 1865	51,559
Pianos, tables, &c., Machine for cutting legs for	A. Stoeckel	New York, N. Y	July 10, 1855	13,236
Pianos, Tension-scale for tuning	C. R. Edwards	Niagara City, N. Y	Aug. 2, 1864	43,674
Pianos, Wrest-pin for	G. Schilling	Hoboken, N. J	July 14, 1857	17,812
Pick	N. Rose	Belmont, N. Y	July 6, 1869	92,370
Pick	J. D. Cochran	Kern River, Cal	July 26, 1864	43,638
Pick	J. C. Reed	Cincinnati, Ohio	June 5, 1860	28,638
Pick and ax combined	H. L. Lowman	New York, N. Y	Oct. 2, 1866	58,438
Pick and pickax	E. P. Dickie	Morristown, N. J	July 14, 1868	79,815
Pick and rammer socket	J. Pearce	Barton, Md	July 16, 1872	129,164
Pickax	L. Baker	Pultney, Vt	Jan. 15, 1822	....
Pickax	J. C. Conklin	Peekskill, N. Y	Dec. 20, 1853	10,356
Pickax	J. C Conklin	Yorktown, N. Y	July 21, 1868	80,146
Pickax	M. Gale	San Antonio, Mexico	Oct. 20, 1868	83,274
Pickax-eye	G. Nimmo	Jersey City, N. J	Nov. 5, 1867	70,459
Pickax, &c., eyes, Machine for forming	H. L. Lowman	New York, N. Y	June 30, 1868	79,364
Pick, Digging and tamping	C. Carroll	North Vernon, Ind	May 17, 1870	103,014
Pick-eyes, Die for forming	H. M. Hamilton	New York, N. Y	Aug. 27, 1867	68,187
Pick-handle	F. Bellinger	Lockport, N. Y	Mar. 29, 1870	101,214
Pick-handle	W. Blay	Helena, Mont	June 9, 1868	78,718
Pick-handle	J. E. Emerson	Sacramento, Cal	Mar. 29, 1859	23,358
Pick-handles, Attaching	R. C. Grover and C. W. Randall	Newton, Mass	June 1, 1869	90,742
Pick-handles, Attaching	P. J. Hogan	Cincinnati, Ohio	Nov. 3, 1868	83,710
Pick-handles, Attaching	H. W. Knight	Columbus, Ohio	Nov. 19, 1867	71,184
Pick-handles, Attaching	W. H. Livingston	New York, N. Y	Oct. 1, 1861	33,394
Pick-handles, Attaching	T. H. Neal	Allegheny, Pa	June 8, 1869	91,154
Picks to handles, Attaching	J. P. Walsh	Helena, Mont	Dec. 17, 1867	72,245
Pick-handles, Fastening	J. H. Fisher	Placerville, Cal	July 17, 1860	29,155
Pick handles, Fastening mining	J. B. Pressey and D. Sheets	Buffalo, N. Y	Oct. 8, 1861	33,454
Pick, Miner's	H. M. Hamilton	New York, N. Y	Aug. 20, 1867	67,875
Pick or ax	A. Moore	San Francisco, Cal	Mar. 24, 1863	37,970
Pick-pocket alarm	I. T. Dyer	Quincy, Ill	Jan. 5, 1869	85,649
Pick-pocket detector	S. W. Ruggles	Fitchburgh, Mass	July 29, 1856	15,440
Picks and rammers, Manufacture of	D. McNally	Mount Savage, Md	Nov. 7, 1871	120,658
Picks and similar tools, Machine for making	A. Buerkle	Pittsburgh, Pa	Apr. 25, 1871	113,977
Picks, Apparatus for opening the eyes of	R. Blake	Scranton, Pa	Jan. 31, 1871	111,306
Picks, &c., Machine for forming the eyes of	H. M. Hamilton	New York, N. Y	Sept. 24, 1867	69,090
Picks, &c., Making mining	E. Hughes	McCartysville, Cal	May 7, 1861	32,249
Picks, Manufacturing	D. Collins	Newark, N. J	June 3, 1873	139,455

Index of patents issued from the United States Patent Office from 1790 *to* 1873, *inclusive*—Continued.

Invention.	Inventor.	Residence.	Date.	No.
Picker: *See* Clover-picker. Cotton-picker. Cotton-seed picker. Cranberry-picker. Fruit-picker. Grain-picker. Hair-picker. Hop-picker. Loom-picker. Nut-picker. Pea-picker. Pea-nut picker. Waste-picker. Wool-picker.				
Picker	A. Holbrook	Providence, R. I	July 9, 1861	32, 770
Picker	A. Holbrook	Providence, R. I	July 9, 1861	32, 771
Picker-collar	J. M. Gotham	Blackstone, Mass	Sept. 27, 1870	107, 772
Picker-cylinders, Tooth-clothing for	R. Heneage	Lowell, Mass	Oct. 31, 1854	11, 861
Picker-house for opening and cleaning cotton, &c	R. Kitson	Lowell, Mass	May 29, 1866	55, 118
Picker-motion	W. Nugent	Chicopee, Mass	May 14, 1861	32, 308
Picker-staff check	E. Robbins	Worcester, Mass	Jan. 7, 1868	73, 045
Picker-staff motion	N. S. Bean	Manchester, N. H	Jan. 22, 1861	31, 205
Picker-staff motion	E. H. Graham	Manchester, N. H	Oct. 16, 1860	30, 441
Picker-staff, Mounting	W. Mason	Taunton, Mass	June 2, 1868	78, 463
Picker-staff, Stop for	J. H. Knowles	Lawrence, Mass	Sept. 17, 1872	131, 354
Picker to picker-staff, Attaching	J. D. Barrie	Lawrence, Mass	May 3, 1870	102, 475
Picket-cutter	J. W. Clark	Outville, Ohio	Apr. 4, 1871	113, 629
Picket for fences and walls	A. Dabb	Elizabethport, N. J	Apr. 2, 1867	63, 482
Picket-heading machine	H. D. Heiser, and H. F. and G. F. Snyder.	Williamsport, Pa	May 14, 1872	126, 634
Picket-holding clamp	P. McCarthy	Rochester, N. Y	Apr. 14, 1868	76, 644
Picket-making machine, Wooden	T. J. Jolly	Versailles, Ind	Jan. 1, 1867	60, 900
Picket-pointer	J. W. Minor	Middleborough, Mass	Sept. 12, 1871	118, 964
Picket-pointing machine	H. Balcom	Ann Arbor, Mich	Mar. 18, 1873	136, 807
Picket-pointing machine	A. B. Corby	Binghamton, N. Y	Feb. 4, 1873	135, 467
Picket-pointing machine	W. W. Johnson	Nashville, Tenn	Jan. 14, 1868	73, 339
Picket-shaping machine	A. J. Sutherland	Ann Arbor, Mich	Dec. 23, 1873	145, 766
Picking, carding, &c., machines, Means of feeding wool, &c., to.	S. R. Parkhurst	Bloomfield, N. J	May 30, 1865	47, 976
Pickle and cruet stand	T. Leach	Taunton, Mass	May 21, 1872	127, 073
Pickle-fork	H. Laurence	New Orleans, La	Aug. 27, 1872	130, 923
Picnic or excursion seat	J. M. Perkins	Chicago, Ill	Oct. 9, 1860	30, 344
Picture and card case, Metallic	H., J., T., and H. Hall, jr	Philadelphia, Pa	Feb. 25, 1862	34, 489
Picture and card exhibiter	B. F. Bostian	New Columbia, Pa	Sept. 2, 1873	142, 370
Picture and curtain knob	J. Gardner	New Haven, Conn	Jan. 7, 1868	73, 088
Picture and looking-glass frames, Ornamenting and constructing.	J. Parker and L. P. Clover	New York, N. Y	Mar. 23, 1829	....
Picture and mirror cord retainer	W. Read, jr	Boston, Mass	Feb. 25, 1868	74, 771
Picture and mirror hanging device	A. A. Fielding	Boston, Mass	Aug. 1, 1871	117, 529
Picture and other frames, Device for driving nails, &c., into.	W. Bascom	New York, N. Y	Sept. 15, 1863	39, 985
Picture and other frames, Device for hanging	C. B. Davies	Dayton, Ohio	Nov. 24, 1868	84, 266
Picture-card frame	R. W. Potter	New York, N. Y	Mar. 7, 1865	46, 699
Picture-case	G. Brownlee	Princeton, Ind	Nov. 23, 1869	97, 034
Picture-case	A. R. Stone	Baldwinsville, N. Y	July 23, 1872	129, 764
Picture-case and cane, Combination	H. D. McGeorge	Rochester, N. Y	Feb. 16, 1869	86, 991
Picture cord	T. Kohn	Hartford, Conn	Apr. 5, 1870	101, 629
Picture-cord, Attachment for adjusting	R. D'Heureuse	New York, N. Y	Mar. 30, 1869	88, 373
Picture-frame	E. W. Appleton	Cincinnati, Ohio	July 16, 1872	129, 450
Picture-frame	J. A. Burch	Buffalo, N. Y	Nov. 18, 1873	144, 599
Picture-frame	J. S. Cannon	New Haven, Conn	June 21, 1864	43, 255
Picture-frame	L. S. Chase	New York, N. Y	May 7, 1867	64, 488
Picture-frame	G. E. and W. C. Collins	Bucksport, Me	Dec. 26, 1871	122, 158
Picture-frame	J. D. Crocker	Norwich, Conn	Aug. 16, 1870	106, 332
Picture-frame	L. M. B. Gray	Boston, Mass	May 14, 1872	126, 633
Picture-frame	H. S. Hale	Philadelphia, Pa	May 13, 1873	138, 884
Picture-frame	C. A. Hodgman	Tuckahoe, N. Y	Mar. 30, 1869	88, 482
Picture-frame	A. J. Holmes	Saratoga Springs, N. Y	Nov. 13, 1866	59, 719
Picture-frame	C. H. Hutchinson	Philadelphia, Pa	Dec. 10, 1872	133, 707
Picture-frame	D. W. Marshall	Pawtucket, R. I	Apr. 6, 1869	88, 651
Picture-frame	A. A. Murfey	Montville, Conn	Sept. 7, 1869	94, 632
Picture-frame	M. Ormsbee	New York, N. Y	Feb. 14, 1865	46, 382
Picture-frame	C. P. Poinier	Boston, Mass	Oct. 22, 1867	70, 116
Picture-frame	F. Reifschneider	New York, N. Y	Oct. 7, 1873	143, 464
Picture-frame	J. E. Rice	Boston, Mass	Apr. 24, 1866	54, 208
Picture-frame	J. T. Schmitt	Brooklyn, N. Y	Nov. 3, 1868	83, 797
Picture-frame	G. Schneider	Buffalo, N. Y	June 22, 1869	91, 673
Picture-frame	J. Sperry	New York, N. Y	Jan. 18, 1870	99, 022
Picture-frame	J. L. Warren and J. W. Adderly	Detroit, Mich	Oct. 21, 1873	143, 861
Picture-frame and exhibiter, Combined	B. Anyan	Fitchville, Ohio	Nov. 18, 1873	144, 648
Picture-frame-beveling machine	O. F. Bedell	New York, N. Y	Sept. 19, 1865	49, 965
Picture-frame clamp	J. A. H. Dunne	Boston, Mass	May 1, 1866	54, 463
Picture-frame clamp	L. A. Johnson	Candor, N. Y	Sept. 19, 1871	119, 151
Picture-frame, Easel	R. Wright	Brooklyn, N. Y	July 18, 1871	117, 133
Picture frame fastening	J. D. Crocker	Norwich, Conn	Feb. 15, 1870	99, 854
Picture-frame hanger	C. H. Moulton	Des Moines, Iowa	Dec. 5, 1871	121, 652
Picture-frame-hanging device	T. C. Barnard	West Meriden, Conn	Sept. 23, 1873	142, 981
Picture-frame, &c., hanging device	J. N. Gillespie	Waltham, Mass	May 10, 1870	102, 805
Picture-frame support	L. L. Hodges	Boston, Mass	Jan. 28, 1868	73, 808
Picture-frames and mirrors, Device for suspending	R. E. Bean	Franklin, N. H	July 27, 1868	93, 040
Picture-frames, Composition for molding for	H. Wurtz	New York, N. Y	Feb. 8, 1870	99, 739
Picture-frames, &c., Corner-piece for	T. Callanan	Brooklyn, N. Y	May 6, 1873	138, 611
Picture-frames, Machine for enameling	J. Sperry and C. W. Sherwood	Brooklyn, N. Y	Apr. 2, 1861	31, 913
Picture-frames, Machine for enameling oval	F. Smith	Allegheny City, Pa	Aug. 2, 1864	43, 715
Picture-frames, Machine for preparing moldings for	E. S. Gardner	New York, N. Y	Mar. 29, 1859	23, 426
Picture-frames, Machinery for making oval	J. E. Sweet	Syracuse, N. Y	Feb. 12, 1867	62, 086
Picture-frames, Machinery for preparing oval	W. Gardner	New York, N. Y	Aug. 17, 1858	21, 192
Picture-frames, &c., Manufacture of moldings for	A. C. Engert	Middlesex County, England.	May 7, 1872	126, 385
Picture-frames, Wooden mat for	H. W. Curtis	Philadelphia, Pa	Oct. 24, 1871	120, 250
Picture-hanger	F. W. Ely	Du Luth, Minn	Aug. 26, 1873	142, 156
Picture-hanger	G. Lamb	Boston, Mass	May 31, 1870	103, 627
Picture-hanging device	C. W. Simpson	Bangor, Me	Apr. 19, 1870	102, 167
Picture-holder	A. Barber	Port Byron, Ill	June 25, 1867	66, 115

Index of patents issued from the United States Patent Office from 1790 to 1873, inclusive—Continued.

Invention.	Inventor.	Residence.	Date.	No.
Picture-holder	W. Walker	New Haven, Conn	June 12, 1866	55,561
Picture-hook	W. D. Gridley	New Britain, Conn	Jan. 12, 1869	85,738
Picture hook and cornice	W. Potts	Handsworth, England	Dec. 1, 1868	84,645
Picture medal, button, &c	H. C. Griggs	Waterbury, Conn	Jan. 3, 1865	45,710
Picture-mount	J. H. Fitzgibbon	Saint Louis, Mo	Sept. 10, 1872	131,260
Picture-nail	J. W. Bishop	New Haven, Conn	May 11, 1869	89,970
Picture-nail	B. H. Bradley	Waterbury, Conn	Nov. 10, 1868	89,970
Picture-nail	S. C. Cary	Brooklyn, N. Y	Mar. 26, 1872	124,933
Picture-nail	H. Hickman	Omaha, Nebr	June 22, 1869	91,629
Picture-nail	C. B. Jenkins	New York, N. Y	Dec. 10, 1872	133,778
Picture-nail	A. D. Judd	New Haven, Conn	Nov. 6, 1866	59,404
Picture-nail	H. L. Judd	Brooklyn, N. Y	June 14, 1870	104,163
Picture-nail	H. L. Judd	Brooklyn, N. Y	Sept. 26, 1871	119,273
Picture-nail	T. C. Richards	New York, N. Y	Dec. 27, 1870	110,500
Picture-nail	T. C. Richards	New York, N. Y	Jan. 10, 1871	110,867
Picture-nail	T. C. Richards	New York, N. Y	Oct. 3, 1871	119,648
Picture-nail	F. J. Seymour	Wolcottville, Conn	June 26, 1866	55,917
Picture-nail	F. W. Smith, jr	Bridgeport, Conn	Dec. 28, 1869	98,438
Picture-nail	J. H. Squier and E. J. Warner	Newark, N. J	Apr. 12, 1870	101,781
Picture-nail	J. Uster	Brooklyn, N. Y	Apr. 16, 1872	125,859
Picture-nail	G. Vintschger	New York, N. Y	Mar. 4, 1873	136,566
Picture-nail head	H. P. Brooks	Wolcottville, Conn	Feb. 20, 1866	52,781
Picture-nail head	J. B. Sargent	New Britain, Conn	Aug. 21, 1860	29,720
Picture-nail knob	W. E. Sparks	New Haven, Conn	Aug. 22, 1871	118,400
Picture-nail, Ornamental-headed	T. C. Richards	New York, N. Y	Dec. 20, 1870	110,287
Picture-nails, Attaching ornamental heads to	A. D. Judd	New Haven, Conn	Mar. 28, 1865	47,023
Picture-nails, Mechanism for manufacturing heads for.	L. Wolf	Meriden, Conn	Mar. 21, 1871	112,880
Picture-poise	J. J. Glover	Quincy, Mass	May 24, 1870	103,445
Picture-stand, Revolving	E. B. Sturdevant	Germantown, Ohio	June 4, 1867	65,446
Picture-stretcher	J. D. Crocker and J. A. Brand	Norwich, Conn	Oct. 3, 1871	119,507
Picture-type	J. McElheran	Brooklyn, N. Y	Mar. 16, 1858	19,645
Picture with looking-glass back-ground	J. Jennings	New York	Nov. 11, 1831	....
Pictures, Apparatus for composing and exhibiting group of card.	S. F. Mills	San Francisco, Cal	Mar. 30, 1869	88,321
Pictures, &c., Apparatus for mounting and exhibiting.	J. W. Hardie	New York, N. Y	Mar. 26, 1872	124,953
Pictures by metallic leaves, Shading	E. Harmon	Cleveland, Ohio	Mar. 27, 1849	6,241
Pictures, Coloring	C. J. Mettais	New York, N. Y	Jan. 7, 1873	134,693
Pictures, looking-glasses, &c., Method of hanging	W. A. Foster	Chester, Conn	July 12, 1859	24,729
Pictures, Machine for mounting	E. P. Christmas	Mansfield, Ohio	Apr. 8, 1873	137,656
Pictures, Molding for hanging	H. Hochstrasser	Philadelphia, Pa	May 8, 1860	28,174
Pictures, Mounting	C. J. Billinghurst	McArthur, Ohio	July 20, 1869	92,694
Pictures, Mounting	E. L. Courtney	Philadelphia, Pa	June 18, 1872	128,125
Pictures on glass, Backing up	T. Fry and C. A. Seely	Brooklyn and New York, N. Y.	Jan. 3, 1860	26,733
Pictures, Preparation of transparent	G. Wedekind	Philadelphia, Pa	Nov. 13, 1860	30,649
Pictures, Screw-support for hanging	M. Judd	New Britain, Conn	Apr. 1, 1862	34,837
Pictures to frames, Attaching	S. W. Hanks	Lowell, Mass	Nov. 20, 1866	59,836
Pictures to translucent surfaces, Transferring	J. T. Foley	New York, N. Y	Aug. 20, 1872	130,630
Pictures transparent, Compound for making	S. D. McPherson	Normal, Ill	Mar. 31, 1868	76,225
Pie-case	H. H. Olds	New Haven, Conn	Mar. 21, 1871	112,953
Pie-crimper	C. A. Shaw	Biddeford, Me	Nov. 6, 1860	30,592
Pie cutter and crimper	H. Mathews	Cambridge, Mass	Apr. 6, 1869	88,726
Pie-marker	T. S. Macomber	Hamilton, N. Y	Dec. 23, 1873	145,808
Pie-rimmer	N. N. Brown	Reading, Pa	Nov. 26, 1867	71,274
Pie-rimmer	J. Stephen and W. Zeller	Womelsdorf, Pa	Sept. 11, 1866	57,990
Pie stamper and cutter	A. Hawkes	Providence, R. I	Apr. 19, 1870	102,117
Pie-tube	N. M. Selden	Chatham, Conn	Oct. 20, 1868	83,327
Pier and bridge	W. H. Wood	Hudson, N. J	Oct. 21, 1862	36,747
Pier and bulk-head	S. J. Seely	New York, N. Y	Jan. 31, 1865	46,144
Pier, dock, and wharf	R. A. Gilpin	Chester County, Pa	June 4, 1867	65,484
Pier-glass frame and cornice	D. A. Hall and D. Garrison	Philadelphia, Pa	Mar. 28, 1871	113,046
Pier, Iron	W. B. Porter	Plattsmouth, Nebr	Sept. 1, 1868	81,682
Pier or breakwater	C. T. Harvey	Marquette, Mich	Apr. 12, 1859	23,574
Pier, Sectional	O. A. Howe	Jersey City, N. J	Feb. 23, 1869	87,171
Piers, &c., Construction of	G. A. Parker	Lancaster, Mass	Mar. 15, 1864	41,935
Piers, Crib for laying foundation for	J. Parish	Chicago, Ill	Mar. 5, 1872	124,284
Piers for bridges, &c., Building	J. Du Bois	Williamsport, Pa	Sept. 23, 1862	36,512
Piers into the water, Lowering	C. Van Horn	Springfield, Mass	Oct. 29, 1861	33,615
Piers, Iron front for stone	J. Abbott	Canton, Ohio	Apr. 29, 1873	138,222
Pig-singeing apparatus	A. and E. M. Denny	Waterford, Ireland	Oct. 23, 1860	30,467
Pigeon-hole	C. J. Clements and E. P. Fowler	New York, N. Y	Apr. 15, 1873	137,892
Pigeon-trap	W. F. Parker	Meriden, Conn	July 1, 1873	140,532
Pigeon-trap	N. B. Tyler	Warren, Ohio	June 10, 1873	139,836
Pigment and vehicle for mixing paints	J. M. Merrymon	Indianapolis, Ind	Dec. 13, 1864	45,426
Pigment from antharcite coal, Black	N. Chater	New York, N. Y	Oct. 5, 1839	1,358
Pigments from iron-ore, Manufacturing	J. H. Davis	Morristown, N. J	Aug. 8, 1854	11,476
Pigments from sulphurets of zinc and lead, Manufacture of.	N. Bartlett	Centreville, N. J	Oct. 27, 1868	83,357
Pigments, Manufacture of	J. Dale	Manchester, England	Aug. 14, 1866	57,264
Pigments, Manufacture of	H. L. Pattinson	Scott's House, England	Aug. 12, 1851	8,292
Pigments, Manufacture of opaque	S. Gwynn	New York, N. Y	Dec. 19, 1865	51,584
Pigments, oils, and gums from cloth used by engravers, Process for recovering.	H. M. Baker	New York, N. Y	Nov. 24, 1868	84,404
Pigments, Producing carbon	A. Farrar	Boston, Mass	Nov. 2, 1869	96,409
Pigments to fibrous and textile material, Mixing	A. Paraf	New York, N. Y	Jan. 25, 1870	99,105
Pike and revolving fire-arm combined	J. C. Campbell	New York, N. Y	June 30, 1863	39,032
Pile	S. B. Cushing	Providence, R. I	Mar. 23, 1869	88,141
Pile and post driver	N. Havermale	Canton, Ill	Dec. 20, 1870	110,359
Pile-driver	S. E. Baker	Chicago, Ill	June 17, 1873	139,853
Pile-driver	J. T. Baldwin	Petaluma, Cal	July 16, 1872	129,083
Pile-driver	J. A. Borgort	Hudson City, N. J	Nov. 3, 1868	83,687
Pile-driver	W. P. Craig	Newport, Ky	Mar. 29, 1859	23,354
Pile-driver	H. Flad	Saint Louis, Mo	Oct. 12, 1869	95,786
Pile-driver	C. D. Hall and E. Badger	New Haven, Conn., and Waukegan, Ill.	Dec. 17, 1872	133,936
Pile-driver	J. W. Hoard	Providence, R. I	Mar. 25, 1856	14,502

Index of patents issued from the United States Patent Office from 1790 *to* 1873, *inclusive*—Continued.

Invention.	Inventor.	Residence.	Date.	No.
Pile-driver	J. Huy	Whistler, Ala	Nov. 29, 1870	109, 623
Pile-driver	H. Roehrs	Washington, D. C	July 25, 1871	117, 465
Pile-driver	T. Shaw	Philadelphia, Pa	Nov. 24, 1868	84, 383
Pile-driver	T. Shaw	Philadelphia, Pa	June 14, 1870	104, 215
Pile-driver	J. J. Simons	East Saint Louis, Mo	Jan. 5, 1869	85, 703
Pile-driver	A. Smith and J. W. Galbraith	Sedalia, Mo	Feb. 16, 1869	87, 010
Pile-driver	C. H. Smith	Bloomer, Wis	July 22, 1873	141, 086
Pile-driver	J. J. Studer	Richmond, Ind	Nov. 6, 1866	59, 476
Pile-driver	J. Wood	Brooklyn, N. Y	Mar. 22, 1864	42, 038
Pile-driver, Adjustable	T. W. Loveless	Corning, N. Y	July 13, 1858	20, 883
Pile-driver, Adjustable	T. Place	Alfred Centre, N. Y	May 3, 1859	23, 858
Pile-driver, Explosive	H. Vogler	Baltimore, Md	Apr. 1, 1873	137, 514
Pile-driver, Steam	J. Nasmyth	Patricroft, England	June 26, 1847	5, 172
Piles, Driving	J. McClay	Hartford, Conn	June 12, 1866	55, 586
Piles, Driving	J. A. Whipple	Boston, Mass	Nov. 8, 1859	26, 073
Pile-driving, Apparatus for atmospheric	A. Holmstrom	New York, N. Y	Jan. 2, 1855	12, 130
Pile-driving machine	R. Benson	New Orleans, La	Sept. 18, 1841	2, 264
Pile-driving machine	J. Du Bois	Williamsport, Pa	May 19, 1863	38, 570
Pile-driving machine by the aid of shifting-wheels, Mode of progressing with.	E. P. Williams	Utica, N. Y	July 22, 1839	1, 253
Pile-fabric	G. Crompton	Worcester, Mass	Mar. 26, 1872	125, 026
Pile-fabric	G. Crompton	Worcester, Mass	June 25, 1872	128, 286
Pile-fabrics, Apparatus for cutting pile of	J. Johnson	Troy, N. Y	Jan. 13, 1852	8, 649
Pile-fabrics, Cutting	E. B. Bigelow	Boston, Mass	Dec. 4, 1855	13, 862
Pile-fabrics, Machinery for manufacturing	C. Miller	New York, N. Y	Mar. 27, 1860	27, 671
Pile-fabrics, Making	J. W. Crossley	Bridgeport, Conn	May 22, 1866	54, 870
Pile-fabrics, Manufacture of	N. Hill	Caton, N. Y	May 29, 1860	28, 484
Pile-fabrics, Weaving	J. Shinn	Philadelphia, Pa	Dec. 3, 1872	133, 676
Pile-fabrics, Weaving	W. Webster	Morrisania, N. Y	May 19, 1868	78, 163
Pile for beams	J. Stokes	Trenton, N. J	Dec. 6, 1870	109, 850
Pile for corrugated beams	R. Montgomery	New York, N. Y	Oct. 11, 1870	108, 165
Pile for engineering purposes	T. W. H. Moseley	Boston, Mass	Nov. 1, 1870	108, 814
Pile for girder-irons	A. Kloman	Pittsburgh, Pa	Mar. 24, 1868	75, 769
Pile for railway-chairs	D. Eynon	Richmond, Va	Sept. 21, 1869	95, 099
Pile for rolling beams	J. Griffen	Phœnixville, Pa	Dec. 1, 1857	18, 738
Pile for wrought-iron beams or girders	G. Walters and T. Shaffer	Phœnixville, Pa	Aug. 27, 1867	68, 266
Pile for wrought-iron beams or girders	G. Walters and T. Shaffer	Phœnixville, Pa	Aug. 27, 1867	68, 267
Pile-hook, Revolving	J. D. Leach and S. Hutchings	Penobscot, Me	Dec. 15, 1868	84, 886
Pile or fagot	J. Barker	Covington, Ky	Sept. 9, 1873	142, 666
Pile-sawing attachment for boats	H. Vogler	Baltimore, Md	Oct. 15, 1872	131, 337
Pile, Secondary electric	G. G. Percival	Brooklyn, N. Y	Apr. 3, 1866	53, 668
Piles, Apparatus for sinking pneumatic	F. E. Sickels	Kennett's Square, Pa	Feb. 1, 1870	99, 359
Piles, Apparatus for sinking pneumatic	F. E. Sickels	Chicago, Ill	Feb. 1, 1870	99, 360
Piles by atmospheric pressure, Driving	W. S. Smith	Trenton, N. J	Sept. 4, 1860	29, 921
Piles, &c., by exhausting the air from the same, Method of sinking hollow.	L. H. Potts	London, England	Jan. 29, 1850	7, 060
Piles for beams, Method of constructing	C. Hewitt	Trenton, N. J	May 4, 1869	89, 763
Piles for forming axles, &c., Method of constructing.	P. Roberts	Philadelphia, Pa	June 15, 1869	91, 486
Piles for girder-beams, Making	A. Kloman	Pittsburgh, Pa	Mar. 24, 1868	75, 770
Piles for wharves, &c., Construction of	C. Walton	Washington, D. C	June 28, 1864	43, 357
Piles for wharves, piers, &c., Mode of staying	E. H. Angamar	New Orleans, La	July 12, 1859	24, 708
Piles, Instrument for curing	J. P. Gilbert	Long Island City, N. Y	May 2, 1865	47, 540
Piles, Instrument for treating	H. Schevenell and S. S. Rembert.	Memphis, Tenn	May 5, 1868	77, 539
Piles, Machine for sinking hollow	W. S. Smith	Noyesville, Ill	Sept. 4, 1866	57, 784
Piles or caissons, Sinking pneumatic	F. E. Sickels	Marshall, Tex	May 13, 1873	138, 945
Piles or fagots, Mode of securing bars of	G. Walters and T. Shaffer	Phœnixville, Pa	Oct. 8, 1867	69, 597
Piles, Preserving wooden	W. H. Smith	Memphis, Tenn	July 30, 1867	67, 366
Piles, Protecting	W. J. L. Moulton	San Francisco, Cal	June 19, 1866	55, 694
Piles, Protecting the surface of wooden	L. J. Henry	San Francisco, Cal	Jan. 24, 1865	46, 049
Piles, Sinking	L. J. Seely	New York, N. Y	Oct. 18, 1864	44, 751
Piles, Sinking broken	E. C. Boobar	San Francisco, Cal	Dec. 16, 1873	145, 483
Piles, Sinking metallic	J. Du Bois	Williamsport, Pa	Oct. 1, 1872	131, 746
Piles-supporter	C. West	Pittsburgh, Pa	Sept. 16, 1873	142, 972
Piles under water, Machine for sawing off	J. Fleming	Portsmouth, Va	May 1, 1855	12, 789
Piles under water, Mechanism for sawing off	V. Palen	Portsmouth, Va	Oct. 31, 1854	11, 872
Pill-box	N. and E. P. Cary	Albany County, N. Y	Apr. 6, 1830	....
Pill, &c., box	G. H. Hawkins	New York, N. Y	Dec. 3, 1867	71, 751
Pill-box	B. F. Stephens	Brooklyn, N. Y	Dec. 15, 1868	84, 972
Pill-boxes, Machine for making	N. D. White	Winchendon, Mass	Dec. 10, 1850	7, 829
Pill-boxes, Machinery for making	A. Fessenden	Baldwinsville, Mass	Mar. 12, 1850	7, 162
Pill cutting and shaping machine	J. Cooper	Philadelphia, Pa	Oct. 15, 1872	132, 200
Pill-machine	J. C. Ayer	Lowell, Mass	July 28, 1857	17, 865
Pill-machine	P. Cauhape	New York, N. Y	Jan. 3, 1871	110, 630
Pill-machine	H. E. Chapman	Albany, N. Y	May 20, 1856	14, 904
Pill-machine	L. G. Merrell	New Bedford, Pa	Jan. 17, 1854	10, 431
Pill-machine	J. H. Shaw	Saco, Me	May 28, 1867	65, 288
Pill-machine	D. J. Tittle	Albany, N. Y	Oct. 3, 1865	50, 288
Pill-machine	W. V. V. Wilson	Savannah, Ga	June 25, 1867	66, 199
Pill-machine	A. H. Wirz	Philadelphia, Pa	Oct. 1, 1867	69, 379
Pill-making machine	F. Bushby	Manchester, England	Dec. 29, 1868	85, 277
Pill-making machine	N. W. Kumler	Cincinnati, Ohio	Jan. 29, 1856	14, 161
Pill-making machine	W. Lewis	....	July 10, 1807	....
Pill-making machine	C. B. Littlefield	Manchester, N. H	Mar. 18, 1873	137, 608
Pill-making machine	E. A. Pond	Rutland, Vt	Dec. 7, 1852	9, 455
Pill-making machine	E. A. Pond	Rutland, Vt	May 29, 1855	12, 960
Pills	D. Coit	New York	June 17, 1820	....
Pills, Anti-bilious	B. Duval	....	May 3, 1797	....
Pills, Anti-bilious	T. H. Rawson	....	July 24, 1802	....
Pills, Anti-bilious	T. H. Rawson	....	Apr. 4, 1803	....
Pills, Anti-dysenteric	J. C. M. Brockway	Lyme, Conn	Mar. 13, 1822	....
Pills, Anti-dyspeptic	G. Smith	New York	Aug. 7, 1821	....
Pills, Apparatus for sugar-coating	F. Marriott	Detroit, Mich	June 20, 1871	116, 204
Pills, Bilious	S. Lee	New London, Conn	May 24, 1810	....
Pills, Bilious	S. H. P. Lee	....	June 26, 1799	....
Pills, Bilious	S. H. P. Lee	New London, Conn	Feb. 8, 1814	....
Pills, Composition for	S. Cooley	....	June 6, 1798	....

Index of patents issued from the United States Patent Office from 1790 to 1873, inclusive—Continued.

Invention.	Inventor.	Residence.	Date.	No.
Pills, Composition of bilious	S. Lee, jr		Apr. 30, 1796	
Pills, Device for administering	J. Sullivan	Thornton, Canada	Dec. 5, 1871	121, 684
Pills, Family	D. Coit		Oct. 5, 1803	
Pills, Hawks's	J. Hawks		Dec. 14, 1799	
Pills, Mode of counting	J. P. Peters	New York, N. Y	June 27, 1839	1, 202
Pills, Rheumatic	E. Deane	Biddeford, Mass	Jan. 13, 1814	
Pills, Rheumatic	E. Deane	Biddeford, Me	Feb. 2, 1828	
Pills, Rheumatic	G. B. Dexter	Boston, Mass	Dec. 18, 1805	
Pillow, Adjustable	H. Eilers	Cincinnati, Ohio	Oct. 11, 1870	108, 122
Pillow and bolster	T. S. Sperry	Chicago, Ill	Apr. 16, 1872	125, 851
Pillow-block	W. R. Manley	New York, N. Y	Dec. 22, 1868	85, 115
Pillow-spread frame	P. W. Van Est	Rochester, N. Y	Jan. 28, 1873	135, 389
Pillow-support	E. T. Annis	Mount Morris, N. Y	Feb. 25, 1868	74, 786
Pillows and bolsters into cases, Putting	D. B. Tiffany	Xenia, Ohio	Sept. 30, 1856	15, 828
Pillows, Formers for making frames for spring	T. S. Sperry	Chicago, Ill	Feb. 13, 1872	123, 738
Pin:				
See Bench-pin. Linchpin.				
Breast-pin. Nine-pin.				
Car-coupling pin. Ox-bow pin.				
Clothes-pin. Piano-pin.				
Crank-pin. Plow-clevis pin.				
Crimping-pin. Rafting-pin.				
Curling-pin. Rolling-pin.				
Dental pin. Safety-pin.				
Diaper-pin. Shawl-pin.				
Dowel pin. Shawl and dress pin.				
Dress-pin. Shielded pin.				
Flask-pin. Swivel-pin.				
Gage-pin. Ten-pin.				
Hair-pin. Tidy-pin.				
Hair-curling pin. Toggle-pin.				
Hook-pin. Tooth-pin.				
Husking-pin. Tuning-pin.				
Iron pin. Wooden pin.				
Jewelry-pin. Wrist-pin.				
Pin	D. R. Pratt	Worcester, Mass	Feb. 26, 1867	62, 361
Pin	W. H. Towers	New York, N. Y	Aug. 26, 1862	36, 313
Pin and needle making machine	W. F. Hill	New York	Oct. 15, 1814	
Pin-book	T. Piper	Birmingham, Conn	Apr. 14, 1868	76, 809
Pin-case, Toilet	T. R. Timby	Saratoga Springs, N. Y	May 25, 1869	90, 608
Pin-case, Toilet	T. R. Timby	Saratoga, N. Y	Aug. 10, 1869	93, 499
Pin-cushion	L. J. Atwood	Waterbury, Conn	Nov. 10, 1868	83, 819
Pin-cushion	T. Earle	Valley Falls, R. I	June 9, 1863	38, 817
Pin-cushion	M. Ormsbee	Brooklyn, N. Y	Nov. 8, 1870	109, 044
Pin, Dress	W. Hunt	New York, N. Y	Apr. 10, 1849	6, 281
Pin-drill	T. Prosser	New York, N. Y	Apr. 27, 1869	89, 502
Pin-fastening	D. Brown	Norfolk, Va	Mar. 20, 1860	27, 519
Pin for attaching wearing-apparel	R. J. Nunn	Savannah, Ga	Dec. 11, 1866	60, 411
Pin for attachment of bows and rosettes	L. H. Foy	Worcester, Mass	Apr. 9, 1867	63, 625
Pin-lock	A. Herman and W. H. Taylor	Stamford, Conn	Apr. 4, 1871	113, 295
Pin-machine	J. I. Howe	Derby, Conn	Mar. 24, 1841	2, 013
Pin-machine	M. L. Morse	Boston, Mass	Aug. 22, 1814	
Pin-machine	H. Weston	Philadelphia, Pa	July 23, 1823	
Pin machine, Wooden	N. W. Brewer	Williamsport, Pa	Mar. 27, 1866	53, 402
Pin-machines, Head-roll for	E. F. Bradley	Birmingham, Conn	Dec. 31, 1867	72, 718
Pin-package	C. O. Crosby	Milford, Conn	Dec. 5, 1871	121, 491
Pin-package	G. Fowler	Seymour, Conn	Dec. 24, 1872	134, 136
Pin-package case	C. O. Crosby	Milford, Conn	Dec. 5, 1871	121, 492
Pin-sticking machine	C. Atwood	New York, N. Y	Dec. 9, 1856	16, 199
Pin-sticking machine	C. O. Crosby	New Haven, Conn	Nov. 1, 1853	10, 180
Pin-sticking machine	C. O. Crosby	New Haven, Conn	Nov. 1, 1853	10, 181
Pin-sticking machine	C. O. Crosby	New Haven, Conn	Nov. 1, 1853	10, 182
Pin-sticking machine	C. O. Crosby	New Haven, Conn	Dec. 5, 1871	121, 493
Pin-sticking machine	T. Fowler	Waterbury, Conn	Aug. 25, 1857	18, 043
Pin-sticking machine	T. Fowler	Waterbury, Conn	Mar. 9, 1858	19, 556
Pin-sticking machine	T. W. Harvey	New York, N. Y	Jan. 3, 1854	10, 385
Pin-sticking machine	J. J. Howe	Derby, Conn	Feb. 24, 1843	2, 970
Pin-sticking machine	J. J. Howe and T. Piper	Derby, Conn	June 10, 1856	15, 112
Pin-sticking machine	J. W. Naramore	Derby, Conn	July 5, 1859	24, 654
Pin-sticking machine	T. Piper	Birmingham, Conn	Jan. 26, 1869	86, 317
Pin-sticking machine	T. Piper and E. F. Bradley	Birmingham, Conn	May 12, 1868	77, 912
Pin-sticking machine	J. B. Terry	Hartford, Conn	Jan. 3, 1854	10, 372
Pin-sticking machine	J. B. Terry	Hartford, Conn	May 29, 1855	12, 975
Pin-sticking machine	J. B. Terry	Hartford, Conn	June 10, 1856	15, 091
Pin-sticking machine	C. W. Van Vliet	Winsted, Conn	Sept. 14, 1858	21, 541
Pins, Device for coating	T. Fowler	Seymour, Conn	Mar. 19, 1861	31, 708
Pins, Device for feeding	C. L. Topliff	New York, N. Y	Mar. 13, 1866	53, 199
Pins in papers, Machine for arranging and sticking	D. Fowler	North Branford, Conn	Sept. 20, 1844	3, 751
Pins in papers, Machine for sticking	C. O. Crosby	New Haven, Conn	Apr. 1, 1851	8, 007
Pins in papers, Machine for sticking	T. Fowler	Waterbury, Conn	Dec. 8, 1857	18, 831
Pins in papers, Machine for sticking	S. Slocum	Poughkeepsie, N. Y	Sept. 30, 1841	2, 275
Pins in papers, Sticking	C. Atwood	New York, N. Y	Oct. 14, 1856	15, 871
Pins in papers, Sticking	W. B. Bartram	Waterbury, Conn	Oct. 14, 1856	15, 877
Pins in papers, Sticking	T. Fowler	Waterbury, Conn	Feb. 12, 1856	14, 234
Pins, Machine for manufacture of	T. and D. Fowler	Northford, Conn	July 31, 1860	29, 431
Pins, Machine for crimping paper for sticking	J. B. Terry	Hartford, Conn	Sept. 11, 1855	13, 553
Pins, &c., Machinery for heading and pointing	S. G. Reynolds	Bristol, R. I	Dec. 31, 1845	4, 346
Pins, Making	L. W. Wright	Manchester, England	Mar. 12, 1825	
Pins, Manufacturing	J. J. Howe	North Salem, N. Y	June 22, 1832	
Pins, Papering	C. Atwood	New York, N. Y	Oct. 14, 1856	15, 874
Pins, Papering	C. O. Crosby	New Haven, Conn	July 8, 1851	8, 202
Pins, Papering	C. O. Crosby	New Haven, Conn	Dec. 2, 1851	8, 564
Pins, Papering	C. O. Crosby	New Haven, Conn	Jan. 11, 1870	98, 745
Pins, Papering	G. Fowler	Seymour, Conn	Jan. 17, 1871	111, 051
Pins, Papering	G. L. Turney	London, England	Sept. 10, 1867	68, 809
Pins, Papering-machine for	T. Fowler	Seymour, Conn	Mar. 17, 1868	75, 540
Pins, rivets, &c., Cutting off	A. Stephens	Andover, Mass	Nov. 11, 1828	
Pins, Separating	T. Fowler	Waterbury, Conn	Nov. 13, 1855	13, 785

Index of patents issued from the United States Patent Office from 1790 *to* 1873, *inclusive*—Continued.

Invention.	Inventor.	Residence.	Date.	No.
Pins upon paper or any other material, Machine for sewing.	E. S. Woodford	Winchester, Conn	July 22, 1856	15, 404
Pinch-bar	A. Reese	Pittsburgh, Pa	May 30, 1871	115, 358
Pinch-bar	P. Spalding	Chicago, Ill	Sept. 21, 1869	95, 159
Pinch-bar for moving heavy weights	W. Siefert	New York, N. Y	Feb. 26, 1867	62, 374
Pinchers	D. Feiler	Marlborough Township, Pa	Oct. 17, 1871	120, 144
Pinchers and grappling-tool, Combined adjustable.	S. B. Dexter	Mason City, Iowa	Mar. 18, 1873	136, 907
Pinchers and nailer, Lasting	F. O. Claflin	Brooklyn, N. Y	Apr. 27, 1869	89, 467
Pinchers, Cutting	P. Broadbooks	Batavia, N. Y	Jan. 25, 1870	99, 057
Pinchers, Die for the manufacture of	W. E. Snediker	Charlotteburgh, N. J	Sept. 16, 1873	142, 955
Pinchers, Double	W. L. Maund	Milan, Tex	Mar. 29, 1870	101, 371
Pinchers for lasting boots and shoes	B. F. Sturtevant	Skowhegan, Me	Oct. 7, 1856	15, 866
Pinchers for operating pile-wires	A. Faulkner	Walpole, N. H	Nov. 23, 1852	9, 417
Pinchers, Hog-ringing	J. C. Schoettle	Collinsville, Ill	Nov. 25, 1873	144, 867
Pinchers, Lasting	G. Kump	Xenia, Ohio	Jan. 16, 1866	52, 055
Pinchers, Lasting	L. B. Richardson	Athol, Mass	Oct. 11, 1859	25, 762
Pinchers, Lasting	B. F. Sturtevant	Skowhegan, Me	Oct. 13, 1857	18, 427
Pinchers, Lasting	A. West	Canton, Ohio	Apr. 19, 1864	42, 451
Pinchers, nail maker and driver for lasting boots	L. R. Blake	Boston, Mass	Feb. 2, 1869	86, 353
Pinchers, Shoemaker's	A. Clarke	Boston, Mass	Oct. 3, 1871	119, 447
Pinchers, Shoemaker's	W. A. Hanna	Chico, Cal	July 22, 1873	141, 047
Pinchers, Shoemaker's	T. Pilote	Marlborough, Mass	Jan. 11, 1870	98, 796
Pinchers, Shoemaker's	T. B. Shelly	Chicago, Ill	Oct. 14, 1873	143, 594
Pinchers, Shoemaker's lasting	A. Dufoult	Spencer, Mass	Dec. 3, 1872	133, 521
Pine leaves for use in the arts, Treating	C. E. Ramus	Lawrence, Kans	Mar. 11, 1873	136, 671
Pine trees, Collecting products of	J. Johnson	Saco, Me	Feb. 2, 1869	86, 553
Pinion	V. W. Blanchard	Bridgeport, Vt	Mar. 9, 1869	87, 623
Pinion	V. W. Blanchard	Bridgeport, Vt	Apr. 26, 1870	102, 363
Pinking-machine	F. L. Hagadom	Baltimore, Md	June 8, 1869	90, 943
Pinking-machine	T. Hagerty	Richmond, Va	Dec. 19, 1871	122, 008
Pinking-machine	W. C. Hooper	Philadelphia, Pa	Oct. 29, 1872	132, 532
Pinking-tool	J. L. McIntosh	Boston, Mass	Sept. 22, 1868	82, 335
Pipe:				
See Aqueduct-pipe. Asphalt pipe. Bit-pipe. Blast-oven pipe. Blow-pipe. Blow-off pipe. Breast-pipe. Building service-pipe. Car-brake pipe. Casting pipe. Cement-pipe. Cement-lined pipe. Chain-locker pipe. Cigar-shaped pipe. Clay pipe. Clay and cement pipe. Composition pipe. Concrete pipe. Conduit-pipe. Confluent-pipe. Copper-coated iron pipe. Coupling-pipe. Dip-pipe. Discharge-pipe. Distributing-pipe. Draft-pipe. Drain-pipe. Drop-pipe. Earthen-pipe. Earthenware-pipe. Escape-pipe. Fire-pipe. Flexible pipe. Fluid-pipe. Furnace air-pipe. Gas-pipe. Gas and water pipe. Gas service-pipe. Glass pipe. Glass-smelting pipe. Hawse pipe. Heating-pipe. Hose-pipe. Hot-air pipe. Hot-blast pipe. Injection-pipe. Iron pipe. Jointed pipe. Lead pipe. Locomotive exhaust-pipe. Locomotive water-supply pipe. Metallic pipe. Molding pipe. Musical-pitch pipe. Organ-pipe. Oven-pipe. Pump-pipe. Sewer-pipe. Sheet-metal pipe. Smoke-pipe. Smoking-pipe. Stand-pipe. Steam-pipe. Steam and water pipe. Steam-boiler-feed pipe. Steam-conveying pipe. Stove-pipe. Straddle-pipe. Tobacco-pipe. Vapor-bath pipe. Waste-pipe. Waste-water pipe. Water-pipe. Water and air proof pipe. Water-circulating pipe. Wind-pipe. Wooden pipe.				
Pipe and bolt cutter	G. R. Kirk	Newark, N. J	Jan. 28, 1868	73, 813
Pipe and bolt cutter	C. C. Parsons	Boston, Mass	Oct. 1, 1867	69, 473
Pipe and bolt wrench	C. Frey and G. Macardle	Newark, N. J., and Brooklyn, N. Y.	Apr. 26, 1870	102, 390
Pipe and cigar-holder	V. A. Bond	Cotton Gin, Tex	Dec. 5, 1871	121, 483
Pipe and cylinder flanges, Machine for drilling and boring.	H. S. Fairbanks	Central Falls, R. I	Dec. 29, 1868	85, 293
Pipe and faucet clamp	J. Elliott	New York, N. Y	Dec. 29, 1868	85, 372
Pipe and hose coupling	A. H. Brown	Albany, N. Y	Apr. 30, 1850	7, 318
Pipe and other boxes, Machine for fitting	W. Thurber	Olean, N. Y	Dec. 1, 1863	40, 779
Pipe and pump for conveying water	M. Reeve		Dec. 14, 1798	
Pipe and rod cutter	D. F. Hartford	Boston, Mass	Jan. 21, 1868	73, 528
Pipe and stud wrench	J. B. Barnes	Fort Wayne, Ind	June 4, 1867	65, 465
Pipe and tobacco box	L. G. Carr	Philadelphia, Pa	June 22, 1867	91, 713
Pipe, artificial stone, &c., Machine for the manufacture of.	J. W. Stockwell	Portland, Me	Aug. 6, 1872	130, 252
Pipe-attachment in plumbing	A. Barrows	Philadelphia, Pa	May 9, 1871	114, 635
Pipe-bending machine	T. J. Harrison	New York, N. Y	July 21, 1868	80, 069
Pipe-boring machine	T. W. Purcell	Fond du Lac, Wis	Aug. 2, 1870	106, 077
Pipe-butt	L. S. Bunnell	Troy, N. Y	June 4, 1861	32, 464
Pipe, Casting	S. Fulton	Conshohocken, Pa	May 21, 1861	32, 359
Pipe cleaner, Cast	H. Davies	Newport, Ky	May 24, 1870	103, 305
Pipe-cleaner, Pocket	G. E. Matile	Washington, D. C	Nov. 29, 1864	45, 258
Pipe-cleansing apparatus	J. Van Slooten, C. S. Hunt, and W. McCulloch.	New Orleans, La	July 5, 1870	105, 015

Index of patents issued from the United States Patent Office from 1790 *to* 1873, *inclusive*—Continued.

Invention.	Inventor.	Residence.	Date.	No.
Pipe-clearing device	W. Young	Easton, Pa	May 12, 1868	77, 857
Pipe-connection	J. M. Morehead	Brooklyn, N. Y	Mar. 14, 1871	112, 618
Pipe-connection	A. Weyermann	St. Gall, Switzerland	Oct. 5, 1869	95, 621
Pipe-connections, Clamped molds for making lead joints in.	E. Groyn	Tiffin, Ohio	Mar. 1, 1870	100, 394
Pipe-cores, Device for supporting	C. J. C. Petersen	Port Chester, N. Y	Dec. 19, 1871	121, 961
Pipe-coupling	L. Abbott	Lewiston, Me	Apr. 27, 1869	89, 373
Pipe-coupling	W. N. Abbott	Boston, Mass	Feb. 28, 1865	46, 633
Pipe-coupling	W. D. Alford	Cuyahoga Falls, Ohio	May 9, 1871	114, 511
Pipe-coupling	W. D. Alford and J. H. Pitkin	Cuyahoga Falls, Ohio	Feb. 15, 1870	99, 744
Pipe-coupling	W. C. Allison	Philadelphia, Pa	July 12, 1870	105, 290
Pipe-coupling	V. D. Anderson	Kewanee, Ill	May 16, 1871	114, 903
Pipe-coupling	J. F. Andrews	Nashua, N. H	Aug. 15, 1871	117, 960
Pipe-coupling	D. Ashworth	Wappinger's Falls, N. Y	Nov. 25, 1873	144, 943
Pipe-coupling	E. Barbaroux	Louisville, Ky	Jan. 9, 1866	51, 910
Pipe-coupling	A. E. Barnard	Cleveland, Ohio	June 6, 1865	48, 038
Pipe-coupling	J. R. Brown	New Haven, Conn	Aug. 24, 1869	94, 069
Pipe-coupling	J. R. Brown	New Haven, Conn	May 24, 1870	103, 423
Pipe-coupling	C. Burger	Reading, Pa	Dec. 19, 1871	121, 929
Pipe-coupling	J. Chambers	Boston, Mass	July 4, 1865	48, 517
Pipe-coupling	J. Conner	Philadelphia, Pa	Mar. 4, 1873	136, 365
Pipe-coupling	W. Craig	Newark, N. J	Jan. 29, 1867	61, 607
Pipe-coupling	E. Duffee	Haverhill, Mass	Nov. 27, 1866	59, 982
Pipe-coupling	W. B. Dunning	Geneva, N. Y	Sept. 30, 1873	143, 229
Pipe-coupling	W. Dutemple	Malden, Mass	July 18, 1865	48, 797
Pipe-coupling	C. W. Emery	Dorchester, Mass	July 11, 1865	48, 769
Pipe-coupling	L. A. Farjon	Brussels, Belgium	Aug. 2, 1870	106, 041
Pipe-coupling	G. Fetter and J. S. McClintock	Philadelphia, Pa., and Libertyville, Ill.	Aug. 19, 1856	15, 560
Pipe-coupling	J. J. Fifield	East Boston, Mass	June 15, 1869	91, 319
Pipe-coupling	J. B. Fink	Mechanicsville, Pa	Jan. 24, 1871	111, 187
Pipe-coupling	B. Fitts	Newark, N. J	July 16, 1867	66, 820
Pipe-coupling	H. P. Garabedian	Philadelphia, Pa	Dec. 15, 1868	84, 875
Pipe-coupling	A. M. George	Nashua, N. H	July 11, 1865	48, 674
Pipe-coupling	G. C. Germain	Cuyahoga Falls, Ohio	Dec. 13, 1870	110, 028
Pipe-coupling	A. Gwynne	New York, N. Y	Dec. 4, 1866	60, 178
Pipe-coupling	J. M. Hale	Georgia Plains, Vt	Nov. 9, 1869	96, 585
Pipe-coupling	R. Hale	Roxbury, Mass	Sept. 3, 1861	33, 193
Pipe-coupling	R. Hill	East Boston, Mass	Nov. 16, 1869	96, 914
Pipe-coupling	R. Hoskin	Dutch Flat, Cal	July 19, 1870	105, 456
Pipe-coupling	J. V. Jepson	Brooklyn, N. Y	Jan. 4, 1870	98, 597
Pipe-coupling	D. Kahnweiler	Wilmington, N. C	June 29, 1858	20, 717
Pipe-coupling	W. Kearney	Bergen County, N. J	Jan. 9, 1872	122, 614
Pipe-coupling	D. C. Kellam	Pontiac, Mich	May 7, 1872	126, 553
Pipe-coupling	W. A. Lighthall	New York, N. Y	July 26, 1870	105, 817
Pipe-coupling	J. Mathews, jr	New York, N. Y	Jan. 29, 1867	61, 626
Pipe-coupling	J. H. McGowan	Cincinnati Ohio	Nov. 45, 1870	109, 332
Pipe-coupling	D. Myers	Chicago, Ill	May 6, 1873	138, 515
Pipe-coupling	J. Old	Pittsburgh, Pa	Aug. 1, 1865	49, 142
Pipe-coupling	J. Old	Pittsburgh, Pa	Oct. 24, 1865	50, 619
Pipe-coupling	C. G. Page and R. J. Falconer	Washington, D. C	Dec. 20, 1859	26, 517
Pipe-coupling	H. D. Parker	Geneseo, N. Y	Feb. 20, 1866	52, 793
Pipe-coupling	J. B. Ramp	Cuyahoga Falls, Ohio	Mar. 21, 1871	112, 958
Pipe-coupling	L. W. Reed	East Cambridge, Mass	Dec. 14, 1869	97, 963
Pipe-coupling	L. T. Scofield	Cleveland, Ohio	Aug. 19, 1873	141, 893
Pipe-coupling	G. Shone	Carondelet, Mo	Jan. 23, 1866	52, 214
Pipe-coupling	E. Smith	Hamburg, Germany	Jan. 11, 1870	98, 717
Pipe-coupling	T. Smith	Baltimore, Md	Oct. 5, 1869	95, 528
Pipe-coupling	R. B. Smith and W. H. Whitney.	Cambridge, Mass	Oct. 29, 1872	132, 604
Pipe-coupling	M. Stephens	Brooklyn, N. Y	Oct. 29, 1872	132, 699
Pipe-coupling	A. C. Sweetland	North Attleborough, Mass	Aug. 12, 1873	141, 830
Pipe-coupling	S. P. M. Tasker	Philadelphia, Pa	July 23, 1872	129, 691
Pipe-coupling	N. Thompson	St. John's Wood, England	Sept. 24, 1867	69, 142
Pipe-coupling	N. Thompson	Brooklyn, N. Y	Dec. 24, 1867	72, 566
Pipe-coupling	J. F. Ward	Phillipsburgh, N. J	Aug. 25, 1863	39, 691
Pipe-coupling	J. D. Ware	Savannah, Ga	Dec. 21, 1869	98, 131
Pipe-coupling	C. Warner	Louisville, Ky	Apr. 30, 1850	7, 332
Pipe-coupling	S. R. Warner	New Haven, Conn	Mar. 17, 1863	37, 929
Pipe-coupling	F. R. Wegman	Hartford, Conn	June 15, 1869	91, 291
Pipe-coupling	N. West and N. Thompson	New York	June 27, 1848	5, 651
Pipe-coupling	L. Wilson	Ovid, N. Y	Jan. 14, 1868	73, 422
Pipe-coupling	E. Wright	Boston, Mass	Sept. 1, 1857	18, 116
Pipe-coupling	W. H. Yeaton	Philadelphia, Pa	Oct. 13, 1868	83, 118
Pipe-coupling, Extension	J. McCarthy	New York, N. Y	May 5, 1868	77, 503
Pipe-coupling, Flexible	Q. Rice	Nevada, Cal	Dec. 7, 1869	97, 699
Pipe coupling for car-heaters	T. S. Speakman	Camden, N. J	Apr. 28, 1868	77, 331
Pipe-coupling for connecting inlet-pipes and retorts of gas-machines.	J. Butler	New York, N. Y	Mar. 29, 1870	101, 224
Pipe coupling, Gas and water	J. Holmes	Evanston, Ill	Sept. 30, 1873	143, 348
Pipe-coupling holder	B. Vitalis	Athens, Greece	Apr. 23, 1872	125, 916
Pipe coupling, Lead	J. Hoyt	New York, N. Y	Jan. 4, 1870	98, 501
Pipe coupling, Leader	J. Demarest	Mott Haven, N. Y	Sept. 19, 1871	119, 127
Pipe-coupling lock	G. A. McIlhenny	Washington, D. C	July 5, 1870	105, 105
Pipe-coupling lock	G. A. McIlhenny	Washington, D. C	Jan. 2, 1872	122, 396
Pipe-coupling, Railway-car	M. Henszey, jr	Philadelphia, Pa	May 20, 1873	139, 150
Pipe-coupling, Union	R. M. Potter	Jersey City, N. J	Mar. 8, 1870	100, 665
Pipe couplings, Molding clay	J. Putnam	Salem, Mass	May 16, 1854	10, 911
Pipe-couplings, Wire joint for combined rubber and copper.	F. Kibler	Baltimore, Md	Aug. 1, 1871	117, 546
Pipe-cutter	J. Balmore	New York, N. Y	Feb. 19, 1867	62, 177
Pipe cutter	J. R. Brown	Boston, Mass	Dec. 20, 1859	26, 544
Pipe-cutter	E. Clarkson	Carbondale, Pa	July 13, 1869	92, 516
Pipe-cutter	T. S. Foster	Fitchburgh, Mass	May 28, 1867	65, 066
Pipe-cutter	D. Harrigan	North Somerville, Mass	May 16, 1871	114, 939
Pipe-cutter	J. V. Jepson	Brooklyn, N. Y	Apr. 2, 1867	63, 524
Pipe-cutter	E. S. Moulton	Chelsea, Mass	June 20, 1871	116, 210
Pipe-cutter	J. Peace	Camden, N. J	Aug. 25, 1868	81, 402
Pipe-cutter	R. S. Sanborn	Rockford, Ill	Dec. 17, 1872	134, 007

Index of patents issued from the United States Patent Office from 1790 *to* 1873, *inclusive*—Continued.

Invention.	Inventor.	Residence.	Date.	No.
Pipe-cutter	G. W. Sower	Cambridge, Mass	July 11, 1871	116, 890
Pipe-cutter	A. G. Wilder	Cohoes, N. Y	May 11, 1869	89, 959
Pipe cutter, Gas	W. Kenyon	Steubenville, Ohio	Aug. 14, 1860	29, 602
Pipe-cutter, Self-feeding	M. H. Freeman	Somerville, Mass	Feb. 18, 1868	74, 527
Pipe cutting and threading machine	W. D. Chase	New York, N. Y	Apr. 27, 1869	89, 466
Pipe cutting and threading machine	W. D. Chase	Marion, N. J	Sept. 30, 1873	143, 273
Pipe-cutting device	A. J. Pennock and T. A. Chandler.	Rockford, Ill	Feb. 18, 1873	136, 004
Pipe-cutting machine	M. Bowes	Charlotte, N. C	May 25, 1858	20, 387
Pipe-cutting tool	C. E. Haynes	Boston, Mass	Feb. 11, 1873	135, 644
Pipe-drilling machine	J. Peace and W. P. Cox	Merchantville and Camden, N. J.	Dec. 24, 1872	134, 307
Pipe-driver	L. M. Rumsey and W. P. Smith	Saint Louis, Mo	June 1, 1869	90, 786
Pipe-elbow	W. Austin and W. Opdyke	Philadelphia, Pa	Apr. 4, 1871	113, 614
Pipe-elbow	A. E. Chamberlain and J. B. Crowley.	Cincinnati, Ohio	Sept. 24, 1872	131, 499
Pipe-elbow	I. Leas and W. H. France	Terre Haute, Ind	Oct. 3, 1871	119, 620
Pipe-elbow pattern	J. K. Home	Almonte, Canada	Aug. 12, 1873	141, 645
Pipe elbow, Sheet-metal	F. Dieckmann	Cincinnati, Ohio	June 4, 1872	127, 583
Pipe-elbows, Machine for crimping	C. E. Bell	Greenfield, Ohio	July 9, 1872	128, 781
Pipe-elbows, Machine for forming	C. Hoeller	Cincinnati, Ohio	Mar. 28, 1871	113, 167
Pipe-elbows, Machine for making	L. C. Goodale	Cincinnati, Ohio	May 21, 1872	126, 884
Pipe-elbows, Machine for making	L. C. Goodale	Cincinnati, Ohio	Dec. 24, 1872	134, 267
Pipe elbows, Manufacture of cast	J. G. Weaver, jr	Cincinnati, Ohio	June 28, 1870	104, 908
Pipe-elbows, Molding and casting	G. Ross	Newport, Ky	Feb. 28, 1871	112, 286
Pipe expansion-joint	J. E. Jones	Tidioute, Pa	Apr. 19, 1870	102, 010
Pipe-fittings, Cutting screw-threads on	C. C. Walworth	Boston, Mass	Oct. 11, 1859	25, 779
Pipe-fittings, Machine for tapping	J. L. Pope	Cleveland, Ohio	June 3, 1873	139, 476
Pipe, Flask for casting	J. Demarest	Mott Haven, N. Y	Apr. 25, 1871	113, 986
Pipe, Flask for molding	J. Farrar and L. F. Whiting	Providence, R. I., and Boston, Mass.	Jan. 26, 1869	86, 144
Pipe-joint	P. Ball	Worcester, Mass	Sept. 26, 1865	50, 090
Pipe-joint	P. Clark	Rahway, N. J	Aug. 9, 1870	106, 122
Pipe-joint	R. B. Coar	Jersey City, N. J	Mar. 15, 1870	100, 726
Pipe-joint	R. B. Coar	Jersey City, N. J	May 31, 1870	103, 567
Pipe-joint	J. Demarest	Mott Haven, N. Y	Nov. 18, 1873	144, 663
Pipe-joint	B. Garvin and R. J. Pettibone	Oshkosh, Wis	Mar. 10, 1868	75, 407
Pipe-joint	W. H. Hammond	Tipton, Iowa	June 25, 1872	128, 303
Pipe-joint	A. C. Jones	Philadelphia, Pa	Apr. 23, 1861	32, 135
Pipe-joint	W. R. Maffit	Wilkesbarre, Pa	Mar. 14, 1865	46, 808
Pipe-joint	T. T. Prosser and G. C. Morgan	Chicago, Ill	Apr. 8, 1873	137, 717
Pipe-joint	J. H. Rhodes	Brooklyn, N. Y	Sept. 3, 1867	68, 388
Pipe-joint	H. Smith	Norwalk, Ohio	July 12, 1870	105, 268
Pipe-joint	H. E. Towle	New York, N. Y	June 30, 1863	39, 081
Pipe-joint	S. Trumbore	Easton, Pa	Feb. 20, 1872	123, 954
Pipe-joint	S. R. Warner	New Haven, Conn	Sept. 10, 1861	33, 275
Pipe joint, Adjustable	J. H. Rhodes	Brooklyn, N. Y	Aug. 27, 1867	68, 112
Pipe joint, Cement	D. G. Phipps	New Haven, Conn	June 14, 1870	104, 348
Pipe-joint coupling, Flexible	S. Ainsworth	Pittsburgh, Pa	Sept. 1, 1868	81, 572
Pipe-joint couplings, Sealing-apparatus for	C. Perks	Philadelphia, Pa	Apr. 5, 1870	101, 501
Pipe-joint, Flexible	R. M. Shurtleff	Hartford, Conn	July 9, 1872	128, 760
Pipe joint, Gas	R. C. Robbins	Jersey City, N. J	Jan. 13, 1863	37, 425
Pipe-joint, Metal	W. H. Harrison	Philadelphia, Pa	Jan. 18, 1870	98, 960
Pipe-joint-sealing apparatus	W. Cassidy	New Bedford, Mass	Apr. 12, 1870	101, 708
Pipe joint, Swing	J. R. Worswick and E. Lewis.	Cleveland, Ohio	Mar. 4, 1873	136, 577
Pipe joints, Connection for lead	I. Smith	New York, N. Y	Apr. 26, 1870	102, 443
Pipe-joints, &c., Machinery for forging	J. A. Shipton and R. Mitchell.	Wolverhampton, England.	Oct. 16, 1866	58, 963
Pipe kiln, Clay	J. Dimelow	Philadelphia, Pa	Sept. 10, 1867	68, 716
Pipe-leak stopper	S. Moore	Sudbury, Mass	July 6, 1869	92, 338
Pipe machinery, Socket	F. M. Mattice	Cleveland, Ohio	June 7, 1870	104, 046
Pipe-mold	G. W. Cornell and B. B. Quinn	New York, N. Y	Mar. 6, 1866	52, 971
Pipe-mold	E. Walsh	Rochester, N. Y	Apr. 30, 1872	126, 354
Pipe, Mold for casting	H. M. Bird	Cambridgeport, Mass	Apr. 16, 1867	63, 781
Pipe, Mold for casting	J. Edson	Boston, Mass	Mar. 23, 1869	88, 020
Pipe, Mold for casting	J. Edson	Boston, Mass	Apr. 12, 1870	101, 842
Pipe, Mold for casting	S. Fulton	Conshohocken, Pa	Apr. 16, 1861	32, 058
Pipe, Molding	W. D. Alford	Cuyahoga Falls, Ohio	Sept. 26, 1871	119, 259
Pipe, Molding	J. Aston	Pittsburgh, Pa	Sept. 15, 1868	82, 065
Pipe, Molding	J. Firth and J. Ingham	Phillipsburgh, N. Y	Dec. 20, 1859	26, 486
Pipe, Molding	G. H. Garrett	Saint Louis, Mo	Dec. 29, 1868	85, 440
Pipe, Molding	J. D. Johnson	Hainesport, N. Y	Dec. 3, 1872	133, 584
Pipe, Molding	A. W. A. Logen	Rochdale, England	July 22, 1873	141, 150
Pipe, Molding	F. Shickle	Saint Louis, Mo	Sept. 28, 1869	95, 388
Pipe, Molding	W. Smith	Pittsburgh, Pa	Oct. 25, 1870	108, 647
Pipe, Molding and casting	B. S. Benson	Baltimore, Md	Feb. 17, 1863	37, 670
Pipe, Molding and casting	I. M. Kenworthy	Carbondale, Pa	July 2, 1872	128, 493
Pipe, Molding and casting	L. Martaresche	Pittsburgh, Pa	Oct. 26, 1869	96, 131
Pipe molding and casting device	J. McClelland	Washington, D. C	June 21, 1870	104, 479
Pipe-molding apparatus	F. Fuller	Providence, R. I	Apr. 12, 1870	101, 853
Pipe-molding apparatus	F. Shickle	Saint Louis, Mo	May 25, 1869	90, 591
Pipe-molding machine	B. S. Benson	Baltimore, Md	July 30, 1861	32, 956
Pipe-molding machine	B. S. Benson	Baltimore, Md	Sept. 3, 1861	33, 178
Pipe-molding machine	B. S. Benson	Baltimore, Md	Oct. 13, 1868	83, 028
Pipe-molding machine	B. S. Benson	Baltimore, Md	June 20, 1871	116, 141
Pipe-molding machine	J. T. Glass and L. E. Morrell	Columbus, Ohio	July 16, 1872	129, 547
Pipe-molding machine	D. Long	Louisville, Ky	Dec. 3, 1872	133, 652
Pipe-molding machine	G. Richardson	Milwaukee, Wis	Mar. 8, 1871	112, 495
Pipe molding machine	W. Smith	Pittsburgh, Pa	Oct. 15, 1867	69, 854
Pipe-molding machine	W. Smith	Allegheny City, Pa	Nov. 3, 1868	83, 668
Pipe-molding pattern	G. Ross	Newport, Ky	Mar. 24, 1863	37, 981
Pipe-nippers	G. Smith	New York, N. Y	Dec. 20, 1859	26, 530
Pipe or car coupling	A. G. Story	Waterloo, N. Y	May 13, 1873	138, 822
Pipe or grate boiler and heater	S. Bolton	Philadelphia, Pa	July 7, 1809	....
Pipe or smoking-apparatus	E. S. Franks	Boston, Mass	Sept. 27, 1864	44, 414
Pipe patterns, Mold for plaster	N. P. Bowler	Cleveland, Ohio	Sept. 16, 1873	142, 765
Pipe reaming and squaring tool	W. Barry	Chicago, Ill	June 15, 1869	91, 201
Pipe-stem	J. Davis	Saint Paul, Minn	June 17, 1873	139, 880
Pipe-stem	S. S. Scranton	New Haven, Conn	Apr. 14, 1868	76, 827

Index of patents issued from the United States Patent Office from 1790 *to* 1873, *inclusive*—Continued.

Invention.	Inventor.	Residence.	Date.	No.
Pipe-stem-bending machine	J. Harvey	Philadelphia, Pa	May 20, 1873	139, 148
Pipe-tips, Machine for milling	A. R. Weiss	Brooklyn, N. Y	May 13, 1873	138, 825
Pipe-trap	C. H. Burleigh	Worcester, Mass	Nov. 23, 1869	97, 160
Pipe-trap, Cement	A. A. Lovell	Worcester, Mass	Nov. 4, 1873	144, 277
Pipe-welding furnace	J. Fieldhouse	Taunton, Mass	Aug. 7, 1866	56, 919
Pipe-wrench	W. C. Abbo	Petroleum Centre, Pa	Aug. 20, 1867	67, 937
Pipe-wrench	L. Anderson	Washington, D. C	Mar. 19, 1872	124, 654
Pipe-wrench	R. Bain	Brooklyn, N. Y	Apr. 9, 1867	63, 689
Pipe-wrench	O. G. Barrett	Boston Highlands, Mass	Jan. 31, 1871	111, 422
Pipe-wrench	W. H. Barwick	Montreal, Canada	June 6, 1871	115, 678
Pipe-wrench	W. F. Beecher	Chicago, Ill	May 1, 1860	28, 053
Pipe-wrench	J. R. Brown	Boston, Mass	Oct. 18, 1864	44, 703
Pipe-wrench	J. R. Brown	Boston, Mass	May 28, 1867	65, 162
Pipe-wrench	W. H. Burton	Selma, Ala	Feb. 11, 1873	135, 773
Pipe-wrench	A. Call	Springfield, Mass	Aug. 28, 1866	57, 621
Pipe-wrench	J. W. Close	Buffalo, N. Y	Feb. 5, 1867	61, 715
Pipe-wrench	D. W. Coggeshall	Providence, R. I	June 12, 1866	55, 578
Pipe-wrench	J. Craig	Detroit, Mich	Sept. 10, 1872	131, 282
Pipe-wrench	I. H. Doolittle	Ansonia, Conn	Mar. 13, 1860	27, 622
Pipe-wrench	W. H. Downing	Pioneer, Pa	Aug. 17, 1869	93, 688
Pipe-wrench	D. E. Eaton	Boston, Mass	Nov. 5, 1872	132, 819
Pipe-wrench	H. V. Faries	Indianapolis, Ind	Aug. 26, 1862	36, 289
Pipe-wrench	D. Frank and T. Snyder	Allentown, Pa	Oct. 10, 1871	119, 838
Pipe-wrench	M. H. Freeman	Somerville, Mass	May 14, 1867	64, 656
Pipe-wrench	A. Gauntt	Chagrin Falls, Ohio	Oct. 29, 1872	132, 571
Pipe-wrench	I. S. Hamilton	Hamilton, Ohio	Nov. 7, 1871	120, 641
Pipe-wrench	M. Hastings	Brooklyn, N. Y	Aug. 24, 1869	94, 107
Pipe-wrench	H. A. Hyle	Shamburgh, Pa	Nov. 28, 1871	121, 372
Pipe-wrench	D. D. Ingram	Manistee, Mich	Oct. 21, 1873	143, 908
Pipe-wrench	J. C. Johnson	Oil City, Pa	Apr. 26, 1864	42, 489
Pipe-wrench	R. H. Lecky	Allegheny City, Pa	June 30, 1868	79, 361
Pipe-wrench	N. F. Loi	New York, N. Y	May 26, 1868	78, 299
Pipe-wrench	G. P. Moloney	Detroit, Mich	May 20, 1873	139, 177
Pipe-wrench	E. S. Moulton	Chelsea, Mass	Feb. 18, 1873	136, 089
Pipe-wrench	C. Neames	New Orleans, La	Jan. 9, 1872	122, 638
Pipe-wrench	J. L. Ordner	Cleveland, Ohio	May 28, 1867	65, 111
Pipe-wrench	W. Pearson	Windsor's Locks, Conn	Aug. 23, 1864	43, 926
Pipe-wrench	C. Pomeroy	Mattoon, Ill	Nov. 16, 1869	96, 957
Pipe-wrench	H. F. Read	Brooklyn, N. Y	July 22, 1862	35, 950
Pipe-wrench	J. Rigg	Cleveland, Ohio	June 3, 1873	139, 520
Pipe-wrench	E. H. Robbins	Pittsfield, Mass	May 2, 1871	114, 318
Pipe-wrench	A. Secor	North Adams, Mass	Dec. 10, 1872	133, 804
Pipe-wrench	D. C. Stillson	Charlestown, Mass	Apr. 30, 1872	126, 161
Pipe-wrench	T. Stone	Carbondale, Ill	Dec. 10, 1872	133, 902
Pipe-wrench	G. Warsop	Nottingham, England	July 1, 1873	140, 561
Pipe-wrench	W. C. Westerfield	Fairbury, Ill	May 7, 1872	126, 506
Pipe-wrench	H. B. Wheatcroft	New York, N. Y	Dec. 2, 1873	145, 263
Pipe-wrench	J. A. Wilcox	Rocky Hill, Conn	Sept. 3, 1861	33, 223
Pipe-wrench	J. A. Wilcox	Rocky Hill, Conn	June 28, 1870	104, 913
Pipe-wrench	T. Williams	Providence, R. I	Feb. 25, 1873	136, 350
Pipe-wrench	W. W. Wills	Janesville, Wis	May 29, 1866	55, 190
Pipe-wrench	H. Wilson	Tarr Farm, Pa	Sept. 12, 1871	118, 997
Pipe wrench and cutter	J. L. Brierly	Auburn, Mass	May 19, 1868	78, 047
Pipe wrench and cutter	R. A. Copeland	Baltimore, Md	Jan. 26, 1869	86, 138
Pipe wrench and cutter	A. Robes and J. C. Chapman	Somerville and Cambridgeport, Mass.	Aug. 31, 1869	94, 241
Pipe-wrench attachment for monkey-wrench	A. H. Woodruff	Lansing, Iowa	Oct. 31, 1871	120, 475
Pipe-wrench, Self-adjusting	T. Williams	Providence, R. I	Feb. 25, 1873	136, 351
Pipes across rivers, Mode of laying	J. F. Ward	Jersey City, N. J	Jan. 31, 1871	111, 498
Pipes and hose, Apparatus for transporting, extending, and elevating.	L. Day	Buffalo, N. Y	Oct. 27, 1868	83, 475
Pipes and manifolds, Arrangement of conducting	C. C. Walworth	Boston, Mass	May 5, 1863	38, 433
Pipes and tubes while being screw-threaded, Chuck for holding.	W. T. Cole	New York, N. Y	Aug. 23, 1870	106, 551
Pipes, Construction and joining of	T. L. Truss	Darlington, England	Aug. 21, 1860	29, 731
Pipes, Cutting threads on	J. M. Carpenter	Florence, Mass	Mar. 9, 1869	87, 632
Pipes, Device for cutting off	W. H. Barwick and W. T. Farre	Montreal, Canada	Mar. 19, 1872	124, 659
Pipes, Elastic water-stop for checking the force or momentum of water in.	H. Allen	New York, N. Y	Oct. 25, 1843	3, 315
Pipes from percussion, Regulator to protect	E. A. Chameroy	Paris, France	May 7, 1872	126, 444
Pipes, Hydrostatic apparatus for testing	T. J. McGowan	Cincinnati, Ohio	Sept. 17, 1872	131, 405
Pipes, joints, bottles, casks, &c., Making and coating.	B. Rhodes	Bow, England	Jan. 19, 1864	41, 351
Pipes, Lug and link for connecting	C. Warner	Louisville, Ky	May 8, 1849	6, 428
Pipes, Machine for screwing together	R. T. Crane	Chicago, Ill	Feb. 6, 1872	123, 382
Pipes, Machinery for riveting	J. Ball	New York, N. Y	Dec. 20, 1845	4, 319
Pipes, Manufacture of	G. L. Eagan	San Francisco, Cal	Sept. 2, 1873	142, 448
Pipes, Method and apparatus for casting	A. T. Brodie, R. R. Smith, and J. J. Tyler.	Pittsburgh, Pa	Nov. 21, 1871	121, 151
Pipes of plastic materials, Mold for molding	J. Chilver	Jersey City, N. J	July 9, 1861	32, 754
Pipes or tubes, Casting	W. Badger	Boston, Mass	Dec. 26, 1808	
Pipes, Stool for casting	C. J. Ellis	Louisville, Ky	Apr. 23, 1872	126, 043
Pipes, tiles, sidewalks, &c., Compound for	W. A. Battersby	Williamsburgh, N. Y	Nov. 8, 1870	108, 957
Pipes, Tool for cutting off	H. Collinson	Boston, Mass	Apr. 9, 1872	125, 380
Pipes, tubes, barrels, &c., Process and apparatus for manufacturing composition.	J. K. Mayo	Williamsburgh, N. Y	July 6, 1869	92, 332
Pipes, tubes, &c., Mode of making	J. Putman	Salem, Mass	Jan. 17, 1827	
Pipes, tubes, &c., with silver or other metal by the electro-depositing process, Coating the interior of.	D. D. Parmelee	New York, N. Y	Oct. 18, 1870	108, 510
Pipes with tin, Coating	T. Ewbank	New York, N. Y	May 16, 1832	
Piping, Machine for making	J. Ayers	Chicago, Ill	May 21, 1872	126, 864
Pisciculture	W. H. Furman	Maspeth, N. Y	June 16, 1868	78, 952
Pistol	W. S. Butler	Rockey Hill, Conn	Feb. 3, 1857	16, 571
Pistol	B. and B. M. Darling	Bellingham, Mass	Apr. 23, 1836	
Pistol, Air	R. Brooks, jr	Rockport, Mass	Feb. 15, 1870	99, 754
Pistol, Air	G. H. Snow and E. H. Hawley	New Haven, Conn., and Kalamazoo, Mich.	Sept. 12, 1871	118, 886
Pistol and dirk, Combined	F. Wesson	Worcester, Mass	July 20, 1869	92, 918

Index of patents issued from the United States Patent Office from 1790 *to* 1873, *inclusive*—Continued.

Invention.	Inventor.	Residence.	Date.	No.
Pistol and other fire-arms	E. Allen	Norwich, Conn	Apr. 16, 1845	3,998
Pistol and pocket-knife, Combined	A. J. Beavey	South Montville, Me	Mar. 27, 1866	53,473
Pistol and sword, Combined	C. E. Billings	Springfield, Mass	Sept. 22, 1868	82,279
Pistol and sword, Combined	D. A. Courter	Beloit, Wis	Mar. 11, 1862	34,625
Pistol, Breech-loading	S. M. Perry	Brooklyn, N. Y	June 21, 1864	43,259
Pistol, Breech-loading	S. M. Perry	Brooklyn, N. Y	June 21, 1864	43,260
Pistol, Breech-loading	S. M. Perry and E. Goddard	Plainfield, N. J., and Brooklyn, N. Y.	Apr. 26, 1870	102,429
Pistol, Breech-loading	D. Werner	Saint Louis, Mo	Oct. 6, 1868	82,908
Pistol-breech pins, Device for moving and holding	W. W. Marston	New York, N. Y	June 18, 1850	7,443
Pistol, Burglar-alarm	J. G. Clark	Augusta, Ga	June 7, 1859	24,349
Pistol-frames, Die for forging and shaping	S. P. Legg	Springfield, Mass	Dec. 5, 1865	51,327
Pistol-frames, Die for swaging	C. E. Billings	Windsor, Vt	Aug. 7, 1866	56,885
Pistol-gallery, Portable	F. A. Spofford and M. G. Raffington.	Columbus, Ohio	Feb. 12, 1867	61,960
Pistol-gallery, Portable	F. A. Spofford and M. G. Raffington.	Columbus, Ohio	Feb. 4, 1868	74,163
Pistol-holster	A. A. Bennett	Cincinnati, Ohio	Apr. 21, 1863	28,213
Pistol knife or cutlass	G. Elgin	New York, N. Y	July 5, 1837	254
Pistol, Parlor air	B. Haviland and G. P. Gunn	Hudson and Ilion, N. Y	Apr. 18, 1871	113,766
Pistol, Pocket	V. M. Wallace	West Topham, Vt	Aug. 17, 1835	....
Pistol, Repeating	G. Tigneres	Covington, La	Dec. 20, 1859	29,538
Pistol, Revolving-breech	J. Stevens	Chicopee, Mass	Oct. 7, 1851	8,412
Pistol-saber	R. B. Lawton	Newport, R. I	Nov. 23, 1837	481
Pistol safety-guard	R. D. Hay	Crooked Creek, N. C	Nov. 22, 1870	109,513
Pistols, Base-pin and rammer of revolving	W. H. Elliot	Plattsburgh, N. Y	Dec. 17, 1861	33,932
Pistols, Mode of attaching knives or dirks to	R. W. Andrews	Stafford, Conn	July 31, 1837	328
Pistols, Tool for manufacturing	A. Rebetey	Norwich, Conn	May 10, 1859	23,990
Piston	C. B. Allen	Philadelphia, Pa	Sept. 9, 1873	142,659
Piston	P. A. Brown	Philadelphia, Pa	Oct. 9, 1816	....
Piston	I. D. Buck	Conshohocken, Pa	Oct. 25, 1870	108,682
Piston	S. E. and I. Y. Chubbuck	Roxbury, Mass	Feb. 25, 1868	74,750
Piston	W. H. Holland	Boston, Mass	Nov. 21, 1871	121,103
Piston	H. S. Hopkins	Boston, Mass	May 7, 1872	126,548
Piston	L. Huntoon	Milford, Mass	July 17, 1866	56,420
Piston	A. H. Smith	New York, N. Y	Jan. 10, 1871	110,874
Piston	W. Swayne	United States Army	Mar. 31, 1868	76,268
Piston	J. Wheelock	Worcester, Mass	Mar. 17, 1868	75,714
Piston and packing	A. K. Rider	New York, N. Y	Sept. 30, 1873	143,255
Piston and piston-packing	O. Collier	Sacramento, Cal	Dec. 28, 1869	98,232
Piston and piston-packing	G. H. Reynolds	New York, N. Y	May 16, 1871	114,969
Piston and piston-packing	E. Sullivan	Mount Washington, Pa	Sept. 26, 1871	119,247
Piston and piston-rod connections	A. P. Samuel	New York, N. Y	Mar. 23, 1858	19,722
Piston and piston-valve, Steam-engine	T. S. Davis	Jersey City, N. J	Feb. 5, 1861	31,303
Piston and slide-valve, Horizontal steam-engine	B. H. Brown	Philadelphia, Pa	Sept. 1, 1843	3,241
Piston and valve-cock	J. Lacey	Philadelphia, Pa	May 6, 1823	....
Piston, Balancing	L. Findlay	Saint Louis, Mo	Aug. 22, 1871	118,227
Piston-blowers	C. W. Isbell	New York, N. Y	Apr. 8, 1873	137,553
Piston-connection	H. F. Pease	Brooklyn, N. Y	May 3, 1870	102,703
Piston, Engine	T. T. Dwelley	Charlestown, Mass	May 12, 1863	38,474
Piston, Engine	M. Hunt	Salem, Ohio	Sept. 17, 1867	68,989
Piston, Engine	W. Keemle	Philadelphia, Pa	June 11, 1867	65,753
Piston-engine and stamping-machine, Spring and weight.	E. F. and J. McFarland	Worcester, Mass	Dec. 26, 1865	51,737
Piston engine, Double	A. W. Morrell	Niles, Mich	Sept. 20, 1870	107,524
Piston engine, Oscillating	J. B. and S. M. Davis	Harrisonville, Mo	Nov. 29, 1870	109,721
Piston engine, Vibrating	J. B. Root	New York, N. Y	Apr. 25, 1865	47,459
Piston-facing machine	W. W. Loweree and G. A. Sanderson.	Albany, N. Y	July 4, 1871	116,730
Pistons for steam-engines, Arrangement of passages and valves for cushioning.	N. W. Wheeler	New York, N. Y	Jan. 26, 1858	19,220
Piston head and valve	G. Weinman	Columbia, Ohio	Jan. 28, 1870	104,909
Piston-head-centering device	O. S. Howard	Bangor, Me	July 29, 1873	141,221
Piston head, Steam-engine	O. S. Howard	Bangor, Me	Nov. 7, 1871	120,586
Piston head, Steam-engine	E. Sullivan	Pittsburgh, Pa	June 5, 1866	55,387
Piston, Hydraulic	T. Critchlow	Baldwin, Pa	Oct. 22, 1872	132,445
Piston of pneumatic spring for railway-car, &c	A. Connison	Belleville, N. J	Dec. 23, 1841	2,395
Piston, Printing-press	C. B. Cottrell	Westerly, R. I	Dec. 12, 1871	121,759
Piston-ring and method of deriving motion therefrom in rotary engines.	J. Tremper	Little Britain, N. Y	Apr. 17, 1849	6,359
Piston-rod adjuster	D. Bly	Macon, Ga	May 19, 1868	78,045
Piston-rod, Compound	J. R. St. John	Cleveland, Ohio	Apr. 27, 1841	2,070
Piston-rod guide	E. Dunscomb	Boston, Mass	Apr. 18, 1865	47,283
Piston rod, Steam-engine	J. F. Carl	Pleasantville, Pa	May 21, 1872	127,025
Piston-rods, &c., Bushing for	G. W. Welch	Fort Wayne, Ind	Jan. 14, 1873	134,955
Piston-rods for steam and other powers, Connecting	E. P. Curtis	Madison, Wis	Mar. 12, 1867	62,825
Piston-rods, Machine for	W. B. Davis	Brooklyn, N. Y	Nov. 21, 1865	51,027
Piston-rods with pistons of steam-cylinders, Mode of connecting.	M. W. Baldwin	Philadelphia, Pa	Dec. 17, 1840	1,904
Piston, Rotative	J. M. Cooper	Guildhall, Vt	July 16, 1827	....
Piston-spring	W. R. Brown	Bath, N. Y	Aug. 3, 1869	93,273
Piston, Steam	E. L. Horton	Hartford, Conn	Sept. 5, 1838	905
Piston, Steam	E. Merriman	Allegheny, Pa	July 11, 1871	116,853
Piston, Steam	J. Richards	New York	Oct. 3, 1846	4,781
Piston, Steam-cylinder	M. B. Mason	Aurora, Ind	Sept. 4, 1866	57,823
Piston, Steam-engine	L. B. Batcheller	Rochester, N. Y	Apr. 30, 1861	32,174
Piston, Steam-engine	C. Bollinger	Harrisburgh, Pa	May 26, 1868	78,179
Piston, Steam-engine	G. Bower and J. Qualter	Ashton-under-Lyne and Barnsley, England.	Aug. 28, 1866	57,648
Piston, Steam-engine	S. G. Cabell	Quincy, Ill	Aug. 7, 1866	56,899
Piston, Steam-engine	H. D. Dunbar	Memphis, Tenn	Aug. 14, 1860	29,576
Piston, Steam-engine	H. D. Dunbar	Hartland, Vt	June 16, 1863	38,890
Piston, Steam-engine	H. D. Dunbar	Hartland Four Corners, Vt	Jan. 12, 1864	41,205
Piston, Steam-engine	L. B. Flanders	Philadelphia, Pa	Feb. 5, 1867	61,728
Piston, Steam-engine	F. H. Furniss and J. Hovey	Cleveland, Ohio	Nov. 10, 1863	40,559
Piston, Steam-engine	J. O. Haight	Albany, N. Y	Apr. 9, 1861	31,971
Piston, Steam-engine	J. and H. Hall, sr	Philadelphia, Pa	Mar. 15, 1864	41,961
Piston, Steam-engine	B. F. Hidden	Norwich, Conn	Aug. 4, 1863	39,397
Piston, Steam-engine	J. Hinds and G. H. Lodge	Salem and Swampscott, Mass.	Apr. 7, 1868	76,448

Index of patents issued from the United States Patent Office from 1790 *to* 1873, *inclusive*—Continued.

Invention.	Inventor.	Residence.	Date.	No.
Piston, Steam-engine	T. S. La France	Elmira, N. Y	Oct. 9, 1860	30, 328
Piston, Steam-engine	H. N. J. Mansfield	Malone, N. Y	Oct. 20, 1868	83, 188
Piston, Steam-engine	A. Nadon	Springfield, Mass	Jan. 21, 1868	73, 456
Piston, Steam-engine	W. Ord	Brooklyn, Ohio	Apr. 27, 1869	89, 428
Piston, Steam-engine	J. K. Robinson and J. M. Clark	Bellaire, Ohio	Aug. 14, 1860	29, 621
Piston, Steam-engine	J. K. Robinson and J. M. Clark	Bellaire, Ohio	Oct. 9, 1860	30, 351
Piston, Steam-engine	F. J. Roth and D. R. Gamble	Newark, Ohio	Dec. 19, 1865	51, 622
Piston, Steam-engine	A. J. Scoville and A. H. De Clercq.	Bloomington, Ill	Feb. 25, 1862	34, 552
Piston, Steam-engine	S. M. Seley and P. Hopkins	Peoria, Ill	June 24, 1862	35, 711
Piston, Steam-engine	J. Y. Smith	Pittsburgh, Pa	Apr. 23, 1867	64, 043
Piston, Steam-engine	W. G. Snook	Corning, N. Y	June 1, 1869	90, 884
Piston, Steam-engine	A. M. Sprague	Mobile, Ala	May 9, 1854	10, 898
Piston, Steam-engine	N. P. Stevens	Boston, Mass	Aug. 4, 1863	39, 431
Piston, Steam-engine	N. P. Stevens	Boston, Mass	Oct. 11, 1864	44, 674
Piston, Steam-engine	N. P. Stevens	Boston, Mass	July 18, 1865	48, 851
Piston, Steam-engine	J. Wheelock	Worcester, Mass	Apr. 23, 1867	64, 052
Piston, Steam-engine	W. D. Whitmore	Boston, Mass	Jan. 22, 1867	61, 373
Piston, Steam-engine	R. Winans	Baltimore, Md	Apr. 6, 1858	19, 888
Piston, Steam-engine	R. Witty	Chicago, Ill	Aug. 31, 1869	94, 371
Piston, Steam-engine	Wright and Ketchum	Calhoun County, Mont	June 19, 1835	....
Piston, Steam-engine and pump	J. A. Huss	Bowling Green, Ky	June 13, 1871	115, 858
Piston, Steam-engine metal-packed	G. W. Cotton	Saint Louis, Mo	May 5, 1857	17, 200
Piston, Steam-engine, pump, &c., Metallic	J. Swainson	New York, N. Y	Mar. 30, 1837	159
Pistons and pitmen of steam-engines, Mode of connecting.	R. S. Edwards	Savannah, Mo	Feb. 13, 1872	123, 623
Pistons, Arrangement for lubricating	B. Garvey	New York, N. Y	July 17, 1860	29, 216
Pistons, cylinders, and safety-valves, Steam apparatus substituted for.	A. Patterson	Rush, Pa	Feb. 26, 1839	1, 091
Pistons, Lubricating rotary steam-engine	S. W. Wicks	Seneca Falls, N. Y	Aug. 12, 1873	141, 845
Pistons, Lubricating steam-engine	S. Hall	Basford, England	Oct. 30, 1834	....
Pistons, Method of expanding metallic	J. Touchstone and J. H. Clark	Philadelphia, Pa	Apr. 17, 1849	6, 318
Pistons, Mode of cushioning steam	G. F. Blake	Boston, Mass	Mar. 29, 1870	101, 217
Pistons to piston-rods, Device for fastening	A. T. Norgan	Palo Alto, Pa	Dec. 14, 1869	97, 797
Pitch-board	J. R. Drew	San Francisco, Cal	Nov. 30, 1869	97, 281
Pitch, Composition	H. Ruggles	New York, N. Y	Dec. 2, 1834	....
Pitch, Composition	T. A. Sherman	Scriba, N. Y	Mar. 4, 1836	....
Pitch, Making	H. Ruggles	New York, N. Y	Mar. 19, 1836	....
Pitch, Manufacture of	A. H. Emery	New York, N. Y	Aug. 8, 1865	49, 248
Pitcher	J. H. Hobbs	Wheeling, W. Va	May 11, 1869	89, 996
Pitcher	N. T. Whiting	Lawrence, Mass	Aug. 7, 1866	57, 025
Pitcher and coffee-pot, Revolving ice	J. P. Adams	New York, N. Y	Aug. 10, 1869	93, 575
Pitcher, Beer	W. P. Eayrs	Nashua, N. H	May 29, 1866	55, 076
Pitcher, Beer	L. Hermance	Saratoga Springs, N. Y	Apr. 17, 1860	27, 904
Pitcher, Beer	W. S. Mathews	Meriden, Conn	Nov. 15, 1859	26, 115
Pitcher, Beer	O. F. Pelton	Middletown, Conn	June 19, 1860	28, 771
Pitcher, Beer	D. W. S. Rawson	Peru, Ill	Feb. 2, 1864	41, 447
Pitcher, coffee-pot, &c., Device for tilting	J. Gibson, jr	Albany, N. Y	Dec. 28, 1869	98, 245
Pitcher-cover	W. Bradley	Providence, R. I	Jan. 7, 1873	134, 586
Pitcher, Double-walled	G. Jepson	Chelsea, Mass	Nov. 10, 1868	83, 969
Pitcher, Enameled metallic ice	C. C. Foote	West Meriden, Conn	June 30, 1868	79, 335
Pitcher for cooling liquids	H. Pietsch	New York, N. Y	July 27, 1869	93, 001
Pitcher-holder	B. T. Currier	Boston, Mass	Nov. 5, 1872	132, 808
Pitcher, Ice	J. B. Bailey	New York, N. Y	Apr. 7, 1868	76, 378
Pitcher, Ice	G. H. Bechtel	Philadelphia, Pa	Apr. 23, 1867	63, 986
Pitcher, Ice	W. Bellamy	Newark, N. J	July 16, 1867	66, 779
Pitcher, Ice	W. Bellamy	Newark, N. J	Feb. 11, 1868	74, 285
Pitcher, Ice	W. Bellamy	Newark, N. J	Sept. 7, 1869	94, 550
Pitcher, Ice	C. Conradt	Philadelphia, Pa	Sept. 5, 1865	49, 724
Pitcher, Ice	S. Curtis, jr	New York, N. Y	Jan. 21, 1862	34, 228
Pitcher, Ice	J. Dawson	New York, N. Y	Jan. 9, 1872	122, 576
Pitcher, Ice	C. Dickinson and W. Bellamy	Newark, N. J	Apr. 5, 1859	23, 455
Pitcher, Ice	E. A. Dodge	Washington, D. C	Dec. 10, 1872	133, 837
Pitcher, Ice	S. Eakins	Philadelphia, Pa	June 26, 1855	13, 125
Pitcher, Ice	C. Englebert and J. S. Von Nieda.	Philadelphia, Pa	Dec. 5, 1871	121, 500
Pitcher, Ice	G. A. Eno	Philadelphia, Pa	Feb. 12, 1867	61, 021
Pitcher, Ice	T. B. Fitts and A. D. Cook	New York, N. Y	Apr. 12, 1870	101, 850
Pitcher, Ice	N. F. Griswold	Meriden, Conn	May 1, 1860	28, 076
Pitcher, Ice	R. Holmes	Middletown, Conn	Apr. 14, 1868	76, 761
Pitcher, Ice	E. Kauffman	Philadelphia, Pa	Apr. 6, 1858	19, 855
Pitcher, Ice	G. Lane	New York, N. Y	Apr. 14, 1868	76, 638
Pitcher, Ice	N. Lawrence	Taunton, Mass	Nov. 5, 1867	70, 582
Pitcher, Ice	T. Leach	Taunton, Mass	Aug. 4, 1868	80, 748
Pitcher, Ice	E. B. Manning	Middletown, Conn	Feb. 5, 1867	61, 747
Pitcher, Ice	F. C. Meyer	Philadelphia, Pa	Aug. 8, 1865	49, 351
Pitcher, Ice	F. Miller	Wallingford, Conn	Aug. 29, 1871	118, 630
Pitcher, Ice	F. J. Miller	Brooklyn, N. Y	Apr. 21, 1868	76, 934
Pitcher, Ice	B. Morahan	Brooklyn, N. Y	Aug. 17, 1869	93, 734
Pitcher, Ice	E. A. Parker	West Meriden, Conn	July 9, 1872	128, 810
Pitcher, Ice	E. A. Parker	West Meriden, Conn	July 9, 1872	128, 811
Pitcher, Ice	E. A. Parker	West Meriden, Conn	May 6, 1873	138, 686
Pitcher, Ice	M. Simons	Middletown, Conn	June 18, 1867	65, 958
Pitcher, Ice	G. W. Smith	Hartford, Conn	June 15, 1858	20, 592
Pitcher, Ice	J. Stimpson	Baltimore, Md	Oct. 5, 1858	21, 717
Pitcher, Ice	J. H. Stimpson	Baltimore, Md	Mar. 8, 1859	23, 209
Pitcher, Ice	W. C. Wood	Washington, D. C	Nov. 24, 1868	84, 458
Pitcher-lid lifter	F. W. L. Knuschke	Providence, R. I	June 11, 1867	65, 757
Pitcher, Molasses	T. B. Atterbury	Pittsburgh, Pa	June 24, 1873	130, 236
Pitcher, Molasses	H. W. Goodrich	Boston, Mass	June 17, 1856	15, 128
Pitcher, Molasses	W. C. King	Pittsburgh, Pa	May 6, 1873	138, 501
Pitcher, Molasses	E. B. Manning	West Meriden, Conn	July 1, 1873	140, 515
Pitcher, Molasses, &c	E. Mingay	Boston, Mass	Nov. 17, 1857	18, 645
Pitcher, Molasses	E. Page	Worcester, Mass	Aug. 7, 1855	13, 400
Pitcher, Molasses	J. A. Parise and C. W. MacCord.	New York, N. Y	Oct. 24, 1865	50, 620
Pitcher, Molasses	C. Reistle	Brooklyn, N. Y	Feb. 23, 1869	87, 201
Pitcher, Molasses	D. C. Ripley	Pittsburgh, Pa	July 15, 1873	140, 793

Index of patents issued from the United States Patent Office from 1790 *to* 1873, *inclusive*—Continued.

Invention.	Inventor.	Residence.	Date.	No.
Pitcher, Molasses	M. Robbins	Cincinnati, Ohio	Aug. 7, 1866	56, 991
Pitcher, Molasses	H. Wright	Pittsburgh, Pa	June 10, 1873	139, 698
Pitcher, Refrigerating	F. D. Hall	Philadelphia, Pa	July 1, 1856	15, 231
Pitcher, Refrigerating	W. W. Lyman	West Meriden, Conn	June 8, 1858	20, 499
Pitcher-shield	F. B. Carleton	Cambridge, Vt	Mar. 1, 1870	100, 374
Pitcher, Sirup	H. Bullard	Middletown, Conn	Feb. 18, 1868	74, 663
Pitcher, Sirup	J. H. Hobbs	Wheeling, W. Va	Apr. 19, 1870	102, 124
Pitcher, Sirup	J. Hyslop and C. E. Phillips	Abington, Mass	July 2, 1867	66, 349
Pitcher, Sirup	D. Scanlin	Rochester, N. Y	Oct. 29, 1687	70, 367
Pitcher, Sirup	J. P. Whipple	Woonsocket, R. I	Apr. 13, 1869	88, 999
Pitcher spout and lid	D. Baker	Harwich, Mass	June 19, 1860	28, 726
Pitcher, Water-cooling	A. Hebbard	New York, N. Y	Oct. 3, 1857	18, 546
Pitchers and other vessels, Mode of constructing water.	K. Goddard	Richmond, N. Y	Nov. 30, 1869	97, 390
Pitchers, Lining metal ice	L. W. Wilbur and A. Tyler	Fitchburgh, Mass	July 2, 1872	128, 690
Pitchers, Manufacture of double-walled ice	D. C. Wilcox	Meriden, Conn	Nov. 3, 1868	83, 747
Pitchers, Method of securing cover to glass	C. Ballinger	Pittsburgh, Pa	Mar. 2, 1869	87, 318
Pitchers, Silvering glass	J. W. Haines	Somerville, Mass	Apr. 4, 1865	47, 101
Pitchers, Top for molasses	G. M. Irwin	Birmingham, Pa	Nov. 7, 1871	120, 748
Pitchers, Top for molasses	G. M. Irwin	Birmingham, Pa	Nov. 26, 1872	133, 317
Pitchers, Top for molasses	G. P. Lang, jr	Allegheny, Pa	Feb. 18, 1873	136, 076
Pitchers, Top for molasses	G. P. Lang, jr	Allegheny City, Pa	Mar. 25, 1873	137, 139
Pitchers, Top for sirup	G. P. Lang and P. Lauster	Allegheny City, Pa	Oct. 31, 1871	120, 387
Pitchers, Top for sirup	T. Smith, jr	Boston, Mass	Nov. 16, 1869	96, 987
Pitchfork	J. Byington	Hinesborough, Vt	Jan. 12, 1813	....
Pitchfork	J. Herald and C. B. Tompkins	Trumansburgh, N. Y	Dec. 6, 1859	26, 354
Pitchfork	I. D. Peck	South Bristol, N. Y	Oct. 11, 1864	44, 652
Pitchfork	C. Perkins	Waterbury, Conn	Jan. 27, 1829	....
Pitchfork	R. Schuyler and W. Crowninshield.	Seneca Falls, N. Y	Jan. 5, 1869	85, 699
Pitchfork	G. Van Sickle	Shorterville, N. Y	Jan. 4, 1870	98, 648
Pitchfork, Horse	W. H. Palmer and W. Crumb	Orleans, N. Y	Feb. 4, 1862	34, 317
Pitchfork, Horse	S. Raymond	Venice, N. Y	Nov. 11, 1862	36, 915
Pitchfork, Horse	S. Raymond	Genoa, N. Y	Nov. 24, 1863	40, 709
Pitman	H. Aufdembrinke	Cincinnati, Ohio	Mar. 11, 1873	136, 696
Pitman	E C. Bacon	New York, N. Y	Dec. 29, 1868	85, 266
Pitman	A. Blaikie and W. Clark	Saint Clair, Mich	Feb. 26, 1856	14, 302
Pitman	J. Butler	Buffalo, N. Y	Nov. 26, 1867	71, 450
Pitman	J. H. Carothers	Lewisburgh, Pa	Sept. 9, 1873	142, 611
Pitman	M. D. Drake	Scituate, R. I	June 22, 1869	91, 613
Pitman	J. H. Holly and J. F. Robertson	Warwick, N. Y	Dec. 30, 1873	145, 947
Pitman	M. Grover	Clyde, Ohio	Jan. 19, 1869	85, 926
Pitman	F. Graham	Rockford, Ill	Jan. 20, 1863	37, 473
Pitman	G. W. Jason	Lodi, Ohio	Mar. 7, 1871	112, 463
Pitman	T. Kealy	Lewisville, Tex	May 17, 1870	103, 196
Pitman	J. H. McGowan	Cincinnati, Ohio	Nov. 8, 1870	109, 138
Pitman	C. H. Perkins	Providence, R. I	Nov. 8, 1870	109, 047
Pitman	J. Swan	Richmond, Ohio	Oct. 7, 1873	143, 475
Pitman	O. F. Thomas	Barnes's Corners, N. Y	Aug. 19, 1873	141, 902
Pitman	T. Welch	Churchville, N. Y	Dec. 17, 1867	72, 344
Pitman	S. Wheeler	Albany, N. Y	Dec. 17, 1872	134, 117
Pitman and cutter-bar connection	W. Green	Ashland, Ohio	Feb. 8, 1870	99, 563
Pitman and mode of attachment to band-wheel	L. N. Rouse	Covington, Ky	Feb. 6, 1872	123, 363
Pitman-box	A. S. Acker	Albion, N. Y	Aug. 14, 1866	57, 061
Pitman-connection	M. L. Ballard	Canton, Ohio	Oct. 30, 1866	59, 310
Pitman-connection	A. Ketchum	Estherville, Iowa	Dec. 12, 1871	121, 789
Pitman-connection	S. T. Lamb	Albany, Ind	Feb. 22, 1870	100, 157
Pitman-connection	W. K. Miller	Canton, Ohio	July 13, 1869	92, 628
Pitman-connection	S. Mahoney and J. Geller	Terre Haute, Ind	Dec. 10, 1872	133, 869
Pitman-connection	A. Neer	Bellefontaine, Ohio	Jan. 30, 1872	123, 285
Pitman-connection	J. A. Shephard	Portsmouth, Ohio	June 24, 1873	140, 312
Pitman-connection for operating cranks	C. W. Reed	Chagrin Falls, Ohio	Aug. 20, 1872	130, 596
Pitman-connection for steam-engines	S. Van Emon	Covington, Ky	Oct. 24, 1871	120, 222
Pitman-coupling	G. W. Clark	Manchester, Ind	Dec. 10, 1867	71, 853
Pitman-coupling	W. J. Keeney	Florence, Ind	Oct. 29, 1867	70, 224
Pitman-coupling	W. J. Keeney	Florence, Ind	Oct. 29, 1867	70, 225
Pitman for presses, punches, &c., Adjustable	N. C. Stiles	Meriden. Conn	July 24, 1866	56, 631
Pitman-head	J. L. Krester	Tustin, Wis	Sept. 27, 1870	107, 788
Pitman-head and crank-wrist connection	W. N. Whiteley	Springfield, Ohio	Nov. 5, 1867	70, 661
Pitman-head and wrist-pin	W. N. Whiteley	Springfield, Ohio	Oct. 22, 1867	70, 143
Pitman-heads, sickles of harvesters, &c., Joint for	I. Poffenberger	Champaign County, Ohio	Apr. 19, 1870	102, 151
Pitman, Reverse-lever	J. and W. W. Porter	Wauconda, Ill	Sept. 17, 1867	68, 900
Pitman-rod	E. S. Blake	Pittsburgh, Pa	July 4, 1871	116, 542
Pitman-rod	S. N. Wate, jr	Danville, Pa	Oct. 14, 1873	143, 648
Pitman-rod connection	E. G. Fish	Colfax, Iowa	Mar. 28, 1871	113, 039
Pitman-rod connection	E. G. Fish	Colfax, Iowa	Jan. 16, 1872	122, 715
Pitmen, Adjusting length of	N. C. Stiles	Middletown, Conn	Jan. 28, 1873	135, 380
Pitmen, Attaching	A. J. Sweeney	Wheeling, W. Va	Sept. 3, 1872	131, 132
Pitmen, Connecting	T. Hall	Northampton, Mass	Apr. 23, 1872	126, 051
Pitmen, &c., Hanging	E. H. Craige	Brooklyn, N. Y	Aug. 21, 1866	57, 292
Pitmen, Securing close joints in	A. Harroun	Perryville, N. Y	July 16, 1872	128, 959
Pitmen with shafts, Connecting	R. Cleaveland	Covington, Pa	Sept. 9, 1873	142, 616
Pivot-bearings	F. C. Lowthorp	Trenton, N. J	Jan. 3, 1860	26, 689
Pivot for seat	J. J. Wilson	New York, N. Y	Apr. 4, 1871	113, 606
Pivoted stump-joint	A. Searls	New York, N. Y	Feb. 11, 1868	74, 434
Placards, &c., Device for suspending and exhibiting	E. Worrell and B. C. Moore	Philadelphia, Pa	Dec. 16, 1873	145, 709
Plain and figured fabrics, Manufacture of	F. W. Norton	Lasswade, Great Britain	Sept. 13, 1853	10, 019
Plaiting-device	W. Walker	Brooklyn, N. Y	Nov. 1, 1870	108, 854
Plaiting-device	W. Walker	Brooklyn, N. Y	Nov. 1, 1870	108, 855
Plaiting-machine	J. F. McKenney	Baltimore, Md	Feb. 2, 1864	41, 441
Plaiting-machine	G. C. Nelson	Petersham, Mass	Oct. 27, 1863	40, 424
Plaiting-machine	J. E. Wheat	Rochester, N. Y	July 16, 1872	129, [illegible]04
Plaiting machine, Box	O. M. Chamberlain	New York, N. Y	Mar 4, 1873	136, 362
Plane	L. A. Alexander	Pittsfield, Mass	July 16, 1872	129, 508
Plane	V. Bitsch	Saint Louis, Mo	Sept. 15, 1868	82, 074
Plane	G. Buckel	Detroit, Mich	Aug. 25, 1868	81, 335
Plane	L. Bundy	Mooer's Forks, N. Y	Nov. 15, 1870	109, 174
Plane	T. O. Callahan	Boston, Mass	Sept. 27, 1870	107, 757

Index of patents issued from the United States Patent Office from 1790 *to* 1873, *inclusive*—Continued.

Invention.	Inventor.	Residence.	Date.	No.
Plane	A. H. Comp	Mount Joy, Pa	Sept. 8, 1868	81, 879
Plane	A. S. Cross	Ripon, Wis	Sept. 10, 1861	33, 240
Plane	W. B. Glover	Boston, Mass	Oct. 25, 1870	108, 586
Plane	C. Jensen	Boston, Mass	May 14, 1872	126, 707
Plane	J. T. Jones	Philadelphia, Pa	May 14, 1836	....
Plane	E. H. Morris	Salem, Ohio	Nov. 8, 1870	109, 037
Plane	E. Odell	Winterset, Iowa	Jan. 19, 1864	41, 317
Plane	F. Smith and I. Carpenter	Lancaster, Pa	Aug. 25, 1868	81, 425
Plane	J. K. P. Smith	Jeffersonville, Ind	May 9, 1871	114, 613
Plane	J. C. Spencer	Phelps, N. Y	May 6, 1873	138, 591
Plane	E. G. Storke	Auburn, N. Y	Oct. 19, 1869	96, 052
Plane	J. Vendrand	Paris, France	June 24, 1862	35, 719
Plane, Accelerating	J. W. Newberry	Avon, N. Y	Nov. 15, 1831	....
Plane and manner of fastening the iron, Carpenter's	P. Meigs	Madison, Wis	Feb. 9, 1831	....
Plane, Bench	L. C. Ashley	Troy, N. Y	Mar. 18, 1856	14, 436
Plane, Bench	J. R. Bailey	Woonsocket, R. I	July 26, 1870	105, 767
Plane, Bench	J. R. Bailey	Woonsocket, R. I	Mar. 14, 1871	112, 675
Plane, Bench	L. Bailey	Boston, Mass	Aug. 6, 1867	67, 398
Plane, Bench	C. S. Beardsley and S. Wood	Auburn, N. Y	May 22, 1849	6, 459
Plane, Bench	B. A. Blandin	Charlestown, Mass	May 7, 1867	64, 477
Plane, Bench	L. C. Bliss	Richmond, Ind	Oct. 17, 1865	50, 530
Plane, Bench	J. Brooks	Boston, Mass	June 11, 1872	127, 842
Plane, Bench	W. H. Brown and D. F. Williams.	Worcester, Mass., and Woonsocket, R. I.	Nov. 11, 1873	144, 381
Plane, Bench	M. Chittenden	Danbury, Iowa	May 21, 1872	127, 026
Plane, Bench	P. S. Foster	Richmond, Me	Jan. 26, 1869	86, 295
Plane, Bench	C. H. Hardy	Boston, Mass	Sept. 24, 1872	131, 544
Plane, Bench	C. H. Hardy	Boston, Mass	Sept. 23, 1873	143, 072
Plane, Bench	W. C. Hopper	Pittsburgh, Pa	Jan. 16, 1855	12, 234
Plane, Bench	S. C. Howes	South Chatham, Mass	Feb. 17, 1863	37, 694
Plane, Bench	G. W. Huber and A. E. Flickinger.	Norwalk, Ohio	Dec. 2, 1873	145, 106
Plane, Bench	H. C. Hunt	Ottumwa, Iowa	Apr. 24, 1860	27, 983
Plane, Bench	H. L. Kendall	Baltimore, Md	June 8, 1858	20, 493
Plane, Bench	J. Lehner	Galena, Ill	Nov. 19, 1872	133, 162
Plane, Bench	W. S. Loughborough	Rochester, N. Y	May 10, 1859	23, 928
Plane, Bench	E. Mathew	Morgantown, Va	Mar. 4, 1856	14, 363
Plane, Bench	G. Müllear	San Francisco, Cal	Oct. 10, 1865	50, 378
Plane, Bench	N. Palmer	Auburn, N. Y	May 14, 1867	64, 790
Plane, Bench	S. W. Palmer and E. G. Storke	Auburn, N. Y	Nov. 28, 1871	121, 406
Plane, Bench	C. H. Sawyer	Hollis, Me	Apr. 16, 1867	63, 948
Plane, Bench	W. H. Taber	Lowell, Mass	Feb. 28, 1865	46, 614
Plane, Bench	C. E. Torrance	Holyoke, Mass	Jan. 2, 1872	122, 339
Plane, Bench	L. D. Tredway	Saint Louis, Mo	Jan. 25, 1870	99, 275
Plane, Bench	T. Vaughan	Boston, Mass	July 23, 1872	129, 695
Plane, Bench	R. Washburn	Ramapo, N. Y	June 7, 1864	43, 053
Plane, Bench	S. Williams	Philadelphia, Pa	June 28, 1864	43, 360
Plane, Bench	T. D. Worrall	Lowell, Mass	June 23, 1857	17, 657
Plane, Beveling	H. W. Lewis	Bath, N. Y	Jan. 13, 1852	8, 655
Plane, Beveling	M. J. Wheeler, G. W. Rogers, H. W. Pierce, and M. B. Tidey.	Dundee, N. Y	July 4, 1854	11, 235
Plane-bit	H. Harris	Gorham, N. Y	Sept. 18, 1855	13, 575
Plane-bits, Device for securing	T. M. Richardson	Stockton, Me	Oct. 2, 1860	30, 248
Plane-bits, Machine for whetting	J. M. Gilstrap	Washington County, Ark	Aug. 11, 1857	17, 965
Plane-bits, Securing	T. D. Worrall	Boston, Mass	May 27, 1856	14, 979
Plane, Boot	A. V. Hill and R. Arnold	Hamburgh, N. Y	Apr. 11, 1848	5, 501
Plane, Carpenter's	L. Bailey	Boston, Mass	Dec. 24, 1867	72, 443
Plane, Carpenter's	J. R. and W. Brown	Boston, Mass	Feb. 28, 1871	112, 218
Plane, Carpenter's	E. M. Chapin and S. Rust	Pine Meadow, Conn	Mar. 31, 1868	76, 051
Plane, Carpenter's	O. R. Chaplin	Boston, Mass	May 7, 1872	126, 519
Plane, Carpenter's	H. N. Frederick	Hancock, N. Y	Sept. 19, 1871	119, 133
Plane, Carpenter's	H. A. Holt	Wilton, N. H	Jan. 9, 1872	122, 609
Plane, Carpenter's	C. G. Miller	Brattleborough, Vt	June 28, 1870	104, 753
Plane, Carpenter's	W. Miller	Boston, Mass	Feb. 21, 1871	112, 062
Plane, Carpenter's	G. Müller	New York, N. Y	May 29, 1866	55, 207
Plane, Carpenter's	G. Müller	New York, N. Y	Oct. 26, 1869	96, 258
Plane, Carpenter's	O. Nichols	Lowell, Mass	Mar. 10, 1857	16, 805
Plane, Carpenter's	R. Phillips	Gardiner, Me	Aug. 13, 1867	67, 671
Plane, Carpenter's	R. Phillips	Boston, Mass	Aug. 30, 1870	106, 868
Plane, Carpenter's	R. Phillips	Boston, Mass	Oct. 24, 1871	120, 212
Plane, Carpenter's	Z. Phillips	Dixon, Ill	May 10, 1870	102, 966
Plane, Carpenter's	G. D. Spooner and L. N. Johnson	Rutland and Brandon, Vt	Aug. 6, 1867	67, 458
Plane, Carpenter's	J. B. Tarr	Chicago, Ill	Sept. 22, 1868	82, 450
Plane, Carpenter's	M. B. Tidey	Ithaca, N. Y	Mar. 2, 1857	16, 889
Plane, Carpenter's	J. A. Trant	New Britain, Conn	Mar. 4, 1873	136, 469
Plane, Carpenter's	C. E. Tucker	Boston, Mass	July 26, 1870	105, 869
Plane, Carpenter's	R. Wellford and J. H. Deas	Philadelphia, Pa	May 14, 1821	....
Plane, Carpenter's	J. Woodville	Cincinnati, Ohio	Nov. 6, 1866	59, 498
Plane, Carpenter's	S. W. Woodward	Buffalo, N. Y	Apr. 27, 1869	89, 369
Plane, Carpenter's grooving	T. Duval	Hartford, Conn	Nov. 23, 1869	97, 177
Plane, Combination	A. Johnson	San Francisco, Cal	Dec. 30, 1873	146, 004
Plane, Cooper's crozing	H. and J. C. Taylor	Cincinnati, Ohio	Oct. 2, 1855	13, 626
Plane, Core-box	E. W. Lewis	Philadelphia, Pa	Dec. 4, 1866	60, 206
Plane, Crozing	S. G. Crane	Rochester, N. Y	Jan. 19, 1858	19, 130
Plane, Crozing and chamfering	A. M. Strattan	Ladoga, Ind	Aug. 12, 1873	141, 828
Plane, Dado	R. H. Dorn	Port Henry, N. Y	July 16, 1872	129, 010
Plane, Double-iron trying	W. B. Reynolds	Saint Clairsville, Ohio	July 7, 1832	....
Plane, Floor	C. E. Barlow	Philadelphia, Pa	Mar. 9, 1858	19, 539
Plane for bevel edges	W. W. Blye	De Ruyter, N. Y	Apr. 10, 1849	6, 304
Plane for boards, Bevel wheel	J. Newhall	Washington, Me	Jan. 25, 1831	....
Plane for crossing railways, Revolving	J. P. Fairlamb	New Castle, Del	July 7, 1830	....
Plane for cutting blind-slats	J. L. Bess and A. Hagny	Keokuk, Iowa	July 30, 1867	67, 157
Plane for cutting blind-slats	C. Kupfer	Madison, Wis	Sept. 1, 1868	81, 795
Plane for cutting blind-slats	R. E. Lowe	Upper Alton, Ill	Sept. 22, 1868	82, 424
Plane for cutting blind-slats	J. H. Miller	Oskaloosa, Iowa	Sept. 8, 1868	81, 929
Plane for finishing grooves in patterns	J. P. Robinson	Matteawan, N. Y	Dec. 18, 1855	13, 957
Plane for jointing table-leaves	T. P. Granger	Pocatonica, Ill	Jan. 12, 1864	41, 258
Plane for making blind-slats	E. K. Thomas and H. H. Andresen.	Rock Island, Ill., and Davenport, Iowa.	Apr. 10, 1866	53, 899

Index of patents issued from the United States Patent Office from 1790 *to* 1873, *inclusive*—Continued.

Invention.	Inventor.	Residence.	Date.	No.
Plane for scraping	J. Jones	Newark, N. J	Jan. 28, 1873	135, 341
Plane for shaving leather	F. Woodward	New Lisbon, N. Y	July 27, 1812	....
Plane for shaving whalebone	J. A. Sevey	Boston, Mass	Dec. 28, 1869	98, 305
Plane for tonguing and grooving boards	J. A. Woodbury	Boston, Mass	June 11, 1850	7, 432
Plane, Groove	E. W. Carpenter	Lancaster, Pa	Jan. 30, 1830	....
Plane, Groove	J. Longanecker and C. Myers	Lancaster County, Pa	Feb. 7, 1829	....
Plane, Grooving	J. Herman	Lancaster, Ohio	Aug. 27, 1835	....
Plane-guide	M. Garland	West Eau Claire, Wis	Aug. 30, 1870	106, 808
Plane-guide	J. Woodville	Cincinnati, Ohio	Mar. 23, 1869	88, 109
Plane, Hand	B. F. Bee	Harwich, Mass	Nov. 11, 1851	8, 503
Plane, Hand	S. S. Dodge	Sunapee, N. H	May 10, 1859	23, 978
Plane, Hand	B. Holly	Seneca Falls, N. Y	July 6, 1852	9, 094
Plane, Horizontal knife and roller	U. Emmons	New York	Mar. 6, 1832	....
Plane, Inclined circular	M. Isaacs	New York	Nov. 17, 1819	....
Plane-iron	I. Almy and S. A. Drake	Covert, N. Y	Apr. 25, 1871	114, 085
Plane-iron	W. H. Eckert	Syracuse, N. Y	Apr. 23, 1867	64, 001
Plane-iron	W. Hovey	Boston, Mass	Mar. 10, 1830	....
Plane-iron	S. Markee	Auburn, N. Y	Apr. 30, 1867	64, 341
Plane-iron	E. Quast	Jerseyville, Ill	Dec. 9, 1873	145, 311
Plane-iron	P. Williamson	Baltimore, Md	Sept. 9, 1825	....
Plane iron, Bench	I. H. A. Bleckman	Ronsdorf, Prussia	Nov. 6, 1855	13, 745
Plane iron, Carpenter's	A. M. Cross	Necedah, Wis	Dec. 3, 1872	133, 632
Plane-iron, Double	F. Beals	Pittsfield, Mass	Mar. 16, 1852	8, 796
Plane-iron roller	A. R. Reynolds	Auburn, N. Y	Nov. 9, 1869	96, 728
Plane-iron, scythe, &c	D. Pettibone	Philadelphia, Pa	May 6, 1813	....
Plane-iron sharpener	J. Turner	Cambridgeport, Mass	June 26, 1860	28, 946
Plane-irons and regulating the throats of planes, Adjusting the position of.	E. W. Carpenter	Lancaster, Pa	Mar. 27, 1849	6, 226
Plane-irons, Cap for	C. N. Tuttle	Auburn, N. Y	Nov. 1, 1870	108, 946
Plane-irons, Device for adjusting	L. Bailey	Winchester, Mass	June 22, 1858	20, 615
Plane-irons, Fastening and adjusting	L. Bliss	Truxton, N. Y	June 16, 1846	4, 576
Plane-irons, Forming projections on caps of	N. B. Reynolds	Auburn, N. Y	June 11, 1867	65, 604
Plane-irons in their sockets, Securing and adjusting.	W. Stoddard	Lowell, Mass	June 23, 1857	17, 645
Plane-irons in their stocks, Holding and adjusting.	W. W. Chipman	Lowell, Mass	June 23, 1857	17, 618
Plane-irons to the stock of bench-planes, Securing	L. Bailey	Winchester, Mass	Aug. 31, 1858	21, 311
Plane-irons to their stocks, Securing	P. A. Gladwin	Boston, Mass	Feb. 16, 1858	19, 359
Plane, Joiner's	G. C. Beckwith	Boston, Mass	Jan. 25, 1870	99, 137
Plane, Joiner's	W. A. Cole	New York, N. Y	June 6, 1848	5, 629
Plane, Joiner's	A. Gray	Naples, Me	June 11, 1867	65, 562
Plane, Joiner's	B. I. Lane	Newburyport, Mass	May 12, 1857	17, 286
Plane, Joiner's	J. Lashbrooks	Owensborough, Ky	May 19, 1857	17, 332
Plane, Joiner's	J. F. Palmer	Auburn, N. Y	Feb. 3, 1857	16, 569
Plane, Joiner's	B. F. Shellabarger	Mifflintown, Pa	Mar. 28, 1848	5, 486
Plane, Joiner's	G. A. Warren	North Bridgewater, Mass	Feb. 14, 1871	111, 890
Plane, Joiner's	T. D. Worrall	Lowell, Mass	Aug. 4, 1857	17, 951
Plane, Joiner's beveling	T. A. Chandler	Rockport, Ill	Mar. 16, 1858	19, 620
Plane, Match	J. Edwards	New York, N. Y	June 10, 1873	139, 710
Plane, Match	C. E. Marshall	Boston, Mass	Oct. 8, 1872	131, 959
Plane, Match	C. G. Miller	New Britain, Conn	Aug. 19, 1873	142, 037
Plane, Metallic	J. A. Baines	New York, N. Y	Aug. 5, 1873	151, 535
Plane, Metallic	J. F. Baldwin	Boston, Mass	Nov. 25, 1873	144, 823
Plane, Miter	J. Sawyer	Moravia, N. Y	Dec. 4, 1866	60, 265
Plane, Molding	J. E. Donaldson	Montezuma, Ind	May 16, 1871	114, 779
Plane, Molding	C. Fleming	Ypsilanti, Mich	Sept. 11, 1860	29, 962
Plane, Molding	A. W. Maxwell	Milton, Pa	Feb. 9, 1869	86, 851
Plane, Molding	E. H. Morris	Canton, Ohio	Mar. 21, 1871	112, 949
Plane, Molding	A. S. Robertson	Boston, Mass	Jan. 21, 1873	135, 046
Plane, Multiform-molding	T. Worrall	Mount Holly, N. J	Aug. 29, 1854	11, 635
Plane or scraper, Box	A. F. Cushman	Hartford, Conn	Feb. 9, 1864	41, 571
Plane, Plow	I. White	Philadelphia, Pa	Jan. 9, 1834	....
Plane, Rabbet	G. M. Darley	Nebraska City, Nebr	July 2, 1872	128, 470
Plane, Rabbet	F. Smith	Boston, Mass	Sept. 23, 1873	143, 101
Plane, Revolving	L. Hedge	Brattleborough, Vt	June 28, 1836	....
Plane, Revolving timber	D. N. Smith	Warwick, Mass	Dec. 28, 1826	....
Plane, Saw rabbet	D. D Whitker	Hudson, N. Y	Feb. 6, 1866	52, 478
Plane, Shoemaker's edge	F. Killbrith	Pembroke, Mass	July 13, 1858	20, 882
Plane, Splint	D. E. and A. A. Aiken	Adrian, Mich	May 5, 1868	77, 434
Plane, Splint	J. Dempsey	Richmond, Ind	Nov. 28, 1865	51, 153
Plane, Splint	P. N. Drake and D. Drummond	McGregor, Iowa	Sept. 27, 1870	107, 765
Plane, Splint	H. Ogborn	Richmond, Ind	Nov. 14, 1865	50, 947
Plane, Splint	H. L. Weagant	Morrisburg, Canada	Oct. 14, 1873	143, 737
Plane-stock	S. M. Adams	Fitchburgh, Mass	June 4, 1872	127, 541
Plane-stock	L. Bailey	New Britain, Conn	Mar. 28, 1871	113, 003
Plane-stock	G. F Evans	Norway, Me	Jan. 28, 1862	34, 248
Plane-stock	G. F. Evans	Norway, Me	Mar. 22, 1864	41, 983
Plane-stock	J. Katz	Cincinnati, Ohio	Apr. 26, 1870	102, 406
Plane-stock	J. B. Thomas	Cincinnati, Ohio	Mar. 11, 1856	14, 423
Plane stock, Bench	J. Bryant	Brooklyn, N. Y	June 16, 1857	17, 553
Plane stock, Bench	G. E. Davis	Lowell, Mass	May 1, 1855	12, 787
Plane stock, Bench	J. Gorham	Bairdstown, Ga	Apr. 19, 1859	23, 678
Plane-stock, Cast-iron	W. Foster	Washington, D. C	Nov. 24, 1843	3, 355
Plane-stock, Cast-iron	H. Knowles	Colchester, Conn	Aug. 24, 1827	....
Plane-stocks and mouth-pieces, Hanging	M. G. Hubbard	New York, N. Y	July 3, 1855	13, 174
Plane-stocks, Hanging	M. G. Hubbard	New York, N. Y	Sept. 5, 1854	11, 648
Plane-stocks, Machine for cutting throats of carpenter's.	H. S. Dewey	Bethel, Vt	Mar. 31, 1857	16, 954
Plane-stocks, Machine for dressing throats in	F. B. Marble	Columbus, Ohio	Feb. 14, 1865	46, 372
Plane-stocks, Machine for mortising	J. Richards	Columbus, Ohio	Feb. 14, 1865	46, 391
Plane-stocks, Machine for mortising	J. Richards	Columbus, Ohio	Feb. 14, 1865	46, 392
Plane, Surfacing	G. E. Franklin	Natick, Mass	May 6, 1873	138, 625
Plane turner, Sliding	J. Sparrow	Portland, Me	Dec. 26, 1827	....
Plane, Wood	P. Viccellio	New Orleans, La	June 29, 1869	91, 990
Planes, Adjusting bits of carpenter's	T. D. Worrall	Lowell, Mass	Dec. 23, 1856	16, 309
Planes, Adjusting size of mouth in	T. J. Tolman	South Scituate, Mass	Jan. 13, 1857	16, 412
Planes, Attaching adjustable handles to joiner's	T. D. Worrall	Lowell, Mass	Sept. 29, 1857	18, 312
Planes, Bevel-attachment for bench	L. O. Fairbanks	Nashua, N. H	Mar. 19, 1861	31, 707
Planes, Constructing screw-arms of	E. W. Carpenter	Lancaster, Pa	Feb. 6, 1838	594

Index of patents issued from the United States Patent Office from 1790 to 1873, inclusive—Continued.

Invention.	Inventor.	Residence.	Date.	No.
Planes or jointers, Making irons for	C. E. West	Colchester, Conn	Jan. 10, 1827	....
Planes, Setting the bit in bench	L. Sanford	East Solon, N. Y	Nov. 26, 1844	3,838
Planes, Stock for smoothing	J. F. W. Erdmann	Philadelphia, Pa	Aug. 4, 1857	17,921
Planes, Tonguing and grooving hand	P. A. Gladwin	Boston, Mass	June 9, 1857	17,541
Planer, Crank	W. H. Warren	Worcester, Mass	Jan. 14, 1868	73,412
Planer cutter-stock, Metal	J. Mason	Paterson, N. J	July 22, 1856	15,379
Planer for fellies, Rotary	C. H. Denison	Green River, Vt	Feb. 12, 1856	14,228
Planer, Iron	C. Carr	Boston, Mass	Jan. 19, 1869	85,904
Planer-knife grinder	J. Grant	Northampton, Mass	Mar. 10, 1868	75,411
Planer, Metal	E. H. Cleveland	Worcester, Mass	July 29, 1856	15,413
Planer, Metal	J. H. Sternbergh	Reading, Pa	Nov. 5, 1867	70,479
Planer, Metal	W. H. Warren	Worcester, Mass	June 20, 1871	116,243
Planer, Nut	J. T. Campbell	Altoona, Pa	Nov. 24, 1868	84,410
Planer, Veneer	G. Williamson	Newark, N. J	Jan. 1, 1861	31,070
Planers and lathes, Belt-shifting mechanism for	T. Reeve	Brooklyn, N. Y	Sept. 16, 1873	142,814
Planers, Arrangement of parts in rotary	A. J. Kramer	Marion, Iowa	Dec. 11, 1860	30,903
Planers, Attachment for	A. S. Hewlett	Sebastopol, Cal	Nov. 28, 1871	121,248
Planers, Cutter-head for	J. More	New York, N. Y	Oct. 4, 1870	107,943
Planers, Cutter-head for rotary	U. Nixon	Adrian, Mich	Nov. 6, 1855	13,757
Planers, Feed for iron	H. B. Weaver	Hartford, Conn	Apr. 5, 1870	101,550
Planers, Feeding-roller for rotary	A. T. Serrell	New York, N. Y	Mar. 12, 1861	31,678
Planers, Gearing for metal	W. M. Barr	Williamsport, Pa	Nov. 8, 1870	108,955
Planers, Holder for lathe	J. P. Manton	Providence, R. I	May 26, 1868	78,387
Planers, Shipping-apparatus for metal	D. Slate	Hartford, Conn	Apr. 20, 1869	89,177
Planers, Tool-holder for	J. S. Eltenborough	Easton, Pa	Dec. 24, 1872	134,198
Planetarium	L. Allen	Pekin, Ill	Sept. 20, 1859	25,476
Planetarium	J. Davis	Allegheny City, Pa	Dec. 24, 1867	72,612
Planetarium	B. O. Swain	Annisquam, Mass	Aug. 21, 1849	6,656
Planing and joining, tonguing and grooving, &c	J. McGregor, jr	Wilton, N. Y	Aug. 28, 1833	....
Planing and matching machine	L. B. Grover and D. Lawton	Springwater, N. Y	Dec. 23, 1834	....
Planing and matching machine, Combined	W. H. Doane and W. E. London	Cincinnati, Ohio	Nov. 11, 1862	36,901
Planing and matching machine, Rotary	C. B. Morse	Rhinebeck, N. Y	Jan. 9, 1855	12,211
Planing and milling machine, Cutting-tool for	E. T. Prindle	Aurora, Ill	Apr. 28, 1868	77,211
Planing and molding machine	F. Douglas	Norwich, Conn	July 26, 1870	105,657
Planing and molding machine	J. P. Grosvenor	Lowell, Mass	Sept. 15, 1868	82,143
Planing and other machines, Device for changing speed and reversing motion of	W. Heckert	Newcastle, Pa	Apr. 16, 1872	125,676
Planing and sawing machines, Dust and shaving conveyer for	D. R. Miller	Harrisburgh, Pa	Oct. 2, 1866	58,456
Planing and slotting machine	C. A. Meinhard	Fort Wayne, Ind	Nov. 26, 1867	71,401
Planing boards	E. Lane	Cincinnati, Ohio	Oct. 26, 1831	....
Planing boards	J. Percival	Philadelphia, Pa	July 20, 1831	....
Planing boards and planks and shaving shingles, Machine for	W. Badger	Madison County, Miss	Mar. 2, 1813	....
Planing boards, grinding and polishing marble, and sawing tenons with circular saws	M. Lancaster	Philadelphia, Pa	Apr. 5, 1832	....
Planing boards, Machine for	C. Ritter	Beaver, Pa	Aug. 11, 1809	....
Planing by machinery	J. Bennock	Boston, Mass	June 1, 1805	....
Planing curved surfaces, Machine for	J. P. Grosvenor	Lowell, Mass	Dec. 27, 1859	26,584
Planing-cutter head	J. M. Patton and W. F. Fergus	Philadelphia, Pa	July 6, 1852	9,100
Planing-cutter head, Rotary	H. D. Stover	New York, N. Y	Aug. 14, 1860	29,633
Planing-cutter, Rotary	H. H. Baker	New Market, N. J	Aug. 18, 1857	18,007
Planing-cutter, Rotary	J. Sperry	New York, N. Y	Oct. 12, 1858	21,782
Planing-cutter, Rotary	H. D. Stover	Boston, Mass	May 19, 1857	17,343
Planing-cutter, Rotary	H. D. Stover and J. W. Bicknell	Boston, Mass	Aug. 30, 1859	25,286
Planing, edging, and grooving boards	J. Tees	Kensington, Pa	June 13, 1831	....
Planing-engine, Circular	U. Emmons	New York	Sept. 10, 1829	....
Planing irregular forms	A. Goodman and J. W. Gibbs	Dana, Mass	Nov. 21, 1848	5,926
Planing irregular surfaces, Machine for	J. H. Nelson	Oskaloosa, Iowa	Sept. 28, 1858	21,618
Planing-knife, Rotary	E. Webber	Gardiner, Me	May 2, 1854	10,871
Planing-knives, Arrangement of rotary	D. N. Hurlbut	Utica, N. Y	Mar. 18, 1856	14,455
Planing-knives while grinding, Holder for	L. Jennings	Ewing, Mass	Mar. 16, 1858	19,641
Planing lumber, Fastening the cutter in machine for	B. Becknell	Cincinnati, Ohio	Mar. 21, 1845	3,961
Planing lumber, Machine for	F. Walcott and J. H. Hutchinson	Boston, Mass	July 16, 1839	1,244
Planing lumber "out of wind," Machine for	S. S. Gray	South Boston, Mass	Aug. 22, 1854	11,582
Planing-machine	E. G. Allen	Boston, Mass	Apr. 17, 1849	6,365
Planing-machine	E. G. Allen	Boston, Mass	Oct. 23, 1849	6,809
Planing-machine	M. L. Andrew	Cincinnati, Ohio	Feb. 15, 1870	99,808
Planing-machine	J. E. Andrews	Boston, Mass	Nov. 21, 1845	4,283
Planing-machine	J. Atkins	Augusta, Me	June 24, 1873	140,235
Planing-machine	N. Barlow	Saint Louis, Mo	May 27, 1851	8,125
Planing-machine	D. Barnum and T. J. Wells	New York, N. Y	Mar. 13, 1849	6,185
Planing-machine	C. B. Beall and J. K. Leach	Hamilton, Ohio	June 18, 1872	128,007
Planing-machine	G. W. Beardslee	Buffalo, N. Y	May 20, 1851	8,098
Planing-machine	G. W. Beardslee	Albany, N. Y	Nov. 4, 1851	8,497
Planing-machine	J. D. Beers and I. Winslow	Philadelphia, Pa	Feb. 25, 1851	7,949
Planing-machine	R. and C. S. Bixby and J. Garst	Dayton, Ohio	May 13, 1851	8,086
Planing-machine	A. T. Boon and J. Collins	Galesburgh, Ill	June 7, 1864	43,001
Planing-machine	O. Bridgeman	Addison, N. Y	Dec. 11, 1866	60,332
Planing-machine	A. L. Brooks	Lowell, Mass	Jan. 7, 1835	....
Planing-machine	B. Brown	Burlington, Vt	Oct. 9, 1845	4,225
Planing-machine	J. B. Brown	Lowell, Mass	Dec. 7, 1869	97,476
Planing-machine	W. W. Carey and G. W. Harris	Lowell, Mass	Jan. 25, 1870	99,157
Planing-machine	T. L. Carley and M. Broughton	Homer, N. Y	Apr. 26, 1870	102,219
Planing-machine	J. Casson	Sheffield Parish, England	May 4, 1869	89,556
Planing-machine	H. Climer and J. D. Riley	Cincinnati, Ohio	Nov. 27, 1866	59,966
Planing-machine	J. Close	Brooklyn, N. Y	Sept. 19, 1865	49,981
Planing-machine	G. W. Cole	Canton, Ill	Dec. 28, 1869	98,231
Planing-machine	J. Coppinger	Baltimore, Md	Oct. 31, 1809	....
Planing-machine	L. Curtis	Sherburne, N. Y	June 16, 1836	....
Planing-machine	J. D. Dale	Philadelphia, Pa	Dec. 8, 1857	18,806
Planing-machine	T. E. Daniels	Worcester, Mass	Dec. 23, 1834	....
Planing-machine	J. Dixon	New York, N. Y	May 8, 1866	54,511
Planing-machine	W. H. Doane, G. V. Orton, and W. L. London	Cincinnati, Ohio	May 21, 1867	64,849
Planing-machine	L. Doolittle	Bushville, N. Y	June 30, 1863	39,037

Index of patents issued from the United States Patent Office from 1790 *to* 1873, *inclusive*—Continued.

Invention.	Inventor.	Residence.	Date.	No.
Planing-machine	F. Douglas	Norwich, Conn	Oct. 12, 1869	95, 782
Planing-machine	F. Douglas	Norwich, Conn	Aug. 9, 1870	106, 139
Planing-machine	J. H. Draper	Mooresville, Ind	Jan. 5, 1869	85, 648
Planing-machine	G. B. Durkee	Chicago, Ill	May 7, 1872	126, 381
Planing-machine	C. Emmons	New York	June 27, 1848	5, 648
Planing-machine	B. Fitts	Worcester, Mass	July 31, 1860	29, 369
Planing-machine	B. Fitts	Newark, N. J	Dec. 24, 1867	72, 618
Planing-machine	B. Fitts	Newark, N. J	Dec. 24, 1867	72, 619
Planing-machine	M. R. Fletcher	Cambridge, Mass	Feb. 11, 1868	74, 333
Planing-machine	A. H. Frank and G. Spire	Buffalo, N. Y	Sept. 30, 1873	143, 232
Planing-machine	A. Fuller	Milford, N. H	June 5, 1866	55, 272
Planing-machine	T. F. Fuller	Bristol, Pa	Oct. 25, 1832	....
Planing-machine	J. Garfield	Groton, Mass	July 24, 1866	56, 548
Planing-machine	I. Gay	Dunstable, N. H	June 25, 1836	....
Planing-machine	J. Goodrich and H. J. Colburn	Fitchburgh, Mass	Feb. 7, 1871	111, 632
Planing-machine	L. Gould	Norwich, Conn	Apr. 29, 1873	138, 244
Planing-machine	L. Gould	Norwich, Conn	Nov. 4, 1873	144, 199
Planing-machine	S. S. Gray	Boston, Mass	Jan. 24, 1860	26, 902
Planing-machine	W. H. Gray	Boston, Mass	Sept. 2, 1873	142, 460
Planing-machine	J. Griffen	Phœnixville, Pa	Mar. 19, 1872	124, 813
Planing-machine	J. S Graham	Rochester, N. Y	Jan. 9, 1872	122, 599
Planing-machine	E. P. Halsted	Worcester, Mass	May 11, 1869	89, 990
Planing-machine	S. M. Hamilton	Baltimore, Md	Sept. 1, 1868	81, 776
Planing-machine	S. M. Hamilton	Baltimore, Md	Sept. 8, 1868	81, 898
Planing-machine	W. D. Hatch	South Antrim, N. H	Oct. 29, 1867	70, 208
Planing-machine	J. Hawkins	Philadelphia, Pa	Feb. 19, 1806	....
Planing-machine	A. M. Hills	Lowell, Mass	Apr. 5, 1870	101, 462
Planing-machine	J. Howarth	Salem, Mass	Mar. 23, 1852	8, 823
Planing-machine	C. L. and L. P. Hoyt	Aurora, Ill	Oct. 29, 1872	132, 663
Planing-machine	A. G. Hull	Brooklyn, N. Y	Sept. 20, 1837	385
Planing-machine	H. Jeter	Lexington, Ky	Oct. 30, 1849	6, 829
Planing-machine	A. Judson	Brooklyn, N. Y	Sept. 20, 1870	104, 502
Planing-machine	S. U. King	Windsor, Vt	Nov. 1, 1864	44, 870
Planing-machine	H. Knowles	Washington, D. C	Apr. 10, 1849	6, 294
Planing-machine	B. Kugler	Philadelphia, Pa	June 3, 1833	....
Planing-machine	H. Law	Wilmington, N. C	Apr. 10, 1849	6, 309
Planing-machine	H. A. Lee	Worcester, Mass	June 13, 1865	48, 185
Planing-machine	H. A. Lee	Worcester, Mass	Nov. 7, 1871	120, 589
Planing-machine	C. Levey	Toronto, Canada	Mar. 21, 1871	112, 821
Planing-machine	T. Llewellyn	London, England	Nov. 1, 1870	108, 801
Planing-machine	A. Lockwood	Chicago, Ill	July 22, 1856	15, 403
Planing-machine	W. N. Manning	Salem, Mass	July 10, 1860	29, 089
Planing-machine	G. H. Mansfield	Concord, N. H	May 10, 1870	102, 842
Planing-machine	G. H. Mansfield	Concord, N. H	Sept. 19, 1871	119, 163
Planing-machine	P. M. Martz	Marion County, Ind	June 25, 1836	....
Planing-machine	C. E. McBeth, F. Bentel, W. C. Margedant, and H. Climer.	Hamilton, Ohio, and Muscatine, Iowa.	Oct. 31, 1871	120, 448
Planing-machine	H G. McDuffe	Bradford, Vt	Apr. 21, 1868	77, 067
Planing-machine	J. McGregor, jr	Wilton, N. Y	Jan. 9, 1838	557
Planing-machine	J. McGregor, jr	Savage Factory, Md	July 15, 1840	1, 690
Planing-machine	W. B. McIver	Worcester, Mass	Mar. 27, 1866	53, 466
Planing-machine	McLaughlin and Hill	Sunderland, Vt	Oct. 28, 1835	....
Planing-machine	R. N. Meriam	Worcester, Mass	Nov. 5, 1867	70, 592
Planing-machine	F. B. Miles	Philadelphia, Pa	Sept. 2, 1873	142, 493
Planing-machine	J. Miller	New York, N. Y	Sept. 2, 1873	142, 495
Planing-machine	C. B. Morse	Rhinebeck, N. Y	May 13, 1856	14, 880
Planing-machine	E. Myers	Cincinnati, Ohio	Nov. 26, 1867	71, 403
Planing-machine	S. W. Nelson	Worcester, Mass	Sept. 16, 1873	142, 936
Planing-machine	D. Niles	Fly Creek, N. Y	July 10, 1866	56, 254
Planing-machine	N. G. Norcross	Middlesex County, Mass	Feb. 12, 1850	7, 087
Planing-machine	N. G. Norcross	Lowell, Mass	June 22, 1852	9, 058
Planing-machine	H. Osgood	Waterville, Me	Dec. 9, 1856	16, 185
Planing-machine	J. F. Ostrander	New York, N. Y	Sept. 3, 1850	7, 621
Planing-machine	G. T. Pearsall	Apalachin, N. Y	Aug. 9, 1870	106, 281
Planing-machine	C. H. Peck and C. Hicks	Saint Louis, Mo	Apr. 24, 1849	6, 392
Planing-machine	C. R. Penfield	Lockport, N. Y	Oct. 18, 1864	44, 774
Planing-machine	R. H. Pindell	Fayette County, Ky	Oct. 4, 1853	10, 100
Planing-machine	F. K. Plumbly	Buffalo, N. Y	Oct. 13, 1868	83, 089
Planing-machine	F. J. Plummer	Worcester, Mass	July 9, 1867	66, 519
Planing-machine	J. Rankin	Binghamton, N. Y	Nov. 25, 1873	145, 012
Planing-machine	L. Read	North Brookfield, N. Y	May 14, 1867	64, 706
Planing-machine	J. Reihm	Savage Factory, Md	Nov. 1, 1827	....
Planing-machine	J. Richards	Philadelphia, Pa	May 17, 1870	103, 080
Planing-machine	S. M. Richardson	Worcester, Mass	Jan. 1, 1867	60, 787
Planing-machine	G. J. Rugg	Worcester, Mass	Jan. 15, 1867	61, 264
Planing-machine	J. J. Russ	Worcester, Mass	Aug. 14, 1866	57, 191
Planing-machine	J. J. Russ	Worcester, Mass	June 27, 1871	116, 496
Planing-machine	J. B. Schenck	Matteawan, N. Y	Feb. 2, 1869	86, 460
Planing-machine	J. B. Schenck	Matteawan, N. Y	Feb. 2, 1869	86, 461
Planing-machine	J. B. Schenck	Matteawan, N. Y	Jan. 18, 1870	99, 008
Planing-machine	J. B. Schenck	Matteawan, N. Y	Feb. 8, 1870	99, 599
Planing-machine	W. Sellers	Philadelphia, Pa	June 19, 1866	55, 723
Planing-machine	H. F. Shaw	West Roxbury, Mass	Jan. 29, 1867	61, 573
Planing-machine	T. Shaw	Philadelphia, Pa	Feb. 18, 1873	135, 941
Planing-machine	J. Sheldon	New Haven, Conn	Jan. 26, 1847	4, 941
Planing-machine	J. Sheldon and J. S. Barden	New Haven, Conn	Apr. 17, 1849	6, 339
Planing-machine	A. H. Shipman	Albany, N. Y	May 6, 1873	138, 703
Planing-machine	L. T. Smart	Centre Ossipee, N. H	Sept. 5, 1871	118, 754
Planing-machine	H. B. Smith	Lowell, Mass	Sept. 26, 1865	50, 178
Planing-machine	H. B. Smith	Lowell, Mass	Oct. 24, 1865	50, 637
Planing-machine	D. H. Southworth	New York, N. Y	Dec. 10, 1850	7, 827
Planing-machine	D. H. Southworth	New York, N. Y	Feb. 18, 1851	7, 942
Planing-machine	C. A. Spring and W. H. Derick	Kensington, Pa	Apr. 3, 1849	6, 249
Planing-machine	D. Stearns	Rome, N. Y	Mar. 16, 1852	8, 808
Planing-machine	F. Stedman	Aquackanckin, N. J	Aug. 17, 1835	....
Planing-machine	H. D. Stover	New York, N. Y	Aug. 21, 1860	29, 728
Planing-machine	H. D. Stover	New York, N. Y	Sept. 4, 1860	29, 923
Planing-machine	H. D. Stover	New York, N. Y	July 23, 1861	32, 904

Index of patents issued from the United States Patent Office from 1790 *to* 1873, *inclusive*—Continued.

Invention.	Inventor.	Residence.	Date.	No.
Planing machine	H. D. Stover	New York, N. Y	Dec. 8, 1868	84,847
Planing-machine	T. J. Talbot	Milford, N. H.	May 1, 1866	54,441
Planing-machine	J. B. Tarr	Chicago, Ill	June 5, 1866	55,389
Planing-machine	W. Teal	Rochester, N. Y	Aug. 2, 1870	105,997
Planing-machine	J. Tesseyman	Dayton, Ohio	Apr. 9, 1867	63,673
Planing-machine	R. R. Throckmorton	New York	Oct. 6, 1835	....
Planing-machine	R. R. Throckmorton	New York	Oct. 22, 1835	....
Planing-machine	R. R. Throckmorton	Brooklyn, N. Y	Aug. 28, 1849	6,667
Planing-machine	G. W. Tolhurst	Cleveland, Ohio	Jan. 20, 1852	8,666
Planing-machine	M. Tuells	Milo, N. Y	Feb. 10, 1836	....
Planing-machine	J. Utter	Central Falls, R. I	Oct. 14, 1873	143,734
Planing-machine	D. Van Fleet	Sandusky, Ohio	Nov. 28, 1854	12,009
Planing-machine	A. Van Haagen	Philadelphia, Pa	June 18, 1872	128,190
Planing-machine	L. B. Walker	Chicago, Ill	Mar. 8, 1870	100,695
Planing-machine	C. P. S. Wardwell	Lake Village, N. H	Jan. 14, 1862	34,174
Planing-machine	A. J. Watson	Lyons Township, Mich	Nov. 14, 1871	121,028
Planing-machine	W. Watson	Chicago, Ill	July 6, 1852	9,108
Planing-machine	T. J. Wells	New York, N. Y	Mar. 20, 1849	6,195
Planing-machine	A. Whitcomb	Worcester, Mass	Sept. 1, 1868	81,854
Planing-machine	S. Whitesides	De Pere, Wis	May 30, 1871	115,401
Planing-machine	B. D. Whitney	Winchendon, Mass	Jan. 1, 1867	60,812
Planing-machine	A. A. Wilder	Detroit, Mich	Dec. 21, 1852	9,496
Planing-machine	A. A. Wilder	Detroit, Mich	May 7, 1867	64,609
Planing-machine	S. Wilmarth	Malden, Mass	June 1, 1869	90,907
Planing-machine	G. E. Woodbury	East Cambridge, Mass	Aug. 6, 1867	67,623
Planing-machine	G. E. Wooobury	Cambridge, Mass	Mar. 22, 1870	101,072
Planing-machine	J. A. Woodbury	Winchester, Mass	Feb. 7, 1854	10,512
Planing-machine	J. A. Woodbury	Winchester, Mass	Nov. 13, 1855	13,808
Planing-machine	J. A. Woodbury	Winchester, Mass	June 8, 1858	20,527
Planing-machine	J. A. Woodbury	Boston, Mass	July 24, 1866	56,656
Planing-machine	J. P. Woodbury	Boston, Mass	Mar. 20, 1849	6,211
Planing-machine	J. P. Woodbury	Boston, Mass	Apr. 20, 1873	138,462
Planing-machine	S. A. Woods	Boston, Mass	Feb. 14, 1871	111,894
Planing-machine	S. A. Woods	Boston, Mass	Nov. 11, 1873	144,588
Planing-machine	S. A. Woods and G. E. Woodbury	Boston and Cambridge, Mass.	July 18, 1871	117,230
Planing-machine	W. Woodworth	New York, N. Y	Nov. 15, 1836	....
Planing-machine	M. Wright	Montpelier, Vt	July 16, 1872	128,995
Planing-machine attachment	L. C. Brastow and A. M. Zwiebel	Wilkesbarre, Pa	Nov. 28, 1871	121,230
Planing machine, Board	H. Allen	Randolph County, N. C	Dec. 22, 1826	....
Planing machine, Board	J. Lombard	Boston, Mass	Oct. 13, 1838	979
Planing machine, Board	R. Luscomb	Penn Yan, N. Y	June 12, 1838	779
Planing machine, Board	S. Whitney	Dunstable, N. H	June 3, 1837	223
Planing machine, Board	W. Woodworth	Hudson, N. Y	Dec. 27, 1828	....
Planing machine, Board	W. Woodworth	New York, N. Y	Nov. 15, 1836	80
Planing machine, Board and timber	H. Law	Wilmington, N. C	Sept. 30, 1841	2,272
Planing-machine cutter	E. G. Allen and C. Briggs	Boston and New Bedford, Mass.	Nov. 26, 1850	7,793
Planing-machine cutter	J. M. Patton and W. F. Fergus	Philadelphia, Pa	Dec. 23, 1851	8,612
Planing-machine-cutter head	B. J. Barber	Ballston Spa, N. Y	Apr. 12, 1870	101,701
Planing-machine-cutter head	L. M. Berry	Boston, Mass	July 22, 1856	15,365
Planing-machine-cutter head	A. Caldwell	Lexington, Ky	Oct. 16, 1847	5,334
Planing-machine-cutter head	W. H. Christie	Albany, N. Y	Nov. 26, 1867	71,455
Planing-machine-cutter head	H. Climer and C. E. McBeth	Hamilton, Ohio	Apr. 4, 1871	113,262
Planing-machine-cutter head	M. F. Connett	Ladoga, Ind	Aug. 27, 1867	68,048
Planing-machine-cutter head	J. D. Dale	Philadelphia, Pa	Oct. 10, 1854	11,778
Planing-machine-cutter head	J. Kuehnle	Cincinnati, Ohio	May 17, 1870	103,200
Planing machine-cutter head	H. A. Lee	Worcester, Mass	July 16, 1867	66,719
Planing-machine-cutter head	E. G. Richards	Detroit, Mich	Sept. 9, 1873	142,731
Planing-machine-cutter head	A. T. Stearns	Dorchester, Mass	June 2, 1868	78,616
Planing-machine-cutter head	D. Stevens	Danbury, Conn	Aug. 16, 1870	106,421
Planing-machine-cutter head	L. D. Towne	Worcester, Mass	Feb. 12, 1856	14,263
Planing-machine-cutter head, Wood	T. Tostevin	Council Bluffs, Iowa	Apr. 7, 1868	76,556
Planing-machine-cutter holder	W. Walkington	Leeds, England	Jan. 21, 1873	135,182
Planing-machine-cutter-setting gage	A. Evans	Cincinnati, Ohio	Apr. 27, 1869	89,394
Planing-machine cutter, Wood	C. Livingston	Redwood City, Cal	July 17, 1866	56,424
Planing-machine, Cylindrical	U. Emmons	New York	Apr. 25, 1829	....
Planing machine feed-motion	S. C. and W. W. Hurlbut	Boonville, N. Y	Oct. 2, 1855	13,618
Planing-machine feed-roll	S. Inman	Rockford, Ill	Jan. 28, 1873	135,224
Planing-machine, &c., feed-roller	J. Hall	Worcester, Mass	July 21, 1857	17,833
Planing-machine feed-roller	F. J. Plummer	Worcester, Mass	Nov. 6, 1866	59,514
Planing-machine feeder	J. Cumberland	Mobile, Ala	Feb. 3, 1852	8,702
Planing-machine-feeding device	C. B. Cottrell	Westerly, R. I	May 17, 1859	24,078
Planing-machine for cutting blind-slats	M. Free	Philadelphia, Pa	June 5, 1866	55,415
Planing-machine for dressing the edges of boards	W. E. Cornell	Boston, Mass	Jan. 1, 1851	7,868
Planing-machine for sawing bellows boards, &c	J. Hinman	New Haven, Conn	June 1, 1805	....
Planing-machine guide	F. Keagey	Chambersburgh, Pa	Jan. 4, 1870	98,600
Planing-machine, Hand	T. P. Butterfield	Indianapolis, Ind	May 17, 1859	24,076
Planing-machine hood	B. D. Mott	Green Island, N. Y	Aug. 30, 1870	106,856
Planing machine, Iron	A. B. Couch	Worcester, Mass	Feb. 1, 1870	99,409
Planing machine, Iron	H. Gastel	Philadelphia, Pa	Mar. 19, 1872	124,676
Planing machine, Iron	W. Sellers	Philadelphia, Pa	May 17, 1864	42,797
Planing machine, Metal	M. Allan	Utica, N. Y	Nov. 22, 1859	26,151
Planing machine, Metal	R. A. Belden and E. H. Cutler	New Haven, Conn	Sept. 14, 1869	94,860
Planing machine, Metal	J. Carhart	New York, N. Y	July 10, 1859	24,792
Planing machine, Metal	R. Harper	Glen Mills, Pa	July 22, 1873	141,049
Planing machine, Metal	F. W. Howe	Windsor, Vt	June 21, 1853	9,797
Planing machine, Metal	J. T. Kichner and W. H. Odenatt	Philadelphia, Pa	Sept. 23, 1873	143,080
Planing machine, Metal	M. Kremser	Indianapolis, Ind	Sept. 19, 1871	119,156
Planing machine, Metal	G. C. Larned	Worcester, Mass	July 22, 1873	141,006
Planing machine, Metal	W. McPherson	New York, N. Y	Nov. 12, 1867	70,876
Planing machine, Metal	D. F. Mellen	Wentworth, N. H	Aug. 1, 1854	11,463
Planing machine, Metal	T. F. Rowland	Green Point, N. Y	Oct. 7, 1862	36,624
Planing machine, Metal	J. H. Thompson	Paterson, N. J	Apr. 24, 1855	12,767
Planing-machine presser bar	H. Snow	Dubuque, Iowa	Nov. 21, 1854	11,984
Planing-machine, Rotary	H. C. and H. S. Ingraham	Guilford and Granger, Ohio	Mar. 15, 1859	23,2[illegible]0
Planing-machine, Rotary	W. H. Smith	Newport, R. I	May 3, 1859	23,8[illegible]0
Planing-machine, Shell-roller-bed	G. Darby and J. E. Young	Augusta, Me	July 28, 1857	17,873

Index of patents issued from the United States Patent Office from 1790 *to* 1873, *inclusive*—Continued.

Invention.	Inventor.	Residence.	Date.	No.
Planing machine, Stave	J. De Libatcheff	Yardslawl, Russia	Oct. 2, 1860	30, 210
Planing machine suction-tube	G. C. Westover	Vernon, Ind	Dec. 30, 1873	146, 115
Planing-machine-tool head	S. W. Paine	Williamsport, Pa	Sept. 26, 1871	119, 395
Planing-machine-tool holder	C. Hall	New York, N. Y	Jan. 22, 1867	61, 420
Planing machine, Wood	C. B. Cottrell	Westerly, R. I	Oct. 5, 1858	21, 720
Planing machine, Wood	W. H. Doane and W. E. London	Cincinnati, Ohio	June 18, 1867	65, 796
Planing machine, Wood	F. Douglas	Norwich, Conn	Apr. 21, 1868	77, 013
Planing machine, Wood	F. Douglas	Norwich, Conn	June 9, 1868	78, 728
Planing machine, Wood	G. B. Durkee and W. H. Murray	Chicago, Ill	Dec. 10, 1867	71, 862
Planing machine, Wood	N. C. Freck and S. Strock	Millersburgh, Pa	Oct. 31, 1871	120, 426
Planing machine, Wood	O. P. Furman	Addison, N. Y	Apr. 14, 1868	76, 740
Planing machine, Wood	D. A. Harris	Ithaca, N. Y	Feb. 11, 1868	74, 218
Planing machine, Wood	R. N. Meriam	Worcester, Mass	Apr. 14, 1868	76, 791
Planing machine, Wood	G. H. Ober	Newbury, Ohio	July 23, 1867	66, 988
Planing machine, Wood	G. Rinewalt, jr	Pendleton, Ind	July 30, 1861	32, 926
Planing machine, Wood	J. J. Russ	Worcester, Mass	Mar. 24, 1868	75, 984
Planing machine, Wood	F. Schmidt	Cincinnati, Ohio	June 18, 1867	65, 954
Planing machine, Wood	P. T. Smith	Salem, Ohio	May 28, 1867	65, 128
Planing machine, Wood	H. D. Stover	New York, N. Y	Dec. 18, 1860	30, 693
Planing machine, Wood	A. and H. Streit	Cincinnati, Ohio	Dec. 3, 1867	71, 814
Planing machine, Wood	W. Tucker	Paris, Ill	May 28, 1867	65, 308
Planing-machines, Adjusting cutter-heads to	C. R. Tompkins	Rochester, N. Y	Nov. 26, 1867	71, 341
Planing-machines, Adjusting stuff in	W. B. Emery	Albany, N. Y	Feb. 27, 1855	12, 438
Planing-machines, Arrangement of feed-rollers for	B. Fitts	Worcester, Mass	Aug. 11, 1857	17, 963
Planing-machines, Arrangement of feed-rollers for	H. C. Wight	Worcester, Mass	Jan. 1, 1856	14, 038
Planing machines, Arrangement of feed-rollers in wood.	H. H. Baker	New Market, N. J	Feb. 15, 1859	22, 926
Planing-machines, Arrangement of pressure and feed rollers in.	C. A. Spring and P. Boon	Kensington, Pa	July 30, 1850	7, 536
Planing-machines, Certain devices in	V. Houck	Buffalo, N. Y	June 17, 1856	15, 129
Planing-machines, Clamping cutters in cutter-heads for.	J. P. Grosvenor	Lowell, Mass	Dec. 2, 1856	16, 144
Planing-machines, Clamping polygonal pieces in	J. W. Killam	East Wilton, N. H	Mar. 23, 1858	19, 702
Planing-machines, Cutter-arm for	R. N. Meriam	Worcester, Mass	Sept. 20, 1864	44, 333
Planing-machines, Device for dogging lumber in	D. N. B. Coffin and H. D. Stover	Newton and Boston, Mass	May 19, 1857	17, 315
Planing-machines, Device for operating cutter-heads of.	T. F. Taft	Worcester, Mass	June 6, 1874	11, 017
Planing-machines, Device for securing cutters in rotary.	S. F. Forman	New York, N. Y	June 29, 1858	20, 762
Planing-machines, Feed apparatus of	J. Whitney	Winchester, Mass	Apr. 13, 1852	8, 881
Planing-machines, Feeding planks to	N. Barlow	Newark, N. J	July 31, 1855	13, 345
Planing-machines, Gearing for feed-roller of	C. Burleigh	Fitchburgh, Mass	Feb. 12, 1856	14, 272
Planing-machines, Hanging knives in	M. G. Hubbard	New York, N. Y	Jan. 2, 1855	12, 162
Planing-machines, Mechanism for operating bed of	A. F. Shaw	West Roxbury, Mass	Sept. 8, 1868	81, 953
Planing machines, Molding	T. Dickinson	Newark, N. J	June 26, 1866	55, 833
Planing-machines, Operating feed-roller for	B. Fitts	Worcester, Mass	Mar. 15, 1859	23, 240
Planing-machines, Pressure-block for	J. Lawrence	East Morrisania, N. Y	Sept. 15, 1863	39, 996
Planing-machines, Preventing back-lash in feed-motion of.	T. H. Burridge	Jersey City, N. J	Dec. 24, 1850	7, 852
Planing-machines, Rotary cutter for	D. McKinley	Philadelphia, Pa	May 26, 1863	38, 712
Planing-machines, Rotary cutter-head for	T. Christian	New York, N. Y	Aug. 28, 1860	29, 768
Planing-machines, Shaving-conductor for	W. Weaver	Burlington, Vt	Dec. 9, 1873	145, 375
Planing-machines, Stock for holding cutters in rotary.	I. Gibbs	Worcester, Mass	July 27, 1858	20, 999
Planing-machines, Universal dog for	S. S. Gray	South Boston, Mass	Sept. 4, 1855	13, 536
Planing-machinery, Device for adjusting	L. Tilton	New York, N. Y	May 15, 1855	12, 880
Planing machinery, Metal	W. Sellers	Philadelphia, Pa	Aug. 5, 1862	36, 113
Planing machinery, Metal	W. Sellers	Philadelphia, Pa	June 14, 1864	43, 137
Planing machinery, Metal	W. W. Spofford	Boston, Mass	Sept. 6, 1853	9, 996
Planing machinery, Portable metal	A. C. Jones	New Orleans, La	Sept. 18, 1847	5, 301
Planing-mechanism	C. Van Haagen	Philadelphia, Pa	Oct. 11, 1870	108, 305
Planing metal	C. Van Horn	Springfield, Mass	Aug. 12, 1856	15, 538
Planing metals, &c., Mounting the cutters of machines for.	P. Saulnier	New York, N. Y	Dec. 28, 1852	9, 504
Planing moldings and smooth-planing	C. Thompson	Poughkeepsie, N. Y	Jan. 30, 1834	....
Planing moldings, Cutter for	E. J. Arens	Boston, Mass	July 2, 1867	66, 280
Planing moldings, Machine for	J. D. Dale	Philadelphia, Pa	Jan. 4, 1853	9, 515
Planing moldings, Machine for	J. D. Dale	Philadelphia, Pa	Jan. 4, 1853	9, 516
Planing moldings, Machine for	W. Rankin	Richmond, Va	Sept. 11, 1860	29, 992
Planing or molding machines, Cutter-head for	J. D. Kimball	Reading, Mass	Sept. 16, 1873	142, 794
Planing planks, boards, and clapboards, Machine for.	B. Langdon	Troy, N. Y	Jan. 9, 1838	555
Planing shavings for upholsterers, Machine for	S. A. and W. H. Post	Durham, N. Y	Sept. 1, 1863	39, 747
Planing facets of polygonal bars, nuts, &c., Machine for.	W. F. Batho	Birmingham, England	May 28, 1872	127, 142
Planing, tonguing, and grooving boards	J. Drummond	Whitestown, N. Y	Apr. 27, 1832	....
Planing, tonguing, and grooving machine	J. Powell, N. Barlow, and E. Holden.	Saint Louis, Mo	Feb. 27, 1847	4, 983
Planing, tonguing, and grooving, &c., planks and other lumber.	R. R. Throckmorton	Brooklyn, N. Y	May 1, 1845	4, 017
Planing-tool	N. Russell	Plymouth, Mass	Apr. 18, 1871	113, 800
Planing valve-seats of steam-engines	C. B. Long	Worcester, Mass	Oct. 2, 1860	30, 234
Planing warped surfaces, Machine for	J. Green	Brooklyn, N. Y	June 19, 1860	28, 808
Planing-wheel, Bramah	E. Jones	Greenfield, Mass	Dec. 2, 1856	16, 163
Planing wood, Adjusting knives of rotary cutter-heads for.	B. Fitts	Worcester, Mass	Apr. 26, 1859	23, 763
Planishing and hammering machine, Metal	C. D. Goodrich	Ann Arbor, Mich	Oct. 3, 1848	5, 818
Plank and timber dresser	E. Baily	Cincinnati, Ohio	Nov. 3, 1868	83, 684
Plank, Gang	F. G. Johnson	Brooklyn, N. Y	Apr. 15, 1873	137, 776
Planking-clamp	W. Glenn	Philadelphia, Pa	May 1, 1866	54, 327
Planking-clamp	J. Macotter	Baltimore, Md	Nov. 14, 1865	50, 941
Planking-clamp	J. C. Thomas	Kennebunk, Me	June 12, 1866	55, 554
Planking-clamp, Floor	H. B. Gregg	Camden, Ohio	Sept. 18, 1866	58, 167
Planking-clamp for ships	S. Staples	Bath, Me	Mar. 25, 1856	14, 522
Planking-clamp for ship's sides or floors	J. J. Hill	Sodus Point, N. Y	Nov. 26, 1867	71, 387
Planking-set	E. H. Parker	Bucksport, Me	July 26, 1870	105, 719
Plant-fender	C. Barber	Lostant, Ill	July 28, 1863	39, 333
Plant-grower	B. B. Chadwick	Buffalo, N. Y	May 23, 1871	115, 162

Index of patents issued from the United States Patent Office from 1790 *to* 1873, *inclusive*—Continued.

Invention.	Inventor.	Residence.	Date.	No.
Plant-protector	R. M. Bartlett	Storrs Township, Ohio	Feb. 25, 1868	74, 879
Plant-protector	J. M. Hurt	Blacks and Whites, Va	Oct. 27, 1868	83, 386
Plant-protector	I. Mayfield	Mayfield, Ky	July 11, 1871	116, 974
Plant-protector	E. Mosher	Flushing, Mich	May 10, 1859	23, 935
Plant-protector	H. K. Nelson	Penn Yan, N. Y	Nov. 2, 1869	96, 468
Plant-protector	W. N. Sprague	Keene, N. H	Jan. 18, 1870	98, 892
Plant-protector	T. R. Timby	Tarrytown, N. Y	Jan. 21, 1873	135, 173
Plant-protector	A. Tuttle	Mannsville, N. Y	Jan. 4, 1870	98, 645
Plant-protector	J. M. Watson	Sharon, Mass	June 30, 1868	79, 422
Plant-protector	J. M. Watson	Sharon, Mass	Nov. 23, 1869	97, 259
Plant-protector	J. Weed	Muscatine, Iowa	Dec. 8, 1863	40, 873
Plant-protector	W. B. Wickes	Sharon, Mass	Feb. 23, 1869	87, 313
Plant-setting device	J. C. Fuller	Phœnix, N. Y	Feb. 16, 1869	87, 037
Plant-stand, Folding	B. B. Nourse	Westborough, Mass	Dec. 5, 1871	121, 541
Plants and animals, Accelerating growth of	A. I. Pleasonton	Philadelphia, Pa	Sept. 26, 1871	119, 242
Plants and fruits, Growing	A. C. Chamberlain	Newport, R. I	Nov. 26, 1861	33, 774
Plants and trees, Compound for destroying vermin on.	F. S. Townsend	South Seaville, N. J	Jan. 26, 1869	86, 259
Plants, Apparatus for fumigating	C. C. Preston	Bayland, Tex	Dec. 17, 1867	72, 419
Plants, Box for propagating	E. B. Newman, jr	Watkins, N. Y	Apr. 30, 1867	64, 245
Plants, Box for transplanting	J. Jenkins	Salem, Ohio	Mar. 12, 1867	62, 751
Plants, Box for transporting	M. L. Thompson	Brooklyn, N. Y	June 21, 1864	43, 242
Plants, Box for transporting	T. F. Wardwell	Penn Yan, N. Y	Apr. 11, 1865	47, 234
Plants, Compound for destroying worms in	J. Bingaman	Jersey Shore, Pa	Mar. 16, 1869	87, 820
Plants, Extracting saline matters from marine	C. F. Brackett and G. L. Goodale.	Brunswick and Saco, Me	Oct. 29, 1867	70, 317
Plants for useful purposes, Preparation of the roots of.	J. M. Katzenmayer	New York, N. Y	Sept. 29, 1863	40, 143
Plants, Forcing	C. Reese	Baltimore, Md	Apr. 2, 1872	125, 332
Plants from frost, Preservation of	G. Morris	....	July 10, 1792	....
Plants from hoes, Implement for shielding	A. Kirkpatrick	Newark, N. J	June 23, 1868	79, 133
Plants from worms, Device for protecting young	J. W. Colburn	Rose, N. Y	Mar. 16, 1869	87, 911
Plants, Hand-lights for protecting	O. R. L. and M. P. A. Crozier	Paris, Mich	May 26, 1863	38, 656
Plants, Implement for setting out	E. W. Packer	Paulsborough, N. J	Apr. 3, 1866	53, 662
Plants, Instrument for transplanting	W. C. S. Ellerbe	Camden, S. C	Oct. 16, 1866	58, 794
Plants, Machine for treating	A. Bouchard	New Orleans, La	May 2, 1871	114, 400
Plants, Machine for treating fibrous	W. B. Shedd	Boston, Mass	Apr. 30, 1872	126, 337
Plants, Machinery for treating fibrous	G. E. Hopkins and W. B. Shedd	Boston, Mass	Jan. 30, 1872	123, 103
Plants, Package-case for	P. L. Suine	Shirleysburgh, Pa	July 10, 1866	56, 290
Plants, Potting and packing	B. L. Ryder	Chambersburgh, Pa	Nov. 10, 1868	84, 002
Plants, Propagation and treatment of house	W. Lilley	Washington, D. C	Mar. 17, 1868	75, 555
Plants, &c., Protecting	J. Shepard	Bristol, Conn	Sept. 1, 1868	81, 693
Plants to prevent the approach of worms, Implement for trenching around.	W. H. Halleck	Ann Arbor, Mich	Dec. 8, 1868	84, 692
Planter	W. Beall	Hainesville, W. Va	May 7, 1872	126, 438
Planter	W. F. Caldwell	Oxford, Me	Sept. 18, 1866	58, 060
Planter and broad-cast seeder, Combined corn	A. J. Edgett	Hornellsville, N. Y	Sept. 18, 1866	58, 166
Planter and chopper, Combined cotton	A. R. Nixon	Polo, Ill	July 4, 1871	116, 741
Planter and cultivator	S. Bacher	Edina, Mo	Dec. 20, 1870	110, 187
Planter and cultivator	N. Earlywine	Centreville, Iowa	Sept. 13, 1870	107, 235
Planter and cultivator	A. Little	Jamestown, Ohio	Apr. 30, 1872	126, 147
Planter and cultivator	A. J. Misenhimer	Oskaloosa, Ill	Sept. 14, 1869	94, 835
Planter and cultivator	C. Rubbles	Lowville, N. Y	May 5, 1868	77, 535
Planter and cultivator	N. Whitehall	Newton, Ind	Aug. 31, 1869	94, 365
Planter and cultivator	A. Q. Withers	Holly Springs, Miss	Jan. 17, 1871	111, 101
Planter and cultivator combined	J. Adams	Clarksville, Tex	Aug. 20, 1867	68, 026
Planter and cultivator, Combined	T. M. and J. Brooks	Cincinnati, Ohio	Apr. 30, 1872	126, 258
Planter and cultivator combined	J. N. Burton	Griffin, Ga	Mar. 26, 1872	125, 018
Planter and cultivator combined	I. H. Chappell	Lawrence, Kans	Mar. 19, 1867	63, 019
Planter and cultivator combined	I. H. Chappell	Decatur, Ill	Apr. 16, 1867	63, 858
Planter and cultivator, Combined	I. H. Chappell	Decatur, Ill	Nov. 19, 1867	71, 134
Planter and cultivator, Combined	J. A. Currie	Xenia, Ohio	Aug. 31, 1869	94, 289
Planter and cultivator, Combined	W. L. Gebby	New Richland, Ohio	May 7, 1867	64, 519
Planter and cultivator combined	G. W. Kinzer	Linden Station, Ohio	Aug. 18, 1868	81, 177
Planter and cultivator combined	E. B. and J. F. McClellan	Alexandria, Ala	Jan. 30, 1872	123, 186
Planter and cultivator, Combined	I. W. McGaffey	Chicago, Ill	Mar. 28, 1865	47, 029
Planter and cultivator, Combined	A. G. Perry	Union County, Miss	Oct. 8, 1872	131, 969
Planter and cultivator, Combined	A. W. Putnam	Suisun, Cal	Apr. 2, 1867	63, 424
Planter and cultivator, Combined	J. A. Rockwood	Kinderhook, Ill	Nov. 23, 1869	97, 121
Planter and cultivator, Combined	W. C. Switzer	Nelsonville, Tex	Mar. 16, 1869	87, 885
Planter and cultivator, Combined	J. Vaughn	College Grove, Tenn	Dec. 24, 1867	72, 703
Planter and cultivator combined	C. L. Wilcox	Wayne, Ohio	July 16, 1872	129, 443
Planter and cultivator, Combined	T. Wilson	Garton, England	Nov. 30, 1869	97, 467
Planter and cultivator, Combined corn	A. G. Aikin	Somerton, Ohio	Aug. 10, 1869	93, 510
Planter and cultivator, Combined corn	E. Braggins	Mount Vernon, Ohio	July 4, 1871	116, 678
Planter and cultivator, Combined corn	J. Campbell	London, Ohio	Aug. 3, 1869	93, 172
Planter and cultivator, Combined corn	G. De Vany, jr	Darbyville, Ohio	June 24, 1873	140, 254
Planter and cultivator, Combined corn	C. Dyer	Coal Run, Ohio	Oct. 27, 1868	83, 478
Planter and cultivator, Combined corn	A. N. Gow	Mount Vernon, Ohio	Feb. 19, 1867	62, 263
Planter and cultivator, Combined corn	J. C. Hazen	West Independence, Ohio	Nov. 2, 1869	96, 319
Planter and cultivator, Combined corn	J. W. Hollaman	Maple Grove, Ill	Jan. 12, 1869	85, 826
Planter and cultivator, Combined corn	M. J. Hunt	Rising Sun, Md	Jan. 8, 1867	61, 071
Planter and cultivator, Combined corn	D. W. Jacoby	Shelbyville, Ill	Aug. 27, 1867	68, 201
Planter and cultivator, Combined corn	M. J. Kavanagh and M. Gregg	Joliet and Chicago, Ill	July 20, 1869	92, 837
Planter and cultivator, Combined corn	J. S. Mason	Coal Run, Ohio	Sept. 15, 1868	82, 137
Planter and cultivator, Combined corn	J. Palmer	Oriskany, N. Y.	July 25, 1865	48, 991
Planter and cultivator, Combined corn	W. Scott	High Banks, Ind	Jan. 19, 1869	86, 105
Planter and cultivator, Combined corn	J. F. Sterett and C. M. J. Reynolds.	Ottumwa, Iowa	Oct. 1, 1867	69, 502
Planter and cultivator, Combined corn	S. J. Taylor	Rome, N. Y	July 16, 1867	66, 910
Planter and cultivator, Combined corn	F. Underwood	South Rutland, N. Y	Mar. 7, 1871	112, 516
Planter and cultivator, Combined cotton-seed	J. A. Wright	Marietta, Ga	Apr. 4, 1871	113, 383
Planter and cultivator, Combined potato	C. F. Hoffman	New Orleans, La	Aug. 25, 1868	81, 371
Planter and cultivator, Combined seed	G. W. Carpenter	Butler, Ind	Nov. 16, 1869	96, 774
Planter and cultivator, Combined seed	P. Limhold	Saint Louis, Mo	Aug. 29, 1865	49, 661
Planter and cultivator, Convertible	S. E. Harrington	Greenfield, Mass	Apr. 19, 1864	42, 370
Planter and cultivator, Corn	N. Breed	Jeffersonville, Ind	Jan. 11, 1870	98, 663
Planter and cultivator, Corn	A. Edmister	Westfield, Ohio	Nov. 10, 1868	83, 943

Index of patents issued from the United States Patent Office from 1790 *to* 1873, *inclusive*—Continued.

Invention.	Inventor.	Residence.	Date.	No.
Planter and cultivator, Corn	J. Jenkins	Andrew County, Mo	Oct. 19, 1869	95, 906
Planter and cultivator, Seed	D. W. Shares	Hamden, Conn	Dec. 12, 1854	12, 075
Planter and cultivator, Seed	M. R. Snodgrass	Jamestown, Ohio	Dec. 24, 1867	72, 556
Planter and cultivator, Seed	W. D. Stoud	Oshkosh, Wis	Apr. 12, 1870	101, 939
Planter and digger, Potato	J. C. Walker	Farmington, Mich	Mar. 21, 1871	112, 870
Planter and distributer	D. McKellar	Selma, Ala	Sept. 27, 1870	107, 796
Planter and drill, Cotton-seed	W. J. Arrington	Jefferson County, Ga	July 6, 1869	92, 144
Planter and earth-excavator, Potato	L. Rice	Robbinston, Me	June 29, 1833	....
Planter and fertilizer	M. Barbour	Oxford, Ohio	Apr. 30, 1872	126, 250
Planter and fertilizer, Combined corn	B. F. Grimes	Dawsonville, Md	May 28, 1867	65, 073
Planter and fertilizer, Corn	A. L. Holcomb	Hopewell, N. J	June 21, 1870	104, 591
Planter and fertilizer, Corn	H. K. Roberts	Jefferson County, Ky	Jan. 18, 1870	98, 886
Planter and fertilizer, Corn	M. M. Sprinkle	Rochelle, Va	June 1, 1869	90, 698
Planter and fertilizer-distributer, Combined corn	C. W. Barrick	Walkersville, Md	July 25, 1871	117, 369
Planter and fertilizer-distributer, Combined cotton-seed.	J. Johnson	Northampton County, N. C	July 16, 1867	66, 713
Planter and fertilizer-distributer, Combined cotton-seed.	W. M. McLendon	Greenville, Ga	May 17, 1870	103, 067
Planter and fertilizer-distributer, Cotton	B. H. Washington	Columbia County, Ga	Oct. 18, 1870	108, 415
Planter and fertilizer-distributer, Cotton-seed	H. C. Harris	Fort Valley, Ga	Mar. 8, 1870	100, 526
Planter and fertilizer-distributer, Seed	J. Arrington	Livingston, Ala	Feb. 22, 1870	100, 001
Planter and fertilizer-distributer, Seed	E. P. Jones	Greensborough, N. C	June 17, 1873	140, 046
Planter and fertilizer-distributer, Seed	R. Montfort	Butler, Ga	Oct. 21, 1873	143, 832
Planter and fertilizer-distributer, Seed	J. Sample	Franklin County, Miss	Sept. 26, 1871	119, 411
Planter and fertilizer, Seed	D. P. Webster	Boston, Mass	Aug. 23, 1870	106, 751
Planter and grain-drill, Corn	G. W. Dickinson	Charleston, Ill	Aug. 31, 1869	94, 404
Planter and guano-distributer, Combined cotton	J. H. Nicholes	Sumter, S. C	Feb. 28, 1871	112, 169
Planter and guano-distributer, Combined cotton	J. E. Ross	Greensborough, Ga	Dec. 20, 1870	110, 290
Planter and guano-distributer, Combined cotton and seed.	S. L. Donnell	Humboldt, Tenn	Mar. 22, 1870	100, 990
Planter and guano-distributer, Combined cotton-seed.	L. Garnett	Camilla, Ga	Jan. 3, 1871	110, 644
Planter and guano-distributer, Cotton-seed	W. W. Croom	Montgomery, Ala	Jan. 10, 1871	110, 832
Planter and guano-distributer, Cotton-seed	J. Rafter	Winona, Miss	May 9, 1871	114, 600
Planter and guano-distributer, Cotton-seed	C. G. Wilson	Milledgeville, Ga	Dec 10, 1872	133, 911
Planter and guano-distributer, Cotton-seed	C. G. Wilson	Milledgeville, Ga	Sept. 23, 1873	143, 211
Planter and guano-distributer, Seed	H. L. Tillery	Halifax, N. C	Jan. 3, 1871	110, 694
Planter and guano-distributer, Seed	W. L. Traynham	Warrenton, Ga	Oct. 11, 1870	108, 213
Planter and guano-sower, Corn	J. B. Gemmill	Strawbridge, Pa	July 23, 1867	66, 959
Planter and hoe, Combined corn	H. Soggs	Columbus, Pa	Feb. 26, 1867	62, 376
Planter and lime-spreader, Combined corn	M. M. Mackerly	South Salem, Ohio	Feb. 18, 1862	34, 438
Planter and manure-distributer	B. F. Whitner	Madison, Fla	May 21, 1867	64, 926
Planter and manure-distributer, Combined	B. F. Whitner	Madison, Mich	Apr. 14, 1868	76, 865
Planter and manure-distributer, Cotton	N. Donaldson	Line Creek, S. C	Jan. 3, 1871	110, 636
Planter and marker, Corn	F. Ogden	Fisherville, Ky	Oct. 4, 1870	108, 044
Planter and marker, Corn	T. Ryan	Saint Martin's, Ohio	Feb. 6, 1872	123, 514
Planter and plaster-dropper, Combined corn	H. Rhodes	Clarence Centre, N. Y	Oct. 1, 1867	69, 486
Planter and plow, Combined corn	J. T. McElhiney	Moberly, Mo	Apr. 1, 1873	137, 309
Planter and plow, Combined corn	P. McIsaac	Waterloo, Iowa	Jan. 7, 1868	73, 022
Planter and plow, Combined corn	S. M. Swartz	Millheim, Pa	Sept. 11, 1866	57, 994
Planter and plow, Combined seed	E. Seibel	Witenburgh, Mo	Feb. 2, 1869	86, 457
Planter and plow, Combined sulky corn	R. and A. S. Robins	Dundas, Ill	June 29, 1869	91, 874
Planter and plow, Corn	J. H. Frampton	Hopewell, Ohio	Apr. 14, 1868	76, 738
Planter and roller, Combined	J. A. Comstock	Bowling Green, Mo	Aug. 22, 1871	118, 342
Planter and seed-drill, Corn	J. Weaver, jr	Elizabethville, Pa	June 16, 1868	79, 038
Planter and seeder	H. H. Webster	Claremont, N. H	Feb. 6, 1866	52, 500
Planter and seeder, Combined	W. E. Fricke	Mexico, Mo	Aug. 30, 1870	106, 806
Planter and seeder, Combined	S. Hiestand	Hillsborough, Ohio	Apr. 4, 1871	113, 718
Planter and seeder, Combined corn	R. Cowan	Bloomfield, Ill	Jan. 14, 1873	134, 855
Planter and seeding-machine	S. S. Kimball	Laconia, N. H	Mar. 16, 1869	87, 855
Planter and shovel-plow, Combined corn	A. M. Franklin, W. J. Hastings, and J. A. Holford.	Rising Sun, Ind	Oct. 20, 1868	83, 271
Planter, Automatic check-row corn	J. L. Kreider	Chesnut Level, Pa	July 13, 1869	92, 616
Planter, Automatic seed	D. Lorriaux	Ottawa, Ill	May 14, 1872	126, 721
Planter, Broom-corn-seed	J. Bush	Darien, Wis	May 24, 1870	103, 427
Planter, Combined corn and cotton	W. B. Townes	Grenada, Miss	May 9, 1871	114, 625
Planter, Combined corn and cotton	A. H. Wooton	Bartow, Ga	Sept. 21, 1869	95, 062
Planter, Combined cotton and corn	L. A. Perrault	Natchez, Miss	Mar. 7, 1871	112, 486
Planter, Combined hoe and hand	H. Fessler and I. E. Betz	Canton, Ohio	May 14, 1867	64, 755
Planter, Combined seed and potato	O. N. Chase	Boston, Mass	Jan. 24, 1865	45, 976
Planter, Convertible corn	A. Runstetler	Peoria, Ill	June 27, 1871	116, 493
Planter, Corn	D. B. Abbey	Horseheads, N. Y	Nov. 12, 1861	33, 688
Planter, Corn	G. Abbott	White's Corners, N. Y	Oct. 15, 1867	69, 739
Planter, Corn	H. Ackerman	Pittsburgh, Pa	Aug. 11, 1868	80, 892
Planter, Corn	M. Ackerman	Steamboat Rock, Iowa	Sept. 6, 1870	106, 981
Planter, Corn	J. C. Adams	Greensburgh, Ind	Oct. 4, 1859	25, 613
Planter, Corn	J. W. Adams	Birmingham, Iowa	May 9, 1865	47, 603
Planter, Corn	S. W. Adams	Moultrie County, Ill	Dec. 11, 1860	30, 861
Planter, Corn	W. H. Adle, P. D. Miles, and G. Custer.	Norristown, Pa	Nov. 27, 1860	30, 708
Planter, Corn	J. Agnew	Bath, Pa	Apr. 5, 1864	42, 151
Planter, Corn	J. M. Aitchison	Omar, N. Y	Feb. 21, 1871	112, 006
Planter, Corn	D. R. Alden	Unionville, Ohio	Sept. 8, 1857	18, 127
Planter, Corn	T. K. Alexander	Decatur, Ill	July 19, 1864	43, 559
Planter, Corn	E. C. Allen	Le Roy, N. Y	Nov. 8, 1859	26, 006
Planter, Corn	E. W. Allen	Auburn, N. Y	Apr. 14, 1868	76, 689
Planter, Corn	H. R. Allen	Athens, Ohio	Sept. 8, 1857	18, 128
Planter, Corn	T. Allen	Arrow Rock, Mo	Oct. 1, 1867	69, 301
Planter, Corn	J. M. Allison	Cranberry, Pa	May 19, 1868	78, 039
Planter, Corn	J. M. Allison	Cranberry, Pa	Dec. 1, 1868	84, 604
Planter, Corn	D. Altman	Nashville, Tenn	Nov. 14, 1871	120, 926
Planter, Corn	D. S. Ames	Laclede, Mo	Mar. 24, 1868	75, 825
Planter, Corn	D. W. Amos	Bedford, Pa	Nov. 1, 1864	44, 845
Planter, Corn	A. Anable	Middlesex, N. Y	Jan. 3, 1860	26, 642
Planter, Corn	J. K. Andrews and J. D. Green	Antrim, Ohio	Sept. 3, 1867	68, 514
Planter, Corn	A. Armstrong	Gillespie, Ill	Apr. 21, 1868	76, 971
Planter, Corn	J. Armstrong, jr	Elmira, Ill	July 22, 1862	35, 914
Planter, Corn	J. Armstrong, jr	Elmira, Ill	Sept. 20, 1864	44, 273

Index of patents issued from the United States Patent Office from 1790 *to* 1873, *inclusive*—Continued.

Invention.	Inventor.	Residence.	Date.	No.
Planter, Corn	J. Armstrong, jr	Elmira, Ill	Sept. 21, 1869	95, 068
Planter, Corn	J. N. Arvin and J. M. Whitmore	Valparaiso, Ind	Feb. 12, 1867	61, 981
Planter, Corn	J. N. Arvin and J. M. Whitmore	Valparaiso, Ind	Sept. 29, 1868	82, 578
Planter, Corn	F. J. Ashburn	West Union, W. Va	Jan. 12, 1869	85, 721
Planter, Corn	S. Avery	Pisgah, Mo	July 10, 1860	29, 046
Planter, Corn	A. G. Babcock	Galesburgh, Ill	Sept. 7, 1858	21, 404
Planter, Corn	W. R. Baldwin	Philadelphia, Pa	Nov. 13, 1866	59, 543
Planter, Corn	T. H. Ballard	Elk Grove, Wis	Dec. 3, 1872	133, 618
Planter, Corn	W. C. Banks	Como Depot, Miss	May 1, 1860	28, 051
Planter, Corn	W. C. Banks	Como Depot, Miss	May 1, 1860	28, 052
Planter, Corn	A. Barber	Port Byron, Ill	Feb. 5, 1867	61, 701
Planter, Corn	H. F. and T. R. Bargar	Border Plains, Iowa	Nov. 17, 1863	40, 595
Planter, Corn	A. Barnes	Bloomington, Ill	Nov. 5, 1872	132, 792
Planter, Corn	A. J. Barnhart	Schoolcraft, Mich	Aug. 14, 1855	13, 419
Planter, Corn	H. F. Batcheller	Sterling, Ill	Oct. 13, 1863	40, 233
Planter, Corn	M. Bates	Carlton, N. Y	Sept. 23, 1856	15, 755
Planter, Corn	H. Baughman	Columbus, Ohio	Sept. 7, 1869	94, 546
Planter, Corn	H. Baughman	Sandusky, Ohio	May 2, 1871	114, 390
Planter, Corn	L. Becker	Jackson Township, Pa	Feb. 7, 1871	111, 510
Planter, Corn	T. M. Bedgood	Cleveland, Ind	Aug. 17, 1858	21, 180
Planter, Corn	U. Beebee	Oakland, Mich	June 19, 1855	13, 077
Planter, Corn	W. T. Beckman	Petersburgh, Ill	Feb. 23, 1869	87, 237
Planter, Corn	M. Bemiss	Lyme, Ohio	July 15, 1856	15, 322
Planter, Corn	G. I. Bergen	Galesburgh, Ill	Dec. 1, 1863	40, 789
Planter, Corn	G. I. Bergen	Galesburgh, Ill	Mar. 7, 1865	46, 629
Planter, Corn	G. I. Bergen	Galesburgh, Ill	Aug. 29, 1865	49, 601
Planter, Corn	E. L. Bergstresser	Hublersburgh, Pa	Sept. 24, 1867	69, 296
Planter, Corn	H. C. Beshler	Berrysburgh, Pa	July 20, 1869	92, 780
Planter, Corn	H. C. Beshler	Berrysburgh, Pa	Oct. 4, 1870	107, 857
Planter, Corn	C. F. Bilhimer	Irwin's Station, Pa	Feb. 14, 1871	111, 808
Planter, Corn	O. Billings	La Grange, Ohio	Dec. 3, 1867	71, 572
Planter, Corn	L. F. Bingham and N. O. Pierce	Chicago, Ill	May 10, 1859	23, 895
Planter, Corn	W. F. Blandin	Macomb, Ill	Dec. 15, 1863	40, 900
Planter, Corn	J. M. Blessing	Jeffersonville, Ohio	Apr. 28, 1868	77, 251
Planter, Corn	W. Blessing	Jeffersonville, Ohio	June 13, 1865	48, 150
Planter, Corn	D. G. Boardman	Albia, Iowa	Mar. 18, 1873	136, 961
Planter, Corn	R. Bocklen	Jersey City, N. J	May 6, 1856	14, 801
Planter, Corn	J. L. Bond	Marshalltown, Iowa	Sept. 6, 1870	107, 156
Planter, Corn	J. H. Bonham	Elizabethtown, Ohio	Dec. 8, 1857	18, 798
Planter, Corn	J. H. Bonham	Elizabethtown, Ohio	May 22, 1860	28, 343
Planter, Corn	R. A. Boulware	Doniphan, Kans	Mar. 21, 1871	112, 894
Planter, Corn	B. Bower	Millersburgh, Ohio	Nov. 27, 1860	30, 713
Planter, Corn	R. M. and W. H. Bowman	London, Ohio	July 30, 1872	129, 924
Planter, Corn	W. C. Bowman	Clarksburgh District, Md	May 16, 1871	114, 908
Planter, Corn	W. H. Boyle	Cazenovia, N. Y	May 9, 1865	47, 614
Planter, Corn	H. Bradt and J. Otis	Schenectady, N. Y	Jan. 30, 1872	123, 080
Planter, Corn	N. Breed	Jeffersonville, Ind	Apr. 6, 1869	88, 604
Planter, Corn	T. R. Brent	Muscatine, Iowa	Aug. 5, 1862	36, 069
Planter, Corn	A. Brightbill	Bethel Township, Pa	Aug. 6, 1861	33, 015
Planter, Corn	A. W. Brinkerhoff	Upper Sandusky, Ohio	May 10, 1859	23, 896
Planter, Corn	A. W. Brinkerhoff	Upper Sandusky, Ohio	Jan. 8, 1861	31, 072
Planter, Corn	A. W. Brinkerhoff	Upper Sandusky, Ohio	Sept. 2, 1862	36, 335
Planter, Corn	J. H. Broad	Lodi, N. Y	Aug. 30, 1864	43, 968
Planter, Corn	T. Brockway	Chatsworth, Ill	Mar. 29, 1870	101, 349
Planter, Corn	J. Broughton	New York, N. Y	May 12, 1857	17, 258
Planter, Corn	E. C. Brown	Crawfordsville, Ind	Sept. 9, 1873	142, 675
Planter, Corn	G. W. Brown	Galesburgh, Ill	Apr. 5, 1864	42, 177
Planter, Corn	G. W. Brown	Galesburgh, Ill	Apr. 5, 1864	42, 178
Planter, Corn	G. W. Brown	Galesburgh, Ill	Apr. 5, 1864	42, 179
Planter, Corn	G. A. Bruce	Mechanicsburgh, Ill	Oct. 4, 1853	10, 064
Planter, Corn	A. B. Bruen	Trenton, Mo	Dec. 31, 1872	134, 356
Planter, Corn	J. T. Bryan	Lebanon, Ind	May 2, 1865	47, 514
Planter, Corn	S. B. Buck	Elyria, Ohio	Mar. 8, 1870	100, 495
Planter, Corn	G. Bunch and J. A. Price	Grand River Township, and Breckinridge, Mo.	Jan. 24, 1865	45, 971
Planter, Corn	G. W. Bunker	Saint Anthony, Minn	June 29, 1869	91, 904
Planter, Corn	J. A. Burchard	Beloit, Wis	Oct. 13, 1868	82, 918
Planter, Corn	A. C. and G. F. Burgner	Charleston, Ill	Feb. 4, 1873	135, 521
Planter, Corn	J. Burns	Elyria, Ohio	Apr. 16, 1867	63, 851
Planter, Corn	R. Burns	New York, N Y	July 11, 1865	48, 654
Planter, Corn	T. D. Burrall	Geneva, N. Y	June 26, 1835	....
Planter, Corn	C. Busch	Brooklyn, Iowa	Dec. 17, 1872	134, 032
Planter, Corn	W. R. Butler	Greenbush, Ill	Nov. 20, 1866	59, 742
Planter, Corn	E. M. Butz	Allegheny City, Pa	Feb. 11, 1868	74, 298
Planter, Corn	S. J. Bye	Bluff Point, Ind	July 4, 1871	116, 552
Planter, Corn	W. Caldwell	Bryan, Ohio	July 6, 1869	92, 159
Planter, Corn	H. W. Camp and A. W. Fox	Owego, N. Y	June 25, 1867	66, 002
Planter, Corn	A., W., and J. Campbell	Harrison, Ohio	June 28, 1859	24, 537
Planter, Corn	J. Campbell	Harrison, Ohio	May 8, 1866	54, 642
Planter, Corn	J. Campbell	Harrison, Ohio	Nov. 19, 1872	133, 199
Planter, Corn	J. P. Campbell	Danville, Pa	Nov. 16, 1869	96, 883
Planter, Corn	S. O. Campbell	Leavenworth, Kans	Sept. 1, 1868	81, 597
Planter, Corn	A. C. Carey	Ipswich, Mass	June 1, 1858	20, 467
Planter, Corn	J. H. Carothers	Davidsburgh, Pa	July 26, 1853	9, 882
Planter, Corn	G. J. Carpenter	Adrian, Mich	June 24, 1873	140, 114
Planter, Corn	J. T. Carter and M. B. Williamson.	Xenia, Ill	Dec. 20, 1870	110, 341
Planter, Corn	J. Case	La Fayette, Ind	Aug. 18, 1868	81, 066
Planter, Corn	J. Case	Springfield, Ohio	Dec. 9, 1873	145, 396
Planter, Corn	R. and J. L. Cassady	Hardingsville, N. J	Sept. 17, 1867	68, 841
Planter, Corn	A. Cassidy	Putnam, Ohio	Feb. 3, 1863	37, 564
Planter, Corn	W. R. Center	Athens, Ill	Dec. 4, 1860	30, 799
Planter, Corn	J. D. Chambers	Carthage, Mo	Dec. 15, 1868	84, 936
Planter, Corn	W. B. Chambers	Decatur, Ill	Aug. 15, 1871	117, 981
Planter, Corn	J. F. Champlin	Aurora, N. Y	July 24, 1866	56, 667
Planter, Corn	J. B. Chapman	Morrison, Ill	Aug. 24, 1869	94, 077
Planter, Corn	E. C. Chesney	Abingdon, Ill	June 21, 1864	43, 181
Planter, Corn	W. E. Chesney	Abingdon, Ill	Mar. 7, 1865	46, 637

Index of patents issued from the United States Patent Office from 1790 *to* 1873, *inclusive*—Continued.

Invention.	Inventor.	Residence.	Date.	No.
Planter, Corn	Z. T. Clagett	Washington, D. C	July 21, 1868	80, 058
Planter, Corn	B. Clark	Mackinaw, Ill	Sept. 4, 1866	57, 676
Planter, Corn	N. Clark	Rochester, Mass	Oct. 22, 1872	132, 355
Planter, Corn	W. R. Clark	Indianola, Ill	May 26, 1868	78, 262
Planter, Corn	J. Clarridge	Pancoastburgh, Ohio	Apr. 23, 1867	64, 076
Planter, Corn	R. J. Clay	Saint Louis, Mo	May 11, 1858	20, 193
Planter, Corn	J. D. Cochran	Milford, N. H	May 8, 1866	54, 507
Planter, Corn	J. S. Coen	Attica, Ind	Jan. 5, 1869	85, 563
Planter, Corn	J. S. Coen	Attica, Ind	Sept. 21, 1869	95, 085
Planter, Corn	W. Cogswell	Ottawa, Ill	July 10, 1866	56, 183
Planter, Corn	A. Colton	Attica, N. Y	May 28, 1861	32, 412
Planter, Corn	W. Combs	Duquoin, Ill	Jan. 8, 1861	31, 076
Planter, Corn	J. B. Conaty and M. Catt	Indianapolis, Ind	Mar. 28, 1871	113, 018
Planter, Corn	J. Conrad	Centralia, Ill	Nov. 13, 1866	59, 559
Planter, Corn	L. H. Converse	Springfield, Ill	July 16, 1872	129, 320
Planter, Corn	G. T. Cooley	Wooster, Ohio	Feb. 16, 1869	87, 028
Planter, Corn	J. Cooley	Tafton, Wis	Mar. 19, 1861	31, 700
Planter, Corn	J. P. Coonley	Farmington, Mich	Aug. 9, 1859	24, 993
Planter, Corn	A. M. Corbit	Bethlehem, Iowa	Mar. 26, 1867	63, 219
Planter, Corn	R. B. and H. Corson	Henry County, Mo	Dec. 20, 1870	110, 207
Planter, Corn	J. Cosand	Russiaville, Ind	Sept. 20, 1870	107, 457
Planter, Corn	W. H. Cox	Virden, Ill	Oct. 23, 1866	58, 988
Planter, Corn	W. H. Cox	Virden, Ill	Mar. 16, 1869	87, 825
Planter, Corn	W. Craig	Urbana, Ill	Dec. 8, 1863	40, 822
Planter, Corn	W. J. C. Crandall	Steele's, Ind	Jan. 7, 1873	134, 521
Planter, Corn	W. H. Crosby	Parish, N. Y	May 14, 1872	126, 788
Planter, Corn	A. L. Crow	Pennville, Ind	June 1, 1869	90, 640
Planter, Corn	I. Crum	Port Union, Ohio	Dec. 18, 1866	60, 483
Planter, Corn	J. W. Crume	Troy, Mo	Aug. 2, 1870	106, 034
Planter, Corn	W. Daggett, 4th	Cordova, Ill	Mar. 10, 1868	75, 380
Planter, Corn	T. Dale	Russellville, Ky	July 12, 1870	105, 314
Planter, Corn	J. Davis	Allegheny City, Pa	Nov. 14, 1865	50, 915
Planter, Corn	J. A. Davis	Verona, Miss	July 12, 1870	105, 318
Planter, Corn	J. R. Davis	Bloomfield, Iowa	Nov. 17, 1863	40, 610
Planter, Corn	J. R. Davis	Bloomfield, Iowa	May 9, 1865	47, 622
Planter, Corn	J. S. Davis	Arcadia, Ohio	Aug. 17, 1852	21 187
Planter, Corn	J. S. Davis	Ithaca, N. Y	Dec. 23, 1873	145, 728
Planter, Corn	N. C. Davis	Worthington, Ohio	June 14, 1870	104, 280
Planter, Corn	F. Dean	Beloit, Wis	July 4, 1865	48, 528
Planter, Corn	W. A. Dean	New Lexington, Ohio	July 16, 1872	129, 008
Planter, Corn	J. D. Delap	Tyrone Township, Pa	Apr. 18, 1871	113, 747
Planter, Corn	S. G. Dentler	Orangeville, Ill	July 2, 1867	66, 308
Planter, Corn	G. Dickerson	Harveysburgh, Ohio	July 14, 1868	79, 814
Planter, Corn	S. Dixson	Roseville, Ill	Jan. 21, 1873	135, 098
Planter, Corn	J. Doak	Keithsburgh, Ill	Dec. 13, 1864	45, 395
Planter, Corn	S. L. Donnell	Humboldt, Tenn	Mar. 19, 1872	124, 728
Planter, Corn	W. A. Donnell	Greensburgh, Ind	Aug. 13, 1867	67, 735
Planter, Corn	N. Drake	Newton, N. J	Feb. 2, 1858	19, 242
Planter, Corn	W. Drummond	Woodbridge, N. J	July 6, 1858	20, 781
Planter, Corn	T. Duncanson	Buford, Ohio	Jan. 5, 1869	85, 516
Planter, Corn	D. Dutcher	Blue Grass, Iowa	June 26, 1860	28, 847
Planter, Corn	J. W. Eardly	Cascade, Mich	Mar. 9, 1869	87, 651
Planter, Corn	W. A. Eastin	Henderson County, Ky	Jan. 21, 1873	134, 984
Planter, Corn	T. M. Edgar	Paris, Tenn	Oct. 31, 1871	120, 424
Planter, Corn	A. Edwards	Cuba, Mo	Feb. 14, 1871	111, 826
Planter, Corn	R. S. Edwards	Savannah, Mo	Mar. 24, 1868	75, 883
Planter, Corn	Y. I. Edwards	Trenton, Tenn	Jan. 10, 1871	110, 837
Planter, Corn	P. Eidmann	Pekin, Ill	Feb. 11, 1868	74, 326
Planter, Corn	J. Elbertson	Kirksville, Mo	May 12, 1868	77, 807
Planter, Corn	J. Elbertson	Kirksville, Mo	Jan. 5, 1869	85, 652
Planter, Corn	S. Elliott	Washington, Ind	Mar. 15, 1859	23, 235
Planter, Corn	B. H. Elmore	Richmond, Ind	Feb. 19, 1861	31, 448
Planter, Corn	J. H. Elward	Ottawa, Ill	Jan. 5, 1864	41, 176
Planter, Corn	D. J. Ely	Acton, Ind	Dec. 19, 1865	51, 575
Planter, Corn	R. Erdley	Selm's Grove, Pa	Oct. 18, 1870	108, 468
Planter, Corn	J. H. Ernst	Millerstown, Pa	Jan. 8, 1869	91, 106
Planter, Corn	J. W. Fawkes	Decatur, Ill	Aug. 8, 1865	49, 250
Planter, Corn	R. R. Fenner	Paxton, Ill	May 19, 1863	38, 627
Planter, Corn	R. W. Fenwick and R. Bocklen	Brooklyn, N. Y., and Jersey City, N. Y.	Aug. 7, 1855	13, 387
Planter, Corn	J. G. Fetzer	Brunswick, Mo	May 24, 1870	103, 315
Planter, Corn	S. Filson and W. E. Kinert	Bluffton, Ind	Nov. 17, 1868	84, 178
Planter, Corn	W. H. Fish, jr	Scarsdale, N. Y	May 12, 1868	77, 808
Planter, Corn	D. S. Fisher	Cedar Spring, Ind	May 28, 1867	65, 197
Planter, Corn	D. Fitzpatrick and J. Knull	Saint Paris, Ohio	Dec. 21, 1869	98, 157
Planter, Corn	F. G. and E. A. Floyd	Macomb, Ill	Aug. 28, 1860	29, 777
Planter, Corn	M. C. Floyd	Bloomfield, Iowa	Jan. 30, 1866	52, 281
Planter, Corn	W. C. Ford	West Salem, Ohio	Apr. 9, 1861	31, 966
Planter, Corn	J. T. Force	Henry County, Ky	July 11, 1871	111, 827
Planter, Corn	R. Forman	Normal, Ill	July 6, 1869	92, 297
Planter, Corn	C. Foster	Prairie City, Ill	Nov. 24, 1863	40, 681
Planter, Corn	J. S. Fowler	Peoria, Ill	Aug. 14, 1860	29, 578
Planter, Corn	J. M. Foy	Fountain Green, Ill	Apr. 16, 1861	32, 056
Planter, Corn	A. Franklin	Genoa Cross-Roads, Ohio	July 14, 1857	17, 786
Planter, Corn	J. A. Freese	Hanover, Ohio	July 8, 1862	35, 814
Planter, Corn	P. A. Freylinghousen and J. G. Heilman.	Johnstown, Pa	June 21, 1859	24, 452
Planter, Corn	J. P. Fulghum and L. L. Lawrence.	Dublin, Ind	July 1, 1873	140, 493
Planter, Corn	E. C. Gage	Witoka, Minn	Dec. 30, 1873	146, 060
Planter, Corn	T. A. Galt and G. S. Tracy	Sterling, Ill	June 17, 1873	140, 031
Planter, Corn	W. M. Garce	Granville, Ohio	Mar. 6, 1860	27, 358
Planter, Corn	W. C. Gardner	Pokagon, Mich	Feb. 2, 1869	86, 386
Planter, Corn	J. Gilbert	Wyalusing, Wis	Jan. 28, 1868	73, 797
Planter, Corn	R. B. Gilbert	Sutherland Springs, Tex	May 10, 1859	23, 915
Planter, Corn	W. Gilman	Ottawa, Ill	Mar. 15, 1870	100, 748
Planter, Corn	D. H. Gobin	Springfield, Ill	July 2, 1872	128, 613
Planter, Corn	F. A. Goddard	Lexington, Ill	July 31, 1860	29, 433

Index of patents issued from the United States Patent Office from 1790 *to* 1873, *inclusive*—Continued.

Invention.	Inventor.	Residence.	Date.	No.
Planter, Corn	A. J. Going	Clinton, La	May 12, 1868	77, 811
Planter, Corn	J. Golder	Sterling, Ill	Aug. 2, 1864	43, 682
Planter, Corn	W. B. Goodwin	Ettingham, Ill	Sept. 29, 1868	82, 515
Planter, Corn	W. B. Goodwin	Kinmundy, Ill	Jan. 5, 1869	85, 581
Planter, Corn	J. M. Gordon and E. Christianson.	Saint Joseph, Mo	Nov. 26, 1867	71, 379
Planter, Corn	H. Gortner	Washport, Ohio	Dec. 14, 1869	97, 906
Planter, Corn	H. Gortner	Washport, Ohio	Nov. 28, 1871	131, 357
Planter, Corn	H. Gortner	Frazeyburgh, Ohio	Nov. 5, 1872	132, 824
Planter, Corn	L. Graham	Plymouth, Ill	Apr. 4, 1871	113, 422
Planter, Corn	J. B. Greely	Summit, Iowa	Oct. 21, 1862	36, 753
Planter, Corn	H. P. Gregg	Cincinnati, Ohio	Mar. 24, 1868	75, 749
Planter, Corn	J. Gross	Decatur, Ill	Jan. 3, 1860	26, 670
Planter, Corn	J. Gross	Decatur, Ill	June 6, 1865	48, 130
Planter, Corn	R. Hackman	Clearfield, Pa	Sept. 3, 1872	131, 097
Planter, Corn	S. H. Hamilton	Deerfield, N. J	June 11, 1872	127, 871
Planter, Corn	J. A. Hamrick	Parnassus, Va	Nov. 9, 1869	96, 586
Planter, Corn	J. W. Harbin	Delaware Station, Ind	Oct. 16, 1860	30, 445
Planter, Corn	E. E. Hardy and W. Dubrul	Joliet, Ill	June 14, 1870	104, 303
Planter, Corn	J. J. Harpel	Lebanon, Pa	Oct. 12, 1869	95, 681
Planter, Corn	S. Harpster	Centre Hall, Pa	Aug. 27, 1867	68, 070
Planter, Corn	W. N. and J. J. Harrison	Hornby, N. Y	June 9, 1868	78, 664
Planter, Corn	S. E. Hartwell	New York, N. Y	May 10, 1859	23, 919
Planter, Corn	C. A. Haskell	Galena, Ill	Dec. 27, 1870	110, 458
Planter, Corn	J. J. S. Hassler	Ripley, Va	Dec. 15, 1857	18, 846
Planter, Corn	P. Hatch	Norwich, Vt	June 22, 1858	20, 639
Planter, Corn	L. L. Haworth	Decatur, Ill	Sept. 19, 1871	119, 142
Planter, Corn	G. D. Haworth	Decatur, Ill	Sept. 3, 1861	33, 196
Planter, Corn	G. D. Haworth	Springfield, Ill	Oct. 18, 1864	44, 725
Planter, Corn	G. D. Haworth	Decatur, Ill	Feb. 22, 1870	100, 032
Planter, Corn	G. D. Haworth	Decatur, Ill	Jan. 14, 1873	134, 747
Planter, Corn	A. Hayes and J. Vancuren	Chenoa, Ill	May 22, 1860	28, 367
Planter, Corn	G. Hayes	Lawrenceburgh, Ind	June 10, 1873	139, 672
Planter, Corn	J. Haynes	Cameron, Ill	Mar. 29, 1859	23, 371
Planter, Corn	A. A. Hazard	New York, N. Y	Aug. 8, 1865	49, 265
Planter, Corn	A. Hearst	Peoria, Ill	Jan. 26, 1869	86, 226
Planter, Corn	A. Hearst	Peoria, Ill	Nov. 7, 1871	120, 643
Planter, Corn	C. W. S. Heaton	Belleville, Ill	Feb. 24, 1863	37, 754
Planter, Corn	G. W. Hendricks	Rushville, Ind	Aug. 6, 1872	130, 220
Planter, Corn	E. E. Henegan	Downsville, Wis	Oct. 7, 1873	143, 450
Planter, Corn	C. W. Henkle	Washington Court-House, Ohio.	Mar. 19, 1867	63, 047
Planter, Corn	C. W. Henkle	Washington Court-House, Ohio.	Feb. 11, 1868	74, 357
Planter, Corn	D., J., and J. F. Herr	Lancaster, Pa	Oct. 9, 1860	30, 316
Planter, Corn	A. E. Herrington and J. D. Richards.	Big Prairie Ronde, Mich	Aug. 25, 1868	81, 502
Planter, Corn	J. A. C. and A. S. Hickman	Summerfield, Ill	Sept. 3, 1861	33, 197
Planter, Corn	W. Hickok	Pharisburgh, Ohio	Oct. 1, 1872	131, 756
Planter, Corn	J. Hildebrand	Taneytown, Md	Dec. 24, 1872	134, 279
Planter, Corn	J. H. Hill and J. T. Hammond	Clinton, Ill	June 23, 1868	79, 120
Planter, Corn	T. M. Hill	Eaton, Ohio	Feb. 6, 1866	52, 413
Planter, Corn	A. F. Hines	Washington, D. C	Oct. 3, 1865	50, 246
Planter, Corn	I. P. Hines	Independence, Iowa	Jan. 5, 1864	41, 120
Planter, Corn	C. Hippensteel	Lee's Cross-Roads, Pa	Apr. 19, 1870	102, 004
Planter, Corn	J. W. Hitchcock and J. K. Deyo	Morrisville, N. Y	Nov. 19, 1867	71, 172
Planter, Corn	J. L. Hoag	Genoa, Ill	Jan. 18, 1859	22, 650
Planter, Corn	W. J. Hobson	Savannah, Mo	Feb. 18, 1867	62, 200
Planter, Corn	A. and E. Hodgson	El Paso, Ill	July 20, 1869	92, 826
Planter, Corn	E. R. Holford	Westford, Wis	Oct. 9, 1866	58, 724
Planter, Corn	H. R. Holland	Wilmington, Va	Oct. 4, 1870	107, 908
Planter, Corn	T. H. Hoskings	Crawfordsville, Ind	Jan. 20, 1844	3, 413
Planter, Corn	G. O. Houck	Springfield, Ohio	Apr. 22, 1873	138, 158
Planter, Corn	W. House	Aurora, Ind	Oct. 21, 1873	143, 905
Planter, Corn	H. J. Howe	Onargo, Ill	May 14, 1861	32, 294
Planter, Corn	M. A. Howell, jr	Ottawa, Ill	Apr. 10, 1860	27, 807
Planter, Corn	W. W. Hubbard	Edinburgh, Ind	Apr. 24, 1866	54, 166
Planter, Corn	W. W. Hubbard	Edinburgh, Ind	July 23, 1867	67, 115
Planter, Corn	D. W. Hughes	New London, Mo	Sept. 8, 1857	18, 148
Planter, Corn	J. Hughes and N. Stonecipher	Cambridge, Ind	Jan. 18, 1859	22, 652
Planter, Corn	N. G. Hughes	Waynesburgh, Pa	Feb. 4, 1868	74, 092
Planter, Corn	D. Humphreys	Cincinnati, Ohio	Apr. 16, 1861	32, 065
Planter, Corn	W. Hunter	Hastings, Minn	Apr. 9, 1867	63, 724
Planter, Corn	W. H. Hunter	Ridgefield, Ill	May 16, 1865	47, 728
Planter, Corn	C. Hutchins	Aubrey, Kans	Oct. 21, 1873	143, 826
Planter, Corn	B. Ingebrightson	Cambridge, Wis	Nov. 7, 1871	120, 649
Planter, Corn	J. Ingels and W. R. Mount	Milton, Ind	Feb. 11, 1873	135, 813
Planter, Corn	H. Ingraham	Naples, N. Y	Oct. 27, 1857	18, 508
Planter, Corn	J. Jackson, jr	Chatsworth, Ill	Aug. 15, 1871	118, 134
Planter, Corn	D. W. Jacoby	Shelbyville, Ill	Sept. 3, 1867	68, 443
Planter, Corn	J. D. Jeffers, J. Sparks, and J. H. Jeffers.	Philadelphia, Pa	July 29, 1856	15, 426
Planter, Corn	L. K. Jenne	Grand Rapids, Mich	Feb. 12, 1861	31, 427
Planter, Corn	J. John	Massillon, Ohio	Nov. 26, 1867	71, 493
Planter, Corn	H. J. Johnson	Saint Peter, Minn	Dec. 24, 1867	72, 501
Planter, Corn	J. Johnson	Naples, Ill	May 29, 1860	28, 499
Planter, Corn	J. A. Johnson	Pendleton, Ind	Aug. 31, 1869	94, 417
Planter, Corn	J. B. Johnson	Rock Island, Ill	July 20, 1869	92, 728
Planter, Corn	J. J. Jones and S. H. Dwight	Decatur, Ill	July 9, 1872	128, 888
Planter, Corn	S. F. Jones	Saint Paul, Ind	Apr. 5, 1864	42, 196
Planter, Corn	S. W. Jones	Bluffton, Ind	Sept. 29, 1868	82, 531
Planter, Corn	H. Jordan	Milton, Ohio	Jan. 3, 1865	45, 721
Planter, Corn	W. H. Karicofe	Harrisonburgh, Va	June 19, 1866	55, 668
Planter, Corn	H. Keeney and C. H. See	Danville and New Florence, Mo.	July 7, 1868	79, 576
Planter, Corn	D. Keethler	Mount Oreb, Ohio	Nov. 23, 1869	97, 200
Planter, Corn	D. Keethler	Mount Oreb, Ohio	July 5, 1870	105, 090
Planter, Corn	C. A. Kellog	Elyria, Ohio	Nov. 24, 1868	84, 426

Index of patents issued from the United States Patent Office from 1790 *to* 1873, *inclusive*—Continued.

Invention.	Inventor.	Residence.	Date.	No.
Planter, Corn	W. G. Kenedy	Greenfield, Ind	Nov. 8, 1864	44, 955
Planter, Corn	D. A. Kershner	Elliottstown, Ill	Apr. 21, 1868	77, 050
Planter, Corn	I. J. Kidd	Young Settlement, Tex	Sept. 8, 1868	81, 911
Planter, Corn	E. W. Kimball	Ottawa, Ill	Jan. 8, 1861	31, 085
Planter, Corn	P. H. Kimball	Prophetstown, Ill	Aug. 1, 1865	49, 121
Planter, Corn	W. H. King	Charleston, Ill	May 10, 1859	23, 980
Planter, Corn	J. M. Kiracofe	Mount Solon, Va	Apr. 19, 1870	102, 017
Planter, Corn	A. Kirlin	New Boston, Ill	Oct. 18, 1859	25, 833
Planter, Corn	M. L. and J. B. Kissell	Springfield, Ohio	Sept. 20, 1870	107, 508
Planter, Corn	J. Klar	Shelbyville, Ill	Dec. 30, 1873	145, 952
Planter, Corn	G. W. Knapp	Corning, N. Y	June 11, 1867	65, 755
Planter, Corn	J. A. Knetzer	Fillmore, Ind	Aug. 1, 1871	117, 646
Planter, Corn	M. Knickerbocker	New Lenox, Ill	Sept. 16, 1873	142, 922
Planter, Corn	J. I. Knight	Philadelphia, Pa	Jan. 10, 1860	26, 819
Planter, Corn	J. Knull and J. P. Pence	Saint Paris, Ohio	Oct. 3, 1871	119, 467
Planter, Corn	H. Koeller	Camp Point, Ill	Nov. 15, 1870	109, 225
Planter, Corn	H. Koeller and W. Uecke	Camp Point, Ill	July 16, 1872	129, 144
Planter, Corn	J. Krebs and A. Johns	Massillon, Ohio	Oct. 1, 1867	69, 449
Planter, Corn	J. L. Kreider	Chesnut Level, Pa	Jan. 16, 1872	122, 770
Planter, Corn	R. Kuschke and P. Merkel	Saint Louis, Mo	May 26, 1857	17, 380
Planter, Corn	E. P. Lacey	Rochester, N. Y	Apr. 8, 1856	14, 631
Planter, Corn	L. Lacey	Bath, Ill	Apr. 29, 1873	138, 259
Planter, Corn	A. Ladd	Saint Lawrence, N. Y	Sept 4, 1866	57, 734
Planter, Corn	D. Ladd	Dearborn, Mich	Nov. 30, 1858	22, 183
Planter, Corn	J. G. La Ponte	Dayton, Ohio	Nov. 25, 1873	144, 989
Planter, Corn	N. C. Lamb	Wyandotte, Mich	Sept. 9, 1873	142, 706
Planter, Corn	R. B. Lanum	Washington, Ohio	May 17, 1864	42, 781
Planter, Corn	H. B. Lansing	Hudson, Mich	Feb. 27, 1866	52, 861
Planter, Corn	H. B. Lansing and H. W. Grenell	Hudson, Mich	Sept. 30, 1862	36, 569
Planter, Corn	L. Larchar	Utica, N. Y	Aug. 27, 1867	68, 091
Planter, Corn	E. B. Lawrence and C. Quick	Lakeville, Ohio	Nov. 10, 1868	83, 979
Planter, Corn	D. P. Leach	Franklin, Ind	Mar. 23, 1869	88, 048
Planter, Corn	J. L. Leas	York Sulphur Springs, Pa	Sept. 22, 1868	82, 422
Planter, Corn	J. L. Leas	Hampton, Pa	Aug. 13, 1872	130, 378
Planter, Corn	J. Lee	Galesburgh, Ill	May 22, 1860	28, 383
Planter, Corn	W. Lees	Germantown, Ohio	Sept. 13, 1859	25, 425
Planter, Corn	J. C. Leffel	Shelbina, Mo	Dec. 1, 1863	40, 760
Planter, Corn	G. W. Lewis	Winchester, Ky	Apr. 4, 1871	113, 717
Planter, Corn	N. A. Lewis	Glen's Falls, N. Y	July 21, 1857	17, 841
Planter, Corn	N. Liddell	La Fayette, N. Y	Apr. 2, 1867	63, 401
Planter, Corn	N. and M. Liddell	La Fayette, N. Y	Mar. 10, 1868	75, 435
Planter, Corn	O. Lippincott	Camden, N. J	Apr. 27, 1858	20, 074
Planter, Corn	W. H. Littel	Prairie du Chien, Wis	July 26, 1870	105, 702
Planter, Corn	H. C. Locke	Somerville, Tenn	Apr. 21, 1868	77, 057
Planter, Corn	C. Long	Paris, Ill	Jan. 1, 1867	60, 914
Planter, Corn	F. Loos	Denver City, Colo	Aug. 9, 1864	43, 782
Planter, Corn	J. T. Lowrey, J. A. Case, and R. Chew.	High Banks, Ind	Mar. 3, 1868	75, 174
Planter, Corn	D. and D. F. Luse	Spring Mills, Pa	Mar. 27, 1866	53, 459
Planter, Corn	J. W. Magers	Reinersville, Ohio	May 3, 1870	102, 564
Planter, Corn	N. Maisonneuve	Kankakee, Ill	May 11, 1869	90, 008
Planter, Corn	S. Malone	Tremont, Ill	Jan. 3, 1854	10, 393
Planter, Corn	W. H. Maple	Chariton, Iowa	Apr. 28, 1863	38, 313
Planter, Corn	T. Marcus	New York, N. Y	Dec. 1, 1857	18, 756
Planter, Corn	A. S. Markham	Monmouth, Ill	Oct. 27, 1863	40, 418
Planter, Corn	A. S. and D. Markham	Monmouth, Ill	Dec. 11, 1860	30, 883
Planter, Corn	T. R. Markillie	Winchester, Ill	Nov. 13, 1855	13, 794
Planter, Corn	A. C. Martin	Waddam's Centre, Ill	Apr. 8, 1873	137, 700
Planter, Corn	A. C. Martin and R. Ferguson	Lena, Ill	Dec. 2, 1873	145, 115
Planter, Corn	J. Matthews	Lincoln, Ill	Nov. 21, 1871	121, 116
Planter, Corn	H. Maxell	Canton, Ohio	Feb. 19, 1867	62, 280
Planter, Corn	J. A. McClure	Mount Carroll, Ill	July 19, 1870	105, 473
Planter, Corn	T. B. McConaughey	Newark, Del	Dec. 4, 1866	60, 218
Planter, Corn	R. McCorkell	Warsaw, Minn	Oct. 25, 1864	44, 810
Planter, Corn	W. H. McCormick	Muncie, Ind	July 6, 1869	92, 260
Planter, Corn	D. McCullough	Oxford Township, Canada	July 27, 1869	92, 984
Planter, Corn	O. C. McCune	Darby Creek, Ohio	Dec. 13, 1859	26, 442
Planter, Corn	O. C. McCune	Pleasant Valley, Ohio	Aug. 2, 1864	43, 699
Planter, Corn	O. C. McCune	Darby Creek, Ohio	Aug. 7, 1866	56, 969
Planter, Corn	I. W. McGaffey	Buffalo, N. Y	June 16, 1857	17, 566
Planter, Corn	J. McGinnis	Frazeysburgh, Ohio	Dec. 10, 1872	133, 789
Planter, Corn	W. McLucas	Reinersville, Ohio	Feb. 4, 1868	74, 112
Planter, Corn	W. McLucas	Reinersville, Ohio	July 20, 1869	92, 861
Planter, Corn	N. Mendenhall	Greensburgh, Ind	Dec. 6, 1870	109, 923
Planter, Corn	N. Mendenhall	Greensburgh, Ind	Nov. 19, 1872	133, 109
Planter, Corn	C. T. Merry and M. A. Dunton	Norwalk, Ohio	Aug. 8, 1871	117, 910
Planter, Corn	C. T. Merry and M. A. Danton	Norwalk, Ohio	Oct. 15, 1872	132, 306
Planter, Corn	A. Merwin and C. H. Hobart	Padua, Ill	Nov. 3, 1863	40, 492
Planter, Corn	S. P. Metz and M. Rohrer	McDonaldsville, Ohio	Sept. 25, 1866	58, 277
Planter, Corn	W. M. Meyers	New Brunswick, N. J	Sept. 20, 1870	107, 520
Planter, Corn	S. Mickley	York, Pa	Sept. 13, 1870	107, 280
Planter, Corn	G. G. J. Millar	Lockbourne, Ohio	Aug. 20, 1872	130, 653
Planter, Corn	A. Miller	Brockport, N. Y	Aug. 14, 1860	29, 653
Planter, Corn	J. Miller	Bucyrus, Ohio	Mar. 31, 1857	16, 930
Planter, Corn	P. Mills	Ridge Farm, Ill	Oct. 27, 1863	40, 423
Planter, Corn	T. S. Mills	Iberia, Ohio	June 26, 1860	28, 886
Planter, Corn	H. Mitchell	Osborn, Ohio	July 7, 1868	79, 588
Planter, Corn	H. S. Mitchell and C. Search	Hublersburgh, Pa	Aug. 6, 1867	67, 444
Planter, Corn	J. G. Mitchell	Collington, Md	Mar. 29, 1859	23, 382
Planter, Corn	M. Mitchell	Altona, Ill	Aug. 9, 1859	25, 033
Planter, Corn	N. B. Moody	Woodman, Wis	Oct. 3, 1871	119, 470
Planter, Corn	J. Moore	Bushnell, Ill	Feb. 18, 1868	74, 710
Planter, Corn	J. C. Moore	Peoria, Ill	May 29, 1860	28, 498
Planter, Corn	J. C. Moore	Peoria, Ill	July 8, 1862	35, 831
Planter, Corn	Q. R. and P. Moor and E. L. Patrick.	Forest Hill, Ind	Dec. 1, 1868	84, 568
Planter, Corn	W. H. Moore	Blooming Grove, Ind	Feb. 18, 1868	74, 711
Planter, Corn	O. P. Moran	Hainesville, Mo	Nov. 8, 1859	26, 044

Index of patents issued from the United States Patent Office from 1790 *to* 1873, *inclusive*—Continued.

Invention.	Inventor.	Residence.	Date.	No.
Planter, Corn	R. W. Moran	Chicago, Ill	Dec. 24, 1867	72, 525
Planter, Corn	L. Morris	Woodbury, Ill	June 26, 1860	28, 891
Planter, Corn	W. Morrison	Carlisle, Pa	Sept. 13, 1859	25, 435
Planter, Corn	W. Morrison	Carlisle, Pa	Apr. 4, 1871	113, 550
Planter, Corn	P. C. Mosier	Homer, Ill	Jan. 26, 1858	19, 198
Planter, Corn	S. Mowry and E. Deppen	Womelsdorf, Pa	Dec. 11, 1860	30, 885
Planter, Corn	D. Moyer	New Hamburgh, Pa	June 5, 1860	28, 595
Planter, Corn	G. W. Moyers	Gordonsville, Va	June 29, 1869	92, 084
Planter, Corn	W. Mull	Rantoul, Ill	Aug. 12, 1873	141, 723
Planter, Corn	J. Y. D. Murphy	Half Moon, Pa	Jan. 22, 1861	31, 186
Planter, Corn	J. Y. D. Murphy	Half Moon, Pa	Mar. 8, 1864	41, 854
Planter, Corn	C. C. Myers	Sterling, Ill	Oct. 16, 1866	58, 876
Planter, Corn	C. K. Myers	Pekin, Ill	Apr. 23, 1861	32, 168
Planter, Corn	D. C. Myers	Richmondale, Ohio	June 12, 1860	28, 684
Planter, Corn	E. J. Myers	Onawa, Iowa	Dec. 5, 1871	121, 649
Planter, Corn	D. Nichols	Onargo, Ill	Jan. 3, 1860	26, 700
Planter, Corn	H. A. Nicholls	Saint Louis, Mo	Sept. 13, 1870	107, 402
Planter, Corn	J. Nitkey	Minneapolis, Minn	Apr. 8, 1873	137, 563
Planter, Corn	A. J. Noe	Mitchell, Ind	Nov. 5, 1872	132, 770
Planter, Corn	R. F. Norwood	Charlotte, N. C	Feb. 28, 1871	112, 171
Planter, Corn	J. Olmsted	Knoxville, Ill	Jan. 19, 1864	41, 345
Planter, Corn	J. Olmsted	Knoxville, Ill	Nov. 1, 1864	44, 920
Planter, Corn	S. Y. Orr	Morning Sun, Iowa	June 29, 1869	91, 961
Planter, Corn	W. F. Osgood	Lowell, Mass	Jan. 12, 1864	41, 263
Planter, Corn	G. Gaddington	Waubeck, Iowa	Dec. 6, 1870	109, 837
Planter, Corn	A. G. Parker	North Gage, N. Y	July 29, 1862	36, 020
Planter, Corn	J. B. Parker	Knobnoster, Mo	Dec. 14, 1869	97, 957
Planter, Corn	T. H. Parker and D. Kellison	Parkersburgh, Ill	Oct. 15, 1867	69, 836
Planter, Corn	E. Parmentier	Clifton, Ill	July 1, 1873	140, 432
Planter, Corn	J. M. Parsons	Charles City, Iowa	Feb. 14, 1871	111, 774
Planter, Corn	B. F. Partridge	Syracuse, N. Y	Jan. 9, 1849	6, 016
Planter, Corn	G. F. Partridge	Adrian, Mich	Sept. 15, 1868	82, 155
Planter, Corn	J. I. Patton	Tiffin, Ohio	July 27, 1869	93, 116
Planter, Corn	E. Peck	Middleport, Ill	June 23, 1863	38, 979
Planter, Corn	G. H. Peck	East Hamburgh, N. Y	Nov. 22, 1870	109, 543
Planter, Corn	L. G. Peel	Webster County, Ga	Aug. 9, 1859	25, 041
Planter, Corn	W. T. Pepper	Rising Sun, Ind	June 16, 1857	17, 582
Planter, Corn	S. M. Perkins	Fort Hill, Ill	Feb. 3, 1857	16, 551
Planter, Corn	L. B. Phelps	Geneva, Ohio	May 18, 1858	20, 297
Planter, Corn	G. W. Phillips and B. C. Richardson.	Oconomowoc, Wis	Feb. 16, 1869	86, 941
Planter, Corn	D. Pierson, J. W. Macy, and J. D. Moore.	Grinnell, Iowa	Feb. 25, 1873	136, 337
Planter, Corn	J. Pies	Sparland, Ill	Dec. 26, 1871	122, 191
Planter, Corn	J. M. Piper	Harrison City, Pa	July 12, 1870	105, 365
Planter, Corn	G. W. Pittman	Winona, Miss	Aug. 9, 1870	106, 202
Planter, Corn	P. Platter	Moore's Hill, Ind	July 5, 1859	24, 703
Planter, Corn	F. W. Poe, jr	Vandalia, Ill	Jan. 3, 1871	110, 786
Planter, Corn	W. R. Pomeroy	Millersburgh, Ohio	Apr. 8, 1862	34, 902
Planter, Corn	B. Porter	Ossian, N. Y	May 24, 1870	103, 499
Planter, Corn	H. G. Porter	Hopkinton, Iowa	Dec. 6, 1870	109, 936
Planter, Corn	N. M. Powers	Kirksville, Mo	Mar. 30, 1869	88, 508
Planter, Corn	F. B. Preston	Fayette, Mo	Feb. 26, 1861	31, 579
Planter, Corn	J. Price	Harrison, Ohio	July 10, 1860	29, 100
Planter, Corn	B. L. Prime	Hamilton, Ohio	Dec. 1, 1857	18, 768
Planter, Corn	W. S. Purdey	Butler, Ind	Sept. 21, 1869	95, 139
Planter, Corn	A. Putnam	Owego, N. Y	Sept. 18, 1866	58, 135
Planter, Corn	J. Rader	Daleville, Ind	July 20, 1869	92, 882
Planter, Corn	J. B. Raines	Tremont, Iowa	Apr. 23, 1867	64, 143
Planter, Corn	P. Raines	London, Ohio	May 1, 1855	12, 783
Planter, Corn	F. A. Ramey and R. R. Cross	Woodstock, Va	Oct. 17, 1871	119, 950
Planter, Corn	B. Randall	Adams, N. Y	Apr. 14, 1868	76, 658
Planter, Corn	J. R. Randall	Camargo, Ill	Feb. 2, 1869	86, 589
Planter, Corn	J. R. Randall	Camargo, Ill	Jan. 3, 1871	110, 788
Planter, Corn	S. G. Randall	Rockford, Ill	Apr. 29, 1856	14, 776
Planter, Corn	J. H. Rankin	Versailles, Mo	Dec. 4, 1860	30, 836
Planter, Corn	W. Rayhill	Pana, Ill	Jan. 21, 1868	73, 649
Planter, Corn	L. M. Reamy	Kokomo, Ind	Mar. 12, 1867	62, 886
Planter, Corn	H. Reed and W. P. Pennewell	Middletown, Mo	Aug. 14, 1866	57, 259
Planter, Corn	S. B. Reeder	Meacham, Ill	Oct. 6, 1868	82, 875
Planter, Corn	J. M. Reeds	Millwood Township, Mo	Feb. 12, 1867	62, 067
Planter, Corn	J. Rice	Prairie Creek, Ind	Nov. 12, 1872	133, 055
Planter, Corn	W. E. Rich	New Providence, Iowa	June 26, 1866	55, 908
Planter, Corn	S. Richardson	Jericho, Vt	June 16, 1857	17, 584
Planter, Corn	J. Richelderfer	Cridersville, Ohio	Jan. 3, 1871	110, 679
Planter, Corn	J. S. Rickel	Genesco, Ill	Sept. 25, 1866	58, 295
Planter, Corn	J. S. Rickel	Geneseo, Ill	Mar. 12, 1867	62, 888
Planter, Corn	J. W. Ricketts	Charlestown, Ill	Nov. 23, 1869	97, 229
Planter, Corn	J. J. Rider	Wilton Junction, Iowa	Oct. 13, 1863	40, 281
Planter, Corn	P. Ritchey	Hamilton, Ill	Apr. 16, 1861	32, 084
Planter, Corn	M. Robbins	Cincinnati, Ohio	Feb. 10, 1857	16, 611
Planter, Corn	B. F. Robertson	Cap Au Gris, Mo	May 28, 1867	65, 278
Planter, Corn	P. Rogers	Sharon, Ohio	Mar. 30, 1869	88, 415
Planter, Corn	A. Röhmér	Newport, Ky	June 3, 1873	139, 617
Planter, Corn	C. Ropp	McLean County, Ill	Oct. 4, 1859	25, 575
Planter, Corn	L. Ruf	Rock Falls, Ill	July 16, 1872	129, 595
Planter, Corn	A. Runstetler	Peoria, Ill	Apr. 13, 1869	88, 812
Planter, Corn	D. Ruppart	Nimisilla, Ohio	Feb. 19, 1867	62, 290
Planter, Corn	C. B. Ruth	Doylestown, Pa	Dec. 21, 1869	98, 193
Planter, Corn	H. Ruth	Summerfield, Ill	June 10, 1862	35, 545
Planter, Corn	J. B. Ryder	Wapello, Iowa	Apr. 21, 1863	38, 214
Planter, Corn	W. H. Sabins	Rockford, Ill	Feb. 2, 1869	86, 595
Planter, Corn	W. J. Sager	Milesburgh, Pa	Apr. 18, 1871	113, 926
Planter, Corn	J. M. Sampson	Waynesville, Ill	Nov. 19, 1867	71, 070
Planter, Corn	O. Sampson	Petersburgh, Ill	May 26, 1868	78, 237
Planter, Corn	W. G. Savage	Clinton, Ill	Sept. 25, 1860	30, 158
Planter, Corn	M. Saviers and W. N. Ayres	Wyandot and Bristolville, Ohio.	May 30, 1865	47, [illegible]87

Index of patents issued from the United States Patent Office from 1790 *to* 1873, *inclusive*—Continued.

Invention.	Inventor.	Residence.	Date.	No.
Planter, Corn	M. Schnapp and W. J. Hollis	De Witt, Mo	Apr. 16, 1872	125, 847
Planter, Corn	C. Schnepf	Lancaster, Pa	July 28, 1857	17, 898
Planter, Corn	J. Schott	Peoria, Ill	Feb. 21, 1871	111, 979
Planter, Corn	L. Scofield	Watertown, Wis	Feb. 7, 1871	111, 687
Planter, Corn	L. Scofield	Watertown, Wis	Nov. 21, 1871	121, 204
Planter, Corn	A. Seaman	Winnebago County, Ill	May 1, 1860	28, 106
Planter, Corn	J. Seibel	Manlius, Ill	May 30, 1865	47, 990
Planter, Corn	J. Selby	Peoria, Ill	Aug. 30, 1864	44, 019
Planter, Corn	J. Selby	Peoria, Ill	Sept. 15, 1868	82, 162
Planter, Corn	J. Selby	Peoria, Ill	July 13, 1869	92, 659
Planter, Corn	J. Selby	Peoria, Ill	May 14, 1872	126, 751
Planter, Corn	W. G. Selby and J. Bowman	Princeville, Ill	June 4, 1872	127, 648
Planter, Corn	J. P. Selsor	Cherry Box, Mo	June 11, 1867	65, 771
Planter, Corn	H. C. Shafer	Petersburgh, Ind	June 15, 1869	91, 489
Planter, Corn	H. A. Sharp	Harlem, Mo	May 27, 1873	139, 269
Planter, Corn	T. T. Shawcross	Allisonville, Ind	Oct. 2, 1866	58, 557
Planter, Corn	J. Sheer	Lincoln, Ill	Feb. 18, 1873	135, 942
Planter, Corn	D. J. Sheers	Belmont, Wis	Nov. 14, 1871	121, 012
Planter, Corn	A. Shellabarger	Miami County, Ohio	July 20, 1869	92, 889
Planter, Corn	P. Shellenberger	Millerstown, Pa	Aug. 3, 1869	93, 237
Planter, Corn	W. H. Shepherd	College Corner, Ohio	Nov. 12, 1867	70, 753
Planter, Corn	L. Sipe	Keezletown, Va	Nov. 18, 1873	144, 708
Planter, Corn	E. and E. C. Slosson	Vienna, Ill	Apr. 27, 1869	89, 512
Planter, Corn	F. J. Smiley	Marshall, Mich	Dec. 10, 1867	71, 917
Planter, Corn	A. D. Smith	Cincinnati, Ohio	Jan. 14, 1873	134, 942
Planter, Corn	A. F. Smith	Montrose, Iowa	Sept. 30, 1873	143, 262
Planter, Corn	A. N. Smith	Chillicothe, Mo	Sept. 23, 1873	143, 036
Planter, Corn	C. Smith	Nauvoo, Ill	July 31, 1860	29, 412
Planter, Corn	E. F. Smith	Orangeville, Ill	Jan. 15, 1867	61, 273
Planter, Corn	F. J. Smith	Four Corners, Ohio	May 26, 1857	17, 397
Planter, Corn	J. D. Smith	Muskegon, Mich	Feb. 25, 1873	136, 341
Planter, Corn	J. L. Smith	Neoga, Ill	June 26, 1860	28, 943
Planter, Corn	J. L. Smith	Neoga, Ill	Oct. 6, 1863	40, 194
Planter, Corn	J. M. Smith	Galva, Ill	Feb. 14, 1865	46, 398
Planter, Corn	W. T. F. Smith	Lexington, Ill	Dec. 10, 1872	133, 808
Planter, Corn	J. H. Sorey	Xenia, Ill	May 12, 1863	38, 510
Planter, Corn	P. Soule	Windsor, N. Y	Sept. 13, 1870	107, 302
Planter, Corn	A. Springsteen	Oquawka, Ill	Nov. 18, 1873	144, 711
Planter, Corn	A. H. Stark	Nevada, Iowa	May 20, 1873	139, 205
Planter, Corn	A. H. Stark and J. C. Mitchell	Nevada, Iowa	Oct. 24, 1871	120, 340
Planter, Corn	P. S. Stames	Majority Point, Ill	Mar. 22, 1870	101, 176
Planter, Corn	G. W. Starrett	Dublin, Ohio	Sept. 9, 1873	142, 656
Planter, Corn	J. Statz	Cross Plains, Wis	Nov. 4, 1873	144, 233
Planter, Corn	S. Stevens	North Fryeburgh, Me	Nov. 7, 1871	120, 793
Planter, Corn	W. Stinson	Georgetown, Pa	July 3, 1855	13, 185
Planter, Corn	V. D. Stoddard	Muscatine, Iowa	Nov. 22, 1864	45, 192
Planter, Corn	C. S. Stone	Greenwood, Ind	July 30, 1872	130, 082
Planter, Corn	J. Stone and J. T. Archibald	Wapello, Iowa	Apr. 16, 1861	32, 091
Planter, Corn	P. Stover	West Alexandria, Ohio	June 5, 1860	28, 615
Planter, Corn	B. T. Stowell	Quincy, Ill	May 22, 1860	28, 418
Planter, Corn	L. F. Straight	Fairbury, Ill	Sept. 11, 1860	30, 008
Planter, Corn	A. Streeter	Adrian, Mich	May 13, 1862	35, 280
Planter, Corn	L Study	Plum Hollow, Iowa	Oct. 15, 1867	69, 862
Planter, Corn	S. L. Sweeney	Morrison, Ill	Oct. 13, 1868	83, 105
Planter, Corn	N. Swigart	Richfield, Ohio	Oct. 4, 1870	108, 065
Planter, Corn	D. F. Taft	New Bedford, Mass	Oct. 20, 1868	83, 338
Planter, Corn	D. F. Taft	New Bedford, Mass	May 27, 1873	139, 310
Planter, Corn	J. O. Talmage	Cardington, Ohio	July 16, 1872	129, 619
Planter, Corn	L. B. Tarbox	Colliersville, N. Y	May 3, 1870	102, 618
Planter, Corn	G. Taylor	Richmond, Indiana	Apr. 20, 1858	20, 024
Planter, Corn	R. Taylor and R. Sprague	Prairie City, Ill	June 26, 1860	28, 917
Planter, Corn	C. W. Thiessero	Effingham, Ill	Oct. 27, 1868	83, 567
Planter, Corn	J. C. Thomas	Redpoint, Md	Apr. 25, 1865	47, 468
Planter, Corn	A. G. and A. J. Thompson	Belleville, Ohio	Mar. 1, 1859	23, 126
Planter, Corn	G. Thomson	Springfield, Ill	June 11, 1872	127, 938
Planter, Corn	J. Thompson	Aledo, Ill	July 16, 1872	129, 624
Planter, Corn	J. Thomson and J. Ramsey	Aledo, Ill	Sept. 27, 1864	44, 472
Planter, Corn	G. W. Thrall and W. L. Rayment.	Burlington, Mich	Dec. 29, 1868	85, 346
Planter, Corn	E. A. Thrush	New Kingston, Pa	Feb. 22, 1870	10[illegible], 215
Planter, Corn	J. S. Toan	Venice, N. Y	Jan. 13, 1857	16, 410
Planter, Corn	W. S. Todd	Mechanicsville, Iowa	Mar. 3, 1863	37, 834
Planter, Corn	B. F. Tomb	Tiffin, Ohio	June 21, 1870	104, 664
Planter, Corn	W. F. Tunnard	East Baton Rouge Parish, La.	July 27, 1869	93, 140
Planter, Corn	C. G. Udell	Morris, Ill	June 28, 1859	24, 594
Planter, Corn	F. Underwood	South Rutland, N. Y	Sept. 17, 1872	131, 482
Planter, Corn	J. Underwood	Cameron, Ill	Sept. 25, 1860	30, 169
Planter, Corn	H. Upjohn	Richland, Mich	Sept. 20, 1864	44, 355
Planter, Corn	A. Upson	Randolph, Ohio	Apr. 17, 1866	54, 042
Planter, Corn	J. M. E. Valk	Baltimore, Md	Mar. 12, 1872	124, 643
Planter, Corn	A. J. Van Boekél	Uniontown, N. J	Feb. 6, 1866	52, 471
Planter, Corn	J. W. Vandiver	Shelbina, Mo	Oct. 6, 1863	40, 2[illegible]2
Planter, Corn	F. Van Doren	Adrian, Mich	Dec. 1, 1868	84, 594
Planter, Corn	C. Van Honten	Sunbury, Ohio	Sept. 21, 1858	21, 583
Planter, Corn	S. and H. Vannuys	Bethel, Ind	Jan. 21, 1873	135, 178
Planter, Corn	R. M. Varner	Oxford, Miss	Oct. 4, 1859	25, 690
Planter, Corn	G. B. Vaughn	Marshall, Mo	Oct. 25, 1870	108, 656
Planter, Corn	G. J. Vaught	Hanley, Ky	Apr. 18, 1871	113, 819
Planter, Corn	W. E. Vernon	Franklin County, Mo	May 17, 1870	103, 259
Planter, Corn	C. L. Waffle	Sharon, Ohio	Aug. 21, 1860	29, 734
Planter, Corn	D. F. Wagner	West Hanover, Pa	Nov. 23, 1869	97, 139
Planter, Corn	E. M. Walker	Gallatin, Mo	May 1, 1866	54, 451
Planter, Corn	J. F. Walker	Murrayville, Ill	Nov. 3, 1868	83, 677
Planter, Corn	J. C. Walkinshaw	Leavenworth, Kans	Dec. 11, 1866	60, 447
Planter, Corn	W. J. Wallingford	Portland, Ill	Nov. 5, 1872	132, 783

Index of patents issued from the United States Patent Office from 1790 *to* 1873, *inclusive*—Continued.

Invention.	Inventor.	Residence.	Date.	No.
Planter, Corn	H. W. Wansbrough and H. M. Diggins.	Cincinnati, Ohio	Sept. 20, 1864	44, 357
Planter, Corn	J. P. Ware	Montgomery City, Mo	Mar. 23, 1869	88, 244
Planter, Corn	D. R. Warfield	Muscatine, Iowa	July 31, 1856	56, 839
Planter, Corn	J. Waterman	Keokuk, Iowa	May 3, 1864	42, 612
Planter, Corn	Z. D. Waters	Brookville, Md	July 1, 1873	140, 562
Planter, Corn	J. R. Weber	Bourbon, Ind	Dec. 10, 1867	71, 932
Planter, Corn	R. Welch	Worth, Ill	Oct. 1, 1861	33, 410
Planter, Corn	J. D. Wells	Franklin County, Ohio	Apr. 30, 1867	64, 390
Planter, Corn	D. F. Welsh	Nevada, Ohio	Jan. 3, 1871	110, 807
Planter, Corn	J. K. Welter	Springfield, Ill	July 16, 1872	129, 380
Planter, Corn	J. K. Welter	Springfield, Ill	Dec. 24, 1872	134, 336
Planter, Corn	J. E. West	Georgetown, Ky	Feb. 26, 1867	62, 458
Planter, Corn	J. W. West	Hillsborough, Ohio	July 5, 1859	24, 683
Planter, Corn	L. West	Georgetown, Ky	Aug. 30, 1870	106, 898
Planter, Corn	C. L. Westbrook	New York, N. Y	Mar. 7, 1865	46, 738
Planter, Corn	F. F. Westerfield	Fort Dodge, Iowa	Feb. 5, 1867	61, 783
Planter, Corn	C. Whitaker	Davenport, Iowa	Sept. 13, 1859	25, 461
Planter, Corn	F. W. White	Worcester, Mass	Aug. 31, 1858	21, 393
Planter, Corn	H. Whitman	Kingsville, Ohio	Aug. 24, 1858	21, 287
Planter, Corn	J. M. Whitmore and J. N. Arvin	Valparaiso, Ind	Jan. 3, 1871	110, 811
Planter, Corn	T. H. Wible	Quincy, Ill	Mar. 31, 1868	76, 282
Planter, Corn	B. Wieland	Orangeville, Ill	July 16, 1867	66, 923
Planter, Corn	L. Wright and O. G. Ewings	Whitewater and La Grange, Wis.	July 16, 1872	129, 383
Planter, Corn	H. Wiley	Frankfort, Ohio	July 5, 1859	24, 687
Planter, Corn	F. L. Wilkens	Saint Mary's, Ohio	Dec. 7, 1869	97, 739
Planter, Corn	M. Wilkinson	Burlington, Mich	May 8, 1866	54, 633
Planter, Corn	W. C. Willey	Princeton, Ill	Jan. 22, 1861	31, 201
Planter, Corn	T. C. Williams	Warrenton, Mo	July 16, 1872	129, 446
Planter, Corn	U. T. Wilson	De Soto City, Miss	Mar. 28, 1871	113, 233
Planter, Corn	W. W. Wilson	Collins Station, Ill	Jan. 14, 1862	34, 176
Planter, Corn	A. Windeck	Peoria, Ill	Dec. 1, 1868	84, 666
Planter, Corn	A. Windeck	Peoria, Ill	Mar. 26, 1872	124, 926
Planter, Corn	A. Windeck and A. Runstetler	Peoria, Ill	Dec. 18, 1866	60, 604
Planter, Corn	L. Winston	Pontiac, Ill	Aug. 30, 1870	106, 976
Planter, Corn	C. Wisdom	Flat Rock, Mich	Nov. 1, 1870	108, 865
Planter, Corn	S. Witt and G. W. Albaugh	Greencastle, Pa	Apr. 29, 1856	14, 785
Planter, Corn	D. and H. Wolf	Lebanon, Pa	Apr. 28, 1860	29, 839
Planter, Corn	D. D. Wood	Paris, Ill	Feb. 16, 1869	87, 018
Planter, Corn	G. H. Wood	Cambridge City, Ind	June 15, 1869	91, 399
Planter, Corn	J. B. Woolsey	Bloomfield, Iowa	Oct. 18, 1864	44, 762
Planter, Corn	E. M. Wright	Wilmington, Ohio	June 13, 1865	48, 229
Planter, Corn	J. J. Wright and J. H. Penny	Harrison, Ohio	May 19, 1868	78, 171
Planter, Corn	L. D. Wyatt	Castleton, Ind	Nov. 17, 1868	84, 075
Planter, Corn	H. M. Wyeth	Bloomfield, Iowa	Sept. 15, 1863	39, 982
Planter, Corn	J. Yoder, J. Gilliford, and E. Gouver.	Juniata, Pa	Oct. 3, 1848	5, 834
Planter, Corn	R. M. Yorks	Schoolcraft, Mich	Nov. 13, 1866	59, 703
Planter, Corn	G. W. N. Yost	Yellow Springs, Ohio	Apr. 3, 1860	27, 764
Planter, Corn and cane	J. M. Fate	Boonesborough, Iowa	Oct. 16, 1866	58, 798
Planter, Corn and cotton	G. C. Boyd	New London, Pa	Apr. 11, 1836	....
Planter, Corn and cotton	J. W. Savage and F. M. Doty	Parmitchie, Miss	May 20, 1873	139, 195
Planter, Corn and cotton	Z. B. Sims	Bonham, Tex	June 3, 1873	139, 620
Planter, Corn and cotton	B. J. Svenson	Manor Station, Tex	Aug. 20, 1872	130, 764
Planter, Corn and cotton-seed	C. C. Garrett	Dayton, Ala	Mar. 12, 1867	62, 834
Planter, Corn and cotton-seed	D. E. Holt	Wilkinson County, Miss	Nov. 2, 1869	96, 323
Planter, Corn and cotton-seed	A. Ingalis	Memphis, Tenn	Sept. 10, 1872	131, 274
Planter, Corn and cotton-seed	R. R. McGregor	Covington, Tenn	Feb. 16, 1869	86, 992
Planter, Corn and cotton-seed	M. L. and R. W. Thornton	Lumpkin, Ga	July 16, 1867	66, 912
Planter, Corn and seed	E. C. Brown	Crawfordsville, Ind	Mar. 24, 1868	75, 850
Planter, Corn and seed	D. D. Franklin	Flora, Ill	July 27, 1869	92, 951
Planter, Corn and seed	J. S. Hersey	Elizabethtown, Pa	May 22, 1866	54, 901
Planter, Corn and seed	W. C. Lockwood	Spring Mills, Mich	July 26, 1870	105, 703
Planter, Corn and seed	F. W. Marriott	Richwood, Ohio	May 26, 1868	78, 303
Planter, Corn and seed	R. H. Mathews	Nebraska City, Nebr	June 6, 1871	115, 755
Planter, Corn and seed	N. Mendenhall	Greensburgh, Ind	Dec. 29, 1868	85, 466
Planter, Corn and seed	M. M. Rutt and A. B. Baer	East Hempfield, Pa	Nov. 26, 1867	71, 540
Planter, Cotton	P. H. Altstatt	Clarke County, Ind	June 20, 1871	116, 003
Planter, Cotton	E. L. Barnett	El Dorado, Ark	June 30, 1868	79, 433
Planter, Cotton	M. Beam	Buffalo, N. C	Feb. 13, 1835	....
Planter, Cotton	M. Beam	Buffalo, N. C	Mar. 30, 1836	....
Planter, Cotton	H. Blair	Glencoss, Md	Aug. 31, 1836	....
Planter, Cotton	H. Blair	United States	Aug. 31, 1836	15
Planter, Cotton	R. E. Bowen	George's Creek, S. C	Aug. 12, 1873	141, 753
Planter, Cotton	J. M. Brooks	Woodbury, Ga	Feb. 11, 1873	135, 767
Planter, Cotton	J. P. Clopton	Terry, Tenn	July 4, 1871	116, 556
Planter, Cotton	W. J. Cox and W. T. Smith	Lilesville, N. C	July 29, 1873	141, 328
Planter, Cotton	J. S. Dickason	Sulphur Well, Tenn	Nov. 22, 1870	109, 499
Planter, Cotton	J. L. A. Edwards	New Orleans, La	Apr. 9, 1867	63, 623
Planter, Cotton	D. C. Ellis and G. N. Deming	Rochester, N. Y	Mar. 14, 1871	112, 567
Planter, Cotton	A. G. W. Foster	Franklin, Ga	Oct. 25, 1870	108, 580
Planter, Cotton	J. Gatling	Murfreesborough, N. C	June 20, 1835	....
Planter, Cotton	W. L. Gebby	New Richland, Ohio	Dec. 4, 1866	60, 170
Planter, Cotton	J. G. Goodman and W. S. McDonald.	Wilson County, Tenn	June 29, 1833	....
Planter, Cotton	R. T. Goodman	Ballsville, Va	Sept. 18, 1835	....
Planter, Cotton	R. G. Hobson	Houlka, Miss	Nov. 7, 1871	120, 742
Planter, Cotton	J. Hughes	New Berne, N. C	Apr. 4, 1871	113, 671
Planter, Cotton	F. E. Moran	Millburn, Ill	Feb. 26, 1867	62, 356
Planter, Cotton	J. H. Nale	Memphis, Tenn	Apr. 30, 1872	126, 320
Planter, Cotton	C. H. Nixon	Polo, Ill	Nov. 18, 1873	144, 784
Planter, Cotton	G. Paterson	Waynesborough, Ga	June 10, 1873	139, 732
Planter, Cotton	J. A. and W. L. D. Pope	Charlotte, N. C	May 7, 1872	126, 487
Planter, Cotton	W. Price	Mount Olive, N. C	Apr. 22, 1873	138, 043
Planter, Cotton	B. A. Ramsay	Trenton, Tenn	Dec. 20, 1870	110, 393
Planter, Cotton	W. E. Rhodes	Darlington Court-House, S. C.	Jan. 28, 1873	135, 290
Planter, Cotton	C. Richmond	Memphis, Tenn	May 12, 1868	77, 761

Index of patents issued from the United States Patent Office from 1790 *to* 1873, *inclusive*—Continued.

Invention.	Inventor.	Residence.	Date.	No.
Planter, Cotton	H. A. Ridley	Jacksonport, Ark	Dec. 12, 1871	121, 815
Planter, Cotton	H. A. Ridley	Jacksonport, Ark	Nov. 12, 1872	132, 984
Planter, Cotton	N. F. Sandelin	New York, N. Y	Mar. 12, 1872	124, 450
Planter, Cotton	J. P. Selsor	Shelbyville, Mo	Feb. 26, 1867	62, 447
Planter, Cotton	P. Seymour	East Bloomfield, N. Y	Oct. 28, 1873	144, 037
Planter, Cotton	O. L. Slater	Wall Hill, Miss	Sept. 2, 1873	142, 355
Planter, Cotton	E. L. Sykes	Okolona, Miss	Jan. 18, 1870	99, 031
Planter, Cotton	T. T. and G. T. Thorne	Whitaker's Station, N. C	Feb. 1, 1870	99, 499
Planter, Cotton	A. R. Wiggs	Iuka, Miss	May 11, 1869	89, 958
Planter, Cotton	J. Wood	Wedowee, Ala	May 28, 1872	127, 399
Planter, Cotton and corn	A. Pennington, jr	Brenham, Tex	Nov. 11, 1873	144, 562
Planter, Cotton and corn	E. Wagner	Westminster, Md	Nov. 1, 1870	108, 947
Planter, Cotton and corn	A. R. Wiggs	Iuka, Miss	Nov. 23, 1869	97, 257
Planter, Cotton-seed	L. Acree	Taliaferro County, Ga	Feb. 14, 1860	27, 091
Planter, Cotton-seed	H. P. Allen	Bowling Green, Ky	Aug. 31, 1858	21, 308
Planter, Cotton-seed	J. P. Allen	Dawson, Ga	Mar. 29, 1870	101, 206
Planter, Cotton-seed	F. M. Bacon	Ripon, Wis	Sept. 26, 1865	50, 089
Planter, Cotton-seed	W. Badger	Memphis, Tenn	Jan. 13, 1857	16, 368
Planter, Cotton-seed	N. E. Badgley	Gadsden, Ala	May 1, 1860	28, 049
Planter, Cotton-seed	N. E. Badgley	New York, N. Y	Nov. 6, 1866	59, 339
Planter, Cotton-seed	P. B. Baker	Wall Hill, Miss	Dec. 13, 1859	26, 404
Planter, Cotton-seed	W. C. Banks	Como Depot, Miss	Aug. 6, 1867	67, 481
Planter, Cotton-seed	C. Battle	Warrenton, Ga	Mar. 13, 1860	27, 421
Planter, Cotton-seed	E. P. Beauchamp	Preston, Ga	Aug. 9, 1859	24, 984
Planter, Cotton-seed	D. I. Beecher	Greenville, Miss	Sept. 2, 1856	15, 640
Planter, Cotton-seed	C. R. Belt	Near Washington, D. C	Oct. 21, 1856	15, 9 8
Planter, Cotton-seed	E. T. Bastrom	Newnan, Ga	June 29, 1858	20, 694
Planter, Cotton-seed	N. Breed	Jeffersonville, Ind	June 1, 1869	90, 722
Planter, Cotton-seed	A. W. Brian	Ouachita County, Ark	June 15, 1869	91, 206
Planter, Cotton-seed	T. E. C. Brinly	Louisville, Ky	June 13, 1871	115, 929
Planter, Cotton-seed	R. M. Brooks	Greenville, Ga	Dec. 27, 1859	26, 562
Planter, Cotton-seed	A. D. Brown, sr	Columbus, Ga	Feb. 23, 1869	87, 089
Planter, Cotton-seed	F. H. Brown	Chicago, Ill	June 26, 1866	55, 811
Planter, Cotton-seed	F. H. Brown	Auburn, N. Y	Oct. 2, 1866	58, 372
Planter, Cotton-seed	L. B. Brown	Scriven County, Ga	Jan. 15, 1861	31, 107
Planter, Cotton-seed	M. S. Burns	Memphis, Tenn	Oct. 18, 1870	108, 445
Planter, Cotton-seed	A. Carey	Rome, Ga	Feb. 14, 1860	27, 107
Planter, Cotton-seed	F. F. Carroll	Midway, S. C	Apr. 4, 1871	113, 626
Planter, Cotton-seed	Z. Carter	Line Creek, S. C	Dec. 16, 1873	145, 554
Planter, Cotton-seed	J. G. Clark	Middletown, Ohio	Oct. 2, 1866	58, 381
Planter, Cotton-seed	T. T. and H. W. S. Collier	Lavernia, Tex	Sept. 13, 1859	25, 388
Planter, Cotton-seed	J. M. Cooper	Palmyra, Ga	Mar. 7, 1854	10, 595
Planter, Cotton-seed	J. A. Cox	Humboldt, Tenn	June 11, 1867	65, 647
Planter, Cotton-seed	W. W. Croom	Gainesville, Ala	Mar. 29, 1870	101, 232
Planter, Cotton-seed	W. M. Croom	Opelika, Ala	Oct. 22, 1872	132, 388
Planter, Cotton-seed	J. P. Crutcher	Silver Spring, Tenn	Apr. 12, 1859	23, 554
Planter, Cotton-seed	I. T. Donovan and W. J. Fowler	Seguin, Tex	Apr. 27, 1858	20, 049
Planter, Cotton-seed	Z. Doolittle	Perry, Ga	July 10, 1860	29, 065
Planter, Cotton-seed	Z. Doolittle and A. M. Crowder	Houston Factory, Ga	Feb. 18, 1868	74, 518
Planter, Cotton-seed	J. M. Elliott	Winnsborough, S. C	July 20, 1869	92, 804
Planter, Cotton-seed	H. R. Fell and E. Phifer	Trenton, N. J	Nov. 20, 1866	59, 765
Planter, Cotton-seed	D. P. Ferguson	Jonesborough, Ga	Nov. 18, 1873	144, 607
Planter, Cotton-seed	T. C. Garlington	Chambers Court-House, Ala	Aug. 2, 1870	105, 935
Planter, Cotton-seed	C. C. Garrett	Spring Hill, Ala	Mar. 8, 1859	23, 164
Planter, Cotton-seed	W. A. Gates	Mount Comfort, Tenn	Nov. 16, 1852	9, 403
Planter, Cotton-seed	W. Gessner	Cape Girardeau, Mo	July 30, 1872	129, 943
Planter, Cotton-seed	O. L. Gibson	Fort Bend County, Tex	Sept. 11, 1860	29, 964
Planter, Cotton-seed	A. J. Going	Clinton, La	Mar. 10, 1868	75, 410
Planter, Cotton-seed	A. J. Going	Clinton, La	Dec. 29, 1868	85, 379
Planter, Cotton-seed	A. D. Griswold	Rocky Hill, Conn	May 6, 1873	138, 497
Planter, Cotton-seed	F. E. Habersham	Old Church, Va	Dec. 17, 1872	133, 983
Planter, Cotton-seed	J. T. Ham	Senatobia, Miss	Oct. 9, 1860	30, 313
Planter, Cotton-seed	D. Herlong	Sandy Ridge, Ala	Jan. 15, 1861	31, 120
Planter, Cotton-seed	J. L. Horn	Edgecombe County, N. C	Feb. 12, 1856	14, 240
Planter, Cotton-seed	W. A. Horrall	Washington, Ind	Oct. 30, 1866	59, 314
Planter, Cotton-seed	E. J. Hudson	Golconda, Ill	June 14, 1870	104, 157
Planter, Cotton-seed	J. S. Huggins and R. Chapman	Darlington District, S. C	June 1, 1858	20, 432
Planter, Cotton-seed	O. P. Humber	Greenville, N. C	Nov. 16, 1869	96, 919
Planter, Cotton-seed	J. W. Huntley	Lane's Creek, N. C	Aug. 16, 1859	25, 119
Planter, Cotton-seed	G. Jessup	Shortsville, N. Y	Sept. 8, 1868	81, 906
Planter, Cotton-seed	J. M. Jones	Palmyra, N. Y	Jan. 22, 1856	14, 134
Planter, Cotton-seed	H. L. Justice and J. H. Galbreath	Goodlettsville, Tenn	May 5, 1857	17, 221
Planter, Cotton-seed	C. Kesler and F. Reinhard	Columbus, Tex	Dec. 27, 1859	26, 596
Planter, Cotton-seed	J. C. King	Spring Place, Ga	Nov. 15, 1870	109, 222
Planter, Cotton-seed	L. D. Law	Henderson, Ga	Dec. 22, 1857	18, 913
Planter, Cotton-seed	J. Lytch	Laurinburgh, N. C	May 31, 1870	103, 759
Planter, Cotton-seed	C. W. McClanahan	Victoria, Tex	Nov. 27, 1860	30, 743
Planter, Cotton-seed	A. McDonald	Salem, Miss	June 15, 1858	20, 572
Planter, Cotton-seed	M. McMillian	Caney, Ark	Nov. 30, 1869	97, 424
Planter, Cotton-seed	J. M. Merrymon and W. M. Dunn	Indianapolis, Ind, and Gurleysville, Ala.	Mar. 5, 1867	62, 664
Planter, Cotton-seed	E. H. Metcalf	Montgomery, Ala	Aug. 13, 1872	130, 521
Planter, Cotton-seed	S. Miller	Washington College, Tenn	Apr. 19, 1853	9, 676
Planter, Cotton-seed	A. H. Morrel	Marlin, Tex	May 15, 1855	12, 867
Planter, Cotton-seed	I. Myers and M. D. Wellman	Pittsburgh, Pa	Jan. 31, 1865	46, 130
Planter, Cotton-seed	R. C. Nash	Somerville, Tenn	Dec. 18, 1860	30, 981
Planter, Cotton-seed	D. B. Neal	Mount Gilead, Ohio	Feb. 23, 1858	19, 438
Planter, Cotton-seed	H. Nicholls	Fairfield, Ky	Aug. 30, 1870	106, 859
Planter, Cotton-seed	A. R. Nixon	Memphis, Tenn	Dec. 21, 1869	98, 183
Planter, Cotton-seed	J. F. Orr	Orville, Ala	Feb. 3, 1857	16, 550
Planter, Cotton-seed	B. Owen	Dayton, Ohio	June 26, 1860	28, 896
Planter, Cotton-seed	B. Owen	Dayton, Ohio	Mar. 20, 1866	53, 377
Planter, Cotton-seed	A. Packham	Prestonville, Ky	May 22, 1866	54, 942
Planter, Cotton-seed	B. and N. Platt	Pana, Ill., and Saint Louis, Mo.	Aug. 1, 1865	49, 150
Planter, Cotton-seed	J. Price	New Harrisburgh, Ohio	Jan. 30, 1866	52, 518
Planter, Cotton-seed	O. Richardson	Boston, Mass	Dec. 31, 1867	72, 907

Index of patents issued from the United States Patent Office from 1790 *to* 1873, *inclusive*—Continued.

Invention.	Inventor.	Residence.	Date.	No.
Planter, Cotton-seed	J. Riggsbee	Chapel Hill, N. C	July 5, 1870	105, 000
Planter, Cotton-seed	T. J. Rogers	Cassville, Ga	Sept. 1, 1857	18, 104
Planter, Cotton-seed	C. A. Rose	Columbia, Ala	Dec. 4, 1860	30, 839
Planter, Cotton-seed	J. Ross	Midway, Ala	Apr. 6, 1858	19, 874
Planter, Cotton-seed	J. L. Russell	Pella, Iowa	July 10, 1866	56, 274
Planter, Cotton-seed	D. H. A. Sanders	Senatobia, Miss	Apr. 6, 1869	88, 670
Planter, Cotton-seed	J. Shearer and M. B. Armstrong	Timbersville, Ill	July 13, 1869	92, 481
Planter, Cotton-seed	N. B. Sherwood	Millville, N. Y	June 5, 1866	55, 372
Planter, Cotton-seed	N. B. Sherwood	Millville, N. Y	June 5, 1866	55, 373
Planter, Cotton-seed	N. B. Sherwood	Millville, N. Y	Mar. 3, 1868	75, 207
Planter, Cotton-seed	F. Sloan	Bolivar, Tenn	Sept. 27, 1870	107, 730
Planter, Cotton-seed	J. L. Slocumb	Raymond, Miss	July 22, 1873	141, 085
Planter, Cotton-seed	A. C. Smith	Roaring Falls, Tenn	Oct. 25, 1870	108, 643
Planter, Cotton-seed	B. Smith	Falkland, N. C	June 2, 1868	78, 614
Planter, Cotton-seed	B. Smith	Falkland, N. C	Feb. 1, 1870	99, 361
Planter, Cotton-seed	B. Smith	Hood Swamp, N. C	July 5, 1870	105, 006
Plantor, Cotton-seed	P. E. Smith	Scotland Neck, N. C	May 23, 1871	115, 123
Planter, Cotton-seed	P. M. Smith and T. T. Collier	Lavernia, Tex	Sept. 13, 1859	25, 449
Planter, Cotton-seed	J. H. Sorey	Flora, Ill	Oct. 31, 1871	120, 544
Planter, Cotton-seed	B. Spencer	Lewisburgh, Pa	Nov. 8, 1864	44, 984
Planter, Cotton-seed	A. V. M. Sprague and R. F. Osgood.	Rochester, N. Y	July 6, 1869	92, 389
Planter, Cotton-seed	J. A. Stewart	Franklin, Ky	July 1, 1856	15, 260
Planter, Cotton-seed	W. A. and J. F. Suddith	Charlestown, Va	June 26, 1860	28, 913
Planter, Cotton-seed	L. B. Sutton	Windsor, N. C	Feb. 25, 1873	136, 342
Planter, Cotton-seed	S. P. Sweeny	Columbia, Tex	Apr. 3, 1860	27, 748
Planter, Cotton-seed	J. C. Tobias	Helena, Ark	Aug. 27, 1867	68, 262
Planter, Cotton-seed	J. Trump	Springfield, Ohio	Apr. 12, 1870	101, 947
Planter, Cotton-seed	W. F. Tunnard	East Baton Rouge Parish, La.	Sept. 7, 1869	94, 671
Planter, Cotton-seed	A. W. Washburn	Yazoo City, Miss	Mar. 26, 1856	14, 529
Planter, Cotton-seed	M. D. Wells	Morgantown, W. Va	Nov. 6, 1866	59, 490
Planter, Cotton-seed	T. W. White	Milledgeville, Ga	Oct. 20, 1857	18, 482
Planter, Cotton-seed	L. T. Wilcox and W. G. Caldwell.	Three Rivers, Mich	July 16, 1867	66, 924
Planter, Cotton-seed	I. Williams and I. W. Bausman.	Allegheny County, Pa	Jan. 23, 1855	12, 294
Planter, Cotton-seed	W. R. Wright	Allendale, S. C	May 28, 1872	127, 400
Planter, Cotton-seed and corn	N. Foster	Palmyra, N. Y	Feb. 11, 1868	74, 334
Planter, Cotton-seed and corn	J. G. B. Gill	Chesnut Grove, S. C	Aug. 31, 1869	94, 301
Planter, Cotton-seed and corn	J. B. Godwin	Williamston, N. C	Mar. 29, 1870	101, 257
Planter, Cotton-seed and corn	A. Runstetler and A. Windeck	Peoria, Ill	Mar. 10, 1868	75, 466
Planter, Cotton-seed and corn	I. A. Towers	Quincy, Fla	Dec. 20, 1870	110, 311
Planter, cultivator, and digger, Potato	C. J. C. Peterson	Port Chester, N. Y	Feb. 2, 1869	86, 442
Planter cultivator, Corn	M. H. Lineback	Greenfield, Ind	Mar. 13, 1866	53, 156
Planter, Double corn	W. McLucas	Riuersville, Ohio	July 14, 1868	79, 998
Planter drill, Seed	J. F. Keller	Hagerstown, Md	July 26, 1870	105, 810
Planter, dropper, and cultivator, Combined seed	C. M. Cornell	Ionia, Mich	July 6, 1869	92, 171
Planter, fertilizer, and plow, Combined seed	H. C. Eves	Orangeville, Pa	Aug. 3, 1869	93, 239
Planter, fertilizer, distributer, cotton-chopper, and cultivator, Combined corn and cotton.	J. F. Tucker	Monticello, Fla	May 31, 1870	103, 688
Planter, Foot corn	G. B. Roe	Paine's Point, Ill	Mar. 17, 1863	37, 922
Planter, Foot corn	N. Silvester	Granger, Ohio	May 20, 1862	35, 330
Planter gearing, Seed	A. Sandoe	Mifflintown, Pa	Apr. 16, 1850	7, 294
Planter, Grain	I. Bogart	Newport, Ind	May 24, 1870	103, 289
Planter, Grain	N. H. Purcell	Avon, N. Y	Nov. 14, 1865	50, 954
Planter, grain-drill, and harrow, Combined corn	H. Smith	Kirksville, Mo	Mar. 2, 1869	87, 376
Planter, Hand	R. Arey	Rock Falls, Ill	Mar. 11, 1873	136, 688
Planter, Hand	J. Jeffcoat	Onawa, Iowa	Mar. 2, 1869	87, 414
Planter, Hand	J. H. Latimer	Crystal Lake, Ill	Dec. 26, 1871	122, 121
Planter, Hand	F. H. Roberts	Wilmington, Ind	Mar. 1, 1864	41, 784
Planter, Hand	N. Safford	Pleasant Valley, Vt	Oct. 2, 1866	58, 486
Planter, Hand	D. B. Seely	Sterling, Ill	Mar. 25, 1873	137, 246
Planter, Hand	S. S. Stults	Cedar Bluffs, Nebr	Jan. 28, 1873	135, 294
Planter, Hand	G. Windle	New Cambria, Mo	Feb. 25, 1873	136, 294
Planter, Hand	Y. F. Wright	Hannahatchee, Ga	Jan. 23, 1872	122, 980
Planter, Hand corn	G. Atkins	Pittsburgh, Pa	June 17, 1856	15, 114
Planter, Hand corn	H. F. Batcheller	Sterling, Ill	Mar. 9, 1858	19, 540
Planter, Hand corn	O. Billings	La Grange, Ohio	May 22, 1866	54, 999
Planter, Hand corn	S. P. Briggs	Saratoga Springs, N. Y	Apr. 30, 1861	32, 178
Planter, Hand corn	D. Broy	Canton, Mo	Apr. 14, 1868	76, 709
Planter, Hand corn	G. Burson	East Palestine, Ohio	Apr. 6, 1869	88, 607
Planter, Hand corn	G. Burson	East Palestine, Ohio	Sept. 20, 1870	107, 594
Planter, Hand corn	G. Burson	East Palestine, Ohio	Sept. 20, 1870	107, 595
Planter, Hand corn	M. Case	Kasoag, N. Y	Aug. 18, 1863	39, 548
Planter, Hand corn	T. A. Chandler	Rockford, Ill	Nov. 25, 1856	16, 135
Planter, Hand corn	J. W. Chappell	Loomisville, Mich	Aug. 6, 1861	32, 975
Planter, Hand corn	D. G. Choppin	Cincinnati, Ohio	Apr. 6, 1858	19, 833
Planter, Hand corn	R. L. Cleveland	Durand, Ill	Sept. 11, 1866	57, 868
Planter, Hand corn	J. W. Coleman	Bridgeville, Mich	June 24, 1873	140, 186
Planter, Hand corn	A. C. L. Davis	Saint Louis, Mo	May 17, 1870	103, 153
Planter, Hand corn	S. L. Denney	Lancaster, Pa	June 3, 1856	15, 035
Planter, Hand corn	W. Douglass	Westport, Mo	July 30, 1861	32, 937
Planter, Hand corn	H. Dyer	Fort Scott, Kans	Aug. 23, 1870	106, 568
Planter, Hand corn	J. Fairbank and E. C. Durfee	Leon, N. Y	Feb. 9, 1858	19, 329
Planter, Hand corn	W. Gilbert	Bardstown, Ky	June 26, 1866	55, 850
Planter, Hand corn	H. Gortner	Nashport, Ohio	Apr. 29, 1873	138, 392
Planter, Hand corn	H. B. Hammon	Bristolville, Ohio	Sept. 9, 1856	15, 696
Planter, Hand corn	H. B. Hammon	Bristolville, Ohio	Mar. 6, 1860	27, 365
Planter, Hand corn	J. M. Harrison	Spartanburgh, Ind	Jan. 24, 1871	111, 202
Planter, Hand corn	E. W. Haven	Brandon, Vt	Oct 12, 1869	95, 683
Planter, Hand corn	F. A. Hawkins	Noblesville, Ind	Jan. 14, 1873	134, 884
Planter, Hand corn	L. O. Hayworth	New Cumberland, Ind	May 18, 1869	90, 169
Planter, Hand corn	H. Hickman	Omaha, Nebr	May 10, 1870	102, 818
Planter, Hand corn	D. H. Howell	Independence, Iowa	Jan. 23, 1866	52, 171
Planter, Hand corn	D. Humphreys	Oskalcosa, Iowa	July 3, 1866	56, 053
Planter, Hand corn	W. Jenks	Alexandria, Va	Mar. 25, 1856	14, 504
Planter, Hand corn	S. G. Jones	Philadelphia, Pa	Dec. 10, 1872	133, 709

Index of patents issued from the United States Patent Office from 1790 *to* 1873, *inclusive*—Continued.

Invention.	Inventor.	Residence.	Date.	No.
Planter, Hand corn	I. A. Keeler	Middleville, Mich	July 19, 1870	105, 578
Planter, Hand corn	C. A. Kellogg	Elyria, Ohio	July 4, 1865	48, 562
Planter, Hand corn	A. C. Kent	Janesville, Wis	Apr. 2, 1872	125, 306
Planter, Hand corn	J. F. Klinglesmith	Hardin County, Ky	Aug. 4, 1868	80, 746
Planter, Hand corn	R. Lambert	Cortland Village, N. Y	Apr. 10, 1866	53, 838
Planter, Hand corn	J. S. Lawson	Disco, Mich	Feb. 11, 1868	74, 231
Planter, Hand corn	W. C. Lewman	Kansas, Ohio	Jan. 8, 1867	61, 076
Planter, Hand corn	C. Martratt	Albany, N. Y	Aug. 26, 1856	15, 616
Planter, Hand corn	P. McCollum	Fayette, Mo	Nov. 22, 1870	109, 437
Planter, Hand corn	D. McKanna	Madison, Wis	June 12, 1866	55, 517
Planter, Hand corn	J. Morris	Auburn, Mo	June 13, 1865	48, 197
Planter, Hand corn	J. H. Morris and T. B. Harrison	Maquoketa, Iowa	Nov. 24, 1868	84, 435
Planter, Hand corn	M. P. Nemmers	Saint Donatus, Iowa	Nov. 25, 1873	144, 919
Planter, Hand corn	M. S. Orton	Galesburgh, Ill	Mar 21, 1865	46, 928
Planter, Hand corn	W. S. Pelham	Kirkville, Iowa	Sept. 14, 1869	94, 910
Planter, Hand corn	L. H. Richards	Rising Sun, Md	Feb. 4, 1873	135, 489
Planter, Hand corn	E. Rogers	Rochester, Ind	Nov. 18, 1873	144, 701
Planter, Hand corn	M. C. Root	Janesville, Wis	Nov. 11, 1873	144, 566
Planter, Hand corn	M. Russell and A. G. Burdick	Mill Rock, Iowa	Nov. 12, 1867	70, 747
Planter, Hand corn	H. Sage	Omaha City, Nebr	June 11, 1872	127, 696
Planter, Hand corn	S. Z. Shores	Towanda, Pa	July 2, 1861	32, 725
Planter, Hand corn	S. S. Smith	North Fairfield, Ohio	June 14, 1864	43, 139
Planter, Hand corn	O. Stoddard	Busti, N. Y	June 26, 1855	13, 151
Planter, Hand corn	W. D. Stroud	Oshkosh, Wis	May 3, 1870	102, 611
Planter, Hand corn	J. O. Talmage	Cardington, Ohio	Nov. 11, 1873	144, 485
Planter, Hand corn	J. D. Tracy and J. F. Platt	Sterling, Ill	Dec. 16, 1873	145, 698
Planter, Hand corn	S. E. Tyler and R. Tattershall	Beloit, Wis	Apr. 3, 1866	53, 752
Planter, Hand corn	J. P. Van Vleck	Rock County, Wis	Feb. 26, 1867	62, 380
Planter, Hand corn	C. A. Wakefield	Pittsfield, Mass	June 5, 1866	55, 399
Planter, Hand corn	J. M. and W. C. Wallis	Milton, Iowa	Oct. 28, 1862	36, 815
Planter, Hand corn	L. Weaver	Canton, Ohio	Jan. 22, 1867	61, 490
Planter, Hand corn	H. B. West and C. A. Kellogg	Elyria, Ohio	Oct. 13, 1863	40, 299
Planter, Hand seed	S. P. Briggs	Saratoga Springs, N. Y	May 26, 1857	17, 362
Planter, Hand seed	J. H. Bruen	Penn Yan, N. Y	Feb. 24, 1857	16, 677
Planter, Hand seed	W. Bullack	Philadelphia, Pa	Nov. 2, 1852	9, 381
Planter, Hand seed	T. Crane	Fort Atkinson, Wis	Apr. 21, 1857	17, 080
Planter, Hand seed	H. V. Davis	Amherst, N. H	Aug. 27, 1867	68, 050
Planter, Hand seed	J. Decker	Sparta, N. J	Apr. 21, 1857	17, 081
Planter, Hand seed	J. Dyson	Philadelphia, Pa	Aug. 17, 1869	93, 690
Planter, Hand seed	P. B. Green	Chicago, Ill	Apr. 21, 1857	17, 089
Planter, Hand seed	J. Haines	West Middleburgh, Ohio	Sept. 8, 1859	18, 145
Planter, Hand seed	J. Heberling	Mount Pleasant, Ohio	June 7, 1870	104, 028
Planter, Hand seed	E. Hopkins	Cincinnati, Ohio	Apr. 29, 1856	14, 767
Planter, Hand seed	D. W. Hughes	New London, Mo	Nov. 20, 1855	13, 820
Planter, Hand seed	J. H. Jones	Rockton, Ill	Aug. 26, 1856	15, 610
Planter, Hand seed	J. H. Jones	Rockford, Ill	Apr. 9, 1867	63, 643
Planter, Hand seed	H. Koeller and W. Uecke	Camp Point, Ill	Feb. 11, 1868	74, 383
Planter, Hand seed	A. T. Large	Tomah, Wis	Nov. 27, 1866	60, 019
Planter, Hand seed	J. H. Latimer	Crystal Lake, Ill	Dec. 4, 1866	60, 200
Planter, Hand seed	J. T. Macomber	Grand Isle, Vt	Apr. 5, 1870	101, 480
Planter, Hand seed	A. C. Miller	Morgantown, Va	July 29, 1856	15, 431
Planter, Hand seed	A. More	Moresville, N. Y	July 13, 1869	92, 632
Planter, Hand seed	G. Sanford	Ellenville, N. Y	June 15, 1852	9, 037
Planter, Hand seed	A. Stickney	Concord, N. H	Dec. 25, 1855	14, 003
Planter, Hand seed	F. Van Doren	Adrian, Mich	Feb. 14, 1860	27, 172
Planter, harrow, and cultivator, Combined	D. D. Stelle	New Brunswick, N. J	July 16, 1867	66, 904
Planter, Hedge	N. L. Griffith	South Plymouth, Ohio	Aug. 1, 1871	117, 622
Planter, Hedge	J. J. Tucker	Eugene, Ill	June 1, 1869	90, 703
Planter, Hedge	J. J. Tucker	Albia, Iowa	Aug. 3, 1869	93, 250
Planter, Hedge	W. Young	Bloomington, Ill	Dec. 22, 1868	85, 195
Planter, hoe, and potato-digger, Combined potato	F. B. Marden	Bangor, Me	Jan. 21, 1868	73, 618
Planter, Potato	J. R. Albertson	Allegheny, Pa	Jan. 26, 1858	19, 178
Planter, Potato	L. A. Aspinwall	Watervliet, N. Y	Nov. 14, 1865	50, 890
Planter, Potato	L. A. Aspinwall	Albany, N. Y	Nov. 30, 1869	97, 339
Planter, Potato	L. A. Aspinwall	Albany, N. Y	Oct. 10, 1871	119, 687
Planter, Potato	J. E. Bendix and M. Deitsch	New York and West Chester, N. Y.	Nov. 12, 1867	70, 786
Planter, Potato	H. Farmer	Pontiac, Mich	Jan. 7, 1868	73, 086
Planter, Potato	S. T. Godfrey	Seaville, N. J	Jan. 11, 1870	98, 687
Planter, Potato	E. E. Hawley	New Haven, Conn	Feb. 9, 1858	19, 294
Planter, Potato	T. Herbert	Philadelphia, Pa	Oct. 25, 1870	108, 591
Planter, Potato	L. J. Holcomb	Nunda, Ill	June 19, 1866	55, 657
Planter, Potato	H. J. Kent	Palmyra, N. Y	Jan. 24, 1871	111, 217
Planter, Potato	G. W. and J. J. Kersey	Beartown, Pa	Sept. 25, 1860	30, 144
Planter, Potato	G. Knowlton	Johnstown, Pa	Mar. 7, 1871	112, 470
Planter, Potato	J. Krehbril	Clarence Centre, N. Y	May 17, 1870	103, 058
Planter, Potato	F. S. McWhorter	Smyrna, Del	Apr. 20, 1858	20, 001
Planter, Potato	J. C. Mills	Rochester, N. Y	Mar. 14, 1871	112, 617
Planter, Potato	J. Moore	Quincy Point, Mass	July 29, 1856	15, 433
Planter, Potato	C. Morgan	Philadelphia, Pa	Feb. 12, 1856	14, 270
Planter, Potato	W. Nevins	Irving, N. Y	Mar. 22, 1864	42, 014
Planter, Potato	A. E. Payne	Jonesville, Mich	Dec. 19, 1871	121, 957
Planter, Potato	A. Randel	Verona, N. Y	Feb. 15, 1848	5, 446
Planter, Potato	J. P. Scudder	Lawrenceville, N. J	June 11, 1867	65, 695
Planter, Potato	N. B. Sherwood	Millville, N. Y	Nov. 24, 1868	84, 309
Planter, Potato	J. C. Stoddard	Worcester, Mass	Mar. 17, 1859	24, 065
Planter, Potato	S. H. Strong	Brunswick, Ohio	Dec. 8, 1857	18, 827
Planter, Potato	C. Svendsen	Blair, Nebr	July 15, 1873	140, 965
Planter, Potato	J. L. True	Benton, Me	Jan. 5, 1869	85, 547
Planter, Potato	J. L. True	Benton, Me	Apr. 16, 1872	125, 705
Planter, Potato	L. Van Wie	Bethlehem, N. Y	June 22, 1869	91, 798
Planter, Potato	H. Wainwright and S. T. Williams.	Farmingdale, N. J	Jan. 5, 1858	19, 054
Planter, Potato	F. W. Worstell	Saint Louis, Mo	June 24, 1873	140, 178
Planter, Potato and corn	A. J. Taylor	Manchester, Ind	June 15, 1869	91, 494
Planter, Rice	E. Wagoner	Westminster, Md	Feb. 25, 1868	74, 957
Planter, Seed	M. Adams	Chilmark, Mass	June 8, 1869	90, 911
Planter, Seed	C. C. Aldrich	Faribault, Minn	Sept. 8, 1857	18, 126

Index of patents issued from the United States Patent Office from 1790 *to* 1873, *inclusive*—Continued.

Invention.	Inventor.	Residence.	Date.	No.
Planter, Seed	T. K. Alexander	Decatur, Ill	Feb. 13, 1866	52,635
Planter, Seed	J. P. Allen	Midville, Ga	Oct. 4, 1859	25,616
Planter, Seed	J. P. Allen	Dover, Ga	Aug. 14, 1860	29,555
Planter, Seed	P. Alling	Norwalk, Ohio	July 26, 1870	105,621
Planter, Seed	A. Anderson	Markham, Canada	Jan. 2, 1855	12,152
Planter, Seed	C. F. Anderson	Charlestown, N. H	May 31, 1859	24,185
Planter, Seed	J. Andrews	Winchester, Mass	Dec. 12, 1854	12,084
Planter, Seed	L. Arnold	Janesville, Wis	Feb. 17, 1857	16,636
Planter, Seed	M. Atwood	New Sharon, Iowa	Aug. 4, 1868	80,585
Planter, Seed	L. and S. H. Bachelder	Hampstead, N. H., and Haverhill, Mass.	Apr. 30, 1840	1,577
Planter, Seed	B. Baker	Hopkinton, Iowa	Aug. 1, 1871	117,502
Planter, Seed	H. F. Baker	Centreville, Ind	Jan. 5, 1858	19,010
Planter, Seed	J. C. Baker	Mechanicsburgh, Ohio	Mar. 1, 1859	23,069
Planter, Seed	S. Baker	Mount Pulaski, Ill	Jan. 19, 1858	19,122
Planter, Seed	E. F. Ballard	Lincoln, Nebr	Jan. 30, 1872	123,216
Planter, Seed	G. Banister	Hartford, Vt	June 22, 1869	91,595
Planter, Seed	W. H. Barber	Wolcottville, Conn	Sept. 4, 1860	29,853
Planter, Seed	A. J. Barnhart	Schoolcraft, Mich	Feb. 27, 1855	12,465
Planter, Seed	J. Barnhill	Circleville, Ohio	May 27, 1851	8,116
Planter, Seed	J. F. Beckwith and A. G. Gage	Alabama, N. Y	Nov. 30, 1858	22,156
Planter, Seed	J. F. Beckwith and A. G. Gage	South Alabama, N. Y	Jan. 18, 1859	22,617
Planter, Seed	L. Beemer	Libertyville, N. J	Feb. 3, 1857	16,522
Planter, Seed	H. Bell	Clinton, Ill	Dec. 13, 1859	26,456
Planter, Seed	A. Bennett	Rockford, Ill	Mar. 5, 1867	62,595
Planter, Seed	G. F. Bennett	Mount Olive, N. C	July 17, 1860	29,134
Planter, Seed	I. C. Benthall	Oakland, Tex	Mar. 1, 1859	23,071
Planter, Seed	A. Berdan	Macon, Mich	Aug. 10, 1858	21,112
Planter, Seed	C. W. Billings	South Deerfield, Mass	Mar. 14, 1854	10,632
Planter, Seed	J. Blackwood	Franklin County, Ohio	Jan. 30, 1855	12,307
Planter, Seed	W. Blessing	Jeffersonville, Ohio	Dec. 13, 1859	26,410
Planter, Seed	R. Boeklen	Jersey City, N. J	Feb. 10, 1857	16,592
Planter, Seed	C. B. and B. S. Borden and A. McLean	West Dresden, N. Y	Apr. 3, 1855	12,618
Planter, Seed	C. P. Brown	Shortsville, N. Y	Oct. 9, 1866	58,588
Planter, Seed	G. W. Brown	Tylerville, Ill	Aug. 2, 1853	9,893
Planter, Seed	G. W. Brown	Galesburgh, Ill	May 8, 1855	12,811
Planter, Seed	G. W. Brown	Galesburgh, Ill	Feb. 28, 1865	46,615
Planter, Seed	G. W. Brown	Galesburgh, Ill	Feb. 1, 1870	99,286
Planter, Seed	J. Brown	Lawn Ridge, Ill	Jan. 30, 1855	12,308
Planter, Seed	J. A. Brown	Richmond, Ind	Jan. 19, 1858	19,126
Planter, Seed	Z. B. Brown and M. C. Godard	Granby, Conn	Sept. 13, 1859	25,380
Planter, Seed	D. Broy	Canton, Mo	Apr. 23, 1861	32,117
Planter, Seed	J. H. Bruen	Penn Yan, N. Y	May 12, 1857	17,260
Planter, Seed	J. Bryant	Brooklyn, N. Y	Jan. 4, 1859	22,484
Planter, Seed	A. Bugbee	Elkhart, Ind	Sept. 19, 1865	50,065
Planter, Seed	W. Bullock	Red Falls, N. Y	Aug. 1, 1854	11,464
Planter, Seed	H. Bundel and J. Williams	Dayton, Ohio	Jan. 7, 1868	72,971
Planter, Seed	M. S. Burdick	Milton, Wis	Dec. 10, 1867	71,972
Planter, Seed	J. W. Butterick	Farmington, Wis	Jan. 15, 1867	61,153
Planter, Seed	L. A. Butts	Cuba, N. Y	June 5, 1855	12,990
Planter, Seed	L. A. Butts	Cuba, N. Y	Feb. 23, 1858	19,404
Planter, Seed	L. A. Butts	Ripon, Wis	Aug. 6, 1867	67,409
Planter, Seed	W. Call, jr	Haverstraw, N. Y	June 17, 1873	139,943
Planter, Seed	J. Campbell	Harrison, Ohio	Nov. 9, 1869	96,546
Planter, Seed	S. Cannon	New Richmond, Pa	Nov. 12, 1850	7,770
Planter, Seed	L. R. Carpenter	Lancaster, Ohio	Sept. 27, 1859	25,556
Planter, Seed	N. R. Carrington	Coldwater, Miss	Mar. 6, 1860	27,350
Planter, Seed	J. Carroll	La Porte, Ohio	Dec. 8, 1857	18,802
Planter, Seed	T. Carter	Laurens District, S. C	Apr. 11, 1854	10,755
Planter, Seed	J. Case	Springfield, Ohio	Jan. 16, 1855	12,231
Planter, Seed	J. Case	Bloomington, Ill	Dec. 7, 1858	22,228
Planter, Seed	L. Caswell	Harrison, Me	Aug. 2, 1853	9,894
Planter, Seed	J. Charlton	Allegheny City, Pa	May 4, 1858	20,143
Planter, Seed	E. E. Chesney	Abingdon, Ill	Apr. 16, 1867	63,859
Planter, Seed	E. E. Chesney	Bushnell, Ill	Oct. 24, 1871	120,241
Planter, Seed	A. Chrisman and M. Whitmer	Sugar Creek, Iowa	Nov. 22, 1864	45,139
Planter, Seed	W. Clark	Palmyra, Ill	Aug. 9, 1859	24,991
Planter, Seed	N. Clute and O. W. Marshall	Schenectady, N. Y., and Columbus, Ohio.	Sept. 24, 1867	69,076
Planter, Seed	L. W. Colver	Louisville, Ky	Feb. 28 1854	10,565
Planter, Seed	L. W. Colver	Louisville, Ky	May 22, 1855	12,895
Planter, Seed	L. W. Colver	Louisville, Ky	Dec. 7, 1852	9,439
Planter, Seed	J. W. Corey	Crawfordsville, Ind	Apr. 10, 1855	15,672
Planter, Seed	W. M. Cory	Jerseyville, Ill	Oct. 28, 1851	8,463
Planter, Seed	E. Cox	Point Pleasant, Ohio	July 7, 1863	39,122
Planter, Seed	G. Crampton	Marshall, Mich	June 28, 1859	24,541
Planter, Seed	W. Cressler	Shippensburgh, Pa	May 17, 1853	9,740
Planter, Seed	C. H. Dana	West Lebanon, N. H	Sept. 5, 1854	11,641
Planter, Seed	C. H. Dana	West Lebanon, N. H	June 5, 1855	12,995
Planter, Seed	J. H. Dancy	Dancyville, Tenn	Oct. 29, 1872	132,564
Planter, Seed	L. Daser	Washington, D. C	Sept. 5, 1854	11,642
Planter, Seed	G. Davis	Mumford, Ala	June 25, 1872	128,366
Planter, Seed	H. V. Davis	Amherst, N. H	Oct. 24, 1865	50,657
Planter, Seed	L. H. Davis and S. and M. Pennock.	Kennett's Square, Pa	Nov. 16, 1852	9,399
Planter, Seed	N. C. Davis	West Jefferson, Ohio	Oct. 25, 1853	10,169
Planter, Seed	W. Davis	Morgantown, Va	June 6, 1854	11,032
Planter, Seed	D. Deihl	Hanover, Pa	June 12, 1849	6,516
Planter, Seed	E. I. Dickey	Chester County, Pa	Jan. 23, 1849	6,049
Planter, Seed	W. C. Doss	Texana, Tex	Mar. 9, 1858	19,549
Planter, Seed	J. K. Dugdale	Richmond, Ind	Aug. 4, 1863	39,386
Planter, Seed	A. B. Earle	Franklin, N. Y	Oct. 24, 1854	11,829
Planter, Seed	D. Eberly	Strasburgh, Pa	Oct. 8, 1850	7,698
Planter, Seed	J. Edge	Acquackanonck, N. J	Mar. 26, 1867	63,147
Planter, Seed	C. R. Edwards	Bowling Green, Ky	Sept. 20, 1870	107,467
Planter, Seed	J. M. Elliott	Winsborough, S. C	Nov. 8, 1870	108,988
Planter, Seed	J. W. Ells and J. Charlton	Pittsburgh, Pa	Sept. 8, 1859	18,140

Index of patents issued from the United States Patent Office from 1790 *to* 1873, *inclusive*—Continued.

Invention.	Inventor.	Residence.	Date.	No.
Planter, Seed	F. E. A. Engelman	Cheektowaga, N. Y	June 8, 1869	91,002
Planter, Seed	G. T. Enoch and D. Wissinger	Springfield, Ohio	May 9, 1854	10,881
Planter, Seed	G. M. Evans	Pittsburgh, Pa	Apr. 28, 1857	17,145
Planter, Seed	G. M. Evans	Pittsburgh, Pa	Oct. 25, 1859	25,889
Planter, Seed	H. C. Fairchild	Brooklyn, Pa	Aug. 10, 1858	21,127
Planter, Seed	H. C. Fairchild	Brooklyn, Pa	June 5, 1860	28,567
Planter, Seed	H. C. Fairchild	Brooklyn, Pa	Nov. 5, 1867	70,541
Planter, Seed	J. W. Fawkes	Bart Township, Pa	Dec. 17, 1850	7,837
Planter, Seed	D. S. Fisher	Mauckport, Ind	May 10, 1859	23,913
Planter, Seed	L. B. Fisher	Cold Water, Mich	Mar. 7, 1854	10,607
Planter, Seed	L. B. Fisher	Cold Water, Mich	Mar. 28, 1854	10,694
Planter, Seed	G. Fletcher, sr., and T. Barnes	Greensburgh, Ind	Mar. 26, 1850	7,213
Planter, Seed	J. Fordyce	Morgantown, Va	Sept. 9, 1856	15,691
Planter, Seed	N. Foster, G. Jessup, and H. L. and C. P. Brown.	Palmyra, N. Y	Nov. 4, 1851	8,484
Planter, Seed	J. K. Frantz	Goodville, Pa	Dec. 24, 1867	72,472
Planter, Seed	R. Friday	Crockett, Tex	Feb. 25, 1873	136,231
Planter, Seed	T. C. Carlington	La Fayette, Ala	Feb. 11, 1873	135,704
Planter, Seed	I. H. Garretson	Clay, Iowa	Mar. 29, 1853	9,636
Planter, Seed	C. C. Garrett	Spring Hill, Ala	Feb. 12, 1861	31,381
Planter, Seed	J. C. Gaston	Reading, Ohio	Mar. 7, 1854	10,608
Planter, Seed	R. J. Gatling	Murfreesborough, N. C	May 10, 1844	3,581
Planter, Seed	R. and W. L. Gebby	New Richland, Ohio	Feb. 12, 1856	14,235
Planter, Seed	W. L. Gebby	New Richland, Ohio	Aug. 14, 1860	29,581
Planter, Seed	J. Gibbons	Adrian, Mich	Aug. 25, 1840	1,731
Planter, Seed	W. Y. Gill	Henderson, Ky	Oct. 6, 1857	18,333
Planter, Seed	R. Gillaspie	New Richmond, Ohio	Mar. 6, 1866	52,989
Planter, Seed	J. M. Gitchell	Haverhill, N. H	Oct. 20, 1868	83,151
Planter, Seed	W. W. Golsan	Autaugaville, Ala	Sept. 4, 1860	29,878
Planter, Seed	F. Goodwin	Astoria, N. Y	Mar. 3, 1857	16,729
Planter, Seed	A. M. Gould and A. Flanders	Cambria, N. Y	Oct. 6, 1857	18,334
Planter, Seed	J. Green	Kennett's Square, Pa	June 5, 1860	28,572
Planter, Seed	J. D. Green	Antrim, Ohio	Mar. 19, 1867	63,038
Planter, Seed	P. B. Green and E. A. Kenedy	Chicago and Newark, Ill	June 10, 1856	15,101
Planter, Seed	T. M. Green	Milledgeville, Ga	Aug. 21, 1860	29,689
Planter, Seed	J. P. Groshon	Yonkers, N. Y	Mar. 19, 1850	7,187
Planter, Seed	R. B. Ground	Marine Town, Ill	June 29, 1858	20,709
Planter, Seed	D. Haldeman	Morgantown, Va	Oct. 5, 1852	9,296
Planter, Seed	G. Hall	Morgantown, Va	June 24, 1856	15,182
Planter, Seed	J. Hall	Honey Cut, Ala	Oct. 13, 1857	18,393
Planter, Seed	L. Hariman	Anderson, Ind	Aug. 21, 1860	26,691
Planter, Seed	W. D. Harrah and B. S. Baldwin	Davenport, Iowa	Aug. 9, 1859	25,011
Planter, Seed	E. P. Harris	Conneautville, Pa	May 26, 1868	78,281
Planter, Seed	E. Hart	New Albany, Ind	Aug. 6, 1850	7,544
Planter, Seed	E. Hart	New Albany, Ind	Oct. 19, 1852	9,343
Planter, Seed	J. Haselton	Oxford, N. H	May 12, 1857	17,275
Planter, Seed	W. W. Haupt	Mountain City, Tex	Apr. 27, 1869	89,404
Planter, Seed	J. D. Havis	Perry, Ga	Oct. 28, 1856	15,974
Planter, Seed	M. Hayden	Rochester, Mich	May 9, 1865	47,637
Planter, Seed	A. Hays	Guy's Mills, Pa	June 9, 1868	78,666
Planter, Seed	T. D. Henson and G. Rohr	Charlestown, Va	Feb. 28, 1854	10,572
Planter, Seed	G. E. Herrick	Lynn, Mass	July 2, 1867	66,236
Planter, Seed	I. F. Herrin	San Antonio, Tex	Apr. 13, 1869	88,870
Planter, Seed	G. Hetrick	Reidsburgh, Pa	July 10, 1860	29,078
Planter, Seed	J. Hildebrand	East Berlin, Pa	Feb. 10, 1857	16,590
Planter, Seed	D. Hill	Bartonia, Ind	June 27, 1854	11,159
Planter, Seed	G. A. Hill and C. Lohnes	Springfield, Ohio	May 14, 1867	64,670
Planter, Seed	A. F. Hines	Washington, D. C	Sept. 4, 1860	29,937
Planter, Seed	P. Hinkley	Charleston, Ill	Oct. 20, 1857	18,450
Planter, Seed	S. M. Hockman	Tom's Brook, Va	Aug. 29, 1854	11,609
Planter, Seed	C. T. Holman	Conneautville, Pa	June 18, 1867	65,810
Planter, Seed	D. S. Holman	Conneautville, Pa	Jan. 22, 1867	61,431
Planter, Seed	S. T. Holly	Rockford, Ill	June 16, 1857	17,568
Planter, Seed	A. J. Holt	Peru, Ind	Sept. 29, 1868	82,524
Planter, Seed	P. Horn	Hagerstown, Md	Aug. 23, 1853	9,955
Planter, Seed	T. B. Houghton	Bloomington, Ill	Feb. 10, 1857	16,585
Planter, Seed	J. S. Huggins	Timmonsville, S. C	Feb. 28, 1860	27,291
Planter, Seed	D. H. Hull	Plantsville, Conn	May 21, 1867	64,877
Planter, Seed	M. J. Hunt and J. H. Haines	Rising Sun, Md	Jan. 5, 1858	19,026
Planter, Seed	S. C. Hunter	East Hickory, Pa	Nov. 26, 1867	71,488
Planter, Seed	O. Hyde	Oakland, Cal	Sept. 17, 1872	131,400
Planter, Seed	S. Ide	East Shelby, N. Y	July 4, 1854	11,226
Planter, Seed	S. C. Ives	Land of Promise, Va	Aug. 27, 1872	130,920
Planter, Seed	R. M. Jackson	Penningtonville, Pa	Oct. 5, 1852	9,298
Planter, Seed	S. Jenkins	Portsmouth, Pa	Sept. 20, 1873	10,032
Planter, Seed	S. Johnson	Cold Spring, N. Y	Feb. 5, 1861	31,354
Planter, Seed	W. B. Johnson	Staunton, Va	June 6, 1854	11,029
Planter, Seed	W. D. Johnson	Raleigh, N. C	Aug. 13, 1867	67,656
Planter, Seed	J. J. Johnston	Allegheny, Pa	May 4, 1858	20,158
Planter, Seed	J. F. Keller	Greencastle, Pa	Feb. 14, 1865	46,364
Planter, Seed	J. F. Keller	Greencastle, Pa	Jan. 29, 1867	61,544
Planter, Seed	J. F. Keller	Greencastle, Pa	Jan. 29, 1867	61,545
Planter, Seed	J. M. Kelley	Clinton, Ill	Sept. 15, 1863	39,931
Planter, Seed	R. Ketcham	South Dansville, N. Y	Mar. 20, 1866	53,304
Planter, Seed	C. Ketchum	Penn Yan, N. Y	May 12, 1857	17,305
Planter, Seed	W. Kilburn	Lawrenceville, Pa	June 11, 1842	2,663
Planter, Seed	W. Kilburn and F. Haines	Lawrenceville and Marietta, Pa.	Dec. 31, 1844	3,870
Planter, Seed	E. W. Kimball	Ottawa, Ill	Aug. 10, 1858	21,137
Planter, Seed	J. H. King, jr	Georgetown, D. C	Aug. 29, 1854	11,611
Planter, Seed	A. Klaus	Belleville, Ill	Oct. 18, 1859	25,835
Planter, Seed	W. Knowland and K. Collings	Henryville, Ind	June 27, 1871	116,327
Planter, Seed	A. Kraber	York, Pa	July 27, 1852	9,151
Planter, Seed	B. Kuhns	Dayton, Ohio	Dec. 29, 1868	85,455
Planter, Seed	B. Kuhns and M. J. Haines	Dayton, Ohio, and Delaware City, Del.	Sept. 30, 1856	15,810
Planter, Seed	L. L. Lancaster	Rocky Mount, N. C	Sept. 20, 1859	25,513
Planter, Seed	J. Landes	Selma, Ohio	Feb. 10, 1857	16,597

Index of patents issued from the United States Patent Office from 1790 *to* 1873, *inclusive*—Continued.

Invention.	Inventor.	Residence.	Date.	No.
Planter, Seed	J. N. Lane	Bethel, Ky	Apr. 3, 1866	53, 635
Planter, Seed	G. W. Lee	Ercildown, Pa	Nov. 21, 1854	11, 967
Planter, Seed	J. Lee	Galesburgh, Ill	Dec. 8, 1857	18, 821
Planter, Seed	J. H. Lee	Camanche, Iowa	Dec. 13, 1859	26, 439
Planter, Seed	J. S. Lewis	Elkader, Iowa	Sept. 28, 1869	95, 241
Planter, Seed	T. J. and G. F. Lewis	Boston, Mass	June 27, 1840	1, 657
Planter, Seed	T. J. and G. F. Lewis	Boston, Mass	June 27, 1840	1, 661
Planter, Seed	H. C. Locke	Somerville, Tenn	Sept. 21, 1869	95, 121
Planter, Seed	C. O. Luce	Brandon, Vt	Oct. 6, 1857	18, 344
Planter, Seed	I. G. McFarlane	Perry County, Pa	Mar. 14, 1854	10, 655
Planter, Seed	W. A. Mahaffy	Carimona, Minn	Aug. 31, 1858	21, 397
Planter, Seed	J. L. Manlove	Connersville, Ind	Oct. 22, 1867	70, 100
Planter, Seed	E. Marshall	Clinton, N. J	Apr. 11, 1854	10, 753
Planter, Seed	F. M. Marshall	Seguin, Tex	Dec. 28, 1858	22, 438
Planter, Seed	J. W. Masten	Utica, Mich	May 22, 1860	28, 386
Planter, Seed	E. G. Mathews	Newton, Mass	June 8, 1869	91, 143
Planter, Seed	A. Maurer	New Carlisle, Ind	Oct. 11, 1859	25, 746
Planter, Seed	J. M. Maxwell	Cape Elizabeth, Me	Nov. 29, 1864	45, 257
Planter, Seed	T. B. McConaughey	Newark, Del	Mar. 27, 1860	27, 644
Planter, Seed	E. McCormick	Connellsville, Pa	Oct. 16, 1855	13, 683
Planter, Seed	I. W. McGaffey	Syracuse, N. Y	Apr. 3, 1855	12, 641
Planter, Seed	I. W. McGaffey	Chicago, Ill	July 1, 1862	35, 771
Planter, Seed	I. W. McGaffey	Chicago, Ill	July 3, 1866	56, 076
Planter, Seed	J. McKown	Gardstown, Va	June 22, 1858	20, 651
Planter, Seed	J. McKown	Gardstown, Va	May 24, 1859	24, 135
Planter, Seed	J. McLaughlin	Duncannon, Pa	Aug. 28, 1860	29, 803
Planter, Seed	J. B. McMillan	Tipton, Ind	Aug. 9, 1859	25, 026
Planter, Seed	G. A. Meacham	New York, N. Y	June 10, 1856	15, 106
Planter, Seed	G. A. Meacham	New York, N. Y	Mar. 31, 1857	16, 929
Planter, Seed	N. Mendenhall	Greensburgh, Ind	Jan. 25, 1870	99, 217
Planter, Seed	J. T. Mercer	Seneca Township, Ohio	Dec. 27, 1859	26, 605
Planter, Seed	J. Miller	Russellville, Ky	Sept. 26, 1865	50, 213
Planter, Seed	J. C. Miller	Marietta, Pa	Jan. 23, 1849	6, 050
Planter, Seed	J. R. Mills	Bloomfield, Iowa	Aug. 28, 1860	29, 807
Planter, Seed	J. B. Miner	Groton, Conn	June 22, 1869	91, 655
Planter, Seed	T. S. Minniss	Meadville, Pa	May 29, 1855	12, 958
Planter, Seed	E. Moore	Avon, N. Y	Dec. 9, 1856	16, 212
Planter, Seed	H. Moore	Clinay, Mich	Mar. 27, 1855	12, 603
Planter, Seed	L. Moore	Bart Township, Lancaster County, Pa.	Apr. 18, 1848	5, 522
Planter, Seed	E. Morgan	Morgantown, Va	Jan. 16, 1855	12, 256
Planter, Seed	E. Morgan	Morgantown, Va	Mar. 18, 1856	14, 465
Planter, Seed	E. Morse	Walpole, N. H	Mar. 20, 1855	12, 554
Planter, Seed	G. Mottmiller	Columbus, Ohio	Sept. 1, 1843	3, 247
Planter, Seed	W. R. Mozier	Higginsville, Ill	Dec. 17, 1867	72, 318
Planter, Seed	W. G. Murphy	Seguin, Tex	Aug. 16, 1859	25, 134
Planter, Seed	J. Musgrave	New Cumberland, W. Va	Dec. 8, 1868	84, 751
Planter, Seed	E. Myers	Carroll County, Md	June 19, 1849	6, 542
Planter, Seed	D. B. Neal	Mount Gilead, Ohio	Oct. 23, 1855	13, 706
Planter, Seed	D. B. Neal	Mount Gilead, Ohio	Dec. 1, 1857	18, 762
Planter, Seed	H. Nycum	Uniontown, Pa	Dec. 14, 1852	9, 468
Planter, Seed	R. F. Osgood	Rochester, N. Y	Sept. 15, 1868	82, 153
Planter, Seed	B. Owen	Dayton, Ohio	Sept. 7, 1858	21, 440
Planter, Seed	G. Page	Baltimore, Md	May 25, 1840	1, 617
Planter, Seed	E. Parker	Baltimore, Md	June 1, 1858	20, 440
Planter, Seed	J. Peirson	Wilmington, Del	July 5, 1848	5, 655
Planter, Seed	J. Peirson	Wilmington, Del	Dec. 25, 1849	6, 976
Planter, Seed	M. and S. Pennock	East Marlborough, Pa	Mar. 12, 1841	1, 999
Planter, Seed	S. and M. Pennock	Kennett's Square, Pa	Dec. 10, 1850	7, 822
Planter, Seed	H. Perrin and W. Rudduck	Wilmington, Ohio	Sept. 20, 1853	10, 037
Planter, Seed	D. H. Phillips	Greenville, Ill	Mar. 20, 1855	12, 557
Planter, Seed	W. C. Pitts	Austin, Texas	May 15, 1860	28, 301
Planter, Seed	F. Plummer	Manchester, Ind	Jan. 22, 1856	14, 144
Planter, Seed	F. Plummer and G. B. Rollins	Manchester, Ind	Sept. 11, 1855	13, 551
Planter, Seed	W. G. Pollock and J. W. Sener	Fredericksburgh, Va	Oct. 9, 1860	30, 318
Planter, Seed	J. F. Pond	Cleveland, Ohio	Aug. 20, 1861	33, 103
Planter, Seed	L. Pratt	Amherst, N. H	Apr. 25, 1844	3, 562
Planter, Seed	J. A. Preston	Greensborough, Ga	Jan. 16, 1872	122, 736
Planter, Seed	D. R. Prindle	Bethany, New York	Mar. 29, 1859	23, 391
Planter, Seed	J. Putnam	Hamilton, N. Y	Mar. 7, 1846	4, 405
Planter, Seed	B. A. Ramsey	Trenton, Tenn	Oct. 4, 1870	108, 051
Planter, Seed	C. Randall	Palmyra, Ga	Nov. 2, 1852	9, 370
Planter, Seed	S. G. Randall	Rockford, Ill	Feb. 10, 1857	16, 610
Planter, Seed	S. G. Randall	Dixon, Ill	June 2, 1857	17, 450
Planter, Seed	S. G. Randall and J. H. Jones	Rockton, Ill	Aug. 7, 1855	13, 401
Planter, Seed	J. Redhead	Woodville, Miss	Mar. 9, 1858	19, 579
Planter, Seed	W. Redick	Uniontown, Pa	Nov. 18, 1851	8, 532
Planter, Seed	W. Redick	Uniontown, Pa	Aug. 29, 1854	11, 617
Planter, Seed	I. Rexford	Malone, N. Y	Dec. 8, 1868	84, 762
Planter, Seed	I. Reynolds	Republic, Ohio	Mar. 9, 1852	8, 790
Planter, Seed	D. B. Rhodes	Concord, N. Y	Dec. 10, 1850	7, 823
Planter, Seed	A. Richards	Anderson, Tex	May 21, 1872	126, 985
Planter, Seed	J. Robb	Lewistown, Pa	Oct. 12, 1852	9, 333
Planter, Seed	J. S. Robb and S. P. Allison	New Cumberland, W. Va	Jan. 12, 1869	85, 854
Planter, Seed	J. Robinson	Sharpstown, Md	Dec. 1, 1857	18, 772
Planter, Seed	J. Robinson	Sharpstown, Md	Apr. 17, 1860	27, 929
Planter, Seed	T. B. and R. N. Rockwell	Batavia, Ill	Aug. 13, 1861	33, 053
Planter, Seed	A. J. Rogers	Stephentown, N. Y	June 19, 1860	28, 777
Planter, Seed	T. B. Rogers	Wethersfield, Conn	Aug. 9, 1859	25, 047
Planter, Seed	G. Rohr	Charlestown, Va	June 21, 1853	9, 803
Planter, Seed	R. Romaine	Montreal, Canada	Feb. 20, 1855	12, 447
Planter, Seed	J. H. Rose	Versailles, Ill	Aug. 17, 1858	21, 217
Planter, Seed	J. P. Ross	Lewisburgh, Pa	Sept. 25, 1849	6, 743
Planter, Seed	J. P. Ross	Lewisburgh, Pa	Jan. 1, 1851	7, 877
Planter, Seed	J. P. Ross	Lewisburgh, Pa	June 8, 1852	9, 004
Planter, Seed	E. Russell	Coatesville, Pa	May 15, 1860	28, 330
Planter, Seed	T. Russell	Waldoborough, Me	Apr. 13, 1858	19, 953
Planter, Seed	W. J. Saffrey	Bremen, Ohio	July 8, 1873	140, 648

Index of patents issued from the United States Patent Office from 1790 *to* 1873, *inclusive*—Continued.

Invention.	Inventor.	Residence.	Date.	No.
Planter, Seed	B. D. Sanders	Holliday's Cove, Va	June 8, 1852	9,006
Planter, Seed	C. R. Sargent	Newburyport, Mass	Apr. 25, 1871	114,047
Planter, Seed	M. Satterlee	Louisa, Ill	July 26, 1853	9,879
Planter, Seed	B. Saunders	Claverack, N. Y	July 4, 1871	116,633
Planter, Seed	W. F. Schroeder	La Porte, Ind	Aug. 14, 1860	29,624
Planter, Seed	J. F. Seaman	Walcott, N. Y	Sept. 30, 1856	15,822
Planter, Seed	G. M. and S. H. Seward	Guilford, Conn	June 6, 1865	48,104
Planter, Seed	J. B. Seymour	Pittsburgh, Pa	Sept. 24, 1867	69,130
Planter, Seed	P. Seymour	East Bloomfield, N. Y	July 24, 1855	13,326
Planter, Seed	J. Shattuck	Waterloo, N. Y	Jan. 29, 1867	61,689
Planter, Seed	J. M. Shaw	Water Valley, Miss	Nov. 30, 1869	97,446
Planter, Seed	J. W. Sherman	Ontario, N. Y	Nov. 6, 1849	6,853
Planter, Seed	N. C. Sherman and J. Mason	Hazel Green, Wis	Dec. 23, 1856	16,314
Planter, Seed	J. H. Shireman	East Berlin, Pa	Oct. 21, 1856	15,955
Planter, Seed	J. Signer and T. N. Shipton	Kishacoquillas Valley, Pa	Dec. 10, 1850	7,831
Planter, Seed	H. Sloan	Franklin, Ind	June 26, 1860	28,909
Planter, Seed	D. M. Smith	Springfield, Vt	May 10, 1859	23,955
Planter, Seed	G. Smith and A. G. Perry	Clyde, Ohio	June 29, 1858	20,738
Planter, Seed	H. W. Smith	Paradise, Pa	Nov. 4, 1846	4,833
Planter, Seed	J. D. Smith	Lancaster, Ohio	Oct. 27, 1857	18,524
Planter, Seed	J. D. Smith	Peoria, Ill	June 24, 1862	35,713
Planter, Seed	L. Smith	Chester Centre, Mass	Aug. 23, 1870	106,631
Planter, Seed	T. G. Smith	Canton, Miss	Apr. 13, 1869	88,989
Planter, Seed	T. H. Smith	Clyde, N. Y	Aug. 25, 1868	81,548
Planter, Seed	T. H. Smith	Clyde, N. Y	May 24, 1870	103,382
Planter, Seed	B. M. Snell	Hancock, Md	Mar. 20, 1855	12,561
Planter, Seed	T. Snow	Social Circle, Ga	Jan. 2, 1872	122,412
Planter, Seed	J. G. Snyder and J. Young	Wheatfield Township, Pa	Feb. 28, 1854	10,583
Planter, Seed	W. Sprague	Ellicottsville, N. Y	Mar. 14, 1854	10,644
Planter, Seed	R. H. Springstead	Wooster, Ohio	July 24, 1849	6,605
Planter, Seed	J. Stark	Thomasville, Ga	Dec. 17, 1867	72,333
Planter, Seed	E. M. Stevens, J. B. Crosby, and J. W. Pearson.	Boston and Winchester, Mass.	May 22, 1855	12,924
Planter, Seed	S. L. Stockstill	Medway, Ohio	May 24, 1859	24,158
Planter, Seed	S. L. Stockstill and P. H. Humes	Brandt, Ohio	Jan. 16, 1855	12,260
Planter, Seed	B. T. Stowell and A. Marcellus	Waddam's Grove, Ill	Apr. 13, 1852	8,877
Planter, Seed	J. L. Strait	Cooksville, Miss	Dec. 28, 1869	98,313
Planter, Seed	W. H. Stuart	Millington, Md	Oct. 4, 1859	25,685
Planter, Seed	I. T. Suggs	Greene Hill, Tex	Apr. 22, 1873	138,105
Planter, Seed	J. F. Tannehill	Staunton, Va	Aug. 14, 1860	29,636
Planter, Seed	J. H. Thomas and P. P. Mast	Springfield, Ohio	July 27, 1858	21,034
Planter, Seed	H. Thomason	La Fayette, Ind	Feb. 10, 1857	16,617
Planter, Seed	J. Thompson	Durhamville, N. Y	Jan. 13, 1857	16,409
Planter, Seed	S. Thompson	Hopedale, Ohio	Mar. 30, 1858	19,818
Planter, Seed	D. L. Tilton	Mount Carmel, Ill	Feb. 23, 1858	19,456
Planter, Seed	F. W. Tilton	Moline, Ill	Mar. 17, 1868	75,600
Planter, Seed	H. Todd	Oxford, N. H	Dec. 15, 1843	3,381
Planter, Seed	F. Townsend	Cambria, N. Y	Nov. 2, 1852	9,372
Planter, Seed	H. M. Tremble	Mattoon, Ill	Oct. 18, 1864	44,758
Planter, Seed	C. S. Trevitt	Ellicottsville, N. Y	Nov. 2, 1852	9,373
Planter, Seed	C. J. Turner and M. L. Wilkinson.	Olean, N. Y	Apr. 25, 1871	114,066
Planter, Seed	J. Urmy	Wilmington, Del	Apr. 6, 1852	8,866
Planter, Seed	F. Vandoren	Adrian, Mich	Apr. 13, 1852	8,879
Planter, Seed	T. G. S. Vaniz	Canton, Miss	Aug. 8, 1871	117,835
Planter, Seed	H. Vermillion	Rising Sun, Md	Nov. 2, 1852	9,374
Planter, Seed	J. T. and L. J. Wait	Waterloo, S. C	Aug. 15, 1854	11,537
Planter, Seed	C. A. Wakefield	Essex County, N. Y	Feb. 19, 1850	7,110
Planter, Seed	C. A. Wakefield	Plainfield, Mass	July 25, 1854	11,395
Planter, Seed	A. Wales	Pontiac, Ill	June 29, 1858	20,749
Planter, Seed	L. F. Ward	Marathon, N. Y	Nov. 24, 1857	18,716
Planter, Seed	L. F. Ward	Marathon, N. Y	Dec. 29, 1857	18,999
Planter, Seed	M. Ward	Owego, N. Y	Mar. 27, 1855	12,608
Planter, Seed	S. B. Ward	Auburn, Ind	Jan. 5, 1869	85,711
Planter, Seed	M. Warner	West Middleburgh, Ohio	Aug. 21, 1860	29,735
Planter, Seed	D. Warren	Gettysburgh, Pa	June 26, 1860	28,926
Planter, Seed	S. J. Wasterbarg	Altona, Ill	Sept. 27, 1859	25,595
Planter, Seed	M. Waterbury	Cuba, N. Y	Nov. 21, 1854	11,980
Planter, Seed	G. Watt	Richmond, Va	Mar. 8, 1859	23,206
Planter, Seed	M. D. Wells	Morgantown, Va	Dec. 14, 1852	9,475
Planter, Seed	W. F. West	Haverstraw, N. Y	Jan. 3, 1871	110,706
Planter, Seed	J. E. White	Clinton, Ill	June 7, 1870	104,088
Planter, Seed	T. W. White	Milledgeville, Ga	Apr. 16, 1861	32,098
Planter, Seed	J. Whitehead	Manchester, Va	May 26, 1857	17,402
Planter, Seed	E. Wicks	Bart, Pa	Feb. 10, 1852	8,728
Planter, Seed	A. Wieting	Middletown, Pa	Apr. 1, 1851	8,018
Planter, Seed	J. H. Wiggin	Boston, Mass	Feb. 2, 1858	19,274
Planter, Seed	S. C. Wilder	Sardinia, Ohio	Mar. 17, 1868	75,613
Planter, Seed	H. Willard	Vergenes, Vt	Oct. 6, 1857	18,366
Planter, Seed	W. B. Willis	Charlestown, Va	Jan. 22, 1850	7,044
Planter, Seed	J. D. Willoughby	Chambersburgh, Pa	June 5, 1849	6,504
Planter, Seed	J. D. Willoughby	Pleasant Hall, Pa	Jan. 26, 1858	19,222
Planter, Seed	J. D. Willoughby	Carlisle, Pa	Aug. 3, 1858	21,102
Planter, Seed	C. B. Winder	North Lewisburgh, Ohio	Nov. 24, 1857	18,717
Planter, Seed	S. and W. H. Witherow	Gettysburgh, Pa	Jan. 18, 1853	9,551
Planter, Seed	D. and H. Wolf	Lebanon, Pa	Feb. 15, 1853	9,589
Planter, Seed	D. and H. Wolf	Lebanon, Pa	Mar. 21, 1854	10,682
Planter, Seed	L. Woodruff	Ann Arbor, Mich	May 2, 1865	47,593
Planter, Seed	D. C. Woods	Waxahatchie, Tex	Dec. 21, 1869	98,136
Planter, Seed	J. Woodward	Haverhill, N. H	July 13, 1852	9,125
Planter, Seed	A. R. Worth	Nantucket, Mass	Apr. 9, 1867	63,685
Planter, Seed	W. H. Worth and L. Finlay	Canton, Mo	Dec. 13, 1859	26,455
Planter, Seed	R. C. Wrenn	Mount Gilead, Ohio	Nov. 29, 1863	10,278
Planter, Seed	L. R. Wright	Cohoes, N. Y	Apr. 16, 1861	32,108
Planter, Seed	L. R. Wright	Cohoes, N. Y	Oct. 23, 1866	59,115
Planter, Seed	R. B. Wright	Vermillion, Ill	Mar. 17, 1868	75,616
Planter, Seed	H. Wyant	Knox County, Ind	Dec. 9, 1856	16,198
Planter, Seed	F. G. Wynkoop	Corning, N. Y	Oct. 16, 1855	13,694
Planter, Seed	E. Young	Fayetteville, Mo	June 26, 1860	28,936

Index of patents issued from the United States Patent Office from 1790 *to* 1873, *inclusive*—Continued.

Invention.	Inventor.	Residence.	Date.	No.
Planter, Seed and guano	T. W. White	Milledgeville, Ga	June 18, 1867	65, 976
Planter, seed-sower, and cultivator, Corn	D. P. Leach	Franklin, Ind	Oct. 19, 1869	96, 015
Planter, seeder, and cultivator, Combined potato	B. F. Field	Sheboygan Falls, Wis	Sept. 26, 1865	50, 202
Planter, Self-working rotary corn	J. W. Simonton	Taylorsville, Ind	Feb. 15, 1870	99, 792
Planter shield, Corn	R. T. Taylor	Everton, Ind	Oct. 20, 1868	83, 340
Planter, Single-row corn	J. Clarridge	Pancoastburgh, Ohio	Oct. 9, 1866	58, 598
Planter, sower, revolving harrow, and cultivator, Combined.	W. P. Byler	Leavenworth, Kans	Feb. 18, 1868	74, 496
Planter, Sugar-cane	E. Cortés	Sagua la Grande, Cuba	Dec. 18, 1866	60, 620
Planter, Walking	A. W. Dunlevy	Fair Play, Ohio	Oct. 31, 1871	120, 507
Planter, Walking	N. Earlywine	Centerville, Iowa	Oct. 31, 1871	120, 423
Planter, Walking	G. W. Heath	Burlington, Pa	Apr. 9, 1872	125, 568
Planter, Walking	L. D. Noble	Cerro Gordo, Ill	Jan. 30, 1872	123, 121
Planter, Walking	M. W. Stephenson	Pickensville, Ala	Jan. 9, 1872	122, 675
Planters, Actuating the feeding-apparatus of seed	J. K. Dugdale	Richmond, Ind	Apr. 7, 1863	38, 099
Planters, Check-row attachment for corn	W. C. Grimes	Decatur, Ill	Apr. 18, 1871	113, 761
Planters, Cultivating seed	W. Flory and G. A. Grove	Chambersburgh, Pa	Mar. 12, 1850	7, 163
Planters, Device for sowing in seed	W. P. Clements	Ellerslie, Ga	Oct. 7, 1851	8, 408
Planters, Draft-apparatus of seed	J. Mumma	Mount Joy, Pa	Aug. 16, 1853	9, 940
Planters, Drill-teeth in seed	L. Haverstick	Manor Top, Pa	Apr. 2, 1850	7, 251
Planters, Dropper for seed	H. H. Koeller	Camp Point, Ill	Dec. 16, 1873	145, 575
Planters, Gearing for seed	M. J. Hunt	Rising Sun, Md	June 3, 1851	8, 138
Planters, Gearing for seed	J. Pierson	Wilmington, Del	Apr. 9, 1850	7, 268
Planters, Guide for corn and seed	E. C. Brown	Crawfordsville, Ind	Mar. 24, 1868	75, 851
Planters, Marking-attachment to corn	E. Sawyer	Madison, Iowa	Sept. 13, 1870	107, 294
Planters, Seed-distributer for seed	D. and H. Wolf	Lebanon, Pa	June 3, 1851	8, 132
Planter, Seed-roller for seed	E. Wicks	Bart Township, Pa	Mar. 26, 1850	7, 227
Planters, Seed-tube for corn	L. Scofield	Watertown, Wis	Mar. 14, 1871	112, 741
Planters, Seed-tube for corn	L. Scofield	Watertown, Wis	July 11, 1871	116, 998
Planters, Seeding-apparatus for seed	G. S. Gardner	Charlestown, Va	Oct. 8, 1850	7, 699
Planters, Seeding-apparatus for seed	D. Horner	Knox County, Ohio	July 29, 1851	8, 264
Planters, Seeding-apparatus for seed	L. Moore	Bart, Pa	July 2, 1850	7, 479
Planters, Seeding-apparatus for seed	S. and M. Pennock	Kennett's Square, Pa	July 9, 1850	7, 495
Planters, Seeding-apparatus for seed	S. and M. Pennock	Kennett's Square, Pa	Feb. 11, 1851	7, 930
Planters, Seeding-apparatus for seed	G. Rohr	Charlestown, Va	July 16, 1850	7, 513
Planters, Seeding-apparatus for seed	C. C. Van Enery	Victor, N. Y	Oct. 21, 1851	8, 459
Planters, Seeding-roller for seed	A. Palmer	Brockport, N. Y	Sept. 10, 1850	7, 642
Planters, Shoe for seed	A. W. Brinkerhoff	Upper Sandusky, Ohio	July 16, 1861	32, 819
Planters, Slide for seed	R. I. Colvin	Lancaster, Pa	Oct. 1, 1850	7, 680
Planters, Tube for seed	J. C. Haines	Dublin, Ind	Dec. 15, 1857	18, 843
Planting and cultivating machine, Cotton	R. Herbert	Williamson County, Tenn	Apr. 19, 1828	....
Planting and cultivating machine, Cotton	F. H. Smith	Richmond, Va	Feb. 15, 1826	....
Planting and digging potatoes, plowing corn, &c	P. Meigs and M. C. Arnold	Madison, Conn	Nov. 3, 1830	....
Planting and fertilizing machine	S. L. Allen	Cinnaminson, N. J	Oct. 25, 1870	108, 672
Planting and preparing seeds, Machine for	Z. Lane	Harrisburgh, N. Y	Sept. 21, 1829	....
Planting harrow, Seed	C. A. Wakefield	Essex County, N. Y	Feb. 19, 1850	7, 109
Planting, Corn	C. R. Belt	Washington, D. C	Jan. 15, 1836	....
Planting corn, &c	H. Todd	Pembroke, N. H	Oct. 1, 1831	....
Planting corn and pease	R. Coffey	Burke County, N. C	Dec. 14, 1830	....
Planting, Cotton	H. Allen	Fayetteville, Tenn	June 16, 1836	....
Planting cotton	C. Ford	Paineville, Va	May 26, 1825	....
Planting, hoeing, and digging potatoes, Combined machine for.	J. C. Clement	Kenduskeag, Me	Dec. 19, 1865	51, 560
Planting Indian corn in rows, Machine for	J. Blecker	Lycoming County, Pa	Mar. 18, 1809	....
Planting-machine	S. L. Allen	Cinnaminson, N. J	Nov. 24, 1868	84, 247
Planting-machine	M. Atwood, jr	Hampstead, N. H	June 24, 1839	1, 188
Planting-machine	J. M. Bright and E. Standt	Bernville, Pa	Aug. 5, 1873	141, 538
Planting-machine	R. Craggs and O. Reynolds	Williamson and Webster, N. Y.	Aug. 14, 1847	5, 237
Planting-machine	J. M. Forrest	Princess Ann Court-House, Va.	June 25, 1839	1, 194
Planting-machine	J. Jones	Newton, N. J	Oct. 11, 1841	2, 303
Planting-machine	M. Nichols	Clearfield, Pa	Oct. 16, 1840	1, 830
Planting-machine	G. Ray	Kinderhook, N. Y	Oct. 23, 1866	59, 071
Planting-machine	J. L. Sater	Cincinnati, Ohio	June 10, 1862	35, 546
Planting-machine	E. Woods	Beloit, Wis	Jan. 10, 1845	3, 879
Planting-machine	R. B. Wright	Vermillion, Ill	Feb. 26, 1867	62, 461
Planting machine, Corn	D. and W. F. Chipman	Mount Carmel, Ill	Apr. 28, 1868	77, 255
Planting machine, Corn	A. Cock	Flushing, N. Y	Mar. 20, 1821	....
Planting machine, Corn, &c	S. W. Cole	Chelsea, Mass	Aug. 25, 1840	1, 729
Planting machine, Corn	M. Lennox, W. Croft, and H. Pitner.	Steubenville, Ohio	Jan. 26, 1829	....
Planting machine, Corn	N. R. and O. G. Merchant	Gilford, N. Y.	Oct. 12, 1849	1, 366
Planting machine, Corn and bean	E. Spooner	....	Jan. 25, 1799	....
Planting machine, Corn and other seed	D. S. Rockwell	New Canaan, Conn	Mar. 12, 1839	1, 097
Planting machine, Cotton, corn, &c	J. Lobdell	Warren County, Miss	Dec. 26, 1826	....
Planting machine, Cotton-seed	J. Armstrong	Bucyrus, Ohio	Nov. 26, 1867	71, 263
Planting machine, Cotton-seed	Z. N. Morrel	Cameron, Tex	July 5, 1859	24, 652
Planting machine, Cotton-seed	W. Price	Mount Olive, N. C	June 5, 1860	28, 602
Planting machine, Cotton-seed	R. S. Thomas	Bennetsville, S. C	July 30, 1841	2, 205
Planting, Machine for making holes for	E. R. Ensign	East Hartford, Conn	Sept. 3, 1867	68, 424
Planting, Machine for marking ground for	P. McQuaid	Wenona, Ill	Aug. 7, 1866	56, 970
Planting machine, Grain	O. Starr	Richmond, N. Y	Aug. 22, 1828	....
Planting machine, Potato	L. A. Aspinwall	Ireland's Corners, N. Y	Oct 14, 1862	36, 634
Planting machine, Potato	G. J. Bundy	Lyndon, Vt	July 21, 1857	17, 827
Planting machine, Potato	T. Marcus	Washington, D. C	Sept. 6, 1864	44, 103
Planting machine, Potato	J. W. Pelletreau	East Moriches, N. Y	Apr. 6, 1858	19, 869
Planting machine, Potato	J. L. True	Garland, Me	Feb. 7, 1865	46, 281
Planting machine, Potato	T. B. Whyte	Greenwich, N. Y	Feb. 9, 1858	19, 322
Planting machine, Seed	D. Brener	Petersburgh, Tenn	Apr. 4, 1844	3, 525
Planting machine, Seed	W. Buckminster	Framingham, Mass	Oct. 8, 1838	969
Planting machine, Seed	A. H. and L. Robbins, jr	Denmark, N. Y	Aug. 28, 1828	....
Planting machine, Seed	J. Woodward	Haverhill, N. H	Feb. 29, 1848	5, 460
Planting onion-seed, beans, &c	P. Smith and T. H. Arnold	Haddam, Conn	Jan. 20, 1831	....
Planting or sowing, Toothed roller for preparing the earth for.	A. Densmore	Le Roy, N. Y	July 29, 1815	....
Plas er	M. C. Borgia and H. B. Taylor	Philadelphia, Pa	July 17, 1866	56, 353
Plaster	G. E. Mitchell	Lowell, Mass	Sept. 12, 1871	118, 872

Index of patents issued from the United States Patent Office from 1790 *to* 1873, *inclusive*—Continued.

Invention.	Inventor.	Residence.	Date.	No.
Plaster, Adhesive	J. Lynch	Columbia, S. C	Sept. 1, 1868	81, 657
Plaster, Adhesive	J. Melvin	Lowell, Mass	Feb. 3, 1863	37, 604
Plaster, Adhesive	W. H. Sherut and H. H. Day	New York, N. Y	Mar. 26, 1845	3, 965
Plaster, Adhesive compound and	J. Hirsh	Chicago, Ill	Feb. 2, 1869	86, 398
Plaster and seed sower	D. Dick and O. W. Preston, jr	Corning, N. Y	Feb. 11, 1868	74, 320
Plaster, Blister	E. Perkins	Baltimore, Md	Jan. 15, 1830	....
Plaster-crushing mill	E. Kent	....	Sept. 14, 1804	....
Plaster for walls	B. R. Smith and J. C. Harris	Philadelphia, Pa	Jan. 24, 1871	111, 267
Plaster, Healing	W. Kramer	New York, N. Y	May 10, 1870	102, 836
Plaster, lime, &c	A. and J. Krauss	Upper Milford, Pa	Oct. 16, 1830	....
Plaster, lime, &c., Machine for spreading	G. U. Relyea	Watkins, N. Y	Nov. 3, 1868	83, 816
Plaster, Medical	S. I. Merrill	Falmouth, Me	Nov. 10, 1868	83, 872
Plaster, Medicated	W. S. Bright and J. G. Morey	New Orleans, La	Jan. 8, 1867	61, 045
Plaster-molds, Process of reducing the size of	N. A. Downer	Canandaigua, Mich	Jan. 8, 1869	91, 100
Plaster, Mustard	B. I. Crew	Philadelphia, Pa	July 14, 1868	79, 811
Plaster of Paris, Breaking	P. Geiger	Robeson, Pa	May 12, 1812	....
Plaster of Paris, Mill for breaking	W. Brown	Baltimore, Md	Dec. 8, 1808	....
Plaster on cloth, Machine for spreading adhesive	A. Mouson	Milton, Pa	Aug. 8, 1837	341
Plaster-patterns, Manufacture of	J. Semple	Chicago, Ill	Sept. 30, 1873	143, 384
Plaster-sowing machine	A. Bugbee	Elkhart, Ind	Sept. 19, 1865	49, 973
Plaster-spreading apparatus	N. Wood	Portland, Me	Oct. 29, 1872	132, 614
Plaster-spreading gage	J. M. Keep	Bath, Me	Apr. 30, 1850	7, 323
Plaster-spreading machine	W. N. Reed	Arlington, Va	June 4, 1872	127, 517
Plaster, Sticking or adhesive	J. C. Battersby	New York, N. Y	May 16, 1871	114, 750
Plaster, Whooping-cough	F. Hower	Brooklyn, N. Y	Nov. 23, 1869	97, 089
Plasters, Apparatus for making medical	A. D. Richards	Lowell, Mass	Apr. 16, 1867	63, 814
Plasters, Apparatus for spreading medical	A. J. Schafhirt	Washington, D. C	Apr. 4, 1871	113, 453
Plasters, Composition for healing	D. G. Williams	Boston, Mass	July 11, 1871	116, 906
Plasterer's float	L. A. Goodsell	New Haven, Conn	May 27, 1873	139, 383
Plasterer's hawk	T. J. McGeary	Newark, N. J	Oct. 19, 1869	96, 022
Plasterer's head-support	T. T. Wright	Fayetteville, N. C	Apr. 29, 1873	138, 463
Plastering, Corner-bead for	E. F. Rice	Worcester, Mass	Apr. 16, 1872	125, 844
Plastering, Foundation for	J. John	Chicago, Ill	Oct. 3, 1871	119, 615
Plastering-machine	J. L. Coburn	Mineral Point, Wis	Mar. 23, 1869	88, 135
Plastering-machine	I. Hussey	Harveysburgh, Ohio	Mar. 7, 1854	10, 590
Plastering-machine	J. Keene	Washington, D. C	July 9, 1867	66, 664
Plastering-machine	T. McKinley	New York, N. Y	July 27, 1869	92, 985
Plastering-machine	G. Stevens and J. H. Watson	Tawas City, Mich	Dec. 9, 1873	145, 459
Plastering surfaces	J. J. Althause	New York, N. Y	Jan. 7, 1862	34, 099
Plastering walls, Composition for	D. Young	Kokomo, Ind	Aug. 17, 1869	93, 941
Plastic articles, Composition for molding	C. Legg	Malcom, Iowa	Mar. 11, 1873	136, 658
Plastic composition	H. C. Gaskins	Union Vale, N. Y	Oct. 20, 1868	83, 149
Plastic composition	T. B. Gunning	New York, N. Y	Mar. 19, 1872	124, 736
Plastic compound	F. Baschnagel	Beverly, Mass	Mar. 6, 1860	27, 343
Plastic materials, Method of molding	W. B. Gleason	Boston, Mass	Dec. 10, 1867	72, 017
Plastic materials, Mold for	J. J. Wiggin	Cincinnati, Ohio	Dec. 4, 1866	60, 310
Plate	N. D. Stevens	Westbrook, Me	Oct. 12, 1869	95, 742
Plate and tumbler lock, Revolving	L. Jennings	New York, N. Y	Apr. 2, 1850	7, 244
Plate, Anti-slipping	W. B. Gould and W. H. Harris	Boston and Taunton, Mass	Sept. 8, 1868	81, 890
Plate, Army	W. Terneson	Philadelphia, Pa	Feb. 16, 1864	41, 652
Plate, Coated metal	E. Morewood and G. Rogers	Enfield, England	Apr. 6, 1858	19, 866
Plate cover, Dinner	M. M. J. O'Sullivan	New Haven, Conn	Nov. 14, 1871	120, 995
Plate, Dinner	J. K. Andrews	Antrim, Ohio	Nov. 19, 1867	71, 115
Plate handling-fork, Heated	G. W. Hyatt	Auburn, N. Y	Nov. 11, 1856	16, 058
Plate-holder	J. Carlin	New York, N. Y	Jan. 18, 1870	98, 923
Plate-holder	W. T. Stantenborough	Brooklyn, N. Y	Jan. 31, 1871	111, 399
Plate holder, Camera	A. S. Southworth	Boston, Mass	Apr. 10, 1855	12, 709
Plate holder, Focusing	S. W. Burcaw	Allentown, Pa	Oct. 24, 1865	50, 555
Plate holder, Glass	T. Jones	New York, N. Y	Mar. 30, 1869	88, 487
Plate-holder, Non-corrosive metal-coated	J. F. Ryder	Cleveland, Ohio	May 16, 1871	114, 864
Plate, Hollow pressure	J. B. Fontaine	Philadelphia, Pa	Nov. 13, 1866	59, 716
Plate-lifter	D. B. Beaty	Aurora, Ill	Dec. 10, 1867	71, 954
Plate-lifter	C. F. Bosworth	Milford, Conn	Aug. 6, 1867	67, 402
Plate-lifter	H. P. Brooks	Waterbury, Conn	Nov. 29, 1870	109, 715
Plate-lifter	J. A. Burns	New Haven, Conn	Sept. 3, 1867	68, 557
Plate-lifter	E. Gibbs	Richland Centre, Wis	June 14, 1870	104, 298
Plate-lifter	D. E. Roe	Elmira, N. Y	Jan. 7, 1868	73, 199
Plate-lifter	G. O. Roe	Conantsville, Conn	Feb. 18, 1868	74, 598
Plate-lifter	D. M. Skinner	Sandwich Center, N. H	June 11, 1867	65, 616
Plate-lifter	J. M. Smith	Centre Sandwich, N. H	Nov. 26, 1867	71, 336
Plate-lifter	S. J. Talbott	Milford, N. H	Apr. 14, 1868	76, 649
Plate-lifter	D. Welch	Lowell, Mass	Mar. 5, 1867	62, 716
Plate-lifter	J. B. Willett	West Meriden, Conn	June 4, 1867	65, 460
Plate-lifter and bread-toaster	T. D. Keith	Mayville, Wis	Nov. 23, 1869	97, 093
Plate-lifter, Hot	C. L. Merrill	Saint Louis, Mo	Nov. 5, 1872	132, 848
Plate-metal along curved or straight lines, Machine for dividing.	N. Waterman and A. T. Perkins.	Toledo, Ohio	Apr. 30, 1872	126, 357
Plate or salver	H. McManus and J. B. Hatting	New York, N. Y	Dec. 15, 1868	84, 894
Plate, Pie	J. F. Kohler and S. B. Conover	New York, N. Y	July 23, 1867	67, 061
Plate spring, Circular metallic	J. W. Adams	New York, N. Y	May 15, 1855	12, 849
Plate, Table	D. H. Shirley	Boston, Mass	June 12, 1860	28, 693
Plate-warmer	J. C. Palmer	New York, N. Y	Aug. 27, 1867	68, 229
Plates and glass, Composition for cleaning	H. Teats	Ann Arbor, Mich	Apr. 5, 1870	101, 679
Plates and sheets, Manufacturing into	E. J. Ellicott	Baltimore, Md	May 26, 1818	....
Plates, Machine for folding tinned	O. W. Stow	Plantsville, Conn	Dec. 24, 1867	72, 561
Plates or bars, Machine for making holes in horizontal.	B. King	Washington, D. C	May 29, 1817	....
Plates, Wiping and polishing the exterior surface of punched or etched.	S. Conillard, jr	Boston, Mass	Oct. 9, 1830	....
Plated articles, Nickel	J. A. Whitman and T. M. Neal	Auburn, Me	Apr. 16, 1872	125, 808
Plated goods, Finishing the surface of	J. Rogers	Newark, N. J	Nov. 1, 1870	108, 940
Plated metal, Manufacture of	J. D. Grüneberg	Philadelphia, Pa	Dec. 26, 1865	51, 714
Plated ware	J. C. Blackman	West Meriden, Conn	Oct. 29, 1867	70, 156
Plated-ware-burnishing machine	S. A. Chapman	Waterbury, Conn	Feb. 19, 1867	62, 251
Plated-ware handles, Fastening	H. G. Smith and E. M. Pomeroy	West Meriden, Conn	May 8, 1866	54, 629
Platen-gage	H. Byxbe	Williamsport, Pa	Jan. 30, 1872	123, 233
Plating and coating metals	E. E. De Lobstein	Paris, France	Aug. 13, 1872	130, 352
Plating and gilding, Cleaning metallic articles for	G. J. Sturdy and S. W. Young	Providence, R. I	Mar. 2, 1869	87, 442

Index of patents issued from the United States Patent Office from 1790 *to* 1873, *inclusive*—Continued.

Invention.	Inventor.	Residence.	Date.	No.
Plating and polishing compound	M. J. A. Keane	New York, N. Y	Feb. 21, 1871	111, 945
Plating, Compound for silver	E. Hunter	Cleveland, Ohio	May 28, 1867	65, 084
Plating iron and steel	E. Savage	West Meriden, Conn	Dec. 26, 1865	51, 754
Plating metals	R. G. Pine	Newark, N. J	July 11, 1854	11, 256
Plating metals	H. Tucker	Newton, Mass	Apr. 27, 1869	89, 522
Plating, Nickel	I. Adams, jr	Boston, Mass	July 4, 1871	116, 658
Plating, Nickel	I. Adams, jr	Boston, Mass	Mar. 11, 1873	136, 634
Plating, Nickel	M. G. Farmer	Salem, Mass	July 4, 1871	116, 579
Plating spoons, &c	M. L. Forbes	West Meriden, Conn	Mar. 10, 1868	75, 258
Platform: *See* Boat-builder's platform. Car-platform. Dumping-platform. Elevating-platform. Extension-platform. Ferry-track platform. Harvester-platform. Harvester-binders' platform. Harvester dropping-platform. Harvester dumping-platform. Harvester grain-platform. Ladder-platform. Moving platform. Oyster-platform. Railway-platform. Railway-ferry-track platform. Railway-scale platform. Revolving platform. Scale-platform. Stove-platform. Strainer-platform. Turning platform. Wagon-platform. Window-platform.				
Platform and windlass, Combined	T. I. Burbyte	Fond du Lac, Wis	Nov. 7, 1865	50, 787
Platform-elevator, Hydraulic telescopic	J. Parker and I. Cook	San Francisco, Cal	July 16, 1872	129, 421
Platform-elevator, Safety-catch for	B. Tatham and J. W. Brittin	New York and Brooklyn, N. Y.	Mar. 19, 1872	124, 864
Platform for steamboats and other vessels, Landing	N. W. Wheeler	Brooklyn, N. Y	Apr. 25, 1865	47, 482
Platform spring-coupling	B. T. Parsels and J. L. Hedges	Hanover, N. J	Apr. 4, 1871	113, 335
Platform-supporter	C. E. Flagg	Sherburne, Mass	June 10, 1856	15, 099
Platform-switch	G. M. Guerrero	New Orleans, La	Nov. 5, 1872	132, 721
Platoon-battery	W. Billinghurst and J. Regna	Rochester, N. Y	Sept. 16, 1862	36, 448
Playing-table	H. Seher	New York, N. Y	Jan. 5, 1869	85, 539
Pleating-machine	S. Sterns	New York, N. Y	Dec. 7, 1869	97, 721
Plettro-lira	P. Trajetta	Philadelphia, Pa	July 22, 1833	....
Plier	J. Bounds	Bridgeport, Conn	Jan. 14, 1868	73, 289
Plier	G. W. Moore	Newark, N. J	June 27, 1871	116, 473
Plier	A. M. Olds	New York, N. Y	June 11, 1867	65, 589
Plier	N. Thompson	Brooklyn, N. Y	Oct. 10, 1871	119, 726
Plier	S. Walker	New York, N. Y	Jan. 8, 1867	61, 032
Plier-joint die	G. R. Andrus	East Berlin, Conn	Jan. 5, 1869	85, 503
Plier-making die	S. H. Wood	East Berlin, Conn	Apr. 7, 1868	76, 574
Pliers and scissors, Combined	J. H. Price	Boston, Mass	Oct. 12, 1869	95, 834
Pliers, Cutting	R. Curtis	Springfield, Mass	Oct. 25, 1832	....
Pliers, Cutting	W. L. Truland	Waterford, N. Y	May 18, 1869	90, 136
Pliers, Making	C. W. Sykes	New York, N. Y	Sept. 14, 1858	21, 525
Pliers, Manufacture of	H. Wilkinson	Collinsville, Conn	June 1, 1858	20, 460
Plotting-instrument	S. Barnett	Washington, Ga	July 17, 1860	29, 133
Plotting instrument	J. E. Crupper	Berlin, Ky	June 20, 1871	116, 163
Plotting-instrument	H. S. Herney	Windsor, Conn	Dec. 22, 1863	41, 005
Plotting-instrument	T. Hinkley	Hallowell, Me	Oct. 18, 1853	10, 133
Plotting-instrument	C. R. Iliff	Falmouth, Ky	June 24, 1856	15, 183
Plotting-instrument	C. R. Iliff	Falmouth, Ky	Jan. 12, 1858	19, 091
Plotting-instrument, Surveyor's	W. J. Card	Lancaster, Ohio	Apr. 16, 1842	2, 561
Plow	J. Adams	Eatonton, Ga	Dec. 23, 1856	16, 260
Plow	S. Aland	Rome, N. Y	Apr. 12, 1864	42, 264
Plow	E. Albert	East Germantown, Ind	Mar. 6, 1847	4, 995
Plow	D. M. Allen	Jeffersonville, Ind	Nov. 5, 1872	132, 707
Plow	R. S. Allen	New York, N. Y	Mar. 6, 1866	52, 048
Plow	D. Almy	Tiverton Four Corners, R. I	May 14, 1867	64, 619
Plow	T. G. Andrews and A. Riviere	Barnesville, Ga	Aug. 5, 1873	141, 533
Plow	E. Andrus	Geneva, N. Y	Mar. 31, 1857	16, 901
Plow	A. Anschutz, A. Seidel, and M. Weber.	Livingston County, Mo	Nov. 4, 1873	144, 308
Plow	C. C. Ansley	Americus, Ga	Nov. 30, 1869	97, 337
Plow	D. Anthony, sr	Union Spring, N. Y	May 30, 1846	4, 549
Plow	J. Anthony	Zanesville, Ohio	Feb. 1, 1831	....
Plow	J. Archer	Springfield, Wis	Dec. 27, 1870	110, 417
Plow	W. J. Arrington	Jefferson County, Ga	July 6, 1869	92, 143
Plow	G. M. Atherton	Friendsville, Ill	Apr. 20, 1869	89, 115
Plow	J. B. Atwater	Chicago, Ill	Mar. 14, 1865	46, 768
Plow	S. Austin	Sempronius, N. Y	Feb. 27, 1824	....
Plow	B. F. Avery	Louisville, Ky	Jan. 8, 1856	14, 044
Plow	B. F. Avery	Louisville, Ky	July 16, 1867	66, 774
Plow	G. D. Avery	Wood County, Va	Dec. 28, 1818	....
Plow	G. D. Avery	Georgetown, D. C	Feb. 15, 1825	....
Plow	G. D. Avery	Georgetown, D. C	Mar. 25, 1825	....
Plow	J. Bader, sr	Perrysburgh, Ohio	Feb. 16, 1869	86, 896
Plow	B. F. Baker	Ballston Spa, N. Y	May 21, 1872	126, 865
Plow	H. F. Baker	Centreville, Ind	Oct. 24, 1854	11, 821
Plow	H. H. Baker	New Market, N. J	Dec. 11, 1860	30, 863
Plow	N. Baker	Penn Township, Mont	Mar. 24, 1835	....
Plow	E. Ball	Greentown, Ohio	Feb. 20, 1845	3, 918
Plow	E. Ball	Greentown, Ohio	Mar. 23, 1852	8, 819
Plow	E. Ball	North Manchester, Ind	Feb. 14, 1865	46, 321
Plow	E. Ball, jr	Canton, Ohio	June 11, 1867	65, 529
Plow	E. Ball, jr	Canton, Ohio	Aug. 17, 1869	93, 853
Plow	J. Ball	Greentown, Ohio	Nov. 8, 1845	4, 263
Plow	J. Ball	Canton, Ohio	Sept. 1, 1868	81, 730
Plow	J. Ball	Canton, Ohio	July 27, 1869	93, 036
Plow	J. Banks	Dadeville, Ala	Dec. 1, 1857	18, 726
Plow	J. Banks	Dadeville, Ala	Jan. 26, 1858	19, 179
Plow	A. Barnaby	Ithaca, N. Y	Sept. 11, 1839	1, 320
Plow	A. P. Barry	Ashland, Miss	July 16, 1872	129, 265
Plow	G. Bartlett	Smithfield, R. I	Feb. 20, 1847	4, 976

Index of patents issued from the United States Patent Office from 1790 *to* 1873, *inclusive*—Continued.

Invention.	Inventor.	Residence.	Date.	No.
Plow	I. W. Bartlett	Otter Creek, Ill	Dec. 18, 1866	60, 461
Plow	A. Barton	Syracuse, N. Y	July 15, 1856	15, 321
Plow	E. Bass	Pachitta, Ga	Dec. 4, 1860	30, 793
Plow	W. J. M. Batchelder and C. Leiber.	Dayton and Harrisburgh, Ohio.	Mar. 13, 1866	53, 100
Plow	N. Batchly	Windsor, N. Y	Dec. 19, 1848	5, 973
Plow	C. Bates	Warsaw, Ill	Aug. 6, 1872	130, 180
Plow	L. M. Bates	Newark, Ohio	Apr. 24, 1866	54, 095
Plow	S. C. Baughn	Calhoun, Ky	Nov. 19, 1872	133, 187
Plow	W. Beach	Philadelphia	June 27, 1827	....
Plow	L. W. Beal	Dixon, Ill	June 16, 1868	78, 856
Plow	Z. M. Beall	Russellville, Ky	June 4, 1861	32, 459
Plow	J. S. Beals	Alabama Centre, N. Y	Apr. 10, 1866	53, 773
Plow	J. S. Beals	Alabama Centre, N. Y	Aug. 27, 1867	68, 152
Plow	H. T. Beam	Palestine, Ill	July 21, 1868	80, 118
Plow	C. A. Beard and E. E. Evans	Zanesville, Ohio	Apr. 18, 1871	113, 960
Plow	S. Beckett	Olive Branch, Ohio	July 3, 1866	55, 984
Plow	J. R. Beggs	New Albany, N. Y	June 9, 1863	38, 803
Plow	C. Beidler	Allentown, Pa	Jan. 21, 1862	34, 191
Plow	C. Beidler	Allentown, Pa	July 17, 1866	56, 350
Plow	J. C. Bell	Lebanon, Ind	Aug. 31, 1869	94, 383
Plow	A. C. Belt	Goresville, Va	Aug. 27, 1867	68, 032
Plow	H. F. Bemendefer and G. Smith	Attica, Ohio	July 2, 1867	66, 284
Plow	E. D. Benjamin	Old Town, Ill	Nov. 24, 1868	84, 252
Plow	A. Benkelmann	Langford, N. Y	Nov. 27, 1860	30, 712
Plow	J. F. Benton	Penn Yan, N. Y	Mar. 9, 1869	87, 533
Plow	J. F. Benton	Penn Yan, N. Y	Aug. 26, 1873	142, 074
Plow	M. Berdan	Maumee City, Ohio	Oct. 13, 1868	83, 030
Plow	C. Bergen	Brooklyn, N. Y	Nov. 11, 1819	....
Plow	C. Bergen	Brooklyn, N. Y	May 16, 1842	2, 626
Plow	E. L. Bergstresser	Hublersburgh, Pa	Sept. 3, 1867	68, 550
Plow	T. F. Bertrand and P. Sanes	Rockford, Ill	Jan. 29, 1867	61, 508
Plow	R. L. and A. C. Betts	Brunswick, N. Y	May 17, 1864	42, 813
Plow	R. W. Biggs	Jacksonville, Fla	May 28, 1867	65, 050
Plow	J. O. Billing	Halcyon Dale, Ga	Dec. 20, 1870	110, 336
Plow	C. Billups	Norfolk, Va	Feb. 20, 1872	123, 858
Plow	W. Black	Anne Arundel County, Md	Oct. 4, 1831	....
Plow	W. Black	Manchester, Pa	Apr. 17, 1858	21, 182
Plow	W. Black	Scott County, Ill	Sept. 14, 1843	3, 259
Plow	W. Blackstone	Shelby County, Ill	Oct. 14, 1873	143, 660
Plow	J. Blanchard	Iowa Falls, Iowa	Apr. 4, 1871	113, 484
Plow	J. Blanchard	East Saginaw, Mich	Nov. 11, 1873	144, 433
Plow	N. Blatchly	Windsor, N. Y	July 20, 1852	9, 129
Plow	S. R. Bliven	McDonough, N. Y	July 27, 1858	20, 984
Plow	C. Blodgett	Morrison, Ill	July 9, 1861	32, 746
Plow	C. A. Blodgett	Columbus, Nebr	June 4, 1872	127, 549
Plow	B. C. Blomsten	Waupaca, Wis	Oct. 11, 1870	108, 095
Plow	E. H. Bloodworth	Thomaston, Ga	Feb. 16, 1858	19, 401
Plow	J. Boatwright	Columbia, S. C	June 11, 1829	....
Plow	A. Boles	Kinder, Ind	Sept. 7, 1869	94, 554
Plow	A. J. Bonander	Rockford, Ill	Sept. 10, 1872	131, 243
Plow	J. C. Bond	Greenwood, S. C	Aug. 14, 1860	29, 564
Plow	A. T. Boon	Galesburgh, Ill	Apr. 3, 1866	53, 559
Plow	T. Borden	Newport, R. I	Jan. 13, 1830	....
Plow	E. Bourne	New Iberia, La	Nov. 18, 1873	144, 653
Plow	H. R. Bowen and L. D. Robnett	New Washington, Ind	Jan. 17, 1871	111, 033
Plow	J. D. Bowen	Roseburgh, Oreg	July 23, 1867	66, 939
Plow	W. B. Bradford	Charlotte, N. C	Apr. 23, 1872	125, 930
Plow	J. Bradley	Owatonna, Minn	Feb. 25, 1868	74, 885
Plow	F. P. Brannan	Richmond, Va	Sept. 24, 1872	131, 596
Plow	H. Briggs	Smithland, Ky	Dec. 16, 1867	71, 966
Plow	C. R. Brinkerhoff	Batavia, N. Y	Oct. 11, 1853	10, 101
Plow	T. E. C. Brinly	Simpsonville, Ky	Apr. 13, 1858	19, 909
Plow	T. E. C. Brinly	Louisville, Ky	Sept. 4, 1860	29, 858
Plow	T. E. C. Brinly	Louisville, Ky	July 3, 1866	55, 999
Plow	T. E. C. Brinly	Louisville, Ky	July 16, 1867	66, 787
Plow	T. E. C. Brinly	Louisville, Ky	Dec. 10, 1867	71, 968
Plow	T. E. C. Brinly	Louisville, Ky	May 18, 1869	90, 232
Plow	T. E. C. Brinly	Louisville, Ky	July 20, 1869	92, 698
Plow	T. E. C. Brinly	Louisville, Ky	Feb. 15, 1870	99, 830
Plow	T. E. C. Brinly	Louisville, Ky	Nov. 15, 1870	109, 290
Plow	T. E. C. Brinly	Louisville, Ky	Nov. 15, 1870	109, 291
Plow	T. E. C. Brinly	Louisville, Ky	June 17, 1873	140, 004
Plow	T. E. C. Brinly and J. G. Dodge	Louisville, Ky	Aug. 12, 1862	36, 136
Plow	W. Britton	Truro, Ill	Apr. 18, 1871	113, 733
Plow	R. H. Brooks	Greenville, Ga	Feb. 14, 1860	27, 099
Plow	R. M. Brooks	Greenville, Ga	Dec. 27, 1859	26, 563
Plow	C. Brown and L. H. Gerth	Peoria, Ill	June 9, 1868	78, 785
Plow	R. A. Brown	Oakland, Miss	Dec. 2, 1873	145, 088
Plow	S. R. Brown and W. McClean	Norfolk, Va	Jan. 19, 1858	19, 125
Plow	T. Brown	New York	Sept. 11, 1829	....
Plow	C. M. Bryan	Wright City, Mo	May 10, 1859	23, 898
Plow	W. Bryant	Davidson County, Tenn	Mar. 31, 1840	1, 527
Plow	J. Buch	East Rushville, Ohio	Dec. 28, 1832	....
Plow	W. Bullock	Jersey City, N. J	July 30, 1845	4, 127
Plow	W. T. Bunn	Humboldt, Tenn	Oct. 18, 1870	108, 444
Plow	L. D. Burch	Sherburne, N. Y	Aug. 14, 1860	29, 567
Plow	L. D. Burch	Sherburne, N. Y	Mar. 12, 1861	31, 654
Plow	W. Burch	North Fairfield, Ohio	Mar. 4, 1873	136, 361
Plow	L. E. Burdin	Paris, Ky	June 28, 1859	24, 536
Plow	L. E. Burdin	Lexington, Ky	Nov. 21, 1871	121, 153
Plow	T. J. Burgess	Kingston, N. Y	Dec. 3, 1872	133, 517
Plow	W. D. Burgess and G. W. Zeigler.	Maumee, Ohio	July 7, 1868	79, 547
Plow	J. M. Burke	Dansville, N Y	Nov. 9, 1858	22, 013
Plow	P. Burns	Indiana, Pa	Nov. 30, 1869	97, 352
Plow	N. F. Burton	Plymouth, Ill	Oct. 29, 1861	33, 568

Index of patents issued from the United States Patent Office from 1790 *to* 1873, *inclusive*—Continued.

Invention.	Inventor.	Residence.	Date.	No.
Plow	O. F. Burton and L. B. Hoit	New York, N. Y., and Cedar Falls, Iowa	Jan. 9, 1866	51, 917
Plow	W. V. Burton	Orange, Ohio	July 26, 1853	9, 875
Plow	E. T. Bussell	Indianapolis, Ind	July 2, 1872	128, 588
Plow	J. Butler	Huff Township, Ind	Dec. 5, 1871	121, 582
Plow	M. Butler	Vernon, Ind	Oct. 4, 1870	108, 004
Plow	M. Butler	Vernon, Ind	Aug. 20, 1872	130, 697
Plow	J. J. Cadenhead	Macon County, Ala	Mar. 4, 1856	14, 346
Plow	F. M. Caldwell	New York, N. Y	Nov. 23, 1869	97, 162
Plow	J. Campbell	Newtown, Ill	Oct. 13, 1868	83, 036
Plow	W. P. Cannon	Monroe County, Tenn	Mar. 4, 1836	....
Plow	R. Cantleton	Montgomery, Ala	Apr. 9, 1867	63, 610
Plow	S. Canterberry	Holmes County, Miss	Aug. 14, 1860	29, 569
Plow	J. Carothers	Shirleysburgh, Pa	Dec. 10, 1831	....
Plow	A. Carson	Memphis, Tenn	May 14, 1867	64, 747
Plow	T. Carter	Laurens District, S. C	Dec. 16, 1833	....
Plow	E. Cartwright	De Witt, Nebr	Nov. 4, 1873	144, 255
Plow	F. F. Cary	New York, N. Y	May 3, 1864	42, 631
Plow	J. Case	Springfield, Ill	Dec. 8, 1857	18, 803
Plow	V. M. Chafee	Xenia, Ill	Jan. 17, 1860	26, 833
Plow	C. F. Chambers	Huttonsville, Ill	June 6, 1871	115, 701
Plow	J. J. Chandler and P. Ranger	Wilton, Me	Jan. 20, 1836	....
Plow	T. F. Chapin	Walpole, N. H	July 25, 1854	11, 356
Plow	A. B. Chapman	Pittsfield, Mass	Dec. 1, 1863	40, 733
Plow	L. Chapman	Collinsville, Conn	Feb. 6, 1862	123, 330
Plow	R. B. Chenoweth	Baltimore, Md	Nov. 25, 1808	....
Plow	H. S. Chichester	Brunswick, N. Y	Aug. 6, 1861	32, 976
Plow	C. M. Clark	Seward, Nebr	Dec. 16, 1873	145, 627
Plow	E. B. Clark	Tallahassee, Fla	Feb. 14, 1860	27, 109
Plow	E. B. Clark	Tallahassee, Fla	July 17, 1860	29, 139
Plow	G. B. Clarke	Leonardsville, N. Y	May 24, 1864	42, 838
Plow	J. Clark and G. W. N. Yost	Washington, D. C., and Pittsburgh, Pa	Feb. 12, 1856	14, 224
Plow	T. J. and G. M. Clark	Higganum, Conn	Mar. 25, 1873	137, 060
Plow	G. Clezy	Baltimore, Md	Sept. 14, 1843	3, 266
Plow	J. Clifton	Georgetown, Tex	Dec. 27, 1870	110, 550
Plow	S. Cline	Plumstead, Pa	July 17, 1835	....
Plow	J. C. Cloud	May's Landing, N. J	Feb. 6, 1849	6, 100
Plow	H. C. Cloyd	West Alexandria, Ohio	June 7, 1870	103, 982
Plow	P. J. Clute	Schenectady, N. Y	Nov. 16, 1822	....
Plow	J. M. Cobb	Jackson, Tenn	June 26, 1860	28, 836
Plow	J. M. Cobb	Jackson, Tenn	Apr. 22, 1873	138, 131
Plow	W. Cock	Luzerne Township, Pa	Apr. 28, 1826	....
Plow	D. Cockley	Lancaster, Pa	Sept. 28, 1858	21, 598
Plow	R. Coddington and D. McCall	Leonidas and Kalamazoo, Mich.	Oct. 29, 1861	33, 573
Plow	O. Coe	Port Washington, Wis	Feb. 16, 1864	41, 603
Plow	W. Coggeshall	Finley, Ohio	May 27, 1862	35, 415
Plow	G. W. Coles	Canton, Ill	Apr. 14, 1868	76, 714
Plow	G. D. Colton	Galesburgh, Ill	Jan. 18, 1859	22, 629
Plow	W. S. Colwell	Pittsburgh, Pa	Jan. 21, 1868	73, 504
Plow	W. S. Colwell	Allegheny, Pa	Apr. 13, 1869	88, 851
Plow	W. H. Conaway	Dillsborough, Ind	Dec. 3, 1872	133, 631
Plow	J. H. Conklin	Peekskill, N. Y	Feb. 27, 1847	4, 980
Plow	E. S. Cook	Laurel Grove, Va	Aug. 6, 1872	130, 196
Plow	I. Cook and J. T. Bever	Hainesville, Mo	July 5, 1859	24, 617
Plow	S. Cooley	Clarkston, Mich	Nov. 7, 1871	120, 718
Plow	W. Cooley	Bunker Hill, Wis	Feb. 5, 1867	61, 809
Plow	W. G. Coombs	New Gloucester, Me	Aug. 23, 1870	106, 556
Plow	G. W. Cooper	Ogeechee, Ga	Jan. 1, 1856	14, 013
Plow	G. W. Cooper	Palmyra, N. Y	Apr. 16, 1861	32, 052
Plow	J. B. Cooper	Brooklyn, N. Y	Apr. 9, 1861	31, 955
Plow	T. J. Cornell	Decatur, Ill	Feb. 20, 1866	52, 807
Plow	T. R. Cornick	Cap Au Gris, Mo	Nov. 27, 1860	30, 723
Plow	J. Coston	Bowden, Ga	June 6, 1871	115, 710
Plow	T. L. Cotten	Water Valley, Miss	Apr. 15, 1873	137, 894
Plow	T. Cottman	Cincinnati, Ohio	Feb. 13, 1866	52, 539
Plow	G. D. Cotten	Galesburgh, Ill	July 20, 1858	20, 935
Plow	M. C. Cox	Bennettsville, S. C	Nov. 2, 1869	96, 399
Plow	F. Cremer	Elmwood, Ill	Feb. 21, 1871	111, 911
Plow	R. S. Crockett	Rossville, S. C	Nov. 8, 1870	108, 975
Plow	H. F. Cromwell	Cynthiana, Ky	Jan. 3, 1860	27, 655
Plow	J. Cromwell	Cynthiana, Ky	Oct. 9, 1834	....
Plow	C. Crow	Covington, Ind	Dec. 7, 1869	97, 609
Plow	G. Crowl	Sleepy Creek, Va	June 27, 1832	....
Plow	J. M. Cullen	Benton, Miss	May 30, 1846	4, 553
Plow	H. Culver	Dansville, N. Y	Mar. 9, 1869	87, 641
Plow	T. Cuming, jr	Brookhaven, Miss	Oct. 24, 1871	120, 248
Plow	J. G. Cummings	Columbus, Miss	Nov. 24, 1857	18, 682
Plow	W. H. Cummings and H. L. Childs.	Boonesborough, Iowa	June 22, 1869	91, 721
Plow	A. G. and J. R. Cummins	McKinney, Tex	July 6, 1869	92, 277
Plow	S. A. Cummins	Vienna, N. J	Aug. 1, 1871	117, 608
Plow	G. W. Cunningham	Paris, Mo	Sept. 25, 1860	30, 125
Plow	S. Curtis	Fitchburgh, Wis	Feb. 23, 1869	87, 149
Plow	A. K. Dahl	Fox Lake, Wis	Nov. 25, 1873	144, 960
Plow	M. K. Dahl	Waupun, Wis	Mar. 15, 1870	100, 869
Plow	A. A. Dailey	Wilson, N. Y	Sept. 13, 1870	107, 228
Plow	J. Dalkener	Canton, Ohio	Jan. 15, 1836	....
Plow	E. Davidson	Batesville, Ark	May 10, 1859	23, 904
Plow	A. B. Davis	Catahoula Parish, La	Aug. 3, 1869	93, 281
Plow	B. Davis and J. M. Scroggin	La Grange, Ga	Mar. 13, 1860	27, 619
Plow	G. Davis	Sandy Springs, Md	May 26, 1818	....
Plow	G. Davis	Georgetown, D. C	Oct. 1, 1825	....
Plow	J. R. Davis	McKay, Ohio	Nov. 19, 1867	70, 972
Plow	A. Day	Crystal Springs, Miss	Nov. 18, 1870	108, 979
Plow	J. Deats	Middletown, Pa	Apr. 26, 1828	....
Plow	J. Deats	Roxburgh, N. J	Dec. 28, 1831	....

Index of patents issued from the United States Patent Office from 1790 *to* 1873, *inclusive*—Continued.

Invention.	Inventor.	Residence.	Date.	No.
Plow	J. Deats	Roxburgh, N. J	July 31, 1837	327
Plow	J. Deats	Roxburgh, N. J	Nov. 25, 1838	1, 019
Plow	J. Deaver	....	July 12, 1804	....
Plow	J. Deere	Moline, Ill	Feb. 21, 1865	46, 454
Plow	R. Deighton, jr	Fairweather, Ill	Aug. 15, 1865	49, 389
Plow	N. Demary, jr	Attica, N. Y	Dec. 15, 1857	18, 882
Plow	J. Dement	Dixon, Ill	July 30, 1861	32, 966
Plow	J. Dement	Dixon, Ill	Aug. 30, 1864	43, 978
Plow	S. T. Demise	Red Bank, N. J	Dec. 3, 1867	71, 715
Plow	S. T. Demise	Red Bank, N. J	Oct. 6, 1868	82, 809
Plow	L. B. Dennett	Portland, Me	May 29, 1866	55, 069
Plow	G. W. Depew	Peekskill, N. Y	Mar. 19, 1861	31, 761
Plow	T. J. De Yampert	Shohola, Pa	May 10, 1859	23, 974
Plow	J. M. Dick	Buffalo, N. Y	Feb. 24, 1863	37, 740
Plow	R. Dickie and H. K. Johnston	Bunker Hill, Ill	Aug. 23, 1870	106, 559
Plow	A. Dickson	Hillsborough, N. C	June 22, 1858	20, 633
Plow	A. A. Dickson	Anderson, S. C	Nov. 29, 1859	26, 259
Plow	J. Dickson	Newcastle, Pa	Oct. 19, 1858	21, 824
Plow	J. H. Dickson	Alford, Ind	Aug. 18, 1868	81, 148
Plow	E. Dietsch	Findley, Ohio	Nov. 1, 1870	108, 892
Plow	O. P. Dills	Falmouth, Ky	Sept. 5, 1865	49, 733
Plow	J. Dodge	Adrian, Mich	Jan. 9, 1872	122, 582
Plow	J. G. Dodge	Louisville, Ky	Oct. 8, 1867	69, 643
Plow	A. Doe	Concord, N. H	Jan. 30, 1855	12, 310
Plow	L. Donnell	Jefferson Township, Ohio	July 23, 1833	....
Plow	W. Donnelly	Calverton, N. Y	Aug. 12, 1873	141, 705
Plow	J. M. Dormon	Claiborne Parish, La	May 4, 1869	89, 565
Plow	W. H. H. Doty	Sonora, Ohio	Dec. 26, 1871	122, 162
Plow	J. W. Downs	Bowdon, Ga	July 16, 1872	129, 112
Plow	V. C. Duclos	New Harmony, Ind	Nov. 17, 1868	84, 097
Plow	J. Dudley	Fleming County, Ky	May 8, 1832	....
Plow	J. C. Duncan	Olney, Ill	May 14, 1867	64, 647
Plow	H. B. Durfee	Decatur, Ill	May 18, 1869	90, 246
Plow	H. B. Durfee	Dacatur, Ill	Nov. 9, 1869	96, 680
Plow	J. Dutcher	New York, N. Y	Oct. 9, 1839	1, 360
Plow	J. L. Dutton, jr	Cherry Lake, Fla	Jan. 31, 1860	26, 976
Plow	I. Eastwood	Lanark, Ill	Aug. 16, 1870	106, 344
Plow	I. Eastwood	Lanark, Ill	Oct. 11, 1870	108, 243
Plow	H. H. Ebaugh	Hereford, Md	July 9, 1867	66, 477
Plow	J. B. Edgell and J. W. and E. A. Alexander.	Independence, Iowa	Oct. 2, 1866	58, 395
Plow	T. Edmunds	Talcott, Va	Apr. 15, 1871	118, 119
Plow	A. N. Edwards	Greenville, Ala	July 20, 1869	92, 803
Plow	D. Eldred	Monmouth, Ill	Oct. 18, 1859	25, 816
Plow	G. Ellington	Madison County, Ala	Dec. 25, 1824	....
Plow	J. W. Elliott	Locust Mount, Va	Aug. 6, 1872	130, 115
Plow	C. A. Elton	Hillsborough, Ohio	July 16, 1867	66, 691
Plow	J. H. Elward	Polo, Ill	Nov. 3, 1868	83, 768
Plow	G. Emery and A. C. Wilson	Newfield, Me	Oct. 18, 1859	25, 817
Plow	A. F. Eppes	Stony Creek, Va	May 10, 1870	102, 789
Plow	G. Esterly	Heart Prairie, Wis	Feb. 13, 1855	12, 381
Plow	W. A. Estes	China, Me	Sept. 30, 1873	143, 279
Plow	G. Evans	Pittsburgh, Pa	Sept. 18, 1833	....
Plow	J. D. Evans	Pleasant Hill, Ga	Sept. 24, 1867	69, 083
Plow	L. G. Evans	Spring Hill, Ala	June 19, 1855	13, 082
Plow	W. C. Evarts	Danby, N. Y	July 23, 1867	66, 958
Plow	W. P. Everdon	Leavenworth, Ind	Feb. 11, 1868	74, 327
Plow	O. G. Ewings	Heart Prairie, Wis	Aug. 29, 1854	11, 601
Plow	P. Falker	Lanesville, Ill	July 20, 1869	92, 806
Plow	S. A. Fanning	Jacksonville, Ill	Apr. 2, 1872	125, 185
Plow	J. Farlee	Mercer County, Ky	Apr. 21, 1836	....
Plow	A. B. Farquhar	York, Pa	June 18, 1872	127, 970
Plow	N. Faught	Pittsborough, Ind	Jan. 3, 1871	110, 758
Plow	V. Felker	Carmel, Me	Apr. 30, 1861	39, 231
Plow	V. Felker	Carmel, Me	June 27, 1865	48, 387
Plow	R. R. Fenner	Urbana, Ill	Aug. 27, 1872	130, 797
Plow	J. B. Ferguson and S. M. White	Big Lick, Va	Feb. 28, 1871	112, 234
Plow	J. G. Fetzer	Brunswick, Mo	Aug. 11, 1868	80, 859
Plow	J. Fish	Smelser, Wis	Apr. 2, 1867	63, 377
Plow	D. S. Fisher	Cedar Spring, Ind	May 28, 1867	65, 198
Plow	J. Fisher	Middletown, Pa	Aug. 4, 1868	80, 715
Plow	S. Fisher	West Windsor, N. J	Nov. 27, 1860	30, 727
Plow	C. L. Fleischmann	New York, N. Y	Feb. 26, 1867	62, 325
Plow	F. Fogelgesang	Canton, Ohio	Nov. 13, 1866	59, 577
Plow	S. Folton	Huntingdon, Pa	Nov. 1, 1830	....
Plow	G. G. Foreman	Stockton, Ga	June 18, 1872	128, 032
Plow	C. Forster	Lebanon, Pa	Nov. 26, 1867	71, 376
Plow	M. Fort	Talbotton, Ga	Mar. 15, 1870	100, 743
Plow	L. Fosdick	Tiskilwa, Ill	Nov. 3, 1868	93, 703
Plow	J. Fowler	Rahway, N. J	Feb. 6, 1866	52, 407
Plow	J. Fowler	Hartland, Wis	Mar. 26, 1867	63, 151
Plow	J. Fox	Oskaloosa, Iowa	June 25, 1872	128, 295
Plow	W. Frank	Saint Louis, Mo	Aug. 25, 1863	39, 639
Plow	A. Franklin	Springfield, Ohio	Nov. 16, 1869	96, 907
Plow	A. and F. M. Franklin	Springfield, Ohio	Oct. 11, 1870	108, 247
Plow	F. M. Franklin	Springfield, Ohio	June 8, 1869	91, 109
Plow	F. M. Franklin	Springfield, Ohio	Sept. 30, 1870	107, 606
Plow	E. J. Faser	Kansas, Mo	Apr. 23, 1861	32, 129
Plow	L. F. Frazee	Jersey City, N. J	Mar. 21, 1871	112, 913
Plow	E. J. Frazer	Kansas City, Mo	Sept. 25, 1860	30, 136
Plow	O. M. French	Rochester, Pa	June 16, 1868	78, 869
Plow	A. Friberg	Moline, Ill	Oct. 13, 1868	82, 934
Plow	L. C. Frost	Fredericksburgh, Va	Oct. 14, 1873	143, 620
Plow	D. Fulton	Saint Helena, Cal	Sept. 27, 1870	107, 676
Plow	W. Furnas	Ononwa, Iowa	Dec. 3, 1861	33, 834
Plow	H. Gale	Albion, Mich	May 17, 1870	103, 038
Plow	H. Gale	Albion, Mich	Sept. 6, 1870	117, 033
Plow	W. Gallagher	Shullsbury, Wis	Apr. 28, 1868	77, 272

Index of patents issued from the United States Patent Office from 1790 *to* 1873, *inclusive*—Continued.

Invention.	Inventor.	Residence.	Date.	No.
Plow	J. Gallatin	Manchester, Ohio	Apr. 15, 1843	3, 052
Plow	A. Gardner	Cincinnati, Ohio	Oct. 26, 1852	9, 362
Plow	T. C. Garlington	La Fayette, Ala	May 5, 1857	17, 211
Plow	W. Garrell	Trenton, Tenn	Jan. 22, 1861	31, 172
Plow	W. A. Gates	Mount Comfort, Tenn	Dec. 21, 1852	9, 443
Plow	J. Gehr	College of Saint James, Md	Nov. 2, 1858	21, 953
Plow	J. George	Greene, Mo	Apr. 18, 1865	47, 294
Plow	D. Ghormley	Wayne Township, Ohio	Feb. 13, 1835	....
Plow	G. Gibbs	Canton, Ohio	Apr. 30, 1867	64, 213
Plow	G. Gibbs	Canton, Ohio	Dec. 3, 1867	71, 734
Plow	J. Gibbs	Canton, Ohio	June 16, 1836	....
Plow	J. Gibbs	Canton, Ohio	Aug. 15, 1854	11, 523
Plow	L. Gibbs	Canton, Ohio	Mar. 12, 1867	62, 835
Plow	L. Gibbs	Canton, Ohio	June 25, 1867	66, 016
Plow	L. Gibbs	Canton, Ohio	Aug. 6, 1872	130, 292
Plow	M. L. Gibbs	Canton, Ohio	Oct. 25, 1870	108, 699
Plow	M. L. Gibbs	Canton, Ohio	June 20, 1871	116, 048
Plow	R. Gibbs	Brunswick, Mo	Sept. 6, 1870	107, 037
Plow	W. and G. Gibbs and L. P. Wikidal.	Canton, Ohio	Dec. 3, 1867	71, 735
Plow	J. Gibson	Montgomery County, N. Y	May 2, 1822	....
Plow	B. Giger	Sangamon County, Ill	June 24, 1851	8, 170
Plow	J. R. Gilbert	Wootens, Ga	Nov. 30, 1869	97, 388
Plow	P. M. Gilbert	Kewanee, Ill	Jan. 1, 1867	60, 875
Plow	C. G. and G. Gill	Chester, S. C	Mar. 25, 1822	....
Plow	J. L. Gill	Columbus, Ohio	Aug. 28, 1855	13, 493
Plow	H. Gillette	Millville, N. Y	June 4, 1872	127, 475
Plow	J. W. Gilliam	Elkton, Ky	Aug. 10, 1869	93, 430
Plow	J. Gilmer	Monticello, Fla	Sept. 17, 1872	131, 434
Plow	A. Gilmore	Phœnixville, Pa	Dec. 17, 1867	72, 386
Plow	T. Gilson and N. Martin	Port Washington, Wis	June 5, 1866	55, 279
Plow	J. K. Gingrich	North Annville, Pa	Feb. 19, 1861	31, 452
Plow	N. T. Glenn	Watkinsville, Ga	Jan. 23, 1872	122, 885
Plow	C. Glidden	Milwaukee, Wis	Dec. 4, 1866	60, 172
Plow	J. S. Godfrey	Leslie, Mich	Mar. 29, 1870	101, 256
Plow	J. S. Godfrey	Leslie, Mich	July 19, 1870	105, 446
Plow	S. T. Godfrey	Seaville, N. J	June 15, 1869	91, 328
Plow	H. Goldson	Greensborough, Miss	Nov. 18, 1851	8, 524
Plow	E. Goldthait	Fort Wayne, Ind	Nov. 25, 1851	8, 544
Plow	J. H. Gooch	Oxford, N. C	May 8, 1860	28, 169
Plow	C. M. Gordon	La Porte, Ind	Sept. 20, 1870	107, 481
Plow	J. Gorham	Bairdstown, Ga	May 5, 1857	17, 212
Plow	J. Gorham	Bairdstown, Ga	Dec. 6, 1859	26, 349
Plow	J. L. Graham	Carthage, Ill	Dec. 23, 1873	145, 862
Plow	R. A. Graham	New Paris, Ohio	Oct. 4, 1853	10, 069
Plow	C. W. Grant	Iona Island, N. Y	Aug. 20, 1867	67, 976
Plow	L. Green	Great Bend, Pa	July 17, 1860	29, 162
Plow	L. Green	Great Bend, Pa	Mar. 19, 1861	31, 712
Plow	L. Green	Great Bend, Pa	May 19, 1863	38, 581
Plow	L. Green	Great Bend, Pa	Nov. 15, 1864	45, 036
Plow	W. J. Griffies	Marietta, Ga	Mar. 29, 1859	23, 369
Plow	W. Griffin	Bennettsville, S. C	July 3, 1860	28, 981
Plow	C. T. Grimes	Gerrard County, Ky	Sept. 8, 1868	81, 996
Plow	R. B. Ground	Edwardsville, Ill	Apr. 15, 1873	137, 915
Plow	M. Grover	Clyde, Ohio	Oct. 6, 1857	18, 335
Plow	M. Grover	Clyde Village, Ohio	Jan. 7, 1862	34, 092
Plow	J. Hage	Shiloh, Ill	May 12, 1863	38, 478
Plow	J. Hage	Shiloh, Ill	Apr. 2, 1867	63, 381
Plow	D. B. Haight	Perryville, N. Y	Dec. 26, 1848	5, 981
Plow	E. Haiman	Columbus, Ga	Jan. 17, 1871	111, 055
Plow	H. B. Hakes	Worcester, Mass	Sept. 24, 1872	131, 679
Plow	H. G. and E. L. Hall	Putnam, Ohio	May 21, 1867	64, 974
Plow	J. D. Hall	Canton, Ohio	July 2[illegible], 1870	105, 673
Plow	J. M. Hall	Warrenton, Ga	May 18, 1858	20, 269
Plow	J. M. Hall	Warrenton, Ga	Feb. 22, 1859	23, 023
Plow	J. S. Hall	Manchester, Pa	Feb. 7, 1854	10, 505
Plow	J. S. Hall	Manchester, Pa	Aug. 1, 1854	11, 456
Plow	J. S. Hall	West Manchester, Pa	June 2, 1857	17, 430
Plow	J. S. Hall	West Manchester, Pa	Aug. 14, 1860	29, 590
Plow	J. S. Hall	West Manchester, Pa	Aug. 14, 1860	29, 591
Plow	J. S. Hall	Pittsburgh, Pa	Apr. 28, 1868	77, 188
Plow	J. S. Hall	Pittsburgh, Pa	May 21, 1872	126, 952
Plow	J. S. Hall	Pittsburgh, Pa	Nov. 18, 1873	144, 760
Plow	S. Hall	Pittsburgh, Pa	Apr. 25, 1848	5, 529
Plow	T. J. Hall	Tawakana Hills, Tex	Apr. 3, 1855	12, 627
Plow	J. Hanes	Polkville, Ky	Jan. 24, 1865	45, 995
Plow	J. Hapgood	Shrewsbury, Mass	Sept. 12, 1871	118, 932
Plow	J. Hardey	Moline, Ill	Aug. 28, 1860	29, 843
Plow	W. C. Hardin	Bowling Green, Mo	Mar. 24, 1868	75, 901
Plow	T. Harding	La Fayette, Ind	Jan. 17, 1871	111, 056
Plow	W. E. Harding	Bowling Green, Mo	Aug. 13, 1867	67, 756
Plow	D. Harper	Crystal Lake, Ill	Oct. 21, 1862	36, 711
Plow	W. K. Harrell	Clarinda, Iowa	June 18, 1872	128, 141
Plow	H. Harris	Bullitt, Ky	Feb. 24, 1808	....
Plow	I. P. Harris	Byhalia, Miss	July 6, 1858	20, 790
Plow	I. P. Harris	Byhalia, Miss	Nov. 8, 1859	26, 033
Plow	J. Harris	Kansas, Ill	Oct. 23, 1866	59, 010
Plow	I. P. Harris	Byhalia, Miss	Sept. 2, 1856	15, 649
Plow	J. R. Harris	Hazlehurst, Miss	June 25, 1872	128, 390
Plow	J. R. Harris	Hazlehurst, Miss	Nov. 5, 1872	132, 828
Plow	J. R. Harris	Hazlehurst, Miss	Dec. 24, 1872	134, 275
Plow	R. N. Harrison	New York	Dec. 19, 1818	....
Plow	S. Hartpence and J. D. Bowne	Kingwood, N. J	July 5, 1837	259
Plow	C. Hartzell	Saint Joseph, Mo	June 1, 1869	90, 747
Plow	C. Hartzell	Saint Joseph, Mo	Dec. 31, 1872	134, 377
Plow	W. Haslup	Sidney, Ohio	June 11, 1872	127, 878
Plow	P. Hastings	Dagsborough, Del	Dec. 19, 1833	....
Plow	J. M. Hawley	Holton, Ind	July 2, 1867	66, 335

Index of patents issued from the United States Patent Office from 1790 *to* 1873, *inclusive*—Continued.

Invention.	Inventor.	Residence.	Date.	No.
Plow	J. Heckendorn	Elkton, Md	Dec. 23, 1856	16, 277
Plow	J. Heckendorn	Reading, Pa	June 8, 1869	91, 129
Plow	D. Heiges	Cashtown, Pa	Mar. 8, 1870	100, 624
Plow	J. C. Henry	Point Douglass, Minn	Aug. 6, 1867	67, 542
Plow	W. Henry	Wyoming, Pa	Mar. 29, 1864	42, 086
Plow	W. Hess	Lower Sancon, Pa	Mar. 27, 1835	....
Plow	J. Hibbs	Tullytown, Pa	Nov. 14, 1854	11, 935
Plow	B. S. Higgins	Olney, Ill	Oct. 13, 1868	83, 061
Plow	D. H. Hill	Union Springs, Ala	June 22, 1869	91, 631
Plow	W. Hinds	Little Falls N. Y	Mar. 22, 1864	41, 997
Plow	D. Hitchcock	New York	July 16, 1823	....
Plow	R. G. Hobson	Houlka, Miss	Jan. 7, 1873	134, 670
Plow	D. Hoke	Byhalia, Miss	Mar. 9, 1858	19, 563
Plow	Z. S. Holdridge and H. L. Lawson.	Berne, N. Y	Aug. 9, 1826	....
Plow	W. C. Holmes	Barnesville, Ga	Apr. 12, 1859	23, 580
Plow	B. R. Hood	Clinton, N. C	Oct. 11, 1859	25, 738
Plow	W. U. Hoover	Daysville, Ky	Dec. 15, 1868	84, 946
Plow	S. Horney, jr	Richmond, Ind	Apr. 5, 1853	9, 646
Plow	J. T. Hovis	Clintonville, Pa	Aug. 20, 1872	130, 717
Plow	W. S. Howell	Alfred, N. Y	Jan. 21, 1868	73, 606
Plow	B. C. Hoyt	Port Washington, Wis	Sept. 2, 1856	15, 654
Plow	J. S. Huggins	Timmonsville, S. C	May 22, 1860	28, 372
Plow	D. W. Hughes	Quincy, Ill	Jan. 21, 1868	73, 607
Plow	S. Hulbert	Ogdensburgh, N. Y	Sept. 20, 1853	10, 031
Plow	S. Hulbert	Ogdensburgh, N. Y	Sept. 7, 1858	21, 423
Plow	S. Hulbert	Ogdensburgh, N. Y	June 2, 1868	78, 457
Plow	N. Hull	De Kalb, Miss	May 12, 1842	2, 620
Plow	H. A. Hummer	Franklin Township, N. J	Sept. 4, 1866	57, 724
Plow	D. F. Humphrey	Saline, Mich	July 29, 1862	36, 004
Plow	D. F. Humphrey	Saline, Mich	Feb. 14, 1865	46, 362
Plow	G. W. Hunt	Muscatine, Iowa	Mar. 27, 1860	27, 634
Plow	G. W. Hunt	Muscatine, Iowa	Sept. 15, 1863	39, 929
Plow	J. Y. Hunt	Tunbridge, Vt	Dec. 24, 1824	....
Plow	L. Hunt	Weathersfield, Vt	June 15, 1869	91, 445
Plow	L. Hunt	Weathersfield, Vt	Aug. 24, 1869	93, 994
Plow	J. T. Hunter and D. L. H. Mitchell.	Forest, Miss	Feb. 25, 1873	136, 245
Plow	C. B. Hunting	Clinton, Ill	Sept. 11, 1866	57, 909
Plow	W. S. Huntington	Byron, Mich	Jan. 15, 1867	61, 203
Plow	A. P. Ingalls	Shelbyville, Ill	Feb. 9, 1869	86, 839
Plow	C. B. Ingersoll	Morris, Ill	June 16, 1857	17, 577
Plow	W. D. Ivey	Milford, Ga	Apr. 10, 1860	27, 808
Plow	W. B. Jackson and J. M. and O. J. Childs.	Utica, N. Y	May 13, 1873	138, 895
Plow	J. Jacobs	Maysville, Ky	July 8, 1834	....
Plow	J. Jacobs	Maysville, Ky	July 8, 1834	....
Plow	D. Jacoby	Mendota, Ill	Aug. 14, 1866	57, 147
Plow	A. C. Jaques	Leavenworth, Kans	July 14, 1868	79, 834
Plow	J. W. Jenkins	Hudson, N. Y	July 6, 1820	....
Plow	J. R. P. Jett	Knoxville, Tenn	Oct. 11, 1870	108, 149
Plow	J. E. Jinkins	Milton, Fla	Oct. 22, 1867	69, 999
Plow	J. Johnson	Wooster, Ohio	May 30, 1846	4, 552
Plow	M. Johnson	Three Rivers, Mich	Dec. 29, 1868	85, 451
Plow	R. Johnson	Frederick, Md	Apr. 29, 1862	35, 098
Plow	R. Johnson	Lawrence, Kans	Sept. 12, 1871	118, 945
Plow	W. H. Johnson	Richmond, Ark	Feb. 14, 1860	27, 188
Plow	J. Jones	New Castle, Del	Aug. 10, 1858	21, 167
Plow	J. H. and H. P. Jones	Herndon, Ga	Jan. 19, 1869	86, 079
Plow	R. Jones	Waynesburgh, Ohio	Feb. 10, 1863	37, 626
Plow	W. Jones	Wilson, Minn	Nov. 25, 1862	36, 999
Plow	W. T. Jones	Joliet, Ill	Nov. 15, 1859	26, 111
Plow	A. C. Judson	Grand Rapids, Ohio	Sept. 7, 1869	94, 609
Plow	A. C. Judson	Grand Rapids, Ohio	Feb. 22, 1870	100, 042
Plow	C. T. Kee	Chester Court-House, S. C	July 30, 1872	129, 964
Plow	G. W. Keeler	New Haven, Ohio	Oct. 13, 1868	82, 959
Plow	H. M. Keith	Commerce, Mich	Oct. 18, 1870	108, 488
Plow	H. M. Keith	Commerce, Mich	Nov. 28, 1871	121, 382
Plow	E. Kelley	Bainbridge, Ohio	Aug. 26, 1833	....
Plow	I. Kennedy	Ithaca, N. Y	Oct. 2, 1866	58, 431
Plow	I. Kennedy	Binghamton, N. Y	May 18, 1869	90, 271
Plow	M. Kennedy	Boston, Mass	June 5, 1866	55, 312
Plow	H. Killam	Mendon, Mich	Oct. 5, 1869	95, 487
Plow	J. Kilmer	Barnerville, N. Y	Mar. 7, 1865	46, 755
Plow	J. D. Kincaid	Bowling Green, Mo	Aug. 13, 1867	67, 772
Plow	S. B. King	Starkville, Ga	Feb. 7, 1871	111, 651
Plow	C. Kinkel	New York, N. Y	Mar. 19, 1867	63, 059
Plow	D. J. Kirkman and E. H. Gray	Winchester, Ill	June 25, 1867	66, 031
Plow	J. Klay	Frederick County, Md	Jan. 11, 1812	....
Plow	S. A. Knox	Worcester, Mass	Oct. 14, 1856	15, 887
Plow	J. Koffend	Appleton, Wis	May 26, 1868	78, 293
Plow	M. Laflin and E. Slosson	Chicago, Ill	June 14, 1870	104, 166
Plow	D. M. Lamb	Strathroy, Canada	Dec. 26, 1871	122, 261
Plow	J. Lane, jr	Lockport, Ill	Dec. 1, 1857	18, 750
Plow	J. Lane	Chicago, Ill	Mar. 31, 1868	76, 208
Plow	J. Lane	Chicago, Ill	July 21, 1868	80, 189
Plow	J. Lane	Chicago, Ill	July 5, 1870	104, 964
Plow	J. Lane	Chicago, Ill	Feb. 14, 1871	111, 854
Plow	J. Lane, jr	Chicago, Ill	Apr. 4, 1871	113, 436
Plow	M. W. Lane	Hillsborough, Ohio	June 4, 1872	127, 613
Plow	B. Langdon	Troy, N. Y	June 22, 1842	2, 689
Plow	W. M. Lanham	Noblesville, Ind	May 3, 1870	102, 687
Plow	W. M. Lanham	Noblesville, Ind	July 20, 1869	92, 845
Plow	W. Lape	Troy, N. Y	June 25, 1861	32, 666
Plow	R. K. and J. Laraway	Battle Creek, Mich	Mar. 5, 1867	62, 641
Plow	J. S. Lash	Carlisle, Pa	Oct. 20, 1857	18, 459
Plow	J. B. Latimer	Stewart County, Ga	Feb. 27, 1872	124, 143
Plow	J. L. Laughlin	Peru, Ill	Nov. 11, 1873	144, 462

Index of patents issued from the United States Patent Office from 1790 *to* 1873, *inclusive*—Continued.

Invention.	Inventor.	Residence.	Date.	No.
Plow	J. Layman	Lebanon, Ohio	Jan. 2, 1849	5, 999
Plow	S. J. Leach	Tuscaloosa, Ala	Dec. 17, 1867	72, 305
Plow	S. J. Leach	Tuscaloosa, Ala	Oct. 15, 1872	132, 295
Plow	A. Lee	West Chester, N. Y	July 22, 1822	....
Plow	E. D. and Z. W. Lee	Blakely, Ga	Oct. 4, 1859	25, 654
Plow	H. A. Lee	Parkman, Ohio	Oct. 29, 1872	132, 677
Plow	J. Lee	Galesburgh, Ill	Dec. 8, 1857	18, 820
Plow	Z. W. Lee	Blakely, Ga	Aug. 21, 1866	57, 341
Plow	E. D. and L. W. Legg	Speedsville, N. Y	June 16, 1857	17, 579
Plow	J. M. Leonard	Marshall, Mich	Jan. 10, 1871	110, 924
Plow	W. E. Levoy	Cincinnati, Ohio	Aug. 20, 1867	67, 890
Plow	C. C. Lewis	Gainesville, Ala	Aug. 13, 1872	130, 435
Plow	J. W. Lewis	Fetterman, W. Va	Jan. 8, 1867	61, 075
Plow	T. S. and J. A. Lockhart	Wellington, Mo	Nov. 27, 1860	30, 740
Plow	N. Locklin	Sparta, N. Y	Apr. 28, 1836	....
Plow	N. S. Lockwood and J. D. Winn	Dayton, Ohio	June 17, 1856	15, 137
Plow	P. Loeb	Dayton, Ohio	June 17, 1873	140, 053
Plow	I. Long	Terre Haute, Ind	Jan. 22, 1867	61, 470
Plow	J. Long	Leavenworth, Ind	June 15, 1869	91, 354
Plow	J. Love	Cusseta, Tex	Feb. 11, 1873	135, 829
Plow	R. Loveridge	Knox County, Ohio	Mar. 8, 1828	....
Plow	C. M. Lufkin	Acworth, N. H	Aug. 4, 1863	39, 411
Plow	C. M. Lufkin	Claremont, N. H	June 19, 1866	55, 684
Plow	C. B. Magruder	Thomasville, Ga	Oct. 20, 1857	18, 463
Plow	D. H. Maloy	Temperance, Ga	Feb. 26, 1861	31, 549
Plow	A. S. Mann	Saint Louis, Mo	July 8, 1873	140, 716
Plow	H. F. Mann	La Porte, Ind	Apr. 16, 1861	32, 073
Plow	D. A. Manuel	Napa City, Cal	Jan. 12, 1869	85, 746
Plow	S. March	Norfolk, Va	Mar. 26, 1867	63, 276
Plow	T. March	Dallas, Mich	Sept. 10, 1867	68, 635
Plow	T. R. Markillie	Winchester, Ill	June 19, 1860	28, 759
Plow	J. Marr	Simcoe, Canada	Aug. 19, 1873	141, 881
Plow	C. Marsh, 2d	Natchez, Miss	Sept. 16, 1873	142, 800
Plow	G. W. Marsh	Clinton, N. C	Dec. 1, 1868	84, 563
Plow	J. S. Marsh	Lewisburgh, Pa	Oct. 16, 1866	58, 855
Plow	H. Marshall	Atlanta, Ga	June 30, 1868	79, 486
Plow	P. Marshall and J. B. Smith	Lambertsville, N. J	Mar. 3, 1820	....
Plow	H. A. Martin	Jeffersonville, Ind	July 24, 1866	56, 584
Plow	H. D. Martin	Ypsilanti, Mich	Sept. 10, 1861	33, 256
Plow	J. R. Mason	Elgin, Ill	Feb. 11, 1862	34, 371
Plow	B. F. Masters	Middleport, Ill	Mar. 16, 1869	88, 794
Plow	D. Mater	Bellmore, Ind	Oct. 6, 1868	82, 858
Plow	R. G. Matheny and L. R. Barnes	De Kalb, Miss	Nov. 27, 1860	30, 741
Plow	J. T. Mathis and G. W. Harrison.	Kosciusko, Miss	July 5, 1870	105, 103
Plow	D. Matteson and T. P. Williamson.	Stockton, Cal	Mar. 12, 1867	62, 766
Plow	E. G. Matthews	Worcester, Mass	Feb. 23, 1841	1, 993
Plow	E. G. Matthews	South Natick, Mass	Nov. 26, 1867	71, 513
Plow	E. G. Matthews	Boston, Mass	May 19, 1868	77, 996
Plow	E. G. Matthews	Newton, Mass	Mar. 2, 1869	87, 420
Plow	E. G. Matthews	Oakham, Mass	Jan. 30, 1872	123, 272
Plow	T. F. Mattox	Griffin, Ga	Mar. 15, 1870	100, 780
Plow	H. H. May	Galesburgh, Ill	May 2, 1843	3, 069
Plow	H. H. May	Galesburgh, Ill	Apr. 25, 1846	4, 482
Plow	J. M. May	Philadelphia, Pa	May 2, 1846	4, 488
Plow	J. M. May	Philadelphia, Pa	May 2, 1846	4, 493
Plow	T. McConaughy	Burnsville, Ala	Mar. 23, 1858	19, 706
Plow	W. C. McCool	Guthrie Centre, Iowa	Oct. 8, 1872	131, 961
Plow	W. C. McCool	Guthrie Centre, Iowa	Nov. 11, 1873	144, 552
Plow	S. McCormick	Fauquier, Va	Jan. 28, 1826	......
Plow	S. McCormick	Fauquier County, Va	Oct. 22, 1828	....
Plow	S. McCormick	Auburn, Va	Dec. 1, 1837	501
Plow	M. C. McCullers	Herndon, Ga	Apr. 24, 1860	28, 000
Plow	M. C. McCullers	Herndon, Ga	July 17, 1860	29, 184
Plow	J. McKinley	Bethesda, Ohio	Sept. 4, 1866	57, 748
Plow	A. A. McMahen	Oxford, Miss	Nov. 2, 1858	21, 975
Plow	F. W. McMeekin	Morrison's Mill, Fla	Sept. 18, 1866	58, 119
Plow	F. M. McMeekin	Putnam County, Fla	Feb. 13, 1872	123, 717
Plow	F. M. McMeekin	Orange Springs, Fla	Oct. 29, 1872	132, 679
Plow	J. C. McVutt and A. B. Furman	Strattonville, Pa	Sept. 28, 1869	95, 251
Plow	S. Mead	New Haven, Conn	Sept. 15, 1863	39, 943
Plow	J. B. Mell	Riceborough, Ga	Feb. 19, 1856	14, 288
Plow	W. W. Metcalfe	York Springs, Pa	Oct. 17, 1848	5, 855
Plow	J. and E. P. Miles	Bloomingdale, Ind	Jan. 8, 1867	61, 083
Plow	L. Miller and H. Kaller	Camp Point, Ill	Dec. 13, 1864	45, 427
Plow	P. Miller	Washington County, N. Y	Oct. 26, 1818	....
Plow	R. J. Miller	Sherman, Iowa	Jan. 23, 1872	122, 956
Plow	R. M. Miller	Springfield, Ohio	July 16, 1872	129, 418
Plow	T. Miller	Pittsburgh, Pa	Oct. 23, 1829	....
Plow	T. Miller	Pittsburgh, Pa	Feb. 15, 1831	....
Plow	T. Miller	Pittsburgh, Pa	July 17, 1835	....
Plow	T. Miller	Pittsburgh, Pa	July 2, 1836	....
Plow	W. D. Miller	Enon, Ohio	June 15, 1869	91, 472
Plow	M. and S. J. Mims	Starkville, Miss	Dec. 23, 1841	2, 399
Plow	J. G. Miner	Nashville, Tenn	June 6, 1871	115, 629
Plow	J. G. Miner	Nashville, Tenn	Aug. 1, 1871	117, 662
Plow	H. Mitchell	Racine, Wis	Sept. 1, 1863	39, 741
Plow	J. J. Mitchell	Hopkinsville, Ky	Dec. 30, 1873	146, 088
Plow	J. C. Moltrup	Bucyrus, Ohio	May 10, 1859	23, 933
Plow	J. Mooers	Hazleton, Pa	July 1, 1844	3, 644
Plow	A. N. Moore	North Cohocton, N. Y	Dec. 31, 1867	72, 880
Plow	E. Moore	Slabtown, S. C	June 14, 1859	24, 399
Plow	G. Moore	Moline, Ill	Aug. 27, 1867	68, 102
Plow	G. Moore	Moline, Ill	Nov. 26, 1867	71, 507
Plow	G. H. Moore	Rochester, N. Y	Jan. 1, 1861	31, 028
Plow	J. Moore	Cynthiana, Ky	Aug. 17, 1832	....

Index of patents issued from the United States Patent Office from 1790 *to* 1873, *inclusive*—Continued.

Invention.	Inventor.	Residence.	Date.	No.
Plow	J. Moore	Lexington, Ky	July 22, 1843	3,193
Plow	J. Moore	Bushnell, Ill	Oct. 29, 1867	70,243
Plow	J. A. Moore	Louisville, Ky	Nov. 6, 1866	59,437
Plow	J. B. Moore	Wilmington, Del	Nov. 24, 1843	3,352
Plow	J. Morford	Maysville, Ky	Apr. 8, 1834	....
Plow	J. Morgan and J. B. Harris	Scipio, N. Y	Oct. 11, 1814	....
Plow	D. Morris	Bunker Hill, Ill	Sept. 27, 1870	107,705
Plow	S. D., D. A., and J. B. Morrison	Fort Madison, Iowa	Aug. 8, 1871	117,801
Plow	J. Mott	Danville, Cal	May 15, 1866	54,756
Plow	J. M. and G. W. Moyers	Gordonsville, Va	Apr. 9, 1872	125,478
Plow	M. Murray	Middletown, Md	Aug. 7, 1813	....
Plow	S. Myers	Marion, Ohio	July 11, 1842	2,712
Plow	T. Nabers	Elyton, Ala	Apr. 23, 1872	126,077
Plow	C. Nash	Bacon, Ill	Oct. 22, 1872	132,406
Plow	J. Nash	Middlebury, Ohio	May 26, 1843	3,110
Plow	C. O. Nason	Moline, Ill	July 29, 1873	141,375
Plow	H. W. Neal	Sidney, Ohio	May 26, 1868	78,390
Plow	H. W. Neal	Sidney, Ohio	Aug. 10, 1869	93,549
Plow	D. Nelson	Port Washington, Ohio	Sept. 27, 1864	44,442
Plow	C. Newbold	....	June 26, 1797	....
Plow	J. R. Nichols	Bastrop, Tex	Feb. 25, 1873	136,257
Plow	W. Nichols	Floyd County, Ga	May 10, 1859	23,938
Plow	W. D. Nichols and N. W. Clark	Chicago, Ill	Sept. 24, 1867	69,117
Plow	S. Nisbet	Toboyne Township, Pa	May 25, 1830	....
Plow	G. Nixon	New York	June 3, 1823	....
Plow	W. Noble	New Haven, Conn	July 27, 1869	92,992
Plow	H. Nolte	Lincoln, Ill	Sept. 7, 1869	94,634
Plow	J. B. Norris	Richmond, Va	Dec. 2, 1873	145,120
Plow	H. Norton	Farmington, Me	Oct. 9, 1855	13,653
Plow	H. L. Norton	Granville, N. Y	Sept. 23, 1842	2,784
Plow	J. B. Norton	Philadelphia, Pa	July 17, 1837	273
Plow	C. O'Bryan and H. Kreps	Minerva, Ohio	Aug. 20, 1861	33,090
Plow	S. Ogle	Adams County, Pa	Feb. 28, 1832	....
Plow	W. Ogle	Frederick, Md	Apr. 6, 1843	3,034
Plow	J. Oliver	South Bend, Ind	Feb. 21, 1871	111,965
Plow	J. Oliver	South Bend, Ind	June 18, 1872	128,061
Plow	J. Oliver	South Bend, Ind	Nov. 18, 1873	144,785
Plow	S. J. Olmsted	Binghamton, N. Y	Mar. 17, 1863	37,939
Plow	S. J. Olmsted	Binghamton, N. Y	Sept. 13, 1864	44,215
Plow	W. O'Neill	Pine Level, Ala	Sept. 13, 1859	25,436
Plow	W. O'Neill	Pine Level, Ala	Sept. 13, 1859	25,437
Plow	W. O'Neill	Pine Level, Ala	June 8, 1869	91,157
Plow	J. Ormiston	Waterford, Ohio	Sept. 5, 1848	5,750
Plow	J. Ormiston	Centre Township, Ohio	June 2, 1857	17,476
Plow	O. Osborn	Trumansburgh, N. Y	Feb. 23, 1869	87,284
Plow	O. Osborn	Trumansburgh, N. Y	Sept. 7, 1869	94,636
Plow	O. Osborn	Erie, Pa	Nov. 15, 1870	109,340
Plow	G. Page	Washington, D. C	Aug. 7, 1847	5,218
Plow	W. F. Pagett	Springfield, Ohio	Sept. 7, 1869	94,637
Plow	J. Parker	Milroy, Ind	Feb. 26, 1867	62,497
Plow	W. F. Parker	Troy, Ala	May 2, 1871	114,334
Plow	J. A. Partlett and J. Thompson	Elmira, N. Y	Apr. 2, 1861	31,903
Plow	G. W. Parmer	Jefferson Township, Ohio	Aug. 17, 1832	....
Plow	M. Patrick	Scipio, N. Y	June 2, 1813	....
Plow	T. Patterson	Rush, Ill	May 14, 1861	32,311
Plow	C. Pawling	Gregg Township, Pa	May 21, 1830	....
Plow	H. Peachey	Philadelphia, Pa	July 25, 1834	....
Plow	D. Peacock	Burlington, N. J	Apr. 1, 1807	....
Plow	D. Peacock	New Mills, N. J	May 29, 1817	....
Plow	D. Peacock	Northampton, N. J	Jan. 21, 1822	....
Plow	H. Pease	Enfield, Conn	Aug. 28, 1813	....
Plow	E. Peck	Chicago, Ill	Sept. 15, 1868	82,157
Plow	J. D. Peck	Triangle, N. Y	Mar. 10, 1868	75,455
Plow	S. T. Peck	Penfield, Ga	Nov. 29, 1859	26,289
Plow	M. Penning	Leavenworth City, Kans	Mar. 31, 1868	76,243
Plow	D. Peters and J. W. Pauley	Keokuk, Iowa	Mar. 12, 1867	62,881
Plow	P. Pettis	Portage County, Ohio	Feb. 15, 1825	....
Plow	J. P. Pettit	Cold Spring, Ky	Nov. 27, 1860	30,756
Plow	O. Phelps and G. Morehouse	Lansing, N. Y	Nov. 17, 1821	....
Plow	J. Pierpont	La Harpe, Ill	Oct. 18, 1864	44,745
Plow	J. Pinkham	New Market, N. H	Jan. 24, 1871	111,244
Plow	W. C. Pitts	Austin, Tex	May 15, 1860	28,329
Plow	J. Plank	Carlisle, Pa	June 2, 1836	....
Plow	H. M. Platt	Darien, Conn	June 22, 1858	20,659
Plow	N. Platt	Saint Louis, Mo	Mar. 21, 1865	46,937
Plow	F. Poindexter	Franklin, N. C	July 16, 1872	129,054
Plow	H. A. G. Pomeroy and R. F. Hudson	Providence, R. I., and Hartford, Conn.	June 12, 1860	28,687
Plow	J. N. Pond	Wakefield, Va	Nov. 6, 1866	59,515
Plow	W. R. Pool	Havanna, Ala	Nov. 9, 1869	96,614
Plow	S. W. Pope	Louisville, Ky	May 10, 1870	102,860
Plow	S. W. Pope	Louisville, Ky	July 26, 1870	105,843
Plow	R. D. Porter	Zanesville, Ohio	July 2, 1872	128,505
Plow	J. C. Potter	Saint Helena, Cal	June 24, 1873	140,302
Plow	M. Prentiss	Cambridge, N. Y	Oct. 1, 1867	69,478
Plow	S. Prentiss and G. Flint	De Soto, Mo	Jan. 12, 1869	85,851
Plow	B. Price	Leesville, Ohio	July 31, 1866	56,798
Plow	J. Price	Greenfield, Ind	Oct. 30, 1866	59,262
Plow	W. Price	Wayne County, N. C	June 5, 1860	28,601
Plow	W. Price	Mount Olive, N. C	Apr. 8, 1873	137,716
Plow	J. P. Pritchard	Conn Valley, Cal	Jan. 24, 1871	111,247
Plow	D. Prouty	Dorchester, Mass	Jan. 13, 1847	4,928
Plow	D. Prouty and J. Mears	Boston, Mass	Mar. 4, 1836	....
Plow	D. Prouty and J. Mears	Boston, Mass	Sept. 15, 1838	922
Plow	D. Prouty and J. Mears	Boston and Dorchester, Mass.	June 16, 1841	2,132
Plow	T. E. Putnam	Winneconne, Wis	July 16, 1872	129,296
Plow	J. A. Quick	South Dansville, N. Y	Sept. 18, 1866	58,177

Index of patents issued from the United States Patent Office from 1790 *to* 1873, *inclusive*—Continued.

Invention.	Inventor.	Residence.	Date.	No.
Plow	T B. Quigley	Mansfield, Ohio	June 14, 1843	3, 137
Plow	W. S. Rabb	Winnsborough, S. C	Apr. 20, 1869	89, 242
Plow	W. S. Rabb	Winnsborough, S. C	May 31, 1870	103, 777
Plow	Y. Rakestrow	White House, Ohio	Oct. 20, 1868	83, 206
Plow	W. Rall	South Bend, Ind	July 16, 1872	129, 364
Plow	J. O. Ramage	La Fayette, Ala	Feb. 2, 1858	19, 262
Plow	T. Rams	Keokuk, Iowa	Sept. 11, 1866	57, 968
Plow	I. N. Rankin	Middletown, Iowa	May 22, 1860	28, 407
Plow	T. S. and T. W. Rappleye	Farmer, N. Y	July 2, 1861	32, 714
Plow	G. W. Ream	Canton, Ohio	Oct. 25, 1870	108, 730
Plow	W. Reaney	Berzelia, Ga	Oct. 19, 1858	21, 846
Plow	J. W. Reed	Careyville, Mo	Jan. 28, 1873	135, 289
Plow	R. Reeder and S. D. Ashley	Madisonville, Ohio	Oct. 21, 1829	......
Plow	J. Reedy	Toledo, Iowa	May 14, 1867	64, 707
Plow	E. Reese	Eutaw, Ala	Jan. 28, 1873	135, 363
Plow	F. Reese	Elyton, Ala	Oct. 13, 1868	82, 990
Plow	F. Reese	Wilsonville, Ala	Dec. 10, 1872	133, 722
Plow	F. F. Reynolds	Burke County, Ga	Jan. 1, 1867	60, 938
Plow	I. Reynolds	Republic, Ohio	Feb. 13, 1855	12, 398
Plow	T. H. Reynolds	Rome, Ga	May 16, 1871	114, 855
Plow	D. Rhodes	Pawtuxet, R. I	Apr. 19, 1864	42, 401
Plow	J. Rhodes	Urbana, Ohio	June 11, 1829	......
Plow	M. G. Rhodes and J. M. Skaggs	Talladega, Ala	May 22, 1860	28, 408
Plow	R. Rhodes	Charlotteville, Va	Feb. 20, 1827	......
Plow	R. Rhodes	Albemarle County, Va	Feb. 25, 1823	......
Plow	J. Rich	Troy, N. Y	July 31, 1849	6, 611
Plow	J. Rich	Kingsbury, N. Y	July 15, 1856	15, 344
Plow	M. Richards and J. Vandegrift	Princeton, Ill	Feb. 5, 1867	61, 762
Plow	F. E. Richardson	Hicksford, Va	Nov. 30, 1852	9, 433
Plow	L. W. Richardson	Roscoe, Ill	Nov. 15, 1870	109, 250
Plow	L. B. Richardson	Carrollton, Ill	July 1, 1873	140, 434
Plow	W. Richardson	Hookstown, Md	Sept. 17, 1867	68, 901
Plow	C. F. Richter	Columbia, S. C	June 5, 1860	28, 605
Plow	W. Richter	Williamsburgh, Ind	Jan. 9, 1849	6, 007
Plow	A. Rickard	Schoharie, N. Y	Feb. 21, 1871	111, 975
Plow	A. Rickard	Schoharie, N. Y	July 16, 1872	129, 057
Plow	M. Rigell	Newton, Ala	Mar. 1, 1870	100, 325
Plow	M. Rigell	Newton, Ala	Mar. 1, 1870	100, 326
Plow	G. Ringen	Smith City, Mo	Oct. 29, 1867	70, 267
Plow	A. W. L. Rivers	Midway, S. C	Nov. 27, 1860	30, 762
Plow	A. Riviere	Barnesville, Ga	Nov. 11, 1873	144, 477
Plow	J. L. Roberts	Brunswick, Ga	Oct. 23, 1866	59, 078
Plow	M. L. Roberts	Smithville, Canada	Feb. 26, 1867	62, 367
Plow	S. I. Roberts	Jeffersonville, Pa	Feb. 12, 1845	3, 913
Plow	H. H. Robertson	Kingston, Mo	July 10, 1860	29, 104
Plow	J Robinson	Hobart, N. Y	Sept. 23, 1862	36, 532
Plow	J. G. Robinson	Biddeford, Me	Dec. 4, 1860	30, 837
Plow	N. Robinson	Sacket's Harbor, N. Y	Feb. 13, 1835	......
Plow	N. Robinson	Patchogue, N. Y	Sept. 14, 1869	94, 841
Plow	A. Roden	Talladega, Ala	Aug. 28, 1860	29, 823
Plow	A. Roden	Mumford, Ala	June 27, 1871	116, 355
Plow	J. M. Rodman	South Union, Ky	Feb. 26, 1861	31, 559
Plow	H. D. Rogers	Grafton, Ohio	Feb. 12, 1861	31, 407
Plow	R. Roles	Cary, N. C	May 31, 1870	103, 663
Plow	L. Ronat	Carondelet, Mo	Jan. 5, 1869	85, 696
Plow	G. W. Roney	Bayley's Mill, Fla	Dec. 27, 1859	26, 620
Plow	H. Roney	Dayton, Ohio	Mar. 28, 1871	113, 098
Plow	E. Rood	Darien, Wis	Dec. 24, 1872	134, 316
Plow	J. Roop	Clinton, Tenn	Apr. 22, 1873	138, 197
Plow	M. P. Rose	Napa, Cal	July 25, 1871	117, 335
Plow	B. C. Rouse	Morris, Ill	Aug. 27, 1867	68, 313
Plow	I. Rulofson	Penn Yan, N. Y	Mar. 1, 1859	23, 116
Plow	I. Rulofson and D. De Garmo	Rochester, N. Y	Mar. 12, 1861	31, 677
Plow	J. Runnyon	Marshall Township, Mich	Aug. 20, 1872	130, 751
Plow	J. Runnyon and G. Ingersoll	Marshall Township and Marshall, Mich.	Aug. 10, 1869	93, 558
Plow	G. W. Russell	Rockford, Ill	July 6, 1869	92, 371
Plow	G. M. and G. S. Salsbury	Wilson and Clarendon, N. Y.	Aug. 4, 1863	39, 425
Plow	A. Sanborn	Saint Johnsbury, Vt	May 28, 1872	127, 372
Plow	A. Sanborn	Saint Johnsbury, Vt	Dec. 10, 1872	133, 802
Plow	T. Sanford	Redding Ridge, Conn	Feb. 23, 1858	19, 455
Plow	W. R. Saunders	Buena Vista, Miss	Mar. 27, 1860	27, 651
Plow	W. Savage and H. Davidson	Horse Well, Ky	Aug. 24, 1831	......
Plow	S. D. Sayre	Rockford, Ill	Apr. 19, 1870	102, 162
Plow	B. B. Scofield	Andover, Ill	Sept. 28, 1858	21, 630
Plow	T. S. Scoville	Elmira, N. Y	Apr. 6, 1858	19, 878
Plow	S. F. Seely	Sylvania, Ohio	Apr. 10, 1866	53, 891
Plow	G. and J. Seibert	Ashley, Ill	Feb. 26, 1861	31, 563
Plow	H. Selick	Lewistown, Pa	Sept. 28, 1869	95, 386
Plow	H. Selick	Lewistown, Pa	Feb. 11, 1873	135, 850
Plow	E. Seward	Hamilton, Ohio	Aug. 6, 1867	67, 456
Plow	B. Seyler	Franklin County, Pa	Oct. 16, 1849	6, 788
Plow	L. W. Shaffar	Shelbyville, Ky	Dec. 10, 1861	33, 906
Plow	M. R. Shalters	Alliance, Ohio	May 9, 1871	114, 611
Plow	J. P. Sharp	Geneseo, N. Y	Dec. 20, 1833	......
Plow	T. Sharp	Nashville, Tenn	Oct. 6, 1857	18, 355
Plow	S. Shearer	Big Prairie, Ohio	July 14, 1845	4, 112
Plow	W. F. Shedd	Ripley, Ohio	Feb. 26, 1861	31, 565
Plow	T. Sheehan	Dunkirk, N. Y	July 5, 1870	105, 134
Plow	G. Shelton	Normal, Ill	June 4, 1872	127, 650
Plow	J. Shepard	Newport, Me	June 29, 1869	91, 972
Plow	J. W. Shipp and C. W. Crenshaw.	La Grange, Tenn	June 26, 1860	28, 907
Plow	W. T. Shipp, C. J. Peterson, and R. L. McLurd.	Brevard Station, N. C	Nov. 18, 1873	144, 707
Plow	C. I. Shiver	Camden, S. C	May 22, 1860	28, 416
Plow	W. G. Shuart	Orange County, N. Y	Feb. 5, 1822	......

Index of patents issued from the United States Patent Office from 1790 *to* 1873, *inclusive*—Continued.

Invention.	Inventor.	Residence.	Date.	No.
Plow	H. Shultz	Lancaster, Pa	Dec. 17, 1814	....
Plow	A. Shunk, sr	Bucyrus, Ohio	Nov. 19, 1867	71, 234
Plow	J. Sietz	Strasburgh, Pa	Feb. 8, 1813	....
Plow	H. B. Sinclear	Lyndonville, N. Y	Jan. 9, 1849	6, 020
Plow	D. T. Singleton	Eatonton, Ga	Mar. 22, 1870	101, 056
Plow	H. M. Skinner	Rockford, Ill	Sept. 10, 1872	131, 309
Plow	J. B. Skinner	Rockford, Ill	Aug. 14, 1866	57, 200
Plow	J. B. Skinner	Rockford, Ill	Apr. 30, 1867	64, 259
Plow	J. B. Skinner	Rockford, Ill	July 2, 1867	66, 259
Plow	J. B. Skinner	Rockford, Ill	July 2, 1867	66, 260
Plow	S. F. Skinner	Jacksonville, Mo	Jan. 14, 1868	73, 264
Plow	W. W. Skinner	Davenport, Iowa	Dec. 1, 1857	18, 776
Plow	M. G. Slemmons	Cadiz, Ohio	Oct. 9, 1860	30, 357
Plow	L. L. Sloss	Auburn, Ky	Jan. 4, 1870	98, 640
Plow	W. Small	North Argyle, N. Y	Apr. 22, 1839	1, 133
Plow	W. H. Simley	Bentonville, Ark	Sept. 24, 1872	131, 636
Plow	A. Smith	Bloomfield, Mich	May 6, 1844	3, 576
Plow	A. Smith	Macoupin County, Ill	Jan. 16, 1855	12, 241
Plow	A. B. Smith	Rochester, Pa	Nov. 5, 1872	132, 779
Plow	A. C. Smith	Joyner's Depot, N. C	Dec. 27, 1870	110, 506
Plow	D. Smith	Cedar Falls, Iowa	July 28, 1868	80, 314
Plow	D. H. and E. E. Smith	Glenn Springs, S. C	Apr. 3, 1860	27, 745
Plow	D. W. Smith	Dooly County, Ga	Mar. 19, 1861	31, 749
Plow	F. C. Smith	Harper's Ferry, Va	Apr. 25, 1848	5, 526
Plow	G. K. Smith and J. Strasser	Allegheny City, Pa	Jan. 21, 1868	73, 550
Plow	H. Smith	Moline, Ill	Nov. 15, 1870	109, 352
Plow	H. B. Smith	Tremont, Ill	May 21, 1872	127, 110
Plow	J. Smith	Norfolk, Va	Aug. 14, 1860	29, 629
Plow	J. C. Smith	Kingwood, N. J	July 11, 1837	260
Plow	J. M. Smith	Haddam, Conn	Aug. 30, 1870	106, 966
Plow	J. M. Smith	Haddam Neck, Conn	Dec. 5, 1871	121, 676
Plow	M. T. Smith	Keeler, Mich	Nov. 6, 1866	59, 468
Plow	C. W. Snead	Milledgeville, Ga	Oct. 4, 1870	107, 973
Plow	I. Snider	Mount Pleasant, Pa	July 2, 1836	....
Plow	C. Snyder	Middletown, Ill	Sept. 27, 1870	107, 828
Plow	S. W. Soule	New York, N. Y	Sept. 30, 1873	143, 263
Plow	O. Sparks	Shelbina, Mo	Dec. 11, 1860	30, 892
Plow	H. Speihlman and D. Miller	Lancaster, Pa	Mar. 25, 1833	....
Plow	G. Spiehlman	Strasburgh, Pa	Oct. 15, 1867	69, 943
Plow	W. S. Spratt	West Manchester, Pa	Sept. 5, 1865	49, 799
Plow	W. T. Sprouse	Sangamon, Ill	Feb. 15, 1838	604
Plow	W. T. Sprouse	Petersburgh, Ill	Mar. 13, 1849	6, 179
Plow	W. T. Sprouse	Chandlerville, Ill	Aug. 27, 1867	68, 253
Plow	D. Staebler	Lancaster, Pa	Mar. 15, 1834	....
Plow	P. Stahl and T. Diffenbasher	Turbut, Pa	Sept. 18, 1835	....
Plow	S. W. Standart	Bellevue, Ohio	Feb. 18, 1868	74, 729
Plow	H. Stanley	Erie County, Pa	Nov. 10, 1857	18, 609
Plow	T. A. Stansbury	Saybrook, Ill	Feb. 27, 1866	52, 903
Plow	P. H. Starke	Richmond, Va	Aug. 21, 1860	29, 726
Plow	P. H. Starke	Richmond, Va	May 5, 1868	77, 547
Plow	P. H. Starke	Richmond, Va	Dec. 29, 1868	85, 342
Plow	P. H. Starke	Richmond, Va	Aug. 1, 1871	117, 574
Plow	P. H. Starke	Richmond, Va	Aug. 19, 1873	141, 960
Plow	S. S. Starns	Macomb, Ill	Nov. 26, 1867	71, 419
Plow	J. L. Stearns	Mahomet, Ill	Dec. 15, 1868	84, 970
Plow	L. M. Stearns	Cardiff, N. Y	Jan. 4, 1861	32, 489
Plow	C. Steen	Athens, Ohio	Oct. 2, 1866	58, 500
Plow	H. Stein	Mifflinburgh, Pa	July 20, 1869	92, 898
Plow	C. P. Steinmetz	Madison, Wis	June 2, 1868	78, 492
Plow	R. S. Stenton	New York, N. Y	Dec. 8, 1857	18, 825
Plow	J. Stephens	Agency City, Iowa	Sept. 7, 1869	94, 663
Plow	R. L. and E. A. Stevens	Hoboken, N. J	Sept. 13, 1820	....
Plow	J. A. Stewart	Philadelphia, Pa	Oct. 2, 1860	30, 263
Plow	S. M. Stewart	New Harrisburgh, Ohio	Mar. 7, 1871	112, 508
Plow	C. St. John	Keosauqua, Iowa	Oct. 16, 1866	58, 911
Plow	S. L. Stockstill and A. D. Kutez	Medway, Ohio, and Harrisburgh, Pa.	Apr. 15, 1873	137, 870
Plow	A. W. Stoker	Petersburgh, Ill	Sept. 11, 1866	57, 991
Plow	J. T. Story	Magnolia, Ark	June 13, 1871	115, 907
Plow	R. E. Strait	Galesburgh, Mich	Dec. 21, 1869	98, 120
Plow	A. L. W. Stroud	Munford, Ala	Apr. 25, 1871	114, 221
Plow	D. A. Stubblefield and W. H. Luse	Yazoo County, Miss	Aug. 24, 1869	94, 041
Plow	Z. W. Sturtevant	Dunstable, Mass	Dec. 14, 1869	97, 828
Plow	R. B. Summers and S. Dement	San José, Ill	Sept. 25, 1866	58, 318
Plow	J. Swan	Scipio, N. Y	July 5, 1814	....
Plow	D. Swartz	Thomas Brook, Va	June 22, 1852	9, 061
Plow	H. H. Sweetland	Centreton, Ohio	Aug. 27, 1872	130, 882
Plow	J. Sweitzer	Mishawaka, Ind	Jan. 22, 1861	31, 194
Plow	C. W. Sykes	Suffield, Conn	Oct. 31, 1865	50, 749
Plow	N. S. Tallmadge	Fond du Lac, Wis	Sept. 1, 1863	39, 761
Plow	E. C. Tavener and O. Nesmith	Hamilton, Va	May 8, 1855	12, 838
Plow	A. Taylor	New Garden, Ohio	Dec. 19, 1844	3, 864
Plow	C. B. Taylor	Bainbridge, Ohio	Aug. 23, 1833	....
Plow	H. Taylor	Montague, Mass	May 17, 1838	743
Plow	J. V. Taylor	Dixon, Ill	Jan. 3, 1860	26, 718
Plow	A. Teague	Jackson, Tenn	Mar. 10, 1843	2, 998
Plow	A. Teague	Madisonville, Ky	Sept. 21, 1869	95, 163
Plow	J. S. Tefft	Amherst, N. Y	Oct. 17, 1835	....
Plow	J. S. Tefft	Amherst, N. Y	Aug. 25, 1842	2, 762
Plow	M. Tessier	Cairo, Ill	Sept. 7, 1869	94, 667
Plow	B. Thacker, jr	Barnstable, Mass	July 20, 1831	....
Plow	D. K. Thom	Farmington, Tenn	Oct. 20, 1857	18, 475
Plow	G. W. Thompson	Ripley, Ohio	Sept. 4, 1866	57, 796
Plow	J. Thompson	Ripley, Ohio	Apr. 17, 1844	3, 542
Plow	J. T. Thompson	Jackson, Tenn	June 26, 1860	28, 919
Plow	S. R. Thompson	New Market, N. H	May 4, 1869	89, 608
Plow	T. Thompson	Thompsonville, N. C	Jan. 19, 1858	19, 163

Index of patents issued from the United States Patent Office from 1790 *to* 1873, *inclusive*—Continued.

Invention.	Inventor.	Residence.	Date.	No.
Plow	I. P. Tice	New York, N. Y	Sept. 5, 1871	118,758
Plow	J. Tietz	Baltimore, Md	July 23, 1867	67,002
Plow	J. M. Tilford	Murfreesborough, Tenn	July 1, 1836	....
Plow	D. L. Tilton	Mount Carmel, Ill	Mar. 23, 1858	19,725
Plow	J. Tinkler	Warwick Township, Ohio	Mar. 2, 1835	....
Plow	W. D. Titus	Brooklyn, N. Y	Dec. 24, 1867	72,568
Plow	J. Tomlinson	Racine, Wis	Jan. 26, 1864	41,407
Plow	J. T. Townsend	Brenham, Tex	Nov. 15, 1859	26,133
Plow	A. J. Traver	Lisburn, Pa	Mar. 23, 1869	88,233
Plow	F. Traxler	Salem, Mich	Apr. 16, 1861	32,094
Plow	M. Turley	Galesburgh, Ill	Feb. 16, 1858	19,388
Plow	B. W. Tuttle	Galena, Ill	May 14, 1872	126,656
Plow	W. H. Tyler	Conneautville, Pa	Jan. 1, 1869	90,895
Plow	C. Urie	Evansville, Ind	Aug. 20, 1872	130,773
Plow	J. Urie	Evansville, Ind	Jan. 7, 1868	73,209
Plow	G. Utley	Louisburgh, N. C	Mar. 16, 1858	19,658
Plow	G. Utley	Chapel Hill, N. C	Feb. 12, 1861	31,419
Plow	G. Utley	Chapel Hill, N. C	May 26, 1868	78,339
Plow	A. Vail	Henry, Ill	Aug. 25, 1868	81,571
Plow	J. Vandegrift	Princeton, Ill	Oct. 15, 1867	69,867
Plow	W. W. Van Loan	Catskill, N. Y	Feb. 16, 1858	19,391
Plow	G. B. Vaughan	Peoria, Ill	May 20, 1873	139,036
Plow	S. O. Vaughan	De Kalb, Ill	Feb. 28, 1860	27,322
Plow	W. Veber, jr	Shingle Creek, N. Y	Sept. 25, 1866	58,323
Plow	J. C. Vertrees	Gallatin, Tenn	Sept. 26, 1871	119,433
Plow	R. Vincent	Rockton, Ill	Dec. 21, 1858	22,389
Plow	B. Volkmann	New York, N. Y	Nov. 27, 1866	60,095
Plow	F. Volkman	Hoboken, N. J	June 11, 1867	65,709
Plow	G. A. Walker	Annville, Pa	Mar. 12, 1861	31,684
Plow	J. Walker	Bellefontaine, Ohio	June 19, 1847	5,168
Plow	R. Walker	Washingtonville, Pa	Mar. 10, 1830	....
Plow	R. Walker	Savannah, Mo	Aug. 6, 1872	130,256
Plow	S. Walker	Kingston, Ga	Dec. 27, 1859	26,633
Plow	W. Walker	Washingtonville, Pa	Oct. 6, 1835	....
Plow	W. and M. C. Walker	Washingtonville, Pa	Oct. 20, 1843	3,311
Plow	J. Wallace	Berks County, Pa	Nov. 28, 1865	51,245
Plow	J. Wallace	Sheridan, Pa	June 25, 1867	66,109
Plow	J. Wallace	Sheridan, Pa	Feb. 20, 1872	123,957
Plow	W. R. Walpole	Chicago, Ill	Apr. 2, 1867	63,586
Plow	N. Warlick	La Fayette, Ala	Apr. 3, 1855	12,650
Plow	N. Warlick	La Fayette, Ala	Oct. 20, 1857	18,480
Plow	J. Warren	Warren County, N. Y	July 31, 1849	6,620
Plow	T. P. Warren	Norfolk, Va	May 12, 1868	77,855
Plow	W. Warren	Penn Yan, N. Y	July 20, 1858	20,968
Plow	W. Warren	Penn Yan, N. Y	Sept. 25, 1860	30,170
Plow	H. Washburn	Pultney, N. Y	Feb. 28, 1871	112,302
Plow	F. Watson	Harrison, Ill	Oct. 15, 1867	69,874
Plow	W. Watson	Bishopville, S. C	Mar. 13, 1860	27,490
Plow	W. M. Watson	Tonica, Ill	Jan. 7, 1873	134,716
Plow	G. Watt	Gainesville, Ala	Apr. 11, 1842	2,548
Plow	G. Watt	Richmond, Va	Dec. 9, 1856	16,218
Plow	G. Watt	Richmond, Va	Feb. 9, 1858	19,321
Plow	G. Watt	Richmond, Va	July 10, 1866	56,298
Plow	G. Watt	Richmond, Va	Nov. 26, 1867	71,560
Plow	G. Watt	Richmond, Va	July 6, 1869	92,408
Plow	A. Weaber	Lebanon, Pa	Apr. 25, 1871	114,237
Plow	J. Weaver	Brownsville, Pa	Aug. 17, 1832	....
Plow	J. W. Webb	Cotton Valley, Ala	Nov. 1, 1870	108,860
Plow	A. P. Webber	Saratoga Township, Ill	Aug. 27, 1872	130,832
Plow	L. T. Webster	Northfield, Mass	Mar. 8, 1870	100,696
Plow	L. Weeks and J. S. Trimble	Shelby, Ohio	May 14, 1872	126,856
Plow	W. B. West	Utica, Wis	Oct. 19, 1869	95,960
Plow	G. Wharton	Jerseyville, Ill	July 19, 1870	105,531
Plow	R. J. Wheatley	Saint John, Ill	Feb. 19, 1867	62,243
Plow	R. J. Wheatley	Duquoin, Ill	Mar. 15, 1870	100,957
Plow	C. White	Bladensburgh, Md	Feb. 4, 1868	74,183
Plow	E. J. White	Locke, N. Y	May 13, 1862	35,272
Plow	L. B. White	Norfolk, Va	July 16, 1872	129,076
Plow	L. B. White	Norfolk, Va	July 8, 1873	140,749
Plow	N. Whitehall	Attica, Ind	July 3, 1855	13,191
Plow	W. Whiteley	Springfield, Ohio	June 2, 1868	78,501
Plow	T. A. Whities	Bellefontaine, Ohio	Jan. 14, 1832	....
Plow	E. G. Whiting	Racine, Wis	July 11, 1839	1,232
Plow	E. G. Whiting	Northfield, Minn	Sept. 5, 1865	49,816
Plow	S. Whitley	Tallahassee, Fla	Dec. 4, 1860	30,849
Plow	S. W. Whitman	Harpersfield, Ohio	May 7, 1825	....
Plow	J. M. Whitney	Bolton, Mass	Mar. 1, 1859	23,134
Plow	A. H. Whittick	Clarksville, Ind	Sept. 12, 1871	118,890
Plow	E. Wiard	Louisville, Ky	Dec. 22, 1868	85,152
Plow	E. Wiard	Louisville, Ky	May 25, 1869	90,475
Plow	E. Wiard	Louisville, Ky	Aug. 31, 1869	94,366
Plow	E. Wiard	Louisville, Ky	Aug. 31, 1869	94,367
Plow	E. Wiard	Louisville, Ky	July 11, 1871	117,024
Plow	E. Wiard	Louisville, Ky	Aug. 26, 1873	142,136
Plow	E. Wiard	Louisville, Ky	Nov. 11, 1873	144,584
Plow	T. Wiard	Avon, N. Y	Apr. 16, 1842	2,557
Plow	T. Wiard	Louisville, Ky	Dec. 14, 1858	22,332
Plow	W. Wiard	Avon, N. Y	Jan. 26, 1828	....
Plow	J. Widman and F. Millica	El Paso, Ill	Feb. 13, 1866	52,633
Plow	C. Wilkin	Dundas, Ill	Apr. 14, 1868	76,681
Plow	R. S. Williams	Bairdstown, Ga	June 26, 1868	28,930
Plow	S. Williams, jr	Hume, N. Y	Mar. 8, 1859	23,211
Plow	W. B. Williams	Warrenton, N. C	Sept. 13, 1859	25,463
Plow	W. B. Williams	Warrenton, N. C	Sept. 13, 1859	25,464
Plow	J. C. Williamson	Washington, Ga	Apr. 6, 1858	19,886
Plow	J. C. Williamson	Washington, Ga	July 16, 1872	129,196
Plow	F. R. Willson	Columbus, Ohio	May 7, 1872	126,507
Plow	J. M. Wilson	Lexington, Miss	June 23, 1868	79,091

Index of patents issued from the United States Patent Office from 1790 *to* 1873, *inclusive*—Continued.

Invention.	Inventor.	Residence.	Date.	No.
Plow	J. S. Wilson	Waynesborough, Ga	May 8, 1860	28,227
Plow	L. S. Wilson	Wabash, Ind	Sept. 5, 1871	118,769
Plow	W. H. Wilson	Summerfield, Ohio	Apr. 12, 1859	23,636
Plow	G. A. Wing	Albany, N. Y	Feb. 2, 1869	86,615
Plow	T. Winslow	Cleveland, Ohio	Apr. 25, 1865	47,486
Plow	R. B. Winston	Richmond, Va	Dec. 1, 1857	18,783
Plow	S. Witherow	Gettysburgh, Pa	Jan. 15, 1836	....
Plow	S. Witherow	Gettysburgh, Pa	Apr. 18, 1846	4,465
Plow	G. Wolf	Fairfield County, Ohio	Jan. 27, 1832	
Plow	L. Wolf	Hamburgh, Mo	Jan. 15, 1861	31,136
Plow	L. Wolf	Saint Louis, Mo	Aug. 5, 1862	36,122
Plow	C. Wood and G. Brundage	Blooming Grove, N. Y	Nov. 9, 1820	....
Plow	D. Wood and A. Byington	Byron, Ill	May 15, 1860	28,323
Plow	J. Wood	Scipio, N. Y	July 1, 1814	....
Plow	J. Wood	Poplar Ridge, N. Y	Sept. 1, 1819	....
Plow	B. Woodcock	Mount Pleasant, Pa	Nov. 23, 1836	....
Plow	B. Woodcock	Mount Pleasant, Pa	Nov. 23, 1836	85
Plow	B. Woodcock	Mount Pleasant, Pa	June 14, 1837	233
Plow	B. Woodcock	Mount Pleasant, Pa	Nov. 23, 1837	476
Plow	B. Woodcock	Wheeling, Va	Jan. 31, 1845	3,898
Plow	B. Woodcock	Williamsburgh, Pa	Aug. 21, 1860	29,741
Plow	J. Woodward	Haverhill, N. C	Mar. 9, 1852	8,794
Plow	J. Woolley	Troy, N. Y	Sept. 14, 1822	....
Plow	W. E. Wormell	Germantown, Tenn	Aug. 28, 1860	29,841
Plow	A. Wright	Allegheny City, Pa	Jan. 21, 1868	73,685
Plow	A. Wright	Allegheny City, Pa	Dec. 7, 1869	97,746
Plow	J. S. Wright	Greenwich, N. Y	July 28, 1820	....
Plow	G. D. Wyman	Oshkosh, Wis	Feb. 18, 1873	136,120
Plow	W. F. Yeager	Starkville, Miss	Dec. 6, 1859	26,390
Plow	G. W. N. Yost	Nashville, Tenn	Aug. 11, 1863	39,536
Plow	W. Yost	Goshen, Ohio	Dec. 5, 1871	121,567
Plow	W. B. Young	Chicago, Ill	Feb. 14, 1865	46,418
Plow	W. B. Young	Chicago, Ill	Feb. 11, 1868	74,474
Plow	J. Yuger	Harrisburgh, Pa	June 9, 1830	....
Plow	G. W. Zeigler	Tiffin City, Ohio	May 27, 1856	14,989
Plow	W. Zeller and R. Lechner	Lebanon County and Berks County, Pa.	Apr. 16, 1867	63,838
Plow	C. Zimmerman	Collinsville, Ohio	Oct. 2, 1866	58,534
Plow	W. T. Zollickoffer	Shelbyville, Tenn	Sept. 25, 1860	30,188
Plow	S. Zook	Armagh Township, Pa	June 18, 1834	....
Plow, Adjustable shovel	T. Donnely and J. B. Cressler	Southampton, Pa	Jan. 28, 1868	73,699
Plow, Adjustable shovel	G. W. Morter and E. Berry	Hartville, Ohio	Dec. 29, 1868	85,469
Plow, Adjusting	W. K. Allan	Brownsborough, Ky	Jan. 31, 1844	3,416
Plow, Agricultural	A. D. Armstrong	Springfield, Ohio	Jan. 15, 1830	....
Plow and action of the mold-board	H. Peachey	Washington, D. C	Apr. 5, 1833	
Plow and clevis	I. Reynolds	West Liberty, Ohio	Mar. 26, 1850	7,223
Plow and cotton-scraper	J. V. Greif	Paducah, Ky	May 28, 1867	65,212
Plow and cultivator	J. Lane	Chicago, Ill	Sept. 15, 1868	82,130
Plow and cultivator, Combined	S. Huber	Danville, Pa	Aug. 3, 1869	93,203
Plow and cultivator, Combined	J. M. Landes	Souders, Pa	Aug. 30, 1870	106,837
Plow and cultivator, Combined	A. Romann and J. Peterka	Wilton, Iowa	Aug. 27, 1867	68,114
Plow and cultivator, Combined	J. Woipert	Louisville, Ky	Mar. 22, 1870	101,200
Plow and cultivator, Combined	C. J. Woods and J. A. Phillips	Centreville, Ind	Nov. 9, 1869	96,757
Plow and cultivator, Combined gang	J. A. Medaris	Sullivan, Ind	Jan. 24, 1871	111,228
Plow and cultivator, Combined gang	O. Sampson	Petersburgh, Ill	Feb. 21, 1871	112,079
Plow and cultivator, Combined gang	W. W. St. John	Saint Louis, Mo	Dec. 18, 1866	60,589
Plow and cultivator, Combined gang	S. C. Thornton	Macomb, Tex	Aug. 23, 1870	106,743
Plow and cultivator, Combined shovel	A. Smith	Providence Township, Pa	Sept. 9, 1873	142,655
Plow and cultivator, Combined sulky	W. G. Crossley	Shullsburgh, Wis	Jan. 7, 1868	72,981
Plow and cultivator, Convertible	J. Orr and H. H. Martin	Oxford, Ohio	Oct. 27, 1868	83,404
Plow and cultivator, Convertible	C. L. Reid	Louisville, Ky	Mar. 29, 1870	101,380
Plow and cultivator, Corn	I. B. Arthur	Sidonsburgh, Pa	Feb. 4, 1868	74,032
Plow and cultivator, Cotton	A. Runstetler and A. Windeck	Peoria, Ill	Feb. 4, 1868	74,005
Plow and cultivator fender	J. C. and W. F. Curryer	Thorntown, Ind	July 5, 1870	105,047
Plow and cultivator, Gang	S. H. Mitchell	El Paso, Ill	Feb. 14, 1865	46,378
Plow and cultivator, Gang	G. W. Price	Bloomington, Ill	Jan. 29, 1867	61,566
Plow and cultivator, Shovel	B. F. Ream	Ada, Ohio	Feb. 18, 1868	74,721
Plow and cultivator, Shovel	E. G. and A. D. Rowell, J. Rice, and S. M. Seeley	Hartford, Wis	Nov. 1, 1870	108,834
Plow and cultivator, Steam	P. H. Standish	Martinez, Cal	Mar. 10, 1868	75,310
Plow and cultivator, Steam-propeller	J. Marquis	San Francisco, Cal	Sept. 29, 1868	82,538
Plow and grain-drill, Combined corn	C. Hollar	Abington, Ind	July 19, 1870	105,575
Plow and gun, Combined	C. M. French and W. H. Fancher	Waterloo, N. Y	June 17, 1862	35,600
Plow and harrow	J. Haessel	Saint Louis, Mo	Oct. 20, 1868	83,154
Plow and harrow	H. Higley	Canaan, Conn	Dec. 11, 1832	
Plow and harrow	P. Wonsey	Ogden, N. Y	Feb. 4, 1868	74,024
Plow and harrow, Combined	J. F. Braucher	Lincoln, Ill	Oct. 4, 1870	107,862
Plow and harrow, Combined	J. Harsha	Circleville, Ohio	Jan. 7, 1868	73,002
Plow and harrow, Combined	A. Moore and F. Wendel	Chillicothe, Ohio	Oct. 19, 1869	95,925
Plow and harrow, Sulky	C. N. Bakewell	Normal, Ill	Mar. 24, 1868	75,832
Plow and harrow, Sulky	J. E. Cheasebro	Buffalo, N. Y	Oct. 16, 1866	58,773
Plow and hedge-grader, Ditching	D. Harmon	Coles County, Ill	Apr. 12, 1870	101,729
Plow and hoe, Combined	D. W. Colburn	Loami, Ill	Sept. 3, 1867	68,489
Plow and marker, Combined	O. M. Pond	Independence, Iowa	May 2, 1871	114,472
Plow and marker, Combined corn	D. Wyant	West Lodi, Ohio	Jan. 19, 1869	85,982
Plow and marker, Corn	G. W. Meixell	Hecktown, Pa	Oct. 22, 1872	132,405
Plow and marker, Garden	H. Haynsworth	Sumter, S. C	Dec. 7, 1869	97,508
Plow and planter	G. C. Avery	Waldron, Ind	June 9, 1868	78,638
Plow and planter	P. Kling	Springfield, Ill	Oct. 4, 1870	108,062
Plow and planter, Combined	J. D. Marshall	Renick, Mo	Oct. 29, 1867	70,237
Plow and planter, Combined	M. Shackleford	Montgomery, Ala	Feb. 2, 1869	86,458
Plow and planter, Combined	I. H. Walker	Newton, Ill	Aug. 4, 1868	80,686
Plow and planter, Corn	J. H. Hannon	Halifax, N. C	Oct. 18, 1870	108,476
Plow and planter, Corn	H. C. Osborn	Clarke County, Ohio	Sept. 27, 1870	107,709
Plow and planter, Corn	C. Sintz	Clarke County, Ohio	Sept. 6, 1870	107,110
Plow and planter, Cotton	Z. B. Sims	Bonham, Tex	Aug. 31, 1869	94,349
Plow and planter, Seeding	J. B. and J. C. Watt	Greene County, Ohio	Aug. 30, 1870	106,896
Plow and rail-cleaner, Snow	L. J. Woodruff	Mohawk, N. Y	May 9, 1871	114,633

Index of patents issued from the United States Patent Office from 1790 *to* 1873, *inclusive*—Continued.

Invention.	Inventor.	Residence.	Date.	No.
Plow and road-scraper, Combined	T. B. Parker	Linesville, Pa	Oct. 11, 1870	108, 176
Plow and roller, Combined	J. A. Alley	Clifty, Ind	Feb. 11, 1868	74, 267
Plow and scraper, Combined	J. C. Cameron	Madison Station, Miss	Apr. 4, 1871	113, 625
Plow and scraper, Combined	R. Elliot	Plainfield, N. J	Jan. 2, 1866	51, 817
Plow and scraper, Combined	J. Reynolds	Crystal Springs, Miss	Sept. 14, 1869	94, 774
Plow and scraper, Combined cotton	T. P. Warren	Norfolk, Va	June 18, 1867	65, 847
Plow and scraper, Combined subsoil	J. Reynolds	Crystal Springs, Miss	Feb. 22, 1870	100, 188
Plow and scraper, Grading	A. P. Hopkins	Bentleyville, Pa	Jan. 26, 1869	86, 229
Plow and scraper, Railway snow	E. Trenholm	Washington, D. C	June 23, 1863	38, 996
Plow and seed-planter, Combined	W. Croasdale	Hartsville, Pa	Nov. 27, 1849	6, 908
Plow and seeder, Combined	W. C. Bibb	Madison, Ga	Apr. 4, 1871	113, 482
Plow and seeder, Gang	D. Stone	Pittsford, N. Y	Aug. 9, 1864	43, 804
Plow and stock for cultivator, Combined	M. Kennedy	Chicago, Ill	Oct. 3, 1871	119, 618
Plow and subsoiler	G. Robinson	Plymouth, Ohio	Mar. 9, 1869	87, 594
Plow and tobacco-hiller attachment, Sulky	J. L. Spencer	Wellville, Va	July 16, 1867	66, 748
Plow, Angular	J. Lupton	Virginia	July 31, 1817	....
Plow-attachment	W. Bennett	Rushville, Ind	Jan. 7, 1868	73, 158
Plow-attachment	A. E. Cruttenden	Canasaragra, N. Y	Feb. 2, 1869	86, 370
Plow-attachment	W. J. Martin	Catawissa, Pa	Feb. 11, 1868	74, 398
Plow-attachment	J. W. Monical	Mooresville, Ind	Jan. 5, 1869	85, 601
Plow-attachment	M. Roberts	Saint Paul, Minn	Apr. 3, 1866	53, 749
Plow-attachment	H. B. Smith	Eureka, Ill	Feb. 5, 1867	61, 885
Plow-attachment	C. E. Wilson	Palmyra, Me	Sept. 15, 1868	82, 189
Plow-attachment, (double-tree)	L. E. Morey	Vandalia, Ill	Dec. 8, 1868	84, 750
Plow-attachment for cutting stubble	J. T. Hovis	Clintonville, Pa	June 27, 1871	116, 446
Plow-attachments, Making blanks for	O. A. Anthony	Mayfield, N. Y	Aug. 13, 1872	130, 403
Plow, Auburn	W. F. Richardson	Auburn, N. Y	June 28, 1825	....
Plow, Automatic spading	F. L. Cagwin	Joliet, Ill	Oct. 27, 1868	83, 456
Plow, Bar-share	J. and J. Butler	Brandywine and Chester County, Pa.	Mar. 1, 1814	....
Plow, Bar-share	J. Deakyne	Petersburgh, Va	Nov. 18, 1828	....
Plow, Bar-share	M. Huffman	Elizabeth, Tenn	Dec. 28, 1832	....
Plow, Bar-share	A. Mitchell	Washington County, Tenn	Oct. 1, 1830	....
Plow, Bar-share	E. Pugh	Chatham County, N. C	Dec. 24, 1827	....
Plow, Bar-share	E. M. Waggenet	Adair County, Ky	Jan. 12, 1831	....
Plow, Bar-share or fallow	G. Davis	Georgetown, D. C	Mar. 6, 1833	....
Plow-beam	E. Bement	Fostoria, Ohio	Sept. 16, 1862	36, 446
Plow-beam	W. Eddy	Greenwich, N. Y	June 4, 1867	65, 357
Plow-beam	E. T. Ford	Stillwater, N. Y	May 6, 1862	35, 147
Plow-beam	W. Gilman	Ottawa, Ill	July 30, 1867	67, 188
Plow-beam	L. B. Griffith	Honey Brook, Pa	Oct. 4, 1853	10, 085
Plow-beam	J. Houghton	Ogden, N. Y	June 22, 1842	2, 691
Plow-beam	S. H. Miller	Hamilton, Ill	Feb. 14, 1871	111, 764
Plow-beam blanks, Rolled bar for	G. W. Jope and W. Bunton	Pittsburgh, Pa	Dec. 20, 1870	110, 371
Plow-beam-clevis attachment	J. L. Baldwin	Troy, Pa	July 20, 1869	92, 775
Plow beam, Gang	J. W. Sursa	San Francisco, Cal	Sept. 20, 1870	107, 560
Plow-beam handle	M. C. Brelsford	Girard, Ill	Jan. 5, 1864	41, 056
Plow-beam, Iron	A. Ball	Canton, Ohio	Jan. 3, 1871	110, 722
Plow-beam-rolling machine	W. Gilman	Ottawa, Ill	July 26, 1870	105, 668
Plow-beams, Attaching colters to	J. H. Hall	Maysville, Ky	Mar. 10, 1868	75, 414
Plow-beams, carriage-shafts, &c., Machine for hewing.	E. G. Matthews	Worcester, Mass	Oct. 22, 1840	1, 836
Plow-beams, Metallic fastening for	L. Gibbs	Canton, Ohio	Apr. 10, 1866	53, 918
Plow, Bogging and ditching	J. Gordon	Copake, N. Y	Jan. 13, 1829	....
Plow-bolt	C. A. Harper	Burlington, N. J	Mar. 4, 1873	136, 508
Plow-brace die	G. Moore	Moline, Ill	July 23, 1867	67, 131
Plow, Bull	B. Johnson	Hickory Grove, Ill	Feb. 20, 1835	....
Plow, Bull	R. Tousley and J. Swan	....	Nov. 9, 1814	....
Plow, Bull	S. Tousley	Onondaga County, N. Y	May 28, 1813	....
Plow, Carpenter's	C. G. Miller	New Britain, Conn	Sept. 17, 1872	131, 367
Plow, Carpenter's	R. B. Rice	Williamsburgh, Mass	Dec. 21, 1869	98, 108
Plow, Carpenter's	H. Vanbuskirk	Vienna, Mich	Nov. 30, 1869	97, 328
Plow, Carriage	H. M. Bullitt	Louisville, Ky	Oct. 7, 1873	143, 434
Plow, Carriage	J. M. Hammitt and H. T. Miller	Toledo, Iowa	Jan. 1, 1867	60, 721
Plow, Carriage	A. Ingalls	Independence, Iowa	May 29, 1866	55, 108
Plow, Carriage	W. L. Jeffries	Lancaster, Ohio	Mar. 30, 1869	88, 486
Plow, Carriage	H. Minuse	Milan, Ohio	Sept. 10, 1867	68, 777
Plow, Carriage	G. F. Willey	Laconia, N. H	Mar. 31, 1868	76, 283
Plow-carrier and cotton-chopper, Combined	F. L. Kirtley	Cleyborne, Tex	May 18, 1869	90, 274
Plow, Cast-iron	T. Altason	Newton Township, N. J	Apr. 16, 1824	....
Plow, Cast-iron	D. Barnard	Washington, Del	Apr. 21, 1830	....
Plow, Cast-iron	D. Chase and A. and J. Gregg	Palmyra, N. Y	Aug. 4, 1824	....
Plow, Cast-iron	J. Dutcher	Durham, N. Y	Feb. 12, 1822	....
Plow, Cast-iron	T. Fairbanks	Saint Johnsbury, Vt	Apr. 19, 1826	....
Plow, Cast-iron	W. Falconner	New York	Sept. 13, 1821	....
Plow, Cast-iron	G. W. Hawkins and H. Emery	Windsor County, Vt	Feb. 16, 1822	....
Plow, Cast-iron	D. Hitchcock	New York	Dec. 29, 1821	....
Plow, Cast-iron	R. McMillen	Middlebury, Ohio	Dec. 14, 1841	2, 389
Plow, Cast-iron	J. Minturn	Urbana, Ohio	June 11, 1829	....
Plow, Cast-iron	J. K. Odell and W. S. Little	Deckertown, N. J	Dec. 6, 1870	109, 929
Plow, Cast-iron	J. J. Schoonmaker and J. Dolson.	Ulster County and Dutchess, N. Y.	Dec. 7, 1822	....
Plow, Cast-iron	J. Seaver and J. Fay	Rockingham, Vt	Dec. 14, 1820	....
Plow, Cast-iron	O. Seely	Pottstown, N. Y	Aug. 27, 1821	....
Plow, Cast-iron	R. L. and E. A. Stevens	Hoboken, N. J	Apr. 23, 1821	....
Plow, Cast-iron	R. Sweeny	Warren County, Ohio	May 18, 1827	....
Plow, Cast-iron	A. Taylor	Green Township, Ohio	Dec. 26, 1833	....
Plow, Cast-iron	R. Winans	Vernon, N. Y	May 11, 1824	....
Plow-castings	F. F. Smith	Collinsville, Conn	May 16, 1865	47, 753
Plow-castings, Chill for	J. Oliver	South Bend, Ind	May 2, 1871	114, 479
Plow-castings, Machine for grinding	J. Gibbs	Canton, Ohio	Oct. 4, 1853	10, 068
Plow-cleaner	G. W. Burr	East Line, N. Y	Dec. 26, 1871	122, 155
Plow-cleaner	C. P. Devereaux	North Newburgh, Mich	July 16, 1867	66, 809
Plow-cleaner	D. D. Gitt	Butler Township, Pa	May 7, 1850	7, 344
Plow-cleaner	J. and E. P. Miles	Pleasant Hill, Ohio	June 29, 1869	91, 957
Plow-cleaner	J. A. B. Patterson	Springfield, Ill	Apr. 20, 1869	89, 237
Plow-cleaner	J. B. Place	Auburn Township, Pa	Nov. 5, 1872	132, 857
Plow-cleaner	J. F. Reasin	Darlington, Md	Apr. 9, 1850	7, 274
Plow-cleaner	D. Warren	Gettysburgh, Pa	Sept. 10, 1850	7, 640

Index of patents issued from the United States Patent Office from 1790 *to* 1873, *inclusive*—Continued.

Invention.	Inventor.	Residence.	Date.	No.
Plow-cleaner	O. F. Warren	Pembroke, N. Y	July 10, 1866	56, 296
Plow-cleaning attachment	T. Terrel	Spring Hill, Ohio	Aug. 14, 1866	57, 216
Plow-clearer	A. C. Black	Kaukauna, Wis	May 25, 1869	90, 488
Plow-clevis	C. Adams	Pittsburgh, Pa	May 22, 1860	28, 337
Plow-clevis	L. W. Alden	Fosterville, N. Y	Jan. 23, 1866	52, 123
Plow-clevis	H. W. Austin and W. Schaw	Kalamazoo, Mich	Sept. 3, 1867	68, 545
Plow-clevis	W. Axford	Carrollton, Ill	Mar. 5, 1872	124, 313
Plow-clevis	J. Ball	Canton, Ohio	Sept. 1, 1868	81, 731
Plow-clevis	J. A. Bilz	Pleasanton, Cal	May 18, 1869	90, 068
Plow-clevis	R. A. Blair and J. B. Reed	New Philadelphia, Ohio	Aug. 20, 1861	33, 066
Plow-clevis	T. E. C. Brinly	Louisville, Ky	May 3, 1870	102, 652
Plow-clevis	J. Brison	Competine, Iowa	Oct. 3, 1871	119, 502
Plow-clevis	A. Carman	Columbus, N. J	June 20, 1838	791
Plow-clevis	A. A. Dailey	Wilson, N. Y	Mar. 5, 1872	124, 339
Plow-clevis	G. P. Darrow	Cincinnati, Ohio	Aug. 21, 1866	57, 436
Plow-clevis	T. Don	Yorktown, Ill	Aug. 31, 1869	94, 296
Plow-clevis	G. Erkson	Hobart, N. Y	Sept. 17, 1850	7, 651
Plow-clevis	P. Gallagher	Chambersburgh, Pa	Nov. 26, 1845	4, 292
Plow-clevis	J. S. Hall	Manchester, Pa	Dec. 4, 1860	30, 873
Plow-clevis	H. Ingraham	Naples, N. Y	Aug. 27, 1867	68, 200
Plow-clevis	A. Kaufman	Davisville, Cal	Mar. 25, 1873	137, 078
Plow-clevis	J. Keech	Waterloo, N. Y	May 5, 1863	38, 394
Plow-clevis	W. Lahman	Wellsville, Ohio	July 8, 1873	140, 715
Plow-clevis	E. G. Matthews	Newton, Mass	Mar. 2, 1869	87, 421
Plow-clevis	A. McCollam	New Orleans, La	July 18, 1871	117, 187
Plow-clevis	C. L. Meech	Preston, Conn	May 28, 1846	4, 544
Plow-clevis	J. Napier	Martin's Ferry, Ohio	Aug. 6, 1872	130, 145
Plow-clevis	J. Newhart	Terre Haute, Ind	Oct. 29, 1867	70, 351
Plow-clevis	F. Reese	Wilsonville, Ala	Dec. 10, 1872	133, 799
Plow-clevis	R. Sandiford	Joliet, Ill	June 2, 1868	78, 542
Plow-clevis	A. Shogren	Sandwich, Ill	July 18, 1865	48, 844
Plow-clevis	D. Stewart	Corinna, Me	Oct. 6, 1868	82, 764
Plow-clevis	E. Stewart	Fort Madison, Iowa	July 16, 1872	129, 614
Plow-clevis	J. B. Stoner	Southampton, Pa	Sept. 10, 1850	7, 638
Plow-clevis	J. H. Sarpley	Greensborough, N. C	Dec. 14, 1869	97, 990
Plow-clevis	J. Van Brocklin	Middleport, N. Y	Mar. 13, 1847	5, 016
Plow-clevis	J. D. Willoughby	Pleasant Hall, Pa	June 2, 1857	17, 462
Plow-clevis	S. Wilt	Hagerstown, Md	Sept. 3, 1846	4, 730
Plow-clevis	E. A. Wright	Fort Madison, Iowa	June 18, 1872	128, 087
Plow-clevis attachment	H. C. Sieverling	Carrollton, Ill	Dec. 13, 1870	110, 080
Plow-clevis attachment	G. Watt	Richmond, Va	July 6, 1869	92, 409
Plow-clevis pin	H. McKiney	Earlham, Iowa	July 16, 1872	129, 355
Plow-clevises, Machine for making	T. Meikle	Louisville, Ky	Mar. 3, 1868	75, 179
Plow-clevises, Spring-link for	J. Ball	Canton, Ohio	Mar. 2, 1869	87, 458
Plow clod-fender	J. and J. W. Higgins	Orth, Ind	Oct. 26, 1869	96, 234
Plow clod-fender	R. A. Kelly	Hope, Ind	July 6, 1869	92, 318
Plow clod-fender	W. B. Kidder	Pike Township, Ind	Aug. 2, 1870	106, 064
Plow-colter	S. Casebeer	Roseburgh, Oreg	Aug. 21, 1866	57, 286
Plow-colter	G. Curkendall	Dixon, Ill	May 17, 1870	103, 148
Plow-colter	J. Custer and C. Rowland	Clinton, Ill	June 12, 1866	55, 472
Plow-colter	C. Furst	Chicago, Ill	Nov. 9, 1869	96, 575
Plow-colter	D. D. Gibson	Springfield, Iowa	Oct. 17, 1871	120, 055
Plow-colter	C. M. Gordon	La Porte, Ind	Apr. 4, 1871	113, 421
Plow-colter	E. C. Hodge	Oneonta, N. Y	Feb. 21, 1871	112, 039
Plow-colter	W. H. Hoefelman	Columbus, Nebr	Aug. 20, 1872	130, 638
Plow-colter	H. S. Hoxie	Adrian, Mich	May 5, 1868	77, 616
Plow-colter	H. S. Hoxie	Raisin Centre, Mich	May 16, 1871	114, 822
Plow-colter	J. Lane	Chicago, Ill	Mar. 30, 1869	88, 309
Plow-colter	J. Pinkham	New Market, N. H	Feb. 20, 1872	123, 838
Plow-colter	F. F. Smith	Collinsville, Conn	Jan. 19, 1869	85, 971
Plow-colter	W. H. Stone	Lebanon, Mich	Apr. 25, 1871	114, 220
Plow-colter	B. W. Tuttle	Galena, Ill	May 14, 1872	126, 655
Plow-colter	S. Way	La Porte, Ind	Jan. 24, 1871	111, 158
Plow-colter	E. Wiard	Louisville, Ky	Aug. 26, 1873	142, 310
Plow colter and share	S. A. Sperry	Ann Arbor Village, Mich	Oct. 27, 1835	....
Plow-colter, Revolving	H. Delano	Mottsville, N. Y	Apr. 1, 1842	2, 529
Plow-colter, Revolving	M. Richards	Princeton, Ill	Apr. 5, 1870	101, 509
Plow-colter, Revolving	M. Sattley	Taylorsville, Ill	Nov. 24, 1868	84, 380
Plow-colters, Apparatus for cleaning	E. C. Bills, jr	Perry, N. Y	Oct. 21, 1856	15, 919
Plow-colters, Attaching	D. W. Hughes	Palmyra, Mo	Aug. 17, 1869	93, 888
Plow, Combined iron and steel	W. Morrisson	Chadd's Ford, Pa	Jan. 28, 1862	34, 262
Plow, Combined single and double shovel	J. W. Nicholson	Indianapolis, Ind	Feb. 2, 1869	86, 577
Plow, Combined subsoil, drill, and side-hill	W. W. Speer	Allegheny City, Pa	Sept. 26, 1871	119, 423
Plow, Combined turn and subsoil	J. C. Gross	Goshen Hill, S. C	Aug. 31, 1869	94, 307
Plow, Common	N. Turbutt	Fredericktown, Md	Sept. 7, 1811	....
Plow, Corn	P. Barnhart	Chillicothe, Ohio	Oct. 23, 1866	58, 973
Plow, Corn	W. H. Bott	York, Pa	Nov. 9, 1869	96, 667
Plow, Corn	W. B. Broadwell	Springfield, Ill	June 3, 1862	35, 434
Plow, Corn	M. C. Buffington	La Harpe, Ill	Sept. 24, 1867	69, 071
Plow, Corn	M. C. Buffington	La Harpe, Ill	Apr. 12, 1870	101, 706
Plow, Corn	A. Canfield	Lyons, Iowa	Aug. 20, 1867	67, 843
Plow, Corn	A. Canfield	Lyons, Iowa	Aug. 20, 1867	67, 844
Plow, Corn	L. H. Castor	Edington, Ill	June 28, 1864	43, 288
Plow, Corn	J. M. Clark	Somerville, Ohio	June 11, 1867	65, 725
Plow, Corn	W. C. Clifton	Elk River Township, Iowa	Oct. 11, 1870	108, 239
Plow, Corn	S. Coats	La Fayette, Wis	June 5, 1849	6, 501
Plow, Corn	S. H. Cox and W. H. Pence	Mattoon, Ill	Jan. 7, 1868	73, 166
Plow, Corn	T. Dillon	Highland, Ohio	Nov. 17, 1868	84, 094
Plow, Corn	N. Etheridge, G. Heath, and I. G. Glynn	Little Falls, N. Y	July 23, 1831	....
Plow, Corn	W. and J. C. French	Keokuk, Iowa	Nov. 22, 1870	109, 403
Plow, Corn	M. B. Goff	Delavan, Wis	June 28, 1870	104, 729
Plow, Corn	L. Guthrie	Waterloo, Ind	Sept. 7, 1869	94, 489
Plow, Corn	T. W. Hammon	Montfort, Wis	Mar. 7, 1865	46, 752
Plow, Corn	H. Harrier	Indianapolis, Ind	Aug. 23, 1870	106, 693
Plow, Corn	J. Hernly	East Hempfield, Pa	Dec. 10, 1838	1, 031
Plow, Corn	J. Hindmarsh	Henry, Ill	Jan. 15, 1867	61, 195
Plow, Corn	R. C. Howard	Lena, Ill	Oct. 23, 1866	59, 022

Index of patents issued from the United States Patent Office from 1790 to 1873, inclusive—Continued.

Invention.	Inventor.	Residence.	Date.	No.
Plow, Corn	R. J. King	Lancaster, Pa	Mar. 5, 1850	7, 141
Plow, Corn	J. Marsh	Seneca, Ill	Sept. 24, 1867	69, 109
Plow, Corn	A. McCreight	Tranquility, Ohio	Sept. 1, 1868	81, 660
Plow, Corn	A. D. Michener and J. W. Steigmeyer.	Attica, Ohio	June 15, 1869	91, 471
Plow, Corn	S. J. Miller and L. Wright	Economy, Ind	Dec. 22, 1868	85, 118
Plow, Corn	F. G. Mourning	Bascon, Ill	Sept. 13, 1870	107, 400
Plow, Corn	C. J. Paine	Young America, Ill	Oct. 27, 1868	83, 537
Plow, Corn	C. E. Paxson	Salem, Ohio	Nov. 26, 1861	33, 793
Plow, Corn	E. Phillips	Centreville, Ind	Dec. 6, 1870	109, 934
Plow, Corn	J. Preston	Plymouth, Ohio	May 13, 1873	138, 818
Plow, Corn	W. B. Raper	Carthage, Ill	Jan. 11, 1870	98, 708
Plow, Corn	M. Redlinger	Freeport, Ill	July 2, 1867	66, 253
Plow, Corn	B. W. and N. T. Remy	Brookville, Ind	Mar. 28, 1871	113, 206
Plow, Corn	W. C. Rhinehart and R. Gaston.	Oskaloosa, Iowa	Dec. 8, 1868	84, 763
Plow, Corn	J. Snyder	Williamsfield, Ohio	Dec. 10, 1867	72, 104
Plow, Corn	J. R. Thomas	Mifflintown, Pa	Sept. 10, 1867	68, 670
Plow, Corn	J. L. Walter and E. Bushman	White House, Pa	Nov. 1, 1870	108, 857
Plow, Corn	W. S. Weir, jr	Monmouth, Ill	Feb. 7, 1865	46, 285
Plow, Corn	D. Wilde	Washington, Iowa	May 15, 1866	54, 860
Plow, Corn	J. Wolf	Young America, Ill	Apr. 21, 1868	76, 964
Plow, Corn and cotton	J. W. Milroy	Galveston, Ind	Dec. 29, 1868	85, 324
Plow, Corn and cotton cultivating	W. R. Blanchard	Hertford, N. C	Aug. 2, 1870	105, 892
Plow, Corn and potato	W. Notman	Deerfield, Ohio	Nov. 24, 1868	84, 437
Plow, Corn-dressing	E. Bean	Massachusetts	Mar. 4, 1816	....
Plow, Cotton	H. C. Godfrey	Elizabeth City, N. C	Dec. 10, 1872	133, 850
Plow, Cotton	W. W. Graves	Fort Adams, Miss	Nov. 27, 1860	30, 729
Plow, Cotton	T. Guice	Mount Andrew, Ala	Dec. 6, 1870	109, 893
Plow, Cotton, &c	H. W. Pitts	Wilsonville, Ala	Mar. 31, 1836	....
Plow, Cotton	Z. B. Sims	Bonham, Tex	Aug. 31, 1869	94, 350
Plow, Cotton	Z. B. Sims	Bonham, Tex	Aug. 31, 1869	94, 351
Plow, Cotton	T. A. Wainwright	Wilson, N. C	July 7, 1868	79, 706
Plow, Cotton	J. Wiley	Warsaw, N. C	Dec. 20, 1870	110, 411
Plow, Cotton	W. B. Williams	Warrenton, N. C	Aug. 13, 1867	67, 831
Plow, Cotton and corn	D. C. Richardson	Weldon, N. C	Oct. 22, 1867	70, 120
Plow, Cotton-scraping	J. M. Cobb	Jackson, Tenn	Nov. 11, 1873	144, 509
Plow, Cotton-thinning	W. Crichton	Diamond Grove, Va	Oct. 23, 1847	5, 340
Plow, Cotton-thinning	L. B. Joyner	Hilliardston, N. C	Apr. 24, 1860	27, 987
Plow-coupling	G. Owen	Jacksonville, Ill	Feb. 28, 1871	112, 273
Plow-coupling	J. K. Pitts	McLean, Ill	Jan. 28, 1873	135, 360
Plow-coupling	W. Reck	Mendota, Ill	Aug. 31, 1869	94, 339
Plow-coupling	T. L. Thrasher	Paris, Tex	Aug. 19, 1873	141, 964
Plow coupling, Double	G. Owen	Jacksonville, Ill	Feb. 4, 1862	34, 316
Plow, Covering	W. Roberts	Rocheport, Mo	Jan. 15, 1861	31, 130
Plow, Cultivating	W. Bagnall	Otsego, Ohio	Nov. 4, 1873	144, 247
Plow, Cultivating	A. Hughes	Gratiot, Ohio	May 29, 1860	28, 487
Plow, Cultivator	S. B. Forbes	Steubenville, Ohio	Mar. 23, 1869	88, 025
Plow, Cultivator	B. F. and J. V. Guy	Macomb, Ill	Oct. 13, 1868	82, 938
Plow, Cultivator	T. F. Jones	Hicksford, Va	Nov. 29, 1870	109, 629
Plow, cultivator, and marker, Combined	J. Dooley	Saint Paul, Minn	July 13, 1869	92, 590
Plow, cultivator, and planter, Combined	J. Rider	Woodburn, Ill	Mar. 12, 1841	2, 002
Plow, cultivator, and potato-digger, Combined	H. B. Smith	Tremont, Ill	July 27, 1869	93, 154
Plow, Cultivator gang	H. P. McCleave	Tomales, Cal	Nov. 2, 1869	96, 334
Plow, Cultivator gang	T. T. Prosser and M. C. and K. A. Darling.	Chicago, Ill., and Fond du Lac, Wis.	Jan. 3, 1865	45, 750
Plow, cultivator, &c., Shovel	P. A. Ross	Harvey's, Pa	Dec. 10, 1867	72, 085
Plow-cutter	G. F. Pykiet	Fairfield, Ill	Dec. 20, 1870	110, 283
Plow cutter and point	D. H. Jones and W. F. Richardson.	Palmyra, N. Y	July 29, 1823	....
Plow, Digging	E. Peck	Deer Park, N. Y	Nov. 10, 1857	18, 600
Plow, Ditching	E. S. Bartlett	Romulus, N. Y	Sept. 13, 1859	25, 377
Plow, Ditching	J. Brooks	Romulus, N. Y	July 31, 1860	29, 357
Plow, Ditching	T. I. Burhyte	Fond du Lac, Wis	Nov. 21, 1865	51, 014
Plow, Ditching	W. Burton	Lake, Ind	Feb. 13, 1872	123, 613
Plow, Ditching	G. Chamberlin	Olean, N. Y	Sept. 6, 1870	107, 162
Plow, Ditching	W. R. Clark	Indianola, Ill	Sept. 10, 1867	68, 697
Plow, Ditching	E. L. Foreman	Rantoul, Ill	Jan. 26, 1869	86, 294
Plow, Ditching	J. Kelley and W. H. Hennis	Winomac, Ind	Jan. 30, 1872	123, 262
Plow, Ditching	J. Lyon	Harrisburgh, Iowa	July 25, 1854	11, 382
Plow, Ditching	J. T. Miller	Iowa Falls, Iowa	Feb. 19, 1867	62, 215
Plow, Ditching	F. Pimmer	Grand Junction, Tenn	Feb. 12, 1861	31, 399
Plow, Ditching	R. M. Primmer	Vinton, Iowa	Aug. 8, 1871	117, 813
Plow, Ditching	I. S. Sheets	Troy, Ohio	Feb. 16, 1869	86, 947
Plow, Ditching	I. S. Sheets	Troy, Ohio	May 3, 1870	102, 599
Plow, Ditching	D. S. Stafford	Rochester, Ill	July 10, 1845	4, 107
Plow, Ditching	J. C. Tiffany	Coxsackie, N. Y	Apr. 25, 1854	10, 815
Plow, Ditching	W. West	Pecksburgh, Ind	Mar. 16, 1869	87, 892
Plow, Ditching	H. D. Williams	Fair View, Iowa	Nov. 7, 1871	120, 690
Plow, Ditching	J. L. Wilson and J. R. Haworth	Iowa Falls, Iowa	Jan. 28, 1868	73, 858
Plow, Ditching	S. S. Wood	Brooklyn, N. Y	Aug. 16, 1870	106, 443
Plow, Ditching and mole	E. H. Morton	Oxford, Iowa	Jan. 3, 1865	45, 735
Plow, Double	P. and B. Altenderfer	Coots Town, Pa	June 11, 1829	....
Plow, Double	J. Cromwell	Roystertown, Md	Oct. 23, 1816	....
Plow, Double	P. Eichar	Wooster, Ohio	Dec. 5, 1843	3, 372
Plow, Double	A. Peebles	Henry County, Ga	July 20, 1831	....
Plow, Double	A. Smith	Bloomfield, Mich	May 10, 1844	3, 579
Plow, Double	G. W. N. Yost	Nashville, Tenn	Aug. 11, 1863	39, 537
Plow, Double corn	L. Guthrie	Waterloo, Ind	Sept. 7, 1869	94, 490
Plow, Double-mold-board	S. Gregory	Sawpitts, N. Y	Nov. 14, 1838	1, 008
Plow, Double revolving	J. Mott	Danville, Cal	Aug. 14, 1866	57, 171
Plow, Double-shovel	J. Balltlirope	Loudoun County, Va	Sept. 20, 1820	....
Plow, Double-shovel	J. Clarridge	Pancoastville, Ohio	Oct. 9, 1866	58, 597
Plow, Double-shovel	A. J. Craig	Ashmore, Ill	Apr. 28, 1868	77, 260
Plow, Double-shovel	J. M. Eby	Warren, Ill	Apr. 9, 1867	63, 711
Plow, Double-shovel	I. G. Flisher and E. M. Bates	Stark County, Ohio	Apr. 14, 1868	76, 736
Plow, Double-shovel	G. W. Lawbaugh	Geneseo, Ill	Mar. 1, 1870	100, 298
Plow, Double-shovel	G. Perry	Granville, Ill	Apr. 19, 1870	102, 150
Plow, Double-shovel	S. G. Rayl	Agency City, Iowa	Sept. 13, 1870	107, 409

Index of patents issued from the United States Patent Office from 1790 to 1873, inclusive—Continued.

Invention.	Inventor.	Residence.	Date.	No.
Plow, Double-shovel	M. R. Shalters and S. Ray	Alliance, Ohio	Mar. 24, 1868	75, 988
Plow, Double-shovel	L. L. Sloss	South Union, Ky	Aug. 27, 1867	68, 247
Plow, Double-shovel	H. Stephens	Mount Vernon, Ohio	May 7, 1867	64, 592
Plow, Double-shovel	R. P. Van Horne	Gratiot, Ohio	Feb. 14, 1871	111, 793
Plow, Double-shovel	C. I. Voigt	West Salem, Ill	Oct. 5, 1869	95, 541
Plow, Drain	D. Ballard	Texas, Ohio	July 21, 1868	80, 113
Plow, Drain	M. Barrowman	Buffalo, N. Y	June 29, 1858	20, 689
Plow, Drain	R. W. Downman	Georgetown, D. C	June 25, 1867	66, 133
Plow, Drain	C. G. Grabo	Greenfield, Mich	May 26, 1863	38, 677
Plow, Drain	J. Hanon, jr	Taylorsville, Ill	Mar. 27, 1860	27, 630
Plow, Draining	S. Hall	Boston, Mass	July 13, 1810	....
Plow, Draining and ditching	H. B. Smawley	Greensburgh, Ind	Apr. 16, 1867	63, 952
Plow-drawing machine	L. Johnson	Leominster, Mass	Aug. 6, 1816	....
Plow drill and cotton-thinner	W. Willis	Edgefield District, S. C	May 7, 1822	....
Plow, Expanding	A. W. Wilkins and S. T. Eskridge.	Rome, Ga	Nov. 30, 1869	97, 465
Plow, Expanding double-shovel	E. Wiard	Louisville, Ky	Aug. 10, 1869	93, 507
Plow, Expanding triple-shovel	E. Wiard	Louisville, Ky	Dec. 21, 1869	98, 214
Plow-fastening	D. and S. Swartz	Tom's Brook, Va	Aug. 22, 1854	11, 575
Plow-fastening device	J. Robb	Lewistown, Pa	Oct. 12, 1852	9, 332
Plow-fender	B. F. Neely	Yorktown, Ind	July 25, 1871	117, 317
Plow-fender	S. J. Reed	Middletown, Ohio	Nov. 10, 1868	83, 899
Plow-fender	A. P. Webber	Saratoga Springs, Ill	Dec. 19, 1871	122, 087
Plow fender, Corn	R. Cook	Franklin, Ohio	Mar. 26, 1867	63, 140
Plow fender, Corn	A. B. Thornton	Berlin, Ill	Sept. 28, 1869	95, 287
Plow fender, Corn	L. C. Witt and W. F. Jones	Boston, Mass	Oct. 26, 1869	96, 293
Plow for breaking and cultivating sward-ground	G. Gray	Industry, Me	Sept. 18, 1835	....
Plow for cutting bogs	J. Coffee	Monroe, N. Y	Nov. 13, 1866	59, 557
Plow for cutting potato-roots	H. M. Keith	Commerce, Mich	Sept. 6, 1870	107, 063
Plow for draining and subsoiling	L. Green	Great Bend, Pa	July 23, 1861	32, 866
Plow for excavating ditches	J. Herbert	La Grange, Ind	Apr. 13, 1844	3, 538
Plow for listing and preparing land	J. Weaver	Washington, D. C	May 8, 1823	....
Plow for planting corn	H. Russell	Litchfield, Me	Jan. 16, 1827	....
Plow for planting cotton	B. Murphy	Union District, S. C	Dec. 31, 1827	....
Plow for planting potatoes	W. Price	Goldsborough, N. C	Aug. 29, 1854	11, 616
Plow, Four-wheel	N. B. Norton	Burlington, Wis	Sept. 15, 1868	82, 241
Plow-frame	S. March	Norfolk, Va	Apr. 28, 1868	77, 203
Plow, Furrowing	S. Perry	Troy, N. Y	July 10, 1866	56, 327
Plow, Furrowing	A. P. and C. F. Rohrer and J. H. Blose	Clarke County, Ohio	Dec. 26, 1871	122, 284
Plow-gage	J. Cluckner	Arcadia, Ind	Mar. 22, 1870	100, 984
Plow-gage	H. B. Spedden	Baltimore, Md	Aug. 3, 1869	93, 362
Plow-gage wheel	P. D. Beckwith	Dowagiac, Mich	Jan. 21, 1868	73, 564
Plow, Gang	S. H. Adams	Coultervillle, Ill	Nov. 8, 1864	44, 924
Plow, Gang	J. Alloways	Decatur, Ill	Aug. 2, 1870	105, 879
Plow, Gang	J. H. Andrews	Benicia, Cal	Sept. 1, 1868	81, 724
Plow, Gang	J. H. Andrews	Benicia, Cal	Sept. 6, 1870	106, 982
Plow, Gang	J. and G. Armstrong	Elmira, N. Y	Aug. 6, 1872	130, 177
Plow, Gang	C. Atwood	Lebanon, Ill	Dec. 19, 1865	51, 536
Plow, Gang	C. Atwood	Lebanon, Ill	Apr. 24, 1866	54, 089
Plow, Gang	C. Atwood	Lebanon, Ill	Oct. 18, 1870	108, 311
Plow, Gang	G. C. Avery	Cone's Creek, Ind	Aug. 27, 1867	68, 277
Plow, Gang	W. Battell	Quincy, Ill	Aug. 7, 1866	56, 878
Plow, Gang	R. Baxter	French Camp, Cal	Aug. 6, 1867	67, 483
Plow, Gang	D. Bequeret and E. Dumoulin	Jamestown, Ill	Nov. 20, 1866	59, 813
Plow, Gang	J. C. Bethea	Blakely, Ga	Feb. 5, 1867	61, 796
Plow, Gang	C. Bishop	Norwalk, Ohio	Oct. 12, 1852	9, 314
Plow, Gang	J. F. and W. L. Black	Lancaster, Ill	July 30, 1861	32, 958
Plow, Gang	J. F. and W. L. Black	Lancaster, Ill	Dec. 19, 1865	51, 543
Plow, Gang	J. Blackwood	Madison Township, Ill	Aug. 15, 1871	117, 970
Plow, Gang	S. B. Bowen and A. M. Abbot	Stockton, Cal	Mar. 12, 1872	124, 480
Plow, Gang	W. J. Boyce and G. W. Haines	Maine Prairie, Cal	Nov. 16, 1869	96, 875
Plow, Gang	G. T. Brewer	Prairie du Rocher, Ill	Oct. 29, 1867	70, 159
Plow, Gang	J. C. Brown and G. H. Slimpert	Pinckneyville, Ill	June 6, 1865	48, 049
Plow, Gang	A. H. Burlingame	Sparta, Ill	Aug. 14, 1866	57, 081
Plow, Gang	R. Carson	Meredosia, Ill	Sept. 21, 1869	95, 082
Plow, Gang	G. R. Carter	New York, N. Y	May 18, 1869	90, 237
Plow, Gang	L. Chapman	Collinsville, Conn	June 29, 1869	92, 016
Plow, Gang	L. Chapman	Collinsville, Conn	Apr. 4, 1871	113, 627
Plow, Gang	L. Chapman	Collinsville, Conn	Nov. 7, 1871	120, 572
Plow, Gang	F. G. Charles	Galesburgh, Ill	Feb. 20, 1872	123, 869
Plow, Gang	P. Conrath	Freeburgh, Ill	Aug. 17, 1869	93, 680
Plow, Gang	R. Coreth	West Belleville, Ill	Aug. 26, 1873	142, 084
Plow, Gang	T. J. Cornell	Decatur, Ill	May 15, 1866	54, 693
Plow, Gang	T. J. Cornell	Decatur, Ill	Oct. 2, 1866	58, 384
Plow, Gang	C. W. Corr	Carlinville, Ill	Aug. 1, 1865	49, 087
Plow, Gang	A. T. Covell	San Leandro, Cal	Aug. 6, 1867	67, 501
Plow, Gang	H. Cowing	New Orleans, La	Nov. 26, 1850	7, 795
Plow, Gang	J. and S. Cox	Eugene City, Oreg	Feb. 8, 1870	99, 538
Plow, Gang	M. A. Cravath	Lodi, Ill	Jan. 12, 1858	19, 077
Plow, Gang	F. R. Crothers	Sparta, Ill	Feb. 23, 1864	41, 686
Plow, Gang	F. R. Crothers	Sparta, Ill	June 29, 1869	92, 019
Plow, Gang	M. S. Curtiss	Bradford, Ill	Nov. 19, 1872	133, 206
Plow, Gang	M. S. Curtiss	Bradford, Ill	July 1, 1873	140, 480
Plow, Gang	G. J. Dahl	Stockton, Cal	Mar. 24, 1868	75, 871
Plow, Gang	H. N. Dalton	Pacheco, Cal	Oct. 2, 1866	58, 386
Plow, Gang	F. S. Davenport	Jerseyville, Ill	Feb. 9, 1864	41, 491
Plow, Gang	F. S. Davenport	Jerseyville, Ill	Oct. 9, 1866	58, 612
Plow, Gang	A. Davison	San Leandro, Cal	Mar. 30, 1869	88, 280
Plow, Gang	G. A. Davison	San Leandro, Cal	July 7, 1868	79, 639
Plow, Gang	J. N. Davison and N. Spencer, jr.	Buffalo, Ill	Nov. 3, 1868	83, 698
Plow, Gang	J. W. Donaldson, D. Sheets, and A. C. Miller.	Suisun, Cal	Sept. 4, 1866	57, 688
Plow, Gang	J. H. Douthit	Albany, Oreg	Jan. 22, 1867	61, 408
Plow, Gang	A. P. Durant	Atlanta, Ill	Feb. 6, 1866	52, 401
Plow, Gang	G. R. Duval	Salem, Oreg	Mar. 1, 1870	100, 383
Plow, Gang	C. L. Eastham	Rhodes Point, Ill	June 18, 1867	65, 798

Index of patents issued from the United States Patent Office from 1790 *to* 1873, *inclusive*—Continued.

Invention.	Inventor.	Residence.	Date.	No.
Plow, Gang	J. H. Eckert	Lebanon, Ill	Mar. 27, 1866	53, 424
Plow, Gang	A. Ellison	Marysville, Cal	Aug. 30, 1870	106, 799
Plow, Gang	C. A. Fargo	Soquel, Cal	June 1, 1869	90, 831
Plow, Gang	A. Farrow	Carrollton, Ill	Jan. 28, 1868	73, 707
Plow, Gang	D. M. Fay	Monmouth, Ill	Mar. 19, 1872	124, 802
Plow, Gang	P. M. Flansburgh	Hayward's, Cal	May 10, 1870	102, 801
Plow, Gang	M. Flinn	Saint Louis, Mo	Apr. 14, 1868	76, 735
Plow, Gang	W. Foster	Greenfield, Ind	Dec. 31, 1867	72, 730
Plow, Gang	A. Freeman	Homer, Ill	Apr. 2, 1872	125, 289
Plow, Gang	W. H. Freeman	Bloomfield, Iowa	Jan. 24, 1865	45, 988
Plow, Gang	J. Frye	Springfield, Ill	Mar. 31, 1857	16, 912
Plow, Gang	J. Frye	Mendota, Ill	May 25, 1858	20, 342
Plow, Gang	W. J. Funk	Portland, Oreg	Oct. 19, 1869	95, 997
Plow, Gang	T. E. Gardiner, jr	Bryantown, Md	May 21, 1867	64, 967
Plow, Gang	R. R. Gaskill	El Paso, Ill	Feb. 13, 1866	52, 558
Plow, Gang	W. S. Gatlin and B. R. Hubbard.	Green Top, Mo	Mar. 24, 1868	75, 895
Plow, Gang	C. F. Gay	Albany, Oreg	July 27, 1869	93, 077
Plow, Gang	P. M. Gilbert	Kewanee, Ill	May 30, 1865	47, 942
Plow, Gang	S. I. and G. M. Gillham	Carlisle, Ill	Aug. 27, 1867	68, 065
Plow, Gang	S. J. Gillham, W. C. Taylor, and J. W. Stolle.	Vandalia, Ill	June 27, 1871	116, 428
Plow, Gang	A. L. and B. F. Gilliland	Littleton, Ill	Mar. 6, 1866	52, 990
Plow, Gang	J. H. Glass	McGregor, Iowa	July 25, 1871	117, 404
Plow, Gang	D. H. Gleeson	San Leandro, Cal	Dec. 7, 1869	97, 500
Plow, Gang	E. E. Gore	Phœnix, Oreg	Dec. 28, 1869	98, 370
Plow, Gang	S. Graham	Fennimore, Wis	July 28, 1868	80, 405
Plow, Gang	R. R. Graves	Montgomery, Ala	July 9, 1867	66, 583
Plow, Gang	G. A. Groves	East Clarkson, N. Y	Jan. 7, 1873	134, 540
Plow, Gang	J. Haege	Shiloh, Ill	Dec. 18, 1860	30, 967
Plow, Gang	J. Haege	Shiloh, Ill	Feb. 24, 1863	37, 750
Plow, Gang	J. Haege	Shiloh, Ill	Apr. 14, 1863	38, 161
Plow, Gang	G. W. Haines	Maine Prairie, Cal	Mar. 12, 1872	124, 571
Plow, Gang	T. J. Hall	Tawakana Falls, Tex	May 1, 1855	12, 791
Plow, Gang	T. J. Hall	Bryan, Tex	Apr. 20, 1869	89, 144
Plow, Gang	A. Hammond	Jacksonville, Ill	July 11, 1865	48, 679
Plow, Gang	A. Hammond	Jacksonville, Ill	Mar. 27, 1866	53, 443
Plow, Gang	J. Harris	Santa Clara County, Cal	Nov. 26, 1867	71, 301
Plow, Gang	J. Harris	San Francisco, Cal	Dec. 20, 1870	110, 356
Plow, Gang	M. W. Harris	Des Moines, Iowa	July 2, 1872	128, 482
Plow, Gang	W. Hay and T. B. Freeman	Hillsborough, Oreg	July 11, 1871	116, 956
Plow, Gang	T. S. Heptinstall	Mendota, Ill	Dec. 27, 1859	26, 587
Plow, Gang	T. S. Heptinstall	Mendota, Ill	July 17, 1860	29, 169
Plow, Gang	P. Herbert	Saint Louis, Mo	Aug. 10, 1869	93, 618
Plow, Gang	C. Hess	Lyons City, Iowa	Mar. 10, 1868	75, 268
Plow, Gang	W. F. Higgins and J. Perry	Watsonville, Cal	Apr. 7, 1868	76, 447
Plow, Gang	G. W. Hildreth	Lockport, N. Y	Oct. 13, 1857	18, 397
Plow, Gang	F. A. Hill	Maysville, Cal	Sept. 7, 1869	94, 491
Plow, Gang	L. Holloway	Gilroy, Cal	Mar. 21, 1865	46, 903
Plow, Gang	L. Holloway	San Francisco, Cal	July 27, 1869	92, 968
Plow, Gang	J. D. Hope	Philadelphia, Pa	June 4, 1850	7, 415
Plow, Gang	C. L. Horn, jr., and L. Mancy	Saint Morgan, Ill	Sept. 15, 1868	82, 223
Plow, Gang	A. W. Hoyt	Denver, Ill	Sept. 10, 1872	131, 218
Plow, Gang	H. R. Hine	Hayward's, Cal	Aug. 11, 1863	39, 483
Plow, Gang	H. R. Huie	Hayward's, Cal	Oct. 20, 1868	83, 283
Plow, Gang	H. R. Huie	Hayward's, Cal	Sept. 28, 1869	95, 229
Plow, Gang	J. M. Huie and E. Card	San Francisco, Cal	Oct. 31, 1871	120, 384
Plow, Gang	A. M. Humphrey	Broadhead, Wis	Feb. 18, 1873	136, 062
Plow, Gang	J. B. Hunter	Ashley, Ill	June 19, 1866	55, 660
Plow, Gang	J. B. Hunter	Ashley, Ill	Nov. 9, 1869	96, 593
Plow, Gang	S. Hutchinson	Griggsville, Ill	Aug. 7, 1866	56, 942
Plow, Gang	J. and J. Ingham	San José, Cal	May 28, 1867	65, 087
Plow, Gang	B. Jennings	Gilroy, Cal	Dec. 7, 1869	97, 646
Plow, Gang	E. C. Jones	Pittsburgh, Pa	Dec. 1, 1857	18, 749
Plow, Gang	J. L. Keasor	Laconia, N. H	Mar. 24, 1868	75, 925
Plow, Gang	C. Kewin	San Francisco, Cal	Nov. 12, 1872	132, 910
Plow, Gang	H. Killam and G. Valleau	Scottsville, N. Y	Mar. 30, 1852	8, 844
Plow, Gang	S. L. Kingston and D. Gore	Plain View, Ill	Oct. 6, 1857	18, 343
Plow, Gang	J. D. Kneedler	Collinsville, Ill	Nov. 3, 1868	83, 641
Plow, Gang	W. Kuehn	Lively, Ill	Nov. 17, 1863	40, 629
Plow, Gang	H. P. Kynett	Lisbon, Iowa	Mar. 26, 1867	63, 260
Plow, Gang	J. H. La Boyteaux and C. A. Ashton.	Jacksonville, Ill	July 11, 1865	48, 696
Plow, Gang	J. Lane	Eugene, Ind	Nov. 5, 1872	132, 842
Plow, Gang	L. B. Lathrop	San José, Cal	May 28, 1867	65, 094
Plow, Gang	J. Lee	Galesburgh, Ill	Nov. 24, 1857	18, 698
Plow, Gang	J. T. Legg	Lewis County, Mo	June 25, 1867	66, 155
Plow, Gang	J. W. Lewis	Oregon City, Oreg	May 18, 1869	90, 178
Plow, Gang	J. W. Lewis	Oregon City, Oreg	June 22, 1869	91, 643
Plow, Gang	M. Likes	Mansfield, Ohio	June 27, 1871	116, 332
Plow, Gang	J. J. Lindly	Lebanon, Ill	Aug. 3, 1869	93, 318
Plow, Gang	J. B. Logan	Richview, Ill	Dec. 1, 1868	84, 634
Plow, Gang	D. A. Manuel	Napa, Cal	June 4, 1872	127, 495
Plow, Gang	G. W. Manuel	San Francisco, Cal	May 19, 1868	78, 111
Plow, Gang	G. W. Manuel	Napa, Cal	Apr. 8, 1873	137, 697
Plow, Gang	T. R. Markillie	Winchester, Ill	Sept. 13, 1864	44, 206
Plow, Gang	W. Mason	Independence, Oreg	Jan. 12, 1869	85, 838
Plow, Gang	W. Mason	Independence, Oreg	Nov. 8, 1870	109, 136
Plow, Gang	W. W. Mathews	Yates City, Ill	Feb. 11, 1868	74, 238
Plow, Gang	D. C. Matteson	Stockton, Cal	June 22, 1858	20, 647
Plow, Gang	D. C. Matteson	Stockton, Cal	June 2, 1868	78, 464
Plow, Gang	J. R. McConnell	Marengo, Iowa	Jan. 24, 1871	111, 226
Plow, Gang	J. K. McLennan	Elmira, Ill	Feb. 11, 1873	135, 654
Plow, Gang	F. McFarnahan	Santa Clara, Cal	Dec. 8, 1868	84, 748
Plow, Gang	P. Merkel	Saint Louis, Mo	Nov. 20, 1866	59, 855
Plow, Gang	M. Murphy	Vacaville, Cal	May 4, 1869	89, 788
Plow, Gang	J. Murry	Silveyville, Cal	Mar. 28, 1871	113, 081
Plow, Gang	R. Nation	Chebanse, Ill	Apr. 5, 1864	42, 258

Index of patents issued from the United States Patent Office from 1790 *to* 1873, *inclusive*—Continued.

Invention.	Inventor.	Residence.	Date.	No.
Plow, Gang	W. Nelson	Cacheville, Cal	Mar. 17, 1868	75, 567
Plow, Gang	W. Newlin	Attica, Ind	Oct. 25, 1870	108, 724
Plow, Gang	J. Oler	Eagle Point, Ill	Apr. 25, 1871	114, 033
Plow, Gang	J. S. Padon	Lebanon, Ill	Sept. 1, 1863	39, 743
Plow, Gang	J. S. Padon	Summerfield, Ill	Sept. 12, 1865	49, 911
Plow, Gang	S. E. Parr	Smithville, Ill	Jan. 28, 1873	135, 357
Plow, Gang	W. Parrish	Dayton, Oreg	June 13, 1871	115, 980
Plow, Gang	W. Parrish	Dayton, Oreg	June 25, 1872	128, 245
Plow, Gang	L. M. Patterson	Jordon's Grove, Ill	Mar. 20, 1866	53, 328
Plow, Gang	T. Peppler	Hightstown, N. J	Dec. 20, 1870	110, 281
Plow, Gang	H. L. Perry	Aurora, N. Y	Apr. 30, 1867	64, 358
Plow, Gang	H. L. Perry	Aurora, N. Y	Nov. 19, 1867	71, 053
Plow, Gang	D. Petticrew	Westville, Ohio	Jan. 21, 1868	73, 644
Plow, Gang	J. C. Pfeil	Arenzville, Ill	Jan. 31, 1865	46, 137
Plow, Gang	J. C. Pfeil	Arenzville, Ill	June 25, 1867	66, 039
Plow, Gang	J. F. Porter and A. Norton	Tidioute, Pa	July 14, 1868	79, 917
Plow, Gang	J. Price	San Leandro, Cal	Mar. 22, 1870	101, 038
Plow, Gang	J. L. Purcell	Thompson, Ill	Oct. 18, 1870	108, 516
Plow, Gang	W. B. Quick	Belleville, Ill	June 6, 1871	115, 639
Plow, Gang	G. Quirin and L. Berkel	Smithton, Ill	Nov. 8, 1864	44, 971
Plow, Gang	W. B. Ready	Sacramento, Cal	Dec. 3, 1861	33, 851
Plow, Gang	F. F. Reynolds	Bethany, Ga	Aug. 24, 1869	94, 133
Plow, Gang	W. B. Rice	Oakland, Oreg	June 29, 1869	92, 099
Plow, Gang	D. C. Riggs	Saint Joseph, Mo	July 30, 1867	67, 351
Plow, Gang	L. Roach	Covington, Ky	Mar. 16, 1858	19, 652
Plow, Gang	L. O. Rockwood	Ottawa, Ill	May 21, 1867	64, 906
Plow, Gang	W. T. Rogers	Quincy, Ill	July 24, 1866	56, 613
Plow, Gang	J. L. Runk	Nashville, Ill	Sept. 15, 1863	39, 961
Plow, Gang	J. L. Runk	Nashville, Ill	Feb. 16, 1864	41, 643
Plow, Gang	J. L. Runk, J. H. Brown, and E. M. Morgan	Nashville, Ill	May 22, 1866	54, 963
Plow, Gang	M. Sattley	Taylorsville, Ill	Feb. 2, 1864	41, 449
Plow, Gang	M. Sattley	Taylorsville, Ill	Dec. 5, 1865	51, 358
Plow, Gang	D. A. Sears	Rockford, Ill	July 19, 1870	105, 600
Plow, Gang	G. and J. Seibert	Ashley, Ill	May 4, 1869	89, 797
Plow, Gang	J. Seibel	Manlius, Ill	May 30, 1865	47, 989
Plow, Gang	E. Sexton	Munson, Mass	Apr. 23, 1867	64, 152
Plow, Gang	T. Short	Fairmount, Ill	Jan. 10, 1865	45, 897
Plow, Gang	J. B. Skinner	Rockford, Ill	July 18, 1865	48, 846
Plow, Gang	J. B. Skinner	Rockford, Ill	Apr. 17, 1866	54, 029
Plow, Gang	A. Smith	Portland, Me	Dec. 1, 1868	84, 652
Plow, Gang	A. Smith and W. P. Watson	Portland, Oreg	Aug. 11, 1868	80, 838
Plow, Gang	A. and T. S. Smith	Troy, Ill	Mar. 4, 1856	14, 373
Plow, Gang	F. P. Smith	Petaluma, Cal	Sept. 15, 1868	82, 165
Plow, Gang	F. P. Smith	Petaluma, Cal	Sept. 15, 1868	82, 166
Plow, Gang	H. Smith	San Lorenzo, Cal	Aug. 22, 1865	49, 564
Plow, Gang	H. B. Smith	Tremont, Ill	Sept. 19, 1871	119, 192
Plow, Gang	H. C. Smith	Ridge Farm, Ill	Jan. 2, 1866	51, 875
Plow, Gang	J. D. Smith	Peoria, Ill	June 26, 1866	55, 921
Plow, Gang	N. Spencer, jr	Eagle Point, Ill	Sept. 27, 1870	107, 829
Plow, Gang	H. P. Stafford	Decatur, Ill	Aug. 13, 1867	67, 814
Plow, Gang	P. H. Standish	Martinez, Cal	Sept. 1, 1868	81, 700
Plow, Gang	G. Steinegger	Highland, Ill	Mar. 3, 1868	75, 069
Plow, Gang	J. Stone	Plattsburgh, Mo	Sept. 20, 1864	44, 351
Plow, Gang	J. W. Sursa	San Leandro, Cal	May 21, 1867	64, 923
Plow, Gang	J. W. Sursa	San Leandro, Cal	Apr. 5, 1870	101, 539
Plow, Gang	J. W. Sursa	San Leandro, Cal	Mar. 7, 1871	112, 394
Plow, Gang	J. Sutter	Saint Louis County, Mo	June 16, 1857	17, 591
Plow, Gang	Z. T. Sweet	Eugene City, Oreg	Jan. 5, 1869	85, 621
Plow, Gang	J. N. Thompson and W. Kenady	Belpassi, Oreg	Oct. 5, 1869	95, 539
Plow, Gang	J. Totten	Adams, Ill	Feb. 2, 1869	86, 472
Plow, Gang	J. E. Travis	Greenville, Ill	May 9, 1865	47, 686
Plow, Gang	J. E. Travis	Greenville, Ill	Feb. 13, 1866	52, 651
Plow, Gang	J. W. Treadway	Crown Point Centre, N. Y	Oct. 11, 1870	108, 214
Plow, Gang	J. Tustin	Portland, Oreg	Feb. 2, 1869	86, 608
Plow, Gang	A. Walker	Claremont, N. H	July 28, 1868	80, 521
Plow, Gang	E. W. Walton	San Leandro, Cal	July 19, 1870	105, 528
Plow, Gang	J. T. Watkins	Santa Clara, Cal	July 14, 1868	79, 880
Plow, Gang	W. M. Watson	Tonica, Ill	May 1, 1866	54, 453
Plow, Gang	H. J. Wattles	Rockford, Ill	Sept. 10, 1867	68, 673
Plow, Gang	S. Way	La Porte, Ind	Jan. 21, 1868	73, 480
Plow, Gang	T. U. Webb	Springfield, Ill	Dec. 29, 1868	85, 496
Plow, Gang	H. Webster	Beetown, Wis	Jan. 31, 1865	46, 164
Plow, Gang	L. T. Webster	Northfield, Mass	Aug. 1, 1871	117, 707
Plow, Gang	S. Welch	Belpassi, Oreg	Feb. 1, 1870	99, 379
Plow, Gang	G. Wharton	Jerseyville, Ill	July 14, 1868	80, 039
Plow, Gang	G. Wharton	Jerseyville, Ill	Apr. 13, 1869	88, 998
Plow, Gang	T. Wiard	East Avon, N. Y	Nov. 24, 1843	3, 356
Plow, Gang	J. C. Wilson	Cedar Hill, Tex	July 3, 1860	29, 027
Plow, Gang	L. Wolf	Saint Louis, Mo	Nov. 24, 1863	40, 721
Plow, Gang	S. A. Worthen	Thompson, Ill	Aug. 1, 1871	117, 711
Plow, Gang	G. W. N. Yost	Cincinnati, Ohio	Apr. 27, 1858	20, 122
Plow, Gang and subsoil	J. L. Bond	Saint Louis, Mo	Jan. 31, 1871	111, 428
Plow, Gang and subsoil	J. L. Bond	Marshalltown, Iowa	Apr. 4, 1871	113, 390
Plow, Gang and subsoil	R. L. Dodge and E. M. Walker	Gallatin, Mo	Oct. 16, 1866	58, 788
Plow, Gang and trench	J. G. Robinson	Springfield, Ill	Mar. 30, 1869	88, 413
Plow, Garden	G. W. Cole	Farmington, Ill	Aug. 3, 1869	93, 277
Plow, Garden	J. M. Crawford	Newcastle, Ky	Sept. 11, 1866	57, 871
Plow, Garden	W. F. Pagett and S. H. Gard	Springfield, Ohio	Aug. 31, 1869	94, 434
Plow, Garden	R. Scott	La Porte, Ind	Sept. 7, 1869	94, 657
Plow, Garden hand	J. Starr	Grand Rapids, Mich	Aug. 11, 1868	80, 842
Plow, Grading	A. P. Hopkins	Bentleyville, Pa	May 28, 1872	127, 239
Plow, Grading and ditching	S. N. Caldwell	Pilot Grove, Ind	Aug. 2, 1870	105, 902
Plow, Grape	R. Hardenbrook	Bath, N. Y	Aug. 27, 1867	68, 190
Plow grinder and polisher	M. Devault	Charleston, Ill	May 3, 1870	102, 508
Plow, Grooving or panel	W. S. Loughborough	Rochester, N. Y	May 3, 1864	42, 585
Plow guard, Shovel	W. J. M. Batchelder	Dayton, Ohio	Mar. 26, 1867	63, 134
Plow, Hand	W. Gowen	Bartlett, Tenn	Aug. 10, 1869	93, 431

Index of patents issued from the United States Patent Office from 1790 *to* 1873, *inclusive*—Continued.

Invention.	Inventor.	Residence.	Date.	No.
Plow, Hand	F. Keefer	Greenfield, Ind	May 14, 1867	64,771
Plow, Hand	L. McWhinney	Winterset, Iowa	Sept. 28, 1869	95,367
Plow, Hand	N. Rue	Harrodsburgh, Ky	Mar. 8, 1871	112,497
Plow, Hand	J. Winecoff	Berlin, Pa	Nov. 24, 1868	84,331
Plow, Hand	W. B. Winton	Marion, Iowa	Sept. 1, 1868	81,857
Plow, Hand garden	W. D. Smith	Homerville, Ga	Aug. 22, 1871	118,290
Plow, Hand garden	J. Von Achen	Bloomfield, Iowa	Mar. 6, 1866	53,061
Plow-handle	S. J. Olmsted	Binghamton, N. Y	Apr. 5, 1864	42,220
Plow-handle	J. T. Raftery	El Dara, Ill	Apr. 29, 1873	138,434
Plow-handle	G. Watt	Richmond, Va	May 7, 1867	64,464
Plow-handle	W. E. Wyche	Brookville, N. C	Oct. 25, 1870	108,671
Plow-handles, Device for bending	W. Heylman	Peoria, Ill	May 4, 1869	89,764
Plow-handles, Forming	G. W. Matthews	York, Pa	May 24, 1859	24,131
Plow-handles, Machine for bending	B. F. Avery	Louisville, Ky	Jan. 22, 1856	14,130
Plow-handles, Machine for bending	J. G. Ernst	York, Pa	Aug. 2, 1859	24,971
Plow-handles, Machine for bending	G. V. Griffith	Fort Wayne, Ind	July 13, 1869	92,604
Plow-handles, Machine for bending	J. Woodburn and S. F. Smith	Saint Louis, Mo., and Indianapolis, Ind.	Dec. 13, 1870	110,184
Plow-handles, Machine for forming	W. A. Ellis	Ashtabula, Ohio	July 31, 1866	56,736
Plow-handles, Machine for forming	G. V. Griffith	Fort Wayne, Ind	Aug. 17, 1869	93,710
Plow-handles, Machine for forming	E. G. Matthews	Oakham, Mass	Apr. 25, 1871	114,170
Plow-handles, &c., Machine for polishing	T. Blanchard	Boston, Mass	Feb. 7, 1854	10,497
Plow-handles, Machine for turning	F. C. Coppage	Terre Haute, Ind	July 23, 1861	32,910
Plow-handles, Machinery for making	A. B. Farquhar	York, Pa	July 1, 1873	140,355
Plow-handles, Manufacture of	T. E. C. Brinly	Louisville, Ky	Dec. 24, 1867	72,596
Plow, harrow, cultivator, and roller, Combined	J. Johnston	Pemberton, Ohio	July 2, 1867	66,240
Plow, Hilling	G. Notman	Deerfield, Ohio	Mar. 2, 1869	87,356
Plow, Hill-side	H. B. Abbott	Felicity, Ohio	Mar. 3, 1868	75,104
Plow, Hill side	H. S. Akins	Speedsville, N. Y	Aug. 31, 1858	21,306
Plow, Hill-side	W. H. Babbit	Greene County, Pa	Nov. 6, 1847	5,363
Plow, Hill-side	W. H. Babbit	Waynesburgh, Pa	Aug. 16, 1853	9,944
Plow, Hill-side	F. G. Bakes	Vevay, Ind	Jan. 1, 1867	60,820
Plow, Hill-side	I. Brewster	Stamford, N. Y	Nov. 14, 1848	5,922
Plow, Hill-side	M. L. Chase	Frankfort, Me	July 16, 1850	7,505
Plow, Hill-side	J. P. Cobbs	Nelson County, Va	Oct. 1, 1830	....
Plow, Hill-side	H. Cox	Peach Bottom, Va	Nov. 7, 1848	5,909
Plow, Hill-side	S. Dennis, jr	Jasper, N. Y	Mar. 2, 1858	19,496
Plow, Hill-side	A. Eldred	Little Falls, N. Y	July 24, 1849	6,606
Plow, Hill-side	R. H. Ewing	Clives, Ohio	Nov. 27, 1860	30,726
Plow, Hill-side	F. Feldhaus	Baltimore, Md	Oct. 8, 1867	69,555
Plow, Hill-side	D. Gochnour, jr	Conemaugh, Pa	June 12, 1840	1,632
Plow, Hill-side	S. Hall	Manchester, Pa	Jan. 4, 1853	9,519
Plow, Hill-side	A. J. Hardin	Shelby, N. C	Oct. 6, 1857	18,336
Plow, Hill-side	N. Harrison and J. W. H. Metcalf.	Ridgeville, Va	Oct. 11, 1853	10,107
Plow, Hill-side	C. H. McCormick	Rockbridge County, Va	June 13, 1831	....
Plow, Hill-side	M. Merk	Rochester, N. Y	July 6, 1858	20,812
Plow, Hill-side	G. C. Miller and R. Henry	Cincinnati, Ohio	Aug. 21, 1860	29,708
Plow, Hill-side	D. H. B. Newcomb	Conewango, N. Y	June 21, 1853	9,801
Plow, Hill-side	D. Robb	Sangamon County, Ill	June 26, 1849	6,553
Plow, Hill-side	J. Rorabaugh	Luney's Creek, Va	Dec. 3, 1846	4,870
Plow, Hill-side	A. Sanborn	Glover, Vt	June 18, 1861	32,587
Plow, Hill-side	H. Sloop	Mount Healthy, Ohio	Jan. 28, 1868	73,933
Plow, Hill-side	E. J. Smith and H. Griswold	Delhi, N. Y	July 25, 1848	5,677
Plow, Hill-side	N. Staples	Patrick County, Va	Nov. 1, 1828	....
Plow, Hill-side	I. Teeter	Johnstown, Pa	Oct. 3, 1838	959
Plow, Hill-side	J. W. Thurman	Buchanan, Va	Aug. 28, 1849	6,677
Plow, Hill-side	J. Trump	Connellsville, Pa	Sept. 9, 1845	4,186
Plow, Hill-side	E. Van Camp	Readington, N. J	May 10, 1859	23,964
Plow, Hill-side	J. B. Wilder	Belfast, Me	June 21, 1853	9,808
Plow, Hill-side double-nosed cast-iron	J. Shepherd	De Ruyter, N. Y	Apr. 12, 1826	....
Plow, Hinged-wing hill-side	W. B Donald	Fairfield, Va	June 29, 1833	....
Plow-holder, Self-acting	J. L. Keasor	Laconia, N. H	Oct. 8, 1867	69,678
Plow, Ice	J. H. Cutter	Cambridge, Mass	Feb. 20, 1872	123,873
Plow, Ice	G. B. Gruman	Bridgefield, Conn	Apr. 2, 1872	125,131
Plow, Iron	J. Boynton	South Coventry, Conn	July 10, 1829	....
Plow-irons, Machine for making	J. M. McGinty and T. Nolan	Moulton, Tex	Mar. 4, 1873	136,444
Plow-irons, Machine for welding	C. G. Cross	Chicago, Ill	Apr. 30, 1872	126,188
Plow-jointer	D. Dillenback	Galesburgh, Mich	Nov. 1, 1870	108,771
Plow-jointer, Adjustable	J. M. Leonard	Marshall City, Mich	Dec. 20, 1870	110,251
Plow jointer, Prairie	A. J. Spicer	Galesburgh, Mich	Feb. 14, 1871	111,786
Plow-landside	J. Bacon	Medina, Wis	Apr. 28, 1868	77,242
Plow-landside	A. Christ	Unity, Ohio	Sept. 18, 1849	6,724
Plow-landsides, Manufacture of	J. Duff, H. Roberts, and G. D. Nourse.	Peoria, Ill., and Saint Louis, Mo.	Sept. 12, 1871	118,848
Plow-lays, Forming	J. Lane	Chicago, Ill	Feb. 4, 1868	73,983
Plow-marking attachment	C. Morris	Stockton Township, N. J	Apr. 2, 1867	63,551
Plow, Metallic double-shovel	W. W. Love	Athens, Ohio	Oct. 19, 1869	96,018
Plow mold-board	E. M. Bard	Philadelphia, Pa	Mar. 14, 1854	10,629
Plow mold-board	G. A. Beard	Washington County, Md	Aug. 28, 1866	57,463
Plow mold-board	S. H. Dwight and C. Wells	Decatur, Ill., and Pittsburgh, Pa.	May 25, 1869	90,512
Plow mold-board	I. T. Dyer	Macon, Ga	Apr. 4, 1871	113,642
Plow mold-board	R. Gaines and M. Scott	Fairfield, Iowa	Feb. 18, 1868	74,679
Plow mold-board	J. Lane	Chicago, Ill	Oct. 4, 1870	107,925
Plow mold-board	J. Oliver	South Bend, Ind	Apr. 14, 1868	76,652
Plow mold board	G. Peacock	Selma, Ala	Aug. 29, 1871	118,551
Plow mold-board	L. P. Rider	Munson, Ohio	Oct. 30, 1866	59,267
Plow mold-board	L. P. Rider	Pittsburgh, Pa	Apr. 25, 1871	114,044
Plow mold-board	T. E. Session and S. A. King	Worcester, Mass	Feb. 1, 1870	99,516
Plow mold-board	J. Seymour	Coventry, N. Y	Mar. 24, 1868	75,987
Plow mold-board	R. Smith	....	May 19, 1800	....
Plow mold-board	L. Witherow and D. Pierce	Philadelphia, Pa	Oct. 5, 1839	1,357
Plow mold-board, Reversible	H. S. Akins	Berkshire, N. Y	June 16, 1857	17,547
Plow mold-boards, Attaching handles to	C. Williams	Jackson, Miss	June 30, 1868	79,532
Plow mold-boards, Die for making blanks for	M. W. Watson	Tonica, Ill	Mar. 15, 1870	100,825
Plow mold-boards, Machine for bending	W. W. Skinner	Davenport, Iowa	Feb. 7, 1860	27,078
Plow mold-boards, Machinery for swaging	H. H. May	Galesburgh, Ill	June 27, 1846	4,609

Index of patents issued from the United States Patent Office from 1790 *to* 1873, *inclusive*—Continued.

Invention.	Inventor.	Residence.	Date.	No.
Plow mold-boards, Making	T. A. Chandler	Rockford, Ill	Dec. 4, 1855	13, 865
Plow, Molding	B. F. Avery	Louisville, Ky	Oct. 25, 1859	25, 873
Plow, Mole	J. Adair	Mendota, Ill	Mar. 27, 1860	27, 606
Plow, Mole	S. Adams	Toulon, Ill	Feb. 28, 1860	27, 283
Plow, Mole	W. B. Atkinson	Plymouth, Ill	Sept. 18, 1860	30, 036
Plow, Mole	H. Bagley	Tipton, Iowa	Sept. 18, 1860	30, 041
Plow, Mole	H. F. Baker	Centreville, Ind	Oct. 4, 1859	25, 618
Plow, Mole	M. Bales	Big Plain, Ohio	Feb. 15, 1859	22, 928
Plow, Mole	H. H. Ballard and H. McClure	Mount Pleasant, Iowa	Mar. 26, 1861	31, 771
Plow, Mole	J. R. Barnett	Galesburgh, Ill	July 9, 1872	128, 842
Plow, Mole	A. Bowers, J. H. Griggs, and J. Wilson.	Monmouth, Ill	Nov. 15, 1859	26, 082
Plow, Mole	J. Carrington	Avoca, N. Y	Mar. 29, 1859	23, 348
Plow, Mole	J. Case	Bloomington, Ill	Jan. 25, 1859	22, 701
Plow, Mole	S. A. Clemens	Rockford, Ill	July 7, 1863	39, 118
Plow, Mole	C. U. and J. H. Crandall and H. N. Hawkins.	Cameron, Ill	Aug. 23, 1859	25, 178
Plow, Mole	J. Creamer	Jeffersonville, Ohio	Feb. 19, 1867	62, 116
Plow, Mole	J. Creamer and T. W. Ricards	London, Ohio	Apr. 5, 1859	23, 452
Plow, Mole	A. Defenbaugh	Walnut Run, Ohio	Sept. 14, 1858	21, 491
Plow, Mole	A. Elmer	Shabbona Grove, Ill	Aug. 16, 1859	25, 105
Plow, Mole	J. H. Edward	Ottawa, Fla	Nov. 13, 1860	30, 625
Plow, Mole	A. Gillet	Lyndon, Ill	Jan. 15, 1861	31, 117
Plow, Mole	W. P. Goolman	Dublin, Ind	Mar. 22, 1859	23, 334
Plow, Mole	W. P. Goolman	Dublin, Ind	Dec. 13, 1859	26, 426
Plow, Mole	G. L. Griffin and J. H. Carper	Dallas City, Ill	Feb. 28, 1860	27, 285
Plow, Mole	J. A. Hammer and J. P. Gordon	Lisbon, Iowa	Feb. 5, 1861	31, 313
Plow, Mole	A. Hammond	Jacksonville, Ill	Aug. 16, 1859	25, 114
Plow, Mole	A. Hammond	Jacksonville, Ill	Apr. 10, 1860	27, 796
Plow, Mole	F. E. Hinckley	Galesburgh, Ill	Dec. 6, 1859	26, 355
Plow, Mole	I. Hobson	Stout's Grove, Ill	Sept. 6, 1859	25, 334
Plow, Mole	M. A. Howell, jr	Ottawa, Ill	Feb. 5, 1861	31, 317
Plow, Mole	R. Hussey and U. Thornburgh, sr.	Walnut Run, Ohio	Oct. 4, 1859	25, 649
Plow, Mole	H. R. Jerome	Monroeville, Ohio	Aug. 16, 1859	25, 121
Plow, Mole	S. F. Jones	Saint Paul, Ind	Oct. 25, 1859	25, 902
Plow, Mole	S. F. Jones	Saint Paul, Ind	Aug. 27, 1861	33, 149
Plow, Mole	A. M. Karr	Mount Pleasant, Iowa	July 24, 1860	29, 285
Plow, Mole	J. Lee	Galesburgh, Ill	Aug. 16, 1859	25, 127
Plow, Mole	A. Miller	Mount Pleasant, Iowa	Oct. 18, 1859	25, 845
Plow, Mole	A. Miller	Chicago, Ill	Oct. 23, 1866	59, 049
Plow, Mole	J. Morrison	De Witt, Ill	Oct. 18, 1859	25, 846
Plow, Mole	E. and W. Parish, jr	Galesburgh, Ill	Feb. 21, 1860	27, 233
Plow, Mole	D. F. Robbins and S. Morrison	De Witt, Ill	Apr. 12, 1859	23, 609
Plow, Mole	H. W. Rowland and E. Forbis	Newport, Ohio	Apr. 19, 1859	23, 745
Plow, Mole	R. P. Smith and J. R. Gates	Louisville, Ky	Jan. 7, 1862	34, 074
Plow, Mole	C. W. Stafford	Burlington, Iowa	July 17, 1860	29, 201
Plow, Mole	O. Sturdevant	Maquon, Ill	Nov. 13, 1860	30, 659
Plow, Mole	A. L. O. Wall, G. Roberts, and M. S. Cartter.	Decatur, Ill	Apr. 3, 1860	27, 751
Plow, Mole	A. Watson	Walnut Run, Ohio	Aug. 2, 1859	24, 969
Plow, Mole	G. Whitcomb	Springfield, Ohio	Nov. 1, 1859	25, 988
Plow or cultivator, Cotton	F. M. Shields	Macon, Miss	July 16, 1867	66, 895
Plow, planter, and cultivator	E. Bourne	New Iberia, La	Sept. 3, 1870	107, 218
Plow, planter, and cultivator, Combined	T. J. Smith	Holly Springs, Miss	Jan. 25, 1870	99, 118
Plow, planter, and cultivator, Combined corn	I. A. Arthur	Sidonsburgh, Pa	Apr. 28, 1868	77, 238
Plow, Planting	A. B. Earle	Colesville, N. Y	Oct. 17, 1848	5, 858
Plow, Planting	B. Hussey	Jonesborough, Tenn	Apr. 1, 1834	....
Plow, Planting	F. Woodward	New Lisbon, N. Y	July 18, 1809	....
Plow plate and point, Shovel	H. Miller	Roadside, Va	Nov. 23, 1869	97, 213
Plow-plates, Cooling and tempering cast-steel	F. F. Smith	Collinsville, Conn	Feb. 2, 1864	41, 464
Plow-plates from molten steel, Making	F. F. Smith	Momence, Ill	Nov. 20, 1860	30, 691
Plow-point	E. Bement	Fostoria, Ohio	Sept. 16, 1862	36, 447
Plow-point	L. D. Burch	Sherburne, N. Y	Oct. 20, 1868	83, 130
Plow-point	E. C. Gero and J. N. Cooley	Kalamazoo, Mich	Mar. 2, 1869	87, 484
Plow-point	M. L. Gibbs	Canton, Ohio	June 25, 1872	128, 221
Plow-point	C. B. Kerr	Columbus, Ind	Apr. 27, 1869	89, 486
Plow-point	H. Kniphals	Davenport, Iowa	Sept. 10, 1867	68, 755
Plow-point	C. W. Saladee and T. Simpson	Newark, Ohio	Mar. 27, 1866	53, 491
Plow-point	D. J. Selden	Mount Vernon, Ohio	May 21, 1867	64, 914
Plow-point	O. O. Storle	Norway, Wis	May 25, 1869	90, 604
Plow-point, Reversible	R. M. Pattills	Cartersville, Ga	Sept. 16, 1873	142, 864
Plow-points, Attaching	G. W. Hildreth	Lockport, N. Y	Apr. 4, 1871	113, 665
Plow-points, Mold for casting	H. J. Brunner	Nazareth, Pa	Apr. 23, 1872	126, 015
Plow, Potato	H. T. Basye	Dyersburgh, Tenn	June 24, 1873	140, 240
Plow, Potato	A. Eldred	Oppenheim, N. Y	July 24, 1846	4, 656
Plow, Potato	A. Leonard	Newell's Run, Ohio	Aug. 23, 1870	106, 702
Plow, Potato	J. P. Stanton	Pedricktown, N. J	Sept. 3, 1867	68, 392
Plow, Potato	M. Stoll	Conestoga Township, Pa	Sept. 8, 1868	82, 043
Plow, Potato	R. P. Terhune and B. J. Romaine.	Hackensack, N. J	Jan. 26, 1869	86, 190
Plow, Potato and corn	C. F. Noftz	Toledo, Ohio	Nov. 10, 1868	83, 992
Plow, potato-planter, and seeder, Combined	S. Shetter	New Cumberland, W. Va	Dec. 31, 1867	72, 922
Plow, Prairie	C. K. Bartlett	Geneseo, Ill	Mar. 28, 1842	2, 515
Plow, Prairie	J. Frye	Springfield, Ill	Mar. 31, 1857	16, 913
Plow, Prairie	M. Turley	Galesburgh, Ill	Dec. 9, 1856	16, 216
Plow-press and drill	T. E. C. Brinly	Simpsonville, Ky	Sept. 21, 1858	21, 547
Plow, Railway	C. Medbury and T. Wyatt	Providence, R. I	Jan. 2, 1866	51, 850
Plow, Railway snow	J. R. Adams	Cisco, Cal	Jan. 14, 1868	73, 276
Plow, Railway snow	J. W. Arnold	Boston, Mass	May 13, 1873	138, 836
Plow, Railway snow	J. K. Babcock	Honeoye Falls, N. Y	Feb. 16, 1858	19, 339
Plow, Railway snow	O. D. Baird	Newark, N. Y	Apr. 22, 1873	138, 116
Plow, Railway snow	P. Boyden	Sandy Creek, N. Y	Apr. 29, 1862	35, 074
Plow, Railway snow	J. C. Carncross	Philadelphia, Pa	Jan. 16, 1866	52, 028
Plow, Railway snow	H. H. Clemons	Oshkosh, Wis	Sept. 17, 1867	68, 845
Plow, Railway snow	T. A. Davies	New York, N. Y	Mar. 11, 1873	136, 709
Plow, Railway snow	W. Davis	Five Corners, N. Y	Mar. 18, 1873	136, 975
Plow, Railway snow	S. L. Denney	Christiana, Pa	Sept. 15, 1863	39, 894

Index of patents issued from the United States Patent Office from 1790 *to* 1873, *inclusive*—Continued.

Invention.	Inventor.	Residence.	Date.	No.
Plow, Railway snow	T. Dougherty	Philadelphia, Pa	Dec. 27, 1870	110, 446
Plow, Railway snow	C. L. Garfield	Albany, N. Y	Aug. 24, 1869	94, 102
Plow, Railway snow	J. A. Gregg	Derry, N. H	June 13, 1846	4, 568
Plow, Railway snow	S. G. Hadley	Cape Vincent, N. Y	Dec. 3, 1867	71, 607
Plow, Railway snow	A. Hall and S. Sturtevant	South Paris, Me	Feb. 28, 1854	10, 568
Plow, Railway snow	H. T. Hartman	Lexington, Va	Feb. 16, 1858	19, 361
Plow, Railway snow	S. W. Hemenway	Lansing, Iowa	Dec. 24, 1872	134, 202
Plow, Railway snow	C. L. Heywood	Boston, Mass	Jan. 2, 1866	51, 829
Plow, Railway snow	A. Hotchkiss	Sharon Valley, Conn	Dec. 22, 1857	18, 903
Plow, Railway snow	J. B. Hulbert and J. Anderson	Hermon, N. Y	Nov. 5, 1872	132, 834
Plow, Railway snow	W. S. Huntington	Andrusville, N. Y	June 21, 1859	24, 463
Plow, Railway snow	B. A. Johnson	North Auburn, Me	Jan. 13, 1863	37, 420
Plow, Railway snow	J. Jones and T. G. Eiswald	Providence, R. I	Sept. 15, 1868	82, 226
Plow, Railway snow	J. S. Munson	Jamaica Plain, Mass	Dec. 10, 1872	133, 792
Plow, Railway snow	W. Rhoads	Baltimore, Md	Apr. 19, 1859	23, 709
Plow, Railway snow	S. Richards	Philadelphia, Pa	May 13, 1856	14, 886
Plow, Railway snow	E. Robbins	Worcester, Mass	Mar. 24, 1868	75, 979
Plow, Railway snow	T. L. Shaw	Omaha, Nebr	Dec. 14, 1869	97, 971
Plow, Railway snow	P. A. Smith	New York, N. Y	Feb. 25, 1873	136, 272
Plow, Railway snow	W. H. Van Gieson	Whitewater, Wis	Feb. 25, 1873	136, 346
Plow, Railway snow	W. Walker	Fort Bridger, Wyo	Feb. 4, 1873	135, 499
Plow, Railway snow	W. Wheelock	Decorah, Iowa	July 8, 1873	140, 663
Plow, Railway snow	R. Wilson	Des Moines, Iowa	Apr. 5, 1870	101, 693
Plow, Railway snow	C. L. Wood	Calais, Me	Nov. 4, 1873	144, 377
Plow-regulator	H. Sprague	Riga, N. Y	Dec. 14, 1852	9, 473
Plow, Reversible	C. F. Barager	Candor, N. Y	June 3, 1862	35, 432
Plow, Reversible	E C. Hodge	Oneonta, N. Y	Nov. 1, 1870	108, 907
Plow, Reversible	G. W. Howe	Vineland, N. J	July 29, 1873	141, 222
Plow, Reversible	E. Jennings	Candor, N. Y	June 27, 1871	116, 314
Plow, Reversible	J. W. Jones	Thompson, Ill	Aug. 24, 1869	94, 001
Plow, Reversible	G. W. Thompson	Ripley, Ohio	Jan. 3, 1871	110, 692
Plow, Reversible side-hill	C. B. Pettingill	Hebron, Me	Aug. 15, 1871	118, 849
Plow, Reversing hill-side	C. Delano	Livermore, Me	July 5, 1833	....
Plow, Revolving	M. A. and I. A. Cravath	Bloomington, Ill	July 16, 1867	66, 802
Plow, Revolving	W. J. Dawson	Brookfield, Mo	Sept. 21, 1869	95, 005
Plow, Revolving spade	J. W. Milroy	Galveston, Ind	May 12, 1868	77, 830
Plow, Ridge	W. Atwood and D. Hamblet	Cornish, N. H	Dec. 24, 1833	....
Plow, Riding	B. C. Hoyt	Fort Atkinson, Wis	May 23, 1871	115, 057
Plow, Right and left	G. Dofler	Fredericktown, Md	Aug. 20, 1827	....
Plow roller-attachment	C. M. Young	Meadville, Pa	June 27, 1871	116, 388
Plow roller-cutter	J. W. Lewis	Oregon City, Oreg	June 22, 1869	91, 644
Plow roller-cutter	R. Newton	Jerseyville, Ill	Aug. 10, 1869	93, 636
Plow roller-cutter	J. H. Sherman	Galesburgh, Ill	July 30, 1867	67, 222
Plow roller-wheel	L. B. Pitcher	Salina, N. Y	Nov. 19, 1867	71, 057
Plow root-cutter	J. W. Baker	Elkton, Md	May 23, 1871	115, 013
Plow, Rotary	H. Berkstresser	Quaker Bottom, Ohio	Sept. 10, 1867	68, 689
Plow, Rotary	W. E. Bleecker	Brooklyn, N. Y	July 16, 1872	129, 206
Plow, Rotary	E. T. Russell	Indianapolis, Ind	Sept. 3, 1867	68, 410
Plow, Rotary	L. H. Colborn	Chicago, Ill	Apr. 3, 1866	53, 577
Plow, Rotary	C. Comstock	Milwaukee, Wis	May 13, 1862	35, 218
Plow, Rotary	C. T. Elliston	Clinton, Mo	Aug 26, 1873	142, 094
Plow, Rotary	J. W. Haggard and G. Bull	Bloomington, Ill	Feb. 13, 1855	12, 387
Plow, Rotary	A. B. Hoffmeyer and J. Schmidt	Copenhagen, Denmark	Dec. 2, 1873	145, 177
Plow, Rotary	N. T. Judd	Washington, D. C	Nov. 29, 1870	109, 741
Plow, Rotary	D. Myers	Chicago, Ill	Feb. 6, 1866	52, 496
Plow, Rotary	J. Young	Jefferson, Me	July 11, 1848	5, 665
Plow, Rotary-cutter	J. C. Pfeil	Argenzville, Ill	Apr. 7, 1868	76, 343
Plow, Rotary-cutter	T. J. Tuthill	Elmira, N. Y	Feb. 6, 1849	6, 091
Plow, Rotary power	J. Tranter, J. Kinsey, and J. M. Carr.	Cincinnati, Ohio	Sept. 20, 1870	107, 639
Plow, Rotary steam	J. T. Wilson	Rochester, N. Y	Aug. 16, 1870	106, 441
Plow, Rotary subsoil	J. R. Morris	Houston, Tex	Mar. 28, 1871	113, 190
Plow, Rotating	J. Hoadley	Zanesville, Ohio	Aug. 23, 1864	43, 913
Plow-scraper	H. Leach	Camanche, Iowa	Aug. 26, 1873	142, 113
Plow, seeder, and roller, Combined	O. B. Cheatham	Henderson, Ky	Jan. 9, 1872	122, 567
Plow, Seeding	C. Atkinson	Vermont, Ill	Apr. 10, 1860	27, 766
Plow, Seeding	J. Peeler	Tallahassee, Fla	Apr. 24, 1860	28, 009
Plow, Seeding	W. P. Penn	Belleville, Ill	Jan. 15, 1861	31, 129
Plow, Seeding	W. P. Penn	Belleville, Ill	July 30, 1861	32, 924
Plow, Seeding	J. S. Snider	Lancaster, Ohio	Aug. 2, 1859	24, 959
Plow, Self-sharpening	W. Beach	Philadelphia, Pa	Aug. 26, 1823	....
Plow, Self-sharpening	W. Beach	Philadelphia, Pa	Apr. 9, 1824	....
Plow, Self-sharpening	W. Beach	Philadelphia, Pa	Dec. 28, 1832	....
Plow, Self-sharpening	R. B. Chenoweth	Baltimore, Md	Mar. 17, 1834	....
Plow, Self-sharpening	C. and O. Evans	Philadelphia, Pa	Apr. 14, 1825	....
Plow, Self-sharpening	J. B. Norton	Utica, N. Y	Apr. 27, 1832	....
Plow, Self-sharpening	J. Ormiston	Centre, Ohio	Mar. 17, 1838	638
Plow, Self-sharpening	J. W. Post	Baltimore, Md	Oct. 8, 1838	970
Plow, Self-sharpening	I. Snider	Mount Pleasant, Pa	July 29, 1837	295
Plow, Self-sharpening	B. Woodcock	Mount Pleasant, Pa	Jan. 26, 1832	....
Plow, Self-sharpening horizontal	C. H. McCormick	Rockbridge County, Va	Nov. 19, 1833	....
Plowshare	T. K. Bronson	East Avon, N. Y	Apr. 3, 1866	53, 565
Plowshare	B. Carpenter	Somer's Town, N. Y	July 1, 1822	....
Plowshare	G. W. Cooper	Ogeechee, Ga	Dec. 8, 1868	84, 798
Plowshare	J. S. Hall	Pittsburgh, Pa	July 8, 1862	35, 819
Plowshare	B. Harvey	West Bloomfield, N. Y	Dec. 10, 1872	133, 854
Plowshare	J. Lane	Chicago, Ill	Jan. 10, 1871	110, 860
Plowshare	A. Maschka	Chicago, Ill	Nov. 14, 1865	50, 942
Plowshare	W. D. Mendenhall	Farmington, Ill	June 4, 1867	65, 412
Plowshare	L. M. Reed	Troy, Ohio	Jan. 5, 1869	85, 533
Plowshare-blanks, Plate for	W. Parlin	Canton, Ill	July 6, 1869	92, 349
Plowshare, Cast-iron	J. H. Conklin	Peekskill, N. Y	Jan. 13, 1830	....
Plowshare, Chilling	J. Oliver and H. Little	South Bend, Ind	June 30, 1857	17, 694
Plowshare, colter, and mold-board	W. Holt	Buffalo, N. Y	Aug. 27, 1835	....
Plowshare, Extension	G. W. Thorp	Columbus, Kans	Jan. 3, 1871	110, 693
Plowshare forging, bending, and shaping machine.	J. S. Hall	Pittsburgh, Pa	Oct. 14, 1862	36, 691
Plowshare, Shifting	J. Wood	Castleton, N. Y	Feb. 1, 1821	....
Plowshares and land-sides, Blank for	D. H. Rowe	Martinsville, Ill	Apr. 23, 1872	125, 909

Index of patents issued from the United States Patent Office from 1790 *to* 1873, *inclusive*—Continued.

Invention.	Inventor.	Residence.	Date.	No.
Plowshares, Apparatus for forming	E. Ball, jr	Canton, Ohio	Feb. 20, 1872	123, 805
Plowshares by rolling out, Making	J. Potts	Montgomery County, Pa	Nov. 7, 1818	....
Plowshares, Casting	J. Hunton	Haversack, N. J	Mar. 24, 1868	75, 919
Plowshares, Hardening edge of	R. L. and E. A. Stevens	Hoboken, N. J	Apr. 23, 1821	....
Plowshares, Machine for manufacturing	D. H. Rowe	Martinsville, Ill	Apr. 23, 1872	125, 908
Plowshares, Manufacture of	J. Lane	Chicago, Ill	Aug. 30, 1870	106, 838
Plowshares, Mold for casting	C. H. Brady	Mount Joy, Pa	July 1, 1862	35, 744
Plowshares to plow-stock, Fastening	W. G. Beckwith	Lowndesborough, Ala	Nov. 2, 1869	96, 383
Plow, Shield	M. Kirkham	Eminence Post-Office, Ind	May 19, 1868	78, 100
Plow shield, Corn	D. F. Brown and E. C. Brown	Champaign, Ill., and Crawfordsville, Ind.	Sept. 21, 1869	95, 079
Plow shield, Corn	J. Fox	Homer, Ind	Feb. 2, 1869	86, 521
Plow, Shifting-shovel	A. Snyder	Packard, Ohio	Dec. 6, 1870	109, 960
Plow, Shovel	I. A. Benedict	West Springfield, Pa	Dec. 13, 1870	109, 999
Plow, Shovel	F. H. Bowlds	Fairfield, Ky	Feb. 28, 1871	112, 212
Plow, Shovel	J. C. Boyd	Milroy, Ind	Oct. 9, 1866	58, 581
Plow, Shovel	J. F. Cameron	Livingston County, Mo	Mar. 6, 1860	27, 347
Plow, Shovel	H. C. Chandler	Erie Township, Ind	Jan. 14, 1868	73, 231
Plow, Shovel	J. C. Daugherty	Bridgeport, Ky	Oct. 30, 1866	59, 191
Plow, Shovel	P. Dennis	Bemus Heights, N. Y	Feb. 23, 1858	19, 412
Plow, Shovel	P. Dennis	Schuylerville, N. Y	June 19, 1866	55, 630
Plow, Shovel	D. Eberly	Waynesville, Ohio	Dec. 22, 1857	18, 894
Plow, Shovel	W. B. Evans	Bracken County, Ky	Oct. 27, 1868	83, 481
Plow, Shovel	C. Ford	Forest City, Ill	May 2, 1865	47, 536
Plow, Shovel	J. H. Forman	Sharon, Ala	Feb. 10, 1852	8, 721
Plow, Shovel	R. J. Gatling	Murfreesborough, N. C	May 29, 1847	5, 130
Plow, Shovel	D. Gilbert	Carbondale, Ill	June 25, 1867	66, 144
Plow, Shovel	F. Goss	Wexford, Pa	Dec. 28, 1869	98, 251
Plow, Shovel	W. R. Harmon	Unionport, Ohio	Sept. 17, 1867	68, 982
Plow, Shovel	S. W. Jackson	Baldville, Ohio	Feb. 19, 1867	62, 272
Plow, Shovel	A. Jennings	West Cairo, Ohio	Aug. 18, 1868	81, 173
Plow, Shovel	G. Jennings	West Cairo, Ohio	Aug. 24, 1869	93, 997
Plow, Shovel	W. Johnson and M. Ranney	Northfield, Ohio	May 6, 1873	138, 656
Plow, Shovel	E. H. Kamerer	Greentown, Ohio	Oct. 19, 1869	96, 010
Plow, Shovel	E. Knepper	Columbus, Ohio	Apr. 6, 1869	88, 723
Plow, Shovel	J. Lattimer	Chattoogaville, Ga	Mar. 16, 1852	8, 802
Plow, Shovel	W. H. Luce	Hampton, Ill	Sept. 18, 1866	58, 112
Plow, Shovel	L. Luppen	Pekin, Ill	Oct. 11, 1870	108, 272
Plow, Shovel	L. Luppen	Pekin, Ill	Oct. 11, 1870	108, 273
Plow, Shovel	L. Luppen	Pekin, Ill	Oct. 11, 1870	108, 274
Plow, Shovel	L. Luppen	Pekin, Ill	Oct. 11, 1870	108, 275
Plow, Shovel	L. Luppen	Pekin, Ill	Oct. 11, 1870	108, 276
Plow, Shovel	B. F. McCollester	California, Mo	Aug. 18, 1868	81, 188
Plow, Shovel	T. Meikle	Louisville, Ky	Dec. 5, 1871	121, 535
Plow, Shovel	J. Meyer	Bloom Township, Ohio	Aug. 18, 1868	81, 189
Plow, Shovel	J. J. Mitchell	Livingston, Ala	Aug. 5, 1873	141, 583
Plow, Shovel	I. Mosher and W. Eddy	Mosherville and Union Village, N. Y.	Mar. 31, 1863	38, 056
Plow, Shovel	W. C. Pagett	Greene County, Ohio	Oct. 17, 1842	2, 818
Plow, Shovel	W. F. Pagett	Stone Bridge, Va	Mar. 30, 1852	8, 842
Plow, Shovel	P. B. Parcell	Ashmore, Ill	Oct. 4, 1870	108, 047
Plow, Shovel	S. W. Pope	Louisville, Ky	Apr. 4, 1871	113, 341
Plow, Shovel	S. Riley	Northcutt's Store, Ky	June 8, 1869	91, 164
Plow, Shovel	G. H. Smith	Des Moines, Iowa	Apr. 9, 1872	125, 494
Plow, Shovel	M. E. Stanger	Wheeling, Ill	Sept. 24, 1867	69, 137
Plow, Shovel	U. T. Stewart	Fayette County, Tenn	Mar. 30, 1869	88, 344
Plow, Shovel	T. Terrell	Springhill, Ohio	May 8, 1866	54, 623
Plow, Shovel	R. Trowbridge	Waterloo, Iowa	July 2, 1867	66, 267
Plow, Shovel	M. D. Wells	Morgantown, Va	May 9, 1846	4, 500
Plow, Shovel	A. Wilcox	Maquoketa, Iowa	May 21, 1867	64, 927
Plow, Shovel	S. H. Wooldridge	Venice, Ill	Dec. 27, 1864	45, 664
Plow, Shovel-cutter	S. Wilson	Darlington, S. C	Feb. 6, 1830	....
Plow, Side-hill	D. C. Day	San José, Cal	July 18, 1871	117, 054
Plow, Side-hill	P. H. Flansburgh	Eden Township, Cal	June 25, 1867	66, 012
Plow, Side-hill	G. W. Leonard	Middle Valley, Pa	June 20, 1871	116, 070
Plow, Side-hill	D. A. Manuel	Napa City, Cal	May 5, 1868	77, 630
Plow, Side-hill	E. McKesson	Philip's Mills, Pa	Jan. 17, 1865	45, 929
Plow, Side-hill	J. Rich, jr	Troy, N. Y	Feb. 10, 1825	....
Plow, Side-hill	I. and H. H. Scoville	Oakland, Cal	May 4, 1869	89, 796
Plow, Side-hill	H. B. Smith	Springfield, Mass	Mar. 7, 1865	46, 716
Plow, Side-hill	N. Vars	New Market, N. J	Jan. 10, 1865	45, 882
Plow, Snow	W. N. Ball	La Porte, Ind	Feb. 14, 1865	46, 322
Plow, Snow	A. L. Bausman	Minneapolis, Minn	Aug. 8, 1865	49, 216
Plow, Snow	G. Beer	Grafton, Wis	Dec. 7, 1869	97, 474
Plow, Snow	T. S. Brown	Greenfield, Mich	Sept. 12, 1871	118, 784
Plow, Snow	R. Bustin	Saint John, New Brunswick.	May 4, 1869	89, 554
Plow, Snow	A. M. Butts	Waterbury, Conn	Oct. 25, 1870	108, 565
Plow, Snow	R. Cassaday and D. Clark	Buffalo, N. Y	Feb. 26, 1861	31, 529
Plow, Snow	T. C. Churchman	Sacramento, Cal	June 14, 1870	104, 113
Plow, Snow	M. A. and I. M. Cravath	Bloomington, Ill	Oct. 26, 1869	96, 205
Plow, Snow	J. N. Drake	Liberty, Mich	July 12, 1870	105, 185
Plow, Snow	A. Dunbar	New York, N. Y	Oct. 18, 1870	108, 338
Plow, Snow	C. L. Ericzon	Salt Lake City, Utah	Oct. 5, 1869	95, 449
Plow, Snow	N. S. Green	Weelaunee Post-Office, Wis	Aug. 18, 1868	81, 081
Plow, Snow	H. Harris	Circleville, Ohio	Nov. 24, 1868	84, 357
Plow, Snow	R. C. Harris	Maple Green, New Brunswick.	Sept. 20, 1870	107, 485
Plow, Snow	R. S. Harris	Dubuque, Iowa	July 23, 1867	67, 049
Plow, Snow	O. Heggem	Chicago, Ill	Mar. 19, 1872	124, 739
Plow, Snow	C. F. Hornbeck and W. J. Carns	Slaterville, N. Y	Dec. 6, 1870	109, 819
Plow, Snow	A. H. Jackson	Bear Valley, Cal	Oct. 19, 1869	96, 006
Plow, Snow	F. L. Knapp	Gosport, N. Y	Feb. 23, 1858	19, 426
Plow, Snow	S. Lewis	Brooklyn, N. Y	July 14, 1868	79, 913
Plow, Snow	C. Lusted	New York, N. Y	Jan. 7, 1868	73, 189
Plow, Snow	R. B. Nevens	Lowell, Mass	Sept. 18, 1866	58, 124
Plow, Snow	A. Nutting	Quincy, Mass	Dec. 10, 1867	71, 902
Plow, Snow	J. H. Pawling	Philadelphia, Pa	Mar. 9, 1858	19, 577

Index of patents issued from the United States Patent Office from 1790 *to* 1873, *inclusive*—Continued.

Invention.	Inventor.	Residence.	Date.	No.
Plow, Snow	G. Place	New York, N. Y	Feb. 25, 1868	74, 844
Plow, Snow	W. A. Plantz	Iowa Falls, Iowa	Jan. 5, 1869	85, 691
Plow, Snow	E. A. Putnam	Oakfield, Wis	Sept. 19, 1865	50, 031
Plow, Snow	S. Richards	Philadelphia, Pa	Apr. 13, 1858	19, 950
Plow, Snow	S. Richards	Philadelphia, Pa	June 11, 1867	65, 605
Plow, Snow	S. Richards	Philadelphia, Pa	Dec. 3, 1867	71, 642
Plow, Snow	W. J. Roberts	Cold Spring, N. Y	Sept. 23, 1873	143, 095
Plow, Snow	J. Sheridan	Saint Louis, Mo	Apr. 22, 1862	35, 066
Plow, Snow	J. Sheridan	Saint Louis, Mo	Apr. 19, 1864	42, 411
Plow, Snow	M. B. Spofford	Warsaw, N. Y	July 19, 1859	24, 829
Plow, Snow	F. J. Steinhauser	Lancaster, Pa	Feb. 19, 1861	31, 491
Plow, Snow	D. D. Stilwell	Philadelphia, Pa	Dec. 22, 1846	4, 898
Plow, Snow	A. Stutzman	Somerset, Pa	May 9, 1871	114, 623
Plow, Snow	B. B. Sweet and J. Noble	Wilmington, Del., and Philadelphia, Pa.	Oct. 14, 1873	143, 731
Plow, Snow	C. W. Tierney	Altoona, Pa	Mar. 16, 1869	87, 989
Plow, Snow	P. Vonlackum	Elba, Minn	Nov. 19, 1867	71, 249
Plow, Snow	W. Y. Warner	Wilmington, Del	May 19, 1868	78, 161
Plow, Snow	J. B. Williams	Madison, Wis	Sept. 20, 1864	44, 360
Plow, Snow	D. L. Winsor	Cambridge, Mass	July 16, 1867	66, 926
Plow, Snow	J. S. Zane	Pleasant Plains, Ill	Nov. 19, 1867	71, 111
Plow, Spade	E. Harris	Princeton, Ill	May 22, 1860	28, 368
Plow, Spade	D. Russell	Drewersburgh, Ind	Feb. 27, 1855	12, 466
Plow spring, Gang	H. N. Dalton	Pacheco, Cal	Oct. 5, 1869	95, 437
Plow, Steam	I. S. Allen, M. P. Browen, and C. W. Moulthrop.	San Francisco, Cal	Nov. 25, 1873	144, 820
Plow, Steam	S. K. Bassett	Galesburgh, Ill	Feb. 8, 1859	22, 848
Plow, Steam	W. Beckett	Kingston, Jamaica	May 31, 1870	103, 704
Plow, Steam	A. Bigelow	Hamilton, Canada	June 19, 1860	28, 732
Plow, Steam	T. H. Burridge	Saint Louis, Mo	July 31, 1860	29, 358
Plow, Steam	J. Curtis	Chicago, Ill	Sept. 6, 1864	44, 077
Plow, Steam	J. C. Delavigne	New Orleans, La	Mar. 31, 1868	76, 060
Plow, Steam	J. W. Evans	New York, N. Y	Oct. 5, 1858	21, 661
Plow, Steam	J. W. Fawkes	Christiana, Pa	Dec. 13, 1859	26, 422
Plow, Steam	J. W. Fawkes	Decatur, Ill	Dec. 10, 1861	33, 882
Plow, Steam	J. Fowler, jr., D. Greig, and R. Noddings.	London and Leeds, England.	Aug. 28, 1866	57, 653
Plow, Steam	J. Fowler, jr., W. Worby, and D. Greig.	Cornhill, Ipswich, and New Cross, England.	Aug. 28, 1866	57, 652
Plow, Steam	J. R. Gray	Fair Play, Wis	Oct. 20, 1857	18, 446
Plow, Steam	N. A. Gray	Cleveland, Ohio	Oct. 27, 1863	40, 403
Plow, Steam	N. A. Gray	Cleveland, Ohio	Apr. 21, 1868	77, 031
Plow, Steam	A. W. Hall	Saint Louis, Mo	Apr. 21, 1863	38, 260
Plow, Steam	J. Hawkins	Wilkins Township, Pa	Oct. 18, 1859	25, 826
Plow, Steam	O. Hyde	Oakland, Cal	Oct. 17, 1871	120, 071
Plow, Steam	C. F. Johnson, jr	Owego N. Y	Sept. 5, 1865	49, 761
Plow, Steam	P. Klingle	Linæon Hill, D. C	Feb. 23, 1858	19, 427
Plow, Steam	J. G. Knapp	Madison, Wis	Nov. 30, 1869	97, 299
Plow, Steam	M. N. Lynn	New Albany, Ind	May 31, 1870	103, 635
Plow, Steam	M. N. Lynn	New Albany, Ind	July 4, 1871	116, 610
Plow, Steam	P. S. McDonald	San Francisco, Cal	June 17, 1873	139, 966
Plow, Steam	A. E. McGaughey	East Minneapolis, Minn	Jan. 14, 1873	134, 913
Plow, Steam	A. E. and S. N. McGaughey	Wastedo, Minn	Nov. 29, 1859	26, 279
Plow, Steam	J. W. McLean	Indianapolis, Ind	Dec. 6, 1859	26, 397
Plow, Steam	W. H. H. Millen	Littleton, N. H	Dec. 11, 1860	30, 884
Plow, Steam	H. Miller	Bellville, Tex	Aug. 15, 1871	118, 143
Plow, Steam	J. H. Northcott	Mechanicsburgh, Ill	June 22, 1869	91, 558
Plow, Steam	E. G. Otis	Yonkers, N. Y	Oct. 20, 1857	18, 468
Plow, Steam	H. E. Paine	Milwaukee, Wis	Aug. 27, 1867	68, 310
Plow, Steam	G. W. Ramsey	New York, N. Y	Feb. 21, 1860	27, 242
Plow, Steam	O. Redmond	Rochester, N. Y	Oct. 23, 1866	59, 073
Plow, Steam	J. Reynolds	New York, N. Y	Dec. 18, 1860	30, 986
Plow, Steam	C. W. Saladee	Jefferson County, Tex	June 25, 1861	32, 652
Plow, Steam	S. L. Shotwell and S. R. Hicks	Ottawa, Ill., and North Hempstead, N. Y.	May 28, 1861	32, 437
Plow, Steam	G. Simonson	Mount Carmel, Ind	Apr. 24, 1866	54, 224
Plow, Steam	J. R. Smith	Trenton, N. J	Apr. 23, 1861	33, 153
Plow, Steam	D. B. Spencer	Parkersburgh, Va	Mar. 31, 1857	16, 937
Plow, Steam	L. Stewart	San Francisco, Cal	June 15, 1869	91, 383
Plow, Steam	A. L. Savean	Chaptico, Md	Mar. 29, 1870	101, 395
Plow, Steam	A. P. Thayer	Syracuse, N. Y	Nov. 24, 1863	40, 717
Plow, Steam	E. A. Tounley and E. S. Friedrich.	Washington, D. C	Apr. 27, 1869	89, 361
Plow, Steam	S. B. Wilkins	Milton, Pa	June 1, 1869	90, 799
Plow, Steam	G. Willard	New York, N. Y	Mar. 26, 1867	63, 349
Plow, Steam	L. B. Woolfolk	Nashville, Tenn	June 19, 1860	28, 801
Plow, Steam	L. B. Woolfolk	Nashville, Tenn	June 26, 1860	28, 933
Plow, Steam gang	W. H. H. Heydrick	Chestnut Hill, Pa	Mar. 26, 1867	63, 247
Plow-stock	J. A. Byrd	Jackson County, Fla	May 1, 1860	28, 064
Plow-stock	L. G. Peel	Preston, Ga	Mar. 2, 1869	87, 362
Plow-stock, Convertible	E. T. Parker	Berkley, Ala	Jan. 21, 1851	7, 910
Plow-stocks, Double-footed	J. L. Hood	Floyd County, Ga	July 13, 1869	92, 448
Plow, Stubble and subsoil	R. R. Fenner	Urbana, Ill	June 22, 1869	91, 617
Plow stubble-attachment	J. Kinney	London, Canada	Feb. 28, 1871	112, 152
Plow stubble-attachment	A. Osborn and E. Wulzen	Streator, Ill	Jan. 26, 1869	86, 245
Plow stubble-cleaner	J. Lacy and G. Watkins	Bristol, Wis	Oct. 4, 1864	44, 535
Plow stubble-turner	G. B. Smith	Shopiere, Wis	Nov. 21, 1871	121, 207
Plow, Submarine	E. T. Ligon	Demopolis, Ala	Sept. 17, 1867	69, 000
Plow, Submarine	A. Morrison and R. E. Rose	New Orleans, La	July 5, 1870	104, 981
Plow, Subsoil	G. B. Birmingham	Trenton, Tenn	July 1, 1873	140, 460
Plow, Subsoil	T. E. C. Brinly	Louisville, Ky	Sept. 17, 1872	131, 391
Plow, Subsoil	W. H. H. Burnham and S. B. Pierce.	Homer, N. Y	Aug. 27, 1861	33, 131
Plow, Subsoil	J. F. Cameron	Livingston County, Mo	Jan. 15, 1861	31, 108
Plow, Subsoil	L. Carleton	Pomeroy, Ohio	Oct. 18, 1870	108, 329
Plow, Subsoil	T. L. Cotten	Madison County, Miss	July 19, 1870	105, 551
Plow, Subsoil	J. Custer	Corsica, Ohio	Oct. 27, 1868	83, 472
Plow, Subsoil	S. E. Fletcher	Ballstown, Ind	Nov. 19, 1872	133, 216

Index of patents issued from the United States Patent Office from 1790 to 1873, inclusive—Continued.

Invention.	Inventor.	Residence.	Date.	No.
Plow, Subsoil	C. Garrett and T. Cottman	Cincinnati, Ohio	June 3, 1856	15, 039
Plow, Subsoil	E. Gross	Goshen Hill, S. C	Mar. 13, 1860	27, 626
Plow, Subsoil	C. K. Hartman	Vincennes, Ind	Apr. 14, 1868	76, 627
Plow, Subsoil	J. W. Howard	Greenville, Ala	Mar. 11, 1873	136, 726
Plow, Subsoil	J. H. Johnson	Bentonville, Ark	Oct. 18, 1870	108, 485
Plow, Subsoil	M. R. Jones	Bradford, Wis	Sept. 24, 1867	69, 099
Plow, Subsoil	M. R. Jones	Walworth County, Wis	Jan. 3, 1871	110, 660
Plow, Subsoil	P. Manny	Waddam's Grove, Ill	Apr. 22, 1856	14, 726
Plow, Subsoil	L. V. B. Marting	Tuscaloosa, Ala	Oct. 11, 1870	108, 164
Plow, Subsoil	J. McCollum	Brownsville, Ala	Apr. 23, 1861	32, 139
Plow, Subsoil	J. W. Murfee	Havanna, Ala	June 22, 1869	91, 657
Plow, Subsoil	C. Myers and W. Gummow	Marysville, Cal	Nov. 19, 1872	133, 167
Plow, Subsoil	W. C. Pagett	Xenia, Ohio	Oct. 22, 1850	7, 732
Plow, Subsoil	R. Peet	Castile, N. Y	June 4, 1867	65, 426
Plow, Subsoil	S. T. W. Potter	Scott, N. Y	July 15, 1862	35, 890
Plow, Subsoil	J. B. Pullman	Los Angeles, Cal	Apr. 27, 1869	89, 432
Plow, Subsoil	E. M. Duery	Harris Depot, N. C	July 26, 1870	105, 844
Plow, Subsoil	R. Themar and B. Brothers	Sheboygan, Wis	May 30, 1871	115, 543
Plow, Subsoil	J. R. Turner and J. Jacobs	Fredericktown, Mo	Nov. 11, 1873	144, 487
Plow, Subsoil	S. D. Tuttle	Eaton, Ohio	Mar. 31, 1868	76, 275
Plow, Subsoil	A. L. P. Vairin	Ripley, Miss	July 8, 1873	140, 660
Plow, Subsoil	J. Vaughn and E. Chamness	Miami County and Grant County, Ind.	May 19, 1868	78, 031
Plow, Subsoil	W. Watkins	Joliet, Ill	Feb. 1, 1870	99, 377
Plow, Subsoil	T. G. Wilder	Camden, Miss	Dec. 27, 1870	110, 525
Plow, Subsoil	J. Wood and R. North	Rochester, Wis	Nov. 10, 1857	18, 619
Plow, Subsoil	G. W. N. Yost	Nashville, Tenn	Sept. 15, 1863	40, 003
Plow subsoil-attachment	E. Cutcliffe	East Bethany, N. Y	Sept. 30, 1873	143, 335
Plow subsoil-attachment	C. Hayden	Collinsville, Conn	Mar. 10, 1868	75, 419
Plow subsoil-attachment	R. Johnson	Lawrence, Kans	May 10, 1870	102, 825
Plow subsoil-attachment	R. Johnson	Lawrence, Kans	Feb. 14, 1871	111, 852
Plow subsoil-attachment	J. A. Krake	Alden, N. Y	July 9, 1867	66, 597
Plow subsoil-attachment	J. C. Leonard and J. J. Gobar	Clinton, Mo	July 28, 1868	80, 356
Plow subsoil-attachment	A. F. Roberts	Lexington, Ky	Nov. 1, 1870	108, 939
Plow, Subsoil corn	H. Bacon	Tecumseh, Mich	June 5, 1849	6, 508
Plow, Subsoil gang	C. Myers	Marysville, Cal	Dec. 9, 1873	145, 361
Plow, Sulky	A. Q. Allis	Dayton, Ohio	Feb. 11, 1868	74, 268
Plow, Sulky	J. G. Boyd	Decatur, Tex	Apr. 5, 1870	101, 574
Plow, Sulky	J. Brewer	Albany, Ill	July 4, 1865	48, 512
Plow, Sulky	G. Burket and S. M. Gaskill	Bluffton, Ohio	Jan. 15, 1867	61, 151
Plow, Sulky	D. W. Colburn	Loami, Ill	Apr. 2, 1867	63, 473
Plow, Sulky	J. H. Cole	Vacaville, Cal	Apr. 5, 1870	101, 435
Plow, Sulky	W. B. Cummins	Leon, Iowa	Feb. 13, 1872	123, 616
Plow, Sulky	O. P. Dils	Falmouth, Ky	Mar. 5, 1867	62, 615
Plow, Sulky	I. Donaldson	Toledo, Iowa	Dec. 17, 1867	72, 375
Plow, Sulky	C. A. Edwards	Chatfield, Minn	Dec. 7, 1869	97, 490
Plow, Sulky	M. A. Elliott	Stratford Hollow, N. H	Mar. 7, 1871	112, 434
Plow, Sulky	R. R. Gaskill	Mendota, Ill	Sept. 11, 1866	57, 889
Plow, Sulky	A. Hale	Eureka, Ill	Nov. 6, 1866	59, 390
Plow, Sulky	B. R. Hubbard	Hillsborough, Ill	Sept. 6, 1870	107, 052
Plow, Sulky	D. W. Hughes	Mexico, Mo	July 1, 1873	140, 505
Plow, Sulky	L. Hunt	Weathersfield, Vt	May 5, 1868	77, 617
Plow, Sulky	W. H. Isaacs and G. E. Banner	Terre Haute, Ind., and Newark, N. J.	Jan. 26, 1869	86, 231
Plow, Sulky	J. R. Jackson	Pelahatchee Depot, Miss	Mar. 9, 1869	87, 577
Plow, Sulky	A. Keith	Lisbon, Ill	Nov. 7, 1865	50, 828
Plow, Sulky	G. Knight	Boone, Iowa	Oct. 30, 1866	59, 232
Plow, Sulky	E. Levee	West Point, Iowa	Feb. 18, 1868	74, 554
Plow, Sulky	J. B. Lewis and J. E. Udall	Concord, Ill	Sept. 1, 1868	81, 799
Plow, Sulky	C. H. Littlefield	Turner, Me	Feb. 26, 1867	62, 430
Plow, Sulky	S. M. Lockwood	Chicago, Ill	Mar. 27, 1866	53, 457
Plow, Sulky	H. W. Mason	Hagerstown, Md	Oct. 4, 1870	107, 938
Plow, Sulky	J. R. McConnell	Marengo, Iowa	Sept. 15, 1868	82, 140
Plow, Sulky	E. Meloy and A. R. Stanley	Shullsburgh, Wis	Jan. 31, 1871	111, 366
Plow, Sulky	W. Ough	Orion, Ill	Feb. 18, 1873	136, 003
Plow, Sulky	C. N. Owen	Salem, Ohio	Nov. 5, 1872	132, 772
Plow, Sulky	I. C. Pratt	Morton, Ill	Sept. 5, 1865	49, 833
Plow, Sulky	J. H. Rankin	Versailles, Ill	Nov. 5, 1867	70, 615
Plow, Sulky	J. J. Reed	Polo, Ill	Nov. 13, 1866	59, 654
Plow, Sulky	S. J. Reed	Camden, Ohio	July 21, 1868	80, 221
Plow, Sulky	J. C. Rogers	Alden, N. Y	June 18, 1867	65, 834
Plow, Sulky	J. Root	Hartland, N. Y	Jan. 12, 1869	85, 768
Plow, Sulky	B. Slusser	Sidney, Ohio	July 28, 1868	80, 427
Plow, Sulky	B. Slusser	Sidney, Ohio	Aug. 3, 1869	93, 358
Plow, Sulky	A. R. Stanley and H. W. Ensign	Shullsburgh, Wis	Sept. 1, 1868	81, 701
Plow, Sulky	S. Stout	Tremont, Ill	Sept. 11, 1866	57, 992
Plow, Sulky	B. W. Sutherlen	Freesoil, Minn	Apr. 14, 1868	76, 648
Plow, Sulky	J. L. Van Gorder	Sidney, Ohio	Aug. 2, 1870	106, 003
Plow, Sulky	J. L. Van Gorder	Sidney, Ohio	Aug. 29, 1871	118, 563
Plow, Sulky	J. Warwick	Springborough, Ohio	Jan. 7, 1868	73, 143
Plow, Sulky	I. Wing	Earlville, Iowa	June 25, 1867	66, 200
Plow, Sulky	T. Wolfe	Girard, Ill	Jan. 16, 1866	52, 104
Plow, Sulky	J. Worrell	Clayton, Ind	Dec. 17, 1872	134, 121
Plow, Sulky	J. Worrell and J. H. Rynerson	Clayton, Ind	June 4, 1872	127, 538
Plow, Sulky	P. Young	El Paso, Ill	July 24, 1866	56, 662
Plow sulky-attachment	J. H. Rynerson and J. Worrell	Clayton, Ind	July 26, 1870	105, 849
Plow sulky-attachment	G. M. Smith	Pittsburgh, Ind	May 30, 1871	115, 537
Plow sulky-attachment	J. Worrell and J. H. Rynerson	Clayton, Ind	Mar. 28, 1871	113, 234
Plow sulky-attachment	J. Worrell and J. H. Rynerson	Clayton, Ind	Mar. 28, 1871	113, 235
Plow sulky-attachment	J. Worrell and J. H. Rynerson	Clayton, Ind	Oct. 31, 1871	120, 560
Plow, Sulky gang	I. C. Pratt	Morton, Ill	Mar. 21, 1865	46, 974
Plow, Swivel	F. Holbrook, J. A. Howe, and J. Nourse.	Brattleborough, Vt., and Boston, Mass.	May 17, 1870	103, 187
Plow, Swivel	E. G. Mathews	Oakham, Mass	Nov. 1, 1870	108, 919
Plow, Trench	J. G. Robinson	Springfield, Ill	Mar. 30, 1869	88, 414
Plow, Trenching	W. Wise	Washington, D. C	Mar. 9, 1858	19, 597
Plow-truck	J. Clees	Darbyville, Ohio	Mar. 2, 1869	87, 398
Plow-truck	J. Flanagin	Pawnee City, Nebr	June 3, 1873	139, 564

Index of patents issued from the United States Patent Office from 1790 to 1873, inclusive—Continued.

Invention.	Inventor.	Residence.	Date.	No.
Plow-truck	J. G. Moore	Kingston, Ohio	Mar. 16, 1869	87, 866
Plow, Truck	M. Mickelson	Ashland Mills, Oreg	Aug. 10, 1869	93, 464
Plow, Turf	H. B. Sommers	Greenfield, Ind	June 5, 1847	5, 148
Plow, Twin	N. G. Cryer	Wentworth, N. C	Mar. 24, 1827	....
Plow, Under-drain	J. and E. Nevison	Morgan, Ohio	Nov. 30, 1858	22, 194
Plow, Underground-drain	A. Watson	Walnut Run, Ohio	Feb. 8, 1859	22, 906
Plow-wheel	A. J. Borland	Charlestown, Iowa	Aug. 8, 1871	117, 734
Plow-wheel	A. J. Borland	Donaldson, Iowa	Apr. 30, 1872	126, 255
Plow-wheel	L. C. Bristol	Holly, Mich	June 11, 1872	127, 734
Plow-wheel	E. A. Chubb	Ionia, Mich	Sept. 28, 1869	95, 194
Plow-wheel	G. Dodge	Kalamazoo, Mich	Aug. 6, 1867	67, 513
Plow-wheel	H. Galentine	Greece, N. Y	Nov. 7, 1871	120, 581
Plow-wheel	L. E. Palmer	Le Roy, Pa	Oct. 15, 1867	69, 834
Plow-wheel	M. Potter	Mendon, Mich	Feb. 25, 1873	136, 179
Plow-wheel	E. S. Rice	Paw Paw, Mich	Mar. 3, 1868	75, 054
Plow-wheel	J. B. Webster and R. Baxter	Stockton, Cal	Dec. 3, 1867	71, 827
Plow, Wheel	J. Bingham and O. M. Pond	Waterloo, Iowa	Dec. 2, 1873	145, 147
Plow, Wheel	T. D. Burrall	Geneva, N. Y	Oct. 28, 1843	3, 320
Plow, Wheel	H. C. Carr	Bordentown, N. J	Aug. 23, 1870	106, 548
Plow, Wheel	E. A. Chace	Rosemond, Ill	May 21, 1867	64, 839
Plow, Wheel	J. Cochrane	Indianola, Iowa	May 7, 1872	126, 447
Plow, Wheel	W. J. Connor	Benton, Ill	Sept. 12, 1871	118, 792
Plow, Wheel	R. Coreth	New Braunfels, Tex	Sept. 5, 1871	118, 694
Plow, Wheel	B. J. Crane	Ripon, Wis	July 16, 1872	128, 952
Plow, Wheel	W. G. Crossley	Apple River, Ill	Jan. 2, 1872	122, 442
Plow, Wheel	O. P. Dills	Falmouth, Ky	Sept. 11, 1866	57, 878
Plow, Wheel	J. C. Ferguson	Hydersville, Miss	June 22, 1836	....
Plow, Wheel	H. M. Freeman, J. Lowe, and J. F. Stevens.	Lathrop, Mo	Jan. 14, 1873	134, 878
Plow, Wheel	I. B. Green	Gillespie, Ill	May 20, 1873	139, 059
Plow, Wheel	W. S. Grimshaw	Pittsfield, Ill	Feb. 11, 1873	135, 800
Plow, Wheel	J. D. Harrison	Middletown, Ohio	Sept. 23, 1873	143, 147
Plow, Wheel	F. Hasbrook	Stokes' Mound, Mo	Apr. 29, 1873	138, 329
Plow, Wheel	F. Hasbrook	Stokes' Mound, Mo	Nov. 11, 1873	144, 453
Plow, Wheel	J. H. Janes, H. Tucker, and T. D. Terry.	Knobnoster, Mo	Apr. 8, 1873	137, 611
Plow, Wheel	J. Kay	Salem, Ind	Oct. 27, 1868	83, 507
Plow, Wheel	I. Long	Bucyrus, Ohio	Mar. 9, 1844	3, 465
Plow, Wheel	W. Mason	Monmouth, Oreg	Sept. 3, 1872	131, 063
Plow, Wheel	W. C. McCool	Guthrie Centre, Iowa	May 21, 1872	126, 976
Plow, Wheel	M. A. Melvin	Washington Court-House, Ohio.	Feb. 6, 1872	123, 499
Plow, Wheel	H. W. Neal	Wellsville, Pa	July 16, 1872	129, 161
Plow, Wheel	I. F. Nutting	Palmer, Mass	Nov. 7, 1865	50, 837
Plow, Wheel	C. N. Owen	Salem, Ohio	Feb. 6, 1872	123, 505
Plow, Wheel	W. H. and J. G. Parish	Portland, Oreg	Jan. 30, 1872	123, 194
Plow, Wheel	J. C. Pearl	Mendota, Ill	July 22, 1873	141, 073
Plow, Wheel	J. Pierce	Wyanet, Ill	Dec. 20, 1864	45, 520
Plow, Wheel	T. B. Quigley and H. Hall	Mansfield, Ohio	Oct. 7, 1845	4, 222
Plow, Wheel	L. W. Richardson	Roscoe, Ill	Aug. 19, 1873	141, 951
Plow, Wheel	J. H. and S. Robbins	Bethel, Oreg	Jan. 30, 1872	123, 294
Plow, Wheel	L. Sachse	Monmouth, Oreg	Mar. 11, 1873	136, 674
Plow, Wheel	M. Sattley	Taylorville, Ill	Feb. 4, 1873	135, 594
Plow, Wheel	H. M. Skinner	Rockford, Ill	July 1, 1873	140, 551
Plow, Wheel	C. B. Stevens	Donelson, Iowa	May 20, 1873	139, 032
Plow, Wheel	J. H. Suydam	Atwater, Minn	Dec. 20, 1870	110, 401
Plow, Wheel	J. E. Swallow	Hagerstown, Md	Aug. 9, 1870	106, 294
Plow, Wheel	J. E. Swallow	Hagerstown, Md	Nov. 29, 1870	109, 684
Plow, Wheel	W. P. Sweeny	Marine, Ill	Apr. 15, 1873	137, 975
Plow, Wheel	T. D. Terry, A. Case, and C. Larkin.	Knobnoster, Mo	Sept. 16, 1873	142, 964
Plow, Wheel	G. Tozer	Jackson, Mo	July 9, 1872	128, 829
Plow, Wheel	J. Worrell	Belleville, Ind	Mar. 18, 1873	137, 044
Plow, Wheeled	R. Wilson	Fowler, N. Y	Feb. 14, 1865	46, 412
Plow with double-pointed wings, Cast-iron	W. C. Pancost	Addison, Vt	Jan. 29, 1824	....
Plow, Wrought-iron	J. and H. F. Cromwell	Cynthiana, Ky	Sept. 30, 1841	2, 274
Plows, Adjustable draft-device for	A. J. Jackson	Millersburgh, Ill	Nov. 15, 1870	109, 215
Plows, Adjustable land-side for	G. Heffley, S. Conrad, and J. Wigle.	Berlin, Pa	Mar. 25, 1851	7, 994
Plows, Adjustable mole for Mole	L. Kazar	Leroy, Ill	Aug. 14, 1860	29, 601
Plows, Adjustable share for corn	D. Wolf	North Lebanon, Pa	Apr. 23, 1850	7, 315
Plows, and converting the same into two single plows, Constructing double hill-side.	M. Rich	Ithaca, N. Y	Mar. 24, 1838	653
Plows, Apparatus for cleaning colters of	A. Reeder	Wrightstown, Pa	May 18, 1858	20, 300
Plows, Apparatus for welding together the lay and land-side of.	J. Lane	Chicago, Ill	Aug. 4, 1868	80, 552
Plows, Attaching cast points to steel mold-boards of	A. and S. Peacock	Cincinnati, Ohio	July 10, 1855	13, 228
Plows, Attaching draft to	G. W. Kidwell	Elwood, Ind	Sept. 27, 1870	107, 692
Plows, Attaching handles to	J. L. Laughlin	Peru, Ill	May 23, 1871	115, 069
Plows, Attaching handles to joiners'	C. H. Weigle	York, Pa	May 3, 1870	102, 630
Plows, Attaching points to the shanks of subsoil	J. W. Murfee	Havanna, Ala	Aug. 9, 1870	106, 193
Plows, Attachment for revolving mold-board for	J. S. Godfrey	Leslie, Mich	Mar. 7, 1871	112, 333
Plows, Attachment for revolving mold-board for	J. S. Godfrey	Rochester, Pa	Apr. 25, 1871	114, 002
Plows, Attachment to	J. W. Murfee	Havanna, Ala	Aug. 9, 1870	106, 192
Plows, Bending mold-boards for	B. Pitcher	Peoria, Ill	Apr. 27, 1858	20, 128
Plows by attaching the mold-boards and sheaths, or cast-iron standards together by rivets.	B. F. Jewett	Springfield, Ill	Feb. 12, 1841	1, 976
Plows, Capstan for	H. C. Drew	Stockbridge, Mich	Feb. 19, 1861	31, 446
Plows, Capstan for ditching	E. Forbis	London, Ohio	July 24, 1860	29, 258
Plows, Carriage-attachment for	J. Zoeberlein	Baltimore, Md	Nov. 22, 1870	110, 567
Plows, Cast-iron cutter for	J. Swan	Cayuga County, N. Y	Apr. 24, 1824	....
Plows, Clod-fender and pulverizing-attachment for	E. Lanny	Mowrytown, Ohio	May 17, 1870	103, 203
Plows, Combination	B. F. McCarty and J. W. and R. J. Orr.	Florence, Ga	Aug. 10, 1869	93, 543
Plows, Combined	H. Brown	Payson, Ill	Mar. 9, 1844	3, 474
Plows, Combined	S. Cline	New Britain, Pa	Nov. 12, 1830	....
Plows, Combined	S. Cline	Berks County, Pa	Oct. 15, 1836	....
Plows, Combined	S. Cline	Bucks County, Pa	Oct. 15, 1836	53

Index of patents issued from the United States Patent Office from 1790 *to* 1873, *inclusive*—Continued.

Invention.	Inventor.	Residence.	Date.	No.
Plows, Combined	J. Knodle	Bakersville, Md	Apr. 8, 1840	1,543
Plows, Combined	J. and C. Krauser	Reading, Pa	Sept. 5, 1846	4,745
Plows, Combined	A. Leland	Milton, Pa	Jan. 2, 1849	5,998
Plows, Combined	J. Parsons	Dublin, Ind	Dec. 12, 1842	2,879
Plows, Composition-metal for casting	G. K. Smith	Waterloo, Iowa	Oct. 7, 1873	143,539
Plows, Constructing and fastening shares and cutters of.	M. Smith	Tinicum, Pa	Jan. 28, 1840	1,482
Plows, Corn-planting attachment for	J. F. Byland	Walton, Ky	Mar. 22, 1870	100,974
Plows, cultivator-teeth, &c., Plate and bar for construction of.	W. H. Singer	Pittsburgh, Pa	Oct. 25, 1870	108,641
Plows, Cutter-attachment for	J. Long	Leavenworth, Ind	Apr. 8, 1862	34,893
Plows, Cutter-attachment for	T. E. Marable	Petersburgh, Va	Aug. 18, 1868	81,187
Plows, Cutter-stock for swivel	E. G. Mathews	Oakham, Mass	May 17, 1870	103,215
Plows, Device for cleaning	R. Groom	Albany, N. Y	July 27, 1869	92,959
Plows, Device for cleaning weeds from	J. Jameson	Philadelphia, Pa	Aug. 6, 1867	67,550
Plows, Device for operating	J. O. Potter	Rouseville, Pa	Nov. 8, 1870	109,048
Plows, Device for securing clevis to	R. B. Prindle	Coventry, N. Y	June 14, 1859	24,403
Plows, Device for securing shield to	J. F. Cameron	Bedford, Mo	Jan. 15, 1861	31,109
Plows, Device for securing stock to guide-rod of joiners'.	S. Going	New York, N. Y	July 7, 1857	17,735
Plows, Draft-attachment for	S. H. Dailey	Olcott, N. Y	May 21, 1872	126,934
Plows, Draft-attachment for	N. Westcott	Nelson, N. Y	Nov. 18, 1873	144,811
Plows, Draft-regulator for	M. Prillaman	Tipton, Ind	Dec. 7, 1869	97,690
Plows, Draft-regulator for	M. Prillaman	Tipton, Ind	Jan. 16, 1872	122,853
Plows, Equalizing-attachment for	D. H. King and W. M. Hulse	Palmyra, Ill	July 8, 1873	140,632
Plows, Fastening colters to	A. and A. K. Whittlesy	Springport, N. Y	Oct. 22, 1850	7,736
Plows, Fastening shoe of hill-side	W. L. Chase	Boston, Mass	July 22, 1850	7,518
Plows, Fertilizer-attachment for	J. I. Boswell	Christiansville, Va	Nov. 2, 1869	96,388
Plows for canals, &c., Excavating	J. Richmond	Hudson, N. Y	June 28, 1817	....
Plows from choking, Device for preventing	W. M. Eccles	Saint Louis, Mo	July 26, 1870	105,660
Plows, Gage for constructing mold-boards of	E. Vaughn	Marion, N. Y	Feb. 5, 1861	31,362
Plows, Handle-seat for	E. Wiard	Louisville, Ky	Apr. 19, 1870	102,071
Plows, &c., Hardening and uniting iron and steel in the manufacture of.	W. Howell and N. W. Browning.	Webster City, Iowa	Nov. 24, 1868	84,420
Plows, Inverting hill-side and horizontal	J. W. Jordan	Lexington, Va	Oct. 28, 1835	....
Plows, Inverting hill-side and horizontal	J. W. Jordan	Lexington, Va	Apr. 19, 1839	1,128
Plows, Locomotive machine for propelling	W. P. Miller	Marysville, Cal	May 3, 1859	23,853
Plows, Machine for making	J. Urie	Evansville, Ind	May 3, 1870	102,731
Plows, Machine for making mold-boards for	H. H. Scoville	Syracuse, N. Y	May 29, 1860	28,507
Plows, Making	J. Deere	Moline, Ill	Apr. 2, 1867	63,369
Plows, Making	W. P. Long	Wheatland, Ind	Feb. 19, 1867	62,211
Plows, Manure-rake attachment for	M. Van Duine and J. De Jonge	Zeeland, Mich	Jan. 24, 1871	111,155
Plows, Mode of coupling and draft of	J. Card and G. Newell	Mentor, Ohio	Nov. 9, 1839	1,401
Plows, Mole for drain	T. S. Cox	La Fayette, Ind	Aug. 16, 1859	25,098
Plows, Mole for drain	A. B. Hawkins and J. Puntenney.	Cameron, Ill	June 12, 1860	28,667
Plows, Mole for drain	J. Lane	Lockport, Ill	Jan. 10, 1860	26,771
Plows, Mole for drain	A., E., and C. Marquiss and C. Emerson.	Monticello and Decatur, Ill.	Feb. 19, 1856	14,287
Plows, Mounting the cutters of rotary	P. H. Standish	Martinez, Cal	May 26, 1868	78,400
Plows of malleable cast-iron, Manufacture of	C. Alger	Boston, Mass	Aug. 3, 1838	869
Plows, Operating cutter for steam	O. Hyde	Oakland, Cal	Jan. 24, 1871	111,212
Plows, Operating drain	D. Watson	Newport, Ohio	Jan. 4, 1859	22,522
Plows, Plant-protector attachment to	J. Ahearn	Baltimore, Md	July 20, 1869	92,772
Plows, Plate for making blanks for mold-boards and shares for.	H. Barnes	Franklin Township, Iowa	Aug. 3, 1869	93,162
Plows, Poly-share for	F. Brewster	Burlington, Vt	Jan. 24, 1833	....
Plows, Portable capstan for mole	A. L. O. Wall, G. Roberts, and M. S. Cartter.	Decatur, Ill	Sept. 11, 1860	30,015
Plows, Portable crab for mole	E. Thorn	Selma, Ohio	Oct. 18, 1859	25,855
Plows, Pulverizing-attachment for	A. A. Rhoads and W. Tash	Berlin, Ill	Jan. 25, 1870	99,236
Plows, Regulating draft of	J. M. C. Arnsby	Worcester, Mass	Apr. 18, 1846	4,466
Plows, Revolving mold-board for	J. S. Godfrey	Rochester, Pa	Apr. 18, 1871	113,760
Plows, Riding-attachment for	A. H. Ballagh	Bowensburgh, Ill	Dec. 2, 1873	145,083
Plows, Riding-attachment for	A. E. and A. L. Porter	Lamoille, Ill	Mar. 15, 1870	100,800
Plows, Riding-attachment for gang	L. Dominy	Ottawa, Ill	Sept. 10, 1867	68,718
Plows, Scraper-attachment for	J. P. Swofford	Brownville, Miss	Feb. 11, 1873	135,863
Plows, Securing point to	H. D. Rogers	Grafton, Ohio	Oct. 9, 1860	30,352
Plows, Seed-sower and harrow-attachment for gang	J. B. Webster	Stockton, Cal	Apr. 4, 1868	77,141
Plows, &c., Self-adjusting equalizer for	J. T. Hagerty	Camp Point, Ill	June 29, 1869	92,042
Plows, Sower-attachment for	J. Brown	Southampton, Ill	Dec. 2, 1873	145,148
Plows, Spring-beam for	W. Morrison	Carlisle, Pa	Sept. 17, 1850	7,656
Plows, Steam-carriage for	M. N. Lynn	New Albany, Ind	July 25, 1871	117,307
Plows, Stop for carpenters'	T. Nichols	Vicksburgh, Miss	May 9, 1871	114,590
Plows, stoves, pipes, levees, dams, &c., Finishing iron-work of.	S. P. Townsend	Union County, N. J	May 28, 1867	65,306
Plows, Subsoil-pulverizing attachment for	L. E. Burdin	Paris, Ky	Oct. 4, 1870	108,002
Plows, Sulky-attachment for breaking	A. H. Allison	Charlottesville, Ind	Feb. 14, 1871	111,799
Plows, Swivel-clevis for	J. B. Small, E. G. Matthews, and F. F. Holbrook.	Boston, Oakham, and Boston, Mass.	Apr. 25, 1871	114,212
Plows to traction-engines, Attaching	W. H. H. Heydrick	Chestnut Hill, Pa	Aug. 15, 1871	118,015
Plows to wheeled carriages, Coupling	M. S. Curtiss	Bradford, Ill	Oct. 1, 1867	69,323
Plows, Traction-machine for	S. S. Stuntz	Jamestown, N. Y	May 17, 1870	103,102
Plows, Traction-wheel for rotary	L. S. Fithian	Rahway, N. J	Mar. 28, 1865	47,005
Plows, Varying the motion of	J. Sanford	Sharon, Conn	Aug. 20, 1811	....
Plows, Vibrating-colter for	J. S. Johnston	Rockford, Ill	Apr. 12, 1870	101,739
Plows, Weed-gatherer for	L. M. Doddridge	Mount Pleasant, Ind	Sept. 7, 1869	94,578
Plows, Weed-gathering device for	D. Hills	East Hartford, Conn	Oct. 7, 1844	3,777
Plows, Wheel-gage for	G. S. Deane	Grand Rapids, Mich	June 16, 1868	78,934
Plows, Whiffletree-attachment for	J. B. Ripsom	Kendall, N. Y	Jan. 15, 1867	61,260
Plower	W. Moore and W. Pruett	Kokomo, Ind	May 28, 1872	127,358
Plowing and breaking up ground, Machine for	J. Bryan	Lebanon, Ill	Jan. 8, 1869	90,921
Plowing and harrowing machine	A. Cunningham	Cincinnati, Ohio	Jan. 4, 1870	98,569
Plowing and tilling land by steam, Drum or pulley to prevent ropes from slipping in machinery for.	J. Fowler, jr., R. Burton, D. Greig, and J. Head.	Cornhill, Kingsland, New Cross, and Newcastle-upon-Tyne, England.	July 23, 1861	39,912
Plowing and tilling land by steam, Machinery for	J. Fowler, jr	Leeds, England	Apr. 9, 1861	32,026
Plowing and tilling land, Machinery for	J. Fowler, jr	Leeds, England	Apr. 9, 1861	32,025

Index of patents issued from the United States Patent Office from 1790 *to* 1873, *inclusive*—Continued.

Invention.	Inventor.	Residence.	Date.	No.
Plowing and tilling land, Machinery for	J. Fowler, jr	London, England	July 9, 1861	32, 809
Plowing and tilling land, Machinery for	J. Fowler, jr., and D. Greig	Leeds, England	Apr. 9, 1861	32, 027
Plowing, Device for burying weeds and stubble while.	J. W. Beer and J. B. Wampler	Shelbyville, Ill	Aug. 21, 1866	57, 279
Plowing-in stubble, Device for	J. Kilmer	Cobleskill, N. Y	Dec. 13, 1864	45, 454
Plowing-machine	A. Bondeli	Philadelphia, Pa	Dec. 21, 1869	98, 017
Plowing-machine	J. W. Fawkes	Christiana, Pa	Jan. 26, 1858	19, 189
Plowing-machine	D. D. Foley	Washington, D. C	Jan. 26, 1864	41, 371
Plowing-machine	H. Moeser	Pittsburgh, Pa	Nov. 10, 1857	18, 596
Plowing-machine	J. Pierce	Wyanet, Ill	Nov. 8, 1864	44, 969
Plowing-machine	W. Stoddard	Lowell, Mass	Jan. 26, 1858	19, 215
Plowing-machine, Steam	Z. Rider	Painesville, Ohio	July 18, 1871	117, 113
Plug-cutter	M. Jincks	Wallace, N. Y	Aug. 31, 1869	94, 416
Plug-cutter for wood-workers	D. L. Kniffen	West Camden, N. Y	Aug. 8, 1871	117, 786
Plug-cutting machines, Feed-motion in	S. Ingersoll	New York, N. Y	June 21, 1853	9, 799
Plug, Fusible safety	F. Curtis	Newburyport, Mass	Nov. 24, 1863	40, 677
Plumb and level	S. A. Bostwick	Laconia, N. H	Dec. 17, 1867	72, 159
Plumb and level	S. Sanderson	New Bedford, Mass	Oct. 21, 1873	143, 933
Plumb and level, Combined	F. A. Trout	New Britain, Conn	Apr. 21, 1863	38, 252
Plumb and level indicators, Method of attaching the plumb-line to.	J. L. Rowe	New York, N. Y	Mar. 30, 1858	19, 817
Plumb and square, Reversible	M. V. Nobles	Saint Anthony Falls, Minn	Jan. 19, 1864	41, 344
Plumb-bob	B. F. Chappell	Norwich, Conn	Nov. 13, 1860	30, 612
Plumb-bob	G. B. Cowles	Bridgeport, Conn	May 13, 1873	138, 790
Plumb, square, and level, Combined	A. T. Ward	Marietta, Ohio	July 21, 1868	80, 252
Plumbago-press	J. C. Ford	Cambridgeport, Mass	Mar. 29, 1870	101, 358
Plumbers' clamp	F. R. Langwith	New York, N. Y	July 29, 1856	15, 427
Plumbers' hook blank	B. F. Gladding	Providence, R. I	Aug. 8, 1865	49, 343
Plumbers' joints, Making	J. P. Hayes	Philadelphia, Pa	June 19, 1860	28, 750
Plush, paper, &c., Manufacture of frosted	H. V Edmund	Norwich, Conn	June 7, 1870	103, 999
Pneumatic and rubber spring, Combined	P. S. Devlan	Jersey City, N. J	Jan. 31, 1871	111, 438
Pneumatic brake and car-starter	C. A. Haskins	Chicago, Ill	May 28, 1872	127, 164
Pneumatic brake for railway-cars	C. R. Peddle	Terre Haute, Ind	Feb. 5, 1867	61, 860
Pneumatic dispatch-apparatus	C. W. Siemens	Westminster, England	July 29, 1873	141, 294
Pneumatic drill	S. Gwynn	New York, N. Y	Oct. 18, 1864	44, 722
Pneumatic drill	S. Gwynn	New York, N. Y	Nov. 1, 1864	44, 862
Pneumatic drill	H. Haupt	Cambridge, Mass	Mar. 7, 1865	46, 668
Pneumatic elevator for water	J. M. Blanchard and W. E. Prall	Washington, D. C	July 15, 1873	140, 879
Pneumatic elevator for wool	J. Penman	Paris, Canada	Mar. 19, 1872	124, 851
Pneumatic engine	J. F. Haskins	Fitchburgh, Mass	May 23, 1871	115, 198
Pneumatic engine	J. R. Remington	Aberfoil, Ala	May 26, 1842	2, 637
Pneumatic engine	R. Spear	New Haven, Conn	Dec. 14, 1869	97, 822
Pneumatic machine	D. Lloyd	Cambria County, Pa	Jan. 7, 1816	....
Pneumatic motor	T. B. Jeffrey	Chicago, Ill	Dec. 5, 1871	121, 626
Pneumatic signal-apparatus	W. Foster, jr	New York, N. Y	Aug. 26, 1873	142, 224
Pneumatic signaling-apparatus	W. Foster, jr	New York, N. Y	Apr. 4, 1871	113, 646
Pneumatic signaling-apparatus	W. Foster, jr	New York, N. Y	Apr. 4, 1871	113, 647
Pneumatic spring	J. Bevan and B. W. Hitchcock	Port Richmond and Flushing, N. Y.	Jan. 31, 1871	111, 303
Pneumatic spring	W. A. Dripps	Fort Wayne, Ind	Oct. 29, 1867	70, 177
Pneumatic spring	W. R. Feé	Cincinnati, Ohio	Mar. 30, 1858	19, 764
Pneumatic spring	J. F. Heyward	Wilmington, Del	July 10, 1855	13, 248
Pneumatic spring	I. W. Hoagland	New Brunswick, N. J	July 16, 1861	32, 848
Pneumatic spring	J. Lewis	New Haven, Conn	Feb. 10, 1847	4, 965
Pneumatic spring	M. F. Maury	Richmond, Va	Apr. 16, 1872	125, 749
Pneumatic spring	L. Sterne	London, England	Feb. 23, 1869	87, 307
Pneumatic spring	D. B. Strope	Fort Wayne, Ind	May 28, 1867	65, 135
Pneumatic spring	E. Ware	Roxbury, Mass	July 6, 1852	9, 107
Pneumatic tube	A. E. Beach	Stratford, Conn	May 7, 1867	64, 400
Pneumatic tube for transporting goods	A. Brisbane	New York, N. Y	June 22, 1869	91, 513
Pneumatic ways for transmission of parcels, &c.	E. P. Needham	New York, N. Y	Sept. 6, 1864	44, 107
Pneumo-hydraulic engine	R. Willcox	New York	Nov. 1, 1823	....
Pocket-alarm	I. Goodspeed	Norwich, Conn	Nov. 1, 1859	26, 001
Pocket-alarm	C. W. Simonds	Boscawen, N. H	Dec. 6, 1870	109, 952
Pocket-book	A. L. Adams	Philadelphia, Pa	Oct. 13, 1863	40, 223
Pocket-book	F. Busch	New York, N. Y	Jan. 30, 1872	123, 232
Pocket-book	A. Button	Dunkirk, N. Y	Dec. 21, 1869	98, 145
Pocket-book	S. C. Currie	New York, N. Y	June 8, 1869	90, 997
Pocket-book	J. C. Dickinson and R. Bate	Hudson, Mich	June 10, 1856	15, 068
Pocket-book	J. F. Dubber	Brooklyn, N. Y	Jan. 10, 1865	45, 819
Pocket-book	W. T. Fry	New York, N. Y	Feb. 20, 1866	52, 702
Pocket-book	E. Hassenpflug	Boston, Mass	June 2, 1863	38, 741
Pocket-book	F. Heiles	New York, N. Y	Oct. 12, 1869	95, 797
Pocket-book	W. H. King	Westbury, England	Feb. 18, 1873	135, 990
Pocket-book	J. Lehman	New York, N. Y	Dec. 6, 1870	109, 827
Pocket-book	L. Saarback	Philadelphia, Pa	July 11, 1865	48, 727
Pocket-book	L. Saarback	Philadelphia, Pa	Aug. 8, 1865	49, 307
Pocket-book	L. Wendt	Philadelphia, Pa	Apr. 22, 1873	138, 217
Pocket-book alarm	T. Blodgett and W. S. Weatherby	Belchertown and Granby, Mass.	May 18, 1869	90, 070
Pocket-book alarm	W. Stoddard	Lowell, Mass	Dec. 8, 1857	18, 826
Pocket-book and fan combined	O. Brück	New York, N. Y	Apr. 21, 1868	76, 988
Pocket-book and watch safe	J. Colton	Amherst, Mass	Apr. 18, 1846	4, 462
Pocket-book clasp	J. W. Brown and E. Rice	New York, N. Y	Mar. 27, 1866	53, 520
Pocket-book clasp	J. F. Dubber	Brooklyn, N. Y	Sept. 2, 1873	142, 447
Pocket-book clasp	D. E. Fisk	Springfield. Mass	Nov. 19, 1867	71, 154
Pocket-book clasp	H. Ropes	Brooklyn, N. Y	Mar. 5, 1867	62, 689
Pocket-book, &c., fastening	J. C. Arms	Northampton, Mass	June 16, 1868	78, 914
Pocket-book fastening	J. C. Arms	Northampton, Mass	Sept. 29, 1868	82, 577
Pocket-book fastening	F. Heiles	New York, N. Y	Oct. 28, 1873	144, 094
Pocket-book, Fire-proof	E. W. Glover and G. L. Damon	Malden and Cambridge, Mass.	Oct. 8, 1872	132, 068
Pocket-book guard	C. G. Cahoone and A. W. Robinson.	Providence, R. I	Dec. 26, 1871	122, 223
Pocket-book, porte-monnaie, &c	B. F. Cowan	New York, N. Y	Nov. 14, 1865	50, 909
Pocket-book protector	A Drummond	Newark, N. J	June 26, 1866	55, 838
Pocket-book safety-attachment	J. Brown, jr	New York, N. Y	May 2, 1871	114, 404
Pocket-book safety-attachment	W. H. Ferguson	Rochester, N. Y	Nov. 19, 1867	70, 982

Index of patents issued from the United States Patent Office from 1790 *to* 1873, *inclusive*—Continued.

Invention.	Inventor.	Residence.	Date.	No.
Pocket-book safety-attachment	A. McNeill	Cincinnati, Ohio	Apr. 21, 1868	77, 068
Pocket-book safety-attachment	S. B. Parker	New York, N. Y	Mar. 26, 1867	63, 291
Pocket-book safety-attachment	T. A. Weber	New York, N. Y	Apr. 2, 1867	63, 590
Pocket-book safety-guard	T. Potter	Jackson, Miss	May 6, 1873	138, 581
Pocket-book safety-hook	F. C. Likar	Baltimore, Md	May 6, 1873	138, 667
Pocket-books, Manufacture of	J. C. Arms	Northampton, Mass	Aug. 16, 1870	106, 450
Pocket-books, &c., Method of securing	O. Cox	Alexandria County, Va	Apr. 6, 1858	19, 832
Pocket-case for railway-schedule	S. E. Allen	Company Shops, N. C	May 25, 1869	90, 418
Pocket-clasp, Safety	A. T. Large	Tomah, Wis	Nov. 6, 1866	59, 415
Pocket-comb	W. J. Thorn	Westbrook, Me	May 17, 1853	9, 732
Pocket-flask	A. T. Becker	Cohoes, N. Y	Jan. 25, 1870	99, 051
Pocket-flask	R. George	New York, N. Y	Mar. 19, 1872	124, 733
Pocket-flask	R. George	New York, N. Y	July 16, 1872	129, 024
Pocket-flask, Self-heating	S. Hughes	Hudson, N. Y	Apr. 25, 1871	114, 143
Pocket for attachment to suspenders, Safety	J. Summers	Winchester, Va	June 29, 1869	92, 123
Pocket for billet-strap	H. H. Smith	Wilmington, Del	Feb. 25, 1868	74, 859
Pocket for garments	D. Harrington	Philadelphia, Pa	Oct. 11, 1841	2, 309
Pocket for garments, Money	E. Schnopp	East New York, N. Y	Apr. 9, 1872	125, 489
Pocket-hook, Safety	P. E. Malmström and P. E. Dummer.	New York, N. Y	Mar. 11, 1873	136, 742
Pocket-implement	J. A. Russ	Springfield, Mass	Mar. 19, 1867	63, 102
Pocket induction-apparatus	C. W. Meyer	New York, N. Y	Apr. 26, 1870	102, 295
Pocket-knife	F. H. Barnard and W. L. Brace	Hartford, Conn	Nov. 23, 1869	97, 154
Pocket-knife	G. H. Gardner	Philadelphia, Pa	July 5, 1864	43, 398
Pocket-knife	G. and J. Kay	Esopus, N. Y	Jan. 1, 1867	60, 905
Pocket-knife	S. Mason	Northfield, Conn	Aug. 26, 1862	36, 321
Pocket-knife	J. E. McBeth	New Orleans, La	June 23, 1868	79, 139
Pocket-knife	T. D. Miller	Kankakee, Ill	Mar. 13, 1866	53, 166
Pocket-knife	R. B. Milliken	Springfield, Vt	Jan. 15, 1867	61, 229
Pocket-knife	J. Mosley	New Haven, Conn	Mar. 24, 1868	75, 955
Pocket-knife	C. H. Palmer	Newark, N. J	Apr. 30, 1867	64, 248
Pocket-knife	J. G. Perry	South Kingston, R. I	Oct. 7, 1862	36, 620
Pocket-knife	H. Staude	Troy, N. Y	Apr. 22, 1873	138, 052
Pocket-knife	H. Twitchell	Naugatuck, Conn	Mar. 22, 1864	42, 031
Pocket-knife	A. G. Waterhouse	San Francisco, Cal	June 22, 1869	91, 693
Pocket-knife and door-fastener	J. Armstrong and O. Dame	Bucyrus and Wyandot County, Ohio.	Feb. 9, 1869	86, 808
Pocket-knife and envelope-opener, Combined	A. S. Pennington	Paterson, N. J	Jan. 23, 1872	122, 960
Pocket-knife and pistol, Combined	A. J. Peavey	South Montville, Me	Sept. 5, 1865	49, 784
Pocket-knife handle	J. W. Ayers	West Meriden, Conn	Apr. 8, 1873	137, 648
Pocket-knife handle	C. Beckham	Alexandria, Va	Sept. 5, 1871	118, 675
Pocket-lock, Burglar-proof	W. J. Scott	Albany, N. Y	Sept. 25, 1860	30, 160
Pocket of wearing-apparel	F. Darling	North Blackstone, Mass	Jan. 29, 1861	31, 227
Pocket of wearing-apparel, Secret	J. C. Howells	Washington, D. C	Aug. 11, 1863	39, 482
Pocket-opening, Fastening	J. W. Davis	Reno, Nev	May 20, 1873	139, 121
Pocket-register of count	P. D. Richards and F. N. Thayer.	New Orleans, La	July 12, 1859	24, 757
Pocket safe or fastening	G. R. McIlroy	Covington, Ky	May 5, 1857	17, 225
Pocket, Safety	C. H. Bagley	Elgin, Ill	July 3, 1866	56, 139
Pocket, Safety	J. A. Bell	Lacon, Ill	Aug. 14, 1866	57, 072
Pocket, Safety	J. T. Chambers	Utica, N. Y	Sept. 18, 1866	58, 062
Pocket, Safety	S. Chittenden	Willimantic, Conn	June 11, 1872	127, 845
Pocket, Safety	J. Colton	New Orleans, La	Dec. 3, 1867	71, 706
Pocket, Safety	S. Engel	Chicago, Ill	Mar. 18, 1873	136, 983
Pocket, Safety	R. W. Fisk	Olney, Ill	July 16, 1867	66, 698
Pocket, Safety	J. Foster	Sandwich, Mass	Apr. 20, 1869	89, 034
Pocket, Safety	S. French	Boston, Mass	June 24, 1862	35, 731
Pocket, Safety	S. French	Boston, Mass	June 9, 1863	38, 860
Pocket, Safety	L. Goodyear	Trumansburgh, N. Y	May 28, 1872	127, 339
Pocket, Safety	G. Hamel	Abington, Pa	Jan. 1, 1867	60, 720
Pocket, Safety	H. Harris	Newark, N. J	June 16, 1857	17, 573
Pocket, Safety	G. G. Hickman	Downingtown, Pa	Oct. 24, 1865	50, 582
Pocket, Safety	W. Hindhaugh	New York, N. Y	July 3, 1866	56, 049
Pocket, Safety	H. L. Hopkins	Morrisville, N. Y	Dec. 2, 1873	145, 178
Pocket, Safety	J. C. Iden	Buckingham Township, Pa	Apr. 28, 1868	77, 192
Pocket, Safety	E. Jones	Grand River, Iowa	June 27, 1871	116, 318
Pocket, Safety	G. A. Mitchell	Turner, Me	July 10, 1866	56, 248
Pocket, Safety	A. Q. Ross	Cleves, Ohio	Jan. 21, 1868	73, 655
Pocket, Safety	F. Russell	Cambridge, Mass	July 9, 1867	66, 638
Pocket, Safety	C. Sanborn	Chichester, N. H	Feb. 4, 1868	74, 008
Pocket, Safety	H. D. Synder	Carbondale, Pa	Oct. 29, 1867	70, 372
Pocket, Safety	P. A. Stecher	New York, N. Y	July 31, 1866	56, 821
Pocket, Safety	A. Wilmot	New Haven, Conn	Nov. 3, 1863	40, 528
Pocket safety-attachment	C. V. Boughton	Titusville, Pa	June 1, 1869	90, 632
Pocket safety-attachment	J. Brosius	Ranch's Gap, Pa	Apr. 7, 1868	76, 297
Pocket safety-attachment	A. W. Browne	Brooklyn, N. Y	Aug. 17, 1869	93, 856
Pocket safety-attachment	H. R. S. Colton	Houghton, Mich	Mar. 15, 1870	100, 728
Pocket safety-attachment	J. Metzendorf	New York, N.Y	Sept. 18, 1866	58, 120
Pocket safety-attachment	F. L. Roell	Northampton, Mass	Mar. 15, 1870	100, 805
Pocket safety-attachment	R. L. Russell	Brooklyn, N. Y	Nov. 18, 1873	144, 702
Pocket safety-attachment	E. Williams	New York, N. Y	Nov. 19, 1867	71, 106
Pocket safety-clasp	D. C. Tuck	Farmington, Me	Dec. 26, 1871	122, 140
Pocket, Safety watch	W. O. Sumner	Brooklyn, N. Y	Aug. 3, 1869	93, 246
Pocket, Seamless	G. Köttgen	Barmen, Prussia	Sept. 11, 1866	58, 035
Pocket, Suspender	J. F. Clark	Baltimore, Md	Nov. 28, 1865	51, 146
Poison, Precautionary attachment for bottles containing.	J. Harrison	Philadelphia, Pa	Jan. 3, 1871	110, 760
Poke	A. E. Cruttenden	Canaseraga, N. Y	May 10, 1870	102, 778
Poke, Animal	A. W. Bishop	York, Ohio	Aug. 1, 1865	49, 071
Poke, Animal	B. E. Blakslee	Medina, Ohio	Mar. 12, 1872	124, 478
Poke, Animal	G. W. Carpender	Jarvis, Ind	Dec. 17, 1867	72, 265
Poke, Animal	H. F. Chapin	Rochester, N. Y	Mar. 14, 1871	112, 546
Poke, Animal	H. F. Chapin	Rochester, N. Y	July 4, 1871	116, 683
Poke, Animal	N. Denny	Saranac, Mich	Aug. 8, 1871	117, 751
Poke, Animal	T. J. Dickerson	Auburn, N. Y	Dec. 10, 1872	133, 765
Poke, Animal	J. Green	Helmick, Ohio	Feb. 13, 1872	123, 692
Poke, Animal	C. Hinman	Akron, Ohio	Jan. 21, 1873	134, 988
Poke, Animal	C. Hinman	Akron, Ohio	June 17, 1873	139, 893

Index of patents issued from the United States Patent Office from 1790 *to* 1873, *inclusive*—Continued.

Invention.	Inventor.	Residence.	Date.	No.
Poke, Animal	J. Hopkins	Akron, Ohio	May 30, 1871	115, 319
Poke, Animal	J. Hopkins	Akron, Ohio	Aug. 5, 1873	141, 438
Poke, Animal	L. Kelley	Saranac, Mich	Feb. 28, 1871	112, 150
Poke, Animal	W. Kelly	Saranac, Mich	June 22, 1869	91, 639
Poke, Animal	W. Kelly	Saranac, Mich	July 19, 1870	105, 579
Poke, Animal	S. C. Leonard	Rushville, N. Y	Jan. 2, 1872	122, 392
Poke, Animal	R. McIntyre	Boston, Mass	June 25, 1872	128, 236
Poke, Animal	F. M. Moulton	Ferrisburgh, Vt	June 13, 1871	115, 884
Poke, Animal	I. N. Peck	Cohocton, N. Y	Oct. 1, 1872	131, 901
Poke, Animal	I. W. Sherwood	Mount Morris, N. Y	May 21, 1872	127, 107
Poke, Animal	J. N. Wallis	Fleming, N. Y	Dec. 10, 1872	133, 815
Poke, Cattle	S. P. Clemons	Dansville, N. Y	May 3, 1870	102, 497
Poke, Cattle	A. H. and M. P. Kennedy	Brunswick, Ohio	Nov. 6, 1866	59, 408
Poke, Cattle	O. and C. H. Sweet	South Glen's Falls, N. Y	Apr. 30, 1872	126, 239
Poke, Horse	M. J. Van Aken	Sturgis, Mich	Feb. 22, 1870	100, 088
Poke, Horse	G. Whitbeck	Phelps, N. Y	Mar. 15, 1870	100, 827
Poke, &c., Horse and cattle	G. W. Bell and G. W. Fulmer	Hinckley, Ohio	Apr. 21, 1868	76, 881
Poke, Horse and cattle	W. Miller	Granger, Ohio	June 14, 1870	104, 185
Poke, Horse and cattle	B. Page	Saint Joseph County, Ind	Nov. 26, 1872	133, 334
Poke, Horse and cattle	N. Silvester	Weymouth, Ohio	June 4, 1867	65, 440
Poke, Horse and cattle	N. Silvester	Weymouth, Ohio	Dec. 10, 1867	71, 916
Poker	J. F. Brener	Plantsville, Conn	Dec. 29, 1868	85, 425
Poker and lid-lifter combined	W. Clegg	Philadelphia, Pa	Feb. 2, 1869	86, 505
Poker and tongs, Combined	J. Blair	Saint Louis, Mo	Nov. 13, 1866	59, 547
Poker, Fire	J. C. Klein	Birmingham, Pa	Sept. 24, 1872	131, 618
Poker, Fire	G. R. Moore	Mount Joy, Pa	Feb. 5, 1856	14, 200
Poker, lifter, and tongs, Combined	P. Sander and J. Manthorn	Philadelphia, Pa	Feb. 6, 1872	123, 349
Poker, tongs, wrench, &c	J. Richards	Washington, D. C	Dec. 18, 1866	60, 556
Pole and hound coupling	F. Bremerman	Indianapolis, Ind	Dec. 1, 1868	84, 531
Pole and post puller	I. Holmes	South New Berlin, N. Y	Sept. 11, 1866	57, 905
Pole and thill coupling	H. L. Taylor	Fredonia, N. Y	Feb. 13, 1866	52, 621
Pole-ascending apparatus	G. Fleming	New York, N. Y	Mar. 8, 1870	100, 613
Pole-attachment	G. N. Compton	Canton, Ohio	Jan. 14, 1868	73, 234
Pole-clamp	H. Haering	New York, N. Y	Jan. 7, 1873	134, 598
Pole-socket	C. W. and J. D. Wilcox	Kingston, R. I	Apr. 26, 1870	102, 456
Pole-straps to neck-yokes, Attaching	N. Irish	Rochester, Minn	Aug. 18, 1868	81, 092
Poles, &c., Evener for wagon	H. F. Willson	Elyria, Ohio	Sept. 4, 1866	57, 806
Police-club	J. L. Rowe	New York, N. Y	Feb. 28, 1860	27, 335
Policeman's rattle	J. McCord	Philadelphia, Pa	Nov. 20, 1855	13, 823
Policeman's rattle	F. P. Smiley, C. K. Becker, and Q. Townsend.	Philadelphia, Pa	Nov. 27, 1866	60, 072
Polisher and grinder, Knife	G. L. Witsil	Philadelphia, Pa	July 11, 1865	48, 755
Polisher, Cylinder	G. Cowing	Seneca Falls, N. Y	Apr. 14, 1863	38, 153
Polisher, Cylindric	E. Stowell	Middlebury, Vt	Aug. 20, 1813	....
Polishing and cleaning metals, &c., Compound for	J. B. Emerson	Rochester, N. Y	Aug. 27, 1872	130, 907
Polishing and cleaning powder	J. W. Bates	Saint Paul, Minn	June 8, 1869	91, 070
Polishing and scouring composition	J. B. Rand	Concord, N. H	Oct. 31, 1871	120, 397
Polishing, Apparatus for holding plates for	A. S. Southworth and J. J. Hawes.	Boston, Mass	June 13, 1846	4, 573
Polishing, Artificial tripoli for	T. J. Platt	Newark, N. J	Jan. 8, 1867	61, 094
Polishing-box	W. A. Morse	Philadelphia, Pa	July 31, 1866	56, 781
Polishing buff and brush plates by revolving cylinder.	W. J. Stone	Washington, D. C	Apr. 30, 1831	....
Polishing-composition	C. J. H. F. Kleemann	New York, N. Y	Feb. 28, 1871	112, 154
Polishing-compound	E. J. Combs	Prattsburgh, N. Y	Nov. 11, 1873	144, 444
Polishing-device	J. C. Morrison	Chicago, Ill	May 20, 1873	139, 178
Polishing-fabric and flexible abrader	J. H. Crane	Charlestown, Mass	Sept. 8, 1868	81, 986
Polishing hard and soft substances, Machine for	B. Green	Hartford, Vt	Mar. 27, 1827	....
Polishing-iron	G. J. Prentiss	Fall River, Mass	Oct. 18, 1859	25, 850
Polishing knives, Composition for	A. W. Stewart	Middletown, Ohio	Dec. 3, 1867	71, 810
Polishing knives, &c., Machinery for	W. Vine	New York	Dec. 5, 1846	4, 875
Polishing-machine	A. Ball	Worcester, Mass	Nov. 10, 1863	40, 546
Polishing-machine	T. Beran	Philadelphia, Pa	Apr. 25, 1871	114, 098
Polishing-machine	J. Gooden	Lockport, N. Y	Nov. 29, 1870	109, 610
Polishing-machine	J. Moore	Gardiner, Me	July 22, 1856	15, 382
Polishing-machine	F. Pederson and R. Olsen	Racine, Wis	July 16, 1872	128, 975
Polishing-machine	P. F. Randolph	Jerseyville, Ill	Sept. 28, 1869	95, 265
Polishing-machine	P. F. Randolph	Jerseyville, Ill	May 2, 1871	114, 341
Polishing-machine	J. Stackhouse	West Pittsburgh, Pa	May 5, 1868	77, 546
Polishing-machine	H. Volkening	New York, N. Y	Oct. 10, 1854	11, 797
Polishing machine, Sheet-metal	H. Todd	Bridgeport, Conn	Mar. 5, 1867	62, 703
Polishing-machine, Traverse-motion for	J. Miller	New York, N. Y	Oct. 22, 1872	132, 481
Polishing metallic surfaces, Preparation for	C. M. Rowley	Chicago, Ill	Mar. 16, 1869	87, 793
Polishing metals, Composition of matter for	A. M. Sawyer	Athol, Mass	Dec. 4, 1866	60, 264
Polishing metals, glass, &c., Compound for	M. C. Bland	Chicago, Ill	July 23, 1872	129, 708
Polishing metals, Paste for	P. McManus and G. W. Latimer	Detroit, Mich	June 7, 1870	103, 909
Polishing plates used in taking likeness, &c., Apparatus for.	J. Johnson	New York	Dec. 14, 1841	2, 391
Polishing-powder	A. Hamilton and J. De Gray	Broad Brook, Conn., and Brooklyn, N. Y.	Oct. 24, 1865	50, 578
Polishing-powder	T. R. Hubbard	Brooklyn, N. Y	July 16, 1872	129, 029
Polishing-powder, Manufacture of	G. Wheelock	Washington, D. C	July 30, 1872	129, 999
Polishing powder, Metal	H. Greenwood	Providence, R. I	Jan. 28, 1818	
Polishing-tool	H. Cottrell	Newark, N. J	Nov. 18, 1873	144, 744
Polishing-tool	T. J. Mayall	Roxbury, Mass	Jan. 22, 1861	31, 180
Polishing-tools to shafts, Connecting rotary	H. F. Wheeler	Boston, Mass	Sept. 3, 1872	131, 140
Polishing turned articles, Machine for	W. Wadleigh	Sanbornton Bridge, N. H	July 21, 1863	39, 315
Polishing-wheel	A. Broad	Litchfield, Conn	June 6, 1814	....
Polishing-wheel	L. Holtzscheiter	Philadelphia, Pa	Apr. 10, 1866	53, 825
Polishing-wheel	A. M. Sawyer	Athol, Mass	Jan. 5, 1869	85, 538
Polishing-wheel	B. Webb	Unadilla Forks, N. Y	July 25, 1854	11, 396
Polishing-wheel, Elastic	L. Hale	Milford, Mass	Apr. 12, 1859	23, 571
Polishing-wheels, Attachment for balancing	H. K. Jones	Kensington, Conn	July 14, 1868	79, 984
Polishing-wheels, Balancing	H. K. Jones	Kensington, Conn	June 30, 1868	79, 475
Polishing-wheels, Balancing	A. W. Stephenson	Kensington, Conn	June 30, 1868	79, 405
Polishing-wheels, Holder for	E. W. Nichols	Worcester, Mass	Aug. 7, 1860	29, 512
Polygraph	N. Ames	Saugus, Mass	Dec. 12, 1854	12, 048
Polygraph	N. Ames	Saugus, Mass	Dec. 12, 1854	12, 049

Index of patents issued from the United States Patent Office from 1790 *to* 1873, *inclusive*—Continued.

Invention.	Inventor.	Residence.	Date.	No.
Pomatum, Putting up	E. Ludde	New York, N. Y	Dec. 6, 1870	109, 917
Poncho	P. W. Neefus	New York, N. Y	Mar. 17, 1863	37, 919
Ponton	H. Carberry and W. Tatham	Washington, D. C	July 19, 1816	....
Ponton	T. J. Mayall	Roxbury, Mass	Oct. 27, 1863	40, 421
Ponton-boat	E. Harrison	New York, N. Y	Nov. 3, 1863	40, 539
Ponton-boat	J. Hegeman	Vischer's Ferry, N. Y	Apr. 23, 1867	64, 103
Ponton-boat	R. Pallett	New York, N. Y	Nov. 24, 1863	40, 706
Ponton-equipage for military and other operations	J. F. Lane	Terre Haute, Ind	June 17, 1840	1, 634
Ponton wagon-boat, Metallic	J. Francis	New York, N. Y	Jan. 19, 1864	41, 288
Pool-ball rack	H. W. Collender	New York, N. Y	July 4, 1871	116, 561
Pool-board	H. W. Collender	New York, N. Y	June 20, 1871	116, 159
Pop-corn balls, Machine for forming	W. N. Thompson	Moline, Ill	July 16, 1872	129, 623
Porcelain, Architectural	W. J. Cheyney and E. F. Dieterichs.	Wallingford and Philadelphia, Pa.	Oct. 1, 1867	69, 315
Porcelain for use in filtering, &c., Porous	A. Vidal	Paris, France	May 18, 1869	90, 324
Porcelain, glass, &c., by the use of flurosilicate, Manufacture of.	T. Cobley and J. C. Coombe	Hahl, Bavaria, and Hoxton, England.	Apr. 28, 1863	38, 286
Porcelain, Manufacture of	W. J. Cheyney	Wallingford, Pa	Apr. 30, 1867	64, 196
Porcelain or pottery to receive designs, &c., Method of preparing the surface of.	J. Cartisser	New York, N. Y	Oct. 24, 1865	50, 561
Porcelain-paste, Apparatus for pulverizing	J. R. Alsing	Newcastle-upon-Tyne, England.	Jan. 25, 1870	99, 128
Porcelain, Plate	W. J. Cheyney and E. F. Dieterichs.	Wallingford and Philadelphia, Pa.	Oct. 1, 1867	69, 316
Porcelain printing-frame	C. L. Lochman	Carlisle, Pa	July 24, 1866	56, 579
Pork, Mode of curing	M. H. Smith	Brooklyn, N. Y	Aug. 14, 1866	57, 203
Porous ware, Manufacture of	B. S. and M. R. Pierce	New Bedford and Mansfield, Mass.	Dec. 27, 1859	26, 613
Port-hole closer	J. V. Murray and C. Borst	Brooklyn and New York, N. Y.	Apr. 5, 1864	42, 215
Port-hole for directing ordnance, Adjustable	P. G. Peltz	United States Navy	May 5, 1863	38, 407
Port-light for vessels	G. C. Gourlay	New York, N. Y	Oct. 23, 1860	30, 519
Port, Ship's	C. E. Marshall	Chicago, Ill	June 21, 1870	104, 474
Port-stopper fastening	J. Carver	Lincolnville, Me	July 22, 1873	141, 113
Portable battery or platoon-gun	C. Stauf and C. J. Steinbach	Saint Louis, Mo	Dec. 24, 1861	34, 017
Portable bath	N. Bartlett	Belvidere, Ill	Sept. 23, 1845	4, 203
Portable bath	J. W. Warren, jr	Boston, Mass	July 31, 1840	1, 710
Portable bellows, Construction of	A. Ewing	New York, N. Y	Oct. 31, 1839	1, 393
Portable boat	R. C. Buchanan	Baltimore, Md	Mar. 31, 1857	16, 904
Portable-boat	R. Knudsen and W. T. Lassoe	Brooklyn, N. Y	June 19, 1866	55, 675
Portable body-battery	S. J. Seeley	Brooklyn, N. Y	Dec. 3, 1861	33, 854
Portable boiler for pitch, &c	L. S. Mills	Brooklyn, N. Y	Mar. 26, 1867	63, 286
Portable box	A. Dreyspring	Montgomery, Ala	Oct. 12, 1858	21, 748
Portable-box alarm	W. Shepard and A. T. Murden	Brooklyn, N. Y	Jan. 9, 1872	122, 540
Portable chair	G. Gardner	Clarksville Post-Office, N. J.	Jan. 17, 1871	111, 053
Portable chair	Z. Lyford	Lowell, Mass	May 13, 1856	14, 877
Portable clamp for school-books	T. H. Dennison	Baltimore, Md	Aug. 3, 1869	93, 287
Portable elevator and conveyor	H. C. Crosby	Chicago, Ill	Nov. 8, 1870	108, 976
Portable evaporator	H. T. Douglas and J. Cooper	Zanesville and Mount Vernon, Ohio.	Sept. 6, 1859	25, 323
Portable evaporator	H. L. Plumb	Hamer, Ohio	Dec. 17, 1867	72, 326
Portable fountain	T. H. Bonham	Elizabethtown, Ohio	May 8, 1866	54, 494
Portable fountain	J. Hegarty	Jersey City, N. J	Oct. 22, 1867	69, 994
Portable fountain and lawn-sprinkler	R. Brusie	New York, N. Y	Mar. 25, 1873	137, 175
Portable furnace	H. Beatty	Massillon, Ohio	Aug. 15, 1871	118, 095
Portable furnace	T. Chadwick	Newton, Iowa	Feb. 22, 1870	100, 117
Portable furnace	B. G. Fitzhugh	Frederick, Md	Oct. 17, 1871	119, 977
Portable furnace	S. Gregory	Galesburgh, Mich	May 31, 1870	103, 736
Portable furnace	J. H. Harper	Pittsburgh, Pa	Dec. 7, 1869	97, 634
Portable furnace	J. P. Hayes	Boston, Mass	Nov. 5, 1850	7, 765
Portable furnace	J. Hulbert, jr	Richmond, Ind	Apr. 5, 1870	101, 621
Portable furnace	W. J. Keep	Troy, N. Y	Feb. 13, 1872	123, 706
Portable furnace	J. Lewis	Derby, Conn	July 8, 1834	....
Portable furnace	H. J. Lingenfelter	Glen, N. Y	Dec. 30, 1873	145, 954
Portable furnace	J. H. Lyon	Pittsburgh, Pa	Apr. 12, 1870	101, 893
Portable furnace	A. Magee	Lawrence, Mass	Nov. 10, 1863	40, 570
Portable furnace	J. B. Marvin	New York, N. Y	Mar. 20, 1860	27, 552
Portable furnace	D. B. Montague	Springfield, Mass	Feb. 28, 1871	112, 164
Portable furnace	J. L. Mott	New York, N. Y	Oct. 19, 1838	983
Portable furnace	M. F. Potter	Charlemont, Mass	Jan. 22, 1850	7, 039
Portable furnace	J. Reynolds	Philadelphia, Pa	Mar. 21, 1871	112, 961
Portable furnace	W. D. Robertson	Knoxville, Tenn	Mar. 21, 1871	112, 965
Portable furnace	M. Saulson	Troy, N. Y	Sept. 3, 1872	131, 030
Portable furnace	T. Stead and J. S. Oram	Cleveland, Ohio	July 30, 1872	130, 081
Portable furnace	C. Van De Mark	Phelps, N. Y	Nov. 29, 1870	109, 780
Portable furnace	G. E. Waring	Stamford, Conn	Mar. 16, 1844	3, 491
Portable furnace and kettle	F. Sapaire and F. X. Jacquot	Wilmington, Del	Nov. 19, 1872	133, 231
Portable furnace and wash-boiler, Combined	H. and T. Humphreville	Lancaster, Pa	June 18, 1872	127, 977
Portable furnace for casting brass, &c	J. Share	Baltimore, Md	Mar. 30, 1813	....
Portable furnace for heating water in bathing-tubs	R. Densmore	Hopewell, N. Y	Nov. 26, 1840	1, 866
Portable heater	A. T. Hutson	Richmond, Ind	Nov. 6, 1866	59, 397
Portable heater	W. Shaw	Albany, N. Y	Nov. 14, 1871	120, 905
Portable house	J. Burnside	Washington, D. C	Oct. 21, 1862	36, 701
Portable house	D. Fitzgerald	New York, N. Y	Mar. 4, 1856	14, 355
Portable house	D. Fitzgerald	New York, N. Y	Mar. 12, 1867	62, 831
Portable house	H. W. Forman	Centralia, Kans	July 23, 1872	129, 805
Portable house	A. E. Hotchkiss	Hubbardsville, N. Y	Oct. 29, 1872	132, 533
Portable house	W. R. Mears	Grafton, Ill	Apr. 6, 1869	88, 657
Portable house	D. N. Skillings	Boston, Mass	Nov. 19, 1861	33, 758
Portable house, Sectional	J. Montgomery	New York, N. Y	Feb. 2, 1869	86, 572
Portable houses, Constructing	D. Fitzgerald	New York, N. Y	May 27, 1856	14, 952
Portable houses, Constructing	D. Fitzgerald	New York, N. Y	May 12, 1857	17, 270
Portable houses, Constructing frame for	J. Morgan	Dundee, Ohio	Jan. 3, 1865	45, 734
Portable houses for emigrants, Constructing	F. S. Barnard	Philadelphia, Pa	Dec. 21, 1839	1, 439
Portable lock	A. D. Perry	New York, N. Y	Sept. 5, 1848	5, 751
Portable lubricator	H. Hammond	Hartford, Conn	Jan. 25, 1870	99, 083
Portable mill	N. Burr	Batavia, Ill	Aug. 7, 1860	29, 462

Index of patents issued from the United States Patent Office from 1790 to 1873, inclusive—Continued.

Invention.	Inventor.	Residence.	Date.	No.
Portable mill	S. Dodson	Jersey City, N. J	Dec. 31, 1867	72, 822
Portable mill	B. Harnish and R. J. King	Lancaster, Pa	Mar. 9, 1869	87, 667
Portable press	T. L. Chase	Philadelphia, Pa	Nov. 21, 1865	51, 019
Portable refreshment-fountain	A. J. Ohmer	Hamilton, Ohio	Dec. 19, 1865	51, 612
Portable register	C. S. Watson	Philadelphia, Pa	Dec. 27, 1859	26, 639
Portable roller for moving heavy bodies	J. R. Anderson	New York, N. Y	Apr. 18, 1871	113, 830
Portable switch	B. C. Galvin	New York, N. Y	Dec. 17, 1867	72, 184
Portable sprinkler	W. L. Fish	Springfield, Mass	Aug. 30, 1870	106, 802
Porte-monnaie	E. Lindner and C. Hoffman	New York, N. Y	Feb. 12, 1856	14, 246
Porte-monnaie	J. L. Mason	Germantown, Pa	Oct. 14, 1856	15, 891
Porte-monnaie and pocket-book fastening	J. Hewson	Newark, N. J	May 13, 1856	14, 867
Porte-monnaie, Armlet	A. W. Pratt	Pultneyville, N. Y	Mar. 10, 1868	75, 458
Porte-monnaie lock and clasp	D. C. Smith	Tecumseh, Mich	Apr. 14, 1857	17, 057
Porte-monnaie, Machine for manufacturing	B. S. Stedman	West Meriden, Conn	Sept. 14, 1852	9, 264
Porters' box	I. Barman	Portland, Oreg	Sept. 16, 1873	142, 888
Portfolio	R. Arthur	Philadelphia, Pa	May 19, 1857	17, 311
Portfolio	G. and E. Ashworth	Manchester, Great Britain	Sept. 8, 1868	81, 864
Portfolio	G. Harvey	New York, N. Y	Jan. 2, 1872	122, 461
Portfolio	L. Heyl	Philadelphia, Pa	Sept. 6, 1864	44, 094
Portfolio	L. Heyl	Philadelphia, Pa	Mar. 20, 1866	53, 299
Portfolio	C. W. Palmer	Cleveland, Ohio	July 8, 1873	147, 725
Portfolio	W. H. Pratt	Davenport, Iowa	July 16, 1872	129, 589
Portfolio	L. F. Raymond	Bordeaux, France	Dec. 16, 1873	145, 683
Portfolio	J. Shaw	Providence, R. I	Dec. 24, 1850	7, 860
Portfolio	J. Shaw	Providence, R. I	June 17, 1856	15, 150
Portfolio	H. T. Sisson	Providence, R. I	Dec. 29, 1857	18, 994
Portfolio	H. T. Sisson	Providence, R. I	Apr. 5, 1859	23, 506
Portfolio	D. Slote	New York, N. Y	Sept. 17, 1872	131, 473
Portfolio	S. J. H. Smith	Boston, Mass	June 5, 1855	13, 014
Portfolio and writing-tablet	G. C. Hathaway	Plymouth, Mass	Dec. 24, 1861	33, 991
Portfolio, Commercial	S. E. Mandlebaum	Saint Louis, Mo	Feb. 22, 1870	100, 050
Portfolio for filing music, printed matter, &c	J. C. Koch	New York, N. Y	May 21, 1861	32, 374
Portfolio-holder	J. B. Aiken	Franklin, N. H	Sept. 16, 1873	142, 884
Portfolio or music stand	A. Eliaers	Boston, Mass	Sept. 22, 1857	18, 239
Portfolio-stand	L. Dubernet	New York, N. Y	May 9, 1865	47, 624
Post:				
See Arbor or fence post. / Hitching and sign post.				
Clothes-post. / Iron post.				
Clothes-line post. / Iron gate or fence post.				
Composition-post. / Jack-post.				
Fence-post. / Lamp-post.				
Fence and trellis post. / Newel-post.				
Gas-lamp post. / Sewing-machine post.				
Gate-post. / Stone post.				
Guide-post. / Telegraph-post.				
Hitching-post.				
Post and pile driver	O. Hyde	Benicia, Cal	June 15, 1858	20, 563
Post-boring and rail-pointing machine	R. Martin and M. Harner	Taneytown, Md	June 21, 1870	104, 476
Post-boring and rail-pointing machine	J. Young	Middletown, Md	July 10, 1855	13, 243
Post-boring and rail-pointing machine	J. Young and C. I. Grumbine	Fairview and Frederick, Md.	May 25, 1869	90, 622
Post-boring machine	A. Butler	Vermont	Nov. 7, 1811	
Post-boring machine	W. F. Cline and P. D. Weaver	Bendersville, Pa	Nov. 14, 1871	120, 941
Post-boring machine	J. R. Group	Idaville, Pa	Feb. 7, 1871	111, 532
Post-boring machine	B. F. Mohr	Mifflinburgh, Pa	Mar. 3, 1868	75, 041
Post-driver	I. V. Adair	Romulus, N. Y	Mar. 25, 1873	137, 120
Post-driver	A. B. Clark	Richmond, Ind	Aug. 4, 1868	80, 600
Post-driver	J. Ellenberger	Easton, Ohio	Nov. 17, 1868	84, 100
Post-driver	C. T. Fitch	Harbor Creek, Pa	Nov. 26, 1867	71, 372
Post-driver	D. D. Foley	Washington, D. C	Nov. 27, 1866	59, 990
Post-driver	A. McClara	Whitney's Point, N. Y	Nov. 3, 1868	83, 692
Post-driver	S. McCullough and A. Robins	Buffalo, Ohio	Dec. 24, 1867	72, 518
Post-driver	J. M. Sampson	Waynesville, Ill	Mar. 25, 1856	14, 520
Post-driver	C. H. Williamson and J. H. Allyn.	Whitestown, N. Y	Dec. 31, 1872	134, 412
Post-driving machine	W. Altick	Dayton, Ohio	Aug. 10, 1869	93, 576
Post-driving machine	C. T. Fitch	Harbor Creek, Pa	May 7, 1867	64, 514
Post frost-protector	S. T. Varian	Plainfield, N. J	Feb. 16, 1869	87, 015
Post, Hitching	D. C. Robie	Springfield, Mass	Sept. 17, 1867	69, 027
Post-hole borer	J. Cothron	Illiopolis, Ill	Sept. 22, 1868	82, 293
Post-hole borer	J. K. Miller	New York, N. Y	Mar. 24, 1868	75, 947
Post-hole borer	J. M. Walter and S. Shank	Springfield, Ohio	Sept. 15, 1868	82, 185
Post-hole-boring machine	A. Q. Allis	Dayton, Ohio	Feb. 25, 1868	74, 785
Post-hole-boring machine	J. Corbett	Hanover, Pa	Dec. 21, 1819	
Post-hole-boring machine	W. I. Hale	Ashley, Ill	Apr. 14, 1868	76, 752
Post-hole-boring machine	W. R. Iles	West Rushville, Ohio	Dec. 24, 1867	72, 497
Post-hole-cutting instrument	N. S. Vance and E. Watkins	Decatur, Ill	Oct. 22, 1867	70, 137
Post-hole digger	W. E. Ball	Bethesda, Ohio	May 14, 1872	126, 773
Post-hole digger	J. Boone	Clintonville, Ky	June 14, 1864	43, 085
Post-hole digger	J. W. Harper	Xenia, Ohio	Feb. 19, 1867	62, 131
Post-hole digger	B. B. Herrick	Decatur, Mich	Mar. 7, 1871	112, 454
Post-hole digger	B. B. Herrick and C. W. Wicker.	Duquoin, Ill	Nov. 23, 1869	97, 086
Post-hole digger	J. Lee	Bolivar, Ohio	June 26, 1860	28, 874
Post-hole digger	A. E. Lindsley	Paw Paw, Mich	Feb. 14, 1871	111, 757
Post-hole digger	A. E. Lindsley	Paw Paw, Mich	May 20, 1873	139, 169
Post-hole digger	E. R. Sumner	Freeport, Ill	Aug. 29, 1871	118, 560
Post-hole digger	E. R. Sumner	Freeport, Ill	Oct. 29, 1872	132, 607
Post-hole digger and ditch	R. L. Jones	Cleveland, Ohio	July 25, 1871	117, 297
Post-hole-digging machine	A. W. Porter	Saint Johnsville, N. Y	Aug. 7, 1860	29, 518
Post-hole-digging machine	J. Thompson and B. B. Herrick.	Edgewood, Ill	Nov. 19, 1867	71, 086
Post-hole-excavating machine	W. K. Johnston	Rock Island, Ill	Mar. 9, 1858	19, 565
Post-hole excavator	H. Sutliff	Waverly, N. Y	Sept. 14, 1869	94, 790
Post-hole-making machine	J. M. Kirkpatrick	Utica, Ohio	Dec. 20, 1870	110, 372
Post-mold for fences, Stone	E. Boston	Gloucester, N. J	Oct. 14, 1815	
Posts, Setting	P. Burke	Helena, N. Y	Sept. 24, 1861	33, 335
Posts, Setting	U. B. Stribling	Madison, Ind	Sept. 20, 1870	107, 637

Index of patents issued from the United States Patent Office from 1790 *to* 1873, *inclusive*—Continued.

Invention.	Inventor.	Residence.	Date.	No.
Posts, timber, &c., Composition for setting	A. D. Purinton	Dover, Mass	June 9, 1868	78, 691
Post-office box	E. Stockwell and W. H. Taylor.	Stamford, Conn	June 17, 1873	140, 092
Post-office-box front	C. J. Clements	New York, N. Y	June 18, 1872	128, 116
Post-office-box lock	L. Yale, jr	Bernardston, Mass	Oct. 24, 1871	120, 177
Post-office cabinet for Sunday-school and other rooms.	W. M. and A. Clark	Pittsburgh, Pa	Aug. 16, 1870	106, 325
Post-office-delivery box	J. H. Beidler	Lincoln, Ill	Aug. 28, 1866	57, 464
Post-office distributing-table	C. H. Bradley	West Chester, Pa	July 9, 1861	32, 748
Post-office letter-box	C. H. Bradley	West Chester, Pa	Sept. 15, 1863	39, 879
Post-office letter-box	B. C. Davis	Binghamton, N. Y	Dec. 7, 1869	97, 483
Post-office letter-case	A. F. Miller	Galva, Ill	Apr. 9, 1872	125, 606
Post-office, &c., pigeon-hole	H. G. Pearson	New York, N. Y	Nov. 5, 1867	70, 674
Postal card	H. M. Johnston	New York, N. Y	Nov. 18, 1873	144, 677
Postal-card case	C. H. Townsend, T. E. Hughes, and A. A. Keith.	Minneapolis, Minn	Nov. 11, 1873	144, 423
Postal-card machine	J. M. D. Keating	New York, N. Y	Apr. 22, 1873	138, 028
Postal-card screen	L. Conroy	New York, N. Y	Sept. 16, 1873	142, 899
Postal wrapper	E. Reed	Joliet, Ill	Nov. 6, 1866	59, 451
Pot:				
See Annealing-pot. Batter-pot. Bean-pot. Bone burning or charring pot. Chamber-pot. Coffee-pot. Culinary-pot. Dinner-pot. Eel-pot. Fire-pot. Flower-pot. Glass-pot. Glass-house pot. Glass-maker's pot. Glass-making pot. Glass-melting pot. Glue-pot. Lead-pot. Marking-pot. Molder's pot. Mucilage-pot. Mustard-pot. Paint-pot. Painter's pot. Salinometer-pot. Sprinkling-pot. Sugar-drip pot. Tea-pot. Tea and coffee pot.				
Pot and kettle cover	S. Van Horn	Chicopee, Mass	Aug. 8, 1871	117, 834
Pot and kettle tipping attachment	C. Coester, jr., and W. L. Dewey.	Bridgeport, Conn	Jan. 29, 1867	61, 654
Pot and kettle tilting device	A. J. Redway	Cincinnati, Ohio	Sept. 17, 1872	131, 410
Pot-cover	W. H. Barker	Windsor, Canada	Sept. 12, 1871	118, 896
Pot, kettle, &c., cover	G. Reuben	San Francisco, Cal	Dec. 29, 1868	85, 398
Pot, kettle, &c., elevator	J. Edwards	New York	Mar. 19, 1825	....
Potash and leaching ashes, Manufacturing	E. Parce	Lincklaen, N. Y	July 20, 1831	....
Potash and leaching ashes, Manufacturing	E. Parce	Pitcher, N. Y	Jan. 23, 1834	....
Potash and pearlash	E. Williams	Erie, Pa	Mar. 8, 1836	....
Potash and pearlash furnace	E. Craft, jr	....	May 12, 1804	....
Potash and pearlash furnace	E. Ryan	....	Apr. 29, 1793	....
Potash and pearlash, Making	S. Hopkins	....	July 31, 1790	....
Potash and pearlash, Manufacture of	B. F. Jewett	Malone, N. Y	Sept. 4, 1866	57, 728
Potash and pearlash, Manufacturing	J. Latting	New York	Apr. 26, 1816	....
Potash and phosphate of lime, Manufacture of	M. B. Manwaring and R. D. Birch.	New York, N. Y., and Philadelphia, Pa.	Mar. 26, 1872	124, 964
Potash and preparing and leaching ashes, Manufacture of.	E. Williams	Harbor Creek, Pa	Aug. 16, 1835	....
Potash and soda, Manufacture of the prussiate of	M. Kalbfleisch	Bushwick, N. Y	Jan. 25, 1848	5, 419
Potash and soda, Preparing	G. Thompson	East Tarentum, Pa	July 24, 1855	13, 325
Potash and soda, Preparing hydrated silicates of	J. M. Ordway	Manchester, N. H	May 5, 1863	38, 449
Potash feldspar for obtaining certain salts, Decomposing.	R. A. Tilghman	Philadelphia, Pa	Dec. 4, 1847	5, 384
Potash into sal polychrestum, &c., Separating, &c., sulphate of.	A. McNitt	Geneva, N. Y	June 15, 1805	....
Potash, Manufacture of	Hartsuff and French	Aurelius, N. Y	July 17, 1835	....
Potash, Manufacture of chromate of	I. Tyson, jr	Baltimore, Md	Oct. 9, 1845	4, 224
Potash, Manufacture of nitrate of	J. C. Pennington	Paterson, N. J	July 22, 1862	35, 946
Potash, Manufacture of pure carbonate of	C. F. Moll	Kenton, Ohio	Aug. 2, 1870	105, 968
Potash, Manufacturing	W. Frobisher	....	Nov. 17, 1796	....
Potash, Manufacturing	T. Power	....	Sept. 19, 1801	....
Potash, Manufacturing	E. Ryan	....	Nov. 16, 1796	....
Potash, Manufacturing	T. H. Sherman	Hastings, N. Y	Feb. 2, 1831	....
Potash, Extracting alkali from ashes in the manufacture of.	J. Osborn	Elyria, Ohio	July 18, 1840	1, 697
Potash, Leaching ashes in the manufacture of	J. H. Ward	Randolph, Ohio	July 16, 1841	2, 178
Potash or soda from alkaline silicate, Liberating	F. O. Ward	London, England	Mar. 21, 1865	46, 979
Potash or soda, Making stannate of	J. Young	Manchester, England	Aug. 20, 1850	7, 588
Potash, Retort for the manufacture of prussiate of	D. B. and A. D. Coles	Newark, N. J	Nov. 14, 1865	50, 907
Potash, soda, &c., Manufacture of sulphates of	H. Deacon	Warrington, England	Feb. 20, 1872	123, 875
Potassa and soda, Preparing stannate of	J. Greenwood, J. Mercer, and J. Barnes.	England	Aug. 12, 1846	4, 688
Potassa and soda, Putting up and preserving caustic.	T. C. Taylor	Philadelphia, Pa	Feb. 6, 1866	52, 466
Potassa, Manufacture of bitartrate of	L. Adler	Baltimore, Md	Dec. 16, 1873	145, 607
Potassa, Manufacture of bitartrate of	G. Bourgade	Jersey City, N. J	Aug. 13, 1872	130, 407
Potassa, Preparation of nitrate of	H. S. Osborn	Belvidere, N. J	May 2, 1865	47, 562
Potassa, Retort for the manufacture of prussiate of	J. Clark	Newark, N. J	Aug. 6, 1861	32, 977
Potato and drill machine	S. L. Spencer	Hopewell X Roads, Md	Aug. 14, 1866	57, 207
Potato and pea-nut digger	C. E. Peirce	Centreville, Mich	Mar. 18, 1873	136, 860
Potato and root washer	W. Ellis	Waterville, Me	Feb. 25, 1835	....
Potato assorter and sifter	L. Perrine and P. G. Conover	Freehold, N. J	Feb. 16, 1869	87, 001
Potato-assorting apparatus	J. F. Unlish	Webster, N. Y	May 10, 1870	102, 886
Potato-assorting machine	D. M. King	Garrettsville, Ohio	Sept. 30, 1873	143, 357
Potato-assorting machine	B. D. and D. A. Vandeveer and T. Denise.	Freehold, N. J	Oct. 25, 1870	108, 655
Potato-baker	A. Reid	Buffalo, N. Y	Oct. 1, 1867	69, 480
Potato-bug cleaner	J. Corbeil	Lind, Wis	July 28, 1868	80, 458
Potato-bug collector	J. Corbeil	Lind, Wis	Jan. 14, 1873	134, 854
Potato-bug collector	R. P. Main	Oregon, Wis	May 21, 1872	127, 079

Index of patents issued from the United States Patent Office from 1790 *to* 1873, *inclusive*—Continued.

Invention.	Inventor.	Residence.	Date.	No.
Potato-bug destroyer	A. A. Mixer	Hamilton, Ohio	Sept. 30, 1873	143, 245
Potato-bug destroyer	W. H. Myers	Oregon, Wis	Sept. 23, 1873	143, 178
Potato-bug destroyer	J. B. Turney	Inkster, Mich	July 11, 1871	116, 894
Potato-bug exterminator	C. Cole	Chatfield, Minn	Oct. 28, 1873	144, 065
Potato-bugs, Compound for destroying	L. Pagin	Niles, Mich	Mar. 14, 1871	112, 732
Potato-bugs, Machine for destroying	H. Pitchforth and W. Benson	Muscatine, Iowa	June 4, 1867	65, 427
Potato cultivator and digger	W. J. Avery and T. Laberteaux	Marshall Township, Mich	Feb. 22, 1870	100, 102
Potato cutter and planter	L. J. Newborn	Kinston, N. C	Oct. 14, 1873	143, 707
Potato-cutting machine	A. Williams	Ashfield, Mass	Mar. 12, 1839	1, 101
Potato-digger	G. W. Adams and J. R. Hopper	Rochester, N. Y	Sept. 14, 1869	94, 693
Potato-digger	J. Albaugh	Lyons, N. Y	June 21, 1870	104, 401
Potato-digger	R. L. Allen	New York, N. Y	Jan. 18, 1859	22, 612
Potato-digger	A. Q. Allis	Rochester, N. Y	Oct. 22, 1872	132, 427
Potato-digger	A. Anderson	Markham, Canada	Dec. 29, 1857	18, 009
Potato-digger	W. F. Andrews	Watson, N. Y	June 29, 1869	91, 897
Potato-digger	L. Annis	Quincy, Mich	Dec. 29, 1868	85, 418
Potato-digger	S. E. Anthony	Stillwater, N. Y	Sept. 20, 1870	107, 433
Potato-digger	S. E. Anthony	Stillwater, N. Y	Apr. 25, 1871	114, 088
Potato-digger	P. Antonides	Freehold, N. J	Feb. 23, 1864	41, 739
Potato-digger	P. Antonides	Freehold, N. J	Dec. 31, 1867	72, 773
Potato-digger	L. A. Aspinwall	Watervliet, N. Y	Nov. 14, 1865	50, 889
Potato-digger	L. A. Aspinwall	Albany, N. Y	June 25, 1867	66, 064
Potato-digger	B. Avis	Penn's Grove, N. J	May 14, 1872	126, 619
Potato-digger	J. B. Baker	Syracuse, N. Y	Sept. 12, 1871	118, 895
Potato-digger	T. Baker	Stillwater, N. Y	July 7, 1863	39, 108
Potato-digger	O. W. Baldwin	Greenfield, Ohio	May 18, 1869	90, 065
Potato-digger	O. W. Baldwin and F. H. Pope	Greenfield, Ohio	Apr. 9, 1867	63, 604
Potato-digger	H. G. Barrows	Benton, Me	Jan. 28, 1873	135, 307
Potato-digger	G. C. Bartlett	Paris, N. Y	July 24, 1860	29, 339
Potato-digger	A. L. Bausman	Minneapolis, Minn	Oct. 11, 1870	108, 093
Potato-digger	J. Bawden	Freehold, N. J	May 8, 1860	28, 232
Potato-digger	W. Beaty	Pontiac, Mich	Mar. 16, 1869	87, 901
Potato-digger	J. Belknap	Adrian, Mich	Nov. 16, 1869	96, 770
Potato-digger	D. D. Bell	Wawarsing, N. Y	Dec. 9, 1851	8, 574
Potato-digger	E. Bennett	Nankin, Mich	June 9, 1868	78, 641
Potato-digger	L. Berry	Clyde, N. Y	Sept. 13, 1870	107, 327
Potato-digger	L. L. Bettys	Ontario, N. Y	May 4, 1869	89, 547
Potato-digger	D. Bibbee and W. Rand	Letart Falls, Ohio	July 14, 1870	104, 251
Potato-digger	S. Bigelow and L. C. Bigelow	Newburgh and Kirtland, Ohio.	May 6, 1873	138, 470
Potato-digger	J. H. Billmeyer	Raisin, Mich	May 3, 1870	102, 479
Potato-digger	J. W. Blodgett	Three Rivers, Mich	June 30, 1868	79, 439
Potato-digger	C. Blood	Malta, N. Y	Sept. 1, 1863	39, 712
Potato-digger	J. I. Brinkerhoff	Auburn, N. Y	Mar. 9, 1869	87, 536
Potato-digger	D. D. T. Brown	Mumford, N. Y	June 11, 1872	127, 676
Potato-digger	J. Brown	Hamlin, Mich	Jan. 19, 1869	85, 994
Potato-digger	A. G. Brush	Susquehanna Depot, Pa	Feb. 18, 1873	136, 028
Potato-digger	G. J. Bundy	Lyndon, Vt	July 4, 1854	11, 206
Potato-digger	I. S. Bunnell	Montrose, Pa	Dec. 29, 1857	18, 954
Potato-digger	T. and G. B. Burditt	Dansville, N. Y	May 14, 1867	64, 627
Potato-digger	A. Burhaus	Albany, N. Y	Sept. 8, 1868	81, 980
Potato-digger	J. M. Burke	Dansville, N. Y	July 27, 1869	93, 050
Potato-digger	J. W. Burnham and W. Conlon	Middletown Point, N. J	May 26, 1868	78, 257
Potato-digger	W. H. Burridge	Cleveland, Ohio	Sept. 25, 1866	58, 334
Potato-digger	J. Burt	Sturgis, Mich	Sept. 29, 1868	82, 592
Potato-digger	W. F. Caldwell	Oxford, Me	July 17, 1866	56, 366
Potato-digger	C. B. Cannon	Keokuk, Iowa	Dec. 18, 1866	60, 618
Potato-digger	A. S. Capron and D. S. Davis	Grass Lake, Mich	Oct. 11, 1859	25, 721
Potato-digger	H. C. Carr	Bordentown, N. J	Mar. 15, 1870	100, 854
Potato-digger	H. Carrier	Kirtland, Ohio	June 15, 1869	91, 415
Potato-digger	W. H. Chamberlin	Medina, N. Y	Mar. 3, 1868	74, 991
Potato-digger	A. F. Chandler	Winthrop, Me	Apr. 9, 1867	63, 612
Potato-digger	O. M. Chase	Boston, Mass	Feb. 2, 1864	41, 424
Potato-digger	J. C. Clark	Elmira, N. Y	Mar. 3, 1868	75, 123
Potato-digger	L. Cochran	Penn's Grove, N. J	Nov. 8, 1870	109, 111
Potato-digger	W. W. Cole and T. McGhee	Eudora, Kans	July 13, 1869	92, 518
Potato-digger	S. B. Conover	New York, N. Y	June 3, 1862	35, 435
Potato-digger	S. B. Conover	New York, N. Y	Jan. 5, 1864	41, 058
Potato-digger	S B. Conover	New York, N. Y	May 2, 1867	63, 475
Potato-digger	C. Cooney	Columbus, Nebr	May 9, 1871	114, 531
Potato-digger	I. Cooper	Middlefield, Ohio	May 25, 1869	90, 341
Potato-digger	V. P. Corbett	Alexandria County, Va	Nov. 10, 1868	83, 833
Potato-digger	V. P. Corbett	Alexandria County, Va	Aug. 20, 1872	130, 569
Potato-digger	J. W. Corwin	Lebanon, Ohio	Feb. 2, 1869	86, 509
Potato-digger	J. W. Corwin	Lebanon, Ohio	Nov. 22, 1870	109, 389
Potato-digger	W. Cousens	Orono, Me	Oct. 3, 1871	119, 577
Potato-digger	W. J. Cowan	Cortland, N. Y	Oct. 6, 1868	82, 806
Potato-digger	I. Curtis	Des Moines, Iowa	Apr. 6, 1869	88, 613
Potato-digger	C. H. Dana	West Lebanon, N. H	May 16, 1854	10, 926
Potato-digger	J. P. Davison	Rome, N. Y	Aug. 4, 1868	80, 611
Potato-digger	C. Darling	Utica, N. Y	Sept. 13, 1870	107, 229
Potato-digger	J. Davis	Saratoga, N. Y	Jan. 31, 1871	111, 437
Potato-digger	R. G. Dayton	North Granville, N. Y	Mar. 19, 1872	124, 726
Potato-digger	R. G. Dayton	North Granville, N. Y	Nov. 5, 1872	132, 814
Potato-digger	D. De Garmo	Rochester, N. Y	July 24, 1860	29, 252
Potato-digger	D. N. Denman	Milburn, N. J	Sept. 20, 1864	44, 288
Potato-digger	P. Dennis	Stillwater, N. Y	Mar. 3, 1857	16, 722
Potato-digger	W. Dillon	Sonoma, Cal	Oct. 11, 1870	108, 117
Potato-digger	A. Dolloff	Lake Village, N. H	Feb. 27, 1866	52, 830
Potato-digger	E. O. Doud and W. F. Beardsley.	Penfield, N. Y	Apr. 30, 1867	64, 291
Potato-digger	H. M. and W. W. Dowd, jr	Saratoga Springs and North Granville, N. Y.	June 10, 1873	139, 709
Potato-digger	M. F. Drake	Pleasant Ridge, Ohio	Dec. 11, 1866	60, 347
Potato-digger	C. Dunham	Batavia, N. Y	May 22, 1860	28, 356
Potato-digger	R. Dunlap, 1st	South Lyons, Mich	June 29, 1869	92, 024
Potato-digger	W. H. Elliot	New York, N. Y	Jan. 8, 1867	61, 058
Potato-digger	J. Farlow	Harrison, Ind	July 16, 1872	129, 120

Index of patents issued from the United States Patent Office from 1790 *to* 1873, *inclusive*—Continued.

Invention.	Inventor.	Residence.	Date.	No.
Potato-digger	H. Farmer	Pontiac, Mich	July 20, 1869	92, 712
Potato-digger	W. A. Field	Schuylkill Haven, Pa	May 4, 1869	89, 571
Potato-digger	E. Finch	Albion, N. Y	July 12, 1870	105, 324
Potato-digger	L. Finch	Vienna, Va	Apr. 7, 1868	76, 427
Potato-digger	E. T. Ford	Stillwater, N. Y	Aug. 11, 1863	39, 475
Potato-digger	E. T. Ford	Stillwater, N. Y	Oct. 18, 1864	44, 716
Potato-digger	E. T. Ford	Stillwater, N. Y	Sept. 1, 1868	81, 765
Potato-digger	E. T. Ford	Stillwater, N. Y	July 13, 1869	92, 437
Potato-digger	E. T. Ford	Stillwater, N. Y	Oct. 28, 1873	144, 083
Potato-digger	E. G. Ford and J. F. Penquite	Delphos, Ohio	Mar. 6, 1866	52, 984
Potato-digger	H. France	Hinmansville, N. Y	Nov. 17, 1868	84, 056
Potato-digger	J. E. Giles and W. Ferry	Mead's Mills, Mich	Dec. 8, 1868	84, 741
Potato-digger	J. E. Giles and C. S. McRobert	Mead's Mills, Mich	June 25, 1867	66, 145
Potato-digger	A. Gilmore	Fort Atkinson, Wis	Dec. 18, 1866	60, 501
Potato-digger	D. Gorman	Hornellsville, N. Y	Oct. 19, 1869	95, 896
Potato-digger	C. G. Grabo	Detroit, Mich	July 23, 1867	67, 040
Potato-digger	A. Graves	Marcellus Falls, N. Y	Apr. 7, 1868	76, 438
Potato-digger	A. Graves	Marcellus Falls, N. Y	Apr. 28, 1868	77, 279
Potato-digger	J. H. Gray and C. W. Calhoun	Florence Township, Mich	Jan. 14, 1868	73, 319
Potato-digger	B. F. Green	Syracuse, N. Y	Aug. 6, 1872	130, 125
Potato-digger	J. M. Green	West Bloomfield, N. Y	July 16, 1867	66, 704
Potato-digger	W. Green	Holly, Mich	Dec. 7, 1869	97, 630
Potato-digger	I. Griffen	Quaker Springs, N. Y	June 2, 1857	17, 428
Potato-digger	E. V. W. Griffith	Utica, N. Y	Dec. 17, 1867	72, 289
Potato-digger	A. L. Grinnell and J. Z. Williams.	Willet, Wis	June 10, 1856	15, 100
Potato-digger	L. B. Griswold	Pennfield, N. Y	Aug. 9, 1859	25, 010
Potato-digger	I. C. Groom	Albany, N. Y	July 6, 1869	92, 304
Potato-digger	G. W. Haag	Cairo, Pa	May 6, 1873	138, 561
Potato-digger	A. Hadwen	Rochester, N. Y	Mar. 17, 1868	75, 543
Potato-digger	A. M. Hall	Falmouth, Me	Nov. 11, 1868	83, 847
Potato-digger	J. Hall, jr	Temperanceville, Pa	May 10, 1870	102, 812
Potato-digger	J. Hall, jr	Temperanceville, Pa	May 10, 1870	102, 814
Potato-digger	J. Hall, jr., and O. Flanigan	Temperanceville, Pa	May 10, 1870	102, 811
Potato-digger	M. Hall	Livonia, Mich	Apr. 23, 1867	64, 097
Potato-digger	J. E. Hardenbergh	Fultonville, N. Y	Dec. 22, 1857	18, 899
Potato-digger	U. R. Harlow	Farmersville, Cal	Aug. 2, 1870	106, 058
Potato-digger	W. D. Harrell	Moore's Hill, Ind	Mar. 14, 1871	112, 592
Potato-digger	L. W. Harris	Waterville, N. Y	Apr. 6, 1858	19, 849
Potato-digger	D. B. Hart	Mentor, Ohio	Apr. 30, 1867	64, 218
Potato-digger	R. A. Haw	Bucksport, Cal	July 25, 1871	117, 415
Potato-digger	L. Henderson	Manson, N. C	Aug. 23, 1870	106, 583
Potato-digger	M. Henderson	Detroit, Mich	July 21, 1868	80, 070
Potato-digger	T. N. Henderson	Jackson, Mich	May 4, 1869	89, 661
Potato-digger	A. Heulings	Philadelphia, Pa	May 6, 1856	14, 810
Potato-digger	A. Heulings	Philadelphia, Pa	May 28, 1872	127, 345
Potato-digger	S. S. Hickok	Marlborough, N. J	Feb. 17, 1863	37, 691
Potato-digger	I. Hicks	Hartford, Wis	Oct. 3, 1871	119, 461
Potato-digger	J. J. Hill	Xenia, Ohio	Aug. 22, 1865	49, 521
Potato-digger	A. P. Hinz	Dunton, Ill	Sept. 1, 1868	81, 636
Potato-digger	G. N. Hoag	Muscatine, Iowa	Nov. 15, 1870	109, 211
Potato-digger	H. Holcroft and C. S. Smith	Media, Pa	Mar. 24, 1863	39, 964
Potato-digger	J. R. Hopper	Rochester, N. Y	Mar. 30, 1869	88, 295
Potato-digger	J. H. Hotchkin	Prattsburgh, N. Y	July 13, 1869	92, 450
Potato-digger	W. H. Hutchins	Lockport, Ill	Nov. 19, 1867	71, 177
Potato-digger	J. W. Innis	Newburgh, N. Y	Feb. 19, 1867	62, 271
Potato-digger	J. W. Innis	Newburgh, N. Y	Feb. 4, 1873	135, 558
Potato-digger	J. O. Ives	Saint Louis, Mo	Aug. 15, 1865	49, 413
Potato-digger	M. Jacobs	Sturgis, Mich	Dec. 8, 1868	84, 828
Potato-digger	M. Johnson	Three Rivers, Mich	June 30, 1868	79, 473
Potato-digger	M. Johnson	Three Rivers, Mich	May 17, 1870	103, 194
Potato-digger	M. Johnson	Three Rivers, Mich	May 24, 1870	103, 469
Potato-digger	M. Johnson	Three Rivers, Mich	Nov. 28, 1871	121, 379
Potato-digger	M. Johnson	Three Rivers, Mich	Mar. 12, 1872	124, 585
Potato-digger	M. Johnson	Three Rivers, Mich	Apr. 2, 1872	125, 303
Potato-digger	M. Johnson	Three Rivers, Mich	Oct. 22, 1872	132, 364
Potato-digger	L. Johonnott	Burlington, Vt	June 18, 1872	128, 150
Potato-digger	F. Jones	Terre Haute, Ind	Sept. 19, 1871	119, 152
Potato-digger	R. D. Jones and T. Purcell	Rochester, N. Y	Mar. 29, 1864	42, 088
Potato-digger	W. Jones	Saint Louis, Mo	Dec. 8, 1863	40, 842
Potato-digger	W. Joseph	Quincy, Mich	Nov. 15, 1870	109, 217
Potato-digger	A. C. Ketchum	Schenectady, N. Y	Feb. 20, 1844	3, 442
Potato-digger	D. M. and G. E. King	Mantua Station, Ohio	Apr. 15, 1873	137, 931
Potato-digger	S. Kinman	Eureka, Cal	Jan. 12, 1869	85, 833
Potato-digger	G. W. Kintz	West Henrietta, N. Y	June 25, 1867	66, 029
Potato-digger	G. W. Kintz	West Henrietta, N. Y	Mar. 31, 1868	76, 200
Potato-digger	H. J. Kintz	Greece, N. Y	Mar. 12, 1867	62, 858
Potato-digger	G. W. Knight	Camden, N. J	Feb. 8, 1870	99, 575
Potato-digger	M. W. Knox	Sheridan, N. Y	Apr. 5, 1870	101, 474
Potato-digger	M. W. Knox	Sheridan, N. Y	Oct. 18, 1870	108, 490
Potato-digger	G. Koenig and G. Otto	Plymouth, Mich	Feb. 18, 1868	74, 699
Potato-digger	J. P. Lafetra	Shrewsbury, N. J	Apr. 4, 1871	113, 308
Potato-digger	D. M. Lamb	Strathroy, Canada	Dec. 10, 1872	133, 710
Potato-digger	T. Lane	San Francisco, Cal	Dec. 9, 1862	37, 100
Potato-digger	E. S. Lenox	New York, N. Y	Dec. 26, 1865	51, 731
Potato-digger	E. S. Lenox and E. Spaulding	New York, N. Y	Apr. 23, 1867	64, 114
Potato-digger	J. J. Lingley	La Fayette, Ind	Oct. 26, 1869	96, 153
Potato-digger	D. Locke	Geneva, Wis	Sept. 28, 1869	95, 242
Potato-digger	G. C. Love	English Centre, Pa	Feb. 7, 1871	111, 656
Potato-digger	J. M. Lumbard	Decatur, Mich	Apr. 23, 1872	125, 968
Potato-digger	O. E. Mallory	Batavia, N. Y	Feb. 22, 1870	100, 168
Potato-digger	A. Marcellus	Pittsford, N. Y	Feb. 6, 1866	52, 428
Potato-digger	A. Marcellus	Pittsford, N. Y	Sept. 17, 1867	69, 0[illegible]5
Potato-digger	A. A. Marcellus	New York, N. Y	Nov. 20, 1855	13, 822
Potato-digger	P. Marcy	Tunkhannock, Pa	June 21, 1859	24, 474
Potato-digger	P. Marcy	Tunkhannock, Pa	Aug. 20, 1861	33, 093
Potato-digger	G. M. Marks	Half Moon, Pa	Sept. 27, 1870	107, 698
Potato-digger	I. W. McGaffey	Philadelphia, Pa	Nov. 7, 1854	11, 899

Index of patents issued from the United States Patent Office from 1790 *to* 1873, *inclusive*—Continued.

Invention.	Inventor.	Residence.	Date.	No.
Potato-digger	P. C. McManus	Troy, N. Y	Dec. 14, 1869	97, 790
Potato-digger	G. Meader	Prairie Centre, Ill	June 11, 1867	65, 583
Potato-digger	W. D. Miller	Enon, Ohio	June 5, 1866	55, 337
Potato-digger	A. Minnis	Canton Township, Mich	May 21, 1867	64, 996
Potato-digger	H. J. Moore	Topeka, Kans	Aug. 8, 1871	117, 913
Potato-digger	F. A. Morley	Sodus Point, N. Y	July 9, 1861	32, 781
Potato-digger	F. A. Morley	New York, N. Y	May 29, 1866	55, 206
Potato-digger	F. A. Morley	Syracuse, N. Y	Nov. 24, 1868	84, 297
Potato-digger	F. A. Morley	Syracuse, N. Y	June 8, 1869	91, 036
Potato-digger	H. B. Morton	Rochester, N. Y	June 2, 1868	78, 473
Potato-digger	J. Munson and J. R. Lyon	Burlington and Shelburne, Vt.	Feb. 25, 1862	34, 519
Potato-digger	W. Musseld	New York, N. Y	Dec. 9, 1856	16, 184
Potato-digger	J. J. Naylon	Chicago, Ill	Mar. 19, 1872	124, 693
Potato-digger	D. Niven	Rochester, N. Y	May 15, 1860	28, 294
Potato-digger	R. Niven	Gates, N. Y	Mar. 1, 1859	23, 106
Potato-digger	H. B. Norton	Rochester, N. Y	Oct. 22, 1867	70, 108
Potato-digger	H. B. Norton	Rochester, N. Y	Mar. 4, 1873	136, 536
Potato-digger	N. S. Noyes	Plymouth, Mich	Aug. 27, 1867	68, 309
Potato-digger	J. D. C. Outwater	Newark, N. J	July 23, 1867	67, 068
Potato-digger	C. Palmer	Willoughby, Ohio	Mar. 23, 1869	88, 199
Potato-digger	O. Patterson	South Dansville, N. Y	Apr. 5, 1870	101, 499
Potato-digger	W. Peacock	New York, N. Y	June 4, 1872	127, 640
Potato-digger	E. Phillips	North Beaver Township, Pa.	Feb. 23, 1869	87, 197
Potato-digger	C. H. Pickering	Indianapolis, Ind	July 2, 1867	66, 252
Potato-digger	W. R. Prince	Parkersburgh, W. Va	Mar. 8, 1870	100, 667
Potato-digger	W. Proctor and D. C. Payne	Elkhart, Ind	Sept. 29, 1863	40, 120
Potato-digger	J. P. Radley	Albany, N. Y	Apr. 4, 1871	113, 449
Potato-digger	J. P. Radley	Albany, N. Y	Nov. 12, 1872	132, 923
Potato-digger	S. Ransom	Kingsville, Ohio	Apr. 25, 1871	114, 041
Potato-digger	G. Ray	Kinderhook, N. Y	Aug. 28, 1866	57, 568
Potato-digger	B. Reamer and C. Van Derzee	Albany, N. Y	July 20, 1869	92, 749
Potato-digger	J. D. Rensink	Bridgeport, N. J	May 13, 1873	138, 934
Potato-digger	E. Rexford	Mentor, Ohio	May 14, 1867	64, 708
Potato-digger	J. C. Richardson	Benton, Me	June 4, 1867	65, 434
Potato-digger	S. Richardson and J. S. Adams	Jericho and Burlington, Vt	Jan. 15, 1867	61, 259
Potato-digger	R. B. Robbins	Adrian, Mich	Feb. 7, 1871	111, 571
Potato-digger	R. B. Robbins	Adrian, Mich	Nov. 7, 1871	120, 775
Potato-digger	F. A. Roberts	North Vassalborough, Me	Dec. 22, 1868	85, 182
Potato-digger	J. Roberts	White Pigeon, Mich	June 15, 1869	91, 487
Potato-digger	E. Robertson	Hartford, Conn	May 8, 1860	28, 197
Potato-digger	E. Ross	Washington, Iowa	Apr. 12, 1870	101, 921
Potato-digger	W. B. Ryan	East Pembroke, N. Y	May 20, 1862	35, 326
Potato-digger	F. C. Shaffer	Brooklyn, N. Y	Jan. 4, 1853	9, 523
Potato-digger	L. Shafer, jr	Mount Carmel, Ill	May 13, 1873	138, 942
Potato-digger	L. F. Schanck	Holmdel, N. J	Mar. 20, 1866	53, 349
Potato-digger	J. P. Scudder	Hightstown, N. J	Oct. 2, 1860	30, 253
Potato-digger	J. N. Seeley	Forsyth, Ga	Dec. 21, 1852	9, 489
Potato-digger	I. Shaw	Four Corners, Md	Mar. 23, 1869	88, 088
Potato-digger	I. Shaw	Four Corners, Md	Apr. 13, 1869	88, 916
Potato-digger	T. W. Shepard	Hennepin, Ill	Dec. 24, 1867	72, 552
Potato-digger	J. Sherwood	Ottumwa, Iowa	July 27, 1869	93, 015
Potato-digger	P. Sisson	Brant, N. Y	Apr. 9, 1867	63, 659
Potato-digger	J. Slocum	Syracuse, N. Y	Mar. 25, 1862	34, 787
Potato-digger	E. Smith	West Milton, N. Y	Feb. 11, 1868	74, 438
Potato-digger	H. M. Smith	Long Branch, N. J	Aug. 31, 1869	94, 356
Potato-digger	H. P. Smith	Denton, Mich	July 23, 1867	67, 076
Potato-digger	J. Smith	Ridgeville, Ohio	Nov. 14, 1871	120, 832
Potato-digger	J. V. Smith	Hudson, N. Y	Dec. 4, 1866	60, 273
Potato-digger	S. Soules	Dowagiac, Mich	June 5, 1866	55, 384
Potato-digger	S. Soules	Cresco, Iowa	Sept. 24, 1867	69, 136
Potato-digger	S. Soules	Dowagiac, Mich	May 25, 1869	90, 404
Potato-digger	W. W. Speer	Pittsburgh, Pa	Feb. 20, 1872	123, 846
Potato-digger	W. Stark	White Pigeon, Mich	Aug. 11, 1868	80, 841
Potato-digger	W. Starkey	Cedarville, N. J	Mar. 26, 1872	125, 094
Potato-digger	J. Stewart	Jackson, Mich	Nov. 3, 1868	83, 670
Potato-digger	M. Stewart	Southfield, Mich	Dec. 12, 1871	121, 906
Potato-digger	T. B. Stout	Key Port, N. J	Apr. 26, 1853	9, 689
Potato-digger	A. Snyder	Jackson, Mich	Jan. 29, 1867	61, 574
Potato-digger	J. Taggart	Roxbury, Mass	Apr. 21, 1857	17, 129
Potato-digger	D. C. Thomas	Easton, N. Y	June 8, 1869	91, 182
Potato-digger	D. C. Thomas	Easton, N. Y	June 11, 1872	127, 937
Potato-digger	G. S. Tiffany	Palmyra, Mich	June 21, 1859	24, 505
Potato-digger	D. J. Tittle	Albany, N. Y	Mar. 20, 1866	53, 363
Potato-digger	W. Tripp	Mechanicsville, N. Y	Dec. 27, 1870	110, 514
Potato-digger	S. W. Tyler	Troy, N. Y	Aug. 19, 1873	141, 903
Potato-digger	N. Vandenburgh	Schuylersville, N. Y	Oct. 18, 1870	108, 537
Potato-digger	N. Vandenburgh	Schuylersville, N. Y	Dec. 6, 1870	109, 974
Potato-digger	B. D. Vanderveer and D. Riddel.	Freehold, N. J	Aug. 18, 1868	81, 230
Potato-digger	C. Van Derzee and B. Reamer	Albany, N. Y	May 17, 1870	103, 258
Potato-digger	A. Van Norman	Detroit, Mich	July 21, 1868	80, 247
Potato-digger	G. Vowles	New Hudson, Mich	May 26, 1868	78, 345
Potato-digger	J. A. Wadhams	Blue Island, Ill	Apr. 4, 1871	113, 373
Potato-digger	A. Walker	Claremont, N. H	Oct. 2, 1866	58, 560
Potato-digger	W. H. Wall	Elbridge, N. Y	Jan. 21, 1873	135, 183
Potato-digger	J. Walmsley	Buffalo, N. Y	July 23, 1867	67, 004
Potato-digger	H. Webster and C. Powers	East Pembroke, N. Y	Apr. 20, 1869	89, 069
Potato-digger	D. B. Westfall	Lyons, N. Y	Jan. 21, 1870	104, 518
Potato-digger	A. Wheeler	Mallory, Iowa	Dec. 1, 1863	40, 786
Potato-digger	J. Wheeler	Dowagiac, Mich	Sept. 21, 1869	95, 175
Potato-digger	S. H. Wheeler and L. V. Rouse	Dowagiac, Mich	May 18, 1869	90, 142
Potato-digger	J. M. Wilcox	Albany, N. Y	July 23, 1867	67, 148
Potato-digger	M. H. Wiley	Boston, Mass	Mar. 31, 1868	76, 127
Potato-digger	A. Willmann	Mount Carmel, Ill	Nov. 25, 1873	145, 037
Potato-digger	D. S. Wing and H. Green	Rome, N. Y	Feb. 20, 1872	123, 966
Potato-digger	A. R. Wixom	Farmington, Mich	Nov. 8, 1870	109, 093

Index of patents issued from the United States Patent Office from 1790 *to* 1873, *inclusive*—Continued.

Invention.	Inventor.	Residence.	Date.	No.
Potato-digger	J. W. Woolsey	Niles, Mich	Sept. 9, 1872	36, 434
Potato-digger	S. Woolson	Moodna, N. Y	Aug. 26, 1856	15, 628
Potato-digger	B. P. Wright	San Francisco, Cal	Apr. 7, 1868	76, 371
Potato-digger	J. S. Wright and I. Benedict	Palmyra, N. Y	Feb. 4, 1873	135, 458
Potato-digger	G. W. B. Yocom, R. J. Walker, and E. Sharp.	Arcata, Cal	Nov. 2, 1869	96, 370
Potato-digger	T. C. Zulick	Schuylkill Haven, Pa	Mar. 5, 1861	31, 641
Potato digger and cultivator	M. Johnson	Three Rivers, Mich	Oct. 19, 1869	96, 007
Potato digger and cultivator, Combined	M. Johnson	Three Rivers, Mich	Aug. 3, 1869	93, 307
Potato digger and picker	W. J. Thompson	Normal, Ill	July 6, 1869	92, 402
Potato digger and separator	J. W. Bartlett	Harmar, Ohio	May 30, 1865	47, 916
Potato digger and separator	W. Green	Holly, Mich	Sept. 1, 1868	81, 772
Potato-digger and shovel-plow frame	J. M. Dick	Buffalo, N. Y	Aug. 2, 1870	105, 922
Potato-digger and stone-gatherer	J. T. Foster	New York, N. Y	June 29, 1852	9, 069
Potato-digger and vine-puller	J. W. Newton	Geneva, Wis	Nov. 3, 1868	83, 785
Potato-digger and weeder	G. W. Hall	Triangle, N. Y	Oct. 1, 1867	69, 339
Potato digger, Sweet	R. Darlington and E. L. Watson	Auburn, N. Y	Feb. 15, 1870	99, 859
Potato-digger, Agitator-wheel for	J. Hall, jr	Temperanceville, Pa	May 10, 1870	102, 813
Potato-diggers, Vine-puller for	L. Berry	Clyde, N. Y	Sept. 13, 1870	107, 328
Potato digging and gathering machine	J. B. Parvin	Heightstown, N. J	May 10, 1859	23, 988
Potato-digging fork	H. Doolittle	East Cleveland, Ohio	Oct. 2, 1866	58, 390
Potato-digging machine	J. W. Bartlett	Harmar, Ohio	Feb. 19, 1867	62, 246
Potato-digging machine	D. Clint and I. Lynd	Poestenkill, N. Y	Feb. 5, 1861	31, 299
Potato-digging machine	S. B. Conover and M. Spring	New York, N. Y	Apr. 30, 1861	32, 183
Potato-digging machine	N. Gear	Zanesville, Ohio	Sept. 7, 1858	21, 413
Potato-digging machine	J. Heulings	Philadelphia, Pa	July 21, 1857	17, 856
Potato-digging machine	H. Kewley	Perry, Ohio	July 2, 1867	66, 358
Potato-digging machine	M. Little	Clyde, N. Y	July 20, 1858	20, 949
Potato-digging machine	H. M. Peavey	Swanville, Me	July 10, 1866	56, 260
Potato-digging machine	J. Pettingill	Carroll, N. H	Oct. 14, 1862	36, 669
Potato-digging machine	J. C. Stoddard	Worcester, Mass	Mar. 29, 1859	23, 408
Potato-digging machine	A. Wells	Brooklyn, N. Y	Aug. 17, 1858	21, 225
Potato-digging machine	L. White	Essex, Vt	Aug. 17, 1858	21, 226
Potato-drill	J. Croco	Holmesville, Ohio	Oct. 2, 1866	58, 385
Potato-drill	J. Croco	Holmesville, Ohio	Apr. 23, 1867	64, 080
Potato-dropper	J. Gibbs	Canton, Ohio	June 25, 1872	128, 384
Potatoes, &c., Bin for preserving sweet	H. T. Basye	Dyersburgh, Pa	June 3, 1873	139, 532
Potato-fork	J. S. Patterson	Whitney's Point, N. Y	Oct. 23, 1866	59, 061
Potato-grinding machine	J. King	Florida, N. Y	Mar. 1, 1814	....
Potato-harvesting machine	O. S. and T. C. St. John	Willoughby, Ohio	Apr. 21, 1868	76, 953
Potato-masher	W. Ball	Peru, Mass	Dec. 13, 1864	45, 382
Potato-masher	V. Fountain, jr	Castleton, N. Y	Dec. 11, 1866	60, 355
Potato-masher	C. A. Frederick	San Francisco, Cal	Mar. 14, 1871	112, 577
Potato-masher	A. H. Lorton	New York, N. Y	Nov. 5, 1861	33, 654
Potato-masher	M. H. Monroe	Rochester, N. Y	Jan. 28, 1868	73, 743
Potato masher	A. L. Whitney	Brooklyn, N. Y	May 22, 1866	54, 987
Potato-masher	W. Zeiger	Elmore, Ohio	Mar. 23, 1869	88, 111
Potato masher and strainer	R. Sebille	Baltimore, Md	June 24, 1873	140, 169
Potato-masher, steak-pounder, and ice-breaker, Combined.	H. P. Stichter	Pottsville, Pa	Nov. 28, 1871	121, 434
Potato-probe	J. A. Beal	Waterford, N. Y	Mar. 7, 1871	112, 312
Potato rot, Compound for treating	L. Reed	Baltimore, Md	July 27, 1858	21, 023
Potato-rot, Mode of preventing	C. Cory	Lima, Ind	June 28, 1864	43, 292
Potato-rot, Treating potatoes to prevent	A. Mallet	Erie, Pa	Mar. 13, 1866	53, 159
Potato-sacking machine	E. Seely	Elkhart, Ind	Dec. 8, 1868	84, 713
Potato-screen	G. Claflin	Miller's Corners, N. Y	July 15, 1873	140, 886
Patato-seedlings, Instrument for cutting	W. P. L. Herr	Brooklyn, N. Y	Apr. 4, 1865	47, 106
Potato separator	S. Harrison	Saint Michael's, Md	Oct. 13, 1863	40, 258
Potato separator	R. Haviland	North Branch, Md	Apr. 12, 1870	101, 731
Potato-starch, Manufacture of	E. M., J. S., and W. H. Gordon	Woodstock, N. H	Nov. 28, 1848	5, 940
Potato vine cutter and potato-digger	T. E. C. Brinly	Louisville, Ky	Aug. 29, 1871	118, 585
Potato-washer	A. Bentley	Honesdale, Pa	May 4, 1852	8, 918
Potato-washer	O. H. Cooke	Morrisville, Vt	May 5, 1868	77, 461
Potato-washer	I. J. Currier and J. Hawse	Wolcott, Vt	July 6, 1869	92, 279
Potato-washer	D. K. Hickok	Morrisville, Vt	Oct. 13, 1868	82, 950
Potato-washer	J. Lawton, S. Hibbert, and J. Rhodes.	Manchester, England	Nov. 17, 1863	40, 631
Potato-washer	E. N. Porter and P. P. Roberts	Morrisville, Vt	Feb. 11, 1868	74, 420
Potato-washer	L. B. Sherwin	Hyde Park, Vt	Dec. 29, 1868	85, 405
Potato-washer	S. H. Tift	Morrisville, Vt	May 5, 1868	77, 679
Potato-washer	J. H. Williams	East Craftsbury, Vt	July 24, 1866	56, 652
Potato washer and pan, Combined	J. M. Chaplin	Morrisville, Vt	Apr. 7, 1868	76, 399
Potato-washing machine	J. H. Fairchild and S. Richardson.	Jericho, Vt	Apr. 18, 1854	10, 796
Potatoes and carrots, Machine for cutting	J. Clarke	Hampton, Conn	July 7, 1835	....
Potatoes, coal, &c., Machine for assorting	M. D. Dickinson	Pilesgrove, N. J	Sept. 25, 1866	58, 229
Potatoes, Cutting and planting	S. Hutchinson	Rockport, Ind	Oct. 25, 1853	10, 154
Potatoes, Keeping sweet	W. and J. Davis	Jefferson County, Iowa	Sept. 8, 1863	39, 797
Potatoes, Keeping sweet	A. H. Vestal	Cambridge City, Ind	Aug. 16, 1844	3, 708
Potatoes, Machine for covering	H. L. Bennett	Long Branch, N. J	July 31, 1860	29, 351
Potatoes, pumpkins, &c., for cattle or swine by steam, Machine for boiling.	H. E. Potter	Otsego, N. Y	Nov. 20, 1810	....
Potters' clay, Purifying	A. Weber	Womelsdorf, Pa	Sept. 9, 1835	....
Potters' ware, Machine for molding	E. W. Blackmer	McGrawville, N. Y	Nov. 7, 1865	50, 791
Pottery-kiln	G. R. Booth	Hanley, England	Aug. 31, 1852	9, 230
Pottery-machine	S. R. Thompson	Portsmouth, N. H	June 28, 1870	104, 795
Pottery, Machine for making	A. M. Cheeseman	Trenton, N. J	Aug. 24, 1869	93, 966
Pottery, &c., Manufacture of	T. L. and R. Boote	Burslem, England	Nov. 21, 1865	51, 123
Pottery, &c., Manufacture of	P. Marquardt	Buffalo, N. Y	Mar. 1, 1870	100, 304
Pottery, Molding	J. C. Mendall and R. B. Ricketts.	Masonville, Ky	June 30, 1836	....
Pottery-molding machine	A. Keil and J. Tresch	New York, N. Y	Apr. 2, 1867	63, 529
Pottery-molding machine	W. Linton	Baltimore, Md	Feb. 12, 1861	31, 394
Pottery-molding machine	J. Tresch	New York, N. Y	May 5, 1863	38, 430
Pottery-stock, Machine for preparing	S. R. Thompson	Portsmouth, N. H	May 13, 1873	138, 824
Pottery-ware and crucible kiln	J. Dixon	Jersey City, N. J	Mar. 5, 1850	7, 136
Pottery-ware, Machine for making	H. R. Bodine	Falls Township, Ohio	Jan. 30, 1866	52, 261
Pottery-ware, Machine for making	S. R. Thompson	Portsmouth, N. H	Apr. 18, 1871	113, 816

Index of patents issued from the United States Patent Office from 1790 *to* 1873, *inclusive*—Continued.

Invention.	Inventor.	Residence.	Date.	No.
Pottery-ware, Machine for manufacturing	P. Pointon	Baraboo, Wis	Sept. 29, 1857	18, 298
Pottery-ware, Manufacture of	J. Farnam	Stillwater, N. Y	Oct. 25, 1845	4, 242
Pottery-ware, Manufacture of	I. W. Knowles	East Liverpool, Ohio	Oct. 11, 1870	108, 157
Pottery-ware, Molding	R. J. Marcher	New York, N. Y	June 16, 1863	38, 905
Pottery-ware, Molding and finishing	F. Durand	Paris, France	Sept. 13, 1864	44, 270
Pottery-ware, Burning	G. C. Hicks	Baltimore, Md	Mar. 26, 1872	125, 048
Pottery-ware, Safeguard for protecting	B. Jackson	Trenton, N. J	Jan. 31, 1865	46, 109
Pottery-ware, Working clay for	J. Akrill	Williamsburgh, N. Y	Aug. 5, 1851	8, 280
Poudrette-drying apparatus	L. Duvall	Philadelphia, Pa	June 3, 1873	139, 555
Poultice-cloth	M. L. J. Chollet and C. H. E. Hamilton.	Paris, France	Dec. 8, 1868	84, 731
Poultice, Composition for	J. P. Scott	Newport, Ky	Oct. 6, 1863	40, 189
Poultry-fountain	J. S. Orndorff	Virginia City, Nev	Aug. 8, 1871	117, 807
Poultry-hatching apparatus	G. F. Quick	Moorestown, N. J	Oct. 13, 1863	40, 277
Poultry, Machine for curing croup in	T. V. Bush	Clark, Ky	June 29, 1832	....
Pounce-holder	R. Cushman and J. R. Dennis	Pawtucket and Central Falls, R. I.	Nov. 24, 1868	84, 346
Pouncing-machine	J. Rosenkranz	Boston, Mass	Aug. 16, 1870	106, 409
Pounding and chopping block, Spring	F. Köhler and A. J. Alsing	New York, N. Y	Oct. 19, 1869	95, 914
Powder:				
See Abrasive powder. Baking-powder. Blasting-powder. Bronze-powder. Cattle-powder. Condition-powder. Core-powder. Disinfecting-powder. Explosive powder. Fulminating-powder. Gunpowder. Gun and blasting powder. Horse-powder. Horse and cattle powder. Ink-powder. Insect-powder. Lime powder. Metal-powder. Metal-cleaning powder Percussion-powder. Polishing-powder. Preserving powder. Toilet-powder. Tooth-powder. Welding-powder. Yeast-powder.				
Powder-box	B. F. Brown	Boston, Mass	Aug. 1, 1871	117, 510
Powder-boxes used for domestic purposes, Cap for	H. A. Bartlett	Philadelphia, Pa	Feb. 22, 1870	100, 234
Powder by hydrogen gas and platina, Igniting	A. Jones	Mobile, Ala	Apr. 22, 1834	....
Powder, Exploding compressed	E. Gomez	New York, N. Y	Feb. 8, 1870	99, 665
Powder-flask	J. H. Breckenridge	Meriden, Conn	Feb. 16, 1858	19, 342
Powder-flask	G. A. and G. D. Capewell	Woodbury, Conn	Sept. 24, 1872	131, 497
Powder-flask	G. A., G. D., and J. T. Capewell	Woodbury, Conn	Aug. 15, 1871	118, 104
Powder-flask	J. T. and G. D. Capewell	Woodbury, Conn	Mar. 14, 1871	112, 542
Powder-flask	M. Cilik	Chicago, Ill	Sept. 9, 1873	142, 614
Powder-flask	F. E. Darrow	Bristol, Conn	Dec. 20, 1870	110, 210
Powder-flask	A. Drizel	Omaha, Nebr	Oct. 17, 1871	120, 047
Powder-flask	T. L. Sturtevant	Framingham, Mass	July 2, 1872	128, 673
Powder-flask	T. L. Sturtevant	Framingham, Mass	Apr. 1, 1873	137, 332
Powder-flask	G. W. Whipple	West Acton, Mass	May 1, 1860	28, 134
Powder-flask charger	H. C. Bascom	La Crosse, Wis	Mar. 24, 1868	75, 839
Powder-flask charger	C. C. Dickey	Philadelphia, Pa	Jan. 23, 1866	52, 147
Powder-flask for breech-loading gun	D. Moore	Williamsburgh, N. Y	Oct. 31, 1854	11, 871
Powder-keg	L. Austin and J. E. Hall	Cleveland, Ohio	Feb. 1, 1870	99, 280
Powder-keg	J. B. and D. J. Fleming	Xenia, Ohio	Oct. 6, 1868	82, 818
Powder-keg	C. Green and W. Wilson, jr	Wilmington, Del	July 15, 1873	140, 913
Powder-keg safety-conductor	M. Ward	Mount Carmel, Pa	Jan. 9, 1872	122, 688
Powder-magazine, Fire-proof	R. S. Sanborn	Ripon, Wis	July 30, 1867	67, 220
Powder-mill	G. Keyser	....	July 27, 1803	....
Powder-mixer	J. Burns	New York, N. Y	Mar. 26, 1867	63, 208
Powder-packing apparatus	K. F. Knowles	New York, N. Y	Aug. 13, 1872	130, 433
Powder pressing and graining machine	P. A. Oliver	New York, N. Y	Mar. 22, 1870	101, 032
Powder-proof lock	W. Hall	Boston, Mass	Aug. 1, 1848	5, 686
Powder-proof lock	E. Kershaw	Boston, Mass	Aug. 1, 1848	5, 685
Powders, Apparatus for dividing	G. P. Allen	Woodbury, Conn	June 27, 1871	116, 253
Powders, Apparatus for dividing	F. Schaeffer	Philadelphia, Pa	Oct. 19, 1869	96, 038
Powders, Instrument for measuring	W. Thomson	Madison, Wis	Oct. 31, 1871	120, 404
Powders, Machine for folding	J. W. Maxwell	Louisville, Ky	May 3, 1870	102, 568
Powders, &c., Putting up	H. Sawyer	Roxbury, Mass	Jan. 5, 1864	41, 097
Powders, Putting up	H. Sawyer	Chelsea, Mass	Sept. 12, 1871	118, 819
Powders, Putting up and using	A. P. Willard	Battle Creek, Mich	Aug. 31, 1869	94, 263
Power:				
See Animal-power. Atmospheric power. Atmospheric motive-power. Churn-power. Dog-power. Fan-power. Foot-power. Hammer-power. Hoisting-power. Hop-power. Horse-power. Human power. Hydraulic-power. Lever-power. Loom-power. Maintaining power. Man-power. Manual power. Mechanical power. Metal expansion and contraction power. Motive-power. Multiplying-power. Pendulum-power. Propelling-power. Railway-power. Ratchet-power. Regulating-power. Screw-power. Sewing-machine power. Spring-power. Steam water-power. Thrashing-power. Tide-power. Transmitting power. Tread-power. Washing-machine power. Water-power. Wave-power. Wind-power. Wind-wheel power. Windlass-crank power.				
Power-accumulating apparatus	J. F. Latimer	Detroit, Mich	Oct. 15, 1867	69, 821
Power-accumulating apparatus	A. C. Platt	Sandusky, Ohio	July 25, 1871	117, 454
Power-accumulating device	G. H. Becker	Memphis, Tenn	Dec. 17, 1867	72, 156
Power and hand drill	H. Woodman	Biddeford, Me	May 25, 1858	20, 385
Power-apparatus	G. W. Vosburgh	Sun Prairie, Wis	Oct. 8, 1872	131, 981
Power by certain fluids, Obtaining	M. J. Brunel	London, England	Mar. 30, 1827	....

Index of patents issued from the United States Patent Office from 1790 *to* 1873, *inclusive*—Continued.

Invention.	Inventor.	Residence.	Date.	No.
Power-generating apparatus	P. Shearer	Reading, Pa	Mar. 5, 1861	31, 631
Power for small machinery	C. L. Johnson	Omaha, Nebr	June 13, 1871	115, 864
Power-indicator	N. P. Bowsher	Ligonier, Ind	May 12, 1868	77, 864
Power-machine	W. Hosford	Washington Township, Ohio.	Dec. 31, 1829	......
Power, Machine for gaining	H. Bickel	Elizabeth City, N. J	June 6, 1865	48, 042
Power-machinery	R. Mitchell	Cynthiana, Ky	Oct. 10, 1829	......
Power, Machinery for accumulating and transmitting.	E. Stevens	Barnet, Vt	Apr. 26, 1859	23, 794
Power, Machinery for gaining	C. Redheffer	Philadelphia, Pa	July 11, 1820	......
Power, Mechanism for transmitting	J. Rankin	Binghamton, N. Y	Oct. 14, 1873	143, 640
Power-meter	J. A. Bradshaw, W. H. Brown, and D. Whithead.	Lowell, Mass	Sept. 13, 1870	107, 331
Power, Obtaining and transmitting	J. J. Gorman	Cincinnati, Ohio	Aug. 9, 1854	43, 768
Power-press	C. J. Beasley	Petersburgh, Va	Apr. 4, 1871	113, 246
Power-press	E. W. Bliss	Brooklyn, N. Y	July 26, 1870	105, 632
Power-press	D. Campbell	Elizabeth, N. J	July 6, 1869	92, 263
Power-press	R. H. Fisher	Beaver Falls, Pa	Jan. 26, 1869	86, 292
Power-press	A. D. Hamlin	Brooklyn, N. Y	Dec. 27, 1870	110, 455
Power-press	C. W. Johnson	Waterbury, Conn	Nov. 7, 1865	50, 826
Power-press	C. Oysten	Little Falls, N. Y	May 22, 1860	28, 401
Power-press	W. F. Parker	Meriden, Conn	Apr. 6, 1869	88, 732
Power-press	C. M. and O. G. Stratton	Greenfield, Mass	July 19, 1870	105, 605
Power-press	J. Weed	Muscatine, Iowa	May 22, 1860	28, 429
Power-press	M. G. Wilder	Meriden, Conn	Aug. 9, 1864	43, 818
Power-press	M. G. Wilder	West Meriden, Conn	June 27, 1871	116, 383
Power-press, Centripetal	E. S. Scripture	Cazenovia, N. Y	Sept. 26, 1834	......
Power-press-feed device	P. E. Austin	New Haven, Conn	Oct. 18, 1870	108, 312
Power-press for punching copper, &c	G. Darracott	Boston, Mass	July 21, 1829	......
Power-press, Hay, &c	J. H. Wittmer	Manor, Pa	May 11, 1869	90, 039
Power-press, Portable	T. B. Wait	Zena, Oreg	June 27, 1871	116, 376
Power-press, Self-acting	E. Davis	Montpelier, Vt	Mar. 23, 1854	10, 960
Power-press, Spring	R. McOmber	Galway, N. Y	Oct. 27, 1829	......
Power-rake	H. N. Tripp	Alfred, Me	Dec. 20, 1853	10, 327
Power-regulator for friction-surfaces	B. Tatham	New York, N. Y	Apr. 12, 1870	101, 940
Power to machinery, Applying	D. Durfey	Fort Seneca, Ohio	Sept. 6, 1859	25, 361
Power to machinery, Applying	G. Page	Keene, N. H	Sept. 28, 1831	......
Power to machinery, Mechanism for applying	L. Curdts	New York, N. Y	Aug. 13, 1867	67, 730
Power-transferring apparatus	E. B. Clark	Cohoes, N. Y	Dec. 23, 1873	145, 721
Power-transmitting apparatus	T. Damon	Thompsonville, Conn	Dec. 12, 1871	121, 854
Preservation of flesh for food	N. B. Marsh	Cincinnati, Ohio	Nov. 30, 1858	22, 185
Preservation of meat and other articles of food	F. Cassel	Cologne, Prussia	Aug. 16, 1870	106, 465
Preservative materials, Impregnating substances with.	E. Sabathé and L. Jourdan	Paris, France	Sept. 11, 1866	58, 036
Preservatory	A. Smith	Perrysburgh, Ohio	Feb. 13, 1866	52, 650
Preserve-can	B. L. Agnew	Indiana, Pa	Oct. 18, 1859	25, 863
Preserve-can	J. K. Chace	Brooklyn, N. Y	Aug. 19, 1873	141, 994
Preserve-can	P. H. Cotton	Demopolis, Ala	Dec. 21, 1858	22, 351
Preserve-can	H. G. Dayton	Maysville, Ky	June 8, 1858	20, 485
Preserve-can	H. S. Fisher	Newburgh, Pa	Aug. 19, 1862	36, 264
Preserve-can	W. Fridley and F. Cornman	Carlisle, Pa	Oct. 25, 1859	25, 894
Preserve-can	N. S. Gilbert	Lockport, N. Y	Jan. 29, 1861	31, 235
Preserve-can	E. W. Gilmore	North Easton, Mass	May 11, 1858	20, 203
Preserve-can	J. L. Gray	Baltimore, Md	Apr. 2, 1867	63, 505
Preserve-can	J. F. Griffen	New York, N. Y	Apr. 5, 1864	42, 186
Preserve-can	W. H. Harn	Carlisle, Pa	May 22, 1860	28, 366
Preserve-can	W. D. Ludlow	New York, N. Y	June 28, 1859	24, 566
Preserve-can	E. Manley	Marion, N. Y	Aug. 3, 1858	21, 078
Preserve-can	J. F. Martin and H. C. Nicholson.	Mount Washington, Ohio	Feb. 15, 1859	22, 962
Preserve-can	F. O. More	Bellefontaine, Ohio	July 12, 1859	24, 751
Preserve-can	S. Morrett	West Pennsborough, Pa	Mar. 29, 1859	23, 384
Preserve-can	C. Newman	Birmingham, Pa	Dec. 20, 1859	26, 515
Preserve-can	P. H. Niles	Boston, Mass	Nov. 28, 1865	51, 264
Preserve-can	J. Roussel, L. Delaugre, and L. L. Robin.	Nantes, France	July 5, 1864	43, 463
Preserve-can	T. Sellers	East Birmingham, Pa	May 22, 1860	28, 413
Preserve-can	C. F. Stilz and J. C. Knoepke	Philadelphia, Pa	Aug. 14, 1866	57, 212
Preserve-can	A. T. Twing, E. Wood, and W. Elderhorst.	Louisburgh and Troy, N. Y	May 22, 1860	28, 424
Preserve-can cover	C. L. Kelling	Mechanicsburgh, Pa	Jan. 24, 1860	26, 907
Preserve can, jar, &c	T. Earle	Smithfield, R. I	Nov. 10, 1863	40, 556
Preserve-can, Self-sealing	R. Arthur	Washington, D. C	Jan. 2, 1855	12, 153
Preserve-can stopper	V. P. Corbett	Washington, D. C	Sept. 20, 1859	25, 490
Preserve-can stopper	W. P. Patton	Harrisburgh, Pa	Jan. 10, 1860	26, 820
Preserve-cans, Apparatus for exhausting air from	D. N. Phelps	San Leandro, Cal	Feb. 6, 1872	123, 418
Preserve-cans, Closing	T. Earle	Valley Falls, R. I	Feb. 2, 1864	41, 425
Preserve-cans, Hermetically-sealing	C. Branwhite	New York, N. Y	Mar. 18, 1856	14, 439
Preserve-cans, Hermetically-sealing	C. E. Russell	Saint Louis, Mo	June 10, 1856	15, 088
Preserve-cans, Sealing	H. S. Fisher	Newburgh, Pa	Nov. 12, 1861	33, 729
Preserve-cans, Sealing	R. W. Lewis	Honesdale, Pa	Feb. 12, 1856	14, 245
Preserve-cans, Sealing	S. Lutz	Philadelphia, Pa	Oct. 23, 1855	13, 707
Preserve-cans, Sealing	W. W. Lyman	West Meriden, Conn	Aug. 31, 1858	21, 348
Preserve-cans, Sealing	H. C. Nicholson and J. Spratt	Cincinnati, Ohio	Jan. 3, 1854	10, 396
Preserve-cans, Sealing	W. W. Paddock	Cincinnati, Ohio	Nov. 6, 1860	30, 585
Preserve-cans, Sealing	A. Taylor	Baltimore, Md	Dec. 7, 1858	22, 247
Preserve-cans, Sealing-device for	W. H. Elliot	Plattsburgh, N. Y	July 17, 1855	13, 291
Preserve canisters, Sealing	H. Hunt	B ooklyn, N. Y	Sept. 6, 1853	9, 989
Preserve-jar	T. Earle	Valley Falls, Smithfield Township, R. I.	Dec. 27, 1864	45, 594
Preserve-jar	T. Earle	Valley Falls, Smithfield Township, R. I.	Mar. 21, 1865	46, 887
Preserve-jar	N. S. Gilbert	Lockport, N. Y	Dec. 17, 1861	33, 938
Preserve-jar	F. H. Lauterbach	Boston, Mass	June 12, 1866	55, 585
Preserve-jar	T. G. Otterson	Camden, N. J	Aug. 31, 1869	93, 236
Preserve-jar	P. W. Reid	Birmingham, Pa	July 21, 1863	39, 327
Preserve-jar	F. J. Shefferly	Detroit, Mich	Jan. 7, 1868	73, 202
Preserve-jar	C. F. Spencer	Rochester, N. Y	Feb. 10, 1863	37, 647
Preserve-jar	J. J. Squire	Windsor Locks, Conn	Oct. 18, 1864	44, 752

Index of patents issued from the United States Patent Office from 1790 *to* 1873, *inclusive*—Continued.

Invention.	Inventor.	Residence.	Date.	No.
Preserve-jar	N. Thompson	Brooklyn, N. Y	Mar. 9, 1869	87, 730
Preserve-jar	J. M. Whitall	Philadelphia, Pa	June 18, 1861	32, 594
Preserve-jar	A. Q. Withers	Holly Springs, Miss	Feb. 20, 1872	123, 851
Preserve-jar cap	L. R. Boyd	New York, N. Y	May 11, 1869	89, 845
Preserve-jar cap	S. B. Rowley	Philadelphia, Pa	Feb. 11, 1868	74, 249
Preserve-jar cap	W. Taylor and C. Hodgetts	Brooklyn, N. Y	July 18, 1871	117, 236
Preserve-jar stand	K. E. Ashley	Williamsburgh, N. Y	Oct. 31, 1865	50, 674
Preserve-jars, Clasp for closing	J. A. Cowles	Chicago, Ill	Apr. 28, 1863	38, 288
Preserve-jars, Composition for sealing	J. Beckley	Cincinnati, Ohio	July 7, 1863	39, 111
Preserve-jars, Sealing	R. M. Dalbey	Mount Washington, Ohio	Nov. 16, 1858	22, 066
Preserve-vessels	J. Adams	Pittsburgh, Pa	May 20, 1862	35, 286
Preserve-vessel, Self-sealing	W. J. Stevenson	New York, N. Y	May 27, 1856	14, 974
Preserves, Packing-apparatus for	C. T. Provost	New York, N. Y	June 7, 1870	103, 927
Preserving and canning meat, fruit, vegetables, &c., Apparatus for.	N. S. Shipley	Baltimore, Md	Jan. 24, 1871	111, 264
Preserving and condensing milk	G. Borden and J. G. Borden	White Plains and South East, N. Y.	Nov. 4, 1873	144, 310
Preserving and curing meat	J. Tilton	New York, N. Y	Oct. 2, 1866	58, 511
Preserving and drying green corn on the cob, Method of.	D. Parker	Shaker Village, N. H	Apr. 10, 1866	53, 861
Preserving and freezing	D. E. Somes	Washington, D. C	Jan. 25, 1870	99, 254
Preserving and freezing fish, meats, &c., Apparatus for.	P. Nunan	Sandusky, Ohio	Oct. 27, 1868	83, 533
Preserving and packing box for perishable articles	B. Yaw	New Concord, Ohio	Nov. 12, 1872	133, 000
Preserving and packing compound for eggs, wood, &c.	J. Connell, jr	Collingwood, Canada	May 20, 1873	139, 116
Preserving and packing meat	C. E. Richardson	Cambridge, Mass	Feb. 11, 1868	74, 247
Preserving and smoking meat	A. Robertson	Seymour, Ind	May 16, 1871	114, 972
Preserving and transporting fish, Apparatus for	W. Davis	Detroit, Mich	Jan. 19, 1869	85, 914
Preserving and transporting fresh meat	W. Dugan	Chicago, Ill	Apr. 4, 1871	113, 505
Preserving animal and vegetable substances	J. Bruckner	New Orleans, La	Aug. 13, 1872	130, 474
Preserving animal and vegetable substances	D. Cooley	New York	Oct. 24, 1814	
Preserving animal and vegetable substances	J. Coppinger	Beaufort, S. C	Nov. 23, 1809	
Preserving animal and vegetable substances	J. Gamgee and A. Gamgee	Bayswater, England, and Edinburgh, Scotland.	Nov. 26, 1867	71, 377
Preserving animal and vegetable substances	R. Jones	London, England	Nov. 28, 1865	51, 280
Preserving animal and vegetable substances	L. H. Spear	Braintree, Vt	Mar. 22, 1864	42, 025
Preserving animal and vegetable substances	L. H. Spear	Braintree, Vt	Aug. 20, 1867	67, 921
Preserving animal and vegetable substances	F. Stabler	Baltimore, Md	Nov. 14, 1865	50, 965
Preserving animal and vegetable substances	A. Fryer	Manchester, England	Oct. 8, 1872	132, 002
Preserving animal and vegetable substances, Apparatus for.	E. Piper	Camden, Me	Aug. 5, 1862	36, 107
Preserving animal and vegetable substances during transportation, Case for.	J. G. Staunton	Buffalo, N. Y	Jan. 3, 1865	45, 764
Preserving animal and vegetable substances in transit.	E. R. Morney	McDonough, Del	Sept. 14, 1869	94, 908
Preserving animal and vegetable substances on ship-board.	J. F. Baldwin	Provincetown, Mass	July 6, 1869	92, 246
Preserving animal and vegetable substances, Refrigerating and condensing apparatus for.	E. D. Brainard	Albany, N. Y	Jan. 14, 1868	73, 292
Preserving animal matters, Process and apparatus for.	A. Rocke	New Orleans, La	Feb. 27, 1872	124, 161
Preserving animal substances	E. Daggett and T. Kensett	New York	Jan. 19, 1825	
Preserving animal substances	O. Lugo	New York, N. Y	Aug. 18, 1868	81, 185
Preserving animal substances	H. Medlock and W. Bailey	London and Wolverhampton, England.	Aug. 1, 1871	117, 660
Preserving apples and other fruit, beets, sweet potatoes, &c., Mode of.	A. Hart	Wharton Township, Pa	Aug. 10, 1829	
Preserving articles of food	J. McCall and B. G. Sloper	London and Walthamstow, England.	Dec. 9, 1862	37, 137
Preserving beer and other liquids, Apparatus for	T. Byrne	New York, N. Y	Sept. 19, 1865	50, 085
Preserving box, Fruit	O. E. Doolittle	Boston, Mass	June 2, 1868	78, 437
Preserving bread	J. J. A. Mouries	Phalsbourg, France	Sept. 6, 1870	107, 088
Preserving butter	P. Giraud	New York, N. Y	Oct. 15, 1861	33, 482
Preserving butter	M. Johnson		June 30, 1797	
Preserving butter	J. F. Saiger	Shelby, Ohio	Jan. 25, 1870	99, 240
Preserving butter	H. D. Wetmore	Cleveland, Ohio	Mar. 5, 1861	31, 638
Preserving butter, Can for	W. C. Reutgen	Vicksburgh, Miss	Dec. 5, 1865	51, 352
Preserving butter, Composition for	L. De Corn	Cincinnati, Ohio	Aug. 3, 1852	9, 167
Preserving butter, meat, &c	W. Ross	Day's Store, Pa	Jan. 1, 1867	60, 942
Preserving butter, &c., Vessel for	J. G. Staunton	Buffalo, N. Y	Jan. 3, 1865	45, 763
Preserving-can	A. R. Davis	East Cambridge, Mass	July 16, 1872	129, 214
Preserving-can	O. N. Weaver	Dover, Ky	July 12, 1859	24, 770
Preserving-can for food	A. S. Lyman	New York, N. Y	May 11, 1858	20, 209
Preserving-can for food	A. S. Lyman	New York, N. Y	June 29, 1858	20, 722
Preserving can, Fruit, &c	G. W. Griswold	Logansport, Ind	July 22, 1862	35, 933
Preserving case, Butter, cheese, &c	T. R. Timby	Saratoga, N. Y	Mar. 2, 1869	87, 446
Preserving chopped meat, &c	W. C. Marshall	New York, N. Y	Feb. 23, 1864	41, 712
Preserving coffee	B. T. Babbitt	New York, N. Y	Mar. 24, 1868	75, 828
Preserving coffee	A. Eikerenkotter and F. Silver	Searsville, Cal	Sept. 3, 1867	68, 422
Preserving corn	L. McMurray	Baltimore, Md	June 3, 1873	139, 595
Preserving cranberries	J. W. Campbell	Philadelphia, Pa	July 25, 1871	117, 256
Preserving cranberries	L. Kniffen	Worcester, Mass	Mar. 19, 1872	124, 827
Preserving dead bodies	J. A. Gaussardia	Washington, D. C	Oct. 28, 1856	15, 972
Preserving dead bodies, &c	J. Morgan	Dublin, Ireland	Sept. 27, 1864	44, 495
Preserving dead bodies	G. W. Scollay	Saint Louis, Mo	Oct. 19, 1869	95, 939
Preserving dead bodies	C. C. St. Clair	Washington, D. C	July 23, 1867	67, 145
Preserving eggs	A. M. Blinval	New York, N. Y	Jan. 5, 1864	41, 053
Preserving eggs, &c	C. Boize	New York, N. Y	Dec. 17, 1867	72, 158
Preserving eggs	L. H. Boole	New York, N. Y	Sept. 17, 1867	68, 834
Preserving eggs	J. Brakeley	Bordentown, N. J	Mar. 10, 1868	75, 242
Preserving eggs	A. R. Davis	East Cambridge, Mass	Sept. 24, 1872	131, 507
Preserving eggs	P. Gaughran and L. Sweeney	San Francisco, Cal	Aug. 6, 1867	67, 427
Preserving eggs	J. H. Hall	New York, N. Y	Mar. 9, 1869	87, 562
Preserving eggs	W. Hansford	San Francisco, Cal	Apr. 11, 1865	47, 202
Preserving eggs, &c	W. B. Ingersoll	New York, N. Y	May 31, 1864	42, 948
Preserving eggs	J. Kaye, jr	Setzler's Store, Pa	Nov. 22, 1870	109, 420
Preserving eggs	J. Knapp	Prattsburgh, N. Y	July 12, 1870	105, 342

Index of patents issued from the United States Patent Office from 1790 *to* 1873, *inclusive*—Continued.

Invention.	Inventor.	Residence.	Date.	No.
Preserving eggs	J. Knapp	Prattsburgh, N. Y	Dec. 26, 1871	122, 258
Preserving eggs	A. G. and E. E. Kyle	Newville, Pa	June 12, 1866	55, 502
Preserving eggs	C. A. La Mont	New York, N. Y	Nov. 28, 1865	51, 263
Preserving eggs	J. Longmaid	New York, N. Y	Sept. 21, 1869	95, 028
Preserving eggs	J. K. Marsh	Terre Haute, Ind	Feb. 26, 1867	62, 432
Preserving eggs	W. S. Marsh	Raymond, Wis	Dec. 5, 1871	121, 640
Preserving eggs	N. Patton	Coles County, Ill	Jan. 16, 1866	52, 072
Preserving eggs	R. S. Rhodes and E. Whyte	Chicago, Ill	Dec. 12, 1865	51, 480
Preserving eggs	A. Van Camp	Washington, D. C	Feb. 18, 1868	74, 733
Preserving eggs and other articles	C. Bruner	Marshall, Mo	Nov. 30, 1869	97, 349
Preserving eggs, butter, &c	T. Edmundson	Pipe Creek, Md	Apr. 26, 1827	....
Preserving eggs, Composition for	W. C. Bruson	Chicago, Ill	Apr. 27, 1869	89, 285
Preserving eggs, Composition for	J. T. McKim	Remington, Ind	July 11, 1871	116, 976
Preserving eggs, Composition for	N. Patton	Kansas, Ill	Dec. 29, 1868	85, 328
Preserving eggs, Composition for	P. J. Stouffer	Uniontown, Pa	Apr. 23, 1867	64, 047
Preserving eggs, Compound for	J. B. Patterson	Portage City, Wis	Aug. 18, 1868	81, 104
Preserving eggs, meats, &c	J. Perkins	Newark, N. J	Apr. 19, 1864	42, 396
Preserving fish	R. A. Adams	Cambridge, Mass	Sept. 28, 1869	95, 179
Preserving fish	E. Piper	Camden, Me	Mar. 19, 1861	31, 736
Preserving fish by freezing	S. H. and D. W. Davis	Detroit, Mich	Feb. 28, 1871	112, 129
Preserving fish, meat, &c., Compound for	T. Sim	Charleston, S. C	Jan. 19, 1869	86, 040
Preserving fish and meat	M. Gross	Washington, D. C	Dec. 13, 1859	26, 427
Preserving flour, grain, &c	J. Carlin	Cincinnati, Ohio	Oct. 2, 1866	58, 374
Preserving flowers	E. M. Stigale	Philadelphia, Pa	Apr. 27, 1869	89, 515
Preserving flowers and other vegetable forms	P. T. Vining	New York, N. Y	June 11, 1867	65, 777
Preserving food	W. A. Keeler	New York, N. Y	July 31, 1860	29, 383
Preserving food for transportation, Apparatus for	M. Brune	New York, N. Y	July 25, 1865	48, 898
Preserving fruit, &c	W. D. Brooks	Baltimore, Md	July 2, 1867	66, 208
Preserving fruit	G. Jáques	Boston, Mass	Sept 27, 1870	107, 689
Preserving fruit	J. K. Jenkins	Kingston, Pa	Dec. 28, 1858	22, 433
Preserving fruit	D. M. Mefford	Norwalk, Ohio	Sept. 22, 1868	82, 429
Preserving fruit	D. M. Mefford	Norwalk, Ohio	Feb. 2, 1869	86, 433
Preserving fruit	E. R. Norny	McDonough, Del	June 22, 1869	91, 557
Preserving fruit, &c	E. C. Roberts	Salem, Mich	Mar. 7, 1865	46, 707
Preserving fruit, &c	N. S. Shaler	Newport, Ky	Oct. 11, 1864	44, 664
Preserving fruit, &c	H. C. Smith	Chicago, Ill	June 21, 1864	43, 232
Preserving fruit	W. H. Trissler	Cleveland, Ohio	June 18, 1867	65, 844
Preserving fruit and vegetables	W. G. Barbee	High Point, N. C	Nov. 18, 1873	144, 592
Preserving fruit and vegetables	J. M. Blanco y Nuño	Havana, Cuba	Sept. 29, 1868	82, 588
Preserving fruit and vegetables	E. G. Holden	Covington, Ky	Nov. 6, 1866	59, 396
Preserving fruit and vegetables	F. C. Hooten	West Chester, Pa	Aug. 8, 1871	117, 884
Preserving fruit and vegetables	E. C. Roberts	Salem, Mich	June 17, 1862	35, 626
Preserving fruit and vegetables and compounds therefor.	E. R. Norny	McDonough, Del	Apr. 27, 1869	89, 330
Preserving fruit, Apparatus for	L. C. Cooley	New York, N. Y	June 3, 1873	139, 547
Preserving fruit, Apparatus for	J. Cope	East Fairfield, Ohio	Nov. 26, 1872	133, 413
Preserving fruit, Apparatus for	W. W. Lyman	Meriden, Conn	May 10, 1870	102, 951
Preserving fruit, Apparatus for	C. C. Williams	Brooklyn, N. Y	Oct. 26, 1869	96, 179
Preserving fruit, Building for	B. M. Nyce	Kingston, Ind	Mar. 19, 1861	31, 734
Preserving fruit from decay, Composition for	G. Jáques	Boston, Mass	Sept. 27, 1870	107, 690
Preserving fruit in jars, &c	S. J. Parker	Ithaca, N. Y	Jan. 26, 1864	41, 392
Preserving fruit, &c., in sealed cans	H. B. Slaughter	Crumpton, Md	Nov. 11, 1862	36, 921
Preserving fruit, meat, &c	E. S. Bartholomew	Westfield, N. Y	Dec. 18, 1866	60, 462
Preserving fruit, meat, &c	V. W. Blanchard	Bridport, Vt	Nov. 16, 1869	96, 871
Preserving fruit, meat, and vegetables	J. B. Hyde	New York, N. Y	Oct. 6, 1868	82, 719
Preserving fruit, meat, &c., Apparatus for	J. G. McMillan	Baltimore, Md	Oct. 15, 1867	69, 924
Preserving fruit, meat, fish, &c	J. G. Staunton	Buffalo, N. Y	Jan. 3, 1865	45, 765
Preserving fruit, vegetables, &c	L. Bradley and T. D. Phillips	Buffalo, N. Y	Feb. 8, 1870	99, 628
Preserving fruit, vegetables. &c	A. T. Jones and L. C. Cooley	New York, N. Y	June 3, 1873	139, 581
Preserving fruit, vegetables, &c	D. M. Mefford	Huron County, Ohio	Dec. 31, 1872	134, 390
Preserving fruit, vegetables, &c., Process and apparatus for.	J. C. Wrenshall	Baltimore, Md	June 21, 1870	104, 680
Preserving green corn	S. J. Carrol	Baltimore, Md	Jan. 22, 1867	61, 316
Preserving green corn	W. L. Gilroy	Philadelphia, Pa	Jan. 11, 1870	98, 758
Preserving green corn	D. Rowe	Baltimore County, Md	June 30, 1857	17, 697
Preserving green corn	I. Winslow	Philadelphia, Pa	May 13, 1862	35, 274
Preserving green corn	I. Winslow	Philadelphia, Pa	May 20, 1862	35, 346
Preserving green corn and other vegetables	G. L. Merrill and O. F. Soule	Syracuse, N. Y	Dec. 16, 1873	145, 581
Preserving grapes and other fruit	E. A. Wible	Georgetown, Cal	June 10, 1862	35, 556
Preserving grated horse-radish	J. G. Lunz	Chicago, Ill	May 7, 1872	126, 557
Preserving hops	J. C. and M. D. Baldwin and R. Brayton.	Watertown, N. Y., Brantford, Canada, and Buffalo, N. Y.	Sept. 25, 1860	30, 113
Preserving-house	H. Blackburn	Bedford County, Pa	Dec. 7, 1869	97, 475
Preserving-house	J. C. Gove	Cleveland, Ohio	Jan. 1, 1867	60, 716
Preserving-house	W. S. Mabbett	Calverton Mills, Md	July 22, 1862	35, 943
Preserving house, Fruit, &c	N. Hellings	Philadelphia, Pa	Oct. 15, 1867	69, 806
Preserving house, Fruit, &c	E. F. Olds	Hudson, Mich	July 18, 1865	48, 833
Preserving house, Fruit	J. Prior	Adrian, Mich	Feb. 1, 1870	99, 349
Preserving-houses, Ice-floor or	T. S. Rankin	Quenemo, Kansas	Mar. 5, 1872	124, 386
Preserving-jar	J. Borden	Bridgeton, N. J	Apr. 13, 1858	19, 964
Preserving jar and can	G. S. G. Spence	Salem, Mass	Nov. 18, 1862	36, 970
Preserving-jar cap	J. Borden	Bridgeton, N. J	Feb. 12, 1867	61, 921
Preserving liquid and other substances	R. d'Heureuse	New York, N. Y	Aug. 3, 1869	93, 182
Preserving malt-liquor, Apparatus for	J. Keane	New York, N. Y	Oct. 12, 1858	21, 761
Preserving meat, &c	W. Davis	Detroit, Mich	June 16, 1868	78, 932
Preserving meat, &c	E. De La Granja	Boston, Mass	Sept. 17, 1867	68, 850
Preserving meat	J. E. Dotch	Washington, D. C	Dec. 1, 1868	84, 481
Preserving meat	J. E. Dotch	Washington, D. C	June 25, 1872	128, 372
Preserving meat	J. E. Dotch and O. Loew	Washington, D. C., and New York, N. Y.	June 25, 1872	128, 371
Preserving meat	H. Gahn	Upsala, Sweden	Jan. 25, 1870	99, 180
Preserving meat, &c	W. Gillespie	Louisa Court-House, Va	Apr. 2, 1872	125, 285
Preserving meat	F. H. Hatch and B. R. Hawley	New Orleans, La., and Normal, Ill.	Sept. 13, 1870	107, 367
Preserving meat	W. C. Marshall	New York, N. Y	July 12, 1864	43, 516
Preserving meat	H. B. Meech	Fort Edward, N. Y	Oct. 30, 1866	59, 247
Preserving meat, &c	J. F. Meiners	New York, N. Y	June 4, 1872	127, 627

Index of patents issued from the United States Patent Office from 1790 *to* 1873, *inclusive*—Continued.

Invention.	Inventor.	Residence.	Date.	No.
Preserving meat	G. W. Oliver	New York, N. Y	June 19, 1860	28, 812
Preserving meat	M. Perl	Houston, Tex	Oct. 6, 1868	82, 871
Preserving meat	D. E. Somes	Biddeford, Me	Apr. 9, 1861	32, [illegible]09
Preserving meat	L. H. Spear	Peekskill, N. Y	Aug. 23, 1870	106, 632
Preserving meat	W. Wiesmann	Bonn, Prussia	Nov. 10, 1868	84, 038
Preserving meat and fruit	J. Hampton	Lexington, Ky	Mar. 12, 1830	......
Preserving meat and other perishable articles, Process and apparatus for.	T. Sim	Charleston, S. C	Dec. 22, 1868	85, 184
Preserving meat and vegetables	A. S. Lyman	New York, N. Y	Apr. 30, 1872	126, 148
Preserving meat and vegetables, Apparatus for	W. Maxwell	Henderson, Tex	Jan. 30, 1872	123, 273
Preserving meat, Apparatus for	C. F. Carr	Norwich, N. Y	Apr. 9, 1872	125, 540
Preserving meat, Composition for	J. J. Harrison	Saint Michael's, Md	July 10, 1866	56, 217
Preserving meat, fish, &c	B. Mosquera	Santiago, Chili	June 25, 1872	128, 320
Preserving meat, fish, &c., and making meat extracts.	T. F. Henley	Pimlico, England	Oct. 1, 1872	131, 820
Preserving meat, fish, &c., Apparatus for	C. F. Pike	Providence, R. I	Dec. 10, 1867	71, 910
Preserving fish, oysters, &c	J. E. Dotch	Washington, D. C	Nov. 8, 1870	108, 983
Preserving meat for food	W. Adamson and C. F. A. Simonin.	Philadelphia, Pa	Apr. 2, 1872	125, 248
Preserving meat for pastry purposes	G. H. Munroe	New York, N. Y	Oct. 12, 1869	95, 715
Preserving meat, fowls, &c	C. Havard and M. X. Harmony	London, England	June 8, 1869	90, 944
Preserving meat, poultry, fish, &c	J. E. Dotch	Washington, D. C	Aug. 3, 1869	93, 183
Preserving meat, poultry, fish, &c	A. Vazquez and J. E. Rosenberg.	Santiago, Chili	Mar. 26, 1872	125, 102
Preserving milk	J. L. Granger	New York, N. Y	Mar. 19, 1836	......
Preserving milk	L. F. Kirchhofer	Union Hill, N. J	Aug. 19, 1873	141, 878
Preserving milk, Apparatus for	N. P. Holmes	Indianapolis, Ind	Sept. 4, 1866	57, 721
Preserving milk, Apparatus for	N. P. Holmes	Indianapolis, Ind	Dec. 4, 1866	60, 190
Preserving milk, Apparatus for	E. Rönning	Brodhead, Wis	Jan. 31, 1871	111, 476
Preserving organic substances	J. G. Staunton	Buffalo, N. Y	Jan. 3, 1865	45, 762
Preserving oysters	E. T. Gilmore	Springfield, Mass	Nov. 20, 1866	59, 833
Preserving potatoes	C. S. Edwards	Rushville, Ill	Dec. 31, 1845	4, 337
Preserving potatoes	J. Mumford	Clarksburgh, Ohio	Oct. 20, 1868	83, 299
Preserving-powder	G. A. Mariner	Chicago, Ill	May 19, 1868	78, 112
Preserving, Preparing fruit and vegetables for	A. Eckert	Trenton, Ohio	Mar. 6, 1866	52, 980
Preserving provisions, Apparatus for	C. Winship	New Haven, Conn	Oct. 11, 1864	44, 683
Preserving provisions, Cooling-room for	D. E. Somes	Biddeford, Me	Sept. 15, 1863	39, 967
Preserving, refrigerating, and transporting perishable articles.	C. F. Pike	Providence, R. I	Dec. 31, 1867	72, 894
Preserving, storing, and transporting fruits, vegetables, and other perishable articles.	J. Rutter	West Chester, Pa	July 9, 1867	66, 666
Preserving sugar-cane	F. O. Darbey	Attakapas, La	Jan. 23, 1872	122, 936
Preserving sweet potatoes	H. T. Basye	Dyersburgh, Tenn	Dec. 13, 1870	110, 108
Preserving sweet potatoes	J. C. Tilton	Pittsburgh, Pa	Mar. 26, 1872	124, 987
Preserving tomatoes	F. G. Ford	Bridgeton, N. J	Oct. 15, 1872	132, 272
Preserving tomatoes, Compound for	D. C. Yates	Big Lick, Va	Sept. 19, 1871	119, 214
Preserving the color in dried fruit	D. Ackart	Schaghticoke, N. Y	Oct. 3, 1871	119, 442
Preserving various articles of food	H. L. Palmer	Stillwater, N. Y	July 19, 1870	105, 484
Preserving vegetable extracts	H. L. Palmer	Stillwater, N. Y	July 19, 1870	105, 485
Preserving vegetables	F. H. Smith	Baltimore, Md	Aug. 17, 1869	93, 841
Preserving vegetables, fruit, &c	M. D. Mefferd	Norwalk, Ohio	Mar. 29, 1870	101, 373
Preserving vegetables in hermetically sealed cans.	I. Winslow	Philadelphia, Pa	Aug. 26, 1862	36, 326
Preserving-vessel	J. B. Wilson	Williamstown, N. J	Dec. 3, 1861	33, 870
Preserving-vessels, Closing	A. Warner	Williamsburgh, N. Y	Mar. 24, 1868	76, 009
Preserving-vessels, Cover for	I. Stratton	Philadelphia, Pa	Feb. 3, 1863	37, 595
Preserving-vessels, Stopper for	J. B. Wilson	Williamstown, N. J	July 9, 1861	32, 805
Press:				

See Anti-friction press.
Back-pad press.
Baling-press.
Beater-press.
Book-holding press.
Brick-press.
Brick and tile press.
Bundling-press.
Butter-print press.
Card-press.
Centripetal press.
Cheese-press.
Cheese and fruit press.
Cider-press.
Cider and wine press.
Cigar-mold press.
Cloth-press.
Clothes-press.
Comb-molding press.
Compressing-press.
Compressing and beating press.
Compression and baling press.
Concrete-building-block press.
Copying-press.
Copying and folding press.
Cotton-press.
Crushing-press.
Cutting-press.
Cylinder-press.
Double-cylinder press.
Drop-press.
Embossing-press.
Field-press.
Filter-press.
Filtering-press.
Fly-wheel press.
Foot-press.
Foundry-press.
Fruit-press.
Fruit and lard press.
Fruit-packing press.
Fuel-press.
Furnace-block press.
Garment-press.
Gear-press.
Glass-press.
Glass-ware press.
Glass-ware-molding press.
Gluing-press.
Gunpowder-press.
Harness-pad press.
Hat-press.
Hat-finishing press.
Hay-press.
Hay and cotton press.
Hay and hemp press.
Hay or cotton press.
Heel-press.
Hop-press.
Hop and hay press.
Hydraulic press.
Hydraulic-wheel press.
Hydro-mechanical press.
Hydrostatic press.
Ink-press.
Jeweler's hand-press.
Lard-press.
Lever-press.
Lever-bottom press.
Liquid-removing press.
Lithographic press.
Meat-press.

Index of patents issued from the United States Patent Office from 1790 *to* 1873, *inclusive*—Continued.

Press—Continued.

See Meat or fruit press.
Metallic press.
Mold-board press.
Mop-press.
Oil-press.
Oil and cider press.
Oil or blubber press.
Oil, tobacco, and cotton press.
Pad-crimp press.
Pen-press.
Photograph-burnishing press.
Photographic-card press.
Plumbago-press.
Power-press.
Printing-press.
Propelling-press.
Provision-press.
Pulley-press.
Punching-press.
Punching and stamping press.
Rafter-press.
Revolving-screw press.
Rim-press.
Rocket-press.
Rocket-filling press.
Rolling-press.
Screw-press.
Screw-die press.
Sector-press.
Self-acting press.
Self adjusting press.
Shearing and punching press.
Stamping-press.
Standing press.
Steam-press.
Steam and hydraulic press.
Stencil-printing press.
Sugar-press.
Sugar-cane press.
Tailor's press.
Tile-press.
Tobacco-press.
Toggle-joint press.
Upsetting-press.
Veneering-press.
Water-power press.
Wedge-press.
Windlass-lever press.
Wine-press.
Wool-press.
Yarn-press.

Invention.	Inventor.	Residence.	Date.	No.
Press	J. E. Alford	Franklinville, N. C	Oct. 15, 1872	132, 229
Press	J. P. Arnold	Louisville, Ky	Apr. 3, 1855	12, 615
Press	S. J. Austin	Freeport, Me	Feb. 7, 1865	46, 205
Press	T. and J. P. Bakewell	Pittsburgh, Pa	Jan. 14, 1829	......
Press	H. B. Barber	Scott, N. Y	June 4, 1867	65, 464
Press	E. C. Betts	Huntsville, Ala	Feb. 21, 1860	27, 198
Press	L. Boudreaux	Thibodeaux, La	Apr. 25, 1865	47, 387
Press	T. G. Brooke	Newark, Ohio	June 4, 1872	127, 557
Press	M. D. Cheek	Memphis, Tenn	May 23, 1848	5, 597
Press	G. C. Chesley	Rocky Mount, Va	June 27, 1838	812
Press	J. Christison	New York, N. Y	Sept. 15, 1863	39, 885
Press	J. E. Conor	Brooklyn, N. Y	July 7, 1868	79, 637
Press	J. Cramton	New York	Feb. 18, 1834	......
Press	D. Curtis	Madison, Wis	Mar. 8, 1870	100, 607
Press	R. V. De Guinon and G. W. Barclay.	Hudson, N. J., and Brooklyn, N. Y.	Apr. 15, 1862	34, 996
Press	D. Dick	Meadville, Pa	Oct. 10, 1848	5, 840
Press	D. Dick	Meadville, Pa	Oct. 17, 1848	5, 856
Press	H. Dodge	Albany, N. Y	Apr. 11, 1865	47, 192
Press	S. S Edmonston	New York	Sept. 1, 1813	......
Press	R. Esmond	New York, N. Y	Aug. 26, 1873	142, 220
Press	P. G. Gardner	New York, N. Y	Feb. 22, 1848	5, 453
Press	J. A. McGillirrae	Dyer, Ind	Jan. 10, 1865	45, 844
Press	J. B. Godwin	Washington, D. C	June 17, 1873	140, 033
Press	G. E. Harding	Bath, Me	May 17, 1864	42, 763
Press	G. E. Harding	Bath, Me	Jan. 3, 1865	45, 711
Press	L. W. Harris	Waterville, N. Y	Aug. 28, 1860	29, 783
Press	T. W. Harvey	New York	Feb. 4, 1843	2, 941
Press	H. Henley	New Garden, Ind	June 9, 1868	78, 738
Press	H. F Hicks	Grand View, Ind	Sept. 20, 1859	25, 546
Press	S. Hitchcock	Vernon, N. Y	June 25, 1813	......
Press	E. Huddleston and B. M. Harrison.	Lawrence, Kans., and Terre Haute, Ind.	Apr. 9, 1861	31, 979
Press	P. C. Ingersoll	Green Point, N. Y	Jan. 24, 1865	46, 051
Press	S. Ingersoll	Brooklyn, N. Y	Mar. 30, 1858	19, 811
Press	A. Jackson	Liberty, Ill	June 24, 1844	3, 637
Press	G. D. Jones	New York, N. Y	June 11, 1867	65, 752
Press	D. King	Aberdeen, Ohio	Mar. 26, 1867	63, 258
Press	R. Kingsley	Springfield, Mass	Jan. 13, 1857	16, 389
Press	J. Huebler	Belleville, Ill	Jan. 20, 1863	37, 475
Press	A. Lasswell	Springfield, Tex	Sept. 24, 1872	131, 619
Press	W. B. Leonard	Matteawan Works, N. Y	Feb. 20, 1843	2, 960
Press	J. Lewis	Wellington, Ohio	Dec. 26, 1865	51, 732
Press	G. Matthewman	Brooklyn, N. Y	Feb. 13, 1866	52, 584
Press	G. Matthewman	Brooklyn, N. Y	Sept. 15, 1868	82, 139
Press	D. McComb	Port Gibson, Miss	Feb. 27, 1849	6, 141
Press	T. H. McCray	Tellico, Tex	Jan. 17, 1860	26, 855
Press	W. J. McDermott	Covington, Tenn	Sept. 28, 1869	95, 249
Press	J. A. McGillerrae	Dyer, Ind	July 24, 1866	56, 588
Press	W. Norman	New York, N. Y	Apr. 3, 1866	53, 657
Press	B. S. Norris	Ripley, Ohio	May 26, 1868	78, 318
Press	A. Norton	Litchfield, Conn	Feb. 11, 1812	......
Press	I. Peck and W. H. H. Glover	Southold and New York, N. Y.	May 19, 1863	38, 599
Press	G. W. Penniston	North Vernon, Ind	Jan. 3, 1860	26, 706
Press	E. B. Platt	Florence, Ind	Apr. 29, 1873	138, 432
Press	W. H. Ragan	Fillmore, Ind	Jan. 30, 1872	123, 125
Press	A. Randel	New York N. Y	Oct. 23, 1860	30, 500
Press	W. Randle	Hadensville Station, Ky	Feb. 13, 1872	123, 644
Press	W. C. Ray	Redington, N. J	Apr. 28, 1863	38, 333
Press	C. O. and J. H. Ritchie	North Madison, Ind	Mar. 31, 1868	76, 101
Press	C. H. Robinson	Bath, Me	Jan. 3, 1865	45, 754
Press	A. Roden	Talladega, Ala	Aug. 21, 1860	29, 719
Press	C. E. Rymes	Somerville, Mass	Aug. 27, 1867	63, 117
Press	U. Schegg	Nauvoo, Ill	Nov. 19, 1867	71, 071
Press	J. Seitz, sr	Bloom, Ohio	Mar. 5, 1861	31, 630
Press	A. C. Semple	Cincinnati, Ohio	June 28, 1853	9, 824

Index of patents issued from the United States Patent Office from 1790 *to* 1873, *inclusive*—Continued.

Invention.	Inventor.	Residence.	Date.	No.
Press	W. I. Tate	Philadelphia, Pa	Oct. 11, 1870	108, 208
Press	E. Thomas	Beverly, Va	Feb. 19, 1861	31, 495
Press	I. H. Trabue	Livingston County, Ky	Aug. 31, 1869	94, 257
Press	C. P. Wagner	New York, N. Y	Sept. 10, 1867	68, 812
Press	R. Wakeman	Port Deposit, Md	Mar. 26, 1861	31, 841
Press	M. Wappich	Sacramento, Cal	Sept. 4, 1860	29, 928
Press	W. E. Warner	Syracuse, N. Y	Apr. 8, 1873	137, 641
Press	T. B. Webster	New York, N. Y	July 25, 1865	49, 046
Press	J. P. White	New York, N. Y	Oct 25, 1864	44, 842
Press	J. J. Wise	Baltimore, Md	May 16, 1839	1, 152
Press and strainer, Combined	J. H. Littlefield	Cambridge, Mass	June 16, 1868	78, 981
Press and strainer, Combined	J. H. Littlefield	Cambridge, Mass	Jan. 5, 1869	85, 530
Press and strainer, Combined	W. D. Medbury	Bangor, Wis	Mar. 29, 1870	101, 294
Press-board, Tailor's	W. Frizzell	Boston, Mass	Apr. 23, 1872	125, 889
Press-board, Tailor's	A. M. Thompson and C. W. Burbank.	Alfred, Me	Sept. 8, 1868	81, 959
Press frame, Brown's standing	R. Hoe	New York	Apr. 22, 1828	....
Press-plate	A. Smith	Cincinnati, Ohio	Feb. 27, 1866	52, 899
Press-spindle adjustment	A. H. Merriman	New Britain, Conn	May 26, 1868	78, 225
Press-tub for pressing oleaginous seeds	F. Follet	Petersburgh, Va	Aug 14, 1834	....
Presses, Machine for removing molded forms from	G. Pattern	Chester, Pa	Dec. 24, 1867	72, 533
Presses, Operating	O. T. Earle and T. J. Rider	Norwalk, Conn., and New York, N. Y.	Aug. 26, 1873	142, 093
Presses, Operating	D. L. Miller	Madison, N. J	Aug. 3, 1858	21, 079
Presses, &c., Trip-motion for	A. Hamlin	Brooklyn, N. Y	Sept. 13, 1870	107, 363
Pressing	J. Tinkler	Massillon, Ohio	Aug. 31, 1831	....
Pressing and boring machine	R. Ramsey	Hanover, N. H	July 9, 1808	....
Pressing and molding pliable materials, Machine for.	G. C. Howard	Philadelphia, Pa	June 19, 1866	55, 658
Pressing and stuffing machine	E. S. Holloway	Columbiana, Ohio	July 4, 1871	116, 709
Pressing, hoisting, and lowering machine	P. Freeman	Perth Amboy, N. J	Feb. 20, 1822	....
Pressing, ironing, and smoothing machine	H. Hamill	New York, N. Y	June 6, 1871	115, 604
Pressing-machine	R. P. Cunningham	Pomfret, Conn	May 21, 1808	....
Pressing-machine	L. E. Dennison	Saybrook, Conn	Oct. 1, 1830	....
Pressing-machine	T. Puffer	Paris, N. Y	May 23, 1815	....
Pressing-machine	R. V. W. and J. Thorne	New York	Mar. 26, 1810	....
Pressing-machine, Tailor's	J. W. Thorp	Hillsborough Bridge, N. H	May 12, 1868	77, 934
Pressure alarm, Hydraulic	J. L. Pillsbury and J. S. Shorb	Columbus and Canton, Ohio	May 16, 1871	114, 849
Pressure and gravitation machine	C. Monson	New Haven, Conn	Sept. 26, 1865	50, 151
Pressure and water indicator, Steam	W. C. Grimes	Philadelphia, Pa	Sept. 7, 1858	21, 468
Pressure-bag for separating elaine and stearine	M. H. Shepard	Cincinnati, Ohio	May 26, 1843	3, 109
Pressure engine, Regulated	D. Livermore	Blairsville, Pa	July 24, 1834	....
Pressure-gage	A. Allan	Perth, Scotland	Nov. 15, 1864	45, 010
Pressure-gage	E. H. Ashcroft	Boston, Mass	July 12, 1853	9, 836
Pressure-gage	R. C. Blake	Cincinnati, Ohio	May 30, 1871	115, 420
Pressure-gage	E. Bourdon	Paris, France	Aug. 3, 1852	9, 163
Pressure-gage	J. Brown, jr	New York, N. Y	Jan. 16, 1872	122, 700
Pressure-gage	W. Burnett	Boston, Mass	Aug. 13, 1861	33, 062
Pressure-gage	H. Chwatal	New York, N. Y	Apr. 12, 1870	101, 832
Pressure-gage	B. Crawford	Allegheny City, Pa	Mar. 16, 1852	8, 797
Pressure-gage	G. H. Crosby	East Sommerville, Mass	Dec. 23, 1873	145, 726
Pressure-gage	C. M. Daboll	New London, Conn	Dec. 4, 1860	30, 803
Pressure-gage	J. B. Eads and H. Flad	Saint Louis, Mo	July 25, 1871	117, 394
Pressure-gage	J. B. Eads and H. Flad	Saint Louis, Mo	Aug. 20, 1872	130, 705
Pressure-gage	H. W. Farley	Hannibal, Mo	Jan. 24, 1860	26, 900
Pressure-gage	R. Finnegan and A. F. W Schulte.	New York, N. Y	June 11, 1861	32, 514
Pressure-gage	W. C. Grimes	Philadelphia, Pa	July 6, 1858	20, 848
Pressure-gage	T. C. Hargrave	Boston, Mass	June 11, 1872	127, 874
Pressure-gage	H. J. H. King	Glasgow, Great Britain	Nov. 9, 1869	96, 705
Pressure-gage	J. Leavens	Brooklyn, N. Y	Mar. 19, 1861	31, 721
Pressure-gage	C. Liedke	Sandusky, Ohio	Aug. 9, 1870	106, 180
Pressure-gage	J. W. Maloy	Boston, Mass	Dec. 3, 1867	71, 773
Pressure-gage	J. W. Maloy	Boston, Mass	Apr. 5, 1870	101, 636
Pressure-gage	J. Matthews, jr	New York, N. Y	Aug. 21, 1855	13, 468
Pressure-gage	J. Matthews, jr	New York, N. Y	Apr. 9, 1867	63, 648
Pressure-gage	G. Sewell	Brooklyn, N. Y	Aug. 26, 1873	142, 128
Pressure-gage	J. W. Stiles	New York, N. Y	Apr. 1, 1873	137, 331
Pressure-gage	E. A. Wood	Utica, N. Y	Apr. 26, 1870	102, 350
Pressure gage and alarm	M. Toulmin	New Orleans, La	Sept. 16, 1873	142, 969
Pressure-gage and safety-valve	A. A. Kent	Lyons, Iowa	June 7, 1870	103, 893
Pressure-gage dial	J. Annin	New York, N. Y	July 15, 1873	140, 867
Pressure-gage, Diaphragm	R. C. Robbins	New York, N. Y	July 18, 1865	48, 877
Pressure-gage, Duplex	P. B. Hovey	New London, Conn	June 21, 1870	104, 593
Pressure-gage for hydrostatic presses	T. Harbottle	Brooklyn, N. Y	May 11, 1869	89, 992
Pressure-gage for steam-boilers	E. H. Ashcroft	Boston, Mass	Mar. 6, 1860	27, 341
Pressure-gage for water-backs, stoves, ranges, &c	J. Anderson	Allegheny, Pa	Aug. 29, 1871	118, 506
Pressure gage, Hot-blast	J. Storer	New York, N. Y	Aug. 3, 1869	93, 366
Pressure-gage, Recording	J. B. Edson	Brooklyn, N. Y	Aug. 16, 1870	106, 345
Pressure-gage, Self-recording	G. P. Clarke and M. B. and J. B. Edson.	New York, N. Y	Aug. 3, 1869	93, 275
Pressure-gages, Adjustment of mercurial	M. E. Campfield	Newark, N. J	Apr. 26, 1870	102, 464
Pressure-indicator	F. T. Riegel	Philadelphia, Pa	Aug. 11, 1868	81, 006
Pressure indicator, Hydrostatic	J. E. Wootten	Philadelphia, Pa	May 14, 1861	32, 329
Pressure-regulator for steam, water, or gas, Balance	E. Osgood	Boston, Mass	Apr. 29, 1862	35, 104
Pressure-regulator for water and steam apparatus	J. C. Hagan	Nashville, Tenn	Nov. 1, 1870	108, 904
Pressure-regulator for water-works	B. Holly	Lockport, N. Y	May 9, 1871	114, 683
Pressure regulator, Hydraulic	T. J. McGowan	Cincinnati, Ohio	Dec. 29, 1868	85, 322
Pressure regulator, Steam	G. H. Corliss	Providence, R. I	Jan. 5, 1869	85, 566
Pressure regulator, Steam and water	N. C. Locke	Salem, Mass	Mar. 28, 1871	113, 069
Pressure regulator, Steam and water supply and	J. C. Hagan	Nashville, Tenn	Nov. 1, 1870	108, 905
Pressure regulator, Water-pipe	J. Johnson	Lowell, Mass	July 2, 1872	128, 492
Preventing a person while asleep from turning on his back, Apparatus for.	R. T. Sullivan	Needham, Mass	Oct. 22, 1872	132, 500
Price-current sheets, Producing	W. Smith and H. D. Rogers	Cincinnati, Ohio	July 12, 1870	105, 380
Price-indicator	P. Topham	Newark, N. J	Feb. 7, 1860	27, 034
Printer and stamp, Electro-magnetic date-time	J. C. Hinchman	Summit, N. J	July 29, 1873	141, 3[illegible]2
Printer, Consecutive number	R. M. Evans	Buffalo, N. Y	Nov. 4, 1873	144, 196

Index of patents issued from the United States Patent Office from 1790 *to* 1873, *inclusive*—Continued.

Invention.	Inventor.	Residence.	Date.	No.
Printer's ball-stock	J. Babcock	New Haven, Conn	Mar. 18, 1818	....
Printer's blanket	J. Taylor	Lawrence, Mass	Sept. 5, 1865	49, 804
Printer's blocks, Use of caseine for making	T. J. Denne	London, England	May 21, 1872	127, 032
Printer's bodkin and tweezers, Pocket-case for	W. Quail	New York, N. Y	Oct. 29, 1867	70, 261
Printer's case	A. T. De Puy	New York, N. Y	May 2, 1871	114, 273
Printer's case	W. H. A. Gresham	Atlanta, Ga	Mar. 7, 1871	112, 445
Printer's case-stand	R. Chapman	Dublin, Ireland	Apr. 23, 1872	126, 017
Printer's case-stand	P. S. Hoe	New York, N. Y	July 16, 1872	129, 281
Printer's chair	A. J. Finch	New York, N. Y	Feb. 1, 1870	99, 302
Printer's chase	E. Allen	Norwich, Conn	July 16, 1872	129, 079
Printer's chase	A. M. Blanchard	Saint Louis, Mo	Feb. 20, 1866	52, 662
Printer's chase	L. G. Chaput	New York, N. Y	June 3, 1873	139, 542
Printer's chase	T. A. Clements	Little Rock, Ark	Nov. 1, 1870	108, 759
Printer's chase	T. A. Clements	Little Rock, Ark	Nov. 1, 1870	108, 760
Printer's chase	R. Dick	Buffalo, N. Y	Apr. 14, 1868	76, 722
Printer's chase	J. N. Murray	Chicago, Ill	July 9, 1867	66, 514
Printer's chase	W. O. Stoddard	Champaign, Ill	July 9, 1861	32, 798
Printer's chase	E. A. Warren	Brooklyn, N. Y	May 13, 1873	138, 967
Printer's cloth, &c., Process for cleaning plate	H. M. Baker	Washington, D. C	June 15, 1869	91, 408
Printer's forms, Locking up	E. H. Sprague	Zanesville, Ohio	June 13, 1854	11, 091
Printer's frisket	A. Overend	Philadelphia, Pa	June 13, 1854	11, 072
Printer's furniture	B. B. Blackwell	Jamaica, N. Y	Aug. 29, 1871	118, 425
Printer's furniture	F. T. M. A. Guyon	Saint Brieux, France	Apr. 7, 1868	76, 440
Printer's furniture	H. A. Hempel	Saint Joseph, Mo	Dec. 23, 1873	145, 800
Printer's furniture	A. N. Kellogg and J. J. Schock	Chicago, Ill	Apr. 19, 1870	102, 013
Printer's furniture	H. P. Montague	Belchertown, Mass	Aug. 5, 1873	141, 450
Printer's furniture	R. B. Topham	Washington, D. C	Jan. 3, 1871	110, 695
Printer's furniture	J. F. Uhlhorn	Sacramento, Cal	May 23, 1871	115, 136
Printer's furniture	W. H. Windsor	Little Rock, Ark	May 30, 1871	115, 404
Printer's galley	J. W. Baker	Warsaw, Ind	July 16, 1867	66, 776
Printer's galley	W. G. Blymyer	Findlay, Ohio	Mar. 17, 1868	75, 619
Printer's galley	J. F. Bronson	Waterbury, Conn	Nov. 21, 1871	121, 041
Printer's galley	A. T. De Puy	New York, N. Y	Dec. 4, 1866	60, 151
Printer's galley	A. T. De Puy	New York, N. Y	Aug. 10, 1869	93, 422
Printer's galley	A. T. De Puy	New York, N. Y	July 16, 1872	128, 954
Printer's galley	E. Hutchings	Hartford, Conn	Mar. 23, 1869	88, 041
Printer's galley	C. H. Lawrence	New York, N. Y	July 14, 1868	79, 988
Printer's galley	H. E. Long	Plymouth, Ind	Mar. 24, 1868	75, 935
Printer's galley	P. G. Meek	Bellefonte, Pa	Nov. 19, 1867	71, 200
Printer's galley	W. Quail	New York, N. Y	Oct. 26, 1869	96, 145
Printer's galley	J. W. Skinner	Quincy, Ill	Sept. 24, 1872	131, 713
Printer's galley	J. Snyder	Burlington, Iowa	Jan. 2, 1866	51, 877
Printer's galley	J. Wilson, jr	New York, N. Y	Aug. 24, 1869	94, 053
Printer's galley-rest	J. A. Doyle and J. A. Murphy	Lancaster, Pa	Sept. 30, 1873	143, 228
Printer's galley-rest	J. M. Murphy	Olympia, Wash	Aug. 3, 1869	93, 216
Printer's galleys, Side-stick for	R. B. Hindle	Saint Louis, Mo	July 1, 1873	140, 368
Printer's ink-roller	G. Littleton	Cleveland, Ohio	Apr. 7, 1863	38, 113
Printer's ink-roller	G. Russell and T. B. Hull	New York and Brooklyn, N. Y.	Jan. 1, 1867	60, 792
Printer's ink-rollers, Machine for washing	O. H. Reed and A. L. Carrier	Washington, D. C	Dec. 15, 1868	84, 904
Printer's inking-roller	L. Francis	New York, N. Y	Mar. 8, 1864	41, 887
Printer's inking-roller	C. S. Westcott	Elizabeth, N. J	Dec. 27, 1870	110, 522
Printer's inking-roller from rubber sponge	S. Moulton	Bradford-on-Avon, England.	Sept. 7, 1869	94, 631
Printer's inking-rollers, Composition for	W. Harvey	Portland, Me	Nov. 13, 1866	59, 597
Printer's inking-rollers, Composition for making	L. Francis and F. W. Letmate	New York, N. Y	May 6, 1862	35, 149
Printer's lapping	S. W. Baker	Providence, R. I	May 6, 1862	35, 133
Printer's lead-casting machine	J. I. Sturgis	Milwaukee, Wis	Jan. 28, 1873	135, 384
Printer's lead-rack	O. A. Dearing	San Francisco, Cal	Feb. 18, 1873	135, 894
Printer's leads and rules, Machine for bending	F. C. Smith and H. McCollum	New York and Long Island City, N. Y.	Mar. 5, 1872	124, 295
Printer's leads, Apparatus for casting	W. Filmer	San Francisco, Cal	Sept. 16, 1873	142, 783
Printer's leads, Cutting	J. W. H. Cheney	Hartford, Conn	Aug. 25, 1868	81, 476
Printer's leads, Cutting	W. E. Clark	Boston, Mass	Aug. 18, 1868	81, 140
Printer's leads, Machine for making	K. M. Klees	New York, N. Y	July 19, 1870	105, 463
Printer's leads, Machine for making	I. Schoenberg, jr	New York, N. Y	Mar. 7, 1871	112, 384
Printer's metallic plates, Preparation of	S. W. Lowe	Philadelphia, Pa	Sept. 18, 1855	13, 587
Printer's miter-machine	J. A. Stansbury	Salem, Ohio	Apr. 8, 1873	137, 575
Printer's quoin	C. W. Ames	New York, N. Y	May 27, 1873	139, 351
Printer's quoin	D. Dorrity	Pont-Audemere, France	Aug. 10, 1869	93, 603
Printer's quoin	H. A. Marinoni and F. N. Chandrè.	Paris, France	Sept. 22, 1863	40, 081
Printer's rollers, Apparatus for cleaning and preserving.	S. Crump	Brooklyn, N. Y	Oct. 17, 1871	120, 043
Printer's rollers, Composition for	L. K. Bingham	New York, N. Y	Sept. 11, 1866	57, 849
Printer's rollers, Diamond-holder for engraving	J. Hope	Providence, R. I	Nov. 9, 1869	96, 591
Printer's rollers, Recovering the materials of worn out.	J. H. Osgood	Peabody, Mass	Nov. 3, 1868	83, 786
Printer's rolls, Renewing the surface of	C. Sentell	Waterloo, N. Y	June 27, 1865	48, 453
Printer's rule	C. Reuter	Cincinnati, Ohio	Sept. 2, 1873	142, 512
Printer's rules, Machine for cutting and mitering	F. H. Aiken	Franklin, N. H	Feb. 21, 1871	111, 896
Printer's rules, Machine for making	S. D. Tucker	New York, N. Y	Apr. 20, 1869	89, 183
Printer's rules, Machine for mitering	W. E. Cameron and A. A. Dettlaff.	Green Island, N. Y	Sept. 28, 1869	95, 193
Printer's rules, Machine for preparing	J. B. Bancroft	Hopedale, Mass	Dec. 16, 1873	145, 549
Printer's shears	C. E. Fisk	New York, N. Y	May 25, 1869	90, 518
Printer's shooting-stick	B. B. Blackwell	New York, N. Y	Sept. 6, 1870	107, 154
Printer's side-stick and quoin	F. Keehn	Milwaukee, Wis	Dec. 16, 1873	145, 574
Printer's use, Combination-rack for	R. Yeomans	Chicago, Ill	Oct. 24, 1865	50, 650
Printing	B. Beniowski	London, England	Oct. 29, 1850	7, 738
Printing	W. M. Dodge	New York, N. Y	Mar. 6, 1823	....
Printing	C. N. Morris	Cincinnati, Ohio	Oct. 24, 1865	50, 612
Printing, Anastatic	C. F. Baldamus and F. W. Siemens.	Berlin, Prussia	Oct. 25, 1845	4, 239
Printing and drawing checks to prevent counterfeiting.	J. Dainty	Philadelphia, Pa	July 31, 1837	320
Printing and dyeing fabrics	T. Crossley	Bridgeport, Conn	June 10, 1873	139, 705
Printing and dyeing woolen cloth	W. Duncan	Belleville, N. J	Sept. 18, 1835	....

Index of patents issued from the United States Patent Office from 1790 *to* 1873, *inclusive*—Continued.

Invention.	Inventor.	Residence.	Date.	No.
Printing and embossing skirts, Machine for	H. J. Davies	Brooklyn, N. Y	Mar. 25, 1873	137, 184
Printing and finishing, Copper-coated iron-rolls for	A. Paraf	Mulhouse, France	Apr. 23, 1867	64, 135
Printing and mode of preventing counterfeits, Copper-plate.	J. Perkins	Boston, Mass	June 16, 1810	....
Printing and numbering press	G. I. Hill	Buffalo, N. Y	Sept. 7, 1858	21, 418
Printing and ornamenting	J. L. Wells	Philadelphia, Pa	Sept. 12, 1871	118, 992
Printing and registering apparatus, Ticket	J. Dyer	Chicago, Ill	July 29, 1873	141, 336
Printing-apparatus	F. A. De Mey	Brooklyn, N. Y	June 9, 1863	38, 815
Printing-apparatus	J. Warren	Warwick Township, Ohio	May 2, 1835	....
Printing-apparatus	J. M. Wilbur	Cleveland, Ohio	Nov. 19, 1867	71, 103
Printing-apparatus, Attaching blocks to belts of	A. S. Davis	Boston, Mass	Nov. 26, 1861	33, 775
Printing-apparatus, Elastic	J. McDermott	Frederick, Md	Jan. 14, 1868	73, 359
Printing-apparatus, Register-points for	J. F. Shearman	Brooklyn, N. Y	May 28, 1867	65, 127
Printing, Applying dry metallic and colored powder to type.	G. J. Newberry	New York	Feb. 1, 1821	....
Printing bank-notes	A. Tichenor	Newark, N. J	June 7, 1859	24, 341
Printing bank-notes, Apparatus for	J. W. Hayes	Newark, N. J	Oct. 22, 1861	33, 526
Printing bank-notes, &c	A. Sellers	New York, N. Y	Feb. 23, 1864	41, 724
Printing, Biting the figures on steel cylinders for calico.	D. H. Mason and M. W. Baldwin.	Philadelphia, Pa	Oct. 30, 1827	....
Printing-blanket	S. W. Baker	Providence, R. I	Dec. 20, 1859	26, 467
Printing-blocks, Constructing	L. Jarosson	Jersey City, N. J	Nov. 14, 1854	11, 954
Printing-blocks, Constructing molds for making	J. Berry	Roxbury, Mass	Mar. 14, 1854	10, 630
Printing-blocks, Construction of	B. Underwood	Brooklyn, N. Y	Jan. 31, 1854	10, 483
Printing-blocks, Making	T. Crossley	Boston, Mass	July 18, 1854	11, 321
Printing books	S. Randall	Warren, R. I	Jan. 4, 1816	....
Printing by means of rollers	B. Allison and W. Elliott	Washington, D. C	Mar. 11, 1819	....
Printing calico	R. Ferguson and J. Clark	Glasgow, Scotland	Apr. 25, 1846	4, 476
Printing, Calico	R. Prince and A. Lovis	Lowell and Boston, Mass	Dec. 11, 1855	13, 915
Printing calico	B. Woodcroft	Manchester, England	Apr. 17, 1847	5, 064
Printing calico and paper, Machine for	L. Beatty	Wilkesbarre, Pa	Dec. 28, 1810	....
Printing calico, Machine for	A. Sibley	Pawtucket, Mass	July 9, 1838	823
Printing, Card-board for	A. R. Davis	East Cambridge, Mass	Mar. 3, 1868	75, 130
Printing checks, Machine for	J. Pollak	Chicago, Ill	July 4, 1865	48, 589
Printing, Cleansing and drying gum-elastic or cloth bands in calico.	J. Hunter	Blockley Township, Pa	Aug. 27, 1850	7, 602
Printing colored figures on hard substances	J. Edwards	New York	Aug. 22, 1816	....
Printing colors on calico, &c	B. Woodcroft	Ardwich, England	Apr. 5, 1838	678
Printing colors on silk, linen, &c., Forming blocks for.	J. Crabtree	New York, N. Y	Dec. 7, 1837	507
Printing colors on textile material	A. Paraf	New York, N. Y	Sept. 21, 1869	95, 040
Printing, Composition for producing elastic forms for.	I. L. Miles	Philadelphia, Pa	Oct. 15, 1867	69, 926
Printing, Composition to be used instead of gum-senegal in calico.	W. Leversidge	Dorchester, Mass	Aug. 8, 1837	344
Printing, Copper-plate	W. Conisbee	Lambeth, Great Britain	July 2, 1872	128, 595
Printing, Cylinder	J. Webster and S. H. Folsom	Boston and Lowell, Mass	Mar. 25, 1856	14, 542
Printing, Cylinder-roller for distributing ink on copper plates for.	A. Maverick	New York	Apr. 17, 1810	....
Printing-cylinders, Attaching blankets to	A. Overend	Philadelphia, Pa	Apr. 30, 1872	126, 228
Printing cylinders, Method of operating the doctors of calico.	J. Baxendale	Fall River, Mass	Mar. 28, 1854	10, 719
Printing, Die-sinking press for obtaining relief-plates for surface.	W. H. Oakes	New York, N. Y	Jan. 5, 1864	41, 088
Printing-disks, Pattern for casting	J. Goldsborough	Philadelphia, Pa	Mar. 5, 1872	124, 350
Printing, Elastic transfer	D. W. Bowdoin	Salem, Mass	July 12, 1870	105, 300
Printing, Electographic	A. W. Thompson	Philadelphia, Pa	Apr. 26, 1845	4, 012
Printing, Etching relief-plates for surface	C. Henry and J. and E. McLoughlin.	Brooklyn, Morrisania, and New York, N. Y.	Nov. 17, 1868	84, 187
Printing fabrics, books, &c., Manufacture of India-rubber blanket or apron for.	C. McBurney	Boston, Mass	June 7, 1859	24, 321
Printing fabrics, Constructing roller or cylinder for.	R. F. Sturges	Birmingham, England	May 5, 1857	17, 239
Printing fabrics	E. A. D. Guichard	Paris, France	June 25, 1872	128, 302
Printing fabrics, Press for	H. H. Townsend	Milton Mills, N. H	Apr. 8, 1873	137, 658
Printing, Facilitating	S. N. Dickenson	Boston, Mass	Nov. 19, 1833	....
Printing-fluid	E. Whitefield	Buffalo, N. Y	June 27, 1865	48, 471
Printing for colors, Plate	J. Enthoffer	Washington, D. C	Nov. 30, 1869	97, 287
Printing-form	E. Edwards	Middlesex County, England.	Dec. 31, 1872	134, 470
Printing, Forming plates for polychromatic	J. Donlevy	New York, N. Y	Jan. 3, 1854	10, 363
Printing-frame	O. H. Burdick	Auburn, N. Y	Apr. 19, 1864	42, 344
Printing-frame	S. F. Conant and H. A. Manley	Skowhegan, Me	Feb. 11, 1868	74, 311
Printing from engraved plates	J. Meyer	Bay Ridge, N. Y	July 5, 1864	43, 421
Printing from engraved plates, Machine for	R. Neale	Clermont County, Ohio	Jan. 9, 1855	12, 213
Printing from engraved plates, Machine for	J. F. Starrett	New York, N. Y	Feb. 19, 1856	14, 295
Printing from engraved plates, Machine for	L. Stewart and J. McClelland	Washington, D. C	Mar. 31, 1857	16, 952
Printing in colors	W. Croome	Brooklyn, N. Y	May 19, 1857	17, 319
Printing in colors	A. D. McKinzie	Philadelphia, Pa	Nov. 6, 1846	4, 838
Printing in colors	J. Rennie	Lodi, N. J	Aug. 9, 1834	....
Printing in colors	H. F. Smart	Worcester, Mass	Nov. 17, 1868	84, 225
Printing in colors for electro-chemical voting apparatus.	A. N. Henderson	Buffalo, N. Y	July 22, 1850	7, 521
Printing in colors, Machine for	T. F. Adams	Philadelphia, Pa	Sept. 17, 1844	3, 744
Printing in colors, Machine for	R. S. Weaver	Maysville, Ky	Oct. 28, 1851	8, 475
Printing in different colors, Machine for	J. R. Wright	Philadelphia, Pa	Aug. 9, 1859	25, 075
Printing-ink roller	E. Pratt	Salem, Mass	Aug. 10, 1858	21, 148
Printing, Inking-apparatus for color	T. L. Bayliss	Richmond, Ind	July 14, 1868	79, 888
Printing, Inking-apparatus for color	L. Biddle	Knoxville, Iowa	Apr. 16, 1872	125, 655
Printing, Inking-apparatus for color	J. Hunt	Richmond, Ind	July 14, 1868	79, 910
Printing, Inking-apparatus for color	T. Moore and P. H. Day	Bloomington, Ill	Apr. 13, 1869	88, 976
Printing, Inking-apparatus for color	I. L. G. Rice	Cambridge, Mass	Apr. 19, 1870	102, 157
Printing, Inking-apparatus for color	I. L. G. Rice	Cambridge, Mass	Oct. 25, 1870	108, 63[illegible]
Printing, Inking apparatus for color	I. L. G. Rice	Cambridge, Mass	Aug. 27, 1872	130, 822
Printing, Inking-apparatus for color	L. B. Waterman	Indianapolis, Ind	Mar. 10, 1868	75, 499
Printing, Inking apparatus for color	G. W. Wood	Richmond, Ind	June 16, 1868	79, 043
Printing, Inking-attachment for chromatic	E. A. Howitt	Whitewater, Wis	May 20, 1873	139, 155
Printing labels on spools, Machine for	G. Hall, jr	South Wilmington, Conn	July 5, 1870	104, 951

Index of patents issued from the United States Patent Office from 1790 *to* 1873, *inclusive*—Continued.

Invention.	Inventor.	Residence.	Date.	No.
Printing long-napped fabrics	W. A. White	Roxbury, Mass	Aug. 15, 1854	11, 540
Printing-machine	M. Bebro	Manchester, Great Britain	Aug. 29, 1871	118, 509
Printing-machine	W. Bullock	Pittsburgh, Pa	Apr. 14, 1863	38, 200
Printing-machine	W. A. Burt	Detroit, Mich	July 23, 1829	....
Printing-machine	G. W. Cartwright	Mount Pleasant, N. Y	Apr. 29, 1828	....
Printing-machine	J. H. Cooper	Philadelphia, Pa	May 20, 1856	14, 907
Printing-machine	O. T. Eddy	Baltimore, Md	Nov. 12, 1850	7, 771
Printing-machine	J. B. Fairbank	Leon, N. Y	Sept. 17, 1850	7, 652
Printing-machine	S. W. Francis	New York, N. Y	Oct. 27, 1857	18, 504
Printing-machine	R. M. Hoe	New York, N. Y	July 26, 1859	24, 875
Printing-machine	J. M. Jones	Palmyra, N. Y	May 20, 1856	14, 919
Printing-machine	S. D. Tucker	New York, N. Y	July 9, 1867	66, 654
Printing-machine	C. Thurber	Worcester, Mass	Aug. 26, 1843	3, 228
Printing-machine	C. F. Voorhies	Newark, N. J	Apr. 8, 1834	....
Printing machine, Card	J. S. Moody	Cincinnati, Ohio	Dec. 1, 1857	18, 795
Printing machine, Electrical	T. A. Edison	Newark, N. J	Nov. 12, 1872	133, 019
Printing-machine for paper-hangings, cloth, books, &c.	P. Force	Washington, D. C	Aug. 22, 1822	....
Printing machine, Stencil	A. J. Fullam	Springfield, Vt	Oct. 2, 1860	30, 216
Printing machines, Blanket for calico	J. Fallon	Lawrence, Mass	May 19, 1857	17, 322
Printing-machines, Inking-apparatus for	J. K. Lowe	Cleveland, Ohio	Apr. 2, 1867	63, 540
Printing-machines, Movement for doctors of calico	J. Standing	Fall River, Mass	Feb. 5, 1856	14, 214
Printing-machines, Washing blankets of	T. W. Clarke	Manchester, N. H	June 6, 1865	48, 053
Printing-machinery	J. Hatch	Boston, Mass	Mar. 21, 1832	....
Printing-machinery	S. Savage	Fredonia, N. Y	July 16, 1872	129, 176
Printing-machinery, Cloth-feeding apparatus for	E. O. Potter	Pawtucket, R. I	Dec. 29, 1868	85, 475
Printing machinery, Cylindrical-plate	G. F. Lewis and F. D. Stuart	Philadelphia, Pa., and Washington, D. C.	May 12, 1868	77, 896
Printing machinery, Plate	I. T. Robertson	New York, N. Y	Feb. 20, 1872	123, 933
Printing, Making mill and skeleton dies for	T. Paton	Providence, R. I	Oct. 24, 1848	5, 871
Printing names or directions on packages, &c., Machine for.	J. Spencer	Toronto, Canada	Nov. 23, 1858	22, 136
Printing numbers, Machine for	C. L. Sholes	Milwaukee, Wis	Apr. 30, 1867	64, 375
Printing on bottles, Apparatus for	I. L. Miles	Charlestown, Mass	July 24, 1866	56, 593
Printing on and decorating enameled surfaces, Apparatus for.	S. J. Hoggson	New Haven, Conn	Nov. 25, 1873	144, 979
Printing on enameled or vitreous surfaces	S. J. Hoggson	New Haven, Conn	Dec. 26, 1871	122, 170
Printing on fabrics, Machinery for	C. Holliday	Huddersfield, England	May 26, 1868	78, 288
Printing on glass	I. L. Miles	Charlestown, Mass	Jan. 22, 1867	61, 350
Printing on glass, Apparatus for	I. L. Miles	Charlestown, Mass	Mar. 5, 1867	62, 666
Printing on glass, Apparatus for	D. B. Wingate	Natick, Mass	July 7, 1868	79, 713
Printing on glass, porcelain, &c	E. C. Jayne	Philadelphia, Pa	Feb. 6, 1866	52, 418
Printing on paper, calico, &c., Forming raised surfaces for.	G. Wocne	London, England	Apr. 2, 1838	670
Printing on solid or inflexible surfaces, Apparatus for.	A. Wilbaux	Paris, France	Sept. 2, 1873	142, 541
Printing on spools, Machine for	G. Hall, jr. and G. W. Averell	South Wilmington, Conn., and New York, N. Y.	Apr. 26, 1870	102, 257
Printing on tin-foil, Apparatus for	J. Polhemus and C. H. Lilienthal.	Jersey City, N. J., and Yonkers, N. Y.	Nov. 19, 1867	71, 058
Printing on the back or reverse face of bank-notes to prevent counterfeits.	J. Kneass	....	Apr. 28, 1815	....
Printing on uneven surfaces, Apparatus for	A. Leighton	London, England	Mar. 5, 1867	62, 646
Printing on woven fabrics of cotton, linen, &c., Block.	R. Hampson	Manchester, Great Britain	June 7, 1841	2, 116
Printing or painting on surfaces, Composition for	A. Farrar	Brookline, Mass	Mar. 15, 1870	100, 877
Printing or painting on surfaces, Composition for	A. Farrar	Brookline, Mass	Apr. 18, 1871	113, 755
Printing, painting, &c., Preparing canvas for	E. Lee	Baltimore, Md	May 12, 1857	17, 308
Printing paper, cloth, &c., Machine for	G. S. Barton	Worcester, Mass	Oct. 12, 1869	95, 758
Printing paper-hangings, Machine for	F. S. Munroe, jr., and T. Mason	Grantville and Boston, Mass.	June 13, 1865	48, 199
Printing paper-hangings, Roller for	T. Van Deventer	New Brunswick, N. J	Dec. 6, 1859	26, 387
Printing paper, leather, &c., Machine for	J. Dixey	....	Jan. 24, 1798	....
Printing paper on both sides	H. Betts	Norwalk, Conn	Sept. 14, 1833	....
Printing, Photo-mechanical	E. Edwards	Middlesex County, England	Dec. 10, 1872	133, 701
Printing, Plate	W. H. Oakes	New York, N. Y	Oct. 23, 1860	30, 495
Printing plate, Relief	B. Day	Hoboken, N. J	Apr. 26, 1864	42, 530
Printing plate, &c., Relief	D. C. Hitchcock and E. B. and E. M. Larchar.	New York, N. Y	Nov. 13, 1860	30, 630
Printing-plates, Removing ink from engraved	J. S. Ives	New York, N. Y	Nov. 11, 1873	144, 544
Printing, Preparation of engraved metal plates for	J. M. Batchelder and L. L. Smith.	Cambridge, Mass., and New York, N. Y.	Nov. 24, 1857	18, 668
Printing, Preparing fabrics for	J. Coates	Manchester, England	Mar. 21, 1848	5, 480
Printing-press	A. S. Adams	Chelsea, Mass	Mar. 19, 1861	31, 760
Printing-press	S. Adams	Boston, Mass	Sept. 27, 1844	3, 769
Printing-press	S. Adams	Boston, Mass	Mar. 8, 1853	9, 606
Printing-press	J. B. Adt	Baltimore, Md	Oct. 24, 1871	120, 225
Printing-press	E. Allen	Norwich, Conn	Nov. 12, 1867	70, 773
Printing-press	E. Allen	Norwich, Conn	Feb. 4, 1868	73, 943
Printing-press	J. F. Allen and R. W. McGowan	New York, N. Y	Dec. 15, 1863	40, 892
Printing-press	F. J. Austin	New York, N. Y	May 7, 1850	7, 335
Printing-press	G. H. Babcock	Westerly, R. I	Dec. 23, 1856	16, 263
Printing-press	W. H. Babcock	Homer, N. Y	Oct. 23, 1860	30, 452
Printing-press	F. L. Bailey	Boston, Mass	Nov. 25, 1856	16, 109
Printing-press	F. L. Bailey	Boston, Mass	June 22, 1857	17, 615
Printing-press	F. L. Bailey	Boston, Mass	Feb. 21, 1860	27, 197
Printing-press	F. L. Bailey	Boston, Mass	Apr. 30, 1861	32, 172
Printing-press	F. L. Bailey	Boston, Mass	July 5, 1864	43, 383
Printing-press	F. L. Bailey	Boston, Mass	Dec. 13, 1864	45, 461
Printing-press	F. L. Bailey	Boston, Mass	Sept. 12, 1871	118, 835
Printing-press	F. L. Bailey and J. Watson	Boston and Everett, Mass	Feb. 11, 1873	135, 618
Printing-press	S. J. Baird	Staunton, Va	Nov. 24, 1868	84, 403
Printing-press	W. H. Baker and G. J. Hill	Buffalo, N. Y	June 2, 1863	38, 781
Printing-press	W. L. Balch	Boston, Mass	Aug. 22, 1871	118, 182
Printing-press	A. H. Bangle	Brooklyn, Cal	June 7, 1870	103, 961
Printing-press	A. H. Bangle	Brooklyn, Cal	Feb. 14, 1871	111, 805
Printing-press	A. H. Bangle	Brooklyn, Cal	July 16, 1872	129, 084
Printing-press	H. Barth	Cincinnati, Ohio	Oct. 22, 1867	70, 067

Index of patents issued from the United States Patent Office from 1790 *to* 1873, *inclusive*—Continued.

Invention.	Inventor.	Residence.	Date.	No.
Printing-press	H. Barth	Cincinnati, Ohio	Feb. 22, 1870	100, 002
Printing-press	C. F. Batchelder and W. E. Gard	Chicago, Ill	Oct. 29, 1872	132, 517
Printing-press	M. S. Beach	Brooklyn, N. Y	Nov. 9, 1858	22, 010
Printing-press	M. S. Beach	Brooklyn, N. Y	June 26, 1860	28, 823
Printing-press	V. Beaumont	New York, N. Y	Sept. 6, 1853	9, 987
Printing-press	H. Betts	Norwalk, Conn	Apr. 13, 1869	88, 835
Printing-press	H. A. Bills and S. W. Wood	West Winsted, Conn., and Cornwall, N. Y.	Mar. 23, 1858	19, 672
Printing-press	T. W. Bracher	New York, N. Y	Feb. 1, 1870	99, 395
Printing-press	W. Braidwood	Mount Vernon, N. Y	May 5, 1868	77, 577
Printing-press	W. Braidwood	New York, N. Y	July 28, 1868	80, 444
Printing-press	J. B. Brown	Nashville, Tenn	May 24, 1870	103, 422
Printing-press	J. M. Brownson	Brooklyn, N. Y	Mar. 8, 1870	100, 494
Printing-press	W. Bullock	Philadelphia, Pa	Feb. 12, 1867	61, 996
Printing-press	J. S. Burdick	Norwich, N. Y	Mar. 27, 1849	6, 236
Printing-press	J. S. Burdick	New York, N. Y	June 2, 1857	17, 418
Printing-press	T. H. Burridge	Saint Louis, Mo	Dec. 20, 1859	26, 545
Printing-press	H. Camp	Trenton, Ohio	Mar. 4, 1836	....
Printing-press	A. Campbell	Brooklyn, N. Y	July 31, 1866	56, 701
Printing-press	A. Campbell	Brooklyn, N. Y	Mar. 8, 1870	100, 598
Printing-press	A. Campbell	Brooklyn, N. Y	Apr. 25, 1871	114, 105
Printing-press	A. Campbell	Brooklyn, N. Y	Apr. 25, 1871	114, 106
Printing-press	J. A. Campbell	New Orleans, La	Sept. 14, 1858	21, 484
Printing-press	V. M. Chafee	Xenia, Ill	Jan. 17, 1860	26, 834
Printing-press	W. H. Chandler	Winchester, Mass	Mar. 30, 1869	88, 448
Printing-press	W. R. Collier	Washington, D. C	Apr. 11, 1834	....
Printing-press	S. Conillard	Boston, Mass	Oct. 5, 1827	....
Printing-press	R. J. Coons	Greensburgh, Pa	Oct. 17, 1871	120, 040
Printing-press	S. R. Cotton	Green Bay, Wis	Dec. 28, 1858	22, 414
Printing-press	C. B. Cottrell	Westerly, R. I	May 2, 1871	114, 268
Printing-press	C. B. Cottrell	Westerly, R. I	Oct. 17, 1871	120, 041
Printing-press	C. B. Cottrell	Westerly, R. I	July 15, 1873	140, 813
Printing-press	A. H. Cragin, M. and J. H. Buck, and F. A. Temrey.	Lebanon, N. H	Nov. 23, 1852	9, 426
Printing-press	R. Cummings	Newport, Vt	Oct. 27, 1868	83, 471
Printing-press	W. H. Danforth	Salem, Mass	Aug. 5, 1856	15, 477
Printing-press	G. W. Davis	Seneca Falls, N. Y	Mar. 30, 1858	19, 758
Printing-press	M. Davis	New York, N. Y	July 24, 1855	13, 335
Printing-press	M. Davis	New York, N. Y	Nov. 3, 1857	18, 567
Printing-press	G. R. Dean	Mayville, N. Y	May 7, 1861	32, 242
Printing-press	F. O. Degener	New York, N. Y	Jan. 11, 1859	22, 611
Printing-press	F. O. Degener	New York, N. Y	Apr. 24, 1860	27, 973
Printing-press	F. O. Degener	Brooklyn, N. Y	May 28, 1872	127, 316
Printing-press	H. B. Denny	Washington, D. C	Jan. 5, 1869	85, 515
Printing-press	J. Densmore	Blooming Valley, Pa	Nov. 9, 1852	9, 383
Printing-press	T. H. Dodge	Nashua, N. H	Nov. 18, 1851	8, 521
Printing-press	A. Dougherty	Brooklyn, N. Y	Aug. 9, 1859	25, 000
Printing-press	A. Dougherty	Brooklyn, N. Y	Aug. 28, 1866	57, 486
Printing-press	A. A. Dunk	Philadelphia, Pa	Mar. 10, 1868	75, 394
Printing-press	S. Fairlamb	New York	Nov. 4, 1826	....
Printing-press	G. F. Folsom	Roxbury, Mass	Apr. 1, 1856	14, 558
Printing-press	W. H. Forbush	Buffalo, N. Y	Jan. 19, 1869	86, 064
Printing-press	J. H. Frey and W. Heckert	Sharon, Pa	Jan. 15, 1867	61, 186
Printing-press	M. Gally	Rye, N. Y	Nov. 9, 1869	96, 577
Printing-press	M. Gally	Rye, N. Y	Nov. 9, 1869	96, 578
Printing-press	M. Gally	Rye, N. Y	Nov. 9, 1869	96, 579
Printing-press	M. Gally	Rochester, N. Y	Nov. 23, 1869	97, 185
Printing-press	M. Gally	Rochester, N. Y	Feb. 8, 1870	99, 551
Printing-press	M. Gally	Rochester, N. Y	Mar. 29, 1870	101, 254
Printing-press	M. Gally	Rochester, N. Y	May 2, 1871	114, 285
Printing-press	G. and S. P. Gary	Oshkosh, Wis	Feb. 26, 1861	31, 573
Printing-press	A. Gilman	Troy, N. Y	Aug. 23, 1844	3, 716
Printing-press	W. H. Golding	Chelsea, Mass	Dec. 2, 1873	145, 101
Printing-press	H. Goodrich	Stoneham, Mass	Oct. 1, 1861	33, 387
Printing-press	G. P. Gordon	New York, N. Y	Mar. 26, 1850	7, 215
Printing-press	G. P. Gordon	New York, N. Y	Aug. 5, 1851	8, 285
Printing-press	G. P. Gordon	New York, N. Y	Aug. 31, 1852	9, 234
Printing-press	G. P. Gordon	New York, N. Y	June 13, 1854	11, 064
Printing-press	G. P. Gordon	New York, N. Y	Jan. 1, 1856	14, 016
Printing-press	G. P. Gordon	New York, N. Y	Oct. 3, 1857	18, 545
Printing-press	G. P. Gordon	New York, N. Y	July 13, 1858	20, 874
Printing-press	G. P. Gordon	New York, N. Y	Apr. 19, 1859	23, 677
Printing-press	G. P. Gordon	New York, N. Y	July 5, 1859	24, 626
Printing-press	G. P. Gordon	New York, N. Y	Jan. 10, 1860	26, 761
Printing-press	G. P. Gordon	Brooklyn, N. Y	Apr. 23, 1861	32, 130
Printing-press	G. P. Gordon	Brooklyn, N. Y	Jan. 7, 1862	34, 111
Printing-press	G. P. Gordon	Brooklyn, N. Y	Oct. 28, 1862	36, 840
Printing-press	G. P. Gordon	Brooklyn, N. Y	Sept. 29, 1863	40, 099
Printing-press	G. P. Gordon	Brooklyn, N. Y	Mar. 8, 1864	41, 841
Printing-press	G. P. Gordon	Brooklyn, N. Y	June 2, 1864	43, 196
Printing-press	G. P. Gordon	Brooklyn, N. Y	Dec. 18, 1866	60, 504
Printing-press	G. P. Gordon	Rahway, N. J	May 18, 1869	90, 091
Printing-press	G. P. Gordon	Rahway, N. J	Oct. 31, 1871	120, 377
Printing-press	G. P. Gordon and F. O. Degener	New York, N. Y	May 11, 1858	20, 204
Printing-press	G. P. Gordon and F. O. Degener	New York, N. Y	Aug. 9, 1859	25, 008
Printing-press	J. Gordon	Caledonia, N. Y	Oct. 28, 1862	36, 839
Printing-press	J. Gough	London, England	Apr. 25, 1871	114, 130
Printing-press	F. J. Grace	Fort Lee, N. J	Jan. 1, 1867	60, 717
Printing-press	S. T. Guernsey	Montpelier, Vt	Oct. 19, 1852	9, 342
Printing-press	H. M. Hall and G. W. Espey	Moore's Hill, Ind	July 7, 1868	79, 751
Printing-press	C. W. Hawkes	Boston, Mass	June 4, 1850	7, 413
Printing-press	J. R. Hathaway and J. P. Strippel.	Norfolk, Va	Oct. 21, 1851	8, 445
Printing-press	C. G. Havens and F. C. Penfield	West Meriden, Conn	Oct. 8, 1872	131, 950
Printing-press	C. W. Hawkes	Boston, Mass	Dec. 24, 1850	7, 855
Printing-press	C. W. Hawkes	Boston, Mass	Sept. 14, 1852	9, 259
Printing-press	C. W. Hawkes	Boston, Mass	Dec. 8, 1857	18, 812
Printing-press	G. F. Hebard, G. J. Hill, and S. D. Rockwell.	Buffalo, N. Y	Aug. 7, 1860	29, 547

Index of patents issued from the United States Patent Office from 1790 *to* 1873, *inclusive*—Continued.

Invention.	Inventor.	Residence.	Date.	No.
Printing-press	W. Heckert	Newcastle, Pa	Apr. 16, 1872	125, 678
Printing-press	J. Henry	Vevay, Ind	Dec. 1, 1857	18, 744
Printing-press	J. Henry	Jersey City, N. J	Feb. 12, 1867	61, 032
Printing-press	J. Henry	Millburn, N. J	May 9, 1871	114, 557
Printing-press	G. J. Hill and S. Greene	Philadelphia, Pa	Aug. 28, 1866	57, 634
Printing-press	R. M. Hoe	New York, N. Y	Apr. 17, 1844	3, 551
Printing-press	R. M. Hoe	New York, N. Y	July 30, 1844	3, 687
Printing-press	R. M. Hoe	New York, N. Y	May 1, 1845	4, 025
Printing-press	R. M. Hoe	New York, N. Y	July 10, 1847	5, 188
Printing-press	R. M. Hoe	New York, N. Y	July 24, 1847	5, 200
Printing-press	R. M. Hoe	New York, N. Y	Apr. 18, 1871	113, 769
Printing-press	R. M. Hoe and S. D. Tucker	New York, N. Y	Dec. 1, 1868	84, 627
Printing-press	R. M. Hoe and S. D. Tucker	New York, N. Y	June 29, 1869	92, 050
Printing-press	R. M. Hoe and S. D. Tucker	New York, N. Y	Nov. 1, 1870	108, 785
Printing-press	R. M. Hoe and S. D. Tucker	West Farms and New York, N. Y.	Sept. 10, 1872	131, 217
Printing-press	J. C. Holbrook and E. H. Thomas.	Brattleborough, Vt	Feb. 7, 1828	....
Printing-press	J. C. Holbrook and E. H. Thomas.	Brattleborough, Vt	Feb. 7, 1828	....
Printing-press	H. Holt	Winchester, Mass	Mar. 17, 1857	16, 837
Printing-press	G. C. Howard	Philadelphia, Pa	June 26, 1860	28, 864
Printing-press	B. Huber	Williamsburgh, N. Y	Nov. 7, 1871	120, 646
Printing-press	B. Huber	Williamsburgh, N. Y	Mar. 19, 1872	124, 822
Printing-press	G. A. Hunt	Newburgh, N. Y	May 20, 1873	139, 156
Printing-press	G. W. Hunt	New York, N. Y	Feb. 28, 1871	112, 249
Printing-press	T. H. Ide	Claremont, N. H	June 20, 1871	116, 060
Printing-press	M. G. Imbach	New York, N. Y	May 1, 1866	54, 357
Printing-press	D. E. James	Utica, N. Y	Nov. 30, 1858	22, 181
Printing-press	W. Johnson	Lambertville, N. J	July 9, 1872	128, 731
Printing-press	J. M. Jones	Palmyra, N. Y	Aug. 11, 1868	80, 865
Printing-press	J. M. Jones	Palmyra, N. Y	Nov. 19, 1872	133, 156
Printing-press	A. Judson	Brooklyn, N. Y	July 30, 1867	67, 198
Printing-press	J. W. Kelberg	Philadelphia, Pa	Dec. 20, 1870	110, 244
Printing-press	A. N. Kellogg	Baraboo, Wis	Jan. 6, 1863	37, 293
Printing-press	S. Kelsey	Erie, Pa	Jan. 2, 1855	12, 183
Printing-press	W. A. Kelsey	Meriden, Conn	May 6, 1873	138, 660
Printing-press	W. A. Kerr	Easton, Pa	June 7, 1870	103, 894
Printing-press	J. L. Kingsley	New York	Apr. 22, 1835	....
Printing-press	J. L. Kingsley	New York, N. Y	Oct. 31, 1839	1, 394
Printing-press	J. L. Kingsley	New York	Jan. 4, 1845	3, 874
Printing-press	S. Kingsley	New York, N. Y	Mar. 2, 1836	....
Printing-press	A. Kinsley	....	Nov. 16, 1796	....
Printing-press	J. C. Kneeland	Troy, N. Y	Feb. 20, 1845	3, 917
Printing-press	M. Laemmel	Bay Ridge, N. Y	Jan. 30, 1872	123, 266
Printing-press	J. W. Latcher	Northville, N. Y	Jan. 31, 1860	26, 999
Printing-press	S. D. Learned	Boston, Mass	May 26, 1857	17, 405
Printing-press	T. Leavitt	Everett, Mass	July 18, 1871	117, 088
Printing-press	J. Lewis	Buffalo, N. Y	Aug. 9, 1853	9, 923
Printing-press	J. Lewis	Prattville, N. Y	Jan. 2, 1855	12, 185
Printing-press	J. C. MacDonald and J. Calverly.	Waddon and Camberwell. England.	June 1, 1869	90, 858
Printing-press	J. C. MacDonald and J. Calverly.	Waddon and Camberwell, England.	Apr. 25, 1871	114, 020
Printing-press	J. M. Marsh	New York, N. Y	Oct. 3, 1848	5, 819
Printing-press	W. W. Marston	New York	Sept. 12, 1846	4, 756
Printing-press	C. C. Maurice	New York, N. Y	May 24, 1870	103, 351
Printing-press	J. W. McDonald	Osgood, Ind	Dec. 24, 1867	72, 660
Printing-press	G. McKay	Boston, Mass	Feb. 15, 1859	22, 968
Printing-press	Z. Mills	Hartford, Conn	Feb. 27, 1813	....
Printing-press	C. Montague	Pittsfield, Mass	Nov. 23, 1852	9, 424
Printing-press	C. Montague	Pittsfield, Mass	Sept. 6, 1853	9, 992
Printing-press	C. Montague	Pittsfield, Mass	Sept. 6, 1853	9, 993
Printing-press	C. Montague	Hartford, Conn	Nov. 9, 1858	22, 027
Printing-press	C. Montague	Hartford, Conn	Dec. 10, 1861	33, 897
Printing-press	C. Montague	Boston, Mass	Dec. 21, 1869	98, 087
Printing-press	C. Montague	Boston, Mass	Dec. 21, 1869	98, 088
Printing-press	R. W. Moran	Saint Louis, Mo	Apr. 18, 1865	47, 319
Printing-press	W. T. Morgans	Youngsville, N. Y	Jan. 25, 1870	99, 101
Printing-press	T. N. Morse	Fairhaven, Mass	Mar. 26, 1872	124, 907
Printing-press	D. Neale	Philadelphia, Pa	Feb. 13, 1834	....
Printing-press	D. Neale	Philadelphia, Pa	Nov. 15, 1825	....
Printing-press	A. and B. Newbury	Windham Centre, N. Y	Sept. 16, 1856	15, 740
Printing-press	F. B. Nichols	Morrisania, N. Y	Aug. 3, 1858	21, 080
Printing-press	J. G. Nicolay	Pittsfield, Ill	Oct. 5, 1852	9, 305
Printing-press	J. G. Northrup	Cortlandville, N. Y	Apr. 1, 1845	3, 985
Printing-press	J. G. Northrup	Syracuse, N. Y	Aug. 9, 1853	9, 925
Printing-press	J. G. Northrup	Syracuse, N. Y	June 12, 1855	13. 069
Printing-press	J. G. Northrup	Cortlandville, N. Y	Sept. 30, 1842	2, 793
Printing-press	T. and A. Parkes	Brooklyn, N. Y	July 29, 1856	15, 437
Printing-press	F. C. Penfield	West Meriden, Conn	May 6, 1873	138, 689
Printing-press	J. G. Peterson	Morgantown, N. C	Sept. 9, 1873	142 648
Printing-press	D. Phelps	Boston, Mass	Sept. 15, 1826	....
Printing-press	J. N. Phelps	Brooklyn, N. Y	Nov. 2, 1869	96, 480
Printing-press	D. Pierson	Newburyport, Mass	July 16, 1813	....
Printing-press	C. Potter, jr	Westerly, R. I	Mar. 5, 1861	31, 621
Printing-press	C. Potter, jr	Westerly, R. I	Aug. 20, 1867	68, 001
Printing-press	C. Potter, jr	Westerly, R. I	Jan. 7, 1868	73, 195
Printing-press	C. Potter, jr	Plainfield, N. J	Sept 24, 1872	131, 702
Printing-press	J. E. Priest	Saint Louis, Mo.	Aug. 28, 1860	29, 816
Printing-press	G. W. Prouty	Charlestown, Mass	Nov. 28, 1871	121, 256
Printing-press	G. W. Prouty	Charlestown, Mass	Oct. 29, 1872	132, 599
Printing-press	A. Ramage	Philadelphia, Pa	May 28, 1818	....
Printing-press	A. Ramage	Philadelphia, Pa	Nov. 19, 1834	....
Printing-press	H. Redlich	Chicago. Ill	Apr. 25, 1865	47, 454
Printing-press	L. and H. Reese	Philadelphia, Pa	Nov. 24, 1868	84, 440
Printing-press	T. S. Reynolds	Athens, Ga	Apr. 27, 1858	20, 090
Printing-press	J. T. Robertson	New York, N. Y	Apr. 4, 1871	113, 346

Index of patents issued from the United States Patent Office from 1790 to 1873, inclusive—Continued.

Invention.	Inventor.	Residence.	Date.	No.
Printing-press	J. T. Robertson	New York, N. Y	Aug. 6, 1872	130, 153
Printing-press	L. Rodney	New York, N. Y	Aug. 6, 1867	67, 586
Printing-press	S. P. Ruggles	Boston, Mass	Nov. 10, 1840	1, 851
Printing-press	S. P. Ruggles	Boston, Mass	Jan. 1, 1851	7, 878
Printing-press	S. P. Ruggles	Boston, Mass	Aug. 2, 1853	9, 904
Printing-press	S. P. Ruggles	Boston, Mass	Nov. 16, 1852	9, 410
Printing-press	S. P. Ruggles	Boston, Mass	Feb. 28, 1854	10, 588
Printing-press	S. P. Ruggles	Boston, Mass	May 10, 1859	23, 951
Printing-press	S. Rust	New York	May 13, 1821	....
Printing-press	S. Rust	New York	Apr. 17, 1829	....
Printing-press	J. Sangster	Buffalo, N. Y	June 27, 1865	48, 493
Printing-press	C. G. Sargent and A. Keach	Lowell and Boston, Mass	Dec. 9, 1856	16, 221
Printing-press	J. Saxton	Washington, D. C	Mar. 21, 1845	3, 958
Printing-press	W. and T. Schnebly	Hagerstown, Md	Sept. 7, 1839	1, 315
Printing-press	A. Sherman	New York	Feb. 26, 1831	....
Printing-press	E. H. Smith	Bergen, N. J	Feb. 7, 1871	111, 581
Printing-press	E. H. Smith	New York, N. Y	Mar. 19, 1872	124, 701
Printing-press	J. A. Smith and L. M. Orris	Fond du Lac and Oakfield, Wis.	June 26, 1860	28, 948
Printing-press	P. Smith	New York	Apr. 6, 1822	....
Printing-press	W. W. Smith	New York	Feb. 1, 1842	2, 439
Printing-press	E. E. Sneider	New York, N. Y	Aug. 10, 1858	21, 154
Printing-press	A. O. Stansbury	New York	Apr. 7, 1821	....
Printing-press	R. G. Stuart	Yonkers, N. Y	July 18, 1871	117, 219
Printing-press	C. A. Swett	Boston, Mass	Nov. 27, 1855	13, 857
Printing-press	A. B. Taylor	Newark, N. J	June 26, 1860	28, 915
Printing-press	W. H. R. Toye	Philadelphia, Pa	May 30, 1871	115, 391
Printing-press	E. B. Tripp	Concord, N. H	Oct. 3, 1854	11, 765
Printing-press	E. B. Tripp	New York, N. Y	Sept. 14, 1858	21, 528
Printing-press	S. D. Tucker	New York, N. Y	June 28, 1864	43, 349
Printing-press	S. D. Tucker	New York, N. Y	June 28, 1864	43, 350
Printing-press	S. D. Tucker	New York, N. Y	Dec. 29, 1868	85, 493
Printing-press	S. D. Tucker	New York, N. Y	Mar. 12, 1872	124, 460
Printing-press	O. Tufts	Boston, Mass	July 30, 1831	....
Printing-press	O. Tufts	Boston, Mass	Aug. 22, 1834	....
Printing-press	J. H. Utter	New York, N. Y	Oct. 20, 1857	18, 477
Printing-press	L. T. Verney	Paris, France	Sept. 3, 1867	68, 471
Printing-press	E. A. Warren	New York, N. Y	June 11, 1872	127, 815
Printing-press	R. C. Warwick	New York, N. Y	Jan. 31, 1871	111, 407
Printing-press	J. Watson	Everett, Mass	Dec. 17, 1872	133, 959
Printing-press	C. Wells and H. Barth	Cincinnati, Ohio	Aug. 7, 1860	29, 554
Printing-press	J. J. Wells	Hartford, Conn	Feb. 8, 1819	....
Printing-press	J. J. Wells	Hartford, Conn	June 29, 1829	....
Printing-press	L. T. Wells	Cincinnati, Ohio	Mar. 20, 1855	12, 568
Printing-press	L. T. Wells	Cincinnati, Ohio	Sept. 6, 1859	25, 357
Printing-press	L. T. Wells	Saint Louis, Mo	July 7, 1868	79, 617
Printing-press	O. E. Weston	Roxbury, Mass	Jan. 17, 1860	26, 869
Printing-press	S. Wilcox, jr	Westerly, R. I	Nov. 10, 1857	18, 618
Printing-press	J. A. Wilkinson	Fireplace, N. Y	Jan. 4, 1853	9, 525
Printing-press	D. K. Winder	Cincinnati, Ohio	June 2, 1857	17, 463
Printing-press	D. Wolfe	Dixon, Ohio	Aug. 17, 1858	21, 228
Printing-press	B. O. Woods and W. S. Tuttle	Boston, Mass	Aug. 6, 1867	67, 475
Printing-press	J. Worms	Paris, France	Sept. 23, 1851	8, 386
Printing-press	J. K. Wright	Philadelphia, Pa	June 8, 1869	91, 191
Printing-press	C. D. Wrightington	Fairhaven, Mass	Sept. 20, 1870	107, 583
Printing-press	J. Young	Philadelphia, Pa	May 10, 1853	9, 721
Printing-press and folding-machine, Combined	E. L. Ford	New York, N. Y	Dec. 17, 1872	133, 976
Printing-press and mode of inking type	W. Elliot	New York	Feb. 17, 1813	....
Printing-press apron	C. McBurney	Roxbury, Mass	Jan. 17, 1860	26, 854
Printing-press attachment, Chromatic	E. A. Howitt	Whitewater, Wis	Jan. 7, 1873	134, 602
Printing press, Card	F. L. Bailey	Boston, Mass	June 16, 1857	17, 549
Printing press, Card	W. W. Clarkson	Baltimore, Md	Apr. 27, 1858	20, 039
Printing press, Card	C. E. Emery	Canandaigua, N. Y	June 9, 1857	17, 499
Printing press, Card	T. Harsha	West Union, Ohio	Oct. 23, 1855	13, 703
Printing press, Card	S. Orcutt	Boston, Mass	Aug. 25, 1840	1, 732
Printing press, Card	D. K. Winder	Cincinnati, Ohio	Sept. 18, 1855	13, 576
Printing press, Card	D. K. Winder	Cincinnati, Ohio	Oct. 9, 1855	13, 671
Printing press, Card and ticket	G. E. Peck	Duquoin, Ill	Oct. 21, 1873	143, 780
Printing press, Chromatic	T. H. Burridge and J. M. Kershaw.	Saint Louis, Mo	June 24, 1873	140, 242
Printing press, Chromatic	F. L. Heughes	Rochester, N. Y	Dec. 31, 1872	134, 380
Printing press, Chromatic	G. W. Woodside	Philadelphia, Pa	Jan. 7, 1873	134, 579
Printing-press, Circular	J. P. Sawin and T. B. Wait	Boston, Mass	Feb. 1, 1810	....
Printing press, Color	A. M. and G. H. Babcock	Westerly, R. I	Oct. 31, 1854	11, 853
Printing press, Color	G. W. Wood	Richmond, Ind	Feb. 18, 1868	74, 741
Printing press, Copper and steel plate	E. C. Middleton, E. Nevers, and R. Neale.	Cincinnati and Mount Carmel, Ohio.	Nov. 19, 1850	7, 786
Printing press, Copper and steel plate	J. Perkins	Newburyport, Mass	June 29, 1813	....
Printing press, Copper-plate	T. S. Bates	Boston, Mass	Dec. 16, 1873	145, 610
Printing press, Copper-plate	C. Durand	New York	May 22, 1828	....
Printing-press, Cylinder	B. Beniowski	London, England	Aug. 13, 1850	7, 558
Printing-press, Cylinder	H. C. Chandler	Indianapolis, Ind	Nov. 5, 1867	70, 524
Printing-press, Cylinder	F. O. Degener	New York, N. Y	Nov. 5, 1861	33, 676
Printing-press, Cylinder	J. G. Northrup	Syracuse, N. Y	Nov. 16, 1852	9, 408
Printing-press, Cylindrical	J. P. Sawin and T. B. Wait	Roxbury, Mass	Jan. 28, 1811	....
Printing-press, Cylindrical	C. G. Williams	New York	May 29, 1828	....
Printing, Device for hand	B. Livermore	Hartland, Vt	July 21, 1863	39, 296
Printing press, Double	J. Booth, sr., and J., T., J., and J. Booth, jr.	New York	Sept. 1, 1829	....
Printing-press, (Napier's,) Double	S. Newton	New York, N. Y	Feb. 26, 1833	....
Printing-press, Double-cylinder	R. M. Hoe	New York	May 20, 1842	2, 629
Printing-press, Engraved-plate	P. C. Andrews and J. Shinn	Leverington, Pa	Oct. 21, 1862	36, 759
Printing-press, Engraved-plate	M. C. Gritzner	Washington, D. C	Apr. 14, 1857	17, 036
Printing-press, Faustus	S. Adams	Boston, Mass	May 23, 1832	....
Printing-press feed-board	E. Allen	Norwich, Conn	Nov. 12, 1872	133, 002
Printing-press feed-gage	H. Barth and R. J. Morgan	Cincinnati, Ohio	Apr. 9, 1872	125, 432
Printing-press feed-gage	E. N. Maxwell	Louisville, Ky	Jan. 16, 1872	122, 842
Printing-press feed-gage	C. N. Morris	Cincinnati, Ohio	Dec. 23, 1873	145, 890

Index of patents issued from the United States Patent Office from 1790 *to* 1873, *inclusive*—Continued.

Invention.	Inventor.	Residence.	Date.	No.
Printing-press feed-gage	C. Potter, jr	Westerly, R. I	Jan. 7, 1868	73, 196
Printing-press feed-gage	G. Wilcox	Norwich, Conn	Oct. 14, 1873	143, 652
Printing-press feeder	H. Barth	Cincinnati, Ohio	Oct. 23, 1860	30, 456
Printing-press feeder	F. L. Heughes	Rochester, N. Y	Aug. 12, 1873	141, 713
Printing-press feeder	R. J. Stuart	Yonkers, N. Y	Mar. 25, 1873	137, 156
Printing-press-fly frame	T. H. Mead	Boston, Mass	Mar. 9, 1869	87, 695
Printing-press-fly frame	T. H. Mead	Boston, Mass	May 18, 1869	90, 114
Printing-press-fly frame	A. Overend	Philadelphia, Pa	Apr. 30, 1872	126, 227
Printing-press for cards and bill-heads, Self-feeding	N. Ames	Saugus, Mass	Mar. 29, 1859	23, 421
Printing-press frisket	T. W. M. Castle and J. B. Conner.	Adriance, Ind	June 23, 1868	79, 053
Printing-press gage	A. Campbell	Brooklyn, N. Y	July 26, 1870	105, 638
Printing-press griper	T. J. Plunket	New York, N. Y	Aug 1, 1871	117, 566
Printing-press-griper and gage	H. Barth	Cincinnati, Ohio	Feb. 15, 1870	99, 749
Printing-press guide	A. L. Bevans	Flushing, N. Y	Jan. 31, 1871	111, 304
Printing-press-guide gage	O. H. Reed	Washington, D. C	May 5, 1868	77, 527
Printing press, Hand	F. J. Austin	New York, N. Y	Oct. 8, 1836	41
Printing press, Hand	D. H. Chamberlain	West Roxbury, Mass	Aug. 14, 1855	13, 423
Printing press, Hand	N. L. Chamberlin	West Roxbury, Mass	Mar. 3, 1857	16, 718
Printing press, Hand	F. S. Coburn	Ipswich, Mass	Mar. 17, 1857	16, 861
Printing press, Hand	S. Conillard, jr	Boston	July 14, 1827	
Printing press, Hand	W. and W. H. Dunkerly, jr	Woonsocket, R. I	Apr. 8, 1873	137, 664
Printing press, Hand	P. Evens, jr	Cincinnati, Ohio	Dec. 23, 1856	16, 270
Printing press, Hand	C. Foster	Cincinnati, Ohio	Oct. 5, 1852	9, 295
Printing press, Hand	C. A. Haskins	New York, N. Y	June 15, 1858	20, 556
Printing press, Hand	J. M. Jones	Palmyra, N. Y	Dec. 22, 1857	18, 907
Printing press, Hand	C. Keniston	Boston, Mass	Mar. 20, 1855	12, 553
Printing press, Hand	H. Moeser	Pittsburgh, Pa	Jan. 20, 1852	8, 674
Printing press, Hand	J. Morse	Canton, Mass	Oct. 27, 1857	18, 527
Printing press, Hand	A. and B. Newbury	Windham Centre, N. Y	July 5, 1859	24, 655
Printing press, Hand	J. N. Phelps	New York, N. Y	Nov. 2, 1858	21, 980
Printing press, Hand	N. Ross	Washington, D. C	Jan. 7, 1873	134, 705
Printing press, Hand	J. Schaffer and E. Spencer	Saint Louis, Mo	Jan. 14, 1862	34, 166
Printing press, Hand	S. I. Smith	New York, N. Y	Nov. 3, 1857	18, 557
Printing press, Hand	H. Underhill	Canandaigua, N. Y	Mar. 28, 1854	10, 717
Printing press, Hand	C. Whipple	Providence, R. I	Nov. 8, 1864	45, 000
Printing press, Hand	D. K. Winder	Cincinnati, Ohio	July 8, 1856	15, 312
Printing-press hand-roller	B. F. Allen	Boston, Mass	Aug. 6, 1872	130, 175
Printing-press ink-fountain	H. B. Allen	Brooklyn, N. Y	Aug. 6, 1872	130, 176
Printing-press ink-fountain	W. V. Wallace	New York, N. Y	July 4, 1871	116, 778
Printing-press inking-apparatus	T. L. Baylies	Richmond, Ind	May 26, 1868	78, 358
Printing-press inking-apparatus	T. L. Baylies	Richmond, Ind	July 7, 1868	79, 722
Printing-press inking-apparatus	F. O. Degener	Brooklyn, E. D., N. Y	Dec. 13, 1870	110, 018
Printing-press inking-apparatus	M. England	New York, N. Y	Dec. 9, 1873	145, 341
Printing-press inking-apparatus	G. K. Farrington and B. S. Potter.	Bloomington, Ill	Oct. 8, 1872	132, 000
Printing-press inking-apparatus	R. M. Hoe	New York, N. Y	July 31, 1847	5, 212
Printing-press inking-apparatus	J. Prince	New York	Dec. 3, 1829	
Printing-press inking-apparatus	I. L. G. Rice	Cambridge, Mass	Apr 20, 1869	89, 244
Printing-press inking-apparatus	I. L. G. Rice	Cambridge, Mass	June 14, 1870	104, 203
Printing-press inking-apparatus	N. F. Turner	Williamsburgh, N. Y	July 11, 1871	116, 892
Printing-press inking-apparatus	C. Wells	Cincinnati, Ohio	Jan. 26, 1869	86, 263
Printing-press inking-apparatus	G. W. Wood	Richmond, Ind	June 18, 1867	65, 984
Printing-press inking-apparatus	R. Wood	New York	Nov. 4, 1830	
Printing-press inking-apparatus, Card	D. K. Winder	Cincinnati, Ohio	Oct. 16, 1855	13, 689
Printing-press inking-apparatus, Oscillating	G. W. Prouty	Charlestown, Mass	July 22, 1873	141, 077
Printing-press inking-apparatus, Rotary	E. Allen	Norwich, Conn	Sept. 9, 1873	142, 607
Printing-press inking-roller	E. E. Barrett	Chicago, Ill	July 21, 1857	17, 824
Printing press, Job and card	J. A. Campbell	New Orleans, La	June 28, 1859	24, 538
Printing press, Lithographic	A. G. Shackford	Malden, Mass	Aug. 4, 1868	80, 771
Printing-press, Locomotive-cylinder	C. J. Carr and A. Smith	Belper, England	Sept. 10, 1840	1, 775
Printing-press, Oscillating	C. Potter, jr	Westerly, R. I	June 2, 1857	17, 449
Printing-press paper-gage	J. W. Thyng	Salem, Mass	July 3, 1866	56, 123
Printing-press, Plate	G. C. Howard and J. Dainty	Philadelphia, Pa	May 27, 1873	139, 392
Printing-press, Portable	S. W. Lowe	Philadelphia, Pa	July 29, 1856	15, 428
Printing press, Power	I. Adams	Boston, Mass	Oct. 4, 1830	
Printing press, Power	I. Adams	Boston, Mass	Mar. 2, 1836	
Printing press, Power	W. H. Danforth	Salem, Mass	June 21, 1853	9, 794
Printing press, Power	J. Morse	Canton, Mass	June 9, 1857	17, 543
Printing press, Power	J. Morse	Canton, Mass	June 7, 1859	24, 357
Printing press, Power	D. Treadwell	Boston, Mass	Mar. 2, 1826	
Printing press, Railway-ticket	W. H. Forbush	Buffalo, N. Y	Nov. 6, 1866	59, 533
Printing press, Railway-ticket	S. Greene and W. H. Forbush	Philadelphia, Pa., and Buffalo, N. Y.	Aug. 13, 1867	67, 748
Printing-press, Register for Napier's	M. Caton and J. C. Rives	Washington, D. C	Mar. 11, 1835	
Printing-press register-point	N. Hopkins	New York, N. Y	Nov. 5, 1867	70, 568
Printing-press register-point	R. W. Macgowan	New York, N. Y	Dec. 10, 1867	72, 059
Printing-press register-point	A. Overend	Philadelphia, Pa	Mar. 9, 1869	87, 587
Printing-press registering-apparatus	J. Kirk	Dover, Del	June 19, 1866	55, 774
Printing-press registering-apparatus	G. McKay	Boston, Mass	May 26, 1857	17, 388
Printing-press registering-apparatus	J. W. Richards	Hoboken, N. J	May 10, 1853	9, 711
Printing-press registering-attachment	A. Hilgenreiner	New York, N. Y	Oct. 21, 1873	143, 904
Printing press, Roller	J. Laird	Pittsburgh, Pa	Aug. 16, 1828	
Printing-press, Rotary	S. H. Bingham	Philadelphia, Pa	May 27, 1873	139, 360
Printing-press, Rotary	W. Braidwood and H. J. Hewitt.	New York, N. Y	July 29, 1873	141, 314
Printing-press, Rotary	W. Braidwood and H. J. Hewitt.	New York, N. Y	July 29, 1873	141, 315
Printing-press, Rotary	C. C. Child	Boston, Mass	Nov. 18, 1873	144, 742
Printing-press, Rotary	C. B. Cottrell	Westerly, R. I	Apr. 1, 1873	137, 347
Printing-press, Rotary	J. C. Davis and W. Miller	Elizabeth, N. J	Mar. 17, 1857	16, 826
Printing-press, Rotary	R. M. Hoe	New York, N. Y	July 24, 1847	5, 199
Printing press, Self-feeding card	N. Ames	Saugus Centre, Mass	Apr. 22, 1862	35, 055
Printing-press sheet-gage	H. Barth	Cincinnati, Ohio	Jan. 25, 18[illegible]0	99, 133
Printing-press sheet-gage	G. B. Cummings	Cambridge, Mass	Jan. 25, 1870	99, 167
Printing press, Stencil	H. W. Rudolf	Louisville, Ky	Aug. 30, 1870	106, 872
Printing-press, Stop-cylinder	H. Barth	Cincinnati, Ohio	May 27, 1873	139, 229
Printing press, Zincographic	G. H. Korff	Hoboken, N. J	May 18, 1858	20, 276

Index of patents issued from the United States Patent Office from 1790 to 1873, inclusive—Continued.

Invention.	Inventor.	Residence.	Date.	No.
Printing-presses, Adjustable attachment of cams for fliers of.	C. Potter, jr	Plainfield, N. J	Feb. 14, 1871	111, 776
Printing-presses and paper-machines, Apparatus for receiving and transferring to the pile sheets of paper from.	I. Adams	Boston, Mass	Mar. 26, 1850	7, 205
Printing-presses and ruling-machines, Apparatus for feeding paper to.	D. Baldwin	Godwinville, N. J	Jan. 2, 1855	12, 118
Printing-presses, Apparatus for delivering paper from.	C. O. Furbush	Machias, Me	Apr. 25, 1865	47, 411
Printing-presses, Apparatus for delivering sheets from.	T. J. Mayall	Boston, Mass	Dec. 30, 1873	146, 085
Printing-presses, Apparatus for feeding paper to	J. B. Hall	New York, N. Y	Apr. 10, 1855	12, 702
Printing-presses, Apparatus for feeding paper to	W. Kuhlenshmidt and W. Hauff.	New York, N. Y	Apr. 25, 1854	10, 828
Printing-presses, Apparatus for feeding paper to	R. Larter, jr	Newark, N. J	Jan. 10, 1860	26, 772
Printing-presses, Apparatus for feeding paper to hand.	E. Mathers and W. D. Siegfried.	Morgantown, Pa	Apr. 3, 1855	12, 634
Printing-presses, Attaching blankets to cylinders of	W. H. Street	New York, N. Y	May 16, 1854	10, 935
Printing-presses, Attachment for	H. Willson	Chicago, Ill	May 17, 1870	103, 267
Printing-presses, Automatic paper-feeder for	W. Bullock	Newark, N. J	Sept. 21, 1858	21, 591
Printing-presses, Bed-recoil spring for	A. Campbell	Brooklyn, N. Y	July 31, 1866	56, 700
Printing-presses, Checking momentum of	A. B. Taylor	New York, N. Y	Apr. 4, 1846	4, 442
Printing-presses, Driving	C. M. Langley	Lowell, Mass	Aug. 27, 1867	68, 212
Printing-presses, Feed-table for	H. Barth	Cincinnati, Ohio	May 25, 1869	90, 482
Printing-presses, Feeding-apparatus for	T. Hall	Saint Louis, Mo	Aug. 7, 1860	29, 485
Printing-presses, Feeding out paper from	M. S. Beach	Brooklyn, N. Y	Nov. 9, 1858	22, 011
Printing-presses, Feeding paper to	M. S. Beach	Brooklyn, N. Y	Dec. 2, 1856	16, 138
Printing-presses, Feeding paper to	H. W. Dickinson	Rochester, N. Y	Oct. 30, 1855	13, 737
Printing-presses, Feeding paper to	G. H. and S. Ferguson	Malden Bridge, N. Y	July 26, 1859	24, 861
Printing-presses, Feeding paper to	R. M. Hoe	New York, N. Y	Nov. 10, 1857	18, 589
Printing-presses, Feeding paper to	R. M. Hoe	New York, N. Y	Aug. 23, 1859	25, 199
Printing-presses, Feeding paper to	J. Hunt	New York, N. Y	Nov. 1, 1864	44, 917
Printing-presses, Feeding paper to	C. H. Stoddard and E. D. Norton.	Ithaca, N. Y	July 15, 1873	140, 962
Printing-presses, Feeding paper to and from	C. Potter, jr., and C. B. Cottrell	Westerly, R. I	Sept. 20, 1859	25, 524
Printing-presses, Feeding paper to rotary	J. T. Ashley	Brooklyn, N. Y	Nov. 25, 1873	144, 822
Printing-presses, Griper for cylinder	T. E. Manger	New York, N. Y	July 4, 1871	116, 616
Printing-presses, Machine for feeding paper to	D. Baldwin	Godwinville, N. J	Dec. 9, 1856	16, 162
Printing-presses, Machine for feeding paper to	M. S. Beach	Brooklyn, N. Y	Dec. 23, 1856	16, 311
Printing-presses, Machine for feeding paper to	A. Campbell	Newark, N. J	July 24, 1855	13, 333
Printing-presses, Machine for feeding paper to	S. I. Chapman	Charleston, S. C	Jan. 15, 1856	14, 084
Printing-presses, Machine for feeding paper to	A. B. Childs and H. W. Dickinson.	Rochester, N. Y	Feb. 20, 1855	12, 401
Printing-presses, Machine for feeding paper to	W. F. Collier	Worcester, Mass	June 27, 1854	11, 188
Printing-presses, Machine for feeding paper to	I. B. Livingston and M. Waterhouse.	Barnet, Vt	Apr. 24, 1855	12, 761
Printing-presses, Machine for feeding paper to	B. A. Rugg and E. H. Benjamin	Oak Hill, N. Y	Sept. 5, 1854	11, 653
Printing-presses, Machine for feeding sheets of paper to.	D. Babson	Groton, Conn	Sept. 2, 1856	15, 639
Printing-presses, Machine for feeding sheets of paper to.	H. Clark	New Orleans, La	Apr. 25, 1854	10, 824
Printing-presses, Machine for feeding sheets of paper to.	A. H. Rowand	Allegheny City, Pa	July 3, 1855	13, 183
Printing-presses, Machine for wetting and cutting paper for.	M. S. Beach	Brooklyn, N. Y	Aug. 25, 1857	18, 032
Printing-presses, Machinery to feed sheets of paper to.	M. S. Beach	Brooklyn, N. Y	Aug. 9, 1859	24, 982
Printing-presses, Nippers for	C. W. Hawkes	Boston, Mass	Mar. 21, 1854	10, 670
Printing-presses, Operating feed-table of	G. Little	Utica, N. Y	Apr. 25, 1854	10, 827
Printing-presses, Operating fingers of	S. D. Tucker	New York, N. Y	Sept. 6, 1859	25, 356
Printing-presses, Operating fly-frames of	R. M. Hoe	New York, N. Y	Nov. 17, 1857	18, 640
Printing-presses, Paper-feeder for	L. T. Wells	Cincinnati, Ohio	Oct. 19, 1858	21, 859
Printing presses, Self-inking machine for hand	W. J. Spence	New York	Apr. 30, 1834	....
Printing-presses, Straightening toggle-joint in hand	O. Tufts	Boston, Mass	Nov. 7, 1831	....
Printing-presses, Tympan-frame for	D. U. Stoner	Mount Joy, Pa	Mar. 9, 1869	87, 723
Printing-presses, Tympans for	L. T. Wells	Cincinnati, Ohio	May 4, 1858	20, 179
Printing-presses, Tympan-sheet for	J. Gorman	Portland, Me	Dec. 13, 1870	110, 032
Printing, Producing from printed paper new plates for re-.	C. and C. Vogt	New York, N. Y	May 4, 1869	89, 715
Printing, Production of plates for	J. Ramage and T. Nelson	Edinburgh, Scotland	Mar. 17, 1868	75, 700
Printing railway and other tickets, Machine for	R. M. Hoe	New York, N. Y	Mar. 8, 1859	23, 172
Printing railway-tickets, Machine for	G. Bailey	Buffalo, N. Y	Mar. 20, 1860	27, 510
Printing railway-tickets, Machinery for	W. T. Cushing	New York, N. Y	May 29, 1866	55, 201
Printing register, Plate	J. L. Harley	Washington, D. C	Mar. 1, 1870	100, 399
Printing-rollers, Composition for	I. G. Husted	Brooklyn, N. Y	Oct. 11, 1870	108, 145
Printing, Self-acting feeder for calico	A. Boyd	Providence, R. I	Apr. 22, 1845	4, 010
Printing silks and other textile fabrics	L. Prang	Boston, Mass	Apr. 4, 1871	113, 343
Printing, Stencil-plate	S. T. Sanford	Fall River, Mass	May 19, 1857	17, 340
Printing-surfaces, Mode of obtaining curved	W. H. Elliot	Plattsburgh, N. Y	Mar. 15, 1859	23, 236
Printing-surfaces, Process of obtaining	P. Schulze	Brooklyn, N. Y	Dec. 2, 1862	37, 078
Printing textile fabrics, Machine for	J. Melville and J. Burch	Roebank Works and Crag Hall, Great Britain.	Aug. 7, 1855	13, 376
Printing, Typographic	J. Donlevy	New York, N. Y	Apr. 3, 1866	53, 587
Printing-wheel, &c	T. S. Hudson	East Cambridge, Mass	June 5, 1866	55, 300
Printing woolen and other goods, Machine for	T. Crossley	Boston, Mass	June 20, 1854	11, 118
Prints, &c., Process for transferring	H. Loewenberg	New York, N. Y	Nov. 3, 1863	40, 489
Prism, Water	B. Kane	Cincinnati, Ohio	Nov. 26, 1867	71, 309
Prison-alarm apparatus	W. O. Hills	Nottingham, N. H	Aug. 31, 1858	21, 339
Prison-bar	J. A. Marden	Boston, Mass	July 13, 1869	92, 626
Prison-door bolt	B. F. Haugh	Indianapolis, Ind	Aug. 18, 1868	81, 165
Prison-door bolt	T. Lalor	Toronto, Canada	Nov. 4, 1873	144, 210
Prison-door lock	C. E. Felton	Buffalo, N. Y	Aug. 13, 1867	67, 641
Prison-grating	T. F. Baker	Cincinnati, Ohio	May 18, 1869	90, 219
Prison-grating	E. May	Indianapolis, Ind	May 9, 1871	114, 584
Prison, Iron	E. Jacobs	Cincinnati, Ohio	July 24, 1860	29, 282
Prison-lock	S. A. Denio	Boston, Mass	Sept. 13, 1859	25, 394
Prison-lock	H. R. Towne	Stamford, Conn	Feb. 7, 1871	111, 586
Prison-lock attachment	E. Kershaw	Boston, Mass	July 30, 1850	7, 525

Index of patents issued from the United States Patent Office from 1790 to 1873, inclusive—Continued.

Invention.	Inventor.	Residence.	Date.	No.
Prisons, Construction of	I. Hodgson	Indianapolis, Ind	June 16, 1868	78, 966
Prisons, Construction of	E. Jacobs	Cincinnati, Ohio	June 7, 1859	24, 307
Prisons, Construction of	E. May	Indianapolis, Ind	Oct. 4, 1859	25, 662
Prisons, Construction of	E. May	Indianapolis, Ind	Dec. 27, 1870	110, 483
Privy	E. E. Clarke	New Haven, Conn	Oct. 24, 1865	50, 654
Privy	M. M. Duplat	Charleston, S. C	Mar. 16, 1820	......
Privy	D. S. Forney	Wytheville, Va	Oct. 2, 1866	58, 400
Privy	J. Ingram	New York, N. Y	Mar. 17, 1868	75, 549
Privy	F. Riedel	San Francisco, Cal	Feb. 21, 1871	111, 976
Privy	F. Riedel	San Francisco, Cal	May 23, 1871	115, 239
Privy	W. J. Warren	Philadelphia, Pa	Nov. 8, 1870	109, 160
Privy and sink cleaner	R. R. Retowsky	Baltimore, Md	Oct. 7, 1873	143, 420
Privy and sink cleaning apparatus	A. M. Hobbs	Washington, D. C	Apr. 8, 1873	137, 680
Privy-bowls, Detachable flange for	R. Diven	Brooklyn, N. Y	Dec. 4, 1866	60, 153
Privy-indicator	W. C. Duffield	Wheeling, W. Va	Feb. 25, 1873	136, 224
Privy-seat	H. W. Carpenter	Madison County, N. Y	June 18, 1872	128, 111
Privy-seat	W. P. Hastings	Maryville, Tenn	July 16, 1872	129, 223
Privy-seat	E. Mangeon	Paris, France	Oct. 23, 1866	59, 150
Privy-seat	F. Peters and G. Clem	Cincinnati, Ohio	May 18, 1869	90, 298
Privy-seat	F. Reed	Fitchburgh, Mass	Dec. 7, 1869	97, 694
Privy-seat cover	W. Beach	Philadelphia, Pa	Aug. 7, 1866	56, 881
Privy-seat cover	W. Street	New York, N. Y	Oct. 17, 1871	120, 121
Privy, Portable	W. H. Pulver	Schuylersville, N. Y	Oct. 25, 1870	108, 728
Privy-vaults, Apparatus for emptying	J. G. Berger	Nuremberg, Bavaria	Nov. 2, 1869	96, 385
Privies, Apparatus for cleaning	H. C. Bull	New Orleans, La	June 6, 1871	115, 565
Privies, Apparatus for emptying	F. Datichiy	New York, N. Y	Dec. 17, 1850	7, 834
Privies, Signal for	I. H. Doughty	New York, N. Y	Sept. 18, 1849	6, 722
Probangs or instruments for treating diseased orifices.	J. S. and M. F. Lovell	Philadelphia, Pa	July 27, 1869	92, 980
Progressive kiln	H. Dueberg	Baltimore, Md	Sept. 7, 1869	94, 580
Projectile	J. Absterdam	New York, N. Y	Apr. 6, 1869	88, 689
Projectile	H. Bates	New London, Conn	Nov. 10, 1857	18, 568
Projectile	C. C. Brand	Norwich, Conn	May 19, 1857	17, 312
Projectile	C. F. Brown	Warren, R. I	Sept. 22, 1868	82, 284
Projectile	J. G. Butler	Fortress Monroe, Va	Sept. 26, 1871	119, 313
Projectile	W. A. Cobb	Orange, Mass	Mar. 23, 1869	88, 012
Projectile	J. W. Cochran	New York, N. Y	Sept. 25, 1860	30, 123
Projectile	J. Curtis	Chicago, Ill	June 1, 1869	90, 732
Projectile	E. Drake	Stoughton, Mass	Nov. 29, 1870	109, 600
Projectile	E. C. Hancock	New Orleans, La	May 16, 1871	114, 807
Projectile	S. H. Haycock	Ottawa, Canada	June 19, 1866	55, 796
Projectile	S. H. Haycock	Ottawa, Canada	Apr. 1, 1873	137, 442
Projectile	J. H. Helm	Pittsburgh, Pa	Dec. 19, 1871	122, 014
Projectile	J. W. Hill	Jefferson, Iowa	June 22, 1869	91, 633
Projectile	T. Hill	Vallejo, Cal	July 18, 1871	117, 172
Projectile	J. G. Hope	Topeka, Kans	Oct. 4, 1870	107, 909
Projectile	J. G. Hope	Wichita, Kans	Sept. 2, 1873	142, 394
Projectile	C. T. James	Providence, R. I	Feb. 26, 1856	14, 315
Projectile	R. H. Jones	San Francisco, Cal	Mar. 2, 1869	87, 498
Projectile	J. Link	United States Army	May 24, 1870	103, 477
Projectile	T. G. Orwig	New York, N. Y	Feb. 21, 1865	46, 490
Projectile	A. F. Potter	San Francisco, Cal	Sept. 21, 1869	95, 137
Projectile	J. D. Richards	Muscatine, Iowa	Sept. 7, 1869	94, 647
Projectile	J. Rigney	West Point, N. Y	June 11, 1872	127, 796
Projectile	N. Scholfield	Norwich, Conn	Aug. 19, 1856	15, 577
Projectile	M. Shaw	Sandwich, Mass	Mar. 3, 1857	16, 753
Projectile	W. H. Shock	Washington, D. C	May 24, 1870	103, 514
Projectile	C. W. Stafford	Burlington, Iowa	July 7, 1863	39, 180
Projectile	W. J. Von Hammerhuber	Washington, D. C	May 1, 1855	12, 801
Projectile	R. Weir	Philadelphia, Pa	July 1, 1862	35, 795
Projectile, Compound	J. E. Crane and J. Fox	Lowell, Mass	Sept. 11, 1866	57, 870
Projectile, Compound	S. Sawyer	Fitchburgh, Mass	Nov. 13, 1855	13, 799
Projectile, Compound explosive	M. L. Callender	New York, N. Y	Oct. 14, 1862	26, 686
Projectile, Compound explosive	F. A. Markley	Waynesborough, Va	May 13, 1873	138, 907
Projectile, Compound sub-caliber	C. Arick	Saint Clairsville, Ohio	July 28, 1863	39, 369
Projectile, Enameled	J. F. Clew	New York, N. Y	Nov. 15, 1864	45, 022
Projectile, Explosive	T. H. Burrowes	Chicopee, Mass	Oct. 27, 1863	40, 396
Projectile, Explosive	J. M. Connel	Newark, Ohio	Feb. 3, 1863	37, 565
Projectile, Explosive	A. O. H. Hardenstein	Clinton, Miss	Oct. 6, 1868	82, 714
Projectile, Explosive	B. B. Hotchkiss	Sharon, N. Y	May 6, 1862	35, 153
Projectile, Explosive	C. W. Isbell	New York, N. Y	May 13, 1862	35, 277
Projectile, Explosive	J. Jobson	Derby, England	Nov. 9, 1869	96, 595
Projectile, Explosive	J. Johnson	Brooklyn, N. Y	May 12, 1863	38, 491
Projectile, Explosive	J. Long	Shaver's Creek, Pa	Dec. 1, 1868	84, 635
Projectile, Explosive	J. N. Smith	Jersey City, N. J	Dec. 8, 1863	40, 888
Projectile for breech-loading ordnance	W. W. Hubbell	Philadelphia, Pa	May 1, 1860	28, 084
Projectile for fire-arms	W. H. Arnold	Washington, D. C	Apr. 12, 1859	23, 538
Projectile for fire-arms	J. F. Clew	New York, N. Y	Nov. 15, 1864	45, 021
Projectile for fire-arms	R. Cranston and H. Bates	New London, Conn	Aug. 14, 1860	29, 573
Projectile for fire-arms	A. H. Emery	New York, N. Y	Mar. 17, 1863	37, 906
Projectile for fire-arms	J. Holroyd	Washington, D. C	Nov. 1, 1859	25, 967
Projectile for fire-arms	E. Hoyt, jr	Chelsea, Mass	May 1, 1855	12, 795
Projectile for fire-arms	W. W. Hubbell	Philadelphia, Pa	Jan. 24, 1860	26, 904
Projectile for fire-arms	S. E. Jones	Santa Fé, N. Mex	Feb. 20, 1872	123, 828
Projectile for fire-arms	E. Matteson	Brooklyn, N. Y	Nov. 19, 1861	33, 746
Projectile for fire-arms	D. M. Mefford	Cincinnati, Ohio	Jan. 22, 1862	34, 285
Projectile for fire-arms	J. B. Read	Tuscaloosa, Ala	Nov. 24, 1857	18, 707
Projectile for fire-arms	R. Shaler	Madison, Conn	July 16, 1861	32, 844
Projectile for fire-arms	T. Smith	Pittsburgh, Pa	Apr. 22, 1856	14, 742
Projectile for fire-arms	B. Swain	Washington, D. C	Feb. 21, 1860	27, 245
Projectile for fire-arms	W. Taggart	Haverhill, Mass	Nov. 11, 1856	16, 076
Projectile for fire-arms	O. G. Warren	New York, N. Y	Oct. 20, 1863	40, 385
Projectile for fire-arms	C. A. Waterbury	New York, N. Y	Sept. 15, 1863	39, 977
Projectile for fire-arms, Elongated	J. N. Smith	New York, N. Y	Aug. 18, 1863	39, 593
Projectile for fire-arms, Hollow	T. D. Gibson	Wilmington, Del	May 18, 1869	90, 164
Projectile for killing whales	R. Brown	New London, Conn	June 14, 1859	24, 371
Projectile for killing whales	R. Sibley	Norwich, Conn	May 26, 1857	17, 407
Projectile for many-chambered guns	W. McCord	Sing Sing, N. Y	Sept. 15, 1863	39, 940

Index of patents issued from the United States Patent Office from 1790 *to* 1873, *inclusive*—Continued.

Invention.	Inventor.	Residence.	Date.	No.
Projectile for ordnance	S. C. Abbott	Zanesville, Ohio	Jan. 15, 1861	31, 099
Projectile for ordnance	C. F. Brown	Warren, R. I	Oct. 1, 1861	33, 378
Projectile for ordnance	J. G. Butler	Fort Leavenworth, Kans	June 14, 1870	104, 108
Projectile for ordnance	J. G. Butler	United States Army	Feb. 28, 1871	112, 120
Projectile for ordnance	J. W. Cochran	New York, N. Y	Nov. 1, 1859	25, 951
Projectile for ordnance	J. M. Currie	Washington, Iowa	Oct. 16, 1866	58, 783
Projectile for ordnance	J. Gault	Boston, Mass	July 2, 1861	32, 689
Projectile for ordnance	H. Alvorson	Cambridge, Mass	Sept. 29, 1863	40, 142
Projectile for ordnance	A. Hotchkiss	Sharon, Conn	Oct. 16, 1855	13, 679
Projectile for ordnance	B. B. Hotchkiss	New York, N. Y	July 27, 1869	93, 589
Projectile for ordnance	B. B. Hotchkiss	New York, N. Y	Feb. 28, 1871	112, 144
Projectile for ordnance	O. Lugo	New York, N. Y	Sept. 5, 1865	49, 773
Projectile for ordnance, &c	W. Mansfield, J. Morse, and H. H. Mansfield.	Canton, Mass	Apr. 29, 1862	35, 103
Projectile for ordnance	J. B. Read	Tuscaloosa, Ala	Oct. 28, 1856	15, 999
Projectile for ordnance	M. Ritner	Vincennes, Ind	Jan. 6, 1863	37, 361
Projectile for ordnance	D. E. Somes	Washington, D. C	Dec. 15, 1863	40, 958
Projectile for ordnance	C. W. Stafford	Burlington, Iowa	Jan. 27, 1863	37, 557
Projectile for ordnance, Arrow	W. Cousins	New York, N. Y	June 27, 1865	48, 371
Projectile for ordnance, Compound	L. E. Reynolds	Mendon, Ill	Dec. 8, 1863	40, 856
Projectile for ordnance, Explosive	W. E. Brown	Valley Falls, R. I	June 10, 1862	35, 503
Projectile for ordnance, Explosive	L. D. Gerarden	Jersey City, N. J	Feb. 10, 1863	37, 661
Projectile for ordnance, Explosive	W. W. Hanes	Covington, Ky	Aug. 26, 1862	36, 295
Projectile for ordnance, Explosive	J. Johnson	Brooklyn, N. Y	June 7, 1864	43, 029
Projectile for ordnance, Explosive	W. F. Rippon	Providence, R. I	Oct. 28, 1862	36, 858
Projectile for ordnance, Hot	C. T. James	Providence, R. I	Jan. 21, 1862	34, 207
Projectile for ordnance, Rifled	W. Page	New York, N. Y	Nov. 5, 1861	36, 660
Projectile for rifled cannon	T. T. S. Laidley	United States Army	Aug. 4, 1857	17, 935
Projectile for rifled cannon	T. T. S. Laidley	United States Army	Aug. 25, 1857	18, 049
Projectile for rifled cannon	J. M. Sigourney	Watertown, N. Y	Mar. 3, 1857	16, 755
Projectile for rifled fire-arm	C. Sharps	Philadelphia, Pa	July 11, 1865	48, 729
Projectile for rifled ordnance	B. S. Alexander	United States Army	Jan. 7, 1862	34, 091
Projectile for rifled ordnance	J. B. Atwater	Chicago, Ill	Mar. 17, 1863	37, 891
Projectile for rifled ordnance	A. Berney	Jersey City, N. J	July 7, 1863	39, 112
Projectile for rifled ordnance	J. W. Cochran	New York, N. Y	Nov. 8, 1859	26, 016
Projectile for rifled ordnance	J. W. Cochran	New York, N. Y	Nov. 8, 1859	26, 017
Projectile for rifled ordnance	J. W. Cochran	New York, N. Y	Dec. 6, 1859	26, 337
Projectile for rifled ordnance	B. C. Converse	Cincinnati, Ohio	Aug. 16, 1864	43, 835
Projectile for rifled ordnance	J. A. Dahlgren	Philadelphia, Pa	Aug. 6, 1861	32, 986
Projectile for rifled ordnance	H. E. Dunick	Saint Louis, Mo	July 14, 1863	39, 216
Projectile for rifled ordnance	A. H. Emery	New York, N. Y	Oct. 28, 1862	36, 773
Projectile for rifled ordnance	W. H. Havens	Paterson, N. J	Feb. 25, 1862	34, 493
Projectile for rifled ordnance	J. L. Henry	District of Columbia	July 8, 1862	35, 824
Projectile for rifled ordnance	B. B. Hotchkiss	Sharon, Conn	July 24, 1860	29, 272
Projectile for rifled ordnance	B. B. Hotchkiss	Sharon, Conn	May 14, 1861	32, 293
Projectile for rifled ordnance	B. B. Hotchkiss	New York, N. Y	June 7, 1864	43, 027
Projectile for rifled ordnance	W. W. Hubbell	Philadelphia, Pa	July 8, 1862	35, 825
Projectile for rifled ordnance	J. H. Merrill	Baltimore, Md	Oct. 13, 1857	18, 401
Projectile for rifled ordnance	R. P. Parrott	Cold Spring, N. Y	Aug. 20, 1861	33, 099
Projectile for rifled ordnance	R. P. Parrott	Cold Spring, N. Y	Aug. 20, 1861	33, 100
Projectile for rifled ordnance	B. S. Roberts	United States Army	Mar. 24, 1863	37, 979
Projectile for rifled ordnance	C. H. Sayre	Utica, N. Y	Dec. 24, 1861	34, 031
Projectile for rifled ordnance	C. W. Small	Bangor, Me	Mar. 4, 1862	34, 596
Projectile for rifled ordnance	W. H. Smith	Birmingham, Conn	Dec. 23, 1862	37, 260
Projectile for rifled ordnance	W. H. Smith	Birmingham, Conn	Jan. 5, 1864	41, 127
Projectile for rifled ordnance	B. F. Sturtevant	Boston, Mass	Aug. 5, 1862	36, 116
Projectile for rifled ordnance	I. P. Tice	New York, N. Y	Nov. 12, 1861	33, 718
Projectile for rifled ordnance	W. S. Williams	Canton, Ohio	May 3, 1864	42, 614
Projectile for rifled ordnance, Explosive	C. W. Smith, G. H. Babcock, and B. B. and C. A. Hotchkiss.	New York, N. Y., and Sharon, Conn.	Apr. 28, 1863	38, 359
Projectile for small-arms	C. Madnell	New Orleans, La	Jan. 9, 1872	122, 620
Projectile for smooth-bored guns	J. L. McConnel	Jacksonville, Ill	July 28, 1857	17, 886
Projectile for smooth-bored ordnance	J. A. Woodbury	Boston, Mass	Dec. 3, 1861	33, 863
Projectile for smooth-bored ordnance, Rotating	T. F. Reilly	New York, N. Y	Mar. 18, 1862	34, 693
Projectile, Hollow	S. Gardiner, jr	New York, N. Y	Nov. 3, 1863	40, 468
Projectile, Incendiary subcaliber	C. W. Stafford	New York, N. Y	Aug. 4, 1863	39, 427
Projectile, Percussion	A. McBurth	Elizabeth, N. J	Aug. 21, 1855	13, 469
Projectile, Percussion	J. Lippincott	Philadelphia, Pa	Mar. 18, 1856	14, 460
Projectile, Rifled	J. McMurtry	Lexington, Ky	Sept. 15, 1863	39, 942
Projectile, Subcaliber	E. A. Dana	Boston, Mass	Nov. 12, 1872	132, 903
Projectile, Subcaliber rifled	J. L. Henry	Kentucky	Dec. 13, 1864	45, 462
Projectile, Submarine explosive	M. L. Callender	New York, N. Y	Aug. 18, 1863	39, 612
Projectile, Submarine explosive	J. D. Willoughby	Washington, D. C	Jan. 5, 1864	41, 112
Projectiles, Alloy for sabots of	T. Taylor	Washington, D. C	Jan. 8, 1867	61, 117
Projectiles, Apparatus for throwing	S. D. Wollison	Pittsfield, Mass	Sept. 8, 1863	39, 858
Projectiles, Application of soft-metal packing to	W. Boekel	Philadelphia, Pa	Sept. 16, 1862	36, 449
Projectiles, Banding and covering	J. Absterdam	New York, N. Y	Nov. 7, 1865	50, 783
Projectiles by pressure, Press for making cylindro-conical hollow.	W. M. B. Hartley	New York, N. Y	Mar. 20, 1855	12, 574
Projectiles, Composition for covering	J. Absterdam	New York, N. Y	Aug. 12, 1862	36, 132
Projectiles, Composition for filling shrapnell and other similar.	A. Berney	Jersey City, N. J	June 17, 1862	35, 659
Projectiles, Composition for packing	P. S. Devlan	Jersey City, N. J	Sept. 15, 1863	39, 897
Projectiles, Directing	S. H. Noble	Prairie du Chien, Wis	Nov. 3, 1863	40, 543
Projectiles, Discharging	A. B. Cooley	Philadelphia, Pa	June 17, 1862	35, 587
Projectiles for fire-arms, Casting	B. A. Mason	Newport, R. I	Jan. 14, 1862	34, 153
Projectiles for ordnance, Attaching sabots to spherical.	G. P. Gauster	New York, N. Y	June 7, 1864	43, 017
Projectiles for ordnance, Mode of preparing	E. Lindner	New York, N. Y	July 30, 1861	32, 949
Projectiles for rifled ordnance, Packing	J. Absterdam	New York, N. Y	Feb. 23, 1864	41, 668
Projectiles for rifled ordnance, Packing	C. Arick	Saint Clairsville, Ohio	Mar. 28, 1865	47, 078
Projectiles for rifled ordnance, Packing	W. Boekel	Philadelphia, Pa	Apr. 12, 1864	42, 268
Projectiles for rifled ordnance, Packing	L. W. Broadwell	New Orleans, La	June 19, 1866	55, 761
Projectiles for rifled ordnance, Packing	J. W. Cochran	New York, N. Y	July 31, 1866	56, 712
Projectiles for rifled ordnance, Packing	B. B. Hotchkiss	New York, N. Y	May 16, 1865	47, 725
Projectiles for rifled ordnance, Packing	B. B. Hotchkiss	New York, N. Y	Oct. 10, 1865	50, 357
Projectiles for rifled ordnance, Packing	A. J. S. Molinard	Baltimore, Md	Apr. 11, 1865	47, 213
Projectiles for rifled ordnance, Packing	W. H. Smith	Birmingham, Conn	Oct. 11, 1864	44, 669

Index of patents issued from the United States Patent Office from 1790 *to* 1873, *inclusive*—Continued.

Invention.	Inventor.	Residence.	Date.	No.
Projectiles for rifled ordnance, Packing	W. H. Smith	Birmingham, Conn	May 16, 1865	47, 754
Projectiles for rifled ordnance, Sabot for	M. Ritner	Vincennes, Ind	June 10, 1862	35, 544
Projectiles for rifled ordnance, Sabot for	C. W. Stafford	New York, N. Y	Dec. 20, 1864	45, 567
Projectiles from smooth-bored ordnance, Rotating	T. Goodrem and C. Jackson	Providence, R. I	July 29, 1862	36, 001
Projectiles, Giving rotation to ordnance	H. K. Kenyon	Steubenville, Ohio	Nov. 18, 1862	36, 959
Projectiles, Igniting explosive	E. Pertuiset, A. Mundel, and J. E. A. De Fleron.	Paris, France	May 26, 1868	78, 322
Projectiles, India-rubber wad for	J. N. Bird	Trenton, N. J	Oct. 28, 1862	36, 879
Projectiles, Inflammable composition for filling	E. Harrison	New York, N. Y	Feb. 9, 1864	41, 577
Projectiles, Jacket for subcaliber	W. H. Smith	Birmingham, Conn	Oct. 11, 1864	44, 670
Projectiles, Machine for throwing	A. B. Smith and W. Weaver	Clinton, Pa	Aug. 12, 1856	15, 529
Projectiles, Mold for casting	H. Conant	Hartford, Conn	Jan. 16, 1855	12, 258
Projectiles, Mold for hollow	E. Allen	Worcester, Mass	July 29, 1856	15, 454
Projectiles, Molding chilled-iron	W. Boekel	Philadelphia, Pa	Jan. 6, 1863	37, 272
Projectiles, Sabot for	E. A. Dana	Brookline, Mass	Oct. 31, 1865	50, 692
Projectiles, Sabot for	C. W. Stafford	Burlington, Iowa	July 7, 1863	39, 179
Projectiles, Sabot for feathered	J. P. Woodbury and S. S. Gray	West Roxbury and Boston, Mass.	July 22, 1862	35, 965
Projectiles, Sabot for ordnance	J. F. Clew	New York, N. Y	Nov. 15, 1864	45, 023
Projector	W. Farini	London, England	June 13, 1871	115, 837
Proof meter and register for alcoholic spirits and other liquids	W. M. Storm	New York, N. Y	May 7, 1867	64, 456
Propagating-bed, Apparatus for heating	W. Chalmers	Philadelphia, Pa	Feb. 28, 1871	112, 220
Propagating-box	A. D. Manchester	Westport, Mass	Dec. 15, 1868	84, 955
Propagating-box for plants	C. Jillson	Worcester, Mass	Aug. 24, 1869	94, 169
Propagating-box for plants and vines	F. L. Perry	Canandaigua, N. Y	Mar. 27, 1866	53, 477
Propeller. (*See also* Canal-boat.)				
Propeller	J. F. Allen	New York, N. Y	May 8, 1866	54, 481
Propeller	R. Atkin	Brooklyn, N. Y	June 23, 1868	79, 183
Propeller	E. Beard	New Sharon, Me	Oct. 18, 1853	10, 124
Propeller	T. Bell	Bellport, N. Y	Oct. 26, 1869	96, 075
Propeller	J. Busser	Philadelphia, Pa	Jan. 3, 1865	45, 697
Propeller	R. D. Chatterton	Bath, England	Mar. 5, 1867	62, 723
Propeller	W. A. Cobb	Orange, Mass	July 16, 1867	66, 797
Propeller	A. Cotton	Sycamore, Ill	Jan. 25, 1870	99, 063
Propeller	G. R. Comstock	Little Falls, N. Y	Nov. 9, 1858	22, 016
Propeller	J. Covert	New York, N. Y	Feb. 19, 1867	62, 255
Propeller	E. Dalman y Sala	New York, N. Y	Nov. 21, 1865	51, 025
Propeller	C. De Berque	London, England	Jan. 9, 1855	12, 190
Propeller	W. H. Degges	Washington, D. C	Apr. 28, 1863	38, 292
Propeller	A. Duboce	Brooklyn, N. Y	July 24, 1855	13, 336
Propeller	E. Etinge and J. C. Broadhead	Kingston, N. Y	Jan. 10, 1871	110, 908
Propeller	W. Fields	Wilmington, Del	Nov. 19, 1867	71, 153
Propeller	J. D. Ford	Baltimore, Md	Sept. 20, 1870	107, 471
Propeller	R. Forward	Cincinnati, Ohio	Jan. 16, 1872	122, 823
Propeller	F. G. Fowler	Springfield, Ill	May 28, 1867	65, 202
Propeller	H. Fromm	East New York, N. Y	June 4, 1867	65, 370
Propeller	C. F. Gardiner	East Boston, Mass	June 15, 1858	20, 606
Propeller	B. Gilbert	Pittsburgh, Pa	Nov. 15, 1853	10, 229
Propeller	R. Griffiths	London, England	Apr. 7, 1857	•17, 016
Propeller	B. Handforth	Chicago, Ill	Jan. 1, 1867	60, 887
Propeller	H. W. Hewett	New York, N. Y	June 7, 1853	9, 673
Propeller	D. H. Heyen	New York, N. Y	June 9, 1868	78, 739
Propeller	J. O. Heyworth	Chicago, Ill	June 29, 1869	91, 845
Propeller	R. Hines and L. Beyer	Washington, D. C	July 16, 1872	129, 028
Propeller	L. and C. Howard	Watkins, N. Y	Aug. 30, 1870	106, 824
Propeller	D. Hughes	Rochester, N. Y	Nov. 15, 1859	26, 106
Propeller	R. Hunter	New York, N. Y	July 7, 1868	79, 657
Propeller	R. Hunter	New York, N. Y	Aug. 11, 1868	80, 960
Propeller	R. Hunter	New York, N. Y	Sept. 13, 1870	107, 376
Propeller	P. M. Hutton and W. Huston	Troy, N. Y., and Wilmington, Del.	June 6, 1848	5, 623
Propeller	F. Jacob	New York, N. Y	Oct. 31, 1865	50, 714
Propeller	A. Johnson	Buffalo, N. Y	Dec. 8, 1857	18, 814
Propeller	W. D. Jones	Poughkeepsie, N. Y	Apr. 17, 1855	12, 752
Propeller	C. Kinzler	New York, N. Y	Apr. 26, 1870	102, 276
Propeller	W. Lawton	Green Point, N. Y	Oct. 1, 1867	69, 451
Propeller	H. Leach	Boston, Mass	Jan. 31, 1854	10, 474
Propeller	H. Link	Little Falls, N. Y	May 8, 1855	12, 823
Propeller	A. C. Loud	San Francisco, Cal	Jan. 19, 1869	86, 086
Propeller	A. C. Loud	San Francisco, Cal	May 4, 1869	89, 777
Propeller	L. H. Markley	Line Lexington, Pa	May 24, 1859	24, 130
Propeller	A. J. Marshall	Warrenton, Va	Mar. 7, 1871	112, 366
Propeller	J. T. Martin	New York, N. Y	Jan. 9, 1867	51, 955
Propeller	G. L. McKay	Linwood Station, Pa	Jan. 30, 1872	123, 274
Propeller	G. Meader	Ottawa, Ill	Jan. 30, 1866	52, 308
Propeller	C. G. Meinhardt	Altoona, Pa	July 28, 1868	80, 421
Propeller	D. E. Merick	Cleveland, Ohio	July 20, 1858	20, 953
Propeller	C. F. Miller and M. Priester	New York, N. Y	June 11, 1872	127, 909
Propeller	S. B. Morey	San Francisco, Cal	Dec. 14, 1869	97, 952
Propeller	N. Nolan	New York, N. Y	Dec. 17, 1867	72, 319
Propeller	C. O'Hara	London, Great Britain	Apr. 22, 1862	35, 035
Propeller	C. M. O'Hara	New York, N. Y	Apr. 14, 1868	76, 651
Propeller	G. Parker	Worcester, Mass	May 24, 1864	42, 870
Propeller	J. Patch	Boston, Mass	Nov. 27, 1849	6, 914
Propeller	F. Peale	Philadelphia, Pa	Feb. 20, 1855	12, 446
Propeller	J. H. Penny and T. B. Rogers	New York, N. Y	June 12, 1855	13, 041
Propeller	O. C. Phelps	New York, N. Y	July 12, 1864	43, 522
Propeller	J. H. Phillips	Saint Louis, Mo	Feb. 5, 1867	61, 861
Propeller	G. R. Pierce	Grand Rapids, Mich	Feb. 27, 1872	124, 009
Propeller	E. S. Renwick	New York, N. Y	Sept. 24, 1861	33, 364
Propeller	H. F. Roberts	Pittsburgh, Pa	Dec. 29, 1868	85, 403
Propeller	H. Rolle	Boston, Mass	July 16, 1867	66, 742
Propeller	T. W. Roys	Southampton, N. Y	June 3, 1862	35, 475
Propeller	J. Salter	Williamsburgh, N. Y	Nov. 29, 1870	109, 673
Propeller	R. Schmidt	Cincinnati, Ohio	Feb. 13, 1872	123, 733
Propeller	C. J. Schveder	Astoria, N. Y	Oct. 9, 1860	30, 355
Propeller	C. Sharps	Philadelphia, Pa	May 25, 1869	90, 590

Index of patents issued from the United States Patent Office from 1790 *to* 1873, *inclusive*—Continued.

Invention.	Inventor.	Residence.	Date.	No.
Propeller	C. Sharps	Philadelphia, Pa	Sept. 21, 1869	95, 150
Propeller	S. P. Snyder and G. W. Cook	Minneapolis, Minn	Mar. 8, 1859	23, 199
Propeller	A. W. Thompson	Philadelphia, Pa	Jan. 21, 1851	7, 907
Propeller	W. Thurber	Olean, N. Y	June 29, 1858	20, 744
Propeller	C. E. Toley	Brooklyn, N. Y	July 30, 1867	67, 376
Propeller	J. Trees	Salem, Pa	Oct. 25, 1853	10, 164
Propeller	W. Van Dusen	Philadelphia, Pa	Nov. 30, 1858	22, 209
Propeller	S. B. Wait	Mariner's Harbor, N. Y	Feb. 19, 1867	62, 238
Propeller	C. R. M. Wall	New York, N. Y	May 17, 1859	24, 069
Propeller	C. Ward	Detroit, Mich	Apr. 6, 1869	88, 595
Propeller	C. T. P. Ware	New York, N. Y	Oct. 4, 1853	10, 091
Propeller	J. White	Mauch Chunk, Pa	June 11, 1829	
Propeller	A. A. Wilder and W. Gooding	Detroit, Mich	May 15, 1866	54, 801
Propeller	W. L. and T. Winans	London, England, and Baltimore, Md.	Oct. 9, 1866	58, 744
Propeller, Adjustable and reversible	A. B. Cooley	Philadelphia, Pa	May 13, 1862	35, 219
Propeller, Adjustable screw	C. F. Brown	Warren, R. I	July 12, 1853	9, 849
Propeller and fire-extinguisher, Combined	A. Turner	Bronson, Mich	Apr. 2, 1872	125, 232
Propeller and paddle-wheel shaft	W. Peters	Baltimore, Md	Nov. 29, 1859	26, 290
Propeller and steerer for boats, Portable	T. Reece	Philadelphia, Pa	Oct. 23, 1866	59, 074
Propeller and steering-apparatus	H. Ressel	Vienna, Austria	Oct. 11, 1864	44, 696
Propeller-blade	C. H. Burton	Cleveland, Ohio	Apr. 21, 1863	28, 216
Propeller-blade	G. Hibsch	Buffalo, N. Y	May 12, 1857	17, 276
Propeller-blade	W. G. Oliver	Buffalo, N. Y	Apr. 7, 1863	38, 121
Propeller-blade	G. W. Swartz	Buffalo, N. Y	Aug. 4, 1857	17, 943
Propeller blades, Screw	C. C. Gates	Albany, N. Y	Dec. 12, 1865	51, 446
Propeller-blades to their shafts, Means of affixing	L. B. Flanders	Philadelphia, Pa	Aug. 16, 1864	43, 842
Propeller, Buoyant	W. Lansdell	Memphis, Tenn	May 29, 1855	12, 955
Propeller, Buoyant	E. Matteson	Brooklyn, N. Y	Oct. 13, 1868	82, 968
Propeller, Buoyant	J. Montgomery	New York, N. Y	Dec. 21, 1858	22, 373
Propeller, Buoyant spiral	T. J. Wells	New York	Dec. 23, 1841	2, 400
Propeller, Cantering	P. W. Mackenzie	Jersey City, N. J	Aug. 12, 1862	36, 161
Propeller, Centripetal	J. W. Nystrom	Stockholm, Sweden	Mar. 19, 1850	7, 194
Propeller, Chain	J. Neumann	Brooklyn, N. Y	Dec. 10, 1872	133, 794
Propeller, Chain	E. P. Russell	Manlius, N. Y	Sept. 9, 1873	142, 653
Propeller, Child's horse and self	J. H. Nolan	Waterville, N. Y	Mar. 1, 1870	100, 316
Propeller, Conical screw	H. Hubbell	Moyamensing, Pa	July 14, 1845	4, 111
Propeller-cranks, Apparatus for oiling	J. Davis	New York, N. Y	Oct. 8, 1867	69, 638
Propeller, Double-cone marine	J. R. Baylis	Baltimore, Md	Dec. 10, 1861	33, 874
Propeller, Duck's-foot	G. Meader	Ottawa, Ill	Sept. 1, 1863	39, 738
Propeller, Duck's-foot	G. Seibert	Hagerstown, Md	Nov. 7, 1848	5, 911
Propeller, Endless-chain	C. C. Bisbee	Rochester, N. Y	Sept. 24, 1872	131, 494
Propeller, Endless-chain	W. W. Bowman	Graves County, Ky	Jan. 26, 1864	41, 360
Propeller, Endless-chain	L. Chevrier	Philadelphia, Pa	June 11, 1836	
Propeller, Endless-chain	C. F. Fisher	New Orleans, La	Oct. 7, 1851	8, 407
Propeller, Endless-chain	W. Keeler	Towanda, Pa	Sept. 17, 1872	131, 450
Propeller, Endless-chain	A. McDonald	Mattawan, Mich	July 4, 1865	48, 574
Propeller, Endless-chain	A. B. Smith	Yankton, Dak	Oct. 22, 1867	70, 036
Propeller, Endless-chain	T. Teed	Alexandria, Va	Oct. 21, 1873	143, 793
Propeller, Endless-chain	J. W. Whinyates	Liverpool, England	Mar. 25, 1873	137, 269
Propeller-engine	E. D. Ashe	Brompton, England	Apr. 19, 1864	42, 336
Propeller-engine	H. Whitaker	New York, N. Y	July 23, 1872	129, 700
Propeller, Flexible	R. Hunter	Cincinnati, Ohio	May 16, 1871	114, 944
Propeller for boats	J. Hamilton	New York, N. Y	Dec. 28, 1858	22, 422
Propeller for boats	M. Nelson	New York, N. Y	July 13, 1858	20, 889
Propeller for boats	L. C. St. John	Buffalo, N. Y	Aug. 31, 1858	21, 378
Propeller for boats, &c., Oblique-paddle	R. Bulkley	New York, N. Y	Mar. 13, 1844	3, 476
Propeller for boats or carriages	B. Taylor	New York	Apr. 25, 1808	
Propeller for boats, Spiral	J. Copley	Warrior Marsh, Pa	May 22, 1830	
Propeller for boats, Steam	A. W. Von Schmidt	San Francisco, Cal	Nov. 13, 1866	59, 686
Propeller for canal-boats	E. Backus	Rochester, N. Y	Aug. 21, 1860	29, 660
Propeller for canal-boats	A. Burbank	Buffalo, N. Y	July 13, 1858	20, 862
Propeller for canal-boats	R. Cartwright	Ithaca, N. Y	July 19, 1859	24, 794
Propeller for canal-boats	S. D. Gilson	Oswego Falls, N. Y	Nov. 17, 1868	84, 108
Propeller for canal-boats	E. Raynale	Birmingham, Mich	Aug. 28, 1866	57, 569
Propeller for canal-boats	O. Plumb	Millport, N. Y	Sept. 23, 1862	36, 526
Propeller for canal-boats	G. W. Swartz	Buffalo, N. Y	June 18, 1857	17, 592
Propeller for canal-boats	G. W. Swartz	Buffalo, N. Y	Mar. 16, 1858	19, 656
Propeller for canal-boats	G. G. Wyland and T. M. Rathmell.	Williamsport, Pa	May 2, 1871	114, 382
Propeller for canal-boats, Chain	E. T. Ligon	Demopolis, Ala	June 25, 1872	128, 319
Propeller for canal-navigation	W. F. Tyson	Owingsburgh, Pa	June 21, 1853	9, 810
Propeller for canals	B. Burling	Buffalo, N. Y	Feb. 22, 1859	23, 010
Propeller for canals	N. P. Otis	Yonkers, N. Y	May 9, 1865	47, 657
Propeller for life-boats	M. M. Camp	New Haven, Conn	Dec. 21, 1858	22, 346
Propeller for life-boats	G. W. La Baw	Jersey City, N. J	May 6, 1856	14, 843
Propeller for shallow water, Wave	J. B. Eads	Saint Louis, Mo	Dec. 16, 1862	37, 157
Propeller for small boats, Hand	J. Fehrenbatch	Indianapolis, Ind	Sept. 18, 1866	58, 083
Propeller for spindle-shaped ships	W. L. and T. Winans	London, England	Sept. 4, 1866	57, 836
Propeller for steamships	C. Kinkel and M. Hubbe	New York, N. Y	June 19, 1866	55, 773
Propeller for steam-vessels	R. Sutton	New Castle, Del	Sept. 6, 1870	107, 120
Propeller for vessels	A. Aubert	Nogent-le-Rotrou, France	Sept. 7, 1869	94, 463
Propeller for vessels	L. De Lill	Phœnix, N. Y	Nov. 21, 1871	121, 163
Propeller for vessels	H. Everett	Windsor, Vt	Aug. 7, 1847	5, 219
Propeller for vessels	S. J. Gold	Cornwall, Conn	Mar. 6, 1847	5, 000
Propeller for vessels	C. A. S. Harris	Saint Louis, Mo	Mar. 28, 1871	113, 163
Propeller for vessels	E. C. Hubbard	Green Bay, Wis	Feb. 4, 1873	135, 555
Propeller for vessels	L. W. Langdon	Northampton, Mass	July 16, 1872	129, 571
Propeller for vessels	W. P. McConnell	Washington, D. C	May 16, 1846	4, 529
Propeller for vessels	C. M. O'Hara	London, England	Nov. 5, 1867	70, 460
Propeller for vessels	L. Phleger	Wilmington, Del	Sept. 13, 1845	4, 193
Propeller for vessels	A. Raplin	France	Apr. 3, 1866	53, 762
Propeller for vessels	C. M. Raynale	Birmingham, Mich	Jan. 8, 1867	61, 020
Propeller for vessels	J. E. Smith	New York, N. Y	Mar. 14, 1848	5, 468
Propeller for vessels	H. Stanley	Troy, N. Y	Oct. 9, 1860	30, 360
Propeller for vessels	C. E. Tripler	New York, N. Y	June 27, 1871	116, 513
Propeller for vessels, Chain	T. W. Carter	New Orleans, La	July 1, 1873	140, 468
Propeller for vessels, Form of screw	E. Beard	New Sharon, Me	Apr. 10, 1841	2, 045

Index of patents issued from the United States Patent Office from 1790 *to* 1873, *inclusive*—Continued.

Invention.	Inventor.	Residence.	Date.	No.
Propeller for vessels, Rotary inclined	R. F. Loper	Philadelphia, Pa	Feb. 28, 1844	3, 459
Propeller for vessels, Vibrating	D. Bainbridge	Philadelphia, Pa	Jan. 28, 1873	135, 257
Propeller for vessels, Vibrating	J. D. Fraser	Pictou, Canada	Dec. 9, 1873	145, 342
Propeller, Four-wheel	R. G. Eider	New York, N. Y	Feb. 1, 1870	99, 421
Propeller, Hand	J. G. Ross	New York, N. Y	May 27, 1856	14, 973
Propeller, Hydraulic	J. B. Booth	Portsmouth, Va	June 2, 1863	38, 724
Propeller, Hydraulic	A. Pagenstecher	Valparaiso, Chili	Oct. 4, 1864	44, 584
Propeller, Jointed scull	C. Dann	Rushford, Minn	Mar. 10, 1863	37, 856
Propeller, Marine	W. Aitken	Newark, Ill	Sept. 15, 1863	39, 984
Propeller, Marine	B. T. Babbitt	New York, N. Y	May 5, 1863	38, 368
Propeller, Marine	G. L. Carver	Brazil, South America	Aug. 7, 1860	29, 464
Propeller, Marine	G. B. Cocks	Cincinnati, Ohio	Mar. 28, 1871	113, 143
Propeller, Marine	R. Covington	Washington, D. C	Nov. 15, 1864	45, 127
Propeller, Marine	J. Eaton	Belleville, Canada	Oct. 19, 1858	21, 825
Propeller, Marine	A. E. Harding	Middletown, Ohio	Sept. 11, 1860	29, 969
Propeller, Marine	A. L. Hatch	Addison, N. Y	Apr. 25, 1865	47, 419
Propeller, Marine	H. W. Herbert	Herbertsville, Va	Mar. 27, 1860	27, 629
Propeller, Marine	H. Hirsch	Berlin, Prussia	Aug. 23, 1859	25, 197
Propeller, Marine	L. Holtslander	Oberlin, Ohio	Sept. 6, 1859	65, 336
Propeller, Marine	W. H. Johnson	Springfield, Mass	June 3, 1862	35, 451
Propeller, Marine	E. Lacroix, jr	Rouen, France	Dec. 10, 1861	33, 893
Propeller, Marine	T. Mason	Boston, Mass	Oct. 11, 1864	44, 641
Propeller, Marine	D. D. Porter	United States Navy	June 12, 1860	28, 688
Propeller, Marine	H. D. J. Pratt	Washington, D. C	Nov. 27, 1860	30, 757
Propeller, Marine	J. Reynolds	Providence, R. I	Feb. 14, 1860	27, 156
Propeller, Marine	J. B. Root	Brooklyn, N. Y	May 5, 1863	38, 415
Propeller, Marine	J. Taggart	Roxbury, Mass	Mar. 29, 1859	23, 432
Propeller, Marine	R. C. Vernol and J. T. Martin	New York, N. Y	Jan. 5, 1864	41, 109
Propeller, Marine	M. Wappich	Sacramento, Cal	Oct. 8, 1861	33, 459
Propeller, Marine	E. Webber	Gardiner, Me	May 29, 1860	28, 542
Propeller, Marine	W. H. Willard	Cleveland, Ohio	Sept. 16, 1862	36, 492
Propeller, Marine	W. D. Wilson	Newark, N. J	Dec. 19, 1865	51, 640
Propeller, Marine hand	E. C. Brackett	Newton Corner, Mass	June 14, 1859	24, 368
Propeller, Minnow	W. D. Chapman	Theresa, N. Y	July 5, 1870	104, 930
Propeller, Minnow	W. D. Chapman	Theresa, N. Y	May 30, 1871	115, 434
Propeller operating as air-pump and condenser, Submerged.	W. W. Hunter	Gosport, Va	Mar. 12, 1842	2, 487
Propeller, Paddle-wheel	J. May	Columbus, Ga	Sept. 7, 1858	21, 432
Propeller, Reciprocating	M. Depuy	Pittsburgh, Pa	Aug. 1, 1865	49, 095
Propeller, Reciprocating	J. Galt	Philadelphia, Pa	June 28, 1859	24, 552
Propeller, Reciprocating	H. W. Hewet	New York, N. Y	Oct. 9, 1849	6, 782
Propeller, Reversible hydraulic	R. G. Dunbar	Logansport, Ind	Oct. 15, 1872	132, 262
Propeller, Rocking	E. Landis	Baltimore, Md	July 24, 1860	29, 288
Propeller-ropes, Preventing fouling of	R. H. Hudson	Glasgow, Scotland	Aug. 20, 1867	67, 982
Propeller, Screw	C. Arlen and C. Gautschi	New York, N. Y	May 3, 1870	102, 645
Propeller, Screw	A. Aubert	Nogent-le-Rotrou, France	June 25, 1872	128, 203
Propeller, Screw	B. F. Bee	Harwich, Mass	Jan. 4, 1859	22, 525
Propeller, Screw	D. Bell	Buffalo, N. Y	June 25, 1861	32, 609
Propeller, Screw	O. Byrne and J. G. Elliott	New York, N. Y	Oct. 5, 1858	21, 650
Propeller, Screw	J. Cochrane	Wall Township, N. J	June 27, 1871	116, 414
Propeller, Screw	L. H. Colborn	Chicago, Ill	Dec. 5, 1865	51, 295
Propeller, Screw	W. J. Curtis	Holloway, England	May 19, 1863	38, 566
Propeller, Screw	W. E. Davis	Jersey City, N. J	Aug. 7, 1866	56, 907
Propeller, Screw	H. Elliott	Boston, Mass	Aug. 20, 1861	33, 077
Propeller, Screw	J. Ericsson	New York, N. Y	Sept. 9, 1845	4, 181
Propeller, Screw	D. Freed	Philadelphia, Pa	Aug. 19, 1873	141, 871
Propeller, Screw	J. Hartman, jr	Philadelphia, Pa	Apr. 1, 1873	137, 365
Propeller, Screw	L. and C. Howard	Watkins, N. Y	Sept. 3, 1872	131, 099
Propeller, Screw	H. Hirsch	Paris, France	Apr. 26, 1870	102, 399
Propeller, Screw	F. Jacob	New York, N. Y	Jan. 24, 1865	46, 004
Propeller, Screw	A. Jouan	San Francisco, Cal	Jan. 25, 1859	22, 731
Propeller, Screw	J. E. Kennedy	New Orleans, La	May 12, 1868	77, 888
Propeller, Screw	M. Kolb	New York, N. Y	Sept. 19, 1871	119, 155
Propeller, Screw	G. Malo	Dunkirk, France	Nov. 18, 1851	8, 529
Propeller, Screw	J. Montgomery	Memphis, Tenn	Aug. 7, 1847	5, 224
Propeller, Screw	J. Montgomery	Baltimore, Md	Apr. 12, 1859	23, 598
Propeller, Screw	N. A. Patterson	Athens, Tenn	Jan. 21, 1873	135, 149
Propeller, Screw	N. A. Patterson	Cleveland, Tenn	Aug. 26 1873	142, 269
Propeller, Screw	S. W. Rowell	Rutland, Vt	Oct. 12, 1869	95, 731
Propeller, Screw	J. B. Root	New York, N. Y	May 23, 1865	47, 864
Propeller, Screw	G. Safford	New York, N. Y	Mar. 1, 1859	23, 117
Propeller, Screw	C. G. Toense	Jersey City, N. J	Feb. 22, 1870	100, 216
Propeller, Screw	A. G. Tompkins	New York, N. Y	Mar. 26, 1861	31, 838
Propeller, Screw	H. Vansittart	Richmond, England	May 4, 1869	89, 712
Propeller, Screw	J. J. B. Vergne	Paris, France	June 21, 1859	24, 508
Propeller, Screw	M. M. Wilson	Honey Grove, Tex	Feb. 18, 1873	136, 015
Propeller, Screw	F. Wittram	San Francisco, Cal	Aug. 17, 1869	93, 848
Propeller, Screw	H. Zahn	San Francisco, Cal	Aug. 22, 1871	118, 325
Propeller-screws, Mode of driving	R. C. Jackson	Cleveland, Ohio	Apr. 21, 1863	38, 228
Propeller, Sculler	A. Bond	Philadelphia, Pa	June 19, 1849	6, 533
Propeller-shaft	W. Burtis	New York, N. Y	Nov. 3, 1868	83, 601
Propeller-shaft	A. Jouan	San Francisco, Cal	Nov. 18, 1856	16, 091
Propeller-shaft coupling, Ship	W. L. and T. Winans	London, England	Sept. 4, 1866	57, 835
Propeller-shafts and receiving rudders of stern-propellers, Means of supporting.	J. Beattie	Liverpool, England	Jan. 15, 1856	14, 081
Propeller-shafts, Anti-friction roller for	D. S. Chase	Augusta, Ga	Aug. 14, 1860	29, 570
Propeller-shafts in keels, Inclosing	A. Arnold	Troy, N. Y	Apr. 8, 1856	14, 589
Propeller-shafts, Stern-bearing for	R. E. Campbell	New York, N. Y	Dec. 20, 1864	45, 473
Propeller, Shell	J. Trees	Salem, Pa	May 14, 1850	7, 371
Propeller-shield	J. Carhart	New York, N. Y	July 9, 1861	32, 751
Propeller, Shifting-bucket	J. Busser	Philadelphia, Pa	July 2, 1867	66, 214
Propeller, Shifting-bucket	J. Busser	Philadelphia, Pa	Apr. 14, 1868	76, 711
Propeller, Ship's	J. K. Peters	New York, N. Y	Mar. 1, 1859	23, 112
Propeller, Spiral	B. Woodcroft	Manchester, England	Apr. 4, 1846	4, 441
Propeller, Spring and valve marine	W. Willis	Charleston, S. C	May 2, 1828	
Propeller, Steam land	G. W. N. Yost	Pittsburgh, Pa	June 3, 1856	15, 050
Propeller, Steamship	A. Doyle	New York, N. Y	Sept. 4, 1866	57, 690
Propeller, Steering	F. G. Fowler	Norwalk, Conn	Jan. 4, 1870	98, 483

Index of patents issued from the United States Patent Office from 1790 *to* 1873, *inclusive*—Continued.

Invention.	Inventor.	Residence.	Date.	No.
Propeller, Steering	F. G. Fowler	Bridgeport, Conn	Oct. 21, 1873	143, 896
Propeller, Steering	H. E. Fessel	Chicago, Ill	Dec. 28, 1858	22, 417
Propeller, Steering	W. Harsen	Green Point, N. Y	Nov. 11, 1873	144, 538
Propeller, Submarine	J. L. Sullivan	Boston, Mass	Mar. 24, 1817	....
Propeller, Submerged	H. Everett	Windsor, Vt	June 9, 1843	3, 129
Propeller, Submerged	R. Vaile	Westbourne Green, England	June 30, 1863	39, 083
Propeller, Submerged	P. Von Schmidt	Washington, D. C	May 30, 1844	3, 606
Propeller, Valve	A. Colton	Le Roy, N. Y	May 25, 1858	20, 332
Propeller, Vibrating	F. Kellsey	Middletown, Conn	Nov. 2, 1852	9, 366
Propeller, Vibrating	C. P. Macowitzky	Corpus Christi, Tex	Feb. 18, 1873	135, 994
Propeller-wheel	C. T. Finlayson	Albany, Oreg	Sept. 28, 1869	95, 335
Propeller-wheel	J. Harrington and F. Caffrey	New Haven, Conn	Mar. 20, 1866	53, 297
Propeller-wheel	R. Johnson	San Francisco, Cal	Apr. 9, 1872	125, 458
Propeller-wheel	J. A. Joyner	New York, N. Y	Dec. 28, 1869	98, 268
Propeller-wheel	T. Tripp	Buffalo, N. Y	Nov. 22, 1859	26, 213
Propeller-wheel	T. Tripp	Chicago, Ill	Apr. 10, 1866	53, 902
Propeller-wheels, Mold-box for casting	W. Lechler and J. K. Schupp	Cleveland, Ohio	Jan. 6, 1863	37, 295
Propellers and chimneys for canal-boats, Arrangement of.	B. M. Smith	Ridgway, N. Y	Apr. 23, 1850	7, 310
Propellers, Application of high-pressure engine to screw.	H. Whittaker	Buffalo, N. Y	Oct. 18, 1853	10, 145
Propellers, Application of steam to	W. Spalding	Port Clinton, Ohio	July 7, 1868	79, 695
Propellers, Applying steam-power to	A. M. Sawyer	Athol, Mass	May 8, 1860	28, 200
Propellers, Arrangement and connection of screw	P. S. Devlan	Reading, Pa	May 21, 1850	7, 376
Propellers, Attaching and housing	W. Webster	Jefferson County, Wash	June 29, 1858	20, 751
Propellers, Elevating and depressing	S. G. Hill	Muscatine, Iowa	Aug. 5, 1873	141, 558
Propellers for boats and other vessels, Segmental spiral.	J. Laing	Ellicott's Mills, Md	Oct. 22, 1842	2, 825
Propellers for steamboats, &c., Coupling the shafts of submerged.	R. F. Loper	Philadelphia, Pa	Oct. 9, 1844	3, 786
Propellers, Journal for oscillating	M. A. Crooker	New York, N. Y	Oct. 16, 1849	6, 793
Propellers, Manufacture of	J. Sutherland	New York, N. Y	Apr. 4, 1865	47, 137
Propellers, Mechanism for operating screw	F. H. Tobias	New York, N. Y	Dec. 16, 1873	145, 697
Propellers, Mechanism for operating screw	J. Wilcoxen	Russellville, Ill	May 20, 1873	139, 098
Propellers, Molding screw	A. K. Rider	Elizabeth, N. J	Aug. 31, 1869	94, 442
Propellers of machinery to be used in currents	J. Hardie	Victoria, Tex	Nov. 18, 1851	8, 525
Propellers in steam-vessels, Device for regulating the revolution of.	H. Dale	Boston, Mass	Feb. 26, 1867	62, 399
Propellers of steam-vessels, Elevating and depressing.	R. F. Soper	Philadelphia, Pa	Nov. 26, 1845	4, 285
Propellers, Operating screw	A. Lee	Whitemarsh, Pa	Oct. 29, 1872	132, 588
Propellers, Operating screw	A. McDonald	Cumberland, Md	Nov. 25, 1873	145, 000
Propellers, Operating vibrating	T. Spiller and A. Crowhurst	London, England	Nov. 8, 1853	10, 215
Propellers relatively to the draught of water, Apparatus for adjusting.	W. H. Willard	Cleveland, Ohio	June 10, 1862	35, 573
Propellers, Shafting-coupling for	S. Wilmarth, S. L. Hay, and D. N. B. Coffin, jr.	Charlestown, Reading, and Newton, Mass.	Apr. 6, 1858	19, 887
Propellers, Shipping and unshipping	S. R. Parkhurst	New York, N. Y	Nov. 26, 1845	4, 290
Propellers, Shipping and unshipping screw	G. Leach	Leeds, England	Nov. 1, 1870	108, 800
Propellers, Stern bearing for	J. Boiles	New York, N. Y	July 18, 1871	117, 039
Propellers to driving-shafts, Attaching	J. L. Cathcart	Washington, D. C	Apr. 18, 1854	10, 790
Propellers to vessels, Attaching	G. Smith	Philadelphia, Pa	Feb. 27, 1872	124, 170
Propelling and navigating boats, &c	A. G. D. Tuthill	New York	Mar. 22, 1810	....
Propelling and steering apparatus	S. and S. Huse, jr	Chicago, Ill	Dec. 28, 1858	22, 431
Propelling and steering apparatus	R. H. Lecky	McClure, Pa	Sept. 15, 1863	39, 936
Propelling and steering apparatus	M. Lytle	Allegheny, Pa	May 24, 1859	24, 129
Propelling and steering apparatus	J. C. Salomon, jr	Cincinnati, Ohio	Dec. 2, 1851	8, 558
Propelling and steering apparatus for vessels	J. Jopling	Bishopwearmouth, England	Apr. 3, 1866	53, 759
Propelling and steering boats, &c	T. L. Jones	New York	July 16, 1842	2, 727
Propelling and steering canal-boats	S. D. Bennett	Butternuts, N. Y	Jan. 30, 1829	....
Propelling and steering canal-boats	P. Roberts	New York, N. Y	June 18, 1872	128, 069
Propelling-apparatus	J. F. Alexander	New York, N. Y	Nov. 23, 1869	97, 146
Propelling-apparatus	E. Averill	Sacramento, Cal	Dec. 28, 1869	98, 217
Propelling-apparatus	E. S. Barnes	Nebraska City, Nebr	Oct. 6, 1868	82, 787
Propelling-apparatus	J. Bourne	London, England	Dec. 28, 1869	98, 222
Propelling-apparatus	S. D. Carpenter	Madison, Wis	Jan. 12, 1869	85, 789
Propelling-apparatus	J. S. Cunningham	New York, N. Y	Sept. 7, 1869	94, 573
Propelling-apparatus	A. Gemünder	New York, N. Y	Oct. 24, 1865	50, 574
Propelling-apparatus	J. Granger	Zanesville, Ohio	Sept. 29, 1868	82, 516
Propelling-apparatus	R. Hunter	New York, N. Y	Dec. 14, 1869	97, 924
Propelling-apparatus	R. Kearney	Havre de Grace, Newfoundland.	Sept. 13, 1870	107, 384
Propelling-apparatus	E. S. Ligon	Demopolis, Ala	July 5, 1870	105, 098
Propelling-apparatus	D. S. Merrit	Mount Morris, Mich	Apr. 20, 1869	89, 231
Propelling-apparatus	L. Moody	Malden, Mass	June 22, 1869	91, 760
Propelling-apparatus	H. Mulford	Philadelphia, Pa	July 20, 1869	92, 741
Propelling-apparatus	E. Offhaus	Newark, N. J	July 28, 1868	80, 497
Propelling-apparatus	J. Paradis	Brooklyn, N. Y	Aug. 3, 1869	93, 337
Propelling-apparatus	F. R. Pike	New York, N. Y	Aug. 17, 1869	93, 902
Propelling-apparatus	P. Roberts	New York, N. Y	Apr. 27, 1869	89, 435
Propelling-apparatus	C. Schilling	New York, N. Y	Aug. 16, 1870	106, 512
Propelling-apparatus	C. T. Smith	Nyack, N. Y	Mar. 2, 1869	87, 375
Propelling-apparatus	C. Wolf	Washington, D. C	Aug. 3, 1869	93, 384
Propelling-apparatus	A. Zink	Lancaster, Ohio	June 8, 1869	91, 196
Propelling-apparatus for land-conveyance	J. Wolf	Burr Oak, Mich	June 20, 1871	116, 130
Propelling-apparatus, Marine	E. Campbell	Boston, Mass	Sept. 29, 1857	18, 314
Propelling boats	J. S. Anderson	Flintville, Wis	Mar. 25, 1873	137, 164
Propelling boats	A. Bartholomew	Albany, N. Y	Dec. 1, 1818	....
Propelling boats	T. Beach	Wilmington, Ohio	Apr. 2, 1830	....
Propelling boats	A. Beekman	New York, N. Y	Sept. 30, 1873	143, 270
Propelling boats	J. Black and D. W. Jones	Philadelphia, Pa	Nov. 18, 1873	144, 732
Propelling boats	W. Burk	Whitemarsh, Pa	Dec. 2, 1834	....
Propelling boats	C. Dancker	Hoboken, N. J	Feb. 6, 1872	123, 383
Propelling boats, &c	T. Davis	Lawrence, Ind	Mar. 15, 1828	....
Propelling boats	J. P. Durand	New York	Jan. 17, 1818	....
Propelling boats	J. J. Giraud	Baltimore, Md	Apr. 10, 1818	....
Propelling boats	J. Gray	Peekskill, N. Y	Nov. 18, 1818	....
Propelling boats	P. Heaton, M. B. Boyes, and P. Heaton, jr.	Plymouth, Mass	Mar. 4, 1808	....

Index of patents issued from the United States Patent Office from 1790 to 1873, inclusive—Continued.

Invention.	Inventor.	Residence.	Date.	No.
Propelling boats	J. Heavin	Montgomery County, Va	Apr. 17, 1806	
Propelling boats	J. Heroux	Saint Paul, Minn	May 4, 1869	89, 663
Propelling boats	G. Hotchkiss	Windsor, N. Y	June 2, 1836	
Propelling boats	D. Keller		May 25, 1795	
Propelling boats	J. Lyon	Statesburgh	Apr. 3, 1819	
Propelling boats	M. Mallory	Urbanna, N. Y	Aug. 4, 1841	2, 210
Propelling boats	G. Manny	Montgomery, N. Y	June 8, 1815	
Propelling boats	P. Noble	Westfield, Mass	Jan. 20, 1836	
Propelling boats	J. Ring	Ogden, N. Y	Oct. 25, 1832	
Propelling boats	D. J. Ross	Havre de Grace, Md	Dec. 28, 1869	98, 302
Propelling boats, &c	P. Sear	Boston, Mass	Dec. 30, 1841	2, 410
Propelling boats	J. Shackford		Mar. 21, 1799	
Propelling boats	I. Silliman	Philadelphia, Pa	Dec. 17, 1812	
Propelling boats	R. Steel	Philadelphia, Pa	June 25, 1872	128, 258
Propelling boats	J. H. Street	Philadelphia, Pa	Dec. 27, 1843	3, 393
Propelling boats	A. Taylor	New York	Nov. 18, 1818	
Propelling boats	I. Theal	New York	Dec. 24, 1834	
Propelling boats, &c	P. C. Traver	Rhinebeck, N. Y	Apr. 30, 1842	2, 598
Propelling boats	E. Williams	Marianna, Fla	Apr. 30, 1867	64, 392
Propelling boats	S. W. Wood	Washington, D. C	Apr. 29, 1856	14, 786
Propelling boats across ferries	S. Ross	Troy, N. Y	Oct. 30, 1828	
Propelling boats and extinguishing fires, Machinery for	S. Bates and G. Titcomb	Boston, Mass	Mar. 4, 1842	2, 477
Propelling boats and other vessels	G. Erkson	Pittsburgh, Pa	Mar. 26, 1845	3, 967
Propelling boats and vessels	A. Anderson	Philadelphia, Pa	Aug. 21, 1812	
Propelling boats and vessels	J. J. Giraud	Baltimore, Md	Dec. 12, 1826	
Propelling boats, Apparatus for	J. Repetti	Philadelphia, Pa	Feb. 7, 1871	111, 681
Propelling boats, &c., Apparatus for	H. Ronalds	Albion, Ill	Dec. 14, 1840	1, 899
Propelling boats, Apparatus for	S. Stillwell and D. T. Wandall	New York	June 3, 1808	
Propelling boats, Apparatus for	M. Vergnes	New York, N. Y	Sept. 4, 1866	57, 798
Propelling boats by air, Apparatus for	J. Black	Orangeville, Pa	July 24, 1838	850
Propelling boats by condensed and compressed air	J. L. Sullivan	Boston, Mass	Mar. 24, 1817	
Propelling boats by animal-power	M. Rogers	New York	Feb. 9, 1814	
Propelling boats by cattle	H. Voight		Aug. 10, 1791	
Propelling boats by endless chain of paddles	A. Bragg	New York	Jan. 24, 1842	2, 434
Propelling boats by horse-wheels	R. Browne and J. Murphy	New York	Nov. 18, 1818	
Propelling boats, &c., by horses	W. P. Sprague		June 19, 1795	
Propelling boats by horses, Machine for	M. Crafts	Troy, N. Y	Feb. 14, 1809	
Propelling boats by horses, mules, &c	R. Bowne	New York	July 18, 1816	
Propelling boats by levers and weights	J. T. Schenck	New York	Oct. 23, 1813	
Propelling boats, &c., by man-power	A. Renoir	New Orleans, La	July 20, 1831	
Propelling boats by means of jets of water	T. W. Reiley	McMinn County, Tenn	Jan. 24, 1842	2, 430
Propelling boats by oars	T. G. Pringle	New York, N. Y	June 20, 1871	116, 095
Propelling boats, &c., by oars, &c., Machine for	J. A. Pearce	Louisville, Ohio	Apr. 7, 1807	
Propelling boats by power of weight	M. Martin	Woodbridge, N. J	July 8, 1816	
Propelling boats by screws, &c., in still water	J. L. Smith	Charleston, S. C	Sept. 18, 1835	
Propelling boats by steam	J. Evelett	Boston, Mass	Apr. 23, 1812	
Propelling boats, &c., by steam, &c	J. Fitch		Aug. 26, 1791	
Propelling boats by steam	J. Rodgers	Albany, N. Y	Sept. 27, 1811	
Propelling boats, &c., by steam	N. J. Roosevelt	New Jersey	Dec. 1, 1814	
Propelling boats, &c., by steam	W. Fatham	Washington, D. C	Apr. 18, 1816	
Propelling boats by steam, Apparatus for	R. Letton	Albany, N. Y	Aug. 29, 1810	
Propelling boats by steam, Compound wheel for	P. Antrim and S. Bolton	Philadelphia, Pa	Nov. 1, 1810	
Propelling boats, &c., by steam-engines	J. Starr		Apr. 28, 1797	
Propelling boats by step-wheels	F. Waterman, 3d	Washington, D. C	May 9, 1823	
Propelling boats by wheels and by animal-power	A. Olney	Canton, Mass	Dec. 11, 1817	
Propelling boats, Chain-paddle for	A. Pierson	Cumberland County, N. J	Mar. 2, 1811	
Propelling boats, Elastic floor for	O. Phelps	Lansing, N. Y	Nov. 17, 1821	
Propelling boats, Elbow-wheel for	R. Bulkley	New York	Apr. 24, 1824	
Propelling boats for inland navigation	W. Bell		Nov. 25, 1803	
Propelling boats in shallow water, Method of	J. Dougherty	Mount Union, Pa	Aug. 20, 1850	7, 574
Propelling boats, &c., Lever-purchase machine for	S. Kimball	Rhinebeck, N. Y	Apr. 24, 1811	
Propelling boats, Machine for	T. Kearsing	New York	June 6, 1818	
Propelling boats, Machine for	J. Martin		Nov. 21, 1797	
Propelling boats, Machine for	W. Wadsworth	Hartford, Conn	Mar. 15, 1808	
Propelling boats, Machinery for	J. Chittenden and A. W. Delano	Stevenson, N. Y	May 1, 1823	
Propelling boats, Machinery for	C. Craft	Woodbury, Conn	Mar. 29, 1826	
Propelling boats, Machinery for	J. J. Giraud	Baltimore, Md	Jan. 3, 1821	
Propelling boats, Machinery for	E. Savage	Albany, N. Y	May 21, 1872	126, 906
Propelling boats, Machinery for	H. R. Teller	Schenectady, N. Y	Sept. 28, 1825	
Propelling boats, Mechanism for	P. Rippingham	Virginia City, Mo	July 16, 1872	129, 366
Propelling boats on canals	W. F. Goodwin	Metuchen, N. J	Jan. 11, 1870	98, 760
Propelling boats on canals and in shoal water	L. Burnell	Elyria, N. Y	May 21, 1834	
Propelling boats on canals by steam	M. Battel	Albany, N. Y	Oct. 14, 1840	1, 820
Propelling boats on canals or rivers	J. Finlay	Baltimore, Md	Apr. 17, 1837	165
Propelling boats or vessels	W. J. Lewis	Virginia	Feb. 3, 1819	
Propelling boats or vessels	J. Rumsay		Aug. 26, 1791	
Propelling boats or vessels by horses	S. Fuller	Clark County, Ind	Feb. 19, 1811	
Propelling boats or vessels by steam-engines	R. Fulton	New York	Feb. 9, 1811	
Propelling boats, &c., Polygonal wheel and chain-paddle for	T. Elmer and A. Pierson	Bridgetown, N. J	Jan. 31, 1822	
Propelling boats, Revolving lever for	L. Merchand	Blakely, Ala	July 10, 1822	
Propelling boats, Screw for	E. P. Fitzpatrick	Mount Morris, N. Y	Nov. 23, 1835	
Propelling boats, Screw-power applied to	H. William	Wheatley, N. Y	Dec. 30, 1818	
Propelling boats, Screw-wheel for	J. Widdifield	Philadelphia, Pa	Oct. 11, 1815	
Propelling boats up rapids, Machinery for	L. Thwing	Dedham, Mass	Aug. 12, 1820	
Propelling boats, Vertical bucket in	S. Costill	Philadelphia, Pa	Oct. 17, 1827	
Propelling boats, vessels, &c	D. Fitzgerald	New York	Oct. 9, 1841	2, 292
Propelling boats, Water-wheel for	P. Palmiter, jr	Jamestown, N. Y	Apr. 14, 1831	
Propelling boats, Water-wheel for	B. Raymond, jr	Beverly, Mass	Nov. 13, 1826	
Propelling boats, Water-wheel for	A. Waters	Millbury, Mass	Apr. 30, 1831	
Propelling boats, Wheels of elastic oars for	L. M. Latour		Feb. 16, 1813	
Propelling boats, working mills, &c	W. Shaw	Otsego County, N. Y	June 5, 1812	
Propelling canal and other boats	H. R. Worthington	New York, N. Y	Feb. 2, 1844	3, 424
Propelling canal and other vessels	J. Jochum	Brooklyn, N. Y	Aug. 22, 1871	118, 244
Propelling canal-boats	A. Ames	Natchez, Miss	Jan. 7, 1873	134, 504
Propelling canal-boats	H. H. Baker	New Market, N. J	June 17, 1873	139, 996
Propelling canal-boats	G. A. Beidler	Philadelphia, Pa	Jan. 2, 1872	122, 429

Index of patents issued from the United States Patent Office from 1790 *to* 1873, *inclusive*—Continued.

Invention.	Inventor.	Residence.	Date.	No.
Propelling canal-boats	N. Bingham and H. Warner	Rochester, N. Y	Dec. 15, 1828	......
Propelling canal-boats	H. Camp	Dunkirk, N. Y	Feb. 16, 1858	19, 346
Propelling canal-boats	J. Echols	Columbus, Ga	Nov. 26, 1845	4, 293
Propelling canal-boats	W. F. Goodwin	Metuchen, N. J	Mar. 22, 1870	101, 004
Propelling canal-boats	J. Hough	Buckingham, Pa	Oct. 15, 1872	132, 288
Propelling canal-boats	C. H. Jenner	Rockport, N. Y	May 28, 1872	127, 401
Propelling canal-boats	A. Johnson	Brooklyn, N. Y	Feb. 18, 1873	136, 064
Propelling canal-boats	W. Leavenworth	New York, N. Y	Apr. 18, 1839	1, 126
Propelling canal-boats	H. B. Meech	Fort Edwards, N. Y	Mar. 7, 1871	112, 369
Propelling canal-boats	W. F. Miller	East Walpole, Mass	June 10, 1873	139, 727
Propelling canal-boats	W. P. More	Moresville, N. Y	Apr. 2, 1872	125, 207
Propelling canal-boats	J. Reid	Catskill, N. Y	Nov. 15, 1870	109, 248
Propelling canal-boats	P. Rippingham	Virginia City, Nev	Dec. 10, 1872	133, 891
Propelling canal-boats	J. Smith and J. Brown	New York	Nov. 19, 1833	......
Propelling canal-boats	O. C. Smith	Ipswich, Mass	Sept. 23, 1873	143, 038
Propelling canal-boats	G. Stackpole	New York, N. Y	Mar. 26, 1867	63, 184
Propelling canal-boats	J. A. Sterling	New York, N. Y	Nov. 26, 1872	133, 495
Propelling canal-boats	J. T. Teasdill	Lockport, N. Y	July 16, 1872	129, 192
Propelling canal-boats	W. W. Virdin	Baltimore, Md	Apr. 18, 1871	113, 951
Propelling canal-boats	E. K. Watson	Shokan, N. Y	Sept. 20, 1870	107, 573
Propelling canal boats and other vessels	L. Bastel	New York, N. Y	Nov. 4, 1873	144, 248
Propelling canal-boats by lock-paddle wheels	H. L. B. Lewis	Buffalo, N. Y	Nov. 2, 1830	......
Propelling canal-boats by steam	F. A. Dorman	Norfolk, Va	Mar. 7, 1833	......
Propelling canal-boats, Device for	J. B. Calnan	New Haven, Conn	Aug. 23, 1870	106, 547
Propelling canal-boats, Device for	T. K. McDowell	San Francisco, Cal	Apr. 2, 1872	125, 205
Propelling canal-boats, Machinery for	D. W. Horton	Petersburgh, Ind	July 16, 1872	129, 228
Propelling canal-boats, Mode of applying wheels for	T. Jackson	Reading, Pa	Jan. 18, 1839	1, 070
Propelling canal-boats, Water-wheel for	H. Averill	Richland, N. Y	Oct. 1, 1830	......
Propelling engine, Self	N. S. Bean	Manchester, N. H	Mar. 10, 1868	75, 348
Propelling-hoop	J. Faye	Philadelphia, Pa	July 16, 1867	66, 697
Propelling horse-boats	B. Langdon	Troy, N. Y	May 1, 1821	......
Propelling-mechanism	C. Sharps	Philadelphia, Pa	Nov. 22, 1870	109, 458
Propelling pleasure-boats	J. O. Belknap	Mobile, Ala	May 19, 1866	78, 043
Propelling-power	N. B. Baldwin	Chicago, Ill	Nov. 14, 1871	120, 846
Propelling ships	J. Ericsson	New York	Dec. 31, 1844	3, 869
Propelling ships, &c	D. Rudd	Bozrah, Conn	Mar. 23, 1842	2, 509
Propelling ships, Apparatus for	A. E. Harding	Middletown, Ohio	Apr. 5, 1859	23, 463
Propelling ships, boats, &c	E. F. Aldrich	New York	July 30, 1841	2, 204
Propelling ships, &c., Machinery for	A. Hermange	Baltimore, Md	May 31, 1828	......
Propelling ships, Machinery for	A. Hermange	Baltimore, Md	Nov. 26, 1828	......
Propelling ships, Machinery for	A. Hermange	Baltimore, Md	Nov. 26, 1828	......
Propelling small boats	V. Dresser	Leavenworth, Kans	Nov. 2, 1869	96, 404
Propelling steam and other boats	W. Scarborough	Savannah, Ga	Apr. 8, 1835	......
Propelling steam and other vessels	G. H. Moreau	France	Jan. 26, 1844	3, 414
Propelling steam and other vessels	F. P. Smith	London, England	Nov. 12, 1841	2, 353
Propelling steam and other vessels, Operating-paddle for.	S. Swett, jr	Chelsea, Mass	May 11, 1841	2, 089
Propelling steamboats, &c	T. Cook	New York	May 12, 1842	2, 623
Propelling steamboats	J. Hunt	East Chester, N. Y	Mar. 11, 1812	......
Propelling steamboats, &c	J. McCurdy	Norwich, Conn	Sept. 18, 1833	......
Propelling steamboats, &c	J. Sheffield and J. Ingraham	Erie County, N. Y	May 17, 1832	......
Propelling steamboats and other vessels	A. Planton	Philadelphia, Pa	Apr. 23, 1834	......
Propelling steam-vessels	J. Ericsson	Sweden	Feb. 1, 1838	588
Propelling vessels, Constructing and	W. W. Hunter and B. Harris	United States Navy and Norfolk, Va.	Mar. 12, 1841	2, 004
Propelling vessels	E. F. Aldrich	New York	June 13, 1848	5, 634
Propelling vessels	L. Alexander	New York, N. Y	Mar. 25, 1856	14, 487
Propelling vessels	J. L. Arman	Bordeaux, France	Oct. 11, 1870	108, 230
Propelling vessels	B. T. Babbitt	New York, N. Y	July 14, 1868	79, 937
Propelling vessels	C. W. Cahoon	Portland, Me	Dec. 18, 1866	60, 471
Propelling vessels	H. C. Camp	Saint Paul, Minn	Jan. 14, 1873	134, 730
Propelling vessels	H. Case	Norwalk, Ohio	Aug. 13, 1828	......
Propelling vessels	F. P. Dimpfel	Philadelphia, Pa	Oct. 25, 1853	10, 149
Propelling vessels	G. W. Dow	Brooklyn, N. Y	July 22, 1873	141, 039
Propelling vessels	J. Echols	Columbus, Ga	May 9, 1846	4, 503
Propelling vessels	J. W. Evans	New York, N. Y	July 29, 1873	141, 340
Propelling vessels	E. Fuller	Providence, R. I	Mar. 2, 1827	......
Propelling vessels	G. W. Fulton	Brazoria, Tex	Aug. 26, 1846	4, 716
Propelling vessels	S. Gerhard	Camden, N. J	Apr. 1, 1873	137, 359
Propelling vessels	J. Ghisi	Genoa, Italy	May 17, 1870	103, 040
Propelling vessels	J. J. Giraud	Baltimore, Md	Dec. 1, 1829	......
Propelling vessels	G. A. Milani	Frankfort, Ind	Oct. 19, 1869	95, 922
Propelling vessels	T. L. Mitchell	Birkenhead, England	July 5, 1853	9, 831
Propelling vessels	J. O'Neil	New York, N. Y	Nov. 11, 1873	144, 474
Propelling vessels	F. Peltier	New York	Oct. 1, 1830	......
Propelling vessels	C. A. Read	Bridgeport, Conn	July 2, 1872	128, 657
Propelling vessels	W. Shepard	Saint Louis, Mo	Aug. 13, 1872	130, 391
Propelling vessels	S. D. Sillman	Jersey City, N. J	June 21, 1870	104, 512
Propelling vessels	R. R. Spedden and D. F. Stafford	Astoria, Oreg	Dec. 10, 1867	72, 106
Propelling vessels	R. R. Stevens	Mokelumne Hill, Cal	Oct. 22, 1867	70, 130
Propelling vessels	O. Tenny	Littleton, Mass	July 21, 1868	80, 239
Propelling vessels	G. Titcomb	Chelsea, Mass	Aug. 13, 1872	130, 455
Propelling vessels	M. K. Wildman	Brooklyn, N. Y	Dec. 26, 1871	122, 301
Propelling vessels	B. Wynkoop		Sept. 13, 1794	......
Propelling vessels, Apparatus for	M. A. Crooker	New York, N. Y	June 29, 1852	9, 066
Propelling vessels, Apparatus for	A. D. Owen and J. D. Sherman	Thorntown, Ind., and Paw Paw, Mich.	Nov. 16, 1869	96, 931
Propelling vessels, Apparatus for	B. Phillips	Philadelphia, Pa	Aug. 16, 1828	......
Propelling vessels, Apparatus for	D. Winer	Lockport, N. Y	Aug. 7, 1866	57, 033
Propelling vessels at sea, Machine for	E. Bryan	New York	Dec. 22, 1827	......
Propelling vessels by screw or spiral levers	J. J. Giraud	Baltimore, Md	Apr. 29, 1831	......
Propelling vessels, &c., by animal-power	J. P. Duraud	New York	Oct. 3, 1817	......
Propelling vessels by means of continuous streams of water.	R. Schuyler	New York	Mar. 17, 1843	3, 003
Propelling vessels, &c., by pendulum-power	J. Cerneau	New York	Aug. 28, 1817	......
Propelling vessels by reaction	M. W. Ruthven	New York, N. Y	May 22, 1849	6, 468
Propelling vessels by the buoyancy of air in water acting against inclined surfaces.	S. W. Hall	Troy, N. Y	Nov. 6, 1843	3, 332

Index of patents issued from the United States Patent Office from 1790 *to* 1873, *inclusive*—Continued.

Invention.	Inventor.	Residence.	Date.	No.
Propelling vessels by the direct action of steam on the water.	W. J. McIntire	New York, N. Y	Aug. 7, 1855	13,394
Propelling vessels, Cylinder for	B. Phillips	Philadelphia, Pa	Jan. 23, 1830	....
Propelling vessels, Device for	B. T. Babbitt	New York, N. Y	June 17, 1873	139,994
Propelling vessels, Device for	S. P. Harbaugh	Alleghany County, Md	July 30, 1872	130,042
Propelling vessels, Device for	G. R. Moore	Philadelphia, Pa	July 16, 1872	129,488
Propelling vessels, Endless-chain paddle for	B. Douglas	Albany, N. Y	Sept. 28, 1843	3,285
Propelling vessels in a calm	S. Rudd	New York	June 14, 1816	....
Propelling vessels in shoal water	J. W. Wetmore	Erie, Pa	June 16, 1857	17,598
Propelling vessels, &c., Machine for	A. Babcock	....	Dec. 2, 1793	....
Propelling vessels, Machine for	J. J. Giraud	Baltimore, Md	Nov. 22, 1820	....
Propelling vessels, Machine for	C. Stondinger	....	June 2, 1798	....
Propelling vessels, Machinery for	S. J. Gold	Cornwall, Conn	July 16, 1867	66,830
Propelling vessels, Mechanism for	L. Murdoch	Marble Hill, Mo	Sept. 23, 1873	143,091
Propelling vessels, Mechanism for	E. Raynale	Birmingham, Mich	Aug. 15, 1871	118,151
Propelling vessels, Mechanism for	J. Weich	New York, N. Y	Mar. 4, 1873	136,475
Propelling vessels on canals	B. O. Mesnil and M. Eyth	Brussels, Belgium, and Stutgart, Wurtemberg.	Feb. 9, 1869	86,737
Propelling vessels, Sluice for	M. W. Ruthven	Middlesex, England	Apr. 10, 1866	53,937
Propelling-wheel	C. Comstock and C. Glidden	Milwaukee, Wis	Apr. 30, 1861	32,182
Propelling-wheel	J. L. Husband	Philadelphia, Pa	Nov. 22, 1859	26,187
Propelling-wheel	A. C. Loud	San Francisco, Cal	May 3, 1870	102,747
Propelling-wheel	C. Seymour	La Porte, Ind	Dec. 22, 1868	85,248
Propelling-wheel	A. A. Wilder	Warsaw, N. Y	Mar. 8, 1836	....
Propelling-wheel for boats	W. Kelly	Pittsburgh, Pa	July 12, 1834	....
Propelling-wheel for boats	W. Kelly	Pittsburgh, Pa	Nov. 17, 1834	....
Propelling-wheel for boats, &c	J. Prince	New York, N. Y	Oct. 2, 1820	....
Propelling-wheel for river and canal boats	H. Fenton	Cleveland, Ohio	July 4, 1865	48,541
Propelling-wheel for steamboats	J. J. Giraud	Baltimore, Md	Sept. 15, 1826	....
Propelling-wheel for steamships, also employed as wind or water wheel for mills.	J. Hobday and W. I. Cocke	Portsmouth, Va	Mar. 29, 1841	2,023
Propelling-wheel, Screw	B. M. Smith	Rochester, N. Y	Nov. 20, 1829	....
Propelling-wheel, Submerged	T. Kendall, sr	San Francisco, Cal	Apr. 28, 1857	17,156
Propulsion, Apparatus for marine	F. W. Harris	Montreal, Canada	Aug. 4, 1863	39,394
Propulsion by gas-explosion	W. A. Leggo	Montreal, Canada	May 6, 1373	138,665
Propulsion, Canal	J. L. Nicolai	Chicago Ill	Feb. 6, 1872	123,504
Propulsion, Device for canal	G. Stackpole	New York, N. Y	Mar. 6, 1866	53,088
Propulsion for tug-boats, Screw	E. D. Gird	Syracuse, N. Y	Apr. 19, 1870	102,111
Propulsion, Marine	L. B. Flanders	Cleveland, Ohio	Sept. 18, 1860	30,057
Propulsion, Marine	B. S. Heath	Philadelphia, Pa	Aug. 15, 1871	118,127
Propulsion, Marine	A. J. Reynolds	Sturgis, Michigan	Mar. 17, 1868	75,648
Propulsion, Means of	S. R. Foster	Saint John, Canada	Oct. 29, 1872	132,568
Propulsion of boats	L. H. Watson	Pittsburgh, Pa	Apr. 30, 1872	126,168
Propulsion of canal-boats	H. H. Baker	New Market, N. J	Oct. 1, 1872	131,839
Propulsion of canal-boats	W. C. Bibb	Madison, Ga	Aug. 22, 1871	118,329
Propulsion of canal-boats	G. W. Bishop	Brooklyn, N. Y	Sept. 24, 1872	131,652
Propulsion of canal-boats	J. T. Burke	Virginia, Ill	Dec. 12, 1871	121,749
Propulsion of canal-boats	O. Coogan	Pittsfield, Mass	Oct. 10, 1871	119,744
Propulsion of canal boats	M. N. Cumminskey	Paterson, N. J	Aug. 8, 1871	117,869
Propulsion of canal-boats	W. T. Duvall	Georgetown, D. C	June 27, 1871	116,281
Propulsion of canal-boats	N. T. Edson	New Orleans, La	Oct. 31, 1871	120,371
Propulsion of canal-boats	P. H. Fontaine	Reidsville, N. C	Oct. 17, 1871	119,923
Propulsion of canal-boats	H. Fowler	Washington, D. C	Dec. 12, 1871	121,712
Propulsion of canal-boats	H. Fowler	Washington, D. C	Dec. 12, 1871	121,713
Propulsion of canal-boats	H. Fowler	Washington, D. C	Dec. 12, 1871	121,714
Propulsion of canal-boats	I. J. Hilgerd	New York, N. Y	May 14, 1872	126,701
Propulsion of canal-boats	J. L. Hornig	Chicago, Ill	Dec. 27, 1870	110,466
Propulsion of canal-boats	J. L. Hornig	Chicago, Ill	May 9, 1871	114,561
Propulsion of canal-boats	F. M. Mahan	Memphis, Tenn	July 25, 1871	117,437
Propulsion of canal-boats	J. M. McMaster	Rochester, N. Y	Nov. 14, 1871	120,823
Propulsion of canal-boats	T. Main	Green Point, N. Y	Jan. 31, 1871	111,462
Propulsion of canal-boats	M. P. Muller, A. Schwaab, and F. Krenson.	Savannah, Ga	June 11, 1872	127,911
Propulsion of canal-boats	W. E. Prall and J. D. Defrees	Washington, D. C	Dec. 5, 1871	121,469
Propulsion of canal-boats	C. Schilling	New York, N. Y	Oct. 10, 1871	119,792
Propulsion of canal-boats	T. Smith	Green Island, N. Y	June 4, 1872	127,437
Propulsion of canal-boats	H. W. Trackman	Philadelphia, Pa	Mar. 5, 1872	124,348
Propulsion of navigable vessels	C. C. Brand and C. C. Moore	Norwich, Conn	July 9, 1872	128,846
Propulsion of steamboats	A. Fellows	Magnoketa, Iowa	Mar. 28, 1865	47,003
Propulsion of vessels	G. A. Beidler	Philadelphia, Pa	May 9, 1871	114,637
Propulsion of vessels	R. Bayman	London, England	May 30, 1871	115,425
Propulsion of vessels	J. P. Bruce	Brooklyn, N. Y	Oct. 31, 1871	120,414
Propulsion of vessels	P. Burke	Boston, Mass	May 16, 1871	114,914
Propulsion of vessels	S. Chase	Portland, Me	Dec. 10, 1872	133,758
Propulsion of vessels	K. P. Cool	Glen's Falls, N. Y	July 18, 1871	117,051
Propulsion of vessels	G. B. De Boucherville	Quebec, Canada	Oct. 7, 1873	143,440
Propulsion of vessels	S. F. Gard	New Orleans, La	Nov. 7, 1871	120,735
Propulsion of vessels	J. S. Godfrey	Leslie, Mich	Oct. 24, 1871	120,264
Propulsion of vessels	W. M. Harding	Chelsea, Mass	Aug. 22, 1871	118,363
Propulsion of vessels	E. Harriott	Plainfield, N. J	Aug. 29, 1871	118,607
Propulsion of vessels	R. W. Heywood	Baltimore, Md	May 23, 1871	115,204
Propulsion of vessels	H. Jackson	Brooklyn, N. Y	Sept. 26, 1871	119,363
Propulsion of vessels	C. P. Macowitzky	Corpus Christi, Tex	Dec. 30, 1873	146,009
Propulsion of vessels	E. Matteson	Norwich, Conn	Jan. 9, 1872	122,624
Propulsion of vessels	C. Mickle	Guelph, Canada	Dec. 19, 1871	122,046
Propulsion of vessels	H. P. Miller	Paducah, Ky	Oct. 1, 1872	131,770
Propulsion of vessels	H. Niles	Chicago, Ill	Jan. 9, 1872	122,640
Propulsion of vessels	H. H. Olds	New Haven, Conn	Dec. 12, 1854	12,067
Propulsion of vessels	T. B. Raymond	Winona, Mich	Jan. 9, 1872	122,650
Propulsion of vessels	W. Richards	Rochester, N. Y	Sept. 3, 1872	131,118
Propulsion of vessels	J. S. Stites	Baltimore, Md	June 20, 1871	116,236
Propulsion of vessels	H. Waterman	Brooklyn, N. Y	Sept. 26, 1871	119 255
Propulsion of vessels	W. Wells	Salem, Mass	Nov. 19, 1872	133,275
Propulsion of vessels	E. Whitehead	Cincinnati, Ohio	Dec. 12, 1871	121,738
Propulsion of vessels	A. F. Yardell	San Francisco, Cal	Dec. 1, 1868	84,602
Propulsion of vessels, Apparatus for	J. Briggs	Louisville, Ky	Oct. 27, 1863	40,393
Propulsion of vessels, Apparatus for water	J. Watt	Buffalo, N. Y	May 12, 1863	38,522
Propulsion of vessels by direct action of steam	G. W. Jones	Philadelphia, Pa	Dec. 2, 1873	145,210

Index of patents issued from the United States Patent Office from 1790 *to* 1873, *inclusive*—Continued.

Invention.	Inventor.	Residence.	Date.	No.
Propulsion of vessels in shoal water, Wheel	O. Olds	Aurora, N. Y	July 11, 1865	48, 715
Propulsion of vessels, Speed in	J. F. Reigart	Washington, D. C	June 30, 1863	39, 067
Propulsion, Screw	E. C. Gregg	Trumansburgh, N. Y	Mar. 7, 1871	112, 335
Propulsion, Wave-power	C. W. Cahoon	Portland, Me	Apr. 24, 1866	54, 107
Propulsion, Wheel for marine	B. Reed	Allegheny City, Pa	Aug. 4, 1863	39, 423
Protector:				
See Bee-hive protector. Body-protector. Book-cover protector. Bottle-protector. Breast-protector. Breast-strap protector. Broom-protector. Car-window protector. Carriage-protector. Carriage-top protector. Check-protector. Chest-protector. Collar-edge protector. Corner-protector. Diamond-protector. Dress-protector. Ear-protector. Eave-protector. Eye-protector. Eye and lung protector. Fly-protector. Frost-protector. Furniture-protector. Grape-vine protector. Gun-nipple protector. Hat-protector. Hatchway-protector. Hay and grain protector. Hay-cock protector. Heel-protector. Hoisting-protector. Horse-head protector. Horse-neck protector. Horse-tail protector. Hose-protector. Joist-protector. Key-hole protector. King-bolt protector. Lamp-chimney protector. Lamp-shade protector. Omnibus-step protector. Pencil-point protector. Plant-protector. Rope-protector. Safe-protector. Sap-protector. Shaft-protector. Skirt-protector. Spool-protector. Stocking-heel protector. Sun-stroke protector. Tree-protector. Wall-protector.				
Protracting-table	W. T. Steiger	Washington, D. C	Sept. 7, 1838	907
Protractor	C. Gordon	Washington, D. C	Mar. 29, 1859	23, 365
Protractor	J. Leyman	Lenox, Mass	May 25, 1858	20, 356
Protractor and parallel ruler	W. L. Apthorp	Tallahassee, Fla	May 28, 1872	127, 210
Protractor, Compound	W. Rutherford	Athens, Ga	Oct. 15, 1867	69, 844
Protractor, Facilitating	E. Brown	Seneca County, Ohio	Aug. 1, 1833	……
Protractor, Rectangular	N. Goodwin	Hartford, Conn	Dec. 24, 1823	……
Protractor, Universal	E. Williams	Philadelphia, Pa	Jan. 19, 1810	……
Provision-cooler	J. W. Bartlett	New York, N. Y	Apr. 14, 1863	38, 142
Provision-cutter	W. Smith	Cincinnati, Ohio	June 8, 1858	20, 517
Provision-jar	W. Zettle	Cincinnati, Ohio	Aug. 5, 1862	36, 131
Provision-press	F. Potter	Rushville, Ill	June 4, 1872	127, 427
Provisions, Preparing	D. Lardner and J. Davison	New York, N. Y	Aug. 9, 1845	4, 144
Prow	J. Revercomb	Botetourt Springs, Va	June 27, 1854	11, 201
Prunes, Curing	I. Reckhow	Brooklyn, N. Y	May 22, 1860	28, 441
Pruning and hedge shears	E. K. Bigelow	Litchfield, Mich	May 10, 1870	102, 752
Pruning and hedge shears	L. Campbell	Marengo, Mich	Sept. 22, 1868	82, 290
Pruning-hatchet	J. White	Providence, Pa	Jan. 17, 1871	111, 024
Pruning-hook	J. Christy	Clyde, Ohio	Oct. 28, 1873	143, 962
Pruning-hook	W. M. Doty	Woodbridge, N. J	Jan. 4, 1870	98, 575
Pruning-hook	A. Downer	Hammondsville, Ohio	May 3, 1870	102, 513
Pruning-hook	M. Gates	Paw Paw, Mich	June 12, 1866	55, 483
Pruning-hook	L. M. Harris	Mattawan, Mich	Mar. 14, 1865	46, 797
Pruning hook	W. S. McLean	Williamsburgh, Ohio	Jan. 2, 1867	51, 899
Pruning-hook	B. M. Parks	Saint Louis, Mo	Oct. 6, 1868	82, 868
Pruning-hook	D. B. Seeley	Sterling, Ill	July 19, 1870	105, 601
Pruning-hook	J. Stark	Thomasville, Ga	July 7, 1868	79, 700
Pruning-hook	A. Travis	Peekskill, N. Y	Dec. 8, 1863	40, 869
Pruning-hook, Extension	J. Stark	Thomasville, Ga	Mar. 9, 1869	87, 719
Pruning-implement	A. P. Bettersworth	Carlinville, Ill	Oct. 14, 1873	143, 659
Pruning-implement	J. Fasig	Congress, Ohio	Mar. 13, 1860	27, 437
Pruning-implement	A. C. Hulse and J. S. Crum	Palmyra, Ill	Sept. 16, 1873	142, 915
Pruning-implement	J. Reynolds	Washington, D. C	Sept. 24, 1872	131, 703
Pruning-implement	G. F. Waters	Boston, Mass	May 25, 1869	90, 618
Pruning-implement	J. R. Woodworth	Nunda, N. Y	Oct. 26, 1869	96, 294
Pruning-implements, Combined	C. W. Dawson	Paynesville, Mo	Nov. 22, 1870	109, 498
Pruning-instrument	R. Collier	Springfield, Ohio	June 19, 1866	55, 623
Pruning-instrument	W. H. Callings	Town, Mo	Sept. 23, 1873	143, 064
Pruning-instrument	J. Evans and R. H. Seymour	Newark and Bloomfield, N. J.	Sept. 4, 1866	57, 817
Pruning-instrument	J. Schroy	Fortville, Ind	Jan. 12, 1869	85, 861
Pruning-instrument	G. F. Waters	Waterville, Me	Sept. 25, 1866	58, 326
Pruning-knife	H. Alter	Lakeport, Cal	Feb. 1, 1870	99, 390
Pruning-knife	G. G. Belcher	Worcester, Mass	May 10, 1859	23, 975
Pruning-knife	J. S. Crum	Palmyra, Ill	May 20, 1873	139, 119
Pruning-knife	J. Fasig	West Salem, Ohio	May 28, 1872	127, 229
Pruning-knife	F. P. Goodall	Deering, N. H	Oct. 18, 1859	25, 818
Pruning-knife	M. Grover	Clyde, Ohio	June 7, 1870	103, 873
Pruning-knife	A. C. Hulse	Palmyra, Ill	Aug. 26, 1873	142, 236
Pruning-knife	C. McCrea	Leavenworth, Kans	Oct. 9, 1866	58, 659
Pruning-knife	D. Morris	Bartlett, Ohio	Jan. 3, 1872	122, 633
Pruning-knife	L. B. Morris	Hopkinsville, Ky	Dec. 4, 1866	60, 225
Pruning-knife	W. Richard	Clyde, Ohio	June 29, 1869	92, 100
Pruning-knife	W. Richard	Clyde, Ohio	Sept. 20, 1870	107, 542
Pruning-knife	J. Surerus	Newark, N. J	Jan. 30, 1866	52, 337
Pruning-knife	S. O. and P. W. T. Vaughn	De Kalb, Ill	Dec. 12, 1865	51, 497

Index of patents issued from the United States Patent Office from 1790 *to* 1873, *inclusive*—Continued.

Invention.	Inventor.	Residence.	Date.	No.
Pruning knife, hook, and saw	R. W. Porter	Nashua, N. H	Aug. 27, 1867	68, 109
Pruning-saw	P. Eberle	Freetown, Ind	June 13, 1871	115, 834
Pruning-saw	A. Travis	Peekskill, N. Y	July 16, 1872	129, 071
Pruning-shears	G. W. Anesley	Waungo Township, Mich	Aug. 1, 1871	117, 590
Pruning-shears	W. E. Ball	Bethesda, Ohio	Sept. 23, 1873	143, 114
Pruning-shears	W. Baker	Ilion, N. Y	Sept. 16, 1873	142, 886
Pruning-shears	A. Barling	West Chester, Pa	May 30, 1871	115, 415
Pruning-shears	A. Baumann	Philadelphia, Pa	May 29, 1866	55, 043
Pruning-shears	S. J. Beigh and E. F. Beard	Republic, Ohio	Jan. 16, 1872	122, 753
Pruning-shears	S. J. Beigh and E. F. Beard	Republic, Ohio	Oct. 1, 1872	131, 808
Pruning-shears	T. S. Bell	Wapello, Iowa	June 16, 1868	78, 918
Pruning-shears	G. Bergner	Washington, Mo	Apr. 5, 1870	101, 571
Pruning-shears	G. Bergner	Washington, Mo	Aug. 23, 1870	106, 538
Pruning-shears	G. Bergner	Washington, Mo	Nov. 28, 1871	121, 319
Pruning-shears	F. Binder	Baltimore, Md	Oct. 30, 1866	59, 168
Pruning-shears	H. W. Black	Cecilton, Md	July 26, 1870	105, 629
Pruning-shears	T. Borden	Squancum, N. J	Feb. 25, 1873	136, 209
Pruning-shears	G. Brewster	Otsego, Mich	Aug. 6, 1872	130, 274
Pruning-shears	P. Broadbooks	Batavia, N. Y	June 11, 1872	127, 735
Pruning-shears	P. Broadbooks	Batavia, N. Y	Apr. 15, 1873	137, 821
Pruning-shears	R. Bullard	Litchfield, Mich	May 7, 1872	126, 443
Pruning-shears	R. Bullard and G. E. Mills	Litchfield, Mich	Aug. 9, 1870	106, 253
Pruning-shears	J. Calder	Macedon, N. Y	Jan. 24, 1871	111, 106
Pruning-shears	D. Campbell	Elizabeth, N. J	Feb. 19, 1867	62, 180
Pruning-shears	D. Campbell	Elizabeth, N. J	Aug. 4, 1868	80, 595
Pruning-shears	T. Campbell	East Palestine, Ohio	Jan. 30, 1872	123, 234
Pruning-shears	S. P. Carpenter	Milford, Mass	Dec. 10, 1867	71, 975
Pruning-shears	A. B. Chapman	Clyde, Ohio	July 12, 1870	105, 173
Pruning-shears	O. Chase	Rutland, Ohio	Jan. 7, 1873	134, 591
Pruning-shears	W. F. and G. H. Clemmer	Alexandria, Ohio	Sept. 5, 1871	118, 690
Pruning-shears	G. H. Clinton and D. H. Harris	New Haven, Conn	Feb. 5, 1867	61, 806
Pruning-shears	J. Christy	Clyde, Ohio	June 6, 1871	115, 571
Pruning-shears	J. F. Creighton	Placerville, Cal	Oct. 3, 1871	119, 452
Pruning-shears	M. De Groodt	Eaton, N. Y	Mar. 18, 1873	136, 820
Pruning-shears	A. J. Doolittle	Hamden, Conn	Mar. 12, 1867	62, 737
Pruning-shears	J. Evans	San José, Cal	July 16, 1861	32, 827
Pruning-shears	J. Evans	Newark, N. J	July 18, 1865	48, 803
Pruning-shears	C. H. Eggleston	Marshall, Mich	Aug. 1, 1871	117, 525
Pruning-shears	E. D. Gaines and J. A. McMillen	Chicago, Ill	Feb. 14, 1871	111, 738
Pruning-shears	P. M. Gilbert	Camanche, Iowa	Nov. 19, 1872	133, 220
Pruning-shears	I. Grass	Sandusky, Ohio	Sept. 26, 1871	119. 221
Pruning-shears	M. Grover	Clyde, Ohio	Sept. 3, 1872	130, 990
Pruning-shears	R. Hall	Clyde, Ohio	Jan. 9, 1872	122, 603
Pruning-shears	A. L. Hatfield	Clyde, Ohio	May 24, 1870	103, 327
Pruning-shears	V. Hayes and L. Cuatt	Tekonsha and Eckford, Mich.	Oct. 26, 1869	96, 232
Pruning-shears	J. M. Heath	Allegan, Mich	Aug. 13, 1872	130, 495
Pruning-shears	H. C. Hills	Bickney, Mich	July 29, 1873	141, 273
Pruning-shears	W. E. Hughes	Aylmer, Canada	June 6, 1871	115, 735
Pruning-shears	H. H. Ingalsbe and G. A. Prescott.	South Hartford and Sandy Hill, N. Y.	Aug. 8, 1871	117, 888
Pruning-shears	S. Jocelin	New Haven, Conn	July 13, 1809	....
Pruning-shears	S. W. Jones	Bluffton, Ind	July 16, 1867	66, 850
Pruning-shears	P. Keck	Zanesville, Ohio	May 21, 1867	64, 882
Pruning-shears	D. Keethler	Mount Orab, Ohio	May 21, 1872	126, 888
Pruning-shears	C. A. Kellogg	Elyria, Ohio	May 17, 1870	103, 197
Pruning-shears	M. C. Malone	Palmyra, Ill	Apr. 29, 1873	138, 338
Pruning-shears	E. J. Marsters	Shaw's Flat, Cal	Aug. 23, 1870	106, 705
Pruning-shears	W. McCray	Black Oak, Mo	Apr. 29, 1873	138, 340
Pruning-shears	S. A. McFarlane	Grand Rapids, Mich	July 6, 1869	92, 202
Pruning-shears	J. W. Mix	Batavia, N. Y	Sept. 19, 1871	119, 041
Pruning-shears	W. C. Moreau and H. J. Waldo	Batavia and Elba, N. Y	Nov. 5, 1872	132, 850
Pruning-shears	A. R. Moulton	Fall's Branch, Tenn	Jan. 14, 1873	134, 922
Pruning-shears	J. Neff, jr	Pultney, N. Y	May 14, 1867	64, 692
Pruning-shears	W. Richard	Clyde, Ohio	Sept. 7, 1869	94, 514
Pruning-shears	J. G. Rogers	Dowagiac, Mich	July 12, 1870	105, 255
Pruning-shears	J. G. Rogers	Berrien County, Mich	July 1, 1873	140, 388
Pruning-shears	O. L. Samson and J. R. Dill	Crawfordsville, Iowa	Oct. 1, 1872	131, 830
Pruning-shears	D. B. Seely	Portland, Ill	Mar. 2, 1869	87, 436
Pruning-shears	D. B. Seely	Sterling, Ill	June 6, 1871	115, 773
Pruning-shears	J. H. Shehan	Lima, Ind	Dec. 14, 1869	97, 973
Pruning-shears	J. Sill	Montoursville, Pa	Dec. 31, 1867	72, 923
Pruning-shears	F. Smiley	Batavia, N. Y	Mar. 7, 1871	112, 392
Pruning-shears	F. Smiley	Batavia, N. Y	Aug. 29, 1871	118, 492
Pruning-shears	O. Snell	Williamsburgh, Ohio	Mar. 12, 1872	124, 455
Pruning-shears	G. D. Spielman	Lancaster, Ohio	May 28, 1872	127, 279
Pruning-shears	G. Vanfossen	Bethesda, Ohio	July 15, 1873	140, 974
Pruning-shears	W. A. Wallet and I. Harbaugh	West Salem, Ohio	June 13, 1871	115, 995
Pruning-shears	P. R. Walsh and G. C. Eaton	Rochester, N. Y	Jan. 18, 1870	99, 038
Pruning-shears	G. F. Waters	Waterville, Me	Nov. 22, 1864	45, 196
Pruning-shears	H. Wendt	Elizabeth, N. J	Jan. 11, 1870	98, 826
Pruning-shears	F. A. Will	San Francisco, Cal	Sept. 20, 1870	107, 577
Pruning-shears	E. L. Yancey	Batavia, N. Y	June 27, 1871	116, 387
Pruning-shears	E. L. Yancey	Batavia, N. Y	Jan. 30, 1872	123, 321
Pruning-shears	E. L. Yancey	Batavia, N. Y	Dec. 24, 1872	134, 185
Pruning-shears and fruit-picker, Combined	M. Grover	Clyde, Ohio	July 30, 1872	129, 946
Pruning shears and knife	J. Spear and J. A. Hull	Carbondale, Ill	June 2, 1868	78, 615
Pruning-tool	S. Gamwell	Wayland, Mich	Sept. 6, 1870	107, 172
Pruning trees, Implement for	W. W. Harvey	Saltville, Va	Jan. 15, 1856	14, 097
Prussian blue, Manufacture of	J. Clark	Newark, N. J	June 12, 1860	28, 651
Prussian blue, Manufacture of	T. A. Helwig	Minersville, Pa	Apr. 3, 1860	27, 716
Pudding-mold	J. Musgrove	East Newark, N. J	Jan. 5, 1869	85, 605
Puddle-ball squeezer	S. Wiler	Newcastle, Pa	Apr. 15, 1856	14, 701
Puddle-balls, Machine for compressing	J. F. Winslow	Troy, N. Y	Jan. 14, 1862	34, 177
Puddler, Rotary	W. Sellers and G. H. Sellers	Philadelphia, Pa., and Wilmington, Del.	Nov. 11, 1873	144, 416
Puddler's ball-elevator	S. S. Jackman	Lock Haven, Pa	Feb. 12, 1856	14, 242
Puddler's ball-roller	R. M. Bassett	Birmingham, Conn	May 27, 1873	139, 356

Index of patents issued from the United States Patent Office from 1790 to 1873, inclusive—Continued.

Invention.	Inventor.	Residence.	Date.	No.
Puddler's ball-squeezer	F. Danks	Cincinnati, Ohio	July 29, 1873	141,207
Puddler's ball-squeezing machine	S. Danks	Cincinnati, Ohio	Nov. 15, 1870	109,186
Puddlers' balls and other masses of iron, Machine for compressing.	S. S. Jackman	Lock Haven, Pa	July 31, 1855	13,355
Puddlers' balls, Machine for compressing	J. Dangerfield	New Albany, Ind	Apr. 2, 1872	125,181
Puddlers' balls, Machinery for rolling	J. F. Winslow and J. Snider	Troy, N. Y	July 5, 1848	5,660
Puddlers' balls, Pressing out instead of hammering	W. Jones	Haverstraw, N. Y	Apr. 18, 1834	....
Puddlers' balls, Rolling and compressing	J. F. Winslow	Troy, N. Y	Dec. 18, 1847	5,399
Puddling and boiling furnace	A. and J. Reese	Pittsburgh, Pa	Apr. 23, 1872	125,987
Puddling and boiling furnace	W. Swindell	Allegheny City, Pa	May 12, 1868	77,932
Puddling and boiling furnace door and bit	J. G. Haberfield	Wheeling, W. Va	May 27, 1873	139,244
Puddling and boiling furnaces, Fettlings for	J. C. McAninch and G. Atkins	Youngstown, Ohio	Feb. 18, 1873	136,083
Puddling and boiling furnaces, Lining or "fix" for	S. Danks	Cincinnati, Ohio	Sept. 10, 1867	68,715
Puddling and heating furnace	S. Cuddick	Pembroke, Me	Oct. 29, 1867	70,322
Puddling and heating furnace	O. F. Mayhew	Indianapolis, Ind	July 17, 1866	56,429
Puddling and other furnace	J. A. Jones, R. Howson, and J. Gjers.	Middlesborough-on-Tees, England.	Apr. 25, 1871	114,151
Puddling and other furnace	T. E. Purchase	Reading, Pa	Apr. 5, 1870	101,658
Puddling and other furnace	J. Russell	Washington, D. C	Sept. 19, 1871	119,182
Puddling and other furnace door	G. E. Reynolds	Philadelphia, Pa	Sept. 24, 1872	131,562
Puddling and other furnace door	N. L. Sibley and B. Shiverick	Weston and Waltham, Mass	Apr. 10, 1866	53,894
Puddling and other furnaces, "Fix" for	A. Reese	Pittsburgh, Pa	May 16, 1871	114,854
Puddling and refining furnace	J. Heatley	Etna, Pa	Apr. 27, 1869	89,310
Puddling and refining furnace, Iron	J. S. Gustin	New York	Aug. 2, 1842	2,743
Puddling and reheating furnaces, Combination of ash-trap with.	L. Scofield and E. Cooper	South Trenton, N. J., and New York, N. Y.	Apr. 3, 1849	6,279
Puddling and reverberatory furnace, Iron and steel	I. Hersey	Fort Edward, N. Y	May 7, 1872	126,546
Puddling, boiling, and other furnaces, Stack for	W. Swindell	Allegheny City, Pa	Aug. 23, 1870	106,636
Puddling-furnace	C. D. Baker	Wheeling, W. Va	Feb. 27, 1866	52,813
Puddling-furnace	S. Bunn	Belleville, Ill	May 9, 1871	114,641
Puddling-furnace	M. J. Davies	Zanesville, Ohio	Aug. 16, 1864	43,880
Puddling-furnace	J. Green	Philadelphia, Pa	Feb. 3, 1857	16,541
Puddling-furnace	J. P. and J. Grove	Montour County, Pa	Apr. 6, 1858	19,843
Puddling-furnace	W. and J. Groves	Providence, R. I	June 27, 1865	48,390
Puddling-furnace	D. and J. Hall	Wheeling, W. Va	Nov. 14, 1865	50,929
Puddling-furnace	D. and J. Hall	West Wheeling, Ohio	Oct 30, 1866	59,214
Puddling-furnace	G. E. Harding	New York, N. Y	July 30, 1872	130,044
Puddling-furnace	J. Heatley	Harrisburgh, Pa	June 10, 1873	139,788
Puddling-furnace	W. Jeffries	West Bromwick, England	Oct. 1, 1867	69,345
Puddling-furnace	C. Jones	Saint Louis, Mo	June 24, 1873	140,140
Puddling-furnace	T. J. Jones	Scranton, Pa	Feb. 11, 1868	74,374
Puddling-furnace	P. Keenan and E. O. Connor	Chartier Township and West Pittsburgh, Pa.	Nov. 14, 1865	50,937
Puddling-furnace	E. Lloyd	Pittsburgh, Pa	Feb. 11, 1873	135,650
Puddling-furnace	H. McDonald	Pittsburgh, Pa	Mar. 9, 1869	87,578
Puddling-furnace	H. McDonald	Pittsburgh, Pa	July 16, 1872	129,153
Puddling-furnace	S. McLaughlin	Philadelphia, Pa	Mar. 1, 1870	100,308
Puddling-furnace	D. Morgan	Pittsburgh, Pa	Jan. 16, 1872	122,731
Puddling-furnace	J. Neville	Brooklyn, N. Y	May 2, 1871	114,465
Puddling-furnace	E. Riley	Pontnewynydd, England	Sept. 30, 1873	143,301
Puddling-furnace	J. B. Robinson	Duncansville, Pa	Aug. 31, 1869	94,342
Puddling-furnace	H. Ross	Pittsburgh, Pa	Jan. 28, 1868	73,931
Puddling-furnace	F. M. Ruschhaupt	Philadelphia, Pa	Apr. 19, 1864	42,446
Puddling-furnace	J. Snyder	Wheeling, W. Va	Oct. 16, 1866	58,904
Puddling-furnace	J. A. Stearns	Rolla, Mo	July 23, 1872	129,762
Puddling-furnace	W. Stevenson	Allegheny City, Pa	Jan. 21, 1868	73,665
Puddling-furnace	G. A. Whipple	West Pittsburgh, Pa	July 10, 1866	56,333
Puddling-furnace	J. Williams	Montreal, Canada	Oct. 3, 1865	50,319
Puddling-furnace door	J. S. Rees	Phillipsburgh, N. J	July 8, 1873	147,730
Puddling-furnace door or frame	P. E. Shear	Saugerties, N. Y	Mar. 2, 1869	87,438
Puddling-furnace for manufacturing iron with anthracite coal.	T. Cooper	New York, N. Y	Aug. 25, 1840	1,733
Puddling-furnace for treating iron and steel, Revolving.	S. Danks	Cincinnati, Ohio	Nov. 24, 1868	84,347
Puddling-furnace, Gas-heated	J. G. Blunt	Leavenworth, Kans	July 16, 1872	129,268
Puddling furnace, Iron	R. Jenkins	Newark, Ohio	July 4, 1871	116,598
Puddling furnace, Iron	N. S. Snedeker	Philadelphia, Pa	Aug. 22, 1865	49,565
Puddling-furnace, Mechanical	R. G. Wood and J. R. Jackson	McKeesport, Pa	July 16, 1872	129,199
Puddling-furnace, Revolving	L. S. Goodrich	Waverly, Tenn	Dec. 24, 1872	134,268
Puddling-furnace, Rotary	W. Baynton	Pottsville, Pa	Sept. 5, 1871	118,674
Puddling-furnace, Rotary	S. Danks	Cincinnati, Ohio	Oct. 26, 1869	96,206
Puddling-furnace, Rotary	J. Davies	Knoxville, Tenn	May 21, 1872	126,937
Puddling-furnace, Rotary	L. S. Goodrich	Waverly, Tenn	Dec. 24, 1872	134,138
Puddling-furnace, Rotary	T. and J. W. Richardson and A. Spencer.	West Hartlepool, Great Britain.	Oct. 15, 1872	132,180
Puddling-furnace, Rotary	E. Wood	Pittsburgh, Pa	Nov. 15, 1870	109,284
Puddling-furnace with a steam-generator, Combination of.	G. W. Hawksley and M. Wild	Sheffield, England	May 26, 1868	78,370
Puddling-furnaces, Apparatus for heating	S. A. Hill and C. F. Thumm	Oil City, Pa	Oct. 12, 1869	95,686
Puddling-furnaces, Application of hot-blast to	J. Reese	Pittsburgh, Pa	Dec. 6, 1864	45,343
Puddling-furnaces, Composition for lining	J. Williams	Montreal, Canada	June 14, 1864	43,168
Puddling-furnaces, Fix for	C. S. Lynch	Boston, Mass	Nov. 1, 1870	108,802
Puddling-furnaces, Fixing for	H. McDonald	Pittsburgh, Pa	Oct. 17, 1865	50,483
Puddling-furnaces, &c., in generating steam, Utilizing the waste heat of.	J. Watt	Buffalo, N. Y	May 12, 1863	38,521
Puddling-furnaces, Lining rotary	S. Danks	Cincinnati, Ohio	Mar. 4, 1873	136,421
Puddling-furnaces, Lining rotary	G. H. Sellers	Wilmington, Del	July 23, 1872	129,685
Puddling-furnaces, Shield for	H. McDonald	Pittsburgh, Pa	Dec. 29, 1868	85,320
Puddling-furnaces, Water-bosh for	J. Stokes and J. Brough	Trenton, N. J	Aug. 4, 1868	80,573
Puddling, melting, and heating furnace, Rotary	S. Danks	Cincinnati, Ohio	Aug. 26, 1873	142,152
Puddling, steam-boiler, and other furnaces, Fire-chamber for.	W. F. Beecher	Pittsburgh, Pa	Apr. 4, 1871	113,481
Puffing-iron	M. Parry	Rochester, N. Y	Feb. 27, 1872	124,079
Pug-mill	D. H. Gage	Dover, N. H	May 6, 1873	138,628
Pug-mill	J. A. Hamer	Reading, Pa	Jan. 26, 1858	19,194
Pug-mill	J. C. McKenzie	Adrian, Mich	Sept. 29, 1868	82,626
Pug-mill	J. E. Noyes	Washington, D. C	Mar. 22, 1870	101,154
Pug-mill for mixing clay	C. F. Schlickeysen	Berlin, Prussia	June 24, 1856	15,197

Index of patents issued from the United States Patent Office from 1790 *to* 1873, *inclusive*—Continued.

Invention.	Inventor.	Residence.	Date.	No.
Pug-mill grinding-attachment	D. H. Cage	Dover, N. H	Mar. 23, 1858	19, 696
Puller: *See* Bean-puller. Cork-puller. Flax-puller. Hoop-pole puller. Stalk and cane puller. Stone-puller. Weed-puller.				
Pulley	A. S. Blake	Waterbury, Conn	Oct. 30, 1866	59, 172
Pulley	J. A. Burnap	Albany, N. Y	Sept. 1, 1868	81, 860
Pulley	J. J. Cowell	Newark, N. J	May 7, 1872	126, 524
Pulley	G. B. Cowles	Bridgeport, Conn	June 14, 1870	104, 119
Pulley	F. Hewitt	Newark, N. J	Apr. 28, 1868	77, 284
Pulley	C. C. Moore	New York, N. Y	Feb. 11, 1873	135, 722
Pulley	R. W. Parker	Woburn, Mass	Oct. 30, 1866	59, 256
Pulley	J. P. Smith	Oshawa, Canada	Aug. 4, 1868	80, 777
Pulley	S. A. Smith	Philadelphia, Pa	Feb. 7, 1871	111, 582
Pulley	J. S. Tibbets	Terre Haute, Ind	Nov. 20, 1866	59, 874
Pulley	M. Ward	Mount Carmel, Pa	Sept. 5, 1871	118, 761
Pulley	T. A. Weston	Birmingham, England	Aug. 6, 1867	67, 470
Pulley	M. K. Whipple	Warren, Mass	Mar. 25, 1873	137, 270
Pulley and block	S. and J. Roebuck	New York, N. Y	Jan. 21, 1868	73, 653
Pulley and cable combined	R. Heneage, G. Milsom, and H. Spendelow.	Buffalo, N. Y	Jan. 29, 1867	61, 554
Pulley and clamp, Combined	M. W. Clark	Worcester, Mass	Apr. 12, 1870	101, 824
Pulley and gearing for machinery	S. Wheeler	Albany, N. Y	Apr. 14, 1868	76, 680
Pulley and spring cord, Calisthenic	J. Wood	New York, N. Y	June 14, 1864	43, 149
Pulley and wheel fastener	A. Newell	Chicago, Ill	Feb. 6, 1872	123, 413
Pulley-attachment for raising weights	G. W. Gregory	Binghamton, N. Y	Aug. 14, 1866	57, 125
Pulley attachment, Loose	C. F. Thayer	Linesville Station, Pa	Apr. 9, 1872	125, 503
Pulley, Band	C. H. Perkins	Providence, R. I	Jan. 21, 1873	135, 151
Pulley, Band	J. Simpson	Decatur, Ga	July 7, 1846	4, 618
Pulley, Band	E. P. West	Jersey City, N. J	Jan 14, 1873	134, 778
Pulley-banding	R. W. Parker	Roxbury, Mass	Feb. 17, 1852	8, 745
Pulley-belting	E. C. Thayer	Providence, R. I	Dec. 13, 1859	26, 451
Pulley-belts. Machine for measuring the strain on	J. B. Duff and T. W. Keating	New York, N. Y	June 19, 1860	28, 743
Pulley-block	E. A. Barrett	New York, N. Y	Jan. 21, 1873	135, 066
Pulley-block	L. A. Beardsley	South Edmeston, N. Y	July 23, 1861	32, 856
Pulley-block	J. Boylo	New York, N. Y	Sept. 24, 1872	131, 594
Pulley-block	G. C. Brown	Brooklyn, N. Y	May 23, 1871	115, 022
Pulley-block	T. B. Brown	Fairfield, Me	May 7, 1872	126, 376
Pulley-block	J. A. Burnap	Albany, N. Y	Dec. 10, 1867	71, 973
Pulley-block	H. Cherry	Birmingham, England	Nov. 14, 1871	120, 940
Pulley-block	J. C. Cottingham	Philadelphia, Pa	Mar. 5, 1872	124, 253
Pulley-block	A. S. Dickinson	Brooklyn, N. Y	Oct. 7, 1873	143, 441
Pulley-block	J. J. Doyle	Brooklyn, N. Y	Sept. 27, 1864	44, 409
Pulley-block	J. M. Drake	Amityville, N. Y	Mar. 10, 1863	37, 859
Pulley-block	G. A. Ford	Cleveland, Ohio	Aug. 26, 1873	142, 095
Pulley-block	R. P. Fuller	Machias, Me	May 2, 1865	47, 598
Pulley-block	J. L. Hovey	Lockport, N. Y	Mar. 27, 1860	27, 633
Pulley-block	C. H. Knapp	Lawrenceville, Pa	Dec. 28, 1869	98, 275
Pulley-block	S. F. Lewis	San Francisco, Cal	Nov. 8, 1859	26, 038
Pulley-block	P. Luck	Williamsburgh, N. Y	May 2, 1865	47, 558
Pulley-block	R. Marsden	Sheffield, England	July 4, 1871	116, 613
Pulley-block	J. W. Norcross	Boston, Mass	Jan. 7, 1868	73, 030
Pulley-block	J. W. Norcross	Boston, Mass	Apr. 4, 1871	113, 688
Pulley-block	H. F. Shaw	West Roxbury, Mass	June 23, 1868	79, 264
Pulley-block	H. Smith	Providence, R. I	May 23, 1871	115, 248
Pulley-block	S. Van Henneck and T. Allen	New York, N. Y	Apr. 25, 1865	47, 471
Pulley-block	N. C. Whitcomb and W. Paddock.	Oak Hill, N. Y	July 12, 1864	43, 541
Pulley-block, Differential	G. F. and M. Clemens	Springfield and Boston, Mass.	Apr. 23, 1872	125, 882
Pulley-block, Differential	C. Hall	New York, N. Y	Oct. 17, 1871	119, 981
Pulley-block, Differential	C. Hall	New York, N. Y	Feb. 6, 1872	123, 342
Pulley-block, Differential	C. Hall	New York, N. Y	June 11, 1872	127, 689
Pulley-block, Differential	C. Hall and E. Hubner	New York, N. Y	May 7, 1872	126, 391
Pulley-block, Differential	R. A. Hardcastle	Newcastle-upon-Tyne, England.	May 7, 1867	64, 527
Pulley-blocks, Metal	C. W. Gregory	Watertown, N. Y	May 25, 1869	90, 525
Pulley-blocks. Molding and casting the sheaves of	A. Worrall	New York	June 12, 1816	....
Pulley box, Loose	C. Purdy	Bedford, Ohio	Nov. 26, 1867	71, 535
Pulley-brake, Self-acting	J. Jochum	Brooklyn, N. Y	Nov. 29, 1864	45, 248
Pulley bush, Wood	T. B. Stout	Keyport, N. J	Apr. 6, 1869	88, 751
Pulley case, Sash	C B. Clark	Buffalo, N. Y	Apr. 23, 1872	126, 019
Pulley casing, Window	N. Thompson	Brooklyn, N. Y	Sept. 19, 1871	119, 199
Pulley, Cast-iron chain	J. Bird	New York, N. Y	Dec. 11, 1866	60, 329
Pulley, Cast-metal	J. A. Evarts	West Meriden, Conn	Nov. 1, 1859	25, 998
Pulley, Clamp	R. Chandler	Springfield, Mass	Apr. 3, 1866	53, 730
Pulley, Clothes-line	G. H. Ryer	New York, N. Y	May 20, 1873	139, 024
Pulley, Clutch	J. Shinn	Leverington, Pa	Sept. 17, 1861	33, 321
Pulley-coupling	J. E. Atwood	Mansfield, Conn	Nov. 29, 1870	109, 709
Pulley-couplings, Clutch for	J. Knickerbacker	Stockport, N. Y	Dec. 6, 1859	26, 369
Pulley, Differential	T. A. Weston	Ridgewood, N. J	Dec. 24, 1872	134, 337
Pulley, Door	J. Reiser	Trenton, N. J	June 8, 1869	91, 163
Pulley, Door	P. J. Tillman	Trenton, N. J	Apr. 6, 1869	88, 592
Pulley, Dynamometer	I. P. Tice	New York, N. Y	Apr. 1, 1873	137, 393
Pulley, Expanding	G. S. Barton	Worcester, Mass	Apr. 18, 1871	113, 723
Pulley, Expanding	J. E. Fales	Worcester, Mass	Apr. 1, 1873	137, 352
Pulley, Expanding	T. H. Savery	Wilmington, Del	June 9, 1868	78, 763
Pulley, Expansible belt	H. A. Hummer and J. H. Stover	Frenchtown, N. J	Dec. 17, 1872	131, 061
Pulley, Extension	W. Onions and I. Bagnall	Saint Louis, Mo	Apr. 18, 1871	113, 789
Pulley-facing machine	T. Blume	Cincinnati, Ohio	Jan. 10, 1860	26, 810
Pulley, Fast and loose	C. H. Brown	Fitchburgh, Mass	Mar. 8, 1864	41, 829
Pulley for band-saws	W. H. Doane	Cincinnati, Ohio	May 21, 1872	127, 033
Pulley for belting	M. Lewis and S. Miller	Greenville, Conn	Mar. 26, 1867	63, 266
Pulley for belts and brakes, Binder	M. C. Bryant	Lowell, Mass	Nov. 13, 1849	6, 864
Pulley for blind-cords, Ratchet	J. B. Holmes, jr	New York, N. Y	Dec. 13, 1859	26, 433
Pulley for fishing-boats, Self-adjusting	W. Woodbury	Chelsea, Mass	Feb. 13, 1866	52, 634
Pulley for hoisting-apparatus	D. L. Miller	Madison, N. J	Mar. 24, 1868	75, 946

Index of patents issued from the United States Patent Office from 1790 *to* 1873, *inclusive*—Continued.

Invention.	Inventor.	Residence.	Date.	No.
Pulley for window-sashes, Spring	D. Bickford	Westerly, R. I	July 13, 1858	20, 857
Pulley for window-sashes, Spring	J. Shopland	Honesdale, Pa	Aug. 12, 1856	15, 528
Pulley, Friction	E. F. Allen	Providence, R. I	Nov. 1, 1870	108, 749
Pulley, Friction	C. Burleigh	Fitchburgh, Mass	July 8, 1862	35, 853
Pulley, Friction	W. Ebbitt	New York, N. Y	Jan. 16, 1872	122, 818
Pulley, Friction	J. Kaceroosky	Bridgeport, Conn	Sept. 17, 1872	131, 449
Pulley, Friction	L. G. Mason	Worcester, Mass	Aug. 21, 1866	57, 352
Pulley, Friction	C. B. Smith	Newark, N. J	Dec. 4, 1866	60, 269
Pulley, Friction	E. Spaulding	Westborough, Mass	May 24, 1859	24, 156
Pulley, Friction	J. Steger	New York, N. Y	Oct. 18, 1870	108, 533
Pulley, Friction	J. Steger	New York, N. Y	July 11, 1871	117, 010
Pulley, Friction-clutch	J. W. Bishop	Stamford, Conn	Oct. 16, 1866	58, 765
Pulley, Friction-clutch	A. B. Clemons	Ansonia, Conn	Dec. 8, 1868	84, 681
Pulley, Friction-clutch	J. D. Crocker	Norwich, Conn	June 3, 1873	139, 456
Pulley, Friction-clutch	D. Harrington	Worcester, Mass	June 16, 1868	78, 961
Pulley, Friction-clutch	C. D. Palmiter	Oswego, N. Y	Nov. 12, 1867	70, 888
Pulley, Friction-clutch	H. K. Smith	Norwich, Conn	July 14, 1868	80, 024
Pulley, Friction-clutch	C. Wright	Newark, N. J	July 14, 1868	80, 045
Pulley, Gate	E. S. Axtell	Macomb, Mich	Sept. 27, 1870	107, 646
Pulley, Gate	D. Bedell	Seneca Falls, N. Y	May 8, 1860	28, 146
Pulley, Grip	A. S. Hallidie	San Francisco, Cal	Feb. 22, 1870	100, 140
Pulley, Griping	A. S. Hallidie	San Francisco, Cal	June 11, 1872	127, 690
Pulley, Grooved	L. Planer	New York, N. Y	Jan. 3, 1860	26, 707
Pulley-hanger	J. Kent	New York, N. Y	June 17, 1873	139, 898
Pulley, Hoisting and transferring	A. D. Manley	Washington, Mich	May 5, 1868	77, 629
Pulley, Loose	E. S. Capen	Worcester, Mass	July 7, 1868	79, 634
Pulley, Loose	W. W. Carey	Lowell, Mass	Dec. 24, 1872	134, 250
Pulley, Loose	J. P. Grosvenor	Lowell, Mass	April 9, 1872	125, 388
Pulley, Loose	D. Harrington	Worcester, Mass	Aug. 11, 1868	80, 948
Pulley, Loose	L. Knight	Salem, N. Y	June 17, 1873	140, 048
Pulley, Loose	N. P. Otis	Yonkers, N. Y	Nov. 21, 1865	51, 077
Pulley, Loose	J. J. Ralya	Cleveland, Ohio	Dec. 31, 1872	134, 487
Pulley, Lubricating	J. H. Gray	Boston, Mass	Sept. 15, 1868	82, 110
Pulley, Lubricating	D. Harrington	Worcester, Mass	July 7, 1868	79, 752
Pulley, Lubricating	J. K. McLanahan	Hollidaysburgh, Pa	Oct. 21, 1873	143, 830
Pulley lubricator, Loose	J. W. Brockway	New York, N. Y	Apr. 26, 1870	102, 215
Pulley lubricator, Loose	F. Keifel	Cincinnati, Ohio	Mar. 12, 1872	124, 589
Pulley lubricator, Loose	C. A. King	Springfield, Mass	Oct. 19, 1869	95, 911
Pulley lubricator, Loose	C. A. King	Springfield, Mass	Oct. 19, 1869	95, 912
Pulley lubricator, Loose	G. M. Morris and J. McCreary	Cohoes, N. Y	Feb. 4, 1868	74, 116
Pulley lubricator, Loose	S. Ustick	Philadelphia, Pa	Feb. 7, 1871	111, 590
Pulley lubricator, Loose	J. P. Wendell and S. P. M. Tasker.	Philadelphia, Pa	May 3, 1870	102, 739
Pulley, Machine	C. S. Hunt	Bridgewater, Mass	July 13, 1858	20, 881
Pulley-mechanism	W. H. Brown	Bangor, Me	Sept. 6, 1870	106, 993
Pulley-molding apparatus	G. L. Scott	Manchester, England	May 18, 1869	90, 126
Pulley-molding apparatus	J. Yocom, jr	Philadelphia, Pa	Apr. 19, 1864	42, 426
Pulley molding machine	T. R. and S. Knowles	Jersey City, N. J	Nov. 12, 1867	70, 864
Pulley press, Power	W. and R. Skene	Louisville, Ky	June 28, 1859	24, 584
Pulley, Sash	J. Andrews	Marlborough, Mass	Jan. 30, 1872	123, 075
Pulley, Sash	J. D. Browne	Cincinnati, Ohio	Sept. 22, 1868	82, 287
Pulley, Sash	H. Cash	Newport, Ky	Aug. 6, 1867	67, 412
Pulley, Sash	C. C. Clark	Buffalo, N. Y	Sept. 21, 1869	94, 945
Pulley, Sash	S. Drum	Allegheny City, Pa	May 21, 1867	64, 957
Pulley, Sash	O. S. Garretson	Buffalo, N. Y	Apr. 12, 1870	101, 854
Pulley, Sash	A. Le Page	Woodhaven, N. Y	Nov. 21, 1871	121, 178
Pulley, Sash	B. Roux	Cincinnati, Ohio	Nov. 16, 1869	96, 969
Pulley, Sash	Z. E. Sawtell	Boston, Mass	May 31, 1870	103, 783
Pulley, Sash	A. P. Seymour, jr., and W. R. Goodrich.	Hecla Works and Whitestown, N. Y.	Aug. 18, 1868	81, 218
Pulley, Sash-rope	J. C. Price	New Philadelphia, Ohio	Sept. 21, 1869	95, 138
Pulley, Self-detaching	J. E. Gustin	Elmira, N. Y	July 7, 1868	79, 653
Pulley, Self-lubricating	W. W. Crane	Auburn, N. Y	Nov. 26, 1872	133, 416
Pulley, Self-oiling	J. Goodrich and H. J. Colburn	Fitchburgh, Mass	May 4, 1869	89, 652
Pulley, Self-oiling	J. Goodrich and H. J. Colburn	Fitchburgh, Mass	Apr. 25, 1871	114, 129
Pulley, Self-stopping	T. C. Richards	New York, N. Y	Mar. 20, 1866	53, 341
Pulley shade-fixture, Rack	W. Johnson	Waterbury, Conn	May 16, 1871	114, 825
Pulley, Sheave	V. Knecht	Cincinnati, Ohio	May 24, 1870	103, 474
Pulley, Sheave	J. B. Vannan and N. P. Cramer	Carbondale, Pa	Aug. 3, 1869	93, 252
Pulley, sheave, and wheel	N. Thompson	Brooklyn, N. Y	Sept. 26, 1871	119, 430
Pulley, Suspension	J. A. Evarts	West Meriden, Conn	Oct. 28, 1873	144, 078
Pulley-suspension hook	D. B. Baker and P. S. Miller	Rollersville, Ohio	Nov. 13, 1866	59, 542
Pulley, Tension	A. B. Nimbs	Buffalo, N. Y	Apr. 18, 1865	47, 323
Pulley, Tight and loose	J. P. Gates	Lincoln, Ill	Dec. 10, 1867	72, 014
Pulley, Tight and loose	J. G. McCormick	Louisville, Ky	Sept. 13, 1870	107, 275
Pulley to ascertain ship's way, &c., Loxodromick	J. S. Donlery	New York	Sept. 21, 1815	....
Pulley-turning machine	G. A. Gray, jr	Hamilton, Ohio	Apr. 29, 1873	138, 394
Pulley-wheels, Mode of constructing wooden	A. Newell	Erie, Pa	May 6, 1873	138, 518
Pulley, Window	J. M. Ford	Brooklyn, N. Y	June 29, 1869	91, 836
Pulley, Window	O. S. Garretson	Buffalo, N. Y	Sept. 17, 1867	68, 868
Pulley, Window-cord	M. C. Ames	Hartford, Conn	Apr. 4, 1865	47, 150
Pulley, Window-cord	J. W. Bliss	Hartford, Conn	Feb. 21, 1854	10, 540
Pulley, Window-sash	O. P. Briggs	Chicago, Ill	Oct. 15, 1872	132, 135
Pulley, Window-sash	O. S. Judd	New Britain, Conn	June 13, 1848	5, 626
Pulleys and journals, Lubricating loose	S. Ustick	Philadelphia, Pa	Sept. 19, 1871	119, 065
Pulleys and wheel-hubs to shafts, Fastening	E. G. Shortt	Carthage, N. Y	June 20, 1871	116, 228
Pulleys and wheel to shafting, Attaching	C. L. Smith	Rahway, N. J	Oct. 8, 1872	132, 0[illegible]9
Pulleys, Balancing	E. W. Phelps	Elizabeth, N. J	Nov. 28, 1871	121, 411
Pulleys, &c., Anti-friction bush for	J. Palmer	Saint Catharine's, Canada	July 16, 1872	129, 585
Pulleys, Bearing for loose	W. Campbell	West Philadelphia, Pa	Oct. 24, 1854	11, 848
Pulleys by friction, Tightening	F. Skinner	New Haven, Conn	Aug. 8, 1865	49, 312
Pulleys, Casting metallic boxes or centers of sheaves or trundles for.	L. Larrabee	Nantucket, Mass	Nov. 4, 1811	....
Pulleys, Cog for	W. Shotwell		June 24, 1800	....
Pulleys, Compound to increase the friction between belts and.	L. F. Robertson	New York, N. Y	June 14, 1870	104, 3[illegible]6
Pulleys, constructing self-lubricating	W. W. Crane	Auburn, N. Y	June 4, 1867	65, 349
Pulleys, Construction of	A. Warth	Stapleton, N. Y	Apr. 3, 1866	53, 711

Index of patents issued from the United States Patent Office from 1790 *to* 1873, *inclusive*—Continued.

Invention.	Inventor.	Residence.	Date.	No.
Pulleys, Device for locking loose	W. J. Linton	Detroit, Mich	Aug. 25, 1868	81, 384
Pulleys, Device for oiling fast and loose	W. Hamilton	Chicopee, Mass	Mar. 24, 1868	75, 753
Pulleys, Lubricating loose	I. F. Brown	New London, Conn	Apr. 30, 1872	126, 261
Pulleys, Lubricating loose	S. Ristick	Philadelphia, Pa	Sept. 26, 1871	119, 287
Pulleys, Lubricating-sleeve for loose	O. E. Greene	Lawrence, Mass	Apr. 20, 1869	89, 038
Pulleys, Lubrication of clutch	A. Canis and F. Higgins	Manchester, N. H	Mar. 11, 1873	136, 699
Pulleys, Mold for casting	W. D. Rinehart	Pittsburgh, Pa	June 19, 1866	55, 713
Pulleys, Molding	W. Neemes	Pittsburgh, Pa	Mar. 5, 1867	62, 668
Pulleys or sheaves, Bushing and pin for	M. H. Marshall	Gloucester, Mass	July 10, 1834	....
Pulleys, Rubber cover for flat-belt	J. W. Sutton	Akron, Ohio	Apr. 2, 1872	125, 228
Pulleys, Securing	J. W. Reid	New York, N. Y	Nov. 28, 1865	51, 218
Pulleys, Shipper-gear for	A. Betteley	Boston, Mass	Aug. 23, 1859	25, 169
Pulleys to shafts, Attaching	C. Clareni	New York, N. Y	Sept. 5, 1854	11, 639
Pulleys to shafts, Device for securing	J. H. Buckman	Cincinnati, Ohio	Dec. 21, 1869	98, 144
Pulleys to shafts, Friction-attachment for securing.	H. Cox	Peterborough, Canada	June 10, 1873	139, 704
Pulleys to shafts, Securing	D. K. Overhiser	Williamsport, Pa	Nov. 19, 1872	133, 246
Pulleys to shafts, Securing	W. C. Van Voorhis	New York, N. Y	May 27, 1873	139, 282
Pulleys with shafts, Coupling and uncoupling	P. Yates	Milwaukee, Wis	Apr. 25, 1843	3, 063
Pulmonometer	A. Eckert	Dayton, Ohio	Jan. 10, 1860	26, 754
Pulp-box drier	S. Wheeler and E. Jerome	Albany, N. Y	July 16, 1867	66, 918
Pulp-box, Finishing	S. Wheeler and E. Jerome	Albany, N. Y	July 16, 1867	66, 919
Pulp-dresser	C. Sellers	Philadelphia, Pa	June 6, 1832	....
Pulp-dresser	E. H. Thomas and N. Woodcock.	Brattleborough, Vt	Aug. 11, 1830	....
Pulp-engine or crusher	M. R. Fletcher	Boston, Mass	June 20, 1871	116, 039
Pulp from wood, straw, and other matters, Making.	J. W. Dixon	Philadelphia, Pa	June 26, 1866	55, 836
Pulp-machine	G. Sweetland	New Haven, Conn	Sept. 5, 1848	5, 756
Pulp machine, Wood	S. C. Taft	Mendon, Mass	Apr. 12, 1870	101, 785
Pulp, Making boxes of paper	A. French and C. Frost	Waterbury, Conn	July 1, 1856	15, 228
Pulp, paste, or slip pigments, Preservation of	P. C. Tiemann	New York, N. Y	Sept. 23, 1873	143, 106
Pulp, Preparing and dressing	J. Ames	Springfield, Mass	Sept. 1, 1832	....
Pulp, Process for reducing fibrous substances to	H. B. Meech	Fort Edward, N. Y	Aug. 7, 1866	56, 971
Pulp-screen	G. West	Tyringham, Mass	Aug. 19, 1851	8, [illegible]
Pulp-strainer	H. H. Olds	New Haven, Conn	Feb. 18, 1873	136, 002
Pulp-strainer	J. Sawyer	Newbury, Vt	Jan. 21, 1832	....
Pulp-washer, Submerged centrifugal	R. R. Sylands	Millburn, N. J	May 25, 1869	90, 472
Pulp-washing machine	F. Goucher	Chester County, Pa	Apr. 12, 1833	....
Pulp, Wood-grinding machine for making	S. B. Zimmer	Elkhart, Ind	Aug. 19, 1873	141, 976
Pulpit, Apparatus for ventilating	J. P. Herron	Huntsville, Ohio	Jan. 12, 1858	19, 089
Pulverized material, Can for	H. Everett	Philadelphia, Pa	Sept. 24, 1872	131, 511
Pulverizer	I. N. Jennings	Danbury, Conn	Jan. 5, 1869	85, 591
Pulverizer, Agricultural	B. F. Stickney	Vistula, Mich	Mar. 1, 1834	....
Pulverizer and amalgamator, Quartz	W. H. and I. Scoville	Chicago, Ill	Sept. 25, 1860	30, 162
Pulverizer and corn-marker, Combined	T. B. Jones	Hiawatha, Kans	Dec. 5, 1871	121, 629
Pulverizer and harrow, Earth	J. Lefeber and G. W. Shults	Cambridge City, Ind	Oct. 31, 1871	120, 446
Pulverizer and seed-sower	G. W. Bonham	Henry, Ill	Jan. 7, 1862	34, 049
Pulverizer and seeder, Combined earth	C. Wilson	Stamford, Conn	Sept. 3, 1867	68, 403
Pulverizer, Clay	G. C. Bovey	Cincinnati, Ohio	Aug. 15, 1871	117, 973
Pulverizer, Clay	I. Hersey and J. H. Van Riper.	New York, N. Y	June 9, 1857	17, 505
Pulverizer, Earth	W. Elwell	Gardiner, Me	Aug. 22, 1865	49, 513
Pulverizer, Earth	J. Johnson	Mount Washington, Ohio	Nov. 29, 1864	45, 249
Pulverizer, Earth	J. Prutzman	Hancock County, Ill	Aug. 27, 1867	68, 110
Pulverizer for reducing ligneous substances to powder.	N. Jones	Boston, Mass	July 23, 1816	....
Pulverizer, Ground	E. A. Olleman	Mooresville, Ind	Aug. 16, 1870	106, 397
Pulverizer, Land	T. R. Denby	Carlinville, Ill	Apr. 15, 1873	137, 764
Pulverizer, leveler, and marker	L. Jones	Funk's Grove, Ill	Dec. 15, 1868	85, 010
Pulverizer, Rotary	J. Thompson	Louisville, Ky	Nov. 15, 1870	109, 273
Pulverizer, Soil	C. Berninger	Mier, Ill	Aug. 25, 1868	81, 333
Pulverizer, Soil	D. Osborn	Paoli, Ind	Jan. 9, 1872	122, 643
Pulverizer, Soil	W. Shumard	Richmond, Ind	Oct. 19, 1869	95, 940
Pulverizer, Subsoil	G. S. Newson	Nashville, Tenn	Aug. 9, 1870	106, 195
Pulverizing and crushing vegetable and mineral matter, Machine for.	M. Smith	Pittsburgh, Pa	Oct. 1, 1861	32, 495
Pulverizing and furrowing device	C. Shabley	Brooklyn, N. Y	Dec. 26, 1865	51, 757
Pulverizing and grinding machine	J. J. Webster	Magog, Canada	Jan. 16, 1872	122, 748
Pulverizing and levigating substances, Machine for.	O. E. Noble	Penn Yan, N. Y	Sept. 27, 1864	44, 444
Pulverizing-barrel, Self-discharging	A. B. Paul	San Francisco, Cal	May 10, 1870	102, 857
Pulverizing-chaser	T. Pugh	Chicago, Ill	Dec. 28, 1869	98, 297
Pulverizing-machine	W. Adamson	Philadelphia, Pa	Mar. 29, 1864	42, 060
Pulverizing machine, Mineral	S. and G. E. Mills	New York, N. Y	June 28, 1859	24, 570
Pulverizing machine, Rock, ore, &c	W. F. Goodwin and C. R. Squire	East New York and New York, N. Y.	Oct. 8, 1867	59, 655
Pulverizing machine, Sand, &c	J. G. Savage	South Reading, Mass	Mar. 7, 1865	46, 713
Pulverizing machine, Soil	G. P. De Yo	Groton Township, Ohio	June 27, 1871	116, 278
Pulverizing machine, Soil	L. S. Fithian	Absecom, N. J	Jan. 24, 1865	46, 048
Pulverizing machine, Soil	A. C. Tower	Mendota, Ill	Apr. 29, 1873	138, 301
Pulverizing machine, Wood	A. Tolson	Georgetown, D. C	June [illegible], 1821	....
Pulverizing soil and clods, Roller for	J. Custer	Corsica, Ohio	Mar. 26, 1867	63, 224
Pulverizing soil, Device for	G. Shelton	Normal, Ill	July 25, 1871	117, 473
Pulverizing tailings from gold-washers	J. H. Hanchett	Beloit, Wis	May 23, 1865	47, 818
Pump	W. Adair	Liverpool, England	June 16, 1868	78, 854
Pump	C. C. Alexander	Denver, Colo	Sept. 22, 1863	40, 006
Pump	M. J. Althouse	Waupun, Wis	Jan. 12, 1864	41, 186
Pump	M. J. Althouse	Waupun, Wis	Aug. 7, 1866	56, 871
Pump	M. J. Althouse	Waupun, Wis	Nov. 13, 1866	59, 536
Pump	M. J. Althouse	Waupun, Wis	July 2, 1872	128, 576
Pump	R. H. Andrews	Elizabethtown, Pa	May 8, 1866	54, 483
Pump	A. Arnold	Heidelberg, Pa	Apr. 21, 1863	28, 209
Pump	J. B. Atwater	Brooklyn, N. Y	Mar. 23, 1858	19, 671
Pump	J. E. Atwood	Bucksport, Me	May 8, 1860	28, 140
Pump	S. L. Avery	Norwich, N. Y	Jan. 1, 1867	60, 818
Pump	H. Bachman	Lancaster, Pa	Apr. 2, 1835	....
Pump	B. C. Bailey	Constitution, Ohio	June 7, 1870	103, 960
Pump	C. Baker	Weymouth, Mass	May 14, 1867	64, 735

Index of patents issued from the United States Patent Office from 1790 *to* 1873, *inclusive*—Continued.

Invention.	Inventor.	Residence.	Date.	No.
Pump	A. Balding	Wheeling, W. Va	June 28, 1870	104, 688
Pump	E. Balding	Memphis, Tenn	Aug. 17, 1869	93, 794
Pump	A. C. Baldwin	Washington, D. C	May 7, 1872	126, 509
Pump	O. Baldwin	Summitville, Iowa	Aug. 16, 1864	43, 826
Pump	J. S. Barden	Providence, R. I	Mar. 3, 1868	75, 112
Pump	J. S. Barden	Providence, R. I	Feb. 8, 1870	99, 521
Pump	J. S. Barden	Providence, R. I	Jan. 7, 1873	134, 626
Pump	A. Barker	Honesdale, Pa	Feb. 17, 1852	8, 733
Pump	W. C. Barker	Ypsilanti, Mich	Nov. 8, 1870	108, 954
Pump	A. B. Barlow	Ripon, Wis	Oct. 23, 1866	58, 972
Pump	E. Barlow	Marietta, Ohio	Feb. 29, 1848	5, 462
Pump	W. Barnes	Maquoketa, Iowa	July 23, 1872	129, 779
Pump	W. A. Barnes	Decatur, Ill	Feb. 6, 1866	52, 377
Pump	W. T. and J. Barnes	Buffalo, N. Y., and Oakville, Canada.	Oct. 14, 1856	15, 878
Pump	N. Barrett	New York, N. Y	June 25, 1861	32, 607
Pump	W. S. Bartle	Newark, N. Y	Jan. 7, 1862	34, 101
Pump	J. R. Bassett	Cincinnati, Ohio	Jan. 3, 1871	110, 622
Pump	A. Batby	New York	May 18, 1830	....
Pump	F. Bauschtleker and B. Vanfleet	Washington, D. C	Oct. 22, 1867	70, 068
Pump	W. D. Baxter	New York, N. Y	Oct. 13, 1868	83, 027
Pump	W. D. Baxter	New York, N. Y	Apr. 8, 1873	137, 527
Pump	J. Bean	Hudson, Mich	Dec. 20, 1864	45, 467
Pump	J. Bean	Hudson, Mich	Feb. 20, 1866	52, 660
Pump	J. Bean	Hudson, Mich	Feb. 19, 1867	62, 247
Pump	J. Bean	Hudson, Mich	Nov. 19, 1867	71, 122
Pump	J. Bean	Hudson, Mich	Feb. 9, 1869	86, 804
Pump	J. Bean	Hudson, Mich	Mar. 26, 1872	125, 008
Pump	J. Bean	Hudson, Mich	Dec. 17, 1872	134, 025
Pump	N. S. Bean	Manchester, Pa	June 12, 1860	28, 644
Pump	A. Beeler and J. B. Christian	Mount Carroll, Ill	Apr. 12, 1859	23, 544
Pump	L. Beemer	Libertyville, N. J	Feb. 6, 1866	52, 485
Pump	S. Belden	Visalia, Cal	Apr. 27, 1869	89, 378
Pump	H. Belfield	Philadelphia, Pa	Apr. 17, 1860	27, 945
Pump	A. Bellingrath	Atlanta, Ga	Jan. 22, 1861	31, 206
Pump	C. F. Bellows	Seneca Falls, N. Y	Feb. 1, 1859	22, 777
Pump	C. Bemis	Mishawaka, Ind	June 11, 1867	65, 718
Pump	W. A. Bemis	Lyndon Centre, Vt	Aug. 9, 1864	43, 755
Pump	S. Benson	Allegheny City, Pa	Dec. 29, 1868	85, 270
Pump	J. A. Bloom	Philadelphia, Pa	Dec. 20, 1865	51, 686
Pump	J. Boley	Baldwinsville, N. Y	July 11, 1865	48, 647
Pump	C. Bollinger	Glen Rock, Pa	Oct. 1, 1861	33, 375
Pump	C. Bollinger	Harrisburgh, Pa	Jan. 5, 1864	41, 133
Pump	C. Bollinger	Harrisburgh, Pa	Jan. 5, 1864	41, 174
Pump	T. Bourke		Nov. 16, 1796	....
Pump	W. Boyers	Mount Carroll, Ill	Feb. 9, 1858	19, 286
Pump	H. E. Braunfeld	Philadelphia, Pa	Dec. 12, 1871	121, 843
Pump	J. F. Brickley	Winchester, Ind	Dec. 15, 1868	84, 991
Pump	S. Brillinger	Clarence Centre, N. Y	May 11, 1869	89, 849
Pump	J. Brokenshire	Oswego, N. Y	Aug. 11, 1868	80, 905
Pump	T. Brooks	Rutland, N. Y	Apr. 13, 1826	....
Pump	A. P. Brown	New York, N. Y	May 2, 1871	114, 403
Pump	W. R. Brown	Cleveland, Ohio	Apr. 5, 1859	23, 443
Pump	G. Bruce	Corydon, Ind	Oct. 15, 1867	69, 758
Pump	J. Bryan	New York, N. Y	Mar. 8, 1870	100, 595
Pump	M. Buchlin	Grafton, N. H	Oct. 25, 1832	....
Pump	J. Budd	Sandy Hill, N. Y	Apr. 29, 1862	35, 077
Pump	J. and D. Budd	Albany, N. Y	Oct. 13, 1863	40, 239
Pump	J. S. Burnham	Yorkville, N. Y	Jan. 13, 1857	16, 373
Pump	J. H. Burns	Clinton Station, N. J	Jan. 31, 1865	46, 073
Pump	A. Burr	Middlesex, Conn	May 1, 1866	54, 289
Pump	F. S. Burt	Mount Pleasant, Iowa	Jan. 13, 1863	37, 383
Pump	L. Button and R. Blake	Waterford, N. Y	Apr. 29, 1862	35, 079
Pump	J. W. Cahill	Madison, Ind	Nov. 13, 1866	59, 708
Pump	J. Camack	Dane, Wis	Sept. 5, 1865	49, 716
Pump	A. S. Cameron	New York, N. Y	May 1, 1866	54, 291
Pump	H. Camp	Rouseville, Pa	Jan. 31, 1871	111, 316
Pump	J. Canfield	Morristown, N. J	June 28, 1809	....
Pump	J. F. Carll	Pleasantville, Pa	May 6, 1873	138, 477
Pump	W. S. Carr	New York, N. Y	Jan. 11, 1870	98, 740
Pump	W. S. Carr	New York, N. Y	Sept. 24, 1872	131, 663
Pump	I. Carter	Champlain, N. Y	June 27, 1854	11, 192
Pump	A. Carver	Little Falls, N. Y	July 11, 1865	48, 707
Pump	A. Carver	Little Falls, N. Y	Apr. 26, 1870	102, 369
Pump	S. Caswell	Harrison, Me	May 19, 1840	1, 610
Pump	F. Catudal	Webster, Mass	Jan. 14, 1873	134, 785
Pump	E. S. Cavnah and D. Yeagley	Bourbon, Ind	Nov. 30, 1869	97, 274
Pump	W. N. Chamberlain	Van Buren, Mich	Nov. 2, 1869	96, 307
Pump	T. Chamberlin and T. E. Garrett	Philadelphia, Pa	Feb. 18, 1868	74, 500
Pump	T. Chambers	Saint Louis, Mo	Nov. 3, 1868	83, 691
Pump	J. B. Christian and A. Beeler	Mount Carroll, Ill	Apr. 5, 1859	23, 447
Pump	D. W. Christman	New York, N. Y	Feb. 8, 1817	....
Pump	M. R. Clapp	New York, N. Y	Mar. 7, 1865	46, 638
Pump	W. H. T. Clark	San Francisco, Cal	May 18, 1869	90, 080
Pump	R. Cochran	Morrison, Ill	June 26, 1866	55, 821
Pump	N. T. Coffin	Knightstown, Ind	Feb. 21, 1871	112, 018
Pump	N. T. Coffin	Knightstown, Ind	Mar. 28, 1871	113, 144
Pump	J. K. Cohick and J. Fesher	Mountville, Pa	Sept. 20, 1864	44, 282
Pump	D. G. Colburn	Wilmington, Vt	Apr. 25, 1846	4, 467
Pump	G. M. Cole	Folsom City, Cal	Sept. 24, 1867	69, 183
Pump	J. W. Cole	Mount Pleasant, Iowa	May 11, 1869	89, 918
Pump	H. T. Coleman	Williamsburgh, N. Y	June 17, 1873	140, 016
Pump	G. Collins and E. Piper	Camden, Me	Jan. 7, 1862	34, 105
Pump	J. F. Collins	New York, N. Y	Jan. 26, 1869	86, 213
Pump	H. Comstock	Seneca Falls, N. Y	Jan. 22, 1867	61, 319
Pump	A. Conant and I. F. Brown	New London, Conn	Nov. 6, 1866	59, 364
Pump	J. Coney	Boston, Mass	Dec. 16, 1856	16, 229

Index of patents issued from the United States Patent Office from 1790 *to* 1873, *inclusive*—Continued.

Invention.	Inventor.	Residence.	Date.	No.
Pump	G. W. Cook and Z. E. B. Nash	Saint Paul, Minn	Mar. 11, 1862	34, 624
Pump	A. Cooley	Springfield, Ill	Nov. 30, 1858	22, 165
Pump	A. Cooley	Chicago, Ill	June 10, 1862	35, 508
Pump	P. Cope	Oakfield, N. Y	June 18, 1872	128, 123
Pump	C. A. Cowell	Newark, N. J	Sept. 21, 1866	95, 088
Pump	G. Cowing	Seneca Falls, N. Y	Mar. 16, 1869	87, 912
Pump	J. P. Cowing	Seneca Falls, N. Y	Oct. 21, 1856	15, 922
Pump	J. C. Crawford	Clintonville, Ill	Apr. 6, 1869	88, 552
Pump	S. B. Crittenden	Cincinnati, Ohio	Oct. 20, 1863	40, 328
Pump	J. E. Cronk	Poughkeepsie, N. Y	Apr. 5, 1859	23, 453
Pump	J. D. Cross	Petaluma, Cal	Dec. 31, 1872	134, 466
Pump	R. W. Crouse	Westminster, Md	July 14, 1868	79, 959
Pump	R. W. Crouse	Westminster, Md	Aug. 17, 1869	93, 810
Pump	E. S. Crowell	Augusta, Me	Aug. 22, 1871	118, 206
Pump	H. G. Crowell	Roxbury, Mass	May 21, 1861	32, 356
Pump	W. H. Culp	Hammondsville, Ohio	Nov. 14, 1865	50, 911
Pump	C. G. Curtis	Springfield, Mass	Mar. 20, 1855	12, 544
Pump	P. C. Curtis	Utica, N. Y	May 22, 1835	....
Pump	L. H. Davis	West Chester, Pa	Apr. 28, 1863	38, 291
Pump	L. H. Davis	Newark, Del	Mar. 16, 1869	87, 828
Pump	J. M. Dearborn	Boston, Mass	May 3, 1817	....
Pump	J. N. Dennisson	Newark, N. J	Oct. 23, 1866	59, 125
Pump	T. De Witt	Detroit, Mich	May 3, 1870	102, 511
Pump	E. N. Dickerson and E. K. Root	New York, N. Y., and Hartford, Conn.	Feb. 5, 1856	14, 186
Pump	I. Dillinhgam	Rockbottom, Mass	Mar. 14, 1871	112, 561
Pump	A. Dixon	Aurora, Ill	Aug. 31, 1869	94, 190
Pump	G. H. Dodge	Camden, N. J	Apr. 1, 1862	34, 819
Pump	L. P. Dodge	Newburgh, N. Y	May 13, 1862	35, 222
Pump	L. P. and W. F. Dodge	Newburgh, N. Y	June 7, 1853	9, 777
Pump	W. F. Dodge	New York, N. Y	Apr. 11, 1865	47, 193
Pump	B. Douglas	Middletown, Conn	Nov. 6, 1860	30, 603
Pump	B. Douglas	Middletown, Conn	Sept. 17, 1861	33, 299
Pump	B. Douglas	Middletown, Conn	July 1, 1862	35, 802
Pump	B. and W. Douglas	Middletown, Conn	Apr. 12, 1859	23, 649
Pump	J. W. Douglas	Middletown, Conn	Apr. 22, 1862	35, 059
Pump	J. W. Douglas	Middletown, Conn	May 5, 1863	38, 444
Pump	J. W. Douglas	Middletown, Conn	July 24, 1866	56, 671
Pump	J. W. Douglas	Middletown, Conn	Feb. 11, 1868	74, 321
Pump	J. W. Douglas	Middletown, Conn	Dec. 14, 1869	97, 767
Pump	J. W. Douglas	Middletown, Conn	Nov. 25, 1873	144, 965
Pump	W. and B. Douglas	Middletown, Conn	Dec. 31, 1842	2, 895
Pump	L. Drescher	Matanzas, Cuba	Sept. 25, 1866	58, 360
Pump	C. H. Dreyer	Nashville, Tenn	Oct. 13, 1868	83, 051
Pump	S. S. Durbon	Lebanon, Ind	July 17, 1866	56, 387
Pump	A. Duvall	Baltimore, Md	Dec. 13, 1864	45, 398
Pump	J. B. Eads	Saint Louis, Mo	July 5, 1870	105, 056
Pump	C. F. Eastlack	Mantua, N. J	June 23, 1868	79, 217
Pump	J. Edson	Boston, Mass	Apr. 4, 1854	10, 746
Pump	J. Edson	Boston, Mass	May 9, 1854	10, 885
Pump	J. Edson	Boston, Mass	Nov. 8, 1859	26, 025
Pump	J. Edson	Boston, Mass	Oct. 21, 1873	143, 751
Pump	J. Edson and P. Noyes	Boston and Lowell, Mass	Aug. 22, 1871	118, 222
Pump	E. Edwards	Westminster, England	June 14, 1870	104, 289
Pump	J. Eldridge	West Buxton, Me	Nov. 21, 1865	51, 028
Pump	E. Elliott	Petaluma, Cal	Apr. 14, 1863	38, 157
Pump	D. S. Evans	Brockway, Mich	Jan. 5, 1869	85, 517
Pump	J. Evans	Whitestown, N. Y	Jan. 6, 1812	....
Pump	J. Evens	Lebanon, Ohio	Nov. 9, 1838	1, 002
Pump	J. Evens	Lebanon, Ohio	May 25, 1840	1, 615
Pump	J. L. Fagan	Anaqua, Tex	Feb. 22, 1859	23, 019
Pump	J. K. Fairbank	Waupun, Wis	Mar. 1, 1864	41, 764
Pump	D. L. Farnam	Philadelphia, Pa	Jan. 24, 1834	....
Pump	G. B. Farnam	Meriden, Conn	July 6, 1858	20, 783
Pump	J. Farnam	Stillwater, N. Y	Nov. 16, 1841	2, 370
Pump	B. W. Felthousen	Milwaukee, Wis	May 24, 1870	103, 435
Pump	A. Fisher	Nashua, N. H	Apr. 7, 1868	76, 318
Pump	E. Fitzgerald	New York, N. Y	Dec. 17, 1861	33, 936
Pump	A. Fitzpatrick	New York, N. Y	June 2, 1863	38, 736
Pump	E. Flanegin and A. B. Smith	Pittsburgh, Pa	Mar. 26, 1867	63, 235
Pump	A. F. Fletcher	Athol, Mass	Sept. 18, 1866	58, 085
Pump	A. F. Fletcher	Athol, Mass	Oct. 16, 1866	58, 801
Pump	A. V. and A. F. Fletcher	Athol, Mass	Oct. 25, 1864	44, 792
Pump	N. T. Fogg	Lewiston, Me	May 22, 1866	54, 881
Pump	D. Foley	Adams, N. Y	Mar. 12, 1872	124, 563
Pump	I. N. Forrester	Bridgeport, Conn	June 8, 1869	90, 940
Pump	I. N. Forrester	Bridgeport, Conn	Dec. 28, 1869	98, 366
Pump	I. N. Forrester	Bridgeport, Conn	Dec. 28, 1869	98, 367
Pump	I. N. Forrester	Bridgeport, Conn	Dec. 28, 1869	98, 368
Pump	I. N. Forrester and J. H. Luddington.	Bridgeport, Conn	Jan. 5, 1869	85, 577
Pump	G. R. Forsyth	Pemberton, Ohio	Dec. 24, 1867	72, 620
Pump	A. D. Foster	Jordan, N. Y	June 13, 1865	48, 165
Pump	T. Foulds, jr	Treverton, Pa	Nov. 5, 1872	132, 718
Pump	B. Frazee	Belleville, N. J	Dec. 5, 1865	51, 302
Pump	W. A. Fry	Worcester, Pa	Oct. 15, 1867	69, 792
Pump	R. M. Fryer	New York, N. Y	Jan. 14, 1868	73, 316
Pump	A. Fuller	Marietta, Pa	Nov. 30, 1869	97, 382
Pump	A. Fuller and F. J. Bray	Buffalo, N. Y	Mar. 26, 1867	63, 152
Pump	G. W. Fulton	Baltimore, Md	May 29, 1849	6, 486
Pump	J. P. Gallagher	Saint Louis, Mo	May 2, 1871	114, 284
Pump	G. W. Gardner and O. Higgins	Napoleon, Ohio	Mar. 31, 1868	76, 072
Pump	J. G. Garretson	Salem, Iowa	Jan. 19, 1864	41, 291
Pump	A. D. Gates	Surrey, Wis	Oct. 26, 1869	96, 099
Pump	G. W. B. Gedney	New York, N. Y	Jan. 6, 1857	16, 366
Pump	A. A. Genung	Painesville, Ohio	July 6, 1858	20, 787
Pump	H. Getty	Brooklyn, N. Y	July 23, 1867	67, 037

Index of patents issued from the United States Patent Office from 1790 *to* 1873, *inclusive*—Continued.

Invention.	Inventor.	Residence.	Date.	No.
Pump	R. Gilliland and R. H. Armstrong.	Hudson, Mich	Mar. 24, 1868	75, 747
Pump	K. Goddard	Philadelphia, Pa	Apr. 11, 1865	47, 200
Pump	J. Goland	Batavia, Ill	Aug. 4, 1863	39, 391
Pump	J. Goodyear	Carlisle Borough, Pa	Apr. 10, 1866	53, 810
Pump	D. J. Gorton	West Eau Claire, Wis	Mar. 15, 1870	100, 885
Pump	D. J. Gorton	Quincy, Ill	July 4, 1871	116, 703
Pump	L. A. Gould	Santa Clara, Cal	Mar. 6, 1866	52, 991
Pump	R. J. Gould	Newark, N. J	Feb. 15, 1870	99, 877
Pump	T. B. Goulding	Columbus, Ga	May 9, 1871	114, 668
Pump	E. Graser	Union City, Pa	Mar. 12, 1872	124, 434
Pump	A. G. Gray	London, England	Sept. 1, 1863	39, 728
Pump	S. H. Gray	Bridgeport, Conn	Aug. 21, 1855	13, 459
Pump	S. H. Gray	Bridgeport, Conn	Sept. 14, 1858	21, 560
Pump	J. Greaves	Utica, N. Y	June 24, 1862	35, 677
Pump	J. Greaves	Utica, N. Y	Oct. 21, 1862	36, 708
Pump	W. W. Green	Cambridge, Pa	Oct. 22, 1872	132, 462
Pump	M. T. Greenleaf	Quincy, Ill	July 16, 1861	32, 831
Pump	C. L. and J. P. Griscom	Mahanoy Plane, Pa	Sept. 28, 1869	95, 222
Pump	J. P. Gruber	New York, N. Y	Aug. 28, 1866	57, 499
Pump	B. F. Gustin	Middletown, Ind	Apr. 19, 1870	102, 199
Pump	E. Hale	Terre Haute, Ind	Jan. 5, 1864	41, 154
Pump	E. J. Hall	Indianapolis, Ind	Nov. 24, 1868	84, 274
Pump	M. P. Hall	Gayville, Ill	Feb. 15, 1870	99, 884
Pump	W. M. Hamilton	Jacksonville, Ill	Oct. 19, 1869	96, 000
Pump	W. M. Hamilton	Jacksonville, Ill	May 31, 1870	103, 738
Pump	E. C. Hammond	Oswego, N. Y	Mar. 29, 1870	101, 261
Pump	J. Hampson and G. Ladue	Newburgh, N. Y	Apr. 18, 1865	47, 299
Pump	T. Hansbrow	Sacramento, Cal	Feb. 5, 1861	31, 314
Pump	T. Hansbrow	Sacramento, Cal	Dec. 15, 1863	40, 927
Pump	T. Hansbrow	Sacramento, Cal	May 15, 1866	54, 719
Pump	T. Hansbrow	Sacramento, Cal	Nov. 24, 1868	81, 355
Pump	A. S. Hanson	Milan, Mich	Sept. 11, 1866	57, 897
Pump	M. Hanstine	Waynesborough, Pa	Dec. 7, 1869	97, 633
Pump	H. A. M. Harris	Philadelphia, Pa	Feb. 20, 1866	52, 709
Pump	H. A. M. Harris	Philadelphia, Pa	Sept. 24, 1867	69, 092
Pump	W. H. Harrison	Philadelphia, Pa	June 23, 1857	17, 625
Pump	R. Hartley	Pittsburgh, Pa	July 10, 1866	56, 218
Pump	W. Hartley	Durand, Ill	Dec. 6, 1870	109, 896
Pump	E. Hartzler	Orville, Ohio	July 26, 1859	24, 868
Pump	P. Harvey	Chicago, Ill	Jan. 24, 1871	111; 120
Pump	A. L. Hatfield	Clyde, Ohio	Oct. 21, 1873	143, 759
Pump	M. C. Hawkins	Edinborough, Pa	Aug. 24, 1869	93, 990
Pump	D. Hayes	New York, N. Y	Mar. 26, 1861	31, 804
Pump	W. M. Henderson	Baltimore, Md	Oct. 4, 1859	25, 642
Pump	W. M. Henderson	Baltimore, Md	Apr. 28, 1863	38, 308
Pump	W. Hendrick	New York	June 12, 1818	......
Pump	D. P. Henry	Windsor, Ill	May 4, 1869	89, 662
Pump	R. Henry	Morrisania, N. Y	June 17, 1862	35, 609
Pump	F. Henshaw	Washington, D. C	Sept. 14, 1858	21, 561
Pump	S. Hewit	Seneca Falls, N. Y	May 19, 1857	17, 327
Pump	S. Hewit	Seneca Falls, N. Y	Oct. 25, 1859	25, 898
Pump	G. Hibberd	Wheeling, W. Va	Oct. 25, 1870	108, 705
Pump	G. Hibsch	Buffalo, N. Y	July 13, 1858	20, 880
Pump	S. C. Higbie	Openheim, N. Y	Mar. 21, 1843	3, 010
Pump	O. Higgins	Napoleon, Ohio	Feb. 7, 1871	111, 645
Pump	B. S. Hill	New York, N. Y	June 27, 1865	48, 401
Pump	J. M. Hirlinger	Red Rock, Pa	Nov. 12, 1867	70, 842
Pump	S. S. Hogle	Lansingburgh, N. Y	May 29, 1841	2, 112
Pump	B. Holly	Seneca Falls, N. Y	June 5, 1849	6, 500
Pump	B. Holly	Seneca Falls, N. Y	July 14, 1857	17, 820
Pump	B. Holly	Lockport, N. Y	Feb. 14, 1860	27, 128
Pump	B. Holly	Lockport, N. Y	July 24, 1860	29, 266
Pump	B. Holly	Lockport, N. Y	July 14, 1863	39, 259
Pump	B. Holly	Lockport, N. Y	Nov. 15, 1864	45, 040
Pump	D. O. Holman	Adams, N. Y	Oct. 11, 1870	108, 264
Pump	D. O. Holman	Adams, N. Y	Oct. 11, 1870	108, 265
Pump	J. Holmes	Saint Clair, Pa	Oct. 2, 1860	30, 221
Pump	C. Hood	Seneca Falls, N. Y	Apr. 17, 1860	27, 905
Pump	W. D. Hooker	Stockton, Cal	Aug. 15, 1865	49, 408
Pump	J. W. Hopkins	New York, N. Y	Jan. 30, 1872	123, 175
Pump	A. Horn	Easton, Pa	Jan. 6, 1809	......
Pump	H. Hosick	Paris, Pa	Jan. 19, 1864	41, 298
Pump	N. Hotz	Brooklyn, N. Y	Feb. 14, 1865	46, 360
Pump	J. G. Hovey	Waverly, Iowa	Nov. 28, 1865	51, 183
Pump	B. J. C. Howe	Syracuse, N. Y	July 12, 1864	43, 502
Pump	B. J. C. Howe	Syracuse, N. Y	Oct. 11, 1864	44, 627
Pump	C. W. Hoyt	South Norwalk, Conn	Oct. 13, 1868	83, 065
Pump	H. S. Huges	Warren County, Va	June 29, 1839	1, 207
Pump	A. J. Hull	Sterling, Ill	Nov. 14, 1871	120, 973
Pump	J. Humphrey	Keene, N. H	Feb. 14, 1871	111, 748
Pump	H. C. Hunt and G. W. Devin	Ottumwa, Iowa	Nov. 4, 1862	36, 837
Pump	J. I. Hurlbutt	Norwalk, Ohio	July 12, 1870	105, 337
Pump	J. Icard	Donaldsonville, La	Nov. 29, 1870	109, 624
Pump	W. L. Jacobs	Lancaster, Pa	Sept. 23, 1842	2, 785
Pump	D. L. Jaques	Hudson, Mich	Oct. 23, 1866	59, 030
Pump	E. A. Jeffery and J. D. Quackenbush.	Corning, N. Y	June 10, 1862	35, 522
Pump	E. S. Jenkins	Brooklyn, N. Y	Jan. 17, 1871	110, 976
Pump	J. Johnson and C. W. Singer	Saginaw, Mich., and Anderson's Store, Va.	Mar. 6, 1866	53, 007
Pump	J. B. Johnson	Boston, Mass	Dec. 4, 1860	30, 825
Pump	N. Johnson	Ripon, Wis	Dec. 26, 1865	51, 726
Pump	N. Johnson	Ripon, Wis	Aug. 28, 1866	57, 516
Pump	W. J. Johnson	Newton, Mass	Oct. 23, 1860	30, 480
Pump	T. O. Jones	Galesburgh, Ill	Jan. 31, 1871	111, 347
Pump	W. F. Jones	Syracuse, N. Y	June 22, 1869	91, 750

Index of patents issued from the United States Patent Office from 1790 to 1873, inclusive—Continued.

Invention.	Inventor.	Residence.	Date.	No.
Pump	J. Jonson	Baltimore, Md	Sept. 12, 1871	118, 859
Pump	J. O. Joyce	Cincinnati, Ohio	Mar. 23, 1858	19, 699
Pump	J. O. Joyce	Dayton, Ohio	Nov. 10, 1868	83, 858
Pump	E. B. Jucket	Pawtucket, R. I	Jan. 3, 1865	45, 722
Pump	E. B. Jucket	Roxbury, Mass	Aug. 28, 1866	57, 635
Pump	W. S. Judd	Clanhassen, Minn	Aug. 18, 1863	39, 577
Pump	A. C. Judson	Grand Rapids, Ohio	Dec. 21, 1869	98, 067
Pump	A. Jusberg	Galva, Ill	Oct. 22, 1867	70, 000
Pump	A. B. Keeley and J. S. Bleck	Philadelphia, Pa	May 17, 1859	24, 032
Pump	W. H. Keek	Stockton, Cal	Sept. 1, 1868	81, 648
Pump	A. L. Keeports and G. Palmer	Littlestown, Pa	Nov. 30, 1858	22, 182
Pump	H. M. Keith	Commerce, Mich	July 11, 1865	48, 691
Pump	L. L. Kellogg	Leon Centre, N. Y	Aug. 10, 1869	93, 537
Pump	W. S. Kelly	Schenectady, N. Y	Feb. 25, 1862	34, 505
Pump	W. S. Kelly	Schenectady, N. Y	Sept. 13, 1864	44, 272
Pump	W. S. Kelly	Schenectady, N. Y	Jan. 15, 1867	61, 209
Pump	H. K. Kenyon	Steubenville, Ohio	Nov. 29, 1870	109, 631
Pump	J. T. Kimbel	Vernon, Ind	Nov. 5, 1867	70, 443
Pump	A. Knecht	Ilchester, Md	Oct. 18, 1870	108, 365
Pump	J. Knibbs	Troy, N. Y	May 24, 1864	42, 920
Pump	L. J. Knowles	Worcester, Mass	June 27, 1871	116, 454
Pump	T. C. H. Kraus	Fishkill, N. Y	Aug. 19, 1873	142, 028
Pump	F. Krieg	Baltimore, Md	Feb. 6, 1833	....
Pump	R. M. Lafferty	Three Rivers, Mich	June 21, 1870	104, 604
Pump	R. M. Lafferty	Toledo, Ohio	May 28, 1872	127, 173
Pump	L. M. Laighton	Portsmouth, N. H	Sept. 1, 1817	....
Pump	E. Lamphear	Stepney, Conn	Mar. 22, 1870	101, 138
Pump	J. W. Lane	Newton, N. J	Nov. 19, 1861	33, 765
Pump	S. Lane, jr	Englewood, N. J	Nov. 26, 1872	133, 373
Pump	S. Lane, jr	Englewood, N. J	Oct. 21, 1873	143, 769
Pump	A. C. Lanning	Wilkesbarre, Pa	Apr. 19, 1859	23, 690
Pump	T. J. Lapsley	Nashville, Tenn	Jan. 31, 1871	111, 354
Pump	G. H. Laub	West Lebanon, Ind	Jan. 24, 1871	111, 219
Pump	E. Lawrence	Antrim, N. H	Nov. 8, 1864	44, 963
Pump	E. Lawrence and R. Lafley, 2d	Waterford, N. Y	Mar. 1, 1859	23, 094
Pump	A. D. Laws and J. C. Cooke	Bridgeport, Conn	June 6, 1871	115, 745
Pump	A. Leuchtweiss	Cincinnati, Ohio	Apr. 10, 1867	53, 939
Pump	W. B. Le Van	Philadelphia, Pa	May 16, 1865	47, 737
Pump	C. N. Lewis	Seneca Falls, N. Y	Jan. 1, 1856	14, 024
Pump	D. W. Lewis	Janesville, Wis	July 23, 1861	32, 881
Pump	E. T. Ligon	Richmond, Va	Oct. 14, 1856	15, 888
Pump	G. Lindsay	Petersburgh, Va	May 1, 1860	28, 090
Pump	H. Lindsey	Asheville, N. C	Dec. 22, 1857	18, 916
Pump	H. Lindsey	Asheville, N. C	Dec. 4, 1855	13, 881
Pump	T. Ling	New York, N. Y	Apr. 30, 1867	64, 337
Pump	T. J. Linton	Providence, R. I	Mar. 7, 1865	46, 683
Pump	A. W. Lloyd	Otis, Mass	May 3, 1859	23, 849
Pump	H. Locke	South Boston, Mass	Apr. 30, 1861	32, 196
Pump	F. R. Lockling	Hannibal, Mo	Aug. 20, 1872	130, 647
Pump	D. Loomis, J. Winters, and A. Stark.	Clyde, Ohio	Feb. 28, 1871	112, 157
Pump	G. W. Low	Ravenna, Ohio	Feb. 22, 1870	100, 162
Pump	J. H. Luddington	Bridgeport, Conn	May 4, 1869	89, 672
Pump	J. M. Lunquest	Griffin, Ga	Mar. 1, 1859	23, 100
Pump	R. M. Marchant	London, England	Dec. 20, 1870	110, 380
Pump	C. Markley	New York, N. Y	Nov. 29, 1870	109, 639
Pump	J. Marquis	San Francisco, Cal	Nov. 7, 1871	120, 655
Pump	G. Marshall	New York, N. Y	May 9, 1865	47, 652
Pump	G. Marshall	Brooklyn, N. Y	July 28, 1868	80, 418
Pump	G. W. Martin	Morrisania, N. Y	Apr. 9, 1861	32, 030
Pump	S. G. Mason and C. B. Gill	Rochester, N. Y	Feb. 6, 1866	52, 431
Pump	L. Matthews	Antrim, Ohio	Apr. 24, 1860	27, 996
Pump	J. M. May	Janesville, Wis	Aug. 23, 1859	25, 207
Pump	J. M. May	Janesville, Wis	Aug. 23, 1859	25, 208
Pump	J. M. May	Janesville, Wis	May 7, 1861	32, 251
Pump	J. Mayher	East Hampton, Mass	May 14, 1872	126, 643
Pump	T. Mayhew	New York, N. Y	Jan. 30, 1866	52, 364
Pump	J. McBride	Flint, Mich	Jan. 14, 1868	73, 253
Pump	A. McCarter	Norristown, Pa	Apr. 21, 1863	38, 234
Pump	R. A. McCauley	Baltimore, Md	Aug. 29, 1865	49, 640
Pump	R. A. McCauley	Baltimore, Md	Dec. 26, 1865	51, 736
Pump	R. A. McCauley	Baltimore, Md	Jan. 1, 1867	60, 919
Pump	A. Y. McDonald	Dubuque, Iowa	May 24, 1870	103, 481
Pump	J. H. McGowan	Cincinnati, Ohio	Nov. 22, 1870	109, 534
Pump	J. H. McGowan	Cincinnati, Ohio	Jan. 24, 1871	111, 227
Pump	J. H. McGowan	Cincinnati, Ohio	Jan. 2, 1872	122, 475
Pump	J. H. and T. J. McGowan	Cincinnati, Ohio	Nov. 12, 1867	70, 733
Pump	T. J. McGowan	Cincinnati, Ohio	Oct. 28, 1862	36, 844
Pump	T. J. McGowan	Cincinnati, Ohio	May 26, 1863	38, 689
Pump	T. J. McGowan	Cincinnati, Ohio	Sept. 1, 1868	81, 806
Pump	T. J. McGowan	Cincinnati, Ohio	Feb. 13, 1872	123, 636
Pump	B. F. McKeehan	Clarksburgh, W. Va	Mar. 14, 1871	112, 730
Pump	C. S. McMahan	Centreville, Ind	June 2, 1868	78, 602
Pump	J. L. McPherson	Sacramento City, Cal	July 8, 1862	35, 829
Pump	C. L. Merrill	Watertown, N. Y	Dec. 7, 1869	97, 670
Pump	C. L. Merrill	Watertown, N. Y	Aug. 2, 1870	106, 072
Pump	C. L. Merrill	Watertown, N. Y	Feb. 21, 1871	111, 961
Pump	J. W. Merrill and E. H. Lawrence.	Berlin, Mass	Aug. 13, 1867	67, 666
Pump	M. Metee	Baltimore, Md	May 20, 1830	....
Pump	M. Mettee	Baltimore, Md	Mar. 8, 1833	....
Pump	J. M. P. Metivier	Paris, France	Apr. 7, 1863	38, 117
Pump	T. Metzler	Wooster, Ohio	Oct. 26, 1869	96, 136
Pump	J. Michel	Rochester, N. Y	Mar. 27, 1866	53, 467
Pump	N. Miller	Finley, Ohio	June 18, 1861	32, 577
Pump	O. Miller	Salem, Ohio	Dec. 18, 1866	60, 535
Pump	G. E. Mills	New York, N. Y	July 18, 1865	48, 826
Pump	G. H. Mills	East Boston, Mass	May 22, 1860	28, 439

Index of patents issued from the United States Patent Office from 1790 *to* 1873, *inclusive*—Continued.

Invention.	Inventor.	Residence.	Date.	No.
Pump	J. R. Mills	Bloomfield, Iowa	Oct. 14, 1862	36, 665
Pump	A. Miner	Jordan, N. Y	July 7, 1835	....
Pump	A. T. Mixsell	Oxford, N. J	Sept. 28, 1839	1, 346
Pump	A. T. Mixwell	Oxford, N. J	June 30, 1836	....
Pump	A. Moon	Maquoketa, Iowa	June 11, 1867	65, 684
Pump	J. A. Morrell	Saint Charles, Mo	Jan. 7, 1862	34, 068
Pump	J. A. Morrell	Chicago, Ill	Oct. 29, 1867	70, 244
Pump	J. A. Morrell	New York, N. Y	June 8, 1869	91, 151
Pump	W. Mullally	Saint Paul, Minn	Oct. 28, 1862	36, 835
Pump	W. B. Munger	Hillsdale, Mich	Nov. 1, 1864	44, 879
Pump	A. Munson	Peterborough, N. H	June 3, 1862	35, 465
Pump	J. Munson	San José, Cal	June 14, 1864	43, 124
Pump	J. L. Munson	New Haven, Conn	Apr. 26, 1819	....
Pump	J. Neal and C. W. Emery	Boston, Mass	Jan. 1, 1856	14, 029
Pump	J. P. Nevens	Starks, Mo	May 10, 1864	42, 681
Pump	N. Newman	Cincinnati, Ohio	May 6, 1851	8, 078
Pump	J. Nicholson	Allegheny City, Pa	Jan. 22, 1867	61, 447
Pump	T. Nicholson	Washington, D. C	Apr. 29, 1817	....
Pump	T. Nicholson and S. Gum	Genoa, N. Y	Aug. 6, 1813	....
Pump	H. Norman and C. F. Dietrich	New Orleans, La	Feb. 25, 1873	136, 258
Pump	S. B. B. Nowlan	New York, N. Y	Nov. 16, 1869	96, 947
Pump	W. H. Noyes	Franklin, Pa	June 4, 1867	65, 419
Pump	G. F. Nutting	Randolph, Vt	June 23, 1868	79, 143
Pump	J. K. O'Neil	Kingston, N. Y	Apr. 5, 1859	23, 485
Pump	M. S. Orton	Galesburgh, Ill	Mar. 29, 1870	101, 304
Pump	D. C. Owen	Adams County, Ill	May 11, 1869	89, 885
Pump	N. Page, jr	Danvers, Mass	Oct. 8, 1867	69, 578
Pump	A. Palmer	Hudson, Mich	Dec. 8, 1868	84, 754
Pump	G. Palmer	Littlestown, Pa	June 5, 1860	28, 599
Pump	G. Palmer	Littlestown, Pa	July 15, 1862	35, 888
Pump	H. H. Palmer	Rockford, Ill	Apr. 29, 1862	35, 108
Pump	A. N. Parkhurst	Peoria, Ill	Feb. 10, 1863	37, 638
Pump	A. N. Parkhurst	Peoria, Ill	Jan. 16, 1866	52, 070
Pump	A. N. Parkhurst	Knoxville, Ill	Nov. 2, 1869	96, 475
Pump	H. Parks	Athens, N. Y	June 11, 1867	65, 690
Pump	J. S. Patric	Victor, N. Y	May 1, 1866	54, 399
Pump	J. S. Patric	Rochester, N. Y	Feb. 28, 1871	112, 177
Pump	J. Patrick	Pitt County, N. C	Sept. 7, 1869	94, 507
Pump	T. Patterson	New York, N. Y	June 4, 1867	65, 424
Pump	J. Peabody	Dixmont Centre, Me	June 13, 1865	48, 202
Pump	F. S. Pease	Buffalo, N. Y	Dec. 12, 1865	51, 474
Pump	F. S. Pease	Buffalo, N. Y	Jan. 9, 1867	51, 965
Pump	H. Pease	Rockport, N. Y	July 7, 1857	17, 768
Pump	H. Pease	Rockport, N. Y	Sept. 11, 1860	29, 989
Pump	W. Peck	Rockford, Ill	Jan. 25, 1859	22, 743
Pump	W. Peck	Rockford, Ill	Jan. 3, 1860	26, 705
Pump	W. Peck	Rockford, Ill	May 8, 1860	28, 193
Pump	J. B. Pell	New York	Oct. 31, 1829	....
Pump	A. M. Perkins	Springfield, Mass	Apr. 2, 1861	31, 904
Pump	J. Perkins		July 9, 1802	....
Pump	J. Perkins	Newburyport, Mass	Mar. 23, 1813	....
Pump	A. Perry and M. C. Hawkins	Edinborough, Pa	Apr. 16, 1867	63, 937
Pump	E. Perry	Baldwinsville, N. Y	Sept. 20, 1864	44, 337
Pump	N. Phelps	Turin, N. Y	Oct. 20, 1831	....
Pump	J. Pike and W. Fisk	Cumberland, R. I	Aug. 13, 1839	1, 285
Pump	J. H. Plank	Pulaski, Iowa	Nov. 24, 1863	40, 708
Pump	J. R. Pomroy and L. J. Walter	Lockport, N. Y	Apr. 27, 1869	89, 338
Pump	J. Poppe	Green Point, N. Y	Sept. 1, 1868	81, 681
Pump	A. F. Porter	Philadelphia, Pa	Apr. 25, 1865	47, 452
Pump	C. Powell	Birmingham, England	Aug. 24, 1869	94, 027
Pump	J. Powers	New York, N. Y	Apr. 5, 1859	23, 489
Pump	G. W. Preston	Clyde, Ohio	May 7, 1872	126, 574
Pump	O. W. Preston, jr	Corning, N. Y	Nov. 30, 1858	22, 201
Pump	A. J. Pritchard	Liverpool, Ohio	May 25, 1869	90, 576
Pump	F. F. Prud'homme	Paris, France	Sept. 22, 1863	40, 062
Pump	W. Purcell	New York	Nov. 1, 1817	....
Pump	A. M. Putnam	Antrim, N. H	May 13, 1862	35, 258
Pump	A. M. Putnam	Antrim, N. H	Aug. 31, 1869	94, 338
Pump	S. S. Putnam	Boston, Mass	June 1, 1858	20, 442
Pump	W. Race	Seneca Falls, N. Y	May 22, 1860	28, 405
Pump	W. Race	Lockport, N. Y	June 30, 1863	39, 065
Pump	R. Ramsden	South Easton, Pa	Apr. 24, 1860	28, 012
Pump	S. G. Randall	Worcester, Mass	Feb. 28, 1860	27, 308
Pump	F. Ransom	Buffalo, N. Y	July 11, 1865	48, 720
Pump	C. D. Rathbone	Belpre, Ohio	July 8, 1873	140, 643
Pump	J. W. Redding	Belleville, Ohio	July 2, 1861	32, 716
Pump	J. Redeslperger	Mansfield, N. J	Oct. 31, 1835	....
Pump	J. Reed	Marshfield, Mass	Aug. 5, 1831	....
Pump	J. Reed	Marshfield, Mass	July 24, 1838	853
Pump	J. Reed	Marshfield, Mass	Apr. 16, 1841	2, 055
Pump	K. Reed	New York, N. Y	Feb. 6, 1866	52, 446
Pump	H. W. Regan	Cressona, Pa	Oct. 12, 1858	21, 801
Pump	A. Richard	Washington, Mo	Feb. 7, 1860	27, 070
Pump	J. W. Reid	New York, N. Y	Jan. 5, 1864	41, 091
Pump	A. Rewerick	San Francisco, Cal	Mar. 13, 1866	53, 185
Pump	A. J. Reynolds	Dayton, Ohio	Mar. 3, 1863	37, 826
Pump	A. J. Reynolds	Sturgis, Mich	May 10, 1864	42, 686
Pump	A. J. Reynolds	Sturgis, Mich	Sept. 24, 1867	69, 250
Pump	G. A. Reynolds and G. H. Babcock.	Mystic Bridge, Conn	Dec. 15, 1863	40, 945
Pump	E. Rhoads, sr	Clyde, Ohio	Oct. 7, 1862	36, 622
Pump	F. Richter	Milwaukee, Wis	June 4, 1867	65, 435
Pump	T. Rider	Valparaiso, Chili	Mar. 28, 1865	47, 075
Pump	J. Roberts	New Madison, Ohio	Oct. 3, 1871	119, 650
Pump	J. Robinson and L. Shaw	Bath, Me	Feb. 1, 1827	....
Pump	W. F. Robinson	Bellaire, Ohio	July 25, 1871	117, 463
Pump	W. W. Robinson	Ripon, Wis	June 4, 1861	32, 485
Pump	J. M. Roebuck and W. R. Reece	Donaldson, Pa	July 3, 1866	56, 101

Index of patents issued from the United States Patent Office from 1790 *to* 1873, *inclusive*—Continued.

Invention.	Inventor.	Residence.	Date.	No.
Pump	H. Roesen	Elkhart, Ind	May 5, 1863	38, 416
Pump	C. Rogers	Allegheny City, Pa	Feb. 11, 1868	74, 431
Pump	J. Ross	Greenville, Mich	Jan. 22, 1867	61, 468
Pump	P. C. Rowe	Boston, Mass	Nov. 8, 1864	44, 978
Pump	M. V. Rowley	Worcester, N. Y	Nov. 8, 1870	109, 057
Pump	M. E. Rudasill	Shelby, N. C	Oct. 9, 1860	30, 354
Pump	A. Russell	Newburyport, Mass	Aug. 6, 1861	33, 008
Pump	E. G. Russell	Ravenna, Ohio	Aug. 26, 1873	142, 123
Pump	S. Russell	Olean, N. Y	Feb. 4. 1826	....
Pump	E. B. Scammon	Scipio, N. Y	June 20. 1816	....
Pump	L. B. Schafer	Baltimore, Md	Mar. 22, 1859	23, 314
Pump	C. Schmidt	Rock Island, Ill	July 6, 1869	92, 373
Pump	A.H. Scholfield and D.C. Sterry	Worcester, Mass	June 23, 1868	79, 150
Pump	J. Scott	Earlville, Iowa	Dec. 28, 1869	98, 422
Pump	J. and J. Seeberger	West Troy, N. Y	May 11, 1869	89, 892
Pump	G. C. Selfridge	North Greenfield, N. Y	Sept. 4, 1860	29, 918
Pump	J. Selser	Williamsport, Pa	May 10, 1859	23, 953
Pump	G. F. and H. F. Shaw	West Roxbury, Mass	Feb. 21, 1871	111, 983
Pump	J. Shaw	Bridgeport, Conn	Aug. 25, 1868	81, 417
Pump	J. W. Sheaffer	Sterling, Ill	May 1, 1866	54, 420
Pump	W. Shearer	Atlanta, Ga	Aug. 21, 1860	29, 723
Pump	W. Shearer	Atlanta, Ga	Nov. 29, 1870	109, 678
Pump	H. A. Sheldon	Middlebury, Vt	Dec. 15, 1857	18, 870
Pump	N. P. Sheldon	San Francisce, Cal	Dec. 20, 1870	110, 297
Pump	N. P. Sheldon	San Francisco, Cal	Jan. 10, 1871	110, 936
Pump	A. Sherman and J. W. Sheaffer	Sterling, Ill	Nov. 15, 1870	109, 349
Pump	G. Shield	Cincinnati, Ohio	Nov. 5, 1867	70, 634
Pump	C. W. Sholes and H. C. Kelly	Morrison, Ill	Dec. 2, 1873	145, 247
Pump	W. Shoup	Saltsburgh, Pa	Dec. 27, 1864	45, 647
Pump	J. A. Sinclair	Woodsfield, Ohio	Apr. 26, 1870	102, 442
Pump	J. A. Sinclair	Woodsfield, Ohio	June 6, 1871	115, 776
Pump	J. A. Sinclair	Woodsfield, Ohio	Sept. 17, 1872	131, 469
Pump	S. B. Sivertson	Chicago, Ill	Dec. 6, 1870	109, 955
Pump	W. N. Slason	South Reading, Mass	Mar. 20, 1860	27, 576
Pump	A. Sluthour	Cleveland, Ohio	Oct. 19, 1869	95, 943
Pump	J. Smart	Easton, Pa	Feb. 3, 1838	589
Pump	J. Smart	Northern Liberties, Pa	Mar. 28, 1854	10, 708
Pump	R. T. Smart and R. T. Smart, jr.	Troy, N. Y	Jan. 30, 1872	123, 203
Pump	J. A. C. J. Smith	Philadelphia, Pa	July 10, 1860	29, 110
Pump	O. Snell	Williamsburgh, Ohio	Dec. 8, 1868	84, 846
Pump	P. M. and O. Snell	Williamsburgh, Ohio	Jan. 28, 1870	104, 786
Pump	P. M. and O. Snell	Williamsburgh, Ohio	Aug. 9, 1870	106, 226
Pump	T. J. Southard	Richmond, Me	July 2, 1861	32, 724
Pump	C. M. Soward and J. A. Dykeman.	Dyer County, Tenn	Dec. 17, 1872	133, 955
Pump	H. Spear	Portland, Me	Nov. 21, 1871	121, 209
Pump	H. Spear	Cape Elizabeth, Me	Nov. 19, 1872	133, 266
Pump	W. P. Squire	Paris, Ill	Nov. 20, 1866	59, 869
Pump	J. A. Stansbury	Baldwinsville, N. Y	July 24, 1866	56, 627
Pump	E. L. Staples	Nashville, Tenn	May 22, 1866	55, 022
Pump	G. Steck and F. Arnold	Hughesville, Pa	Nov. 14, 1871	121, 016
Pump	N. Stedman	Aurora, Ind	Apr. 28, 1863	38, 360
Pump	N. Stedman	Aurora, Ind	Sept. 8, 1863	39, 842
Pump	J. M. Stephenson	Anderson, Ind	May 22, 1860	28, 417
Pump	C. A. Stillman	Westerly, R. I	Dec. 12, 1865	51, 522
Pump	H. M. Stoker	Watson, Ill	Apr. 3, 1866	53, 750
Pump	S. D. Stout	Charleston, Tenn	Oct. 9, 1860	30, 363
Pump	R. E. Strait	West Oneonta, N. Y	Aug. 28, 1866	57, 595
Pump	R. E. Strait	Battle Creek, Mich	Oct. 29, 1867	70, 285
Pump	N. Sutton	New York, N. Y	Nov. 3, 1857	18, 559
Pump	N. Sutton	Detroit, Mich	Apr. 18, 1865	47, 344
Pump	Z. T. Sweet	Tiskilwa, Ill	May 16, 1871	114, 989
Pump	C. O. Sylvester	Attica, Ind	July 9, 1872	128, 921
Pump	J. Tapley	Frankfort, Me	Sept. 19, 1854	11, 704
Pump	C. L. Taverdon and J. Moret	Paris and Sevres, France	Sept. 29, 1868	82, 659
Pump	A. Thayer	Chatham, N. Y	Feb. 1, 1842	2, 438
Pump	D. M. Thomas	Dowagiac, Mich	July 24, 1866	56, 637
Pump	S. Thomas	Boston, Mass	Oct. 10, 1817	....
Pump	W. H. Thomas	Sacramento, Cal	Mar. 29, 1864	42, 128
Pump	J. E. Thorp	New York, N. Y	July 29, 1862	36, 042
Pump	W. P. Tilton	Boston, Mass	May 6, 1817	....
Pump	J. Tomlinson	Racine, Wis	Apr. 19, 1864	42, 417
Pump	A. Tower	New York, N. Y	Feb. 3, 1857	16, 558
Pump	F. W. Tulley and T. Reece	Philadelphia, Pa	July 17, 1866	56, 468
Pump	H. Tyler	Gaines, N. Y	Sept. 10, 1867	68, 810
Pump	H. Tyler	Gaines, N. Y	Oct. 29, 1867	70, 379
Pump	F. J. Underwood	Rock Island, Ill	Oct. 8, 1867	69, 595
Pump	J. Underwood	Mason County, Ill	Jan. 5, 1869	85, 548
Pump	D. Updike	New York	Feb. 19, 1818	....
Pump	C. D. Van Allen	Petersburgh, Pa	July 8, 1841	2, 160
Pump	G. Van Camp	Somerset County, N. J	Sept. 17, 1861	33, 323
Pump	S. Vance	Newburyport Mass	Jan. 30, 1866	52, 346
Pump	H. Van Keuren	Jersey City, N. J	Aug. 16, 1870	106, 433
Pump	J. Vaughn and J. McGee	Galena, Ill	Feb. 18, 1868	74, 638
Pump	C. Verniand and D. J. Lucie	Quincy, Ill	Aug. 20, 1867	68, 014
Pump	W. W. Virdin	Baltimore, Md	Sept. 15, 1863	39, 974
Pump	W. T. Vose	Newtonville, Mass	Mar. 20, 1855	12, 566
Pump	E. Wade	Norwich, Conn	June 12, 1860	28, 704
Pump	T. J. Wadleigh	Sutton, N. H	Nov. 5, 1861	33, 671
Pump	H. Wadsworth	Duxbury, Mass	Apr. 26, 1870	102, 452
Pump	J. F. Walker	Easton, Pa	May 6, 1836	....
Pump	F. Walther	Winchester, Va	May 1, 1845	4, 026
Pump	J. J. Walton	New York, N. Y	Dec. 23, 1873	145, 918
Pump	R. Ward	Newark, N. J	Feb. 27, 1872	124, 024
Pump	S. B. Ward	Drummondtown, Va	Feb. 2, 1869	86, 612
Pump	W. G. Ward	Fayette, N. Y	Dec. 13, 1870	110, 177
Pump	C. Warner	Lexington, Ky	Nov. 10, 1841	2, 350
Pump	C. Warner	Brooklyn, N. Y	July 31, 1860	29, 424

Index of patents issued from the United States Patent Office from 1790 *to* 1873, *inclusive*—Continued.

Invention.	Inventor.	Residence.	Date.	No.
Pump	Z. Waters and S. Bradley	Bloomington, Ill	Dec. 12, 1871	121, 918
Pump	W. E. Watters	East Bend, Ky	May 7, 1861	32, 260
Pump	W. E. Watters	East Bend, Ky	Jan. 3, 1865	45, 799
Pump	J. H. Webster	Saint Louis, Mo	Feb. 28, 1845	3, 933
Pump	J. R. Weisiger	Danville, Ky	Jan. 22, 1867	61, 492
Pump	J. R. Weisiger	Danville, Ky	Mar. 3, 1868	75, 088
Pump	V. Weitz	Cleveland, Ohio	June 18, 1861	32, 592
Pump	J. D. West	New York, N. Y	Sept. 29, 1857	18, 309
Pump	E. C. Wharton	Philadelphia, Pa	Oct. 28, 1873	144, 171
Pump	E. C. Wharton	Philadelphia, Pa	Dec. 23, 1873	145, 921
Pump	N. W. Wheeler	Morristown, N. J	Nov. 14, 1871	120, 841
Pump	W. M. Wheeler	Liberty, Mo	May 15, 1841	2, 097
Pump	J. F. Whipple	Chicago, Ill	May 3, 1870	102, 634
Pump	J. W. Whitaker	Kenosha, Wis	Nov. 19, 1872	133, 278
Pump	J. F. White	Keene, N. H	Apr. 4, 1865	47, 146
Pump	P. B. and H. M. C. White	Dowagiac, Mich	Mar. 3, 1868	75, 228
Pump	R. White and D. Moritz	New York, N. Y	Sept. 24, 1872	131, 644
Pump	S. White	Providence, R. I	Mar. 24, 1868	75, 819
Pump	J. Wilcke	Newark, N. J	May 5, 1868	77, 694
Pump	M. Wilcox	Sacramento, Cal	Aug. 30, 1870	106, 904
Pump	R. Wileman	East Hartford, Conn	July 9, 1829	....
Pump	T. J. Willett	Nunda, N. Y	Dec. 10, 1861	33, 914
Pump	C. Wilson	Bridgeport, Conn	May 28, 1872	127, 284
Pump	C. A. Wilson and W. R. Dunlap	Cincinnati, Ohio	Sept. 29, 1868	82, 670
Pump	J. E. Wilson	Bridgeport, Conn	Apr. 26, 1870	102, 458
Pump	T. C. Wilson	Watkins, N. Y	May 3, 1870	102, 743
Pump	P. Wineman	Joliet, Ill	May 18, 1869	90, 143
Pump	E. B. Winship	Racine, Wis	Dec. 18, 1866	60, 605
Pump	H. E. Wolcott	Elbridge, N. Y	Dec. 13, 1870	110, 182
Pump	T. J. Wolfe	Baltimore, Md	Apr. 10, 1844	3, 529
Pump	J. Wolford and R. Conner	Knightsville, Ind	Aug. 5, 1873	141, 514
Pump	D. S. Wood	Delavan, Wis	May 8, 1866	54, 636
Pump	S. Woodruff and H. B. Beach	Hartford, Conn	Sept. 22, 1868	82, 371
Pump	R. L. Wright	West Nantmeal, Pa	Apr. 12, 1870	101, 960
Pump	W. Wright	Hartford, Conn	Nov. 15, 1859	26, 139
Pump	W. Wright	New York, N. Y	Mar. 8, 1870	100, 702
Pump	H. M. Wyeth	Pulaski, Iowa	Mar. 18, 1862	34, 711
Pump	H. M. Wyeth	Bloomfield, Iowa	Nov. 25, 1862	37, 024
Pump	H. M. Wyeth	Newark, Ohio	July 12, 1870	105, 401
Pump	F. G. Wynkoop	Corning, N. Y	Dec. 29, 1857	19, 003
Pump	F. G. Wynkoop	Corning, N. Y	Dec. 20, 1864	45, 551
Pump	F. Yeiser	Danville, Ky	Jan. 7, 1868	73, 153
Pump	J. H. Young	Saint Louis, Mo	May 17, 1859	24, 073
Pump	S. M. Young and P. Brand	Jacksonville, Ill	Apr. 2, 1872	125, 243
Pump	H. Zeng	Elizabethport, N. J	July 27, 1858	21, 043
Pump	M. W. and J. Zimmerman	Earl Township, Pa	May 3, 1864	42, 617
Pump	O. Zwietusch	Milwaukee, Wis	Nov. 9, 1869	96, 760
Pump, Acid	L. H. Fisher	Oil City, Pa	Dec. 3, 1872	133, 639
Pump, Acid	F. Nichols	New London, Conn	Mar. 18, 1873	136, 934
Pump, Air	G. E. Barker and S. F. Mack	Waverly, N. Y	Dec. 30, 1873	145, 926
Pump, Air	N. H. Borgfeldt	New York, N. Y	Nov. 29, 1864	45, 220
Pump, Air	D. Carpenter	Peekskill, N. Y	Mar. 26, 1867	63, 209
Pump, Air	T. Claxton	Boston, Mass	June 7, 1834	....
Pump, Air	T. Doane	Boston, Mass	Dec. 31, 1867	72, 820
Pump, Air	W. H. Flanigan	Philadelphia, Pa	Mar. 26, 1872	125, 037
Pump, Air	H. Getty	Brooklyn, N. Y	Oct. 1, 1867	69, 423
Pump, Air	C. Goodyear	Philadelphia, Pa	Mar. 30, 1835	....
Pump, Air	C. B. Hardick	Brooklyn, N. Y	Feb. 14, 1871	111, 742
Pump, Air	J. F. Haskins	Fitchburgh, Mass	Nov. 29, 1870	109, 732
Pump, Air	H. H. Hendrick	Cincinnati, Ohio	Apr. 19, 1864	42, 437
Pump, Air	J. W. Hopkins	Brooklyn, N. Y	Jan. 25, 1870	99, 193
Pump, Air	A. S. Lyman	New York, N. Y	Jan. 31, 1865	46, 122
Pump, Air	R. M. Marchant	London, England	Feb. 21, 1871	112, 060
Pump, Air	J. Molyneaux	Bordentown, N. J	Apr. 11, 1865	47, 243
Pump, Air	G. M. Mowbray	Titusville, Pa	Nov. 22, 1864	45, 168
Pump, Air	R. Porter	Melrose, Mass	May 26, 1863	38, 715
Pump, Air	F. Ransom	Buffalo, N. Y	Aug. 8, 1865	49, 301
Pump, Air	E. S. Ritchie	Brookline, Mass	Nov. 19, 1867	71, 218
Pump, Air	D. E. Somes	Washington, D. C	Oct. 5, 1869	95, 613
Pump, Air	J. M. Wightman	Boston, Mass	Nov. 10, 1841	2, 338
Pump, Air	J. H. Wilhelm	Chicago, Ill	Dec. 26, 1865	51, 769
Pump, Air	G. M. Woodward	New York, N. Y	May 30, 1865	48, 009
Pump, Air-compressing	J. N. Dennison	Newark, N. J	Oct. 23, 1866	59, 124
Pump, Air-compressing	B. Holly	Lockport, N. Y	May 22, 1866	54, 905
Pump, Air-compressing	C. W. Wailey	New Orleans, La	Nov. 10, 1868	84, 030
Pump, Air or water	J. Barron	Hampton, Va	Feb. 20, 1819	....
Pump and condenser, Steam air	S. Gibbons	Freedom, Pa	Sept. 2, 1873	142, 333
Pump and cupping-instrument, Stomach and breast	L. B. White	New York	July 22, 1833	....
Pump and cupping-instrument, Stomach and breast	L. B. White	New York	Dec. 23, 1834	....
Pump and engine, Air	T. Beach	Freeport, Pa	Mar. 25, 1873	137, 123
Pump and engine piston	D. Hinman	Brunswick, Ohio	Mar. 7, 1846	4, 396
Pump and engine, Rotary	D. D. Hardy	Cincinnati, Ohio	Dec. 11, 1866	60, 365
Pump and engine, Rotary	D. D. Hardy and J. J. Morris	Cincinnati, Ohio	Dec. 11, 1866	60, 366
Pump and fire-engine	B. T. Babbitt, S. C. Higbee, and P. W. Plantz.	Little Falls and Oppenheim, N. Y.	Oct. 7, 1842	2, 801
Pump and fire-engine	S. Weirick and H. Lathrope	Warsaw, Ind	June 10, 1873	139, 840
Pump and fire-engine, Steam	J. W. Whitaker	Kenosha, Wis	June 13, 1871	115, 918
Pump and fire-engine, Suction	I. Jennings	Philadelphia, Pa	Apr. 1, 1812	....
Pump and fountain	J. J. Sink	Philadelphia, Pa	June 4, 1872	127, 651
Pump and funnel, Combined atmospheric exhaust	G. H. Randall	Forestville, Md	Feb. 25, 1873	136, 264
Pump and gasometer, Compound air	S. Chichester	Poughkeepsie, N. Y	Apr. 20, 1858	19, 981
Pump and measure, Combined	J. M. Brooks and P. Munson	Independence, Iowa	Oct. 15, 1867	69, 756
Pump and nursing-bottle, Combined breast	O. H. Needham	Needham, N. Y	Sept. 19, 1871	119, 044
Pump and other oscillating rods, Protector for	L. Wilson	Middletown, Conn	July 18, 1865	48, 880
Pump and pump-box	J. C. Helme	Newport, R. I	Aug. 1, 1817	....
Pump and reel for artesian wells, Sand	J. A. Fleming	Shamburgh, Pa	Jan. 2, 1872	122, 449
Pump and reservoir, Combined	E. Nickerson	Provincetown, Mass	Nov. 19, 1867	71, 044
Pump and siphon, Combined	J. D. Averill, G. A. Higgins, and T. Gordon.	New York, N. Y., and Shrewsbury, N. J.	Oct. 11, 1870	108, 087

Index of patents issued from the United States Patent Office from 1790 *to* 1873, *inclusive*—Continued.

Invention.	Inventor.	Residence.	Date.	No.
Pump and siphon, Combined	W. Pitman	New York, N. Y	June 21, 1870	104, 641
Pump, Artesian	L. Jennings	Brooklyn, N. Y	Aug. 4, 1868	80, 631
Pump, Artesian-well steam	H. W. Faucett and A. T. Comer	Titusville, Pa	Oct. 29, 1872	132, 567
Pump, Atmospheric	L. Keiler	Catawissa, Pa	Apr. 28, 1857	17, 154
Pump-attachment	T. J. Jones	Summit, N. J	Sept. 18, 1866	58, 170
Pump, Automatic railway	J. B. Atwater	Chicago, Ill	Jan. 5, 1864	41, 047
Pump, Automatic steam	W. E. Prall	Washington, D. C	July 4, 1871	116, 751
Pump, Automatic steam	E. Thayer	Worcester, Mass	Mar. 28, 1865	47, 051
Pump, Balance-wheel	L. Schrack	Norristown, Pa	Dec. 8, 1809	
Pump, Balance-wheel uniform-working	J. Josslyn	Hanover, Mass	July 25, 1809	
Pump, Barge	C. P. Gaver	Ashland, Ky	Sept. 23, 1873	143, 069
Pump, Barrel	J. W. Kelly	Hartford City, W. Va	July 16, 1872	129, 141
Pump-barrels, Device for escape of waste water from.	J. E. Cronk	Poughkeepsie, N. Y	July 10, 1855	13, 214
Pump-barrels to their bases, Adjustable attachment of.	J. W. Douglas	Middletown, Conn	Jan. 11, 1870	98, 674
Pump, Beer, &c	L. Pitkins	Rochester, N. Y	Sept. 11, 1828	
Pump, Beer and cider	R. Sealy	New York	May 6, 1825	
Pump, Bellows	E. Parker	Beverly, N. Y	Oct. 11, 1864	44, 649
Pump, Bellows	J. M. Wallis and E. P. Swearingen.	Milton, Iowa	Nov. 27, 1866	60, 098
Pump, Belt	S. C. Murdock and T. H. Borland.	Pittsburgh, Pa	Dec. 26, 1871	122, 274
Pump, Boring	C. Leach		Apr. 13, 1797	
Pump, Bottle	G. Asmus	Houghton, Mich	June 11, 1867	65, 527
Pump-box	J. Munson	San José, Cal	May 10, 1859	23, 937
Pump-box	W. A. Spooner	Springfield, Mass	Dec. 2, 1873	145, 133
Pump-box	F. and J. Stock	San José, Cal	Apr. 26, 1859	23, 795
Pump-box	A. S. Whitehouse	Ashland, Mass	Dec. 20, 1870	110, 319
Pump-box	I. P. Williams and S. G. Rea	Salem, Mass	Dec. 9, 1828	
Pump-box, Elliptic-valve	J. Baker	Charlestown, Mass	May 9, 1817	
Pump-box, Metallic single-valve	J. Scott	Boston, Mass	May 2, 1817	
Pump-box or pump for raising water on ships of war, &c.	J. Stickney		May 1, 1804	
Pump-bracket	S. Selden and O. R. Hanchett	Erie, Pa	Aug. 26, 1873	142, 286
Pump, Breast	J. H. Beadle	New York, N. Y	Aug. 21, 1860	29, 662
Pump, Breast	J. H. Beadle	New York, N. Y	Oct. 20, 1863	40, 319
Pump, Breast	E. Blake	New York	Apr. 24, 1830	
Pump, Breast	J. Cole	Brooklyn, N. Y	Dec. 24, 1867	72, 604
Pump, Breast	L. O. Colvin	Philadelphia, Pa	Feb. 17, 1863	37, 677
Pump, Breast	L. O. Colvin	Philadelphia, Pa	Oct. 17, 1865	50, 457
Pump, Breast	L. O. Colvin	Newark, N. J	Nov. 1, 1870	108, 882
Pump, Breast	W. T. Fry	New York, N. Y	Nov. 19, 1867	70, 988
Pump, Breast	W. T. Fry	New York, N. Y	June 9, 1868	78, 731
Pump, Breast	M. Mattson	New York, N. Y	Sept. 8, 1868	82, 015
Pump, Breast	M. Mattson	New York, N. Y	Dec. 29, 1868	85, 818
Pump, Breast	O. H. Needham	New York, N. Y	June 20, 1854	11, 135
Pump, Breast	C. H. Wilder	Argyle, Wis	May 16, 1865	47, 760
Pump-bucket	D. Coon	Mendon, N. Y	Nov. 28, 1829	
Pump-bucket	W. F. Horton	Lockport, N. Y	Jan. 19, 1858	19, 173
Pump-bucket	E. Quinn	Trenton, N. J	June 1, 1858	20, 443
Pump bucket, Chain	W. C. Barker	Ypsilanti, Mich	June 20, 1871	116, 138
Pump bucket, Chain	A. Bauman	Toledo, Ohio	Nov. 21, 1871	121, 148
Pump bucket, Chain	J. D. Clark	Leicester, Mass	Dec. 23, 1862	37, 218
Pump bucket, Chain	F. P. Hunt	Northborough, Mass	May 31, 1870	103, 619
Pump bucket, Chain	W. Hutchinson	Belleville, Mich	Mar. 26, 1872	125, 051
Pump bucket, Chain	E. Morris	Burlington, N. J	Jan. 23, 1855	12, 296
Pump bucket, Chain	J. S. Wilcox	Ypsilanti, Mich	Oct. 3, 1871	119, 679
Pump bucket, Endless-chain	C. Polley	May's Landing, N. J	Dec. 14, 1852	9, 466
Pump-buckets, Machine for making	J. F. Semple	Chicago, Ill	Mar. 22, 1870	101, 182
Pump, Cast-iron	J. and J. Reed	Plymouth County, Mass	Nov. 19, 1833	
Pump, Cattle	J. B. Atwater	Chicago, Ill	Sept. 20, 1864	44, 274
Pump, Cattle	J. B. Atwater	Chicago, Ill	Jan. 16, 1866	52, 011
Pump, Cattle	J. Augsperger	Trenton, Ohio	Dec. 20, 1859	26, 465
Pump, Cattle	A. Bailey	Jefferson, Ohio	Aug. 3, 1839	1, 274
Pump, Cattle	J. F. Black and A. Buracker	Lancaster, Ill	Aug. 19, 1862	36, 203
Pump, Cattle	J. A. Dickson	Sandwich, Ill	Sept. 20, 1864	44, 289
Pump, Cattle	D. P. Farnham	Johnstown Centre, Wis	Feb. 8, 1859	22, 862
Pump, Cattle	P. Fraer	West Springfield, Pa	Sept. 10, 1861	33, 243
Pump, Cattle	A. Gerred	Sparta, Ill	Jan. 18, 1859	22, 640
Pump, Cattle	J. B. Ghormley	Bellefontaine, Ohio	May 19, 1863	38, 578
Pump, Cattle	J. H. Irwin	Carlinville, Ill	Jan. 18, 1859	22, 653
Pump, Cattle	D. J. Keller	Kane, Ill	Jan. 12, 1869	85, 832
Pump, Cattle	E. C. Kellogg	Rome, N. Y	Sept. 3, 1867	68, 514
Pump, Cattle	W. Lindsey	Oberlin, Ohio	Apr. 9, 1867	63, 732
Pump, Cattle	E. A. and S. Moore and D. Mooney.	Findley, Ohio	Dec. 3, 1861	33, 848
Pump, Cattle	W. Nichols	Lima, Ohio	Dec. 6, 1859	26, 399
Pump, Cattle	J. S. Patrie	Rochester, N. Y	June 5, 1866	55, 356
Pump, Cattle	T. H. Powers	Wyocena, Wis	May 13, 1856	14, 884
Pump, Cattle	G. B. Roe	Paine's Point, Ill	Mar. 17, 1863	37, 923
Pump, Cattle	G. B. Roe	Ogle County, Ill	Dec. 12, 1871	121, 731
Pump, Cattle	S. Stadon	Greenwood, Pa	Apr. 2, 1841	2, 031
Pump, Cattle	H. W. Thornburg	Morristown, Ind	May 21, 1867	65, 026
Pump, Cattle	M. D. Wilder	La Porte, Ind	Apr. 16, 1867	63, 972
Pump, Cattle	W. H. Wright	Rochester, N. Y	Mar. 1, 1870	100, 485
Pump, Centrifugal	W. D. Andrews	New York, N. Y	Mar. 14, 1846	4, 418
Pump, Centrifugal	W. D. Andrews	New York, N. Y	Aug. 22, 1854	11, 544
Pump, Centrifugal	W. D. Andrews	New York, N. Y	Mar. 15, 1870	100, 838
Pump, Centrifugal	E. Brazelton	Saint Louis, Mo	May 19, 1863	38, 553
Pump, Centrifugal	E. H. Covel	New York, N. Y	Dec. 4, 1866	60, 145
Pump, Centrifugal	H. A. Due	Charleston, S. C	Jan. 12, 1869	85, 802
Pump, Centrifugal	W. C. Hibbard	West Roxbury, Mass	Oct. 12, 1858	21, 756
Pump, Centrifugal	E. M. Jones	New Orleans, La	Jan. 28, 1868	73, 898
Pump, Centrifugal	N. H. Lebby	Charleston, S. C	Aug. 27, 1867	68, 214
Pump, Centrifugal	N. H. Lebby	Charleston, S. C	Dec. 7, 1869	97, 655
Pump, Centrifugal	C. F. Martorana	Baltimore, Md	May 28, 1867	65, 099
Pump, Centrifugal	S. P. Thayer	Baldwinsville, N. Y	Oct. 12, 1869	95, 749

Index of patents issued from the United States Patent Office from 1790 *to* 1873, *inclusive*—Continued.

Invention.	Inventor.	Residence.	Date.	No.
Pump, Centrifugal	A. W. and J. H. Von Schmidt	Washington, D. C	July 24, 1847	5, 203
Pump, Centrifugal-screw	J. H. White and W. S. Henson	Lima, South America, and Newark, N. J.	Jan. 26, 1869	86, 264
Pump, Chain	A. G. Bradford	Freeport, Ill	June 10, 1862	35, 499
Pump, Chain	A. Coates	Marlborough, Ohio	Aug. 7, 1860	29, 466
Pump, Chain	J. M. Connel	Newark, Oho	June 21, 1864	43, 253
Pump, Chain	W. Cooper	Ypsilanti, Mich	June 4, 1872	127, 576
Pump, Chain	D. Du Pré	Raleigh, N. C	June 7, 1859	24, 288
Pump, Chain	L A. Fisher	Chicago, Ill	Dec. 20, 1864	45, 484
Pump, Chain	J. Harrison, jr	New York, N. Y	May 5, 1857	17, 217
Pump, Chain	E. Morris	Burlington, N. J	Feb. 10, 1857	16, 603
Pump, Chain	J. Robingson	New Brighton, Pa	Nov. 4, 1856	16, 024
Pump, Chain	A. Wyckoff	Columbus, Ohio	Apr. 3, 1855	12, 645
Pump, Chinese chain	J. Black	Northumberland, Pa	Aug. 2, 1810	....
Pump, Cog and pendulum	B. Lies	New York, N. Y	Aug. 6, 1812	....
Pump, Common suction	D. Watson	Plymouth, Mass	Mar. 4, 1808	....
Pump, Compound high and low pressure steam	A. J. L. Loretz	Brooklyn, N. Y	May 20, 1873	139, 071
Pump, Compound hydraulic	H. Rodgers	Auburn, N. Y	Oct. 9, 1841	2, 279
Pump, Compound propeller	Y. Shaw	Philadelphia, Pa	Feb. 15, 1870	99, 791
Pump, Compound steam	A. J. L. Loretz	Brooklyn, N. Y	May 20, 1873	139, 070
Pump-coupling	S. H. Gray	Bridgeport, Conn	Apr. 27, 1858	20, 060
Pump, Cream	T. A. Case	Ellington, N. Y	May 25, 1869	90, 424
Pump, Cream	M. A. Richardson	Sherman, N. Y	Sept. 23, 1862	36, 530
Pump-cylinders, Lining for	W. S. Owen	Oskaloosa, Iowa	Dec. 23, 1873	145, 895
Pump, Deep-well	J. A. Bloom	Philadelphia, Pa	Jan. 16, 1866	52, 019
Pump, Deep-well	R. Boeklen	Brooklyn, N. Y	Nov. 21, 1865	51, 007
Pump, Deep-well	A. Carver	Little Falls, N. Y	Aug. 14, 1866	57, 086
Pump, Deep-well	J. J. G. Collins	Philadelphia, Pa	Nov. 28, 1865	51, 148
Pump, Deep-well	R. Cornelius	Philadelphia, Pa	Mar. 13, 1866	53, 117
Pump, Deep-well	R. Cornelius	Philadelphia, Pa	May 1, 1866	54, 300
Pump, Deep-well	R. Cornelius	Philadelphia, Pa	June 26, 1866	55, 822
Pump, Deep-well	B. Crawford	Allegheny, Pa	July 17, 1866	56, 380
Pump, Deep-well	J. H. Davis	Allegheny City, Pa	June 19, 1866	55, 626
Pump, Deep-well	N. Dodge	New York, N. Y	Mar. 12, 1850	7, 161
Pump, Deep-well	N. Dodge	New York, N. Y	June 27, 1865	48, 378
Pump, Deep-well	S. H. Early	Lynchburgh, Va	Mar. 13, 1866	53, 128
Pump, Deep-well	W. H. Genung	Madison, Ohio	July 16, 1872	129, 220
Pump, Deep-well	J. Harrison	Jamestown, N. Y	Apr. 17, 1866	53, 974
Pump, Deep-well	S. E. Hewes	Albany, N. Y	Nov. 14, 1865	50, 932
Pump, Deep-well	W. S. Kelly	Schenectady, N. Y	Apr. 17, 1866	53, 990
Pump, Deep-well	H. K. Kenyon	Steubenville, Ohio	June 28, 1870	104, 855
Pump, Deep-well	T. J. Lovegrove	Philadelphia, Pa	Dec. 19, 1865	51, 602
Pump, Deep-well	W. E. Morrison and W. L. Betts	Funkville, Pa	July 17, 1866	56, 435
Pump, Deep-well	J. Old	Pittsburgh, Pa	Feb. 18, 1862	34, 444
Pump, Deep-well	J. Old	Pittsburgh, Pa	June 6, 1865	48, 089
Pump, Deep-well	J. Sheffield	Pultneyville, N. Y	June 6, 1865	48, 105
Pump, Deep-well	W. B. Snow	Titusville, Pa	Dec. 18, 1866	60, 586
Pump, Deep-well	J. W. Summers	Tarr Farm, Pa	Sept. 18, 1866	58, 150
Pump, Deep-well	T. J. Trapp	Williamsport, Pa	July 11, 1871	117, 019
Pump, Deep-well	J. Warren	Buffalo, N. Y	May 10, 1864	42, 705
Pump, Deep-well	J. T. Whipple	Chicago, Ill	Feb. 18, 1868	74, 738
Pump-device for steam and other enginery	C. E. Blake	San Francisco, Cal	Jan. 26, 1869	86, 273
Pump device, Steam	M. Wilcox	Sacramento, Cal	Mar. 8, 1870	100, 576
Pump, Diaphragm	D. F. Hitchcock	Warren, Mass	Jan. 17, 1854	10, 428
Pump, Diaphragm	J. L. McPherson and J. O. Joice	New Vienna and Cincinnati, Ohio.	Feb. 12, 1856	14, 247
Pump, Diaphragm	J. A. Pease	New York, N. Y	Aug. 22, 1854	11, 559
Pump, Direct-acting hydraulic steam	R. B. Gorsuch	New York, N. Y	Apr. 3, 1855	12, 625
Pump, Direct-acting hydraulic steam	H. R. Worthington	Brooklyn, N. Y	July 31, 1855	13, 370
Pump, Direct-acting steam	O. T. Earle	Norwalk, Conn	Aug. 26, 1873	142, 091
Pump, Direct-acting steam	J. A. Reed	Jersey City, N. J	Apr. 28, 1863	38, 334
Pump, Direct-acting steam	W. Sewell and A. S. Cameron	New York, N. Y	May 10, 1864	42, 694
Pump, Double-acting	P. M. Barker	Grand Haven, Mich	June 10, 1873	139, 650
Pump, Double-acting	M. Braun	Cape Vincent, N. Y	July 5, 1870	105, 035
Pump, Double-acting	D. W. Clark	Bridgeport, Conn	Nov. 20, 1855	13, 816
Pump, Double-acting	W. H. Davis	Austin, Ind	July 19, 1859	24, 797
Pump, Double-acting	P. Foley	Nineveh, N. Y	Aug. 4, 1868	80, 617
Pump, Double-acting	T. N. Henderson	Jackson, Mich	Oct. 8, 1872	131, 952
Pump, Double acting	B. J. C. Howe	Syracuse, N. Y	May 17, 1864	42, 773
Pump, Double-acting	W. H. Ivens	Trenton, N. J	Feb. 8, 1870	99, 680
Pump, Double-acting	E. A. Jeffery	Corning, N. Y	June 5, 1855	13, 007
Pump, Double-acting	B. F. Joslyn	Worcester, Mass	Apr. 3, 1855	12, 631
Pump, Double-acting	W. S. Kelly	Schenectady, N. Y	Aug. 23, 1864	43, 917
Pump, Double-acting	F. A. Mack	Niles, Mich	June 22, 1869	91, 648
Pump, Double-acting	K. Nolan	Galesburgh, Ill	Jan. 21, 1873	135, 146
Pump, Double-acting cylinder	P. Mix	Germantown, Pa	Jan. 9, 1835	
Pump, Double-acting force	A. Bagley	Ypsilanti, Mich	July 12, 1870	105, 162
Pump, Double-acting force	J. Barney	Perry's Mills, N. Y	Mar. 28, 1871	113, 132
Pump, Double-acting force	E. Bellamy	Saint Louis, Mo	Apr. 13, 1858	19, 907
Pump, Double-acting force	W. C. and I. S. Burnham	New York, N. Y	Feb. 6, 1855	12, 342
Pump Double-acting force	D. W. Clark and S. H. Gray	Bridgeport, Conn	Dec. 19, 1854	12, 092
Pump, Double-acting force	D. L. Farnam	New York	Oct. 23, 1837	437
Pump, Double-acting force	J. P. Flanders	Vergennes, Vt	July 15, 1873	140, 819
Pump, Double-acting force	G. Fowler	Northford, Conn	Jan. 30, 1855	12, 312
Pump, Double-acting force	J. H. McGowan, jr	Cincinnati, Ohio	June 27, 1854	11, 169
Pump, Double-acting force	T. J. D. Yampert	Mobile, Ala	Sept. 11, 1855	13, 559
Pump, Double-acting hollow-piston	F. A. Chase	Jordan, N. Y	July 26, 1864	43, 637
Pump, Double-acting submerged	A. J. Reynolds	Sturgis, Mich	Nov. 1, 1864	44, 891
Pump, Double-acting suction and force	J. Farnam	Stillwater, N. Y	Dec. 14, 1841	2, 387
Pump, Double-acting water	J. Farnam	Stillwater, N. Y	Nov. 10, 1841	2, 345
Pump, Double air and water	B. Smallman	Philadelphia, Pa	Mar. 27, 1815	
Pump, Double-bellows	D. L. Deshon, 2d, and E. L. Webster.	New London, Conn	Sept. 12, 1848	5, 770
Pump, Double-bored pendulum	A. Farra	Montgomery County, Pa	Dec. 5, 1809	....
Pump, Double-cylinder force	W. H. Ivens	Trenton, N. J	Feb. 8, 1870	99, 681
Pump, Double-cylinder suction and force	J. J. Rice	Salina, N. Y	Mar. 10, 1838	625
Pump, Double forcing	W. Baker	Springfield, N. Y	Dec. 31, 1819	....
Pump, Double forcing	W. Beach	Franklin, N. Y	Feb. 19, 1811	....

Index of patents issued from the United States Patent Office from 1790 *to* 1873, *inclusive*—Continued.

Invention.	Inventor.	Residence.	Date.	No.
Pump, Double forcing	C. Colver	Champlain, N. Y	Apr. 1, 1807	....
Pump, Double forcing	W. Douglass	Middletown, Conn	Aug. 20, 1835	....
Pump, Double forcing	T. Ferris	Dutchess County, N. Y	Feb. 20, 1811	....
Pump, Double forcing	S. Gillett	Windsor, Conn	Apr. 8, 1820	....
Pump, Double forcing	L. Gray	Otsego County, N. Y	Mar. 7, 1811	....
Pump, Double forcing	L. Newton	Alexander, N. Y	Mar. 8, 1836	....
Pump, Double forcing	J. G. White	Dryden, N. Y	July 1, 1836	....
Pump, Double piston	S. W. Allen	Geneva, N. Y	July 21, 1831	....
Pump, Draining	W. S. Nelson	Saint Louis, Mo	Jan. 2, 1866	51, 854
Pump-dredge, Submarine	D. Quinn	Chicago, Ill	Apr. 20, 1869	89, 073
Pump, Driven	S. Waite	New Bedford, Mass	Sept. 17, 1867	68, 917
Pump, Driven-well	J. W. Harrington and J. A. Yingling.	Cincinnati and Hamilton, Ohio.	Sept. 30, 1873	143, 345
Pump, Driving	S. H. Rhoades	Clyde, Ohio	Oct. 2, 1866	58, 479
Pump, Dry-bellows	E. Alden		Nov. 15, 1803	
Pump, Elastic-tube	R. Porter and J. D. Bradley	Washington, D. C., and Brattleborough, Vt.	Apr. 17, 1855	12, 753
Pump, Electro-magnetic	C. H. Rudd and G. W. Shawk	Cleveland, Ohio	Sept. 3, 1867	68, 577
Pump, Elliptic-valve	J. Baker	Charlestown, Mass	Jan. 16, 1817	....
Pump, Elliptical rotary	B. Holly	Seneca Falls, N. Y	Feb. 6, 1855	12, 350
Pump, Endless-chain	J. A. and H. A. Pitts	Winthrop and Livermore, Me.	Apr. 28, 1834	....
Pump engine, Steam	J. Sutherland and F. Moakly	East Hampton, Mass	Mar. 24, 1868	75, 998
Pump for agitating the surface of water in wells, Attachment to.	W. D. Mayfield	Elkton, Ky	Dec. 24, 1850	7, 857
Pump for artesian wells, Sand	T. J. Lovegrove	Philadelphia, Pa	Mar. 7, 1865	46, 756
Pump for boxed wells	J. Suggett	Cortland, N. Y	Mar. 29, 1864	42, 126
Pump for continuous discharge	S. D. Lount	Rahtbone, W. Va	July 13, 1869	92, 621
Pump for deep wells, Tubular	C. H. Duncan	Pithole City, Pa	Apr. 20, 1869	89, 134
Pump for diving-bells, Hydropneumatic	G. Williamson	Brooklyn, N. Y	Jan. 1, 1856	14, 039
Pump for elevating meal, grain, &c	J. Ewing and D. Dickey	Oxford, Pa	Oct. 2, 1813	....
Pump for elevating water mixed with mineral substances.	W. Ball	Chicopee, Mass	Dec. 23, 1851	8, 602
Pump for excavating wells, &c., Gravel	J. I. and E. Rice	Salina, N. Y	Aug. 15, 1838	883
Pump for exhausting and sealing cans, Air	W. Y. Gill	Henderson, Ky	Aug. 14, 1860	29, 582
Pump for fire-engines	J. N. Dennisson and R. J. Gould	Newark, N. J	Nov. 26, 1867	71, 287
Pump for fire-engines	H. Gates	Northampton, Mass	June 12, 1835	....
Pump for fire-engines	J. Newman	Baltimore, Md	Apr. 14, 1838	691
Pump for fire-engines, Force	J. N. Dennisson	Newark, N. J	July 13, 1858	20, 867
Pump for fire-engines, Horizontal double-forcing	E. Smith	Palmer, Mass	Nov. 6, 1810	....
Pump for hydraulic presses, Force	W. P. Callahan	Dayton, Ohio	July 14, 1868	79, 949
Pump for liquid freight, Ship's	S. F. Paullin	Baltimore, Md	May 28, 1872	127, 364
Pump for locomotive-engines	C. L. Rice	Milwaukee, Wis	July 17, 1860	29, 195
Pump for locomotive-tenders, &c	A. W. Todd	Chicago, Ill	Dec. 6, 1864	45, 360
Pump for low-pressure steam-engine	J. Vial	Cleveland, Ohio	Sept. 9, 1862	36, 431
Pump for machinery	L. Knap	Brighton, N. Y	May 22, 1826	....
Pump for marine-alarms, Air	S. G. Cabell	Quincy, Ill	June 18, 1867	65, 872
Pump for mines, Portable	W. E. Sidney	Knightsville, Ind	Oct. 14, 1873	143, 724
Pump for oil-wells, Sand	J. F. Carll	Brooklyn, N. Y	Jan. 21, 1868	73, 577
Pump for railway-stations	H. S. Lansdell	New York, N. Y	Feb. 20, 1866	52, 721
Pump for railway-stations	A. W. Todd	Chicago, Ill	July 17, 1866	56, 467
Pump for railway-stations, Steam	W. H. Butler	Chicago, Ill	Mar. 20, 1866	53, 269
Pump for raising and measuring liquids	D. Hunsicker	Hartleton, Pa	Mar. 7, 1846	4, 409
Pump for raising water, Air or bellows	D. M. Miller		Mar. 5, 1804	....
Pump for raising water, Spring	W. Finn	Baths, Md	Feb. 13, 1806	....
Pump for ships, mines, &c	J. Stickney		Nov. 29, 1799	....
Pump for steam-boilers	A. Judson	Sweden, N. Y	Feb. 23, 1827	....
Pump for steam-boilers, Feed-water	E. W. Ellsworth	East Windsor Hill, Conn	Dec. 9, 1856	16, 206
Pump for steam-engines	J. F. Hamilton	Pittsburgh, Pa	July 24, 1860	29, 271
Pump for steam-engines, Air	T. B. Stillman	New York, N. Y	May 15, 1837	193
Pump for steam-engines, Air	D. A. Woodbury	Rochester, N. N	May 8, 1860	28, 229
Pump for steam-engines, Double air	J. Smallman	Philadelphia, Pa	Aug. 29, 1815	....
Pump for vessels, Alarm	E. Haskell	Dover, N. H	Oct. 24, 1871	120, 194
Pump, Force	W. Beers and W. Raynor	Milan, Ohio	Jan. 11, 1870	98, 374
Pump, Force	B. Brundage	Sumpter, Mich	Aug. 20, 1872	130, 620
Pump, Force	A. Carver	Little Falls, N. Y	June 21, 1864	43, 180
Pump, Force	N. Chapin	Penn Yan, N. Y	Mar. 30, 1836	....
Pump, Force	N. Clute	Schenectady, N. Y	Feb. 9, 1869	86, 645
Pump, Force	W. G. Crutchfield	Dayton, Ohio	Dec. 4, 1866	60, 146
Pump, Force	S. Currey	Williamson City, Tenn	Jan. 3, 1825	....
Pump, Force	J. De Long	Upper Sandusky, Ohio	Oct. 25, 1864	44, 790
Pump, Force	B. Egbert	Lansing, N. Y	Jan. 23, 1836	....
Pump, Force	J. D. Egbert	Lansing, N. Y	Sept. 22, 1834	....
Pump, Force	J. Eveleth		June 13, 1801	....
Pump, Force	G. B. Farnam	New York, N. Y	Jan. 9, 1855	12, 199
Pump, Force	J. Farnam	Stillwater, N. Y	Oct. 22, 1834	....
Pump, Force	S. D. Gilson	Syracuse, N. Y	Nov. 24, 1863	40, 723
Pump, Force	C. B. Guy	Lybrand, Iowa	Aug. 2, 1864	43, 684
Pump, Force	E. B. Harris	Wilmington, Ill	July 24, 1866	56, 556
Pump, Force	F. Kettler	Milwaukee, Wis	July 26, 1859	24, 876
Pump, Force	F. Kettler	Milwaukee, Wis	July 26, 1859	24, 877
Pump, Force	J. L. Kitchin	Fayetteville, N. C	July 2, 1872	128, 635
Pump, Force	B. Sapham	Waterford, N. Y	Nov. 19, 1833	....
Pump, Force	W. W. Lesner	Venice, N. Y	Mar. 12, 1836	....
Pump, Force	J. Paddleford	Lebanon, N. Y	May 6, 1823	....
Pump, Force	R. Poole	Baltimore, Md	Sept. 6, 1859	25, 366
Pump, Force	H. Rogers	Ferrisburgh, Vt	Jan. 30, 1855	12, 326
Pump, Force	J. F. Rogers	Waterford, N. Y	Mar. 30, 1836	....
Pump, Force	O. Snell	Williamsburgh, Ohio	July 29, 1873	141, 394
Pump, Force and suction	A. Bailey	Jefferson, Ohio	May 4, 1838	722
Pump, Force and suction	J. Clark	Westminster, Vt	Sept. 11, 1823	....
Pump, Force and suction	A. Kasslar	Canajoharie, N. Y	Oct. 28, 1837	443
Pump, Force and suction	J. F. Rogers	Waterford, N. Y	Feb. 27, 1833	....
Pump, Force and vacuum	J. O. Joyce	Dayton, Ohio	July 2, 1872	128, 631
Pump, Frictionless	C. V. Card	New Bedford, Mass	Nov. 26, 1836	87
Pump, Frictionless	E. Whitfield	New York, N. Y	June 2, 1836	....
Pump, Gage	J. D. Woodside	Washington, D. C	July 14, 1834	....
Pump, Garden	C. G. Korth	Carlstadt, N. J	Mar. 12, 1872	124, 593

Index of patents issued from the United States Patent Office from 1790 *to* 1873, *inclusive*—Continued.

Invention.	Inventor.	Residence.	Date.	No.
Pump, Gas	W. Foster	Brooklyn, N. Y	Sept. 5, 1871	118, 709
Pump, Gas	W. H. Pollard	Seneca Falls, N. Y	Aug. 8, 1871	117, 925
Pump-gear	H. and F. J. L. Blandy	Zanesville, Ohio	July 5, 1864	43, 327
Pump-gearing	J. P. Carr	Mattapoisett, Mass	May 31, 1859	24, 193
Pump, Hand	T. J. Mayall	Boston, Mass	Apr. 16, 1872	125, 824
Pump-handle	I. M. Frazer	Elmer, N. J	Jan. 28, 1873	135, 269
Pump-handle bracket	D. S. Messler	Philadelphia, Pa	May 21, 1872	127, 087
Pump, Horizontal	L. Holland	Belchertown, Mass	Mar. 28, 1810	....
Pump, Horizontal	L. Holland	Belchertown, Mass	Sept. 10, 1811	....
Pump, Hydraulic	E. Squire	Cold Spring, N. Y	Oct. 25, 1864	44, 829
Pump, Hydraulic steam	H. N. Black	Philadelphia, Pa	Mar. 22, 1853	9, 622
Pump, Hydropneumatic force	A. B. Latta	Cincinnati, Ohio	June 27, 1854	11, 165
Pump, Impellent	S. Thayer	....	June 9, 1801	....
Pump into a suction and forcing pump, or vice versa, Converting a lifting.	J. F. Brickley	Winchester, Ind	Mar. 10, 1857	16, 785
Pump, Labor-saving	D. Smith	China, N. Y	June 13, 1831	....
Pump, Lift and force	A. H. Powell	Philadelphia, Pa	Jan. 14, 1873	134, 927
Pump, Lifting	F. Crocker	Titusville, Pa	Oct. 11, 1864	44, 610
Pump, Lifting	F. Raymond and A. Miller	Cleveland, Ohio	May 24, 1864	42, 874
Pump, Liquor	M. Cavanaugh	Philadelphia, Pa	July 26, 1870	105, 776
Pump, Liquor	H. Rodgers	Moravia, N. Y	Sept. 17, 1842	2, 779
Pump-lock, Ship's	A. G. Zeising	Weehawken, N. J	Mar. 15, 1870	100, 833
Pump, Locomotive-engine-feed	P. Cheswell	Manchester, N. H	June 15, 1869	91, 211
Pump-log-boring machine	S. Draper	Camillus, N. Y	Mar. 31, 1828	....
Pump, Measuring	S. H. Wheeler	Dowagiac, Mich	June 5, 1866	55, 433
Pump, Mechanical	W. Shultz	Baltimore, Md	June 24, 1809	....
Pump, Mercury-packed	P. Marcelin	Chambery, France	Aug. 23, 1864	43, 960
Pump, Metallic	J. Beckwith	Saratoga, N. Y	Dec. 27, 1833	....
Pump, Mine	T. M. Fell	New York, N. Y	Apr. 18, 1865	47, 289
Pump, Mining	T. J. Chubb	Williamsburgh, N. Y	Oct. 22, 1872	132, 384
Pump, Mining	G. E. Mills	New York, N. Y	Apr. 22, 1873	138, 037
Pump, Mining	H. Wolf	Saint Louis, Mo	Sept. 21, 1869	95, 061
Pump, Non-corrosive cast-iron	J. A. Rumsey	Seneca Falls, N. Y	May 25, 1869	90, 586
Pump, Oblique	J. Hunt	....	June 7, 1798	....
Pump, Oil	W. H. Elliot	Plattsburgh, N. Y	May 2, 1865	47, 530
Pump, Oil	L. S. Lapham	Providence, R. I	June 14, 1870	104, 168
Pump, Oil	A. P. Odell	Shamburgh, Pa	Dec. 17, 1872	134, 093
Pump, Oil-well	E. H. Ashcroft	Lynn, Mass	Dec. 27, 1864	45, 573
Pump, Oil-well	W. H. Elliot	Plattsburgh, N. Y	Nov. 19, 1861	33, 736
Pump, Oil-well	M. D. Fenner	Rochester, N. Y	June 30, 1868	79, 333
Pump, Oil-well	J. S. Fish	Cleveland, Ohio	Apr. 26, 1870	102, 241
Pump, Oil-well	A. S. Hill	Pleasantville, Pa	June 11, 1872	127, 880
Pump, Oil-well	J. B. Root	New York, N. Y	Apr. 4, 1865	47, 133
Pump, Oil-well	J. B. Root	New York, N. Y	Apr. 11, 1865	47, 224
Pump, Oil-well	T. Rose	Cortlandville, N. Y	Nov. 7, 1865	50, 846
Pump, Oil-well	H. Searl	Rochester, N. Y	July 25, 1865	48, 983
Pump, Oil-well	J. Thomas, jr	Cleveland, Ohio	Sept. 13, 1864	44, 264
Pump, Oil-well	N. Weare	Marietta, Ohio	Oct. 1, 1872	131, 800
Pump, Oil-well	W. S. Wilkinson	Baltimore, Md	Feb. 27, 1866	52, 947
Pump operated by exhaust steam, Steam vacuum	W. Burdon	Brooklyn, N. Y	Jan. 28, 1873	135, 203
Pump or blower, Rotary	A. Brear	Saugatuck, Conn	Dec. 16, 1873	145, 617
Pump or engine for raising water and ejecting air	J. Beckwith	Saratoga, N. Y	Apr. 16, 1831	....
Pump or engine, Rotary	W. P. Maxson	Elmira, N. Y	July 22, 1873	141; 155
Pump or syringe, Hand	W. B. Robins	Shepherd's Bush, England	May 27, 1873	139, 263
Pump or water-elevator, Rotary	T. J. Lovegrove	Philadelphia, Pa	Dec. 12, 1871	121, 793
Pump, Oscillating	E. Cope and I. W. Bragg	Cincinnati, Ohio	Mar. 23, 1858	19, 680
Pump, Oscillating	J. E. Hallett	Greenbush, N. Y	Mar. 20, 1866	53, 373
Pump, Oscillating	W. Painter	Baltimore, Md	Nov. 12, 1872	133, 048
Pump, Oscillating lifting	E. Horsey	Kingston, Canada	Jan. 16, 1872	122, 722
Pump, Package	J. A. Smith	Windham, Conn	Feb. 3, 1830	....
Pump, pan, &c., Vacuum	E. Dunscomb	Boston, Mass	Mar. 12, 1867	62, 739
Pump, Petroleum	J. Sparks	Rouseville, Pa	Aug. 29, 1871	118, 496
Pump-pipe	J. P. Cowing	Seneca Falls, N. Y	Aug. 30, 1864	43, 975
Pump-pipe joint	D. Knowlton	Camden, Me	July 19, 1859	24, 809
Pump-piston	J. D. Alvord	Bridgeport, Conn	Dec. 13, 1870	109, 994
Pump-piston	D. W. Bell	Saint Louis, Mo	Mar. 21, 1871	112, 768
Pump-piston	F. A. Cramblitt	Petroleum Centre, Pa	Aug. 6, 1867	67, 417
Pump-piston	S. Crowell	Syracuse, N. Y	Feb. 23, 1869	87, 148
Pump-piston	W. C. Culbertson	Girard, Pa	Aug. 19, 1873	142, 004
Pump-piston	W. E. Derrick	Jordan, N. Y	Sept. 17, 1867	68, 852
Pump-piston	W. F. Dodge	New York, N. Y	Apr. 4, 1865	47, 095
Pump-piston	A. H. Foe	Strathroy, Canada	Oct. 17, 1871	120, 054
Pump-piston	J. T. Follansbee and G. Doolittle.	Bridgeport, Conn	Mar. 7, 1871	112, 438
Pump-piston	P. Giffard	Paris, France	Jan. 16, 1872	122, 825
Pump-piston	S. P. Gilbert	Racine, Wis	Feb. 4, 1868	73, 969
Pump-piston	C. B. Gill	Weedsport, N. Y	June 4, 1872	127, 594
Pump-piston	C. B. Gill	Rochester, N. Y	Feb. 18, 1873	135, 905
Pump-piston	A. Griffin	Shamburgh, Pa	June 4, 1872	127, 480
Pump-piston	H. S. Hopkins	Boston, Mass	Apr. 14, 1868	76, 635
Pump-piston	J. Humphrey	Keene, N. H	Oct. 17, 1871	120, 070
Pump-piston	D. Johnson	Ashland, Ohio	Oct. 10, 1871	119, 852
Pump-piston	S. G. Mason	Elbridge, N. Y	Aug. 28, 1866	57, 531
Pump-piston	S. G. Mason	Rochester, N. Y	Apr. 14, 1868	76, 643
Pump-piston	S. G. Mason	Rochester, N. Y	Apr. 20, 1869	89, 060
Pump-piston	F. P. Michel	Rochester, N. Y	Apr. 21, 1868	76, 933
Pump-piston	B. H. Naves	Philadelphia, Pa	Dec. 31, 1867	72, 884
Pump piston	W. Newcomb	Baltimore, Md	Nov. 7, 1871	120, 767
Pump piston	G. Page	Washington, D. C	Apr. 17, 1847	5, 071
Pump-piston	J. Patterson	Monroe, N. Y	Nov. 11, 1862	36, 912
Pump-piston	B. and E. Pickering and B. Pickering.	West Milton and Montgomery County, Ohio.	Sept. 4, 1866	57, 760
Pump-piston	R. Poole	Baltimore, Md	Sept. 6, 1859	25, 367
Pump-piston	P. C. Rowe	Boston, Mass	June 6, 1855	48, 098
Pump-piston	O. Salgee	New York, N. Y	May 5, 1868	77, 538
Pump-piston	J. Van Tassel	Toledo, Ohio	Dec. 19, 1871	122, 082
Pump-piston	J. D. Westcott	Waterford, Pa	Aug. 25, 1868	81, 565
Pump-piston	D. S. Wood	Delaware, Wis	July 10, 1866	56, 314

Index of patents issued from the United States Patent Office from 1790 to 1873, inclusive—Continued.

Invention.	Inventor.	Residence.	Date.	No.
Pump-piston	J. Wood	Franklin, Pa	Aug. 25, 1868	81, 451
Pump-piston	P. Zeiher	Pomeroy, Ohio	June 18, 1872	128, 196
Pump piston, Deep-well	R. N. Allen	Cleveland, Ohio	Sept. 11, 1866	57, 840
Pump piston, Deep-well	J. W. Hoagland	New Brunswick, N. J	July 17, 1866	56, 414
Pump piston, Deep-well	C. Jarecki	Erie, Pa	Nov. 10, 1868	83, 968
Pump piston, Deep-well	E. Y. Kneeland	Buffalo, N. Y	Apr. 3, 1866	53, 630
Pump-piston, Double-acting	C. B. Gill	Rochester, N. Y	July 30, 1867	67, 291
Pump piston, Force	H. H. Cooley	Battle Creek, Mich	Oct. 21, 1862	36, 702
Pump piston and valve, Rotary	J. Gatley	Rome, N. Y	Sept. 12, 1854	11, 666
Pump piston, Steam	L. J. Knowles	Worcester, Mass	July 19, 1870	105, 465
Pump-pistons, engines, &c., Material for	D. Bickford	Boston, Mass	Dec. 24, 1867	72, 580
Pump-pistons, Operating	S. Wood	Worcester, Mass	Feb. 2, 1858	19, 276
Pump-pistons, Working	D. Whittier	Belfast, Me	Apr. 14, 1838	697
Pump-plunger and pump	N. P. Sheldon	San Francisco, Cal	July 15, 1873	140, 795
Pump, Pneumatic	J. A. Bailey	Detroit, Mich	May 11, 1869	89, 909
Pump, Pneumatic	H. Gottfried	New York, N. Y	May 1, 1866	54, 328
Pump, Pneumatic	W. S. Guild	Williamsburgh, N. Y	Nov. 28, 1865	51, 175
Pump, Pneumatic	J. Powell	Cincinnati, Ohio	Oct. 3, 1871	119, 641
Pump, Pneumatic breast	J. P. Stabler	Sandy Spring, Md	June 19, 1834	....
Pump, Pneumatic forcing	P. Upham	Brookfield, Mass	Jan. 6, 1809	....
Pump, Portable	J. A. Cheron	New York, N. Y	July 13, 1869	92, 584
Pump, Portable	W. and B. Douglas	Middletown, Tex	Apr. 6, 1858	19, 834
Pump, Portable	F. Eichler	New Lisbon, Wis	Dec. 15, 1868	85, 002
Pump, Portable	W. T. Vose	Newtonville, Mass	Nov. 15, 1859	26, 135
Pump, Portable sprinkling	W. Servant	Providence, R. I	Sept. 20, 1870	107, 633
Pump, Portable steam	W. Boardman, jr	New York, N. Y	Nov. 6, 1847	5, 352
Pump, Quadruple forcing	I. Ferris	Ellicott, N. Y	Aug. 12, 1828	....
Pump, Quicksilver	M. P. Boss	Bullionville, Nev	Dec. 23, 1873	145, 718
Pump, Railway	J. B. Atwater	Chicago, Ill	Oct. 4, 1864	44, 501
Pump, Reaction	E. Rhodes, jr	Clyde, Ohio	Jan. 14, 1873	134, 770
Pump, Reciprocating	W. D. Andrews	New York, N. Y	June 17, 1862	35, 577
Pump, Reciprocating	D. M. Weston	Boston, Mass	July 3, 1866	56, 129
Pump reel, Sand	H. T. Hunt	Titusville, Pa	May 21, 1872	127, 058
Pump reel, Sand	M. A. Richardson	Sherman, N. Y	Sept. 24, 1872	131, 706
Pump-regulator	S. H. Wheeler	Dowagiac, Mich	Aug. 29, 1871	118, 659
Pump-rod adjustment	B. Densmore	New York, N. Y	Aug. 30, 1870	106, 791
Pump-rod attachment	H. H. Locke	Pleasantville, Pa	Mar. 31, 1870	103, 632
Pump-rod connection for deep wells	H. De Zavala	New York, N. Y	Aug. 23, 1870	106, 673
Pump-rod coupling	I. A. Dewar, D. S. Smith, and R. A. Brashear.	Franklin, Pa	June 8, 1869	90, 935
Pump-rod coupling	H. T. Hunt	Titusville, Pa	Mar. 16, 1869	87, 778
Pump-rods, Connecting	H. F. Purmort	Saginaw City, Mich	Feb 4, 1868	74, 133
Pump, Rotary	C. L. Adancourt	Troy, N. Y	Sept. 22, 1863	40, 008
Pump, Rotary	W. B. Allyn	Boston, Mass	Jan. 21, 1870	104, 403
Pump, Rotary	M. L. Andrew	Cincinnati, Ohio	Apr. 27, 1869	89, 268
Pump, Rotary	M. L. Andrew	Cincinnati, Ohio	Apr. 27, 1869	89, 269
Pump, Rotary	W. D. Andrews	Brook Haven, N. Y	Dec. 20, 1870	110, 330
Pump, Rotary	J. Banks	New York, N. Y	Sept. 23, 1862	36, 500
Pump, Rotary	J. Banks	New York, N. Y	Aug. 18, 1863	39, 540
Pump, Rotary	J. S. Barden	New Haven, Conn	Jan. 26, 1858	19, 180
Pump, Rotary	A. Barker	Honesdale, Pa	Mar. 27, 1855	12, 579
Pump, Rotary	A. Barker	Honesdale, Pa	Mar. 24, 1857	16, 869
Pump, Rotary	J. A. Bazin	Canton, Mass	July 8, 1856	15, 274
Pump, Rotary	J. A. Bazin	Canton, Mass	June 25, 1867	65, 992
Pump, Rotary	P. Bennett	New York, N. Y	Jan. 7, 1851	7, 883
Pump, Rotary	J. D. Birch	Philadelphia, Pa	July 16, 1867	66, 671
Pump, Rotary	J. Broughton	Chicago, Ill	June 10, 1856	15, 059
Pump, Rotary	R. Burdine	Washington, D. C	Apr. 25, 1854	10, 822
Pump, Rotary	L. Burnell	Milwaukee, Wis	Aug. 31, 1858	21, 318
Pump, Rotary	R. Bush	Brooklyn, N. Y	May 10, 1870	102, 768
Pump, Rotary	W. Butterfield	Madison, Wis	Jan. 8, 1867	60, 995
Pump, Rotary	S. D. Carpenter	Madison, Wis	Oct. 10, 1854	11, 776
Pump, Rotary	S. D. Carpenter	Madison, Wis	June 24, 1856	15, 173
Pump, Rotary	L. Chapman	Collinsville, Conn	Sept. 19, 1871	119, 011
Pump, Rotary	L. Chapman	Collinsville, Conn	Mar. 25, 1873	137, 054
Pump, Rotary	L. Chapman	Collinsville, Conn	Mar. 25, 1873	137, 055
Pump, Rotary	L. Chapman	Collinsville, Conn	Mar. 25, 1873	137, 057
Pump, Rotary	L. Chapman	Collinsville, Conn	Mar. 25, 1873	137, 058
Pump, Rotary	M. R. Clapp	Seneca Falls, N. Y	Sept. 21, 1858	21, 550
Pump, Rotary	C. N. Clow	Port Byron, N. Y	July 1, 1856	15, 221
Pump, Rotary	C. N. Clow	Port Byron, N. Y	July 8, 1856	15, 280
Pump, Rotary	T. Crane	Fort Atkinson, Wis	Dec. 25, 1855	13, 979
Pump, Rotary	T. Crane	Fort Atkinson, Wis	Apr. 8, 1856	14, 599
Pump, Rotary	W. H. Davis	Maysville, Ky	Nov. 5, 1850	7, 751
Pump, Rotary	R. Defrees	Washington, D. C	July 31, 1866	56, 850
Pump, Rotary	F. O. Deschamps	Philadelphia, Pa	Mar. 7, 1871	112, 429
Pump, Rotary	J. Doyle and T. A. Martin	Hoboken, N. J., and New York, N. Y.	Dec. 10, 1867	72, 001
Pump, Rotary	J. Doyle and T. A. Martin	Hoboken, N. J., and New York, N. Y.	Nov. 8, 1870	109, 117
Pump, Rotary	C. I. Du Pont, jr	Wilmington, Del	Aug. 6, 1872	130, 113
Pump, Rotary	J. L. Fagan	Anaqua, Tex	Mar. 8, 1859	23, 162
Pump, Rotary	J. J. Flanders	Manchester, N. H	Nov. 1, 1859	25, 999
Pump, Rotary	G. S. Follensbee	Philadelphia, Pa	Apr. 23, 1872	126, 046
Pump, Rotary	J. T. Foster	New York, N. Y	Jan. 4, 1870	98, 482
Pump, Rotary	R. S. Foster	New York, N. Y	Dec. 27, 1864	45, 596
Pump, Rotary	W. and R. Foster	Brooklyn, N. Y	Jan. 9, 1866	51, 936
Pump, Rotary	T. Freeman, jr	Providence, R. I	Apr. 26, 1859	23, 764
Pump, Rotary	C. V. Genung	Duquoin, Ill	Nov. 17, 1868	84, 106
Pump, Rotary	R. Gilbert	Rochester, N. Y	Apr. 7, 1857	16, 974
Pump, Rotary	R. Gilbert	Rochester, N. Y	Mar. 29, 1864	42, 083
Pump, Rotary	J. E. Gillespie	Norwich, Conn	July 22, 1873	141, 000
Pump, Rotary	J. S. Godfrey	Leslie, Mich	May 7, 1872	126, 540
Pump, Rotary	J. Gray	Boston, Mass	Dec. 5, 1854	12, 022
Pump, Rotary	G. W. Griswold	Carbondale, Pa	Mar. 24, 1857	16, 875
Pump, Rotary	R. C. Grover and J. Nickelson	Newton, Mass	Dec. 26, 1865	51, 713
Pump, Rotary	W. H. Guild	Brooklyn, E. D., N. Y	June 17, 1862	35, 604
Pump, Rotary	J. S. Groynne	New York, N. Y	Jan. 14, 1851	7, 901

Index of patents issued from the United States Patent Office from 1790 *to* 1873, *inclusive*—Continued.

Invention.	Inventor.	Residence.	Date.	No.
Pump, Rotary	E. Hale	Hyde Park, N. Y	Aug. 3, 1833	....
Pump, Rotary	E. R. Hale	Hyde Park, N. Y	July 14, 1830	....
Pump, Rotary	I. Hall	Poughkeepsie, N. Y	Apr. 22, 1835	....
Pump, Rotary	D. S. Hamilton	Elmira, N. Y	Nov. 5, 1861	33, 643
Pump, Rotary	D. D. Hardy and J. J. Morris	Cincinnati, Ohio	Mar. 5, 1861	31, 597
Pump, Rotary	C. H. Harrison	San Francisco, Cal	June 21, 1864	43, 204
Pump, Rotary	G. W. Heald and L. D. Cisco	Baldwinsville, N. Y	July 25, 1865	48, 938
Pump, Rotary	G. W. Heald and L. D. Cisco	Baldwinsville, N. Y	Aug. 23, 1870	106, 581
Pump, Rotary	C. H. Hersey	Boston, Mass	July 26, 1859	24, 872
Pump, Rotary	C. H. Hersey	Boston, Mass	Oct. 6, 1868	82, 833
Pump, Rotary	W. Hinds	Little Falls, N. Y	Aug. 16, 1864	43, 849
Pump, Rotary	A. P. Holly	Seneca Falls, N. Y	July 6, 1858	20, 796
Pump, Rotary	C. P. Holmes	Havana, N. Y	Mar. 19, 1872	124, 820
Pump, Rotary	C. P. Holmes	Havana, N. Y	Sept. 9, 1873	142, 696
Pump, Rotary	R. A. Horning	Lenark, Ill	Aug. 2, 1870	106, 061
Pump, Rotary	C. E. Hutson	Saint Louis, Mo	Aug. 29, 1871	118, 455
Pump, Rotary	C. W. Isbell	New York, N. Y	Jan. 11, 1870	98, 772
Pump, Rotary	C. W. Isbell	New York, N. Y	June 6, 1871	115, 613
Pump, Rotary	C. L. Johnson	Little Falls, N. Y	Apr. 9, 1861	32, 029
Pump, Rotary	W. H. Johnson	Springfield, Mass	Mar. 21, 1846	4, 426
Pump, Rotary	D. L. Jones	Nebraska City, Nebr	Apr. 30, 1872	126, 306
Pump, Rotary	J. Jones and A. K. Rider	New York, N. Y., and Poultney, Vt.	May 21, 1861	32, 372
Pump, Rotary	S. W. Kelly	Nashville, Tenn	Nov. 14, 1871	120, 819
Pump, Rotary	F. Kettler	Milwaukee, Wis	June 3, 1862	35, 452
Pump, Rotary	A. H. Knapp	Newton Centre, Mass	Nov. 17, 1863	40, 660
Pump, Rotary	A. H. Knapp	Needham, Mass	July 20, 1869	92, 842
Pump, Rotary	J. C. Lamb and E. Lavens	Middletown, Conn	Dec. 24, 1872	134, 294
Pump, Rotary	E. Lavens and J. C. Lamb	Middletown, Conn	Apr. 19, 1870	102, 136
Pump, Rotary	H. B. Leach	Boston, Mass	May 8, 1866	54, 565
Pump, Rotary	A. Leuchtweiss	Cincinnati, Ohio	June 21, 1870	104, 469
Pump, Rotary	H. Maas	Homestead, Iowa	June 5, 1866	55, 318
Pump, Rotary	H. W. Mather	New York, N. Y	Apr. 19, 1870	102, 140
Pump, Rotary	J. Megaw	Wilmington, Del	July 5, 1859	24, 650
Pump, Rotary	J. J. Morris	Cincinnati, Ohio	May 8, 1866	54, 581
Pump, Rotary	N. E. Nash	Providence, R. I	Aug. 6, 1872	130, 146
Pump, Rotary	J. Naughton	Buffalo, N. Y	Feb. 23, 1869	87, 108
Pump, Rotary	G. W. Nichols	Wheatland, Iowa	June 3, 1862	35, 446
Pump, Rotary	O. Palmer	Buffalo, N. Y	Dec. 29, 1857	18, 986
Pump, Rotary	O. Palmer	Cincinnati, Ohio	Dec. 5, 1865	51, 347
Pump, Rotary	H. Pease	Brockport, N. Y	Oct. 20, 1857	18, 488
Pump, Rotary	H. Pease	Brockport, N. Y	Mar. 31, 1863	38, 058
Pump, Rotary	W. Peirce	New Orleans, La	Mar. 9, 1858	19, 581
Pump, Rotary	E. Perry	Baldwinsville, N. Y	Sept. 30, 1862	36, 588
Pump, Rotary	Peters and Dean	Poughkeepsie, N. Y	Oct. 31, 1835	....
Pump, Rotary	F. B. Pierce	Brockport, N. Y	May 27, 1862	35, 388
Pump, Rotary	J. Poppe	Green Point, N. Y	Dec. 3, 1867	71, 785
Pump, Rotary	G. W Putnam	South Glen's Falls, N. Y	Mar. 7, 1871	112, 491
Pump, Rotary	R. Ramsden	South Easton, Pa	June 9, 1857	17, 516
Pump, Rotary	F. Rochow	New York, N. Y	Dec. 17, 1861	33, 961
Pump, Rotary	G. W. Rogers	Philadelphia, Pa	Apr. 20, 1869	89, 080
Pump, Rotary	G. W. Rogers	Brooklyn, N. Y	July 23, 1872	129, 864
Pump, Rotary	F. B. Scott	Buffalo, N. Y	Sept. 30, 1862	36, 577
Pump, Rotary	F. B. Shannon	Port Washington, Ohio	Feb. 5, 1867	61, 882
Pump, Rotary	J. H. Shedd and W. Edson	Boston, Mass	Jan. 19, 1864	41, 327
Pump, Rotary	N. P. Sheldon	San Francisco, Cal	Jan. 30, 1872	123, 130
Pump, Rotary	A. Sluthour	Cleveland, Ohio	Apr. 19, 1870	102, 057
Pump, Rotary	H. C. Spalding and G. Stickney	Hartford, Conn	Apr. 20, 1852	8, 893
Pump, Rotary	M. Stannard	New Britain, Conn	Aug. 20, 1861	33, 126
Pump, Rotary	T. Sutton	Norwich, Conn	Jan. 28, 1831	....
Pump, Rotary	T. Swan	Manlius, N. Y	May 31, 1870	103, 684
Pump, Rotary	P. Sweeney	Buffalo, N. Y	Nov 27, 1849	6, 906
Pump, Rotary	B. F. Taber	Buffalo, N. Y	Oct. 22, 1861	33, 550
Pump, Rotary	J. A. Talpey	Somerville, Mass	Mar. 24, 1868	75, 811
Pump, Rotary	H. P. Tenant	East Germantown, Ind	May 23, 1871	115, 254
Pump, Rotary	B. T. Trimmer	Rochester, N. Y	Sept. 28, 1858	21, 632
Pump, Rotary	W. C. Trowbridge	Southeast, N. Y	Jan. 27, 1835	....
Pump, Rotary	P. Ulmholtz	Tremont, Pa	July 4, 1865	48, 604
Pump, Rotary	A. Walker	Claremont, N. H	Nov. 15, 1864	45, 096
Pump, Rotary	D. M. Walker	Cavendish, Vt	Aug. 15, 1835	....
Pump, Rotary	W. H. Ward	Auburn, N. Y	Oct. 3, 1871	119, 482
Pump, Rotary	S. A. West and L. Goodwin	San Francisco, Cal., and Virginia City, Nev.	Oct. 1, 1872	131, 801
Pump, Rotary	I. Williams	Baldwinsville, N. Y	Jan. 17, 1871	111, 026
Pump, Rotary	G. Wingate	Philadelphia, Pa	Jan. 25, 1859	22, 762
Pump, Rotary	C. D. Wright	Fort Atkinson, Wis	Dec. 11, 1855	13, 930
Pump, Rotary	W. A. Young	Charlotte, N. C	Dec. 29, 1857	19, 004
Pump, Rotary air	G. B. Hill	New York, N. Y	July 11, 1865	48, 770
Pump, Rotary forcing	G. W. Billings	Saint Paul, Minn	Nov. 11, 1862	36, 888
Pump, Rotary hydropneumatic	H. M. Paine	Newark, N. J	Mar. 29, 1870	101, 306
Pump, Rotary steam	H. O. Ames	New Orleans, La	Jan. 29, 1867	61, 650
Pump, Rotary water	S. A. Lee	Boston, Mass	Sept. 4, 1841	2, 236
Pump, Rotary wrecking	O. Palmer	Buffalo, N. Y	July 17, 1855	13, 273
Pump, Saliva	F. Davison	Liberty, Va	Jan. 17, 1854	10, 426
Pump, Saliva	A. M. Shurtleff	Boston, Mass	Dec. 3, 1867	71, 799
Pump, Saliva	A. M. Shurtleff	Boston, Mass	Mar. 31, 1868	76, 257
Pump, Sand	J. Benson	Bellaire, Ohio	Oct. 9, 1866	58, 579
Pump, Sand	R. Cornelius	Philadelphia, Pa	May 22, 1866	54, 867
Pump, Sand	J. B. Kibler	Girard, Pa	May 15, 1866	54, 739
Pump, Sand	O. B. Latham	Seneca Falls, N. Y	Dec. 26, 1865	51, 730
Pump, Sand	C. Mather	Manchester, England	Sept. 4, 1866	57, 830
Pump, Sand	B. Morahan	Brooklyn, N. Y	Oct. 2, 1866	58, 457
Pump, Sand	E. A. L. Roberts	Titusville, Pa	May 28, 1867	65, 121
Pump, Sand	J. Smith	Franklin, Pa	June 4, 1872	127, 655
Pump, Sea-motion	B. Lies	New York	Aug. 19, 1811	....
Pump, Segmental	J. Robertson	Brooklyn, N. Y	Feb. 13, 1872	123, 579
Pump, Self adjustable or anchoring	T. Ling	Shelby, Ohio	Mar. 27, 1855	12, 601
Pump, Ship's	T. Bell	Bellport, N. Y	June 18, 1872	128, 060

Index of patents issued from the United States Patent Office from 1790 *to* 1873, *inclusive*—Continued.

Invention.	Inventor.	Residence.	Date.	No.
Pump, Ship's	R. Bulkley	New York, N. Y	Dec. 20, 1847	5, 030
Pump, Ship's	G. Clymer		Dec. 22, 1801	....
Pump, Ship's	A. Coates and S. M. Perry	New York and Brooklyn, N. Y.	Sept. 15, 1857	18, 192
Pump, Ship's	D. A. Dunham	Pilatka, Fla	Apr. 4, 1871	113, 641
Pump, Ship's	J. Edson	Boston, Mass	Aug. 15, 1865	49, 393
Pump, Ship's	J. J. Heard	Boston, Mass	July 10, 1855	13, 219
Pump, Ship's	G. Moulton, jr	Bath, Me	May 29, 1866	55, 147
Pump, Ship's, &c	T. Odiorne	Portsmouth, N. H	Aug. 27, 1835	....
Pump, Ship's	S. F. Paullin	Philadelphia, Pa	Dec. 17, 1872	133, 950
Pump, Ship's	S. Pearn	New York, N. Y	Oct. 2, 1855	13, 622
Pump, Ship's	F. Ransom	New York, N. Y	Mar. 19, 1850	7, 198
Pump, Ship's	B. K. Rogers	Jonesport, Me	Dec. 28, 1869	98, 419
Pump, Ship's	T. S. Speakman and N. Hand	Camden, N. J	Nov. 21, 1865	51, 094
Pump, Siphon	J. M. Bois	Salamanca, N. Y	Mar. 18, 1873	136, 809
Pump, Siphon	T. J. Trapp	Williamsport, N. Y	Oct. 25, 1870	108, 652
Pump, Siphon steam	T. O'Rorke	Pittsburgh, Pa	July 25, 1871	117, 450
Pump-spear, Horizontal circular-valve	E. Leman	Boston, Mass	May 23, 1817	....
Pump, Spiral	C. Foss	Perry, Ohio	Aug. 4, 1832	....
Pump-spout	J. H. Gleim	Tipton, Mo	Apr. 19, 1870	101, 995
Pump, Spring or drop valve	E. Kendall	Ashbury, Mass	June 22, 1832	....
Pump-sprinkler	W. F. Class	Cleveland, Ohio	May 15, 1866	54, 685
Pump, Steam	W. Aldrich	Dayton, Ohio	May 20, 1873	139, 103
Pump, Steam	L. H. Allen and W. Barton	Tamaqua, Pa	Feb. 13, 1872	123, 659
Pump, Steam	L. H. Allen and W. Barton	Tamaqua, Pa	Feb. 13, 1872	123, 660
Pump, Steam	L. H. Allen and W. Barton	Tamaqua, Pa	Feb. 13, 1872	123, 661
Pump, Steam	W. Arthur	Newport, Ky	Aug. 27, 1872	130, 890
Pump, Steam	W. Arthur	Newport, Ky	Jan. 21, 1873	135, 061
Pump, Steam	J. B. Atwater	Chicago, Ill	Aug. 1, 1865	49, 063
Pump, Steam	J. S. Barden	New Haven, Conn	Mar. 27, 1860	27, 610
Pump, Steam	J. S. Barden	Providence, R. I	Nov. 24, 1868	84, 251
Pump, Steam	W. Baxter, jr	Newark, N. J	Apr. 18, 1871	113, 725
Pump, Steam	P. Bernard and T. Soper	Whitestown, N. Y	Jan. 16, 1811	....
Pump, Steam	G. F. Blake	Boston, Mass	Oct. 25, 1870	108, 554
Pump, Steam	G. F. Blake	Boston, Mass	Feb. 20, 1872	123, 766
Pump, Steam	A. and F. Brown	New York, N. Y	Mar. 13, 1860	27, 426
Pump, Steam	F. Brown	New York, N. Y	Dec. 5, 1865	51, 291
Pump, Steam	L. and T. E. Button	Waterford, N. Y	Sept. 26, 1871	119, 314
Pump, Steam	A. S. Cameron	New York, N. Y	Dec. 17, 1867	72, 363
Pump, Steam	L. Carricaburn	Havana, Cuba	Nov. 6, 1866	59, 519
Pump, Steam	J. C. and E. H. Dean	Indianapolis, Ind	Nov. 26, 1872	133, 419
Pump, Steam	R. Eickemeyer	Yonkers, N. Y	May 6, 1873	138, 622
Pump, Steam	H. Epping	Pittsburgh, Pa	May 7, 1872	126, 536
Pump, Steam	A. Friedmann	Vienna, Austria	Apr. 6, 1869	88, 620
Pump, Steam	A. Friedmann	Vienna, Austria	Aug. 23, 1870	106, 683
Pump, Steam	J. B. Gardiner	Springfield, Mass	Feb. 4, 1868	74, 075
Pump, Steam	J. B. Gardiner and E. H. Hyde	Springfield, Mass	Dec. 17, 1867	72, 383
Pump, Steam	J. W. Gardner, T. W. Ranson, and E. Martin.	Cleveland, Ohio	Dec. 10, 1872	133, 847
Pump, Steam	L. P. Garner	Ashland, Pa	Aug. 17, 1869	93, 696
Pump, Steam	W. W. Gilbert	New York, N. Y	Apr. 27, 1869	89, 398
Pump, Steam	W. W. Gilbert	New York, N. Y	Nov. 30, 1869	97, 389
Pump, Steam	L. Griscom	Port Carbon, Pa	Mar. 19, 1872	124, 679
Pump, Steam	J. W. Handren	Jersey City, N. J	Jan. 14, 1873	134, 746
Pump, Steam	J. W. Handren	Jersey City, N. J	Feb. 4, 1873	135, 426
Pump, Steam	T. Harrington	Pittsburgh, Pa	Jan. 10, 1871	110, 850
Pump, Steam	W. Harsen	Green Point, N. Y	Feb. 4, 1868	74, 084
Pump, Steam	W. M. Henderson	Philadelphia, Pa	June 18, 1867	65, 911
Pump, Steam	W. L. Horne	Chicago, Ill	July 30, 1872	129, 891
Pump, Steam	I. Illofsky	Pesth, Hungary	Oct. 4, 1864	44, 532
Pump, Steam	J. Jordan	Wyandotte, Kans	Nov. 6, 1866	59, 400
Pump, Steam	W. E. Kelly	New Brunswick, N. J	May 14, 1872	126, 814
Pump, Steam	L. J. Knowles	Worcester, Mass	Apr. 18, 1871	113, 897
Pump, Steam	L. J. Knowles	Worcester, Mass	May 2, 1871	114, 507
Pump, Steam	B. S. Lawson	New York, N. Y	Dec. 23, 1873	145, 806
Pump, Steam	J. A. Lidback	Portland, Me	Dec. 28, 1869	98, 279
Pump, Steam	W. Livingstone	New York, N. Y	May 31, 1870	103, 756
Pump, Steam	G. W. Long	Old Point Comfort, Va	Feb. 16, 1826	....
Pump, Steam	J. R. Maxwell and E. Cope	Cincinnati, Ohio	June 21, 1870	104, 616
Pump, Steam	J. R. Maxwell and E. Cope	Cincinnati, Ohio, and Covington, Ky.	Jan. 30, 1872	123, 185
Pump, Steam	J. R. Maxwell and E. Cope	Cincinnati, Ohio, and Covington, Ky.	Oct. 1, 1872	131, 765
Pump, Steam	J. Mayher	East Hampton, Mass	Oct. 8, 1872	131, 960
Pump, Steam	J. McCloskey	New York, N. Y	Apr. 20, 1869	89, 158
Pump, Steam	G. McFeely	Steubenville, Ohio	Aug. 31, 1869	94, 426
Pump, Steam	J. J. Miller	Chicago, Ill	Aug. 23, 1864	43, 952
Pump, Steam	J. C. Morgan	Alliance, Ohio	Mar. 13, 1866	53, 170
Pump, Steam	J. North	New York, N. Y	Oct. 15, 1872	132, 315
Pump, Steam	J. North	New York, N. Y	Dec. 31, 1872	134, 481
Pump, Steam	J. V. Pangburn	Galesburgh, Ill	Nov. 5, 1872	132, 735
Pump, Steam	A. and G. W. Perry	New Philadelphia and Mahanoy City, Pa.	Dec. 5, 1871	121, 658
Pump, Steam	W. Porter	Brooklyn, N. Y	Apr. 23, 1872	126, 083
Pump, Steam	J. B. Pottmeyer	Pittsburgh, Pa	Feb. 26, 1867	62, 360
Pump, Steam	A. J. Reynolds	Chicago, Ill	Oct. 27, 1868	83, 549
Pump, Steam	G. J. Roberts	Dayton, Ohio	May 14, 1872	126, 834
Pump, Steam	G. J. Roberts	Dayton, Ohio	May 6, 1873	138, 695
Pump, Steam	W. H. Roberts	Mauch Chunk, Pa	Mar. 30, 1869	88, 512
Pump, Steam	L. C. Rodier	Springfield, Mass	Aug. 14, 1866	57, 188
Pump, Steam	G. W. Rogers	Brooklyn, N. Y	June 25, 1872	128, 426
Pump, Steam	R. Schmidt	New York, N. Y	Aug. 6, 1867	67, 592
Pump, Steam	H. C. Sergeant	New York, N. Y	Oct. 7, 1862	36, 623
Pump, Steam	H. C. Sergeant	New York, N. Y	June 22, 1869	91, 676
Pump, Steam	W. Sewell	New York, N. Y	Oct. 28, 1862	36, 862
Pump, Steam	T. Simmons	New York, N. Y	Aug. 13, 1867	67, 812
Pump, Steam	T. and G. T. Snowdon and I. V. Lynn.	Pittsburgh, Pa	Jan. 11, 1870	98, 810

Index of patents issued from the United States Patent Office from 1790 to 1873, inclusive—Continued.

Invention.	Inventor.	Residence.	Date.	No.
Pump, Steam	S. Stanton	New York, N. Y	Feb. 27, 1872	124,095
Pump, Steam	C. L. Stevens	Galesburgh, Ill	Oct. 4, 1870	107,977
Pump, Steam	C. L. Stevens and A. A. Denton	Galesburgh, Ill	Jan. 19, 1869	86,111
Pump, Steam	C. Swinscoe	Boston, Mass	Mar. 12, 1872	124,639
Pump, Steam	W. R. Thomas	Catasauqua, Pa	July 14, 1868	80,030
Pump, Steam	J. E. Thorpe	Erie, Pa	Apr. 23, 1867	64,166
Pump, Steam	A. W. Todd	Chicago, Ill	July 18, 1865	48,853
Pump, Steam	L. W. Turrell	Newburgh, N. Y	Jan. 24, 1865	46,036
Pump, Steam	W. Watts	Newark, N. J	Sept. 8, 1863	39,852
Pump, Steam	G. M. Weinman	Columbus, Ohio	Dec. 6, 1870	109,980
Pump, Steam	J. C. Wightman	Newton, Mass	Apr. 27, 1869	89,534
Pump, Steam	M. Wilcox	Middlebury, Ohio	Apr. 2, 1861	31,923
Pump, Steam	M. Wilcox	Sacramento, Cal	Aug. 15, 1865	49,461
Pump, Steam	C. and G. M. Woodward	New York, N. Y	July 22, 1862	35,966
Pump, Steam and vacuum	A. J. Simmons	Indianapolis, Ind	Oct. 14, 1873	143,725
Pump, Steam-engine air	A. Hartupee	Pittsburgh, Pa	Feb. 5, 1867	61,735
Pump, Steam-jet	C. Barnes	Cincinnati, Ohio	Aug. 7, 1866	57,035
Pump, Steam-jet	A. J. Blakslee and G. C. Williams	Duquoin, Ill	Dec. 13, 1870	110,110
Pump, Steam-jet	J. D. Toppin	Newark, N. J	Dec. 17, 1872	134,018
Pump, Steam siphon	T. O'Rorke	Pittsburgh, Pa	June 18, 1872	127,984
Pump, Steam vacuum	I. Barnum	New York	Apr. 16, 1834	....
Pump, Steam vacuum	W. Burdon	Brooklyn, N. Y	July 23, 1872	129,647
Pump, Steam vacuum	W. Burdon	Brooklyn, N. Y	Jan. 28, 1873	135,198
Pump, Steam vacuum	W. Burdon	Brooklyn, N. Y	Jan. 28, 1873	135,199
Pump, Steam vacuum	W. Burdon	Brooklyn, N. Y	Jan. 28, 1873	135,200
Pump, Steam vacuum	W. Burdon	Brooklyn, N. Y	Jan. 28, 1873	135,201
Pump, Steam vacuum	W. Burdon	Brooklyn, N. Y	Jan. 28, 1873	135,202
Pump, Steam vacuum	W. Burdon	Brooklyn, N. Y	May 27, 1873	139,235
Pump, Steam vacuum	W. Burdon	Brooklyn, N. Y	Aug. 5, 1873	141,540
Pump, Steam vacuum	C. H. Hall	New York, N. Y	Sept. 24, 1872	131,515
Pump, Steam vacuum	C. H. Hall	New York, N. Y	Sept. 24, 1872	131,516
Pump, Steam vacuum	C. H. Hall	New York, N. Y	Sept. 24, 1872	131,517
Pump, Steam vacuum	C. H. Hall	New York, N. Y	Sept. 24, 1872	131,518
Pump, Steam vacuum	C. H. Hall	New York, N. Y	Sept. 24, 1872	131,519
Pump, Steam vacuum	C. H. Hall	New York, N. Y	Sept. 24, 1872	131,520
Pump, Steam vacuum	C. H. Hall	New York, N. Y	Sept. 24, 1872	131,521
Pump, Steam vacuum	C. H. Hall	New York, N. Y	Sept. 24, 1872	131,522
Pump, Steam vacuum	C. H. Hall	New York, N. Y	Sept. 24, 1872	131,523
Pump, Steam vacuum	C. H. Hall	New York, N. Y	Sept. 24, 1872	131,524
Pump, Steam vacuum	C. H. Hall	New York, N. Y	Sept. 24, 1872	131,525
Pump, Steam vacuum	C. H. Hall	New York, N. Y	Sept. 24, 1872	131,526
Pump, Steam vacuum	C. H. Hall	New York, N. Y	Sept. 24, 1872	131,527
Pump, Steam vacuum	C. H. Hall	New York, N. Y	Sept. 24, 1872	131,528
Pump, Steam vacuum	C. H. Hall	New York, N. Y	Sept. 24, 1872	131,529
Pump, Steam vacuum	C. H. Hall	New York, N. Y	Sept. 24, 1872	131,530
Pump, Steam vacuum	C. H. Hall	New York, N. Y	Sept. 24, 1872	131,531
Pump, Steam vacuum	C. H. Hall	New York, N. Y	Sept. 24, 1872	131,532
Pump, Steam vacuum	C. H. Hall	New York, N. Y	Sept. 24, 1872	131,533
Pump, Steam vacuum	C. H. Hall	New York, N. Y	Sept. 24, 1872	121,534
Pump, Steam vacuum	C. H. Hall	New York, N. Y	Sept. 24, 1872	131,535
Pump, Steam vacuum	C. H. Hall	New York, N. Y	Sept. 24, 1872	131,536
Pump, Steam vacuum	C. H. Hall	New York, N. Y	Sept. 24, 1872	131,537
Pump, Steam vacuum	C. H. Hall	New York, N. Y	Sept. 24, 1872	131,538
Pump, Steam vacuum	C. H. Hall	New York, N. Y	Sept. 24, 1872	131,539
Pump, Steam vacuum	C. H. Hall	New York, N. Y	Sept. 24, 1872	131,540
Pump, Steam vacuum	C. H. Hall	New York, N. Y	Sept. 24, 1872	131,541
Pump, Steam vacuum	C. H. Hall	New York, N. Y	Sept. 24, 1872	131,542
Pump, Steam vacuum	C. H. Hall	New York, N. Y	Sept. 24, 1872	131,543
Pump, Steam vacuum	J. R. Little	Monmouth, Ill	Mar. 4, 1873	136,441
Pump, Steam vacuum	J. M. Morehead	Brooklyn, N. Y	Jan. 30, 1872	123,117
Pump, Steam vacuum	J. M. Morehead	Brooklyn, N. Y	May 21, 1872	126,898
Pump, Steam vacuum	G. H. Nye	Monmouth, Ill	Nov. 5, 1872	132,731
Pump, Steam vacuum	G. H. Nye	Monmouth, Ill	Nov. 5, 1872	132,732
Pump, Steam vacuum	J. H. Pattee	Monmouth, Ill	July 16, 1872	129,051
Pump, Steam vacuum	J. H. Pattee	Monmouth, Ill	Aug. 6, 1872	130,309
Pump, Steam vacuum	J. H. Pattee and H. J. Graham	Monmouth, Ill	Jan. 10, 1871	110,933
Pump, Steam vacuum	J. H. Pattee and G. H. Nye	Monmouth, Ill	May 17, 1870	103,076
Pump, Steam vacuum	W. E. Prall	Washington, D. C	Jan. 30, 1872	123,292
Pump, Steam vacuum	W. E. Prall and D. A. Burr	Washington, D. C	Dec. 30, 1873	146,095
Pump, Steam vacuum	C. L. Riker	Schraalenburgh, N. J	Apr. 9, 1872	125,410
Pump, Steam vacuum	J. K. Simpson and S. N. Drake	New York, N. Y	Jan. 7, 1873	134,568
Pump, Steam vacuum	I. P. Tice	New York, N. Y	Feb. 25, 1873	136,344
Pump, Stock	W. T. Armstrong	Freeland, Ill	Sept. 15, 1868	82,064
Pump, Stock	W. A. Durrin	Milledgeville, Ill	Feb. 24, 1863	37,744
Pump, Stock	A. H. Piland and A. H. Turner	Indianapolis, Ind	Aug. 18, 1868	81,287
Pump, Stock	A. H. Russell	Adrian, Mich	June 23, 1868	79,146
Pump-stocks, Machine for securing sheet-metal lining in.	N. T. Coffin	Knightstown, Ind	Aug. 2, 1870	105,912
Pump, Stomach	J. M. Youngblood	Saint Louis, Mo	July 20, 1869	92,769
Pump-strainer	W. S. Blunt	Brooklyn, N. Y	Aug. 20, 1872	130,560
Pump, Submerged	A. C. Blethen	Lynn, Mass	Apr. 6, 1869	88,538
Pump, Submerged	B. F. Brown	Woburn, Mass	Dec. 22, 1868	85,063
Pump, Submerged	P. H. Cardwell	Knoxville, Tenn	Feb. 1, 1870	99,403
Pump, Submerged	J. Haney	Yorktown, Ind	July 9, 1872	128,877
Pump, Submerged	D. G. Hussey	Nantucket, Mass	Aug. 20, 1872	130,642
Pump, Submerged	H. Lindsey	Asheville, N. C	Sept. 6, 1859	25,340
Pump, Submerged	J. A. Morrell and J. Brewer	New York, N. Y	Aug. 12, 1873	141,807
Pump, Submerged	H. M. Stoker	Watson, Ill	June 13, 1865	48,219
Pump, Submerged	H. M. Stoker	Watson, Ill	June 13, 1865	48,220
Pump, Submerged	J. H. Williamson	Branchville, N. J	Mar. 7, 1865	49,764
Pump, Submerged force	A. Balding	Flora, Ill	Aug. 8, 1865	49,212
Pump, Submerged force	A. T. Hafford	Saint Louis, Mo	May 20, 1873	139,003
Pump, Submerged force	C. C. Hiatt	Ridgeville, Ind	Jan. 19, 1869	86,071
Pump, Submerged rotary	D. D. Hardy	Cincinnati, Ohio	Sept. 1, 1868	81,778
Pump-sucker	J. Weis	Bordentown, N. J	Dec. 18, 1855	13,967
Pump, Suction	T. C. Barton	Washington, N. Y	Feb. 20, 1836	....
Pump, Suction	A. H. Mathes	Madisonville, Tenn	Dec. 31, 1833	....

Index of patents issued from the United States Patent Office from 1790 *to* 1873, *inclusive*—Continued.

Invention.	Inventor.	Residence.	Date.	No.
Pump, Suction	E. Whiton	Groton, Mass	Oct. 2, 1834	....
Pump, Suction and force	J. Stevens	Newark, N. J	Dec. 20, 1837	523
Pump, Suction and lifting	E. Tolles	Litchfield, Conn	Sept. 28, 1831	....
Pump, Supply	J. L. Sullivan	New York	Dec. 28, 1833	....
Pump, Test	T. Shaw	Philadelphia, Pa	Oct. 4, 1864	44, 559
Pump to raise water, Force	A. Taylor	....	Oct. 4, 1804	....
Pump, Tube-well	J. H. Devirs and D. Gravatt	Pleasantville, Pa	May 18, 1869	90, 157
Pump-tubes, Wrench for elevating	J. A. Fleming	Shamburgh, Pa	Sept. 7, 1869	94, 484
Pump, Turbinate force	J. H. A. Gericke	New Orleans, La	May 29, 1866	55, 082
Pump, Universal	A. Kinsley	....	Oct. 28, 1799	....
Pump used in low-pressure or condensing steam-engines, Air.	C. Reeder	Baltimore, Md	Nov. 15, 1843	3, 334
Pump, Vacuum	E. Dunscomb	Boston, Mass	Mar. 12, 1867	62, 738
Pump-valve chamber	M. C. Taylor	Grass Valley, Cal	Dec. 1, 1868	84, 516
Pump ventilator for ships, mines, &c., Air	R. Robotham	....	Oct. 10, 1801	....
Pump, Vibrating	S. Davis	Derby, Vt	Feb. 25, 1836	....
Pump, Vibrating	E. Garrette	Wilmington, Del	Oct. 9, 1855	13, 643
Pump, Water	J. Babcock, jr	Edmiston, N. Y	May 6, 1825	....
Pump, Water	A. H. Fitch	Cuylersville, N. Y	May 23, 1848	5, 591
Pump, Water	J. F. Flanders	Newburyport, Mass	July 22, 1851	8, 239
Pump, Water	J. W. Hillias	Baltimore, Md	Sept. 16, 1829	....
Pump, Water	D. Hinman	Brunswick, Ohio	Oct. 24, 1846	4, 823
Pump, Water	F. B. Hyatt	Maysville, Ky	June 16, 1846	4, 583
Pump, Water	G. Ketchum	Marshfield, Mich	Dec. 18, 1847	5, 401
Pump, Water	J. B. Read	Tuscaloosa, Ala	Sept. 11, 1849	6, 714
Pump, Water	J. Renfrew	Fayetteville, Pa	June 16, 1846	4, 580
Pump, Water	A. Stiven	New York, N. Y	Dec. 4, 1849	6, 927
Pump, Water	P. Upham	Brookfield, Mass	Mar. 9, 1811	....
Pump, Water	C. Warner	Louisville, Ky	July 8, 1846	4, 619
Pump, Wind-power	H. Bronson	Kent, Conn	Mar. 8, 1820	....
Pump, Wind-wheel	L. D. Parsons	Tremont, N. Y	Oct. 19, 1869	95, 930
Pump, Wind-wheel	E. Pepple	NewCarlislePost-Office,Ind	Apr. 26, 1864	42, 503
Pump with concentric piston, Force	P. M. Kafer	Trenton, N. J	Sept. 7, 1869	94, 611
Pump with double action for engines, Forcing	E. Cady, jr	Chatham, N. Y	Mar. 9, 1808	....
Pumps, Air-chamber for	R. H. Hilton	New Berne, N. C	Nov. 19, 1867	71, 007
Pumps, Air-escape for	H. Smith	Norwalk, Ohio	Mar. 4, 1856	14, 372
Pumps and steam-engines, Construction of cylinders and pistons for.	W. Wells	New York, N. Y	Oct. 12, 1858	21, 789
Pumps, Anti freezing device for	J. G. Hanning	Indianapolis, Ind	May 24, 1870	103, 325
Pumps, Application of power to	N. Underwood	Baltimore County, Md	July 17, 1828	....
Pumps, Arrangement of	L. H. Mosby	Powhatan County, Va	Feb. 15, 1823	....
Pumps, Attaching air-chambers to	C. N. Lewis	Seneca Falls, N. Y	June 23, 1857	17, 653
Pumps, Attachment for direct-acting steam	O. T. Earle	Norwalk, Conn	Aug. 26, 1873	142, 092
Pumps, Basket-attachment for pistons of deep-well	E. D. Brown	Buffalo, N. Y	Jan. 22, 1867	61, 311
Pumps by wind, Operating	H. Moore	Charleston, Mich	Mar. 13, 1855	12, 517
Pumps by wind-wheels, Regulating	J. W. Goodwin and M. C. Hawkins.	Edinborough, Pa	Apr. 8, 1856	14, 626
Pumps, Case for rotary	P. H. and F. M. Roots	Connersville, Ind	Aug. 11, 1868	81, 010
Pumps, &c., Communicating reciprocating motion to	I. Van Olinda	Brooklyn, N. Y	Oct. 16, 1866	58, 918
Pumps, Constructing and working ships'	A. Henshaw and N. Harlow, jr	Bangor, Mass	Aug. 30, 1811	....
Pumps, Construction of	W. Rhoades	New York	May 16, 1808	....
Pumps, Cylinder for lining wooden	R. W. Wheeler	Bushnell, Ill	July 15, 1873	140, 802
Pumps, Débris-check for	S. O. Blanding	Smithfield, R. I	Nov. 17, 1868	84, 081
Pumps, Device for air-chambers of	J. P. Cowing	Seneca Falls, N. Y	Mar. 20, 1855	12, 542
Pumps, &c., Device for drawing water from and supplying air to air-vessels of.	A. H. Ranch	Bethlehem, Pa	June 5, 1860	28, 604
Pumps, &c., Double-reciprocating split piston-rod for.	J. A. Burnap	Albany, N. Y	July 24, 1855	13, 299
Pumps for ships, Arrangement of means in pendulum.	J. Stever	Bristol, Conn	July 22, 1856	15, 395
Pumps for steam-engines, Method of varying the stroke of feeding.	J. R. Sees	New York, N. Y	Apr. 1, 1856	14, 576
Pumps for wells, &c., Constructing force	T. W. H. Moseley	Paris, Ky	Sept. 30, 1839	1, 351
Pumps, Glass cylinder for	J. Bryan	New York, N. Y	Aug. 27, 1872	130, 896
Pumps, Glass piston and packing for	A. E. Gay	Nashua, N. H	July 8, 1873	140, 697
Pumps, Hand-lever attachment for steam	O. T. Earle	Norwalk, Conn	Sept. 9, 1873	142, 562
Pumps, Handle for ships'	K. Markuson	Gloucester, Mass	Nov. 5, 1867	70, 587
Pumps, Hoisting water by wind	J. Kerr	Maury County, Tenn	Sept. 6, 1833	....
Pumps in deep wells, Apparatus for protecting	F. Armstrong	Pittsburgh, Pa	May 8, 1866	54, 484
Pumps in propelling boats, Application of	M. Ulman and M. Isaacs	Philadelphia, Pa	Mar. 17, 1818	....
Pumps, Instrument for reaming out barrels of ships'	S. H. Sugett	Eden, Me	Oct. 7, 1862	36, 626
Pumps, Lifting-clamp for	W. H. Downing	Pioneer, Pa	Apr. 5, 1870	101, 595
Pumps, Link for connecting handle and pitman of.	L. W. and C. Olds	Erie, Pa	Aug. 13, 1872	130, 527
Pumps, Link for connecting handle and valve-rod of	S. Selden and M. Griswold	Erie, Pa	Oct. 4, 1870	107, 965
Pumps, Lubricating pistons of air	A. Barker	Honesdale, Pa	Apr. 3, 1855	12, 617
Pumps, Machine for operating	R. E. Moore	Navasota, Tex	Dec. 7, 1869	97, 672
Pumps, Making suction	D. Johnson	Boston, Mass	Apr. 21, 1824	....
Pumps, Mechanism for operating	J. A. Whipple	Boston, Mass	June 6, 1854	11, 020
Pumps, Mechanism for operating valve of force	J. C. King	New York, N. Y	Sept. 17, 1867	68, 994
Pumps, Metallic mold for casting	H. Post	New York, N. Y	Mar. 25, 1862	34, 774
Pumps, Method of effecting uniform pressure upon the pumping-piston of double-acting steam.	R. B. Gorsuch	New York, N. Y	June 10, 1856	15, 070
Pumps on railways, Operating	G. M. Cole	Folsom City, Cal	Dec. 16, 1862	37, 152
Pumps on railway-stations, Operating	H. H. Call	Rohrerstown, Pa	Dec. 3, 1867	71, 695
Pumps on railway-stations, Operating	W. McVeigh	Boone, Ill	Feb. 9, 1858	19, 304
Pumps, Operating	J. Armstrong	Dobbinsville, N. C	Apr. 16, 1861	32, 038
Pumps, Operating	E. Borton	Morris, Ill	Jan. 28, 1868	73, 771
Pumps, Operating	D. J. Rogers	Magnolia, N. C	Oct. 26, 1858	21, 904
Pumps, Operating ships'	S. R. Deverell	Mount Gambier, South Australia.	Jan. 21, 1873	135, 094
Pumps, Operating ships'	A. Roff	Southport, Conn	June 8, 1869	91, 045
Pumps, Preparing molds for casting double	F. Henshaw	Brookfield, Mass	June 29, 1839	1, 206
Pumps, Pressure-head for siphon and force	D. M. Shapley	Saint Louis; Mo	May 8, 1866	54, 615
Pumps, Protection against freezing in	S. Adams	Townsend, Mass	Mar. 11, 1837	141
Pumps, Regulating discharge of water from	J. Mayher	East Hampton, Mass	Jan. 10, 1871	110, 928
Pumps, Return-drip for	J. B. Stevenson	Bloomington, Ill	Nov. 10, 1868	83, 887
Pumps, Rivet-pocket for oil-well	W. H. Birge	Franklin, Pa	Jan. 21, 1873	134, 970
Pumps, Sand-chamber for well	E. C. Johnson	Williamsport, Pa	Jan. 4, 1870	98, 503

Index of patents issued from the United States Patent Office from 1790 *to* 1873, *inclusive*—Continued.

Invention.	Inventor.	Residence.	Date.	No.
Pumps, Strainer for ship's bilge	S. J. Chapman	Baltimore, Md	Sept. 2, 1873	142, 438
Pumps, Stuffing-box for deep-well	J. B. Pettey and J. Fredricks	Conneaut, Ohio	Apr. 9, 1867	63, 747
Pumps, Sucker-rod joint for	A. M. Williams	Oil City, Pa	Nov. 18, 1873	144, 813
Pumps, Supplying water to air	J. F. Haskins	Fitchburgh, Mass	Aug. 1, 1871	117, 626
Pumps to barrels, Device for attaching	J. F. de Navarro	New York, N. Y	June 16, 1868	78, 865
Pumps to bungs of barrels, Apparatus for attaching.	F. A. Pratt	Hartford, Conn	Sept. 22, 1863	40, 082
Pumps to bungs of barrels, Apparatus for attaching.	F. A. Pratt	Hartford, Conn	Oct. 13, 1863	40, 276
Pumps, Valve-arrangement for	W. Sewell and A. S. Cameron	New York, N. Y	Dec. 27, 1864	45, 644
Pumps, Valvular arrangement for diaphragm	S. A. Clemens	Springfield, Mass	July 11, 1854	11, 300
Pumps, Vetilating-attachment for	C. N. Lewis	Seneca Falls, N. Y	Nov. 17, 1857	18, 660
Pumps, Vulcanized-rubber sack for stock	J. S. Patrie	Rochester, N. Y	Oct. 17, 1871	120, 093
Pumps, Water-charger for	T. Dutton and T. Maguire	Port Jervis, N. Y	Oct. 20, 1868	83, 268
Pumps, Water-meter for rotary	E. Keith	Buffalo, N. Y	Sept. 14, 1869	94, 892
Pumps water-tight, Making joints of stone	A. Van Vorhes	Hebardville, Ohio	May 3, 1839	1, 139
Pumps with hydraulic press or ram, Connecting	R. Dillon	New York, N. Y	Feb. 1, 1848	5, 423
Pumps, Working	E. Rigg		July 29, 1794	
Pumps, Working	B. Robbins	Machias, Me	July 19, 1859	24, 822
Pumps, Working	W. Wright	Hartford, Conn	Apr. 7, 1857	17, 009
Pumps, Working ships'	F. E. Boettner	Chicago, Ill	May 12, 1863	38, 460
Pumps, Working ships'	A. Cain	Holyoke, Mass	Jan. 24, 1865	45, 973
Pumps, Working ships'	M. Francis	Boston, Mass	July 7, 1829	
Pumps, Working ships'	D. Gay	Bath, Me	Sept. 8, 1837	373
Pumps, Working ships'	L. Simpkins	Brooklyn, N. Y	June 5, 1866	55, 377
Pumps, Working condenser of steam diaphragm	J. Black and O. Beecher	Philadelphia, Pa	Jan. 25, 1853	9, 553
Pumping and condensing apparatus	G. J. Washburn	Worcester, Mass	Mar. 1, 1864	41, 799
Pumping and measuring apparatus, Liquid	E. F. Wilder	Lowell, Mass	May 7, 1872	126, 604
Pumping and raising water	B. Folger		July 7, 1804	
Pumping-apparatus	D. Stoner	Canton, Ohio	June 14, 1870	104, 374
Pumping-apparatus, Bellows	S. Motte	Paris, France	Sept. 8, 1868	81, 930
Pumping-apparatus for supplying finings and other liquids to casks.	A. M. Dix	Shelton, England	Jan. 26, 1869	86, 287
Pumping-apparatus, Steam	G. H. Corliss	Providence, R. I	June 2, 1857	17, 423
Pumping by the motion of the cars, Device for	R. R. and E. C. Spedden	Astoria, Ga	Sept. 7, 1869	94, 660
Pumping-device	J. Johnson	Saco, Me	Dec. 15, 1868	84, 948
Pumping-engine	R. Allison	Port Carbon, Pa	Sept. 29, 1868	82, 475
Pumping-engine	E. D. and F. W. Eames	Grand Rapids, Mich	May 28, 1872	127, 322
Pumping-engine	C. E. Emery	Brooklyn, N. Y	Nov. 17, 1868	84, 176
Pumping-engine	S. Harrison	Pottsville, Pa	Aug. 5, 1862	36, 089
Pumping-engine	J. L. Lowry	Pittsburgh, Pa	Mar. 10, 1868	75, 284
Pumping-engine	L. C. Rodier	Springfield, Mass	Aug. 23, 1870	106, 727
Pumping-engine	T. Shaw	Philadelphia, Pa	Dec. 15, 1868	84, 912
Pumping-engine	W. J. Stevens	New York, N. Y	Sept. 13, 1864	44, 232
Pumping-engine	H. R. Worthington	Brooklyn, N. Y	July 19, 1859	24, 838
Pumping-engine, Steam	J. S. Barden	New Haven, Conn	Nov. 24, 1857	18, 718
Pumping-engine, Steam	T. E. Blunt	Brookfield, Ohio	Aug. 16, 1870	106, 312
Pumping-engine, Steam	E. Cope	Cincinnati, Ohio	Oct. 26, 1858	21, 873
Pumping-engine, Steam	G. H. and C. P. Deane	Springfield, Mass	May 28, 1872	127, 153
Pumping-engine, Steam	R. J. Gould	Newark, N. J	July 27, 1869	92, 957
Pumping-enging, Steam	W. C. Hicks	Summit, N. J	Oct. 21, 1873	143, 824
Pumping-engine, Steam	L. J. Knowles	Worcester, Mass	May 2, 1871	114, 506
Pumping-engine, Steam	E. D. Leavitt, jr	Cambridge, Mass	July 16, 1872	129, 240
Pumping-engine, Steam	A. J. L. Loretz	Brooklyn, N. Y	Jan. 7, 1873	134, 689
Pumping-engine, Steam	J. H. McConnell	Harmar, Ohio	Dec. 24, 1872	134, 212
Pumping-engine, Steam	G. W. Perry	Shenandoah City, Pa	Sept. 29, 1868	82, 548
Pumping-engine, Steam	W. H. Roberts	Mauch Chunk, Pa	Feb. 15, 1870	99, 953
Pumping-engine, Steam	J. Tufts	Athens, Ohio	Apr. 28, 1868	77, 336
Pumping-engine, Steam	N. W. Wheeler	Morristown, N. J	June 6, 1871	115, 670
Pumping-engine, Steam	H. R. Worthington	New York, N. Y	June 20, 1871	116, 131
Pumping-engine, Steam	W. Wright	Hartford, Conn	Nov. 15, 1859	26, 138
Pumping-engines, Device for securing uniform motion in.	W. H. Roberts	Mauch Chunk, Pa	Mar. 1, 1870	100, 449
Pumping-machine	J. Clark		May 19, 1803	
Pumping-machine	R. Yeamans	Springfield, Mass	May 28, 1810	
Pumping water out of vessels, Method of	A. Kirkwood	Jackson County, Miss	Feb. 27, 1855	12, 458
Pumping water out of vessels by wind	T. Brownell	New York, N. Y	Mar. 23, 1827	
Pumping water, Wind-machine for	F. G. Johnson	Bellwood, Sag Harbor, N.Y	Apr. 3, 1860	27, 758
Punch	M. J. Fitzpatrick and B. Barker	New York, N. Y	Aug. 29, 1865	49, 616
Punch	A. J. Fullam	Springfield, Vt	Apr. 10, 1860	27, 793
Punch	C. B. Holden	Worcester, Mass	Aug. 14, 1866	57, 136
Punch	R. Hughes	Virginia City, Nev	Jan. 8, 1867	61, 069
Punch	R. Humphrey	Unionville, Conn	Oct. 16, 1860	30, 405
Punch	R. J. Mullen	Providence, R. I	Oct. 24, 1871	120, 162
Punch	W. A. Rex	Newville, Ind	Nov. 6, 1866	59, 453
Punch	I. P. Richards	Whitinsville, Mass	June 28, 1870	104, 769
Punch	E. Schlindler and C. H. Metzger	Easton, Pa	Sept. 17, 1867	69, 031
Punch	G. C. Wilder	Lawrence, Kans	Nov. 10, 1868	84, 039
Punch	L. Wolf	Saint Jacob, Ill	July 2, 1867	66, 436
Punch	J. Wright	Middleport, Ohio	June 15, 1869	91, 400
Punch, Adjustable center	E. E. Safford and S. Sawyer	Fitchburgh, Mass	Aug. 22, 1865	49, 553
Punch, Adjustable gang	C. E. Howard	West Bridgewater, Mass	Sept. 13, 1864	44, 191
Punch and awl, Combined	F. P. Pflegbar	Whitneyville, Conn	Jan. 25, 1859	22, 744
Punch and cutter, Leather	H. S. Fickett	Yarmouth, Me	Jan. 14, 1873	134, 794
Punch and cutting-nippers, Hand	J. O. Reilley	Jersey City, N. J	May 20, 1873	139, 079
Punch and die	W. K. Lewis	Boston, Mass	Mar. 7, 1865	46, 681
Punch and die for punching watch-hands	A. L. Dennison	Roxbury, Mass	Aug. 15, 1854	11, 522
Punch and edger combined	G. E. Hernig	Louisville, Ky	Mar. 19, 1872	124, 740
Punch and eyeleting-machine	E. Shaw	Milwaukee, Wis	Mar. 28, 1871	113, 211
Punch and press	W. S. Lord	Crawford, Miss	Feb. 22, 1870	100, 166
Punch and screw-driver, Combined	S. D. Tuttle	Eaton, Ohio	June 1, 1869	90, 704
Punch and screw-driver, Combined band	G. W. Schofield	United States Army	Dec. 26, 1871	122, 286
Punch and shears	D. D. Robinson	Niles, Mich	July 30, 1867	67, 352
Punch and shears, Combined power	A. C. Stanard	Milton, Wis	Sept. 6, 1870	107, 116
Punch, awl, and knife, Combined	H. S. Holmes and D. Williams	South Boston, Mass	May 16, 1871	114, 820
Punch, Belt	J. T. Carson	Greensborough, N. C	Feb. 11, 1868	74, 197
Punch, Belt	M. D. Drake	Scituate, R. I	Jan. 5, 1869	85, 575
Punch, Belt	J. E. Gates	Lowell, Mass	Apr. 30, 1872	126, 141
Punch, Belt	E. Hester	Suffield, Conn	Aug. 4, 1868	80, 734

Index of patents issued from the United States Patent Office from 1790 *to* 1873, *inclusive*—Continued.

Invention.	Inventor.	Residence.	Date.	No.
Punch, Belt	J. P. Jubb	Namaha, Mich	Feb. 2, 1869	86, 412
Punch, Belt	J. Mulchahey	Springfield, Mass	May 7, 1867	64, 556
Punch, Belt	A. Simpson	Worcester, Mass	Sept. 23, 1856	15, 795
Punch, Belt	A. M. Southwick	Boston, Mass	Feb. 11, 1873	135, 734
Punch, Belt	D. M. Weston	Boston, Mass	Sept. 1, 1868	81, 717
Punch, Belt	C. D. Wheeler	New York, N. Y	Jan. 24, 1860	26, 940
Punch-block	J. D. and J. M. Filkins	Johnson Township, Ind	Sept. 22, 1863	40, 030
Punch, Brad	R. F. Cook	Potsdam, N. Y	May 24, 1870	103, 429
Punch, Brad	J. Thorndike	North Weare, N. H	June 29, 1858	20, 743
Punch, Canceling	M. E. Berolzheimer	New York, N. Y	Nov. 30, 1869	97, 344
Punch, Car-ticket	R. J. Kellett	San Francisco, Cal	May 28, 1867	65, 080
Punch, Center	M. Bowker	Fitchburgh, Mass	Oct. 17, 1865	50, 442
Punch, Check	J. R. Mesa	Brooklyn, N. Y	Oct. 29, 1872	132, 593
Punch, Conductor's	J. and G. D. Friese	Baltimore, Md	Dec. 8, 1868	84, 739
Punch, Conductor's registering	A. D. Hoffman	Chicago, Ill	Feb. 22, 1870	100, 036
Punch, Conductor's ticket	J. Beck	Philadelphia, Pa	Sept. 10, 1867	68, 687
Punch, Conductor's ticket	W. J. Phelps	Springfield, Mass	Sept. 17, 1867	69, 019
Punch, Drop	S. Andrews	Perth Amboy, N. J	Apr. 13, 1852	8, 868
Punch for attaching buttons by rivets	E. Pincus and G. Rehfuss	Philadelphia, Pa	Oct. 18, 1864	44, 746
Punch for cutting out welts of boots and shoes	J. H. Walker	Worcester, Mass	June 28, 1864	43, 356
Punch for forming clasps	C. D. Flesche	New York, N. Y	Dec. 10, 1867	72, 007
Punch, Gang	J. H. Keating	Marblehead, Mass	Oct. 30, 1866	59, 230
Punch, Gang	B. G. Welch	Danville, Pa	Feb. 18, 1868	74, 646
Punch, Gun-wad	R. Rathbone	New York, N. Y	Dec. 31, 1867	72, 903
Punch, Hand	A. L. Eckert	Newark, N. J	Mar. 20, 1866	53, 282
Punch, Hand	O. C. Ford	Burlington, Conn	Sept. 13, 1870	107, 242
Punch, Hand	J. D. Higgins	Greenville, Conn	July 28, 1868	80, 349
Punch, Hand	W. Nash	Watertown, N. Y	Jan. 3, 1865	45, 738
Punch, Hand	R. Wood	Grand Ledge, Mich	Sept. 27, 1859	25, 599
Punch, Horseshoe	C. Huil	Lockport, N. Y	Oct. 9, 1866	58, 645
Punch, Leather	J. Lyle	Newark, N. J	Nov. 2, 1869	96, 453
Punch, Metal	W. I. Granger	Chicago, Ill	July 6, 1858	20, 846
Punch, Metal hand	S. F. Leach	Chelsea, Mass	Dec. 16, 1873	145, 578
Punch-mixer and egg-beater	T. Fisler	Camden, N. J	Apr. 17, 1866	53, 965
Punch-operating mechanism	L. W. Holmes	Grand Ledge, Mich	Sept. 22, 1868	82, 317
Punch, Paper-collar-button-hole	S. S. Stone	Troy, N. Y	Feb. 7, 1865	46, 279
Punch, Portable	J. J. Lafely	Ottumwa, Iowa	Apr. 22, 1873	138, 201
Punch, Portable metal	S. McKenna	Cincinnati, Ohio	Mar. 21, 1854	10, 681
Punch, Registering ticket	A. D. Hoffman	Chicago, Ill	Jan. 31, 1871	111, 345
Punch, Registering ticket	J. H. Small	Buffalo, N. Y	Jan. 31, 1871	111, 391
Punch, Registering ticket	J. H. Small	Buffalo, N. Y	May 23, 1871	115, 119
Punch, Revolving spring	S. Merrick	Springfield, Mass	Mar. 17, 1838	636
Punch, Revolving spring	S. Merrick	Springfield, Mass	Feb. 8, 1848	5, 435
Punch, rivet, and screw cutter	L. C. Judson	Waterville, N. Y	Nov. 18, 1825	......
Punch, Rotary	A. Bedford	Cold Water, Mich	Aug. 25, 1868	81, 465
Punch, Self-centering	S. Z. Hall	Camden, N. J	May 30, 1867	48, 018
Punch, shears, and iron-shrinker, Combined	D. C. Burdick	Milton, Wis	Mar. 2, 1869	87, 464
Punch, shears, &c., Operating hand	W. B. Mason	Boston, Mass	July 31, 1866	56, 772
Punch, Spring	P. Bauer	Newark, N. J	Mar. 29, 1864	42, 064
Punch, Spring	P. Bauer	Newark, N. J	Aug. 15, 1865	49, 364
Punch, Spring	M. V. Bryant	North Plains, Mich	July 28, 1868	80, 449
Punch, Spring	J. Lyle	Newark, N. J	Mar. 27, 1866	53, 537
Punch, Spring	A. U. Noble	Kalamazoo, Mich	Feb. 11, 1868	74, 241
Punch, Ticket	J. Chapin	Chicopee, Mass	May 4, 1869	89, 828
Punch, Ticket	R. Engels	Philadelphia, Pa	Apr. 5, 1870	101, 598
Punch, Ticket	H. W. Hewet	Cincinnati, Ohio	Feb. 4, 1873	135, 519
Punch, Ticket	W. Hill	Springfield, Mass	June 30, 1868	79, 470
Punch, Ticket	R. J. Kellett	San Francisco, Cal	June 23, 1868	79, 232
Punch, Ticket	C. H. Palmer and T. K. Leslie	Brooklyn, N. Y	Feb. 2, 1869	86, 438
Punch, Ticket	W. J. Phelps	Springfield, Mass	June 30, 1868	79, 498
Punches to stocks, Device for securing	J. Johnson	New York, N. Y	Apr. 2, 1872	125, 302
Punching and eyeleting machine, Shoe	J. Keith	New Bedford, Mass	Dec. 16, 1862	37, 170
Punching and eyeleting machines, Mechanism for operating.	C. H. Morse	Boston, Mass	Feb. 14, 1871	111, 865
Punching and raising disks from metal	C. Diedrichs	Philadelphia, Pa	Jan. 28, 1868	73, 784
Punching and shaping metal, Machine for	G. Haseltine	Washington, D. C	July 28, 1857	17, 876
Punching and shearing apparatus	S. Renfrew	Marseilles, Ill	Mar. 8, 1864	41, 861
Punching and shearing machine	W. Culver	Vineland, N. J	Jan. 25, 1870	99, 065
Punching and shearing machine	S. R. Houser	Sandusky, Ohio	Oct. 7, 1873	143, 510
Punching and shearing machine	W. H. Ivens and W. E. Brooke	Trenton, N. J	Nov. 9, 1869	96, 594
Punching and shearing machine	J. Kissel	Cassopolis, Mich	Oct. 28, 1873	143, 985
Punching and shearing machine	M. A. Lanagan	Brooklyn, N. Y	July 14, 1868	79, 912
Punching and shearing machine	P. C. Perkins	Waterford, N. Y	Mar. 12, 1861	31, 674
Punching and shearing machine, Combined	W. H. Ivens and W. E. Brooks	Trenton, N. J	May 2, 1871	114, 297
Punching and shearing machine, Combined	W. Lyon	Mamaroneck, N. Y	Sept. 16, 1873	142, 926
Punching and shearing machine, Combined	C. Swanson	Newton, Iowa	Dec. 16, 1873	145, 694
Punching and shearing machine, Iron	L. T. Pope	Boston, Mass	Mar. 17, 1838	639
Punching and shearing metal, Machine for	W. H. Ivens and W. E. Brooke	Trenton, N. J	Aug. 24, 1869	94, 117
Punching and shearing metallic plates, Machine for	J. M. Riter	Pittsburgh, Pa	June 25, 1872	128, 250
Punching and shearing tool	J. C. Jordan	Watertown, Wis	Nov. 12, 1867	70, 857
Punching and stamping machine	F. M. Huntington	Chicago, Ill	Sept. 3, 1872	131, 004
Punching and stamping machine, Metal*	J. Van Hagen	Chicago, Ill	Apr. 25, 1871	114, 068
Punching and stamping press	S. P. Ruggles	Boston, Mass	May 3, 1839	23, 864
Punching and upsetting apparatus	S. E. Lockwood	Westbury, New York	Mar. 3, 1868	75, 032
Punching-apparatus	J. F. Allen	New York, N. Y	Feb. 27, 1866	52, 810
Punching-apparatus	F. De Witt	Detroit, Mich	May 3, 1870	102, 510
Punching-apparatus	T. E. Harris	Green Bay, Wis	Aug. 6, 1867	67, 429
Punching-apparatus	R. Kent	Brooklyn, N. Y	Apr. 20, 1869	89, 226
Punching-apparatus	I. Lamplugh	Peoria, Ill	Feb. 18, 1868	74, 700
Punching-apparatus	P. L. Weimer	Lebanon, Pa	Oct. 31, 1865	50, 754
Punching between rollers	R. Montgomery	New York, N. Y	Feb. 5, 1850	7, 073
Punching corrugated metal, Apparatus for	M. J. Montgomery	New York, N. Y	May 8, 1866	54, 580
Punching countersunk holes	J. V. Westlake	Saint Louis, Mo	Dec. 16, 1862	37, 189
Punching designs in sheet-metal, Apparatus for	W. T. Rudd	Amsterdam, Va	July 8, 1851	8, 211
Punching-die	O. C. Burdict	New Haven, Conn	Apr. 10, 1866	53, 781
Punching-dies, Manufacture of	R. J. Mullin	Providence, R. I	Sept. 2, 1873	142, 407
Punching holes in iron or other metallic plates, Machine for.	W. Adams and A. Hammond	Boston, Mass	May 8, 1843	3, 077

*Ordered to be canceled by decree of circuit court of United States northern district of Illinois, October 2, 1873.

Index of patents issued from the United States Patent Office from 1790 *to* 1873, *inclusive*—Continued.

Invention.	Inventor.	Residence.	Date.	No.
Punching holes in leather, Machine for	G. L. Bailey	Portland, Me	Aug. 16, 1859	25, 083
Punching holes in metal, &c., Machine for	J. Sarchet	Philadelphia, Pa	Apr. 2, 1822	....
Punching linchpin holes and cutting off the journals of axles for wagons, &c., Machine for.	S. H. Hartman	Pittsburgh, Pa	Aug. 26, 1862	36, 298
Punching-machine	G. A. Alger	Manchester, N. H	May 22, 1866	54, 997
Punching-machine	R. Baird	Sterling, Scotland	Mar. 18, 1873	136, 954
Punching-machine	H. W. Bill	Cuyahoga Falls, Ohio	July 14, 1863	39, 209
Punching-machine	J. A. Bradshaw and A. F. Colby	Lowell, Mass	July 24, 1860	29, 231
Punching-machine	E. R. Brown and J. Long	Mauch Chunk and Packerton, Pa.	May 23, 1871	115, 156
Punching-machine	W. Churchill	Saint Louis, Mo	May 11, 1869	89, 851
Punching-machine	J. Duff	Peoria, Ill	June 27, 1871	116, 420
Punching-machine	D. A. Faulkner	Centreville, Cal	June 27, 1871	116, 288
Punching-machine	E. Heath	Fowlersville, N. Y	May 13, 1856	14, 866
Punching-machine	C. Hughes	New Orleans, La	July 24, 1860	29, 343
Punching-machine	S. Kendall	Kalamazoo, Mich	Apr. 3, 1849	6, 277
Punching-machine	C. Keniston	Somerville, Mass	June 27, 1871	116, 323
Punching-machine	J. M. Laughlin	Boston, Mass	Sept. 2, 1873	142, 403
Punching-machine	G. W. Lewis	Dansville, N. Y	July 4, 1871	116, 726
Punching-machine	J. M. Long	Hamilton, Ohio	Feb. 6, 1872	123, 407
Punching-machine	W. Lyon	New York, N. Y	Nov. 10, 1868	83, 981
Punching-machine	C. S. Moseley	Elgin, Ill	June 28, 1870	104, 755
Punching-machine	S. W. Murray	Milton, Pa	Feb. 25, 1873	136, 172
Punching-machine	F. Nelson	Wyandotte, Mich	Mar. 24, 1868	75, 782
Punching-machine	R. Porter	Washington, D. C	Jan. 29, 1856	14, 166
Punching-machine	H. Powers	Florence, Italy	Feb. 19, 1861	31, 476
Punching-machine	J. C. Rhodes	South Abington, Mass	Mar. 7, 1871	112, 381
Punching-machine	W. T. Richards	Bridgeport, Conn	Oct. 13, 1868	81, 993
Punching-machine	J. F. Sargent and E. Townsend	Boston, Mass	Mar. 29, 1864	42, 146
Punching-machine	W. E. Scott and D. L. Wood	Terre Haute, Ind	Jan. 26, 1869	86, 324
Punching-machine	D. S. Sherman	Lowell, Mass	June 8, 1858	20, 516
Punching-machine	O. Smith	Bridgeton, N. J	May 21, 1872	127, 112
Punching-machine	O. Snow	West Meriden, Conn	Mar. 12, 1867	62, 897
Punching-machine	D. Sprague	Elizabethport, N. J	Mar. 12, 1861	31, 681
Punching-machine	J. Steadman	Pecatonica, Ill	Dec. 19, 1865	51, 630
Punching-machine	C. Steinbach	Lima, Mich	Mar. 3, 1868	75, 068
Punching-machine	W. H. Van Cleve	Ypsilanti, Mich	Oct. 10, 1871	119, 900
Punching-machine	W. Welch	Bridgeport, Conn	Oct. 15, 1867	69, 877
Punching-machine	W. Woiceski	Bloomington, Ill	Aug. 26, 1873	142, 317
Punching-machine, Automatic	J. E. Wiggin	Stoneham, Mass	Oct. 13, 1868	83, 014
Punching machine, Copper	W. Ballard	Boston	Dec. 6, 1827	....
Punching machine, Copper or brass	J. Shugert	Pittsburgh, Pa	July 20, 1831	....
Punching-machine feeding-device	W. B. Bement	Philadelphia, Pa	Apr. 1, 1873	137, 407
Punching-machine, Hand	G. C. Miller	Rockford, Ill	May 31, 1870	103, 764
Punching-machine, Hydraulic	J. B. Barnes	Fort Wayne, Ind	Feb. 12, 1867	61 987
Punching-machine, Hydraulic	R. Livingston	Albany, N. Y	Oct. 24, 1871	120, 203
Punching machine, Leather	J. H. Haskell	Baltimore, Md	Oct. 22, 1867	69, 991
Punching machine, Metal	J. Allonas	Mansfield, Ohio	Sept. 14, 1869	94, 857
Punching machine, Metal	J. H. Brown	Grand Ledge, Mich	May 31, 1859	24, 191
Punching machine, Metal	S. H. Brown	Wheeling, Va	Aug. 14, 1839	1, 288
Punching machine, Metal	J. Clark	Haverstraw, N. Y	Aug. 12, 1873	141, 629
Punching machine, Metal	E. Craddock	London, England	Apr. 2, 1872	125. 121
Punching machine, Metal	O. J. Davis and T. W. Stephens	Erie, Pa	Oct. 4, 1853	10, 098
Punching machine, Metal	E. Doty	Jaynesville, Wis	May 10, 1870	102, 924
Punching machine, Metal	E. R. Hollands	Northampton Square, England.	Jan. 24, 1865	46, 056
Punching machine, Metal	H. A. Shipp	London, England	Nov. 2, 1869	96, 492
Punching machine, Metal	N. C. Stiles	Middletown, Conn	Aug. 27, 1872	130, 878
Punching machine, Metal	N. Thompson	Brooklyn, N. Y	Sept. 10, 1872	131, 193
Punching machine, Metal	G. Zender	Caledonia, Minn	Feb. 6, 1872	123, 435
Punching machine, Sheet-metal	M. Stevens	Brooklyn, N. Y	Oct. 7, 1873	143, 473
Punching-machine, Steam	J. Sparrow	Portland, Me	Sept. 27, 1859	25, 588
Punching machine, Tube	J. T. Brigden	Hornellsville, N. Y	Sept. 29, 1868	82, 591
Punching-machine with a combination of adjustable gages.	R. S. Tilden	Saint Louis, Mo	Mar. 10, 1849	6, 154
Punching machinery, Metal	D. T. Walker	Brooklyn, E. D., N. Y	Sept. 12, 1865	49, 937
Punching metal	L. C. Bunnel	Otsego, N. Y	Oct. 10, 1834	....
Punching metal	P. Koch	New Haven, Conn	Oct. 4, 1859	25, 653
Punching metallic sheathing, Machine for	A. H. Teeple	New York	Aug. 28, 1846	4, 727
Punching nail-holes, Machine for	C. C. Crosby	Nantucket, Mass	Oct. 9, 1860	30, 374
Punching or cutting iron or steel by machinery	J. Richards	Elbridge, N. Y	Sept. 23, 1831	....
Punching-press	G. H. Corliss and E. Harris	Providence, R. I	Mar. 25, 1856	14, 493
Punching-press	N. C. Stiles	West Meriden, Conn	Jan. 26, 1864	41, 403
Punching-press	M. G. Wilder	West Meriden, Conn	May 28, 1867	65, 143
Punching-presses, Conveyer to transfer blanks from	J. H. Baird	Oakville, Conn	June 8, 1869	90, 914
Punching-press rod	J. V. C. Crate	Waterbury, Vt	Nov. 20, 1866	59, 823
Punching, pressing, cutting, and slitting machine for iron plates, saws, &c.	J. Bennett	Brutus, N. Y	July 25, 1832	....
Punching rivet-holes in hose, Machinery for	J. R. Hague	Pittsburgh, Pa	July 4, 1854	11, 222
Punching sheet-metal	M. Seiferth	Morristown, N. J	Nov. 12, 1867	70, 905
Punching sheets of metal, Machine for	J. M. Riter and L. J. Farquhar	Pittsburgh, Pa	Oct. 9, 1866	58, 676
Punching sheets of metal, Machinery for	S. T. Sanford	Fall River, Mass	Jan. 13, 1852	8, 660
Purchase, Suspended	W. H. Brown	Erie, Pa	Jan. 2, 1855	12, 155
Purifying and refining spirits, Apparatus for	T. Thompson	Baltimore, Md	June 14, 1864	43, 141
Purifying fatty bodies	R. A. Tilghman	Philadelphia, Pa	Oct. 3, 1854	11, 766
Purifying liquids by galvanism	A. Crosse	Broomfield, England	Jan. 6, 1848	5, 409
Purifying spirits and other liquids, Compound for	P. J. Badoux	New York, N. Y	Aug. 27, 1867	68, 028
Pushing-jack	A. Freeman	Peoria, Ill	June 29, 1869	91, 837
Putty, Composition for	J. and W. H. Lucas	Philadelphia, Pa	Aug. 28, 1866	57, 528
Putty-knife	S. W. Gerelds	Worcester, Mass	Jan. 18, 1870	98, 949
Putty-knife and sprig-box, Combined	J. H. Gaches	Atlanta, Ga	Mar. 4, 1873	136, 499
Putty, Manufacture of	G. W. Hatfield	Dayton, Ohio	Oct. 26, 1869	96, 231
Puzzle	J. W. Mueller	Detroit, Mich	Feb. 24, 1863	37, 763
Puzzle, Alphabetical-instruction	E. F. Gilbert	Lyons, N. Y	June 22, 1869	91, 737
Puzzle, Child's	W. B. Rice	Feltonville, Mass	Feb. 9, 1864	41, 536
Pyrite-burning furnace for manufacture of sulphuric acid, &c.	J. Hughes	Edgewater, Mass	June 18, 1867	65, 914
Pyrites, Treating auriferous and argentiferous	L. Solomon	New York, N. Y	Jan. 11, 1859	22, 587

Index of patents issued from the United States Patent Office from 1790 to 1873, inclusive—Continued.

Invention.	Inventor.	Residence.	Date.	No.
Pyrometer	W. H. Bailey	Albion Works, Salford, Great Britain.	Feb. 28, 1871	112, 106
Pyrometer	E. Brown	Philadelphia, Pa	June 1, 1869	90, 815
Pyrometer	E. Brown	Philadelphia, Pa	May 3, 1870	102, 654
Pyrometer	E. Brown	Philadelphia, Pa	Aug. 27, 1872	130, 894
Pyrometer	E. Brown	Philadelphia, Pa	Aug. 27, 1872	130, 895
Pyrotechnic night-signal	B. F. Coston	Washington, D. C	Apr. 5, 1859	23, 536
Pyrotechnic night-signal	M. J. Coston	Washington, D. C	June 13, 1871	115, 935
Pyrotechnic night-signal	G. A. Lilliendahl	New York, N. Y	Apr. 5, 1859	23, 529
Pyrotechnic signal	H. J. Harris	Shreveport, La	Sept. 29, 1868	82, 619
Pyrotechnic signal	A. Lamarre	Paris, France	Oct. 28, 1873	144, 030
Pyroxile, pyroxyline, &c., for forming useful and ornamental articles, Treating.	L. R. Streeter	Chelsea, Mass	Apr. 20, 1869	89, 254
Pyroxyline and articles therefrom, Manufacture of	S. Deitz and B. P. Wayne	Albany, N. Y	Dec. 17, 1872	133, 969
Pyroxyline, Apparatus for compressing and molding plastic.	J. Brockway	Albany, N. Y	Apr. 18, 1871	113, 735
Pyroxyline articles, Manufacturing	J. W. Hyatt	Albany, N. Y	Apr. 29, 1873	138, 254
Pyroxyline, Process and apparatus for manufacturing.	J. S. and J. W. Hyatt	Albany, N. Y	Nov. 19, 1872	133, 229
Pyroxyline, Treating and molding	J. W., jr., and J. S. Hyatt	Albany, N. Y	July 12, 1870	105, 338

AMERICA IN TWO CENTURIES:
An Inventory

An Arno Press Collection

American Association of Museums. **A Statistical Survey of Museums in the United States and Canada.** 1965

Andrews, Israel D. **On the Trade and Commerce of the British North American Colonies, and Upon the Trade of the Great Lakes and Rivers.** 1853

Audit Bureau of Circulations. **Scientific Space Selection.** 1921

Austin, E. L. and Odell Hauser. **The Sesqui-Centennial International Exposition.** 1929

Barnett, James H. **The American Christmas.** 1954

Barton, L[eslie] M. **A Study of 81 Principal American Markets.** 1925

Bennitt, Mark, comp. **History of the Louisiana Purchase Exposition.** 1905

Bowen, Eli. **The United States Post-Office Guide.** 1851

Bureau of Applied Social Research, Columbia University. **The People Look at Radio.** 1946

Burlingame, Roger. **Engines of Democracy:** Inventions and Society in Mature America. 1940

Burlingame, Roger. **March of the Iron Men:** A Social History of Union Through Invention. 1938

Burnham, W. Dean. **Presidential Ballots, 1836-1892.** 1955

Cochrane, Rexmond C. **Measures for Progress:** A History of the National Bureau of Standards. 1966

Cohn, David L. **The Good Old Days.** 1940

Cozens, Frederick W. and Florence Scovil Stumpf. **Sports in American Life.** 1953

Day, Edmund E. and Woodlief Thomas. **The Growth of Manufactures, 1899 to 1923.** 1928

Edwards, Richard Henry. **Popular Amusements.** 1915

Evans, Charles H., comp. **Exports, Domestic and Foreign, From the American Colonies to Great Britain, From 1697 to 1789, Inclusive;** Exports, Domestic, From the U.S. to All Countries, From 1789 to 1883, Inclusive. 1884

Federal Reserve System, Board of Governors. **All-Bank Statistics, United States, 1896-1955.** 1959

Flexner, Abraham. **Funds and Foundations:** Their Policies, Past and Present. 1952

Flint, Henry M. **The Railroads of the United States.** 1868

Folger, John K. and Charles B. Nam. **Education of the American Population.** 1967

Handel, Leo A. **Hollywood Looks At Its Audience:** A Report of Film Audience Research. 1950

Harlow, Alvin F. **Old Waybills:** The Romance of the Express Companies. 1934

Harrison, Shelby M. **Social Conditions in an American City:** A Summary of the Findings of the Springfield Survey. 1920

Homans, J. Smith, comp. **An Historical and Statistical Account of the Foreign Commerce of the United States.** 1857

Ingram, J. S. **The Centennial Exposition.** 1876

Institute of American Meat Packers and the School of Commerce and Administration of the University of Chicago. **The Packing Industry:** A Series of Lectures. 1924

Leech, D[aniel] D. T[ompkins]. **The Post Office Department of the United States of America.** 1879

Leggett, M. D., comp. **Subject-Matter Index of Patents for Inventions Issued by the United States Patent Office From 1790 to 1873, Inclusive.** 1874. Three vols.

Magazine Marketing Service. **M.M.S. County Buying Power Index.** 1942

Martin, Robert F. **National Income in the United States, 1799-1938.** 1939

McCullough, Edo. **World's Fair Midways.** 1966

Melish, John. **Surveys for Travellers, Emigrants and Others.** 1976

National Advertising Company. **America's Advertisers.** 1893

Peters, Harry T. **America On Stone:** The Other Printmakers to the American People. 1931

Peters, Harry T. **California On Stone.** 1935

Peters, Harry T. **Currier & Ives:** Printmakers to the American People. 1929/1931. Two vols.

Pownall, T[homas]. **A Topographical Description of the Dominions of the United States of America.** Edited by Lois Mulkearn. 1949

Reed, Alfred Zantzinger. **Present-Day Law Schools in the United States and Canada.** 1928

Reed, Alfred Zantzinger. **Training for the Public Profession of the Law.** 1921

Rogers, Meyric R. **American Interior Design.** 1947

Romaine, Lawrence B. **A Guide to American Trade Catalogs, 1744-1900.** 1960

Scammon, Richard M., comp. **America at the Polls:** A Handbook of American Presidential Election Statistics, 1920-1964. 1965

Smillie, Wilson G. **Public Health:** Its Promise for the Future. 1955

Thompson, Warren S. **Population: The Growth of Metropolitan Districts in the United States, 1900-1940.** 1947

Thorndike, E[dward] L. **Your City.** 1939

Truman, Ben[jamin] C. **History of the World's Fair.** 1893

U.S. Bureau of the Census, Department of Commerce. **Housing Construction Statistics: 1889 to 1964.** 1966

U.S. Census Office (12th Census). **Street and Electric Railways.** 1903

Urban Statistical Surveys. 1976

Wayland, Sloan and Edmund de S. Brunner. **The Educational Characteristics of the American People.** 1958

Woytinsky, W. S. **Employment and Wages in the United States.** 1953

U.S. Census Office (1st Census, 1790). **Return of the Whole Number of Persons Within the Several Districts of the United States.** 1802

U.S. Census Office (2nd Census, 1800). **Return of the Whole Number of Persons Within the Several Districts of the United States.** 1802

U.S. Census Office (3rd Census, 1810). **Aggregate Amount of Each Description of Persons Within the United States of America.** 1811

U.S. Census Office (4th Census, 1820). **Census for 1820.** 1821

U.S. Census Office (5th Census, 1830). **Abstract of the Returns of the Fifth Census.** 1832

U.S. Census Office (6th Census, 1840). **Compendium of the Enumeration of the Inhabitants and Statistics of the United States.** 1841

U.S. Census Office (7th Census, 1850). **The Seventh Census of the United States.** 1853

U.S. Census Office (8th Census, 1860). **Statistics of the United States in 1860.** 1866

U.S. Census Office (9th Census, 1870). **A Compendium of the Ninth Census.** 1872

U.S. Census Office (10th Census, 1880). **Compendium of the Tenth Census.** Parts I and II. 1883. Two vols.

U.S. Census Office (11th Census, 1890). **Abstract of the Eleventh Census.** 1894

U.S. Bureau of the Census (12th Census, 1900). **Abstract of the Twelfth Census of the United States.** 1904

U.S. Bureau of the Census (13th Census, 1910). **Thirteenth Census of the United States: Abstract of the Census.** 1913

U.S. Bureau of the Census (14th Census, 1920). **Abstract of the Fourteenth Census of the United States.** 1923

U.S. Bureau of the Census (15th Census, 1930). **Fifteenth Census of the United States: Abstract of the Census.** 1933

U.S. Bureau of the Census (16th Census, 1940). **Sixteenth Census of the United States: United States Summary.** 1943

U.S. Bureau of the Census (17th Census, 1950). **A Report of the Seventeenth Decennial Census of the United States: United States Summary.** 1953

U.S. Bureau of the Census (18th Census, 1960). **The Eighteenth Decennial Census of the United States: United States Summary.** 1964

U.S. Bureau of the Census (19th Census, 1970). **1970 Census of Population: United States Summary.** 1973. Two vols.